FUNGI

ON PLANTS AND PLANT PRODUCTS IN THE UNITED STATES

David F. Farr   Gerald F. Bills   George P. Chamuris   Amy Y. Rossman

APS PRESS

The American Phytopathological Society
St. Paul, Minnesota USA

Authors:

David F. Farr
Gerald F. Bills
George P. Chamuris
Amy Y. Rossman

Systematic Botany and Mycology Laboratory
Beltsville Agricultural Research Center
USDA-Agricultural Research Service
Beltsville, MD 20705

This publication is No. 5 in the series Contributions from the U.S. National Fungus Collections.

This book has been reproduced directly from computer-generated copy submitted in final form to APS Press by the authors of the volume. No editing or proofreading has been done by APS Press.

Library of Congress Catalog Card Number: 89-84541
International Standard Book Number: 0-89054-099-3

Published 1989 by The American Phytopathological Society
Second Printing, 1995

Printed in the United States of America on acid-free paper

The American Phytopathological Society
3340 Pilot Knob Road
St. Paul, Minnesota 55121, USA

This book is dedicated to

Muriel J. O'Brien

1915-1985

Muriel J. O'Brien was born on April 10, 1915, in Brooklyn, New York, where she spent her childhood and school years. She earned a B.A. degree from Columbia University in 1940 and an M.S. degree from Iowa State University in 1947. Miss O'Brien then moved to Washington, D.C., to begin her 38-year career with the U.S.D.A. as a Research Plant Pathologist.

Miss O'Brien's research accomplishments are many and varied: she discovered, described, and named several disease-causing fungi of important crop plants; her evaluation of 473 plant introduction accessions of eggplant for wilt-resistance produced five resistant lines currently being used in commercial breeding programs; the pioneer work by Miss O'Brien and W.B. Raymer on potato spindle tuber disease opened up a whole new research area on the characterization and identity of the low-molecular weight entity now called a "viroid."

Miss O'Brien was invited to participate in plant pathology symposia in India on potato diseases and production. She also spent several months in Copenhaagen, Denmark, with the Institute of Seed Pathology for Developing Countries, working on species concepts in the genus *Alternaria* to provide direction for indexing literature and other material on seedborne diseases. She was a frequent contributor and presenter at national and international meetings, technical conferences, and workshops.

In her advisory and consultant activities, she assisted the Animal and Plant Health Inspection Service (Plant Protection and Quarantine) in identification of foreign disease organisms and provided technical guidance and information on host range and taxonomy of crop diseases to researchers both in the United States and abroad.

During her career Miss O'Brien authored or coauthored 57 publications; three of these extensive works still serve as reference guides for plant pathologists, mycologists, botanists, bacteriologists, nematologists, and virologists. *Bibliography of Potato Disease Literature through 1945: with Common and Scientific Names* (1970. Misc. Publication 1162. United States Department of Agriculture, Agriculture Research Service) is a guide to potato disease citations existing before the *Bibliography of Agriculture* and other indexes were started. *Potato Diseases* (1976. Agriculture Handbook 474. United States Department of Agriculture, Agriculture Research Service), a 79-page handbook with keys, discusses the causal agents and control measures of 62 disease problems of potato. This publication was awarded the Society of Technical Communications Award of Excellence in the category of "Bulletin." The award was conferred for superior writing, total integration of subject matter, editing, and graphics.

Although the book was developed by a consortium of scientists, Miss O'Brien was instrumental in the completion of the *Index of Plant Diseases in the United States,* Agriculture Handbook 165. This work, which was first published as a series in *Plant Disease Reporter* in 1953, and then as one volume in 1960, has become the standard reference on diseases found on economic crops, ornamentals, and grasses in the United States. The book has received worldwide distribution and acceptance. Miss O'Brien had begun work on the revision of this valuable handbook when she was stricken with leukemia and died August 3, 1985.

Muriel O'Brien's passing was a great loss to the scientific community, but her research continues to benefit people throughout the world. Her friends and colleagues will always remember Muriel's concern for others, her positive outlook, and her sense of humor, traits she retained to the end.

CONTENTS

Preface	vii
Acknowledgments	viii
Introduction	1
Host-Fungus List	7
Fungus List	555
Literature Cited	1029
Host Index	1095
Host Common Name Index	1101
Fungus Index	1139
Fungal Author Names	1241

PREFACE

This book provides an up-to-date, accurate, and comprehensive account of the fungi that occur on plants and plant products in the United States. Knowledge of these fungi has relevance to a wide variety of scientists and administrators at a national and international level, including agricultural and biological research scientists, plant pathologists, mycologists, crop-oriented workers, extension agents, post-harvest pathologists, forest product scientists, agronomists, horticulturists, plant breeders, foresters, and other plant scientists as well as those who teach in these areas.

Before this publication, such information was available to a limited extent in the *Index of Plant Diseases in the United States* published from 1950 to 1953 by the United States Department of Agriculture and later revised as Handbook 165,* which provided a listing of organisms causing plant diseases and their hosts, based on information accumulated to the later 1940s. The publication of several state and regional lists since the appearance of Handbook 165 demonstrates the continuing interest in comprehensive accounts of plant disease organisms. Over the past 35-40 years there has been a continued demand for identification of fungi causing plant diseases. Availability of information about these fungi has increased, and many new fungal diseases have been described. In addition, the scientific names of certain fungi have changed in accordance with the International Rules of Botanical Nomenclature (1987). *Fungi on Plants and Plant Products in the United States* incorporates this new information in providing both an update of Handbook 165 and important additional information to meet the current needs of today's scientific community.

In comparing *Fungi on Plants and Plant Products in the United States* with the Handbook 165, several similarities are apparent. Both publications share the basic objective of providing a listing of fungi that occur on vascular plant hosts in each state. The most significant difference between the two books is the inclusion of a section on the taxonomy and nomenclature of fungi in the new publication. This is the first time detailed information covering such a large number of fungi has been brought together in one publication. It also provides users of information on fungi with an important tool for evaluating the fungal literature. Secondly, *Fungi on Plants and Plant Products in the United States* provides the literature references on which each report is based. These references were not included in Handbook 165; thus the value of the information was diminished by the inability to investigate the origin of a report. Using the Index to Fungal Names in the current book, it is possible to determine the geographic distribution, host range, and other relevant information about a particular fungus, which was not possible using Handbook 165.

Handbook 165 included nematodes, virus, and bacteria, all of which are not covered in the current book. A current comprehensive treatment is available for the nematodes (Anonymous. 1985. *Distribution of Plant-Parasitic Nematode Species in North America.* Society of Nematologists). However, no similar treatments are available for viruses and bacteria.

This project utilized a PRIME minicomputer and PRIME Information software. This software is a variant of the Pick operating system and is similar to the Revelation software which runs on an IBM Personal Computer. The photo-ready copy, including headers and page numbers, was produced in its entirety directly from the computer databases. No intermediate print files or word processing files were created. A Hewlett Packard LaserJet Series II was used in conjunction with SoftCraft Fontware soft fonts. The photo-ready copy was reduced about 20% for publication.

The data included in this book are part of an ongoing database project of the U.S.D.A./A.R.S. This project will continue to provide current up-to-date information on fungi in the United States, and it is hoped that it will stimulate future work to clarify and revise our knowledge of the distribution and host relationships of fungi.

* The *Index to Plant Diseases in the United States* first appeared in five parts in the *Plant Disease Survey Special Publications as Special Publications* Number 1. Part 1 was published in 1950 and part 5 in 1952. This information was reprinted without change in a single volume: Anonymous. 1960. *Index to Plant Diseases in the United States*. U.S. Dept. of Agric. Handbook. No. 165. Washington D.C.

ACKNOWLEDGMENTS

With any publication of this size, there are always some unsung heroes. In this respect two individuals must be cited: Maryann Loftus and Lois Brako. Maryann worked with Muriel O'Brien in the initial stages and continued to be an important part of the project. By being involved in the project from the beginning, she provided needed continuity as the project evolved. The project would not have been completed without Lois Brako's significant contribution. She happily took on those tasks for which the rest of us had little enthusiasm. Her willingness to concentrate fully on all aspects of the project has greatly improved the final product.

Many mycologists assisted with the review of fungal names and their help is most gratefully acknowledged. Without their willingness to contribute time and expertise, a synopsis such as this one would never have been possible. At the same time, it must be clearly stated that they are in no way responsible for the content.

M.E. Palm (Maryland) assisted in many areas of the project and was a constant source of valuable suggestions and comments. H. Burdsall (Wisconsin) provided many of the references for the wood decay fungi. J. McCain (Minnesota) worked through many of the more difficult rust names. G. Samuels (New York) provided extensive help with the ascomycete names. K. Vanky (West Germany) made extensive annotations to the smut names.

Other mycologists making valuable contributions were: J. Alcorn (Australia); H.A. van der Aa (Holland); G. Boerema (Holland); M. Barr Bigelow (Massachusetts); J. Bissett (Canada); P. Cannon (United Kingdom); S. Carpenter (Oregon); M. Corlett (Canada); M. Dunn (Washington); G. De Hoog (Netherlands); M.L. Farr (Maryland); L. Gilbertson (Arizona); J. Ginns (Canada); J. Haines (New York); D. Hawksworth (United Kingdom); J. Hennen (Indiana); H.H. Ho (New York); S. Hughes (Canada); T. Iturriaga (Venezuela); P. Kirk (United Kingdom); R.P. Korf (New York); R. Lowen (New York); G. Morgan-Jones (Alabama); T. Nag Raj (Canada); P. Nelson (Pennsylvania); D. Pfister (Massachusetts); F. Pollack (Florida); E. Punithalingam (United Kingdom); S. Redhead (Canada); J. Rogers (Washington), C.T. Rogerson (New York); M. Sherwood-Pike (Oregon); R. Shoemaker (Canada); E. Simmons (Indiana); A. Sivanesan (United Kingdom); B. Sutton (United Kingdom); T. Toussoun (Pennsylvania); M. Wingfield (South Africa); and Wen-ying Zhueng (China).

Jacqueline Kallunki (New York) is gratefully acknowledged for reviewing the vascular plant names and providing much-needed updating of the nomenclature. Other vascular plant taxonomists making valuable contributions were W. W. Thomas (New York) and J. Wiersema (Maryland).

Several people assisted in providing the computer databases that were used for checking the vascular plant names: E. Bird and J. Mowder (Germplasm Services Laboratory, U.S.D.A./A.R.S), L. Morse (Nature Conservancy, Washington, D.C.), B. Magill (Missouri Botanical Garden) and E.R. Farr (Smithsonian Institution).

Technical support was cheerfully provided by our laboratory staff: Sharey Biles, Harriet Gladish, Marjorie Richeson, Loretta Alessandrini, Dulcie Powell, June McDaniels and Eleanor Edinger.

The Committee established to oversee this project is acknowledged for their input of ideas and encouragement, as are the administrators in the Agricultural Research Service who approved and funded this project: T. Barksdale, W. Dowler, E. Dutkey, E. Finney, A. Grybauskas, W. Klassen, L. Knutson, H. Moline, N. O'Neil, A. Stoner, and H. Waterworth.

FUNGI

ON PLANTS AND PLANT PRODUCTS IN THE UNITED STATES

INTRODUCTION

This introduction consists of three sections: 1) a discussion of the general purpose and information provided in the book, 2) instructions for using the book, and 3) guidelines followed in compiling and presenting the information.

I. PURPOSE AND CONTENT

Fungi on Plants and Plant Products in the United States provides information that serves four major functions:

1. To provide a comprehensive account of fungi reported on plants and plant products on a state or regional basis, with brief notes on the diseases they cause. Each of the more than 78,000 unique fungus-host combinations is documented by the literature source(s).

2. To serve as a reference for accepted names of the fungi reported on vascular plant material in the United States. Each of the 13,000 accepted fungal names is listed by its accurate scientific name based on the latest mycological literature. Relevant nomenclatural information is included, specifically, the correct author citation and some widely used synonyms. Additional information includes the related anamorph or teleomorph (if known), worldwide distribution, diseases caused by the fungus, general host range, relevant taxonomic information, and important literature references about the fungus. The host genera from which the fungus is reported in the United States are listed.

3. To serve as a guide to the literature reports of fungi on plants and plant products. Over 4,000 articles, books, monographs, and herbarium records are cited by reference number in the Host-Fungus List. Complete citations to these references are presented in the Literature Cited section.

4. To aid in the identification of fungi on plants and plant products in the United States. By surveying a large number of articles and condensing the information in these articles, by presenting this information according to host and by providing supplementary information for each fungus, the authors have created a body of material that will facilitate identification of fungi. While this publication can be a powerful tool to assist in the identification of these fungi, it cannot substitute for an identification based on direct examination of the relevant morphological features of the fungus. Rossman, A. Y. et al. (1987. *A Literature Guide for the Identification of Plant Pathogenic Fungi*. APS Press, St. Paul Minnesota) provide information on literature useful in the identification of fungi.

To facilitate access to the information four indexes are included: 1) an index to the genera and families of the vascular plant hosts; 2) an index to the common names of the vascular plant hosts; 3) an index to all fungal names; 4) a list of authors of fungal taxa and their abbreviations.

Specific applications of the information provided by this publication include the following:

1. Agricultural scientists may use this book as a resource in their research programs and training activities. The concise information on the taxonomy and nomenclature of fungi may assist communication among scientists, clarify relationships between groups of fungi, and provide editors with a standardized reference to fungal names.
2. Regulatory agencies may rely on this book for information on which to base their import/export and other regulatory decisions involving fungal plant pathogens both throughout the United States and internationally. As the book provides an up-to-date, comprehensive list of fungi present in the United States, it may be especially helpful in avoiding unnecessary restriction of germplasm by such agencies.
3. Agronomists may use the information provided on host-fungus distributions to locate areas of potential threat of disease or for identification of local crop resistance or susceptibility. Potential disease threats to newly introduced crop species can also be ascertained.
4. Plant breeders may use this information in the development of disease-resistant germplasm lines, since it provides a comprehensive list of the fungi found on the plant of interest as well as on related plants.
5. This reference may serve as an important resource for plant disease clinics in the diagnosing of plant diseases and assist local and state extension agents in serving the concerns of citizens about plant diseases.

II. HOW TO USE THIS BOOK

The book is divided into three major sections, followed by three indexes and a list of the authors of fungal names.

Host-Fungus List

Format. This section of the book is arranged in alphabetical order by vascular plant host family to facilitate comparison of fungi found on related hosts. Host family names are printed in bold roman type. Host genera within each family are in alphabetical order. Each genus is followed by two parts (Figure 1).

The first part lists the host species. Each species has a number that is used to identify this host in the second section. Each host species name is followed by its author(s), appropriate synonyms, and relevant common name(s) (in brackets).

The second part contains the fungi reported for each host genus, listed according to the major fungal groups (e.g., Ascomycotina, Basidiomycotina; Figure 2). Each entry contains the currently accepted name of the fungus, comments about the host-fungus combination, one or more host species numbers (from above) in boldface, the states or regions from which the particular fungus has been reported, and the literature source(s) for each report.

Comments. The placement of plant genera within families follows Mabberley D.J. (1987. The Plant-book, Cambridge University Press, Cambridge).

This section contains only accepted fungal names. Reports of fungi on a particular host not in agreement with the reports in the fungal taxonomic literature are enclosed in brackets.

The comments represent a compilation of all information found in the literature concerning the part of the host affected by the fungus or the symptoms produced by the fungus. If two reports stated different symptoms, both statements were included. A comment enclosed in single quotes indicates a designated common name as published in *Plant Disease* 69:649-676, 1985.

Localities are reported as they appeared in the literature; no attempt was made to consolidate localities or to present a generalized disease distribution. Locality abbreviations and a map of the United States with the state abbreviations are presented at the end of the introduction.

Fungus List

Format. Fungi reported in the Host-Fungus List are arranged alphabetically in this section by genus (Figure 3). The genus name (in boldface) is followed by the author(s) of the genus, the class and order in which the genus is placed, and pertinent comments. The accepted species name (in boldface) is listed with its correct author citation. Pertinent synonyms are listed below the accepted name. Teleomorphs, anamorphs, or synanamorphs are listed along with the synonyms and are so indicated. Following the synonymy, comments on the general distribution, taxonomy, and biology may be included. Following the comments is a list of the associated host genera reported in the Host-Fungus List for that fungus.

Comments. The use of the word "type" in brackets in the distribution field indicates that the only reference found was the type description. Reports of fungi on a particular host not in agreement with the reports in the fungal taxonomic literature are enclosed in brackets. The list of vascular plant genera following "Host:" are those reported in the Host-Fungus List and are thus only those reported in the surveyed literature.

Literature Cited

All references used in compiling the Host-Fungus List are cited in full in alphabetical order by author.

Host Index

All host genera and families in the Host-Fungus List are listed alphabetically. Each is followed by the page number on which the entry begins.

Host Common Name Index

Common names of the hosts listed in the Host-Fungus List are given here. Each is followed by the page number on which it is found.

Fungus Index

All fungal names in this publication are listed by specific epithet. The accepted name is printed in boldface. Synonyms are cross referenced. Since only accepted fungal names are provided in the Host-Fungus List, this index serves as an entry point into the Host-Fungus List for those fungal names now considered synonyms. Two sample entries from the Fungus Index followed by their interpretations are listed below:

detospora, (Septoria, Titaeospora) = **Titaeospora equiseti**
accepted name = Titaeospora equiseti
synonyms = Septoria detospora and Titaeospora detospora

donacis, Pseudoseptoria (Selenophoma, Septoria)
accepted name = Pseudoseptoria donacis
synonyms = Selenophoma donacis and Septoria donacis

Authors of Fungal Names List

The authorities are listed in alphabetical order by the full name, which is followed by the abbreviation.

III. GUIDELINES USED IN COMPILING THIS INFORMATION

The various decisions made regarding the content of the book and its presentation are outlined in this section.

Criteria for Inclusion

1. The report must be verifiable; i.e., a published account or a herbarium specimen.
2. A definitive reference must be made to the fungus occurring in one of the 50 United States, or a region in the United States.
3. A definitive reference must be made to the fungus occurring on a particular vascular plant host. If the host is identified by the common name, determination of the scientific name of that host must be unequivocal.
4. Only fungi reported to occur on the host under natural conditions are included. Fungi reported to be pathogenic in laboratory or greenhouse studies after inoculation with pure fungal cultures are not included.
5. The host must be a vascular plant, including ferns and fern allies. Fungi reported to occur on mosses, algae, and lichens are not included.
6. Bacteria, viruses, nematodes and fungi that are mycoparasitic, mycorrhizal, or lichenized are not included.

Sources of Data

Reports of fungi were derived from the literature and, to a limited extent, from herbarium specimens. Handbook 165 was the largest source of reports, covering the literature up to 1950. Additional information for the current work was derived primarily from the mycological and plant pathological literature published during the years 1950-1987. Five major journals were systematically searched: *Plant Disease* 1982-1987 and its predecessor *Plant Disease Reporter* 1950-1981, *Phytopathology* 1950-1987, *Canadian Journal of Plant Pathology* 1975(year initiated)-1987, *Mycologia* 1950-1987, and *Mycotaxon* 1971(year initiated)-1987. Some articles appearing in these and other journals before 1950 were added; however, the pre-1950 literature was not searched systematically.

Additional information was obtained from the collection of reprints, books and monographs in the John A. Stevenson Reference Room of the U.S. National Fungus Collections (U.S.D.A./A.R.S.) The reprints containing reports of fungi occurring on vascular plant hosts span a number of international and regional journals. Although not systematically searched, a serious effort was made to review reprints by important mycologists and plant pathologists for reports of fungi in the United States.

Other sources of information were:

1. Published checklists of fungi reported on plants in one particular state. Lists of fungi included in mimeographed or locally distributed reports from plant disease clinics were not included.
2. Specimens filed in the U.S. National Fungus Collections. Records derived from these specimens were included if the identity of the fungus was assured through the deposition or annotation of the specimen by an established mycologist.
3. Fungal specimens issued in multiple copies to major fungal herbaria as exsiccati.

Several difficulties were encountered while surveying the literature such as references that listed all hosts for a particular fungus followed by a list of localities that did not specify the relationship between host and locality and reports that failed to provide a clear or complete reference to the locality. Unless the locality, host and fungus information could be related, the information was not included. Because of these difficulties, some fungal species known to occur on vascular plants in the United States may not be included.

Compiling the Host-Fungus List

Host-fungus relationships covered. Host plants and their associated fungi occur in a number of ecological relationships. Some fungi are parasitic on living plant parts, while others attack dying tissues or dead parts of living plants. Certain fungi cause deterioration of stored fruits, vegetables, and grains. Still others decay and discolor wood, wood products, and other plant materials. All the above types of fungi are included in this book. Reports of fungal-like organisms are not included such as *Rhodochytrium*, an organism only recently determined to be a parasitic red alga, or *Cephaleuros*, a plant-pathogenic alga.

The database includes 77,830 different host-fungus combinations. Of these combinations, 32% are Basidiomycotina, 24% are Ascomycotina, and 38% are Deuteromycotina. The Zygomycotina, Mastigomycotina, and Myxomycota were present in 6% of these combinations. The more common vascular plant families were the Poaceae, Pinaceae, Rosaceae, Asteraceae and Fabaceae. Members of the Poaceae were present in 13% of these combinations. For the other families, the values were Pinaceae (9%), Rosaceae (6%), Asteraceae (6%), and Fabaceae (6%).

Sources of information on vascular plant names. The scientific names of the vascular plant hosts included here were initially reviewed electronically for accuracy and acceptability using the following vascular plant databases:

1. Anonymous. 1982. *National List of Scientific Plant Names*. United States Department of Agriculture, Soil Conservation Service, SCS-TP-159.
2. Terrell E.E., S.R. Hill, J.H. Wiersema, and W.E. Rice. 1986. *A*

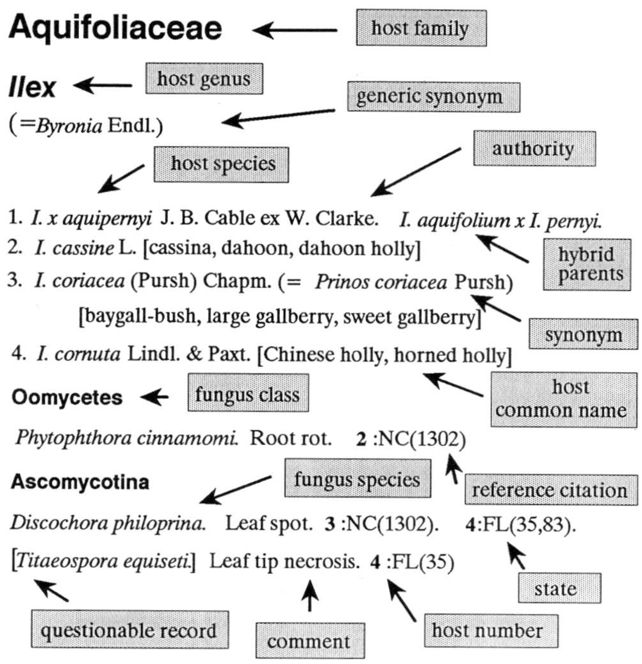

Figure 1. Components of a record in the Host-Fungus List.

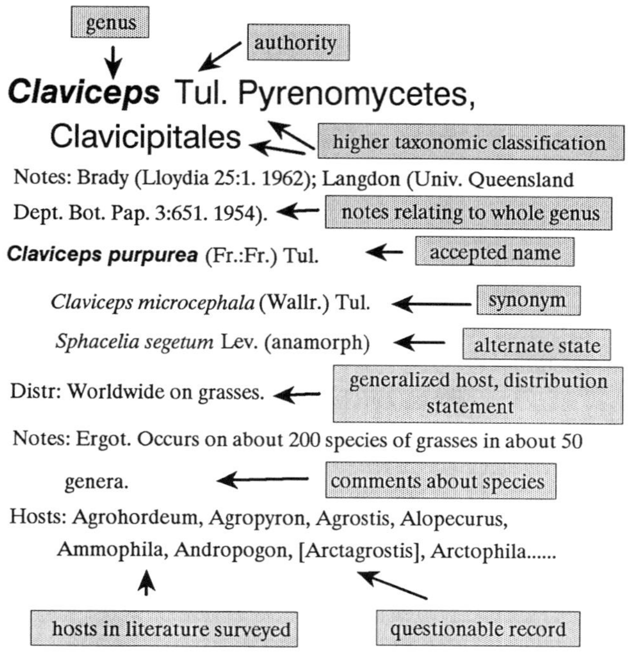

Figure 3. Components of a record in the Fungus List.

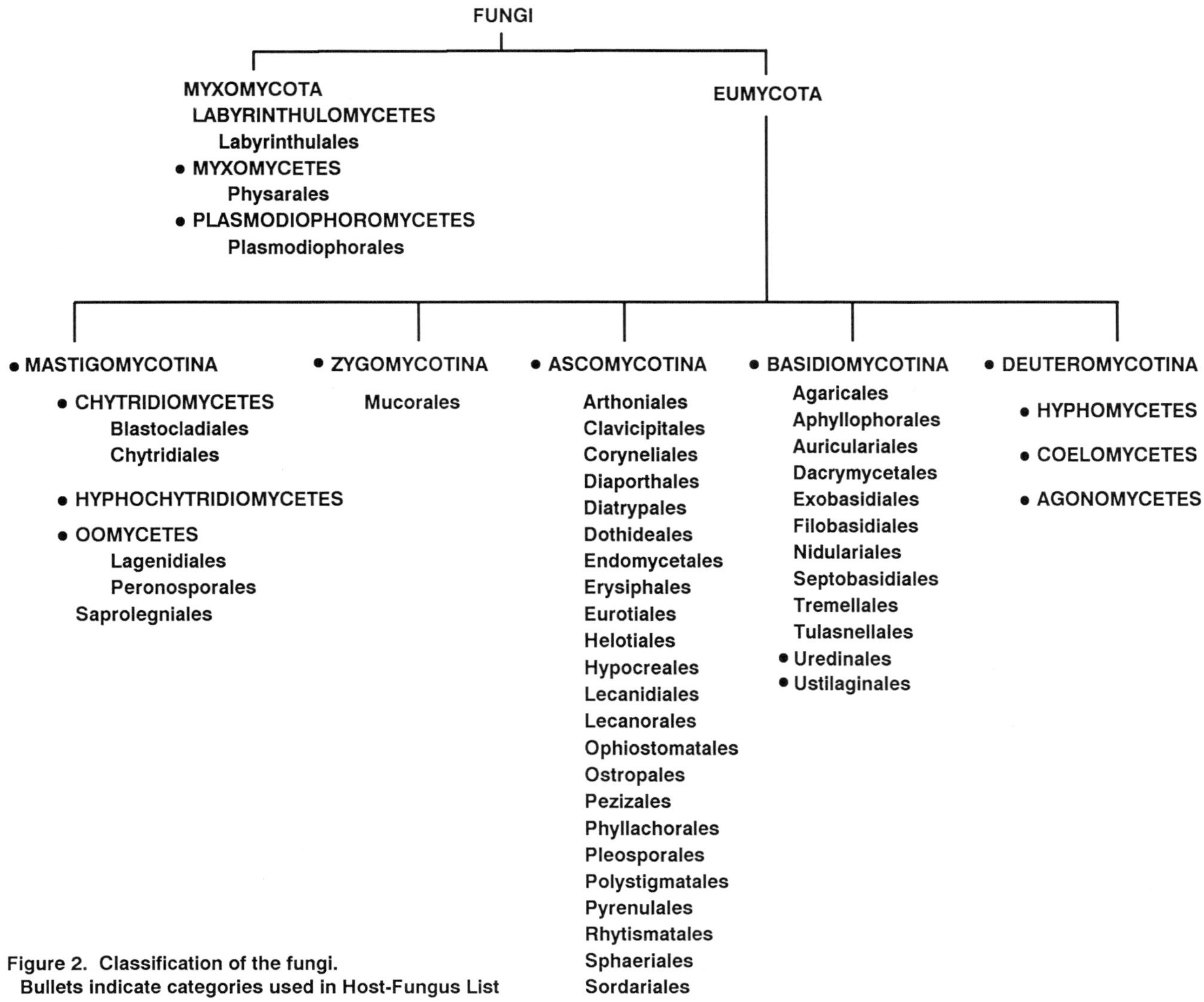

Figure 2. Classification of the fungi.
Bullets indicate categories used in Host-Fungus List

Checklist of Names for 3,000 Vascular Plants of Economic Importance. United States Department of Agriculture Handbook 505 (Revised Edition).
3. Germplasm Resources Information Network, nomenclature file. United States Department of Agriculture.

To complete the review of vascular plant host names, the following literature was consulted:

1. Bailey, L.H., and E.Z. Bailey. 1976. *Hortus Third*. Macmillian Pub. Co., Inc., New York.
2. *Index Kewensis*. 1893-date. Vol. 1-2, supplements 1-18. Clarendon Press, Oxford.
3. Fernald, M.L. 1950. *Gray's Manual of Botany*. American Book, New York.
4. Hitchcock, A.S. 1935. *Manual of the Grasses of the United States*, U.S.D.A., Misc. Publ. 200, Washington D.C.

In addition, several regional floristic studies were consulted to clarify problems not covered by the above-mentioned sources.

This publication is not intended as an authoritative reference on the nomenclature and synonymy of the vascular plant hosts. An attempt was made to find the correct current names for all host plants. The extent of synonomy goes only as far as the names appearing in the fungal-host records. Additional synonymy is beyond the scope of this manual. The common names listed are mostly taken from Terrell et al. (1986), Bailey and Bailey (1976) and the literature searched. No attempt has been made to include all common names for each host.

Fungi with anamorph and teleomorph states. The Host-Fungus List reports the morph cited in the literature. Thus, it is possible that the same fungus will be reported from an individual host species under both the anamorph and teleomorph name. These relationships are not indicated in the Host-Fungus List. The anamorph-teleomorph relationships may be clarified through reference to the Fungus List.

Compiling the Fungus List

The rapidly appearing nomenclatural and taxonomic changes occurring as a result of current research in fungal systematics has created a need for a comprehensive reference concerning the nomenclature and taxonomy of fungi reported on vascular plants. This publication provides a major step toward this goal. However, with the large gaps in the mycological literature and the large number of taxa involved, the full realization of this goal requires additional work. The diffuse nature of much of the systematic mycological information makes it very difficult to review all relevant material concerning a specific fungus. The willingness of many mycologists to annotate the list of fungi in their specific groups has been extremely helpful.

Fungi included and their taxonomic status. The 16,500 fungal names occurring in the literature surveyed formed the basis of the Fungus List. The Fungus List contains 12,666 accepted taxa in 1,988 genera. In addition there are 10,142 synonyms, 864 anamorphs, 543 teleomorphs, and 88 synanamorphs. The synonymy for each species is not intended to be comprehensive but does include those names likely to used in the 20th century North American literature. Decisions concerning nomenclature and synonymy are based upon a review of the relevant mycological literature. For groups where comprehensive monographic studies are available, this was a reasonably straight forward procedure. Frequently, the status and placement of species may be the subject of conflict among authorities, and therefore, choice of placement was based on our interpretation. In these cases, problems associated with the name are indicated in the notes. In many groups, particularly the Ascomycotina and Deuteromycotina, generic concepts have been drastically limited without a systematic revision of all the species named in that genus. As a result, many species are listed in genera that by modern standards would be inappropriate.

Varieties presented a special problem. When a variety is described, two distinct taxa are formed. One is the new variety and the second is the morphological form represented by the original species description (the type variety). *Ascochyta lophanti* var. *lophanti* is the type variety in contrast to *Ascochyta lophanti* var. *lycopina*. If a report gives only the species name, it is generally impossible to assign this fungus to one of the two varieties without examining the material used as a basis for the report. Thus, where a variety or varieties are accepted, entries under the species name alone may include fungi referable to any of the varieties.

For Ascomycotina, the class and order are those presented in Eriksson (*Systema Ascomycetum* 6:3-109. 1987); all other fungi follow Hawksworth et al. (1983. *Dictionary of Fungi*. Commonwealth Mycological Institute, England).

Geographic distribution. The worldwide geographic distribution of each accepted fungal name is given. For well-studied fungi, the distributions listed should be indicative of presently known ranges. In other cases the range of the species may be broadly disjunct, but these gaps in ranges may likely be an artifact of the small number of collections. Geographic data is presented in the following order: United States and Puerto Rico listed by postal codes; Canada including the province where provided; Mexico; West Indies; Central America; South America; all other regions arranged alphabetically.

Host ranges. At the end of each fungus species record there is a generalized statement regarding the host range of the fungus extracted from the mycological literature and a list of the hosts reported in the surveyed literature. Sometimes there is an apparent conflict between the distribution statement or comments and the list of hosts from the United States. This discrepancy may occur for several reasons. Since we did not systematically survey the world's mycological and phytopathological literature, the reports in this book may be only a subset of the distribution for a particular fungus. In some cases, the reports cited may be in error and require reexamination. Clarification of each of these discrepancies would require detailed studies, including examination of the specimens, a task clearly beyond the scope of this book. Indeed one of the principal values of this book is to condense a myriad of literature reports into a form that will stimulate future work to increase our knowledge of fungal distributions and host associations.

Author Abbreviations

Author citations are abbreviated so that each abbreviation is unique within the disciplines of botany and mycology. A list of the fungal authors is provided with the full name and the abbreviation. These author abbreviations may be different from those used in the Handbook 165 because those abbreviations may no longer provide an unambiguous reference to one author. The following guidelines were proposed for determining the appropriate abbreviations and are in agreement with the guidelines established by the International Commission on Botanical Author Abbreviations at the International Botanical Congress (1987):

a. If an author is listed in Stafleu, F. A. & R. S. Cowen (1976-1988. *Taxonomic Literature*. Vol. 1-7. Second ed. Bohn, Scheltema & Holkema. Urecht.) that abbreviation was followed. Of the 2,485 fungal authors used in this book, 27% were included in *Taxonomic Literature*.

b. If the author was not in *Taxonomic Literature*, then the name listed in Meikle, R. D. (1980. *Draft Index of Author Abbreviations compiled at the Herbarium Royal Botanic Gardens, Kew*. Her Majesty's Stationery Office.) was used.

c. If the author was not cited in either of the above references, the last name was generally written out in full with one or two initials included, depending on the perceived frequency of the name. The abbreviation "Hort." is not an author abbreviation. This term indicates a plant name of no botanical standing used by horticulturists. When followed by a personal name (e.g. Hort. Lemoine), it indicates a plant name published in a catalogue of a nursery firm. The notation "of Authors" indicates conventional use of a name that is not necessarily taxonomically correct.

Abbreviations of Journals and Books

Abbreviations of journals are cited according to Lawrence, G. H. M. et al. (1968. *Botanicum-Periodicum-Huntianum*. Hunt Botanical Library, Pittsburgh, PA). Book titles are abbreviated using the guidelines for abbreviating words in *Botanicum-Periodicum-Huntianum* or else they are cited in full.

Glossary of Nomenclatural Terms

An outline of the principles and techniques involved in the naming and describing of fungi can be found in Hawksworth, D. L. (1974. *Mycologist's Handbook*. Commonwealth Mycological Institute, England). Also included in this reference is an explanation of the rules controlling the nomenclature of fungi (International Code of Botanical Nomenclature) and a glossary of nomenclatural terms. Specialized terms used in this book are briefly defined below.

anamorph - asexual form of a fungus.

basionym - the name on which a new transfer or new combination is based.

hom. illeg. (homonym illegitimum) - a validly published name spelled exactly like another validly published name in the same rank but based on a different type. For example, five scientists, using different types, have published the name *Septoria longispora*. Only the earliest name is accepted. The others need to be renamed. Where homonyms exist, we have noted them using the word "non." *Septoria longispora* Shear non Overholts hom. illeg., means we are using the name based on the type published by Shear and not that of Overholts, which was published at a later date.

ICBN - Internation Code of Botanical Nomenclature.

nom. ambig. (nomen ambiguum) - a name used in different senses (e.g., applied to different taxa) that has become a persistent source of error.

nom. conf. (nomen confusum) - a name based on a type specimen consisting of discordant elements.

nom. cons. (nomen conservandum) - a validly published name whose retention has been authorized by an International Botanical Congress and that would have to be replaced by a less well known name by the strict application of the Code.

nom. dub. (nomen dubium) - a name of uncertain application (e.g., because it cannot be satisfactorily typified).

nom. illeg. (nomen illegitimum) - a name validly published but for some other reason is not acceptable. A name contrary to the rules of nomenclature.

nom. inval. (nomen invalidum) - a name not validly published.

nom. nud. (nomen nudum) - a name published without a description or diagnosis and without any reference to a previously published description or diagnosis; consequently not a valid name because it was not validly published.

pro parte - in part.

synanamorph - the name refering to any one of two or more anamorphs that have the same teleomorph.

teleomorph - the sexual form of a fungus.

type - the single specimen of a taxon to which the name is permanently attached.

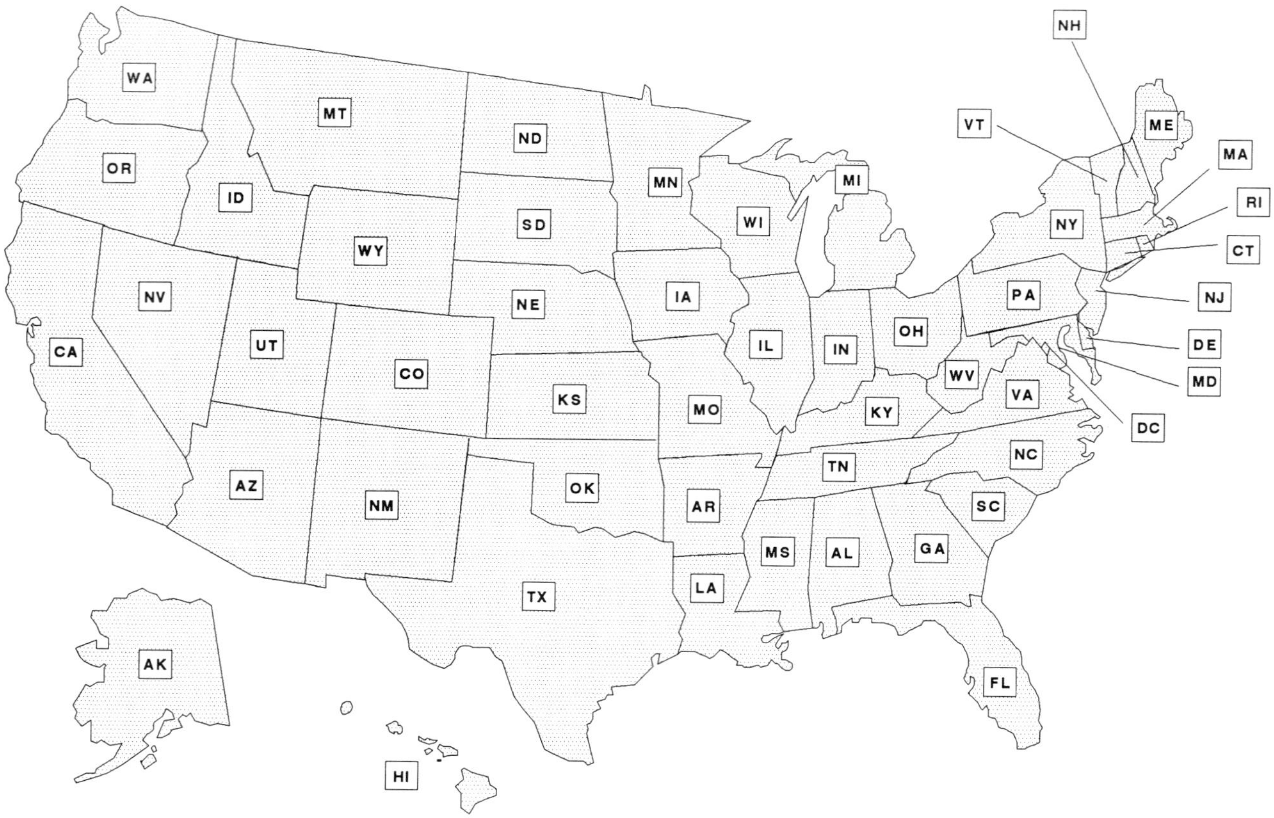

States

AL	Alabama	IL	Illinois	MT	Montana	RI	Rhode Island
AK	Alaska	IN	Indiana	NE	Nebraska	SC	South Carolina
AZ	Arizona	IA	Iowa	NV	Nevada	SD	South Dakota
AR	Arkansas	KS	Kansas	NH	New Hampshire	TN	Tennessee
CA	California	KY	Kentucky	NJ	New Jersey	TX	Texas
CO	Colorado	LA	Louisiana	NM	New Mexico	UT	Utah
CT	Connecticut	ME	Maine	NY	New York	VT	Vermont
DE	Delaware	MD	Maryland	NC	North Carolina	VA	Virginia
DC	District of Columbia	MA	Massachusetts	ND	North Dakota	WA	Washington
FL	Florida	MI	Michigan	OH	Ohio	WV	West Virginia
GA	Georgia	MN	Minnesota	OK	Oklahoma	WI	Wisconsin
HI	Hawaii	MS	Mississippi	OR	Oregon	WY	Wyoming
ID	Idaho	MO	Missouri	PA	Pennsylvania		

Regions

E states	Eastern states	NW states	Northwestern states
NE states	Northeastern states	SW states	Southwestern states
SE states	Southeastern states	E & C states	Eastern and central states
S states	Southern states	C & W states	Central and western states
N states	Northern states	Gulf states	Gulf states
C states	Central states	Lake states	Great Lake states
NC states	North central states	NGP	Northern Great Plains states
SC states	South central states	Unk	Unknown or unspecified locality in the U.S.
W states	Western states	Range of Host	Cosmopolitan, general, widespread, etc.

Map of the United States with state and locality abbreviations.

Host-Fungus List

Acanthaceae

Anisacanthus Nees
1. *A. thurberi* (Torr.) A. Gray [chuparosa, desert-honeysuckle]
2. *Anisacanthus* sp.

Basidiomycotina - Uredinales
Puccinia anisacanthi. Autoecious. **1**:AZ(747,1195,1203,2058,3981). **2**:AZ(2058).

Aphelandra R. Br.
1. *A. squarrosa* Nees [saffron-spike, zebraplant]

Oomycetes
Phytophthora nicotianae var. *parasitica*. Crown, root and stem rot. **1**:FL(35,556).
Pythium aphanidermatum. Root rot. **1**:FL(35).
P. splendens. Root rot. **1**:FL(35).
Pythium sp. Root rot. **1**:HI(2903).

Ascomycotina
Nectriella pironii. Occasionally on leaves. 'Stem gall.' **1**:FL(35).

Deuteromycotina - Hyphomycetes
Alternaria sp. Leaf spot. **1**:FL(35).
Cercospora sp. Leaf spot. **1**:FL(35).
Corynespora cassiicola. 'Corynespora leaf spot.' **1**:FL(35).
Kutilakesa pironii. Leaf and stem galls. **1**:FL(34).
Myrothecium sp. Leaf spot. **1**:FL(35).

Deuteromycotina - Coelomycetes
Colletotrichum sp. Anthracnose. **1**:FL(35).

Deuteromycotina - Other
Rhizoctonia solani. 'Rhizoctonia stem rot.' **1**:FL(35).
Sclerotium rolfsii. 'Southern blight.' **1**:FL(35).

Barleria L.
1. *B. cristata* L. [Philippine violet]
2. *Barleria* sp.

Ascomycotina
Nectriella pironii. Canker, stem gall. **1**:FL(35).

Basidiomycotina - Uredinales
Puccinia sp. **1**:FL(35).

Deuteromycotina - Hyphomycetes
Corynespora sp. Leaf spot. **1**:FL(35).

Deuteromycotina - Other
Sclerotium rolfsii. **2**:HI(2903).

Beloperone Nees in Wallich
This genus is considered a synonym of *Justicia* but the appropriate combination could not be located.
1. *B. californica* Benth.

Basidiomycotina - Uredinales
Uromyces beloperones. Autoecious. **1**:AZ(747,2058,3981), CA(747,2058).
U. ruelliae. Autoecious. **1**:AZ(82,1203), CA(81,82), NV(82).

Blechum P. Br.
1. *Blechum* sp.

Oomycetes
Pythium sp. Root rot. **1**:FL(35).

Ascomycotina
Mycosphaerella pteridis. Leaf spot. **1**:FL(35).

Basidiomycotina - Uredinales
Puccinia lateripes. **1**:FL(35).

Carlowrightia A. Gray
1. *C. arizonica* A. Gray
2. *C. cordifolia* A. Gray
3. *C. glabrata* Fernald

Basidiomycotina - Uredinales
Puccinia anisacanthi. Autoecious. **1**:AZ(747,3981). **2**:AZ(747). **3**:AZ(747).

Crossandra Salisb.
1. *C. infundibuliformis* (L.) Nees [firecracker-flower]
2. *Crossandra* sp.

Oomycetes
Phytophthora sp. Root and stem rot. **1**:FL(35).
Pythium sp. Root rot. **1**:FL(35).

Deuteromycotina - Hyphomycetes
Alternaria sp. Flower spot. **1**:FL(35).
Botrytis cinerea. Gray leaf mold. **2**:FL(35).
Helminthosporium sp. Flower spot. **1**:FL(35).

Deuteromycotina - Other
Rhizoctonia solani. Root and stem rot. **1**:FL(35).

Dianthera L.
This genus is considered a synonym of *Justicia*, but the appropriate combination could not be located.
1. *D. ovata* (Lindau) Walter

Deuteromycotina - Hyphomycetes
Cercospora diantherae. Leaf spot. **1**:OK(2872).

Dicliptera Juss.
1. *D. resupinata* (Vahl) Juss.

Basidiomycotina - Uredinales
Puccinia tetramerii. Autoecious. **1**:AZ(747).

Dyschoriste Nees in Wallich
1. *D. oblongifolia* (Michx.) Kuntze

Basidiomycotina - Uredinales
Aecidium tracyanum. **1**:FL(35,82,2058).

Deuteromycotina - Hyphomycetes
Cercospora consociata. Leaf spot. **1**:FL(35).

Deuteromycotina - Other
Sclerotium rolfsii. Southern blight. **1**:FL(35).

Eranthemum L.
1. *E. pulchellum* Andr. (=*E. nervosum* (Vahl) R. Br. ex Roem. & Schult.) [blue sage]

Deuteromycotina - Coelomycetes
Lasiodiplodia theobromae. Twig dieback. **1**:FL(35).
Phyllosticta sp. Leaf spot. **1**:FL(35), NJ(82).

Deuteromycotina - Other
Rhizoctonia solani. Web blight. **1**:FL(35).

Fittonia E. Coemans
1. *F. verschaffeltii* (Hort. ex Lem.) Coem. [mosaic-plant, nerveplant, silver fittonia, silver nerveplant, silver-threads, silvernet-plant, white-leaved fittonia]
2. *F. verschaffeltii* (Hort. ex Lem.) Coem. var. *argyroneura* (Coem.) Regel
3. *Fittonia* sp.

Oomycetes
Phytophthora sp. Root rot. **1**:FL(35).
Pythium splendens. Root rot. **1**:FL(35). **3**:FL(35).
Deuteromycotina - Hyphomycetes
Alternaria sp. Leaf spot. **3**:FL(35).
Cladosporium sp. Leaf spot. **3**:FL(35).
Curvularia sp. Leaf spot. **1**:FL(35).
Fusarium sp. Root rot. **1**:FL(35).
Deuteromycotina - Coelomycetes
Colletotrichum sp. Anthracnose. **3**:FL(35).
Phyllosticta sp. Leaf spot. **3**:FL(35).
Deuteromycotina - Other
Rhizoctonia solani. Leaf and stem blight, crown and root rot. **1**:'Rhizoctonia stem rot.' FL(35). **2**:OH(943). **3**:FL(35).
Sclerotium rolfsii. Southern blight. **1**:'Southern blight.' FL(35). **3**:FL(35).

Hemigraphis Nees in A. DC.

1. *H. alternata* (N. L. Burm.) T. Anderson (=*H. colorata* (Blume) H. Hallier) [red-ivy, redflame-ivy]
2. *Hemigraphis* sp.

Oomycetes
Pythium sp. Root rot. **1**:FL(35).
Ascomycotina
Sclerotinia sclerotiorum. Leaf and stem rot. **1**:FL(35).
Deuteromycotina - Hyphomycetes
Alternaria sp. Leaf spot. **1**:FL(35).
Corynespora sp. **1**:HI(2903).
Curvularia sp. Leaf spot. **1**:FL(35).
Deuteromycotina - Coelomycetes
Colletotrichum sp. Leaf spot. **1**:FL(35).
Deuteromycotina - Other
Rhizoctonia solani. Root and stem rot. **1**:FL(35).
Sclerotium rolfsii. Southern blight, basal stem rot. **1**:FL(35). **2**:HI(2903).

Hygrophila R. Br.

1. *H. angustifolia* R. Br.
2. *H. difformis* (L. fil.) Blume [water-wisteria]
3. *H. stricta* (Vahl) Nees

Oomycetes
Aphanomyces sp. Leaf rot and spot. **3**:FL(35).
Pythium sp. Black rot of stems, leaf spot. **1**:FL(35). **2**:FL(35). **3**:FL(35).
Deuteromycotina - Other
Rhizoctonia solani. Basal stem rot. **1**:FL(35).

Justicia L.

(= *Beloperone* Nees in Wallich, *Dianthera* L., *Jacobinia* Nees in Moric.)

1. *J. americana* (L.) Vahl (=*Dianthera americana* L.)
2. *J. brandegeana* Wassh. & L. B. Sm. [false hop-plant, junta-de-cobra-pintada, shrimp-bush, shrimp-plant]
3. *J. carnea* Lindl. (=*Jacobinia carnea* (Lindl.) G. Nichols.) [balsamo-cor-de-carne, Brazilian plume, flamingo-plant, king's-crown, paradise-plant, plume-flower, plume-plant]
4. *J. mexicana* (Seem.) J. Rose (=*Jacobinia mexicana* Seem.)
5. *J. runyonii* Small
6. *Justicia* sp.

Oomycetes
Phytophthora sp. Root rot. **6**:FL(35).
Pythium sp. Root and stem rot. **2**:FL(35). **3**:FL(35).
Ascomycotina
Dimerium langloisii. Black mildew. **1**:LA(82), MS(82).
Dothideovalsa diantherae. Black knot. **1**:TX(82).
Basidiomycotina - Uredinales
Puccinia justiciae. Autoecious. **5**:TX(2058).
P. lateripes. **6**:FL(35).
P. tetramerii. **4**:AZ(747).
Uromyces ruelliae. Autoecious. **2**:FL(35).
Deuteromycotina - Hyphomycetes
Alternaria sp. Leaf spot. **2**:FL(35).
Botrytis sp. Leaf blight. **3**:NH(1118).
Cercospora diantherae. Leaf spot. **1**:DE(3543), KS(82), MD(82), OK(2872), TX(82), WV(3907). **3**:FL(35).
C. justiciae. Leaf spot. **6**:FL(35).
C. justiciicola. Leaf spot. **6**:FL(35).
Corynespora cassiicola. Leaf spot. **2**:FL(35). **6**:FL(35).
Fusarium sp. Brown spot on flowers and leaves. **2**:FL(35).
Phymatotrichopsis omnivora. Root rot. **1**:TX(82).
Deuteromycotina - Coelomycetes
Colletotrichum sp. Anthracnose. **2**:FL(35). **6**:FL(35).

Pachystachys Nees in Mart.

1. *P. lutea* Nees

Oomycetes
Pythium sp. Root rot. **1**:FL(35).
Deuteromycotina - Hyphomycetes
Alternaria sp. Leaf spot. **1**:FL(35).
Cercospora justiciicola. Leaf spot. **1**:FL(35).
Corynespora cassiicola. Leaf spot. **1**:FL(35).

Pseuderanthemum Radlk.

1. *P. reticulatum* (J. D. Hook.) Radlk.

Deuteromycotina - Coelomycetes
Phyllosticta sp. Leaf spot. **1**:FL(35).

Ruellia L.

1. *R. ciliosa* Pursh
2. *R. drummondiana* A. Gray
3. *R. humilis* Nutt.
4. *R. makoyana* Hort. Makay ex J. Closon [monkey-plant, trailing velvet-plant]
5. *R. nudiflora* (A. Gray) Urb.
6. *R. parviflora* Britton
7. *R. pedunculata* Torr. ex A. Gray
8. *R. strepens* L.
9. *R. tuberosa* L. [meadow-weed, menow-weed]
10. *Ruellia* sp.

Oomycetes
Pythium sp. Root rot. **4**:FL(35). **10**:FL(35).
Basidiomycotina - Uredinales
Puccinia lantanae. Autoecious. **5**:TX(2058).
P. lateripes. Autoecious. **1**:GA(1439), IL(2058), IN(2058), KS(2058), MS(2720). **3**:E states(82), FL(35), KS(82), TX(82). **5**:TX(82). **7**:MO(2058). **8**:IL(2058), IN(2058), KS(82,2058), MO(2058), MS(2720), OH(2058), TX(82). **9**:TX(82). **10**:FL(35), MO(82), TX(82,652).
P. longiana. Autoecious. **2**:TX(2058). **5**:TX(2058). **10**:TX(2058).
P. ruelliae-parviflorae. Autoecious. **6**:TN(2058).
Deuteromycotina - Hyphomycetes
Botrytis cinerea. Gray mold. **4**:FL(35).
Cercospora consociata. Leaf spot. **1**:AL(82), MS(2720). **8**:OK(2872). **10**:FL(35), IA(82).
Deuteromycotina - Coelomycetes
Ascochyta sp. **1**:WI(1359).
Deuteromycotina - Other
Rhizoctonia solani. Root and stem rot. **3**:FL(35). **4**:FL(35). **10**:FL(35).

Sanchezia Ruiz & Pav.

1. *S. nobilis* J. D. Hook.

Basidiomycotina - Other
Armillaria tabescens. Root rot. **1**:FL(35,82).
Deuteromycotina - Other
Rhizoctonia solani. Root rot. **1**:FL(35).

Strobilanthes Blume

1. *Strobilanthes* sp.

Oomycetes
Phytophthora sp. Damping-off. **1**:FL(35).
Pythium sp. Damping-off. **1**:MD(2852).

Tetramerium Nees in Benth.

1. *T. hispidum* Nees

Basidiomycotina - Uredinales
Puccinia tetramerii. Autoecious. **1**:AZ(1195,1203,2058,3379,3981).

Thunbergia Retz.

1. *T. fragrans* Roxb.
2. *T. grandiflora* (Rottler) Roxb.
3. *Thunbergia* sp.

Basidiomycotina - Uredinales
Puccinia sp. **3:**FL(35).
Basidiomycotina - Other
Armillaria tabescens. Root rot. **2:**FL(35).
Deuteromycotina - Hyphomycetes
Alternaria sp. Leaf spot. **3:**FL(35).
Cercospora sp. Leaf spot. **1:**HI(2899,2903,3074). **3:**FL(35).
Oidium sp. Powdery mildew. **3:**FL(35).
Deuteromycotina - Coelomycetes
Colletotrichum gloeosporioides. Anthracnose. **2:**FL(35). **3:**FL(35).
Diplodia sp. Stem gall. **3:**FL(35).
Deuteromycotina - Other
Sclerotium rolfsii. Southern blight. **3:**FL(35).

Xantheranthemum Lindau in Engl. & Prantl

1. *X. igneum* (Linden) Lindau

Deuteromycotina - Other
Sclerotium rolfsii. Southern blight. **1:**FL(35).

Aceraceae

Acer L.

1. *A. barbatum* Michx. (=*A. floridanum* (Chapm.) Pax) [Florida maple, southern sugar maple, sugar-tree]
2. *A. campestre* L. [field maple, hedge maple]
3. *A. circinatum* Pursh [vine maple]
4. *A. concolor* (Gordon) Hoppe
5. *A. ginnala* Maxim. [Amur maple]
6. *A. glabrum* Torr. [dwarf maple, Rocky Mountain maple]
7. *A. glabrum* Torr. subsp. *douglasii* (Hook.) Wesm. (=*A. douglasii* Hook.) [Douglas maple]
8. *A. glabrum* Torr. var. *neomexicanum* (Greene) Kearn. & Peebles (=*A. neomexicanum* Greene)
9. *A. macrophyllum* Pursh [bigleaf maple, canyon maple, Oregon maple]
10. *A. negundo* L. [ash-leaved maple, box elder]
11. *A. negundo* L. subsp. *californicum* (Torr. & A. Gray) Wesm. (=*A. negundo* L. var. *californicum* (Torr. & A. Gray) Sarg.) [California box elder]
12. *A. nigrum* F. Michx. (=*A. saccharum* Marsh. subsp. *nigrum* (F. Michx.) Y. Desmarais, *A. saccharum* Marsh. var. *nigrum* (F. Michx.) Britton) [black maple, black sugar maple, rock maple]
13. *A. oblongum* Wallich ex DC.
14. *A. palmatum* Thunb. [Japanese maple]
15. *A. pensylvanicum* L. [moosewood, Pennsylvania maple, striped maple, whistlewood]
16. *A. platanoides* L. (=*A. platanoides* L. f. *schwedleri* K. Koch) [Norway maple]
17. *A. pseudoplatanus* L. [mock-plane, sycamore maple]
18. *A. rubrum* L. [red maple, scarlet maple, soft maple, swamp maple]
19. *A. rubrum* L. var. *rubrum* [red maple]
20. *A. rubrum* L. var. *trilobum* Torr. & A. Gray ex K. Koch [trident maple]
21. *A. saccharinum* L. (=*A. dasycarpum* Ehrh.) [river maple, silver maple, soft maple, white maple]
22. *A. saccharum* Marsh. (=*A. saccharophorum* K. Koch) [hard maple, rock maple, sugar maple]
23. *A. saccharum* Marsh. subsp. *grandidentatum* (Nutt. ex Torr. & A. Gray) Y. Desmarais (=*A. grandidentatum* Nutt. ex Torr. & A. Gray) [bigtooth maple , Rocky Mountain maple, sugar maple]
24. *A. saccharum* Marsh. subsp. *leucoderme* (Small) Y. Desmarais (=*A. leucoderme* Small) [chalk maple, whitebark maple]
25. *A. spicatum* Lam. [moose maple, mountain maple]
26. *A. truncatum* Bunge [Shantung maple]
27. *A. truncatum* Bunge subsp. *mono* (Maxim.) A. E. Murray (=*A. mono* Maxim.) [mono maple]
28. *Acer* sp.

Myxomycetes
Comatricha typhoides. **28:**NC(3944).
Oomycetes
Phytophthora cactorum. Bleeding canker. **12:**MA(82), NJ(82), RI(82). **16:**CT(82), MA(82), NJ(82), NY(82), RI(82), VA(3493). **17:**RI(82). **19:**MS(82), NC(1303,3657), NJ(82), RI(82), TN(3657). **22:**MA(82), NC(3657), NJ(82), RI(82), TN(3657).
P. cambivora. Basal canker. **16:**NJ(82).
Phytophthora sp. On submerged leaves. **21:**IL(3218).
Pythium sp. Root rot. **16:**HI(2903). **19:**FL(35).
Zygomycotina
Mortierella isabellina. Isolated from fallen leaves. **22:**WI(2005).
Mortierella sp. Isolated from fallen leaves and wood. **19:**ME(2860), NH(2860). **22:**WI(2005).
Mucor sp. Isolated from wood. **19:**ME(2860), NH(2860).
Ascomycotina
Acarosporina microspora. Trunk canker. **19:**Lake States(802), MI(82), MN(82), WI(82). **22:**Lake States(802), MI(82), MN(82,2874), WI(82). **28:**NJ(3225), NY(3225).
Acrospermum foliicola. **10:**GA(1438).
Amphisphaeria sp. **6:**ID(3202).
Anthostoma saprophilum. **10:**GA(1438).
Asteridiella negundinis. On living twigs and small limbs. **10:**FL(3483).
Bertia moriformis. On wood. **9:**OR(3202). **21:**OH(684). **28:**OR(3202).
Biscogniauxia nummularia. On wood. **10:**IA(1225).
Bisporella citrina. **3:**WA(3202). **6:**ID(3202), WA(638,3202).
Botryosphaeria dothidea. **9:**WA(3202). **10:**GA(1438). **19:**GA(1438). **21:**GA(1438).
B. obtusa. Twig canker. **10:**GA(82), IA(82), IL(82), MD(82), MN(209), OK(82,2873), PA(82), TN(82). **19:**GA(82), MD(82), VA(82). **21:**IN(82).
B. ribis. On twigs. **17:**NE(82).
B. stevensii. **28:**FL(3213), NC(3213), SC(3213).
B. trames. **9:**WA(3202).
Botryosphaeria sp. Associated with dieback. **14:**NC(1303). **21:**FL(35).
Brunnipila clandestina. Dieback, leaf spot. **3:**WA(1834,3202).
Bulgaria inquinans. On fallen trunk. **28:**NY(2463).
Byssochlamys nivea. Isolated from wood. **28:**IA(1228).
Calosporella sp. Dead limbs. **19:**NC(1303).
Camarops polysperma. **10:**GA(1438).
C. pugillus. **28:**NC(2561), NY(2561).
Capitotricha bicolor. On dead limbs and twigs. **6:**CA(642), MT(3202).
Capnodium salicinum. On dead twigs and limbs. **10:**MT(3202).
Cenangium episphaeria. **19:**GA(1438).
C. griseum. **19:**PA(2677).
Ceratocystis coerulescens. Sap streak. **22:**MI(1881), NC(82,1303,1527,1533,3047), NY(236), VT(1646).
Ceratocystis sp. Isolated from wood. **19:**ME(2860), NH(2860).
Chlorociboria aeruginosa. Green stain of wood. **6:**ID(3202), WA(3202). **15:**VT(82). **22:**Range of host(82), ME(82), MN(2874), VT(82).
Ciboria acerina. Inflorescence blight. **19:**MA(82), NY(82). **21:**NY(82,1960).
Cistella dentata. **28:**WA(3202).
Claussenomyces prasinula. On wood. **28:**NJ(850).
Coniochaeta ligniaria. **9:**OR(3202).
C. pulveracea. **26:**GA(1438).
Crocicreas cyathoideum var. *olympianum*. On petioles. **3:**WA(505).
Cryptendoxyla hypophloia. Under bark. **22:**NY(1188,2550).
Cryptodiaporthe acerinum. On limbs. **25:**MA(2179), NJ(2179).
C. lebiseyi. On limbs. **11:**CA(82).
C. petiolophila. On limbs, petioles and small twigs. **15:**NY(82). **22:**GA(445,1438). **25:**NH(82), NY(82).
Cryptodiscus pallidus. On wood. **28:**NY(3225).
Cryptosporella acerina. On limbs. **6:**CA(82,639,642).
C. hypodermia. On decorticated wood. **21:**IA(1227).
Cucurbitaria ailanthi. **10:**ND(377).
Daldinia concentrica. White rot. **10:**VA(82). **19:**NE states(82), MI(82). **21:**VA(82). **24:**GA(82,1438). **28:**IA(1225).
D. grandis. On wood. **9:**ID(82), WA(82,3202).
D. vernicosa. White rot. **19:**Range of host(82), MD(82), MO(82).
Dasyscyphus acerinus. On leaves. **19:**GA(1438). **28:**NJ(849).
Dasyscyphus sp. **15:**GA(1438).
Dermea acerina. On limbs. **19:**GA(82), ME(82), MI(82), NH(82,2673), NJ(82), NY(82), PA(82,2673), VA(82), WV(3907). **21:**NY(82). **22:**NY(82,1376). **28:**MA(664), NY(1376).

Diaporthe acerina. On dead limbs and trunks. **21:**IA(82,1227). **22:**IA(1227). **25:**IA(82), NC(1303,1307), NH(82), NY(82), VA(82).
D. dubia. On limbs. **10:**GA(1438). **21:**MI(82), MO(82), NH(82), NY(82), OH(82), VT(82). **22:**IA(1227), MI(82), MO(82), NH(82), NY(82), OH(82), VT(82). **24:**GA(82,1438).
D. eres. **19:**FL(2782). **21:**IA(82). **28:**IA(1227).
Diaporthella sphendamnina. On limbs. **19:**PA(3807).
Diatrype albopruinosa. On dead limbs. **8:**NM(1013). **12:**IA(3629). **21:**IA(3629). **28:**IA(3629).
D. asterostoma. On dead limbs. **19:**NC(1303).
D. bullata. On dead limbs. **6:**ID(3202).
D. flavovirens. **9:**OR(3202), WA(3202).
D. oregonensis. **28:**OR(3202).
D. stigma. On dead wood. **6:**WA(3202). **12:**IA(3629). **19:**GA(1438), NC(1303,1307), SC(1307). **28:**IA(3629), NC(1303,3944).
Diatrypella annulans. On dead wood. **19:**GA(1438).
D. favacea. **10:**GA(1438). **19:**GA(1438).
D. frostii. **12:**IA(3629). **28:**IA(3629), IL(1243).
D. quercina. **19:**GA(1438).
D. subfulva. **19:**GA(1438).
Dictyoporthe acerophila. On limbs. **22:**NH(184).
Discohainesia oenotherae. Leaf spot. **19:**NC(82,1303), VA(82). **22:**NY(82).
Discosphaerina fagi. On overwintered leaves. **22:**MA(179).
Discostroma muricata. **9:**WA(3202).
Dothidea puccinioides. **10:**GA(1438).
Endothia gyrosa. **21:**VA(2365,2996).
Endoxylina allantospora. On limbs. **10:**CO(82,3254), KS(82), OK(2872).
Eutypa lata var. *aceri*. **10:**GA(1438).
E. limaeformis. **28:**MI(2925).
E. maura. **6:**ID(3202).
Eutypa sp. **21:**WA(3202).
Eutypella angulosa f. *negundinis*. **10:**ND(377).
E. dissepta. **21:**IA(3629).
E. juglandicola. **10:**GA(1438). **19:**GA(1438). **24:**GA(1438).
E. leprosa. Bark canker. **10:**IA(3629), NC(1303), ND(377). **12:**IA(3629). **16:**IA(3629). **21:**IA(3629), MN(2874), NY(82). **22:**MN(82,2874). **28:**IA(3629).
E. parasitica. Canker. **10:**MN(2874). **12:**NY(2457). **17:**NY(1117). **19:**Lake States(802), WI(1773). **21:**MN(2874), NY(1117,2457). **22:**Lake States(802), MI(82), MN(82,2874), NH(82,802), NY(82,802), VT(82,802), WI(82,1773). **28:**MN(1242), WI(1921).
E. scoparia. Canker. **19:**SC(49,2925).
Glonium stellatum. **28:**NY(2456).
Glyphium corrugatum. **6:**ID(3202), OR(3202), WA(3202). **9:**WA(3202).
G. elatum. **3:**WA(3202). **6:**ID(3202).
Godronia urceolus. **3:**WA(1834).
Gorgoniceps iowensis. On dead limbs. **19:**NC(1303,1307).
Graphostroma platystoma. On limbs. **19:**GA(1438).
Hainesia lythri. Isolated from fallen leaves. **22:**WI(2005).
Herpotrichia macrotricha. **25:**NY(189).
Holwaya mucida. **28:**NY(2456).
Hyalinia crenato-marginata. **3:**WA(1834).
Hymenoscyphus caudatus. On decaying leaves. **28:**Unk.(3874).
H. epiphyllus. **28:**NC(1307), VA(2130).
Hypocrea citrina. **19:**NC(3944).
H. schweinitzii. **19:**GA(1438).
Hypoderma pacificensis. **9:**OR(3202,3586).
H. rufilabrum. **28:**NE states(3586).
Hypoxylon albocinctum. **19:**MI(2862).
H. atropunctatum. **19:**GA(1438).
H. caries. **28:**NJ(2403).
H. cinereo-lilacinum. On wood. **10:**IA(1225).
H. crocopeplum. **19:**GA(1438).
H. fuscum. On wood. **9:**WA(1811,3202). **19:**MI(2862). **28:**IA(1225).
H. grenadense var. *macrospora*. On bark. **3:**WA(3202). **9:**WA(1811,3013,3202).
H. howeianum. On wood. **19:**NC(1303).
H. hypophlaeum. **19:**GA(1438).
H. investiens. **9:**WA(3202). **10:**GA(1438).
H. jecorinum. On bark. **19:**SC(128).
H. mammatum. Trunk canker. **19:**MI(82), MN(82). **22:**MI(82), MN(82,2874).
H. mediterraneum. On wood. **3:**WA(1811,3202). **12:**IA(1225). **28:**IA(1225).
H. multiforme. On bark. **3:**WA(3202). **9:**WA(1811,3202).
H. regale. On wood. **28:**OH(2403).
H. rubiginosum. **15:**GA(1438). **19:**GA(1438,2399). **28:**NC(3944).
H. rubiginosum var. *ferrugineum*. **9:**WA(1811,3202).
H. rubrostromaticum. **10:**GA(1438).
H. serpens. **9:**WA(2786). **19:**GA(1438).
H. serpens var. *macrospora*. On wood. **6:**ID(3202). **7:**ID(1811). **9:**WA(3202). **28:**ID(2786).
H. tinctor. **1:**GA(1438). **10:**GA(1438). **19:**GA(1438). **24:**GA(1438).
H. truncatum. **19:**GA(1438).
H. vogesiacum. On living and dead limbs. **6:**ID(1811,3202). **9:**WA(1811,3202).
Hypoxylon sp. On living and dead limbs. **10:**Unk.(82). **12:**Unk.(82). **22:**NY(2456).
Hysterographium flexuosum. **19:**GA(1438).
H. fraxini. On decorticated wood. **6:**CO(3549).
Hysterographium sp. **6:**ID(3202), WA(3202). **9:**WA(3202).
Hysteropatella sp. **6:**ID(3202).
Kirschsteiniothelia acerina. Associated with absorbing, mycorrhizal rootlets. **22:**NY(3043).
K. recessa. On wood. **19:**NY(195).
Laestadia brunnea. Leaf spot. **19:**NC(82,1303,3944), SC(82,634).
L. glaucescens. Leaf spot. **19:**SC(82).
Lanzia luteovirescens. On dead leaves and petioles. **28:**NY(1960).
L. rufescens. On leaves. **3:**WA(911). **28:**WA(911,1833).
Lasiosphaeria spermoides. **9:**WA(3202).
L. strigosa. **9:**WA(3202).
Lophiostoma angustilabrum var. *crenatum*. **10:**GA(1438).
Lophodermium punctiforme. On fallen leaves. **21:**NY(82).
Lulworthia grandispora. On wood submerged in sea water. **28:**NC(1791).
L. submersa. On panels submerged in sea water. **28:**NC(1791).
Macrodiaporthe everhartii. On dead limbs. **19:**KS(82), MA(82), PA(82). **21:**IA(1227), KS(82), MA(82), PA(82). **28:**IA(1227), NJ(82).
Massaria inquinans. Leaf spot. **9:**WA(3202). **19:**GA(1438). **24:**GA(1438). **28:**MS(2720).
M. lantanae. **10:**GA(1438).
Massarinula lignorum. On limbs. **6:**CA(639,642).
Massariovalsa sudans. On dead limbs. **16:**IA(1227). **19:**GA(82,1438), MA(82), MI(82), NJ(82), NY(82), SC(82). **21:**GA(82), IA(1227), MA(82), MI(82), NJ(82), NY(82), SC(82). **22:**IA(1227). **28:**IA(1227).
Melanconis sp. On dead limbs. **10:**Unk.(82). **12:**Unk.(82). **19:**Unk.(82). **21:**Unk.(82). **22:**Unk.(82).
Meliola amphitricha. **10:**GA(1438).
Microsphaera penicillata. **9:**OR(82,3202), WA(3202).
Moellerodiscus lentus. On leaves. **28:**NY(850).
Mollisia cinerea. **28:**VT(1224).
Mycosphaerella punctiformis. **19:**GA(1438). **22:**GA(82), NY(82). **24:**GA(1438).
M. septorioides. **22:**NY(82).
M. sphaerellula. On twigs. **15:**NY(82,179).
Mycosphaerella sp. On leaves. **3:**CA(82). **22:**GA(1438).
Mycothyridium lividum. **9:**WA(3202).
Myriangium duriaei. **19:**NC(3944).
Nectria cinnabarina. Dieback, canker. **6:**AZ(1195). **9:**CA(82), OR(82), WA(82,3202). **10:**E & C states(82), CO(1195), GA(1438). **12:**Range of host(82). **14:**NE states(82). **16:**NE states(82). **17:**NE states(82). **19:**Range of host(82), GA(1438). **21:**Range of host(82). **22:**NC states(82), NE states(82), MA(1781).
N. coccinea. Secondary canker. **15:**VT(82). **19:**NE states(82), NY(82). **22:**NE states(82), NY(82).
N. flammea. **28:**Unk.(3166).
[*N. galligena*]. Trunk canker. **3:**OR(82). **9:**OR(82), WA(82). **19:**Lake states(82), NE states(82), MN(2240). **22:**Lake states(82), NE states(82).
N. ochroleuca. Bark canker; fungus infects through unprotected wounds, small injuries and leaf scars. Can result in significant reduction in volume of merchantable timber. **3:**WI(3085). **28:**PA(3085).

N. oropensoides. **19:**NY(3085).
N. peziza. On dead stump. **17:**NY(3085). **22:**NH(3085), NY(3085).
N. purtonii. On dead stumps. **22:**NY(3085). **28:**NY(891,3085).
N. verrucosa. **10:**ND(377).
N. vulpina. **28:**NJ(3085).
Niesslia pulchriseta. On wood chips. **28:**NY(195).
Nummularia sp. **19:**GA(1438).
Ophiostoma piceae. Isolated from wood chips. **22:**ME(1005).
O. piliferum. On bark. **28:**NJ(82), OH(82).
O. rostrocoronatum. On wood. **22:**ME(3714).
Patellaria lecideola. **3:**WA(1834,3202).
Pezicula acericola. On limbs. **15:**PA(82). **22:**ME(82), VT(82). **25:**NH(82), NY(82,1376), VA(82).
P. carnea. On twigs. **19:**GA(82,1438), MD(82), NC(3482), NH(82), NJ(82,1378), NY(82,1378), PA(82).
P. spiculata. **25:**NY(3167,3168).
P. subcarnea. **15:**ME(82,1378).
Peziza repanda. **28:**NY(2456).
Pezizella vulgaris. **28:**NY(3874).
Phacidium negundinis. On twigs. **10:**IL(3589).
[*Phyllactinia guttata*]. Powdery mildew. **3:**ID(3202). **6:**ID(3202). **9:**CA(82), OR(82). **10:**CO(1195), NM(1003), SD(82), WY(1195). **21:**NC(82,1303), WI(82). **22:**IA(82), VT(82).
Physalospora abdita. Powdery mildew. **19:**FL(82), GA(82), SC(82).
Physalospora sp. On branches. **12:**NC(1303). **15:**NC(82).
Plagiostoma acerophilum. On overwintered petioles. **15:**NY(269). **19:**NY(269). **28:**NJ(269), NY(269).
Platystomum compressum. On overwintered petioles. **6:**WA(3202).
P. compressum var. *pseudomacrostomum*. On wood. **28:**NJ(959).
Pleomassaria acericola. On limbs. **28:**PA(2779).
Pleosphaeria echinata. **19:**GA(1438).
Pleospora sp. Leaf spot. **10:**FL(35).
Pleuroceras tenellum. On fallen leaves. **1:**GA(445,1438). **19:**GA(82,1438), NC(82,1303), NJ(82). **22:**GA(82,1438). **24:**GA(445,1438).
Poculum fraternum. On decaying petioles. **28:**NY(3874).
P. sydowiana. **28:**OR(3202).
Propolomyces versicolor. On dead limbs. **3:**OR(3202). **19:**NC(1303,1307).
Prosthecium appendiculatum. On dead limbs. **10:**GA(82), IA(1227). **19:**GA(82,1438).
P. stylosporum. On limbs. **25:**NH(82,3809), NY(82,3808,3809).
Pseudotrichia mutabilis. On dead limbs. **19:**NH(3296). **28:**Unk.(3166).
Pycnopeziza sympodialis. On buds, flowers and decaying leaves. **19:**IA(3549), NY(35,3549,3875), WV(3549).
Pyrenopeziza leucodermis. On fallen leaves. **1:**GA(1438). **19:**GA(82,1438). **24:**GA(82,1438).
Rebentischia massalongii. On limbs. **15:**ME(186).
Rhizodiscina lignyota. **28:**WA(1834,3202).
Rhynchostoma tinctum. **9:**OR(3202), WA(3202).
Rhytisma acerinum. Tar spot on leaves. **1:**GA(1438). **11:**OR(82), WA(82). **16:**NY(82), OH(82), VA(3493). **19:**Range of host(82), GA(1438), MS(2720), NC(1303,3944), NY(2456), OK(2873), VA(3493). **21:**Range of host(82), MS(2720), NC(1303), NH(3780), SD(2243), VA(3493). **28:**IL(1052), MI(2862), MS(2720), OK(2872).
R. punctatum. Speckled tar spot on leaves. **3:**OR(82). **6:**ID(82), OR(82), WA(82). **9:**CA(81,82), OR(82), WA(82). **10:**NY(82). **11:**CA(82). **15:**Range of host(82), VA(3493). **19:**NC(1303), PA(82). **21:**IN(82), NC(82,1303), WI(82). **22:**IN(82), NC(1303), PA(82), VA(82,3493). **23:**UT(1195). **25:**Range of host(82).
Rhytisma sp. Tar spot. **16:**NJ(1667), NY(1667), OH(3780).
Rosellinia aquila. On dead limbs. **19:**NC(1303). **28:**NC(3944).
R. glandiformis. On dead limbs. **10:**GA(1438).
Rutstroemia sp. **28:**NY(2456).
Schizothyrium perexiguum. On twigs. **22:**IN(82).
S. pomi. Fly speck. **9:**CA(924).
Schizoxylon compositum. Fly speck. **28:**NY(3225).
Scutellinia scutellata. **28:**NY(2456).
Sphaerognomonia carpinea. On fallen leaves. **19:**GA(82,1438,2410).
Sphaerotheca fuliginea. Powdery mildew. **2:**CA(1155). **3:**CA(1155). **9:**CA(1155). **23:**Unk.(3976), CA(1155).
Splanchnonema pupula. On limbs. **16:**Range of host(82). **17:**Range of host(82). **28:**PA(188).
Stictis radiata. **28:**CA(3225).
Tapesia fusca. On dead limbs. **25:**NC(1303,1307).
Taphrina aceris. Leaf blister. **23:**UT(82,831,2453,2454).
T. carveri. Leaf blister. **21:**AL(82,1750,2454), KS(3361), KY(82), MI(82,1750), MO(82,1740,1750,1967,2454).
T. darkeri. **3:**OR(82,2454,3202). **6:**ID(3202).
T. dearnessii. Black leaf blister. **19:**AL(1750), GA(82,1438), MI(82,1750,2454,3397), MN(82,2454,2874), NC(82,1303,1750,2454), NY(82,1750,2454,3361), OK(82,1750,2872), PA(82,1750), VA(82,2454), WI(1356). **25:**NY(82,1750).
T. letifera. Leaf blight. **19:**NC(1303). **25:**MA(82,2454), MN(82,2454), NC(82,2454), NY(82,1750,2454,3361).
T. sacchari. Brown leaf blister. **12:**OH(82,1750). **22:**AR(82,1750,2454), GA(82,1438,1750), IN(82,1750), KS(82,2454,3361), ME(82,1750,2454), MI(82,1750), MO(82,1750), NH(82,1750), NY(82,1743,1750), OH(82,1750), PA(82,1750,2454), TN(82,1750), WI(82,1329,1342,1750), WV(82,1750). **28:**UT(3620).
T. wiesneri. Leaf blister. **16:**MS(2720).
Taphrina sp. Leaf blister. **16:**MS(82). **19:**MD(3780), VA(3780). **22:**VA(3780). **28:**NC(1303).
Teichospora megastega. **6:**MT(3202).
T. negundinis. **6:**CO(1195).
Thyridaria rubronotatum. On bark. **28:**VT(272).
Thyronectria pyrrhochlora. On limbs. **22:**GA(1438). **28:**Unk.(3166).
Triblidium caliciiforme. **28:**WA(3202).
Tubeufia pezizula. **19:**GA(1438), NC(3944).
Tympanis truncatula. Erumpent through bark of twigs. **19:**NC(1307). **28:**NH(1380).
Uncinula bicornis. **19:**GA(1438).
U. circinata. Powdery mildew. **12:**IN(82). **15:**NY(82). **19:**E states(82), GA(1438), MI(82), MO(82), MS(2720), NC(1303,3944). **21:**E & C states(82), IA(82). **22:**E & C states(82), IA(82), NC(1303). **25:**MI(82), NY(82), PA(82), WI(1345).
Ustulina deusta. White heart and butt rot. **10:**NC states(82), NE states(82). **12:**NC states(82), NE states(82). **19:**NC states(82), NE states(82). **21:**NC states(82), NE states(82). **22:**NC states(82), NE states(82).
Valsa ambiens. On dead limbs. **19:**GA(82,1438), NJ(3415), WV(3415). **21:**IA(1226), WV(3415). **22:**IA(1227). **25:**NJ(82), NY(82), WV(82). **28:**NJ(3415).
V. ambiens subsp. *leucostomoides*. Decay and discoloration of sapwood after tapping. **10:**CO(3415), IL(3415), KS(3415), MI(3415), NE(3415), NY(3415), SD(3415). **21:**IA(1226), NY(3415). **22:**NY(3415), VT(3452). **28:**NY(3415).
V. ceratosperma. On dead limbs. **9:**WA(831,3202). **16:**IA(1226). **19:**GA(1438), NJ(3415), PA(2673), WV(3415,3907). **22:**IA(1226).
V. sordida. **6:**ID(3202), MT(3202), WA(3202). **28:**Unk.(82).
Valsaria exasperans. **28:**IA(1226).
V. moroides var. *acerina*. **10:**ND(377).
V. rubricosa. **9:**WA(3202). **22:**GA(1438). **24:**GA(1438).
Venturia acerina. Scab, leaf spot. **19:**NY(82), VA(82), WV(82,3907). **21:**NC(1303), WI(82). **22:**MN(2874), PA(82), WI(82).
Vibrissea truncorum. On samaras. **28:**NY(638).
Xylaria acuta. On fallen trunk. **28:**NY(3017).
X. bulbosa. On wood. **19:**NY(3015). **22:**VT(3015).
X. cornu-damae. On wood. **28:**NH(3017), NY(3017), WI(3017).
X. cubensis. On wood. **19:**GA(1438). **28:**NC(3016).
X. hypoxylon. **3:**WA(3202). **9:**WA(3202). **19:**GA(1438).
X. longipes. On wood. **28:**ME(3015).
X. mali. Root rot. **16:**VA(82).
X. polymorpha. **22:**NY(3018). **28:**MD(3018), NY(2456).

Basidiomycotina - Other

Abortiporus biennis. White rot. **10:**MT(3202). **21:**IA(1004).
A. fractipes. White rot. **28:**Unk.(1204).
[*Acanthophysium lividocaeruleum*]. **6:**NM(128,1190).
Agaricus subrufescens. **21:**IA(1004).
Amphinema byssoides. White rot. **6:**AZ(1202), NM(1190).
Amylocorticium subincarnatum. **9:**WA(3202).
[*Amylostereum chailletii*]. Usually reported on conifers. **9:**WA(3202).
Antrodia albida. Brown rot; often on fence posts and structural timbers. **10:**Range of host(82), MT(128). **12:**Range of host(82). **19:**Range of host(82). **21:**Range of host(82). **22:**Range of

host(82).
A. malicola. Brown rot. **10:**Range of host(82). **12:**Range of host(82). **19:**Range of host(82). **21:**Range of host(82). **22:**Range of host(82). **28:**MN(1197).
Antrodiella semisupina. On wood. **6:**MT(3202,3832).
Armillaria mellea. White rot. **12:**Range of host(82). **16:**NJ(82), PA(82). **19:**NC states(82), NE states(3240), Range of host(82), MA(59), NC(3944). **21:**NC states(82), NE states(82), Range of host(82). **22:**Range of host(82), CT(59), NY(59), VT(59). **28:**CA(2898), NY(59), OK(2872).
A. tabescens. Root rot. **10:**GA(2963). **19:**FL(35). **21:**GA(2684), MO(82).
Athelia coprophila. **28:**MN(1197).
A. salicum. On wood. **3:**WA(128).
Auricularia auricula. On wood. **6:**CA(642). **10:**MO(914).
A. fuscosuccinea. On wood. **10:**MD(128).
Auriporia aurea. On wood. **28:**NY(128).
Bjerkandera adusta. White rot. **10:**Range of host(82), ND(378). **12:**Range of host(82). **19:**Range of host(82). **21:**Range of host(82). **22:**Range of host(82). **28:**NC(3944).
B. fumosa. White rot. **19:**MD(82).
Calocera cornea. On fallen log. **9:**WA(3202). **28:**NH(2671).
Candelabrochaete septocystidia. White rot. **22:**MI(432). **28:**TN(432).
Ceraceomyces serpens. **6:**ID(128).
Ceriporia purpurea. **9:**WA(3202). **19:**NY(128).
C. spissa. **9:**WA(3202).
Cerrena unicolor. **3:**WA(82). **6:**MT(82). **10:**E & C states(82). **12:**E & C states(82). **19:**E & C states(82), NE states(3240), MA(1394). **21:**E & C states(82). **22:**E & C states(82). **23:**MT(82). **28:**NY(2456).
Cheimonophyllum candidissimum. On wood. **6:**WA(128).
Chondrostereum purpureum. On wood. **9:**WA(543,638). **10:**Range of host(82), IN(543). **12:**Range of host(82). **19:**Range of host(82). **21:**Range of host(82). **22:**Range of host(82), NY(543). **25:**VT(543). **28:**IN(543), NY(543).
Claudopus corticincinctus. **3:**WA(3202).
Clavicorona pyxidata. **21:**NY(884). **28:**TN(884).
Climacocystis borealis. **22:**VT(2536).
Climacodon pulcherrimus. On wood. **10:**OH(82).
C. septentrionalis. White trunk rot of living trees. **19:**Range of host(82). **21:**Range of host(82), IA(1004). **22:**MN(2874). **28:**PA(128).
Collybia sp. White trunk rot of living trees and wood. **21:**IA(1004).
Confertobasidium olivaceo-album. White rot, on wood. **6:**AZ(1178,1202).
Conocybe blattaria. White rot. **9:**OR(3202).
Coprinus atramentarius. **21:**IA(1004).
C. micaceus. Root and butt rot of living trees. **21:**WV(3717).
Coprinus sp. On wood. **22:**IL(3786).
Coriolopsis gallica. White rot. **10:**Range of host(82). **12:**Range of host(82). **19:**Range of host(82). **21:**Range of host(82). **22:**Range of host(82).
C. rigida. On wood. **28:**Unk.(2681).
Coronicium alboglaucum. **28:**Unk.(2115).
Crepidotus contortus. On decaying twigs. **3:**WA(1544).
C. hamulatus. On wood. **28:**MI(1544).
C. versutus. **3:**WA(3202). **9:**WA(3202).
Cyathus striatus. **28:**WA(3202).
Cylindrobasidium corrugum. On wood. **6:**AZ(1178), NM(128,1190).
C. laeve. On bark. **15:**VT(128). **25:**NY(128). **28:**NY(543).
C. torrendii. **9:**WA(3202).
Cyphellopsis anomala. **19:**NH(128).
Cystostereum murrayi. Canker, heart rot. **15:**NH(543), NY(543), VT(543). **19:**NE states(82), NH(543). **21:**NH(543). **22:**NE states(82), NY(543). **25:**NH(543). **28:**IA(543).
C. pini-canadense subsp. *subabruptum*. White rot. **9:**WA(128,3202).
Cytidia patelliformis. **9:**WA(3202).
Dacrymyces capitatus. **6:**ID(3202).
Dacryopinax elegans. On wood. **28:**IA(128).
D. spathularia. On stumps. **28:**AR(128).
Daedalea quercina. **21:**VT(82).
Daedaleopsis confragosa. White rot. **10:**E & C states(82). **12:**E & C states(82). **15:**VT(82). **19:**E & C states(82), NC(3944). **21:**E & C states(82), NC(1303). **22:**E & C states(82).
Datronia mollis. White rot. **10:**Range of host(82). **12:**Range of host(82). **19:**Range of host(82). **21:**Range of host(82), IA(1004). **22:**Range of host(82).
D. scutellata. White rot. **10:**ND(82,378). **19:**ME(82), VA(82).
D. stereoides. On dead limbs. **28:**Unk.(2681).
Dendrocorticium roseocarneum. White rot. **10:**PA(82), TX(82). **12:**PA(82), TX(82). **19:**CT(128), PA(82), TX(82). **21:**PA(82), TX(82). **22:**PA(82), TX(82). **28:**NH(2671).
Dendrophora albobadia. White rot. **6:**AZ(1202). **10:**AZ(1202), IA(543), IN(543).
D. erumpens. White rot. **19:**VA(543).
D. versiformis. White rot. **10:**Range of host(82). **12:**Range of host(82). **19:**Range of host(82), CT(543), DC(543), GA(491,543,1439). **21:**Range of host(82). **22:**Range of host(82). **28:**TN(543).
Dendrothele acerina. On bark. **22:**ME(2084). **28:**IA(2084), MA(82), VT(82).
D. alliacea. On bark. **22:**ME(2084). **28:**NY(2084).
D. candida. On bark. **9:**CA(2084). **19:**NC(1303,1307). **22:**MO(2084). **28:**IN(2084), WV(2084).
D. dryina. On bark. **28:**NH(2084).
D. griseo-cana. On bark. **2:**FL(2084), MI(2084).
D. microspora. On bark. **10:**NC(2084). **16:**NJ(2084). **19:**VT(2084). **22:**MI(2084). **28:**NH(2084), NY(2084).
Dentipellis sp. **22:**NY(128).
Dentocorticium sulphurellum. White rot. **25:**NY(2043). **28:**NY(1173).
Dichostereum effuscatum. **28:**NY(128).
Diplomitoporus linbladii. **28:**WA(3202).
Exidia recisa. **28:**NC(3944).
Exidiopsis calcea. **9:**WA(3202).
E. molybdea. **28:**VT(1224).
Flammulina velutipes. On wood. **10:**MS(2720). **22:**CT(82), MA(82).
Flammulina sp. **21:**IA(1004).
Fomes fasciatus. On wood. **1:**FL(82). **19:**GA(1439). **28:**GA(491).
F. fomentarius. White mottled rot. **10:**NC states(82), NE states(82). **12:**NC states(82), NE states(82). **19:**NE states(82), NW states(82). **21:**NC states(82), NE states(82). **22:**NC states(82), NE states(82). **28:**WI(1921).
Fomitopsis cajanderi. Brown cubical rot. **9:**WA(3202). **19:**MD(82).
F. durescens. On logs and stumps. **21:**IA(1004), MS(82), OH(82).
F. meliae. Brown cubical rot. **28:**Unk.(2681).
F. pinicola. Brown cubical rot. **6:**ID(82). **19:**MI(2862). **22:**NY(82).
F. spraguei. Brown cubical rot. **9:**OR(82), WA(82,3202). **12:**MI(82). **28:**OR(2180,3202), WA(2180).
Fuscocerrena portoricensis. White rot. **10:**MN(2874). **28:**IA(531).
Ganoderma applanatum. White rot. **10:**Range of host(82). **12:**Range of host(82). **19:**NE states(3240), Range of host(82), NC(1303). **21:**Range of host(82), IA(1004), MN(2874). **22:**Range of host(82). **25:**Range of host(82).
G. lobatum. White rot of dead logs and stumps. **28:**VA(2535).
G. lucidum. White root and butt rot. **10:**RI(17,18). **21:**IA(1004). **22:**CT(82), MN(2874), OH(17,18). **28:**LA(17,18).
Globifomes graveolens. White heart rot. **19:**TN(1513,1538).
Gloeocystidiellum clavuligerum. White rot. **22:**MI(2553), NY(2553).
G. porosum. White rot. **25:**MI(2553).
Gloeodontia columbiensis. On wood. **6:**MT(440).
Gloeophyllum sepiarium. **28:**NY(2962).
G. trabeum. Brown cubical rot. **10:**Range of host(82). **12:**Range of host(82). **19:**Range of host(82). **21:**Range of host(82). **22:**Range of host(82).
Gloeoporus dichrous. White rot. **10:**Range of host(82). **12:**Range of host(82). **19:**Range of host(82). **21:**Range of host(82). **22:**Range of host(82).
Grandinia granulosa. **22:**MA(656).
G. nespori. White rot. **23:**AZ(1202).
G. pruni. White rot. **6:**AZ(1202).
G. subalutacea. **22:**MN(128,1197).
Grifola frondosa. Butt rot. **12:**IA(82).
Guepiniopsis buccina. **6:**ID(3202).
Hapalopilus rutilans. White rot. **28:**Unk.(2681).
Hericium americanum. White rot. **10:**Range of host(82). **12:**Range of host(82). **19:**Range of host(82). **21:**Range of host(82). **22:**Range of host(82).
H. coralloides. On wood. **19:**NC(1303). **28:**NY(128).
H. erinaceus. Spongy white rot. **12:**NE states(82), MI(82), MN(82). **21:**IA(1004), KS(82), MD(82), MN(2874), NJ(82). **22:**NE

states(82), MI(82), MN(82). **28:**Unk.(1006).
Heterobasidion annosum. Butt and root rot. **9:**OR(82,3202). **19:**MI(2862).
Hohenbuehelia nigra. Butt and root rot. **9:**WA(3202).
Hymenochaete agglutinans. Bark canker. **19:**CT(82,1312), MD(82), MI(82,2862). **25:**VT(128).
H. cinnamomea. On wood. **9:**WA(3202).
H. corrugata. On wood. **10:**Range of host(82). **12:**Range of host(82). **19:**Range of host(82), MA(1394). **21:**Range of host(82). **22:**Range of host(82).
H. episphaeria. White rot. **28:**NY(2456).
H. fusca. **28:**NY(2456).
H. spreta. **9:**WA(3202).
H. tabacina. White rot. **10:**Range of host(82). **12:**Range of host(82). **19:**Range of host(82). **21:**Range of host(82). **22:**Range of host(82).
Hyphoderma heterocystidia. **28:**NY(2456).
H. leoninum. White rot. **28:**NC(444,1303), OH(444).
H. puberum. **22:**MA(656).
H. rimosum. White rot. **28:**MD(444).
H. setigerum. White rot. **22:**MN(1197). **28:**NY(1173).
Hyphodontia stipata. White rot. **28:**NY().
Hypholoma fasciculare. **9:**OR(3202).
H. sublateritium. Wood decay. **28:**NC(1303).
Hypochnicium punctulatum. **9:**WA(3202).
H. sphaerosporum. White rot. **10:**AZ(1202).
H. vellereum. **28:**WI(1921).
Hypsizygus tessellatus. On wood. **19:**SE states(2421), MD(2421). **28:**MA(2935).
H. ulmarius. Sapwood decay. **10:**CO(49,82), IA(82), KS(82), MN(209), WI(128). **19:**CT(82), MA(2935), NC(82,3944).
Inonotus cuticularis. White rot. **10:**Range of host(82), AZ(1181). **11:**WA(82). **12:**Range of host(82). **19:**Range of host(82). **21:**Range of host(82). **22:**Range of host(82). **28:**AR(2180,2675), CA(2180), LA(2675), NM(2180), UT(2180), WA(2180).
I. dryadeus. White root and butt rot. **9:**OR(82,3202,4006).
I. dryophilus. White heart rot. **28:**Unk.(2681).
I. glomeratus. White heart rot. **10:**MT(82). **19:**NE states(3240,3241), MI(82), MN(82), NY(82), OH(82), VT(82). **22:**NE states(3241), MI(82), MN(82), NY(82), OH(82). **28:**MI(2862), WI(1921).
I. hispidus. White heart rot. **10:**Range of host(82), AZ(1181). **12:**Range of host(82). **19:**Range of host(82). **21:**Range of host(82). **22:**Range of host(82). **28:**Range of host(82).
Irpex lacteus. White rot. **10:**Range of host(82). **12:**Range of host(82). **19:**Range of host(82). **21:**Range of host(82). **22:**Range of host(82), MN(1197). **28:**NC(3944).
Ischnoderma resinosum. Yellowish white rot. **28:**Unk.(2681).
Junghuhnia collabens. **9:**OR(3202).
J. fimbriatella. **28:**NY(2538).
J. nitida. On dead limbs. **3:**ID(128). **19:**NH(2168), PA(128,2168), VT(2168). **22:**NH(128,2168), NY(2168). **25:**NH(128,2168). **28:**Range of host(82), ME(128,2168), NH(2671), VT(128).
J. separabilima. White rot. **28:**Unk.(1204).
Laetiporus sulphureus. Brown cubical rot. **12:**NC states(82), NE states(82). **19:**NC states(82), NE states(82). **22:**NC states(82), NE states(82).
Laxitextum bicolor. White rot. **10:**Range of host(82). **12:**Range of host(82). **19:**Range of host(82). **21:**Range of host(82), MD(543). **22:**Range of host(82). **28:**Range of host(82), AR(2675), LA(2675).
Lentaria byssiseda. Isolated from living tree. **28:**TN(2751).
Lentinellus micheneri. **19:**Unk.(2423).
L. ursinus. On wood. **10:**SW states(2423). **19:**MD(2423).
Lentinus strigosus. On wood. **1:**MS(128). **19:**CT(128), ME(128), NH(128), NY(128), VT(128).
L. tigrinus. **21:**IA(1004).
Lenzites betulina. White rot. **10:**Range of host(82). **12:**Range of host(82). **19:**NC states(82), NE states(82), GA(1439), NC(1303). **21:**Range of host(82). **22:**Range of host(82). **28:**NC(3944).
Leptosporomyces ovoideus. Isolated from mine timbers. **28:**Unk.(1006).
Licrostroma subgiganteum. On dead limbs. **19:**ME(2084), NH(2084). **28:**NH(2671).
Lopharia cinerascens. White rot. **10:**DC(543), GA(543), IA(543), KS(543), MA(543), MO(543), ND(543), NY(543), OH(543), SD(543).
Marasmiellus opacus. On dead limbs. **28:**WV(3717).
Marasmius rotula. On wood. **10:**ND(378). **21:**DC(128).
Meruliopsis albostramineus. **4:**AZ(128).
M. corium. On wood. **3:**WA(128). **9:**WA(3202). **19:**GA(491,1439).
M. hirtellus. **28:**NY(128).
Micromphale foetidum. On dead limbs and trunks. **25:**NC(1303,1307).
Microstroma sp. Frosty mildew. **16:**IL(82), OH(82).
Mycena radicatella. On wood. **28:**WV(3717).
Mycoacia fuscoatra. On wood. **28:**NY(1173).
Mycogloea carnosa. On bark. **19:**SC(2643).
Oxyporus latemarginatus. **19:**NY(2166). **28:**Range of host(82), NY(2166).
O. populinus. Heart rot. **15:**NE states(82). **16:**CT(82). **19:**E states(82), NE states(3240), GA(491,1439), NC(1303). **22:**E states(82), MN(82), NC(1303,1307), NH(3575). **25:**NE states(82). **28:**NC(1303), NY(2456), WI(1921).
Pachykytospora tuberculosa. On wood. **19:**MI(2862).
Panellus pusillus. On wood. **28:**Unk.(2419).
P. stipticus. On wood. **22:**MN(2874), WI(128).
Pellicularia koleroga. **21:**LA(3640).
Peniophora cinerea. On twigs. **9:**WA(3202). **10:**MN(209). **24:**GA(1439).
P. decorticans. **9:**WA(3202).
P. incarnata. **3:**WA(128). **9:**WA(3202).
P. malenconii subsp. *americana*. On wood. **6:**AZ(542).
P. polygonia. **6:**ID(3202).
Perenniporia compacta. Butt rot. **12:**MI(82).
P. fissiliformis. White rot. **28:**MO(2682).
P. fraxinea. **10:**NY(82). **22:**NY(82).
P. fraxinophila. **10:**ND(128), OR(82,3202), WY(82). **19:**CT(82), MS(82). **20:**CT(82), MS(82).
P. medulla-panis. **28:**Range of host(82).
P. narymica. On wood. **28:**PA(2678).
P. ohiensis. **10:**WY(128). **28:**MI(1853).
P. robiniophila. On trunks. **28:**Unk.(2681).
P. subacida. On fallen trunk. **19:**NH(2671).
P. tenuis. White rot. **28:**Range of host(82), NJ(2534).
Phanerochaete affinis. White rot. **19:**CT(128), NH(128), NY(128). **25:**NH(128). **28:**NH(128), NY(128,433), TN(433).
P. carnosa. On dead log. **25:**NH(128). **28:**NH(128,2671).
P. chrysorhiza. White rot. **28:**TN(433).
P. ericina. White rot. **28:**TN(433).
P. filamentosa. White rot. **28:**NC(433).
P. ravenelii. **19:**AR(2675), LA(2675).
P. rimosa. **28:**NY(2456), VT(1224).
P. sanguinea. White rot. **6:**MT(433), WA(3202). **19:**NY(128).
P. sordida. White rot. **19:**NH(128), NY(128,433). **22:**MA(656), MI(433). **28:**MT(433).
P. velutina. White rot. **22:**MI(433). **28:**NY(128), TN(433).
P. viticola. White rot. **22:**NH(128). **25:**NH(128,433), NY(128). **28:**NH(128).
Phellinus conchatus. On wood. **19:**NH(2671). **22:**MN(2874).
P. ferreus. **3:**WA(3202). **6:**ID(3202). **9:**OR(82,3202), WA(3202).
P. ferruginosus. White laminated rot. **19:**MI(2862). **28:**Range of host(82).
P. fragrans. On wood. **22:**MI(2047). **28:**NY(2047).
P. gilvus. White rot. **10:**Range of host(82). **12:**Range of host(82). **19:**Range of host(82). **21:**Range of host(82), IA(1004). **22:**Range of host(82). **28:**OK(2872).
P. igniarius. White heart rot. **11:**ID(82), OR(82), WA(82). **15:**NE states(82). **19:**NE states(3240), Range of host(82), NC(1303). **22:**Range of host(82). **25:**NE states(82).
P. melleoporus. On wood. **28:**Gulf states(1205).
P. pini. **28:**VT(128).
P. punctatus. White rot. **28:**Range of host(82).
P. robustus. On wood. **9:**WA(3202). **10:**MS(2720).
P. viticola. White rot. **3:**OR(3202). **9:**OR(4006). **28:**Range of host(82), Unk.(2681).
Phlebia concentrica. White rot. **3:**WA(3202).
P. ludoviciana. White rot. **19:**FL(2556).
P. merismoides. **9:**WA(3202).
P. ochraceofulva. **21:**NY(128).
P. rufa. **22:**NY(1238).
P. subserialis. White rot. **6:**MT(2556).

P. tremellosus. **21:**IA(1004).
Phlebiella ralla. **28:**Unk.(2115).
P. tulasnelloidea. **28:**Unk.(2115).
P. vaga. **28:**MA(128).
Pholiota adiposa. Butt and root rot. **9:**WA(3202). **10:**MT(82). **21:**DE(82), IA(1004), MD(82), MN(2874).
P. albocrenulata. Butt rot, brown-mottled heart rot and root rot. **28:**NC(1540), NY(2456).
P. limonella. On wood. **10:**MI(3332), WA(3202).
P. squarrosoides. On wood. **10:**NY(82).
P. stratosa. On wood. **19:**MI(3332).
Pholiota sp. **28:**VA(1017).
Phyllotopsis nidulans. White rot. **19:**ME(128), VT(128). **22:**NH(128).
Piloderma sphaerosporum. On wood. **3:**WA(1707).
Pleurotellus hypnophilus. Wound rot. **19:**NC(3944). **28:**NC(82,1303), WA(3202).
Pleurotus cystidiosus. White rot. **19:**IN(2418).
P. levis. Wound decay. **19:**DC(128). **22:**MA(656), NC(1303).
P. ostreatus. White flaky rot of heartwood and sapwood. **10:**E & C states(82). **12:**E & C states(82). **21:**E & C states(82). **22:**E & C states(82), CT(82), MI(63), MN(63), MS(82), NC(1303). **28:**NC(3944).
Pleurotus sp. Sapwood rot, wound rot. **19:**NC(1303).
Plicaturopsis crispa. On wood. **15:**VT(128). **28:**MI(2945).
Pluteus aurantiorugosus. **19:**OH(229).
Polyporus admirabilis. White heart rot. **28:**Unk.(2681).
P. arcularius. On wood. **9:**WA(3202). **21:**IA(1004).
P. badius. On wood. **28:**Unk.(2681), MI(2862).
P. brumalis. **10:**Range of host(82). **12:**Range of host(82). **19:**Range of host(82). **21:**Range of host(82). **22:**Range of host(82).
P. melanopus. **3:**WA(3202). **6:**WA(3202).
P. mori. On dead limbs. **6:**ID(3202). **19:**NC(3944). **28:**Unk.(2681).
P. squamosus. On wood. **6:**MT(82). **10:**NC states(82), NE states(82). **22:**NC states(82), NE states(82).
P. varius. **3:**OR(3202), WA(3202). **6:**ID(3202), WA(3202). **19:**NC(3944). **28:**MI(2862).
Porotheleum perenne. On wood. **22:**NY(645).
Postia balsamea. Isolated from mine timbers. On wood. **28:**Unk.(1006), WA(3202).
P. caesia. Brown rot. **6:**ID(3202), WA(3202). **9:**OR(3202).
P. floriformis. **9:**OR(3202). **28:**OR(4006).
P. guttulata. Usually reported on conifers. **19:**NE states(795).
P. perdelicata. **6:**WA(3202).
P. stiptica. On wood. **28:**Unk.(2681).
P. tephroleuca. Brown cubical rot. **28:**Unk.(2681).
P. undosa. **21:**MA(397). **28:**NY(397).
Psathyrella incerta. On roots around stump. **22:**NY(3491).
Pseudotomentella flavovirens. On wood. **28:**NY(2032).
P. griseopergamacea. On wood. **28:**NY(2037).
Pulcherricium caeruleum. **28:**NC(3944).
Pycnoporellus fulgens. On roots of overturned tree. **28:**ID(3202,3831).
Pycnoporus cinnabarinus. White rot. **6:**ID(3202). **19:**NC(3944). **21:**IA(1004). **22:**NH(2671). **28:**NH(2671).
P. sanguineus. **19:**NC(3944).
Rhodotus palmatus. On wood. **19:**MI(2422).
Rigidoporus ulmarius. Root and butt rot. **10:**FL(35), WV(3907). **19:**Gulf States(82), FL(35). **28:**LA(2675), MS(2675).
Schizophyllum commune. Canker. **21:**IA(1004). **28:**Range of host(82), OK(2872).
Schizopora paradoxa. **28:**Range of host(82).
Scytinostroma galactinum. On fallen trunks. **28:**NH(2671).
S. portentosum. **19:**ME(128).
Sebacina incrustans. At base of living trees. **28:**MI(128).
Septobasidium castaneum. Parasitic on scale insects. **10:**FL(35), WV(3907). **19:**FL(35,82). **21:**FL(35). **28:**FL(35).
S. cokeri. Parasitic on scale insects. **10:**NC(82,1303).
S. fumigatum. Parasitic on scale insects. **19:**AL(82), FL(35,82), LA(82), NC(82,1303), SC(82), VA(82).
S. pseudopedicellatum. Parasitic on scale insects. **10:**FL(35,82), WV(3907). **19:**FL(35,82). **21:**FL(35), KY(82). **22:**KY(82), TN(82). **28:**FL(35).
Sistotrema brinkmannii. **9:**OR(3202).
Skeletocutis nivea. White rot. **6:**MT(3202).
Spongipellis delectans. White mottled rot. **28:**NY(82).
S. spumeus. White rot. **28:**Unk.(2681).
S. unicolor. On trunks and limbs of living trees. **12:**MI(82). **21:**MD(82).
Steccherinum fimbriatum. **9:**WA(3202). **22:**MN(128).
S. laeticolor. On wood. **21:**IA(1004), NY(128,1175). **28:**NY(128).
S. ochraceum. White rot. **3:**OR(82). **19:**MD(82), NH(82), VA(82). **28:**MN(128).
S. peckii. On wood. **28:**NY(159).
S. subcrinale. **28:**NY(128).
S. tenue. **28:**TN(443).
[*Stereum gausapatum*]. On exposed roots. **28:**NJ(128).
S. hirsutum. White rot. **3:**OR(543). **9:**WA(543). **19:**CT(543), IN(543), MA(543), ME(543), NH(2671), VT(543). **21:**CT(543). **22:**MA(543), NH(543), NY(543), VT(543). **25:**NY(543). **28:**Range of host(82), CT(543), NY(543).
S. ostrea. White rot. **9:**WA(82). **19:**MD(543), VA(543). **21:**CT(543). **28:**MD(543), NY(543).
Thelephora albidobrunnea. Encircling stems of seedlings or young trees. **19:**NC(82,1303,3944), VA(82).
Tomentella angulospora. On wood. **28:**NY(2040).
T. atrorubra. **28:**MI(2039), NY(2039).
T. bicolor. On wood. **28:**NY(2039).
T. botryoides. **19:**NY(128). **28:**NY(2039).
T. bresadolae. On wood. **28:**NY(2035,2039).
T. bryophila. **28:**NC(2039), NY(2039).
T. calcicola. **28:**NY(2039).
T. chlorina. **28:**NM(1190), NY(2039).
T. coerulea. **9:**WA(2036,3202). **28:**WA(2039).
T. crinalis. **28:**MT(3202), NY(2039).
T. epigaea. On wood. **28:**NY(2039).
T. ferruginea. **28:**MI(2039), NM(1190,2039), NY(2030,2039).
T. fuscoferruginosa. **28:**NY(2039).
T. lateritia. **28:**MI(2039), NY(2039).
T. neobourdotii. On wood. **28:**MI(2039), NC(2039), NY(2033,2039).
T. ochracea. **28:**NY(2036,2039).
T. olivascens. **28:**NY(2039).
T. pilosa. **28:**NY(2039).
T. punicea. **28:**NC(2039), NY(128,2030,2039).
T. ramosissima. **28:**NY(2039).
T. ruttneri. On wood. **22:**NY(2035). **28:**NY(2035).
T. subcinerascens. On wood. **28:**NC(2039).
T. sublilacina. **28:**MI(2039), NY(2039), WV(2039).
T. subvinosa. **28:**NJ(2039), NY(2039).
T. terrestris. **28:**NC(2039).
T. umbrinospora. **22:**NY(2030). **28:**NY(2039).
T. violaceofusca. **28:**NY(2039).
T. viridescens. On wood. **28:**NM(2039), NY(2032,2039).
T. viridula. On wood. **28:**NM(2039), NY(2032,2039).
Trametes conchifer. On wood. **19:**NC(1303). **28:**Unk.(2681).
T. ectypus. On wood. **28:**Unk.(2681).
T. elegans. On wood. **28:**WV(3907).
T. hirsuta. White rot. **3:**Range of host(82). **6:**Range of host(82). **7:**Range of host(82). **9:**Range of host(82). **10:**Range of host(82). **11:**Range of host(82). **12:**Range of host(82). **19:**Range of host(82), MI(2862). **21:**Range of host(82), IA(1004). **22:**Range of host(82). **23:**Range of host(82).
T. pubescens. On wood. **10:**ID(3202). **28:**Unk.(2681).
T. trogii. On wood. **28:**Unk.(2681).
T. versicolor. White rot. **3:**Range of host(82). **6:**Range of host(82). **7:**Range of host(82). **9:**Range of host(82). **10:**Range of host(82). **11:**Range of host(82). **12:**Range of host(82). **19:**Range of host(82), NC(1303,3944). **21:**Range of host(82), IA(1004). **22:**Range of host(82), VT(3768). **23:**Range of host(82). **28:**Unk.(1006).
T. villosa. On wood. **28:**Unk.(2681).
Trechispora farinacea. **10:**NM(128). **28:**MN(1197).
T. mollusca. **9:**OR(3202). **28:**ID(3202).
Tremella fimbriata. On wood. **19:**NC(1303).
T. mesenterica. On dead limbs. **19:**MA(1394).
Trichaptum biforme. On wood. **10:**Range of host(82). **12:**Range of host(82). **19:**Range of host(82), NC(1303,1307,3944), SC(1307). **21:**Range of host(82), IA(1004). **22:**Range of host(82), NC(1303,1307), SC(1307).
T. sector. White rot. **6:**MT(3832).
Tulasnella bifrons. **19:**MA(128).
T. pruinosa. On bark. **28:**NH(128).

T. violea. On bark. **19:**NH(128). **28:**IA(128), PA(2673).
Typhula phacorrhiza. On leaves. **28:**NY(2955).
T. setipes. **9:**WA(3202).
T. variabilis. On leaves. **28:**NY(2955).
Tyromyces chioneus. White stringy rot. **28:**OR(3202,4006).
T. fissilis. White rot. **10:**MS(2720). **19:**NC(1303). **28:**Unk.(2681).
T. fumidiceps. On wood. **28:**Unk.(2681).
T. galactinus. On wood. **21:**IA(1004). **28:**Unk.(2681).
Vararia investiens. On wood. **19:**CT(128), MA(656), VT(128). **22:**MA(656). **25:**NH(128).
Volvariella bombycina. On living or dead trees. **10:**AR(82), LA(82), MI(82), NY(82), OH(82). **12:**AR(82), LA(82), MI(82), NY(82), OH(82). **19:**AR(82), LA(82,2675), MI(82), NY(82), OH(82). **21:**AR(82), IA(1004), LA(82), MI(82), NY(82), OH(82). **22:**AR(82), LA(82), MI(82), NC(1303), NY(82), OH(82).
Wolfiporia cocos. Brown cubical butt and root rot. **28:**PA(793), SC(1241).
Xenasma praeteritum. **28:**Unk.(2115).
X. rimicola. On wood. **9:**OR(2115). **28:**WA(3202).

Deuteromycotina - Hyphomycetes

Acremonium bacillisporum. Isolated from fallen leaves. **22:**WI(2005).
Acrodictys atroapicula. On rotten wood. **19:**NY(3771).
Alatospora acuminata. On submerged leaves. **21:**IL(3218).
Alternaria alternata. Isolated from fallen leaves. **22:**WI(2005).
A. negundinicola. On living leaves, possibly a secondary invader. **10:**KS(82).
Alternaria sp. Leaf spot. **19:**FL(35). **21:**IL(3218). **28:**WI(1921).
Ambrosiella brunnea. In tunnels of *Monarthrum mali* (Coleoptera). **22:**WV(203).
A. xylebori. In tunnels of *Xyleborus compactus* (Coleoptera). **19:**GA(203).
Anguillospora longissima. On submerged leaves. **21:**IL(3218).
Aspergillus sp. Isolated from wood. **19:**NH(1645). **28:**WI(1921).
Aureobasidium apocryptum. Anthracnose, leaf blight, leaf spot. **10:**E & C states(82), IL(1052), MN(82), SD(2243). **14:**NC(1303). **15:**PA(82). **16:**E states(82), CT(82), MN(82), MO(82). **19:**MD(82), MI(82), NJ(82). **21:**CT(82), IA(82), IL(82), NC(1303), WI(82). **22:**E states(82), CT(82), IN(82), ME(82), MI(3780), MN(82), NC(1303), NY(82), OH(82), VA(3780), VT(3780), WI(82). **28:**CT(3770), IL(1052), MS(2720), OK(2872).
A. pullulans. Isolated from fallen leaves. **13:**AZ(662). **22:**WI(2005).
Bacillispora aquatica. On submerged leaves. **21:**IL(3218).
Berkleasmium moriforme. **28:**NC(3944).
Bispora effusa. On bark. **22:**NY(3546).
Brachysporium nigrum. **28:**VT(1224).
Cacumisporium capitulatum. **28:**NY(1264).
Candida sp. **28:**WI(1921).
Cephalosporium sp. Isolated from wood. **19:**ME(1319,2860), NH(2860). **28:**WI(1921).
Cercospora negundinis. Leaf spot. **10:**KS(82), NC(1303), NE(82), WI(82). **19:**FL(35).
C. saccharini. Leaf spot. **21:**IL(2119).
Cercospora sp. Leaf spot. **10:**FL(35). **19:**FL(35). **28:**OK(2872).
Cladosporium cladosporioides. Isolated from fallen leaves. **22:**WI(2005).
C. herbarum. **16:**WA(3202).
C. humile. Scab, leaf spot. **10:**Unk.(82). **12:**Unk.(82). **19:**NY(2822), WI(2822). **21:**NC(1303), WI(1350,2822).
Cladosporium sp. Isolated from wood. **19:**ME(2860), NH(2860). **28:**WI(1921).
Clavariopsis aquatica. On submerged leaves. **21:**IL(3218).
Coniosporium olivaceum. On decorticated wood. **19:**PA(2679).
Coniothecium applanatum. **28:**WA(3202).
Conoplea globosa. Trunk canker. **19:**PA(82). **28:**NH(1672).
Crinula caliciiformis. **28:**VA(2130).
Cristulariella depraedans. Leaf spot. **10:**CT(82). **14:**CT(82), NJ(82), WV(82). **16:**CT(82), NJ(82). **19:**CT(82). **21:**CT(2929). **22:**CT(82,2929), NY(2929). **25:**NY(82).
C. moricola. Zonate leaf spot, premature defoliation. **10:**CT(3783), WV(3907). **12:**WV(815,816,3907). **16:**NC(261,1303), VA(3493), WV(816,3907). **17:**NY(82), PA(82,3783). **19:**FL(35). **21:**FL(35), WV(816). **22:**FL(35), NC(1303), NY(82,3783), PA(82), WV(3907). **28:**FL(35).
Cryptostroma corticale. Sooty bark on stored pulpwood. **17:**WA(1276,3202,3657). **22:**MI(2637), WI(2637). **28:**WI(1358).
Curvularia sociata. On leaves. **19:**LA(2500).
Curvularia sp. **28:**WI(1921).
Dendrodochium nigrescens. **10:**MT(3202).
Dimorphospora foliicola. On submerged leaves. **21:**IL(3218).
Epicoccum nigrum. Isolated from fallen leaves. **10:**DE(3543). **22:**WI(2005).
Epicoccum sp. **28:**WI(1921).
Filosporella annelidica. On submerged leaves. **21:**IL(3216,3218).
Flagellospora curvula. On submerged leaves. **21:**IL(3218).
Fusarium lateritium. Red stain of wood. **10:**ND(378).
F. pallidoroseum. On samaras. **10:**GA(659).
F. sambucinum. On samaras. **10:**WI(1341).
F. solani. Canker. **16:**NJ(82). **22:**PA(3821).
Fusarium sp. On leaves and wood. **10:**NC(1303). **19:**ME(2860), NH(1645,2860). **22:**WI(2005). **28:**WI(1921).
Geotrichum sp. Canker. **28:**WI(1921).
Gliocladium roseum. Isolated from fallen leaves. **19:**ME(2860), NH(2860). **22:**WI(2005).
Gliocladium sp. Branch stubs. **19:**ME(1319).
Graphium sp. Isolated from wood. **19:**ME(2860), NH(1645,2860). **22:**ME(1005). **28:**WI(1921).
Helicoma ambiens. On bark. **28:**OH(1271).
H. polysporum. On bark. **21:**OH(1271).
H. repens. On inner bark. **21:**OH(1271).
Helicomyces roseus. On stumps. **28:**IA(1270).
Helicosporium viride. On wood. **15:**ME(1266). **19:**MA(1266), RI(1266). **28:**CT(1266), MA(1266).
Helminthosporium velutinum. On dead twigs and limbs. **3:**WA(3202). **10:**NC(1303,1307).
Isariopsis sp. Leaf spot. **28:**FL(35).
Itersonilia perplexans. **9:**WA(2618,3202,3396).
Lemonniera aquatica. On submerged leaves. **19:**PA(2676). **21:**IL(3218).
L. filiforme. On submerged leaves. **19:**NY(2750).
L. pseudofloscula. On submerged leaves. **21:**IL(3218).
L. terrestris. On submerged leaves. **21:**IL(3218).
Mycopappus aceris. Leaf spot. **9:**OR(82,2940), WA(82,831,2940,3202).
Nigrospora sp. Leaf spot. **28:**WI(1921).
Nodulisporium sp. Isolated from branch stubs. **19:**ME(1319). **28:**WI(1921).
Paecilomyces carneus. Isolated from fallen leaves. **22:**WI(2005).
P. farinosus. Isolated from fallen leaves. **22:**WI(2005).
P. marquandii. Isolated from fallen leaves. **22:**WI(2005).
Papularia sp. **28:**WI(1921).
Penicillium brevicompactum. Isolated from fallen leaves. **22:**WI(2005).
P. citrinum. Isolated from fallen leaves. **22:**WI(2005).
P. glabrum. Isolated from fallen leaves. **22:**WI(2005).
P. granulatum. Isolated from fallen leaves. **22:**WI(2005).
P. herquei. Isolated from fallen leaves. **22:**WI(2005).
P. implicatum. Isolated from fallen leaves. **22:**WI(2005).
P. lividum. Isolated from fallen leaves. **22:**WI(2005).
P. sclerotiorum. Isolated from fallen leaves. **22:**WI(2005).
P. thomii. Isolated from fallen leaves. **22:**WI(2005).
Penicillium sp. Isolated from wood. **19:**ME(1319,2860), NH(1645,2860). **28:**WI(1921).
Petrakia echinata. **21:**WI(1354).
Phaeostalagmus altissimus. On rotten wood. **28:**NY(3771).
P. arbusculus. On rotten wood. **22:**NY(3771).
Phialophora melinii. Isolated from wood. **19:**ME(2860), NH(2860).
Phialophora sp. Isolated from wood. **19:**ME(1319), NH(1645). **22:**VT(3768). **28:**WI(1921).
Phymatotrichopsis omnivora. Root rot. **10:**TX(82). **21:**TX(82). **22:**TX(82).
Ramularia lethalis. Leaf spot. **19:**NY(82), WI(82).
Rhodotorula sp. **28:**WI(1921).
Sclerotiomyces colchicus. **21:**WI(1355). **22:**WI(1356).
Spadicoides catenulata. On rotten wood. **22:**NY(3771).
Spegazzinia parkeri. Black mold on leaves. **28:**FL(35).
Sporidesmium hormiscioides. **28:**NJ(1686).
Sporobolomyces roseus. On leaves. **9:**CA(2879).
Stachycoremium parvulum. On wood. **19:**DE(3543). **28:**NJ(3177).
Stigmina negundinis. On living limbs. **10:**WY(834). **22:**NY(82).
S. robusta. On twigs and limbs. **28:**NJ(1678).
Stilbella acerina. Leaf spot. **19:**TN(82,2679).
S. aleuriata. On bark. **10:**TX(3178).

Stromatocrea cerebriforme. **6:**ID(3202), WA(641,3202). **7:**ID(641), WA(641).
Taeniolella breviuscula. On bark. **28:**NC(255).
T. muricata. On bark. **23:**UT(831).
Tetracladium marchalianum. On submerged leaves. **21:**IL(3218).
Tilletiopsis minor. On leaves. **9:**WA(2619).
T. washingtonensis. On leaves. **9:**WA(2619).
Trichocladium sp. Isolated from branch stubs. **19:**ME(1319). **28:**WI(1921).
Trichoderma hamatum. Isolated from fallen leaves. **22:**WI(2005).
T. koningii. Isolated from fallen leaves. **22:**WI(2005).
T. viride. Isolated from wood. **19:**ME(2860), NH(2860).
Trichoderma sp. Isolated from wood. **19:**NH(1645). **28:**WI(1921).
Trichothecium sp. **28:**WI(1921).
Tripospermum myrti. Isolated from fallen leaves. **22:**WI(2005).
Tubercularia smaragdina. On wood. **28:**NJ(3178).
T. vulgaris. On bark. **10:**NM(1003).
Umbelopsis versiformis. Isolated from roots of living tree. **19:**WV(51). **22:**WV(51).
Vargamyces aquaticus. On submerged leaves. **21:**IL(3218).
Verticillium albo-atrum. Wilt. **2:**IL(1558). **5:**IL(1558). **11:**CA(82), OR(82), WA(82). **12:**IL(1558). **14:**Range of host(82), CA(3896), VA(3493). **16:**NW states(1558), Range of host(82), IL(992,1558), NC(1533,1558), VA(1533,1558,3493). **17:**C states(82), NE states(82). **19:**C states(82), E states(82), IL(82,1558), TN(82), VA(3493). **21:**E states(82), NC states(82), IL(82,1558), TN(82), VA(3493). **22:**E & C states(82), IL(82,992,1558), MA(1558), NC(1533,1558), NY(1558), TN(82), VA(3493). **27:**TN(1029). **28:**CA(1558).
V. lecanii. Isolated from fallen leaves. **22:**WI(2005).
V. psalliotae. Isolated from fallen leaves. **22:**WI(2005).
Verticillium sp. Wilt. **22:**WI(2005). **28:**AL(285), KY(1780), WI(1921).

Deuteromycotina - Coelomycetes

Aplosporella clintonii. Dieback. **10:**IA(82), KS(82), ND(82,378), SD(82,2243), WI(82). **12:**Unk.(82). **16:**NJ(82), NY(82). **19:**NY(82). **21:**IL(82,2738). **22:**IL(82). **25:**Unk.(82). **28:**IL(971).
A. negundinis. Dieback. **10:**KS(82).
Ascochyta negundinis. On leaves. **10:**IL(3585).
Ascochyta sp. Leaf spot. **28:**FL(35).
Asteromella sp. Isolated from fallen leaves. **22:**WI(2005).
Botryodiplodia acerina. On limbs. **15:**WV(82,3907).
Chaetophoma pellicula. **3:**CA(81).
Colletotrichum gloeosporioides. Anthracnose, leaf blight. **10:**NJ(82), TX(82).
Colletotrichum sp. Anthracnose. **28:**FL(35).
Coniothyrium fuckelii. Isolated from fallen leaves. **22:**WI(2005).
C. negundinis. Canker, leaf spot. **10:**IL(82,3589), OK(82,2873). **19:**NC(1303). **21:**NC(1303).
Coniothyrium sp. Canker. **10:**WI(1344,1360). **19:**FL(35).
Cryptocoryneum condensatum. On bark. **28:**NY(3149).
Cryptosporiopsis sp. Bleeding trunk canker; associated with tree crickets. **19:**CT(3579).
Cylindrosporium acerinum. On leaves. **3:**CA(81), OR(4006). **6:**UT(128). **15:**NC(3482). **25:**NY(836).
C. consociatum. Leaf spot. **6:**CA(81), ID(832), MT(1195), WA(832), WY(1195,3812). **23:**UT(1195).
C. negundinis. Leaf spot. **10:**MS(2720).
C. saccharinum. Leaf spot. **19:**MS(2720). **21:**WI(966).
Cylindrosporium sp. On leaves. **10:**IL(1052). **19:**FL(35). **28:**FL(35).
Cytospora annulata. Leaf spot. **10:**CO(49), NE(82), SD(82).
C. chrysosperma. Canker. **3:**MT(82). **6:**MT(82), WA(82). **7:**ID(82), MT(82). **14:**CT(82), MA(82), NJ(82), NY(82). **16:**CT(82), MA(82), NJ(82), NY(82). **17:**CT(82), MA(82), NJ(82), NY(82). **19:**E states(82), SC states(82), MN(82). **21:**E states(82), SC states(82), MN(82). **28:**MI(2862), OK(2872), WA(3202).
Cytospora sp. Cytospora canker. **16:**MI(3319). **19:**FL(35), ME(1319,2860), NH(2860). **22:**VT(82). **28:**CO(49), WI(1921).
Cytosporella sp. Canker. **28:**UT(3620).
Dendrophoma sp. Leaf spot. **28:**FL(35).
Dichomera juglandis. **19:**MD(2783).
Didymosporina aceris. Leaf blight, leaf spot. **9:**OR(82,3202). **16:**CT(951), NY(82).
Diplodia acericola. On leaves. **22:**IL(82,3591).
D. atrata. On twigs. **10:**NE(82), WV(82,3907). **16:**PA(82).
D. sarmentorum. On limbs. **6:**WA(82).
D. subtecta. On limbs. **9:**WA(82,3202). **28:**MI(82).
Diplodina acerina. On leaves and twigs. **10:**Unk.(82), MI(82). **19:**FL(35), MI(82), NC(82), NY(82), SC(82), WV(82). **21:**IN(82), MO(82,962), WV(82,3907).
D. fusispora. On twigs. **21:**IL(2738), MN(2874).
Discella pilosula. **6:**CO(1195).
Discosia sp. Leaf spot. **16:**NC(1303). **21:**IL(3218). **22:**NC(1303).
Discula sassafras. On leaves. **22:**IL(3587).
Dothiorella sp. Leaf spot. **22:**IL(2567). **28:**FL(35).
Endophragmiella biseptata. **28:**NY(1677).
Fusicoccum sp. Stem canker. **28:**FL(35).
Gloeodes pomigena. On twigs. **22:**IN(82).
Gloeosporium aceris. Leaf blight, leaf spot. **19:**NC(1303).
G. decolorans. On living leaves. **19:**MA(1390).
Gyrostroma missouriense. **28:**CT(3173).
[*Kirschsteiniella applanata*]. On dead limbs. **25:**NC(1303,1307).
Leptothyrella aceris. On twigs. **25:**NY(82).
Leptothyrium maximum. On twigs. **10:**IL(82,3589). **12:**IL(82,3589).
Libertella acerina. **15:**NY(82).
Macrophoma negundinis. Leaf spot. **10:**CO(82), OK(82,2872).
Macrophoma sp. Twig blight, canker. **21:**FL(35). **28:**AR(82), NY(82).
Macrophomina phaseolina. Charcoal rot of seedlings. **21:**IL(82).
Marssonina castagnei. Leaf spot. **21:**OR(82).
Marssonina sp. On gall spots on living leaves. **10:**WI(1354).
Melanconium atrum. **10:**WA(3202).
Melasmia acerina. Tar spot. **28:**GA(3549), MA(3549), MD(3549), WI(3549), WV(3549).
Microdiplodia perpusila. On twigs. **21:**NY(82).
Monochaetia kansensis. Zonate leaf spot. **19:**GA(82,1509), NC(82,1509), OK(82), TN(82,1509).
M. monochaeta. Leaf spot. **19:**NC(1303), OK(2873).
Monostichella hysteroidea. Leaf spot. **22:**OH(832).
Myxosporium acerinum. On twigs. **10:**NE(82). **21:**Range of host(82), IL(2738), WI(82).
M. negundicola. On dead shoots. **10:**KS(834).
M. seriatum. On twigs. **15:**NY(82).
Myxosporium sp. Twig canker. **14:**PA(82).
Pestalotia adusta. On dead leaves. **16:**NJ(1393).
P. zahlbruckneriana. On dead, weathered leaves. **10:**MD(1393).
Pestalotia sp. **6:**ID(3202).
Phloeospora pseudoplatani. **10:**CO(1195).
Phoma cava. Isolated from fallen leaves. **22:**WI(2005).
P. exigua. Isolated from fallen leaves. **22:**WI(2005).
P. negundicola var. *ramicola*. On twigs. **10:**WV(3907).
P. negundinicola. On twigs. **10:**IA(82), NJ(82).
P. pennsylvanica. On twigs and dead limbs. **15:**WV(82,3907).
P. pezizoides. On decorticated wood. **28:**WV(307).
Phoma sp. Canker. **21:**FL(35). **22:**WI(2005). **28:**WI(1921).
Phomopsis acerina. Isolated from trunk. **16:**NJ(2814).
P. lebiseyi. On seedlings. **28:**IA(82).
Phomopsis sp. Canker, dieback. **6:**MT(3202). **14:**E states(82), NC(1303), RI(3780). **16:**NJ(82). **19:**NC(1303). **21:**FL(35). **22:**OH(82).
Phyllosticta arida. Dieback, canker. **10:**AL(64).
P. hesleri. **12:**TN(82). **22:**TN(82,837).
P. minima. Purple-bordered leaf spot. **6:**ID(82), MO(82). **10:**Range of host(82), NC(1303). **12:**Range of host(82). **14:**CT(82), NC(1303). **16:**E states(82), MO(82), NC(1303), WI(1352). **17:**CT(82), PA(82). **19:**Range of host(82), LA(2675), MA(656), MS(2720), NC(1303,1307,3944), NJ(2), NY(2456), OK(2872), PA(1039), SD(2243), WI(1360). **21:**Range of host(82), MS(2720), NC(1303,3944), OK(2873), WI(64). **22:**Range of host(82), MS(2720), PA(1039), WI(64). **25:**NY(82), WI(64,82). **28:**AL(285), FL(35), MS(2720), NJ(2).
P. minutissima. On leaves. **6:**CO(1195), ID(3202), MT(1195,3202), OR(3202), WA(3202), WY(1195). **10:**Unk.(82). **19:**Unk.(82), WI(1334). **22:**WI(1348,1355). **23:**UT(1195). **25:**WI(1356). **28:**VA(2130).
P. negundinis. Leaf spot. **10:**E states(82), SD(2243), TX(82), WI(82). **28:**NC(3944).
P. platanoidis. **10:**CO(1195), NM(1195). **11:**CA(81).
Phyllosticta sp. On leaves. **19:**WI(1352). **22:**WI(1357). **28:**CO(49), IL(1052).
Piggotia negundinis. Leaf blotch. **10:**WI(82,1359), WY(82).

Pilidium acerinum. Leaf spot. **19:**FL(35). **22:**OK(82,2872).
P. concavum. On leaves. **19:**Unk.(3212).
Pseudolachnea hispidula. **10:**ND(378).
Selenophoma sp. Leaf spot. **19:**FL(35).
Septoria aceris. Leaf spot. **6:**CO(649). **7:**AK(526). **10:**CO(1195), IL(2568), NC(1303), OK(2873), UT(1195), WY(1195). **21:**KS(966), NC(1303), OK(2872).
S. aceris-macrophylli. Leaf spot. **9:**OR(4007).
S. circinata. **9:**CA(81).
S. curvispora. On leaves. **6:**CA(642), CO(82,649).
S. marginata. **6:**CA(638,642).
S. saccharina. Leaf spot. **22:**MS(2720).
S. samarae-macrophylli. **9:**WA(831).
Septoria sp. On leaves. **28:**Unk.(82).
Sphaceloma sp. Leaf spot. **19:**FL(35).
Sphaeronaema acerinum. On dead limbs and trunks. **19:**NY(2736), PA(2672). **28:**MD(2783), MI(2862).
S. canum. On dead limbs. **10:**KS(966), PA(82).
Sphaeropsis amplispora. On dead limbs. **21:**KS(833).
S. negundinis. Canker. **10:**IL(3589).
Sphaeropsis sp. Twig canker. **14:**MD(82), NJ(82). **19:**FL(35). **28:**FL(35).
Stegonsporium acerinum. On twigs. **17:**MA(82). **22:**NY(1434), PA(2674).
S. pyriforme. Dieback. **9:**WA(3202). **10:**Unk.(82). **12:**Unk.(82). **16:**MN(2874). **19:**WV(3907). **21:**Unk.(82). **22:**MN(2874), NC(1303), WV(3910). **28:**WV(2467).
Tubakia dryina. Leaf spot. **19:**FL(35). **22:**IL(82,1246,3587). **28:**NC(1303).
Deuteromycotina - Other
Rhizoctonia crocorum. Root rot. **10:**TX(82).
R. solani. Damping-off. **10:**Range of host(82). **12:**Range of host(82). **19:**Range of host(82), FL(35). **21:**Range of host(82), FL(35), OK(2872). **22:**Range of host(82). **28:**FL(35).

Actinidiaceae

Actinidia Lindl.

1. *A. chinensis* Planch. [Chinese gooseberry, kiwi]
Oomycetes
Phytophthora citrophthora. **1:**CA(1110).
P. cryptogea. **1:**CA(1110).
P. megasperma. **1:**CA(1110).
Phytophthora sp. **1:**CA(1110).
Pythium sp. Seedling root rot. **1:**CA(1110), FL(35).
Ascomycotina
Diaporthe actinidiae. Stem-end rot; isolated from fruits. **1:**CA(3383), IL(3383).
Eutypa lata. Stem end rot. **1:**CA(1110).
Basidiomycotina - Other
Armillaria mellea. Root rot. **1:**CA(1110,3347).
Trametes versicolor. **1:**CA(1110).
Deuteromycotina - Hyphomycetes
Alternaria sp. **1:**CA(1110).
Botrytis cinerea. **1:**FL(35).
Botrytis sp. **1:**CA(1110).
Cercospora sp. Leaf spot. **1:**FL(35).
Cladosporium sp. **1:**CA(1110).
Cylindrocladium crotalariae. Root rot. **1:**SC(1972).
Fusarium solani. **1:**CA(1110).
Fusarium sp. **1:**CA(1110).
Penicillium sp. **1:**CA(1110).
Pseudocercospora handelii. **1:**CA(1110).
Verticillium dahliae. **1:**CA(1110).
Deuteromycotina - Coelomycetes
Lasiodiplodia theobromae. Postharvest decay of fruits. **1:**CA(939).
Phyllosticta sp. Leaf spot. **1:**FL(35).
Deuteromycotina - Other
Rhizoctonia solani. Seedling root rot. **1:**CA(1110), FL(35).

Adoxaceae

Adoxa L.

1. *A. moschatellina* L. [moschatel, muskroot, townhall-clock]
Chytridiomycetes
Synchytrium anomalum. Leaf gall. **1:**IA(82).
Basidiomycotina - Uredinales
Puccinia adoxae. Autoecious. **1:**AK(60,526), CO(82,1195), MN(2874), UT(82), WY(82).
P. argentata. Heteroecious: aecial host. **1:**IA(82), MN(82,2874), WI(82).
Deuteromycotina - Hyphomycetes
Ramularia adoxae. On leaves. **1:**CO(1195,3378).
Deuteromycotina - Coelomycetes
Phyllosticta adoxae. Leaf spot. **1:**CO(82).

Agavaceae

Agave L.

1. *A. americana* L. [American century-plant, century-plant, maguey]
2. *A. attenuata* Salm-Dyck
3. *A. desertii* Engelm.
4. *A. palmeri* Engelm.
5. *A. parryi* Engelm. [mescal]
6. *A. sisalana* Perrine [hemp-plant, sisal, sisal hemp]
7. *A. victoriae-reginae* T. Moore
8. *Agave* sp.
Oomycetes
Pythium splendens. Root rot. **8:**FL(35).
Pythium sp. Root rot. **7:**HI(2903).
Ascomycotina
Dothidella parryi. Leaf spot. **8:**CA(82).
Glomerella cingulata. Anthracnose. **1:**FL(82). **6:**Unk.(82).
Pleospora bataanensis. **8:**CA(2783).
Basidiomycotina - Other
Armillaria mellea. Root rot. **2:**CA(2895).
Hyphoderma amoenum. **5:**AZ(1202).
Laetisaria agavei. On leaves. **8:**AZ(438).
Megalocystidium lactescens. **4:**AZ(1202).
[*Phanerochaete crassa*]. **4:**AZ(128,1202).
P. tuberculata. **4:**AZ(1202). **5:**AZ(1202).
Schizopora apacheriensis. **4:**AZ(1202). **8:**AZ(1193).
Deuteromycotina - Hyphomycetes
Botrytis cinerea. Gray mold blight. **1:**Unk.(82). **8:**AK(526).
Coniosporium diversum. **1:**GA(3149).
Coniothecium agaves. On decaying leaves. **8:**HI(2784).
Exosporium sp. Leaf spot. **8:**TX(82).
Fusarium moniliforme. Leaf spot. **8:**FL(35).
Fusarium sp. Leaf spot. **1:**FL(35).
Deuteromycotina - Coelomycetes
Colletotrichum crassipes. Black rot following mechanical injury. **8:**Unk.(82).
C. gloeosporioides. Anthracnose. **1:**FL(2782), MS(2720). **8:**Unk.(82), FL(2782).
Colletotrichum sp. Anthracnose. **8:**FL(82).
Dothiorella agavicola. On petioles. **6:**HI(2784).
Melanconium americanum. On leaves. **1:**NY(82).
Microsphaeropsis concentrica. Leaf spot. **1:**Unk.(82), FL(35). **8:**Unk.(82), FL(35).
Phyllosticta sp. Leaf spot. **8:**FL(35).
Piptarthron macrosporum. On leaves. **8:**AZ(3549), TX(3549).
P. yuccae. **8:**AZ(662).
Stagonospora gigantea. Leaf blight, leaf spot. **1:**CA(81), TX(82). **3:**CA(81). **8:**CA(82), NM(82).

Cordyline Commerson ex Juss.

1. *C. terminalis* (L.) Kunth [good-luck-plant, ti, ti-palm, tree-of-kings]
2. *Cordyline* sp.
Oomycetes
Phytophthora nicotianae var. *parasitica*. Leaf spot. **1:**HI(2903,3684).
Phytophthora sp. Root rot. **1:**FL(35), NJ(82). **2:**NJ(82).
Pythium aphanidermatum. Root rot. **1:**HI(2903).
Ascomycotina
Capnodiopsis sp. **1:**HI(2903).

Glomerella cingulata. Leaf spot. 'Anthracnose.' **1**:FL(35), NJ(82). **2**:NJ(82).
Deuteromycotina - Hyphomycetes
Alternaria sp. Leaf spot. **1**:FL(35).
Cercospora cordylines. Leaf spot. **1**:FL(35), HI(2903). **2**:FL(35).
Cercospora sp. Leaf spot. **1**:HI(2299).
Cladosporium dracaenatum. Leaf spot. **2**:FL(35), SC(82).
Fusarium moniliforme. Leaf spot. **1**:FL(35). **2**:FL(35).
Deuteromycotina - Coelomycetes
Colletotrichum sp. Anthracnose. **1**:FL(35).
Coniothyrium sp. Leaf spot. **1**:FL(35).
Macrophoma sp. Leaf spot. **1**:FL(35).
Phyllosticta cordylinophila. On living leaves. **1**:HI(2903).
P. dracaenae. Leaf spot. **1**:FL(35), OH(82), PA(82). **2**:FL(35), OH(82), PA(82).
P. maculicola. **2**:NJ(82).
Phyllosticta sp. Leaf spot. **1**:HI(2903).
Deuteromycotina - Other
Rhizoctonia solani. Root rot. **1**:FL(35). **2**:FL(35).

Dasylirion Zucc.

1. *D. wheeleri* S. Wats. [spoonflower]
2. *Dasylirion* sp.

Ascomycotina
Asterina sp. Leaf spot. **2**:FL(35).
Botryosphaeria dasylirii. On leaves. **1**:AZ(82,195), NM(82), TX(82).
Basidiomycotina - Other
Athelia arachnoidea. **1**:AZ(1202).
[*Phanerochaete crassa*]. **1**:AZ(128,1202).
Deuteromycotina - Coelomycetes
Actinopelte sp. Leaf spot. **2**:FL(35).
Coniothyrium sp. Stem canker, leaf spot. **2**:FL(35).
Microsphaeropsis concentrica. On leaves. **2**:AZ(2783), TX(82).

Dracaena Vand. ex L.

1. *D. aurea* H. Mann [golden dracaena, halapepe]
2. *D. deremensis* Engl.
3. *D. draco* (L.) L. (=*Yucca draconis* L.) [dragon-tree]
4. *D. fragrans* (L.) Ker-Gawl.
5. *D. fragrans* (L.) Ker-Gawl. cv. 'massangeana' [corn-plant]
6. *D. marginata* Lam.
7. *D. reflexa* Lam.
8. *D. sanderana* Sander ex Mast. [Belgian evergreen]
9. *D. surculosa* Lindl. (=*D. godseffiana* Hort. ex Baker) [gold-dust dracaena, spotted dracaena]
10. *Dracaena* sp.

Oomycetes
Phytophthora cinnamomi. Basal stem rot, root rot. **6**:FL(35).
P. nicotianae var. *parasitica*. Cutting, leaf and stem rot. **2**:FL(35).
Phytophthora sp. Root rot. **4**:FL(35). **9**:FL(35). **10**:FL(35).
Pythium aphanidermatum. Root rot. **2**:HI(2903).
P. splendens. Root rot. **2**:FL(35). **4**:FL(35). **6**:FL(35). **10**:FL(35).
Pythium sp. Root rot. **3**:FL(35). **5**:HI(2903). **8**:FL(35). **9**:FL(35).
Ascomycotina
Botryosphaeria rhodina. Root rot. **3**:MD(82). **10**:MD(82).
Botryosphaeria sp. Dieback, leaf spot. **2**:FL(35). **8**:FL(35).
Glomerella cingulata. Tip blight, leaf spot. 'Anthracnose.' **2**:FL(35). **3**:Unk.(82), FL(82), LA(82), NJ(82), PA(82), WV(82). **4**:FL(35), WV(3202,3907). **8**:WA(3202). **9**:WA(3202). **10**:FL(82), LA(82), NJ(82), PA(82), WV(82).
Guignardia sp. Brown tips. **2**:FL(35).
Leptosphaeria dracaenae. **1**:HI(2903).
Lophodermium dracaenae. Anthracnose. **3**:CA(82). **10**:CA(82).
Meliola dracaenae. Black mildew. **1**:HI(1272,1452,2903). **3**:HI(82). **10**:HI(82).
M. subdentata. **1**:HI(1452).
Nectriella pironii. Stem gall. **6**:FL(35).
Pleospora sp. Leaf spot. **4**:FL(35).
Basidiomycotina - Other
Auricularia cornea. **1**:HI(3948).
Sphaerobolus stellatus. On mulch shot. **5**:HI(2903).
Deuteromycotina - Hyphomycetes
Alternaria sp. Bud rot, leaf spot. **2**:FL(35). **4**:FL(35). **6**:FL(35). **7**:FL(35). **8**:FL(35). **9**:FL(35). **10**:FL(35).
Aspergillus niger. Wilt. **8**:'Stem rot.' FL(2564). **10**:FL(35).
Botrytis cinerea. 'Gray mold.' **9**:FL(35).
Cercospora sp. Leaf spot. **3**:FL(35). **4**:FL(35). **6**:FL(35). **7**:FL(35). **8**:FL(35). **10**:FL(35).
Cladosporium dracaenatum. On leaves. **3**:SC(82). **10**:OR(), SC(82).
Corynespora cassiicola. Leaf spot. **10**:FL(35).
Curvularia sp. Leaf spot. **10**:FL(35).
Fusarium moniliforme. 'Fusarium leaf spot.' **3**:FL(35). **4**:FL(35). **6**:HI(2903). **7**:FL(35). **10**:FL(35).
F. oxysporum. Crown and root rot, leaf spot. **2**:FL(35). **6**:FL(35). **7**:FL(35). **10**:FL(35).
F. solani. Cane and stem rot. **2**:FL(35). **6**:FL(35).
Fusarium sp. Cutting, leaf, root and basal stem rot. **2**:FL(35). **3**:FL(35). **4**:FL(35). **8**:FL(35). **9**:FL(35).
Helminthosporium sp. Leaf spot. **2**:FL(35). **4**:FL(35). **6**:FL(35).
Myrothecium sp. Leaf spot. **2**:FL(35).
Deuteromycotina - Coelomycetes
Colletotrichum gloeosporioides. Anthracnose, leaf spot. **2**:FL(35). **4**:FL(35), OK(2872). **6**:FL(35). **8**:FL(35). **9**:FL(35).
Colletotrichum sp. Anthracnose. **5**:WA(1300). **8**:WA(1300). **9**:WA(1300).
Coniothyrium dracaenae. On leaves. **1**:HI(2903). **3**:HI(82). **10**:HI(82,2784).
Coniothyrium sp. Leaf spot. **3**:FL(35).
Diplodia sp. Leaf spot; dieback. **2**:FL(35). **4**:FL(35).
Gloeosporium sp. Anthracnose, dieback, leaf spot. **5**:WA(1300).
Kellermania anomala. On leaves. **3**:CA(2502).
Lasiodiplodia theobromae. On leaves. **3**:Unk.(82). **10**:Unk.(82).
Macrophoma sp. Cutting rot, leaf spot. **2**:FL(35). **8**:FL(35).
Phyllosticta dracaenae. Leaf spot. **2**:FL(35). **3**:FL(35), HI(2903), NJ(82). **4**:FL(35). **10**:FL(35), NJ(82).
P. draconis. **3**:PA(82). **10**:PA(82).
P. maculicola. **3**:MA(82), NJ(82). **10**:MA(82), NJ(64,82).
Phyllosticta sp. Leaf spot. **6**:FL(35).
Deuteromycotina - Other
Rhizoctonia solani. Cutting, leaf, root and stem rot. **2**:FL(35). **4**:FL(35). **6**:FL(35).
Rhizoctonia sp. Root, basal stem rot. **2**:HI(2903). **8**:FL(35). **10**:FL(35).
Sclerotium rolfsii. 'Southern blight.' **10**:FL(35).

Furcraea Venten.

1. *F. selloa* K. Koch [maguey]

Deuteromycotina - Coelomycetes
Microsphaeropsis concentrica. Leaf spot. **1**:FL(35).

Manfreda Salisb.

1. *M. maculosa* (Hook.) J. Rose [spice-lily, wild-tuberose]
2. *M. virginica* (L.) Salisb. ex J. Rose (=*Agave virginica* L.) [false aloe, rattlesnake-master]

Ascomycotina
Diaporthe arctii. **2**:GA(1438).
Basidiomycotina - Uredinales
Aecidium modestum. **1**:TX(82).
Deuteromycotina - Hyphomycetes
Cercospora amaryllidis. **1**:TX(82).
Deuteromycotina - Coelomycetes
Phyllosticta hymenocallidis. **1**:TX(82).

Nolina Michx.

(= *Beaucarnea* Lem.)

1. *N. microcarpa* S. Wats. [sacahuista]
2. *N. texana* S. Wats.
3. *N. tuberculata* Hort. (=*Beaucarnea recurvata* Lem.) [elephant-foot-tree, ponytail]
4. *Nolina* sp.

Oomycetes
Phytophthora palmivora. Bud and stem rot. **3**:FL(35).
Pythium splendens. Root rot. **3**:FL(35).
Basidiomycotina - Uredinales
[*Uromyces hordeinus*]. **1**:KS(82), OK(82), TX(82).
U. primaverilis. Autoecious. **1**:MO(82), TX(82).
Basidiomycotina - Ustilaginales
Clintamra nolinae. Smut. **1**:SW states(673), AZ(82,922,1077,3602), NM(82,1077), TX(82). **2**:TX(1077,3602). **4**:AZ(3602).

Deuteromycotina - Hyphomycetes
Fusarium moniliforme. Leaf spot. **3**:FL(35).
F. oxysporum. Crown rot. **3**:FL(35).
F. solani. Root rot. **3**:FL(35).
Fusarium sp. Root and stem rot. **3**:FL(35).
Stigmina beaucarneae. **3**:TX(2843).
Deuteromycotina - Coelomycetes
Alpakesa nolinae. On leaves. **1**:AZ(2840,3549).
A. uniseptata. On leaves. **1**:AZ(2501).
Colletotrichum circinans. Anthracnose. **1**:OK(82).
Phyllosticta sp. Leaf spot. **3**:FL(35).
Deuteromycotina - Other
Rhizoctonia sp. Stem rot. **3**:FL(35).

Phormium J. R. Forster & G. Forster

1. *P. tenax* J. R. Forster & G. Forster [harakeke, korari, New Zealand flax, New Zealand hemp]

Oomycetes
Pythium sp. Crown rot. **1**:FL(35).
Ascomycotina
Glomerella cingulata. **1**:CA(81).
Basidiomycotina - Other
Armillaria mellea. Root rot. **1**:CA(2900).
Deuteromycotina - Hyphomycetes
Cercospora phormii. Leaf spot. **1**:FL(35).
Pyricularia sp. Leaf spot. **1**:FL(35).
Deuteromycotina - Coelomycetes
Colletotrichum sp. Anthracnose. **1**:FL(35).
Macrophoma sp. Leaf spot. **1**:FL(35).
Phyllosticta sp. Leaf spot. **1**:FL(35).
Deuteromycotina - Other
Rhizoctonia sp. Crown rot. **1**:FL(35).

Polianthes L.

1. *P. tuberosa* L. [tuberose]

Oomycetes
Pythium debaryanum. Root rot. **1**:TX(82).
Pythium sp. Root rot. **1**:FL(35).
Deuteromycotina - Hyphomycetes
Botrytis cinerea. Blossom blight. **1**:HI(2903).
B. elliptica. Leaf spot. **1**:CA(97).
Cercospora sp. Leaf spot. **1**:TX(82).
Curvularia sp. Leaf spot. **1**:FL(35).
Fusarium sp. Leaf spot. **1**:FL(35).
Helminthosporium sp. Leaf spot, stem spot. **1**:TX(82).
Deuteromycotina - Other
Rhizoctonia solani. Root rot. **1**:FL(35), NC(82,1303).
Rhizoctonia sp. Root rot. **1**:HI(2903).
Sclerotium rolfsii. Southern blight. **1**:FL(35), HI(2903).

Sansevieria Thunb.

1. *S. trifasciata* Prain var. *laurentii* (De Wild.) N. E. Brown [mother-in-law's-tongue, snake-plant]
2. *S. zeylanica* (L.) Willd. [Ceylon bowstring hemp]
3. *Sansevieria* sp.

Oomycetes
Phytophthora sp. Root and stem rot. **1**:FL(35). **3**:FL(35).
Pythium sp. Root rot, leaf blight. **2**:FL(35), MS(2720). **3**:FL(35).
Ascomycotina
Glomerella cingulata. Leaf blight, root rot. **2**:WA(3202). **3**:TX(82).
Deuteromycotina - Hyphomycetes
Fusarium moniliforme. 'Red leaf spot.' **2**:MS(2720), WA(3202). **3**:FL(35,82), MO(82), WA(82).
Kutilakesopsis sp. Cutting rot. **3**:FL(35).
Deuteromycotina - Coelomycetes
Colletotrichum gloeosporioides. Leaf spot. **1**:FL(35). **2**:FL(35), MS(2720). **3**:FL(82).
Colletotrichum sp. Anthracnose. **3**:FL(35).
Phyllosticta sp. Leaf spot. **3**:FL(35).
Deuteromycotina - Other
Rhizoctonia solani. Root and stem rot. **1**:FL(35). **3**:FL(35).
Sclerotium rolfsii. 'Southern blight.' **2**:MS(2720).

Yucca L.

1. *Y. aloifolia* L. [aloe yucca, dagger-plant, Spanish-bayonet]
2. *Y. angustifolia* Pursh
3. *Y. baccata* Torr. [banana yucca, blue yucca, datil, Spanish-bayonet]
4. *Y. brevifolia* Engelm. (=*Y. arborescens* (Torr.) Trel.) [Joshua-tree]
5. *Y. elata* Engelm. (=*Y. radiosa* Trel. in L. H. Bailey) [palmella, soaptree, soapweed]
6. *Y. elephantipes* Regel [izote, spineless yucca]
7. *Y. filamentosa* L. [Adam's-needle, bear-grass, needle-palm, silk-grass, spoonleaf yucca]
8. *Y. flaccida* Haw. [weakleaf yucca]
9. *Y. glauca* Nutt. [soapweed, soapwell]
10. *Y. gloriosa* L. [Lord's-candlestick, palm-lily, Roman-candle, Spanish-dagger]
11. *Y. recurvifolia* Salisb.
12. *Y. rupicola* Scheele [twisted-leaf yucca]
13. *Y. schidigera* Roezl ex K. E. Ortgies (=*Y. mohavensis* Sarg.)
14. *Y. schottii* Engelm. (=*Y. macrocarpa* Engelm.)
15. *Y. smalliana* Fernald [Adam's-needle, bear-grass]
16. *Y. whipplei* Torr. [our-lord's-candle]
17. *Yucca* sp.

Oomycetes
Phytophthora sp. Basal stem rot. **6**:FL(35).
Pythium splendens. Root rot. **1**:FL(35).
Pythium sp. Basal stem rot; root rot. **6**:FL(35).
Ascomycotina
Allantonectria yuccae. Root rot, basal stem rot. **13**:CA(81).
Anthostomella nigro-annulata. On dead leaves. **1**:SC(82). **7**:FL(82), GA(82), NC(82).
Apodothina pringlei. Leaf spot. **1**:FL(35,82). **13**:CA(81). **14**:AZ(82), CA(82,195), NM(82). **16**:CA(81).
Botryosphaeria disrupta. Leaf spot. **7**:MS(82).
Botryosphaeria sp. Leaf spot. **1**:FL(35).
Cenangium yuccae. On dead leaves. **13**:CA(81).
Diaporthe arctii. On leaves. **7**:GA(82,1438).
Didymosphaeria clementsii. On dead leaves. **9**:CO(82), WY(82,1195).
D. yuccogena. On dead leaves. **9**:CA(82).
Discochora philoprina. **17**:IL(650).
D. yuccae. **6**:FL(281). **17**:FL(281).
Dothidea conspicua. On leaves. **2**:CO(970). **9**:CO(82,1195), MT(82,3202). **17**:AZ(662), CO(1003), FL(2782).
Glomerella cingulata. On stems. **17**:TX(82).
Guignardia sp. Leaf spot. **17**:FL(35).
Hyponectria mohavensis. **4**:CA(81).
Leptosphaeria clavigera. **7**:GA(1438).
L. filamentosa. Leaf spot. **1**:NC(1303). **4**:NC(1303). **7**:NJ(82). **17**:TX(82).
L. pacifica. Secondary invader in leaf spots. **13**:CA(81).
Lojkania decorticata. On old leaf tips. **17**:CA(189).
Lophiostoma caulium. **7**:GA(1438).
Montagnula thuemeniana. On leaves. **1**:GA(82,1438), MS(82), SC(82).
Mycosphaerella acervata. **9**:OK(660).
M. tassiana. Leaf spot. On leaves. **9**:OK(82,2872), WA(82,3202).
M. yuccae. Leaf spot. **7**:AL(525). **9**:ND(377).
Mycosphaerella sp. On leaves. **7**:AL(82). **9**:ND(82).
Paraphaeosphaeria obtusispora. Leaf spot. **1**:FL(35,3369). **6**:FL(35). **10**:AL(82), MS(82,2720,3251). **15**:OR(3202), WA(3202). **17**:FL(35).
Phomatospora argyrostigma. On leaves. **7**:AL(82), GA(1438), VA(82). **17**:OH(82).
Phragmodothis circumscissa. On leaves. **10**:FL(82). **17**:FL(35).
Physalospora sp. Anthracnose. **1**:FL(35).
Pleospora ellisii. Secondary invader in leaf spots. **4**:NC(1303).
Stomatogene agaves. On leaves. **13**:CA(81,2434). **16**:CA(81). **17**:CA(82).
Systremma sp. On leaves. **9**:CO(82).
Teichospora aridophila. On dead leaves. **4**:CA(81).
Venturia weiriana. On leaves. **9**:WY(82).
Basidiomycotina - Uredinales
Puccinia sporoboli var. *robusta*. **9**:NE(82).
Basidiomycotina - Other
Ceriporia tarda. **5**:AZ(1202). **17**:AZ(128).
Erythricium chaparralum. On dead flower stalks. **17**:AZ(438).
Phanerochaete tuberculata. **4**:AZ(1202).
Tomentella ferruginea. On dead stem. **17**:AZ(2030,2039).

Deuteromycotina - Hyphomycetes
Alternaria alternata. **7**:WV(82,3907).
Alternaria sp. Leaf spot. **1**:FL(35). **6**:FL(35). **17**:CO(49), FL(35).
Cercospora floricola. Flower blight. **12**:TX(82,581).
C. yuccae. Leaf spot. **17**:GA(581).
Cercospora sp. Leaf spot. **1**:FL(35). **17**:FL(35).
Cladosporium atriellum. On decayed fruits. **1**:SC(634).
Dicyma pulvinata. On leaves. **10**:AL(2482).
Epicoccum nigrum. Leaf spot. **17**:TX(82).
Fusarium lateritium. Leaf spot. **1**:FL(35). **6**:FL(35).
Heterosporium sp. **17**:WA(3202).
Periconia maculans. On leaves. **4**:CA(82). **17**:AZ(82), CA(82), SC(82,721), TX(82), WY(82).
Stemphylium botryosum. On panicles. **7**:SC(634).
Stigmina concentrica. Leaf spot. **4**:TX(82). **7**:CT(82), IL(312), NJ(82), OK(82,2872), TX(82). **9**:KS(82), MS(82,2720), OK(82,2872). **10**:IA(82). **17**:GA(82), TX(82).
Torula herbarum. On leaves. **1**:LA(82). **17**:CO(1195,3378).
Zygosporium oscheoides. On leaves. **10**:AL(2482).
Deuteromycotina - Coelomycetes
Alpakesa uniseptata. On leaves. **12**:TX(2501,3549).
A. yuccifolia. On dead leaves. **7**:GA(2501), KS(2501), NC(1303). **17**:AL(3549), GA(82,2501,3549), IA(82), KS(3549), WA(82,2501,3202,3549).
Botryodiplodia circinans. Leaf spot. **1**:FL(35,2782). **17**:TX(82).
Catenophora yuccae. On leaves. **3**:CO(2545). **17**:AZ(2545), CO(2545).
Colletotrichum dematium. On leaves. **7**:NJ(82), PA(82), WV(82,3907).
C. gloeosporioides. Leaf spot. **7**:MO(82,962). **17**:TX(82).
Coniothyrium bartholomaei. Leaf spot. **13**:CA(81). **16**:CA(81).
Cylindrosporium sp. Leaf spot. **1**:NC(1303). **10**:NC(1303).
Cytosporina sp. Leaf necrosis, gray leaf spot. **1**:FL(35). **17**:FL(35).
Hendersonia sp. Tip dieback. **1**:FL(35).
Kellermania anomala. Secondary leaf blight. **1**:FL(35,82,2502). **4**:AZ(2502), CA(82,2502), NV(1195). **5**:NM(82,2502). **7**:KS(82), NE(82), OK(82), TX(82,2502). **9**:Range of host(82), CO(1195,3378). **10**:MS(82,2720). **16**:CA(2502). **17**:AZ(3549), CA(82,2502,3549), FL(35,3549), NM(3549), TX(82,2502,3549).
K. yuccigena. On leaves. **2**:CO(2502), KS(2502), MT(2502). **5**:NM(1003,1195,2502). **7**:KS(2502), OK(2872), TX(2502). **9**:CO(2502), MO(2502), MT(2502,3202), NE(2502), OK(2872), SD(2243), TX(2502), WA(3202), WY(1195,2502). **16**:CA(2502). **17**:CA(3549), CO(2502), KS(2502), MO(2502), ND(2502,3549), NE(2502), NM(3549), NV(3549), SD(2502), WY(3549).
Macrophoma smilacina. Leaf spot. **1**:FL(35). **6**:FL(35). **17**:AL(82), FL(82,963).
Macrophoma sp. Leaf spot. **1**:FL(35). **6**:FL(35). **17**:FL(35).
Microsphaeropsis concentrica. Secondary invader in leaf spots. **1**:Range of host(82), FL(35,3369), NC(1303). **4**:NC(1303). **6**:FL(35). **7**:Range of host(82), AL(2506), AZ(3984), FL(3369), MA(1390), MS(2720), WI(1337), WV(3907). **8**:Range of host(82), WV(3907). **9**:Range of host(82), AZ(3984), CA(1003), CO(1195), NC(1303), NM(1003). **10**:Range of host(82). **11**:AL(2506). **15**:FL(3369). **17**:FL(35), HI(2903).
Pestalozziella yuccae. Secondary leaf spot. **17**:TX(82).
Phyllosticta maculicola. **5**:NM(82).
Phyllosticta sp. Leaf spot. **4**:TX(82).
Piptarthron macrosporum. On leaves. **4**:AZ(2502). **14**:TX(2502). **17**:CA(2783), CO(82).
P. yuccae. On leaves. **4**:NV(2502). **17**:AZ(662), CO(2502,3549), FL(2502,3549), NV(3549).
Septoria yuccae. Leaf spot. **6**:FL(35). **10**:PA(82).
Septoria sp. Leaf spot. **17**:GA(82), NC(82,1303), TX(82).
Sphaeropsis filamentosa. On leaves. **7**:SC(82).
Stagonospora gigantea. Leaf spot. **3**:CA(81). **4**:NC(1303). **13**:CA(81). **16**:CA(81). **17**:CA(82), FL(35).
Deuteromycotina - Other
Sclerotium rolfsii. Southern blight, stem rot. **7**:MD(82). **17**:FL(35).

Aizoaceae

Carpobrotus N. E. Brown
1. *C. chilensis* (Molina) N. E. Brown [sea-fig]
2. *C. edulis* (L.) L. Bolus [hottentot-fig]
Oomycetes
Phytophthora cryptogea. Root rot. **1**:CA(2223). **2**:CA(2223).
Pythium aphanidermatum. Root rot. **1**:CA(2223). **2**:CA(2223).

Lampranthus N. E. Brown
1. *L. aureus* (L.) N. E. Brown
2. *L. glomeratus* N. E. Brown [akulikuli]
Oomycetes
Pythium sp. Stem rot. **1**:FL(35).
Deuteromycotina - Other
Rhizoctonia sp. Root rot. **2**:HI(2903).

Lithops N. E. Brown
1. *Lithops* sp.
Deuteromycotina - Hyphomycetes
Bipolaris cactivora. Stem rot. **1**:FL(35).
Fusarium oxysporum. Isolated from rotted leaves. **1**:Unk.(1165).
Fusarium sp. Stem rot. **1**:FL(35).

Mesembryanthemum L.
1. *M. crystallinum* L. [ice-plant]
2. *Mesembryanthemum* sp.
Ascomycotina
Gibberella cyanogena. On stems. **2**:CA(82).
Deuteromycotina - Hyphomycetes
Torula herbarum. Sooty mold. **2**:CA(82).
Deuteromycotina - Coelomycetes
Camarosporium ellipticum. On stems. **2**:CA(82).
Diplodia sp. Leaf spot. **1**:FL(35).
Pestalotia sp. Leaf spot. **1**:FL(35).

Sesuvium L.
1. *S. portulacastrum* (L.) L. [cenicilla, sea purslane]
Oomycetes
Albugo trianthemae. White blister. **1**:TX(82).
Basidiomycotina - Uredinales
[*Puccinia aristidae*]. **1**:TX(82).
Deuteromycotina - Hyphomycetes
Phymatotrichopsis omnivora. Root rot. **1**:TX(82).

Tetragonia L.
1. *T. tetragonioides* (Pall.) Kuntze (=*T. expansa* Thunb. ex J. A. Murray) [New Zealand spinach, tsuru-na, warrigal-cabbage]
Oomycetes
Peronospora effusa. Leaf spot. **1**:FL(35).
Zygomycotina
Cunninghamella elegans. **1**:WA(3202).
Deuteromycotina - Hyphomycetes
Alternaria sp. Leaf spot. **1**:OK(2872).
Cercospora sp. Leaf spot. **1**:FL(35), IN(82), MA(82).
Helminthosporium sp. Leaf spot. **1**:TX(82).
Ulocladium consortiale. **1**:WA(3202).
Verticillium albo-atrum. **1**:CA(3896), WA(3202).

Trianthema L.
1. *T. portulacastrum* L. [horse purslane]
Oomycetes
Albugo trianthemae. White rust. **1**:NM(82).
Deuteromycotina - Hyphomycetes
Gibbago trianthemae. On leaves. **1**:TX(3273).

Alismataceae

Alisma L.
1. *A. geyeri* Torr.
2. *A. gramineum* Lej.
3. *A. plantago-aquatica* L. (=*A. triviale* Pursh, *A. plantago* L.) [large-flowered water-plantain]
4. *A. subcordatum* Raf.

5. *Alisma* sp.
Plasmodiophoromycetes
Ligniera junci. On roots. **3:**NY(82).
Chytridiomycetes
Physoderma maculare. Leaf gall. **3:**ID(3202), MD(3202), WA(3202). **5:**MT(82), NY(82), WI(82).
Ascomycotina
Massariosphaeria rubelloides. On dead stems. **3:**SD(970).
Basidiomycotina - Ustilaginales
Doassansia alismatis. Leaf smut, yellow spot. **1:**UT(82). **2:**UT(1077). **3:**C states(82), NE states(82), CA(82,1077), IA(1077), ID(1195,3202), KS(1077), MI(1077), MN(1077,2874), MO(1077), MT(1077,3202), ND(82,378,1077,3212), NE(1077), NY(1077), WA(82,1077,3202), WI(1077).
Deuteromycotina - Hyphomycetes
Cercospora alismatis. Leaf spot. **3:**MN(2874), ND(378). **4:**MN(2874). **5:**DE(82), IA(82), MT(82), ND(82), NY(82), WI(82).
C. callae. **3:**DE(3543).
Ramularia alismatis. Leaf spot. **3:**IA(82), MN(2874), MT(82), ND(82), NY(82), WA(82,3202), WI(82). **4:**Unk.(82), IA(82), MN(2874), NY(82).
Deuteromycotina - Other
Sclerotium hydrophilum. **5:**MN(2888).

Echinodorus Rich. ex Engelm. in A. Gray
1. *E. brevipedicellatus* (Kuntze) Buchenau
2. *E. cordifolius* (L.) Griseb. [Texas mud-baby]
3. *E. radicans* Engelm. in A. Gray
4. *E. rostratus* (Nutt.) Engelm. in A. Gray
5. *Echinodorus* sp.
Oomycetes
Aphanomyces euteiches. Blight, petiole necrosis. **1:**FL(35,2978).
Pythium sp. Root rot. **5:**FL(35).
Basidiomycotina - Ustilaginales
Burrillia echinodori. Leaf smut. **2:**CA(82,1077), FL(35,1077), TX(82,1077). **4:**CA(82), FL(1077), TX(82).
Doassansia alismatis. **2:**FL(35,1077). **4:**CA(1077).
Deuteromycotina - Hyphomycetes
Cercospora sp. Leaf spot. **3:**TX(82).
Deuteromycotina - Other
Sclerotium sp. Leaf rot, petiole rot. **5:**FL(35).

Sagittaria L.
(= *Lophotocarpus* Th. Durand.)
1. *S. calycinus* Engelm. ex Torr. (=*Lophotocarpus calycinus* (Engelm. ex Torr.) J. G. Sm.)
2. *S. cuneata* E. Sheld. [wapato]
3. *S. graminea* Michx. [coastal arrowhead, slender arrowhead]
4. *S. lancifolia* L. [arrowhead]
5. *S. latifolia* Willd. (=*S. chinensis* Pursh) [common arrowhead, duck-potato, wapato]
6. *S. montevidensis* Cham. & Schlechtend. [giant arrowhead]
7. *S. rigida* Pursh
8. *S. sagittifolia* L. (=*S. arifolia* Nutt. ex J. G. Sm., *S. variabilis* Engelm. in A. Gray) [Chinese arrowhead, old-world arrowhead, swan-potato]
9. *Sagittaria* sp.
Basidiomycotina - Uredinales
Puccinia sp. **9:**FL(35).
Basidiomycotina - Ustilaginales
Burrillia limnanthemi. Leaf rot, petiole rot. **5:**IL(1077), NE(1077), WI(1077).
B. pustulata. **5:**IL(82,1077,3724), NE(82,1077), WI(82,1077).
Doassansia alismatis. **5:**MO(1077).
D. deformans. Leaf gall. **4:**FL(1077). **5:**NE states(82), CT(1077), FL(1077), IA(1077), IL(1077), IN(1077), MA(1077), ME(1077), MN(1077), MO(82,1077), NE(1077), NJ(1077), NY(1077), RI(1077), SD(82,1077), WI(1077). **9:**FL(35,82,1077), TX(82,1077).
D. opaca. Blister, leaf smut. **5:**NE states(82), CT(1077), DE(1077), IL(82,1077), IN(1077), MA(1077), NY(1077), WI(82,1077). **8:**CT(1077), DE(1077), IL(1077), IN(1077), MA(1077), NY(1077), RI(1077).
D. sagittariae. Blister, leaf smut. **1:**WI(82,1077). **2:**IL(1077), WI(1077), WY(1077). **3:**DE(1077), IL(1077). **4:**FL(1077). **5:**C states(82), NE states(82), CT(1077), IN(1077), KS(1077), MO(1077), MT(82,1077,3202), ND(378,1077), OH(1077), TX(82,1077), WI(1077). **6:**AR(3593). **7:**IN(82,1077), WI(82,1077). **8:**IN(1077), KS(1077), MO(1077), NY(1077), OH(1077), WI(1077). **9:**AR(82,1077), DE(82), IL(82), ME(1077), MO(82,1077), SD(1077), VT(82,1077), WI(82), WY(82).
Doassansiopsis furva. Leaf spot. **7:**WI(82,3724).
D. intermedia. Leaf smut. **5:**IA(82,1077), MN(82,1077,2874), ND(82,1077), NH(82,1077,3724), WI(82,1077). **7:**WI(1077). **9:**LA(1077).
Pseudodoassansia obscura. Leaf smut. **5:**CT(82,1077), MA(82,1077,3724), WI(1077).
Deuteromycotina - Hyphomycetes
Cercospora alismatis. Leaf spot. **1:**MN(2874). **2:**MN(2874). **7:**MN(2874), WI(82). **9:**TX(82).
C. sagittariae. Leaf spot. **5:**OK(2873).
Cercospora sp. Leaf spot. **9:**FL(35).
Ramularia alismatis. Leaf spot. **7:**WI(82).
Deuteromycotina - Coelomycetes
Colletotrichum sp. Anthracnose. **8:**HI(2903).
Gloeosporium confluens. Leaf spot. **5:**IA(82), MA(82), TX(82), WI(82).
Marssonina sp. Leaf spot. **7:**LA(82).
Pestalotia sp. Leaf spot. **8:**HI(2903).
Deuteromycotina - Other
Sclerotium hydrophilum. **8:**WI(2888).

Amaranthaceae

Acnida L.
This genus is considered a synonym of *Amaranthus*, but appropriate combinations could not be located.
1. *A. cuspidata* Bertol. ex Spreng.
2. *A. tamariscina* (Nutt.) A. Wood [water-hemp]
3. *Acnida* sp.
Oomycetes
Albugo amaranthi. White rust. **1:**FL(35).
A. bliti. White rust. **2:**MN(2874), OK(2872).
Deuteromycotina - Hyphomycetes
Alternaria sp. Root rot. **3:**FL(35).
Phymatotrichopsis omnivora. Root rot. **2:**TX(82).
Deuteromycotina - Other
Sclerotium rolfsii. Southern blight. **3:**FL(35).

Alternanthera Forssk.
1. *A. achyrantha* R. Br.
2. *A. ficoidea* (L.) R. Br. ex Roem. & Schult. cv. 'bettzichiana' (=*A. bettzichiana* (Regel) G. Nichols.) [calico-plant]
3. *A. ficoidea* (L.) R. Br. ex Roem. & Schult. cv. 'amoena' (=*A. amoena* (Lem.) Voss) [parrotleaf, shoofly joyweed]
4. *A. philoxeroides* (Mart.) Griseb. [alligator-weed, periquito-saracura]
5. *Alternanthera* sp.
Oomycetes
Pythium sp. Root rot. **5:**FL(35).
Basidiomycotina - Uredinales
Uredo maculans. **5:**FL(35).
Basidiomycotina - Other
Thanatephorus cucumeris. **4:**LA(3524).
Deuteromycotina - Hyphomycetes
Alternaria alternantherae. Root rot, leaf and stem spot. **4:**LA(1605), NC(1303). **5:**FL(35).
Cercospora alternantherae. Leaf spot. **1:**LA(82,581).
Fusarium oxysporum. Root rot, wilt. **2:**NJ(82).
Fusarium sp. Root and stem rot. **5:**FL(35).
Deuteromycotina - Coelomycetes
Colletotrichum sp. Anthracnose. **5:**FL(35).
Phyllosticta amaranthi. Leaf spot. **5:**NJ(82).
Deuteromycotina - Other
Rhizoctonia solani. Root rot. **2:**IL(82), NJ(82). **4:**LA(82). **5:**FL(35).
Rhizoctonia sp. Root rot. **3:**HI(2903).
Sclerotium rolfsii. Southern blight. **5:**HI(2903).

Amaranthus L.

1. *A. albus* L. [prostrate amaranth, tumble pigweed, tumbleweed]
2. *A. blitoides* S. Wats. [prostrate amaranth, prostrate pigweed]
3. *A. cannabinus* (L.) Sauer (=*Acnida cannabina* L.)
4. *A. caudatus* L. (=*A. edulis* Speg.) [Inca-wheat, love-lies-bleeding, tassel-flower]
5. *A. gracilis* Desf. [pakai, slender amaranth]
6. *A. graecizans* L. Specimens of *Amaranthus blitoides* are sometimes misidentified as this species. [prostrate amaranth, spreading pigweed, tumbleweed]
7. *A. hybridus* L. [green amaranth, pigweed, smooth pigweed, wild-beet]
8. *A. powellii* S. Wats.
9. *A. retroflexus* L. [green amaranth, pigweed, redroot, redroot pigweed, rough pigweed, wild-beet]
10. *A. spinosus* L. [spiny amaranth, spiny pigweed]
11. *A. tricolor* L. (=*A. gangeticus* L.) [Chinese amaranth, Ganges amaranth, Joseph's-coat, tampala]
12. *A. tuberculatus* (Moq.) Sauer (=*Acnida tuberculata* Moq.)
13. *A. viridis* L. [green amaranth, slender amaranth]
14. *Amaranthus* sp.

Plasmodiophoromycetes

Ligniera junci. On roots. **7**:NC(1303).
Polymyxa betae. On roots. **9**:NE(2025).

Oomycetes

Albugo bliti. White rust. **2**:Range of Host(82), ND(377). **6**:Range of host(82), MN(2874), MT(3202), ND(377), SD(2243), WA(3202), WI(1343), WY(1195). **7**:E states(82), FL(35,1231), NE(82), TX(82). **8**:WI(1354). **9**:ID(3202), MA(1390), MN(2874), MS(2720), MT(3202), NC(1303), ND(377), NY(2456), OK(2872), SD(2243), WA(3202), WI(1345), WV(3907), WY(1195). **10**:Range of host(82), NC(3944), OK(2872). **12**:IA(82), KS(82), WI(82). **13**:MN(2874). **14**:Range of host(82), FL(35), NY(2456).
Peronospora farinosa. Downy mildew. **14**:NJ(82), NY(82).
Pythium aphanidermatum. Root, seedling and stem rot. **11**:FL(35).
P. debaryanum. Damping-off. **4**:CT(82).

Zygomycotina

Choanephora cucurbitarum. Leaf blight. **9**:FL(35). **10**:FL(35).

Ascomycotina

Leptospora rubella. **9**:PA(3248). **14**:PA(3248).
Lophiostoma caulium. **6**:ND(377).
Sclerotinia sclerotiorum. Stem rot. **9**:CO(447), MT(82,3202).

Basidiomycotina - Uredinales

[*Puccinia aristidae*]. **2**:CO(82). **6**:CO(82), TX(82). **9**:CO(82), TX(82). **14**:FL(35).

Basidiomycotina - Other

Thanatephorus cucumeris. **6**:WA(3202). **9**:WA(3202).

Deuteromycotina - Hyphomycetes

Alternaria amaranthi. Leaf spot. **9**:IN(82), KS(82), NY(82), TX(82), UT(82), WA(82). **11**:FL(35).
A. solani. Early blight. **10**:HI(2903). **11**:HI(2903).
A. tenuissima. Leaf spot. **9**:NC(1303).
Alternaria sp. **9**:OK(2872).
Cercospora acnidae. Leaf spot. **3**:DE(82,3543). **12**:WI(82,1335).
C. brachiata. Leaf spot. **1**:WI(1349). **2**:DE(3543), WI(1349). **9**:DE(82,3543), FL(82), WI(1349). **10**:TX(82).
C. canescens. Leaf spot. **14**:AL(82), MD(82).
Cercospora sp. Leaf spot. **9**:HI(2903).
Phymatotrichopsis omnivora. Root rot. **4**:TX(82). **6**:TX(82). **9**:TX(82). **10**:TX(82). **14**:TX(82).
Ramularia sp. Leaf spot. **9**:IA(82).
Verticillium albo-atrum. **9**:CA(422). **14**:OR(3202).

Deuteromycotina - Coelomycetes

Colletotrichum gloeosporioides. Anthracnose. **10**:AL(82).
C. truncatum. Isolated from leaves. **9**:IL(1482).
Gloeosporium amaranthicola. Anthracnose. **11**:NC(1303).
Macrophomina phaseolina. Charcoal rot. **10**:HI(2903).
Phoma amaranthi. On stems. **7**:NJ(82).
P. amaranthicola. On inflorescences and stems. **6**:OR(82,3202).
P. longissima. On stems. **9**:OH(82), TX(82).
Phoma sp. Stem necrosis. **14**:FL(35).
Phyllosticta amaranthi. Leaf spot. **8**:WI(1349). **9**:AL(82), KS(82), KY(82), MS(82), NJ(82), TX(82), WI(1358). **12**:NY(82), WI(1357).
Phyllosticta sp. Leaf spot. **10**:FL(35).
Septoria smardosii. Leaf spot. **14**:FL(35).
Septoria sp. On leaves. **9**:WI(1345).

Deuteromycotina - Other

Rhizoctonia solani. Root and stem lesions, root rot. **2**:IL(82), WA(82). **6**:TX(82), WA(82). **9**:NE(336), TX(82), WA(82).
Rhizoctonia sp. Root rot. **5**:HI(2903).
Sclerotium rolfsii. Stem rot. **9**:FL(82), LA(82). **10**:FL(82).

Celosia L.

1. *C. argentea* L. [veludo-branco, woolflower]
2. *C. argentea* L. var. *cristata* (L.) Kuntze (=*C. cristata* L.) [cockscomb, crista-de-galo]
3. *Celosia* sp.

Oomycetes

Phytophthora nicotianae var. *parasitica*. Damping-off. **2**:TN(1027).
Pythium sp. Root rot. **2**:HI(2903).

Deuteromycotina - Hyphomycetes

Alternaria sp. Leaf spot. **1**:NJ(82). **2**:FL(35).
Cercospora celosiae. Leaf spot. **1**:AL(82), OK(82,2872). **2**:FL(35).
Fusarium sp. Stem rot, wilt. **2**:FL(35). **3**:OK(2872).

Deuteromycotina - Coelomycetes

Macrophomina phaseolina. Charcoal rot. **1**:TX(82).
Phyllosticta sp. Leaf spot. **1**:NJ(82).

Deuteromycotina - Other

Rhizoctonia solani. Damping-off, root and stem rot. **1**:CT(82), NC(1303).
Rhizoctonia sp. Root rot. **2**:HI(2903).

Froelichia Moench

1. *F. floridana* (Nutt.) Moq.
2. *F. floridana* (Nutt.) Moq. var. *campestris* (Small) Fernald
3. *Froelichia* sp.

Oomycetes

Albugo bliti. White rust. **3**:NE(82), TX(82).

Deuteromycotina - Hyphomycetes

Alternaria amaranthi. Leaf spot. **1**:FL(35).
Cercospora crassoides. Leaf spot. **1**:OK(2873), WI(1340). **2**:FL(35). **3**:OK(82), TX(82), WI(82).
Helminthosporium sp. Leaf spot. **1**:FL(35).
Phymatotrichopsis omnivora. Root rot. **3**:TX(82).

Gomphrena L.

1. *G. globosa* L. [globe amaranth, perpetua]
2. *Gomphrena* sp.

Oomycetes

Albugo bliti. White rust. **2**:NM(82).

Deuteromycotina - Hyphomycetes

Alternaria gomphrenae. Leaf spot. **1**:HI(2903).
Alternaria sp. Leaf spot. **1**:OK(2872). **2**:OK(2872).
Cercospora gomphrenae. Leaf spot. **1**:GA(82), OK(82,581,2872), TX(82).

Deuteromycotina - Other

Rhizoctonia sp. Root rot. **1**:HI(2903).

Guilleminea Kunth

(= *Brayulinea* Small.)

1. *G. illecebroides* H. B. K. (=*Brayulinea densa* Small)

Basidiomycotina - Uredinales

Puccinia guillemineae. **1**:AZ(747,1195,1203,3981).

Iresine P. Br.

1. *I. celosioides* L.
2. *I. herbstii* Hook. [beefplant, beefsteak-plant, chicken-gizzard]
3. *I. rhizomatosa* Standl.
4. *Iresine* sp.

Oomycetes

Pythium sp. Root rot. **2**:FL(35).

Basidiomycotina - Ustilaginales

Thecaphora iresine. Inflorescence smut. **1**:IN(82,1077). **3**:IN(1077).

Deuteromycotina - Coelomycetes

Colletotrichum sp. Leaf necrosis. **2**:FL(35).
Septoria iresines. Leaf spot. **1**:OK(82,2873).

Deuteromycotina - Other

Rhizoctonia crocorum. Root rot. **4**:TX(82).
R. solani. Root and stem rot. **2**:FL(35), IL(82).

Tidestromia Standl.

1. *T. lanuginosa* (Nutt.) Standl. (=*Cladothrix lanuginosa* (Nutt.) Nutt. ex S. Wats.)

Basidiomycotina - Ustilaginales

Thecaphora thornberi. **1**:AZ(1077).

Anacardiaceae

Anacardium L.

1. *A. occidentale* L. [caju, cashew, maranon]

Oomycetes

Pythium splendens. Root rot. **1**:FL(35).

Deuteromycotina - Other

Sclerotium rolfsii. Seedling blight. **1**:FL(35,82).

Bouea Meisn.

1. *B. macrophylla* Griff.

Oomycetes

Pythium sp. Root rot. **1**:FL(35).

Cotinus Mill.

1. *C. coggygria* Scop. (=*Rhus cotinus* L.) [common smoke-tree, smokebush, Venetian-sumac, wigtree]
2. *C. obovatus* Raf. (=*C. americanus* Nutt.) [American smoke-tree, chittamwood]
3. *Cotinus* sp.

Oomycetes

Pythium sp. Root rot. **2**:FL(35).

Ascomycotina

Botryosphaeria dothidea. **1**:GA(1438).
B. obtusa. Dieback. **1**:GA(82).
Discohainesia oenotherae. Leaf spot. **1**:MD(82).
Massaria sp. **1**:GA(1438).
Physalospora sp. Twig dieback. **2**:FL(35).

Basidiomycotina - Uredinales

Pileolaria cotini-coggygriae. Autoecious. **1**:GA(3615,3616). **2**:AR(3594,3656).

Basidiomycotina - Other

Armillaria mellea. Root rot. **1**:CA(2897).

Deuteromycotina - Hyphomycetes

Aspergillus flavus. **2**:GA(1438).
A. tamarii. **2**:GA(1438).
Cercospora sp. Leaf spot. **1**:FL(35).
Phymatotrichopsis omnivora. Root rot. **1**:TX(82).
Pseudocercospora rhoina. Leaf spot. **1**:AL(82), TX(82).
Verticillium albo-atrum. Wilt. **1**:CT(82), IL(82,992,1558), NC(1303), NE(82), NJ(82), NY(82), OH(946,1558), RI(1558).
Verticillium sp. Verticillium wilt. **3**:CO(49).

Deuteromycotina - Coelomycetes

Macrophoma sp. Dieback. **1**:OK(2872).
Septoria rhoina. Leaf spot. **1**:CT(82), MA(82), NY(82), VA(82). **2**:Unk.(82).

Mangifera L.

1. *M. indica* L. [mango]
2. *Mangifera* sp.

Oomycetes

Pythium sp. Root rot. **1**:FL(35).

Ascomycotina

Antennellina hawaiiensis. **1**:HI(2903).
Botryosphaeria disrupta. Dieback. **1**:FL(35).
B. quercuum. Dieback. **1**:FL(35).
B. rhodina. On limbs. **1**:FL(82).
B. ribis. On limbs. **1**:HI(82,2903).
Capnodium sp. Sooty mold. **1**:FL(82).
Chaetothyrium mangiferae. **1**:HI(2903).
Diaporthe sp. On twigs and limbs. **1**:HI(82,2903).
Elsinoe mangiferae. Spot anthracnose. **1**:FL(35,82).
Erysiphe cichoracearum. **1**:CA(1110).
Glomerella cingulata. Anthracnose, flower and twig blight, withertip. **1**:FL(35,82), HI(82,2903), TX(82).
Guignardia citricarpa. Anthracnose. **1**:FL(2328).
Hypoxylon serpens var. *effusum*. **1**:HI(2903).
Lachnum abnorme. **1**:HI(2903).
Leptosphaeria sp. Leaf spot. **1**:FL(35).
Nummularia broomeiana. On trunks and larger limbs. **2**:SE states(1810).
Orbilia epipora. On logs. **1**:HI(2903).
Phaeosphaerella mangiferae. Leaf spot. **1**:HI(82,2903).
Physalospora abdita. On limbs. **1**:FL(82).
Rhytidhysteron rufulum. On dead limbs. **1**:FL(82).
Scutellinia hirta. **2**:HI(2784).
Vialaea minutella. On decaying and dead leaves. **1**:HI(2781).

Basidiomycotina - Other

Crepidotus bakerae. On dead stump. **1**:HI(3706).
Earliella scabrosa. On wood. **1**:HI(82,2903).
Hexagonia hydnoides. On wood. **1**:FL(82).
Phellinus gilvus. Sapwood rot. **1**:HI(82,2903).
Pycnoporus sanguineus. On wood. **1**:FL(35,82).
Septobasidium pilosum. Parasitic on scale insects. **1**:FL(35,82).
S. pseudopedicellatum. Brown felt. **1**:FL(35,82).

Deuteromycotina - Hyphomycetes

Alternaria sp. Leaf spot. **1**:FL(35).
Botrytis cinerea. Blossom blight. **1**:HI(2903).
Cercospora sp. Leaf spot. **1**:FL(35).
Curvularia sp. Leaf spot. **1**:FL(35).
Fusarium moniliforme. Witches' brooms. **1**:FL(35).
Johncouchia mangiferae. Parasitic on scale insects. **1**:FL(1683).
Oidium sp. Powdery mildew. **1**:CA(82), FL(35,82).
Phymatotrichopsis omnivora. Root rot. **1**:TX(82).
Verticillium albo-atrum. Verticillium wilt. **1**:FL(35,2256).
V. lecanii. Leaf spot. **1**:FL(35).

Deuteromycotina - Coelomycetes

Colletotrichum gloeosporioides. Anthracnose, blossom and peduncle blight, leaf spot, fruit stain and rot. **1**:HI(90,2903).
Diplodia sp. Twig blight. **1**:FL(35).
Macrophoma sp. Leaf spot. **1**:FL(35).
Pestalotiopsis mangiferae. Leaf spot. **1**:FL(35,82).
Phoma sp. Stem canker. **1**:FL(35).
Phomopsis sp. Dieback. **1**:FL(35,82).
Phyllosticta anacardiacearum. Leaf spot. **1**:FL(2,35).
P. mortonii. Leaf spot. **1**:FL(35,82), TX(82).
Septoria sp. Leaf spot. **1**:FL(35,82).
Sphaceloma mangiferae. On leaves. **1**:FL(1745).
Sphaeropsis sp. Dieback, stem gall. **1**:FL(35).

Deuteromycotina - Other

Rhizoctonia solani. Root rot; leaf necrosis. **1**:FL(35).

Pistacia L.

1. *P. chinensis* Bunge [Chinese pistachio]
2. *P. vera* L. [green-almond, pistachio-nut]
3. *Pistacia* sp.

Oomycetes

Phytophthora nicotianae var. *parasitica*. Leaf necrosis, root rot. **2**:CA(1110).
Phytophthora sp. **2**:CA(1110).
Pythium sp. **2**:CA(1110).

Zygomycotina

Rhizopus sp. **2**:CA(1110).

Ascomycotina

Botryosphaeria dothidea. **1**:GA(1438).
B. obtusa. On twigs. **1**:GA(82).
Rhytidhysteron rufulum. On dead limbs. **1**:FL(82).

Basidiomycotina - Other

Armillaria mellea. **2**:CA(1110).
Pellicularia koleroga. Thread blight. **1**:FL(82), TX(82). **2**:TX(82). **3**:TX(82).
Pleurotus ostreatus. Sapwood rot. **2**:CA(82,1110).
Schizophyllum commune. Sapwood rot. **2**:CA(1110). **3**:CA(82).

Deuteromycotina - Hyphomycetes

Alternaria sp. Leaf spot. **1**:FL(35). **2**:CA(1110). **3**:FL(35).
Aspergillus flavus. Isolated from nuts. **3**:CA(3621).
Botrytis cinerea. Shoot blight. **2**:CA(323,1110).
Fusarium oxysporum. **2**:CA(1110).
Fusarium sp. **2**:CA(1110).
Penicillium sp. **2**:CA(1110).
Phymatotrichopsis omnivora. Root rot. **1**:TX(82). **2**:CA(1110), TX(82). **3**:AZ(82), CA(82), TX(82).
Verticillium albo-atrum. Hadromycosis, wilt. **2**:CA(115,3366).

V. dahliae. Wilt. **2:**CA(114,1110). **3:**CA(2881).
Verticillium sp. Wilt. **2:**CA(115).

Deuteromycotina - Coelomycetes

Hyalotia pistacina. On leaves. **2:**MD(2215).
Phomopsis sp. **2:**CA(1110).
Phyllosticta lentisci. Leaf spot. **2:**TX(82,128).
Septoria sp. Leaf spot. **3:**MD(2221).

Deuteromycotina - Other

Rhizoctonia solani. Thread blight. **1:**FL(35). **3:**FL(35).

Rhus L.

(= *Toxicodendron* Mill.)

1. *R. aromatica* Aiton [fragrant sumac, polecat-bush, sweet-scented sumac]
2. *R. chinensis* Mill. [Chinese sumac, nutgall-tree]
3. *R. choriophylla* Woot. & Standl.
4. *R. copallina* L. [dwarf sumac, flame-tree sumac, mountain sumac, shining sumac, wing-rib sumac, winged sumac]
5. *R. diversiloba* Torr. & A. Gray (=*Toxicodendron diversilobum* (Torr. & A. Gray) Greene) [Pacific poison-oak, Western poison-oak]
6. *R. floridana* Mearns
7. *R. glabra* L. [scarlet sumac, smooth sumac, vinegar-tree]
8. *R. glabra* L. var. *cismontana* (Greene) Cockerell
9. *R. integrifolia* (Nutt.) Benth. & J. D. Hook. ex S. Wats. in L. C. Wheeler (=*Neostyphonia integrifolia* (Nutt.) Shafer in Britton) [lemonade sumac, lemonade-berry, sourberry]
10. *R. lancea* L. fil. [bastard-willow, karee, karoo-tree]
11. *R. laurina* Nutt. [laurel sumac]
12. *R. ovata* S. Wats. [sugar sumac, sugarbush sumac]
13. *R. radicans* L. (=*Toxicodendron radicans* (L.) Kuntze) [cow-itch, markry, mercury, poison-ivy]
14. *R. radicans* L. var. *rydbergii* (Small) Rehd.
15. *R. rhomboidea* Small
16. *R. sempervirens* Scheele [evergreen sumac]
17. *R. toxicodendron* L. (=*R. diversifolia* Engl. in DC.) [hiedra, poison-ivy, poison-oak]
18. *R. trilobata* Nutt. ex Torr. & A. Gray [skunkbush]
19. *R. typhina* L. [staghorn sumac, velvet sumac, Virginian sumac]
20. *R. venenata* DC.
21. *R. vernix* L. [poison sumac, poison-dogwood, poison-elder, swamp sumac]
22. *R. virens* Lindh. ex A. Gray [evergreen sumac, lentisco, tobacco sumac]
23. *Rhus* sp.

Chytridiomycetes

Synchytrium sp. On leaves and petioles. **13:**LA(1847).

Ascomycotina

Amphiporthe aculeans. Canker and dieback. **4:**E states(82), GA(1438), IA(82), MO(82), NC(1303), OK(82), WV(3907). **7:**E states(82), GA(1438), IA(82), OK(82). **19:**E states(82), IA(82), OK(82), WV(3907). **21:**E states(82), IA(82), OK(82). **23:**GA(1438), SC(634).
Botryosphaeria dothidea. Canker, dieback. **4:**GA(1438). **7:**GA(1438), OK(660).
B. obtusa. Inflorescence blight; possibly causing canker on limbs. **4:**Unk.(82). **7:**E states(82), MN(82), NE(82). **19:**E states(82), KS(82), MI(82), WV(3907). **21:**E states(82), KS(82), MI(82).
B. quercuum. Dieback. **7:**FL(35). **23:**FL(634).
B. rhodina. Inflorescence blight; possibly causing canker on limbs. **7:**FL(82).
B. ribis. Canker, dieback, umbrella disease. **4:**E states(82), MN(82), NE(82). **7:**E states(82), MN(82,2874), NE(82). **19:**WI(1342).
B. stevensii. On limbs. **18:**CA(82).
Botryosphaeria sp. **17:**Unk.(82).
Caliciopsis sp. Dead limbs. **4:**NC(1303).
Dasyscyphus citrinoalbus. On dead limbs. **2:**HI(2903).
D. corticalis. On roots. **5:**WA(3202). **17:**WA(849).
Diaporthe peckii. On stems. **17:**NJ(82), NY(82,195). **21:**MI(82).
D. spiculosa. On dead limbs. **4:**AL(82). **17:**IL(82). **19:**IL(82), MA(82), NJ(82), NY(82). **21:**IL(82), MA(82), NJ(82), NY(82).
Diatrype albopruinosa. **23:**IA(3629).
Diatrypella rhois. **5:**WA(3202).
Discohainesia oenotherae. Leaf spot. **1:**NJ(82). **4:**E states(82), WV(82,3907). **7:**E states(82), GA(1438), WV(82,3907). **17:**NY(82), TN(82), VA(82). **19:**E states(82), WV(82,3907). **21:**E states(82), WV(82,3907).
Dothidea puccinioides. **7:**GA(1438). **17:**GA(1438).
Erysiphe cichoracearum. Powdery mildew. **11:**CA(1155).
Eutypella elevans. **7:**PA(2925).
E. juglandicola. **7:**GA(1438).
Glyphium corrugatum. **7:**WA(3202).
Godronia rhois. **19:**NY(1382).
Hainesia lythri. **1:**WI(1342). **4:**WI(1347). **7:**WI(1342). **13:**WI(1342). **19:**WI(1342).
Hypoxylon rubiginosum. **4:**GA(1438). **7:**GA(1438). **17:**GA(1438).
Hysterographium flexuosum. On dead twigs. **11:**CA(2439).
Hysterographium sp. **7:**WA(3202).
Lachnellula theiodea. **20:**NJ(849).
Lophodermium petiolicola. **7:**OK(2873).
Metasphaeria anisometra. On twigs. **11:**CA(2439).
Mycosphaerella nigredo. On leaves. **7:**PA(82,967).
Nectria aureo-fulva. **20:**NJ(3085).
N. cinnabarina. Canker, dieback. **5:**CA(81), WA(3202). **19:**NY(82), OK(82,2873), WV(3907).
N. galligena. Trunk canker. **19:**NY(82), PA(82), VA(82).
Perrotia flammea. **18:**CO(849).
Phaeangium sphaeroides. On bleached wood. **11:**CA(2439).
Phomatospora leptasca. On dead limbs. **19:**NY(82,195).
Phomatospora sp. **4:**GA(1438).
Phyllactinia guttata. Powdery mildew. **13:**WA(3202). **17:**MI(82), WA(82).
Physalospora abdita. Inflorescence blight; possibly causing canker on limbs. **7:**FL(82).
Rhytidhysteron rufulum. On dead limbs. **4:**AL(82), FL(82), GA(82). **5:**CA(82). **7:**GA(1438). **17:**GA(82,1438), MS(82,2720).
Sphaerotheca macularis. Powdery mildew. **1:**CT(82), IN(82), PA(82). **4:**E & C states(82), FL(35), WI(1352). **7:**E & C states(82), GA(1438). **19:**E & C states(82), OH(950). **21:**E & C states(82).
S. pannosa. **7:**GA(1438).
Sphaerulina tarda. Leaf spot. On leaves. **5:**CA(81,82,1468).
Stictis radiata. Leaf spot. **5:**OR(3202). **23:**OR(3225).
[*Taphrina caerulescens*]. **4:**KS(2458).
T. purpurescens. Blister, leaf curl. **4:**AR(2454), GA(1438), KS(2454), MA(2454), MS(2720), NC(3944), NJ(2454), VA(2454). **7:**E states(82), SC states(82), CT(2454). **19:**E states(82), CT(82,2454), KS(82), MD(82), TX(82).
Telimena rhoina. Black spot. **9:**CA(82).
Thyronectria pyrrhochlora. **7:**GA(1438).
T. xanthoxyli. On limbs. **7:**CT(82), GA(82), MS(82). **9:**CA(82).
Valsa ceratosperma. **4:**GA(1438). **7:**GA(1438). **17:**GA(1438). **19:**NY(3415). **20:**NJ(3415).
Xylogramma nigerrima. **11:**CA(2439).

Basidiomycotina - Uredinales

Pileolaria brevipes. Autoecious. **6:**MS(2720). **13:**AZ(1195,1203), FL(35), IN(647), MI(2862), MN(2874), MT(3202), NC(1303), ND(378), SD(2243), WA(3202), WV(3907). **14:**AZ(3981). **23:**SC(634).
P. effusa. Autoecious. **7:**AZ(1203). **8:**Unk.(82). **14:**AZ(1203,3981). **15:**OK(2872). **17:**AZ(82), CO(82).
P. patzcuarensis. Autoecious. **18:**AZ(1203,3981), CO(82), NM(82,1195), OK(82,2872).
Uredo mexicana. **3:**AZ(1194,1203,3981). **22:**AZ(1203), TX(742).

Basidiomycotina - Other

Antrodia albida. On wood. **23:**MN(2874).
A. sinuosa. On wood. **11:**CA(2439).
Armillaria mellea. Root rot. **1:**CA(2897). **4:**CA(2897). **11:**CA(82). **12:**CA(2897).
A. tabescens. Root rot. **4:**FL(35,82).
Bjerkandera adusta. On trunks. **4:**Range of host(82). **7:**Range of host(82). **17:**MD(82), WV(82). **19:**Range of host(82). **21:**Range of host(82).
Ceratobasidium cornigerum. On bark. **7:**IA(128).
Corticium roseum. On wood. **7:**GA(491,1439).
Cytidia patelliformis. **5:**WA(3202).
Daedaleopsis confragosa. On wood. **13:**NC(1303).
Dendrophora albobadia. **21:**GA(543).
D. versiformis. **19:**NY(543). **23:**AL(543).
Ganoderma lucidum. **10:**AZ(1202).
Gloeoporus dichrous. **7:**ID(3202).
Hyphoderma setigerum. **5:**WA(3202).

Irpex lacteus. On trunks. **4:**Range of host(82), IN(654). **7:**Range of host(82), IN(654). **19:**Range of host(82). **21:**Range of host(82).
Peniophora cinerea. **5:**WA(3202).
P. decorticans. **5:**WA(3202).
Perenniporia medulla-panis. On wood. **5:**CA(82). **17:**MI(82).
Phellinus gilvus. On dead or injured stems. **17:**MD(82).
P. punctatus. On wood. **21:**MI(82).
Porodisculus pendulus. White rot. **4:**FL(35,82), MD(82), NC(1303), WV(82,3907). **7:**FL(82), MD(82), WV(82,3907). **19:**FL(82), MD(82), WV(82,3907). **21:**FL(82), MD(82), WV(82,3907).
Schizophyllum commune. On wood. **4:**Range of host(82), FL(35). **7:**Range of host(82), FL(35), IN(654). **17:**VA(82). **19:**Range of host(82). **21:**Range of host(82). **23:**FL(35).
Scytinostroma galactinum. White root rot. **7:**MD(82).
Steccherinum ochraceum. On wood. **4:**PA(82). **7:**PA(82). **19:**PA(82). **21:**PA(82).
Stereum hirsutum. **4:**WV(543). **17:**NC(3944).
S. ochraceo-flavum. **4:**FL(543). **21:**MD(543).
S. striatum. **4:**WV(3907).
Stilbum rhois. On dead limbs. **4:**NC(1303). **7:**DE(3543).
Trametes hirsuta. On wood. **4:**Range of host(82). **7:**Range of host(82). **17:**MD(82). **19:**Range of host(82). **21:**Range of host(82).
T. versicolor. On trunks. **4:**Range of host(82). **7:**Range of host(82). **17:**MD(651), NC(3944). **19:**Range of host(82). **21:**Range of host(82).
Trichaptum biforme. On trunks. **4:**Range of host(82). **7:**Range of host(82). **19:**Range of host(82). **21:**Range of host(82).
Wolfiporia cocos. On roots. **4:**FL(35,82).

Deuteromycotina - Hyphomycetes

Alternaria alternata. Leaf spot. **19:**CO(3611).
Alternaria sp. Leaf spot. **13:**WV(3907). **23:**FL(35).
Cercospora toxicodendri. Leaf spot. **7:**IN(82). **13:**NC(1303). **17:**IA(82), IN(82), KS(82), MA(82), MD(82), NJ(82), TX(82), WI(82,1330). **20:**DE(3543). **21:**DE(82). **23:**FL(35).
Cercospora sp. Leaf spot. **13:**FL(35).
Cercosporella californica. Leaf spot. **5:**CA(81,82,325).
Cladosporium aromaticum. On leaves. **1:**CA(82), WI(1352). **4:**WI(82,1340). **7:**NE(82), WI(82,1330). **19:**MS(2720), NY(82), WI(82), WV(82).
Conoplea globosa. **23:**WA(3202).
Cristulariella moricola. On leaves. **17:**AL(2056), GA(2056).
Cylindrocladium sp. Root rot. **23:**NC(1303).
Fusarium oxysporum f. sp. *rhois*. Wilt. **19:**CT(82,1533), VA(82,1533).
Fusarium sp. Wilt. **19:**CO(3611). **23:**CO(49).
Oidium sp. Powdery mildew. **1:**IL(2567), WI(82). **7:**IL(1052).
Phymatotrichopsis omnivora. Root rot. **4:**TX(82). **7:**TX(82). **16:**TX(82). **17:**TX(82). **18:**TX(82). **19:**TX(82). **21:**TX(82).
Podosporium rigidum. **17:**DE(3543), MS(2720).
Pseudocercospora rhoina. Dieback, leaf spot. **1:**AL(82), CA(82), MO(82), OK(2872), WI(1344). **7:**Range of host(82), DE(3543), FL(35), IL(2568), MN(2874), MS(2720), OK(2872). **16:**TX(82). **18:**OK(2872). **19:**Unk.(82), MN(2874), WI(1339). **21:**Unk.(82), WI(1332). **23:**FL(35).
Sarcinella copallina. On leaves. **4:**AL(82,1681), DC(1681). **7:**MS(1681).
Sporidesmium inflatum. On dead limbs. **4:**NC(1303).
Stigmina pallida. Leaf spot. **7:**ID(3202). **17:**CO(82,3547), TX(82).
Verticillium albo-atrum. Wilt. **1:**IL(1558), MA(82). **7:**IA(82), IL(82,992,1558), MN(82,2874). **18:**IA(82). **19:**IA(82), IL(1558), MN(2874).
Verticillium sp. Verticillium wilt. **23:**CO(49).

Deuteromycotina - Coelomycetes

Botryodiplodia compressa f. *toxicodendri*. On stems. **17:**NY(82).
Colletogloeum rhois. On leaves. **17:**NJ(3549).
Colletotrichum sp. Leaf spot. **13:**FL(35).
Coniothyrium rhois-radicantis. Leaf spot. **12:**CA(81). **16:**TX(82). **22:**TX(3597).
Coniothyrium sp. Canker. **23:**FL(35).
Cytospora rhuina. Dieback. **5:**WA(3202). **7:**ME(82), MI(82), NY(82). **19:**WV(82,3907).
C. sacculus. Dieback. **19:**AL(82), NY(82).
Cytospora sp. Dieback. **23:**CO(49).
Dendrophoma nigrescens. On dead limbs. **7:**NY(82).
Diplodia resurgens. On limbs. **18:**CA(82,637).
Discosia artocreas. Leaf spot. **17:**MS(2720).
Dothiorella radicans. On stems. **17:**CO(82), IL(82), NJ(82,970).
D. rhoina. On stems. **4:**MD(2783). **17:**CO(82), IL(82), NJ(82).
D. toxica. On stems. **17:**CO(82), IL(82,971), NJ(82).
Dothiorella sp. On dead limbs. **4:**NC(1303).
Gloeodes pomigena. Sooty blotch. **23:**NC(1303).
Gloeosporium toxicodendri. Leaf spot. **5:**CA(82), OR(82,3202), WA(82,3202). **7:**IL(2568). **17:**E states(82), KS(82), MS(82,2720), MT(82), OK(2872). **18:**CA(81), MT(82). **23:**Unk.(82).
Harknessia rhoina. Leaf spot. **9:**CA(82,971,2548,3549).
Hendersonula toruloidea. Leaf spot. **19:**WV(3907).
Libertella olivacea. On limbs. **7:**NE(82).
Myxosporium rhois. On twigs. **7:**WV(82,3907). **19:**WV(82,3907).
Phlyctema albocincta. On stems. **5:**NJ(82).
Phoma pulchella. On twigs. **4:**SC(82). **19:**ME(82).
Phoma sp. Leaf spot. **13:**NC(1303).
Phomopsis sp. **23:**FL(35).
Phyllosticta rhoicola. Leaf spot. **5:**OR(82,3202). **17:**MS(2720), NY(82), TX(82), WA(82), WV(82).
P. rhoiseda. Leaf spot. **9:**CA(82,2439).
P. toxica. On leaves. **13:**IA(2), WI(1341). **17:**IA(82,959), IL(82), MS(82,2720), SC(82), WV(82).
P. toxicodendri. Leaf spot. **17:**SC(82).
Phyllosticta sp. Leaf spot. **1:**WI(1352). **4:**FL(35). **7:**WI(1358). **13:**WI(1347). **23:**FL(35), MN(2874).
Septoria rhoina. Leaf spot. **4:**Range of host(82), FL(35), WI(1340). **7:**Range of host(82), MN(2874), ND(378). **13:**WI(1343), WV(3907). **17:**ID(638). **19:**Range of host(82), MS(2720). **21:**Range of host(82). **23:**Unk.(82).
Sphaeronaema rhoidis. On limbs. **4:**SC(634).
Sphaeropsis sumachi. Inflorescence blight; possibly causing canker on limbs. **4:**E states(82), KS(82), MI(82). **7:**KS(82), MI(82), NJ(82), WV(82,3907). **19:**KS(82), MI(82), NJ(82), WV(82,3907).
Tubakia dryina. Leaf spot. **13:**IL(1246). **17:**IL(82).
Zythia rhoina. On dead stems. **13:**NJ(970).

Deuteromycotina - Other

Rhizoctonia crocorum. Root rot. **17:**TX(82).

Schinus L.

1. *S. molle* L. [Australian-pepper, California pepper-tree, molle, Peruvian mastic-tree, Peruvian pepper-tree, pirul]
2. *S. terebinthifolius* Raddi [Brazilian pepper-tree, Christmas-berry-tree]
3. *Schinus* sp.

Ascomycotina

Botryosphaeria ribis. On twigs. **1:**HI(82,2903). **2:**HI(82,2903).
Diaporthe sp. On twigs and limbs. **1:**HI(82,2903). **2:**HI(82,2903).
Nectria cinnabarina. Canker, dieback. **2:**FL(35).
Seuratia millardetii. On leaves. **2:**HI(2337).

Basidiomycotina - Other

Armillaria mellea. Root rot. **1:**CA(82,2898), TX(82). **2:**TX(82).
A. tabescens. Root rot. **2:**FL(35).
Coriolopsis gallica. Sapwood rot. **1:**CA(82).
Ganoderma applanatum. Heart rot. **1:**CA(82).
G. brownii. On wood. **3:**CA(1204).
G. lucidum. On wood. **1:**AZ(1202), CA(82).
Inonotus cuticularis. **1:**AZ(1181).
I. dryophilus. Heart rot. **1:**CA(82).
I. farlowii. Heart rot. **1:**AZ(82), CA(82).
I. munzii. **1:**AZ(1202).
Laetiporus sulphureus. Heart rot. **1:**CA(82).
Schizophyllum commune. Wound rot. **1:**CA(82).
Stereum hirsutum. On wood. **1:**CA(82).
Trametes versicolor. Sapwood rot. **1:**CA(82).

Deuteromycotina - Hyphomycetes

Alternaria sp. Leaf spot. **2:**FL(35).
Cercospora schinii. Leaf spot. **2:**FL(35).
Corynespora sp. Leaf spot. **2:**FL(35).
Helminthosporium sp. Leaf spot. **2:**FL(35).
Phymatotrichopsis omnivora. Root rot. **1:**AZ(82), TX(82). **2:**TX(82).
Verticillium albo-atrum. Wilt. **2:**CA(82).

Deuteromycotina - Coelomycetes

Diplodia sp. Twig blight. **2:**FL(35).
Macrophoma sp. Stem gall. **2:**FL(35).

Phyllosticta sp. Leaf spot. **2**:FL(35).
Sphaeropsis tumefaciens. Gall. **2**:FL(35,2260).

Spondias L.

1. *S. purpurea* L. [hog-plum, jocote, purple mombin, red mombin, Spanish plum]

Deuteromycotina - Coelomycetes
Sphaceloma spondiadis. Spot anthracnose. **1**:FL(35,82).

Annonaceae

Annona L.

1. *A. cherimola* Mill. [cherimalla, cherimoya, chirimoya, custard-apple]
2. *A. reticulata* L. [bullock's-heart, corazon, custard-apple]
3. *A. squamosa* L. [custard-apple, sugar-apple, sweetsop]
4. *Annona* sp.

Oomycetes
Phytophthora sp. **1**:CA(1110).
Pythium sp. Root rot. **1**:CA(1110). **4**:FL(35).
Ascomycotina
Glomerella cingulata. Anthracnose, fruit rot. **1**:FL(82).
Mycosphaerella sp. Leaf spot. **3**:TX(82).
Basidiomycotina - Uredinales
Phakopsora cherimoliae. **1**:FL(35,82). **3**:FL(35). **4**:FL(35).
Basidiomycotina - Other
Armillaria mellea. Root rot. **1**:CA(82).
A. tabescens. Root rot. **1**:FL(35,82).
Deuteromycotina - Hyphomycetes
Alternaria sp. Leaf spot. **1**:CA(1110). **3**:FL(35).
Cercospora sp. Leaf spot. **3**:FL(35). **4**:FL(35).
Helminthosporium sp. Leaf spot. **3**:FL(35).
Penicillium sp. **1**:CA(1110).
Phymatotrichopsis omnivora. Root rot. **3**:TX(82).
Verticillium dahliae. **1**:CA(1110).
Deuteromycotina - Coelomycetes
Colletotrichum gloeosporioides. Anthracnose, fruit rot. **1**:FL(35), HI(2903). **3**:FL(35). **4**:FL(35).
Colletotrichum sp. Anthracnose, fruit rot. **1**:CA(1110).
Dothiorella sp. **1**:CA(1110).
Lasiodiplodia theobromae. Stem blight. **3**:TX(82).
Phomopsis sp. Leaf and fruit spot. **1**:CA(1110), FL(35). **3**:FL(35).
Phyllosticta sp. Leaf spot. **2**:FL(35). **3**:FL(35).
Deuteromycotina - Other
Rhizoctonia solani. **1**:CA(1110).

Asimina Adans.

1. *A. parviflora* (Michx.) Dunal
2. *A. triloba* (L.) Dunal [pawpaw]
3. *Asimina* sp.

Ascomycotina
Dothidea puccinioides. **2**:GA(1438).
Glomerella sp. Leaf spot. **3**:FL(35).
Karstenula sp. **2**:GA(1438).
Mycosphaerella asiminae. On living leaves. **2**:OH(967).
M. punctiformis. **1**:GA(1438).
Mycosphaerella sp. Leaf spot. **2**:IN(82), OH(82), WV(82).
Nectria cinnabarina. **2**:IN(82), WV(82,3907).
Valsa ambiens. On dead twigs. **2**:VA(82), WV(82,3907).
Basidiomycotina - Other
Phellinus punctatus. **2**:VA(82).
P. viticola. Sapwood rot. **2**:VA(82), WV(82,3907).
Phylloporia frutica. On wood. **2**:FL(35), GA(491,1439). **3**:SE states(2681), FL(35,82), GA(82).
Deuteromycotina - Hyphomycetes
Conoplea olivacea. Canker. **3**:OH(1672).
Cristulariella moricola. Leaf spot. **2**:WV(82,3680,3907).
Cylindrocladium scoparium. On fallen leaves. **2**:WV(82,3907).
Mycocentrospora asiminae. Leaf spot. **2**:AL(82), IL(312), KS(82,973), MD(82), MS(82,2720), TX(82). **3**:AL(82), FL(35), MS(82).
Myrothecium roridum. On leaves. **2**:WV(82,3907).
Phymatotrichopsis omnivora. Root rot. **2**:TX(82).
Rhopaloconidium asiminae. Leaf blotch. **2**:IL(82), IN(654,987), KS(82), MO(82), OH(82), WV(82,3907).
Deuteromycotina - Coelomycetes
Aplosporella asiminae. **2**:MD(82), NC(2783), WV(82,3907).
Diplodia sp. Leaf spot. **3**:FL(35).
Dothiorella aberrans. On limbs. **2**:OH(82).
D. asiminae. On fallen leaves. **2**:WV(82,3907).
Gloeodes pomigena. On twigs. **2**:IN(82).
Phoma microsporella. On leaves. **2**:IN(82), OH(82).
Phyllosticta asiminae. Eye spot, bordered leaf spot. **2**:SE states(82), FL(82), IL(650), IN(82), MD(82), MI(82), MS(2720), TX(82), WV(3907). **3**:FL(35).
Septoria asiminae. Leaf spot. **2**:TX(82).

Monodora Dunal

1. *M. myristica* (Gaertn.) Dunal [African nutmeg, Calabash nutmeg, Jamaica nutmeg]

Oomycetes
Pythium sp. Root rot. **1**:FL(35).

Rollinia St. Hil.

1. *R. pulchrinervis* A. DC. (=*R. deliciosa* Saff.) [biriba]

Ascomycotina
Glomerella cingulata. Dieback, fruit rot. **1**:FL(35,82).

Stelechocarpus (Blume) J. D. Hook. & Th. Thomson

1. *S. burakoi* J. D. Hook. & Th. Thomson

Deuteromycotina - Hyphomycetes
Cylindrocladium sp. Root rot. **1**:HI(2903).

Apiaceae

Anethum L.

1. *A. graveolens* L. [dill]

Ascomycotina
Heterosphaeria patella. **1**:MN(2874).
Sclerotinia sclerotiorum. Stem rot. **1**:TX(82).
Deuteromycotina - Hyphomycetes
Cercosporella anethi. **1**:ND(378).
Cercosporidium punctum. Leaf and stem spot. **1**:FL(35), ND(82), OR(82), SD(2243), TX(82).
Didymaria anethiana. Leaf spot. **1**:OR(666,3202).
Fusarium sp. Root rot, wilt. **1**:OH(1725).
Phymatotrichopsis omnivora. Root rot. **1**:TX(82).
Deuteromycotina - Coelomycetes
Phoma anethi. Leaf and stem spot. **1**:CT(82), IA(82), IN(82).
Deuteromycotina - Other
Rhizoctonia solani. Damping-off. **1**:GA(82).
Rhizoctonia sp. Root rot. **1**:HI(2903).

Angelica L.

1. *A. ampla* A. Nels.
2. *A. arguta* Nutt. ex Torr. & A. Gray (=*A. lyallii* S. Wats.)
3. *A. atropurpurea* L. [alexanders, great angelica, masterwort, purple-stemmed angelica]
4. *A. genuflexa* Nutt.
5. *A. grayi* (J. M. Coulter & J. Rose) J. M. Coulter & J. Rose
6. *A. hendersonii* J. M. Coulter & J. Rose
7. *A. lucida* L.
8. *A. pinnata* S. Wats.
9. *A. tomentosa* S. Wats.
10. *A. triquinata* Michx. [filmy angelica]
11. *Angelica* sp.

Chytridiomycetes
Physoderma pluriannulatum. **5**:WY(2924). **8**:WY(1195).
Oomycetes
Plasmopara crustosa. **4**:WA(3202).
Basidiomycotina - Uredinales
Puccinia angelicae. Autoecious. **2**:OR(82,3202), PA(82), WA(82,3202). **3**:NY(82). **4**:AK(60,526), WA(3202). **7**:AK(3431). **11**:OR(82), PA(82), TX(82), WA(82).
P. bistortae. Heteroecious: aecial host. **7**:AK(60,526). **11**:AK(82).
P. coelopleuri. Autoecious. **7**:AK(60,526), WA(3202). **11**:AK(82).
P. ellisii. Autoecious. **2**:CA(82), WY(1195). **4**:ID(3202), OR(3202). **6**:CA(81). **9**:CA(81). **11**:CA(82), ID(82), NV(82), OR(82), WA(82).

P. ligustici. Autoecious. **5:**WY(2924). **8:**UT(1195). **11:**AK(82), CO(82).
P. poromera. Autoecious. **11:**UT(82).
Deuteromycotina - Hyphomycetes
Botrytis cinerea. **2:**ID(3202).
Cercospora angelicae. Leaf spot. **11:**AK(82,526).
C. thaspii. Leaf spot. **11:**AL(82), TX(82).
Cercosporidium angelicae. Leaf spot. **1:**WY(839). **2:**Range of host(82). **3:**Range of host(82), IA(839), VT(839), WI(839). **10:**Range of host(82). **11:**Range of host(82).
C. depressum. **1:**WY(1195). **2:**ID(3202), MT(3202), WA(3202). **8:**WY(1195). **11:**AK(526).
Heterosporium sp. On stems. **11:**WY(82).
Phymatotrichopsis omnivora. Root rot. **11:**TX(82).
Ramularia archangelicae. Leaf spot. **2:**WY(1195). **3:**WI(1361). **4:**AK(526), OR(3202), WA(3202). **11:**AK(82), CO(82), WA(82).
Deuteromycotina - Coelomycetes
Colletotrichum dematium. On stems. **3:**NY(82).
Leptostromella angelicae. On stems. **3:**NY(82).
Phoma complanata. On stems. **11:**WY(82).
Phyllosticta angelicae. Leaf spot. **1:**WY(1195). **2:**CA(81), WA(3202). **3:**WI(1337). **4:**CA(81). **6:**WA(3202). **9:**CA(81). **11:**CA(82), WY(82).
Phyllosticta sp. **3:**WI(1356).
Piggotia depressa. Leaf spot. **2:**MT(82).
Plenodomus sp. **11:**WA(2219).
Septogloeum angelicae. Leaf spot. **11:**SC(82).
Septoria dearnessii. Leaf spot. **3:**WI(82,1330). **10:**NC(82,1303). **11:**AK(526), TX(82).

Apium L.

1. *A. graveolens* L. [celery]
2. *A. graveolens* L. var. *dulce* (Mill.) Pers. [celery]
3. *A. graveolens* L. var. *rapaceum* (Mill.) Gaudin [celeriac, knob celery, turnip-rooted celery]

Myxomycetes
Physarum sp. Slime mold. **2:**CA(82), FL(82), NY(82), OR(82).
Oomycetes
Aphanomyces euteiches. Damping-off. **2:**MA(82).
Pythium debaryanum. Damping-off, root rot. **1:**FL(35). **2:**'Pythium damping-off. Pythium root rot.' CO(82), CT(82), FL(82), NH(82).
P. irregulare. Damping-off, root rot. **1:**CA(2371).
P. ultimum. Damping-off, root rot. **1:**CA(2371). **2:**OH(82).
Pythium sp. Seedling blight, damping-off. **2:**CO(49), HI(2903).
Ascomycotina
Chaetomium succineum. Isolated from seeds. **2:**CA(3306).
Sclerotinia minor. Damping-off, pink rot, stem rot, watery rot. **2:**'Sclerotinia foot rot.' Range of host(82), FL(35). **3:**Range of host(82).
S. sclerotiorum. Damping-off, crown rot, pink rot, stem rot, storage rot, watery rot. **1:**CA(2371), ID(3202), MT(3202), NY(2593), OR(3202). **2:**'Sclerotinia pink rot.' Range of host(82), FL(35), HI(2903). **3:**CA(82).
Sclerotinia sp. Damping-off, pink rot, stem rot, storage rot and watery rot. **2:**Range of host(82). **3:**CA(82).
Basidiomycotina - Other
Thanatephorus cucumeris. Damping-off, crater rot, root rot. **1:**CA(2371), WA(3202).
Typhula variabilis. Storage rot. **1:**NY(82). **2:**NY(2955).
Deuteromycotina - Hyphomycetes
Alternaria ramulosa. Leaf spot. 'Stemphylium leaf spot.' **2:**FL(35,82).
Alternaria sp. Leaf spot. **2:**FL(82), NC(1303).
Botrytis cinerea. Soft rot. **1:**AK(526), CA(2371), ID(3202), NY(2593), OR(3202), WA(3202). **2:**'Gray mold blight.' AK(82), CA(82), FL(35), MA(82), WA(82).
Cephalosporium sp. Black spot. **1:**CT(2971), NY(3174).
Cercospora apii. Early blight, leaf spot. **1:**CA(2371), FL(3500), MS(2720), WA(3202), WI(1334). **2:**'Early blight.' Range of host(82), CO(49), FL(35), HI(2903), MN(2874), NC(1303), OK(2872), WV(3907). **3:**CT(82), IL(82), NY(82).
Fusarium oxysporum. Wilt. 'Fusarium yellows.' **2:**E & C states(82), CA(82,985), CO(82), MI(985), MO(82), MT(82).
F. oxysporum f. sp. *apii*. Wilt, yellows. **1:**CA(2371), FL(35), MT(3202), UT(3620). **2:**CO(49), FL(35), MI(984), MN(2874), TX(2275).
Fusarium sp. Root rot. **2:**CA(82), CO(82), FL(82), MI(82), NJ(82), OH(82).
Mycocentrospora acerina. Storage rot. **1:**NY(2576,2593). **2:**NY(82).
Stemphylium botryosum. On seeds. **1:**Unk.(82).
Trichothecium roseum. Pink mold in storage. **2:**NY(82).
Verticillium albo-atrum. Wilt. **2:**CA(82).
Deuteromycotina - Coelomycetes
Colletotrichum sp. **2:**FL(35,707).
Phoma apiicola. Root and storage rot. **1:**CA(2371). **2:**'Phoma root rot. Phoma crown rot.' CA(82), MI(82), NY(82,2593), OH(82). **3:**NY(82), WI(3908).
Phyllosticta apii. Leaf spot. **2:**DE(82), NJ(82).
Septoria apiicola. Late blight, foliage blight, large leaf spot. **1:**AK(2163), CA(2371), CO(1195), ID(3202), OR(3202), SD(2243), WA(3202). **3:**Range of host(82), CA(82), MI(82).
Deuteromycotina - Other
Rhizoctonia solani. Damping-off, root and stem rot. **1:**FL(3500), MS(2720). **2:**'Rhizoctonia crater rot. Rhizoctonia damping-off.' Range of host(82), FL(35).
Rhizoctonia sp. Damping-off, root rot. **1:**FL(3644). **2:**HI(2903), NC(1303).
Sclerotium rolfsii. Southern blight. **2:**FL(35).

Berula Besser ex W. Koch in Mert. & W. Koch

1. *B. erecta* (Huds.) Cav. (=*B. pusilla* (Nutt.) Fernald)

Deuteromycotina - Coelomycetes
Septoria sii. **1:**WI(1357).

Bupleurum L.

1. *B. americanum* J. M. Coulter & J. Rose

Ascomycotina
Lewia scrophulariae. **1:**WY(3810).
Pleospora chlamydospora. On dead stem and leaves. **1:**AK(1938).
P. helvetica. **1:**AK(526).
P. tragacanthae. On dead stem. **1:**AK(1938).
Basidiomycotina - Uredinales
Puccinia bupleuri. **1:**AK(60,526).

Carum L.

1. *C. carvii* L. (=*C. velenovskyi* Rohlena) [caraway]

Ascomycotina
Comoclathris permunda. **1:**WY(3810).
Leptosphaeria erigerontis. **1:**WY(3810).
Mycosphaerella dolichospora. On stems. **1:**WY(3811).
Nodulosphaeria olivacea. **1:**WY(3250).
Sclerotinia sclerotiorum. Stem rot. **1:**WA(82).
Deuteromycotina - Coelomycetes
Septoria sp. **1:**WI(1352).

Centella L.

1. *C. asiatica* (L.) Urb. (=*Hydrocotyle asiatica* L.) [Asiatic pennywort]
2. *C. erecta* (L. fil.) Fernald

Deuteromycotina - Coelomycetes
Septoria asiatica. Leaf spot. **1:**FL(35).
S. hydrocotyles. Leaf spot. **1:**FL(35). **2:**FL(82).
Septoria sp. Leaf spot. **2:**HI(2903).

Chaerophyllum L.

1. *C. procumbens* (L.) Crantz

Deuteromycotina - Coelomycetes
Septoria sii. **1:**WI(1338).

Cicuta L.

1. *C. bulbifera* L.
2. *C. douglasii* (DC.) J. M. Coulter & J. Rose (=*C. occidentalis* Douglas ex Heynh.)
3. *C. maculata* L. [musquash-root, spotted cowbane, spotted water-hemlock, water-hemlock]
4. *Cicuta* sp.

Oomycetes
Plasmopara crustosa. Downy mildew. **3:**WI(82).
Ascomycotina
Protomyces macrosporus. Leaf gall. **3:**WI(82).
Basidiomycotina - Uredinales
Nyssopsora echinata. Autoecious. **2:**CA(2461), ID(2461).
Puccinia cicutae. Autoecious. **2:**W states(82), AK(60,526),

AZ(1195,1203), CA(81,82,642), ID(3202), OR(82,3202), WA(3202), WY(1195). **3**:AZ(3981), ME(82), NE(82), NY(82), WI(82), WV(3907). **4**:AK(82), CA(82), FL(35,82), NV(82).
Uromyces lineolatus subsp. *nearcticus*. Heteroecious: aecial host. **1**:NE(82), WI(82). **3**:SE states(82), CO(82), FL(35,82), NE(3101), WI(82,3101).
Deuteromycotina - Hyphomycetes
Alternaria sp. **3**:WV(3907).
Botrytis cinerea. **3**:WV(3907).
Dendryphiella vinosum. **3**:PA(82), TX(82).
Phymatotrichopsis omnivora. Root rot. **3**:TX(82).
Stemphylium botryosum. **3**:OK(2872).
Deuteromycotina - Coelomycetes
Coniothyrium sp. **3**:OK(2872).
Cylindrosporium cicutae. Leaf spot. **3**:TX(82), WI(82).
Neottiospora umbelliferarum. **3**:WI(1361).
Phyllosticta sp. Leaf spot. **3**:OK(2872).
Septoria sii. Leaf spot. **1**:WI(82). **2**:WY(82,1195). **3**:KS(82), WI(82). **4**:FL(35).
Stagonospora thaspii. Leaf spot. **3**:WI(82).

Coelopleurum Ledeb.
1. *C. gmelinii* (DC.) Ledeb.
Basidiomycotina - Uredinales
Nyssopsora echinata. **1**:AK(2461).

Conioselinum Hoffm.
1. *C. chinense* (L.) B. S. P. (=*C. benthamii* (S. Wats.) Fernald)
2. *C. pacificum* (S. Wats.) J. M. Coulter & J. Rose (=*Selinum pacificum* S. Wats.)
3. *C. scopulorum* (A. Gray) J. M. Coulter & J. Rose
Basidiomycotina - Uredinales
Nyssopsora echinata. Autoecious. **1**:AK(526). **2**:CA(2461). **3**:CO(2461).
Puccinia bistortae. Heteroecious: aecial host. **1**:AK(526).
P. coelopleuri. Autoecious. **1**:AK(526).
P. ligustici. Autoecious. **1**:AK(60,526). **3**:WY(1195).
[*P. longipes*]. **1**:AK(526).
Deuteromycotina - Hyphomycetes
Botrytis cinerea. **1**:AK(526).
Cercospora selini-gmelini. **1**:AK(526).
Deuteromycotina - Coelomycetes
Septoria petroselini. **3**:WY(1195).
S. petroselini subsp. *treleasiana*. **1**:AK(526).

Conium L.
1. *C. maculatum* L. [California-fern, Nebraska-fern, poison-hemlock, spotted-hemlock, winter-fern]
Ascomycotina
Massarina microcarpa. On dead stems. **1**:CA(81).
Deuteromycotina - Coelomycetes
Phomopsis asteriscus. On dead stems. **1**:CA(81).

Coriandrum L.
1. *C. sativum* L. [Chinese parsley, cilantro, coriander, parsley]
Oomycetes
Phytophthora nicotianae var. *parasitica*. Root rot. **1**:HI(2903).
Deuteromycotina - Hyphomycetes
Alternaria sp. Leaf spot. **1**:HI(2899,2903,3074).
Deuteromycotina - Other
Rhizoctonia sp. Root rot. **1**:HI(2903).

Cryptotaenia DC.
1. *C. canadensis* (L.) DC. (=*Deringa canadensis* (L.) Kuntze) [honewort, white chervil]
Ascomycotina
Leptospora rubella. **1**:DE(3248).
Basidiomycotina - Uredinales
Puccinia cryptotaeniae. Autoecious. **1**:MN(2874).
Basidiomycotina - Other
Ceratobasidium anceps. **1**:WI(1355).
Deuteromycotina - Hyphomycetes
Cercospora cryptotaeniae. Leaf spot. **1**:NY(581).
Deuteromycotina - Coelomycetes
Colletotrichum sp. On leaves. **1**:WI(1357).
Septoria cryptotaeniae. Leaf spot. **1**:MN(2874).
Stagonospora cryptotaeniae. On leaves. **1**:WI(1337).

Cymopterus Raf.
1. *C. anisatus* A. Gray (=*Pseudocymopterus anisatus* J. M. Coulter & J. Rose)
2. *C. bipinnatus* S. Wats. (=*Pseudocymopterus bipinnatus* J. M. Coulter & J. Rose)
3. *C. littoralis* A. Gray
4. *C. terebinthinus* Torr. & A. Gray (=*Pteryxia californica* J. M. Coulter & J. Rose)
Ascomycotina
Comoclathris permunda. **4**:WA(3202).
Mycosphaerella sp. On stems. **1**:CO(82).
Pleospora herbarum. **2**:MT(3202).
Basidiomycotina - Uredinales
Puccinia musenii. Autoecious. **1**:CO(82), MT(82), WY(82). **2**:MT(3202).
P. pseudocymopteri. Autoecious. **1**:UT(82). **2**:MT(3202). **4**:CA(642).
Deuteromycotina - Coelomycetes
Phyllosticta angelicae. **3**:CA(81).
Selenophoma maculicola. On stems. **1**:WY(3812).

Daucus L.
1. *D. carota* L. [devil's-plague, queen's-lace, Queen-Anne's-lace, wild carrot]
2. *D. carota* L. subsp. *sativus* (Hoffm.) Arcang. (=*D. carota* L. var. *sativa* Hoffm.) [carrot]
3. *Daucus* sp.
Oomycetes
Phytophthora megasperma. Root rot. **2**:CA(82).
Pythium irregulare. Forking, brown root. **1**:WI(2377).
P. paroecandrum. Forking, brown root. **1**:WI(2377).
P. sylvaticum. Forking, brown root. **1**:WI(2377).
Pythium sp. Seedling blight, damping-off, forking, brown root. **1**:ID(3202), WI(2377). **2**:CO(49), FL(35), ID(82).
Zygomycotina
Cunninghamella elegans. On seeds. **2**:MN(3202).
Rhizopus arrhizus. Black-mold rot in transit and storage. **2**:Range of host(82).
R. stolonifer. Black-mold rot in transit and storage. **2**:Range of host(82).
Rhizopus sp. Black-mold rot in transit and storage. **2**:Range of host(82), FL(35).
Ascomycotina
Botryotinia fuckeliana. **2**:Unk.(82).
Chaetomium megalocarpum. Isolated from seeds. **2**:MN(3306).
Crocicreas nigrescens. **1**:NJ(505).
Diaporthe arctii. On stems. **1**:GA(1438). **2**:GA(82).
Erysiphe heraclei. Powdery mildew. **2**:NC(1754).
E. polygoni. Powdery mildew. **2**:CA(7).
Leptosphaeria longipedicellata. **1**:GA(1438).
Leptosphaeria sp. **1**:GA(1438).
Massarina gloeospora. **1**:GA(1438).
Metasphaeria sp. **1**:GA(1438).
Ophiobolus nigro-clypeata. **1**:GA(1438).
Pleospora penicillus. **1**:GA(1438).
Sclerotinia minor. Cottony rot, watery soft rot. **2**:Unk.(82).
S. sclerotiorum. Watery soft rot; cottony rot. **2**:Range of host(82), FL(35), GA(1438).
Basidiomycotina - Uredinales
Uromyces lineolatus. Heteroecious: aecial host. **1**:OR(3202).
U. lineolatus subsp. *nearcticus*. Heteroecious: aecial host. **2**:OR(82).
Basidiomycotina - Other
Athelia rolfsii. Cottony rot, watery soft rot. **1**:OR(3202).
Helicobasidium brebissonii. **2**:Unk.(82).
Thanatephorus cucumeris. Storage rot. **1**:ID(3202), OR(3202), WA(3202). **2**:NY(82).
Typhula variabilis. Leaf rot. **1**:ID(3202). **2**:ID(82).
Typhula sp. Storage rot. **2**:IL(82), NY(82).
Deuteromycotina - Hyphomycetes
Alternaria alternata. Seed mold, secondary leaf spot. **2**:Range of host(82), FL(35).
A. dauci. Leaf blight, leaf spot. **1**:ID(3202), NC(1303,3931). **2**:Range of host(82), CO(49), FL(35), HI(2903), IL(312), MS(2720),

NC(1303), NM(82), WV(3907).
[*A. longipes*]. Leaf blight, late blight, leaf spot. **3**:GA(659).
A. radicina. Black rot of crown. **1**:AK(2163), CA(1368), ID(3202), OH(653), OR(3202), WA(3202). **2**:ID(82), MA(82), NY(82), PA(82), WA(82), WI(1343).
Alternaria sp. Isolated from inflorescences. **1**:MD(651).
Botrytis cinerea. Gray-mold rot. **1**:AK(2163), ID(3202), WA(3202). **2**:Range of host(82), FL(35).
Cercospora carotae. Leaf spot. **2**:C states(82), NE states(82), Range of host(82), CO(49), FL(35), HI(82), MN(3202), MS(2720), SD(2243), WI(1330), WV(3907).
Chalara thielavioides. On roots. **1**:TX(2550).
Fusarium avenaceum. On roots. **2**:NY(82).
F. moniliforme. Isolated from seed. **2**:Unk.(82).
Fusarium sp. Scab, dry rot, stem rot, wilt. **1**:ID(3202), NC(1303). **2**:AZ(82), CA(82), FL(35), ID(82), NC(1303), NH(82), NM(82).
Geotrichum candidum. Sour rot. **1**:CA(3960), IL(3960).
Gliocladium roseum. In stored carrot roots. **2**:NY(82).
Mycocentrospora acerina. Black crown, side rot, storage rot. **2**:NY(82).
Penicillium expansum. Bule mold rot in transit and storage. **2**:Unk.(82).
Penicillium sp. **2**:Unk.(82).
Phymatotrichopsis omnivora. Root rot. **2**:AZ(82), LA(82), TX(82).
Ramularia sp. Leaf spot. **2**:KS(82).
Stemphylium botryosum. On seeds. **2**:Unk.(82).
Trichothecium roseum. Pink-mold rot. **2**:IN(82).
Ulocladium consortiale. On seeds. **2**:Unk.(82).
Deuteromycotina - Coelomycetes
Septoria sp. On leaves. **1**:OR(3202). **2**:NY(82).
Deuteromycotina - Other
Rhizoctonia carotae. Crater rot. **2**:IL(82), NY(82).
R. crocorum. Violet root rot. **1**:OR(3202), WA(3202). **2**:OR(82), TX(82), WA(82).
R. solani. Canker, damping-off. **2**:Range of host(82), FL(35), MN(1365), TX(82).
R. zeae. Root rot. **2**:NC(1303).
Rhizoctonia sp. Root canker, lesions, hairy root, crown rot, root rot, cavity spot. **2**:CO(49), HI(2903), MI(82), NC(1303), NY(82), OR(82), WA(82), WI(2378).
Sclerotium rolfsii. Southern blight, root and stem blight. **1**:SE states(2884), GA(2885), NC(2885,3931). **2**:SE states(82), CA(82), FL(35), HI(2903), NC(1303), OR(82).
S. rolfsii var. *delphinii*. Southern blight. **2**:NY(82).

Erigenia Nutt.

1. *E. bulbosa* (Michx.) Nutt. [harbinger-of-spring, pepper-and-salt]
Basidiomycotina - Uredinales
Puccinia erigeniae. Autoecious. **1**:OH(82).

Eryngium L.

1. *E. aquaticum* L. [button-snakeroot, eryngo]
2. *E. leavenworthii* Torr. & A. Gray
3. *E. yuccifolium* Michx. [button snake-root, rattlesnake-master]
4. *Eryngium* sp.
Basidiomycotina - Ustilaginales
Entyloma eryngii. White smut. **1**:IA(82). **3**:IA(1077), WI(1077,1330).
Deuteromycotina - Hyphomycetes
Phymatotrichopsis omnivora. Root rot. **4**:TX(82).
Deuteromycotina - Coelomycetes
Cylindrosporium eryngii. Leaf spot. **1**:IA(82), KS(82), WI(82). **3**:KS(973). **4**:TX(82).
Macrophomina phaseolina. Charcoal rot, stem rot. **2**:OK(2873). **4**:OK(82).
Phoma sp. On living stems. **3**:WI(1352).
Septoria eryngicola. Leaf spot. **1**:WI(82). **3**:WI(1330,1332).

Foeniculum Mill.

1. *F. vulgare* Mill. [fennel, finocchio, Florence fennel]
2. *F. vulgare* Mill. var. *dulce* Battand. & Trabut
3. *Foeniculum* sp.
Ascomycotina
Botryosphaeria dothidea. **1**:GA(1438).
Dothidea puccinioides. **1**:GA(1438).
Leptosphaeria doliolum. **1**:GA(1438).
Massarina gloeospora. **1**:GA(1438).
Pleospora herbarum. **1**:CA(2439).
P. penicillus. **1**:GA(1438).
Sclerotinia sclerotiorum. Stem rot. **2**:IL(82), NJ(82), TX(82).
Basidiomycotina - Other
Calyptella capula. On dead stems. **3**:NY(456), SC(456).
Deuteromycotina - Hyphomycetes
Botrytis cinerea. Gray mold. **1**:NY(82).
Cladosporium herbarum. On stems. **1**:CA(2439).
Heterosporium sp. **1**:WA(3202).
Phymatotrichopsis omnivora. Root rot. **1**:TX(82).
Deuteromycotina - Other
Rhizoctonia solani. Damping-off, stem pitting. **1**:GA(82), NJ(82).

Glehnia F. W. Schmidt ex Miq.

1. *G. leiocarpa* Mathias
Deuteromycotina - Hyphomycetes
Pollaccia peucedani. Damping-off, stem pitting. **1**:WA(3202).

Heracleum L.

1. *H. sphondylium* L. subsp. *montanum* (Schleich. ex Gaudin) Briq. (=*H. maximum* Bartram, *H. lanatum* Michx.) [American cow-parsnip, masterwort]
2. *Heracleum* sp.
Ascomycotina
Crocicreas cyathoideum. On dead stems. **1**:AZ(662), NY(505).
C. nigrofuscum. **1**:CA(661).
Diaporthe arctii. On stems. **1**:CA(82).
Didymella exigua. On stems. **1**:MT(82,3202).
Heterosphaeria patella. **2**:AK(526).
Hyponectria sceptri. On dead stems. **1**:CO(649,3084).
Leptosphaeria asparagina. **1**:WA(3202,3814).
L. doliolum. **1**:AK(526). **2**:AK(526).
L. erigerontis. **2**:WY(3810).
L. simmonsii. **1**:WA(3202).
Leptosphaeria sp. **1**:WA(3202).
Naevia stenospora. On dead stems. **1**:WA(82,3202).
[*Ophiobolus acuminatus*]. **1**:WA(3202,3815).
Phyllachora heraclei. **1**:AK(526), ID(3202), WA(3202), WY(1195).
Trichopezizella relicina var. *macrospora*. **1**:CO(1416).
Deuteromycotina - Hyphomycetes
[*Cercosporidium angelicae*]. Leaf spot. **1**:WI(82).
Ostracoderma sp. **1**:OR(648,3202).
Phymatotrichopsis omnivora. Root rot. **1**:TX(82).
Ramularia heraclei. Leaf spot. **1**:Range of host(82), AK(526,3431), CA(81), CO(649), DE(3543), MT(3202), ND(378), NY(82), SD(2243), WA(3202), WI(1358), WY(1195). **2**:AK(526).
Stemphylium botryosum. **1**:WA(3202).
Deuteromycotina - Coelomycetes
Colletotrichum dematium. **1**:ND(378).
Heteropatella umbilicata. **1**:CO(649).
Leptothyrella graminis. **1**:WA(3816).
Phoma complanata. On stem. **1**:AK(82), CA(82), CO(82), NM(82), UT(82). **2**:AK(526), AK(526).
P. heraclei. On stems. **1**:CO(82), TX(82).
P. herbarum. **1**:WA(3202).
P. minuta. **1**:WA(3202).
Phoma sp. On stems. **1**:Unk.(82).
Phomopsis asteriscus. On stems. **1**:ME(82).
Phyllosticta heraclei. Leaf spot. **1**:AK(82), OR(82,3202), TX(82). **2**:AK(526).
Rhabdospora aristata. **1**:MT(3202).
R. heraclei. On stems. **1**:NV(82).
R. pleosporoides. **1**:CA(81).
Septoria pastinacae. On stems and seeds. **1**:AK(82). **2**:AK(526).
Sirococcus americanus. **1**:WA(3202).
Sphaerographium abditum. **2**:AK(526).
Deuteromycotina - Other
Sclerotium varium. **2**:AK(526).

Hydrocotyle L.

1. *H. americana* L.
2. *H. interrupta* Muhl.
3. *H. ranunculoides* L. fil.
4. *H. umbellata* L. [water pennywort]
5. *H. verticillata* Thunb.

6. *Hydrocotyle* sp.

Chytridiomycetes

Physoderma hydrocotylidis. On leaves. **3:**CA(81,3401), FL(3401). **4:**CA(3401). **6:**NJ(3401).

Synchytrium aureum. Leaf gall. **1:**WI(82). **4:**MA(1394).

S. bonaerense. Leaf gall. **6:**LA(82).

Basidiomycotina - Uredinales

Puccinia hydrocotyles. Autoecious. **1:**GA(1439). **3:**SE states(82), AZ(82), CA(82), NV(82), TX(82), UT(82). **4:**SE states(82), AZ(82), CA(82), FL(35), MS(2720), NV(82), TX(82), UT(82). **5:**AZ(747,1203,3981), CA(81), HI(2903). **6:**FL(35), LA(658,2675).

Uromyces americanus. Heteroecious: aecial host. **4:**TX(740,3101).

U. lineolatus subsp. *nearcticus*. Heteroecious: aecial host. **4:**DE(82), FL(35,82), MA(82).

Basidiomycotina - Ustilaginales

Entyloma fimbriata. **3:**CA(1077).

E. hydrocotyles. **3:**CA(82).

Deuteromycotina - Hyphomycetes

Cercospora hydrocotyles. Leaf spot. **1:**SE states(82), TX(82). **2:**LA(962). **3:**SE states(82), TX(82). **4:**SE states(82), FL(35), NJ(3543), TX(82). **5:**MS(2720). **6:**FL(35).

Deuteromycotina - Coelomycetes

Ascochyta sp. Leaf spot. **4:**FL(35).

Septoria hydrocotyles. Leaf spot. **1:**MN(2874), NY(82), WI(82). **6:**FL(35), MS(82,2720).

S. pallidula. Leaf spot. **1:**NY(82).

Ligusticum L.

1. *L. actaeifolium* Michx. (=*L. canadensis* (L.) Vail ex Small & Vail)
2. *L. apiifolium* (Nutt. ex Torr. & A. Gray) A. Gray (=*L. apiodorum* J. M. Coulter & J. Rose)
3. *L. canbyi* J. M. Coulter & J. Rose (=*L. leibergii* J. M. Coulter & J. Rose)
4. *L. filicinum* S. Wats.
5. *L. grayi* J. M. Coulter & J. Rose (=*L. cusickii* J. M. Coulter & J. Rose, *L. purpureum* J. M. Coulter & J. Rose)
6. *L. hultenii* Porta
7. *L. porteri* J. M. Coulter & J. Rose (=*L. simulans* J. M. Coulter & J. Rose)
8. *L. scothicum* L. [sea-lovage]
9. *Ligusticum* sp.

Chytridiomycetes

Physoderma pluriannulatum. Leaf gall. **5:**WA(3202). **7:**WY(1195). **9:**OR(82), WA(82), WY(82).

Physoderma sp. **5:**WA(1844).

Synchytrium aureum. **5:**CA(642).

Synchytrium sp. **5:**WA(3202).

Oomycetes

Plasmopara crustosa. Downy mildew. **6:**AK(526). **8:**AK(82). **9:**AK(526).

Ascomycotina

Diaporthe arctii. On stems. **1:**GA(82,1438).

Entodesmium niessleanum. **5:**WA(3202,3814).

Hyponectria sceptri. On dead stems. **4:**UT(3084).

Leptosphaeria erigerontis. **5:**WA(3202,3814).

Mycosphaerella delphiniicola. **5:**WA(3202,3815).

Mycosphaerella sp. On stems. **1:**GA(82,1438).

Nodulosphaeria olivacea. On dead stems. **7:**CO(128), WY(3250).

Ophiobolus nigro-clypeata. **1:**GA(1438).

Pleospora helvetica. **5:**WA(3202,3814).

Basidiomycotina - Uredinales

Nyssopsora echinata. Autoecious. **2:**CA(81,2461). **3:**ID(2461,3202). **4:**UT(2461). **5:**CA(638,642,2461), OR(2461,3202), WA(3202). **7:**NM(2461), WY(1195,2461). **9:**CA(82), CO(2461), ID(82), NM(82), OR(82), UT(82), WA(82).

Puccinia bistortae. Heteroecious: aecial host. **2:**WA(3202).

P. jonesii. Autoecious. **7:**WY(1195).

P. ligustici. Autoecious. **2:**OR(3202), WA(3202). **3:**ID(3202). **5:**CA(638,642), WA(3202). **7:**WY(1195). **9:**CA(82), CO(82), ID(82), NV(82), OR(82), UT(82), WA(82), WY(82,2924).

Uromyces lineolatus subsp. *nearcticus*. **8:**ME(82).

Deuteromycotina - Hyphomycetes

Cercospora ligusticicola. Leaf spot. **5:**CA(581).

Ramularia reticulata. **9:**AK(526).

Ramularia sp. **5:**WA(3202).

Deuteromycotina - Coelomycetes

Cylindrosporium umbelliferarum. Leaf spot. **7:**WY(1195). **9:**CA(82), WY(82).

Heteropatella umbilicata. **5:**CA(642), WA(3202,3816).

Kellermania alpina. On stems. **5:**CA(638). **9:**CA(82).

Phyllosticta angelicae. **2:**CA(81).

Rhabdospora conii. **5:**WA(3202,3816).

Septoria aromatica. **4:**CO(649). **5:**CA(639,642).

S. ligustici. Leaf spot. **8:**MA(82).

Septoria sp. Leaf spot. **9:**CA(82).

Lomatium Raf.

(= *Cogswellia* Roem. & Schult.)

1. *L. ambiguum* (Nutt.) J. M. Coulter & J. Rose
2. *L. brandegei* (J. M. Coulter & J. Rose) Macbr.
3. *L. californicum* (Nutt.) Mathias & Constance
4. *L. columbianum* Mathias & Constance
5. *L. cous* (S. Wats.) J. M. Coulter & J. Rose (=*L. montanum* J. M. Coulter & J. Rose)
6. *L. cusickii* (S. Wats.) J. M. Coulter & J. Rose
7. *L. dissectum* (Nutt. ex Torr. & A. Gray) Mathias & Constance
8. *L. dissectum* (Nutt. ex Torr. & A. Gray) Mathias & Constance var. *dissectum* (=*Leptotaenia dissecta* Nutt.)
9. *L. dissectum* (Nutt. ex Torr. & A. Gray) Mathias & Constance var. *multifidum* (Nutt.) Mathias & Constance (=*Leptotaenia multifida* Nutt., *Ferula multifida* (Nutt.) A. Gray)
10. *L. foeniculaceum* (Nutt. ex Torr. & A. Gray) J. M. Coulter & J. Rose (=*Peucedanum foeniculaceum* Nutt. ex Torr. & A. Gray, *L. daucifolium* J. M. Coulter & J. Rose)
11. *L. geyeri* (S. Wats.) J. M. Coulter & J. Rose
12. *L. gormanii* (Howell) J. M. Coulter & J. Rose
13. *L. grayi* (J. M. Coulter & J. Rose) J. M. Coulter & J. Rose
14. *L. leptocarpum* (Torr. & A. Gray) J. M. Coulter & J. Rose [slender-fruited lomatium]
15. *L. macrocarpum* (Nutt. ex Torr. & A. Gray) J. M. Coulter & J. Rose
16. *L. martindalei* J. M. Coulter & J. Rose (=*L. angustatum* (J. M. Coulter & J. Rose) St. John)
17. *L. nudicaule* (Pursh) J. M. Coulter & J. Rose [pestle parsnip]
18. *L. orientale* J. M. Coulter & J. Rose (=*Cogswellia orientalis* (J. M. Coulter & J. Rose) M. E. Jones)
19. *L. parryi* (S. Wats.) Macbr. (=*Cynomarathrum parryi* (S. Wats.) J. M. Coulter & J. Rose)
20. *L. salmoniflorum* (J. M. Coulter & J. Rose) Mathias & Constance
21. *L. suksdorfii* (S. Wats.) J. M. Coulter & J. Rose
22. *L. triternatum* (Pursh) J. M. Coulter & J. Rose [buck parsnip]
23. *L. utriculatum* (Nutt. ex Torr. & A. Gray) J. M. Coulter & J. Rose [bladder parsnip, spring-gold]
24. *Lomatium* sp.

Oomycetes

Peronospora parasitica. Downy mildew. **5:**MT(3205). **14:**MT(3205).

Plasmopara crustosa. Downy mildew. **24:**ID(3202), MT(82,3202).

Ascomycotina

Comoclathris permunda. **19:**WY(3810). **21:**WA(3202).

Crocicreas cyathoideum var. *cacaliae*. On dead stems. **7:**UT(505).

[*Guignardia reticulata*]. **21:**WA(3202).

Hyponectria sceptri. **24:**OR(3202).

Lewia scrophulariae. **13:**WA(3202).

Mycosphaerella tassiana. **5:**WA(3202). **24:**OR(3202).

Nodulosphaeria olivacea. **24:**OR(3202).

Phyllachora sp. **7:**WA(3202). **13:**ID(3202), WA(3202). **16:**WA(3202).

Sphaerella sp. **17:**WA(3202).

Basidiomycotina - Uredinales

Nyssopsora echinata. **9:**UT(2461).

Puccinia asperior. Autoecious. **7:**CA(82), OR(82), WA(82). **8:**CA(81).

P. jonesii. Autoecious. **1:**WA(3202). **4:**WA(3202). **5:**MT(3202), WA(3202). **6:**OR(3202). **7:**AZ(1203), ID(3202), MT(3202), OR(3202), WA(3202), WY(1195). **9:**AZ(3981), ID(638), MT(1195). **10:**KS(82), NE(82), OK(2872), SD(2243), WA(3202). **11:**ID(3202), WA(3202). **12:**WA(3202). **13:**ID(638,3202), UT(82), WA(82,3202). **15:**CA(81), ID(638,3202), MT(3202), OR(3202), WA(3202), WY(2924). **16:**ID(3202), OR(82,3202). **17:**CA(81). **18:**AZ(82,1203), CO(82), NE(82), OK(82,2872),

WY(82). **20:**WA(3202). **21:**WA(3202). **22:**OR(82,3202), WA(82,3202). **24:**W states(82), CA(82), WA(82).
P. ligustici. Autoecious. **11:**ID(3202). **12:**WA(3202). **24:**ID(82), WA(82).
P. musenii. Autoecious. **5:**WY(749,1195).
Deuteromycotina - Hyphomycetes
Cercospora sp. **5:**OR(3202).
Pollaccia peucedani. **2:**WA(3202). **17:**ID(1195,3202), OR(3202), WA(3202). **22:**OR(3202).
Pseudocercosporella herpotrichoides. On stems. **22:**ID(82), OR(82), WA(82).
Ramularia heraclei. **6:**OR(3202). **13:**WA(3202). **14:**ID(3202).
R. reticulata. **3:**CA(638).
Deuteromycotina - Coelomycetes
Ascochyta lomatii. On leaves. **13:**WA(641,3202).
Heteropatella umbilicata. **24:**OR(3202), WA(3202).
Phoma karstenii. **23:**CA(81).
P. lomatiae. On leaves. **22:**OR(82,3202).
Phyllosticta sp. **14:**OR(3202). **17:**WA(3202).

Musineon Raf.

1. *M. divaricatum* (Pursh) Nutt. ex Torr. & A. Gray

Basidiomycotina - Uredinales
Puccinia jonesii. Autoecious. **1:**MT(3202), SD(2243).

Oenanthe L.

1. *O. sarmentosa* K. Presl ex DC. (=*O. sarmentosa* K. Presl ex DC. var. *californica* J. M. Coulter & J. Rose, *O. californica* S. Wats.)

Basidiomycotina - Uredinales
Nyssopsora echinata. Autoecious. **1:**AK(60,526), CA(81,2461), WA(2461,3202).
Uromyces americanus. **1:**CA(3101).
U. lineolatus subsp. *nearcticus*. Heteroecious: aecial host. **1:**CA(81).
Deuteromycotina - Coelomycetes
Septoria oenanthis. **1:**WA(3202).

Orogenia S. Wats.

1. *O. linearifolia* S. Wats.

Deuteromycotina - Hyphomycetes
Ramularia orogeniae. Leaf spot. **1:**WA(666,3202).

Osmorhiza Raf.

1. *O. chilensis* (Molina) Hook. & Arnott (=*O. brevipes* Suksd. in P. Allorge, *O. divaricata* (Britton) Piper in Piper & Beattie, *O. intermedia* (Rydb.) Blank., *O. nuda* Torr.)
2. *O. claytonii* (Michx.) C. B. Clarke [hairy sweet cicely, sweet jarvil, woolly sweet cicely]
3. *O. depauperata* Phil. (=*O. obtusa* Fernald)
4. *O. longistylis* (Torr.) DC. [aniseroot]
5. *O. occidentalis* (Nutt. ex Torr. & A. Gray) Torr. (=*Glycosma occidentalis* Nutt. ex Torr. & A. Gray)
6. *Osmorhiza* sp.

Chytridiomycetes
Physoderma pluriannulatum. Leaf gall. **4:**MT(82,3202).
Ascomycotina
Crocicreas cyathoideum var. *cacaliae*. On dead stems. **6:**UT(505).
C. nigrofuscum var. *allantosporum*. On stems. **5:**UT(505).
Mycosphaerella glycosmae. **5:**CO(82).
Mycosphaerella sp. On stems. **4:**MN(2874). **5:**NV(82).
Pleospora richtophensis. **5:**WY(3810).
Pleospora sp. **5:**OR(3202).
Basidiomycotina - Uredinales
Puccinia pimpinellae. Autoecious. **2:**Range of host(82), MN(2874), TN(1539), WV(3907). **3:**MT(3202), WA(3202), WY(1195). **4:**Range of host(82), AK(526), MN(2874), SD(2243). **5:**Range of host(82), CA(81,661), ID(3202), MT(3202), OR(3202), WA(3202), WY(1195). **6:**MN(2874).
Deuteromycotina - Hyphomycetes
Cercospora osmorrhizae. Leaf spot. **2:**IL(82), MN(2874), OH(82), WI(82). **4:**DE(581,3543), MD(82), MN(2874), OH(82), WI(82).
[*Cercosporidium angelicae*]. Leaf spot. **6:**WA(82).
C. depressum. **1:**WA(3202).
Heterosporium sp. **5:**ID(3202).
Mycocentrospora acerina. On living leaves. **1:**OR(3202,3421). **6:**OR(82).
Ramularia reticulata. Leaf spot. **1:**WA(3202). **2:**WI(82). **4:**VA(82), WI(82). **6:**WY(82).
Deuteromycotina - Coelomycetes
Colletotrichum dematium. On stems. **2:**PA(82). **4:**NE(82), VA(82).
Phloeospora osmorrhizae. Leaf spot. **1:**ID(3202), WA(3202), WY(1195,2924). **2:**NY(82). **3:**CO(1195), ID(3202), MT(3202), WY(1195,2924). **4:**IA(82), NY(82), WI(82). **5:**CA(81), MT(82), OR(3202), WY(82,1195). **6:**CA(82), CO(82), ID(82), WA(82), WY(82).
Phyllosticta sp. **2:**WI(1340).
Septoria aegopodii. Leaf spot. **2:**OH(82). **4:**ND(82), NY(82), WI(82).
S. micropuncta. **2:**VA(82).
S. occidentalis. **5:**WA(82). **6:**WA(82).
Stagonospora thaspii. **2:**WI(1350).

Oxypolis Raf.

1. *O. fendleri* (A. Gray) A. Heller
2. *O. occidentalis* J. M. Coulter & J. Rose
3. *O. rigidior* (L.) J. M. Coulter & J. Rose [cowbane]

Ascomycotina
Ophiobolus nigro-clypeata. **3:**GA(1438).
Basidiomycotina - Uredinales
Puccinia ligustici. Autoecious. **1:**CO(1195,3379).
Deuteromycotina - Hyphomycetes
Cercospora sp. **3:**WI(1333).
Cercosporidium depressum. **3:**WI(1332).
C. punctiforme. **3:**WI(1335).
Ramularia heraclei. **2:**CA(81).
Deuteromycotina - Coelomycetes
Neottiospora umbelliferarum. On living leaves. **3:**WI(1352).
Stagonospora thaspii. **3:**WI(1343).

Pastinaca L.

1. *P. sativa* L. [parsnip, wild parsnip]
2. *Pastinaca* sp.

Oomycetes
Phytophthora nicotianae var. *parasitica*. Dry rot. **1:**PA(82).
Zygomycotina
Rhizopus sp. Black-mold rot in storage and markets. **1:**Unk.(82).
Ascomycotina
Chaetomium globosum. Isolated from seeds. **1:**CA(3306).
C. murorum. Isolated from seeds. **1:**MI(3306).
Diaporthe arctii. On stems. **1:**NY(82).
Pleospora lactucicola. **1:**NJ(82).
Sclerotinia sclerotiorum. Watery soft rot. **1:**CA(82), ID(82), IN(82), LA(82), MA(82), NY(82), OR(82), TX(82).
Basidiomycotina - Other
Thanatephorus cucumeris. **1:**WA(3202).
Deuteromycotina - Hyphomycetes
Alternaria alternata. On seeds. **1:**Unk.(82).
Alternaria sp. Leaf spot. **1:**NJ(82).
Botrytis cinerea. Gray-mold rot in transit and storage. **1:**Unk.(82).
Cercospora pastinacae. Leaf spot. **1:**MD(82), NE(82), NY(82), TX(82), WV(82,3907).
Phymatotrichopsis omnivora. Root rot. **1:**AZ(82), TX(82).
Stilbella annulata. **2:**SC(3178).
Deuteromycotina - Coelomycetes
Cylindrosporium crescentum. **1:**TX(82).
Phoma nebulosa. Root rot. **1:**PA(82), TX(82).
Phomopsis diachenii. On fruits, leaves, seeds and stems. **1:**IL(82), NY(82,3081), WI(82,1337,1357).
Septoria sp. On leaves. **1:**WI(1348).
Stagonospora polytaeniae. **1:**WI(1342).
Deuteromycotina - Other
Rhizoctonia solani. Black scurf. **1:**NC(82,1303), TX(82), WA(82).

Perideridia Rchb.

1. *P. gairdneri* (Hook. & Arnott) Mathias (=*Carum gairdneri* (Hook. & Arnott) A. Gray) [false caraway, Indian caraway, squawroot]

Basidiomycotina - Uredinales
Puccinia cicutae. **1:**CA(81).
Deuteromycotina - Hyphomycetes
Cercosporidium depressum. **1:**WY(1195).
C. punctiforme. **1:**NW states(3202).

Petroselinum J. Hill
1. *P. crispum* (Mill.) Nym. ex A. W. Hill (=*P. hortense* Hoffm., *P. sativum* Hoffm., *P. vulgare* Lag., *Carum petroselinum* Benth. & J. D. Hook., *Apium petroselinum* L.) [parsley]

Oomycetes
Phytophthora nicotianae var. *parasitica*. Root rot. **1**:HI(2903).
Pythium aphanidermatum. **1**:HI(2903).
P. debaryanum. Root rot. **1**:MI(4030).
Pythium sp. Damping-off. **1**:NJ(82).

Ascomycotina
Chaetomium murorum. Isolated from seeds. **1**:OH(3306).
C. succineum. Isolated from seeds. **1**:CA(3306).
Comoclathris permunda. On overwintered stems. **1**:CA(661).
Hyponectria sceptri. On dead stems. **1**:CA(661).
Sclerotinia sclerotiorum. Stem rot. **1**:CT(82), GA(82), HI(2903), LA(82), NY(82), PA(82), TX(82), VA(82).

Deuteromycotina - Hyphomycetes
Alternaria alternata. On seeds. **1**:Unk.(82).
A. dauci. **1**:CT(82).
A. radicina. On seeds. **1**:Unk.(82).
Alternaria sp. **1**:NJ(82), TX(82).
Botrytis cinerea. Gray mold. **1**:AK(82,526).
Cercospora sp. Leaf spot. **1**:HI(2903).
Cercosporidium punctum. Leaf spot. **1**:NJ(82).
Fusarium oxysporum f. sp. *apii*. Wilt. **1**:OR(1423,3202).
Fusarium sp. Wilt. **1**:NJ(82,3748), VA(82).
Phymatotrichopsis omnivora. Root rot. **1**:TX(82).
Stemphylium botryosum. On seeds. **1**:Unk.(82).
Stemphylium sp. Leaf spot. **1**:NJ(2104).

Deuteromycotina - Coelomycetes
Phyllosticta petroselini. **1**:NY(82).
Septoria petroselini. Leaf spot. **1**:CA(82), CT(82), HI(2903), NJ(82), NY(82), OK(2872), TX(82).

Deuteromycotina - Other
Rhizoctonia solani. Damping-off, root rot. **1**:FL(35), NJ(82), NY(82), TX(82).
Rhizoctonia sp. Root rot. **1**:HI(2903).

Pimpinella L.
1. *P. anisum* L. (=*Anisum vulgare* Gaertn.) [anise]
2. *P. integerrima* (DC.) Benth. & J.D. Hook. (=*Zizia integerrima* DC.)

Oomycetes
Pythium sp. Root rot. **1**:FL(35).

Ascomycotina
Sclerotinia sclerotiorum. Stem rot. **1**:CA(82), TX(82).

Basidiomycotina - Uredinales
Puccinia pimpinellae. Autoecious. **1**:CA(82).

Deuteromycotina - Hyphomycetes
Cercospora malkoffii. Leaf spot. **1**:VA(82).
Cercospora sp. Leaf spot. **1**:FL(35).
Cercosporidium punctiforme. On leaves. **2**:Unk.(962).
Phymatotrichopsis omnivora. Root rot. **1**:TX(82).

Deuteromycotina - Coelomycetes
Colletotrichum sp. Leaf spot. **1**:FL(35).

Polytaenia DC.
1. *P. nuttallii* DC.

Deuteromycotina - Hyphomycetes
Cercospora polytaeniae. On leaves. **1**:KS(973).

Deuteromycotina - Coelomycetes
Phomopsis pimpinellae. On immature fruits. **1**:WI(1344).
Stagonospora polytaeniae. On living leaves. **1**:WI(1338,1342,1344).

Pseudocymopterus J. M. Coulter & J. Rose
1. *P. montanus* (A. Gray) J. M. Coulter & J. Rose (=*P. sylvaticus* A. Nels.)
2. *Pseudocymopterus* sp.

Basidiomycotina - Uredinales
Puccinia pseudocymopteri. Autoecious. **1**:AZ(1195,1203,3981), CO(82,1195), NM(82), UT(82), WY(82,1195). **2**:WY(2924).

Deuteromycotina - Coelomycetes
Phloeospora osmorrhizae. **1**:WY(1195).

Ptilimnium Raf.
1. *P. capillaceum* (Michx.) Raf.

Deuteromycotina - Other
Rhizoctonia solani. Stem necrosis. **1**:FL(35).

Sanicula L.
1. *S. bipinnatifida* Douglas ex Hook. [purple sanicle]
2. *S. canadensis* L.
3. *S. crassicaulis* Poepp. ex DC. (=*S. menziesii* Hook. & Arnott)
4. *S. gregaria* C. Bicknell
5. *S. marilandica* L.
6. *Sanicula* sp.

Chytridiomycetes
Physoderma pluriannulatum. Leaf gall. **1**:CA(81). **2**:MS(82,2720). **3**:CA(81), OR(3202), WA(3202). **5**:IA(82), IL(82), MI(82), MS(2720), NY(82), WI(82,1344). **6**:AL(82), CA(82), OR(82), WA(82), WI(82).

Ascomycotina
Crocicreas fuscum. On stems. **3**:CA(505,2801).
Ophiobolus nigro-clypeata. On stems. **2**:GA(1438).

Basidiomycotina - Uredinales
Puccinia marylandica. Autoecious. **2**:GA(1439), IN(987), MO(82), TN(82), WV(82,3907). **5**:CO(82), IA(82), IN(82), NC(82,1303), PA(82), TN(82), WI(82). **6**:E states(82), FL(35), IN(647), MO(82), SD(82).

Basidiomycotina - Ustilaginales
Entyloma eryngii. White smut. **2**:IA(1077), NY(1077). **3**:CA(1077). **4**:WI(1077). **5**:IA(82,1077), IL(82,1077), IN(82), MD(82,1077), NY(82,1077), VA(82,1077), WI(82,1077). **6**:AL(82,1077), CA(82), OR(82,1077,3202), WI(82).

Deuteromycotina - Hyphomycetes
Cercospora saniculae. Leaf spot. **4**:WI(82,581).
Cercosporidium depressum. **6**:OR(3202).

Deuteromycotina - Coelomycetes
Septoria saniculae. Leaf spot. **2**:NE(82). **5**:IA(82), WI(82).
Stagonospora thaspii. Leaf spot. **5**:WI(82,1341). **6**:WI(82).

Sium L.
1. *S. suave* Walter (=*S. cicutifolium* Schrank) [water-parsnip]

Chytridiomycetes
Physoderma palustris. On leaves and stems. **1**:MI(3400).
P. vagans. Leaf spot. **1**:MI(3398), WI(82).
Physoderma sp. On leaves and stems. **1**:MI(3406).

Basidiomycotina - Uredinales
Uromyces lineolatus subsp. *nearcticus*. Heteroecious: aecial host. **1**:ND(82,378), NY(82), WI(82).

Deuteromycotina - Hyphomycetes
Cercosporidium sii. Leaf spot. **1**:IA(82), WI(82,1350).

Deuteromycotina - Coelomycetes
Phyllosticta sp. Leaf spot. **1**:ND(82,378).
Septoria sii. Leaf spot. **1**:ME(82), ND(82), NY(82), WI(82).
Stagonospora sp. On leaves. **1**:WI(1341).

Sphenosciadium A. Gray
1. *S. capitellatum* A. Gray

Ascomycotina
Comoclathris permunda. On dead stems. **1**:CA(81).

Taenidia (Torr. & A. Gray) Drude in Engl. & Prantl
1. *T. integerrima* (L.) Drude [yellow pimpernel]

Basidiomycotina - Uredinales
Puccinia angelicae. Autoecious. **1**:IN(82), MI(82), MO(82), NY(82), WI(82).

Deuteromycotina - Hyphomycetes
[*Cercosporidium angelicae*]. Leaf spot. **1**:WI(82).
C. punctiforme. On leaves. **1**:MO(839).

Deuteromycotina - Coelomycetes
Phomopsis pimpinellae. Leaf spot. **1**:MN(82,2874), WI(1342,1347).

Tauschia Schlechtend.
1. *T. hartwegii* (A. Gray) Macbr. (=*Drudeophytum hartwegii* (A. Gray) J. M. Coulter & J. Rose)

Basidiomycotina - Uredinales
Puccinia jonesii. **1**:CA(81).

Thaspium Nutt.
1. *T. trifoliatum* (L.) A. Gray (=*T. aureum* Nutt., *Zizia cordata* K. Koch ex DC.)
2. *T. trifoliatum* (L.) A. Gray var. *flavum* S. F. Blake
Deuteromycotina - Hyphomycetes
Cercospora thaspii. Leaf spot. **1**:DE(3543).
C. thaspiicola. Leaf spot. **1**:WI(581,1334,1336). **2**:WI(1338).
C. ziziae. On leaves. **1**:PA(962). **2**:WI(1341).
Deuteromycotina - Coelomycetes
Septoria ziziae. **1**:ND(378).
Stagonospora thaspii. **1**:WI(1331). **2**:WI(1341).
Stagonospora sp. On leaves. **1**:WI(1340).

Torilis Adans.
1. *T. japonica* (Houtt.) DC.
Ascomycotina
Protomyces macrosporus. Leaf gall. **1**:AR(3594,3722).

Trachymene Rudge
1. *T. coerulea* Graham [blue laceflower]
Oomycetes
Phytophthora cactorum. Wilt. **1**:Unk.(3887).
Deuteromycotina - Hyphomycetes
Fusarium sp. Root rot. **1**:CT(82), NJ(82).
Deuteromycotina - Other
Rhizoctonia solani. Stem rot. **1**:NJ(82).

Zizia W. Koch
1. *Z. aptera* (A. Gray) Fernald
2. *Z. aurea* (L.) W. Koch [golden alexanders, meadow-parsnip]
Chytridiomycetes
Physoderma pluriannulatum. Leaf gall. **1**:WI(1351). **2**:IA(82), WI(82,1330).
Ascomycotina
Erysiphe polygoni. Powdery mildew. **1**:PA(82), WV(82,3907). **2**:PA(82), WV(82).
Basidiomycotina - Uredinales
Puccinia ziziae. Autoecious. **1**:WA(82,3202).
Basidiomycotina - Other
Postia placenta. Heart rot. **1**:CO(82).
Deuteromycotina - Hyphomycetes
Cercospora thaspiicola. Leaf spot. **1**:MN(2874), ND(82), WA(82,3202), WI(82). **2**:MN(2874), WI(1336).
C. ziziae. Leaf spot. **1**:PA(82), SD(2243), WI(1340). **2**:WI(82,1329).
Deuteromycotina - Coelomycetes
Phyllosticta sp. On leaves. **1**:WI(1339,1348).
Septoria ziziae. Leaf spot. **1**:ND(82).
Stagonospora thaspii. Leaf spot. **1**:WI(82,1343). **2**:WI(82).

Apocynaceae

Acokanthera G. Don
1. *A. oppositifolia* (Lam.) Codd [bushman's-poison]
Basidiomycotina - Other
Armillaria tabescens. Root rot. **1**:FL(35).

Adenium Roem. & Schult.
1. *A. obesum* (Forssk.) Roem. & Schult. (=*A. coetaneum* Stapf in Dyer) [desert-rose]
Oomycetes
Phytophthora sp. Basal stem rot. **1**:FL(35).
Deuteromycotina - Hyphomycetes
Cercospora sp. Leaf spot. **1**:FL(35).
Fusarium sp. Basal stem rot. **1**:FL(35).
Deuteromycotina - Other
Rhizoctonia sp. Crown and root rot. **1**:HI(2903).

Allamanda L.
1. *A. cathartica* L. [allamanda, golden-trumpet]
2. *A. neriifolia* Hook. [bush allamanda]
3. *Allamanda* sp.
Oomycetes
Pythium sp. Root rot, basal stem rot. **3**:FL(35).
Basidiomycotina - Uredinales
Coleosporium sp. **2**:FL(35).
Deuteromycotina - Hyphomycetes
Alternaria sp. Leaf spot. **3**:FL(35).
Botrytis sp. Leaf tip rot. **3**:FL(35).
Cercospora sp. Leaf spot. **1**:FL(35). **2**:FL(35). **3**:FL(35).
Corynespora cassiicola. Leaf spot. **1**:FL(35).
Deuteromycotina - Coelomycetes
Colletotrichum gloeosporioides. Anthracnose. **2**:FL(35). **3**:FL(35).
Diplodia sp. Dieback. **2**:FL(35).
Phoma sp. Stem lesions. **1**:FL(35).
Phyllosticta sp. **1**:FL(35). **2**:FL(35). **3**:FL(35).
Deuteromycotina - Other
Rhizoctonia solani. Root and stem rot. **1**:FL(35). **2**:FL(35).

Alstonia R. Br.
1. *A. macrophylla* Wallich ex G. Don [batino]
Deuteromycotina - Hyphomycetes
Cercospora sp. Leaf spot. **1**:FL(35).

Alyxia Banks ex R. Br.
1. *A. oliviformis* Gaud.
Ascomycotina
Amazonia alyxiae. Black mildew. **1**:HI(1272,1452,2903).
Elsinoe alyxiae. Spot anthracnose. **1**:HI(3788).
Guignardia alyxiae. **1**:HI(2903).
Meliola moerenhoutiana. Black mildew. **1**:HI(1272,1452,2903).
Trichopeltis rhyacoides. **1**:HI(2903).
Trichothyrium reptans. **1**:HI(2903).
Basidiomycotina - Uredinales
Uromyces alyxiae. Autoecious. **1**:HI(2903).
Deuteromycotina - Hyphomycetes
Trichothallus hawaiiensis. **1**:HI(2903).

Amsonia Walter
1. *A. ciliata* Walter
2. *A. tabernaemontana* Walter
Ascomycotina
Hypoderma commune. **2**:GA(1438).
Mycosphaerella sp. **2**:GA(82,1438).
Ophiobolus nigro-clypeata. **2**:GA(1438).
Basidiomycotina - Uredinales
Coleosporium apocynaceum. Heteroecious: telial host. **1**:FL(35,82), GA(82), SC(82,634). **2**:AL(82), GA(82), NC(1303), SC(82).
Puccinia seymouriana. Heteroecious: aecial host. **2**:Unk.(82).
Deuteromycotina - Coelomycetes
Phyllosticta sp. **2**:WI(1349).

Apocynum L.
1. *A. androsaemifolium* L. [common dogbane, spreading dogbane]
2. *A. androsaemifolium* L. var. *pumilum* A. Gray
3. *A. cannabinum* L. [hemp dogbane, Indian hemp]
4. *A. cannabinum* L. var. *glaberrimum* A. DC.
5. *A. medium* Greene
6. *A. sibiricum* Jacq. (=*A. hypericifolium* Aiton) [prairie dogbane]
Ascomycotina
Botryosphaeria apocyni. On stems. **6**:ND(82,377).
Didymosphaeria futilis. **6**:ND(377).
Nectria apocyni. **3**:NY(3085).
Schizoxylon berkeleyanum. **6**:ND(377,3225).
Basidiomycotina - Uredinales
Puccinia seymouriana. Heteroecious: aecial host. **3**:IN(82), KS(82), NJ(82), NY(82). **6**:IL(82), KS(82), NE(82), OK(82,2872), SD(82,2243).
P. smilacis. Heteroecious: aecial host. **1**:DE(82), IL(82), KS(82), MD(82), NC(82,1303), TN(82), VA(82), WI(82). **3**:IL(82), KS(82), MD(82), NC(82,3944), TN(82), VA(82), WI(82).
Basidiomycotina - Other
Thanatephorus cucumeris. **3**:WA(3202).
Deuteromycotina - Hyphomycetes
Cercospora apocyni. Leaf spot. **1**:NC states(82), TX(82), VA(82). **3**:ID(3202), MT(3202). **5**:MN(2874). **6**:MN(2874).
Phymatotrichopsis omnivora. Root rot. **1**:TX(82). **3**:TX(82).
Deuteromycotina - Coelomycetes
Colletotrichum truncatum. Isolated from stems. **3**:IL(1482).

Cylindrosporium apocyni. Leaf spot. **1**:DE(82,962), IA(82), MN(2874), OK(2872), TX(82), WI(82). **3**:OK(2872), VA(82), WI(1350), WV(3907).
Macrophomina phaseolina. Charcoal rot. **3**:IL(311).
Pestalozziella andersonii. On leaves. **2**:CA(2549). **3**:MT(3202).
Phyllosticta apocyni. Leaf spot. **1**:IA(82), NY(64,82), OR(82,3202), PA(82), WI(82,1355). **3**:FL(35), MS(82,2720), NJ(82), NY(82), OK(2872), WI(82,1329).
Septoria littorea. Leaf spot. **1**:MI(82). **3**:KS(82), MS(2720), ND(82,378), NE(82), OH(82). **6**:OH(82).
Stagonospora apocyni. Leaf spot. **1**:IA(82), IL(82), MN(2874), NY(82), WI(82). **3**:IN(82), MN(2874), NY(82), SD(2243), VA(82), WI(82,1334). **4**:MN(2874). **5**:WI(1342). **6**:MN(2874), WI(1359).

Carissa L.

1. *C. carandas* L. [karanda, perunkila]
2. *C. macrocarpa* (Ecklon) A. DC. (=*C. grandiflora* (E. Mey.) A. DC.) [amatungulu, big num-num, natal-plum]
3. *Carissa* sp.

Oomycetes
Phytophthora nicotianae var. *parasitica*. Root rot. **2**:FL(35). **3**:FL(35).
Pythium sp. Root rot. **2**:FL(35). **3**:FL(35).
Ascomycotina
Botryosphaeria obtusa. **2**:FL(82).
B. quercuum. Leaf spot. **2**:FL(35).
B. rhodina. On twigs. **2**:FL(82).
Nectria sp. Stem gall. **2**:FL(35).
Deuteromycotina - Hyphomycetes
Alternaria sp. Leaf spot. **2**:FL(35).
Fusarium sp. Root rot, leaf spot. **2**:FL(35). **3**:FL(35).
Oidium sp. Powdery mildew. **2**:FL(35).
Phymatotrichopsis omnivora. Root rot. **2**:TX(82).
Deuteromycotina - Coelomycetes
Colletotrichum gloeosporioides. Anthracnose, leaf spot. **1**:FL(82). **2**:FL(35). **3**:FL(35).
Diplodia sp. Dieback. **2**:FL(82).
Lasiodiplodia theobromae. Stem canker, dieback. **1**:FL(35). **2**:FL(35). **3**:FL(35).
Phoma sp. Stem gall. **2**:FL(35).
Phomopsis sp. Stem gall. **2**:FL(35).
Phyllosticta sp. Leaf spot. **2**:FL(35).
Septoria sp. Leaf spot. **2**:CA(82).
Sphaeropsis tumefaciens. Canker, stem gall. **2**:FL(35,2259). **3**:FL(35).
Sphaeropsis sp. Dieback of twigs. **1**:FL(82).
Deuteromycotina - Other
Rhizoctonia ramicola. Thread blight, dieback. **2**:FL(35). **3**:FL(35).
R. solani. Leaf blight, dieback. **2**:FL(35,40). **3**:FL(35).
Rhizoctonia sp. Root rot. **2**:HI(2903).

Catharanthus G. Don

1. *C. roseus* (L.) G. Don (=*Vinca rosea* L., *Lochnera rosea* (L.) Rchb., *Ammocallis rosea* (L.) Small) [bright-eyes, Madagascar periwinkle, rose periwinkle]
2. *Catharanthus* sp.

Oomycetes
Phytophthora colocasiae. Leaf rot. **1**:HI(82).
P. nicotianae var. *parasitica*. Foliage blight. **1**:CA(1219,1862). **2**:HI(2903).
P. palmivora. Root rot. **1**:FL(35).
Pythium sp. Stem rot. **1**:FL(35).
Basidiomycotina - Uredinales
Puccinia vincae. Autoecious. **1**:FL(35), Fl(35).
Deuteromycotina - Hyphomycetes
Fusarium sp. Stem rot. **1**:FL(35).
Deuteromycotina - Coelomycetes
Colletotrichum sp. Anthracnose. **1**:FL(35).
Phyllosticta sp. Leaf spot. **1**:FL(35).
Deuteromycotina - Other
Rhizoctonia solani. Leaf blight, root rot. **1**:FL(35).

Cycladenia Benth.

1. *C. humilis* Benth.

Ascomycotina
Comoclathris permunda. On dead stems. **1**:CA(642).

Kopsia Blume

1. *K. fruticosa* (Ker-Gawl.) A. DC.

Oomycetes
Pythium sp. Root rot. **1**:FL(35).
Deuteromycotina - Other
Rhizoctonia solani. Root rot. **1**:FL(35).

Mandevilla Lindl.

1. *Mandevilla* sp.

Oomycetes
Pythium sp. Root rot. **1**:Fl(35).
Ascomycotina
Glomerella sp. Leaf spot. **1**:Fl(35).
Deuteromycotina - Hyphomycetes
Botrytis cinerea. Gray mold. **1**:Fl(35).
Cercospora sp. Leaf spot. **1**:Fl(35).
Corynespora sp. Leaf spot. **1**:Fl(35).
Fusarium oxysporum. Root rot. **1**:Fl(35).
Deuteromycotina - Coelomycetes
Colletotrichum sp. Anthracnose. **1**:Fl(35).
Phyllosticta sp. Leaf spot. **1**:Fl(35).
Deuteromycotina - Other
Rhizoctonia solani. Root and stem rot. **1**:Fl(35).

Nerium L.

1. *N. oleander* L. [oleander, rosebay]
2. *Nerium* sp.

Oomycetes
Phytophthora sp. Root rot. **1**:FL(35).
Pythium sp. Root rot. **1**:HI(2903).
Ascomycotina
Botryosphaeria dothidea. **1**:GA(1438).
B. obtusa. On twigs. **1**:GA(82), HI(82,2903).
Botryosphaeria sp. Leaf spot. **1**:FL(35).
Calonectria crotalariae. Basal stem rot. **1**:Unk.(2807).
Capnodium elongatum. Sooty mold. **1**:AL(82), FL(82), OK(2873). **2**:FL(634).
Diaporthe sp. Dieback. **1**:FL(35), HI(82,2903).
Basidiomycotina - Other
Armillaria mellea. Dieback. **1**:CA(2898).
A. tabescens. Root rot. **1**:FL(35,82).
Ceriporia xylostromatoides. **1**:AZ(1202).
Meruliopsis corium. **1**:FL(128).
Scytinostroma galactinum. Root rot. **1**:MD(82).
Deuteromycotina - Hyphomycetes
Alternaria sp. Leaf spot. **1**:AL(82), CA(82), FL(35,82), GA(82), MS(82,2720).
Aspergillus versicolor. On stems. **1**:CA(2740).
Cercospora repens. **1**:TX(82).
Dicranidion fragile. On wood. **1**:CA(2740).
Fusarium solani. Leaf spot. **1**:FL(35).
Fusarium sp. Leaf spot, stem rot. **1**:FL(35).
Helminthosporium sp. Leaf spot. **1**:FL(35).
Macrosporium nerii. Leaf spot. **1**:AL(285).
Phymatotrichopsis omnivora. Root rot. **1**:TX(82).
Deuteromycotina - Coelomycetes
Aplosporella nerii. On limbs. **1**:FL(35,82).
Botryodiplodia nerii. **1**:FL(2782).
Botryodiplodia sp. Dieback. **1**:FL(35).
Colletotrichum gloeosporioides. Leaf spot. **1**:MS(2720).
Colletotrichum sp. Leaf spot. **1**:HI(2903).
Diplodia nerii. On twigs. **1**:CA(82).
Dothiorella sp. Dieback. **1**:FL(35).
Pestalotia versicolor. Leaf spot. **1**:FL(35).
Phoma exigua. Dieback. **1**:CA(1863).
Phoma sp. Dieback. **1**:FL(35).
Phomopsis sp. Canker, dieback. **1**:FL(35).
Phyllosticta nerii. Leaf spot. **1**:FL(35,82), HI(82,2903), LA(82), MI(82), MS(82,2720), NC(1303).
Septoria oleandrina. Leaf spot. **1**:CA(82), FL(35,82), TX(82).
Sphaceloma oleandri. Scab. **1**:FL(35,2524), LA(1745).
Sphaeropsis gouldiae. Canker. **1**:HI(2903).
S. tumefaciens. Witches' broom. **1**:FL(35).
Sphaeropsis sp. Canker, witches' broom. **1**:FL(82), NH(82).

Deuteromycotina - Other
Rhizoctonia solani. Root rot. **1**:FL(35).
Rhizoctonia sp. Root rot. **1**:HI(2903).

Ochrosia Juss.

1. *O. elliptica* Labill.

Deuteromycotina - Other
Rhizoctonia solani. Root rot. **1**:FL(35).

Pachypodium Lindl.

1. *Pachypodium* sp.

Oomycetes
Pythium sp. Root rot. **1**:FL(35).
Deuteromycotina - Hyphomycetes
Alternaria sp. Leaf spot. **1**:FL(35).
Corynespora sp. Leaf spot. **1**:FL(35).
Deuteromycotina - Coelomycetes
Phyllosticta sp. Leaf spot. **1**:FL(35).

Plumeria L.

1. *P. rubra* L. [nosegay]
2. *P. rubra* L. f. *acutifolia* (Poir.) Woodson (=*P. acuminata* Aiton, *P. acutifolia* Poir.) [pagoda-tree]
3. *Plumeria* sp.

Oomycetes
Pythium sp. Root rot. **2**:HI(2903).
Ascomycotina
[*Apiosporina morbosa*]. Black knot. **3**:FL(35).
Basidiomycotina - Uredinales
Coleosporium plumierae. **1**:Fl(35). **3**:FL(35,82).
Deuteromycotina - Hyphomycetes
Botrytis sp. Blossom blight. **2**:HI(2903).
Cercospora plumeriae. Leaf spot. **1**:FL(35). **2**:FL(3065).
Cercospora sp. Leaf spot. **3**:Fl(35).
Chalara paradoxa. **3**:HI(2903).
Phymatotrichopsis omnivora. Root rot. **3**:TX(82).
Deuteromycotina - Coelomycetes
Colletotrichum gloeosporioides. Anthracnose. **3**:HI(2903).
Macrophoma sp. Dieback. **1**:Fl(35).
Phomopsis sp. Leaf spot. **3**:FL(35), HI(2903).
Deuteromycotina - Other
Rhizoctonia sp. Root rot. **2**:HI(2903).

Rhabdadenia Müll. Arg. in Mart.

1. *R. biflora* (Jacq.) Müll. Arg.

Deuteromycotina - Coelomycetes
Phyllosticta sp. Leaf spot. **1**:FL(35).

Tabernaemontana L.

1. *T. divaricata* (L.) R. Br. (=*Ervatamia coronaria* (Jacq.) Stapf, *T. coronaria* (Jacq.) Willd., *Ervatamia divaricata* (L.) Burkill) [Adam's apple, broadleaved rosebay, crape-gardenia, crape-jasmine, East Indian rosebay, flowers-of-love, moonbeam, pinwheel-flower, waxflower]
2. *Tabernaemontana* sp.

Oomycetes
Pythium sp. Root rot. **1**:HI(2903).
Zygomycotina
Choanephora cucurbitarum. Flower blight. **1**:Fl(35).
Ascomycotina
Mycosphaerella sp. Leaf spot. **1**:Fl(35).
Basidiomycotina - Other
Armillaria tabescens. Root rot. **1**:FL(35), Fl(82).
Deuteromycotina - Hyphomycetes
Cercospora peregrina. Leaf spot. **1**:Fl(35). **2**:FL(35).
Cladosporium sp. On leaves. **1**:FL(82), Fl(35).
Pseudocercospora tabernaemontanae. Leaf spot. **1**:Fl(35).
Deuteromycotina - Coelomycetes
Colletotrichum gloeosporioides. Leaf spot. **1**:Fl(35). **2**:FL(35,82).
Colletotrichum sp. Leaf spot. **2**:FL(35).

Thevetia L.

1. *T. peruviana* (Pers.) K. Schum. [be-still-tree, luckynut, yellow oleander]

Basidiomycotina - Other
Coriolopsis gallica. **1**:AZ(1202).

Trachelospermum Lem.

1. *T. difforme* (Walter) A. Gray
2. *T. jasminoides* Lem. [confederate-jasmine, star-jasmine]

Chytridiomycetes
Synchytrium trachelospermi. Leaf gall. **1**:LA(627).
Oomycetes
Pythium sp. Root rot. **2**:FL(35).
Ascomycotina
Botryosphaeria sp. Leaf spot. **2**:FL(35).
Schiffnerula pulchra. Sooty mold. **2**:LA(82).
Basidiomycotina - Other
Armillaria tabescens. Root rot. **2**:FL(35,82).
Deuteromycotina - Hyphomycetes
Cercospora repens. Leaf spot. **1**:LA(962). **2**:LA(82).
Cercospora sp. Leaf spot. **2**:FL(35).
Corynespora cassiicola. Leaf spot. **2**:FL(35).
Cylindrocladium sp. Leaf spot. **2**:FL(35).
Fusarium sp. Basal stem rot. **2**:FL(35).
Helminthosporium sp. Leaf spot. **2**:FL(35).
Deuteromycotina - Coelomycetes
Colletotrichum sp. Anthracnose. **2**:FL(35).
Coniothyrium sp. Leaf spot. **2**:FL(35).
Phyllosticta sp. Leaf spot. **2**:FL(35).
Deuteromycotina - Other
Rhizoctonia solani. Root rot. **2**:FL(35).

Vinca L.

1. *V. major* L. [band-plant, bigleaf periwinkle, blue-buttons, greater periwinkle]
2. *V. minor* L. [common periwinkle, lesser periwinkle, running myrtle]
3. *Vinca* sp.

Oomycetes
Phytophthora nicotianae var. *parasitica*. Stem and leaf blight, root and stem rot. **1**:HI(2903). **2**:HI(2903), MD(82), NC(1303). **3**:HI(2903).
Phytophthora sp. Root rot. **2**:FL(35).
Pythium sp. Seedling blight, damping-off, root rot. **2**:CO(49), NC(1303). **3**:HI(2903).
Ascomycotina
Botryosphaeria obtusa. On stems. **2**:NY(82), VA(82).
Stictis pustulata. **3**:OH(3225).
Basidiomycotina - Uredinales
Coleosporium sp. **2**:FL(35).
Puccinia vincae. Autoecious. **1**:ID(3202), IL(312), MA(82), MI(82), NY(82), WA(82,3202).
Deuteromycotina - Hyphomycetes
Alternaria sp. Stem rot, leaf spot. **2**:FL(35), PA(82), TX(82).
Botrytis cinerea. Gray mold blight. **1**:HI(2903). **2**:CT(82), FL(35), WA(82).
Botrytis sp. Gray mold blight, tip rot. **2**:WA(3202). **3**:WA(3202).
Cladosporium herbarum. On leaves. **3**:MD(82), NY(82), PA(82).
Fusarium sp. Root rot. **2**:CO(49).
Stemphylium botryosum. Leaf spot. **2**:FL(35).
Volutella vincae. Leaf spot. **2**:NY(82), PA(82).
Deuteromycotina - Coelomycetes
Colletotrichum sp. Anthracnose, leaf spot. **1**:FL(35). **2**:FL(35,82), NC(1303).
Diplodia vincae. On leaves and stems. **2**:MD(82), OH(82).
Macrophoma vincae. Dieback, leaf spot. **1**:IL(82). **2**:NC(1303), NY(82).
Phoma exigua. Dieback. **2**:NC(1303).
P. exigua var. *inoxydabilis*. **3**:CA(82).
Phoma sp. **1**:OR(1423,3202).
Phomopsis lirella. Canker, dieback. **2**:MD(82), NJ(82), OH(82), PA(82), VA(82).
Phyllosticta minor. Leaf spot. **2**:MD(82), NJ(82), NY(82), VA(82), WI(1354).
Phyllosticta sp. Stem rot, leaf spot. **1**:GA(82), NJ(82). **2**:CT(82), NC(1303).
Septoria vincae. Leaf spot. **1**:NJ(82). **2**:NY(82).
Deuteromycotina - Other
Rhizoctonia solani. Damping-off, root and stem rot. **1**:FL(35), IL(82).

2:FL(35), MD(82), NC(1303), NJ(82), PA(82), VA(82).
Rhizoctonia sp. Root and stem rot. **2**:CO(49). **3**:HI(2903).
Sclerotium rolfsii. Southern blight, root and stem rot. **2**:IL(312), NC(1303).

Aponogetonaceae

Aponogeton L. fil.

1. *A. ulvaceus* Baker

Oomycetes

Pythium myriotylum. Basal leaf rot, petiole rot. **1**:FL(35).

Aquifoliaceae

Ilex L.

(= *Byronia* Endl.)
1. *I. ambigua* (Michx.) Torr.
2. *I. anomala* Hook. & Arnott f. *sandwicensis* (Endl.) Loes. [kawa'ii]
3. *I. aquifolium* L. [English holly, European holly, Oregon holly]
4. *I.* × *aquipernyi* J. B. Gable ex W. Clarke. *I. aquifolium* × *I. pernyi*.
5. *I. arenicola* Ashe
6. *I. cassine* L. [cassina, dahoon, dahoon holly, yaupon]
7. *I. cassine* L. var. *myrtifolia* (Walter) Sarg.
8. *I. coriacea* (Pursh) Chapm. (=*Prinos coriacea* Pursh) [baygall-bush, large gallberry, sweet gallberry]
9. *I. cornuta* Lindl. & Paxt. [Chinese holly, horned holly]
10. *I. cornuta* Lindl. & Paxt. cv. 'burfordii'
11. *I. crenata* Thunb. [box-leaved holly, Japanese holly]
12. *I. crenata* Thunb. cv. 'rotundifolia'
13. *I. decidua* Walter [possum haw]
14. *I. glabra* (L.) A. Gray [Appalachian-tea, gallberry, inkberry, winterberry]
15. *I. helleri* Hort. A listed name of no botanical standing for *I. crenata* cultivars.
16. *I. laevigata* (Pursh) A. Gray [smooth winterberry]
17. *I. montana* Torr. & A. Gray ex A. Gray (=*I. ambigua* (Michx.) Torr. var. *montana* (Torr. & A. Gray) W. Ahles) [large-leaved holly, mountain holly, mountain winterberry]
18. *I. opaca* Aiton [American holly]
19. *I. verticillata* (L.) A. Gray [black-alder, common winterberry]
20. *I. vomitoria* Aiton [cassina, yaupon]
21. *Ilex* sp. (=*Byronia sp.*)

Chytridiomycetes

Synchytrium ilicicola. Leaf gall. **13**:LA(627). **20**:LA(627).

Oomycetes

Phytophthora cinnamomi. Root rot. **6**:NC(1303). **11**:MD(1729,3624), NC(1303). **14**:NC(1303).
P. ilicis. Leaf disease, twig disease. **3**:NW states(1389), NW states(423), OR(3202), WA(3202).
P. nicotianae var. *parasitica*. Root decay. **11**:NC(1303).
Phytophthora sp. Leaf and twig blight, root rot. **3**:OR(424). **9**:FL(35). **11**:FL(35). **18**:FL(35). **21**:FL(35).
Pythium sp. Root rot; some associated with nematodes. **6**:FL(35). **9**:FL(35). **11**:FL(35), GA(1208), NC(1303). **18**:FL(35). **21**:FL(35).

Ascomycotina

Anthostomella nigrotecta. On leaves. **21**:SC(1389).
Asterina lepiniana. Black mildew. **8**:FL(82,634), GA(82,1438), MS(82,2720). **18**:AL(82), MS(82,2720), TX(82).
A. orbicularis. Black spot, on leaves. **6**:E states(1389), Gulf states(1389), FL(35,82). **8**:E states(1389), Gulf states(1389), AL(82), FL(82), GA(1438), MS(82,2720), SC(82). **14**:E states(1389), Gulf states(1389), FL(35,82), GA(82,1438). **18**:E states(1389), Gulf states(1389), SE states(82), FL(35,82,634), GA(1438), TX(82). **21**:FL(35).
A. tantalina. Black mildew, black spot. **21**:HI(2784).
Asterina sp. Black mildew. **14**:NJ(82). **18**:GA(82), MS(82).
Asterinella humiriae. **2**:HI(2903).
Aulographum hederae. On fallen leaves. **18**:NJ(1389).
Belonium consanguineum. On dead twigs. **21**:AL(1389).
Botryosphaeria dothidea. Limb canker. **1**:GA(1438). **11**:NC(1303). **14**:GA(1438). **18**:GA(1438).
B. obtusa. On limbs. **17**:GA(82). **18**:MD(82). **21**:CA(82).
B. quercuum. Leaf spot, on twigs. **18**:FL(35), TX(82,1389).
B. rhodina. Leaf spot. **18**:FL(82), SC(82).
B. ribis. On limbs. **18**:GA(82,1389), MD(1389).
Botryosphaeria sp. Die-back. **9**:NC(1303). **21**:E states(2761), FL(35), NC(1303).
Calosphaeria subcuticularis. Canker, dieback, gall, leaf spot. **18**:NJ(1389).
Capnodium citri. On leaves. **14**:S states(1389).
C. elongatum. Sooty mold. **6**:Gulf States(82). **8**:Gulf States(82,1389). **14**:Gulf States(82,1389). **18**:Gulf States(82). **20**:Gulf States(82).
Capnodium sp. Black mildew. **20**:FL(35).
Cenangium tuberculiforme. **14**:NJ(1389).
Daldinia vernicosa. On wood. **18**:MD(82,1389).
Dermea olivacea. On twigs. **14**:NJ(82).
D. peckiana. On twigs. **14**:MI(1389), NJ(82,1389). **19**:MI(82,1389), MS(2720), NJ(82,1389).
Dermea sp. **18**:WA(82).
Diapleella coniothyrium. Leaf spot. **18**:FL(35).
Diaporthe eres. Canker, dieback. **3**:OR(82,1389,3202), WA(1389,3202).
D. ilicis. On limbs and twigs. **19**:MA(1389), MI(82,959,1389,3807). **21**:NC(1307).
D. oxyspora. On limbs. **1**:GA(1438). **17**:GA(82). **18**:MD(82), NC(82), NJ(82), NY(82), OK(82,2873), WV(82,3907). **19**:NJ(82). **21**:E states(1389), OK(1389).
Diaporthe sp. On dead limbs. **17**:GA(1438).
Diatrype friabilis. **13**:NC(1389), PA(1389), SC(1389).
Diatrypella favacea. **6**:GA(1438). **13**:GA(1438).
D. nigro-angulata. On dead limbs. **18**:MS(1389), TX(1389).
D. opaca. On dead limbs and bark. **18**:FL(82,634,1389), GA(82,1389), MS(82,1389), TX(82,1389), WV(82,1389,3907).
D. quercina. **18**:GA(82,1389,1438), WV(82,1389,3907).
Dimerium nigrosporum. Black mildew. **8**:MS(1389).
Dimerosporium ilicinum. On leaves. **7**:NC(1389), SC(1389).
Discochora philoprina. Leaf spot. **3**:CA(82), NJ(82), WA(82,3202). **8**:FL(35,82). **9**:NC(1303). **11**:NC(1303), NJ(82). **12**:GA(899). **18**:NC(82,1303), NJ(82), NY(82), SC(82), TX(82), WV(82,3907). **19**:NY(82). **21**:CA(82), FL(35).
Echidnodella angustiformis. Sooty mold, black scurf, black spot. **6**:MS(2720). **8**:Gulf states(1389), MS(82,2720). **21**:FL(35).
E. prinoides. Sooty mold, black scurf. **8**:MS(82,1389,2720).
Echidnodella sp. Black scurf. **8**:Gulf States(82), Gulf states(82). **14**:Gulf States(82), Gulf states(82).
Elsinoe ilicis. Spot anthracnose on leaves, scab. **9**:Gulf states(1389), FL(35), LA(1745,2826). **21**:FL(35).
Elsinoe sp. Leaf spot. **21**:NC(1303).
Endothia gyrosa. Orange hobnail canker. **18**:TX(108).
Eutryblidiella viburni. On twigs. **13**:Gulf states(1389), AL(82), LA(82). **16**:Gulf states(1389). **19**:Gulf states(1389), AL(82), LA(82). **21**:LA(2675).
Eutypella juglandicola. **18**:GA(1438).
Glomerella sp. Leaf spot. **18**:FL(35). **21**:Fl(35).
Godroniopsis nemopanthis. On twigs. **19**:MI(82,1389), NY(82,1389).
Guignardia sp. Leaf spot. **9**:FL(35). **21**:Fl(35).
Hypoxylon fuscum. On dead limbs. **18**:E & C states(1389).
H. punctulatum. **20**:TX(652).
H. rubiginosum. **18**:E & C states(1389). **20**:TX(652).
Lembosia brevis. Black mildew. **21**:FL(82).
Macroderma curtisii. Tar spot. **6**:FL(35). **21**:E states(82), FL(35), TX(82).
Metasphaeria ilicis. On leaves. **3**:CA(82).
Microsphaera penicillata. Powdery mildew. **13**:AL(82), IL(82). **14**:FL(35). **17**:AL(82), NC(82,1303), PA(82). **19**:WI(82). **21**:FL(35,82).
Microthyriella cuticulosa. Black speck. **18**:GA(82,1438), MS(82,2720).
Mycosphaerella ilicella. On leaves. **18**:NY(82).
M. ilicis. Leaf spot. **6**:FL(35). **14**:FL(82), MD(82), MS(82,2720), NJ(82), SC(82). **21**:FL(35).
Mycosphaerella sp. Leaf spot. **13**:GA(1438).
Myriangium superficialis. Parasitic on scale insects. **13**:GA(82,1438). **19**:GA(82).
Nectria aquifolii. **14**:NY(891).
N. coccinea. On limbs. **18**:CT(82), WV(82,3907).
N. galligena. **3**:OR(3202).

N. rubricarpa. **14:**NJ(82).
Nectria sp. Canker, dieback. **21:**E states(2761).
Phacidium elegantissimum. **18:**NC(3944).
P. sphaeroideum. On leaves. **14:**NJ(82,864).
Phacidium sp. Tar spot. **21:**E states(2761).
Phillipsiella ilicis. Black mildew. **14:**Gulf states(1389), NJ(82,1389). **18:**Gulf states(1389), GA(82,1438), MS(82,2720), NJ(1389).
Phyllactinia guttata. Powdery mildew. **18:**TN(82).
Physalospora sp. Twig blight. **11:**NC(1303).
Pseudonectria aquifoliae. On leaves. **18:**NJ(850,1389).
Rebentischia massalongii. On limbs. **19:**MA(186).
Rhytidhysteron hysterinum. On twigs. **6:**AL(82), LA(82), NC(82,1303). **20:**TX(652).
R. rufulum. On twigs. **6:**AL(82), FL(35,82). **18:**FL(35).
Rhytisma concavum. Tar spot. **19:**MA(1390), OH(82), WI(82).
R. ilicincola. Tar spot. **6:**FL(35), TX(82). **13:**NC(3944). **18:**FL(35). **20:**VA(82). **21:**AL(285), FL(35).
R. prini. Tar spot. **13:**GA(82,1438). **17:**NC(82,1303). **19:**E states(82), E states(82), IL(82), MS(82,2720), WI(82,1358), WV(3907).
R. velatum. Tar spot. **13:**MS(82,2720), NC(82,1303).
Rhytisma sp. Tar spot. **6:**GA(1438). **13:**OK(2872).
Schizoxylon insigne. **18:**GA(1438).
Trochila ilicina. On leaves. **3:**CA(81,82), WA(82,3202).
Ustulina deusta. On wood. **18:**MD(82,1389).
Valsa ceratosperma. On dead wood. **14:**GA(1438).
Vialaea insculpta. Canker, dieback. **3:**NW states(1389), OR(82,3202), WA(82,3202).
Winterina lobata. On leaves. **8:**FL(82), GA(82), MS(82). **21:**FL(35).
Xylaria filiformis. Leaf spot. **18:**GA(1438).

Basidiomycotina - Uredinales

Chrysomyxa ilicina. Heteroecious: telial host. **18:**NC(82,1307), TN(82), WV(82,3907).

Basidiomycotina - Other

Armillaria mellea. Blight, root rot. **9:**CA(2897). **11:**NC(1303). **4:**CA(2897).
A. tabescens. Root rot. **18:**FL(35,82). **21:**FL(35).
Bjerkandera adusta. White rot. **3:**OR(82). **18:**GA(1439), MD(82).
Corticium stevensii. Limb blight. **18:**FL(35).
Erythricium salmonicolor. Limb blight. **20:**FL(35).
Ganoderma australe. **18:**FL(35,82).
Hyphoderma rimosum. White rot. **20:**MS(444).
Meruliopsis corium. **5:**FL(128).
Pellicularia koleroga. Thread blight. **8:**LA(82). **14:**LA(82).
Phanerochaete crassa. On dead limbs. **18:**MD(82).
Phellinus inermis. On wood. **14:**LA(82). **17:**MD(82), MI(82), PA(82). **18:**MD(82), PA(82). **21:**LA(2675).
Pleurotus ostreatus. On wood. **18:**MD(82).
Rigidoporus crocatus. On wood. **18:**PA(82).
Schizophyllum commune. On wood. **18:**MD(82).
Schizopora paradoxa. On wood. **14:**AR(82), LA(82). **18:**MD(82), PA(82). **21:**AR(2675), LA(2675).
Scytinostroma galactinum. Root rot. **21:**MD(669,2090).
Septobasidium castaneum. Brown felt. **18:**FL(35,82), NC(82).
S. pseudopedicellatum. Brown felt. **18:**FL(82), NC(82,1303). **21:**FL(35), MS(2720).
Stereum hirsutum. On dead limbs. **18:**MD(82).
S. striatum. On dead limbs. **18:**MD(82,543).
Trametes hirsuta. On wood. **3:**OR(82,3202). **18:**MD(82).
T. versicolor. White rot. **18:**GA(1439).
Trichaptum biforme. On wood. **18:**AL(82).

Deuteromycotina - Hyphomycetes

Alternaria alternata. Leaf spot. **18:**MS(2720).
Alternaria sp. Fruit and leaf spot. **6:**FL(35). **9:**FL(35). **11:**FL(35). **20:**FL(35). **21:**Fl(35).
Botrytis sp. Inflorescence blight, gray mold. **18:**NJ(132). **21:**E states(2761).
Cercospora ilicis. Leaf spot. **6:**FL(35). **9:**FL(35). **14:**AL(82), FL(35,82), MS(82,2720), NJ(82). **18:**FL(35), TX(3597). **21:**FL(35).
C. ilicis-opacae. Leaf spot. **18:**FL(35), GA(581), SC(581).
C. pulvinula. Leaf spot. **6:**FL(35,82). **18:**MD(82), MS(2720), NJ(82), TX(82).
Cercospora sp. Edema, leaf spot. **11:**NC(1303). **21:**E states(2761), FL(35), NC(1303).
Chalara thielavioides. Black mold. **18:**NJ(1545).
Cladosporium sp. **3:**OR(3202).
Cylindrocladium avesiculatum. Defoliation, twig dieback, leaf spot. **9:**Unk.(1213), GA(1216). **11:**Unk.(1213), FL(35), GA(1216). **18:**Unk.(1213), GA(1216). **20:**FL(35,1213,1216), GA(1213,1216). **21:**FL(1694), Fl(35).
C. scoparium. Fruit and leaf spot. **9:**FL(35). **20:**NC(1303).
Cylindrocladium sp. Cutting rot. **21:**Fl(35).
Fumago vagans. Sooty mold. **3:**OR(82,3202), WA(82,3202).
Fumago sp. Sooty mold. **21:**E states(2761).
Fusarium solani. **18:**NJ(82).
Fusarium sp. Leaf and stem necrosis, root rot, stem rot. **6:**FL(35). **9:**FL(35). **11:**FL(35). **14:**FL(35). **18:**FL(35). **20:**FL(35). **21:**FL(35), NC(1303).
Mycoleptodiscus indicus. Leaf spot. **20:**FL(35).
M. minimus. **18:**AL(82).
Phymatotrichopsis omnivora. Root rot. **6:**Unk.(82). **13:**TX(82). **20:**Unk.(82).
Ramularia prini. On leaves. **19:**MI(82), NY(82), WI(82).
Thielaviopsis basicola. Black root rot. **9:**NC(1303). **11:**AL(2016), FL(35,2016), NC(1303,2016), VA(2015,2016). **15:**KY(1777). **20:**NC(1303).
Trichothecium roseum. Dieback, black root rot. **19:**WV(3907).
Triscelophorus monosporus. Leaf litter. **2:**HI(2903).
Verticillium sp. Wilt. **21:**FL(35).

Deuteromycotina - Coelomycetes

Amerosporium ilicinum. On leaves. **13:**LA(82).
Colletotrichum sp. Anthracnose, fruit rot. **6:**FL(35). **9:**NC(1303). **20:**FL(35). **21:**Fl(35).
Coniothyrium ilicinum. On leaves. **18:**MD(82), NY(82).
Coniothyrium sp. Canker. **18:**FL(35).
Cytosporella sp. Twig blight. **18:**FL(35).
Dendrophoma nigrescens. On twigs. **19:**NY(82).
Diplodia sp. Dieback, leaf spot. **3:**MD(82). **9:**FL(35). **18:**FL(35,82), NJ(82), NY(82), WV(82). **20:**FL(35). **21:**Fl(35), NC(1303).
Discosia artocreas. Secondary leaf spot. **18:**AL(82), MD(82).
D. deflectens. **18:**SC(82).
Discosia sp. **18:**NC(1307).
Dothiorella sp. Leaf spot. **21:**Fl(35).
Entomosporium sp. Leaf spot. **6:**FL(35). **18:**FL(35).
Fusicoccum ilicinum. Leaf spot. **6:**FL(35). **18:**WV(82,3907).
Fusicoccum sp. Leaf spot. **9:**FL(35).
Gloeosporium aquifolii. Leaf spot. **3:**NJ(82), TX(82), WA(82,3202).
G. ilicis. Leaf spot. **6:**FL(35). **9:**FL(35). **18:**FL(35), TN(82). **21:**Fl(35).
G. niveum. Leaf spot. **19:**WI(82).
Gloeosporium sp. Canker. **9:**AL(3530).
Leptothyrium foraminulatum. Leaf spot. **18:**FL(35,82), MD(82), NC(82), TN(82), WV(3907), WY(82). **21:**Fl(35).
Macrophoma ilicella. Dieback, leaf spot. **9:**FL(35). **20:**FL(35).
Macrophoma sp. Dieback, leaf spot. **9:**FL(35). **11:**FL(35). **18:**FL(35). **21:**E states(2761), NC(1303).
Melanconium sp. Dieback. **21:**Fl(35).
Microdiplodia ilicigena. On twigs. **21:**NY(82).
Pestalotia stellata. Secondary leaf spot. **18:**MD(82), MS(82,2720), NJ(82), SC(82), TX(82).
Pestalotia sp. Secondary leaf spot. **11:**NC(1303).
Pestalotiopsis annulata. Leaf spot. **6:**FL(35,82). **18:**AL(82), MS(82,2720), NJ(82).
Phoma ilicina. On leaves and twigs. **3:**CA(82), ID(3202), MT(3202), OR(3202), WA(3202). **18:**MD(82).
P. phacidiella. Leaf spot. **3:**WA(82,3202). **18:**NJ(82), TN(82).
Phoma sp. Dieback. **20:**FL(35). **21:**FL(35).
Phomopsis ilicis. On leaves. **18:**NJ(82).
P. oxyspora. **19:**MD(2783).
Phomopsis sp. Canker, dieback. **6:**FL(35). **9:**FL(35). **11:**NC(1303). **18:**FL(35). **21:**E states(2761), FL(35), NC(1303), NJ(82).
Phyllosticta concentrica. Leaf spot. **18:**OK(2872). **20:**FL(35).
P. concomitans. Leaf spot. **21:**LA(82).
P. haynaldii. Leaf spot. **10:**GA(898). **19:**WV(82,3907).
P. opaca. Leaf spot. **18:**MS(2720), OK(2872), SC(82), TX(82), WV(82,3907). **21:**CA(82).
P. terminalis. Leaf spot. **6:**FL(35,82). **9:**FL(35). **18:**FL(35), MS(82,2720). **21:**FL(35).
Phyllosticta sp. Scald of cuttings, leaf spot. **3:**CA(82), NJ(82), WA(82,3202). **11:**IL(312), NC(1303). **14:**FL(35). **19:**WI(1349).

21:E states(2761), IL(1052), NC(1303).
Rhabdospora ilicigena. On twigs. **21**:NY(82).
Sclerophoma sp. Scab. **3**:NW states(410), OR(410,3202).
Septoria ilicifolia. Leaf spot. **9**:FL(35). **21**:FL(35).
Sphaceloma sp. Spot anthracnose, leaf spot. **9**:LA(82,2827).
Sphaeropsis tumefaciens. Stem gall. **6**:FL(35). **9**:FL(35). **11**:FL(35). **18**:FL(35). **20**:FL(35). **21**:FL(35).
Sporonema ilicis. On leaves. **18**:AL(82).
[*Titaeospora equiseti*]. Leaf tip necrosis. **9**:FL(35).
Deuteromycotina - Other
Rhizoctonia ramicola. Thread blight. **9**:FL(35). **11**:FL(35). **20**:FL(35).
R. solani. Web blight, leaf rot of cuttings, root rot. **3**:MD(82). **6**:FL(35). **9**:FL(35). **11**:FL(35), NC(1303). **14**:FL(35). **18**:FL(35), MD(82). **20**:FL(35). **21**:FL(35).
Rhizoctonia sp. Damping-off, root rot. **11**:NC(1303). **21**:E states(2761).

Nemopanthus Raf.
1. *N. mucronatus* (L.) Trel. [catberry]
2. *Nemopanthus* sp.

Ascomycotina
Curreya peckiana. On dead twigs. **1**:NY(3081).
Dermea peckiana. On twigs. **1**:MI(82), NY(82).
Durandiella nemopanthis. On dead limbs. **1**:NH(128), NY(82,128,1381), PA(128), VT(82,128,1381).
Godroniopsis nemopanthis. On twigs. **1**:NY(82).
Microsphaera penicillata. Powdery mildew. **1**:NY(82), OH(948,950), WI(82).
Protoventuria curviseta. On leaves. **1**:NY(82).
Rhytisma prini. Tar spot. **1**:NC states(82), NE states(82), MI(82,2862), WI(1330), WV(3907).
Basidiomycotina - Other
Phellinus inermis. On wood. **1**:PA(82).
Deuteromycotina - Hyphomycetes
Conoplea olivacea. **2**:NY(1672).
Ramularia nemopanthis. Leaf spot. **1**:NY(82).

Araceae

Acorus L.
1. *A. calamus* L. [calamus, flagroot, sweet-flag]
2. *Acorus* sp.

Oomycetes
Pythium sp. Root rot. **2**:FL(35).
Ascomycotina
Phaeosphaeria acori. On living leaves. **1**:WI(1341).
Sphaerulina acori. On leaves. **1**:NY(82).
Basidiomycotina - Uredinales
[*Uromyces sparganii*]. **1**:E states(82), IA(2712), IL(82), IN(2712), MA(82), MI(2712), MN(2874), MS(82,2720), NY(2712), WI(1350,2712).
Deuteromycotina - Hyphomycetes
Septocylindrium aromaticum. Leaf spot. **1**:NE states(82), CT(82), IN(82), KS(82), MN(2874), WI(82).
Septocylindrium sp. Leaf spot. **1**:NY(82).
Deuteromycotina - Coelomycetes
Colletotrichum sp. Anthracnose. **1**:FL(35). **2**:FL(35).
Deuteromycotina - Other
Rhizoctonia solani. Root rot. **1**:FL(35). **2**:FL(35).

Aglaonema Schott
1. *A. commutatum* Schott
2. *A. modestum* Schott ex Engl. [Chinese evergreen]
3. *A. pictum* (Roxb.) Kunth
4. *A. simplex* Blume [Chinese evergreen]
5. *A. treubii* Engl.
6. *Aglaonema* sp.

Oomycetes
Phytophthora sp. Root and stem rot. **1**:FL(35). **2**:FL(35). **6**:FL(35).
Pythium splendens. Root rot. **2**:FL(35). **4**:FL(35,82). **5**:HI(2903).
P. ultimum. Root rot. **6**:FL(35).
Pythium sp. Root rot. **6**:FL(35), HI(2903).
Ascomycotina
Glomerella cingulata. Anthracnose. **1**:WA(3202). **2**:FL(35).
Glomerella sp. Leaf spot. **1**:FL(35). **4**:FL(35). **6**:FL(35).
Basidiomycotina - Other
Sphaerobolus stellatus. Glebal deposits on leaves. **4**:FL(276).
Deuteromycotina - Hyphomycetes
Cephalosporium sp. Leaf spot. **1**:FL(35). **4**:FL(35).
Cercospora sp. Leaf spot. **6**:FL(35).
Cladosporium sp. Leaf spot. **6**:FL(35).
Corynespora sp. Leaf spot. **6**:FL(35).
Cylindrocladiella peruviana. Basal stem rot, root rot. **1**:FL(35). **4**:FL(35). **6**:FL(35).
Cylindrocladium sp. Stem rot. **2**:FL(35).
Fusarium oxysporum. Root and stem rot. **2**:FL(35).
Fusarium sp. Root and stem rot. **4**:FL(35).
Myrothecium roridum. 'Myrothecium leaf spot.' **1**:FL(35). **2**:FL(35). **6**:FL(35).
Myrothecium sp. Leaf spot. **4**:FL(35).
Deuteromycotina - Coelomycetes
Colletotrichum sp. Anthracnose, leaf spot. **1**:Unk.(1300). **2**:FL(35). **3**:FL(35). **4**:CO(49). **6**:FL(35).
Deuteromycotina - Other
Rhizoctonia solani. Root rot. 'Rhizoctonia aerial blight.' **1**:FL(35). **2**:FL(35). **4**:FL(35). **6**:FL(35).
Rhizoctonia sp. Root rot. **1**:HI(2903). **6**:HI(2903).
Sclerotium rolfsii. 'Southern blight.' **1**:FL(35). **2**:FL(35). **6**:FL(35).

Alocasia (Schott) G. Don
1. *A. cucullata* (Lour.) G. Don [Chinese taro]
2. *Alocasia* sp.

Oomycetes
Phytophthora colocasiae. **2**:HI(2903).
Pythium sp. Root rot. **2**:FL(35).
Deuteromycotina - Hyphomycetes
Alternaria sp. Leaf spot. **2**:FL(35).
Cercospora sp. Leaf spot. **1**:HI(2899,3074). **2**:HI(2903).
Deuteromycotina - Coelomycetes
Colletotrichum sp. Anthracnose. **2**:FL(35).
Deuteromycotina - Other
Rhizoctonia solani. Root rot. **2**:FL(35).

Amorphophallus Blume ex Decne.
1. *Amorphophallus* sp.

Deuteromycotina - Hyphomycetes
Botrytis cinerea. **1**:AK(526).
Fusarium lateritium. Stem canker. **1**:FL(35).
Thielaviopsis sp. Leaf rot. **1**:FL(35).
Deuteromycotina - Coelomycetes
Colletotrichum gloeosporioides. Leaf spot. **1**:FL(35).

Anthurium Schott
1. *A. andraeanum* Linden [flamingo-lily, oilcloth-flower]
2. *A.* × *cultorum* Birdsey (=*A. andraeanum* Linden var. *giganteum* Hort.). A name used for complex hybrids involving *A. andraeanum*, *A. linderianum*, *A. hymphaeifolium* and *A. ornatum*.
3. *Anthurium* sp.

Oomycetes
Phytophthora sp. Root rot. **3**:FL(35).
Pythium splendens. Root rot. **1**:HI(2903). **3**:FL(35).
Ascomycotina
Botryosphaeria sp. Leaf spot. **1**:FL(35).
Calonectria sp. Root rot. **3**:HI(2903).
Deuteromycotina - Hyphomycetes
Alternaria sp. Leaf spot. **3**:FL(35).
Botrytis cinerea. Leaf rot. **3**:FL(35).
Cercospora richardiaecola. Leaf spot. **3**:FL(35).
Fusarium sp. Root rot. **3**:FL(35).
Deuteromycotina - Coelomycetes
Ascochyta minima. Anthracnose. **1**:OK(2872). **2**:OK(2872).
Colletotrichum gloeosporioides. Anthracnose. 'Colletotrichum spadix blight.' **1**:HI(91). **3**:FL(35), HI(2903).
Diplodia sp. Cane rot, leaf spot. **1**:FL(35).
Phomopsis sp. Leaf spot. **1**:FL(35).
Phyllosticta sp. Leaf spot. **1**:FL(35).
Septoria sp. Leaf spot. **3**:FL(35).

Deuteromycotina - Other
Rhizoctonia solani. Root and stem rot. 'Rhizoctonia aerial blight.' **3:**FL(35).
Rhizoctonia sp. Root rot. **1:**HI(2903).

Arisaema Mart.

1. *A. dracontium* (L.) Schott [dragon arum, dragonroot, green-dragon]
2. *A. triphyllum* (L.) Torr. (=*A. atrorubens* (Aiton) Blume) [dragonroot, Indian turnip, jack-in-the-pulpit]
3. *Arisaema* sp.

Ascomycotina
Streptotinia arisaemae. On leaves and stalks. **2:**NC(1307), NY(3865).
Basidiomycotina - Uredinales
Uromyces ari-triphylli. Autoecious. **1:**E states(82), FL(35), MS(2720), NC(1303), NE(82), OK(2872), TX(82), WV(3907). **3:**FL(35).
Uromyces sp. **2:**CO(49).
Deuteromycotina - Hyphomycetes
Botrytis sp. On leaflets. **2:**WI(1349).
Cladosporium sp. On leaves. **2:**VA(82).
Ramularia arisaematis. **2:**WI(1338).
Streptobotrys streptothrix. Leaf blight. **2:**DE(3543). **3:**IA(82), IL(82), MD(82), NY(82), PA(82), WI(82).
Volutella sp. Leaf spot. **2:**IN(82).
Deuteromycotina - Coelomycetes
Colletotrichum arisaematis. On living leaves. **2:**WI(1349).
Phyllosticta sp. On leaves. **2:**WI(1353).

Caladium Venten.

1. *C. bicolor* (Aiton) Venten. [elephant's-ear, heart-of-Jesus]
2. *C.* × *hortulanum* Hort. Of mixed hybrid origin. [fancy-leaved caladium]
3. *Caladium* sp.

Oomycetes
Phytophthora sp. Root rot. **3:**FL(35).
Pythium irregulare. Root rot. **2:**FL(35). **3:**FL(35).
P. myriotylum. Root rot. **2:**FL(35). **3:**FL(35).
P. splendens. Root rot. **2:**FL(35). **3:**FL(35).
Zygomycotina
Rhizopus stolonifer. Corm rot. **2:**FL(35). **3:**FL(35).
Ascomycotina
Botryotinia ricini. Rot. **1:**FL(1451).
Deuteromycotina - Hyphomycetes
Botrytis sp. Leaf blotch and petiole rot. **1:**NC(1303).
Cercospora sp. Leaf spot. **3:**FL(35).
Fusarium oxysporum. Corm rot. **3:**FL(35).
F. solani. Corm rot. **2:**FL(35,1932). **3:**FL(35).
Helminthosporium sp. Leaf spot. **3:**FL(35).
Penicillium sp. Corm rot. **3:**FL(35).
Deuteromycotina - Coelomycetes
Colletotrichum sp. Anthracnose, leaf spot. **2:**FL(35). **3:**FL(35).
Phyllosticta sp. Leaf spot. **1:**NC(1303).
Deuteromycotina - Other
Rhizoctonia solani. Petiole rot, tuber rot. **2:**FL(35). **3:**FL(35).
Sclerotium rolfsii. Southern blight. **1:**FL(82). **2:**FL(35). **3:**FL(35,82).

Calla L.

1. *C. palustris* L. [water arum, water-dragon, wild calla]

Deuteromycotina - Hyphomycetes
Cercospora callae. Leaf spot. **1:**MN(2874), MS(82), NY(82), WI(82).
Deuteromycotina - Coelomycetes
Ascochyta pellucida. **1:**WI(1359).
Marssonia callae. Leaf spot. **1:**NY(82).

Colocasia Schott in Schott & Endl.

1. *C. esculenta* (L.) Schott (=*C. antiquorum* Schott, *Arum esculentum* L.) [coco-yam, dasheen, eddo, elephant's-ear-plant, kalo, taro]
2. *C. esculenta* (L.) Schott var. *globulifera* (Engl. & K. Krause) R. A. Young
3. *Colocasia* sp.

Chytridiomycetes
Allomyces arbuscula. Southern blight, corm and stem rot. **1:**HI(2903).
Oomycetes
Phytophthora colocasiae. Leaf blight. **1:**HI(82,2903).
Pythium aphanidermatum. Crown and root rot. **1:**HI(2903).
P. carolinianum. Corm and root rot. **1:**HI(2903).
P. debaryanum. Corm and root rot. **1:**HI(2903).
P. graminicola. Corm and root rot. **1:**HI(2903).
Zygomycotina
Rhizopus stolonifer. Corm rot in storage. **1:**HI(2903).
Deuteromycotina - Hyphomycetes
Cercospora sp. Leaf spot. **1:**HI(82,2903). **3:**FL(35).
Cladosporium sp. On leaves. **1:**HI(82,2903).
Fusarium roseum. Corm rot. **1:**HI(2903).
F. solani. Corm rot, powdery gray rot. **1:**FL(35,82), HI(82,2903), TX(82). **3:**FL(35).
Fusarium sp. Root rot. **1:**NC(1303). **2:**HI(2903).
Helminthosporium sp. Leaf spot. **3:**FL(35).
Macrosporium sp. **1:**HI(2903).
Deuteromycotina - Coelomycetes
Diplodia sp. Black rot of tubers. **1:**FL(35,82), SC(82), TX(82).
Phyllosticta colocasiophila. Leaf spot. **1:**HI(82,2903).
Deuteromycotina - Other
Rhizoctonia sp. Pocket rot. **2:**HI(2903).
Sclerotium rolfsii. Southern blight, corm and stem rot. **1:**FL(35,82), HI(2903), NY(82). **3:**FL(35).

Cryptocoryne Fisch. ex Wydl.

1. *Cryptocoryne* sp.

Oomycetes
Pythium sp. Leaf spot. **1:**FL(35).

Dieffenbachia Schott

1. *D. amoena* W. Bull
2. *D. maculata* (Lodd.) G. Don (=*D. picta* (Lodd.) Schott, *D. picta* (Lodd.) Schott cv. 'rudolph') [spotted dumbcane]
3. *D. seguine* (Jacq.) Schott [dumbcane, mother-in-law plant]
4. *Dieffenbachia* sp.

Oomycetes
Phytophthora palmivora. Root and basal stem rot. **1:**FL(35). **2:**CA(82), FL(35). **4:**FL(35).
Pythium splendens. Root and stem rot. **1:**FL(35). **2:**FL(35,82). **4:**FL(35), HI(2903).
Ascomycotina
Glomerella cingulata. Anthracnose, leaf spot. **1:**FL(35), WA(3202). **2:**FL(35), NJ(82), WA(3202). **3:**NJ(82). **4:**FL(35).
Glomerella sp. Anthracnose. **1:**WA(1300). **4:**WA(1300).
Mycosphaerella sp. Leaf spot. **4:**FL(35).
Nectriella pironii. Stem gall. **4:**FL(35).
Phaeosphaeria eustoma. Brown leaf spot. **1:**FL(35). **4:**FL(35).
Deuteromycotina - Hyphomycetes
Alternaria sp. Leaf spot. **2:**FL(35).
Cephalosporium dieffenbachiae. Leaf spot. **2:**NY(82).
Cephalosporium sp. Leaf spot. **1:**FL(35). **2:**FL(35). **4:**FL(35).
Cercospora sp. Leaf spot. **1:**FL(35). **4:**FL(35).
Fusarium moniliforme. Leaf margin blight. **4:**FL(35).
F. oxysporum. Root and stem rot. **1:**FL(35). **4:**FL(35).
F. solani. Cutting rot, leaf spot. 'Fusarium stem rot.' **2:**FL(35,562). **4:**FL(35).
Myrothecium roridum. 'Myrothecium leaf spot.' **2:**NC(1303). **4:**FL(35), NC(1303).
Volutella sp. Leaf spot. **4:**FL(35).
Deuteromycotina - Coelomycetes
Ascochyta minima. Anthracnose. **2:**OK(2872).
Colletotrichum gloeosporioides. Anthracnose. **1:**FL(35). **2:**FL(35). **4:**FL(35), HI(2903).
Colletotrichum sp. Anthracnose. **1:**WA(1300). **2:**Unk.(82), WA(1300). **4:**NC(1303).
Gloeosporium sp. **1:**WA(1300). **2:**WA(1300).
Phyllosticta sp. Leaf spot. **1:**FL(35). **2:**FL(35). **4:**FL(35).
Deuteromycotina - Other
Rhizoctonia solani. Root and basal stem rot. 'Rhizoctonia aerial blight.' **1:**FL(35). **2:**FL(35). **4:**FL(35).
Sclerotium rolfsii. 'Southern blight.' **1:**FL(35). **4:**FL(35).

Epipremnum Schott

1. *E. aureum* (Linden ex André) Bunting (=*Pothos aureus* Linden ex André, *Scindapsus aureus* (Linden ex André) Engl.) [devil's ivy, golden Ceylon creeper, hunter's-robe, ivy arum, pothos, Solomon Island ivy, taro-vine, variegated philodendron]

Oomycetes
Phytophthora sp. Root and stem rot. **1**:FL(35).
Pythium splendens. Root rot. **1**:FL(82,1930).
Pythium sp. Root rot. **1**:FL(35), HI(2903).
Ascomycotina
Glomerella cingulata. Anthracnose. **1**:FL(35).
Sclerotinia sclerotiorum. Cottony blight. **1**:FL(35).
Basidiomycotina - Other
Crepidotus mollis. **1**:HI(3706).
Melanotus flavolivens. **1**:HI(2903).
Deuteromycotina - Hyphomycetes
Alternaria sp. Leaf spot. **1**:FL(35).
Botrytis cinerea. Gray mold. **1**:FL(35).
Cercospora richardiaecola. Leaf spot. **1**:FL(35).
Corynespora sp. Leaf spot. **1**:FL(35).
Cylindrocladium sp. Root rot. **1**:FL(35).
Fusarium sp. Root and stem rot. **1**:FL(35).
Myrothecium sp. Leaf spot. **1**:FL(35).
Thielaviopsis basicola. Root rot. **1**:MD(1867).
Deuteromycotina - Coelomycetes
Colletotrichum sp. Anthracnose. **1**:FL(35).
Phyllosticta aricola. Leaf spot. **1**:HI(82).
P. pothicola. Leaf spot. **1**:HI(82).
Phyllosticta sp. Leaf spot. **1**:FL(35).
Deuteromycotina - Other
Rhizoctonia solani. Leaf and stem rot. 'Rhizoctonia aerial blight.' **1**:FL(35).
Rhizoctonia sp. Crown, foot and root rot. **1**:HI(2903), MD(1867), MO(2438).
Sclerotium rolfsii. Root and stem rot. 'Southern blight.' **1**:FL(35).

Homalomena Schott ex Schott & Endl.
1. *H. wallisii* Regel
Ascomycotina
Glomerella cingulata. Leaf spot. **1**:NJ(82).

Lysichiton Schott
1. *L. americanum* Hultén & St. John [western skunk-cabbage, yellow skunk-cabbage]
Ascomycotina
Ombrophila lysichitonis. **1**:WA(3202).
Deuteromycotina - Hyphomycetes
Stemphylium sp. **1**:WA(3202).
Deuteromycotina - Coelomycetes
Septoria sp. **1**:OR(3202).

Monstera Adans.
1. *M. deliciosa* Liebm. (=*Philodendron pertusum* Kunth & Bouché) [breadfruit-vine, ceriman, cutleaf philodendron, fruit-salad-plant, hurricane-plant, monstera, Swiss-cheese-plant]
2. *M. punctulata* Schott ex Engl.
Oomycetes
Phytophthora sp. Leaf spot, root rot. **1**:FL(35).
Pythium sp. Root rot. **1**:FL(35).
Ascomycotina
Botryosphaeria sp. Marginal leaf necrosis. **1**:FL(35).
Glomerella cingulata. Anthracnose. **1**:FL(35), WA(3202).
Leptosphaeria sp. Leaf spot. **1**:FL(35).
Basidiomycotina - Other
Armillaria tabescens. Root rot. **1**:FL(35).
Deuteromycotina - Hyphomycetes
Alternaria sp. Leaf spot. **1**:FL(35).
Cercospora sp. Leaf spot. **1**:FL(35).
Corynespora sp. Leaf spot. **1**:FL(35).
Fusarium sp. Leaf necrosis, tip rot. **1**:FL(35).
Deuteromycotina - Coelomycetes
Colletotrichum sp. **1**:Unk.(1300), WA(1300). **2**:WA(1300).
Diplodia sp. Leaf necrosis. **1**:FL(35).
Gloeosporium sp. **1**:Unk.(1300).
Macrophoma philodendri. Leaf spot. **1**:FL(35,82), MI(82).
Phomopsis sp. Leaf necrosis. **1**:FL(35).
Phyllosticta sp. Leaf spot. **1**:FL(35).
Deuteromycotina - Other
Rhizoctonia solani. Root and stem rot. 'Rhizoctonia aerial blight.' **1**:FL(35).

Nephthytis Schott
1. *N. afzelii* Schott
Oomycetes
Pythium splendens. Root rot. **1**:FL(82).
Deuteromycotina - Hyphomycetes
Cephalosporium sp. Leaf spot. **1**:MD(82).

Orontium L.
1. *O. aquaticum* L. [goldenclub]
2. *Orontium* sp.
Ascomycotina
Mycosphaerella sp. Leaf spot. **1**:NJ(82).
Physalospora orontii. On leaves. **1**:NJ(82).
Deuteromycotina - Hyphomycetes
Ramularia orontii. Leaf spot. **1**:DE(3543), NJ(82).
Stilbella aciculosa. **2**:NJ(3178).
Streptobotrys streptothrix. Leaf blight. **1**:DE(3543), NJ(82).
Volutella diaphana. Leaf spot. **1**:NJ(82).
Deuteromycotina - Coelomycetes
Phyllosticta orontii. Leaf spot. **1**:NJ(82), TX(82).

Peltandra Raf.
1. *P. glauca* (Elliott) A. Wood
2. *P. sagittifolia* (Michx.) Morong [spoonflower]
3. *P. virginica* (L.) Schott & Endl. [arrow arum, tuckahoe, Virginian wake-robin]
4. *Peltandra* sp.
Basidiomycotina - Uredinales
Uromyces ari-triphylli. Autoecious. **1**:GA(82), NC(82). **2**:NC(1303). **3**:E states(82), FL(35), GA(1439), IA(82), IL(82), IN(82), NC(1303), TN(1539).
Deuteromycotina - Hyphomycetes
Cercospora callae. Leaf spot. **1**:AL(82). **3**:DE(82,3543), FL(35,82), NY(82).
Ramularia sp. Leaf spot. **3**:MI(82).
Deuteromycotina - Coelomycetes
Colletotrichum paludosum. Leaf spot. **3**:DE(82), FL(35), IN(82), MA(82), NY(82).
Colletotrichum sp. Leaf spot. **1**:AL(82).
Pestalotiopsis aquatica. Secondary leaf spot. **3**:NJ(82,966), NY(82).
Deuteromycotina - Other
Sclerotium caladii. On leaves. **4**:NC(1303).

Philodendron Schott
1. *P. cordatum* (Vell.) Kunth [heart-leaf philodendron]
2. *P. hastatum* K. Koch & H. Sello
3. *P. panduriforme* (H. B. K.) Kunth
4. *P. scandens* K. Koch & H. Sello [heart-leaf philodendron]
5. *P. scandens* K. Koch & H. Sello f. *micans* (Klotzsch ex K. Koch) Bunting (=*P. micans* Klotzsch ex K. Koch) [velvet-leaf philodendron]
6. *P. selloum* K. Koch
7. *Philodendron* sp.
Oomycetes
Phytophthora nicotianae. Brown leaf spot. **2**:FL(35). **4**:FL(35).
P. nicotianae var. *parasitica*. Brown leaf spot. **3**:FL(35). **6**:FL(35). **7**:FL(35).
Pythium splendens. Root rot. **3**:FL(1930). **6**:FL(35).
Pythium sp. Root rot. **2**:FL(35). **3**:FL(35,1930). **4**:FL(35). **5**:HI(2903). **7**:FL(35), HI(2903).
Ascomycotina
Glomerella cingulata. **1**:WA(3202). **7**:WA(3202).
Nectriella pironii. Stem gall. **4**:FL(35).
Sclerotinia sclerotiorum. Cottony blight. **4**:FL(35). **6**:FL(35).
Deuteromycotina - Hyphomycetes
Acremonium crotocinigenum. Leaf spot. **4**:FL(35). **7**:FL(35).
Alternaria sp. Leaf spot. **6**:FL(35). **7**:FL(35).
Botryosporium sp. **7**:WA(3202).
Botrytis cinerea. 'Gray mold.' **4**:FL(35). **7**:FL(35).
Cercospora sp. Leaf spot. **6**:FL(35).
Curvularia sp. Leaf spot. **4**:FL(35).
Cylindrocarpon sp. Root rot. **7**:FL(35).
Cylindrocladium sp. Root rot. **4**:FL(35).
Dactylaria humicola. 'Dactylaria leaf spot.' **2**:FL(35). **4**:FL(35). **6**:FL(35). **7**:FL(35).

Fusarium moniliforme. Leaf spot. **7**:FL(35).
Fusarium sp. Cutting and root rot, leaf spot. **2**:FL(35). **4**:FL(35). **7**:FL(35).
Deuteromycotina - Coelomycetes
Colletotrichum gloeosporioides. Anthracnose, leaf spot. **2**:FL(35). **3**:FL(35). **6**:FL(35). **7**:FL(35), HI(2903), NJ(82).
Colletotrichum sp. Anthracnose. **7**:HI(2903).
Diplodia sp. Leaf necrosis. **6**:FL(35).
Gloeosporium sp. **1**:Unk.(1300). **2**:Unk.(1300). **3**:Unk.(1300).
Phyllosticta sp. Leaf spot. **2**:FL(35). **4**:FL(35). **6**:FL(35). **7**:FL(35).
Deuteromycotina - Other
Rhizoctonia solani. Basal stem rot, root rot. 'Rhizoctonia aerial blight.' **2**:FL(35). **4**:FL(35). **6**:FL(35). **7**:FL(35).
Rhizoctonia sp. Root rot. **2**:HI(2903).
Sclerotium rolfsii. 'Southern blight.' **1**:CA(1111). **4**:FL(35). **7**:FL(35).

Pothos L.
1. *Pothos* sp.
Deuteromycotina - Coelomycetes
Leptothyrium pothi. **1**:HI(2903).
Phyllosticta aricola. **1**:HI(2903).
P. pothicola. **1**:HI(2903).

Spathiphyllum Schott
1. *Spathiphyllum* sp.
Oomycetes
Phytophthora nicotianae var. *parasitica*. Leaf blight. **1**:FL(35).
Pythium splendens. Root rot. **1**:FL(35).
Pythium sp. **1**:HI(2903).
Ascomycotina
Botryosphaeria sp. Leaf spot. **1**:FL(35).
Deuteromycotina - Hyphomycetes
Alternaria sp. Leaf spot. **1**:FL(35).
Cercospora sp. Leaf spot. **1**:FL(35).
Cylindrocladium spathiphylli. Petiole and root rot, leaf spot. **1**:FL(35,565,3150).
Fusarium sp. Root and spathe rot. **1**:FL(35).
Myrothecium sp. Leaf spot. **1**:FL(35).
Deuteromycotina - Coelomycetes
Ascochyta sp. Leaf spot. **1**:FL(35).
Colletotrichum gloeosporioides. Anthracnose. **1**:FL(35).
Colletotrichum sp. Anthracnose. **1**:HI(2903).
Phyllosticta sp. Leaf spot. **1**:FL(35).
Deuteromycotina - Other
Rhizoctonia solani. Root rot. **1**:FL(35).
Rhizoctonia sp. Crown and root rot. **1**:HI(2903).
Sclerotium rolfsii. Southern blight. **1**:FL(35), HI(2903).

Symplocarpus Salisb. ex Nutt.
1. *S. foetidus* (L.) Nutt. [polecat-weed, skunk-cabbage]
Deuteromycotina - Hyphomycetes
Botrytis sp. Leaf blight. **1**:CT(82), IL(82), NJ(82), NY(82).
Cercospora symplocarpi. Leaf spot. **1**:E states(82), DE(3543), IN(82), MN(2874), WI(82).
Streptobotrys streptothrix. **1**:DE(3543).
Deuteromycotina - Coelomycetes
Ascochyta sp. Leaf spot. **1**:WI(1361).
Septoria spiculosa. Leaf spot. **1**:MD(82), MN(2874), NY(82), PA(82), WI(82).

Syngonium Schott
1. *S. podophyllum* Schott [African evergreen, arrowhead-vine, nephthytis]
2. *Syngonium* sp.
Oomycetes
Phytophthora sp. Root rot. **2**:FL(35).
Pythium splendens. Root rot. **1**:FL(35). **2**:FL(35).
Ascomycotina
Botryosphaeria sp. Leaf spot. **2**:FL(35).
Ceratocystis fimbriata. Blight, chlorosis, leaf spot. 'Black cane rot.' **1**:HI(2903,3701).
Glomerella cingulata. Blight, chlorosis, leaf spot. **1**:WA(3202).
Basidiomycotina - Uredinales
Uromyces sp. **1**:FL(35).
Deuteromycotina - Hyphomycetes
Acremonium crotocinigenum. Leaf blight, leaf spot. 'Cephalosporium leaf spot.' **1**:FL(35), HI(3702). **2**:FL(35).
Alternaria sp. Leaf spot. **1**:FL(35). **2**:FL(35).
Cephalosporium cinnamomeum. Leaf spot. **1**:FL(41), NY(82).
Cercospora sp. Leaf spot. **1**:FL(35).
Cercosporidium sp. Leaf spot. **1**:FL(35).
Fusarium oxysporum. Cutting rot. **2**:FL(35).
F. solani. Cutting rot. **1**:FL(35).
Myrothecium sp. Leaf spot. **1**:FL(35).
Thielaviopsis sp. Black root. **1**:HI(2903).
Deuteromycotina - Coelomycetes
Colletotrichum falcatum. Anthracnose. **2**:HI(2903).
C. gloeosporioides. Anthracnose. **1**:FL(35). **2**:FL(35).
Colletotrichum sp. **1**:Unk.(1300).
Diplodia sp. Leaf necrosis. **1**:FL(35).
Gloeosporium sp. **1**:Unk.(1300).
Phyllosticta sp. Leaf spot. **1**:FL(35).
Deuteromycotina - Other
Rhizoctonia solani. Leaf blight, root and stem rot. 'Rhizoctonia aerial blight.' **1**:FL(35). **2**:FL(35).
Sclerotium rolfsii. Stem rot of cuttings. 'Southern blight.' **1**:FL(35). **2**:FL(35).

Thaumatophyllum Schott
This genus is considered a synonym of *Philodendron*, but the appropriate combination could not be located.
1. *T. spruceanum* Schott
Deuteromycotina - Coelomycetes
Diplodia sp. Leaf spot. **1**:FL(35).

Xanthosoma Schott in Schott & Endl.
1. *X. brasiliense* Engl.
2. *X. sagittifolium* (L.) Schott
Oomycetes
Pythium sp. Root rot. **1**:HI(2903).
Basidiomycotina - Other
Armillaria mellea. Root rot. **2**:CA(82).
Deuteromycotina - Hyphomycetes
Fusarium solani. Powdery gray rot. **2**:FL(35,82).
Stemphylium sp. Leaf spot. **2**:FL(35).
Deuteromycotina - Coelomycetes
Colletotrichum sp. Anthracnose. **2**:FL(35).
Deuteromycotina - Other
Sclerotium rolfsii. Southern blight. **2**:FL(35).

Zantedeschia Spreng.
1. *Z. aethiopica* (L.) Spreng. [arum-lily, common calla-lily, florist's calla, garden calla, pig-lily, trumpet-lily, white calla-lily]
2. *Z. elliottiana* (Wm. Wats.(Kew)) Engl. [golden calla-lily, yellow calla-lily]
3. *Z. rehmannii* Engl. [pink calla-lily, red calla-lily]
4. *Zantedeschia* sp.
Oomycetes
Phytophthora erythroseptica. Rhizome rot. **1**:CA(82,3654). **3**:CA(82).
P. richardiae. Root rot. **1**:CA(82), FL(35,82), IL(82), IN(82), MA(82), NJ(82), NY(82), OH(82), OR(82), PA(82), WA(82). **4**:FL(35).
Phytophthora sp. **4**:WA(3202).
Ascomycotina
Glomerella cingulata. **1**:WA(3202).
Basidiomycotina - Uredinales
Uromyces ari-triphylli. **4**:FL(35).
Basidiomycotina - Other
Armillaria mellea. Root rot. **1**:CA(82,2898).
Deuteromycotina - Hyphomycetes
Alternaria sp. Secondary leaf spot. **1**:CA(82), NJ(82), NY(82), OR(82).
Botrytis cinerea. Gray mold blight. **1**:AK(82), NJ(82). **4**:AK(526).
Cercospora callae. Leaf spot. **1**:FL(35).
C. richardiaecola. Leaf spot. **1**:AL(82), FL(35), MS(82,2720). **4**:FL(35).
Coniothecium richardiae. Leaf spot. **1**:CA(82), FL(35,82), HI(82,2903), MA(82), NJ(82), NY(82), OR(82).
Corynespora cassiicola. Leaf spot. **1**:HI(2903).
Corynespora sp. Leaf spot. **1**:FL(35).

Fusarium sp. Bulb rot. **4:**FL(35).
Deuteromycotina - Coelomycetes
Colletotrichum gloeosporioides. Leaf spot. **1:**WA(82).
Phoma sp. Dry rot of rhizomes. **1:**CA(82), OR(82).
Deuteromycotina - Other
Rhizoctonia solani. Root rot. **2:**CA(82).
Sclerotium rolfsii. Southern blight. **1:**CA(82), FL(35,82), OR(82).

Araliaceae

Aralia L.

1. *A. californica* S. Wats. [elk-clover]
2. *A. cordata* Thunb. (=*A. edulis* Siebold & Zucc.) [oudo, spikenard, udo]
3. *A. hispida* Venten. [bristly sarsaparilla]
4. *A. nudicaulis* L. [wild sarsaparilla]
5. *A. racemosa* L. [American spikenard, life-of-man, petty-morel]
6. *A. spinosa* L. [angelica-tree, devil's-walkingstick, Hercules-club, prickly-ash]
7. *Aralia* sp.

Oomycetes
Phytophthora sp. Root rot. **7:**FL(35).
Pythium splendens. Root rot. **7:**FL(35).
Ascomycotina
Acrospermum album. On dead stems. **5:**NY(195).
Botryosphaeria dothidea. **6:**GA(1438).
B. ribis. Limb canker, dieback. **6:**GA(82), VA(82), WV(82,3907).
Dasyscyphus setigerus. On stems. **7:**CA(849).
Diaporthe araliae. On limbs. **6:**GA(82,1438), NC(1303), WV(82,3907).
Leptosphaeria collinsoniae. **5:**NY(3249).
L. conoidea. **5:**NY(3249).
L. doliolum. **5:**NY(3249).
Nectria cinnabarina. Canker, dieback. **6:**MN(2874), VA(82), WV(82,3907).
Ophiobolus niesslii. **5:**NY(3248).
Phyllactinia guttata. Powdery mildew. **4:**MI(82), NE(82).
Sclerotinia sclerotiorum. Stem rot. **2:**MD(82).
Basidiomycotina - Uredinales
Nyssopsora clavellosa. Autoecious. **1:**CA(81,2461). **3:**MI(82), NY(82), TX(82). **4:**CA(81,82,2461), ID(2461,3202), ME(2461), MI(2461), MN(82,2461,2874), MT(2461,3202), NH(2461), NY(2461), OR(82,2461,3202), VT(2461), WA(2461,3202), WI(2461). **5:**AZ(1195,1203,2461,3379,3981), OR(82,2461,3202).
Basidiomycotina - Ustilaginales
Mundkurella mossii. Systemic; mainly fructicolous. **4:**SD(3109).
Basidiomycotina - Other
Ceratobasidium anceps. **4:**WI(1343).
Exidia nucleata. **6:**WV(128).
Irpex lacteus. Sapwood rot. **6:**MD(82).
Oxyporus latemarginatus. **6:**VA(2166).
Deuteromycotina - Hyphomycetes
Alternaria araliae. On living leaves. **5:**WI(1347).
A. panax. Leaf spot. **7:**FL(35).
Alternaria sp. Blight, leaf spot. **2:**DE(82), NJ(82). **3:**NY(82). **5:**NY(82). **6:**FL().
Cercospora leptosperma. Leaf spot. **3:**MI(82), WI(82). **4:**DE(3543), IA(82), MI(82), MN(2874), NY(82), WI(82,1335).
Cercospora sp. Leaf spot. **5:**WI(1354). **7:**FL(35).
Phaeoisariopsis nigricans. Leaf spot. **6:**LA(82), TX(82), WV(3907).
Phymatotrichopsis omnivora. Root rot. **6:**TX(82).
Ramularia repens. Leaf spot. **4:**WI(82). **5:**WI(82).
Verticillium albo-atrum. Wilt. **5:**NY(82).
Xenosporium berkeleyi. **6:**FL(634).
Deuteromycotina - Coelomycetes
Aplosporella araliae. On limbs. **6:**WV(82,3907).
Ascochyta marginata. Leaf spot. **4:**WI(82).
Colletotrichum sp. Anthracnose. **7:**FL(35).
Diplodia sp. Dieback. **7:**FL(35).
Hendersonia petiolorum. Leaf spot. **6:**WV(3907).
Macrophoma sp. Leaf spot. **7:**FL(35).
Pestalotia brevipes. On petioles. **6:**FL(634).
Phoma exigua. Leaf spot. **5:**WI(64,82,1333).
P. melaleuca. On twigs. **6:**AL(82), WV(82,3907).
Phomopsis sp. Stem gall. **7:**FL(35).
Phyllosticta everhartii. Leaf spot. **6:**TX(82), WV(82,3907). **7:**FL(35).
Phyllosticta sp. On leaves. **5:**WI(1353).
Septoria clementsii. On leaves. **4:**CO(82), WI(1361).
Septoria sp. On leaflets. **4:**WI(1358). **5:**WI(1354).
Sphaceloma araliae. Spot anthracnose, scab, on leaves. **6:**MD(82,1736), MO(82,1745).
Stagonospora sp. Leaf spot. **6:**WV(82).
Deuteromycotina - Other
Rhizoctonia solani. Root rot. **6:**TX(82). **7:**FL(35).
Sclerotium deciduum. Leaf spot. **4:**WI(82,1342).

Cheirodendron Nutt. ex Seem.

1. *C. trigynum* (Gaud.) A. Heller (=*C. trigynum* (Gaud.) A. Heller var. *trigynum*)
2. *Cheirodendron* sp.

Oomycetes
Phytophthora cinnamomi. Root rot. **1:**HI(232). **2:**HI(2903).
Zygomycotina
Mucor sp. On leaves. **1:**HI(144,2903).
Ascomycotina
Asteridiella cheirodendronis. Black mildew. **1:**HI(1268,2903).
Coniochaeta sp. On leaves. **1:**HI(144,2903).
Glomerella pezizoidea. On decaying leaves. **2:**HI(2781).
Nectria sp. On leaves. **1:**HI(144,2903).
Preussia vulgare. On leaf surface. **1:**HI(144).
Trichothyrium reptans. On leaves. **1:**HI(144,2903).
Venturia carpophila. On leaf surface. **1:**HI(144).
Deuteromycotina - Hyphomycetes
Acremonium curvulum. On leaves. **2:**HI(2903).
Acremonium sp. On leaves. **1:**HI(144,2903).
Acrostaphylus sp. On leaves. **1:**HI(144,2903).
Aspergillus niger. On leaves. **1:**HI(144,2903).
Aureobasidium pullulans. On leaves. **1:**HI(144,2903).
Cercospora sp. On leaves. **1:**HI(144,2903).
Chalara sp. On leaves. **1:**HI(2903).
Cladosporium carpophilum. On leaves. **1:**HI(2903).
C. cladosporioides. On leaves. **1:**HI(144,2903).
Cladosporium sp. On leaves. **1:**HI(144,2903).
Cylindrocephalum sp. On leaves. **2:**HI(2903).
Fusarium roseum. On leaves. **1:**HI(144,2903).
Geotrichum sp. On leaves. **1:**HI(144,2903).
Monocillium sp. On leaves. **1:**HI(144,2903).
Nodulisporium gregarium. On leaves. **1:**HI(144,2903).
Oidiodendron sp. On leaves. **1:**HI(144,2903).
Paecilomyces sp. On leaves. **1:**HI(144).
Penicillium canescens. On leaves. **1:**HI(144,2903).
P. glabrum. On leaves. **1:**HI(144,2903).
P. simplicissimum. On leaves. **1:**HI(144,2903).
Penicillium sp. On leaves. **1:**HI(144,2903).
Periconia byssoides. On leaves. **1:**HI(144,2903).
Rhinocladiella sp. On leaves. **1:**HI(144,2903).
Scolecobasidium terreum. On leaves. **1:**HI(144,2903).
Scopulariopsis sp. On leaves. **1:**HI(144,2903).
Trichoderma viride. On leaves. **1:**HI(144,2903).
Tritirachium sp. On leaves. **1:**HI(144,2903).
Zasmidium cellare. On leaves. **1:**HI(144,2903).
Deuteromycotina - Coelomycetes
Melanconium sp. On leaves. **1:**HI(144,2903).
Pestalotia sp. On leaves. **1:**HI(144,2903).
Phoma sp. On leaves. **2:**HI(2903).
Sphaeropsis sp. On leaves. **1:**HI(144,2903).

Cussonia Thunb.

1. *C. spicata* Thunb. [cabbage-tree]

Oomycetes
Phytophthora sp. Root and stem rot. **1:**FL(35).

Didymopanax Decne. & Planch.

1. *D. morototonii* (Aubl.) Decne. & Planch.

Deuteromycotina - Hyphomycetes
Fusarium sp. Root and basal stem rot. **1:**FL(35).

Dizygotheca N. E. Brown

This genus is considered a synonym of *Schefflera*, but the appropriate combination could not be located.

1. *D. elegantissima* (J. H. Veitch) R. Vig. & Guillaumin

Oomycetes

Pythium irregulare. Root rot. **1**:FL(35).
P. splendens. Root rot. **1**:FL(35).

Basidiomycotina - Other

Trametes hirsuta. White rot. **1**:FL(35).

Deuteromycotina - Hyphomycetes

Alternaria panax. Leaf spot. **1**:CA(3703), FL(35).
Cercospora sp. Leaf spot. **1**:FL(35).
Fusarium oxysporum. Tip necrosis. **1**:FL(35).

Deuteromycotina - Coelomycetes

Colletotrichum sp. Anthracnose. **1**:FL(35).
Diplodia sp. Dieback. **1**:FL(35).
Phyllosticta sp. Root rot, leaf spot. **1**:FL(35).
Sphaceloma araliae. Spot anthracnose. **1**:FL(35).

Deuteromycotina - Other

Rhizoctonia solani. Root rot. **1**:FL(35).

Eleutherococcus Maxim.

1. *E. sieboldianus* (Makino) Koidz. (=*Acanthopanax sieboldianus* Makino)

Deuteromycotina - Hyphomycetes

Alternaria panax. Leaf spot. **1**:MO(82).
Alternaria sp. Leaf spot. **1**:MO(82).
Phymatotrichopsis omnivora. Root rot. **1**:TX(82).

×*Fatshedera* Guillaumin

Fatsia × *Hedera*.

1. × *F. lizei* (Hort. ex C.-Cochet) Guillaumin. *Fatsia japonica* cv. '*Moseri*' × *Hedera helix* var. *hibernica*.

Oomycetes

Phytophthora nicotianae var. *parasitica*. Damping-off. **1**:FL(35).
Pythium splendens. Root rot. **1**:FL(35).

Ascomycotina

Glomerella cingulata. Leaf spot. **1**:FL(35).
Guignardia sp. Leaf spot. **1**:FL(35).

Deuteromycotina - Hyphomycetes

Alternaria sp. Leaf spot. **1**:FL(35).
Cercospora sp. Leaf spot. **1**:FL(35).
Corynespora cassiicola. Leaf spot. **1**:FL(35).
Fusarium sp. Cutting rot. **1**:FL(35).

Deuteromycotina - Coelomycetes

Colletotrichum sp. Anthracnose. **1**:FL(35).
Phyllosticta concentrica. Leaf spot. **1**:FL(35).

Deuteromycotina - Other

Rhizoctonia solani. Root and stem rot. 'Rhizoctonia aerial blight. Rhizoctonia stem rot.' **1**:FL(35).

Fatsia Decne. & Planch.

1. *F. japonica* (Thunb.) Decne. & Planch. [Formosa rice-tree, glossy-leaved paperplant, Japanese fatsia, paperplant]

Oomycetes

Phytophthora nicotianae var. *parasitica*. Basal stem rot. **1**:FL(35).
Pythium splendens. Root rot. **1**:FL(35).

Ascomycotina

Nectriella pironii. Stem gall. **1**:FL(35).

Deuteromycotina - Hyphomycetes

Alternaria sp. Leaf spot. **1**:FL(35).
Cercospora sp. Leaf spot. **1**:FL(35).
Fusarium sp. Root rot. **1**:FL(35).

Deuteromycotina - Coelomycetes

Colletotrichum sp. Anthracnose. **1**:FL(35).
Macrophoma sp. Leaf spot. **1**:FL(35).

Deuteromycotina - Other

Rhizoctonia solani. Root rot. 'Rhizoctonia aerial blight. Rhizoctonia stem rot.' **1**:FL(35).

Hedera L.

1. *H. helix* L. [English ivy]
2. *Hedera* sp.

Oomycetes

Phytophthora nicotianae var. *parasitica*. Root and stem rot. **1**:NC(1303).
P. palmivora. Stem blight, root rot. 'Phytophthora leaf spot.' **1**:CA(1866), FL(35), HI(3700).
Phytophthora sp. Stem blight, root rot. **1**:CA(881).
Pythium splendens. Root rot. **1**:FL(35).
Pythium sp. Root rot. **2**:HI(2903).

Ascomycotina

Botryosphaeria obtusa. On stems. **1**:MD(82).
Elsinoe sp. Scab. **1**:MD(1736).
Erysiphe cichoracearum. Powdery mildew. **1**:CT(3770), FL(35), OK(82,2872).
Glomerella cingulata. Dieback. **1**:CT(82), MD(82), NY(82), TX(82).
Nectriella pironii. Stem gall. **1**:FL(35).
Sclerotinia sclerotiorum. Stem rot. **1**:FL(35).

Deuteromycotina - Hyphomycetes

Alternaria alternata. Leaf spot. **1**:FL(35).
Alternaria sp. Associated with leaf spot. **1**:NC(1303).
Botrytis sp. Gray mold. **1**:FL(35).
Cercospora hederae. Leaf spot. **1**:FL(35).
Cercospora sp. Leaf spot. **1**:NC(1303).
Cladosporium brunneolum. On leaves. **1**:CA(82).
C. punctulatum. **1**:CA(81,82).
Corynespora sp. Leaf spot. **1**:FL(35).
Fusarium oxysporum. Root rot. **1**:FL(35).
F. solani. Root rot. **1**:FL(35).
Fusarium sp. Leaf spot. **1**:FL(35).
Phymatotrichopsis omnivora. Root rot. **1**:TX(82).
Ramularia hedericola. Leaf spot. **1**:TX(82).
Stemphylium sp. Leaf spot. **1**:FL(35).

Deuteromycotina - Coelomycetes

Aplosporella hedericola. Gall, leaf and stem spot. **1**:FL(35), NY(82), WV(82,3907).
Colletotrichum trichellum. Anthracnose, leaf and stem spot. **1**:'Anthracnose.' E states(82), Unk.(82), FL(35), GA(659), IL(312), IN(663), MN(2874), NC(1303), OK(82,2872), OR(82,3202), TX(82), WA(82,3202). **2**:CA(3549), FL(3549).
Colletotrichum sp. Anthracnose, leaf and stem spot. **1**:OR(3202).
Coniothyrium sp. Leaf spot. **1**:FL(35).
Diplodia hederae. On dead stems. **1**:CT(82), SC(82), VA(82).
Diplodia sp. Dieback. **1**:FL(35).
Macrophoma sp. Leaf necrosis, spot. **1**:FL(35), TX(82).
Macrophomina phaseolina. Root and stem rot. **1**:NC(1303).
Phoma sp. Gall. **1**:FL(35).
Phyllosticta concentrica. 'Phyllosticta leaf spot.' **1**:E states(82), CA(81,82), FL(35), MS(2720), NC(1303), NE(82), NJ(2), TX(82), WA(82,3202), WV(3907).
P. hederae. **1**:DE(82), NY(82).
P. hedericola. Leaf spot. **1**:CA(82), CT(82), MS(2720), NJ(82), NY(82), OR(82,3202).
Pyrenochaeta sp. Root rot. **1**:FL(35).
Sphaceloma hederae. Spot anthracnose and scab on leaves. **1**:CA(82,1745,1747), MD(1729,1736,1737,1745,1747), NC(82,1303,1745,1747), VA(82,1745,1747).

Deuteromycotina - Other

Rhizoctonia solani. Leaf blight, root rot. 'Rhizoctonia aerial blight. Rhizoctonia stem rot.' **1**:CT(82), FL(35), NC(1303).
Rhizoctonia sp. Root rot. **2**:HI(2903).

Meryta J. R. Forster & G. Forster

1. *Meryta* sp.

Deuteromycotina - Coelomycetes

Phyllosticta sp. Leaf spot. **1**:FL(35).

Oplopanax (Torr. & A. Gray) Miq.

1. *O. horridum* (Sm.) Miq. (=*Fatsia horrida* (Sm.) Benth. & J. D. Hook. ex Brewer & S. Wats.) [devil's-club]
2. *Oplopanax* sp.

Ascomycotina

Coniochaeta ligniaria. **1**:WA(3202,3815).
Diaporthe araliae. **1**:WA(3202,3815).
Nummularia subconcava. **1**:ID(3202).
Trichopezizella barbata. **1**:WA(1416).
T. nidulus var. *setigera*. **1**:WA(3202).
Zignoella ovoidea. **1**:WA(3202,3815).

Deuteromycotina - Hyphomycetes
Botrytis cinerea. Gray mold blight. **1**:AK(82). **2**:AK(526).
Cercospora daemonicola. On living leaves. **1**:OR(82,3202,3421).
Ramularia sp. **1**:WA(3202).
Deuteromycotina - Coelomycetes
Ascochyta sp. **1**:WA(3202).

Panax L.

1. *P. gaudichaudii* DC. (=*Cheirodendron gaudichaudii* Seem.)
2. *P. quinquefolius* L. [American ginseng, ginseng]
3. *P. trifolius* L. [dwarf ginseng, groundnut]

Oomycetes
Phytophthora cactorum. Root and stem rot. **2**:E states(82), IA(82), MI(82), WA(82,3202).
Pythium debaryanum. Damping-off. **2**:NY(82).
Ascomycotina
Asteridiella cheirodendronis. Black mildew. **1**:HI(1272,1452).
Calonectria sp. **2**:NC(1303).
Sclerotinia sclerotiorum. White rot of rhizomes. **2**:MI(82), NY(82), OH(82), PA(82), WI(82).
Stromatinia panacis. Black rot of rhizomes. **2**:MI(82), MN(82,2874), NY(82,3167), WI(82).
Basidiomycotina - Uredinales
Puccinia araliae. Autoecious. **3**:MA(82), PA(82).
Basidiomycotina - Other
Thanatephorus cucumeris. **2**:WA(3202).
Deuteromycotina - Hyphomycetes
Alternaria panax. Blight, stem rot. **2**:E states(82), MN(82), MO(82), WI(1344).
Botrytis cinerea. Gray mold blight, root, seed and stem rot. **2**:E states(82), MI(82), WA(82,3202).
Cylindrocarpon destructans. Superficial discoloration of rhizomes, **2**:MI(82), NY(82), OR(82,3202), WA(82,3202), WI(82).
Fusarium acuminatum. Root rot. **2**:E states(82), MO(82), WI(82).
F. equiseti. Root rot, wilt. **2**:E states(82), MO(82), NC(1303), WA(3202), WI(82).
Thielaviopsis basicola. Black root. **2**:IL(82), MI(82), NJ(82), NY(82), OH(82).
Verticillium albo-atrum. Wilt. **2**:IN(82), KY(82), MI(82), NJ(82), NY(82), OH(82), OR(3202), PA(82), TN(82), WI(82).
Deuteromycotina - Coelomycetes
Colletotrichum dematium. Secondary leaf spot. **2**:E states(82), MN(82), MO(82).
Septoria araliae. **3**:WI(82).
Septoria sp. Leaf spot. **2**:MI(82), MN(2874).
Deuteromycotina - Other
Rhizoctonia solani. Damping-off. **2**:AR(82), IN(82), MI(82), NJ(82), NY(82), WA(82).

Polyscias J. R. Forster & G. Forster

1. *P. balfouriana* (Hort. ex André) L. H. Bailey [Balfour aralia]
2. *P. fruticosa* (L.) Jaeger (=*Nothopanax fruticosus* (L.) Miq.) [ming aralia]
3. *P. guilfoylei* (W. Bull) L. H. Bailey [coffee-tree, geranium-leaf aralia, wild-coffee]
4. *Polyscias* sp.

Oomycetes
Phytophthora sp. Root rot. **4**:FL(35).
Pythium splendens. Root rot. **1**:FL(35). **2**:FL(35). **3**:HI(2903). **4**:FL(35).
Pythium sp. **2**:HI(2903).
Ascomycotina
Diaporthe sp. On limbs. **3**:HI(2903).
Elsinoe mulleri. Spot anthracnose. **3**:HI(1746).
Nectriella pironii. Stem gall. **1**:FL(35). **2**:FL(35).
Deuteromycotina - Hyphomycetes
Alternaria panax. 'Alternaria leaf spot.' **1**:FL(35). **2**:FL(118). **4**:FL(35).
Cercospora sp. Leaf spot. **1**:FL(35). **4**:FL(35).
Corynespora cassiicola. Leaf spot. **1**:FL(35).
Fusarium oxysporum. Root rot. **4**:FL(35).
Fusarium sp. Basal stem rot, root rot, leaf spot. **1**:FL(35). **2**:FL(35).
Deuteromycotina - Coelomycetes
Colletotrichum dematium. Anthracnose. **2**:FL(35).
C. gloeosporioides. Anthracnose, leaf spot. **1**:FL(35). **2**:HI(2903). **4**:FL(35).
Coniothyrium sp. Leaf spot. **1**:FL(35).
Diplodia sp. Dieback. **4**:FL(35).
Lasiodiplodia theobromae. **2**:HI(2903).
Phyllosticta sp. Leaf spot. **1**:FL(35). **4**:FL(35).
Deuteromycotina - Other
Rhizoctonia solani. Leaf, petiole and root rot. 'Rhizoctonia aerial blight. Rhizoctonia stem rot.' **1**:FL(35). **2**:FL(35). **3**:HI(2899,3074). **4**:FL(35).
Rhizoctonia sp. Root rot. **3**:HI(2903).

Pseudopanax K. Koch

1. *P. crassifolia* (Sol.) K. Koch [lancewood]

Ascomycotina
Glomerella sp. Anthracnose. **1**:HI(2903).

Schefflera J. R. Forster & G. Forster

(= *Brassaia* Endl., *Dizygotheca* N. E. Brown.)
1. *S. actinophylla* (Endl.) Jaeger (=*Brassaia actinophylla* Endl.) [Australian umbrella-tree, octopus-tree, Queensland umbrella-tree, rubber-tree, starleaf]
2. *S. arboricola* Hayata
3. *Schefflera* sp. (=*Brassaia sp.*)

Oomycetes
Phytophthora nicotianae var. *parasitica*. Leaf spot. **1**:FL(35). **2**:FL(35).
Pythium splendens. Seedling blight, root rot. **1**:FL(35), HI(3967). **2**:FL(35).
Pythium sp. Root rot. **3**:HI(2903).
Ascomycotina
Botryosphaeria sp. Canker. **1**:FL(35).
Sclerotinia sclerotiorum. Leaf and stem rot. **2**:FL(35).
Deuteromycotina - Hyphomycetes
Alternaria panax. Leaf spot. **1**:CA(3703), FL(35), HI(2903). **2**:'Alternaria leaf spot.' FL(35). **3**:FL(35).
Alternaria sp. Leaf spot. **1**:NC(1303). **2**:FL(117).
Cercospora sp. Edema, leaf spot. **1**:FL(35). **2**:FL(35). **3**:FL(35).
Cladosporium sp. Leaf spot. **2**:FL(35).
Corynespora sp. Leaf spot. **1**:FL(35).
Fusarium solani. Damping-off. **1**:FL(35).
Fusarium sp. Damping-off, stem rot. **2**:FL(35).
Wiesneriomyces laurinus. On leaf litter. **3**:HI(2903).
Deuteromycotina - Coelomycetes
Colletotrichum sp. Anthracnose. **1**:FL(35). **3**:FL(35).
Coniothyrium sp. Stem canker, leaf spot. **1**:FL(35). **2**:FL(35).
Diplodia sp. Stem necrosis. **1**:FL(35).
Macrophoma sp. Leaf spot. **1**:FL(35).
Macrophomina phaseolina. Charcoal root rot. **1**:FL(35).
Phomopsis sp. Leaf spot. **1**:FL(35).
Phyllosticta sp. Leaf spot. **1**:FL(35).
Deuteromycotina - Other
Rhizoctonia solani. Root rot. **1**:FL(35). **2**:'Rhizoctonia aerial blight. Rhizoctonia stem rot.' FL(35).
Rhizoctonia sp. Root rot. **2**:HI(2903).
Sclerotium rolfsii. Southern blight. **1**:FL(33,35). **2**:'Southern blight.' FL(35).

Tetrapanax (K. Koch) K. Koch

1. *T. papyriferum* (Hook.) K. Koch [Chinese rice-paper-plant]

Deuteromycotina - Hyphomycetes
Cercospora sp. Leaf spot. **1**:FL(35).
Fusarium sp. Leaf necrosis. **1**:FL(35).
Deuteromycotina - Coelomycetes
Phomopsis sp. Leaf and petiole spot. **1**:FL(35).

Tetraplasandra A. Gray

This genus is considered a synonym of *Gastonia*, but the appropriate combination could not be located.
1. *T. hawaiensis* A. Gray
2. *T. meiandra* (Hildebr.) Harms

Ascomycotina
Seynesiopeltis tetraplasandrae. On leaves. **1**:HI(2903). **2**:HI(2903).

Trevesia Vis.

1. *T. palmata* (Roxb.) Vis.

Oomycetes

Phytophthora sp. Root rot. **1**:FL(35).
Pythium splendens. Root rot. **1**:FL(35).

Ascomycotina

Nectriella pironii. Stem gall. **1**:FL(35).

Deuteromycotina - Hyphomycetes

Alternaria sp. Leaf spot. **1**:FL(35).
Botrytis sp. Gray mold. **1**:FL(35).
Fusarium solani. Root rot. **1**:FL(35).
Fusarium sp. Root rot. **1**:FL(35).

Deuteromycotina - Coelomycetes

Colletotrichum sp. Leaf spot. **1**:FL(35).
Phyllosticta sp. Leaf spot. **1**:FL(35).

Tupidanthus J. D. Hook. & Th. Thomson

1. *T. calyptratus* J. D. Hook. & Th. Thomson

Oomycetes

Pythium sp. Root rot. **1**:FL(35).

Deuteromycotina - Hyphomycetes

Alternaria panax. Leaf spot. **1**:CA(3703), FL(35).

Araucariaceae

Araucaria Juss.

1. *A. angustifolia* (Bertol.) Kuntze (=*A. brasiliana* A. Rich.) [Brazilian pine, Paraná pine]
2. *A. araucana* K. Koch (=*A. imbricata* Pav.) [Chilean pine, monkey-puzzle-tree]
3. *A. columnaris* (G. Forster) Hook. (=*A. excelsa* (Lamb.) R. Br.) [New Caledonia pine]
4. *A. heterophylla* (Salisb.) Franco [Australian pine, house pine, Norfolk Island pine]
5. *Araucaria* sp.

Oomycetes

Phytophthora cinnamomi. Tree decline, seedling root rot. **3**:HI(2299). **4**:HI(2903).
Phytophthora sp. Root rot. **4**:FL(35).
Pythium splendens. Root rot. **4**:FL(35).
Pythium sp. Root rot. **2**:FL(35). **5**:FL(35).

Ascomycotina

Botryosphaeria sp. Dieback. **2**:FL(35). **5**:FL(35).
Calonectria kyotensis. Collar rot. **4**:HI(2903).
Cryptosporella araucariae. **2**:WA(3202).
Glomerella sp. Needle necrosis. **4**:FL(35). **5**:FL(35).
Leptosphaeria californica. **2**:CA(82).
Leptosphaeria sp. Needle blight. **4**:FL(35). **5**:FL(35).
Physalospora sp. Dieback. **5**:FL(35).
Servazziella longispora. Needle blight. **4**:FL(35).
Stictis monilifera. On leaves. **2**:CA(82). **5**:CA(3225).

Basidiomycotina - Other

Sebacina sp. **4**:HI(2903).

Deuteromycotina - Hyphomycetes

Alternaria sp. Needle necrosis. **2**:FL(35).
Corynespora sp. Needle browning. **5**:FL(35).
Curvularia sp. Needle necrosis. **4**:FL(35).
Cylindrocladium clavatum. Root rot. **4**:FL(35).
C. ilicicola. Twig blight. **4**:FL(35).
Fusarium decemcellulare. Stem necrosis. **4**:FL(35).
F. lateritium. Stem necrosis. **4**:FL(35).
F. oxysporum. Root rot. **4**:FL(35).
F. solani. Root and stem rot. **4**:FL(35).
Helminthosporium sp. Needle blight. **4**:FL(35).
Mycoenterolobium platysporum. On wood. **4**:HI(2903). **5**:HI(1265).
Phymatotrichopsis omnivora. Root rot. **1**:TX(82).

Deuteromycotina - Coelomycetes

Botryodiplodia sp. Canker, dieback. **2**:FL(35). **4**:FL(35).
Colletotrichum sp. Needle browning, needle tip necrosis. **2**:FL(35). **4**:FL(35).
Diplodia sp. Dieback. **4**:FL(35).
Dothiorella sp. Needle browning, bleeding canker. **4**:FL(35), HI(1915,2903). **5**:FL(35).
Lasiodiplodia theobromae. Bleeding canker, dieback. **2**:FL(35). **4**:'Bleeding canker.' HI(1915,2903).
Macrophoma sp. Dieback. **2**:FL(35). **4**:FL(35). **5**:FL(35).
Pestalotia sp. Needle browning. **2**:FL(35). **4**:FL(35). **5**:FL(35).
Pestalotiopsis dichaeta. On leaves. **2**:PA(82).
P. funerea. On leaves. **2**:CA(81,82).
Phoma sp. Dieback. **4**:FL(35).
Phomopsis sp. Dieback. **4**:FL(35).
Phyllosticta araucariae. Needle blight, needle browning. **2**:FL(35). **4**:FL(35). **5**:FL(35).
Sphaeropsis sp. Dieback. **4**:FL(35). **5**:FL(35).

Deuteromycotina - Other

Rhizoctonia solani. Damping-off, root rot. **3**:HI(2899,3074). **4**:FL(35).
Sclerotium rolfsii. Southern blight. **4**:FL(35).

Arecaceae

Records with host identified to family only.

1. *Arecaceae* sp.

Oomycetes

Phytophthora palmivora. Wilt, bud rot. **1**:FL(35,82).
Phytophthora sp. Root rot. **1**:FL(35).
Pythium splendens. Dieback, root rot. **1**:FL(35).

Ascomycotina

Ceratocystis paradoxa. Root rot, stem bleeding. **1**:FL(35).
Didymella phacidiomorpha. Leaf spot. **1**:FL(35).
Glomerella cingulata. Canker, leaf spot. **1**:Range of host(82).
Leptosphaeria sp. Bud necrosis. **1**:FL(35).
Meliola palmicola. **1**:HI(82).
Nectria ochroleuca. On petioles. **1**:HI(891).
Phyllachora palmicola. Black spot. **1**:FL(35).
Phyllachora sp. Tar spot. **1**:FL(35).
Physalospora sp. Leaf spot. **1**:FL(35).
Tubeufia pezizula. On petioles. **1**:HI(1269).

Basidiomycotina - Ustilaginales

Graphiola phoenicis. **1**:Range of host(82), FL(35).

Basidiomycotina - Other

Armillaria tabescens. Root rot. **1**:FL(35).
Ganoderma zonatum. **1**:FL(35).
Phellinus palmicolus. On wood. **1**:FL(1205).
Stilbotulasnella conidiophora. On leaves. **1**:HI(157).
Tyromyces palmarum. **1**:FL(2179).

Deuteromycotina - Hyphomycetes

Alternaria sp. Leaf spot. **1**:FL(35).
Cercospora sp. Leaf spot. **1**:FL(35).
Chalara paradoxa. Root rot. **1**:FL(82).
Cylindrocladium scoparium. Leaf spot. **1**:FL(35).
Delortia palmicola. On dead petioles and fronds. **1**:HI(1269).
Fusarium sp. Root rot. **1**:FL(35).
Gliocladium sp. Stem rot. **1**:FL(35).
Helicoma muelleri. On petioles. **1**:HI(1269).
H. palmigenum. On dead leaves. **1**:HI(1269).
Helicomyces roseus. On dead petioles. **1**:HI(1269).
Helminthosporium sp. Leaf spot. **1**:FL(35).
Hobsonia mirabilis. On petioles and fronds. **1**:HI(1269).
Phaeotrichoconis sp. Leaf spot. **1**:FL(35).
Stigmina palmivora. Leaf spot. **1**:FL(35), NY(82), OH(82), TX(82).
Xenosporium berkeleyi. On leaves. **1**:HI(1269).

Deuteromycotina - Coelomycetes

Colletotrichum gloeosporioides. Leaf blight. Anthracnose. **1**:Range of host(82), FL(35).
Coniothyrium sp. Leaf spot. **1**:FL(35).
Cytospora palmarum. Wilt, bud rot. **1**:FL(35,82).
Cytospora sp. Leaf spot. **1**:FL(35).
Cytosporina sp. Leaf spot. **1**:FL(35).
Diplodia sp. Leaf spot. **1**:FL(35).
Lasiodiplodia theobromae. Dieback. **1**:Range of host(82).
Pestalotiopsis palmarum. On leaves. Leaf spot. **1**:Range of host(82), FL(35).
Phoma maculata. On leaves. **1**:CA(82).
Phomopsis sp. Leaf spot. **1**:FL(35).
Phyllosticta sp. Leaf spot. **1**:FL(35,82).
Septoria sp. Leaf spot. **1**:FL(35).

Deuteromycotina - Other
Rhizoctonia solani. Root and stem rot. **1**:FL(35).

Acoelorrhaphe H. Wendl.
1. *A. wrightii* (Griseb. & H. Wendl. ex Griseb.) H. Wendl. ex Becc. [everglades palm, Paurotis palm, saw cabbage palm, silver saw palm]

Ascomycotina
Botryosphaeria sp. Anthracnose. **1**:FL(35).
Catacauma sabal. Black spot. **1**:FL(35).
Basidiomycotina - Ustilaginales
Graphiola phoenicis. False smut. **1**:FL(35).
Deuteromycotina - Hyphomycetes
Cercospora sp. Leaf spot. **1**:FL(35).
Fusarium sp. Leaf spot. **1**:FL(35).
Stigmina palmivora. Leaf spot. **1**:FL(35).
Deuteromycotina - Coelomycetes
Sphaeropsis sp. Leaf spot. **1**:FL(35).
Deuteromycotina - Other
Rhizoctonia solani. Root rot. **1**:FL(35).

Archontophoenix H. Wendl. & Drude
1. *A. alexandrae* (F. v. Muell.) H. Wendl. & Drude [Alexandra palm, kingpalm, northern Bangalow palm]
2. *Archontophoenix* sp.

Deuteromycotina - Hyphomycetes
Alternaria sp. Leaf spot. **1**:FL(35).
Cercospora sp. Leaf spot. **1**:FL(35). **2**:FL(35).
Fusarium sp. Bud rot. **1**:FL(35).
Helminthosporium sp. Leaf spot. **1**:FL(35).
Deuteromycotina - Coelomycetes
Pestalotia sp. Leaf spot. **1**:FL(35).
Phomopsis sp. Leaf spot. **1**:FL(35).
Phyllosticta sp. Leaf spot. **1**:FL(35).
Deuteromycotina - Other
Rhizoctonia sp. Root rot. **1**:FL(35).

Arenga Labill. in DC.
1. *A. pinnata* (F. Wurmb) Merr. (=*A. saccharifera* Labill. ex DC.) [areng palm, black-fiber palm, gomuti palm, sugar palm]

Basidiomycotina - Ustilaginales
Graphiola phoenicis. False smut. **1**:Unk.(82).

Bactris Jacq. ex Scop.
1. *B. porschiana* Burret

Oomycetes
Phytophthora sp. Root rot. **1**:FL(35).

Balaka Becc.
1. *Balaka* sp.

Oomycetes
Pythium sp. Root rot. **1**:FL(35).

Bismarckia Hildebr. & H. Wendl.
1. *B. nobilis* Hildebr. & Wendl.

Deuteromycotina - Coelomycetes
Pestalotiopsis palmarum. Leaf spot. **1**:FL(35).
Phyllosticta sp. Leaf spot. **1**:FL(35).

Butia (Becc.) Becc.
1. *B. capitata* (Mart.) Becc. [South American jelly palm]

Oomycetes
Phytophthora sp. Bud rot. **1**:FL(35).
Pythium sp. Root rot. **1**:FL(35).
Basidiomycotina - Ustilaginales
Graphiola phoenicis. False smut. **1**:FL(35).
Basidiomycotina - Other
Armillaria tabescens. Root rot. **1**:FL(35).
Ganoderma zonatum. Trunk rot. **1**:FL(35).
Deuteromycotina - Hyphomycetes
Alternaria sp. Leaf spot. **1**:FL(35).
Cercospora sp. Leaf spot. **1**:FL(35).
Fusarium sp. Bud rot. **1**:FL(35).
Stigmina palmivora. Leaf spot. **1**:FL(35).
Deuteromycotina - Coelomycetes
Colletotrichum sp. Leaf spot. **1**:FL(35).
Diplodia sp. Leaf stalk lesions. **1**:FL(35).
Pestalotiopsis palmarum. Leaf spot. **1**:FL(35).
Phyllosticta sp. Leaf spot. **1**:FL(35).
Deuteromycotina - Other
Rhizoctonia sp. Bud rot. **1**:FL(35).

Calamus L.
1. *Calamus* sp.

Oomycetes
Pythium sp. Root rot. **1**:FL(35).

Calyptrocalyx Blume
(= *Paralinospadix* Burret.)
1. *Calyptrocalyx* sp.

Oomycetes
Pythium sp. Root rot. **1**:FL(35).

Caryota L.
1. *C. mitis* Lour. [Burmese fishtail palm, clustered fishtail palm, tufted fishtail palm]
2. *C. urens* L. [fishtail palm, jaggery palm, kittul-tree, sago palm, toddy palm, wine palm]
3. *Caryota* sp.

Oomycetes
Phytophthora sp. Root rot. **3**:FL(35).
Pythium sp. Root rot. **1**:FL(35). **2**:FL(35). **3**:FL(35).
Ascomycotina
Glomerella cingulata. Anthracnose, leaf blight. **1**:FL(35). **3**:Unk.(82), MS(2720).
Deuteromycotina - Hyphomycetes
Alternaria sp. Leaf spot. **1**:FL(35).
Bipolaris setariae. 'Helminthosporium leaf spot.' **1**:FL(35).
Cercospora sp. Leaf spot. **1**:FL(35). **2**:FL(35).
Curvularia lunata. Leaf spot. **1**:FL(35).
[*Exserohilum rostratum*]. 'Helminthosporium leaf spot.' **1**:FL(35).
Fusarium sp. Root rot, leaf spot. **1**:FL(35). **2**:FL(35).
Phaeotrichoconis crotalariae. Leaf spot. **1**:FL(35).
Stigmina palmivora. Leaf spot. **1**:FL(35). **2**:FL(35).
Deuteromycotina - Coelomycetes
Macrophoma sp. Leaf spot. **1**:FL(35).
Pestalotiopsis palmarum. Leaf spot. **1**:FL(35). **2**:FL(35). **3**:FL(35).
Phomopsis sp. Leaf spot. **1**:FL(35). **2**:FL(35).
Phyllosticta sp. Leaf spot. **2**:FL(35).
Deuteromycotina - Other
Rhizoctonia solani. Basal stem rot. **1**:FL(35). **2**:FL(35).

Chamaedorea Willd.
1. *C. elegans* Mart. [good-luck palm, pacaya, parlor palm]
2. *C. erumpens* H. Moore [bamboo palm]
3. *C. seifrizii* Burret
4. *Chamaedorea* sp.

Oomycetes
Phytophthora palmivora. Root rot. **1**:FL(35). **2**:FL(35).
Phytophthora sp. Root rot. **3**:FL(35).
Pythium splendens. Root and basal stem rot. **1**:FL(35). **2**:FL(35). **3**:FL(35).
Pythium sp. Root rot. **1**:HI(2903). **4**:FL(35).
Ascomycotina
Botryosphaeria sp. Stem necrosis. **2**:FL(35).
Deuteromycotina - Hyphomycetes
Alternaria sp. Leaf spot. **1**:FL(35). **2**:FL(35). **3**:FL(35). **4**:FL(35).
Bipolaris setariae. 'Helminthosporium leaf spot.' **1**:FL(35). **3**:FL(558).
Botrytis cinerea. On fronds. **1**:FL(35).
Cercospora sp. Leaf spot. **2**:FL(35). **4**:FL(35).
Curvularia sp. Leaf spot. **1**:FL(35).
Cylindrocladium pteridis. Leaf spot. **1**:FL(35).
[*Exserohilum rostratum*]. 'Helminthosporium leaf spot.' **1**:FL(35). **3**:FL(35,558).
Fusarium moniliforme. Seed rot, leaf spot. **1**:FL(35).
F. oxysporum. Basal stem rot, leaf spot. **1**:FL(35). **2**:FL(35).
F. roseum. Seed rot. **1**:FL(35).
F. solani. Root and stem rot. **1**:FL(35). **2**:FL(35). **3**:FL(35).

4:FL(35).
Fusarium sp. Root rot, basal stem rot. **1**:FL(35). **3**:FL(35). **4**:FL(35).
Gliocladium vermoeseni. Bleeding canker, stem rot. 'Gliocladium stem rot.' **1**:FL(35). **2**:FL(35). **3**:FL(35,2960). **4**:FL(35), HI(2903).
Helminthosporium sp. Leaf spot. **2**:FL(35). **3**:FL(35). **4**:FL(35).
Phaeotrichoconis crotalariae. Leaf spot. **1**:FL(35).
Stemphylium sp. Leaf spot. **2**:FL(35). **3**:FL(35).
Verticillium sp. Wilt. **1**:FL(35).
Deuteromycotina - Coelomycetes
Colletotrichum sp. Anthracnose. **1**:FL(35). **2**:FL(35). **4**:FL(35).
Macrophoma sp. Leaf spot. **2**:FL(35).
Pestalotiopsis palmarum. Leaf spot. **1**:FL(35).
Phoma sp. Stem necrosis. **3**:FL(35).
Phomopsis sp. Leaf spot. **1**:FL(35). **2**:FL(35). **3**:FL(35). **4**:FL(35).
Phyllosticta sp. Leaf spot. **1**:FL(35). **2**:FL(35). **4**:FL(35).
Deuteromycotina - Other
Rhizoctonia solani. Root and stem rot. **1**:FL(35). **2**:FL(35). **3**:FL(35). **4**:FL(35).

Chambeyronia Vieill.

1. *C. macrocarpa* (Brongn.) Vieill. ex Becc.
Oomycetes
Pythium sp. Root rot. **1**:FL(35).

Chrysalidocarpus H. Wendl.

1. *C. lutescens* H. Wendl. [areca palm, butterfly palm, cane palm, golden cane palm, golden feather palm, Madagascar palm, yellow butterfly palm, yellow palm]
2. *Chrysalidocarpus* sp.
Oomycetes
Phytophthora sp. Root rot. **1**:FL(35).
Pythium splendens. Root rot. **1**:FL(35).
Pythium sp. Root rot. **1**:HI(2903).
Basidiomycotina - Ustilaginales
Graphiola phoenicis. False smut. **1**:FL(35).
Basidiomycotina - Other
Ganoderma applanatum. On wood. **2**:FL(35).
G. zonatum. On wood. **1**:FL(35).
Deuteromycotina - Hyphomycetes
Alternaria alternata. Leaf spot. **1**:FL(35).
Bipolaris setariae. Leaf spot. 'Helminthosporium leaf spot.' **1**:FL(35,558).
Cercospora sp. Leaf spot. **1**:FL(35).
Corynespora sp. Leaf spot. **1**:FL(35).
Cylindrocladium sp. Leaf spot. **2**:FL(35).
Exserohilum rostratum. 'Helminthosporium leaf spot.' **1**:FL(35,558).
Exserohilum sp. Leaf spot. **1**:HI(2903).
Fusarium sp. Root rot, leaf spot. **1**:FL(35).
Gliocladium vermoeseni. Stem rot. **1**:'Gliocladium stem rot.' FL(35). **2**:FL(35).
Monilochaetes sp. Leaf spot. **1**:FL(35).
Phaeotrichoconis crotalariae. 'Helminthosporium leaf spot.' **1**:FL(35).
Deuteromycotina - Coelomycetes
Colletotrichum gloeosporioides. Anthracnose. **1**:FL(35).
Dothiorella sp. On fronds. **1**:FL(35).
Macrophoma sp. Frond and stem necrosis. **1**:FL(35).
Pestalotiopsis palmarum. Leaf spot. **1**:FL(35).
Phomopsis sp. Leaf spot. **1**:FL(35).
Phyllosticta sp. Leaf spot. **1**:FL(35).
Deuteromycotina - Other
Rhizoctonia solani. Root and stem rot. **1**:FL(35).
Rhizoctonia sp. Root rot. **1**:HI(2903).

Coccothrinax Sarg.

1. *Coccothrinax* sp.
Ascomycotina
Didymella phacidiomorpha. On leaves. **1**:FL(82).

Cocos L.

1. *C. nucifera* L. [coconut palm]
Oomycetes
Phytophthora katsurae. **1**:HI(2903).
P. palmivora. Bud rot, wilt. **1**:FL(35,82), HI(2903).
Phytophthora sp. **1**:HI(2903).
Pythium sp. Root rot, wilt. **1**:FL(35,82).
Ascomycotina
Botryosphaeria disrupta. Leaf spot. **1**:FL(35).
B. quercuum. On roots and trunk. **1**:FL(35).
B. rhodina. On roots and trunk. **1**:FL(82).
Ceratocystis paradoxa. Stem bleeding, bleeding canker, leaf-bitten disease. **1**:FL(35,82), HI(2903).
Halosphaeria salina. On fruit mesocarp washed up on shore. **1**:NC(1943).
Mycosphaerella sp. Leaf spot. **1**:FL(35).
Physalospora abdita. On leaves. **1**:FL(82).
Torpedospora radiata. On fruit exocarp washed up on shore. **1**:NC(1943).
Valsa chlorina. **1**:FL(2782).
Basidiomycotina - Ustilaginales
Graphiola phoenicis. False smut. **1**:FL(35).
Basidiomycotina - Other
Ganoderma zonatum. Butt rot. **1**:FL(35).
Pellicularia koleroga. Thread blight. **1**:FL(82).
Deuteromycotina - Hyphomycetes
Alternaria sp. Leaf spot. **1**:FL(35).
Bipolaris incurvata. Leaf spot. **1**:FL(84).
Cercospora sp. Leaf spot. **1**:FL(35).
Curvularia lunata. Leaf spot. **1**:FL(35).
Cylindrocladium pteridis. Leaf spot. **1**:FL(35).
Endocalyx melanoxanthus. On petioles. **1**:HI(82,2903).
Epicoccum nigrum. Leaf spot. **1**:FL(35).
Fusarium moniliforme. Bud rot. **1**:FL(35).
F. oxysporum. Petiole and stem rot. **1**:FL(35).
F. solani. Bud and root rot. **1**:FL(35).
Graphium sp. Bud rot. **1**:HI(2899,2903,3074).
Helminthosporium sp. Leaf spot. **1**:FL(35).
Macrosporium cocos. Leaf spot. **1**:FL(35).
Oidium sp. Powdery mildew. **1**:FL(35).
Ramularia necator. Leaf spot. **1**:FL(35).
Deuteromycotina - Coelomycetes
Colletotrichum sp. Anthracnose. **1**:FL(35).
Cytospora palmarum. Leaf blight. **1**:FL(35).
Diplodia epicocos. Leaf and stalk rot. **1**:FL(35).
Lasiodiplodia theobromae. Nut rot. **1**:FL(35), HI(2903).
Lasmeniella cocoes. On leaves. **1**:FL(3549).
Macrophoma sp. Leaf spot. **1**:FL(35).
Melanconium sp. Leaf spot. **1**:FL(35), HI(2903).
Pestalotia sp. **1**:HI(2903).
Pestalotiopsis palmarum. Gray leaf spot. **1**:FL(35,82).
Phomopsis sp. Leaf spot. **1**:FL(35).
Phyllosticta palmetto. Leaf spot. **1**:FL(35).
Deuteromycotina - Other
Rhizoctonia solani. Root rot. **1**:FL(35).

Elaeis Jacq.

Oomycetes

Euterpe Mart.

1. *E. edulis* Mart. [assai palm]
Oomycetes
Phytophthora sp. Root rot. **1**:FL(35).
Pythium sp. Root rot. **1**:FL(35).
Deuteromycotina - Hyphomycetes
Corynespora sp. Leaf spot. **1**:FL(35).
Fusarium sp. Leaf spot. **1**:FL(35).
Helminthosporium sp. Leaf spot. **1**:FL(35).
Deuteromycotina - Coelomycetes
Colletotrichum sp. Leaf spot. **1**:FL(35).
Diplodia sp. On seeds. **1**:FL(35).

Geonoma Willd.

1. *G. interrupta* Mart.
Deuteromycotina - Hyphomycetes
Exserohilum rostratum. Seedling necrosis. **1**:FL(35).

Gronophyllum Scheff.

1. *Gronophyllum* sp.
Deuteromycotina - Hyphomycetes
Phaeotrichoconis crotalariae. Leaf spot. **1**:FL(558).

Howeia Becc.

1. *H. belmoreana* (C. Moore & F. v. Muell.) Becc. [Belmore palm, curly palm]
2. *H. forsteriana* (C. Moore & F. v. Muell.) Becc. [Forster palm, Forster sentry palm, kentia palm, thatch palm, thatch-leaf palm]
3. *Howeia* sp.

Oomycetes
Phytophthora sp. Root rot. **2:**FL(35).
Pythium splendens. Root rot. **2:**FL(35).
Ascomycotina
Nectria depauperata. **3:**OR(3202).
Deuteromycotina - Hyphomycetes
Alternaria sp. Leaf spot. **1:**FL(35). **2:**FL(35).
Cercospora sp. Leaf spot. **2:**FL(35).
Curvularia sp. Leaf spot. **2:**FL(35).
Cylindrocladium pteridis. Leaf spot. **2:**FL(35).
C. scoparium. Leaf spot. **2:**FL(35).
Fusarium sp. Leaf spot. **2:**FL(35).
Gliocladium vermoeseni. Stem rot. **2:**FL(35).
Helminthosporium sp. Leaf spot. **1:**FL(35). **2:**FL(35).
Stigmina palmivora. Leaf spot. **2:**FL(35).
Deuteromycotina - Coelomycetes
Colletotrichum sp. Leaf spot. **2:**FL(35).
Macrophoma sp. Leaf spot. **2:**FL(35).
Phyllosticta sp. Leaf spot. **2:**FL(35).
Deuteromycotina - Other
Rhizoctonia solani. Crown rot. **1:**FL(35). **2:**FL(35).

Hyophorbe Gaertn.

1. *Hyophorbe* sp.

Oomycetes
Pythium sp. Root rot. **1:**FL(35).
Deuteromycotina - Hyphomycetes
Cercospora sp. Leaf spot. **1:**FL(35).
Deuteromycotina - Other
Rhizoctonia sp. Root rot. **1:**FL(35).

Hyphaene Gaertn.

1. *H. thebaica* (L.) Mart. [doum palm, Egyptian doum, ginberbread-tree, gingerbread palm]

Deuteromycotina - Coelomycetes
Protostroma hyphaenea. On bark. **1:**FL(200).

Laccospadix Drude & H. Wendl.

1. *L. australasica* H. Wendl. & Drude (=*Calyptrocalyx australasicus* (H. Wendl. & Drude) Scheff.)

Oomycetes
Pythium sp. Root rot. **1:**FL(35).

Livistona R. Br.

1. *L. carinensis* (Chiov.) Dransf. & N. Uhl (=*Wissmannia carinensis* (Chiov.) Burret)
2. *Livistona* sp.

Oomycetes
Pythium sp. **1:**FL(35).
Deuteromycotina - Coelomycetes
Phomopsis arecae. Dieback. **2:**MD(82), NJ(82).
P. palmicola. Dieback. **2:**MD(82), NJ(82).

Nengella Becc.

1. *Nengella* sp.

Oomycetes
Pythium sp. Root rot. **1:**FL(35).

Paralinospadix Burret

This genus is considered a synonym of *Calyptrocalyx*, but the appropriate combination could not be located.

1. *P. hollrungii* (Becc.) Burret

Oomycetes
Pythium sp. Root rot. **1:**FL(35).

Phoenix L.

1. *P. canariensis* Hort. ex J. B. Chabaud [Canary Island date palm]
2. *P. dactylifera* L. [date palm]
3. *P. reclinata* Jacq. [Senegal date palm, wild date palm]
4. *P. roebelenii* O'Brien [dwarf date palm, miniature date palm, pygmy date palm, Roebelin palm]
5. *Phoenix* sp.

Oomycetes
Phytophthora palmivora. Bud rot. **1:**FL(35).
Pythium sp. Root rot. **1:**FL(35). **2:**FL(35). **4:**FL(35).
Zygomycotina
Rhizopus sp. **2:**CA(1110).
Ascomycotina
Anthostomella contaminans. **2:**CA(1110).
Botryosphaeria quercuum. Leaf spot. **1:**FL(35).
B. rhodina. On fronds. **1:**FL(82).
Ceratocystis paradoxa. Black scorch. **2:**AZ(82), CA(82,1110).
C. radicicola. Root rot, wilt. **2:**CA(82,302,1110,1691,3714).
Didymopleella sp. Leaf spot. **3:**FL(35).
Myxotrichum setosum. **2:**CA(1110).
Pleospora herbarum. Fruit rot. **2:**CA(82,1110).
Basidiomycotina - Ustilaginales
Graphiola phoenicis. False smut, leaf spot. **1:**Range of host(82), FL(35), GA(2405). **2:**Range of host(82), CA(1110), FL(35), HI(2903), MS(2720), TX(82). **3:**FL(35). **4:**FL(35). **5:**AL(285), WV(3907).
Basidiomycotina - Other
Armillaria mellea. Root rot. **1:**CA(82).
A. tabescens. Root rot. **1:**FL(35,82). **2:**FL(35). **3:**FL(35).
Atractogloea stillata. On spathes. **1:**CA(2627).
Ceriporia xylostromatoides. **1:**AZ(1202). **2:**AZ(1202).
Ganoderma zonatum. Root rot. **1:**FL(35). **3:**FL(35).
Grammothele fuligo. **2:**AZ(82), CA(82).
Marasmiellus pigmentatus. Decline disease. **2:**CA(82,1110).
Omphalia tralucida. Decline disease. **2:**CA(82,1110).
Oxyporus latemarginatus. Wood rot. **2:**AZ(82), CA(82).
Schizopora paradoxa. Wood rot. **2:**AZ(82), CA(82).
Trechispora farinacea. **1:**AZ(1202).
Tubulicium capitatum. On leaves. **1:**FL(341,444).
Deuteromycotina - Hyphomycetes
Alternaria citri. Brown spot of fruit. **2:**AZ(82), CA(82,1110), TX(82).
A. stemphylioides. Fruit spoilage. **2:**CA(82,1110).
Alternaria sp. Leaf spot. **1:**FL(35,82). **2:**AZ(82), CA(82,1110), TX(82). **3:**FL(35). **4:**FL(35).
Annellophora phoenicis. Leaf spot. **1:**TX(3725).
Aspergillus flavus. **2:**CA(1110).
A. niger. Calyx-end rot. **2:**CA(82,1110), WA(82).
Bipolaris setariae. Leaf spot. **4:**FL(35).
Catenularia fuliginea. Fruit rot. **2:**AZ(82), CA(82,1110).
Cephalosporium sp. **2:**CA(1110).
Cercospora sp. Leaf spot. **3:**FL(35). **4:**FL(35).
Chalaropsis sp. Rachis necrosis. **1:**FL(35).
Cladosporium sp. **2:**CA(1110).
Cylindrocladiella peruviana. Root rot. **4:**FL(35).
Dendryphiella vinosum. Brown spot of fruit. **2:**AZ(82), CA(82,1110).
Fusarium moniliforme. Inflorescence blight. **2:**AZ(82).
F. pallidoroseum. Inflorescence blight. **2:**AZ(82).
F. solani. Root rot. **4:**FL(35).
Fusarium sp. Crown and root rot, leaf spot. **1:**FL(35). **2:**CA(1110), FL(35). **3:**FL(35). **4:**FL(35).
Gliocladium roseum. Fruit rot. **2:**AZ(82), CA(82,1110).
G. vermoeseni. Canker. **1:**CA(82). **2:**CA(1110).
Helminthosporium sp. Leaf spot. **1:**FL(35). **3:**FL(35).
Phymatotrichopsis omnivora. Root rot. **2:**AZ(82), CA(82,1110).
Stemphylium sp. **2:**CA(1110).
Stigmina palmivora. Leaf spot. **1:**FL(35,82), LA(82), MS(2720), TX(82). **2:**Gulf States(82), FL(35), MS(2720). **3:**FL(35). **4:**FL(35).
Deuteromycotina - Coelomycetes
Colletotrichum gloeosporioides. Anthracnose, leaf spot. **2:**TX(82). **4:**FL(35).
Diplodia phoenicum. Shoot blight, stalk rot. **1:**FL(35). **2:**AZ(82), CA(82,1110).
Diplodia sp. Rachis lesions, crown rot. **2:**FL(35). **4:**FL(35).
Leptothyrium sp. Rachis spot. **4:**FL(35).
Macrophoma sp. Leaf spot. **4:**FL(35).
Pestalotia sp. Leaf spot. **1:**MS(2720). **2:**CA(82,1110), TX(82).
Pestalotiopsis palmarum. Leaf spot. **1:**FL(35,82). **2:**CA(1110), FL(35). **3:**FL(35). **4:**FL(35).

Phomopsis phoenicola. Fruit rot. **2**:CA(82,1110).
Phomopsis sp. Rachis spot. **1**:FL(35). **2**:FL(35). **3**:FL(35). **4**:FL(35).
Phyllosticta sp. Leaf and rachis spot. **1**:FL(35). **4**:FL(35).
Septoria sp. Leaf spot. **4**:FL(35).
Deuteromycotina - Other
Rhizoctonia solani. Seedling dieback, root and basal stem rot. **1**:FL(35). **3**:FL(35). **4**:FL(35).

Pholidocarpus Blume in Schult. & J. H. Schult.

1. *P. majadum* Becc.

Oomycetes
Pythium sp. Root rot. **1**:FL(35).

Pritchardia Seem. & H. Wendl. ex Wendl.

1. *P. martii* (Gaud.) H. Wendl.
2. *Pritchardia* sp.

Ascomycotina
Meliola lyoni. **1**:HI(2903).
M. palmicola. Black mildew. **2**:HI(1268,2903).
Mollisia cinerea. **2**:HI(2903).
Oxydothis pandanicola. On dead leaves. **2**:HI(2781).
Deuteromycotina - Hyphomycetes
Melanographium cookei. **2**:HI(2903).
Deuteromycotina - Coelomycetes
Phomopsis pritchardiae. On leaves. **2**:CA(82).

Ptychosperma Labill.

1. *P. macarthurii* (H. Wendl.) G. Nichols. [cluster palm, hurricane palm, Macarthur feather palm]

Deuteromycotina - Other
Rhizoctonia sp. Root rot. **1**:HI(2903).

Rhapis L. fil. ex Aiton

1. *R. excelsa* (Thunb.) A. Henry ex Rehd. [bamboo palm, dwarf ground rattan, fern rhapis, lady palm, miniature fan palm, rattan palm, slender lady palm]
2. *Rhapis* sp.

Oomycetes
Pythium sp. Root rot. **1**:FL(35). **2**:FL(35).
Deuteromycotina - Hyphomycetes
Cercospora sp. Leaf spot. **1**:FL(35). **2**:FL(35), HI(2903).
Cylindrocladiella parva. Root rot. **1**:FL(35).
Exserohilum rostratum. 'Helminthosporium leaf spot.' **1**:FL(35).
Fusarium moniliforme. Root rot. **1**:FL(35).
Fusarium sp. Leaf spot, wilt. **1**:FL(35). **2**:FL(35).
Helminthosporium sp. Leaf spot. **1**:FL(35), HI(2903). **2**:FL(35).
Phaeotrichoconis crotalariae. Leaf spot. **1**:FL(35).
Stigmina palmivora. Leaf spot. **1**:FL(35). **2**:FL(35).
Deuteromycotina - Coelomycetes
Pestalotiopsis palmarum. Leaf spot. **1**:FL(35). **2**:FL(35).
Phomopsis sp. Leaf spot. **2**:FL(35).
Phyllosticta sp. Leaf spot. **1**:FL(35).
Deuteromycotina - Other
Rhizoctonia solani. Crown rot. **1**:FL(35).

Roystonea Cook

1. *R. elata* (W. Bartram) F. Harper [Florida royal palm]
2. *Roystonea* sp.

Oomycetes
Phytophthora palmivora. Bud rot, wilt. **1**:FL(35). **2**:FL(82).
Pythium sp. Root rot. **1**:FL(35).
Ascomycotina
Glomerella sp. Leaf spot. **1**:FL(35).
Linocarpon versisporum. On leaf stalks. **1**:FL(35). **2**:FL(82).
Nectria cinnabarina. On bark. **1**:FL(35). **2**:FL(82).
Physalospora sp. On fronds. **1**:FL(35).
Basidiomycotina - Ustilaginales
Graphiola phoenicis. False smut. **1**:FL(35). **2**:FL(35,82).
Basidiomycotina - Other
Ganoderma zonatum. **1**:FL(35).
Deuteromycotina - Hyphomycetes
Alternaria sp. Leaf spot. **1**:FL(35). **2**:FL(35,82).
Cercospora sp. **1**:FL(35).
Epicoccum nigrum. Leaf spot. **1**:FL(35).
[*Exserohilum rostratum*]. Leaf spot. **1**:FL(35).
Fusarium oxysporum. Bud and heart rot. **1**:FL(35).
Helicoma palmigenum. On inflorescences and leaves. **2**:HI(1271).
Helminthosporium sp. Leaf stripe. **1**:FL(35). **2**:FL(35,82).
Septonema sp. Leaf spot. **1**:FL(35).
Stigmina palmivora. Leaf spot. **1**:FL(35).
Thielaviopsis sp. Black root rot. **1**:FL(35), HI(2903).
Deuteromycotina - Coelomycetes
Colletotrichum gloeosporioides. Anthracnose. **1**:FL(35). **2**:FL(82), TX(82).
Diplodia sp. Leaf spot. **1**:FL(35). **2**:FL(82).
Macrophomina phaseolina. Charcoal rot. **1**:HI(2903).
Melanconium palmarum. On leaf stalks. **1**:FL(35). **2**:FL(82).
Pestalotiopsis palmarum. Leaf spot. **1**:FL(35).
Phyllosticta sp. Leaf spot. **1**:FL(35).
Deuteromycotina - Other
Rhizoctonia solani. Root rot. **1**:FL(35).

Sabal Adans.

1. *S. etonia* Swingle ex Nash [Etonia palmetto, scrub palmetto]
2. *S. megacarpa* Small
3. *S. mexicana* Mart. (=*S. texana* (J. Cook) Becc.) [Mexican palmetto, Rio Grande palmetto, Texas palmetto, Victoria palmetto]
4. *S. minor* (Jacq.) Pers. [bluestem palmetto, bush palmetto, dwarf palmetto, scrub palmetto]
5. *S. palmetto* (Walter) Schult. & J. H. Schult. [cabbage palmetto, Florida cabbage palm]
6. *Sabal* sp.

Oomycetes
Pythium sp. Root rot. **5**:FL(35).
Ascomycotina
Acanthorhynchus sp. **4**:GA(1438). **5**:GA(1438).
Anthostomella leucobasis. On dead leaf blades and stalks. **4**:GA(1438). **5**:GA(1438). **6**:SE states(82), LA(82).
A. melanosticta. On dead leaf blades and stalks. **6**:SE states(82), LA(82).
A. minor. On dead leaf blades and stalks. **4**:GA(1438). **5**:GA(1438). **6**:SE states(82), FL(35), LA(82).
A. sabalensioides. On dead leaf blades and stalks. **6**:SE states(82), LA(82).
Asterina sabalicola. Black mildew, leaf spot. **4**:GA(1438). **5**:FL(35). **6**:FL(35,82), GA(82).
Catacauma sabal. Black spot. **1**:FL(82), GA(82). **3**:TX(82). **4**:FL(82), GA(82). **5**:FL(35,82), GA(82). **6**:FL(35).
Ceratocystis paradoxa. Stem bleeding. **5**:FL(35).
Diatrypella nitschkei. **5**:FL(2782).
Ellisiodothis inquinans. On dead leaves. **5**:FL(2782), GA(1438). **6**:FL(82), LA(82), MS(82).
Eutypella sabalina. On leaf stalks. **6**:AL(82), FL(35,82), GA(82).
Gloniopsis praelonga. **5**:FL(2782).
Hypoxylon nummularium var. *exutans*. **5**:FL(2782).
Hysterostomella sparsa. On leaf stalks. **5**:FL(35,195). **6**:FL(82).
Linocarpon palmetto. On leaf stalks. **5**:LA(184).
L. versisporum. On leaf stalks. **5**:FL(35). **6**:FL(35,82).
Massarina floridana. On leaf stalks. **5**:FL(2777).
Meliola amphitricha. Sooty mold. **5**:MS(2720). **6**:FL(634), MS(82).
M. bidentata. Sooty mold. **6**:FL(82).
M. furcata. Black mildew. **6**:FL(82).
M. palmicola. Sooty mold. **4**:GA(1438). **5**:Unk.(1452), MS(2720). **6**:Gulf States(82), HI(82).
M. sabalidis. Black mildew, sooty mold. **6**:FL(82).
Metasphaeria apiosporoides. On petioles. **5**:FL(2782).
M. palmetta. On petioles. **4**:GA(1438). **5**:FL(2782), GA(1438). **6**:GA(82).
Mollisia sabalidis. On petioles. **5**:FL(35). **6**:FL(82).
Mycothyridium deviatum. On petioles. **5**:FL(2777).
Myriangium sabaleos. Black speck, on leaves. **1**:FL(82,1745), GA(82). **4**:FL(82), GA(82). **5**:FL(35,82), GA(82). **6**:FL(35).
Oxydothis sabalensis. Black speck. **4**:GA(1438).
Venturia sabalicola. On dead leaves. **6**:LA(82).
Xylaria poitei. **3**:TX(3016).
Basidiomycotina - Ustilaginales
Graphiola congesta. False smut. **5**:FL(35,1074). **6**:AL(82), FL(35,82), SC(82).
G. phoenicis. False smut. **4**:MS(2720), TX(595). **5**:FL(35), MS(2720).

6:FL(35,82), MS(82).
G. thaxteri. False smut. **2**:FL(1074). **6**:FL(35,82).
Basidiomycotina - Other
Epithele sulphurea. On bark. **5**:FL(128), HI(341).
Favolaschia sabalensis. On dead leaves. **5**:FL(550).
Ganoderma zonatum. Butt and root rot. **5**:FL(35). **6**:FL(82), GA(82).
Hydropus sabalis. Butt and root rot. **5**:FL(3289).
Hyphoderma echinocystis. **5**:FL(443).
Hyphodontia lanata. On inflorescence. **5**:MS(443).
Irpex lacteus. On dead stems. **6**:FL(82).
Mycena sabali. On petioles. **6**:LA(3326).
Phanerochaete burtii. White rot. **5**:FL(433).
Rectipilus fasciculatus. **5**:FL(128).
Schizophyllum commune. On dead trunks. **6**:SC(82).
Tomentella botryoides. **6**:FL(2039).
Trametes cubensis. On dead trunks. **6**:FL(82).
Wolfiporia cocos. Root rot. **5**:FL(35). **6**:FL(35,82).
Wrightoporia avellanea. On wood. **5**:FL(35,128), SC(128). **6**:FL(82).
Deuteromycotina - Hyphomycetes
Alternaria sp. Root rot. **5**:FL(35).
Cercospora sp. Leaf spot. **6**:FL(35).
Curvularia lunata. Leaf blotch. **6**:FL(35).
Endocalyx melanoxanthus. **5**:FL(2782).
Fusarium solani. Root rot. **5**:FL(35).
Fusarium sp. Bud rot, leaf spot. **5**:FL(35). **6**:FL(35).
Helminthosporium spiculiferum. Bud rot, leaf spot. **4**:MS(2720). **5**:MS(2720). **6**:LA(82), MS(82).
Helminthosporium sp. Leaf spot. **5**:FL(35).
Mycoleptodiscus indicus. Frond necrosis, leaf spot. **5**:FL(35).
Stigmina palmivora. Frond necrosis, leaf spot. **5**:FL(35). **6**:FL(35).
Deuteromycotina - Coelomycetes
Amerosporium sabalinum. On dead leaves. **6**:LA(82).
Ascochyta sp. Leaf spot. **5**:FL(35).
Colletotrichum sp. Leaf spot. **5**:FL(35).
Coniothyrium sp. Leaf spot. **5**:FL(35).
Hendersonia sabaleos. On leaves. **4**:MS(2720). **5**:MS(2720). **6**:MS(82).
Leptothyrium sp. Leaf spot. **5**:FL(35).
Melanconium palmarum. On leaf stalks. **5**:FL(35). **6**:FL(82), GA(82).
Pestalotiopsis palmarum. On leaf stalks. **5**:FL(35). **6**:FL(35).
Phyllosticta palmetto. Leaf spot. **4**:MS(2720). **5**:LA(966), MS(2720). **6**:LA(82), MS(82).
Phyllosticta sp. Leaf spot. **5**:FL(35).
Rhabdospora sabalensis. On leafstalks. **6**:LA(82), SC(82).
Septoria sp. **4**:GA(82).
Sphaeropsis sabalicola. On leafstalks. **6**:AL(82).
Deuteromycotina - Other
Rhizoctonia solani. Root rot. **5**:FL(35).

Serenoa J. D. Hook. in Benth. & J. D. Hook.
1. *S. repens* (W. Bartram) Small (=*Sabal serrulata* (Michx.) Nutt. ex Schult. & J. H. Schult., *S. serrulata* (Michx.) G. Nichols.) [saw palmetto, scrub palmetto]
2. *Serenoa* sp.
Ascomycotina
Anthostomella consanguinea. **1**:FL(1098).
Cryptovalsa deusta. On petioles. **1**:FL(2782).
Gloniopsis lineolata. On living leaves. **1**:MS(2720).
Linocarpon versisporum. On dead petioles. **1**:FL(974).
Meliola palmicola. Black mildew. **1**:FL(1452), HI(1268), MS(2720). **2**:HI(2903).
Mycosphaerella serrulata. Leaf spot. **1**:FL(35).
Seynesia serrulata. On petioles. **1**:FL(2782).
Stictis serenoae. On leaves. **1**:FL(522).
Basidiomycotina - Other
Acanthophysium cerussatum. On wood. **1**:FL(2083).
Lazulinospora cinnamomea. **1**:FL(443).
Deuteromycotina - Hyphomycetes
Chalara longipes. On leaves. **1**:FL(3548).
Corynespora obclavata. On dead petioles. **2**:FL(928).
Endocalyx melanoxanthus. On petioles. **1**:FL(3548).
Gyrothrix microsperma. On dead leaves. **1**:FL(3548).
Menispora uncinata. On leaves. **1**:FL(3548).
Mycoleptodiscus indicus. On dead leaves. **1**:FL(3548).
Oncopodium simmonsii. On dead rachises. **1**:FL(3548).
Deuteromycotina - Coelomycetes
Botryodiplodia palmarum. **1**:FL(2782).
Endophragmiella boothii. On dead rachises. **1**:FL(3548).
E. dimorphospora. On leaves. **1**:FL(3548).

Syagrus Mart.
(= *Arecastrum* (Drude) Becc.)
1. *S. inajai* (Spruce) Becc. [pupunha palm]
2. *S. romanzoffianum* (Cham.) Glassman (=*Arecastrum romanzoffianum* (Cham.) Becc.) [queen palm]
3. *Syagrus* sp. (=*Arecastrum sp.*)
Oomycetes
Phytophthora palmivora. Bud rot, wilt. **2**:FL(35,82).
Pythium sp. Root rot. **1**:FL(35). **2**:FL(82).
Ascomycotina
Catacauma sp. Root rot, wilt. **2**:FL(35).
Glomerella cingulata. Leaf spot. **2**:NJ(82).
Glomerella sp. Leaf spot. **2**:FL(35).
Basidiomycotina - Ustilaginales
Graphiola phoenicis. False smut, leaf spot. **2**:Unk.(82), FL(35,82).
Basidiomycotina - Other
Ganoderma zonatum. White root and butt rot. **2**:FL(575).
Deuteromycotina - Hyphomycetes
Alternaria alternata. Leaf spot. **2**:FL(35).
Cercospora sp. Leaf spot. **2**:FL(35).
Chalara sp. Trunk rot. **2**:FL(35).
Cylindrocladium pteridis. Leaf spot. **2**:FL(35).
Fusarium moniliforme. Bud and leaf blight. **2**:FL(35). **3**:HI(2903).
Fusarium sp. Root rot. **2**:FL(35).
Gliocladium vermoeseni. Canker. **2**:CA(82).
Helicoma palmigenum. On inflorescence. **2**:HI(1268,1269).
Helicosporium panacheum. On wood. **3**:HI(2466).
Helminthosporium sp. Leaf spot. **2**:FL(35).
Stigmina palmivora. Leaf spot. **2**:FL(82).
Deuteromycotina - Coelomycetes
Colletotrichum sp. Anthracnose. **2**:FL(35).
Diplodia sp. Leaf spot. **2**:FL(35).
Dothiorella gregaria. Gummosis. **2**:CA(82).
Melanconium sp. Leaf spot. **2**:FL(35).
Pestalotiopsis palmarum. Leaf spot. **2**:FL(35).
Phomopsis sp. Leaf spot. **2**:FL(35).
Phyllosticta sp. Leaf spot. **2**:FL(35,82).
Septoria cocoina. Leaf spot. **2**:MO(82).

Thrinax Sw.
1. *Thrinax* sp.
Ascomycotina
Meliola palmicola. Black mold. **1**:HI(2903).
Deuteromycotina - Hyphomycetes
Cercospora sp. Leaf spot. **1**:FL(35).
Fusarium sp. Heart rot. **1**:FL(35).
Stigmina palmivora. Leaf spot. **1**:FL(35).
Deuteromycotina - Coelomycetes
Phomopsis sp. Leaf spot. **1**:FL(35).
Phyllosticta sp. Leaf spot. **1**:FL(35).

Trachycarpus H. Wendl. in J. Gay
1. *T. fortunei* (Hook.) H. Wendl. [Chinese windmill palm, hemp palm, windmill palm]
Oomycetes
Pythium sp. Root rot. **1**:FL(35).
Ascomycotina
Botryosphaeria sp. Leaf spot. **1**:FL(35).
Glomerella sp. Leaf spot. **1**:FL(35).
Deuteromycotina - Hyphomycetes
Alternaria sp. Leaf spot. **1**:FL(35).
Cercospora sp. Leaf spot. **1**:FL(35).
Fusarium sp. Root rot. **1**:FL(35).
Deuteromycotina - Coelomycetes
Macrophoma sp. Leaf necrosis. **1**:FL(35).
Pestalotia sp. Leaf spot. **1**:FL(35).
Phomopsis arecae. On dead leaves. **1**:CA(81).
Phyllosticta sp. Leaf spot. **1**:FL(35).

Veitchia H. Wendl. in Seem.

1. *V. merrillii* (Becc.) H. Moore [Christmas palm, Manila palm, Merrill palm]
2. *Veitchia* sp.

Oomycetes
Pythium sp. Root rot. **1**:FL(35).
Deuteromycotina - Hyphomycetes
Alternaria sp. Leaf spot. **1**:FL(35).
Cercospora sp. Leaf spot. **1**:FL(35).
Fusarium sp. Bud rot. **1**:FL(35).
Helminthosporium sp. Leaf spot. **1**:FL(35). **2**:FL(35).
Stigmina palmivora. Leaf spot. **1**:FL(35). **2**:FL(35).
Deuteromycotina - Coelomycetes
Colletotrichum sp. Leaf spot. **1**:FL(35).
Pestalotiopsis palmarum. Leaf spot. **1**:FL(35). **2**:FL(35).
Phomopsis sp. Leaf spot. **1**:FL(35).
Phyllosticta sp. Leaf spot. **1**:FL(35). **2**:FL(35).
Deuteromycotina - Other
Rhizoctonia solani. Root and stem rot. **1**:FL(35). **2**:FL(35).
Rhizoctonia sp. Root rot. **1**:HI(2903).

Washingtonia H. Wendl.

1. *W. filifera* (Linden ex André) H. Wendl. [California fan palm, California palm, California Washington palm, desert fan palm, petticoat palm]
2. *W. robusta* H. Wendl. [desert palm, Mexican Washington palm, thread palm]
3. *Washingtonia* sp.

Oomycetes
Phytophthora nicotianae var. *parasitica*. Trunk rot. **1**:CA(775), FL(775). **3**:AZ(3864).
P. palmivora. Bud and trunk rot. **2**:FL(35,116). **3**:AZ(3864), FL(35).
Phytophthora sp. Collar rot, stem rot. **2**:CA(1864). **3**:AZ(82).
Pythium sp. Bud, collar and stem rot. **2**:FL(35). **3**:FL(35,82).
Ascomycotina
Auerswaldia sp. Leaf spot. **3**:CA(82).
Euryachora neowashingtonianae. On leaf stalks. **2**:FL(35). **3**:FL(35,82).
Metasphaeria washingtoniae. On dead leaves. **3**:CA(82).
Mycosphaerella tassiana. On leaves. **3**:CA(82).
Sphaerodothis neowashingtoniae. On leaves and leafstalks. **1**:CA(81). **3**:CA(82).
Basidiomycotina - Ustilaginales
Graphiola phoenicis. On leaves. **2**:FL(35). **3**:FL(35,82), TX(82).
Basidiomycotina - Other
Armillaria tabescens. Root rot. **2**:FL(35). **3**:FL(35,82).
Ceriporia xylostromatoides. **1**:AZ(1202).
Ganoderma zonatum. On wood. **2**:FL(35). **3**:FL(35).
Henningsomyces candidus. **1**:AZ(1202).
Marasmiellus pigmentatus. On decaying leaf bases. **3**:CA(82).
Omphalia tralucida. On decaying leaf bases. **3**:CA(82).
Deuteromycotina - Hyphomycetes
Cercospora sp. Leaf spot. **3**:FL(35,82).
Cylindrocladium pteridis. Leaf spot. **2**:FL(35,1694). **3**:FL(35,82,3221).
[*Exserohilum rostratum*]. Leaf spot. **3**:FL(35).
Gliocladium vermoeseni. Trunk rot. **2**:CA(506).
Phymatotrichopsis omnivora. Root rot. **3**:TX(82).
Stigmina palmivora. Leaf spot. **2**:FL(35).
Deuteromycotina - Coelomycetes
Colletotrichum sp. Anthracnose, leaf spot. **3**:FL(35,82).
Pestalotiopsis palmarum. Anthracnose. **3**:FL(35,82).
Phomopsis palmicola. Leaf spot. **2**:FL(35). **3**:TX(82).
P. pritchardiae. Leaf spot. **1**:WA(3202). **3**:CA(82).
Phomopsis sp. On leaves. **3**:TX(82).
Deuteromycotina - Other
Rhizoctonia sp. Seedling root rot. **2**:FL(35).
Sclerotium rolfsii. Southern blight of seedlings. **2**:FL(35).

Aristolochiaceae

Aristolochia L.

1. *A. clematitis* L.
2. *A. durior* J. Hill [Dutchman's-pipe, pipevine]
3. *A. macrophylla* Lam.
4. *A. serpentaria* L. [Virginia snakeroot]
5. *Aristolochia* sp.

Oomycetes
Pythium splendens. Root rot. **5**:FL(35).
Ascomycotina
Botryosphaeria obtusa. On stems. **1**:OH(82).
Basidiomycotina - Other
Hymenochaete tabacina. On dead limbs. **5**:CA(82).
Deuteromycotina - Hyphomycetes
Botrytis cinerea. Gray mold blight. **2**:CT(82), MD(82).
Cercospora guttulata. Gray mold blight. **2**:WV(82). **3**:WV(3907).
C. serpentaria. Leaf spot. **2**:WV(82). **4**:AL(82), CT(82), DE(82,581,962,3543).
Ovularia aristolochiae. Leaf spot. **2**:WV(82,128). **3**:WV(3907).
Deuteromycotina - Coelomycetes
Aplosporella squieriae. On stems. **5**:MA(2783), NY(82).
Colletotrichum gloeosporioides. Leaf spot. **2**:MA(82).
Diplodia radicicola. Root rot. **4**:Unk.(82).
Phyllosticta aristolochiae. Leaf spot. **2**:NJ(82).
Deuteromycotina - Other
Sclerotium rolfsii. Stem rot. **5**:FL(35).

Asarum L.

1. *A. canadense* L. [snakeroot]
2. *A. caudatum* Lindl.
3. *A. hartwegii* S. Wats.
4. *A. lemmonii* S. Wats.
5. *A. virginicum* L.
6. *Asarum* sp.

Chytridiomycetes
Synchytrium asari. Leaf gall. **1**:CA(82), MI(82), MN(2874), WI(82). **2**:CA(82). **3**:CA(638). **5**:CA(82).
Ascomycotina
Laestadia asarifolia. Leaf spot. **6**:SC(82).
Mycosphaerella sp. On leaves. **5**:AL(82).
Sclerotinia sclerotiorum. Rhizome rot. **1**:NY(82).
Basidiomycotina - Uredinales
Puccinia asarina. Autoecious. **2**:CA(81,82), ID(82,3202), OR(82,3202), WA(82,3202). **3**:CA(642). **4**:CA(642).
Deuteromycotina - Coelomycetes
Ascochyta versicolor. Leaf spot. **2**:ID(82,3202).
Ascochyta sp. Leaf spot. **1**:WI(1361).
Colletotrichum sp. **1**:WI(1360).
Septoria asaricola. **2**:ID(3202), WA(3202).

Asclepiadaceae

Asclepias L.

1. *A. amplexicaulis* Sm.
2. *A. angustifolia* Sessé & Moc.
3. *A. asperula* (Decne.) Woodson subsp. *capricornu* (Woodson) Woodson (=*Asclepiodora decumbens* (Nutt.) A. Gray)
4. *A. brachystephana* Engelm.
5. *A. curassavica* L. [bloodflower milkweed, false ipecac, redhead]
6. *A. eriocarpa* Benth.
7. *A. erosa* Torr.
8. *A. exaltata* L. (=*A. phytolaccoides* Pursh)
9. *A. hirtella* (Pennell) Woodson (=*Acerates hirtella* Pennell)
10. *A. incarnata* L. [swamp milkweed]
11. *A. lanuginosa* Nutt.
12. *A. mexicana* Cav.
13. *A. obtusifolia* Michx.
14. *A. ovalifolia* Decne.
15. *A. pumila* (A. Gray) Vail
16. *A. purpurascens* L. [purple milkweed]
17. *A. rubra* L.
18. *A. speciosa* Torr.
19. *A. subverticillata* (A. Gray) Vail
20. *A. sullivantii* Engelm.
21. *A. syriaca* L. (=*A. cornuti* Decne.) [common milkweed]
22. *A. tuberosa* L. [butterfly-weed, chigger-flower, Indian paintbrush, pleurisy root, tuber-root]
23. *A. variegata* L.
24. *A. verticillata* L. (=*A. galioides* H. B. K.) [eastern whorled

milkweed, horsetail milkweed]
25. *A. viridiflora* Raf. (=*Acerates viridiflora* (Raf.) Eaton, *Acerates viridiflora* (Raf.) Eaton var. *lanceolata* (Ives) Torr.)
26. *A. viridis* Walter (=*Asclepiodora viridis* (Walter) A. Gray)
27. *Asclepias* sp.

Chytridiomycetes

Physoderma sp. On stems. **10**:MI(3400,3409).

Ascomycotina

Comoclathris permunda. On stems. **10**:KS(82). **27**:CA(82).
Diaporthe arctii. On stems. **21**:IA(82), WV(82,3907). **22**:NJ(82). **23**:GA(1438). **27**:GA(82).
Didymella cornuta. On stems. **21**:NY(82).
Ditopella asclepiadea. **12**:CA(81).
Erysiphe cichoracearum. Powdery mildew. **21**:OH(82). **22**:WI(1343). **27**:IL(82).
Glomerella cingulata. Anthracnose, stem blight, leaf spot. **21**:NC(1303).
Guignardia sp. Anthracnose, stem blight, leaf spot. **27**:GA(1438).
Leptosphaeria asclepiadis. **13**:NY(3249).
L. dearnessii. **21**:NY(3249).
L. doliolum. **10**:NY(3249).
Microsphaera diffusa. **22**:WI(1343).
Phyllactinia guttata. Powdery mildew. **21**:NY(82).
Rosellinia aquila. Root rot. **27**:TX(82).
Sphaerotheca sparsa. On leaves. **27**:AZ(370).

Basidiomycotina - Uredinales

Puccinia chloridis. Heteroecious: aecial host. **3**:AZ(3981), OK(82,2872), TX(82). **4**:AZ(1203,3981). **9**:WI(1352). **15**:AZ(1203,3981). **18**:C & W states(82), SD(2243). **19**:AZ(1203,3981). **21**:C & W states(82), MN(2874), SD(2243). **22**:C & W states(82), OK(2872). **24**:C & W states(82), IL(312), SD(2243). **26**:KS(82), OK(82,2872). **27**:C & W states(82), FL(35).
P. obliqua. **7**:AZ(3981).
P. seymouriana. **16**:WI(1353). **18**:ND(82,378), SD(82,2243). **21**:MO(82), WI(82). **23**:NC(3944). **27**:NY(82).
P. vexans. Heteroecious: aecial host. **22**:WI(82). **24**:WI(1343). **25**:WI(1332,1344,1348).
Uromyces asclepiadis. Autoecious. **3**:NM(82), TX(82). **5**:FL(35). **10**:Unk.(82). **18**:Unk.(82). **21**:Unk.(82), MN(2874), MS(2720), WI(1335), WV(3907). **22**:Unk.(82). **24**:Unk.(82). **26**:KS(82). **27**:Unk.(82), FL(35).

Deuteromycotina - Hyphomycetes

Alternaria alternata. Leaf spot. **26**:OK(82,2872).
Alternaria sp. Leaf spot. **10**:VT(82). **21**:MI(82), WV(82,3907). **27**:FL(35), SC(82).
Asperisporium asclepiadis. On leaves. **6**:CA(81,82). **21**:MI(82).
Cercospora asclepiadis. On leaves. **2**:AZ(1194). **21**:DE(3543), KY(581). **22**:WI(1338). **25**:DE(3543).
C. asclepiodorae. Leaf spot. **26**:AL(581), KS(82,581).
C. briareus. **9**:WI(1338). **11**:WI(1331). **25**:MD(3543).
C. clavata. Leaf spot. **10**:C & W states(82), DE(3543), MS(2720), OH(82), OK(2873), WV(3907). **12**:OR(3202), WA(3202). **14**:WI(1338,1342). **16**:WI(1343). **18**:C & W states(82), CA(81), OK(2872), WY(1195). **21**:C & W states(82), CA(642), MA(1390), MS(2720), WV(3907). **22**:C states(82), W states(82). **24**:WI(1331). **27**:FL(35), WI(82).
C. elaeochroma. Leaf spot. **1**:WI(1334). **18**:ND(82,378). **20**:WI(1358). **21**:ND(82). **27**:WI(82).
C. hansenii. **21**:CA(82). **27**:CA(82).
C. venturioides. Leaf spot. **5**:FL(35). **18**:WA(82). **21**:IA(82), IL(82), IN(82), MA(82), MO(82), NY(82), TX(82). **27**:FL(82), NC(82), NY(82), TX(82).
Cladosporium herbarum. **21**:DE(3543).
C. molle. On leaves. **27**:SC(634).
Macrosporium sp. **21**:DE(3543).
Phymatotrichopsis omnivora. Root rot. **21**:TX(82). **22**:TX(82). **27**:TX(82).

Deuteromycotina - Coelomycetes

Ascochyta asclepiadicola. **22**:WI(1342).
A. asclepiadis. Leaf spot. **11**:WI(1357). **20**:WI(1344). **21**:DE(82), WI(1356).
Colletotrichum fusarioides. Anthracnose. **1**:WI(1357). **18**:NM(82), UT(82). **20**:WI(1339). **21**:KS(82), MA(82), MT(82), NJ(82), NY(82), PA(82), TX(82).
C. truncatum. Isolated from leaves and stems. **21**:IL(1482).
Colletotrichum sp. **8**:WI(1360).
Diplodia asclepiadea. On stems. **21**:NJ(82).
Dothiorella hawaiiensis. On dead inflorescences. **5**:HI(2784).
Macrophomina phaseolina. Charcoal rot. **21**:IL(311).
Phoma asclepiadea. Stem blight. **18**:UT(82). **21**:NY(82), WV(82,128,3907). **27**:TX(82).
Phoma sp. On dead stems. **21**:WV(3907).
Phomopsis missouriensis. On stems. **27**:MO(82).
Phyllosticta cornuti. Leaf spot. **21**:IL(82), KS(82), MI(82), WV(3907).
P. tuberosa. Leaf spot. **3**:OK(82,2872). **10**:DE(82). **22**:IL(82), NJ(82), WI(1354).
Phyllosticta sp. Leaf spot. **5**:FL(35). **8**:WI(1355). **10**:WI(1353).
Septoria asclepiadicola. Leaf spot. **8**:WI(1338). **10**:NE states(82), OK(82,2873), WI(82), WV(3907). **17**:MO(966). **21**:MD(82), MO(82), PA(82).
S. asclepiadicola f. *syriaca*. Leaf spot. **21**:MN(2874).
S. cryptotaeniae. **10**:IA(82). **21**:IA(82), TX(82). **27**:TX(82).
S. incarnata. **10**:KS(82). **21**:IN(82).
Sphaeropsis sphaerospora. On stems. **21**:NY(82).
Stagonospora zonata. Leaf spot. **8**:WI(1343,1357). **9**:WI(1341). **14**:WI(1338). **21**:WI(82). **22**:WI(1340). **24**:WI(1344). **25**:WI(1338).
Stagonospora sp. **11**:WI(1344).

Deuteromycotina - Other

Rhizoctonia solani. Root rot. **21**:TX(82). **27**:TX(82).

Calotropis R. Br.

1. *C. gigantea* (L.) Dryander ex W. T. Aiton [bowstring hemp, crownflower, crownplant, giant milkweed, madar, mudar]
2. *C. procera* (Aiton) Dryander ex W. T. Aiton [auricula-tree, dumb-cotton, madar, small crownflower, small mudar]

Basidiomycotina - Uredinales

Puccinia obliqua. **2**:AZ(3981).
Uromyces asclepiadis. Autoecious. **1**:FL(35).

Cynanchum L.

1. *C. arizonicum* (A. Gray) Shinners (=*Metastelma arizonicum* A. Gray)
2. *C. barbigerum* (Scheele) Shinners (=*Metastelma barbigerum* Scheele)
3. *C. laeve* (Michx.) Pers. (=*Ampelamus albidus* (Nutt.) Britton, *Gonolobus laevis* Michx.) [angle-pod, bluevine, honeyvine, sandvine]
4. *C. palustre* (Pursh) A. Heller (=*Seutera palustris* (Pursh) Vail, *Vincetoxicum palustre* (Pursh) A. Gray, *Seutera maritima* Decne. in DC.)
5. *Cynanchum* sp.

Oomycetes

Plasmopara gonolobi. Downy mildew. **3**:SC(82).

Ascomycotina

Meliola bidentata. Black mildew. **3**:NC(82).

Basidiomycotina - Uredinales

Puccinia obliqua. Autoecious. **1**:AZ(1194,1203,3981). **2**:TX(742). **3**:OK(82,2872), TX(82). **5**:FL(35).

Deuteromycotina - Hyphomycetes

Cercospora gonolobi. Leaf spot. **3**:IN(663), OK(82,2872).

Deuteromycotina - Coelomycetes

Phyllosticta sp. Leaf spot. **4**:FL(35).
Septoria sp. Leaf spot. **3**:LA(82).

Edithcolea N. E. Brown

1. *E. grandis* N. E. Brown

Deuteromycotina - Hyphomycetes

Cercospora sp. Basal stem rot. **1**:FL(35).

Deuteromycotina - Coelomycetes

Phomopsis sp. Basal stem rot. **1**:FL(35).

Gonolobus Michx.

1. *Gonolobus* sp.

Chytridiomycetes

Synchytrium gonolobi. Leaf gall. **1**:LA(627).

Hoya R. Br.

1. *H. carnosa* (L. fil.) R. Br. [honeyplant, waxplant]
2. *Hoya* sp.

Oomycetes

Phytophthora sp. Cutting and root rot. **1**:FL(35). **2**:FL(35).
Pythium splendens. Root rot. **2**:FL(35).

Deuteromycotina - Hyphomycetes

Alternaria sp. Leaf spot. **1**:FL(35). **2**:FL(35).
Cercospora sp. Leaf spot. **1**:FL(35).
Corynespora sp. Leaf spot. **1**:FL(35). **2**:FL(35).
Cylindrocladium sp. Leaf spot. **1**:FL(35). **2**:FL(35).
Fusarium sp. Root rot. **1**:FL(35).
Myrothecium sp. Leaf cutting rot. **2**:FL(35).

Deuteromycotina - Coelomycetes

Colletotrichum gloeosporioides. Anthracnose. **1**:FL(35). **2**:FL(35).
Macrophoma sp. Leaf spot. **2**:FL(35).
Phyllosticta sp. Leaf spot. **1**:FL(35). **2**:FL(35).
Pyrenochaeta sp. Leaf spot. **1**:FL(35).

Deuteromycotina - Other

Rhizoctonia solani. Leaf blight, root rot. **1**:'Rhizoctonia aerial blight.' FL(35). **2**:FL(35).

Huernia R. Br.

1. *Huernia* sp.

Oomycetes

Pythium sp. Basal stem rot. **1**:FL(35).

Matelea Aubl.

1. *M. carolinensis* (Jacq.) Woodson (=*Gonolobus carolinensis* (Jacq.) R. Br., *Gonolobus hirsutus* Michx., *Vincetoxicum hirsutum* (Michx.) Britton)
2. *M. gonocarpos* (Walter) Shinners (=*Gonolobus gonocarpos* (Walter) L. M. Perry, *Vincetoxicum gonocarpos* Walter)
3. *M. producta* (Torr.) Woodson (=*Gonolobus productus* Torr.)
4. *M. suberosa* (L.) Shinners

Oomycetes

Plasmopara gonolobi. Downy mildew. **2**:MS(82,2720). **4**:FL(35).

Basidiomycotina - Uredinales

Puccinia obliqua. Autoecious. **1**:MS(2720). **2**:MS(82,2720), SC(82). **3**:AZ(1203,3981). **4**:FL(35).
Uromyces asclepiadis. **2**:WV(82).

Deuteromycotina - Hyphomycetes

Cercospora punctiformis. Leaf spot. **1**:AL(82,581), TX(82).

Morrenia Lindl.

1. *M. odorata* (Hook. & Arnott) Lindl. [strangler-vine, tasi]

Oomycetes

Phytophthora palmivora. Leaf blight, root rot, girdled stem. **1**:FL(35,452,1025).

Basidiomycotina - Uredinales

Puccinia obliqua. Autoecious. **1**:FL(35).

Deuteromycotina - Hyphomycetes

Cercospora punctiformis. Leaf blight, root rot, girdled stem. **1**:FL(35).
Fusarium sp. Stem necrosis. **1**:FL(35).
Oidium sp. Powdery mildew. **1**:FL(35).

Deuteromycotina - Coelomycetes

Phyllosticta sp. Leaf spot. **1**:FL(35).

Deuteromycotina - Other

Rhizoctonia sp. Leaf necrosis. **1**:FL(35).

Pachycarpus E. Mey.

1. *P. dealbatus* E. Mey.

Ascomycotina

Erysiphe polygoni. Powdery mildew. **1**:CA(1155).

Philibertia Kunth

1. *Philibertia* sp.

Ascomycotina

Phyllactinia guttata. Powdery mildew. **1**:FL(82).

Basidiomycotina - Uredinales

Puccinia chloridis. **1**:AZ(82).
P. obliqua. Autoecious. **1**:AZ(82), CA(82), NM(82), TX(82).

Sarcostemma R. Br.

1. *S. clausum* (Jacq.) Roem. & Schult. (=*Philibertia clausa* (Jacq.) K. Schum. in Engl. & Prantl)
2. *S. cynanchoides* Decne. (=*Funastrum cynanchoides* (Decne.) Schlechtend., *Philibertia cynanchoides* (Decne.) A. Gray)
3. *S. cynanchoides* Decne. var. *hartwegii* (Vail) Shinners
4. *S. heterophyllum* Engelm. ex Torr. (=*Funastrum heterophyllum* (Engelm. ex Torr.) Standl.)
5. *S. hirtellum* (A. Gray) R. Holm

Ascomycotina

Phyllactinia guttata. Powdery mildew. **1**:FL(35).

Basidiomycotina - Uredinales

Puccinia chloridis. **3**:AZ(3981). **4**:AZ(1203).
P. obliqua. Autoecious. **1**:FL(35,82). **2**:AZ(1203,3981), OK(2872). **3**:AZ(3981). **4**:AZ(1203). **5**:AZ(747,3981), CA(747).

Deuteromycotina - Coelomycetes

Colletotrichum sp. Leaf spot. **1**:FL(35).
Phomopsis sp. Leaf spot. **1**:FL(35).

Stapelia L.

1. *Stapelia* sp.

Oomycetes

Phytophthora sp. Root rot. **1**:FL(35).
Pythium sp. Root rot. **1**:FL(35).

Deuteromycotina - Hyphomycetes

Bipolaris cactivora. Stem rot. **1**:FL(35).
Fusarium oxysporum. Stem rot. **1**:FL(35).

Deuteromycotina - Coelomycetes

Colletotrichum gloeosporioides. Leaf spot. **1**:FL(35).
Diplodia sp. **1**:OK(2872).
Phomopsis sp. Stem necrosis. **1**:FL(35).
Phyllosticta sp. Stem rot. **1**:FL(35).

Stephanotis Thouars

1. *S. floribunda* Brongn.

Ascomycotina

Glomerella cingulata. **1**:WA(3202).

Deuteromycotina - Hyphomycetes

Botrytis elliptica. Flower blight. **1**:CA(3652).
Oidium sp. Powdery mildew. **1**:HI(2903).

Deuteromycotina - Coelomycetes

Gloeosporium sp. **1**:Unk.(1300).

Deuteromycotina - Other

Rhizoctonia sp. Crown and root rot. **1**:HI(2903).

Vincetoxicum C. Wolf

1. *Vincetoxicum* sp.

Oomycetes

Plasmopara gonolobi. Downy mildew. **1**:E states(82), TX(82).

Basidiomycotina - Uredinales

Puccinia obliqua. Autoecious. **1**:E states(82), AZ(82), OK(82,2872).
Uromyces asclepiadis. Autoecious. **1**:IN(82).

Deuteromycotina - Hyphomycetes

Cercospora bellynckii. Leaf spot. **1**:TX(82).
Phymatotrichopsis omnivora. Root rot. **1**:TX(82).

Asteraceae

Acanthospermum Schrank

1. *A. australe* (Loefl.) Kuntze

Oomycetes

Bremia sp. Downy mildew. **1**:FL(35).
Plasmopara halstedii. Downy mildew. **1**:FL(35).

Deuteromycotina - Coelomycetes

Phyllosticta acanthospermi. Leaf spot. **1**:FL(35).

Achillea L.

1. *A. borealis* Bong.
2. *A. lanulosa* Nutt. (=*A. millefolium* L. subsp. *lanulosa* (Nutt.) Piper)
3. *A. millefolium* L. [common yarrow, milfoil, nose-bleed, sanguinary, thousand-seal]
4. *A. ptarmica* L. [sneezeweed, sneezewort]
5. *A. sibirica* Ledeb. (=*A. multiflora* Hook.)
6. *Achillea* sp.

Ascomycotina
Crocicreas coronatum. On dead stems. **4:**NY(505).
Erysiphe cichoracearum. Powdery mildew. **3:**MO(82), MT(82,3202), PA(82), SD(82), VT(82), WI(82). **6:**AK(82).
E. communis. **1:**AK(526). **3:**WY(2924).
Leptosphaeria millefolii. **3:**MI(3249), NY(3249).
L. multiseptata. **3:**WA(3202,3814).
Lophiostoma herbarum. **3:**WA(3202,3815).
Ophiobolus nigro-clypeata. **3:**GA(1438).
Pleospora helvetica. **3:**WA(3202,3814).
P. richtophensis. **3:**WY(3810).
Basidiomycotina - Uredinales
Puccinia cnici-oleracei. **1:**AK(60,526). **2:**CA(81,82), CO(82), ID(82), MT(82), NM(82), OR(82), TX(82), UT(82), WA(82), WY(82). **3:**CA(81,82), CO(82), ID(82,3202), MT(82,3202), NM(82), OR(82,3202), TX(82), UT(82), WA(82,3202), WY(82,1195). **5:**AK(526).
Basidiomycotina - Ustilaginales
Entyloma compositarum. **2:**WA(1077). **3:**ID(3202), WA(3202).
Deuteromycotina - Hyphomycetes
Alternaria sp. Leaf spot. **3:**FL(35), MT(3202), WA(3202).
Cercosporella sp. **3:**WA(3202).
Cladosporium sp. **3:**WI(1347).
Oidium sp. Powdery mildew. **3:**FL(35).
Phymatotrichopsis omnivora. Root rot. **3:**TX(82).
Deuteromycotina - Coelomycetes
Camarosporium compositarum. On stems. **3:**CA(82). **6:**CA(82).
C. kriegeri. **3:**ID(3202), WA(3202,3816).
Colletotrichum sp. Anthracnose. **3:**FL(35).
Heteropatella umbilicata. **3:**WA(3202).
Phoma erysiphoides. On stems. **3:**CO(82).
Septoria achilleae. On living leaves. **3:**WI(1346).
Deuteromycotina - Other
Rhizoctonia solani. Root rot. **3:**Range of host(82).

Adenocaulon Hook.

1. *A. bicolor* Hook.

Basidiomycotina - Uredinales
Coleosporium adenocaulonis. **1:**ID(3202), OR(746,3202), WA(3202).
Deuteromycotina - Hyphomycetes
Ramularia sp. **1:**ID(3202).
Deuteromycotina - Coelomycetes
Septoria adenocauli. **1:**ID(3202).

Ageratina Spach

1. *A. aromatica* (L.) Spach
2. *A. jucunda* (Greene) Clewell & Wooten

Basidiomycotina - Uredinales
Aecidium ageratinae. **1:**MS(3101), TN(3101). **2:**FL(3101).

Ageratum L.

1. *A. conyzoides* L. [whiteweed]
2. *A. houstonianum* Mill. [ageratum]
3. *Ageratum* sp.

Oomycetes
Pythium mamillatum. Root rot. **2:**Unk.(82).
Pythium sp. Root rot. **3:**FL(35).
Ascomycotina
Erysiphe cichoracearum. Powdery mildew. **1:**MD(82).
Basidiomycotina - Uredinales
Puccinia conoclinii. Autoecious. **1:**FL(35), GA(82,752), HI(2903), MS(82,2720), NC(82,1303). **2:**TN(752). **3:**GA(82), HI(82), MS(82), NC(82), OH(946), PA(1875).
Deuteromycotina - Hyphomycetes
Alternaria sp. Collar rot. **3:**FL(35).
Cercospora aciculina. Leaf spot. **1:**HI(2903).
Mycovellosiella perfoliati. Leaf spot. **1:**HI(82,2903).
Deuteromycotina - Other
Rhizoctonia solani. Root rot. **1:**IL(82), NJ(82), NM(82,2107). **3:**Il(82), NJ(82), NM(82).
Rhizoctonia sp. Root and stem rot. **3:**HI(2903).
Sclerotium rolfsii. Southern blight. **1:**NC(82,1303), NJ(82). **2:**NC(82,1303), NJ(82). **3:**FL(35), MS(2720).

Agoseris Raf.

1. *A. apargioides* (Less.) Greene
2. *A. aurantiaca* (Hook.) Greene (=*A. gracilens* (A. Gray) Kuntze)
3. *A. cuspidata* (Pursh) Raf.
4. *A. elata* (Nutt.) Greene
5. *A. glauca* (Pursh) Raf. (=*A. laciniata* (Nutt.) Greene, *A. parviflora* (Nutt.) Greene, *A. turbinata* Rydb., *A. scorzoneraefolia* (Schrad.) Greene)
6. *A. grandiflora* (Nutt.) Greene
7. *A. heterophylla* (Nutt.) Greene (=*A. major* Jeps. ex Greene)
8. *A. plebia* Greene
9. *A. purpurea* Greene
10. *Agoseris* sp.

Oomycetes
Bremia lactucae. Southern blight. **5:**OR(3202), WA(3202). **6:**ID(3202), WA(3202). **7:**ID(3202), WA(3202). **10:**CO(82), WA(82).
Ascomycotina
Leptosphaeria erigerontis. **2:**WY(2924).
Sphaerotheca fuliginea. Powdery mildew. **10:**CA(82), CO(82), MT(82), NE(82), UT(82), WA(82), WY(82).
S. macularis. Powdery mildew. **2:**ID(3202), WA(3202). **4:**WA(3202). **5:**MT(3202), WY(1195). **7:**CA(81), ID(3202), OR(3202), WA(3202). **10:**CA(82), CO(82), MT(82), NE(82), UT(82), WA(82), WY(82).
Basidiomycotina - Uredinales
Puccinia agoseridis. **2:**CO(2713), WA(2713). **6:**CA(2713). **8:**CA(2713).
P. cnici-oleracei. Autoecious. **5:**MT(3202), OR(3202), WA(3202). **10:**WA(82,2713), WY(82).
P. crepidis-montanae. Autoecious. **5:**ID(3202), MT(3202), WA(3202).
P. dioicae. Heteroecious: aecial host. **2:**CO(649). **5:**MT(1195,3202), WY(1195). **10:**CO(82), MT(82), WA(3202), WY(82).
P. hieracii. Autoecious. **2:**CO(649), ID(3202), MT(3202), WA(3202), WY(2924). **3:**WI(1334). **4:**WA(3202). **6:**CA(638,642), MT(3202), WA(3202). **7:**CA(81,642), WA(3202). **10:**C states(82), W states(82), AZ(82), CA(82), WA(82).
P. stipae. Heteroecious: aecial host. **10:**CO(82).
P. suksdorfii. Autoecious. **5:**AZ(1203,2713,3981), CO(1195,2713), ID(3202), MT(1195,2713,3202), WA(746,3202), WY(1195,2713). **6:**MT(3202). **10:**W states(82), NE(82), WA(82).
P. troximontis. Autoecious. **1:**CA(81). **2:**UT(2713). **4:**WA(2713). **5:**MT(2713), WY(2713). **7:**CA(81). **8:**CA(81).
Basidiomycotina - Ustilaginales
Entyloma compositarum. White smut. **10:**CO(82).
E. polysporum. **2:**ID(3202), WA(3202), WY(1077). **5:**ID(3202). **9:**CO(1077).
Deuteromycotina - Hyphomycetes
Cladosporium sp. **5:**WA(3202).
Ovularia compacta. Leaf spot. **2:**MT(3202). **5:**ID(3202), MT(3202), OR(3202), WA(3202), WY(1195). **6:**OR(3202), WA(3202). **10:**CO(82).
Ramularia agoseridis. Leaf spot. **5:**MT(3202), WA(666,3202). **6:**ID(666,3202), WA(666,3202). **7:**WA(666,3202). **10:**CA(82), CO(82), OR(3202), WA(82), WY(82).
Deuteromycotina - Coelomycetes
Heteropatella umbilicata. **2:**CA(642).
Phyllosticta sp. **2:**WY(2924).
Septoria lactucae. On leaves. **3:**WI(1334).

Ambrosia L.

(= *Franseria* Cav.)

1. *A. acanthicarpa* Hook.
2. *A. ambrosioides* (Cav.) Payne (=*Franseria ambrosioides* Cav.)
3. *A. artemisiifolia* L. [common ragweed]
4. *A. artemisiifolia* L. var. *elatior* (L.) Descourt.
5. *A. bidentata* Michx.
6. *A. confertifolia* DC.
7. *A. deltoidea* (Torr.) Payne (=*Franseria deltoidea* Torr.)
8. *A. dumosa* (A. Gray) Payne (=*Franseria dumosa* A. Gray ex Torr.)
9. *A. hispida* Pursh
10. *A. psilostachya* DC. [western ragweed]
11. *A. tomentosa* Nutt. (=*Franseria discolor* Nutt.) [skeleton-leaf bur, skeleton-leaf bur-sage]
12. *A. trifida* L. (=*A. aptera* DC.) [buffaloweed, giant ragweed]

13. *Ambrosia* sp. (=*Franseria sp.*)

Chytridiomycetes

Monochytrium stevensianum. In cells of leaves and petioles. **3:**NC(82,1303).

Oomycetes

Albugo tragopogonis. White rust. **1:**WA(3202). **3:**E states(82), FL(35), MS(2720), NM(82), WA(3202), WI(82). **6:**AZ(1195). **10:**C states(82), ND(82,377), UT(82). **12:**C states(82), MS(2720), ND(82,377), UT(82), WI(1361). **13:**TX(82).

Plasmopara halstedii. White rust. **3:**E states(82), CO(82), NC(1303,3944), WI(82). **10:**C states(82), MT(82,3202), ND(377). **11:**WY(1195). **12:**NE states(82), MN(82), OK(2872), TX(82).

Pythium palingenes. Rootlet rot. **12:**VA(82).

Ascomycotina

Calycina herbarum. **12:**ND(377).

Diaporthe arctii. On stems. **3:**GA(82,1438). **12:**GA(82,1438), LA(82).

Didymella prominens. On stems. **12:**LA(82).

Erysiphe cichoracearum. Powdery mildew. **3:**Range of host(82), FL(35), GA(1307,1438), MS(2720), NC(1303,1307,3944), OK(2873), SD(2243). **5:**Range of host(82). **10:**Range of host(82), MT(3202), OK(2873), WI(1351). **12:**Range of host(82), MS(2720), MT(3202), NC(1303,3944), ND(377), OK(2872), SD(2243), WY(1195). **13:**AZ(82), FL(35).

Hyponectria dakotensis. **12:**ND(377).

Leptosphaeria complanata. **3:**GA(1438).

L. conoidea. **12:**NY(3249).

L. consessa. **10:**ND(377). **12:**ND(377).

L. doliolum. **12:**GA(1438), ND(3249).

L. fulgida. **12:**NY(3248).

Macroventuria wentii. **13:**NV(1).

Metasphaeria ambrosiaecola. **3:**GA(1438). **10:**ND(377).

M. ambrosiaecola var. *praetans*. **12:**ND(377).

Mollisia atrata. On stems. **3:**GA(1438). **12:**ND(82,377), PA(82).

Ophiobolus acuminatus. On stems, probably saprophytic. **3:**NY(82).

O. anguillidus. On stems, probably saprophytic. **3:**GA(1438), MS(3248), NY(3248). **10:**ND(377). **12:**GA(82,1438), IN(82), ND(82,377). **13:**MI(3248).

O. drechsleri. **12:**DE(3248), IA(3248), IL(3248), KY(3248), NE(3248), NY(3248).

O. nigro-clypeata. **3:**GA(1438).

Phyllachora ambrosiae. Leaf spot. **3:**Unk.(82), AL(82), FL(35,82), GA(1438), MS(82,2720), SC(82), VA(82), WI(82). **10:**NE(82), WI(82). **12:**AL(82), IN(82), WI(82).

Pleospora herbarum. **3:**GA(1438).

P. lactucicola. **3:**GA(1438).

Protomyces andinus. Stem gall. **3:**OR(3202), WI(82). **12:**WI(82).

P. gravidus. Hypertrophy of aerial parts. **3:**MA(807).

Sclerotinia sclerotiorum. Stem rot. **3:**FL(35,82). **9:**FL(35).

Basidiomycotina - Uredinales

Puccinia canaliculata. Heteroecious: aecial host. **3:**NC(82,1303), ND(82). **6:**AZ(3981). **12:**CT(82), IL(82), IN(82), MO(82).

P. conoclinii. Autoecious. **3:**NC(1303).

P. franseriae. Autoecious. **7:**AZ(744,747,1195,3379,3981), CA(747). **8:**AZ(747), CA(747).

P. splendens. **2:**AZ(1203). **7:**AZ(1203). **8:**AZ(1203), CA(81). **13:**AZ(82), CA(82).

P. xanthii. Autoecious. **3:**SE states(82), FL(35), NC(1303), NE(82), OK(2872), TX(82). **6:**AZ(1195,3379). **9:**FL(35). **10:**C states(82), AZ(1195,3379), CO(82), MT(82,3202), SD(2243), WI(1330). **12:**E states(82), MT(82,3202), OK(2872), SD(2243), TX(82), WA(3202). **13:**FL(35).

Uromyces junci. Heteroecious: aecial host. **10:**CA(82), CO(82), ND(82,378), NE(82).

Basidiomycotina - Ustilaginales

Entyloma compositarum. White smut. **3:**E states(82), FL(35,1077), IA(82,1077), IL(1077), IN(1077), MA(1077), ME(1077), MI(1077), NC(1303), NH(1077), NY(1077), PA(1077), SD(1077,2243), TX(82,1077). **4:**FL(1077). **5:**IL(1077), KS(82). **10:**KS(82,1077), WI(82,1077). **12:**NE states(82), CO(1077), CT(1077), DC(1077), DE(1077), IA(1077), IL(1077), LA(1077), MD(1077), MO(82,1077), ND(1077), NY(1077), PA(1077), WI(82,1077).

E. polysporum. On leaves. **3:**E states(82), CT(1077), IA(82,1077), IL(82,1077), IN(1077), MI(1077), NC(1303), NH(1077), NY(1077), PA(1077), TX(82), WI(1077). **4:**IL(1077), MI(1077), NY(1077). **5:**IL(1077). **10:**KS(82), WI(82). **12:**NE states(82), CO(1077), IA(1077), IL(1077), IN(1077), LA(1077), MO(82), NY(1077), PA(1077), WI(82,1077). **13:**FL(35).

Basidiomycotina - Other

Thanatephorus cucumeris. White smut. **3:**LA(3524). **12:**LA(3524).

Deuteromycotina - Hyphomycetes

Cercospora arcti-ambrosiae. Leaf spot. **3:**FL(35). **12:**HI(82,2903), NJ(82), OK(82,2872), WI(1358).

Fusarium oxysporum. Root rot. **13:**FL(35).

F. solani. Root rot. **13:**FL(35).

Mycovellosiella ambrosiae. **3:**WI(82,1339). **10:**WI(82).

M. ferruginea. **12:**DE(3543).

M. teucrii. Leaf spot. **12:**KS(962). **13:**TX(82).

Oidium ambrosiae. On leaves. **6:**AZ(1195,3378).

Oidium sp. Powdery mildew. **13:**FL(35).

Phymatotrichopsis omnivora. Root rot. **3:**TX(82). **5:**TX(82). **10:**TX(82). **12:**TX(82). **13:**TX(82).

Deuteromycotina - Coelomycetes

Ascochyta compositarum. **12:**WI(1360).

Ascochyta sp. On leaves. **12:**WI(1343).

Colletotrichum truncatum. Isolated from leaves. **12:**IL(1482).

Macrophomina phaseolina. Charcoal rot. **3:**IL(311). **5:**IL(311), TX(82).

Phyllosticta ambrosiae. Leaf spot. **12:**WI(82).

Phyllosticta sp. **10:**WI(1352).

Septoria ambrosiicola. Leaf spot. **12:**TX(82,3597).

S. bacilligera. Leaf spot, on cotyledonary leaves. **12:**IL(82), KS(82), MO(82), MS(2720), OK(82,2873), WI(82,1341,1345).

Septoria sp. Leaf spot. **10:**WI(1345,1361).

Stagonospora ambrosiae. On leaves. **12:**ND(82).

S. trifidae. On living leaves. **12:**WI(1360).

Deuteromycotina - Other

Rhizoctonia crocorum. Root rot. **12:**TX(82).

R. solani. Root rot. **3:**TX(82). **5:**TX(82). **10:**NE(82), TX(82). **12:**TX(82). **13:**FL(35).

Sclerotium rolfsii. Southern blight. **3:**FL(35,82), NC(1303), TX(82). **5:**TX(82). **10:**TX(82). **12:**MS(2720), TX(82).

Anaphalis DC.

1. *A. margaritacea* (L.) Benth. & J. D. Hook. (=*A. margaritacea* (L.) Benth. & J. D. Hook. var. *occidentalis* Greene) [pearly everlasting]

Ascomycotina

Mycosphaerella tassiana. On fading leaves. **1:**NY(82).

Basidiomycotina - Uredinales

Uromyces amoenus. Autoecious. **1:**CA(81,82,638,642), ID(82,3202), MI(82), MT(82,1195,3202), OR(82,3202), SD(2243), WA(82,3202), WY(82).

Deuteromycotina - Hyphomycetes

Ramularia sp. **1:**ID(3202), WI(1344).

Deuteromycotina - Coelomycetes

Platycarpium caulicola. **1:**WA(3202,3816).

Septoria margaritaceae. Leaf spot. **1:**NY(82), OR(82,3202), SD(2243), WA(3202,3816), WI(82).

Antennaria Gaertn.

1. *A. anaphaloides* Rydb.
2. *A. flagellaris* (A. Gray) A. Gray
3. *A. luzuloides* Torr. & A. Gray
4. *A. neglecta* Greene var. *attenuata* (Fernald) Cronq. (=*A. petaloidea* Fernald)
5. *A. plantaginifolia* (L.) Richardson
6. *A. plantaginifolia* (L.) Richardson var. *ambigens* (Greene) Cronq. (=*A. fallax* Greene)
7. *A. plantaginifolia* (L.) Richardson var. *arnoglossa* (Greene) Cronq. (=*A. parlinii* Fernald)
8. *A. pulcherrima* (Hook.) Greene
9. *A. racemosa* Hook.
10. *A. rosea* Greene
11. *Antennaria* sp.

Oomycetes

Albugo tragopogonis. **5:**OK(2872), WI(1355). **11:**IL(82), NE(82).

Ascomycotina

Crocicreas phaeoconium. On leaves. **5:**NY(505,850).

Erysiphe cichoracearum. **2:**ID(3202).

Globulina antennariae. On leaves. **11:**NY(82).

Lasiostemma fimbriatum. On leaves. **11**:NY(82).
Mycosphaerella tassiana. On dead stems and leaves. **11**:AK(1938).
Nematostoma occidentalis. **6**:WI(1344). **7**:WI(1339).
Ophiobolus gnaphalii var. *lanaria*. **4**:WI(1355).
Pleospora penicillus. On dead stems. **11**:AK(1938).
Basidiomycotina - Ustilaginales
Entyloma calendulae. **3**:WA(1077,3202).
Entyloma sp. **1**:ID(3202), MT(3202). **3**:ID(3202). **8**:MT(3202). **9**:WA(3202).
Deuteromycotina - Coelomycetes
Ascochyta sp. **7**:WI(1360).
Phoma antennariae. Leaf spot. **11**:CO(82).
Phyllosticta antennariae. Leaf spot. **5**:WI(1355). **6**:WI(1332). **11**:DE(82), WI(82).
P. ragnhildae. On living leaves. **8**:WY(1195,3376).
Phyllosticta sp. **10**:MT(3202).
Septoria lanaria. Leaf spot. **4**:WI(1355). **7**:WI(1333). **11**:NY(82).
S. mollisia. **5**:WI(1333). **11**:NY(82).

Anthemis L.

1. *A. arvensis* L. [corn chamomile]
2. *A. cotula* L. [common dogfennel, mayweed, stinking chamomile]

Ascomycotina
Sphaerotheca macularis. **2**:WA(3202).
Basidiomycotina - Other
Thanatephorus cucumeris. **1**:WA(3202). **2**:WA(3202).
Deuteromycotina - Hyphomycetes
Phymatotrichopsis omnivora. Root rot. **1**:TX(82). **2**:TX(82).
Deuteromycotina - Coelomycetes
Septoria matricariae. **2**:WI(1360).
Deuteromycotina - Other
Rhizoctonia solani. Damping-off. **1**:WA(82).

Apargidium Torr. & A. Gray

1. *A. boreale* (Bong.) Torr. & A. Gray (=*Scorzonella borealis* (Bong.) Greene)

Ascomycotina
Sphaerotheca fuliginea. **1**:AK(526).
Basidiomycotina - Uredinales
Puccinia dioicae. **1**:AK(526).

Arctium L.

1. *A. lappa* L. (=*Lappa major* Gaertn.) [cuckold, edible burdock, gobo, great burdock, harlock]
2. *A. minus* Bernh. [common burdock, lesser burdock]
3. *Arctium* sp.

Ascomycotina
Crocicreas cyathoideum. **2**:SD(505).
Diaporthe arctii. On stems. **1**:TX(82). **2**:MI(82), TX(82).
Erysiphe cichoracearum. Powdery mildew. **1**:TX(82). **2**:IA(82), KS(82), TX(82).
Leptosphaeria collinsoniae. **3**:NY(3249).
Leptospora rubella. **3**:NY(3248).
Ophiobolus anguillidus. **3**:MI(3248).
Basidiomycotina - Uredinales
[*Puccinia calcitrapae*]. **2**:WY(1195).
P. calcitrapae var. *bardanae*. Autoecious. **1**:E states(82), MO(82), NC(1303), ND(82). **2**:E states(82), MA(1390), MO(82), MS(2720), ND(82), UT(82), WA(3202).
Deuteromycotina - Hyphomycetes
Cercospora arcti-ambrosiae. Leaf spot. **1**:HI(82,2903), NC(1303), NJ(82), NY(82), TX(82), WI(82). **2**:WI(1332).
Phymatotrichopsis omnivora. Root rot. **1**:TX(82). **2**:TX(82).
Deuteromycotina - Coelomycetes
Ascochyta doronici. Leaf spot. **1**:TX(82). **2**:IN(82), NY(82), TX(82).
Colletotrichum fuscum. Anthracnose. **1**:OR(82), PA(82).
Phlyctema arcuata. Leaf spot. **1**:TX(82). **2**:TX(82).
Phyllosticta lappae. Leaf spot. **1**:E states(82), NC(1303), TX(82), WI(82).
Phyllosticta sp. Leaf spot. **2**:WI(1360).
Rhabdospora arctii. On stems. **2**:MI(82).
Septoria lapparum. Leaf spot. **1**:IN(82), MA(82), MI(82). **2**:WI(82).
Sphaeropsis lappae. Leaf spot. **1**:TX(82). **2**:TX(82).

Arctotis L.

1. *A. stoechadifolia* Bergius [blue-eyed African daisy]

Oomycetes
Plasmopara halstedii. Downy mildew. **1**:FL(35).
Deuteromycotina - Hyphomycetes
Cercospora sp. Leaf blotch. **1**:FL(35,82).
Phymatotrichopsis omnivora. Root rot. **1**:TX(82).
Deuteromycotina - Other
Sclerotium rolfsii. Southern blight. **1**:FL(35).

Argyroxiphium DC.

1. *A. gymnoxiphium* (A. Gray) D. Keck (=*Wilkesia gymnoxiphium* A. Gray)
2. *A. sandwicense* DC.

Deuteromycotina - Hyphomycetes
Acrodictys fimicola. **2**:HI(2903).
Aspergillus niger. On leaves. **2**:HI(2903).
A. terricola var. *indicus*. **2**:HI(2903).
Cladosporium sp. Leaf spot. **1**:HI(2903).
Pithomyces chartarum. **2**:HI(2903).
Deuteromycotina - Coelomycetes
Plenodomus sp. On leaves. **2**:HI(2903).

Arnica L.

1. *A. alpina* (L.) Olin
2. *A. amplexicaulis* Nutt.
3. *A. chamissonis* Less.
4. *A. chamissonis* Less. subsp. *foliosa* (Nutt.) Maguire (=*A. foliosa* Nutt.)
5. *A. cordifolia* Hook. (=*A. pumila* Rydb.)
6. *A. discoidea* Benth.
7. *A. diversifolia* Greene
8. *A. fulgens* Pursh
9. *A. latifolia* Bong.
10. *A. latifolia* Bong. var. *gracilis* (Rydb.) Cronq. (=*A. gracilis* Rydb.)
11. *A. longifolia* D. C. Eaton
12. *A. longifolia* D. C. Eaton subsp. *myriadenia* (Piper) Maguire
13. *A. louiseana* Farr var. *frigida* (C. A. Mey.) Maguire (=*A. frigida* Mey.)
14. *A. mollis* Hook. (=*A. subplumosa* Greene)
15. *A. parryi* A. Gray
16. *A. sororia* Greene
17. *A. viscosa* A. Gray
18. *Arnica* sp.

Ascomycotina
Apiosporella alpina. **12**:CA(642).
Crocicreas cyathoideum var. *cacaliae*. On stems. **17**:CA(505).
Didymella delphinii. On stems. **9**:NV(681). **18**:CA(661).
Erysiphe cichoracearum. Powdery mildew. **18**:CO(82).
Hysteropezizella sp. On overwintered herbage. **17**:CA(642).
Leptosphaeria doliolum. **17**:CA(642).
Phaeosphaeria heterospora. On overwintered herbage. **17**:CA(642).
Pleospora anthyllidis. **17**:CA(642).
P. helvetica. **15**:WA(3202,3814).
P. herbarum. On overwintered herbage. **17**:CA(642).
P. tragacanthae. On dead stem. **13**:AK(1938).
Sphaerotheca fuliginea. Powdery mildew. **9**:AK(526). **14**:CA(638,642). **18**:AK(82), CA(82), WA(82), WY(82).
S. macularis. **1**:WA(3202). **2**:ID(3202). **3**:WA(3202). **4**:WY(1195). **8**:WA(3202). **9**:WA(3202).
Trichopezizella nidulus var. *setigera*. On bases of overwintered stems. **17**:CA(638,642).
T. relicina var. *macrospora*. **17**:CA(1416).
Basidiomycotina - Uredinales
Coleosporium pacificum. Heteroecious: telial host. **3**:WA(3202). **8**:WA(3202).
Puccinia arnicalis. **3**:WY(1195). **5**:CO(1195), MT(3202), OR(3202), WY(1195,2924). **7**:OR(3202). **8**:WA(3202). **9**:MT(3202). **13**:AK(60,526,1938,3113). **14**:CO(1195), WY(1195). **18**:Unk.(82).
Uromyces junci. Heteroecious: aecial host. **3**:WA(3202), WY(1195). **5**:OR(3202). **14**:MT(3202). **18**:CA(82), CO(82), MT(82), OR(82), SD(82), WY(82).
Basidiomycotina - Ustilaginales
Entyloma arnicale. White smut. **3**:WY(1195). **5**:WY(1195,2924).

9:CO(649), WY(2924). **14:**WY(1195). **18:**AK(82), CO(82), ID(82), MT(82), UT(82), WA(82), WY(82).

E. calendulae. **3:**AK(526), WA(1077,3202), WY(1077,1195). **4:**WY(1077). **5:**CO(1077), ID(1077,1195,3202), MT(1077,3202), OR(1077,3202), WA(1077,3202), WY(1077,1195). **8:**CO(1077). **9:**ID(3202), MT(3202), WA(1077,3202), WY(1077). **10:**WY(1077). **14:**CO(1077), UT(1077), WA(1077,3202), WY(1077,1195). **18:**AK(1077), WA(3202).

E. compositarum. **2:**AK(3431), WA(3202). **5:**ID(3202), WA(1077,3202).

Entyloma sp. **3:**ID(3202), OR(3202), WA(3202). **6:**OR(3202). **7:**OR(3202). **8:**ID(3202), WA(3202). **9:**OR(3202), WA(3202). **11:**ID(3202). **14:**ID(3202), MT(3202), OR(3202), WA(3202). **15:**MT(3202), WA(3202). **16:**WA(3202).

Thecaphora californica. **5:**CO(920,1077).

Deuteromycotina - Hyphomycetes

Ovularia hughesiana. Leaf spot. **18:**MT(82,3202).

Deuteromycotina - Coelomycetes

Phoma sp. **17:**CA(642).

Phyllosticta arnicae. Leaf spot. **5:**MT(3202), WY(1195,2924). **18:**CO(82), ID(3202), MT(82), UT(82), WY(82).

Artemisia L.

1. *A. abrotanum* L. [old-man, southern wormwood, southernwood]
2. *A. absinthium* L. [absinthe, absinthium, common wormwood]
3. *A. arbuscula* Nutt. [low sagebrush]
4. *A. biennis* Willd.
5. *A. californica* Less. [California sagebrush]
6. *A. campestris* L. (=*A. canadensis* Michx.)
7. *A. campestris* L. subsp. *caudata* (Michx.) H. Hall & Clements (=*A. caudata* Michx.)
8. *A. cana* Pursh [silver sagebrush]
9. *A. carruthii* A. Wood (=*A. wrightii* A. Gray)
10. *A. douglasiana* Besser ex Hook. (=*A. vulgaris* L. var. *heterophylla* (Nutt.) Jeps., *A. heterophylla* Nutt.)
11. *A. dracunculus* L. (=*A. aromatica* A. Nels., *A. dracunculoides* Pursh, *A. glauca* Pall. ex Willd.) [estragon, tarragon]
12. *A. frigida* Willd. [estafiata, fringed sagebrush]
13. *A. globularia* Cham. ex W. S. Vesser
14. *A. gnaphalodes* Nutt. [cudweed, western mugwort, white sage]
15. *A. longifolia* Nutt.
16. *A. ludoviciana* Nutt. (=*A. purshiana* Besser) [Louisiana sagebrush, Louisiana wormwood, western mugwort, white sage]
17. *A. ludoviciana* Nutt. var. *pabularis* (A. Nels.) Fernald (=*A. pabularis* (A. Nels.) Rydb.)
18. *A. macrobotrys* Ledeb. (=*A. laciniata* Willd.)
19. *A. michauxiana* Besser
20. *A. norvegica* Fr. var. *comata* (Rydb.) Welsh (=*A. arctica* Less. var. *comata* (Rydb.) Hultén)
21. *A. norwegica* Fr. var. *saxatilis* Jeps. (=*A. arctica* Less.)
22. *A. nova* A. Nels.
23. *A. palustris* L.
24. *A. pycnocephala* DC.
25. *A. rigida* (Nutt.) A. Gray
26. *A. scopulorum* A. Gray
27. *A. serrata* Nutt.
28. *A. spinescens* D. C. Eaton
29. *A. suksdorfii* Piper
30. *A. tilesii* Ledeb. (=*A. tilesii* Ledeb. var. *elatior* Torr. & A. Gray)
31. *A. tridentata* Nutt. [basin sagebrush, big sagebrush, common sagebrush]
32. *A. trifurcata* Stephenson (=*A. tacomensis* Rydb.)
33. *A. tripartita* Rydb.
34. *A. vulgaris* L. [felonherb, mugwort]
35. *A. vulgaris* L. var. *littoralis* Suksd.
36. *Artemisia* sp.

Chytridiomycetes

Synchytrium aureum. Leaf gall. **4:**TX(82), WI(82).

Oomycetes

Albugo tragopogonis. White rust. **4:**IA(82), MT(82,3202), ND(82,377), SD(82), TX(82), WI(82).

Peronospora leptosperma. Downy mildew. **4:**ND(377), SD(2243). **16:**CA(82), IA(82), KS(82), ND(82,377), OR(3202), WI(82). **36:**WA(3202), WI(82).

Ascomycotina

Botryosphaeria obtusa. On stems. **34:**SD(82).

Crocicreas cyathoideum var. *cacaliae*. **10:**CA(661).

Cucurbitaria obducens. **31:**OR(3202), WA(3202). **36:**ID(3202).

Didymella effusa. On stems. **4:**ND(82), UT(82).

Didymosphaeria catalina. On dead stems. **5:**CA(2439).

Dothidea puccinioides. Black pustule. **1:**PA(82).

D. tuberculiformis. On stems. **36:**CA(82).

Erysiphe cichoracearum. Powdery mildew. **2:**WA(3202). **4:**CA(82), IA(82), SD(82), TX(82). **10:**CA(81). **11:**MT(3202), WI(1344). **15:**ND(377). **16:**C & W states(82), AK(82), CA(82), MT(3202), WA(82,3202). **19:**MT(3202). **29:**WA(3202). **34:**CA(82). **36:**C & W states(82), AK(82), CA(82), OR(3202), SD(2243), WA(82,3202).

E. communis. **21:**AK(526). **30:**AK(526). **36:**AK(526).

Glyphium corrugatum. **31:**ID(3202).

Graphyllium dakotense. **12:**ND(377).

Guignardia caryophyllea. **32:**WA(3202,3815).

Lachnella villosa. On dead stems. **36:**Unk.(456).

Lasiosphaeria sp. Leaf spot. **16:**OK(2872).

Leptosphaeria artemisiae. On stems. **7:**ND(377). **8:**MT(3202). **12:**ND(3249), NE(82). **16:**ND(82,377). **36:**ID(3202), MT(82), ND(82).

L. millefolii. On dead stems. **36:**AK(1938).

L. preandina. On stems. **31:**NV(128).

L. tetonensis. **8:**MT(3202). **12:**ND(377,3249). **16:**ND(3249). **36:**MT(82), ND(82).

L. tumefaciens. Stem gall. **5:**CA(3249). **31:**NV(82), UT(82), WY(1195,3249,3378). **36:**CA(82), ID(3202), MT(82,3202).

Mycosphaerella eriophila. On leaves. **36:**AK(82).

M. tassiana. On dead stems. **13:**AK(1938). **36:**AK(526).

Mycosphaerella sp. On dead leaves. **27:**WI(1352).

Nematostoma occidentalis. On leaves. **16:**IA(82), IL(82), WI(82).

Odontotrema oregonense. On wood. **31:**OR(3232).

Ophiobolus claviger. **4:**ND(377).

O. instabilis. **4:**WI(1357).

O. tanaceti. **4:**ND(3248).

Pleospora chlamydospora. On dead stem. **36:**AK(1938).

P. helvetica. **20:**AK(526). **32:**WA(3202).

P. herbarum. **32:**WA(3202).

P. herbarum var. *occidentalis*. **32:**WA(3814).

P. penicillus. **32:**WA(3814).

Pleospora sp. On leaves. **7:**WI(1332).

Pyrenopeziza artemisiae. On dead stems. **30:**AK(1938). **31:**ND(82).

Rosellinia ovalis. On stems. **36:**UT(82).

Stictis radiata. **35:**CA(81).

Teichospora solitaria. **6:**ND(377).

T. variabilis. **36:**MT(3202).

Tympanis montanensis. On stems. **36:**MT(3167,3202).

Basidiomycotina - Uredinales

Puccinia artemisiae-norvegicae. Autoecious. **21:**AK(60,526).

P. atrofusca. Heteroecious: aecial host. **4:**ND(82,378). **7:**WI(1354). **8:**MT(3202), WY(2924). **11:**AZ(1203), MT(1195,3202). **12:**C states(82), CA(82), MT(3202), ND(378), WA(82). **14:**C states(82), CA(82), WA(82). **15:**MT(3202), ND(378). **16:**C states(82), CA(82), MT(3202), ND(378), WA(82,3202). **31:**C states(82), CA(82), WA(82). **33:**MT(3202). **36:**C states(82), CA(82), WA(82).

P. cnici-oleracei. Autoecious. **11:**NM(1195). **12:**W states(82), AK(82), CA(82), MT(3202), WA(82), WI(1336). **14:**W states(82), AK(82), CA(82), WA(82), WY(1195). **16:**W states(82), AK(82), AZ(747,1203,3981), CA(82), MT(3202), OK(2872), OR(3202), TX(742), WA(82,3202), WY(1195). **18:**AK(60,526). **29:**WA(3202). **30:**AK(60,526,1938). **31:**W states(82), AK(82), CA(82), WA(82). **34:**CA(82). **36:**W states(82), AK(82,1938), CA(82), WA(82).

P. dioicae. Heteroecious: aecial host. **9:**AZ(1195,1203,3981).

P. ludovicianae. Autoecious. **9:**AZ(3981). **16:**AZ(3981), MT(746).

P. similis. Autoecious. **22:**AZ(3981). **31:**AZ(3981), WY(744).

[*P. tanaceti*]. **3:**ID(3202), OR(3202). **5:**CA(81). **8:**CO(638), MT(3202), ND(378), OR(3202), SD(2243), WY(1195). **9:**AZ(1203). **12:**C states(82), CA(82), HI(82), ND(378), OR(3202), WA(82). **14:**C states(82), CA(82), HI(82), OK(2872), WA(82). **15:**ND(378). **16:**C states(82), AZ(1203), CA(82), HI(82), MT(3202), ND(378), NM(1195), OR(3202), SD(2243),

TX(742), WA(82,3202), WY(1195). **17**:SD(2243). **22**:AZ(1203,3981). **23**:SD(2243). **24**:CA(81). **25**:ID(3202), OR(3202), WA(3202). **28**:ID(3202). **31**:C states(82), AZ(1195,1203,3981), CA(82), CO(1195), HI(82), ID(1195,3202), MT(1195,3202), NM(1195), OR(3202), UT(1195), WA(82,3202), WY(2924). **33**:ID(3202), WY(1195). **34**:CA(82). **36**:C states(82), CA(82), HI(82), SD(2243), WA(82).
P. tanaceti var. *dracunculina*. **11**:AZ(3981), WI(1352). **16**:AZ(3981).
Uromyces oblongisporus. Autoecious. **31**:WY(82,746).

Basidiomycotina - Other

Guepiniopsis buccina. **31**:OR(3202).
Heliocybe sulcata. **31**:AZ(1202).
Heterobasidion annosum. Root rot. **31**:CA(3349).
Hyphodermella corrugata. **34**:CA(128).

Deuteromycotina - Hyphomycetes

Botrytis cinerea. Gray mold blight. **36**:AK(82).
Cladosporium artemisiae. On living leaves. **7**:WI(1344).
Epicoccum nigrum. **31**:WA(3202).
Epochnium isthmophorum. **36**:WA(3202).
Heterosporium sp. Leaf spot. **36**:AK(82).
Mycovellosiella ferruginea. Leaf spot. **2**:ND(378), NY(82), WI(82). **16**:WI(82). **29**:OR(3202). **36**:NY(82).
Phymatotrichopsis omnivora. Root rot. **4**:TX(82). **36**:TX(82).
Ramularia artemisiae. Leaf spot. **16**:ID(3202). **36**:NY(82), WI(82).
Stigmina dothideoides. On stems. **8**:MT(82,3202,3253).
S. sycina. **31**:MT(3202).

Deuteromycotina - Coelomycetes

Camarosporium compositarum. On stems. **12**:NM(82). **36**:CA(82), WA(82,3202).
Chaetosticta perforata. On leaves. **16**:IA(718).
Colletotrichum phyllachoroides. On leaves. **10**:CA(971). **34**:CA(82,3549). **36**:CA(3549).
Cylindrosporium artemisiae. Leaf spot. **16**:WI(82,1334). **29**:WA(3202). **36**:WA(82), WI(82).
Diplodia abrotani. On stems. **1**:ND(82).
Diplodina tridentatae. On leaves. **31**:UT(666).
Heteropatella umbilicata. **32**:WA(3202,3816).
Phoma terrestris. Isolated from rhizosphere. **31**:OR(2791).
Phomopsis oblita. On dead stems. **10**:CA(2439).
Phyllosticta raui. Leaf spot. **16**:MT(82,3202). **26**:NY(64). **31**:WY(82). **36**:CO(82).
Placosphaeria sp. On stems. **36**:CA(82).
Rhabdospora ellisii. On stems. **36**:CO(82).
Septoria artemisiae. Leaf spot. **36**:WA(82,3202).
S. fusca. On leaves. **34**:NY(82).

Deuteromycotina - Other

Sclerotium sp. Stem blight. **36**:OR(82).

Aster L.

(= *Sericocarpus* Nees.)
1. *A. acuminatus* Michx.
2. *A. alpigenus* (Torr. & A. Gray) A. Gray
3. *A. alpinus* L. [rock aster]
4. *A. azureus* Lindl.
5. *A. borealis* Prov. (=*A. junciformis* Rydb.)
6. *A. bracteolatus* Nutt. (=*A. eatonii* (A. Gray) Howell)
7. *A. bractyactis* S. F. Blake (=*A. angustus* (L.) Torr. & A. Gray)
8. *A. campestris* Nutt.
9. *A. chilensis* Nees (=*A. menziesii* Lindl.)
10. *A. chilensis* Nees subsp. *adscendens* (Lindl. ex DC.) Cronq. (=*A. adscendens* Lindl. ex DC.)
11. *A. ciliolatus* Lindl. ex Hook. (=*A. lindleyanus* Torr. & A. Gray)
12. *A. conspicuus* Lindl. ex DC.
13. *A. cordifolius* L. [bluewood aster]
14. *A. divaricatus* L. [whitewood aster]
15. *A. drummondii* Lindl.
16. *A. dumosus* L.
17. *A. engelmannii* (D. C. Eaton) A. Gray
18. *A. ericoides* L. (=*A. multiflorus* Aiton) [heath aster]
19. *A. ericoides* L. var. *platyphyllus* Torr. & A. Gray
20. *A. falcatus* Lindl. (=*A. adsurgens* Greene, *A. commutatus* (Torr. & A. Gray) A. Gray)
21. *A. foliaceus* Lindl. ex DC.
22. *A. foliaceus* Lindl. var. *canbyi* A. Gray
23. *A. foliaceus* Lindl. var. *parryi* (D. C. Eaton) Cronq. (=*A. frondeus* Greene)
24. *A. frondosus* (Nutt.) Torr. & A. Gray
25. *A. glaucescens* (A. Gray) S. F. Blake
26. *A. glaucodes* S. F. Blake (=*A. glaucus* Torr. & A. Gray)
27. *A. greatae* Parish
28. *A. hesperius* A. Gray (=*A. coerulescens* DC., *A. fluvialis* Osterh.)
29. *A. integrifolius* Nutt.
30. *A. jessicae* Piper
31. *A. laevis* L.
32. *A. lateriflorus* (L.) Britton
33. *A. ledophyllus* A. Gray (=*A. engelmannii* (D. C. Eaton) A. Gray var. *ledophyllus* A. Gray)
34. *A. linariifolius* L.
35. *A. lucidulus* (A. Gray) Wiegland
36. *A. macrophyllus* L.
37. *A. modestus* Lindl.
38. *A. novae-angliae* L. [New England aster]
39. *A. novi-belgii* L. (=*A. longifolius* Lam.)
40. *A. oblongifolius* Nutt.
41. *A. oblongifolius* Nutt. var. *angustatus* Shinners
42. *A. occidentalis* (Nutt.) Torr. & A. Gray (=*A. fremontii* A. Gray)
43. *A. pansus* (S. F. Blake) Cronq.
44. *A. parryi* A. Gray (=*Xylorhiza parryi* (A. Gray) Greene)
45. *A. patens* Dryander in Aiton
46. *A. paternus* Cronq. (=*Sericocarpus asteroides* (L.) B. S. P.)
47. *A. pilosus* Willd.
48. *A. praealtus* Poir. (=*A. salicifolius* Aiton)
49. *A. prenanthoides* Muhl. ex Willd.
50. *A. ptarmicoides* (Nees) Torr. & A. Gray [white upland aster]
51. *A. puniceus* L.
52. *A. sagittifolius* Willd.
53. *A. sagittifolius* Willd. var. *dissitiflorus* Burgess
54. *A. sericeus* Venten.
55. *A. shortii* Lindl.
56. *A. sibiricus* L.
57. *A. simmondsii* Small
58. *A. simplex* Willd. (=*A. paniculatus* Lam. var. *simplex* (Willd.) Burgess)
59. *A. simplex* Willd. var. *ramosissimus* (Torr. & A. Gray) Cronq. (=*A. paniculatus* Lam.)
60. *A. solidagineus* Michx. (=*Sericocarpus linifolius* B. S. P., *Sericocarpus solidagineus* (Michx.) Nees)
61. *A. spinosus* Benth.
62. *A. stenomeres* A. Gray
63. *A. subspicatus* Nees
64. *A. subulatus* Michx. (=*A. exilis* Elliott)
65. *A. subulatus* Michx. var. *ligulatus* Shinners
66. *A. tradescantii* L.
67. *A. umbellatus* Mill.
68. *A. undulatus* L.
69. *A. vimineus* Lam.
70. *Aster* sp.

Chytridiomycetes

Synchytrium asterum. Leaf gall. **70**:LA(627).
S. nigrescens. Leaf gall. **70**:WI(82).

Oomycetes

Basidiophora entospora. On leaves. **9**:MT(3202), WY(1195). **10**:WY(3378). **31**:WI(82,1332). **38**:IL(82), IN(82), MD(82), MO(82), WI(82). **41**:WI(1339). **70**:FL(35), NE(82), WI(82).
Pythium sp. Downy mildew. **70**:FL(35).

Ascomycotina

Acrospermum foliicola. **68**:GA(1438).
Anthostomella sp. **19**:GA(1438).
Clypeoporthella brencklei. On stems. **70**:ND(82).
Comoclathris permunda. **70**:OR(3202).
Crocicreas cyathoideum var. *cacaliae*. On dead stems. **17**:UT(505).
Diaporthe linearis. On stems. **14**:GA(1438). **16**:GA(1438). **19**:GA(1438). **32**:GA(1438). **46**:GA(82,1438). **51**:GA(1438). **53**:GA(1438). **68**:GA(1438). **70**:GA(82), MI(82), NJ(82), NY(82).
Diaporthopsis sp. **14**:GA(1438).
Diatrypella sp. **19**:GA(1438).
Didymosphaeria futilis. **19**:GA(1438).
Elsinoe solidaginis. **34**:WI(1348).
Erysiphe cichoracearum. Powdery mildew. **3**:Range of host(82).

4:WI(1335). **6:**CA(1155). **9:**CA(81), MT(3202). **10:**WY(1195,3378). **12:**MT(3202). **13:**Range of host(82). **18:**Range of host(82), ND(377), OK(2872). **19:**GA(1438). **20:**MT(3202), WY(1195). **21:**MT(3202). **25:**WA(3202). **27:**CA(1155). **28:**WY(1195). **29:**WY(1195). **30:**WA(3202). **31:**Range of host(82), MT(3202). **33:**Unk.(3976), CA(1155). **35:**WI(1333). **36:**WI(1330). **38:**Range of host(82), WI(1335). **39:**Range of host(82), ND(377), WI(1330). **40:**WI(1332). **42:**ID(3202), WA(3202). **43:**MT(3202). **47:**WI(1332). **48:**OK(2872). **49:**GA(1307), NC(1307), WI(1340). **50:**Range of host(82). **53:**GA(1438). **54:**WI(1346). **57:**FL(35). **59:**ND(377), SD(2243). **70:**GA(1438), HI(2899), ID(3202), SD(2243).

Eutypella goniostoma. **13:**OH(2925).

Guignardia haydenii. Black spot. **4:**WI(1349). **7:**WI(1340). **35:**WI(1353). **38:**NY(82). **47:**WI(1355). **51:**WI(1359). **52:**WI(1360). **59:**ND(377). **70:**IA(82), ND(82), NJ(82).

G. scabiosae. **18:**ND(377).

Heptameria obesa. **19:**GA(1438).

Hymenoscyphus scutula. **70:**OR(3202).

Hypoderma commune. **70:**GA(1438).

Leptosphaeria astericola. On dead stems. **18:**KS(970).

L. complanata. **68:**GA(1438).

L. conoidea. **70:**MA(3249), MI(3249).

L. doliolum. **16:**GA(1438). **19:**GA(1438). **70:**IL(3249).

L. erigerontis. **39:**ND(377).

L. heliopsidis. Black knot. **5:**WI(1332). **18:**ND(377). **20:**MT(3202). **35:**WI(1332). **38:**NY(82). **47:**WI(1353). **70:**ND(82), NE(82), WI(82).

L. licatensis. Black knot. **70:**IL(650).

L. ogilviensis. **19:**GA(1438). **45:**GA(1438). **70:**AK(3249).

L. planiuscula. **52:**GA(3249). **53:**GA(1438). **67:**NY(3249).

Leptosphaeria sp. **53:**GA(1438).

Lophiostoma caulium. **18:**ND(377). **46:**GA(1438).

Massarina gloeospora. **19:**GA(1438).

Mollisia atrata. **19:**GA(1438).

M. cinerea. **51:**GA(1438).

Mycosphaerella aronici. **21:**WA(3202,3815).

Mycosphaerella sp. On leaves and stems. **46:**GA(82,1438). **67:**WI(1353).

Ophiobolus anguillidus. **19:**GA(1438).

O. crassus. **51:**NC(3248).

Phyllachora asterigena. Anthracnose. **70:**KS(82), NE(82).

Pleospora helvetica. **21:**WA(3202,3814).

P. penicillus. On stems. **70:**CA(82).

P. richtophensis. **21:**ID(3202), WA(3202).

P. vulgaris f. *astericola*. **18:**ND(377).

Pyrenopeziza sp. On stems. **46:**GA(82). **60:**GA(1438).

Pyrenophora leucelenes. On leaves and stems. **70:**NM(82).

Rhytisma asteris. **34:**WI(1350). **47:**WI(1350). **52:**WI(1350).

Sclerotinia sclerotiorum. Stem rot. **70:**CT(82), FL(35).

Basidiomycotina - Uredinales

Coleosporium asterum. Heteroecious: telial host. **1:**MA(664). **3:**Unk.(82). **4:**WI(1333). **5:**WI(1333). **6:**MT(3202), WA(3202). **9:**CA(81), ID(3202), OR(3202). **11:**WI(1333), WY(1195). **12:**ID(3202), MT(1195,3202), OR(3202), WA(3202). **13:**Unk.(82), WI(1333). **16:**GA(1439). **18:**Unk.(82), MA(1391), ND(378), WI(1332,1333). **21:**ID(3202), MT(3202), OR(3202), WA(3202). **24:**WA(3202). **29:**ID(3202). **31:**Unk.(82), ID(3202), MT(3202), ND(378), WA(3202), WI(1333). **32:**GA(1439), WI(1333). **33:**CA(81), WA(3202). **35:**WI(1333). **36:**NY(2456), WI(1333). **37:**MT(3202). **38:**Unk.(82), WI(1333). **39:**Unk.(82), MA(1391), WI(1332,1333). **40:**WI(1332,1333). **41:**WI(1332). **42:**ID(3202), MT(3202), OR(3202), WA(3202). **43:**ID(3202), WA(3202). **46:**CT(82), GA(1439). **47:**WI(1330,1333). **49:**WI(1333). **50:**Unk.(82), WI(1349). **51:**GA(1439), WI(1333,1354). **52:**MS(2720), WI(1333). **55:**WI(1341). **56:**MT(3202). **57:**FL(35). **58:**SD(2243), WI(1333). **59:**ND(378). **63:**AK(60), MT(3202), OR(3202), WA(3202). **67:**WI(1333). **70:**Unk.(82), FL(35), OR(657), SD(2243).

Puccinia atrofusca. Heteroecious: aecial host. **32:**WI(1354).

P. cnici-oleracei. Autoecious. **4:**WI(1343). **8:**ID(3202). **9:**CA(81), MT(3202). **11:**WI(1330). **12:**ID(3202), MT(3202), OR(3202), WA(3202), WY(1195). **13:**Unk.(82), MS(2720). **15:**OK(2872). **18:**Unk.(82), ND(378), WI(1332). **19:**GA(1439). **20:**AZ(3981), MT(1195,3202). **21:**AK(526), ID(3202), MT(3202), WA(3202). **26:**WY(1195). **28:**AZ(1203,3981). **29:**WY(2768). **31:**Unk.(82), ID(3202), MT(3202), WA(3202), WI(1335). **32:**GA(1439). **35:**WI(1330). **38:**Unk.(82), ND(378). **39:**Unk.(82), ND(378). **40:**WI(1330,1340). **41:**WI(1337). **42:**ID(3202), WA(3202). **43:**MT(3202). **47:**WI(1355). **50:**Unk.(82). **51:**GA(1439), WI(1360). **55:**WI(1344). **58:**SD(2243). **59:**ND(378). **62:**ID(3202). **64:**AZ(1203). **65:**AZ(3981). **70:**FL(35), SD(2243).

P. dioicae. Heteroecious: aecial host. **4:**WI(1337). **9:**CA(81), MT(3202), WA(3202). **12:**MT(3202). **13:**Unk.(82), MS(2720). **18:**ND(378), WI(1354). **20:**WA(3202). **21:**AK(526), ID(1195,3202), MT(1195,3202), WA(3202), WY(1195). **22:**WY(3379). **23:**WY(1195). **28:**ID(3202). **31:**Unk.(82). **33:**WA(3202). **36:**VT(1224). **39:**ND(378). **41:**WI(1338,1341). **42:**CA(638,642), ID(3202), MT(3202), OR(3202), WA(3202). **47:**WI(1342,1344). **55:**WI(1341,1348). **58:**WI(1334). **59:**ND(378). **70:**SD(2243), WY(1195).

P. dulichii. Heteroecious: aecial host. **46:**TN(82). **70:**NC(1303).

P. grindeliae. Autoecious. **44:**WY(1195). **70:**Unk.(82).

P. stipae. Heteroecious: aecial host. **18:**CO(82), IA(82), KS(82), ND(82), NE(82), SD(82). **31:**WI(1346). **40:**WI(1357). **41:**WI(1344). **70:**SD(2243).

Uromyces compactus. Autoecious. **61:**Unk.(82), AZ(1203,3981), TX(742).

U. junci. Heteroecious: aecial host. **36:**Unk.(82).

U. silphii. Heteroecious: aecial host. **36:**WI(1344).

Basidiomycotina - Ustilaginales

Entyloma compositarum. **10:**WY(1077). **13:**MA(82). **21:**ID(1195,3202), OR(3202), WA(3202), WY(1195). **29:**WA(1077,3202). **32:**WI(1344). **39:**MA(82,1077), ME(82,1077). **47:**WI(1355). **49:**WI(1342). **51:**MA(1077), NH(1077), WI(1077). **59:**WI(1077). **70:**DE(1077), PA(1077), SD(82), WA(3202), WI(82).

E. polysporum. White smut. **36:**WI(1077). **54:**WI(1077,1330). **70:**WI(82).

Basidiomycotina - Other

Ceratobasidium anceps. **49:**WI(1355). **59:**WI(1355).

Sebacina incrustans. Encrusting living leaves and stems. **70:**WV(128).

Thanatephorus cucumeris. **70:**WA(3202).

Typhula phacorrhiza. **70:**WA(3202).

Deuteromycotina - Hyphomycetes

Alternaria sp. Leaf spot. **34:**WI(1346). **57:**FL(35). **70:**FL(35), MI(82), TX(82), VT(82).

Botrytis cinerea. Gray mold blight. **70:**AK(82), CT(82), FL(35), NJ(82).

Cercospora asterata. Leaf spot. **3:**TX(82). **35:**WI(1333). **70:**AL(82), TX(82).

C. quarta. On leaves. **67:**WI(82,581,1335).

Cercosporella virgaureae. Leaf spot. **6:**MT(3202). **12:**ID(3202). **17:**MT(3202). **21:**WA(3202). **29:**ID(3202), MT(3202), WA(3202). **31:**WI(82,1337). **35:**WI(1359). **37:**ID(3202). **50:**WI(1334). **67:**WI(1335). **69:**IL(82,581). **70:**CO(82), MS(82,2720), OR(82), WI(82).

Cladosporium astericola. On leaves. **4:**WI(1346). **36:**WI(1361). **41:**WI(1339). **52:**WI(1335). **70:**IA(82), WI(82).

Coniothecium mollerianum. **18:**ND(378).

Denticularia tertia. On leaves. **50:**WI(82,1335).

Fusarium sp. Root and stem rot, wilt. **70:**FL(35), NC(1303).

Oidium sp. Powdery mildew. **70:**HI(2903).

Ovularia asteris. Leaf spot. **17:**WY(1195). **70:**WY(82).

Ovularia sp. **17:**ID(3202), MT(3202), WA(3202). **33:**OR(3202), WA(3202).

Ramularia asteris. Leaf spot. **21:**WY(1195). **31:**TX(82). **38:**IA(82), MI(82), NE(82), WI(82). **42:**WY(1195). **47:**WI(1332). **49:**WI(1358). **50:**IA(82), MI(82), NE(82), WI(82,1332). **70:**IA(82), MI(82), NE(82), TX(82), WI(82), WY(82).

R. filaris. **70:**CO(82), NH(82).

R. macrospora. **2:**WA(3202). **24:**WA(3202).

R. macrospora var. *asteris*. **38:**IA(82).

Septocylindrium sp. **52:**WI(1358).

Stemphylium botryosum. Leaf spot. **70:**FL(35).

Verticillium albo-atrum. Wilt. **70:**CT(82), OR(3202).

Deuteromycotina - Coelomycetes

Ascochyta compositarum. Leaf spot. **4:**WI(1354). **13:**WI(1361). **32:**WI(1356). **36:**WI(1358). **49:**WI(1358). **52:**WI(1358). **55:**WI(1354). **67:**WI(1349). **70:**WI(82).

Ascochyta sp. On leaves. **54:**WI(1346).

Asteromella asteris. On leaves. **70:**KS(82), NE(82).
Colletotrichum dematium. **34:**WI(1341).
Exophoma astericola. On leaves. **70:**IL(82).
Hendersonia leucelenes. On leaves and stems. **70:**NM(82).
Heteropatella umbilicata. **21:**WA(3202,3816).
Kellermania alpina. On stems. **70:**CA(82).
Leptothyrella graminis. **21:**WA(3816).
Leptothyrium astericola. Leaf spot. **3:**TX(82). **67:**WI(1358). **70:**TX(82), WI(82).
L. doellingeriae. Leaf spot. **70:**NY(82).
L. similisporum. **36:**WI(1355).
Macrophoma astericola. On leaves and stems. **70:**AL(82).
Microdiplodia leucelenes. On leaves and stems. **70:**NM(82).
Phoma iowana. On leaves. **35:**WI(1330). **50:**IA(82), WI(82,1332,1342). **70:**NH(82), WI(82).
P. pedicularis. **21:**WA(3202).
Phyllosticta sp. On living leaves. **33:**OR(3202). **34:**WI(1343). **36:**WI(1355). **38:**WI(1352). **40:**WI(1346). **47:**WI(1355). **49:**WI(1360). **51:**WI(1359). **52:**WI(1356,1357).
Placosphaeria decipiens. Leaf spot. **70:**NM(82).
P. haydenii. Black spot (tar spot) on bracts, leaves and stems. **18:**NE(82). **38:**IA(82), KS(82), ND(82), NJ(82), WI(82), WV(82).
Septoria angularis. On living leaves. **15:**TX(3597). **70:**TX(82).
S. astericola. Leaf spot. **13:**DE(82,966), MA(82), MI(82), OH(82), WI(82). **47:**WI(1332). **49:**WI(1340). **54:**WI(1355). **70:**WI(82).
S. asterina. On leaves. **15:**OK(2872), TX(3597). **70:**TX(82), WY(82).
S. atropurpurea. On leaves. **4:**WI(1344). **5:**WI(1355). **11:**WI(1351,1359). **13:**NE states(82), IA(82), WI(82,1343). **31:**NE states(82), IA(82), WI(82,1351). **35:**WI(1332). **36:**WI(1337,1351). **38:**NE states(82), IA(82), WI(82,1330). **49:**WI(1350). **51:**WI(1332). **52:**WI(1340). **54:**WI(1344). **70:**NE states(82), IA(82), MA(1390), MI(82), NY(82), OR(82), PA(82), TX(82), WI(82), WY(82).
S. fumosa. **52:**WI(1341). **66:**WI(1337).
S. punicea. **70:**MN(82), NY(82).
S. shastensis. **70:**CA(82).
S. solidaginicola. **13:**WI(1342). **58:**WI(1332). **67:**WI(1332).
S. solidaginis. **70:**IL(82), TX(82), WI(82).
Septoria sp. On leaves. **7:**WI(1357). **21:**WA(3202). **31:**WI(1354). **32:**WI(1358). **37:**WA(3202). **47:**WI(1344,1348). **55:**WI(1351). **59:**WI(1352).
Stagonospora astericola. On leaves. **18:**WI(1331,1355). **32:**WI(1355). **59:**WI(1359).
S. herbarum. On stems. **70:**CA(82).

Deuteromycotina - Other

Rhizoctonia solani. **70:**CA(12).
Rhizoctonia sp. On foliage, root rot, sore shin. **70:**HI(2903), NC(1303).
Sclerotium deciduum. Root rot, sore shin. **59:**WI(1342).
S. rolfsii. Southern stem rot. **70:**NC(1303).
Sclerotium sp. On leaves. **36:**WI(1344).

Baccharis L.

1. *B. emoryi* A. Gray
2. *B. glutinosa* Pers. [seep-willow, sticky baccharis, water-motie, water-willow]
3. *B. halimifolia* L. [consumption-weed, groundsel-bush, groundsel-tree, sea-myrtle, silverling]
4. *B. pilularis* DC. [chaparral-broom, coyote-brush, dwarf baccharis, kidneywort]
5. *B. pilularis* DC. subsp. *consanguinea* (DC.) C. Wolf
6. *B. sarothroides* A. Gray [rosin-brush]
7. *B. thesioides* H. B. K.
8. *B. vaccinioides* H. B. K.
9. *B. viminea* DC. [mule-fat]
10. *Baccharis* sp.

Ascomycotina

Botryosphaeria dothidea. Stem dieback. **3:**GA(1438). **10:**FL(35).
B. obtusa. **3:**GA(82).
B. rhodina. **10:**SC(82).
Capnodium salicinum. **4:**CA(81).
Dasyscyphus corticalis. **4:**CA(81).
Diaporthe sarothamni var. *baccharidis*. On stems. **3:**GA(82), NY(82).
Diaporthopsis sclerophila. Dieback, witches' broom. **5:**CA(81,330).
Diatrype prominens. On limbs. **4:**CA(81). **5:**CA(81).
Diatrypella favacea. **3:**GA(1438).
Dothidea puccinioides. Black pustule. **3:**FL(35,82), GA(1438), MD(82), NC(82,1303), NY(82), VA(82). **10:**CA(82), FL(35).
Erysiphe cichoracearum. Powdery mildew. **4:**CA(3975). **8:**CA(1155).
Eutypella goniostoma. on stems. (?1). **2:**CA(82).
Hypoxylon hypophlaeum. **3:**GA(1438).
Lasiostemma melioloides. Black mildew. **3:**FL(82), SC(82).
Leptosphaeria bicuspidata. On stems. **4:**CA(82).
L. conoidea. **10:**NY(3249).
Metasphaeria anisometra. On dead twigs. **9:**CA(2439).
Phaeosaccardinula dematia. **4:**CA(81,2434), WA(3202). **5:**CA(2434).
Phyllachora ambrosiae. Leaf spot. **10:**MS(2720).
Physalospora abdita. On stems. **10:**SC(82).
Pleosphaeria salicina. On bark. **4:**CA(81). **5:**CA(2434).
Stictis radiata. **10:**CA(3225).

Basidiomycotina - Uredinales

Puccinia baccharidis. Autoecious. **1:**AZ(1203,3981). **2:**AZ(82,747,1194,1195,1203,3981), CA(82), NM(82), TX(82,742). **3:**FL(35). **9:**AZ(3981), CA(81). **10:**AZ(82), CA(82).
P. evadens. Autoecious. **1:**AZ(747,1203,3981), CA(81). **3:**SE states(82), FL(35), TX(82). **4:**CA(81,82,638), OR(638,3202). **6:**AZ(747,1195,1203,3981). **7:**AZ(1203,3981). **10:**SW states(82), AZ(1203), FL(35), UT(82).
P. melampodii. **2:**TX(82).
P. pistorica. Autoecious. **3:**FL(35). **10:**FL(35,82).
Puccinia sp. **3:**NC(1303).
Uromyces compactus. Autoecious. **10:**Unk.(82).

Basidiomycotina - Other

Dacrymyces minor. **6:**AZ(1202).
Dendrophora albobadia. **3:**FL(543), VA(543). **6:**AZ(1202).
Ganoderma applanatum. Wood rot. **3:**NY(82).
Grandinia pruni. **6:**AZ(1202).
Hyphodermella corrugata. **4:**CA(81).
Irpex lacteus. Wood rot. **3:**MD(82).
Meruliopsis corium. **6:**AZ(1202).
Oxyporus latemarginatus. **6:**AZ(1202).
Peniophora incarnata. On branches. **10:**SC(634).
P. malenconii subsp. *americana*. On wood. **4:**CA(542).
P. tamaricicola. On wood. **6:**AZ(437,1180,1202).
Phanerochaete omnivora. White rot. **10:**AZ(433).
P. xerophila. **10:**AZ(433).
Phellinus conchatus. **3:**NY(82).
Phlebia ludoviciana. **6:**AZ(1202).
Schizophyllum commune. On wood. **3:**FL(35,82), NY(82). **10:**FL(35).
[*Stereum gausapatum*]. Wood rot. **4:**CA(82).
Stilbum sphaerocephalum. **4:**CA(81).

Deuteromycotina - Hyphomycetes

Alternaria sp. Leaf spot. **3:**FL(35). **10:**FL(35).
Cercospora baccharidis. Leaf spot. **2:**CA(82). **10:**CA(82)
C. palustrium. Leaf spot. **10:**FL(35).
Oidium sp. Powdery mildew. **3:**FL(35).
Phymatotrichopsis omnivora. Root rot. **10:**TX(82).

Deuteromycotina - Coelomycetes

Aplosporella baccharidicola. On stems. **3:**NY(82).
Phyllosticta baccharidis. Leaf spot. **3:**FL(35), NY(64,82).
Placosphaeria baccharidis. On stems. **10:**NY(82).
Stagonospora baccharidis. Leaf spot. **3:**FL(35). **10:**FL(35), TX(82), VA(82).
Stevensonula ciliata. On twigs. **3:**FL(2778).

Bahia Lag.

1. *B. dissecta* (A. Gray) Britton

Basidiomycotina - Uredinales

Puccinia canaliculata. **1:**AZ(3981).
Uromyces junci. Heteroecious: aecial host. **1:**AZ(1203).

Baileya Harv. & A. Gray ex A. Gray

1. *B. multiradiata* Harv. & A. Gray ex Torr.

Basidiomycotina - Uredinales

Puccinia grindeliae. Autoecious. **1:**AZ(1195,3379).
P. grindeliae subsp. *riddellii*. **1:**AZ(3981).

Balsamorhiza Nutt.

1. *B. careyana* A. Gray
2. *B. deltoidea* Nutt.
3. *B. hookeri* Nutt. var. *hirsuta* (Nutt.) A. Nels. (=*B. hirsuta* Nutt.)

4. *B. incana* Nutt.
5. *B. macrophylla* Nutt.
6. *B. sagittata* (Pursh) Nutt. [Oregon sunflower]
7. *Balsamorhiza* sp.

Ascomycotina

Comoclathris permunda. **6**:WY(3810).
Erysiphe cichoracearum. Powdery mildew. **6**:WY(82).
Mycosphaerella balsamorhizae. On stems. **7**:NV(82).
Pleospora chlamydospora. On stems. **2**:CO(82). **6**:CO(82), ID(3202), WA(3202).
P. coloradensis. **6**:MT(3202).
P. herbarum. **6**:MT(3202).

Basidiomycotina - Uredinales

Puccinia balsamorhizae. Autoecious. **1**:OR(3202). **2**:Unk.(82), CA(81), ID(3202), OR(3202), WA(3202). **3**:ID(3202). **4**:MT(3202). **5**:Unk.(82), UT(2701). **6**:W states(82), CA(82), ID(638,1195,3202), MT(1195,3202), OR(3202), WA(82,3202), WY(1195,2924). **7**:Unk.(82).

Basidiomycotina - Other

Typhula sp. **6**:ID(3202).

Deuteromycotina - Hyphomycetes

Pseudocercosporella herpotrichoides. On roots and stems. **7**:ID(82), OR(82), WA(82).

Deuteromycotina - Coelomycetes

Cylindrosporium balsamorhizae. Leaf spot. **6**:WY(128,1195).
Phloeospora balsamorhizae. **6**:WY(1195).
Phyllosticta ambrosiae. **2**:CA(81).
Septoria sp. Leaf spot. **2**:WA(82,3202).

Bellis L.

1. *B. perennis* L. [English daisy]

Oomycetes

Pythium mastophorum. Rootlet rot. **1**:MD(82).

Ascomycotina

Sclerotinia sclerotiorum. Crown rot. **1**:NJ(82).

Deuteromycotina - Hyphomycetes

Botrytis cinerea. Gray mold blight. **1**:AK(82).
Cercospora sp. Leaf spot. **1**:MN(82).
Phymatotrichopsis omnivora. Root rot. **1**:TX(82).

Bidens L.

1. *B. aristosa* (Michx.) Britton
2. *B. bipinnata* L. [beggarticks, Spanish-needles]
3. *B. cernua* L. [nodding beggarticks]
4. *B. coronata* (L.) Britton
5. *B. frondosa* L. [devil's beggarticks]
6. *B. laevis* (L.) B. S. P. (=*B. nashii* Small)
7. *B. pilosa* L. (=*B. leucantha* Willd.) [bur-marigold, hairy beggarticks, Spanish-needles]
8. *B. pinnata* Noronha
9. *B. tenuisecta* A. Gray
10. *B. vulgata* Greene [tall beggarticks]
11. *Bidens* sp.

Oomycetes

Phytophthora nicotianae var. *parasitica*. Root rot. **7**:HI(2903).
Plasmopara halstedii. Downy mildew. **5**:MS(2720). **7**:FL(35). **11**:E states(82), ND(82), TX(82).
Pythium oedochilum. Root rot. **11**:MD(82).

Ascomycotina

Diaporthe arctii. **2**:GA(1438).
Erysiphe cichoracearum. Powdery mildew. **5**:CO(82), IN(82), MA(82). **11**:FL(35), TN(82).
Hypoderma commune. **5**:GA(1438).
Leptosphaeria corallorhizae. **2**:NY(3249).
Massarina gloeospora. **2**:GA(1438).
Mollisia atrata. **5**:GA(1438).
Protomyces andinus. Leaf and stem gall. **2**:FL(35). **5**:MA(82), WI(82). **11**:MA(82), WI(82).
P. gravidus. Leaf and stem galls. **3**:WI(807). **5**:WI(807).
Sphaerotheca fuliginea. **10**:ND(377).
S. macularis. Powdery mildew. **1**:GA(1307). **2**:FL(35). **3**:WA(3202). **5**:Range of host(82), MS(2720), OR(3202), SD(2243), WA(3202). **6**:Range of host(82). **7**:Range of host(82), FL(35). **8**:Range of host(82). **10**:WA(3202). **11**:Range of host(82).

Basidiomycotina - Uredinales

Puccinia obtecta. Heteroecious: aecial host. **2**:FL(35). **3**:CO(3101). **5**:IA(82), IL(82), IN(82,3101), NE(82), NJ(82), WI(82). **11**:CO(82), KS(82), NE(82), NM(82), TX(82), WI(82).
Uromyces bidenticola. Autoecious. **2**:FL(35). **7**:CA(82), FL(35), Fl(82), HI(82,2903), NM(82). **11**:CA(82), FL(35), Fl(82), HI(82,2903), NM(82).
U. bidentis. Autoecious. **7**:CA(128), FL(128).
[*U. martinii*]. **2**:FL(35). **7**:FL(35).

Basidiomycotina - Ustilaginales

Entyloma compositarum. White smut. **2**:FL(35). **3**:WI(1353). **5**:IA(82,1077). **6**:KS(82,1077), WI(1077). **7**:FL(35), HI(2903). **9**:NM(1077). **10**:WI(1077). **11**:FL(35), NM(82), TX(82).
E. polysporum. **2**:FL(35). **5**:MA(82,1077), NJ(82,1077). **7**:FL(1077). **11**:FL(82).

Deuteromycotina - Hyphomycetes

Alternaria sp. **2**:FL(35).
Cercospora bidentis. Leaf spot. **2**:FL(35). **3**:WI(1332). **4**:WI(1334). **6**:TX(3597). **7**:FL(35). **11**:FL(35), TX(82), WI(82).
C. megalopotamica. Leaf spot. **7**:HI(2903). **11**:WI(82).
C. umbrata. Leaf spot. **3**:WI(1341). **4**:WI(1360). **5**:C states(82), SE states(82), MS(2720), TX(82), WI(82,1341). **11**:C states(82), SE states(82), TX(82), WI(82).
Corynespora sp. Leaf spot. **2**:FL(35).
Cristulariella moricola. On leaves. **5**:IL(2572).
Fusarium moniliforme. Leaf spot. **7**:FL(35).
Oidium sp. Powdery mildew. **2**:FL(35).
Phymatotrichopsis omnivora. Root rot. **11**:Unk.(82).
Septocylindrium concomitans. Leaf spot. **5**:IA(82), IN(82), NJ(82), WI(82). **6**:TX(82). **11**:IA(82), NC(82), WI(82).

Deuteromycotina - Coelomycetes

Colletotrichum sp. Anthracnose. **7**:FL(35).
Macrophomina phaseolina. Charcoal rot. **2**:IL(311).
Phoma exigua. On leaves. **5**:WI(64,82,1333).
Septoria bidentis. Leaf spot. **5**:WI(82). **8**:MO(82).

Deuteromycotina - Other

Sclerotium deciduum. **5**:WI(1342).

Bigelowia DC.

1. *Bigelowia* sp.

Ascomycotina

Schizoxylon ligustri. **1**:CO(3225).

Boltonia L'Hér.

1. *B. asteroides* (L.) L'Hér.
2. *B. asteroides* (L.) L'Hér. var. *latisquama* (A. Gray) Cronq. (=*B. latisquama* A. Gray)
3. *B. diffusa* Elliott

Ascomycotina

Discosphaerina boltoniae. Anthracnose, on stems. **3**:OK(82,179,2872).
Erysiphe cichoracearum. Powdery mildew. **1**:NE(82), SD(82).

Basidiomycotina - Uredinales

Puccinia dioicae. **1**:IA(82), ND(82,378), NE(82), SD(82,2243).
Uromyces compactus. Autoecious. **3**:TX(82).

Basidiomycotina - Ustilaginales

Entyloma compositarum. **1**:WI(82,1077).

Deuteromycotina - Hyphomycetes

Didymaria sp. **1**:WI(1340).

Deuteromycotina - Coelomycetes

Phoma boltoniae. **3**:OK(179,2872).
Selenophoma boltoniae. On stems. **3**:OK(82,179,2872).
Septoria erigerontis f. *boltoniae*. Leaf spot. **1**:IA(82), WI(82). **2**:OK(2872). **3**:OK(2872).

Borrichia Adans.

1. *B. arborescens* (L.) DC.
2. *B. frutescens* (L.) DC.

Basidiomycotina - Uredinales

Aecidium borrichiae. **2**:AL(746).
[*Puccinia balsamorhizae*]. **1**:FL(35).
P. mirifica. Autoecious. **1**:FL(35). **2**:TX(740,746).

Deuteromycotina - Hyphomycetes

Cercospora sp. Leaf spot. **1**:FL(35).
Cercosporidium sp. Leaf spot. **2**:FL(35).
Ramularia sp. Leaf spot. **1**:FL(35).

Brickellia Elliott

(= *Kuhnia* L.)
1. *B. amplexicaulis* B. L. Robinson
2. *B. betonicaefolia* A. Gray
3. *B. californica* (Torr. & A. Gray) A. Gray
4. *B. chlorolepis* (Woot. & Standl.) Shinners (=*Kuhnia chlorolepis* Woot. & Standl.)
5. *B. cordifolia* Elliott
6. *B. coulteri* A. Gray
7. *B. floribunda* A. Gray
8. *B. grandiflora* (Hook.) Nutt. [tassel-flower]
9. *B. lemmonii* A. Gray
10. *B. rusbyi* A. Gray
11. *B. simplex* A. Gray
12. *B. umbellata* (Greene) Rydb.
13. *Brickellia* sp.

Ascomycotina

Comoclathris permunda. **8:**WY(3810).

Basidiomycotina - Uredinales

Aecidium arcularium. **13:**AZ(82), CO(82), NM(82).
Coleosporium arizonicum. Telial host. **3:**AZ(745,746,3981), CA(745). **13:**CA(82).
C. steviae. Telial host. **3:**AZ(1195,1203).
Puccinia kuhniae. Autoecious. **1:**AZ(1194,1195,3981). **2:**AZ(1203,3981). **4:**AZ(3981). **5:**FL(35). **6:**AZ(1203). **9:**AZ(1195,1203). **11:**AZ(752,1203,3981). **13:**AZ(82), FL(82).
P. kuhniae var. *brickelliae*. **6:**AZ(745,746,3981).
P. kuhniae var. *robusta*. **4:**AZ(752), CO(752), NM(752), TX(752). **9:**AZ(752,3981).
P. poarum. **3:**AZ(1203,3981). **7:**AZ(1203,3981). **8:**AZ(1195,1203,3981). **10:**AZ(3981). **13:**AZ(1203).
P. subdecora. Autoecious. **8:**CA(642,752), CO(752), UT(752), WY(1195,3379). **13:**AZ(82), CO(82), NM(82), UT(82).

Deuteromycotina - Hyphomycetes

Cercospora coleosanthi. Leaf spot. **3:**CA(81,82,581). **12:**CO(82,581). **13:**CA(82), CO(82).
Phymatotrichopsis omnivora. Root rot. **13:**TX(82).

Cacalia L.

1. *C. atriplicifolia* L. [pale Indian plantain]
2. *C. muhlenbergii* (Schultz-Bip.) Fernald (=*C. reniformis* Muhl.) [great Indian plantain]
3. *C. plantaginea* (Raf.) Shinners (=*C. tuberosa* Nutt.)
4. *C. suaveolens* L.
5. *Cacalia* sp. (unnamed hybrids)

Oomycetes

Plasmopara halstedii. **4:**WI(1348).

Ascomycotina

Erysiphe cichoracearum. Powdery mildew. **1:**MS(2720).

Basidiomycotina - Uredinales

Aecidium mesadeniae. **2:**WI(746).

Deuteromycotina - Coelomycetes

Ascochyta compositarum. **2:**WI(1359).
Phyllosticta cacaliae. **1:**WI(1349). **3:**WI(1347). **4:**WI(1357). **5:** WI(1351).
P. pentstemonicola. On living leaves. **3:**WI(1338).
Septoria cacaliae. Leaf spot. **1:**FL(35), MS(2720). **3:**FL(35), MS(2720), OK(2872), WI(1337). **4:**WI(1331).

Calendula L.

1. *C. officinalis* L. [pot marigold]

Oomycetes

Pythium splendens. Root rot. **1:**FL(35).
P. ultimum. Root rot. **1:**CA(82).

Ascomycotina

Erysiphe cichoracearum. Powdery mildew. **1:**CA(82), FL(35).
E. polygoni. Powdery mildew. **1:**FL(35), PA(82).
Sclerotinia sclerotiorum. Stem rot. **1:**CA(82), FL(35,82), LA(82), MO(82), OH(82), TX(82).

Basidiomycotina - Uredinales

Puccinia melampodii. Autoecious. **1:**FL(35), GA(1439), IA(82), IL(82), IN(82), KS(82), MO(82), NE(82), OH(946), TX(82).

Basidiomycotina - Ustilaginales

Entyloma calendulae. White smut. **1:**CA(82,1077), NH(82,1077), NY(1077), OR(82,1077,3202).
E. compositarum. **1:**NY(1077), WA(82,1077,3202).

Deuteromycotina - Hyphomycetes

Alternaria sp. Leaf and stem spot. **1:**CA(149), FL(35), NY(82).
Botrytis cinerea. Gray mold blight. **1:**AK(82,526), FL(35), MO(82), NJ(82), NY(82).
Cercospora calendulae. Leaf spot. **1:**FL(35), PA(82), TX(82), VA(82).
Phymatotrichopsis omnivora. Root rot. **1:**TX(82).

Deuteromycotina - Coelomycetes

Colletotrichum gloeosporioides. Leaf spot. **1:**VA(82).

Deuteromycotina - Other

Rhizoctonia solani. Stem rot. **1:**FL(35), IN(82), NC(82), NJ(82), TX(82).
Sclerotium rolfsii. Southern blight. **1:**FL(35), TX(82).

Callistephus Cass.

1. *C. chinensis* (L.) Nees (=*Aster chinensis* L., *C. hortensis* (Cass.) Cass.) [annual aster, China aster]
2. *Callistephus* sp.

Oomycetes

Basidiophora entospora. Downy mildew. **1:**FL(35,82), TX(82).
Phytophthora cryptogea. **1:**CA(82), IL(82).
Phytophthora sp. Blackleg. **1:**MD(82).
Pythium splendens. Root rot. **1:**HI(2903).
P. ultimum. Root rot. **1:**CA(82), ND(82).
Pythium sp. Stem rot, wilt. **1:**FL(35).

Ascomycotina

Diaporthe arctii. On stems. **1:**NJ(82).
Erysiphe cichoracearum. Powdery mildew. **1:**DE(82), FL(35), MN(82), NC(82,1303), NE(82), VT(82), WA(82,3202).
E. polygoni. **1:**NJ(82).
Pleospora herbarum. Seed mold. **1:**Range of host(82).

Deuteromycotina - Hyphomycetes

Alternaria alternata. Flower, leaf and stem blight. **1:**FL(35), WI(1347). **2:**AK(526).
Alternaria sp. Leaf spot. **1:**Range of host(82), CA(149).
Botrytis cinerea. Flower, seedling and stem blight. **1:**AK(82,526), CA(82), CT(82), FL(35), IL(82), MA(1390), NJ(82), NY(82), OK(946), PA(82), WI(82).
Cercospora sp. On leaves. **1:**WI(1355).
Fusarium culmorum. Secondary stem rot following wilt. **1:**Unk.(82).
F. oxysporum f. sp. *callistephi*. Stem rot, wilt, yellows. **1:**Range of host(82), AL(285), FL(35), ID(3202), MS(2720), MT(3202), OR(3202), WA(3202).
F. oxysporum f. sp. *conglutinans*. Wilt. **1:**AL(285).
Fusarium sp. Wilt. **1:**HI(2903).
Phymatotrichopsis omnivora. Root rot. **1:**TX(82).
Stemphylium callistephi. Leaf spot. **1:**CA(148).
S. lycopersici. Leaf spot. **1:**FL(35).
Verticillium albo-atrum. Wilt. **1:**FL(35).
Verticillium sp. Wilt. **1:**CA(82), CT(82), IL(82), MA(82), NY(82).

Deuteromycotina - Coelomycetes

Ascochyta sp. Leaf spot. **1:**FL(35).
Camarosporium eriogoni. Leaf spot. **1:**FL(35).
Colletotrichum gloeosporioides. Anthracnose. **1:**FL(35).
Colletotrichum sp. Anthracnose. **1:**FL(82).
Phoma exigua. Leaf spot. **1:**CA(82), ND(82), NY(82), OH(82).
Phomopsis callistephi. Stem canker. **1:**HI(2903), IL(82), WI(82).
Septoria atropurpurea. Leaf spot. **1:**FL(35).
S. callistephi. Leaf spot. **1:**AL(82), DE(82), HI(2903), IL(82), MI(82), MO(82), NJ(82), NY(82), OH(82), PA(82), WA(3202).

Deuteromycotina - Other

Rhizoctonia solani. Dry rot, root rot. **1:**Unk.(82), FL(35), OK(2872).
Rhizoctonia sp. **2:**HI(2903).
Sclerotium rolfsii. Southern blight. **1:**MS(82,2720), NC(82,1303).

Calycadenia DC.

1. *C. truncata* DC. (=*Hemizonia truncata* (DC.) A. Gray)

Basidiomycotina - Uredinales

Puccinia nuda. **1:**OR(3202).

Carduus L.

1. *C. acanthoides* L. [plumeless thistle]
2. *C. tenuiflorus* Curtis

Basidiomycotina - Uredinales

Puccinia carduorum. **2:**CA(2839,3789).

Deuteromycotina - Coelomycetes
Stagonospora cirsii. **1**:WI(1350).

Carthamus L.

1. *C. tinctorius* L. [bastard saffron, false saffron, safflower, strawberry clover]

Oomycetes
Phytophthora cryptogea. Root and stem rot. **1**:CA(1924).
P. drechsleri. Root and stem rot. **1**:AZ(4016), CA(1924,4016), MT(4016), ND(4016), NE(999,4016).
P. nicotianae var. *parasitica*. Root and stem rot. **1**:CA(1924).
Plasmopara halstedii. **1**:SD(2243).

Ascomycotina
Erysiphe cichoracearum. **1**:SD(2243).
Erysiphe sp. Powdery mildew. **1**:UT(4013).
Sclerotinia sclerotiorum. Stem rot. **1**:CA(4016), IN(82), ND(82), VA(82), WA(3202).

Basidiomycotina - Uredinales
Puccinia calcitrapae var. *centaureae*. Autoecious. **1**:W states(1923), AZ(408,1195,1203,3981,4016), CA(4016), FL(35), MA(82), MT(4016), ND(4016), NE(82,3159,4016), OR(3202), SD(2243), WA(3202).

Deuteromycotina - Hyphomycetes
Alternaria sp. Leaf spot. **1**:AZ(4016), CA(4016), ID(3202), IN(82), MT(4016), ND(4016), NE(82,4016), SD(2243).
Botrytis cinerea. Head blight. **1**:CA(4016), ID(4016).
Botrytis sp. **1**:ID(3202).
Cercospora sp. **1**:SD(2243).
Fusarium oxysporum. On roots. **1**:MD(3604).
F. oxysporum f. sp. *carthami*. Wilt. **1**:CA(4016).
Oidium sp. Powdery mildew. **1**:AL(1411), AZ(257).
Stemphylium sp. Leaf spot. **1**:FL(35).
Verticillium albo-atrum. Wilt. **1**:AZ(4016), CA(4016), MT(4016), ND(4016), NE(4016).

Deuteromycotina - Coelomycetes
Gloeosporium carthami. Anthracnose. **1**:IN(82), SD(82,2243), TX(82), VA(82).
Phoma sp. On leaves. **1**:IN(82).
Pyrenochaeta lycopersici. **1**:CA(1371).
Septoria carthami. Leaf spot. **1**:IN(82), SD(2243), TX(82).
Septoria sp. Leaf spot. **1**:OK(2872).

Deuteromycotina - Other
Rhizoctonia solani. Blight, stem canker. **1**:NM(4017).

Centaurea L.

1. *C. americana* Nutt. [basket-flower, cardo-del-valle, thornless thistle]
2. *C. cineraria* L. [dusty-miller]
3. *C. cyanus* L. [bachelor's-button, bluebottle, cornflower]
4. *C. macrocephala* Mussin-Puschk.
5. *C. montana* L. [mountain bluet]
6. *C. repens* L. (=*C. picris* Pall.) [creeping knapweed, Russian knapweed]
7. *Centaurea* sp.

Oomycetes
Albugo tragopogonis. White rust. **7**:TX(82).
Bremia lactucae. Downy mildew. **3**:CA(82), OR(3202).
Phytophthora cactorum. Stem rot. **1**:NY(82). **3**:NY(82).
Plasmopara halstedii. Downy mildew. **7**:IA(82), TX(82).
Pythium debaryanum. Root rot. **3**:FL(35).
Pythium sp. Root rot. **3**:NJ(82).

Ascomycotina
Erysiphe cichoracearum. Powdery mildew. **1**:CA(82). **2**:CT(82). **7**:CA(82).
Sclerotinia sclerotiorum. Stem rot, wilt. **3**:FL(35), IL(312), MO(82), MS(82,2720). **5**:IN(82). **7**:CA(82), MS(82,2720), TX(82).

Basidiomycotina - Uredinales
Puccinia acroptili. Autoecious. **6**:AZ(747,3981).
P. calcitrapae var. *centaureae*. **1**:OK(2872), TX(82,740).
P. cyani. Autoecious. **3**:E states(82), CA(81,82), ID(3202), IL(312), IN(82), NC(1303,3944), OR(82,3202), WA(82,3202). **7**:C states(82), SE states(82), TX(82).

Deuteromycotina - Hyphomycetes
Cercospora centaureae. Stem rot, wilt. **3**:FL(35).
Cercospora sp. **4**:WI(1355).
Fusarium oxysporum f. sp. *callistephi*. Wilt. **3**:MI(82).
Phymatotrichopsis omnivora. Root rot. **1**:TX(82). **7**:TX(82).
Verticillium albo-atrum. Wilt. **3**:NY(82).

Deuteromycotina - Coelomycetes
Septoria centaureicola. **3**:OR(3202).
S. centaureicola var. *brevispora*. Leaf spot. **1**:OK(2872). **3**:IL(312), NY(82).
S. cyani. Leaf spot. **3**:OK(2872).
Septoria sp. On leaves and stems. **3**:OR(82).

Deuteromycotina - Other
Rhizoctonia solani. Thread blight, root rot. **3**:FL(35), NJ(82), NY(82), TX(82). **7**:IL(82), TX(82).
Sclerotium rolfsii. Southern blight. **3**:CT(82), FL(35), MD(82), TX(82). **7**:NJ(82).

Chaenactis DC.

1. *C. douglasii* (Hook.) Hook. & Arnott

Ascomycotina
Mycosphaerella minor. **1**:WA(3202).
M. tassiana. **1**:WA(3202).

Chondrilla L.

1. *C. juncea* L.

Basidiomycotina - Uredinales
Puccinia chondrillina. Autoecious. **1**:WA(11).

Chrysanthemum L.

1. *C. balsamita* L. var. *tanacetoides* (Boiss.) Boivin
2. *C. cinerariifolium* (Trevir.) Vis. (=*Tanacetum cinerariifolium* (Trevir.) Schultz-Bip., *Pyrethrum cinerariifolium* Trevir.) [Dalmatian insect-flower, Dalmatian pyrethrum, pyrethrum]
3. *C. coccineum* Willd. (=*C. roseum* Adams) [common pyrethrum, painted daisy, Persian insect-flower]
4. *C. frutescens* L. [marguerite, Paris daisy, white daisy, white marguerite, whiteweed]
5. *C. indicum* L.
6. *C. leucanthemum* L. [oxeye daisy]
7. *C. leucanthemum* L. var. *pinnatifidum* Lecoq & Lamotte
8. *C. maximum* Ramond [daisy chrysanthemum, Max chrysanthemum, Shasta daisy]
9. *C.* × *morifolium* Ramat. [florist's chrysanthemum, mum]
10. *C. parthenium* (L.) Bernh. (=*Tanacetum parthenium* (L.) Schultz-Bip.) [feverfew]
11. *C. segetum* L. [corn chrysanthemum, corn marigold]
12. *Chrysanthemum* sp.

Oomycetes
Peronospora leptosperma. Downy mildew. **9**:FL(35).
Phytophthora cryptogea. **9**:CA(1580).
P. nicotianae var. *parasitica*. Basal stem rot. **9**:FL(35).
Phytophthora sp. Root rot. **9**:NC(1303).
Pythium aphanidermatum. Damping-off, root and stem rot. **9**:FL(35), NC(1303).
P. debaryanum. Root and stem rot. **9**:FL(35).
P. irregulare. Root rot. **9**:FL(35).
P. megalacanthum. Root rot. **9**:FL(35).
P. paroecandrum. Root and stem rot. **9**:NC(1303).
P. splendens. Root rot. **9**:FL(35).
P. ultimum. Root rot. 'Pythium root rot.' **9**:FL(35), MD(82).
Pythium sp. Root rot. **3**:CO(82). **8**:NJ(82). **9**:HI(2903), NC(1303), ND(82), TN(82), TX(82).

Zygomycotina
Choanephora sp. Blossom blight. **9**:NC(1303).
Rhizopus stolonifer. Flower and leaf rot. **9**:FL(35).
Rhizopus sp. Following injury. **9**:NC(1303).

Ascomycotina
Diaporthe arctii. On stems. **4**:GA(82). **6**:GA(1438).
Diaporthe sp. **12**:WA(3202).
Didymella ligulicola. Spur blight, stem canker. **9**:'Ascochyta blight.' SE states(82), CA(82), MS(82), NC(1303), OH(82), SD(2243). **12**:MS(2720).
Erysiphe cichoracearum. Powdery mildew. **4**:NJ(82). **10**:NY(82). **9**:'Powdery mildew.' Range of host(82), FL(35), IL(312), NC(1303), SD(2243). **12**:FL(35), WA(3202).
Mycosphaerella tassiana. **6**:WA(3202,3815).
Nectria cinnabarina. Stem rot. **9**:FL(35). **12**:FL(35).
Nectriella pironii. Stem gall. **9**:FL(35).

Pleospora herbarum. **6:**WA(3202).
P. penicillus. **6:**WA(3202,3814).
Sclerotinia sclerotiorum. Stem rot. **3:**VA(82). **6:**TX(82). **8:**MT(82,3202), WA(82,3202). **9:**'Sclerotinia rot.' AZ(82), FL(35), HI(2903), LA(82), MI(82), NC(1303), VA(82). **12:**WA(3202).

Basidiomycotina - Uredinales

Puccinia horiana. Autoecious. **12:**NJ(2763).
P. tanaceti. Autoecious. **5:**CA(81), GA(1439), HI(2903). **8:**FL(35). **9:**'Rust.' Range of host(82), AZ(1203,3981), FL(35), HI(2903), IL(312), MT(3202), NC(1303), OK(2872), OR(3202). **12:**ID(3202), MS(2720), OK(2872), OR(3202), WA(3202), WI(1335).

Basidiomycotina - Other

Thanatephorus cucumeris. **10:**WA(3202).

Deuteromycotina - Hyphomycetes

Acremonium strictum. Vascular wilt. **8:**CA(555,564).
Alternaria alternata. Leaf spot. **9:**FL(35).
A. leucanthemi. Leaf spot. **8:**FL(35,3272). **9:**FL(35). **12:**FL(35).
Alternaria sp. On leaves and flowers. **9:**NC(1303), OK(82), TX(82). **12:**OK(2872).
Aspergillus sp. Secondary. **9:**NC(1303).
Bipolaris setariae. Flower petal spot. 'Bipolaris leaf spot.' **9:**FL(35).
Botryosporium sp. **9:**WA(3202).
Botrytis cinerea. Flower blight, gray mold blight. **3:**PA(82). **9:**'Gray mold.' Range of host(82), FL(35), HI(2903), NC(1303). **12:**FL(35), MS(2720), WA(3202).
Cercospora chrysanthemi. Leaf spot. **8:**FL(35), OK(82,2872). **9:**'Cercospora leaf spot.' AL(82), FL(35), LA(82), MD(82), NC(1303), PA(82), TX(82).
Cylindrocladium scoparium. Cutting rot. **9:**FL(35).
Fusarium moniliforme. Petiole rot, stem rot. **9:**FL(35).
F. oxysporum. Stem rot. 'Fusarium wilt.' **9:**FL(35).
F. oxysporum f. sp. *callistephi*. Wilt. **9:**OK(82).
F. oxysporum f. sp. *chrysanthemi*. Wilt. **9:**FL(35), NC(1303,3513).
[*F. oxysporum* f. sp. *conglutinans*]. Wilt. **12:**OK(2872).
[*F. oxysporum* f. sp. *tracheiphilum*]. Wilt. **9:**NC(3513).
F. poae. Flower rot. **12:**NJ(2762).
F. roseum. Root and stem rot. **6:**TX(82). **8:**TX(82). **9:**FL(35). **12:**MD(1729).
F. solani. Stem rot. **6:**TX(82). **8:**TX(82). **9:**'Stem rot.' FL(35,993). **12:**FL(35).
Fusarium sp. Root and stem rot, wilt. **8:**MS(2720). **9:**AL(82), FL(82), IL(82), NC(1303), NH(82), NJ(82), OH(82), OK(82), TX(82), VA(82). **12:**AL(285), MS(2720), OK(2872).
Itersonilia perplexans. Petal blight. **9:**'Itersonilia petal blight.' FL(35). **12:**FL(2334).
Oidium sp. Powdery mildew. **9:**HI(2903).
Phymatotrichopsis omnivora. Root rot. **2:**TX(82). **6:**TX(82). **9:**'Phymatotrichum root rot.' AZ(82), TX(82).
Sporobolomyces albo-rubescens. On leaves. **12:**KS(2686).
Stemphylium lycopersici. Flower rot, ray speck, leaf spot. **9:**'Ray speck.' FL(35,1711). **12:**FL(35).
Tilletiopsis minor. On leaves. **12:**KS(2686).
T. washingtonensis. On leaves. **12:**KS(2686).
Verticillium albo-atrum. Wilt. **4:**NJ(82). **6:**OR(3202). **9:**'Verticillium wilt.' FL(35), IL(82), IN(82), MA(82), MD(82), MI(82), MN(82), NC(1303), NJ(82), NY(82), OH(82), PA(82), WA(82,3202).

Deuteromycotina - Coelomycetes

Ascochyta sp. Leaf spot. **1:**WI(1339,1345).
Colletotrichum sp. Anthracnose. **3:**NC(1303).
Cylindrosporium chrysanthemi. Leaf spot. **6:**TX(82). **9:**'Cylindrosporium leaf spot.' E states(82), KS(82), NC(1303). **12:**AL(285), MS(2720), WI(1340).
Cylindrosporium sp. Leaf spot. **9:**FL(35).
Diplodia sp. Leaf spot. **9:**FL(35).
Macrophoma sp. Stem necrosis. **9:**FL(35).
Macrophomina phaseolina. Charcoal stem rot. **9:**'Charcoal stem rot.' OK(82). **12:**OK(2872).
Melanconium sp. Black stem. **9:**FL(35).
Phoma chrysanthemi. Stem and ray blight, stem canker. **9:**FL(35), HI(2903), NC(1303). **12:**FL(35).
Phomopsis sp. Petiole rot, stem rot. **9:**FL(35).
Phyllosticta chrysanthemi. Leaf spot. **8:**FL(35). **9:**FL(35,82), MA(82), MS(82), NC(1303), VA(82).
Septoria chrysanthemella. Leaf spot. **4:**HI(2903). **7:**WI(1332). **12:**OK(2872), WI(1351).
S. chrysanthemi. Leaf spot. **6:**CT(82), IA(82), NY(82), TX(82), VT(82), WA(82). **11:**NY(82). **9:**'Septoria leaf spot.' Unk.(82), FL(35), NC(1303), SD(2243). **12:**MS(2720), WI(1351).
S. leucanthemi. Leaf blotch. **6:**CT(82), NY(82), WI(1340). **8:**CA(82), HI(2899,2903), OR(82,3202), WA(3202). **9:**CT(82), IL(312), MD(82), NJ(82), NY(82), TX(82), WA(82,3202). **12:**CA(81), OK(2872), TX(3597).
S. obesa. Leaf blotch, spot. **9:**FL(35).
S. socia. **1:**WI(1341).
Septoria sp. On leaves. **3:**WI(1355). **4:**OK(2872). **7:**WI(1332). **9:**HI(2899,2903), OR(3202).

Deuteromycotina - Other

Rhizoctonia solani. Damping-off, root and stem rot. **3:**NJ(82). **8:**MD(82). **10:**WA(82). **9:**'Rhizoctonia stem rot.' Range of host(82), FL(35), NC(1303). **12:**MS(2720).
Rhizoctonia sp. Root rot. **9:**HI(2903).
Sclerotium rolfsii. Southern blight. **6:**TX(82). **9:**'Southern blight.' FL(35,82), HI(2903), NC(1303), VA(82). **12:**MS(2720).

Chrysogonum L.

1. *C. virginianum* L.

Deuteromycotina - Other

Sclerotium rolfsii. Southern blight. **1:**NC(1303).

Chrysopsis (Nutt.) Elliott

1. *C. asprella* Greene
2. *C. breweri* A. Gray
3. *C. graminifolia* Elliott [silk-grass]
4. *C. mariana* (L.) Elliott
5. *C. oregana* A. Gray
6. *C. villosa* (Pursh) Nutt. (=*C. berlandieri* Greene, *C. foliosa* Nutt., *C. fulcrata* Greene)
7. *Chrysopsis* sp.

Oomycetes

Basidiophora entospora. Downy mildew. **7:**FL(35).

Ascomycotina

Erysiphe cichoracearum. Powdery mildew. **6:**MT(3202). **7:**MT(82), WY(82).
Leptosphaeria doliolum. **4:**GA(1438).
Massarina gloeospora. **3:**GA(1438).
Pyrenopeziza polygoni. **3:**GA(1438).

Basidiomycotina - Uredinales

Coleosporium asterum. Heteroecious: telial host. **5:**CA(81). **7:**CO(82), FL(35,82), NE(82).
Puccinia cyperi. Heteroecious: aecial host. **7:**FL(35).
P. grindeliae. Autoecious. **7:**CA(82), CO(82), OK(82), UT(82), WY(82).
P. marianae. Autoecious. **7:**FL(35).
P. stipae. Heteroecious: aecial host. **6:**AZ(1203,3981), MT(1195,3202). **7:**AZ(82), CO(82), FL(35,82), MT(82), NE(82), WY(82).

Basidiomycotina - Ustilaginales

Entyloma compositarum. **1:**CO(1077). **6:**CO(1077).

Deuteromycotina - Hyphomycetes

Cercospora macroguttata. Leaf spot. **3:**AL(82,581), MS(82,581,2720). **7:**FL(35).
Ovularia asteris. **2:**CA(81).
Ramularia chrysopsidis. On living leaves. **4:**NY(835). **7:**NY(82).

Chrysothamnus Nutt.

1. *C. nauseosus* (Pall. ex Pursh) Britton [rubber rabbitbrush]
2. *C. nauseosus* (Pall. ex Pursh) Britton subsp. *albicaulis* (Nutt.) H. Hall & Clements (=*C. nauseosus* (Pall. ex Pursh) Britton var. *occidentalis* H. Hall)
3. *C. nauseosus* (Pall. ex Pursh) Britton subsp. *graveolens* (Nutt.) Piper (=*C. graveolens* (Nutt.) Greene)
4. *C. nauseosus* (Pall. ex Pursh) Britton subsp. *nauseosus* (=*C. plattensis* Greene)
5. *C. parryi* (A. Gray) Greene
6. *C. pulchellus* (A. Gray) Greene (=*C. baileyi* Woot. & Standl.)
7. *C. viscidiflorus* (Hook.) Nutt. [Douglas rabbitbrush]
8. *C. viscidiflorus* (Hook.) Nutt. subsp. *lanceolatus* (Nutt.) Greene
9. *C. viscidiflorus* (Hook.) Nutt. subsp. *pumilus* (Nutt.) H. Hall & Clements (=*C. pumilus* Nutt.)

10. *Chrysothamnus* sp.

Ascomycotina

Cucurbitaria umbilicata. Leaf spot. **1:**ND(82). **3:**ND(377).
Dothidea bigeloviae. On stems. **10:**MT(82,3202).
Erysiphe cichoracearum. Powdery mildew. **1:**MT(3202). **7:**MT(82,3202), NM(82), WY(82). **9:**WY(1195). **10:**CA(82), WA(3202).
E. polygoni. **1:**MT(3202).
E. pseudosepulta. Powdery mildew. **7:**UT(369).
Leptosphaeria arthrophyma. On stems. **1:**NM(82,128,1013).
Lophiostoma caulium. **1:**ID(3202).
L. caulium var. *congregatum*. **2:**CA(661).
Melanomma occidentale. **3:**ND(377).
Rosellinia ovalis. **3:**ND(377).
Xylopezia biseptata. On decorticated wood. **1:**OR(3233).

Basidiomycotina - Uredinales

Puccinia grindeliae. Autoecious. **1:**CO(1195), MT(1195,3202), NM(1195). **4:**MT(3202). **6:**TX(742). **7:**ID(3202), MT(3202), WA(3202), WY(1195).
P. stipae. Heteroecious: aecial host. **1:**WA(3202). **7:**MT(3202), WY(1195). **8:**WY(3379). **10:**ID(3202).

Basidiomycotina - Other

Peniophora malenconii subsp. *americana*. **1:**WA(3202).
Peniophora sp. **1:**WA(3202).

Deuteromycotina - Hyphomycetes

Cylindrocolla bigeloviae. On stems. **10:**CO(82).
Epochnium isthmophorum. On limbs and stems. **1:**ID(3202), NV(82), UT(82), WA(82). **2:**CA(638,642,661).
Stemphylium botryosum. **10:**MT(3202).
Stigmina ellisii. On stems. **10:**CO(82).
S. utahensis. On dead limbs. **1:**UT(128,3082).

Deuteromycotina - Coelomycetes

Camarosporium rosellinioides. On stems. **10:**CO(82).
Diplodia sp. On leaves. **3:**WI(1357).
Macrophoma chrysothamni. On leaves. **2:**CA(642,666).
Marssonina chrysothamni. **5:**WY(1195).
Phoma terrestris. Isolated from rhizosphere. **1:**OR(2791).

Cichorium L.

1. *C. endivia* L. [endive, escarole]
2. *C. intybus* L. [barbe-de-capuchin, blue-sailors, common chicory, succory, witloof]

Oomycetes

Bremia lactucae. Downy mildew. **2:**FL(35,82), PA(82).
Phytophthora cryptogea. Root rot. **2:**AZ(3467).
Pythium debaryanum. Root rot. **1:**CT(82), FL(35,82).

Ascomycotina

Erysiphe cichoracearum. Powdery mildew. **1:**FL(35). **2:**ID(82,3202), NJ(82).
Sclerotinia sclerotiorum. Drop, watery soft rot. **1:**AZ(1203), FL(35), MT(3202). **2:**TX(82), WA(82,3202).
Sclerotinia sp. Leaf burn. **1:**NC(1303).

Basidiomycotina - Uredinales

Puccinia hieracii. Autoecious. **1:**CA(82), CT(82), MA(82), NY(82). **2:**CA(81,82), CT(82), ID(3202), MA(2713), NY(82), OR(82,3202), PA(82), VT(82,2713), WA(82,3202), WI(82).

Deuteromycotina - Hyphomycetes

Alternaria sonchi. Leaf spot. **1:**FL(35). **2:**FL(35).
A. tenuissima. Leaf spot. **2:**FL(35).
Alternaria sp. Leaf spot. **1:**CT(82), FL(82), NY(82). **2:**FL(1203), WI(1203).
Botrytis cinerea. Gray mold rot. **1:**CA(82), FL(35,82), NY(82), PA(82).
Cercospora cichorii. Leaf spot. **1:**TX(82). **2:**MI(1203), NY(1203), SD(2243), TX(1203), WI(581,1203).
Fusarium sp. **1:**NC(1303).
Microdochium panattonianum. Leaf spot. **1:**TX(82).
Phymatotrichopsis omnivora. Root rot. **1:**AZ(82), TX(82). **2:**AZ(3467), TX(82).
Ramularia cichorii. Leaf spot. **2:**NY(82).

Deuteromycotina - Coelomycetes

Phoma sp. Root rot. **1:**NY(82).

Deuteromycotina - Other

Rhizoctonia solani. Damping-off, bottom rot. **1:**CT(1203), FL(35,1203), NY(1203), TX(1203). **2:**FL(35), TX(82).
Sclerotium rolfsii. Southern blight. **1:**FL(35,82), TX(82). **2:**FL(35,82), TX(82).

Cirsium Mill.

1. *C. altissimum* (L.) Spreng. (=*Cnicus altissimus* (L.) Willd.)
2. *C. arizonicum* (A. Gray) Petr. (=*C. nidulum* (M. E. Jones) Petr.)
3. *C. arvense* (L.) Scop. (=*Cnicus arvensis* (L.) Roth) [Canada thistle]
4. *C. brevifolium* Nutt.
5. *C. brevistylum* Cronq.
6. *C. calcareum* (M. E. Jones) Woot. & Standl. (=*C. pulchellum* (Greene) Woot. & Standl.)
7. *C. callilepis* (Greene) Jeps.
8. *C. canescens* Nutt.
9. *C. canovirens* (Rydb.) Petr.
10. *C. centaureae* (Rydb.) K. Schum. (=*Carduus centaureae* Rydb., *C. americanum* (A. Gray) F. P. Daniels)
11. *C. discolor* (Muhl. ex Willd.) Spreng.
12. *C. douglasii* DC. (=*C. breweri* (A. Gray) Jeps.)
13. *C. douglasii* DC. var. *canescens* (Petr.) J. T. Howell
14. *C. drummondii* Torr. & A. Gray
15. *C. eatonii* (A. Gray) B. L. Robinson (=*Carduus leiocephalus* (D. C. Eaton) A. Heller)
16. *C. edule* Nutt.
17. *C. engelmannii* Rydb.
18. *C. flodmanii* (Rydb.) Arth.
19. *C. foliosum* (Hook.) DC.
20. *C. hillii* (Canby) Fernald
21. *C. hookerianum* Nutt.
22. *C. horridulum* Michx.
23. *C. muticum* Michx.
24. *C. neomexicanum* A. Gray
25. *C. occidentale* (Nutt.) Jeps.
26. *C. occidentale* (Nutt.) Jeps. var. *coulteri* Jeps. (=*C. coulteri* (Jeps.) Harv. & A. Gray)
27. *C. ochrocentrum* A. Gray
28. *C. parryi* (A. Gray) Petr.
29. *C. pulcherrimum* (Rydb.) K. Schum.
30. *C. remotifolium* (Hook.) DC. (=*Cnicus remotifolius* (Hook.) A. Gray)
31. *C. scariosum* Nutt.
32. *C. scopulorum* (Greene) Cockerell (=*Carduus scopulorum* Greene)
33. *C. subcoriaceum* Schultz-Bip. ex Seem.
34. *C. subniveum* Rydb. (=*C. wallowense* Peck)
35. *C. undulatum* (Nutt.) Spreng. (=*Carduus undulatus* Nutt., *Cnicus undulatus* (Nutt.) A. Gray)
36. *C. undulatum* (Nutt.) Spreng. var. *megacephalum* (A. Gray) Fernald
37. *C. vulgare* (Savi) Ten. (=*C. lanceolatum* Scop.) [bull thistle]
38. *C. wrightii* A. Gray
39. *Cirsium* sp.

Oomycetes

Albugo tragopogonis. **22:**FL(35). **36:**OK(2872).

Ascomycotina

Acrospermum foliicola. **1:**GA(1438).
Botryosphaeria dothidea. **1:**GA(1438).
Calycina herbarum. **37:**ID(3202).
Diaporthe arctii. **1:**GA(1438).
Erysiphe cichoracearum. Powdery mildew. **1:**SD(2243), WI(1330). **18:**ND(377). **22:**FL(35). **33:**CA(1155). **35:**MT(3202).
Hyponectria sceptri. **10:**WY(3084).
Leptosphaeria doliolum. **39:**NY(3249).
Metasphaeria sp. **1:**GA(1438).
Mycosphaerella tassiana. **37:**WA(3202).
Nematostoma occidentalis. **1:**WI(1347). **11:**WI(1335).
Nodulosphaeria modesta. **39:**OR(657).
Ophiobolus acuminatus. **3:**NY(3248). **18:**ND(377). **37:**ID(3202), WA(3815).
O. anguillidus. **1:**GA(1438). **39:**IL(3248).
O. cirsii f. *furcatus*. **18:**ND(377).
O. cirsii-altissima. **1:**GA(1438).
O. ophioboloides. **32:**CO(3248).
Phyllactinia guttata. **37:**WA(3202).
Pleospora herbarum. On dead stems. **25:**CA(2439).
P. richtophensis var. *pallida*. **39:**OR(657).
Pyrenopeziza carduorum. **37:**ID(3202).
Sclerotinia sclerotiorum. **3:**MT(396,3202), WA(3202).

Basidiomycotina - Uredinales
Puccinia altissimorum. Autoecious. **11:**DE(746).
P. altissimorum var. *horriduli*. Autoecious. **22:**LA(746).
[*P. calcitrapae*]. **10:**WY(1195). **13:**CA(661). **35:**WY(1195).
P. calcitrapae var. *centaureae*. Autoecious. **1:**E states(82), CA(82), OR(82), SD(2243), TX(82). **4:**ID(3202), OR(3202), WA(3202). **5:**ID(3202), WA(3202). **6:**AZ(1203,3981). **8:**WY(1195). **9:**OR(3202). **10:**OR(3202), WA(3202), WY(1195). **11:**OK(2872). **12:**CA(81), WA(3202). **14:**AZ(3981). **16:**ID(3202), MT(1195,3202), OR(3202), WA(3202). **17:**WY(1195). **18:**MT(3202), ND(378). **19:**ID(3202), MT(3202), OR(3202), WY(1195). **21:**ID(3202), MT(3202). **22:**FL(35). **25:**CA(81). **26:**CA(81). **27:**AZ(1203,3981). **28:**AZ(1203,3981). **29:**ID(3202). **30:**CA(81), WA(3202). **32:**MT(3202). **35:**MT(1195,3202), OK(2872), OR(3202), SD(2243), WA(3202). **37:**CA(81), ID(638), MS(2720). **39:**AZ(1195,1203,3379), WY(2924).
P. californica. Autoecious. **2:**AZ(747,3981). **12:**CA(746). **38:**AZ(3981).
P. californica var. *deserticola*. **24:**AZ(3981).
P. cnici. Autoecious. **7:**OR(3202). **31:**UT(1195). **37:**CA(81), ID(3202), MT(3202), OR(3202), WA(3202).
P. inclusa. Autoecious. **3:**OR(3202). **4:**ID(3202), WA(3202). **5:**OR(3202). **16:**WA(3202). **18:**MT(3202). **30:**WA(3202). **34:**OR(3202). **35:**KS(746), MT(3202), OR(3202), WA(3202).
P. punctiformis. Autoecious. **3:**CA(81), ID(3202), MA(1394), MT(1195,3202,3696), OR(3202), SD(2243), WA(3202), WY(1195,3379). **35:**OK(2872).
Uromyces junci. Heteroecious: aecial host. **18:**MT(3202), ND(378). **21:**MT(3202). **22:**FL(35).
Basidiomycotina - Ustilaginales
Thecaphora trailii. **10:**CO(920). **15:**UT(1077). **27:**CO(920,1077). **35:**CO(920,1077). **39:**UT(1077).
Basidiomycotina - Other
Thanatephorus cucumeris. **3:**WA(3202).
Deuteromycotina - Hyphomycetes
Cercospora cirsii. Leaf spot. **30:**WA(581,3202).
C. kansensis. Leaf spot. **1:**KS(581).
C. obesa. Leaf spot. **35:**KS(581).
Fusarium sp. **3:**WA(3202).
Ramularia cynarae. **3:**WA(3202). **37:**WA(3202).
Deuteromycotina - Coelomycetes
Chaetosticta perforata. On leaves. **1:**IA(718), IL(718), KS(718). **11:**IL(718), WI(1355). **23:**WI(1355).
Macrophomina phaseolina. Charcoal rot. **3:**IL(311).
Phyllosticta cacaliae. **11:**WI(1349).
P. cirsii. **23:**WI(1359). **35:**WA(3202), WI(1340). **37:**WI(1335).
Septoria cirsii. Leaf spot. **3:**MS(2720). **20:**WI(1338).
S. commonsii. On leaves. **1:**DE(966), WI(1337).
Stagonospora cirsii. Leaf spot. **1:**WI(82). **11:**WI(1352).

Cnicus L.

1. *C. benedictus* L. [blessed thistle]
2. *Cnicus* sp.

Ascomycotina
Sclerotinia sclerotiorum. Stem rot. **2:**FL(35).
Basidiomycotina - Uredinales
Puccinia hieracii. **2:**ND(378).
Deuteromycotina - Other
Sclerotium rolfsii. Southern blight. **1:**GA(82).

Conyza Less.

1. *C. bonariensis* (L.) Cronq. (=*Erigeron bonariensis* L.)
2. *C. canadensis* (L.) Cronq. (=*Erigeron canadensis* L., *Leptilon canadense* (L.) Britton & A. Brown) [butterweed, hogweed, horseweed]

Oomycetes
Basidiophora entospora. Downy mildew. **2:**IL(82), NC(3944), TX(82).
Ascomycotina
Crocicreas nigrescens. **2:**NJ(505).
Diaporthe arctii. On stems. **2:**NJ(82).
Diaporthopsis apiculosa. On stems. **2:**NJ(82).
Didymosphaeria futilis. **2:**GA(1438).
Erysiphe cichoracearum. Powdery mildew. **1:**CA(1155). **2:**MT(82,3202), TX(82).
Heptameria obesa. **2:**GA(1438).
Hypoderma commune. **2:**GA(1438).
Leptosphaeria erigerontis. **2:**NJ(3249).
L. longipedicellata. **2:**GA(1438).
L. planiuscula. **2:**ND(377).
Mollisia atrata. **2:**GA(1438).
M. exigua. On old stems. **2:**NJ(850).
Sphaerotheca fuliginea. Powdery mildew. **2:**IA(82), WI(1336).
S. macularis. Powdery mildew. **2:**IA(82), ID(3202), MD(82), MT(82,3202), WA(3202).
Stictis radiata. **2:**GA(1438).
Basidiomycotina - Uredinales
Puccinia cyperi. **2:**NE states(82), OK(2872), TX(82), WI(82).
P. dioicae. **2:**E states(82), FL(35), MS(82), SD(82), TX(82).
Deuteromycotina - Hyphomycetes
Phymatotrichopsis omnivora. Root rot. **2:**TX(82).
Ramularia macrospora. Leaf spot. **2:**ID(3202), OR(3202), TX(82), WA(3202).
Deuteromycotina - Coelomycetes
Colletotrichum sp. Dead stems. **2:**NC(1303).
Macrophomina phaseolina. Charcoal stem rot. **2:**IL(82,311).
Septoria erigerontis. Leaf spot. **2:**E states(82), OK(2872), TX(82), WI(82).
Deuteromycotina - Other
Sclerotium rolfsii. Southern blight. **2:**FL(35,82).

Coreopsis L.

1. *C. grandiflora* Hogg ex Sweet
2. *C. lanceolata* L.
3. *C. major* Walter (=*C. oemleri* Elliott, *C. major* Walter var. *oemleri* (Elliott) Britton)
4. *C. palmata* Nutt.
5. *C. tinctoria* Nutt. [calliopsis, plains coreopsis]
6. *Coreopsis* sp.

Oomycetes
Plasmopara halstedii. Downy mildew. **6:**FL(35).
Ascomycotina
Acrospermum foliicola. **3:**GA(1438).
Erysiphe cichoracearum. Powdery mildew. **1:**IL(312), NY(82). **4:**WI(1333). **5:**MD(82), MN(82), WI(82).
Massarina gloeospora. **3:**GA(1438).
Ophiobolus anguillidus. **3:**GA(1438).
O. nigro-clypeata. **3:**GA(1438).
Sclerotinia sclerotiorum. Stem rot. **1:**OH(1203), WA(1203). **5:**OH(82), WA(82). **6:**WA(3202).
Sphaerotheca macularis. **6:**WA(3202).
Basidiomycotina - Uredinales
Coleosporium inconspicuum. Heteroecious: telial host. **1:**NC(82,3944). **2:**NC(1203), TN(82,1203). **3:**GA(1439). **5:**GA(1203), MD(1203), NC(1203), OH(1203), SC(1203), TN(82), VA(1203), WV(1203).
Deuteromycotina - Hyphomycetes
Botrytis cinerea. Gray mold blight. **1:**AK(82). **6:**AK(526).
Cercospora bidentis. Leaf spot. **6:**FL(35).
C. coreopsidis. Leaf spot. **1:**OK(82,581,2872).
Cercospora sp. Leaf spot. **6:**MS(2720).
Cladosporium coreopsidis. On living leaves. **4:**WI(1350).
Cladosporium sp. On living leaves. **4:**WI(1348).
Oidium sp. Powdery mildew. **6:**HI(2903).
Phymatotrichopsis omnivora. Root rot. **6:**TX(82).
Verticillium albo-atrum. Wilt. **2:**NY(82).
Deuteromycotina - Coelomycetes
Phyllosticta coreopsidis. On leaves and stems. **4:**WI(1335). **5:**WI(82).
Septoria coreopsidis. Leaf spot. **5:**IA(82), WI(82). **6:**TX(82).
Deuteromycotina - Other
Rhizoctonia solani. Root and stem rot. **2:**NY(82). **5:**MN(1203), TX(1203). **6:**TX(82).
Rhizoctonia sp. Root rot. **2:**HI(2903).
Sclerotium rolfsii. Southern blight. **2:**FL(82). **5:**AZ(1203). **6:**FL(35), HI(2903).

Cosmos Cav.

1. *C. bipinnatus* Cav. [common cosmos]
2. *C. parviflorus* (Jacq.) Pers.
3. *C. sulphureus* Cav. [orange cosmos, yellow cosmos]

4. *Cosmos* sp.

Oomycetes

Pythium sp. Root rot. **1:**MD(82).

Ascomycotina

Diaporthe stewartii. Blight, stem canker. **1:**NE states(82), CA(82), KS(82), SD(82), TX(82). **3:**TX(82).

Erysiphe cichoracearum. Powdery mildew. **1:**CA(82), IA(82), MD(82), NC(82), NE(82), TX(82). **3:**TX(82).

Basidiomycotina - Uredinales

Puccinia canaliculata. **2:**AZ(3981).

Basidiomycotina - Ustilaginales

Entyloma calendulae. **3:**LA(1077).

Deuteromycotina - Hyphomycetes

Cercospora zinniae. Leaf spot. **4:**FL(35).

Cercospora sp. Leaf spot. **1:**TX(82).

Fusarium sp. Wilt. **1:**NJ(82). **4:**FL(35).

Oidium sp. Powdery mildew. **1:**HI(2903).

Phymatotrichopsis omnivora. Root rot. **1:**AZ(82), TX(82). **3:**AZ(82), TX(82).

Deuteromycotina - Coelomycetes

Septoria sp. Leaf spot. **1:**CT(82).

Deuteromycotina - Other

Rhizoctonia solani. Root and stem rot. **1:**CT(82), MD(82), TX(82). **3:**TX(82).

Sclerotium rolfsii. Southern blight. **1:**MS(82,2720).

Crepis L.

1. *C. acuminata* Nutt.
2. *C. atribarba* A. Heller
3. *C. bakeri* Greene
4. *C. barbigera* Leiberg ex Coville
5. *C. capillaris* (L.) Wallr. [smooth hawksbeard]
6. *C. intermedia* A. Gray
7. *C. modocensis* Greene
8. *C. occidentalis* Nutt. [Western hawksbeard]
9. *C. pleurocarpa* A. Gray (=*C. acuminata* Nutt. var. *pleurocarpa* (A. Gray) Jeps.)
10. *C. pygmaea* L.
11. *C. runcinata* (E. James) Torr. & A. Gray (=*C. glauca* Torr. & A. Gray, *C. riparia* A. Nels.)
12. *C. tectorum* L. [narrowleaf hawksbeard]
13. *Crepis* sp.

Ascomycotina

Erysiphe cichoracearum. Powdery mildew. **2:**WA(3202). **11:**PA(82). **13:**NE(82).

Pleospora herbarum. **10:**AK(526).

Sphaerotheca macularis. Powdery mildew. **11:**WY(82).

Basidiomycotina - Uredinales

Puccinia crepidis-montanae. Autoecious. **1:**WY(82). **2:**WA(3202). **4:**WA(2713,3202). **6:**ID(3202), MT(1872), OR(3202), WA(1872). **8:**MT(82,3202), OR(3202). **11:**CO(82), MT(1195,3202), WY(1195,2713). **13:**CO(82), ID(82), OR(82), UT(82), WA(82).

P. dioicae. Heteroecious: aecial host. **6:**WA(3202). **11:**MT(82,3202), NE(82). **13:**MT(82), NE(82), NV(82), WA(82).

P. hieracii. Autoecious. **1:**W states(82), CA(82,2713), CO(1195,3379), ID(3202), MT(3202), OR(3202), WA(82,3202). **2:**ID(3202). **3:**OR(2713). **6:**WA(3202), WY(2713). **7:**ID(3202). **8:**W states(82), CA(82), UT(2713), WA(82). **9:**CA(81). **11:**W states(82), CA(82), MT(3202), OR(3202), WA(82,3202). **13:**W states(82), CA(82), WA(82).

P. stipae. Heteroecious: aecial host. **1:**MT(82,3202). **13:**UT(82), WA(82).

Basidiomycotina - Ustilaginales

Entyloma sp. **11:**WA(3202).

Deuteromycotina - Hyphomycetes

Cercoseptoria crepidis. **11:**WY(1195).

Cercospora stromatis. Leaf spot. **1:**CO(82).

Ramularia crepidis. Leaf spot. **11:**NM(82,964). **13:**NM(82).

Deuteromycotina - Coelomycetes

Phyllosticta eximia. Leaf spot. **1:**CO(82). **5:**WA(3202). **8:**ID(3202).

Septoria crepidis. **12:**WI(1341).

Cynara L.

1. *C. cardunculus* L. (=*C. scolymus* L.) [artichoke, cardoon, globe artichoke]

Oomycetes

Phytophthora megasperma. Root rot. **1:**CA(82).

Ascomycotina

Erysiphe cichoracearum. Powdery mildew. **1:**CA(82).

Sclerotinia sclerotiorum. Stem rot. **1:**OR(82,3202), WA(3202).

Deuteromycotina - Hyphomycetes

Botrytis cinerea. Gray mold blight. **1:**CA(82), NY(82).

Cercospora obscura. Leaf spot. **1:**CA(82), TX(82,581).

Cladosporium sp. Secondary leaf spot. **1:**AZ(1203).

Phymatotrichopsis omnivora. Root rot. **1:**TX(82).

Ramularia cynarae. Leaf spot. **1:**CA(82).

Deuteromycotina - Other

Rhizoctonia solani. Root and stem rot. **1:**MS(82,2720), TX(82).

Sclerotium rolfsii. Southern blight. **1:**GA(82), NC(1303), TX(82).

Dahlia Cav.

1. *D. pinnata* Cav. (=*D. variabilis* (Willd.) Desf.). The different cultivars of garden dahlias were apparently derived from hybrids between *D. pinnata* and *D. coccinea.*
2. *Dahlia* sp.

Oomycetes

Pythium debaryanum. Root rot and stem rot of cuttings, watery rot of tubers. **1:**CT(82), MD(82), WA(82,3202).

P. oedochilum. Root and stem rot of cuttings. **1:**MD(82).

P. ultimum. Root rot and stem rot of cuttings, watery rot of tubers. **1:**MD(82).

Pythium sp. Root rot. **1:**NC(1303).

Zygomycotina

Choanephora cucurbitarum. Blossom blight. **1:**FL(35,82).

Ascomycotina

Botryosphaeria obtusa. On stems. **1:**TX(82).

Erysiphe cichoracearum. Powdery mildew. **1:**Range of host(82), FL(35), MS(2720), SD(2243). **2:**GA(1438), WA(3202).

E. communis. **2:**GA(1438).

E. polygoni. Powdery mildew. **1:**CA(82), DE(82), FL(35), GA(82), IA(82), MO(82), NJ(82), PA(82), VA(82).

Mycosphaerella dahliae. On stems. **1:**MI(82), NJ(82), VT(82).

Sclerotinia sclerotiorum. Stem rot. **1:**CA(82), FL(35), ME(82), NY(82).

Uncinula sp. Powdery mildew. **1:**NC(82).

Basidiomycotina - Ustilaginales

Entyloma dahliae. Leaf smut. **1:**CA(82,1077), NC(1303), NJ(82,1077), NY(1077), OR(82,1077,3202). **2:**OR(3202), WA(1077,3202).

Basidiomycotina - Other

Armillaria mellea. Root rot. **1:**CA(82).

Thanatephorus cucumeris. **1:**WA(3202).

Deuteromycotina - Hyphomycetes

Alternaria alternata. Leaf spot. **1:**MS(2720).

Alternaria sp. Leaf spot. **1:**E states(82), FL(35), MI(82), MO(82), NC(1303), WA(82,3202).

Botrytis cinerea. Gray mold blight of buds, flowers and stems, storage rot of tubers. **1:**Range of host(82), FL(35), NC(1303), WA(3202).

Cercospora sp. Leaf spot. **1:**FL(35,82), MS(82,2720). **2:**OK(2872).

Fusarium oxysporum. Wilt. **1:**FL(35).

Fusarium sp. Root, stem and tuber rot, wilt. **1:**AZ(82), MD(82), MO(82), MS(2720), NC(1303), NH(82), NJ(82), NY(82), OH(82). **2:**OK(2872).

Oidium sp. On cultivated plants. **1:**CA(661). **2:**HI(2903).

Penicillium sp. Powdery mildew. **2:**OK(2872).

Phymatotrichopsis omnivora. Root rot. **1:**TX(82).

Verticillium albo-atrum. Wilt. **1:**IL(82), MI(82), MO(82), NJ(82), OH(82), TX(82).

Deuteromycotina - Coelomycetes

Colletotrichum coccodes. Root and stem rot. **1:**FL(35).

C. dematium. Anthracnose. On dead stems. **2:**CA(81), OK(2872).

Diplodina dahliae. Anthracnose. **1:**WA(3202).

Macrophomina phaseolina. Charcoal rot. **1:**OK(82), SC(82). **2:**OK(2873).

Phoma dahliae. On leaves. **1:**NJ(82), OH(82).

Phomopsis achilleae. On dead plants. **2:**HI(2903).

Phyllosticta dahliaecola. Leaf spot. **1:**FL(35).

Phyllosticta sp. On flowers, stems. **1:**NC(1303).

Deuteromycotina - Other
Rhizoctonia solani. Thread blight, stem and tuber rot. **1**:FL(35), MD(82), MS(82,2720), NC(82,1303), NJ(82), TX(82), WA(82).
Rhizoctonia sp. Stem and tuber rot. **1**:NC(1303).
Sclerotium rolfsii. Southern blight. **1**:FL(35,1203), KS(1203), MS(1203,2720), NC(1203,1303), NJ(1203), TX(1203).

Dimorphotheca Moench
1. *D. annua* Less.
2. *D. aurantiaca* DC.
3. *D. sinuata* DC. [blue-eyed cape marigold]
4. *Dimorphotheca* sp.

Oomycetes
Plasmopara halstedii. **3**:IA(2101).
Pythium ultimum. Root rot. **2**:CA(82). **4**:CA(2101).
Zygomycotina
Choanephora sp. **4**:GA(2101).
Ascomycotina
Sclerotinia sclerotiorum. **4**:CA(2101).
Basidiomycotina - Uredinales
Puccinia melampodii. **2**:IL(82), NJ(82). **4**:IA(2101), IL(2101), NJ(2101).
Deuteromycotina - Hyphomycetes
Alternaria alternata. **4**:IA(2101).
Alternaria sp. **4**:GA(2101).
Botrytis cinerea. Gray mold blight. **1**:AK(82), CT(82). **2**:AK(82,526), CT(82). **4**:CT(2101), IA(2101), NE(2101).
Cephalosporium sp. **4**:CA(2101).
Curvularia sp. **4**:GA(2101).
Fusarium sp. Wilt. **1**:FL(82). **4**:CA(2101), FL(2101), GA(2101), IA(2101).
Verticillium albo-atrum. Wilt. **2**:IA(82). **4**:IA(2101), MD(2101), NY(2101).
Verticillium sp. **4**:CA(2101).
Deuteromycotina - Coelomycetes
Sphaeropsis sp. **4**:GA(2101).
Deuteromycotina - Other
Rhizoctonia solani. Root rot. **1**:IL(82). **4**:IL(2101).
Sclerotium rolfsii. Southern stem rot. **4**:NC(1303).

Doronicum L.
1. *D. plantagineum* L. [leopard's-bane]

Ascomycotina
Erysiphe cichoracearum. Powdery mildew. **1**:CA(82).

Dracopis Cass.
1. *D. amplexicaulis* (Vahl) Cass. (=*Rudbeckia amplexicaulis* Vahl) [coneflower]

Ascomycotina
Erysiphe cichoracearum. Powdery mildew. **1**:OK(2872).

Dubautia Gaud. in Freyc.
1. *D. laxa* Hook. & Arnott
2. *Dubautia* sp.

Ascomycotina
Echidnodella raillardiae. **2**:HI(2903).
Deuteromycotina - Hyphomycetes
Graphium dubautiae. **1**:HI(2903).

Dyssodia Cav.
1. *D. pentachaeta* (DC.) B. L. Robinson
2. *D. tenuiloba* (DC.) B. L. Robinson [Dahlberg daisy, golden-fleece]

Ascomycotina
Erysiphe cichoracearum. Powdery mildew. **2**:CA(1155).
Basidiomycotina - Uredinales
Puccinia dyssodiae. Autoecious. **1**:AZ(745,3981).
P. grindeliae. Autoecious. **1**:AZ(1203).

Echinacea Moench
1. *E. pallida* (Nutt.) Nutt.
2. *E. pallida* (Nutt.) Nutt. var. *angustifolia* (DC.) Cronq. (=*E. angustifolia* DC.)
3. *E. purpurea* (L.) Moench (=*Brauneria purpurea* Britton)

Ascomycotina
Metasphaeria sp. **2**:OK(2872).
Basidiomycotina - Ustilaginales
Entyloma compositarum. **3**:WI(1077).
E. polysporum. **1**:WI(1077,1338).
Deuteromycotina - Hyphomycetes
Cercospora tabacina. Leaf spot. **3**:IA(82,581).
Phymatotrichopsis omnivora. Root rot. **3**:TX(82).
Deuteromycotina - Coelomycetes
Ascochyta compositarum. **3**:WI(1338).
Phyllosticta sp. **1**:WI(1352).
[*Septoria helianthi*]. **1**:WI(1341).
S. lepachydis. On leaves. Leaf spot. **3**:WI(82,1332,1347).

Echinops L.
1. *E. sphaerocephalus* L.
2. *Echinops* sp.

Deuteromycotina - Coelomycetes
Phyllosticta sp. Leaf spot. **1**:WI(1344).
Deuteromycotina - Other
Sclerotium rolfsii. Crown rot. **2**:CT(82).

Eclipta L.
1. *E. prostrata* (L.) L. (=*E. alba* (L.) Hassk.)

Deuteromycotina - Hyphomycetes
Cercospora sp. Leaf spot. **1**:FL(35).

Elephantopus L.
1. *E. carolinianus* Raeusch.
2. *E. elatus* Bertol.
3. *E. nudatus* A. Gray
4. *E. tomentosus* L. [devil's-grandmother, tobacco-weed]
5. *Elephantopus* sp.

Ascomycotina
Diaporthe arctii. **4**:GA(1438).
Basidiomycotina - Uredinales
Coleosporium vernoniae. Heteroecious: telial host. **1**:MS(2720), OK(2872), TX(652). **2**:FL(35), MS(2720). **3**:MS(2720). **4**:GA(1439), MS(2720), NC(3944). **5**:FL(35).
Basidiomycotina - Other
Athelia arachnoidea. Thread blight. **5**:FL(35).
Sebacina incrustans. Encrusting leaves and stems. **1**:MD(128).
Deuteromycotina - Hyphomycetes
Cercospora elephantopi. On leaves. **1**:DE(962,3543), MO(962), OK(1022). **5**:FL(35).

Emilia Cass.
1. *E. javanica* (J. Burm.) Merr. (=*E. sagittata* (Vahl) DC.) [Flora's paintbrush, tassel-flower]
2. *E. sonchifolia* (L.) DC. ex Wight [red tassel-flower]
3. *Emilia* sp.

Basidiomycotina - Uredinales
Puccinia emiliae. Autoecious. **1**:FL(82). **2**:FL(82).
P. melampodii. Autoecious. **1**:FL(35). **3**:FL(35).
Deuteromycotina - Hyphomycetes
Alternaria sonchi. Leaf spot. **2**:HI(2903).
Oidium sp. Powdery mildew. **2**:HI(2299,2903).
Deuteromycotina - Other
Sclerotium rolfsii. Southern blight, stem rot. **2**:HI(82,2903).

Encelia Adans.
1. *E. californica* Nutt.
2. *Encelia* sp.

Ascomycotina
Lachnella villosa. On dead twigs. **1**:CA(2439).
Basidiomycotina - Uredinales
Puccinia enceliae. Autoecious. **2**:CA(82).
Basidiomycotina - Other
Meruliopsis corium. On dead twigs. **1**:CA(2439).
Peniophora malenconii subsp. *americana*. On wood. **1**:CA(542).

Engelmannia A. Gray ex Nutt.
1. *E. pinnatifida* Torr. & A. Gray ex Nutt. [Engelmann daisy]

Chytridiomycetes
Synchytrium taraxaci. Leaf gall. **1**:TX(82).
Synchytrium sp. On leaves. **1**:TX(1848).

Erechtites Raf.

1. *E. hieracifolia* (L.) Raf. ex DC. [American burnweed, burnweed, fireweed]
2. *E. prenanthoides* Greenm. & Hieron.
3. *E. valerianifolia* (Wolf) DC.
4. *Erechtites* sp.

Chytridiomycetes

Synchytrium erechtitis. Leaf gall. **1:**LA(627).

Oomycetes

Bremia lactucae. Downy mildew. **1:**PA(82).
Plasmopara halstedii. On leaves and stems. Downy mildew. **1:**IL(82), MA(82,1390), MD(82), NJ(82), TX(82), WI(82).

Ascomycotina

Crocicreas cyathoideum. Downy mildew. **2:**CA(505).
Erysiphe cichoracearum. Powdery mildew. **1:**MD(82). **2:**CA(1155).
E. polygoni. **1:**NY(82).
Leptospora rubella. In culm tissue. **2:**CA(3754).
Sclerotinia sclerotiorum. Stem rot. **1:**FL(35,82).
Sphaerotheca fuliginea. Powdery mildew. **1:**IA(82), MI(82), PA(82), WI(82).
S. macularis. Powdery mildew. **1:**E states(82), IL(82), MN(82), MT(82,3202).
Stictis radiata. **4:**CA(3225).

Deuteromycotina - Hyphomycetes

Cercospora erechtitis. Leaf spot. **1:**AL(82), FL(35,82), TX(82).
Phymatotrichopsis omnivora. Root rot. **1:**TX(82).
Ramularia wisconsina. On living leaves. **1:**WI(1342).
Torula granulosa. **4:**CA(81).

Deuteromycotina - Coelomycetes

Ascochyta compositarum. **1:**WI(1352).
Phyllosticta erechtitis. Leaf spot. **3:**HI(82). **4:**HI(2903).
Septoria erechtitis. Leaf spot. **1:**AK(82), DE(82), TX(82), WI(82).

Deuteromycotina - Other

Sclerotium rolfsii. Stem rot. **1:**FL(35,82).

Erigeron L.

1. *E. acris* L.
2. *E. annuus* (L.) Pers. [annual fleabane, daisy fleabane, sweet scabious, whitetop]
3. *E. caespitosus* Nutt.
4. *E. compositus* Pursh
5. *E. corymbosus* Nutt.
6. *E. coulteri* Porter
7. *E. divaricatus* Michx.
8. *E. divergens* Torr. & A. Gray
9. *E. eatonii* A. Gray
10. *E. elatior* (A. Gray) Greene
11. *E. filifolius* (Hook.) Nutt.
12. *E. glabellus* Nutt.
13. *E. humilis* Graham
14. *E. inornatus* (A. Gray) A. Gray
15. *E. lonchophyllus* Hook.
16. *E. nematophyllus* Rydb.
17. *E. peregrinus* (Banks ex Pursh) Greene
18. *E. peregrinus* (Banks ex Pursh) Greene subsp. *callianthemus* (Greene) Cronq. (=*E. salsuginosus* A. Gray)
19. *E. philadelphicus* L.
20. *E. poliospermus* A. Gray
21. *E. pulchellus* Michx. [robin's plantain]
22. *E. pumilus* Nutt.
23. *E. simplex* Greene
24. *E. speciosus* (Lindl.) DC.
25. *E. speciosus* (Lindl.) DC. var. *macranthus* (Nutt.) Cronq. (=*E. macranthus* Nutt.)
26. *E. strigosus* Muhl. ex Willd. (=*E. ramosus* (Walter) B. S. P.)
27. *E. subtrinervis* Rydb. ex Porter & Britton
28. *E. uniflorus* L. subsp. *eriocephalus* (J. Vahl) Cronq.
29. *Erigeron* sp.

Chytridiomycetes

Synchytrium erigerontis. Leaf gall. **19:**LA(82).

Oomycetes

Basidiophora entospora. Downy mildew. **2:**IN(82), TX(82). **19:**LA(82). **26:**IL(82), MO(82), NC(82,1303), WI(82). **29:**LA(82).
Plasmopara halstedii. Downy mildew. **2:**IA(82), OK(946), TX(82), WI(82). **19:**MD(82). **29:**IA(82).

Ascomycotina

Crocicreas nigrescens. On stems. **29:**NJ(505,850).
C. nigrofuscum. **24:**UT(505).
Diaporthe arctii. On stems. **2:**NJ(82). **29:**NJ(82).
Erysiphe cichoracearum. Powdery mildew. **2:**MD(82), TX(82). **5:**MT(3202), WA(3202). **12:**MT(82,3202). **14:**CA(638,642). **15:**MT(3202). **24:**ID(3202), MT(3202). **26:**MT(82,3202), NC(82,1303,3944). **27:**CO(1195), SD(2243). **29:**CO(82), MI(82), MT(82), NM(82), PA(82), SD(82), WY(82).
Guignardia sp. **11:**WA(3202).
Leptosphaeria compressa. On dead stems. **29:**IL(3754).
L. erigerontis. **18:**WY(3810).
Massarina gloeospora. **26:**GA(1438).
Mycosphaerella subcongregata. **17:**WA(967,3202).
M. taraxaci. **4:**WA(3202).
Mycosphaerella sp. On stems. **18:**WA(82).
Nodulosphaeria olivacea. **18:**WY(3250).
Ophiobolus anguillidus. **26:**GA(1438).
Phyllactinia guttata. Powdery mildew. **22:**WA(3202). **29:**WA(82).
Pleospora allii. **11:**WA(3202).
P. comata. On dead stem. **13:**AK(1938). **28:**AK(1938).
P. penicillus. **4:**WA(3202).
P. richtophensis. **18:**WY(3810).
Sclerotinia sclerotiorum. **2:**OH(946).
Sphaerotheca fuliginea. Powdery mildew. **2:**MI(82), PA(82), WI(1341). **7:**WI(1357). **17:**AK(526). **26:**WI(1361). **29:**Ak(82), IN(82), NY(82).
S. macularis. Powdery mildew. **2:**PA(82). **17:**WA(3202).
Wettsteinina duplex. **29:**WA(3252).
W. macrospora. **29:**WA(3202,3252).

Basidiomycotina - Uredinales

Coleosporium asterum. Heteroecious: telial host. **17:**AK(60,526). **18:**WY(2924). **29:**AK(82), CA(82).
Puccinia cnici-oleracei. Autoecious. **10:**WY(1195).
P. cyperi. Heteroecious: aecial host. **2:**NE states(82), TX(82), WI(82). **21:**MO(82). **26:**NE states(82), OK(2872), TX(82), WI(82).
P. dioicae. Heteroecious: aecial host. **2:**E states(82), MS(82), NC(1303), SD(82), TX(82). **18:**W states(82), WY(1195). **19:**E & C states(82), MS(2720). **21:**E & C states(82). **24:**W states(82), CO(1195), MT(3202), OR(3202). **26:**E states(82), MS(82), NC(3944), OK(2872), SD(82,2243), TX(82). **29:**W states(82).
P. dovrensis var. *leptoticha*. Autoecious. **23:**CO(746,1520), WY(746). **29:**WY(1520).
P. dovrensis var. *russa*. Autoecious. **18:**CA(746).
P. excursionis. Autoecious. **17:**CO(649,1520).
P. grindeliae. Autoecious. **3:**CO(1195), ID(3202). **9:**WY(1195,3379). **16:**WY(1195,3379). **17:**WA(3202). **18:**UT(82). **20:**WA(3202). **29:**CO(82), NE(82), WY(82).
P. stipae. Heteroecious: aecial host. **5:**MT(1195,3202). **19:**MT(3202). **22:**MT(3202). **29:**CO(82), WY(82).

Basidiomycotina - Ustilaginales

Entyloma compositarum. White smut, leaf spot. **1:**WA(3202). **6:**ID(3202), OR(3202), UT(1077). **10:**WA(1077). **12:**CO(1077). **17:**ID(3202), MT(3202), OR(3202), WA(1077,3202), WY(1077). **19:**ID(3202), MI(82,1077), ND(82,378,1077), WI(82,1077). **22:**OR(3202). **24:**ID(3202), MT(3202), OR(3202), WA(3202), WY(1195,3378). **25:**CO(1077). **27:**CO(1077), MT(3202). **29:**IL(1077).

Basidiomycotina - Other

Sebacina incrustans. Encrusting leaves and stems. **2:**DC(128).
Thanatephorus cucumeris. **19:**LA(3524). **21:**WI(1349).

Deuteromycotina - Hyphomycetes

Botrytis cinerea. Gray mold blight. **29:**AK(82).
Cercosporella virgaureae. On lower leaves. Leaf spot. **1:**WA(3202). **2:**C states(82), E states(82), DE(3543), NE(82), TX(82), WI(1357). **12:**WI(1341). **19:**LA(82), MS(2720), WI(1334). **26:**C states(82), E states(82), NC(3944), NE(82), OK(2872), TX(82). **29:**DE(3543).
Oidium sp. Powdery mildew. **29:**HI(2903).
Phymatotrichopsis omnivora. Root rot. **2:**TX(82).
Ramularia macrospora. Leaf spot. **2:**IL(82), MD(82), NC(82,3944), NJ(82), TN(82), TX(82). **4:**MT(3202). **5:**ID(3202). **8:**MT(3202), OR(3202), WA(3202). **15:**MT(3202), OR(3202). **19:**MT(3202). **22:**ID(3202), MT(3202), WA(3202). **24:**ID(3202).

Verticillium albo-atrum. Wilt. **29:**MA(82).

Deuteromycotina - Coelomycetes

Leptothyrium dearnessii. **2:**TX(82).

L. punctiforme. On leaves. **2:**NY(82), WI(82,1332). **19:**NY(82), WI(1341). **26:**WI(82,1334), WV(82).

Macrophomina phaseolina. Charcoal rot. **26:**IL(311).

Phoma selenophomoides. On stems. **18:**WY(3812).

Phyllosticta sp. Leaf spot. **2:**WI(1361). **12:**WI(1352).

Rhabdospora sp. Leaf spot. **2:**WI(1351).

Septoria erigerontis. Leaf spot. **2:**E states(82), MS(2720), TX(82), WA(3202), WI(82). **19:**IA(82), IN(82), KS(82), NY(82), WA(3202). **21:**NY(82), OH(82), WI(1338). **22:**ID(3202). **26:**E states(82), NC(1303), OK(2872), SD(2243), TX(82), WI(82). **29:**NE states(82), MI(82), NE(82).

Deuteromycotina - Other

Sclerotium rolfsii. Stem rot. **12:**IL(82).

Eriophyllum Lag.

1. *E. lanatum* (Pursh) J. Forbes
2. *E. lanatum* (Pursh) J. Forbes var. *grandiflorum* (A. Gray) Jeps.
3. *E. nevinii* A. Gray
4. *Eriophyllum* sp.

Ascomycotina

Sphaerotheca macularis. **2:**CA(81).

Basidiomycotina - Uredinales

Puccinia hieracii. **1:**ID(1195,3202), OR(3202). **3:**CA(2439). **4:**CA(82), OR(82), WY(82).

Uromyces junci. Heteroecious: aecial host. **4:**CA(82).

Deuteromycotina - Hyphomycetes

Cercosporella virgaureae. **1:**ID(3202).

Deuteromycotina - Other

Rhizoctonia solani. Brown patch. **4:**FL(35).

Eupatorium L.

1. *E. album* L.
2. *E. altissimum* L.
3. *E. capillifolium* (Lam.) Small [dogfennel]
4. *E. coelestinum* L. [blue boneset, hardy ageratum, mistflower]
5. *E. compositifolium* Walter (=*E. coronopifolium* Willd.)
6. *E. fistulosum* J. Barratt [trumpetweed]
7. *E. greggii* A. Gray
8. *E. herbaceum* (A. Gray) Greene
9. *E. hyssopifolium* L.
10. *E. incarnatum* Walter
11. *E. maculatum* L. [Joe-Pye-weed, smokeweed]
12. *E. occidentale* Hook.
13. *E. perfoliatum* L. [common boneset]
14. *E. purpureum* L. [green-stemmed Joe-Pye-weed, Joe-Pye-weed, sweet Joe-Pye-weed]
15. *E. pycnocephalum* Less.
16. *E. rotundifolium* L. (=*E. rotundifolium* L. var. *rotundifolium*)
17. *E. rugosum* Houtt. (=*E. urticifolium* Reichard, *E. ageratoides* (L.) L. fil.) [white sanicle, white snakeroot]
18. *E. serotinum* Michx.
19. *E. sessilifolium* L. [upland boneset]
20. *E. solidaginifolium* A. Gray
21. *Eupatorium* sp.

Plasmodiophoromycetes

Ligniera junci. On roots. **3:**NC(1303).

Oomycetes

Plasmopara halstedii. Downy mildew. **14:**NE states(82), MO(82), WI(82).

Ascomycotina

Acrospermum foliicola. **14:**GA(1438).

Anthostomella tomicoides. On stems. **14:**NJ(1098).

Belonidium eupatorii. On dead stems. **14:**PA(3167).

Crocicreas coronatum. On stems. **17:**NY(505).

C. cyathoideum. On stems. **17:**NY(505).

C. helios. On dead plants. **21:**HI(2903).

Dasyscyphus solenia. On stems. **17:**NY(849).

Diaporthe arctii. On stems. **3:**GA(1438). **9:**GA(1438). **11:**GA(1438). **14:**GA(82,1438). **16:**GA(1438). **17:**GA(82).

Diatrypella favacea. **3:**GA(1438).

Erysiphe cichoracearum. Powdery mildew. **13:**Range of host(82). **14:**Range of host(82). **17:**Range of host(82). **21:**FL(35).

[*E. galeopsidis*]. **17:**WI(1339).

Hypoderma commune. **14:**GA(1438).

Leptosphaeria doliolum. **14:**GA(1438).

L. helminthospora. **3:**GA(1438).

L. macrospora. **6:**VT(272).

L. ogilviensis. **17:**NY(3249).

L. planiuscula. **3:**GA(1438).

Leptosphaeria sp. **3:**GA(1438).

Massarina gloeospora. **3:**GA(1438). **9:**GA(1438).

Mollisia atrata. **14:**GA(1438).

Mycosphaerella sp. On stems. **14:**GA(82,1438).

Nectriella pironii. Stem gall. **21:**FL(35).

Pleospora herbarum. On overwintered herbage. **12:**CA(642).

P. penicillus. **3:**GA(1438).

P. tragacanthae. **12:**OR(3202).

Pleospora sp. **3:**GA(1438).

Rosellinia umbrinella. On stems. **5:**SC(967).

Trichopezizella relicina. **11:**NY(1416).

Tryblidaria sp. **3:**GA(1438).

Basidiomycotina - Uredinales

Aecidium eupatorii-rotundifolii. **16:**AL(3101), DE(3101), TN(3101).

Puccinia concinna. **7:**AZ(3981).

P. conoclinii. Autoecious. **4:**E states(82), IL(82,752), MS(2720), OK(2872), TX(752), WI(82), WV(752). **7:**AZ(1194,1203). **8:**AZ(752,1194,1203,3981). **10:**IN(752). **15:**AZ(741,752,1195,1203,3379,3981). **20:**AZ(752,1203,3981). **21:**AZ(82), FL(35,82), IL(82), IN(82), LA(82).

P. eleocharidis. Heteroecious: aecial host. **2:**WI(1350). **11:**IA(3101), NE(3101), WI(1337,3101). **13:**E & C states(82), AL(82), TX(82), WI(1351), WY(82). **14:**E & C states(82), AL(82), TX(82), WY(82). **16:**MS(2720), NC(3944). **17:**NC(1303), VA(82), WV(82). **18:**NC(1303), OK(2872), TX(740).

P. inanipes. Autoecious. **20:**AZ(752,1195,1203,3379,3981).

P. tenuis. Autoecious. **17:**E states(82), IN(987), MN(82), MS(2720), NE(82), NY(836). **21:**FL(35).

P. tolimensis. Autoecious. **17:**NY(82,752).

Basidiomycotina - Ustilaginales

Entyloma compositarum. White smut. **12:**OR(3202). **17:**IA(82,1077), IL(82,1077), MS(82,1077,2720), OR(1077), WI(82,1077), WV(82,1077).

Basidiomycotina - Other

Ceratobasidium anceps. **11:**WI(1350).

Deuteromycotina - Hyphomycetes

Alternaria zinniae. Leaf spot. **21:**FL(35).

Botrytis cinerea. Gray mold blight. **4:**NJ(82). **21:**NJ(82).

Cercospora ageratoides. Leaf spot. **16:**MS(2720). **17:**DE(3543), MS(2720), NJ(82), WI(82), WV(82). **21:**AL(82), FL(35), MS(82), NJ(82), TX(82).

C. eupatorii. Leaf spot. **1:**AL(82), NY(82,581). **13:**IN(82). **14:**IN(82). **17:**NJ(82). **21:**FL(35).

Cercosporella sp. **2:**WI(1354).

Conoplea fusca. **21:**NJ(1672).

Cristulariella moricola. On leaves. **4:**IL(2572). **17:**IL(2572).

Fusarium sp. Wilt. **21:**NJ(82).

Mycovellosiella perfoliati. On leaves. **13:**MI(82), WI(82). **14:**WI(82). **17:**IL(82), WI(82). **19:**WI(1361). **21:**FL(35).

Phymatotrichopsis omnivora. Leaf spot. **4:**TX(82). **17:**TX(82). **21:**TX(82).

Ramularia dispar. On old leaves. **11:**WI(1350). **14:**TX(82), WI(82). **17:**TX(82), WI(82). **21:**IL(82), VA(82).

Sporidesmium irregularis. On stems. **21:**SC(634).

Deuteromycotina - Coelomycetes

Ascochyta compositarum. Leaf spot. **13:**WI(1350). **14:**WI(82), WV(82). **17:**WI(82).

Macrophomina phaseolina. Charcoal rot. **4:**IL(311). **18:**IL(311).

Phoma exigua. Leaf spot. **13:**WI(64,82,1333). **19:**WI(1333).

Phyllosticta umbrino-fumosa. On leaves. **14:**WI(1342). **17:**WI(1337).

Phyllosticta sp. **13:**WI(1352). **17:**WI(1358).

Septoria eupatorii. Leaf spot. **2:**WI(1338). **4:**IL(82). **11:**WI(1342). **13:**MS(82,2720), WI(1334,1351). **17:**IA(82), TX(82). **21:**LA(82), MS(82), TX(82), WI(82).

S. eupatoriicola. **13:**IL(82).

Septoria sp. **19:**WI(1361).

Deuteromycotina - Other

Rhizoctonia solani. Thread blight, root rot. **21:**FL(35), NJ(82).

Sclerotium rolfsii. Root and stem rot. **4:**CT(82), IL(312), NJ(82). **17:**IL(82), TX(82). **21:**IL(82), MD(82), NJ(82).

Euryops (Cass.) Cass.

1. *E. speciosissimus* DC. (=*E. athanasiae* (L. fil.) Less. ex Harv.) [Clanwilliam daisy]

Basidiomycotina - Other

Armillaria mellea. Root and stem rot. **1:**CA(2897).

Euthamia Nutt.

1. *E. graminifolia* (L.) Nutt. (=*Solidago graminifolia* (L.) Salisb.)
2. *E. occidentalis* (Torr. & A. Gray) Nutt. (=*Solidago occidentalis* Torr. & A. Gray)
3. *E. tenuifolia* (Pursh) Greene (=*Solidago tenuifolia* Pursh)

Ascomycotina

Diaporthe linearis. On stems. **1:**GA(82), MI(82), ND(82), NJ(82).
Erysiphe cichoracearum. **1:**WI(1349). **2:**MT(3202).
Phyllachora solidaginum. **1:**Unk.(82).
Rhytisma solidaginis. On leaves. **1:**Unk.(82), WI(1346,1350).

Basidiomycotina - Uredinales

Coleosporium asterum. **1:**MA(1390). **3:**MA(1390).
C. delicatulum. Heteroecious: telial host. **1:**Unk.(82), IN(987), WI(1332,1358).
Puccinia dioicae. **1:**NE states(82), WI(82). **2:**OR(3202), WA(2219,3202).

Deuteromycotina - Hyphomycetes

Cladosporium astericola. **1:**WI(1344).
Didymaria spissa. On living leaves. **2:**CA(1468).
Ramularia serotina. Leaf spot. **1:**CO(82), IL(82), TX(82), WI(82), WY(82).

Deuteromycotina - Coelomycetes

Septoria angularis. **1:**WI(1330).
S. fumosa. **1:**WI(82).

Felicia Cass.

1. *F. bergeriana* (Spreng.) O. Hoffm. ex Zahlbr. [kingfisher daisy]

Deuteromycotina - Hyphomycetes

Botrytis cinerea. Gray mold. **1:**FL(35).

Filago L.

1. *F. gallica* L.

Deuteromycotina - Hyphomycetes

Phaeoisariopsis gnaphaliaceae. **1:**CA(81).

Gaillardia Foug.

1. *G. aristata* Pursh
2. *G. pulchella* Foug. [firewheel]
3. *G. pulchella* Foug. var. *picta* (Sweet) A. Gray
4. *Gaillardia* sp.

Oomycetes

Pythium ultimum. Root rot. **1:**CA(82).

Ascomycotina

Erysiphe cichoracearum. Powdery mildew. **1:**MT(3202), OK(2873).
Sphaerotheca fuliginea. Powdery mildew. **1:**MT(82), WA(82), WY(82).
S. macularis. **1:**MT(3202), OR(3202), WA(3202).

Basidiomycotina - Uredinales

Coleosporium asterum. Heteroecious: telial host. **1:**CA(82).
Puccinia helianthellae. **1:**CA(82).

Basidiomycotina - Ustilaginales

Entyloma calendulae. **1:**WA(1077,3202).
E. compositarum. **1:**WA(3202). **2:**NE(1077).
E. polysporum. White smut. **1:**CO(1077), MN(82). **2:**KS(82,1077), NE(82). **4:**MN(1077).
Entyloma sp. **1:**ID(3202), MT(3202), WA(3202).

Deuteromycotina - Hyphomycetes

Cladosporium sp. **4:**AK(526).
Phymatotrichopsis omnivora. Root rot. **1:**TX(82). **2:**TX(82).
Ramularia sp. **1:**MT(3202), WA(3202).

Deuteromycotina - Coelomycetes

Phyllosticta sp. Leaf spot. **1:**WA(3202). **2:**FL(35). **3:**WI(1343).
Septoria gaillardiae. Leaf spot. **1:**IA(82), IL(312), OK(82,2872), TX(82), WI(1359). **2:**IA(82), KS(82), MO(82), OK(82,2872), TX(82). **3:**WI(1338). **4:**OK(2872).

Deuteromycotina - Other

Rhizoctonia solani. Root rot. **2:**FL(35).

Galinsoga Ruiz & Pav.

1. *G. parviflora* Cav. [quickweed, small-flowered galinsoga]

Ascomycotina

Sclerotinia sclerotiorum. **1:**MD(14).

Deuteromycotina - Hyphomycetes

Alternaria solani. Early blight. **1:**HI(2903).

Deuteromycotina - Other

Rhizoctonia sp. Root rot. **1:**HI(2903).

Gazania Gaertn.

1. *G. rigens* (L.) Gaertn. [treasure-flower]
2. *Gazania* sp.

Oomycetes

Pythium sp. Root rot. **2:**FL(35).

Deuteromycotina - Hyphomycetes

Alternaria sp. Leaf spot. **2:**FL(35).
Corynespora sp. Leaf spot. **2:**FL(35).
Fusarium sp. Root rot. **2:**FL(35).

Deuteromycotina - Other

Rhizoctonia solani. Crown rot, root rot. **1:**CA(82). **2:**FL(35).
Rhizoctonia sp. Root rot. **1:**HI(2903).

Geraea Torr. & A. Gray

1. *G. viscida* (A. Gray) S. F. Blake

Basidiomycotina - Uredinales

Puccinia abramsii. Autoecious. **1:**CA(746).

Gerbera L.

1. *G. jamesonii* H. Bolus ex J. D. Hook. [African daisy, Barberton daisy, Transvaal daisy, veldt daisy]

Oomycetes

Phytophthora cryptogea. Crown and root rot. **1:**FL(35).
Phytophthora sp. Blight, crown and root rot. **1:**CA(82), FL(35), NC(1303), NJ(82), NY(82).
Plasmopara sp. Downy mildew. **1:**FL(35).
Pythium sp. Root rot. **1:**FL(35), HI(2903).

Ascomycotina

Erysiphe cichoracearum. Powdery mildew. **1:**CA(82), FL(35), HI(2899), NC(1303), OK(82,2872).
Sclerotinia sclerotiorum. Crown rot. **1:**MD(82), NY(82).

Deuteromycotina - Hyphomycetes

Alternaria gerberae. Leaf spot. **1:**HI(2903).
[*A. porri*]. Leaf spot. **1:**CA(149), HI(2899,2903).
Alternaria sp. Leaf spot. **1:**FL(35), HI(2903).
Botrytis cinerea. Gray mold blight. **1:**FL(35,82), NY(82), WA(3202).
Cercospora gerberae. Leaf spot. **1:**FL(35), HI(2903).
Fusarium oxysporum. Damping-off, crown rot. **1:**FL(35).
F. solani. Crown rot. **1:**FL(35).
Oidium sp. Powdery mildew. **1:**HI(2903).
Thielaviopsis basicola. Root rot. **1:**MD(1867).
Verticillium albo-atrum. Hadromycosis. **1:**CA(3366).

Deuteromycotina - Coelomycetes

Ascochyta doronici. Leaf spot. **1:**FL(35,3500).
Phyllosticta sp. Leaf spot. **1:**FL(35).
Septoria sp. Leaf spot. **1:**FL(35).

Deuteromycotina - Other

Rhizoctonia solani. Damping-off, root rot. **1:**FL(35).
Sclerotium rolfsii. Southern blight, stem rot. **1:**FL(35).

Gnaphalium L.

1. *G. chilense* Spreng.
2. *G. macounii* Greene
3. *G. microcephalum* Nutt.
4. *G. obtusifolium* L. [everlasting, fragrant cudweed, rabbit tobacco, sweet everlasting]
5. *G. palustre* Nutt.
6. *G. polycephalum* Michx.
7. *G. purpureum* L.
8. *G. purpureum* L. var. *spathulatum* (Lam.) W. Ahles [cudweed]
9. *G. uliginosum* L. [low cudweed]
10. *G. viscosum* H. B. K. (=*G. decurrens* Ives)
11. *Gnaphalium* sp.

Oomycetes

Bremia lactucae. Downy mildew. **8:**FL(35).
Plasmopara halstedii. Downy mildew. **7:**AL(82), MS(82,2720).

8:FL(35). 11:MS(82).
Ascomycotina
Diaporthopsis trinucleata. 4:GA(1438).
Heptameria obesa. 4:GA(1438).
Leptosphaeria galiorum var. *gnaphaliana*. On dead stems. 11:CA(2439).
Mycosphaerella taraxaci. 11:WA(3202).
M. tassiana. On stems. 11:CA(82).
Sclerotinia sclerotiorum. Crown rot. 8:FL(35).
Basidiomycotina - Uredinales
Puccinia gnaphaliicola. Autoecious. 4:NC(1303). 7:AL(82), GA(1439), LA(82), SC(82). 8:FL(35). 11:AL(82), NC(82), VA(638).
P. investita. Autoecious. 2:AZ(1203,3981). 4:OH(946). 6:MA(1390). 10:AZ(82), NY(82), VT(82), WI(82). 11:CA(82), CT(82), MA(82), NY(82), PA(82), TN(82), WV(82).
Basidiomycotina - Ustilaginales
Entyloma compositarum. White smut. 11:AK(1077), AL(82), MD(1077).
Deuteromycotina - Hyphomycetes
Alternaria sp. 5:ID(3202).
Cercospora gnaphalii. On living leaves. 3:ID(3202). 7:OR(3202), WA(3202). 10:CA(82,1468). 11:TX(82).
Fusicladium gnaphaliatum. On leaves. 1:CA(81).
Heterosporium sp. 1:WA(3202).
Ovularia sp. 9:WA(3202).
Phaeoisariopsis gnaphaliaceae. Leaf spot. 7:AL(2484), DE(3543), FL(35), MS(82,2720). 9:WI(1340). 11:KS(82), MS(82), TX(82), WI(82).
Phymatotrichopsis omnivora. Root rot. 11:TX(82).
Stemphylium botryosum. Leaf spot. 8:FL(35).
Deuteromycotina - Coelomycetes
Cylindrosporium gnaphalicola. Leaf spot. 11:AL(82), TX(82).
Heteropatella cercosperma. Leaf spot. 11:TX(82).
Phoma erysiphoides. Stem canker. 7:TX(82). 10:TX(82). 11:TX(82).
Deuteromycotina - Other
Rhizoctonia solani. Leaf necrosis. 7:FL(35).
Sclerotium rolfsii. Stem rot. 7:FL(35).

Grindelia Willd.

1. *G. camporum* Greene
2. *G. erecta* A. Nels.
3. *G. humilis* Hook. & Arnott
4. *G. inornata* Greene
5. *G. integrifolia* DC.
6. *G. maritima* (Greene) Steyerm. (=*G. robusta* Nutt. var. *maritima* (Greene) Jeps.)
7. *G. nana* Nutt.
8. *G. perennis* A. Nels.
9. *G. robusta* Nutt. (=*G. cuneifolia* Nutt.)
10. *G. squarrosa* (Pursh) Dunal [curly-cup gumweed]
11. *G. stricta* DC. (=*G. oregana* A. Gray)
12. *Grindelia* sp.

Ascomycotina
Erysiphe cichoracearum. Powdery mildew. 1:CA(1155). 3:CA(1155). 7:WA(3202). 8:WY(1195). 9:CA(1155). 10:C & W states(82), CA(82), MT(82,3202), ND(377), SD(2243). 12:C & W states(82), CA(82), MT(82).
Leptosphaeria nigricans var. *inculta*. 10:ND(377).
Basidiomycotina - Uredinales
Coleosporium asterum. Heteroecious: telial host. 10:CO(82), WI(82). 11:CA(81). 12:CA(82), WA(82).
Puccinia dulichii. Heteroecious: aecial host. 10:KS(82), NE(82), TX(82).
P. grindeliae. Autoecious. 6:CA(81). 7:OR(3202), WA(3202). 10:C states(82), CA(82), MT(82,3202), NM(1195). 12:C states(82), CA(82), MT(82), UT(1195).
P. stipae. Heteroecious: aecial host. 10:CO(82), KS(82), ND(82,378), NE(82), SD(82).
Uromyces junci. Heteroecious: aecial host. 2:WY(2924). 12:CA(82).
Basidiomycotina - Ustilaginales
Thecaphora californica. 9:CA(1077), OR(920). 10:CO(1077), NY(1077), UT(82,1077). 12:CA(82), OR(920,1077,3202).
T. cuneata. 4:CO(920,1077). 10:CA(920,1077), CO(82,920,1077), KS(82,920,1077), NE(82,920,1077), NM(82,920,1077), UT(920,1077), WY(920). 12:NE(82), WY(1077).
Deuteromycotina - Hyphomycetes
Cercospora grindeliae. Leaf spot. 10:TX(82). 12:CA(82), WI(82).
Cercosporella virgaureae. 5:WA(3202).
Phymatotrichopsis omnivora. Root rot. 12:TX(82).
Ramularia grindeliae. On old leaves. 10:KS(82), TX(82), WI(1349). 12:KS(82), NM(82).
Deuteromycotina - Coelomycetes
Camarosporium parasiticum. On living leaves. 10:WI(1349).
Phoma leptospora. On stems. 10:ND(378), TX(82).
Phyllosticta sp. On leaves. 10:WI(1338).
Septoria grindeliae. Leaf spot. 5:OR(3202). 10:CO(82), KS(82), TX(82). 12:OR(82).
S. grindeliicola. On leaves. 10:WI(82,1335).

Guardiola Cerv. ex Humb. & Bonpl.

1. *G. platyphylla* A. Gray

Ascomycotina
Comoclathris permunda. 1:AZ(662).
Lophiostoma caulium. 1:AZ(662).
Deuteromycotina - Coelomycetes
Colletotrichum gloeosporioides. 1:AZ(662).

Gutierrezia Lag.

1. *G. longifolia* Greene
2. *G. sarothrae* (Pursh) Britton & Rusby [broom snakeroot, broom snakeweed]
3. *Gutierrezia* sp.

Ascomycotina
Erysiphe cichoracearum. Powdery mildew. 2:CO(82,1195), MT(82,3202), TX(82). 3:CO(82), MT(82), TX(82).
E. polygoni. 2:MT(82).
Basidiomycotina - Uredinales
Coleosporium asterum. Heteroecious: telial host. 3:TX(82).
Puccinia grindeliae. Autoecious. 1:AZ(1203,3981). 2:W states(82), AZ(1203,3981), CA(82), MT(3202), ND(378), UT(82), WY(1195,3379). 3:W states(82), CA(82), UT(82).
P. stipae. Heteroecious: aecial host. 2:CO(1195), MT(82,3202), NE(82), NM(82), TX(82). 3:TX(82).
Deuteromycotina - Hyphomycetes
Phymatotrichopsis omnivora. Root rot. 3:TX(82).
Deuteromycotina - Coelomycetes
Diplodina coloradensis. On stems. 2:CO(82), TX(82).
Rhabdospora gutierreziae. On stems. 3:CO(82).
Septoria helenii. Leaf spot. 2:TX(82), WI(82). 3:TX(82).

Gynura Cass.

1. *G. aurantiaca* (Blume) DC. [purple-passionvine, royal-vineplant, velvetplant]
2. *Gynura* sp.

Oomycetes
Phytophthora sp. Leaf rot. 1:FL(35).
Pythium sp. Root rot. 1:FL(35). 2:FL(35).
Ascomycotina
Sclerotinia sclerotiorum. Cottony rot. 2:FL(35).
Deuteromycotina - Hyphomycetes
Alternaria sp. Leaf spot. 2:FL(35).
Botrytis cinerea. Leaf and stem blight. 2:FL(35).
Corynespora sp. Leaf necrosis. 2:FL(35).
Fusarium oxysporum. Root and stem rot. 1:FL(35). 2:FL(35).
Deuteromycotina - Coelomycetes
Colletotrichum gloeosporioides. Anthracnose. 1:FL(35).
Deuteromycotina - Other
Rhizoctonia solani. Root and stem rot. 1:FL(35). 2:FL(35).

Haplopappus Cass.

1. *H. acaulis* (Nutt.) A. Gray (=*Stenotus acaulis* Nutt.)
2. *H. bloomeri* A. Gray
3. *H. bloomeri* A. Gray var. *angustatus* A. Gray (=*Chrysothamnus bloomeri* (A. Gray) Greene var. *angustatus* A. Gray)
4. *H. canus* (A. Gray) S. F. Blake
5. *H. ciliatus* (Nutt.) DC.
6. *H. divaricatus* A. Gray (=*Isopappus divaricatus* Torr. & A. Gray)
7. *H. glutinosus* Cass. (=*H. coronopifolius* DC.)
8. *H. gracilis* (Nutt.) A. Gray (=*Machaeranthera gracilis* (Nutt.)

Shinners)
9. *H. heterophyllus* (A. Gray) S. F. Blake (=*Isocoma heterophylla* (A. Gray) Greene)
10. *H. integrifolius* Porter ex A. Gray
11. *H. lanceolatus* (Hook.) Torr. & A. Gray
12. *H. laricifolius* A. Gray (=*Ericameria laricifolia* (A. Gray) Shinners)
13. *H. racemosus* (Nutt.) Torr. subsp. *congesta* (Greene) H. Hall (=*H. racemosus* (Nutt.) Torr. var. *congesta* (Greene) Peck)
14. *H. ravenii* R. Jackson
15. *H. squarrosa* Hook. & Arnott (=*Hazardia squarrosa* (Hook. & Arnott) Greene)
16. *H. tenuisectus* (Greene) S. F. Blake ex L. Benson [burro-weed]
17. *H. venetus* (H. B. K.) S. F. Blake
18. *Haplopappus* sp.

Ascomycotina

Dasyscyphus aplopappi. On twigs. On dead limbs. **3:**CA(524,642).
D. salmoneus. On dead limbs. **3:**CA(524,642).
Dasyscyphus sp. **2:**WA(3202).
Erysiphe cichoracearum. Powdery mildew. **5:**OK(2872). **18:**CO(82), WY(82).
Hyponectria gregaria. On leaves. **3:**CA(183).
Leptosphaeria ogilviensis. **6:**GA(1438).
L. tumefaciens. Stem gall. **18:**CO(82).
Lophiostoma caulium. **2:**CA(661).
L. caulium var. *congregatum*. **3:**CA(642).
Massarina gloeospora. **6:**GA(1438).
Mollisia hysteroides. On dead limbs and stems. **3:**CA(524,642).
Perrotia flammea. On dead limbs. **3:**CA(642).
Pleospora herbarum. **10:**MT(3202).
Rosellinia riminicola. On dead limbs. **3:**CA(638,642).
Sphaerotheca fuliginea. Powdery mildew. **4:**CA(1155).
Teichospora megastega. On woody limbs. **3:**CA(642).
T. pomiformis. On woody stems. **2:**CA(661).
T. solitaria. On dead stems. **2:**CA(661).

Basidiomycotina - Uredinales

Coleosporium asterum. Heteroecious: telial host. **11:**MT(3202). **13:**OR(81). **18:**MT(82).
C. pacificum. Heteroecious: telial host. **13:**CA(81).
Puccinia grindeliae. Autoecious. **1:**WY(1195). **3:**CA(638,642). **8:**AZ(1203,3981). **12:**AZ(1203,3981). **14:**AZ(3981). **15:**CA(2439). **18:**Unk.(82), AZ(3981), WA(3202).
P. marianae. **18:**Unk.(82).
P. stipae. **18:**CO(82).

Basidiomycotina - Ustilaginales

Thecaphora pilulaeformis. Inflorescence smut. **7:**AZ(1077). **17:**AZ(920), CA(920,1077). **18:**AZ(82), CA(82).

Basidiomycotina - Other

Dendrophora albobadia. **12:**AZ(1202).
Lentinus sp. On dead limbs. **3:**CA(642).
Phanerochaete chrysorhiza. **12:**AZ(1202).
Resupinatus applicatus. **16:**AZ(1202).
Schizopora apacheriensis. On roots of living bush. **12:**AZ(1193,1202).

Deuteromycotina - Hyphomycetes

Phymatotrichopsis omnivora. Root rot. **18:**TX(82).

Deuteromycotina - Coelomycetes

Phoma estrelti. **9:**NM(2399).
Septoria sp. Leaf spot. **5:**OK(2872).

Helenium L.

1. *H. amarum* (Raf.) H. Rock (=*H. tenuifolium* Nutt.) [bitter sneezeweed, bitterweed]
2. *H. autumnale* L. [sneezeweed]
3. *H. curtisii* A. Gray
4. *H. hoopesii* A. Gray
5. *H. microcephalum* DC. [sneezeweed]
6. *Helenium* sp.

Chytridiomycetes

Synchytrium sp. On leaves. **4:**CO(1849).

Ascomycotina

Erysiphe cichoracearum. Powdery mildew. **1:**TX(82). **2:**CO(82), MT(82), OR(82), UT(82), WA(82), WI(82), WY(82). **6:**CO(82), MT(82), OR(82), UT(82), WA(82), WI(82), WY(82).
Hypoderma commune. **3:**GA(1438).
Nodulosphaeria modesta. **4:**AZ(662).
Pleospora scrophulariae var. *compositarum*. **1:**GA(1438).

Basidiomycotina - Uredinales

Puccinia dulichii. Heteroecious: aecial host. **2:**NE(82), TX(82). **6:**CO(82), NE(82).
P. poarum. Heteroecious: aecial host. **4:**AZ(741,1195,1203,1857,3981), UT(1195,3379). **6:**AZ(82), CO(82), NM(82).

Basidiomycotina - Ustilaginales

Entyloma compositarum. Leaf smut. **2:**MT(1077), TX(82), WI(82,1077).
E. polysporum. **2:**MT(82).

Deuteromycotina - Hyphomycetes

Cercospora helenii. Leaf spot. **1:**AL(82). **5:**TX(82,581,3597).
Cladosporium aecidiicola. Leaf spot. **4:**AZ(1857).
Phymatotrichopsis omnivora. Root rot. **1:**TX(82). **2:**TX(82).

Deuteromycotina - Coelomycetes

Ascochyta compositarum. **2:**WI(1354).
Septoria helenii. Leaf spot. **6:**IA(82), PA(82).

Helianthella Torr. & A. Gray

1. *H. californica* A. Gray
2. *H. californica* A. Gray var. *nevadensis* (Greene) Jeps.
3. *H. quinquenervis* (Hook.) A. Gray
4. *H. uniflora* (Nutt.) Torr. & A. Gray (=*Helianthus uniflorus* Nutt.)
5. *H. uniflora* (Nutt.) Torr. & A. Gray var. *douglasii* (Torr. & A. Gray) W. A. Weber

Ascomycotina

Hyponectria sceptri. **4:**UT(3084).
[*Pyrenophora schroeteri*]. On dead stems. **2:**CA(81).

Basidiomycotina - Uredinales

Puccinia helianthellae. Autoecious. **1:**CA(81). **2:**CA(81). **3:**WY(1195,3379). **4:**WY(1195). **5:**WA(638).

Basidiomycotina - Ustilaginales

Entyloma compositarum. **4:**WA(1077).

Deuteromycotina - Hyphomycetes

Cercosporella helianthellae. **5:**WA(638).
Didymaria wyethiae. **5:**WA(638).
Ramularia helianthi. **3:**WY(1195).

Helianthus L.

1. *H. angustifolius* L. [narrow-leaf sunflower, swamp sunflower]
2. *H. annuus* L. (=*H. lenticularis* Douglas) [common sunflower, Hopi sunflower, mirasol]
3. *H. argophyllus* Torr. & A. Gray [silver-leaved sunflower]
4. *H. atrorubens* L. [dark-eye sunflower, dark-red sunflower, hairywood sunflower]
5. *H. bracteatus* E. Wats.
6. *H. debilis* Nutt. [beach sunflower, weak sunflower]
7. *H. decapetalus* L. [giant sunflower, ten-petals sunflower, thin-leaf sunflower, woodland sunflower]
8. *H. divaricatus* L. [divergent sunflower]
9. *H. giganteus* L. [giant sunflower]
10. *H. grosseserratus* G. Martens [Sawtooth sunflower]
11. *H. heterophyllus* Nutt. [different-leaf sunflower, hairy sunflower]
12. *H. hirsutus* Raf. [rough sunflower, stiff-haired sunflower]
13. *H. kellermanii* Britton
14. *H.* × *laetiflorus* Pers. (=*H. scaberrimus* Elliott). *H. rigidus* × *H. tuberosus* var. *rigidus*. [cheerful sunflower, showy sunflower]
15. *H. maximilianii* Schrad. [Maximilian's sunflower]
16. *H. microcephalus* Torr. & A. Gray [small-headed sunflower, smallwood]
17. *H. mollis* Lam. [ashy sunflower, soft sunflower]
18. *H.* × *multiflorus* L. *H. annuus* × *H. decapetalus*.
19. *H. nitidus* Lunell
20. *H. nuttallii* Torr. & A. Gray (=*H. fascicularis* Greene) [Nuttall's sunflower]
21. *H. occidentalis* Riddell [western sunflower]
22. *H. petiolaris* Nutt. [Kansas sunflower, petioled sunflower, prairie sunflower]
23. *H. rigidus* (Cass.) Desf. [stiff sunflower]
24. *H. strumosus* L. [swollen sunflower, woodland sunflower]
25. *H. subrhomboideus* Rydb.
26. *H. tuberosus* L. (=*H. tomentosus* Michx.) [girasole, Jerusalem-artichoke, sunchoke, woodland sunflower]
27. *Helianthus* sp.

Oomycetes

Albugo tragopogonis. White rust. **2:**WI(82). **27:**IL(82), MO(82).

Phytophthora cryptogea. Stem rot. **2:**CA(1925).
Plasmopara halstedii. Basal gall, downy mildew. **2:**NE states(82), KS(82), MN(4014,4015), MT(82,3202), ND(377,4014,4015), SD(511,2243), WY(1195). **3:**TX(82). **9:**C states(82), ND(82). **10:**MT(3202). **15:**C states(82), ND(82,377). **20:**MT(3202), WY(1195). **23:**C states(82), ND(82), WI(1332). **24:**WI(1349). **25:**ND(377). **26:**C states(82), NE states(82), SD(82). **27:**C states(82), MT(82), ND(82), NY(82), TX(82), WY(82).
Pythium debaryanum. Root rot. **2:**ND(82).
P. irregulare. Root rot. **2:**HI(82,2903).
P. rostratum. **2:**HI(2903).
P. splendens. Root rot. **2:**HI(82,2903).
Pythium sp. Root rot. **27:**FL(35).

Zygomycotina

Mucor fragilis. Isolated from achenes. **2:**GA(3007).
M. hiemalis. Isolated from achenes. **2:**GA(3007).
Mucor sp. Isolated from achenes. **2:**GA(3007).
Rhizomucor miehei. Isolated from achenes. **2:**GA(3007).
Rhizopus arrhizus. Head rot. Isolated from achenes. **2:**CA(1925), GA(3007), KS(3010), OK(3010), TX(3010,3970).
R. stolonifer. Head rot, soft rot of tubers. **2:**TX(3970). **26:**MN(82).
Rhizopus sp. Isolated from achenes. **2:**GA(3007).

Ascomycotina

Apiospora montagnei. Isolated from achenes. **2:**GA(3007).
Ascotricha xylina. Isolated from achenes. **2:**GA(3007), ND(3006).
Chaetomium globosum. Isolated from achenes. **2:**GA(3007).
C. incomptum. Isolated from achenes. **2:**GA(3007).
C. indicum. Isolated from achenes. **2:**GA(3007).
C. spirale. Isolated from achenes. **2:**GA(3007).
Chaetomium sp. Isolated from achenes. **2:**GA(3007).
Diaporthe arctii. On stems. **2:**NE(82). **4:**GA(82,1438). **8:**GA(1438). **16:**GA(1438). **27:**GA(82).
D. helianthi. Stem canker, leaf spot. **2:**TX(3972).
Diaporthe sp. Stem canker. **2:**OH(1537).
Emericella nidulans. Isolated from achenes. **2:**GA(3007).
Erysiphe cichoracearum. Powdery mildew. **2:**Range of host(82), CA(1925), FL(35), GA(1438), ID(3202), MT(3202), ND(377), OK(2872), SD(2243), WA(3202), WI(1340). **4:**E states(82), MT(82), TX(82). **6:**IA(82). **7:**E states(82), MT(82), TX(82). **9:**E states(82), MT(82), TX(82). **10:**MT(3202), SD(2243). **12:**MS(2720). **13:**WI(1330). **15:**E states(82), MT(82,3202), ND(377), SD(2243), TX(82). **17:**E states(82), MT(82), TX(82). **20:**MT(3202). **21:**WI(1332). **22:**IA(82), MI(82). **23:**E states(82), MT(82), SD(2243), TX(82), WI(1344). **26:**Range of host(82), GA(1438,2301), MA(1390), MS(2720), NC(1303), ND(377). **14:**GA(1438). **27:**FL(35), MS(2720).
E. cichoracearum var. *latispora*. Powdery mildew. **10:**WI(367).
E. polygoni. Powdery mildew. **2:**FL(35).
Hypoderma commune. **16:**GA(1438).
Lachnella villosa. On dead stems. **27:**Unk.(456).
Leptosphaeria consessa. **25:**ND(377).
L. helianthi. **9:**ND(377).
L. heliopsidis. Canker, black patch on leaves and stems. **8:**GA(1438), MS(2720), NC(3944). **20:**MT(3202). **23:**WI(82,1339,1344). **24:**WI(2277). **27:**GA(82), IL(82), MO(82), MS(82,2720), NC(82,1303), VA(82), WI(82,1335).
L. lindquistii. On overwintered stalks. **2:**MN(886), ND(886).
Lophiostoma caulium. **16:**GA(1438).
Massarina gloeospora. **16:**GA(1438).
Microascus cinereus. Isolated from achenes. **2:**GA(3007).
M. cirrosus. Isolated from achenes. **2:**GA(3007).
M. senegalensis. Isolated from achenes. **2:**GA(3007).
M. trigonosporus. Isolated from achenes. **2:**GA(3007).
Monascus ruber. Isolated from achenes. **2:**GA(3007).
Mycosphaerella sp. On leaves and stems. **16:**GA(1438). **24:**WI(1354). **27:**GA(82).
Naevia sp. **27:**GA(1438).
Ophiobolus anguillidus. **8:**GA(1438).
O. drechsleri. **2:**KS(3248). **10:**KS(3248).
O. nigro-clypeata. **16:**GA(1438).
Ophioceras sp. **2:**GA(1438).
Phyllachora ambrosiae. Black pustule. **15:**NM(82). **27:**CA(82), NM(82), WY(82).
Pithoascus intermedius. Isolated from achenes. **2:**GA(3007).
P. nidicola. Isolated from achenes. **2:**GA(3007).
Pleospora lactucicola. **15:**ND(377).
Schizoxylon berkeleyanum. **15:**ND(377).
Sclerotinia minor. **2:**NY(1946).
S. sclerotiorum. Head rot, root rot, basal stem rot, wilt. **2:**NC states(82), NE states(82), W states(82), CA(1925), HI(1402,2903), MT(3202), ND(1611), OR(3202), WA(3202). **26:**MA(82), MN(82), OR(3202), WA(82,3202). **27:**NC(1303).
Sordaria fimicola. Isolated from achenes. **2:**GA(3007).
Thermoascus crustaceus. Isolated from achenes. **2:**GA(3007).

Basidiomycotina - Uredinales

Coleosporium helianthi. Heteroecious: telial host. **2:**CA(3974). **7:**E states(82), IN(82), LA(82). **9:**E states(82), IN(82), LA(82), MS(2720). **16:**GA(1307), NC(1303,1307). **23:**OK(82,2872). **26:**AL(82), GA(1307), IL(82), NC(82,1303,1307), NY(82), OK(82), PA(82), SC(82), TN(82), VA(82). **27:**FL(35).
C. pacificum. **2:**CA(3974).
Puccinia helianthi. Autoecious. **1:**MS(2720), NC(3944). **2:**Range of host(82), AZ(747,1203,3981), CA(81,747), FL(35), GA(1439), ID(3202), MD(2655), MS(2720), MT(3202), ND(378), OK(2872), OR(3202), SD(2243), TX(742,3969), WA(3202), WY(1195). **4:**Range of host(82), GA(1439), MS(2720). **6:**CA(82), FL(82), IA(82), TX(82). **7:**Range of host(82), WI(1357). **8:**IN(987), NC(3944). **9:**Range of host(82), SD(2243), WI(1334). **10:**MT(3202), SD(2243), WI(1334). **11:**SC(634). **12:**MS(2720), OK(2872), WI(1352). **15:**Range of host(82), MT(3202), ND(378), SD(2243). **16:**NC(1303). **17:**Range of host(82), MS(2720), OK(2872). **19:**ND(378). **20:**MT(3202), WY(1195). **22:**Range of host(82), AZ(747,3981), CA(747), MT(3202), OK(660,2872), WI(1346). **23:**Range of host(82), NC(3944), SD(2243), WI(1331). **24:**GA(1439), NC(1303,3944), WI(1334). **25:**ND(378), SD(2243). **14:**IN(987), NC(3944), WI(1330). **18:**WI(1330). **27:**Range of host(82), FL(35), OK(2872), SD(2243).
P. massalis. Autoecious. **27:**NM(82), TX(82).
Uromyces junci. Heteroecious: aecial host. **2:**NE(82). **7:**NE(82). **15:**ND(82,378), NE(82). **20:**WY(1195). **22:**OK(82,2872). **23:**CA(82), KS(82), NE(82), SD(82), WI(1350), WY(82). **24:**WI(1344). **26:**ND(82,378), NE(82).
U. silphii. Heteroecious: aecial host. **9:**NY(82). **24:**WI(1333). **26:**WI(1360). **27:**IL(82), MO(82), NY(82), TN(82), WI(82).

Basidiomycotina - Ustilaginales

Entyloma polysporum. Leaf smut. **2:**MT(82,1077,3202).
Sporidiobolus pararoseus. On rusted leaves. **27:**KS(2686).

Basidiomycotina - Other

Armillaria mellea. Root rot. **2:**OR(82).
Schizophyllum commune. Isolated from achenes. **2:**GA(3007).
Thanatephorus cucumeris. **2:**WA(3202).

Deuteromycotina - Hyphomycetes

Acremonium implicatum. Isolated from achenes. **2:**GA(3007).
A. sclerotigenum. Isolated from achenes. **2:**GA(3007).
Acremonium sp. Isolated from achenes. **2:**GA(3007).
Acrodontium sp. Isolated from achenes. **2:**GA(3007).
Alternaria alternata. Leaf spot. **2:**FL(35), GA(3007), MT(3202). **27:**FL(35).
A. helianthi. Seedling blight, leaf spot, stem spot. **2:**HI(1402), MN(3198), NC(1303), OH(1536).
A. helianthinficiens. **2:**ND(3272).
A. zinniae. Leaf spot. **2:**Range of host(500).
Alternaria sp. Isolated from achenes. **2:**GA(3007).
Aspergillus candidus. Isolated from achenes. **2:**GA(3007).
A. flavipes. Isolated from achenes. **2:**GA(3007).
A. flavus. Isolated from achenes. **2:**GA(3007).
A. fumigatus. Isolated from achenes. **2:**GA(3007).
A. glaucus. Isolated from achenes. **2:**GA(3007).
A. hollandicus. Isolated from achenes. **2:**GA(3007).
A. niger. Isolated from achenes. **2:**GA(3007).
A. niger var. *awamorii*. Isolated from achenes. **2:**GA(3007).
A. ochraceus. Isolated from achenes. **2:**GA(3007).
A. parasiticus. Isolated from achenes. **2:**GA(3007).
A. penicillioides. Isolated from achenes. **2:**GA(3007).
A. tamarii. Isolated from achenes. **2:**GA(3007).
A. terreus. Isolated from achenes. **2:**GA(3007).
A. terricola. Isolated from achenes. **2:**GA(3007).
A. versicolor. Isolated from achenes. **2:**GA(3007).
A. wentii. Isolated from achenes. **2:**GA(3007).
Aureobasidium sp. Isolated from achenes. **2:**GA(3007).

Beauveria sp. Isolated from achenes. **2:**GA(3007).
Botrytis cinerea. Gray mold blight. **2:**CA(82), GA(3007), OR(82).
Botrytis sp. **24:**WI(1361).
Cephaliophora tropica. Isolated from achenes. **2:**GA(3007).
Cephalosporium sp. Wilt. **2:**IN(2976).
Cercospora helianthi. Leaf spot. **15:**TX(82). **23:**IL(82), KS(82), WI(1350). **26:**KS(82). **27:**FL(35), KS(82), MO(82), OH(82), WI(82).
C. pachypus. Leaf spot. **2:**Al(82), KS(82,973), OK(82,2872), TX(82). **22:**KS(82).
Cladosporium cladosporioides. Isolated from achenes. **2:**GA(3007).
C. cucumerinum. Isolated from achenes. **2:**GA(3007).
C. herbarum. Isolated from achenes. **2:**GA(3007).
C. oxysporum. Isolated from achenes. **2:**GA(3007).
C. sphaerospermum. Isolated from achenes. **2:**GA(3007).
Drechslera sp. Isolated from achenes. **2:**GA(3007).
Epicoccum nigrum. Isolated from achenes. **2:**GA(3007).
Eupenicillium cinnamopurpureum. Isolated from achenes. **2:**GA(3007).
Eurotium amstelodami. Isolated from achenes. **2:**GA(3007).
E. chevalieri. Isolated from achenes. **2:**GA(3007).
E. echinulatum. Isolated from achenes. **2:**GA(3007).
E. herbariorum. Isolated from achenes. **2:**GA(3007).
E. repens. Isolated from achenes. **2:**GA(3007).
E. rubrum. Isolated from achenes. **2:**GA(3007).
E. tonophilum. Isolated from achenes. **2:**GA(3007).
Fusarium chlamydosporum. Isolated from achenes. **2:**GA(3007).
F. concolor. Isolated from achenes. **2:**GA(3007).
F. equiseti. Isolated from achenes. **2:**GA(3007).
F. lateritium. Isolated from achenes. **2:**GA(3007).
F. moniliforme. Isolated from achenes. **2:**GA(3007).
F. oxysporum. Isolated from achenes. **2:**GA(3007).
F. pallidoroseum. Isolated from achenes. **2:**GA(3007).
F. poae. Isolated from achenes. **2:**GA(3007).
F. solani. Isolated from achenes. **2:**GA(3007).
F. sporotrichioides. Isolated from achenes. **2:**GA(3007).
F. subglutinans. Isolated from achenes. **2:**GA(3007).
F. ventricosum. Isolated from achenes. **2:**GA(3007).
Fusarium sp. Basal stem rot, storage rot, wilt. **2:**TX(2653). **5:**WA(3202). **26:**GA(2301). **27:**FL(35).
Malbranchea sulfurea. Isolated from achenes. **2:**GA(3007).
Microdochium nivale. Isolated from achenes. **2:**GA(3007).
Nigrospora oryzae. Isolated from achenes. **2:**GA(3007).
N. sphaerica. Isolated from achenes. **2:**GA(3007).
Oidium sp. Powdery mildew. **26:**HI(2903).
Paecilomyces variotii. Isolated from achenes. **2:**GA(3007).
Paecilomyces sp. Isolated from achenes. **2:**GA(3007).
Papulaspora sp. Isolated from achenes. **2:**GA(3007).
Penicillium sp. Isolated from achenes. **2:**GA(3007).
Phymatotrichopsis omnivora. Root rot, wilt. **2:**TX(82,2653). **15:**TX(82). **22:**TX(82). **26:**TX(82). **27:**TX(82).
Ramularia helianthi. Root rot, wilt. **27:**CA(82), WA(82).
Scopulariopsis brevicaulis. Isolated from achenes. **2:**GA(3007).
S. brumptii. Isolated from achenes. **2:**GA(3007).
S. candida. Isolated from achenes. **2:**GA(3007).
Scopulariopsis sp. Isolated from achenes. **2:**GA(3007).
Stemphylium botryosum. Isolated from achenes. **2:**GA(3007).
Thermomyces lanuginosus. Isolated from achenes. **2:**GA(3007).
T. stellatus. Isolated from achenes. **2:**GA(3007).
Tilletiopsis washingtonensis. On rusted leaves. **27:**KS(2686).
Trichothecium roseum. Isolated from achenes. **2:**GA(3007).
Verticillium dahliae. Wilt. Isolated from achenes. **2:**CA(1925), GA(3007).
V. nigrescens. Wilt. **2:**GA(3007).
Verticillium sp. Isolated from achenes. **2:**GA(3007).

Deuteromycotina - Coelomycetes

Ascochyta compositarum. Leaf spot. **26:**WI(1361). **27:**WI(82).
A. doronici. **24:**WI(1343).
Colletotrichum helianthi. Leaf spot. **26:**WI(1360). **27:**WI(82).
C. helianthi var. *macromaculans*. On living leaves. **24:**WI(1356).
Colletotrichum sp. **24:**WI(1346).
Macrophomina phaseolina. Charcoal stem rot, wilt. **2:**CA(1925), HI(1402), MD(82), TX(2653,3971).
Macrophomopsis sp. Isolated from achenes. **2:**GA(3007).
Phoma macdonaldii. Black stem. Isolated from achenes. **2:**GA(3007), HI(1402).
Phoma sp. Black stem. **2:**GA(3007). **10:**WI(1358).
Phomopsis sp. Stem canker. Isolated from achenes. **2:**GA(3007), OH(1537).
Phyllosticta helianthi. Stem canker. **23:**WI(1343). **27:**WI(82).
P. wisconsinensis. On living leaves. **21:**WI(1359,1360). **23:**WI(1360).
Phyllosticta sp. Leaf spot. **9:**WI(1360). **21:**WI(1339). **24:**WI(1354).
Septoria helianthi. Leaf spot. **2:**E states(82), Range of host(500), W states(82), HI(1402), MS(82,2720), MT(82,3202), ND(378), NM(82), OK(2872), WA(3202), WY(1195). **3:**TX(82). **7:**WV(82). **9:**WI(1356), WV(82). **22:**MN(82), MT(82,3202), WI(1349). **23:**C states(82), NE states(82), MN(82). **26:**IA(82), IL(82), WI(82). **27:**C states(82), FL(35), MN(82), NC(1303), OK(2872).
S. helianthicola. On stems. **2:**CA(82).
S. inulae. **17:**IL(82).
S. paupera. Leaf spot. **8:**NJ(959).
Sphaceloma sp. **1:**FL(2525).
Sphaerophoma brencklei. On stems. **26:**ND(82).

Deuteromycotina - Other

Rhizoctonia crocorum. Violet root rot. **2:**TX(82).
R. solani. Root rot. **2:**FL(35), IL(82), MD(82), NE(82), NY(82), WI(82).
R. zeae. Isolated from achenes. **2:**GA(3007).
Rhizoctonia sp. Root rot. Isolated from achenes. **2:**GA(3007), HI(2903).
Sclerotium rolfsii. Southern blight, tuber rot. **2:**CA(2886), FL(35), GA(2886), LA(82), TX(82). **26:**FL(82), GA(2301), LA(82), MS(82,2720), NC(1303), SC(82), TX(82). **27:**FL(35), NC(1303).

Helichrysum Mill.

1. *H. bracteatum* (Venten.) Andr. [strawflower]

Oomycetes

Bremia lactucae. Downy mildew. **1:**FL(35).
Pythium sp. Root rot. **1:**FL(35).

Deuteromycotina - Hyphomycetes

Fusarium sp. Stem rot. **1:**WA(82).
Verticillium albo-atrum. Wilt. **1:**CA(82).

Deuteromycotina - Other

Sclerotium rolfsii. Southern blight. **1:**FL(35).

Heliopsis Pers.

1. *H. helianthoides* (L.) Sweet (=*Buphthalmum helianthoides* L.)
2. *H. helianthoides* (L.) Sweet cv. 'pitcherana' (=*H. pitcherana* Hort.)
3. *H. helianthoides* (L.) Sweet var. *scabra* (Dunal) Fernald (=*H. scabra* Dunal)
4. *H. parvifolia* A. Gray
5. *Heliopsis* sp.

Ascomycotina

Erysiphe cichoracearum. Powdery mildew. **3:**IA(82), MN(82), ND(82), NE(82), NM(82), WI(82).
E. cumminsiana. Powdery mildew. **3:**CO(369), NM(369).
[*E. galeopsidis*]. **3:**WI(1349).
Leptosphaeria heliopsidis. Black patch on leaves and stems. **1:**MS(82). **3:**MS(82).
Leveillula taurica. Powdery mildew. **3:**IA(82).
Mycosphaerella sp. On leaves. **1:**WI(1360,1361).
Ophiobolus drechsleri. **5:**IL(3248).

Basidiomycotina - Uredinales

Coleosporium helianthi. Heteroecious: telial host. **1:**NC(1303).
Puccinia batesiana. Autoecious. **1:**DE(82), PA(82). **3:**IA(82), KS(82), MD(82), MN(82), NE(82), WI(82).
P. canaliculata. Heteroecious: aecial host. **4:**AZ(1195,1203,3981).
P. helianthi. Autoecious. **1:**NC(1303), WI(1355). **3:**IA(82), IN(82), MN(82).

Deuteromycotina - Hyphomycetes

Phymatotrichopsis omnivora. Root rot. **1:**TX(82). **3:**TX(82).

Deuteromycotina - Coelomycetes

Ascochyta compositarum. On ray flowers. **1:**WI(1360). **3:**WI(1345).
Phyllosticta heliopsidis. On living leaves. **1:**WI(1355).
P. pitcheriana. Leaf spot. **1:**NY(82). **2:**NY(64).
[*Septoria helianthi*]. Leaf spot. **1:**IN(82).
S. heliopsidis. **3:**WI(82).
Septoria sp. **3:**WI(1344).

Helipterum DC.

1. *H. roseum* (Hook.) Benth.

Deuteromycotina - Other

Rhizoctonia sp. Damping-off. **1:**FL(35).

Hemizonia DC.

1. *H. luzulaefolia* DC.
2. *H. multicaulis* Hook. & Arnott (=*H. citrina* Greene)

Basidiomycotina - Uredinales

Puccinia nuda. Autoecious. **1:**CA(81). **2:**CA(81).

Heterotheca Cass.

1. *H. subaxillaris* (Lam.) Britton & Rusby

Basidiomycotina - Uredinales

Coleosporium asterum. Heteroecious: telial host. **1:**FL(35).
Puccinia cyperi. Heteroecious: aecial host. **1:**FL(35), TX(652).

Basidiomycotina - Ustilaginales

Entyloma compositarum. **1:**KS(1077).

Deuteromycotina - Hyphomycetes

Alternaria sp. Leaf spot. **1:**FL(35).

Hieracium L.

1. *H. albertinum* Farr
2. *H. albiflorum* Hook.
3. *H. aurantiacum* L. [devil's-paintbrush, king's-devil, orange hawkweed]
4. *H. canadense* Michx.
5. *H. cusickii* Gand.
6. *H. cynoglossoides* Arv.-Touv.
7. *H. cynoglossoides* Arv.-Touv. var. *nudicaule* A. Gray
8. *H. fendleri* Schultz-Bip.
9. *H. gracile* Hook.
10. *H. longipilum* Torr.
11. *H. piloselloides* Vill. (=*H. florentinum* All.) [king-devil hawkweed]
12. *H. scabrum* Michx.
13. *H. scouleri* Hook.
14. *H. umbellatum* L. [narrow-leaved hawkweed]
15. *H. venosum* L. [poor-Robin's plantain, rattlesnake-weed]
16. *Hieracium* sp.

Oomycetes

Bremia lactucae. Downy mildew. **3:**WI(82).

Ascomycotina

Entodesmium niessleanum. **2:**WA(3814).
Erysiphe cichoracearum. Powdery mildew. **2:**WA(3202). **3:**TX(82). **4:**TX(82). **16:**PA(82), WA(82).
Leptosphaeria salsolae. **2:**WA(3202,3814).
Pleospora helvetica. **16:**WA(3202,3814).
P. lactucicola. **1:**WA(3202).
Sphaerotheca fuliginea. Powdery mildew. **16:**IL(82).
S. macularis. **2:**WA(3202).

Basidiomycotina - Uredinales

Aecidium columbiense. **1:**WA(3202). **2:**CA(81), OR(3202), WA(3202). **15:**NY(82). **16:**CA(82), ID(82), OR(82), WA(82).
Puccinia atrofusca. **14:**ND(378).
P. cnici-oleracei. Autoecious. **1:**ID(3202), WA(638). **6:**WA(3202). **13:**OR(3202), WA(3202). **16:**OR(82,2713), TX(82), WA(82).
P. dioicae. Heteroecious: aecial host. **2:**OR(3202). **4:**IL(82), WI(82). **5:**WY(1195). **8:**AZ(3981). **9:**MT(3202). **10:**WI(1337). **12:**WI(1341). **13:**OR(3202). **16:**NC states(82), NE states(82), CA(82), MT(82), OR(82).
P. fraseri. Autoecious. **1:**ID(3202). **2:**MT(1195,2713,3202,3379). **12:**NY(2713). **15:**GA(1439), PA(82), TN(82), VA(82), WV(82). **16:**MT(82), NE(82), NH(82), NY(82), PA(82).
P. hieracii. Autoecious. **1:**ID(3202), MT(1195,2713,3202), WA(3202). **2:**AK(60,526), CA(81,638,642,661), ID(3202), MT(3202), OR(2713,3202), WA(3202), WY(1195). **3:**TX(82). **4:**ID(3202), MT(3202), SD(2243), WA(3202), WI(1337,2713). **8:**AZ(1203,3981). **9:**CA(638,642), OR(3202), WA(2713,3202). **12:**GA(1439), WI(1332,2713). **13:**OR(3202), WA(2713,3202). **14:**ND(378), SD(2243). **15:**NC(82,3944), NY(82). **16:**CA(642,661).

Basidiomycotina - Ustilaginales

Entyloma calendulae. **16:**WA(1077,3202).
E. compositarum. **1:**ID(3202), OR(3202), WA(1077,3202). **2:**ID(1077,3202), WA(1077,3202). **7:**CA(638,642,1077).

Deuteromycotina - Hyphomycetes

Cercospora erechtitis. **3:**WI(1342).
C. hieracii. Leaf spot. **15:**AL(82), NC(82,3944).
Cercospora sp. **10:**WI(1340).
Phymatotrichopsis omnivora. Root rot. **3:**TX(82). **16:**TX(82).
Ramularia macrospora. **2:**WA(3202).
R. parva. On living leaves. **10:**WI(1352).
R. taraxaci. **10:**WI(1342).
Ramularia sp. On leaves. **10:**WI(1340).

Deuteromycotina - Coelomycetes

Ascochyta sp. Leaf spot. **3:**WI(1361).
Heteropatella cercosperma. Leaf spot. **3:**TX(82). **16:**TX(82).
Phoma exigua. Leaf spot. **3:**WI(64,82,1333).
P. hieracii. Stem blight. **3:**TX(82).
Phomopsis hieracii. On living leaves. **3:**WI(1356). **10:**WI(1346).
Phyllosticta hieracii. **10:**WI(1340).
Phyllosticta sp. **9:**WA(3202).
Septoria hieracicola. **11:**WI(1356). **12:**WI(1356). **16:**NY(82), WA(3202).

Hymenoclea Torr. & A. Gray in A. Gray

1. *H. monogyra* Torr. & A. Gray
2. *H. salsola* Torr. & A. Gray (=*H. fasciculata* A. Nels.)
3. *H. salsola* Torr. & A. Gray var. *pentalepis* (Rydb.) L. Benson (=*H. pentalepis* Rydb.)

Basidiomycotina - Uredinales

Puccinia franseriae. Autoecious. **1:**AZ(1195,3379). **3:**AZ(1195,3379,3981).
P. splendens. Autoecious. **1:**AZ(747,1195,1203,3981), CA(747), TX(742). **2:**AZ(747,3981), CA(81,747), NV(1195).

Hymenopappus L'Hér.

1. *H. scabiosaeus* L'Hér. (=*H. caroliniensis* Porter)

Oomycetes

Plasmopara halstedii. Downy mildew. **1:**OK(82,2872).

Basidiomycotina - Uredinales

Puccinia grindeliae. Autoecious. **1:**OK(82,2872).

Deuteromycotina - Hyphomycetes

Phymatotrichopsis omnivora. Root rot. **1:**TX(82).

Hymenoxys Cass.

1. *H. acaulis* (Pursh) K. Parker
2. *H. grandiflora* (Torr. & A. Gray) K. Parker (=*Rydbergia grandiflora* (Torr. & A. Gray) Greene)

Ascomycotina

Pleospora stenospora. **2:**WY(3810).

Basidiomycotina - Uredinales

Puccinia grindeliae. Autoecious. **1:**MT(3202).

Hypochaeris L.

1. *H. glabra* L. [smooth cat's-ear]
2. *H. radicata* L. [false dandelion, spotted cat's-ear]

Basidiomycotina - Uredinales

Puccinia hieracii. **1:**CA(82), WA(3202). **2:**CA(82), OR(82,3202).
P. hypochaeridis. **1:**CA(81,2713), WA(2713).

Deuteromycotina - Hyphomycetes

Phymatotrichopsis omnivora. Root rot. **2:**TX(82).
Ramularia hypochoeridis. **2:**WA(3202).

Inula L.

1. *I. helenium* L. [elecampane]

Ascomycotina

Anthostomella rostraspora. On stems. **1:**NY(82).
Erysiphe cichoracearum. Powdery mildew. **1:**NY(82), SD(2243), WI(82).

Basidiomycotina - Uredinales

Puccinia hieracii. Autoecious. **1:**WI(82).

Deuteromycotina - Hyphomycetes

Ramularia sp. Leaf spot. **1:**MI(82).

Deuteromycotina - Coelomycetes

Phyllosticta sp. **1:**WI(1355).

Iva L.

1. *I. axillaris* Pursh [poverty-weed]
2. *I. frutescens* L.
3. *I. xanthifolia* Nutt. [marsh-elder]
4. *Iva* sp.

Oomycetes

Albugo tragopogonis. White rust. **3:**WI(82). **4:**AZ(82), IL(82), KS(82), MO(82), NM(82).

Basidiophora kellermanii. Downy mildew. **3:**C states(82), CO(82), MT(82,3202), ND(377), SD(2243).

Plasmopara halstedii. Downy mildew. **3:**SD(82).

Ascomycotina

Crocicreas cyathoideum. **3:**MN(505), ND(377).

Dasyscyphus ivae. **3:**ND(377,3167).

Erysiphe cichoracearum. Powdery mildew. **2:**NJ(82). **3:**CO(82), ND(82,377).

Lewia scrophulariae. **1:**WA(3202).

Metasphaeria ambrosiaecola. **3:**ND(377).

Ophiobolus anthrisci. **3:**ND(3248).

O. claviger. **3:**ND(377).

Phyllachora ambrosiae. Leaf spot. **2:**FL(35), GA(82), SC(82). **3:**Unk.(82), CO(82), IA(82), KS(82), MT(82,3202), TX(82). **4:**GA(82).

Pyrenopeziza absinthii. On stems. **3:**NC(82), ND(377).

Sclerotinia sclerotiorum. **4:**MT(3202).

Basidiomycotina - Uredinales

Aecidium ivae. **2:**DE(82,746), FL(35,82), LA(82), MD(82), VA(82).

[*Coleosporium helianthi*]. Heteroecious: telial host. **1:**NC(82).

Puccinia helianthi. Autoecious. **3:**C states(82), W states(82), AZ(1203,3981), ID(3202).

P. intermixta. Autoecious. **1:**W states(82), CA(81), ID(3202), MT(3202), OR(3202), SD(2243), WA(3202), WY(1195).

Deuteromycotina - Hyphomycetes

Botrytis cinerea. **4:**WA(3202).

Heterosporium tuberculans. Leaf spot. **3:**ND(82,378), TX(82).

Phymatotrichopsis omnivora. Root rot. **3:**TX(82).

Ramularia ivae. On leaves. **1:**WY(82).

Deuteromycotina - Coelomycetes

Microdiplodia ivicola. On dead stems. **3:**SD(970).

Phyllosticta ivaecola. Leaf spot. **2:**LA(82). **3:**ND(64,82,378), TX(82), WI(82).

Septoria ivicola. **3:**TX(82).

S. xanthifolia. Leaf spot. **3:**KS(82), ND(82,378), WI(82). **4:**GA(82), KS(82), ND(82), WI(82).

Jaumea Pers.

1. *J. carnosa* (Less.) A. Gray

Basidiomycotina - Uredinales

Uromyces junci. Heteroecious: aecial host. **1:**WA(3202).

Krigia Schreb.

1. *K. biflora* (Walter) S. F. Blake
2. *K. dandelion* (L.) Nutt.
3. *K. virginica* (L.) Willd. (=*Serinia oppositifolia* Kuntze)
4. *Krigia* sp.

Oomycetes

Bremia lactucae. Downy mildew. **1:**WI(82). **2:**MS(2720). **3:**WI(1333). **4:**MO(82), MS(82), OK(82,2872).

Ascomycotina

Mycosphaerella krigiae. Leaf spot. **1:**IL(82), WI(82).

Basidiomycotina - Uredinales

Puccinia cnici-oleracei. **1:**IL(82), MI(82,2713), MO(82,2713), MS(2713), PA(82,2713), TN(82,2713). **4:**MO(82), MS(82), PA(82).

P. dioicae. **1:**IA(82), IL(82), WI(82).

P. helianthi. Autoecious. **1:**NC(1303).

P. hieracii. Autoecious. **1:**NC(82).

Deuteromycotina - Hyphomycetes

Phymatotrichopsis omnivora. Root rot. **4:**TX(82).

Deuteromycotina - Coelomycetes

Phoma krigiae. On stems and bracts. **4:**Unk.(82).

Septoria krigiae. Leaf spot. **1:**KY(82), NY(82), WI(82). **3:**WI(1334). **4:**WI(82).

Kuhnia L.

This genus is considered a synonym of *Brickellia*, but the appropriate combinations could not be located.

1. *K. eupatorioides* L. (=*K. glutinosa* Elliott) [false boneset]
2. *K. eupatorioides* L. var. *corymbulosa* Torr. & A. Gray (=*K. hitchcockii* A. Nels., *K. reticulata* A. Nels.)
3. *K. leptophylla* Scheele (=*K. rosmarinifolia* Venten.)
4. *Kuhnia* sp.

Ascomycotina

Pleospora scrophulariae var. *compositarum*. Leaf spot. **4:**NM(82), TX(82).

Basidiomycotina - Uredinales

Puccinia kuhniae. Autoecious. **1:**AL(752), FL(35), IA(752), IL(752), IN(752), MI(665,752), MO(752), MS(2720), MT(3202), ND(378,752), NE(752), SD(752,2243), WI(1358), WY(752). **3:**AZ(1195,1203). **4:**FL(82), MT(82), TX(82).

P. kuhniae var. *robusta*. **2:**CO(752), MT(752), WY(752). **3:**AZ(752), CO(752), NM(752), TX(752).

Deuteromycotina - Hyphomycetes

Phymatotrichopsis omnivora. Root rot. **4:**TX(82).

Deuteromycotina - Coelomycetes

Ascochyta doronici. On living leaves. **1:**WI(1358).

Ascochyta sp. **1:**WI(1347).

Septoria parvimaculans. On living leaves. **1:**WI(1338).

Stagonospora sp. On leaves. **1:**WI(1358).

Lactuca L.

1. *L. biennis* (Moench) Fernald (=*L. leucophaea* A. Gray, *L. spicata* (Lam.) Hitchc.)
2. *L. canadensis* L. [Canada wild lettuce, tall lettuce]
3. *L. graminifolia* Michx. [grassleaf lettuce]
4. *L. ludoviciana* (Nutt.) Riddell
5. *L. muralis* (L.) Fresen. (=*Prenanthes vulgaris* Gueldenst.)
6. *L. sativa* L. [garden lettuce, lettuce]
7. *L. sativa* L. var. *capitata* L. [head lettuce]
8. *L. sativa* L. var. *longifolia* Lam. [Cos lettuce, Romaine lettuce]
9. *L. serriola* L. (=*L. scariola* L.) [prickly lettuce]
10. *L. tatarica* (L.) C. A. Mey. subsp. *pulchella* (Pursh) Stebb. (=*L. oblongifolia* Nutt., *L. pulchella* (Pursh) DC.) [blue lettuce]
11. *Lactuca* sp.

Myxomycetes

Physarum cinereum. Slime mold. **6:**Unk.(82).

Chytridiomycetes

Olpidium brassicae. Big-vein. **6:**AZ(174).

Synchytrium aureum. Leaf gall. **11:**LA(82).

Oomycetes

Albugo sp. White rust. **6:**TX(82).

Bremia lactucae. Downy mildew. **1:**Unk.(82), WA(3202). **2:**Unk.(82), MS(2720), NC(1303,3944). **4:**MT(3202), WI(1334). **6:**'Downy mildew.' Range of host(82), AK(2163), CA(2797), CT(2974), FL(35,705), ID(3202), MS(2720), MT(3202), NC(1303), NY(3993), OR(3202), SD(2243), WA(3202), WI(1332). **7:**HI(2903). **9:**Unk.(82), MT(3202), NC(3944), WA(3202). **10:**Unk.(82), MT(3202), ND(377). **11:**Unk.(82), FL(35), OK(2872).

Plasmopara halstedii. Downy mildew. **2:**MD(82).

Pythium catenulatum. Isolated from roots of hydroponically grown plants. **6:**AZ(3508).

P. debaryanum. Damping-off. **6:**NC(1303), WA(3202).

P. dissotocum. Root pathogen of hydroponically grown plants. **6:**AZ(3508).

P. irregulare. Isolated from roots of hydroponically grown plants. **6:**AZ(3508).

P. polymastum. Root rot. **6:**OH(672).

P. rostratum. Isolated from roots of hydroponically grown plants. **6:**AZ(3508).

P. spinosum. Damping-off. **6:**NC(1303).

P. sylvaticum. Isolated from roots of hydroponically grown plants. **6:**AZ(3508).

P. tracheiphilum. 'Wilt and leaf blight.' **6:**WI(3669).

P. uncinulatum. Isolated from roots of hydroponically grown plants. **6:**AZ(3508).

P. violae. Isolated from roots of hydroponically grown plants. **6:**AZ(3508).

Pythium sp. Damping-off, root rot. **6:**Unk.(82), FL(35), MT(3202),

OK(2872). **7**:HI(2903).

Zygomycotina

Mucor sp. Damping-off, root rot. **6**:NY(2805).

Ascomycotina

Calosphaeria herbicola. Leaf spot. **9**:TX(82). **11**:TX(82).

Diaporthe arctii. On stems. **2**:GA(82,1438). **11**:LA(82).

Erysiphe cichoracearum. Powdery mildew. **1**:Range of host(82), ND(377). **2**:Range of host(82), NC(1303). **6**:'Powdery mildew.' CA(82,3365), FL(35), MI(82). **9**:Range of host(82), ID(3202), NC(1303,3944), ND(377), WA(3202), WY(1195). **10**:Range of host(82), MT(3202). **11**:FL(35).

Hyponectria depressa. On stems. **2**:KS(82), NY(82). **11**:NE states(183), MI(82).

Leptosphaeria agnita. On overwintered stems. **11**:NC(1303).

L. conoidea. **1**:NY(3249).

L. doliolum. **11**:NY(3249).

L. heliopsidis. Black knot on stems. **11**:CT(82).

Leptosphaerulina sp. On leaves. **1**:WI(1355).

Leptospora rubella. **11**:NY(3248).

Melanopsamma sp. **2**:GA(1438).

Mycosphaerella lactucae. On leaves. **2**:KS(525,967). **9**:ID(3202).

Phyllactinia guttata. **9**:WA(3202).

Pleospora herbarum. Seed mold. **6**:FL(35,82), KY(82), NY(82).

P. lactucicola. On decaying stems. **2**:NJ(967).

Sclerotinia minor. 'Sclerotinia drop rot.' **6**:CA(876,1946,2727), NY(1946,3965), OR(3202), WA(3202).

S. sclerotiorum. Drop, watery soft rot, stem rot. **6**:'Sclerotinia drop rot.' Range of host(82), AL(285), AZ(1946), CA(2727), FL(35), GA(1438), ID(3202), KY(1776), MS(2720), MT(3202), NC(1303), NY(3965), OK(2872), OR(3202), WA(3202). **7**:HI(2903). **9**:CA(82). **11**:MD(82), MS(82).

Sphaerotheca fuliginea. Powdery mildew. **1**:MI(82). **2**:MD(82), WI(82). **5**:MS(2720). **11**:AL(82), IN(82), MD(82), MI(82), NJ(82), WI(82).

S. macularis. Powdery mildew. **5**:MS(2720).

Basidiomycotina - Uredinales

Puccinia dioicae. Heteroecious: aecial host. **1**:NC states(82), NE states(82), Range of host(82), CO(82), ID(82). **3**:NC states(82), NE states(82), Range of host(82), CO(82), ID(82). **4**:WI(1341). **6**:'Rust.' IN(82), MA(82), MN(82), ND(82), SD(2243), WI(82). **9**:NC states(82), NE states(82), Range of host(82), CO(82), ID(82), ND(378). **10**:NC states(82), NE states(82), Range of host(82), CO(82), ID(82), ND(378), SD(2243). **11**:NC states(82), NE states(82), Range of host(82), CO(82), ID(82,3202).

P. hieracii. Autoecious. **6**:'Rust.' CA(82). **9**:CA(81).

P. minussensis. Autoecious. **2**:MI(82), NY(82). **10**:W states(82), AZ(3981), ID(3202), MT(1195,2713,3202), ND(378,2713), WA(2713,3202), WI(2713), WY(1195,2713,3379).

Puccinia sp. **11**:FL(35).

Basidiomycotina - Ustilaginales

Entyloma compositarum. Leaf smut. **2**:MN(82,1077), SD(1077), TX(82,1077). **9**:TX(82,1077). **11**:TX(82).

Basidiomycotina - Other

Thanatephorus cucumeris. **6**:OR(3202), WA(3202).

Deuteromycotina - Hyphomycetes

Alternaria sonchi. Leaf spot. **2**:IL(82), TX(82). **6**:'Alternaria leaf spot.' HI(2903). **9**:TX(82). **11**:TX(82).

Alternaria sp. Bottom rot, leaf spot. **6**:FL(35,82), LA(82), NY(2805), OH(82), OK(82,2872), TX(82), WA(82,3202). **11**:FL(35).

Botrytis cinerea. Gray mold, crown rot, head rot. **6**:'Gray mold.' Range of host(82), AK(526), FL(35), ID(3202), KY(1776), OR(3202), WA(3202). **7**:HI(2903).

Botrytis sp. Leaf blight, damping-off. **6**:AK(2162). **7**:HI(2903).

Cercospora apii. Leaf spot. **2**:NC(1303).

C. longissima. Leaf spot. **6**:'Cercospora leaf spot.' FL(35,82), HI(2899), IL(82), IN(82), TX(82), VA(82), WI(82). **7**:HI(2903). **8**:HI(2899,2903). **9**:WI(1332). **11**:LA(82).

Cercospora sp. Leaf spot. **6**:NC(1303). **7**:HI(2903).

Fusarium oxysporum. Bottom rot. **6**:NY(2805).

Fusarium sp. Seed blast, damping-off, root rot. **6**:AK(2163), FL(35), KY(82), OH(82).

Microdochium panattonianum. Anthracnose. **6**:'Anthracnose.' E states(82), AK(526), CA(82), FL(35), ID(3202), MI(82), NC(1303), OR(82,3202), TX(82), WA(82,3202). **9**:ID(82,3202), WA(82,3202). **11**:ID(3202), WA(3202).

Oidium sp. Powdery mildew. **6**:HI(2903). **7**:HI(2299).

Ovularia carletonii. Leaf spot. **10**:CO(82), KS(82), NE(82), SD(82). **11**:KS(82).

Phymatotrichopsis omnivora. Root rot. **6**:'Phymatotrichum root rot.' AZ(82), TX(82). **9**:TX(82).

Stemphylium botryosum. 'Stemphylium leaf spot.' **6**:FL(35).

Stemphylium sp. Leaf spot. **7**:HI(2903).

Trichoderma sp. Bottom rot. **6**:NY(2805).

Deuteromycotina - Coelomycetes

Ascochyta sp. On leaves. **2**:WI(1352).

Asteroma lactucae. On leaves. **2**:TX(82), WI(82). **11**:TX(82), WI(82).

Cylindrosporium lactucicola. Leaf spot. **9**:NC(82,1303,3944). **11**:CO(82).

Leptothyrium lactucae. On leaves. **9**:KS(82).

Macrophomina phaseolina. Charcoal rot. Charcoal stem rot. **9**:IL(82,311).

Phoma exigua. **1**:WI(64,82). **2**:WI(64,82,1333).

P. pulchellicola. On stems. **10**:WY(3812).

Phyllosticta lactucae. **2**:AL(82), TX(82). **9**:TX(82).

P. scariolicola. **9**:IL(82).

Phyllosticta sp. On leaves. **1**:WI(1346). **2**:Unk.(82).

Rhabdospora lactucarum. Leaf spot. **10**:MT(3202).

R. solidaginis. On stems. **11**:CO(82).

Septoria brunellae. Leaf spot. **5**:MS(2720).

S. lactucae. Leaf spot. **2**:C states(82), E states(82). **3**:FL(82). **6**:'Septoria leaf spot.' E & C states(82), Unk.(82), W states(82), CO(82), FL(35), HI(82,2903), MS(2720), NC(1303), SD(82). **7**:HI(2903). **9**:C states(82), E states(82), CA(82), GA(82), LA(82). **10**:WI(82). **11**:FL(35).

S. lactucicola. Leaf spot. **1**:WI(1337). **2**:E states(82), MS(82,2720), OK(82,2872), WI(82). **4**:WI(1332). **9**:MI(82), OH(82), WI(82). **11**:E states(82), CO(82), FL(35), SD(82).

S. unicolor. Leaf spot. **2**:KS(82), TX(82). **9**:TX(82). **11**:CO(82), IL(82), MO(82), TX(82).

Stagonospora lactucicola. On living leaves. **1**:WI(1354).

Stagonospora sp. **1**:WI(1351).

Deuteromycotina - Other

Rhizoctonia solani. Root rot. **6**:'Bottom rot.' Range of host(82), FL(35), NC(1303), NY(2805), WI(2229). **9**:NY(82).

Rhizoctonia sp. Bottom rot, root rot. **7**:HI(2903).

Sclerotium rolfsii. Southern blight. **2**:TX(82). **6**:'Southern blight.' CA(82), FL(35,82), NC(82,1303), SC(82), TX(82), VA(82). **9**:TX(82). **11**:TX(82).

Lagophylla Nutt.

1. *L. ramosissima* Nutt.

Basidiomycotina - Uredinales

Puccinia nuda. Autoecious. **1**:CA(81), OR(3202), WA(3202).

Basidiomycotina - Ustilaginales

Entyloma compositarum. **1**:WA(1077,3202).

Lapsana L.

1. *L. communis* L.
2. *Lapsana* sp.

Basidiomycotina - Uredinales

Puccinia lapsanae. Autoecious. **1**:OR(3202), WA(3202), WI(1330). **2**:NY(2713), OR(2713), WA(2713).

Lasianthaea DC.

1. *L. podocephala* (A. Gray) K. Becker (=*Zexmenia podocephala* A. Gray)

Basidiomycotina - Uredinales

Puccinia zexmeniae. **1**:AZ(1194,1203,3981).

Lasthenia Cass.

(= *Baeria* Fisch. & C. A. Mey.)

1. *L. chrysostoma* (Fisch. & C. A. Mey.) Greene (=*Baeria chrysostoma* Fisch. & C. A. Mey. subsp. *gracilis* (DC.) Ferris) [goldfields]
2. *L. macrantha* (A. Gray) Greene (=*Baeria macrantha* A. Gray)
3. *L. minor* (DC.) Ornd. (=*Baeria minor* (DC.) Ferris)

Chytridiomycetes

Synchytrium sp. On flower bracts, leaves, petioles and stems. **1**:CA(1847).

Ascomycotina

Sphaerotheca fuliginea. Powdery mildew. **2**:CA(1155). **3**:CA(1155).

Layia Hook. & Arnott

1. *Layia* sp.

Ascomycotina

Erysiphe cichoracearum. Powdery mildew. **1:**CA(82).

Leontodon L.

1. *L. autumnalis* L. [fall-dandelion, hawkbit]

Basidiomycotina - Uredinales

Puccinia hieracii. Autoecious. **1:**ME(82), NH(82).

Deuteromycotina - Coelomycetes

Septoria sp. **1:**WI(1340).

Liatris Gaertn. ex Schreb.

1. *L. aspera* Michx. var. *intermedia* (Lunell) Gaiser
2. *L. cylindracea* Michx.
3. *L. ligulistylis* (A. Nels.) K. Schum.
4. *L. punctata* Hook.
5. *L. pycnostachya* Michx.
6. *L. scariosa* (L.) Willd. (=*L. spheroides* Michx.)
7. *L. spicata* (L.) Willd. [blazing-star]
8. *L. squarrosa* (L.) Michx.
9. *L. squarrosa* (L.) Michx. var. *compacta* Torr. & A. Gray (=*L. compacta* (Torr. & A. Gray) Rydb.)
10. *Liatris* sp.

Ascomycotina

Erysiphe cichoracearum. Powdery mildew. **6:**OK(82,2872).
Leptosphaeria trimeroides. On stems. **6:**ND(377), OK(2873). **7:**ND(82), OK(82).
Sclerotinia sclerotiorum. Stem rot. **10:**NY(82).

Basidiomycotina - Uredinales

Coleosporium laciniariae. Heteroecious: telial host. **5:**TX(82). **6:**AR(82), FL(82), GA(82), TN(82). **10:**E states(82), AR(82), FL(35).
Puccinia poarum. Heteroecious: aecial host. **1:**WI(1352). **2:**WI(1331). **3:**WI(1331). **4:**MT(3202), ND(378), WI(1331). **5:**IN(82), WI(82,1330,1331,1334). **7:**NE(82), WI(82,1331,1350). **8:**WI(1331). **10:**NC states(82), W states(82), MT(82).

Deuteromycotina - Hyphomycetes

Phymatotrichopsis omnivora. Root rot. **5:**TX(82).
Verticillium albo-atrum. Wilt. **5:**NJ(82). **6:**NJ(82). **10:**NJ(82).

Deuteromycotina - Coelomycetes

Leptothyrium sp. On leaves. **1:**WI(1350).
Phomopsis liatridis. On living leaves. **3:**WI(1341).
Phyllosticta liatridis. Leaf spot. **2:**WI(1338,1343). **5:**WI(1342). **6:**OK(2873), WI(1334,1337). **7:**OK(82), WI(64,82). **9:**WI(1341). **10:**WI(1343).
Septoria liatridis. Leaf spot. **1:**WI(1352). **2:**WI(1335). **3:**WI(1332). **5:**TX(82), WI(82,1330). **6:**ND(82,378), WI(82). **7:**WI(82,970). **8:**WI(1332). **9:**WI(1335). **10:**IA(82), MN(82), WI(82).

Deuteromycotina - Other

Sclerotium rolfsii. Leaf spot. **10:**FL(35).

Ligularia Cass.

1. *L. tussilaginea* (N. L. Burm.) Makino (=*L. kaempferi* Siebold & Zucc.)

Deuteromycotina - Hyphomycetes

Cercospora sp. Leaf spot. **1:**HI(2899,2903).

Deuteromycotina - Other

Rhizoctonia sp. Root rot. **1:**HI(2903).
Sclerotium rolfsii. Southern blight. **1:**HI(2299,2903).

Luina Benth.

1. *L. nardosmia* (A. Gray) Cronq.

Deuteromycotina - Hyphomycetes

Ramularia variegata. **1:**WA(3202).

Deuteromycotina - Coelomycetes

Phyllosticta sp. **1:**WA(3202).

Lygodesmia D. Don

1. *L. grandiflora* (Nutt.) Torr. & A. Gray
2. *L. juncea* (Pursh) D. Don ex Hook.
3. *L. spinosa* Nutt.
4. *L. texana* (Torr. & A. Gray) Greene

Basidiomycotina - Uredinales

Puccinia grindeliae. Autoecious. **2:**MT(3202), OR(3202).
P. hieracii var. *harknessii*. Autoecious. **1:**AZ(1203,3981). **2:**OR(3202), WY(1195,2713,3379). **3:**CA(2713), NV(2713), OR(2713,3202). **4:**TX(2713).
P. stipae. Heteroecious: aecial host. **2:**MT(3202), SD(2243).

Machaeranthera Nees

1. *M. canescens* (Pursh) A. Gray (=*Aster canescens* Pursh)
2. *M. canescens* (Pursh) A. Gray var. *shastensis* (A. Gray) B. L. Turner (=*Aster shastensis* A. Gray)
3. *M. glabriuscula* (Nutt.) Cronq. & D. Keck (=*Xylorhiza glabriuscula* Nutt.)
4. *M. pinnatifida* (Hook.) Shinners (=*Haplopappus spinulosus* (Pursh) DC., *Haplopappus junceus* Greene)

Ascomycotina

Erysiphe cichoracearum. **1:**MT(3202).
Mycosphaerella taraxaci. **1:**WA(3202).
Pleospora penicillus. **2:**CA(642).
Pleospora sp. On overwintered herbage. **2:**CA(642).

Basidiomycotina - Uredinales

Coleosporium asterum. **2:**CA(638,642).
Puccinia grindeliae. **3:**WY(1195).

Deuteromycotina - Hyphomycetes

Alternaria sp. **1:**WA(3202).
Cladosporium astericola. **1:**WA(3202).

Deuteromycotina - Coelomycetes

Heteropatella umbilicata. **2:**CA(642).
Septoria shastensis. On leaves and stems. **2:**CA(638,642).
Septoria sp. **1:**OR(3202), WA(3202). **2:**OR(3202).

Madia Molina

1. *M. anomala* (A. Gray) Greene (=*M. dissitiflora* Torr. & A. Gray)
2. *M. citriodora* Greene
3. *M. elegans* D. Don [common madia, common tarweed]
4. *M. exigua* (Sm.) Greene
5. *M. glomerata* Hook.
6. *M. gracilis* (Sm.) D. Keck & J. Clausen ex Appleg.
7. *M. madioides* (Nutt.) Greene (=*M. nuttallii* A. Gray)
8. *M. minima* (A. Gray) D. Keck
9. *M. sativa* Molina [Chilean tarweed, coast tarweed, madia-oilplant]
10. *Madia* sp.

Oomycetes

Plasmopara halstedii. Downy mildew. **9:**CA(82).

Ascomycotina

Erysiphe cichoracearum. Powdery mildew. **3:**CA(82). **5:**ID(3202), MT(3202). **10:**CA(82), MT(82), NM(82).
Sphaerotheca macularis. Powdery mildew. **4:**WA(3202). **6:**CA(81,1155). **9:**CA(82).

Basidiomycotina - Uredinales

Coleosporium helianthi. Heteroecious: telial host. **9:**CA(3974).
C. pacificum. Heteroecious: telial host. **1:**CA(81). **2:**OR(3202). **3:**CA(82), OR(82,3202), WA(82). **4:**OR(3202). **5:**OR(3202), WA(3202). **6:**CA(642), OR(3202), WA(3202). **7:**CA(81). **8:**WA(3202). **9:**CA(81,82,745,3974), OR(82,3202), WA(82,3202). **10:**CA(82), OR(82), WA(82).
Puccinia nuda. Autoecious. **3:**OR(82,3202). **9:**WA(3202). **10:**OR(3202), WA(82).

Basidiomycotina - Ustilaginales

Entyloma compositarum. White smut. **4:**ID(1077,3202), WA(1077,3202). **5:**ID(3202), ND(1077). **10:**CA(1077), ND(82), OR(1077,3202).

Malacothrix DC.

1. *M. saxatilis* (Nutt.) Torr. & A. Gray var. *arachnoidea* (McGregor) B. S. Williams (=*M. arachnoidea* McGregor)
2. *Malacothrix* sp.

Chytridiomycetes

Synchytrium innominatum. Leaf gall. **2:**CA(82,1842).

Basidiomycotina - Uredinales

Puccinia hieracii var. *harknessii*. Autoecious. **1:**CA(82). **2:**CA(82).
P. nuda. Autoecious. **1:**OR(82).

Matricaria L.

1. *M. matricarioides* (Less.) C. L. Porter [pineapple-weed]

Oomycetes

Albugo tragopogonis. White rust. **1:**CA(82), ND(82), OR(82,3202),

WA(82,3202), WY(1195).
Ascomycotina
Erysiphe cichoracearum. Powdery mildew. **1**:WA(82,3202), WI(1351).
Sphaerotheca fuliginea. Powdery mildew. **1**:AK(526), WA(82), WI(1351).
S. macularis. **1**:ID(3202), WA(3202).
Sphaerotheca sp. **1**:AK(526).

Melanthera Rohr
1. *M. angustifolia* A. Rich.
2. *M. nivea* (L.) Small (=*M. hastata* (Walter) Michx.)
3. *M. parvifolia* Small
Oomycetes
Pythium sp. Root rot. **2**:FL(35).
Basidiomycotina - Uredinales
Uromyces columbianus. Autoecious. **1**:FL(35,128).
U. martinii. Autoecious. **2**:FL(35,746). **3**:FL(746).

Microseris D. Don
1. *M. cuspidata* (Pursh) Schultz-Bip.
2. *M. laciniata* (Hook.) Schultz-Bip. (=*Scorzonella laciniata* L., *Scorzonella procera* Greene)
3. *M. nutans* (Geyer ex Hook.) Schultz-Bip. (=*Scorzonella nutans* Geyer ex Hook.)
4. *M. troximoides* A. Gray
5. *Microseris* sp.
Chytridiomycetes
Synchytrium innominatum. **2**:CA(81,1842).
Ascomycotina
Sphaerotheca macularis. **4**:WA(3202).
Basidiomycotina - Uredinales
Puccinia dioicae. **3**:CA(81).
P. hieracii. Autoecious. **1**:WI(2713). **3**:CA(2713), CO(2713), ID(3202), UT(2713). **5**:WA(3202).
P. suksdorfii. Autoecious. **1**:MT(3202).
P. troximontis. **2**:CA(81).
Deuteromycotina - Coelomycetes
Septoria scorzonellae. On leaves and scapes. **3**:WA(666,3202).

Mikania Willd.
1. *M. cordifolia* (L. fil.) Willd.
2. *M. scandens* (L.) Willd. [climbing hempweed]
3. *Mikania* sp.
Oomycetes
Pythium sp. Root rot. **3**:FL(35).
Basidiomycotina - Uredinales
Puccinia spegazzinii. Autoecious. **1**:FL(752). **2**:AL(82), FL(35,82), GA(1439), MS(82,2720), NC(82,1303,3944), TX(752). **3**:FL(35,82).
Deuteromycotina - Hyphomycetes
Cercospora mikaniacola. Leaf spot. **2**:FL(35).
C. mikaniae. Leaf spot. **2**:MS(82,2720).
Deuteromycotina - Coelomycetes
Septoria mikanii. Leaf spot. **2**:CT(82), TX(82).
Deuteromycotina - Other
Rhizoctonia solani. Leaf rot, root rot, stem rot. **3**:FL(35).

Osteospermum L.
1. *O. ecklonis* (DC.) Norl. [Vanstadens River daisy]
2. *O. fruticosum* (L.) Norl.
3. *Osteospermum* sp.
Oomycetes
Phytophthora cryptogea. Isolated from crowns, roots and stems. **3**:CA(1220).
Pythium sp. Root rot. **1**:FL(35).
Ascomycotina
Sclerotinia sclerotiorum. On stems, on leaves. **1**:CA(1217). **2**:CA(1217).

Parthenice A. Gray
1. *P. mollis* A. Gray
Basidiomycotina - Uredinales
Puccinia invelata. Autoecious. **1**:AZ(747,1203,3981).

Parthenium L.
1. *P. argentatum* A. Gray [guayule]
2. *P. hysterophorus* L.
3. *P. integrifolium* L. [American feverfew]
4. *Parthenium* sp.
Oomycetes
Albugo tragopogonis. White rust. **3**:IA(82). **4**:KS(82).
Phytophthora drechsleri. Root rot. **1**:AZ(82), CA(82), NM(82), TX(82).
Pythium aphanidermatum. Isolated roots. **1**:AZ(2376).
P. ultimum. Damping-off. **1**:CA(82).
Ascomycotina
Erysiphe cichoracearum. **3**:WI(1341).
Sclerotinia minor. Stem rot. **1**:CA(82).
S. sclerotiorum. Stem rot. **1**:CA(82).
Basidiomycotina - Uredinales
Coleosporium helianthi. Heteroecious: telial host. **3**:AL(82), GA(82,1439), LA(82), NC(82,1303), TN(82), VA(82). **4**:GA(82), NC(82), TN(82), VA(82).
Puccinia melampodii. Autoecious. **4**:TX(82).
P. parthenii. Autoecious. **4**:TX(82).
Deuteromycotina - Hyphomycetes
Alternaria sp. Secondary leaf spot. **1**:CA(82).
Botrytis cinerea. Seedling blight. **1**:CA(82).
Cercospora partheniphila. On leaves. **2**:WI(1335). **3**:WI(82,1335). **4**:WI(82).
Fusarium solani. Leaf spot. **1**:TX(2612).
Phymatotrichopsis omnivora. Root rot. **1**:AZ(82), TX(82). **4**:TX(82).
Verticillium albo-atrum. Wilt. **1**:AZ(82), CA(82), NM(82), TX(82).
Deuteromycotina - Coelomycetes
Ascochyta compositarum. **3**:WI(1359).
Lasiodiplodia theobromae. Dieback. **1**:CA(82), TX(82).
Macrophomina phaseolina. Charcoal rot. Charcoal stem rot. **1**:CA(82), TX(82,2614).
Septoria inclusa. Charcoal stem rot. **3**:WI(1342).
Deuteromycotina - Other
Rhizoctonia solani. Damping-off. **1**:CA(82).
Sclerotium rolfsii. Southern blight. **1**:AZ(82).

Petasites Mill.
1. *P. frigidus* (L.) Fr.
2. *P. frigidus* (L.) Fr. var. *nivalis* (Greene) Cronq. (=*P. corymbosus* Rydb., *P. hyperboreus* Rydb.)
3. *P. frigidus* (L.) Fr. var. *palmatus* (Aiton) Cronq. (=*P. palmatus* (Aiton) A. Gray)
4. *P. sagittatus* (Banks ex Pursh) A. Gray
Chytridiomycetes
Synchytrium aureum. Leaf gall. **3**:WI(82).
Ascomycotina
Leptosphaeria doliolum. On dead petiole. **1**:AK(1938).
Nodulosphaeria dolioloides. On dead petioles and stems. **1**:AK(1938).
Pleospora penicillus. On dead stem. **1**:AK(1938).
Basidiomycotina - Uredinales
Puccinia conglomerata. Autoecious. **1**:AK(60,82,526,2059,3112), MI(82), MN(82), NY(82), WI(82). **3**:AK(60), NY(3112).
P. poarum. Heteroecious: aecial host. **1**:AK(60,82,526). **2**:AK(60,526). **4**:AK(60,526,1904).
Deuteromycotina - Hyphomycetes
Ramularia variegata. On dead stems. **3**:MN(82).
Deuteromycotina - Coelomycetes
Stagonospora petasitidis. Leaf spot. **1**:WA(3202). **3**:WI(82).

Petradoria Greene
1. *P. pumila* (Nutt.) Greene (=*Solidago petradoria* S. F. Blake)
Basidiomycotina - Uredinales
Puccinia grindeliae. **1**:CO(1195).
Deuteromycotina - Coelomycetes
Phyllosticta solidaginis. **1**:CO(1195).

Pinaropappus Less.
1. *P. roseus* (Less.) Less.
Basidiomycotina - Uredinales
Puccinia pinaropappi. Autoecious. **1**:TX(746,2713).

Piqueria Cav.

1. *P. trinervia* Cav. [stevia]

Oomycetes

Pythium sp. Root rot. **1**:OK(2872).

Ascomycotina

Erysiphe cichoracearum. Powdery mildew. **1**:IL(82).

Deuteromycotina - Other

Rhizoctonia solani. Damping-off. **1**:IL(82).
Sclerotium rolfsii var. *delphinii*. Stem rot. **1**:NJ(82).

Pluchea Cass.

1. *P. camphorata* (L.) DC. [marsh fleabane]
2. *P.* × *fosbergii* Cooperr. & Galang
3. *Pluchea* sp.

Ascomycotina

Didymosphaeria futilis. On stems. **1**:CA(2439).
Halosphaeria cucullata. On submerged limbs. **2**:HI(1942,2903).
Lignincola laevis. On submerged limb. **2**:HI(2903).

Basidiomycotina - Uredinales

Puccinia biocellata. **1**:FL(35,82). **3**:FL(35,82).

Basidiomycotina - Other

Crepidotus appalachianensis. **3**:HI(3706).

Deuteromycotina - Hyphomycetes

Alternaria sp. Leaf spot. **1**:FL(35).
Humicola alopallonella. On limbs. **2**:HI(1942).
Periconia prolifica. On submerged limbs. **2**:HI(1942,2903).

Deuteromycotina - Coelomycetes

Phloeospora plucheae. Leaf spot. **1**:FL(35). **3**:TX(82).
Septoria plucheae. Leaf spot. **1**:FL(35), MA(82).

Polymnia L.

1. *P. canadensis* L.

Ascomycotina

Acrospermum foliicola. **1**:GA(1438).
Diaporthe arctii. **1**:GA(1438).
Leptosphaeria longipedicellata. **1**:GA(1438).
Leptosphaeria sp. **1**:GA(1438).

Deuteromycotina - Hyphomycetes

Sclerotiomyces colchicus. On leaves. **1**:WI(1354).

Porophyllum Adans.

1. *P. scoparium* A. Gray

Basidiomycotina - Uredinales

Puccinia gallula. Autoecious. **1**:TX(746).

Prenanthes L.

1. *P. alata* (Hook.) D. Dietr.
2. *P. alba* L.
3. *P. altissima* L.
4. *P. aspera* Michx.
5. *P. boottii* D. Dietr.
6. *P. crepidinea* Michx.
7. *P. racemosa* Michx. (=*Nabalus racemosus* Hook.)
8. *P. serpentaria* Pursh
9. *P. trifoliata* (Cass.) Fernald
10. *Prenanthes* sp.

Chytridiomycetes

Synchytrium aureum. Leaf gall. **2**:WI(82).

Oomycetes

Bremia lactucae. Downy mildew. **2**:IA(82), MA(82), MN(82). **10**:MA(82).

Ascomycotina

Erysiphe cichoracearum. Powdery mildew. **2**:NY(82). **10**:IL(82), MD(82), NC(82,1303,3944), PA(82).
Mycosphaerella prenanthis. Lef spot. **6**:AL(179). **10**:AL(82).
Sphaerotheca fuliginea. Leaf spot. **1**:AK(526). **2**:OH(82), PA(82), WI(82). **10**:E states(82), AK(82,526), IL(82), MN(82).

Basidiomycotina - Uredinales

Puccinia atropuncta. Heteroecious: aecial host. **9**:PA(82), VA(82).
P. dioicae. **2**:IL(82), MN(82), NY(82), WI(82). **7**:ND(378). **9**:VA(82). **10**:IL(82), MN(82), NY(82), WI(82).
P. orbicula. Autoecious. **2**:E states(82), GA(1439), MI(2713), ND(82), NY(2713). **3**:MA(2713). **5**:NH(2713), SC(1307). **8**:GA(1439). **9**:E states(82), ME(2713), ND(82), NY(2713). **10**:E states(82), DE(2713), MI(2713), ND(82), NY(2713), PA(2713).
P. variabilis var. *insperata*. Autoecious. **1**:AK(60,526,2713), OR(745,2713,3202). **10**:AK(82), OR(82).

Basidiomycotina - Other

Thanatephorus cucumeris. **7**:WI(1361).

Deuteromycotina - Hyphomycetes

Cercospora prenanthis. Leaf spot. **2**:IN(82), WI(82). **3**:NC(3944). **4**:KS(82,973). **10**:AL(82), AR(82), NC(82,1303).
C. tabacina. Leaf spot. **2**:WI(82).
C. wisconsinensis. On leaves. **2**:WI(1334). **4**:WI(1341).

Deuteromycotina - Coelomycetes

Ascochyta compositarum. **2**:WI(1354). **7**:WI(1349).
Ascochyta sp. **7**:WI(1344).
Lasiodiplodia theobromae. On stems. **10**:AL(82).
Phlyctema simulans. On stems. **10**:NH(82).
Phyllosticta cacaliae. **2**:WI(1361).
Phyllosticta sp. **2**:WI(1354).
Septoria nabali. Leaf spot. **2**:NC states(82), NE states(82), MS(2720), WI(82). **7**:WI(1335). **9**:VA(82). **10**:NE states(82), OH(82), WI(82).
S. prenanthis. On leaves. **10**:WI(966).

Psilostrophe DC.

1. *P. cooperi* (A. Gray) Greene
2. *P. sparsiflora* (A. Gray) A. Nels.
3. *P. tagetina* (Nutt.) Greene

Basidiomycotina - Uredinales

Puccinia grindeliae. Autoecious. **1**:AZ(1195,1203,3379). **2**:AZ(1203). **3**:AZ(1203).
P. grindeliae subsp. *riddellii*. **1**:AZ(3981). **2**:AZ(3981). **3**:AZ(3981).

Pyrrhopappus DC.

1. *P. carolinianus* (Walter) DC. [Carolina false dandelion]
2. *P. grandiflorus* (Nutt.) Nutt.
3. *P. multicaulis* DC.
4. *Pyrrhopappus* sp.

Chytridiomycetes

Synchytrium sp. On leaves. **4**:OR(1849).

Basidiomycotina - Uredinales

Puccinia hieracii. Autoecious. **1**:FL(35,2713), GA(1439), MS(2720), NC(1303,3944), OK(2872). **2**:OK(2872). **3**:OK(2872).

Deuteromycotina - Coelomycetes

Septoria sp. Leaf spot. **1**:OK(2872). **3**:OK(2872).

Rafinesquia Nutt.

(= *Nemoseris* Greene.)

1. *R. californica* Nutt. (=*Nemoseris californica* Greene)
2. *R. neomexicana* A. Gray
3. *Rafinesquia* sp. (=*Nemoseris sp.*)

Ascomycotina

Mycosphaerella nemoseridis. On smaller limbs of stems. **1**:CA(2439).
Pleospora herbarum. **3**:CA(2439).

Basidiomycotina - Uredinales

Puccinia hieracii var. *harknessii*. **2**:AZ(747,3981).

Deuteromycotina - Hyphomycetes

Cercospora rafinesquiae. On leaves. **1**:CA(581,1468).

Railliardia Gaud. in Freyc.

1. *Railliardia* sp.

Ascomycotina

Echidnodella raillardiae. **1**:HI(2903).

Ratibida Raf.

1. *R. columnifera* (Nutt.) Woot. & Standl. (=*R. columnaris* (Sims) D. Don, *Lepachys columnaris* (Sims) Torr. & A. Gray)
2. *R. pinnata* (Venten.) Barnhart (=*Lepachys pinnata* (Venten.) Torr. & A. Gray)
3. *R. tagetes* (E. James) Barnhart

Oomycetes

Plasmopara halstedii. Downy mildew. **2**:IA(82), WI(1338).

Ascomycotina

Anthostomella ratibidae. On leaves and stems. **3**:NM(1195,3376).
Erysiphe cichoracearum. Powdery mildew. **1**:ND(82,377), TX(82).
Mycosphaerella sp. On leaves. **2**:WI(1353).
Phyllachora ambrosiae. On living leaves. Leaf spot. **1**:MT(82,970,3202).

Basidiomycotina - Uredinales
Uromyces perigynius. Heteroecious: aecial host. **1**:ND(82), TX(82).
Basidiomycotina - Ustilaginales
Entyloma compositarum. Leaf spot. **1**:NE(82,1077), SD(1077,2243). **2**:C states(82), IA(1077), IL(1077), IN(1077), KS(82,1077), MI(1077), MN(82,1077), WI(1348), WI(1077).
E. polysporum. **1**:SD(1077).
Deuteromycotina - Hyphomycetes
Cercospora ratibidae. Leaf spot. **1**:KS(82). **2**:WI(82).
Phymatotrichopsis omnivora. Root rot. **1**:TX(82).
Ramularia rudbeckiae. Leaf spot. **1**:ID(82,3202).
Deuteromycotina - Coelomycetes
Ascochyta compositarum. **2**:WI(1346).
Phoma exigua. **1**:WI(1341).
Phyllosticta entylomicola. **2**:WI(1348).
Phyllosticta sp. On living leaves. **2**:WI(1339).
Septoria infuscata. Leaf spot. **2**:MI(82), MO(82), WI(82).
S. lepachydis. **1**:KS(82). **2**:WI(82,1332).
S. rudbeckiae. **1**:ND(82), NM(82,378).
Deuteromycotina - Other
Rhizoctonia crocorum. Violet root rot. **1**:TX(82).

Rudbeckia L.

1. *R. hirta* L. [black-eyed-Susan]
2. *R. hirta* L. var. *angustifolia* (T. Moore) Perdue (=*R. bicolor* Nutt.)
3. *R. hirta* L. var. *pulcherrima* Farw. (=*R. serotina* Nutt.)
4. *R. laciniata* L. [coneflower, golden-glow]
5. *R. occidentalis* Nutt.
6. *R. subtomentosa* Pursh
7. *R. triloba* L. [brown-eyed-Susan]
8. *Rudbeckia* sp.

Chytridiomycetes
Synchytrium aureum. Leaf gall. **4**:IL(82), WI(82).
Oomycetes
Plasmopara halstedii. Downy mildew. **1**:ND(82,377), NY(82), OK(2872). **4**:IA(82), NC(82), ND(82), NE(82), TX(82), WI(82). **7**:IA(82).
Ascomycotina
Crocicreas cyathoideum var. *cacaliae*. On dead stems. **5**:UT(505).
C. nigrofuscum var. *allantosporum*. On dead stems. **5**:ID(505).
Diaporthe arctii. On stems. **4**:GA(82,1438).
Erysiphe cichoracearum. Powdery mildew. **1**:IA(82), MA(82,1390), SD(82), VA(82), WI(1343). **2**:CA(1155). **4**:Range of host(82), UT(82). **5**:ID(3202). **8**:FL(35).
Hypoderma commune. **4**:GA(1438).
Leptosphaeria artemisiae. **4**:MO(3249).
L. complanata. **4**:GA(1438).
Massarina gloeospora. **4**:GA(1438).
Mycosphaerella rudbeckiae. On stems. **1**:CO(82).
Ophiobolus anguillidus. **1**:GA(1438). **5**:UT(3248).
Phyllactinia guttata. **4**:NM(1195).
Pleospora penicillus. **1**:GA(1438).
P. scrophulariae var. *compositarum*. **5**:WY(3810).
P. tomentosa. **5**:WY(3810).
Sclerotinia sclerotiorum. Stem rot. **4**:CT(82).
Basidiomycotina - Uredinales
Aecidium batesii. **1**:NE(82).
Puccinia cnici-oleracei. **8**:TX(82).
P. dioicae. Heteroecious: aecial host. **4**:MD(82), SD(82).
Uromyces perigynius. Heteroecious: aecial host. **4**:C states(82), NE states(82), MS(2720), MT(82,3202). **8**:ID(3202).
U. rudbeckiae. Autoecious. **4**:SE states(82), IA(746), MS(82,2720), MT(82,3202), ND(378), NM(82,1195), SD(2768), WY(1195). **5**:ID(3202). **7**:MS(82,2720). **8**:FL(35), TX(82).
Basidiomycotina - Ustilaginales
Entyloma compositarum. White smut. **4**:IA(82,1077), IL(1077), MO(82,1077), OH(82,1077), WI(82,1077). **8**:IA(82).
E. davisii. **1**:WI(1077).
E. polysporum. **1**:WI(82).
Basidiomycotina - Other
Ceratobasidium anceps. **4**:WI(1355).
Deuteromycotina - Hyphomycetes
Alternaria sp. Leaf spot. **8**:FL(35).
Cercospora tabacina. Leaf spot. **1**:WI(82). **4**:IL(82), NY(82), TX(82), WI(82). **7**:IA(82). **8**:FL(35).
Cercospora sp. **1**:WI(1352). **4**:WI(1351). **7**:WI(1354).
Myrothecium roridum. Leaf spot. **8**:FL(35).
Phymatotrichopsis omnivora. Root rot. **7**:TX(82). **8**:TX(82).
Ramularia rudbeckiae. Leaf spot. **4**:E states(82), CO(82), ID(82,3202), MS(82,2720), MT(3202), WY(1195). **5**:MT(3202), WY(1195). **8**:MT(82), VA(82), VT(82), WV(82).
Verticillium albo-atrum. Wilt. **1**:NY(82).
Deuteromycotina - Coelomycetes
Ascochyta compositarum. **4**:WI(1358).
A. doronici. **4**:WI(1341).
Colletotrichum rudbeckiae. On stems. **4**:NY(82).
Macrophomina phaseolina. Charcoal rot. **1**:IL(311).
Phoma exigua. **1**:WI(1333).
Phyllosticta rudbeckiae. Leaf spot. **4**:IA(82), NY(82), TX(82), WI(82).
Septoria rudbeckiae. Leaf spot. **1**:DE(82), OK(2872), WI(82,1332,1344). **3**:WI(1361). **4**:IA(82), KS(82), NE(82), TX(82), WI(82). **5**:ID(3202). **6**:WI(1332). **7**:MS(2720). **8**:CO(82), ID(3202), IL(82), MA(82), MS(82,2720), WA(82,3202), WI(82).
S. rudbeckiae var. *oaklandica*. Leaf spot. **1**:NY(82).
Deuteromycotina - Other
Sclerotium rolfsii. Southern blight. **4**:FL(82), NJ(82). **8**:FL(35).

Santolina L.

1. *S. chamaecyparissus* L. [lavender-cotton]
2. *S. virens* Mill. [green santolina]
3. *Santolina* sp.

Deuteromycotina - Hyphomycetes
Alternaria tenuissima. Foliar blight. **1**:AL(3535). **2**:AL(3535).
Deuteromycotina - Coelomycetes
Macrophomina phaseolina. On bark and wood. **3**:NC(1303).
Deuteromycotina - Other
Rhizoctonia sp. Isolated from blighted foliage. **3**:NC(1303).

Saussurea DC.

1. *S. americana* Eaton
2. *S. angustifolia* (Willd.) DC.

Ascomycotina
Pleospora helvetica. On dead stem. **2**:AK(1938).
Deuteromycotina - Coelomycetes
Phyllosticta barssii. On living leaves. **1**:OR(3202,3421).

Scorzonera L.

1. *S. hispanica* L. [black oysterplant, black salsify, Spanish oysterplant, Spanish salsify]

Oomycetes
Albugo tragopogonis. White rust. **1**:CA(82).

Senecio L.

1. *S. amplectens* A. Gray
2. *S. ampullaceus* Hook.
3. *S. arnicoides* DC.
4. *S. aronicoides* DC.
5. *S. aureus* L.
6. *S. bigelowii* A. Gray
7. *S. bolanderi* A. Gray (=*S. harfordii* Greenm.)
8. *S. chenopodioides* Kunth (=*S. confusus* Britton) [Mexican flamevine]
9. *S. crassulus* A. Gray
10. *S. cruentus* (Masson ex L'Hér.) DC. [florist's cineraria]
11. *S. cymbalarioides* H. Buek (=*S. subnudus* DC.)
12. *S. douglasii* DC. [bush senecio]
13. *S. fendleri* A. Gray
14. *S. foetidus* Howell
15. *S. foetidus* Howell var. *hydrophiloides* (Rydb.) T. Barkley (=*S. hydrophiloides* Rydb.)
16. *S. fremontii* Torr. & A. Gray
17. *S. hartwegii* Benth. (=*S. seemannii* Schultz-Bip.)
18. *S. hydrophilus* Nutt.
19. *S. integerrimus* Nutt. (=*S. columbianus* Greene)
20. *S. integerrimus* Nutt. von *exaltatus* (Nutt.) Cronq. (=*S. dispar* A. Nels., *S. perplexus* A. Nels.)
21. *S. jacobaea* L. [ragwort, tansy ragwort]
22. *S. lugens* Richardson
23. *S. mendocinensis* A. Gray

24. *S. mikanioides* Otto ex Walp. [cape ivy, German ivy, parlor ivy, water ivy]
25. *S. pauciflorus* Pursh
26. *S. pauperculus* Michx. (=*S. balsamitae* Muhl. ex Willd., *S. pauperculus* Michx. var. *balsamitae* (Muhl. ex Willd.) Fernald) [balsam groundsel]
27. *S. pseudaureus* Rydb.
28. *S. rapifolius* Nutt.
29. *S. robbinsii* Oakes ex Rusby
30. *S. rydbergii* A. Nels.
31. *S. serra* Hook.
32. *S. spartioides* Torr. & A. Gray
33. *S. streptanthifolius* Greene (=*S. platylobus* Rydb.)
34. *S. triangularis* Hook.
35. *S. vulgaris* L. [common groundsel]
36. *Senecio* sp.

Chytridiomycetes

Synchytrium aureum. Leaf gall. **5:**WI(82).
Synchytrium sp. On leaves. **11:**WY(1847). **34:**WA(3202).

Oomycetes

Albugo tragopogonis. White rust. **5:**IN(82), MO(82), MT(82,3202), WI(82). **11:**MT(3202). **19:**NV(82). **23:**CA(81). **31:**MT(3202), WA(3202). **36:**CA(82), CO(82), MO(82), MT(82), TX(82), UT(82), WA(82).
Phytophthora sp. Root and stem rot. **10:**NJ(82). **36:**FL(35), NJ(82).
Plasmopara halstedii. Downy mildew. **10:**NY(82). **36:**OR(3202).
Pythium ultimum. **10:**CA(82).
Pythium sp. Root rot. **10:**MD(82). **36:**FL(35).

Ascomycotina

Bisporella drosodes. **36:**ID(3202).
Comoclathris permunda. **36:**WY(3810).
Crocicreas coronatum. On stems. **36:**MI(505).
C. cyathoideum var. *cacaliae*. On dead stems. **31:**UT(505). **33:**UT(505).
Didymolepta mimuli. On overwintered herbage. **34:**CA(638,661).
Erysiphe cichoracearum. Powdery mildew. **5:**MN(82), VA(82). **10:**MA(82), WA(3202). **18:**WA(3202). **36:**ID(82,3202), NE(82), WA(82,3202).
E. cumminsiana. Powdery mildew. **17:**AZ(367).
Hyponectria sceptri. On stems. **31:**UT(3084).
Leptosphaeria doliolum. **34:**WA(3202,3814).
L. ogilviensis. **21:**OR(3249).
Lewia scrophulariae. **30:**WY(3810).
Mycosphaerella tassiana. On overwintered herbage. **19:**ID(3202). **34:**CA(642), WY(2924).
Nodulosphaeria octoseptata. **9:**WY(3250).
N. olivacea. **9:**WY(3810).
Pleospora herbarum var. *occidentalis*. **30:**WY(3810).
P. richtophensis. **9:**WY(3810).
P. stenospora. **36:**WY(3810).
Sclerotinia sclerotiorum. Stem rot. **10:**WA(82,3202). **36:**FL(35), LA(82).
Sphaerotheca fuliginea. On leaves and petioles. **22:**AK(1938). **34:**CA(81,82), CO(82), MT(82), WY(82). **36:**UT(82).
S. macularis. Powdery mildew. **32:**UT(1195). **34:**MT(3202), WA(3202), WY(82,1195). **36:**CA(82).
Systremma sp. Black scab. **8:**FL(35). **36:**FL(35).

Basidiomycotina - Uredinales

Baeodromus californicus. Autoecious. **12:**CA(746).
[*Coleosporium tussilaginis*]. **14:**WA(3202). **34:**ID(82,3202), MT(82,3202), OR(82,3202), WA(82,3202). **36:**CA(82), CO(82), MT(82), UT(82), WA(82), WY(82).
Puccinia angustata. Heteroecious: aecial host. **16:**OR(3202).
P. canaliculata. Heteroecious: aecial host. **2:**TX(740).
P. dioicae. Heteroecious: aecial host. **19:**ID(3202), NE(82). **36:**NM(82).
P. eriophori. Heteroecious: aecial host. **5:**NC states(82), NE states(82), MN(82), NY(3101). **36:**NH(82), OR(82), UT(82), VT(82).
P. glomerata. Autoecious. **19:**WA(3202), WY(1195). **36:**CA(82), UT(82), WA(82), WY(82,2924).
P. grindeliae. Autoecious. **19:**WA(3202).
P. recedens. Autoecious. **5:**NE states(82), IA(82), NY(836), TN(82), WI(82). **7:**OR(3202). **25:**ID(3202). **27:**ID(3202), MT(3202). **36:**NW states(82), NC(82), WA(82).
P. senecionis. Autoecious. **1:**CO(1520). **5:**SD(2243). **6:**CO(1520). **9:**CO(1195,3379), WY(1195). **11:**WA(3202). **14:**WA(3202). **18:**WA(3202). **19:**NC(1303), ND(82), NE(82), NV(82), WA(3202). **23:**CA(81). **34:**ID(82,3202), MT(3202), UT(82), WA(82,3202). **36:**W states(82), CA(82), WA(82).
P. stipae. Heteroecious: aecial host. **19:**CO(82), WY(82,1195). **20:**WY(1195).

Basidiomycotina - Ustilaginales

Entyloma calendulae. **27:**ID(3202), MT(3202). **34:**WY(1195). **36:**WA(1077,3202).
E. compositarum. White smut. **3:**CA(1077). **4:**CA(81). **5:**IN(1077), KS(82,1077), MD(82,1077), NE(82,1077), PA(82,1077), TX(82), WI(1077,1334). **13:**CO(1077). **25:**WA(1077,3202). **26:**KS(1077). **29:**NY(1077). **36:**KS(82), PA(82), TX(82,1077).

Basidiomycotina - Other

Armillaria tabescens. Root rot. **36:**FL(35).
Ceratobasidium anceps. **5:**WI(1355).

Deuteromycotina - Hyphomycetes

Alternaria cinerariae. Leaf spot. **8:**FL(35). **10:**CA(149). **36:**FL(35).
Botrytis cinerea. Damping-off, leaf spot. **36:**AK(82), IN(82), MO(82), NJ(82), PA(82).
Cercospora jacquiniana. Leaf spot. **5:**DE(3543). **8:**FL(35). **35:**CO(82), RI(82). **36:**TX(82).
C. senecionicola. Leaf spot. **5:**WI(82). **26:**WI(1333).
Corynespora sp. Leaf spot. **8:**FL(35).
Cylindrocladium sp. Collar and root rot. **24:**HI(2903).
Fusarium sp. Root and stem rot, wilt. **10:**PA(82). **36:**FL(35), NJ(82).
Helminthosporium sp. Leaf spot. **36:**FL(35).
Oidium sp. Powdery mildew. **36:**FL(35).
Phymatotrichopsis omnivora. Root rot. **36:**TX(82).
Ramularia filaris. Leaf spot. **36:**CO(82), MT(82,3202).
R. pruinosa. **28:**WY(1195). **31:**WY(1195). **34:**WY(82,1195). **36:**CO(82), WY(82).
R. senecionis. **3:**CA(638). **14:**ID(3202), WA(3202). **15:**WY(1195). **18:**OR(3202). **19:**ID(3202), WA(3202), WY(1195). **34:**ID(3202), MT(3202), OR(3202), WY(2924). **36:**CA(82), CO(82).
R. senecionis var. *carniolica*. On leaves. **3:**CA(642). **36:**CA(82).
Stemphylium sp. Leaf spot. **8:**FL(35).
Thielaviopsis basicola. Root rot. **10:**MA(82), MD(1867).
Verticillium albo-atrum. Wilt. **10:**NJ(82), NY(82), WA(82,3202). **35:**OR(3202). **36:**WA(82).

Deuteromycotina - Coelomycetes

Ascochyta compositarum. **5:**WI(1355).
Ascochyta sp. On leaves. **26:**WI(1341).
Asteroma sp. **19:**WA(3202).
Colletotrichum sp. Anthracnose. **36:**FL(35).
Gloeosporium senecionis. Leaf spot. **4:**CA(82).
Heteropatella umbilicata. **19:**WA(3202). **34:**OR(3202).
Phoma bacilliformis. On stems. **36:**WY(3812).
Phyllosticta cacaliae. **5:**WI(1347).
P. garrettii. Leaf spot. **14:**WA(3202). **19:**ID(3202). **20:**UT(64). **27:**MT(3202). **34:**AK(82,526), OR(82,3202), WY(82,1195). **36:**UT(82), WY(82).
Phyllosticta sp. **36:**NJ(82).
Septoria cacaliae. Leaf spot. **36:**AL(82), IN(82), TX(82).
S. senecionis. **36:**CA(82).
S. senecionis-aurei. **5:**WI(82,1342). **36:**WI(82).

Deuteromycotina - Other

Rhizoctonia solani. Damping-off, root rot. **10:**IL(82), NJ(82). **36:**FL(35), IL(82), NJ(82).
Sclerotium rolfsii. Southern blight. **36:**FL(35).

Sericocarpus Nees

This genus is considered a synonym of *Aster*, but the appropriate combinations could not be located.

1. *S. rigidus* Lindl.
2. *Sericocarpus* sp.

Basidiomycotina - Uredinales

Coleosporium asterum. Heteroecious: telial host. **1:**CA(81).
Puccinia dulichii. Heteroecious: aecial host. **2:**IN(82), NC(82).

Silphium L.

1. *S. asperrimum* Hook.
2. *S. asteriscus* L. (=*S. dentatum* Elliott, *S. elliottii* Small)
3. *S. compositum* Michx.
4. *S. compositum* Michx. subsp. *venosum* (Small) C. R. Sweeney & T.

Fisher (=*S. lapsuum* Small)
5. *S. integrifolium* Michx.
6. *S. laciniatum* L. [compass-plant]
7. *S. perfoliatum* L. [Indian-cup]
8. *S. terebinthinaceum* Jacq. [prairie-dock]
9. *S. trifoliatum* L.
10. *Silphium* × sp. (unnamed hybrids)
11. *Silphium* sp.

Oomycetes

Plasmopara halstedii. Downy mildew. **6:**IA(82), IL(82). **7:**C states(82), KS(82), MN(82), SD(2243). **8:**C states(82), KS(82), MN(82), WI(1350). **10:** On *S. terebinthinaceum* x *S. laciniatum*. WI(1355). **11:**KS(82), NE(82), TX(82), WI(82), WV(82).

Ascomycotina

Diapleella coniothyrium. Downy mildew. **2:**GA(1438). **11:**GA(82).
Diaporthe arctii. On stems. **2:**GA(1438). **4:**GA(1438). **11:**GA(82).
Erysiphe cichoracearum. Powdery mildew. **8:**CT(82), WI(1341). **11:**MD(82).
Hymenoscyphus epiphyllus. On dead stems. **3:**NC(1303,1307).
Hypoderma commune. **2:**GA(1438). **4:**GA(1438).
Mycosphaerella sp. On leaves. **7:**WI(82).

Basidiomycotina - Uredinales

Coleosporium helianthi. Heteroecious: telial host. **1:**MS(2720). **2:**GA(1439), NC(3944). **5:**MS(2720), NC(3944). **7:**WI(1334). **8:**WI(1334). **11:**E states(82), FL(35), KS(82), TX(82).
Puccinia silphii. Autoecious. **5:**MS(2720), WI(1344). **6:**SE states(82), ND(82), TX(82), WI(1344,1348). **7:**SE states(82), MS(2720), ND(82,378), SD(2243), TX(82). **8:**SE states(82), ND(82), TX(82), WI(1332,1338). **9:**NC(1303). **10:** On *S. terebinthinaceum* x *S. laciniatum* and *S. integrifolium* x *S. perfoliatum*. WI(1347,1349). **11:**SE states(82), ND(82), TX(82).
Uromyces silphii. Heteroecious: aecial host. **5:**WI(1334). **6:**WI(1330). **7:**GA(1439), WI(1338). **8:**WI(1342). **10:** *S. laciniatum* x *S. terebinthinaceum* WI(1352). **11:**C states(82), KS(82), WI(82).

Basidiomycotina - Ustilaginales

Entyloma compositarum. White smut. **5:**WI(1077). **11:**TX(82,1077), WI(82).

Deuteromycotina - Hyphomycetes

Cercospora silphii. Leaf spot. **5:**MS(2720), WI(1349). **6:**IL(82), WI(82,1332,1353). **7:**IL(82). **8:**AL(82), KS(82), TX(82), WI(1332), WV(82).
Hyalodendron sp. **11:**GA(659).
Periconia byssoides. **11:**GA(659).
Phymatotrichopsis omnivora. Root rot. **11:**TX(82).

Deuteromycotina - Coelomycetes

Ascochyta compositarum. Leaf spot. **7:**WI(82,1335).
A. treleasei. **11:**WI(82).
Colletotrichum dematium. Leaf spot. **5:**WI(1332). **7:**WI(82,1332). **11:**FL(35), WI(82).
C. graminicola. On leaves. **11:**MO(82).
Phyllosticta cacaliae. **7:**WI(1342,1347). **8:**WI(1344,1347).
P. favillensis. On living leaves. **5:**WI(1343).
Phyllosticta sp. Leaf spot. **7:**WI(1357,1360,1361).
Septoria alba. Leaf spot. **11:**IL(82), KS(82).
S. silphii. On leaves. **5:**WI(1334). **7:**IA(82), WI(82). **11:**TX(82), WI(82).

Deuteromycotina - Other

Rhizoctonia solani. Root rot. **11:**ME(82).
Sclerotium deciduum. **8:**WI(1342).

Silybum Adans.

1. *S. marianum* (L.) Gaertn. [blessed thistle, holy thistle, milk thistle, St.-Mary's thistle]

Deuteromycotina - Coelomycetes

Septoria silybi. **1:**CA(81).

Smallanthus Mackenz. in Sm.

1. *S. uvedalia* (L.) Mackenz. ex Small

Ascomycotina

Diaporthe arctii. **1:**GA(1438).
Leptosphaeria longipedicellata. **1:**GA(1438).
Leptosphaeria sp. **1:**GA(1438).
Metasphaeria sp. **1:**GA(1438).

Solidago L.

(= *Chrysoma* Nutt.)
1. *S. angusta* Torr. & A. Gray
2. *S. bicolor* L. [silver-rod, white goldenrod]
3. *S. boottii* Hook.
4. *S. brachyphylla* Chapm.
5. *S. caesia* L. [blue-stem goldenrod, wreath goldenrod]
6. *S. canadensis* L. [Canada goldenrod]
7. *S. canadensis* L. var. *gilvocanescens* Rydb. (=*S. gilvocanescens* (Rydb.) Smyth)
8. *S. canadensis* L. var. *lepida* DC. (=*S. lepida* DC.)
9. *S. canadensis* L. var. *salebrosa* (Piper) M. E. Jones (=*S. elongata* Nutt., *S. serotina* Aiton var. *salebrosa* Piper)
10. *S. canadensis* L. var. *scabra* (Muhl.) Torr. & A. Gray (=*S. altissima* L.) [tall goldenrod]
11. *S. castrensis* Steele
12. *S. curtisii* Torr. & A. Gray
13. *S. elliottii* Torr. & A. Gray (=*S. edisoniana* Mackenz.)
14. *S. erecta* Pursh
15. *S. fistulosa* Mill.
16. *S. flexicaulis* L. (=*S. latifolia* L.)
17. *S. gigantea* Aiton (=*S. gigantea* Aiton var. *leiophylla* Fernald, *S. serotina* Aiton, *S. pitcheri* Nutt.)
18. *S. hispida* Muhl. ex Willd. (=*S. monticola* Torr. & A. Gray)
19. *S. juncea* Aiton
20. *S. leavenworthii* Torr. & A. Gray [Leavenworth goldenrod]
21. *S. missouriensis* Nutt.
22. *S. missouriensis* Nutt. var. *fasciculata* Holz. (=*S. glaberrima* G. Martens)
23. *S. mollis* Bartl.
24. *S. multiradiata* Aiton
25. *S. nana* Nutt.
26. *S. nemoralis* Aiton
27. *S. odora* Aiton [sweet goldenrod]
28. *S. ohioensis* Riddell
29. *S. patula* Muhl. ex Willd.
30. *S. pauciflosculosa* Michx. (=*Chrysoma pauciflosculosa* (Michx.) Greene)
31. *S. petiolaris* Aiton (=*S. milleriana* Mackenz.)
32. *S. pinetorum* Small
33. *S. riddellii* J. Frank ex Riddell
34. *S. rigida* L.
35. *S. roanensis* Porter
36. *S. rugosa* Mill.
37. *S. sciaphila* Steele
38. *S. sempervirens* L. [seaside goldenrod]
39. *S. sparsiflora* A. Gray
40. *S. spathulata* DC.
41. *S. speciosa* Nutt. (=*S. conferta* Mackenz.)
42. *S. uliginosa* Nutt.
43. *S. ulmifolia* Muhl. ex Willd.
44. *Solidago* × sp. (unnamed hybrids)
45. *Solidago* sp.

Chytridiomycetes

Synchytrium sp. **45:**LA(1847).

Oomycetes

Basidiophora entospora. Downy mildew. **17:**WI(1361). **45:**IL(82), WI(82).
Plasmopara halstedii. Downy mildew. **45:**IL(82), WI(82).

Ascomycotina

Acrospermum foliicola. **10:**GA(1438). **17:**GA(1438).
Anthostomella sp. **10:**GA(1438). **27:**GA(1438). **45:**GA(1438).
Botryosphaeria dothidea. **27:**GA(1438).
B. obtusa. On stems. **45:**GA(82).
Crocicreas coronatum. On stems. **21:**WA(3202). **45:**NY(505).
Diaporthe arctii. **41:**GA(1438).
D. leariformis. On stems. **21:**ND(3807).
D. linearis. On stems. **2:**GA(82), MI(82), ND(82), NJ(82). **3:**GA(1438). **4:**GA(1438). **5:**GA(82,1438), MI(82), ND(82), NJ(82). **6:**GA(82), MI(82), ND(82), NJ(82). **10:**GA(82,1438), MI(82), ND(82), NJ(82). **14:**GA(1438). **17:**GA(1438). **26:**GA(1438). **27:**GA(1438). **31:**GA(1438). **36:**GA(82), MI(82), ND(82), NJ(82). **45:**GA(82), MI(82), ND(82), NJ(82).
Diatrypella herbacea. On stems. **45:**FL(2782).
Dothidea puccinioides. **45:**GA(1438).

D. solidaginis. On leaves. **45**:SC(634).
Elsinoe solidaginis. Spot anthracnose on leaves and stems. **2**:FL(82), GA(82). **5**:FL(82), GA(82). **6**:FL(82), GA(82). **10**:FL(82), GA(82), SC(82). **13**:FL(1745), GA(1438,1745). **15**:GA(1438). **17**:GA(1438,1745). **20**:GA(1438). **36**:FL(82), GA(82). **38**:FL(1745), GA(1438). **45**:FL(35,82,2525), GA(82), SC(82).
Erysiphe cichoracearum. Powdery mildew. **2**:Range of host(82). **5**:Range of host(82). **6**:Range of host(82), Unk.(3976), ND(377), SD(2243), WA(3202). **10**:Range of host(82). **16**:WI(1349). **17**:MT(3202), WA(3202), WI(1335). **21**:MT(3202). **23**:ND(377). **25**:MT(3202). **33**:WI(1335). **34**:MT(3202), WI(1340). **36**:Range of host(82). **45**:FL(35), MS(2720), WY(1195).
E. communis. **45**:AK(526).
Helotium sp. **10**:GA(1438).
Heptameria obesa. **10**:GA(1438). **17**:GA(1438).
Herpotrichia macrotricha. On stalks. **45**:NY(189).
Hypoderma commune. **5**:GA(1438). **10**:GA(1438). **14**:GA(1438). **17**:GA(1438). **27**:GA(1438).
H. nitidum. On dead stems. **45**:MS(2720).
Lachnella villosa. On dead stems. **45**:Unk.(456).
Leptosphaeria doliolum. **4**:GA(1438). **6**:NY(3249). **9**:WA(3249). **10**:GA(1438). **12**:GA(1438). **17**:NY(3249). **42**:NY(3249). **45**:NY(3249).
L. heliopsidis. Black knot. **45**:CT(82), MO(82), ND(82), NY(82).
L. longipedicellata. **5**:GA(1438).
L. macrospora. **5**:GA(3249).
L. noveboracensis. **45**:NY(3249).
L. planiuscula. **41**:GA(1438). **45**:MA(3249).
L. senecionis. **36**:NY(3249). **45**:MI(3249), MO(3249).
Leptosphaeria sp. **5**:GA(1438). **26**:GA(1438).
Lophiostoma caulium. **10**:GA(1438). **17**:GA(1438).
L. vagabundum. **5**:GA(1438).
Massarina gloeospora. **4**:GA(1438). **14**:GA(1438). **26**:GA(1438). **31**:GA(1438).
Mycosphaerella sp. On leaves. **3**:GA(1438). **5**:GA(82,1438). **6**:KS(82). **45**:FL(35,82), GA(82).
Ophiobolus nigro-clypeata. **17**:GA(1438).
Phyllachora solidaginum. **5**:Unk.(82). **6**:MT(3202). **36**:Unk.(82). **45**:NC(3944).
Phyllactinia guttata. Powdery mildew. **17**:WA(3202). **45**:WA(82).
Plagiostoma solidaginis. On overwintered stems. **38**:CT(633).
Pyrenopeziza artemisiae var. *solidaginis*. On dead stems. **45**:NY(82).
Rhytisma solidaginis. **26**:WI(1350). **29**:WI(1350).
Schizoxylon insigne. **10**:GA(1438).
Sphaerotheca fuliginea. Powdery mildew. **45**:IN(82).
Stictis radiata. **45**:GA(1438).
Uncinula sp. Powdery mildew. **45**:NY(82).

Basidiomycotina - Uredinales

Coleosporium asterum. Heteroecious: telial host. **2**:Range of host(82). **3**:GA(1439). **4**:GA(1439). **5**:Range of host(82), GA(1439), WI(1361). **6**:ID(3202), IN(987), MS(2720), MT(3202), NC(3944), ND(378), OR(3202), SD(2243), WA(3202), WI(1333). **7**:ND(378). **9**:CA(81). **10**:Range of host(82), GA(1439), IN(987), WI(1333,1342,1355). **11**:WI(1333). **12**:GA(1307,1439), NC(1307). **14**:GA(1307), NC(1303,1307). **15**:MS(2720). **16**:NY(2456), WI(1333). **18**:GA(1439), WI(1333). **19**:IN(647), MD(651), NY(2456), WI(1335). **21**:AZ(3981), ID(3202), MT(1195,3202), OR(3202), WA(3202), WI(1352). **22**:WI(1333). **23**:ND(378). **24**:AK(60,526), MT(3202), WA(3202). **26**:GA(1439), WI(1333). **29**:WI(1333). **30**:MS(2720). **32**:NC(1303). **33**:WI(1330,1333). **34**:WI(1337). **35**:GA(1307), NC(1303,1307). **36**:Range of host(82), GA(1439), MA(1390), MD(651), NC(1303), NY(2456). **37**:WI(1333). **38**:MA(1390), NC(1303). **40**:CA(81), MT(3202). **41**:WI(1333). **42**:WI(1333). **43**:IN(987), WI(1333). **44**: On *S. rugosa* x *S. sempervirens*. Range of host(82), AK(60,526), LA(2675), MA(1390), NY(2456), OK(2872). **45**:CO(649), MI(665), MS(2720), NC(1303), OR(657).
C. delicatulum. Heteroecious: telial host. **10**:NC(1303). **38**:NJ(1252). **45**:FL(35), NC(1303).
Puccinia dioicae. Heteroecious: aecial host. **6**:MS(2720), MT(3202), ND(378). **17**:ID(3202), MT(3202), ND(378). **21**:AZ(1195,1203,3981), MT(1195,3202), ND(378), WI(1350). **24**:MT(3202). **28**:WI(1348). **29**:WI(1332,1360). **34**:MT(3202). **39**:AZ(1203,3981). **41**:WI(1350). **42**:WI(1360). **43**:WI(1335).
P. dulichii. Heteroecious: aecial host. **2**:Unk.(82). **5**:Unk.(82), NC(1303). **6**:Unk.(82), GA(1439), IN(3101), NC(3944). **9**:CA(81). **10**:Unk.(82), NC(3944). **17**:GA(1439). **18**:OK(2873). **36**:Unk.(82). **45**:FL(35), VT(1224).
P. grindeliae. Autoecious. **9**:WY(1195). **21**:WA(3202). **24**:WY(1195). **26**:WI(1349). **45**:C states(82), W states(82), WA(82).
P. stipae. Heteroecious: aecial host. **6**:MT(3202). **10**:AZ(1203,3981). **21**:ND(378). **23**:MT(3202), ND(378). **25**:AZ(1195,1203,3981). **45**:CO(82), FL(35), MT(82), ND(82), NE(82), NM(82), SD(82).
P. virgae-aurea. **10**:IL(82). **36**:NH(82). **45**:AK(526), MA(82), MI(82), NY(82).
Uromyces perigynius. Heteroecious: aecial host. **36**:ME(82).
U. sommerfeltii. Autoecious. **17**:ID(3202), MT(3202), OR(3202), SD(2768), WA(3202). **45**:CO(82), ID(82), MS(2720), MT(82), OR(82), WA(82), WY(82).

Basidiomycotina - Ustilaginales

Thecaphora cuneata. Inflorescence smut. **21**:KS(920,1077). **45**:KS(82).

Basidiomycotina - Other

Ceratobasidium anceps. On leaves. **10**:WI(1350). **17**:WI(128,1342,1343).
Pellicularia koleroga. Thread blight. **10**:LA(3640). **45**:LA(82).
Thanatephorus cucumeris. **45**:LA(3524).
Typhula phacorrhiza. On stems. **45**:NY(2955).

Deuteromycotina - Hyphomycetes

Cercospora solidaginis. On leaves. **19**:WI(82,581,1334). **22**:WI(1342).
C. stomatica. Leaf spot. **10**:WI(1334,1343). **16**:WI(581). **17**:WI(82,581,1343). **19**:WI(1337,1343). **21**:SD(2243). **33**:WI(1332,1343). **34**:WI(1332,1343). **42**:WI(1332,1343). **45**:FL(35).
Cercosporella virgaureae. On basal rosette leaves. **2**:E states(82), KS(82), WI(82). **6**:E states(82), KS(82), WI(82,1348). **10**:E states(82), KS(82), WI(82). **18**:WI(1355). **21**:ID(3202). **26**:WI(1357). **29**:WI(1332). **33**:WI(1339). **37**:WI(1340). **45**:E states(82), FL(35), KS(82), MT(82), WI(82).
Cladosporium astericola. Leaf spot. **10**:WI(82,1335). **11**:WI(1333). **26**:WI(1335). **41**:WI(1332,1334). **42**:WI(1338). **43**:WI(1335). **45**:WI(82).
Cristulariella moricola. Leaf mold. **6**:IL(2572).
Mycovellosiella minax. On leaves. **10**:WI(1349). **17**:WI(1354). **45**:TX(82), WI(82).
Phymatotrichopsis omnivora. Root rot. **10**:TX(82). **45**:TX(82).
Pseudocercosporella nivea. **17**:WI(1335). **21**:WI(1352). **37**:WI(1343). **41**:WI(1340).
Ramularia serotina. Leaf spot. **2**:CO(82), IL(82), TX(82), WI(82), WY(82). **5**:CO(82), IL(82), TX(82), WI(82), WY(82). **6**:CO(82), IL(82), TX(82), WI(82), WY(82,1195,3378). **8**:WY(1195). **10**:CO(82), IL(82), TX(82), WI(82), WY(82). **17**:WY(1195). **21**:WY(1195). **36**:CO(82), IL(82), TX(82), WI(82), WY(82). **45**:CO(82), IL(82), TX(82), WI(82), WY(82).
R. tenuis. Leaf spot. **1**:KS(840). **16**:WI(811,840).
Sclerotiomyces colchicus. On leaves. **26**:WI(1348).

Deuteromycotina - Coelomycetes

Ascochyta compositarum. Leaf spot. **10**:WI(82). **16**:WI(1358). **17**:WI(1343). **43**:WI(1354,1357).
A. greenei. On living leaves. **10**:WI(1354).
Asteroma solidaginis. Black scurf on leaves. **6**:IA(82).
Botryodiplodia gallicola. On blister-galls. **45**:ND(82).
Colletotrichum dematium. Leaf spot. **10**:NE(82), WI(82,1338). **19**:WI(1334). **22**:WI(1343). **26**:WI(1330). **33**:WI(1330). **37**:WI(1340). **45**:KS(82), WI(82).
Leptothyrium similisporum. On leaves. **10**:WI(1336). **26**:WI(1332). **34**:WI(1361). **42**:WI(1359). **45**:IL(82), NY(82), WI(82).
Macrophoma sphaeropsispora. On leaves and stems. **45**:CA(82).
Macrophomina phaseolina. Charcoal rot. **10**:IL(311).
Phoma exigua. **17**:WI(1357).
P. solidaginis var. *longispora*. On stems. **45**:NY(82).
Phomopsis sp. Leaf spot. **6**:WI(1357). **45**:FL(35).
Phyllosticta cacaliae. **34**:WI(1347).
P. gallicola. On blister-galls. **21**:ID(3202). **45**:CO(82).
P. siccata. On leaves. **17**:WI(1334). **26**:WI(1343).
P. solidaginicola. Leaf spot. **45**:IL(82,3590), WI(82).
P. solidaginis. **6**:WY(82,1195). **17**:WI(1331). **37**:WI(1354). **45**:WI(82).

P. solidaginis-serotinae. On leaves. **17**:WI(1331).
Phyllosticta sp. Leaf spot. **6**:WI(1360). **10**:WI(1350). **16**:WI(1349). **17**:WI(1350,1355). **29**:WI(1361). **33**:WI(1361). **41**:WI(1342).
Placosphaeria haydenii. Black spot on leaves and stems. **6**:IA(82).
Rhabdospora solidaginis. On stems. **6**:CO(82), IN(82). **45**:NJ(82).
R. subgrisea. On stems and galls. **45**:MI(82), NY(82).
Septoria angularis. **16**:WI(1337).
S. atropurpurea. **10**:WI(1336). **16**:WI(1336). **21**:WI(1342). **41**:WI(1350). **42**:WI(1343). **45**:IL(82), WI(82,1351).
S. bacilligera. Leaf spot. **45**:OK(2873).
S. brevis. **45**:NY(82).
S. davisii. Leaf spot. **6**:TX(82). **10**:WI(1334). **45**:CO(82), FL(35,82), TX(82), WI(82).
S. fumosa. **6**:NY(82). **10**:WI(82). **19**:WI(1338). **21**:WI(1338). **34**:WI(1340). **43**:WI(1335). **45**:IA(82), IL(82), MO(82), NY(82), WI(82).
S. gallarum. On old galls. **45**:NY(965).
S. intermedia. **45**:WI(966).
S. solidaginicola. **5**:SD(82), WI(82). **6**:SD(2243), WI(82,1350). **10**:WI(82). **16**:NY(2456). **26**:WI(1332). **29**:WI(1330). **34**:WI(1347). **45**:KS(82), NY(82), WI(82).
S. solidaginis. **6**:IA(82). **45**:SC(82).
S. virgaureae. **6**:IA(82).
Septoria sp. On leaves. **17**:WI(1333). **34**:WI(1347). **41**:WI(1359). **45**:OK(2872).
Stagonospora astericola. Leaf spot. **10**:WI(82). **45**:WI(82).
Stagonospora sp. On insect galls. **17**:WI(1352).
Deuteromycotina - Other
Sclerotium deciduum. **6**:WI(1342).
S. mendax. On leaves. **6**:WI(82). **10**:NY(82), WI(82,1332). **45**:WI(82).

Sonchus L.

1. *S. arvensis* L. [perennial sowthistle]
2. *S. asper* (L.) J. Hill [spiny sowthistle, spiny-leaved sowthistle]
3. *S. oleraceus* L. [annual sowthistle, common sowthistle]
4. *Sonchus* sp.

Oomycetes
Bremia lactucae. Downy mildew. **1**:MD(82), NY(82), OR(82). **2**:OR(3202). **3**:FL(82), NY(82), SC(82), TX(82). **4**:FL(35).
Ascomycotina
Diaporthe arctii. On stems. **3**:GA(82).
Erysiphe cichoracearum. Powdery mildew. **2**:OK(2872). **3**:FL(82), PA(82). **4**:FL(35).
Leptosphaeria fulgida. **3**:GA(1438).
Mollisia atrata. **3**:GA(1438).
Sclerotinia sclerotiorum. Stem rot. **1**:TX(82). **2**:TX(82). **3**:FL(82), TX(82). **4**:FL(35).
Sphaerotheca fuliginea. Powdery mildew. **3**:IA(82), PA(82).
Basidiomycotina - Uredinales
[*Coleosporium tussilaginis*]. **1**:WA(82,3202). **2**:WI(82).
Deuteromycotina - Hyphomycetes
Alternaria sonchi. Blight, leaf spot. **1**:MA(82,1394). **2**:LA(82), NY(82), OK(2872), TX(82), WI(82). **3**:FL(82), HI(2903), TX(82). **4**:FL(35).
Cercospora sp. Leaf spot. **4**:FL(35).
Phymatotrichopsis omnivora. Root rot. **2**:TX(82). **3**:TX(82).
Deuteromycotina - Coelomycetes
Phoma exigua. Leaf spot. **1**:ND(82), WI(1352). **2**:WI(1347).
Septoria sonchifolia. Leaf spot. **1**:WA(82), WI(1334). **2**:ME(82), SC(82,634), TX(82), WI(82). **3**:HI(2903), IL(82), TX(82), WI(1349). **4**:WA(3202).

Steirodiscus Less.

(= *Gamolepis* Less.)
1. *Steirodiscus* sp. (=*Gamolepis sp.*)

Oomycetes
Pythium sp. Leaf spot. **1**:FL(35).
Deuteromycotina - Hyphomycetes
Fusarium sp. Basal stem rot. **1**:FL(35).
Deuteromycotina - Coelomycetes
Colletotrichum sp. Rotted cuttings. **1**:FL(35).

Stephanomeria Nutt.

(= *Ptiloria* Raf.)
1. *S. cichoriacea* A. Gray
2. *S. lactucina* A. Gray
3. *S. paniculata* Nutt.
4. *S. pauciflora* (Torr.) A. Nels.
5. *S. runcinata* Nutt.
6. *S. tenuifolia* (Raf.) H. Hall
7. *S. thurberi* A. Gray
8. *Stephanomeria* sp. (=*Ptiloria sp.*)

Ascomycotina
Comoclathris permunda. **6**:WY(3810).
Pleospora herbarum. On dead stems. **8**:CA(2439).
Basidiomycotina - Uredinales
Puccinia hieracii var. *stephanomeriae*. Autoecious. **1**:CA(82,2713). **2**:CA(81,638,642,661,2713). **3**:OR(3202), WA(3202). **4**:AZ(747,1203,2713,3981). **5**:MT(3202). **6**:AZ(2713), CO(2713), ID(3202), MT(2713,3202), UT(1195,2713), WA(2713,3202), WY(2713). **7**:AZ(3981). **8**:W states(82), CA(82).
Deuteromycotina - Hyphomycetes
Cercospora clavicarpa. Leaf spot. **1**:CA(82).
Deuteromycotina - Coelomycetes
Discula runcinata. On dead stems. **5**:MT(966,3202).

Stokesia L'Hér.

1. *S. laevis* (J. Hill) Greene [Stokes' aster]

Ascomycotina
Erysiphe cichoracearum. Powdery mildew. **1**:MD(1902).
Sclerotinia sclerotiorum. Basal stem rot. **1**:FL(35).
Deuteromycotina - Hyphomycetes
Alternaria sp. Leaf spot. **1**:FL(35).
Botrytis cinerea. Head blight. **1**:NY(82).
Cercospora sp. Leaf spot. **1**:FL(35).
Stemphylium lycopersici. Leaf spot. **1**:FL(35).
Deuteromycotina - Coelomycetes
Ascochyta sp. Leaf spot. **1**:IA(82), PA(82).
Phyllosticta sp. Leaf spot. **1**:FL(35).
Deuteromycotina - Other
Rhizoctonia sp. Flower and leaf blight. **1**:FL(35).
Sclerotium rolfsii. Southern blight. **1**:FL(35).

Tagetes L.

1. *T. erecta* L. [African marigold, Aztec marigold, big marigold]
2. *T. patula* L. [French marigold]
3. *Tagetes* sp.

Oomycetes
Phytophthora cryptogea. Stem rot, wilt. **3**:FL(35), NY(82).
Pythium ultimum. Root rot. **2**:CA(82).
Pythium sp. Damping-off, root rot. **3**:FL(35), HI(2903).
Zygomycotina
Choanephora cucurbitarum. Flower blight. **3**:FL(35).
Ascomycotina
Sclerotinia sclerotiorum. Stem rot, wilt. **3**:FL(35), NY(82).
Basidiomycotina - Uredinales
Coleosporium pacificum. Heteroecious: telial host. **1**:CA(82).
Puccinia tageticola. Telial host; aecial host unknown. **3**:TX(82).
Puccinia sp. **3**:HI(2903).
Deuteromycotina - Hyphomycetes
Alternaria tagetica. Leaf and petal blight, leaf spot. **1**:NJ(692), SC(1641). **2**:NJ(692).
Alternaria sp. Leaf and petal blight, leaf spot. **3**:FL(35), HI(2903), NC(1303), NJ(82).
Botrytis cinerea. Head blight, flower and leaf rot. **1**:AK(82,526), CT(82), NJ(82), PA(82). **2**:AK(82), CT(82), NJ(82), PA(82). **3**:AK(82), CT(82), FL(35), NJ(82), PA(82).
Cercospora tageticola. Leaf spot. **2**:FL(82). **3**:FL(35).
Cercospora sp. Leaf spot. **3**:CT(82), NC(1303).
Fusarium oxysporum. Stem rot, wilt. **3**:FL(35).
Fusarium sp. Stem rot, wilt. **1**:CA(82), NJ(82), NY(82). **2**:NJ(82), NY(82). **3**:NC(1303).
Helminthosporium sp. Flower spot. **1**:TX(82).
Verticillium albo-atrum. Wilt. **3**:NY(82).
V. dahliae. Wilt. **3**:CA(1573).

Deuteromycotina - Coelomycetes
Ascochyta sp. Leaf spot. **3:**FL(35).
Colletotrichum gloeosporioides. Anthracnose. **3:**NC(1303).
Macrophomina phaseolina. Charcoal stem rot. **1:**OK(82).
Phyllosticta sp. Leaf spot. **3:**FL(35).
Septoria tageticola. Leaf spot. **1:**FL(546). **3:**FL(35).
Deuteromycotina - Other
Rhizoctonia solani. Root rot. **1:**TX(82). **2:**TX(82). **3:**FL(35), NC(1303).
Rhizoctonia sp. Damping-off, bottom rot, stem rot. **1:**HI(2903).
Sclerotium rolfsii. Southern blight. **3:**FL(35,82), NC(1303), NJ(82), VA(82).

Tanacetum L.

1. *T. camphoratum* Less.
2. *T. capitatum* Torr. & A. Gray
3. *T. douglasii* DC.
4. *T. vulgare* L. (=*Chrysanthemum vulgare* (L.) Bernh.) [common tansy, golden-buttons]

Ascomycotina
Erysiphe cichoracearum. Powdery mildew. **1:**CA(1155). **3:**CA(1155). **4:**PA(82), SD(2243), WA(3202).
Deuteromycotina - Hyphomycetes
Oidium sp. Powdery mildew. **4:**MS(2720).
Ramularia tanaceti. Leaf spot. **2:**WI(82).

Taraxacum Wigg.

1. *T. erythrospermum* Andrzeiovski ex Besser [redseed dandelion]
2. *T. kok-saghyz* Rodin [Russian dandelion]
3. *T. laevigatum* (Willd.) DC.
4. *T. lyratum* (Ledeb.) DC.
5. *T. mutilum* Greene
6. *T. officinale* Wigg. (=*T. vulgare* Schrank) [common dandelion]
7. *Taraxacum* sp.

Myxomycetes
Physarum cinereum. Slime mold. **6:**MS(2720).
Chytridiomycetes
Synchytrium taraxaci. Leaf gall. **6:**MN(82).
Ascomycotina
Erysiphe cichoracearum. Powdery mildew. **6:**IA(82), IN(82), NE(82), NJ(82), NY(82), OH(82), PA(82), SD(2243).
Phyllactinia guttata. Powdery mildew. **6:**WA(82,3202).
Protomyces pachydermus. Leaf gall. **6:**UT(82).
Sclerotinia sclerotiorum. Crown, peduncle, and stem rot. **6:**FL(35), PA(82), WA(2323,3202).
Sphaerotheca fuliginea. Powdery mildew. **6:**Unk.(82), AK(526), CA(642), ND(377).
S. macularis. Powdery mildew. **6:**Unk.(3976), CO(1195), GA(1438), ID(3202), MT(3202), NC(1303,3944), SD(2243), WA(3202), WY(1195).
Basidiomycotina - Uredinales
Puccinia hieracii. Autoecious. **1:**WI(1335). **4:**AK(526). **5:**AK(526). **6:**Range of host(82), AK(526), AZ(1203,2713,3981), CA(81,642,2713), CO(1195,2713), FL(35), GA(1439), HI(2903), ID(1195,2713,3202), IL(2713), IN(2713), KY(2713), MA(2713), MI(665,2713), MS(2720), MT(1195,2713,3202), NC(1303,1307), ND(378), NJ(2713), OH(2713), OK(2872), OR(3202), SD(2243), WA(3202), WY(1195,2713,2924). **7:**AK(60,526).
P. variabilis. Autoecious. **6:**FL(35), SD(2243).
Basidiomycotina - Ustilaginales
Sporidiobolus pararoseus. On mildewed leaves. **7:**KS(2686).
Basidiomycotina - Other
Thanatephorus cucumeris. **3:**WA(3202).
Deuteromycotina - Hyphomycetes
Botrytis cinerea. Gray mold blight. **6:**NY(82).
Fusarium equiseti. **2:**MT(3202).
F. sambucinum. **2:**MT(3202).
Phymatotrichopsis omnivora. Root rot. **6:**TX(82).
Ramularia lineola. Leaf spot. **6:**IN(82), NC(82,3944), ND(82), NY(82), TX(82).
R. taraxaci. Leaf spot. **1:**WI(1334). **6:**E & C states(82), CO(1195), ID(3202), MS(2720), NC(1303), ND(378), OR(3202), WA(82,2323,3202), WY(1195). **7:**AK(82,526), IA(82), MI(82), WA(82).
Ramularia sp. Leaf spot. **2:**CT(82).
Tilletiopsis minor. On mildewed leaves. **7:**KS(2686).
Deuteromycotina - Coelomycetes
Ascochyta doronici. Leaf spot. **6:**WI(1354). **7:**AK(82,526).
Septoria britannica. **6:**WI(1343).
S. unicolor. Leaf spot. **6:**WA(82,3202).
Vermicularia dematium f. *minor*. **6:**ND(378).
Deuteromycotina - Other
Rhizoctonia solani. Root rot. **2:**CT(82). **6:**CT(82), NJ(82), WA(82). **7:**WA(82).
Sclerotium rolfsii. Southern blight. **6:**FL(35).

Tetradymia DC.

1. *T. canescens* DC.

Basidiomycotina - Uredinales
Puccinia grindeliae. Autoecious. **1:**WA(3202).
Basidiomycotina - Other
Crucibulum parvulum. **1:**ID(3202).

Tithonia Desf. ex Juss.

1. *T. rotundifolia* (Mill.) S. F. Blake (=*T. speciosa* (Hook.) Griseb.)
2. *T. thurberi* A. Gray

Basidiomycotina - Uredinales
Puccinia enceliae. Autoecious. **2:**AZ(747,1194,1203,3981).
Basidiomycotina - Other
Armillaria tabescens. Root rot. **1:**FL(35).
Deuteromycotina - Hyphomycetes
Alternaria sp. Leaf spot. **1:**OK(2872).
Cercospora pachypus. Leaf spot. **1:**FL(35).
Deuteromycotina - Coelomycetes
Ascochyta compositarum. **1:**WI(1358).
Deuteromycotina - Other
Sclerotium rolfsii. Southern blight. **1:**FL(35).

Tragopogon L.

1. *T. dubius* Scop. (=*T. major* Jacq.) [western salsify]
2. *T. mirus* Ownbey
3. *T. miscellus* F. Ownbey
4. *T. porrifolius* L. [oyster-plant, salsify, vegetable-oyster]
5. *T. pratensis* L. [Jack-go-to-bed-at-noon, John-go-to-bed-at-noon, meadow salsify]
6. *Tragopogon* sp.

Oomycetes
Albugo tragopogonis. White rust. **1:**ID(3202), WA(3202). **3:**WA(3202). **4:**Range of host(82), MT(3202), ND(377), SD(2243), WA(3202), WY(1195). **5:**WA(3202).
Ascomycotina
Chaetomium convolutum. **1:**ID(3202).
Erysiphe cichoracearum. Powdery mildew. **4:**Range of host(82), OK(2872), WI(1330). **6:**WA(3202).
Sclerotinia minor. Crown rot. **4:**IL(82,1946).
S. sclerotiorum. Crown rot. **4:**IL(82).
Basidiomycotina - Other
Thanatephorus cucumeris. **4:**WA(3202).
Deuteromycotina - Hyphomycetes
Alternaria alternata. **1:**ID(3202), WA(3202). **2:**ID(3202), WA(3202). **4:**ID(3202), WA(3202).
A. scorzonerae. Leaf blight. **4:**AL(82), MD(82), NY(82), PA(82), VA(82), WV(82).
A. tenuissima. **1:**ID(3202).
Cercospora tragopogonis. Leaf spot. **4:**MT(82,3202), OK(82,2872). **5:**WI(1332).
Cladosporium cladosporioides. **1:**ID(3202).
C. herbarum. **3:**ID(3202). **4:**ID(3202), WA(3202). **5:**ID(3202).
Cladosporium sp. **1:**WA(3202).
Hormodendrum viride. **4:**WA(3202). **5:**ID(3202).
Memnoniella echinata. **1:**WA(3202).
Phymatotrichopsis omnivora. Root rot. **4:**AZ(82), TX(82).
Stachybotrys chartarum. **1:**ID(3202). **5:**ID(3202).
Stemphylium botryosum var. *tragopogonis*. Leaf spot. **4:**NY(82).
Verticillium albo-atrum. Wilt. **4:**NY(82).
Deuteromycotina - Coelomycetes
Phyllosticta sp. On leaves. **5:**WI(1352).
Deuteromycotina - Other
Rhizoctonia solani. Root rot. **4:**TX(82), WA(82).
Sclerotium rolfsii. Southern blight. **4:**TX(82).

Tridax L.
1. *T. procumbens* L.
Basidiomycotina - Uredinales
Puccinia melampodii. Autoecious. **1**:FL(35).
Deuteromycotina - Hyphomycetes
Cercospora sp. Leaf spot. **1**:FL(35).

Tussilago L.
1. *T. farfara* L. [coltsfoot]
Ascomycotina
Mycosphaerella tussilaginis. Leaf spot. **1**:NY(82).
Deuteromycotina - Hyphomycetes
Ramularia brunnea. **1**:NY(2456).
Deuteromycotina - Coelomycetes
Septoria farfaricola. Leaf spot. **1**:TX(82).

Verbesina L.
1. *V. alternifolia* (L.) Britton ex Kearn. (=*Actinomeris alternifolia* (L.) DC.) [wingstem, yellow-ironweed]
2. *V. encelioides* (Cav.) Benth. & J. D. Hook. ex A. Gray [butter daisy, goldencrown daisy]
3. *V. holwayi* B. L. Robinson
4. *V. longifolia* A. Gray
5. *V. occidentalis* (L.) Walter
6. *V. rothrockii* B. L. Robinson & Greenm.
7. *V. virginica* L. (=*V. texana* Buckley)
8. *Verbesina* sp.
Oomycetes
Plasmopara halstedii. Downy mildew. **2**:NM(82), TX(82).
Ascomycotina
Diaporthe arctii var. *achilleae*. On stems. **1**:IN(82).
Erysiphe cichoracearum. Powdery mildew. **1**:DE(82), KS(82), MD(82), PA(82), VA(82), WV(82). **5**:SC(82), TX(82), VA(82). **8**:FL(35).
Heptameria obesa. **5**:GA(1438). **7**:GA(1438).
Leptosphaeria doliolum. **1**:GA(1438). **7**:GA(1438).
Leptospora rubella. **7**:NC(3248).
Massarina gloeospora. **5**:GA(1438).
Ophiobolus crassus. **7**:NC(3248).
O. drechsleri. **7**:NC(3248).
Pleospora penicillus. **5**:GA(1438). **7**:GA(1438).
Sphaerodothis pirifera. **8**:FL(35).
Basidiomycotina - Uredinales
Coleosporium viguierae. Autoecious. **2**:AZ(82,1203,3981), TX(742). **7**:FL(82), NC(82,3944). **8**:FL(35,82), TX(82).
[*Puccinia abrupta*]. **2**:TX(82). **6**:AZ(1203). **8**:CA(82).
P. cognata. Autoecious. **2**:AZ(1203,3981), TX(82). **3**:AZ(1203). **4**:AZ(1203,3981). **7**:AR(82), LA(82), MS(2720), TN(82), TX(82). **8**:FL(35), TX(82).
P. verbesinae. Autoecious. **1**:MD(82), ME(82), NC(3944), TN(82). **5**:SE states(82), GA(1439), IN(82), LA(82), NC(1303,3944).
P. ximenesiae. Autoecious. **2**:TX(746). **6**:AZ(1203,3981).
Deuteromycotina - Hyphomycetes
Cercospora anomala. Leaf spot. **1**:IA(82).
C. fulvella. Leaf spot. **7**:TX(82,581).
Cercospora sp. Leaf spot. **8**:FL(35).
Phymatotrichopsis omnivora. Root rot. **2**:TX(82).
Stachybotryella repens. On leaves. **7**:TX(82).
Deuteromycotina - Coelomycetes
Colletotrichum sp. Leaf spot. **2**:AZ(82).
Phyllosticta verbesinae. Leaf spot. **7**:TX(64). **8**:TX(82).
Deuteromycotina - Other
Rhizoctonia crocorum. Root rot. **7**:TX(82).
Sclerotium rolfsii. Southern blight. **8**:FL(35).

Vernonia Schreb.
1. *V. acaulis* (Walter) Gleason
2. *V. arkansana* DC. (=*V. crinita* Raf.)
3. *V. baldwinii* Torr. [western ironweed]
4. *V. baldwinii* Torr. var. *interior* (Small) W. Z. Faust (=*V. interior* Small)
5. *V. fasciculata* Michx.
6. *V. gigantea* (Walter) Trel. ex F. V. Branner & Coville (=*V. altissima* Nutt.) [tall ironweed]
7. *V. glauca* (L.) Willd.
8. *V. missurica* Raf.
9. *V. noveboracensis* (L.) Michx. (=*V. ovalifolia* Torr. & A. Gray) [ironweed]
10. *V. serrata* Less. (=*V. palustris* Gardner)
11. *Vernonia* sp.
Oomycetes
Plasmopara halstedii. Downy mildew. **9**:KS(82), MO(82).
Ascomycotina
Diaporthe arctii. **1**:GA(1438). **9**:GA(1438).
Erysiphe cichoracearum. Powdery mildew. **3**:OK(2873). **5**:Range of host(82). **6**:Range of host(82), FL(35). **9**:Range of host(82). **11**:Range of host(82), FL(35).
Hypoderma commune. **9**:GA(1438).
Lasiosphaeria sp. **7**:GA(1438).
Leptosphaeria doliolum. **6**:TN(3249).
L. fulgida. **9**:GA(1438).
Leptosphaeria sp. **7**:GA(1438).
Metasphaeria sp. **7**:GA(1438).
Mycosphaerella decidua. On leaves. **11**:KS(82,967).
Sphaerotheca fuliginea. Powdery mildew. **9**:MO(82).
Basidiomycotina - Uredinales
Coleosporium vernoniae. Heteroecious: telial host. **2**:OK(2872). **3**:MS(2720), OK(2872). **5**:Range of host(82), MS(2720), OK(2872). **8**:OK(2873). **9**:Range of host(82), GA(1439). **10**:SE states(82), KY(82), La(82), MS(82,2720), SC(1533), TX(82). **11**:FL(35), LA(2675), OH(653).
Puccinia longipes. Autoecious. **2**:OK(2872). **3**:IL(3715), MS(2720), OK(2872). **4**:OK(2872). **5**:Range of host(82), IN(3715), ND(378), SD(2243). **6**:Range of host(82), FL(35), MI(3715). **7**:VA(3715). **9**:Range of host(82), DC(3715), DE(3715), IL(3715), VA(3715). **11**:Range of host(82), FL(35), WV(3715).
P. longipes var. *brevipes*. Autoecious. **3**:IA(3715), KS(3715), MO(3715), MO(746), NE(3715), OK(3715), TX(3715). **5**:IA(3715), ND(3715), NE(3715), SD(3715). **6**:TX(3715).
Deuteromycotina - Hyphomycetes
Cercospora noveboracensis. On leaves. **9**:MO(82,962).
C. oculata. Leaf spot. **3**:OK(2872). **5**:SE states(82), NE(82), TX(82), WI(82). **6**:SE states(82), NE(82), TX(82), WI(82). **9**:SE states(82), DE(3543), NE(82), TX(82), WI(82). **11**:SE states(82), NE(82), TX(82), WI(82).
C. vernoniae. Leaf spot. **3**:MS(2720). **5**:SE states(82), NE(82), TX(82), WI(82). **6**:SE states(82), FL(35), NE(82), TX(82), WI(82). **8**:OK(2872). **9**:SE states(82), NE(82), TX(82), WI(82). **11**:SE states(82), FL(35), NE(82), TX(82), WI(82).
Phymatotrichopsis omnivora. Root rot. **11**:TX(82).
Ramularia sp. Leaf spot. **11**:FL(35).
[*Sarcinella heterospora*]. On living leaves. **9**:FL(1681).
S. vernoniae. On living leaves. Black mildew. **11**:MO(82), NC(1681).
Deuteromycotina - Coelomycetes
Ascochyta treleasei. Black mildew. **9**:WI(82).
Phoma exigua. **5**:WI(1333).
Septoria sp. Leaf spot. **11**:MO(82).
Deuteromycotina - Other
Rhizoctonia solani. Thread blight. **11**:FL(35).
Sclerotium rolfsii. Southern blight. **11**:FL(35).

Viguiera Kunth in Humb., Bonpl. & Kunth
1. *V. annua* (Jones) S. F. Blake [goldeneye]
2. *V. cordifolia* A. Gray
3. *V. deltoidea* A. Gray
4. *V. deltoidea* A. Gray var. *parishii* (Greene) Vasey & J. Rose
5. *V. dentata* (Cav.) Spreng.
6. *V. dentata* (Cav.) Spreng. var. *lancifolia* S. F. Blake
7. *V. longifolia* (B. L. Robinson & Greenm.) S. F. Blake
8. *V. multiflora* (Nutt.) S. F. Blake
9. *V. stenoloba* S. F. Blake [resin-bush]
10. *Viguiera* sp.
Ascomycotina
Erysiphe cichoracearum. Powdery mildew. **10**:UT(82).
Basidiomycotina - Uredinales
Coleosporium viguierae. Autoecious. **2**:AZ(3981). **5**:AZ(1194,1195,1203,3379,3981), TX(742). **6**:AZ(741,1194,1203,3981). **8**:AZ(1203).
Puccinia abrupta. Autoecious. **1**:AZ(82,1203,3981). **2**:AZ(1203). **5**:AZ(1194). **6**:AZ(1194,1203,3379,3981). **7**:AZ(3981). **8**:AZ(1203,3981). **10**:AZ(1203), TX(82).

P. calanticariae. Autoecious. **2:**AZ(1203,3981). **5:**AZ(1195,1203,3981).
P. caleae. Autoecious. **9:**TX(742).
P. chloracae. **9:**AZ(1203).
P. enceliae. Autoecious. **1:**AZ(1203,3981). **8:**AZ(1195,1203,3981), CO(1195), WY(1195).
P. turgidipes. Autoecious. **3:**AZ(747,1203). **4:**AZ(3981). **10:**AZ(82), CA(82).
Basidiomycotina - Ustilaginales
Entyloma compositarum. White smut. **10:**UT(82,1077).
Deuteromycotina - Hyphomycetes
Phymatotrichopsis omnivora. Root rot. **10:**TX(82).
Deuteromycotina - Other
Rhizoctonia crocorum. Root rot. **10:**TX(82).

Wedelia Jacq.

1. *W. trilobata* (L.) Hitchc. [creeping-Charlie, porch-vine, trailing wedelia]
2. *Wedelia* sp.

Deuteromycotina - Hyphomycetes
Alternaria sp. Leaf spot. **1:**HI(2903).
Cercospora wulffiae. Leaf spot. **1:**FL(35). **2:**FL(35).
Cercospora sp. Leaf spot. **1:**HI(2903).
Oidium sp. Powdery mildew. **1:**FL(35).
Deuteromycotina - Coelomycetes
Colletotrichum sp. Anthracnose. **1:**FL(35).
Deuteromycotina - Other
Rhizoctonia sp. Root rot. **1:**HI(2903).

Wyethia Nutt.

1. *W. amplexicaulis* (Nutt.) Nutt.
2. *W. angustifolia* (DC.) Nutt.
3. *W. arizonica* A. Gray
4. *W. helianthoides* Nutt.
5. *W. mollis* A. Gray
6. *Wyethia* sp.

Ascomycotina
Clathrospora pentamera. **5:**CA(661).
Pleospora sp. On stems. **5:**CA(661).
Sphaerotheca macularis. **4:**WY(1195).
Basidiomycotina - Uredinales
Puccinia balsamorhizae. Autoecious. **1:**CO(82), UT(82). **2:**CA(81,82), UT(82). **3:**AZ(1203,3981), CO(1195). **5:**CA(661). **6:**WA(3202).
Deuteromycotina - Hyphomycetes
Didymaria conferta. Leaf spot. **1:**ID(3202), OR(82,3202), UT(82), WA(3202), WY(82,1195). **2:**OR(3202).
D. wyethiae. Leaf spot. **1:**ID(1195,3202), WA(82,3202). **6:**CA(82).
Deuteromycotina - Coelomycetes
Phyllosticta ambrosiae. On leaves. **5:**CA(81).
Septoria wyethiae. On living leaves. **5:**CA(81,661,1468). **6:**CA(82), UT(82).

Xanthium L.

1. *X. saccharatum* Wallr.
2. *X. spinosum* L. [spiny cocklebur]
3. *X. strumarium* L. (=*X. pensylvanicum* Wallr., *X. chinense* Mill.) [common cocklebur]
4. *X. strumarium* L. var. *canadense* (Mill.) Torr. & A. Gray (=*X. canadense* Mill., *X. commune* Britton, *X. echinatum* Andr. Murray, *X. italicum* Moretti, *X. oviforme* Wallr., *X. speciosum* Kearn., *X. orientale* L.) [cocklebur]
5. *X. strumarium* L. var. *glabratum* (DC.) Cronq. (=*X. inflexum* Mackenz. & Bush)
6. *Xanthium* sp.

Chytridiomycetes
Synchytrium macrosporum. Red galls on leaves and petioles. **3:**TX(1848).
Oomycetes
Albugo tragopogonis. White rust. **1:**OR(82). **3:**OR(3202).
Plasmopara halstedii. Downy mildew. **1:**IA(82).
Ascomycotina
Diaporthe arctii. On stems. **1:**GA(82). **4:**GA(1438).
Erysiphe cichoracearum. Powdery mildew. **1:**SE states(82), TX(82), WI(82). **3:**FL(35), GA(1438), MS(2720). **4:**GA(1438), WI(1335).
Leptosphaeria conoidea. **3:**NY(3249).
Massarina gloeospora. **4:**GA(1438).
Mycosphaerella xanthicola. On stems. **1:**CA(82).
Ophiobolus acuminatus. **4:**GA(1438).
O. anguillidus. **4:**GA(1438).
Sclerotinia sp. On seeds and stems. **3:**MD(15).
Basidiomycotina - Uredinales
Puccinia canaliculata. Heteroecious: aecial host. **1:**E states(82), SC states(82), AZ(1203), NE(82), NM(82). **3:**AZ(3981). **4:**OK(2873). **6:**OK(2872).
P. xanthii. Autoecious. **1:**Range of host(82), AZ(1203), HI(2903), TX(742). **2:**IN(82), MT(82), WA(3202). **3:**AZ(747), FL(35), ID(3202), MT(3202), NC(1303), OK(2872), WA(3202). **5:**OK(2872). **6:**MS(2675), NC(2702), OK(2872), PA(2702).
Basidiomycotina - Other
Thanatephorus cucumeris. **3:**LA(3524).
Deuteromycotina - Hyphomycetes
Botrytis cinerea. Leaf spot. **1:**WV(82).
Cercospora xanthicola. Leaf spot. **1:**TX(82), WI(82). **4:**WI(1335).
Oidium sp. Powdery mildew. **1:**HI(2903). **4:**HI(2903).
Phymatotrichopsis omnivora. Root knot. **1:**TX(82).
Verticillium albo-atrum. **4:**NM(2109).
Deuteromycotina - Coelomycetes
Colletotrichum gloeosporioides. On stems. **1:**NJ(82), TX(82).
C. truncatum. Isolated from leaves. **3:**Unk.(3059), IL(1482).
Lasiodiplodia theobromae. On stems. **1:**AL(82).
Rhabdospora xanthii. On stems. **1:**NY(82), TX(82).
Septoria xanthii. Leaf spot. **1:**CA(82), DE(82), ND(82).

Youngia Cass.

1. *Y. japonica* (L.) DC. (=*Crepis japonica* (L.) Benth.)

Ascomycotina
Sclerotinia sclerotiorum. Cottony rot. **1:**FL(35).
Deuteromycotina - Hyphomycetes
Oidium sp. Powdery mildew. **1:**FL(35).
Stemphylium sp. Leaf spot. **1:**FL(35).
Deuteromycotina - Coelomycetes
Phyllosticta sp. Leaf spot. **1:**FL(35).

Zexmenia Llave & Lex.

1. *Z. brevifolia* A. Gray

Basidiomycotina - Uredinales
Puccinia electrae. Autoecious. **1:**TX(742).
P. electrae var. *depressiporosa*. Autoecious. **1:**TX(746).
P. electrae var. *robusta*. Autoecious. **1:**TX(746).

Zinnia L.

1. *Z. peruviana* (L.) L. (=*Z. multiflora* L.)
2. *Z. violacea* Cav. (=*Z. elegans* Jacq.) [zinnia]
3. *Zinnia* sp.

Oomycetes
Phytophthora cryptogea. Stem rot. **2:**NJ(82).
Phytophthora sp. Stem rot. **2:**NC(1303).
Pythium sp. Root rot. **2:**HI(2903).
Zygomycotina
Choanephora cucurbitarum. Blossom rot. **2:**FL(35).
Choanephora sp. Blossom rot. **2:**FL(82).
Ascomycotina
Erysiphe cichoracearum. Powdery mildew. **2:**Range of host(82), FL(35), GA(1438), HI(2899), MA(1390), MS(2720), NC(1303), OK(2872), WI(1330).
E. graminis f. sp. *avenae*. Powdery mildew. **2:**AR(3029).
Sclerotinia sclerotiorum. Stem rot. **2:**CA(82), CO(82), MA(82), MT(82,3202), OR(82), PA(82), WA(82). **3:**OR(3202), WA(3202).
Sclerotinia sp. Cottony rot. **2:**FL(35).
Basidiomycotina - Other
Athelia rolfsii. **2:**WA(3202).
Deuteromycotina - Hyphomycetes
Alternaria zinniae. Alternaria blight, leaf spot. **2:**CA(149), FL(35), IL(312), NC(1303), WI(1357). **3:**CT(82), NJ(82), NY(82), PA(82), SC(82).
Alternaria sp. Blight, stem canker, secondary leaf spot. **2:**HI(2903), NC(1303), OK(2872), OR(3202).

Botrytis cinerea. Head and leaf blight. **2**:AK(82,526), CA(82), CT(82), NC(1303), NJ(82), OR(82), PA(82). **3**:WA(3202).
Cercospora zinniae. Leaf spot. **1**:AL(82,581), FL(82,581). **2**:SE states(82), CO(82), FL(35), IN(82), MS(2720), NC(1303), PA(82), TX(82).
Corynespora sp. Leaf spot. **2**:FL(35).
Fusarium oxysporum. Wilt. **2**:HI(2903).
Fusarium sp. Root and stem rot. **2**:CO(82), FL(35), IA(82), MO(82), NC(1303), NY(82), TX(82), WA(82,3202). **3**:WA(3202).
Oidium sp. Powdery mildew. **2**:HI(2903). **3**:HI(2903).
Phymatotrichopsis omnivora. Root rot. **2**:TX(82).
Deuteromycotina - Coelomycetes
Macrophomina phaseolina. Charcoal rot. Charcoal stem rot. **2**:OK(2872), TX(82).
Deuteromycotina - Other
Rhizoctonia solani. Thread blight, damping-off. **2**:CA(82), FL(35), NJ(82), TX(82).
Rhizoctonia sp. Root and stem blight. **2**:HI(2903), NC(1303).
Sclerotium rolfsii. Southern blight, southern stem rot. **2**:FL(35,82), NC(1303), NJ(82).

Balsaminaceae

Impatiens L.

1. *I. balsamina* L. [garden balsam, rose balsam]
2. *I. capensis* Meerb. (=*I. biflora* Walter, *I. fulva* Nutt.) [jewelweed, lady's-earrings, spotted touch-me-not]
3. *I. noli-tangere* L. [touch-me-not]
4. *I. oliveri* Wm. Wats. (Kew)
5. *I. pallida* Nutt. [pale snapdragon, pale touch-me-not]
6. *I. wallerana* J. D. Hook. (=*I. sultanii* J. D. Hook.) [busy-lizzy, patience-plant, patient-lucy, sultana, Zanzibar balsam]
7. *Impatiens* sp.

Chytridiomycetes
Synchytrium impatientis. Galls of leaf blades, petioles and stems. **2**:LA(625). **5**:LA(625).
Oomycetes
Phytophthora sp. Root rot. **7**:FL(35).
Plasmopara obducens. Downy mildew. **2**:NE states(82), IN(82), MN(82,2874), MS(2720). **3**:MT(3202). **5**:E states(82), MO(82), WI(82,1330). **7**:MN(2874), WV(3907).
Pythium paroecandrum. Rootlet rot. **5**:MD(82).
Pythium sp. Damping-off, root rot. **1**:FL(35). **6**:WA(82,3202). **7**:FL(35), HI(2903).
Ascomycotina
Crocicreas cyathoideum. On stems. **2**:NY(505). **5**:NY(505). **7**:ID(505), MI(505), NY(505), VA(505).
Hymenoscyphus scutula. **2**:GA(1438).
Leptosphaeria aulica. **2**:GA(1438).
L. conoidea. **2**:NY(3249).
Metasphaeria sp. **2**:GA(1438).
Mycosphaerella impatiens. On leaves. **2**:IA(82), NY(82), VT(82), WI(82,1330). **5**:IA(82), VA(82). **7**:WV(3907).
Phialea sp. **2**:IN(647).
Basidiomycotina - Uredinales
Puccinia argentata. Heteroecious: telial host. **2**:E states(82), GA(1439), IA(82), MN(82,2874), MS(82,2720), NC(1303,1307,3944). **5**:IA(82), IL(82), IN(82), MS(2720), NY(82), TN(82).
P. recondita. Heteroecious: aecial host. **3**:WA(3202). **5**:Range of host(82), GA(1439), MN(2874), MS(2720), NC(1303), SD(2243), WV(3907). **7**:FL(35), MN(2874).
Basidiomycotina - Other
Armillaria mellea. Root rot. **4**:CA(2893,2898).
Deuteromycotina - Hyphomycetes
Alternaria sp. Leaf spot. **1**:FL(35).
Botrytis cinerea. Gray mold. **7**:FL(35).
Botrytis sp. **2**:WI(1343).
Bullera alba. On rusted leaves. **7**:KS(2686).
Cercospora fukushiana. Leaf blight, leaf spot. **1**:FL(35), HI(2899,2903), KS(82), WI(1357). **7**:FL(35).
Cristulariella moricola. Leaf spot. **5**:WV(3680,3907).
Fusarium sp. Root rot. **1**:FL(35).
Periconia byssoides. **2**:DE(3543).
Ramularia impatientis. Leaf spot. **2**:NE states(82), MN(82,2874), NE(82). **3**:WA(3202). **5**:NE states(82), MN(82,2874), NE(82). **7**:MN(2874).
Stemphylium botryosum. Leaf spot. **1**:NJ(1817).
Torula herbarum. **2**:DE(3543).
Verticillium albo-atrum. Wilt. **1**:NY(82).
Deuteromycotina - Coelomycetes
Discosia maculicola. Leaf spot. **2**:NY(82).
Gloeosporium impatientis. Leaf spot. **2**:IL(82), IN(82), WV(3907).
Phomopsis impatientis. On stems. **2**:NY(82).
Phyllosticta sp. Leaf spot. **1**:WI(1357). **6**:NJ(82).
Septoria noli-tangeris. Leaf spot. **1**:OH(82). **2**:NE states(82), IL(82), MN(2874), WI(82). **5**:IA(82), WI(1333), WV(82,3907). **7**:WI(1344).
Deuteromycotina - Other
Rhizoctonia solani. Damping-off, root rot. **1**:FL(35,82). **7**:FL(35).
Rhizoctonia sp. Root rot. **1**:HI(2903).
Sclerotium rolfsii. Southern blight, stem rot. **6**:IL(82). **7**:HI(2903).

Basellaceae

Anredera Juss.

(= *Boussingaultia* Kunth.)
1. *Anredera* sp. (=*Boussingaultia sp.*)

Ascomycotina
Schizoxylon compositum. **1**:LA(3225).

Basella L.

1. *B. alba* L. (=*B. rubra* L.) [country-spinach, Indian spinach, Malabar-spinach, vine-spinach]

Deuteromycotina - Hyphomycetes
Alternaria sp. Leaf spot. **1**:FL(35).
Cercospora sp. Leaf spot. **1**:HI(2903).

Begoniaceae

Begonia L.

1. *B. cucullata* Willd.
2. *B. cucullata* Willd. var. *hookeri* (A. DC.) L. B. Sm. & Schub. (=*B. semperflorens* Link & Otto)
3. *B. maculata* Raddi
4. *B.* × *rex-cultorum* L. H. Bailey. A group of interrelated cultivars of complex hybrid origin derived mostly from *B. rex* and related rhizomatous spp. [beefsteak-geranium, rex begonia]
5. *B.* × *tuberhybrida* Voss. A group of cultivars of complex hybrid origin derived through hybridization and selection from several Andean spp. [hybrid tuberous begonia]
6. *Begonia* sp.

Oomycetes
Phytophthora sp. Root rot. **4**:FL(35). **6**:FL(35).
Pythium debaryanum. Damping-off, leaf, root and stem rot. **2**:MO(82), NY(82). **5**:MO(82), NY(82). **6**:RI(82).
P. intermedium. Damping-off, leaf, root and stem rot. **5**:CA(82), NJ(82), NY(82).
P. irregulare. Root rot. **6**:FL(35).
P. splendens. Damping-off, leaf, root and stem rot. **2**:MO(82). **4**:FL(35). **6**:FL(35).
P. ultimum. Damping-off, leaf, root and stem rot. **1**:FL(35). **2**:CA(82), MO(82), NJ(82), NY(82). **4**:FL(35). **5**:CA(82), MO(82), NJ(82), NY(82).
Pythium sp. Crown, root and stem rot. **6**:CO(49), HI(2903), NC(1303), UT(3620), WA(3202).
Ascomycotina
Botryosphaeria sp. Leaf spot. **6**:FL(35).
Erysiphe cichoracearum. Powdery mildew. **4**:FL(35), HI(2899). **5**:CA(82). **6**:CO(49), FL(35), NH(3620), UT(3620).
E. polygoni. Powdery mildew. **3**:CA(1155).
Glomerella cingulata. Anthracnose. **4**:FL(35). **6**:FL(35), WA(3202).
Sclerotinia sclerotiorum. Stem rot. **5**:CA(82).
Talaromyces bacillisporus. Leaf spot. **6**:NY(82).
Basidiomycotina - Other
Armillaria mellea. Root rot. **5**:CA(82).

Deuteromycotina - Hyphomycetes
Alternaria sp. Leaf spot. **6:**FL(35), NC(1303).
Botrytis cinerea. Gray mold. **1:**FL(35). **2:**Unk.(82), NC(1303). **4:**Unk.(82), FL(35), NC(1303). **6:**Unk.(82), CO(49), FL(35), WA(3202).
Cercospora begoniae. Gray mold blight, leaf blight, blotch. **4:**FL(35). **6:**FL(35).
Cercospora sp. Leaf spot. **6:**MS(2720).
Corynespora cassiicola. Leaf spot. **6:**FL(35).
Fusarium sp. Basal rot, stem rot. **6:**FL(35), NC(1303).
Helminthosporium sp. Leaf spot. **4:**FL(35). **6:**FL(35).
Myrothecium roridum. Leaf spot. **4:**FL(35). **6:**FL(35).
Oidium sp. Powdery mildew. **4:**FL(35), HI(2903). **6:**FL(35,82), NC(82,1303).
Septonema sp. Leaf spot. **6:**FL(35).
Stemphylium sp. Leaf spot. **6:**FL(35).
Thielaviopsis basicola. Root rot. **2:**MA(82), MD(1867), OH(82).
Verticillium albo-atrum. Wilt. **6:**CT(82), NY(82).
Deuteromycotina - Coelomycetes
Colletotrichum sp. Anthracnose. **4:**FL(35). **6:**FL(35).
Macrophoma sp. Leaf spot. **6:**FL(35).
Macrophomina phaseolina. Leaf necrosis. **6:**FL(35).
Phomopsis sp. Leaf spot. **6:**FL(35).
Phyllosticta sp. Leaf spot. **4:**NJ(82). **6:**FL(35), NJ(82), PA(82).
Sphaeropsis begoniicola. On leaves. **6:**AL(82).
Deuteromycotina - Other
Rhizoctonia solani. Damping-off, leaf and root rot. **1:**FL(35). **2:**Range of host(82). **4:**Range of host(82), FL(35). **5:**Range of host(82). **6:**Range of host(82), FL(35), MS(2720).
Rhizoctonia sp. Leaf, root and stem rot. **6:**CO(49), HI(2903), NC(1303).
Sclerotium rolfsii. Stem rot. **6:**IL(82), NC(1303).

Berberidaceae

Achlys DC.

1. *A. triphylla* (Sm.) DC. [deerfoot, vanillaleaf]

Deuteromycotina - Coelomycetes
Ascochyta achlyicola. Leaf spot. **1:**WA(82,3202).
Stagonospora achlydis. Leaf spot. **1:**CA(81), OR(82,3202,3421).

Berberis L.

1. *B. aquifolium* Pursh (=*Mahonia aquifolium* (Pursh) Nutt.) [blue barberry, holly barberry, holly mahonia, mountain-grape, Oregon-grape]
2. *B. atrocarpa* C. K. Schneid.
3. *B. bealei* Fortune (=*Mahonia bealei* (Fortune) Carrière)
4. *B. canadensis* Mill. [Allegheny barberry, American barberry]
5. *B. dictyota* Jeps. (=*Mahonia dictyota* (Jeps.) Fedde)
6. *B. fendleri* A. Gray [Colorado barberry]
7. *B. fortunei* Lindl. (=*Mahonia fortunei* (Lindl.) Fedde)
8. *B. fremontii* Torr. (=*Mahonia fremontii* (Torr.) Fedde) [Fremont mahonia]
9. *B. haematocarpa* Woot. (=*Mahonia haematocarpa* (Woot.) Fedde) [red barberry]
10. *B.* × *mentorensis* L. Ames. *B. julianae* × *B. thunbergii.*
11. *B. nervosa* Pursh (=*Mahonia nervosa* (Pursh) Nutt.) [Cascades mahonia]
12. *B. pinnata* Lag. (=*Mahonia pinnata* (Lag.) Fedde) [cluster mahonia]
13. *B. piperana* (Abrams) McMinn (=*Mahonia piperana* Abrams)
14. *B. pumila* Greene (=*Mahonia pumila* (Greene) Fedde)
15. *B. repens* Lindl. (=*Mahonia repens* (Lindl.) G. Don) [creeping barberry]
16. *B. thunbergii* DC. [Japanese barberry]
17. *B. trifoliolata* Moric. (=*Mahonia trifoliata* (Moric.) Fedde)
18. *B. vulgaris* L. [common barberry, European barberry, jaundice-berry, piprage]
19. *Berberis* sp.

Oomycetes
Pythium debaryanum. Damping-off. **19:**CA(82).
P. splendens. Root rot. **3:**FL(35).
Ascomycotina
Botryosphaeria dothidea. Canker, dieback. **16:**FL(35). **19:**FL(35).
B. ribis. Canker, dieback. **18:**FL(82).
Coccomyces coronatus. On leaves. **1:**WA(3202). **11:**OR(3202,3229), WA(3202,3229). **12:**CA(82). **15:**ID(3202), WA(3202).
C. dentatus. On leaves. **11:**OR(638). **15:**ID(3229).
Cucurbitaria berberidis. On limbs and twigs. **1:**CA(82), OR(82,3202). **18:**ME(82), NY(82), SC(82). **19:**WA(82).
Diaporthe detrusa. On limbs. **18:**MA(82). **19:**MD(82).
Dothidea berberidis. On limbs. **1:**OR(82,3202). **18:**MA(82), WA(3202). **19:**WA(82).
Eutypella aequilinearis. **4:**NJ(2925).
Hymenoscyphus caudatus. **15:**ID(3202).
Hypoxylon serpens. Endophyte. **11:**OR(2786).
Laestadia berberidis. Leaf spot. **1:**MD(82).
Leptosphaeria berberidis. Canker, leaf blotch. **1:**ID(82), MD(82). **18:**IA(82).
Mycosphaerella berberidis. On twigs. **11:**WA(82,3202). **16:**NY(82).
Phyllactinia guttata. Powdery mildew. **18:**MA(82), VT(82).
Thyridium vestitum. **16:**IA(1226).
Togninia minima. On limbs. **18:**MA(190).
Basidiomycotina - Uredinales
Cumminsiella mirabilissima. Autoecious. **2:**OR(218,3202). **5:**CA(218). **7:**CA(2300). **11:**CA(81,82,218), ID(82,3202), OR(2300), WA(82,3202). **12:**CA(81,82,218,2300). **13:**CA(81,218,638,642). **14:**CA(218), OR(218,3202). **15:**W states(82), W states(218), AZ(1195,1203,3379,3981), CA(82), CO(2300), ID(2300,3202), MT(638,1195,3202), SD(2243), UT(218,1195), WA(82,3202), WY(1195,2300,2924,3379). **19:**OR(82).
C. texana. Autoecious. **17:**TX(218,742,2300).
C. wootoniana. Autoecious. **8:**AZ(218,1203,3981). **9:**AZ(218,1203,3981), NM(218,2300), TX(742,2300).
Puccinia brachypodii var. *arrhenatheri.* **1:**ID(82,3202), OR(82,3202), WA(82,3202). **6:**CO(82), NM(82). **15:**W states(82), ID(82,3202), MT(3202), WY(1195).
P. graminis. Heteroecious: aecial host. **4:**MA(82), NY(82), VA(82), WV(82). **6:**CO(82), MN(2874), NM(82). **12:**CA(82). **18:**Range of host(82), ID(3202), MN(2874), SD(2243), WA(3202). **19:**CO(49), FL(35).
P. montanensis. Heteroecious: aecial host. **6:**CO(82), NM(82). **15:**WY(1195). **19:**CO(49).
P. oxalidis. Heteroecious: aecial host. **15:**NM(82), TX(82).
P. pygmaea. Heteroecious: aecial host. **15:**AZ(3981), CO(1195).
Basidiomycotina - Other
Marasmius copelandii. **11:**WA(2932).
M. salalis. **11:**WA(855).
Phellinus punctatus. Heart rot. **18:**MD(82).
Deuteromycotina - Hyphomycetes
Botrytis cinerea. Gray mold on flowers or leaves. **18:**CA(82). **19:**MO(82).
Cercospora boutelouae. Leaf spot. **18:**OK(2873).
Cercospora sp. Leaf spot. **16:**FL(35). **19:**FL(35).
Cladosporium sp. **18:**WA(3202).
Corynespora sp. Leaf spot. **19:**FL(35).
Cylindrocladium ellipticum. Leaf necrosis. **3:**FL(35).
C. scoparium. Leaf spot. **3:**FL(35).
Kabatiella berberidis. Leaf spot. **11:**WA(3202). **15:**ID(3202). **18:**MN(2874).
Oidium sp. Powdery mildew. **1:**IL(312).
Phymatotrichopsis omnivora. Root rot. **4:**TX(82). **16:**TX(82). **18:**TX(82).
Verticillium albo-atrum. Wilt. **16:**E states(82), IL(82,992,1558), MA(1558), MI(82), NJ(1558), OR(3202). **18:**CT(82).
Deuteromycotina - Coelomycetes
Coniothyrium sp. Leaf spot. **1:**OK(2872).
Cytoplea insitiva. On stems and twigs. **18:**NE(82), WI(82,1333).
Diplodia berberidina. On twigs. **18:**NY(82).
Pestalotia pestalozzioides. **19:**MA(1393).
Pestalotia sp. On leaves. **16:**NC(1303).
Phomopsis berberina. On berries and twigs. **16:**MA(82). **18:**IA(82).
Phomopsis sp. Stem gall, leaf spot. **1:**OR(3202). **3:**FL(35). **11:**WA(82). **19:**NY(82).
Phyllosticta berberidis. Leaf spot. **15:**WA(82,3202). **18:**KY(82). **10:**OK(2872).
P. japonica. Leaf spot. **16:**AK(82,526), OK(2872).
P. mahoniicola. **1:**AL(82). **11:**OR(3202,4007), WA(82,3202).

15:ID(3202).
Phyllosticta sp. Leaf spot. **1:**NJ(82), OK(2872). **3:**FL(35). **16:**WI(1343). **19:**FL(35).
Pyrenochaeta berberidis. On limbs. **1:**CA(82).
Stagonospora berberidina. On leaves. **18:**NY(82).
Deuteromycotina - Other
Rhizoctonia solani. Root rot. **18:**CT(82).
Sclerotium rolfsii. Southern blight. **19:**FL(35).

Caulophyllum Michx.

1. *C. thalictroides* (L.) Michx.

Ascomycotina
Laetinaevia caulophylli. On stems. **1:**NY(82).
Deuteromycotina - Hyphomycetes
Botrytis sp. Leaf blight. **1:**NJ(82), NY(82).
Cercospora caulophylli. Leaf spot. **1:**E states(82), MO(82), MS(2720), WI(82).
Cylindrocephalum sp. On bark, leaves and wood. **1:**WI(1334,1340).
Drechslera gigantea. **1:**IL(650).
Deuteromycotina - Coelomycetes
Colletotrichum dematium. **1:**IL(650).
C. hysteriiformis. On stems. **1:**NY(82), WI(82,1329,1342).
Phyllosticta sp. Leaf spot. **1:**WI(1360).

Jeffersonia Barton

1. *J. diphylla* (L.) Pers. [twinleaf]

Deuteromycotina - Coelomycetes
Colletotrichum sp. On leaves. **1:**WI(1358).

Mahonia Nutt.

1. *Mahonia* sp.

Oomycetes
Pythium sp. Root rot. **1:**FL(35).
Ascomycotina
Leptosphaeria berberidis. Leaf blotch. **1:**FL(35).
Nectriella pironii. Stem gall. **1:**FL(35).
Basidiomycotina - Uredinales
Cumminsiella texana. Autoecious. **1:**Unk.(82).
C. wootoniana. Autoecious. **1:**Unk.(82).
Puccinia graminis. Heteroecious: aecial host. **1:**MI(82), ND(82).
Deuteromycotina - Hyphomycetes
Alternaria sp. Leaf spot. **1:**FL(35).
Cercospora sp. Leaf spot. **1:**LA(82).
Deuteromycotina - Coelomycetes
Macrophoma sp. Leaf spot. **1:**FL(35), MS(2720).
Phomopsis sp. Stem galls. **1:**FL(35).
Phyllosticta japonica. **1:**CT(82).
P. mahoniicola. Leaf spot. **1:**AL(82), MS(2720), NC(1303), NY(64).
Phyllosticta sp. Leaf spot. **1:**FL(35).

Nandina Thunb.

1. *N. domestica* Thunb. [heavenly-bamboo, sacred-bamboo]

Oomycetes
Phytophthora sp. Root rot. **1:**FL(35).
Pythium splendens. Root rot. **1:**FL(35).
Pythium sp. Root rot. **1:**FL(35).
Ascomycotina
Glomerella cingulata. Anthracnose. **1:**OK(2872), TX(82).
Basidiomycotina - Other
Ceriporia xylostromatoides. **1:**AZ(1202).
Deuteromycotina - Hyphomycetes
Alternaria sp. Leaf spot. **1:**FL(35).
Cercospora nandinae. Leaf spot. **1:**AL(82), FL(35), OK(2872), SC(82).
Cylindrocladium sp. Leaf spot. **1:**FL(35).
Phymatotrichopsis omnivora. Root rot. **1:**TX(82).
Deuteromycotina - Coelomycetes
Macrophoma sp. Leaf spot. **1:**FL(35).
Phyllosticta nandinae. Leaf spot. **1:**OK(2872).
Phyllosticta sp. Leaf spot. **1:**FL(35).
Deuteromycotina - Other
Rhizoctonia solani. Leaf rot. **1:**FL(35).

Podophyllum L.

1. *P. peltatum* L. [mandrake, mayapple, raccoonberry, wild-jalap, wild-lemon]

Ascomycotina
Discohainesia oenotherae. Leaf spot. **1:**VA(82).
Glomerella cingulata. Leaf spot. **1:**DE(82), TX(82).
Septotinia podophyllina. Leaf blight. **1:**DE(82), MD(82), MO(82), NJ(82), NY(82,1960), VA(82), WV(82,3907).
Basidiomycotina - Uredinales
Puccinia podophylli. Autoecious. **1:**Range of host(82), GA(1439), MS(2720), NC(1303,3944), OK(2872), WI(1361), WV(3907).
Deuteromycotina - Hyphomycetes
Botrytis cinerea. Gray mold. **1:**NJ(82).
Cercospora podophylli. Leaf spot. **1:**IL(82,581).
Deuteromycotina - Coelomycetes
Phyllosticta podophylli. Leaf spot. **1:**E states(82), AR(82), MD(2783), MS(2720), OH(2), OK(2872), WI(82,1329).
P. podophyllina. **1:**IL(82).
Phyllosticta sp. **1:**WI(1356).
Septoria podophyllina. Leaf spot. **1:**E states(82), MO(82), MS(82,2720), WI(82), WV(3907).
Septotis podophyllina. Leaf blight. **1:**MO(965,3551), NJ(3551), NY(3551), PA(3551).
Deuteromycotina - Other
Rhizoctonia sp. Leaf blight. **1:**MO(82).

Vancouveria C. Morr. & Decne.

1. *V. hexandra* (Hook.) C. Morr. & Decne.
2. *V. planipetala* Calloni (=*V. parviflora* Greene) [inside-out-flower]

Ascomycotina
Phragmodothis berberidis. Leaf spot. **1:**CA(81). **2:**CA(81,82).
Deuteromycotina - Hyphomycetes
Ramularia vancouveriae. Leaf spot. **1:**CA(82), OR(82,3202,3421), WA(3202). **2:**CA(81).

Betulaceae

Alnus Mill.

1. *A. crispa* (Aiton) Pursh [American green alder, green alder, mountain alder]
2. *A. crispa* (Aiton) Pursh var. *mollis* (Fernald) Fernald
3. *A. glutinosa* (L.) Gaertn. (=*A. vulgaris* Pers.) [black alder, European alder]
4. *A. incana* (L.) Moench [speckled alder, white alder]
5. *A. japonica* (Thunb.) Steud. [Japanese alder]
6. *A. oblongifolia* Torr. [New Mexican alder]
7. *A. pumila* Steud.
8. *A. rhombifolia* Nutt. [Sierra alder, white alder]
9. *A. rubra* Bong. (=*A. oregona* Nutt.) [Oregon alder, red alder]
10. *A. rugosa* (Du Roi) Spreng. [hazel alder, smooth alder, speckled alder]
11. *A. serrulata* (Aiton) Willd. [hazel alder]
12. *A. sinuata* (Regel) Rydb. (=*A. fruticosa* Rupr. var. *sinuata* (Regel) Rydb.) [green alder, Sitka alder]
13. *A. tenuifolia* Nutt. [mountain alder, thinleaf alder]
14. *A. viridis* (Chaix) DC. [European green alder]
15. *Alnus* sp.

Zygomycotina
Mortierella ramanniana var. *angulispora*. On leaf surface. **9:**OR(3042).
Ascomycotina
Acrospermum compressum. **10:**GA(1438).
Atopospora betulina. **15:**AK(526).
Bertia moriformis. On wood. **4:**ID(3202). **9:**OR(3202). **15:**WA(684,3202).
Biscogniauxia nummularia. **9:**CA(81).
Bisporella citrina. **4:**ID(3202). **9:**OR(3202), WA(3202). **12:**AK(3431), ID(3202).
Botryosphaeria obtusa. **10:**SC(82), VA(82).
B. ribis. On seedlings. **3:**FL(2505).
Byssosphaeria alnea. On limbs. **12:**AK(82). **15:**NY(82,189).
Calycellina leucella. **12:**AK(3431).
Camarops microspora. On limbs. **4:**ID(3202), MI(2561,2862), NY(2561). **15:**AK(526,2561), ID(2561), MA(2561), ME(2561), NH(2561), NY(2561).

Capitotricha bicolor. On dead twigs. **9:**WA(3202). **12:**AK(3431). **13:**CA(661). **15:**WA(3202).
Cenangium sp. **12:**WA(3202).
Ceratocystis sp. **9:**WA(3202).
Chaetomium funicola. On leaf surface. **9:**OR(3042).
C. globosum. On leaf surface. **9:**OR(3042).
Chlorociboria aeruginosa. **9:**OR(3202). **12:**AK(3431). **15:**WA(3202).
Ciboria alni. On overwintered seeds. **9:**OR(82,3202).
C. caucus. On fallen catkins. **9:**OR(82,3202). **15:**ID(3202).
Cistella xylita. **9:**WA(3202).
Coniochaeta ligniaria. **4:**ID(3202). **15:**OR(3202).
C. pulveracea. **9:**WA(3202). **13:**CA(661).
Cryptodiaporthe pyrrhocystis. **12:**ID(3202).
Cryptosporella paucispora. On twigs. **15:**NY(2949).
Cucurbitaria conglobata. **12:**AK(526). **15:**AK(526).
C. fraxini. **10:**GA(1438).
Daldinia concentrica. Wood rot. **4:**MT(3202). **10:**ME(82), MI(82), NH(82). **15:**ID(3202), MI(2862).
D. occidentalis. On wood. **4:**ID(3202), WA(3202). **8:**ID(82), MT(82), WA(82,3202). **9:**CA(81). **13:**ID(82), MT(82), WA(82). **15:**MT(3202).
D. vernicosa. **4:**ID(3202), MT(3202), WA(3202). **13:**ID(82), MT(82), WA(82).
Dasyscyphella dryinus. **4:**WA(3202).
Dasyscyphus scabrovillosus. **12:**WA(3202).
Diatrype bullata. **4:**ID(3202). **9:**WA(3202). **15:**ID(3202), OR(3202).
D. disciformis. **15:**AK(526).
D. flavovirens. **9:**WA(3202). **15:**ID(3202).
D. stigma. On limbs. **4:**MN(2874). **15:**AK(526), ID(3202).
Diatrypella decorata. **9:**OR(3202). **11:**SC(634). **12:**OR(3202). **15:**WA(3202,3815).
D. favacea. On limbs. **4:**MN(2874). **9:**WA(1248). **10:**GA(1438). **15:**ID(3202), WA(3202).
D. placenta. On dead limbs. **4:**MT(3202), OR(3202). **10:**GA(1438), NC(3944). **13:**CA(661).
D. rimosa. **15:**OR(3202,3209).
Didymosphaeria nana. **15:**AK(526).
D. nana var. *brachyspora*. On leaves. **12:**AK(82).
D. oregonensis. Branch and trunk canker. **4:**ID(3202), OR(3202). **9:**ID(82), OR(82,3202), WA(82,3202). **12:**ID(82), OR(82,3202), WA(82). **13:**ID(82), OR(82), WA(82). **15:**AK(526).
Ditopellopsis alni. On fallen leaves. **10:**GA(82,1438).
Dothidea puccinioides. **10:**GA(1438).
Dothivalsaria megalospora. **9:**OR(870).
Emericella nidulans. On leaf surface. **9:**OR(3042).
Encoelia furfuracea. On twigs. **4:**ID(3202), ME(82), MI(82), MT(3202), NY(82). **8:**CA(82). **9:**WA(82,3202). **10:**ME(82), MI(82), NY(82). **13:**ID(82), MT(82).
Endoxylina moroides. **15:**IA(1226).
Erysiphe aggregata. Powdery mildew on fertile catkins. **4:**NE states(82), MN(82,2874). **8:**CA(82), WA(82). **9:**CA(81). **10:**NE states(82), MI(82). **15:**WV(3907).
Eutypella cerviculata. Powdery mildew on fertile catkins. **4:**CO(2925), ID(3202), MN(2874), WA(3202). **9:**Range of host(82), CA(81), OR(3202), WA(82,3202). **10:**GA(1438), MS(2720), NC(3944). **11:**IA(3629). **12:**Range of host(82), AK(526), ID(3202). **13:**Range of host(82), CO(82), UT(2925). **15:**IA(3629), ID(1249), MS(2720), NY(2925), WA(3202,3815).
E. glandulosa. On bark of dead limbs and trunks. **15:**GA(82), TN(82).
E. stellulata. **10:**NC(3944). **15:**AK(526), ID(3202).
Eutypella sp. **12:**OR(3202).
Gelasinospora tetrasperma. On leaf surface. **9:**OR(3042).
Gloniella sp. **8:**WA(3202).
Glyphium corrugatum. **12:**OR(3202), WA(3202).
G. elatum. **9:**WA(3202).
Gnomonia alni. Leaf spot. **8:**CA(81,82), OR(81,3202). **9:**CA(82), OR(82,3202). **13:**TX(82).
G. gnomon. **10:**GA(1438).
G. setacea. **10:**GA(1438).
Gnomoniella tubiformis. On dead leaves. **1:**NC(82). **4:**E states(82), OR(3202), WA(3202), WI(82). **9:**OR(82,3202). **10:**E states(82), GA(1438), WI(82). **11:**WV(3907). **12:**AK(82,526). **13:**CA(81), CO(82), OR(82). **15:**AK(526).
Helotium fastidiosum. Leaf spot. **15:**OR(3202).
Hyalinia crenato-marginata. **9:**WA(3202).
Hyaloscypha alniseda. **15:**WA(3202).
H. hyalina. **15:**WA(3202).
Hymenoscyphus alniellus. **15:**OR(3202).
H. calyculus. On twigs. **4:**WA(3202). **9:**WA(3202). **12:**AK(3431), ID(3202). **13:**CA(661). **15:**ID(3202).
H. caudatus. **12:**AK(3431).
H. epiphyllus. **15:**WA(3202).
H. imberbis. **15:**WA(3202).
Hypocrea patella. **12:**AK(3431).
Hypoxylon bartholomaei. **9:**WA(3202).
H. fragiforme. On wood. **12:**AK(526,1904).
H. fuscum. On wood. **4:**ID(3202), MT(3202), WA(3202). **9:**CA(81), ID(3202), WA(1811,3202). **10:**GA(1438), NC(3944). **11:**SC(634), WV(3907). **12:**WA(3202). **15:**Unk.(82), AK(526), MI(2862), MT(3202), OR(3202).
H. hypophlaeum. On wood. **10:**GA(1438).
H. mammatum. On limbs and trunks. **4:**Unk.(82). **13:**CO(82).
H. mediterraneum. On dead limbs and trunks, occasionally on living trees. **9:**WA(3202).
H. mediterraneum var. *microspora*. **9:**WA(1811).
H. multiforme. On dead limbs and trunks. **4:**ID(3202). **8:**WA(3202). **9:**CA(81), WA(3202). **12:**AK(3431). **15:**Unk.(82), AK(526), ID(3202), OR(3202).
H. rubiginosum. **10:**GA(1438), NC(3944). **15:**WA(3202,3815).
H. vogesiacum. **4:**ID(3202).
Hysterographium fraxini. **15:**ID(3202).
Karstenula alnicola. **15:**MD(2780).
Lachnum alneum. **15:**WA(3202).
L. hyalinellum. **15:**WA(3202).
L. virgineum. On twigs. **10:**GA(1438). **13:**CA(661).
Lanzia rufescens. On dead leaves. **9:**OR(82,3202). **15:**OR(911,3202), WA(911).
Lasiosphaeria chrysentera. On leaves. **4:**ID(3202).
L. strigosa. **15:**OR(3202).
Lopadostoma gastrinum. **15:**OR(3202).
Lophiosphaera sp. **12:**OR(3202).
Melanconis alni. **12:**ID(3202). **15:**CO(649).
M. marginalis. On limbs. **1:**NE states(82), MI(1759), WI(82). **4:**NE states(82), MN(2874), WI(82). **9:**CA(82), OR(82). **10:**NE states(82), MS(2720), WI(82). **12:**AK(82,526), ID(82). **13:**CO(82), ID(1759). **14:**NY(1759).
Melanomma pulvis-pyrius. Twig blight. **8:**ID(3202). **9:**WA(3202). **15:**AK(526), WA(3202,3815).
Melanomma sp. **4:**ID(3202).
Microsphaera penicillata. Powdery mildew. **4:**E states(82), MT(3202), TX(82), WI(82). **9:**CA(81), OR(82,3202), WA(82,3202). **10:**E states(82), GA(1438), TX(82), WI(82). **12:**AK(82,526), MT(82), WA(3202). **13:**TX(82), UT(82). **15:**MN(2874).
Moellerodiscus tenuistipes. **15:**WA(3202).
Mollisia amenticola. **15:**WA(3202).
M. benesuada. **4:**WA(3202).
M. caespiticia. **15:**WA(3202).
Mycosphaerella alnicola. On fallen leaves. **1:**NY(82).
M. borealis. On leaves. **15:**NH(179).
M. harthensis. **4:**NY(82).
Mycosphaerella sp. On leaves. **4:**ID(3202). **6:**CA(82). **8:**CA(82). **9:**CA(82). **12:**AK(82), CA(82). **13:**CA(82).
Nectria cinnabarina. On twigs. **8:**WA(82,3202). **12:**AK(82,526). **15:**AK(526).
N. coccinea. On limbs. **4:**NY(82), PA(82).
N. ditissima. **9:**OR(3202), WA(3202).
N. episphaeria. **4:**ID(3202). **15:**AK(526), WA(3202).
N. galligena. Trunk canker. **4:**NH(82). **9:**OR(82).
N. pithoides. **9:**WA(82,3202).
Nectria sp. **12:**WA(3202).
Neurospora sitophila. On leaf surface. **9:**OR(3042).
Nitschkia grevillii. **13:**CA(661).
Orbilia paradoxa. **15:**WA(3202).
Patinella flavobrunnea. **4:**WA(3202).
Perrotia flammea. **13:**CA(661).
Pezicula alni. On dead limbs. **4:**FL(2874), MI(82), MN(82), NC(82). **10:**MI(82), MN(82,2874), NC(82). **15:**MN(2874).
Pezicula sp. **10:**GA(1438). **15:**ID(3202).
Phaeosphaerella sp. **10:**GA(1438).
Phragmodiaporthe tiliacea. **9:**CA(81).

Phyllactinia guttata. Powdery mildew. **4:**NE states(82), FL(2874), ID(3202), MT(3202), WA(3202), WI(82). **8:**ID(3202). **9:**WA(3202). **10:**AL(82), FL(35), OH(82,950), SC(82). **13:**NM(1195), WY(1195).
Physalospora abdita. On limbs. **10:**NC(82).
Plagiostoma alneum. On leaves. **9:**OR(82,3202), WA(82,3202).
Platychora alni. Black leaf spot. **1:**NY(82). **15:**AK(526).
Propolomyces versicolor. **10:**GA(1438).
Pseudociboria umbrina. On fallen leaves. **15:**WA(910,3202).
Pseudovalsella thelebola. On limbs. **4:**GA(82), MT(3202), NY(82), PA(82). **9:**CA(82), WA(82,3202). **10:**GA(82,1438), NY(82), PA(82). **13:**MT(82). **15:**AK(526).
Pyrenopeziza foliicola. **10:**GA(1438).
Rhytisma salicinum. **12:**AK(3431).
Rosellinia aquila. **9:**CA(81).
Scopinella caulincola. **15:**OR(648,3202).
Scorias spongiosa. Sooty mold. **4:**E states(82), WV(82). **10:**GA(1438), NC(3944), WV(3907).
Scutellinia hirta. Sooty mold. **4:**ID(3202), WA(3202).
S. scutellata. On wood. **9:**WA(3202). **15:**CA(81), WA(3202).
Sphaerella alni-viridis. **15:**AK(526).
Sphaerognomonia carpinea. On dead leaves. **10:**GA(1438).
Stictis radiata. **15:**CA(3225).
Talaromyces luteus. On leaf surface. **9:**OR(3042).
Tapesia fusca. **12:**AK(3431).
T. mollisioides. On branches. **13:**CA(661).
Taphrina alni. Hypertrophy of catkins. **9:**AK(82,526,2454). **10:**GA(1438).
T. japonica. Leaf curl. **5:**CA(3361). **9:**AK(82,526,2454), CA(81,82,2454), OR(82,2454,3202).
T. occidentalis. On cones and trees. **4:**ID(3202), MT(3202), OR(3202), WA(3202). **8:**CA(81,82,2454), OR(82,3202). **9:**CA(82,2454), OR(82,638,2454,3202), WA(3202). **12:**ID(3202), MT(3202), OR(82,2454,3202), WA(3202). **13:**CA(81,82,661), ID(82,638,2454,2457), MT(82,2454), OR(82,2454), WA(82,2454).
T. populina. **10:**NY(3361).
T. robinsoniana. Catkin hypertrophy. **4:**E states(82), AL(285), MN(82,2874), MS(2720), OK(2872), TX(82). **10:**E states(82), AR(82), CT(2454), GA(1438), MA(2454), ME(638,2454), MI(2454), NC(2454), NH(2454), NY(2454,3361), VA(638), VT(2454), WI(1348). **11:**AL(2454), AR(2454), CT(2454), GA(2454), MA(2454), NY(2454), TN(2454), VA(2454). **15:**MA(2454).
T. tosquinetii. Catkin hypertrophy. **1:**NH(82). **2:**NH(2454).
Tympanis alnea. On dead limbs. **4:**ID(3202), MT(3202), NY(82), PA(82), WA(3202). **13:**CA(82), MT(82). **15:**AK(526), ME(2670), MN(2670), ND(2670), NY(2670), WV(2670).
Ustulina deusta. **9:**OR(648,3202).
Valsa ceratosperma. On dead limbs. **6:**Range of host(82). **8:**Range of host(82). **9:**Range of host(82). **10:**GA(1438). **11:**SC(634). **12:**Range of host(82). **13:**Range of host(82). **15:**CO(49).
V. melanodiscus. Canker. **10:**NY(3415). **15:**AK(526), NY(3415).
Valsella alnicola. **9:**CA(81).
V. furva. **12:**WA(3202,3815).
Winterella aurantiaca. **10:**MI(2948).
W. femoralis. On twigs. **4:**GA(82), NJ(82), NY(82), WV(82). **10:**GA(82,1438), MI(2948), NJ(82), NY(82), WV(82,2948,3907). **11:**WV(3907). **15:**NY(2948), VA(2948), WV(2948).
W. suffusa. On twigs. **13:**CO(82). **15:**CO(2948).
Xylaria cornu-damae. **12:**AK(3431).
X. subterranea. **12:**AK(3431).
Xylaria sp. **15:**WA(3202).

Basidiomycotina - Uredinales

Melampsoridium betulinum. Heteroecious: telial host. **8:**CA(82). **9:**CA(81,82).

Basidiomycotina - Other

[*Anomoporia myceliosa*]. **15:**WA(128).
[*Antrodia crassa*]. **9:**WA(3202).
A. heteromorpha. **4:**ID(3202), MT(3202). **9:**ID(3202). **13:**MT(3832). **15:**MT(128), WA(3202).
A. sinuosa. On wood. **9:**OR(3202). **10:**MS(2720).
Antrodiella semisupina. **4:**MT(3202). **13:**MT(3832).
Aporpium caryae. On wood. **15:**WA(3202).
Armillaria mellea. Root rot. **8:**CA(82). **9:**CA(59). **15:**CA(2898).
Arrhytidia sp. **12:**AK(3431).
Athelia decipiens. **6:**AZ(1202).
A. epiphylla. **6:**AZ(1178,1202).
Basidioradulum radula. **4:**WA(3202). **12:**ID(128).
Bjerkandera adusta. On wood. **6:**AZ(1202). **8:**WA(3202). **9:**CA(81), ID(3202), OR(3202), WA(3202). **12:**AK(3431). **15:**Range of host(82), WA(3202).
Botryobasidium pruinatum. **15:**AK(128).
B. subcoronatum. **6:**AZ(1202).
Calocera cornea. **9:**WA(3202).
Cellypha subgelatinosa. **11:**SC(456).
Ceraceomyces tessulatus. **12:**ID(3202).
Ceriporia purpurea. **9:**OR(3202). **15:**WA(128).
C. reticulata. **4:**ID(3202), WA(3202). **6:**AZ(1202). **9:**WA(128). **15:**WA(128).
C. spissa. **9:**CA(128).
C. viridans. **6:**AZ(1202). **9:**OR(3202), WA(3202). **15:**WA(128).
Ceriporiopsis gilvescens. **15:**WA(3202).
C. mucida. **9:**CA(81).
Cerocorticium confluens. **6:**AZ(1202).
Cerrena unicolor. On wood. **4:**ID(3202), ME(82), MT(3202), VA(82), WA(3202). **9:**OR(3202), WA(82,3202). **10:**ME(82), VA(82). **13:**ID(82), MT(82). **15:**ID(3202), MT(3202).
Chondrostereum purpureum. Wood rot. **9:**WA(3202). **10:**ME(543), VT(543). **12:**AK(526,1904). **13:**AK(1904). **15:**Unk.(82), AK(526,543), ID(3202), MT(3202), WA(543,3202).
Clavicorona pyxidata. **15:**CO(884).
Coniophora puteana. Brown cubical rot. **4:**ID(3202). **6:**AZ(1202). **13:**ID(82,128).
Coprinus alnivorus. **15:**WA(314).
Corticium roseum. **6:**AZ(128,1202).
Crepidotus alnicola. On bark. **4:**MI(1544).
C. mollis var. *cystidiosus*. **15:**WA(1544).
C. occidentalis. On wood. **15:**WA(1544).
C. ochraceus. On wood. **15:**OR(1544).
C. pseudoflammeus. On fallen twigs. **15:**WA(1544).
C. submollis. **15:**WA(3202).
C. subverrucisporus var. *megalosporus*. On wood. **15:**CO(1544).
Cyphellopsis anomala. **10:**NC(3944).
C. confusa. **15:**MI(2945).
Cyptotrama asprata. **6:**AZ(1202).
Cystostereum pini-canadense. **15:**OR(3202).
C. pini-canadense subsp. *subabruptum*. **15:**OR(128).
Cytidia flocculenta. **12:**AK(526).
Dacrymyces capitatus. On wood. **10:**MI(3176). **12:**OR(3202).
D. chrysospermus. **9:**ID(3202).
D. stillatus. **9:**WA(3202).
Daedaleopsis confragosa. On wood. **4:**IN(82). **6:**AZ(1202). **15:**ID(3202), OR(3202).
Datronia mollis. On wood. **9:**WA(3202). **13:**AK(1904). **15:**WA(3202).
D. scutellata. On wood. **4:**E states(82), ID(3202), MI(82,2862), MN(2874). **10:**E states(82), MI(82). **12:**AK(3431). **15:**AZ(2180), ID(2180), MN(1197), MT(2180), OR(2180), WA(2180,3202).
D. stereoides. **15:**AK(526).
Dendrocorticium roseocarneum. **4:**VT(128).
Dendrophora albobadia. On wood. **6:**AZ(1202). **15:**Unk.(82).
D. erumpens. **9:**OR(543). **13:**MT(543). **15:**GA(543), MD(543), WA(543).
D. versiformis. **7:**AZ(128). **15:**MA(543), NY(543), PA(543).
Dichomitus campestris. **15:**ID(3202).
Dichostereum effuscatum. **15:**MA(128).
D. pallescens. **6:**AZ(1202). **15:**WA(128,3202).
Diplomitoporus lenis. **4:**ID(3202). **9:**OR(3202).
D. linbladii. **9:**WA(128,3202).
Erythricium laetum. **13:**MT(128).
Exidia glandulosa. On wood. **15:**MI(2862).
E. repanda. **4:**ID(3202).
Exidia sp. **12:**AK(3431).
Exidiopsis grisea. **15:**WA(3853).
[*Fibricium rude*]. **6:**AZ(1202).
Fomes fomentarius. **9:**OR(82,3202), WA(82,3202). **15:**ID(3202).
Fomitopsis cajanderi. **15:**ID(3202).
F. pinicola. **4:**MT(3202). **9:**ID(3202), OR(3202), WA(3202). **13:**MT(3832). **15:**WA(3202).

Fuscocerrena portoricensis. **8:**WA(3202).
Galerina pallidispora. On logs. **15:**WA(3329).
Ganoderma applanatum. On wood. **9:**AK(1904), WA(82,3202). **15:**OR(3202).
Gloeocystidiellum porosum. White rot. **15:**OR(2553), WA(2553).
Gloeodontia columbiensis. On wood. **15:**MT(440).
Gloeophyllum odoratum. **15:**WA(128,3202).
G. sepiarium. On wood. **4:**ID(3202), WA(3202). **13:**ID(82), WA(82).
G. trabeum. On wood. **15:**Range of host(82).
Gloeoporus dichrous. On wood. **1:**Range of host(82). **4:**Range of host(82), ID(3202), MT(3202). **10:**Range of host(82). **15:**AK(2419).
Grandinia arguta. **6:**AZ(1202).
G. crustosa. **4:**ID(3202). **6:**AZ(1202). **12:**AK(3431). **15:**WA(3202).
G. pallidula. **6:**AZ(1202).
G. quercina. **15:**ID(3202).
G. spathulata. **6:**AZ(1202).
Hapalopilus rutilans. White rot. **12:**AK(3431). **15:**Unk.(2681).
Hericium coralloides. On wood. **13:**AK(1904).
Heterobasidion annosum. **4:**WA(3202).
Hydnochaete olivacea. Wood rot. **1:**VA(82). **4:**VA(82). **9:**WA(82,3202). **10:**VA(82).
Hymenochaete agglutinans. Canker. **4:**CT(1312), ID(3202), MI(82), NY(82), PA(82), VA(82). **10:**FL(35). **13:**ID(82,128). **15:**FL(35), NY(128).
H. corrugata. On wood. **4:**ID(3202). **10:**GA(491,1439). **11:**SC(634).
H. pinnatifida. On wood. **10:**GA(491,1439).
H. spreta. **15:**OR(3202).
H. tabacina. On wood. **4:**ID(3202). **9:**WA(3202). **12:**AK(526). **13:**AK(1904), CA(661).
Hyphoderma amoenum. **15:**GA(3315), VT(3315).
H. cristulatum. **6:**AZ(1202). **12:**AK(526).
H. guttuliferum. **6:**AZ(1202).
H. litschaueri. **15:**OR(3202).
[*H. pilosum*]. **6:**AZ(1202).
H. praetermissum. **6:**AZ(1202).
H. puberum. **6:**AZ(1202).
H. sambuci. **15:**ID(3202).
H. setigerum. **12:**AK(526).
Hyphodontia rimosissima. **6:**AZ(1202).
H. stipata. **13:**WA(128).
Hypochnicium bombycinum. **6:**AZ(1202).
H. geogenium. **4:**ID(3202).
H. punctulatum. **6:**AZ(128,1202).
H. sphaerosporum. **6:**AZ(1202).
Hypochnopsis mustialaensis. **13:**ID(128).
Hypsizygus tessellatus. On wood. **15:**WA(2421,3330).
Inonotus obliquus. On wood. **9:**ID(3202). **13:**MT(128). **15:**Range of host(82).
I. radiatus. On wood. **4:**MT(3202). **12:**AK(526). **13:**AK(1904), MT(3832). **15:**Range of host(82), AK(526), OR(3202).
Irpex lacteus. On wood. **4:**ID(3202), MI(2862). **9:**ID(3202), WA(3202). **15:**Range of host(82), ID(3202), OR(3202).
Junghuhnia fimbriatella. **15:**WA(128).
J. nitida. **4:**ME(128). **9:**WA(128,2168,3202). **10:**ME(2168). **15:**ME(2168), NY(128,2168), WA(2168).
Kuehneromyces mutabilis. On wood. **9:**AK(1904), WA(3202). **15:**OR(3202).
Laxitextum bicolor. On wood. **6:**Range of host(82). **15:**Range of host(82), OR(543,3202).
Lentinellus cochleatus. Decay of dead wood. **6:**AZ(1202).
L. flabelliformis. **15:**Unk.(2423).
L. micheneri. **3:**Unk.(2423). **4:**ID(3202), MT(3202), OR(3202), WA(3202). **13:**Unk.(2423).
Lenzites betulina. **4:**ID(3202). **10:**GA(1439).
Leptosporomyces galzinii. **6:**AZ(1202).
Leucogyrophana pinastri. **15:**ID(3202).
Megacollybia platyphylla. **6:**AZ(1202).
Megalocystidium lactescens. **6:**AZ(128,1202).
M. leucoxanthum. **12:**AK(3431).
M. luridum. **9:**ID(3202). **12:**AK(3431).
Merismodes fasciculatus. On limbs and twigs. **9:**OR(3202), WA(3202). **10:**GA(1439). **11:**SC(634). **15:**MA(2945), MI(2945), NY(2456), OR(4007).
M. ochraceus. On bark. **15:**Range of host(82).
Meruliopsis corium. On wood. **6:**AZ(1202). **9:**WA(3202). **10:**GA(491,1439). **12:**AK(3431). **15:**MD(128).
Mycena alnicola. On wood. **15:**WA(3326).
M. haematopoda var. *cuspidata*. On wood. **15:**CO(2450).
M. rugulosiceps. **15:**OR(3202), WA(3202).
Mycoacia uda. **15:**WA(128).
Osteina obducta. **15:**ID(3202).
Oxyporus corticola. **9:**OR(3202).
Panellus longinquus subsp. *pacificus*. On wood. **9:**OR(2120), WA(2120).
P. serotinus. Sapwood rot. **4:**ID(3202). **13:**ID(82).
Peniophora aurantiaca. On wood. **9:**ID(3202). **10:**MS(2720), NC(3944). **15:**AK(526), ID(3202), MT(3202), OR(3202), WA(3202), WI(128).
P. cinerea. **10:**GA(1439). **15:**OR(3202), WA(3202).
P. erikssonii. On dead, still-attached limbs. **15:**AL(128), IN(128), NY(3315), PA(128), WA(128).
P. incarnata. On wood. **10:**GA(491,1439).
Perenniporia amylodextrinoidea. On wood. **15:**AZ(1205).
P. ellipsospora. **6:**AZ(1202).
P. subacida. On wood. **9:**OR(3202), WA(3202). **15:**Range of host(82).
P. tenuis. **6:**AZ(1202), MT(128). **10:**GA(1439). **15:**MT(128).
Phaeomarasmius erinaceus. **15:**WA(2934).
Phanerochaete affinis. **4:**NH(128), NY(128). **12:**AK(526). **15:**WA(3202).
P. avellanea. White rot. **15:**OH(433).
P. burtii. **6:**AZ(128).
P. carnosa. **6:**AZ(1202). **15:**ID(128).
P. ravenelii. On wood. **10:**GA(491,1439). **15:**GA(128).
P. sanguinea. White rot. **15:**AL(433).
P. sordida. White rot. **6:**AZ(1202). **12:**AK(3431). **13:**FL(433), ID(128), WA(128,433). **15:**WA(128).
P. velutina. White rot. **6:**AZ(1202). **13:**ID(433).
Phellinus conchatus. On wood. **4:**ID(3202). **15:**MN(2874).
P. ferreus. **4:**AK(526), ID(3202). **9:**OR(128,3202), WA(128,3202). **12:**AK(526). **15:**Range of host(82), AK(526).
P. ferruginosus. **4:**ID(3202), MT(3202). **9:**OR(3202), WA(3202). **12:**AK(526). **15:**Range of host(82), MT(128).
P. gilvus. On wood. **4:**MT(3202). **6:**AZ(1202). **15:**Range of host(82), MT(3202,3832).
P. igniarius. White heart rot. **4:**ID(3202), MT(3202), OR(3202), VT(82), WA(3202). **8:**ID(82,3202), MT(82), WA(3202). **9:**ID(3202), OR(82,3202), WA(3202). **12:**WA(3202). **13:**AK(1904), ID(82), MT(82,3832), TX(82). **15:**ID(3202).
P. laevigatus. **4:**ID(3202). **13:**ID(128). **15:**ID(128).
P. lundellii. On wood. **15:**E states(1205).
P. nigricans. **12:**AK(526).
P. pini. White pocket heart rot. **9:**OR(82). **13:**ID(82), MT(82).
P. punctatus. **4:**MT(3202). **12:**AK(526). **13:**MT(3832). **15:**Range of host(82).
P. viticola. **4:**ID(3202).
Phlebia coccineo-fulva. **4:**ID(3202).
P. concentrica. **15:**MT(3202), OR(3202), WA(128,3202).
P. ludoviciana. White rot. **15:**WI(2556).
P. merismoides. On wood. **4:**MT(3202). **6:**AZ(1202). **9:**CA(81,643), OR(3202). **12:**WA(3202). **15:**OR(3202,4007).
P. tremellosus. On wood. **11:**WV(3907).
Phlebiella vaga. **6:**AZ(1202).
Pholiota adiposa. Butt rot. **4:**ID(3202), WA(3202). **13:**ID(82), WA(82). **15:**ID(1017), WA(3202).
P. alnicola. Butt rot. **4:**ID(3202). **9:**OR(82,3202). **13:**ID(82).
P. alniphila. On wood. **15:**OR(3202,3332).
P. aurivelloides. On dead wood. **13:**CO(82), NM(82), WY(82).
Pleurotus ostreatus. **15:**OR(3202).
Plicatura nivea. **4:**ID(3202), MT(3202). **13:**CA(661), MT(3832). **15:**AK(526), ID(3202), MI(2862), OR(3202,4007).
Plicaturopsis crispa. **15:**MN(1197).
Pluteus latifolius. On wood. **15:**WA(3202,3336).
Polyporus arcularius. **4:**MT(3202). **8:**ID(3202).
P. badius. **4:**ID(3202). **9:**OR(3202), WA(3202). **12:**AK(3431). **15:**WA(3202).
P. brumalis. **4:**WA(3202). **12:**WA(3202). **15:**ID(3202).
P. melanopus. **4:**ID(3202).
P. squamosus. **4:**ID(3202).

P. varius. On wood. **4:**ID(3202), MT(3202), WA(3202). **9:**ID(3202), WA(3202). **12:**AK(3431). **13:**CA(661). **15:**MI(2862).
Porotheleum fimbriatum. **4:**ID(3202).
Postia caesia. **12:**AK(3431). **15:**WA(3202).
P. stiptica. **6:**AZ(1202).
P. tephroleuca. **6:**AZ(1202).
Psathyrella alboalutacea. On dead wood. **15:**WA(3335).
P. candidissima. **15:**WA(3335).
P. maculata. **15:**WA(3202).
P. naucoria. On logs. **15:**WA(3335).
Pycnoporellus fulgens. **13:**MT(3832).
Pycnoporus cinnabarinus. **9:**WA(3202). **10:**NC(3944). **15:**OR(3202).
Rectipilus fasciculatus. **15:**GA(128).
Resinomycena saccharifera subsp. *kalalochensis*. On wood. **9:**WA(3326).
Schizophyllum commune. Wood rot. **4:**ID(3202). **8:**WA(82,3202). **9:**WA(3202).
Schizopora paradoxa. **4:**ID(3202). **9:**OR(3202), WA(128,3202). **12:**WA(3202). **13:**ID(128). **15:**Range of host(82), WA(128).
Scytinostroma portentosum. **15:**VT(128).
Sistotrema brinkmannii. **13:**ID(128).
Skeletocutis alutacea. **15:**MT(3202), WA(128).
S. nivea. **15:**ID(3202), WA(3202).
Spongipellis delectans. **4:**MT(3202).
Steccherinum alaskense. On wood. **4:**AK(128,2143).
S. fimbriatum. **9:**WA(128). **15:**MT(128).
S. ochraceum. On wood. **4:**ME(82), MI(82), NY(82). **6:**AZ(1202). **9:**CA(81), WA(82,3202). **10:**ME(82), MI(82), NY(82). **12:**AK(3431). **13:**AK(1904). **15:**CA(81), MI(2862).
[*Stereum gausapatum*]. White heart rot. **12:**AK(526,1904).
S. hirsutum. On wood. **4:**WA(3202). **9:**OR(3202), WA(543,3202). **10:**NC(3944), NY(543), PA(543), RI(543). **13:**AK(543,1904), CA(661), CO(543,1195), MT(543), WY(1195). **15:**Range of host(82), AK(526), ID(3202), MT(3202).
S. ochraceo-flavum. On wood. **11:**DE(543), NY(543). **15:**MD(543), MS(2720), NH(543).
S. ostrea. On wood. **4:**ID(3202). **15:**Unk.(82).
S. rugosum. **1:**AK(543). **15:**AK(526,543).
S. sanguinolentum. On dead limbs. **13:**CA(661). **15:**WA(3202).
S. striatum. **15:**MD(543).
Stereum sp. **12:**WA(3202).
Stromatocyphella conglobata. On limbs. **15:**MI(2945), NH(456), NY(456).
Subulicystidium longisporum. On limbs. **6:**AZ(1202). **15:**AZ(128).
Thanatephorus cucumeris. **4:**ID(3202).
Tomentella bryophila. **4:**ID(3202). **13:**ID(128). **15:**ID(2039).
T. calcicola. **4:**WA(3202). **13:**WA(128). **15:**WA(2039).
T. chlorina. **6:**AZ(1202).
T. cinerascens. **4:**MT(3202). **13:**MT(128). **15:**MT(2039).
T. crinalis. **13:**ID(128). **15:**WA(128,2039,3202).
T. ferruginea. **15:**WA(2030,2039,3202).
T. neobourdotii. **15:**MT(2039,3202).
T. ramosissima. **4:**ID(3202). **15:**ID(2039), MT(2039,3202).
T. sublilacina. **15:**AZ(2039).
Tomentella sp. **13:**ID(128).
Tomentellastrum badium. On wood. **4:**ID(3202). **13:**ID(2032).
T. montanensis. On wood. **15:**MT(128,2031,2034,3202).
Trametes hirsuta. On wood. **4:**ID(3202), MN(2874), MT(3202), WA(3202). **6:**AZ(1202). **8:**ID(3202), WA(3202). **9:**OR(3202), WA(3202). **15:**Range of host(82), ID(3202), WA(3202).
T. pubescens. On wood. **4:**ID(3202), MT(3202). **9:**WA(3202). **13:**MT(3832). **15:**Range of host(82).
T. versicolor. On wood. **1:**Range of host(82). **3:**VA(82). **4:**Range of host(82), ID(3202), MT(3202), WA(3202). **6:**AZ(1202). **9:**AK(1904), ID(3202), OR(3202), WA(3202). **10:**Range of host(82). **12:**AK(3431). **15:**AK(2419), CA(81), WA(3202).
Trechispora cohaerens. On wood. **6:**AZ(1202).
T. mollusca. **15:**OR(3202).
Tremella fimbriata. **4:**ID(3202).
T. mesenterica. **9:**ID(3202). **12:**OR(3202).
Trichaptum biforme. On wood. **4:**MT(3202), WA(3202). **10:**Range of host(82). **13:**MT(3832). **15:**Range of host(82).
Tulasnella bifrons. **15:**MA(128).
Typhula phacorrhiza. On leaves. **4:**NY(2955).
T. phacorrhiza var. *heterogenea*. **13:**ID(264).
Typhula sp. On leaves. **12:**AK(3431).
Tyromyces chioneus. On wood. **4:**ID(3202), MT(3202), WA(3202). **9:**OR(3202). **12:**AK(526,1904), ID(3202). **15:**Range of host(82), MI(2862), WA(3202).
T. destructor. **6:**AZ(1202).
T. galactinus. **4:**MT(3202).
Vesiculomyces citrinus. **6:**AZ(1202).

Deuteromycotina - Hyphomycetes

Acremonium strictum. On leaf surface. **9:**OR(3042).
Alternaria alternata. On leaf surface. **9:**OR(82).
Anguillospora vermiformis. Leaf spot. **1:**WI(82). **10:**WI(2940).
Arthrinium arundinis. On leaf surface. **9:**OR(3042).
Aspergillus niger. On leaf surface. **9:**OR(3042).
Aureobasidium pullulans. On leaf surface. **9:**OR(3042).
Botrytis cinerea. Gray mold, on leaf surface. **9:**OR(3042), WA(82).
Cladosporium cladosporioides. Gray mold. **9:**OR(3042).
C. herbarum. On leaf surface. **9:**OR(3042), WA(3202).
C. macrocarpum. On leaf surface. **9:**OR(3042).
C. oxysporum. On leaf surface. **9:**OR(3042).
C. sphaerospermum. On leaf surface. **9:**OR(3042).
Curvularia lunata. On leaf surface. **9:**OR(3042).
Epicoccum nigrum. On leaf surface. **9:**OR(3042).
Fusarium arthrosporioides. On leaf surface. **9:**OR(3042).
Fusicladium poricola. On wood. **15:**CA(81).
Geniculosporium sp. On leaf surface. **9:**OR(3042).
Helicoma microscopicum. On decaying catkins. **11:**NJ(1271).
Helminthosporium macrocarpum. **9:**WA(3202).
Humicola fuscoatra. On leaf surface. **9:**OR(3042).
Memnoniella echinata. On leaf surface. **9:**OR(3042).
Monocillium sp. On leaf surface. **9:**OR(3042).
Monodictys sp. On leaf surface. **9:**OR(3042).
Mycopappus alni. Leaf spot. **9:**OR(82,2940,3202), WA(82,2940,3202). **12:**AK(82,526,2940).
Nematogonum ferrugineum. **15:**WA(3202).
Nodulisporium sp. On leaf surface. **9:**OR(3042).
Passalora alni. On leaves. **1:**WI(82,1331,1350). **3:**FL(35). **12:**AK(526).
Penicillium sp. Leaf spot. **9:**OR(3042).
Phialophora hoffmannii. On leaf surface. **9:**OR(3042).
P. verrucosa. On leaf surface. **9:**OR(3042).
Phymatotrichopsis omnivora. Root rot. **3:**TX(82). **9:**TX(82). **13:**TX(82).
Pseudospiropes nodosus. **15:**WA(1685).
Ramularia alnicola. **15:**ID(3202).
Scopulariopsis asperula. On leaf surface. **9:**OR(3042).
S. brevicaulis. On leaf surface. **9:**OR(3042).
S. brumptii. On leaf surface. **9:**OR(3042).
Scytalidium lignicola. On leaf surface. **9:**OR(3042).
Septonema subramosum. On dead twigs. **13:**CA(81).
Stilbella flavipes. **15:**NH(3178).
Taeniolella alta. **15:**MN(1679), NY(1679).
Taeniolella sp. On dead limbs. **13:**CA(661).
Torula herbarum. **15:**ID(3202).
Trichocladium asperum. On leaf surface. **9:**OR(3042).
Trichoderma viride. On leaf surface. **9:**OR(3042).
Trichothecium roseum. On leaf surface. **9:**OR(3042).
Tubercularia vulgaris. **15:**CO(649).
Ulocladium botrytis. On leaf surface. **9:**OR(3042).
U. chartarum. On leaf surface. **9:**OR(3042).

Deuteromycotina - Coelomycetes

Aplosporella alnicola. On twigs. **3:**NY(82). **4:**NY(82).
Asteroma alneum. On leaves. **9:**CA(81). **10:**NY(82).
Cylindrosporium alni. Leaf spot. **4:**MT(3202). **9:**AK(82,526). **13:**MT(82), UT(128).
Cytospora umbrina. **12:**Unk.(82).
Cytospora sp. **12:**AK(526).
Diplodia sarmentorum. On twigs. **9:**WA(82,3202).
Dothiorella inversa. **12:**AK(526).
[*Kirschsteiniella applanata*]. **15:**ID(3202), WA(3202,3815).
Leptothyrium alneum. On leaves. **12:**AK(526).
Melanconium apiocarpum. **4:**ID(3202). **12:**AK(526).
M. candidum. On limbs. **9:**OR(82,3202), WA(82,3202).
M. sphaeroideum. **15:**WI(1329).
Melanconium sp. **12:**AK(3431).
Monostichella alni. Leaf spot. **10:**WV(82). **11:**WV(3907).

15:AK(526).
Pestalotia microspora. On leaf surface. **9:**OR(3042).
Phomopsis alnea. Basal stem cankers. **3:**KY(2625).
Phomopsis sp. Canker, dieback. **3:**FL(35).
Phyllosticta alnea. Leaf spot. **12:**AK(82). **15:**AK(526).
P. carpini. **6:**NM(1195).
P. maculiformis. **10:**NY(64). **12:**ID(3202). **15:**WA(3202).
Rhabdospora maculans. On twigs. **10:**SC(82).
Septogloeum variegatum. Leaf spot. **9:**CA(82).
Septoria alni. Leaf spot. **1:**NY(82), WI(82). **3:**WI(82,607,1335). **4:**NY(82), WI(82,607). **9:**OR(4007). **10:**NY(82), WI(82). **13:**MT(607).
S. alnifolia. Leaf spot. **4:**ID(3202), MT(3202), WA(3202). **8:**CA(82,607). **9:**CA(81,82,607), ID(3202), OR(82,3202,4007), WA(82,607,3202). **12:**AK(82,526), ID(3202), WA(3202).
S. ostryae. **6:**NM(1195).
S. weiriana. Leaf spot. **1:**WI(607). **4:**WA(3202). **13:**ID(607).
Sphaeropsis alni. On twigs. **10:**SC(82). **15:**WA(3202).

Betula L.

1. *B. alleghaniensis* Britton [gray birch, yellow birch]
2. *B. glandulosa* Michx. [bog birch, dwarf birch]
3. *B. lenta* L. [black birch, cherry birch, mahogany birch, mountain-mahogany, sweet birch]
4. *B. microphylla* Bunge
5. *B. nigra* L. (=*B. rubra* F. Michx.) [black birch, red birch, river birch]
6. *B. occidentalis* Hook. (=*B. papyrifera* Marsh. var. *occidentalis* (Hook.) Sarg., *B. fontinalis* Sarg.) [mountain birch, water birch]
7. *B. papyrifera* Marsh. [canoe birch, paper birch, white birch]
8. *B. papyrifera* Marsh. var. *cordifolia* (Regel) Fernald
9. *B. papyrifera* Marsh. var. *humilis* (Regel) Fernald (=*B. neoalaskana* Sarg.)
10. *B. papyrifera* Marsh. var. *kenaica* (W. H. Evans) A. Henry (=*B. kenaica* W. H. Evans) [Kenai birch]
11. *B. papyrifera* Marsh. var. *minor* (Tuck.) S. Wats. & J. M. Coulter
12. *B. pendula* Roth (=*B. alba* L.) [European white birch]
13. *B. populifolia* Marsh. [fire birch, gray birch, old-field birch, white birch]
14. *B. pumila* L. (=*B. lutea* F. Michx.) [low birch, swamp birch]
15. *B. pumila* L. var. *glandulifera* Regel
16. *B. resinifera* Britton
17. *B.* × *sandbergii* Britton [Sandberg's birch]
18. *Betula* sp.

Chytridiomycetes

Blastocladia ramosa. **7:**MT(3202).

Oomycetes

Phytophthora cactorum. Bleeding canker. **14:**NJ(82).
P. cinnamomi. **7:**SE states(3624). **12:**SE states(3624).
Pythium sp. Root rot. **5:**FL(35).

Ascomycotina

Acrospermum compressum. **5:**GA(1438).
Anisogramma virgultorum. On limbs. **18:**NH(184).
Arthothelium lichenalis. On bark. **18:**VT(195).
Atopospora betulina. Black-speckled leaf spot. **2:**AK(526), ME(82). **6:**ID(82), NM(82). **14:**MN(2874), WI(82). **15:**WI(1330). **18:**MT(3202), WA(3202).
Bertia moriformis. Black-speckled leaf spot. **14:**NC(1303).
Bisporella citrina. Rotted limbs. **7:**ID(3202), WA(3202). **18:**NC(1303,1307).
Botryosphaeria dothidea. Canker. **5:**GA(1438). **12:**IL(727).
B. obtusa. On dead limbs. **5:**GA(82).
Botryosphaeria sp. Canker, dieback. **5:**NC(1303).
Calycellina populina. On fallen leaves. **18:**NH(2181).
Chaetomium sp. **1:**MI(1882).
Chaetosphaeria pulviscula. **7:**ID(3202).
Chapeckia nigrospora. **7:**MI(82). **14:**NY(82).
Chlorociboria aeruginosa. Green stain of dead wood. **3:**GA(1307), NC(1307). **14:**NC(1303). **18:**Range of host(82), OR(3202).
Ciboria betulae. Green stain of sapwood. **5:**MD(82).
Coccomyces coronatus. On dead leaves. **18:**ME(3229), NH(3229).
Coniochaeta sp. **1:**MI(1882).
Coronophora annexa. **5:**GA(1438). **18:**MD(2783).
Dactylospora bloxamii. **3:**NC(1307).
D. stygia. **7:**MN(2874).
Daldinia concentrica. On wood. **3:**NC(1303,1307). **5:**GA(1438), NC(1303). **7:**ME(82), VT(82). **13:**CT(82). **14:**NC(1303,1307). **18:**Range of host(82), WV(3907).
D. occidentalis. Wood rot of stumps and logs. **2:**MT(82,3202). **6:**AK(526), ID(82), WA(82). **7:**AK(1904), ID(3202), WA(3202). **18:**AK(526).
D. vernicosa. Wood rot. **18:**NY(82).
Daldinia sp. **7:**ID(3202).
Dermea molliuscula. On dead twigs. **3:**NE states(82), MI(82). **14:**NE states(82), MI(82), NC(1307).
Diaporthe alleghaniensis. Canker. **1:**MI(1882).
Diaporthella aristata. On limbs. **18:**NY(195).
Diatrype albopruinosa. **7:**IA(3629).
D. disciformis. On dead limbs. **14:**NC(1303).
D. stigma. On dead limbs and logs. **3:**NC(1303,1307), SC(1307). **6:**ID(82). **7:**IA(3629), ID(3202), MA(82), ME(82), VT(82), WA(3202). **14:**GA(1438), NC(1303). **16:**AK(526). **18:**GA(1438), MN(2874).
D. undulata. On dead limbs and logs. **18:**NJ(2925).
Diatrypella betulina. On dead limbs. **1:**WV(3907). **3:**NY(82), WI(82), WV(3907). **6:**ID(82). **7:**ID(82,3202), ME(82), MI(82), MN(82,2874), NY(82). **13:**ME(82). **14:**GA(1438), MN(2874), NY(82), WI(82). **18:**VT(1224).
D. decorata. On limbs. **3:**GA(82), MS(82), NY(82), WV(82). **13:**NY(82). **14:**GA(82,1438), MS(82), NY(82), WV(82). **16:**AK(526).
D. favacea. On dead limbs. **5:**GA(82,1438), KY(82). **7:**IA(3629), MN(1248). **13:**NY(82). **14:**PA(82). **18:**IA(3629), NC(1303).
Didymosphaeria sp. **5:**GA(1438).
Discohainesia oenotherae. On leaves. **5:**NC(82).
Durandiella seriata. On dead limbs. **6:**CO(82), UT(128). **14:**NJ(82), NY(82), PA(82).
Eudimeriolum septentrionale. On leaves and twigs. **5:**GA(1020).
Eutypa sp. On wood. **5:**NC(1303).
Eutypella cerviculata. **5:**MO(2925). **7:**MN(2874), SD(82,2243).
E. leprosa. **7:**IA(3629). **18:**IA(3629).
E. scoparia. **7:**IA(3629).
[*E. sorbi*]. On dead limbs. **6:**MT(82). **7:**ID(3202), MT(3202), OR(3202). **13:**NY(82). **18:**WA(1250).
Glomerella cingulata. Defoliation, sever leaf spot. **12:**VA(3494).
Glonium stellatum. On decorticated wood. **3:**NC(1303).
Gnomonia setacea. On fallen leaves. **5:**GA(82,1438). **7:**ID(3202). **12:**MD(82). **14:**NY(82).
Hypocrea gelatinosa. **18:**NY(2456).
H. schweinitzii. **5:**GA(1438).
Hyponectria betulina. On overwintered leaves. **18:**MA(183), ME(183).
Hypoxylon albocinctum. **18:**MI(2862).
H. fuscum. **5:**GA(1438). **7:**ID(3202). **18:**ID(3202).
H. howeianum. On wood. **3:**NC(1303). **5:**GA(1438). **18:**NC(1303).
H. mammatum. On wood. **14:**MI(82), MN(82,2874).
H. mediterraneum. On wood. **14:**IA(1225).
H. multiforme. On wood. **1:**MI(2862). **6:**AK(82), ID(82), WA(82). **7:**ID(1811,3202), MI(2862), MN(2874), WA(3202). **10:**AK(526). **12:**IA(1225). **14:**NC(1303,1307). **18:**MN(2874), VT(1224), WV(3907).
H. papillatum. On wood. **5:**GA(1438).
H. rubiginosum. On decorticated logs. **3:**NC(1303,1307). **5:**GA(1438).
H. truncatum. **5:**GA(1438).
Hysterium angustatum. **18:**MN(2874).
H. pulicare. **5:**GA(1438), NC(3944). **7:**ID(3202).
Incrucipulum sulphurellum. **3:**NC(1307).
Lachnellula cervina. On limbs. **18:**CO(849).
Lamprospora crec'hqueraultii. On logs. **7:**MI(2862).
Lopadostoma gastrinum. **18:**MA(664).
Melanconis decorahensis. Canker, dieback. **5:**GA(82,1438), NC(1303). **7:**MA(82). **12:**MA(82). **13:**NY(82), PA(82). **14:**IA(82), MA(82).
M. stilbostoma. Twig blight, dieback. **1:**WV(3907). **3:**MS(82,2720). **7:**NC states(82), NE states(82), MN(2874). **9:**AK(82). **12:**NC states(82), NE states(82). **13:**ME(82), NY(82). **14:**MN(2874), NY(82), VT(82), WV(82). **16:**AK(526). **18:**MN(2874), WA(3202).
Melanconis sp. **14:**Unk.(82). **18:**MA(664).
Melanomma pulvis-pyrius. **14:**NC(1307).
Microsphaera penicillata. Powdery mildew. **3:**NH(82), OH(950), PA(82), WV(3907). **14:**MA(82), MI(82), MN(2874), WI(82).

17:WI(1334). **18**:MN(2874), WV(3907).
Mycosphaerella harthensis. **2**:AK(526).
M. punctiformis. On dead leaves. **2**:AK(1938).
Nectria cinnabarina. On twigs. **3**:E states(82), MI(82). **5**:E states(82), MI(82). **7**:MN(2874). **14**:E states(82), GA(1438), MI(82). **18**:CO(49).
N. coccinea. Canker. **3**:CT(82), MA(82), MI(82), NY(82). **14**:CT(82), MA(82), MI(82), NY(82).
N. coryli. Canker. **14**:NC(1303).
N. discophora. Canker. **14**:VT(82).
N. galligena. Canker. **3**:CT(82), MA(3567), NY(82), PA(82). **5**:CT(82), NY(82), PA(82). **7**:MN(2240). **13**:MA(3567). **14**:CT(82), NY(82), PA(82). **18**:CO(49).
N. peziza. Canker. **18**:MI(3085), NY(3085).
Nectria sp. **7**:ID(3202).
Ophiostoma rostrocoronatum. Isolated from wood chips. **18**:ME(1005,3714).
O. stenoceras. Isolated from heartwood. **13**:NJ(778).
Peziza repanda. **7**:MI(2862).
Phaeangium magnisporum. **5**:GA(1438).
Phyllactinia guttata. Powdery mildew. **3**:MA(82), NY(82), VA(82). **5**:IL(82), MO(82), NC(1303), TX(82), WI(82). **6**:WA(82). **7**:MT(3202). **14**:VT(82).
Plagiostoma campylostylum. On fallen leaves. **6**:CO(82).
P. campylostylum var. *mirabilis*. On overwintered leaves. **14**:NY(195).
Pleomassaria siparia. On limbs. **18**:ME(82).
Propolomyces versicolor. On dead twigs. **3**:NC(1303,1307).
Prosthecium acrocystis. On limbs. **3**:MI(82), NJ(82), NY(82,184), VT(82). **14**:MI(82), NJ(82), NY(82), VT(82).
Pseudovalsa lanciformis. On limbs and twigs. **3**:NC(1303). **5**:GA(1438). **7**:IL(82). **12**:MA(82), NY(82).
Rhizodiscina lignyota. **18**:MA(664).
Rhytisma sp. **7**:WA(3202).
Rosellinia desmazieresii. **3**:NC(1307).
Scutellinia scutellata. On wood. **1**:NY(848). **14**:NC(1307).
Spathulariopsis velutipes. On rotten log. **18**:MI(2862).
Sphaerognomonia carpinea. On fallen leaves. **5**:GA(82,1438).
Splanchnonema argus. On limbs. **3**:GA(1438). **5**:GA(1438). **18**:MA(188), PA(188).
S. scoriadeum. On limbs. **5**:NC(188). **13**:MA(188).
Tapesia fusca. On limbs and logs. **3**:NC(1303,1307). **14**:NC(1303,1307).
Taphrina americana. Witches'-broom. **6**:CO(82,2454,3361). **7**:NH(82,2454), VT(82), WI(82,1348,2454). **10**:AK(2457). **14**:NH(82,2454), VT(82). **18**:CO(49).
T. bacteriosperma. Witches' broom. **2**:NH(82,2454,3361). **7**:ID(3202). **18**:AK(82).
T. boycei. **6**:NV(2457).
T. carnea. Red leaf blister, leaf curl. **2**:NH(82,2454), NY(82,2454). **5**:NH(82). **7**:MI(2457), NH(82,2454), VT(3361). **8**:MI(2457). **11**:NH(2454). **14**:ME(82,2454), NH(82), VT(2457). **18**:MI(2457), NH(82,2454).
T. flava. Yellow leaf blister. **7**:NH(82), WI(82). **13**:MA(82,2454), ME(82,2454), NH(82,2454).
T. nana. **2**:WY(2457).
Taphrina sp. Leaf blister. **14**:ME(82). **18**:MA(82), NH(82).
Valsa ambiens. Canker. **1**:MI(1882). **18**:IA(1226).
V. betulina. Dieback. **13**:MA(82), NY(82).
V. ceratosperma. Dieback. **5**:GA(1438).
V. melanodiscus. On dead twigs. **7**:NY(82).
V. polyspora. **7**:WA(3202).
V. sordida. On dead twigs. **6**:MT(82). **7**:MT(3202).
Venturia ditricha. On fallen leaves. **13**:ME(82), NY(82).
Winterella betulae. On dead limbs. **3**:NJ(82), NY(82). **5**:GA(82,1438). **7**:MA(82). **13**:NY(82,2948). **18**:MA(2948), MI(2948), NJ(2948), VT(2948).
W. femoralis. **3**:NY(82,2948).
Xylaria cornu-damae. On wood. **1**:NH(3017).
X. curta. On wood. **18**:NH(3015).
X. polymorpha. On wood. **18**:NC(1303).
Xylaria sp. **1**:MI(1882).

Basidiomycotina - Uredinales

Melampsoridium betulinum. Heteroecious: telial host. **2**:AK(60,526,1904), ID(82,3202), MT(82,3202), WA(82,3202). **6**:AK(60,526), CT(82), ID(82,1575), IL(82), ME(82), MI(82), MN(82), MT(82,1575), NH(82), WA(82,1575). **7**:ID(3202), MN(2874), MT(3202), WA(3202). **9**:AK(82,1575). **10**:AK(60,526). **13**:NE states(82), CT(1575), MA(1575), ME(1575), NH(1575), NJ(1575), NY(1575). **14**:IN(82), MA(82,1575), ME(82), MI(82,1575), MN(82,1575,2874), NH(82), VT(82), WI(82,1575). **16**:AK(60,526,1575). **18**:AK(82,526), CA(82).

Basidiomycotina - Other

Abortiporus fractipes. White rot. **18**:Unk.(1204), Unk.(2681).
Aleurodiscus oakesii. On bark. **14**:VT(82).
Amphinema byssoides. **7**:MN(1197). **18**:WA(3202).
[*Anomoporia myceliosa*]. **7**:WA(3202).
[*Antrodia crassa*]. **7**:ID(3202).
A. heteromorpha. **7**:ID(3202), MT(3202).
A. malicola. **7**:MN(2874).
A. variiformis. **7**:WA(3202).
Antrodiella semisupina. **18**:NY(2456).
Armillaria mellea. Root rot. **7**:MN(82,2874). **12**:CA(2897), MN(2874). **14**:NE states(82), NY(59), VT(59). **18**:CA(2898), VT(59).
Arrhytidia involuta. On wood. **5**:NH(128).
Asterodon ferruginosum. On logs. **18**:MI(2862).
Athelia salicum. On wood. **6**:ID(128).
Auricularia auricula. On wood. **3**:NC(1307). **14**:NC(1303).
Bjerkandera adusta. On wood. **5**:GA(1439). **7**:ID(3202), MN(1197), MT(3202). **18**:Range of host(82), MN(2874).
B. fumosa. **7**:MT(3202). **18**:MN(2874).
Ceraceomyces borealis. **14**:NY(128).
Ceriporia reticulata. **18**:MT(128).
C. spissa. **7**:MN(2874). **18**:MN(2874).
Ceriporiopsis aneirina. **18**:WA(3202).
C. pannocincta. **18**:MT(128).
Cerrena unicolor. On wood. **2**:ID(82,3202). **5**:NC(1303). **6**:ID(82), MT(82), WA(82). **7**:Lake states(82), NE states(82), AK(1904), ID(3202), MN(294,1197), MT(3202), WA(3202), WI(294). **9**:AK(82). **13**:ME(82). **14**:NE states(82). **16**:AK(526). **18**:MT(128,3832), NY(2456).
Chondrostereum purpureum. Wood rot. **1**:NH(543), NY(543), VT(543). **6**:AK(526), MT(543). **7**:AK(543,1904), ID(3202), ME(82,543), MI(82), MT(3202), NY(543), VT(543), WA(3202). **12**:MA(543). **13**:CT(543), MA(543). **14**:VT(82). **18**:ID(3202), MA(543), ME(543), NY(543).
Clavicorona pyxidata. On wood. **1**:MI(884). **18**:CT(884), NC(884).
Climacodon pulcherrimus. On wood. **5**:IN(82,128).
Coniophora arida. **13**:NY(128).
C. puteana. Brown cubical rot. **18**:Range of host(82).
Coprinus disseminatus. **5**:NC(3944).
Crepidotus albescens. **18**:NY(2934).
C. cinnamomeus. On dead limbs. **18**:ID(1544).
C. latifolius. **7**:ID(3202).
C. montanensis. **18**:MT(1544).
Crinipellis zonata. **18**:NC(594).
Cylindrobasidium laeve. **18**:AK(526).
Cyphellopsis anomala. **13**:NH(128).
Cystostereum murrayi. Canker, heart rot. **1**:E states(82), MA(543), MI(2862), MN(82), NY(128,543), VT(543). **3**:E states(82), MN(82). **5**:E states(82), MN(82). **6**:ID(82,543), OR(82). **7**:ID(3202), ME(82,543), MN(128,543,1197), NY(82,543), OR(3202), PA(82). **14**:MN(2874). **18**:MI(543), NH(543), NY(543), VA(543).
C. pini-canadense. White rot. **18**:NC(541).
Cytidia lanata. On limbs. **18**:ID(640,3202).
Daedalea quercina. On wood. **14**:CT(82).
Daedaleopsis confragosa. On wood. **3**:E states(82), NC(1303,1307), TN(82), WI(82). **5**:E states(82), GA(1439), NC(3944), TN(82), WI(82). **6**:WA(82). **7**:ID(3202), MI(82,2862), MN(128,1197), MT(3202), VT(82). **14**:E states(82), NC(1303), TN(82), WI(82). **18**:ID(3202), MA(1781), MN(2874), MT(128), NY(2456), WV(3907).
Datronia mollis. On wood. **6**:ID(82), MT(82,3832). **7**:MT(3202). **18**:MN(2874).
D. stereoides. On wood. **6**:AK(82), MT(82), WA(82). **7**:MN(1197). **16**:AK(526). **18**:AK(82).
Dendrophora erumpens. **7**:SD(543).
D. versiformis. White rot. **1**:NY(543), VA(543). **3**:Range of host(82).

Dichostereum pallescens. **7:**ME(128).
Diplomitoporus linbladii. **7:**ID(3202).
Erythricium laetum. **13:**NY(128).
Exidia glandulosa. **3:**NC(1307).
E. recisa. On wood. **5:**GA(491,1439).
Exidiopsis fuliginea. On wood. **4:**ID(128).
E. grisea. **18:**WA(3853).
Fomes fomentarius. On wood. **1:**MI(2862), WV(3907). **6:**NW states(82), AK(526,2419), MT(3832), WA(82). **7:**NE states(82), NW states(82,2862), AK(82,1904), ID(3202), MI(2862), MN(294,1197,2874), MT(2045,3202), WA(3202), WI(294). **9:**AK(82). **12:**NY(2456). **13:**ME(82). **14:**NH(638), NY(2456). **16:**AK(526). **18:**Range of host(82), AK(526), MA(1781), ND(378).
Fomitopsis cajanderi. Brown cubical rot. **6:**MT(82). **7:**MT(3202). **18:**MN(2874).
F. pinicola. Brown cubical rot. **3:**NH(82), VT(82), WI(82). **6:**ID(82), MT(82,3832), WA(82). **7:**ID(3202), MN(1197,2874), MT(3202), VT(82), WA(3202), WI(82). **13:**NY(82). **14:**NH(82), VT(82), WI(82). **18:**MN(2874).
F. spraguei. On wood. **18:**Range of host(82).
Ganoderma applanatum. White rot. **1:**MI(2862). **6:**MT(82,3832). **7:**NC states(82), NE states(82), AK(1904), ID(3202), MI(2862), MN(294,2874), MT(3202), WA(3202), WI(294). **13:**NY(82). **18:**Range of host(82), MN(2874).
G. lucidum. White spongy rot. **14:**NY(82).
G. tsugae. **14:**NY(2962).
Gloeophyllum sepiarium. On wood. **6:**ID(82), WA(82). **7:**ID(3202), WA(3202). **14:**NY(2962). **18:**MN(2874).
G. trabeum. **6:**MT(82,3832). **7:**MN(2874), MT(3202).
Gloeoporus dichrous. On wood. **7:**AK(1904), ID(3202), MT(3202), WA(3202). **14:**NC(1303,1307), SC(1307). **18:**MN(2874).
Grandinia crustosa. **7:**MA(656).
Hapalopilus rutilans. White rot. **7:**ID(3202), MN(2874). **18:**Unk.(2681), MA(1781).
Helicogloea terminalis. On limbs. **18:**NC(2645).
Henningsomyces candidus. **7:**ID(3202). **18:**NY(2456).
Hericium coralloides. On wood. **7:**AK(1904). **18:**MI(1237).
Hydnochaete olivacea. On wood. **3:**NC(1307). **14:**VA(82). **18:**MA(656), NY(430).
Hymenochaete agglutinans. Trunk canker. **1:**MI(82,2862), PA(82), VA(82). **3:**MI(82), PA(82), VA(82). **14:**MI(2862). **18:**VA(128).
H. badio-ferruginea. On wood. **14:**NC(1303).
H. tabacina. On wood. **3:**NC(1303,1307). **18:**ID(3202).
Hymenochaete sp. **7:**ID(3202).
Hyphoderma clavigerum. **7:**MN(1197).
H. setigerum. On dead twigs. **18:**MI(128).
Hypholoma sublateritium. **1:**MI(2862). **14:**MI(2862).
Hypochnicium bombycinum. **18:**AK(128).
Inonotus glomeratus. **6:**MT(3832). **7:**MI(2862), MT(3202). **18:**MI(2862).
I. hispidus. White spongy heart rot. **18:**Range of host(82).
I. obliquus. Trunk canker, white heart rot. **1:**NE states(3241), NY(3994), WV(3907). **3:**Lake States(82), NE States(82). **6:**WA(82). **7:**NE states(3241), ID(3202), MI(82), WA(82). **13:**MA(82), NH(82), PA(82), VT(82). **14:**Lake States(82), NE States(82), NY(2456). **18:**MT(128).
I. radiatus. On wood. **7:**MN(2874). **10:**AK(526). **14:**NC(1303,1307). **18:**AK(82).
Irpex lacteus. On wood. **7:**ID(3202), WA(3202). **18:**Range of host(82), MA(656).
Ischnoderma resinosum. On wood. **1:**MI(2862). **7:**AK(1904), MI(2862), MN(294,2874), WI(294). **16:**AK(526). **18:**AK(82), GA(491,1439), MN(2874).
Laetiporus sulphureus. Brown cubical rot. **7:**MN(2874).
Laxitextum bicolor. **1:**NH(543). **3:**WV(543).
Lentinellus cochleatus. **18:**Unk.(2423).
L. flabelliformis. **18:**Unk.(2423).
L. vulpinus. **18:**Unk.(2423).
Lentinus strigosus. On wood. **5:**GA(1439). **7:**AK(1904). **18:**MN(2874).
Lenzites betulina. White rot. **5:**FL(35), GA(1439), NC(3944). **6:**AK(526), ID(82), MT(82,3832). **7:**AK(1904), ID(3202), MI(2862), MN(82,1197,2874), MT(3202). **16:**AK(526). **18:**Range of host(82), AK(526), FL(35), MN(2874), NC(1303), ND(378).
Leucogyrophana mollusca. **18:**MI(2862).
Marasmius rotula. On logs. **18:**MI(2862).
Merismodes ochraceus. Bark patch. **18:**Range of host(82).
Meruliopsis corium. **16:**AK(526).
Mucronella aggregata. **18:**OR(3202).
Osteina obducta. **6:**MT(3832). **7:**MT(3202), WA(3202).
Oxyporus populinus. **3:**NE states(82). **13:**ME(82). **14:**NE states(82).
Panellus serotinus. Wound rot. **1:**MI(2862), WI(82).
P. stipticus. On dead limbs and logs. **3:**NC(1303). **14:**NC(1303,1307). **18:**ID(3202).
Peniophora aurantiaca. **18:**AK(526).
P. cinerea. **7:**MN(2874). **18:**MI(2862).
P. nuda. **18:**NY(2456).
Perenniporia medulla-panis. On wood. **7:**MN(294), WI(294). **18:**Range of host(82).
P. subacida. White rot. **7:**NY(82), WA(3202). **14:**VT(82). **18:**MI(2862).
P. tenuis. **7:**MT(128).
Phaeolus schweinitzii. **7:**ID(3202).
Phanerochaete affinis. **3:**CT(128). **13:**NH(128). **18:**TN(128).
P. carnosa. **14:**NH(128).
P. filamentosa. **1:**MI(2862).
P. magnoliae. White rot. **1:**NY(433). **14:**NY(128).
P. sanguinea. White rot. **7:**NH(433).
P. sordida. **6:**MT(128). **7:**MA(656).
P. tuberculata. White rot. **7:**MN(433,1197).
P. viticola. White rot. **18:**TN(128,433).
Phellinus conchatus. On wood. **6:**ID(82), MT(82,3832). **7:**MT(3202).
P. everhartii. White heart rot. **7:**ID(3202), MN(82,2874), WI(82). **14:**NC(1307), NJ(82), SC(1307), WI(82).
P. ferreus. **6:**MT(82,3832). **7:**ID(3202), MT(3202). **18:**WA(128,3202).
P. ferruginosus. On wood. **6:**WA(82). **7:**ME(82), MT(3202), PA(82), WA(3202). **13:**ME(82), PA(82). **16:**AK(526).
P. gilvus. On wood. **3:**NC(1303). **5:**GA(1439), NC(3944). **7:**WA(3202). **14:**NC(1303). **18:**Range of host(82).
P. igniarius. White heart rot. **6:**ID(82), MT(82,3202), WA(82). **7:**NC states(82), NE states(82), AK(1904), ID(3202), MI(2862), MN(1197), MT(3202), OR(3202), WA(3202). **9:**AK(82). **13:**ME(82). **14:**MN(2874). **16:**AK(526). **18:**MN(2874), NY(2456), OR(3202), WV(3907).
P. laevigatus. White laminated rot. **3:**Lake States(82), NE states(82), NC(1303,1307). **6:**ID(128), WA(82). **7:**ME(82), MN(128,1197), NY(82), PA(82), WA(3202). **13:**ME(82), NY(82), PA(82). **14:**Lake States(82), NE states(82), Range of host(82). **18:**Range of host(82), ID(128), MI(2862), MT(3202), NY(2456).
P. lundellii. On wood. **18:**E states(1205).
P. nigricans. **7:**MN(2874). **10:**AK(526). **16:**AK(526).
P. pini. On wood. **6:**ID(82). **7:**ID(3202).
P. punctatus. On wood. **6:**MT(82,3832). **7:**MT(3202).
P. robustus. **3:**PA(82). **5:**IA(82), MD(82), PA(82), VA(82), WI(82). **7:**WI(82).
P. viticola. **18:**MI(2862).
Phlebia chrysocrea. On wood. **18:**NC(1303,2165).
P. coccineo-fulva. **18:**AL(128).
P. ludoviciana. White rot. **5:**WI(2556). **18:**WI(2556).
P. merismoides. **7:**ID(3202). **14:**NY(128).
P. tremellosus. On wood. **6:**MT(3832). **7:**ID(3202), MI(2862), MN(2874), MT(3202), WA(3202).
Phlebiella vaga. Open canker face. **7:**MN(1197). **14:**NC(1303,1307). **18:**NY(2456).
Pholiota squarrosa. **14:**NY(1017).
P. squarrosoides. Wood rot. **1:**MI(82,2862). **6:**AK(526). **7:**AK(1904). **14:**MI(2862).
Pholiota sp. Wood rot. **14:**NH(1017).
Piptoporus betulinus. On wood. **6:**AK(526), MT(3832). **7:**AK(1904), ID(3202), MI(2862), MN(2874), MT(3202), WA(3202). **14:**NC(1303). **16:**AK(526). **18:**Range of host(82), AK(526), MA(656,1781), MN(2874), NC(1303), NY(2456), WA(3202).
Pleurotus levis. Heart rot. **7:**WA(3202). **14:**MI(82).
P. minutus. **5:**NC(3944).
P. ostreatus. On wood. **7:**AK(1904).
Plicatura nivea. **6:**MT(3832). **7:**MT(3202).
Plicaturopsis crispa. On wood. **1:**MI(2945). **3:**NC(1303,1307).

6:AK(526). **7:**AK(1904), MN(1197). **14:**NC(1303,1307).
Pluteus atricapillus. On wood. **7:**ID(3202). **18:**MI(2862).
Polyporus admirabilis. On wood. **14:**MN(2874).
P. badius. On wood. **2:**ID(3202). **3:**NC(1303). **7:**ID(3202).
P. brumalis. **7:**MI(2862).
P. mori. On wood. **3:**NC(1303).
P. varius. On wood. **7:**ID(3202), MI(2862), WA(3202). **14:**NC(1303). **18:**MN(2874).
Porotheleum fimbriatum. **7:**WA(3202).
Postia caesia. On wood. **7:**WA(3202). **18:**Range of host(82), ID(3202), NC(1303).
P. undosa. **1:**NH(397), VT(397).
Pseudotomentella flavovirens. On wood. **1:**NY(2032).
Pycnoporus cinnabarinus. White rot. **2:**ID(3202). **5:**NC(1303,3944). **6:**MT(3832). **7:**ID(3202), MN(2874), MT(3202), OR(3202), WA(3202). **18:**AK(82).
P. sanguineus. On wood. **5:**NC(1303).
Radulodon americanus. **7:**MN(1197).
Rectipilus fasciculatus. **18:**NY(128,2456).
Rigidoporus crocatus. **7:**MN(128).
Schizophyllum commune. On wood. **3:**NC(1303). **5:**FL(35). **7:**MN(2874). **18:**Range of host(82), FL(35), MN(2874).
Schizopora paradoxa. White rot. **14:**Lake States(82).
Scytinostroma galactinum. On dead wood. **7:**NH(82,128). **13:**NH(82).
S. portentosum. **14:**VT(128).
Sebacina epigaea. On wood. **18:**GA(2642).
Septobasidium pseudopedicellatum. Brown felt, parasitic on scale insects. **18:**FL(35).
Sistotrema brinkmannii. On wood. **5:**GA(491,1439).
Skeletocutis nivea. **7:**ID(3202).
Spongipellis unicolor. **18:**MN(2874).
Steccherinum ochraceum. White spongy sapwood rot. **1:**WV(3907). **7:**IA(82), MI(82), NJ(82), WI(82). **12:**MI(2862). **14:**NY(82,128), VA(82), WV(82). **18:**MI(2862).
Stereum hirsutum. White rot. **1:**CT(543), ME(543), NY(543). **3:**CT(543), MA(543), NY(543). **5:**OK(2872). **6:**AK(526), MT(543). **7:**AK(1904), CT(543), ID(3202), MA(543), NH(543), NY(543), VT(82,543), WA(3202). **13:**CT(543), MA(543), ME(543), NY(543). **18:**Range of host(82), MA(543,656), ME(543), NY(543,2456), PA(543), SD(543).
S. ochraceo-flavum. White rot. **3:**CT(543). **13:**ME(82), NY(543).
S. ostrea. White rot. **6:**ID(82), MT(82), NM(82), WA(82). **7:**ID(3202), MT(3202), SD(82), WA(3202). **18:**Range of host(82), ID(82), MA(1781), MI(2862), MT(82), WA(82).
S. striatum. On wood. **3:**WV(3907). **5:**MD(82), NC(82,3944). **7:**WV(3907). **18:**OH(543).
Stereum sp. White rot. **18:**OR(3202).
Thelephora caryophyllea. **5:**GA(1439).
Tomentella botryoides. **18:**MI(2039), NY(2039).
T. bryophila. **18:**MA(2039), NY(2039), OH(2039).
T. cinerascens. **18:**MT(2039).
T. ferruginea. **18:**NY(2039).
T. jaapii. On wood. **18:**MN(3746).
T. lateritia. On wood. **18:**NY(2032,2039).
T. neobourdotii. On wood. **18:**MI(2039), NY(2033,2039).
T. olivascens. **18:**NY(2039).
T. pilosa. **18:**MT(2039,3202).
T. punicea. **18:**MA(2030,2039).
T. ruttneri. On wood. **1:**NY(2035).
T. subclavigera. On wood. **18:**MT(2039).
T. sublilacina. **1:**NY(128). **18:**MA(2039), MI(2039), NY(2039), WV(2039).
T. subvinosa. **18:**MN(2039).
T. violaceofusca. **18:**FL(2039).
Tomentellastrum badium. On wood. **1:**NY(2032).
Trametes hirsuta. On wood. **1:**WV(3907). **5:**GA(1439), OK(2872). **6:**AK(526). **7:**AK(1904), ID(3202), MN(1197), MT(3202), WA(3202). **18:**Range of host(82), ID(3202), MN(2874), WA(3202).
T. pubescens. On dead tree. **1:**MI(2862). **6:**MT(3832). **7:**ID(3202), MI(2862), MN(2874), MT(3202), WA(3202). **18:**MN(2874), WV(3907).
T. versicolor. On wood. **3:**NC(1307), SC(1307). **5:**NC(3944). **6:**AK(526). **7:**AK(1904), ID(3202), MT(3202), WA(3202). **18:**Range of host(82), MI(2862), WV(3907).
Trichaptum biforme. White rot. **3:**NC(1303). **5:**GA(1439), NC(3944). **6:**AK(526), MT(3832). **7:**AK(1904), ID(3202), MI(2862), MN(1197,2874), MT(3202). **16:**AK(526). **18:**AK(82,526), MA(656), MN(2874).
T. subchartaceum. On wood. **7:**ID(3202), MT(3202).
Tulasnella bifrons. **18:**MA(128).
T. eichleriana. On wood. **18:**PA(2673).
T. pinicola. On wood. **5:**GA(2642).
T. pruinosa. On bark. **5:**IA(128).
Tyromyces chioneus. On wood. **6:**AK(526). **7:**AK(1904), ID(3202), MI(2862), MT(3202), WA(3202). **9:**AK(82). **14:**NC(1303,1307). **16:**AK(526). **18:**Range of host(82), MN(2874), NY(2456), OR(3202).
Wolfiporia cocos. On wood. **1:**WI(793). **7:**MA(793).
Xeromphalina kauffmanii. On stumps. **18:**MA(270).

Deuteromycotina - Hyphomycetes

Acrodictys atroapicula. On rotten wood. **7:**MA(3771).
Aspergillus sp. Isolated from wood. **1:**NH(1645).
Cercoseptoria septoriopsidis. Leaf spot. **3:**PA(607).
Cercospora sp. Leaf spot. **18:**FL(35).
Cladosporium caducum. Leaf mold. **5:**WI(82).
Cladosporium sp. On leaves. **1:**MI(1882). **7:**WI(1348).
Conoplea globosa. **18:**MA(1672).
Coryne sarcoides var. *viridescens*. On stumps. **7:**MI(2862).
Dictyocatenulata alba. On bark. **18:**NY(3181).
Epicoccum sp. **1:**MI(1882).
Fusarium sp. Isolated from wood. **1:**NH(1645). **18:**NE states(1415).
Graphium sp. Isolated from wood. **1:**NH(1645).
Haplotrichum conspersum. On logs. **18:**MA(2131).
Penicillium sp. Isolated from wood. **1:**NH(1645).
Phialophora sp. Isolated from wood. **1:**NH(1645).
Phymatotrichopsis omnivora. Root rot. **5:**TX(82).
Sesquicillium candelabrum. **18:**WA(3202).
Sporidesmium macrosporoides. **18:**MT(3202).
Sporotrichum sp. **1:**MI(1882).
Stilbella fusca. On stump sap. **18:**NY(3178).
Torula ligniperda. Red stain of heartwood. **3:**NE states(82). **7:**NE states(82).
Trichoderma viride. **14:**NY(2456).
Trichoderma sp. Isolated from wood. **1:**NH(1645).
Xenosporium berkeleyi. **5:**NC(3944).

Deuteromycotina - Coelomycetes

Aplosporella alnicola. **12:**NY(82).
Asteroma microspermum. Leaf spot. **3:**NY(82,607), PA(82), WV(82,607,3907). **7:**WI(82). **14:**NY(82).
Asterosporium betulinum. On limbs. **6:**WA(128).
Coryneum betulinum. **5:**NC(3547). **18:**NJ(3547).
[*C. umbonatum*]. On dead limbs and twigs. **3:**NC(1303,1307).
Cryptocline betularum. Leaf spot. **3:**IL(82), NJ(82), PA(82), TX(82), WI(1356). **5:**E states(82), FL(35), NC(1303), OK(2872), TX(82), WI(82), WV(3907). **7:**CO(82), OK(2872), WI(82). **18:**CO(49), FL(35).
Cylindrosporella betulae-papyriferae. Anthracnose, leaf spot. **7:**NY(82), PA(82).
C. leptothyrioides. **17:**WI(1334).
Cylindrosporium betulae. Leaf spot. **7:**DE(82), NY(82), WI(82,1345). **13:**WI(1355). **14:**WI(82,607,1341). **17:**WI(1334). **18:**WA(3202).
Cytospora chrysosperma. On twigs. **6:**WA(82). **7:**MN(2874). **18:**UT(3620).
C. decipiens. Perennial canker. **7:**MN(2874).
Cytospora sp. Canker. **7:**ID(3202). **18:**CO(49).
Discosia artocreas. On fallen leaves. **13:**ME(82).
Discula betulina. Anthracnose, leaf spot. **5:**NC(1303). **6:**UT(1195). **13:**NY(82).
Gelatinosporium betulinum. On limbs. **3:**NY(861). **14:**NY(3549).
G. fulvum. On limbs. **18:**NE states(1415).
Gelatinosporium sp. Canker. **1:**MI(1882).
Melanconium betulinum. Canker, dieback. **7:**NE states(1415), MN(2874). **18:**CO(49).
M. bicolor. Canker, dieback. **5:**NC(1303), WV(3907). **18:**PA(3780).
Myxocyclus polycistis. Canker, dieback. **18:**AL(3547,3549).
Phomopsis sp. Limb canker, limb and twig dieback. **1:**NE states(1415). **7:**NE states(1415).
Phyllosticta betulina. Leaf spot. **5:**OK(82,2872).
Septoria betulae. Leaf spot. **1:**NY(607), WI(607). **3:**MA(607), VT(82).

6:OR(82), WA(82), WI(82). **7:**MA(607), NY(607), OR(3202), WA(3202), WI(82,607). **12:**IA(82). **13:**KS(607), VT(82), WI(607,1352). **14:**NY(82), WI(82,607). **17:**WI(1335). **18:**MN(2874), WI(82).
S. betulae-odoratae. Leaf spot. **6:**MT(607), UT(607,1195), WA(607). **7:**WI(607).
S. weiriana. Leaf spot. **6:**OR(607). **7:**OR(3202).
Septoria sp. Leaf spot. **13:**NC(1303). **18:**FL(35).
Sphaeropsis betulae. On twigs. **13:**NY(82). **14:**NY(82).
S. betulae var. *foliicola*. On leaves. **7:**WI(82).
Sphaeropsis sp. Gall. **18:**FL(35).
Stegonsporium pyriforme. On twigs. **3:**VT(82).

Carpinus L.

1. *C. betulus* L. [European hornbeam]
2. *C. caroliniana* Walter [American hornbeam, blue beech, musclewood, water beech]
3. *Carpinus* sp.

Chytridiomycetes
Synchytrium carpini. Leaf gall. **3:**LA(627).
Ascomycotina
Aglaospora profusa. **2:**WV(3907).
Botryosphaeria obtusa. On dead limbs. **2:**Unk.(82). **3:**Unk.(82).
Calosphaeria microsperma. **2:**MA(190), NY(190).
Daldinia concentrica. Wood rot. **2:**MD(82). **3:**IA(1225).
D. loculata. Wood rot. **2:**MN(82,2874).
Diaporthe bakeri. Wood rot. **2:**GA(82,1438), NY(82,3807), WI(3807).
Diatrype albopruinosa. **2:**IA(3629).
D. stigma. **2:**IA(3629).
Diatrypella decorata. **2:**GA(1438).
D. eutypaeformis. **2:**IA(3629).
D. nitschkei. **2:**FL(2782).
Discostroma corticola. **2:**GA(1438).
Enchnoa subcorticalis. On limbs. **2:**MA(272), NY(195,272).
Endoxylina moroides. **3:**IA(1226).
Eutypella cerviculata. On dead limbs. **2:**IA(3629), MI(82), MS(82,2720), NY(82).
E. scoparia. On dead limbs. **2:**GA(82,1438), NC(1303).
Gibberella baccata. **2:**AL(82).
Graphostroma platystoma. On limbs. **2:**GA(1438). **3:**SC(634).
Hypoxylon fuscum. **3:**MO(82).
H. hypophlaeum. **2:**GA(1438).
H. mammatum. **3:**IA(82,1225).
H. rubiginosum. **2:**GA(1438). **3:**Range of host(82).
H. tinctor. **2:**GA(1438). **3:**GA(82).
Mamiania fimbriata. On leaves. **2:**MI(82), MN(2874), NY(82), VT(82), WI(82), WV(82).
Massaria sp. **2:**GA(1438).
Melanconis chrysostroma. **2:**GA(1438).
M. chrysostroma var. *ellisii*. On dead limbs. **1:**MD(82), NY(82). **2:**GA(82), IA(82), MD(82,2780), MI(82), NY(82), PA(82), SC(82), VA(82).
M. platystroma. On limbs. **2:**MI(82,3809).
Melogramma campylosporum. On dead limbs. **2:**GA(82,1438).
M. patens. On dead limbs. **2:**IN(82), OH(82).
Microsphaera penicillata. Powdery mildew. **2:**IA(82), IN(82), MA(82), MI(82), TX(82), WI(82).
Mycosphaerella punctiformis. On leaves. **3:**Unk.(3943).
Nectria galligena. Trunk canker. **2:**CT(82), NY(82).
Nitschkia callista. On dead limbs. **2:**GA(1438), TX(82).
N. confertula. **2:**GA(1438).
Ostreichnion curtisii. **2:**GA(1438).
Pezicula carpinea. Limb and trunk canker. **2:**E states(82), GA(1438), NC(3944), OK(82,2872).
Phyllactinia guttata. Powdery mildew. **2:**AL(82,285), FL(35), IN(82), OH(82), TX(82), WI(82).
Pleomassaria carpini. On limbs. **2:**GA(82,1438), NY(188).
Rosellinia aquila. On wood. **2:**GA(1438). **3:**GA(82).
Rutstroemia bolaris. On twigs. **2:**NY(1960).
Sphaerognomonia carpinea. Leaf spot. **2:**GA(82,1438), NY(82), PA(82), WI(82), WV(82,3907).
Taphrina australis. Leaf blister. **2:**AL(82,2454), CT(82,2454), KY(82).
Urnula craterium. On wood. **2:**NC(1303).
Valsa ceratosperma. **2:**GA(1438).
Valsaria insitiva. **2:**GA(1438).
V. rubricosa. **2:**GA(1438).
Valsella laschii. **2:**GA(1438).
Xylaria corniformis. On wood. **2:**NC(1303).
Basidiomycotina - Other
Aleurodiscus oakesii. On bark. **2:**NY(82).
Armillaria mellea. Root rot. **2:**FL(35,82).
Bjerkandera adusta. **2:**Range of host(82). **3:**Range of host(82).
Chionosphaera apobasidialis. On limbs. **2:**IA(128), IL(700), PA(128).
Chondrostereum purpureum. **2:**CT(543), DE(543).
Cystostereum murrayi. White rot. **2:**VA(82,543).
Daedaleopsis confragosa. On wood. **2:**NC(82,3944).
Datronia mollis. On trunks. **3:**FL(634).
Dendrophora albobadia. On wood. **2:**GA(491,1439).
D. erumpens. **2:**IN(543).
D. versiformis. **2:**CT(543).
Fomitella supina. **3:**FL(82).
Gloeocystidiellum fimbriatum. On wood. **2:**FL(436).
Hydnochaete olivacea. **2:**GA(1439).
Hymenochaete corrugata. On wood. **2:**GA(491,1439).
H. curtisii. On wood. **2:**GA(491,1439).
Hyphoderma heterocystidia. **2:**MI(2321). **3:**MI(2321).
Irpex lacteus. On wood. **2:**NC(1303). **3:**Range of host(82).
Mycoacia meridionalis. **3:**TX(443).
Mycorrhaphium adustum. On wood. **2:**CT(82,128).
Panellus pusillus. On wood. **2:**Unk.(2681).
Peniophora cinerea. On wood. **2:**MS(2720).
Phanerochaete affinis. **3:**NY(128).
P. chrysorhiza. White rot. **2:**FL(433).
P. crassa. White rot. **2:**VA(82).
P. flavido-alba. White rot. **2:**FL(433).
Phellinus gilvus. **3:**Range of host(82).
P. igniarius. White spongy heart rot. **2:**VT(82).
P. punctatus. On wood. **3:**AR(82,2675), LA(82,2675).
Phleogena faginea. **2:**NY(2456).
Physisporinus vitreus. On wood. **2:**FL(35). **3:**FL(82).
Pleurotus ostreatus. Wood rot. **2:**NC(1303,3944).
Pycnoporus sanguineus. On wood. **2:**MS(2319).
Schizophyllum commune. On wood. **3:**Range of host(82).
Schizopora paradoxa. On wood. **3:**MI(82).
Septobasidium pseudopedicellatum. Brown felt. **2:**FL(35,82), NC(82,3944), TN(82), VA(82).
Steccherinum ochraceum. **2:**GA(1439), VA(82).
[*Stereum gausapatum*]. White heart rot. **2:**Range of host(82).
S. hirsutum. White rot. **2:**Range of host(82), CT(543), NC(3944), NY(543). **3:**NY(2456).
S. ostrea. White rot. **2:**Range of host(82), IN(543).
S. striatum. On wood. **2:**Range of host(82), CT(543), DC(543), FL(543), IA(543), IN(543,654), LA(543), MA(543), MD(543), MN(2874), NY(543), OH(543), PA(543), VA(543), WV(543). **3:**NY(2456).
Thelephora albidobrunnea. **2:**NC(3944).
Tomentella neobourdotii. **3:**NY(2039).
Trametes hirsuta. **3:**Range of host(82).
T. pubescens. **3:**Range of host(82).
T. versicolor. **3:**Range of host(82).
Tulasnella pruinosa. **2:**NY(128).
Deuteromycotina - Hyphomycetes
Cladosporium caryigenum. **2:**GA(1281).
C. caryigenum var. *carpinum*. Leaf spot. **2:**WI(82).
Conoplea globosa. **3:**CT(1672).
Fusarium carpineum. On leaves. **2:**WI(808).
Phymatotrichopsis omnivora. Root rot. **2:**TX(82).
Pseudocercosporella caryigena. Leaf spot. **2:**WI(82).
Teratosperma cornigerum. On bark and leaves. **2:**MD(82), NY(82), WI(82).
Xenosporium berkeleyi. **3:**FL(634).
Deuteromycotina - Coelomycetes
Cryptosporiopsis fasciculata. Parasitic on twigs. **2:**CT(1377), NY(1377), WI(1377).
Depazea carpinea. On dead leaves. **2:**NC(82).
Monostichella robergei. Defoliation, leaf spot. **2:**NY(3283), WI(1358).
Myxosporium carpini. On twigs. **2:**NY(82), RI(82).
Phyllosticta hesperidearum. **2:**CO(82). **3:**CO(82).
Phyllosticta sp. Leaf spot. **2:**OK(2872). **3:**OK(82).
Septoria betulae. Leaf spot. **2:**WI(607).

S. ostryae. Leaf spot. **2:**MI(82).

Corylus L.

1. *C. americana* Marsh. [American filbert, American hazelnut]
2. *C. avellana* L. [European filbert, European hazelnut]
3. *C. cornuta* Marsh. (=*C. cornuta* Marsh. var. *cornuta* , *C. rostrata* Aiton) [beaked filbert, beaked hazelnut]
4. *C. cornuta* Marsh. var. *californica* (A. DC.) W. Sharp (=*C. californica* (A. DC.) J. Rose, *C. rostrata* Aiton var. *californica* A. DC.) [California hazelnut, western hazel]
5. *C. maxima* Mill. [giant filbert]
6. *Corylus* × sp. (unnamed hybrids)
7. *Corylus* sp.

Ascomycotina

Allantoporthe decedens. On dead limbs and nuts. **1:**IA(82), MI(82), NY(82). **2:**NY(82). **3:**IA(82), MI(82), NY(82), OR(3202).
Anisogramma anomala. Blight, canker, dieback. **1:**NC states(82), NE states(82), WI(82,1334). **2:**CT(82), DE(82), IL(82), MA(82), WA(1282). **3:**NY(82). **5:**NC(1303). **6:** MD(3780), NC(3780), WA(821), WI(1355). **7:**WA(3202).
Botryosphaeria obtusa. On dead limbs. **1:**NJ(82). **2:**CA(82,1110). **4:**CA(82,1110).
Cucurbitaria conglobata. On dead limbs. **1:**NY(82), PA(82). **3:**NY(82), PA(82).
Diatrype albopruinosa. **1:**IA(3629).
Diatrype sp. **7:**OR(3202).
Diatrypella decipiens. **3:**OR(3202).
D. favacea. **1:**IA(3629).
D. frostii. On dead limbs. **1:**ND(82,377).
D. minutispora. **3:**NY(82).
D. missouriensis. **1:**MO(82), NY(82).
Encoelia furfuracea. On dead limbs. **1:**ND(82,377).
Endothia gyrosa. **2:**GA(1438).
Endotryblidium insculptum. On limbs. **4:**CA(82,1110).
Eutypella cerviculata. **7:**OR(3202).
E. tetraploa. **3:**WA(3202).
Glyphium corrugatum. **7:**WA(3202).
Gnomonia gnomon. **1:**GA(82,1438), NY(82), WI(1358). **3:**WA(3202), WI(1358).
G. setacea. Leaf spot. **4:**CA(1110).
Hypoxylon cohaerens. **3:**OR(3202).
H. fuscum. On dead limbs. **1:**MI(82). **3:**OR(3202).
H. rubiginosum. On wood. **3:**WA(3202). **7:**IA(1225).
Mamianiella coryli. Leaf spot. **1:**Range of host(82), MN(2874), WV(3907). **3:**Range of host(82), MN(2874), OR(3202), WA(3202). **4:**Range of host(82), CA(81). **7:**MN(2874), OR(3202).
M. coryli var. *spiralis.* Leaf spot. **1:**IL(82). **3:**MI(82).
Melanconis decorahensis. **7:**OR(3202).
M. flavovirens. On dead limbs. **1:**IA(82). **3:**IA(82).
Microsphaera penicillata. Powdery mildew. **1:**NC states(82), NE states(82), MN(2874), OH(950), SD(2243), WI(1354). **3:**NC states(82), NE states(82). **4:**Unk.(3976), CA(81). **7:**ND(377).
Mollisia melaleuca. **5:**GA(1438).
Mollisia sp. **5:**GA(1438).
Mycosphaerella punctiformis. **5:**GA(1438).
Nectria coryli. On dead limbs. **4:**NJ(82), NY(82).
N. episphaeria. **7:**OR(3202).
Nematospora coryli. Kernel-spot. **2:**OR(3202). **7:**OR(1424).
Pezicula corylina. On dead limbs and twigs. **2:**PA(82). **3:**NH(82), NY(82), WA(3202).
Phyllactinia guttata. Powdery mildew. **1:**NC states(82), NE states(82). **2:**OR(3202). **3:**NC states(82), NE states(82), OR(3202), WA(3202), WI(1354). **4:**Unk.(3976), CA(81,1110), OR(82), WA(82). **5:**OR(82,3202), WA(82). **7:**MN(2874), OR(2424,2425,2426,2427,2428,2429,2430,2431).
Schizoxylon insigne. **1:**GA(1438).
Scorias spongiosa. On leaves and twigs following aphid or scale infestation. **1:**MD(82).
Sphaerognomonia carpinea. Leaf spot. **5:**GA(82,1438).
Splanchnonema pupula. **3:**OR(3202).
Stictis chrysopsis. **7:**OR(3225), WA(3225).
S. radiata. **7:**CA(3225).
Taphrina coryli. Leaf blister. **1:**CT(82,2454), MA(82,2454), WI(82,2454,3361). **2:**PA(82). **3:**CT(82,2454), WA(3202). **4:**WA(2457).
Valsa ambiens. On dead limbs. **1:**NY(82). **3:**OR(3202). **4:**OR(82).
Winterella alnicola. **7:**NY(2948).
Xylaria hypoxylon. On dead wood. **3:**WA(3202). **4:**WA(82).

Basidiomycotina - Other

Armillaria mellea. Root rot. **2:**OR(82,3202), WA(82,3202). **4:**CA(82,1110). **7:**OR(2430).
Cerocorticium cremoricolor. **1:**ND(378).
Cyphellopsis anomala. **3:**WA(3202).
Cytidia patelliformis. **3:**WA(3202).
Dacryobolus karstenii. **3:**WA(3202).
Datronia stereoides. White rot. **1:**Unk.(82). **3:**Unk.(82).
Exidia candida. **3:**WA(3202).
Hymenochaete agglutinans. At base of limbs. **3:**ME(82). **7:**ME(128).
H. cinnamomea. On dead limbs. **1:**ND(82,378).
Hyphoderma mutatum. **3:**WA(3202).
H. setigerum. **3:**WA(3202).
Inonotus radiatus. On wood. **1:**Unk.(82). **3:**Unk.(82).
Junghuhnia collabens. **3:**OR(3202).
Peniophora cinerea. **3:**WA(3202).
P. nuda. **3:**WA(3202).
Phanerochaete velutina. **3:**WA(3202).
Phellinus ferreus. White rot. **3:**OR(3202). **4:**CA(82,1110). **7:**OR(3202).
P. ferruginosus. **3:**OR(3202). **4:**CA(81).
Phlebia concentrica. **3:**WA(3202).
P. merismoides. On dead limbs. **4:**CA(81).
Polyporus varius. On wood. **1:**Unk.(82). **3:**Unk.(82).
Sebacina incrustans. **1:**WI(128).
Stereum hirsutum. White rot. **2:**CT(543). **3:**WA(3202). **4:**WA(82).
S. ochraceo-flavum. **1:**MO(543).
Trametes hirsuta. **3:**OR(3202).
T. versicolor. **3:**OR(3202).
Tremella mesenterica. **3:**WA(3202).
Tremella sp. **3:**WA(3202).
Tyromyces chioneus. White rot. **1:**Unk.(82). **3:**Unk.(82).

Deuteromycotina - Hyphomycetes

Anguillospora coryli. Leaf spot. **1:**WI(2940).
A. vermiformis. Leaf spot. **1:**WI(82). **3:**WA(3202), WI(1358). **4:**CA(81).
Cercospora corylina. Leaf spot. **1:**OK(2872). **3:**OK(82,2872).
Phymatotrichopsis omnivora. Root rot. **1:**TX(82). **3:**TX(82).
[*Sarcinella heterospora*]. **1:**WI(1359).
Sclerotiomyces colchicus. **1:**WI(1356).
Troposporium album. On wood. **3:**CA(1468,2740). **7:**CA(2740).

Deuteromycotina - Coelomycetes

Colletotrichum sp. On leaf midribs. **1:**WI(1360).
Diplodia coryli. On dead twigs. **1:**MI(82).
D. sarmentorum. On twigs. **2:**OR(82,3202).
Phyllosticta coryli. Leaf spot. **1:**IA(82), IN(82), MO(82), TX(82). **3:**WA(3202). **4:**WA(82).
Phyllosticta sp. On leaves. **1:**WI(1351).
Piggotia coryli. Leaf spot. **1:**MI(82), OK(82,2873), WI(82). **2:**NJ(82). **3:**NJ(82), NY(82), OR(3202), VT(82), WA(3202), WI(82). **4:**OR(82), WA(82). **7:**Unk.(2087).
Pilidium concavum. **1:**WI(1357).
Scolicosporium coryli. On leaves. **1:**NY(82).
Septogloeum profusum. Leaf spot. **1:**IN(82), MA(82), MS(82,965).
Septoria ostryae. Leaf spot. **1:**MI(82,607), MN(2874), NE(82,607), NY(607), WI(82,607,1332). **3:**MN(2874), NY(607), OR(82,3202), WA(82,3202), WI(607). **4:**CA(81,607,1110), MA(607), NJ(607), NY(607), OR(607,4007), WA(607).
Sphaeropsis coryli. On dead limbs. **1:**NJ(82), NY(82). **2:**NJ(82). **3:**NJ(82), NY(82).

Ostrya Scop.

1. *O. virginiana* (Mill.) K. Koch [American hophornbeam, leverwood]
2. *Ostrya* sp.

Oomycetes

Pythium sp. Root rot. **1:**FL(35).

Ascomycotina

Acrospermum foliicola. **1:**GA(1438).
Botryosphaeria dothidea. **1:**GA(1438).
B. obtusa. On limbs. **1:**GA(82), NY(82).
B. ribis. On limbs. **1:**GA(82).

Chaetosphaeria pannicola. **1**:GA(1438).
Coniochaeta ligniaria. **2**:IN(82).
Daldinia loculata. On dead limbs. **1**:GA(82).
Diaporthe eres. On dead limbs. **1**:NE(82), NY(82).
Diatrype albopruinosa. On dead limbs. **1**:IA(3629), MS(82), ND(2925). **2**:IA(3629).
D. americana f. *ostryae*. On dead limbs. **1**:MN(2874), ND(82,377).
D. stigma. On dead limbs. **1**:IA(3629), NY(82).
Eutypella cerviculata. **1**:IA(3629).
Gnomonia gnomon. **1**:GA(82,1438), NY(82). **2**:GA(82), NY(82).
Godronia sp. **1**:GA(1438).
Graphostroma platystoma. On dead limbs. **1**:AL(82).
Hypoxylon albocinctum. On wood. **1**:IA(1225). **2**:IA(1225).
H. atropunctatum. On wood. **1**:GA(82,1438).
H. fuscum. On wood. **1**:AL(82), IN(82). **2**:IA(1225).
H. howeianum. On wood. **1**:NY(82). **2**:IA(1225).
H. hypophlaeum. On wood. **1**:GA(82,1438).
H. mediterraneum. On wood. **2**:IA(1225).
H. rubiginosum. On wood. **1**:GA(82,1438).
H. truncatum. On wood. **1**:AL(82).
Mamiania fimbriata. On leaves. **1**:WI(1359). **2**:MA(82), ME(82), NY(82).
Melanconis ostryae. On limbs. **1**:IA(82), MI(82), NE(82), NY(82).
Microsphaera penicillata. Powdery mildew. **1**:NC states(82), NE states(82), WI(82).
Mycosphaerella punctiformis. **1**:GA(1438).
Nectria sp. Canker. **1**:NY(82).
Phyllactinia guttata. Powdery mildew. **1**:E states(82), FL(35), MS(2720), TX(82), WI(82).
Rosellinia aquila. On dead limbs. **1**:IA(82), IN(82). **2**:IA(1225).
Sphaerognomonia carpinea. On decaying wood. **1**:GA(82,1438).
[*Taphrina caerulescens*]. **1**:KS(2458).
T. virginica. Leaf curl. **1**:E states(82), FL(35), GA(1438), KS(1967,2454,3361), MA(2454), NH(2454), TX(82), VT(2454), WI(82,2454).
Uncinula macrospora. Powdery mildew. **1**:MI(82), WI(82).
Valsa ambiens. On dead limbs. **1**:MS(82).
V. ceratosperma. **1**:GA(1438).
Basidiomycotina - Uredinales
[*Melampsoridium betulinum*]. **1**:NY(82).
[*M. carpini*]. Telial host. **1**:NY(1575).
Basidiomycotina - Other
Aleurobotrys botryosus. On dead stems. **1**:CT(2083).
Aleurodiscus oakesii. On limbs. **1**:AL(82), GA(1439,2083), IA(82,2083), MI(2083), MN(2874), MO(82), MS(2083), NH(2083), NY(82,2083), OH(82), TN(2083), VT(2083), WI(82), WV(82,2083,3907). **2**:NY(2456).
Armillaria mellea. Root rot. **2**:Unk.(82).
A. tabescens. Root rot. **1**:FL(35,82).
Bjerkandera adusta. On wood. **1**:Unk.(82).
Cerrena unicolor. Wood rot. **1**:VT(82).
Chondrostereum purpureum. **1**:NY(543).
Cystostereum murrayi. **1**:MN(128,543,1197). **2**:PA(82).
Dacrymyces minor. On wood. **1**:IA(128).
Daedaleopsis confragosa. On wood. **1**:VT(82).
Datronia mollis. Wood rot. **1**:NY(82), VT(82).
Dendrophora albobadia. White rot. **2**:MD(82).
Dendrothele griseo-cana. On wood. **1**:IA(82), MO(82).
D. strumosa. On wood. **1**:FL(35,82).
D. subfusispora. On wood. **1**:FL(444).
Ganoderma applanatum. On wood. **2**:Range of host(82).
Gloeocystidiellum clavuligerum. White rot. **1**:MI(2553).
G. fimbriatum. On wood. **1**:FL(436).
G. turpe. On wood. **1**:FL(436).
Inonotus cuticularis. On wood. **1**:IN(82).
I. obliquus. On wood. **1**:PA(82).
Junghuhnia nitida. **1**:NY(2168).
Kavinia himantia. On wood. **1**:MA(82).
Meruliopsis corium. **2**:MD(128).
Mycorrhaphium adustum. On dead limbs. **1**:NC(1303).
Oxyporus latemarginatus. On wood. **1**:Unk.(82).
O. populinus. On wood. **2**:MA(82), VT(82).
Panellus pusillus. On wood. **1**:FL(441).
Perenniporia ohiensis. On wood. **2**:IA(82).
Phanerochaete chrysorhiza. White rot. **1**:FL(433).
Phellinus conchatus. On wood. **2**:VT(82).
P. everhartii. On wood. **2**:WI(82).
P. ferruginosus. On wood. **1**:Unk.(82).
P. igniarius. White spongy heart rot. **1**:SD(378). **2**:MI(82), ND(82), SD(82), VT(82).
P. punctatus. On wood. **1**:OH(82), PA(82), VT(82).
Pleurotus ostreatus. **1**:NY(82,2936).
Polyporus melanopus. On wood. **1**:Unk.(2681).
Rectipilus fasciculatus. **1**:LA(128).
Sebacina helvelloides. At base of living trees. **1**:IA(128).
Septobasidium pseudopedicellatum. Brown felt, parasitic on scale insects. **1**:FL(35,82).
Steccherinum fimbriatum. **1**:MN(128), NY(128).
S. ochraceum. **2**:NY(128).
Trametes conchifer. On wood. **2**:ND(378).
T. hirsuta. On wood. **1**:VT(82).
T. versicolor. On wood. **1**:VT(82).
Deuteromycotina - Hyphomycetes
Cercospora sp. Leaf spot. **1**:FL(35).
Conoplea fusca. **1**:MN(2874).
C. globosa. Trunk canker. **1**:MD(82), PA(82), WV(82,3907).
C. olivacea. **2**:NY(1672).
Fusarium lateritium. On twigs. **1**:OH(82).
Fusarium sp. Root rot. **1**:FL(35).
Haplotrichum curtisii. On bark. **1**:NY(2131).
Phymatotrichopsis omnivora. Root rot. **1**:TX(82).
Deuteromycotina - Coelomycetes
Monostichella robergei. Defoliation, leaf spot. **1**:NJ(82), NY(82,3283), OK(82,2873), PA(82), WI(82,1358).
Septoria ostryae. Leaf spot. **1**:IA(82), NY(82,607), VA(82), WI(82,607).
Deuteromycotina - Other
Rhizoctonia solani. Damping-off. **1**:MO(82). **2**:MO(82).

Bignoniaceae

Bignonia L.

1. *B. capreolata* L. [cross-vine, quarter-vine]
2. *Bignonia* sp.

Chytridiomycetes
Synchytrium bignoniae. Leaf gall. **1**:LA(627).
Ascomycotina
Asterella bignoniae. Black mildew. **1**:LA(82).
Botryosphaeria obtusa. On stems. **1**:AL(82).
Capnodium elongatum. Sooty mold. **1**:AL(82), MS(82,2720).
Dothidea puccinioides. **1**:GA(82,1438).
Hypoderma commune. **1**:GA(1438).
Meliola bidentata. Black mildew. **1**:Gulf States(82), S states(1452), Unk.(1452), FL(35), GA(1438), MS(2720).
M. bignoniacearum. **2**:FL(1452).
M. furcata. Black mildew. **1**:FL(35,82).
Valsa ceratosperma. **2**:PA(3415).
Deuteromycotina - Hyphomycetes
Cercospora duplicata. Leaf spot. **1**:AL(82,285), FL(35), MS(82,2720).
Cercospora sp. Leaf spot. **1**:IL(312).
Deuteromycotina - Coelomycetes
Leptostromella bignoniae. Leaf spot. **1**:TN(82).
Phyllosticta sp. Leaf spot. **1**:IL(312).
Sphaceloma sp. Spot anthracnose. **1**:LA(82).

Campsis Lour.

1. *C. radicans* (L.) Seem. ex Bureau (=*Tecoma radicans* (L.) Juss.) [cow-itch, trumpet honeysuckle, trumpet-creeper, trumpetvine]

Chytridiomycetes
Synchytrium tecomae. Leaf gall. **1**:LA(627).
Ascomycotina
Botryosphaeria dothidea. **1**:GA(1438).
B. obtusa. **1**:GA(82).
Ellisiodothis inquinans. On stems. **1**:GA(82,1438).
Erysiphe cichoracearum. Powdery mildew. **1**:IL(82).
Gloniopsis curvata. **1**:FL(2782).
Microsphaera penicillata. Powdery mildew. **1**:SE states(82), GA(1438), IN(82), MS(2720), OH(950), TX(82).
Mycosphaerella tecomae. Leaf spot, on wood. **1**:Range of host(82),

MS(2720), WV(3907).
Physalospora abdita. Leaf spot. **1**:NJ(82), VA(82).
Physalospora sp. **1**:GA(1438).
Valsa ceratosperma. **1**:GA(1438).
Valsaria bignoniae. On stems. **1**:PA(82).
Basidiomycotina - Other
Plicaturopsis crispa. **1**:NC(3944).
Pycnoporus sanguineus. On dead stems. **1**:LA(2319).
Typhula elegans. On stems. **1**:IN(264).
Deuteromycotina - Hyphomycetes
Cercospora ampelopsidis. **1**:DE(3543).
Cristulariella moricola. Leaf spot. **1**:AL(2056), GA(2056), IL(2572).
Myrothecium roridum. Leaf spot. **1**:TX(82).
Myrothecium sp. Leaf spot. **1**:FL(35).
Phymatotrichopsis omnivora. Root rot. **1**:TX(82).
Pseudocercospora sordida. Blight, leaf spot. **1**:MS(2720), NC(3944), OK(2873).
Deuteromycotina - Coelomycetes
Aplosporella tecomae. On stems. **1**:NY(832), OK(82), TX(82), WV(82).
Ascochyta tecomae. On stems. **1**:NY(82).
Diplodiella tecomae. On wood. On stems. **1**:LA(82), SC(634).
Microdiplodia minuta. On wood. **1**:MS(82,2720).
Phomopsis majuscula. On stems. **1**:NY(82).
Phyllosticta tecomae. Leaf spot. **1**:MS(82,2720).
Rhabdospora translucens. On stems. **1**:NY(82).
Seiridium breviaristata. On stems. **1**:MS(82,2720).
Septoria tecomae. Leaf spot. **1**:OK(82,2873), TX(82), WV(82,3907).
S. tecomaxochitl. **1**:IL(82).

Catalpa Scop.

1. *C. bignonioides* Walter (=*C. cordifolia* Moench) [catawba, common catalpa, Indian bean catalpa, southern catalpa]
2. *C. speciosa* (Warder ex E. E. Barney) Warder ex Engelm. [catawba-tree, cigar-tree, Indian bean, northern catalpa, western catalpa]
3. *Catalpa* sp.

Oomycetes
Pythium ultimum. Damping-off. **3**:KS(82).
Ascomycotina
Botryosphaeria obtusa. On limbs. **3**:AL(82), MD(82).
B. rhodina. **3**:AL(82).
B. ribis. On twigs. **1**:MD(82).
Dasyscyphus sp. **1**:GA(1438).
Didymosphaeria catalpae. On leaves. **1**:OH(82).
Eutypella juglandicola. On dead limbs. **1**:GA(82,1438).
Hypoxylon rubiginosum. On dead limbs. **1**:GA(82,1438), PA(82). **2**:GA(82), PA(82).
Hysterographium mori. On limbs. **1**:GA(1438), MN(2874).
Microsphaera elevata. Powdery mildew. **1**:E states(82), NE(82), TX(82).
M. penicillata. Powdery mildew. **1**:GA(1438), MA(1390), OH(950). **2**:GA(1438), OH(950). **3**:FL(35).
[*M. vaccinii*]. Powdery mildew. **1**:E states(82), NE(82), TX(82), WV(3907). **2**:E states(82), IL(2568), MS(2720), NE(82), TX(82). **3**:NJ(82).
Mycosphaerella sp. **1**:GA(1438).
Phyllactinia guttata. Powdery mildew. On leaves. **1**:IL(82), IN(82), OH(82,950), SC(82,634), VA(82). **2**:IN(82), KY(82), MA(82).
Propolomyces versicolor. Powdery mildew. **1**:GA(1438).
Basidiomycotina - Other
Antrodia albida. Brown crumbly sapwood rot. **3**:KS(82).
Armillaria mellea. Root rot. **1**:CA(2900). **2**:WA(82,3202).
Bjerkandera adusta. On wood. **2**:OH(82).
Dendrophora albobadia. White rot. **3**:Unk.(82).
D. erumpens. **1**:RI(543).
Flammulina velutipes. Wound rot. **2**:IN(82).
Hohenbuehelia nigra. **1**:GA(1439).
Schizophyllum commune. Sapwood rot. **1**:Range of host(82). **2**:Range of host(82). **3**:Range of host(82).
Stereum hirsutum. White rot. **3**:Range of host(82), IN(82), MO(82), OH(82).
S. ostrea. White rot. **3**:Unk.(82).
Trametes hirsuta. White rot. **2**:ID(3202), OH(82).
T. versicolor. Spongy yellow heart rot. **1**:NC states(82), NE states(82), KS(82), WI(82). **2**:NC states(82), NE states(82), KS(82), WI(82).
Deuteromycotina - Hyphomycetes
Alternaria catalpae. Leaf spot. **1**:MS(2720). **2**:OK(2872). **3**:NC(1303).
Alternaria sp. Leaf spot. **1**:E states(82), FL(35), IL(82), NC(1303), TX(82). **2**:E states(82), CO(49), IL(82), TX(82).
Cercospora catalpae. Leaf spot. **1**:E states(82), FL(35), IA(82), OK(2872), TX(82). **3**:FL(35).
Cristulariella moricola. Leaf spot. **2**:WV(3680,3907). **3**:FL(35).
Epicoccum nigrum. Zonate leaf spot. **1**:WV(3907).
Hyalotrochophora lignitalis. On wood. **3**:IA(2510).
Leptoxyphium axillatum. On leaves. **1**:SC(634). **2**:LA(82), SC(82).
Phymatotrichopsis omnivora. Sooty mold. **1**:TX(82). **2**:OK(82), TX(82).
Sporidesmium capsularum. **2**:OK(2872).
Thielaviopsis basicola. Black root rot (seedlings). **2**:OH(82).
Verticillium albo-atrum. Wilt. **1**:IL(82), NJ(82), NY(82). **2**:IL(82,992,1558), KS(1558,3022), NJ(82), NY(82). **3**:CA(3896), UT(3620).
Verticillium sp. Wilt. **3**:CO(49).
Deuteromycotina - Coelomycetes
Ascochyta catalpae. On leaves. **2**:IL(82), PA(82).
Colletotrichum gloeosporioides. Anthracnose. **1**:MA(82), MD(82), NJ(82), NY(82), PA(82). **2**:MA(82), MD(82), NC(1303), NJ(82), NY(82), PA(82). **3**:FL(35), NC(1303).
Macrophomina phaseolina. Anthracnose, leaf spot. **2**:OK(2872).
Phoma sp. Stem gall. **3**:FL(35).
Phomopsis sp. Twig dieback. **1**:FL(35,82), OH(82).
Phyllosticta catalpae. Leaf spot. **1**:Range of host(82), NC(1303), WV(3907). **2**:Range of host(82), MS(2720), OK(2872). **3**:NC(1303).
Rhabdospora baculum. On limbs. **3**:NJ(82), NY(82).
Sphaceloma sp. Spot anthracnose. **2**:LA(82). **3**:LA(82).
Deuteromycotina - Other
Rhizoctonia crocorum. Root rot of seedlings. **3**:OH(82).
R. solani. Damping-off, root and stem rot. **1**:FL(35). **3**:NE(82).
Sclerotium rolfsii. Southern blight of seedlings. **2**:TX(82).

Chilopsis D. Don

1. *C. linearis* (Cav.) Sweet [desert-willow, flowering-willow]

Oomycetes
Pythium ultimum. Damping-off. **1**:NE(82).
Pythium sp. **1**:OR(3202).
Basidiomycotina - Other
Ceriporia tarda. **1**:AZ(1202).
Coniophora arida. **1**:AZ(1202).
Coriolopsis gallica. On wood. **1**:OK(618).
Gloeophyllum trabeum. **1**:AZ(1202).
Grandinia pruni. **1**:AZ(1202).
Heliocybe sulcata. **1**:AZ(1202).
Hyphoderma clavigerum. **1**:AZ(1202).
H. setigerum. **1**:AZ(1202).
Perenniporia medulla-panis. **1**:AZ(1202).
Phanerochaete omnivora. White rot. **1**:AZ(433).
P. tuberculata. **1**:AZ(1202).
Phellinus badius. **1**:AZ(128,1202).
Schizopora apacheriensis. **1**:AZ(1193,1202).
Thanatephorus cucumeris. **1**:OR(3202).
Deuteromycotina - Hyphomycetes
Fusarium sp. **1**:OR(3202).
Phymatotrichopsis omnivora. Root rot. **1**:OK(2873), TX(82).
Deuteromycotina - Coelomycetes
Phyllosticta erysiphoides. Leaf spot. **1**:Unk.(82).
Deuteromycotina - Other
Rhizoctonia solani. Damping-off. **1**:NE(82), TX(82).

Crescentia L.

1. *C. cujete* L. [calabash]

Deuteromycotina - Hyphomycetes
Cercospora sp. Leaf spot. **1**:FL(35).

Cydista Miers

1. *C. aequinoctialis* (L.) Miers [garlicvine]
2. *Cydista* sp.

Deuteromycotina - Hyphomycetes
Cercospora sp. Leaf spot. **1**:FL(35). **2**:FL(35).
Oidium sp. Powdery mildew. **1**:FL(35).
Deuteromycotina - Coelomycetes
Colletotrichum sp. Anthracnose. **1**:FL(35). **2**:FL(35).
Macrophoma sp. Leaf spot. **2**:FL(35).
Phyllosticta sp. Leaf spot. **2**:FL(35).

Jacaranda Juss.
1. *J. acutifolia* Humb. & Bonpl.
Oomycetes
Pythium sp. Root rot. **1**:FL(35).
Basidiomycotina - Other
Armillaria mellea. Root rot. **1**:CA(82).
A. tabescens. Root rot. **1**:FL(35).
Deuteromycotina - Hyphomycetes
Phymatotrichopsis omnivora. Root rot. **1**:TX(82).
Deuteromycotina - Coelomycetes
Phomopsis sp. Leaf spot. **1**:FL(35).
Phyllosticta sp. Leaf spot. **1**:FL(35).

Kigelia DC.
1. *K. africana* (Lam.) Benth. (=*K. pinnata* (Jacq.) DC.) [sausage-tree]
Ascomycotina
Phyllachora sp. Tar spot. **1**:FL(35).
Deuteromycotina - Hyphomycetes
Fusarium sp. Dieback. **1**:FL(35).
Deuteromycotina - Coelomycetes
Colletotrichum sp. Anthracnose. **1**:FL(35).
Sphaeropsis sp. Leaf spot. **1**:FL(35).

Macfadyena DC.
1. *M. unguis-cati* (L.) A. Gentry (=*Doxantha unguis-cati* (L.) Miers) [cat's-claw trumpet, cat's-claw-vine, funnel-creeper]
Deuteromycotina - Hyphomycetes
Phymatotrichopsis omnivora. Root rot. **1**:TX(82).

Pandorea (Endl.) Spach
1. *Pandorea* sp.
Deuteromycotina - Hyphomycetes
Oidium sp. Powdery mildew. **1**:FL(35).

Pyrostegia K. Presl
1. *P. venusta* (Ker-Gawl.) Miers [flameflower, flamevine, flaming-trumpet, golden-shower]
2. *Pyrostegia* sp.
Deuteromycotina - Hyphomycetes
Cercospora pyrostegiae. Leaf spot. **1**:FL(35). **2**:FL(35).
Deuteromycotina - Coelomycetes
Colletotrichum sp. Anthracnose. **2**:FL(35).
Phyllosticta sp. Leaf spot. **1**:FL(35). **2**:FL(35).

Spathodea P. Beauv.
1. *S. campanulata* P. Beauv. [African tuliptree, flame-of-the-forest]
Deuteromycotina - Hyphomycetes
Graphium sp. Leaf blister. **1**:FL(35).
Helicomyces roseus. On pods. **1**:HI(1269).

Tabebuia B. A. Gomes ex DC.
1. *T. caraiba* (Mart.) Bureau (=*T. argentea* (Bureau & K. Schum.) Britton) [Paraguayan trumpet-tree, silver trumpet-tree, tree-of-gold]
2. *T. impetiginosa* (Mart. ex DC.) Tol.
3. *T. riparia* (Raf.) Sandw. (=*T. pentaphylla* (L.) Hemsl.) [whitewood]
4. *Tabebuia* sp.
Oomycetes
Pythium sp. Damping-off, root rot. **1**:FL(35). **3**:HI(2903).
Ascomycotina
Botryosphaeria obtusa. On limbs. **4**:AL(82), MD(82).
Basidiomycotina - Uredinales
Prospodium plagiopus. Autoecious. **1**:FL(35). **4**:FL(82).
Sphaerophragmium sp. **3**:FL(35).
Deuteromycotina - Hyphomycetes
Alternaria sp. Leaf spot. **1**:FL(35). **3**:FL(35).
Cercospora jahnii. Leaf spot. **1**:FL(35). **3**:FL(35).
Corynespora cassiicola. Leaf spot. **1**:FL(35). **3**:FL(35).
Fusarium sp. Root and stem rot. **1**:FL(35).
Deuteromycotina - Coelomycetes
Colletotrichum sp. Anthracnose. **1**:FL(35).
Phoma sp. Dieback. **1**:FL(35).
Phomopsis sp. Twig dieback. **1**:FL(35). **3**:FL(35).
Phyllosticta sp. Leaf spot. **3**:FL(35).
Deuteromycotina - Other
Rhizoctonia solani. Leaf necrosis. **1**:FL(35).
Rhizoctonia sp. Crown and root rot. **2**:HI(2903).

Tecoma Juss.
1. *T. stans* (L.) H. B. K. (=*Stenolobium stans* (L.) Seem.) [trumpet-flower, yellow elder, yellow-bells]
Basidiomycotina - Uredinales
Prospodium appendiculatum. Autoecious. **1**:FL(35,82), TX(82).
Basidiomycotina - Other
Armillaria tabescens. Root rot. **1**:FL(35,82).
Deuteromycotina - Hyphomycetes
Cercospora sp. Leaf spot. **1**:FL(35).
Phymatotrichopsis omnivora. Root rot. **1**:TX(82).
Deuteromycotina - Coelomycetes
Phomopsis sp. Stem canker. **1**:FL(35).

Tecomaria (Endl.) Spach
1. *T. capensis* (Thunb.) Spach [Cape honeysuckle]
Oomycetes
Pythium splendens. Root rot. **1**:FL(35).
Basidiomycotina - Other
Armillaria mellea. Root rot. **1**:CA(82).
A. tabescens. Root rot. **1**:FL(35,82).
Deuteromycotina - Hyphomycetes
Cercospora sp. Leaf spot. **1**:FL(35).
Oidium sp. Powdery mildew. **1**:FL(35).
Deuteromycotina - Coelomycetes
Colletotrichum gloeosporioides. Anthracnose. **1**:FL(35), TX(82).
Deuteromycotina - Other
Rhizoctonia solani. Root rot. **1**:FL(35).
Rhizoctonia sp. Crown and root rot. **1**:HI(2903).

Bixaceae

Bixa L.
1. *B. orellana* L. [achiote, annatto, lipstick-tree]
Oomycetes
Pythium sp. Root rot. **1**:FL(35).
Ascomycotina
Erysiphe sp. Powdery mildew. **1**:FL(35).
Basidiomycotina - Uredinales
Crossopsora bixae. **1**:FL(35).
Deuteromycotina - Hyphomycetes
Cercospora bixae. Leaf spot. **1**:FL(35).
Oidium bixae. Powdery mildew. **1**:FL(35).
Deuteromycotina - Coelomycetes
Phyllosticta sp. Leaf spot. **1**:FL(35).

Cochlospermum Kunth in H. B. K.
1. *C. vitifolium* (Willd.) Spreng.
Basidiomycotina - Other
Armillaria tabescens. Root rot. **1**:FL(35).
Deuteromycotina - Hyphomycetes
Cercospora cochlospermi. Leaf spot. **1**:FL(35).

Bombacaceae

Bombax L.
1. *B. ceiba* L. (=*Salmalia malabarica* (DC.) Schott & Endl., *B. malabaricum* DC.) [red silk-cotton-tree]
Oomycetes
Pythium sp. Root rot. **1**:FL(35).
Deuteromycotina - Hyphomycetes
Corynespora sp. Leaf spot. **1**:FL(35).

Ceiba Mill.
1. *C. pentandra* (L.) Gaertn. [ceiba, kapok, silk-cotton-tree, white silk-cotton-tree]

Deuteromycotina - Hyphomycetes
Cercospora sp. Leaf spot. **1**:FL(35).
Corynespora cassiicola. Leaf spot. **1**:FL(35).
Deuteromycotina - Coelomycetes
Colletotrichum sp. Anthracnose. **1**:FL(35).

Chorisia Kunth
1. *C. speciosa* St. Hil. [floss-silk-tree]

Oomycetes
Pythium splendens. Root rot. **1**:FL(35).
Deuteromycotina - Hyphomycetes
Corynespora sp. Leaf spot. **1**:FL(35).
Deuteromycotina - Coelomycetes
Colletotrichum sp. Leaf spot. **1**:FL(35).
Macrophoma sp. Leaf necrosis. **1**:FL(35).
Phyllosticta sp. Leaf spot. **1**:FL(35).

Pachira Aubl.
1. *P. aquatica* Aubl. [Guiana-chestnut, provision-tree, water-chestnut, wild-cocoa]

Deuteromycotina - Hyphomycetes
Curvularia sp. Leaf spot. **1**:FL(35).

Pseudobombax Dugand
1. *P. ellipticum* (Kunth) Dugand [shaving-brush-tree]

Ascomycotina
Leptosphaeria sp. Leaf spot. **1**:FL(35).
Deuteromycotina - Hyphomycetes
Cercospora sp. Leaf spot. **1**:FL(35).

Boraginaceae

Amsinckia Lehm.
1. *A. douglasiana* A. DC.
2. *A. intermedia* Fisch. & C. A. Mey.
3. *A. lycopsoides* Lehm.
4. *A. menziesii* (Lehm.) A. Nels. & Macbr.
5. *A. retrorsa* Suksd.
6. *A. tessellata* A. Gray

Oomycetes
Peronospora echinospermi. On leaves. **2**:AZ(1195,3381).
P. myosotidis. **2**:OR(3202). **3**:ID(3202). **4**:ID(3202), OR(3202), WA(3202). **5**:ID(3202), OR(3202), WA(3202).
Ascomycotina
Erysiphe cichoracearum. Powdery mildew. **2**:ID(3202), OR(3202), WA(3202). **3**:WA(3202). **5**:ID(3202), OR(3202), WA(3202). **6**:WA(3202).
E. cynoglossi. **2**:AZ(1195).
Basidiomycotina - Ustilaginales
[*Entyloma compositarum*]. **2**:WA(1077).
E. serotinum. **1**:CA(1077). **2**:WA(1077,3202). **6**:OR(3202).

Anchusa L.
1. *A. officinalis* L. [alkanet, bugloss]

Basidiomycotina - Uredinales
Puccinia recondita. Heteroecious: aecial host. **1**:IN(82), MI(82), WA(3202).
Deuteromycotina - Other
Rhizoctonia solani. Damping-off. **1**:CT(82).

Borago L.
1. *B. officinalis* L. [borage, cool-tankard, talewort]

Deuteromycotina - Hyphomycetes
Cercospora agnostoica. Leaf spot. **1**:SD(2243).
Ramularia sp. Leaf spot. **1**:CA(82).

Cordia L.
1. *C. boissieri* A. DC. [anacahuita]
2. *Cordia* sp.

Ascomycotina
Elsinoe sp. Scab. **2**:FL(35).
Erysiphe cichoracearum. Powdery mildew. **1**:TX(82).
Basidiomycotina - Uredinales
Alveolaria cordiae. **1**:CA(82).
Puccinia cordiae. **2**:FL(35).
Basidiomycotina - Other
Armillaria tabescens. Root rot. **2**:FL(82).
Deuteromycotina - Hyphomycetes
Phymatotrichopsis omnivora. Root rot. **1**:TX(82).
Deuteromycotina - Coelomycetes
Phyllosticta sp. Leaf spot. **2**:FL(35).

Cryptantha Lehm. ex G. Don
1. *C. affinis* (A. Gray) Greene
2. *C. ambigua* (A. Gray) Greene
3. *C. angustifolia* (Torr.) Greene
4. *C. barbigera* (A. Gray) Greene
5. *C. crassisepala* (Torr. & A. Gray) Greene
6. *C. intermedia* (A. Gray) Greene (=*C. hendersonii* (A. Nels.) J. Nels.)
7. *C. pterocarya* (Torr.) Greene
8. *C. torreyana* (A. Gray) Greene
9. *C. watsonii* (A. Gray) Greene
10. *Cryptantha* sp.

Chytridiomycetes
Synchytrium myosotidis. Leaf gall. **3**:AZ(1846). **8**:CA(1846). **10**:AZ(82), CA(82).
Oomycetes
Peronospora myosotidis. **1**:ID(3202), OR(3202), WA(3202). **2**:MT(3202). **8**:ID(3202), MT(3202), OR(3202), WA(3202). **9**:ID(3202).
Ascomycotina
Erysiphe cichoracearum. Powdery mildew. **1**:CA(642), OR(3202), WA(3202). **2**:WA(3202). **6**:CA(81). **8**:CA(81). **10**:CA(82), NV(82).
E. cynoglossi. Powdery mildew. **4**:AZ(1195,3378).
Basidiomycotina - Uredinales
[*Puccinia aristidae*]. **4**:AZ(1203). **5**:AZ(1203). **7**:AZ(1203). **10**:AZ(82).
P. cryptanthes. Autoecious. **8**:WA(3202). **10**:CA(82), OR(3202), WA(82).
P. subnitens. **4**:AZ(3981). **5**:AZ(3981). **7**:AZ(3981).

Cynoglossum L.
1. *C. amabile* Stapf & J. R. Drumm. [Chinese forget-me-not]
2. *C. boreale* Fernald
3. *C. officinale* L. [hound's-tongue]
4. *C. virginianum* L.
5. *C. zeylanicum* (Vahl ex Hornem.) Thunb. ex Lehm.
6. *Cynoglossum* sp.

Oomycetes
Peronospora cynoglossi. Downy mildew. **3**:IL(82), MD(82), TX(82).
Ascomycotina
Erysiphe cichoracearum. Powdery mildew. **2**:MN(2874). **3**:OR(3202), UT(82), WY(82,1195). **4**:VA(82).
Sclerotinia sclerotiorum. Stem rot. **1**:WA(82,3202).
Deuteromycotina - Hyphomycetes
Botrytis cinerea. **6**:WA(3202).
Cercospora cynoglossi. Leaf spot. **3**:IN(82). **6**:WI(1335).
Phymatotrichopsis omnivora. Root rot. **3**:TX(82).
Ramularia lappulae. Leaf spot. **3**:TX(82), WI(82).
Deuteromycotina - Coelomycetes
Ascochyta sp. On leaves. **3**:WI(1338).
Phoma cynoglossi. On stems. **3**:NY(82,832), TX(82).
P. exigua. Leaf spot. **3**:MN(2874), TX(82), WI(64,82,1333).
Deuteromycotina - Other
Rhizoctonia solani. Root and stem rot. **1**:FL(35). **5**:FL(35).
Sclerotium rolfsii. Southern blight. **1**:FL(35,82). **5**:FL(35).

Echium L.
1. *E. vulgare* L. [blue-devil, blueweed, viper's bugloss]

Ascomycotina
Rosellinia sp. Root rot. **1**:CA(82).
Deuteromycotina - Hyphomycetes
Cercospora echii. Leaf spot. **1**:Unk.(82), MS(2720), WI(1344).

Hackelia Opiz in Bercht. & Opiz
1. *H. californica* (A. Gray) I. M. Johnst.
2. *H. cinerea* (Piper) I. M. Johnst.
3. *H. floribunda* (Lehm.) I. M. Johnst. (=*Lappula floribunda* (Lehm.) Greene)
4. *H. virginiana* (L.) I. M. Johnst. (=*Lappula virginiana* (L.) Greene)
5. *Hackelia* sp.

Oomycetes
Peronospora echinospermi. Downy mildew. **2**:WA(3202). **3**:MT(3202). **4**:IA(82), WI(1342).
P. myosotidis. **3**:MT(82).

Ascomycotina
Erysiphe cichoracearum. Powdery mildew. **1**:CA(81). **4**:IA(82), MI(82), MN(82), MO(82), OH(82,950), TX(82), WI(82), WV(3907). **5**:MT(3202).
Microsphaera sp. Powdery mildew. **4**:IL(82).

Basidiomycotina - Uredinales
Puccinia hydrophylli. **3**:UT(1520).
P. hydrophylli subsp. *mertensiae*. Autoecious. **3**:UT(82).

Basidiomycotina - Ustilaginales
Entyloma serotinum. Leaf smut. **3**:UT(82,1077).

Deuteromycotina - Hyphomycetes
Cercospora cynoglossi. Leaf spot. **4**:WI(82,1335).
Ovularia asperifolii var. *lappulae*. Leaf spot. **4**:WI(82).
Ramularia lappulae. On leaves. **3**:CA(661).

Deuteromycotina - Coelomycetes
Phoma exigua. Leaf spot. **4**:WI(64,82,1333).

Heliotropium L.
1. *H. arborescens* L. (=*H. peruvianum* L.) [cherry-pie, heliotrope]
2. *H. convolvulaceum* (Nutt.) A. Gray (=*Euploca convolvulacea* Nutt.)
3. *H. curassavicum* L. [salt heliotrope, seaside heliotrope]

Basidiomycotina - Uredinales
Puccinia aristidae. Heteroecious: aecial host. **3**:AZ(82), CA(82), NE(82), NM(82), TX(82), UT(82).

Deuteromycotina - Hyphomycetes
Botrytis cinerea. Leaf spot. **1**:Range of host(82), AK(526).
Cercospora heliotropii. Leaf spot. **2**:OK(2873). **3**:NM(82), OK(2872), OR(82,3202).
Cladosporium heliotropii. **1**:AK(526).
Verticillium albo-atrum. Wilt. **1**:MD(82).

Lappula Moench
1. *L. redowskii* (Hornem.) Greene (=*L. texana* (Scheele) Britton) [stickseed]
2. *L. squarrosa* (Retz.) Dumort. (=*L. echinata* Gilib.)
3. *Lappula* sp.

Oomycetes
Peronospora echinospermi. Downy mildew. **1**:AZ(1195,3378), ID(3202), KS(82), MT(82,3202), NE(82). **2**:ND(82), TX(82). **3**:ND(377), TX(82).

Ascomycotina
Erysiphe cichoracearum. Powdery mildew. **1**:CA(82), MT(82,3202), TX(82), WY(82). **2**:MN(2874). **3**:MN(2874).
E. cynoglossi. Powdery mildew on stems and leaves. **1**:AZ(1195,3378).
Leptospora rubella. **3**:OH(3248).
Ophiobolus collapsus. **3**:ND(377).

Basidiomycotina - Uredinales
[*Puccinia aristidae*]. **1**:AZ(82,1203), CO(82), UT(82).
P. hydrophylli subsp. *mertensiae*. **3**:TX(82).
P. subnitens. **1**:AZ(3981).

Deuteromycotina - Hyphomycetes
Phymatotrichopsis omnivora. Root rot. **2**:TX(82). **3**:TX(82).

Lithospermum L.
1. *L. arvense* L.
2. *L. canescens* (Michx.) Lehm. [hoary puccoon, Indian-paint]
3. *L. caroliniense* (J. F. Gmelin) MacMill. (=*L. croceum* Fernald)
4. *L. incisum* Lehm. (=*L. angustifolium* Michx.)
5. *L. ruderale* Douglas ex Lehm. [western gromwell]
6. *L. tuberosum* Rugel ex DC.

Ascomycotina
Calloria lithospermi. **2**:MT(3202).
Erysiphe cichoracearum. Powdery mildew. **1**:PA(82), TX(82). **2**:CO(82), MT(82), NM(82), NV(82), UT(82), WY(82).

Basidiomycotina - Uredinales
Aecidium hesleri. **6**:TN(82).
[*Puccinia montanensis*]. **4**:ND(378).
P. recondita. Heteroecious: aecial host. **1**:TX(82). **2**:ND(82). **4**:ND(82), NE(82), SD(82,2243).

Deuteromycotina - Hyphomycetes
Botrytis cinerea. **5**:ID(3202).
Cercospora lithospermi. On living leaves. **2**:WI(1340). **3**:WI(1339).
Phymatotrichopsis omnivora. Root rot. **1**:TX(82).
Pseudocercosporella herpotrichoides. On roots. **5**:ID(82), OR(82), WA(82).

Deuteromycotina - Coelomycetes
Septoria lithospermi. On leaves. **2**:WI(82,1337).

Mertensia Roth
1. *M. alpina* (Torr.) G. Don
2. *M. arizonica* Greene
3. *M. bella* Piper
4. *M. ciliata* (E. James ex Torr.) G. Don
5. *M. lanceolata* (Pursh) DC.
6. *M. lanceolata* (Pursh) DC. var. *brachyloba* (Greene) A. Nels. (=*M. brachyloba* Greene)
7. *M. maritima* (L.) S. F. Gray
8. *M. oblongifolia* (Nutt.) G. Don (=*M. oblongifolia* (Nutt.) G. Don var. *nevadensis* (A. Nels.) L. O. Williams)
9. *M. paniculata* (Aiton) G. Don
10. *M. platyphylla* A. Heller
11. *M. virginica* (L.) Pers. [Roanoke bells, Virginia bluebells, Virginia cowslip]
12. *M. viridis* A. Nels. (=*M. bakeri* Greene)
13. *Mertensia* sp.

Chytridiomycetes
Synchytrium sp. **4**:WY(1847).

Oomycetes
Peronospora myosotidis. **9**:OR(3202). **13**:MT(3202).
Peronospora sp. Downy mildew. **4**:MT(82).

Ascomycotina
Crocicreas nigrofuscum var. *allantosporum*. **13**:UT(505).
Didymosphaeria johansenii. **7**:AK(526).
Erysiphe cichoracearum. Powdery mildew. **4**:CO(649), MT(3202), WY(1195). **8**:WA(526). **9**:MT(3202). **11**:PA(82). **13**:CO(82), MN(2874), MT(82), NM(82), NV(82), OR(3202), UT(82), WY(82,1195).
E. communis. **13**:AK(526).
Lachnella cenangioides. On stems. **9**:WA(849).
Leptosphaeria erigerontis. **9**:WA(3202).
Nodulosphaeria olivacea. **4**:WY(3250).
Sclerotinia sclerotiorum. Stem rot. **5**:CO(82).

Basidiomycotina - Uredinales
Puccinia hydrophylli subsp. *mertensiae*. Autoecious. **1**:MT(1520), WY(1520). **2**:UT(1520). **4**:CO(82,1520), NV(82), UT(82,1520), WY(82,1195,1520,3379). **8**:UT(1520), WA(526), WY(1520). **9**:AK(60,526,1520). **12**:CO(1520), UT(1520). **13**:CO(82), NV(82), UT(82), WY(82).
P. recondita. Heteroecious: aecial host. **3**:ID(3202). **9**:ID(82,3202), MT(82,3202), OR(3202), WA(3202). **10**:ID(3202). **13**:OR(82).

Basidiomycotina - Ustilaginales
Entyloma serotinum. Leaf smut. **3**:OR(3202). **4**:CA(1077), ID(3202), MT(3202), WY(82,1077,1195). **6**:CO(1077). **8**:ID(526), OR(526). **9**:ID(3202), OR(3202), WA(1077,3202). **11**:DC(1077), IA(82,1077), IN(82,1077), MD(82,1077), VA(82,1077), WY(1077).

Deuteromycotina - Coelomycetes
Ascochyta madisonensis. On living leaves. **11**:WI(1352).
Septoria poseyi. **9**:OR(3202).

Myosotis L.
1. *M. alpestris* F. W. Schmidt
2. *M. alpestris* F. W. Schmidt subsp. *asiatica* Vestergr. ex Hultén
3. *M. discolor* Pers.
4. *M. laxa* Lehm.
5. *M. macrosperma* Engelm.
6. *M. micrantha* Pall. ex Lehm.
7. *M. scorpioides* L. [forget-me-not]
8. *M. sylvatica* Ehrh. ex Hoffm. [garden forget-me-not]

9. *M. verna* Nutt. (=*M. virginica* B. S. P.)
10. *Myosotis* sp.

Oomycetes

Peronospora myosotidis. Downy mildew. **3:**OR(3202). **4:**WI(82). **5:**MS(2720). **6:**ID(3202), WA(3202). **9:**IL(82), MI(82), MS(82,2720), OR(3202), WA(3202), WI(82).
Plasmopara myosotidis. On leaves. **4:**WI(1344,3201).

Ascomycotina

Erysiphe cichoracearum. Powdery mildew. **8:**CA(1155).
Pleospora chlamydospora. On dead stem. **2:**AK(1938).
Sclerotinia sclerotiorum. Crown rot. **7:**IL(82), WA(82). **8:**IL(82), WA(82). **10:**WA(3202).

Basidiomycotina - Uredinales

Puccinia eatoniae. Heteroecious: aecial host. **5:**MS(2720). **9:**IL(82), IN(82), MO(82), NC(82,3944), WI(82).
P. hydrophylli. **1:**CO(1520).
P. hydrophylli subsp. *mertensiae*. Autoecious. **4:**CO(82). **8:**MT(3202).

Basidiomycotina - Ustilaginales

Entyloma serotinum. **8:**MT(3202).

Deuteromycotina - Hyphomycetes

Botrytis cinerea. Gray mold blight. **7:**Range of host(82). **8:**Range of host(82).
Cercospora sp. **9:**WI(1354).
Oidium sp. **8:**OH(950).

Deuteromycotina - Coelomycetes

Phoma exigua. **7:**WI(1356).

Omphalodes Mill.

1. *Omphalodes* sp.

Deuteromycotina - Hyphomycetes

Botrytis cinerea. **1:**WA(3202).

Onosmodium Michx.

1. *O. molle* Michx. [marbleseed]
2. *O. molle* Michx. var. *occidentale* (Mackenz.) I. M. Johnst.

Basidiomycotina - Uredinales

[*Puccinia montanensis*]. **1:**ND(378).
P. recondita. Heteroecious: aecial host. **2:**CO(82), KS(82), ND(82), NE(82).

Deuteromycotina - Hyphomycetes

Phymatotrichopsis omnivora. Root rot. **2:**Unk.(82).

Deuteromycotina - Coelomycetes

Phyllosticta sp. **2:**WI(1344).
Septoria lithospermi. **2:**WI(1341).

Pectocarya DC. ex Meisn.

1. *P. linearis* (Ruiz & Pav.) DC.
2. *P. linearis* (Ruiz & Pav.) DC. var. *ferocula* I. M. Johnst.
3. *P. penicillata* (Hook. & Arnott) A. DC.

Chytridiomycetes

Synchytrium myosotidis. Leaf gall. **1:**AZ(1846). **3:**CA(1846).

Ascomycotina

Erysiphe cichoracearum. **1:**WA(3202).

Basidiomycotina - Uredinales

[*Puccinia aristidae*]. **1:**AZ(1203).
P. subnitens. **2:**AZ(3981).

Plagiobothrys Fisch. & C. A. Mey.

1. *P. scouleri* (Hook. & Arnott) I. M. Johnst.
2. *P. scouleri* (Hook. & Arnott) I. M. Johnst. var. *penicillatus* (Greene) Cronq. (=*P. ramosus* Piper)
3. *P. tenellus* (Nutt. ex Hook.) A. Gray

Oomycetes

Peronospora myosotidis. **1:**ID(3202), OR(3202), WA(3202). **2:**WA(3202). **3:**WA(3202).

Ascomycotina

Erysiphe cichoracearum. **1:**OR(3202), WA(3202).

Symphytum L.

1. *S. officinale* L. [boneset, common comfrey, healing herb]
2. *S.* × *uplandicum* Nym. (=*S. peregrinum* Ledeb.). *S. asperum* × *S. officinale*. [blue comfrey, Quaker comfrey, Russian comfrey]

Deuteromycotina - Hyphomycetes

Stemphylium sp. Leaf spot. **2:**NY(372).

Deuteromycotina - Other

Rhizoctonia sp. Root rot. **1:**HI(2903).

Tournefortia L.

1. *T. argentea* L. (=*Messerschmidia argentea* (L.) I. M. Johnst.)

Deuteromycotina - Other

Rhizoctonia sp. Root rot. **1:**HI(2903).

Brassicaceae

Alyssum L.

1. *A. argenteum* Vitm.
2. *A. saxatile* L. [basket-of-gold, gold-dust, golden-tuft alyssum, golden-tuft madwort, rock madwort]
3. *Alyssum* sp.

Plasmodiophoromycetes

Plasmodiophora brassicae. Clubroot. **2:**NJ(82). **3:**WA(3202).

Oomycetes

Pythium sp. Basal stem rot. **3:**FL(35).

Deuteromycotina - Other

Rhizoctonia solani. Damping-off. **1:**NJ(82). **2:**NJ(82). **3:**FL(35).

Arabidopsis Heynh. in Holl & Heynh.

1. *A. thaliana* (L.) Heynh. (=*Sisymbrium thalianum* (L.) Gay & Monnard) [mouse-ear cress]

Oomycetes

Albugo candida. **1:**NC(3944).
Peronospora parasitica. **1:**NC(3944).

Ascomycotina

Sclerotinia sp. Leaf and stem blight. **1:**MD(2474).

Arabis L.

1. *A. blepharophylla* Hook. & Arnott
2. *A. canadensis* L.
3. *A. cusickii* S. Wats.
4. *A. divaricarpa* A. Nels.
5. *A. drummondii* A. Gray
6. *A. furcata* S. Wats.
7. *A. hirsuta* (L.) Scop.
8. *A. holboellii* Hornem.
9. *A. holboellii* Hornem. var. *retrofracta* (R. J. Graham) Rydb. (=*A. exilis* A. Nels., *A. lignipes* A. Nels. in J. M. Coulter)
10. *A. laevigata* (Muhl. ex Willd.) Poir.
11. *A. lignifera* A. Nels.
12. *A. lyallii* S. Wats.
13. *A. lyatra* L. subsp. *occidentalis* (S. Wats.) Piper
14. *A. lyrata* L. var. *kamchatica* Fisch. ex DC.
15. *A. lyrata* L. var. *kamtschatica* (Fisch.) Hultén
16. *A. microphylla* Nutt.
17. *A. nuttallii* B. L. Robinson
18. *A. perennans* S. Wats.
19. *A. platysperma* A. Gray
20. *A. puberula* Nutt.
21. *A. shortii* (Fisch.) Gleason
22. *A. sparsiflora* Nutt.
23. *A. suffrutescens* S. Wats.
24. *A. virginica* (L.) Trel. [rock cress]
25. *A. whitedii* Piper
26. *Arabis* sp.

Chytridiomycetes

Synchytrium lacunosum. **26:**MT(1848).

Oomycetes

Albugo candida. White rust. **4:**MN(2874). **6:**MT(3202), WA(3202). **7:**ID(3202). **10:**OK(2872). **24:**MS(2720). **26:**NE states(82), CO(82), NC(1303), TX(82), WA(82).
Peronospora parasitica. On leaves. **2:**WI(1361). **5:**AZ(1195), MT(3202), WY(3378). **7:**AK(526). **8:**ID(1195,3202), WA(3202). **21:**WI(1356). **22:**OR(3202). **26:**AK(82), AL(82), CO(82), IN(82), MI(82), NC(1303), TX(82), WI(82).

Ascomycotina

Belonium arabicola. **6:**WA(3202).
Comoclathris permunda. On stems. **19:**CA(642). **26:**CA(82), WA(3202).
Erysiphe polygoni. Powdery mildew. **1:**CA(3975). **19:**CA(638,642).

Heptameria tenera. **5:**WA(3202,3814).
Microthyrium microscopicum. On dead stems. **19:**CA(642).
Mycosphaerella cruciferarum. On capsules. **26:**Unk.(82).
M. tassiana. On overwintered stems. **6:**WA(3202). **19:**CA(642).
M. tassiana var. *arctica*. On dead stem. **26:**AK(1938).
Pleospora coloradensis. On dead stem. **26:**AK(1938).
P. pluriseptata. **26:**WA(3202).
Basidiomycotina - Uredinales
Puccinia monoica. Heteroecious: aecial host. **3:**WA(3202). **4:**MN(2874). **5:**MT(3202), WY(1195). **8:**ID(3202), MT(1195,3202), WA(3202), WY(1195). **9:**MT(3202). **12:**MT(3202), OR(3202), WA(3202). **18:**AZ(3981). **19:**CA(638,642,661). **20:**WA(3202). **23:**OR(3202). **25:**WA(3202). **26:**C states(82), W states(82), AZ(1203,3981), CA(82,638,642), CO(1195), WA(82).
P. thlaspeos. Autoecious. **4:**AK(526), ID(3202). **5:**CA(81), MT(3202), WY(1195). **6:**WA(3202). **8:**AK(60,526), MT(1195,3202), WA(3202), WY(1195). **11:**WY(1195). **12:**MT(1195,3202), WA(3202). **13:**AK(526). **14:**AK(526). **15:**AK(60,526). **16:**ID(3202), MT(1195,3202). **17:**ID(3202), MT(3202). **22:**WA(3202). **26:**W states(82), AK(526), AZ(3981), ID(3202).
Deuteromycotina - Hyphomycetes
Botrytis cinerea. Gray mold blight. **26:**WA(82).
Cercospora barbarea. **2:**WI(1343).
Phymatotrichopsis omnivora. Root rot. **26:**TX(82).
Ramularia sp. **7:**WA(3202).
Deuteromycotina - Coelomycetes
Cornularia sphaeroidea. On stems. **12:**WA(3202). **23:**WA(3202). **26:**UT(82).
Heteropatella umbilicata. **5:**WA(3202,3816).
Phlyctema andersonii. On stems. **8:**MT(3202). **26:**MT(82).
Phoma nebulosa. **20:**WA(3202).
Phoma sp. On capsules and stems. **26:**TX(82).
Phyllosticta sp. **8:**WA(3202).
Septoria arabidis. Leaf spot. **9:**CO(1195), WY(1195). **26:**TX(82).
S. arabidis var. *macrospora*. Leaf spot. **26:**CO(82).
Sirexcipula wyomingensis. **19:**CA(642).
Deuteromycotina - Other
Rhizoctonia solani. Damping-off. **26:**NC(1303), NJ(82).

Armoracia P. Gaertn., B. Mey. & Scherb.

1. *A. rusticana* P. Gaertn., B. Mey. & Scherb. (=*Rorippa armoracia* (L.) Hitchc., *Nasturtium armoracia* (L.) Fr., *Cochlearia armoracia* L., *A. lapathifolia* Gilib. ex Usteri, *Radicula armoracia* (L.) B. L. Robinson) [horseradish, red-cole]

Plasmodiophoromycetes
Plasmodiophora brassicae. Clubroot. **1:**IL(82), NC(1303).
Oomycetes
Peronospora parasitica. Downy mildew. **1:**AL(82), IL(82), NC(1303), NJ(82), SD(82), WI(82).
Ascomycotina
Erysiphe polygoni. Powdery mildew. **1:**CA(82), FL(3500).
Deuteromycotina - Hyphomycetes
Alternaria brassicae. Leaf spot. **1:**NE states(82), IA(82), NC(1303), NJ(3543), TX(82).
A. brassicicola. Anthracnose. **1:**NE states(82), NE(82), TX(82), UT(82).
Cercospora armoraciae. Leaf spot. **1:**Range of host(82), FL(35), MN(2874), OK(2873), WV(3907).
Penicillium hirsutum. Root pitting. **1:**IL(82), NJ(82).
Phymatotrichopsis omnivora. Root rot. **1:**TX(82).
Thielaviopsis basicola. Root rot. **1:**KS(82), NJ(82).
Verticillium albo-atrum. Wilt. **1:**MI(82), WA(82,3202).
Deuteromycotina - Coelomycetes
Ascochyta armoraciae. **1:**WA(3202).
Phyllosticta orbicula. **1:**NY(82).
Deuteromycotina - Other
Rhizoctonia solani. Damping-off. **1:**Unk.(82), CT(2970), NC(1303).

Barbarea R. Br. in Aiton & W. T. Aiton

1. *B. orthoceras* Ledeb.
2. *B. vulgaris* R. Br. [upland cress, winter cress, yellow-rocket]

Oomycetes
Albugo candida. White rust. **1:**WA(3202). **2:**CA(81,82), TX(82).
Peronospora parasitica. Downy mildew. **2:**TX(82).
Ascomycotina
Sclerotinia sp. Leaf and stem blight. **2:**MD(2474).
Deuteromycotina - Hyphomycetes
Alternaria sp. Leaf spot. **2:**PA(82).
Cercospora barbarea. Leaf spot. **2:**WI(82).
Cladosporium sp. **2:**DE(3543).
Ramularia barbareae. Leaf spot. **2:**NE states(82), DE(3543), OH(82), TX(82), WI(82).
Deuteromycotina - Coelomycetes
Ascochyta sp. On leaves. **2:**WI(1353).
Deuteromycotina - Other
Sclerotium rolfsii. Stem rot. **2:**TX(82).

Berteroa DC.

1. *B. incana* (L.) DC. [hoary alyssum]

Deuteromycotina - Hyphomycetes
Cercospora nasturtii. **1:**WI(1333,1340).

Brassica L.

1. *B. chinensis* L. (=*B. rapa* L. var. *chinensis* (L.) Kitam., *B. oleracea* L. var. *chinensis* (L.) Prain) [pak-choi]
2. *B. geniculata* (Desf.) J. Ball
3. *B. juncea* (L.) Czernj. & Coss., [brown mustard, Indian mustard, mustard-greens]
4. *B. juncea* (L.) Czernj. & J. M. Coulter var. *crispifolia* L. H. Bailey [brown mustard, curled mustard, Indian mustard, leaf mustard, mustard-greens, ostrich-plume, southern cole]
5. *B. napus* L. [colza, rape, rutabaga]
6. *B. napus* L. var. *napobrassica* (L.) Rchb. (=*B. napobrassica* (L.) Mill., *B. campestris* L. var. *napobrassica* (L.) DC.) [rutabaga, swede, swedish turnip]
7. *B. napus* L. var. *napus* [colza, rape]
8. *B. nigra* (L.) W. Koch [black mustard]
9. *B. oleracea* L. (=*B. caulorapa* Pasq.) [cabbage, kale, kohlrabi, wild cabbage]
10. *B. oleracea* L. var. *acephala* DC. [borecole, braschette, cole, colewort, collards, flowering cabbage, kale]
11. *B. oleracea* L. var. *botrytis* L. [broccoli, cauliflower]
12. *B. oleracea* L. var. *capitata* L. [cabbage, savoy]
13. *B. oleracea* L. var. *gemmifera* DC. [brussel sprouts]
14. *B. oleracea* L. var. *gongylodes* L. [kohlrabi]
15. *B. oleracea* L. var. *italica* Plenck [asparagus broccoli, Italian broccoli, sprouting broccoli]
16. *B. oleracea* L. var. *viridis* L.
17. *B. pekinensis* (Lour.) Rupr. (=*B. pe-tsai* L. H. Bailey) [celery cabbage, Chinese cabbage, pe-tsai, Shantung cabbage]
18. *B. rapa* L. (=*B. campestris* L.) [bird rape, field mustard, turnip]
19. *Brassica* sp.

Plasmodiophoromycetes
Plasmodiophora brassicae. 'Clubroot.' **3:**CA(82), CT(82), NC(1303), OH(82), TX(82), WA(82). **5:**CT(82), MN(82), NC(1303), ND(82), NJ(82), NY(82), WA(82). **6:**CT(82), LA(3278), MN(82,2874), ND(82), NJ(82), NY(82), WA(82,3202). **8:**CA(82), NC(1303). **9:**AK(526), NC(3931), OR(3202), WA(3202). **11:**Range of host(82), CO(49), NC(1303). **12:**Range of host(82), FL(35), HI(2903), MN(2874), NC(1303,3928), OK(2872), WV(3907). **13:**Range of host(82), NC(1303). **14:**Range of host(82). **15:**CA(486). **16:**Range of host(82), NC(1303). **17:**CT(82), FL(35), HI(2903), MA(82), NJ(82), OH(82), PA(82). **18:**E states(82), AK(526), CA(82), CO(49,82), FL(35), LA(2680), MN(82,2874), NC(1303), TX(82), WA(82,3202).
Chytridiomycetes
Olpidium brassicae. On roots. **12:**MN(82,2874), MS(82,2720), NY(82), WI(82).
Oomycetes
Albugo candida. Leaf spot. 'White rust. Staghead.' **1:**HI(2903). **3:**Range of host(82), FL(35), HI(2903), NC(1303), OK(2872). **4:**MS(2720). **5:**MN(82,2874). **6:**HI(82,2903), MA(82), TX(82), VT(82). **8:**Range of host(82), FL(35), HI(2903), ID(3202), MN(2874), MS(2720), NC(1303), WA(3202), WV(3907). **9:**ID(3202). **10:**NC(1303). **11:**CA(82), FL(35), HI(2899,2903), NC(1303). **12:**CA(82), FL(35), ID(82), KY(82), NC(82,1303), NE(82), OH(82), TX(82). **13:**NC(1303). **16:**TX(82). **17:**HI(2903). **18:**Range of host(82), AL(285), CA(485), FL(35), HI(2899,2903), ID(3202), NC(1303), OK(2872). **19:**AL(82),

CO(49), OK(2872).
Peronospora parasitica. Leaf spot. 'Downy mildew. Staghead.' **3:**CT(82), FL(35,82), IA(82), NC(1303), TX(82). **5:**CT(82), NC(1303). **6:**CT(82), FL(35,82), IA(82), NC(1303), NJ(82), OR(82,3202), TX(82). **8:**FL(35,82), IA(82), MN(2874), NC(1303), NY(82), OH(946), SD(82), WI(82). **9:**AL(285), FL(3500), GA(342), MT(3202), NC(3931), OR(3202), WA(3202). **10:**Range of host(82), FL(35), MS(2720), NC(1303). **11:**Range of host(82), CA(1955), FL(35), HI(2903), MN(2874), NC(1303). **12:**Range of host(82), CA(1955), FL(35), HI(2903), MN(2874), MS(2720), NC(1303), OK(2872). **13:**Range of host(82), FL(35), NC(1303). **14:**Range of host(82). **17:**Unk.(82), FL(35), HI(2903). **18:**E states(82), CA(485), FL(35), IL(82), MS(2720), NC(1303,3944), TN(1028), TX(82), WA(3202), WV(3907).
Phytophthora megasperma. 'Phytophthora root rot.' **11:**CA(82), OR(1432). **12:**CA(82), OR(1432). **13:**CA(82).
Phytophthora sp. Root rot. **3:**FL(35). **5:**NC(1303).
Pseudoperonospora cubensis. Downy mildew. **11:**FL(35).
Pythium polymastum. Damping-off. **12:**FL(35).
P. ultimum. Damping-off, head rot. **12:**Unk.(82). **18:**WI(82).
Pythium sp. Seedling blight, damping-off, root rot. **9:**MN(2874). **11:**Range of host(82), CO(49), FL(35). **12:**Range of host(82), FL(35), HI(2903), NC(1303). **13:**Range of host(82), NC(1303). **14:**Range of host(82). **16:**Range of host(82), FL(35). **17:**FL(35), HI(2903). **18:**FL(35), NY(82).

Zygomycotina

Blakeslea trispora. Dieback of seedlings. **12:**FL(35).
Rhizopus fusiformis. Soft rot. **6:**IL(82).
R. stolonifer. 'Black mold rot.' **6:**Range of host(82), NC(1303). **12:**Unk.(82).
Rhizopus sp. Black mold rot. **18:**HI(2903).

Ascomycotina

Chaetomium globosum. Isolated from seeds. **18:**MN(3306).
Erysiphe polygoni. 'Powdery mildew.' **3:**AZ(82), CA(82), FL(35,82), NC(1303), TX(82). **5:**NC(1303), WA(3202), WV(3907). **6:**E states(82), CA(82), FL(35), HI(82), NC(1303), TX(82), WA(82,3202). **8:**CA(82), IA(82), IN(82), NC(1303). **9:**OR(3202). **11:**CA(82), FL(35). **12:**AZ(82), CA(82), CT(82), FL(35,82), MD(82), MN(82), NC(1303), NE(82), OR(82), TX(82). **13:**NC(1303). **16:**CA(82), CT(82), FL(35), MD(82), NC(1303), VA(82). **17:**AZ(82), CA(82), ID(3202), MA(82), NJ(82). **18:**E states(82), W states(82), AZ(82), FL(35), IA(82), ID(3202), NC(1303), NV(82), OK(2872), TX(82), UT(82), WA(3202).
Leptosphaeria maculans. Canker. 'Black leg rot. Phoma root rot.' **9:**OR(3202), WA(3202). **12:**WI(333). **18:**WA(3202).
L. olericola. On stems. **9:**Unk.(82).
Lewia scrophulariae. **19:**WA(3202).
Mycosphaerella brassicicola. 'Ring spot.' **6:**CA(82), WA(3202). **9:**OR(3202), WA(3202). **11:**CA(82), NY(82), OR(82), WA(82). **12:**CA(82), HI(2899,2903), IL(82), OR(82), TX(82), WA(82). **13:**CA(82). **16:**AL(82), CA(82), WA(82). **17:**WA(3202). **18:**OR(82), WA(3202).
Pseudombrophila deerata. On overwintered stalks. **9:**MA(272), ND(377).
Sclerotinia minor. **9:**PA(82), WA(3202). **16:**PA(82).
S. sclerotiorum. White blight, watery soft rot. 'Sclerotinia stem rot. Watery soft rot.' **5:**CA(82), FL(35), ID(82,3202), MS(82,2720). **6:**ME(82), MS(2720), MT(3202), WA(82,3202). **8:**CA(82), TX(82). **9:**CA(1016), OR(3202), WA(3202). **11:**Range of host(82), CA(82), NC(1303), TX(82). **12:**Range of host(82), FL(35), GA(1438), HI(2903), NC(1303), NY(877). **13:**CA(82), LA(82), NC(1303), TX(82). **14:**IN(82). **16:**GA(82,1438), NC(82,1303), OR(82), SC(82), VA(82), WA(82). **17:**AZ(82), HI(2899,2903), MA(82), NY(82), TX(82). **18:**CA(82), CT(82), ID(3202), MD(82,1729), MS(82,2720), TX(82), WA(82,3202). **19:**CA(1217).
Sclerotinia sp. **9:**ID(3202).

Basidiomycotina - Uredinales

[*Puccinia aristidae*]. **3:**CO(82). **8:**CO(82).

Basidiomycotina - Other

Dendrophora albobadia. **9:**NY(543).
Marasmius sp. On living leaves. **19:**HI(2903).
Thanatephorus cucumeris. Wire stem. **6:**WA(3202). **9:**WA(3202). **11:**MN(2874). **12:**MN(2874). **18:**WA(3202).

Deuteromycotina - Hyphomycetes

Alternaria alternata. Seed mold, secondary leaf spot. **11:**Range of host(82). **12:**Range of host(82), FL(35), WA(129). **13:**WA(129). **18:**Range of host(82), FL(35).
A. brassicae. Leaf blight, black mold, gray leaf spot. 'Black spot.' **3:**HI(2903), NC(1303). **6:**FL(35), MI(82), MN(82), PA(82). **8:**IA(82), MS(2720), NC(1303), WI(1345). **9:**AK(526), AL(285), AZ(693), HI(2903), OR(3202), WA(3202). **10:**FL(35). **11:**CA(82), CT(82), FL(35), HI(2903), NY(82), OR(82), TX(82). **12:**CA(82), CT(82), DE(3543), FL(35), HI(2903), MS(2720), NC(1303), NY(82), OR(82), TX(82), WA(129), WV(3907). **13:**FL(35), NC(1303), WA(129). **17:**CA(82), CT(82), FL(35,82), HI(2903), IN(82), MD(82).
A. brassicicola. Anthracnose, stem lesions, brown head rot, black spot of leaves and pods, wilt. 'Black spot.' **3:**NC(1303). **6:**CT(82), MA(82), MN(82,2874), NC(1303), NJ(82), PA(82), TX(82), VA(82). **8:**CA(82), IA(82). **9:**HI(2903), OR(3202), WA(3202). **10:**NC(1303). **11:**Range of host(82), FL(35), NC(1303), OK(2872). **12:**Range of host(82), FL(35), HI(2903), MN(2874), NC(1303), OK(2872), WA(129). **13:**CA(82), IN(82), NC(1303), NY(82), TX(82), WA(129). **16:**Range of host(82). **17:**CA(82), CT(82), FL(35,82), MA(82), NH(82), TX(82). **18:**CT(82), FL(35,82), HI(82,2903), MA(82), MD(82), NC(82), NJ(82), TX(82).
A. raphani. 'Black spot.' **18:**AZ(694).
Alternaria sp. Head blight, leaf spot. **3:**FL(35). **11:**CO(49). **18:**FL(35).
Botrytis cinerea. 'Gray mold.' **6:**Unk.(82). **9:**AK(526), WA(3202). **10:**WV(3907). **11:**CA(82), FL(35,82). **12:**W states(82), FL(35). **13:**CA(82), FL(35).
Botrytis sp. **9:**WA(3202).
Cercospora brassicicola. 'Cercospora leaf spot.' **3:**IN(82), LA(82), NJ(82). **5:**MS(2720), NC(1303). **6:**MS(82,2720), NJ(82), VA(82). **8:**NJ(82). **11:**HI(2903). **12:**CA(82), DE(82), FL(35,82), IL(82), MS(82,2720), NJ(82), OK(82), TX(82). **16:**AL(82). **17:**FL(35,82), GA(82), HI(2903), NH(82), NJ(82). **18:**AL(82), FL(35,82), GA(82,3540), MS(82,2720).
Cladosporium brassicae. **9:**KS(525,959).
[*C. variabile*]. Leaf mold. **9:**MT(3202). **12:**MT(82), NY(82).
Cladosporium sp. Following tip burn. **12:**NC(1303).
Curvularia geniculata. Seed mold in commercial samples. **9:**Unk.(82).
C. inaequalis. Seed mold. **18:**Unk.(82).
Entylomella armoraciae. Leaf spot. **8:**MI(82).
Fusarium oxysporum. 'Yellows.' **5:**NC(1303). **13:**NC(1303).
F. oxysporum f. sp. *conglutinans*. Stem rot, wilt, yellows. **9:**AL(285), FL(935,3644), NC(3931), NM(2107), OR(3202), VA(3644). **10:**AL(82,285), CA(82), IL(82), IN(82), MD(82), MS(82,2720), NC(82,1303), NY(82), SC(82), TX(82), VA(82). **11:**CA(82), DE(82), IN(82), LA(82), MI(82), NC(1303), NJ(82), NM(82), NY(82), TX(82), WI(82). **12:**Range of host(82), CA(2912), FL(35), MN(2874), MS(2720), OK(2872), WV(82). **13:**CA(82), WI(82). **14:**IN(82), WI(82). **16:**AL(82), CA(82), IL(82), IN(82), MD(82), MS(82), NC(82), NY(82), SC(82), TX(82), VA(82). **18:**IN(82), MS(82,2720), TX(82).
Fusarium sp. Damping-off, dry rot, wilt, yellows. **3:**FL(35), ND(82). **4:**MS(2720). **6:**MN(82,2874). **9:**MS(2720). **10:**NC(1303).
Hobsonia mirabilis. On petioles. **19:**HI(1269).
Hormodendrum sp. On leaves. **9:**AK(2163).
Oidium sp. Powdery mildew, leaf spot. **5:**OH(950). **6:**HI(2903). **18:**CA(485), OH(950).
Phymatotrichopsis omnivora. 'Phymatotrichum root rot.' **6:**TX(82). **11:**TX(82). **12:**TX(82). **14:**TX(82). **16:**TX(82). **18:**TX(82).
Pseudocercosporella capsellae. Leaf spot, white spot. 'White leaf spot. Gray stem.' **1:**CA(82,485). **3:**CA(82), FL(35), NC(1303), VA(82), WI(1341). **5:**NC(1303). **6:**CA(82), LA(82), MA(82), MN(82,2874), NC(1303), WV(3907). **8:**FL(35), MA(82), NC(1303). **9:**AK(526), OR(3202). **10:**NC(1303). **11:**IN(82), NC(1303), OR(82), PA(82). **12:**OR(82). **17:**AL(82), CA(82), CT(82), FL(35,82), IN(82), MA(82), MD(82), MO(82), PA(82), VA(82).
Ramularia sp. Leaf spot. **18:**FL(35,82), WA(82,3202).
Stemphylium botryosum. Seed mold. **11:**Unk.(82). **12:**Unk.(82). **18:**Unk.(82).
Stemphylium sp. Leaf spot. **12:**FL(35), WA(129). **13:**WA(129).
Stilbella annulata. **19:**SC(3178).

Ulocladium consortiale. Seed mold. **9**:Unk.(82).
Ulocladium sp. **12**:WA(129). **13**:WA(129).
Verticillium albo-atrum. Hadromycosis. 'Verticillium wilt.' **6**:CA(82). **12**:CA(3366). **13**:CA(3366). **18**:KS(3022).
Deuteromycotina - Coelomycetes
Ascochyta armoraciae. On leaves and siliques. **3**:WI(1341).
Asteromella brassicae. On leaves. **9**:CA(64). **12**:CA(82).
Colletotrichum gloeosporioides. Anthracnose. **3**:NC(1303). **5**:NC(1303). **8**:NC(1303). **10**:NC(1303).
C. higginsianum. Leaf spot. 'Anthracnose.' **6**:GA(82). **8**:NC(82,3944). **12**:FL(35). **17**:FL(35,82). **18**:E states(82), AL(285), FL(35), GA(3540), MS(2720), OK(2872), TX(82).
Macrophomina phaseolina. **2**:CA(1171). **18**:CA(1171).
Phoma lingam. Black leg, leaf, root and stem rot, leaf spot. **5**:NC(1303). **6**:CT(82), GA(82), HI(82,2903), MA(82), MN(82,2874), NC(1303), WA(82). **8**:CA(82), NC(1303). **9**:AL(285), AZ(1194), MN(2874). **11**:Gulf States(82), NE states(82), FL(35). **12**:Range of host(82), CT(3770), FL(35), HI(2903), MS(2720), NC(1303), OK(2872), OR(82,334), WA(82), WV(3907). **13**:Gulf States(82), NE states(82), CA(82), NC(1303). **16**:Gulf States(82), NE states(82), NC(1303). **17**:WA(82). **18**:CT(82), MA(82), WA(82).
Phyllosticta sp. Leaf spot. **18**:TX(82), WA(82), WV(82,3907).
Septomyxa affinis. Leaf spot. **18**:AL(82).
Septoria brassicae. Leaf spot. **8**:WV(82,3907).
Deuteromycotina - Other
Rhizoctonia solani. Foliar blight, soreshin. 'Bottom rot. Damping-off. Headrot. Seedling root rot. Wire stem. Basal stem rot.' **3**:FL(35), NC(1303). **4**:MS(2720). **5**:NC(1303). **6**:FL(35,82), HI(82), MS(2720), WA(82). **8**:Range of host(82), FL(35), NC(1303). **9**:AK(2163), NC(3931). **11**:Range of host(82), FL(35), MS(2720), NC(1303). **12**:Range of host(82), FL(35), HI(2899), NC(1303), NY(4). **13**:Range of host(82), FL(35), NC(1303). **14**:Range of host(82). **16**:Range of host(82), FL(35), NC(1303). **18**:NE states(82), FL(35,82), HI(82), TX(82), VA(82), WA(82). **19**:FL(35).
Rhizoctonia sp. Root rot. **6**:HI(2903). **12**:HI(2903). **17**:HI(2903). **18**:HI(2903).
Sclerotium rolfsii. 'Southern blight.' **5**:NC(1303). **9**:AL(285). **11**:SE states(82), NC(1303), TX(82). **12**:SE states(82), FL(35), TX(82). **13**:SE states(82), NC(1303), TX(82). **14**:SE states(82), TX(82). **16**:SE states(82), AL(285), NC(1303), TX(82). **17**:TX(82). **18**:FL(35,82), TX(82).

Cakile Mill.

1. *C. edentula* (Bigel.) Hook. [American sea-rocket]

Oomycetes
Peronospora sp. Defoliation. **1**:NJ(1252).
Deuteromycotina - Coelomycetes
Phoma sp. Stem canker, black root. **1**:NJ(1252).

Camelina Crantz

1. *C. microcarpa* Andrzeiovski ex DC.
2. *C. sativa* (L.) Crantz

Plasmodiophoromycetes
Plasmodiophora brassicae. Clubroot. **1**:NJ(82). **2**:NJ(82).
Oomycetes
Albugo candida. White rust. **1**:ID(3202), MT(3202), OH(82), OK(82,2872), PA(82), UT(82), VA(82), WA(3202). **2**:MT(82,3202), NE(82), OH(82), OK(82,2872), OR(3202), TX(82), UT(82), VA(82).
Basidiomycotina - Uredinales
[*Puccinia aristidae*]. **1**:CO(82), UT(82).

Capsella Medik.

1. *C. bursa-pastoris* (L.) Medik. [shepherd's-purse]
2. *C. procumbens* Fr. (=*Hutchinsia procumbens* Desv.)
3. *Capsella* sp.

Plasmodiophoromycetes
Plasmodiophora brassicae. Clubroot. **1**:CA(82), NJ(82), NY(82), WA(82,3202).
Oomycetes
Albugo candida. White rust. **1**:Range of host(82), ID(3202), MN(2874), MS(2720), NC(3944), ND(377), OK(2872), WA(3202), WY(1195).
Peronospora parasitica. Downy mildew. **1**:NC states(82), NE states(82), Range of host(82), AK(526), ID(3202), KY(82), MN(2874), OK(2872), OR(3202), TX(82), WA(82,3202).
Ascomycotina
Sclerotinia sp. Leaf and stem blight. **1**:MD(2474).
Basidiomycotina - Uredinales
[*Puccinia aristidae*]. **1**:CO(82), DE(82), NE(82), NV(82), UT(82). **2**:OR(3202).
Basidiomycotina - Other
Thanatephorus cucumeris. **1**:WA(3202).
Deuteromycotina - Hyphomycetes
Entylomella armoraciae. Leaf spot. **1**:AK(82). **3**:AK(526).
Pseudocercosporella capsellae. Leaf spot. **1**:IA(82), IN(82), MO(82,963), TX(82), WI(82).
Deuteromycotina - Other
Rhizoctonia solani. Root rot. **1**:WA(82).

Cardamine L.

1. *C. bellidifolia* L.
2. *C. bellidifolia* L. var. *pachyphylla* Crép. & Leiberg
3. *C. bulbosa* (Schreb. ex Muhl.) B. S. P.
4. *C. cordifolia* A. Gray
5. *C. douglasii* (Torr.) Britton
6. *C. hirsuta* L.
7. *C. occidentalis* (S. Wats.) Howell
8. *C. oligosperma* Nutt.
9. *C. parviflora* L.
10. *C. parviflora* L. var. *arenicola* (Britton) O. E. Schulz
11. *C. pensylvanica* Muhl. ex Willd.
12. *C. purpurea* Cham. & Schlechtend.

Chytridiomycetes
Synchytrium aureum. **3**:WI(1338).
Synchytrium sp. On leaves and stems. **4**:CO(1849).
Oomycetes
Albugo candida. White rust. **3**:WI(82). **5**:WI(82). **9**:GA(82). **11**:NC(82,3944).
Peronospora parasitica. Downy mildew. **3**:DE(82), NY(82), WI(82). **5**:WI(82). **6**:IL(82), MD(82), WV(82). **7**:ID(3202). **8**:OR(3202). **9**:NE(82), VA(82), WI(82). **10**:WI(1337). **11**:ID(3202), NC(82), WA(3202), WI(82).
Ascomycotina
Pleospora comata. On dead stem. **12**:AK(1938).
Basidiomycotina - Uredinales
Puccinia cruciferarum. **1**:AK(60,82,526,1938), CA(82), OR(3202). **2**:CA(642).
P. cruciferarum subsp. *borealis*. **1**:AK(3099).
P. cruciferarum subsp. *wyomingensis*. **4**:UT(82), WY(82,1195,3099).
Deuteromycotina - Coelomycetes
Phoma herbarum. **1**:AK(526).
P. lingam. On stems. **1**:AK(82).
Septoria dentariae. Leaf spot. **3**:DE(82), NY(82), WI(82). **5**:WI(82).

Cheiranthus L.

1. *C. cheiri* L. [English wallflower, wallflower]

Oomycetes
Albugo candida. White rust. **1**:MN(82).
Ascomycotina
Mycosphaerella cruciferarum. On pods. **1**:PA(82).
Deuteromycotina - Hyphomycetes
Botrytis cinerea. Gray mold blight. **1**:AK(82,526), WA(82,3202).
Heterosporium sp. Leaf spot. **1**:OK(82,2872).
Deuteromycotina - Coelomycetes
Ascochyta cheiranthi. On leaves. **1**:AK(82,526).
Deuteromycotina - Other
Rhizoctonia solani. Crown and root rot. **1**:FL(35), NJ(82).

Cochlearia L.

1. *C. officinalis* L. (=*C. officinalis* L. var. *arctica* (Schlechtend.) Gelert ex G. Anmdersson & H. Hesselman) [scurvygrass]

Oomycetes
Albugo candida. **1**:NE(82), NH(82).
Peronospora cochleariae. **1**:AK(1938).

Conringia Heist. ex Fabr.
1. *C. orientalis* (L.) Dumort. [hare's-ear mustard]
Oomycetes
Albugo candida. **1**:WI(1353).
Deuteromycotina - Hyphomycetes
Cercospora sp. **1**:WI(1353).
Deuteromycotina - Coelomycetes
Phyllosticta sp. On leaves. **1**:WI(1353).
Septoria peregrina. **1**:ND(378).

Coronopus Zinn
1. *C. didymus* (L.) Sm.
Oomycetes
Peronospora parasitica. Downy mildew. **1**:NC(82).

Crambe L.
1. *C. abyssinica* Hochst. ex R. E. Fr. [colewort, crambe]
2. *C. maritima* L. [scurvygrass, sea kale]
Ascomycotina
Sclerotinia sclerotiorum. Root rot. **1**:LA(1609).
Deuteromycotina - Hyphomycetes
Alternaria brassicicola. Leaf spot, on fruits, leaves and stems. **1**:LA(1609). **2**:VA(82).
Chromelosporium fulvum. Stem rot on peat cubes. **1**:MD(1896).
Fusarium oxysporum f. sp. *conglutinans*. Yellows. **2**:IN(82).
Fusarium sp. Root rot. **1**:LA(1609).

Dentaria L.
1. *D. diphylla* Michx.
2. *D. laciniata* Muhl.
3. *D. tenella* Pursh
4. *Dentaria* sp.
Oomycetes
Albugo candida. White rust. **1**:WV(3907). **4**:E states(82), CA(82), IN(82), WI(82).
Peronospora parasitica. Downy mildew. **1**:C states(82), NE states(82). **2**:C states(82), NE states(82). **3**:WA(3202).
Basidiomycotina - Uredinales
Puccinia dentariae. Autoecious. **3**:OR(82,3202).
Deuteromycotina - Coelomycetes
Phyllosticta sp. Leaf spot. **3**:OR(82,3202).
Septoria dentariae. Leaf spot. **1**:MI(82), NY(82), WI(1354). **2**:IL(82), IN(82), WI(82).

Descurainia P. Webb & Berthel.
1. *D. pinnata* (Walter) Britton [tansy mustard]
2. *D. pinnata* (Walter) Britton var. *brachycarpa* (Rich.) Fernald (=*Sisymbrium canescens* Nutt. var. *brachycarpon* S. Wats.)
3. *D. richardsonii* (Sweet) O. E. Schulz
4. *D. richardsonii* (Sweet) O. E. Schulz subsp. *incisa* (Engelm.) Detl. (=*D. incisa* (Engelm.) Britton)
5. *D. richardsonii* (Sweet) O. E. Schulz subsp. *procera* (Greene) Detl. (=*Sophia procera* Greene)
6. *D. sophia* (L.) P. Webb ex Prantl [flixweed, herb-sophia, tansy mustard]
7. *Descurainia* sp.
Oomycetes
Albugo candida. **1**:AZ(82,1195,3378), IA(82), IN(82), KS(82), MT(82,3202), NV(82), OK(82,2872), OR(3202), WA(3202). **3**:MT(3202). **4**:CO(82), KS(82), MT(82), NM(82), WA(82), WY(82). **6**:WY(1195). **7**:MT(82).
Peronospora parasitica. Downy mildew. **1**:AL(82), AZ(1195,3378,3381), FL(35), IA(82), KS(82), MT(3202), NE(82), NM(82), SD(82), TX(82), WA(3202), WI(82). **2**:WI(1330). **3**:ID(3202), MT(3202), OR(3202), WY(1195). **4**:CO(82), KS(82), MN(82,2874), MT(82), NE(82), NM(82), OK(82,2872). **5**:WY(1195,3378). **6**:WA(3202). **7**:NV(82), WI(82).
Basidiomycotina - Uredinales
[*Puccinia aristidae*]. **1**:AZ(82), CO(82), NM(82). **4**:CO(82), NE(82), NM(82), UT(82).
P. monoica. Heteroecious: aecial host. **1**:CO(82), NM(82). **4**:CO(82), NM(82), UT(82). **7**:UT(82).
P. thlaspeos. Autoecious. **1**:TX(82).
Basidiomycotina - Ustilaginales
Urocystis coralloides. Root gall. **1**:AZ(82,1077).
Deuteromycotina - Hyphomycetes
Cercospora nasturtii. Leaf spot. **1**:TX(82).
Phymatotrichopsis omnivora. Root rot. **1**:TX(82).
Deuteromycotina - Coelomycetes
Septoria sisymbrii. Leaf spot. **1**:KS(82). **4**:OK(82,2872).
Deuteromycotina - Other
Sclerotium rolfsii. Southern blight. **1**:FL(35).

Draba L.
1. *D. alpina* L.
2. *D. aurea* Vahl ex Hornem. (=*D. luteola* Greene)
3. *D. brachycarpa* Nutt. in Torr. & A. Gray
4. *D. caesia* Adams
5. *D. cuneifolia* Nutt.
6. *D. incerta* Payson
7. *D. nemorosa* L.
8. *D. nivalis* Lilijebl. var. *kamtschatica* (Ledeb.) Pohl
9. *D. platycarpa* Torr. & A. Gray
10. *D. reptans* (Lam.) Fernald (=*D. caroliniana* Walter)
11. *D. sphaerocarpa* Macbr. & Payson
12. *D. stenoloba* Ledeb.
13. *D. verna* L.
14. *Draba* sp.
Chytridiomycetes
Synchytrium lacunosum. On leaves, petioles, and stems. **13**:WY(1848).
Synchytrium sp. **13**:WY(2924).
Oomycetes
Albugo candida. White rust. **7**:WY(82,1195). **10**:KS(82).
Peronospora parasitica. Downy mildew. **5**:AZ(1195,3378), TX(82). **7**:CO(82), SD(82,2243), WI(1336), WY(82,1195). **9**:TX(82). **10**:IA(82), IL(82), KS(82), NE(82), OR(3202), WI(82,1334). **12**:ID(3202), MT(3202), WA(3202). **13**:WY(2924).
Ascomycotina
Leptosphaeria doliolum. On dead stem. **4**:AK(1938). **12**:AK(1938).
Lewia scrophulariae. **2**:WY(3810).
Mycosphaerella pachyasca. On stems. **14**:AK(82).
M. pyrenaica. On dead leaves and stems. **14**:AK(1938).
M. tassiana. On dead stem. **1**:AK(526). **4**:AK(1938). **8**:AK(1938).
[*Phaeosphaeria eustoma*]. **2**:WY(3810).
Pleospora comata. On dead stems and leaves. **14**:AK(1938).
P. penicillus. On dead stems. **8**:AK(1938). **13**:AK(526). **14**:AK(82,526).
Basidiomycotina - Uredinales
[*Puccinia aristidae*]. **5**:AZ(1203). **14**:AZ(82).
P. drabae. Autoecious. **2**:AK(60,526). **6**:WY(749,1195). **11**:WY(749). **14**:AK(82,526), UT(82), WY(82).
P. monoica. Heteroecious: aecial host. **3**:MN(2874). **14**:CA(82), NM(82).
P. subnitens. **5**:AZ(3981).
P. thlaspeos. Autoecious. **14**:CA(82), CO(82), UT(82).
Deuteromycotina - Coelomycetes
Phlyctema andersonii. On stems. **14**:MT(82,3202).
Phoma jejuna. On stems. **2**:WY(3812).
Rhabdospora pleosporoides var. *drabae*. On stems. **2**:WY(3812).

Erysimum L.
1. *E. asperum* (Nutt.) DC. [prairie rocket, western wallflower]
2. *E. capitatum* (Douglas ex Hook.) Greene [coast wallflower]
3. *E. cheiranthoides* L. [wormseed mustard]
4. *E. inconspicuum* (S. Wats.) MacMill.
5. *E. pallasii* (Pursh) Fernald
6. *E. repandum* L. [bushy wallflower, treacle mustard]
7. *Erysimum* sp.
Plasmodiophoromycetes
Plasmodiophora brassicae. Clubroot. **1**:NJ(82). **7**:NJ(82).
Oomycetes
Albugo candida. White rust. **1**:OR(82), TX(82). **2**:ID(3202), OR(3202), WA(3202). **3**:WI(1355). **7**:ID(82).
Peronospora parasitica. Downy mildew. **1**:ID(82). **2**:ID(3202), MT(3202), WA(3202). **3**:WA(3202), WI(1357). **7**:CO(82), IA(82), WY(82).
Ascomycotina
Clathrospora pentamera. On dead stem. **5**:AK(1938).
Erysiphe polygoni. Powdery mildew. **7**:CA(82).
Lewia scrophulariae. On stems. **7**:CA(82).

Basidiomycotina - Uredinales
[*Puccinia aristidae*]. **1**:CO(82), ND(82). **2**:ID(3202). **7**:AZ(82), NE(82), UT(82).
P. consimilis. Autoecious. **4**:MT(3202). **7**:MT(82).
P. thlaspeos. Autoecious. **1**:CO(82).
Deuteromycotina - Hyphomycetes
Cercospora erysimi. Leaf spot. **3**:WI(82,581).
Deuteromycotina - Coelomycetes
Phyllosticta erysimi. **4**:WI(1354).
Septoria repanda. **6**:WA(3202).
Septoria sp. **2**:OR(3202).
Deuteromycotina - Other
Rhizoctonia solani. Root rot. **1**:TX(82).

Hesperis L.

1. *H. matronalis* L. [dame's-rocket, dame's-violet, sweet-rocket]

Oomycetes
Albugo candida. White rust. **1**:NY(82).
Peronospora parasitica. Downy mildew. **1**:NY(82), PA(82), WI(1350).
Plasmopara brassicae. Clubroot. **1**:NJ(82).

Iberis L.

1. *I. amara* L. [rocket candytuft]
2. *I. umbellata* L. [globe candytuft]
3. *Iberis* sp.

Plasmodiophoromycetes
Plasmodiophora brassicae. Clubroot. **2**:MA(82), NJ(82).
Oomycetes
Albugo candida. White rust. **1**:CA(82).
Peronospora parasitica. Downy mildew. **1**:CA(82).
Pythium oligandrum. Root rot. **1**:ME(82). **2**:ME(82).
Ascomycotina
Erysiphe polygoni. Powdery mildew. **1**:CA(82).
Deuteromycotina - Hyphomycetes
Alternaria sp. Leaf spot. **1**:FL(35). **2**:FL(35).
Botrytis cinerea. Blight, gray mold. **2**:AK(82,526). **3**:NC(1303).
Fusarium sp. Stem rot. **3**:NC(1303).
Deuteromycotina - Coelomycetes
Colletotrichum gloeosporioides. Anthracnose. **2**:NC(1303).
Phoma lingam. Root rot. **2**:CA(82).
Deuteromycotina - Other
Rhizoctonia solani. Damping-off, stem rot. **1**:Range of host(82), FL(35). **2**:FL(35). **3**:NC(1303).
Sclerotium rolfsii. Southern blight. **1**:FL(35). **2**:FL(35).

Lepidium L.

1. *L. campestre* (L.) R. Br. [field cress, field pepperweed]
2. *L. densiflorum* Schrad. (=*L. apetalum* Willd.)
3. *L. lasiocarpum* Nutt. ex. Torr. & A. Gray
4. *L. lasiocarpum* Nutt. ex Torr. & A. Gray var. *lasiocarpum*
5. *L. perfoliatum* L.
6. *L. ramosissimum* A. Nels.
7. *L. sativum* L. [garden cress]
8. *L. thurberi* Woot.
9. *L. virginicum* L. [peppergrass, Virginia pepperweed]
10. *L. virginicum* L. var. *medium* (Greene) C. Hitchc. (=*L. medium* Greene, *L. intermedium* A. Gray)
11. *Lepidium* sp.

Plasmodiophoromycetes
Plasmodiophora brassicae. Clubroot. **1**:NJ(82). **9**:NJ(82).
Chytridiomycetes
Synchytrium lepidii. Leaf gall. **9**:LA(82).
Oomycetes
Albugo candida. White rust. **1**:IN(82), NY(82), WI(1334), WV(3907). **5**:WA(3202). **6**:MT(3202). **7**:FL(35), NE(82), TX(82). **9**:Range of host(82), FL(35), MS(2720), WA(3202), WV(3907). **11**:HI(2903).
Peronospora parasitica. Downy mildew. **2**:IA(82), ID(3202), KS(82), MN(2874), ND(82), TX(82), WA(3202), WI(82). **4**:AZ(1195,3378). **6**:MT(3202). **8**:AZ(1195,3378). **10**:AZ(1195,2750,3378). **11**:ID(3202), ND(377,2243).
Pythium debaryanum. Damping-off. **7**:TX(82).
Ascomycotina
Leptosphaeria maculans. **9**:NY(3249).
L. planiuscula. On stems. **9**:GA(82,1438), NJ(967), NY(82). **10**:KS(967).
Pleospora herbarum var. *occidentalis*. On stems. **2**:CO(82), TX(82).
Sclerotinia sclerotiorum. Crown rot. **7**:MA(82).
Sclerotinia sp. Leaf and stem blight. **9**:MD(2474).
Basidiomycotina - Uredinales
[*Puccinia aristidae*]. **2**:CO(82), KY(82), MT(82,3202), ND(82), NE(82), TX(82), UT(82). **4**:AZ(1203). **5**:WA(3202). **7**:TX(82). **9**:TX(82). **11**:AZ(82), DE(82), IN(82), KS(82), NM(82), NV(82).
P. monoica. Heteroecious: aecial host. **1**:WY(82).
P. subnitens. **4**:AZ(3981).
P. thlaspeos. Autoecious. **1**:MT(82), WY(82).
Basidiomycotina - Other
Thanatephorus cucumeris. **1**:WI(1335). **9**:LA(3524).
Deuteromycotina - Hyphomycetes
Alternaria brassicae. Leaf spot. **1**:NY(82), WI(1340). **9**:NY(82).
Cercospora sp. Leaf spot. **1**:NJ(82). **9**:NJ(82).
Phymatotrichopsis omnivora. Root rot. **1**:TX(82). **2**:TX(82). **7**:TX(82). **9**:TX(82).
Pseudocercosporella capsellae. Leaf spot. **9**:IN(82), WI(82).
Deuteromycotina - Coelomycetes
Phoma terrestris. Pink root. **1**:ND(82), SD(82). **7**:ND(82), SD(82).
Phyllosticta sp. **9**:WA(3202).
Septoria lepidiicola. Leaf spot. **2**:NE(82), TX(82,3597), WI(82). **9**:IL(82), MD(82), MS(82,2720), NY(82), PA(82), TX(82), WI(82).
Deuteromycotina - Other
Rhizoctonia solani. Damping-off. **7**:TX(82).

Lesquerella S. Wats.

1. *L. gordonii* (A. Gray) S. Wats. [Gordon bladderpod]
2. *L. purpurea* (A. Gray) S. Wats.
3. *Lesquerella* sp.

Oomycetes
Peronospora parasitica. Downy mildew. **1**:AZ(1195,3381). **2**:AZ(1195,3378).
Basidiomycotina - Uredinales
[*Puccinia aristidae*]. **1**:AZ(1203). **3**:AZ(82).
P. subnitens. **1**:AZ(3981).
Deuteromycotina - Hyphomycetes
Dendryphion nanum. Leaf spot. **3**:TX(82).
Phymatotrichopsis omnivora. Root rot. **3**:TX(82).
Deuteromycotina - Coelomycetes
Phoma punctiformis. Stem blight. **3**:TX(82).

Lobularia Desv.

1. *L. maritima* (L.) Desv. [sweet alyssum]

Plasmodiophoromycetes
Plasmodiophora brassicae. Clubroot. **1**:NJ(82).
Oomycetes
Peronospora parasitica. Downy mildew. **1**:CA(82), NJ(82).
Pythium sp. Seedling blight, damping-off. **1**:CO(49).
Ascomycotina
Erysiphe polygoni. Powdery mildew. **1**:CA(82).
Microsphaera penicillata. Powdery mildew. **1**:FL(35).
Sclerotinia sclerotiorum. Crown rot. **1**:VA(82).
Deuteromycotina - Coelomycetes
Phoma lingam. Root rot. **1**:CA(82).
Deuteromycotina - Other
Rhizoctonia solani. Damping-off, stem rot. **1**:CO(49), NJ(82), NY(82), VA(82).

Lunaria L.

1. *L. annua* L. [bolbonac, honesty-plant, money-plant, pennyflower, silver-dollar]

Plasmodiophoromycetes
Plasmodiophora brassicae. Clubroot. **1**:NJ(82).
Deuteromycotina - Hyphomycetes
Alternaria brassicae. Leaf spot. **1**:CA(149).
A. brassicicola. Black spot of leaves and pods. **1**:MA(82).
Helminthosporium lunariae. Leaf spot. **1**:MA(82).
Deuteromycotina - Coelomycetes
Leptothyrium lunariae. On stems. **1**:MA(82).

Matthiola R. Br. in Aiton & W. T. Aiton

1. *M. bicornis* (Sibth. & Sm.) DC. [evening-scented stock]
2. *M. incana* (L.) R. Br. in W. T. Aiton [Brampton's stock, common stock, gillyflower, imperial stock]
3. *M. incana* (L.) R. Br. in W. T. Aiton var. *annua* (Sweet) Voss [ten-weeks stock]
4. *Matthiola* sp.

Plasmodiophoromycetes
Plasmodiophora brassicae. Clubroot. **1**:NJ(82).
Oomycetes
Peronospora parasitica. Downy mildew. **2**:IL(82).
Phytophthora cryptogea. Stem rot. **2**:CA(82).
P. megasperma. Root rot. **1**:CA(82). **2**:CA(82).
Pythium hydnosporum. Damping-off. **2**:NH(82).
P. oedochilum. Damping-off. **2**:CA(82).
P. vexans. Damping-off. **2**:CA(82).
Pythium sp. Damping-off. **2**:CA(82).
Ascomycotina
Sclerotinia sclerotiorum. Crown rot. **2**:CA(82), MI(82), PA(82). **4**:WA(3202).
Basidiomycotina - Uredinales
Puccinia sp. **2**:FL(35).
Deuteromycotina - Hyphomycetes
Alternaria raphani. Leaf spot. **2**:CA(82).
Alternaria sp. Leaf spot. **2**:FL(35).
Botrytis cinerea. Gray mold blight. **2**:AK(82), FL(35), IL(312), TX(82). **3**:AK(526).
Fusarium avenaceum. Root rot. **2**:DE(82), NJ(82).
F. oxysporum f. sp. *mathioli*. Wilt. **2**:AZ(82), CA(82).
Myrothecium roridum. Leaf spot. **2**:TX(82).
Phymatotrichopsis omnivora. Root rot. **2**:TX(82).
Verticillium albo-atrum. Wilt of garden stock. **2**:CA(2908,3366).
Verticillium sp. Wilt. **1**:NY(82). **2**:NY(82).
Deuteromycotina - Coelomycetes
Colletotrichum gloeosporioides. Leaf spot. **1**:TX(82). **2**:TX(82).
Colletotrichum sp. Anthracnose. **2**:FL(35). **4**:FL(35).
Deuteromycotina - Other
Rhizoctonia solani. Damping-off. **1**:Range of host(82). **2**:Range of host(82), FL(35). **4**:FL(35).
Sclerotium rolfsii. Southern blight. **2**:MS(2720).

Nasturtium R. Br.

1. *N. officinale* R. Br. (=*Rorippa nasturtium-aquaticum* (L.) Hayek, *N. nasturtium-aquaticum* (L.) Karsten) [watercress]

Plasmodiophoromycetes
Plasmodiophora brassicae. On leaves and shoots. **1**:FL(35).
Spongospora subterranea f. sp. *nasturtii*. Crook root. **1**:FL(35,3507), PA(3507).
Oomycetes
Albugo candida. White rust. **1**:MN(82,2874).
Peronospora parasitica. **1**:OR(3202), WA(3202).
Phytophthora cryptogea. Root, seedling and stem rot. **1**:FL(35,2451).
Pythium debaryanum. Root rot. **1**:TX(82).
Zygomycotina
Choanephora cucurbitarum. Leaf and stem blight. **1**:FL(35).
Ascomycotina
Sclerotinia sclerotiorum. Leaf and stem blight. **1**:FL(35).
Basidiomycotina - Uredinales
[*Puccinia aristidae*]. **1**:AZ(82,1203), CO(82), TX(82).
P. subnitens. **1**:AZ(3981).
Deuteromycotina - Hyphomycetes
Alternaria sp. Leaf spot. **1**:FL(35), NC(1303).
Cercospora nasturtii. Leaf spot. **1**:CA(82), CO(82), FL(35,82), HI(82,2903), IN(82), NH(82), OK(2872), TX(82), WI(82), WV(3911).
Fusarium sp. Stem rot. **1**:FL(35).
Deuteromycotina - Coelomycetes
Colletotrichum sp. Anthracnose. **1**:NC(1303).
Sphaceloma sp. Spot anthracnose. **1**:FL(35).
Deuteromycotina - Other
Rhizoctonia solani. Damping-off, root and stem rot. **1**:FL(35), NC(1303), TX(82).
Sclerotium rolfsii. Southern blight, stem rot. **1**:FL(35).

Parrya R. Br.

1. *P. nudicaulis* (L.) Regel

Ascomycotina
Pleospora helvetica. **1**:AK(526).
Basidiomycotina - Uredinales
Puccinia oudemansii. Autoecious. **1**:AK(60,526).
Deuteromycotina - Coelomycetes
Leptothyrium vulgare f. *parryae*. **1**:AK(526), AK(526).

Phoenicaulis Nutt. in Torr. & A. Gray

1. *P. cheiranthoides* Nutt.

Basidiomycotina - Uredinales
Puccinia monoica. Heteroecious: aecial host. **1**:WA(3202).

Raphanus L.

1. *R. raphanistrum* L. [jointed charlock, wild radish]
2. *R. sativus* L. [radish]
3. *R. sativus* L. cv. 'longipinnatus' [Chinese radish, daikon]
4. *Raphanus* sp.

Plasmodiophoromycetes
Plasmodiophora brassicae. 'Clubroot.' **2**:NE states(82), AK(82,526), MN(82,2874), NC(1303), WA(82,3202). **3**:HI(2899,2903).
Oomycetes
Albugo candida. 'White rust. Staghead.' **1**:Unk.(82). **2**:Range of host(82), FL(35,3500), HI(2903), MN(2874), NC(1303), OK(2872), WV(3907), WY(819,1195). **3**:HI(2899,2903).
Aphanomyces raphani. 'Black root.' **2**:E states(82), CA(82), CT(2973), FL(35), IA(82), NC(1303), OK(82,2872), OR(1638,3202).
Aphanomyces sp. Root rot. **2**:HI(2903).
Peronospora parasitica. 'Downy mildew.' **2**:C states(82), NE states(82), CA(82), CT(2969), FL(35), MN(2874), MS(2720), NC(1303), TX(82), WA(3202), WI(1339). **3**:HI(2899,2903).
Pythium aphanidermatum. Black root. **2**:OK(2872). **4**:IN(82), KS(82), MA(82), MI(82), MS(82), NC(1303), NY(82), OH(82), OK(82), PA(82), SC(82), WI(82).
P. debaryanum. Damping-off. **2**:MA(82), MN(82,2874), NJ(82), VA(82).
Pythium sp. Damping-off, root rot. **2**:FL(35), HI(2903).
Ascomycotina
Erysiphe polygoni. 'Powdery mildew.' **2**:CA(82), FL(35), MO(82), TX(82).
Sclerotinia sclerotiorum. Crown rot. 'Sclerotinia stem rot. Watery soft rot.' **1**:CA(82). **2**:CA(82), IN(82), MN(82,2874), NJ(82), TX(82).
Basidiomycotina - Uredinales
[*Puccinia aristidae*]. **2**:AZ(82,1203), CO(82).
P. subnitens. **2**:AZ(3981).
Basidiomycotina - Other
Thanatephorus cucumeris. **2**:WA(3202).
Deuteromycotina - Hyphomycetes
Alternaria brassicae. 'Black spot.' **2**:CA(82), CT(82), FL(35), HI(2903), MI(82), NC(1303), WI(1340).
A. brassicicola. 'Black spot.' **2**:CT(82), NJ(82).
A. raphani. Black pod blotch. 'Black spot.' **2**:CA(82), MI(82), MN(82,2874), NJ(82), OH(82), PA(82).
Cercospora atrogrisea. Leaf spot. **2**:NJ(82).
C. cruciferarum. Leaf spot. **2**:AL(82), FL(35,82), IL(82), MO(82), TX(82).
Cercospora sp. Leaf spot. **4**:HI(2903).
Fusarium oxysporum. Wilt. 'Yellows.' **2**:CO(49).
F. oxysporum f. sp. *raphani*. Wilt. **2**:CA(82,1872), WI(2861).
Heterosporium sp. **2**:WA(3202).
Phymatotrichopsis omnivora. 'Phymatotrichum root rot.' **2**:TX(82).
Stemphylium botryosum. On seeds. **2**:Range of host(82).
Verticillium albo-atrum. Hadromycosis. 'Verticillium wilt.' **2**:CA(3366).
Deuteromycotina - Coelomycetes
Ascochyta sp. Root rot. **2**:MA(82).
Phoma lingam. Pod and root rot. **1**:CA(82). **2**:FL(35,82).
Deuteromycotina - Other
Rhizoctonia solani. 'Black rot. Damping-off. Headrot. Seedling root rot. Wire stem. Basal stem rot.' **2**:Range of host(82), AK(2163), FL(35), OK(2872).
R. zeae. Root rot. **2**:NC(1303).

Rorippa Scop.

1. *R. curvisiliqua* (Hook.) Bessey ex Britton
2. *R. islandica* (Oed. ex J. A. Murray) Borbás [marsh cress]
3. *R. islandica* (Oed. ex J. A. Murray) Borbás var. *hispida* (Desv.) F. K. Butters & Abbe
4. *R. palustris* (L.) Besser
5. *R. sylvestris* (L.) Besser [creeping yellowcress, yellow fieldcress]
6. *Rorippa* sp.

Plasmodiophoromycetes
Plasmodiophora brassicae. Clubroot. **2**:NJ(82). **5**:NJ(82).
Chytridiomycetes
Physoderma sp. On stems. **3**:MI(3400).
Oomycetes
Albugo candida. White rust. **2**:IL(82), MN(2874), ND(82), OR(82), SD(82), WI(82). **4**:OR(3202). **6**:FL(35).
Peronospora parasitica. Downy mildew. **1**:ID(3202). **2**:IA(82), MI(82), WI(82). **4**:MT(3202).
Ascomycotina
Pyrenopeziza sp. **4**:AK(3431).
Sclerotinia sclerotiorum. Crown rot. **5**:FL(35). **6**:FL(35).
Zignoella roripae. **6**:ND(377).
Basidiomycotina - Uredinales
[*Puccinia aristidae*]. **2**:CO(82).
Deuteromycotina - Hyphomycetes
Cercospora nasturtii. Leaf spot. **2**:WI(82). **5**:FL(35), WI(82). **6**:FL(35).
Deuteromycotina - Coelomycetes
Septoria radiculae. Leaf spot. **2**:NY(82).

Sibara Greene

1. *S. virginica* (L.) Rollins

Oomycetes
Albugo candida. **1**:MS(2720).

Sinapis L.

1. *S. alba* L. (=*Brassica alba* (L.) Rabenh., *Brassica hirta* Moench) [white mustard]
2. *S. arvensis* L. (=*Brassica arvensis* (L.) Rabenh., *Brassica kaber* (DC.) L. C. Wheeler) [California rape, charlock]

Plasmodiophoromycetes
Plasmodiophora brassicae. **1**:CA(82). **2**:NJ(82).
Oomycetes
Albugo candida. **2**:Range of host(82), ND(377), WI(1353).
Peronospora parasitica. **1**:NC(1303), SD(82), TX(82). **2**:IA(82).
Ascomycotina
Erysiphe polygoni. Powdery mildew. **1**:FL(35), TX(82). **2**:CA(82), NJ(82).
Leptospora rubella. **2**:ND(377).
Sclerotinia sclerotiorum. Stem rot. **1**:TX(82). **2**:MT(82).
Basidiomycotina - Uredinales
[*Puccinia aristidae*]. **2**:CO(82).
Basidiomycotina - Other
Thanatephorus cucumeris. Damping-off. **2**:MN(2874).
Deuteromycotina - Hyphomycetes
Alternaria sp. Leaf spot. **1**:VT(82).
Cercospora brassicicola. **2**:TX(82).
Pseudocercosporella capsellae. **1**:FL(35).
Deuteromycotina - Other
Rhizoctonia solani. **1**:FL(35), NC(1303). **2**:MN(82), TX(82).

Sisymbrium L.

1. *S. altissimum* L. [hedge mustard, tumble mustard]
2. *S. irio* L. [London-rocket]
3. *S. linifolium* Nutt. ex Torr. & A. Gray (=*Schoenocrambe linifolia* (Nutt. ex Torr. & A. Gray) Greene)
4. *S. officinale* (L.) Scop. [hedge mustard]
5. *Sisymbrium* sp.

Plasmodiophoromycetes
Plasmodiophora brassicae. Clubroot. **1**:NJ(82). **4**:NJ(82), WA(82,3202).
Oomycetes
Albugo candida. White rust. **2**:AZ(1195), NM(82,2107), TX(82). **3**:MT(3202), WA(3202), WY(1195). **4**:Range of host(82), MN(2874), WA(3202), WV(3907).
Peronospora parasitica. Downy mildew. **1**:WA(3202), WI(82). **4**:IA(82). **5**:KS(82), MD(82).
Basidiomycotina - Uredinales
[*Puccinia aristidae*]. **1**:CO(82).
P. consimilis. Autoecious. **3**:MT(3202), WY(1195).
P. monoica. Heteroecious: aecial host. **3**:AZ(3981), MT(3202), WY(1195).
P. thlaspeos. Autoecious. **3**:CO(1195), MT(3202), WY(1195).
Basidiomycotina - Other
Typhula sp. On stems. **1**:MT(82), WA(3202).
Deuteromycotina - Hyphomycetes
[*Cercospora cruciferarum*]. **4**:DE(3543).
C. nasturtii. Leaf spot. **4**:DE(82,962).
Cercospora sp. Leaf spot. **4**:DE(82).
Heterosporium sp. **1**:WA(3202).
Pseudocercosporella capsellae. **4**:WI(1343).
Verticillium albo-atrum. **2**:NM(2109).
Deuteromycotina - Coelomycetes
Septoria sisymbrii. Leaf spot. **4**:ID(82), KY(82), MI(82), NJ(82), WV(82,3907). **5**:ID(3202).
Deuteromycotina - Other
Sclerotium fulvum. **1**:MT(3202).

Smelowskia C. A. Mey. in Ledeb.

1. *S. calycina* (Stephenson ex Willd.) C. A. Mey.
2. *Smelowskia* sp.

Basidiomycotina - Uredinales
Puccinia aberrans. Autoecious. **1**:MT(3202), WA(3202), WY(1195,1520). **2**:CO(82), MT(82), NV(82), UT(82), WA(82).
P. monoica. Heteroecious: aecial host. **2**:CO(82), WY(82).
P. thlaspeos. Autoecious. **2**:NV(82).

Stanleya Nutt.

1. *S. pinnata* (Pursh) Britton [desert-plume]
2. *S. viridiflora* Nutt.

Oomycetes
Albugo candida. **2**:OR(3202).
Basidiomycotina - Uredinales
[*Puccinia aristidae*]. **1**:CO(82), NV(82).
Deuteromycotina - Hyphomycetes
Cercospora nasturtii. Leaf spot. **1**:KS(82).

Streptanthus Nutt.

1. *S. arizonicus* S. Wats.
2. *S. orbiculatus* Greene

Oomycetes
Peronospora parasitica. **1**:AZ(1195).
Deuteromycotina - Coelomycetes
Naemosphaera shastensis. **2**:CA(638,642).

Thelypodium Endl.

1. *T. howellii* S. Wats.
2. *T. integrifolium* (Nutt.) Endl.
3. *T. laciniatum* (Hook.) Endl.
4. *T. lasiophyllum* Greene
5. *T. wrightii* A. Gray (=*Stanleyella wrightii* (A. Gray) Rydb.)

Oomycetes
Albugo candida. Downy mildew. **3**:WA(3202). **5**:UT(1336).
Peronospora parasitica. Downy mildew. **4**:AZ(1195,3381).
Basidiomycotina - Uredinales
[*Puccinia aristidae*]. **3**:WA(3202).
P. thelypodii. Autoecious. **4**:CA(81).
Deuteromycotina - Coelomycetes
Hendersonia stanleyellae. On stems. **5**:NM(2399).
Heteropatella umbilicata. **1**:OR(3202).
Septoria sp. **2**:WA(3202). **3**:ID(3202), WA(3202).

Thlaspi L.

1. *T. alpestre* L.
2. *T. arvense* L. [fanweed, field penny cress, frenchweed, mithridate mustard, stinkweed]
3. *T. montanum* L. (=*T. montanum* L. var. *fendleri* (A. Gray) P. Holmgren, *T. fendleri* A. Gray var. *glaucum* (A. Nels.) C. Hitchc., *T. glaucum* A. Nels., *T. purpurascens* Rydb., *T. fendleri* A. Gray)
4. *Thlaspi* sp.

Plasmodiophoromycetes
Plasmodiophora brassicae. Clubroot. **2:**NJ(82).
Chytridiomycetes
Synchytrium sp. On leaves and stems. **1:**CO(1849).
Oomycetes
Albugo candida. White rust. **2:**CO(82), MT(82,3202), WY(82). **3:**MT(3202). **4:**CO(82), MT(82), WY(82).
Ascomycotina
Mycosphaerella tassiana. **3:**WA(3202).
Sclerotinia sclerotiorum. Crown rot. **2:**MT(82,3202).
Basidiomycotina - Uredinales
[*Puccinia aristidae*]. **2:**CO(82), TX(82).
P. monoica. Heteroecious: aecial host. **3:**CO(1195,3379).
P. oudemansii. Autoecious. **4:**UT(82).
P. thlaspeos. Autoecious. **3:**AZ(1203,3981), ID(3202), OR(3202), WY(3103). **4:**CO(82), ID(82), UT(82), WY(82).
Deuteromycotina - Hyphomycetes
Cercospora thlaspiae. On silicles. **2:**WI(1334).
Entylomella armoraciae. Leaf spot. **4:**TX(82).
Deuteromycotina - Coelomycetes
Phyllosticta sp. On ripened silicles. **2:**WI(1352).
Septoria thlaspii. On silicles. **2:**WI(1331).
Septoria sp. **2:**WI(1351).

Thysanocarpus Hook.
1. *T. curvipes* Hook.
Oomycetes
Albugo candida. **1:**ID(3202).
Peronospora parasitica. **1:**WA(3202).

Bromeliaceae

Aechmea Ruiz & Pav.
1. *A. chantinii* (Carrière) Baker
2. *A. fasciata* (Lindl.) Baker [urn plant]
3. *A. orlandiana* L. B. Sm. [finger-of-God]
4. *Aechmea* sp.
Oomycetes
Phytophthora sp. Root rot, basal stem rot. **1:**FL(35). **4:**FL(35).
Pythium sp. Crown and root rot. **2:**FL(35). **3:**FL(35). **4:**FL(35).
Ascomycotina
Glomerella sp. Anthracnose. **4:**FL(35).
Deuteromycotina - Hyphomycetes
Alternaria sp. Leaf spot. **4:**FL(35).
Botrytis cinerea. Gray mold. **2:**FL(35).
Cercospora sp. Leaf spot. **4:**FL(35).
Exserohilum rostratum. Leaf spot. **2:**FL(35,2254,2255).
Fusarium sp. Leaf rot, root rot, basal stem rot. **1:**FL(35). **4:**FL(35).
Deuteromycotina - Coelomycetes
Colletotrichum gloeosporioides. Anthracnose. **1:**FL(35). **2:**FL(35). **4:**FL(35).
Phomopsis sp. Leaf spot. **2:**FL(35).
Phyllosticta sp. Leaf spot. **1:**FL(35). **3:**FL(35). **4:**FL(35).
Deuteromycotina - Other
Rhizoctonia solani. Basal rot. **1:**FL(35). **2:**FL(35).
Rhizoctonia sp. Crown and root rot. **2:**HI(2903).

Ananas Mill.
1. *A. comosus* (L.) Merr. (=*Bromelia comosa* L.) [pineapple]
Chytridiomycetes
Rhizidiocystis ananasi. In root hairs. **1:**HI(82,2903).
Oomycetes
Aphanomyces sp. Root rot. **1:**HI(82,2903).
Phytophthora cinnamomi. Heart rot. **1:**HI(82,1914,2903,3624).
P. citrophthora. **1:**HI(82,2903).
P. drechsleri. **1:**HI(82,2903).
P. nicotianae var. *parasitica*. Heart rot. **1:**HI(82,1914,2903).
P. palmivora. Heart rot of buds and stems. **1:**HI(82,2903).
Pythium acanthophoron. Isolated from diseased leaves; weakly pathogenic. **1:**HI(82,2821,2903).
P. aphanidermatum. Root rot. **1:**HI(82,2903).
P. arrhenomanes. Root rot. **1:**HI(82,1914,2903).
P. debaryanum. **1:**HI(82,2903).
P. graminicola. Root rot. **1:**HI(82,1914).
P. hydnosporum. **1:**HI(82,2903).
P. indigoferae. Root rot. **1:**HI(82,2903).
P. intermedium. **1:**HI(2903).
P. irregulare. Root rot. **1:**HI(82,1914,2903).
P. mamillatum. **1:**HI(82).
P. megalacanthum. **1:**HI(82,2903).
P. rostratum. **1:**HI(82,2903).
P. splendens. Root rot. **1:**HI(82,1914,2903).
P. torulosum. Root rot. **1:**HI(1914).
P. vexans. **1:**HI(82,2903).
Pythium sp. Root rot. **1:**FL(35), HI(82,2903).
Zygomycotina
Rhizopus stolonifer. Fruit rot. **1:**HI(82,2903).
Rhizopus sp. **1:**HI(2903).
Ascomycotina
Ceratocystis paradoxa. Leaf base rot, soft rot of fruit. **1:**CA(3714), FL(35,82), HI(82,2903).
Mollisia sp. **1:**HI(2903).
Saccharomyces sp. **1:**HI(2903).
Deuteromycotina - Hyphomycetes
Alternaria sp. **1:**HI(2903).
Aspergillus sp. **1:**HI(2903).
Fusarium subglutinans. Fruitlet core rot. **1:**HI(2903).
Fusarium sp. Leaf necrosis, basal rot, brown rot of fruit. **1:**FL(35), HI(82,2903).
Hymenula affinis. Root rot. **1:**HI(82,2903).
Monilia sp. **1:**HI(2903).
Penicillium funiculosum. Interfruitlet corking, leather pocket, fruitlet core rot. **1:**HI(1534,2903).
P. pinophilum. **1:**HI(2903).
Penicillium sp. Fruit rot. **1:**Range of host(82), FL(35).
Trichoderma viride. Secondary fruit mold. **1:**HI(82,2903).
Trichoderma sp. **1:**HI(2903).
Deuteromycotina - Coelomycetes
Colletotrichum gloeosporioides. Anthracnose. **1:**FL(35).
Coniothyrium sp. Basal rot. **1:**FL(35).
Diplodia sp. Basal rot. **1:**FL(35).
Pestalotia sp. **1:**HI(2903).
Phoma sp. **1:**HI(2903).
Phyllosticta sp. Leaf spot. **1:**FL(35).
Deuteromycotina - Other
Rhizoctonia sp. Root rot. **1:**FL(35), HI(2903).

Billbergia Thunb.
1. *Billbergia* sp.
Oomycetes
Pythium arrhenomanes. **1:**HI(2903).
Pythium sp. Root rot. **1:**FL(35).
Deuteromycotina - Hyphomycetes
Alternaria sp. Leaf spot. **1:**FL(35).
Botrytis cinerea. Gray mold. **1:**FL(35).
Fusarium solani. Root rot. **1:**FL(35).
Deuteromycotina - Coelomycetes
Colletotrichum bromeliacearum. Anthracnose. **1:**FL(35).
C. gloeosporioides. Anthracnose. **1:**FL(35).
Phomopsis sp. Leaf spot. **1:**FL(35).
Phyllosticta sp. Leaf spot. **1:**FL(35).

Bromelia L.
1. *Bromelia* × sp. (unnamed hybrids)
Deuteromycotina - Hyphomycetes
[*Exserohilum rostratum*]. Leaf spot. **1:** FL(35).

Canistrum E. Morr.
1. *C. fosterianum* L. B. Sm.
Oomycetes
Pythium sp. Stem rot. **1:**FL(35).

Cryptanthus Otto & A. Dietr.
1. *C. acaulis* (Lindl.) Beer [starfish-plant]
2. *C. bromelioides* Otto & A. Dietr. [pink crypthanthus]
3. *C. bromelioides* Otto & A. Dietr. var. *tricolor* M. B. Foster [rainbow-star]
4. *Cryptanthus* sp.

Oomycetes
Phytophthora sp. Root rot. **4:**FL(35).
Pythium sp. Root rot. **4:**FL(35).
Ascomycotina
Glomerella cingulata. **1:**WA(3202).
Deuteromycotina - Hyphomycetes
Botrytis cinerea. Gray mold. **4:**FL(35).
Cephalosporium sp. Stem rot. **4:**FL(35).
Fusarium sp. Root and basal stem rot. **4:**FL(35).
Verticillium sp. Wilt. **4:**FL(35).
Deuteromycotina - Coelomycetes
Colletotrichum gloeosporioides. Anthracnose. **4:**FL(35).
Gloeosporium sp. **2:**Unk.(1300). **3:**Unk.(1300).
Phyllosticta sp. Leaf spot. **4:**FL(35).
Deuteromycotina - Other
Rhizoctonia sp. Root rot. **4:**FL(35).

×***Cryptbergia*** Hort.
Billbergia × *Cryptanthus*.
1. × *Cryptbergia* sp. (unnamed hybrids)
Deuteromycotina - Coelomycetes
Colletotrichum gloeosporioides. Leaf spot. **1:** FL(35).
Diplodia sp. Leaf blister. **1:** FL(35).
Phyllosticta sp. Leaf spot. **1:** FL(35).

Guzmania Ruiz & Pav.
1. *Guzmania* sp.
Oomycetes
Phytophthora sp. Root rot. **1:**FL(35).
Deuteromycotina - Hyphomycetes
Fusarium sp. Leaf rot. **1:**FL(35).
Deuteromycotina - Coelomycetes
Colletotrichum gloeosporioides. Anthracnose. **1:**FL(35).
Phyllosticta capitalensis. Leaf spot. **1:**FL(35).
Phyllosticta sp. Leaf spot. **1:**FL(35).

Hohenbergia Schult. & J. H. Schult.
1. *Hohenbergia* sp.
Deuteromycotina - Coelomycetes
Phyllosticta sp. Leaf spot. **1:**FL(35).

Neoregelia L. B. Sm.
1. *Neoregelia* sp.
Oomycetes
Phytophthora sp. Crown rot. **1:**FL(35).
Pythium sp. Basal rot. **1:**FL(35).
Deuteromycotina - Hyphomycetes
Fusarium sp. Leaf spot. **1:**FL(35).
Deuteromycotina - Coelomycetes
Colletotrichum gloeosporioides. Leaf spot. **1:**FL(35).
Phomopsis sp. Basal leaf necrosis. **1:**FL(35).
Phyllosticta sp. Leaf spot. **1:**FL(35).

Nidularium Lem.
1. *Nidularium* sp.
Oomycetes
Phytophthora sp. Root and basal rot. **1:**FL(35).
Pythium sp. Root rot. **1:**FL(35).
Deuteromycotina - Coelomycetes
Phyllosticta capitalensis. Leaf spot. **1:**FL(35).
Phyllosticta sp. Leaf spot. **1:**FL(35).

Puya Molina
1. *P. chilensis* Molina
Ascomycotina
Metasphaeria puyae. On dead leaves. **1:**CA(81).

Tillandsia L.
1. *T. fasciculata* Sw. [wild-pineapple]
2. *T. recurvata* (L.) L. [ball-moss, bunch-moss]
3. *T. usneoides* (L.) L. [graybeard, Spanish-moss]
4. *Tillandsia* sp.
Oomycetes
Pythium sp. Basal rot, root rot. **3:**HI(2903). **4:**FL(35).
Ascomycotina
Gelasinospora autosteira. **3:**MS(27).
Glyphium tillandsiae. **1:**FL(522).
Basidiomycotina - Uredinales
Puccinia tillandsiae. Autoecious. **1:**FL(35).
Deuteromycotina - Hyphomycetes
Fusarium solani. Leaf and stem blight. **3:**FL(35,3002), GA(3002), MS(3002), SC(3002).
Fusarium sp. Blight, basal rot. **4:**FL(35).
Volutella cylindrospora. On leaves. **3:**LA(82).
Deuteromycotina - Coelomycetes
Colletotrichum bromeliacearum. On stems. **2:**TX(82).
Colletotrichum sp. Anthracnose. **4:**FL(35).
Phyllosticta sp. Leaf spot. **4:**FL(35).

Vriesea Lindl.
1. *Vriesea* sp.
Oomycetes
Phytophthora sp. Basal stem rot. **1:**FL(35).
Pythium sp. Basal stem rot. **1:**FL(35).
Deuteromycotina - Coelomycetes
Colletotrichum gloeosporioides. Anthracnose. **1:**FL(35).
Phyllosticta sp. Leaf spot. **1:**FL(35).
Deuteromycotina - Other
Rhizoctonia solani. Root and stem rot. **1:**FL(35).

Burseraceae

Bursera Jacq. ex L.
1. *B. simaruba* (L.) Sarg. [gum-elemi, gumbo, gumbo-limbo, West Indian birch]
Basidiomycotina - Other
Gloeodontia discolor. On wood. **1:**FL(440).
Oudemansiella canarii. On wood. **1:**FL(3284).
Deuteromycotina - Hyphomycetes
Cercospora sp. Leaf spot. **1:**FL(35).
Oidium sp. Powdery mildew. **1:**FL(35).
Deuteromycotina - Coelomycetes
Diplodia sp. Stem gall. **1:**FL(35).

Butomaceae

Butomus L.
1. *B. umbellatus* L. [flowering-rush, grassy-rush, water gladiolus]
Chytridiomycetes
Physoderma butomi. On flowering stalks and perianths. **1:**MI(3398).

Buxaceae

Buxus L.
1. *B. microphylla* Siebold & Zucc. [littleleaf boxwood]
2. *B. microphylla* Siebold & Zucc. var. *japonica* (Müll. Arg. ex Miq.) Rehd. & E. Wils. [Japanese boxwood]
3. *B. microphylla* Siebold & Zucc. var. *koreana* Nakai ex Rehd. [Korean boxwood]
4. *B. sempervirens* L. [common boxwood, Turkish boxwood]
5. *B. sempervirens* L. cv. 'arborescens'
6. *B. sempervirens* L. cv. 'suffruticosa'
7. *Buxus* sp.
Oomycetes
Phytophthora cinnamomi. Root rot. **4:**FL(35), NC(1303). **5:**NC(1303).
P. nicotianae var. *parasitica*. Foliage blight, stem necrosis, root rot. **1:**FL(35). **4:**NC(1303). **5:**MD(82), NC(1303). **6:**NC(1303).
Phytophthora sp. Basal stem discoloration, root rot. **6:**VA(2014). **7:**FL(35).
Pythium helicoides. Root rot. **6:**NC(1303).
P. irregulare. Root rot. **7:**FL(35).
P. splendens. Root rot. **1:**FL(35).
Pythium sp. Basal stem discoloration, root rot. **2:**HI(2903). **4:**FL(35), NC(1303). **6:**MA(82), MD(82), VA(2014).

Ascomycotina
Glomerella cingulata. Anthracnose. **7:**FL(35).
Hyponectria buxi. Blight. **4:**MD(183), MS(2720). **5:**NY(82).
Nectria cinnabarina. On twigs. **4:**AL(82).
Nectria sp. Twig blight, stem canker. **4:**NC(1303). **5:**Unk.(82). **7:**FL(35).
Pseudonectria rousseliana. On leaves, limbs and twigs. **5:**NY(82). **7:**MS(2720).
Basidiomycotina - Other
Armillaria mellea. Root rot. **4:**NJ(82).
Erythricium salmonicolor. Limb blight. **1:**FL(35).
Pellicularia koleroga. **7:**LA(3640).
Phellinus punctatus. Heart rot. **5:**VA(82).
Deuteromycotina - Hyphomycetes
Cercospora sp. Leaf spot. **4:**FL(35).
Cladosporium herbarum. Sooty mold associated with scale insect infestation. **4:**CA(2362).
Fusarium buxicola. On limbs. **5:**AL(82), MD(82), PA(82).
F. lateritium. On leaves and twigs. **5:**Unk.(82).
F. moniliforme. Twig dieback. **1:**FL(35). **7:**FL(35).
F. solani. Root rot. **1:**FL(35).
Fusarium sp. Canker, root decay, basal stem discoloration, root rot. **4:**NC(1303). **5:**NC(1303). **6:**VA(2014). **7:**FL(35), WV(3907).
Gliocladium roseum. On leaves. **4:**SC(82).
Helminthosporium velutinum. On twigs. **6:**AL(2503).
Olpitrichum macrosporum. On bark. **7:**VA(2131).
Sesquicillium buxi. Canker, dieback, basal stem discoloration, wilt. **1:**FL(35). **4:**Range of host(82), FL(35), MS(2720), NC(1303). **6:**NC(1303), VA(2014). **7:**FL(35).
Sporidesmium carrii. On dead twigs. **6:**AL(2490).
Verticillium albo-atrum. Wilt. **3:**IL(1558).
Volutella buxi. Leaf blight, canker, dieback, basal stem discoloration. **1:**Unk.(82), FL(35). **4:**Unk.(82), FL(35), MS(2720), NC(1303). **5:**Unk.(82), NC(1303). **6:**Unk.(82), NC(1303), VA(2014). **7:**FL(35), NC(1533).
Deuteromycotina - Coelomycetes
Colletotrichum sp. Limb blight, secondary leaf cast. **4:**MD(82). **5:**MD(82), NC(1303).
Diplodia sp. Dieback. **1:**FL(35). **4:**FL(35). **7:**FL(35).
Macrophoma candollei. Leaf blight, leaf spot. **1:**FL(35). **4:**MS(2720), WA(3202). **5:**Range of host(82), NC(1303). **7:**FL(35).
Macrophoma sp. On limbs. **6:**NC(1303).
Phoma glomerata. Leaf tip blight. **4:**OK(2872). **5:**MD(82), NJ(82), NY(82,3564).
Phoma sp. Dieback. **1:**FL(35). **5:**NC(1303).
Phomopsis stictica. On limbs and twigs. **5:**MD(82).
Phomopsis sp. Dieback. **1:**FL(35). **7:**FL(35).
Phyllosticta sp. Leaf spot. **4:**FL(35), NC(1303).
Deuteromycotina - Other
Rhizoctonia ramicola. Thread blight. **1:**FL(35).
R. solani. Twig dieback, basal stem discoloration, root rot. **1:**FL(35). **4:**Unk.(82). **6:**VA(2014). **7:**FL(35).

Pachysandra Michx.
1. *P. procumbens* Michx. [Alleghany pachysandra, Alleghany spurge]
2. *P. terminalis* Siebold & Zucc. [Japanese pachysandra, Japanese spurge]
3. *Pachysandra* sp.

Ascomycotina
Pseudonectria pachysandricola. Dieback. **2:**CT(82), DE(82), NJ(82), NY(82), PA(82).
Deuteromycotina - Hyphomycetes
Fusarium sp. Stem rot. **2:**NC(1303).
Volutella pachysandrae. Leaf and stem blight. **2:**NC(1303), VA(82), WV(3910). **3:**WV(3911).
Deuteromycotina - Coelomycetes
Macrophoma pachysandrae. On leaves. **2:**VA(82).
Phyllosticta pachysandrae. Leaf spot. **1:**NC(64), NY(82), PA(82), VA(82). **2:**NC(82), NY(82), PA(82), VA(82).
Septoria pachysandrae. Leaf spot. **1:**FL(35,82), TN(82).
Sphaeropsis sp. Tip blight. **2:**NJ(82).
Deuteromycotina - Other
Rhizoctonia sp. Stem rot. **2:**NJ(82).

Sarcococca Lindl.
1. *Sarcococca* sp.

Deuteromycotina - Hyphomycetes
Botrytis cinerea. **1:**WA(3202).

Cabombaceae

Brasenia Schreb.
1. *B. schreberi* J. F. Gmelin [purple-wan dock, watershield]

Deuteromycotina - Hyphomycetes
Cercospora sp. Leaf spot. **1:**FL(35).
Dichotomophthoropsis nymphaearum. Edema, leaf spot. **1:**MN(1767).

Cactaceae

Astrophytum Lem.
1. *A. myriostigma* (Salm-Dyck) Lem. [bishop's-cap, bishop's-hood, monk's-hood]
2. *A. ornatum* (DC.) Britton & J. Rose [bishop's-cap, ornamental monk's-hood, star cactus]

Oomycetes
Pythium sp. Basal stem rot. **1:**FL(35).
Deuteromycotina - Hyphomycetes
Bipolaris cactivora. Stem rot in nurseries. **2:**CA(923).
Bispora sp. Scab-like lesions. **1:**FL(35).

Cactus L.
1. *Cactus* sp.

Deuteromycotina - Coelomycetes
Colletotrichum gloeosporioides. Anthracnose. **1:**MS(2720).

Carnegiea Britton & J. Rose
1. *C. gigantea* (Engelm.) Britton & J. Rose (=*Cereus giganteus* Engelm.) [Arizona-giant, giant cactus, giant saguaro, saguaro, sahuaro]

Ascomycotina
Pichia amethionina var. *pachycereana*. Stem decay. **1:**AZ(3470).
P. cactophila. Isolated from decaying tissue. **1:**AZ(3469).
P. heedii. Isolated from necrotic tissues. **1:**AZ(2793).
Basidiomycotina - Other
Athelia decipiens. Uniform white rot. **1:**AZ(1202,2134,2558).
Ceriporiopsis carnegiea. White rot. **1:**AZ(82,128,1192,1202,2134,2140).
Coniophora eremophila. Brown rot. **1:**AZ(2134).
Coprinus papagoensis. Forms sclerotia. **1:**AZ(2134,2558).
Heliocybe sulcata. Brown rot. **1:**AZ(1202,2134,2558).
Hyphoderma clavigerum. White rot. **1:**AZ(2134).
Nidularia griseolazulina. **1:**AZ(2134).
Peniophora tamaricicola. White rot. **1:**AZ(1202,2134,2558).
Phanerochaete chrysorhiza. Uniform white rot. **1:**AZ(1202,2134).
P. omnivora. White rot. **1:**AZ(433).
P. tuberculata. White rot. **1:**AZ(1202,2134,2558).
P. xerophila. **1:**AZ(433).
Phellinus texanus. White heart rot. **1:**AZ(1202,2134).
Schizopora apacheriensis. White rot. **1:**AZ(2134).
Deuteromycotina - Hyphomycetes
Candida deserticola. Isolated from necrotic stems. **1:**AZ(2794).
Fusarium oxysporum. Soft rot of seedlings. **1:**AZ(337).
F. solani. Soft rot of seedlings. **1:**AZ(337).
Deuteromycotina - Coelomycetes
Camarosporium cerei. On cortex. **1:**AZ(82).

Cephalocereus Pfeiff.
1. *C. collinsii* Britton & J. Rose
2. *C. mezcalensis* Bravo
3. *C. sartorianus* J. Rose
4. *C. tetetzo* (A. Weber) Vaup.

Deuteromycotina - Hyphomycetes
Bipolaris cactivora. Stem rot in nurseries. **1:**CA(923). **2:**CA(923). **3:**CA(923). **4:**CA(923).

Cereus Mill.
1. *C. peruvianus* (L.) Mill. [Peruvian apple cactus]
2. *Cereus* sp.

Oomycetes
Pythium sp. Root rot. **2**:FL(35).

Ascomycotina
Diplotheca tunae. Black spot. **2**:TX(82).
Monographella opuntiae. Anthracnose. **2**:TX(82).

Deuteromycotina - Hyphomycetes
Alternaria sp. Leaf spot. **2**:FL(35).
Aspergillus alliaceus. Branch rot. **2**:TX(82).
Bipolaris cactivora. 'Helminthosporium stem rot. Helminthosporium shattering.' **1**:CA(923). **2**:FL(35).
Botrytis cinerea. Gray mold rot. **2**:Unk.(82).
Fusarium oxysporum. Root and stem rot. **2**:MS(82,2720), TX(82).
Fusarium sp. Basal stem rot, leaf spot. **2**:FL(35).
Phymatotrichopsis omnivora. Root rot. **2**:TX(82).

Deuteromycotina - Coelomycetes
Colletotrichum gloeosporioides. Anthracnose. **2**:HI(2903).
Diplodia sp. Dieback, basal black rot. **2**:FL(35).
Dothiorella opuntiae. **2**:HI(2784).
Hendersonia opuntiae. Scald, scorch. **2**:Unk.(82).
Phaeoseptoria sp. Leaf spot. **2**:FL(35).
Phomopsis sp. Leaf spot. **2**:FL(35).
Septoria cacticola. Stem spot. **2**:TX(82).

Deuteromycotina - Other
Rhizoctonia solani. Root and stem rot. **2**:FL(35).

Coryphantha (Engelm.) Lem.
1. *C. vivipara* (Nutt.) Britton & J. Rose (=*Mammillaria vivipara* Nutt.)
2. *Coryphantha* sp.

Deuteromycotina - Hyphomycetes
Cercospora sp. Leaf spot. **2**:FL(35).

Deuteromycotina - Coelomycetes
Coniothyrium sp. Leaf spot. **2**:FL(35).
Phoma mamillariae. On spines. **1**:MT(965,3202).

Cryptocereus E. J. Alexander
1. *C. anthonyanus* E. J. Alexander

Deuteromycotina - Hyphomycetes
Alternaria sp. Leaf spot. **1**:FL(35).
Cercospora sp. Leaf spot. **1**:FL(35).
Fusarium sp. Leaf spot. **1**:FL(35).

Deuteromycotina - Coelomycetes
Colletotrichum sp. Leaf spot. **1**:FL(35).
Cytospora sp. Leaf spot. **1**:FL(35).

Echinocactus Link & Otto
1. *Echinocactus* sp.

Ascomycotina
Diplotheca sp. Black spot. **1**:TX(82).
Monographella opuntiae. Anthracnose. **1**:TX(82).

Deuteromycotina - Hyphomycetes
Aspergillus alliaceus. Stem rot. **1**:TX(82).
Fusarium sp. Stem rot. **1**:FL(35).
Helminthosporium sp. Basal stem rot. **1**:FL(35).
Phymatotrichopsis omnivora. Root rot. **1**:TX(82).

Deuteromycotina - Coelomycetes
Diplodia sp. Black basal stem rot. **1**:FL(35).
Hendersonia opuntiae. Scald. **1**:TX(82).

Echinocereus Engelm. in Wisl.
1. *E. chloranthus* (Engelm.) Engelm. ex Rümpler
2. *Echinocereus* sp.

Deuteromycotina - Hyphomycetes
Cercosporidium sp. Brown stem lesions. **2**:FL(35).
Fusarium sp. Root rot, basal stem rot. **2**:FL(35).
Helminthosporium sp. Stem spot. **1**:TX(82).

Echinopsis Zucc.
1. *Echinopsis* sp.

Oomycetes
Pythium sp. Basal stem rot. **1**:FL(35).

Ascomycotina
Guignardia sp. Stem lesions. **1**:FL(35).

Deuteromycotina - Hyphomycetes
Bipolaris cactivora. Stem rot. **1**:FL(35).
Fusarium oxysporum. Basal stem rot. **1**:FL(35).

Deuteromycotina - Coelomycetes
Colletotrichum gloeosporioides. Anthracnose. **1**:HI(2903).

Epiphyllum Haw.
1. *Epiphyllum* sp.

Oomycetes
Phytophthora sp. Crown and root rot. **1**:FL(35).
Pythium sp. Root rot. **1**:FL(35).

Ascomycotina
Botryosphaeria sp. Dieback. **1**:FL(35).

Deuteromycotina - Hyphomycetes
Alternaria sp. Stem spots. **1**:FL(35).
Cercospora sp. Edema-like lesions. **1**:FL(35).
Fusarium sp. Leaf and stem rot. **1**:FL(35).

Deuteromycotina - Coelomycetes
Colletotrichum gloeosporioides. Anthracnose, leaf spot. **1**:AL(82), FL(35), HI(2903).
Colletotrichum sp. Anthracnose. **1**:FL(35).

Espostoa Britton & J. Rose
1. *E. lanata* (Kunth) Britton & J. Rose [cotton-ball , new old-man cactus, Peruvian old-man cactus, Peruvian snowball cactus]

Deuteromycotina - Hyphomycetes
Bipolaris cactivora. Stem rot in nurseries. **1**:CA(923).

Ferocactus Britton & J. Rose
1. *Ferocactus* sp.

Deuteromycotina - Hyphomycetes
Bipolaris cactivora. Stem rot in nurseries. **1**:CA(923).

Hylocereus A. Berger
1. *H. undatus* (Haw.) Britton & J. Rose [Honolulu-queen, night-blooming cereus, queen-of-the-night]
2. *Hylocereus* sp.

Oomycetes
Pythium sp. Basal rot. **2**:FL(35).

Deuteromycotina - Hyphomycetes
Botrytis cinerea. Tip rot. **2**:FL(35).
Fusarium sp. Basal rot. **2**:FL(35).

Deuteromycotina - Coelomycetes
Colletotrichum gloeosporioides. Anthracnose on cladodes. **1**:FL(35,82).
Colletotrichum sp. Stem rot. **2**:FL(35).

Deuteromycotina - Other
Rhizoctonia sp. Basal rot. **2**:FL(35).

Lemaireocereus Britton & J. Rose
1. *L. dumortieri* (Scheidw.) Britton & J. Rose
2. *L. martinezii* Ort.
3. *L. thurberi* (Engelm.) Britton & J. Rose [organ-pipe cactus]

Basidiomycotina - Other
Phanerochaete arizonica. **3**:AZ(1202).

Deuteromycotina - Hyphomycetes
Bipolaris cactivora. Stem rot in nurseries. **1**:CA(923). **2**:CA(923).

Lophocereus (A. Berger) Britton & J. Rose
1. *L. gatesii* M. E. Jones
2. *L. schottii* (Engelm.) Britton & J. Rose [senita, whisker cactus]

Deuteromycotina - Hyphomycetes
Bipolaris cactivora. Stem rot in nurseries. **1**:CA(923). **2**:CA(923).
Fusarium oxysporum. Root and stem rot. **2**:AZ(82), CA(82).

Mammillaria Haw.
1. *M. hahniana* Werderm. [lady-of-Mexico cactus]
2. *Mammillaria* sp.

Oomycetes
Pythium sp. Root rot. **2**:FL(35).

Ascomycotina
Monographella opuntiae. Anthracnose. **2**:TX(82).

Deuteromycotina - Hyphomycetes
Bipolaris cactivora. Stem rot in nurseries. **1**:CA(923).
Cercospora sp. Leaf spot. **2**:FL(35).
Fusarium oxysporum. Stem rot. **2**:FL(35).

Phymatotrichopsis omnivora. Root rot. **2:**TX(82).
Deuteromycotina - Coelomycetes
Colletotrichum gloeosporioides. Zonate spot. **2:**Unk.(82).
Colletotrichum sp. Anthracnose. **2:**FL(35).
Phoma mamillariae. On spines. **2:**MT(82).

Nopalea Salm-Dyck

1. *N. cochenillifera* (L.) Salm-Dyck [cochineal cactus, cochineal-plant]
Oomycetes
Pythium sp. Root rot. **1:**FL(35).

Opuntia Mill.

1. *O. arbuscula* Engelm.
2. *O. ficus-indica* (L.) Mill. (=*O. engelmannii* Salm-Dyck, *O. megacantha* Salm-Dyck). Plants cultivated as *O. engelmannii* may be *O. phaeacantha* var. *discata*. [Indian fig, spineless cactus]
3. *O. fulgida* Engelm. [jumping cactus]
4. *O. humifusa* (Raf.) Raf. (=*O. compressa* (Salisb.) Macbr., *O. rafinesquii* Engelm.)
5. *O. lindheimeri* Engelm. [Lindheimer prickly-pear, Texas prickly-pear]
6. *O. macrorhiza* Engelm.
7. *O. missouriensis* DC. (=*O. polycantha* Haw.)
8. *O. nitens* Small
9. *O. phaeacantha* Engelm.
10. *O. spinosior* (Engelm.) Toumey
11. *O. stricta* (Haw.) Haw. var. *dillenii* (Ker-Gawl.) L. Benson (=*O. dillenii* (Ker-Gawl.) Haw.) [prickly-pear cactus]
12. *O. versicolor* Engelm. ex J. M. Coulter
13. *Opuntia* sp.

Oomycetes
Phytophthora nicotianae var. *parasitica*. Stem rot. **13:**NY(82).
Pythium aphanidermatum. Root rot. **13:**FL(35).
P. debaryanum. Stem rot. **13:**CA(82).
Ascomycotina
Anthostomella cacti. On cortex. **2:**CA(82). **13:**NJ(82).
Botryosphaeria obtusa. Black rot. **4:**NY(82).
B. quercuum. Black rot. **13:**FL(35).
B. rhodina. **4:**FL(82).
Clavispora opuntiae. Isolated from necrotic tissue. **2:**HI(2792). **5:**TX(2792). **9:**AZ(2792).
Diplotheca tunae. Black spot. **2:**TX(82). **4:**FL(82), WI(1341). **5:**TX(82). **6:**WI(1355). **8:**FL(1745). **11:**FL(1745). **13:**FL(35,82), TX(82).
Discostroma muricata. **13:**AZ(1003).
Hysterographium opuntiae. On wood. **3:**AZ(406).
Lembosia cactorum. Black mildew. **4:**FL(82). **13:**FL(35).
Leptosphaeria opuntiae. On cortex. **2:**AL(82). **5:**TX(82).
Monographella opuntiae. Anthracnose, black rot. **4:**FL(82), NY(82). **5:**TX(82). **11:**FL(82). **13:**AL(82), FL(35), LA(82,965,967), SC(82).
Nectriella cacti. On cortex. **2:**AL(82).
Pichia cactophila. Isolated from decaying tissues. **2:**HI(3469).
P. deserticola. Isolated from necrotic stems. **5:**TX(2794). **9:**AZ(2794).
Teichospora mammoides var. *opuntiae*. On cortex. **13:**CA(82).
T. opuntiae. **13:**CO(82).
Basidiomycotina - Other
Armillaria mellea. Root and stem rot. **2:**CA(2902).
Crustoderma opuntiae. On wood. **3:**AZ(2559).
Perenniporia medulla-panis. **12:**AZ(128,1202).
Phanerochaete arizonica. **12:**AZ(1202).
P. xerophila. **12:**AZ(433).
[*Phellinus robustus*]. **12:**AZ(128).
P. texanus. **1:**AZ(1202). **10:**AZ(1202). **12:**AZ(1202).
Pycnoporus cinnabarinus. **10:**AZ(1202).
Deuteromycotina - Hyphomycetes
Alternaria sp. Leaf spot. **13:**FL(35).
Aspergillus alliaceus. Cladode rot. **2:**TX(82). **4:**TX(82). **5:**TX(82). **7:**TX(82). **11:**TX(82). **13:**TX(82).
Cephalosporium sp. Stem rot. **13:**FL(35).
Curvularia eragrostidis. On stems. **13:**FL(3548).
Fusarium oxysporum. Root rot. **2:**HI(2903).
Fusarium sp. Basal stem rot, leaf spot. **13:**FL(35).
Helminthosporium sp. Leaf spot. **13:**FL(35).
Microdochium lunatum. Anthracnose. **2:**AL(82), FL(82), TX(82). **4:**AL(82), FL(82), TX(82). **5:**AL(82), FL(82), TX(82). **11:**AL(82), FL(82), TX(82). **13:**FL(35).
Oidium sp. Powdery mildew. **13:**FL(35).
Tretopileus sphaerophorus. On cladodes. **13:**FL(35,82).
Deuteromycotina - Coelomycetes
Ascochyta sp. Leaf spot. **13:**FL(35).
Colletotrichum gloeosporioides. Zonate leaf spot. **13:**FL(35,82), MI(82), MO(82), MS(2720), NJ(82).
Colletotrichum sp. On cladodes. **13:**NY(82), TX(82).
Diplodia opuntiae. Cladode rot. **4:**MD(82), PA(82). **13:**HI(82,2903), KS(82).
Hendersonia opuntiae. Black rot, scorch. **2:**AL(82). **4:**NJ(82). **5:**MT(82). **7:**MT(3202). **13:**FL(35), KS(82), TX(82).
Leptodermella opuntiae. **13:**WA(3202).
Macrophoma opunticola. Cladode rot. **13:**HI(82,2903).
Macrophomina phaseolina. Charcoal rot. **13:**TX(82).
Phoma mamillariae. **7:**MT(3202).
Phoma sp. Cladode rot. **4:**OK(82,2872).
Phomopsis cacti. Cladode spot. **13:**NM(82).
Phyllosticta concava. Black spot, leaf spot. **4:**OK(2872). **13:**MO(82), NJ(82), OK(82), TX(82).
Phyllosticta sp. Cladode spot. **13:**CO(49), FL(35).
Septoria fici-indicae. Cladode spot. **13:**TX(82).
Vermicularia cacti. **4:**Unk.(82).
Deuteromycotina - Other
Rhizoctonia solani. Root and stem rot. **13:**FL(35).

Rhipsalidopsis Britton & J. Rose

1. *R. gaertneri* (Regel) Moran [Easter cactus]
Deuteromycotina - Hyphomycetes
Bipolaris cactivora. 'Helminthosporium stem rot. Helminthosporium shattering.' **1:**FL(557).

Schlumbergera Lem.

1. *S. bridgesii* (Lem.) Loefgr. [Christmas cactus]
2. *S. truncata* (Haw.) Moran (=*Zygocactus truncatus* (Haw.) K. Schum.) [claw cactus, crab cactus, linkleaf, Thanksgiving cactus, yoke cactus]
3. *Schlumbergera* sp.

Oomycetes
Phytophthora nicotianae var. *parasitica*. Root and stem rot. **2:**Unk.(38), FL(35).
Phytophthora sp. Basal stem rot. **3:**FL(35).
Pythium aphanidermatum. Crown rot, root rot. **2:**FL(35), HI(2903).
P. irregulare. Root rot. **2:**FL(35).
Pythium sp. Root rot, basal stem rot. **1:**HI(2903). **3:**FL(35).
Ascomycotina
Glomerella sp. Anthracnose. **2:**FL(35).
Basidiomycotina - Other
Armillaria tabescens. Root rot. **2:**FL(35).
Deuteromycotina - Hyphomycetes
Alternaria sp. Leaf spot. **2:**FL(35).
Bipolaris cactivora. Stem rot. **2:**'Helminthosporium stem rot. Helminthosporium shattering.' FL(35). **3:**FL(35).
Cercospora sp. Leaf spot. **2:**FL(35).
Dichotomophthora sp. Tip blight. **2:**FL(35).
Fusarium moniliforme. Basal stem rot, leaf spot. **2:**FL(35).
F. oxysporum. Dieback, root and stem rot. 'Fusarium stem rot.' **2:**FL(35), MA(2471).
Fusarium sp. Basal stem rot, leaf spot. **2:**FL(35). **3:**FL(35).
Deuteromycotina - Coelomycetes
Colletotrichum gloeosporioides. Anthracnose, zonate leaf spot. **2:**FL(35).
Phomopsis sp. Leaf spot. **2:**FL(35).
Deuteromycotina - Other
Rhizoctonia solani. Root and stem rot. **2:**FL(35).
Rhizoctonia sp. Crown and root rot. **1:**HI(2903).

Selenicereus (A. Berger) Britton & J. Rose

1. *Selenicereus* sp.
Deuteromycotina - Hyphomycetes
Bipolaris cactivora. Stem rot in nurseries. **1:**CA(923).

Trichocereus (A. Berger) Riccob.

1. *T. spachianus* (Lem.) Riccob. [golden-column, torch cactus, white torch cactus]

Deuteromycotina - Hyphomycetes

Bipolaris cactivora. Stem rot. **1**:FL(35).

Deuteromycotina - Other

Rhizoctonia solani. Basal rot of seedlings. **1**:FL(35).

Callitrichaceae

Callitriche L.

1. *C. marginata* Torr. var. *longipedunculata* Jeps.

Basidiomycotina - Ustilaginales

Doassansia callitriches. **1**:CA(1077).

Calycanthaceae

Calycanthus K. Schum.

1. *C. floridus* L. [Carolina allspice, pineapple-shrub, strawberry-shrub]
2. *C. floridus* L. var. *laevigatus* (Willd.) Torr. & A. Gray
3. *C. occidentalis* Hook. & Arnott [California sweetshrub]

Ascomycotina

Botryosphaeria calycanthi. On twigs. **1**:NC(82), VA(82).
B. dothidea. **1**:GA(1438).
B. obtusa. On limbs. **1**:GA(82), NY(82), VA(82).
Nectria cinnabarina. On twigs. **1**:MD(82).
Phyllactinia guttata. Powdery mildew. **3**:CA(81,82).

Deuteromycotina - Hyphomycetes

Cercospora sp. Leaf spot. **1**:FL(35).
Fumago sp. **2**:NC(1307).

Deuteromycotina - Coelomycetes

Aplosporella calycanthi. **1**:OK(2873).
Cytospora laxa. On twigs. **1**:MD(82).
Diplodia sp. Dieback. **1**:FL(35).

Chimonanthus Lindl.

1. *C. praecox* (L.) Link [wintersweet]

Basidiomycotina - Other

Armillaria mellea. Root rot. **1**:CA(2897).

Campanulaceae

Campanula L.

1. *C. americana* L. [American bellflower, tall bellflower]
2. *C. aparinoides* Pursh [marsh bellflower]
3. *C. carpatica* Jacq. [tussock bellflower]
4. *C. divaricata* Michx. [southern harebell]
5. *C. lasiocarpa* Cham.
6. *C. linifolia* Scop.
7. *C. medium* L. [Canterbury-bells]
8. *C. persicifolia* L. [peach-bells, willow bellflower]
9. *C. rapunculoides* L. [creeping bellflower, rover bellflower]
10. *C. rotundifolia* L. (=*C. petiolata* A. DC.) [bluebell, common harebell]
11. *C. scouleri* Hook. ex A. DC.
12. *C. uliginosa* Rydb.
13. *Campanula* sp.

Chytridiomycetes

Synchytrium sp. On leaves and stems. **1**:WA(1849). **10**:WA(3202).

Ascomycotina

Erysiphe cichoracearum. Powdery mildew. **13**:PA(82).
Mycosphaerella campanulae. On stems. **1**:OH(82).
M. minor. On dead leaves and stems. **5**:AK(1938).
Sclerotinia sclerotiorum. Stem rot. **7**:WA(82). **8**:MD(82).

Basidiomycotina - Uredinales

Aecidium campanulastri. **1**:IA(82), MN(82,2874), WI(1338,1343).
Coleosporium tussilaginis. Heteroecious: telial host. **1**:E states(82), IL(650), IN(647), MO(82), MS(82,2720), WI(82), WV(3907). **8**:CA(82). **9**:IL(312), WI(1338). **10**:WI(1350). **13**:E states(82), WI(82).
Puccinia campanulae. Autoecious. **10**:MT(82,3202), NY(82). **11**:CA(81), OR(3202), WA(3202). **13**:CA(82), OR(82), WA(82).

Deuteromycotina - Hyphomycetes

Botrytis cinerea. Gray mold blight. **13**:AK(82).
Cercospora minuta. Leaf spot. **13**:WI(82).
Cercospora sp. **9**:WI(1352).
Fusarium sp. Root rot. **1**:NJ(82). **13**:NY(82).
Ramularia macrospora. Leaf spot. **6**:AK(526). **10**:ID(3202), WI(1344). **13**:AK(82).

Deuteromycotina - Coelomycetes

Ascochyta bohemica. Leaf spot. **1**:WI(1342). **9**:WI(1338). **13**:WI(82).
Macrophomina phaseolina. Stem blight. **9**:IL(312).
Phlyctema campanulae. On stems. **10**:CO(82).
Phoma gentianae. On stems. **10**:TX(82).
Phyllosticta alliariifoliae. Leaf spot. **13**:NJ(82), NY(82).
Phyllosticta sp. **10**:WI(1356).
Septoria campanulae. Leaf spot. **1**:IA(82), IL(82), KS(82), MI(82), MO(82), MS(2720), WI(82). **2**:WI(1334). **9**:WI(1333). **10**:WI(1350). **12**:WI(1357). **13**:WI(82).

Deuteromycotina - Other

Rhizoctonia solani. Root rot. **7**:CT(82), TX(82). **13**:IL(82).
Sclerotium rolfsii. Southern blight. **3**:IL(82). **4**:FL(35). **7**:IL(82), NJ(82). **8**:IL(82). **13**:IL(82).

Clermontia Gaud. in Freyc.

1. *C. kakeana* Walp.
2. *C. multiflora* Hildebr.
3. *C. parviflora* Gaud. ex A. Gray
4. *C. persicifolia* Gaud.
5. *Clermontia* sp.

Ascomycotina

Amazonia clermontiae. Black mildew. **2**:HI(1272,1452,2903). **4**:HI(1272,1452,2903). **5**:HI(1272,2903).
A. psychotriae. **2**:HI(2903). **5**:HI(2903).
Asterina clermontiae. **5**:HI(2784,2903).
Calothyriopeltis clermontiae. **3**:HI(2903).
Clypeolella clermontiae. **5**:HI(2903).
Meliola lobeliae. Black mildew. **5**:HI(1272,1452,2784,2903).
Trichothyrium reptans. **2**:HI(2903). **5**:HI(2903).

Deuteromycotina - Hyphomycetes

Trichothallus hawaiiensis. **5**:HI(2903).

Deuteromycotina - Coelomycetes

Coleophoma sandwicensis. On decaying leaves. **5**:HI(2784).
Septoria clermontiae. **1**:HI(2903). **5**:HI(2903).

Cyanea Gaud. in Freyc.

1. *C. angustifolia* (Cham.) Hildebr.
2. *Cyanea* sp.

Ascomycotina

Mycosphaerella cyaneae. On leaves. **1**:HI(2903,3480).
Trichothyrium reptans. **2**:HI(2903).

Cyphia Bergius

1. *C. pinnata* Roem. & Schult. (=*Lobelia pinnata* Lam.)

Deuteromycotina - Hyphomycetes

Ovularia lotophaga. **1**:OR(3202).

Downingia Torr.

1. *D. elegans* (Douglas ex Lindl.) Torr.

Basidiomycotina - Ustilaginales

Doassansia downingiae. **1**:ID(1077,3202).

Heterocodon Nutt.

1. *H. rariflorum* Nutt.

Deuteromycotina - Coelomycetes

Septoria sp. **1**:ID(3202).

Legousia Durande

(= *Specularia* A. DC.)

1. *Legousia* sp. (=*Specularia sp.*)

Deuteromycotina - Coelomycetes

Septoria speculariae. Leaf spot. **1**:IL(82), LA(82), TX(82).
S. specularina. **1**:SC(82).

Lobelia L.

1. *L. cardinalis* L. [cardinal-flower, Indian pink]
2. *L. erinus* L. [edging lobelia]
3. *L. inflata* L. [Indian tobacco]
4. *L. kalmii* L.
5. *L. puberula* Michx.
6. *L. siphilitica* L. [blue lobelia, cardinal-flower, great lobelia]
7. *L. spicata* Lam. [pale-spike lobelia]
8. *Lobelia* sp.

Chytridiomycetes
Physoderma sp. On leaves and stems. **8:**MI(3400).
Oomycetes
Pythium debaryanum. Damping-off. **2:**MA(82), NY(82). **8:**MA(82), NY(82).
Ascomycotina
Amazonia asterinoides. Black mildew. **8:**HI(82,2903).
A. psychotriae. **8:**HI(2903).
Asterina lobeliae. Black mildew. **8:**HI(82,2784,2903).
A. punctiformis var. *fimbriata*. Black mildew. **8:**HI(82,2903).
Calothyriopeltis metrosideri. Black mildew. **8:**HI(82,2903).
Meliola lobeliae. Black mildew. **8:**HI(1272,1452,2903).
Basidiomycotina - Uredinales
Puccinia lobeliae. Autoecious. **1:**MI(82). **6:**NE states(82), AR(82), WI(82). **7:**MS(2720). **8:**NE states(82), AR(82), NC(82), TX(82), WA(82,3202), WI(82).
Basidiomycotina - Ustilaginales
Entyloma lobeliae. Leaf smut. **3:**NE states(82), CT(1077), IL(1077), IN(1077), MA(664,1077), ME(1077), MO(82), NC(1077), NH(1077), NY(1077), PA(1077), VT(1077), WI(82,1077), WV(1077). **8:**PA(82).
Deuteromycotina - Hyphomycetes
Botrytis cinerea. Gray mold. **2:**AK(526). **8:**Unk.(82).
Cercospora effusa. Leaf spot. **6:**DE(3543).
C. lobeliae. Leaf spot. **1:**WI(1342). **3:**MD(82), TX(82). **6:**IN(82), KS(82), TX(82). **8:**AL(82), TX(82).
Cristulariella moricola. On leaves. **3:**IL(2572). **6:**IL(2572).
Mycovellosiella ferruginea. Leaf spot. **5:**DE(3543), OK(2872).
M. lobeliae-cardinalis. Leaf spot. **1:**KS(82,845), MN(2874), PA(845), TN(82), TX(82). **3:**VA(82). **6:**IA(82), IL(82), TX(82), VA(82), WI(82). **8:**PA(82), TX(82).
Phymatotrichopsis omnivora. Root rot. **3:**TX(82). **6:**TX(82). **8:**TX(82).
Deuteromycotina - Coelomycetes
Colletotrichum gloeosporioides. Leaf spot. **8:**HI(82,2903).
Phyllosticta bridgesii. Leaf spot. **6:**IN(82).
Septoria lobeliae. Leaf spot. **1:**MA(1781), OK(2872). **4:**WI(1360). **6:**OK(2872). **7:**MN(2874). **8:**NE states(82), TX(82), WI(82).
S. lobeliae-syphiliticae. Leaf spot. **6:**NY(82).
Deuteromycotina - Other
Rhizoctonia solani. Root rot. **8:**NY(82), OH(82).
Sclerotium rolfsii var. *delphinii*. Stem rot. **8:**NJ(82).

Platycodon A. DC.

1. *P. grandiflorum* A. DC.

Oomycetes
Phytophthora cactorum. Blight. **1:**MN(82).
Phytophthora sp. Blight. **1:**MN(82,2874).
Deuteromycotina - Hyphomycetes
Botrytis sp. Blight. **1:**OK(2872).
Phymatotrichopsis omnivora. Root rot. **1:**TX(82).
Deuteromycotina - Coelomycetes
Colletotrichum sp. On stems. **1:**CT(82).
Deuteromycotina - Other
Rhizoctonia solani. Root rot. **1:**CT(82), PA(82).

Rollandia Gaud. in Freyc.

1. *R. crispa* Gaud.
2. *R. humboldtiana* Gaud.
3. *R. racemosa* Hillebr.

Ascomycotina
Limaciniopsis rollandiae. **2:**HI(2903). **3:**HI(2903).
Deuteromycotina - Coelomycetes
Septoria rollandiae. **1:**HI(2903).

Triodanis Raf. ex Greene

1. *T. biflora* (Ruiz & Pav.) Greene (=*Specularia biflora* (Ruiz & Pav.) Fisch. & C. A. Mey.)
2. *T. leptocarpa* (Nutt.) Nieuwl. (=*Specularia leptocarpa* (Nutt.) A. Gray)
3. *T. perfoliata* (L.) Nieuwl. (=*Specularia perfoliata* (L.) A. DC.)

Chytridiomycetes
Synchytrium nitidum. On leaves and stems. **1:**LA(1846). **3:**TX(1846).
Synchytrium sp. Leaf gall. **3:**TX(82).
Basidiomycotina - Uredinales
Coleosporium tussilaginis. **3:**NC(82), PA(82).
Basidiomycotina - Ustilaginales
Ustilago specularíae. Seed smut. **3:**OK(82,2873), TX(1077).
Deuteromycotina - Hyphomycetes
Botrytis cinerea. Gray mold. **3:**FL(35).
Cercospora speculariae. Leaf spot. **3:**LA(82,581).
Phymatotrichopsis omnivora. Root rot. **1:**TX(82).
Deuteromycotina - Coelomycetes
Septoria speculariae. Leaf spot. **1:**OK(2872), VA(82). **2:**KS(82). **3:**E states(82), MS(2720), OK(2872), TX(82), WI(82).
Septoria sp. **3:**WA(3202).

Cannabidaceae

Cannabis L.

1. *C. sativa* L. (=*C. sativa* L. var. *sativa* , *C. sativa* L. subsp. *sativa*) [gallowgrass, marijuana, soft hemp, true hemp]

Zygomycotina
Mucor sp. **1:**WI(1826).
Ascomycotina
Botryosphaeria marconii. Stem canker, wilt. **1:**MD(82,551), VA(82,551).
Gibberella cyanogena. Stem rot. **1:**IN(82), VA(82).
G. quinqueseptata. Stem canker. **1:**VA(3220).
Nectria cancri. On roots. **1:**MD(82).
[*Ophiobolus anguillidus*]. On stems. **1:**MN(2874).
Sclerotinia sclerotiorum. Stem rot. **1:**MT(82,3202).
Deuteromycotina - Hyphomycetes
Acrosporium sp. Powdery mildew. **1:**IL(2330).
Alternaria alternata. On inflorescences. **1:**IL(2330), KS(1484).
Alternaria sp. On dew-retted stems. **1:**IA(1127).
Aspergillus flavus. On stored marijuana. **1:**MD(2462), VA(2158), WI(1825,1826,2158).
A. niger. On stored marijuana. **1:**MD(2462), WI(1825,1826).
A. parasiticus. **1:**VA(2158).
Aspergillus sp. On dew-retted stems. **1:**IA(1127).
Botrytis cinerea. Gray mold. **1:**OR(82,3202), VA(82).
Cephalosporium sp. On dew-retted stems. **1:**IA(1127).
Cercospora cannabis. Leaf spot. **1:**MO(2239).
Epicoccum sp. On dew-retted stems. **1:**IA(1127).
Fusarium oxysporum f. sp. *cannabis*. Wilt. **1:**IL(2330).
Fusarium sp. Canker. **1:**IA(1127), IL(82), IN(82), VA(82), WI(82).
Hormodendrum sp. On dew-retted stems. **1:**IA(1127).
Monilia sp. On dew-retted stems. **1:**IA(1127).
Penicillium sp. On dew-retted stems. **1:**IA(1127), WI(1826).
Periconia sp. On dew-retted stems. **1:**IA(1127).
Phymatotrichopsis omnivora. Root rot. **1:**AZ(82), TX(82).
Pseudocercospora cannabina. **1:**MS(2091), WI(1334).
Spicaria sp. On dew-retted stems. **1:**IA(1127).
Trichoderma sp. On dew-retted stems. **1:**IA(1127).
Trichothecium roseum. On dew-retted stems. **1:**IA(1128).
Deuteromycotina - Coelomycetes
Ascochyta cannabis. Leaf spot. **1:**SE states(82), FL(35,82), IA(82), IL(2330), MN(82,2874), SD(2243), TX(82), WI(1340).
Coniothyrium sp. On dew-retted stems. **1:**IA(1127).
Cylindrosporium sp. Leaf spot. **1:**MD(82,128).
Dendrophoma marconii. Stem canker, wilt. **1:**MD(551), VA(551).
Macrophoma sp. Stem canker, wilt. **1:**MD(551), VA(551).
Macrophomina phaseolina. Charcoal rot. **1:**IL(82,311,2330,3588).
Phoma sp. On dew-retted stems. **1:**IA(1127).
Phomopsis cannabina. Canker, stem spot. **1:**IL(128).
P. ganjae. On leaves. **1:**IL(2331).
Phomopsis sp. On stems. **1:**IL(82).
Septoria cannabina. **1:**NY(2737).

Deuteromycotina - Other
Sclerotium rolfsii. Southern blight. **1**:SC(82), TX(82).

Humulus L.

1. *H. japonicus* Siebold & Zucc. [Japanese hop]
2. *H. lupulus* L. (=*H. americanus* Nutt.) [bine, common hop, European hop]
3. *Humulus* sp.

Oomycetes
Phytophthora cactorum. Crown and root rot. **2**:CA(3393), OR(3202).
P. citricola. **2**:CA(1110).
Pseudoperonospora humuli. Downy mildew. **2**:CA(82,1110,3025), ID(3202), NY(82,3025), OR(82,3025,3202), WA(82,1765,1770,3025,3202), WI(82).
Ascomycotina
Ceriospora dubyi. On stems. **2**:NM(82).
Erysiphe cichoracearum. Powdery mildew. **2**:NY(82), UT(82).
Glomerella cingulata. Anthracnose. **2**:IN(82), KS(82), MD(82), NY(82), OR(82,3202), WA(82,3202), WI(82).
Mycosphaerella erysiphina. Leaf spot. **2**:CA(82,1110).
Sphaerotheca fuliginea. Powdery mildew. **1**:MN(82). **2**:MN(82,2874).
S. macularis. Powdery mildew. **1**:Range of host(82). **2**:Range of host(82), CA(1110), WI(1337,1351).
Basidiomycotina - Uredinales
Aecidium humuli. **2**:WA(3202).
Aecidium sp. **2**:WA(82).
Basidiomycotina - Other
Armillaria mellea. Root rot. **2**:OR(82,3202).
Deuteromycotina - Hyphomycetes
Cercospora sp. Leaf spot. **2**:NE(82).
Fumago vagans. Sooty mold on cones. **2**:CA(82), OR(82,3202), WA(82,3202).
Fusarium roseum. Canker. **3**:CA(3973).
Fusarium sp. **2**:CA(1110).
Verticillium albo-atrum. Wilt. **2**:ME(82), OH(82), OR(82,3202), WI(82).
V. dahliae. **2**:CA(1110).
Deuteromycotina - Coelomycetes
Colletotrichum sp. Anthracnose. **2**:Unk.(82).
Cylindrosporium humuli. Leaf spot. **2**:E states(82), DE(962), IA(82), WI(82).
Phoma exigua. Leaf spot. **2**:IA(82), WI(82,1333).
Phyllosticta humuli. Leaf spot. **2**:IA(82), MA(82), MI(82). **3**:IA(966).
Septoria humuli. Leaf spot. **2**:NH(82), SD(2243).
S. lupulina. **2**:KS(82).
Stagonospora humuli-americani. **2**:NM(82).

Cannaceae

Canna L.

1. *C. edulis* Ker-Gawl. [achira, edible canna, gruya, Queensland arrowroot, Spanish arrowroot, tous-les-mois]
2. *C. flaccida* Salisb.
3. *C.* × *generalis* L. H. Bailey [common garden canna]
4. *C. glauca* L.
5. *C. indica* L. [Indian-shot]
6. *Canna* sp.

Oomycetes
Pythium sp. Root rot. **6**:FL(35).
Ascomycotina
Botryosphaeria obtusa. On stems. **4**:AL(82). **5**:AL(82).
Physalospora sp. Leaf necrosis. **6**:FL(35).
Basidiomycotina - Uredinales
Puccinia thaliae. Autoecious. **1**:FL(82). **2**:FL(82). **4**:FL(82). **5**:FL(35), HI(2903). **3**:FL(82), HI(2903), OH(82), TX(82,128). **6**:FL(35), HI(1152), TX(128).
Deuteromycotina - Hyphomycetes
Alternaria sp. Leaf spot. **3**:MI(82), SC(82), TX(82). **6**:FL(35).
Fusarium sp. Rhizome rot. **3**:MN(82,2874), MO(82), PA(82).
Deuteromycotina - Coelomycetes
Colletotrichum gloeosporioides. Anthracnose, capsule blight. **5**:HI(2899,2903).
Macrophoma subconica. On stems. **4**:LA(82).
Phoma sp. On stems. **4**:AL(82).

Deuteromycotina - Other
Rhizoctonia sp. Root rot. **6**:FL(35).
Sclerotium rolfsii. Southern blight. **3**:TX(82).

Capparidaceae

Capparis L.

1. *C. cynophallophora* L. (=*C. jamaicensis* Jacq.) [black willow, Jamaica caper-tree]
2. *C. sandwichiana* DC.
3. *Capparis* sp.

Oomycetes
Albugo candida. White blister. **3**:HI(82).
Ascomycotina
Asteridium lepidigenoides. Black spot. **1**:FL(35,82).
Asterina radians. Black spot. **1**:FL(35). **3**:FL(82).
Meliola bidentata. Sooty mold. **2**:HI(2903).
Deuteromycotina - Coelomycetes
Diplodia sp. Dieback. **1**:FL(35).

Cleome L.

(= *Isomeris* Nutt. ex Torr. & A. Gray.)

1. *C. graveolens* Raf. (=*Polanisia graveolens* Raf.)
2. *C. lutea* Hook. [golden cleome, yellow beeplant, yellow cleome]
3. *C. serrulata* Pursh
4. *C. spinosa* Jacq. [spiny spiderflower]
5. *Cleome* sp. (=*Isomeris sp.*)

Oomycetes
Peronospora parasitica. Downy mildew. **4**:LA(82).
Basidiomycotina - Uredinales
[*Puccinia aristidae*]. **2**:AZ(82,1203). **3**:AZ(82,1203), CO(82), IN(82), MT(82,3202), NE(82), NM(82). **4**:IN(82). **5**:CO(82), DE(82).
P. subnitens. **2**:AZ(3981). **3**:AZ(3981).
Deuteromycotina - Hyphomycetes
Alternaria sp. On living leaves. **1**:WI(1349).
Cercospora cleomes. Leaf spot. **1**:WI(1342). **3**:MI(82). **4**:NJ(82,581).
C. uramensis. **4**:WI(1334).
Cladosporium herbarum. On capsules. **5**:CA(2439).
Heterosporium hybridum. Leaf spot. **3**:IA(82), MT(82,3202).
Oidium sp. **4**:OH(950).
Pseudocercospora conspicua. Leaf spot. **2**:OK(82,2872). **3**:OK(82,2872).
Stigmina carpophila. On follicles. **5**:CA(2439).

Caprifoliaceae

Abelia R. Br.

1. *A. chinensis* R. Br.
2. *A.* × *grandiflora* (André) Rehd. *A. chinensis* × *A. uniflora*. [glossy abelia]

Oomycetes
Pythium sp. Root rot. **2**:FL(35), TX(82).
Ascomycotina
Capnodium sp. Sooty mold. **2**:MS(2720).
Basidiomycotina - Other
Armillaria tabescens. Root rot. **2**:FL(35).
Deuteromycotina - Hyphomycetes
Cercospora abeliae. Leaf spot. **1**:IL(312). **2**:FL(35), LA(2830).
C. deutziae. Leaf spot. **1**:IL(313).
Oidium sp. Powdery mildew. **2**:TX(82).
Phymatotrichopsis omnivora. Root rot. **1**:TX(82). **2**:TX(82).
Deuteromycotina - Other
Rhizoctonia solani. Root rot. **2**:TX(82).
Rhizoctonia sp. Root rot. **2**:FL(35).

Diervilla Mill.

1. *D. lonicera* Mill. [bush honeysuckle]
2. *Diervilla* sp.

Ascomycotina
Godronia turbinata. On dead twigs. **1**:ME(82), MI(128), PA(82,1382).
Microsphaera penicillata. Powdery mildew. **2**:NJ(82), WA(82).
Basidiomycotina - Other
Ceratobasidium anceps. **1**:WI(1343).

Deuteromycotina - Hyphomycetes
Cercospora diervillae. Leaf spot. **1:**ME(82). **2:**TX(82).
Phymatotrichopsis omnivora. Root rot. **1:**TX(82). **2:**TX(82).
Ramularia diervillae. Leaf spot. **1:**ME(82), NH(82), NY(82), WI(82,1333).
R. umbrina. On leaves. **1:**WI(809).
Deuteromycotina - Coelomycetes
Microdiplodia diervillae. On twigs. **1:**NY(82).
Phyllosticta diervillae. Leaf spot. **1:**WI(64,82,1360).
Septoria diervillae. Leaf spot. **1:**IA(82), MA(82), MN(82,2874), WI(82). **2:**MS(2720).
Sphaeropsis diervillae. On twigs. **1:**NY(82).

Kolkwitzia Graebn.

1. *K. amabilis* Graebn. [beautybush]

Deuteromycotina - Hyphomycetes
Cercospora kolkwitziae. Leaf spot. **1:**AL(82), OK(82,2872).
Deuteromycotina - Coelomycetes
Dothiorella sp. **1:**OK(2872).

Linnaea L.

1. *L. borealis* L. (=*L. borealis* L. var. *americana* (J. Forbes) Rehd., *L. borealis* L. var. *longiflora* Torr., *L. longiflora* (Torr.) Howell)

Ascomycotina
Gnomonia linnaeae. On overwintered leaves. **1:**ME(128).
Godronia linnaeae. On stems. **1:**ME(82,128,1382).
Halbaniella linnaeae. Black mildew. **1:**NY(82).
Metacoleroa dickiei. Leaf spot. **1:**ID(82,3202), MI(82), MT(82,3202), NM(82), NY(82), OR(82,3202), WA(82,638,3202), WI(82).
Pleospora penicillus. On dead stem. **1:**AK(1938).
Deuteromycotina - Coelomycetes
Rhabdospora breviuscula. Leaf spot. **1:**NY(82,3081).

Lonicera L.

1. *L. albiflora* Torr. & A. Gray
2. *L. canadensis* Marsh. [American honeysuckle, fly honeysuckle]
3. *L. cauriana* Fernald
4. *L. ciliosa* (Pursh) Poir.
5. *L. conjugialis* Kellogg
6. *L. dioica* L. (=*L. dioica* L. var. *glaucescens* (Rydb.) F. K. Butters) [limber honeysuckle]
7. *L. flava* Sims [yellow honeysuckle]
8. *L. heckrottii* Rehd.
9. *L. hirsuta* Eaton [hairy honeysuckle]
10. *L. hispidula* (Lindl.) Douglas ex Torr. & A. Gray [pink honeysuckle]
11. *L. hispidula* (Lindl.) Douglas ex Torr. & A. Gray var. *vacillans* A. Gray (=*L. catalinensis* Millsp.)
12. *L. involucrata* (Richardson) Banks ex Spreng. [twinberry]
13. *L. involucrata* (Richardson) Banks ex Spreng. var. *ledebourii* (Eschsch.) Jeps.
14. *L. japonica* Thunb. [gold-and-silver-flower, Japanese honeysuckle]
15. *L. morrowii* A. Gray [Morrow honeysuckle]
16. *L. oblongifolia* (Goldie) Hook. [swamp-fly honeysuckle]
17. *L. orientalis* Lam. var. *caucasica* (Pall.) Rehd.
18. *L. prolifera* (Kirchn.) Rehd. [grape honeysuckle]
19. *L. sempervirens* L. [coral honeysuckle, trumpet honeysuckle]
20. *L. tatarica* L. [Tatarian honeysuckle]
21. *L. utahensis* S. Wats.
22. *Lonicera* sp.

Oomycetes
Pythium sp. Root rot. **14:**FL(35).
Ascomycotina
Botryosphaeria dothidea. **19:**GA(1438).
B. obtusa. On stems and twigs. **14:**LA(82), MS(82,2720). **19:**GA(82). **22:**AL(82), NJ(82), VA(82).
Cenangium sp. **12:**WA(3202).
Diaporthe eres. On limbs, stems and twigs. **2:**DE(82), NJ(82), WI(82). **8:**OK(2873). **12:**CA(82). **15:**DE(82), NJ(82). **19:**DE(82), MI(82), NJ(82). **22:**OK(82).
Diatrypella favacea. **15:**GA(1438).
Dothidea puccinioides. On stems. **7:**GA(82,1438). **14:**GA(82,1438), NC(82), VA(82). **19:**PA(82).
D. sambuci. On stems. **19:**PA(82).
Dothidotthia andersonii. **12:**CO(649).
Erysiphe polygoni. Powdery mildew. **2:**CA(1155). **3:**CA(1155). **10:**CA(82). **12:**WY(82).
Erysiphe sp. Powdery mildew. **22:**CO(49).
Eutypa lata. On dead stems. **12:**CA(82). **13:**CA(81).
Eutypella juglandicola. **14:**GA(1438).
Glyphium schizosporum. **12:**CO(649). **22:**CO(649).
Hyponectria lonicerae. On leaves. **10:**CA(82,183).
Hypoxylon rubiginosum. **11:**CA(2439).
Hysterium angustatum. **12:**ID(3202).
Julella vitrispora. On leaves. **10:**CA(82). **12:**WA(3202).
Lachnum fuscofloccosum. **22:**WA(3202).
Lachnum sp. **12:**WA(3202).
Lasiobotrys lonicerae. Anthracnose, black leaf spot. **4:**CA(2434), ID(3202), WA(3202). **10:**CA(81,82,1468), OR(3202). **22:**CA(82), ID(82), MT(82,3202), OR(82,3202), WA(3202).
Microsphaera lonicerae. **12:**WY(1195).
M. penicillata. Powdery mildew. **2:**OH(82,950), WI(1344). **4:**ID(3202), WA(3202). **6:**Range of host(82), MN(2874), OH(950). **7:**Range of host(82). **8:**OK(2872). **9:**MN(2874). **10:**Unk.(3976). **12:**WA(3202), WY(82). **14:**FL(35). **15:**NY(82), WI(1360). **16:**OH(82,950), WI(82). **19:**Range of host(82), FL(35), MN(2874). **20:**Range of host(82), MN(2874). **21:**WY(1195). **22:**Range of host(82), ID(3202), MN(2874), NY(82), OK(82), TX(82), WA(3202), WY(82).
Microsphaera sp. Powdery mildew. **22:**CO(49).
Mollisia complicatula. **22:**WA(3202).
Mycosphaerella clymenia. Leaf spot. **7:**IA(82). **14:**VA(82). **22:**CA(2439).
Ophiobolus lonicerae. **10:**CA(3248).
O. nigro-clypeata. **14:**GA(1438).
Pleospora herbarum. **12:**ID(3202).
Propolomyces versicolor. **22:**WA(3202).
Robergea nigra. On decorticated twigs. **12:**OR(3225,3228).
Rosellinia aquila. On roots. **7:**AL(82).
Stictis ostropoides. **22:**CA(3225), OR(3225,3228).
S. radiata. **12:**WA(3202). **22:**CA(3225), WA(3202).
Tapesia lonicerae. On twigs. **12:**CO(521).
Teichospora sp. **22:**ID(3202).
Teichosporella lonicerina. On dead stems. **22:**CA(2439).
Thyronectria lonicerae. On dead limbs and twigs. **12:**CO(82,3173), ND(82).
Valsa ceratosperma. **14:**GA(1438).
Basidiomycotina - Uredinales
Puccinia festucae. Heteroecious: aecial host. **7:**IA(82), IL(82). **22:**IA(82), IL(82).
Basidiomycotina - Other
[*Aleurobotrys botryosus*]. On dead stems. **14:**SC(2083).
Armillaria mellea. Root rot. **15:**NJ(82).
A. tabescens. Root rot. **19:**FL(35).
Dendrothele griseo-cana. On bark. **20:**MI(2084).
Heterobasidion annosum. **22:**ID(3202).
Insolibasidium deformans. Leaf blight. **2:**CT(82), IA(82), IL(312), NY(82,1287), WI(82,1287). **12:**WA(3202), WY(1195). **15:**IA(82). **16:**WI(82,1287). **20:**IA(82,1287,2626), MI(82,1287), NY(82), WI(82). **22:**IA(82,2626), SD(2243).
Merismodes ochraceus. On bark. **14:**LA(82).
Pellicularia koleroga. Thread blight. **14:**LA(82,3640).
Phylloporia ribis. Collar rot of living plants. **12:**MT(82,3202,3832).
Deuteromycotina - Hyphomycetes
Alternaria sp. Leaf spot. **14:**FL(35).
Botrytis cinerea. Gray mold blight. **12:**AK(82,526). **20:**AK(82,526), CT(82).
Cercospora lonicerae. Leaf spot. **14:**FL(35). **19:**FL(35). **22:**HI(2903).
C. varia. Leaf spot. **14:**TX(82). **19:**FL(35). **22:**TX(82).
Cercospora sp. Leaf spot. **6:**MI(82).
Cladosporium herbarum. **21:**MT(3202).
C. trichophilum. On living leaves. **9:**WI(1344).
Corynespora cassiicola. Leaf spot. **14:**FL(35).
Exosporium sp. **4:**MT(3202).
Graphiothecium sp. **20:**WI(1360).
Phaeoramularia antipus. Leaf spot. **4:**ID(3202), WA(3202). **7:**IA(82), IL(82). **14:**FL(35). **19:**FL(35), IA(82). **21:**ID(3202). **22:**IA(82), MT(82,3202), NE(82), TX(82), WI(82).
P. periclymeni. On leaves. **1:**TX(843).
Phymatotrichopsis omnivora. Root rot. **15:**TX(82). **19:**TX(82).

20:TX(82). **22:**TX(82).
Ramularia lonicerae. **21:**WY(1195).
Ulocladium consortiale. **21:**MT(3202).

Deuteromycotina - Coelomycetes

Aplosporella punctum. Secondary stem spot. **14:**MS(2720). **22:**DC(2783).
Ascochyta lonicerae-canadensis. On living leaves. **2:**WI(1355).
A. tenerrima. **18:**WI(1359).
Ascochytulina deflectens. **20:**AK(526).
Diplodia lonicerae. **20:**MO(82).
Diplodia sp. On stems. **14:**VA(82).
Kabatia lonicerae. Leaf spot. **2:**MI(82), NY(82), PA(82), WI(82). **4:**ID(3202), WA(3202). **5:**CA(1468,3549). **12:**CA(82), CO(82), ID(3202), MT(3202), UT(1195), WA(3202), WY(1195). **16:**WI(82). **22:**CA(82), ID(82), WY(82).
K. lonicerae var. *involucratae*. Leaf lesions. **12:**MT(3549).
K. mirabilis. **12:**WY(82,1195). **16:**WI(1346).
K. mirabilis var. *oblongifoliae*. Leaf lesions. **16:**NY(3549), WI(3549).
K. periclymeni. Leaf lesions. **21:**ID(3202), MT(3202), WA(3202), WY(1195). **22:**MT(3549).
Libertella lonicerae. On twigs. **12:**CA(82).
Phoma mariae. On stems. **6:**WI(1341). **14:**NY(82). **15:**WI(1357). **20:**MA(82), WI(1341). **22:**WA(3202), WI(1340).
P. xylostei. On stems. **10:**CA(82).
Phyllosticta vulgaris. Leaf spot. **20:**MN(2874). **22:**OK(82,2872).
Phyllosticta sp. Leaf spot. **14:**FL(35). **17:**OK(2872).
Rhabdospora lonicerae. On stems. **14:**NJ(82). **22:**NJ(82).
Septoria xylostei. Leaf spot. **2:**WI(82). **6:**MN(2874). **16:**WI(1351). **22:**MT(82,3202), WI(82).
Sphaceloma sp. **16:**WI(1351).

Deuteromycotina - Other

Rhizoctonia solani. Leaf spot. **14:**FL(35).

Sambucus L.

1. *S. caerulea* Raf. (=*S. glauca* Nutt.) [blue elder, blue elderberry, western elderberry]
2. *S. callicarpa* Greene [Pacific coast red elder, red elderberry]
3. *S. canadensis* L. [American elder, American elderberry, sweet elder]
4. *S. javanica* Reinw. ex Blume (=*S. chinensis* Lindl.)
5. *S. melanocarpa* A. Gray [blackbead elder]
6. *S. mexicana* K. Presl ex DC.
7. *S. nigra* L. [European elder]
8. *S. racemosa* L. [European red elder]
9. *S. racemosa* L. var. *callicarpa* Jeps.
10. *S. racemosa* L. subsp. *pubens* (Michx.) House (=*S. pubens* Michx., *S. microbotrys* Rydb., *S. racemosa* L. var. *microbotrys* (Rydb.) Kearn. & Peebles) [American red elderberry, red-berried elder]
11. *S. simpsonii* Rehd. [Florida elder]
12. *Sambucus* sp.

Chytridiomycetes

Synchytrium sambuci. Leaf gall. **3:**LA(82).

Ascomycotina

Asteromassaria distincta. On limbs. **10:**PA(188).
Belonidium sulphureum. **12:**WA(3202).
Botryosphaeria dothidea. Canker, dieback. **3:**GA(1438). **12:**FL(35).
B. obtusa. **3:**AL(82), VA(82).
B. rhodina. **3:**NY(82).
B. ribis. Dieback. **3:**FL(82). **7:**GA(82).
Calosphaeria abnormis. **3:**MD(2779).
Chaetoplaca sp. **1:**CA(661).
Claussenomyces atrovirens. On wood. **12:**FL(522).
Comoclathris permunda. **10:**WY(3810).
Coniochaeta ligniaria. **1:**CA(661), ID(3202). **8:**OR(3202). **10:**AK(526).
C. pulveracea. **1:**ID(3202).
C. sordaria. **3:**MD(2783).
Crocicreas cyathoideum. **1:**CA(661), OR(505).
C. cyathoideum var. *cacaliae*. On dead stems. **10:**UT(505).
Cryptodiaporthe calosphaerioides. On limbs. **2:**AK(82). **10:**AK(526).
Cryptodiscus sambuci. On decorticated stems. **12:**FL(522).
Dendropleella hirta. **12:**CO(649).
Diaporthe megalospora. On limbs. **3:**GA(82,1438), MA(82,3807), MI(82,3807), NY(82).
D. sociabilis var. *sambuci*. On limbs. **1:**CA(82), WA(82). **2:**AK(82), CA(82), WA(82). **3:**GA(82,1438), LA(82), MI(82). **5:**ID(82). **8:**WA(3202). **12:**AK(526).
D. spiculosa. **12:**OR(3202).
Diatrype flavovirens. On limbs. **2:**OR(82). **8:**OR(3202).
D. sambucivora. On limbs. **3:**PA(82), VA(82).
Didymosphaeria futilis. **3:**GA(1438).
Dothidea puccinioides. **3:**GA(1438).
D. sambuci. On twigs. **3:**NY(82).
Dothiora sambucina. On limbs. **3:**NY(82).
Eutypella glandulosa. **3:**MD(82).
E. leprosa. **3:**IA(3629), MD(82), NY(82).
E. scoparia. On limbs. **1:**CA(2439). **3:**GA(82,1438). **12:**CA(81).
E. stellulata. On bark of limbs. **1:**CA(2439).
Gibberella pulicaris. **8:**CA(81).
Hymenoscyphus calyculus. **10:**AK(3431).
H. phiala. **8:**WA(3202).
Hypomyces rosellus. On wood. **1:**CA(2439).
Hypoxylon rubiginosum. **3:**GA(1438).
Hysterographium sp. **1:**WA(3202).
Lachnum flavofuligineum. **12:**WA(3202).
Lecanidion atratum. **1:**CA(2439).
Leptosphaeria sambuci. On dead limbs. **1:**OR(82,3202).
L. sambucina. **5:**CO(82).
Lophiosphaera querceti. On dead limbs. **1:**CA(2439).
Lophiostoma caulium var. *congregatum*. On limbs. **8:**CA(1468).
L. hysterioides. **1:**WA(3202).
Massarina corticola. On twigs. **12:**MI(319).
Massarina sp. On dead canes. **1:**CA(661).
Melanomma pulvis-pyrius. **1:**ID(3202).
M. sambuci. **1:**WA(3202). **12:**AK(526).
Microsphaera grossulariae. Powdery mildew. **3:**Unk.(82), GA(1438), OH(950), WV(3907). **8:**MT(3202). **10:**Unk.(82), OH(950). **12:**FL(35), MN(2874).
M. penicillata. Powdery mildew. **3:**Unk.(82), OH(950). **12:**FL(35).
Mollisia ligni. **12:**AK(526).
Mycosphaerella sp. **5:**NM(82).
Nectria cinnabarina. Canker, dieback. **1:**ID(82,3202), MT(3202), WA(82,3202). **2:**AK(526), Ak(82). **3:**E states(82), ND(82), TN(82). **5:**MT(82). **8:**ID(3202), MT(3202). **10:**AK(526), MN(2874), VA(82), WV(3907).
N. coccinea. **1:**ID(82,3202). **2:**WA(82).
N. episphaeria. **8:**CA(81).
Nectriella pironii. Stem gall. **11:**FL(35).
N. sambuci. On twigs. **2:**CA(82).
Nodulosphaeria erythrospora. On dead canes. **1:**CA(661). **12:**OR(657).
Ophiobolus byssicola. On limbs. **1:**CA(1468). **12:**CO(649).
Orbilia chrysocoma. On rotten wood. **1:**CA(2439).
Otthia amica. **1:**WA(3202). **12:**WA(3202).
Phyllactinia guttata. Powdery mildew. **3:**MI(82).
Physalospora abdita. **3:**AL(82), NY(82).
Rosellinia aquila. **1:**CA(2439).
R. rhyncospora. On dead limbs. **1:**CA(1468).
Schizothyrium pomi. On twigs. **3:**VA(82).
Sphaerotheca macularis. On limbs. **3:**MA(82).
Stictis radiata. **3:**GA(1438). **12:**WA(3202).
S. serpentaria. **2:**WA(1277).
Stictis sp. **8:**WA(3202).
Teichospora symphoricarpi. **2:**WA(3815). **8:**WA(3202).
Thyridaria sambucina. On limbs. **1:**WA(82,3202). **3:**NY(82). **12:**CO(82).
Trichopezizella relicina. **8:**OR(1416).
Valsa ambiens. **3:**IL(3415).
V. sordida. **1:**WA(3202).
Xylaria sp. Root rot. **3:**TX(82).

Basidiomycotina - Uredinales

Puccinia bolleyana. Heteroecious: aecial host. **3:**E states(82), GA(1439), MA(1394), MN(82,2874), MS(2720), NC(1303,3944), TX(82), WI(1350), WV(3907). **7:**MA(82). **10:**DE(82), IN(82), MD(82), MI(82). **11:**FL(35). **12:**FL(35).

Basidiomycotina - Other

Armillaria putrida. On base of trees. **1:**CA(2439).
Bjerkandera adusta. **1:**WA(3202). **12:**MT(3202).
Botryobasidium vagum. **1:**WA(3202).
Ceriporia viridans. **1:**CA(2439).
Corticium microsclerotium. Web blight. **12:**FL(82).

Cylindrobasidium corrugum. On bark. **8**:CA(642).
Datronia mollis. **1**:ID(3202), WA(3202).
D. stereoides. On wood. **1**:ID(82), WA(82).
Dendrophora albobadia. On wood. **3**:MD(128,543).
Exidia nucleata. On bark. **1**:CA(2439).
E. zelleri. **1**:OR(3202).
Exidiopsis calcea. On wood. **1**:CA(128,2439), WA(3202).
Hymenochaete agglutinans. Trunk canker. **2**:WY(82).
H. rubiginosa. On dead limbs. **1**:CA(2439).
Hyphoderma sambuci. **8**:WA(3202). **9**:CA(81).
Inonotus cuticularis. **6**:AZ(1181).
I. farlowii. On wood. **1**:AZ(82), CA(82).
Lopharia cinerascens. **3**:IA(543).
Melzericium udicola. **1**:WA(3202).
Oxyporus latemarginatus. **1**:WA(3202).
Pellicularia koleroga. Thread blight. **3**:LA(3640). **12**:LA(82).
Peniophora incarnata. **1**:WA(3202).
Phellinus igniarius. Heart rot. **1**:CA(2439), ID(82), WA(3202).
Polyporus badius. **1**:WA(3202).
P. varius. **10**:AK(526).
Postia caesia. On wood. **1**:CA(82), OR(82). **12**:OR(3202).
Scytinostroma portentosum. **12**:AK(128).
Trametes versicolor. On wood. **1**:WA(82,3202).

Deuteromycotina - Hyphomycetes

Alternaria sp. Leaf spot. **12**:OK(2872).
Botrytis cinerea. **12**:AK(526).
Brachysporium oosporum. On dead canes. **8**:CA(642).
Cercospora depazeoides. Leaf spot. **3**:Range of host(82), MS(2720), WV(3907). **7**:IA(82), ME(82), MS(2720). **12**:FL(35).
C. lateritia. **3**:IA(82), KS(82). **7**:IA(82). **10**:IA(82).
Cercospora sp. **3**:CA(1110).
Cercosporella prolificans. Leaf spot. **1**:CA(81,82), OR(82). **5**:NM(82).
Curvularia geniculata. Isolated from berries. **4**:NY(374).
Didymobotryum sp. On dead twigs. **10**:NC(1303).
Heterosporium interseminatum. On dead twigs. **12**:AL(82), IA(82), NJ(82), OH(82), WV(82,3907).
H. sambuci. **3**:OK(2872).
Macrosporium hybridum. **1**:WA(3202).
Oidium sp. Powdery mildew. **11**:FL(35).
Phaeoramularia catenospora. Leaf spot. **3**:AL(82), KS(82), MS(82,2720), NC(82,3944), TX(82). **12**:FL(35), OK(2872).
Phymatotrichopsis omnivora. Root rot. **3**:TX(82). **10**:TX(82).
Pseudospiropes nodosus. **2**:WA(1685). **8**:WA(3202). **12**:WA(1685).
Ramularia glauca. Leaf spot. **1**:CA(82), ID(3202), WA(82,3202). **5**:ID(666). **8**:ID(3202). **12**:CO(82).
R. sambucina. Leaf spot. **1**:WA(82,3202). **3**:MO(82), MS(2720), NY(82), WV(82,3907). **8**:ID(3202), WA(3202). **10**:NY(82), PA(82), WI(82). **12**:NM(82).
Stigmina compacta. On dead limbs. **1**:CA(81).
S. negundinis. **1**:ID(3202).
S. pedunculata. On dead canes and twigs. **1**:CA(642,661), ID(638,3202), WA(639,3202). **8**:MT(3202), WA(3202). **12**:CA(82,1110), ID(82,3202), MT(3202), WA(82).
S. sambucina. Leaf spot. **1**:CA(638). **5**:CO(82). **12**:CO(49).
Trichoderma viride. On dead leaves. **1**:CA(2439).
Triposporium elegans. **10**:AK(526).
Tubercularia vulgaris. On dead stems. **1**:CA(81). **10**:AK(526). **12**:CO(3178).
Verticillium albo-atrum. Wilt. **12**:MD(82).
V. tenerum. On canes. **1**:WA(638).

Deuteromycotina - Coelomycetes

Aplosporella sambuci. On limbs. **3**:IN(82), MI(82), ND(82), NY(82).
Ascochyta deformis. On limbs. **5**:WA(82). **8**:WA(3202).
A. wisconsina. Leaf spot. **3**:NY(82), WI(82). **10**:WI(82).
Ascochyta sp. **8**:WA(3202).
Coniothyrium fuscidulum. **2**:AK(526).
C. sambucinum. On twigs. **12**:CO(82), MT(3202).
Cryptospora pulviniceps. On limbs. **3**:NY(82).
Cryptospora sp. On limbs and twigs. **3**:Unk.(82).
Cylindrosporium sp. **12**:CA(1110).
Cytospora chrysosperma. On limbs and twigs. **1**:WA(82). **3**:KS(82), NJ(82), NY(82).
C. leucosperma. On limbs and twigs. **3**:IL(82), KS(82), MS(2720), NY(82).
Cytospora sp. Twig canker. **12**:NJ(82).
Diplodia sambuci. On limbs and twigs. **3**:IL(82).
D. sambucina. On limbs and twigs. **3**:MI(82), NJ(82), NY(82).
Dothiorella macrospora. On limbs. **3**:MI(82), PA(82).
Lasmeniella alpina. On dead limbs. **8**:CA(638). **12**:ID(3202), WA(3202).
Microdiplodia paupercula. On limbs. **3**:NY(82). **10**:VA(82).
Microsphaeropsis olivacea. **10**:AK(526).
Myxosporium sambuci. On twigs. **12**:WA(82,3202).
Phlyctema vagabunda. Leaf spot. **3**:MS(82,2720), TX(82).
Phoma exigua. Leaf spot. **1**:MT(82,3202). **3**:IN(82), MO(82), NY(82), WI(82), WV(82,3907). **7**:NJ(82).
Phomopsis sambucina. **1**:WA(3202). **8**:ID(3202). **12**:Unk.(82).
Phomopsis sp. On twigs. **3**:Unk.(82). **10**:Unk.(82). **12**:Unk.(82).
Phyllosticta sambucicola. **3**:IN(82).
Phyllosticta sp. Leaf spot. **3**:NY(82), PA(82). **10**:WI(1352). **11**:FL(35).
Septoria sambucina. Leaf spot. **1**:CA(81,82), ID(3202), OR(82,3202,4007), WA(82,3202). **2**:CA(82), OR(82), WA(82). **3**:E states(82), OK(2872), TX(82), WI(82). **8**:WA(3202). **10**:WI(82). **11**:FL(35). **12**:AK(526), CO(1195), FL(35).
Sphaceloma sp. Spot anthracnose. **3**:LA(82).
Sphaeropsis sambucina. Dieback. **1**:CA(82), MT(82,3202). **3**:IN(82), KS(82), ND(82), NY(82). **5**:CO(82). **7**:CA(82,1110). **12**:NV(82).
Stagonospora samarorum. Leaf spot. **7**:AK(82).
Stegonsporium tuberculiforme. On stems. **1**:CA(81). **8**:CO(3776). **10**:WY(3812).

Deuteromycotina - Other

Rhizoctonia crocorum. Root rot. **3**:TX(82). **12**:WA(3202).
R. solani. Web blight. **12**:FL(35).
Sclerotium sp. **10**:AK(526).

Symphoricarpos Duhamel

1. *S. acutus* (A. Gray) Dieck
2. *S. albus* (L.) S. F. Blake (=*S. racemosus* Michx.) [common snowberry, waxberry]
3. *S. albus* (L.) S. F. Blake var. *laevigatus* (Fernald) S. F. Blake (=*S. rivularis* Suksd.) [common snowberry, waxberry]
4. *S.* × *chenaultii* Rehd. Probably a hybrid between *S. microphyllus* and *S. orbiculatus*. [coralberry, Indian currant]
5. *S. mollis* Nutt.
6. *S. occidentalis* Hook. [wolfberry]
7. *S. orbiculatus* Moench [coralberry, Indian currant]
8. *S. oreophilus* A. Gray (=*S. vaccinioides* Rydb.)
9. *Symphoricarpos* sp.

Ascomycotina

Anthostoma melanotes var. *symphoricarpi*. On dead limbs. **6**:SD(2243).
Didymosphaeria futilis. On dead limbs. **6**:SD(2243).
Dothidotthia andersonii. **2**:ND(3246). **9**:CO(3246).
Dothiora schizospora. On stems. **7**:MO(2202). **9**:OK(660).
Griphosphaerioma kansensis. **6**:KS(3246), ND(3246), SD(3246). **7**:KS(3246).
Herpotrichia symphoricarpi. On twigs. **9**:CO(128,189).
Hymenoscyphus calyculus. **2**:ID(3202), WA(3202).
Hysterographium sp. **2**:ID(3202).
[*Lasiobotrys lonicerae*]. **6**:ID(1195,3202).
Lecanidion atratum. Leaf spot. **6**:SD(2243).
Lophiostoma macrostomoides. **9**:ND(377).
Microsphaera diffusa. Powdery mildew. **2**:CA(81), ID(3202), MN(526), MT(3202), OH(950), OR(3202), WA(3202). **3**:WI(1330). **5**:ID(3202). **6**:MN(2874), MT(3202), OH(950), SD(2243), WA(3202), WI(1333), WY(1195). **7**:MS(2720), OH(950), OK(660,2872), SD(2243). **4**:OH(950). **9**:MN(2874).
M. penicillata. **6**:WI(1351).
M. symphoricarpi. **6**:ND(377).
Mollisia dakotensis. On dead fallen twigs. **6**:ND(377,850).
Nectria cinnabarina. **6**:ID(3202).
N. coryli. **6**:ND(377).
Podosphaera clandestina. **2**:ID(3202), WA(3202).
Pyrenopeziza symphoricarpi. On stems. **9**:CO(849).
Rhizogene impressa. **5**:CA(81).
Rosellinia parasitica. **6**:MT(3202).
Schizoxylon compositum. On bark. **6**:KS(3226). **9**:KS(3225).
S. decipiens var. *symphoricarpi*. **6**:ND(377).
Stictis fusca. **9**:KS(3225).

Teichospora solitaria. On dead stems. **8:**CA(661).
T. umbonata. On dying twigs. **6:**MT(3202), SD(2243).
Thyronectria lonicerae. **6:**SD(3173).
Valsa ambiens. **6:**ND(377,3415).

Basidiomycotina - Uredinales

Puccinia crandallii. Heteroecious: aecial host. **1:**CA(638). **5:**CA(81). **6:**MT(1195,3202), ND(378), WA(3202), WY(1195). **7:**OK(2872). **8:**AZ(1195,1203,3981), WY(1195). **9:**CA(642).
P. symphoricarpi. Autoecious. **2:**AK(526), CA(81), ID(3202), MT(1195,3202), OR(3202), WA(3202). **3:**AK(60). **5:**OR(3202). **6:**MT(1195,3202).

Basidiomycotina - Other

[*Aleurobotrys botryosus*]. On dead stems. **9:**NC(2083).
Hymenochaete tabacina. **2:**ID(3202), NJ(494).
Irpex lacteus. **6:**ND(378).
Peniophora cinerea. **6:**ND(378).
P. incarnata. **6:**ND(128).
Phylloporia ribis. On living plants. **2:**MT(3202,3832).
Piloderma byssinum. **6:**ND(378).
Polyporus varius. **2:**WA(3202).
Stereum ochraceo-flavum. **9:**MO(543).

Deuteromycotina - Hyphomycetes

Alternaria sp. Brown rot of berries. **7:**CO(49).
Botrytis sp. **9:**WA(3202).
Cladosporium sp. On leaves. **6:**WI(1345).
Papulaspora brachiata. On dead limbs. **6:**CO(721).
Phaeoramularia symphoricarpi. Leaf spot. **2:**AK(526), ID(3202), MT(3202), OR(3202), WA(3202). **3:**WI(1340). **5:**ID(3202), WA(3202). **6:**MT(3202). **8:**WY(1195). **9:**SD(2243), WY(1195,3378).

Deuteromycotina - Coelomycetes

Ascochyta tenerrima. **2:**WA(3202).
Camarosporium umbonatum. On dead limbs. **6:**SD(2243).
Labridella cornu-cervae. On dead limbs. **6:**ND(3549), SD(2243,3549).
Phyllosticta symphoricarpi. Leaf spot. **2:**OR(648,3202), WA(3202). **9:**OK(2872).
Septoria oedospora. **8:**WY(1195).
S. signalensis. On leaves. **2:**WY(1195). **3:**WY(3378). **8:**ID(1195,3378), WY(1195).
S. symphoricarpi. **2:**ID(3202), OR(648,3202), WA(3202). **6:**ID(3202), MT(3202), ND(378), WA(3202).
Septoria sp. Leaf spot. **7:**CO(49).
Sphaceloma symphoricarpi. Spot anthracnose, scab. **2:**MN(526). **3:**IA(1745), IN(1745), MA(1745), MD(1736,1745), MI(1745), MN(2874), NY(1745), OH(1745), WI(1329,1745). **7:**CO(49), IA(1745). **9:**OR(1745,3202).

Triosteum L.

1. *T. angustifolium* L.
2. *T. aurantiacum* C. Bicknell [wild-coffee]
3. *T. perfoliatum* L. [feverwort, tinker's-weed, wild-coffee]

Ascomycotina

Gnomonia triostei. On stems. **3:**NY(82).

Basidiomycotina - Uredinales

Aecidium triostei. **1:**MO(82).

Basidiomycotina - Other

Ceratobasidium anceps. **3:**WI(1355).

Deuteromycotina - Hyphomycetes

Cercospora triostei. On leaves. **3:**WI(581,1337).
Cladosporium triostei. Leaf spot. **3:**WV(3907).
Coremium triostei. On living leaves. **3:**WI(1345).

Deuteromycotina - Coelomycetes

Cylindrosporium triostei. Leaf spot. **2:**OK(2873). **3:**OK(2872).
Phyllosticta sp. Leaf spot. **3:**WI(1360).

Viburnum L.

1. *V. acerifolium* L. [arrowwood, dockmackie, maple-leaved viburnum, possum-haw]
2. *V. alnifolium* Marsh. [American wayfaring-tree, devil's-shoestrings, dog-hobble, dogberry, hobblebush, moosewood, tangelfoot, White Mountains dogwood, witch-hobble]
3. *V. × burkwoodii* Hort. Burkw. & Skipw. *V. carlesii × V. utile*.
4. *V. carlesii* Hemsl. ex Forbes & Hemsl.
5. *V. cassinoides* L. [Appalachian-tea, swamp haw, teaberry, wild-raisin, withe-rod]
6. *V. davidii* Franch.
7. *V. dentatum* L. (=*V. pubescens* (Aiton) Pursh) [arrowwood viburnum]
8. *V. dilatatum* Thunb. [Linden viburnum]
9. *V. edule* (Michx.) Raf. [mooseberry, squashberry]
10. *V. lantana* L. [twistwood, wayfaring-tree]
11. *V. lentago* L. [black haw, cowberry, nannyberry, sheepberry, sweet viburnum, tea-plant]
12. *V. nudum* L. [black-alder, naked viburnum, possum haw, smooth withe-rod, swamp haw]
13. *V. obovatum* Walter
14. *V. odoratissimum* Ker-Gawl. [sweet viburnum]
15. *V. opulus* L. [European cranberry-bush, guelder-rose, snowball-bush, whitten-tree]
16. *V. opulus* L. var. *roseum* L. [snowball-bush]
17. *V. orientale* Pall.
18. *V. pauciflorum* Raf. [mooseberry]
19. *V. plicatum* Thunb. var. *tomentosum* Miq. (=*V. tomentosum* Thunb.) [doublefile viburnum]
20. *V. prunifolium* L. [black haw, nannyberry, sheepberry, stagbush, sweet haw]
21. *V. rafinesquianum* Schult. [downy-leaved arrowwood]
22. *V. rafinesquianum* Schult. var. *affine* (Bush ex C. K. Schneid.) House (=*V. affine* Bush ex C. K. Schneid.)
23. *V. recognitum* Fernald [smooth arrowwood]
24. *V. rhytidophyllum* Hemsl.
25. *V. rotundifolia* Raf. A name of uncertain application; material so listed may be a varient of *V. plicatum*.
26. *V. rufidulum* Raf. [rusty black haw, Southern black haw]
27. *V. semitomentosum* (Michx.) Rehd.
28. *V. suspensum* Lindl.
29. *V. tinus* L. [laurustinus]
30. *V. trilobum* Marsh. [crampbark, cranberry-bush, grouseberry, highbush cranberry, pimbina, summerberry]
31. *Viburnum* sp.

Oomycetes

Phytophthora cinnamomi. **31:**MD(1729,3624).
Phytophthora sp. Root rot. **28:**FL(35). **31:**FL(35).
Plasmopara viburni. Downy mildew. **1:**AL(82), MA(82), MD(82), NY(82), WV(82,3907). **7:**AL(82), MA(82,1390), MD(82), NY(82), WV(82,3907). **15:**MN(2874). **28:**FL(35). **29:**GA(82), MS(2720). **30:**MN(82,2874), WI(82). **31:**FL(35), MN(2874).
Pythium splendens. Root rot. **14:**FL(35).
Pythium sp. Root rot. **28:**FL(35). **31:**FL(35).

Ascomycotina

Botryosphaeria dothidea. Twig dieback. **1:**GA(1438). **12:**GA(1438). **13:**GA(1438). **14:**FL(35). **20:**GA(1438). **25:**GA(1438). **28:**FL(35). **31:**FL(35).
B. quercuum. Dieback, leaf spot. **14:**FL(35). **31:**FL(35).
Diaporthe viburni. **26:**GA(1438).
Didymosphaeria futilis. **26:**GA(1438).
Didymosphaeria sp. **20:**GA(1438).
Durandiella viburnicola. On twigs. **5:**NH(82,1381,3167). **7:**NH(82,3167). **31:**MI(1381), NH(1381), PA(1381), SC(1381).
Eutypella juglandicola. **27:**GA(1438).
Glonium sp. **17:**GA(1438).
Hypoxylon rubiginosum. **5:**GA(1438). **26:**GA(1438).
Hysterographium mori. On twigs. **31:**FL(35).
Lambertella viburni. On leaves. **5:**NH(909).
Leptosphaeria tini. Leaf spot. **29:**LA(82).
Leptosphaeria sp. Dieback. **31:**FL(35).
Massaria lantanae. **7:**MA(3257), NY(3257). **11:**NY(3257). **12:**GA(1438). **20:**GA(1438). **26:**GA(1438,3257). **31:**NY(3257).
Massarina microcarpa. **26:**GA(1438).
Microsphaera penicillata. Powdery mildew. **1:**MS(2720), OH(950). **5:**OH(950). **11:**IL(312), MN(2874), ND(377), OH(950). **15:**WI(1343). **20:**GA(1438), MN(2874), NC(1303), ND(377), OH(950). **21:**NC(1303). **22:**MN(2874), NC(3944). **29:**Unk.(3976). **31:**FL(35), MN(2874).
Mycosphaerella punctiformis. **26:**GA(1438).
Mycosphaerella sp. Leaf spot. **14:**FL(35). **28:**FL(35).
Naevia sp. **20:**GA(1438).
Nectria cinnabarina. **31:**AK(526).
Paradidymella sp. Leaf spot. **31:**FL(35).

Phyllachora sp. Leaf spot. **31**:FL(35).
Rosellinia necatrix. Root rot. **1**:CA(82).
Schizoxylon insigne. **20**:GA(1438). **26**:GA(1438).
S. ligustri. **31**:SC(3225).
Valsa ambiens. On dead limbs. **7**:NY(82), PA(3415).

Basidiomycotina - Uredinales

Aecidium rubromaculans. **31**:FL(35).
Coleosporium viburni. Heteroecious: telial host. **10**:WI(1346). **11**:MN(2874), WI(1357). **15**:WI(1339). **20**:WI(1354). **21**:MN(2874). **22**:MN(2874).
Puccinia linkii. Autoecious. **9**:AK(60,526), ID(3202), MT(3202), WA(3202). **18**:AK(82), ID(82), MI(82), MT(82), NH(82), WA(82).

Basidiomycotina - Other

Antrodia albida. Wood rot. **20**:MO(82).
Armillaria mellea. Root rot. **28**:CA(2900). **31**:CA(2898).
A. tabescens. Dieback, root rot. **14**:FL(35). **24**:FL(82). **29**:LA(2815). **31**:FL(35).
Dendrophora albobadia. **31**:WV(543).
Hyphoderma mutatum. **26**:GA(1439).
Junghuhnia nitida. **20**:VA(2168).
Meruliopsis corium. **13**:FL(128).
Phanerochaete sordida. **13**:FL(128).
Scytinostroma galactinum. Root rot. **4**:MD(668).
Thelephora albidobrunnea. **26**:NC(3944).
Typhula elegans. On slender limbs. **15**:SC(264).
T. viburni. On dead leaves. **31**:NH(264,2955).

Deuteromycotina - Hyphomycetes

Alternaria sp. Leaf spot. **14**:FL(35). **28**:FL(35). **31**:FL(35).
Botrytis cinerea. Gray mold blight. **15**:MA(82), WA(82).
Botrytis sp. **15**:WA(3202).
Cercospora opuli. Leaf spot. **15**:IA(82), MS(82).
C. varia. Leaf spot. **2**:VT(1224). **4**:NY(82). **8**:IL(312). **9**:ID(3202). **10**:IL(82), OK(82). **14**:FL(35). **15**:OK(2872). **28**:FL(35). **31**:AK(526), FL(35), WV(3907).
C. viburnicola. Leaf spot. **4**:WI(1354). **14**:FL(35). **15**:OK(82,2872). **19**:OK(2872). **28**:FL(35). **31**:FL(35).
Cladosporium sp. On leaves. **9**:AK(526). **31**:MS(2720), WV(3907).
Cristulariella moricola. Leaf spot. **23**:WV(3680,3907). **31**:MD(2848).
Fusarium sp. Root rot. **28**:FL(35).
Helminthosporium sp. Leaf spot. **28**:FL(35).
Ramularia viburni. Leaf spot. **11**:WI(82). **20**:TN(82).
Sirosporium beaumontii. Leaf spot. **31**:AL(82), TX(82).
Stigmina tinea. **31**:LA(82).
Verticillium albo-atrum. Wilt. **10**:IL(992,1558), IN(1558). **11**:OR(82), WA(82). **19**:IN(1558). **29**:OR(82,3202). **3**:IL(1558).

Deuteromycotina - Coelomycetes

Ascochyta sp. Leaf spot. **11**:WI(1357). **14**:FL(35). **31**:FL(35).
Coniothyrium sp. Leaf spot. **28**:FL(35).
Corniculariella hystricina. **31**:NJ(861).
Cytospora sp. Canker. **15**:MN(2874).
Hendersonia foliorum var. *viburni*. Leaf spot. **31**:FL(35).
H. tini. Leaf spot. **29**:LA(82).
Hypocenia caespitosa. **31**:MD(2785).
Lasiodiplodia theobromae. Dieback, leaf spot. **14**:FL(35). **28**:FL(35). **31**:FL(35).
Monochaetia jefferisii. On dead limbs. **20**:PA(959).
Phoma viburni. Leaf spot. **2**:NY(82). **5**:NY(64,82). **11**:WI(1350). **15**:IN(663), WI(1345). **31**:FL(35).
Phomopsis sp. Twig dieback. **14**:FL(35). **28**:FL(35). **31**:FL(35).
Phyllosticta tinea. Leaf spot. **6**:MD(82). **28**:FL(35).
P. tineola. Leaf spot. **7**:WI(64). **22**:MN(2874).
Phyllosticta sp. Leaf spot. **5**:WI(1360). **11**:WI(1359). **14**:FL(35). **26**:OK(2872). **31**:FL(35), NC(1303).
Rhabdospora interrupta. **15**:AK(526).
Septoria viburni. **15**:WI(1341,1355).
Septoria sp. Leaf spot. **1**:WI(1355). **11**:MN(82,2874). **14**:FL(35).
Sphaceloma viburni. Spot anthracnose. **13**:FL(2525). **15**:MD(1729), WA(3202). **16**:MD(1736,1737,1745,1748), WA(1745). **28**:FL(35).

Deuteromycotina - Other

Rhizoctonia solani. Root rot. **14**:FL(35).

Weigela Thunb.

1. *W. florida* (Bunge) A. DC. (=*Diervilla florida* (Bunge) Siebold & Zucc.) [pink weigela]
2. *Weigela* sp.

Ascomycotina

Mycosphaerella weigelae. On leaves. **1**:GA(1438). **2**:GA(82), MS(82,2720), NY(82).

Deuteromycotina - Hyphomycetes

Phymatotrichopsis omnivora. Root rot. **2**:TX(82).
Pseudocercospora weigeliae. Leaf spot. **2**:FL(35), MD(82), MS(82,2720), NJ(82).
Ramularia diervillae. Leaf spot. **2**:TN(82).

Deuteromycotina - Coelomycetes

Diplodia weigelae. On twigs. **2**:NY(82).
Diplodia sp. Twig dieback. **2**:FL(35).
Phoma weigeliae. Twig blight. **2**:WA(82,3202).

Caricaceae

Carica L.

1. *C. cauliflora* Jacq.
2. *C. papaya* L. (=*Papaya carica* Gaertn.) [melon-tree, papaya, pawpaw]

Oomycetes

Phytophthora capsici. Fruit rot. **2**:HI(94,2903).
P. nicotianae var. *parasitica*. Postharvest decay of fruits, damping-off, collar rot, fruit rot, stem rot. **2**:FL(35), HI(939,2903).
P. palmivora. Fruit, root and stem rot. **2**:FL(35), HI(82,92,2903).
Pythium aphanidermatum. Root rot. **2**:CA(82,1110), FL(82), HI(82,2903).
P. chamaehyphon. **2**:HI(2821).
P. debaryanum. Damping-off. **2**:TX(82).
P. ultimum. Damping-off. **2**:HI(2899,2903).
P. vexans. **2**:HI(2903).
Pythium sp. Damping-off. **2**:FL(35), HI(2903).

Zygomycotina

Choanephora cucurbitarum. Blossom spot, leaf spot. **2**:FL(35,82).
Rhizopus stolonifer. Fruit rot. **2**:HI(2903).

Ascomycotina

Asterina caricarum. Black mildew. **2**:FL(35,82).
Calonectria crotalariae. Collar rot. **1**:HI(2903). **2**:HI(2903).
Calonectria sp. Collar rot. **1**:HI(2903). **2**:HI(2009).
Didymella sp. Petiole spot. **2**:HI(82,2903).
Erysiphe cichoracearum. **2**:CA(1110).
Erysiphe sp. Powdery mildew. **2**:CA(82).
Glomerella cingulata. Anthracnose (fruit, stem, and leaf spot). **2**:Range of host(82).
Mycosphaerella caricae. Fruit rot, leaf spot. **2**:FL(35).
Mycosphaerella sp. Stem-end rot, target spot. **2**:FL(82), HI(2903).
Sphaerotheca sp. Powdery mildew. **2**:HI(2903).

Deuteromycotina - Hyphomycetes

Alternaria alternata. Fruit spot. **2**:HI(2903).
Alternaria sp. Blossom-end rot, leaf spot. **2**:FL(35), HI(82).
Asperisporium caricae. Leaf spot. **2**:FL(35,82), TX(82).
Cercospora mamaonis. Fruit and leaf spot. **2**:HI(2903).
C. papayae. Leaf spot. **2**:HI(2903).
Cercospora sp. Leaf spot. **2**:FL(35).
Cladosporium sp. **2**:HI(2899,2903).
Corynespora cassiicola. Leaf spot. **2**:FL(35).
Fusarium solani. Fruit rot. **2**:HI(2903).
Fusarium sp. Fruit, root and stem rot. **2**:CA(82,1110), FL(35), HI(82,2899,2903), TX(82).
Oidium caricae. Powdery mildew. **2**:FL(35,82), HI(82,2903), TX(82).
Penicillium sp. **2**:HI(2903).
Phymatotrichopsis omnivora. Root rot. **2**:TX(82).
Stemphylium lycopersici. Storage rot. **2**:HI(570).
Stemphylium sp. Leaf spot. **2**:HI(2903).

Deuteromycotina - Coelomycetes

Colletotrichum falcatum. Fruit rot. **2**:TX(82).
C. gloeosporioides. Anthracnose, fruit rot, leaf spot. **2**:FL(35,2782), HI(939,2784,2903,3687), TX(82).
Diplodia sp. Fruit rot. **2**:TX(82).
Lasiodiplodia theobromae. Fruit rot. **2**:HI(2903).
Phoma caricae-papayae. Stem end fruit rot. **2**:HI(939,1620,2903).

Phomopsis caricae-papayae. **2**:HI(2784).
Phomopsis sp. Stem-end fruit rot. **2**:HI(2299,2903).
Phyllosticta caricae-papayae. Target spot. **2**:FL(35,82).
Phyllosticta sp. Leaf spot. **2**:HI(82,2903).
Deuteromycotina - Other
Rhizoctonia solani. Damping-off. **2**:FL(35), HI(82).
Rhizoctonia sp. Root rot. **2**:HI(2903).
Sclerotium rolfsii. Southern blight. **2**:FL(35), TX(82).

Caryophyllaceae

Agrostemma L.

1. *A. githago* L. [corn cockle, purple cockle]

Deuteromycotina - Hyphomycetes
Fusarium sp. Stem rot. **1**:IN(82).
Phymatotrichopsis omnivora. Root rot. **1**:TX(82).
Deuteromycotina - Coelomycetes
Diplosporonema delastrei. Leaf spot. **1**:IL(82), IN(82), MI(82), MO(82), MS(82,2720).
Septoria lychnidis. Leaf spot. **1**:ND(82).

Arenaria L.

1. *A. capillaris* Poir. (=*A. formosa* Fisch. ex DC.)
2. *A. congesta* Nutt.
3. *A. douglasii* Fenzl ex Torr. & A. Gray
4. *A. fendleri* A. Gray
5. *A. franklinii* Douglas ex Hook.
6. *A. groenlandica* (Retz.) Spreng. [mountain daisy, mountain sandwort]
7. *A. lateriflora* L.
8. *A. macrocarpa* Pursh
9. *A. macrophylla* Hook.
10. *A. nuttallii* Pax
11. *A. physodes* Fisch. ex DC.
12. *A. rubella* (Wahlenb.) Sm. (=*A. verna* L.)
13. *A. sajanensis* Willd. ex Schlechtend.
14. *A. stricta* Michx. [rock sandwort]
15. *Arenaria* sp.

Ascomycotina
Clathrospora pentamera. **12**:OR(3202).
Erysiphe polygoni. Powdery mildew. **5**:WA(3202). **9**:CA(81). **14**:CA(82).
Leptosphaeria stellariae. On dead leaves and stems. **15**:AK(1938).
Mycosphaerella stellarinearum. On leaves. **13**:CO(82).
M. tassiana. On dead leaves and stems. **12**:OR(3202), WA(3202). **15**:AK(1938).
M. tingens. On leaves. **4**:NM(82).
Pleospora comata. On stems. **2**:WA(3202). **8**:AK(526). **12**:AK(526). **15**:AK(82).
P. helvetica. On dead leaves and stems. **12**:OR(3202), WA(3202). **15**:AK(1938).
P. montana. **12**:OR(3202).
P. phaeospora. On stems. **13**:WY(82,1195).
P. tragacanthae. **10**:WA(3202).
Pseudopeziza cerastiorum var. *arenariae*. **7**:AK(526).
Basidiomycotina - Uredinales
Puccinia arenariae. Autoecious. **1**:WA(3202). **2**:ID(3202), MT(82,3202). **3**:CA(81). **7**:AK(60,526), MT(3202). **10**:OR(3202). **11**:AK(60,526). **12**:WA(3202). **14**:NY(82), WI(82). **15**:FL(35).
P. tardissima. Autoecious. **2**:CO(82), UT(82), WY(82). **4**:NM(82). **15**:CO(82), ID(3202), OR(82), TX(82), UT(82), WY(82).
[*Uromyces inaequialtus*]. **15**:CO(82), UT(82).
Basidiomycotina - Ustilaginales
Sorosporium saponariae. **7**:WA(3202). **9**:ID(3202).
Ustilago violacea. Anther smut. **1**:WA(1077,3202). **6**:ME(82,1077), NH(82,1077), NY(82,1077), VT(82,1077). **7**:MN(1077), WA(3202), WI(1077). **15**:TX(82,1077).
Deuteromycotina - Hyphomycetes
Phymatotrichopsis omnivora. Root rot. **15**:TX(82).
Deuteromycotina - Coelomycetes
Hendersonia tenella. Leaf spot. **15**:TX(82).
Phoma herbarum. **4**:WY(2924).
Phyllosticta nebulosa. **7**:WI(1336).
Phyllosticta sp. **7**:WI(1340).
Septoria sp. On leaves. **15**:NJ(82).

Cerastium L.

1. *C. alpinum* L.
2. *C. arvense* L. (=*C. arvense* L. var. *viscidulum* J. F. Gmelin, *C. campestre* Greene, *C. occidentale* Greene, *C. oreophilum* Greene, *C. scopulorum* Greene, *C. strictum* L.) [field chickweed, starry glasswort]
3. *C. beeringianum* Cham. & Schlechtend.
4. *C. brachypodum* (Engelm. ex A. Gray) B. L. Robinson
5. *C. maximum* L.
6. *C. nutans* Raf.
7. *C. tomentosum* L. [snow-in-summer]
8. *C. viscosum* L.
9. *C. vulgatum* L. [mouse-ear chickweed]
10. *Cerastium* sp.

Chytridiomycetes
Synchytrium cerastii. **8**:LA(82).
S. stellariae. **9**:MO(82). **10**:CA(82).
Synchytrium sp. Leaf gall. **2**:TX(82).
Oomycetes
Peronospora alsinearum. Downy mildew. **2**:CO(82). **4**:SD(2243). **6**:IA(82), IL(82), MD(82), MN(2874), WI(82). **8**:AL(82), IA(82), MS(82,2720), NC(82,3944). **9**:AL(82), IA(82), MS(82,2720), NC(82).
Ascomycotina
Clathrospora pentamera. **2**:MT(3202).
Leptotrochila cerastiorum. Leaf spot. **2**:OR(82,3202).
Mycosphaerella minor. **2**:WA(3202).
M. tassiana. On dead stems and leaves. **3**:WA(3202). **10**:AK(1938).
Pleospora tragacanthae. On dead stems. **2**:MT(3202). **10**:AK(1938).
Basidiomycotina - Uredinales
Melampsorella caryophyllacearum. Heteroecious: telial host. **1**:ME(82,1575). **3**:AK(60,526,1575), CO(1575), MT(3202), UT(1575), WY(1575). **8**:OR(1575,3202), WA(1575,3202). **9**:ID(3202), NE(82,1575), OR(82,3202). **10**:AK(82), CO(82), MT(82), NM(82), SD(82), UT(82), WA(82), WY(82,1575).
Puccinia arenariae. Autoecious. **1**:MT(3202). **2**:MT(3202). **10**:CO(82), MT(82).
Puccinia sp. **10**:CO(49).
Basidiomycotina - Ustilaginales
Sorosporium saponariae. Flower smut. **2**:CO(82,1077), NY(82,1077).
Ustilago duriaeana. **8**:OR(1077,3202). **10**:WA(1077,3202).
U. violacea. Anther smut. **5**:AK(526,1077). **8**:OR(1077,3202). **10**:AK(82), WA(1077,3202).
Basidiomycotina - Other
Thanatephorus cucumeris. Root and stem rot. **9**:MN(2874).
Deuteromycotina - Hyphomycetes
Cercoseptoria cerastii. On leaves. **9**:WI(1334).
Cladosporium echinulatum. **2**:WY(1195).
Didymobotryopsis cerastii. **2**:WY(1195).
Isariopsis albo-rosella. Leaf spot. **2**:MT(3202). **6**:ID(3202), WI(82). **8**:MA(82), WI(82). **9**:ID(3202), MS(82,2720), NH(82), NY(82), PA(82), WA(3202), WI(82).
Phymatotrichopsis omnivora. Root rot. **8**:TX(82). **9**:TX(82).
Ramularia sp. **9**:ID(3202).
Deuteromycotina - Coelomycetes
Phoma herbarum. On stems. **8**:TX(82). **9**:TX(82).
Septoria cerastii. Leaf spot. **2**:AL(82), ID(3202), KY(82), MS(82,2720), MT(3202), OR(3202), WA(3202). **8**:AL(82), IL(82), IN(82), KY(82), MO(82), MS(82,2720), NJ(82), TX(82). **9**:AL(82), ID(3202), IL(82), IN(82), KY(82), MO(82), MS(82,2720), NJ(82), TX(82), WA(3202). **10**:HI(2903), ID(82), WI(82).
S. commersoniana. **2**:WY(1195).
Deuteromycotina - Other
Rhizoctonia solani. Root rot. **8**:TX(82). **9**:NE states(82), IL(82), MN(82), TX(82).
Rhizoctonia sp. Foliage blight. **7**:PA(2600).
Sclerotium rolfsii. Southern blight. **8**:TX(82). **9**:TX(82). **10**:FL(35).

Dianthus L.

1. *D. armeria* L. [Deptford pink, grass pink]
2. *D. barbatus* L. [sweet-William]
3. *D. caryophyllus* L. [carnation, clove pink, divine-flower]
4. *D. chinensis* L. [rainbow pink]
5. *D. deltoides* L. [maiden pink]
6. *D. heddewigii* Hort. A name of no botanical standing used for *D. chinensis* cultivars.
7. *D. plumarius* L. [cottage pink, garden pink, grass pink]
8. *Dianthus* sp.

Oomycetes

Peronospora dianthicola. 'Downy mildew.' **3:**CA(1153).
Phytophthora nicotianae var. *parasitica*. Root rot. 'Phytophthora wilt.' **3:**HI(2903).
Pythium ultimum. 'Pythium root rot.' **3:**CA(82), IL(82), WI(82).
P. vexans. Root rot. **3:**CA(82).
Pythium sp. Damping-off, root and stem rot. **3:**FL(35), HI(2903).

Ascomycotina

Mycosphaerella caryophylli. On stems. **3:**AK(82,526).
Mycosphaerella sp. On stems. **1:**CA(82).
Schizothyrium perexiguum. Greasy blotch. **3:**HI(2899,2903).

Basidiomycotina - Uredinales

Puccinia arenariae. Autoecious. **2:**AL(82,285), CT(82), MA(82), MS(82,2720), NY(82), PA(82). **3:**HI(2903). **5:**NY(82).
Uromyces dianthi. Heteroecious: telial host. **1:**IL(82), MS(82), NY(82), TN(82), TX(82), WA(82). **2:**FL(35), IA(82), NE(82), TX(82). **3:**'Rust.' Range of host(82), AK(60,526), CA(81), CO(49), FL(35), GA(1439), HI(2903), ID(3202), MN(2874), MS(2720), MT(3202), OK(2872), OR(3202), SD(2243), WA(3202), WV(3907). **4:**NE(82). **5:**IL(82), MS(82), NY(82), TN(82), TX(82), WA(82). **7:**IL(82), MS(82), NY(82), TN(82), TX(82), WA(82). **8:**ID(3202).

Basidiomycotina - Ustilaginales

Ustilago violacea. Anther smut. **2:**NJ(82,1077). **3:**'Anther smut.' DC(1077), MA(82,1077).

Basidiomycotina - Other

Armillaria mellea. Stem rot. **3:**CA(2906).
Corticium stevensii. **3:**Unk.(82).
Pellicularia koleroga. Web blight. **3:**NC(82).
Thanatephorus cucumeris. Root rot. **1:**WI(1357). **3:**MN(2874), WA(3202).

Deuteromycotina - Hyphomycetes

Alternaria dianthicola. Flower blight, damping-off, leaf spot. **3:**CA(149), HI(2903).
A. saponariae. Leaf spot, stem rot, blight. **2:**FL(35), HI(2903). **3:**'Alternaria blight.' Range of host(82), FL(35), MD(82), MI(82), MN(2874), MS(2720), NC(1303), WA(3202). **4:**C states(82), NE states(82), CA(82), CO(82), TX(82). **7:**C states(82), NE states(82), CA(82), CO(82), TX(82). **8:**DE(3543), MN(2874).
Alternaria sp. Isolated from inflorescences. **1:**MD(651). **3:**HI(2903), NC(1303).
Bipolaris maydis. Petal spot. **3:**FL(35).
B. setariae. Flower petal spot. **3:**FL(35).
Botrytis cinerea. Gray mold. **2:**FL(35). **3:**'Gray mold.' Range of host(82), AK(526), CO(49), FL(35), HI(2903), ID(3202), NC(1303). **4:**AK(82). **6:**AK(526). **7:**AK(82).
Botrytis sp. **3:**HI(2903).
Cladosporium echinulatum. Leaf spot. **3:**'Fairy-ring leaf spot.' Unk.(82), CO(49), FL(35), OR(3202), WA(3202). **7:**CA(82), NY(82), OR(82).
C. herbarum. On leaves. **3:**Range of host(82), FL(35), ID(3202), WA(3202).
Curvularia sp. **1:**MD(651).
Cylindrocladium scoparium. Cutting rot. **3:**FL(35).
Fusarium acuminatum. Branch, root and dry stem rot. **3:**Unk.(82).
F. avenaceum. Branch, root and dry stem rot. **3:**Unk.(82).
F. culmorum. Branch, root and stem rot. **3:**Unk.(82).
F. equiseti. Branch, root and dry stem rot. **3:**Unk.(82).
F. oxysporum f. sp. *dianthi*. Wilt. **2:**CA(82), KS(82). **3:**Range of host(3992), CA(82), CO(49,82), HI(2903), MA(82), MI(82), NC(1303), NE(82), NY(82), UT(3620), WA(82,3202). **8:**CO(49).
F. poae. Bud rot. **3:**NE states(82), Unk.(82), KS(82), NE(82), OH(946), WA(82,3202). **7:**KS(82), NY(82), TX(82), WI(82), WV(82).
F. roseum. Stub dieback, basal foot rot, limb, root and stem rot. **3:**CO(49), HI(2903), NC(1303), PA(2583).
Fusarium sp. Dieback, root and basal stem rot. **2:**FL(35). **3:**Range of host(82), FL(35), HI(2903), MS(2720), MT(3202), NC(1303), OK(2872).
Oidium sp. Powdery mildew. **3:**FL(35,82), NC(82,1303).
Phialophora cinerescens. 'Phialophora wilt.' **3:**CO(49).
Phymatotrichopsis omnivora. Root rot. **2:**TX(82). **3:**'Phymatotrichum root rot.' TX(82). **7:**TX(82).
Stemphylium botryosum. **8:**DE(3543).
S. lycopersici. Petal blight. **3:**FL(35,1582).
Verticillium dahliae. Wilt. **3:**NJ(82).
Volutella sp. On stems. **3:**MI(82), NJ(82), NY(82), OR(82), PA(82), WA(3202).
Zygophiala jamaicensis. 'Greasy blotch.' **3:**CA(150).

Deuteromycotina - Coelomycetes

Ascochyta dianthi. Leaf spot. **1:**NY(82). **7:**MS(82).
Asteroma dianthii. On leaves and stems. **7:**CA(82).
Colletotrichum dematium. **3:**Unk.(82). **8:**WA(3202).
Colletotrichum sp. Anthracnose. **3:**NJ(82), NY(82), TX(82).
Heteropatella valtellinensis. On leaves. **3:**WA(82,3202).
Pestalotia sp. **8:**HI(2903).
Phoma sp. On stems. **2:**VA(82). **7:**LA(82).
Phyllosticta silenes. **7:**MO(82).
Phyllosticta sp. Leaf spot. **2:**AK(526), WA(82,3202), WI(1352).
Septoria dianthi. Leaf spot. **1:**OH(946). **2:**AL(82), IA(82), MI(82), NY(82). **3:**E states(82), CA(82), HI(82,2903), MI(82), MS(2720), TX(82). **4:**AL(82), IA(82), MI(82), NY(82). **7:**MS(82,2720), NC(82), NJ(82), NY(82).
S. dianthicola. **2:**WI(1357).

Deuteromycotina - Other

Rhizoctonia solani. Root and stem rot. **2:**CT(82), FL(35), IL(82), KS(82), MA(82), MS(82,2720), NC(1303), NJ(82), NY(82), PA(82), TX(82). **3:**'Rhizoctonia stem rot.' Range of host(82), FL(35), MS(2720), NC(1303). **4:**CT(82), IL(82), KS(82), MA(82), MS(82,2720), NJ(82), NY(82), PA(82), TX(82). **7:**CT(82), IL(82), KS(82), MA(82), MS(82,2720), NJ(82), NY(82), PA(82), TX(82).
Rhizoctonia sp. Root rot. **2:**HI(2903). **3:**HI(2903).
Sclerotium rolfsii. Southern blight, stem rot. **2:**CT(82), FL(35,82), HI(2903), NC(82,1303). **3:**'Southern blight.' FL(35,82), HI(2903), MS(82,2720), TX(82). **7:**IL(82), TX(82).
Sclerotium sp. **3:**HI(2903).

Gypsophila L.

1. *G. elegans* M. Bieb. [baby's-breath]
2. *G. paniculata* L. [baby's-breath]
3. *Gypsophila* sp.

Oomycetes

Phytophthora cactorum. Wilt. **2:**Unk.(3887).
P. nicotianae var. *parasitica*. Crown and root rot, wilt. **2:**FL(35,990,991).
Phytophthora sp. Root rot. **3:**MN(2874).
Pythium aphanidermatum. Root rot. **2:**FL(35).
P. debaryanum. Damping-off. **2:**CT(82).
P. ultimum. Root rot. **2:**FL(35).
Pythium sp. Root rot. **2:**HI(2903).

Deuteromycotina - Hyphomycetes

Alternaria sp. Flower and petal blight. **2:**FL(35), HI(2903).
Botrytis cinerea. Gray mold blight. **2:**NJ(82).
Fusarium oxysporum. Stem decay. **2:**HI(2903).
Fusarium sp. Stem rot, wilt. **2:**FL(35), NC(1303), OK(2873).

Deuteromycotina - Coelomycetes

Phyllosticta sp. Leaf spot. **2:**FL(35).

Deuteromycotina - Other

Rhizoctonia solani. Damping-off, root and stem rot. **1:**IL(82). **2:**CT(82), FL(35), NC(1303), OK(2873). **3:**IL(82).
Rhizoctonia sp. Foot, root and stem rot. **2:**HI(2903), NC(1303).

Honkenya Ehrh.

1. *H. peploides* (L.) Ehrh. (=*Arenaria peploides* L., *Ammodenia peploides* (L.) Rupr., *Arenaria peploides* L. var. *major* Hook.) [sea chickweed, sea-purslane, seabeach sandwort]

Deuteromycotina - Coelomycetes

Ascochyta sp. **1:**AK(526).
Septoria ammodeniae. On leaves. **1:**AK(82,526).

Lychnis L.

1. *L. chalcedonica* L. [Jerusalem-cross, London-pride, Maltese-cross, scarlet lychnis, scarlet-lightening]
2. *L. coronaria* (L.) Desr. [dusty-miller, mullein pink, rose campion]
3. *L. drummondii* (Hook.) S. Wats.
4. *L. flos-cuculi* L. [cuckoo-flower, ragged-robin]
5. *L. viscaria* L. [German catchfly]
6. *Lychnis* sp.

Oomycetes

Phytophthora cactorum. Stem rot. **6:**IN(82).

Ascomycotina

Pleospora herbarum. **2:**WA(3202).

Basidiomycotina - Uredinales

Puccinia arenariae. Autoecious. **1:**PA(82).
Uromyces suksdorfii. Autoecious. **6:**UT(82).

Basidiomycotina - Ustilaginales

Ustilago violacea. Anther smut. **3:**WY(1077,1195). **6:**MI(82), MN(1077,2874), WI(82,1077), WY(82,1077).

Deuteromycotina - Hyphomycetes

Alternaria saponariae. **1:**AK(526).
Alternaria sp. Leaf spot. **1:**AK(82).
Botrytis cinerea. **6:**AK(526).
Cercospora lychnidis. On living leaves. **2:**WI(1352).
Heterosporium sp. On leaves. **1:**AK(82). **6:**AK(82).

Deuteromycotina - Coelomycetes

Leptothyrium lychnidis. Leaf spot. **6:**AL(82).
Phoma punctiformis. **6:**NY(82).
Phyllosticta lychnidis. Leaf spot. **2:**IA(82).
P. nebulosa. **5:**WI(1359). **6:**WI(1360).
Septoria lychnidis. Leaf spot. **1:**MN(2874), WI(1352). **4:**WI(1334).
Septoria sp. Leaf spot. **6:**MA(82).

Deuteromycotina - Other

Rhizoctonia solani. Root rot. **1:**IL(82). **6:**IL(82).

Paronychia Mill.

(= *Anychia* Michx.)

1. *P. canadensis* (L.) A. Wood (=*Anychia canadensis* (L.) B. S. P.) [forked chickweed]
2. *P. fastigiata* (Raf.) Fernald

Deuteromycotina - Coelomycetes

Phoma sp. On stems. **1:**WI(1348). **2:**WI(1357).
Septoria anychiae. On lower leaves. **1:**WI(1343).

Sagina L.

1. *S. procumbens* L. [bird's-eye pearlwort]
2. *S. subulata* (Sw.) K. Presl
3. *Sagina* sp.

Ascomycotina

Mycosphaerella tassiana. On dead stems. **3:**AK(1938).
Pleospora helvetica. On dead stems. **3:**AK(1938).

Deuteromycotina - Other

Rhizoctonia solani. Brown patch. **1:**NC states(82), NE states(82).
Rhizoctonia sp. Foliage blight. **2:**PA(2600).

Saponaria L.

1. *S. officinalis* L. [bouncing-bet, soapwort]

Ascomycotina

Didymella sp. On leaves. **1:**GA(82,1438).
Leptosphaeria doliolum. **1:**GA(1438).
Massarina gloeospora. **1:**GA(1438).

Deuteromycotina - Hyphomycetes

Alternaria saponariae. Leaf and stem spot. **1:**NE states(82), IN(82), MA(1390), MN(82,2874), WV(3907).
Phymatotrichopsis omnivora. Root rot. **1:**TX(82).

Deuteromycotina - Coelomycetes

Colletotrichum sp. On stems. **1:**NY(82), WI(1332).
Cylindrosporium officinale. **1:**IN(82).
Phyllosticta tenerrima. Leaf spot. **1:**NJ(82), TX(82).
Rhabdospora saponariae. **1:**NY(82).
Septoria noctiflorae. Leaf spot. **1:**IL(82).

Silene L.

1. *S. acaulis* L. [cushion pink, moss campion]
2. *S. alba* (Mill.) J. Krause (=*Lychnis alba* Mill.) [evening lychnis, evening-campion, white campion, white cockle]
3. *S. antirrhina* L. [sleepy catchfly]
4. *S. armeria* L. [garden catchfly, none-so-pretty, sweet-William catchfly]
5. *S. californica* E. M. Durand [California Indian pink]
6. *S. campanulata* S. Wats.
7. *S. caroliniana* Walter subsp. *pensylvanica* (Michx.) R. T. Clausen (=*S. caroliniana* Walter var. *pensylvanicum* (Michx.) Fernald) [wild pink]
8. *S. cserei* Baumg.
9. *S. cucubalus* Wibel (=*S. latifolia* (Mill.) Rendle & Britten) [bladder campion]
10. *S. dichotoma* Ehrh.
11. *S. dioica* (L.) Clairv. (=*Lychnis dioica* L.) [morning campion, red campion]
12. *S. douglasii* Hook. (=*S. lyallii* S. Wats.)
13. *S. douglasii* Hook. var. *douglasii*
14. *S. gallica* L. [French catchfly]
15. *S. grayi* S. Wats.
16. *S. hookeri* Nutt.
17. *S. lemmonii* S. Wats.
18. *S. menziesii* Hook.
19. *S. noctiflora* L. [night-flowering catchfly, sticky-cockle]
20. *S. oregana* S. Wats.
21. *S. parryi* (S. Wats.) C. Hitchc. & Maguire (=*S. macounii* S. Wats.)
22. *S. sargentii* S. Wats. (=*S. watsonii* B. L. Robinson & Seat.)
23. *S. scaposa* B. L. Robinson
24. *S. scouleri* Hook.
25. *S. scouleri* Hook. subsp. *grandis* (Eastw.) Hitchc. & Maguire (=*S. pacifica* Eastw. in J. M. Coulter)
26. *S. spaldingii* S. Wats.
27. *S. stellata* (L.) W. T. Aiton [starry campion, widow's-frill]
28. *S. suksdorfii* B. L. Robinson
29. *S. uralensis* (Rupr.) Bocq. subsp. *apetala* (L.) Bocq. (=*Lychnis apetala* L., *Melandrium apetala* (L.) Fenzl)
30. *Silene* sp.

Oomycetes

Peronospora silenes. Downy mildew. **3:**KS(82), NE(82), SD(2243), TX(82), WI(82). **30:**IL(82), TX(82).

Ascomycotina

Crocicreas cyathoideum var. *cacaliae*. On stems. **30:**CO(505).
Erysiphe polygoni. **30:**ID(3202).
Lewia scrophulariae. **29:**AK(526).
Mycosphaerella tassiana. **28:**WA(3202).
M. tassiana var. *arctica*. On dead stems. **29:**AK(1938).
Mycosphaerella sp. **1:**WA(3202).
Pleospora abscondita. **21:**WA(3202,3814). **30:**WA(3817).
P. cerastii. **29:**AK(526).
P. comata. **21:**WA(3202,3814).
P. helvetica. **1:**AK(3431).
P. herbarum. On overwintered herbage. **13:**CA(642).
P. tragacanthae. **21:**WA(3202). **28:**WA(3202).
Pleospora sp. **13:**ID(3202).

Basidiomycotina - Uredinales

[*Puccinia aristidae*]. **3:**AZ(82,1203), TX(82).
P. subnitens. **3:**AZ(3981).
Uromyces inaequialtus. Autoecious. **5:**CA(81). **13:**CA(638). **19:**PA(82). **30:**CA(82), IA(82), KS(82), MT(82), WA(82).
U. suksdorfii. Autoecious. **20:**ID(3202), WA(3202). **24:**AZ(1195,1203,3981), OR(3202), WA(3202). **25:**OR(3202). **30:**CA(82), ID(82), NM(82), UT(82), WA(82).
U. verruculosus. **2:**IN(82), MI(82), NY(82), TX(82), WI(1343). **11:**NY(82).

Basidiomycotina - Ustilaginales

Sorosporium saponariae. Flower smut. **18:**UT(1077). **22:**NV(1077). **30:**CO(82,1077), NV(82), UT(82,1077).
Ustilago clintoniana. **21:**WA(1077,3202). **22:**WA(1077,3202).
U. violacea. Anther smut. **1:**HI(82), NH(1077). **7:**NY(1077), VA(1077). **12:**MT(1077), WA(1077), WY(1077). **13:**CA(638,642), MT(3202), WA(1077,3202). **15:**CA(638,642,1077). **20:**WA(1077,3202). **21:**MT(3202), WA(3202). **22:**CA(1077), WA(82,1077,3202). **30:**CA(82), ID(1077,3202), MT(82,1077,3202), TX(82,1077), VA(82), WA(82), WY(82).

Basidiomycotina - Other

Scytinostroma galactinum. Root rot. **2:**MD(82).

Deuteromycotina - Hyphomycetes
Alternaria saponariae. **27:**WI(1332).
Phymatotrichopsis omnivora. Root rot. **2:**TX(82). **3:**TX(82). **30:**TX(82).
Deuteromycotina - Coelomycetes
Ascochyta silenes. On leaves and stems. **3:**ID(3202), MT(82,3202), OK(82,2872), WA(3202), WI(82,966). **6:**OR(3202). **8:**WI(1354). **9:**WI(1357). **13:**MT(3202), OR(3202), WA(3202). **16:**OR(3202). **26:**WA(3202).
Colletotrichum sp. On stems. **30:**MD(82).
Diplodina silenes. On stems. **13:**CA(642,666).
Diplosporonema delastrei. Leaf spot. **17:**OR(3202). **23:**OR(3202). **26:**OR(3202). **27:**WI(82,1337).
Heteropatella umbilicata. **13:**WA(3202).
Phyllosticta lychnidis. Leaf spot. **2:**TX(82).
P. nebulosa. Leaf spot. **2:**WI(1343). **4:**WI(1357). **8:**WI(1354). **9:**WI(1334). **19:**MT(82,3202), NY(82), WI(64,82).
P. silenes. **3:**NE(82), NY(82), TX(82).
Phyllosticta sp. **10:**WI(1357).
Septoria dimera. Leaf spot. **3:**NE(82), WI(82,1333). **9:**WI(1333). **30:**WI(82).
S. lychnidis. Leaf spot. **2:**MN(2874), WI(82). **10:**WI(1334). **11:**WI(1343).
S. noctiflorae. **3:**IN(82), TX(82). **19:**KS(82), NJ(82), TX(82), WI(82). **30:**WI(82).
S. silenes. **3:**KS(82), OK(82). **13:**MT(3202), WA(3202). **18:**ID(3202), MT(3202), WA(3202). **20:**WA(3202). **21:**WA(3202). **30:**IL(82).
S. silenicola. On leaves. **3:**NE(82). **5:**CA(81). **14:**CA(81). **19:**NE(82), NY(82). **27:**IA(82), MD(82), NE(82), PA(82), WI(82,1329). **30:**ID(82), MT(82), UT(82).
Septoria sp. Leaf spot. **3:**OK(2872).
Deuteromycotina - Other
Rhizoctonia solani. Damping-off. **30:**IL(82).
Sclerotium rolfsii. Southern blight. **2:**TX(82).

Spergula L.
1. *S. arvensis* L. [corn spurry]

Oomycetes
Albugo caryophyllacearum. White rust. **1:**TX(82).
Peronospora alsinearum. **1:**WA(3202).
P. obovata. Downy mildew. **1:**AK(82,526), CA(81,82), ME(82), OR(3202), WA(3202), WI(82), WV(3907).
Basidiomycotina - Uredinales
Puccinia arenariae. Autoecious. **1:**ME(82), NY(82), PA(82), TX(82), WI(1357).
Deuteromycotina - Hyphomycetes
Phymatotrichopsis omnivora. Root rot. **1:**TX(82).

Spergularia (Pers.) J. Presl & K. Presl
1. *S. rubra* (L.) J. Presl & K. Presl [red sandspurry]
2. *Spergularia* sp.

Oomycetes
Albugo caryophyllacearum. **2:**CA(82), MA(82).
Basidiomycotina - Uredinales
[*Puccinia aristidae*]. **2:**CA(82).
Uromyces acuminatus. Heteroecious: aecial host. **2:**CA(82).
Deuteromycotina - Coelomycetes
Phoma sp. On leaves and stems. **1:**WI(1357).

Stellaria L.
1. *S. aquatica* (L.) Scop. (=*Myosoton aquaticum* (L.) Moench)
2. *S. borealis* Bigel.
3. *S. brachypetala* Bunge
4. *S. calycantha* (Ledeb.) Bong.
5. *S. graminea* L. [lesser starwort, lesser stitchwort, little starwort]
6. *S. humifusa* Rottb.
7. *S. jamesiana* Torr.
8. *S. longifolia* Muhl. ex Willd.
9. *S. longipes* Goldie [common starwort]
10. *S. media* (L.) Vill. [common chickweed, stitchwort]
11. *S. nitens* Nutt.
12. *S. praecox* A. Nels.
13. *S. sitchana* Steud. var. *bongardiana* (Fernald) Hultén (=*S. borealis* Bigel. var. *bongardiana* Fernald)
14. *S. umbellata* Turcz. ex Karel. & Kir.
15. *Stellaria* sp.

Plasmodiophoromycetes
Ligniera hypogeae. In roots. **2:**CA(82). **5:**CA(82). **9:**CA(82). **10:**CA(82). **15:**CA(82).
L. junci. In roots. **10:**CA(82).
Chytridiomycetes
Synchytrium stellariae. Leaf gall. **10:**DC(1838), LA(82,1838), MS(82,2720), NC(3944), OK(82,2872), TX(82,1838), VA(1838).
Oomycetes
Peronospora alsinearum. Downy mildew. **10:**NC(82,3944). **15:**VA(82).
Ascomycotina
Erysiphe polygoni. **10:**WA(3202).
Leptosphaeria stellariae. On dead leaves and stems. **9:**AK(3431). **15:**AK(1938).
Mycosphaerella stellarinearum. On leaves. **9:**AK(526). **15:**AK(82).
M. tassiana. **9:**AK(526).
Omphalospora stellariae. On leaves and stems. **15:**AK(179).
Phaeosphaeria silenes-acaulis. **6:**AK(526).
Sclerotinia sclerotiorum. Stem rot. **10:**CA(82), SC(82).
Basidiomycotina - Uredinales
Melampsorella caryophyllacearum. Heteroecious: telial host. **1:**WI(1358). **2:**CA(82), MT(82), NY(82), OR(82), UT(82), WA(82). **3:**WA(1575). **4:**ID(3202), MT(3202), OR(3202), WA(3202), WI(1357). **7:**CA(2769), ID(2769,3202), UT(2769). **8:**MI(1575), NY(1575), PA(1575), WI(1343,1354). **10:**ID(3202), ME(82), NH(82), WA(82,3202). **13:**CA(81). **14:**ID(3202), WY(1575). **15:**AK(82), CA(82), ID(82,3202), MI(82,1575), NY(82), PA(82), WA(82).
Puccinia arenariae. Autoecious. **2:**CA(82), MT(82), WA(82). **4:**MT(3202), WA(3202). **9:**AK(60,82,526), MT(3202). **15:**MT(82), WY(82).
[*P. aristidae*]. **10:**CO(82), TX(82).
Basidiomycotina - Ustilaginales
Entyloma alsines. Leaf smut. **10:**TX(82).
Sorosporium saponariae. Flower smut. **7:**CO(1077), UT(1077). **15:**CO(82), UT(82).
Ustilago alsineae. Seed smut. **11:**ID(3202), WA(1077,3202). **12:**ID(1077). **15:**ID(82), WA(82).
U. duriaeana. **15:**WA(1077,3202).
Basidiomycotina - Other
Thanatephorus cucumeris. **10:**LA(3524).
Deuteromycotina - Hyphomycetes
Isariopsis albo-rosella. Leaf spot. **15:**WI(82).
Phymatotrichopsis omnivora. Root rot. **10:**TX(82).
Deuteromycotina - Coelomycetes
Ascochyta dianthi. **6:**AK(3431).
Septoria jamesii. Leaf spot. **15:**CO(82), MT(3202).
S. jonesii. **9:**CA(81).
S. stellariae. **1:**WI(1341). **7:**WA(3202). **10:**MD(82), MI(82), NY(82), OR(82,3202,4007), TX(82), WI(82). **15:**WI(82).
Deuteromycotina - Other
Rhizoctonia solani. Brown patch. **5:**NE states(82), MI(82).

Vaccaria Wolf
1. *V. hispanica* (Mill.) Rauschert (=*Saponaria vaccaria* L.) [cockle, cow soapwort, cow-herb]

Basidiomycotina - Uredinales
[*Puccinia aristidae*]. **1:**CO(82).
Deuteromycotina - Hyphomycetes
Alternaria saponariae. Leaf spot. **1:**CA(149), MT(2281).

Casuarinaceae

Casuarina L.
1. *C. cunninghamiana* Miq. [river she-oak]
2. *C. equisetifolia* L. [Australian pine, beefwood-tree, horsetail casuarina, horsetail-tree, mile-tree, South Sea ironwood]
3. *C. stricta* Aiton [drooping she-oak]
4. *Casuarina* sp.

Oomycetes
Phytophthora sp. Needle blight. **4:**FL(35).
Pythium sp. Root rot. **2:**FL(35).

Ascomycotina
Botryosphaeria sp. Dieback. **4**:FL(35).
Rhytidhysteron rufulum. **4**:FL(2782).
Basidiomycotina - Other
Armillaria mellea. Root rot. **2**:FL(35). **3**:CA(2900). **4**:CA(82).
A. tabescens. Root rot. **1**:FL(35). **2**:FL(35). **4**:FL(35,82).
Ganoderma annularis. On wood. **1**:FL(35). **2**:FL(35). **4**:FL(35).
G. applanatum. Heart rot. **4**:HI(82,2903).
G. lucidum. **1**:AZ(1202).
Inonotus rickii. On wood. **4**:AZ(1204).
Phellinus kawakamii. On wood. **2**:HI(2050).
Sebacina sp. **4**:HI(2903).
Deuteromycotina - Hyphomycetes
Allescheriella crocea. **4**:HI(2903).
Helicosporium panacheum. On fruits. **2**:HI(1269,2903).
Oidium sp. Powdery mildew. **1**:FL(35).
Deuteromycotina - Coelomycetes
Botryodiplodia sp. Dieback. **4**:FL(35).
Colletotrichum sp. Anthracnose. **2**:FL(35).
Diplodia sp. Dieback. **1**:FL(35).
Phomopsis sp. Dieback. **2**:FL(35).
Sphaeropsis sp. Gall. **2**:FL(35). **4**:FL(35).
Deuteromycotina - Other
Rhizoctonia solani. Root rot. **2**:FL(35). **4**:FL(35).
Sclerotium rolfsii. Southern blight. **1**:FL(35).

Cecropiaceae

Cecropia Loefl.

1. *Cecropia* sp.

Oomycetes
Pythium sp. Root rot. **1**:FL(35).
Ascomycotina
Mycosphaerella sp. Leaf spot. **1**:FL(35).
Basidiomycotina - Other
Mycocalia reticulata. On petioles. **1**:HI(1267,2903).
Deuteromycotina - Hyphomycetes
Alternaria sp. Leaf spot. **1**:FL(35).
Deuteromycotina - Coelomycetes
Diplodia sp. Leaf spot. **1**:FL(35).

Celastraceae

Canotia Torr.

1. *C. holocantha* Torr.

Basidiomycotina - Other
Phanerochaete crassa. **1**:AZ(128,1202).
Phellinus texanus. **1**:AZ(1202).

Catha Forssk. ex Scop.

1. *C. edulis* (Vahl) Forssk. ex Endl. (=*Celastrus edulis* Vahl) [Abyssinian tea, Arabian tea, cafta, chat, kat, khat, qat, Somali tea]

Deuteromycotina - Hyphomycetes
Ovularia sp. On leaves. **1**:FL(35,82).
Deuteromycotina - Coelomycetes
Colletotrichum gloeosporioides. Leaf-tip blight. **1**:FL(35,82).

Celastrus L.

1. *C. orbiculata* Thunb. [Oriental bittersweet]
2. *C. scandens* L. [American bittersweet, climbing bittersweet, false bittersweet, shrubby bittersweet, staff-vine, waxwork]
3. *Celastrus* sp.

Ascomycotina
Botryosphaeria obtusa. On dead stems. **1**:NY(82), VA(82).
B. ribis. On dead stems. **2**:WV(3907). **3**:IL(82), NY(82), WV(82).
Diaporthe celastrina. On dead stems. **2**:KS(82), MI(82), NY(82).
Eutypella leprosa. On dead stems. **1**:KS(82). **2**:KS(2925).
Glomerella cingulata. Canker. **1**:NC(82), NH(82).
Leptosphaeria conoidea. **2**:NY(3249).
Microsphaera penicillata. Powdery mildew. **1**:WI(82).
Microthyriella celastri. Leaf spot. **2**:KS(82), ME(82), MI(82).
Nectria cinnabarina. Canker. **2**:MI(82), MN(2874), MO(82), NE(82), PA(82), WV(82,3907).
Phyllactinia guttata. Powdery mildew. **1**:NE states(82), SD(82), TX(82). **2**:MN(2874), OH(950), SD(2243).
Valsa ambiens. **2**:KS(3415).
Valsaria insitiva. **2**:IA(1226).
Deuteromycotina - Hyphomycetes
Bactrodesmium pithoideum. On dead stems. **1**:NY(82). **2**:NY(1687,3547).
Phaeoisariopsis melanochaeta. Leaf spot. **2**:IA(82), KS(82), MS(2720), ND(82,378), NE(82).
Phymatotrichopsis omnivora. Root rot. **1**:TX(82).
Ramularia celastri. Leaf spot. **1**:E states(82), ND(82), TX(82). **2**:DE(3543), FL(35), MN(2874), MS(2720), NC(3944), ND(378), WV(3907).
Deuteromycotina - Coelomycetes
Ascochyta sp. Leaf spot. **1**:CT(82). **3**:CT(82).
Camarosporium abnormale. On dead stems. **1**:NY(82).
Cytospora celastri. On stems. **2**:KS(82), NE(82).
Cytospora sp. On stems. **1**:VA(82).
Diplodia propullans. On limbs and dead stems. **1**:KS(82), MI(82), NY(82), TX(82), WA(82). **2**:WV(3907).
Dothiorella celastri. On dead stems. **2**:IL(2738).
Gloeodes pomigena. On stems. **1**:IN(82).
Labrella celastri. On dead stems. **2**:MA(82), NY(82).
Marssonina sp. Anthracnose. **2**:NY(338).
Phoma pallens. On stems. **1**:NY(82).
Phyllosticta celastri. Leaf spot. **2**:IL(82), KS(82), MA(82), NY(82), WV(82,3907).
Septogloeum thomasianum. On leaves and stems. **2**:MD(1749), WI(1353).

Crossopetalum P. Br.

1. *C. ilicifolia* (Poir.) Kuntze (=*Rhacoma ilicifolia* (Poir.) Trel. in A. Gray)

Ascomycotina
Irenopsis tehoniana. **1**:FL(1452,3483).

Euonymus L.

1. *E. alatus* (Thunb.) Siebold [winged spindle-tree]
2. *E. americanus* L. [bursting-heart, strawberry-bush]
3. *E. atropurpureus* Jacq. [burning-bush, eastern wahoo, wahoo]
4. *E. europaeus* L. [European spindle-tree]
5. *E. fortunei* (Turcz.) Hand.-Mazz. [climbing euonymus]
6. *E. fortunei* (Turcz.) Hand.-Mazz. var. *vegetus* (Rehd.) Rehd.
7. *E. japonica* Thunb. [evergreen euonymus, Japanese spindle-tree]
8. *E. kiautschovicus* Loes.
9. *E. nanus* M. Bieb.
10. *E. occidentalis* Nutt. ex Torr. [western burning-bush]
11. *Euonymus* sp.

Myxomycetes
Physarum cinereum. **7**:MS(2720).
Oomycetes
Phytophthora sp. Root rot. **7**:FL(35).
Pythium sp. Root rot. **7**:FL(35).
Ascomycotina
Botryosphaeria dothidea. **2**:GA(1438).
B. obtusa. On limbs. **2**:GA(82).
Diaporthe euonymi. On dead stems. **3**:DE(830).
Elsinoe euonymi-japonici. On living leaves and stems. **7**:FL(1747).
Eutypella leprosa. **11**:IA(3629).
Glomerella cingulata. Anthracnose, on leaves and twigs. **7**:CA(81), MS(82,2720), TX(82).
Hypoxylon sp. **3**:IN(654).
Microsphaera euonymi. **3**:IN(654), WI(1352). **4**:WI(1350).
M. penicillata. Powdery mildew. **1**:NE(82). **2**:SD(2243). **3**:C states(82), MS(2720), SD(82), TX(82). **4**:IL(2567). **11**:MN(2874), UT(3620).
M. pusilla. **2**:DE(364), NY(364).
Nectria atrofusca. **11**:NY(271).
N. cinnabarina. On limbs. **3**:NY(82).
Physalospora sp. On twigs. **11**:CA(82).
Sclerotinia sclerotiorum. Dieback. **1**:RI(1718).
Basidiomycotina - Ustilaginales
Sporidiobolus pararoseus. On leaves. **11**:KS(2686).
Basidiomycotina - Other
Armillaria mellea. **11**:CA(2898).

Erythricium salmonicolor. Limb blight. **7**:FL(35).
Lopharia cinerascens. **3**:IN(654). **4**:MA(543).
Pellicularia koleroga. Thread blight. **7**:LA(82). **11**:LA(3640).
Trametes hirsuta. Sapwood rot. **11**:MA(82).
Deuteromycotina - Hyphomycetes
Alternaria sp. Leaf spot. **7**:FL(35).
Botrytis cinerea. On limbs. **1**:NJ(82).
Bullera alba. On leaves. **11**:KS(2686).
Cercospora destructiva. Leaf spot. **7**:AR(82), FL(35), MS(82,2720), SC(82,962), TX(82), VA(82).
C. euonymi. Leaf spot. **2**:KY(581). **7**:FL(35).
Cladosporium herbarum. On leaves. **3**:WI(82,1332). **7**:SC(82).
Cylindrocladium sp. Leaf spot. **7**:FL(35).
Epicoccum nigrum. **2**:DE(3543).
Fusarium equiseti. Root rot. **7**:NJ(82).
F. lateritium. On twigs. **7**:CA(82).
Fusarium sp. Root rot, wilt. **7**:FL(35). **11**:OK(2872).
Mycocentrospora verrucosa. On living leaves. **8**:OH(2842).
Mycocentrospora sp. Leaf spot. **8**:OH(3868).
Oidium euonymi-japonici. Powdery mildew. **4**:WA(3202). **7**:CA(82), FL(35), IA(82), LA(82), MS(82,2720), NC(1303), NJ(82), TX(82), WA(82).
Oidium sp. Powdery mildew. **2**:IA(82). **3**:CA(82), TX(82). **7**:OH(950).
Phymatotrichopsis omnivora. Root rot. **3**:TX(82). **7**:AZ(82), TX(82). **11**:TX(82).
Ramularia euonymi. Leaf spot. **2**:KY(82). **3**:IA(82), IL(82), KS(82), MO(82), NE(82), WI(82). **7**:AR(82), FL(35). **10**:CA(82). **11**:MN(2874).
Tubercularia sp. **1**:IL(2567).
Verticillium heterocladium. Wilt. **7**:FL(35).
Deuteromycotina - Coelomycetes
Aplosporella euonymi. On dead limbs and twigs. **3**:NE(82,966).
Colletotrichum gloeosporioides. Anthracnose. **5**:MA(2228). **11**:AL(285), NJ(82), NY(82).
Coniothyrium fuckelii. On twigs. **11**:CA(82).
Coniothyrium sp. On leaves. **3**:WI(82,1332).
Cytospora euonymi. Twig blight, canker. **5**:WI(1347). **9**:MN(2874). **11**:CA(82).
Diplodia euonymi. On twigs. **4**:AL(82), NY(82).
Discogloeum concentricum. Leaf spot. **7**:AL(82), MS(82,2720), SC(82), TX(82), VA(82). **11**:AL(285).
Macrophoma ramulicola. On twigs. **7**:LA(82). **11**:CA(82).
Pestalotia planimi. On leaves. **7**:CA(82), LA(82), NC(1303), TX(82).
Phlyctema vagabunda. Anthracnose. **7**:AR(82), MS(82,2720).
Phomopsis foveolaris. On twigs. **4**:OH(82). **6**:OK(2872).
Phyllosticta euonymi. Leaf spot. **3**:DE(82), SD(2243), TX(64,3597). **5**:OK(82). **6**:OK(2872). **7**:MS(82,2720), NC(1303), TX(82), VA(82). **11**:OK(2872).
P. pallens. Leaf spot. **2**:AL(82), IN(82), TX(82). **3**:TX(82). **4**:AL(82).
Phyllosticta sp. Leaf spot. **7**:FL(35).
Septogloeum thomasianum. Leaf spot. **3**:IA(82), IL(82,2568), IN(82), MI(82), MO(82), OH(82), WI(82).
Septoria atropurpurea. Leaf spot. **3**:IL(82).
S. evonymella. **7**:NJ(82), SC(82).
S. evonymi. Leaf spot. **2**:SC(82), VA(82). **3**:IN(82). **7**:MS(82,2720).
S. spiculispora. On leaves. **11**:DE(970).
Sirogloea euonymi. On twigs. **11**:CA(82).
Sphaceloma euonymi-japonici. Leaf scab. **7**:FL(35).
Sphaeropsis sp. Dieback. **7**:FL(35).
Deuteromycotina - Other
Rhizoctonia ramicola. Thread blight. **7**:FL(35).
R. solani. Thread blight. **7**:FL(35).

Maytenus Molina
1. *M. boaria* Molina
Basidiomycotina - Other
Armillaria mellea. Root rot. **1**:CA(2895).

Mortonia A. Gray
1. *M. scabrella* A. Gray
Basidiomycotina - Other
Dendrothele acerina. **1**:AZ(128,1202).
Odontia crustula. On dead limbs. **1**:AZ(1202).
Peniophora quercina. **1**:AZ(1202).
Phanerochaete arizonica. White rot. **1**:AZ(433,1202).
Schizopora apacheriensis. On roots of living plant. **1**:AZ(1193,1202).
Scytinostroma portentosum. **1**:AZ(1202).

Paxistima Raf.
1. *P. canbyi* A. Gray [cliff-green, mountain-lover]
2. *P. myrsinites* (Pursh) Raf. [myrtle box-leaf, Oregon boxwood]
Ascomycotina
Lophiostoma quadrinucleatum. **2**:ID(3202).
Mycosphaerella pachystimae. Leaf spot. **2**:ID(82,3202).
Sarcosoma latahensis. **2**:ID(3202).
Basidiomycotina - Other
Heterobasidion annosum. **2**:W states(3584).
Hymenochaete tabacina. **2**:WA(3202).
Vararia investiens. **2**:WA(3202).
Deuteromycotina - Coelomycetes
Coleophoma empetri. On leaves. **1**:VA(82).
Phyllosticta pachystimae. Graying or silvering of leaves. **2**:WY(3812).

Perrottetia Kunth in Humb., Bonpl. & Kunth
1. *P. sandwicensis* A. Gray
Oomycetes
Phytophthora cinnamomi. **1**:HI(2903).
Ascomycotina
Amazonia stevensii. Black mildew. **1**:HI(1272,1452,2903).
Appendiculella kalalauensis. Black mildew. **1**:HI(1272,1452,2903).
Asteridiella perrottetiae. Black mildew. **1**:HI(1272,1452,2903).
A. waimeana. **1**:HI(1452).

Cephalotaxaceae

Cephalotaxus Siebold & Zucc. ex Endl.
1. *C. harringtonia* (D. Don) K. Koch [Harrington plum yew]
2. *C. harringtonia* (D. Don) K. Koch var. *drupacea* (Siebold & Zucc.) Koidz. (=*C. drupacea* Siebold & Zucc.) [cow's-tail pine, Japanese plum yew, plum fruited yew]
Oomycetes
Phytophthora sp. Root rot. **1**:FL(35).
Deuteromycotina - Coelomycetes
Phomopsis juniperovora. Twig blight. **2**:NY(82).

Cercidiphyllaceae

Cercidiphyllum Siebold & Zucc.
1. *C. japonicum* Siebold & Zucc. [katsura-tree]
Deuteromycotina - Hyphomycetes
Tubercularia sp. Canker. **1**:NY(3755).

Cercropiaceae

Pourouma Aubl.
1. *P. cecropiaefolia* Mart.
Oomycetes
Pythium sp. Root rot. **1**:FL(35).
Deuteromycotina - Hyphomycetes
Fusarium sp. Root rot. **1**:FL(35).

Chenopodiaceae

Allenrolfea Kuntze
1. *A. occidentalis* (S. Wats.) Kuntze
Basidiomycotina - Uredinales
Uredo spirostachydis. **1**:AZ(1203,3981).

Atriplex L.
1. *A. canescens* (Pursh) Nutt. [cenizo, chamiza, four-wing saltbush]
2. *A. confertifolia* (Torr. & Frém.) S. Wats. [shadscale, sheep-fat, spiny saltbush]
3. *A. nuttallii* S. Wats.
4. *A. patula* L. [spreading orache]

5. *A. patula* L. subsp. *hastata* (L.) H. Hall & Clements (=*A. hastata* L., *A. patula* L. var. *hastata* (L.) A. Gray) [halberdleaf orache]
6. *A. patula* L. var. *littoralis* (L.) A. Gray
7. *A. polycarpa* (Torr.) S. Wats. [allscale, desert saltbush]
8. *A. rosea* L. [red orache, redscale]
9. *A. semibaccata* R. Br. [Australian saltbush, berry saltbush]
10. *Atriplex* sp.

Chytridiomycetes
Physoderma pulposum. Leaf gall. **5:**ND(82), TX(82).
Oomycetes
Peronospora farinosa. Downy mildew. **5:**MT(82), TX(82).
Ascomycotina
Eutypella goniostoma. On stems. **1:**NM(82).
Lojkania utahensis. On stems. **2:**UT(82).
Otthia sp. Leaf spot. **10:**FL(35).
Basidiomycotina - Uredinales
[*Puccinia aristidae*]. **5:**CO(82), DE(82), ME(82), ND(82), NY(82), TX(82), UT(82). **8:**AZ(82,1203), CO(82), NM(82), NV(82), UT(82). **10:**CO(82), FL(35), TX(82).
P. subnitens. Heteroecious: aecial host. **5:**ND(378). **8:**AZ(3981).
Uromyces peckianus. Heteroecious: aecial host. **4:**WA(3202).
U. shearianus. Autoecious. **1:**CA(82), NM(82). **2:**CO(82), UT(82), WY(82). **3:**CO(3209). **7:**AZ(82,747,1203,3981). **10:**CA(82).
Deuteromycotina - Hyphomycetes
Cercospora sp. Leaf spot. **10:**FL(35).
Oidium sp. Powdery mildew. **10:**FL(35).
Phymatotrichopsis omnivora. Root rot. **5:**TX(82). **10:**TX(82).
Ramularia dubia. Leaf spot. **4:**MT(3202). **5:**IA(82), MT(82), ND(82,378), NY(82), OH(82), TX(82), WI(82). **10:**IA(82), MT(82), ND(82), NY(82), OH(82), TX(82), WI(82).
Deuteromycotina - Coelomycetes
Microsphaeropsis olivacea. On old stems. **9:**CA(2439).
Phyllosticta chenopodii-albi. **4:**WI(1342).
Stagonospora atriplicis. Leaf spot. **5:**NJ(82), NY(82), PA(64,82). **6:**WI(1334). **10:**KS(82), NE(82).

Bassia All.

1. *B. scoparia* (L.) A. J. Scott (=*Kochia scoparia* (L.) Schrad.) [kochia, summer-cypress]

Oomycetes
Pythium debaryanum. Damping-off. **1:**SD(82).
Basidiomycotina - Uredinales
[*Puccinia aristidae*]. **1:**CO(82), NE(82), TX(82).
Deuteromycotina - Hyphomycetes
Cercospora beticola. **1:**WI(1346).
Phymatotrichopsis omnivora. Root rot. **1:**TX(82).
Deuteromycotina - Coelomycetes
Camarosporium roumeguerii. **1:**WI(1346).
Deuteromycotina - Other
Rhizoctonia solani. Root rot. **1:**TX(82).

Beta L.

1. *B. macrorrhiza* Steven (=*B. vulgaris* L. var. *macrorrhiza* Steven) [mangel]
2. *B. vulgaris* L. [beetroot, common beet, garden beet, mangel, mangold, sea beet, sugarbeet, Swiss chard]
3. *B. vulgaris* L. subsp. *cicla* (L.) W. Koch (=*B. vulgaris* L. var. *cicla* L.) [chard, spinach beet, Swiss chard]
4. *B. vulgaris* L. subsp. *vulgaris* [beetroot, garden beet, mangel, mangold, sugar beet]

Plasmodiophoromycetes
Polymyxa betae. Intracellular parasite in roots. **4:**CA(1015), NE(2025).
Chytridiomycetes
Physoderma leproides. Tumors on crowns and leaves. **4:**CA(3882).
Oomycetes
Aphanomyces cochlioides. Black root. **4:**CA(82), CO(82), IA(82), MI(82), MN(82), MT(82,3202), OH(82), SD(82), WA(82), WI(82).
Peronospora farinosa. Downy mildew. **3:**CA(82). **4:**CA(82), MI(82), NJ(82), NY(82), OR(82,3202), WA(82,3202).
Phytophthora drechsleri. Taproot rot. **4:**CA(82), CO(82), ID(82,3202), OR(82), UT(82).
Pythium aphanidermatum. Damping-off, foot and root rot. **1:**CA(82). **3:**CA(82). **4:**AZ(1574), CA(82), CO(82), SD(82), UT(82), WI(82), WY(82).
P. debaryanum. Damping-off, root rot. **1:**CO(82), CT(82), KS(82), MI(82), NJ(82), UT(82), WI(82). **4:**IA(82), OR(3202), WA(3202).
P. deliense. Root rot. **4:**AZ(3465), TX(3074).
P. dissotocum. Root necrosis. **4:**MI(82).
P. oligandrum. Seedling root rot. **4:**MD(82).
P. ultimum. **4:**CA(82), CO(82), OR(3202), SD(82), WY(82).
Pythium sp. Damping-off, root rot. **4:**Range of host(82), AZ(3071), FL(35), MT(3202).
Zygomycotina
Rhizopus arrhizus. Storage rot, wound rot. **4:**Range of host(82), AZ(3466).
Rhizopus sp. Storage rot, wound rot. **4:**Range of host(82).
Ascomycotina
Botryosphaeria quercuum. Leaf spot. **4:**FL(35).
B. rhodina. Root rot. **3:**FL(82). **4:**AL(82).
Chaetomium globosum. Isolated from seeds. **3:**CA(3306).
C. murorum. Isolated from seeds. **3:**CA(3306). **4:**OH(3306).
C. succineum. Isolated from seeds. **4:**CA(3306).
Erysiphe polygoni. Powdery mildew. **4:**AZ(3072), CA(82,1956,3072,3310), CO(3072), ID(3072), KS(3072), MI(3143), MT(3072), NE(3072), NM(3072), OR(3072), TX(3072), UT(3072), WA(3072,3202), WY(3072).
Mycosphaerella tabifica. **4:**ID(3202), MT(3202), OR(3202), WA(3202).
Physalospora abdita. On leaves. **3:**FL(82).
Sclerotinia sclerotiorum. Soft rot. **4:**CT(82), FL(35), IL(82), MT(3202).
Sclerotinia sp. Crown rot. **1:**WA(82). **3:**MS(82,2720).
Basidiomycotina - Uredinales
[*Puccinia aristidae*]. **4:**CO(82), KS(82), NM(82), UT(82), WA(3202).
Uromyces betae. Autoecious. **1:**OR(82), WA(82). **3:**CA(81,82), OR(82). **4:**AZ(82,1203,3981), CA(81,82), NM(82), OR(82,3202), SD(2243), WA(82,3202).
Uromyces sp. **4:**AZ(3071).
Basidiomycotina - Other
Athelia rolfsii. **4:**OR(3202).
Helicobasidium brebissonii. **4:**CO(82), MT(82,3202), OH(82), TX(82), UT(82), WV(82), WY(82).
Thanatephorus cucumeris. Leaf blight, dry rot canker, damping-off. **4:**W states(82), CO(82), IL(82), MI(82), MN(82), NE(82), WA(3202), WI(82).
Deuteromycotina - Hyphomycetes
Alternaria alternata. Seed discoloration, seed mold. **3:**CA(82). **4:**CA(82).
Alternaria sp. Secondary storage rot, leaf spot. **4:**CA(82), CO(82), FL(35), ID(82), LA(82), MA(82), MI(82), NH(82), NJ(82), NY(82), UT(82), WA(82).
Botrytis cinerea. **4:**MN(426), ND(426,428).
Cercospora beticola. Leaf spot. **1:**Range of host(82). **3:**Range of host(82), HI(2903), OK(2872), WI(1330). **4:**Range of host(82), CO(3378), FL(35,3500), HI(2903), MA(1390), MS(2720), MT(3202), NC(1303), NM(109), OK(2872), OR(3202), SD(2243), TX(109), WA(3202).
Cercospora sp. Leaf spot. **4:**AZ(3071).
Clasterosporium putrefaciens. On leaves. **4:**CO(82), MI(82), WA(82,3202).
Cylindrocarpon destructans. Secondary storage rot. **4:**NY(82).
Cylindrocladium scoparium. **4:**MI(2088).
Epicoccum sp. **4:**OR(3202).
Fusarium avenaceum. Storage rot. **4:**Range of host(82).
F. culmorum. Root rot, storage rot. **4:**Range of host(82).
F. equiseti. Root rot, storage rot. **4:**Range of host(82).
F. flocciferum. Storage rot. **4:**Range of host(82).
F. heterosporum. Storage rot. **4:**MT(82).
F. javanicum. Storage rot. **4:**Range of host(82).
F. merismoides. Root rot. **4:**Range of host(82).
F. oxysporum. Stalk blight, storage rot, leaf wilt. **4:**Range of host(82), CO(82), MT(82), NE(82), NM(82), OR(1369), SD(82), WY(82).
[*F. oxysporum* f. sp. *conglutinans*]. **4:**MT(3202).
F. oxysporum f. sp. *spinaciae*. Wilt. **4:**CO(82), MT(82), NE(82), NM(82), SD(82), WY(82).
F. pallidoroseum. Storage rot. **4:**Range of host(82).
F. roseum. **4:**OR(3202).
F. sambucinum. Storage rot. **4:**Range of host(82).
F. semitectum var. *majus*. Storage rot. **4:**Range of host(82).

F. solani. Root rot. **4:**Range of host(82), OR(3202).
F. trichothecioides. Storage rot. **4:**Range of host(82).
F. ventricosum. Storage rot. **4:**Range of host(82).
Fusarium sp. Seedling blight, root rot. **1:**WA(82). **4:**FL(35), MT(3202), WA(3202).
Gliocladium roseum. Storage rot. **4:**Range of host(82).
Heterosporium betae. Leaf spot. **1:**WA(82). **4:**WA(82,3202).
Macrosporium sp. **4:**MT(3202).
Microdochium dimerum. Root rot, storage rot. **4:**Range of host(82).
Nigrospora sphaerica. **4:**WA(3202).
Penicillium aurantiogriseum. **4:**ND(428).
P. claviforme. **4:**MN(426), ND(426).
P. digitatum. Mold, storage rot. **4:**Range of host(82).
P. funiculosum. **4:**ND(428).
P. puberulum. Mold, storage rot. **4:**Range of host(82).
Phymatotrichopsis omnivora. Root rot. **1:**TX(82). **3:**TX(82). **4:**AZ(82), NM(82), TX(82).
Ramularia beticola. Leaf spot. **3:**WA(82). **4:**CA(82), OR(82,3202), WA(82,3202).
Ramularia sp. Leaf spot. **4:**AZ(3071).
Stemphylium botryosum. Seed discoloration, seed mold. **1:**CA(82), WA(82). **3:**CA(82), WA(82). **4:**CA(82).
Verticillium albo-atrum. **4:**CO(82), OR(3202).
V. dahliae. **4:**WA(2881).
Volutella oxyspora. On roots. **4:**AL(82).

Deuteromycotina - Coelomycetes

Macrophomina phaseolina. Charcoal rot. **4:**CA(82).
Phoma betae. Damping-off, seedling root rot. **1:**MN(82), NY(82), VA(82), WA(82), WI(82). **4:**Range of host(82), FL(35), MN(426), MS(82,2720), MT(82), ND(378,426), OK(2872), OR(427).
Phoma sp. Leaf spot. **4:**AZ(3071).
Septoria betae. Leaf spot. **3:**CA(82). **4:**DE(82), IN(82), MA(82), OH(82), OK(2872).

Deuteromycotina - Other

Rhizoctonia solani. Damping-off, black rot, crown, root and stem rot. **1:**WA(82). **3:**NC(1303), NY(82). **4:**CO(12), FL(35), MS(2720), NC(1303), NE(2029), WY(819).
R. zeae. Root rot. **4:**NC(1303).
Rhizoctonia sp. Leaf blight, late canker, damping-off, crown and root rot. **4:**Range of host(82), AZ(3071), HI(2903), MT(3202).
Sclerotium rolfsii. Southern blight, sclerotial rot. **1:**LA(82), TX(82). **3:**LA(82), SC(82). **4:**SE states(82), AZ(82,2886), CA(82,2886), FL(35), GA(82), IA(82), LA(82), NC(1303), TX(82).
Sclerotium sp. Root rot. **4:**AZ(3071).

Ceratoides Gagnebin

This genus is considered a synonym of *Axyris*, but the appropriate combination could not be located.

1. *C. lanata* (Pursh) J. T. Howell (=*Eurotia lanata* (Pursh) Moq.) [winterfat]

Ascomycotina

Didymella eurotiae. On stems. **1:**NM(82).

Basidiomycotina - Uredinales

Puccinia burnettii. **1:**CO(82), MT(82,3202), NM(82), WA(82,3202), WY(82).
P. eurotiae. **1:**CA(82), NV(82), UT(82).

Deuteromycotina - Coelomycetes

Phoma exigua. On stems. **1:**NM(82).

Chenopodium L.

1. *C. album* L. [lamb's-quarters, white goosefoot]
2. *C. ambrosioides* L. (=*C. ambrosioides* L. var. *anthelminticum* (L.) A. Gray, *C. anthelminticum* L.) [American wormseed, Mexican tea, Spanish tea, wormseed]
3. *C. berlandieri* Moq.
4. *C. bonus-henricus* L. [allgood, fat-hen, good-King-Henry, mercury, perennial goosefoot, wild spinach]
5. *C. botrys* L. [feather-geranium, Jerusalem-oak]
6. *C. capitatum* (L.) Aschers. [Indian-paint, strawberry-blite]
7. *C. chenopodioides* (L.) Aellen
8. *C. gigantospermum* Aellen
9. *C. glaucum* L.
10. *C. hians* Standl.
11. *C. hybridum* L.
12. *C. incanum* (S. Wats.) A. Heller
13. *C. leptophyllum* (Moq.) S. Wats.
14. *C. murale* L. [nettleleaf goosefoot]
15. *C. rubrum* L. [red goosefoot]
16. *C. salinum* Standl.
17. *C. sandwiceum* Moq.
18. *C. urbicum* L.
19. *Chenopodium* sp.

Chytridiomycetes

Physoderma pulposum. Leaf gall. **9:**SD(82), TX(82). **15:**SD(2243).

Oomycetes

Albugo bliti. White rust. **1:**MS(82), TX(82).
A. occidentalis. White rust. **2:**MT(82). **6:**CO(82), IA(82), WY(82,1195). **15:**MT(3202).
Aphanomyces cochlioides. Root rot of seedlings. **1:**IA(82).
Peronospora effusa. Downy mildew. **1:**NE states(82), Range of host(82), HI(2899,2903), KS(82), MA(1390), MS(82,2720), ND(377), OK(82,2872), SD(2243), TX(82), WA(82), WY(1195). **11:**IA(82), IL(82), KS(82), NE(82), NJ(82), SD(82), WI(82). **19:**FL(35).
P. farinosa. Downy mildew. **1:**NE states(82), Range of host(82), AZ(1195,3381), ID(3202), KS(82), MS(82), MT(3202), OK(82), OR(3202), TX(82), WA(3202), WA(82), WY(1195). **8:**WA(3202). **9:**KS(82). **11:**IA(82). **13:**KS(82), WA(3202).
Peronospora sp. Downy mildew. **1:**NC(1303). **19:**OK(2872).

Ascomycotina

Diaporthe arctii. On stems. **2:**GA(82,1438).
Dothidella longissima. On stems. **1:**IA(82), KS(82).
Leptosphaeria anthelmintica. On stems. **1:**NY(82). **9:**ND(82,377).
L. macrospora. **1:**NY(3249).
Mycosphaerella chenopodii. On stems. **13:**KS(82).
Ophiobolus collapsus. **1:**ND(377).
Pleospora calvescens. On stems. **1:**NY(195,270). **2:**NJ(82), NY(82), WV(82).
Sclerotinia sclerotiorum. Root rot. **1:**MT(82,3202).

Basidiomycotina - Uredinales

[*Puccinia aristidae*]. **1:**W states(82), AZ(82,1203), IN(82), MT(82,3202), OK(2872), WA(3202). **8:**WA(3202). **9:**AZ(82), CA(82), CO(82), MT(82), NM(82), TX(82). **11:**CA(82), CO(82), NE(82). **12:**AZ(1203). **13:**AZ(82,1203), CA(82), CO(82), NE(82), NM(82), UT(82). **14:**AZ(1203). **15:**AZ(82,1203), CA(82), CO(82), NM(82), WY(82). **16:**MT(3202).
P. subnitens. **1:**AZ(3981). **7:**AZ(3981). **12:**AZ(3981). **13:**AZ(3981). **14:**AZ(3981).
Uromyces giganteus. Autoecious. **19:**FL(35).
U. peckianus. Heteroecious: aecial host. **1:**CA(82). **14:**CA(82).

Basidiomycotina - Ustilaginales

Glomosporium leptideum. **10:**CO(1042).

Basidiomycotina - Other

Thanatephorus cucumeris. **1:**WA(3202).

Deuteromycotina - Hyphomycetes

Alternaria amaranthi. Leaf spot. **1:**TX(82).
Cercospora beticola. Leaf spot. **1:**IA(82). **5:**WI(1356). **11:**OK(2872), WI(82,1335).
C. chenopodii. **1:**DE(3543), NC(3944), ND(378). **15:**ND(378). **18:**DE(3543).
Cercospora sp. Leaf spot. **1:**HI(2903).
Cristulariella moricola. On leaves. **2:**IL(2572).
Fusarium pallidoroseum. On stems. **2:**MD(82).
Fusisporium pallido-roseum. **2:**SC(634).
Hadrotrichum chenopodinum. On leaves. **1:**NY(82).
Helminthosporium sp. Leaf spot. **19:**FL(35).
Phymatotrichopsis omnivora. Root rot. **1:**TX(82). **2:**TX(82).
Ramularia dubia. Leaf spot. **1:**E states(82), MT(3202), NC(1303), ND(82), OK(2872), SD(2243), TX(82), WA(3202). **3:**WI(1356). **6:**TX(82), WI(82). **11:**IA(82), KS(82), OK(82,2872), SD(2243), TX(82), WI(82). **13:**KS(82), NE(82), OK(2873). **15:**ND(82), SD(2243). **19:**FL(35).

Deuteromycotina - Coelomycetes

Ascochyta hyalospora. On stems. **1:**KS(82), NY(82).
Colletotrichum truncatum. Isolated from leaves and stems. **1:**IL(1482).
Colletotrichum sp. Anthracnose. **19:**FL(35).
Hendersonia bliti. Leaf spot. **6:**CO(82).
Lasiodiplodia theobromae. On stems. **1:**AL(82).
Leptostromella chenopodii. On stems. **1:**NY(82).

Leptothyrium chenopodii. On stems. **11**:KS(82).
Macrophomina phaseolina. Charcoal rot. **1**:IL(311).
Phoma exigua. Leaf spot. **1**:PA(82).
Phyllosticta ambrosioides. Leaf spot. **2**:DE(82), IL(82), IN(82), LA(82), MS(2720), NJ(82), NY(82).
P. chenopodii. **1**:Unk.(82), NY(64).
P. chenopodii-albi. **1**:WI(1332). **11**:WI(1350).
Phyllosticta sp. Leaf spot. **19**:FL(35).
Stagonospora atriplicis. Leaf spot. **1**:Range of host(82), ID(3202), NC(3944), WI(64,1355). **2**:OK(2873). **4**:NY(82). **6**:MI(82), MN(82), WI(82). **11**:NE(82), WI(1342). **15**:ND(82). **17**:HI(2784). **19**:FL(35), WI(64).
S. chenopodii. **15**:ND(378).
Deuteromycotina - Other
Rhizoctonia solani. Thread blight, damping-off, root rot. **1**:WA(82). **13**:KS(82). **19**:FL(35).

Cycloloma Moq.
1. *C. atriplicifolium* (Spreng.) J. M. Coulter [winged pigweed]
Deuteromycotina - Hyphomycetes
Cercospora beticola. **1**:WI(1329).

Monolepis Schrad.
1. *M. nuttalliana* (Schutt) Greene
Oomycetes
Peronospora effusa. Downy mildew. **1**:OK(2872).
Basidiomycotina - Uredinales
[*Puccinia aristidae*]. **1**:MT(3202).

Salicornia L.
1. *S. europaea* L. [chicken-claws, pigeon-foot, samphire]
2. *S. perennis* Mill.
3. *S. virginica* L.
4. *Salicornia* sp.
Ascomycotina
Mycosphaerella salicorniae. On stems. **1**:CT(82).
Pleospora gaudefroyi. On dead standing plants. **4**:MA(1170).
P. herbarum. On stems. **1**:MA(82). **2**:TX(82).
Pleospora sp. On dead standing plants. **4**:MA(1170).
Protomyces conglomeratus. Stem gall. **1**:NY(82).
Basidiomycotina - Uredinales
Uromyces peckianus. Heteroecious: aecial host. **2**:CA(82), TX(82), WA(82). **3**:WA(3202).
Deuteromycotina - Hyphomycetes
Alternaria sp. On dead standing plants. **4**:MA(1170).
Deuteromycotina - Coelomycetes
Camarosporium palliatum. On dead leaves and stems. **3**:NC(1945).
C. roumeguerii. On dead standing plants. **4**:MA(1170).

Salsola L.
1. *S. australis* R. Br. (=*S. kali* L. var. *tenuifolia* Tausch) [Russian-thistle, tumbleweed]
2. *S. iberica* Sennen & Pau (=*S. kali* of American authors, *S. kali* L. subsp. *ruthenica* (Iljin) Sóo, *S. pestifer* A. Nels.) [Russian-thistle, saltwort, tumbleweed]
3. *S. tragus* Torner ex L.
4. *Salsola* sp.
Oomycetes
Pythium debaryanum. Root rot. **2**:MN(82).
Ascomycotina
Arnium apiculatum. **2**:SD(3217).
Ciliolea coronata. **4**:ND(3817).
Graphyllium ipomoeae. On stems. **2**:ND(82).
Melanospora zamiae. On stems. **2**:SD(82).
Pleospora lecanora. On stems. **2**:ND(82). **3**:ND(377).
Podospora curvicolla. **2**:MT(3202).
Basidiomycotina - Uredinales
[*Puccinia aristidae*]. **2**:AZ(82,1203), CO(82), NE(82), NM(82), TX(82), UT(82).
P. subnitens. **2**:AZ(3981).
Deuteromycotina - Hyphomycetes
Phymatotrichopsis omnivora. Root rot. **2**:AZ(82), TX(82).
Deuteromycotina - Coelomycetes
Camarosporium roumeguerii. **1**:WI(1344).

Sarcobatus Nees in Wied.-Neuw.
1. *S. vermiculatus* (Hook.) Torr. [greasewood]
Ascomycotina
Eutypella sarcobati. **1**:CO(2925).
Strickeria mammoides. **1**:OR(3202).
Basidiomycotina - Uredinales
[*Puccinia aristidae*]. **1**:ID(3202), MT(3202), OR(3202), WA(3202), WY(1195).
P. cryptandri var. *luxurians*. **1**:CA(753), CO(753).
P. subnitens. **1**:NV(1468).
Deuteromycotina - Coelomycetes
Hendersonia heterophragmia. On dead twigs. **1**:MT(966), OR(3202).
Septoria sarcobati. **1**:MT(3202).

Spinacia L.
1. *S. oleracea* L. [spinach]
Chytridiomycetes
Olpidium brassicae. On roots. **1**:CT(2972).
Oomycetes
Albugo occidentalis. White rust. **1**:AR(82), LA(2825), OK(82,2872), TX(82).
Albugo sp. **1**:HI(2903).
Aphanomyces cladogamus. Root rot. **1**:NJ(82), VA(82).
Peronospora effusa. Downy mildew, powdery mildew. **1**:Range of host(82), CA(4018), FL(35), HI(2903), OK(2872), SD(2243).
P. farinosa. **1**:OR(3202), WA(3202).
Phytophthora drechsleri. Root rot. **1**:HI(82,2903).
P. megasperma. **1**:CA(82), NC(82), NY(82).
Phytophthora sp. Crown rot. **1**:AZ(82), IN(82), NJ(82), NY(82).
Pythium anandrum. Preemergence damping-off, seed decay, root rot. **1**:C states(82), NE states(82), CA(82).
P. aphanidermatum. Preemergence damping-off, seed decay, root rot. **1**:C states(82), NE states(82), AZ(199), CA(82).
P. debaryanum. Preemergence damping-off, seed decay, root rot. **1**:C states(82), NE states(82), CA(82).
P. dissotocum. Damping-off, seed decay, root rot. **1**:AZ(199), VA(82).
P. helicoides. Preemergence damping-off, seed decay, root rot. **1**:C states(82), NE states(82), CA(82).
P. irregulare. Preemergence damping-off, seed decay, root rot. **1**:C states(82), NE states(82), CA(82).
P. polytylum. Preemergence damping-off, seed decay, root rot. **1**:C states(82), NE states(82), CA(82).
P. rostratum. Preemergence damping-off, seed decay, root rot. **1**:CA(82).
P. salpingophorum. Preemergence damping-off, seed decay, root rot. **1**:VA(82).
P. ultimum. Preemergence damping-off, seed decay, root rot. **1**:CA(82).
P. vexans. Preemergence damping-off, seed decay, root rot. **1**:HI(82,2903).
Ascomycotina
Sclerotinia sclerotiorum. Crown rot. **1**:GA(82), NY(82).
Basidiomycotina - Uredinales
[*Puccinia aristidae*]. **1**:AZ(82,1203), CA(82), CO(82,2657,2916), OR(82,3202), WA(82,3202).
P. subnitens. **1**:AZ(3981).
Basidiomycotina - Ustilaginales
Entyloma ellisii. Leaf smut. **1**:NJ(82), WA(3980).
Basidiomycotina - Other
Thanatephorus cucumeris. **1**:WA(3202).
Deuteromycotina - Hyphomycetes
[*Alternaria cheiranthi*]. Leaf spot. **1**:OK(2872).
Alternaria sp. Leaf spot. **1**:CT(82), FL(35), OH(82), SC(82).
Cercospora beticola. Leaf spot. **1**:Unk.(82), FL(35).
Cercospora sp. Leaf spot. **1**:HI(2903), OK(2872).
Cladosporium macrocarpum. On leaves. **1**:DE(82), FL(35), OK(82,2873), PA(82), TX(82).
C. subnodosum. **1**:DE(82), SC(82).
C. variabile. Leaf mold. **1**:Range of host(82), ID(3202), MT(3202), OK(2872), WA(3202).
Curvularia inaequalis. Seed mold. **1**:Unk.(82).
Fusarium oxysporum f. sp. *spinaciae*. Crown and root rot. **1**:AR(198), CA(82), CT(82), ID(82,3202), MD(82), NY(82), TX(82), VA(82), WA(3202).
F. solani. Root rot. **1**:TX(82).

Heterosporium sp. **1**:OR(3202), WA(3202).
Phymatotrichopsis omnivora. Root rot. **1**:TX(82).
Stemphylium botryosum. Seed mold. **1**:Range of host(82), FL(35).
Verticillium sp. Wilt. **1**:NY(82).
Deuteromycotina - Coelomycetes
Colletotrichum dematium f. *spinaciae*. Anthracnose. **1**:CT(82), LA(82), MS(82,2720), NJ(82), NY(82), TX(82), VA(82).
Hendersonia spinaciae. Leaf spot. **1**:SD(82).
Phoma terrestris. Secondary root rot. **1**:IA(82).
Phyllosticta chenopodii. Leaf spot. **1**:DE(82), NJ(82), NY(82), VA(82).
Pyrenochaeta lycopersici. **1**:CA(1371).
Deuteromycotina - Other
Rhizoctonia solani. Damping-off. **1**:Range of host(82), FL(35).
Rhizoctonia sp. Crown and root rot. **1**:HI(2903).

Suaeda Forssk. ex Scop.
1. *S. californica* S. Wats.
2. *S. depressa* (Pursh) S. Wats. (=*Dondia depressa* (Pursh) Britton)
3. *S. occidentalis* S. Wats.
4. *S. suffrutescens* S. Wats.
5. *S. torreyana* S. Wats.
Ascomycotina
Metasphaeria anisometra. **2**:ND(377).
Pleospora calvescens. **2**:ND(377).
Basidiomycotina - Uredinales
Uromyces giganteus. Autoecious. **1**:AZ(1203). **3**:ID(3202). **4**:TX(742). **5**:AZ(747,3981).

Chrysobalanaceae

Chrysobalanus L.
1. *C. icaco* L. [coco-plum, icaco]
Oomycetes
Pythium aphanidermatum. Root rot. **1**:FL(35).
Ascomycotina
Glomerella cingulata. Fruit rot. **1**:FL(35).
Deuteromycotina - Hyphomycetes
Cercospora chrysobalani. Leaf spot. **1**:FL(35,82).
[*Fusarium oxysporum* f. sp. *tracheiphilum*]. Wilt. **1**:FL(35).
Deuteromycotina - Coelomycetes
Colletotrichum gloeosporioides. Anthracnose. **1**:FL(35).
[*Pestalotiopsis funerea*]. Leaf spot. **1**:FL(35).
Phomopsis sp. Leaf spot. **1**:FL(35).
Phyllosticta sp. Leaf spot. **1**:FL(35).

Licania Aubl.
1. *L. michauxii* Prance (=*Chrysobalanus oblongifolius* Michx.)
2. *L. rigida* Benth. [oiticica]
Deuteromycotina - Hyphomycetes
Cercospora sp. Leaf spot. **1**:FL(35).
Deuteromycotina - Coelomycetes
Phyllosticta sp. Leaf spot. **2**:FL(35).

Cistaceae

Cistus L.
1. *C.* × *cyprius* Lam. *C. ladanifer* × *C. laurifolius*.
2. *C. palhinhai* Ingram [St.-Vincent cistus]
3. *C. populifolius* L.
4. *C.* × *purpureus* Lam. *C. ladanifer* × *C. villosus*.
Basidiomycotina - Other
Armillaria mellea. Root rot. **2**:CA(2897). **3**:CA(2897). **1**:CA(2900).
Deuteromycotina - Hyphomycetes
Verticillium albo-atrum. **4**:CA(3896).

Helianthemum Mill.
1. *H. canadense* (L.) Michx. (=*Crocanthemum canadense* Britton) [frostwort]
2. *H. nummularium* (L.) Mill. [sun-rose]
Deuteromycotina - Hyphomycetes
Phymatotrichopsis omnivora. Root rot. **2**:TX(82).
Deuteromycotina - Coelomycetes
Cylindrosporium eminens. Leaf spot. **1**:WI(82). **2**:TX(82), WI(82).
Septoria chamaecisti. Leaf spot. **2**:WA(82).

Hudsonia L.
1. *H. tomentosa* Nutt. [poverty-grass]
Oomycetes
Phytophthora cinnamomi. Root rot. **1**:NC(1303).
Ascomycotina
Melanomma sp. On hairy branchlets. **1**:WI(1357).
Plagiostoma conradii. On stems. **1**:NJ(82).

Lechea L.
1. *L. intermedia* Legg. ex Britton
2. *L. stricta* Legg.
3. *L. tenuifolia* Michx.
Deuteromycotina - Hyphomycetes
Cercospora lecheae. On leaves. **1**:WI(1334). **2**:WI(1357). **3**:WI(1346).

Clethraceae

Clethra L.
1. *C. acuminata* Michx.
2. *C. alnifolia* L. [summer-sweet, sweet pepperbush]
3. *Clethra* sp.
Chytridiomycetes
Synchytrium vaccinii. Red leaf gall. **2**:NJ(82). **3**:NJ(82).
Ascomycotina
Botryosphaeria obtusa. On limbs. **2**:NJ(82).
Cryptosporella sp. On limbs. **2**:MD(82). **3**:MD(82).
Dasyscyphus subauratus. On leaves. **2**:NJ(849).
Dermea crypta. On limbs. **2**:NJ(82).
Diaporthe tuberculosa var. *corymbosa*. On limbs. **2**:MD(82), NY(82).
Ditopellopsis clethrae. On dead limbs. **2**:MD(2947).
Godronia urceolata. **1**:NC(128). **3**:GA(1382), NC(1382), NJ(1382), NY(1382).
G. urceolus. **1**:GA(1438).
Nectria ochroleuca. **2**:NJ(3085).
Valsa ambiens. **2**:NJ(3415).
Basidiomycotina - Other
Scytinostroma galactinum. Root rot. **2**:MD(82).
Deuteromycotina - Hyphomycetes
Phymatotrichopsis omnivora. Root rot. **2**:TX(82).
Deuteromycotina - Coelomycetes
Kaleidosporium fenestratum. On dead twigs. **2**:NJ(3776).
Phyllosticta clethrae. On leaves. **2**:MA(82), NC(1303).
P. clethricola. Leaf spot. **2**:MD(82), NJ(2,82), TX(82).
Sphaeronaema clethrincola. On limbs. **2**:NJ(959). **3**:NY(82).

Clusiaceae

Calophyllum L.
1. *C. inophyllum* L. [Alexandrian laurel, ati, Indian laurel, laurelwood, tamanu, tamono]
2. *Calophyllum* sp.
Deuteromycotina - Coelomycetes
Phyllosticta sp. Leaf spot. **1**:FL(35). **2**:FL(35).

Clusia L.
1. *C. rosea* Jacq. [balsam-apple, copey, cupey, Scotch-attorney]
Oomycetes
Pythium sp. Root rot. **1**:FL(35).
Deuteromycotina - Hyphomycetes
Alternaria sp. Leaf spot. **1**:FL(35).
Deuteromycotina - Coelomycetes
Ascochyta sp. Leaf spot. **1**:FL(35).
Phomopsis sp. Witches' brooms. **1**:FL(35).
Phyllosticta sp. Leaf spot. **1**:FL(35).
Deuteromycotina - Other
Rhizoctonia solani. Root rot. **1**:FL(35).

Garcinia L.
1. *G. mangostana* L. [mangosteen]
2. *Garcinia* sp.
Ascomycotina
Glomerella cingulata. Leaf spot. **1**:MD(82).
Deuteromycotina - Hyphomycetes
Cylindrocladiella camelliae. Root rot. **2**:FL(35).

Hypericum L.
1. *H. adpressum* Barton
2. *H. anagalloides* Cham. & Schlechtend.
3. *H. ascyron* L.
4. *H. calycinum* L. [Aaron's-beard, creeping St.-John's-wort, goldenflower, rose-of-Sharon]
5. *H. canadense* L.
6. *H. formosum* H. B. K.
7. *H. formosum* H. B. K. var. *scouleri* (Hook.) J. M. Coulter (=*H. scouleri* Hook.)
8. *H. frondosum* Michx.
9. *H. gentianoides* (L.) B. S. P. [orangegrass, pineweed]
10. *H. graveolens* Buckley
11. *H. hypericoides* (L.) Crantz (=*Ascyrum hypericoides* L.) [St.-Andrew's-cross]
12. *H. kalmianum* L. [kalms St.-John's-wort]
13. *H. mitchellianum* Rydb. [Blue Ridge St.-John's-wort]
14. *H. mutilum* L.
15. *H. perforatum* L. [common St.-John's-wort, klamathweed]
16. *H. prolificum* L. [broombrush, shrubby St.-John's-wort]
17. *H. punctatum* Lam. [spotted St.-John's-wort]
18. *H. sphaerocarpum* Michx.
19. *H. stans* (Michx.) Adams & E. Robson (=*Ascyrum stans* Michx.) [St.-Peter's-wort]
20. *H. virginicum* L. [marsh St.-John's-wort]
21. *Hypericum* sp.
Ascomycotina
Botryosphaeria ribis. On stems. **16**:PA(82). **20**:Unk.(82).
Didymella prominens. On stems. **17**:GA(1438). **21**:GA(82).
Erysiphe cichoracearum. Powdery mildew. **21**:Unk.(82).
Lachnum morthieri. **14**:GA(1438).
[*Leptosphaeria heliopsidis*]. Black knot on stems. **21**:MD(82).
Lophiostoma caulium. **17**:GA(1438).
Mollisia atrata. **9**:GA(1438).
Mycosphaerella hypericina. On leaves. **16**:NJ(967).
Mycosphaerella sp. On leaves. **12**:NJ(82). **15**:NJ(82). **21**:IL(82).
Rosellinia necatrix. Root rot. **4**:CA(2297).
Basidiomycotina - Uredinales
Melampsora hypericorum. **6**:AZ(3981), MT(3202). **7**:MT(82).
Uromyces triquetrus. Autoecious. **1**:MA(1394). **2**:CA(81). **5**:WI(1357). **6**:OR(3202). **7**:CA(81,638). **8**:NC(82). **11**:MS(82,2720), NJ(82), TX(82). **12**:IL(82), NY(82), WI(82). **14**:E states(82), GA(1439), IA(82), MS(2720), NC(1303). **15**:CA(642), MA(1390), ME(82), NC(1303,1307), NY(82), PA(82). **16**:E states(82), IN(987), NC(1303), TX(82), WI(82). **17**:GA(1439). **18**:WI(1348). **20**:E states(82), FL(35), MA(1394), MO(82), WI(82). **21**:E states(82), CA(82), FL(35), IN(987), MO(82), WI(82).
Basidiomycotina - Other
Sebacina incrustans. Encrusting stems. **16**:VA(128).
Typhula micans. On stems. **15**:CA(81).
Deuteromycotina - Hyphomycetes
Cercospora hyperici. Leaf spot. **1**:IL(82,581). **18**:WI(1344).
Cercospora sp. On leaves. **3**:WI(1349).
Cladosporium gloeosporioides. Leaf spot. **14**:AL(82), NJ(82), WI(82). **19**:AL(82), MS(2720). **20**:NY(82), WI(82).
Fusicladium ascyrinum. On floral bracts. **11**:LA(82).
Deuteromycotina - Coelomycetes
Colletotrichum sp. Stem anthracnose. **10**:NC(1303). **13**:NC(1303). **15**:NC(1303). **17**:NC(1303).
Leptostroma hyperici. On stems. **15**:PA(82).
Pestalotia sp. On leaves. **12**:WI(1333).
Rhabdospora hyperici. **9**:WI(1357).
R. sphaerelloides. On stems. **14**:IA(82). **21**:KS(82), OH(82), WA(82).

Combretaceae

Bucida L.
1. *B. buceras* L. [black-olive]
Oomycetes
Pythium splendens. Root rot. **1**:FL(35).
Ascomycotina
Botryosphaeria sp. Leaf spot. **1**:FL(35).
Basidiomycotina - Other
Armillaria tabescens. Root rot. **1**:FL(35).
Deuteromycotina - Hyphomycetes
Cristulariella depraedans. Leaf spot. **1**:FL(35).
Cylindrocladium scoparium. Leaf spot. **1**:FL(35).
Helminthosporium sp. Leaf spot. **1**:FL(35).
Deuteromycotina - Coelomycetes
Colletotrichum sp. Anthracnose. **1**:FL(35).
Coniothyrium sp. Leaf spot. **1**:FL(35).
Fusicoccum sp. Leaf spot. **1**:FL(35).
Lasiodiplodia theobromae. Dieback, stem gall. **1**:FL(35).
Phomopsis sp. Twig dieback. **1**:FL(35).
Phyllosticta sp. Leaf spot. **1**:FL(35).
Sphaceloma bucidae. Spot anthracnose. **1**:FL(3788).
Sphaceloma sp. Scab. **1**:FL(35).
Sphaeropsis sp. Canker. **1**:FL(35).
Deuteromycotina - Other
Rhizoctonia solani. Root rot, basal stem rot. **1**:FL(35).

Conocarpus L.
1. *C. erectus* L. [button-mangrove, buttonwood]
2. *C. erectus* L. var. *sericeus* DC. [silver buttonwood, silver-tree]
Oomycetes
Pythium sp. Root rot. **1**:FL(35).
Deuteromycotina - Hyphomycetes
Cercospora conocarpi. Leaf spot. **1**:FL(35).
Cylindrocladium scoparium. Leaf blight. **1**:FL(35). **2**:FL(2413).
Deuteromycotina - Coelomycetes
Colletotrichum sp. Anthracnose. **1**:FL(35).
Diplodia sp. Twig dieback. **1**:FL(35).
Phyllosticta sp. Leaf spot. **1**:FL(35).
Deuteromycotina - Other
Rhizoctonia sp. Root rot. **1**:FL(35).

Laguncularia C. F. Gaertn.
1. *L. racemosa* (L.) C. F. Gaertn. [white buttonwood, white mangrove]
Oomycetes
Pythium sp. Seedling root rot. **1**:FL(35).
Deuteromycotina - Hyphomycetes
Cercospora sp. Leaf spot. **1**:FL(35).
Deuteromycotina - Coelomycetes
Colletotrichum sp. Leaf necrosis. **1**:FL(35).
Septoria sp. Leaf spot. **1**:FL(35).

Quisqualis L.
1. *Q. indica* L. [Rangoon creeper]
Deuteromycotina - Hyphomycetes
Alternaria sp. Leaf spot. **1**:FL(35).
Cercospora sp. Leaf spot. **1**:FL(35).

Terminalia L.
1. *T. arjuna* (Roxb. ex DC.) Wight & Arnott
2. *Terminalia* sp.
Ascomycotina
Mollisia petiolorum. **2**:HI(2903).
Deuteromycotina - Hyphomycetes
Cercospora sp. Leaf spot. **1**:FL(35).
Deuteromycotina - Coelomycetes
Pestalotiopsis disseminata. Leaf spot. **1**:FL(35,82).
Phomopsis sp. Leaf spot. **1**:FL(35,82).
Phyllosticta sp. Leaf spot. **1**:FL(35,82).
Septoria sp. Leaf spot. **1**:FL(35).
Deuteromycotina - Other
Rhizoctonia sp. Leaf necrosis. **1**:FL(35).

Commelinaceae

Callisia Loefl.
1. *C. fragrans* (Lindl.) Woodson
2. *C. repens* L.
3. *Callisia* sp.

Oomycetes
Pythium sp. Root rot. **2**:FL(35).
Deuteromycotina - Hyphomycetes
Myrothecium sp. Basal stem rot. **2**:FL(35).
Deuteromycotina - Coelomycetes
Colletotrichum sp. Leaf spot. **1**:FL(35). **2**:FL(35). **3**:FL(35).
Deuteromycotina - Other
Rhizoctonia sp. Leaf necrosis. **2**:FL(35).

Cochliostema Lem.
1. *Cochliostema* sp.

Oomycetes
Phytophthora sp. Throat rot. **1**:FL(35).

Commelina L.
1. *C. communis* L. [dayflower]
2. *C. diffusa* N. L. Burm. (=*C. nudiflora* Davidson) [creeping dayflower]
3. *C. virginica* L. [Virginia dayflower]
4. *Commelina* sp.

Oomycetes
Pythium arrhenomanes. Root rot. **2**:HI(82,2903).
Ascomycotina
Mycosphaerella tetraspora. Leaf spot. **4**:TX(82).
Phyllachora sp. Leaf spot. **4**:FL(35).
Basidiomycotina - Uredinales
Phakopsora tecta. **2**:HI(1145,2903).
Puccinia commelinae. Telial host. **4**:TX(82).
Uromyces commelinae. Heteroecious: telial host. **2**:HI(2903). **4**:FL(35,82), TX(82).
Basidiomycotina - Ustilaginales
Ustilago commelinae. Smut. **1**:NY(922).
Basidiomycotina - Other
Thanatephorus cucumeris. Banded sclerotial disease. **4**:LA(1010).
Deuteromycotina - Hyphomycetes
Alternaria sp. Leaf spot. **4**:FL(35).
Cercospora nudiflorae. Leaf spot. **4**:FL(35).
Cristulariella moricola. On leaves. **2**:IL(2572).
Fusarium sp. Basal stem rot. **4**:FL(35).
Illosporium commelinae. Leaf spot. **3**:TX(82). **4**:TX(82).
Periconia byssoides. **2**:DE(3543).
Pyricularia grisea. Leaf spot. **4**:FL(35).
Stemphylium globuliferum. Leaf spot. **4**:MD(651).
Deuteromycotina - Coelomycetes
Colletotrichum gloeosporioides. Leaf spot. **4**:FL(35,82), TX(82).
Phyllosticta commelinicola. Leaf spot. **2**:OK(82), TX(82).
Phyllosticta sp. Leaf spot. **3**:OK(2872). **4**:FL(35).

Cyanotis D. Don
1. *C. kewensis* C. B. Clarke [teddy-bear-plant, teddy-bear-vine]

Oomycetes
Pythium sp. Root rot. **1**:FL(35).
Deuteromycotina - Hyphomycetes
Alternaria sp. Leaf spot. **1**:FL(35).
Fusarium sp. Basal stem rot. **1**:FL(35).
Deuteromycotina - Other
Rhizoctonia sp. Root rot. **1**:FL(35).

Dichorisandra J. C. Mikan
1. *Dichorisandra* sp.

Deuteromycotina - Hyphomycetes
Cercospora sp. Leaf spot. **1**:FL(35).
Cylindrocladium sp. Crown rot. **1**:FL(35).
Deuteromycotina - Coelomycetes
Colletotrichum falcatum. Anthracnose. **1**:FL(35).
Phyllosticta sp. Leaf spot. **1**:FL(35).
Deuteromycotina - Other
Sclerotium rolfsii. Southern blight. **1**:FL(35).

Siderasis Raf.
1. *S. fuscata* (Lodd.) H. Moore

Deuteromycotina - Hyphomycetes
Alternaria sp. Leaf spot. **1**:FL(35).
Deuteromycotina - Other
Rhizoctonia solani. Damping-off. **1**:FL(35).

Tradescantia L.
(= *Rhoeo* Hance ex Walp., *Setcreasea* K. Schum. & Syd. in K. Schum.)
1. *T. fluminensis* Vell. [wandering-jew]
2. *T. pallida* (J. Rose) D. Hunt (=*Setcreasea pallida* J. Rose)
3. *T. spathacea* Sw. (=*Rhoeo discolor* (L'Hér.) Hance, *Rhoeo spathacea* (Sw.) Stearn) [boat-lily, man-in-a-boat, Moses-in-a-boat, Moses-in-the-bulrushes, Moses-on-a-raft, oysterplant, purple-leaved spiderwort, three-men-in-a-boat]
4. *T. subaspera* Ker-Gawl.
5. *T. virginiana* L. [common spiderwort, widow's-tears]
6. *Tradescantia* sp.

Chytridiomycetes
Cladochytrium replicatum. On leaves. **1**:NY(82).
Oomycetes
Phytophthora sp. Root and stem rot. **3**:FL(35).
Pythium sp. Root rot. **1**:HI(2903). **3**:FL(35). **6**:FL(35).
Basidiomycotina - Uredinales
Uromyces commelinae. Heteroecious: telial host. **6**:TX(82).
Deuteromycotina - Hyphomycetes
Alternaria sp. Leaf spot. **1**:FL(35). **3**:FL(35). **6**:FL(35).
Botrytis cinerea. Gray mold leaf blight. **1**:AK(82). **6**:FL(35).
Cercospora sp. Leaf spot. **1**:FL(35).
Curvularia eragrostidis. Tan leaf spot. **3**:FL(35,2411).
Curvularia sp. Leaf spot. **2**:FL(35).
Fusarium sp. Root rot. **6**:FL(35).
Helminthosporium sp. Leaf spot. **3**:FL(35).
Myrothecium sp. Leaf spot. **6**:FL(35).
Ramularia sp. Leaf spot. **6**:FL(35).
Deuteromycotina - Coelomycetes
Colletotrichum sp. Anthracnose, crown rot, leaf spot. **1**:FL(35), TX(82). **3**:FL(35). **5**:NJ(82). **6**:FL(35).
Phyllosticta sp. Leaf spot. **3**:FL(35). **4**:WI(1354). **6**:FL(35).
Septoria tradescantiae. Leaf spot. **5**:C states(82), SD(82,2243). **6**:C states(82), IA(82), SD(82).
Deuteromycotina - Other
Rhizoctonia solani. Leaf blight, basal stem rot. **3**:FL(35). **6**:FL(35).
Rhizoctonia sp. Root rot. **1**:NC(1303).

Tripogandra Raf.
1. *T. multiflora* (Sw.) Raf. [fern-leaf inchplant]

Oomycetes
Pythium splendens. Root rot. **1**:FL(35).
Deuteromycotina - Other
Rhizoctonia sp. Root rot. **1**:FL(35).

Zebrina Schnizl.
This genus is considered by some to be a synonym of *Tradescantia*, but it is retained here as a separate genus.
1. *Z. pendula* Schnizl. [inchplant, wandering-jew]
2. *Zebrina* sp.

Oomycetes
Pythium aphanidermatum. Root rot. **1**:FL(35).
Zygomycotina
Actinomucor elegans. **1**:MT(3202).
Basidiomycotina - Uredinales
Uromyces commelinae. Heteroecious: telial host. **1**:FL(35).
Deuteromycotina - Hyphomycetes
Cercospora zebrina. Leaf spot. **1**:FL(35).
Deuteromycotina - Coelomycetes
Chaetoseptoria sp. Leaf spot. **2**:FL(35).
Colletotrichum falcatum. Anthracnose. **1**:FL(35).
Colletotrichum sp. Anthracnose. **2**:FL(35).
Deuteromycotina - Other
Rhizoctonia solani. Root and stem rot. **1**:FL(35). **2**:FL(35).
Sclerotium rolfsii. Southern blight. **1**:HI(2903).

Convolvulaceae

Argyreia Lour.

1. *A. nervosa* (N. L. Burm.) Bojer (=*A. speciosa* (L. fil.) Sweet, *Ipomoea speciosa* Pers.) [elephant-climber, woolly morning-glory]

Basidiomycotina - Uredinales

Coleosporium ipomoeae. Heteroecious: telial host. **1**:FL(35), MS(2720).

Deuteromycotina - Hyphomycetes

Phymatotrichopsis omnivora. Root rot. **1**:TX(82).

Bonamia Thouars

(= *Breweria* R. Br.)

1. *B. humistrata* (Walter) A. Gray (=*Breweria humistrata* (Walter) A. Gray)
2. *Bonamia* sp. (=*Breweria sp.*)

Ascomycotina

Meliola malacotricha. **2**:FL(3483).

Deuteromycotina - Hyphomycetes

Cercospora stylismae. Leaf spot. **1**:MS(2720).

Calystegia R. Br.

1. *C. hederacea* Wallich (=*Convolvulus japonicus* Thunb.) [Japanese bindweed]
2. *C. polymorpha* (Greene) Munz (=*Convolvulus polymorphus* Greene)
3. *C. sepium* (L.) R. Br. (=*Convolvulus sepium* L.) [hedge bindweed, wild morning-glory]
4. *C. spithamaea* (L.) Pursh (=*Convolvulus spithamaeus* L.) [low bindweed]

Oomycetes

Albugo ipomoeae-panduratae. White rust. **3**:E states(82), IL(82), MS(2720), TX(82).

Ascomycotina

Botryosphaeria obtusa. On stems. **3**:Unk.(82).
Dothidella calystegiae. On stems. **1**:CA(82).
Mycosphaerella sp. On stems. **1**:CA(82).

Basidiomycotina - Uredinales

Coleosporium ipomoeae. **1**:MS(2720).
Puccinia convolvuli. Autoecious. **1**:CA(82), OR(82). **2**:CA(81), OR(3202). **3**:E states(82), MS(2720), MT(3202), ND(82,378), TX(82).

Basidiomycotina - Other

Thanatephorus cucumeris. **4**:WI(1359).

Deuteromycotina - Hyphomycetes

Cercospora calystegiae. **3**:WI(82).
Cercospora sp. Leaf spot. **3**:KS(82), NC(82).
Phymatotrichopsis omnivora. Root rot. **1**:TX(82). **3**:TX(82).

Deuteromycotina - Coelomycetes

Ascochyta sp. **3**:WI(1360).
Coniothyrium sepium. On stems. **3**:NM(2399).
Cylindrosporium convolvulicola. Leaf spot. **1**:CA(82), WI(82).
Diplodia convolvuli. On stems. **3**:NY(82), TX(82).
Phomopsis ipomeae-batatas. Leaf spot. **1**:SC(82).
Septoria convolvuli. **1**:NY(82), WA(82), WI(82).
Stagonospora calystegiae. Leaf spot. **1**:CA(82). **2**:CA(81). **3**:NY(82).
S. convolvuli. Leaf spot. **1**:NY(82). **3**:WI(1335). **4**:WI(1343).

Convolvulus L.

1. *C. arvensis* L. [field bindweed, small bindweed]
2. *C. cneorum* L. [silverbush]
3. *C. nyctagineus* Greene
4. *C. occidentalis* A. Gray (=*C. luteolus* A. Gray)
5. *Convolvulus* sp.

Oomycetes

Albugo ipomoeae-panduratae. **5**:FL(35).

Ascomycotina

[*Parodiella hedysari*]. Black mildew. **1**:NC(82,3944).

Basidiomycotina - Uredinales

Coleosporium ipomoeae. Heteroecious: telial host. **1**:AL(82), GA(82), LA(82), MD(82), MS(82), NC(82,3944), TN(82), VA(82), WV(82).
Puccinia convolvuli. Autoecious. **1**:CA(81,82), IN(82), MA(82,1394), NE(82), NY(82), OK(82,2872), SD(2243), TN(82), TX(82). **2**:AZ(3981). **3**:OR(3202). **4**:CA(81).

Basidiomycotina - Ustilaginales

Thecaphora seminis-convolvuli. **1**:OR(3202), WA(3202). **5**:CA(1077).

Deuteromycotina - Hyphomycetes

Cercospora calystegiae. Leaf spot. **1**:OK(2872).
Cercospora sp. Leaf spot. **5**:FL(35).
Phymatotrichopsis omnivora. Root rot. **1**:TX(82).

Deuteromycotina - Coelomycetes

Colletotrichum gloeosporioides. Anthracnose. **5**:FL(35).
Cylindrosporium convolvulicola. **1**:WI(1337).
Phyllosticta sp. Leaf spot. **5**:FL(35), WA(3202).
Rhabdospora sp. Blight, stem rot. **1**:ID(82,3202), WA(82,3202).
Septoria convolvuli. Leaf spot. **1**:Unk.(82), ID(3202), IL(82), NV(82), OK(2872), WA(3202), WI(1329), WY(1195).
Stagonospora calystegiae. Leaf spot. **1**:WI(82).
S. convolvuli. **1**:WI(1345).

Deuteromycotina - Other

Rhizoctonia solani. Damping-off. **1**:MN(82), TX(82).

Cressa L.

1. *C. cretica* L.
2. *C. depressa* Goodd. in J. M. Coulter
3. *C. truxillensis* H. B. K.

Basidiomycotina - Uredinales

Puccinia tuyutensis. **1**:CA(81). **2**:AZ(1203). **3**:AZ(747,1203,3981), CA(81,2439).

Cuscuta L.

1. *C. epithymum* (L.) L. [clover dodder]
2. *C. gronovii* Willd. ex Roem. & Schult. [swamp dodder]
3. *C. pentagona* Engelm. [lespedeza dodder]
4. *Cuscuta* sp.

Deuteromycotina - Coelomycetes

Colletotrichum destructivum. **1**:OR(3202). **3**:OR(3202). **4**:OR(2062).
Phoma sp. On living stems. **4**:WI(1346).
Phomopsis cuscutae. On living stems. **2**:WI(1359).

Dichondra J. R. Forster & G. Forster

1. *D. carolinensis* Michx. (=*D. repens* J. R. Forster & G. Forster var. *carolinensis* (Michx.) Choisy in DC., *D. repens* J. R. Forster & G. Forster) [lawnleaf]
2. *Dichondra* sp.

Oomycetes

Pythium sp. Blight. **2**:FL(35).

Basidiomycotina - Uredinales

Puccinia dichondrae. Autoecious. **1**:CA(81,82), GA(1439), LA(82), MS(82,2720), NC(82,1303), TX(82). **2**:FL(35).

Basidiomycotina - Other

Thanatephorus cucumeris. Banded sclerotial disease. **1**:LA(1010).

Deuteromycotina - Hyphomycetes

Alternaria alternata. Leaf spot. **1**:HI(2903).
Alternaria sp. Leaf spot. **2**:FL(35).
Phymatotrichopsis omnivora. Root rot. **1**:TX(82).

Deuteromycotina - Coelomycetes

Colletotrichum gloeosporioides. Anthracnose. **1**:HI(2903).
Macrophomina phaseolina. Charcoal rot. **2**:HI(2903).

Deuteromycotina - Other

Rhizoctonia solani. Root rot. **2**:FL(35).
Rhizoctonia sp. Root rot. **1**:HI(2903).
Sclerotium rolfsii. Southern blight. **1**:HI(2299,2903), NC(1303). **2**:FL(35).

Evolvulus L.

1. *E. alsinoides* (L.) L.
2. *E. alsinoides* (L.) L. var. *acapulcensis* Ooststr.
3. *E. sericeus* Sw.
4. *Evolvulus* sp.

Basidiomycotina - Uredinales

Puccinia lithospermi. Autoecious. **1**:AZ(1203). **2**:AZ(3981). **3**:AZ(1203,3981). **4**:CO(82), KS(82), NE(82), TX(82).

Deuteromycotina - Hyphomycetes

Phymatotrichopsis omnivora. Root rot. **4**:TX(82).

Ipomoea L.

(= *Calonyction* Choisy.)
1. *I. alba* L. (=*I. bona-nox* L., *Calonyction aculeatum* (L.) House) [moonflower]
2. *I. aquatica* Forssk. (=*I. reptans* of authors) [kangkong, swamp morning-glory, water-spinach]
3. *I. barbigera* Sweet
4. *I. batatas* (L.) Lam. [sweetpotato, yam]
5. *I. coccinea* L. (=*Quamoclit coccinea* (L.) Moench) [red morning-glory, star ipomoea]
6. *I. hederacea* (L.) Jacq. [ivy-leaf morning-glory]
7. *I. indica* (J. Burm.) Merr. (=*I. congesta* R. Br.) [blue dawnflower]
8. *I. lacunosa* L.
9. *I. leptophylla* Torr. [bush moonflower, bush morning-glory, man-of-the-earth, manroot]
10. *I. longifolia* Benth.
11. *I. nil* (L.) Roth (=*Pharbitis nil* (L.) Choisy) [white-edge morning-glory]
12. *I. pandurata* (L.) G. Mey. [big-root morning-glory, man-of-the-earth, man-root, wild sweetpotato-vine]
13. *I. pes-caprae* (L.) R. Br. [beach morning-glory, railroad vine]
14. *I. purpurea* (L.) Roth [common morning-glory, tall morning-glory]
15. *I. quamoclit* L. (=*Quamoclit vulgaris* Choisy, *Quamoclit pennata* (Desr.) Bojer) [cardinal-climber, cypress-vine, star-glory]
16. *I. triloba* L.
17. *Ipomoea* sp.

Myxomycetes

Fuligo septica. Epiphytic on sprouts in hotbeds. **4:**AL(82), DE(82), NJ(82), TX(82).
Physarum cinereum. Slime mold. **4:**AL(82), AR(82), KY(82), MS(82,2720), NJ(82), VA(82).
Stemonitis fusca. Slime mold. **3:**MS(2720).

Oomycetes

Albugo ipomoeae-panduratae. White blister, white rust. **1:**FL(82). **2:**HI(2903). **4:**'White rust.' Range of host(82), FL(35), MS(2720), NC(1303). **5:**FL(35), NM(82), OK(2872). **6:**E states(82), AZ(82), MS(2720), NC(1303), NE(82), OK(2872). **7:**HI(2903). **9:**CO(82), KS(82), TX(82). **10:**OK(2873). **11:**E states(82), AZ(82), NC(1303), NE(82). **12:**E states(82), AZ(82), MS(2720), NC(3944), NE(82), OK(2872). **14:**E states(82), AZ(82), MS(2720), NC(1303,3944), NE(82), OK(2872), WI(1354). **15:**MS(82,2720). **17:**FL(35), TX(652).
Phytophthora sp. Root rot. **4:**NJ(82), VA(82).
Pythium aphanidermatum. Leak in storage, mottle necrosis of mature roots. **4:**CA(82), DE(82), IA(82), IN(82), MD(82), MS(82), NC(82), NJ(82), PA(82), VA(82), WA(82).
P. arrhenomanes. Rootlet rot. **4:**HI(82,2903).
P. debaryanum. Mottle necrosis of mature roots. **4:**CA(82), DE(82), HI(2903), IA(82), IN(82), MD(82), MS(82,2720), NC(82), NJ(82), PA(82), VA(82), WA(82).
P. hydnosporum. Rootlet rot. **4:**HI(82,2903).
P. irregulare. Rootlet rot. **4:**HI(82,2903).
P. rostratum. Rootlet rot. **4:**HI(82,2903).
P. scleroteichum. Mottle necrosis of mature roots. **4:**CA(82), DE(82), IA(82), IN(82), MD(82), MS(82,2720), NC(82), NJ(82), PA(82), VA(82), WA(82).
P. splendens. Mottle necrosis of mature roots. **4:**CA(82), DE(82), HI(2903), IA(82), IN(82), MD(82), MS(82), NC(82), NJ(82), PA(82), VA(82), WA(82).
P. ultimum. Leak and ring rot in storage, mottle necrosis of mature roots. **4:**Unk.(82), CA(82), DE(82), IA(82), IN(82), MD(82), MS(82,2720), NC(82), NJ(82), OK(2873), PA(82), VA(82), WA(82).
Pythium sp. Root rot. **4:**Unk.(82), FL(35), OK(2872), WA(3202).

Zygomycotina

Choanephora cucurbitarum. Flower blight, leaf mold. **4:**'Leaf mold.' FL(35,82). **5:**FL(35).
C. infundibulifera. Blossom rot. **14:**GA(82).
Mucor racemosus. 'Storage rot.' **4:**Range of host(82), FL(35).
Rhizopus arrhizus. Soft rot, common during storage at high temperatures. **4:**Range of host(82).
R. stolonifer. Ring rot, soft rot. **3:**MS(2720). **4:**'Soft rot.' Range of host(82), FL(35), HI(2903), MS(2720), OK(2872).
Rhizopus sp. Soft rot. **4:**Range of host(82), HI(2903), OR(3202).

Ascomycotina

Botryosphaeria quercuum. Black rot. **4:**FL(35).
Ceratocystis fimbriata. Black rot of roots and stems. **3:**MS(2720).
Diaporthe arctii. On stems. **14:**WV(82).
D. phaseolorum. Root and stem rot. 'Dry rot.' **4:**E states(82), FL(35), GA(1438), HI(2903), MO(82), OK(2872), TX(82).
Dothidea puccinioides. **4:**GA(1438).
Elsinoe batatas. 'Spot anthracnose.' **4:**HI(2299,2903).
Glomerella cingulata. On stems. **14:**LA(82).
[*Mycosphaerella macrospora*]. Leaf spot. **17:**MS(2720).
Nectria haematococca. **4:**GA(1438).
N. ipomoeae. On roots and stems following wilt, storage rot. **4:**CA(82), DE(82), MD(82), NJ(82), OH(82), SC(82).
Sclerotinia minor. 'Sclerotinia sprout rot.' **4:**NJ(82).
Sclerotinia sp. Storage rot following chilling. **4:**NC(82).
Sphaerulina ipomoeae. On leaves. **1:**HI(82,2903).

Basidiomycotina - Uredinales

Coleosporium ipomoeae. Heteroecious: telial host. **1:**AL(82), NC(82), SC(82), TX(82). **3:**MS(2720). **4:**'Rust.' AL(82,285), MS(82,2720). **5:**FL(35), GA(1439), IL(82), SC(82), TN(82). **6:**E states(82), GA(1439), KS(82), MS(2720), NC(1303,3944), OK(2872), TX(82). **8:**MS(2720). **11:**E states(82), KS(82), NC(1303), TX(82). **12:**GA(1439), MS(2720), NC(3944), OK(2872). **13:**E states(82), KS(82), OK(82). **14:**E states(82), GA(1439), KS(82), MS(2720), NC(1303,3944), OK(2872), TX(82). **15:**MS(82,2720), NC(82,3944). **16:**GA(1439). **17:**E states(82), FL(35,82), KS(82), MS(82), NC(1303), OK(2872), TX(82).
Puccinia convolvuli. Autoecious. **4:**FL(35).
P. crassipes. Autoecious. **14:**OK(2872). **17:**FL(35,82), GA(82), LA(82), SC(82), TX(82).
[*P. iridis*]. **17:**MS(2720).
Uredo laeticolor. **17:**FL(35).

Basidiomycotina - Other

Helicobasidium brebissonii. 'Violet root rot.' **4:**KS(82), TX(82).
Pellicularia koleroga. Thread blight. **17:**FL(82), LA(3640).
Schizophyllum commune. Dry rot of roots. **4:**NC(82).
Schizophyllum sp. **4:**OK(2872).
Thanatephorus cucumeris. Stem rot. **4:**OK(2872).

Deuteromycotina - Hyphomycetes

Alternaria solani. Foot rot. **4:**DE(82), TX(82).
Alternaria sp. Leaf spot. **4:**Range of host(82), OK(2872). **14:**VA(82). **17:**FL(35).
Aspergillus niger. Secondary storage rot. **4:**Unk.(82).
Aspergillus sp. Secondary storage rot. **4:**Unk.(82), OK(2872).
Asperisporium punctulatum. Leaf spot. **17:**MS(2720).
Botrytis cinerea. 'Botrytis storage rot.' **4:**Range of host(82).
Cercospora ipomoeae. Leaf spot. **4:**FL(35). **5:**FL(35). **6:**FL(82), KS(82), MO(82), TX(82). **12:**FL(82), IL(82), MO(82), MS(82,2720). **13:**HI(2903), TX(82). **14:**AL(82), HI(2903), IL(82), IN(82), MO(82), NC(82,3944), NE(82), NJ(82), OH(82), TX(82), VA(82). **17:**FL(35), GA(82).
Cercospora sp. Leaf spot. **1:**HI(2899,2903). **2:**HI(2903). **4:**FL(82), MS(2720), OK(82). **7:**HI(2899).
Corynespora cassiicola. Leaf spot. **13:**HI(2903).
Cristulariella moricola. On leaves. **6:**IL(2572). **8:**IL(2572).
Epicoccum sp. Storage rot. **4:**Unk.(82).
Fusarium acuminatum. Dry rot, end rot, secondary storage rot. **4:**Unk.(82).
F. culmorum. Dry rot, end rot, secondary storage rot. **4:**Unk.(82).
F. javanicum. Secondary storage rot. **4:**Unk.(82), OK(2872).
F. oxysporum. Stem rot, wilt. 'Surface rot. Fusarium stem rot.' **4:**Range of host(82), AL(285), FL(35), NC(1303), OK(2872).
F. oxysporum f. sp. *batatas*. Stem rot, wilt. **4:**Range of host(82), S states(82), W states(82), AL(285), FL(35), MS(2720), NC(1303), NM(2113), OK(2872), TN(72,73), WA(3202). **17:**NM(2112).
F. roseum. Secondary storage rot. **4:**Unk.(82).
F. semitectum var. *majus*. Secondary storage rot. **4:**Unk.(82).
F. solani. Stem canker, root rot. 'Storage rot.' **4:**Unk.(82), CA(2307), HI(2903), LA(583), NC(583,1303,2518,2577).
Fusarium sp. Wilt. **4:**MS(2720). **6:**TX(82). **9:**TX(82).
Monilochaetes infuscans. 'Scurf.' **4:**Range of host(82), FL(35), HI(2903), MS(2720), NC(1303), NM(2109), OK(2872).
Oidium sp. Powdery mildew. **17:**FL(35).
Penicillium sp. Blue mold. **4:**Range of host(82), OK(2872).

Phymatotrichopsis omnivora. Root rot. **1:**TX(82). **4:**'Phymatotrichum root rot.' AZ(82), NM(82), OK(82,2872), TX(82). **6:**TX(82). **11:**TX(82). **14:**TX(82). **15:**TX(82). **17:**TX(82).
Ramularia ipomoeae. Leaf spot. **1:**HI(82,2903).
Trichoderma koningii. Storage rot. **4:**OK(2872).
Trichoderma sp. Storage rot. **4:**DE(82), NC(82), NJ(82).
Verticillium albo-atrum. Wilt. **4:**CA(82).

Deuteromycotina - Coelomycetes

Colletotrichum dematium. Anthracnose. **14:**NC(1303).
C. dematium f. *ipomoeae*. Anthracnose. **17:**NC(1303).
C. truncatum. Isolated from leaves and stems. **17:**IL(1482).
Colletotrichum sp. Leaf spot. **4:**MS(2720).
Disaeta sp. **4:**OK(2872).
Macrophomina phaseolina. Charcoal rot. **4:**'Charcoal rot.' E states(82), AL(285), CA(82), FL(35), HI(2903), KS(82), NC(1303), OK(2872), TX(82). **6:**IL(311).
Pestalotia batatae. On roots. **4:**AL(82).
Phlyctema ipomoeae. On sepals and stems. **12:**WV(82).
Phoma terrestris. 'Pink root.' **4:**CA(82).
Phomopsis ipomeae-batatas. Leaf spot. 'Leaf blight.' **4:**E states(82), AL(285), FL(35), KS(82), MS(2720), NC(1303), OK(2872), TX(82).
Phomopsis sp. Leaf spot. **13:**HI(82,2903).
Phyllosticta ipomoeae. Leaf spot. **12:**KS(973), MS(2720). **13:**FL(82), KS(82), MS(82), WV(82). **17:**FL(35).
Phyllosticta sp. Leaf spot. **1:**NJ(82).
Plenodomus destruens. 'Foot rot.' **4:**E states(82), CA(82), FL(35), IA(82), LA(82), MS(2720), NC(1303), OK(2872).
Septoria bataticola. 'Septoria leaf spot.' **4:**E states(82), FL(35), HI(82,2903), IA(82), NC(1303), OK(2872), TX(82).
S. convolvuli. Leaf spot. **3:**MS(2720). **6:**KS(82), TX(82). **17:**FL(35,82), PA(82), WI(82).
Sphaeropsis ipomoeae. On stems. **12:**WV(82).
Vermicularia ipomoearum. Stem canker. **5:**PA(82). **14:**NY(82), PA(82).

Deuteromycotina - Other

Rhizoctonia solani. Thread blight, root rot, sprout rot, stem rot. **4:**'Rhizoctonia rootlet rot. Rhizoctonia sprout rot. Stem canker.' E states(82), AZ(82), CA(82), FL(35), OH(82), TX(82), WA(82). **17:**FL(35).
Rhizoctonia sp. Root lesions and rot. **4:**HI(2903), MD(1556).
Sclerotium rolfsii. Southern blight, cottony blight. **3:**MS(2720). **4:**'Southern blight. Circular spot.' SC states(82), SE states(82), FL(35), HI(2903), MS(2720), NC(1303), OK(82,2872), TX(82). **14:**TX(82). **17:**MS(2720).

Jacquemontia Choisy

1. *J. sandwicensis* A. Gray
2. *J. tamnifolia* (L.) Griseb.
3. *Jacquemontia* sp.

Oomycetes

Albugo ipomoeae-panduratae. White rust. **3:**AL(82).

Basidiomycotina - Uredinales

Coleosporium ipomoeae. Heteroecious: telial host. **3:**LA(82).

Basidiomycotina - Other

Pellicularia koleroga. **2:**LA(3640).

Deuteromycotina - Hyphomycetes

Cercospora ipomoeae. Leaf spot. **3:**MS(82).
Stemphylium botryosum. Leaf spot. **1:**HI(2903).

Merremia Dennst. & Endl.

1. *M. dissecta* (Jacq.) H. Hallier (=*Ipomoea dissecta* (Jacq.) Pursh)

Oomycetes

Albugo ipomoeae-panduratae. **1:**FL(35).

Basidiomycotina - Uredinales

Uredo laeticolor. **1:**FL(111).

Cornaceae

Aucuba Thunb.

1. *A. japonica* Thunb. [Japanese aucuba, Japanese-laurel]
2. *A. japonica* Thunb. cv. 'crotonifolia'
3. *Aucuba* sp.

Oomycetes

Phytophthora cinnamomi. Root rot. **1:**NC(1303,3414).
P. citricola. Root rot. **1:**NC(1303,3414).
Phytophthora sp. Root rot. **1:**FL(35).
Pythium sp. Root rot. **1:**NC(1303).

Ascomycotina

Glomerella cingulata. **1:**WA(3202).

Basidiomycotina - Other

Corticium stevensii. Thread blight. **1:**FL(35).

Deuteromycotina - Hyphomycetes

Alternaria sp. Leaf spot. **1:**FL(35).
Cercospora sp. Leaf spot. **1:**FL(35). **3:**FL(35).
Fusarium sp. Wilt. **1:**NC(1303).
Verticillium albo-atrum. Wilt. **1:**NJ(82).

Deuteromycotina - Coelomycetes

Colletotrichum gloeosporioides. Anthracnose, leaf spot. **1:**MS(82,2720), NC(1303), NJ(82).
Colletotrichum sp. **2:**WA(1300).
Diplodia sp. Dieback. **1:**FL(35).
Pestalotia subcuticularis. Leaf spot. **1:**NJ(82).
Phomopsis aucubae. Twig dieback. **1:**NC(1303).
Phomopsis sp. Dieback. **1:**FL(35), SC(82), WA(3202).
Phyllosticta aucubae. Leaf spot. **1:**CA(82), MS(82,2720), NC(1303), SC(82). **3:**FL(35).
P. aucubicola. **1:**CT(82).

Deuteromycotina - Other

Rhizoctonia solani. Leaf necrosis. **3:**FL(35).
Sclerotium rolfsii. Root and stem rot. **1:**FL(35), NC(1303).

Cornus L.

1. *C. alternifolia* L. fil. [green osier, pagoda dogwood]
2. *C. amomum* Mill. [red-willow, silky dogwood]
3. *C. asperifolia* Michx. [small roughleaf cornel]
4. *C.* × *californica* C. A. Mey. *C. occidentalis* × *C. sericea*. [creek dogwood]
5. *C. canadensis* L. [bunchberry, crackerberry, dwarf cornel, pudding-berry]
6. *C. drummondii* C. A. Mey. [roughleaf dogwood]
7. *C. florida* L. [flowering dogwood]
8. *C. glabrata* Benth. [brown dogwood]
9. *C. instolonea* A. Nels.
10. *C. nuttallii* Audubon [mountain dogwood, Pacific dogwood]
11. *C. obliqua* Raf. [silky dogwood]
12. *C. occidentalis* Coville [western osier]
13. *C. racemosa* Lam. (=*C. paniculata* L'Hér.) [gray dogwood, panicled dogwood]
14. *C. rugosa* Lam. [round-leaved dogwood]
15. *C. sanguinea* L. [blood-twig dogwood, dogberry, pegwood]
16. *C. sericea* L. (=*C. stolonifera* Michx.) [American dogwood, red osier dogwood]
17. *C. stricta* Lam. (=*C. foemina* Mill.) [stiff dogwood]
18. *C. suecica* L.
19. *Cornus* sp.

Chytridiomycetes

Synchytrium corni. Leaf gall. **6:**LA(627).

Oomycetes

Phytophthora cactorum. Trunk canker, collar rot. **7:**MA(82), MD(82), NJ(82), NY(82). **10:**WA(82,3202,3527).
P. cinnamomi. Root rot, basal stem rot. **7:**FL(35), NC(1303). **19:**NC(1303).
P. nicotianae var. *parasitica*. Leaf blight. **7:**FL(35,39).
P. palmivora. Leaf blight. **7:**FL(35).
Pythium irregulare. Root rot. **7:**FL(35).
P. splendens. Root rot. **7:**FL(35).
Pythium sp. Root rot of nursery plants. **7:**NJ(82).

Ascomycotina

Acrospermum foliicola. **7:**GA(1438).
Anthostomella cornicola. **19:**WA(3202).
Botryosphaeria dothidea. **2:**GA(1438).
B. obtusa. On limbs. **7:**NJ(82), NY(82), VA(82). **12:**CO(82). **19:**GA(82), NY(82), VA(82).
B. rhodina. **7:**LA(82).
Botryosphaeria sp. Canker, dieback. **7:**FL(35), PA(82).
Calosphaeria cornicola. On limbs. **1:**NY(82). **13:**NY(82).
Clypeosphaeria mamillana. **16:**MT(3202).

Cryptodiaporthe corni. On limbs. **1:**IA(82), MI(82), MS(2720), NY(82), WI(82), WV(82). **12:**CO(82). **16:**IA(82), MI(82), NY(82), WI(82), WV(82).
Daldinia vernicosa. Wood rot. **7:**MD(82).
Diaporthe eres. On dead limbs and twigs. **7:**GA(82,1438). **13:**IA(82).
D. pardalota. **13:**CT(82), MA(82), MD(82), MN(82), ND(82), NY(82). **16:**CT(82), MA(82), MD(82), MN(82), ND(82), NY(82). **19:**CT(82), MA(82), MD(82), MN(82), ND(82), NY(82).
Diatrype albopruinosa. **19:**IA(3629).
Didymosphaeria decolorans var. *americana*. **16:**ND(377).
Discohainesia oenotherae. Leaf spot. **5:**ME(82), NH(82).
Elsinoe corni. Spot anthracnose on floral leaves, seeds, stems and twigs, stem canker. **3:**FL(2525). **7:**DE(82), FL(35,2525,3801), GA(82,897,1438,1745,2401,2405), LA(905), MD(82,1729,1731,1732,1734,1735,1736,1742,1745), NC(82,351,1303,1745), SC(82,351), VA(82,1528,1731), WV(954). **19:**MS(2720).
Eutypa lata. On decorticated wood. **19:**NC(1303,1307).
E. maura. **7:**GA(1438).
Eutypella leprosa. **19:**IA(3629).
E. scoparia. **4:**CA(81). **19:**SC(2925).
E. stellulata. On dead twigs. **19:**NC(1303).
Glomerella sp. Twig dieback. **7:**FL(35).
Glyphium corrugatum. **16:**ID(3202), WA(3202).
Godronia sp. **16:**ID(3202).
Hainesia lythri. **5:**WI(1359). **17:**WI(1357).
Hapalocystis corni. On dead limbs and twigs. **7:**GA(82,1438,3809), MD(2779).
Hymenoscyphus serotinus. **16:**ID(3202).
Hypoxylon mediterraneum. On dead limbs. **7:**IN(82).
H. rubiginosum. On dead wood. **7:**GA(82,1438), VA(82).
H. serpens. On dead wood. **7:**GA(82), VA(82).
H. serpens var. *macrospora*. **16:**ID(2786).
Hysterium pulicare. On bark. **7:**MA(1394).
Hysteropatella prostii. **16:**ND(377).
Lachnum corni. On leaves. **2:**GA(128,849,1438,3167).
Leptosphaeria fiedleri. On dead twigs. **19:**NY(82).
Lophiostoma vagabundum. **6:**OK(660).
Massariovalsa sudans. On dead limbs and twigs. **7:**GA(82,1438).
Meliola nidulans. Black mildew. **2:**Unk.(1452). **13:**Gulf states(82), SE states(82), TN(82). **16:**Unk.(1452). **19:**Gulf states(82), SE states(82), Unk.(1452), MS(2720), TN(82).
Microsphaera penicillata. Powdery mildew. **1:**E states(82), GA(1438), IL(82), WI(82). **13:**E states(82), IL(82), WI(82). **14:**WI(1359). **16:**E states(82), ID(3202), IL(82), WI(82). **19:**E states(82), IL(82), NC(1303), WI(82).
M. tortilis. Powdery mildew. **15:**MO(82).
Mollisia propinqua. On dead twigs. **19:**PA(850).
Monilinia corni. Shoot-tip blight. **10:**WA(3202). **19:**NY(82), WI(82).
Mycosphaerella corni. On fallen leaves. **7:**AL(82), GA(82,1438), SC(82).
M. cornicola. On twigs. **16:**IL(82).
Mycosphaerella sp. On overwintered leaves. **5:**WI(1357).
Myrmaecium harperianum. **19:**WA(3202).
Nectria cinnabarina. On limbs and twigs. **10:**WA(3202). **19:**MD(82).
N. coccinea. **10:**OR(3202). **19:**OR(82).
N. galligena. Trunk canker. **10:**OR(82,3202), WA(82).
N. haematococca. Canker. **7:**NC(1303).
Nitschkia callista. **19:**OR(3202).
N. confertula. **7:**MD(2779).
Penzigia frustulosa. On decorticated wood. **19:**NC(1303).
Pezicula corni. On limbs and twigs. **1:**ID(82,3202). **7:**MA(82), PA(82). **16:**CA(82), ID(82,3202), OR(82,3202). **19:**CA(82), OR(82).
P. cornicola. **19:**MA(82), NY(82).
[*P. ocellata*]. **19:**OR(3202).
Pezicula sp. **19:**AL(82), MD(82).
Phyllachora subcuticularis. Tar spot. **19:**NY(82).
Phyllactinia guttata. Powdery mildew. **1:**Range of host(82). **3:**Range of host(82). **5:**ID(3202), WA(82,3202). **7:**SE states(82), GA(1438), MI(82), MO(82), NC(1303). **8:**CA(81). **10:**CA(81,82), ID(82,3202), OR(82,3202), WA(82,3202). **11:**WI(1361). **12:**Range of host(82). **13:**Range of host(82). **16:**Range of host(82), ID(3202), MT(3202), WA(3202). **4:**CA(81). **19:**Range of host(82).
Pleospora laricina. On limbs. **7:**MA(270).
Pseudomassaria corni. On twigs. **9:**NM(175). **16:**WI(175,681).
P. foliicola. On leaves. **5:**AK(82,526), ME(175).
Romellia cornina. **3:**KS(82). **13:**NY(82). **19:**MI(82).
Rosellinia aquila. **7:**GA(1438).
Schiffnerula pulchra. Sooty mold. **13:**E states(82), TX(82), WI(82). **16:**E states(82), TX(82), WI(82). **19:**E states(82), TX(82), WI(82).
Schizoxylon compositum. **19:**KS(3225).
Stictis radiata. **16:**WA(3202).
S. sphaeroboloidea. **19:**PA(3225).
S. stigma. On dead twigs. **7:**NJ(3225). **19:**NJ(3225).
Trematosphaeria cornina. On dead limbs. **7:**KY(82).
Tubeufia pezizula. On decorticated wood. **7:**NC(1303).
Valsa ambiens. On dead limbs and twigs. **3:**KS(3415). **7:**WV(82). **19:**NY(82).
V. ceratosperma. **7:**GA(1438). **16:**NY(3415).
V. coronata. On dead limbs. **7:**WV(82).
V. fallax. **5:**AK(526).
Venturia systema-solare. On overwintered leaves. **14:**WI(1341). **19:**NY(82).

Basidiomycotina - Uredinales

Puccinia porphyrogenita. Autoecious. **5:**NC states(82), NE states(82), NW states(82), AK(60,82,526,3431), CA(81), ID(3202), MT(1195,3202), OR(657,3202), WA(3202). **18:**AK(60,526).

Basidiomycotina - Other

Amphinema arachispora. On charred tree. **7:**GA(443).
Armillaria mellea. Root rot. **7:**NC(1303), NY(59,82), PA(205), WA(3202). **10:**WA(82).
A. tabescens. **7:**FL(35,82), GA(2684).
Bjerkandera adusta. **7:**NY(82).
Ceratobasidium anceps. Leaf rot. **5:**NH(82), WI(1345).
Ceriporia tarda. **7:**MD(82).
Chondrostereum purpureum. On wood. **7:**NY(82,543).
Clavicorona pyxidata. **7:**DE(884).
Coprinus disseminatus. **7:**NC(3944).
Corticium stevensii. Thread blight. **7:**FL(35).
Daedaleopsis confragosa. On wood. **7:**MD(82), NC(82,1303,3944), PA(82), TN(82), VA(82), WV(82).
Datronia scutellata. On dead wood. **7:**MD(82).
Dendrophora versiformis. **7:**NY(82).
Ductifera sucina. On twigs. **7:**NC(2645).
Gloeoporus dichrous. On wood. **7:**MD(82), NC(1303,1307), SC(1307). **10:**MD(82).
Hyphoderma rimosum. White rot. **7:**MS(444).
Irpex lacteus. **7:**MD(82).
Lenzites betulina. Wood rot. **7:**NC(82,3944).
Pellicularia koleroga. Thread blight. **3:**LA(82). **6:**LA(3640). **7:**FL(82).
Perenniporia medulla-panis. **7:**MD(82).
P. tenuis. On wood. **7:**MD(82).
Phanerochaete affinis. White rot. **7:**NY(128,433).
P. chrysorhiza. White rot. **7:**MS(433), NC(433).
Phellinus ferreus. **10:**OR(3202).
P. igniarius. Heart rot. **10:**ID(3202), OR(82,3202).
Pleurotellus hypnophilus. **7:**NC(3944).
Pulcherricium caeruleum. On wood. **7:**GA(491,1439).
Schizopora paradoxa. **7:**MD(82), PA(82).
Scytinostroma galactinum. Root rot. **7:**E & C states(2090), MD(669,2090), VA(82).
Sebacina incrustans. **19:**LA(2675).
Septobasidium castaneum. Parasitic on scale insects. **7:**NC(1303).
S. cokeri. Brown felt. **2:**NC(1303). **7:**SE states(82), FL(35), NC(1303).
S. fumigatum. Brown felt. **7:**SE states(82), FL(35). **19:**FL(82).
S. pseudopedicellatum. Brown felt. **7:**SE states(82), FL(35), NC(1303,3944). **19:**AL(82), GA(82), NC(82).
Septobasidium sp. Brown felt. **7:**FL(35). **19:**Unk.(82).
Trametes versicolor. **7:**GA(1439), MD(82), VA(82).
Tyromyces chioneus. **7:**VA(82).

Deuteromycotina - Hyphomycetes

Alternaria sp. Leaf spot. **3:**OK(2872). **7:**FL(35).
Botrytis cinerea. Bud, flower and shoot tip blight. **7:**CT(3770), MA(82), MD(82), NC(1303), NJ(82), NY(82). **13:**MA(82).
Botrytis sp. Crown rot, root rot. **7:**CT(2974), MD(1729). **19:**WA(3202).

Cercospora corni. Leaf spot. **13:**WI(82). **19:**LA(82), WI(82).
C. cornicola. Leaf spot. **3:**TX(82). **7:**SE states(82), FL(35), MS(2720), NC(3944), TX(82). **19:**AL(285).
Cladosporium sp. **16:**MT(3202).
Coniosporium olivaceum. **7:**DE(3543).
Cristulariella moricola. Leaf spot. **2:**WV(3680). **7:**FL(35).
Fusarium oxysporum. Root rot. **7:**FL(35).
F. solani. Isolated from cankers. **7:**NC(1303).
Fusicladium sp. On flower bracts. **10:**OR(2173).
Gliocephalotrichum simplex. Root rot, basal stem rot. **7:**FL(35).
Glomopsis corni. Leaf spot. **5:**NC states(82), NE states(82), AK(526), ID(3202), MA(1515), NH(1515), NY(1515), OR(82,3202), WA(3202), WI(1515). **19:**MS(2720).
Helminthosporium macrocarpum. **19:**WA(3202).
H. velutinum. **7:**DE(3543).
Oidium sp. Powdery mildew. **7:**FL(35).
Phymatotrichopsis omnivora. Root rot. **3:**TX(82). **7:**TX(82). **19:**TX(82).
Ramularia gracilipes. Leaf spot. **1:**WI(82).
R. stolonifera. **1:**WI(82). **11:**WI(1350). **16:**ID(3202), MT(3202), WA(82,3202), WI(82). **4:**CA(81). **19:**WI(82).
Ramularia sp. Leaf spot. **5:**NY(82).
[*Sarcinella heterospora*]. **16:**WI(1359). **17:**WI(1359).
S. questieri. On living leaves. **13:**IL(1681), NY(1681), WI(1681). **14:**WI(1681). **16:**IL(1681), WI(1681). **17:**WI(1681). **19:**MD(1681), WV(1681).
Stigmina cornicola. On twigs. **13:**MI(82,3547). **19:**MI(82).
Verticicladium pulvereum. **16:**WA(3202).
Xenosporium berkeleyi. On decorticated wood. **7:**NC(1303).
Deuteromycotina - Coelomycetes
Ascochyta cornicola. Leaf shriveling, leaf spot. **7:**NC(82,1528), VA(82,1528). **16:**WI(966). **19:**OH(82), WI(82).
Colletotrichum gloeosporioides. Dieback. **1:**WI(1350). **7:**TX(3665). **17:**WI(1344).
Colletotrichum sp. Anthracnose. **7:**FL(35).
Coniothyrium sp. Dieback. **7:**FL(35). **11:**WI(1356). **19:**NC(1303).
Cryptosporiopsis cornina. On limbs. **1:**Unk.(82). **16:**CO(82), NE(82), NM(82). **19:**MD(82), NY(82), PA(82).
Cryptostictis sp. Canker. **7:**NJ(82).
Cylindrosporium corni. Leaf spot. **16:**IL(312), OR(3202), WY(82,1195,3376).
Cylindrosporium sp. Leaf spot. **1:**NC(1303,1307). **16:**OK(2872).
Cytospora corni. On twigs. **3:**KS(82). **16:**CO(1195,3378), SD(2243), WY(1195,3378). **19:**KS(82), MS(2720).
Cytospora sp. Canker. **7:**NJ(82).
Diplodia mamillana. **16:**WA(3202).
Diplodia sp. Dieback. **7:**FL(35).
Discosia artocreas. On fallen leaves. **7:**VA(82), WV(82).
Discula sp. **7:**CT(776,1550), MA(1550), MD(1550), NJ(776,1550), NY(776,1550), PA(776,1550), RI(1550), VA(1550), WV(1550). **10:**ID(3083), OR(3083).
Gloeodes pomigena. On twigs. **1:**IN(82). **19:**IN(82).
Hendersonia fiedleri. On twigs. **1:**NY(82). **19:**NY(82).
H. pallida. On twigs. **13:**WI(82).
Macrophoma sp. Dieback. **7:**FL(35). **16:**ID(3202), MT(3202).
Macrophomina phaseolina. Charcoal root rot. **7:**FL(35).
Melanconium sp. **16:**WA(3202).
Myxosporium everhartii. On dead twigs. **7:**MI(82), TN(82).
M. nitidum. Twig blight. **19:**Unk.(82).
Phoma florida. On dead twigs. **7:**NY(82).
Phoma sp. Twig dieback. **7:**FL(35).
Phomopsis corni. **1:**Unk.(82).
Phomopsis sp. On bark. **7:**MD(82), TN(82,3780).
Phyllosticta cornicola. Leaf spot. **1:**Range of host(82), WI(1332). **3:**Range of host(82). **7:**FL(35), NC(1303,1307). **12:**Range of host(82). **13:**Range of host(82). **16:**Range of host(82). **19:**Range of host(82).
P. globifera. Leaf spot. **7:**NC(1533).
P. starbaeckii. Leaf spot. **7:**OK(2873). **10:**CA(81).
Phyllosticta sp. Leaf spot. **5:**ID(3202), NY(82). **7:**E states(82), KS(82), TN(82). **11:**WI(1355). **14:**WI(1351).
Placosphaeria cornicola. On leaves. **10:**OR(82,3202).
Septoria canadensis. Leaf spot. **5:**AK(82,526), ID(3202), ME(82), MT(3202), WA(3202), WA(82), WI(1333).
S. cornicola. Leaf spot. **1:**Range of host(82). **3:**Range of host(82), OK(2872). **7:**E states(82), FL(35), GA(2405), IA(82), MO(82), NC(1303), NY(82), OK(2873), WV(82,954). **12:**Range of host(82). **13:**Range of host(82), NY(2456), OK(2873). **16:**Range of host(82), MT(3202), ND(378), OK(2872), SD(2243), WA(3202), WI(1330). **19:**Range of host(82), MS(2720).
S. cornicola var. *ampla*. On living leaves. **11:**WI(1355). **13:**WI(1341). **14:**WI(1356).
S. floridae. **7:**IL(82).
Sphaceloma sp. Leaf spot. **1:**WI(1351). **19:**NC(1303).
Sphaeropsis cornicola. On twigs. **19:**MD(82).
Sphaeropsis sp. Twig blight, canker. **7:**FL(35), NJ(82). **16:**WA(3202).
Zythia aurantiaca. On twigs. **1:**WI(1341,1348).
Deuteromycotina - Other
Rhizoctonia solani. Thread blight, root rot, basal stem rot. **7:**FL(35).
Sclerotium rolfsii. Southern blight. **7:**FL(35).

Crassulaceae

Aeonium P. Webb & Berthel.

1. *Aeonium* sp.

Oomycetes
Pythium sp. Root rot. **1:**FL(35).

Crassula L.

(= *Tillaea* L.)
1. *C. arborescens* Willd. [Chinese jadeplant, silver jadeplant, silver-dollar]
2. *C. argentea* L. fil. [cauliflower-ears, Chinese rubberplant, dollar-plant, dwarf rubberplant, jadetree, Japanese rubberplant]
3. *Crassula* sp.

Oomycetes
Pythium sp. Root rot. **2:**FL(35). **3:**NJ(82), NY(82).
Ascomycotina
Sphaerotheca macularis. Powdery mildew. **1:**FL(35).
Basidiomycotina - Other
Armillaria mellea. Root rot. **2:**CA(82).
Deuteromycotina - Hyphomycetes
Alternaria sp. Leaf spot. **1:**FL(35). **2:**FL(35).
Bipolaris sp. Leaf spot. **1:**FL(35). **2:**FL(35).
Cercospora sp. Leaf spot. **1:**FL(35). **2:**FL(35).
Corynespora sp. Leaf spot. **2:**FL(35).
Cylindrocladium pteridis. Leaf spot. **2:**FL(35).
Fusarium sp. Stem rot. **2:**FL(35).
Stemphylium bolickii. Leaf spot. **1:**FL(35). **2:**FL(35).
Deuteromycotina - Coelomycetes
Ascochyta sp. Leaf spot. **1:**FL(35). **2:**FL(35).
Diplodia crassulae. On stems. **3:**CA(82).
Diplodia sp. Stem rot. **2:**FL(35).
Phomopsis sp. Leaf spot. **3:**CT(82).
Phyllosticta sp. Leaf spot. **1:**FL(35). **2:**FL(35).
Septoria sp. **3:**NC(1303).
Sphaeropsis sp. Stem gall. **1:**FL(35).
Deuteromycotina - Other
Rhizoctonia solani. Root and stem rot. **1:**FL(35). **2:**FL(35).

Echeveria DC.

1. *Echeveria* sp.

Oomycetes
Pythium aphanidermatum. Root rot. **1:**HI(2903).
Pythium sp. Root rot. **1:**FL(35).
Basidiomycotina - Uredinales
Puccinia echeveriae. Autoecious. **1:**CA(82).
Deuteromycotina - Hyphomycetes
Alternaria sp. Leaf spot. **1:**FL(35).
Cercospora sp. Leaf spot. **1:**FL(35).
Fusarium sp. Stem rot. **1:**FL(35).
Stemphylium bolickii. Leaf spot. **1:**FL(35).
Deuteromycotina - Other
Rhizoctonia solani. Leaf and root rot. **1:**FL(35).

Kalanchoe Adans.

1. *K. blossfeldiana* Poelln.
2. *K. fedtschenkoi* Raym.-Hamet & E. Perrier [lavender-scallops, South American airplant]

3. *K. flammea* Stapf
4. *K. gastonisbonnieri* Raym.-Hamet & E. Perrier
5. *K. globosa* Hort. A listed name of no botanical standing.
6. *K. laciniata* (L.) DC. [Christmas-tree kalanchoe, fig-tree kalanchoe]
7. *K. laxiflora* Baker
8. *K. marmorata* Baker (=*K. somaliensis* Baker) [pen-wiper]
9. *K. orgyalis* Baker
10. *K. pinnata* (Lam.) Pers. [airplant, curtain-plant, floppers, good-luck-leaf, life-plant, Mexican love-plant, miracle-leaf, sprouting-leaf]
11. *K. rosei* Raym.-Hamet & E. Perrier
12. *K. tomentosa* Baker [panda-bear-plant, plush-plant, pussy-ears]
13. *K. velutina* Welw. ex Oliver
14. *Kalanchoe* sp.

Oomycetes
Phytophthora sp. Crown and root rot. **6**:NJ(82), NY(82). **14**:FL(35).
Pythium sp. Root rot. **1**:FL(35). **14**:NC(1303).
Ascomycotina
Botryosphaeria ribis. On stems. **10**:HI(82,2903).
Sphaerotheca fuliginea. Powdery mildew. **1**:FL(35). **6**:MD(82), NJ(82). **14**:FL(35), NC(1303).
Deuteromycotina - Hyphomycetes
Botrytis cinerea. Leaf blight, stem rot. **14**:NC(1303), WA(3202).
Cercospora sp. Leaf spot. **1**:MS(3413).
Cladosporium sp. Leaf spot. **14**:NC(1303).
Cylindrocladium sp. Root rot. **1**:FL(35).
Fusarium sp. Stem rot. **6**:NJ(82). **14**:FL(35).
Stemphylium bolickii. Leaf spot. **1**:FL(35,3374). **2**:FL(3374). **3**:FL(3374). **4**:FL(3374). **5**:FL(3374). **6**:FL(3374). **7**:FL(3374). **8**:FL(3374). **9**:FL(3374). **11**:FL(3374). **12**:FL(3374). **13**:FL(3374). **14**:FL(35,3374).
Deuteromycotina - Coelomycetes
Colletotrichum gloeosporioides. Anthracnose. **14**:FL(35).
Lasiodiplodia theobromae. Leaf and stem rot. **10**:AL(82). **14**:FL(35).
Deuteromycotina - Other
Rhizoctonia solani. Damping-off. **1**:FL(35). **14**:FL(35).
Sclerotium rolfsii. Southern blight, stem rot. **2**:FL(35). **10**:FL(35).

Sedum L.
1. *S. acre* L. [gold-moss, golden-carpet]
2. *S. douglasii* Hook.
3. *S. kamtschaticum* Fisch. & C. A. Mey.
4. *S. lineare* Thunb. cv. 'variegatum'
5. *S. nussbaumerianum* Bitter
6. *S. spathulifolium* Hook.
7. *S. spectabile* Boreau
8. *S. telephium* L. [live-forever, orpine]
9. *Sedum* sp.

Chytridiomycetes
Synchytrium aureum. **2**:WA(3202).
Synchytrium sp. **2**:WA(3202).
Oomycetes
Phytophthora sp. Stem rot. **8**:NY(82).
Ascomycotina
Erysiphe polygoni. Powdery mildew. **3**:CA(1155).
Pleospora sp. Leaf spot. **8**:NY(82).
Basidiomycotina - Uredinales
Puccinia rydbergii. Autoecious. **6**:OR(3202). **9**:UT(82).
P. umbilici. **9**:CO(82), WY(82).
Deuteromycotina - Hyphomycetes
Botrytis cinerea. Cutting rot. **9**:FL(35).
Cercospora sedi. On leaves. **9**:AL(82).
Cercospora sp. Leaf spot. **9**:FL(35).
Corynespora cassiicola. Leaf spot. **9**:FL(35).
Fusarium solani. Root rot. **9**:FL(35).
Stemphylium bolickii. Leaf spot. **5**:FL(3374). **9**:FL(35).
Tilletiopsis washingtonensis. On leaves. **9**:KS(2686).
Deuteromycotina - Coelomycetes
Colletotrichum sp. Anthracnose, stem rot. **1**:VA(82). **8**:VA(82). **9**:FL(35).
Diplodia sedicola. On stems. **9**:CA(82).
Phyllosticta sp. Leaf spot. **9**:FL(35).
Septoria sedi. Leaf spot. **1**:ME(82), NY(82). **7**:IA(82), IL(82). **8**:NY(82), VT(82), WI(82).
Septoria sp. Leaf spot. **9**:FL(35).
Sphaeronaema minutulum. On stems. **7**:NY(82).
Vermicularia beneficiens. **1**:Unk.(82). **8**:Unk.(82).
Deuteromycotina - Other
Rhizoctonia solani. Root and stem rot. **7**:IL(82), NJ(82). **9**:FL(35), IL(82), NJ(82).
Rhizoctonia sp. Root rot. **4**:HI(2903).
Sclerotium rolfsii. Southern blight. **1**:KS(82), NJ(82), VA(82). **9**:NC(1303).

Sempervivum L.
1. *S. tectorum* L. [hens-and-chickens, houseleek, live-forever]
2. *Sempervivum* sp.

Oomycetes
Phytophthora nicotianae var. *parasitica*. Stem rot. **2**:NY(82).
Pythium sp. Root rot. **1**:IA(82).
Basidiomycotina - Uredinales
Endophyllum sempervivi. Autoecious. **1**:MA(82), NJ(82), NY(82). **2**:MA(82), NJ(82), NY(82).
Deuteromycotina - Hyphomycetes
Cercospora sp. Leaf spot. **2**:FL(35).

Tillaea L.
This genus is considered a synonym of *Crassula*, but the appropriate combinations could not be located.
1. *T. minima* Miers (=*T. erecta* Hook. & Arnott)
2. *T. truncata* Spach

Chytridiomycetes
Synchytrium tillaeae. Leaf and petiole galls. **1**:CA(1841).
Ascomycotina
Anthostoma simplex. **2**:GA(1438).

Crossosomataceae

Crossosoma Nutt.
1. *C. bigelovii* S. Wats.
2. *C. californicum* Nutt.
3. *Crossosoma* sp.

Ascomycotina
Eutypa lata. On wood. **2**:CA(2439).
Basidiomycotina - Other
Peniophora malenconii subsp. *americana*. On wood. **3**:CA(542).
Phanerochaete velutina. White rot. **2**:CA(2439). **3**:CA(128,433).
Scytinostroma portentosum. **1**:AZ(1202).

Forsellesia Greene
1. *F. nevadensis* (A. Gray) A. Heller (=*Glossopetalon nevadense* A. Gray)

Ascomycotina
Sphaerotheca macularis. **1**:ID(3202).

Cucurbitaceae

Benincasa Savi
1. *B. hispida* (Thunb.) Cogn. (=*B. cerifera* Savi) [ash gourd, Chinese fuzzy gourd, Chinese preserving melon, Chinese watermelon, Chinese winter melon, tunka, wax gourd, waxgourd, white gourd, white pumpkin, zit-kwa]

Oomycetes
Pseudoperonospora cubensis. Downy mildew. **1**:MA(82).
Deuteromycotina - Hyphomycetes
Cercospora echinocystis. Leaf spot. **1**:HI(2903).
Deuteromycotina - Coelomycetes
Colletotrichum orbiculare. Anthracnose. **1**:IN(82).

Bryonopsis Arnott
1. *B. laciniosa* (L.) Naudin

Oomycetes
Pseudoperonospora cubensis. Downy mildew. **1**:MA(82), OH(82).

Citrullus Schrad. ex Ecklon & Zeyh.
1. *C. lanatus* (Thunb.) Matsum. & Nakai (=*C. vulgaris* Schrad.) [watermelon]
2. *Citrullus* sp.

Oomycetes
Phytophthora cactorum. **1**:AZ(82).
P. capsici. **1**:CO(82).
P. citrophthora. **1**:CA(82,1110).
P. nicotianae var. *parasitica*. **1**:TX(2615).
Phytophthora sp. Damping-off, fruit rot, stem rot. **1**:Unk.(82), CA(1110), FL(35), NC(1303), WA(3202).
Pythium acanthicum. Damping-off, blossom-end rot, root rot. **1**:FL(35,82), GA(82), IA(82), IN(82), MD(82), MO(82), VA(82).
P. aphanidermatum. Damping-off, cottony leak, blossom end rot, root rot. **1**:CA(1110), FL(35,82), HI(2903), NC(1303), VA(82).
P. debaryanum. Damping-off, blossom end rot, fruit and root rot. **1**:Unk.(82), FL(35).
P. helicoides. Fruit rot. **1**:Unk.(82).
P. hydnosporum. Fruit rot. **1**:Unk.(82).
P. irregulare. Damping-off. **1**:IA(82).
P. myriotylum. Damping-off, cottony leak, blossom end rot, fruit and root rot. **1**:FL(35,82), GA(82), NC(1303).
P. periplocum. Fruit rot. **1**:MD(82), VA(82).
P. ultimum. Damping-off, cottony leak, blossom end rot, root rot. **1**:CA(82,1110), NC(1303).
Pythium sp. Damping-off, blossom end rot, root rot. **1**:CA(1110), FL(35), NC(1303), OK(2872).
Zygomycotina
Choanephora cucurbitarum. Blossom end rot. 'Choanephora fruit rot.' **1**:NC(1303).
Rhizopus stolonifer. 'Rhizopus soft rot.' **1**:Unk.(82).
Rhizopus sp. Soft rot. **1**:Unk.(82), CA(1110).
Ascomycotina
Botryosphaeria quercuum. Stem end rot. **1**:FL(35).
B. rhodina. On dead stems. **1**:FL(82).
Erysiphe cichoracearum. Powdery mildew. **1**:AZ(82), CA(82,1110), FL(35,82,3127), GA(82), NC(82,1303), NY(82), TX(82), VA(82).
Glomerella lagenaria. Anthracnose, fruit rot. **1**:FL(35), NC(1303).
G. magna. Anthracnose. **1**:NC(1303).
Physalospora abdita. On dead stems. **1**:FL(82).
Sclerotinia sclerotiorum. Cottony rot. 'Sclerotinia stem rot.' **1**:NJ(82), TX(82).
Sphaerotheca fuliginea. 'Powdery mildew.' **1**:CA(1110,1952), NY(1303).
Basidiomycotina - Other
Athelia rolfsii. Fruit rot. **1**:FL(3128), TX(2914).
Thanatephorus cucumeris. **1**:WA(3202).
Deuteromycotina - Hyphomycetes
Acremonium sp. **1**:CA(1110).
Alternaria sp. **1**:CA(1110).
Cercospora citrullina. 'Cercospora leaf spot.' **1**:FL(35,3127), IL(312), MS(2720), NC(3931).
Cladosporium cucumerinum. Crown blight on leaves, petioles and stems. 'Scab.' **1**:MD(82), NC(1303), NE(82), TX(3314).
Cladosporium sp. **1**:CA(1110).
Corynespora cassiicola. 'Target spot.' **1**:NC(1303).
Cylindrocladium scoparium. Fruit rot. **1**:FL(35,3128).
Fusarium equiseti. Fruit rot. **1**:Unk.(82).
F. roseum. Fruit rot. **1**:NC(1303).
F. solani. **1**:CA(1110).
Fusarium sp. Damping-off, stem rot. **1**:NC(1303), OK(2872).
Helminthosporium sp. Fruit rot. **1**:TX(82).
Penicillium sp. **1**:CA(1110).
Phymatotrichopsis omnivora. Root rot. **1**:AZ(82), TX(82).
Thielaviopsis basicola. 'Black root rot.' **1**:OR(82,3202), UT(82).
Verticillium albo-atrum. 'Verticillium wilt.' **1**:CA(82), FL(35), NH(82), OR(82).
V. dahliae. 'Verticillium wilt.' **1**:CA(1110).
Deuteromycotina - Coelomycetes
Ascochyta sp. Stem rot, leaf spot. **1**:FL(35).
Colletotrichum gloeosporioides. On fruit. **1**:NY(82).
Cribropeltis citrullina. Fruit speck. **1**:IL(82).
Diplodia sp. Gray rot, stem end rot. **1**:SE states(82), AZ(82), FL(35), KS(82), OK(2872).
Macrophoma seminalis. On leaves and seeds. **1**:SC(82).
Macrophoma sp. Blossom end rot. **1**:GA(82), MO(82).
Macrophomina phaseolina. 'Charcoal rot.' **1**:CA(1110), FL(35,3128), TX(82).
Marssonina melonis. Anthracnose. **1**:NY(82).
Phoma cucurbitacearum. Gummy stem blight, stem end rot, leaf spot. **1**:FL(3803), NC(1303).
P. exigua. Gummy stem blight, leaf margin blight. 'Phoma blight.' **1**:NC(1303,1752).
Phoma sp. **1**:WA(3202).
Phyllosticta cucurbitacearum. On stems. **1**:FL(35,82), GA(82), KY(82), MO(82).
Septoria citrulli. On leaves. **1**:NJ(82), NY(82). **2**:NJ(965).
S. cucurbitacearum. **1**:WI(82).
Deuteromycotina - Other
Rhizoctonia solani. Leaf blight. 'Belly rot. Damping-off. Web blight.' **1**:E states(82), CA(1110), FL(35), GA(82), IA(82), NC(1303), TX(82).
Rhizoctonia sp. Root rot. **1**:HI(2903).
Sclerotium rolfsii. Fruit rot. 'Southern blight.' **1**:SE states(82), AL(285), CA(1110), FL(35), HI(2903), MO(82), NC(3931), TX(82).

Cucumis L.

1. *C. anguria* L. [bur gherkin, goarberry gourd, gooseberry gourd, West Indian gherkin]
2. *C. melo* L. (=*C. melo* L. var. *melo*) [melon]
3. *C. melo* L. var. *cantalupensis* Naudin [cantaloupe]
4. *C. melo* L. var. *inodorus* Naudin [casaba melon, honeydew melon, winter melon]
5. *C. melo* L. var. *reticulatus* Naudin [muskmelon, netted melon, nutmeg melon, Persian melon]
6. *C. sativus* L. [cucumber, pumpkin]

Oomycetes
Phytophthora cactorum. Fruit rot. **5**:UT(82).
P. capsici. Fruit rot. **2**:CA(1110). **4**:CA(82), CO(82). **5**:CA(82), CO(82). **6**:CO(82).
P. drechsleri. Fruit rot. **2**:CA(1110). **4**:CA(82).
Phytophthora sp. Fruit rot. **2**:Unk.(82), CA(1110), NC(1303,3931).
Pseudoperonospora cubensis. Downy mildew, powdery mildew. **1**:NC(1303), OH(82), TX(82). **3**:HI(2903). **4**:Range of host(82). **5**:Range of host(82).
Pythium anandrum. Fruit rot. **6**:CA(82).
P. aphanidermatum. Damping-off, cottony leak, fruit and root rot. **2**:CA(1110), FL(35), NC(1303). **4**:AZ(82), CA(82). **5**:AZ(82), CA(82), TX(82). **6**:Unk.(82), FL(35), GA(82), HI(2903), LA(82), NC(82,1303,3931), NJ(82), SC(82), VA(82).
P. debaryanum. Seedling blight, root blight, damping-off. **2**:CA(82), CT(82), FL(35), IA(82), NJ(82), NY(82). **6**:C states(82), E states(82), FL(35), NC(1303).
P. myriotylum. Fruit rot, root rot. **6**:NC(1303), SC(82).
P. oligandrum. Fruit rot. **6**:MD(82).
P. periplocum. Root rot. **2**:CA(82,1110).
P. ultimum. Seedling blight, damping-off, root rot. **2**:CA(82,1110), HI(2903). **6**:C states(82), E states(82), FL(35), HI(2903), MD(82), NC(1303), WA(3202).
Pythium sp. Damping-off, fruit and seed rot. **2**:CA(1110), NC(1303,3931). **6**:NC(1303).
Zygomycotina
Choanephora cucurbitarum. Blossom blight. **6**:FL(35,82), GA(82), HI(2903), NC(2928).
Cunninghamella echinulata. **6**:NC(2928).
Mucor hiemalis f. *silvaticus*. **6**:NC(2928).
Mucor sp. Fruit rot. **2**:NY(82). **6**:NC(2928).
Rhizopus arrhizus. **2**:Unk.(82). **6**:NC(2928).
R. stolonifer. Soft rot. **2**:Unk.(82). **6**:Unk.(82), FL(3390), NC(2928).
Rhizopus sp. Fruit rot, Rhizopus rot. **2**:Range of host(82), FL(35). **4**:Range of host(82).
Ascomycotina
Botryosphaeria rhodina. On dead stems. **2**:NC(82).
Chaetomium aureum. Isolated from seeds. **6**:MI(3306).
C. globosum. Isolated from seeds. **2**:MI(3306), NJ(3306). **6**:NC(2928).
Didymella bryoniae. Gummy stem blight, seedling blight, black rot, leaf spot. **2**:FL(35,82), MS(2720), NC(1303,3931), SC(998). **3**:HI(2903). **5**:DE(82), MA(82), NJ(82), NY(82), TX(82).
Erysiphe cichoracearum. Powdery mildew. **1**:MA(82), NC(1303). **2**:AZ(2251), CA(1110), FL(35), GA(1438), MS(2720), NC(1303,3931), SC(998), TX(1253). **4**:AZ(82), CA(82). **5**:Range of host(82), AZ(82), CA(82), UT(82). **6**:FL(35,2683,3500), MA(1390), MS(2720), NC(1303), OK(2872),

SD(2243), WA(3202).
E. polygoni. **2**:GA(1438).
Glomerella lagenaria. Anthracnose. **2**:NC(1303). **6**:NC(1303).
G. magna. Anthracnose. **2**:NC(1303). **6**:NC(1303).
Monosporascus cannonballus. Decay of secondary roots in greenhouse. **2**:AZ(2847).
Pseudonectria sp. **6**:NC(2928).
Sclerotinia sclerotiorum. Fruit, stem and timber rot. **2**:CA(1110). **5**:AR(82), MA(82), TX(82). **6**:E states(82), NW states(82), CA(82), CO(447), FL(35,82), IL(82), MN(82), NC(1303), OR(3202), TX(82), WA(3202).
Sphaerotheca fuliginea. Powdery mildew. **2**:CA(1110), NC(1304), TX(501). **6**:Unk.(3976), CA(83), NC(1303).
Thielavia basicola. **6**:NC(2928).

Basidiomycotina - Other

Thanatephorus cucumeris. **6**:WA(3202).

Deuteromycotina - Hyphomycetes

Acremonium sp. **2**:CA(1110).
Alternaria alternata. Seed mold. **2**:CA(1110). **6**:Range of host(82), FL(35,533), NC(2928), NY(533).
A. cucumerina. Leaf blight, black mold blight, leaf spot. **2**:Range of host(82), Unk.(82), AL(285), FL(35), MS(2720), NC(1303,3931), OK(2872), SC(998), TX(501,1253). **6**:Range of host(82), AZ(693), FL(35), HI(2903), NC(1303), OK(2872), OR(3202), WA(3202).
Alternaria sp. Fruit rot. **2**:Unk.(82), CA(1110), FL(35), NC(1303). **6**:FL(35).
Arthrinium arundinis. **6**:NC(2928).
Aspergillus elegans. **6**:NC(2928).
A. fumigatus. **6**:NC(2928).
A. niger. **6**:NC(2928).
A. sydowii. **6**:NC(2928).
A. terreus. **6**:NC(2928).
A. unguis. **6**:NC(2928).
A. versicolor. **6**:NC(2928).
Aspergillus sp. Fruit rot. **2**:OK(2872).
Aureobasidium pullulans. **6**:NC(2928).
Botrytis cinerea. Gray mold rot. **6**:FL(35,533), IL(82), NY(533), OH(82), WA(82,3202).
Botrytis sp. Fruit rot. **6**:AK(2162).
Cephalosporium sp. **6**:NC(2928).
Cercospora citrullina. Leaf spot. **2**:DE(82), IL(312), IN(82), NC(1303,3931), TX(82). **6**:FL(82), GA(82), HI(2903), IA(82), IL(312), IN(82), OH(82), SD(2243).
Cercospora sp. Leaf spot. **2**:CO(82), GA(82), TX(82). **6**:FL(35,82), GA(82), IA(82), IN(82), NC(2928), OH(82).
Chrysonilia sitophilia. Fruit rot. **2**:IN(82), NY(82).
Cladosporium cladosporioides. **6**:NC(2928).
C. cucumerinum. Crown blight, scab of foliage and fruit. **2**:E & C states(82), NC(1303), SC(3930), TX(3314). **6**:Range of host(82), FL(35), NC(1303,3931), OK(2872), OR(3202), WA(1138,3202).
Corynespora cassiicola. Leaf spot, target spot. 'Cercospora leaf spot.' **2**:FL(35), NC(1303). **6**:FL(35), NC(1303,3929).
Curvularia trifolii. Seed mold. **6**:NC(2928), NJ(82).
Cylindrocladium sp. Leaf spot. **2**:FL(35).
Dactylaria constricta. **6**:NC(2928).
Fusarium acuminatum. Fruit rot. **2**:Range of host(82).
F. culmorum. Fruit rot. **2**:Range of host(82).
F. episphaeria. **6**:NC(2928).
F. graminearum. Fruit rot. **2**:Range of host(82).
F. javanicum. Root rot. **6**:CT(82), OR(82), WA(82,3202).
F. moniliforme. **6**:NC(2928).
F. oxysporum. **6**:NC(2928).
F. oxysporum f. sp. *cucumerinum*. Wilt. **6**:NC(1303,1753).
F. oxysporum f. sp. *melonis*. Wilt. **2**:CA(1110), FL(35), HI(2903), NC(1303). **4**:AZ(82), CA(82), CO(82). **5**:E & C states(82), Range of host(82), S states(82), W states(82), CA(1395), TX(2274), WA(2237).
F. oxysporum f. sp. *niveum*. Wilt. **2**:OK(2872). **6**:CA(82), MI(82), NM(82), TX(82), WI(82).
F. pallidoroseum. Fruit rot. **2**:Range of host(82).
F. roseum. Corky, dry rot, fruit rot. **2**:E & C states(82), AZ(82), CA(82,1110), CO(82), TX(82,519). **6**:NC(2928).
F. solani. Fruit rot. **2**:Range of host(82). **6**:NC(2928).
F. solani f. sp. *cucurbitae*. **2**:CA(1110).
F. subglutinans. Fruit rot. **2**:Range of host(82).
Fusarium sp. Fruit, root and stem rot, wilt. **2**:E & C states(82), AZ(82), CA(82,661), CO(82), FL(35), NC(1303), OK(2872), TX(82). **6**:FL(533,2683), MS(1782), NC(1303), NY(533), OK(2872), TX(82).
Geotrichum candidum. Market decay. **5**:CA(532), NY(532).
Geotrichum sp. **6**:NC(2928).
Gliocladium roseum. **6**:NC(2928).
Helminthosporium sp. **6**:NC(2928).
Humicola fuscoatra. **6**:NC(2928).
Leandria momordicae. Net spot of leaves. **6**:FL(35,300), IN(82), OH(82).
Myrothecium roridum. Fruit and rind rot. **2**:IL(312). **5**:TX(2325). **6**:NC(2928).
M. verrucaria. **6**:NC(2928).
Myrothecium sp. **6**:MS(1782).
Oidium sp. Powdery mildew. **6**:HI(2903).
Penicillium atramentosum. **6**:NC(2928).
P. chermesinum. **6**:NC(2928).
P. citrinum. **6**:NC(2928).
P. duclauxii. **6**:NC(2928).
P. expansum. **6**:NC(2928).
P. funiculosum. **6**:NC(2928).
P. glabrum. **6**:NC(2928).
P. gladioli. **6**:NC(2928).
P. griseofulvum. **6**:NC(2928).
P. herquei. **6**:NC(2928).
P. implicatum. **6**:NC(2928).
P. janczewskii. **6**:NC(2928).
P. janthinellum. **6**:NC(2928).
P. oxalicum. **6**:NC(2928).
P. restrictum. **6**:NC(2928).
P. rugulosum. **6**:NC(2928).
P. simplicissimum. **6**:NC(2928).
P. sublateritium. **6**:NC(2928).
P. thomii. **6**:NC(2928).
P. variabile. **6**:NC(2928).
Penicillium sp. Fruit rot. **2**:Unk.(82), OK(2872). **6**:NC(2928).
Phymatotrichopsis omnivora. Root rot. **2**:TX(82). **6**:TX(82).
Sporotrichum pruinosum. **6**:NC(2928).
Stysanus sp. **6**:NC(2928).
Trichoderma viride. Green mold rot. **2**:MN(82). **6**:NC(2928).
Trichothecium roseum. Pink mold rot. **2**:CA(82,1110), IN(82), NY(82).
Ulocladium consortiale. **6**:WA(3202).
Verticillium albo-atrum. Wilt. **2**:CA(82,3898), FL(35), OR(82,3202), WA(3202). **6**:CA(1396), FL(35), ME(82), OH(82), OR(82,3202), WA(82,3202), WI(82).
V. dahliae. **2**:CA(1110), WA(3308).
V. tricorpus. **2**:WA(3202,3308).

Deuteromycotina - Coelomycetes

Ascochyta sp. Rot inside fruit. **2**:NC(1303). **6**:NC(2928).
Colletotrichum orbiculare. Anthracnose, scab. **2**:E states(82), S states(82), SW states(82), CO(82), FL(35), HI(2903), MS(2720), NC(1303), ND(82), OK(2872), SC(998). **3**:HI(2903). **6**:Range of host(82), SE states(3613), AL(285), FL(35,2683), LA(125,1631), MS(2720), NC(956,1303,2928,3931), OK(2872).
Colletotrichum sp. Needle rust. **2**:NY(82).
Lasiodiplodia theobromae. Blight, stem gumming, gray rot, stem end rot. **2**:TX(82,244). **6**:TX(82).
Macrophomina phaseolina. Vine decline, charcoal rot, fruit and stem rot. **2**:Unk.(82), CA(82,1110), IL(311), OR(82), TX(82,518). **6**:IL(82,311).
Marssonina melonis. Anthracnose. **2**:NY(82).
Pestalotiopsis guepini var. *macrotricha*. **6**:NC(2928).
Phoma cucurbitacearum. Gummy stem blight, stem end rot, leaf spot. **2**:NC(1303). **6**:NC(2928).
P. exigua. Blight, leaf spot. **2**:NC(1752). **6**:NC(1303,1752).
P. terrestris. Secondary root rot. **2**:CO(82), IA(82).
Phoma sp. **6**:NC(2928).
Phyllosticta cucurbitacearum. Leaf spot. **2**:OH(64). **6**:DE(82), OH(64,82), TX(82).
Phyllosticta sp. Leaf spot. **2**:FL(35), GA(82), OH(82), TX(82).
Septoria cucurbitacearum. Leaf spot. **5**:DE(82), MA(82), MI(82), NH(82), NY(82), PA(82), VT(82), WI(82).
Stagonospora sp. **6**:NC(2928).

Truncatella sp. **6:**NC(2928).
Deuteromycotina - Other
Rhizoctonia solani. Foliage blight, damping-off, belly rot of fruit, root and stem rot. **2:**CA(82,1110), FL(35), GA(82), NC(1303,3931), NJ(82). **6:**Unk.(82), FL(35), NC(956,1303,3931).
R. zeae. Root rot. **6:**NC(1303).
Rhizoctonia sp. Root rot. **3:**HI(2903). **6:**HI(2903).
Sclerotium rolfsii. Southern blight, fruit rot. **2:**E states(82), CA(2886), FL(35), MS(2720), NC(1303,3931), TX(82), VA(82). **6:**SE states(82), FL(35), MS(2720), NC(1303), TX(82).

Cucurbita L.

1. *C. foetidissima* H. B. K. [calabazilla, fetid wild pumpkin, Missouri gourd]
2. *C. maxima* Duchesne [autumn squash, winter squash]
3. *C. moschata* (Duchesne) Duchesne ex Poir. [Canada pumpkin, crookneck squash, pumpkin, winter squash]
4. *C. pepo* L. [marrow, pumpkin, summer squash, warron, winter squash]
5. *C. pepo* L. var. *melopepo* (L.) Alef. (=*C. pepo* L. var. *condensa* L. H. Bailey) [bush pumpkin, bush squash]
6. *C. pepo* L. var. *ovifera* (L.) Alef. [yellow-flowered gourd]
7. *C. texana* A. Gray [Texas gourd, wild marrow]
8. *Cucurbita* sp.

Oomycetes
Phytophthora cactorum. Fruit rot. **2:**AZ(82). **4:**AZ(82).
P. capsici. Blight, wilt. **2:**CO(82), NC(2838). **4:**VA(82). **5:**MO(82), NC(733). **8:**NC(3931).
P. citrophthora. **4:**CA(82).
P. nicotianae var. *parasitica*. Foliage and root rot. **2:**FL(35).
Phytophthora sp. Blight, fruit and root rot. **4:**NC(1303). **8:**HI(2903).
Pseudoperonospora cubensis. 'Downy mildew.' **2:**E states(82), CA(82), FL(35), TX(82). **3:**LA(82), NY(82), TX(82). **4:**NE states(82), HI(2903), IA(82), NC(1303), WI(82). **5:**FL(82), MA(82), NY(82), VA(82). **6:**IN(82), NJ(82), OH(82), TX(82). **8:**FL(35), IL(310), NC(1303,3931).
Pythium aphanidermatum. Blossom end rot, root rot. **2:**AZ(82). **4:**CA(82), MD(82), SC(82). **5:**AZ(82), CA(82), MD(82), SC(82).
P. debaryanum. Damping-off. **2:**CT(82), WI(82).
P. ultimum. Damping-off, fruit rot. **2:**CA(82). **4:**TX(82). **5:**TX(82).
Pythium sp. Damping-off, fruit, root and stem rot. **2:**FL(35). **4:**HI(2903), NC(1303). **8:**FL(35), HI(2903), NC(1303).
Zygomycotina
Choanephora cucurbitarum. Blossom blight, blossom end fruit rot, root rot. 'Choanephora fruit rot.' **2:**FL(35), NJ(82), OK(82,2872), RI(82), TX(82). **3:**OK(82,2872), TX(82). **4:**E states(82), FL(2326), HI(2903), MD(82), MI(82), NC(1303), NJ(82), TX(82), WI(1340). **5:**E states(82), MI(82), TX(82). **8:**FL(35), NC(1303,3931), WV(172).
Rhizopus stolonifer. 'Rhizopus soft rot.' **2:**Range of host(82), ID(3202). **4:**CT(82), NC(1303), NY(82), TX(82). **5:**NY(82), TX(82). **8:**NC(1303).
Rhizopus sp. Fruit and storage rot. **2:**Range of host(82), FL(35). **8:**NC(1303).
Ascomycotina
Chaetomium globosum. Isolated from seeds. **2:**CT(3306).
C. murorum. Isolated from seeds. **2:**MI(3306).
Didymella bryoniae. Black rot of fruit. 'Gummy stem blight.' **2:**CT(82), FL(35), MA(82), MI(82), NH(82), NJ(82), NY(82), PA(82). **3:**MA(82). **4:**FL(35), HI(2903), MA(82), MI(82), NJ(82), NY(82). **5:**FL(82), GA(82), MA(82). **6:**NY(82). **8:**NC(1303).
Erysiphe cichoracearum. Powdery mildew. **1:**CA(82), OK(2872). **2:**Range of host(82), FL(35), OK(2872), WI(1343). **3:**AZ(82), MS(82,2720), TX(82). **4:**Range of host(82), GA(1438), NC(1303), OK(2872), OR(3202), SD(2243), WA(3202). **5:**Range of host(82). **6:**CT(82), PA(82), SD(2243). **8:**FL(35), NC(1303,3931).
E. polygoni. **8:**GA(1438).
Glomerella cingulata. Anthracnose. **8:**NC(1303).
G. lagenaria. Anthracnose. **4:**NC(1303).
G. magna. Anthracnose. **4:**NC(1303). **8:**NC(1303).
Sclerotinia sclerotiorum. Fruit rot. 'Sclerotinia stem rot.' **2:**FL(35), ID(82), MA(82), ME(82), MT(82,3202), NH(82), NY(82), PA(82), WA(82,3202). **4:**ID(82), MT(82,3202). **5:**NY(82).
Sphaerotheca fuliginea. 'Powdery mildew.' **2:**NC(1304). **4:**NC(1304). **5:**CA(1305,1954).
Deuteromycotina - Hyphomycetes
Alternaria alternata. Seed mold. **2:**Range of host(82), FL(35). **3:**Range of host(82).
A. cucumerina. 'Alternaria leaf blight.' **2:**FL(35), ID(3202), NH(82), NY(82), PA(82), UT(82). **3:**TX(82). **4:**IA(82), MN(82), NC(82), NJ(82), WV(82). **5:**DE(82), UT(82). **8:**FL(35), NC(1303).
A. radicina. Seed mold. **4:**Unk.(82).
Alternaria sp. Fruit rot, leaf spot. **2:**MA(82), NH(82), OR(82), VT(82), WA(82,3202). **4:**MN(82), NC(1303), NJ(82), WA(82,3202). **5:**NY(82). **6:**NC(1303).
Botrytis cinerea. 'Gray mold.' **2:**FL(35), ID(82), MA(82), NH(82), OR(82), WA(82,3202). **4:**ID(82), TX(82), WA(3202). **5:**NH(82), NY(82), TX(82).
Botrytis sp. Fruit rot in storage. **8:**AK(2162).
Cercospora citrullina. 'Cercospora leaf spot.' **1:**KS(82), OK(2872), TX(82). **2:**DE(82), IA(82), NJ(82), WI(82). **3:**AL(82), TX(82). **4:**AL(82), DE(82), IN(82), WV(82). **5:**AL(82), MD(82).
Cercospora sp. Leaf spot. **4:**IN(82).
Cladosporium cucumerinum. Storage rot. 'Scab.' **2:**CT(82), MA(82), MD(82), NJ(82), NY(82), OR(82,3202), WA(82,3202). **4:**NC(1303). **5:**MA(82), MD(82), NC(956), NY(82), PA(82). **6:**CT(82). **8:**NC(1303,3931).
Corynespora cassiicola. 'Target spot.' **8:**NC(1303).
Curvularia trifolii. Seed mold. **4:**CT(82).
Fusarium avenaceum. **8:**MA(82).
F. oxysporum f. sp. *niveum*. Wilt. **2:**AZ(82), NC(82). **3:**TX(82). **5:**AZ(82), NM(82).
F. roseum. **2:**WA(3202).
F. solani f. sp. *cucurbitae*. Root rot, wilt. **2:**CA(82), NY(82), OR(82), WA(82). **3:**OR(82). **4:**NC(1303), OR(3202). **5:**CA(82), CT(82), NY(82). **6:**NY(82). **7:**AR(3594).
Fusarium sp. Fruit, stem and storage rot, fruit spot, wilt. **2:**CO(82), FL(35), MA(82), NH(82), OR(82), WA(82), WI(82). **4:**MD(82), NJ(82), PA(82), TX(82), VA(82). **5:**CA(82), CO(82), CT(82), MA(82), OR(82), TX(82), WV(82). **6:**CO(82), CT(82), MA(82), NY(82), OR(82), TX(82), WV(82). **8:**FL(35).
Leandria momordicae. Leaf spot. **3:**NY(82). **5:**IN(82), OH(82). **6:**IN(82), OH(82), PA(82).
Oidium sp. Powdery mildew. **2:**HI(2903). **3:**HI(2903). **5:**HI(2903). **6:**HI(2903). **8:**HI(2903).
Phaeoramularia cucurbiticola. Leaf spot. **8:**FL(35).
Phymatotrichopsis omnivora. Root rot. **2:**TX(82). **3:**TX(82). **4:**TX(82). **5:**TX(82).
Stemphylium botryosum. On seeds. **4:**Unk.(82).
Stemphylium sp. Leaf spot. **3:**NY(82).
Trichothecium roseum. 'Pink mold rot.' **2:**MA(82).
Ulocladium consortiale. Leaf spot. **3:**NY(82). **4:**Unk.(82).
Verticillium albo-atrum. 'Verticillium wilt.' **2:**FL(35), OR(82).
Deuteromycotina - Coelomycetes
Ascochyta sp. Leaf spot. **2:**OR(82).
Colletotrichum orbiculare. 'Anthracnose.' **1:**TX(82). **3:**MD(82), TX(82). **4:**NE states(82), KS(82), NC(1303), TX(82). **5:**TX(82). **6:**NE states(82), KS(82), TX(82). **8:**NC(1303).
Lasiodiplodia theobromae. 'Diplodia fruit rot.' **2:**TX(82). **3:**TX(82). **4:**LA(82), TX(82).
Macrophomina phaseolina. 'Charcoal rot.' **2:**IL(311). **3:**TX(82). **4:**OR(82). **5:**OR(82).
Phoma cucurbitacearum. Gummy stem blight, stem end rot, leaf spot. **2:**WI(1355). **8:**NC(1303).
P. lagenariae. On fruit. **4:**NY(82).
P. subvelata. Fruit and leaf spot. **2:**TX(82). **6:**TX(82).
P. terrestris. Secondary root rot. **4:**ND(82).
Phyllosticta cucurbitacearum. Leaf spot. **2:**IN(82).
P. orbicularis. Leaf spot. **4:**DE(82), NY(82).
Septoria cucurbitacearum. Fruit and leaf spot. **2:**MA(82), NH(82), NY(82), WI(82). **4:**MA(82). **5:**MA(82).
S. vestita. On fruit. **2:**PA(82).
Deuteromycotina - Other
Rhizoctonia solani. Root and stem rot. 'Belly rot. Damping-off. Web blight.' **2:**FL(35), TX(82). **3:**TX(82). **4:**FL(35), NC(1303), TX(82). **5:**TX(82). **8:**NC(1303).
Rhizoctonia sp. Root rot. **4:**HI(2903).
Sclerotium rolfsii. 'Southern blight.' **2:**AL(82,285), FL(35), HI(2903). **4:**FL(35,82), GA(82). **5:**FL(82), GA(82).

Echinocystis Torr. & A. Gray

1. *E. lobata* (Michx.) Torr. & A. Gray [mock cucumber, prickly cucumber, wild balsam-apple, wild cucumber]

Oomycetes

Plasmopara australis. Downy mildew. **1:**IA(82), KS(82), MN(82), OH(82), WI(82).
Pseudoperonospora cubensis. Downy mildew. **1:**IA(82), OH(82).

Ascomycotina

Erysiphe cichoracearum. Powdery mildew. **1:**WI(82).
Mycosphaerella sicyicola. **1:**WI(1349).

Deuteromycotina - Hyphomycetes

Alternaria sp. Leaf spot. **1:**FL(35,82).
Cercospora echinocystis. Fruit and leaf spot. **1:**E states(82), FL(35), NE(82), WI(82).
Fusarium sp. Wilt. **1:**FL(35,82).
Leandria momordicae. **1:**WI(1344).

Deuteromycotina - Coelomycetes

Colletotrichum orbiculare. Anthracnose. **1:**FL(35,82).
Phloeospora megarrhizae. **1:**Unk.(82).
Septoria brencklei. Leaf spot. **1:**IA(82), IL(82), IN(82), MI(82), ND(82,378), WI(82), WY(82).
S. echinocystis. Leaf spot. **1:**CA(82).
S. sicyi. **1:**IA(82), IL(82), IN(82), MI(82), ND(82), WI(82), WY(82).

Lagenaria Ser.

1. *L. siceraria* (Molina) Standl. [bottle gourd, calabash gourd, white-flowered gourd]

Oomycetes

Pseudoperonospora cubensis. Downy mildew. **1:**CT(82), FL(35,82), MA(82), OH(82).

Ascomycotina

Erysiphe cichoracearum. Powdery mildew. **1:**WI(82).
Laestadia cucurbitacearum. Fruit rot. **1:**PA(82).
Sphaerotheca fuliginea. Powdery mildew. **1:**NC(1304).

Basidiomycotina - Other

Pellicularia koleroga. Thread blight. **1:**FL(82).

Deuteromycotina - Hyphomycetes

Cercospora citrullina. Leaf spot. **1:**AL(82), FL(35), IN(82).
Oidium sp. Powdery mildew. **1:**HI(2903).
Phymatotrichopsis omnivora. Root rot. **1:**TX(82).
Stemphylium sp. Fruit spot. **1:**NY(82).

Deuteromycotina - Coelomycetes

Colletotrichum orbiculare. Anthracnose. **1:**CT(82), FL(35), IL(82), IN(82), MN(82), NE(82), PA(82).
Colletotrichum sp. Anthracnose. **1:**HI(2903).
Phoma subvelata. Fruit spot. **1:**TX(82).

Deuteromycotina - Other

Rhizoctonia solani. Thread blight. **1:**FL(35).

Luffa Mill.

1. *L. acutangula* (L.) Roxb. [angled luffa, sing-kwa, towel gourd]
2. *L. aegyptiaca* Mill. (=*L. cylindrica* M. Roem.) [dishcloth gourd, loofah, luffa, smooth loofah, sponge gourd, vegetable sponge]

Oomycetes

Pseudoperonospora cubensis. Downy mildew. **1:**CT(82), NC(82), OH(82). **2:**MA(82), NC(1303).

Ascomycotina

Erysiphe cichoracearum. Powdery mildew. **1:**CT(82).

Deuteromycotina - Hyphomycetes

Cercospora citrullina. Leaf spot. **1:**AL(82). **2:**AL(82).
Cercospora sp. Leaf spot. **2:**FL(35).

Deuteromycotina - Coelomycetes

Colletotrichum orbiculare. Anthracnose. **1:**CT(82), IA(82). **2:**CT(82), IA(82).
Phyllosticta sp. Leaf spot. **2:**FL(35).

Marah Kellogg

(= *Megarrhiza* Torr. & A. Gray.)

1. *M. fabacea* (Naudin) Greene (=*Echinocystis fabacea* Naudin, *Megarrhiza californica* Torr.)
2. *M. oregana* (Torr. & A. Gray) Howell (=*Echinocystis oregana* (Torr. & A. Gray) Cogn.)
3. *Marah* sp. (=*Megarrhiza sp.*)

Ascomycotina

Crocicreas cyathoideum. **1:**CA(505), NY(505).
Didymella megarrhizae. **2:**CA(81).
Mycosphaerella sp. **2:**WA(3202).
Stictis monilifera. **3:**CA(3225).

Deuteromycotina - Coelomycetes

Phoma megarrhizae. **1:**CA(2439).
Septoria echinocystis. Leaf spot. **1:**CA(82). **2:**CA(82).
S. megarrhizae. Leaf spot. **1:**CA(81,2439). **2:**CA(82), OR(82,3202), WA(82,3202).
Septoria sp. Leaf spot. **1:**Unk.(82). **2:**Unk.(82).
Stagonospora sp. **2:**WA(3202).

Melothria L.

1. *M. pendula* L.
2. *M. scabra* Naudin
3. *Melothria* sp.

Oomycetes

Pseudoperonospora cubensis. Downy mildew. **1:**FL(35). **2:**GA(82), OH(82). **3:**TX(82).

Ascomycotina

Erysiphe cichoracearum. Powdery mildew. **2:**WI(82).

Basidiomycotina - Uredinales

Uromyces novissimus. Autoecious. **1:**FL(35).

Deuteromycotina - Hyphomycetes

Cercospora citrullina. Leaf spot. **1:**FL(35).

Momordica L.

1. *M. balsamina* L. [balsam-apple]
2. *M. charantia* L. [balsam-pear, bitter cucumber, bitter gourd, la-kwa]

Oomycetes

Pseudoperonospora cubensis. Downy mildew. **1:**IA(82). **2:**IA(82).
Pythium aphanidermatum. Root rot. **2:**HI(2903).

Ascomycotina

Erysiphe cichoracearum. Powdery mildew. **1:**WI(82). **2:**FL(35), HI(2899), WI(82).

Deuteromycotina - Hyphomycetes

Cercospora citrullina. Leaf spot. **2:**FL(35).
Corynespora cassiicola. Leaf spot. **2:**FL(35).
Leandria momordicae. Net spot. **2:**FL(35).
Oidium sp. Powdery mildew. **2:**HI(2903).
Ramularia momordicae. Leaf blight. **1:**TX(82).

Deuteromycotina - Coelomycetes

Colletotrichum orbiculare. Anthracnose. **1:**IN(82). **2:**FL(35).
Colletotrichum sp. Anthracnose. **2:**HI(2903).

Sechium P. Br.

1. *S. edule* (Jacq.) Sw. [chayote, choyote, christophine]

Ascomycotina

Erysiphe cichoracearum. Powdery mildew. **1:**FL(35), HI(2899).
Glomerella cingulata. Fruit rot. **1:**LA(82).

Deuteromycotina - Hyphomycetes

Cercospora citrullina. Leaf spot. **1:**FL(35), TX(82).
C. sechii. Leaf spot. **1:**FL(82), TX(82).
Corynespora cassiicola. Leaf spot. **1:**FL(35).
Oidium sp. Powdery mildew. **1:**HI(2903).

Deuteromycotina - Coelomycetes

Colletotrichum orbiculare. Anthracnose. **1:**FL(35,82), TX(82).

Deuteromycotina - Other

Sclerotium rolfsii. Southern blight. **1:**TX(82).

Sicana Naudin

1. *S. odorifera* (Vell.) Naudin [casabanana, coroa, curua, curubá]

Zygomycotina

Choanephora cucurbitarum. Flower blight, fruit rot. **1:**FL(35).

Deuteromycotina - Coelomycetes

Colletotrichum orbiculare. Anthracnose. **1:**FL(35,82).

Deuteromycotina - Other

Rhizoctonia solani. Root rot. **1:**FL(35).

Sicyos L.

1. *S. angulatus* L. [bur-cucumber, star-cucumber]

Oomycetes

Plasmopara australis. Downy mildew. **1:**C states(82), E states(82), MS(2720), NE(82), WI(82).
Pseudoperonospora cubensis. Downy mildew. **1:**TX(82).

Ascomycotina
Erysiphe cichoracearum. Powdery mildew. **1**:TX(82), WI(82).
Mycosphaerella sicyicola. **1**:MO(525,967), WI(1348).
Mycosphaerella sp. Leaf spot. **1**:MO(82).
Deuteromycotina - Hyphomycetes
Cercospora echinocystis. Leaf spot. **1**:DE(3543), IL(82), NJ(82), TX(82), WI(82).
Deuteromycotina - Coelomycetes
Colletotrichum orbiculare. Anthracnose. **1**:IA(82).
Phyllosticta orbicularis. **1**:WI(1346).
P. sicyna. On leaves. **1**:NY(64,82), TX(82).
Septoria sicyi. Leaf spot. **1**:NY(82), TX(82).

Trichosanthes L.

1. *T. anguina* L. [club gourd, serpent-cucumber, snake gourd, viper's gourd]

Oomycetes
Pseudoperonospora cubensis. Downy mildew. **1**:MA(82), OH(82).
Ascomycotina
Glomerella magna. Anthracnose. **1**:NC(1303).
Deuteromycotina - Coelomycetes
Colletotrichum orbiculare. Anthracnose. **1**:MD(82).
Macrophoma trichosanthis. Fruit spot. **1**:AL(82).

Cupressaceae

Callitris Venten.

1. *Callitris* sp.

Deuteromycotina - Hyphomycetes
Cercosporidium sp. Needle blight. **1**:FL(35).
Deuteromycotina - Coelomycetes
Phoma sp. Twig dieback. **1**:FL(35).

Calocedrus Kurz

1. *C. decurrens* (Torr.) Florin (=*Heyderia decurrens* (Torr.) K. Koch, *Libocedrus decurrens* Torr.) [California incense cedar]

Oomycetes
Phytophthora cinnamomi. Root rot. **1**:CA(3624,4010).
Ascomycotina
Bertia moriformis. **1**:OR(3202).
Chloroscypha alutipes. On foliage. **1**:CA(850), OR(3202).
C. enterochroma. On twigs. **1**:CA(81,82), OR(3202).
Dermea libocedri. On twigs. **1**:CA(1379).
Diaporthe sp. Foliage endophyte. **1**:OR(508,2787).
Eutryblidiella macrospora. On bark and limbs. **1**:CA(81,82,523,642), OR(3202).
Herpotrichia juniperi. Brown felt blight. **1**:CA(82,661).
Hyalopeziza sp. On twigs. **1**:CA(661).
Lophodermium juniperinum. Needle cast (secondary). **1**:CA(82), OR(82,3202).
Parksia libocedri. On branchlets. **1**:CA(81,82,523).
Pezicula sp. On foliage. **1**:OR(2787).
Pithya cupressina. On foliage. **1**:CA(81,82).
Seynesiella sequoiae. On foliage. **1**:CA(81,82,642), OR(3202).
Xylaria sp. On foliage. **1**:OR(2787).
Basidiomycotina - Uredinales
Gymnosporangium biseptatum. **1**:CA(642).
G. libocedri. Heteroecious: telial host. **1**:CA(81,82,638,642,1877,2771), NV(82), OR(82,3202), WA(3202).
Basidiomycotina - Other
Armillaria mellea. Collar and root rot. **1**:OR(1064).
Cryptoporus volvatus. On wood. **1**:CA(82,642).
Dacrymyces stillatus. On rotten wood. **1**:CA(642).
Fomitopsis pinicola. Brown crumbly heart rot. **1**:CA(82), OR(82).
Gloeophyllum sepiarium. Wood rot. **1**:Range of host(82).
Henningsomyces candidus. On wood. **1**:CA(642).
Heterobasidion annosum. Root rot. **1**:CA(3349).
Neolentinus lepideus. Brown cubical rot of dead trunks and timber. **1**:CA(82).
Phellinus pini. Red ring rot. **1**:OR(82,3202).
P. viticola. White pocket rot. **1**:CA(82).
Postia caesia. **1**:OR(3202).
P. fragilis. **1**:OR(3202).
Stereum hirsutum. On dead limbs. **1**:CA(82,543).
Trametes versicolor. On wood. **1**:CA(82,661).
Trichaptum abietinum. On wood. **1**:CA(82).
Deuteromycotina - Hyphomycetes
Alternaria alternata. On foliage. **1**:OR(2787).
Aureobasidium sp. On foliage. **1**:OR(2787).
Capnocybe spongiosa. On twigs, on limbs. **1**:CA(2434).
Geniculosporium sp. Foliage endophyte. **1**:OR(508,2787).
Linodochium sp. Foliage endophyte. **1**:OR(508).
Nodulisporium sp. On foliage. **1**:OR(2787).
Phialophora decumbens. On foliage. **1**:OR(2787).
Phymatotrichopsis omnivora. Root rot. **1**:TX(82).
Deuteromycotina - Coelomycetes
Camaropycnis libocedri. On branchlets, cones, and fallen limbs. **1**:CA(81,82,523,3549).
Coniothyrium sp. On foliage. **1**:OR(2787).
Gelatinosporium sp. On foliage. **1**:OR(2787).
Kabatina juniperi. On foliage. **1**:OR(2787).
Macrophoma magnifructa. On cones. **1**:CA(638,639,642).
Sclerophoma pythiophila. **1**:OR(3202).
Seiridium cardinale. Branch canker. **1**:CA(82).
S. juniperi. On foliage. **1**:OR(2787).
Sirococcus conigenus. **1**:OR(3202).
Sirodothis sp. Foliage endophyte. **1**:OR(508,2787).
Sphaeropsis sp. Canker. **1**:GA(3780).

Chamaecyparis Spach

1. *C. lawsoniana* (Andr. Murray) Parl. [Lawson cypress, Port Orford cedar]
2. *C. lawsoniana* (Andr. Murray) Parl. cv. 'ellwoodii'
3. *C. nootkatensis* (D. Don) Spach [Alaska yellow-cedar, Nootka yellow-cedar]
4. *C. obtusa* (Siebold & Zucc.) Siebold & Zucc. ex Endl. [Hinoki cypress, Hinoki false cypress, Japanese false cypress]
5. *C. pisifera* (Siebold & Zucc.) Endl. [Sawara cypress]
6. *C. pisifera* (Siebold & Zucc.) Endl. cv. 'squarrosa'
7. *C. thyoides* (L.) B. S. P. (=*Cupressus thyoides* L.) [Atlantic white cedar, southern white cedar, swamp white cedar]
8. *Chamaecyparis* sp.

Oomycetes
Phytophthora cinnamomi. Root rot. **1**:LA(492), OR(3202,3989). **2**:CA(4010).
P. lateralis. Root rot, leaf spot. **1**:CA(1916), OR(82,1274,3202,3679,3691), WA(82,3202). **4**:OR(3202), WA(3202).
Pythium ultimum. Rootlet rot. **1**:CA(82).
Pythium sp. Root rot. **4**:FL(35). **8**:FL(35).
Ascomycotina
Anthostomella sp. On foliage. **1**:OR(2787).
Asterina cupressina. On foliage. **3**:AK(82,526). **7**:NJ(82).
Botryosphaeria dothidea. **7**:GA(1438).
B. obtusa. On limbs. **7**:NJ(82).
B. quercuum. **7**:NJ(82).
B. ribis. On limbs. **7**:GA(82), SC(82).
Botryosphaeria sp. Dieback. **8**:FL(35).
Caliciopsis thujina. On dead foliage. **7**:NJ(82).
Chloroscypha alutipes. Foliage endophyte. **1**:OR(508).
Cryptosporella thujina. On needles. **7**:NY(2548).
Cyanodiscus occidentalis. **3**:WA(3202).
Diaporthe sp. On foliage. **1**:OR(2787).
Didymascella chamaecyparidis. On foliage. **7**:NJ(82).
Eriosphaeria vermicularia. On twigs. **3**:WA(82,3202).
Gelatinodiscus flavidus. On limbs, fallen needles and cones. **3**:OR(504), WA(504,3202).
Gibbera sp. **3**:WA(3202).
Hysteromyxa effugiens. On dead foliage. **7**:NJ(82).
Lophodermium juniperinum f. *cupressi-thyoidis*. On foliage. **7**:NJ(82).
Microthyrium thujae. On wood. **1**:CA(81).
Nectria thujana. On dead foliage. **7**:NJ(82).
N. truncata. On dead bark. **7**:NJ(82).
Pezicula sp. On foliage. **1**:OR(508,2787).
Pithya cupressina. On dead foliage. **7**:NJ(82), NY(82).
Sarea difformis. On foliage. **1**:OR(2787).
Trichosphaeria cupressina. On dead foliage. **7**:NJ(82).
Tryblidaria washingtonensis. **3**:WA(3202).
Xylaria hypoxylon. On foliage. **1**:OR(2787).

Xylaria sp. On foliage. **1:**OR(2787).

Basidiomycotina - Uredinales

Gymnosporangium biseptatum. Heteroecious: telial host. **7:**NE states(82), AL(82), NJ(1877).

G. ellisii. Heteroecious: telial host. **7:**E states(82), AL(82), NC(1303), NJ(638,1877). **8:**FL(35).

G. fraternum. Heteroecious: telial host. **5:**FL(35). **7:**MA(82), ME(82), NJ(82,1877).

G. hyalinum. Heteroecious: telial host. **7:**FL(739,1877).

G. nootkatense. Heteroecious: telial host. **3:**AK(60,82,526), OR(82,1877,3202), WA(82,3202).

Basidiomycotina - Other

Amylostereum chailletii. On wood. **7:**NJ(543,638).

Antrodia albobrunnea. **1:**OR(3202).

A. xantha. **1:**OR(3202). **3:**AK(82).

Armillaria mellea. Root rot. **4:**CA(2891). **7:**VA(82).

A. tabescens. Root rot. **1:**FL(82).

Coniophora arida. **7:**NJ(128).

C. olivacea. **7:**NJ(128).

Cystostereum pini-canadense subsp. *subabruptum*. **1:**CA(81).

Dacryobolus karstenii. **3:**AK(526).

Dendrothele nivosa. On bark. **7:**NJ(82).

Diplomitoporus lenis. White pocket rot. **3:**AK(82,526).

Echinodontium ballouii. On wood. **7:**NJ(82).

Fomitopsis cajanderi. Brown crumbly rot. **7:**NC(82), NJ(82).

F. pinicola. **3:**AK(82).

Gloeophyllum carbonarium. **1:**OR(3202).

G. sepiarium. Brown rot. **3:**AK(82). **7:**VA(82).

Gloeoporus dichrous. **1:**OR(3202).

Heterobasidion annosum. Wood rot. **7:**MI(82), NC(1303,3664).

Luellia sp. On foliage. **1:**OR(2787).

Meruliopsis taxicola. **7:**NJ(82).

Mycena subcana. On twigs. **1:**CA(3326).

Pachykytospora papyracea. On dead wood. **7:**NJ(82), NY(82), PA(82).

Phellinus pini. Butt rot. **1:**OR(82,3202). **3:**AK(82), OR(3202).

P. viticola. White pocket rot. **1:**CA(82).

P. weirii. Root and butt rot. **3:**ID(82), WA(82).

Platygloea unispora. On limbs. **7:**GA(2642).

Postia fragilis. **1:**OR(3202).

Septobasidium sp. Parasitic on scale insects. **5:**FL(35).

Trametes villosa. **7:**NC(82), VA(82).

Trichaptum abietinum. Wood rot. **7:**ME(82).

T. sector. **7:**NC(82).

Xenosperma ludibundum. On wood. **7:**Unk.(2115).

Deuteromycotina - Hyphomycetes

Alternaria sp. Needle blight. **5:**FL(35). **8:**FL(35).

Capnocybe spongiosa. On limbs and twigs. **1:**CA(638,2434).

Cercospora sequoiae. Needle blight. **4:**FL(35). **5:**NC(1303,1591). **6:**NC(3931). **8:**FL(35).

Geniculosporium sp. On foliage. **1:**OR(2787).

Nodulisporium sp. On foliage. **1:**OR(508,2787).

Phymatotrichopsis omnivora. Root rot. **7:**TX(82). **8:**TX(82).

Sigmoidea sp. On foliage. **1:**OR(2787).

Stigmina thujina. Needle blight. **1:**HI(1593,2903).

Deuteromycotina - Coelomycetes

Aplosporella juniperi. On twigs. **8:**FL(82).

Chaetomella oblonga. **1:**CA(81).

Coniothyrium sp. On foliage. **1:**OR(2787).

Diplodia thyoidea. On bark. **7:**NJ(82,635).

Gelatinosporium sp. Foliage endophyte. **1:**OR(508,2787).

Harknessia thujina. On needles. **7:**NY(2548).

Hemidothis sp. On foliage. **1:**OR(2787).

Hendersonia thyoides. On foliage. **7:**NJ(82).

Kabatina juniperi. Foliage endophyte. **1:**OR(508,2787).

Lasiodiplodia theobromae. **1:**HI(2903).

Leptostroma sequoiae. **1:**CA(81).

Leptostroma sp. On foliage. **1:**OR(2787).

Macrophoma cupressi. Twig blight. **5:**FL(35).

Macrophomina phaseolina. Charcoal root rot. **4:**FL(35).

Pestalotiopsis funerea. On foliage. **1:**OR(2787). **7:**MI(82), NJ(82). **8:**NJ(82), TX(82).

Phomopsis juniperovora. Twig blight. **1:**KS(82), NC(82). **8:**KS(82), NC(82).

Phyllosticta sp. Needle blight. **1:**OR(2787). **5:**FL(35).

Scolecosporiella sp. Foliage endophyte. **1:**OR(508).

Seiridium cardinale. Branch canker. **1:**CA(82).

S. unicorne. On foliage. **7:**GA(82), NJ(82,3547).

Sirodothis sp. Foliage endophyte. **1:**OR(508).

Deuteromycotina - Other

Rhizoctonia solani. Needle blight. **4:**FL(35). **8:**FL(35).

Cupressus L.

1. *C. arizonica* Greene (=*C. arizonica* Greene var. *arizonica*) [Arizona cypress, rough-barked Arizona cypress]
2. *C. arizonica* Greene cv. 'bonita'
3. *C. arizonica* Greene var. *glabra* (Sudw.) Little
4. *C. bakeri* Jeps. [modoc cypress]
5. *C. goveniana* Gordon var. *pigmaea* Lemmon
6. *C. lusitanica* Mill. [cedar-of-Goa, Mexican cypress, Portuguese cypress]
7. *C. macrocarpa* Hartw. [Monterey cypress]
8. *C. pygmaea* (Lemmon) Sarg. [Mendocino cypress]
9. *C. sargentii* Jeps. [Sargent cypress]
10. *C. sempervirens* L. (=*C. sempervirens* L. var. *stricta* Aiton) [Italian cypress]
11. *C. sempervirens* L. cv. 'glauca'
12. *C. sempervirens* L. cv. 'stricta'
13. *Cupressus* sp.

Oomycetes

Phytophthora cinnamomi. Root rot. **10:**CA(3624), FL(35). **11:**CA(4010). **13:**FL(35).

Pythium splendens. Root rot. **10:**FL(35). **13:**FL(35).

P. ultimum. Root rot. **12:**HI(2903).

Pythium sp. Root rot. **1:**FL(35). **10:**HI(2903).

Ascomycotina

Botryosphaeria sp. **1:**GA(1438).

Diaporthe eres. Dieback. **10:**FL(35), GA(82), TX(82).

Lophodermium sp. On foliage. **10:**NJ(82).

Nectria ochroleuca. **1:**GA(1438).

Physalospora sp. Dieback. **1:**FL(35). **10:**FL(35).

Pithya cupressina. Foliar blight. **9:**CA(638). **13:**FL(35).

Seynesiella sequoiae. On leaves. **7:**CA(82).

Basidiomycotina - Uredinales

Gymnosporangium cunninghamianum. Heteroecious: telial host. **2:**AZ(1877). **3:**AZ(2773).

Uredo cupressicola. **1:**AZ(2773,3981), NM(2773). **5:**CA(2773). **8:**CA(1877,2764).

Basidiomycotina - Other

Acanthophysium lividocaeruleum. **1:**AZ(128,1202).

Antrodia albida. **1:**AZ(128,1202).

Armillaria mellea. **7:**CA(2898).

A. tabescens. Root rot. **1:**FL(35,82). **10:**FL(82). **13:**FL(35).

Athelia epiphylla. **1:**AZ(1202).

Coniophora puteana. Brown cubical rot. **7:**CA(81,82).

Diplomitoporus rimosus. **1:**AZ(1202).

Gloeophyllum sepiarium. Wood rot. **7:**CA(82).

G. striatum. On wood. **10:**FL(35,82).

Marasmius plicatulus. **13:**CA(856).

Perenniporia medulla-panis. **1:**AZ(1202).

P. subacida. Wood rot. **1:**AZ(82).

P. tenuis. **1:**AZ(128,1202).

Phellinus pini. Heart rot. **13:**CA(82).

Postia balsamea. **1:**AZ(1202).

P. stiptica. On wood. **7:**CA(82).

Steccherinum ochraceum. Wood rot. **7:**CA(82).

Trametes versicolor. On wood. **7:**CA(82).

Deuteromycotina - Hyphomycetes

Cercospora sequoiae. Juniper blight, needle blight, needle spot. **1:**AR(3917), FL(35), GA(1591), SC(1591). **6:**GA(1591), HI(2903). **7:**GA(1590,1591). **10:**GA(1590,1591). **13:**FL(35).

C. sequoiae var. *juniperi*. Needle blight. **1:**GA(1591), NC(1591).

Cercospora sp. Leaf and twig blight. **1:**GA(348), SC(348). **10:**LA(82).

Curvularia intermedia. Seedling tip blight. **1:**AZ(817).

Cylindrocladiella parva. **12:**HI(2903).

C. peruviana. Root rot, basal stem rot. **10:**FL(35).

Fusarium solani. Seedling blight, root rot. **10:**FL(35), TX(82). **13:**FL(35).

Fusarium sp. **13:**OR(3202).

Mycoleptodiscus sp. Twig blight. **1:**FL(35).

Phymatotrichopsis omnivora. Root rot. **1:**AZ(82), TX(82).

10:AZ(82), TX(82).
Stigmina asperula. On twigs. **10:**AL(82,3547).
S. glomerulosa. Needle blight. **10:**FL(35).

Deuteromycotina - Coelomycetes

Cytospora cenisia f. *littoralis*. Twig canker. **1:**CA(82). **7:**CA(82). **10:**CA(82).
Diplodia cyparissa. On twigs. **7:**CA(82). **10:**GA(82).
Diplodia sp. Tip blight. **1:**FL(35). **10:**FL(35).
Macrophoma cupressi. Twig blight, canker. **1:**AL(82), FL(35), TX(82). **7:**CA(82). **10:**FL(82), TX(82). **13:**FL(35).
Macrophoma sp. Leaf blight. **1:**AL(3780), TX(3780).
Macrophomina phaseolina. Charcoal root rot. **1:**FL(35). **10:**FL(35).
Monochaetia sp. On bark and wood. **1:**MS(82), TX(82).
Pestalotiopsis funerea. Leaf and twig blight. **1:**TX(82). **7:**CA(82), HI(2903), TX(82). **10:**MS(2720), TX(82).
Phomopsis juniperovora. Needle and twig blight. **1:**FL(35), GA(3618), MD(82), NC(82), VA(82). **7:**NC(82). **10:**NC(82). **13:**FL(35).
P. occulta. On dead or moribund twigs. **10:**CA(82), TX(82).
Phomopsis sp. Leaf and twig blight. **1:**GA(348), SC(348). **7:**HI(2903).
Phyllosticta sp. Needle dieback. **10:**FL(35).
Seimatosporium berckmansii. **10:**OR(82,3202).
Seiridium cardinale. Canker. **7:**CA(82,3547). **10:**CA(82).
S. unicorne. Canker, dieback. **1:**GA(348), MS(2720), SC(348,1311).
Sphaeropsis tumefaciens. Dieback. **10:**FL(35).
Sporonema sp. **7:**HI(2903).

Deuteromycotina - Other

Rhizoctonia solani. Twig dieback. **1:**FL(35). **13:**FL(35).
Rhizoctonia sp. Root rot. **10:**HI(2903).

Juniperus L.

1. *J. ashei* Buchholz [Ashe juniper, Ozark white cedar]
2. *J. californica* Carrière
3. *J. chinensis* L. [Chinese juniper]
4. *J. chinensis* L. cv. 'pfitzeriana'
5. *J. chinensis* L. cv. 'pyramidalis'
6. *J. chinensis* L. f. *globosa*
7. *J. chinensis* L. var. *sargentii* A. Henry [Sargent juniper]
8. *J. communis* L. (=*J. nana* Willd.) [common juniper]
9. *J. communis* L. var. *depressa* Pursh
10. *J. communis* L. var. *saxatilis* Pall. (=*J. communis* L. var. *montana* Aiton, *J. siberica* Burgsd., *J. communis* L. var. *siberica* Burgsd.) [mountain juniper]
11. *J. conferta* Parl. [shore juniper]
12. *J. deppeana* Steud.
13. *J. deppeana* Steud. var. *pachyphloea* (Torr.) Martinez (=*J. pachyphloea* Torr.) [alligator juniper, sweet-fruited juniper]
14. *J. excelsa* M. Bieb. [Greek juniper]
15. *J. excelsa* M. Bieb. cv. 'stricta' [Greek juniper]
16. *J. flaccida* Schlechtend.
17. *J. horizontalis* Moench (=*J. prostrata* Pers.) [creeping cedar, creeping juniper, creeping savin juniper]
18. *J. horizontalis* Moench f. *alpina*
19. *J. mexicana* Schiede (=*J. sabinoides* Sarg.)
20. *J. monosperma* (Engelm.) Sarg. [cherrystone juniper, one-seed juniper]
21. *J. occidentalis* Hook. [California juniper, Sierra juniper]
22. *J. osteosperma* (Torr.) Little (=*J. californica* Carrière var. *utahensis* Engelm., *J. utahensis* (Engelm.) Lemmon) [Utah juniper]
23. *J. oxycedrus* L. (=*J. glauca* Salisb.) [prickly juniper]
24. *J. phoenicea* L.
25. *J. pinchotii* Sudw. [Pinchot juniper, redberry juniper]
26. *J. sabina* L. [savin]
27. *J. sabina* L. cv. 'tamariscifolia' [savin]
28. *J. scopulorum* Sarg. [Colorado red cedar, Rocky Mountain juniper]
29. *J. silicicola* (Small) L. H. Bailey [Southern red cedar]
30. *J. virginiana* L. [eastern red cedar, red cedar]
31. *J. virginiana* L. cv. 'canaertii'
32. *J. virginiana* L. cv. 'elegantissima'
33. *J. virginiana* L. cv. 'glauca'
34. *J. virginiana* L. cv. 'globosa'
35. *J. virginiana* L. cv. 'pyramidalis'
36. *J. virginiana* L. cv. 'pyramidiformis'
37. *Juniperus* sp.

Myxomycetes

Leocarpus fragilis. **30:**NC(3944).

Oomycetes

Phytophthora cactorum. Root rot. **26:**CA(82).
P. cinnamomi. Crown and root rot. **7:**OR(2174). **11:**GA(1211). **27:**CA(3464), OR(2174).
P. citrophthora. Root rot. **26:**CA(82).
P. cryptogea. **3:**CA(1580).
P. nicotianae var. *parasitica*. Crown and root rot. **11:**GA(1211).
Phytophthora sp. Root rot. **37:**FL(35).
Pythium irregulare. Crown and root rot. **11:**GA(1211).
P. splendens. Root rot. **37:**FL(35).
P. sylvaticum. Crown and root rot. **11:**GA(1211).
Pythium sp. Root and stem rot. **37:**FL(35).

Ascomycotina

Actidium nitidum. **17:**WA(3202). **26:**WA(3202).
Asterina cupressina. On foliage. **8:**WI(82).
Botryosphaeria dothidea. Canker. **37:**FL(35).
B. obtusa. **30:**MD(82), NJ(82), NY(82), VA(82).
B. quercuum. **30:**NY(82).
B. ribis. On limbs. **30:**AL(82), NJ(82), VA(82).
Caliciopsis nigra. On limbs associated with galls. **30:**NY(82,967).
Cenangella helotioides. On twigs. **3:**OR(82).
Chaetosphaeria parvicapsa. **30:**GA(1438).
Chloroscypha cedrina. On foliage. **30:**NC(82,3944), NY(82).
C. sabinae. On dead shoots. **3:**OR(3202). **21:**CA(81).
Coccodothis sphaeroidea. On foliage. **29:**NC(82). **30:**GA(82), LA(82), NC(82,1303,3944), SC(82).
Coccomyces petersii. On bark and twigs. **8:**AL(82). **30:**AL(3229), GA(1438), MA(3229), NC(3944), NY(3229).
Cryptosporella thujina. On needles. **30:**TX(2548).
Diatrypella favacea. **30:**GA(1438).
Didymascella tetramicrospora. Needle blight. **28:**CO(82,2694).
Dothidea acerva. **8:**MA(179), NH(179).
Endoxyla inusta. **37:**GA(1438).
Extrawettsteinina minuta. **8:**VT(179).
Godronia ribis. **21:**CA(128).
Herpotrichia juniperi. Brown felt blight. **8:**W states(82), CA(82), MT(3202), OR(82,3202). **10:**CA(642). **13:**W states(82), CA(82), OR(82). **17:**W states(82), CA(82), OR(82). **28:**MT(3202).
Holmiella sabina. On wood. **8:**MT(3202). **28:**CO(82), WY(1195). **29:**KS(82). **30:**KS(82), NY(82). **37:**CO(82), MT(82), WY(82).
Hypocrea ceramica. **30:**NC(3944).
Lachnella alboviolascens. **20:**AZ(1196,1202).
Lecidea sp. **30:**OK(660).
Lophodermium juniperinum. Needle spot. **3:**MA(82), WA(3202). **4:**WI(1348). **8:**NC states(82), NE states(82), W states(82), CA(81), ID(3202), MT(3202), OR(3202), WA(82,3202). **10:**CA(642), WY(2924). **14:**CT(82), OR(82,3202). **17:**W states(82), CA(82), MT(3202), WA(82). **26:**WA(82,3202). **28:**CO(3457), MT(3202), WA(3202). **30:**MA(82,1394), NY(82), WA(82,3202), WI(82,1335). **37:**CO(649), NC(1303).
Melittosporiella macrospora. On wood. **37:**OR(3231).
Murangium sequoiae. On bark. **30:**SC(3167).
Mycosphaerella juniperina. On foliage. **8:**Unk.(82), IA(82). **17:**Unk.(82), ND(82,377). **30:**MD(82), OK(660).
Mytilinidion acicola. **8:**MT(3202), WA(3202). **37:**MT(3202).
M. decipiens. On bark. **29:**KS(82), MS(82), NC(82), NJ(82). **30:**KS(82), MS(82), NC(82), NJ(82).
M. tortile. **30:**GA(1438).
Octospora wrightii. **30:**NC(3944).
Physalospora abdita. On dead limbs. **29:**GA(82), LA(82). **30:**GA(82), LA(82).
Physalospora sp. Twig dieback. **37:**FL(35).
Pithya cupressina. On dead foliage. **3:**MD(82). **17:**CA(81), GA(82), MI(82), WI(82). **30:**MA(82), MD(82), NC(3944), OK(660), VA(82). **37:**GA(1438), WA(82,3202).
Plectania melastoma. **30:**NC(3944).
P. nannfeldtii. **30:**NC(3944).
Pododimeria gelatinosa. **30:**VA(2206).
P. juniperi. **30:**GA(2206), GA(2206), NJ(2206).
Quasiconcha reticulata. On undigested seeds in ringtail dung. **30:**TX(192).
Rhizodiscina lignyota. **30:**GA(1438).
Robergea albicedrae. On limbs and twigs. **1:**TX(82,3225,3228). **37:**OK(2872).
Rosellinia deerata. On wood. **30:**NJ(967).

Sarea difformis. Foliage endophyte. **21:**OR(508,2787).
Seynesiella exigua. On dead leaves. **8:**ME(272), NH(272), VT(272).
S. juniperi. On foliage. **2:**CA(81). **9:**WI(1361). **30:**NJ(82).
S. sequoiae. **21:**CA(81).
Sthughesia juniperi. On foliage and limbs. **21:**CA(2434), OR(3202), WA(3202). **22:**OR(638). **37:**CA(82), OR(3202).
Valsa cenisia. On dead limbs. **30:**DE(82), MI(82), NJ(82), OK(660).
V. friesii. **30:**SC(634).
Xylopezia biseptata. On decorticated wood. **21:**OR(3233).

Basidiomycotina - Uredinales

Gymnosporangium asiaticum. Heteroecious: telial host. **3:**CT(82), OR(82,3202), WA(82,3202). **30:**WI(1361).
G. bermudianum. Autoecious. **29:**AL(82), FL(82), LA(82), MS(82,2720). **30:**AL(82,285), FL(82), LA(82), MS(82,2720). **37:**FL(35).
G. bethelii. Heteroecious: telial host. **17:**WY(82). **21:**OR(3202). **28:**AZ(1203,2771,3981), CO(82,1877,2768), ID(82,3202), MT(82,3202), ND(82,378), OK(2872), WA(82,3202), WY(82). **37:**AZ(1203), OR(82).
G. clavariiforme. Heteroecious: telial host. **8:**CO(1975), ID(1975), MI(1975), MT(1975,3202), SD(2768), UT(1975,2771), WA(3202), WY(1975).
G. clavipes. Heteroecious: telial host. **8:**Range of host(82), MT(3202), WA(3202). **9:**OK(2873). **30:**Range of host(82), GA(1439), MA(1390), MS(2720), NC(1303,3944), NY(1877), OK(2872), VA(3493). **37:**FL(35).
G. confusum. Heteroecious: telial host. **26:**CA(160).
G. cornutum. Heteroecious: telial host. **8:**W states(82), CO(82), ME(82), MI(82), WA(82). **10:**CO(1877). **17:**UT(82).
G. davisii. Heteroecious: telial host. **8:**ME(82), WI(82). **10:**WI(1877). **30:**FL(82). **37:**FL(35).
G. effusum. Heteroecious: telial host. **30:**MD(82), NJ(82), NY(82), SC(82,1877), VA(82). **37:**OK(2872).
G. exiguum. Heteroecious: telial host. **2:**CA(81). **13:**OK(82), TX(82). **15:**CA(81). **19:**OK(2872). **28:**OK(2872). **30:**TX(82,1877). **37:**OK(82), TX(82).
G. exterum. Heteroecious: telial host. **30:**KY(82,1877), VA(3493).
G. floriforme. Heteroecious: telial host. **30:**SE states(82), AL(1877), MS(2720), OK(82,2872), TX(82,652). **34:**OK(2872). **37:**FL(35).
G. globosum. Heteroecious: telial host. **3:**IL(2573). **9:**OK(2873). **17:**IL(2573), ND(82,378), NJ(82), WI(1352). **28:**IL(1559,2573). **29:**AL(82). **30:**Range of host(82), IL(1052,1559,2573), MA(1877), MI(665), MS(2720), NC(3944), OK(2872), VA(3493).
G. gracile. Heteroecious: telial host. **12:**TX(1876). **16:**TX(1876).
G. harknessianum. Heteroecious: telial host. **21:**CA(81,1877,2771), OR(3202). **22:**NV(2771). **37:**CA(82), OR(82).
G. inconspicuum. Heteroecious: telial host. **12:**AZ(2771,3981). **13:**AZ(1203). **20:**AZ(1203,3981), NM(2771). **21:**CA(2771). **22:**AZ(1203,2771,3981), CO(1877), ID(2771), NV(2771), UT(1195), WY(2771). **37:**CA(82), CO(82), UT(82).
G. japonicum. Heteroecious: telial host. **3:**CT(82), MA(82), NJ(82), WA(82,3202).
G. juniperi-virginianae. Heteroecious: telial host. **3:**IL(2573). **4:**OK(2872). **6:**IL(1559). **9:**OK(2873). **17:**IL(2573). **18:**IL(1559). **22:**IL(1559). **25:**IL(1559). **28:**CO(82), IL(82,1559), NE(82). **29:**AL(82). **30:**Range of host(82), CO(3780), GA(1439), IL(1052,1559,2573,3780), IN(987), MD(3780), MS(2720), NC(1303,1877,3944), ND(378), OH(3780), OK(618,2872), SD(2243), VA(3493), WA(3202), WY(3780). **31:**OK(2872). **32:**IL(1559). **33:**OK(2872). **34:**OK(2872). **35:**OK(2872). **37:**AL(82), FL(35).
G. kernianum. Heteroecious: telial host. **2:**CA(2771). **12:**AZ(1195,3379,3981), TX(742). **13:**AZ(82,1203), NM(82). **20:**AZ(1195,1203,3981). **21:**CA(81), OR(3202). **22:**AZ(1203,2771,3981), CO(1877), ID(3202), WY(2771). **37:**AZ(1203), CA(82), CO(82), ID(82), NV(82), OR(82), UT(82).
G. multiporum. Telial host; aecial host unknown. **12:**AZ(3981). **13:**AZ(1203), NM(82). **20:**SW states(1877), AZ(1203,3981). **21:**CA(2771). **22:**AZ(1203,3981). **37:**CA(82), CO(82), NM(82).
G. nelsonii. Heteroecious: telial host. **2:**CA(2771). **12:**AZ(3981). **16:**TX(742). **17:**AK(60), ME(82), MI(82,1877), MT(82,3202), SD(82), WA(3202), WI(82). **20:**AZ(1195,1203,3981), NM(1195). **21:**OR(3202). **22:**AZ(1203,3981), CA(2771). **28:**CA(2771), ID(3202), MT(1195,3202), ND(82,378), OR(3202), UT(1195), WA(3202), WY(1195,1877). **30:**NY(82). **37:**W states(82), OR(82), UT(82).
G. nidus-avis. Heteroecious: telial host. **17:**W states(82), ME(82), MT(82,3202), ND(82,378), UT(82), WA(82), WI(82). **28:**W states(82), AZ(1203,3981), ID(2771,3202), MT(3202), OR(3202), SD(2243), UT(82,2771), WA(82,3202), WY(1195,2771). **29:**FL(82). **30:**E states(82), CT(1877), GA(1439), MA(1390), MS(82,2720), NC(1303,3944), NE(82), OK(2872), SD(2243). **37:**FL(35).
G. speciosum. Heteroecious: telial host. **12:**AZ(1195,3981), TX(742). **13:**AZ(82,1194,1203), CO(82), NM(82), NV(82), UT(82). **20:**AZ(1194,1195,1203,3981). **21:**CO(1877). **22:**AZ(1203,3981), UT(2771). **37:**AZ(82,1203), CO(82), NM(82), NV(82), UT(82).
G. trachysorum. Heteroecious: telial host. **30:**LA(82), MS(82,2720), SC(82,1877). **37:**FL(35).
G. tremelloides. Heteroecious: telial host. **8:**MT(3202), OR(3202), WA(2219,3202), WY(1195).
G. vauqueliniae. Heteroecious: telial host. **20:**AZ(1203,1877,3981). **37:**AZ(82).
Gymnosporangium sp. **20:**OK(2872). **30:**GA(2405), OK(2872). **31:**TN(1029). **33:**TN(1029). **36:**TN(1029). **37:**OK(2872).
Uredo apacheca. **12:**NM(1877). **37:**NM(2771).

Basidiomycotina - Other

Acanthophysium bertii. **20:**AZ(1202).
A. cerussatum. **20:**AZ(1196).
A. diffissum. On dead limbs and twigs. **37:**CO(2083).
A. lividocaeruleum. **12:**AZ(1196,1202).
Aleurobotrys botryosus. On bark. **30:**LA(1185).
Amylostereum chailletii. **30:**CT(543), NY(543).
Anthracophyllum lateritium. On dead limbs. **30:**LA(1185).
Antrodia albida. Wood rot. **12:**AZ(1196,1202), NM(1190). **20:**AZ(1202). **30:**MO(82), PA(2962), VA(82). **37:**NM(82).
A. ferox. On wood. **12:**AZ(1202), NM(1190). **37:**NM(82).
A. heteromorpha. **20:**AZ(1194,1196,1202). **30:**NC(3944).
A. juniperina. Brown cubical heart rot. **12:**AZ(1196,1202). **20:**AZ(1196,1202,2608). **21:**OR(3202). **22:**AZ(1202). **29:**FL(82), LA(82), SC(82). **30:**FL(82), GA(491,1439,2608), KS(2608), NC(1303), OK(618,660,2872), TX(652), VA(2608). **37:**AZ(82,1194), FL(35), SC(2608).
A. serialis. Wood rot. **28:**WY(82).
A. sinuosa. **37:**AZ(1196), NM(1190).
Armillaria mellea. Root rot. **3:**CA(2897,2898). **17:**CA(2891). **28:**CO(1774). **37:**CA(2898), NC(1303).
A. tabescens. Root rot. **29:**FL(82). **37:**FL(35), KS(3193).
Asterostroma muscicola. On bark of living trees. **30:**FL(1186).
Bjerkandera adusta. On wood. **29:**Range of host(82). **30:**Range of host(82), NC(3944), PA(2962).
Bondarzewia berkeleyi. **30:**NC(3944).
Calocera cornea. **30:**IN(654).
Ceriporia purpurea. **30:**IA(82).
C. tarda. **12:**AZ(1196,1202). **37:**NM(1190).
C. xylostromatoides. On bark. **30:**LA(1185).
Coniophora arida. **20:**NM(128,1190).
C. eremophila. **20:**AZ(1196).
C. fusispora. On wood. **30:**AL(1186), FL(1186).
Conohypha albocremea. On wood. **24:**LA(1186).
Coriolopsis rigida. **20:**NM(128).
Crucibulum laeve. **30:**NC(3944). **37:**MT(3202).
Cyathus olla. **30:**NC(3944).
Cylindrobasidium corrugum. On wood. **28:**AZ(82), CO(82), WY(82).
Dacryobolus sudans. On wood. **20:**AZ(1194,1196,1202). **30:**FL(1186).
Dendrophora albobadia. On dead limbs. **30:**LA(1185).
Dendrothele griseo-cana. On bark. **12:**AZ(128,1202). **30:**LA(1185).
D. incrustans. On wood. **12:**AZ(1196). **30:**LA(1186).
D. itihummensis. On limbs. **30:**LA(1185), MS(1185).
D. nivosa. On bark. **29:**LA(1185). **30:**AR(2084), CT(2084), FL(2084), GA(1439), IN(654,2084), KS(2084), LA(1185,2084), MA(2084), MO(2084), MS(1185), NC(1303,3944), NJ(2084), NY(2084), OK(660), PA(2084), SC(2084), TN(2084), TX(2084). **37:**AL(82), FL(35), OR(82), SC(634), TX(82).
D. pachysterigmata. On dead limbs. **30:**LA(1185).
Diplomitoporus lenis. **37:**ID(128).
D. rimosus. **12:**AZ(128,1196,1202). **20:**AZ(128,1196,1202), NM(128). **21:**ID(3202), OR(128). **22:**AZ(1196,1202), UT(128). **28:**NM(128). **30:**OK(128). **37:**ID(3202), OK(2873), OR(3202).
Exidiopsis calcea. On wood. **12:**AZ(1196,1202). **20:**AZ(128).

24:LA(1186). **30:**FL(1186).
Fomitopsis cajanderi. Heart rot. **30:**IN(654), SC(1533), TN(1533).
F. feei. On wood. **30:**FL(1186), LA(1186).
F. rosea. On wood. **13:**W states(82), CA(82). **28:**W states(82), CA(82). **30:**AL(82), FL(82), MD(82), PA(82), VA(82). **37:**W states(82), CA(82), FL(35).
Ganoderma sp. On roots. **30:**FL(1186).
Gloeophyllum odoratum. Wood rot. **30:**NY(82).
G. sepiarium. Wood rot. **12:**AZ(1196,1202). **13:**AZ(82).
G. trabeum. On wood. **12:**AZ(1196,1202). **13:**AZ(82,128). **30:**PA(82,2962).
Gloeoporus dichrous. On wood. **29:**Range of host(82). **30:**Range of host(82).
Grandinia arguta. **12:**AZ(128,1196,1202).
G. crustosa. On wood. **30:**FL(1186).
G. juniperi. On dead limbs. **1:**TX(1185).
G. nespori. On dead limbs. **30:**FL(1186).
G. spathulata. On dead limbs. **12:**AZ(1196,1202). **30:**FL(1186).
Heliocybe sulcata. **12:**AZ(1196,1202).
Hemimycena albida. On wood. **24:**LA(1186). **30:**FL(1186).
Henningsomyces candidus. On dead limbs. **30:**LA(1185).
Heterobasidion annosum. White root and butt rot. **8:**Unk.(82). **11:**NC(1303). **21:**W states(3584), NV(3055). **22:**W states(3584), UT(3055). **30:**DE(82), GA(491,1439,1533), MA(82), MD(82), NC(82,1303,1533,3944), NY(82), SC(1533), TN(1533), VA(1533,3493). **37:**FL(35), MS(2720).
Heterochaete shearii. On dead limbs. **30:**LA(1185).
Hohenbuehelia approximans. On wood. **30:**FL(1186).
H. elegans. On bark of living trees. **30:**NC(2421).
Hyphoderma argillaceum. **12:**AZ(1202).
H. baculorubrense. On wood. **24:**LA(1186). **30:**FL(1186).
H. deserticola. **12:**AZ(1196).
H. obtusiforme. On wood. **30:**FL(1186).
H. pallidum. **12:**AZ(1196).
H. praetermissum. On wood. **30:**LA(1185).
H. sambuci. On dead limbs. **1:**TX(1185).
Hypholoma fasciculare. On wood. **30:**LA(1185).
Hypochnicium bombycinum. On dead limbs. **30:**FL(1186).
H. eichleri. On dead limbs. **30:**LA(1185).
H. lundellii. On dead limbs. **30:**LA(1185).
H. punctulatum. On wood. **30:**FL(1186).
Inonotus crocitinctus. On wood. **37:**TX(1204).
I. juniperinus. On roots. **37:**TX(1204).
Ischnoderma resinosum. **30:**NC(3944).
Kriegeria sp. Needle spot. **37:**OR(197).
Leptosporomyces galzinii. On living limbs. **8:**ME(128).
L. juniperinus. **12:**AZ(128).
Marasmiellus juniperinus. On living limbs. **30:**LA(1185).
Pachykytospora alabamae. On dead limbs. **29:**LA(1185).
Peniophora junipericola. On dead limbs. **30:**LA(1185), MS(1185).
Perenniporia fraxinophila. **13:**AZ(128). **20:**AZ(128,1196).
P. subacida. On wood. **30:**ID(128), NY(82).
Phanerochaete carnosa. White rot. **8:**NY(433). **30:**MA(128).
P. flavido-alba. White rot. **19:**TX(433,2117). **30:**GA(128,433), LA(1185).
P. velutina. **13:**AZ(128).
Phellinus gilvus. On wood. **29:**Range of host(82). **30:**Range of host(82).
P. pini. Red ring rot. **30:**MA(82).
P. texanus. White heart rot. **12:**AZ(1196,1202). **20:**AZ(1196), NM(1190). **28:**AZ(1196), NM(1190). **30:**TX(82). **37:**AZ(82), NM(82), TX(82).
Phlebia subserialis. **12:**AZ(1202).
Platygloea unispora. On dead limbs. **37:**GA(155), NC(155).
Polyporus arcularius. **12:**AZ(1202). **20:**AZ(1196).
Postia caesia. On wood. **29:**Range of host(82). **30:**Range of host(82), PA(2962).
P. sericeomollis. **21:**OR(3202).
Pyrofomes demidoffii. White rot. **8:**MT(82,3202,3832). **12:**AZ(1202). **20:**AZ(1194,1196,1202). **21:**CA(642,661), OR(3202). **22:**AZ(1196,1202), ID(3202). **28:**OR(3202). **29:**FL(82). **30:**KY(82), MD(82), OK(660), TN(82). **37:**AZ(82), CA(82), CO(82), NM(82), TX(82).
Resupinatus applicatus. **30:**NC(3944).
Schizopora apacheriensis. **12:**AZ(1193,1196,1202).
S. paradoxa. On dead limbs. **30:**LA(1185).
Scytinostroma portentosum. On dead limbs. **30:**LA(1185).
Sebacina epigaea. On wood. **30:**GA(2642), LA(1185).
S. incrustans. At base of living trees. **30:**FL(1186).
Steccherinum ochraceum. On wood. **30:**FL(1186).
Stereum hirsutum. **30:**CT(543), NC(3944).
Stigmatolemma taxi. On bark. **30:**FL(1186), GA(456), LA(1186), SC(456).
Thelephora albidobrunnea. **30:**NC(3944).
T. cuticularis. On bark. **30:**FL(1186).
Tomentella asperula. On wood. **30:**FL(1186), LA(1186). **37:**MI(2039).
T. bresadolae. At base of living tree, on exposed heartwood. **30:**OH(128).
T. ferruginea. **37:**MI(2039), NM(1190).
Tomentella sp. **30:**VA(128).
Trametes hirsuta. **30:**NC(3944).
T. versicolor. On wood. **29:**Range of host(82). **30:**Range of host(82), NC(3944), PA(2962).
T. villosa. White rot. **8:**GA(82), TX(82). **29:**MO(82), TX(82). **30:**MO(82), TX(82).
Trechispora farinacea. On dead limbs. **30:**FL(1186).
Trichaptum abietinum. **30:**NC(3944).
T. byssogenum. On wood. **20:**AZ(1196,1202). **37:**S states(2681).
Tubulicium clematidis. On wood. **30:**FL(1186).
Tubulicrinis medius. On dead limbs. **30:**LA(1185).
Vararia gomezii. On dead limbs. **30:**LA(1185).
Wolfiporia cocos. Tuckahoe, on roots. **29:**NC(82). **30:**NC(82).
Xenosperma murrillii. On limbs. **30:**FL(1186).

Deuteromycotina - Hyphomycetes

Alternaria alternata. On foliage. **21:**OR(2787).
Alternaria sp. Needle blight, leaf spot. **8:**TX(82). **37:**FL(35), OK(2872).
Bactrodesmium cedricola. **26:**WA(3202).
Cercospora sequoiae. Needle blight, needle spot. **5:**HI(2903). **20:**IA(2758). **28:**NE(2758). **30:**GA(1590,1591), SC(1590,1591). **37:**FL(35).
C. sequoiae var. *juniperi*. Needle blight, needle cast. **8:**CT(82), WI(82). **9:**WI(1334). **30:**AR(1591), FL(1591), GA(1591), MO(382), NC(1303,1591), NE(2754,2758), OK(2873), SC(1591), VA(1591), WI(82,962,1591). **37:**KY(82), OK(82,2872).
Cercospora sp. **8:**NC(1303).
Chalara thielavioides. Black mold. **30:**NJ(1545).
Conoplea fusca. **30:**NC(3944).
C. juniperi. On dead needles and limbs. **30:**CT(1672), DE(3543), IN(654), MA(1672), MD(1672), MI(1672), NC(1303,1672), NJ(1672), NY(1672), OH(1672), PA(1672). **37:**AL(1672).
Curvularia intermedia. Seedling tip blight. **30:**AZ(817).
Cylindrocarpon sp. **37:**OR(3202).
Cylindrocladium scoparium. Cutting rot. **37:**FL(35).
Cylindrocladium sp. Root rot. **37:**FL(35).
Fusarium oxysporum. Root and stem rot. **37:**FL(35).
Fusarium sp. Root and stem rot. **37:**FL(35).
Hormonema sp. Foliage endophyte. **21:**OR(508).
Hyphosoma piceae. **21:**OR(3202), WA(3202).
Mycoleptodiscus sp. Needle necrosis. **37:**FL(35).
Myxormia sp. On foliage. **21:**OR(2787).
Phialophora decumbens. On foliage. **21:**OR(2787).
P. hoffmannii. On foliage. **21:**OR(2787).
Phymatotrichopsis omnivora. Root rot. **3:**TX(82). **8:**TX(82). **13:**TX(82). **17:**TX(82). **28:**TX(82). **29:**TX(82). **30:**TX(82). **37:**OK(2872), TX(82).
Sporidesmium hysterioideum. **30:**DE(3543).
Stigmina deflectens. **30:**SD(2243).
S. glomerulosa. Needle blight, twig blight. **8:**IA(82,1591), UT(82). **30:**NC(1870), SC(1870), VA(1870). **37:**CA(82), FL(35).
S. juniperina. Needle cast. **8:**MI(1591), WI(1591). **9:**WI(1357).
Streptothrix sp. Saprophytic on living or dead bark. **30:**MA(82), MI(82), NJ(82), NY(82).
Verticillium sp. Wilt. **37:**FL(35).

Deuteromycotina - Coelomycetes

Aplosporella juniperi. Dieback. **29:**KS(82), MI(82), NY(82). **30:**KS(82), MI(82), NY(82). **37:**FL(35), LA(2783).
Botryodiplodia juniperina. On dead limbs. **30:**NY(82).
Cryptocline sp. Foliage endophyte. **21:**OR(508).
Cylindrosporella sp. On foliage. **21:**OR(2787).

Cytospora sp. Needle blight. **30:**OK(2872).
Diplodia virginiana. On twigs. **29:**SC(82). **30:**SC(82,634).
Diplodia sp. Dieback. **29:**Unk.(82). **30:**Unk.(82), NC(1303). **37:**FL(35).
Discosia sp. On dead foliage and stems. **37:**NC(1303).
Harknessia thujina. On needles. **30:**TX(2548).
Kabatina juniperi. Branch tip dieback. **28:**NE(2665). **30:**ME(2663), NE(2664,2665).
Leptothyrium sp. Twig canker, dieback. **30:**WI(1346). **37:**FL(35).
Macrophoma sp. Dieback. **37:**FL(35).
Macrophomina phaseolina. Charcoal rot. **30:**OK(2872).
Microdiplodia cupressina. **37:**WA(3202).
Microdiplodia sp. Dieback; on twigs. **37:**WA(82).
Pestalotiopsis funerea. Needle blight. **3:**NJ(82). **8:**CA(82), GA(82), IL(82), MI(82), NJ(82), SC(82), TX(82). **13:**CA(82), GA(82), IL(82), MI(82), NJ(82), SC(82), TX(82). **14:**NJ(82). **17:**CA(82), GA(82), IL(82), MI(82), NJ(82), SC(82), TX(82). **28:**CA(82), GA(82), IL(82), MI(82), NJ(82), SC(82), TX(82). **30:**MI(82), MS(2720), NJ(82), TN(82), TX(82), WI(82,1330). **37:**CA(82), GA(82), IL(82), MI(82), NJ(82), SC(82), TX(82), WA(3202).
Phoma glomerata. On wood. **13:**NM(82).
P. terrestris. Isolated from rhizosphere. **21:**OR(2791).
Phoma sp. Needle blight. **37:**FL(35).
Phomopsis juniperovora. Nursery blight, seedling blight, twig blight. **3:**WA(3202). **8:**E states(82), KS(82), NE(82). **13:**E states(82), KS(82), NE(82). **14:**NE states(82), CT(82), NE(82), NJ(82), OK(82). **17:**E states(82), KS(82), NE(82), SD(2243). **26:**MA(82), NY(82). **28:**E states(82), KS(82), NE(82,1412). **30:**E states(82), AL(285,1411), AR(1411), CT(1411), IA(1411), IL(1411), IN(1411), KS(82,1411), KY(1411), LA(1411), MA(1411), MD(1411), MI(1411), MN(82,1411), MO(1411), NC(1303,1411,1533), NE(1411), NJ(1411), NY(1411), OH(1411), OK(618,1411,2873), PA(1411), TN(1411), TX(1411), VA(1411,3493), WA(3202), WI(1411). **37:**E states(82), CT(3770), FL(35), KS(82), NE(82), NJ(82), OK(2872), RI(502), WV(954).
P. occulta. Secondary twig blight. **3:**Range of host(82). **14:**Range of host(82). **26:**Range of host(82). **30:**MA(1412), MD(1412).
Phyllosticta sp. Tip dieback. **37:**FL(35).
Sclerophoma pythiophila. Tip dieback. **3:**WI(379). **17:**WI(379). **23:**WI(379). **26:**WI(379). **30:**WI(379).
Seiridium cardinale. Canker. **3:**CA(82). **26:**CA(82).
S. unicorne. Canker. **30:**SC(1311).
Septogloeum sp. On twigs. **8:**PA(82).
Sirodothis sp. Foliage endophyte. **21:**OR(508).
Sphaeropsis sapinea. On twigs. **8:**NJ(82). **13:**NJ(82). **17:**NJ(82). **28:**NJ(82). **37:**NJ(82).
Sphaeropsis sp. Canker. **8:**Unk.(82). **13:**Unk.(82). **17:**Unk.(82). **28:**Unk.(82). **29:**Unk.(82). **30:**Unk.(82), OK(2872). **37:**Unk.(82).
Stagonospora pini. On foliage and twigs. **30:**IL(2568), TX(82). **37:**TX(82).
Stagonospora sp. Twig dieback. **37:**FL(35).

Deuteromycotina - Other

Apiosporium pinophilum. Sooty mold. **8:**OR(82), WA(82).
Rhizoctonia solani. Needle blight, stem blight, damping-off. **28:**TX(82). **37:**FL(35).

Libocedrus Endl.

1. *Libocedrus* sp.

Ascomycotina

Paschelkiella pini. On wood. **1:**OR(3232).

Thuja L.

1. *T. decurrens* Voss
2. *T. occidentalis* L. (=*T. compacta* J. Standish, *T. occidentalis* L. cv. 'compacta') [American arborvitae, northern white-cedar]
3. *T. orientalis* L. (=*Platycladus orientalis* (L.) Franco) [Oriental arborvitae]
4. *T. orientalis* L. cv. 'conspicuus'
5. *T. plicata* Donn ex D. Don [giant arborvitae, giant cedar, western red cedar]
6. *Thuja* sp.

Oomycetes

Phytophthora sp. Twig blight. **2:**NJ(82).
Pythium splendens. Root rot. **6:**HI(2903).
Pythium sp. Root rot. **3:**FL(35).

Ascomycotina

Actidium sp. **5:**ID(3202).
Amphisphaeria sp. **5:**ID(3202).
Bertia moriformis. **5:**OR(3202).
Botryosphaeria dothidea. Branch blight. **2:**GA(1438). **3:**FL(35).
B. obtusa. On dead limbs. **2:**GA(82), PA(82).
Ceratostomella sp. Wood stain. **2:**IL(82).
Chloroscypha seaveri. Foliage endophyte. **5:**ID(82,3202), MT(82,3202), OR(508).
Diaporthe sp. On foliage. **5:**OR(2787).
Didymascella thujina. Leaf blight. **2:**MI(82), NY(82), TX(82), VA(82), VT(82), WI(82). **3:**TX(82). **5:**NW states(82), AK(526), CA(81,82), ID(3202), MT(3202), WA(3202).
Didymascella sp. Needle blight. **6:**HI(2903).
Exarmidium hysteriiforme. **5:**ID(3202).
E. morthieri. **5:**ID(3202).
Herpotrichia juniperi. Snow blight. **5:**ID(82,3202).
H. macrotricha. On leaves. **6:**ME(189).
Hysterium thujae. On twigs. **2:**NY(82).
Kirschsteiniothelia thujina. On decaying leaves. **2:**MI(2862).
Leptosphaeria sp. Twig dieback. **3:**FL(35).
Leucostoma kunzei. On wood. **5:**ID(82,3202).
Lophiostoma thujae. **2:**MI(2862).
Lophodermium thujae. Needle spot. **2:**ME(82), NH(82), NY(82), WI(82). **3:**WA(3202).
Melanomma conjunctum. On limbs. **5:**ID(82,128,3202).
Microthyrium thujae. On living leaves. **5:**OR(82,832,3202).
Mollisia sp. **2:**GA(1438).
Mycosphaerella pinsapo. On twigs. **2:**NY(82).
Mycosphaerella sp. On twigs. **2:**AL(82), PA(82). **3:**LA(82).
Mytilinidion oblongisporum. **5:**OR(3202).
M. thujarum. On bark. **2:**NY(82,195), WI(82).
Pezicula sp. On foliage. **5:**OR(2787).
Phacidium infestans. Snow blight. **2:**ME(82), NY(82).
P. sherwoodiae. On dead leaves. **5:**OR(863).
Pithya cupressina. On dead foliage. **2:**GA(82,1438), MN(82), NY(82). **5:**ID(3202).
Pleiopatella harperi. **5:**WA(3202).
Pseudographis intermedia. **5:**WA(3202).
P. tetraspora. **2:**MT(3202).
Pseudoplectania melaena. **5:**ID(3202).
P. nigrella. **5:**ID(3202).
Quasiconcha reticulata. On roots. **2:**AZ(283).
Scirrhia conigena. Leaf spot, on cones. **2:**NY(82), OK(2872). **5:**CA(81), ID(82,3202), MT(82,3202), WA(3202).
Stamnaria thujae. On foliage. **2:**WI(82,1344).
Urnula craterium. On logs. **3:**NC(1303).
Valsa abietis. On twigs. **2:**IA(1226). **3:**PA(82). **5:**WA(82,3202).
V. thujae. On dead limbs. **2:**NY(82,195).
V. weiriana. On wood. **5:**ID(82,3202), WA(3202).

Basidiomycotina - Other

Acanthophysium cerussatum. On wood. **5:**Unk.(82).
A. lividocaeruleum. On wood. **5:**ID(82,128,2083,3202), MT(82), WA(82,2083).
A. weirii. On bark. **5:**Unk.(82), ID(2083,3202).
Agrocybe aegerita. On wood. **5:**ID(82,3202).
Aleurobotrys botryosus. On dead stems. **2:**MA(2083).
Aleurodiscus amorphus. On suppressed, injured, or recently dead trees. **5:**OR(82,1449,3202).
A. grantii. **5:**ID(128,2083).
A. tsugae. On bark. **2:**MI(2083). **5:**ID(2083,3202), WA(2083,3202).
Aleurodiscus sp. **5:**ID(3202).
Amyloathelia amylacea. On wood. **5:**ID(2083,3202), MT(2083,3202), OR(82).
Amylocystis lapponica. **5:**ID(3202).
Amylostereum chailletii. On wood. **2:**ME(543), NY(543). **5:**ID(82,3202).
A. laevigatum. On bark. **2:**ME(128).
Antrodia albida. **5:**WA(3202).
A. carbonica. **5:**WA(3202).
A. heteromorpha. **5:**WA(3202).
A. serialis. On wood. **5:**ID(82), WA(82,3202).
A. sinuosa. **5:**ID(3202).

A. vaillantii. Isolated from discarded poles. **5:**ID(82,3202,3394).
A. xantha. **5:**ID(3202), WA(3202).
Armillaria mellea. Stem and root rot. **2:**MI(82), NC(1303), NY(82), TX(82). **3:**MS(82,2720), TX(82). **5:**ID(538,1583,3202), WA(82,3202).
A. tabescens. Dieback, root rot. **2:**FL(35,82). **3:**FL(35,82), LA(2815).
Asterodon ferruginosum. **5:**MT(3202).
Auricularia auricula. On wood. **5:**ID(914,3202), WA(3202).
Basidiodendron eyrei. On wood. **2:**NY(2185).
Bjerkandera adusta. White rot. **2:**MN(82), PA(82,2962). **5:**ID(82,3202), MT(3202), WA(3202).
Ceraceomyces serpens. Wood rot. **5:**WA(82,128,3202).
Ceriporiopsis rivulosa. On wood. **5:**ID(128,3202), WA(82).
Cerocorticium hiemale. **5:**OR(2084).
Chaetoderma luna. **5:**ID(3202).
Columnocystis abietina. Brown cubical rot. **5:**ID(82,3202).
Coniophora arida. Brown cubical rot. **5:**ID(82,3394).
C. arida var. *suffocata*. **2:**ME(128).
C. puteana. Brown cubical rot. **5:**ID(82,3394), OR(3202), WA(82).
Coniophora sp. **6:**MI(128).
Corticium sp. **5:**WA(3202).
Cystostereum pini-canadense. White rot. **2:**MI(541), MN(541).
Datronia mollis. **5:**ID(3202), MT(3202), WA(3202).
Dendrothele griseo-cana. On bark. **2:**MI(2084).
D. nivosa. On bark. **2:**MI(82).
D. pachysterigmata. **2:**MN(1197).
Dichomitus squalens. Red ray rot. **5:**ID(82,3202), MT(3202).
Diplomitoporus lenis. **2:**MI(2862). **5:**ID(3202).
D. linbladii. **5:**ID(3202).
Echinodontium tinctorium. Heart rot of living trees. **5:**ID(82,3202).
Fibricium rude. On wood. **5:**ID(82,3202).
Fomitopsis cajanderi. Isolated from utility pole. **2:**MN(1197). **5:**MI(2591), WA(3202).
F. pinicola. Brown trunk rot. **5:**WA(82,3202).
F. rosea. Brown cubical rot. **2:**ME(82). **5:**OR(82).
Galerina thujina. On logs. **2:**MI(3333).
Ganoderma applanatum. White mottled butt rot. **5:**AK(526), WA(82,3202).
Gloeophyllum carbonarium. **2:**MI(2862). **5:**ID(128,3202), WA(3202).
G. odoratum. **5:**ID(3202).
G. sepiarium. Wood rot. **2:**MN(82). **5:**ID(3202), WA(82,3202).
Gloeoporus dichrous. On wood. **5:**ID(82,3202), MT(82,3202,3832), OR(3202), WA(82,3202).
Grandinia alutaria. **5:**ID(3202).
G. arguta. **2:**MN(1197).
G. hastata. **2:**OH(128).
G. pallidula. **5:**ID(3202).
Heterobasidion annosum. Brown butt rot. **2:**MI(82). **5:**ID(82,568,1583,3202), WA(3202).
Hymenochaete corrugata. On wood. **2:**MI(82). **5:**MT(82,3202).
H. spreta. On wood. **5:**ID(82,3202), WA(82,3202).
H. tabacina. On wood. **2:**MI(82,2862). **5:**ID(82,3202), MT(82,3202), WA(82,3202).
H. tenuis. On wood. **2:**MI(82), NY(82).
Hymenochaete sp. On wood. **5:**Unk.(82).
Hyphoderma puberum. **5:**WA(3202).
H. sambuci. **2:**MN(1197).
Hypochnicium geogenium. On wood. **5:**ID(82).
H. punctulatum. **5:**ID(3202).
H. vellereum. On wood. **5:**ID(82,3202), MT(82), WA(82).
Inonotus subiculosus. Timber rot. **2:**MI(82), NY(82), WI(82).
I. tomentosus. **5:**MT(3202).
Ischnoderma resinosum. **5:**ID(3202).
Junghuhnia collabens. **5:**MT(3202).
Laetiporus sulphureus. On wood. **5:**ID(82), OR(3202), WA(82).
Laurilia sulcata. On wood. **5:**AK(82,543).
Leucogyrophana mollusca. **2:**MI(2862).
Lindtneria thujatsugina. On duff. **5:**ID(2042).
Luellia sp. Foliage endophyte. **5:**OR(508,2787).
Mycena albicolor. On twigs. **5:**OR(3326).
M. clavata. **2:**NY(2936).
M. corticalis. On bark. **5:**OR(3326).
Neolentinus lepideus. **5:**ID(3202).
Odonticium romellii. **5:**ID(128,3202), MT(128,3202).
Oxyporus cuneatus. On wood. **5:**AK(526,1904), ID(82,3202), MT(82,3202), OR(82,3202), WA(82,3202).
O. similis. **5:**ID(3202).
Pachykytospora papyracea. Timber rot. **2:**ME(82), MI(82), NJ(82), NY(82), PA(82), VA(82).
Peniophora violaceolivida. **5:**ID(3202).
Perenniporia medulla-panis. **5:**WA(3202).
P. subacida. On wood. **5:**ID(82,1583,3202), WA(82,3202).
Phaeolus schweinitzii. Brown cubical butt and root rot. **2:**NE states(82), MS(2720). **5:**ID(82,1583,3202), WA(82,3202).
Phanerochaete affinis. **5:**ID(3202).
P. carnosa. White rot. **5:**AK(82), ID(82,128,3202), MT(82,433,3202), OR(82,3202).
P. gigantea. On wood. **5:**MT(82,3202).
P. sanguinea. White rot. **2:**ME(433), MI(433).
P. velutina. On wood. **5:**MT(82,3202).
P. viticola. White rot. **2:**NY(128,433).
Phellinus contiguus. **5:**MT(3832).
P. ferreus. **5:**OR(3202).
P. ferrugineofuscus. On wood. **5:**AK(1904).
P. ferruginosus. Timber rot. **2:**MI(82,2862), WI(82). **5:**OR(3202).
P. nigrolimitatus. White pocket rot. **5:**ID(82,128,538,1583,3202).
P. pini. Red ring rot, stem rot. **5:**ID(82,538,1583,3202), MT(82,3202,3832), WA(82,3202).
P. viticola. White rot. **2:**MI(82). **5:**ID(82,3202), OR(3202), WA(82,3202).
P. weirii. Root and butt rot. **2:**WI(82). **5:**Range of host(82), AK(526), ID(128,538,1583,3202), MT(128,3202,3832), OR(3202), WA(3202).
Phlebia livida. On wood. **5:**ID(82), MT(82,3202), WA(82).
P. romellii. **2:**MN(1197).
Phlebiella vaga. On wood. **5:**ID(82), MT(82), WA(82).
Polyporus badius. **5:**WA(3202).
P. melanopus. On wood. **5:**WA(3202).
P. varius. On wood. **5:**ID(82,3202).
Postia balsamea. **5:**WA(3202).
P. caesia. **5:**ID(3202), WA(3202).
P. fragilis. **5:**WA(3202).
P. guttulata. Brown cubical rot. **5:**ID(82,3202).
P. sericeomollis. Brown cubical rot. **5:**ID(128,538,1583,3202), MT(3202), WA(82,128).
P. stiptica. **5:**ID(3202).
Pycnoporellus fulgens. **5:**ID(3202).
Pycnoporus cinnabarinus. **5:**MT(3202).
Ramaricium alboflavescens. **6:**VA(128).
Resinicium bicolor. Root rot. **5:**ID(538,1583).
R. furfuraceum. On wood. **5:**ID(82,3202), MT(82), WA(82).
Rigidoporus crocatus. On wood. **5:**WA(82,128,3202).
Schizophyllum commune. Wood rot. **2:**ME(82).
Schizopora paradoxa. Timber rot. **2:**MN(82).
Scytinostroma galactinum. White root and butt rot. **5:**ID(82,3202), MT(82), WA(82).
Serpula lacrimans. **5:**WA(3202).
Skeletocutis nivea. **5:**ID(3202).
Stereum sanguinolentum. On wood. **5:**ID(1583), MT(82,3202).
Stereum sp. **5:**ID(3202).
Thelephora terrestris. On stems of seedlings. **5:**WA(82,3202).
Tomentella asperula. **6:**MI(2039).
T. botryoides. **6:**NJ(2039).
T. bryophila. **2:**MI(2862). **6:**MT(2039), PA(2039).
T. punicea. **6:**NJ(2039).
T. ramosissima. **6:**NY(2039).
T. rubiginosa. **2:**MI(128).
T. sublilacina. **6:**MI(2039), NY(2039), OR(2039).
T. violaceofusca. **6:**MI(2039).
Trametes hirsuta. On wood. **2:**ME(82). **5:**ID(82,3202), MT(3202), WA(3202).
T. pubescens. **5:**ID(3202).
T. versicolor. On wood. **2:**ME(82). **5:**ID(82,3202), MT(3202), OR(3202), WA(82,3202).
Trechispora mollusca. **5:**ID(3202).
Trichaptum abietinum. On wood. **5:**ID(82,3202), MT(3202), WA(82,3202).
Tubulicrinis glebulosus. On wood. **5:**ID(82,3202), WA(82,3202).
Tyromyces chioneus. On wood. **5:**ID(82,3202), MT(3202,3832), WA(82,3202).

Vararia racemosa. **5**:ID(128,3202).
Vesiculomyces citrinus. **5**:ID(3202).
Xeromphalina campanella. Root rot. **2**:MI(2862). **5**:ID(538,1583).
Deuteromycotina - Hyphomycetes
Alternaria sp. Twig blight, leaf blotch, dieback. **2**:FL(35), OK(2872). **3**:Unk.(82), FL(35), TX(82).
Aureobasidium pullulans. Isolated from discolored roof shingles. **5**:FL(3136), OR(3136).
Beltrania rhombica. Dieback. **3**:FL(82).
Botrytis cinerea. Gray mold blight. **3**:NJ(82).
Cercospora sequoiae. Blight. **2**:FL(35). **4**:TN(2941).
Chalara thielavioides. Black mold. **2**:NJ(1545). **3**:NJ(1545).
Cladosporium sp. Dieback. **3**:Unk.(82).
Fusarium solani. Seedling root rot. **2**:TX(82).
Fusarium sp. Root rot. **3**:FL(35).
Geniculosporium sp. Foliage endophyte. **5**:OR(508,2787).
Heterosporium sp. **6**:OR(3202).
Hormodendrum microsporum. Blue stain of heartwood. **2**:MN(82).
Nodulisporium sp. On foliage. **5**:OR(2787).
Periconia hispidula. On foliage. **1**:OR(2787). **5**:OR(2787).
Phialophora decumbens. On foliage. **5**:OR(2787).
P. hoffmannii. On foliage. **5**:OR(2787).
Phymatotrichopsis omnivora. Root rot. **2**:TX(82). **3**:TX(82).
Ptychogaster rubescens. Isolated from discarded poles. **5**:ID(3394).
Sigmoidea sp. Foliage endophyte. **5**:OR(508,2787).
Stigmina glomerulosa. Twig blight. **5**:ID(82), OR(82,3202), WA(82,3202).
S. thujina. Twig blight. **2**:OK(2872). **5**:ID(82,3202), OR(82,3202), WA(82,3202).
Deuteromycotina - Coelomycetes
Cylindrosporium sp. On dead needles and twigs. **2**:NC(1303).
Cytospora thujae. **5**:WA(3202).
Cytospora sp. Canker. **2**:OK(2872).
Diplodia thujina. On wood. **2**:NY(82).
Diplodia sp. Dieback. **2**:FL(35,82), NC(1303). **3**:AL(82), FL(35,82).
Gelatinosporium sp. On foliage. **5**:OR(2787).
Harknessia thujina. On bark. **2**:NJ(2548).
Hendersonia thyoides. Twig blight (secondary). **5**:OR(82,3202).
Hendersonia sp. Leaf blight. **3**:TX(82).
Kabatina juniperi. On foliage. **5**:OR(2787).
[*Kirschsteiniella applanata*]. **5**:MT(3202).
Leptostroma sp. On foliage. **5**:OR(2787).
Macrophoma sp. Dieback. **3**:FL(35).
Macrophomina phaseolina. Charcoal root rot. **3**:FL(35).
Micropera tenella. On limbs. **2**:NJ(82).
Microperella sp. Foliage endophyte. **5**:OR(508).
Monochaetia sp. Canker. **5**:WA(3202). **6**:WA(821).
Pestalotia thujicola. On cone scales. **2**:NY(2215).
Pestalotiopsis funerea. Twig blight, dieback, foliage endophyte. **2**:E states(82), MS(82,2720), NC(1303), OK(2872), TX(82). **3**:Range of host(82), FL(35). **5**:OR(508,2787), WA(82). **6**:WV(3911).
Phoma sp. Dieback. **2**:NC(1303).
Phomopsis juniperovora. Nursery blight. **2**:IN(82), KY(82), NJ(82), OH(82), OK(2872), PA(82), VA(82). **3**:FL(35), IL(82), IN(82), PA(82).
P. occulta. Leaf blight, secondary twig blight, foliage endophyte. **2**:CT(82), MD(82), MS(82,2720), NY(82), SC(82). **3**:MS(82,2720), NJ(82), SC(82), TX(82). **5**:OR(508).
Phyllosticta sphaeropsoidea. On leaves. **2**:WI(1341).
Seimatosporium berckmansii. Twig blight. **3**:OR(82,3202), WA(82,3202). **4**:OR(3547,3549).
S. caudatum. **2**:NY(3547).
Seiridium cardinale. Twig canker. **3**:CA(82).
S. juniperi. Foliage endophyte. **5**:OR(508,2787).
Sirodothis sp. On foliage. **5**:OR(2787).
Sphaeropsis sp. **3**:PA(82).
Deuteromycotina - Other
Rhizoctonia solani. Damping-off, root and stem rot. **2**:CT(82), FL(35), TX(82). **3**:FL(35), NH(82), NY(82), TX(82), VA(82).
Sclerotium rolfsii. Southern blight. **3**:FL(35).

Thujopsis Siebold & Zucc. ex Endl.

1. *T. dolabrata* (L. fil.) Siebold & Zucc. [false arborvitae, hiba arborvitae]

Deuteromycotina - Coelomycetes
Phomopsis occulta. Twig blight (secondary). **1**:CA(82).

Cycadaceae

Cycas L.

1. *C. circinalis* L. [fern-palm, queen sago, sago-palm]
2. *C. revoluta* Thunb. [Japanese fern-palm, Japanese sago-palm, sago-palm]

Oomycetes
Phytophthora sp. Root rot. **1**:FL(35). **2**:FL(35).
Pythium sp. Root rot. **1**:FL(35). **2**:FL(35).
Ascomycotina
Leptosphaeria sp. Leaf spot. **2**:FL(35).
Basidiomycotina - Other
Armillaria tabescens. Root rot. **1**:FL(35).
Ganoderma sp. Butt rot. **1**:FL(35).
Deuteromycotina - Hyphomycetes
Alternaria sp. Leaf spot. **1**:FL(35). **2**:FL(35,82).
Cercospora sp. Leaf spot. **2**:FL(35).
Fusarium oxysporum. Leaf spot. **2**:FL(35).
Fusarium sp. Root rot, leaf spot. **2**:FL(35).
Deuteromycotina - Coelomycetes
Ascochyta cycadina. Leaf spot. **2**:MO(82), TX(82).
Colletotrichum sp. Leaf spot. **1**:FL(35). **2**:FL(35).
Coniothyrium sp. Leaf spot. **2**:FL(35).
Dendrophoma clypeata. Leaf spot. **2**:FL(35).
Diplodia sp. Leaf spot. **2**:FL(35).
Pestalotia cycadis. Leaf spot. **2**:CT(82), FL(35,82).
Phoma bresadolae. Leaf spot. **2**:FL(82), MS(82,2720), PA(82).
Phoma sp. Leaf spot. **2**:FL(35,82), MS(82), PA(82).
Phyllosticta cycadina. Leaf spot. **1**:FL(35). **2**:FL(35).
Phyllosticta sp. Leaf spot. **1**:FL(35). **2**:FL(35).
Rhabdospora sp. On leaves. **2**:FL(35,82).
Sphaeropsis sp. Leaf spot. **1**:FL(35).
Deuteromycotina - Other
Rhizoctonia solani. Root and stem rot. **1**:FL(35). **2**:FL(35).

Cymodoceaceae

Halodule Endl.

1. *H. wrightii* Aschers.

Plasmodiophoromycetes
Plasmodiophora diplantherae. Forming shoot galls. **1**:FL(362).

Cyperaceae

Baumea Gaud. in Freyc.

1. *B. meyenii* Kunth

Ascomycotina
Meliola argentina. Black mildew. **1**:HI(1272,1452,2903).

Carex L.

1. *C. abrupta* Mackenz.
2. *C. adusta* Boott [browned sedge]
3. *C. aenea* Fernald
4. *C. aestivalis* M. A. Curtis [summer sedge]
5. *C. albursina* E. Sheld. [white bear sedge]
6. *C. alopecoidea* Tuck. [foxtail sedge]
7. *C. amplectens* Mackenz.
8. *C. amplifolia* Boott
9. *C. annectens* (C. Bicknell) C. Bicknell
10. *C. aquatilis* Wahlenb. (=*C. aquatilis* Wahlenb. var. *altior* (Rydb.) Fernald) [water sedge]
11. *C. arapahoensis* Clokey
12. *C. arcta* Boott [northern clustered sedge]
13. *C. arctata* Boott [drooping wood sedge]
14. *C. artitecta* Mackenz.
15. *C. assiniboinensis* Boott
16. *C. atherodes* Spreng.
17. *C. athrostachya* Olney
18. *C. atrata* L.

19. *C. atratiformis* Britton [black sedge]
20. *C. aurea* Nutt. (=*C. garberi* Fernald) [golden-fruited sedge]
21. *C. backii* Boott
22. *C. barbarae* Dewey
23. *C. bebbii* Olney
24. *C. bicknellii* Britton
25. *C. bigelowii* Torr. ex Schwein.
26. *C. bipartita* All. (=*C. lachenalii* Schkuhr)
27. *C. blanda* Dewey
28. *C. brainerdii* Mackenz.
29. *C. brevior* (Dewey) Mackenz. ex Lunell (=*C. merritt-fernaldii* Mackenz.)
30. *C. breweri* Boott
31. *C. bromoides* Schkuhr
32. *C. brunnescens* (Pers.) Poir. [brownish sedge]
33. *C. buxbaumii* Wahlenb.
34. *C. californica* L. H. Bailey
35. *C. canescens* L. [silvery sedge]
36. *C. cephalantha* (L. H. Bailey) C. Bicknell
37. *C. cephaloidea* Dewey [thin leaved sedge]
38. *C. cephalophora* Muhl. ex Willd. [oval-headed sedge]
39. *C. cherokeensis* Schwein.
40. *C. chordorrhiza* (Ehrh.) L. fil. [creeping sedge]
41. *C. circinata* C. A. Mey.
42. *C. communis* L. H. Bailey
43. *C. comosa* Boott [bristly sedge]
44. *C. complanata* Torr. & Hook. (=*C. triceps* Michx.)
45. *C. concinna* R. Br. ex Richards.
46. *C. concinnoides* Mackenz. [low northern sedge]
47. *C. conoidea* Schkuhr [field sedge]
48. *C. crawfordii* Fernald
49. *C. crinita* Lam. [fringe sedge]
50. *C. cristatella* Britton [crested sedge]
51. *C. debilis* Michx.
52. *C. debilis* Michx. var. *rudgei* L. H. Bailey
53. *C. deflexa* Hornem. [northern sedge]
54. *C. deweyana* Schwein. (=*C. bolanderi* Olney, *C. leptopoda* Mackenz.)
55. *C. dioica* L. (=*C. gynocrates* Wormsk. ex Drejer)
56. *C. disperma* Dewey
57. *C. douglasii* Boott
58. *C. ebenea* Rydb.
59. *C. eburnea* Boott
60. *C. elynoides* Holm
61. *C. emmonsii* Dewey
62. *C. emoryi* Dewey
63. *C. enanderi* Hultén
64. *C. eurycarpa* Holm
65. *C. exilis* Dewey
66. *C. festucacea* Schkuhr ex Willd.
67. *C. feta* L. H. Bailey
68. *C. filifolia* Nutt. [threadleaf sedge]
69. *C. flava* L. var. *fertilis* Peck
70. *C. foenea* Willd. [hay sedge]
71. *C. foetida* All. (=*C. vernacula* L. H. Bailey)
72. *C. fracta* Mackenz.
73. *C. frankii* Kunth
74. *C. geophila* Mackenz.
75. *C. geyeri* Boott [elk sedge]
76. *C. gmelinii* Hook. & Arnott
77. *C. gracillima* Schwein. [graceful sedge]
78. *C. gravida* L. H. Bailey [heavy sedge]
79. *C. grayi* J. Carey
80. *C. gynodynama* Olney
81. *C. haydeniana* Olney (=*C. nubicola* Mackenz.)
82. *C. heliophila* Mackenz.
83. *C. hendersonii* L. H. Bailey
84. *C. hindsii* C. B. Clarke
85. *C. hoodii* Boott
86. *C. hookerana* Dewey
87. *C. houghtoniana* Torr.
88. *C. hystricina* Muhl. ex Schkuhr [porcupine sedge]
89. *C. interior* L. H. Bailey [inland sedge]
90. *C. intumescens* Rudge [bladder sedge]
91. *C. jacobi-peteri* Hultén
92. *C. jonesii* L. H. Bailey
93. *C. joorii* L. H. Bailey
94. *C. lacustris* Willd.
95. *C. laeviconica* Dewey
96. *C. laeviculmis* Meinsh.
97. *C. lanuginosa* Michx. [woolly sedge]
98. *C. lasiocarpa* Ehrh.
99. *C. lasiocarpa* Ehrh. var. *americana* Fernald
100. *C. laxiflora* Lam. [loose-flowered sedge]
101. *C. leersii* Willd. var. *angustata* Mackenz.
102. *C. lenticularis* Michx. (=*C. kelloggii* W. Boott) [lenticular sedge]
103. *C. leptalea* Wahlenb. [bristle-stalked sedge]
104. *C. leptonervia* Fernald
105. *C. limosa* L. [mud sedge]
106. *C. livida* (Wahlenb.) Willd.
107. *C. loliacea* L.
108. *C. lonchocarpa* Willd.
109. *C. lupuliformis* Sartwell ex Dewey [hop-like sedge]
110. *C. lupulina* Muhl. ex Willd. [hop sedge]
111. *C. lurida* Wahlenb. [sallow sedge]
112. *C. luzulaefolia* S. Wats.
113. *C. luzulina* Olney
114. *C. lyngbyei* Hornem. (=*C. lyngbyei* Hornem. subsp. *cryptocarpa* (C. A. Mey.) Hultén)
115. *C. macloviana* Dum. d'Urv. (=*C. macloviana* Dum. d'Urv. subsp. *pachystachya* Cham. ex Steud., *C. pachystachya* Cham. ex Steud., *C. platylepis* Mackenz.)
116. *C. macrocephala* Willd.
117. *C. macrochaeta* C. A. Mey.
118. *C. magellanica* Lam.
119. *C. meadii* Dewey
120. *C. mendocinensis* Olney ex Boott
121. *C. mertensii* Prescott
122. *C. microptera* Mackenz. (=*C. festiva* Dewey, *C. festivella* Mackenz.)
123. *C. muhlenbergii* Schkuhr
124. *C. multicaulis* L. H. Bailey
125. *C. multicostata* Mackenz.
126. *C. muricata* L. (=*C. angustior* Mackenz.) [lesser prickly sedge]
127. *C. nardina* Fr. (=*C. hepburnii* Boott)
128. *C. nebrascensis* Dewey
129. *C. nesophila* Holm
130. *C. neurophora* Mackenz.
131. *C. nigra* (L.) Reichard (=*C. goodenovii* J. Gay)
132. *C. nigricans* C. A. Mey.
133. *C. nigromarginata* Schwein. var. *elliptica* (Boott) Gleason (=*C. peckii* E. Howe) [black-edged sedge]
134. *C. normalis* Mackenz.
135. *C. nudata* Boott
136. *C. obnupta* L. H. Bailey
137. *C. obtusata* Lilijebl.
138. *C. occidentalis* L. H. Bailey
139. *C. oederi* Retz. (=*C. viridula* Michx.) [green sedge]
140. *C. oligosperma* Michx. [few-seeded sedge]
141. *C. pansa* L. H. Bailey
142. *C. pauciflora* Lightf. [few-flowered sedge]
143. *C. paupercula* Michx.
144. *C. pedunculata* Muhl. ex Willd. [long-stalked sedge]
145. *C. pensylvanica* Lam.
146. *C. petasata* Dewey
147. *C. petricosa* Dewey
148. *C. phaeocephala* Piper (=*C. eastwoodiana* Stacey)
149. *C. phyllomanica* S. Wats.
150. *C. picta* Steud.
151. *C. plantaginea* Lam. [plantain-leaved sedge]
152. *C. platyphylla* J. Carey [thicket sedge]
153. *C. pluriflora* Hultén
154. *C. podocarpa* R. Br. [long-awned arctic sedge]
155. *C. praegracilis* W. Boott (=*C. camporum* Mackenz.)
156. *C. prairea* Dewey
157. *C. praticola* Rydb.
158. *C. projecta* Mackenz.
159. *C. pseudocyperus* L.
160. *C. pubescens* Muhl. ex Schkuhr
161. *C. pyrenaica* Wahlenb.

162. *C. pyrenaica* Wahlenb. subsp. *micropoda* (C. A. Mey.) Hultén
163. *C. ramenskii* Kom.
164. *C. rariflora* (Wahlenb.) Sm. (=*C. stygia* Fr.) [loose-flowered alpine sedge]
165. *C. raymondii* Calder
166. *C. raynoldsii* Dewey
167. *C. reniformis* (L. H. Bailey) Small
168. *C. retrorsa* Schwein. [retrorse sedge]
169. *C. rhynchophysa* C. A. Mey. (=*C. laevirostris* Anderss.)
170. *C. richardsonii* R. Br.
171. *C. riparia* Curtis [riverbank sedge]
172. *C. rosea* Schkuhr (=*C. convoluta* Mackenz.) [stellate sedge]
173. *C. rossii* Boott (=*C. brevipes* W. Boott)
174. *C. rostrata* Stokes ex With. (=*C. rostrata* Stokes ex With. var. *utriculata* Boott, *C. inflata* Rendle & Britton)
175. *C. sartwellii* Dewey
176. *C. saxatilis* L. (=*C. physocarpa* K. Presl) [russet sedge]
177. *C. scabrata* Schwein. [rough sedge]
178. *C. scirpoidea* Michx. (=*C. pseudoscirpoidea* Rydb.)
179. *C. scoparia* Schkuhr ex Willd. [pointed-broom sedge]
180. *C. scopulorum* Holm
181. *C. senta* Boott
182. *C. siccata* Dewey [dry-spiked sedge, hillside sedge]
183. *C. simulata* Mackenz.
184. *C. sitchensis* Prescott
185. *C. specifica* L. H. Bailey
186. *C. spectabilis* Dewey
187. *C. sprengelii* Dewey ex Spreng.
188. *C. stellulata* Good var. *cephalantha* Fernald
189. *C. stenophylla* Wahlenb. (=*C. eleocharis* L. H. Bailey)
190. *C. stenophylla* Wahlenb. var. *enervis* (C. A. Mey.) Kükenth.
191. *C. sterilis* Willd. [little prickly sedge]
192. *C. stipata* Muhl. ex Willd. [awn-fruited sedge]
193. *C. straminea* Willd. [straw sedge]
194. *C. straminiformis* L. H. Bailey
195. *C. stricta* Lam. [tussock sedge]
196. *C. stricta* Lam. var. *strictior* (Dewey) J. Carey (=*C. strictior* Dewey)
197. *C. subfusca* W. Boott
198. *C. supina* Willd. [weak arctic sedge]
199. *C. tenera* Dewey [marsh straw sedge]
200. *C. tetanica* Schkuhr [wood sedge]
201. *C. tolmiei* Boott
202. *C. torreyi* Tuck. (=*C. abbreviata* Boott)
203. *C. torta* Boott [twisted sedge]
204. *C. tribuloides* Wahlenb. (=*C. cristata* Schwein.) [blunt-broom sedge]
205. *C. trichocarpa* Muhl. ex Willd.
206. *C. triquetra* Boott
207. *C. trisperma* Dewey [three-fruited sedge]
208. *C. tuckermanii* Boott
209. *C. tumulicola* Mackenz.
210. *C. turgescens* Torr.
211. *C. typhina* Michx.
212. *C. umbellata* Schkuhr (=*C. umbellata* Schkuhr var. *tonsa* Fernald, *C. abdita* C. Bicknell)
213. *C. unilateralis* Mackenz.
214. *C. vaginata* Tausch (=*C. saltuensis* L. H. Bailey)
215. *C. vallicola* Dewey (=*C. vallicola* Dewey var. *rusbyi* (Mackenz.) F. J. Herm., *C. rusbyi* Mackenz.)
216. *C. variabilis* L. H. Bailey
217. *C. venusta* Dewey (=*C. oblita* Steud.)
218. *C. venusta* Dewey var. *rudgei* L. H. Bailey
219. *C. vesicaria* L. (=*C. exsiccata* L. H. Bailey)
220. *C. vesicaria* L. var. *monile* (Tuck.) Fernald
221. *C. vestita* Willd. [velvet sedge]
222. *C. vulpinoidea* Michx. [fox sedge]
223. *C. wahuensis* C. A. Mey.
224. *C. whitneyi* Olney
225. *Carex* sp.

Chytridiomycetes

Synchytrium caricis. On leaves. **225:**UT(82).

Ascomycotina

Acrospermum graminum. **49:**GA(1438).
Bricookea sepalorum. **124:**CA(642).
Calloria caricinella. **225:**ID(3202).
Ciboria aschersoniana. On seeds. **225:**CO(82), NY(82).
Clathrospora elynae. On overwintered herbage. **186:**CA(642).
C. heterospora. **113:**WA(3202).
C. heterospora var. *simmonsii*. **217:**WA(3814).
C. pentamera. **10:**AK(526).
Claviceps grohii. Ergot. **204:**MI(82).
Crocicreas panici. **225:**OR(3202).
Didymella proximella. On culms and leaves. **34:**OR(681). **186:**WA(3202,3815). **225:**WA(3815).
Dothidella caricina. On leaves. **225:**NY(82).
Duplicaria acuminata. **30:**CA(661). **225:**WA(3202).
[*Gaeumannomyces graminis*]. Take-all. **54:**WA(3202). **225:**WA(3446).
Heptameria tenera. **225:**WA(3202,3814).
Herpotrichia macrotricha. On culms. **145:**GA(177). **225:**GA(189).
Hysteropezizella diminuens. On culms and leaves. **35:**AK(526). **225:**AK(82,1938), MI(82), NY(82), VT(82).
H. ignobilis. **10:**AK(526).
H. multipuncta. **225:**NY(82).
Lachnum nardi. **225:**WA(3202).
Lasiosphaeria sp. **111:**GA(1438).
Leptosphaeria epicarecta. **111:**GA(1438).
L. folliculata var. *oxyspora*. **225:**WI(82).
Leptosphaeria sp. On dead culms and leaves. **17:**CA(661). **84:**AK(3427,3431). **148:**OR(3202). **195:**AK(82), GA(82), WI(82). **222:**WI(1355). **225:**AK(82), GA(82), NC(1303), WI(82).
Lophodermium caricinum. On culms and leaves. **22:**CA(3443). **75:**CO(82). **225:**CA(82), NJ(82), NY(82).
Melanodothis caricis. On inflorescences. **33:**AK(82,526,869). **63:**AK(526,869). **105:**AK(526,869). **114:**AK(526,869). **142:**AK(526,869).
Microthyrium microscopicum. **225:**OR(657).
Mollisia culmina. **49:**GA(1438).
Mollisia sp. **49:**GA(1438).
Mycosphaerella caricis. On dead culms and leaves. **195:**NY(82).
M. leptospora. **121:**AK(526).
M. lineolata. On dead culms and leaves. **225:**CO(82).
M. recutita. On dead culms and leaves. **225:**AK(526), NY(82).
M. scirrhoides. On leaves. **225:**NY(82).
M. tassiana. On dead culms and leaves. **64:**CA(661). **72:**CA(642). **75:**CO(82). **225:**AK(82,1938).
M. wichuriana. On dead culms and leaves. **225:**AK(82), MT(82,3202).
Mycosphaerella sp. On bracts. **195:**WI(1359).
Myriosclerotinia caricis-ampullaceae. **10:**WY(82,1195,1960). **174:**WY(82). **225:**MI(3331), WA(3331).
M. longisclerotialis. **156:**NY(1960). **195:**ME(82), NY(82). **225:**ME(82), NY(82).
M. sulcata. Inflorescence blight. **49:**DE(82), NY(82). **128:**OR(82,3202). **195:**DE(82), NY(82). **225:**DE(82), NY(82), WA(3202).
Niptera guestphalica. **111:**GA(1438).
N. melanophaea. On dead leaves. **225:**NY(82).
Phaeosphaeria caricinella. On leaves and stems. **20:**AK(3431). **186:**CA(661). **225:**CA(81,82,642,661).
[*P. insignis*]. On dead herbage. **186:**CA(661).
P. nigrans. On dead sheath. **225:**AK(1938).
P. typharum. **225:**WA(3202,3814).
Phialea sp. On leaves. **225:**CA(642).
[*Phomatospora dinemasporium*]. **225:**ID(3202).
P. therophila. **225:**WY(2924).
Phyllachora caricis. On leaves. **225:**IL(82), ME(82).
Pleospora discors. On dead leaves. **225:**CO(82).
P. gigaspora. On dead herbage. **186:**CA(661).
P. longispora. **193:**MT(3202). **225:**MT(3813).
Pseudopeziza caricina. **182:**ND(377).
Pseudosphaeria wallowaensis. On leaves. **166:**OR(3202,3436).
Psilachnum crinellum. On leaves. **49:**PA(849).
Pyrenopeziza karstenii. **225:**CO(82), NY(82).
P. multipunctoidea. On dead leaves. **225:**NY(82).
P. rigidae. **25:**AK(526).
P. yogoensis. **225:**MT(3202).
Sclerotinia sp. Inflorescence blight. **225:**Unk.(82).
Sphaerulina pallens. On leaves. **225:**WI(82).

Basidiomycotina - Uredinales

Puccinia atrofusca. Heteroecious: telial host. **1:**OR(3202). **49:**C & W states(82), AK(82). **57:**AZ(1195,1203,3981), CA(81), ID(3202), MT(1195,3202), OR(3202), WA(3202). **68:**C & W states(82), AK(82), CA(82), MT(3202), SD(2243), WY(1195). **75:**C & W states(82), AK(82), CA(82), MT(3202), WA(3202), WY(1195,2924). **85:**WY(1195). **122:**C & W states(82), AK(82). **124:**CA(81,638,642), OR(3202). **128:**C & W states(82), AK(82), CA(82). **137:**MT(3202), SD(2243). **145:**MT(3202), ND(378), OR(3202). **146:**MT(3202). **151:**C & W states(82), AK(82). **155:**MT(3202), OR(3202). **173:**CA(81), CO(1195), OR(3202), WA(3202). **182:**AZ(1203,3981). **195:**C & W states(82), AK(82), CA(82). **212:**OR(3202). **225:**C & W states(82), AK(82), AZ(1203), CA(82), ND(378).

P. bolleyana. Heteroecious: telial host. **49:**E states(82), TX(82), WI(82). **68:**E states(82), TX(82), WI(82). **73:**MS(2720). **75:**E states(82), TX(82), WI(82). **110:**WI(1333). **111:**GA(1439). **122:**E states(82), TX(82), WI(82). **128:**E states(82), TX(82), WI(82). **151:**E states(82), TX(82), WI(82). **195:**E states(82), TX(82), WI(82). **225:**E states(82), FL(35), NC(1303), TX(82), WI(82).

P. caricina. Heteroecious: telial host. **4:**NC(1303). **7:**CA(642). **8:**ID(3202), WA(3202). **10:**ID(3202), MT(3202), WY(1195,3379). **16:**MT(3202), ND(378), SD(2243), WA(3202), WI(1355). **23:**ND(378). **25:**AK(3431). **43:**WI(1360). **46:**ID(3202). **47:**WI(1352). **49:**NC states(82), NE states(82), Range of host(82). **52:**NC(1307). **54:**CA(81), ID(3202), WA(3202). **57:**MT(3202). **64:**ID(3202). **69:**WI(1345). **76:**AK(526). **77:**NY(2456). **83:**CA(81). **85:**MT(3202), WA(3202). **93:**TX(652). **96:**AK(3427). **97:**MT(3202), WI(1350). **100:**ND(378). **102:**CA(638,642), ID(3202), OR(3202), WA(3202). **113:**ID(3202). **115:**MT(3202), OR(3202). **117:**AK(60,526,3427). **125:**CA(661). **127:**MT(1195,3202). **128:**NC states(82), NE states(82), Range of host(82), CA(661), ID(3202), MT(3202), OR(3202), WA(3202), WY(1195,3379). **130:**ID(3202). **131:**OR(3202). **135:**CA(81). **136:**CA(81), OR(3202), WA(3202). **139:**CA(81), OR(3202). **144:**WI(1355). **145:**MT(3202), OR(3202). **149:**OR(3202). **151:**NC states(82), NE states(82), Range of host(82). **153:**AK(60,526). **154:**ID(3202). **159:**WI(1355). **164:**AK(526). **165:**AK(526). **173:**WA(3202). **174:**CA(81), CO(1195), ID(1195,3202), MT(3202), OR(3202), WA(3202), WY(1195). **180:**MT(3202). **181:**AZ(1195,1203,3981). **185:**CA(661). **186:**AK(3427,3443), OR(3202), WA(3202). **192:**OR(3202), WA(3202). **201:**AK(526). **205:**SD(2243). **211:**WI(1353). **216:**WY(1195). **224:**CA(638,642).

P. caricis-shepherdiae. Heteroecious: telial host. **10:**WY(2924). **225:**NC states(82), NE states(82), NW states(82), AK(526).

P. dioicae. Heteroecious: telial host. **2:**WI(1340). **3:**MT(3202), WI(1333). **9:**WI(1360). **10:**MT(3202). **12:**OR(3202), WI(1358). **15:**WI(1354). **17:**CA(81), OR(3202), WA(3202). **24:**OK(2872), WI(1344). **28:**CA(638,642). **29:**WI(1355). **35:**OR(3202). **37:**ND(378). **38:**WI(1350). **40:**WI(1333). **45:**WI(1355). **48:**WI(1333). **49:**NC(1303). **51:**WI(1356). **54:**CA(81), OR(3202), WA(3202). **56:**ID(3202,3443), WA(3202), WI(1355). **57:**ID(3202). **66:**ND(378). **67:**OR(3202). **68:**Range of host(82), MT(1195,3202). **70:**WI(1355). **71:**WY(1195). **75:**Range of host(82), WY(1195). **80:**CA(81). **81:**CO(649). **85:**MT(3202), WY(1195). **86:**ND(378). **87:**WI(1355). **89:**WI(1335). **96:**WA(3202). **97:**ND(378). **98:**WI(1359). **102:**OR(3202). **103:**CO(649). **115:**CA(81), ID(3202), MT(3202), OR(3202), WA(3202). **122:**ID(3202), MT(3202), WY(1195). **123:**MS(2720). **125:**CA(81). **126:**ID(3202). **128:**MT(3202). **131:**OR(3202). **134:**WI(1355). **139:**ND(378), WI(1333). **142:**WI(1355). **146:**MT(3202), WY(1195). **148:**ID(1195,3202). **149:**OR(3202). **155:**AZ(1195,1203,3981), ND(378), OR(3202). **157:**MT(3202). **158:**WI(1355). **167:**OK(2873). **168:**CA(81). **172:**WI(1333). **173:**MT(3202). **174:**MT(3202), WA(3202). **179:**OR(3202), WI(1333). **182:**AZ(1203,3981), MT(3202), ND(378). **188:**WI(1336). **191:**WI(1355). **192:**ID(3202), MT(3202), OR(3202), WA(3202). **193:**OR(3202). **197:**OR(3202). **199:**WI(1359). **209:**CA(81), OR(3202). **213:**CA(81). **215:**AZ(1195,1203,3981). **224:**OR(3202).

P. dulichii. Heteroecious: telial host. **225:**Range of host(82).

P. eumacrospora. Heteroecious: telial host. **43:**DE(82), NJ(82). **221:**DE(82), NJ(82).

P. karelica. Heteroecious: telial host. **49:**WI(1360). **225:**NC states(82), NE states(82), AK(82).

P. limosae. Heteroecious: telial host. **105:**NC states(82), NE states(82), WI(1355). **225:**NE states(82).

P. microsora. Telial host; aecial host unknown. **68:**LA(82), OR(82), PA(82), VA(82), WI(82), WV(82). **75:**LA(82), OR(82), PA(82), VA(82), WI(82), WV(82). **122:**LA(82), OR(82), PA(82), VA(82), WI(82), WV(82). **128:**LA(82), OR(82), PA(82), VA(82), WI(82), WV(82). **151:**LA(82), OR(82), PA(82), VA(82), WI(82), WV(82). **177:**MA(664). **195:**LA(82), OR(82), PA(82), VA(82), WI(82), WV(82). **219:**OR(3202). **225:**LA(82), OR(82), PA(82), VA(82), WI(82), WV(82).

P. minuta. **225:**AL(82), FL(35,82).

P. minutissima. Heteroecious: telial host. **98:**WI(1348,1359). **225:**NC states(82), NE states(82), MO(82).

P. molokaiensis. Telial host; aecial host unknown. **225:**HI(82,2903).

P. puritanica. **145:**MA(82), WI(1355).

Puccinia sp. **43:**WI(1360).

Uredo hawaiiensis. **223:**HI(2903).

Uromyces minutus. Telial host; aecial host unknown. **44:**MS(2720). **49:**E states(82), TX(82), WI(82). **68:**E states(82), TX(82), WI(82). **75:**E states(82), TX(82), WI(82). **122:**E states(82), TX(82), WI(82). **128:**E states(82), TX(82), WI(82). **151:**E states(82), TX(82), WI(82). **160:**WI(1340). **195:**E states(82), TX(82), WI(82). **225:**E states(82), TX(82), WI(82).

U. perigynius. Heteroecious: telial host. **15:**WI(1354). **17:**OR(3202), WA(3202). **49:**Range of host(82). **68:**Range of host(82). **75:**Range of host(82). **122:**Range of host(82). **128:**Range of host(82). **133:**WI(1344). **151:**Range of host(82). **195:**Range of host(82). **202:**ND(378). **213:**CA(81). **225:**Range of host(82).

U. uniporulus. Telial host. **218:**NC(1303). **225:**E states(82).

U. valens. Telial host. **110:**WI(1342). **225:**IN(82), WI(82).

Uromyces sp. **152:**IN(987). **170:**WI(1355).

Basidiomycotina - Ustilaginales

Anthracoidea americana. **174:**AK(1997), WY(1997).

A. aspera. **40:**AK(526,1997). **225:**WY(1997).

A. atratae. **18:**CO(1077,1996), MT(1077,1996,3093,3202). **33:**AK(3443). **117:**AK(1996). **129:**AK(1996). **154:**AK(1996), WY(1077,1996). **166:**WY(1077,1996). **186:**CA(661,1996), WA(1996).

A. buxbaumii. **33:**AK(1996).

A. calderi. **21:**MN(1997).

A. caricis. Covered seed smut. **10:**AK(526,1077), CO(649), ID(1077,3202), WA(1077,3202), WI(1358). **13:**ID(1077,3202), MI(1077), NH(1077). **14:**MA(1077), MD(1077), OH(1077), PA(1077). **19:**MT(1077,3202). **27:**IN(1077), WI(1357). **31:**NY(1077). **33:**CT(1077), PA(1077). **36:**NH(1077), NY(1077). **40:**AK(1077). **49:**Range of host(82), NY(1077). **53:**CO(1077). **55:**OR(1077). **57:**CA(1077), CO(1077), MT(1077,3202), NE(1077), OR(1077,3202), WA(1077,3202). **60:**CO(1077). **62:**WI(1354). **65:**MA(1077). **68:**Range of host(82), WY(1195). **70:**CO(1077), VT(1077). **74:**CO(1077), UT(1077). **75:**Range of host(82), WA(1077,3202), WY(1077). **76:**AK(526,1077). **82:**CO(1077). **89:**ME(1077). **98:**MT(1077,3202). **101:**NH(1077). **103:**MI(1077). **105:**MA(1077), MI(3879), WI(1077). **106:**AK(526,1077), ND(1077). **108:**MS(1077,2720). **112:**CA(81,1077). **114:**AK(1077). **115:**AK(526). **119:**WI(1360). **122:**Range of host(82), AK(1077). **124:**CA(642,1077,3093). **126:**NH(1077), VT(1077). **128:**Unk.(82), MT(1077,3202). **132:**CA(638,642,1077), UT(1077). **136:**CA(1077). **138:**WY(1077). **140:**NY(1077). **141:**CA(81). **143:**NY(1077). **144:**ME(1077). **145:**CT(1077), IA(1077), IL(1077), MA(1077,1996), ME(1077), MN(1077), MT(1077,3202), ND(1996), NH(1077,1996), NY(1077,1996), PA(1077), RI(1077), WI(1077,1996). **151:**Range of host(82), Unk.(82). **154:**ID(1077,3202). **155:**MT(1077,3202), UT(1077), WY(1077,1195). **164:**AK(526,1077). **166:**WY(1195). **169:**AK(526,1077). **170:**WI(1357). **173:**AK(1996), CO(1077,1996), WY(1996). **174:**AK(526,1077), MI(1077), NY(1077), PA(1077). **178:**WY(1077). **180:**ID(3202), OR(3202). **184:**CA(1077), WA(3202). **186:**CA(638,642,1077), WA(1077,3202). **189:**MT(3202). **190:**CA(1077), CO(1077). **191:**NH(1077), VT(1077). **194:**CA(1077). **195:**Range of host(82), CT(1077), NE(1077), NY(1077), WI(1077).

196:MI(1077). **200:**MA(1077), ND(1077). **212:**DE(1077), IN(1077), ME(1077), NH(1077), NY(1077), WI(1357). **214:**ME(1077). **220:**VT(1077). **225:**Unk.(82), CA(81,82), CO(1195), FL(35), VT(1224).
A. caricis var. *intermedia*. **75:**WY(3093). **76:**AK(526,3093). **114:**AK(526).
A. caryophylleae. **198:**AK(1996).
A. douglasii. **57:**CO(2562).
A. eleocharidis. **183:**WY(1195). **189:**CA(2562).
[*A. elynae*]. **26:**AK(526,3093).
A. externa. **68:**CA(1077), CO(1077,3093), MT(82,1077,3202), ND(82,378,1077,1996,3093), NE(82,1077), WY(82,1077,1996,3093). **186:**CA(642).
A. fischeri. **35:**AK(1997). **97:**WY(1077,3093). **174:**AK(526,1077), WY(1077). **225:**WY(3093).
A. heterospora. **10:**AK(526), CA(661). **26:**AK(1077). **31:**NY(1077). **32:**AK(526,1077,3093), VT(1077,3093). **35:**AK(526,1077,3093). **54:**AK(1077). **55:**AK(1077). **56:**AK(526,1077). **68:**WY(1077). **106:**AK(3093). **107:**AK(526,1077,3093). **114:**AK(526,3093). **142:**AK(526,1077,3093). **147:**AK(526,1077). **155:**CO(1077), MT(1077,3202), WY(1077). **161:**AK(1077). **163:**AK(526,3093). **173:**AK(526,1077,3093). **215:**WY(1195).
A. karii. **26:**AK(2562). **32:**AK(2562), VT(2562). **35:**AK(2562). **36:**VT(2562). **55:**AK(2562). **56:**AK(2562). **89:**ME(2562). **107:**AK(2562). **126:**VT(2562). **207:**MI(2562).
A. lasiocarpae. **99:**AK(1996).
A. limosa. **33:**AK(526). **70:**CO(3093), WY(3093). **99:**AK(526). **105:**AK(1996), MA(1996), VT(1996). **118:**AK(1996). **143:**AK(526). **153:**AK(1996). **164:**AK(1996).
A. limosa var. *gigantissima*. **105:**AK(526,3093). **114:**AK(526). **143:**AK(526,3093). **153:**AK(526,3093).
A. limosa var. *minor*. **57:**CO(3093), NV(3093), OR(3093). **178:**AK(526,3093), WY(3093).
A. paniceae. **106:**AK(1996).
A. scirpoideae. **178:**AK(1996), CO(1996), WY(1996).
A. subinclusa. Covered seed smut. **95:**ND(1077). **97:**NE(1077), NV(1997), WI(1077). **109:**WI(1077,3093). **110:**WI(1345). **124:**NC states(82), NE states(82), NV(82), OR(82). **174:**OR(3202), WY(1077,1195,1997,3093). **176:**AK(526,1077,1997,3093). **189:**CO(1077). **205:**ND(378). **208:**ME(1997). **219:**WI(1346). **225:**ME(1077), OR(3202).
A. verrucosa. **58:**WY(1077,3093). **115:**AK(526,1077,3093). **122:**CO(1077,3093). **197:**CA(661). **225:**CA(661).
Anthracoidea sp. **72:**CA(661). **112:**CA(661). **176:**CA(661). **225:**CA(661).
Cintractia carpophila var. *kenaica*. **162:**AK(526,3093).
Entorrhiza sp. Root gall. **225:**MN(82).
Farysia thuemenii. **108:**MS(1077,2720). **174:**CA(82,1077), LA(1077), MS(2720), MT(1077,3202), PA(82,1077), WA(82,1077,3202), WI(82,1077). **206:**CA(1077). **210:**FL(1077). **219:**CA(81,1077). **225:**FL(35,82), MS(82).
Orphanomyces arcticus. Smut. **59:**CO(1077). **81:**UT(1077). **225:**CO(922), WY(1077).
Schizonella melanogramma. On leaves. **10:**ID(1077,3202), WY(1077,1195). **19:**CO(1077). **27:**WI(1355). **42:**MA(1077), NH(1077), NJ(1077), NY(1077), PA(1077), TN(1077), WI(1359). **49:**E states(82), CA(82). **54:**WI(1353). **68:**E states(82), CA(82). **75:**E states(82), CA(82). **82:**CO(1077). **85:**UT(1077). **100:**IL(1077), IN(1077). **122:**E states(82), CA(82). **128:**E states(82), CA(82), NE(1077), UT(1077). **132:**WY(1077,1195). **144:**PA(1077). **145:**C & W states(82), NE states(82), CO(1077), CT(1077), IA(1077), IL(1077), IN(1077), MA(1077), MI(1077), MN(1077), ND(378,1077), NY(1077), OH(1077), PA(1077), WI(1077). **150:**IN(1077). **151:**E states(82), CA(82). **166:**ID(1077,3202). **172:**IN(1077), WI(1077). **174:**CO(1077). **178:**WY(1077,1195). **195:**E states(82), CA(82). **212:**WI(1357). **219:**MN(1077). **225:**E states(82), CA(82,1077), OR(1077,3202), UT(1077), WY(1077).
Tolyposporium aterrimum. Head smut. **2:**IA(1077). **68:**AK(1077). **127:**UT(1077). **145:**CO(1077), KS(1077). **154:**WA(1077,3202). **225:**CO(82), IA(82), KS(82), ND(1077), WA(82).
Urocystis fischeri. **206:**CA(1077). **225:**CO(1077), MT(3202), WY(1077).

Basidiomycotina - Other

Cellypha goldbachii. On culms and leaves. **225:**MI(2945), NY(82), PA(82).
Coprinus martinii. On dead stems. **10:**AK(2420).
Marasmius caricis. On culms. **225:**MI(82,1223), ND(82,378).
Mycena culmigena. On dead leaves. **225:**OR(2931), WA(2931).
Sparassis crispa. Root decay. **138:**NW states(82), WA(82).
Tomentella rubiginosa. **225:**WV(2039).
Typhula sp. **113:**WA(3202).

Deuteromycotina - Hyphomycetes

Alternaria flagelloideum. **225:**WI(1352).
Arthrinium caricicola. On dead leaves. **225:**PA(82).
A. cuspidatum. On dead leaves. **225:**CA(642).
A. sporophleum. On dead leaves. **225:**CA(81).
Bipolaris sp. **85:**MT(3202).
Cercospora caricis. Leaf spot. **5:**WI(1358). **6:**WI(1360). **12:**WA(3202,3443). **23:**WA(3202). **24:**WI(1344). **31:**WI(1358). **38:**WI(1354). **48:**WI(1333). **50:**WI(1341). **53:**WI(1359). **78:**WI(1358). **89:**WI(1336). **122:**WA(3202). **123:**WI(1340). **134:**WI(1355). **139:**WA(3202). **145:**WI(1349). **160:**WI(1347). **175:**WI(1350). **192:**WI(1340). **195:**WI(1350). **199:**WI(1349). **222:**WI(1349). **225:**IA(82), NY(82), WI(82).
Cercosporella palustris. **126:**WA(3202,3443).
Cladosporium graminum f. *caricicola*. **66:**ND(378).
C. herbarum. Leaf mold. **186:**CA(661). **203:**DE(3543). **225:**Range of host(82), DE(3543).
Clasterosporium caricinum. Leaf mold. **225:**IL(82), IN(82), NY(82), OH(82,638), PA(82).
Fusarium acuminatum. **41:**AK(3431).
F. avenaceum. **186:**WA(3202).
F. pallidoroseum. **186:**WA(3202).
Graphium sp. **49:**NC(1307).
Helicosporium phragmitis. Leaf spot. **225:**ME(82).
Heterosporium avenae. On frost-injured leaves. **75:**ID(3202,3443).
Macrosporium sp. **225:**WA(3202).
Microdochium nivale. On living leaves. **117:**AK(3427,3431). **186:**WA(3202).
Periconia nigriceps. On dead leaves. **186:**AK(3443).
Ramularia canadensis. Leaf spot. **97:**WI(1359). **134:**WI(1360). **175:**WI(1349). **205:**WI(1349). **225:**WI(82).
Volutella caricicola. On leaves. **39:**MS(2720). **225:**MS(82,2720).

Deuteromycotina - Coelomycetes

Amerosporium atrum. On leaves. **171:**MS(82). **225:**MS(82,2720).
Aposphaeria caricicola. On dry leaves. **173:**WA(3202,3443).
Ascochyta caricicola. On leaves. **225:**AK(526), CA(666).
A. socialis. On leaves. **225:**ND(82).
A. teretiuscula. **75:**ID(3202). **225:**ID(3202).
[*A. zeicola*]. **225:**WA(3202,3816).
Ascochyta sp. On leaves. **217:**WA(3436).
Colletotrichum graminicola. Anthracnose. **225:**WI(82).
C. madisonensis. On living leaves. **43:**WI(1360). **62:**WI(1358). **94:**WI(1347). **110:**WI(1359). **205:**WI(1350).
Colletotrichum sp. Anthracnose. **38:**WI(1347). **78:**OK(2872).
Coniothyrium sp. On leaves. **10:**WI(1347).
Cylindrosporella caricina. **94:**WI(1361).
Hendersonia distans. On leaves. **225:**CA(82).
Leptostroma caricinum. **10:**WA(3202).
Leptothyrella caricis. On leaves. **195:**KS(82).
Phaeoseptoria airae. **148:**MT(3202).
P. caricicola. Leaf spot. **16:**WA(3202,3443). **122:**MT(3202). **128:**WA(3202,3436). **148:**MT(3202). **157:**AK(3427,3431). **179:**OR(3202). **225:**MI(82), OR(82,3202).
P. caricis. On leaves. **225:**IL(82).
Phloeospora caricis. Leaf spot. **225:**CT(82), DE(82), NY(82), WI(82).
P. lunelliana. Leaf spot. **10:**MT(3202), OR(3202). **22:**OR(3202). **23:**ID(3202), MT(3202). **57:**ID(3202), WA(3202). **72:**CA(638,639,642,661). **75:**ID(3202). **85:**ID(3202), MT(3202), WA(3202). **92:**WA(3202). **97:**ID(3202), MT(3202). **115:**WY(1195). **122:**ID(3202), MT(3202). **148:**ID(3202), MT(3202), OR(3202,3443), WA(3202). **157:**WA(3202). **173:**ID(3202), WA(3202,3443). **182:**AZ(3443). **197:**ID(3202). **225:**CA(82), ND(82).
P. skagwayensis. **186:**AK(3443).
Phoma caricis. On dead leaves. **30:**CA(642).
P. herbicola. On dead leaves. **30:**CA(642).
Phyllosticta caricicola. Leaf spot. **56:**ID(3202). **96:**ID(3202). **225:**AK(64,82,526).

P. caricis. Leaf spot. **54:**CA(642). **90:**WI(64). **207:**WI(64). **225:**CA(82), NY(82), WI(82).
P. olympica. On dried leaves. **54:**WA(3202,3436).
Phyllosticta sp. **27:**WI(1355,1359). **175:**WI(1355).
Pseudostegia nubilosa. On leaves. **225:**Unk.(82), MT(3202), NJ(966).
Selenophoma drabae. On leaves. **11:**CO(3443). **28:**CA(3443), OR(3202,3436). **41:**AK(3427). **54:**WA(3202,3443). **57:**WA(3202). **75:**ID(3202,3443). **85:**OR(3202,3443). **91:**AK(3427). **97:**WA(3202). **173:**ID(3202), WA(3202).
Septoria baudysiana. Leaf spot. **10:**ID(3202,3436), WY(1195). **103:**ID(3202). **113:**WA(3202). **122:**OR(3202). **173:**WA(3202). **219:**WY(1195). **225:**WY(82).
S. caricina. **144:**WI(1360). **187:**WI(1345).
S. caricinella. Leaf spot. **3:**WI(3443). **29:**WI(1358). **42:**WI(1359). **57:**WA(3202). **61:**WI(1359). **70:**WI(1358). **75:**CO(82), ID(3202). **94:**WI(1360). **123:**WI(1333). **204:**WI(1333). **208:**WI(1359). **212:**WI(1359). **225:**WA(82,3202), WI(82,1343), WY(82).
S. caricis. Leaf spot. **62:**WI(1354). **144:**WI(1358). **195:**WI(1336). **225:**ND(82), WA(82).
S. carnea. **205:**SD(2243).
S. lineolata. Leaf spot. **116:**OR(3202). **209:**CA(81). **225:**CA(82), MD(82), OR(82).
S. nematospora. **17:**AK(3443), ID(3202). **20:**AK(3427,3431), MT(3202). **31:**WI(1343). **46:**ID(3202). **54:**ID(3202), WA(3202), WI(1358). **56:**AK(3427), MA(3427). **61:**WI(1359). **75:**ID(3202). **85:**WA(3202). **89:**WI(1354). **90:**WI(1360). **96:**AK(3427), ID(3202). **116:**AK(526,3427). **122:**ID(3202). **126:**ID(3202). **145:**AK(3427). **157:**AK(3427). **209:**CA(3443).
S. nematospora f. *aggregata*. On dry leaves. **157:**AK(3427,3431).
S. polita. Leaf spot. **102:**ID(3202). **225:**WI(82).
S. punctoidea. **2:**WI(1358). **90:**WI(1359). **172:**WI(1341).
S. riparia. Leaf spot. **225:**CO(82), MS(82,2720), WA(3202,3816), WI(82).
S. tenuis. Leaf spot. **225:**NY(82), WI(82).
Septoria sp. On leaves. **97:**MT(3436).
Stagonospora albescens. **8:**ID(3202), WA(3202,3443). **47:**WI(1355). **79:**WI(1354). **88:**WI(1360). **89:**WI(1355). **94:**WI(1341). **98:**WI(1352). **156:**WI(1355). **205:**WI(1361).
S. anglica. **8:**ID(3202), WA(3202).
S. biseptata. **97:**WI(1355).
S. caricinella. On old leaf tips. **29:**WI(1355). **54:**WA(3202,3443). **120:**OR(3202). **134:**WI(1355). **145:**WI(1334). **173:**WA(3202). **192:**WA(3202). **225:**ID(3202), WI(82).
S. caricis. **52:**MA(3443). **97:**OR(3202).
S. heleocharidis var. *caricina*. **115:**AK(526). **225:**AK(82).
S. luzulae. **145:**WI(1334).
S. paludosa. **54:**ID(3202). **75:**ID(3202).
S. simplicior. On leaf tips. **57:**WA(3202). **113:**WA(3202). **126:**WA(3202,3443). **157:**MT(3202). **173:**WA(3202,3436). **180:**WA(3202). **186:**WA(3202).
S. strictae. **32:**AK(526). **195:**KS(82).
[*S. subseriata*]. **225:**AK(3427,3443), WA(3443).
S. tetramera. **225:**WI(82).
Stagonospora sp. On overwintered leaves. **5:**WI(1341). **97:**WI(1347). **104:**WI(1355). **145:**WI(1355). **205:**WI(1356).
Tiarosporella paludosa. On leaves. **79:**WI(1355). **97:**WI(1360). **123:**WI(1346). **174:**WI(1331,1344). **179:**WI(1355). **225:**WI(82).

Deuteromycotina - Other

Sclerotium sulcatum. **128:**Unk.(82).

Cladium P. Br.

1. *C. angustifolium* Benth. & J. D. Hook.
2. *C. californicum* (S. Wats.) O'Neill
3. *C. germanicum* Schrad. (=*C. effusum* Torr.)
4. *C. jamaicense* Crantz [sawgrass]
5. *C. meyenii* Benth. & J. D. Hook. ex Drake
6. *C. sandwicense* Müll. Arg.
7. *Cladium* sp.

Ascomycotina

Asterina ildefonsiae. **6:**HI(2903).
Aulographella baumeae. **7:**HI(2903).
Chaetothyrium sp. **4:**FL(35).
Lophodermium arundinaceum. **1:**HI(2903).
Meliola amphitricha. Sooty mold. **4:**MS(2720).
M. argentina. Black mildew. **5:**HI(1272,1452). **7:**HI(2903).
M. cyperi. **1:**HI(2903). **7:**HI(2903).
M. intricata. Black mildew. **3:**Unk.(1452), FL(1452).
M. morbosa. **6:**HI(2903).
Trichothyrium reptans. **1:**HI(2903). **7:**HI(2903).

Basidiomycotina - Uredinales

Puccinia cladii. Autoecious. **2:**AZ(1203,3981). **4:**MS(2720), NC(1303).

Deuteromycotina - Hyphomycetes

Pyricularia grisea. Leaf spot. **4:**FL(35).
Trichothallus hawaiiensis. **1:**HI(2903).

Deuteromycotina - Coelomycetes

Pestalotia sp. **1:**HI(2903). **7:**HI(2903).

Cymophyllus Mackenz. ex Britton & A. Brown

1. *C. fraseri* (Andr.) Mackenz. (=*Carex fraseri* Andr.) [Fraser's sedge]

Basidiomycotina - Other

Epidochium sp. On culms. **1:**WV(82).
Tomentella rubiginosa. **1:**WV(3746).

Deuteromycotina - Hyphomycetes

Conoplea fusca. On leaves. **1:**WV(82).

Cyperus L.

(= *Kyllinga* Rottb.)

1. *C. albomarginatus* Mart. & Schrad. (=*C. flavicomus* A. Gray)
2. *C. alternifolius* L. [umbrella sedge, umbrella-palm, umbrella-plant]
3. *C. aristatus* Rottb. (=*C. inflexus* Muhl.) [awned cyperus]
4. *C. articulatus* L.
5. *C. cylindricus* Chapm. in J. M. Coulter [pine-barren cyperus]
6. *C. erythrorhizos* Muhl. [red-rooted cyperus]
7. *C. esculentus* L. [chufa, tigernut, yellow nut sedge, yellow nutgrass]
8. *C. fendlerianus* Boeck.
9. *C. filiculmis* Vahl [slender cyperus]
10. *C. filiculmis* Vahl var. *macilentus* Fernald
11. *C. grayi* Torr.
12. *C. houghtonii* Torr.
13. *C. ligularis* L.
14. *C. ovularis* (Michx.) Torr. [globose cyperus]
15. *C. polystachyos* Rottb.
16. *C. pseudovegetus* Steud. [marsh cyperus]
17. *C. retrorsus* Chapm.
18. *C. rivularis* Kunth [shining cyperus]
19. *C. rotundus* L. [nutgrass, nutsedge, purple nutsedge]
20. *C. rusbyi* Britton
21. *C. schweinitzii* Torr.
22. *C. strigosus* L. [false nut sedge, straw-colored cyperus]
23. *C. tegetiformis* Sendtn.
24. *C. tenuifolius* (Steud.) Dandy (=*Kyllinga pumila* Michx.)
25. *C. virens* Michx.
26. *Cyperus* sp.

Oomycetes

Peronospora cyperi. Downy mildew. **23:**SC(82), TX(82).
Pythium sp. Root rot. **19:**FL(35).

Ascomycotina

Balansia cyperacearum. Black band, black choke. **14:**AL(82). **19:**FL(35). **22:**FL(82), LA(82). **26:**FL(35,82), LA(82).
B. cyperi. On inflorescences. **25:**LA(82), TX(82). **26:**FL(82), LA(82).
Meliola argentina. Black mildew. **26:**HI(82,2903).
M. circinans. Black mildew. **26:**FL(82).
M. intricata. **26:**FL(1452).
Phyllachora canaliculata. **21:**ND(377).
P. cyperi. Tar spot. **2:**FL(35). **7:**SC(82). **19:**FL(82). **22:**WV(82). **26:**FL(35), ND(82), NE(82), WI(82).
P. viequesensis. Forming stromata on living leaves. **13:**TX(82,3484).
Scirrhia ostiolata. On culms. **4:**TX(82).
Scirrhia sp. Stem blight. **2:**FL(35).
Sclerotinia homoeocarpa. Blight. **19:**MS(138).

Basidiomycotina - Uredinales

Puccinia angustata. Heteroecious: telial host. **26:**FL(35).
P. canaliculata. Heteroecious: telial host. **3:**WI(1333). **7:**S states(82), AZ(1203,3981), CA(81,82), IL(312), NC(1303), OK(2872), TX(740), WI(82). **8:**AZ(3981). **9:**S states(82), CA(82), WI(82). **19:**S states(82), CA(82), FL(35), HI(2903), MS(2720), OK(2872), WI(82). **20:**AZ(1195,1203,3981). **22:**S states(82), CA(82), MS(2720), NC(1303), OK(2872), WI(82,1346). **26:**AZ(1203),

FL(35), OK(2872).
P. cyperi. Heteroecious: telial host. **9:**E states(82), ND(82), OK(2872), TX(82). **14:**E states(82), ND(82), TX(82). **16:**OK(2872). **19:**E states(82), FL(35), ND(82), TX(82). **21:**OK(2872). **22:**E states(82), ND(82), TX(82). **26:**FL(35), MS(2720), NC(3944), OK(2872).
P. cyperi-tagetiformis. Autoecious. **1:**FL(111). **26:**FL(35,82), TX(82).
Uredo nociviola. **26:**FL(35,82).
Basidiomycotina - Ustilaginales
Cintractia cyperi. Inflorescence smut. **9:**CT(82,1077,2147), MA(82,1077), MI(82,1077), NJ(2147). **11:**NJ(1077). **17:**NJ(1077). **26:**NJ(82).
C. leucoderma. **15:**FL(1077).
C. limitata. **11:**NJ(2147), NY(2147). **13:**FL(1077). **26:**FL(35,82).
C. peribebuyensis. **9:**MD(82). **11:**DE(1077), MD(1077), NJ(1077), NY(1077). **26:**AL(82), DE(82), MD(82), NJ(82), NY(82).
Testicularia cyperi. **26:**NC(1077,1303), NY(1077), TX(1077).
Basidiomycotina - Other
Thanatephorus cucumeris. **19:**LA(3524).
Deuteromycotina - Hyphomycetes
Cercospora caricis. Leaf spot. **7:**IL(312). **9:**WI(82).
C. cypericola. On living leaves. **12:**WI(1348). **22:**WI(1346).
Cercospora sp. Leaf spot. **2:**FL(35). **26:**FL(35).
Cladosporium sitchense. Culm spot. **26:**WA(82).
Dactylaria higginsii. On living leaves. **19:**GA(2200). **26:**GA(2200).
Fusarium sp. **19:**HI(2903).
Helminthosporium sp. Leaf spot. **24:**MS(2720). **26:**FL(35), NC(1303).
Deuteromycotina - Coelomycetes
Ascochyta sp. Leaf spot. **7:**GA(82). **19:**FL(35). **26:**FL(35).
Colletotrichum truncatum. **19:**Unk.(3059).
Coniothyrium sp. On living leaves. **6:**WI(1343).
Septoria cyperi. Leaf spot. **26:**AL(82), FL(35).
Stagonospora cyperi. On culms. **5:**MS(82). **17:**MS(2720).
S. cypericola. **10:**WI(1354). **18:**WI(1357). **21:**WI(1355).
Stagonospora sp. **10:**WI(1347).
Deuteromycotina - Other
Rhizoctonia solani. Thread blight, culm rot, root rot. **19:**FL(35,82). **26:**FL(35).

Dulichium Pers.

1. *D. arundinaceum* (L.) Britton

Chytridiomycetes
Physoderma dulichii. On leaves. **1:**MI(1760).
Basidiomycotina - Uredinales
Puccinia dulichii. Heteroecious: telial host. **1:**DE(3101), FL(35), MA(3101), NH(3101), WI(3101).

Eleocharis R. Br.

1. *E. acicularis* (L.) Roem. & Schult. [slender spike rush]
2. *E. compressa* Sullivan
3. *E. flavescens* (Poir.) Urb. (=*E. olivacea* Torr.)
4. *E. intermedia* Schult.
5. *E. montevidensis* Kunth
6. *E. ovata* (Roth) Roem. & Schult. (=*E. obtusa* (Willd.) Schult.) [ovoid spike rush]
7. *E. palustris* (L.) Roem. & Schult. (=*E. calva* Torr., *E. smallii* Britton)
8. *E. parishii* Britton
9. *E. rostellata* (Torr.) Torr.
10. *E. tenuis* (Willd.) Schutt (=*E. elliptica* Kunth)
11. *Eleocharis* sp.

Chytridiomycetes
Physoderma heleocharidis. On leaf sheaths. **1:**OR(3202). **11:**CA(3401).
Ascomycotina
Claviceps nigricans. Ergot, brown leaf spot. **6:**MT(3202,3443). **7:**CA(81), HI(2903), IN(82), MT(3202), ND(82,377), OR(3202), SC(82), SD(82), WA(82,3202), WI(82,1340), WY(82). **9:**IN(82), MI(82), ND(82), SC(82), SD(82), WI(82), WY(82).
Mycosphaerella sp. On tips and stems. **1:**WI(1361). **7:**WI(1357).
Phaeosphaeria sp. On stems. **7:**WI(1352).
Pyrenophora scirpi. **11:**SD(3818).
Basidiomycotina - Uredinales
Puccinia eleocharidis. Heteroecious: telial host. **1:**WI(1331,3101). **4:**IA(3101). **5:**TX(3101). **6:**GA(3101), IN(3101), MO(3101), MS(2720,3101), NC(1303,3101), NY(3101), OK(2872,3101). **7:**KS(3101), NE(3101), NY(3101), WI(1333,3101). **10:**ME(3101), NE(3101), OK(2872). **11:**E & C states(82), LA(658), NE(3101), WI(3101).
P. liberta. Autoecious. **7:**CA(81). **8:**CA(81). **11:**CA(82).
Uromyces eleocharidis. Autoecious. **3:**ND(378). **7:**ND(378). **11:**C states(82), ND(82), WA(3202).
Basidiomycotina - Ustilaginales
Entyloma parvum. **1:**MA(82,1077), WI(82,1077).
Deuteromycotina - Hyphomycetes
Cercospora eleocharidis. On culms. **1:**WI(1357). **2:**WI(1346). **6:**ID(3202,3443). **10:**WI(1346). **11:**LA(658).

Eriophorum L.

1. *E. angustifolium* Honck.
2. *E. chamissonis* C. A. Mey.
3. *E. russeolum* Fr. [russet cottongrass]
4. *E. scheuchzeri* Hoppe
5. *E. vaginatum* L. [sheathed cottongrass]
6. *E. virginicum* L. [Virginia cottongrass]
7. *E. viridicarinatum* (Engelm.) Fernald
8. *Eriophorum* sp.

Ascomycotina
Phaeosphaeria caricinella. On dead leaves and sheaths. **1:**AK(1938).
Basidiomycotina - Uredinales
Puccinia angustata. Heteroecious: telial host. **1:**OR(3202). **6:**WI(1333). **7:**OR(3202). **8:**NC states(82), NE states(82), OR(82).
P. eriophori. **6:**MA(3101), WI(3101). **7:**NY(3101).
Basidiomycotina - Other
Coprinus martinii. **1:**AK(2059). **3:**AK(82). **4:**AK(2059). **5:**AK(82).
Deuteromycotina - Coelomycetes
Septoria chamissonis. Leaf spot. **2:**AK(526). **8:**AK(82).
Septoria sp. **6:**WI(1361).
Stagonospora eriophorella. On leaves. **2:**AK(526). **5:**AK(526). **8:**AK(82).

Fimbristylis Vahl

1. *F. autumnalis* (L.) Roem. & Schutt
2. *F. capillaris* (L.) A. Gray (=*Bulbostylis capillaris* (L.) C. B. Clarke)
3. *F. ciliatifolia* (Elliott) Britton ex S. Wats. (=*Bulbostylis ciliatifolia* (Elliott) Fernald)
4. *F. spadicea* (L.) Vahl (=*F. puberula* (Michx.) Vahl)
5. *F. vahlii* (Lam.) Link
6. *Fimbristylis* sp.

Oomycetes
Pythium sp. Root rot. **3:**FL(35).
Basidiomycotina - Uredinales
Puccinia fimbristylidis. Autoecious. **4:**FL(35), TX(82,111).
Basidiomycotina - Ustilaginales
Cintractia axicola. Inflorescence. **1:**AL(82,1077,2147), AR(1077), DE(82,1077), FL(1077), LA(1077), MS(82,1077,2720), NY(1077), OK(82,1077,2147,2872), VA(82,1077,2147). **5:**OK(1077,2872).
C. taubertiana. **2:**AZ(1077).
Deuteromycotina - Coelomycetes
Septoria fimbristylis. Leaf spot. **6:**MO(82).

Fuirena Rottb.

1. *F. simplex* Vahl (=*F. hispida* Elliott)
2. *F. squarrosa* Michx.

Basidiomycotina - Uredinales
Puccinia fuirenae. **1:**GA(1439). **2:**FL(35), SC(634).
Basidiomycotina - Ustilaginales
Cintractia fuirenae. Smut. **1:**NE(1488).

Gahnia J. R. Forster & G. Forster

1. *G. gaudichaudii* Steud.
2. *G. leptostachya* Boeck.

Ascomycotina
Meliola argentina. Black mildew. **1:**HI(1272,1452,2903).
M. argentina var. *hawaiiensis*. Black mildew. **2:**HI(1272,1452,2903).
M. cyperi. **1:**HI(2903). **2:**HI(2903).

Kobresia Willd.

1. *K. bellardii* (All.) Degl. (=*K. myosuroides* (Vill.) Fiori)
2. *K. simpliciuscula* (Wahlenb.) Mackenz.

Basidiomycotina - Ustilaginales

[*Anthracoidea caricis*]. **1:**UT(1077).
A. elynae. **1:**CO(649,1995), WY(1996).
A. lindebergiae. **2:**AK(1995).

Machaerina Vahl

1. *M. angustifolia* (Gaud.) T. Koyama (=*Vincentia angustifolia* Gaud.)
2. *M. mariscoides* (Gaud.) J. Kern

Ascomycotina

Meliola argentina. Black mildew. **1:**HI(1268,1272,1452,2784,2903). **2:**HI(1268,2903).

Deuteromycotina - Hyphomycetes

Helicoma machaerinae. On leaves. **1:**HI(1269,1271,2903).
Hobsonia mirabilis. On leaves. **1:**HI(1269).

Rhynchospora Vahl

1. *R. alba* (L.) Vahl [white beakrush]
2. *R. capitellata* (Michx.) Vahl
3. *R. cephalantha* A. Gray
4. *R. ciliaris* (Michx.) C. Mohr
5. *R. corniculata* (Lam.) A. Gray (=*R. laxa* Vahl) [horned rush]
6. *R. corymbosa* (L.) Britton
7. *R. cymosa* Nutt.
8. *R. cyperoides* (Sw.) Mart.
9. *R. eximia* Boeck.
10. *R. fascicularis* (Michx.) Vahl
11. *R. fusca* (L.) W. T. Aiton [brown beakrush]
12. *R. globularis* (Chapm.) Small var. *recognita* Gale
13. *R. glomerata* (L.) Vahl
14. *R. glomerata* (L.) Vahl var. *minor* Britton
15. *R. inexpansa* (Michx.) Vahl [nodding beakrush]
16. *R. lavarum* Gaud.
17. *R. leptocarpa* Riddell
18. *R. macrostachya* Torr.
19. *R. mariscus* Nees (=*R. stipitata* Chapm.)
20. *R. megalocarpa* A. Gray (=*R. dodecandra* Baldw. ex A. Gray)
21. *R. microcephala* (Britton) Britton
22. *R. miliacea* (Lam.) A. Gray
23. *R. nitens* (Vahl) A. Gray (=*Psilocarya nitens* (Vahl) A. Wood)
24. *R. plumosa* Elliott (=*R. semiplumosa* A. Gray)
25. *R. scirpoides* (Torr.) Griseb. (=*Psilocarya scirpoides* Torr.)
26. *R. thyrsoidea* Nees & Meyen
27. *R. tracyi* Britton
28. *Rhynchospora* sp.

Ascomycotina

Meliola argentina. Black mildew. **26:**HI(1272,1452,2903).
M. circinans. Black mildew. **28:**FL(82).
M. circinans var. *rhynchosporae*. Black mildew. **20:**FL(1452).
M. cyperi. Black mildew. **28:**FL(82).
Phaeosphaeria clavispora. On culms. **13:**GA(1438). **28:**GA(82).

Basidiomycotina - Uredinales

Puccinia angustatoides. Telial host; aecial host unknown. **5:**AL(82,3101), FL(82), LA(82), MO(82,3101), TN(82,3101), TX(82). **28:**FL(35,3101).
P. cladii. Autoecious. **13:**VA(82).
Uromyces lineolatus subsp. *nearcticus*. Heteroecious: telial host. **28:**NC(1303).
U. rhynchosporae. Telial host; aecial host unknown. **10:**MS(2720). **13:**GA(1439), MS(2720). **16:**HI(2903). **28:**NC states(82), NE states(82), FL(35,82), HI(82), LA(82), MI(82).

Basidiomycotina - Ustilaginales

Cintractia affinis. **5:**NY(82). **18:**NY(1077).
C. farlowii. **13:**DE(1077), FL(82,1077), MA(82), NJ(82,1077). **17:**FL(1077). **28:**FL(35,82), MA(82), NJ(82).
C. leucoderma. **5:**FL(1077), NY(1077). **6:**NJ(82). **18:**NY(1077,2148). **27:**FL(1077). **28:**FL(35), SC(2148).
C. pachyderma. **28:**FL(35).
C. psilocaryae. **23:**FL(1077), MS(1077,2720). **25:**IN(1077), MA(1077), MI(1077), RI(1077), WI(1347).
C. striata. **8:**FL(2148), NJ(82). **27:**FL(82,1077). **28:**FL(35).
C. taubertiana. **1:**MA(1077), NJ(1077), NY(1077), PA(1077). **3:**MS(1077,2720). **7:**MD(1077). **10:**FL(1077). **11:**NY(1077). **15:**SC(1077). **21:**AL(1077). **28:**E states(82), E states(82), FL(35), TN(1539), TX(82), VA(1077).
C. utriculicola. **28:**CO(1077).
Sorosporium rhynchosporae. **24:**MS(82,2720).
Testicularia cyperi. **5:**FL(82,1077), LA(82,1077), NY(82,1077), TX(82,1077). **18:**FL(1077), LA(1077), NY(1077), TX(1077). **28:**FL(35,82).
Thecaphora rhynchosporae. **24:**MS(1077,2720).
Ustilago montagnei. Sori in ovaries. **1:**CT(1077), ME(1077), NY(1077), VT(1077). **2:**FL(1077), NY(1077). **4:**FL(1077). **9:**IN(2146). **13:**CT(1077), DE(1077), IN(1077), MA(1077), MS(1077,2720), NJ(1077), NY(1077). **14:**MA(1077). **19:**FL(1077). **22:**FL(1077). **23:**MS(2146). **24:**MS(1077,2720). **28:**E states(82), E states(82), FL(35), ID(82), MS(82), NC(1077,1303), SC(1077).
U. montagnei var. *major*. **4:**FL(2146). **12:**MD(2146). **13:**DE(2146).

Deuteromycotina - Hyphomycetes

Cercospora crinospora. Leaf spot. **13:**AL(82).

Deuteromycotina - Coelomycetes

Amerosporium macrochaeta. On culms and leaves. **5:**LA(82).

Scirpus L.

1. *S. acutus* Muhl. (=*S. occidentalis* S. Wats.) [hard-stem bulrush, tule]
2. *S. americanus* Pers. (=*S. olneyi* A. Gray) [American bulrush, chairmakers rush, common-threesquare]
3. *S. atrocinctus* Fernald
4. *S. atrovirens* Willd. [dark green bulrush]
5. *S. caespitosus* L. (=*S. caespitosus* L. var. *callosus* Bigel., *Trichophorum caespitosum* (L.) Hartm. subsp. *austriacum* (Palla) Hegi) [tufted club rush]
6. *S. californicus* (C. A. Mey.) Steud. [California bulrush, southern bulrush]
7. *S. congdonii* Britton [woolgrass]
8. *S. cyperinus* (L.) Kunth [woolgrass bulrush]
9. *S. cyperinus* (L.) Kunth var. *pelius* Fernald
10. *S. fluviatilis* (Torr.) A. Gray [river bulrush]
11. *S. georgianus* R. M. Harper
12. *S. hudsonianus* (Michx.) Fernald (=*Trichophorum alpinum* (L.) Pers.)
13. *S. lacustris* L. [great bulrush, mat rush]
14. *S. maritimus* L. (=*S. campestris* Britton, *S. paludosus* A. Nels., *S. robustus* Pursh) [bayonet-grass, prairie bayonet-grass]
15. *S. maritimus* L. var. *paludosus* Kükenth.
16. *S. microcarpus* K. Presl
17. *S. pacificus* Britton ex Parish
18. *S. pallidus* (Britton) Fernald
19. *S. pedicellatus* Fernald
20. *S. rubricosus* Fernald (=*S. eriophorum* Michx.)
21. *S. rubrotinctus* Fernald
22. *S. validus* Vahl [American great bulrush]
23. *Scirpus* sp.

Chytridiomycetes

Synchytrium scirpi. On leaves. **4:**WI(82).

Ascomycotina

Belonopsis junciseda. On culms. **4:**GA(82). **23:**MI(82), NY(82).
Hypoderma scirpinum. On dead culms and stems. **1:**CA(81,82), ND(82,377), NE(82), WA(82). **22:**WA(3202). **23:**MS(82,2720), MT(3202), NY(82).
Leptosphaeria norvegica. **16:**WA(3202,3814).
Leptosphaeria sp. Dead leaf tips. **4:**WI(1359). **22:**WA(3202).
Lophiostoma arundinis. **11:**GA(1438).
Meliola argentina. Black mildew. **23:**HI(82,2903).
Mollisia scirpina. **11:**GA(1438).
Mollisia sp. **20:**GA(1438).
[*Ophiobolus acuminatus*]. **11:**GA(1438).
Phaeosphaerella scirpicola. On leaf tips. **23:**CA(82).
Phaeosphaeria juncicola. On leaves. **5:**AK(526). **23:**AK(82).
Pleospora punctata. **22:**WY(3810).
Wettsteinina niesslii. On leaves. **23:**IA(1228).

Basidiomycotina - Uredinales

Puccinia angustata. Heteroecious: telial host. **3:**NY(3101). **4:**DE(3101), IA(3101), IN(3101), MI(3101), ND(378), NY(3101), OH(3101), SD(2243), WI(3101). **7:**CA(81). **8:**GA(1439), IL(3101), MA(3101), ME(3101), MI(3101), MN(3101), NH(3101), NY(3101), OH(3101), WI(1350,1359,3101). **9:**WI(1348).

14:OR(3202). **16:**CA(81), OR(638). **17:**WA(3202). **18:**NE(3101). **19:**MN(3101). **20:**VA(3101). **21:**WI(1342). **23:**NC(3944).
P. congdonii. **7:**CA(3101).
P. eriophori-alpini. **12:**NH(3101).
P. mcclatchieana. Autoecious. **7:**CA(3101). **16:**CA(81,661,3101), ID(3101,3202), MT(3202), OR(648,3101,3202), WA(3101,3202), WY(1195). **21:**CA(3101), MT(3101), NY(3101), OR(3101), WA(3101), WI(3101).
P. obtecta. Heteroecious: telial host. **1:**NE states(82), CA(82,3101), OR(82,3202), WI(1335). **2:**NE states(82), CA(81,82), ID(3202), IN(3101), MT(3202), OK(2872), OR(82,3202), SD(2243), WY(1195). **10:**CA(81). **13:**SD(2243). **22:**NE states(82), CA(82), CO(3101), ID(3202), MI(3101), OR(82,3202), WI(3101). **23:**NE states(82), CA(82), OR(82), WA(3202).
P. osoyoosensis. **2:**CA(3101), CO(3101), DE(3101). **6:**CA(3101).
Uromyces americanus. Heteroecious: telial host. **1:**CA(3101), DE(3101), TX(3101), WI(3101). **2:**AL(82), DE(82), TX(82,740). **6:**CA(81). **22:**AL(82), DE(82), IN(82), NE(82,3101), TX(82), WI(82,1335). **23:**CA(82).
U. americanus var. *californicus*. **6:**CA(3101).
U. lineolatus. Heteroecious: telial host. **14:**HI(2903), OR(3202).
U. lineolatus subsp. *nearcticus*. Heteroecious: telial host. **1:**CA(81). **6:**CA(81). **10:**Range of host(82), DE(3101), IA(3101), IL(3101), IN(3101), MO(3101), ND(378), NE(3101), NY(3101), OH(3101), SD(3101), WI(3101). **14:**Range of host(82), CA(81), MA(3101), ND(378). **15:**CA(3101), CO(3101), SD(3101), WY(3101). **23:**Range of host(82), CA(81), FL(35).

Basidiomycotina - Ustilaginales
[*Anthracoidea caricis*]. Seed smut. **5:**MI(82).
A. scirpi. **5:**AK(1996), MI(1077).
Cintractia axicola. **23:**AL(1077), WI(1077).

Basidiomycotina - Other
Kriegeria eriophori. On stems. **3:**WI(1357). **4:**IA(1835), WI(82,810,1343,1344). **9:**WI(1346). **10:**WI(1350). **19:**WI(1346).

Deuteromycotina - Hyphomycetes
Alternaria flagelloideum. **1:**WI(1352).
Arthrinium cuspidatum. On culms. **23:**CA(82).
Cercoseptoria scirpi. On living stems. **1:**WI(1347).
Cercosporella scirpina. On leaves. **9:**WI(1346). **23:**WI(82).
Cheiromyces stellatus. On culms. **20:**AL(256), PA(256).
Clasterosporium caricinum. Leaf spot. **1:**MN(82).
Epicoccum nigrum. Leaf mold. **13:**DE(3543). **23:**Range of host(82).
Helminthosporium sp. Leaf spot. **10:**MN(82).
Septocylindrium scirpinum. On leaves. **23:**NY(82).
Tetraploa scabra. On culms. **23:**CA(82).

Deuteromycotina - Coelomycetes
Cylindrosporium scirpivora. On living leaves. **16:**WA(3202,3436).
Hendersonia scirpicola. On culms. **23:**CA(82).
Hymenopsis trochiloides. **13:**DE(3543).
Leptostromella scirpina. On leaves. **4:**NE(82). **10:**NY(82). **23:**NY(82).
Pestalotia scirpina. On culms. **14:**FL(82), LA(82), MD(82). **23:**FL(35,82), LA(82), MD(82).
Phaeoseptoria scirpi. On culms. **22:**WY(3812). **23:**SD(2783).
Phyllosticta sp. **4:**WI(1349).
Septoria narvisiana. On culms. **1:**OR(4007). **14:**DE(82). **22:**OR(3202).
S. peckii. On leaves and sheaths. **10:**Unk.(82), NY(82).
Stagonospora scirpi. On leaves and sheaths. **4:**IL(82).
S. scirpini. **23:**LA(82).
Stagonospora sp. **4:**WI(1348).

Scleria Bergius
1. *S. triglomerata* Michx.
2. *Scleria* sp.

Ascomycotina
Phyllachora scleriicola. Tar spot. **1:**MS(2720). **2:**MS(82).

Basidiomycotina - Uredinales
Puccinia scleriicola. Telial host; aecial host unknown. **2:**FL(35,82), GA(82), IL(82).

Deuteromycotina - Hyphomycetes
Cercospora sp. Leaf spot. **2:**FL(35).

Deuteromycotina - Coelomycetes
Stagonospora sp. On leaves. **1:**WI(1353).

Cyrillaceae

Cliftonia Banks ex C. F. Gaertn.
1. *C. monophylla* (Lam.) Britton ex Sarg. [black titi, buckwheat-bush, buckwheat-tree, ironwood]

Ascomycotina
Coccomyces sp. Leaf spot. **1:**GA(82,1438).
Guignardia sp. On leaves. **1:**GA(82,1438).
Hypoxylon rubiginosum. **1:**GA(1438).
Lembosia cliftoniae. Black spot. **1:**MS(82,2720).

Deuteromycotina - Coelomycetes
Pestalotia versicolor. Leaf spot. **1:**MS(82,2720).

Cyrilla Garden ex L.
1. *C. racemiflora* L. [black titi, he huckleberry, ironwood, leatherwood, myrtle, red titi, southern leatherwood]

Ascomycotina
Botryosphaeria abrupta. On twigs. **1:**NC(82).
B. dothidea. **1:**GA(1438).
Diatrypella favacea. **1:**GA(1438).
Lophodermium cyrillicola. On leaves. **1:**AL(82), GA(1438), MS(82,2720).
Myriangium superficialis. On scale insects on limbs. **1:**GA(82).
Ostreichnion curtisii. **1:**GA(1438).
Stictis sp. **1:**GA(1438).

Basidiomycotina - Uredinales
Aecidium cyrillae. **1:**FL(35,82), LA(82), MS(82,2720).

Deuteromycotina - Coelomycetes
Phyllosticta cyrillae. Leaf spot. **1:**FL(35,82).

Datiscaceae

Datisca L.
1. *D. glomerata* (K. Presl) Baill.

Ascomycotina
Phomatospora datiscae. On dead stems. **1:**CA(1468).

Diapensiaceae

Diapensia L.
1. *D. lapponica* L.
2. *D. obovata* (Friedr. Schmidt) Nakai & Koidz. (=*D. lapponica* L. subsp. *obovata* Fr. Schmidt)

Ascomycotina
Clathrospora pentamera. On leaves. **1:**NY(82).
Mycosphaerella pachyasca. On leaves and stems. **1:**ME(82).
M. tassiana var. *arctica*. On dead stems. **2:**AK(1938).

Galax Sims
1. *G. aphylla* L.
2. *Galax* sp.

Ascomycotina
Clypeolella leemingii. Anthracnose, sooty mold. **1:**SE states(82), GA(1438), MS(82,2720), NC(1303,1307). **2:**MS(2720).
Dimerium langloisii. **1:**GA(1438).
Discohainesia oenotherae. Leaf spot. **1:**NC(82).
Guignardia galactina. On leaves. **1:**NC(82,1303,1307).
Mycosphaerella sp. **1:**GA(1438).

Deuteromycotina - Coelomycetes
Coleophoma empetri. On capsules and flowers. **1:**TN(82).
Phyllosticta galactis. Leaf spot. **1:**NC(82), VA(82), WV(82,3907).

Shortia Torr. & A. Gray
1. *S. galacifolia* Torr. & A. Gray [Oconee-bells]

Ascomycotina
Discohainesia oenotherae. Leaf spot. **1:**NC(82), SC(82).

Dioscoreaceae

Dioscorea L.
1. *D. batatas* Decne. [Chinese yam, cinnamon vine]
2. *D. villosa* L. [wild yam]
3. *Dioscorea* sp.

Zygomycotina
Rhizopus stolonifer. **1**:WA(3202).
Ascomycotina
Didymella dioscoreae. On stems. **3**:FL(35,82), SC(82).
Deuteromycotina - Hyphomycetes
Cercospora sp. Leaf spot. **3**:FL(35).
Corynespora sp. Leaf spot. **3**:FL(35).
Didymaria fulva. On leaves. **2**:IN(82).
Fusarium sp. Wilt. **3**:FL(35).
Phaeoramularia dioscoreae. **2**:DE(3543), IA(82), IL(82), MD(82), MI(82), PA(82), WI(82).
Phymatotrichopsis omnivora. Root rot. **3**:TX(82).
Ramularia dioscoreae. Leaf spot. **3**:WI(82).
Deuteromycotina - Coelomycetes
Colletotrichum gloeosporioides. Leaf spot. **2**:IL(82).
Colletotrichum sp. Anthracnose. **1**:FL(35). **3**:FL(35).
Phoma dioscoreae. On stems. **1**:NY(82).
Phyllosticta dioscoreae. Leaf spot. **2**:SC(82), VA(82), WV(82). **3**:SC(634).

Dipsacaceae

Dipsacus L.
1. *D. sylvestris* Huds. (=*D. fullonum* L.) [common teasel, wild teasel]
2. *Dipsacus* sp.

Oomycetes
Peronospora dipsaci. Downy mildew. **1**:MO(82).
Ascomycotina
Didymella exigua. **1**:ID(3202).
Kalmusia clivensis. **2**:MI(3249).
Mycosphaerella asterinoides. On stems. **1**:NY(82). **2**:NY(965,967).
Phyllactinia guttata. Powdery mildew. **1**:WA(82,3202).
Sclerotinia sclerotiorum. Stem rot. **1**:TX(82).
Deuteromycotina - Hyphomycetes
Cercospora elongata. Leaf spot. **1**:E & C states(82), IL(82), MD(82), WA(82,3202), WI(1338).
Phymatotrichopsis omnivora. Root rot. **1**:TX(82).
Ramularia silvestris. **1**:UT(1195).
Deuteromycotina - Coelomycetes
Phoma herbarum. On stems. **1**:TX(82).
Deuteromycotina - Other
Sclerotium rolfsii. Southern blight. **1**:TX(82).

Scabiosa L.
1. *S. atropurpurea* L. [mourning-bride, pincushions, sweet scabious]
2. *Scabiosa* sp.

Oomycetes
Pythium sp. Root rot. **1**:FL(35).
Ascomycotina
Erysiphe polygoni. Powdery mildew. **1**:Unk.(82). **2**:Unk.(82).
Sclerotinia sclerotiorum. Stem rot. **2**:NY(82).
Deuteromycotina - Hyphomycetes
Phymatotrichopsis omnivora. Root rot. **1**:TX(82). **2**:TX(82).
Deuteromycotina - Other
Sclerotium rolfsii. Southern blight. **1**:FL(35,82).

Dipterocarpaceae

Shorea Roxb. ex C. F. Gaertn.
1. *Shorea* sp.

Deuteromycotina - Hyphomycetes
Aureobasidium pullulans. Isolated from discolored roof shingles. **1**:FL(3136), OR(3136).

Droseraceae

Dionaea J. Ellis
1. *D. muscipula* J. Ellis [Venus's-flytrap]

Zygomycotina
Blakeslea trispora. Necrosis of trap. **1**:FL(35).
Deuteromycotina - Coelomycetes
Colletotrichum gloeosporioides. Anthracnose, foliar blight. **1**:FL(35,1931), NC(1303).
Deuteromycotina - Other
Rhizoctonia sp. Stem and foliage blight. **1**:FL(35).

Ebenaceae

Diospyros L.
1. *D. hillebrandii* (Seem.) Fosb.
2. *D. kaki* L. fil. (=*D. chinensis* Blume) [date-plum, Japanese persimmon, kaki persimmon, keg-fig]
3. *D. sandwicensis* (A. DC.) Fosb.
4. *D. texana* Scheele [black persimmon, Mexican persimmon, Texas persimmon]
5. *D. virginiana* L. [American persimmon, common persimmon, date-plum, possum-apple, possumwood]
6. *Diospyros* sp.

Oomycetes
Phytophthora sp. **6**:CA(1110).
Pythium sp. Root rot. **2**:FL(35).
Zygomycotina
Rhizopus stolonifer. Soft rot of fruit. **4**:TX(82).
Ascomycotina
Asterina aspidii. **3**:HI(2903).
Asterinella mabae. **1**:HI(2903). **3**:HI(2903).
Botryosphaeria dothidea. **5**:GA(1438).
B. obtusa. Black rot of fruit. **5**:AL(82), NY(82), TX(82).
B. rhodina. Black rot of fruit. **5**:AL(82), TX(82).
B. ribis. Limb canker, dieback. **4**:AL(82). **5**:AL(82), LA(82), MD(82).
Ceratostomella sp. Blue stain of timber. **5**:IL(82).
Daldinia concentrica. Wood rot. **5**:LA(82), VA(82). **6**:LA(2675).
D. vernicosa. **5**:VA(82).
Echidnodella mabae. **3**:HI(2903).
Elsinoe diospyri. On leaves. **5**:FL(2524).
Eutypella juglandicola. **5**:GA(1438).
Hypoxylon mediterraneum. **5**:GA(1438).
Meliola sp. **3**:HI(2903).
Mycosphaerella punctiformis. **5**:GA(1438).
Myriangium duriaei. On twigs, parasitic on scale insects. **5**:AL(82), FL(82).
Phyllachora orbiculata. Black spot. **5**:SC(82), TN(82).
Physalospora abdita. Black rot of fruit. **5**:AL(82), GA(82).
Physalospora sp. Black rot of fruit. **5**:Unk.(82).
Podosphaera clandestina. Powdery mildew. **5**:TX(82).
Rhytidhysteron rufulum. On dead limbs. **4**:FL(82). **5**:AL(82), GA(82).
Schizothyrium pomi. Fly speck on fruit. **2**:FL(35), OK(2872). **4**:FL(82). **6**:FL(35).
Thyronectria pyrrhochlora. On dead twigs. **5**:GA(82,1438).
Tubeufia pezizula. Gray stain of logs and timber. **5**:IL(82).
Valsa ceratosperma. On dead twigs. **5**:WV(82,3415).
Valsaria diospyri. On dead twigs. **5**:SC(82).
Basidiomycotina - Other
Armillaria mellea. Root rot. **4**:CA(82,1110).
A. tabescens. Root rot. **6**:FL(35).
Crepidotus alabamensis. **6**:AL(1544).
Dendrophora albobadia. **6**:AL(543).
D. versiformis. **5**:AL(543).
Exidiopsis molybdea. On wood. **5**:GA(2642).
Fomes fasciatus. Heart rot. **5**:AR(82), LA(82), MS(82,2720). **6**:LA(2675).
Fomitella supina. On wood. **5**:LA(82), MS(82,2720). **6**:LA(2675).
Fomitopsis spraguei. Butt rot. **5**:IN(82).
Ganoderma australe. Heart rot. **5**:LA(82), MS(82,2720).
G. lucidum. Butt rot. **5**:LA(82). **6**:LA(2675).
Hericium erinaceus. Usually on living trees. **5**:LA(82).
Heterobasidion annosum. Heart rot. **5**:IN(82).
Lentinus tigrinus. Sapwood rot. **5**:LA(82), MS(82,2720).

Microporellus obovatus. On wood. **5**:AR(82), LA(82). **6**:AR(2675), LA(2675).
Mycena meliigena. On bark. **6**:Unk.(634).
Pellicularia koleroga. Thread blight. **4**:FL(82). **5**:LA(3640).
Phanerochaete ravenelii. **6**:AR(2675), LA(2675).
Phellinus gilvus. On wood. **5**:AR(82), LA(82), MS(82,2720). **6**:AR(2675), LA(2675), MS(2675).
Pleurotus ostreatus. Sapwood rot. **5**:LA(82), MS(82,2720).
Rigidoporus ulmarius. Heart rot. **5**:LA(82), MS(82,2720).
Schizophyllum commune. Sapwood rot. **5**:KY(82), TN(82). **6**:AR(2675), LA(2675).
Stilbum rhois. **5**:DE(3543).
Trametes elegans. On wood. **5**:LA(82), MS(82,2720). **6**:AR(2675), LA(2675), MS(2675).
T. versicolor. On wood. **5**:VA(82).
Trichaptum biforme. On wood. **5**:MS(82,2720).
Tyromyces fissilis. On wood. **5**:LA(82).
Wolfiporia cocos. Tuckahoe, on roots. **5**:FL(82).

Deuteromycotina - Hyphomycetes

Acremonium diospyri. Wilt. **5**:SE states(82), AL(3667), AR(3594,3921), FL(3667), GA(3667), MS(2720,3667), NC(82,1303,3667), OK(3667), SC(3667), TN(3667), TX(82,3667).
Alternaria sp. Fruit spot and rot, secondary leaf spot. **4**:FL(82), OH(82). **6**:FL(35).
Botrytis cinerea. Gray mold rot. **4**:CA(82). **5**:CA(1110). **6**:FL(35).
Cephalosporium sp. Wilt. **5**:FL(35). **6**:FL(35).
Cercospora fuliginosa. Anthracnose. **4**:AL(82), FL(82), GA(82), IL(82), MS(82), TX(82). **5**:AL(82), DE(3543), FL(82), GA(82), IL(82), KS(973), MS(82), TX(82).
C. kaki. Leaf spot. **2**:FL(35). **4**:AL(82), FL(82), GA(82), LA(82), MS(82), TX(82). **6**:FL(35).
C. macclatchieana. Anthracnose, leaf spot. **2**:AL(285), OK(2872). **5**:FL(35), NC(3944). **6**:FL(35).
Cladosporium sp. **2**:CA(1110).
Dendrina diospyri. **5**:DE(3543).
Fusarium sp. Root rot, black stem rot. **2**:FL(35). **6**:FL(35).
Fusicladium levieri. Leaf spot. **5**:CT(82), FL(35,82), MS(82,2720).
Isariopsis linderae. On leaves. **5**:WV(82).
Penicillium expansum. Blue mold. **4**:Unk.(82).
Phymatotrichopsis omnivora. Root rot. **4**:AZ(82), TX(82). **5**:TX(82).
Ramularia sp. Leaf spot. **2**:FL(35). **5**:FL(35,82).
Sirosporium diospyri. Leaf spot. **5**:FL(35), IL(82), MD(82), MS(82,2720), NC(1303), OK(2872), SC(82), TX(82), VA(82).
Verticillium albo-atrum. Wilt. **4**:TX(82). **5**:IL(1558).

Deuteromycotina - Coelomycetes

Aplosporella diospyri. Twig blight, black rot of fruit. **2**:FL(35). **5**:LA(82).
Colletotrichum gloeosporioides. Anthracnose. **5**:FL(35). **6**:FL(35).
Colletotrichum sp. Anthracnose. **2**:HI(2903). **5**:NC(1303).
Diplodia sp. **5**:Unk.(82).
Lasiodiplodia theobromae. Stem canker, twig dieback. **2**:FL(35). **5**:FL(35).
Macrophoma diospyri. Twig blight, fruit spot. **2**:FL(35). **5**:AL(82), NC(82), TX(82).
Pestalotiopsis maculans. Leaf spot. **2**:FL(35). **4**:AL(82), FL(82), KS(82), SC(82). **5**:AL(82), FL(82), KS(82), SC(82). **6**:FL(35).
Phomopsis diospyri. Twig dieback. **2**:FL(35). **4**:FL(82). **5**:CA(1110), FL(35,82), SC(82).
Phomopsis sp. Dieback. **2**:FL(35). **6**:FL(35).
Phyllosticta arxii. Anthracnose on fruits and leaves, leaf spot. **2**:FL(35), MS(2720). **4**:Unk.(82), FL(82), MS(82,2720). **5**:Unk.(82), AL(82), DE(2,963), FL(82), KS(82), LA(82), MS(82,2720), TX(82), WV(82). **6**:FL(35).
P. biformis. Leaf spot. **2**:FL(35), TX(82). **4**:TX(64). **5**:TX(82). **6**:FL(35).
Phyllosticta sp. Leaf spot. **5**:FL(82), IN(82).

Deuteromycotina - Other

Rhizoctonia solani. Thread blight, basal stem rot of seedlings. **2**:FL(35). **6**:FL(35).

Elaeagnaceae

Elaeagnus L.

1. *E. angustifolia* L. [oleaster, Russian olive, silverberry, wild olive]
2. *E. commutata* Bernh. ex Rydb. (=*E. argentea* Pursh) [silverberry]
3. *E. pungens* Thunb. [thorny elaeagnus]
4. *E. umbellata* Thunb. [autumn elaeagnus]
5. *Elaeagnus* sp.

Oomycetes

Phytophthora cactorum. Trunk canker. **1**:IL(82), WY(82).
Phytophthora sp. Foot and root rot. **1**:AZ(82). **3**:FL(35).
Pythium sp. Root rot. **3**:FL(35).

Ascomycotina

Botryosphaeria dothidea. **4**:GA(1438).
B. obtusa. On limbs. **5**:AL(82).
Cucurbitaria elaeagni. On dead limbs. **1**:ND(82). **2**:ND(82,377).
Diaporthe beckhausii. On dead limbs. **2**:NY(82).
Didymosphaeria sp. **1**:AK(526).
Eutypella juglandicola. **4**:GA(1438).
E. scoparia. **4**:GA(1438).
Leptosphaeria sp. Dieback. **3**:FL(35).
Mycosphaerella sp. On leaves. **1**:GA(82), WA(3202). **4**:GA(1438).
Nectria cinnabarina. On twigs. **1**:CA(82), IA(82). **4**:GA(1438). **5**:TX(82).
Phyllactinia eleagni. **2**:WY(82).
P. guttata. Powdery mildew. **2**:OR(82).
Rhytidhysteron rufulum. On dead limbs. **5**:FL(82).
Valsa ambiens. On dead twigs. **1**:ND(82). **2**:ND(377,3415).

Basidiomycotina - Uredinales

Puccinia caricis-shepherdiae. **1**:ND(82). **2**:CO(82), MT(82), ND(82,378), NE(82), SD(82), WY(82,2768).
P. coronata. **2**:MT(82,3202), ND(82).

Basidiomycotina - Other

Armillaria tabescens. Dieback, root rot. **1**:LA(2815).
Exidiopsis podlachica. On wood. **2**:GA(2642).
Hymenochaete pinnatifida. On wood. **4**:GA(491,1439).
Lopharia cinerascens. On wood. **4**:GA(491,1439).
Sebacina epigaea. On wood. **2**:GA(2642).

Deuteromycotina - Hyphomycetes

Alternaria sp. **1**:OK(2872).
Cercospora elaeagni. Leaf spot. **1**:FL(35), MS(82,2720), OK(82,2872), TX(82). **2**:MS(2720), OK(2872). **5**:FL(35,82).
Cercosporidium manitobanum. **2**:ID(1195,3202), MT(3202).
Fusarium roseum. Dieback, wilt. **3**:FL(35).
Fusarium sp. Canker, root rot. **1**:OR(197,3202), WY(82,3780).
Phymatotrichopsis omnivora. Root rot. **1**:OK(82), TX(82). **5**:OK(82), TX(82).
Tubercularia ulmea. Isolated from cankers. **1**:ND(1989), SD(1989).
Verticillium albo-atrum. Wilt. **1**:KS(1558,3022), NM(1558,3355).
Verticillium sp. Wilt. **1**:WA(82).

Deuteromycotina - Coelomycetes

Aplosporella elaeagnina. On dead limbs. **1**:ND(82).
Camarosporium elaeagnellum. On dead limbs. **1**:CA(82), ND(82). **5**:NY(82).
Colletotrichum sp. Leaf spot. **1**:FL(35). **5**:FL(35).
Cytospora elaeagni. Canker. **1**:ID(3202), OK(82,2872).
Diplodia sp. Twig blight. **3**:FL(35). **5**:FL(35).
Hendersonia sp. Twig canker, dieback. **1**:WA(82,3202).
Macrophoma sp. Twig dieback. **3**:FL(35).
Phoma glomerata. Isolated from stem canker. **1**:ND(3871).
Phomopsis sp. Dieback. **1**:FL(35).
Phyllosticta argyraea. Leaf spot. **1**:NC(82), OK(2872), SD(2243). **5**:MD(82), MS(2720).
Phyllosticta sp. Leaf spot. **3**:FL(35). **5**:FL(35).
Septoria elaeagni. Leaf spot. **1**:IA(82), KS(82), ND(82), WI(82). **2**:NE(82).
S. shepherdiae. **2**:WI(1332).

Deuteromycotina - Other

Rhizoctonia ramicola. Thread blight. **3**:FL(35,3804). **5**:FL(35,82).
R. solani. Root rot. **3**:FL(35).
Sclerotium rolfsii. Seedling blight. **1**:TX(82).

Shepherdia Nutt.

1. *S. argentea* (Pursh) Nutt. [silver buffaloberry]
2. *S. canadensis* (L.) Nutt. (=*Elaeagnus canadensis* (L.) A. Nels.) [russet buffaloberry, soapberry]
3. *Shepherdia* sp.

Oomycetes

Pythium ultimum. Damping-off. **1:**NE(82).

Ascomycotina

Capitotricha bicolor. **2:**AK(3431), WA(3202).
Cucurbitaria occultata. **1:**ND(82,377).
C. shepherdiae. On dead twigs. **1:**MT(82,3202).
Karstenula shepherdiae. On dead twigs. **1:**ND(82), UT(82). **2:**NY(82,195).
Phyllactinia guttata. Powdery mildew. **1:**OR(3202), UT(82).
Sphaerotheca fuliginea. Powdery mildew. **1:**MT(82), WY(82). **2:**AK(82), TX(82).
S. macularis. Powdery mildew. **1:**CO(82), MT(82,3202). **2:**AK(526). **3:**AK(526).
S. shepherdiae. **1:**MO(371).
Sydowia lepargyrea. **2:**WA(179).
Tapesia fusca. **2:**AK(3431).
Valsa ambiens. On dead twigs. **1:**CO(82), ND(82,377,3415), WA(82).

Basidiomycotina - Uredinales

Puccinia caricis-shepherdiae. Heteroecious: aecial host. **1:**W states(82), MT(3202), SD(2243). **2:**W states(82), AK(82), ID(3202), MI(82), MT(3202), NM(82), NY(82), OR(82,3202), SD(82,2243), WA(82,3202), WY(1195,2924). **3:**CO(649).
P. coronata. **2:**Unk.(82), W states(82), AK(60,82,526,1938), ID(3202), MT(1195,3202), OR(3202), WA(82,3202), WY(1195,2924).

Basidiomycotina - Other

Perenniporia ellisiana. White heart rot. **1:**CO(82), MT(82,638), ND(638), SD(82,378,2243), WY(82). **2:**NM(82). **3:**CO(1205), ID(1205), MT(1205,3832), ND(1205), NM(1205), OR(1205), SD(1205), UT(1205), WA(1205), WY(1205).
P. fraxinophila. **1:**ID(3202), MT(3202), OR(3202). **2:**WA(3202).

Deuteromycotina - Hyphomycetes

Phymatotrichopsis omnivora. Root rot. **1:**TX(82).
Stigmina dothideoides. On dead twigs. **1:**MT(82).

Deuteromycotina - Coelomycetes

Septoria shepherdiae. Leaf spot. **1:**UT(82), WI(82). **2:**AK(82,526,3431), ID(82,3202), MT(82), WA(3202), WI(82).

Deuteromycotina - Other

Rhizoctonia solani. Damping-off. **1:**NE(82).

Elaeocarpaceae

Elaeocarpus L.

1. *E. bancroftii* F. v. Muell. & L. H. Bailey
2. *E. obtusa* Blume
3. *Elaeocarpus* sp.

Chytridiomycetes

Physoderma heleocharidis. **2:**HI(2903).

Oomycetes

Pythium sp. Root rot. **1:**FL(35). **3:**HI(2903).

Ascomycotina

Trichothyrium reptans. **3:**HI(2903).

Muntingia L.

1. *M. calabura* L. [calabur, Jamaica cherry, Panama berry]

Deuteromycotina - Hyphomycetes

Pseudocercospora muntingiae. Leaf spot. **1:**FL(35,581).

Deuteromycotina - Coelomycetes

Phyllosticta sp. Leaf spot. **1:**FL(35).

Empetraceae

Empetrum L.

1. *E. nigrum* L. [black crowberry, crakeberry, curlewberry, monox]

Ascomycotina

Arwidssonia empetri. **1:**AK(526).
Didymosphaeria empetri. On leaves. **1:**AK(82).
Phaeangellina empetri. On dead leaves. **1:**AK(1938).
Physalospora empetri. On leaves. **1:**AK(82,526), NH(178).

Basidiomycotina - Uredinales

Chrysomyxa empetri. Heteroecious: telial host. **1:**AK(60,526,1938), ME(82), NH(82), NY(82), VT(1224).

Deuteromycotina - Coelomycetes

Coleophoma empetri. On leaves. **1:**AK(82).

Epacridaceae

Styphelia Sm.

1. *S. tameiameiae* (Cham.) F. v. Muell.
2. *Styphelia* sp.

Oomycetes

Phytophthora cinnamomi. Root rot. **1:**HI(2903).

Ascomycotina

Capnodium sp. Sooty mold. **2:**HI(2903).

Basidiomycotina - Other

Armillaria mellea. On wood. **1:**HI(2011,2903).

Ephedraceae

Ephedra L.

1. *E. antisyphilitica* Berland. ex C. A. Mey.
2. *E. californica* S. Wats.
3. *E. nevadensis* S. Wats. [gray ephedra, Nevada ephedra]
4. *E. sinica* Stapf [Chinese ephedra]
5. *E. trifurca* Torr. ex S. Wats. [longleaf ephedra]
6. *E. viridis* Coville [green ephedra, Mormon tea]
7. *Ephedra* sp.

Ascomycotina

Stictis schizoxyloides. **7:**AZ(3225).
Teichospora megastega. On dead stems. **3:**NV(82). **6:**UT(82).

Basidiomycotina - Uredinales

[*Peridermium californicum*]. **2:**CA(111).
P. ephedrae. **1:**NM(111), TX(111). **3:**AZ(111,1203,3981), UT(111). **5:**NM(111). **6:**AZ(111,1203,3981). **7:**AZ(1203).

Deuteromycotina - Hyphomycetes

Phymatotrichopsis omnivora. Root rot. **4:**TX(82).

Equisetaceae

Equisetum L.

1. *E. arvense* L. [field horsetail]
2. *E. fluviatile* L.
3. *E. hyemale* L. [common scouring-rush]
4. *E. laevigatum* A. Braun (=*E. kansanum* J. H. Schaffn.)
5. *E. sylvaticum* L. f. *multiramosum* Fernald (=*E. sylvaticum* L. var. *multiramosum* (Fernald) Wherry)
6. *E. telmateia* Ehrh. var. *braunii* (Milde) Milde
7. *E. variegatum* Schleich. ex F. Weber & D. Mohr [variegated scouring-rush]
8. *Equisetum* sp.

Ascomycotina

Arnium apiculatum. On submerged plants. **8:**IL(3217).
Crocicreas cyathoideum. **7:**AK(3431).
Heptameria arvensis. On stems. **1:**CA(82).
Leptosphaeria equiseti. **3:**NY(82).
Leptosphaeria sp. **3:**ID(3202).
Mycosphaerella altera. On stems. **1:**WI(1347). **3:**NY(82).
Nectria sylvana. **1:**NY(3085).
Phialea rhodoleuca. On stems. **8:**CO(82).
Stamnaria americana. On dead stems. **3:**WA(638), WI(1339,1342). **8:**NC states(82), NE states(82), W states(82).
S. persoonii. **3:**ID(666,3202), WA(666,3202). **4:**ID(666), WA(666). **7:**AK(3431). **8:**ID(666), OR(3202), WA(666,3202).

Deuteromycotina - Hyphomycetes

Cladosporium sp. **8:**ID(3202).
Epicoccum sp. **3:**NY(2456).
Heterosporium equiseti. On branchlets. **5:**WI(1342).
Ramularia equiseti. On stems. **4:**WI(82,1331).

Deuteromycotina - Coelomycetes

Ascochyta equiseti. On dead stems. **1:**WI(1342). **2:**WI(1356). **4:**WI(1361). **6:**CA(81).

Stagonospora sp. On stems. **1:**WI(1359). **3:**WI(1349).
Titaeospora equiseti. Blight. **1:**ID(3202), SD(2243), WA(3202), WI(1340). **3:**ID(3202), WA(3202), WI(1340). **4:**WA(3202), WI(1332,1339), WY(1195). **8:**NC states(82), NE states(82), CO(82,649), WA(82).
Titaeospora sp. On leaves and stems. **8:**CA(3549), ND(3549), WY(3549).
Deuteromycotina - Other
Rhizoctonia solani. Root rot. **1:**IL(82), MN(82), TX(82). **3:**IL(82), MN(82), TX(82).

Ericaceae

Andromeda L.

1. *A. glaucophylla* Link
2. *A. polifolia* L.
3. *Andromeda* sp.

Ascomycotina
Asterina clavuligera. Black mildew. **1:**GA(82), NJ(82).
Durandiella andromedae. **3:**NC(1381).
Gibbera andromedae. On leaves. **1:**NY(128).
G. cassandrae. **1:**WI(1343).
Jattaea microtheca. On limbs. **3:**NJ(190).
Lachnum fuscescens. On leaves. **3:**NJ(849).
Microsphaera vaccinii. Powdery mildew. **1:**MA(82), NJ(82).
Mycosphaerella polifoliae. Leaf spot. **1:**NY(179).
Nectria ochroleuca. **3:**NJ(3085).
Protoventuria alpina. Leaf spot or superficial on leaves. **2:**NY(82).
Rhytisma andromedae. Tar spot. **1:**E states(82), IN(82), MI(2862), MN(82), NY(2456), WI(1340). **2:**AK(82,526), MI(82), NY(82).
Basidiomycotina - Uredinales
[*Chrysomyxa ledi* var. *cassandrae*]. **3:**Unk.(82).
Basidiomycotina - Other
Exobasidium vaccinii. Galls of buds, leaves or shoots, leaf spot, witches' broom. **1:**ME(82), NY(82,2456), WI(82). **2:**AK(82,526), NH(457), NY(82).
Stereum ochraceo-flavum. **3:**NJ(543).
Deuteromycotina - Hyphomycetes
Cercospora kalmiae. **1:**WI(1344).
Cladosporium sp. Leaf spot. **1:**WI(1360).

Arbutus L.

1. *A. arizonica* (A. Gray) Sarg.
2. *A. menziesii* Pursh [madrone, Pacific madrone]
3. *A. unedo* L. [cane-apples, strawberry-tree]
4. *Arbutus* sp.

Oomycetes
Phytophthora cactorum. Trunk canker. **2:**CA(82), WA(82,3202,3527).
Ascomycotina
Biscogniauxia nummularia. **2:**OR(3202).
Coccomyces dentatus. On leaves. **2:**CA(3229).
C. leptideus. On leaves and twigs. **2:**WA(936,3202).
Coccomyces sp. On leaves. **2:**CA(82), WA(82).
Diatrypella prominens. On limbs. **2:**CA(82).
Dothidea rugodisca. On leaves. **2:**CA(82,967).
Durella atrocyanea. **2:**OR(3228).
Elsinoe mattirolianum. Spot anthracnose. **2:**CA(82). **3:**CA(82).
Hypoxylon vogesiacum. **2:**OR(3202).
Hysterographium flexuosum. **2:**WA(3202).
Hysterographium sp. **2:**WA(3202).
Massariovalsa californica. On limbs. **2:**CA(82,3809).
Mycosphaerella arbuticola. On leaves. **2:**CA(81,82,959), OR(82,3202), WA(82,936,3202).
Patellea californica. **2:**WA(3202).
Perrotia flammea. **2:**WA(3202).
Pestalopezia brunneopruinosa. On leaves. **2:**WA(936,3202).
Phacidium arbuti. On leaves. **2:**CA(82,864).
Rhytisma arbuti. Tar spot. **2:**CA(81,82,936), OR(82,3202), TX(82), WA(82).
Stictis radiata. On stems. **2:**CA(81,936), OR(936,3202), WA(3202). **4:**PA(3225).
Valsa ceratosperma. On dead limbs. **2:**OR(82,3202), WA(82,936,3202).
Velutarina rufo-olivacea. On leaves. **2:**CA(81).
Basidiomycotina - Uredinales
Pucciniastrum sparsum. Heteroecious: telial host. **2:**CA(81,82,936), OR(82,3202), WA(82,936,3202).
Basidiomycotina - Other
Acanthophysium diffissum. **2:**OR(3202).
A. succineum. On wood. **2:**CA(2083), OR(2083,3202).
Antrodia albida. Wood rot. **1:**AZ(1202). **2:**CA(81,82,128), OR(3202).
Armillaria mellea. **2:**CA(2898).
Botryobasidium subcoronatum. **1:**AZ(1202).
Ceriporia purpurea. **1:**AZ(1202).
Chondrostereum purpureum. **2:**OR(3202).
Dendrothele dryina. **1:**AZ(1202).
Exobasidium vaccinii. Galls of buds, leaves or shoots, leaf spot, witches' broom. **2:**CA(81,82,457,936), OR(82), TX(82), WA(82).
Fomitopsis cajanderi. Heart rot. **2:**OR(82,3202).
Gloeophyllum sepiarium. Wood rot. **2:**OR(82,3202).
Gloeoporus dichrous. Wood rot. **2:**OR(82,3202).
Grandinia arguta. **1:**AZ(1202).
G. breviseta. **1:**AZ(1202).
Heliocybe sulcata. **1:**AZ(1202).
Heterobasidion annosum. **2:**CA(429).
[*Hyphoderma pilosum*]. **1:**AZ(1202).
Hypholoma fasciculare. On wood. **1:**AZ(1202).
Kavinia himantia. **1:**AZ(1202). **2:**OR(3202).
Micromphale arbuticola. On bark. **2:**CA(853).
Mycena madronicola. On bark. **2:**OR(3326).
Perenniporia medulla-panis. **1:**AZ(128,1202).
P. ohiensis. **1:**AZ(128,1202).
Phanerochaete velutina. White rot. **2:**OR(128,433).
Phellinus ferruginosus. **2:**OR(3202).
P. viticola. **2:**OR(3202).
Stereum hirsutum. **2:**OR(543).
S. ostrea. **1:**AZ(1202).
Trametes versicolor. **2:**OR(3202). **3:**TX(82).
Trichaptum biforme. **2:**OR(3202).
Deuteromycotina - Hyphomycetes
Alternaria sp. Leaf spot. **3:**FL(35).
Cercospora molleriana. Leaf spot. **3:**FL(35).
Mycoleptodiscus minimus. On leaves. **2:**CA(82).
Phymatotrichopsis omnivora. Root rot. **2:**TX(82).
Trichothecium roseum. **2:**WA(3202).
Deuteromycotina - Coelomycetes
Ascochyta hansenii. Leaf spot. **2:**CA(82,959), TX(82).
Coniothyrium asterinum. On leaves. **2:**CA(82), WA(936,3202).
Cryptosporium punctiforme. On leaves. **2:**CA(82).
Dendrophoma rhododendri. On limbs. **2:**CA(82), WA(82).
Didymosporium arbuticola. On leaves. **2:**OR(82,3202), WA(936,3202).
Diplodia maculata. On leaves. **2:**CA(81,82,936), TX(82), WA(3202).
Dothiorella sp. Leaf spot. **3:**FL(35).
Harknessia arctostaphyli. On fallen leaves. **2:**CA(82).
Hendersonula toruloidea. Canker. **2:**WA(820,3202).
Macrophoma maculiformis. On leaves. **2:**CA(82).
Phyllosticta fimbriata. Leaf spot. **2:**CA(82), OR(3202).
Seimatosporium arbuti. On leaves. **2:**CA(82), OR(82,3202), WA(936,3202).
Septoria unedinis. Leaf spot. **3:**OR(82,3202).
Sphaceloma mattirolianum. On leaves. **2:**CA(1745).

Arctostaphylos Adans.

(= *Camarostaphylis* Zucc., *Xylococcus* Nutt.)

1. *A. alpina* (L.) Spreng. (=*Arctotheca calendula* (L.) Levyns, *Arctous alpina* (L.) Nied.) [alpine bearberry, black bearberry]
2. *A. bicolor* (Nutt.) A. Gray (=*Xylococcus bicolor* Nutt.)
3. *A. columbiana* Piper [hairy manzanita]
4. *A. columbiana* Piper var. *tracyi* (Eastw.) Adams (=*A. tracyi* Eastw.)
5. *A. diversifolia* (Parry) Parry ex A. Gray
6. *A. glandulosa* Eastw. [Eastwood manzanita]
7. *A. hookeri* G. Don [Monterey manzanita]
8. *A. manzanita* Parry [Parry manzanita]
9. *A. nevadensis* A. Gray [pine-mat manzanita]
10. *A. nevadensis* A. Gray var. *coloradensis* (Rollins) Harrington
11. *A. nummularia* A. Gray [Bragg manzanita]
12. *A. nummularia* A. Gray var. *sensitiva* (Jeps.) McMinn (=*A. sensitiva* Jeps.)
13. *A. patula* Greene [greenleaf manzanita]

14. *A. pumila* Nutt. [dune manzanita]
15. *A. pungens* H. B. K. [Mexican manzanita, paintleaf manzanita]
16. *A. rubra* (Rehd. & E. Wils.) Fernald (=*Arctous rubra* (Rehd. & E. Wils.) Nakai, *A. alpina* (L.) Spreng. var. *rubra* (Rehd. & E. Wils.) Bean)
17. *A. tomentosa* (Pursh) Lindl. [woolly manzanita]
18. *A. uva-ursi* (L.) Spreng. (=*A. uva-ursi* (L.) Spreng. var. *coactilis* Fernald & Macbr.) [bear's-grape, bearberry, common bearberry, creashak, hog cranberry, kinnikinick, mealberry, mountain-box, sandberry]
19. *A. viscida* Parry [whiteleaf manzanita]
20. *Arctostaphylos* sp.

Chytridiomycetes

Synchytrium vaccinii. **11:**CA(936).

Ascomycotina

Arachnopeziza aurata. On decorticated stems. **4:**CA(3167).
Asterina conglobata. On leaves. **18:**ME(82).
Berlesiella sp. **13:**CA(661).
Botryosphaeria arctostaphyli. **9:**OR(936,3202), WA(3202).
Capitotricha bicolor. On dead twigs. **13:**CA(661).
Chlorociboria aeruginosa. **13:**CA(936).
Chlorociboria sp. On wood. **9:**CA(661).
Cistella coarctata. **9:**WA(936,3202).
Coccomyces arctostaphyli. On leaves. **18:**OR(3229).
Dasyscyphus labrosus. On dead leaves. **15:**CA(2801).
Dennisiodiscus crossotus. **9:**WA(849,3202).
Diaporthe vaccinii. On leaves. **18:**WA(936,3202).
Euantennaria rhododendri. **12:**CA(81).
Eupropolella vaccinii. On leaves. **18:**ID(936,3202), OR(3228), WA(3202).
Gloniella lapponica. On limbs and twigs. **9:**CA(638,642,661,936), WA(3202).
Gloniopsis praelonga. **3:**CA(936).
Glyphium elatum. **13:**CA(661).
Godronia ribis. **13:**CA(128).
Haglundia perelegans. On wood. **13:**CA(661).
Herpotrichiella sp. **13:**CA(661).
Hypoderma tunicatum. On twigs. **9:**CA(661), WA(936,3202).
Hysterographium arctostaphyli. **3:**OR(3202).
H. flexuosum. On limbs. **3:**OR(3202). **20:**CA(81,936).
Hysterographium sp. **13:**OR(3202).
Lachnella alboviolascens. **15:**AZ(1202).
Lachnum niveum. On partially buried sticks. **13:**CA(642).
Lophodermium maculare. On leaves. **20:**OR(82,3202).
Melanomma alpestre. **9:**WA(936,3202).
Meliola sp. Black mildew. **20:**CA(82).
Mollisia shastensis. On limbs and stems. **9:**CA(524,642).
Mycosphaerella minor. On dead sheaths. **1:**AK(1938).
M. shawii. **18:**ID(113,3202).
Ophiocapnocoma phloiophilia. **12:**CA(1673).
Patellea californica. On dead limbs. **13:**CA(642).
Pestalopezia brunneopruinosa. **18:**WA(3202).
Phacidium lacerum. **18:**ID(3202).
P. vaccinii. On fallen leaves. **18:**ID(82,3202), ME(82), MT(82,3202), WA(82,3202).
Phaeangium sphaeroides. **5:**CA(2439).
Propolomyces versicolor. **20:**CA(81,936).
Protoventuria alpina. On leaves. **18:**CA(82), MA(82). **20:**CA(82), WA(936,3202).
P. latispora. On leaves. **20:**OR(657).
Pyrenopeziza latebrosa. On leaves. **15:**CA(849).
Rhytisma vaccinii. **8:**CA(81).
Rosellinia riminicola. **13:**CA(638,936).
R. subsimilis. On limbs. **9:**CA(642,936). **13:**CA(642).
Schizothyrium pomi. Black leaf speck. **18:**ID(3202), WI(82).
Schizothyrium sp. **3:**WA(936,3202). **17:**WA(3202).
Sclerópleella hyperborea. **9:**CA(661).
Stictis chrysopsis. **20:**CA(3225).
S. radiata. **20:**CA(81).
Tapesia fusca. **9:**CA(642).
T. melaleucoides var. *vaccinii*. On dry twigs. **9:**CA(642).

Basidiomycotina - Uredinales

Chrysomyxa arctostaphyli. Heteroecious: telial host. **9:**WA(3202). **10:**CO(2767). **13:**AZ(1203,2767,3981), CO(2767), UT(2767). **18:**AK(60,82,526), CO(82), MT(82,3202), SD(2767), UT(82), WI(82), WY(1195,2924,3379).
Pucciniastrum sparsum. Heteroecious: telial host. **1:**AK(60,82,526). **3:**WA(1575). **7:**CA(1575). **8:**CA(81,82,936), OR(82,3202). **9:**CA(81,638,642,936), OR(936,3202). **13:**CA(81,82,638,642,936), OR(82,3202). **18:**WI(82). **20:**CA(82), OR(82).

Basidiomycotina - Other

Acanthophysium diffissum. On dead limbs and twigs. **9:**CA(638,642,661,936,2083). **13:**CA(642,661,936).
A. fruticetorum. On stems and twigs. **13:**CA(642,2083).
A. macrocystidiatum. On recently dead limbs. **20:**CA(2083).
Aleurodiscus aurantius. On twigs and stems. **4:**CA(81,2083).
Ceriporia tarda. **15:**AZ(1202).
Corticium roseum. **13:**CA(642).
Dacrymyces stillatus. **13:**CA(642).
Exidiopsis grisea. **8:**CA(3853).
Exobasidium angustisporum. On leaves. **1:**AK(1938).
E. uvae-ursi. **18:**WY(1195).
E. vaccinii. Galls of buds or leaves, red leaf spot, witches' broom. **3:**NW states(3202). **6:**CA(936). **8:**NW states(3202), CA(82,457), OR(82), WA(82). **9:**CA(642,661,936), OR(3202), WA(457,936,3202). **13:**CA(81,82,661,936), OR(82), WA(82). **15:**AZ(128), CA(82,457), OR(82), WA(82). **18:**AK(936), CO(82,457,649,1195), ID(82,936,3202), MA(82), MT(82,3202), NJ(82), NY(82), VT(82), WA(457,936,3202), WI(82), WY(2924). **20:**CA(82), MT(3202), OR(82,657), WA(82).
E. vaccinii var. *arctostaphyli*. On leaves and young shoots. **15:**CA(1468).
E. vaccinii-uliginosi. Shoot gall. **3:**WA(936,3202). **9:**CA(638,642,936), OR(936,3202). **13:**CA(82,638,642,936). **18:**ID(936,3202). **20:**OR(82).
Grandinia crustosa. On dead limbs. **13:**CA(642).
Heterobasidion annosum. Root rot. **8:**CA(3349). **19:**CA(3349).
Hymenochaete tabacina. On dead limbs. **13:**CA(642).
Lentinellus flabelliformis. **18:**Unk.(2423).
Naucoria sp. On bark. **20:**CA(642).
Omphalotus olivascens. On stumps or buried wood. **20:**CA(274).
Peniophora incarnata. On dead limbs. **13:**CA(661).
Phanerochaete velutina. **15:**AZ(1202).
Phellinus arctostaphyli. Heart rot. **3:**OR(1524). **13:**OR(82,1524). **15:**AZ(1202). **20:**OR(82).
P. ferreus. **3:**CA(81).
P. ferruginosus. White spongy rot. **20:**CA(82).
P. igniarius. **3:**OR(3202). **13:**CA(642), OR(3202).
Postia leucospongia. **13:**CA(642).
Stereum hirsutum. **13:**CA(661).
Stereum sp. On wood. **13:**CA(642).

Deuteromycotina - Hyphomycetes

Cercospora gaultheriae. Leaf spot. **18:**WI(82).
Phymatotrichopsis omnivora. Root rot. **8:**TX(82). **20:**TX(82).
Polydesmus sp. **13:**CA(661).
Sporidesmium induratum. On leaves. **20:**CA(82).
Trichoderma viride. On wood. **13:**CA(642).

Deuteromycotina - Coelomycetes

Coleophoma empetri. On leaves. **18:**WA(936,3202).
Coniothyrium arctostaphyli. On leaves. **9:**CA(642,661). **13:**CA(81,661,936). **15:**CA(82). **18:**OR(82,3202). **19:**CA(81).
Cryptosporium candidum. On leaves. **15:**CA(82). **18:**WA(82,3202). **20:**CA(82).
Gloeosporium arctostaphyli. Leaf spot. **18:**ID(666,936,3202), WA(3202).
Harknessia arctostaphyli. On old fruits, leaves and limbs. **2:**CA(81,936,2548), OR(2548). **14:**CA(2548). **15:**CA(82). **18:**CA(2548). **19:**CA(81). **20:**CA(82).
Phoma terrestris. Isolated from rhizosphere. **13:**OR(2791).
Phyllosticta amicta. On leaves and limbs. **3:**OR(3202). **8:**CA(81,82,936). **9:**CA(638,642,936), WA(936,3202). **18:**MA(1390), WA(936,3202). **19:**OR(3202). **20:**CA(82).
Placosphaeria arctostaphyli. **9:**OR(936,3202), WA(936,3202).
Seimatosporium arbuti. On leaves. **3:**WA(936). **9:**CA(642). **13:**CA(642,661,936). **20:**OR(82), WA(936).
Seimatosporium sp. **3:**WA(3202).

Deuteromycotina - Other

Sclerotium rolfsii. Blight. **1:**CA(572).

Calluna Salisb.

1. *C. vulgaris* (L.) Hull [heather]
2. *Calluna* sp.

Oomycetes

Phytophthora cinnamomi. **1**:OR(3202).
Pythium sp. Root rot. **2**:FL(35).

Deuteromycotina - Hyphomycetes

Botrytis cinerea. Gray mold. **2**:FL(35).

Deuteromycotina - Coelomycetes

Diplodia sp. Stem canker. **2**:FL(35).

Deuteromycotina - Other

Rhizoctonia solani. Thread blight. **2**:FL(35).

Cassiope D. Don

1. *C. hypnoides* (L.) D. Don
2. *C. lycopodioides* (Pall.) D. Don
3. *C. mertensiana* (Bong.) D. Don
4. *C. tetragona* (L.) D. Don
5. *Cassiope* sp.

Ascomycotina

Discocainia arctica. **4**:AK(526).
Extrawettsteinina andromedae. **3**:WA(3202).
Lophiostoma sieversiae. On leaves. **3**:CA(82).
Lophodermium orbiculare. On leaves. **2**:AK(526). **5**:AK(82).
Mycosphaerella immersa. **4**:AK(526).
M. punctiformis. **2**:AK(526).
Trochila craterium. **4**:AK(526).

Basidiomycotina - Other

Exobasidium myrtilli. On leaves. **4**:AK(936,1938).
E. vaccinii. Galls of buds, leaves, or shoots, leaf spot, witches' broom. **3**:AK(82,526), WA(82,457,936). **5**:AK(3431), WA(3202).

Deuteromycotina - Hyphomycetes

Antennaria rectangularis. On leaves. **1**:AK(82). **5**:AK(526).

Deuteromycotina - Coelomycetes

Diplodina andromedae. On stems. **1**:AK(82).

Chamaedaphne Moench

1. *C. calyculata* (L.) Moench (=*Cassandra calyculata* (L.) D. Don) [cassandra, leatherleaf]
2. *Chamaedaphne* sp.

Chytridiomycetes

Synchytrium vaccinii. Red leaf gall. **1**:NJ(82).

Ascomycotina

Cenangium cassandrae. On dead stems. **1**:NY(82).
C. pezizoides. On dead stems. **1**:NY(82).
Cucurbitaria cassandrae. On dead stems. **1**:MI(82), NY(82).
Godronia cassandrae. On dead stems. **1**:MI(1382), NJ(82), NY(82,1382), WI(128).
Hypoderma cassandrae. On dead stems. **1**:MI(82).
Lophodermium maculare. On leaves. **1**:NH(82), NY(82).
L. orbiculare. On leaves. **1**:NH(82), NY(82).
Physalospora vaccinii. On dead leaves. **1**:Unk.(178).
Protoventuria alpina. Leaf spot. **1**:ME(82), NH(82), NY(82).
Psilachnum cassandrae. On fallen leaves. **1**:MI(82,505).
Pyrenobotrys pulchella. On leaves. **1**:ME(82), MI(82), NH(82), NJ(82), NY(82), WI(82).
Sporomega degenerans. On dead stems. **1**:MI(2862).
Venturia sp. On leaves. **1**:WI(1360).

Basidiomycotina - Uredinales

Chrysomyxa ledi var. *cassandrae*. Heteroecious: telial host. **1**:NE states(82), AK(82,526), MA(1394), MI(2862), MN(82). **2**:AK(526).

Basidiomycotina - Other

Exobasidium vaccinii. Galls of buds, leaves or shoots, leaf spot, witches' broom. **1**:NE states(82), MI(457), NY(457), WI(82).

Deuteromycotina - Coelomycetes

Ascochyta cassandrae. Leaf spot. **1**:MI(82), NY(82), WI(82).
Dendrophoma phyllogena. On leaves. **1**:NY(3081).
Discella arida. On dead leaves. **1**:NY(82).
Dothichiza cassandrae. On dead stems. **1**:MI(82).
Monostichella chamaedaphnes. On leaves. **1**:WI(1345).

Epigaea L.

1. *E. repens* L. [mayflower, trailing arbutus]

Ascomycotina

Microsphaera vaccinii. Powdery mildew. **1**:NE states(82), WI(82).

Deuteromycotina - Hyphomycetes

Cercospora epigaeae. Leaf spot. **1**:NC(82), NY(82), WI(82).

Deuteromycotina - Coelomycetes

Discosia maculicola. On old leaves. **1**:MD(82), WV(82).
Phyllosticta epigaeae. Leaf spot. **1**:MA(82), NY(82).

Erica L.

1. *E. australis* L.
2. *E. carnea* L. [snow heather, spring heath]
3. *E. persoluta* L.
4. *E. regerminans* L.
5. *Erica* sp.

Oomycetes

Phytophthora cinnamomi. Collar and root rot. **2**:OR(3202). **4**:CA(3624,4010). **5**:CA(82), NY(82).

Ascomycotina

Erysiphe polygoni. Powdery mildew. **5**:CA(82).

Basidiomycotina - Uredinales

Uredo ericae. **5**:CA(82,1575).

Deuteromycotina - Hyphomycetes

Verticillium albo-atrum. Hadromycosis. **1**:CA(3366). **3**:CA(3366).

Deuteromycotina - Coelomycetes

Ascochyta sp. Stem rot. **5**:NJ(82).

Gaultheria L.

1. *G. hispidula* (L.) Muhl. (=*Chiogenes hispidula* (L.) Torr. & A. Gray ex Torr.) [creeping snowberry, maidenhair-berry, moxie-plum]
2. *G. humifusa* (R. J. Graham) Rydb. [alpine wintergreen]
3. *G. ovatifolia* A. Gray
4. *G. procumbens* L. [checkerberry, mountain-tea, teaberry, wintergreen]
5. *G. shallon* Pursh [salal, shallon, unionleaf]
6. *Gaultheria* sp.

Oomycetes

Phytophthora cinnamomi. Isolated from roots. **5**:OR(2133).

Ascomycotina

Botryosphaeria ribis. On stems. **4**:MA(82).
B. vaccinii. On dead leaves. **4**:E states(178).
Coccomyces coronatus. **5**:WA(3202).
C. dentatus. On leaves. **5**:WA(936,3229).
C. gaultheriae. On stems. **5**:CA(82), OR(82,936,3202), WA(82,936,3202). **6**:AK(526).
C. leptideus. On living and dead twigs. **5**:OR(3228,3229), WA(3229).
Diapleella coniothyrium. On leaves. **5**:OR(936,3202), WA(3202).
Diaporthe vaccinii. On leaves. **5**:WA(936,3202).
Discohainesia oenotherae. Leaf spot. **4**:VA(82).
Elsinoe ledi. Spot anthracnose. **5**:CA(936), OR(936,3202), WA(936,3202). **6**:OR(82), WA(82).
Lachnum gaultheriae. Leaf spot. **5**:CA(936,3167), OR(849,3167,3202), WA(936,3167,3202). **6**:CA(82), OR(82), WA(82).
Leptosphaeria gaultheriae. On dead stems. **5**:WA(936,3202). **6**:OR(82).
Lophodermium exaridum. On dead leaves. **6**:CA(82).
L. schweinitzii. On leaves. **5**:OR(936,3202), WA(936,3202).
Meliola nidulans. On leaves. **5**:CA(81,936,2434), OR(936,3202), WA(3202).
M. niessleana. On leaves. **2**:OR(3202), WA(3202,3815). **5**:CA(936,1452), WA(936,3202).
Meliola sp. Black mildew. **6**:CA(82), CO(82).
Microsphaera penicillata. Powdery mildew. **4**:MD(82). **5**:OR(3202), WA(936,3202). **6**:OR(82).
Microsphaera sp. Powdery mildew. **5**:OR(2012).
Mollisia gaultheriae. **5**:WA(3202).
M. pteridina. **5**:WA(3202).
Mycosphaerella gaultheriae. Leaf spot. **3**:ID(3202), WA(3202). **4**:NE states(82), WV(82). **5**:W states(82), AK(82,526), CA(936), OR(648,936,3202), WA(936,3202).
Pestalopezia brunneopruinosa. Leaf spot. **5**:CA(936), ID(3202), MT(3202), OR(936,3202), WA(936,3202). **6**:CA(82), OR(82), WA(82).
Phacidium sherwoodiae. On dead leaves. **5**:OR(863,864).

Protoventuria alpina. Leaf spot. **4:**MA(82), MD(82), NJ(82), NY(82), VA(82). **6:**MA(82), MD(82), NJ(82), NY(82), VA(82).
Pseudomassaria leucothoes var. *gaultheriae*. On dead leaves and stalks. **4:**MA(175).
Schizothyrium pomi. Black speck. **4:**NE states(82), NY(2456), WI(82).
Stictis radiata. **6:**CA(3225).
Tapesia cinerella. **5:**WA(3202).
Basidiomycotina - Uredinales
Chrysomyxa chiogenis. Heteroecious: telial host. **1:**MI(82), NH(82), NY(82), WI(82).
Basidiomycotina - Other
Exobasidium vaccinii. **3:**ID(936,3202), WA(936,3202).
Marasmius salalis. On branches and leaves. **5:**OR(855), WA(855).
Mycena gaultheri. On dead leaves. **5:**WA(3202,3326).
Phellinus ferreus. **5:**WA(128).
Skeletocutis nivea. **5:**WA(3202).
Tubulicium vermifera. **5:**CA(3859).
Deuteromycotina - Hyphomycetes
Cercospora gaultheriae. Leaf spot. **4:**NJ(82), WI(82).
Cercospora sp. **4:**WI(1360).
Chaetotrichum macrosporum. **5:**CA(81).
Corynespora pruni. **5:**WA(3202).
Deuteromycotina - Coelomycetes
Apostrasseria lunata. On dead leaves. **4:**NY(2546).
Coleophoma empetri. On leaves. **5:**WA(936,3202).
Discosia blumencronii. On leaves. **5:**OR(936,3202), WA(936,3202).
D. maculicola. Secondary leaf spot. **4:**NJ(82).
Gloeodes pomigena. Sooty blotch. **4:**WI(82).
Hendersonia concentrica. On stems. **5:**WA(936,3202).
Pestalotia multiseta. On leaves. **5:**CA(936).
P. sydowiana. On leaves. **5:**CA(936), OR(936,3202).
Phyllosticta gaultheriae. Leaf spot. **2:**WA(2). **4:**Range of host(82), NJ(2,959). **5:**Range of host(82), CA(936), OR(936,2012,3202), WA(936,3202).
Phyllosticta sp. Leaf spot. **4:**WI(1361).
Seimatosporium arbuti. On leaves. **5:**OR(936,3202), WA(936,3202).

Gaylussacia Kunth in Humb., Bonpl. & Kunth

1. *G. baccata* (Wangenh.) K. Koch (=*G. resinosa* Torr. & A. Gray in A. Gray) [black huckleberry, gueules-noires]
2. *G. brachycera* (Michx.) A. Gray [box huckleberry]
3. *G. dumosa* (Andr.) Torr. & A. Gray [dwarf huckleberry]
4. *G. frondosa* (L.) Torr. & A. Gray ex Torr. [blue-tangle, dangleberry, dwarf huckleberry]
5. *Gaylussacia* sp.

Chytridiomycetes
Synchytrium vaccinii. Leaf gall. **1:**NJ(82).
Ascomycotina
Botryosphaeria vaccinii. On leaves. **2:**MD(82).
Dasyscyphus venturioides. On leaves. **1:**NJ(849). **3:**NJ(849,965).
Discohainesia oenotherae. On leaves. **1:**VA(82).
Incrucipulum sulphurellum. On roots. **5:**NC(1303).
Microsphaera penicillata. **4:**GA(1438).
M. vaccinii. Powdery mildew. **1:**NE states(82), IL(82), WI(82). **5:**NY(82).
Monilinia fructicola. Brown rot. **5:**FL(35).
Ophiodothella vaccinii. Leaf spot. **5:**TX(82).
Protoventuria maculosa. Black mildew. **1:**NJ(82). **4:**NJ(82). **5:**MS(82,2720).
Rhytisma vaccinii. Tar spot. **1:**MA(82). **4:**SC(82). **5:**OK(82,2872).
Valsa ceratosperma. On twigs. **1:**MA(82), NJ(82).
Venturia sp. **1:**WI(1349).
Basidiomycotina - Uredinales
Pucciniastrum vaccinii. Heteroecious: telial host. **1:**NE states(82), MI(1575), PA(1575), VA(1575), WI(82). **4:**PA(82).
Basidiomycotina - Other
Armillaria tabescens. Root rot. **5:**FL(35).
Exobasidium vaccinii. Galls of buds, leaves or shoots, leaf spot, witches' broom. **1:**NE states(82), MA(457), MD(457), ME(457), MI(457), VA(457), WI(82,457,1359). **4:**AL(82), FL(82,457), MA(82,457), MD(82), MI(82,457), NY(82,457). **5:**FL(35).
Deuteromycotina - Hyphomycetes
Cercospora sp. Leaf spot. **5:**FL(35).
Helminthosporium sp. Leaf spot. **5:**FL(35).
Monilia peckiana. On leaves. **1:**NY(82).
Ramularia effusa. Leaf spot. **1:**WI(82).
Deuteromycotina - Coelomycetes
Apostrasseria lunata. On leaves. **2:**NC(82).
Coleophoma empetri. On leaves. **2:**MD(82).
Diplodia sp. Twig blight. **5:**FL(35).
Pestalotia vaccinii. Leaf spot. **2:**WV(82).
Phyllosticta sp. Leaf spot. **1:**NJ(82). **2:**PA(82). **5:**CA(1110), FL(35).

Kalmia L.

1. *K. angustifolia* L. [dwarf laurel, lambkill, pig laurel, sheep laurel, wicky]
2. *K. latifolia* L. [calicobush, ivy, ivybush, mountain laurel, spoonwood]
3. *K. polifolia* Wangenh. [bog kalmia, bog laurel, pale laurel]
4. *Kalmia* sp.

Chytridiomycetes
Synchytrium vaccinii. Leaf gall. **1:**NJ(82).
Oomycetes
Phytophthora cinnamomi. Root rot. **2:**NC(1303).
Ascomycotina
Botryosphaeria ribis. On limbs. **2:**NC(82).
B. vaccinii. On dead leaves. On leaves. **2:**E states(178), CT(82), NJ(82), NY(82), VA(82).
Diaporthe kalmiae. Leaf spot. **2:**NC(128,3942).
Gibbera kalmiae. On leaves. **1:**GA(1438), ME(82), NY(82). **2:**NY(82), SC(1307). **3:**AK(82,526), CA(642,936).
Lophodermium exaridum. On dead leaves. **1:**NH(82), NY(82,195). **2:**NJ(82).
Melanomma rhododendri. On twigs. **1:**NY(82).
Microsphaera penicillata. **3:**ID(3202), WA(3202).
M. vaccinii. Powdery mildew. **3:**ID(936), WI(82).
Mycosphaerella colorata. Eye spot, bordered leaf spot. **1:**E states(82), NY(179,3081). **2:**E states(82), CT(3770), GA(1438), MS(2720), NC(1303), OH(82), TX(82).
Ovulinia azaleae. Flower blight. **2:**SC(82).
Phyllachora kalmiae. Witches' broom. **1:**NY(82). **3:**AK(82), ME(82).
Physalospora kalmiae. On leaves. **2:**CT(178,3942), MA(178), RI(178).
Protoventuria rhododendri. Leaf spot. **2:**NC(1303).
Pseudomassaria erumpens. On dead leaves. **3:**ME(175), NH(175).
Rhytisma andromedae. Tar spot. **3:**VT(82).
Valsaria peckii. On dead twigs. **2:**NJ(82).
Basidiomycotina - Other
Antrodia sinuosa. **2:**WV(82).
Armillaria mellea. Root rot. **2:**MD(82).
Exobasidium vaccinii. Leaf gall. **3:**AK(82,526).
Heterobasidion annosum. **2:**NC(680).
Phellinus punctatus. Wood rot. **2:**MD(82).
Scytinostroma galactinum. Root rot. **2:**MD(82). **4:**MD(669,2090).
Sebacina helvelloides. Encrusting living tree. **4:**NC(128).
Septobasidium sp. Brown felt. **2:**TX(82).
Stereum hirsutum. Wood rot. **2:**VA(82).
Trametes versicolor. Wood rot. **2:**GA(1439), VA(82).
Deuteromycotina - Hyphomycetes
Alternaria sp. Leaf spot. **2:**FL(35).
Arthrobotryum atrocephalum. On dead limbs. **2:**NJ(525). **4:**NJ(3178).
Cercospora kalmiae. Leaf spot. **1:**WV(82). **2:**E states(82), FL(35), MA(656), NC(1303), TN(82).
C. sparsa. Leaf spot. **2:**MS(82,2720), PA(82), SC(82), TX(82).
Cercospora sp. Leaf spot. **2:**NC(1303).
Phymatotrichopsis omnivora. Root rot. **2:**TX(82).
Deuteromycotina - Coelomycetes
Amerodothis sp. **3:**AK(526).
Ceuthospora foliicola. On dead leaves. **2:**NJ(82).
Coleophoma empetri. On leaves. **1:**VA(82).
Hendersonia kalmicola. On twigs. **2:**NJ(82).
Micropera sp. On limbs. **2:**NC(1303).
Pestalotia adusta. Leaf spot. **2:**DE(82), NJ(82), PA(82), TX(82). **4:**WA(3202).
P. vaccinii. Secondary leaf spot. **2:**NC(1303).
Phomopsis kalmiae. Leaf blotch. **2:**CT(3770), MA(82), MD(82), NJ(82), NY(82), SC(82).
Phyllosticta kalmicola. Leaf spot. **2:**AL(285), NC(1303).
P. latifolia. Leaf spot. **2:**E states(82), MS(82,2720).
Phyllosticta sp. Leaf spot. **2:**NC(1303).
Rhabdospora kalmiarum. On dead leaves. **2:**NC(82), PA(82), WV(82).
Seimatosporium arbuti. On leaves. **4:**OR(936,3202), WA(3202).

[*S. rhododendri*]. On leaves. **2:**ME(82), NC(1303,1307).
Septoria angustifolia. Leaf spot. **1:**NJ(82). **2:**NC(1303,1307). **3:**ID(936,3202), WI(82), WV(82).

Ledum L.

1. *L. decumbens* (Aiton) Lodd. ex Steud. (=*L. palustre* L. var. *decumbens* Aiton) [crystal-tea, wild rosemary]
2. *L. glandulosum* Nutt.
3. *L. glandulosum* Nutt. var. *columbianum* (Piper) C. Hitchc.
4. *L. groenlandicum* Oed. (=*L. palustre* L. subsp. *groenlandicum* (Oed.) Hultén) [Labrador tea]
5. *Ledum* sp.

Chytridiomycetes
Synchytrium vaccinii. Leaf gall. **4:**ME(82).
Ascomycotina
Coccomyces coronatus. **2:**ID(3202).
Elsinoe ledi. Spot anthracnose. **2:**CA(81,82,1745), OR(82,1745,3202), WA(82,3202). **3:**WA(936). **4:**ME(82), MI(82), MN(82), NY(82,195), PA(82), WA(3202), WI(82,1333).
Lophodermium maculare. On dead stems. **4:**NY(82).
L. orbiculare. **4:**NH(82).
L. sphaerioides. On leaves. **4:**AK(82), ME(82), MI(82,2862), WA(82,936,3202). **5:**AK(526), WA(936).
Melanomma rhododendri. On twigs. **4:**MI(82).
Microsphaera penicillata. **2:**WA(3202).
M. vaccinii. Powdery mildew. **2:**WA(82).
Mycosphaerella punctiformis. **2:**ID(3202).
Niesslia exilis. **2:**ID(3202).
Rhytisma sp. Tar spot. **2:**ID(82,3202).
Basidiomycotina - Uredinales
Chrysomyxa ledi. Heteroecious: telial host. **2:**CA(82), ID(82,3202), MT(82,1195,3202), NV(82), OR(3202), UT(1976), WA(3202), WY(82). **4:**CT(82), MI(82), NH(82), NY(82), WA(3202), WI(82).
C. ledicola. Heteroecious: telial host. **1:**AK(60,526,1938). **2:**AK(82), ME(82), NH(82), NY(82), WA(82,3202), WI(82). **3:**WA(936). **4:**AK(60,526,1904), VT(1224), WA(936,3202). **5:**AK(82).
Basidiomycotina - Other
Daedaleopsis confragosa. On dead limbs. **4:**MN(82).
Exobasidium vaccinii. Leaf gall. **1:**AK(526). **2:**ID(82,3202), OR(82), WA(82). **4:**AK(82,526,936), ID(82), OR(82,3202), WA(82,3202). **5:**AK(82).
Deuteromycotina - Hyphomycetes
Sporocybe sp. On leaves. **4:**AK(82).
Deuteromycotina - Coelomycetes
Ascochyta ledi. Leaf spot, on bud scales. **4:**WA(936,3202), WI(82).
Diplodina fructigena. On fruit capsules. **2:**OR(936,3202), WA(3202). **4:**WA(936,3202).
Pestalotiopsis maculiformans. On leaves. **3:**CA(936).
Seimatosporium arbuti. On leaves. **2:**CA(82), OR(82,936,3202), WA(3202). **3:**WA(936). **4:**WA(936,3202).
S. lichenicola. Leaf spot. **4:**NY(2820).
[*S. rhododendri*]. On leaves. **4:**NY(82).

Leiophyllum R. Hedw.

1. *L. buxifolium* (Bergius) Elliott

Basidiomycotina - Other
Exobasidium vaccinii. Leaf gall. **1:**NC(82), NJ(82).

Leucothoe D. Don

1. *L. axillaris* (Lam.) D. Don (=*L. catesbaei* (Walter) A. Gray). Material offered as *L. catesbaei* is *L. fontanesiana*.
2. *L. axillaris* (Lam.) D. Don var. *editorum* (Fernald & Schub.) W. Ahles (=*L. editorum* Fernald & Schub.) [dog-hobble, drooping laurel, switch ivy]
3. *L. populifolia* (Lam.) Dippel
4. *L. racemosa* (L.) A. Gray [sweet-bells]
5. *Leucothoe* sp.

Ascomycotina
Asterina diplodioides. Black spot. **5:**AL(82).
Clithris andromedae. On dead limbs and twigs. **1:**NC(82). **4:**NJ(82), NY(82), PA(82).
C. leucothoicola. On leaves. **1:**NJ(82), PA(82). **2:**NC(1307).
Colpoma leucothoicola. On leaves. **1:**NC(1303).
Elsinoe ledi. Spot anthracnose. **1:**FL(35). **3:**FL(2524).
Hypoderma variegatum. On dead limbs. **5:**NC(82).
Hysterium macrosporum. **5:**FL(2782).
Hysterographium mori. **5:**FL(2782).
Lachnum niveum. **1:**GA(1438).
Lophodermium orbiculare. On leaves. **1:**NC(82).
Microsphaera penicillata. Powdery mildew. **1:**GA(1212).
Microsphaera sp. Powdery mildew. **1:**NC(1303).
Phacidium nigrum. Tar spot. **3:**FL(35).
Protoventuria alpina. On leaves. **1:**NC(82), NY(82).
P. maculosa. **4:**NJ(82).
Pseudomassaria leucothoes. Leaf spot. **2:**NC(175). **4:**TN(175).
Rhytisma andromedae. Tar spot. **3:**FL(35).
R. decolorans. Tar spot. **1:**FL(35), TN(82). **4:**CT(82), VA(82). **5:**FL(82), TN(82).
Valsa ceratosperma. On stems. **4:**NJ(82).
Basidiomycotina - Other
Exobasidium vaccinii. Leaf gall. **1:**FL(35), MS(82,2720), NC(82,1307). **4:**MA(82).
Mycena carolinensis. On twigs. **1:**NC(3326), TN(3326).
Septobasidium pseudopedicellatum. On scale insects infesting limbs. **1:**FL(35). **5:**FL(82).
Deuteromycotina - Hyphomycetes
Cladosporium herbarum. On leaves. **1:**NC(1303). **2:**NC(1307).
Cylindrocladium avesiculatum. Blight, leaf spot. **1:**GA(1212).
C. scoparium. Stem canker, shoot dieback, leaf spot, wilted leaves. **1:**NC(1303,2441).
C. theae. Stem lesion, wilt. **1:**NC(1303).
Pseudocercospora leucothoes. Leaf spot. **1:**FL(35), NC(1303), NJ(82), NY(82).
Ramularia andromedae. Leaf spot. **4:**NJ(82).
Deuteromycotina - Coelomycetes
Cryptostictis sp. Leaf spot. **1:**NJ(82).
Macrophoma sp. Dieback. **1:**FL(35).
Pestalotia sp. Leaf spot. **1:**NC(1303).
Pestalotiopsis leucothoes. Leaf spot. **1:**NJ(82).
Phyllosticta terminalis. Leaf spot. **1:**FL(35), NJ(82), NY(82), SC(82). **5:**FL(82).
Deuteromycotina - Other
Rhizoctonia solani. Root rot. **1:**FL(35).

Lyonia Nutt.

1. *L. ferruginea* (Walter) Nutt. (=*Andromeda ferruginea* Walter, *Xolisma ferrugineum* Nash)
2. *L. ligustrina* (L.) DC. (=*L. ligustrina* (L.) DC. var. *foliosiflora* (Michx.) Fernald) [he huckleberry, male-blueberry]
3. *L. lucida* (Lam.) K. Koch (=*Pieris nitida* (W. Bartram) Benth. & J. D. Hook.) [fetterbush]
4. *L. mariana* (L.) D. Don (=*Andromeda mariana* L.) [staggerbrush]
5. *L. paniculata* Nutt. (=*Andromeda ligustrina* Muhl.)
6. *Lyonia* sp.

Ascomycotina
Antennularia robinsonii. Leaf spot. **6:**FL(35).
Asteridium lepidigenum. Black spot. **1:**FL(35,82).
Botryosphaeria dothidea. **3:**GA(1438).
B. obtusa. **2:**Unk.(82).
B. vaccinii. On dead leaves. On leaves. **3:**E states(178), MS(3675).
Colpoma azaleae. **2:**GA(1438).
Cryptosphaeria vexata. On limbs. **1:**Unk.(82).
Cucurbitaria longitudinalis. On dead twigs. **2:**ME(82), NJ(82), NY(82,195).
Diaporthe ligustrina. On dead limbs. **2:**NJ(82).
Didymosphaeria grumata. On twigs. **2:**NJ(82).
Eutypella scoparia. On dead limbs. **2:**NC(1303,1307).
Gloniopsis praelonga. **1:**FL(2782).
Hypoderma variegatum. On twigs. **3:**NC(82).
Hysterostoma weberi. Leaf spot. **6:**FL(35).
Lembosiopsis andromedae. Sooty mold. **3:**AL(82), MS(2720).
Maurodothina dothideoides. Anthracnose, black leaf spot. **3:**FL(82). **6:**FL(35).
Microsphaera vaccinii. Powdery mildew. **2:**AL(82).
Mycosphaerella sp. On leaves. **2:**GA(82), MS(82,2720). **3:**MS(82,2720).
Ophiodothella ferruginea. On leaves. **1:**FL(35,82,959,967).
Phacidium nigrum. On dead leaves. **1:**FL(35,82).
P. vaccinii. Leaf spot. **1:**FL(35,82). **6:**FL(35).

Phacidium sp. On dead leaves. **1**:Unk.(82). **2**:Unk.(82). **3**:Unk.(82). **4**:Unk.(82).
Rhytisma andromedae. Large tar spot. **2**:FL(82), NJ(82), NY(82), VA(82). **3**:FL(82). **6**:FL(35).
R. decolorans. Small tar spot. **4**:NC(82). **6**:FL(35).
R. vaccinii. Tar spot. **6**:FL(35).
Schizoxylon microstomum. On twigs. **5**:NJ(3226).
Valsa ambiens. On dead twigs. **3**:GA(82,1438).
V. ceratosperma. **3**:GA(1438). **5**:NJ(3415).
Basidiomycotina - Uredinales
Pucciniastrum vaccinii. Heteroecious: telial host. **1**:FL(35,82). **2**:AL(82), AR(82), VA(82), WA(936,3202). **3**:GA(82). **4**:DE(82).
Basidiomycotina - Other
Exobasidium vaccinii. Galls of buds, leaves or shoots, shoot hypertropy, leaf spot, witches' broom. **1**:FL(35,82,457), GA(82,457,1439). **2**:AL(82), FL(82), MA(82), NJ(82), NY(82). **4**:FL(82,457), MD(82), NJ(457), NY(82,457). **5**:FL(457), MA(457), NJ(457), NY(457).
Schizopora paradoxa. Wood rot. **2**:ME(82).
Deuteromycotina - Hyphomycetes
Cercospora lyoniae. Leaf spot. **6**:FL(35).
Ramularia andromedae. **5**:DE(3543).
R. cylindriopsis. Leaf blight. **4**:NY(82).
Deuteromycotina - Coelomycetes
Coniothyrium sp. Leaf spot. **6**:FL(35).
Melasmia sp. Tar spot. **2**:NC(1303).
Pestalotia vaccinii. Leaf spot. **1**:FL(82). **6**:FL(35).
Phyllosticta sp. Leaf spot. **6**:FL(35).
Septoria pulchella. Leaf spot. **1**:FL(35,82).

Menziesia Sm.

1. *M. ferruginea* Sm. [rustyleaf]
2. *M. ferruginea* Sm. var. *glabella* (A. Gray) Peck (=*M. glabella* A. Gray)
3. *M. pilosa* (Michx. ex Lam.) Juss. ex Pers. [minniebush]

Ascomycotina
Dermea sp. On dead twigs. **3**:VA(82).
Melogramma sp. **1**:AK(526).
Microsphaera vaccinii. Powdery mildew. **1**:AK(82,526). **3**:VA(82).
Rhytisma arbuti. On leaves. **1**:AK(3431), CA(936), ID(638,936,3202), MT(3202), OR(3202), WA(936,3202).
Rhytisma sp. Tar spot. **1**:Unk.(82), AK(526).
Basidiomycotina - Uredinales
Pucciniastrum vaccinii. Heteroecious: telial host. **1**:ID(936,3202). **3**:WV(82).
Basidiomycotina - Other
Exobasidium vaccinii. Galls of buds, leaves or shoots, leaf spot, witches' broom. **1**:NW states(82), AK(82,526,936,3431), ID(936,3202), MT(3202), OR(3202), WA(936,3202), WY(1195). **2**:ID(457). **3**:AL(82), NC(82), VA(82), WV(82).
Peniophora aurantiaca. **1**:ID(3202).
Tremella mesenterica. **1**:ID(3202).
Deuteromycotina - Coelomycetes
Melasmia menziesiae. Tar spot. **1**:W states(82), AK(526,936,1904), CA(936), ID(936), OR(648), WA(936).

Oxydendrum DC.

1. *O. arboreum* (L.) DC. [sorreltree, sourwood, titi]

Ascomycotina
Botryosphaeria dothidea. **1**:GA(1438).
B. obtusa. On limbs. **1**:GA(82).
Coccomyces triangularis. On dead twigs. **1**:NC(1303).
Diaporthe sp. **1**:GA(1438).
Dothiora wolfii. Twig blight. **1**:NC(82).
Durandiella andromedae. On dead limbs. **1**:NC(82).
D. rugosa. Dieback. **1**:AL(82,1381), GA(82,1381,1438), NC(1303).
Hypoderma variegatum. On dead twigs. **1**:NC(82).
Hypoxylon howeianum. **1**:GA(1438).
Lachnum niveum. Herbaceous stems. **1**:GA(1438).
Lecanidion sp. **1**:GA(1438).
Mycosphaerella caroliniana. Leaf spot. **1**:GA(82,1438), NC(82,3944), TX(82), VA(82), WV(82).
Orbilia coccinella. On dead limbs. **1**:NC(1303).
Pachyella clypeata. **1**:GA(1438).
Patellariopsis clavispora. **1**:GA(1438).
Pezicula sp. On dead limbs. **1**:GA(82,1438).
Pseudomassaria oxydendri. On overwintered leaves. **1**:GA(82,1438), NC(82,175).
Rhytidhysteron rufulum. On dead limbs. **1**:GA(82,1438).
Rhytisma vaccinii. **1**:GA(1438).
Tympanis oxydendri. On dead limbs. **1**:GA(1438), WV(3167).
Valsa ceratosperma. **1**:GA(1438).
Basidiomycotina - Other
Calyptella capula. **1**:GA(1439).
Phellinus punctatus. Wood rot. **1**:WV(82).
Resinicium bicolor. On wood. **1**:GA(491,1439).
Trametes versicolor. On wood. **1**:NC(1303).
Deuteromycotina - Hyphomycetes
Alternaria sp. Leaf spot. **1**:FL(35).
Cercospora oxydendri. Leaf spot. **1**:AL(82), FL(35), MS(82,2720), TX(82), WV(82).
Cristulariella moricola. Zonate leaf spot. **1**:NC(1302,1303).
Fusarium aquaeductuum. On twigs. **1**:NC(82), WV(82).
Fusicladium sp. **1**:NC(1303).
Phymatotrichopsis omnivora. Root rot. **1**:TX(82).
Deuteromycotina - Coelomycetes
Phyllosticta oxydendri. **1**:NC(3944), WV(525).
Phyllosticta sp. Leaf spot. **1**:FL(35).

Phyllodoce Salisb.

1. *P. breweri* (A. Gray) A. Heller [red heather]
2. *P. empetriformis* (Sm.) D. Don
3. *P. glanduliflora* (Hook.) Coville
4. *Phyllodoce* sp.

Ascomycotina
Exarmidium diaphanum. **1**:CA(661).
Herpotrichia juniperi. Brown felt blight. **2**:ID(82,3202), OR(82), WA(82).
Mycosphaerella sp. **2**:WA(3202).
Basidiomycotina - Other
Exobasidium vaccinii-uliginosi. Witches' broom. **2**:OR(82), WA(82,936,3202). **4**:WA(936).
Deuteromycotina - Hyphomycetes
Antennaria rectangularis. On leaves. **3**:AK(526). **4**:AK(82).
Deuteromycotina - Coelomycetes
Diplodina andromedae. On stems. **3**:AK(526). **4**:AK(82).

Pieris D. Don

1. *P. floribunda* (Pursh) Benth. & J. D. Hook. [fetterbush, mountain andromeda]
2. *P. japonica* (Thunb.) D. Don ex G. Don [Japanese andromeda, lily-of-the-valley-bush]
3. *P. phillyreifolia* (Hook.) DC.
4. *Pieris* sp.

Oomycetes
Phytophthora cinnamomi. Root rot. **1**:NC(1303). **2**:NC(1303).
P. citricola. Shoot blight, dieback, leaf spot. **2**:NC(1303), OH(1164).
P. citrophthora. Twig blight, leaf spot. **2**:OH(1164).
Phytophthora sp. Dieback. **2**:PA(82).
Ascomycotina
Botryosphaeria dothidea. Canker, dieback. **2**:NC(1303), OH(1164). **4**:NC(1303).
Gloniopsis curvata. **4**:FL(2782).
Rhytisma andromedae. Tar spot. **2**:FL(35). **3**:FL(82), GA(82).
Basidiomycotina - Other
Armillaria mellea. Root rot. **2**:NJ(82).
Exobasidium vaccinii. Leaf and stem gall. **2**:FL(35).
Deuteromycotina - Hyphomycetes
Alternaria alternata. Leaf spot. **2**:RI(1483).
Cercospora sp. Leaf spot. **2**:FL(35).
Deuteromycotina - Coelomycetes
Cytospora sp. **2**:OR(3202).
Pestalotia sp. Leaf spot. **2**:CT(82), NJ(82).
Phomopsis sp. On dead stems. **2**:NC(1303).
Phyllosticta andromedae. Leaf spot. **1**:NJ(82). **2**:NJ(82).
P. maxima. **1**:CT(82).

Rhododendron L.

(= *Azalea* L.)
1. *R. adenogynum* Diels
2. *R. albiflorum* Hook.
3. *R. anthosphaerum* Diels
4. *R. arborescens* (Pursh) Torr. [smooth azalea, sweet azalea]
5. *R. arboreum* Sm. [farkleberry, tree huckleberry, tree rhododendron]
6. *R. brittania* Hort.
7. *R.* × *broughtonii-aureum* Hort. A listed name of no botanical standing for a hybrid of (*R. maximum* × *R. ponticum*) × *R. molle*.
8. *R. calendulaceum* (Michx.) Torr. (=*Azalea calendulacea* Michx.) [flame azalea, yellow azalea]
9. *R. caloxanthum* Balf. & Farrer
10. *R. campylocarpum* J. D. Hook. [honey-bell rhododendron]
11. *R. campylogynum* Franch.
12. *R. canadense* (L.) Torr. [Canadian rhododendron, rhodora]
13. *R. canescens* (Michx.) Sweet (=*Azalea canescens* Michx.) [Florida pinxter, hoary azalea]
14. *R. carolinianum* Rehd. [Carolina rhododendron]
15. *R. catawbiense* Michx. [Catawba rhododendron, mountain rosebay, purple laurel]
16. *R. decorum* Franch.
17. *R. erubescens* Hutch.
18. *R. fictolacteum* I. B. Balf.
19. *R.* × *gandavense* (K. Koch) Rehd. [Ghent hybrid azaleas]
20. *R. grande* Wight [silvery rhododendron]
21. *R. griersonianum* I. B. Balf. & Forrest
22. *R. indicum* (L.) Sweet [macranthum azalea]
23. *R. japonicum* (A. Gray) Suringar [Japanese azalea]
24. *R. javanicum* (Blume) Benn.
25. *R. kaempferi* Planch. (=*R. obtusum* (Lindl.) Planch. var. *kaempferi* (Planch.) W. Wils.) [torch azalea]
26. *R. lapponicum* (L.) Wahlenb. [Lapland rhododendron, Lapland rosebay]
27. *R. lutescens* Franch.
28. *R. luteum* Sweet [pontic azalea]
29. *R. macrophyllum* D. Don ex G. Don (=*R. californicum* Hook.) [California rosebay, West Coast rhododendron]
30. *R. maximum* L. [great laurel, great rhododendron, rosebay rhododendron]
31. *R. minus* Michx. [Piedmont rhododendron]
32. *R. molle* (Blume) G. Don [Chinese azalea]
33. *R. monosematum* Hutch.
34. *R. mucronatum* G. Don [snow azalea]
35. *R. mucronulatum* Turcz.
36. *R. nudiflorum* (L.) Torr. (=*Azalea nudiflora* L.) [pinxterbloom]
37. *R. obtusum* (Lindl.) Planch. [hiryu azalea]
38. *R. occidentale* (Torr. & A. Gray) A. Gray [western azalea]
39. *R. orbiculare* Decne.
40. *R. oreotrephes* Wm. W. Sm.
41. *R. ponticum* L.
42. *R. praevernum* Hutch.
43. *R. prunifolium* (Small) Millais [plum-leaved azalea]
44. *R. radicans* I. B. Balf. & Forrest
45. *R. russatum* I. B. Balf. & Forrest (=*R. cantabile* I. B. Balf. ex Hutch.)
46. *R. smirnowii* Trautv. ex Regel [Smirnow rhododendron]
47. *R. sutchuenense* Franch.
48. *R. thomsonii* J. D. Hook.
49. *R. vaseyi* A. Gray [pink-shell azalea]
50. *R. viscosum* (L.) Torr. (=*Azalea viscosa* L.) [clammy azalea, swamp azalea, swamp honeysuckle, white swamp azalea]
51. *Rhododendron* × sp. (unnamed hybrids) (=*Azalea x sp.*)
52. *Rhododendron* sp. (=*Azalea sp.*)
53. *Rhododendron* sp. (=*Azalea sp.*)

Myxomycetes

Physarum cinereum. Slime mold. **23:**NJ(82). **19:**NJ(82).

Chytridiomycetes

Synchytrium vaccinii. Leaf gall. **50:**NJ(82).

Oomycetes

Phytophthora cactorum. Blight. 'Phytophthora root rot. Phytophthora dieback.' **15:**MA(82), MD(82), NJ(82), NY(82), OH(82), PA(82), RI(82). **30:**MA(82), MD(82), NJ(82), NY(82), OH(82), PA(82), RI(82). **52:**MA(82), MD(82), NC(1303), NJ(82), NY(82), OH(82), PA(82), RI(82), WA(936,3202).

P. cinnamomi. Stem rot. 'Phytophthora root rot. Phytophthora wilt.' **14:**MD(82), NJ(82,3624), NY(82). **15:**NJ(82), NY(82). **29:**NJ(3624). **34:**AL(82), MD(82), MO(82). **35:**SE states(3624). **37:**AL(82), MD(82), MO(82), NC(242). **38:**OR(2133). **41:**NJ(3624), OR(3202). **52:**FL(35), HI(2299,2903), MD(1729,3624), MS(2720), NC(243,1303), NJ(82), NY(82), WA(936,3202).

P. citricola. 'Phytophthora dieback. Phytophthora root rot.' **52:**NC(1303).

P. heveae. Root rot. 'Phytophthora dieback.' **52:**NC(1303).

P. nicotianae var. *parasitica*. Dieback, root rot. **52:**FL(35), NC(1303).

Phytophthora sp. Dieback, root rot. **52:**NC(243).

Pythium debaryanum. Root rot. **52:**FL(35).

P. dimorphum. Isolated from diseased roots. **52:**NY(1579).

P. irregulare. Root rot. 'Pythium damping-off.' **23:**CA(82). **32:**CA(82). **19:**CA(82). **52:**FL(35).

P. splendens. Root rot. **52:**FL(35).

P. ultimum. Root rot. 'Pythium damping-off.' **23:**CA(82). **32:**CA(82). **19:**CA(82). **52:**FL(35).

Pythium sp. Root rot. **22:**HI(2903). **52:**NC(1303).

Ascomycotina

Acrospermum foliicola. **13:**GA(1438).

Bertia moriformis. On decorticated wood. **30:**GA(1307), NC(1303,1307), SC(1307).

Botryosphaeria disrupta. On dead limbs. **52:**NC(1303).

B. dothidea. Branch canker. 'Botryosphaeria dieback.' **52:**NC(1303).

B. obtusa. On leaves and limbs. **23:**MS(82). **30:**OH(82). **19:**MS(82). **52:**MS(2720).

B. rhodina. On dead limbs. **23:**NC(82). **19:**NC(82).

B. rhodorae. Leaf spot. **14:**CT(82), NC(82), NJ(82), NY(82). **15:**CT(82), MA(82), MD(82), NJ(82), NY(82), PA(82), VA(82). **29:**NJ(82), OR(3202), WA(3202). **30:**CT(82), MA(82), MD(82), NJ(82), NY(82), PA(82), VA(82). **52:**CA(81,178), CT(82), FL(35), MA(82,178), MD(82), NC(178,1303), NH(178), NJ(82), NY(82,178), PA(82), VA(82), WA(936,3202).

B. ribis. Limb canker, dieback. **15:**MA(82), MD(82), NJ(82), NY(82). **30:**MA(82), MD(82), NJ(82), NY(82). **52:**AL(2505), OH(3151).

Botryosphaeria sp. Leaf spot. **52:**FL(35).

Calonectria theae. On leaves. **37:**FL(36). **52:**FL(938).

Camarops polysperma. On wood. **52:**TN(128).

C. pugillus. **52:**TN(2561).

Chaetapiospora rhododendri. Leaf spot. **10:**WA(3202). **18:**WA(3202). **29:**WA(3202). **30:**MD(82), VA(82). **39:**WA(3202). **41:**WA(3202). **52:**WA(936).

Chaetosphaeria sp. Tip blight of petals. **52:**FL(35).

Coccomyces coronatus. On leaves and twigs. **15:**GA(82), NC(82), VA(82). **29:**CA(82,936), OR(82), WA(82,936,3202). **30:**GA(82), NC(82), VA(82).

C. dentatus. On leaves. **29:**CA(81,3229). **30:**GA(1438).

C. leptideus. On leaves and twigs. **29:**OR(3229).

C. radiatus. On leaves. **30:**VA(3229). **52:**TN(3229), VA(3229).

Colpoma azaleae. On dead limbs. **13:**GA(1438). **36:**GA(82), NC(82), PA(82), VA(82).

Dermea crypta. **13:**GA(1438).

Diapleella coniothyrium. On buds. **52:**WA(936,3202).

Diaporthe eres. On dead limbs and twigs. **15:**NC(82). **38:**CA(82). **52:**WA(936,3202).

D. vaccinii. On buds, leaves and stems. **52:**OR(936,3202), WA(936,3202).

Diaporthe sp. On stem galls. **52:**WA(936,3202).

Discostroma corticola. On leaves. **52:**WA(936,3202).

Dothidea puccinioides. **13:**GA(1438).

Elsinoe sp. Leaf spot. **52:**NC(1303).

Encoeliopsis rhododendri. On buds, capsules, leaves and twigs. **29:**CA(82), OR(82,3202). **41:**WA(936,3202). **52:**WA(936).

Erysiphe polygoni. 'Powdery mildew.' **6:**CA(1155). **24:**CA(1155). **28:**CA(1155). **36:**VA(82). **38:**CA(82). **48:**CA(1155). **7:**CA(1155).

Eutypella juglandicola. **13:**GA(1438).

Glomerella cingulata. 'Glomerella canker. Glomerella dieback. Anthracnose.' **52:**MD(82), WA(936,3202).

Glomerella sp. Anthracnose. **52:**FL(35).

Godronia cassandrae. On bud scales, leaves and stems. **5:**WA(936,3202). **9:**WA(936,3202). **33:**WA(936,3202).

52:WA(936,3202).
Graphostroma platystoma. On dead limbs. **30:**NC(1303).
Guignardia sp. **52:**WA(3202).
Herpotrichia sp. Tip blight of petals. **52:**FL(35).
Hyalopeziza ciliata. **52:**OR(3202).
Hypocrea rufa. **30:**NC(1307).
Hypoxylon rubiginosum. **30:**GA(1438).
H. serpens. **30:**GA(1438).
Lachnellula ciliata. **52:**OR(3202).
Lachnum virgineum. On dead twigs. **30:**NC(1303).
Leptosphaeria sp. Tip burn. **52:**FL(35), NC(1303).
Lophodermium melaleucum. **15:**NC(82), TN(82).
L. schweinitzii. Leaf spot. 'Lophodermium leaf spot.' **15:**E states(82), PA(82), TX(82), VA(82), WV(82). **29:**OR(82,3202), WA(82,3202). **30:**E states(82), NC(1303,1307), PA(82), VA(82), WV(82). **52:**WA(936,3202).
Melanomma rhododendri. On stems. **29:**WA(936,3202).
Microsphaera penicillata. 'Powdery mildew.' **8:**GA(1438). **15:**MD(82), NJ(82), NY(82). **23:**NC(3512), NJ(82), NY(82), PA(82), RI(82), VA(82). **38:**Unk.(3976), CA(81,936). **41:**NC(3512). **19:**NJ(82), NY(82), PA(82), RI(82), VA(82). **52:**CA(936), FL(35), NC(1303,3512), NJ(82).
M. vaccinii. Powdery mildew. **8:**GA(82). **36:**NY(82), PA(82), VA(82).
Monilinia azaleae. 'Twig blight.' **13:**GA(1438). **36:**GA(82), NY(82).
Mycosphaerella clintoniana. Leaf spot. **15:**NC(82), NJ(82), NY(82). **30:**NC(82), NJ(82), NY(82). **52:**NW states(3202), FL(35), NC(1303), OR(82), WA(82).
M. millepunctata. On leaves. **30:**NC(178), VA(178). **52:**NC(1303).
Mycosphaerella sp. Leaf spot. **14:**NC(82). **29:**WA(82,3202).
Myriophacidium aphyophyllicum. On leaves. **29:**OR(3224).
Nectria cinnabarina. On stems. **52:**WA(936,3202).
N. radicicola. On stems. **52:**WA(936,3202).
Ovulinia azaleae. Limb blight. 'Ovulinia petal blight.' **15:**SC(82). **22:**HI(2903). **34:**Gulf states(82), SE states(82), CA(82). **36:**SC(82). **52:**CA(82), CT(2132), FL(35), HI(2299), MD(1728,1729), MS(2720), NC(1303), NY(2585).
Pestalopezia rhododendri. On leaves. **30:**GA(1307), NC(1303,1307,1958), TN(82), WV(82).
Pestalosphaeria concentrica. 'Pestalosphaeria leaf spot.' **30:**NC(181,1303), VA(181), WV(181).
Pezicula rhododendricola. On dead twigs. **30:**PA(82).
Phacidium vaccinii. On fallen leaves. **29:**OR(82,3202).
Phomatospora gelatinospora. On dead leaves. **30:**NH(178).
Physalospora abdita. On dead limbs. **23:**NC(82). **50:**GA(82). **19:**NC(82).
P. dearnessii. Leaf spot. **30:**NC(178,1303), PA(178), TN(178), VA(178).
P. rhododendri. Leaf spot. **30:**PA(82), TN(82), VA(82).
P. rhododendri var. *albimontana*. On dead leaves. **26:**NH(178).
Physalospora sp. Twig dieback. **52:**FL(35).
Pleospora herbarum. On leaves and stems. **52:**WA(936,3202).
Pleospora sp. **52:**WA(936,3202).
Propolomyces versicolor. **13:**GA(1438).
Pseudomassaria inversa. On leaves. **26:**NH(272).
P. thistletonia. Leaf spot. **30:**MD(272), NH(272).
Rhytisma arbuti. **2:**WA(3202).
Sordaria destruens. On stems. **52:**WA(936,3202).
Sphaerotheca pannosa. Powdery mildew. **52:**GA(2405).
Stictis radiata. **52:**CA(3225).
Strickeria vaga. **52:**WA(936,3202).
Tapesia fusca. On dead roots. **30:**NC(1303,1307).
Valsa ceratosperma. On dead twigs. **50:**NJ(82). **52:**NJ(3415).
V. microstoma. **13:**GA(1438).
Velutarina rufo-olivacea. On bark and twigs. **30:**NC(82), WV(82).

Basidiomycotina - Uredinales

[*Chrysomyxa ledi*]. 'Chrysomyxa leaf rust.' **21:**WA(3202). **52:**OR(3202), WA(3202).
C. ledi var. *rhododendri*. Heteroecious: telial host. **40:**OR(936), WA(936). **52:**NC(1303), OR(936), WA(936,1288).
[*C. ledicola*]. **21:**WA(936,3202). **52:**WA(936,3202).
C. piperiana. Heteroecious: telial host. 'Chrysomyxa leaf rust.' **29:**CA(81,82,739), OR(82,936,3202), WA(82,936,3202). **52:**OR(936,3202), WA(936,3202).
C. roanensis. Heteroecious: telial host. 'Chrysomyxa leaf rust.' **15:**TN(82). **26:**AK(60). **31:**NC(82,1303), TN(82).
Chrysomyxa sp. **52:**NJ(82).
Pucciniastrum vaccinii. Heteroecious: telial host. 'Pucciniastrum leaf rust.' **12:**ME(82), NH(82). **13:**GA(1439), VT(1575). **27:**CT(82), NJ(82), RI(82). **36:**E states(82), NC(1303), TX(82). **41:**CT(1575), NJ(1575), RI(1575). **43:**NC(1303). **50:**E states(82), TX(82). **19:**NJ(82).

Basidiomycotina - Other

Acanthophysium mirabile. **52:**NC(2083).
Aleurodiscus aurantius. On dead stems and twigs. **38:**CA(2083).
Armillaria mellea. On roots. 'Shoestring root rot.' **15:**NJ(82), NY(82). **23:**CA(82), NJ(82), WA(82). **30:**GA(1307), NC(1303,1307). **19:**CA(82), NJ(82), WA(82). **52:**CA(2898), NC(1303), WA(936,3202).
A. tabescens. Dieback 'Clitocybe root rot.' **22:**LA(2815). **34:**FL(82). **37:**FL(82). **52:**FL(35).
Athelia scutellaris. On wood. **13:**GA(491,1439).
Clavicorona pyxidata. **52:**NC(884).
Coniophora arida. On wood. **30:**NC(1303,1307).
Exobasidium burtii. Yellow leaf spot. 'Exobasidium leaf spot.' **2:**ID(82,936,3202), OR(82,3202), WA(82,936,3202). **38:**CA(638,936). **52:**NJ(82).
E. canadense. On leaves. **12:**NH(3096).
E. decolorans. On leaves. **38:**CA(81,457,936,1468,3096).
E. vaccinii. Galls of buds, leaves or shoots, witches' broom. 'Leaf gall. Flower gall.' **2:**ID(3202), WA(457,3202). **4:**E states(82), TX(82). **8:**E states(82), GA(1439), TX(82). **12:**E states(82), TX(82). **14:**TN(82). **15:**E states(82), MS(82), NC(3944). **22:**Range of host(82), HI(2903), MS(2720). **25:**Range of host(82). **29:**CA(936,3096), OR(82,936,3202), WA(936). **30:**E states(82), MS(82), NC(457,1307). **34:**Range of host(82). **36:**AL(457), MA(457), MS(2720), NC(457,3944), OK(2872). **37:**Range of host(82). **50:**E states(82), GA(1439), MA(457,1390), MS(82,457,1439), NJ(457,958,3096), TX(82). **52:**E states(82), AL(285), CA(936), CT(3770), FL(35), HI(2899,2903), MA(457), MS(82,2720), NC(1303), OR(936,3202), TX(82), WA(936,3202).
E. vaccinii-uliginosi. 'Witches' broom.' **29:**CA(81,82,936), OR(82,936,3202), WA(82,936,3202). **52:**NJ(82).
Exobasidium sp. Leaf gall. **52:**KY(1777).
Gymnopilus fuscosquamulosus. On roots. **52:**NC(1542).
Heterobasidion annosum. Isolated from roots. **30:**NC(680). **52:**NC(1303).
Hymenochaete agglutinans. **30:**PA(82), TN(82). **52:**TN(128).
Leucogyrophana mollusca. **52:**TN(128).
Mycena carolinensis. On dead twigs. **52:**NC(3326), TN(3326).
M. cylindrospora. On dead leaves. **52:**TN(3326).
Panellus violaceofulvus. **30:**NC(1307).
Pellicularia koleroga. Thread blight. **34:**LA(82). **52:**LA(3640).
Phellinus laevigatus. On wood. **30:**NC(1303).
P. viticola. White rot. **30:**NC(82,1303).
Postia caesia. Brown rot. **30:**NC(1303,1307), VA(82).
Rhodocybella rhododendri. On living stems. **30:**NC(173).
Stereum rugosum. On dead limbs. **15:**TN(82). **30:**NC(543). **52:**TN(543).
Tomentella rubiginosa. On decaying leaves. **52:**VA(128).
Trametes versicolor. **30:**NC(1307), SC(1307).
Trichaptum biforme. **30:**AL(82).

Deuteromycotina - Hyphomycetes

Alternaria alternata. Flower spot. **22:**SC(82). **25:**SC(82). **34:**SC(82). **37:**SC(82). **19:**SC(82). **52:**FL(35).
Alternaria sp. Damping-off. **15:**CT(82). **52:**WA(936,3202).
Bispora sp. On leaves. **30:**NC(1303,1307).
Botrytis cinerea. 'Botrytis leaf blight. Botrytis petal blight.' **1:**WA(936,3202). **5:**WA(936,3202). **15:**CA(82), NJ(82), NY(82). **16:**WA(3202). **22:**Range of host(82), HI(2903). **23:**Range of host(82). **25:**Range of host(82). **30:**CA(82), NJ(82), NY(82). **34:**Range of host(82). **36:**Unk.(82). **37:**Range of host(82). **41:**WA(3202). **45:**WA(3202). **19:**Range of host(82). **52:**CA(82), NC(1303), NJ(82), NY(82), WA(936,3202).
Botrytis sp. **52:**WA(3202).
Cercospora sp. Leaf spot. **34:**AL(82), GA(82), MD(82), MS(82), PA(82), VA(82). **37:**AL(82), GA(82), MD(82), MS(82), PA(82), VA(82). **52:**MS(2720), NC(1303), OK(2872).
Cladosporium herbarum. Flower spot, on leaves. **22:**Range of host(82). **23:**Range of host(82). **25:**Range of host(82). **34:**Range of host(82). **37:**Range of host(82). **19:**Range of host(82).

52:FL(35), WA(936,3202).
Corynespora cassiicola. Leaf spot. **37**:FL(3367). **52**:FL(35).
Cylindrocarpon destructans. 'Cylindrocarpon canker. Cylindrocarpon wilt.' **52**:MA(701).
Cylindrocarpon sp. Root rot. **52**:FL(35).
Cylindrocladiella peruviana. Root rot. **52**:FL(35).
Cylindrocladium avesiculatum. Root and stem rot, leaf spot. **52**:FL(35).
C. clavatum. Twig blight, root rot. **22**:HI(2903). **52**:FL(35).
C. scoparium. Blight, leaf, root and stem rot. 'Cylindrocladium blight. Cylindrocladium root rot.' **51**: Unk.(3976), FL(1637), OH(1637). **52**:AL(3634), FL(35,708), NC(1303).
C. theae. Stem rot. 'Stem canker. Cylindrocladium blight. Cylindrocladium root rot.' **52**:FL(35), NC(1303,2441).
Dactylaria sp. On roots. **52**:WA(936,3202).
Didymaria sp. Leaf spot. **52**:FL(35).
Epicoccum sp. Secondary flower spot. **34**:NC(82), SC(82). **37**:NC(82), SC(82).
Fumago vagans. **29**:WA(936,3202). **52**:WA(3202).
Fumago sp. Sooty mold on insect secretions. **52**:NC(1303).
Fusarium moniliforme. Root rot, wilt. **52**:FL(35).
F. oxysporum. Root rot, wilt. **52**:FL(35).
F. solani. Dieback, cutting and root rot. **52**:FL(35).
Fusarium sp. On buds and twigs. **52**:WA(936,3202).
Gliocladium sp. Root rot. **41**:WA(936,3202). **52**:WA(936).
Oidium sp. Powdery mildew. **52**:NC(1303).
Phaeoramularia sp. Leaf spot. **52**:FL(35).
Phymatotrichopsis omnivora. 'Phymatotrichum root rot.' **4**:TX(82). **15**:TX(82).
Pseudocercospora handelii. 'Cercospora leaf spot.' **15**:CA(82), FL(82), NC(82), NJ(82), NY(82), VA(82). **30**:FL(82), NC(82), NJ(82), VA(82).
Pycnostysanus azaleae. Flower blight. 'Bud blight. Twig blight.' **4**:MA(82), NH(82), NJ(82). **15**:GA(82,2405), NC(82), NJ(82), NY(82), PA(82), TN(82), VA(82). **29**:CA(81,82,936), OR(82,3202), WA(82,936,3202). **30**:GA(82), NC(82,1307), NJ(82), NY(82), PA(82), TN(82), VA(82). **31**:NC(1303). **36**:MA(82), NH(82), NJ(82). **49**:NC(82). **50**:NC(1303). **52**:MS(2720), NC(1303), NJ(82).
Ramularia angustata. Leaf spot. **23**:MS(82). **32**:MS(82). **36**:NY(82). **50**:MS(2720). **19**:MS(82).
Ramularia sp. Leaf spot. **15**:CA(82). **30**:CA(82). **52**:CA(82).
Septocylindrium sp. On buds. **52**:WA(936,3202).
Trichoderma viride. Secondary flower spot. **34**:NC(82), SC(82). **37**:NC(82), SC(82).
Trichothecium roseum. On buds. **16**:WA(936,3202). **52**:WA(936).
Verticillium albo-atrum. Wilt. **32**:IL(1558). **43**:WA(936,3202). **52**:WA(936,3202).
Verticillium sp. Wilt. **43**:WA(3202). **52**:MA(82).

Deuteromycotina - Coelomycetes

Ascochyta rhododendri. On bud scales. **52**:WA(936,3202).
Ascochyta sp. Leaf spot. **52**:FL(35).
Asterostomella saccardoi. On leaves. **29**:CA(936), WA(936).
Coleophoma empetri. On leaves. **30**:VA(82). **52**:WA(936,3202).
Colletotrichum gloeosporioides. Leaf spot. **50**:FL(82). **52**:FL(35).
Colletotrichum sp. Anthracnose. **52**:MS(2720), NC(1303).
Coniothyrium phyllogenum. **52**:WA(936,3202).
C. rhododendri. On leaves and stems. **44**:WA(936,3202). **46**:WA(936,3202). **52**:WA(936,3202).
Cornicolariella hystricina. **50**:NJ(861).
Dendrophoma azaleae. On dead limbs. **36**:NY(82).
D. rhododendri. On bud scales and leaves. **16**:WA(936,3202). **52**:WA(936,3202).
Diplodia sp. Twig dieback. **52**:FL(35).
Diplodina eurhododendri. On leaves. **52**:CA(82).
Discosia artocreas. Leaf spot. **5**:WA(936,3202). **10**:WA(936,3202). **15**:MD(82), NY(82). **30**:MD(82), NY(82). **36**:MD(82), NY(82). **52**:WA(936), WA(3202).
Gloeosporium ferrugineum. Leaf spot. **14**:NC(82).
Hendersonia concentrica. Leaf spot. **14**:NC(82). **15**:NC(82,966), TX(82). **30**:WV(82).
H. oudemansii. On stems. **52**:WA(936,3202).
H. sarmentorum. On stems. **52**:WA(936,3202).
Leptothyrium parvulum. On leaves. **14**:NC(82).
Macrophoma sp. Dieback. **52**:FL(35).
Macrophomina phaseolina. Root rot. **52**:FL(35).
Melasmia menziesiae. Tar spot. **52**:WA(82).
M. rhododendri. Tar spot. **52**:AK(82).
Monochaetia cryptomeriae. On leaves. **52**:WA(936,3202).
M. monochaeta. On leaves. **52**:WA(936,3202).
Pestalotia baarnensis. On leaves. **3**:WA(936,3202). **9**:WA(936,3202). **16**:WA(936,3202). **47**:WA(936). **52**:WA(936,3202).
P. sydowiana. Secondary flower spot and leaf spot. 'Gray blight.' **11**:WA(936,3202). **15**:WA(936,3202). **16**:WA(936,3202). **17**:WA(936,3202). **30**:GA(1307), NC(1303). **41**:WA(936,3202). **42**:WA(936,3202). **45**:WA(936,3202). **52**:NC(1303), WA(936,3202).
P. vaccinii. On leaves. **15**:DE(82), IN(82), NC(82). **30**:DE(82), IN(82), NC(82,1303).
Pestalotia sp. Damping-off, secondary flower spot, leaf spot. **31**:NC(1303). **34**:Range of host(82). **37**:Range of host(82). **52**:FL(35), NC(1303).
Pestalotiopsis guepini var. *macrotricha*. Gray blight of leaves and twigs, leaf spot. **14**:NC(82), NJ(82), NY(82), TN(82). **15**:Range of host(82). **29**:NC(82), NJ(82), NY(82), TN(82). **30**:Range of host(82), NC(1303). **52**:Range of host(82), CA(81), CT(3770).
P. maculans. Leaf spot. **4**:NY(82). **15**:CT(82), NY(82), PA(82), WV(82). **29**:OR(3202). **30**:CT(82), NC(181), NY(82), PA(82), VA(181), WV(181), WV(82). **50**:WV(82). **52**:WA(936,3202).
P. sydowiana. Leaf spot. **16**:WA(936,3202). **30**:NC(1303,1307), SC(1307). **52**:NC(1303), NJ(82), OK(2872), OR(3202), WA(936,3202).
Phomopsis ericaceana. On leaves. **29**:CA(82).
Phomopsis sp. Dieback, leaf spot. **14**:NJ(82). **15**:CT(82), NJ(82), NY(82), PA(82). **25**:AL(82), MA(82), MD(82), SC(82), TX(82). **34**:AL(82), MA(82), MD(82), SC(82), TX(82). **52**:CT(82), FL(35), NC(1303), NJ(82), NY(82), OR(3202), PA(82), SC(2432), WA(3202).
Phyllosticta berolinensis. **10**:WA(936). **18**:WA(936). **29**:WA(936). **52**:WA(936).
P. cunninghamii. On leaves. **15**:WA(936,3202). **52**:WA(936).
P. maxima. Leaf spot. **52**:OK(2872).
P. occulta. On leaves. **52**:WA(936,3202).
P. rhododendri. Leaf spot. **15**:NJ(82). **52**:WA(936).
P. rhododendricola. **15**:CA(936), WA(936). **20**:CA(936), WA(936). **40**:CA(936), WA(936). **52**:CA(936), WA(936).
P. rhodorae. **46**:WA(936). **52**:WA(936).
Phyllosticta sp. Leaf spot. **8**:WV(954). **31**:NC(1303). **34**:GA(82), LA(82), SC(82). **36**:WV(954). **37**:GA(82), LA(82), SC(82). **50**:WV(954). **52**:CT(3770), FL(35), NC(1303), WA(3202).
Seimatosporium lichenicola. Leaf spot. **15**:TN(2820).
S. mariae. Leaf spot. 'Cryptostictis leaf spot.' **15**:KY(82), NY(82), TN(82), VA(82). **29**:CA(82), OR(82,936,3202), WA(82,936,3202). **30**:KY(82), NY(82), TN(82), VA(82). **52**:OR(3202), WA(936,3202).
S. rhododendri. Leaf spot. 'Coryneum leaf spot.' **14**:TN(82). **15**:NC(82), PA(82), TN(82,3547), VA(82). **29**:OR(82), WA(82). **30**:NC(82), NY(82), PA(82), TN(82), VA(82). **31**:NC(1303). **52**:OR(4007), TN(3547), WA(936).
Septoria albopunctata. Purple bordered leaf spot. **31**:NC(1303). **52**:CA(936).
S. azaleae. 'Leaf scorch.' **23**:Unk.(82). **19**:Unk.(82). **52**:FL(35), HI(2899,2903), NC(1303), WA(936,3202).
S. azaleae-indicae. On leaves. **38**:OR(936,3202).
S. rhododendri. 'Septoria leaf spot.' **15**:ME(82). **30**:NC(82,1303). **52**:WA(3202).
S. solitaris. Leaf spot. **14**:NC(82). **38**:CA(81,82,638), OR(82,3202). **52**:CA(936), TX(82).
Sphaceloma sp. Spot anthracnose. **29**:WA(82,3202).
Sphaeronaema sp. On cuttings. **52**:WA(936,3202).
Sphaeropsis sp. Stem canker, dieback, leaf spot. **15**:NJ(82), PA(82). **23**:NJ(82). **30**:NJ(82), PA(82). **19**:NJ(82). **52**:FL(35).
Truncatella angustata. On leaves and twigs. **52**:WA(936,3202).

Deuteromycotina - Other

Rhizoctonia ramicola. Thread blight. **52**:FL(35).
R. solani. Leaf blight, root and basal stem rot. 'Rhizoctonia damping-off. Web blight.' **14**:NJ(82), TN(82). **15**:CT(82), NJ(82), NY(82). **22**:Range of host(82). **23**:Range of host(82). **25**:Range of host(82). **30**:CT(82), NJ(82), NY(82). **34**:Range of host(82). **37**:Range of host(82). **19**:Range of host(82). **52**:CT(82), FL(35),

NC(1303), NJ(82), NY(82).
Rhizoctonia sp. Damping-off, root rot. **52:**HI(2903), NC(1303).
Sclerotium durum. **52:**WA(936).

Thibaudia Ruiz & Pav. ex J. St. Hil.

1. *T. punctatifolia* Ruiz & Pav. (=*Vaccinium macrophyllum* Dombey ex Dunal)

Basidiomycotina - Uredinales
Pucciniastrum goeppertianum. **1:**OR(1575), WA(1575).
P. vaccinii. **1:**ID(1575), MT(1575), OR(1575).

Vaccinium L.

1. *V. alaskaense* Howell
2. *V. angustifolium* Aiton [late sweet blueberry, low sweet blueberry, lowbush blueberry, sweet-hurts]
3. *V. angustifolium* Aiton var. *laevifolium* House (=*V. pensylvanicum* Lam.)
4. *V. arboreum* Marsh.
5. *V. ashei* Reade [rabbit's-eye blueberry]
6. *V. brittonii* Porter ex C. Bicknell
7. *V. caespitosum* Michx. [dwarf bilberry]
8. *V. calycinum* Sm.
9. *V. candicans* Michx. (=*Polycodium candicans* (Michx.) Small)
10. *V. chandleri* Jeps.
11. *V. corymbosum* L. (=*V. australe* Small, *V. amoenum* Aiton) [highbush blueberry, swamp blueberry, whortleberry]
12. *V. crassifolium* Andr. [creeping blueberry]
13. *V. darrowii* Camp
14. *V. deliciosum* Piper
15. *V. dentatum* Sm.
16. *V. elliottii* Chapm. [Elliott's blueberry]
17. *V. erythrocarpum* Michx.
18. *V. globulare* Rydb.
19. *V.* × *intermedium* Ruthe
20. *V. lamarckii* Camp
21. *V. macrocarpon* Aiton [American cranberry, cultivated cranberry, large cranberry]
22. *V. membranaceum* Douglas [blue huckleberry, mountain blueberry, thin-leaf huckleberry]
23. *V. myrsinites* Lam.
24. *V. myrtilloides* Michx. (=*V. canadense* Kalm ex Richardson) [sourtop blueberry, velvetleaf blueberry]
25. *V. myrtillus* L. [bilberry, whinberry, whortleberry]
26. *V. nivictum* Camp
27. *V. occidentale* A. Gray [western blueberry]
28. *V. oreophilum* Rydb.
29. *V. ovalifolium* Sm. [mathers]
30. *V. ovatum* Pursh [California huckleberry, evergreen huckleberry, shot huckleberry]
31. *V. ovatum* Pursh var. *saporosum* Jeps.
32. *V. oxycoccos* L. [small cranberry]
33. *V. parvifolium* Sm. [red huckleberry]
34. *V. peleanum* Skottsb.
35. *V. reticulatum* Sm.
36. *V. scoparium* Leiberg ex Crép. [grouseberry, little-leaf huckleberry]
37. *V. simulatum* Small
38. *V. stamineum* L. (=*Polycodium stamineum* (L.) Greene, *Polycodium neglectum* Small) [deerberry, squaw huckleberry]
39. *V. stamineum* L. var. *melanocarpum* D. Mohr (=*Polycodium melanocarpum* (Mohr) Small)
40. *V. tenellum* Aiton
41. *V. uliginosum* L. [bog bilberry, moorberry]
42. *V. vacillans* Torr. (=*V. pallidum* Aiton) [early sweet bilberry, low bilberry, low blueberry, sweet blueberry]
43. *V. virgatum* Aiton
44. *V. vitis-idaea* L. [cowberry, cranberry, foxberry]
45. *V. vitis-idaea* L. var. *minus* Lodd. [lingberry, lingen, lingenberry, mountain cranberry, rock cranberry]
46. *Vaccinium* × sp. (unnamed hybrids)
47. *Vaccinium* sp.

Chytridiomycetes
Synchytrium vaccinii. Red leaf gall. **2:**ME(82). **11:**MS(82,2720). **47:**NJ(82).

Oomycetes
Phytophthora cinnamomi. Crown and root rot. **5:**FL(35), NC(1303). **11:**AR(3478), NC(586,2384), NJ(3061). **30:**CA(1110). **47:**WA(936,3202).
Pythium spinosum. Crown and root rot. **5:**FL(35).
Pythium sp. Root rot. **5:**FL(35).

Zygomycotina
Rhizopus stolonifer. Fruit rot. **11:**NJ(537).

Ascomycotina
Acrospermum foliicola. **9:**GA(1438).
Asteridiella cheirodendronis. **35:**HI(2903).
A. exilis. Black mildew. **11:**Unk.(1452). **35:**HI(1272,1452). **47:**HI(2903).
Asterina clavuligera. On leaves. **47:**FL(634).
A. delitescens. **47:**HI(2903).
Bertia moriformis. **47:**WA(3202,3815).
Bifusella vaccinii. On leaves. **2:**NH(82), NJ(82).
Botryosphaeria corticis. Cane and stem canker. **5:**FL(35), GA(82), MS(82,2720), NC(82,1303). **11:**GA(82), MS(82,2720), NC(82,1303). **40:**MS(2720). **43:**MS(2720).
B. dothidea. Stem blight. **5:**FL(35). **11:**NC(1303).
B. obtusa. Secondary berry rot. **4:**AL(82). **11:**NC(82), NJ(82). **47:**MA(82), NJ(82).
B. ribis. On twigs and leaves. **11:**NJ(82).
B. vaccinii. Leaf spot. **5:**FL(35). **16:**FL(35). **21:**E states(178), Range of host(82), OR(936,3202), WA(936,3202). **32:**Unk.(82).
Botryosphaeria sp. Cane blight. **11:**IL(2922), MI(2922), NJ(2922), OH(2922).
Byssosphaeria salebrosa. **47:**NY(195).
Calonectria crotalariae. Root and stem rot, leaf spot. **5:**NC(1303). **11:**NC(1303,2381).
Cistella coarctata. On limbs. **22:**WA(936,3202). **24:**WA(849).
Coccomyces coronatus. **30:**WA(936,3202).
Cryptosphaeria vexata. On dead limbs. **11:**NJ(82).
Dasyscyphus sp. **22:**WA(3202).
Diapleella coniothyrium. Stem canker, fruit rot. **5:**FL(35). **11:**NJ(537), OR(82). **21:**WA(936,3202). **47:**OR(82), WA(936,3202).
Diaporthe vaccinii. Twig blight, canker, dieback, storage rot. **5:**NC(1303). **11:**MA(82), ME(82), MI(3828,3829), NC(1303), NJ(82), OR(3202). **21:**Range of host(82), OR(3202), WA(936,3202), WI(1121). **32:**Range of host(82), WA(3202). **47:**WA(936).
Diatrype concolor. **11:**NJ(2925).
Discohainesia oenotherae. Storage rot, leaf spot. **11:**MA(4022), NJ(82). **21:**OR(3202), WA(3202). **47:**MA(82), NJ(82), OR(82), WA(82), WV(82).
Discostroma corticola. On twigs. **47:**WA(936,3202).
Elsinoe vaccinii. Leaf anthracnose. **35:**HI(1151).
Elsinoe sp. Anthracnose. **35:**HI(2903).
Eupropolella oxycocci. Witches' broom. **21:**ME(82), MI(82), NH(82), NY(82), WA(936,3202). **32:**ME(82), MI(82), NH(82), NY(82).
E. vaccinii. On leaves. **47:**NY(82).
Gibbera myrtilli. On leaves. **21:**MD(128), WI(128).
G. vaccinii. On stems. **7:**WA(3202). **32:**WA(3202). **44:**AK(82,526), WA(3202).
Glomerella cingulata. Bitter rot of berries, fruit and leaf spot. **11:**NC(1303), NJ(82), NY(82). **21:**Range of host(82), OR(3202), WA(936,3202). **32:**Range of host(82).
Gloniella vaccinicola. On dead twigs. **11:**NY(82).
Gloniopsis praelonga. **47:**FL(2782).
Gloniopsis sp. On dead twigs. **11:**VA(82).
Gnomonia setacea. On dead stems. **21:**OR(82,3202), WA(82,936,3202). **32:**OR(82), WA(82).
Godronia cassandrae. Canker, dieback. **11:**MI(3829,3830), WA(3202). **21:**Range of host(82), MA(128), NJ(128), OR(128), WA(128), WA(3202). **30:**WA(3202). **32:**NW states(3202). **47:**MA(128), OR(3202).
G. cassandrae f. *vaccinii*. Stem canker, end rot, on buds, flowers, fruits, leaves and stems. **7:**AK(82), ME(82), NH(128). **11:**MI(3828), NC(1303), OR(1382). **21:**WA(936). **25:**OR(82), WA(82). **30:**WA(936). **32:**Range of host(82). **44:**AK(82), ME(82). **47:**AK(526), WA(936).
G. kalmiae. On dead limbs. **11:**Unk.(82).
G. urceolata. On dead twigs. **47:**OR(82).
Guignardia sp. Stem lesions. **5:**FL(35).
Hysterium angustatum. **4:**GA(1438), **16:**GA(1438).
H. macrosporum. **47:**FL(2782).

Jattaea microtheca. On limbs. **47:**NJ(190).
Lachnum virginellum. On leaves. **4:**GA(2279). **9:**GA(1438). **11:**NY(2279). **16:**GA(1438). **47:**NJ(849).
L. virgineum. On dead stems. **21:**WI(1120).
Lophodermium maculare. Leaf spot. **4:**FL(35). **11:**NJ(82). **30:**OR(82,3202). **33:**WA(936,3202). **41:**ME(82). **42:**GA(82,1438).
L. melaleucum. On leaves and stems. **22:**MT(3202), WA(936,3202). **30:**OR(82,3202), WA(936,3202).
L. oxycocci. On leaves. **21:**AK(82), MA(82), ME(82), MI(82), MN(82), NJ(82), NY(82), OR(82,3202), WA(82,936,3202), WI(82). **32:**AK(526), MI(82,2862), MN(82), NY(82), OR(82), WA(82,936,3202), WI(82). **44:**AK(82,526,936,3431).
L. schweinitzii. On leaves. **30:**WA(936,3202).
Melanospora sp. **11:**MI(3828).
Meliola atkinsonii. **11:**AL(1452). **47:**Unk.(1452).
M. heteroseta. Black mildew. **35:**HI(1272,2903).
[*M. nidulans*]. Black mildew. **12:**GA(1438). **42:**AL(82), GA(82). **43:**GA(1438). **47:**AL(82), GA(82).
M. niessleana. Black mildew. **11:**FL(1452). **43:**Unk.(1452).
M. vaccinii. Black mildew. **15:**HI(1272,2903). **35:**HI(1272,1452,2903).
Microsphaera penicillata. Powdery mildew. **4:**NC(1303). **11:**NC(1303). **28:**WY(1195). **33:**WA(936,3202). **47:**NC(1303).
M. vaccinii. Powdery mildew. **1:**AK(526). **2:**Range of host(82). **4:**IL(312). **5:**Range of host(82), FL(35). **11:**Range of host(82). **25:**WY(82). **33:**AK(526), WA(936). **38:**NC(1307). **40:**NC(3944). **42:**Range of host(82), GA(1438). **47:**Range of host(82), AK(82), AL(82), MD(651), NJ(82), OH(82), OR(82), WA(82), WY(82).
Mollisia melaleuca. **4:**GA(1438).
Monilinia ledi. Twig blight. **41:**NY(82).
M. megalospora. **41:**NH(207).
M. oxycocci. Hard rot. **21:**MA(82), ME(82), OR(82,3202), WA(82,936,3202), WI(82). **32:**MA(82), ME(82), OR(82,3202), WA(82,3202), WI(82). **47:**AK(82).
M. polycodii. Twig blight. **38:**MD(207), NY(82).
M. urnula. **11:**WA(3202). **22:**ID(3202), WA(3202). **47:**WA(936).
M. vaccinii-corymbosi. Leaf and shoot blight; mummified fruits. **2:**ME(82), NJ(207). **5:**FL(35), MS(82). **6:**NJ(207). **16:**NJ(207). **20:**MD(207). **22:**ID(3202). **23:**NJ(207). **24:**NJ(207). **27:**NJ(207). **40:**MS(2720). **43:**MS(2720). **46:** NJ(207).
Montagnula obtusa. On dead stems. **11:**MA(4022).
Mycosphaerella nigromaculans. Black spot. **21:**OR(3202), WA(3202). **32:**OR(82,3202), WA(82,3202). **47:**OR(82), WA(82).
M. perseae. Stem lesion. **5:**FL(35).
M. punctiformis. **4:**GA(1438). **11:**GA(1438). **42:**GA(1438).
M. vaccinii. On leaves. **4:**GA(82). **5:**FL(35). **11:**NY(82). **13:**FL(35). **35:**HI(2784). **41:**NY(82). **47:**MA(82), NJ(82).
Myriangium superficialis. On limbs infested with scale insects. **11:**GA(82).
Nectria cinnabarina. On stems. **47:**WA(936,3202).
Niesslia exilis. On leaves. **30:**OR(2012,3202), WA(82,936,3202).
Ophiodothella vaccinii. Flyspeck leaf spot. **4:**SE states(82), FL(35), GA(1438), IL(82), NC(1303), TX(82). **5:**MS(82,2720). **9:**GA(1438). **11:**GA(659). **39:**GA(1438). **40:**MS(2720). **43:**MS(2720).
Ostreichnion curtisii. **16:**GA(1438).
Pestalopezia brunneopruinosa. On leaves. **30:**WA(936,3202).
Pezicula myrtillina. On leaves. **30:**CA(82,1110).
Phacidium vaccinii. On leaves. **21:**WA(3202). **30:**WA(936,3202).
Phyllactinia guttata. Powdery mildew. **38:**MS(82,2720), WV(82).
Physalospora vaccinii. On leaves. Blotch rot of berries. Blotch rot of berries, leaf spot. **21:**E states(82), NC(1303), OR(82,3202), WA(82,936,3202), WI(82).
Physalospora sp. Dieback. **5:**FL(35).
Pleospora herbarum. On stems. **47:**WA(936,3202).
Podosphaera clandestina. Powdery mildew. **41:**AK(82,526). **47:**AK(82).
Pseudophacidium ledi. On stems. **21:**OR(3202), WA(936,3202). **47:**OR(82), WA(82).
Pyrenobotrys compacta. Leaf smudge. **21:**Range of host(82), OR(936), WA(936). **32:**WA(936). **36:**WA(936).
P. conferta. **21:**OR(3202), WA(3202). **32:**WA(3202). **36:**WA(3202).
Rhizodiscina lignyota. On exfoliating bark. **11:**MA(4022).
Rhytisma vaccinii. Tar spot. **4:**FL(35,82), LA(2675), MS(2675,2720), OK(82,2873), TX(82). **11:**GA(82), MD(82), VA(82), WV(82). **30:**WA(936,3202). **33:**AK(526). **38:**E states(82), FL(82), NC(3944). **40:**MS(2720). **42:**GA(82), MA(1390), MD(82), VA(82), WV(82). **43:**MS(2720). **47:**AK(82), GA(1438), MS(2720), NC(3944), OK(2872).
Rhytisma sp. **29:**WA(3202).
Saccharomyces sp. Fruit rot. **11:**NJ(537).
Schizothyrium pomi. On fruits. **47:**NJ(82).
Sordaria destruens. On roots and stems. **11:**MA(82), NC(82), NJ(82). **21:**WA(3202). **47:**Unk.(82).
Sporomega degenerans. On dead twigs. **11:**NJ(82).
Talaromyces trachyspermus. Fruit rot. **21:**NJ(82), WA(936,3202).
Tapesia fusca. **22:**WA(936,3202).
Triblidium alpinum. **30:**OR(3228).
Trichothyrium reptans. **35:**HI(2903).
Valsa ceratosperma. On dead twigs. **2:**MA(82), NJ(82). **4:**GA(1438). **11:**MA(82), NJ(82). **21:**WA(936,3202). **30:**CA(936). **47:**MA(82), NJ(82,3415), OR(82), WA(82,936).
Valsaria peckii. On dead twigs. **11:**NY(82).
Xylopezia hemisphaerica. On wood. **47:**OR(3233).

Basidiomycotina - Uredinales

Pucciniastrum goeppertianum. Heteroecious: telial host. **2:**NE states(82), MN(82). **3:**MA(1575), ME(1575), MI(1575), NH(1575), VT(1575), WI(1575). **7:**Range of host(82), CA(81,638,642,936), ID(936,3202), MT(3202), OR(936,3202), UT(1575), WA(936,3202), WY(1195,3379). **10:**CA(1575). **11:**MA(1781), OR(82). **14:**WA(936,3202). **22:**Range of host(82), CA(936), ID(936,1195,3202), MT(3202), OR(936,3202), WA(936,3202), WY(1195,2924). **24:**MA(1575), ME(1575), WI(1575). **25:**Range of host(82), CO(1575), ID(936,3202), MT(3202), OR(3202), WA(3202). **26:**CA(661). **28:**CO(1575), MT(1575), WA(1575). **29:**Range of host(82), AK(60,526), CA(1575), OR(1575,3202), WA(1575,3202). **30:**CA(81,82,936,1110), OR(82,638,936,3202), WA(82,936,3202). **31:**CA(638). **33:**AK(526,936), CA(1575), ID(3202), MT(3202), OR(936,3202), WA(936,3202). **36:**CO(649,1575), ID(936,3202), MT(3202), NM(1575), OR(936,3202), UT(1195), WA(936,3202), WY(1195). **41:**Range of host(82), AK(526), ME(1575). **44:**NC states(82), NE states(82), AK(60,82,526). **47:**Range of host(82), CA(936), CO(1195), ID(936), MI(2862), WA(936).
P. vaccinii. Heteroecious: telial host. **1:**AK(60,526). **2:**E states(82), AR(82), CT(1575), ME(1575), MI(1575), NH(1575), PA(1575), WA(3202), WI(1575). **3:**MA(1390), NY(836). **5:**E states(82), AR(82), FL(35), WA(936,3202). **11:**E states(82), AL(1575), AR(82), PA(1575), WA(3202). **14:**WA(3202). **16:**GA(1439). **17:**WV(1575). **18:**ID(1575,3202), MT(1575,3202). **21:**OR(1575,3202), UT(1575), WA(936,1575,3202). **22:**Range of host(82), CA(81,638,936,1575), ID(936,1575,3202), MT(1195,1575,3202), OR(936,1575,3202), WA(936,1575,3202), WY(1195,1575). **23:**FL(82). **24:**WI(1575). **25:**Range of host(82), AK(526), ID(3202), MT(3202). **28:**CO(1575), MT(1575), NM(1575), WY(1195). **29:**Range of host(82), AK(60,526), OR(3202), WA(936,3202). **33:**CA(1575), WA(936,1575,3202). **35:**HI(2903). **36:**CO(649), ID(3202), WY(1195). **38:**PA(82). **41:**Range of host(82), AK(60,526), NH(1575). **42:**E states(82), AR(82,1575), CT(1575), DE(1575), GA(1439), NC(1575), PA(1575). **43:**VA(1575). **44:**NH(82). **45:**VT(1224). **47:**E states(82), Range of host(82), AK(526), AR(82), ID(936), OR(82), UT(82), WA(82,936).
Pucciniastrum sp. **7:**OR(82), WA(82). **22:**OR(82), WA(82).

Basidiomycotina - Other

Armillaria mellea. Root rot. **11:**MI(1129). **47:**WA(936,3202).
Corticium efibulatum. On bark. **47:**MS(2051).
Exobasidium cordilleranum var. *minor*. On leaves. **22:**ID(3096), MT(3096).
E. dimorphosporum. **7:**CO(649).
E. perenne. Red shoot disease. **21:**WI(2605). **32:**WI(2605).
E. vaccinii. Galls of buds, leaves and shoots, leaf spot, witches' broom, red leaf disease. **2:**E states(82), NW states(82), MA(656), MS(82), TX(82), WI(82). **3:**MA(457), MN(457). **4:**AL(82,457), FL(35,82). **5:**FL(35). **7:**AK(526), CA(642). **8:**HI(1148). **11:**E states(82), NW states(82), GA(1439), MA(1390), MS(82), TX(82), WI(82). **14:**WA(457,936,3202). **21:**Range of host(82), MA(457,3096), OR(3202), WA(936,3202). **22:**AK(936), CA(81), ID(457,936,3202), MT(3202), OR(3202), WA(457,936,3202),

WY(457). **25:**NW states(3202). **29:**NW states(82), AK(82,526,936,3431), ID(936,3202), OR(82,3202), WA(82,936,1642,3202). **30:**CA(82,1110,3096), OR(82), WA(936,3202). **32:**AK(526), OR(3202), WA(936,3202), WI(1340). **33:**CA(638,936), OR(657,1642,3202), WA(936,1642,3202). **34:**HI(1148). **35:**HI(1148). **36:**ID(936,3202), MT(3202), OR(936,3202), WA(936,3202), WY(1195). **38:**AL(457), GA(1439), MD(82), NY(82,457). **40:**MS(2720), NC(3944). **41:**NW states(82), AK(82,526), WA(936,3202). **42:**E states(82), NW states(82), DC(457), GA(1439), MA(457), MD(457), MO(457), MS(82), TX(82), WI(82). **44:**AK(82,526,936), ME(82), NY(82). **19:**WA(457). **47:**E states(82), NW states(82), AK(82), MA(656), ME(2314), MI(2862), MS(82), TX(82), WA(936), WI(82).

E. vaccinii-uliginosi. Shoot gall. **22:**OR(82), WA(457,936,3202), WY(1195). **30:**NW states(3202), OR(82). **41:**AK(1938). **45:**AK(1938).

Hymenochaete badio-ferruginea. **33:**OR(3202).

H. tabacina. **33:**WA(638,936,3202).

Mycena cucullata. **11:**NJ(2934), NY(2934).

M. rugulosiceps. **47:**OR(3202), WA(3202).

Peniophora sphaerocystidiata. White rot. **47:**MS(444).

Phellinus ferreus. **30:**CA(81).

P. ferruginosus. On dead limbs. **30:**CA(82,1110).

Psilocybe agrariella var. *vaccinii*. Fairy ring. **47:**MA(82), NJ(82).

Stereum ochraceo-flavum. **47:**NY(543).

Thelephora terrestris. Encircling canes. **11:**NJ(82). **47:**OR(3202).

Deuteromycotina - Hyphomycetes

Alternaria alternata. Fruit rot, leaf spot. **11:**NC(1303), NJ(537).

A. tenuissima. Leaf spot. **11:**MA(4022), NC(2379).

Alternaria sp. Secondary leaf and twig blight, storage rot, leaf spot. **4:**FL(35). **5:**FL(35). **11:**MA(82), MI(3828), NC(82), NJ(82,534). **21:**IN(82), MA(82), NJ(82), OR(82,3202), WA(82), WI(82).

Aspergillus sp. Fruit rot. **11:**NJ(537).

Bispora sp. **11:**MI(3828).

Botrytis cinerea. Blossom, fruit and twig blight, gray mold rot, storage rot of fruits. **1:**AK(526). **2:**MA(82), ME(82), NJ(82). **5:**FL(35). **11:**CA(82,1110), MA(82), ME(82), MI(3828), NC(1303), NJ(82,534,537), OR(82), WA(82,936,3202). **21:**WA(936,3202). **47:**WA(82,936,3202).

Botrytis sp. **2:**WA(3202). **5:**WA(3202). **21:**OR(3202). **27:**WA(3202). **30:**WA(3202). **33:**WA(3202).

Cephalosporium sp. **11:**MI(3828).

Cercospora calotropidis. Leaf spot. **5:**FL(35).

C. sparsa. Tip dieback, leaf spot. **5:**FL(35).

Cercospora sp. Stem blotch. **5:**NC(1303,2382).

Cladosporium herbarum. Fruit rot, on leaves. **11:**NJ(537). **21:**WA(936,3202).

C. oxycocci. Secondary storage rot, on leaves. **21:**WA(936,3202). **47:**NJ(82), WA(82).

Cladosporium sp. On dead stems. **11:**MA(4022).

[*Conoplea juniperi*]. **11:**DE(3543).

Corynespora arctespora. **11:**DE(3543), NJ(507). **47:**NJ(507).

Cristulariella moricola. Leaf spot. **37:**NC(1302). **47:**MD(2848).

Curvularia inaequalis. Mold on leaves and twigs. **11:**NJ(82). **21:**IL(82), NJ(82,3423), WI(82).

Cylindrocarpon sp. **11:**MI(3828).

Cylindrocladium crotalariae. Root rot. **5:**FL(35).

C. ilicicola. Root rot. **5:**FL(35).

Dematium sp. **21:**OR(3202).

Epicoccum sp. On stems. **11:**MA(4022), MI(3828). **21:**WA(3202).

Fusarium sp. Fruit and root rot. **5:**FL(35). **11:**MI(3828), NJ(537). **21:**OR(3202).

Gloeocercospora inconspicua. Leaf spot. **5:**NC(82). **11:**MD(82), NC(82).

Hyphosoma lumbricoideum. **47:**WA(3202).

Monilia peckiana. On leaves. **2:**NY(82).

Monilia sp. **7:**WA(3202). **22:**WA(3202).

Olpitrichum macrosporum. On leaves. **21:**DC(3209). **47:**NJ(82).

Papulaspora sp. **11:**MI(3828).

Penicillium sp. Blue mold fruit rot. **11:**NJ(537). **21:**OR(3202), WA(936,3202). **47:**Unk.(82).

Phymatotrichopsis omnivora. Root rot. **4:**TX(82).

Pullularia sp. **11:**MI(3828).

Ramularia effusa. On leaves. **2:**MI(82), WI(82). **42:**MD(82).

R. multiplex. Leaf spot. **21:**MI(82), NY(82), WI(82,1333). **32:**MI(82), NY(82), WI(82).

R. vaccinii. Brown leaf spot. **2:**FL(82), MD(82), MI(82), NJ(82), NY(82). **5:**FL(35). **11:**FL(82), MD(82), MI(82), NJ(82), NY(82).

Stemphylium sp. **21:**WA(3202).

Trichothecium roseum. On twigs. **47:**WA(936,3202).

Verticillium sp. **11:**MI(3828).

Deuteromycotina - Coelomycetes

Apostrasseria lunata. Black rot of berries, leaf spot. **21:**MA(82,2546), MD(2546), ME(2546), NJ(82,2546), OR(82,3202), WA(82,2546,3202), WI(82,3161). **32:**WA(82,3202).

Camarosporium sp. On twigs. **47:**WA(936,3202).

Ceuthospora latitans. Leaf spot. **35:**HI(2784). **44:**AK(82,526).

C. minima. **30:**CA(82,1110).

Ceuthospora sp. On fruit and leaves. **21:**WA(936).

Chaetomella sp. On leaves and twigs. **47:**WA(936,3202).

Coleophoma empetri. Leaf spot, berry rot. **21:**ID(3202), MA(82), ME(82), NJ(82), OR(82,3202), WA(82,936,3202), WI(82). **30:**WA(936,3202). **32:**NW states(3202), AK(82,526). **47:**WA(936).

Colletotrichum gloeosporioides. Fruit rot. **11:**IL(3510), MI(3510), NC(3510), NJ(3510), NY(3510).

Coniothyrium vaccinicola. On twigs. **11:**MA(82), PA(82).

Coniothyrium sp. Twig dieback. **5:**FL(35). **11:**MI(3828). **47:**NC(1303).

Coryneum sp. **47:**WA(3202).

Cylindrosporium sp. Leaf spot. **4:**TX(82).

Dendrophoma sp. **11:**MI(3828).

Diplodia vaccinii. On twigs. **21:**OR(3202), WA(3202). **47:**MA(82), NJ(82).

Discosia artocreas. On leaves. **21:**Range of host(82), WA(936,3202). **32:**Range of host(82).

Dothichiza caroliniana. Leaf spot. **5:**FL(35). **11:**NC(82).

Fusicoccum putrefaciens. Canker. **11:**MA(4020), MI(169), NC(1303). **21:**Unk.(82).

Gloeosporium minus. Anthracnose, leaf and stem fleck, leaf spot. **11:**NC(1303,2380). **21:**WA(936,3202). **47:**MD(82), NJ(82), WA(936).

Gloeosporium sp. Anthracnose. **47:**FL(82), NJ(82).

Leptothyrium conspicuum. Leaf spot. **11:**GA(82). **29:**AK(82,526). **33:**AK(526). **42:**NY(82). **47:**AK(82).

Leptothyrium sp. Leaf spot. **5:**FL(35).

Macrophoma sp. Stem canker. **5:**FL(35). **11:**MA(4022).

Micropera vaccinii. On stems. **11:**IN(971).

Micropera sp. On twigs. **47:**NC(1303).

Pestalotia adusta. Leaf spot. **4:**FL(82). **5:**FL(35).

P. vaccinii. Fruit rot, leaf spot, on leaves and twigs. **2:**ME(82). **5:**FL(35). **11:**MA(82), NC(82,1303), NJ(82,537). **21:**OR(3202), WA(936,3202). **38:**MS(82,2720). **47:**Range of host(82), WA(936).

Pestalotiopsis maculans. On dead leaves. **21:**NJ(3209).

P. maculiformans. On leaves. **30:**CA(936), OR(82,3202), WA(82,936,3202).

Phoma houseana. On dead twigs. **11:**NJ(82), NY(82,3081).

P. vaccinii. Fruit rot. **11:**NJ(537), NY(82).

Phoma sp. **11:**MI(3828).

Phomopsis vaccinii. Twig blight, stem canker, leaf spot. **5:**FL(35). **11:**IL(2922), IN(2698), MI(2698,2922), NC(2383), NJ(2922), OH(2922).

Phomopsis sp. Stem gall. **11:**MA(82), MI(82), NJ(82), OR(82), WA(82). **47:**WA(3202).

Phyllosticta cyanococci. **11:**NC(82), NY(82).

P. elongata. On leaves. **21:**MA(3820), WI(3820).

P. putrefaciens. Berry rot, leaf spot. **21:**NJ(797), WA(936,3202). **47:**MA(82), NJ(82).

P. vaccinii. Canker, leaf spot. **4:**AL(2,82,3820), FL(35,82), MS(82,2720). **5:**GA(82), MD(82), MS(82,2720), NC(82). **11:**GA(82), MD(82), NC(82). **30:**CA(82,1110), WA(936,3202).

Phyllosticta sp. Leaf spot. **5:**FL(35,82). **11:**AL(82).

Piggotia vaccinii. Leaf spot. **2:**WI(82). **5:**WI(82).

Pyrenochaeta sp. **11:**MI(3828).

Rhabdospora oxycocci. On dead leaves and stems. **11:**MA(4022). **47:**NJ(82).

Seimatosporium lichenicola. Canker. **11:**MA(4021), MI(3828).

[*S. rhododendri*]. On stems. **47:**WA(936).
Septogloeum sp. Leaf spot. **47:**AK(82).
Septoria albopunctata. Canker, eye spot, purple-bordered leaf spot. **4:**FL(35,82), SC(82), TX(82). **5:**FL(82), GA(82), NC(82,1303). **11:**FL(35,82), GA(82), NC(82,1303).
S. difformis. Leaf spot. **2:**ME(82), WI(82).
S. sheareana. On fruits and leaves. **47:**NJ(82).
S. vaccinii. On stems. **11:**MA(4022).
Sphaceloma sp. Spot anthracnose on stems. **5:**FL(35). **47:**FL(2525).
Sphaeronaema pomorum. Berry rot. **47:**NJ(82).
Sphaeronaema sp. **11:**MI(3828).
Sphaeropsis vaccinii. **30:**WA(666,936,3202).
Strasseria geniculata. On leaves. **21:**DC(2546).
S. oxycocci. Fruit rot, on leaves and dead twig tips. **11:**MA(4022). **21:**MA(2710), NJ(3160), WA(936,3202). **47:**MA(82), NJ(82), WA(936).
Deuteromycotina - Other
Rhizoctonia solani. Root rot. **5:**FL(35).
Rhizoctonia sp. Leaf and stem blight, ectotrophic mycorhiza. **2:**IN(82). **47:**NC(1303).

Zenobia D. Don
1. *Z. pulverulenta* (W. Bartram) Pollard
Ascomycotina
Ophiodothella vaccinii. **1:**GA(1438).

Eriocaulaceae

Eriocaulon L.
1. *E. septangulare* With.
2. *Eriocaulon* sp.
Chytridiomycetes
Cladochytrium replicatum. In cells of old leaves. **2:**NY(82).
Entophlyctis texana. In cells of old leaves. **2:**TX(82).
Basidiomycotina - Ustilaginales
Moesziomyces eriocauli. Seed smut in ovaries. **1:**CT(1077), MA(1077), MA(588), NH(1077). **2:**CT(82), MA(82), NH(82).
Ustilago eriocauli. In ovaries. **1:**CT(1077), MA(588,1077), NH(1077).

Euphorbiaceae

Acalypha L.
1. *A. gracilens* A. Gray
2. *A. lindheimeri* Müll. Arg.
3. *A. neomexicana* Müll. Arg.
4. *A. ostryifolia* Riddell
5. *A. virginica* L. [three-seeded-mercury, Virginia copperleaf]
6. *A. wilkesiana* Müll. Arg. [beefsteak-plant, fire-dragon, Jacob's-coat, match-me-if-you-can, painted copperleaf]
7. *Acalypha* sp.
Chytridiomycetes
Synchytrium aureum. Leaf gall. **5:**WI(82).
Oomycetes
Plasmopara acalyphae. Downy mildew. **5:**WI(82).
Pythium sp. Root rot. **6:**FL(35).
Ascomycotina
Botryosphaeria ribis. On stems. **6:**HI(82,2903).
Diaporthe sp. On limbs. **7:**HI(2903).
Erysiphe cichoracearum. Powdery mildew. **5:**WI(82,1335). **6:**FL(35). **7:**FL(35).
E. sparsa. Powdery mildew. **5:**WI(369).
Physalospora abdita. Black canker. **6:**HI(82,2903). **7:**HI(2903).
Basidiomycotina - Uredinales
Puccinia dietelii. Heteroecious: aecial host. **3:**AZ(741,1203,3981). **4:**AZ(741,1203,3981).
Basidiomycotina - Other
Armillaria tabescens. Root rot. **6:**FL(35,82). **7:**FL(35).
Deuteromycotina - Hyphomycetes
Alternaria sp. Leaf spot. **5:**WI(1341).
Cercospora acalyphae. Leaf spot. **1:**OK(2872). **4:**OK(2872), TX(3597). **5:**E states(82), DE(3543), FL(35), OK(2873), TX(82), WI(82). **7:**AL(82), FL(35), NY(82), OK(82), TX(82).
Fusarium sp. Root rot. **6:**FL(35).
Helminthosporium sp. Leaf spot. **6:**FL(35).
Phymatotrichopsis omnivora. Root rot. **7:**TX(82).
Ramularia acalyphae. Leaf spot. **2:**TX(82,3597).
Volutella acalyphae. On stems. **7:**AL(82).
Deuteromycotina - Coelomycetes
Phyllosticta sp. Leaf spot. **6:**FL(35), NJ(82).
Deuteromycotina - Other
Rhizoctonia solani. Root rot. **6:**FL(35). **7:**FL(35).

Aleurites J. R. Forster & G. Forster
1. *A. fordii* Hemsl. [China wood oil-tree, tung oil-tree]
2. *A. moluccana* (L.) Willd. (=*A. javanica* Gand., *A. triloba* J. R. Forster & G. Forster) [candleberry-tree, candlenut-tree, country-walnut, Indian walnut, otaheite-walnut, varnish-tree]
3. *A. montana* (Lour.) E. Wils. [mu, mu-oil-tree, mu-tree, tung]
4. *Aleurites* sp.
Oomycetes
Phytophthora cinnamomi. Root rot. **1:**S states(3624), LA(82).
Ascomycotina
Ascocoryne sarcoides. On wood. **4:**HI(2903).
Bisporella sulfurina. **4:**HI(2903).
Botryosphaeria dothidea. Branch canker, dieback. **1:**FL(35).
B. quercuum. Branch canker, dieback. **1:**FL(35).
B. rhodina. Limb canker. **4:**FL(82), LA(82), MS(82), TX(82).
B. ribis. Limb and trunk canker, dieback, nut rot. **1:**GA(82), LA(82). **2:**HI(2903).
Dactylospora stygia. **4:**HI(2903).
Diaporthe sp. On cut or dying limbs. **2:**HI(2903).
Glomerella cingulata. Anthracnose. **1:**FL(35,82). **4:**FL(35,82).
Mollisia petiolorum. **4:**HI(2903).
Mycosphaerella aleuritis. Leaf spot. **1:**Range of host(82), FL(35), MS(137).
Nectria cinnabarina. On twigs. **4:**MA(82).
N. subquaternata f. *farinosa*. **2:**HI(2903).
Peziza sp. **2:**HI(2903).
Scutellinia scutellata. **2:**HI(2903).
Ustulina deusta. Wood rot. **2:**HI(82,2903).
Xylaria curta. On dead wood. **2:**HI(82,2903). **4:**HI(2903).
X. schweinitzii. On dead wood. **2:**HI(82,2903). **4:**HI(2903).
Basidiomycotina - Other
Armillaria tabescens. Root rot. **1:**FL(35,82), LA(82). **2:**FL(82), LA(82). **4:**FL(82), LA(82).
Auricularia cornea. **2:**HI(3948).
Corticium stevensii. Thread blight. **1:**FL(35,2027), LA(2027), WA(2027). **3:**FL(2027), LA(2027), WA(2027). **4:**FL(35).
Crepidotus alabamensis. **2:**HI(3706).
Earliella scabrosa. Wood rot. **2:**HI(82,2903). **4:**HI(2903).
Erythricium salmonicolor. Limb blight. **1:**FL(35). **4:**FL(35).
Fomes hawaiiensis. On wood. **2:**HI(82,2903).
Odontia sp. **2:**HI(2903).
Pellicularia koleroga. Thread blight. **1:**FL(82), LA(82,3640), MS(82,2720), NC(82). **4:**FL(82).
Phellinus gilvus. Wood rot. **2:**HI(82,2903).
Scytinostroma portentosum. **1:**FL(128).
Septobasidium pseudopedicellatum. Brown felt. **1:**LA(82).
Thanatephorus cucumeris. Web blight. **1:**LA(82), MS(82,2720).
Typhula hyalina. On leaves. **2:**HI(2903).
Deuteromycotina - Hyphomycetes
Alternaria sp. Leaf spot. **4:**FL(35).
Bactridium flavum. **4:**HI(2784).
Cephalosporium sp. Collar girdle. **1:**LA(82). **2:**LA(82). **4:**LA(82).
Cercospora websteri. Ring spot. **1:**MS(4027).
Cristulariella moricola. Leaf spot. **1:**FL(35).
Helminthosporium sp. Leaf spot. **1:**FL(35).
Oidium sp. Powdery mildew. **2:**HI(2903).
Phymatotrichopsis omnivora. Root rot. **1:**TX(82).
Pseudocercospora aleuritis. Leaf spot. **1:**FL(35).
Deuteromycotina - Coelomycetes
Colletotrichum gloeosporioides. Anthracnose, leaf spot, on nuts. **1:**MS(82,2720). **2:**HI(82,2903).
Colletotrichum sp. Leaf spot. **1:**FL(35).
Diplodia sp. **1:**Unk.(82).
Lasiodiplodia theobromae. Dieback. **1:**FL(35).
Phyllosticta sp. Leaf spot. **1:**FL(35,82), GA(82). **2:**FL(82), GA(82). **4:**FL(82), GA(82).

Deuteromycotina - Other
Rhizoctonia solani. Thread blight, root rot. **1**:FL(35). **4**:FL(35).
Sclerotium rolfsii. Seedling blight. **1**:FL(35), TX(82).

Antidesma L.
1. *A. bunius* (L.) Spreng. [bignay, Chinese-laurel, salamander tree]
2. *A. platyphyllum* H. Mann
3. *A. pulvinatum* Hildebr.

Oomycetes
Phytophthora cinnamomi. Root rot. **3**:HI(2903).
Deuteromycotina - Coelomycetes
Diplodia sp. Dieback. **1**:FL(35).
Pestalotia sp. Leaf spot. **2**:HI(2903).
Phyllosticta sp. Leaf spot. **1**:FL(35).

Argythamnia P. Br.
1. *A. mercurialina* (Nutt.) Müll. Arg.

Deuteromycotina - Hyphomycetes
Cercospora argythamniae. Leaf spot. **1**:OK(2873).

Baccaurea Lour.
1. *B. racemosa* (Reinw.) Müll. Arg. [kapundung, menteng]

Deuteromycotina - Coelomycetes
Phomopsis sp. Stem dieback. **1**:FL(35).

Bischofia Blume
1. *B. trifoliata* (Roxb.) Hook. (=*B. javanica* Blume) [toog]

Oomycetes
Pythium sp. Root rot. **1**:FL(35).
Deuteromycotina - Hyphomycetes
Cercospora bischofiae. Leaf spot. **1**:FL(35).
Corynespora sp. Leaf spot. **1**:FL(35).
Fusarium sp. Root rot. **1**:FL(35).
Deuteromycotina - Coelomycetes
Phomopsis sp. Canker. **1**:FL(35).
Phyllosticta sp. Leaf spot. **1**:FL(35).

Breynia J. R. Forster & G. Forster
1. *B. nivosa* (Bull.) Small

Ascomycotina
Xylaria sp. **1**:HI(2903).
Deuteromycotina - Other
Rhizoctonia sp. Root rot. **1**:HI(2903).

Caperonia St. Hil.
1. *C. palustris* (L.) St. Hil.

Deuteromycotina - Hyphomycetes
Amphobotrys ricini. Stem canker. **1**:TX(3885).
Pseudocercospora caperoniae. Leaf spot. **1**:FL(35).

Chamaesyce Raf.
This genus is considered a synonym of *Euphorbia*, but the appropriate combination could not be located.
1. *C. tracyi* Small

Basidiomycotina - Uredinales
Uromyces euphorbiae. Autoecious. **1**:MS(2720).

Claoxylon Adr. Juss.
1. *C. sandwicense* Müll. Arg.

Ascomycotina
Asterina ildefonsiae. **1**:HI(2903).
Meliola morbosa. Black mildew. **1**:HI(1272,1452,2903).

Cnidoscolus Pohl
1. *C. stimulosus* (Michx.) Engelm. & A. Gray [bullnettle]
2. *C. texanus* (Müll. Arg.) Small [Texas bullnettle, tread-softly]

Deuteromycotina - Hyphomycetes
Cercospora jatrophae. Leaf spot. **1**:AL(82), FL(35).
C. jatrophicola. **2**:TX(82).
Phymatotrichopsis omnivora. Root rot. **2**:TX(82).
Deuteromycotina - Coelomycetes
Phoma sp. Stem lesions. **1**:FL(35).
Phomopsis sp. Stem canker. **1**:FL(35).
Septoria jatrophae. Leaf spot. **1**:AL(82), FL(35), TX(82). **2**:OK(82,2872), TX(82).

Codiaeum Adr. Juss.
1. *C. variegatum* (L.) Blume [garden croton]
2. *C. variegatum* (L.) Blume var. *moluccanum* (Decne.) Müll. Arg.
3. *C. variegatum* (L.) Blume var. *pictum* (Lodd.) Müll. Arg.

Oomycetes
Phytophthora sp. Root rot. **1**:FL(35).
Pythium aphanidermatum. Root rot. **1**:HI(2903).
P. splendens. Root rot. **1**:FL(35).
Ascomycotina
Diaporthe sp. Canker. **1**:FL(35).
Leptosphaeria sp. Stem canker. **1**:FL(35).
Nectriella pironii. 'Stem gall. Canker.' **1**:FL(35).
Basidiomycotina - Uredinales
Phakopsora crotonis. **1**:FL(35).
Deuteromycotina - Hyphomycetes
Alternaria sp. Leaf spot. **1**:FL(35).
Cercospora stevensonii. Leaf spot. **1**:FL(35).
Fusarium oxysporum. Root and stem rot. **1**:FL(35).
F. solani. Root and stem rot. **1**:FL(35).
Fusarium sp. Root and stem rot. **1**:FL(35).
Helminthosporium sp. Leaf spot. **1**:FL(35).
Kutilakesa pironii. Leaf and stem galls. **1**:FL(34).
Phymatotrichopsis omnivora. Root rot. **1**:TX(82).
Verticillium sp. **3**:HI(2903).
Deuteromycotina - Coelomycetes
Colletotrichum gloeosporioides. Anthracnose. **1**:FL(35).
Coniothyrium sp. Stem canker. **1**:FL(35).
Diplodia sp. Dieback. **1**:FL(35).
Dothiorella sp. Leaf spot. **1**:FL(35).
Macrophoma sp. Leaf spot. **1**:FL(35).
Phomopsis sp. Leaf spot. **1**:FL(35).
Phyllosticta ghaesembillae. Leaf spot. **1**:Unk.(82), HI(82). **2**:HI(2903).
Phyllosticta sp. Leaf spot. **1**:FL(35).
Sphaceloma venezuelanum. Scab. **1**:FL(35).
Deuteromycotina - Other
Rhizoctonia solani. Thread blight, root rot. **1**:FL(35).
Rhizoctonia sp. Root rot. **1**:HI(2903).
Sclerotium rolfsii. Southern blight, basal stem rot. **1**:FL(35).

Croton L.
1. *C. capitatus* Michx. [woolly croton]
2. *C. dioicus* Cav.
3. *C. fruticulosus* Engelm.
4. *C. glandulosus* L.
5. *C. glandulosus* L. var. *septentrionalis* Müll. Arg.
6. *C. monanthogynus* Michx. [prairie-tea]
7. *C. pottsii* (Klotzsch) Müll. Arg.
8. *C. punctatus* Jacq. (=*C. maritimus* Walter)
9. *C. texensis* (Klotzsch) Müll. Arg.
10. *Croton* sp.

Ascomycotina
Botryosphaeria dothidea. Twig dieback. **10**:FL(35).
B. ribis. **1**:AL(82), AZ(82), NE(82). **6**:AL(82), AZ(82), NE(82). **9**:AL(82), AZ(82), NE(82).
Basidiomycotina - Uredinales
Phakopsora crotonis. Telial host. **1**:LA(2675), MS(2720), OK(2872), TX(652). **2**:TX(742). **6**:AR(3594), MS(2720), OK(2872). **7**:TX(742). **9**:AZ(747,1203,3981), OK(2872). **10**:FL(35).
P. mexicana. Autoecious. **10**:TX(749).
Deuteromycotina - Hyphomycetes
Cercospora capitati. On living leaves. **1**:TX(82,3597).
C. crotonicola. Leaf spot. **3**:TX(82,581).
C. crotonifolia. Leaf spot. **4**:AL(581), SC(82,581). **10**:FL(35).
C. crotonophila. Leaf spot. **5**:WI(82,581,1330).
C. maritima. Leaf spot. **8**:MS(82,581,2720).
Phymatotrichopsis omnivora. Root rot. **6**:TX(82). **9**:TX(82).
Pseudocercospora crotonis. Leaf spot. **1**:AL(82), FL(82), SC(82), TX(82). **9**:KS(82). **10**:FL(35).

Crotonopsis Michx.
1. *C. linearis* Michx.

Basidiomycotina - Uredinales
Phakopsora crotonis. Autoecious. **1**:OK(2873).

Eremocarpus Benth.

1. *E. setigerus* (Hook.) Benth. [doveweed, turkey mullein]
2. *Eremocarpus* sp.

Chytridiomycetes

Synchytrium eremocarpae. On leaves, petioles, and stems. **2:**CA(1847).

Ascomycotina

Erysiphe immersa. Powdery mildew. **1:**CA(369).

Basidiomycotina - Uredinales

Uredo piscariae. **1:**OR(3202).

Deuteromycotina - Hyphomycetes

Macrosporium rosarium. **1:**OR(3202).

Euphorbia L.

(= *Chamaesyce* A. Gray, *Poinsettia* J. Graham.)

1. *E. albomarginata* Torr. & A. Gray
2. *E. antiquorum* L.
3. *E. brachycera* Engelm.
4. *E. capitellata* Engelm.
5. *E. celastroides* Boiss. in A. DC. var. *amplectans* Sherff
6. *E. chamaesula* Boiss.
7. *E. chamaesyce* L.
8. *E. clusiifolia* Hook. & Arnott
9. *E. cordata* Meyen
10. *E. corollata* L. [flowering spurge, tramp's spurge]
11. *E. crenulata* Engelm.
12. *E. cyparissias* L. [cypress spurge]
13. *E. dentata* Michx. [toothed spurge]
14. *E. dentata* Michx. var. *cuphosperma* Engelm.
15. *E. esula* L. [leafy spurge, wolf's-milk]
16. *E. fendleri* Torr. & A. Gray
17. *E. florida* Engelm.
18. *E. geyeri* Engelm.
19. *E. glyptosperma* Engelm.
20. *E. heterophylla* L. [annual poinsettia, Japanese poinsettia, Mexican fire-plant, moleplant, painted spurge, paintleaf]
21. *E. hirta* L. [garden spurge]
22. *E. hookeri* Steud.
23. *E. humistrata* Engelm.
24. *E. hypericifolia* L.
25. *E. hyssopifolia* L.
26. *E. incisa* Engelm.
27. *E. incisa* Engelm. var. *mollis* (Norton) L. C. Wheeler
28. *E. lagascae* Spreng.
29. *E. lathyris* L. (=*E. lathyrus* L.) [caper spurge, gopher spurge, moleplant, myrtle spurge]
30. *E. leptocera* Engelm. (=*Tithymalus leptocerus* (Engelm.) Arth.)
31. *E. lucida* Waldst. & Kit.
32. *E. lurida* Engelm.
33. *E. maculata* L. (=*E. preslii* Guss.) [nodding spurge, spotted spurge]
34. *E. marginata* Pursh [ghostweed, snow-on-the-mountain]
35. *E. micromera* Boiss.
36. *E. milii* Des Moul. [Christ-plant, Christ-thorn, crown-of-thorns]
37. *E. multiformis* Hook. & Arnott
38. *E. nutans* Lag.
39. *E. olowaluana* Sherff var. *gracilis* (Rock) Sherff
40. *E. oregonensis* Millsp.
41. *E. palmeri* Engelm.
42. *E. peplus* L. [petty spurge]
43. *E. plummerae* S. Wats.
44. *E. polycarpa* Benth.
45. *E. polygonifolia* L. [seaside spurge]
46. *E. prostrata* Aiton
47. *E. pulcherrima* Willd. ex Klotzsch (=*Poinsettia pulcherrima* (Willd. ex Klotzsch) J. Graham) [Christmas-flower, Christmas-star, lobsterplant, Mexican flameleaf, painted-leaf, poinsettia]
48. *E. pulcherrima* Willd. ex Klotzsch var. *plenissima* Hort.
49. *E. robusta* (Engelm.) Small
50. *E. serpens* H. B. K. [round-leaved spurge]
51. *E. serpyllifolia* Pers. [thyme-leaved spurge]
52. *E. spathulata* Lam. (=*E. dictyosperma* Fisch. & C. A. Mey., *E. missouriensis* Small in Britton, *Tithymalus missouriensis* Small)
53. *E. stictospora* Engelm.
54. *E. trigona* Haw. [Abyssinian euphorbia, African milk-tree]
55. *Euphorbia* sp. (=*Poinsettia sp.*)

Oomycetes

Peronospora euphorbiae. Downy mildew. **51:**MS(2720), SD(2243).

Peronospora sp. Downy mildew. **33:**E states(82), SD(82), TX(82). **50:**E states(82), SD(82), TX(82). **51:**E states(82), SD(82), TX(82). **55:**E states(82), FL(35), SD(82), TX(82).

Phytophthora drechsleri. Bract, flower and leaf blight. **47:**HI(3982,3983).

P. nicotianae var. *parasitica*. Bract, flower and leaf blight, leaf, root and stem rot. **47:**FL(35), HI(2903,3982,3983). **55:**FL(35).

Phytophthora sp. Root and stem rot. **36:**FL(35). **47:**MI(3124), NC(1303), NJ(82).

Pythium aphanidermatum. Root rot, on seedlings and seeds. **29:**AZ(3985). **47:**FL(35).

P. debaryanum. Stem rot. **47:**OK(82,2872). **55:**Unk.(1868).

P. perniciosum. Root rot. **47:**CA(82).

P. ultimum. Root and stem rot. **47:**CA(82), MI(3124), NJ(82). **55:**Unk.(1868).

Pythium sp. Root rot. **2:**HI(2903). **36:**FL(35). **47:**HI(2903), NC(1303). **55:**FL(35).

Zygomycotina

Mucor sp. **47:**MI(3124).

Rhizopus stolonifer. Soft rot. **54:**CA(2526).

Rhizopus sp. **47:**NC(1303).

Ascomycotina

Amazonia psychotriae. **8:**HI(2903).

A. tetrorchidii. Black mildew. **8:**HI(1272,1452,2903).

Botryosphaeria ribis. On stems. **47:**TX(82).

Botryotinia sp. Stem lesions. **47:**FL(35).

Chaetomium sp. **47:**MI(3124).

Diaporthe arctii. On stems. **10:**GA(82,1438).

Erysiphe euphorbiae. Powdery mildew. **10:**NC(3944). **24:**NY(195,369).

Erysiphe sp. Powdery mildew. **20:**FL(35).

Glomerella sp. Anthracnose. **20:**FL(35). **36:**FL(35).

Leptosphaeria sp. Leaf spot. **55:**FL(35).

Microsphaera euphorbiae. Powdery mildew. **10:**SE states(82), GA(1438), KS(82), MI(82), MS(2720). **13:**Range of host(82). **25:**GA(1438). **33:**MS(2720), WI(1347,1348). **34:**IA(82), IN(82), KS(82), MO(82). **55:**Range of host(82).

Mycosphaerella degeneri. On living leaves. **55:**HI(2784).

Nectriella pironii. Stem gall. **47:**FL(35).

Questieria euphorbiae. **8:**HI(2903).

Sclerotinia sclerotiorum. Stem rot. **13:**CO(447). **47:**FL(35), NC(1303), WA(82,3202). **51:**MI(82). **55:**FL(35).

Syspastospora parasitica. Leaf and stem rot. **47:**FL(35).

Basidiomycotina - Uredinales

Aecidium tithymali. **55:**IA(82), IL(82), IN(82), MD(82), NE(82), NM(82), TX(82), WI(82).

Melampsora euphorbiae. Autoecious. **12:**NE states(82), IN(82), WI(82), WI(1330). **42:**CA(81). **55:**CA(82), IA(82).

M. euphorbiae-gerardianae. Telial host. **11:**CA(642). **42:**WA(3202). **55:**IN(82), MD(82), VA(82), WI(82), WV(82).

M. monticola. Telial host. **3:**AZ(1203,3981), TX(742). **6:**AZ(1203,3981). **11:**CA(81), OR(3202). **26:**AZ(1195,3981). **27:**AZ(1203). **30:**CA(2439). **32:**AZ(3981). **41:**AZ(1203,3981). **49:**AZ(3981), CO(1195), SD(2243). **55:**SC states(82), CA(82), OR(82).

Puccinia emaculata. **10:**E states(82), GA(1439), IL(2568), MN(82), TX(82), WI(1337). **34:**SC states(82), CO(82), MS(2720), OK(2872), SD(82). **55:**FL(35), NE(82), NJ(82).

P. levata. Autoecious. **5:**HI(2903). **9:**HI(1522). **22:**HI(2903). **37:**HI(2903). **39:**HI(2903). **55:**HI(82,2903).

P. vitata. Autoecious. **39:**HI(1522,2903).

Uredo stevensii. **8:**HI(1522,2903). **55:**HI(2903).

Uromyces coordinatus. Autoecious. **55:**CA(82), UT(82).

U. dictyosperma. Autoecious. **52:**MT(3202), ND(378), OK(2872), SD(2243), WA(3202). **55:**W states(82), TX(82).

U. euphorbiae. Autoecious. **1:**AZ(747,1203,3981). **4:**AZ(1203,3981). **6:**AZ(3981). **7:**AZ(747,1203), MS(2720). **11:**CA(642). **13:**SC states(82), SE states(82), OK(2872), TX(652), UT(82). **14:**AZ(1203,3981). **16:**AZ(3981). **17:**AZ(1203,3981). **18:**MS(2720). **19:**AZ(1195,1203,3981), ND(378), OR(3202), WA(3202), WY(1195). **20:**C states(82), SE states(82), FL(35), KS(82), MS(2720), TX(82). **21:**AZ(1194,1203,3981). **23:**MS(2720), OK(2872). **25:**MS(2720). **34:**C states(82), CO(82), NM(1195), OK(2872), SD(2243). **35:**AZ(747). **40:**OR(3202).

41:AZ(1203,3981). **43:**AZ(3981). **44:**AZ(747,3981), TX(742). **45:**MS(2720). **46:**AZ(3981), MS(2720). **47:**FL(35), OK(82,2872), TX(82). **50:**Range of host(82). **51:**Range of host(82), AZ(1203,3981), CA(81,638,642), HI(2903), MS(2720), MT(3202), ND(378), OR(3202), SD(2243), WA(3202), WY(1195). **53:**OK(2872). **55:**Range of host(82), FL(35), ND(378).
U. mayorii. Autoecious. **50:**SC states(82), CA(82). **55:**SC states(82), CA(82).
U. occidentalis. Heteroecious: aecial host. **6:**AZ(3981). **26:**AZ(3981). **49:**AZ(1203,3981), MT(3202). **55:**AZ(82), CO(82), MT(82), NM(82), NV(82), UT(82).
U. tranzschelii. Autoecious. **31:**AZ(1195). **32:**AZ(747,1203,3981). **41:**AZ(747,3981). **49:**AZ(1203,3981), MT(3202), OR(3202), WY(1195). **55:**W states(82), CA(82).
Basidiomycotina - Ustilaginales
Melanotaenium euphorbiae. **13:**LA(1077). **20:**LA(82,1077,2094). **33:**LA(82,1077).
Basidiomycotina - Other
Armillaria tabescens. Root rot. **47:**FL(35,82). **50:**FL(82). **55:**FL(35,82).
Thanatephorus cucumeris. **33:**WI(1352).
Deuteromycotina - Hyphomycetes
Alternaria angustiovoidea. **15:**IA(3271), ND(3271).
A. euphorbiae. Bract, leaf and stem blight. **47:**HI(3983).
A. euphorbiicola. **47:**FL(3271).
A. tenuissima f. sp. *euphorbiae*. **15:**ND(1991).
Alternaria sp. Leaf spot. **15:**WI(1342,1360). **20:**FL(35). **34:**KS(82), TX(82). **47:**MI(3124). **55:**FL(35).
Botrytis cinerea. Gray mold blight, tip blight, scorch. **13:**CT(82). **34:**NJ(82), WI(1346). **36:**FL(35). **47:**AK(82,526), HI(2903), MO(82), NC(1303), TX(82), WA(82,3202). **55:**FL(35).
Cercospora euphorbiae. Leaf spot. **10:**KS(82), TX(82), WI(1346). **12:**TX(82). **13:**TX(82). **20:**FL(35). **33:**TX(82). **36:**FL(35). **55:**FL(35), MD(82).
C. euphorbiaecola. Leaf spot. **34:**NE(82). **55:**AL(82,581).
C. heterospora. Leaf spot. **10:**WI(82,581).
C. pulcherrimae. Leaf spot. **34:**OK(82,2872), SD(2243), WI(1346). **47:**FL(35), TX(82,3597).
Cercospora sp. Leaf spot. **10:**Unk.(82). **12:**Unk.(82). **34:**Unk.(82).
Cercosporidium chaetomium. Leaf spot on flowers, leaves or stems. **24:**IA(839).
Chalara thielavioides. Root rot. **47:**IL(2744).
Corynespora cassiicola. Bract spot. **47:**AZ(566).
Corynespora sp. Leaf spot. **47:**FL(35).
Cylindrocladium scoparium. Cutting rot. **47:**FL(35).
Dendrodochium affine. **55:**DE(3543).
Fusarium moniliforme. Root and stem rot. **36:**FL(35). **55:**FL(35).
F. oxysporum. Basal stem rot. **36:**FL(35). **47:**FL(35).
F. solani. Basal stem canker. **47:**WI(1511).
Fusarium sp. Root and stem rot, wilt. **10:**WI(1352). **47:**FL(82), MI(3124), NC(1303), NJ(82), WA(3202).
Fusicladium fasciculatum. Leaf and stem mold. **10:**WI(1329,1349). **19:**WI(1333,1349). **38:**NJ(839). **51:**WI(82,1349). **55:**KS(82), MD(82), NJ(82).
F. fasciculatum var. *didymum*. Leaf and stem mold. **10:**IA(82,839).
Helminthosporium sp. Leaf spot. **36:**FL(35). **47:**FL(35). **55:**FL(35).
Hormodendrum sp. **47:**MI(3124).
Oidium sp. Powdery mildew. **55:**FL(35), HI(2903).
Penicillium sp. **47:**MI(3124).
Phymatotrichopsis omnivora. Root rot. **10:**TX(82). **12:**TX(82). **13:**TX(82). **33:**TX(82). **34:**TX(82). **36:**AZ(82). **47:**AZ(82), TX(82).
Thielaviopsis basicola. Root rot. **47:**MD(1867), NC(1303), WA(3202). **55:**Unk.(1868).
Trichoderma sp. **47:**MI(3124).
Deuteromycotina - Coelomycetes
Ascochyta sp. Leaf spot. **15:**WI(1353). **47:**FL(35).
Colletotrichum gloeosporioides. On stems. **47:**TX(82).
Colletotrichum sp. Anthracnose. **20:**FL(35). **36:**FL(35). **47:**FL(35). **55:**FL(35).
Gloeosporium intermedium var. *poinsettiae*. On dead twigs. **47:**CA(81).
Lasiodiplodia theobromae. Twig dieback. **47:**FL(35).
Macrophoma sp. Leaf spot. **55:**FL(35).
Macrophomina phaseolina. Charcoal rot, wilt. **29:**AZ(3985,3986).
Phoma euphorbiicola. On stems. **34:**PA(82), TX(82).
Phoma sp. Stem necrosis. **47:**FL(35), MI(3124).
Phomopsis sp. Stem canker. **47:**FL(35). **55:**FL(35).
Phyllosticta ghaesembillae. On stems. **55:**FL(2782).
Phyllosticta sp. Leaf spot. **10:**OK(82,2872). **34:**NJ(82). **36:**FL(35). **47:**FL(35). **55:**FL(35).
Septoria sp. Leaf spot. **20:**FL(35).
Sphaceloma poinsettiae. Spot anthracnose, scab. **20:**FL(35). **47:**FL(35,82), HI(82,2899,2903). **48:**FL(1745). **55:**FL(35).
Stagonospora sp. Leaf spot. **36:**MD(82).
Deuteromycotina - Other
Rhizoctonia solani. Damping-off, root and stem rot, leaf spot. **29:**AZ(3985). **36:**FL(35). **47:**CA(3612), FL(35,82), IL(82), MI(3124), NC(1303), NJ(82), TX(82). **55:**Unk.(1868), FL(35), HI(2899), NY(82).
Rhizoctonia sp. Foliage blight, root rot. **12:**PA(2600). **47:**HI(2903). **55:**HI(2903).
Sclerotium rolfsii. **28:**MD(3784).

Gymnanthes Sw.

1. *G. lucida* Sw. [crabwood, poisonwood]

Basidiomycotina - Uredinales
Uromyces euphorbiae. **1:**Range of host(82).

Hevea Aubl.

1. *H. brasiliensis* (Willd. ex Adr. Juss.) Müll. Arg. (=*Siphonia brasiliensis* Willd. ex Adr. Juss.) [caouthchouc-tree, Pará rubber-tree, rubber-tree]

Ascomycotina
Glomerella cingulata. Leaf spot. **1:**FL(35,82).
Ophiobolus heveae. Leaf spot. **1:**FL(35,82).
Deuteromycotina - Hyphomycetes
Corynespora cassiicola. Leaf spot. **1:**FL(35).
Drechslera heveae. Bird's-eye spot. **1:**FL(35,82).
Deuteromycotina - Coelomycetes
Phomopsis sp. Dieback. **1:**FL(35,82).

Hura L.

1. *H. crepitans* L. [javillo, monkey's-dinner-bell, monkey-pistol, sandbox-tree]

Deuteromycotina - Hyphomycetes
Cercospora huricola. Leaf spot. **1:**FL(35).

Jatropha L.

1. *J. angustidens* (Torr.) Müll. Arg. (=*Cnidoscolus angustidens* Torr.)
2. *J. curcas* L. [Barbados-nut, pysicnut]
3. *J. macrorhiza* Benth. var. *septemfida* Engelm.
4. *J. urens* L. (=*J. stimulosa* Michx.)
5. *Jatropha* sp.

Oomycetes
Pythium sp. Root rot. **5:**FL(35).
Basidiomycotina - Uredinales
Aecidium sp. **3:**AZ(3981).
Phakopsora jatrophicola. **2:**FL(35), TX(82). **5:**FL(35), TX(82).
Uromyces agnatus. **4:**FL(35,82), SC(82).
U. dictyosperma. Autoecious. **5:**FL(35).
U. oaxacanus. **1:**AZ(1203,3981).
Basidiomycotina - Other
Armillaria tabescens. Root rot. **2:**FL(35,82).
Septobasidium sp. Parasitic on scale insects. **5:**FL(35).
Deuteromycotina - Hyphomycetes
Alternaria sp. Storage rot. **5:**FL(35).
Cercospora jatropharum. Leaf spot. **5:**FL(35).
Cercospora sp. Leaf spot. **5:**FL(35).
Fusarium sp. Root rot. **5:**FL(35).
Phymatotrichopsis omnivora. Root rot. **5:**TX(82).
Stemphylium sp. Leaf spot. **5:**FL(35).
Deuteromycotina - Coelomycetes
Colletotrichum gloeosporioides. Anthracnose, leaf spot. **2:**FL(35,82). **5:**FL(35).
Macrophoma sp. Leaf spot. **5:**FL(35).
Phomopsis sp. Twig dieback. **5:**FL(35).
Sphaeropsis tumefaciens. Witches' broom. **5:**FL(35).

Lasiocroton Griseb.

1. *L. bahamensis* Pax & Hoffm. in Engl.

Deuteromycotina - Hyphomycetes

Cercospora sp. Leaf spot. **1**:FL(35).

Mallotus Lour.

1. *M. japonicus* (Thunb.) Müll. Arg.

Deuteromycotina - Hyphomycetes

Cercospora mallotii. Leaf spot. **1**:MS(82,2720).

Manihot Mill.

1. *M. carthaginensis* Müll. Arg. in DC.
2. *M. esculenta* Crantz (=*M. aipi* Pohl, *M. dulcis* (J. F. Gmelin) Pax, *M. melanobasis* Müll. Arg., *M. utilissima* Pohl, *Jatropha manihot* L.) [bitter cassava, cassava, manioc, sweet-potato-tree, tapioca-plant, yuca]

Oomycetes

Phytophthora nicotianae var. *parasitica*. Root rot. **2**:FL(35).
Pythium sp. Tuber rot. **2**:FL(35).

Zygomycotina

Rhizopus sp. **2**:HI(2903).

Ascomycotina

Botryosphaeria disrupta. Dieback. **2**:FL(35).
B. quercuum. Dieback. **2**:FL(35).
B. rhodina. Dieback. **2**:FL(82).
Physalospora abdita. Dieback. **2**:FL(82).

Deuteromycotina - Hyphomycetes

Alternaria sp. Leaf spot. **2**:FL(35).
Cercospora sp. **2**:HI(2903).
Cercosporidium henningsii. Leaf spot. **2**:FL(35,82), HI(82), TX(82).
Fusarium sp. Tuber rot. **2**:FL(35).
Oidium sp. Powdery mildew. **2**:HI(2903).
Phaeoramularia manihotis. White leaf spot. **2**:FL(35), HI(2903).
Phymatotrichopsis omnivora. Root rot. **2**:TX(82).

Deuteromycotina - Coelomycetes

Ascochyta carthaginensis. On twigs. **1**:SC(82).
Colletotrichum gloeosporioides. On twigs. **1**:SC(82).
Colletotrichum sp. Leaf spot. **2**:FL(35).
Diplodia sp. **2**:HI(2903).
Dothiorella botrya. On limbs. **1**:SC(82).
Phomopsis sp. Leaf spot. **2**:FL(35).
Sphaeropsis sp. Dieback. **2**:FL(35).

Deuteromycotina - Other

Rhizoctonia solani. Root rot. **2**:FL(35,82).

Pedilanthus Neck. ex Poit.

1. *P. tithymaloides* (L.) Poit. [devil's-backbone, Japanese poinsettia, redbird-cactus, redbird-flower, ribbon-cactus, slipper-flower, slipper-plant]

Oomycetes

Pythium sp. Root rot. **1**:FL(35).

Ascomycotina

Erysiphe sp. Powdery mildew. **1**:FL(35).

Deuteromycotina - Hyphomycetes

Alternaria sp. Leaf spot. **1**:FL(35).
Oidium sp. Powdery mildew. **1**:FL(35).

Deuteromycotina - Coelomycetes

Colletotrichum sp. Stem lesions. **1**:FL(35).
Phomopsis sp. Stem lesions. **1**:FL(35).
Sphaeropsis sp. Stem lesions. **1**:FL(35).

Ricinus L.

1. *R. communis* L. [castor bean, castor-oil-plant, plama-christi, wonder-tree]

Chytridiomycetes

Synchytrium sp. Stem gall. **1**:TX(2853).

Oomycetes

Phytophthora cinnamomi. Root rot. **1**:HI(2903).
P. nicotianae. Stem rot. **1**:FL(35,82).
P. nicotianae var. *parasitica*. Root rot. **1**:HI(2903).
P. palmivora. Root and stem rot. **1**:HI(82,2903).
Pythium intermedium. Root rot. **1**:CA(82).
P. ultimum. Root rot. **1**:CA(82).
P. vexans. Root rot. **1**:HI(82,2903).
Pythium sp. Root rot. **1**:Unk.(82).

Ascomycotina

Botryosphaeria dothidea. Stem canker. **1**:FL(35).
B. obtusa. On stems. **1**:AL(82).
B. quercuum. Dieback. **1**:FL(35).
B. rhodina. On stems. **1**:FL(82).
B. ribis. On stems. **1**:FL(82), HI(82,2903).
Botryotinia ricini. Leaf and raceme blight, gray mold. **1**:SE states(82), Unk.(82), FL(35,2651,2854), LA(2854), MD(82), MS(2720,3499), TX(82).
Diaporthe arctii. **1**:GA(1438).
Diaporthe sp. On limbs. **1**:HI(2903).
Didymella ricini. On stems. **1**:LA(82).
Leptosphaeria doliolum. **1**:GA(1438).
Massarina gloeospora. **1**:GA(1438).
Physalospora abdita. On stems. **1**:FL(82).
Physalospora sp. On stems. **1**:Unk.(82).
Pleospora sp. On leaves. **1**:FL(621).
Sclerotinia sclerotiorum. Stem rot. **1**:FL(35,623).

Basidiomycotina - Other

Armillaria tabescens. Root rot. **1**:FL(35,82).
Schizophyllum commune. Twig dieback. **1**:FL(35,82).

Deuteromycotina - Hyphomycetes

Alternaria brassicae. On stems. **1**:NY(82).
A. compacta. Leaf spot. **1**:MD(82).
A. ricini. Seedling blight, spot, rot of leaves and seed pods. **1**:MD(82), MS(2720,3499), TX(388,2854), VA(2854).
Alternaria sp. Leaf spot. **1**:FL(35,82), LA(82), NC(1303), TX(82).
Botrytis cinerea. Gray mold. **1**:FL(35).
Cercospora canescens. **1**:KS(82), MO(82), NJ(82).
C. ricinella. Leaf spot. **1**:Gulf States(82), CA(82), FL(35), MO(82), MS(2720,3499), TX(388).
Cercospora sp. Isolated from blasted racemes. **1**:FL(2854).
Fusarium acuminatum. On stems. **1**:LA(82).
Fusarium sp. Seedling blight. **1**:MD(82).
Heterosporium sp. On seed pods and stems. **1**:CA(82).
Phymatotrichopsis omnivora. Root rot. **1**:TX(82,2854).

Deuteromycotina - Coelomycetes

Colletotrichum sp. Stem lesions. **1**:FL(35).
Diplodia sp. Twig dieback, on seed pods and stems. **1**:FL(35,82), LA(82), TN(82).
Lasiodiplodia theobromae. **1**:Unk.(82).
Macrophomina phaseolina. Charcoal rot. **1**:FL(622), MD(622), SC(622), TX(622), VA(622).
Phoma exigua. Leaf blight. **1**:NC(1303).
Phomopsis ricini. On stems. **1**:SC(82).

Deuteromycotina - Other

Rhizoctonia solani. Damping-off, seed rot. **1**:FL(35,82), KS(82), TX(82).
Rhizoctonia sp. Root rot. **1**:HI(2903).
Sclerotium rolfsii. Southern blight. **1**:FL(35,82).

Sapium P. Br.

1. *S. sebiferum* (L.) Roxb. (=*Stillingia sebifera* (L.) Michx.) [Chinese tallow-tree, tallow-tree]

Oomycetes

Pythium sp. Root rot. **1**:FL(35).

Basidiomycotina - Other

Armillaria mellea. Root rot. **1**:CA(2900).
A. tabescens. Root rot. **1**:FL(35,82).

Deuteromycotina - Hyphomycetes

Alternaria sp. Leaf spot. **1**:FL(35).
Phymatotrichopsis omnivora. Root rot. **1**:TX(82).
Pseudocercospora stillingiae. Leaf spot. **1**:FL(35), LA(82,962).

Deuteromycotina - Coelomycetes

Diplodia sp. Dieback. **1**:FL(35).
Phomopsis sp. Leaf tip necrosis. **1**:FL(35).
Phyllosticta stillingiae. Leaf spot. **1**:FL(35), LA(82,963).

Sebastiania Spreng.

1. *S. fruticosa* (W. Bartram) Fernald (=*S. ligustrina* (Michx.) Müll. Arg.)

Basidiomycotina - Uredinales

Uromyces graminicola. Heteroecious: aecial host. **1**:FL(35), MS(2720).

Stillingia Garden ex L.
1. *S. linearifolia* S. Wats.
2. *S. sylvatica* L. [queen's-delight, queen's-root]

Basidiomycotina - Uredinales
Uromyces graminicola. Heteroecious: aecial host. **2:**OK(82,2872).
Deuteromycotina - Hyphomycetes
Phymatotrichopsis omnivora. Root rot. **1:**TX(82).
Pseudocercospora stillingiae. Leaf spot. **1:**TX(82).

Synadenium Boiss. in A. DC.
1. *S. grantii* J. D. Hook. [American milkbush]

Deuteromycotina - Coelomycetes
Colletotrichum sp. Anthracnose. **1:**FL(35).
Deuteromycotina - Other
Rhizoctonia solani. Basal stem rot. **1:**FL(35).

Tragia L.
1. *T. nepetaefolia* Cav.
2. *Tragia* sp.

Deuteromycotina - Hyphomycetes
Cercospora euphorbiaecola var. *tragiae*. Leaf spot. **1:**TX(82,581,3597).
Phymatotrichopsis omnivora. Root rot. **2:**TX(82).
Deuteromycotina - Other
Sclerotium sp. Stem rot. **2:**TX(82).

Fabaceae

Abrus Adans.
1. *A. precatorius* L. [coral-bead-plant, crab's-eye, Indian licorice, jequirity bean, licorice-vine, love pea, prayer-beads, precatory bean, red-bead-plant, rosary pea, weather-plant, weather-vine, wild-licorice]

Ascomycotina
Myiocopron smilacis. **1:**FL(2782).
Schizoxylon ligustri. On twigs. **1:**FL(522).
Deuteromycotina - Other
Rhizoctonia sp. Basal rot. **1:**HI(2903).

Acacia Mill.
1. *A. angustissima* (Mill.) Kuntze [prairie acacia]
2. *A. confusa* Merr.
3. *A. constricta* Benth. [mescat acacia]
4. *A. cyclopis* A. Cunn. [coastal wattle, cultivated acacia]
5. *A. farnesiana* (L.) Willd. (=*Mimosa farnesiana* L.) [cassie, huisache, popinac, spongetree, sweet acacia, West Indian blackthorn]
6. *A. flexicaulis* Benth. (=*Siderocarpos flexicaulis* (Benth.) Small)
7. *A. greggii* A. Gray [cat's-claw acacia, Gregg cat's-claw, long-flowered cat's-claw, Texas mimosa]
8. *A. greggii* A. Gray var. *arizonica* Isely
9. *A. koa* A. Gray (=*A. koa* A. Gray var. *koa*) [koa]
10. *A. koa* A. Gray var. *hawaiiensis* Rock
11. *A. koa* A. Gray var. *latifolia* Benth.
12. *A. koaia* Hildebr.
13. *A. melanoxylon* R. Br. [Australian blackwood, blackwood acacia]
14. *A. roemerana* Scheele
15. *A. suffrutescens* J. Rose
16. *A. tortuosa* (L.) Willd. [catclaw]
17. *A. wrightii* Benth. [Texas catclaw]
18. *Acacia* sp.

Oomycetes
Phytophthora cinnamomi. **9:**HI(2903). **13:**HI(2903).
Pythium sp. Root rot. **2:**HI(2903). **18:**FL(35).
Ascomycotina
Botryosphaeria disrupta. Leaf spot. **18:**FL(35).
Calonectria crotalariae. Collar rot. **2:**HI(2903). **9:**HI(93,2903). **13:**HI(2903).
C. theae. Shoot blight. **9:**HI(2609,2903).
Cucurbitaria arizonica. On dead limbs. **18:**AZ(970).
Diaporthe rudis. On twigs. **18:**CA(82).
D. sheariana. On dead limbs. **9:**HI(128,2781).
Erysiphe polygoni. Powdery mildew. **4:**CA(82).
Hypoxylon truncatum. **9:**HI(2903).
Khuskia oryzae. On leaf surface. **9:**HI(144).
Lophodermium acacicola. Leaf spot. **9:**HI(82).
L. intermissum. **9:**HI(2903).
Meliola bidentata. **9:**HI(2903).
M. koae. Black mildew, sooty mold on leaves. **2:**HI(1272,1452,2903). **9:**HI(144,1272,1452,2903). **10:**HI(2903).
Mycosphaerella acaciae. On leaves. **18:**CA(967).
M. koae. On decaying fruit. **9:**HI(2784).
M. metrosideri. On leaves. **9:**HI(144). **10:**HI(2903).
Nectria ditissima. Twig canker. **18:**CA(82), SC(82).
N. lateritia. **9:**HI(2784).
Nectria sp. Twig canker, on leaves. **9:**HI(144). **10:**HI(2903). **18:**CA(82), SC(82).
Nummularia guaranitica. **9:**HI(2903).
Pestalosphaeria varia. On dry pods. **9:**HI(2547).
Phyllachora acaciae. Leaf spot. **17:**TX(82,3597). **18:**FL(35).
Physalospora abdita. Leaf spot. **18:**FL(82).
Pleospora herbarum. On leaves. **9:**HI(144). **10:**HI(2903).
Preussia vulgare. On leaf surface. **9:**HI(144).
Pyrenophora avenae. On leaves. **9:**HI(144). **10:**HI(2903).
Thyridaria koae. On decaying leaves. **9:**HI(2781).
Xylaria curta. **9:**HI(2903).
Xylaria sp. **13:**HI(2903).
Basidiomycotina - Uredinales
Atelocauda digitata. Autoecious. **9:**HI(1150,1596,2903). **11:**HI(1150,1596,2903). **12:**HI(1150,2903).
A. koae. Autoecious. **9:**HI(1596,2903). **11:**HI(1150,1596,2903).
Endoraecium acaciae. Autoecious. **9:**HI(1150,1596). **12:**HI(1596).
E. hawaiiense. Autoecious. **9:**HI(1150,1596).
Ravenelia australis. Autoecious. **5:**TX(82).
R. expansa. Autoecious. **1:**AZ(1203), TX(82).
R. hieronymii. **5:**TX(82,111).
R. pringlei. Autoecious. **7:**AZ(82,1195,1203), CA(82,746,1195), NM(82), TX(82,742,746). **8:**AZ(3981).
R. roemerianae. Autoecious. **14:**TX(82,742,746).
R. spegazziniana. Autoecious. **5:**FL(35), HI(82,1150,2903), NM(82), TX(82).
R. subtortuosae. Autoecious. **16:**TX(82).
R. texensis. Autoecious. **1:**AZ(1203,3981). **15:**AZ(82).
R. thornberiana. Autoecious. **3:**AZ(82,746,747,1203,3981), TX(82,742,746).
Basidiomycotina - Other
Armillaria mellea. Heart rot, root rot. **9:**HI(234,2011,2903,2907). **18:**Unk.(82).
A. tabescens. Root rot. **5:**FL(35,82). **18:**FL(35,82).
Auricularia cornea. **9:**HI(3948).
Eichleriella leucophaea. **7:**AZ(1202).
Fomes fasciculatus. On wood. **9:**HI(82,2903).
F. fullageri. On wood. **9:**HI(82,2903).
F. hawaiiensis. On wood. **18:**CA(82), HI(82,2903).
Ganoderma applanatum. On wood. **18:**CA(82).
G. australe. On wood. **9:**Unk.(82), HI(2903). **18:**HI(2903).
G. zonatum. Root rot. **5:**TX(2613).
Ganoderma sp. On wood. **9:**HI(234).
Inonotus texanus. **7:**AZ(1181,1202).
Laetiporus sulphureus. Brown cubical rot. **9:**HI(234,2903).
Peniophora malenconii subsp. *americana*. On wood. **18:**CA(542).
Perenniporia medulla-panis. **7:**AZ(1202).
P. ohiensis. **7:**AZ(128,1202).
Phaeolus schweinitzii. Brown cubical butt and root rot. **9:**HI(234,2903).
Phanerochaete crassa. White rot. **18:**AZ(433).
P. omnivora. White rot. **18:**AZ(433).
Phellinus badius. **7:**AZ(128,1202).
P. kawakamii. On wood. **9:**HI(2050). **12:**HI(2050).
Phlebiella tulasnelloidea. **9:**HI(128).
Pleurotus ostreatus. Heart and root rot. **9:**HI(234,2903).
Polyporus arcularius. **7:**AZ(1202).
Schizophyllum commune. Wound rot. **9:**HI(2903). **18:**CA(82).
Scytinostroma portentosum. **6:**TX(128).
Stereum hirsutum. **18:**CA(543).
Deuteromycotina - Hyphomycetes
Acremonium strictum. On leaves. **9:**HI(144). **10:**HI(2903).
Acremonium sp. On leaves. **9:**HI(144). **10:**HI(2903).
Alternaria alternata. On leaves. **9:**HI(144). **10:**HI(2903).
Alternaria sp. Leaf spot. **18:**FL(35).
Aspergillus sp. On leaves. **2:**HI(2903).
Aureobasidium pullulans. On leaves. **9:**HI(144). **10:**HI(2903).

Bipolaris australiensis. On leaves. **9:**HI(144). **10:**HI(2903).
Cercospora sp. **18:**GA(82).
Chalara sp. On leaves. **10:**HI(2903).
Cladosporium carpophilum. On leaves. **10:**HI(2903).
C. cladosporioides. On leaves. **9:**HI(144). **10:**HI(2903).
Cylindrocarpon obtusisporum. On twigs. **18:**CA(82).
Cylindrocarpon sp. On leaves. **9:**HI(144). **10:**HI(2903).
Cylindrocephalum sp. On leaves. **9:**HI(2903).
Cylindrocladium scoparium. Leaf spot. **18:**FL(35).
Cylindrocladium sp. On leaves. **9:**HI(144). **10:**HI(2903).
Dicyma pulvinata. On leaves. **9:**HI(144). **10:**HI(2903).
Drechslera sp. On leaves. **9:**HI(144). **10:**HI(2903).
Epicoccum nigrum. On leaves. **9:**HI(144). **10:**HI(2903).
Exosporium sp. **9:**HI(2903).
Fusarium lateritium. On leaves. **9:**HI(144). **10:**HI(2903).
F. merismoides. On leaves. **9:**HI(144). **10:**HI(2903).
F. oxysporum. Wilt. **9:**HI(1144).
F. oxysporum f. sp. *koae*. **9:**HI(2903).
Fusarium sp. Root and stem rot. **18:**FL(35).
Fusisporium leguminum. On legumes. **18:**SC(634).
Gliocladium roseum. On leaves. **9:**HI(144). **10:**HI(2903).
Gliocladium sp. On leaves. **9:**HI(144). **10:**HI(2903).
Hansfordia sp. On leaves. **9:**HI(144). **10:**HI(2903).
Helminthosporium sp. Leaf spot. **18:**FL(35).
Humicola brevis. On leaves. **9:**HI(2903).
Microdochium dimerum. On leaves. **9:**HI(2903).
Nigrospora oryzae. On leaves. **10:**HI(2903).
N. sphaerica. On leaves. **9:**HI(144). **10:**HI(2903).
Oidium sp. Powdery mildew. **18:**FL(35).
Paecilomyces sp. On leaves. **9:**HI(144). **10:**HI(2903).
Papulaspora sp. On leaves. **9:**HI(144,2903).
Penicillium canescens. On leaves. **9:**HI(144). **10:**HI(2903).
P. glabrum. On leaves. **9:**HI(144). **10:**HI(2903).
P. janthinellum. On leaves. **9:**HI(144). **10:**HI(2903).
P. raistrickii. On leaves. **9:**HI(144). **10:**HI(2903).
Penicillium sp. On leaves. **10:**HI(2903).
Periconia byssoides. On leaves. **9:**HI(144). **10:**HI(2903).
Periconiella angusiana. On leaves. **10:**HI(2903).
Phymatotrichopsis omnivora. Root rot. **5:**TX(82). **7:**TX(82).
Pithomyces atro-olivaceus. On bark and leaves. **9:**HI(144). **10:**HI(2903). **18:**CA(637).
Rhinocladiella sp. On leaves. **9:**HI(144). **10:**HI(2903).
Rhodotorula sp. On leaves. **9:**HI(144). **10:**HI(2903).
Schizotrichella sp. On leaves. **9:**HI(144). **10:**HI(2903).
Scolecobasidium humicola. On leaves. **9:**HI(144). **10:**HI(2903).
S. variabile. On leaves. **9:**HI(144). **10:**HI(2903).
Scopulariopsis sp. On leaves. **9:**HI(144). **10:**HI(2903).
Sporotrichum laxum. On leaves. **9:**HI(2903).
Stachybotrys chartarum. On litter. **9:**HI(2903).
Stemphylium botryosum. On leaves. **10:**HI(2903).
Strumella acaciae. On limbs and twigs. **18:**CA(82).
Trichoderma viride. On leaves. **9:**HI(144). **10:**HI(2903).
Triscelophorus monosporus. On leaf litter. **9:**HI(2903).
Tritirachium sp. On leaves. **9:**HI(144). **10:**HI(2903).
Volutella sp. On leaves. **9:**HI(144). **10:**HI(2903).
Zygosporium gibbum. On leaves. **9:**HI(144). **10:**HI(2903).

Deuteromycotina - Coelomycetes

Colletotrichum sp. Anthracnose. **18:**FL(35).
Coniothyrium decipiens. On limbs. **18:**CA(82).
Cylindrosporium sp. On leaves. **9:**HI(2903).
Diplodia subcuticularis. Canker on limbs and twigs. **18:**NY(82).
Diplodia sp. On leaves, limbs and phyllodes. **18:**CA(82), FL(35).
Gloeocoryneum hawaiiense. On living leaves. **9:**HI(3554).
Macrophomina sp. Leaf spot. **18:**FL(35).
Monochaetia sp. On leaves. **9:**HI(144). **10:**HI(2903).
Pestalotia breviseta. On leaves. **9:**HI(2903).
P. stevensonii. On leaves. **9:**HI(144,2903). **10:**HI(2903).
Pestalotia sp. On leaves. **9:**HI(2903). **10:**HI(2903).
Pestalotiopsis besseyii. On dry pods. **9:**HI(2547).
Phomopsis sp. Dieback. **18:**FL(35).
Phyllosticta sp. Leaf spot. **9:**HI(144). **10:**HI(2903). **18:**FL(35).
Sphaeropsis sp. Canker, stem gall. **18:**FL(35).

Deuteromycotina - Other

Rhizoctonia solani. Root rot. **18:**FL(35).
Rhizoctonia sp. Root rot. **2:**HI(2903). **9:**HI(2903).

Aeschynomene L.

1. *A. americana* L. [joint-vetch]
2. *A. indica* L. [kat-sola]
3. *A. scabra* G. Don
4. *A. virginica* (L.) B. S. P.

Basidiomycotina - Uredinales

[*Uredo posita*]. **4:**FL(35).

Deuteromycotina - Coelomycetes

Colletotrichum gloeosporioides. Stem canker, leaf spot. **1:**FL(35). **2:**AR(3594).
C. gloeosporioides f. sp. *aeschynomene*. **4:**AR(772).
C. truncatum. Anthracnose. **1:**FL(35). **3:**FL(35).

Albizia Durazz.

(= *Samanea* (DC.) Merr.)

1. *A. julibrissin* Durazz. [mimosa, silktree albizia]
2. *A. lebbeck* (L.) Benth. (=*Mimosa lebeck* L.) [lebbek-tree, siris-tree]
3. *A. saman* (Jacq.) F. v. Muell. (=*Pithecellobium saman* (Jacq.) Benth., *Samanea saman* (Jacq.) Merr.) [rain-tree, saman]

Oomycetes

Phytophthora sp. Root rot. **1:**FL(35).
Pythium sp. Root rot. **3:**HI(2903).

Ascomycotina

Botryosphaeria dothidea. **1:**GA(1438).
B. rhodina. On limbs. **1:**FL(82).
B. ribis. **1:**GA(82).
Eutypella scoparia. **1:**GA(82,1438).
E. stellulata. On limbs. **1:**DC(82,1097).
Nectria cinnabarina. Limb canker, dieback. **1:**DC(82,1097), MD(1097), NC(82,1097), OK(2872), OR(3202), SC(1097), VA(82).
N. coccinea. **1:**GA(82,1438).
N. haematococca. On bark. **3:**HI(82,2903).
Nectriella pironii. Stem gall. **2:**FL(35).
Orbilia rubrococcinea. **1:**GA(1438).
Rhytidhysteron rufulum. **1:**NC(82).
Thyronectria austroamericana. **1:**GA(1438).
Valsa sp. Canker. **1:**OK(2872).
Zopfia albiziae. **2:**TX(1493).

Basidiomycotina - Uredinales

Sphaerophragmium acaciae. Autoecious. **2:**FL(35,746,2272).

Basidiomycotina - Other

Armillaria mellea. Root rot. **1:**CA(2897).
A. tabescens. Root rot. **2:**FL(35).
Dendrophora albobadia. On wood. **1:**DC(82,1097), GA(82,491,1439), NC(82).
Ganoderma lucidum. White root and butt rot. **1:**MS(2309,2720).
Irpex lacteus. On limbs. **1:**DC(82,1097).
Oxyporus latemarginatus. **1:**NC(2166).
Phellinus gilvus. On wood. **3:**HI(2903).
Schizophyllum commune. On limbs. **1:**DC(82,1097), GA(1439).
Trametes hirsuta. On limbs. **1:**AZ(1202), DC(82,1097).
T. versicolor. White rot. **1:**DC(82,1097).

Deuteromycotina - Hyphomycetes

Cercospora sp. Leaf spot. **1:**FL(35), NC(1303). **2:**FL(35).
Fusarium lateritium. Dieback, wilt. On twigs. **1:**FL(35), PA(82), SC(82), VA(82).
F. moniliforme. Dieback, wilt. **1:**FL(35).
F. oxysporum. Dieback, wilt. **1:**FL(35).
F. oxysporum f. sp. *perniciosum*. Dieback, wilt. **1:**AL(82), AR(3921), FL(35), GA(82,1207,1533,2405), MD(2475), NC(82,1303,1533), NJ(82,1533), SC(82,1533), VA(82,1533,2804,3493,3497).
Fusarium sp. Canker. **1:**VA(3493).
Stilbella fimetaria. Gall. **2:**FL(35).
Tubercularia vulgaris. Limb canker. **1:**DC(1097), VA(1097).

Deuteromycotina - Coelomycetes

Botryodiplodia sp. Twig dieback. **3:**HI(2903).
Cytoplea insitiva. On twigs. **1:**Range of host(82), Unk.(82).
Diplodia julibrissin. **1:**WA(3202).
Lasiodiplodia theobromae. Twig dieback. **1:**FL(35). **2:**FL(35).
Leptothyrium sp. Leaf spot. **1:**FL(35).
Phoma henningsii. On dead pods. **2:**HI(82,2903).
Phoma sp. Dieback. **1:**FL(35).
Phomopsis sp. Dieback, wilt. **1:**FL(35), SC(82). **2:**FL(35).
Phyllosticta divergens. Pod spot. **2:**FL(35).
P. pithecolobii. **3:**HI(2903).

Deuteromycotina - Other
Rhizoctonia sp. Root rot. **1**:FL(35).

Alysicarpus Neck. ex Desv.

1. *A. rugosus* (Willd.) DC.
2. *A. vaginalis* (L.) DC. [alyce-clover, one-leaf clover]

Ascomycotina
Erysiphe polygoni. Powdery mildew. **2**:MS(2720).
Basidiomycotina - Other
Thanatephorus cucumeris. Leaf blight, banded sclerotial disease. **2**:LA(1010,3524), MS(2720).
Deuteromycotina - Hyphomycetes
Curvularia eragrostidis. Leaf spot. **2**:FL(2658).
Deuteromycotina - Coelomycetes
Colletotrichum gloeosporioides. Stem canker, leaf spot. **1**:FL(35).
C. truncatum. Stem canker, leaf spot. **1**:FL(35).
Phoma sp. Leaf spot. **2**:MS(2720).

Amorpha L.

1. *A. californica* Nutt. ex Torr. & A. Gray
2. *A. canescens* Pursh [leadplant]
3. *A. fruticosa* L. [bastard indigo, false indigo-bush]
4. *A. fruticosa* L. var. *angustifolia* Pursh (=*A. fragrans* Sweet)
5. *A. fruticosa* L. var. *occidentalis* (Abrams) Kearn. & Peebles
6. *A. herbacea* Walter
7. *A. nana* Nutt. [fragrant false indigo]
8. *Amorpha* sp.

Ascomycotina
Botryosphaeria dothidea. **3**:GA(1438).
B. obtusa. On twigs. **3**:OK(660). **8**:GA(82), KS(82), NE(82), NY(82).
Comoclathris permunda. **3**:OK(660).
Cucurbitaria amorphae. On limbs. **8**:KS(82), NM(82).
Diaporthe amorphae. On limbs. **3**:GA(1438). **8**:GA(82), KS(82), ND(82).
Eutypella leprosa. **3**:IA(3629).
Nematostoma occidentalis. **2**:WI(1346).
Valsa ambiens. **3**:KS(82,3415).
V. ceratosperma. **3**:KS(3415).
Valsaria insitiva. **3**:GA(1438), IA(1226). **8**:GA(82).
Basidiomycotina - Uredinales
Uropyxis amorphae. Autoecious. **1**:AZ(1195,1203,3981), CA(81). **2**:Range of host(82), FL(35), ND(378), OK(2872), SD(2243). **3**:Range of host(82), AZ(1195,1203), FL(35), MS(2720), ND(378), OK(2872), SD(2243), WI(1334). **4**:AZ(1195,1203). **5**:AZ(3981). **6**:SC(219). **7**:ND(378), SD(2243). **8**:FL(35), SC(634).
Deuteromycotina - Hyphomycetes
Cercospora passaloroides. Leaf spot. **3**:Range of host(82), FL(35), OK(2872), TX(82).
Phymatotrichopsis omnivora. **8**:TX(82).
Deuteromycotina - Coelomycetes
Cytospora sacculus. On twigs. **3**:SD(2243). **8**:KS(82), SD(82).
Diplodia amorphae. Leaf spot. **3**:MS(2720), SD(2243).
Phyllosticta sp. **3**:WI(1343).
Sphaeropsis amorphae. On limbs and twigs. **3**:KS(82), NY(82).
S. peckii. **3**:NE(82).
Stagonospora cytisporoides. On dead limbs. **3**:KS(82).

Amphicarpaea Elliott ex Nutt.

1. *A. bracteata* (L.) Fernald (=*Falcata comosa* Kuntze, *A. monoica* (L.) Elliott) [hog-peanut]
2. *A. bracteata* (L.) Fernald var. *comosa* (L.) Fernald [pitcher's hog-peanut]
3. *Amphicarpaea* sp.

Chytridiomycetes
Synchytrium decipiens. Leaf gall. **3**:E states(82), KS(82), ND(82).
S. lagerheimii. Leaf gall. **1**:FL(35), MS(2720), NY(2456), SD(2243). **2**:MS(2720).
Ascomycotina
Erysiphe polygoni. Powdery mildew. **1**:Range of host(82), NC(1307). **2**:Range of host(82).
Parodiella hedysari. Black mildew. **3**:MD(82).
Basidiomycotina - Uredinales
Cerotelium tanakae. **1**:IN(654,987).
Puccinia andropogonis. **1**:ND(378). **3**:IA(82), IL(82), ND(82), TN(82), WI(82).
Uromyces appendiculatus. Autoecious. **1**:FL(35).
Deuteromycotina - Hyphomycetes
Cercospora simulans. On leaves. **1**:WI(1347), WV(970). **2**:WV(82).
Epicoccum nigrum. **1**:WI(1348).
Pseudocercospora monoicae. Leaf spot. **1**:DE(3543), IA(82), IL(82), IN(82), KS(82), NE(82), OH(82), WI(82). **2**:IA(82), IL(82), IN(82), KS(82), NE(82), OH(82), WI(82).
Trichosporum falcatae. On leaves. **2**:TX(82).
Deuteromycotina - Coelomycetes
Ascochyta sp. **1**:WI(1357).
Colletotrichum sp. Leaf spot. **3**:WV(82).
Phyllosticta phaseolina. **1**:WI(1359).
Septoria glycines. **1**:WI(1360).
Stagonospora sp. **1**:WI(1343).

Anthyllis L.

1. *A. vulneraria* L. [kidney-vetch, lady's-fingers, woundwort]

Deuteromycotina - Hyphomycetes
Fusarium sp. Leaf blight. **1**:NC(82).
Deuteromycotina - Coelomycetes
Phyllosticta sp. Leaf spot. **1**:NC(82).

Apios Fabr.

1. *A. americana* Medik. (=*A. tuberosa* Moench) [American potato bean, groundnut, wild bean]

Chytridiomycetes
Synchytrium decipiens. **1**:IN(987).
Ascomycotina
Erysiphe polygoni. Powdery mildew. **1**:IN(82).
Metasphaeria sp. **1**:GA(1438).
Microsphaera diffusa. Powdery mildew. **1**:MS(82,2720), NE(82), TN(82).
M. penicillata. Powdery mildew. **1**:AL(82), MS(82,2720), OH(82).
Mycosphaerella sp. On leaves. **1**:WI(1359).
Myiocopron sp. **1**:GA(1438).
Basidiomycotina - Uredinales
Puccinia andropogonis. Heteroecious: aecial host. **1**:FL(35), IA(82), MN(82), NE(82), TX(82).
Deuteromycotina - Hyphomycetes
Alternaria sp. Leaf spot. **1**:NY(82).
Cercospora tuberosa. Leaf spot. **1**:E states(82), DE(3543), FL(35), TX(82), WI(82).
Phymatotrichopsis omnivora. **1**:TX(82).
Deuteromycotina - Coelomycetes
Ascochyta sp. **1**:WI(1360).
Colletotrichum sp. On leaves. **1**:WI(1356).
Phyllosticta phaseolina. **1**:WI(1338).
Stagonospora sp. On leaves. **1**:WI(1352).

Arachis L.

1. *A. hypogaea* L. [common peanut, earthnut, goober, grassnut, groundnut, maní, monkeynut, pindar]

Oomycetes
Phytophthora sp. Stem rot. **1**:NC(1303).
Pythium myriotylum. Seedling blight, peg, pod and root rot. 'Pod rot. Pythium peg rot. Pythium root rot. Pythium wilt.' **1**:FL(35), NC(1303,2859), VA(2859).
P. ultimum. 'Pythium peg rot. Pythium root rot.' **1**:NC(82).
Pythium sp. Peg, pod and root rot. **1**:CA(82), GA(82), NC(82,1303).
Zygomycotina
Rhizomucor pusillus. Isolated from kernels. **1**:OK(3568), TX(3568).
Rhizopus stolonifer. Isolated from endocarps. **1**:VA(1156).
Rhizopus sp. Crown, seed and seedling rot. **1**:FL(35), NC(1303), SC(82).
Ascomycotina
Botryosphaeria quercuum. Pod and stem rot. **1**:FL(35).
B. rhodina. Stem rot. **1**:AL(82), FL(82), GA(82), TX(82).
Calonectria crotalariae. Black rot of pegs, pods and roots, collar rot. **1**:HI(2903), NC(1303).
Chaetomium funicola. On greenhouse plant showing malnutrition (MD) and on seeds (NC). **1**:MD(82), NC(631).
C. globosum. Isolated from endocarps. **1**:VA(1156).
C. trilaterale. On shells. **1**:GA(632).
Chaetomium sp. Isolated from kernels. **1**:TX(3568).
Diaporthe phaseolorum. Stem blight. **1**:GA(1438), VA(82), WV(82).

Leptosphaerulina crassiasca. Pepper scorch, brown leaf spot. 'Pepper spot. Scorch.' **1**:FL(35), GA(1297,1438,2207), NC(1303).
L. trifolii. **1**:GA(1438).
Melanocarpus albomyces. Isolated from kernels. **1**:OK(3568).
Mycosphaerella arachidis. Halo spot, brown leaf spot. **1**:Gulf states(82), GA(1438), MS(2720), NC(1303).
M. berkeleyi. Leaf spot. **1**:Range of host(82), GA(1438), HI(82,2903), MS(2720), WI(82).
Neocosmospora vasinfecta. On pods and roots. **1**:AL(82), GA(82,1438).
Pleospora sp. Leaf and stem spot. **1**:AK(82), OK(82,2873).
Sclerotinia minor. 'Sclerotinia blight.' **1**:LA(3528), NC(1303,2859), OK(3740), TX(3528), VA(1946,2859).
S. sclerotiorum. 'Sclerotinia blight.' **1**:NC(268,2858), OK(3741), VA(268,2858).
Sclerotinia sp. Blight, stem rot. **1**:NC(1303).
Talaromyces thermophilus. Isolated from kernels. **1**:TX(3568).
Thermoascus aurantiacus. Isolated from kernels. **1**:OK(3568), TX(3568).

Basidiomycotina - Uredinales

Puccinia arachidis. Telial host. 'Rust.' **1**:FL(35,82), GA(82,1439,2923), LA(82), NC(1303,3850), TX(82,1477), VA(3318).

Basidiomycotina - Other

Thanatephorus cucumeris. **1**:GA(1439).

Deuteromycotina - Hyphomycetes

Alternaria alternata. Isolated from endocarps. 'Alternaria spot. Alternaria veinal necrosis.' **1**:VA(1156).
Alternaria sp. Leaf spot. **1**:FL(35,82), MO(82), NJ(82), NM(82), SC(82).
Aspergillus candidus. Storage rot. **1**:AL(871).
A. flavus. Seedling blight. 'Aspergillus damping off. Yellow mold.' **1**:FL(35), VA(1156).
A. fumigatus. Isolated from kernels. **1**:OK(3568), TX(3568).
A. glaucus. Storage rot. **1**:AL(871).
A. niger. Seedling blight. 'Aspergillus damping-off. Aspergillus crown rot.' **1**:FL(35), NC(1303).
A. tamarii. Storage rot. **1**:AL(871).
Aspergillus sp. Seed and storage rot. **1**:AL(871), NC(1303).
Botrytis cinerea. Gray mold, leaf, pod and stem rot. 'Botrytis blight.' **1**:CT(82), MD(82,2654), MS(82,2720), NC(1303), TN(82), VA(82,2654).
Botrytis sp. **1**:OK(2872).
Cercospora arachidicola. 'Early leaf spot.' **1**:Range of host(1762), FL(35,1717), HI(2903), NC(1303,2859), OK(2872), TX(1477), VA(2859).
Cercospora sp. Stem canker, petiole canker. **1**:NC(1303).
Cladosporium herbarum. Black mold of pods. **1**:Range of host(82), FL(35).
Cladosporium sp. Storage rot. **1**:AL(871).
Cristulariella moricola. 'Zonate leaf spot.' **1**:GA(3340).
Curvularia inaequalis. **1**:SC(82).
Cylindrocladium crotalariae. Black rot of pegs, pods and roots. 'Cylindrocladium black rot.' **1**:FL(35), GA(239,1694), NC(1303,3055,3056).
Cylindrocladium sp. Black rot. **1**:VA(1157).
Fusarium lateritium. On pods. **1**:SC(82).
F. moniliforme. Seedling root rot, on pods. **1**:AR(82), NC(82), OK(82,2873).
F. oxysporum f. sp. *vasinfectum*. Root rot. **1**:GA(82), NC(82), SC(82), VA(82).
F. solani f. sp. *phaseoli*. Root rot, wilt. **1**:GA(82).
Fusarium sp. Seedling blight, peg, pod, root and stem rot, wilt. **1**:SE states(82), FL(35), MO(82), NC(1303), OK(2872), TX(82), VA(1156).
Helminthosporium sp. Root rot. **1**:OK(82,2873).
Mycoleptodiscus terrestris. Root rot. **1**:FL(35).
Penicillium citrinum. Storage rot. **1**:AL(871).
P. janthinellum. Storage rot. **1**:AL(871).
Penicillium sp. Blue mold on pods and nuts. **1**:Range of host(82), VA(1156).
Phaeoisariopsis personata. 'Late leaf spot.' **1**:Range of host(1762), AL(285), FL(35,1717), HI(2903), NC(1303,2859), OK(2872), TX(1477), VA(2859).
Phymatotrichopsis omnivora. 'Phymatotrichum root rot.' **1**:AZ(82), TX(82).
Stemphylium sp. Leaf spot. **1**:ND(82).
Thermomyces lanuginosus. Isolated from kernels. **1**:OK(3568), TX(3568).
Thielaviopsis basicola. Root rot. 'Blackhull.' **1**:NC(82).
Trichoderma viride. Seed rot. **1**:SC(82).
Trichothecium roseum. Storage rot. **1**:AL(871).
Verticillium albo-atrum. 'Verticillium wilt.' **1**:NC(1303), NM(3357).
Wallemia sebi. Storage rot. **1**:AL(871).

Deuteromycotina - Coelomycetes

Ascochyta sp. Web blotch, leaf spot. **1**:AR(82), FL(35), TX(2788).
Colletotrichum sp. Leaf and stem spot. **1**:AR(82), NC(1303), OK(82,2873).
Diplodia frumenti. Stem rot. **1**:OK(2872).
Diplodia sp. Stem rot. **1**:SE states(82), TX(82).
Phoma arachidicola. Web blight. 'Web blotch.' **1**:NC(1303), VA(2802).
Phoma sp. Root rot. **1**:AL(82), GA(82), MO(82), VA(82).
Phyllosticta arachidis-hypogaea. Leaf spot. 'Phyllosticta leaf spot.' **1**:FL(35).
Phyllosticta sp. Leaf spot. **1**:AL(82), AR(82), HI(82,2903), IA(82), MS(82,2720).
Septogloeum arachidis. **1**:HI(2903).

Deuteromycotina - Other

Rhizoctonia solani. Leaf, nut and seedling blight; peg, pod, root and stem rot. 'Rhizoctonia foliar blight. Rhizoctonia peg blight. Rhizoctonia root rot. Pod rot.' **1**:FL(35), HI(2899), NC(1303), OK(2872), VA(1156).
Rhizoctonia sp. Root rot. **1**:SE states(82), CA(82), HI(2903), OK(82), TX(82).
Sclerotium rolfsii. Southern blight, peg and pod rot, blue stain. 'Stem rot.' **1**:Range of host(82), SE states(82), AL(2886), AZ(82), FL(35), GA(2886), HI(82,2903), MS(2720), NC(1303,2886), NM(2109,2112), OK(82,2872), TX(2886), VA(2886).

Astragalus L.

1. *A. aboriginorum* Richardson in Franklin
2. *A. adsurgens* Pall. (=*A. nitidus* Douglas ex Hook.)
3. *A. adsurgens* Pall. var. *robustior* Hook. (=*A. striatus* Nutt. ex Torr. & A. Gray) [milkvetch]
4. *A. agrestis* Douglas ex G. Don (=*A. goniatus* Nutt.)
5. *A. allochrous* A. Gray
6. *A. alpinus* L.
7. *A. arizonicus* A. Gray
8. *A. atropubescens* J. M. Coulter & E. M. Fisher
9. *A. bisulcatus* (Hook.) A. Gray
10. *A. bourgovii* A. Gray
11. *A. canadensis* L. (=*A. carolinianus* L.)
12. *A. canadensis* L. var. *mortonii* (Nutt.) S. Wats. (=*A. mortonii* Nutt.)
13. *A. chamaeleuce* A. Gray
14. *A. cicer* L. [cicer milkvetch]
15. *A. crassicarpus* Nutt. var. *berlandieri* Barneby (=*A. mexicanus* A. DC.)
16. *A. crassicarpus* Nutt. var. *crassicarpus* (=*A. caryocarpus* Ker-Gawl.)
17. *A. crotalariae* (Benth.) A. Gray
18. *A. diaphanus* Douglas ex Hook.
19. *A. drummondii* Douglas ex Hook.
20. *A. falcatus* Lam. [milkvetch]
21. *A. flexuosus* (Hook.) G. Don [flexile milkvetch]
22. *A. frigidus* (L.) A. Gray
23. *A. gambelianus* E. Sheld.
24. *A. gilensis* Greene
25. *A. gilviflorus* E. Sheld. (=*Orophaca caespitosa* Britton)
26. *A. hamosus* L.
27. *A. humistratus* A. Gray var. *hosackiae* (Greene) M. E. Jones (=*A. hosackiae* Greene)
28. *A. hypoglottis* L.
29. *A. inflexus* Douglas ex Hook.
30. *A. lentiginosus* Douglas ex Hook.
31. *A. lentiginosus* Douglas ex Hook. var. *diphysus* (A. Gray) Jones (=*A. diphysus* A. Gray)
32. *A. lotiflorus* Hook.
33. *A. melilotoides* Pall.
34. *A. miser* Douglas ex Hook.
35. *A. miser* Douglas ex Hook. var. *decumbens* (Nutt.) Cronq. (=*A. decumbens* (Nutt.) A. Gray)

36. *A. missouriensis* Nutt. [milkvetch]
37. *A. mollissimus* Torr. [woolly loco, woolly locoweed]
38. *A. mollissimus* Torr. var. *bigelovii* (A. Gray) Barneby (=*A. bigelovii* A. Gray)
39. *A. mollissimus* Torr. var. *thompsonae* (S. Wats.) Barneby (=*A. thompsonae* S. Wats.)
40. *A. mulfordae* M. E. Jones
41. *A. neglectus* (Torr. & A. Gray) E. Sheld.
42. *A. nuttallianus* DC.
43. *A. nuttallii* (Torr. & A. Gray) J. T. Howell (=*A. menziesii* A. Gray)
44. *A. parryi* A. Gray
45. *A. pectinatus* Douglas ex Hook.
46. *A. purshii* Douglas ex Hook.
47. *A. reventus* A. Gray
48. *A. salmonis* M. E. Jones
49. *A. scopulorum* Porter
50. *A. spaldingii* A. Gray
51. *A. spathulatus* E. Sheld. (=*Homalobus caespitosus* Nutt. ex Torr. & A. Gray)
52. *A. tenellus* Pursh (=*A. multiflorus* A. Gray)
53. *A. thurberi* A. Gray
54. *A. trichopodus* (Nutt.) A. Gray var. *lonchus* (Jones) Barneby (=*A. leucopsis* Torr. & A. Gray)
55. *A. umbellatus* Bunge
56. *A. utahensis* (Torr.) Torr. & A. Gray
57. *A. vexilliflexus* E. Sheld.
58. *A. whitneyi* A. Gray
59. *A. wootonii* E. Sheld.
60. *A. wootonii* E. Sheld. var. *candolleanus* (Kunth) Barneby (=*A. candolleanus* Boiss.)
61. *Astragalus* sp.

Oomycetes

Peronospora parasitica. On leaves. **6:**WY(3378).
P. trifoliorum. Downy mildew. **2:**MT(3202). **6:**AK(3431), WY(1195). **11:**IA(82), KS(82), MN(82), MT(82,3202), ND(82,377), SD(82), WI(82). **23:**OR(3202). **32:**IA(82), KS(82), MN(82), MT(82,3202), ND(82), SD(82), WI(82). **50:**OR(3202), WA(3202). **61:**MT(3202), WA(3202).
Peronospora sp. **33:**WA(82).

Ascomycotina

Clathrospora pentamera. **8:**WA(3202).
Erysiphe polygoni. Powdery mildew. **4:**MT(3202). **6:**AK(3431), WY(1195). **8:**ID(3202). **11:**MT(3202), SD(2243), WA(3202), WY(1195). **20:**WA(3202). **34:**MT(3202). **35:**WY(1195). **52:**MT(3202). **61:**Range of host(82), MT(3202), WA(3202).
Microsphaera euphorbiae. Powdery mildew. **2:**MT(3202). **3:**CO(82), MT(82), NY(82). **41:**CO(82), MT(82), NY(82).
Mycosphaerella pachyasca. On leaves. **6:**SD(82).
M. tassiana. **6:**AK(526). **11:**WA(3202). **47:**WA(3202).
Pleospora comata. On leaves. **3:**WY(82). **10:**MT(3202). **58:**WA(3202).
P. herbarum. **22:**AK(526).
P. penicillus. **8:**ID(3202). **47:**WA(3202). **61:**ID(3202).
P. rudis. **8:**WA(3202). **48:**OR(3202).
P. tragacanthae. **48:**OR(3202).
Polystigma astragali. Leaf spot. **2:**WA(3202). **3:**CO(82), MT(82), ND(82), UT(82), WY(82). **4:**MT(3202), ND(377), WA(3202), WY(1195). **9:**CO(82), MT(82), ND(82,377), UT(82), WY(82,1195). **21:**CO(82), MT(82), ND(82), UT(82), WY(82). **28:**CO(82), MT(82), ND(82), UT(82), WY(82). **45:**CO(82), MT(82,3202), ND(82), UT(82), WY(82). **61:**Unk.(82).
Sphaerotheca fuliginea. Powdery mildew. **61:**AK(82,526).
S. macularis. **6:**WY(1195).

Basidiomycotina - Uredinales

[*Uromyces hedysari-obscuri*]. **2:**ND(378).
U. lapponicus. Autoecious. **6:**NW states(82), AK(82,3431), CO(82), ID(3202), WY(1195). **21:**NW states(82), AK(82), CO(82). **61:**AK(526), AZ(1203), MT(3202), OR(3202).
U. phacae-frigidae. Autoecious. **55:**AK(60,526). **61:**AK(526).
U. punctatus. Heteroecious: telial host. **1:**MT(3202). **2:**MT(3202), ND(378). **4:**WA(3202). **5:**AZ(747,3981), CA(747). **7:**AZ(1203,3981). **11:**SD(2243), WI(1332). **17:**AZ(747,3981), CA(747). **18:**OR(3202). **24:**AZ(1203,3981). **25:**MT(3202), ND(378). **26:**OK(2873). **27:**AZ(1203,3981). **29:**ID(3202). **30:**AZ(747,3379,3981), CA(747). **32:**MT(3202). **36:**MT(3202). **37:**TX(742). **38:**AZ(1203,3981). **43:**CA(81). **46:**ID(3202), MT(3202), OR(3202), WA(3202). **50:**WA(3202). **51:**ND(378). **53:**AZ(1203,3981). **56:**ID(3202). **57:**MT(3202). **59:**AZ(1203,3981), ID(3202). **60:**AZ(1203). **61:**C states(82), AZ(1203), CA(82), TX(82), WA(82).

Basidiomycotina - Ustilaginales

Thecaphora deformans. Seed smut. **5:**AZ(82), CO(82), NM(82), UT(82), WY(82). **9:**AZ(82), CO(82,1077), NM(82,1077), UT(82), WY(82,1077). **12:**CA(1077). **16:**AZ(82), CO(82), NM(82,1077), UT(82), WY(82). **19:**AZ(82), CO(82,1077), NM(82,1077), UT(82), WY(82). **31:**AZ(82), CO(82), NM(82), UT(82), WY(82). **36:**AZ(82), CO(82), NM(82), UT(82,1077), WY(82). **37:**CO(1077). **39:**AZ(82), CO(82), MN(82), UT(82,1077), WY(82). **42:**AZ(82,1077), CO(82), NM(82,1077), UT(82), WY(82). **49:**AZ(82), CA(1077), CO(82,1077), NM(82), UT(82), WY(82). **52:**AZ(82), CO(82), NM(82), UT(82,1077), WY(82). **54:**CA(1077). **59:**AZ(82,1077), CO(82), NM(82,1077), UT(82), WY(82).

Deuteromycotina - Hyphomycetes

Alternaria sp. **20:**NW states(3202).
Cercospora astragali. Leaf spot. **11:**WI(1334). **14:**OK(82,2872). **61:**WI(82).
Cercosporidium sp. Leaf spot. **15:**OK(2872).
Dendryphion nubilosum. On leaves and stems. **11:**MT(3202). **21:**MT(82,3202). **44:**CO(82).
Fusarium moniliforme. Stem lesions, wilt. **11:**WI(1331).
F. tricinctum. Stem lesions, wilt. **11:**WI(1331).
Fusarium sp. Stem rot, wilt. **11:**WI(82). **37:**TX(82).
Fusicladium brevipes. On leaves. **9:**MT(3202). **11:**WA(3202). **28:**CO(82).
Fusicladium sp. **6:**AK(3431).
Napicladium astragali. Leaf spot. **13:**MT(82,3202). **15:**OK(82,2873).
Phymatotrichopsis omnivora. **61:**TX(82).
Ramularia astragali. Leaf spot. **11:**IA(82), ID(3202), IL(82), KS(82), MT(3202), ND(82,378), NE(82), WI(82).
Stemphylium sp. Leaf spot. **20:**WA(82).
Volutella occidentalis. On stems. **19:**MT(82,3202). **21:**MT(82,3202).

Deuteromycotina - Coelomycetes

Ascochyta sp. **40:**ID(3202).
Colletotrichum dematium. On pods. **16:**KS(82).
Cylindrosporium sp. **6:**ID(3202), MT(3202).
Heteropatella umbilicata. **10:**MT(3202).
Phoma astragali. On stems. **4:**CA(82), KS(82), ND(82,378). **9:**CA(82), KS(82), ND(82). **11:**CA(82), KS(82), ND(82). **43:**CA(82,128), KS(82), ND(82,128).
Phyllosticta astragali. Leaf spot. **61:**MN(82), MT(82,3202), WA(82).
P. davisii. On leaves. **11:**WI(82).
Phyllosticta sp. **11:**WI(1357).
Septoria astragalicola. Leaf spot. **2:**MT(3202), WY(1195). **3:**AZ(82), CO(82), MT(82). **19:**AZ(82), CO(82), MT(82). **34:**MT(3202). **57:**MT(3202).
S. astragalina. **3:**CO(649). **52:**CO(649).
S. psammophila. **45:**ND(378).
S. wisconsina. On leaves. **11:**WI(1334).

Baptisia Venten.

1. *B. australis* (L.) R. Br. [blue wild indigo]
2. *B. bracteata* Muhl. ex Elliott
3. *B. lactea* (Raf.) Thieret (=*B. leucantha* Torr. & A. Gray)
4. *B. lanceolata* Walter
5. *B. leucophaea* Nutt.
6. *B. perfoliata* (L.) R. Br.
7. *B. tinctoria* (L.) Venten.
8. *Baptisia* sp.

Ascomycotina

Acrospermum foliicola. **2:**GA(1438).
Anthostomella baptisiae. On stems. **6:**SC(82).
Botryosphaeria dothidea. **6:**GA(1438).
B. obtusa. On stems. **8:**GA(82).
Diapleella coniothyrium. **4:**GA(82,1438).
Diaporthe arctii. On stems. **7:**NJ(82).
Erysiphe polygoni. On leaves and stems. **5:**WI(1344). **7:**MA(1390).
Microsphaera penicillata. Powdery mildew. **2:**WI(82). **5:**WI(1344).
Mycosphaerella baptisiicola. On dead stems. **3:**SC(82). **6:**GA(1438). **7:**NJ(82), SC(82).

M. oenotherae. On stems. **7**:NJ(82).
M. punctiformis. **2**:GA(1438).
Mycosphaerella sp. On leaves. **2**:KS(82).
Basidiomycotina - Uredinales
Puccinia andropogonis. Heteroecious: aecial host. **1**:KS(82), OK(82,2872). **2**:OK(2872). **5**:KS(82), NE(82). **7**:SC(82).
Basidiomycotina - Other
Armillaria mellea. Root rot. **1**:CA(2897).
Scytinostroma galactinum. Root rot. **1**:MD(668).
Deuteromycotina - Hyphomycetes
Alternaria sp. **3**:WI(1342).
Cercospora velutina. Leaf spot. **1**:KS(82). **2**:WI(82). **3**:IL(82), WI(82).
Cladosporium baptisiae. On living leaves. **5**:WI(1338).
Phymatotrichopsis omnivora. Root rot. **8**:TX(82).
Deuteromycotina - Coelomycetes
Aplosporella baptisiae. **8**:NJ(2783).
Coniothyrium baptisiae. On stems. **6**:SC(82).
Phoma baptisiae. On stems. **1**:SC(82).
Septoria baptisiae. On leaves. **6**:SC(82,634), TX(82).
Stagonospora baptisiae. Leaf spot. **3**:IA(82), WI(82). **5**:WI(1341). **7**:WI(1358).

Bauhinia L.

1. *B. galpinii* N. E. Brown (=*B. punctata* Bolle) [nasturtium bauhinia, red bauhinia]
2. *B. hookeri* F. v. Muell. [mountain-ebony]
3. *B. variegata* L. [mountain-ebony, orchidtree]
4. *Bauhinia* sp.

Oomycetes
Pythium sp. Root rot. **1**:FL(35). **4**:FL(35).
Ascomycotina
Hypoxylon sp. On wood. **3**:FL(35).
Microsphaera diffusa. Powdery mildew. **4**:MD(82).
Nectriella pironii. Stem gall. **3**:FL(35). **4**:FL(35).
Basidiomycotina - Other
Armillaria tabescens. Root rot. **3**:FL(35).
Deuteromycotina - Hyphomycetes
Alternaria sp. Leaf spot. **1**:FL(35). **3**:FL(35).
Corynespora sp. Leaf spot. **3**:FL(35). **4**:FL(35).
Fusarium sp. Basal stem rot. **4**:FL(35).
Helminthosporium sp. Leaf spot. **3**:FL(35). **4**:FL(35).
Pseudocercospora bauhiniae. Leaf spot. **3**:FL(35).
Deuteromycotina - Coelomycetes
Botryodiplodia sp. Canker. **4**:FL(35).
Colletotrichum sp. Anthracnose, leaf spot. **1**:FL(35). **3**:FL(35), TX(82). **4**:FL(35).
Diplodia sp. Twig blight, dieback. **3**:FL(35). **4**:FL(35).
Phomopsis sp. Leaf spot. **4**:FL(35).
Phyllosticta sp. Leaf spot. **1**:FL(35). **2**:FL(82). **3**:FL(35).
Sphaeropsis tumefaciens. Stem gall. **3**:FL(35). **4**:FL(35).
Deuteromycotina - Other
Rhizoctonia solani. Basal stem rot. **3**:FL(35).

Biserrula L.

1. *B. pelecinus* L.

Deuteromycotina - Other
Rhizoctonia sp. Crown and root rot. **1**:HI(2903).

Caesalpinia L.

(= *Poinciana* L.)
1. *C. gilliesii* (Wallich ex Hook.) Benth. (=*Poinciana gilliesii* Wallich ex Hook.) [bird-of-paradise]
2. *C. pulcherrima* (L.) Sw. [dwarf poinciana, flowering-fence, paradise-flower, pride-of-Barbados]
3. *Caesalpinia* sp.

Oomycetes
Pythium sp. Seedling root rot. **2**:FL(35).
Ascomycotina
Botryosphaeria ribis. Dieback. **1**:FL(35,82), TX(82).
Nectria cinnabarina. Canker. **1**:OK(2872).
Physalospora sp. Dieback. **2**:FL(35).
Basidiomycotina - Uredinales
Ravenelia humphreyana. Autoecious. **1**:FL(35). **2**:FL(35). **3**:FL(35,82).
Basidiomycotina - Other
Armillaria tabescens. Root rot. **2**:FL(35).
Schizophyllum commune. **1**:CA(82).
Deuteromycotina - Hyphomycetes
Phymatotrichopsis omnivora. Root rot. **1**:TX(82).
Deuteromycotina - Coelomycetes
Colletotrichum sp. Anthracnose. **2**:FL(35).
Cytospora sp. **1**:OK(2872).
Cytosporina sp. Dieback. **1**:OK(2872).
Sphaeropsis sp. Dieback. **2**:FL(35).
Deuteromycotina - Other
Rhizoctonia solani. Dieback, root rot. **2**:FL(35).

Cajanus DC.

1. *C. cajan* (L.) Huth [Angola pea, catjang pea, Congo pea, dahl, no-eye pea, pigeon pea, red gram]

Oomycetes
Phytophthora cinnamomi. Root rot. **1**:HI(2903).
P. drechsleri. **1**:HI(2903).
P. nicotianae var. *parasitica*. **1**:HI(2903).
P. palmivora. **1**:HI(2903).
Pythium arrhenomanes. Root rot. **1**:HI(2903).
P. debaryanum. Root rot. **1**:HI(2903).
P. hydnosporum. Root rot. **1**:HI(2903).
P. irregulare. Root rot. **1**:HI(2903).
P. rostratum. Root rot. **1**:HI(2903).
P. splendens. Root rot. **1**:HI(2903).
Ascomycotina
Botryosphaeria quercuum. Dieback. **1**:FL(35).
Nectria grammicospora. Stem canker. **1**:FL(35,82).
Basidiomycotina - Uredinales
Uromyces dolicholi. Autoecious. **1**:FL(35,82), TX(82).
Deuteromycotina - Hyphomycetes
Alternaria tenuissima. Leaf spot. **1**:TX(82).
Cercospora sp. Leaf spot. **1**:FL(35).
Phymatotrichopsis omnivora. Root rot. **1**:TX(82).
Stemphylium sp. Leaf spot. **1**:HI(2903).
Deuteromycotina - Coelomycetes
Colletotrichum gloeosporioides. Anthracnose. **1**:FL(82).
Diplodia cacaoicola. Stem rot. **1**:FL(35,82).
Macrophomina phaseolina. Charcoal rot. **1**:CA(82).
Phyllosticta cajani. Leaf and pod spot. **1**:FL(35,82).
Deuteromycotina - Other
Rhizoctonia sp. Stem and root rot. **1**:FL(35), HI(82,2903).
Sclerotium rolfsii. Southern blight, root rot. **1**:HI(82,2903).

Calliandra Benth.

1. *C. eriophylla* Benth. [fairy-duster, mesquitilla, mock-mesquite]
2. *C. haematocephala* Hassk. [red powderpuff]
3. *C. humilis* Benth.
4. *C. humilis* Benth. var. *reticulata* (A. Gray) L. Benson (=*C. reticulata* A. Gray)
5. *Calliandra* sp.

Ascomycotina
Leptosphaeria sp. Stem gall. **2**:FL(35).
Basidiomycotina - Uredinales
Ravenelia texensis. Autoecious. **1**:AZ(3981). **2**:FL(35). **3**:AZ(1203,3981). **4**:AZ(1195,1203,3379,3981). **5**:AZ(82).
Basidiomycotina - Other
Armillaria tabescens. Root rot. **5**:FL(35).
Deuteromycotina - Hyphomycetes
Cercospora sp. Leaf spot. **5**:FL(35).
Helminthosporium sp. Leaf spot. **5**:FL(35).

Calopogonium Desv.

1. *Calopogonium* sp.

Deuteromycotina - Other
Rhizoctonia solani. Web blight. **1**:FL(35).

Camoensia Welw. ex Benth. in Benth. & J. D. Hook.

1. *C. scandens* (Welw. ex Benth.) J. B. Gillett (=*C. maxima* Welw. ex Benth.)

Oomycetes
Pythium sp. Root rot. **1**:FL(35).

Canavalia DC.

1. *C. ensiformis* (L.) DC. [giant stock bean, horse bean, jack bean, sword bean, wonder bean]
2. *C. gladiata* (Jacq.) DC. [sword bean]
3. *C. hawaiiensis* Degener & Sauer
4. *C. kauensis* St. John
5. *Canavalia* sp.

Oomycetes

Pythium debaryanum. Root rot. **1**:HI(2903).
P. hydnosporum. Root rot. **1**:HI(2903).
P. irregulare. Root rot. **1**:HI(2903).
P. rostratum. Root rot. **1**:HI(2903).
P. splendens. Root rot. **1**:HI(2903).

Ascomycotina

Enchnoa canavaliae. **5**:HI(2903).

Deuteromycotina - Hyphomycetes

Cercospora ternateae. Leaf spot. **5**:AL(82).
Fusarium sp. **1**:HI(2903).
Stenella canavaliae. Leaf spot. **2**:FL(35).

Deuteromycotina - Coelomycetes

Colletotrichum capsici. Pod spot. **1**:FL(82). **2**:FL(35).
C. gloeosporioides. Anthracnose, stem canker, dieback. **1**:HI(82,2903).
C. truncatum. Pod spot. **1**:AL(82).
Fusicoccum canavaliae. Secondary stem spot. **1**:HI(82,2903).
Septoria canavaliae. Leaf spot. **1**:HI(82,2903). **5**:HI(2903).
S. mollereana. **3**:HI(2903).
Septoria sp. Leaf spot. **4**:HI(1146).

Caragana Fabr.

1. *C. arborescens* Lam. [Siberian pea-tree]
2. *Caragana* sp.

Oomycetes

Phytophthora cactorum. Wilt of seedlings. **1**:MO(82).

Ascomycotina

Cucurbitaria caraganae. On limbs. **1**:OH(82).
Eutypella leprosa. **2**:IA(3629).
Thyridium vestitum. **2**:IA(1226).

Basidiomycotina - Other

Phellinus punctatus. On living trees. **1**:ND(3759).
Thanatephorus cucumeris. Root rot of seedlings. **1**:ND(82).

Deuteromycotina - Hyphomycetes

Phymatotrichopsis omnivora. Root rot. **1**:TX(82).

Deuteromycotina - Coelomycetes

Ascochyta sp. Leaf blight. **1**:OH(82).
Hendersonia septem-septata. On twigs. **1**:KS(82).
Phloeospora caraganae. **1**:WY(1195).
Phomopsis caraganae. On limbs. **1**:OH(82).
P. rudis. Leaf spot, on limbs. **1**:OH(82).
Phyllosticta gallorum. **1**:AK(526).
Septoria caraganae. Defoliation. **1**:IA(1228), WI(1336).
Septoria sp. Leaf blight. **1**:MN(82).
Sphaeropsis sp. On limbs. **1**:IA(82).

Deuteromycotina - Other

Rhizoctonia solani. Damping-off. **1**:ND(82).

Cassia L.

1. *C. artemisioides* Gaud. [feathery cassia, shower-tree, wormwood senna]
2. *C. brachiata* (Pollard) Macbr. (=*Chamaecrista brachiata* Pollard)
3. *C. chamaecrista* L.
4. *C. fasciculata* Michx. [partridge pea]
5. *C. fistula* L. [golden-rain, golden-shower, Indian laburnum, pudding-pipe-tree, purging cassia, purging fistula]
6. *C. javanica* Ghesq. var. *indochinensis* Gagnep. (=*C. nodosa* F. Ham. ex Roxb.) [jointwood, pink-and-white-shower]
7. *C. nictitans* L. [wild sensitive-plant]
8. *C. procumbens* L. [sensitive partridgepea]
9. *C. tora* L. [sickle senna, sicklepod]
10. *Cassia* sp.

Ascomycotina

Botryosphaeria rhodina. On dead limbs. **10**:FL(82), MS(82,2720).
Botryotinia sp. Foliar necrosis. **10**:FL(35).
Diaporthe arctii. On dead stems. **3**:GA(1438). **4**:GA(82).
Diaporthopsis sp. **3**:GA(1438).
Erysiphe polygoni. Powdery mildew. **4**:MS(82,2720), TX(82). **9**:FL(82), MD(82).
Glomerella cingulata. Anthracnose. **4**:AL(82), GA(82). **8**:GA(82).
Leptosphaeria cassiaecola. On dead stems. **10**:NY(82), TX(82).
Myiocopron sp. **3**:GA(1438).
Ophiobolus nigro-clypeata. **3**:GA(1438).
Phyllachora canafistulae. Root rot, tar spot. **5**:FL(35). **6**:TX(82). **10**:FL(35).
Physalospora abdita. On dead limbs. **10**:FL(82), MS(82,2720).
Pleospora cassiae. On dead stems. **10**:TX(82).
Urosporella alabamensis. **3**:GA(1438).

Basidiomycotina - Uredinales

Aecidium chamaecristae. **4**:IA(749), KS(82), NE(82), TX(82).
Ravenelia cassiaecola. Autoecious. **4**:FL(82), GA(82), LA(82), MS(82,2720), OH(82), TX(82). **7**:AL(746), MS(2720). **8**:AL(82), MS(82,2720). **10**:FL(35).
R. mesilliana. Autoecious. **10**:AZ(1203).

Basidiomycotina - Other

Armillaria tabescens. Root rot. **5**:FL(35). **6**:FL(82). **10**:FL(35).
Ganoderma lucidum. Heart rot. **10**:FL(35).

Deuteromycotina - Hyphomycetes

Alternaria sp. Leaf spot. **10**:FL(35).
Botrytis sp. Gray mold. **10**:FL(35).
Cercospora chamaecrista. Leaf spot. **3**:KS(409). **4**:KS(82), NE(82).
C. pinnulaecola. Leaf spot. **8**:AL(82).
Fusarium oxysporum f. sp. *batatas*. Wilt. **10**:SC(82).
Phaeoisariopsis nigricans. Leaf spot. **4**:WI(1361). **5**:FL(35). **9**:AL(82), FL(82), LA(82), MO(82), SC(82), TX(82,3597). **10**:FL(35).
P. simulata. Leaf spot. **10**:FL(35).
Phaeoramularia occidentalis. Leaf spot. **10**:FL(35).
Phymatotrichopsis omnivora. Root rot. **4**:TX(82). **8**:TX(82). **9**:TX(82).

Deuteromycotina - Coelomycetes

Aplosporella chamaecristae. **2**:FL(2782).
Lasiodiplodia theobromae. Dieback. **1**:TX(82).
Macrophomina phaseolina. Charcoal rot. **8**:AL(82).
Phomopsis cassiae. Dieback. **9**:FL(82). **10**:FL(35).
Phomopsis sp. Dieback. **5**:FL(35). **10**:FL(35), SC(82).
Phyllosticta cassiicola. Leaf spot. **10**:KS(82).
Septoria cassiicola. Leaf spot. **4**:KS(82), TX(82), WI(82). **8**:KS(82), TX(82), WI(82). **10**:KS(82), TX(82), WI(82).

Deuteromycotina - Other

Rhizoctonia solani. Seedling blight. **5**:FL(35). **9**:FL(82). **10**:FL(35).

Castanospermum A. Cunn. ex Hook.

1. *C. australe* A. Cunn. & C. Fraser [blackbean, Moreton Bay chestnut]

Ascomycotina

Rhytisma sp. Tar spot. **1**:FL(35).

Deuteromycotina - Coelomycetes

Phomopsis sp. Leaf spot. **1**:FL(35).
Phyllosticta sp. Leaf spot. **1**:FL(35).

Centrosema (DC.) Benth.

1. *C. brasilianum* (L.) Benth.
2. *C. plumieri* (Turpin ex Pers.) Benth.
3. *C. pubescens* Benth.
4. *C. virginianum* (L.) Benth.

Deuteromycotina - Hyphomycetes

Cercospora sp. Leaf spot. **4**:FL(35).
Pseudocercospora clitoriae. **4**:AL(82).

Deuteromycotina - Coelomycetes

Colletotrichum gloeosporioides. Anthracnose, canker, leaf spot. **3**:FL(35).
C. truncatum. Anthracnose. **4**:FL(35).
Colletotrichum sp. On pods. **4**:AL(82).

Deuteromycotina - Other

Rhizoctonia solani. Web blight. **1**:FL(35). **2**:FL(35). **3**:FL(35). **4**:FL(35).

Ceratonia L.

1. *C. siliqua* L. [algarroba bean, carob, locust bean, St.-John's-bread]

Oomycetes

Phytophthora cactorum. Root rot. **1**:CA(82).

Ascomycotina

Botryosphaeria ribis. Canker. **1**:CA(82).

Basidiomycotina - Other
Armillaria mellea. Root rot. **1**:CA(2897).
A. tabescens. Root rot. **1**:FL(35).
Ganoderma sp. Butt rot. **1**:FL(35).
Deuteromycotina - Hyphomycetes
Phymatotrichopsis omnivora. Root rot. **1**:TX(82).
Pseudocercospora ceratoniae. Leaf spot. **1**:FL(35).
Verticillium albo-atrum. Wilt. **1**:CA(82).

Cercidium Tul.
1. *Cercidium* sp.
Basidiomycotina - Other
Inonotus rickii. On wood. **1**:AZ(1204).

Cercis L.
1. *C. canadensis* L. [eastern redbud]
2. *C. chinensis* Bunge [Chinese redbud]
3. *C. occidentalis* Torr. ex A. Gray [western redbud]
4. *C. reniformis* Engelm.
Oomycetes
Phytophthora cinnamomi. Root rot. **1**:NC(1303).
Phytophthora sp. Root rot. **1**:FL(35).
Plasmopara cercidis. On leaves. **1**:TN(3201).
Pythium sp. Root rot. **1**:FL(35).
Ascomycotina
Acrospermum foliicola. **1**:GA(1438).
Botryosphaeria dothidea. Canker, dieback. **1**:NC(1303), VA(3493).
B. obtusa. On dead limbs. **1**:Unk.(82).
B. ribis. Canker, dieback. **1**:DE(82), GA(2405), MD(82), NC(82,1533), NJ(82), TX(82), VA(82), WV(954,3907). **2**:MD(82).
Cryptodiaporthe sp. **1**:GA(82,1438).
Diatrype stigma. **1**:IA(3629).
Diatrypella favacea. **1**:OK(660).
Didymella leguminis-cytisis. On old legumes. **3**:CA(81).
Discohainesia oenotherae. On limbs and twigs. **1**:VA(82).
Endothia viridistroma. **1**:GA(82,1438,2365).
Eutypella juglandicola. **1**:GA(1438).
E. leprosa. **1**:IA(3629).
Hypoxylon hypophlaeum. **1**:GA(1438).
H. rubiginosum. **1**:GA(82,1438).
H. truncatum. **1**:GA(82,1438).
Lecanidion simile. On dead twigs. **1**:VA(82).
Lophiostoma macrostomoides. **3**:OK(660).
Mycosphaerella cercidicola. Leaf spot. **1**:Range of host(82), FL(35), GA(1438), MS(2720), NC(1303), OK(2872), WV(3907). **2**:CT(82), KY(82), MD(82). **3**:TX(82).
Nectria sp. On trunks. **1**:FL(35).
Nitschkia broomeiana. **1**:GA(1438).
Rosellinia medullaris. **1**:IN(82).
Basidiomycotina - Other
Armillaria tabescens. Root rot. **1**:FL(35).
Bjerkandera adusta. Sapwood rot. **1**:Range of host(82).
Ceriporia spissa. **1**:MI(82).
Dendrophora albobadia. On limbs. **1**:MD(82).
Erythricium salmonicolor. Limb blight. **1**:FL(35).
Ganoderma applanatum. On wood. **1**:NC(1303).
Microstroma sp. Leaf spot. **1**:MS(2720).
Omphalotus olivascens. On stumps and buried wood. **3**:CA(274).
Peniophora cinerea. **1**:GA(1439).
Phellinus gilvus. Sapwood rot. **1**:Range of host(82), NC(1303,3944).
P. punctatus. **1**:MI(82).
Schizophyllum commune. On wood. **1**:NC(1303).
Stereum ochraceo-flavum. **1**:MA(543).
Trametes hirsuta. **1**:MD(82).
T. versicolor. Sapwood rot. **1**:Range of host(82).
Deuteromycotina - Hyphomycetes
Alternaria sp. Leaf spot. **1**:FL(35), OK(2872).
Cercospora cercidicola. Leaf spot. **1**:FL(35), NC(1303,3944), WI(1352).
C. chionea. Leaf spot. **1**:FL(35), IL(82), IN(82), KS(82), NC(82), OK(82,2872). **2**:IN(82).
Cylindrocladium scoparium. **1**:LA(1694).
Fusarium solani. Canker. **1**:FL(35).
Ovularia cercidis. On leaves. **1**:MO(82).
Phymatotrichopsis omnivora. Root rot. **1**:AZ(82), TX(82). **3**:AZ(82), TX(82). **4**:AZ(82), TX(82).
Pseudocercospora vitis. Leaf spot. **1**:FL(35).
Verticillium albo-atrum. Wilt. **1**:IL(992), KS(1558,3022), VA(3493).
Verticillium sp. Wilt. **1**:OH(82).
Deuteromycotina - Coelomycetes
Aplosporella cercidis. On dead limbs. **1**:KS(82).
Cytospora cercidis. Root rot. **1**:WV(82,3907).
Diplodia cercidis. On twigs. **1**:NY(82), WV(82,3907). **2**:NY(82).
Diplodia sp. Dieback. **1**:FL(35).
Discosia artocreas. Leaf spot. **1**:OK(2872).
Entomosporium mespili. Leaf spot. **1**:FL(35).
Gloeodes pomigena. On twigs. **1**:IN(82).
Gloeosporidiella cercidis. On living leaves. **1**:WI(1361).
Macrophoma cercis. On leaves. **1**:FL(35), IL(82), NC(1303).
Pestalotia sp. Leaf spot. **1**:OK(2872).
Pestalotiopsis maculans. On leaves. **1**:IL(2568).
Phoma cercidicola. On dead limbs. **2**:NY(82).
Phyllosticta cercidicola. Leaf spot. **1**:FL(35,82), IN(82), OK(2872), WV(82,3907).
P. siliquastri. On pods. **1**:AL(82), TX(82).
Phyllosticta sp. Leaf spot. **1**:NC(1303).
Sphaeropsis cercidis. On dead twigs. **1**:KS(82).
Tubakia dryina. **1**:IL(82,1246).

Cicer L.
1. *C. arietinum* L. [Bengal gram, chickpea, Egyptian-pea, garbanzo, yellow gram]
Oomycetes
Pythium ultimum. Root rot, wilt. **1**:CA(82,2901,3863).
Ascomycotina
Mycosphaerella rabiei. **1**:ID(1831).
Mycosphaerella sp. Stem necrosis. **1**:AL(82).
Basidiomycotina - Uredinales
Uromyces ciceris-arietini. **1**:SD(3731), TX(128).
Deuteromycotina - Hyphomycetes
Fusarium oxysporum f. sp. *ciceris*. Root rot, wilt. **1**:CA(1001,3863).
F. solani f. sp. *pisi*. Root rot, wilt. **1**:CA(3863).
Thielaviopsis basicola. Root rot. **1**:ID(347), WA(347).
Verticillium albo-atrum. **1**:CA(1002).
Deuteromycotina - Coelomycetes
Macrophomina phaseolina. Root rot, wilt. **1**:CA(2901,3863).
Phoma rabiei. Blight. **1**:ID(851).
Deuteromycotina - Other
Rhizoctonia solani. Root rot, stem rot, wilt. **1**:CA(82,2901,3863).

Cladrastis Raf.
1. *C. lutea* (F. Michx.) K. Koch [virgilia, yellow-wood]
2. *Cladrastis* sp.
Ascomycotina
Botryosphaeria ribis. Dieback. **1**:OK(2872), PA(82).
Phyllactinia guttata. Powdery mildew. **1**:OK(82,2872).
Basidiomycotina - Other
Fomitopsis spraguei. Wood rot. **1**:MD(82).
Oxyporus latemarginatus. **2**:DC(2166).
Deuteromycotina - Hyphomycetes
Verticillium albo-atrum. Wilt. **1**:IL(82,1558).

Clitoria L.
1. *C. mariana* L.
2. *C. ternatea* L. [Asian pigeonwings, butterfly pea]
3. *Clitoria* sp.
Deuteromycotina - Hyphomycetes
Cercospora ternateae. Leaf spot. **2**:FL(35).
Phymatotrichopsis omnivora. Root rot. **1**:TX(82).
Pseudocercospora clitoriae. Leaf spot. **1**:AL(82). **2**:FL(35,82).
P. cruenta. Leaf spot. **2**:FL(82).
Deuteromycotina - Coelomycetes
Colletotrichum gloeosporioides. Anthracnose, stem canker, leaf spot. **2**:FL(35). **3**:FL(35).
C. truncatum. Anthracnose, stem canker, leaf spot. **2**:FL(35).
Phyllosticta sojicola. Leaf spot. **3**:FL(35).

Colutea L.

1. *C. arborescens* L. [bladder senna]

Oomycetes

Phytophthora cactorum. Wilt of seedlings. **1:**MO(82).

Ascomycotina

Cucurbitaria elongata. On dead limbs. **1:**Unk.(82).
Dothidea puccinioides. On twigs. **1:**PA(82).
Erysiphe polygoni. Powdery mildew. **1:**Unk.(82).

Basidiomycotina - Uredinales

[*Uromyces coluteae*]. Telial host. **1:**KS(82).

Basidiomycotina - Other

Ganoderma sp. Root rot. **1:**OK(82,2872).

Deuteromycotina - Coelomycetes

Camarosporium coluteae. On limbs. **1:**NY(82).
Diplodia coluteae. Twig blight. **1:**PA(82).
Phomopsis coluteae. On leaves. **1:**NJ(82).

Coronilla L.

1. *C. varia* L. [crownvetch, trailing crownvetch]

Oomycetes

Pythium myriotylum. Damping-off, root rot, wilt. **1:**MD(926).

Ascomycotina

Sclerotinia trifoliorum. **1:**PA(3301).

Deuteromycotina - Hyphomycetes

Cercospora rautensis. Leaf and stem spot. **1:**PA(1296).
Dictyochaeta fertilis. Root rot. **1:**NC(1303).
Fusarium roseum. Storage decay. **1:**PA(596).
Stemphylium sp. On leaves. **1:**PA(1296).

Deuteromycotina - Coelomycetes

Colletotrichum destructivum. Isolated from diseased plants. **1:**PA(1091).
C. trifolii. Anthracnose. **1:**NC(1303).
Colletotrichum sp. Anthracnose of leaves and stems. **1:**PA(1296).

Deuteromycotina - Other

Rhizoctonia sp. Root rot. **1:**HI(2903).
Sclerotium rolfsii. **1:**KY(2886).

Coursetia DC.

1. *C. glandulosa* A. Gray
2. *C. microphylla* A. Gray
3. *Coursetia* sp.

Basidiomycotina - Uredinales

Phragmopyxis acuminata. Autoecious. **1:**AZ(747). **2:**AZ(1195,1203,3981). **3:**AZ(82), CA(2458).
P. noelii. Autoecious. **1:**AZ(747). **2:**AZ(222,1203,3981).

Cracca Benth. in Benth. & Örst.

1. *C. edwardsii* A. Gray

Basidiomycotina - Uredinales

Phragmopyxis deglubens. Autoecious. **1:**AZ(1194,1195,1203,3981).

Crotalaria L.

1. *C. brevidens* Benth. var. *intermedia* (Kotschy) Polh. (=*C. intermedia* Kotschy)
2. *C. incana* L.
3. *C. juncea* L.
4. *C. pallida* Aiton (=*C. mucronata* Desv., *C. striata* DC.) [smooth crotalaria, striped crotalaria]
5. *C. retusa* L.
6. *C. sagittalis* L.
7. *C. spectabilis* Roth [showy crotalaria]
8. *C. verrucosa* L.
9. *C. zanzibarica* Benth. (=*C. usaramoensis* E. G. Baker)
10. *Crotalaria* sp.

Oomycetes

Phytophthora nicotianae var. *parasitica*. Root rot. **2:**HI(2903). **10:**HI(2903).
P. palmivora. Root rot. **2:**HI(2903). **9:**HI(2903).

Ascomycotina

Calonectria crotalariae. Collar rot. **2:**HI(2903). **4:**HI(2903). **5:**HI(2903). **10:**HI(2903).
Diaporthe crotalariae. Stem canker. **7:**FL(35,82).
Erysiphe polygoni. Powdery mildew. **7:**FL(35,82).
Microsphaera diffusa. Powdery mildew. **3:**FL(35), VA(82). **10:**FL(35), MD(82).
Mycosphaerella sp. Leaf spot. **10:**FL(35).
Neocosmospora vasinfecta. On dead stems. **5:**GA(82). **10:**GA(1438).
Parodiella hedysari. Black mildew. **10:**FL(82).

Basidiomycotina - Other

Armillaria tabescens. Root rot. **4:**FL(82). **10:**FL(35).
Thanatephorus cucumeris. Thread blight, stem spot. **1:**FL(82). **2:**SC(82). **7:**FL(35), LA(82), MS(82,2720).

Deuteromycotina - Hyphomycetes

Alternaria sp. Leaf spot. **7:**FL(35,82), TX(82).
Botrytis cinerea. Blight, leaf spot. **1:**NC(82). **7:**FL(35,82), NC(82).
Cercospora cotizensis. Leaf spot. **7:**FL(35). **10:**FL(35).
C. demetrioniana. Leaf spot. **4:**HI(2903). **6:**DE(82,3543), MO(82), TX(82). **10:**FL(35), HI(2903), LA(82).
Cercospora sp. Leaf spot. **4:**FL(82), LA(82), MS(82,2720), NC(82). **7:**FL(82), LA(82), MS(82,2720), NC(82).
Fusarium oxysporum. Wilt. **9:**HI(2903).
F. oxysporum f. sp. *tracheiphilum*. Wilt. **7:**MS(2720).
F. oxysporum f. sp. *vasinfectum*. **7:**HI(2903).
F. udum f. sp. *crotalariae*. Wilt. **1:**MS(2720). **7:**MS(2720). **10:**FL(35,82), LA(82), MS(82).
Fusarium sp. Root rot, wilt. **1:**FL(82), NC(82). **5:**GA(82), TX(82). **7:**LA(82), MS(82), NC(82), OK(82,2873). **10:**OK(2872).
Helminthosporium sp. Pod blight. **7:**FL(35,82).
Oidium erysiphoides f. *crotalariae*. Powdery mildew. **5:**FL(82). **7:**LA(82), MS(82,2720). **10:**FL(35).
Phymatotrichopsis omnivora. Root rot. **1:**TX(82). **4:**TX(82). **7:**TX(82).
Stemphylium sp. On stems. **8:**SC(82).

Deuteromycotina - Coelomycetes

Ascochyta pisi. Anthracnose. **7:**OK(2872).
Chaetopyrena sp. On leaves. **10:**LA(82).
Colletotrichum gloeosporioides. Anthracnose. **4:**FL(82). **7:**FL(35,82). **10:**FL(35).
Diplodia sp. Dieback. **10:**FL(35).
Macrophomina phaseolina. Charcoal root rot. **1:**FL(82). **4:**OK(82), TX(82). **7:**FL(35,82), OK(82,2873), TX(82). **10:**FL(35), OK(2872).
Phoma sp. Leaf spot. **6:**MT(82,3202).
Pyrenochaeta sp. Secondary leaf spot. **7:**LA(82), MS(82,2720).
Septoria sp. Leaf spot. **4:**NC(82).

Deuteromycotina - Other

Rhizoctonia solani. Root and stem rot. **10:**Unk.(82), FL(35,82).
Rhizoctonia sp. Stem canker. **1:**FL(82). **7:**LA(82).
Sclerotium rolfsii. Root and stem rot. **3:**FL(35). **10:**FL(35,82), MS(82,2720), NC(82).

Cyamopsis DC.

1. *C. tetragonoloba* (L.) Taub. (=*C. psoralioides* DC.) [cluster bean, guar]

Ascomycotina

Leveillula taurica. Powdery mildew. **1:**AZ(2374).

Deuteromycotina - Hyphomycetes

Alternaria cucumerina. Leaf spot. **1:**GA(2197).
Alternaria sp. Leaf spot. **1:**CA(82), MS(2720).
Cercospora kikuchii. Purple stain. **1:**MS(1784).
Fusarium sp. Wilt. **1:**AZ(82), OK(82), TX(82).
Phymatotrichopsis omnivora. Root rot. **1:**AZ(82), TX(82).

Deuteromycotina - Coelomycetes

Macrophomina phaseolina. Charcoal rot. **1:**AZ(2375).

Deuteromycotina - Other

Rhizoctonia solani. Black root rot. **1:**AZ(82), OK(2872).
Sclerotium rolfsii. Southern blight, white mold, root rot. **1:**AZ(82), GA(2197), MS(2720), TX(82).

Cytisus L.

1. *C. racemosus* Hort. ex Marnock
2. *C. scoparius* (L.) Link (=*Sarothamnus scoparius* (L.) Wimm. ex W. Koch) [Scotch-broom]
3. *Cytisus* sp.

Oomycetes

Pythium sp. Root rot. **1:**FL(35).

Ascomycotina

Botryosphaeria obtusa. On limbs. **3:**AL(82).
Diaporthe inaequalis. On dead limbs. **3:**NJ(82), NY(82).
D. rudis. On dead limbs. **3:**NJ(82), NY(82).

D. sarothamni. On dead limbs. **3:**NJ(82), NY(82).
Dothidea puccinioides. On dead limbs. **3:**CA(82).
Nectria cinnabarina. On stems. **2:**MA(1390).
N. coccinea. On limbs. **2:**OR(3202). **3:**OR(82).
Thielavia basicola. On dead roots. **2:**WI(82). **3:**WI(82).
Basidiomycotina - Other
Maireina marginata. **2:**OR(3202).
Meruliopsis corium. **2:**OR(3202,4007).
Trichaptum biforme. **2:**OR(3202).
Deuteromycotina - Hyphomycetes
Botrytis cinerea. Gray mold. **2:**FL(35).
Periconia byssoides. **2:**GA(659).
Pleiochaeta setosa. **3:**Unk.(82).
Deuteromycotina - Coelomycetes
Camarosporium laburni. On dead limbs. **2:**CA(81).
Pestalotia sp. **2:**GA(659).
[*Pestalotiopsis funerea*]. On twigs. **3:**CA(82).
Phomopsis sp. On limbs. **3:**MA(82), NJ(82).
Selenophoma juncea. On stems. **2:**MA(1394).
Deuteromycotina - Other
Rhizoctonia sp. Root rot. **1:**FL(35).

Dalbergia L. fil.

1. *D. sissoo* Roxb. ex DC. [Indian rosewood, sissoo, sisu]

Oomycetes
Phytophthora sp. Root rot. **1:**FL(35).
Pythium sp. Root rot. **1:**FL(35).
Basidiomycotina - Other
Armillaria tabescens. Root rot. **1:**FL(35).
Deuteromycotina - Hyphomycetes
Alternaria sp. Leaf spot. **1:**FL(35).
Fusarium sp. Root rot. **1:**FL(35).
Deuteromycotina - Coelomycetes
Colletotrichum gloeosporioides. Anthracnose. **1:**FL(35).
Diplodia sp. Dieback. **1:**FL(35).
Phomopsis sp. Leaf spot. **1:**FL(35).

Dalea L.

(= *Petalostemon* Michx.)
1. *D. albiflora* A. Gray
2. *D. candida* Willd. (=*Kuhnistera candida* (Willd.) Kuntze, *Petalostemon candidum* (Willd.) Michx., *Petalostemon oligophyllum* (Torr.) Rydb.) [white prairie-clover]
3. *D. candida* Michx. ex Willd. var. *oligophylla* (Torr.) Shinners (=*Petalostemon occidentale* (A. Heller) Fernald)
4. *D. cylindriceps* Barneby (=*Petalostemon compactus* Swezey)
5. *D. enneandra* Nutt. (=*D. laxiflora* Pursh)
6. *D. formosa* Torr. [feather-plume]
7. *D. frutescens* A. Gray [black dalea]
8. *D. mollis* Benth.
9. *D. multiflora* (Nutt.) Shinners (=*Petalostemon multiflorus* Nutt.)
10. *D. purpurea* Venten. (=*Petalostemon purpureum* (Venten.) Rydb., *Petalostemon violaceus* Michx.)
11. *D. villosa* (Nutt.) Spreng. (=*Petalostemon villosum* Nutt.)
12. *Dalea* sp. (=*Petalostemon sp.*)

Ascomycotina
Nematostoma occidentalis. **11:**WI(1346).
Pleospora comata. On dead stems. **2:**OK(2872). **9:**OK(2873). **10:**OK(2872). **12:**NM(82), OK(82).
Basidiomycotina - Uredinales
Puccinia andropogonis. **5:**KS(82), NE(82), OK(2872), SD(82). **10:**NM(1195). **12:**W states(82), CO(82), TX(82), WY(82).
P. paroselae. Autoecious. **8:**CA(82,746).
Uropyxis daleae. **1:**AZ(3981). **6:**AZ(3981). **7:**TX(742).
U. daleae var. *eysenhardtiae*. Autoecious. **1:**AZ(219,1203). **6:**AZ(1203).
U. petalostemonis. **2:**AZ(1203), MT(3202), ND(378), NM(111). **3:**AZ(3981). **4:**SD(2243). **12:**CO(82), MN(82), ND(82), NM(82), WI(82,219).
Deuteromycotina - Hyphomycetes
Alternaria sp. On blighted tips of leaflets. **2:**WI(1332).
Cercospora daleae. Leaf spot. **5:**KS(82,581).
Phymatotrichopsis omnivora. Root rot. **12:**TX(82).
Deuteromycotina - Coelomycetes
Hendersonia petalostemonis. On stems. **2:**NM(2399).

Delonix Raf.

1. *D. regia* (Bojer ex Hook.) Raf. (=*Poinciana regia* Bojer ex Hook.) [flamboyant-tree, flame-tree, royal poinciana]

Basidiomycotina - Uredinales
Ravenelia humphreyana. Autoecious. **1:**FL(35).
Basidiomycotina - Other
Armillaria tabescens. Root rot. **1:**FL(35,82).
Hexagonia hydnoides. On wood. **1:**FL(35).
Phanerochaete sordida. **1:**FL(128).
Deuteromycotina - Coelomycetes
Colletotrichum sp. Anthracnose. **1:**FL(35).
Diplodia sp. Dieback. **1:**FL(35).
Phomopsis sp. Twig dieback. **1:**FL(35).
Phyllosticta sp. Leaf spot. **1:**FL(35).
Sphaeropsis sp. Canker. **1:**FL(35).
Deuteromycotina - Other
Rhizoctonia solani. Root rot. **1:**FL(35).

Desmanthus Willd.

1. *D. acuminatus* Benth. in Hook.
2. *D. brachylobus* Benth.
3. *D. cooleyi* (Eaton) Trel.
4. *D. illinoensis* (Michx.) MacMill. ex B. L. Robinson & Fernald [prairie-mimosa, prickleweed]
5. *D. velutinus* Scheele [velvet bundleflower]
6. *Desmanthus* sp.

Ascomycotina
Diaporthe phaseolorum. **4:**OK(660).
Erysiphe polygoni. Powdery mildew. **4:**IL(82), MS(82,2720).
Basidiomycotina - Uredinales
Ravenelia texensis. Autoecious. **3:**AZ(3981), TX(82).
Deuteromycotina - Hyphomycetes
Cercosporidium desmanthi. Leaf spot. **1:**TX(267). **2:**Unk.(962). **4:**KS(82,267,839), LA(82,839), MO(82,839), NM(82,839), OK(267), SD(82,839,2243), TN(82,839). **5:**TX(267).
Deuteromycotina - Coelomycetes
Colletotrichum truncatum. Anthracnose. **6:**FL(35).

Desmodium Desv.

1. *D. acuminatum* DC.
2. *D. canadense* (L.) DC. [beggarweed]
3. *D. canescens* (L.) DC.
4. *D. ciliare* (Muhl. ex Willd.) DC.
5. *D. cuspidatum* (Muhl. ex Willd.) Loud. (=*D. bracteosum* (Michx.) DC., *D. bracteosum* (Michx.) DC. var. *longifolium* (Torr. & A. Gray) B. L. Robinson, *D. longifolium* (Torr. & A. Gray) F. P. Daniels)
6. *D. dillenii* Darl.
7. *D. glabrum* (Mill.) DC. (=*D. molle* (Vahl) DC.)
8. *D. glutinosum* (Muhl. ex Willd.) Schindl.
9. *D. illinoense* A. Gray
10. *D. laevigatum* (Nutt.) DC.
11. *D. lineatum* DC.
12. *D. marilandicum* (L.) DC.
13. *D. neomexicanum* A. Gray (=*D. exiguum* A. Gray)
14. *D. nudiflorum* (L.) DC.
15. *D. obtusum* (Muhl. ex Willd.) DC. (=*D. rigidum* (Elliott) DC.)
16. *D. ochroleucum* M. A. Curtis ex Canby
17. *D. paniculatum* (L.) DC.
18. *D. paniculatum* (L.) DC. var. *angustifolium* Torr. & A. Gray
19. *D. perplexum* B. G. Schubert
20. *D. procumbens* (Mill.) Hitchc.
21. *D. rhombifolium* (Elliott) DC.
22. *D. rosei* Schub.
23. *D. rotundifolium* DC.
24. *D. sessilifolium* (Torr.) Torr. & A. Gray
25. *D. strictum* (Pursh) DC.
26. *D. tortuosum* (Sw.) DC. (=*D. purpureum* (Mill.) Fawc. & Rendle) [beggarweed, Florida beggarweed]
27. *D. viridiflorum* (L.) DC.
28. *Desmodium* sp.

Chytridiomycetes
Synchytrium lagerheimii. Leaf gall. **3:**MS(82,2720), TX(82).
Zygomycotina
Choanephora cucurbitarum. Flower and leaf blight. **28:**FL(35).

Ascomycotina
Diaporthe arctii. On stems. **3:**GA(1438). **18:**GA(1438). **28:**NJ(82), NY(82).
Ellisiodothis inquinans. **28:**GA(1438).
Elsinoe wisconsinensis. Spot anthracnose. **9:**WI(2278).
Elsinoe sp. **9:**WI(1348).
Erysiphe polygoni. Powdery mildew. **3:**OH(82), TX(82). **28:**FL(35).
Hypoderma commune. **3:**GA(1438).
Leptosphaeria distributa. On dead stems. **28:**NJ(967).
Massarina gloeospora. **14:**GA(1438).
Metasphaeria sp. **2:**GA(1438).
Microsphaera diffusa. Powdery mildew. **1:**WI(1336). **5:**WI(1345). **9:**WI(1339). **28:**MS(2720), NC(1303,3944).
M. ludens. Powdery mildew. **2:**WI(82,1332).
Microthyrium sp. **2:**GA(1438).
Mycosphaerella sp. Midribs of living leaflets. **2:**WI(1354). **3:**MO(82).
Myiocopron sp. **2:**GA(1438).
Ophiobolus nigro-clypeata. **2:**GA(1438). **14:**GA(1438). **18:**GA(1438).
Parodiella hedysari. Black mildew, black spot. **1:**WI(1353). **2:**GA(1438), NC(3944). **3:**MS(2720). **12:**OK(2872). **15:**MS(2720). **17:**MS(2720). **25:**SC(3547). **28:**E states(82), MS(82), ND(82).
Phyllachora simillima. Tar spot. **11:**SC(82).
Phyllactinia guttata. Powdery mildew. **2:**IA(82). **8:**IA(82).
Basidiomycotina - Uredinales
Puccinia andropogonis. Heteroecious: aecial host. **8:**MN(82), WI(82). **19:**AL(82).
Uromyces hedysari-paniculati. Autoecious. **2:**GA(1439), MS(2720). **3:**Unk.(634), MS(2720), NC(1303). **5:**MS(2720). **6:**MS(2720). **10:**NC(1303,3944). **12:**MS(2720), NC(3944). **15:**MA(1390), MS(2720), NC(1303). **16:**GA(1439). **17:**GA(1439), MS(2720), NC(1303,3944), OK(2872), TX(652). **23:**IN(987). **24:**OK(2872). **25:**NC(1303). **27:**GA(1439). **28:**E states(82), AZ(1195,1203,3981), FL(35), IN(647), MN(82), NC(1303), TX(82).
U. mexicanus. Autoecious. **13:**AZ(3981). **20:**AZ(1195,1203,3379,3981). **22:**AZ(1195,3379,3981).
Basidiomycotina - Ustilaginales
Thecaphora deformans. On living leaves. **5:**NJ(1077). **8:**NJ(82). **14:**MD(82,1077), PA(82,1077), VA(82,1077), WI(1357).
Deuteromycotina - Hyphomycetes
Cercospora desmodii. Leaf spot. **7:**AL(82). **8:**IL(82), KS(82), MI(82), NY(82), WI(82). **14:**IA(82), MO(82), WI(82). **23:**VA(82). **27:**DE(3543). **28:**FL(35), OK(2872).
C. desmodiicola. Leaf spot. **2:**WI(1334). **3:**OK(82,2872). **9:**WI(1354). **23:**IN(82). **26:**FL(82). **28:**FL(35).
Cladosporium sp. On leaves. **14:**WI(1350).
Cristulariella moricola. On leaves. **3:**IL(2572).
Dendryphiella infuscans. On living leaves. **25:**SC(82).
Fusarium sp. Leaf spot. **28:**FL(35).
Isthmospora glabra. Leaf spot. **28:**TX(82).
Phymatotrichopsis omnivora. Root rot. **28:**TX(82).
Ramularia desmodii. Leaf spot. **2:**IA(82), IL(82), KS(82), MS(82,2720), NE(82), WI(82). **3:**NE(82). **4:**SC(82,634). **5:**KS(82). **9:**KS(82), WI(82). **11:**LA(82), MS(82,2720), SC(82). **17:**WI(82). **19:**AL(82), TX(82). **28:**FL(35).
Deuteromycotina - Coelomycetes
Ascochyta sp. **9:**WI(1345).
Colletotrichum dematium. On stems. **14:**Unk.(82).
C. gloeosporioides. Anthracnose, stem canker, leaf spot. **28:**FL(35).
C. truncatum. Anthracnose, on leaves, pods and stems, seedling blight. **26:**FL(497). **28:**FL(35,2086), GA(3845).
Colletotrichum sp. On dead upper portions of stems. **5:**WI(1360).
Gloeosporidiella desmodii. On living leaves. **2:**WI(1344).
Phyllosticta desmodii. Leaf spot. **2:**IA(82), IL(82), WI(82,1334). **5:**WI(1342). **9:**WI(82). **23:**IN(82). **28:**FL(35), WI(966).
P. macroguttata. Leaf spot. **6:**AL(64). **19:**AL(82). **28:**AL(64).
P. meibomiae. Leaf spot. **19:**IL(82). **21:**FL(82). **28:**FL(35).
Phyllosticta sp. **5:**WI(1343). **9:**WI(1359).
Septoria sp. **14:**WI(1351).
Stagonospora desmodii. On stems. **26:**FL(82). **28:**FL(35).
Deuteromycotina - Other
Rhizoctonia solani. Web blight. **28:**FL(35).
Sclerotium desmodii. On leaves. **23:**SC(82).
S. rolfsii. Southern blight, root and stem rot. **7:**FL(82), TX(82). **26:**FL(82), TX(82). **28:**FL(35).

Dichrostachys (DC.) Wight & Arnott
1. *D. cinerea* (L.) Wight & Arnott (=*D. nutans* (Pers.) Benth.)
Ascomycotina
Hypoxylon rubiginosum. **1:**WA(3202).
Rhytidhysteron rufulum. **1:**WA(3202).

Enterolobium Mart.
1. *Enterolobium* sp.
Deuteromycotina - Hyphomycetes
Alternaria sp. Petiole lesions. **1:**FL(35).
Fusarium sp. On stems. **1:**FL(35).
Deuteromycotina - Coelomycetes
Diplodia sp. Stem lesion. **1:**FL(35).

Erythrina L.
1. *E. crista-galli* L. [cockspur coraltree, cry-baby-tree]
2. *E. flabelliformis* Kearn.
3. *E. herbacea* L. [cardinal-spear, Cherokee bean, coral bean]
4. *E. sandwicensis* Degener
5. *Erythrina* sp.
Ascomycotina
Lecanidion atratum. **4:**HI(2903).
Nectria cinnabarina. On stems. **5:**AL(82).
Basidiomycotina - Other
Armillaria mellea. Root rot. **1:**CA(2900).
A. tabescens. Root rot. **5:**FL(35,82).
Pellicularia koleroga. Thread blight. **5:**FL(82).
Deuteromycotina - Hyphomycetes
Cercospora erythrinae. Leaf spot. **1:**LA(581,962), TX(581). **5:**FL(35), LA(82), TX(82).
C. erythrinicola. On living leaves. **3:**TX(581,3597). **5:**TX(82).
Epicoccum nigrum. **2:**AZ(662).
Oidium sp. Powdery mildew. **4:**HI(2903). **5:**FL(35).
Phymatotrichopsis omnivora. Root rot. **5:**TX(82).
Ramularia sp. Leaf spot. **4:**HI(2899,2903).
Verticillium sp. Wilt. **5:**CA(82).
Deuteromycotina - Coelomycetes
Colletotrichum gloeosporioides. Anthracnose. **5:**FL(35,82).
Phoma erythrinicola. On stems. **5:**MS(82,2720).
Phyllosticta sp. Leaf spot. **5:**FL(35).
Stagonospora erythrinae. On dead plants. **4:**HI(2903).
Deuteromycotina - Other
Rhizoctonia ramicola. Thread blight. **5:**FL(35,82).

Eysenhardtia Kunth in Humb., Bonpl. & Kunth
1. *E. amorphoides* H. B. K. (=*E. orthocarpa* S. Wats.)
2. *E. polystachya* (Ort.) Sarg. [kidneywood]
Basidiomycotina - Uredinales
Uropyxis daleae. Autoecious. **2:**AZ(1195).
U. daleae var. *eysenhardtiae*. Autoecious. **2:**AZ(1203,3981).
U. holwayi. Autoecious. **1:**AZ(219). **2:**AZ(1194,1203,3981).
Deuteromycotina - Hyphomycetes
Phymatotrichopsis omnivora. Root rot. **2:**TX(82).

Galactia P. Br.
1. *G. eggersii* Urb.
2. *G. elliottii* Nutt.
3. *G. mollis* Michx. (=*G. pilosa* Nutt.)
4. *G. striata* (Jacq.) Urb.
5. *G. volubilis* (L.) Britton
Deuteromycotina - Hyphomycetes
Cercospora flagellifera. Leaf spot. **3:**AL(581). **5:**OK(2873).
C. galactiae. Leaf spot. **2:**FL(35).
Deuteromycotina - Coelomycetes
Colletotrichum gloeosporioides. Stem canker, leaf spot. **4:**FL(35).
C. truncatum. Stem canker, leaf spot. **4:**FL(35).
Sphaceloma floridense. Anthracnose, scab. **1:**FL(1745). **2:**FL(35,82).

Genista L.
1. *Genista* sp.
Ascomycotina
Botryosphaeria obtusa. On limbs. **1:**NJ(82), NY(82).
Diaporthe rudis. On limbs. **1:**NJ(82).

Dothidea puccinioides. On limbs. **1**:NJ(82).
Erysiphe polygoni. Powdery mildew. **1**:MO(82).
Deuteromycotina - Coelomycetes
Cytospora sp. On limbs. **1**:NY(82).
Diplodia sp. On limbs. **1**:CA(82), NJ(82).
Phoma genista. On limbs. **1**:CA(82).
Phomopsis rudis. **1**:Unk.(82).

Gleditsia L.

1. *G. japonica* Miq. [Japanese honey locust]
2. *G. triacanthos* L. [honey locust, honeyshuck, sweet locust]
3. *G. triacanthos* L. var. *inermis* (Pursh) C. K. Schneid.
4. *Gleditsia* sp.

Oomycetes
Phytophthora citrophthora. Collar rot. **2**:CA(82).
Ascomycotina
Aglaospora profusa. On limbs and twigs. **2**:PA(82).
Biscogniauxia marginata. Canker. **2**:MS(2720).
Botryosphaeria dothidea. **2**:GA(1438).
B. obtusa. On limbs. **2**:Range of host(82).
B. quercuum. Dieback. **2**:FL(35).
B. rhizogena. **2**:OK(2872).
B. rhodina. On limbs. **2**:AL(82), SC(82).
B. ribis. **2**:GA(82).
Camarops tubulina. **4**:LA(2675).
Crocicreas cyathoideum. On fallen pods. **2**:OH(638).
Cucurbitaria elongata. Canker. **2**:OH(82,878).
C. recuperata. On limbs. **2**:SC(82).
Dothidea puccinioides. **2**:GA(1438).
Eutypella conseptata. **4**:PA(2925).
E. juglandicola. **2**:GA(82,1438).
E. scoparia. **2**:IA(3629).
Glomerella cingulata. On leaves. **2**:MD(82).
Hypoxylon laurus. On dead limbs. **2**:OH(2403), VA(2403).
H. rubiginosum. On wood. **2**:IA(1225).
Hysterographium mori. **2**:GA(1438).
Linospora gleditschiae. Tar spot. **2**:SC states(82), SE states(82), FL(35), GA(184,1438), MS(2720), NC(3944), OK(184).
Microsphaera penicillata. Powdery mildew. **2**:Range of host(82), FL(35), GA(1438), OH(950), OK(2872).
M. ravenelii. Powdery mildew. **2**:Unk.(82), GA(1438), MS(2720). **4**:SC(634).
Mycosphaerella sp. **2**:Unk.(82), GA(1438).
Nectria cinnabarina. Canker. **2**:AL(82), MI(82), MN(230), SC(82).
N. episphaeria. Stem canker. **2**:FL(35).
Nectria sp. Canker. **2**:CO(49).
Nummularia signata. On large limbs and trunks. **4**:SE states(1810).
Parodiella hedysari. On leaves. **2**:MO(82).
Pezicula australis. On bark. **2**:MA(1394).
Thyronectria austroamericana. Twig and trunk canker, wilt. **1**:MA(82,3171,3172). **2**:E states(82), AL(82,1721), CO(49,231,1668,1721), IL(231,1721), KS(735,1721), LA(82), MA(1721,3171,3172), MS(82,1721), NE(82,1721,2983), OK(618,1721), TN(82,1721), WV(3907).
Valsa ceratosperma. **2**:KS(3415).
Valsaria insitiva. On dead limbs. **2**:IA(1226), IL(1245).
V. notarisii. **2**:GA(1438).
Xylaria carpophila. On wood. **2**:OK(618).
X. mali. Root rot. **2**:VA(82).
Basidiomycotina - Uredinales
Ravenelia opaca. Autoecious. **2**:IL(82,746).
Basidiomycotina - Other
Auricularia auricula. **4**:AR(2675), LA(2675).
Bjerkandera adusta. **4**:AR(2675), LA(2675), MS(2675).
Botryobasidium sp. On wood. **2**:OK(618).
Calocera cornea. **4**:LA(2675).
Dacrymyces capitatus. On dead limbs. **2**:KS(3193).
Fomes fasciatus. Heart rot. **2**:FL(35). **4**:LA(2675).
Fomitella supina. Wood rot. **2**:Unk.(82). **4**:LA(2675).
Fomitopsis meliae. Heart rot. **2**:Range of host(82), S states(82), FL(35). **4**:AR(2675), LA(2675).
Ganoderma applanatum. Heart rot. **2**:Range of host(82), FL(35).
G. lucidum. White root and butt rot. **3**:MS(2721). **4**:AR(2675), LA(2675), MS(2675).
Hexagonia hydnoides. Wood rot. **2**:Unk.(82). **4**:LA(2675).
Inonotus porrectus. On wood. **2**:MS(2720).
Irpex lacteus. On wood. **2**:Unk.(82), GA(1439), OK(618).
Laetiporus sulphureus. On wood. **2**:Unk.(82), MS(2720).
Oxyporus latemarginatus. **2**:S states(82).
O. populinus. **2**:Range of host(82).
Perenniporia tenuis. Wood rot. **2**:Unk.(82).
Phanerochaete flavido-alba. **2**:LA(128).
Phellinus conchatus. On wood. **2**:MS(2720).
P. gilvus. On wood. **2**:MS(2720).
P. laevigatus. Heart rot. **2**:Range of host(82), FL(35). **4**:LA(2675).
P. robiniae. Stem canker. **2**:OK(618).
P. spiculosus. On wood. **2**:MS(2720).
Phlebia chrysocrea. On wood. **2**:MS(2720).
Pluteus atricapillus. On wood. **2**:OK(618).
Polyporus arcularius. Wood rot. **2**:Unk.(82). **4**:LA(2675).
Schizophyllum commune. Wood rot. **2**:Range of host(82), FL(35), OK(618). **4**:AR(2675), LA(2675).
Scytinostroma portentosum. **4**:LA(2675).
Trametes elegans. On wood. **2**:S states(82). **4**:AR(2675), LA(2675), MS(2675).
T. hirsuta. **2**:NC(3944).
T. versicolor. Wood rot. **2**:Unk.(82).
T. villosa. **4**:AR(2675), LA(2675).
Trichaptum biforme. White rot. **2**:Unk.(82).
Deuteromycotina - Hyphomycetes
Alternaria sp. Leaf spot. **2**:OK(2872).
Cercospora condensata. Leaf spot. **2**:NC states(82), DE(3543), OK(2872).
C. seymouriana. On leaves. **2**:C states(82), SE states(82), MS(2720), TX(82).
Phymatotrichopsis omnivora. Root rot. **2**:OK(82,2872), TX(82).
Deuteromycotina - Coelomycetes
Cytospora sacculus. Canker. **2**:CO(49), KS(82), OK(618). **3**:IL(3147).
Dothiorella gleditschiae. On limbs. **2**:MS(82,2720), SC(82).
Gyrostroma austroamericana. Bark lesions. **3**:IL(3549).
Hendersonia sp. **2**:Unk.(82).
Libertella gleditschiae. Canker. **2**:MO(82), OK(618).
Macrophoma mamillaris. On limbs. **2**:SC(82).
M. triacanthi. **2**:Unk.(82).
Melasmia hypophylla. Tar spot. **2**:Unk.(82), NC(3944), OK(2872), WI(1333).
Phoma mamillaris. **2**:Unk.(82).
Sphaeropsis gleditschiae. On legumes. **4**:SC(634).

Glycine Willd.

1. *G. max* (L.) Merr. (=*Soja hispida* Moench, *Soja max* (L.) Piper, *Phaseolus max* L., *G. gracilis* Skvortzov) [sojabean, soybean]

Oomycetes
Peronospora manshurica. 'Downy mildew.' **1**:E states(82), S states(82), FL(35), IA(82,713,715,918), MS(1787,1889,2720), NC(1303), OK(2872), SD(2243), WI(82,1335).
Phytophthora megasperma. 'Phytophthora root rot. Phytophthora stem rot.' **1**:WI(1428).
Phytophthora sp. Damping-off. **1**:NC(1303,3307).
Pythium aphanidermatum. 'Pythium damping-off. Pythium root rot.' **1**:NC(1303).
P. debaryanum. Neck rot. 'Pythium damping-off. Pythium root rot.' **1**:IA(82), IL(82), MN(82), MO(82), NC(82), ND(82).
P. myriotylum. Seedling blight. 'Pythium damping-off. Pythium root rot.' **1**:FL(35).
P. ultimum. Stem rot. 'Pythium damping-off. Pythium root rot.' **1**:NC(1303).
Pythium sp. Damping-off. **1**:FL(3131), NC(1303,3307).
Zygomycotina
Mucor sp. Isolated from roots. **1**:IL(2520), MS(2435).
Rhizopus stolonifer. Isolated from vegetative and reproductive parts. **1**:MS(1899,2435).
Rhizopus sp. Isolated from roots. **1**:IL(2520).
Ascomycotina
Calonectria crotalariae. Stem rot. 'Red crown rot. Black root rot.' **1**:LA(1804), NC(1303).
Chaetomium cupreum. Seedborne. **1**:IL(3979).
C. globosum. **1**:MS(2435).
C. trilaterale. **1**:MS(2435).
Chaetomium sp. Isolated from reproductive parts and roots.

1:IL(1814,2520), MS(1899).
Diaporthe phaseolorum. Pod and stem blight, root rot. 'Pod rot. Stem rot. Stem canker.' 1:E states(82), NC states(3053), SE states(3053), AR(1576), FL(35), GA(1438,3053), IA(82,713,715,717,915), IL(82,1162), IN(82), KS(82), LA(82), MD(1999,3053), MI(74,82,1584), MS(1787,1889,2435,2720), NC(1303,2079), NY(2999), OH(1584,1927,1928), OK(82,2872), SC(1973), TX(3886).
Erysiphe polygoni. Powdery mildew. 1:FL(35).
[*Gaeumannomyces graminis*]. 1:IN(3754).
Glomerella cingulata. Anthracnose. 1:Unk.(2238,3059), FL(35), IA(82), MS(2435), NC(82,2082).
G. glycines. 'Anthracnose.' 1:C states(82), SE states(82), Unk.(2238,3059), GA(1438), IA(82), IL(1482), MS(2435,2720), NC(82,2082), NE(82), OK(2872), TX(82).
Leptosphaerulina trifolii. 'Leptosphaerulina leaf spot.' 1:MD(1387).
Melanospora sp. 1:MS(2435).
Microascus trigonosporus. On seeds. 1:AL(82).
Microsphaera diffusa. 'Powdery mildew.' 1:DE(82,2070), IA(82,917), IL(2322,2730), MN(1874), MS(1783), NC(82,1303), SC(82), TX(82), WI(106).
Microsphaera sp. Powdery mildew. 1:NC(82).
Mycosphaerella cruenta. Leaf spot. 1:GA(82,1438), MS(82,2720).
Mycosphaerella sp. 1:MS(2435).
Nematospora coryli. On seeds. 'Yeast spot.' 1:FL(204), IL(1814), MD(204), NC(82), OK(82,2872), SC(82), VA(82).
N. lycopersici. 1:MD(204).
Neocosmospora vasinfecta. Root rot, wilt. 'Neocosmospora stem rot.' 1:AL(1315), FL(35), GA(82,1438), NC(1303).
Ophioceras sp. 1:GA(1438).
Pleosphaerulina sojicola. Leaf spot. 1:MD(82), MS(2720), NJ(82).
Podospora sp. 1:MS(2435).
Sclerotinia sclerotiorum. Blight, white mold disease. 'Sclerotinia stem rot.' 1:FL(35), IA(82), IL(540), MD(82), MN(1308), NH(540), NY(82), VA(82,2803).
Sordaria fimicola. 1:MS(2435).
Sporormiella minima. 1:MS(2435).
Thielavia basicola. On dead roots. 1:NC(82).

Basidiomycotina - Uredinales

[*Uromyces appendiculatus*]. 1:GA(1439).

Basidiomycotina - Other

Stilbum sp. Isolated from roots. 1:IL(2520).
Thanatephorus cucumeris. Root rot. 1:Range of host(82), LA(3524), MS(2720).

Deuteromycotina - Hyphomycetes

Acremonium strictum. 1:MS(2435).
Alternaria alternata. Leaf spot, seed rot. 1:IL(2002), MN(82), MS(82,2435), NC(82).
A. atrans. Leaf spot. 1:AZ(82), NY(82), PA(82).
A. tenuissima. Pod and seed loss, leaf spot. 1:IL(3262), MD(82).
Alternaria sp. Root and basal stem rot, leaf spot. 1:Range of host(82), FL(35), HI(2903), IA(82), IL(1162,1814,2520), MS(1889,1899,2720).
Aspergillus flavus. 1:Range of host(82), IL(82), MN(82).
A. fumigatus. 1:Range of host(82), IL(82), MN(82).
A. glaucus. 1:Range of host(82), IL(82), MN(82).
A. niger. Isolated from reproductive and vegetative parts. 1:Range of host(82), IL(82), MN(82), MS(1899).
A. ochraceus. 1:Range of host(82), IL(82), MN(82).
Aspergillus sp. Root and basal stem rot. 1:Range of host(82), IL(82,1162,1814,2520), MN(82), MS(1899,2435).
Bipolaris victoriae. Blight. 1:IA(3423).
Botrytis cinerea. Gray mold leaf spot. 1:CT(82), OH(82).
Cephalosporiopsis sp. 1:MS(2435).
Cercospora canescens. Leaf spot, on pods and stems. 1:AL(82), FL(35), IL(82), MD(82), MS(82,2720), NC(82), TX(82), WV(82).
C. kikuchii. Blight, seed stain, leaf spot. 'Purple stain.' 1:Unk.(1803), AR(3765), FL(35), IA(715,717), IL(1814), MD(82), MS(1787,1889,1899,2435,2720), NC(82,2530), SD(2243), VA(82).
C. sojina. 'Frogeye leaf spot.' 1:E states(82), FL(35), GA(2799), HI(2903), IA(82), MI(82), MS(2435), MS(1787,1899,2720), NC(1303), OK(82,2872).
Cercospora sp. Leaf, pod and stem spot. 1:MS(1787,1899).
Cladosporium herbarum. 1:MS(2435).
Cladosporium sp. Isolated from reproductive and vegetative parts. 1:IL(1814), MS(1899).
Clasterosporium sp. Isolated from roots. 1:MI(2160).
Corynespora cassiicola. Root and stem rot, pod and target spot. 'Target spot.' 1:Unk.(82), FL(35), MS(1899,2720), NC(82), NE(335), VA(3223).
Curvularia lunata. 1:MS(2435).
C. lunata var. *aeria*. 1:MS(2435).
Curvularia sp. Isolated from reproductive and vegetative parts. 1:MS(1899).
Cylindrocarpon sp. 1:MS(2435).
Cylindrocladium crotalariae. Root and stem rot. 1:NC(1303).
Dendryphion sp. Isolated from vegetative parts. 1:MS(1899).
Dictyochaeta fertilis. Root rot. 1:NC(1303).
Epicoccum nigrum. 1:MS(2435).
Epicoccum sp. Isolated from roots. 1:IL(2520).
Eurotium sp. Isolated from roots. 1:IL(2520).
Fusarium acuminatum. Seed rot. 1:MS(2435), ND(82).
F. equiseti. 1:MS(2435).
F. moniliforme. Root and basal stem rot. 1:IL(1162), MS(2435).
F. oxysporum. Root and basal stem rot, isolated from reproductive and vegetative parts. 'Fusarium blight.' 1:DE(1050), IA(916), IL(1162), MS(1899), NC(1303).
F. oxysporum f. sp. *tracheiphilum*. Wilt. 1:E states(82), CA(82), HI(2903), LA(82), OK(2872), TX(82).
F. pallidoroseum. 1:MS(2435).
F. roseum. Isolated from reproductive and vegetative parts. 1:MS(1899).
F. solani. Seedling blight, root and basal stem rot, isolated from reproductive and vegetative parts. 1:FL(35), IL(1162), MS(1899).
F. tricinctum. 1:MS(2435).
F. udum f. sp. *crotalariae*. Wilt. 1:SC(82).
Fusarium sp. Blight, damping-off, wilt. 1:FL(3131), IA(915), IL(1814,2520), NC(1303,3307).
Geniculosporium sp. 1:MS(2435).
Gliocladium roseum. Isolated from roots. 1:IL(2520,2521).
Gliocladium sp. Root and basal stem rot. 1:IL(1162).
Helminthosporium sp. Isolated from vegetative parts. 1:MS(1899,2435).
Mycoleptodiscus terrestris. Root and basal stem rot, leaf spot. 'Mycoleptodiscus root rot.' 1:FL(35), IL(1162,1318).
Myrothecium roridum. Secondary leaf spot. 1:LA(82), MS(2435).
Myrothecium sp. Isolated from roots. 1:IL(2520).
Nigrospora sp. Isolated from reproductive parts. 1:MS(1899,2435).
Nodulisporium sp. 1:MS(2435).
Oidium sp. Powdery mildew. 1:HI(2903).
Paecilomyces sp. 1:MS(2435).
Papularia sp. 1:MS(2435).
Penicillium sp. Root and basal stem rot. 1:IL(1162,1814,2520), MS(1899,2435), OH(82).
Periconia circinata. 1:MS(2435).
Phialophora gregata. 'Brown stem rot.' 1:FL(2800), IA(6,82,713,715,717,919,2800), IL(6,82), IN(6,82), KY(82), MN(1879), MO(82), NC(1303,2800,3039), OH(82), VA(2800,3039), WI(2347).
Phymatotrichopsis omnivora. 'Phymatotrichum root rot.' 1:OK(82,2872), TX(82).
Pithomyces chartarum. 1:MS(2435).
Pseudocercospora cruenta. Leaf spot. 1:AL(285).
Pyricularia grisea. 1:MS(2435).
Thielaviopsis basicola. 'Thielaviopsis root rot.' 1:MI(2160), NC(1303).
Trichoderma sp. Isolated from the roots and reproductive parts. 1:IL(1162,2520), MS(1899,2435).
Verticillium dahliae. 1:MS(2435).
V. nigrescens. 1:MS(2435).
Volutella sp. 1:MS(2435).

Deuteromycotina - Coelomycetes

Ascochyta sp. On leaves. 1:AR(82), WV(82).
Botryodiplodia pallida. On stems. 1:AL(82).
Chaetomella sp. 1:MS(2435).
Colletotrichum dematium. 'Anthracnose.' 1:NC(1303).
C. destructivum. 1:IL(1482).
C. gloeosporioides. 1:Unk.(3059), MS(2435).
Colletotrichum sp. Anthracnose, leaf and stem spot. 1:IL(1814), LA(82), MS(1787,1899,2435).
Lasiodiplodia theobromae. 1:MS(2435).

Pestalotia sp. **1**:MS(2435).
Phoma sp. On stems. **1**:IL(82,1814), MS(2435).
Phomopsis longicolla. Isolated from pods, seeds and stems. **1**:IA(1585), IL(1585), MS(1585), OH(1585).
P. phaseoli. Pod and stem blight, isolated from reproductive and vegetative parts. **1**:Unk.(82), IL(1585), IN(1585), MD(1585), MS(1899), NC(1303), OH(1585), OK(2873), SC(82).
Phomopsis sp. Isolated from immature, symptomless plants. **1**:IL(1814,2520), OH(1927,1928).
Phyllosticta phaseolina. **1**:WI(1336).
P. sojicola. Pod lesions. 'Phyllosticta leaf spot.' **1**:AR(3766), IA(915,916), IL(82), MD(82), MO(82), NC(82), OK(2873), VA(82).
Phyllosticta sp. Canker, leaf spot, isolated from reproductive and vegetative parts. **1**:MD(1738), MS(1787,1889,1899).
Pseudorobillarda sojae. Isolated from senescent or overwintered stems. **1**:AL(3705), GA(3705), MD(3705).
Septoria glycines. Black spot. 'Brown spot.' **1**:AR(82), DE(82), FL(35), IA(82,713,715,717,915), IL(2124), IN(82), MD(82), MS(2435), NC(82,1303), NJ(495), SD(2243), WI(82,1335).

Deuteromycotina - Other

Rhizoctonia solani. Isolated from reproductive and vegetative parts. 'Rhizoctonia damping-off. Rhizoctonia root rot. Rhizoctonia stem rot. Rhizoctonia aerial blight.' **1**:Unk.(82), FL(35), IA(713,715,717,915,2013,3570), IL(1162), LA(2624), MS(1899,2435), NC(1303), TN(72,73).
R. zeae. Root rot. **1**:NC(1303).
Rhizoctonia sp. Damping-off, root rot. **1**:FL(3131), HI(2903), IL(2520), NC(1303,3307).
Sclerotium rolfsii. Leaf and stem rot. 'Southern blight.' **1**:SE states(82), AL(285), FL(35), HI(2903), IA(82), MS(1787,1889,1899,2720), NC(1303,2081), OK(2873), TX(82).

Glycyrrhiza L.

1. *G. lepidota* Pursh [American licorice, wild licorice]
2. *G. lepidota* Pursh var. *glutinosa* (Nutt.) S. Wats.

Ascomycotina

Erysiphe communis. **1**:ND(377).
E. polygoni. Powdery mildew. **1**:ND(82), NM(82).
Microsphaera diffusa. Powdery mildew. **1**:CO(82), MT(82,3202), WA(82,3202), WY(82).

Basidiomycotina - Uredinales

Uromyces glycyrrhizae. Autoecious. **1**:W states(82), AZ(1203,3981), CA(82), CO(746,1195), ID(3202), MT(1195,3202), ND(378), NM(1195), OR(3202), SD(2243), UT(1195), WA(82,3202), WI(1333), WY(1195). **2**:CA(81).

Deuteromycotina - Coelomycetes

Cylindrosporium glycyrrhizae. On living leaves. **1**:CA(1468).
Septoria glycyrrhizae. Leaf spot. **1**:KS(82), WA(82,3202).

Gymnocladus Lam.

1. *G. dioica* (L.) K. Koch [chicot, Kentucky coffee-tree, nicker-tree]

Basidiomycotina - Other

Ganoderma lucidum. Root rot. **1**:OK(618).
Perenniporia tenuis. Wood rot. **1**:IN(82), MI(82).

Deuteromycotina - Hyphomycetes

Cercospora gymnocladi. Leaf spot. **1**:NC states(82), OK(660,2872).
Fusarium sp. **1**:OK(2873).
Phymatotrichopsis omnivora. Root rot. **1**:OK(82,2872), TX(82).
Verticillium albo-atrum. Wilt. **1**:IL(1558).

Deuteromycotina - Coelomycetes

Amerosporium subclausum. On fallen leaves. **1**:KS(82).
Gloeodes pomigena. On limbs. **1**:IN(82).
Marssonina sp. Leaf spot. **1**:NE(82).
Phoma sp. Twig dieback. **1**:OK(618).
Phyllosticta gymnocladi. Leaf spot. **1**:IL(82,3590).
Phyllosticta sp. On leaflets. **1**:WI(1348).

Hedysarum L.

1. *H. alpinum* L.
2. *H. alpinum* L. subsp. *americanum* Fedch. (=*H. alpinum* L. var. *americanum* Michx.)
3. *H. boreale* Nutt. (=*H. boreale* Nutt. subsp. *boreale* , *H. pabulare* A. Nels.)
4. *H. mackenzii* Richardson
5. *H. marginatum* Greene
6. *H. occidentale* Greene
7. *H. sulphurescens* Rydb.
8. *H. utahense* A. Nels.
9. *Hedysarum* sp.

Oomycetes

Peronospora trifoliorum. **2**:AK(3431).

Ascomycotina

Comoclathris permunda. **8**:WY(3810).
Leptosphaeria oreophila. **8**:WY(3810).
Parodiella hedysari. Black mildew. **9**:NY(82).
Pleospora stenospora. **9**:WY(3810).

Basidiomycotina - Uredinales

Uromyces hedysari-obscuri. Autoecious. **1**:AK(3431). **2**:AK(60,526,1938,3431). **3**:AK(82,526), CO(82), ID(82), MT(82,1195), NM(82), SD(82), UT(82), WY(82,1195). **4**:AK(526). **5**:WY(1195). **6**:WY(1195). **7**:MT(1195), WY(1195). **9**:AK(82,526), CO(82), ID(82), MT(82), NM(82), SD(82), UT(82), WY(82).

Deuteromycotina - Coelomycetes

Phloeosporella hedysari. Leaf spot. **5**:WY(82,1195,3549).
Sirexcipula wyomingensis. On stems. **8**:WY(3812).

Hoffmanseggia Cav.

1. *H. oxycarpa* Benth.

Basidiomycotina - Uredinales

Uredo hoffmanseggiae. Autoecious. **1**:TX(82,746).

Indigofera L.

1. *I. hirsuta* L. [hairy indigo]
2. *I. sphaerocarpa* A. Gray
3. *I. suffruticosa* Mill.
4. *Indigofera* sp.

Ascomycotina

Nectria cinnabarina. On stems. **4**:MD(82).
Nectria sp. Basal stem rot. **1**:FL(35).
Parodiella hedysari. Leaf spot. **4**:FL(35), NM(82).

Basidiomycotina - Uredinales

Ravenelia indigoferae. Autoecious. **2**:AZ(1194,1195,1203,3981). **4**:FL(35).
R. laevis. Autoecious. **4**:TX(82).
Uromyces indigoferae. Autoecious. **2**:AZ(3981). **3**:FL(35). **4**:FL(35,82), TX(82).

Deuteromycotina - Hyphomycetes

Phymatotrichopsis omnivora. Root rot. **4**:TX(82).
Verticillium sp. **1**:FL(35).

Deuteromycotina - Coelomycetes

Colletotrichum gloeosporioides. Stem canker, leaf spot. **4**:FL(35).
C. truncatum. Stem canker, leaf spot. **4**:FL(35).
Lasiodiplodia theobromae. Dieback. **4**:FL(35).

Deuteromycotina - Other

Rhizoctonia solani. Root and stem rot. **1**:FL(35). **4**:FL(35).
Sclerotium rolfsii. Southern stem rot. **4**:NC(1303).

Kummerowia Schindl.

1. *K. stipulacea* (Maxim.) Makino (=*Lespedeza stipulacea* Maxim.) [Korean lespedeza]

Ascomycotina

Erysiphe polygoni. Powdery mildew. **1**:IL(312).
Glomerella cingulata. Leaf spot. **1**:GA(82), NC(82), SC(82), VA(82).
Microsphaera diffusa. Powdery mildew. **1**:FL(82), IL(82).
Phyllachora lespedezae. Tar spot. **1**:IL(312), NC(1303).

Basidiomycotina - Other

Pellicularia koleroga. Brown leaf patch. **1**:NC(82).
Thanatephorus cucumeris. **1**:LA(3524).

Deuteromycotina - Hyphomycetes

Aureobasidium caulivorum. Stem canker. **1**:IL(82).
Cercospora latens. Leaf spot. **1**:IL(82).
Dictyochaeta fertilis. Root rot. **1**:NC(1303).

Deuteromycotina - Coelomycetes

Catosphaeropsis caulivora. Stem canker. **1**:IL(82).
Colletotrichum trifolii. Southern anthracnose. **1**:IL(82).
Colletotrichum sp. Leaf spot. **1**:NC(82), SC(82).
Macrophomina phaseolina. Charcoal rot. **1**:GA(82), IL(82,311).

Deuteromycotina - Other
Rhizoctonia solani. **1**:NC(12).
Sclerotium rolfsii. Southern blight. **1**:AR(82), NC(82), TX(82).

Lablab Adans.

1. *L. purpureus* (L.) Sweet (=*L. niger* Medik., *L. vulgaris* Savi, *Dolichos lablab* L.) [bonavist bean, hyacinth bean, lablab bean]

Ascomycotina
Neocosmospora vasinfecta. On dead stems. **1**:SC(82).
Parodiella hedysari. Black mildew. **1**:NC(82).
Deuteromycotina - Hyphomycetes
Cercospora canescens. Leaf and pod spot. **1**:FL(35,82), HI(2899,2903).
Phymatotrichopsis omnivora. Root rot. **1**:TX(82).
Deuteromycotina - Other
Rhizoctonia solani. Web blight, root and stem rot. **1**:FL(35).
Sclerotium rolfsii. **1**:GA(2886).

Laburnum Medik.

1. *L. anagyroides* Medik. [golden-chain]

Ascomycotina
Botryosphaeria ribis. On limbs. **1**:MD(82).
Cucurbitaria laburni. On limbs. **1**:NY(82), OH(82).
Diaporthe rudis. On limbs. **1**:NE states(82), CA(82), OH(82).
Deuteromycotina - Hyphomycetes
Fusarium lateritium. Twig blight. **1**:NJ(82).
Fusarium sp. Twig blight. **1**:MD(82), OH(82).
Pseudocercospora laburni. Leaf spot. **1**:OK(82,581,2872).
Deuteromycotina - Coelomycetes
Camarosporium laburni. **1**:Unk.(82).
Phomopsis rudis. **1**:Unk.(82).
Rabenhorstia rudis. On limbs. **1**:MD(82), PA(82).
Rhabdospora rudis. On limbs. **1**:OH(82).

Lathyrus L.

1. *L. arizonicus* Britton
2. *L. bijugatus* T. G. White
3. *L. bolanderi* S. Wats.
4. *L. eucosmus* F. K. Butters & St. John
5. *L. hirsutus* L. [Austrian winter pea, Caley pea, rough peavine, singletary pea]
6. *L. japonicus* (Roth) Willd. [beach pea, heath pea, seaside pea]
7. *L. japonicus* (Roth) Willd. subsp. *maritimus* (L.) P. Ball (=*L. japonicus* (Roth) Willd. var. *glaber* (Ser.) Fernald, *L. maritimus* Bigel.)
8. *L. laetiflorus* Greene subsp. *alefeldii* (T. G. White) R. V. Bradshaw (=*L. strictus* Nutt.)
9. *L. laetivirens* Greene
10. *L. lanswertii* Kellogg
11. *L. lanswertii* Kellogg subsp. *aridus* (Piper) R. V. Bradshaw (=*L. lanswertii* Kellogg var. *aridus* (Piper) Jeps.)
12. *L. latifolius* L. [everlasting pea, perennial pea]
13. *L. leucanthus* Rydb.
14. *L. littoralis* (Nutt.) Endl. [beach pea]
15. *L. nevadensis* S. Wats.
16. *L. ochroleucus* Hook.
17. *L. odoratus* L. [sweet pea]
18. *L. palustris* L. [marsh peavine, wing-stemmed wild peavine]
19. *L. palustris* L. var. *myrtifolius* (Muhl. ex Willd.) A. Gray
20. *L. pauciflorus* Fernald
21. *L. polymorphus* L. (=*L. decaphyllus* Pursh, *L. ornatus* Nutt. ex Torr. & A. Gray)
22. *L. polyphyllus* Nutt. ex Torr. & A. Gray
23. *L. sulphureus* W. H. Brewer ex A. Gray
24. *L. sylvestris* L. [everlasting pea, flat pea, perennial pea]
25. *L. tingitanus* L. [Tangier pea]
26. *L. torreyi* A. Gray
27. *L. utahensis* M. E. Jones
28. *L. venosus* Muhl. ex Willd.
29. *L. venosus* Muhl. ex Willd. var. *intonsus* (F. K. Butters & St. John) C. Hitchc.
30. *L. vestitus* Nutt.
31. *Lathyrus* sp.

Chytridiomycetes
Physoderma lathyri. Leaf and stem galls. **23**:OR(82,3202).
Oomycetes
Aphanomyces euteiches. Root rot. **12**:WI(82). **17**:IN(82), MI(82), WI(82).
Peronospora lathyri-palustris. **7**:AK(3431).
P. trifoliorum. Downy mildew. **17**:FL(35,82).
Phytophthora cactorum. Root rot. **17**:CT(82).
Pythium debaryanum. Root rot. **17**:CT(82), NC(82), NJ(82).
P. oligandrum. Root rot. **17**:NY(82).
P. ultimum. Root rot. **17**:MA(82).
Pythium sp. Root rot. **5**:CA(82), CT(82), MD(82), NJ(82). **12**:CA(82), CT(82), MD(82), NJ(82). **17**:CA(82), CT(82), MD(82), NJ(82).
Ascomycotina
Crocicreas cyathoideum var. *cacaliae*. **31**:UT(505).
Erysiphe polygoni. Powdery mildew. **6**:AK(526). **7**:CA(1155), MA(1390), WA(3202). **12**:Unk.(3976), GA(1438), WA(3202), WI(1343). **17**:Unk.(82), FL(35), GA(1438), ID(3202), OK(2872), WA(3202). **18**:AK(82,526). **22**:WA(3202). **23**:OR(3202), WA(3202).
Glomerella cingulata. Anthracnose. **17**:E & C states(82), FL(35), HI(2903), MS(2720).
Melanospora multiformis. White mold of greenhouse crops. **12**:CA(82), WA(82). **17**:MA(82), NJ(82), NY(82), PA(82), TX(82).
Microsphaera diffusa. **30**:OR(81,3202).
M. penicillata. Powdery mildew. **17**:Range of host(82), FL(35,82), SD(2243). **18**:IA(82), IL(82), MI(82), NY(82), OH(82), WI(82). **19**:WI(1336).
M. ravenelii. **17**:ND(377).
Mycosphaerella pinodes. Basal stem rot, leaf and stem spot. **7**:AK(3431). **12**:MT(82). **17**:WI(82). **25**:MS(82,2720).
M. tulasnei. **7**:AK(3431).
Mycosphaerella sp. Stem spot. **5**:AL(82). **12**:NJ(82).
Sclerotinia sp. Stem rot. **17**:MD(82), PA(82).
Basidiomycotina - Uredinales
Uromyces fabae. Autoecious. **1**:AZ(1195,1203,3981). **2**:ID(3202), WA(3202). **3**:CA(81). **4**:AZ(1195,1203,3981). **6**:AK(526). **7**:AK(60,3431), WA(3202). **8**:CA(81). **9**:AZ(1195). **10**:OR(3202), UT(1195), WA(3202). **11**:CA(638,642). **13**:AZ(1203,3981), ID(3202), WY(1195,3379). **14**:CA(81). **15**:CA(81), OR(3202), WA(3202). **16**:ID(3202), SD(2243). **18**:Range of host(82), WA(3202). **20**:ID(3202), MT(3202), OR(3202), WA(3202). **22**:CA(81), OR(3202), WA(3202). **23**:CA(81), OR(3202), WA(3202). **24**:ND(2243). **26**:CA(81), OR(3202). **28**:NC(3944), ND(378). **30**:OR(3202). **31**:Range of host(82), AK(526), CA(81), OR(648).
Uromyces sp. **17**:FL(35).
Basidiomycotina - Ustilaginales
Thecaphora deformans. **27**:WA(1077).
Basidiomycotina - Other
Thanatephorus cucumeris. **17**:WA(3202).
Deuteromycotina - Hyphomycetes
Alternaria alternata. Leaf spot. **18**:VT(82).
Alternaria sp. Secondary leaf spot. **17**:MA(82), MD(82), NJ(82), NY(82), PA(82), TX(82).
Botrytis cinerea. Gray mold. **17**:AK(82,526), MN(82), PA(82), WA(82,3202).
Cercospora lathyri. Leaf spot. **6**:MA(82), NY(82). **7**:MA(1391).
C. lathyrina. Leaf spot. **12**:GA(82), OK(82,2872).
C. zonata. **18**:WI(82).
Cercospora sp. **12**:WI(1350).
Cladosporium herbarum. **12**:WA(3202).
Fusarium oxysporum f. sp. *vasinfectum*. Wilt. **17**:FL(35,82), NY(82).
F. solani. Stem rot, wilt. **17**:CO(82), FL(35,82), IN(82), MS(82), ND(82), NY(82), OH(82), TX(82).
Fusarium sp. Root and stem rot. **17**:HI(2903), NC(1303).
Fusicladium lathyrinum. **16**:MT(3202).
Heterosporium sp. Leaf spot. **17**:OK(2872). **24**:OK(2873). **31**:OK(82).
Ovularia sp. Leaf spot. **12**:CA(82).
Phaeoisariopsis griseola. Leaf spot. **17**:CT(82).
Phymatotrichopsis omnivora. Root rot. **17**:TX(82).
Ramularia alba. Leaf spot. **12**:Unk.(82). **17**:Unk.(82), CA(82,152), MA(82), NJ(82), NY(82), PA(82), TX(82), WA(82).
R. deusta. Leaf spot. **12**:WA(666). **18**:AK(82,526).

R. deusta f. *latifolii*. Leaf spot. **12:**CA(152).
Ramularia sp. Leaf spot. **17:**NY(82).
Stemphylium majusculum. On dead leaves. **7:**NY(3269).
Stigmina sycina. On leaves. **11:**CA(642).
Thielaviopsis basicola. Black root rot. **17:**CA(82), CO(82), CT(82), FL(35,82), IL(82), MN(82), NY(82), OH(82), OR(82), PA(82), WA(82).
Verticillium sp. Wilt. **17:**NJ(82), NY(82).

Deuteromycotina - Coelomycetes

Ascochyta lathyri. Stem spot. **6:**WA(82). **7:**WA(3202). **12:**NJ(82).
A. pisi. Leaf spot. **11:**CA(642). **16:**WI(1355). **18:**CA(82), WI(82). **24:**OK(2873). **29:**WI(1340). **31:**OK(82).
Ascochyta sp. Leaf and stem spot. **5:**LA(82). **17:**KS(82), MN(82), MO(82), WI(82), WV(82). **25:**HI(2903).
Colletotrichum gloeosporioides. Anthracnose, leaf spot. **12:**AL(82), FL(82), GA(82), SC(82). **17:**AL(285), FL(35).
Phyllosticta davisii. Leaf spot. **28:**IA(82).
P. orobella. Leaf spot. **6:**MA(82), NY(82). **7:**NY(64). **12:**TX(82), WA(3202). **16:**MT(3202). **17:**TX(82).
P. suecica. **6:**CT(82).
Septoria astragali. Leaf spot. **6:**MA(82), ME(82), VT(82). **7:**WI(1333). **11:**CA(642). **28:**ND(378).
S. emaculata. Leaf spot. **18:**NY(82), WI(82,1331).
S. lathyri. On leaves. **12:**NJ(82,966).
Septoria sp. **16:**WI(1346).
Stagonospora baptisiae. **7:**WI(1343).

Deuteromycotina - Other

Rhizoctonia solani. Damping-off, root and stem rot. **17:**Range of host(82), AL(285), FL(35), OK(2872). **31:**TX(82).
Sclerotium rolfsii. Southern blight. **17:**FL(35,82), TX(82). **31:**TX(82).

Lens Mill.

1. *L. culinaris* Medik. (=*L. esculenta* Moench, *Ervum lens* L., *Cicer lens* Willd.) [lentil]

Oomycetes

Pythium sp. Isolated from roots and stems of seedlings. **1:**NW states(3202), NW states(3927), NW states(3927).

Ascomycotina

Chaetomium globosum. **1:**WA(3202).
Sclerotinia sclerotiorum. Isolated from leaves, roots and stems of seedlings. **1:**NW states(3202), NW states(3927), NW states(3927).
S. trifoliorum. **1:**WA(3202).

Deuteromycotina - Hyphomycetes

Alternaria alternata. Isolated from leaves, roots and stems of seedlings. **1:**NW states(3927), NW states(3927), WA(3202).
Botrytis cinerea. Isolated from roots of seedlings. **1:**NW states(3927), NW states(3927), WA(3202).
Cladosporium herbarum. Isolated from leaves, roots and stems of seedlings. **1:**NW states(3927), NW states(3927), WA(3202).
Epicoccum nigrum. Isolated from leaves of seedlings. **1:**NW states(3927), NW states(3927), WA(3202).
Fusarium oxysporum. Isolated from leaves, roots and stems of seedlings. **1:**NW states(3927), WA(3202).
F. roseum. Isolated from roots and stems of seedlings. **1:**NW states(3202,3927), WA(2129,3202).
Gilmaniella humicola. **1:**WA(3202).
Paecilomyces sp. Isolated from leaves, roots and stems of seedlings. **1:**NW states(3202), NW states(3927), NW states(3927).
Penicillium sp. Isolated from leaves, roots and stems of seedlings. **1:**NW states(3202), NW states(3927), NW states(3927).
Stemphylium botryosum. Isolated from leaves, roots and stems of seedlings. **1:**NW states(3927), WA(3202).
Thielaviopsis basicola. Root rot. **1:**ID(347), WA(347).
Verticillium albo-atrum. Isolated from roots and stems of seedlings. **1:**NW states(3202), NW states(3927), NW states(3927).

Deuteromycotina - Coelomycetes

Ascochyta lentis. Foliar blight. **1:**WA(1829,1830).
Phoma pinodella. Isolated from leaves, roots and stems of seedlings. **1:**NW states(3927), NW states(3927), WA(3202).

Deuteromycotina - Other

Rhizoctonia sp. Isolated from roots and stems of seedlings. **1:**NW states(3202), NW states(3927), NW states(3927).

Lespedeza Michx.

1. *L. bicolor* Turcz. [bicolor lespedeza, shrub bushclover]
2. *L. capitata* Michx. [roundhead lespedeza]
3. *L. cuneata* (Dum. Cours.) G. Don [Chinese lespedeza, Sericea lespedeza]
4. *L. hirta* (L.) Hornem.
5. *L. intermedia* (S. Wats.) Britton
6. *L. procumbens* Michx.
7. *L. repens* (L.) Barton
8. *L. reticulata* Pers. (=*L. frutescens* DC.)
9. *L. simulata* Mackenz. & Bush
10. *L. striata* (Thunb. ex J. A. Murray) Hook. & Arnott [common lespedeza, Japanese lespedeza, Japanese-clover, striate lespedeza]
11. *L. stuevei* Nutt.
12. *L. violacea* (L.) Pers.
13. *L. virginica* (L.) Britton
14. *Lespedeza* sp.

Ascomycotina

Botryosphaeria dothidea. **13:**GA(1438).
B. obtusa. On stems. **13:**GA(82).
Diaporthe arctii. On stems. **13:**GA(1438), VA(82).
D. phaseolorum. **14:**GA(1438).
Didymella sp. On stems. **8:**GA(82,1438).
Dothidea puccinioides. **1:**GA(1438).
Ellisiodothis inquinans. **1:**GA(1438). **14:**GA(1438).
Erysiphe graminis f. sp. *avenae*. Powdery mildew. **14:**AR(3029).
E. polygoni. Powdery mildew. **10:**ND(2243), SD(82), VA(82).
Eutypella juglandicola. **1:**GA(1438).
Glomerella cingulata. Leaf spot. **10:**GA(82), NC(82), SC(82), VA(82).
Gloniopsis praelonga. **13:**GA(1438).
Hypoderma commune. **8:**GA(1438).
Leptosphaeria clavigera. **13:**GA(1438).
L. distributa. **13:**GA(1438).
Leptosphaerulina trifolii. Leaf spot. **3:**GA(82).
Lophiostoma caulium. **4:**GA(1438).
Lophiotrema sp. **14:**GA(1438).
Massarina gloeospora. **8:**GA(1438). **13:**GA(1438).
Metasphaeria sp. **13:**GA(1438).
Microsphaera diffusa. Powdery mildew. **4:**IL(82), PA(82). **10:**AL(82), FL(82), GA(82,1438), IL(82), MS(82,2720), SC(82), VA(82). **14:**FL(35), MA(82), MS(82,2720), NC(48,1303), VA(82), WV(82).
Microthyrium sp. **8:**GA(1438).
Mycosphaerella sp. Blight. **14:**AL(82).
Myiocopron sp. **8:**GA(1438).
Parodiella hedysari. Black mildew. **14:**KS(82), MO(82), SC(82).
Phyllachora lespedezae. Tar spot. **2:**Range of host(82), WI(1332,1350). **4:**Range of host(82). **6:**GA(1438). **7:**Range of host(82), GA(1438). **11:**Range of host(82). **12:**Range of host(82). **13:**Range of host(82). **14:**Range of host(82), FL(35), IN(987).
Rhytidhysteron rufulum. **1:**GA(1438).
Urosporella alabamensis. **13:**GA(1438).

Basidiomycotina - Uredinales

Uromyces lespedezae-procumbentis. Autoecious. **2:**MS(2720), NC(1303), WI(1343,1358). **4:**GA(1439), MS(2720), NC(3944), OK(2872). **5:**IL(312), MS(2720), NC(1303), OK(2872). **6:**GA(1439), NC(1303), OK(2872). **7:**GA(1439), MS(2675,2720), NC(1303,3944), OK(2872). **8:**GA(1439), NC(1303). **11:**NC(1303), OK(2872). **12:**MS(2720), OK(2872). **13:**GA(1439), MD(651), NC(1303), OK(2872). **14:**Range of host(82), FL(35), IN(647).

Basidiomycotina - Other

Coriolopsis rigida. Wood rot. **11:**Gulf States(82).
Physalacria luttrellii. On dead stems. **1:**GA(143).
Thanatephorus cucumeris. **1:**LA(3524). **3:**LA(3524). **10:**LA(3524).

Deuteromycotina - Hyphomycetes

Alternaria sp. Leaf spot. **3:**DE(82), MD(82).
Cercospora latens. Leaf spot. **2:**IL(581), KS(82,581), WI(82). **4:**IL(82), NY(82). **12:**WI(82,1333). **14:**TX(82).
Dictyochaeta fertilis. Root rot. **3:**NC(1303). **10:**NC(1303).
Phymatotrichopsis omnivora. Root rot. **14:**TX(82).
Stemphylium sp. Leaf spot. **14:**MD(82).

Deuteromycotina - Coelomycetes

Colletotrichum trifolii. Anthracnose. **3:**NC(1303). **10:**IL(312).
Colletotrichum sp. Anthracnose, leaf spot. **3:**NC(82), SC(82). **4:**MS(2720). **14:**NC(1303).

Macrophomina phaseolina. Charcoal rot. **3**:TX(82).
Deuteromycotina - Other
Rhizoctonia solani. Stem blight, summer blight. **10**:NC(1303). **14**:NC(48).
Rhizoctonia sp. Summer blight, damping-off, root rot. **1**:NC(1303). **4**:MS(2720). **14**:GA(82), NC(1303).
Sclerotium rolfsii. Southern blight, root and stem blight. **3**:GA(3849). **9**:FL(35,82), NC(45,48). **10**:NC(82).

Leucaena Benth.

1. *L. leucocephala* (Lam.) de Wit (=*L. glauca* (L.) Benth.) [ipil, jumbie bean, leadtree, leucaena]
2. *L. pulverulenta* (Schlechtend.) Benth.
3. *Leucaena* sp.

Oomycetes
Phytophthora drechsleri. Canker. **1**:HI(2903).
Ascomycotina
Botryosphaeria obtusa. On limbs. **1**:HI(2903). **3**:HI(82).
B. ribis. On limbs. **1**:HI(2903). **3**:HI(82).
Diaporthe sp. On limbs. **1**:HI(2903).
Otthia shearii. On dry twigs. **1**:HI(2784).
Rhytidhysteron rufulum. **1**:HI(2903).
Basidiomycotina - Uredinales
Ravenelia leucaenae. Autoecious. **2**:TX(82).
Basidiomycotina - Other
Ganoderma zonatum. Root rot. **2**:TX(82).
Deuteromycotina - Coelomycetes
Diplodia shearii. On decaying limbs. **1**:HI(2784).
Sphaeropsis gouldiae. Canker. **1**:HI(2903).

Lonchocarpus Kunth

1. *Lonchocarpus* sp.

Basidiomycotina - Other
Ganoderma zonatum. Stem rot. **1**:FL(35).
Deuteromycotina - Coelomycetes
Colletotrichum gloeosporioides. Leaf spot. **1**:FL(35).

Lotus L.

1. *L. aboriginum* Jeps. (=*L. stipularis* (Benth.) Greene var. *subglaber* Ottley)
2. *L. americanus* (Nutt.) Bisch. var. *glaber* Ewan & Jeps.
3. *L. argophyllus* (A. Gray) Greene (=*Syrmatium ornithopus* Greene)
4. *L. corniculatus* L. [bird's-foot trefoil]
5. *L. denticulatus* (E. R. Drew) Greene
6. *L. douglasii* Greene
7. *L. eriophorus* Greene
8. *L. glaber* (Vogel) Greene (=*Hosackia glabra* (Vogel) Torr.)
9. *L. humistratus* Greene
10. *L. micranthus* Benth. (=*Hosackia parviflora* Benth. in Lindl.) [small-flowered birdsfoot trefoil]
11. *L. nevadensis* (S. Wats.) Greene
12. *L. purshianus* (Benth.) Clements & S. E. Clements (=*L. americanus* (Nutt.) Bisch.) [deervetch, prairie trefoil]
13. *L. scoparius* (Nutt.) Ottley
14. *L. stipularis* (Benth.) Greene
15. *L. strigosus* (Nutt.) Greene (=*Hosackia strigosa* Nutt. ex Torr. & A. Gray)
16. *L. subpinnatus* Lag.
17. *L. torreyi* (A. Gray) Greene
18. *L. uliginosus* Schkuhr [big trefoil]
19. *Lotus* sp.

Chytridiomycetes
Physoderma potteri. Wart. **18**:OR(1460,3202).
Oomycetes
Peronospora lotorum. Downy mildew. **4**:OR(82,3202). **12**:OR(3202), WA(3202).
Ascomycotina
Erysiphe communis. **1**:CA(81).
E. polygoni. Powdery mildew. **2**:CA(81). **5**:CA(81), WA(3202). **10**:OR(3202), WA(3202). **12**:CA(81,82), MN(82), OR(82,3202), SD(82), WA(82,3202). **14**:CA(81). **16**:CA(81). **19**:CA(82), MN(82), OR(82), SD(82), WA(82).
Mycosphaerella hosackiae. On stems. **13**:CA(82).
Pleospora herbarum. On stems. **12**:CA(82).
Pyrenopeziza compressula. On stems. **14**:CA(82).
Sclerotinia sclerotiorum. **4**:NY(82), WA(3202).
S. trifoliorum. Crown and stem rot. **4**:NY(1946,2998), VA(892), WA(3202). **19**:NY(2999).
Sclerotinia sp. Crown rot, wilt. **4**:WV(952). **19**:WA(82).
Basidiomycotina - Uredinales
Uromyces euphorbiae-corniculati. Autoecious. **3**:CA(2439). **4**:PA(4003). **19**:CA(82).
U. punctatus. **7**:CA(81).
Basidiomycotina - Ustilaginales
Thecaphora deformans. Seed smut. **9**:AZ(82,1077). **10**:WA(82,1077,3202). **12**:WA(3202).
Basidiomycotina - Other
Thanatephorus cucumeris. **4**:GA(1439), LA(3524).
Deuteromycotina - Hyphomycetes
Alternaria sp. Secondary leaf spot. **4**:VA(82).
Cercospora loti. Leaf spot. **18**:GA(1983), MD(1983).
Cercospora sp. Leaf spot. **4**:MN(2957). **12**:OK(2872).
Epicoccum nigrum. **12**:ND(378).
Fusarium roseum. Blight. **4**:NY(1087).
Fusarium sp. Aerial blight, root rot, wilt. **4**:MD(892), NC(892), NY(82,2998), VA(892).
Mycoleptodiscus sphaericus. Crown and root rot. **4**:MD(2659), VA(2659).
M. terrestris. Crown and root rot, leaf spot. **4**:FL(35), MD(2659), VA(2659).
Myrothecium verrucaria. Leaf spot. **4**:NY(375). **19**:NY(375).
Ovularia lotophaga. Leaf spot. **12**:ND(82,378), OR(3202), WA(3202). **17**:CA(642). **19**:CA(82), OR(82).
Phymatotrichopsis omnivora. Root rot. **12**:TX(82).
Pseudospiropes lotorus. Isolated from roots. **4**:AL(2487).
Ramularia schulzeri. On leaves and seedlings. **4**:NY(1088). **12**:NE(82), TX(82).
Stemphylium loti. Stem canker, target leaf spot. **4**:IA(216,217), KY(892), MD(892), MN(2957), NC(892), NJ(494), NY(2998,3004,3005), TN(892), VA(892), WV(892,952). **19**:ME(1890), NH(1890,1897), NY(2999), RI(1890), VT(1890).
Stemphylium sp. Stem canker, leaf spot. **4**:NE states(1291), NY(1291).
Verticillium sp. Root rot. **4**:MD(892), NC(892), VA(892).
Volutella colletotrichoides. **4**:IA(576).
Deuteromycotina - Coelomycetes
Colletotrichum truncatum. Anthracnose. **18**:GA(3840). **19**:LA(82).
Colletotrichum sp. **4**:MN(2957).
Macrophomina phaseolina. Root rot. **4**:MD(892).
Microdiplodia hosackiae. On stems. **8**:CA(81). **13**:CA(82).
Pestalotia moorei. On dead stems. **8**:CA(1468).
Phoma sp. Canker, stem spot. **4**:MO(892).
Phyllosticta hosackiae. On leaflets. **6**:WA(666). **11**:WA(3202).
Septoria hosackiae. Leaf spot. **15**:CA(82,1468).
Deuteromycotina - Other
Rhizoctonia solani. Foliage blight, summer blight, stem canker, brown patch, crown and root rot. **4**:IA(215,217), KY(892), MD(892), NC(892), TN(892), VA(892), WV(892). **18**:GA(3840), NC(48). **19**:MD(82), NC(45).
Rhizoctonia sp. Root and stem rot. **4**:HI(2903), IA(714).
Sclerotium rolfsii. Southern blight. **4**:VA(893). **18**:GA(3840).

Lupinus L.

1. *L. albicaulis* Douglas ex Hook. (=*L. albicaulis* Douglas ex Hook. var. *albicaulis*) [sickle-keeled lupine]
2. *L. albicaulis* Douglas ex Hook. var. *shastensis* (A. Heller) Ch. P. Sm.
3. *L. albifrons* Benth. ex Lindl.
4. *L. albus* L. (=*L. termis* Forssk.) [Egyptian lupine, field lupine, white lupine, wolf bean]
5. *L. alpestris* A. Nels.
6. *L. angustifolius* L. [European blue lupine]
7. *L. arboreus* Sims [tree lupine]
8. *L. arcticus* S. Wats.
9. *L. argenteus* Pursh
10. *L. argenteus* Pursh var. *tenellus* (Douglas ex G. Don) D. Dunn (=*L. laxiflorus* Douglas)
11. *L. benthamii* A. Heller [Bentham's annual lupine]
12. *L. bicolor* Lindl. [bicolored lupine]
13. *L. blumeri* Greene
14. *L. candicans* Rydb.
15. *L. densiflorus* Benth.

16. *L. diffusus* Nutt.
17. *L. formosus* Greene
18. *L. hallii* Abrams
19. *L. holosericeus* Nutt.
20. *L. ingratus* Greene
21. *L. latifolius* Lindl. ex J. Agardh
22. *L. latifolius* Lindl. ex J. Agardh var. *subalpinus* (Piper & B. L. Robinson) Ch. P. Sm. (=*L. subalpinus* Piper & B. L. Robinson, *L. volcanicus* Greene)
23. *L. lepidus* Douglas
24. *L. leucophyllus* Douglas [woolly-leaved lupine]
25. *L. littoralis* Douglas [chinook licorice, seashore lupine]
26. *L. luteus* L. [European yellow lupine]
27. *L. lyallii* A. Gray
28. *L. micranthus* Guss.
29. *L. microcarpus* Sims
30. *L. mutabilis* Sweet [tarwi]
31. *L. nanus* Douglas ex Benth.
32. *L. nevadensis* A. Heller
33. *L. nootkatensis* Donn ex Sims [Nootka lupine]
34. *L. obtusilobus* A. Heller
35. *L. ornatus* Douglas ex Lindl.
36. *L. palmeri* S. Wats.
37. *L. parviflorus* Nutt. ex Hook. & Arnott
38. *L. perennis* L. [sundial lupine, wild lupine]
39. *L. pilosus* J. A. Murray (=*L. hirsutus* L.) [blue lupine]
40. *L. plattensis* S. Wats.
41. *L. polyphyllus* Lindl. [Washington lupine]
42. *L. polyphyllus* Lindl. var. *burkei* (S. Wats.) C. Hitchc. (=*L. burkei* S. Wats.)
43. *L. rivularis* Douglas ex Lindl. (=*L. lignipes* A. Heller) [riverbank lupine]
44. *L. rubricaulis* Greene
45. *L. sellulus* Kellogg var. *artulus* (Jeps.) Eastw.
46. *L. sericeus* Pursh (=*L. leucopsis* J. Agardh)
47. *L. sp. (fruticose or arborescent)*
48. *L. sp. (herbaceous)*
49. *L. sp. (ornamental)*
50. *L. subcarnosus* Hook. [Texas bluebonnet, Texas lupine]
51. *L. succulentus* Douglas ex K. Koch
52. *L. sulphureus* Douglas ex Hook.
53. *L. texensis* Hook. [Texas bluebonnet]
54. *L. wyethii* S. Wats.
55. *Lupinus* sp.

Chytridiomycetes

Synchytrium sp. On flowers, leaves, petioles and stems. **21:**WA(3202). **46:**WA(1848). **52:**WA(3202).

Oomycetes

Peronospora trifoliorum. Downy mildew. **10:**ID(3202). **23:**OR(3202), WA(3202). **29:**OR(3202). **38:**WI(82,1334). **41:**MT(3202). **46:**WA(3202). **55:**FL(35), WI(82).

Phytophthora megasperma f. sp. *glycinea*. Root and stem rot. **4:**MS(1806). **6:**MS(1806). **26:**MS(1806).

P. nicotianae var. *parasitica*. Stem rot. **4:**FL(82). **55:**FL(35).

Pythium debaryanum. Damping-off, stem rot. **50:**TX(82). **55:**CA(82), TX(82).

P. graminicola. Root rot. **4:**Unk.(82). **26:**Unk.(82).

P. ultimum. Root and stem rot. **4:**GA(3824). **6:**GA(3824). **55:**CA(82).

Pythium sp. Root rot. **11:**AL(82), GA(82). **26:**AL(82), GA(82).

Ascomycotina

Apiosporella alpina. **21:**WA(3202). **22:**WA(3815).

Apiosporella sp. On overwintered herbage. **2:**CA(642).

Clathrospora pentamera. **37:**WY(3810).

Comoclathris permunda. **32:**WA(3202). **37:**WY(3810). **52:**WA(3202).

Dasyscyphus sp. **55:**WA(3202).

Diaporthe lupini. On limbs and stems. **7:**CA(1468). **55:**CA(82).

D. phaseolorum. **39:**GA(1438).

Diatrype prominens. **43:**CA(81).

Didymella lupini. On stems. **55:**CA(82).

Erysiphe communis. **34:**CA(638).

E. polygoni. Powdery mildew. **6:**MS(2720). **9:**MT(3202), WA(3202), WY(1195). **21:**CA(81), WA(3202). **24:**WA(3202). **29:**WA(3202). **33:**AK(526,3431). **38:**NE states(82), IL(82), WI(82). **41:**OR(81,82,3202), WA(3202). **46:**WA(3202). **50:**TX(82). **55:**Range of host(82), FL(35), ID(3202).

Glomerella cingulata. Anthracnose, leaf and pod spot. **4:**Unk.(82). **6:**AL(82), FL(82), GA(82,2749). **39:**GA(1438). **55:**FL(35).

Hysteropezizella sp. On overwintered herbage. **34:**CA(642).

Leptosphaeria agnita. **21:**WA(3202). **22:**WA(3814). **37:**WY(3810).

L. byssincola. **55:**CO(3249).

L. castrensis. **34:**CA(3249).

L. erigerontis. **21:**WA(3202). **22:**WA(3814).

L. foeniculacea subsp. *lupina*. On dead stems. **55:**AK(526), CA(81).

L. galiorum. On dead herbage. **55:**CA(661).

L. lupini. **24:**WA(3202).

L. macrospora. On dead stems. **55:**CA(661).

[*L. maculans*]. On dead herbage, limbs and stems. **1:**CA(642). **55:**CA(661).

[*L. millefolii*]. **34:**CA(638,642).

L. ogilviensis. On stems. **2:**CA(638,3249).

L. salsolae. **21:**WA(3202). **22:**WA(3814).

L. shastensis. **2:**CA(3249).

[*L. tetonensis*]. **21:**WA(3202).

L. wehmeyeri. **44:**WY(3249).

Leptosphaeria sp. On overwintering herbage. **2:**CA(642).

Lewia infectoria. On twigs. **18:**CA(2439).

Lophiostoma herbarum. **23:**WA(3202). **27:**WA(3815).

Microsphaera diffusa. Powdery mildew. **39:**GA(3614).

Microsphaera sp. Powdery mildew. **55:**NJ(82).

Mycosphaerella aquilegiae-jonesii. **34:**CA(642).

M. pinodes. Leaf spot. **55:**CT(82), WI(82).

M. punctiformis. On pods. **55:**CA(81).

M. tassiana. On dead stems. **10:**WA(3202). **21:**WA(3202). **22:**WA(3815). **32:**WA(3202). **34:**CA(661). **52:**WA(3202).

Mycosphaerella sp. Stem necrosis. **11:**AL(82). **26:**AL(82).

Nectria episphaeria. **7:**CA(81).

N. eucalypti. On stems. **55:**CA(82).

N. sanguinea. On stems. **7:**CA(81). **55:**CA(82).

Nodulosphaeria modesta. **55:**OR(657).

Phaeosphaeria heterospora. On dead stems. **34:**CA(661).

Phomatospora berkeleyi. **34:**CA(661).

Pleospora amplispora. On stems. **2:**CA(638). **55:**WA(3202,3813).

P. chlamydospora. On overwintered herbage. **1:**CA(82). **2:**CA(642). **34:**CA(638,642).

P. comata. **52:**WA(3202).

P. gigaspora. On overwintered herbage. **2:**CA(642). **34:**CA(642).

P. helvetica. On stems. **21:**WA(3202). **22:**WA(3814). **55:**CO(82).

P. herbarum. On pods. **55:**CA(81,82).

P. herbarum var. *occidentalis*. **37:**WY(3810).

P. magnifica. **35:**CA(661).

P. montana. **10:**WA(3202). **21:**WA(3202). **23:**MT(3202). **32:**WA(3202). **52:**WA(3202). **55:**CO(3813), WY(3813).

P. njegusensis. On dead stem. **8:**AK(1938).

P. penicillus. **2:**CA(642). **21:**WA(3202). **22:**WA(3814). **23:**WA(3202). **27:**WA(3814).

P. pluriseptata. **35:**CA(661).

P. rainierensis. **21:**WA(3202). **22:**WA(3814).

P. togwotiensis. **14:**WY(3810).

P. tragacanthae. **23:**MT(3202).

Pleospora sp. **2:**CA(642).

Proliferodiscus earoleucus. **55:**AK(526).

Rhynchostoma exasperans. On overwintered herbage. **34:**CA(642).

Sclerotinia sclerotiorum. Stem rot. **6:**FL(82), LA(82), MS(2720). **38:**NY(82), TX(82). **55:**FL(35), LA(82), NY(82).

Septoriella sp. On overwintered plant herbage. **34:**CA(642).

Stictis lupini. **7:**CA(2801). **55:**CA(3225).

Valsa lupini. On stems. **55:**CA(82).

Wettsteinina duplex. **27:**WA(3252).

Basidiomycotina - Uredinales

Puccinia andropogonis. Heteroecious: aecial host. **38:**MI(82), MN(82), NY(82), WI(82). **55:**C & W states(82), MI(82), NY(82), WY(82).

[*P. thelypodii*]. **11:**CA(82,738).

Uromyces lupini. Autoecious. **3:**CA(81). **7:**CA(81). **10:**OR(3202). **13:**AZ(3981). **17:**CA(81). **19:**MT(3202). **21:**CA(81), OR(3202), WA(3202). **24:**MT(3202), WA(3202). **25:**CA(81). **36:**AZ(3981). **40:**WY(1195). **41:**CA(81,82), OR(82,3202), WA(82,3202). **43:**CA(81), OR(3202). **46:**ID(3202), WA(3202). **52:**WA(3202). **55:**W states(82), CA(82), CO(1195), ID(3202),

MT(82).
U. occidentalis. Heteroecious: telial host. **9:**AZ(1203,3981), MT(3202). **10:**MT(3202). **19:**MT(3202). **23:**WA(3202). **46:**MT(3202). **55:**W states(82), AZ(1203), CA(82), OR(82,3202).
Uromyces sp. **23:**OR(3202). **25:**WA(3202). **55:**FL(35).
Basidiomycotina - Ustilaginales
Thecaphora deformans. Seed smut. **40:**WY(1077,1195). **55:**CO(82,1077), WY(82).
Basidiomycotina - Other
Armillaria mellea. Root rot. **55:**CA(82).
Flammulina velutipes. Wound rot. **55:**CA(82).
Pleurotus ostreatus. Wound rot. **15:**CA(82). **55:**CA(82).
Deuteromycotina - Hyphomycetes
Alternaria sp. On stem lesions. **26:**AL(82), GA(82). **41:**MA(82). **55:**FL(35), MA(82).
Botrytis cinerea. Gray mold blight, wilt. **24:**MT(3202). **26:**AL(82), FL(82), GA(82). **41:**MA(82). **55:**FL(35), MT(82), NY(82), WA(3202).
Botrytis sp. Gray mold. **55:**NC(1303).
Cercospora longispora. Leaf spot. **38:**NY(82), TX(82), WI(82). **41:**OR(3202). **55:**FL(35,82), MO(82), NY(82), WI(82).
C. lupinicola. Leaf spot. **50:**TX(82). **53:**TX(3597).
Cladosporium herbarum. **10:**WA(3202). **18:**CA(2439).
Epicoccum nigrum. **10:**WA(3202). **21:**WA(3202). **23:**WA(3202). **28:**OR(3202). **33:**AK(3431). **55:**AK(82,526), ID(3202).
Fusarium moniliforme. Root rot. **4:**AL(82), GA(82). **6:**AL(82), GA(82). **26:**AL(82), GA(82).
F. oxysporum f. sp. *radici-lupini*. Root rot. **4:**AL(82), FL(82), GA(82). **6:**AL(82), FL(82), GA(82). **26:**AL(82), FL(82), GA(82).
F. solani f. sp. *lupini*. Seedling blight, root rot, wilt. **4:**Unk.(82). **6:**FL(82). **26:**Unk.(82). **55:**FL(35).
Fusarium sp. Root and stem rot. **6:**MS(2720). **26:**Gulf states(82), SE states(82), CA(82), HI(82). **39:**HI(2903). **55:**E states(82), CA(82), HI(82,2903), LA(82), NJ(82), VA(82).
Hadrotrichum globiferum. Leaf blight. **5:**WY(1195). **9:**SD(2243), UT(1195,3378), WY(1195). **37:**WY(1195,3378). **55:**CO(1195), WA(3202). **47:**CA(82). **48:**CA(82), CO(82), OR(82), WA(82), WY(82). **49:**CA(82), CO(82), OR(82), WA(82), WY(82).
Ovularia lupini. On leaflets. **55:**CO(666).
O. lupinicola. Leaf spot. **1:**OR(3202). **9:**SD(2243). **10:**MT(3202). **38:**WA(82). **41:**WA(3202). **55:**CO(82).
Ovularia sp. Leaf spot. **55:**WA(82).
Phymatotrichopsis omnivora. Root rot. **38:**TX(82). **50:**TX(82). **55:**TX(82).
Pleiochaeta setosa. Black spot, brown spot. **4:**FL(82), LA(82). **6:**AL(82), FL(82), GA(82), MS(2720). **26:**FL(82).
Pseudocercospora lupini. Leaf spot. **16:**FL(82,581), SC(82,581,634). **41:**OR(82). **55:**FL(35).
Ramularia lupini. Leaf spot. **10:**ID(3202), MT(3202), OR(3202), WA(3202). **19:**ID(3202). **20:**ID(3202). **21:**OR(3202), WA(3202). **41:**CA(661), OR(3202). **52:**WA(3202). **54:**MT(3202). **55:**AK(526), TX(82), WA(82).
Stemphylium botryosum. Leaf spot. **6:**FL(3846,3848), GA(3846).
S. solani. Leaf spot. **6:**FL(3848). **55:**FL(35).
Thielaviopsis basicola. Root rot. **55:**WI(82).
Deuteromycotina - Coelomycetes
[*Ascochyta gossypii*]. Canker. **6:**GA(3823).
A. pisi. **38:**WI(1334).
Ascochyta sp. Blight. **38:**WI(82). **55:**CT(82), ID(3202).
Colletotrichum sp. Stem spot. **11:**MD(82). **26:**MD(82). **55:**MD(82).
Cryptosporium lupini. On stems. **55:**CA(82).
Cylindrosporium burkii. Leaf spot. **41:**ID(3202). **42:**ID(666).
C. lupini. On leaves. **21:**CA(81). **33:**AK(3431). **55:**CA(82).
Cylindrosporium sp. **55:**WY(82).
Diplodia lupini. On leaves and stems. **34:**CA(642). **55:**CA(82).
Diplodia sp. **10:**WA(3202).
Gloeosporium lupinicola. **10:**ID(3202).
Hendersonia lupini. On stems. **35:**CA(661). **55:**CA(82).
Hendersonia sp. On dead stems. **34:**CA(661).
Heteropatella conglutinata. **21:**WA(3202).
Lasiodiplodia theobromae. Powdery mildew. **4:**AL(82), FL(82).
Leptostroma lupini. On stems. **14:**WY(3812). **22:**WA(3816).
Leptothyrella graminis. **22:**WA(3816).
Macrophoma lupini. On stems. **55:**CA(82), WA(3202).
Macrophomina phaseolina. Charcoal root rot. **30:**CA(82). **55:**FL(35).
Pellionella tetonensis. On stems. **37:**WY(3812).
Phoma lupini. Leaf spot. **45:**CA(661). **55:**SW states(82), W states(82), CA(82,661), CO(82), ID(3202), NM(82), OR(3202), WA(82,3202).
P. lupinicola. On stems. **55:**CO(82).
P. minuta. On stems. **21:**WA(3202). **37:**WY(3812).
P. pedicularis. **21:**WA(3202).
Phoma sp. **2:**CA(642).
Phomopsis leptostromiformis. Stem blight. **26:**FL(2662), GA(2662).
Phyllosticta ferax. Leaf spot. **2:**CA(638,642). **5:**WY(1195). **9:**SD(2243), UT(1195), WY(1195). **10:**WA(3202). **12:**WA(3202). **19:**OR(3202). **28:**OR(3202). **34:**CA(638). **46:**WY(1195). **55:**W states(82), CA(82), CO(1195), MT(3202), WA(82,3202), WY(82,2924).
P. lupini. **12:**CA(81). **28:**CA(81,82). **31:**CA(81,82). **51:**CA(81).
Phyllosticta sp. On leaves. **38:**WI(1353).
Placosphaeria lupini. Leaf spot. **46:**WA(3202). **55:**AK(82,526), CA(82), WA(82).
Rhabdospora pleosporoides. **21:**WA(3202). **22:**WA(3816). **23:**WA(3202). **27:**WA(3816).
Septogloeum lupini. Leaf spot. **38:**AK(82), CA(82), MI(82). **55:**AK(82), CA(82), MI(82).
Septoria lupini. On leaves. **12:**WA(3202). **15:**CA(82,1468). **21:**MT(3202). **24:**MT(3202). **55:**CA(82).
S. lupinicola. Leaf spot. **33:**AK(3431). **38:**WI(82,1332). **52:**WA(3202). **55:**WI(82), WY(82).
Sirexcipula wyomingensis. On stems. **14:**WY(3812).
Siroscyphellina lupini. **21:**WA(3202). **22:**WA(3816).
Sphaeropsis lupini. On stems. **55:**CA(82).
Stagonospora sp. **46:**ID(3202).
Deuteromycotina - Other
Rhizoctonia solani. Damping-off, stem rot. **4:**AL(82), FL(82), GA(82). **6:**AL(82), FL(82), GA(82). **26:**AL(82), FL(82), GA(82). **38:**CT(82). **50:**TX(82). **55:**Unk.(82), FL(35).
Rhizoctonia sp. Root rot. **6:**MS(2720).
Sclerotium rolfsii. Southern blight, stem rot. **4:**AL(82), FL(82), GA(82). **6:**AL(82), FL(82), GA(82,2886), MS(2720). **26:**AL(82), FL(82), GA(82). **38:**TX(82). **50:**TX(82). **55:**S states(82), FL(35), GA(82), KS(82), TX(82).

Lysiloma Benth.

1. *L. bahamensis* Benth. [Bahama lysiloma]
2. *L. latisiliqua* (L.) Benth. [sabicu, wild tamarind]
3. *Lysiloma* sp.

Ascomycotina
Nectriella pironii. Stem gall. **1:**FL(35).
Basidiomycotina - Uredinales
Ravenelia annulata. Autoecious. **1:**FL(35,82,746).
R. lysilomae. Autoecious. **1:**FL(35,82).
Basidiomycotina - Other
Phellinus dependens. On wood. **3:**FL(2681).
P. extensus. White pocket heart rot. **1:**FL(82).
Deuteromycotina - Hyphomycetes
Helicomina sp. Leaf splotch. **2:**FL(35).
Deuteromycotina - Coelomycetes
Phomopsis sp. Leaf spot. **1:**FL(35).

Maackia Rupr.

1. *M. amurensis* Rupr. & Maxim. [Amur maackia]

Deuteromycotina - Hyphomycetes
Phymatotrichopsis omnivora. Root rot. **1:**TX(82).

Macroptilium (Benth.) Urb.

1. *M. atropurpureum* (Moc. & Sessé ex DC.) Urb. (=*Phaseolus atropurpureus* Moc. & Sessé ex DC.) [conchito, purple bean, siratro]
2. *M. bracteatum* (Nees & Mart.) Maréchal & Baudet (=*Phaseolus bracteatus* Nees & Mart.)
3. *M. lathyroides* (L.) Urb. [one-leaf-clover]

Basidiomycotina - Uredinales
Uromyces appendiculatus. **1:**AZ(3981), FL(35), IN(82).
Deuteromycotina - Hyphomycetes
Cercospora sp. Leaf spot. **1:**FL(35).
Oidium sp. Powdery mildew. **1:**FL(35), HI(2903).

Deuteromycotina - Coelomycetes
Colletotrichum gloeosporioides. Stem canker, leaf spot. **3:**FL(35).
C. truncatum. Anthracnose. **2:**FL(35).
Deuteromycotina - Other
Rhizoctonia solani. Foliar blight, web blight, stem canker, damping-off. **1:**FL(12,35,3391). **2:**FL(35). **3:**FL(35).

Macrotyloma (Wight & Arnott) Verdc.

1. *M. uniflorum* (Lam.) Verdc. (=*Dolichos biflorus* of authors, *Dolichos uniflorus* Lam.) [horsegram]

Ascomycotina
Microsphaera euphorbiae. Powdery mildew. **1:**IN(82).
Deuteromycotina - Coelomycetes
Colletotrichum truncatum. Anthracnose. **1:**GA(3847).
Deuteromycotina - Other
Rhizoctonia solani. Web blight, root and stem rot. **1:**FL(35).

Marina Liebm.

1. *M. parryi* (Torr. & A. Gray) Barneby (=*Dalea parryi* Torr. & A. Gray)

Basidiomycotina - Uredinales
Uropyxis daleae. **1:**AZ(3981).
U. daleae var. *eysenhardtiae*. **1:**AZ(1203).

Medicago L.

1. *M. arabica* (L.) Huds. [spotted burclover, spotted medic]
2. *M. lupulina* L. [black medic, hop-clover, nonesuch, yellow-trefoil]
3. *M. polymorpha* L. (=*M. polymorpha* L. var. *vulgaris* (Benth.) Shinners, *M. hispida* Gaertn.) [California burclover]
4. *M. sativa* L. [alfalfa, lucerne]
5. *M. sativa* L. subsp. *falcata* (L.) Arcang. (=*M. falcata* L.) [yellow-flowered alfalfa]
6. *Medicago* × sp. (unnamed hybrids)
7. *Medicago* sp.

Myxomycetes
Physarum cinereum. Slime mold. **4:**CA(82), KS(82), NY(82), PA(82).
P. vernum. Slime mold. **4:**CA(82), KS(82), NY(82), PA(82).
Physarum sp. Slime mold. **4:**CA(82), KS(82), NY(82), PA(82).
Chytridiomycetes
Physoderma alfalfae. Crown wart. **3:**CA(82). **4:**'Crown wart.' E & C states(82), W states(82), AL(285), CA(82), HI(2903), ID(3202), KY(1777), MS(2720), MT(3202), OR(3202), PA(2068), WA(82,3202). **5:**CA(82). **7:**KY(3583).
Physoderma sp. On leaves. **1:**MS(1844).
Oomycetes
Aphanomyces euteiches. Root rot. **4:**WI(82,847).
Aphanomyces sp. **4:**CA(1299).
Peronospora trifoliorum. Downy mildew. **2:**IA(82). **4:**'Downy mildew.' NE states(82), Range of host(82), AZ(3381), FL(35), HI(1272), IA(215,217,714,716), ID(3202), IL(309), NC(45,48,1303,3931), NY(2998,2999), OK(2872), OR(3202), SD(2243), VA(82,1034), WA(3202), WI(1454), WV(952), WY(1195).
P. viciae. Downy mildew. **4:**MS(2720), NH(1897).
Peronospora sp. Downy mildew. **4:**ME(1890), NH(1890).
Phytophthora cryptogea. Root rot. **4:**CA(1000), IL(463).
P. megasperma. 'Phytophthora root rot.' **4:**Range of host(1299), Unk.(1578), AZ(1572), CA(1123,2966), IA(1124), IL(1123), MN(1123,1124), MS(1123,1428), NC(1303,3857), NE(1124), NY(3901), OH(1123), PA(2069), SD(1124), WA(3202), WI(1124,1428).
P. megasperma f. sp. *medicaginis*. Root rot. **4:**WY(1313).
Pythium aphanidermatum. **4:**AZ(1299).
P. debaryanum. 'Damping-off.' **4:**CA(82), IA(82), MO(82), SD(2243).
P. dissotocum. Isolated from roots. **4:**CA(1437).
P. irregulare. Isolated from roots. 'Damping-off.' **4:**CA(1437).
P. mamillatum. **4:**CA(82), MO(82).
P. middletonii. **4:**CA(82).
P. paroecandrum. Isolated from roots. **4:**CA(1437).
P. pulchrum. **4:**IA(82).
P. rostratum. **4:**CA(82).
P. splendens. 'Damping-off.' **4:**IA(82).
P. ultimum. 'Damping-off.' **4:**CA(82), SD(2243).
P. vexans. Isolated from roots. **4:**CA(1437).
P. violae. Isolated from roots. **4:**CA(1437).
Pythium sp. Damping-off. **4:**AL(1314), NC(1303).
Zygomycotina
Mucor sp. **4:**OR(3202).
Rhizopus stolonifer. **4:**OR(3202).
Ascomycotina
Byssothecium circinans. Crown and root rot. **4:**SD(317,3183).
Calonectria crotalariae. Crown and root rot. **4:**HI(2649,2903).
Chaetomium sp. **4:**OR(3202).
Erysiphe polygoni. Powdery mildew. **3:**CA(82), MS(82,2720). **4:**MA(82), WY(82).
Leptosphaeria pratensis. Leaf spot, root rot. **4:**CA(82), IA(215), KS(82), MD(82), MO(82), MT(82,3202), NV(82), NY(82,3005), OK(82,2873), PA(82), UT(82), VA(82), WA(3202), WI(82).
Leptosphaerulina trifolii. Pepper scorch, brown leaf spot. **1:**AL(82), GA(82). **2:**GA(1297,1438), NY(82), WI(1353). **3:**GA(1297,1438).
Leptospora rubella. On stems. **4:**VA(82).
Leptotrochila medicaginis. Yellow leaf blotch. **3:**CA(82). **4:**'Yellow leaf blotch.' Range of host(82), IA(215,714,716), ID(3202), IL(309), MN(2957), MT(3202), NC(45), NH(1895), NJ(494), NY(2999), OK(2872), OR(3202), SD(2243), WA(3202), WI(1454), WV(952). **5:**Range of host(82).
Metschnikowia pulcherrima. Isolated from floral nectar. **4:**ID(204).
Nectria pseudotrichia. Canker. **4:**LA(82), TX(82).
Neocosmospora vasinfecta. **4:**MN(1299).
Pleospora herbarum. Leaf spot. **1:**MS(82,2720). **4:**E states(82), NC(45,46,48,1303), NJ(494), NY(2999), SD(82), TX(82).
P. rehmiana. On stems. **4:**ID(82,3202), VA(82).
Pseudopeziza medicaginis. Yellow leaf blotch, leaf spot, black stem. **1:**GA(82,1438). **2:**GA(82), NY(82), VT(82), WI(82,1357), WV(952). **4:**'Common leaf spot.' NE states(1890), Range of host(82), AK(2078), AL(759), AZ(1195,3378), CA(81), CO(1195), FL(3500), GA(1438), HI(2903), IA(215,217,714,716), ID(3202), MN(2957), MS(2720), MT(3202), NC(45,48,1303), ND(377), NH(1897), NJ(494), NY(1890,2998,2999,3004,3005), OK(2872), OR(3202), SD(2243), WA(3202), WI(1454), WV(952), WY(819). **5:**KS(82), WI(82).
P. trifolii f. sp. *medicaginis-sativae*. On leaflets. **4:**Unk.(2363).
Pseudopeziza sp. Leaf spot. **4:**VA(1034).
Pseudoplea sp. Pepper spot. **4:**NY(3005).
Rosellinia necatrix. White root rot. **4:**CA(82).
Sclerotinia sclerotiorum. Collar rot. **3:**CA(82). **4:**CA(82), DE(82), ID(82,3202), NY(82), OR(3202), WA(82,3202).
S. trifoliorum. Root rot. 'Sclerotinia crown rot. Sclerotinia stem rot.' **4:**Range of host(82), AL(759,1314), CT(1890), ID(3202), IL(312), MA(1890), MN(1890), MS(2720), NC(45,48,1303), NE(3626), NJ(494), NY(1890,2998,3005), OR(3202), RI(1890), TN(73), VA(1034), WA(3202), WV(952).
Basidiomycotina - Uredinales
Uromyces anthyllidis. Heteroecious: telial host. **3:**TX(128).
U. striatus. Heteroecious: telial host. **2:**E states(82), CA(81,82), HI(2903), NC(1303), OK(82,2872), OR(82,3202), TX(82), WA(3202), WI(1357), WV(952). **3:**CA(82), LA(82), TX(82). **4:**'Rust.' Range of host(82), AL(759,1314), AZ(1203,3981), CA(81,2439), FL(35), GA(1439), HI(2903), IA(215,217,714,716), MD(1729), MN(2957), MS(2720), MT(3202), NC(45,48,1303), NH(1897), NY(2998,2999), OK(2872), SD(2243), WI(1454), WV(3910). **5:**SD(82). **7:**FL(35).
Basidiomycotina - Other
Armillaria mellea. Mushroom rot. **4:**CA(82).
Helicobasidium brebissonii. 'Violet root rot.' **4:**OK(2872).
Thanatephorus cucumeris. Root rot. **1:**LA(3524). **2:**LA(3524). **3:**LA(3524). **4:**NC states(82), NE states(82), AZ(82), CA(82), MI(82), MN(82), OK(82).
Deuteromycotina - Hyphomycetes
Alternaria alternata. **4:**OR(3202).
Alternaria sp. Secondary leaf spot. **4:**DE(82), OK(82,2872), VA(82).
Aspergillus niger. **4:**GA(1438).
Aspergillus sp. **4:**OR(3202).
Aureobasidium pullulans. **4:**OR(3202).
Bipolaris sorokiniana. Black stem. **4:**MN(1299,2956), WI(2956).
Botrytis cinerea. Leaf and shoot blight. **3:**CA(82). **4:**OR(3202), PA(82), WI(1801).
Botrytis sp. Blight. **4:**WI(1454).
Cercospora medicaginis. Leaf spot. **1:**MS(2720). **3:**AL(285), MS(2720), TX(82). **4:**'Summer black stem. Leaf spot.' E & C

states(82), AL(1314), CA(82), FL(35), GA(986), IA(215,217), MS(2720), SD(2243), UT(82), WI(1335), WV(952). **7:**FL(35).
C. zebrina. Summer blackstem, leaf spot. **1:**AL(82), FL(82), GA(82), LA(82), MS(82), TX(82). **3:**AL(82), GA(82), LA(82), MS(82), SC(82), TX(82). **4:**IA(714,716), KY(1780), MN(2957), NC(1303), OK(2872), WI(1454). **7:**FL(35).
Cercospora sp. Leaf spot, spring black stem. **4:**HI(2903), NC(1303).
Cladosporium cladosporioides. **4:**OR(3202).
C. elatum. **4:**OR(3202).
Curvularia sp. **4:**AL(1314).
Cylindrocarpon magnusianum. Root rot. **4:**Unk.(82).
C. obtusisporum. Root rot. **4:**Unk.(82).
Cylindrocladium clavatum. Crown and root rot. **4:**HI(2649,2903).
C. crotalariae. Root and stem rot. **4:**FL(35).
Cylindrocladium sp. **4:**AL(1314).
Dictyochaeta fertilis. Root rot. **2:**NC(1303). **4:**NC(1303).
Epicoccum nigrum. **4:**OR(3202).
Fusarium acuminatum. 'Damping-off.' **4:**ND(82).
F. avenaceum. Collar rot. **4:**WI(82).
F. graminearum. Collar rot. **4:**KS(82), WI(82).
F. moniliforme. Root rot. **4:**IA(714).
F. oxysporum. Root rot, wilt. **4:**CA(1437), IA(714), NC(1303), OR(2063).
F. oxysporum f. sp. *medicaginis*. Collar rot. 'Fusarium wilt.' **4:**AZ(82), CA(82), IA(217), MN(82,1788), MS(82,2720), MT(3202), NE(82), OR(3202).
F. roseum. Root rot. **4:**CA(1437), IA(714), OR(3202).
F. sambucinum. Collar rot. **4:**OH(82), WI(82).
F. solani. Collar and root rot. **4:**CA(1437), IA(714), KY(82), NM(82), OR(2063,3202).
Fusarium sp. Collar, crown and root rot. **4:**C & W states(82), AL(1314), GA(82), HI(2903), IA(215), MN(2957), NC(45,48,1303), OK(2872), OR(2063), VA(82), WA(3202).
Helminthosporium sp. **4:**WA(3202).
Mycoleptodiscus terrestris. Crown and root rot. **4:**NC(1303).
Oidium sp. Powdery mildew. **4:**HI(2903).
Penicillium sp. **4:**OR(3202).
Periconia circinata. Root rot. **4:**ND(3423).
Phymatotrichopsis omnivora. Root rot. **1:**TX(82). **2:**TX(82). **3:**TX(82). **4:**'Phymatotrichum root rot.' AR(82), AZ(82), CA(82), NM(82), NV(82), OK(82,2872), TX(82).
Scolicotrichum sp. Brown stripe. **4:**NC(1303).
Stemphylium botryosum. Leaf spot, summer black stem. **1:**LA(82), MS(82,2720). **3:**LA(82). **4:**'Stemphylium leaf spot.' NE states(1890), AL(759), CA(698,699), FL(35), HI(2903), IA(215,217), KY(1777,1780), MD(699), MN(2957), NC(1303), NH(1897), NY(2998), OR(3202), PA(699), VA(1034), WI(1454), WV(953).
Thielaviopsis basicola. 'Black root rot.' **4:**MA(82), NJ(82), TX(82).
Trichoderma sp. **4:**OR(3202).
Trichothecium sp. **4:**OR(3202).
Ulocladium consortiale. **4:**OR(3202), WA(3202).
Verticillium albo-atrum. 'Verticillium wilt.' **4:**KS(3529), MD(1388), MN(1125), NY(1832), OR(1298), WA(1298), WI(1309), WY(1316).
Volutella colletotrichoides. **4:**IA(576).
Volutella sp. On foliage. **4:**IA(1299).

Deuteromycotina - Coelomycetes

Ascochyta sp. **4:**AL(759).
Botryodiplodia sp. **4:**MS(1299).
Colletotrichum dematium. **4:**WA(3202).
C. destructivum. Anthracnose, brown leaf spot, black stem. **4:**Unk.(3059), AL(759), GA(1802), IA(215,217,714), KY(3718), MN(2957), NC(1303), WI(1343).
C. gloeosporioides. Anthracnose. **4:**HI(2903).
C. graminicola. Isolated from stem lesions. **4:**Unk.(3059), AL(759), GA(1802), KY(1802).
C. trifolii. Anthracnose. **1:**AL(82), FL(82), GA(82), LA(82), MS(82,2720). **2:**AL(82), TX(82). **3:**AL(82), GA(82), LA(82), MS(82,2720), OK(82,2872). **4:**'Anthracnose' E states(82), AL(1314), AZ(82), CA(82,1000), GA(1802), MD(2660), MN(1126), MS(2720), NC(45,48,1303), NJ(494), NM(82), NY(3000), OK(44,2872), UT(82), WA(3202), WI(1343), WV(3911). **7:**FL(35).
C. truncatum. Anthracnose. **4:**HI(2903), NC(1303).
Colletotrichum sp. Anthracnose. **4:**E states(82), AZ(82), CA(82), NM(82), UT(82), WA(2234).
Diplodia sp. Crown rot. **4:**AZ(1299).
Lasiodiplodia theobromae. Crown rot. **4:**AZ(3312).
Macrophomina phaseolina. Charcoal rot. **4:**AL(1314), IL(82,311), OK(82,2872).
Phoma anceps. Stem canker. **4:**MT(82).
P. medicaginis. Seedling blight, stem canker, root rot, spring black stem, leaf spot. **5:**AK(526), KS(82). **6:** AK(526).
P. pinodella. **4:**OR(3202).
P. sclerotioides. 'Plenodomus root rot.' **4:**AK(1299).
P. terrestris. Isolated from rhizosphere. **4:**OR(2791).
P. vulgaris. Stem canker. **4:**WV(82).
Phoma sp. Stem blight, stem canker, crown and root rot. **2:**UT(82). **4:**Unk.(82), NC(3931).
Placosphaeria medicaginis. Leaf blotch, tar spot. **4:**IL(82,309).
Pyrenochaeta sp. Root rot. **4:**IA(714).
Rhabdospora allantoidea. On stems. **4:**PA(82).
Sporonema phacidioides. Yellow leaf blotch, leaf spot. **4:**C states(3976), HI(2903), NC(1303), OK(2872).
Stagonospora meliloti. Anthracnose, stem canker, crown rot, leaf spot. **1:**FL(82), GA(82), MS(23,82,2720). **2:**NY(82), TX(82), WI(1333). **4:**'Stagonospora leaf spot. Stagonospora root rot.' CA(1000), FL(35), KS(973), NC(3931), NY(2998), OK(2872), VA(1034), WI(1454). **7:**FL(35).
Stagonospora sp. Leaf spot. **4:**NC states(82), NE states(82), MT(82), NH(1897), NY(2999), UT(82). **5:**WI(82).

Deuteromycotina - Other

Rhizoctonia crocorum. Violet root rot. **4:**AZ(82), IA(82,716), KS(82), MI(82), MO(82), NC(1303), NE(82), OK(82), OR(82), TX(82), WA(82), WI(1358,1456).
R. leguminicola. **4:**WV(1299).
R. solani. Aerial and summer blight, stem canker, crown rot. **2:**NC(82). **4:**'Rhizoctonia root rot. Damping-off.' AL(1314), CA(1000,1437), FL(35), IA(716), MN(2957), MS(2720), NC(12,45,48), OK(2872).
Rhizoctonia sp. Summer blight, root and stem rot. **4:**C states(82), Range of host(82), W states(82), HI(2903), IA(714), NC(82,1303), VA(82,1034).
Sclerotium rolfsii. Root rot. 'Southern blight.' **4:**AL(1314), GA(82), HI(82,2903), LA(82), MS(82,2720), NC(45,48,1303), TX(82), VA(82).

Melilotus Mill.

1. *M. alba* Medik. [Bukhara-clover, white melilot, white sweet-clover]
2. *M. alba* Medik. var. *annua* H. S. Coe [Hubam clover]
3. *M. dentata* (Waldst. & Kit.) Pers.
4. *M. indica* (L.) All. [Indian sweet-clover, sour-clover]
5. *M. italica* (L.) Lam.
6. *M. officinalis* (L.) Lam. [melist, yellow melilot, yellow sweet-clover]
7. *M. suaveolens* Ledeb.
8. *Melilotus* sp.

Oomycetes

Aphanomyces euteiches. Seedling root rot. **1:**WI(82).
Peronospora trifoliorum. Downy mildew. **1:**WI(82,1340). **2:**TX(82), WI(82). **6:**IA(82), ID(82,3202), MT(82), OR(3202), WA(3202), WY(2924).
Phytophthora cactorum. Root rot. **1:**WI(82). **6:**SD(2243). **8:**WI(1454).
Pythium debaryanum. Damping-off. **1:**WI(82). **6:**WI(82).
P. ultimum. Damping-off. **1:**WI(82). **6:**WI(82).
Pythium sp. Seedling blight, root rot. **1:**IA(82), IL(82), KY(82), OH(82), WA(82). **4:**IA(82). **8:**WA(3202).

Ascomycotina

Cilioplea kansensis. On dead stems. **1:**KS(82,970,3817).
Didymella lethalis. Black stem. **1:**NE states(82), ND(82), OK(2872), TX(82). **3:**WI(82). **4:**LA(82). **6:**IA(82), NJ(82), NY(82). **8:**WI(1454).
D. obscura. **8:**MD(82).
Erysiphe polygoni. Powdery mildew. **4:**CA(82).
Leptosphaeria pratensis. Leaf spot. **1:**C states(82), NE states(82), W states(82), ID(3202), NH(1895), WA(82,3202). **4:**WV(82). **6:**ID(3202), MT(82,3202), WA(3202), WI(82). **8:**IA(215,217).
Leptosphaeria sp. On dead stems. **1:**NY(82), OH(82), WI(82).
Leptosphaerulina argentinensis. **1:**PA(1297).

Mycosphaerella davisii. Leaf spot. **1**:E states(82), KS(82), MN(82). **6**:IA(82), OK(82,2873), WI(82).
M. linhartiana. On stems. **1**:NJ(82).
Phaeosphaeria viridella. On stems. **1**:NY(195,3249), WI(3249).
Pleospora herbarum. On dead stems. **8**:CA(2439).
P. rehmiana. **8**:ID(3202).
Pseudopeziza trifolii. Leaf spot. **1**:MN(2957), MS(2720). **6**:MN(2957). **8**:WI(1454).
Pseudopeziza sp. Leaf spot. **1**:NH(1895).
Pseudoplea sp. Leaf spot. **1**:MN(2957). **6**:MN(2957).
Sclerotinia minor. Collar rot. **1**:MN(82).
S. sclerotiorum. Stem rot. **1**:CA(82), MT(82,3202). **4**:CA(82). **6**:MT(82,3202), WA(3202).
S. trifoliorum. Root and stem rot. **1**:CA(82), ID(3202), KY(82), MS(2720), OR(82,3202), TX(82), VA(82), WA(82,3202), WV(82).

Basidiomycotina - Uredinales

Uromyces striatus. Heteroecious: telial host. **5**:IA(2100).

Basidiomycotina - Ustilaginales

Entyloma meliloti. White smut. **4**:AL(82,1077), AZ(82,1077), CA(82,1077), LA(82,1077), SC(82), TX(82,1077). **6**:LA(82,1077).

Basidiomycotina - Other

Thanatephorus cucumeris. **8**:WA(3202).

Deuteromycotina - Hyphomycetes

Aureobasidium caulivorum. Northern anthracnose. **1**:IL(82), KS(82).
Botrytis sp. Blight. **8**:WI(1454).
Cercospora davisii. Leaf spot, summer black stem. **1**:MN(2957), OK(2872), WI(581). **6**:HI(2903), OK(2872). **8**:IA(215,217), WI(1454).
C. zebrina. Leaf spot. **1**:CA(82), NY(82). **4**:CA(82). **6**:NY(82), SD(2243).
Cercospora sp. Leaf spot. **1**:HI(2903).
Dictyochaeta fertilis. Root rot. **1**:NC(1303).
Fusarium sp. Root rot, wilt. **1**:KS(82), MN(82), ND(82), NM(82), OK(2872), TN(82). **8**:IA(217).
Mycoleptodiscus terrestris. Root rot, leaf spot. **1**:FL(35).
Phymatotrichopsis omnivora. Root rot. **1**:AZ(82), TX(82). **2**:AZ(82), TX(82). **4**:TX(82). **6**:TX(82).
Ramularia meliloti. Leaf spot. **4**:CA(82).
Ramularia sp. Leaf spot. **1**:NY(82).
Stemphylium botryosum. Leaf spot. **1**:MN(2957). **6**:MN(2957).
S. sarciniforme. Leaf spot. **1**:WA(82,3202).
Stemphylium sp. Leaf spot. **1**:MO(82).

Deuteromycotina - Coelomycetes

Ascochyta lethalis. Stem canker, goose neck, spring black stem, crooked tip. **1**:IA(82), IN(82), KS(82), MN(82,2957), MO(82), ND(82), OK(82), WI(82,1342). **2**:MS(2720). **6**:HI(2903), MN(2957), MT(82), OK(2873), PA(82), SD(2243). **8**:IA(215,217), MT(3202), WI(1454).
A. pisi. Leaf spot, black stem. **1**:OK(2872).
Ascochyta sp. Stem rot. **1**:WA(3202). **4**:LA(82).
Colletotrichum destructivum. Anthracnose. **1**:IA(82), KY(82), MN(2957), WI(82). **6**:MN(2957). **8**:IA(215).
C. gloeosporioides. On leaves. **1**:AL(82).
C. trifolii. Southern anthracnose. **1**:IA(82), IL(82), MS(2720), NC(1303), OK(82,2872), TX(82), WI(1343). **2**:MS(82), TX(82). **4**:LA(82), TX(82).
Colletotrichum sp. **1**:IA(82).
Hendersonia sp. On dying stems. **1**:NY(82), WI(82).
Macrophomina phaseolina. Charcoal rot. **1**:IL(311).
Phoma medicaginis. Black stem. **6**:ID(82,3202), OK(82,2873), WA(3202). **8**:WI(1454).
P. terrestris. Isolated from rhizosphere. **6**:OR(2791).
Sporonema phacidioides. Leaf spot. **1**:NC(1303).
Stagonospora meliloti. Leaf spot. **1**:ID(1195,3202), NH(1895). **6**:SD(2243), WI(23). **7**:WI(23). **8**:WI(1454).

Deuteromycotina - Other

Rhizoctonia solani. Root and stem rot. **1**:E states(82), MS(2720), TX(82), WA(82).
Rhizoctonia sp. Root rot. **4**:LA(82).
Sclerotium rolfsii. Southern blight. **1**:TX(82). **4**:TX(82).

Mimosa L.

1. *M. biuncifera* Benth.
2. *M. dysocarpa* Benth.
3. *M. grahamii* A. Gray
4. *M. laxiflora* Benth.
5. *M. strigillosa* Torr. & A. Gray
6. *Mimosa* sp.

Ascomycotina

Nectria cinnabarina. Canker. **6**:OK(2872).

Basidiomycotina - Uredinales

Ravenelia fragrans. Autoecious. **1**:AZ(1203,3981). **2**:AZ(1203,3981). **3**:AZ(1194,1203,3981). **4**:AZ(3981). **6**:AZ(82), TX(82).
R. fragrans var. *evernia*. Autoecious. **4**:AZ(747). **6**:AZ(82).

Basidiomycotina - Other

Armillaria tabescens. Root rot. **6**:FL(35).
Phanerochaete arizonica. White rot. **1**:AZ(433).
Sebacina epigaea. On wood. **6**:GA(2642).
Sirobasidium brefeldianum. On limbs. **6**:GA(2642).
Tulasnella pinicola. On wood. **6**:GA(2642).

Deuteromycotina - Hyphomycetes

Phymatotrichopsis omnivora. Root rot. **6**:TX(82).

Deuteromycotina - Coelomycetes

Cylindrosporium sp. Leaf spot. **5**:TX(652). **6**:TX(82).
Phyllosticta hawaiiensis. On dead twigs. **6**:HI(2784).

Mucuna Adans.

(= *Stizolobium* P. Br.)

1. *M. gigantea* (Willd.) DC.
2. *M. novaguineensis* Scheff.
3. *M. pruriens* (L.) DC. var. *utilis* (Wallich ex Wight) Baker ex Burck (=*Stizolobium deeringianum* K. S. Bort, *M. deeringiana* (K. S. Bort) Merr.) [Bengal velvetbean, cowage velvetbean, Florida velvetbean, Mauritius velvetbean, velvetbean, Yokohama velvetbean]
4. *Mucuna* sp. (=*Stizolobium sp.*)

Oomycetes

Phytophthora cinnamomi. Root rot. **1**:HI(2903).
P. nicotianae var. *parasitica*. Root rot. **1**:HI(2903). **3**:FL(35,82), HI(2903). **4**:FL(35).
P. palmivora. **1**:HI(2903). **3**:HI(2903).

Ascomycotina

Mycosphaerella cruenta. Leaf spot. **3**:GA(82,1438).

Basidiomycotina - Other

Thanatephorus cucumeris. Stem rot. **4**:AL(35,285).

Deuteromycotina - Hyphomycetes

Fusarium sp. Pod spot. **4**:TX(82).
Phymatotrichopsis omnivora. Root rot. **3**:AZ(82), TX(82).
Pseudocercospora stizolobii. Leaf spot. **3**:FL(35), MS(2720), NC(1303).

Deuteromycotina - Coelomycetes

Colletotrichum gloeosporioides. Anthracnose, cutting rot. **2**:HI(2899). **3**:HI(2903).
Phyllosticta mucunae. Leaf spot. **3**:AL(82), HI(82), TX(82). **4**:AL(285), HI(2903).

Deuteromycotina - Other

Rhizoctonia solani. Stem rot. **3**:GA(82), LA(82).
Sclerotium rolfsii. Southern blight. **3**:AL(82), FL(35,82), GA(82), MS(2720), TX(82). **4**:FL(35).

Neonotonia J. Lackey

1. *N. wightii* (J. Graham ex Arnott) J. Lackey var. *wightii* (=*Glycine javanica* of authors, *Glycine wightii* (J. Graham ex Arnott) Verdc.)

Deuteromycotina - Other

Rhizoctonia solani. Root rot, basal stem rot. **1**:FL(35).

Olneya A. Gray

1. *O. tesota* A. Gray [desert ironweed]

Basidiomycotina - Other

Coniophora submembranacea. **1**:AZ(1202).
Grandinia pruni. **1**:AZ(1202).
Phanerochaete chrysorhiza. **1**:AZ(1202).
Phellinus texanus. **1**:AZ(1202).
Wolfiporia sulphurea. **1**:AZ(1202).

Onobrychis Mill.
1. *O. viciifolia* Scop. [esparcet, holy clover, sainfoin]

Oomycetes

Phytophthora megasperma. Seedling blight. **1**:WY(1317).

Deuteromycotina - Hyphomycetes

Fusarium solani. Crown and root rot. **1**:MT(3165).

Deuteromycotina - Other

Sclerotium rolfsii. Isolated from crown tissue. **1**:TX(3952).

Ornithopus L.
1. *O. sativus* Brot. [serradella]

Deuteromycotina - Other

Rhizoctonia solani. Root rot. **1**:Unk.(82).

Oxytropis DC.
1. *O. arctica* R. Br.
2. *O. besseyi* (Rydb.) Blank.
3. *O. campestris* (L.) DC. (=*Astragalus campestris* L.)
4. *O. campestris* (L.) DC. var. *gracilis* (A. Nels.) Barneby (=*O. gracilis* (A. Nels.) K. Schum.)
5. *O. chartacea* Fassett
6. *O. deflexa* (Pall.) DC.
7. *O. lambertii* Pursh (=*Aragallus lambertii* Greene) [Lambert crazyweed, Lambert loco, locoweed, purple loco]
8. *O. maydelliana* Trautv.
9. *O. nigrescens* (Pall.) Fisch. ex DC.
10. *O. podocarpa* A. Gray
11. *O. sericea* Nutt. (=*O. albiflora* (A. Nels.) K. Schum., *O. saximontana* (A. Nels.) A. Nels.)
12. *O. viscida* Nutt. ex Torr. & A. Gray
13. *Oxytropis* sp.

Ascomycotina

Erysiphe polygoni. Powdery mildew. **4**:AK(3431). **7**:MT(82,3202), SD(82), WY(82).
Mycosphaerella tassiana. On dead stems. **13**:AK(1938).
Phaeosphaeria vagans. **12**:AK(526).
Pleospora helvetica. On dead stems. **9**:AK(526). **13**:AK(82,526).
P. herbarum. Leaf spot. **1**:AK(526). **3**:MT(3202). **13**:AK(82), MT(82).
P. tragacanthae. On dead stems. **8**:AK(1938). **13**:AK(1938).
Sphaerotheca fuliginea. On leaves and stems. **13**:AK(1938).

Basidiomycotina - Uredinales

Uromyces lapponicus. Autoecious. **3**:WY(1195). **4**:AK(60), WY(1195). **9**:AK(60,526). **11**:CO(1195), WY(2924). **13**:CO(82), WY(82).
U. punctatus. Heteroecious: telial host. **2**:ID(3202). **3**:MT(3202), WA(3202). **4**:WY(2924). **5**:WI(82,1332). **7**:W states(82), MT(3202), ND(378), NM(82). **10**:ID(3202). **11**:MT(3202), WY(1195). **13**:W states(82).

Deuteromycotina - Hyphomycetes

Ramularia sp. **6**:MT(3202).

Deuteromycotina - Coelomycetes

Phloeospora oxytropidis. Leaf spot. **7**:MT(82,3202), NM(82).
Phyllosticta sp. **13**:AK(3431).
Septoria sp. **13**:AK(3431).

Pachyrhizus Rich. ex DC.
1. *P. erosus* (L.) Urb. (=*P. angulatus* Rich. ex DC., *P. bulbosus* (L.) Kurz) [jicama, yambean]

Deuteromycotina - Other

Rhizoctonia sp. Root rot. **1**:HI(2903).

Parkinsonia L.
1. *P. aculeata* L. [horse-bean, Jerusalem-thorn, Mexican palo-verde, retaima]
2. *P. florida* (Benth. ex A. Gray) S. Wats. (=*Cercidium floridum* Benth. ex A. Gray) [palo-verde]
3. *P. microphylla* Torr. (=*Cercidium microphyllum* (Torr.) J. Rose & I. M. Johnst.)
4. *Parkinsonia* sp.

Oomycetes

Pythium sp. Seedling root rot. **1**:FL(35).

Ascomycotina

Erysiphe sp. Powdery mildew. **1**:FL(35).
Leptosphaeria sp. Stem canker. **1**:FL(35).
Nectriella pironii. Stem gall. **1**:FL(35).

Basidiomycotina - Other

Armillaria tabescens. Root rot. **1**:FL(35). **4**:FL(82).
Dacrymyces minor. **3**:AZ(1202).
Ganoderma lucidum. **3**:AZ(17).
Inonotus rickii. On wood. **4**:AZ(1204).
Peniophora tamaricicola. White rot. **3**:AZ(1180,1202).
Perenniporia ohiensis. **1**:AZ(128,1202).
Phanerochaete arizonica. White rot. **3**:AZ(433,1202).
Phellinus texanus. **3**:AZ(1202).

Deuteromycotina - Hyphomycetes

Alternaria sp. Leaf spot. **1**:FL(35).
Cylindrocladium clavatum. Root rot. **1**:FL(35).
Fusarium sp. Damping-off. **1**:FL(35).
Phymatotrichopsis omnivora. Root rot. **2**:SW states(82). **4**:AZ(82), TX(82).
Ptychogaster cubensis. **1**:AZ(1202).

Deuteromycotina - Coelomycetes

Asteroma parkinsoniae. On pods. **2**:AZ(82).
Coniothyrium sp. Twig dieback. **1**:FL(35).
Cylindrosporium parkinsoniae. Leaf spot. **4**:TX(82).
Diplodia sp. Dieback. **1**:FL(35).
Macrophoma sp. Twig dieback. **1**:FL(35).
Phoma sp. Twig dieback. **1**:FL(35).
Phomopsis sp. Twig dieback. **1**:FL(35).
Phyllosticta parkinsoniae. Leaf spot. **1**:FL(35). **4**:TX(82).

Parryella Torr. & A. Gray ex A. Gray
1. *P. filifolia* Torr. & A. Gray

Basidiomycotina - Uredinales

Uropyxis amorphae. Autoecious. **1**:NM(1195).

Pediomelum Rydb.
This genus is considered a synonym of *Orbexilum*, but appropriate combinations could not be located.
1. *P. argophyllum* (Pursh) Grimes (=*Psoralea argophylla* Pursh)
2. *P. cuspidatum* (Pursh) Rydb. (=*Psoralea cuspidata* Pursh)
3. *P. digitatum* (Nutt. ex Torr. & A. Gray) Isely (=*Psoralea digitata* Nutt. ex Torr. & A. Gray)
4. *P. esculentum* (Pursh) Rydb. (=*Psoralea esculenta* Pursh) [breadroot, Indian turnip, prairie-potato, prairie-turnip]

Ascomycotina

Parodiella hedysari. Black mildew, black spot. **1**:KS(82), ND(82,377), NE(82), SD(82,2243). **2**:OK(2872). **3**:OK(2873).

Basidiomycotina - Uredinales

Uromyces coloradensis. **1**:ND(378).
U. psoraleae. **4**:W states(82), AZ(82), OR(82).
U. psoraleae var. *argophyllae*. **1**:C states(82), AZ(82), MT(82), SD(2243). **2**:SD(2243). **4**:C states(82), AZ(82), MT(82).

Deuteromycotina - Hyphomycetes

Cercospora latens. Leaf spot. **1**:KS(82).
Cercospora sp. On calyx. **4**:WI(1343).
Cercosporidium cassiocarpum. Brown leaf and stem spot. **3**:OK(605,2872).

Deuteromycotina - Coelomycetes

Actinonema psoraleae. Leaf spot. **3**:KS(82).
Colletotrichum psoraleae. On leaves. **4**:NE(82), WI(1333).
Septoria argophylla. Leaf spot. **1**:KS(82), OR(3202).

Phaseolus L.
1. *P. acutifolius* A. Gray [Tepary bean]
2. *P. coccineus* L. (=*P. multiflorus* Lam.) [Dutch-case-knife bean, scarlet runner bean]
3. *P. lunatus* L. (=*P. inamoenus* L., *P. limensis* Macfady., *P. tunkinensis* Lour.) [butter bean, Carolina bean, civet bean, garden bean, lima bean, sewee bean, sieva bean]
4. *P. vulgaris* L. (=*P. nanus* L.) [French bean, frijol, garden bean, green bean, haricot bean, kidney bean, navy bean, runner bean, salad bean, snap bean, string bean, wax bean]
5. *P. vulgaris* L. var. *humilis* Alef. [bush bean]
6. *Phaseolus* sp.

Myxomycetes

Diachea leucopodia. Slime mold. **4**:NC(1303).

Oomycetes

Aphanomyces euteiches f. sp. *phaseoli*. Hypocotyl and root rot.

4:WI(2789).

Phytophthora nicotianae var. *parasitica*. Pod, root and stem rot. 'Downy mildew.' 4:FL(35), NC(1303).

P. phaseoli. Downy mildew. 3:E & C states(82), DE(734), KY(82), MN(82), OH(82).

Phytophthora sp. Wilt. 4:CO(2656).

Pythium aphanidermatum. Damping-off, cottony leak of market beans, root and stem rot, wilt. 3:CA(82), NC(1303). 4:'Pythium blight.' CA(82), CO(82,2656), FL(35), ID(82,2339,3202), MD(896), NC(1303), NY(82), OK(82,2872), VA(82), WA(3202). 6:ID(3925).

P. aristosporum. Root rot. 4:WI(1939).

P. catenulatum. Root rot. 4:WI(1939).

P. debaryanum. Damping-off, pod and root rot. 4:CT(82), DE(82), FL(82), ID(82), NC(1303), NJ(82), NY(82), OH(82).

P. dissotocum. Root rot. 4:WI(1939).

P. helicoides. Damping-off. 4:FL(35,82).

P. myriotylum. Damping-off, root and stem rot. 4:GA(1159), MD(896,1159), NC(1303).

P. oligandrum. Root rot. 4:FL(35,82), VA(82).

P. rostratum. Root rot. 4:CA(82).

P. ultimum. Damping-off, root and stem rot, wilt. 3:CA(82). 4:CA(82), DE(896), FL(35,82), LA(82), NC(1303,3931), WA(1964,3202).

P. vexans. 4:MD(82), VA(82).

Pythium sp. Stem blight, pod decay, root rot. 3:FL(35), MD(82). 4:CO(1854), ID(2339), MS(2720), NE(3475), WI(1939).

Zygomycotina

Choanephora sp. Pod and seed rot. 3:NC(1303). 4:FL(35).

Rhizopus arrhizus. Soft rot of market beans. 3:Unk.(82). 4:Range of host(82).

R. stolonifer. Soft rot of market beans. 3:Unk.(82). 4:Range of host(82), FL(35), NC(1303).

Rhizopus sp. 4:MT(3202).

Ascomycotina

Botryotinia fuckeliana. Gray mold. 4:NY(2837).

B. ricini. Wilt. 4:Unk.(82).

Chaetomium globosum. Isolated from seeds. 4:CA(3306).

Diaporthe arctii. On stems. 4:GA(82,1438).

D. phaseolorum. Pod blight, leaf spot. 3:E states(82), FL(35), GA(82,1438), LA(82), MS(2720), NC(1303), OH(82), OK(82,2873). 4:GA(82,1438), HI(2903), MS(2720).

Diaporthe sp. Root rot. 4:AL(82).

Ellisiodothis inquinans. 3:GA(1438).

Elsinoe phaseoli. Scab. 3:SC(3295).

Erysiphe graminis f. sp. *avenae*. Powdery mildew. 4:AR(3029).

E. polygoni. Powdery mildew. 1:CA(82). 2:CA(82). 3:CA(82), CT(82), FL(35,82), IA(82), NC(1303), NJ(82), NY(82), TX(82), VA(82). 4:'Powdery mildew.' E states(82), SW states(82), Unk.(3976), FL(35,3500), GA(1438), MS(2720), NC(1303).

Glomerella lindemuthiana. 4:GA(1438).

Leptosphaeria phaseolorum. On stems. 4:NJ(82).

Leptosphaerulina trifolii. 3:GA(1438).

Microsphaera diffusa. Powdery mildew. 3:IL(82), MD(82). 4:GA(82,1438), MD(82).

M. euphorbiae. Powdery mildew. 3:IN(82).

Mycosphaerella cruenta. Leaf blotch. 4:E states(82), AR(82), FL(35), GA(1438), OK(82), TX(82), WI(82).

Nematospora lycopersici. On seeds. 3:FL(204).

N. phaseoli. Yeast spot. 3:SE states(82), IL(82), MS(2720), NC(1303).

Pleospora herbarum. Leaf spot. 4:Unk.(82), NC(1303,3931).

Sclerotinia minor. Wilt. 4:Unk.(82).

S. sclerotiorum. White mold, watery soft rot, stem rot, wilt. 3:CA(82), IN(82), NC(1303), NY(82), TX(82). 4:'White mold.' Range of host(82), AK(2163), CO(447), FL(35,3500), HI(2903), ID(2339,3202), MT(3202), NC(1303), OR(3202), WA(3202), WY(819). 6:ID(3925), NE(624).

Basidiomycotina - Uredinales

Uromyces appendiculatus. Autoecious. 1:CA(82), TX(82). 2:AL(82,285), MA(82), NH(82). 3:E states(82), Range of host(82), CA(82), MS(2720), OH(82), TX(82). 4:'Rust.' Range of host(82), AL(285), AZ(1203,3981), CA(81), FL(35,3473,3500), GA(1439), HI(2781,2903), ID(3202), MD(3473), MI(3473), MN(25), MS(2720), MT(3202), NC(1303,3931), ND(378,3472,3473), NE(3473), NJ(3473), NM(2109), NY(1800,3473), OK(2872), OR(3202), SD(2243), TN(3473), TX(3473), VA(3473), WA(3202), WI(1354), WY(819). 5:MA(1390). 6:NY(448), WA(2349), WY(1195).

Basidiomycotina - Other

Corticium microsclerotium. Web blight. 3:FL(82), LA(1010). 4:FL(82), LA(82,1010), MS(82,2720), NC(82), SC(82).

Thanatephorus cucumeris. Root and stem rot, banded sclerotial disease. 3:Unk.(82), LA(1010). 4:Range of host(82), AK(2163), AL(285), CO(446), ID(3202), LA(1010), MS(2720), NC(1303), OK(2872), OR(3202), WA(3202), WY(446).

Deuteromycotina - Hyphomycetes

Alternaria alternata. Flecking of pods, leaf spot. 3:NC(1303). 4:'Alternaria leaf spot.' FL(35), HI(2903), MT(3202), NC(1303), ND(378), NY(3), OK(2872), WA(3202).

A. tenuissima. Leaf spot. 3:GA(82), MD(82). 4:FL(35), HI(2903), OH(82).

Alternaria sp. Leaf spot, on leaves, roots and stems. 3:KY(82), MD(82), NJ(82), NY(82), PA(82), VA(82). 4:Range of host(82), FL(3563), ID(3202), NE(3475), OR(3202).

Aspergillus sp. On roots and stems. 4:NE(3475).

Aureobasidium pullulans. Seed spot. 4:NY(82).

Botrytis cinerea. Gray mold shoot blight, rot, leaf spot. 3:AK(82), CA(82,1871), FL(35,82), LA(82), ME(82), NC(1303), NY(82), TX(82), WA(82,3202). 4:'Gray mold.' AK(82), CA(82), FL(35,82), IL(312), KY(1776), LA(82), ME(82), NY(82), OR(3202), TX(82), WA(82,3202). 6:AK(526).

Botrytis sp. Pod rot. 4:ID(3202), WA(3202). 6:WA(2349).

Cercospora canescens. Canker, leaf spot, sometimes on pods. 3:MS(2720), NC(1303). 4:AL(82), FL(35,82), KS(82), MO(82), NC(1303), NJ(82), TX(82), WI(1347).

Cercospora sp. Leaf spot. 3:FL(35). 4:NC(1303).

Cladosporium herbarum. Pod and seed spot. 3:CA(82), FL(35,82), GA(82), MD(82), NJ(82), VA(82). 4:FL(35), MT(82,3202).

Cristulariella moricola. 3:WV(158). 4:WV(158).

Dictyochaeta fertilis. Root rot. 4:NC(1303).

Epicoccum nigrum. Secondary leaf spot. 3:MD(82), NJ(82). 4:MD(82), MS(82,2720), NJ(82).

Epicoccum sp. Secondary leaf spot. 3:MD(82), NJ(82).

Fusarium oxysporum. Wilt. 'Fusarium root rot.' 4:HI(2903), NE(3475).

F. oxysporum f. sp. *phaseoli*. Yellows, wilt. 4:CA(82), CO(82), ID(82,3202), MT(82), NC(1303), SC(102).

F. oxysporum f. sp. *vasinfectum*. Root decay, wilt. 4:AL(82), FL(35,82), NC(1303).

F. roseum. On roots and stems. 4:NE(3475).

F. solani. 'Fusarium root rot.' 4:AL(285), NE(3475), OK(2872).

F. solani f. sp. *phaseoli*. Seedling blight, seed decay, pod rot, dry root rot, seed stain. 1:CA(82), TX(82). 2:NY(82), OR(3202). 3:CA(82), FL(35), IL(82), MD(82), MS(82,2720), NC(1303), NY(82), OH(82), PA(82), TX(82), VA(82), WV(82). 4:Range of host(82), FL(35), HI(2781,3683), ID(2339,3202), MS(2720), MT(3202), NC(1303), OR(3202), WA(3202), WY(819). 6:ID(3925), WA(2349).

Fusarium sp. Root rot. 4:NC(3931), OK(2872), OR(3202).

Gliocladium roseum. Root rot. 4:ID(1663,3202).

Gliocladium sp. On roots and stems. 4:NE(3475).

Heterosporium sp. Sooty spot, on old leaves and pods. 3:Unk.(82). 4:Unk.(82), ID(2339,3202).

Mycoleptodiscus terrestris. Root rot, leaf spot. 3:FL(35). 4:FL(35).

Oidium sp. Powdery mildew. 4:HI(2903).

Penicillium sp. On roots and stems. 4:MT(3202), NE(3475).

Phaeoisariopsis griseola. 'Angular leaf spot.' 4:E states(82), FL(3500), HI(82,2903), IN(82), MD(3794), MI(686), MS(2720), NC(1303,3931), NJ(82), OK(82,2873), TX(82), WI(1340,1410).

Phialophora gregata. Brown stem rot. 4:IL(82). 6:IL(82).

Phymatotrichopsis omnivora. Root rot. 1:TX(82). 3:AZ(82), TX(82). 4:'Phymatotrichum root rot.' AZ(82), OK(82,2873), TX(82).

Pseudocercospora cruenta. Leaf spot. 2:AL(82,285). 3:OK(2872). 4:'Cercospora leaf blotch.' AL(285), FL(35), HI(2903), MO(82), MS(2720), SC(82).

Stachybotrys chartarum. 4:ID(3202).

Stemphylium botryosum. Leaf spot. 4:OR(3202), WA(82,3202).

Stemphylium sp. Leaf spot. 4:FL(35).

Thielaviopsis basicola. Black root rot. 3:CA(82). 4:'Black root rot.' AL(82), CA(82), NC(1303), NH(82), NJ(82), NY(82), WA(3202).

Thielaviopsis sp. On roots and stems. 4:NE(3475).

Trichothecium roseum. Pink pod rot. **4:**ND(3690).

Deuteromycotina - Coelomycetes

Aristastoma oeconomicum. Leaf spot. **4:**GA(82).

Colletotrichum lindemuthianum. Anthracnose. **2:**NY(82). **3:**E states(82), FL(35), HI(2899,2903), KS(82), MS(2720), TX(82). **4:**'Anthracnose.' W states(82), AK(526), AL(285), FL(35), HI(2903), ID(3202), MS(2720), NC(1303), ND(378), OK(2872), OR(3202), TX(82), WA(3202).

C. truncatum. Anthracnose, russeting of pods. **4:**SE states(82), TX(82).

Macrophomina phaseolina. Ashy stem blight, root rot, leaf spot. **3:**CA(82), GA(82), NC(82,1303), OK(82,2873), SC(82), TX(82). **4:**'Charcoal rot.' SE states(82), CA(82), CO(82), FL(35), HI(2903), IL(311), MS(2720), NC(1303,3931), NM(2107), OK(2873), SD(2243), TX(82).

Phoma exigua. Leaf and pod spot. **3:**NC(957,1303). **4:**NC(1303,3931), WA(82,3202).

P. terrestris. Secondary root rot. **3:**CA(82).

Phoma sp. On roots and stems. **4:**NE(3475).

Phomopsis phaseoli. Pod blight, leaf spot. **3:**Unk.(82), FL(35), HI(2903).

Phomopsis sp. Stem tip necrosis. **4:**FL(35).

Phyllosticta phaseolina. Leaf and pod spot. **3:**E states(82), IL(312), MI(82), NJ(1278), TX(82). **4:**'Phyllosticta leaf spot.' E states(82), FL(35), IN(82), MI(82), MS(2720), NC(1303,3931), NJ(1278), NY(1278), OK(2872), TX(82), WA(3202), WI(1278).

Phyllosticta sp. Leaf spot. **4:**HI(2903), WI(1349).

Stagonospora phaseoli. Leaf spot. **4:**TN(82).

Stagonosporopsis hortensis. Leaf spot. **2:**OR(82). **3:**HI(2903), NC(82), PA(82). **4:**'Ascochyta leaf spot.' OR(82,3202), WA(3202).

Deuteromycotina - Other

Rhizoctonia solani. Web blight, stem canker, damping-off, root and stem rot, sore shin, brown stem. **3:**Range of host(82), FL(35,3500), NC(1303), NY(875). **4:**'Rhizoctonia damping off. Rhizoctonia stem rot.' CO(1854), FL(35,3500), NC(1303,3931), OK(2872).

R. zeae. Root rot. **3:**NC(1303).

Rhizoctonia sp. Web blight, canker, foot, root, seed and stem rot. **3:**NC(1303), WA(3202). **4:**Range of host(82), HI(2903), KY(1780), NC(1303). **6:**ID(3925).

Sclerotium rolfsii. Southern blight. **1:**AL(82). **3:**SE states(82), FL(35), MS(82,2720), NC(1303). **4:**'Southern blight.' SE states(82), AL(285), AR(82), CA(82,2886), FL(35), HI(2903), MS(2720), NC(1303,3931), TX(82).

Piptanthus Sweet

1. *P. nepalensis* (Hook.) D. Don

Ascomycotina

Mycosphaerella pinodes. **1:**WA(3202).

Piscidia L.

1. *P. piscipula* (L.) Sarg. (=*P. erythrina* L.) [fish-poison-tree, Jamaican dogwood, West Indian dogwood]

Basidiomycotina - Uredinales

Ravenelia piscidiae. Autoecious. **1:**FL(35,82).

Deuteromycotina - Hyphomycetes

Cercospora ichthyomethiae. Leaf spot. **1:**FL(35,832).

Pisum L.

1. *P. sativum* L. [common pea, English pea, garden pea]
2. *P. sativum* L. var. *arvense* (L.) Poir. (=*P. arvense* L.) [field pea]
3. *P. sativum* L. var. *macrocarpon* Ser. [edible-podded pea, snow pea, sugar pea]
4. *Pisum* sp.

Oomycetes

Aphanomyces euteiches. Root rot. **1:**Range of host(82,2105), ID(345,3202), MI(2161), MT(3202), OR(3202), WA(3202), WI(1332). **2:**AL(82), GA(82), ID(82), MS(82).

A. euteiches f. sp. *phaseoli*. Hypocotyl and root rot. **1:**WI(2789).

A. euteiches f. sp. *pisi*. Root rot. **1:**WI(2789).

Aphanomyces sp. Root rot. **2:**ID(1035).

Peronospora pisi. Downy mildew. **1:**Unk.(82), MI(2161), WI(1409). **2:**Unk.(82).

P. viciae. Downy mildew. **1:**'Downy mildew.' Range of host(82), FL(35), ID(3202,3926), MS(2720), MT(3202), NC(1303), OR(3202), WA(3202,3926). **2:**AL(82), GA(82), LA(82), MS(82), OK(2872), OR(82).

Phytophthora sp. Root rot. **1:**CA(82), CT(82), NC(1303).

Pythium aphanidermatum. **1:**UT(82).

P. debaryanum. Root rot. **1:**CA(82), FL(35,82), WI(82).

P. dissotocum. **1:**DE(82), MD(82), NJ(82), WI(82).

P. graminicola. Root rot. **1:**MS(2720). **2:**AL(82), GA(82), MS(82).

P. irregulare. Root rot. **1:**MS(2720). **2:**AL(82), GA(82), MS(82).

P. oligandrum. On fruits and roots. **1:**CT(82), DE(82), ME(82), NJ(82), NY(82), WI(82).

P. salpingophorum. **1:**NY(82).

P. ultimum. Root rot. **1:**CA(82), FL(35), MT(82,3202), NY(82), WA(1964,3202), WI(82,1466).

P. vexans. **1:**DE(82), MD(82), NJ(82).

Pythium sp. Root and stem rot. **1:**HI(2903), ID(3202), NC(1303), OR(3202). **2:**ID(1035). **3:**HI(2903).

Zygomycotina

Choanephora cucurbitarum. Pod rot. **1:**NC(3931).

C. infundibulifera. Leaf rot. **1:**GA(82).

Mucor sp. **1:**OR(3202).

Rhizopus stolonifer. Black mold. **1:**Range of host(82), FL(35), OR(3202).

Ascomycotina

Erysiphe polygoni. Powdery mildew. **1:**Range of host(82), FL(35), GA(1438), ID(3202), MI(2161), MS(2720), MT(3202), NC(1303), OK(2872), OR(3202), SD(2243), WA(3202), WI(1336). **2:**FL(82), GA(82,1438), LA(82), ME(82), MS(82), OK(2872), TX(82).

Erysiphe sp. **2:**ID(1035).

Leptosphaerulina americana. Leaf spot. **1:**MS(82,2720).

Mycosphaerella pinodes. Blight, foot rot, leaf spot. **1:**'Aphanomyces root rot.' Range of host(82), AK(526), FL(35), HI(2903), ID(3202), MS(2720), MT(3202), OR(3202), WA(3202). **2:**AL(82), GA(82,1438), OR(82), TX(82). **3:**HI(2903).

Pleospora herbarum. **1:**WA(3202).

Sclerotinia sclerotiorum. Pod and stem rot, wilt. **1:**'Sclerotinia rot.' CA(82), DE(82), FL(35,82), ID(82,3202), MI(2161), MT(82,3202), NJ(82), OR(3202), PA(82), TX(82), VA(82), WA(82,3202). **2:**GA(82), ID(82).

Sclerotinia sp. White mold, crown and stem rot. **1:**ID(1035). **2:**NC(1303).

Thielavia basicola. On dead roots. **2:**NC(82).

Basidiomycotina - Uredinales

Uromyces fabae. Autoecious. **1:**CA(82), FL(35), ID(82,3202), MA(82), ME(82), MN(82), ND(82), NE(82), OR(2335,3202), WA(82,3202), WY(82). **2:**TX(82).

Basidiomycotina - Other

Thanatephorus cucumeris. Root rot. 'Rhizoctonia seedling rot.' **1:**Range of host(82), ID(3202), MT(3202), OR(3202), WA(3202).

Deuteromycotina - Hyphomycetes

Alternaria alternata. Pod blight, leaf spot. 'Alternaria blight.' **1:**OR(3202), WI(3544).

Alternaria sp. On leaves and stems. **1:**AK(2163), DE(82), NH(82), NY(82). **2:**ID(1035).

Aspergillus sp. **1:**OR(3202).

Aureobasidium pullulans. Seed spot. **1:**NY(82), OR(3202).

Botrytis cinerea. Blight, stem girdling. **1:**'Gray mold.' AK(82), FL(35), ID(82,3202), ME(82), OK(2872), OR(3202), WA(82,1089,3202). **2:**ID(82).

Cercospora lathyrina. Leaf spot. **1:**MS(2720). **2:**GA(82), MS(82). **4:**MS(1467).

C. pisa-sativae. 'Cercospora leaf spot.' **1:**GA(82).

Cladosporium cladosporioides. 'Cladosporium blight.' **1:**OR(3202).

C. cladosporioides f. sp. *pisicola*. Necrotic spotting of aerial parts. **1:**CA(82), ME(82), OR(82,3202), TX(82), UT(82), WA(82,3202).

C. elatum. **1:**OR(3202).

C. herbarum. On leaves, pods, and seeds. **1:**AK(82,526), CA(82), IL(82), MI(82), MT(82,3202), NY(82), WA(3202).

C. myriosporum. **1:**NY(82), WY(82).

C. pisi. **1:**OR(3202).

C. vignae. Spot. **1:**NC(3931).

Corynespora cassiicola. Target spot. **1:**NC(3931).

Dictyochaeta fertilis. Root rot. **1:**NC(1303).

Epicoccum nigrum. **1:**OR(3202).

Fusarium acuminatum. Wilt. **1:**MN(82), WI(82).

F. equiseti. Root rot, wilt. **1:**MN(82).

F. merismoides. Basal stem rot. **1:**AZ(82).

F. moniliforme. Root rot. **1**:MS(82,2720).
F. oxysporum. 'Near-wilt.' **1**:ID(3202).
F. oxysporum f. sp. *medicaginis*. Root rot. **1**:KS(82), MS(82,2720).
F. oxysporum f. sp. *pisi*. Wilt. **1**:E & C states(82), E states(82), CA(82), CO(82), ID(82,3202), IL(82), MI(2161), MN(82), MT(82,3202), NC(1303), OR(82,3202), TX(82), WA(82,1965,3202). **3**:HI(2903).
F. roseum. **1**:OR(3202).
F. solani f. sp. *pisi*. Root rot. **1**:E states(82), AK(2163), CA(82), CO(82), FL(35), ID(82,3202), MI(2161), MN(82), MO(82), MT(82,3202), NC(1303), NH(1907), NM(82), OR(82,3202), WA(82,3202), WY(82).
F. solani var. *coeruleum*. **1**:ID(3202).
F. tricinctum. **1**:OR(3202).
Fusarium sp. Foot and root rot, wilt. **1**:Unk.(82), HI(2903), NC(1303,3931). **2**:AL(82), GA(82), ID(1035), KS(82), MS(82), VA(82), WA(82).
Fusicladium pisicola. 'Black leaf.' **1**:ID(82,3202), UT(82).
Heterosporium sp. Secondary leaf spot. **1**:MT(82).
Mycoleptodiscus terrestris. Root rot, leaf spot. **1**:FL(35).
Oidium sp. Powdery mildew. **1**:HI(2903). **3**:HI(2903).
Penicillium sp. **1**:OR(3202).
Periconia byssoides. On dead stems. **1**:VA(82).
Phymatotrichopsis omnivora. Root rot. **1**:AZ(82), TX(82).
Stemphylium botryosum. **1**:AK(526), OR(3202).
S. polymorphum. Leaf spot. **1**:ME(82).
Stemphylium sp. **2**:ID(1035).
Thielaviopsis basicola. 'Thielaviopsis root rot.' **1**:AR(82), CA(82), CT(82), MD(82), MT(82,3202), NJ(82), NY(82,304), PA(82), WI(82).
Trichoderma sp. **1**:OR(3202).
Trichothecium sp. **1**:OR(3202).
Ulocladium consortiale. **1**:OR(3202).

Deuteromycotina - Coelomycetes

Aristastoma oeconomicum. Leaf spot. **1**:NC(1303). **2**:NC(82).
Ascochyta pinodes. **1**:WI(128).
A. pisi. Anthracnose, canker. **1**:'Ascochyta leaf spot. Ascochyta pod spot.' Range of host(82), AK(526), FL(35), HI(2903), ID(3202), MS(2720), MT(3202), OK(2872), OR(3202), WA(3202). **2**:AZ(82), GA(82), LA(82), MS(82), MT(82), OK(82,2872), OR(82).
Ascochyta sp. Blight. **1**:HI(2903), MI(2161), NC(1303). **2**:NC(1303).
Colletotrichum falcatum. Leaf spot. **3**:HI(2903).
C. gloeosporioides. Anthracnose. **1**:'Anthracnose.' CT(82), FL(35), GA(82), HI(2903), IA(82), ID(3202), ME(82), MN(82), NC(1303), TX(82), WI(82,1409). **2**:LA(82), TX(82). **3**:HI(2903).
Colletotrichum sp. **1**:OR(3202).
Gloeosporium toxicodendri. **1**:ND(378).
Macrophomina phaseolina. Ashy stem blight. **1**:TX(82).
Phoma exigua. Blight. **1**:NC(3931).
P. pinodella. Foot rot, black stem, leaf spot. **1**:E states(82), CA(82), CO(82), FL(35), ID(3202), MN(82), MS(2720), MT(82), OH(82), OR(82,3202), UT(82), WA(82,3202). **2**:AL(82), GA(82), ID(1035), MS(82), SC(82).
P. terrestris. Secondary root rot. **1**:IA(82).
Phoma sp. Root rot. **1**:NJ(82), WI(82).
Phyllosticta sp. Associated with black leaf. **1**:UT(82).
Septoria flagellifera. Leaf spot. **1**:MN(82), ND(82), SD(82), WI(82).
S. pisi. Leaf blotch, leaf spot. **1**:'Septoria blotch.' Range of host(82), FL(35), IA(82), ID(3202), MS(2720), MT(3202), NC(1303), ND(378), OK(2872), OR(3202), SD(2243), TX(82), WA(3202). **2**:AL(82), GA(82), MS(82), SC(82).
Septoria sp. **2**:ID(1035).

Deuteromycotina - Other

Rhizoctonia solani. Stem canker, damping-off, root rot. **1**:FL(35), NC(1303), OK(2872).
Rhizoctonia sp. Stem canker, damping-off, root rot. **1**:HI(2903), MS(2720), NC(1303). **2**:CO(82), GA(82), KS(82). **3**:HI(2903).
Sclerotium rolfsii. Southern blight. **1**:FL(35,82), NC(1303,3931).

Pithecellobium Mart.

1. *P. dulce* (Roxb.) Benth. [guaymochil, huamuchil, manila, opiuma]
2. *P. flexicaule* (Benth.) J. M. Coulter [ebony blackbead, Texas-ebony]
3. *P. guadalupense* Pers. [blackbead]
4. *P. unguis-cati* (L.) Benth. [blackbead, bread-and-cheeses, cat's-claw]
5. *Pithecellobium* sp.

Ascomycotina

Meliola venezuelana var. *floridensis*. **3**:FL(1452).

Basidiomycotina - Uredinales

Ravenelia floridana. Autoecious. **4**:FL(746,750).
R. gracilis. Autoecious. **2**:TX(82).
R. pithecolobii. Autoecious. **2**:TX(82). **5**:FL(35).
R. siderocarpi. Autoecious. **2**:TX(82,746).

Basidiomycotina - Other

Armillaria tabescens. Root rot. **5**:FL(35).
Phellinus gilvus. On wood. **4**:HI(2903).

Deuteromycotina - Hyphomycetes

Camptomeris floridana. On leaves. **4**:FL(267).
Exosporium sp. Leaf spot. **1**:FL(82). **5**:FL(35).
Pseudocercospora spilosticta. Leaf spot. **5**:FL(35).

Deuteromycotina - Coelomycetes

Colletotrichum gloeosporioides. Leaf spot. **2**:TX(82).
Pestalotiopsis funerea. Leaf spot. **2**:TX(82).
Phomopsis sp. Twig blight. **4**:FL(35).

Deuteromycotina - Other

Rhizoctonia sp. Root rot. **4**:HI(2903).

Pongamia Venten.

This genus is considered a synonym of *Millettia*, but the appropriate combination could not be located.

1. *P. pinnata* (L.) Pierre (=*P. glabra* Venten.) [karum-tree, pongam, poonga oil-tree]

Oomycetes

Pythium sp. Root rot. **1**:FL(35).

Deuteromycotina - Hyphomycetes

Alternaria sp. Leaf spot. **1**:FL(35).
Mycovellosiella pongamiae. Leaf spot. **1**:FL(35).

Deuteromycotina - Coelomycetes

Phyllosticta sp. Leaf spot. **1**:FL(35).

Prosopis L.

1. *P. chilensis* (Molina) Stuntz [algarroba, mesquite]
2. *P. glandulosa* Torr. [mesquite]
3. *P. glandulosa* Torr. var. *torreyana* (L. Benson) M. Johnston (=*P. juliflora* var. *torreyana* L. Benson) [mesquite, western honey mesquite]
4. *P. juliflora* (Sw.) DC. [algarroba, mesquite]
5. *P. pallida* (Humb. & Bonpl. ex Willd.) H. B. K. [algarroba]
6. *P. velutina* Woot. (=*P. juliflora* var. *velutina* (Woot.) Sarg.) [velvet mesquite]
7. *Prosopis* sp.

Ascomycotina

Botryosphaeria obtusa. **7**:HI(2903).
B. stevensii. On limbs. **7**:HI(82,2903).
Diaporthe sp. On limbs. **7**:HI(2903).
Diatrypella underwoodii. **2**:TX(195).
Didymosphaeria cryptosphaerioides. On limbs. **7**:TX(82).
Leveillula taurica. Powdery mildew. **1**:TX(3481).
Pleochaeta prosopidis. On living leaves. **2**:TX(3597).
Rhytidhysteron rufulum. **5**:HI(2903).

Basidiomycotina - Uredinales

Ravenelia arizonica. Autoecious. **2**:AZ(747,3981), OK(2873), TX(742). **4**:SW states(746), AZ(1203). **6**:AZ(3981).
R. holwayi. Autoecious. **2**:TX(111,652). **3**:SW states(746). **4**:CA(111), TX(111).

Basidiomycotina - Other

Antrodia heteromorpha. Brown cubical rot. **4**:AZ(1191,1202).
Athelia arachnoidea. **4**:AZ(1202).
A. coprophila. White rot. **4**:AZ(1191).
Bjerkandera adusta. **7**:TX(82).
Ceriporia tarda. White rot. **4**:AZ(1191,1202).
Coniophora eremophila. Brown rot. **4**:AZ(128,1191).
C. submembranacea. **4**:AZ(1202).
Coriolopsis gallica. White rot. **4**:AZ(1191,1202).
Dacrymyces minor. On wood. **4**:AZ(1191,1202).
D. stillatus. **4**:AZ(1202).
Dendrophora albobadia. White rot. **4**:AZ(1191,1202).
Eichleriella leucophaea. White rot. **4**:AZ(1191,1202).
Exidiopsis calcea. White rot. **4**:AZ(1191,1202).
Ganoderma lucidum. White root and butt rot of living plants.

4:AZ(1191).
G. zonatum. Root rot. 4:TX(2613).
Grandinia pruni. White rot. 4:AZ(1191,1202).
G. quercina. White rot. 4:AZ(1191,1202).
Heliocybe sulcata. Brown cubical rot. 4:AZ(1191,1202).
Henningsomyces candidus. White rot. 4:AZ(128,1191,1202).
Hymenochaete arida. White rot. 4:AZ(1191).
H. rubiginosa. White root and butt rot of living plants. 4:AZ(1191).
Hyphoderma amoenum. White rot. 4:AZ(1191,1202).
H. sambuci. White rot. 4:AZ(1191,1202).
Hyphoderma sp. White rot. 4:AZ(1191).
Hypochnicium bombycinum. White rot. 4:AZ(1191,1202).
H. prosopidis. White rot. 4:AZ(1191).
Inonotus texanus. White-mottled heart rot. 4:AZ(1181,1191,1202). 7:AZ(82), TX(82).
Lindtneria baboquivariensis. White rot. 4:AZ(1191,2042).
Meruliopsis corium. White rot. 4:AZ(1191,1202).
Mycoacia austro-occidentale. White rot. 4:AZ(1191).
Oxyporus corticola. 4:AZ(1202).
O. latemarginatus. White rot. 4:AZ(1191).
Peniophora nuda. White rot. 4:AZ(1191,1202).
P. tamaricicola. White rot. 4:AZ(128,1180,1202). 6:AZ(128).
Perenniporia medulla-panis. White rot. 4:AZ(1191,1202).
Phanerochaete allantospora. White rot. 4:AZ(1191,1202). 6:AZ(433).
P. arizonica. White rot. 4:AZ(1191,1202). 6:AZ(433).
P. chrysorhiza. White rot. 4:AZ(1191,1202).
P. crassa. White rot. 4:AZ(128,1191,1202). 6:AZ(433).
P. omnivora. White rot. 6:AZ(433).
P. tuberculata. White rot. 4:AZ(1191,1202).
P. xerophila. 6:AZ(433).
Phellinus badius. White heart rot. 4:AZ(1191,1202).
P. everhartii. White heart rot. 7:AZ(82), TX(82).
P. ferruginosus. White rot. 4:AZ(1191,1202).
P. gilvus. White heart and butt rot of living plants. 4:AZ(1191,1202).
P. robiniae. White heart rot. 4:AZ(128). 7:AZ(82), TX(82).
Phlebia ochraceofulva. White rot. 4:AZ(1191).
Polyporus arcularius. Uniform white rot. 4:AZ(1191).
Schizophyllum commune. Heart rot. 7:TX(82).
Schizopora apacheriensis. White rot. 4:AZ(1191,1193,1202).
Scytinostroma portentosum. 4:TX(128).
Skeletocutis subincarnata. White rot. 4:AZ(128,1191,1202).
Tomentella coerulea. 4:AZ(1202).
Trametes villosa. 7:TX(82).
Trichaptum byssogenum. On wood. 7:S states(2681).
Vararia tropica. White rot. 4:AZ(1191).
Wolfiporia sulphurea. 4:AZ(1202).

Deuteromycotina - Hyphomycetes

Cercospora prosopidis. Leaf blight. 4:TX(82,581).
Napicladium prosopodium. Leaf spot. 2:TX(3597). 7:TX(82).
Oidium sp. Powdery mildew. 4:FL(35).
Phymatotrichopsis omnivora. Root rot. 7:TX(82).

Deuteromycotina - Coelomycetes

Aplosporella prosopidis. On limbs. 7:TX(82).
Colletotrichum gloeosporioides. Pod spot. 7:TX(82).
Discella leguminum. 7:TX(634).
Phoma jolyana. 5:HI(2903).
Phoma sp. Black pod spot. 7:HI(82,2903).
Phyllosticta juliflora. Leaf spot. 2:OK(2873). 4:TX(64). 7:OK(82), TX(82).
Scleropycnium aureum. Leaf blight. 2:AZ(82), TX(82).
Sphaeropsis malorum. 7:HI(2903).

Pseudovigna (Harms) Verdc.

1. *P. argentea* (Willd.) Verdc. (=*Dolichos argenteus* Willd.)

Deuteromycotina - Other

Rhizoctonia solani. Web blight, root and stem rot. 1:FL(35).

Psoralea L.

1. *P. canescens* Michx. [buckroot]
2. *P. douglasii* Greene
3. *P. macrostachya* DC. [leatherroot]
4. *P. onobrychis* Nutt.
5. *P. physodes* Douglas ex Hook.
6. *P. purshii* Vail
7. *P. tenax* Lindl.
8. *P. tenuiflora* Pursh var. *bigelovii* Macbr.
9. *Psoralea* sp.

Ascomycotina

Microsphaera diffusa. Powdery mildew. 5:WA(82,3202). 7:MS(1783).
M. diffusa f. *elongata*. Powdery mildew. 5:WA(369).
Orbilia myriospora. On dead stems. 3:CA(2801,3167).

Basidiomycotina - Uredinales

Puccinia andropogonis. 4:IL(82).
Uromyces abbreviatus. Autoecious. 5:CA(81,82), WA(82,3202). 6:NV(82). 9:ID(82).
U. psoraleae. Autoecious. 8:AZ(1203,3981).

Deuteromycotina - Hyphomycetes

Ramularia psoraleae. Leaf spot. 2:CA(81). 3:CA(81,82). 5:ID(3202), OR(82,3202), WA(3202).

Deuteromycotina - Coelomycetes

Colletotrichum psoraleae. 1:FL(82).
Phyllosticta psoraleae. Leaf spot. 1:SC(82). 5:OR(3202).
Septoria argophylla. 5:ID(3202).

Psoralidium Rydb.

1. *P. lanceolatum* (Pursh) Rydb. (=*Psoralea lanceolata* Pursh, *Psoralea lanceolata* Pursh var. *scabra* (Nutt.) Piper)
2. *P. tenuiflorum* (Pursh) Rydb. (=*Psoralea tenuiflora* Pursh, *Psoralea tenuiflora* Pursh var. *floribunda* (Nutt. ex Torr. & A. Gray) Rydb., *Psoralea floribunda* Nutt. ex Torr. & A. Gray) [scurfy psoralea]

Ascomycotina

Erysiphe polygoni. Powdery mildew. 2:MT(82,3202).

Basidiomycotina - Uredinales

Uromyces psoraleae. 1:W states(82), AZ(82,1195,1203,3981), CO(1195,3379), ID(3202), MT(3202), OR(82), WY(1195). 2:AZ(1203), MT(3202), WY(1195).
U. psoraleae var. *argophyllae*. 2:C states(82), AZ(82), MT(82), OK(2872), SD(2243).

Deuteromycotina - Hyphomycetes

Cercospora latens. Leaf spot. 2:TX(82).
Phymatotrichopsis omnivora. Root rot. 2:TX(82).
Ramularia psoraleae. Leaf spot. 2:TX(82).

Deuteromycotina - Coelomycetes

Diplodina psoraleae. On dead stems. 2:KS(82).
Septoria psammophila. Stem spot. 1:ND(82).
S. umatillensis. On living leaves. 1:OR(82,3202,3421).

Pterocarpus Jacq.

1. *P. indicus* Willd. [Burmese rosewood, padauk, padouk]

Ascomycotina

Phyllachora sp. Tar spot. 1:FL(35).

Pueraria DC.

1. *P. lobata* (Willd.) Ohwi (=*P. hirsuta* (Thunb.) C. K. Schneid., *P. thunbergiana* (Siebold & Zucc.) Benth., *Dolichos lobatus* Willd.) [kudzu]
2. *P. phaseoloides* (Roxb.) Benth. [tropical kudzu]

Ascomycotina

Botryosphaeria dothidea. 1:GA(1438).
B. obtusa. On dead stems. 1:VA(82).
B. ribis. On dead stems. 1:GA(82).
Eutypella scoparia. 1:GA(1438).
Leptosphaerulina argentinensis. 1:GA(1438).
Microthyrium sp. 1:GA(1438).
Mycosphaerella puerariicola. Angular leaf spot. 1:GA(3825).
Stictis radiata. 1:GA(1438).

Basidiomycotina - Other

Corticium microsclerotium. Web blight. 1:MS(2720). 2:GA(82), MS(82).

Deuteromycotina - Hyphomycetes

Alternaria sp. Secondary leaf spot. 1:NJ(82).
Fusarium sp. Stem rot. 2:LA(82).
Phymatotrichopsis omnivora. Root rot. 1:TX(82).
Pseudocercospora puerariicola. Angular leaf spot. 1:FL(35), GA(3825), MS(2720). 2:AL(82), GA(82), MS(82).

Deuteromycotina - Coelomycetes

Diplodia puerariae. On dead stems. 1:MD(82), TX(82). 2:TX(82).
Macrophomina phaseolina. Stem rot. 2:GA(82).
Phoma sp. On stems. 2:LA(82).
Phomopsis sp. Leaf spot. 1:FL(35).

Deuteromycotina - Other
Rhizoctonia solani. Damping-off. **2**:AL(82), GA(82).

Rhynchosia Lour.

1. *R. cinerea* Nash
2. *R. difformis* (Elliott) DC.
3. *R. reniformis* DC. (=*R. simplicifolia* (Walter) A. Wood)
4. *R. senna* Gillies ex Hook. (=*R. texana* Torr. & A. Gray)
5. *R. tomentosa* (L.) Hook. & Arnott (=*R. erecta* DC.)
6. *Rhynchosia* sp.

Ascomycotina
Parodiella hedysari. Black mildew. **1**:FL(82). **2**:NC(82). **3**:SC(82). **4**:TX(82). **5**:MS(2720), NC(3944). **6**:SC(634), TX(82).
Basidiomycotina - Uredinales
Uromyces dolicholi. Autoecious. **4**:TX(82,742).
Deuteromycotina - Hyphomycetes
Cercospora sp. **5**:DE(3543).
Deuteromycotina - Coelomycetes
Ascochyta rhynchosiae. Leaf spot. **3**:SC(82).
Phyllosticta rhynchosiae. Leaf spot. **5**:MS(82,2720).

Robinia L.

1. *R. hispida* L. [bristly locust, moss locust, rose-acacia]
2. *R. neomexicana* A. Gray
3. *R. pseudoacacia* L. [black locust, false acacia, yellow locust]
4. *Robinia* sp.

Oomycetes
Phytophthora cinnamomi. Seedling root rot. **3**:SE states(3624), MD(82).
P. nicotianae var. *parasitica*. Seedling top wilt. **3**:AL(82), NC(82), VA(82).
Pythium myriotylum. Seedling root rot. **3**:NC(82).
Pythium sp. **3**:OR(3202).
Ascomycotina
Aglaospora profusa. Canker. **3**:E states(82), GA(1438), MI(82).
Botryosphaeria dothidea. **3**:GA(1438).
B. obtusa. On limbs. **3**:E states(82), KS(82), MI(82). **4**:PA(82).
B. ribis. On limbs. **3**:GA(82), TN(82), VA(82).
Byssosphaeria diffusa. **3**:GA(1438).
B. rhodomphala. Leaf spot, on decaying wood. **3**:NJ(82), SC(82), VA(82).
Calycellina populina. On petioles and racheae. **3**:MA(2181).
Cryptosphaeria eunomia. On decorticated wood. **3**:NC(1303).
Cucurbitaria elongata. On limbs. **3**:Range of host(82). **4**:NY(82).
Daldinia vernicosa. **3**:GA(1438).
Diaporthe oncostoma. Canker. **3**:E states(82), GA(1438), IL(82), NC(1303,1307).
Diatrype sp. **3**:GA(1438).
Erysiphe polygoni. Powdery mildew. **3**:CA(82).
Erysiphe sp. Powdery mildew. **3**:FL(35).
Eutypella leprosa. **3**:GA(1438), IA(3629). **4**:IA(3629).
E. scoparia. **3**:GA(1438), IA(3629).
Eutypella sp. **3**:GA(1438).
Gibberella baccata. On twigs. **3**:MI(82), SC(82).
Hysterium insidens. **3**:NC(1307).
Lasiosphaeria hispida. **3**:WA(3202).
Leucostoma cincta. **3**:IA(1226).
Microsphaera diffusa. Powdery mildew. **3**:IL(82), NC(82).
Mycosphaerella pseudacaciae. On dead legumes. **3**:CA(81).
Mycosphaerella sp. **3**:GA(1438).
Nectria cinnabarina. Limb and twig canker. **3**:Range of host(82), FL(35), GA(1438), VA(3493).
N. coccinea. On limbs. **3**:WV(82).
N. decora. On limbs. **3**:MO(82).
Nitschkia broomeiana. **3**:GA(1438).
Phyllactinia guttata. Powdery mildew. **2**:AZ(1195). **3**:NM(82). **4**:NM(82).
Rebentischia massalongii. On limbs. **3**:MA(186).
Rhytidhysteron rufulum. On twigs. **1**:AL(82).
Romellia cornina. **3**:GA(1438).
Rosellinia aquila. On decaying wood. **4**:IA(1225).
Thielavia basicola. On dead roots. **3**:WI(82).
Trematosphaeria sp. **3**:ID(3202).
Tryblidiella sp. Twig blight. **3**:FL(35).
Valsa sp. **3**:OK(2872).
Xylaria carpophila. On wood. **3**:OK(618).
X. cornu-damae. On wood. **4**:VA(3017).
X. hypoxylon. On dead wood. **3**:NC(1303).
X. longiana. On stumps and dead trunks. **3**:Range of host(82).
X. polymorpha. **3**:Range of host(82), TN(3018).
Basidiomycotina - Other
Armillaria mellea. Root rot. **3**:Unk.(82).
Cerocorticium confluens. **2**:AZ(1202).
Cerrena unicolor. Wood rot. **3**:WI(82).
Crepidotus sp. On wood. **3**:OK(618).
Crustoderma marianum. Brown rot. **3**:MD(2555).
Cyphellopsis anomala. On wood. **3**:VA(3717).
Exidiopsis calcea. On wood. **3**:GA(491,1439), ID(3202).
Ganoderma applanatum. White-mottled heart rot. **3**:CA(82).
G. lucidum. White root and butt rot. **3**:MS(2309), NC(1303), OR(128,3202).
Grandinia granulosa. **2**:AZ(1202).
G. stenospora. **3**:VA(128).
Hyphoderma heterocystidia. **3**:NY(2321).
H. sambuci. **2**:AZ(1202).
Laetiporus sulphureus. Brown cubical heart rot. **3**:MA(82), NY(82), WV(82).
Lopharia cinerascens. **3**:CA(543).
Marasmius robinianus. On dead leaves. **3**:MI(1223).
Meruliporia incrassata. Wood rot. **3**:Range of host(82).
Omphalotus olearius. Heart rot. **3**:OK(2872).
Oxyporus latemarginatus. Wood rot. **3**:MI(82), MS(2166), VA(2166).
Perenniporia compacta. On wood. **3**:GA(491,1439).
P. medulla-panis. On wood. **3**:ID(3202), VA(82).
P. ohiensis. On logs and fence posts. **3**:VA(3717), WV(3717).
P. robiniophila. White spongy heart rot. **3**:NE states(82), DC(128), IN(663), MI(82), MO(82,128), TN(128), VA(128,3717).
Phanerochaete carnosa. **2**:AZ(128).
P. ericina. White rot. **3**:TN(433).
P. filamentosa. White rot. **3**:TN(433).
P. sanguinea. White rot. **2**:NM(433).
Phellinus ferruginosus. Wood rot. **3**:MD(82), VA(82), WV(82).
P. gilvus. Wood rot. **3**:WI(82), WV(82).
P. igniarius. White spongy heart rot. **3**:IL(82), WV(82).
P. robiniae. White heart rot. **2**:AZ(1202). **3**:E states(82), FL(35,82), GA(491,1439), ID(128,3202), MS(2309), NC(1303,1307), NM(82), OH(653), OK(618,2872,2981), VA(3493), WI(82). **4**:AZ(82,128), LA(2675), NM(82).
P. robustus. **3**:CA(82).
Phlebia chrysocrea. On wood. **3**:NC(1303,2165).
Ramaricium polyporoideum. On wood. **3**:NC(430).
Spongipellis unicolor. **3**:MD(82), TX(82).
Stigmatolemma taxi. **3**:TN(646).
Thanatephorus cucumeris. **3**:OR(3202).
Tomentella brunneorufa. On wood. **4**:NC(2039).
T. bryophila. **4**:NC(2039).
T. lateritia. **4**:NC(2039).
T. punicea. **4**:NC(2039).
Trametes hirsuta. Wood rot. **3**:NY(82).
Trichaptum biforme. Wood rot. **3**:KY(82).
Deuteromycotina - Hyphomycetes
Alternaria alternata. Leaf spot. **1**:ND(82), OK(2872).
Alternaria sp. Seedling leaf blight. **3**:SE states(82), MO(82).
Cladosporium herbarum. Leaf spot. **3**:TN(82), VA(3493), WV(82).
C. nigrellum. On inner bark of railroad ties. **4**:WV(2489).
Coniosporium sp. **3**:NC(1307).
Fusarium avenaceum. On twigs. **3**:PA(82).
F. sambucinum. Twig canker. **3**:IA(82).
Fusarium sp. Seedling root rot. **3**:AL(82), GA(82), OR(3202).
Gyrothrix pediculata. On fallen leaves. **3**:MD(758).
Haplotrichum ramosissimum. On wood. **3**:VT(2131).
Heterosporium robiniae. On leaves. **3**:NY(82).
Isariopsis sp. Leaf spot. **3**:FL(35).
Macrosporium heteronemum. **1**:ND(378).
Phaeoisariopsis robiniae. Seedling leaf blight. **3**:SE states(82), DC(3209), MO(82), WI(82,1335,1350).
Phymatotrichopsis omnivora. Root rot. **1**:TX(82). **3**:NM(82), OK(82,2872), TX(82). **4**:TX(82).
Stigmina trimera. On limbs. **3**:ID(82,3202), OR(3202), WA(82,3202).
Tubercularia vulgaris. **3**:SD(2243).

Verticillium albo-atrum. Wilt. **3:**IL(82,1558).

Deuteromycotina - Coelomycetes

Colletotrichum gloeosporioides. On living leaves. **3:**NJ(966).
Cryptosporium robiniae. On twigs. **3:**NY(82).
Cylindrosporium solitarium. Leaf spot. **3:**VA(1533).
Cytospora coccinea. **3:**Unk.(82).
C. leucosperma. **3:**Unk.(82).
Dothiorella glandulosa. On limbs. **3:**VA(82), WV(82).
Macrophoma numerosa. On limbs. **3:**NY(82), PA(82).
Macrophomina phaseolina. Seedling stem rot. **3:**AL(82), NC(82), TX(82).
Melanconium viscosum. On dead limbs. **3:**MO(82), PA(82). **4:**PA(82).
Phloeospora robiniae. Leaf spot. **3:**NE states(82), WI(82).
Phomopsis oncostoma. On twigs. **4:**NJ(82).
Phyllosticta neomexicana. Leaf spot. **3:**FL(35).
P. robiniae. Leaf spot. **3:**LA(82).
Rhabdospora breviuscula. On limbs. **3:**SC(82).

Deuteromycotina - Other

Rhizoctonia solani. Damping-off. **3:**SE states(82), NE(82), TX(82).

Schrankia Willd.

1. *S. nuttallii* (DC.) Standl. [cat's-claw, sensitive brier]
2. *S. uncinata* Willd.

Ascomycotina

Diaporthe sp. **2:**GA(1438).

Basidiomycotina - Uredinales

Ravenelia texensis var. *morongiae*. Autoecious. **1:**TX(82).

Deuteromycotina - Hyphomycetes

Cercospora morongiae. On stems. **1:**MS(82). **2:**MS(2720).

Senna Mill.

This genus is considered by some to be a synonym of *Cassia*, but following Irwin & Barneby (Mem. New York Botanical Garden 35. 1982) *Senna* is retained as a separate genus.

1. *S. alata* (L.) Roxb. (=*Cassia alata* L.) [candlestick senna, Christmas-candle, empress-candleplant, ringworm-bush]
2. *S. alexandrina* Mill. (=*Cassia acutifolia* Delile, *Cassia angustifolia* Vahl, *Cassia senna* L.) [Alexandrian senna, Indian senna]
3. *S. bauhinioides* (A. Gray) H. Irwin & Barneby (=*Cassia bauhinioides* A. Gray)
4. *S. bicapsularis* (L.) Roxb. (=*Cassia bicapsularis* L.)
5. *S. covesii* (A. Gray) H. Irwin & Barneby (=*Cassia covesii* A. Gray) [desert senna]
6. *S. lindheimeriana* (Scheele) H. Irwin & Barneby (=*Cassia lindheimeriana* Scheele)
7. *S. marilandica* (L.) Link (=*Cassia marilandica* L.) [wild senna]
8. *S. obtusifolia* (L.) H. Irwin & Barneby (=*Cassia obtusifolia* L.) [sicklepod]
9. *S. occidentalis* (L.) H. Irwin & Barneby (=*Cassia occidentalis* L.) [coffee senna, negro-coffee, stypticweed]

Ascomycotina

Amphisphaeria depressa. On dry twigs. **4:**HI(2784).
Diaporthe arctii. On dead stems. **7:**NY(82).
Erysiphe sp. Powdery mildew. **8:**FL(35).
Microsphaera penicillata. Powdery mildew. **1:**FL(35). **2:**MD(82).
Rhytidhysteron rufulum. On dead plants. **4:**HI(2903).

Basidiomycotina - Uredinales

Ravenelia mesilliana. **3:**NM(82,746), TX(82,742). **5:**AZ(747,1195,1203,3981). **6:**AZ(1203).
R. spinulosa var. *papillifera*. **6:**AZ(3981), TX(742,746,751).

Deuteromycotina - Hyphomycetes

Alternaria cassiae. Seedling blight, leaf and stem lesions, leaf spot. **8:**Unk.(3752), AR(3594), MS(3751), NC(1303).
Alternaria sp. Leaf spot. **1:**FL(35). **8:**FL(35).
Cercosporidium cassiae. Leaf spot. **7:**MO(839).
Cladosporium sp. Stem lesions, leaf spot. **8:**FL(35).
Curvularia sp. Leaf and stem lesions. **8:**FL(35).
Exserohilum rostratum. Leaf spot. **8:**NC(1303).
Fusarium oxysporum. Stem rot, wilt. **8:**FL(35).
F. solani. Stem necrosis, wilt. **8:**FL(35).
Macrosporium leguminum. On legumes. **9:**SC(634).
Oidium sp. Powdery mildew. **9:**HI(2903).
Phaeoisariopsis nigricans. Leaf spot; potential biological control agent. **2:**MS(82,2720). **7:**TX(82), WI(1342). **8:**FL(1602).
P. personata. On living leaves. **9:**TX(3597).
P. simulata. Leaf spot. **7:**E states(82), DE(3543), MS(2720), TX(82), WI(82).
Phaeoramularia occidentalis. Leaf spot. **1:**FL(35). **7:**AL(82), MD(82), TX(82). **9:**SE states(82), SC(634), TX(82).
Phymatotrichopsis omnivora. Root rot. **7:**TX(82).

Deuteromycotina - Coelomycetes

Colletotrichum gloeosporioides. Anthracnose. **8:**FL(35,1650).
C. truncatum. Anthracnose. **9:**AL(1400).
Colletotrichum sp. Leaf and stem lesions. **8:**FL(35).
Macrophomina phaseolina. Charcoal rot. **9:**GA(82).
Phomopsis cassiae. On stems. **9:**SC(82).
Phomopsis sp. Dieback. **7:**SC(82).
Septoria cassiicola. Leaf spot. **7:**KS(82), TX(82), WI(82).

Deuteromycotina - Other

Rhizoctonia solani. Seedling blight. **8:**FL(35).
Sclerotium rolfsii. Stem blight. **8:**FL(35).

Sesbania Scop.

1. *S. macrocarpa* Muhl. (=*S. exaltata* (Raf.) Rydb. ex A. W. Hill) [Colorado River hemp, peatree]
2. *S. platycarpa* Pers. (=*Glottidium vesicarium* R. M. Harper)
3. *S. punicea* (Cav.) Benth. (=*Daubentonia tripetii* Poit., *Sesbanis tripetii* (Poit.) Hort. ex F. T. Hubbard) [scarlet wisteria-tree]
4. *S. tomentosa* Hook. & Arnott

Oomycetes

Pythium sp. Root rot. **4:**HI(2903).

Ascomycotina

Botryosphaeria minor. On stems. **1:**LA(82).
Diaporthe arctii. On stems. **1:**LA(82).
Elsinoe sesbaniae. On stems. **1:**SC(1745).
Erysiphe polygoni. Powdery mildew. **2:**FL(35,82), TX(82).
Erysiphe sp. Powdery mildew. **1:**CA(82).
Physalospora sesbaniae. Leaf spot. **1:**LA(82). **2:**TX(82).
Sphaerella sesbaniae. On dead stems. **1:**LA(965).

Basidiomycotina - Other

Armillaria mellea. Root rot. **3:**CA(2897).

Deuteromycotina - Hyphomycetes

Alternaria sp. Leaf spot. **2:**FL(35).
Cercospora glottidiicola. Leaf spot. **2:**FL(35), MS(82,2720), TX(82).
Phymatotrichopsis omnivora. Root rot. **1:**TX(82). **2:**TX(82).
Pseudocercospora glycines. Leaf spot. **2:**FL(35).
Tetraploa aristata. Leaf spot. **2:**FL(35).

Deuteromycotina - Coelomycetes

Colletotrichum gloeosporioides. Anthracnose, leaf spot. **2:**FL(35), TX(82).
Macrophomina phaseolina. **1:**TX(82).
Phoma clitoricarpa. On stems. **2:**TX(82).
Phoma sp. Stem lesions. **1:**FL(35).

Sophora L.

1. *S. affinis* Torr. & A. Gray
2. *S. chrysophylla* (Salisb.) Seem. [mamane]
3. *S. davidii* (Franch.) Skeels [skeels, vetchleaf sophora]
4. *S. japonica* L. [Japanese pagoda-tree]
5. *S. secundiflora* (Ort.) Lag. ex DC. [frijolito, mescal-bean]
6. *S. sericea* Nutt. [silky sophora]
7. *S. stenophylla* A. Gray
8. *S. tomentosa* L. [silverbush]

Oomycetes

Phytophthora cinnamomi. Root rot. **2:**HI(2903).

Ascomycotina

Microsphaera penicillata. Powdery mildew. **4:**CT(82).
Nectria cinnabarina. Twig blight. **2:**HI(2903). **4:**CT(82), NY(82), OK(2872), WV(3911).

Basidiomycotina - Uredinales

Uromyces hyalinus. Autoecious. **6:**W states(82), AZ(82,1203,3981), NM(1195), WY(82). **7:**AZ(3981).

Basidiomycotina - Other

Armillaria mellea. Root rot. **2:**HI(2011). **4:**CA(2897).

Deuteromycotina - Hyphomycetes

Alternaria sp. Leaf spot. **8:**FL(35).
Cercospora sp. Leaf spot. **4:**FL(35).
Fusarium sp. Basal stem rot. **8:**FL(35).
Oidium sp. Powdery mildew. **2:**HI(2903).

Phymatotrichopsis omnivora. Root rot. **3**:TX(82). **4**:TX(82). **5**:TX(82).
Verticillium albo-atrum. Wilt. **4**:IL(1558).
Deuteromycotina - Coelomycetes
Cytospora sophorae. On twigs. **4**:NH(82), NJ(82), NY(82).
Diplodia sophorae. Dieback. **4**:OH(82).
Macrophoma sp. **1**:OK(2872).
Phyllosticta sophorae. Leaf spot. **4**:OK(2873). **5**:OK(82), TX(82).
Deuteromycotina - Other
Rhizoctonia solani. Damping-off. **4**:CT(82).

Spartium L.

1. *S. junceum* L. [Spanish-broom, weaver's-broom]

Deuteromycotina - Coelomycetes
Diplodia sarothamni. On twigs. **1**:CA(82).
[*Pestalotiopsis funerea*]. On twigs. **1**:CA(82).
Phomopsis sarothamni. On twigs. **1**:CA(82).

Strongylodon Vogel

1. *S. macrobotrys* A. Gray [jadevine]

Deuteromycotina - Coelomycetes
Colletotrichum gloeosporioides. Stem blight. **1**:HI(2899,2903).

Strophostyles Elliott

1. *S. helvola* (L.) Elliott (=*S. missouriensis* Small)
2. *S. leiosperma* (Torr. & A. Gray) Piper (=*S. pauciflora* (Benth.) S. Wats. in A. Gray)
3. *S. umbellata* (Muhl. ex Willd.) Britton

Ascomycotina
Diaporthe phaseolorum. **1**:GA(1438).
Erysiphe polygoni. Powdery mildew. **1**:IA(82), MN(82), MO(82), OK(82). **2**:OK(2872).
Parodiella hedysari. Black mildew. **1**:VA(82).
Basidiomycotina - Uredinales
Uromyces appendiculatus. Autoecious. **2**:E states(82), MS(2720), OK(2872,2873), SD(82,2243), TX(82). **3**:IN(82), MO(82), NC(82,3944), PA(82).
Deuteromycotina - Hyphomycetes
Phymatotrichopsis omnivora. Root rot. **1**:TX(82).
Pseudocercospora cruenta. Leaf spot. **1**:IL(82).
Deuteromycotina - Coelomycetes
Ascochyta sp. Leaf spot. **2**:OK(2872).
Macrophomina phaseolina. Stem blight. **1**:GA(82).
Phomopsis phaseoli. Stem blight. **1**:GA(82).
Phyllosticta phaseolina. Leaf spot. **1**:IA(82), KS(82), NJ(82), OH(82), OK(82,2872), WI(82). **2**:KS(82), NE(82), OK(2873), TX(82), WI(1356).

Stylosanthes Sw.

1. *S. biflora* (L.) B. S. P.
2. *S. elatior* Sw.
3. *S. erecta* P. Beauv. (=*S. guineensis* Schumach.) [Nigerian stylo]
4. *S. guianensis* (Aubl.) Sw. (=*S. gracilis* Kunth) [Brazilian lucerne, Brazilian stylo]
5. *S. hamata* (L.) Taub. [Caribbean stylo]
6. *S. humilis* Kunth [Townsville stylo]
7. *S. scabra* Vogel
8. *S. subsericea* S. F. Blake

Deuteromycotina - Hyphomycetes
Cercospora commonsii. Leaf spot. **2**:DE(962).
Deuteromycotina - Coelomycetes
Colletotrichum gloeosporioides. Anthracnose, stem canker, leaf spot. **3**:FL(35). **4**:FL(35). **5**:FL(35). **6**:FL(35). **7**:FL(35). **8**:FL(35).
C. truncatum. Anthracnose, leaf spot, stem canker. **4**:FL(35,2086). **5**:FL(35,2086). **7**:FL(2086). **8**:FL(35,2086).
Colletotrichum sp. Stem canker, leaf spot. **6**:FL(3387).
Phoma gramma. Leaf spot. **1**:MS(2720).

Tephrosia Pers.

1. *T. hispidula* (Michx.) Pers.
2. *T. spicata* (Walter) Torr. & A. Gray
3. *T. tenella* A. Gray
4. *T. virginiana* (L.) Pers. [catgut, goat's-rue, rabbit's pea]
5. *Tephrosia* sp.

Ascomycotina
Helotium consanguineum. On dead stems. **4**:NJ(850).
Parodiella hedysari. Black mildew. **5**:FL(82).
Basidiomycotina - Uredinales
Ravenelia epiphylla. Autoecious. **1**:NC(1303). **2**:AL(82), FL(35,82), SC(82). **3**:AZ(1194,1203,3981). **4**:E states(82), FL(35), LA(82), MI(665), MS(2720), NC(746,1303,3944), OK(2872), WI(82,1333,1340,1350).
Deuteromycotina - Hyphomycetes
Alternaria alternata. **4**:MI(665).
Cercospora tephrosiae. Leaf spot. **2**:AL(82). **4**:WI(1342).
Phymatotrichopsis omnivora. Root rot. **5**:TX(82).
Deuteromycotina - Coelomycetes
Phyllosticta sp. On leaves. **4**:WI(1350).
Stagonospora tephrosiae. On leaves. **4**:WI(1339,1344).
Deuteromycotina - Other
Rhizoctonia solani. Root rot. **2**:FL(35). **4**:FL(35).

Teramnus P. Br.

1. *T. labialis* (L. fil.) Spreng. [blue wiss]
2. *T. uncinatus* (L.) Sw.
3. *T. volubilis* Sw.

Deuteromycotina - Coelomycetes
Colletotrichum dematium. Anthracnose, stem canker, leaf spot. **1**:FL(35). **2**:FL(35).
C. gloeosporioides. Anthracnose, stem canker, leaf spot. **1**:FL(35). **2**:FL(35).
Deuteromycotina - Other
Rhizoctonia solani. Web blight. **2**:FL(35). **3**:FL(35).

Thermopsis R. Br. ex Aiton & W. T. Aiton

1. *T. divaricarpa* A. Nels.
2. *T. macrophylla* Hook. & Arnott (=*T. californica* S. Wats.)
3. *T. montana* Nutt.
4. *T. rhombifolia* (Nutt. ex Pursh) Nutt. ex Richardson
5. *Thermopsis* sp.

Ascomycotina
Didymella castillejae. **3**:ID(3202).
D. stromatica. **3**:UT(194).
Erysiphe polygoni. Powdery mildew. **1**:WY(1195). **3**:ID(3202), WA(3202). **5**:W states(82).
Leptosphaeria wehmeyeri. **3**:MT(3249).
Basidiomycotina - Uredinales
Poliomopsis thermopsidis. Telial host. **4**:CO(2911).
Deuteromycotina - Hyphomycetes
Ramularia sphaerioides. Leaf spot. **3**:WA(969,3202).
Stigmina earlei. Leaf spot. **1**:WY(844,1195). **4**:MT(3202), WY(844,1195,3378). **5**:CO(82), MT(82).
S. thermopsidis. Leaf spot. **2**:CA(82,959,1468). **3**:ID(3202).
Deuteromycotina - Coelomycetes
Apiocarpella thermopsidis. **1**:WY(1195).
Macrophoma thermopsidis. On stems. **5**:MT(82), NM(82).
Phoma thermopsidicola. Leaf spot. **5**:CA(82).
Phyllosticta thermopsidis. **3**:ID(3202).
Sphaeropsis thermopsidis. **4**:MT(3202).

Tipuana (Benth.) Benth.

1. *T. tipu* (Benth.) Kuntze [pride-of-Bolivia, rosewood, tipu-tree]

Deuteromycotina - Coelomycetes
Phomopsis sp. Stem gall. **1**:FL(35).
Phyllosticta sp. Leaf spot. **1**:FL(35).

Trifolium L.

1. *T. agrarium* L. [hop clover, yellow clover]
2. *T. albopurpureum* Torr. & A. Gray
3. *T. anemophilum* Greene
4. *T. arvense* L. [old-field clover, rabbitfoot clover, stone clover]
5. *T. bifidum* A. Gray
6. *T. carolinianum* Michx. [Carolina clover]
7. *T. ciliatum* Nutt.
8. *T. ciliolatum* Benth.
9. *T. cyathiferum* Lindl.
10. *T. dasyphyllum* Torr. & A. Gray
11. *T. eriocephalum* Nutt.
12. *T. fragiferum* L. [strawberry clover]

13. *T. gymnocarpum* Nutt. (=*T. subcaulescens* A. Gray)
14. *T. hybridum* L. [alsike clover]
15. *T. incarnatum* L. [crimson clover, Italian clover]
16. *T. isthmocarpum* Brot.
17. *T. kingii* S. Wats.
18. *T. latifolium* (Hook.) Greene
19. *T. longipes* Nutt.
20. *T. macraei* Hook. & Arnott
21. *T. macrocephalum* (Pursh) Poir.
22. *T. medium* L. [zigzag clover]
23. *T. michelianum* Savi [bigflower clover]
24. *T. microcephalum* Pursh
25. *T. microdon* Hook. & Arnott
26. *T. nanum* Torr.
27. *T. obtusiflorum* Hook.
28. *T. oliganthum* Steud.
29. *T. oreganum* Howell
30. *T. parryi* A. Gray
31. *T. plumosum* Douglas ex Hook.
32. *T. pratense* L. [red clover]
33. *T. procumbens* L. (=*T. dubium* Sibth.) [cow hop clover, Irish shamrock, small hop clover, yellow clover]
34. *T. reflexum* L.
35. *T. repens* L. [ladino clover, white clover]
36. *T. repens* L. f. *gigantea*
37. *T. repens* L. f. *lodigense* Hort. ex Gams
38. *T. resupinatum* L. [Persian clover, reversed clover]
39. *T. rusbyi* Fresen. subsp. *reflexum* (A. Nels.) D. Heller & Zoh. (=*T. rydbergii* Greene)
40. *T. semipilosum* Fresen. [Kenya clover, Kenya wild white clover]
41. *T. subterraneum* L. [subclover, subterranean clover]
42. *T. tridentatum* Lindl.
43. *T. variegatum* Nutt. [whitetip clover]
44. *T. vesiculosum* Savi [arrowleaf clover]
45. *T. wormskioldii* Lehm. (=*T. willdenovii* Spreng., *T. fendleri* Greene, *T. involucratum* Ort., *T. involucratum* Ort. var. *fimbriatum* (Lindl.) McDermott) [seaside clover, Sierra clover]
46. *Trifolium* sp.

Myxomycetes

Mucilago crustacea. Slime mold. **46:**NC(1303).

Physarum cinereum. Slime mold. **6:**IN(82), MD(82), PA(82). **14:**IN(82), MD(82), PA(82). **22:**IN(82), MD(82), PA(82). **32:**IN(82), MD(82), PA(82). **35:**IN(82), MD(82), PA(82).

Physarum sp. Slime mold. **35:**NC(1303). **46:**NC(1303).

Plasmodiophoromycetes

Polymyxa graminis. On roots. **32:**IL(1163).

Chytridiomycetes

Olpidium trifolii. Leaf curl. **35:**LA(82).

Physoderma trifolii. Leaf gall, wart. **6:**LA(82), MD(82), SC(82). **14:**LA(82), MD(82), SC(82). **22:**LA(82), MD(82), SC(82). **32:**'Leaf gall.' LA(82), MD(82), SC(82). **35:**LA(82), MD(82), NC(48), SC(82). **41:**CA(466).

Synchytrium parksii. Leaf and petiole galls. **25:**CA(1845).

Oomycetes

Peronospora trifoliorum. Downy mildew. **32:**'Downy mildew.' NY(82), WY(82). **33:**OR(82,3202). **35:**NE(82).

P. viciae. Downy mildew. **15:**MS(2720).

Phytophthora erythroseptica. Root rot. **44:**MS(2867).

P. megasperma. Root rot. **41:**MS(1785). **44:**MS(2867).

P. nicotianae var. *parasitica*. Root rot. **35:**MS(1786). **44:**MS(2867).

Pythium sp. Crown and root rot. **1:**WI(82). **14:**MN(2957). **32:**MN(2957), WI(1898). **35:**AL(1426), MN(2957), NC(1303), NH(1893), SC(1426).

Zygomycotina

Absidia sp. On roots. **35:**NH(1893).

Mucor sp. On leaves, roots, seeds, and stolons. **14:**OR(3202). **15:**OR(3202). **32:**OR(3202). **35:**NH(1892,1893), OR(3202).

Rhizopus stolonifer. On leaves, roots, seeds, and stolons. **32:**HI(2899), OR(3202). **35:**NH(1892,1893).

Ascomycotina

Chaetomium sp. On leaves, roots, and seeds. **15:**OR(3202). **32:**HI(2899), OR(3202). **35:**NH(1892,1893).

Cymadothea trifolii. Sooty blotch on leaves. **9:**WA(3202). **14:**NE states(1890), Range of host(82), CA(81), MA(1394), MN(2957), NY(1890), OR(3202), WA(3202). **15:**E states(82), GA(1438), LA(82), MS(2720), NC(45,48,1303), ND(82), OR(3202). **18:**ID(3202). **19:**ID(3202). **24:**WA(3202). **28:**OR(3202). **32:**'Sooty blotch.' NE states(82,1890), IA(215), ID(3202), MN(2957), MS(2720), MT(3202), NH(1893,1897), NY(1890,2998,2999,3005), OR(3202), SD(2243), TX(82), WA(82,3202), WI(1454), WV(952). **34:**MS(2720). **35:**NE states(1890), Range of host(82), AK(526), GA(1438), ID(3202), MN(2957), MS(2720), MT(3202), NC(1303,3944), NH(1893,1897), NY(1890), WA(3202), WI(1454), WV(952). **39:**NM(82). **43:**CA(81), OR(3202), WA(3202). **46:**CA(82), ID(3202), ND(82).

Didymella trifolii. On stems. **32:**MS(82,2720).

Erysiphe cichoracearum. **33:**NC(3944).

E. graminis f. sp. *avenae*. Powdery mildew. **46:**AR(3029).

E. polygoni. Powdery mildew. **1:**CT(82), IL(82), LA(82), NY(82), PA(82), TN(82). **14:**E states(82), NE states(1890), NW states(82), ID(3202), IL(82), MN(82,2957), NY(1890), OR(3202), WA(3202). **15:**GA(82,1438), LA(82). **18:**ID(3202). **19:**MT(3202), WA(3202), WY(1195). **22:**KY(82), NY(82), VT(82). **24:**WA(3202). **31:**ID(3202). **32:**'Powdery mildew.' NE states(1890), Range of host(82), Unk.(3976), AL(285,759), GA(1438), IA(215,217,714,716), ID(3202), MA(1390), MI(2607), MN(2957), MS(2720), MT(3202), NC(3944), NH(1893,1897), NJ(494), NY(1890,2998,3005), OR(3202), WA(3202), WI(1454), WV(952). **33:**AR(82), GA(82,1438), LA(82), MD(82), MS(82,2720), NC(82), WA(3202), WI(1340). **35:**MN(2957), MS(2720), NH(1893,1897), WA(3202). **39:**UT(82). **42:**WA(3202). **43:**MT(82,3202). **44:**GA(2398). **45:**CO(82), MT(3202), OR(82,3202). **46:**FL(35), ID(82), NH(1895), OK(2872), OR(82), WA(82).

Erysiphe sp. Powdery mildew. **35:**NC(1303).

Leptosphaeria monticola. On leaves and stems. **17:**CO(82).

L. pratensis. Root rot, leaf spot. **14:**WA(3202). **19:**MT(3202). **32:**WV(952). **35:**ID(3202), NY(3005), OR(3202).

Leptosphaerulina americana. On stolons. **32:**GA(1438). **35:**NC(46).

L. argentinensis. **32:**GA(1297).

L. trifolii. Summer blight, leaf spot on leaves, seeds and stolons. **6:**AL(82). **12:**WA(3202). **14:**IL(82), MN(2957), MO(82), NY(82), WV(952). **15:**GA(1297), IL(82), MO(82), NY(82). **16:**GA(1297,1438). **32:**AL(759), GA(1297,1438), HI(2899), IA(716), IL(82), KS(82), ME(1297), MN(2957), MO(82), NH(1897), NJ(494), NY(82), WV(952). **35:**E states(82), NE states(1890), GA(1297,1438), LA(82), MN(2957), NC(45,48,1303), NH(1892,1893,1897), NJ(494), NY(1890,2998,2999,3004,3005), WI(82,1341,1454), WV(952,1297).

Microsphaera penicillata. Powdery mildew. **32:**NY(82).

M. trifolii. Powdery mildew. **32:**NC(1303). **46:**AZ(662), NC(1303).

Neocosmospora vasinfecta. Isolated from roots. **46:**WI(1345).

Neocosmospora sp. Isolated from roots and stolons. **35:**NC(1303).

Neurospora sp. On leaves and roots. **35:**NH(1893).

Ophiobolus collapsus. **32:**NJ(3248).

Pleospora herbarum. Seed mold, leaf spot. **6:**Range of host(82). **32:**HI(2903). **46:**FL(35), NM(82).

Pseudopeziza trifolii. Leaf spot. **4:**NC(1303). **14:**ID(3202), ME(1890), MN(2957), NC(1303), NY(82), WV(82). **22:**NY(82). **32:**'Common leaf spot.' NE states(82), NW states(82), AK(82,526), CA(82), GA(1438), ME(1890), MI(2607), MN(2957), NC(1303,1307), ND(377), NE(82), NH(1890,1893), NY(1890,2998,2999,3004,3005), OR(3202), RI(1890), SD(2243), WA(3202), WI(1454), WV(953). **35:**NE states(1890), CA(82), MA(82), ME(1890), MN(2957), MS(82,2720), MT(82,3202), NH(1890,1893), NY(82,3005), OR(82,3202), VT(82), WA(3202), WI(1454). **42:**OR(3202), WA(3202). **46:**OK(2872).

Pyrenopeziza sp. Leaf spot. **4:**MD(82).

Sclerotinia sclerotiorum. **15:**OR(3202).

S. spermophila. **32:**OR(3202), WA(3202). **35:**OR(3202).

S. trifoliorum. Stem canker, crown, root and stem rot, Sclerotinia wilt. **14:**ID(3202), MI(82), NC(82), WV(82). **15:**E states(82), KY(3719), MD(1729), MS(2720), NC(45,48,1303), OR(82,3202), TN(82), VA(1034). **22:**OR(82,3202). **32:**'Sclerotinia crown rot. Sclerotinia root rot.' E states(82), HI(2899), ID(3202), LA(82), MI(82,2607), MS(2720), NC(45,48,1303), NH(1890), NJ(494), NY(1946,2998,3004), OR(82,3202), RI(1890), WA(82,3202), WV(952). **35:**NE states(1890), AL(82,759), KS(82), KY(82),

MS(82,2720), NC(45,1303), NH(1893), NY(82), OR(82), PA(82), WV(952). **46:**OK(2872), WA(3202).

Basidiomycotina - Uredinales

Uromyces elegans. Autoecious. **6:**SE states(82), MS(2720), OK(2872), TX(82). **33:**OK(2872). **46:**FL(35).

U. minor. Autoecious. **2:**CA(81). **3:**WY(1195). **5:**OR(3202). **7:**CA(81). **8:**OR(3202), WA(3202). **9:**CA(81). **10:**CO(649,1195), WY(1195). **11:**OR(3202), WA(3202). **13:**AZ(1203,3981). **18:**MT(3202), WA(3202). **19:**MT(3202), UT(1195), WY(1195). **20:**OR(3202). **21:**OR(3202), WA(3202). **24:**CA(81), OR(3202), WA(3202). **25:**OR(3202). **26:**CO(1520), MT(1520,3202). **28:**OR(3202), WA(3202). **30:**CO(82,649,1195), UT(82), WY(82,1195,2924). **32:**MS(2720). **33:**CA(82), MS(2720), OK(82), OR(82,3202), TX(82), WA(82,3202). **39:**CO(82), MT(82). **42:**CA(81), OR(3202), WA(3202). **43:**CA(81,82), NV(82), WA(82,3202). **45:**CA(82), CO(82), NV(82). **46:**CA(82), OR(82), UT(82), WA(82), WY(82).

U. striatus. **33:**CA(81). **46:**MD(1729).

U. trifolii. Autoecious. **6:**GA(1439). **11:**CA(81). **14:**ID(3202), MN(2957), MT(3202), OR(3202), WA(3202). **15:**OR(3202). **21:**WA(3202). **30:**WY(2924). **32:**'Rust.' CA(81), IA(215,217,714,716), ID(3202), KY(1777), MI(2607), MN(2957), MT(3202), ND(378), NY(2998), OR(3202), WA(3202).

U. trifolii-repentis. Autoecious. **1:**OR(82), WI(82). **14:**NE states(1890), CO(82), MI(82), MT(82), NC(3944), NV(82), NY(82,1890), WA(82), WI(82), WV(952). **35:**E states(82), NE states(1890), NW States(82), CA(82), GA(1439), MO(82), NC(1303,3944), NH(1893,1897), NY(1890), TX(82), WI(1454). **46:**FL(35).

U. trifolii-repentis var. *fallens*. Autoecious. **1:**DE(82), SD(82). **14:**SD(2243). **15:**SD(2243). **22:**IN(82), MD(82), NE(82), NY(82), SD(82,2243). **32:**NE states(1890), Range of host(82), CA(81), GA(1439), MA(1390), NC(1303,1307), NH(1893,1897), NY(1890,2999,3005), OK(2872), SD(2243), WI(1454), WV(952).

Basidiomycotina - Ustilaginales

Thecaphora deformans. **42:**CA(82,1077).

Basidiomycotina - Other

Thanatephorus cucumeris. Stem blight. **1:**LA(3524). **32:**IL(312), LA(3524). **35:**LA(3524). **36:**GA(1439).

Typhula idahoensis. With seeds. **32:**IN(2061).

T. variabilis. In stored seed. **46:**MO(2955).

Deuteromycotina - Hyphomycetes

Alternaria alternata. Seed mold, leaf spot, on leaves, roots, stems and stolons. **14:**OR(3202). **15:**OR(3202). **22:**Range of host(82), VT(82). **32:**'Seed mold.' Range of host(82), HI(2899), OR(3202), VT(82). **35:**NH(1893), OR(3202). **46:**FL(35).

Alternaria sp. Isolated from leaves, roots, seeds, stems and stolons. **35:**NC(1303), NH(1892). **46:**NH(1895).

Aspergillus flavus. On roots. **35:**NH(1893).

A. terreus. On roots. **35:**NH(1893).

Aspergillus sp. **14:**OR(3202). **15:**OR(3202). **32:**OR(3202). **35:**OR(3202).

Aureobasidium caulivorum. Northern anthracnose, leaf spot. **15:**NC(82), OR(1535,3202), VA(82). **32:**'Northern anthracnose.' E states(82), IA(215,217,714,716), ID(3202), ME(1890), MI(2607), MN(2957), MO(82), MT(3202), NH(1890,1893,1897), NY(1890,2998,2999,3004,3005), OR(82,3202), RI(1890), WA(3202), WI(1332,1454), WV(952).

A. pullulans. **14:**OR(3202). **15:**OR(3202). **32:**OR(3202). **35:**OR(3202).

Basidiobotrys pallida. On decaying stems. **46:**NJ(965).

Bipolaris sorokiniana. Black stem. **32:**MN(2956). **46:**WI(2956).

B. spicifera. **15:**OR(3202). **32:**OR(3202).

Botrytis anthophila. On seeds. 'Anther mold.' **32:**OR(82), WA(82).

B. cinerea. Gray mold leaf blight, leaf spot. **4:**NH(1895). **14:**OR(3202). **15:**OR(3202). **32:**NY(82), OR(3202), PA(82). **35:**NH(1895), OR(3202). **46:**NH(1895).

Botrytis sp. Gray mold leaf blight. **15:**DE(82), KY(82). **32:**WI(1454). **35:**AL(82).

Candida sp. **32:**OR(3202).

Cephalothecium sp. **15:**OR(3202).

Cercospora medicaginis. Leaf spot, summer black stem. **15:**GA(986).

C. zebrina. Leaf and stem spot. **1:**ID(3202), MT(3202), NJ(82), NY(82), WI(1333). **4:**NH(1895), WI(1352). **14:**E states(82), LA(82), MN(2957), NH(1895), OK(2872), WI(82), WV(952). **15:**MS(2720), WI(82). **22:**NY(82), VT(82). **32:**'Summer black stem.' NE states(82), AL(759), HI(2899), IA(215,217,714,716), MD(82), MI(2607), MN(82,2957), MO(82), MS(2720), NC(1303), NH(1890,1897), NY(1890,2998), RI(1890), SD(2243), WI(1454), WV(952). **33:**WI(82,1341). **35:**NE states(1890), MN(2957), NH(1892,1893,1897), NJ(494), NY(1890,2998,2999,3004,3005), WI(1454), WV(952). **41:**MS(2869). **46:**FL(35), NH(1895).

Cercospora sp. Leaf spot, summer black stem. **14:**NC(1303). **46:**HI(2903).

Cladosporium cladosporioides. **14:**OR(3202). **15:**OR(3202). **32:**OR(3202). **35:**OR(3202).

C. elatum. **14:**OR(3202). **15:**OR(3202). **32:**OR(3202). **35:**OR(3202).

Cladosporium sp. On leaves, roots, seeds, and stolons. **32:**HI(2899). **35:**NH(1892,1893). **46:**NH(1895).

Cryptococcus albidus. On leaves and seeds. **35:**NH(1893).

Curvularia geniculata. Leaf spot. **35:**NC(45).

C. trifolii. Leaf spot, on flowers, leaves and seeds. **32:**'Curvularia leaf blight.' FL(1982), HI(2899), MD(1982). **35:**NE states(1890), AL(1426), FL(1982), IL(312), LA(1426), MD(82,1982), NC(48,1303,2080), NH(1892,1893,1894), NY(1890), SC(1426), VA(82), WV(952). **46:**E states(1891).

Curvularia sp. Leaf spot. **35:**NH(1895), SC(2316).

Cylindrocladium crotalariae. Crown, root and stem rot. **46:**FL(35).

C. scoparium. Crown, root and stem rot. **32:**TN(82). **46:**FL(35).

Dictyochaeta fertilis. Root rot. **14:**NC(1303). **15:**NC(1303). **32:**NC(1303). **35:**NC(482,483,1303). **38:**NC(1303). **41:**NC(1303). **44:**NC(1303).

Epicoccum nigrum. **14:**OR(3202). **15:**OR(3202). **32:**OR(3202). **35:**OR(3202).

Fusarium graminearum. Root rot. **32:**OH(82). **35:**OH(82).

F. moniliforme. Root rot, wilt. **32:**IA(714), KY(82), PA(82), WI(82). **35:**KY(82), PA(82), WI(82).

F. oxysporum. Crown and stolon rot, on leaves, roots, seeds and stolons, vascular wilt. **15:**MS(2868), OR(3202). **32:**'Fusarium root rot.' HI(2899), IA(217,714,716), ID(3202), KY(82), PA(82), WI(82,1898). **35:**AL(1426), LA(1426), NH(1892,1893,1897), SC(1426).

F. roseum. Crown and root rot, on leaves, roots, seeds and stolons, wilt. **14:**OR(3202). **15:**OR(3202). **32:**IA(714), ID(3202), WI(1898). **35:**AL(1426), LA(1426), NH(1892,1893,1897), OR(3202), SC(1426).

F. solani. Crown and root rot, wilt. **32:**'Fusarium root rot.' IA(714), WI(1898). **35:**AL(1426), KY(82), LA(1426), NH(1893), PA(82), SC(1426), WI(82).

F. tricinctum. **15:**OR(3202).

Fusarium sp. Root, stem, and stolon rot. **14:**AL(82), MN(2957). **32:**IA(215,217), MN(2957), NJ(494), NY(2998,3005), OR(3202), WV(952). **35:**AL(1426), LA(1426), MN(2957), MS(2720), NC(1303), NY(2998,3005), SC(1426,2316). **37:**NY(2998).

Geotrichum sp. **32:**OR(3202).

Gliocladium roseum. Crown, root and stolon rot. **32:**WI(1898). **35:**NH(1893,1897).

Helminthosporium sp. Leaf spot, isolated from roots, seeds and stolons. **35:**NC(1303), NH(1892). **46:**FL(35).

Hormiscium sp. **32:**OR(3202).

Macrosporium sp. **15:**DE(3543).

Mycoleptodiscus terrestris. Root rot, leaf spot. **32:**'Leptodiscus crown rot. Leptodiscus root rot.' IL(1161), MD(1161), VA(1161). **35:**AL(1426), LA(1426), NH(1893), SC(1426). **46:**FL(35).

Nigrospora oryzae. **15:**OR(3202).

Nigrospora sp. On roots. **35:**AL(1426), NH(1893), SC(1426).

Oidium sp. **32:**CA(661).

Penicillium sp. On leaves, roots, seeds, and stolons. **14:**OR(3202). **15:**OR(3202). **32:**HI(2899), OR(3202). **35:**NH(1892,1893), OR(3202).

Phymatotrichopsis omnivora. Root rot. **15:**TX(82). **32:**TX(82). **33:**TX(82). **35:**TX(82).

Polythrincium trifolii. Sooty blotch. **9:**CA(81). **14:**WI(1339), WY(1195). **23:**ND(378). **27:**CA(81). **35:**DE(3543), KS(2458), ND(378). **45:**CA(638). **46:**FL(35).

Ramularia trifolii. **19:**WY(1195). **39:**WY(1195).

Ramularia sp. Leaf spot. **11:**OR(3202). **15:**GA(82). **19:**MT(3202), OR(3202).

Rhodotorula glutinis. On leaves and seeds. **32:**NH(1893).

35:NH(1893).
R. rubra. On leaves. **35:**NH(1893).
Scopulariopsis stercoraria. **32:**OR(3202). **35:**OR(3202).
Stemphylium botryosum. **14:**OR(3202). **15:**OR(3202). **32:**'Target spot.' IA(215,217,714,716), OR(3202), WV(952). **35:**OR(3202).
S. globuliferum. **32:**MA(3269).
S. sarciniforme. Target leaf spot. **14:**IN(82), MN(2957), NY(82), VT(82). **22:**VT(82). **32:**'Target spot.' OH(653), WI(1335). **35:**MN(2957), NH(1897).
S. trifolii. On leaves, seeds, and stolons. **35:**NH(1892,1893), OR(3202), PA(1295).
Stemphylium sp. Leaf spot. **1:**VA(82). **32:**MI(2607). **35:**NE states(1890), AL(759), NY(1890). **46:**HI(2903).
Tetracoccosporium paxianum. **32:**ID(3202).
Thielaviopsis basicola. Black root rot. **1:**WI(82). **14:**CT(82), TX(82). **32:**'Black root rot.' CT(82), NC(82), NY(82). **35:**CT(82), TX(82).
Thielaviopsis sp. **35:**NH(1893).
Trichoderma viride. On roots. **35:**NH(1893), SC(1426).
Trichoderma sp. **32:**OR(3202). **35:**SC(2316).
Trichothecium sp. **15:**OR(3202). **32:**OR(3202).
Ulocladium consortiale. **14:**OR(3202). **15:**OR(3202). **32:**OR(3202). **35:**OR(3202).
Verticillium albo-atrum. **14:**OR(3202).
V. dichotomum. On dead stems. **32:**NJ(965).
Volutella colletotrichoides. **14:**IA(576). **32:**IA(576).

Deuteromycotina - Coelomycetes

Ascochyta pisi. **32:**OR(3202).
Ascochyta sp. Leaf spot, black stem. **14:**IL(82). **18:**ID(3202). **32:**IL(82), MI(2607), WA(82). **33:**WI(1343). **35:**AL(759), MA(82), ME(82), VT(82).
Colletotrichum destructivum. Anthracnose, blight, on leaves, roots and stolons. **14:**MN(2957), UT(82), WV(952). **32:**AL(759), HI(2899), IA(215,217,714), KY(82), MN(82,2957), UT(82), WV(952,953). **33:**NJ(82). **35:**AL(1426), LA(1426), MN(2957), NH(1893,1897), SC(1426), WV(952).
C. graminicola. Anthracnose. **32:**KY(82), OH(82). **35:**KY(82).
C. trifolii. Anthracnose. **14:**NC(1303). **15:**KY(82), NC(82,1303). **32:**'Southern anthracnose.' Range of host(82), SE states(82), AL(285), IL(82), MI(82), NC(1303), NH(1893), NY(82), TX(82), WA(82,3202), WI(82). **33:**NC(1303). **35:**E states(82), AL(1426), IL(82), LA(1426), NC(1303), NY(82), SC(1426), TX(82), WA(82). **38:**NC(1303). **44:**NC(1303). **46:**HI(2903), OK(2872).
Colletotrichum sp. Anthracnose, on roots and stolons. **32:**NC(1303). **35:**NC(1303), SC(2316). **40:**HI(2903).
Macrophomina phaseolina. Charcoal rot, isolated from roots and stolons. **32:**KY(82). **35:**AL(1426), LA(1426), NC(1303), NH(1893).
Macrophomina sp. **35:**SC(2316).
Pestalotia sp. On leaves. **32:**NH(1893).
Phoma medicaginis. Black stem, leaf spot. **14:**MN(2957). **32:**MN(2957). **35:**MN(2957).
P. pinodella. Leaf spot. **14:**OR(3202). **15:**OR(3202). **32:**'Spring black stem.' IA(215,217,714,716), NH(1897), NY(2998,3005), OR(3202), WI(1454), WV(952). **35:**OR(3202).
P. sclerotioides. On roots. **14:**AK(307).
Phoma sp. Crown and root rot, leaf spot, black stem. **32:**MI(2607), WI(1898). **35:**NH(1892,1893). **46:**NH(1895).
Phyllosticta alpinicola. On living leaves. **30:**WY(1195,3376).
Phyllosticta sp. Leaf spot. **15:**TN(82).
Septoria trifolii. On dead leaves. **19:**WA(3202). **29:**OR(3202). **32:**NJ(82).
Sporonema trifolii. Leaf spot. **32:**MI(82), NY(82), OH(82), WI(82,1330).
Stagonospora meliloti. Leaf spot. **14:**MT(82), NY(82), WI(82). **32:**NH(1893,1897). **35:**NE states(1890), AK(23,526), AL(23,82), GA(23), ID(82), LA(82), MD(23), NC(48), NH(1892,1893,1897), NJ(82), NY(82), OR(82), WI(82,1338,1454), WV(82,952).
S. recedens. 'Stagonospora leaf spot.' **32:**CA(82), IA(82), WI(82).
Stagonospora sp. Leaf spot. **35:**NH(1895).

Deuteromycotina - Other

Rhizoctonia crocorum. Violet root rot. **32:**OR(82,3202), WI(1456).
R. solani. Summer blight, brown patch, on leaves, roots, seeds, stems and stolons. **14:**MN(82), OH(82). **32:**HI(2899). **35:**AL(1426), LA(1426), NC(45,48,1303), NH(1892,1893,1897), NY(82), OH(82), SC(1426), WI(82).
Rhizoctonia sp. Crown, root and stem rot. **1:**NC(82). **12:**HI(2903). **32:**NC(1303), WI(1898). **35:**IA(714), MS(2720), SC(2316), WI(1341).
Sclerotium rolfsii. Southern blight, southern stem rot. **6:**AL(82). **15:**LA(82). **35:**HI(2903), NC(45,48,1303). **46:**FL(35), NC(1303).

Trigonella L.

(= *Kentia* Adans.)
1. *T. foenum-graecum* L. [fenugreek]
2. *Trigonella* sp. (=*Kentia sp.*)

Zygomycotina

Rhizopus stolonifer. On seed coat. **1:**IA(2099).

Basidiomycotina - Ustilaginales

Entyloma meliloti. Leaf smut. **1:**CA(82,1077).

Deuteromycotina - Hyphomycetes

Alternaria alternata. On seeds. **1:**IA(2099).
Aspergillus sp. On germinating seeds. **1:**IA(2099).
Botrytis cinerea. On leaves and pods. **1:**IA(2099).
Cercospora traversiana. Leaf spot. **1:**IA(2099).
Cladosporium herbarum. On seeds. **1:**IA(2099).
Fusarium sp. On germinating seeds. **1:**IA(2099).
Penicillium sp. On germinating seeds. **1:**IA(2099).
Stemphylium botryosum. On seeds. **1:**IA(2099).

Deuteromycotina - Coelomycetes

Colletotrichum gloeosporioides. **2:**AK(526).

Vicia L.

1. *V. americana* Muhl. ex Willd. (=*V. sparsifolia* Nutt. ex Torr. & A. Gray, *V. trifida* D. Dietr., *V. linearis* Greene, *V. americana* Muhl. ex Willd. var. *truncata* (Nutt.) Brewer)
2. *V. americana* Muhl. ex Willd. var. *americana*
3. *V. americana* Muhl. ex Willd. var. *oregana* (Nutt.) A. Nels. (=*V. oregana* Nutt.)
4. *V. angustifolia* L. (=*V. sativa* subsp. *nigra* (L.) Ehrh.) [blackpod vetch, common vetch]
5. *V. benghalensis* L. (=*V. atropurpurea* Desf.) [purple vetch]
6. *V. caroliniana* Walter
7. *V. cracca* L. [bird vetch, boreal vetch, Canada pea, cow vetch, tufted vetch]
8. *V. exigua* Nutt.
9. *V. faba* L. (=*Faba vulgaris* Moench) [broad bean, fava bean, horse bean, Windsor bean]
10. *V. gigantea* Hook. [Sitka vetch]
11. *V. pannonica* Crantz [Hungarian vetch]
12. *V. pseudo-orobus* Fisch. & C. A. Mey.
13. *V. pulchella* H. B. K.
14. *V. sativa* L. [common vetch, springvetch, tare]
15. *V. tenuifolia* Roth
16. *V. unguiculata* Clavaud
17. *V. villosa* Roth (=*V. villosa* Roth subsp. *villosa*) [hairy vetch, large Russian vetch, winter vetch]
18. *Vicia* sp.

Oomycetes

Aphanomyces euteiches. Root rot. **14:**AL(82), WI(82).
Peronospora mayorii. Downy mildew. **4:**MD(82).
P. narbonensis. Downy mildew. **2:**CO(82), IA(82), ID(3202), KS(82), MT(3202), ND(82), SD(82), WA(3202), WI(82,1338).
P. viciae. Downy mildew. **1:**ND(377). **4:**GA(82), MD(82), MS(2720), NC(82). **14:**FL(35), MA(82), MD(82), MS(82,2720), OR(82,3202), TX(82), VA(82). **17:**IL(312), MS(82,2720), NC(82), SC(82). **18:**MT(3202), OR(3202).
P. viciae var. *americana*. **4:**NC(3944).
P. vicicola. **10:**OR(3202), WA(3202).
Phytophthora cactorum. Red stem canker. **16:**FL(35).
Phytophthora sp. Crown and root rot. **4:**MS(2472).
Pythium arrhenomanes. Root rot. **9:**CA(82), HI(82,2903).
P. debaryanum. Root rot. **9:**CA(82), HI(82,2903).
P. graminicola. Root rot. **9:**CA(82), HI(82).
P. irregulare. Root rot. **9:**CA(82), HI(82,2903).
P. rostratum. Root rot. **9:**CA(82), HI(82,2903).
P. spinosum. **18:**OR(3202).
P. splendens. Damping-off, root rot. **9:**CA(82), HI(82,2903). **16:**FL(35).
Pythium sp. Root rot. **9:**CA(82), HI(82).

Zygomycotina
Choanephora cucurbitarum. Flower blight. **16:**FL(35).
Mucor sp. **14:**OR(3202). **18:**OR(3202).
Rhizopus stolonifer. **14:**OR(3202). **18:**OR(3202).
Ascomycotina
Chaetomium sp. **17:**OR(3202).
Erysiphe polygoni. Powdery mildew. **1:**MT(82), OR(82). **2:**CA(638), MO(82), MT(82,3202), OR(3202), WA(3202). **4:**TX(82). **5:**TX(82). **9:**TX(82). **10:**AK(526), CA(1155). **11:**TX(82). **14:**TX(82). **16:**FL(35). **17:**MS(2720), TX(82).
Leptosphaeria sp. **9:**AK(526).
Leptosphaerulina americana. Leaf spot. **14:**MS(82,2720).
Microsphaera diffusa. **2:**MT(82), SD(82), WY(82). **10:**CA(81).
M. ludens. Powdery mildew. **1:**CO(82), MT(82), SD(82), WY(82). **2:**OR(82), SD(82), WY(82).
M. penicillata. Powdery mildew. **1:**MT(82), WY(1195). **2:**ID(82,3202), IL(82), MT(82,3202), ND(82), OR(82,3202), SD(82), WA(82,3202), WY(82). **16:**FL(35).
M. ravenelii. **1:**ND(377).
[*M. vaccinii*]. Powdery mildew. **2:**WA(82). **7:**WY(82).
Mycosphaerella pinodes. Leaf spot. **2:**NY(82). **4:**GA(1438), WI(82). **6:**NY(82). **11:**OR(82). **14:**GA(1438), NY(82). **17:**GA(82), NY(82), SC(82), WA(82). **18:**WA(3202).
Mycosphaerella sp. Leaf spot. **9:**LA(82).
Neurospora sp. **18:**OR(3202).
Pseudopeziza medicaginis. Leaf spot. **17:**MS(82,2720).
Sclerotinia sclerotiorum. Stem rot. **9:**CA(82). **16:**FL(35). **18:**OR(3202).
S. trifoliorum. Wilt. **5:**OR(82). **14:**CA(82), OR(82).
Sclerotinia sp. Stem rot. **17:**CA(82), MD(2473).
Sphaerotheca macularis. Powdery mildew. **2:**WA(82). **18:**WA(3202).
Basidiomycotina - Uredinales
Uromyces coloradensis. Autoecious. **6:**WI(82). **7:**UT(82), WY(82).
U. fabae. Autoecious. **1:**NC states(82), W states(82), AZ(1203). **7:**CT(82), VT(82). **8:**CA(81). **9:**AL(82), CA(82), KS(82), LA(82), MS(82,2720), NC(82,1303), PA(82), TX(82). **13:**AZ(1203,3981). **14:**CA(81,82), MA(82,1394), MT(82,3202), WA(82,3202). **18:**NH(1897).
[*U. vignae*]. **16:**FL(35). **18:**FL(35).
Basidiomycotina - Ustilaginales
Thecaphora deformans. Seed smut. **1:**UT(1077). **2:**UT(82,1077). **3:**UT(1077). **6:**NY(82,1077). **7:**UT(82).
Basidiomycotina - Other
Thanatephorus cucumeris. **18:**WA(3202).
Deuteromycotina - Hyphomycetes
Alternaria alternata. Leaf spot. **9:**CA(82), ID(82,3202). **14:**OR(3202). **17:**OR(3202).
A. cassiae. Leaf spot. **4:**NC(1303).
Alternaria sp. Leaf spot. **9:**FL(35). **16:**FL(35).
Aspergillus sp. **14:**OR(3202). **17:**OR(3202).
Aureobasidium nigricans. False anthracnose, pod and stem spot. **4:**LA(82), VA(82). **5:**GA(82). **11:**OR(82). **14:**LA(82), NC(82), NY(82). **17:**GA(82), LA(82), MS(82,2720), MT(82,3202), NC(82), NY(82), OH(82), OK(2873), OR(82,3202), SC(82), TN(82), WI(82,1334). **18:**NC(3944).
A. pullulans. **14:**OR(3202). **17:**OR(3202).
Bipolaris spicifera. **14:**OR(3202). **18:**OR(3202).
Botrytis cinerea. Gray mold, black spot, leaf spot. **4:**FL(82), GA(82), NY(82). **9:**AK(82,526), FL(35,82), GA(82), NC(1303), NY(82). **14:**FL(82), GA(82), OR(3202). **16:**FL(35). **17:**OR(3202). **18:**NH(1897).
Botrytis sp. On stems. **17:**WI(1358).
Cercospora canescens. Leaf spot. **16:**FL(35).
C. zonata. Leaf spot. **6:**MI(82), WI(82). **14:**IA(82), KS(82), MD(82).
Cercospora sp. Leaf lesions on the vetch. **17:**MS(1467).
Cladosporium cladosporioides. **14:**OR(3202). **17:**OR(3202).
C. elatum. **17:**OR(3202).
C. vignae. Leaf and pod spot. **16:**FL(35).
Cladosporium sp. **18:**WA(3202).
Corynespora cassiicola. Leaf spot. **16:**FL(35). **18:**FL(35).
Dictyochaeta fertilis. Root rot. **17:**NC(1303).
Epicoccum nigrum. **14:**OR(3202). **17:**OR(3202).
Fusarium culmorum. Root rot. **9:**UT(82).
F. oxysporum f. sp. *medicaginis*. Wilt. **14:**MS(2720). **18:**AL(82), MS(82).
F. oxysporum f. sp. *tracheiphilum*. Wilt. **16:**FL(35).
F. roseum. **14:**OR(3202). **17:**OR(3202).
F. solani f. sp. *phaseoli*. Root rot. **16:**FL(35).
F. trichothecioides. Root rot. **9:**UT(82).
Fusidium sp. **18:**OR(3202).
Hadrotrichum blasdalei. **10:**CA(81).
Harzia acremonioides. **17:**OR(3202).
Ovularia sphaeroidea. Leaf and stem spot. **2:**ID(3202), WA(3202). **5:**WA(82,3202). **6:**WI(1346). **7:**WA(3202), WI(1350). **12:**OR(3202). **14:**WA(82). **15:**WA(3202). **17:**CA(82), ID(82), OR(82), WI(1347). **18:**ID(3202), OR(3202), WA(3202).
Papulaspora sp. **18:**OR(3202).
Penicillium sp. **14:**OR(3202). **18:**OR(3202).
Phymatotrichopsis omnivora. Root rot. **4:**TX(82). **9:**TX(82). **14:**TX(82).
Pseudocercospora cruenta. Leaf spot. **16:**FL(35).
Stemphylium botryosum. **14:**OR(3202). **18:**OR(3202).
Thielaviopsis basicola. Root rot. **18:**MA(82).
Trichoderma sp. **18:**OR(3202).
Trichothecium sp. **18:**OR(3202).
Ulocladium consortiale. **14:**OR(3202). **18:**OR(3202).
Verticillium albo-atrum. Wilt. **16:**FL(35).
Deuteromycotina - Coelomycetes
Aristastoma oeconomicum. Late leaf spot, white leaf spot. **16:**FL(35).
[*Ascochyta pisi*]. Blight, leaf spot. **1:**WI(82). **2:**CA(82), ID(3202), MT(3202), OR(3202), WA(3202), WI(82). **4:**GA(82), MS(82,2720), NC(82), SC(82), WI(82). **5:**MS(2720). **6:**WI(82). **7:**ID(3202). **9:**CA(82), LA(82), OH(82), OR(3202). **14:**DE(82), GA(82), MS(82,2720), NC(82), NY(82), OR(82,3202). **17:**ID(3202), IL(312), MS(2720), OK(2872), OR(3202), TN(47), WA(3202).
A. viciae. **4:**NC(3944). **18:**NC(3944).
Ascochyta sp. Leaf spot, black stem. **14:**AL(82), MI(82). **17:**KY(3719,3721), MS(2720).
Colletotrichum trifolii. Anthracnose. **17:**NC(1303).
C. truncatum. Anthracnose, leaf spot. **4:**FL(82), GA(82), WI(82,1335,1340). **5:**FL(82), GA(82). **9:**FL(35). **11:**FL(82), GA(82). **14:**FL(82), GA(82), LA(82). **17:**FL(82), GA(82), IL(312), LA(82), MD(82), MS(82,1467,2720), OK(82,2872,2873), PA(82), TN(47), WI(82,1340).
Colletotrichum sp. Anthracnose. **14:**NC(1303). **16:**FL(35).
Diplodia pruni. **1:**FL(82).
Macrophomina phaseolina. Charcoal root rot. **16:**FL(35).
Marssonina punctiformis. On leaves. **2:**WI(82).
Microdiplodia viciae. On leaves. **1:**CA(82), KS(82). **2:**KS(82), NM(82).
Phyllosticta davisii. On pods. **2:**WI(82).
P. phaseolina. Leaf spot. **16:**FL(35).
Phyllosticta sp. **9:**HI(2903).
Septoria astragali. On leaves. **2:**WI(82).
S. pisi. Leaf spot. **17:**GA(82).
S. viciicola. Scald. **11:**OR(82,3202). **14:**OR(82,3202).
Stagonospora meliloti. Leaf spot. **17:**OK(2872).
Deuteromycotina - Other
Rhizoctonia solani. Damping-off, stem canker, root rot. **9:**MI(82), WA(82). **16:**FL(35). **18:**FL(35).
Sclerotium rolfsii. Southern blight. **9:**FL(35,82). **16:**FL(35).

Vigna Savi

1. *V. angularis* (Willd.) Ohwi & H. Ohashi (=*Phaseolus angularis* (Willd.) W. Wight, *Dolichos angularis* Willd.) [azuki bean]
2. *V. luteola* (Jacq.) Benth.
3. *V. marina* (J. Burm.) Merr.
4. *V. mungo* (L.) Hepper (=*Phaseolus mungo* L.) [black gram, urd]
5. *V. radiata* (L.) R. Wilcz. (=*Phaseolus aureus* Roxb., *Phaseolus radiatus* L.) [golden gram, green gram, mung bean]
6. *V. unguiculata* (L.) Walp. (=*V. sinensis* (L.) Savi ex Hassk., *V. unguiculata* (L.) Walp. subsp. *unguiculata*) [black-eyed pea, cow pea, cowpea, crowder pea, southern pea, yard-long bean]
7. *V. unguiculata* (L.) Walp. subsp. *cylindrica* (L.) Verdc. (=*V. cylindrica* (L.) Skeels) [catjang, Jerusalem pea, marble pea]
8. *V. unguiculata* (L.) Walp. subsp. *sesquipedalis* (L.) Verdc. (=*V. sesquipedalis* (L.) Fruw.) [asparagus bean, yard-long bean]
9. *Vigna* sp.

Oomycetes
Peronospora trifoliorum. Downy mildew. **6:**NC(1303), NJ(82), VA(82).
Phytophthora cactorum. Red stem canker. **6:**GA(82), SC(100). **7:**GA(82). **8:**GA(82).
Pythium arrhenomanes. Root rot. **5:**CA(82), HI(82,2903).
P. debaryanum. Root rot. **6:**CA(82), HI(82,2903).
P. hydnosporum. Root rot. **5:**HI(2903). **6:**HI(82,2903).
P. irregulare. Root rot. **5:**HI(2903).
P. splendens. Root rot. **5:**HI(2903). **6:**HI(82,2903).
P. ultimum. Root rot. **6:**CA(82).
Pythium sp. Root rot. **6:**AZ(82), CA(82), HI(82), NC(1303).
Zygomycotina
Choanephora cucurbitarum. Pod rot. **6:**GA(82), LA(82), MS(82,2720), NC(82,1303).
Rhizopus sp. On pods. **6:**NC(1303).
Ascomycotina
Diaporthe phaseolorum. Stem rot. **6:**GA(82,1438,3651), MD(82), VA(82).
Erysiphe graminis f. sp. *avenae*. Powdery mildew. **6:**AR(3029).
E. polygoni. Powdery mildew. **4:**VA(82). **5:**CA(82), OK(82,2873). **6:**E states(82), AL(285), CA(82), GA(1438,3651), KS(82), NC(1303), OK(2872), TX(82). **8:**CA(82).
Glomerella cingulata. **6:**GA(82), IL(82). **7:**GA(82). **8:**GA(82).
Leptosphaerulina trifolii. Leaf spot. **6:**IL(82).
Microsphaera penicillata. Powdery mildew. **6:**FL(82), GA(82,3651), IN(82), SC(82).
Mycosphaerella cruenta. **6:**Unk.(82), GA(1438). **7:**Unk.(82). **8:**Unk.(82).
M. pinodes. Leaf spot. **6:**SC(82).
Nematospora coryli. Yeast spot. **5:**OK(2872). **6:**OK(2872).
N. phaseoli. Yeast spot of seed. **6:**IL(82), VA(82).
Sclerotinia sclerotiorum. Stem rot, Sclerotinia wilt. **4:**VA(82). **5:**TX(82), VA(82). **6:**TX(82).
Basidiomycotina - Uredinales
[*Uromyces appendiculatus*]. **1:**IN(82). **6:**MS(2720).
U. vignae. Autoecious. **6:**SE states(82), AL(285), AZ(3981), CA(82), GA(3651), HI(2903), KS(82), NC(1303), OK(2872), SD(2243), TX(82). **7:**HI(82,2903). **8:**HI(82,2903).
Basidiomycotina - Other
Athelia rolfsii. Southern blight. **6:**GA(3651).
Pellicularia koleroga. Leaf blight. **7:**NC(82).
Thanatephorus cucumeris. Stem canker, stem rot. **6:**Range of host(82), AL(285), GA(3651), MS(2720).
Deuteromycotina - Hyphomycetes
Alternaria alternata. Tip burn. **6:**GA(3651), WI(1349).
Alternaria sp. Leaf spot. **6:**AZ(82), DE(82), IL(82).
Botrytis cinerea. Gray mold blight, shoot blight, leaf spot. **5:**WI(82,1332). **6:**CA(82), DE(82), IL(312), MD(82), NC(1303), NJ(82).
Cercospora canescens. Leaf spot, sometimes on pods. **5:**MS(82,2720), VA(82). **6:**IL(82), MS(82,2720), VA(82), WV(82).
C. caracallae. Leaf spot. **5:**WI(82,1332).
Cercospora sp. Leaf and pod blight, spot. **3:**HI(2899,2903). **6:**NC(1303).
Cladosporium vignae. Scab, leaf and pod blight. **6:**SE states(82), AL(285), CA(82), GA(3651), IN(82), MS(2720), NC(1303), TX(82). **8:**IN(82).
Corynespora cassiicola. Target spot. **6:**FL(82), GA(82,3651), LA(82), NC(82), SC(82).
Cylindrocephalum sp. Seedling rot. **5:**WA(590).
Fusarium oxysporum. Wilt. **6:**HI(2903).
F. oxysporum f. sp. *tracheiphilum*. Wilt. **6:**GA(3651), MS(2720), NC(1303), OK(2872). **7:**S states(82), MO(82), TX(82).
F. oxysporum f. sp. *vasinfectum*. Wilt. **1:**AL(82).
F. solani f. sp. *phaseoli*. Root rot. **1:**NY(82). **6:**CA(82), MS(82), NC(1303), NY(82). **7:**CA(82), MS(82), NY(82). **8:**CA(82), MS(82), NY(82).
Fusarium sp. Seedling blight, foot and stem rot, wilt. **5:**NC(82). **6:**NC(1303).
Isariopsis sp. Leaf spot. **8:**HI(2903).
Myrothecium roridum. Leaf spot. **6:**LA(82), MS(82,2720).
Oidium sp. Powdery mildew. **5:**HI(2903).
Phymatotrichopsis omnivora. Root rot. **5:**TX(82). **6:**OK(82,2872), TX(82). **7:**AZ(82).
Pseudocercospora cruenta. Leaf spot. **2:**LA(962). **4:**AL(82), MS(82,2720), SC(82). **5:**MS(2720). **6:**E states(82), GA(3651), KS(82), MS(2720), NC(1303), OK(2872), TX(82). **7:**AR(82), DE(82), FL(82), LA(82), TN(82). **8:**HI(2903), VA(82).
P. dolichi. Leaf spot. **6:**MS(581).
Ramularia sp. Frosty mildew. **6:**MS(2720).
Thielaviopsis basicola. Root rot. **6:**MD(82).
Verticillium albo-atrum. Wilt. **6:**CA(82). **8:**CA(3896).
Deuteromycotina - Coelomycetes
Aristastoma oeconomicum. Late leaf spot, white leaf spot, zonate leaf spot. **1:**VA(82). **4:**VA(82). **6:**E states(82), IL(82), MS(965,2720), NC(1303), OK(2872), TX(82). **7:**MS(82).
Ascochyta pisi. Leaf spot. **6:**TX(82).
Ascochyta sp. Leaf spot. **6:**DE(82), MD(82), NC(1303), TX(82).
Chaetoseptoria vignae. Leaf spot. **6:**IL(82).
Colletotrichum lindemuthianum. Anthracnose. **6:**AL(82,285), DE(82), FL(82), IL(82), OK(82).
C. truncatum. Anthracnose. **6:**LA(82).
Colletotrichum sp. Anthracnose. **6:**NC(1303). **8:**HI(2903).
Lasiodiplodia theobromae. On pods and stems. **6:**TX(82).
Macrophoma subconica. Stem canker. **6:**AL(82), IL(82).
Macrophomina phaseolina. Ashy stem blight, canker, charcoal rot. **5:**CA(82), OK(82,2872). **6:**AR(82), CA(82), GA(82,2209,3651), IL(82,311), MD(82), MS(2720), NC(82), OK(82,2872), TX(82).
Phoma bakeriana. Pod spot. **6:**MS(82,2720).
P. exigua. Leaf spot. **6:**MD(82).
P. terrestris. **6:**OK(2872).
Phyllosticta phaseolina. Leaf spot. **6:**SE states(82), HI(2903), IA(82), MI(82), OK(82,2872).
Septoria melanophthalmi. **6:**NC(82).
Stagonospora phaseoli. Leaf spot. **6:**TX(82).
Deuteromycotina - Other
Rhizoctonia solani. Root rot, sore shin. **1:**IL(82). **6:**NC(1303).
Rhizoctonia sp. Damping-off, pod lesions, root and stem rot. **6:**Unk.(82), HI(2903), NC(1303). **9:**HI(2903).
Sclerotium rolfsii. Southern blight. **5:**OK(82,2873). **6:**Gulf States(82), HI(2903), IL(82), MS(2720), NC(82), OK(2872), TX(82).

Wisteria Nutt.
1. *W. floribunda* (Willd.) DC. [Japanese wisteria]
2. *W. frutescens* (L.) Poir.
3. *W. sinensis* (Sims) Sweet (=*W. chinensis* DC.) [Chinese wisteria]
4. *Wisteria* sp.

Ascomycotina
Botryosphaeria dothidea. **3:**GA(1438).
B. obtusa. On stems. **4:**NY(82), SC(82).
B. wistariae. On stems. **4:**SC(82).
Erysiphe sp. Powdery mildew. **4:**TX(82).
Lepteutypa fusispora. On limbs. **4:**HI(2784).
Mycosphaerella sp. On leaves. **4:**SC(82).
Nectria cinnabarina. On stems. **4:**CT(82).
Nummularia subconcava. On dead stems. **4:**PA(82).
Phomatospora wistariae. **2:**IL(82).
Valsaria rubricosa. **3:**GA(1438).
Basidiomycotina - Other
Armillaria tabescens. Root rot. **3:**FL(35).
Pleurotus ostreatus. **2:**GA(1439).
Pleurotus sp. Heart rot. **4:**WV(82).
Deuteromycotina - Hyphomycetes
Alternaria sp. Leaf spot. **3:**FL(35).
Cercospora wisteriae. Leaf spot. **3:**FL(35).
Corynespora cassiicola. Leaf spot. **3:**FL(35).
Phymatotrichopsis omnivora. Root rot. **4:**TX(82).
Deuteromycotina - Coelomycetes
Aplosporella wistariae. On stems. **4:**KS(82), NY(82).
Camarosporium wistarianum. On stems. **4:**NY(82).
Colletotrichum sp. Leaf spot. **3:**FL(35).
Hendersonia wistariae. On stems. **4:**SC(82).
Pestalotia sp. Leaf spot. **3:**FL(35).
Phomopsis seposita. On dead twigs and stems. **1:**CA(81). **4:**MA(82).
Phyllosticta wistariae. Leaf spot. **3:**FL(35). **4:**MA(82), MO(82), NJ(82), TX(82).
Septoria wistariae. Leaf spot. **2:**TX(82). **3:**TX(3597). **4:**OK(2872).
Sphaeropsis tumefaciens. Stem gall. **3:**FL(35).

Zornia J. F. Gmelin

1. *Zornia* sp.

Basidiomycotina - Uredinales

Puccinia arachidis var. *offuscata*. Autoecious. **1**:FL(35,82).

Deuteromycotina - Coelomycetes

Colletotrichum dematium. Anthracnose. **1**:FL(35).

Fagaceae

Castanea Mill.

1. *C. alnifolia* Nutt.
2. *C. crenata* Siebold & Zucc. (=*C. japonica* Blume, *C. pubinervis* (Hassk.) C. K. Schneid., *C. stricta* Siebold & Zucc.) [Japanese chestnut]
3. *C. dentata* (Marsh.) Borkh. [American chestnut]
4. *C. mollissima* Blume (=*C. bungeana* Blume, *C. formosana* (Hayata) Hayata) [Chinese chestnut]
5. *C. ozarkensis* Ashe
6. *C. pumila* (L.) Mill. [Allegheny chinquapin]
7. *C. sativa* Mill. (=*C. vesca* Gaertn.) [Eurasian chestnut, European chestnut, Spanish chestnut]
8. *Castanea* sp.

Oomycetes

Phytophthora cambivora. Root rot. **6**:OK(2872).

P. cinnamomi. Root rot of nursery plants and forest trees. **2**:SE states(3624), Unk.(82). **3**:E states(82), SE states(3624), AR(82), LA(82), MO(82), OK(82). **4**:SE states(3624), Unk.(82). **5**:SE states(3624). **6**:E states(82), AR(82), LA(82), MO(82), OK(82). **7**:SE states(3624), GA(82), LA(82).

P. nicotianae. Root rot. **3**:FL(35). **8**:FL(35).

Pythium sp. Root rot. **3**:FL(35).

Ascomycotina

Amphiporthe castanea. Canker. **2**:E states(82), CA(82), IA(82), MI(82), OR(82). **3**:NC(1303), OH(82). **4**:E states(82), CA(82), IA(82), MI(82), OR(82,3202), WA(3202). **7**:MD(82).

Botryosphaeria castaneae. **3**:NC(82).

B. dothidea. **1**:GA(1438). **8**:CA(1110).

B. obtusa. On twigs. **3**:OH(82).

B. quercuum. Stem canker. **3**:FL(35). **8**:FL(35).

B. ribis. Twig blight. **2**:MS(82,2720), VA(82). **3**:MS(82,2720), OH(82).

Camarops petersii. **3**:NC(2561).

Cenangium alboatrum. **3**:WV(82).

C. castaneae. On twigs. **3**:PA(82).

Ceratocystis microspora. Canker. **3**:NC(799), PA(82). **8**:PA(778,1691).

Chlorencoelia versiformis. **3**:GA(82,1438).

Chlorociboria aeruginosa. Green stain. **3**:Range of host(82), GA(1438).

Chlorosplenium chlora. On dead wood. **3**:GA(82,1438), TN(82).

Ciboria americana. On fallen burs. **3**:NY(82), PA(82).

Colpoma quercinum. On dead twigs. **3**:GA(82,1438).

Cryphonectria parasitica. Blight, stem canker. **2**:NY(82), PA(82). **3**:Range of host(82), CA(82), FL(35), GA(1307,1438), IA(1226), MA(656,1781,3567), MN(1112), NC(1303,1307), OR(82,3202), VA(638,3493,3495), WA(82,3202), WV(3495). **4**:GA(1438,1799), NC(1533,1799), NY(82,1799), PA(82), TN(1799), VA(1503,1799,3493), WV(1799,3495). **6**:E states(82), GA(1438), IA(82), LA(3764). **7**:CA(82,1110), OR(82,3202), WA(82,3202). **8**:CA(1110), FL(35).

C. radicalis. **3**:E states(82), MS(82), OH(82). **6**:AR(82).

Cryptodiaporthe sp. **3**:GA(1438).

Cryptodiscus foveolaris. On wood. **8**:NY(3225).

C. pallidus. On wood. **8**:NY(3225).

Diaporthe eres. On dead limbs. **3**:NY(82). **8**:CA(1110).

Diatrypella favacea. **4**:GA(1438).

Didymella castanella. On dead twigs. **3**:NJ(82).

Discohainesia oenotherae. On leaves. **3**:MD(82), VA(82). **4**:GA(82,1438). **6**:MD(82), VA(82). **8**:MD(82), VA(82).

Endothia gyrosa. On exposed roots and dead limbs. **3**:E states(82), AL(82).

Fenestella phaeospora. **3**:NJ(82), WA(82,3202).

Glomerella cingulata. **4**:GA(1096).

Gnomonia setacea. On fallen leaves. **3**:GA(82). **6**:GA(82,1438).

Hyponectria rubescens. On overwintered leaves. **3**:MA(183,269), NJ(183,269).

Hypoxylon fuscum. On wood. **3**:GA(1438). **8**:Unk.(82).

H. howeianum. On wood. **3**:GA(1438). **8**:Unk.(82).

H. illitum. On wood. **8**:VA(2403).

H. punctulatum. On wood. **3**:GA(1438). **8**:Unk.(82).

H. rubiginosum. On wood. **3**:GA(1438). **6**:GA(1438). **8**:Unk.(82).

Hysterium magnosporum. **3**:NC(1307).

Hysterographium flexuosum. **6**:GA(1438).

H. subrugosum. **3**:GA(1438).

Lanzia echinophila. **4**:GA(1438).

Lopadostoma sphinctrinum. On dead limbs. **3**:GA(82,1438), VA(82).

Massaria inquinans. **3**:GA(1438).

Microsphaera penicillata. Powdery mildew. **3**:E states(82), OH(82). **6**:NC(82,3944), VA(82). **7**:MA(82), NC(82).

Microthyrium michelianum. Leaf spot. **3**:FL(35).

Mycosphaerella punctiformis. On fading or fallen leaves. **3**:Unk.(3943), GA(1438), MD(82), NY(82), PA(82), WV(82). **4**:GA(1438).

M. tassiana. **3**:WA(3202).

Ophiostoma stenoceras. Canker. **3**:NC(789,3714).

Ostreichnion curtisii. **6**:GA(1438).

Pezicula cinnamomea. **4**:GA(1438).

P. purpurascens. On dead limbs. **3**:MA(82), PA(82,965), VA(82).

Pezizella vulgaris. On twigs. **8**:PA(849).

Phyllactinia guttata. Powdery mildew. **3**:GA(1438). **7**:GA(1438), OH(82). **8**:E states(82), IN(82).

Pseudovalsa modonia. Twig blight. **3**:CA(82), IN(82), OR(82,3202), VA(82). **7**:OR(82).

Pyrenopeziza sp. **6**:GA(1438).

Rosellinia necatrix. Root rot. **3**:CA(1110).

R. subiculata. **3**:GA(1438).

Rutstroemia bolaris. **3**:GA(1438).

Scutellinia scutellata. **3**:GA(1438).

Sphaerognomonia carpinea. On dead leaves. **2**:LA(82). **3**:GA(82,1438), NJ(82).

Valsa castanicola. On dead limbs. **3**:VA(82).

V. ceratosperma. **3**:GA(1438).

Valsaria rubricosa. **3**:GA(1438).

Winterella cinctula. On dead limbs. **3**:NE states(82), CT(2948), GA(1438), MA(2948), MD(2948), NJ(2948), NY(2948), PA(2948), RI(2948), VA(2948), WV(2948). **6**:GA(82,1438).

Xylaria bulbosa. On fence posts. **8**:MD(3015).

X. hypoxylon. On dead wood. **3**:Range of host(82), GA(1438).

Basidiomycotina - Uredinales

Cronartium quercuum. **3**:MD(82), NC(82). **4**:Unk.(82). **6**:MD(82). **7**:Unk.(82).

Basidiomycotina - Other

Antrodia albida. Brown rot. **3**:Range of host(82).

A. xantha. **8**:MA(664).

Armillaria mellea. Root rot. **3**:CA(1110), NC(82), NJ(82), NY(82), TN(82), VA(82). **6**:SE states(82), AR(82), MO(82).

A. tabescens. **3**:VA(82).

Basidioradulum radula. On wood. **8**:Range of host(82).

Bjerkandera adusta. On wood. **8**:Range of host(82).

Bondarzewia berkeleyi. Butt rot, heart rot of living trees. **6**:Range of host(82). **8**:Range of host(82).

Byssocorticium atrovirens. On wood. **3**:CT(128).

Cerocorticium notabile. **3**:CT(1715).

Cerrena unicolor. On dead wood. **3**:VA(82).

Chondrostereum purpureum. On wood. **8**:Range of host(82).

Coniophora puteana. Brown rot. **3**:Unk.(82).

Crepidotus subapplanatus. **3**:NC(1307).

Crustoderma carolinense. Brown rot. **8**:NC(2555).

C. flavescens. On wood. **3**:NC(1303,2559).

Dacryopinax spathularia. On wood. **3**:MA(128).

Daedalea quercina. Brown heart rot. **2**:PA(82). **3**:NE states(82), IN(82).

Daedaleopsis confragosa. On wood. **3**:VA(82).

Datronia scutellata. On wood. **3**:MD(82), VA(82), WV(82).

Dendrophora albobadia. On wood. **8**:Range of host(82).

D. erumpens. **3**:DC(543), NJ(543).

D. versiformis. On wood. **2**:VA(543). **3**:WV(543). **7**:MD(82,543). **8**:Range of host(82).

Dendrothele acerina. On bark. **3**:MI(82).

D. candida. On bark. **8**:NC(2084).

Dichostereum pallescens. **3**:RI(128).
Fistulina hepatica. On wood. **3**:NE states(82), NC(1303,1307), TN(82).
Flammula sp. On decaying stumps and slash. **3**:NY(82), VA(82).
Fomitopsis pinicola. On wood. **3**:PA(82), WV(82).
F. spraguei. Brown cubical rot. **3**:E states(2533), NC(1303,1307). **8**:Range of host(82).
Ganoderma applanatum. On wood. **3**:VA(82).
Gloeophyllum trabeum. Brown cubical rot. **3**:Range of host(82). **6**:Range of host(82).
Grandinia barba-jovis. **3**:MA(128).
Grifola frondosa. White spongy butt rot. **8**:Range of host(82).
Hapalopilus croceus. White heart rot. **3**:Range of host(82), AR(2171). **6**:Range of host(82), AR(2171). **8**:Unk.(1204,2681).
H. mutans. White rot. **8**:E states(1204), Range of host(82), Unk.(1204).
H. rutilans. Sapwood rot. **2**:Unk.(82). **4**:Unk.(82). **7**:Unk.(82). **8**:Range of host(82), Unk.(2681).
Heterobasidion annosum. White root and butt rot. **3**:MI(82).
Hymenochaete corrugata. On wood. **8**:Unk.(82).
H. curtisii. On wood. **3**:GA(491,1439).
H. episphaeria. On wood. **8**:Unk.(82).
H. rubiginosa. On wood. **3**:MD(82), NC(1307), NY(82), SC(1307), TN(82), TX(82).
H. tabacina. On wood. **3**:GA(491). **8**:Unk.(82).
Hyphoderma mutatum. **3**:GA(1439).
Hypholoma sublateritium. On dead wood, usually roots and stumps. **6**:Range of host(82). **8**:Range of host(82).
Inonotus glomeratus. On wood. **8**:Range of host(82).
Irpex lacteus. Sapwood rot. **2**:Unk.(82). **4**:Unk.(82). **7**:Unk.(82).
Ischnoderma resinosum. On wood. **3**:GA(491,1439).
Junghuhnia nitida. On wood. **3**:VA(2168). **8**:Range of host(82).
Laetiporus sulphureus. Brown cubical rot. **3**:CA(1110), GA(491,1439), NC(1303,1307). **8**:Range of host(82).
Laxitextum bicolor. On wood. **8**:Range of host(82).
Lentinellus cochleatus. **3**:Unk.(2423).
Lentinus strigosus. Sapwood rot of dead sprouts and slash. **3**:Range of host(82).
Lenzites betulina. On wood. **3**:Range of host(82). **6**:Range of host(82).
Leucogyrophana mollusca. On decaying wood. **3**:PA(82), TN(82).
Mycoacia fuscoatra. On wood. **8**:Unk.(82).
Mycorrhaphium adustum. On wood. **3**:Range of host(82).
Odontia lateritia. On wood. **3**:MA(128,1175), NC(1175), NY(1175).
Omphalotus olearius. Sapwood rot. **3**:Range of host(82).
Panellus stipticus. Sapwood rot of dead sprouts and slash. **3**:Range of host(82).
Peniophora cinerea. On wood. **8**:Unk.(82).
Perenniporia medulla-panis. On wood. **8**:Range of host(82).
P. ohiensis. On wood. **3**:MD(82), NC(1303), VA(82), WV(82).
P. subacida. On wood. **8**:Range of host(82).
P. tenuis. On wood. **8**:Range of host(82).
Phanerochaete affinis. White rot. **3**:CT(128). **8**:CT(433).
P. crassa. On wood. **8**:Range of host(82).
P. filamentosa. White rot. **3**:VA(433).
P. sanguinea. White rot. **3**:GA(433), NY(128,433), VA(433).
P. velutina. White rot. **8**:Unk.(82).
Phellinus everhartii. On wood. **3**:IN(82), MI(82).
P. ferreus. **8**:OR(3202).
P. ferruginosus. On wood. **6**:Range of host(82). **8**:Range of host(82).
P. gilvus. Sapwood rot. **2**:Unk.(82). **4**:Unk.(82). **7**:Unk.(82).
P. punctatus. On wood. **8**:Range of host(82).
Phlebia chrysocrea. **3**:MD(2165).
P. rufa. **3**:NJ(128).
P. tremellosus. **3**:VA(82).
Pholiota adiposa. On decaying logs and stumps. **3**:NY(82), WI(82).
P. squarrosa. On decaying logs and stumps. **3**:NY(82), WI(82).
Pleurotus ostreatus. Sapwood rot. **3**:SE states(82).
Polyporus badius. On wood. **3**:GA(491,1439).
P. mori. White rot. **3**:Unk.(82).
P. umbellatus. On wood. **3**:Range of host(82).
Porodisculus pendulus. On wood. **8**:Range of host(82).
Postia caesia. On wood. **3**:GA(491,1439).
Pseudofistulina radicata. **3**:PA(82).
Pulcherricium caeruleum. Wood rot of dead limbs. **3**:AL(82), TN(82). **6**:GA(1439).
Pycnoporus cinnabarinus. White rot. **2**:Unk.(82). **4**:Unk.(82). **7**:Unk.(82).
P. sanguineus. On dead twigs. **4**:MD(2319).
Schizophyllum commune. On wood. **3**:OR(3202). **7**:OR(82,3202). **8**:Range of host(82).
Schizopora paradoxa. On wood. **8**:Range of host(82).
Spongipellis spumeus. On wood. **8**:Range of host(82).
Steccherinum fimbriatum. On dead wood. **8**:Unk.(82).
S. ochraceum. On wood. **3**:Range of host(82).
Stereum gausapatum. White heart rot. **3**:Range of host(82), NY(543). **4**:MD(82). **6**:Range of host(82).
S. hirsutum. On wood. **2**:OR(3202). **3**:OR(3202), PA(543). **4**:OR(3202). **7**:Range of host(82), OR(3202). **8**:Range of host(82).
S. ochraceo-flavum. On wood. **3**:CT(543), MA(543), VA(543). **8**:Range of host(82), NY(543).
S. ostrea. On wood. **3**:CT(543), NY(543). **8**:Range of host(82).
S. striatum. On wood. **8**:Range of host(82).
Tomentella atrorubra. **3**:NY(2030). **8**:NY(2039).
T. botryoides. **8**:NY(2039).
T. bryophila. **8**:NC(2039), NY(2039).
T. ochracea. **3**:NY(2003). **8**:NY(2039).
T. olivascens. **8**:NY(2039), PA(128,2039).
T. sublilacina. **8**:NC(2039), NY(2039).
T. terrestris. **8**:NC(2039).
Trametes hirsuta. Sapwood rot. **2**:Unk.(82). **4**:Unk.(82). **7**:Unk.(82).
T. pubescens. On wood. **8**:Range of host(82).
T. versicolor. White rot. **2**:Unk.(82). **3**:OR(3202). **4**:Unk.(82). **6**:Range of host(82). **7**:Unk.(82).
Tremella mesenterica. On wood. **6**:GA(491,1439).
Trichaptum biforme. White rot. **2**:Unk.(82). **4**:Unk.(82). **7**:Unk.(82). **8**:Range of host(82).
Tyromyces chioneus. On wood. **8**:Range of host(82).
T. fissilis. On wood. **3**:GA(491,1439).
T. galactinus. On wood. **6**:Range of host(82). **8**:Range of host(82).
Vararia investiens. **3**:CT(128).
Xylobolus frustulatus. White pocket heart rot of living trees. **3**:Range of host(82), NY(543), RI(543). **6**:Range of host(82).

Deuteromycotina - Hyphomycetes

Articulophora variispora. On wood. **3**:NY(3771).
Ceratophorum epiphyllum. **8**:PA(1671).
Chalara quercina. Wilt. **4**:MO(383,1094).
Confistulina hepatica. **3**:PA(3462). **8**:NC(3462), WI(3462).
Conoplea globosa. Limb canker. **3**:CT(82), PA(82). **8**:PA(1672).
Helicomyces scandens. On bark. **8**:MA(1270).
Helicosporium viride. On wood. **3**:MA(1266).
Helminthosporium velutinum. On dead twigs. **3**:NC(1303). **8**:NC(1307).
Phymatotrichopsis omnivora. Root rot. **6**:TX(82). **7**:TX(82).
Spadicoides obovata. **8**:NY(1675).
Sporidesmium inflatum. On dead limbs. **3**:WV(82).
Verticillium albo-atrum. Wilt. **7**:CT(2296).

Deuteromycotina - Coelomycetes

Ascoconidium castanae. On dead limbs. **3**:PA(82,2550).
Coniella castaneicola. On fallen leaves. **7**:DE(968).
Coryneum castaneicola. Twig blight. **3**:FL(35), OR(3202). **7**:CA(1110). **8**:FL(35,82), GA(3547,3549), OR(82), PA(3547,3549).
C. modonium. **8**:MD(2779).
[*C. umbonatum*]. Canker. **3**:CA(1110), IA(82), NY(82), OH(82), WV(82). **7**:CA(81,1110). **8**:CA(1110).
Cylindrosporium castaneae. Leaf spot. **3**:FL(35), NJ(82), PA(82).
Cytospora sp. Twig blight. **4**:WA(82,3202). **8**:Unk.(82).
Cytosporella carnea. On dead limbs. **8**:VA(82), WV(82).
Diplodia longispora. Twig blight. **3**:NJ(82), NY(82).
Diplodia sp. Twig blight, canker. **4**:NC(1303), OR(82), TN(82). **8**:FL(35), OR(3202).
Discella ochroleuca. Anthracnose, eye spot, brown-bordered leaf spot. **3**:Range of host(82), FL(35), MS(2720). **4**:MD(82), MS(2720). **6**:Range of host(82), OK(2872). **7**:E states(82). **8**:CA(1110), FL(35), OK(2872).
Discosia artocreas. On fall leaves. **3**:WV(82).
Hendersonula toruloidea. Canker. **7**:CA(82).
Leptothyrium castaneae. On leaves. **3**:FL(35,82), WV(82).
Macrophoma sp. Leaf spot. **3**:FL(35).

Melanconium cinctum. On twigs. **3:**GA(82), TN(82). **6:**NC(82).
Monochaetia concentrica. On leaves. **8:**AL(3549).
M. kansensis. Leaf spot. **3:**FL(35,82), NC(82). **4:**MS(4029). **8:**FL(35).
M. monochaeta. Zonate leaf spot. **2:**LA(966). **3:**GA(82,1509), IN(82,1509), NC(82,1509), TN(82,1509), VA(82,1509), WV(82). **6:**AL(82), NC(82).
Myxosporium castaneum. On dead limbs. **3:**NY(82).
Myxosporium sp. On twigs. **2:**OR(82).
Phoma castanea. On dead limbs. **3:**NY(82), PA(82), VA(82).
Phoma sp. Nut rot, on dead limbs. **3:**CA(1110), FL(35).
Phomopsis sp. Twig blight. **2:**MD(82), VA(82). **4:**MD(82), VA(82).
Phyllosticta castaneae. Leaf spot. **3:**E states(82). **6:**WV(82). **7:**TX(82).
P. fusispora. **3:**AL(82), TX(82).
Phyllosticta sp. Leaf spot. **3:**FL(35).
Scolicosporium macrosporium. Leaf spot. **3:**FL(35,82).
Seiridium castaneae. On bark. **8:**VA(3549).
Septoria gilletiana. Leaf spot. **2:**GA(82), LA(82). **4:**GA(82), LA(82).
Sphaeropsis sp. Canker. **3:**FL(35). **8:**FL(35).
Tubakia dryina. Leaf spot. **3:**FL(35,82), MS(82,2720), NC(1303,1307), NJ(82). **4:**LA(1607). **6:**AL(82), LA(1805). **7:**MS(82,2720), NJ(82,965). **8:**FL(35).

Castanopsis (D. Don) Spach

1. *C. chrysophylla* (Douglas ex Hook.) A. DC. (=*Chrysolepis chrysophylla* (Douglas ex Hook.) Hjelmq.) [giant chinquapin, golden chinquapin]
2. *C. sempervirens* (Kellogg) T. Dudley ex Merriam (=*Chrysolepis sempervirens* (Kellogg) Hjelmq.) [bush chinquapin, California chinquapin, Sierra chinquapin]
3. *Castanopsis* sp.

Ascomycotina

Botryosphaeria sp. On twigs. **3:**GA(82).
Capitotricha bicolor. On dead limbs and twigs. **1:**CA(642).
Coccomyces castanopsidis. On limbs. **1:**CA(82,128,638,642,3229). **2:**CA(128).
Coronellaria castanopsidis. **1:**OR(3202).
Cryphonectria parasitica. Blight. **3:**E states(82).
Dasyscyphus echinophilus. On involucre spines. **2:**CA(527).
Helotium sp. On dead limbs. **2:**CA(642).
Lachnum virgineum. On twigs. **2:**CA(661).
Microsphaera penicillata. Powdery mildew. **1:**NC(82). **3:**NC(82).
Microthelia sp. **1:**OR(657).
Mycosphaerella janus. Leaf spot. **1:**CA(81,82,638,642), OR(82,648,657,3202), WA(3202). **2:**CA(81,661). **3:**CA(82).
Myriophacidium aphyophyllicum. On fallen leaves. **1:**OR(3224).
Pezicula cinnamomea. On dead twigs. **1:**OR(82,3202).
Rutstroemia sp. On dead limbs. **2:**CA(642).
Sphaerulina myriadea. On leaves. **1:**CA(82). **2:**CA(81).
Taphrina castanopsidis. Leaf blister. **1:**CA(82,638,642,2454), OR(648,657,3202). **2:**CA(661,2457,3361).
Valsa coronata. On twigs. **1:**OR(82,3202).

Basidiomycotina - Other

Armillaria mellea. Root rot. **1:**OR(82,3202).
Chaetocalathus bicolor. **2:**CA(81).
Ganoderma lucidum. **1:**OR(3202).
G. oregonense. Root rot. **1:**OR(82).
Hymenochaete tabacina. On dead limbs. **1:**CA(642).
Marasmius copelandii. **3:**CA(856).
Phanerochaete sanguinea. Wood rot. **1:**OR(82,3202).
Phellinus ferreus. Wood rot. **1:**CA(82).
P. igniarius. Heart rot. **1:**OR(82,3202).
Polyporus badius. **1:**OR(3202).
Septobasidium pseudopedicellatum. Brown felt on scale-infested limbs. **3:**GA(82).
Trametes hirsuta. Sapwood rot. **1:**OR(82), WA(82,3202).
T. versicolor. Sapwood rot. **1:**OR(82,3202), WA(82).

Deuteromycotina - Coelomycetes

Asteromella castanicola. **2:**CA(81).
Gloeosporium castanopsidis. On leaves. **1:**CA(82).
Melanconium cinctum. On twigs. **3:**GA(82).

Fagus L.

1. *F. grandifolia* Ehrh. (=*F. americana* Sweet, *F. ferruginea* Aiton) [American beech]
2. *F. sylvatica* L. [European beech]
3. *Fagus* sp.

Oomycetes

Phytophthora cactorum. Bleeding canker. **1:**MA(82), NY(82), RI(82). **2:**MA(82), NY(82), RI(82).
Pythium sp. Root rot. **1:**FL(35).

Ascomycotina

Acarosporina monilifera. On bark. **1:**NY(3226). **3:**NY(3225).
Apiognomonia errabunda. On leaves and petioles. **1:**MA(184), NH(184).
Ascodichaena rugosa. On bark. **1:**E states(82), OH(82).
Asteromassaria macrospora. On limbs and twigs. **2:**MA(82). **3:**MA(82,188), ME(188).
Bertia moriformis. On dead limbs. **1:**GA(1307), ME(684), NC(1303,1307), SC(1307).
Botryosphaeria ribis. **3:**NY(2456).
Calosphaeria fagicola. **3:**PA(82).
Camarops pugillus. **3:**VA(2561).
Ceratocystis coerulescens. Wood stain. **1:**SE states(779), VA(82).
Ceratostomella microspora. **1:**NY(82,525).
Coccomyces coronatus. On dead leaves. **1:**NH(3229), NY(82).
C. tumidus. On fallen leaves. **1:**NY(82).
Cryptodiaporthe galericulata. On limbs. **1:**MI(82), NY(82).
Cryptosporella compta. On limbs. **1:**WV(82).
Daldinia concentrica. Sapwood rot of stumps. **1:**ME(82).
D. vernicosa. **1:**MA(82), MD(82), VA(82), WI(82), WV(82).
Diaporthe fagi. On dead limbs. **1:**ME(82), NY(82).
D. fagi var. *longispora*. On dead limbs. **1:**MI(82).
Diatrype disciformis. On dead limbs. **1:**E states(82), IN(82), MI(82).
D. stigma. **3:**NY(2456).
D. virescens. On dead limbs. **1:**E states(82), GA(1438), IN(82), MI(82), NC(1303,1307,3944).
Diatrype sp. On dead limbs. **1:**E states(82), IN(82), MI(82).
Diatrypella favacea. **1:**GA(1438).
Endothia gyrosa. Limb canker, on bark of exposed roots. **1:**E states(82), GA(1438), IN(987), MD(651), NC(1303,3944), OH(82), WV(2365). **2:**MD(82), NY(3283).
Eutypella juglandicola. **1:**GA(1438).
E. stellulata. On dead twigs. **1:**NC(1303).
Holwaya mucida. **3:**NY(2456).
Hymenoscyphus phyllophilus. **1:**GA(1438).
Hypoxylon atropunctatum. On wood. **1:**GA(1438). **2:**Unk.(82). **3:**Unk.(82), LA(2675).
H. cohaerens. On wood. **1:**GA(1438), NH(1811), OH(653). **2:**Unk.(82). **3:**Unk.(82), NY(2456).
H. fragiforme. On wood. **1:**GA(1438), IN(654), MA(1781), NH(1811). **2:**Unk.(82). **3:**Unk.(82).
H. howeianum. On wood. **1:**GA(1438). **2:**Unk.(82). **3:**Unk.(82).
H. illitum. **1:**GA(1438).
H. investiens. On dead limbs. **1:**NC(1303).
H. mediterraneum. On dead limbs. **1:**GA(1438), IN(82), NY(82), OH(82).
H. rubiginosum. On wood. **1:**GA(1438), NH(1811). **2:**Unk.(82). **3:**Unk.(82).
H. serpens. **1:**GA(1438).
H. serpens var. *macrospora*. **1:**GA(1438).
H. sphaeriostomum. **1:**GA(1438).
H. stygium. **3:**TX(652).
H. truncatum. **1:**GA(1438).
Hypoxylon sp. On wood. **2:**Unk.(82).
Lopadostoma turgidum. On dead limbs. **3:**MD(82), NY(82).
Massaria inquinans. **1:**GA(1438).
Microsphaera penicillata. Powdery mildew. **1:**E states(82), AL(82), FL(35), WI(82).
Mycosphaerella fagi. On leaves. **1:**ME(82).
M. punctiformis. On leaves. **1:**GA(1438), MA(82), NY(82). **3:**Unk.(3943).
Nectria cinnabarina. On twigs. **1:**NE states(82). **2:**MA(82).
N. coccinea. Canker. **1:**ME(82), NH(82).
N. coccinea var. *faginata*. Bole canker. **1:**NH(690), WV(3910). **3:**PA(3671,3674).
N. galligena. Bole and trunk canker. **1:**ME(82), NY(690), PA(690),

WV(2373).
N. peziza. **3:**NY(3085).
Ophiostoma grandifoliae. Sapwood stain. **1:**IA(787,3714).
O. piceae. Isolated from wood chips. **1:**ME(1005).
O. piliferum. On dead limbs. **1:**TX(82).
O. rostrocoronatum. On wood. **1:**ME(3714).
Phyllactinia guttata. Powdery mildew. **1:**E states(82), IL(82), MS(2720), NC(1303), WI(82). **2:**MS(2720).
Scorias spongiosa. On limbs. **1:**GA(1438), NC(3944), VA(638).
Scutellinia pennsylvanica. On wood. **3:**NY(848).
S. scutellata. **3:**NY(2456).
Splanchnonema loricatum. On limbs. **1:**VT(188).
Trematosphaeria cariosa. On logs. **3:**NY(318).
T. pertusa. **1:**NC(1307).
Tubeufia pezizula. Sapwood stain. **1:**IL(82).
Ustulina deusta. On wood. **1:**GA(82,1438), MA(656), ME(82), NH(1811), NY(82), RI(82), VA(82). **3:**NY(2456).
U. linearis. **3:**KY(82).
Valsa ambiens. **3:**VT(82).
Vibrissea truncorum. On dead leaves. **3:**NY(638).
Xylaria bulbosa. On wood. **1:**NY(3015). **3:**MD(3015).
X. corniformis. **3:**WV(82).
X. curta. On wood. **3:**NY(3015).
X. digitata. **1:**VT(82).
X. longipes. On wood. **3:**MI(3015).
X. polymorpha. On dead limbs and roots. **1:**NC(1303).

Basidiomycotina - Other

Abortiporus fractipes. White rot. **3:**Unk.(1204), Unk.(2681).
Albatrellus peckianus. On buried wood. **3:**Unk.(1204).
Antrodia albida. **3:**VT(82), WV(82).
Antrodiella americana. On wood. **1:**NC states(82), NE states(82), MI(82). **2:**NC states(82), NE states(82), MI(82). **3:**NC states(82), NE states(82), MI(82).
Aporpium caryae. On wood. **1:**NC states(82), NE states(82), MI(82). **2:**NC states(82), NE states(82), MI(82). **3:**NC states(82), NE states(82), MI(82).
Armillaria mellea. Root rot. **1:**CT(82), NJ(82), NY(82), OH(82), VT(59). **2:**CT(82), NJ(82), NY(82), OH(82).
Basidiodendron cinereum. On wood. **1:**OH(2185).
Bjerkandera adusta. On wood. **1:**Range of host(82). **2:**Range of host(82). **3:**Range of host(82).
Byssocorticium atrovirens. On bark. **1:**MA(128).
Ceriporia tarda. On wood. **1:**NC states(82), NE states(82), MI(82). **2:**NC states(82), NE states(82), MI(82). **3:**NE states(82).
Cerrena unicolor. On wood. **1:**IN(82), MD(82), NC(1303), NH(82), VT(82).
Chondrostereum purpureum. On wood. **1:**ME(82,543), NH(543), VT(82,543). **3:**NY(543), PA(543), VA(543).
Climacodon pulcherrimus. On wood. **3:**IN(128).
Crepidotus cinnabarinus. On wood. **1:**NY(2192).
Crucibulum laeve. On twigs, on leaves. **1:**MA(1391).
Cylindrobasidium laeve. **3:**NY(128).
Cystostereum murrayi. On wood. **1:**MA(543), NY(82,543), VT(82).
Dacryopinax spathularia. On wood. **3:**NY(128).
Daedalea quercina. Brown cubical heart rot. **1:**NE states(82), OH(82).
Daedaleopsis confragosa. On wood. **1:**NC(82,3944), NY(82). **3:**NY(2456).
Datronia mollis. **3:**NH(82), NY(82).
Exidia glandulosa. On dead limbs. **1:**NC(1303,1307).
Fomes fomentarius. On wood. **1:**E states(82), OH(82).
Fomitopsis pinicola. Brown crumbly heart rot. **1:**NY(82), TN(82), VT(82).
F. rosea. On dead wood. **1:**NY(82). **2:**NY(82). **3:**NY(82).
F. spraguei. Butt rot of living trees. **1:**IN(82), MO(82), OH(82). **2:**IN(82), MO(82), OH(82). **3:**IN(82), MO(82), OH(82).
Ganoderma applanatum. White-mottled heart rot. **1:**E states(82), NC(1307), WI(82).
Ganoderma sp. Butt and root rot. **1:**FL(35).
Globifomes graveolens. On wood. **1:**OH(82). **2:**OH(82). **3:**OH(82).
Gloeophyllum trabeum. Dry rot of timber. **3:**Unk.(82).
Gloeoporus dichrous. On wood. **1:**Range of host(82). **2:**Range of host(82). **3:**Range of host(82).
Hapalopilus rutilans. White rot. **3:**Unk.(2681).
Hericium americanum. On wood. **1:**ME(82), NY(82), VT(82), WV(82). **3:**PA(1237).
H. coralloides. On wood. **1:**ME(82), NY(82), VT(82), WV(82).
H. erinaceus. On wood. **1:**PA(1237).
Hohenbuehelia inversa. On logs. **3:**MI(3330).
Hymenochaete agglutinans. **1:**NH(128).
H. cinnamomea. On wood. **3:**Unk.(82).
H. corrugata. On wood. **1:**GA(491,1439), VA(638). **3:**Unk.(82).
H. tabacina. On wood. **3:**Unk.(82).
Hyphoderma heterocystidia. **3:**KY(2321).
Hypholoma fasciculare. On wood. **1:**NC(1303).
Hypochnicium stratosum. On wood. **1:**FL(444).
Inonotus cuticularis. On wood. **1:**Range of host(82). **2:**Range of host(82). **3:**Range of host(82), NY(2456).
I. glomeratus. Butt rot of living trees. **1:**NE states(82,3241), IN(82), NY(82). **2:**NE states(82), IN(82), NY(82). **3:**NE states(82), IN(82), NY(82).
I. hispidus. On wood. **1:**CT(82). **2:**CT(82). **3:**CT(82).
I. obliquus. On wood. **1:**NC states(82), NE states(82), MI(82). **2:**NC states(82), NE states(82), MI(82). **3:**NC states(82), NE states(82), MI(82).
I. radiatus. On wood. **1:**ME(82), NY(82). **2:**Range of host(82). **3:**ME(82), NY(82).
I. rheades. Butt rot. **1:**MI(82), OH(82). **2:**MI(82), OH(82). **3:**MI(82), OH(82).
Irpex lacteus. On wood. **1:**Range of host(82). **2:**Range of host(82). **3:**Range of host(82).
Ischnoderma resinosum. On wood. **1:**IN(82), NY(82), VT(82). **2:**IN(82), NY(82), VT(82). **3:**IN(82), NY(82), VT(82).
Junghuhnia nitida. **1:**NJ(2168), NY(2168), VT(2168).
Laxitextum bicolor. On wood. **1:**NY(82), OH(82), VA(543).
Lentinellus cochleatus. **1:**Unk.(2423).
L. micheneri. **1:**Unk.(2423).
L. ursinus. On wood. **1:**NC(1303). **2:**E states(2423).
L. vulpinus. On dead wood. **3:**Unk.(82).
Lenzites betulina. On wood. **3:**Range of host(82).
Licrostroma subgiganteum. **3:**PA(2084).
Lindtneria flava. On wood. **3:**NY(1204).
Marasmius epifagus. On fallen leaves. **3:**MA(1222,1223), MI(1222,1223), NY(1223).
M. rotula. **3:**NY(2456).
Microstroma sp. Leaf spot. **1:**IL(82), NJ(82).
Mycena filiformis. On fallen leaves. **3:**MI(3326).
M. haematopoda. On wood. **1:**WV(3717).
M. leaiana. On wood. **3:**MI(3328).
M. tenuiceps. On wood. **1:**TN(3326).
Mycorrhaphium adustum. On wood. **1:**OH(128).
Oxyporus latemarginatus. On wood. **1:**NC states(82), NE states(82), MI(82). **2:**NC states(82), NE states(82), MI(82). **3:**NC states(82), NE states(82,2166), NY(2166).
O. populinus. **1:**CT(82), MI(82), WI(82).
Panellus serotinus. On wood. **3:**Unk.(82).
Peniophora cinerea. On wood. **3:**Unk.(82).
P. incarnata. On wood. **3:**Unk.(82).
Perenniporia fraxinophila. On dead wood. **1:**CT(82). **2:**CT(82). **3:**CT(82).
P. medulla-panis. On wood. **1:**NC states(82), NE states(82), MI(82). **2:**NC states(82), NE states(82), MI(82).
P. ohiensis. On dead wood. **1:**IN(82), PA(82). **2:**IN(82), PA(82). **3:**IN(82), PA(82).
P. tenuis. On wood. **1:**NC states(82), NE states(82). **2:**NC states(82), NE states(82). **3:**NC states(82), NE states(82).
Phanerochaete affinis. **3:**NY(128).
P. burtii. White rot. **3:**FL(433).
P. crassa. On wood. **1:**NH(82).
P. ericina. White rot. **1:**TN(433).
P. filamentosa. White rot. **1:**TN(433). **3:**MI(128).
P. flavido-alba. **3:**GA(128).
P. rimosa. White rot. **1:**TN(433).
P. velutina. On wood. **1:**VT(128). **3:**Unk.(82), ME(128).
P. viticola. White rot. **1:**NC(433), NH(128,433), NY(433). **3:**NY(128).
Phellinus conchatus. On dead wood. **1:**ME(82). **2:**ME(82). **3:**ME(82).
P. everhartii. White heart rot. **1:**NJ(82), NY(82).
P. ferruginosus. On wood. **1:**NC states(82), NE states(82), MI(82). **2:**NC states(82), NE states(82), MI(82). **3:**NC states(82), NE states(82), MI(82).

P. gilvus. On wood. **1:**E states(82), IN(82), LA(82), MS(2720), NC(1303,3944). **2:**E states(82), IN(82), LA(82), MS(2720). **3:**E states(82), AR(2675), IN(82), LA(82,2675), MS(2675).
P. igniarius. White spongy heart rot. **1:**NE states(82), NY(82), WI(82).
P. laevigatus. White spongy heart rot. **1:**SC states(82). **3:**LA(2675).
P. punctatus. On wood. **1:**NC states(82), NE states(82), MI(82). **2:**NC states(82), NE states(82), MI(82). **3:**NC states(82), NE states(82), MI(82).
P. viticola. **3:**PA(82).
Pholiota adiposa. On wood. **3:**Unk.(82).
P. granulosa. On wood. **1:**VA(3717).
P. squarrosoides. On wood. **3:**Unk.(82).
Pleurotus ostreatus. On wood. **1:**NC(1303). **3:**Unk.(82).
Plicaturopsis crispa. On wood. **1:**VA(3717), WV(3717).
Polyporus mori. On decaying wood. **1:**Range of host(82). **2:**Range of host(82). **3:**Range of host(82).
P. squamosus. Butt rot of living trees. **1:**ME(82), NY(82). **2:**ME(82), NY(82). **3:**ME(82), NY(82).
P. varius. On wood. **1:**Range of host(82). **2:**Range of host(82). **3:**Range of host(82).
Postia undosa. **1:**NH(397), NY(397).
Pseudoclitocybe cyathiformis. On wood. **3:**NC(273).
Pycnoporus cinnabarinus. On wood. **1:**Range of host(82). **2:**Range of host(82). **3:**Range of host(82).
Rigidoporus crocatus. On wood. **1:**NC states(82), NE states(82), MI(82). **2:**NC states(82), NE states(82), MI(82). **3:**NC states(82), NE states(82), MI(82).
Schizophyllum commune. White rot. **3:**Range of host(82).
Schizopora paradoxa. On wood. **1:**NC states(82), NE states(82), IN(82). **2:**NC states(82), NE states(82), MI(82). **3:**NC states(82), NE states(82), MI(82).
Scytinostroma galactinum. **1:**NY(128).
Sebacina helvelloides. At base of living trees. **1:**IN(128).
Septobasidium castaneum. On scale insects. **3:**MS(2720).
S. cokeri. Parasitic on scale insects. **1:**FL(35).
Skeletocutis nivea. On wood. **1:**NH(82), NY(82). **2:**NH(82), NY(82). **3:**NH(82), NY(82).
Spongipellis delectans. On wood. **1:**NY(82), OH(82). **2:**NY(82), OH(82). **3:**NY(82), OH(82).
S. spumeus. On wood. **1:**NY(82). **2:**NY(82). **3:**NY(82).
S. unicolor. Butt rot of living trees. **3:**MD(82).
Steccherinum ciliolatum. On wood. **1:**NY(1175).
Stereum hirsutum. On wood. **1:**GA(1439), IN(82), MA(543), MD(82), ME(82), NC(3944), NH(543), NY(543), VT(543). **3:**MD(543), PA(543).
S. ochraceo-flavum. **1:**CT(543). **3:**MI(543).
S. ostrea. On wood. **1:**GA(1439), IN(82,543), OH(82), PA(82), VT(82). **3:**NY(543).
S. striatum. **1:**CT(543).
Tomentella botryoides. **3:**MI(2039), NY(2039).
T. bryophila. **3:**MD(2039), NY(2039).
T. ferruginea. **3:**MI(2039).
T. galzinii. On wood. **3:**NY(2032,2039).
T. lateritia. **3:**MI(2039), NY(2039).
T. neobourdotii. **3:**NY(2039).
T. olivascens. **3:**NY(2039).
T. rubiginosa. **3:**NY(128).
T. ruttneri. On wood. **1:**NY(2035).
T. sublilacina. **3:**MA(2039), NY(2039).
T. subvinosa. **3:**NY(2039).
Tomentellastrum badium. On wood. **3:**NY(2032).
Trametes elegans. White rot. **1:**OH(82).
T. hirsuta. On wood. **1:**Range of host(82), NC(1303). **2:**Range of host(82). **3:**Range of host(82).
T. pubescens. On wood. **1:**Range of host(82). **2:**Range of host(82). **3:**Range of host(82).
T. versicolor. White rot. **1:**Range of host(82), MS(2720), NC(3944). **2:**Range of host(82). **3:**Range of host(82).
Trichaptum abietinum. **1:**NY(2456).
T. biforme. On wood. **1:**Range of host(82), NC(1303,3944). **2:**Range of host(82). **3:**Range of host(82), AR(2675), LA(2675), NY(2456).
Tubulicrinis angustus. **3:**MA(3859).
Tulasnella violea. **1:**NH(128), OH(128). **3:**PA(2673).
Tyromyces chioneus. On wood. **1:**Range of host(82). **2:**Range of host(82). **3:**Range of host(82).
Vararia investiens. **3:**MA(128), NY(128,2456).
Wolfiporia cocos. On wood. **1:**PA(793). **3:**WV(1241).

Deuteromycotina - Hyphomycetes

Alternaria sp. Heteroecious: aecial host. **1:**MI(691), NH(691), OH(691), PA(691).
Aureobasidium pullulans. Heteroecious: aecial host. **1:**MI(691), OH(691), PA(691).
Bispora betulina. Heteroecious: aecial host. **1:**NH(691), PA(691).
Botrytis cinerea. **3:**NY(2456).
Brachysporium nigrum. **1:**NY(2456).
Cephalosporium sp. Heteroecious: aecial host. **1:**MI(691), OH(691), PA(691).
Cercospora sp. Leaf spot. **1:**NY(82).
Cladosporium sp. Heteroecious: aecial host. **1:**MI(691), NH(691), OH(691), PA(691).
Conoplea fusca. **1:**NC(3944).
C. olivacea. **3:**IN(1672), NY(1672), PA(1672), VA(1672), WV(1672).
Cylindrocarpon album. **1:**Unk.(82).
Dictyocatenulata alba. On bark. **1:**NH(2510,3181).
Epicoccum nigrum. Heteroecious: aecial host. **1:**MI(691), NH(691), OH(691), PA(691).
Fusarium sp. Heteroecious: aecial host. **1:**MI(691), NH(691), OH(691), PA(691).
Geotrichum sp. Heteroecious: aecial host. **1:**MI(691), NH(691), OH(691), PA(691).
Graphium album. Sapwood stain. **1:**AR(82).
Haplotrichum conspersum. On trunks. **1:**MA(2131).
Humicola sp. Heteroecious: aecial host. **1:**MI(691), NH(691), OH(691), PA(691).
Leptographium microsporum. Wood stain. **3:**MS(777).
Nodulisporium sp. Heteroecious: aecial host. **1:**MI(691), NH(691), OH(691), PA(691).
Penicillium sp. Heteroecious: aecial host. **1:**MI(691), NH(691), OH(691), PA(691).
Phaeostalagmus arbusculus. On rotten wood. **1:**NY(3771).
Phialophora sp. Heteroecious: aecial host. **1:**NH(691), OH(691), PA(691).
Rhinocladiella sp. Heteroecious: aecial host. **1:**NH(691), OH(691), PA(691).
Spadicoides grovei. **1:**NY(1676).
Trichoderma viride. Heteroecious: aecial host. **1:**MI(691), NH(691), OH(691), PA(691).
Trichoderma sp. Heteroecious: aecial host. **1:**MI(691), NH(691), OH(691), PA(691).
Tubercularia vulgaris. **1:**VT(1224). **3:**NY(2456).

Deuteromycotina - Coelomycetes

Asterosporium asterospermum. On twigs and bark. **1:**MA(82,128), MD(82,128), MI(82,128), MS(2720), NJ(82), NY(128), PA(82,128), VT(128). **2:**MA(128).
Camarosporium sp. Heteroecious: aecial host. **1:**NH(691), OH(691), PA(691).
Coniothyrium fagi. On leaves. **1:**IL(82).
Coniothyrium sp. Heteroecious: aecial host. **1:**MI(691), NH(691), OH(691), PA(691).
Cytospora pustulata. **1:**MI(82), VT(82).
Cytospora sp. Heteroecious: aecial host. **1:**CT(82), NH(691), NJ(82), OH(691), PA(691).
Diplodia sp. Heteroecious: aecial host. **1:**NH(691), PA(691).
Discosia artocreas. On fallen leaves. **1:**AL(82), MA(82), NY(82).
Discula umbrinella. Anthracnose, leaf spot. **1:**NE states(82), IL(2567), NC(1303), NJ(3780), NY(82), WI(82). **2:**NC(1303).
Endobotrya elegans. On dead twigs. **1:**MA(82), ME(82).
Libertella faginea. On dead twigs. **1:**NE states(82), MI(82).
Neohendersonia kickxii. On limbs. **2:**MA(128).
Phoma sp. Heteroecious: aecial host. **1:**MI(691), NH(691), OH(691), PA(691).
Phomopsis sp. Dieback, leaf spot. **1:**FL(35), MI(82), NY(82), PA(82). **2:**MI(82), NY(82), PA(82).
Phyllosticta faginea. Leaf spot. **1:**NE states(82), FL(35), MO(82), NC(1303), WI(82).

Lithocarpus Blume

1. *L. densiflora* (Hook. & Arnott) Rehd. (=*Quercus densiflora* Hook. & Arnott, *Pasania densiflora* Örst.) [tanbark oak]
2. *Lithocarpus* sp.

Oomycetes

Phytophthora cinnamomi. **1**:SE states(3624).

Ascomycotina

Brasiliomyces trina. Powdery mildew. **1**:CA(1154).
Bulgaria inquinans. On wood. **1**:CA(81), OR(3202).
Chlorociboria aeruginosa. Sapwood stain. **1**:CA(82).
Coccomyces dentatus. On leaves. **1**:CA(3229).
Cystotheca lanestris. Powdery mildew. **1**:CA(1154,1155).
Diatrype stigma. **1**:CA(81).
Laestadia caelata. On dead leaves. **1**:CA(1468).
Limacinia lithocarpi. On living leaves. **1**:CA(2434).
Limacinula anomala. On leaves. **1**:CA(2434).
Microsphaera penicillata. Powdery mildew. **1**:CA(1154).
Mollisia lithocarpi. On dead leaves. **1**:CA(81,527).
Nectria episphaeria. **1**:CA(81).
Protopeltis lithocarpi. On leaves. **1**:CA(2434).
Sphaerulina conflicta. Leaf spot. **1**:CA(82).

Basidiomycotina - Uredinales

Cronartium quercuum. Heteroecious: telial host. **1**:CA(81,82).

Basidiomycotina - Other

Aleurodiscus aurantius. On dead stems and twigs. **1**:CA(2083).
Ceriporia viridans. Wood rot of dead trunks and logs. **1**:CA(82).
Cystostereum pini-canadense subsp. *subabruptum*. On wood. **1**:CA(81).
Dendrothele candida. On bark. **1**:CA(2084).
Exidiopsis grisea. **1**:CA(3853).
Hymenochaete tabacina. **1**:CA(81).
Junghuhnia nitida. **1**:OR(3202).
Marasmius copelandii. **1**:CA(856).
M. quercophilus. **1**:CA(856).
M. thiersii. **2**:CA(856).
Omphalotus olivascens. On stumps and buried wood. **1**:CA(274).
Phanerochaete ravenelii. **1**:CA(128).
Phellinus ferreus. On wood. **1**:CA(82).
Phlebia concentrica. **1**:CA(128).
Porotheleum fimbriatum. **1**:CA(81).
Schizopora paradoxa. Wood rot of dead trunks and logs. **1**:CA(82), OR(3202).
Steccherinum fimbriatum. On dead wood. **1**:CA(82).
[*Stereum gausapatum*]. **1**:CA(543).
S. hirsutum. Wood rot. **1**:CA(82).

Deuteromycotina - Hyphomycetes

Chaetochalara setosa. On leaves. **1**:CA(2550).
Fusicolla foliicola. On leaves. **1**:CA(81).
Fusidium griseum. On dead leaves. **1**:CA(81).

Deuteromycotina - Coelomycetes

Chaetasbolisia falcata. **1**:CA(81).
Monochaetia hysteriiformis. On dead leaves. **1**:CA(81).
Pestalotiopsis montellica. Leaf blight. **1**:CA(81,82).

Quercus L.

1. *Q. acutissima* Carruth. [sawtooth oak]
2. *Q. agrifolia* Née [California field oak, California live oak, coast live oak]
3. *Q. agrifolia* Née var. *frutescens* Engelm.
4. *Q. agrifolia* Née var. *oxyadenia* (Torr.) J. T. Howell
5. *Q. alba* L. [white oak]
6. *Q. arizonica* Sarg.
7. *Q. austrina* Small
8. *Q. bicolor* Willd. [swamp white oak]
9. *Q. bushii* Sarg.
10. *Q. cerris* L. [European turkey oak]
11. *Q. chrysolepis* Liebm. [canyon oak, maul oak]
12. *Q. coccinea* Münchh. [scarlet oak]
13. *Q. douglasii* Hook. & Arnott [blue oak]
14. *Q. dumosa* Nutt. [California scrub oak]
15. *Q. durandii* Buckley
16. *Q. durata* Jeps. [leather oak]
17. *Q. ellipsoidalis* E. J. Hill [Jack oak, northern pin oak]
18. *Q. emoryi* Torr. [Emory oak]
19. *Q. engelmannii* Greene [Engelmann oak]
20. *Q. falcata* Michx. [southern red oak, Spanish red oak]
21. *Q. falcata* Michx. var. *pagodifolia* Elliott
22. *Q. gambelii* Nutt. (=*Q. gunnisonii* Rydb., *Q. utahensis* (A. DC.) Rydb.) [Gambel oak]
23. *Q. garryana* Douglas [Oregon oak, western oak]
24. *Q. garryana* Douglas var. *semota* Jeps.
25. *Q. georgiana* M. A. Curtis
26. *Q. grisea* Liebm.
27. *Q. hypoleucoides* A. Camus
28. *Q. ilex* L. [holly oak, Holm oak]
29. *Q. ilicifolia* Wangenh. [bear oak, scrub oak]
30. *Q. imbricaria* Michx. [laurel oak, shingle oak]
31. *Q. incana* Roxb. (=*Q. cinerea* Michx.) [bluejack oak, high-ground willow oak, sand-jack, turkey oak]
32. *Q. kelloggii* Newb. [California black oak, Kellogg oak]
33. *Q. laevis* Walter (=*Q. catesbaei* Michx.) [Catesby's oak, turkey oak]
34. *Q. laurifolia* Michx. [laurel-leaved oak]
35. *Q. lobata* Née [California white oak, valley oak]
36. *Q. lyrata* Walter [overcup oak, swamp post oak]
37. *Q. macdonaldii* Greene [MacDonald oak]
38. *Q. macrocarpa* Michx. [bur oak, mossy-cup oak]
39. *Q. margaretta* Ashe ex Small
40. *Q. marilandica* Münchh. [blackjack oak, jack oak]
41. *Q. michauxii* Nutt.
42. *Q. mongolica* Fisch. ex Turcz. [Mongolian oak]
43. *Q. muhlenbergii* Engelm. [yellow chestnut oak]
44. *Q. myrtifolia* Willd.
45. *Q. nigra* L. (=*Q. aquatica* Walter) [possum oak, water oak]
46. *Q. nuttallii* E. J. Palmer
47. *Q. oblongifolia* Torr.
48. *Q. obtusata* Thunb. & Bonpl.
49. *Q. obtusiloba* Michx.
50. *Q. oglethorpensis* W. Duncan
51. *Q. palmeri* Engelm. (=*Q. dunnii* Kellogg)
52. *Q. palustris* Münchh. [pin oak, Spanish oak]
53. *Q. phellos* L. [willow oak]
54. *Q. prinoides* Willd. [chinquapin oak, dwarf chestnut oak]
55. *Q. prinus* L. (=*Q. montana* Willd.) [basket oak, chestnut oak, rock chestnut oak, swamp chestnut oak]
56. *Q. pumila* Walter
57. *Q. reticulata* Humb. & Bonpl.
58. *Q. robur* L. [English oak, truffle oak]
59. *Q. rubra* L. (=*Q. borealis* F. Michx. var. *maxima* (Marsh.) Sarg., *Q. maxima* (Marsh.) Sarg.) [northern red oak]
60. *Q. rubra* L. var. *borealis* (F. Michx.) Farw. (=*Q. borealis* F. Michx.)
61. *Q. rugosa* Née
62. *Q. sadleriana* R. Br. [deer oak]
63. *Q. semicarpifolia* Sm. (=*Q. obtusifolia* D. Don)
64. *Q. shumardii* Buckley [Shumard's red oak]
65. *Q. stellata* Wangenh. [post oak]
66. *Q. suber* L. [cork oak]
67. *Q. texana* Buckley (=*Q. shumardii* Buckley var. *texana* (Buckley) Ashe) [Texas red oak]
68. *Q. tomentella* Engelm. [island oak]
69. *Q. tomentosa* Willd.
70. *Q. toumeyi* Sarg.
71. *Q. turbinella* Greene
72. *Q. undulata* Torr. [Rocky Mountain scrub oak]
73. *Q. vaccinifolia* Kellogg [huckleberry oak]
74. *Q. velutina* Lam. (=*Q. tinctoria* W. Bartram) [black oak, quercitron, yellow-barked oak]
75. *Q. virginiana* Mill. (=*Q. virens* Aiton, *Q. geminata* Small) [southern live oak]
76. *Q. wislizenii* A. DC. [interior live oak]
77. *Quercus* sp.

Myxomycetes

Arcyria nutans. **14**:CA(2439).
Comatricha nigra. **68**:CA(2439).
Hemitrichia clavata. **77**:NC(3944).
H. serpula. **77**:NC(3944).
Leocarpus fragilis. **77**:NC(3944).
Metatrichia vesparium. **77**:NC(3944).
Physarum leucophaeum. **77**:CA(2439).
P. nutans. On bark. **77**:CA(2439).
Stemonitis fusca. **77**:NC(3944).
S. pallida. **68**:CA(2439).

Trichia favoginea. **55**:NC(3944).

Oomycetes

Phytophthora cactorum. Bleeding canker, trunk canker. **2**:CA(82,2444). **20**:FL(82). **66**:CA(2444).

P. cinnamomi. Trunk canker, root rot. **2**:CA(2444). **5**:SE states(3624), MD(82). **55**:SE states(3624), MD(82). **58**:SE states(3624). **60**:SE states(3624), MD(82). **66**:SE states(3624), CA(2444). **77**:CA(3624), FL(35).

P. citricola. Trunk canker. **2**:CA(2444). **66**:CA(2444).

Phytophthora sp. Root rot. **33**:FL(35). **77**:FL(35), IL(3218).

Pythium sp. Root rot. **33**:FL(35). **34**:FL(35). **75**:FL(35). **77**:FL(35).

Zygomycotina

Mortierella sp. On wood. **77**:ME(3242).

Mucor sp. On wood. **77**:ME(3242).

Ascomycotina

Acantharia echinata. Black leaf spot. **11**:CA(82,2434,3675). **73**:CA(81,3297). **76**:CA(82,2434). **77**:CA(82), OR(3202).

Acanthostigma decastylum. On decorticated limbs. **77**:GA(1438), NC(1303,1307).

Acarosporina monilifera. **77**:NJ(3225).

Acrospermum foliicola. **39**:GA(1438).

Amphiporthe raveneliana. Twig dieback. **5**:E states(82), GA(1438), MI(82), MO(82), NC(1303). **8**:IL(82). **12**:MD(82), NJ(82). **52**:IL(82). **60**:IL(82), MA(82). **77**:DC(2949), FL(35).

Amphisphaeria granulosa. On barrel staves. **77**:NY(971).

A. succinea. **55**:GA(1438).

Angelina rufescens. **77**:GA(1438).

Anthostoma quercina. **77**:MD(2780).

Apiognomonia errabunda. Anthracnose, leaf and twig blight. **2**:OR(82). **5**:E states(82), KS(82), MN(82). **8**:NY(82). **23**:OR(3202). **32**:OR(3202). **38**:SC states(82), SE states(82), WI(82). **55**:CT(82), NJ(82), WV(82). **65**:OR(82). **75**:FL(35), MS(2720). **77**:E & C states(82), CA(82), CT(3770), FL(35), OR(82).

A. veneta. Anthracnose. **5**:IL(1052), NC(1533), PA(1036). **38**:IL(1052). **55**:NC(1533). **59**:GA(1438). **65**:GA(1438). **74**:GA(1438). **77**:DC(3780), MD(3780), NJ(3780), OK(2872), PA(3780).

Apioporthe utahensis. On limbs. **22**:UT(3807).

Arachnopeziza aurelia. **5**:GA(1438). **77**:NC(3944).

Ascocoryne sarcoides. On wood. **77**:NC(3944).

Ascocoryne sp. On wood. **77**:ME(3242).

Asteridiella manca. Sooty mold. **31**:MS(2720). **75**:MS(2720).

Asterina subcyanea. On leaves. **34**:FL(35,82).

Bertia moriformis. On decorticated wood. **5**:GA(1307), NC(1307), SC(1307). **77**:GA(1438), NC(1303).

Bertiella rhodospila. On stumps. **77**:LA(195).

Biscogniauxia nummularia. On wood. **74**:VA(638). **77**:IA(1225).

Bisporella citrina. **23**:OR(3202).

Botryosphaeria dothidea. Dieback. **20**:FL(35). **77**:FL(35).

B. melanops. On dead limbs. **17**:MN(1172). **60**:CT(82).

B. obtusa. On limbs and trunks. **45**:AL(82). **74**:NC(82), VA(82). **75**:HI(2903), SC(3294).

B. quercuum. Twig blight, canker, dieback, dry rot of acorns. **2**:Unk.(82). **5**:E states(82), E states(82), GA(1438), IL(82), MN(82), MO(82). **12**:E states(82), NE states(82), GA(1438), IL(82), MI(82), MO(82). **20**:E states(82), FL(35), MO(82). **33**:FL(35). **34**:FL(35). **40**:IL(82). **43**:IL(82). **45**:FL(35), GA(1438). **52**:NE states(82), IL(82), MI(82). **55**:MD(82), NJ(82), VA(82). **59**:GA(1438). **60**:E states(82), NE states(82), IL(82), MI(82), MO(82). **65**:E states(82), IL(82), MO(82). **74**:NE states(82), IL(82), MI(82). **75**:GA(82,1438). **77**:GA(1438), IL(650).

B. rhodina. **52**:E states(82), AR(82), IA(82). **77**:E states(82), AR(82), IA(82).

B. ribis. **5**:MD(82), VA(82).

Brasiliomyces trina. Powdery mildew. **2**:CA(81,82,638,1154,1468,3977). **11**:CA(1154). **27**:AZ(1194).

Bulgaria inquinans. On wood. **2**:CA(82). **5**:Range of host(82). **59**:NC(1307). **60**:Range of host(82), MA(1781). **77**:Range of host(82), GA(1438), MA(1394), VT(1224).

Calosphaeria barbirostris. On wood. **5**:GA(1438). **77**:GA(1438), NC(1303).

C. pulchelloidea. **5**:NJ(190). **12**:NJ(190). **77**:OR(3202).

Calosphaeria sp. **55**:GA(1438).

Calospora sp. **55**:GA(1438).

Calothyrium pustulatum. Leaf spot. **34**:FL(35,82).

C. stomatophorum. On living or fallen leaves. **34**:FL(82).

Calycellina castanea. On leaves. **34**:FL(525,850,959,2181).

C. punctata. On fallen leaves. **77**:NY(2181).

Camarops ohiensis. **77**:GA(2561), WV(2561).

C. petersii. On wood. **77**:IL(1627), KY(2561), MN(1504), OH(2561), PA(1043), VA(2561,3210).

C. polysperma. **77**:GA(1438).

Cenangium tetrasporum. On dead twigs. **12**:NJ(82).

Ceratocystis coerulescens. Wood stain. **77**:SE states(779).

C. fagacearum. Wilt, on wood. **5**:PA(712,1046), WV(604). **12**:NC(1303), OH(361), PA(1046), TN(994,1029). **17**:MN(1172). **20**:TN(1029). **29**:PA(1046). **30**:MO(65,66), PA(1046). **40**:TN(1029). **45**:NC(1303). **52**:PA(1046). **53**:NC(1303). **55**:PA(712,1046). **59**:MO(65). **60**:Unk.(3714), OH(361), PA(1046). **67**:TX(2106). **74**:OH(361), PA(1046). **75**:SC(3572), TX(2106). **77**:AL(2734), AR(888,2734,2958,3733), IA(2958), IL(2958), IN(2958), KS(983,2755,2958), KY(2958), LA(2734), MD(2958), MI(2958), MN(1820,2958), MO(65,2958,3492), MS(2734), NC(352,2958,3931), NE(2755,2958), OH(2958), OK(888,2958,3733), PA(2958), TN(352,2958), TX(887,1632,2734,2958), VA(2958), WI(2958), WV(2958,3239,3492).

C. fimbriata. Sapwood stain. **17**:MN(484). **55**:WV(82,779).

C. microspora. Canker. **75**:VA(799,3714).

Ceratocystis sp. **77**:ME(3242).

Chaetosphaeria atrobarbum. **77**:GA(1438).

C. pulviscula. **74**:GA(1438).

Chaetothyrina asterinoides. On fallen leaves. **20**:GA(82). **33**:GA(1438). **34**:FL(82). **43**:GA(1438). **59**:GA(1438). **60**:GA(82). **74**:GA(82,1438). **77**:GA(82).

Chlorencoelia versiformis. Green stain. **77**:Range of host(82).

Chlorociboria aeruginosa. Green stain. **77**:Range of host(82), OR(3202).

Chlorosplenium chlora. On dead wood. **77**:E states(82), NC(1303), TN(82).

Ciborinia candolleana. Isolated from veins. **59**:PA(210).

Coccomyces castanopsidis. On wood. **11**:CA(81,527,3229).

C. coronatus. Leaf spot. **2**:NC(82). **5**:E states(82). **10**:MD(82). **23**:OR(82,3202). **34**:FL(35). **45**:FL(35,82), GA(82), NC(82). **55**:TN(82). **77**:Range of host(82), FL(2782), MS(2720).

C. dentatus. On leaves. **2**:CA(3229). **5**:VA(3229). **32**:CA(3229). **41**:TN(3229). **45**:VA(3229). **57**:AZ(3229). **75**:VA(3229). **77**:FL(3229), LA(3229), NC(3229), PA(3229).

C. farlowii. On fallen leaves. **59**:MA(3229).

C. triangularis. On twigs. **5**:E states(82,1438), GA(1438), MI(82), OH(82). **20**:VA(82). **60**:VA(82). **77**:NC(1303).

C. tumidus. On leaves. **59**:MA(3228).

Colpoma quercinum. On twigs. **5**:GA(82), IA(82), NC(1303), PA(82). **77**:GA(1438), NC(3944).

Coniochaeta pulveracea. **12**:GA(1438). **77**:OR(3202).

Coniochaeta sp. **77**:WA(3202).

Cookella microscopica. On leaves. **74**:IL(195). **77**:MS(2720).

Cryphonectria parasitica. Canker, dieback. **5**:CT(82), GA(82), MD(82), PA(82), TN(82), VA(82). **55**:MD(82), WV(82). **65**:CT(82), GA(82), MD(82), PA(82), TN(82), VA(82). **74**:CT(82), IN(82). **75**:MS(2732,3495), NC(2795), VA(2291,2732,3493,3495).

C. radicalis. On dead limbs. **5**:SE states(82), MS(82,2720). **12**:E states(82), IN(82), KY(82). **20**:E states(82), IN(82), KY(82). **40**:KY(82). **45**:SC(82). **55**:MD(82), NC(82), VA(82). **60**:E states(82), IN(82), KY(82). **74**:E states(82), IN(82), KY(82).

Cryptodiscus foveolaris. On wood. **77**:NY(3225).

C. niveopurpureus. On decaying post. **5**:NJ(965).

Cryptosporella farinosa. On leaves and twigs. **5**:NJ(2548), PA(2548).

C. paucispora. On twigs. **5**:PA(82).

Cucurbitaria sp. **45**:GA(1438).

Cyanoderma viridulum. **5**:GA(1438). **55**:GA(1438). **56**:GA(1438). **59**:GA(1438).

Cystotheca lanestris. Brown mildew, powdery mildew. **2**:CA(82,1154,1468). **4**:CA(1155). **5**:AL(82), CA(1154), IA(82), IL(82), MS(82,2720). **8**:AL(82), IA(82), IL(82), MS(82). **10**:CA(1154). **11**:CA(1154,1155). **12**:CA(1154). **13**:CA(1154). **19**:CA(82,1154). **20**:MS(82,2720). **23**:CA(1154). **24**:CA(1155). **28**:CA(1154,1155). **32**:Unk.(3976), CA(1154,1155).

34:CA(1154). **35:**CA(82,1154). **38:**AL(82), CA(1154), IA(82), IL(82), MS(82). **42:**CA(1154). **45:**CA(1154,1155). **48:**CA(1154). **51:**CA(1154). **52:**CA(1154). **53:**CA(1154,1155). **55:**MS(82,2720). **58:**CA(82,1154). **59:**MS(2720). **60:**CA(81,1154), NC(82). **62:**CA(1155). **65:**MS(82,2720). **66:**CA(82). **74:**CA(1154). **76:**CA(1154,1155). **77:**CA(81,82).

Daldinia concentrica. Sapwood rot of stumps and logs. **5:**E states(82), NC states(82), LA(82). **23:**CA(82), OR(82,3202). **77:**Range of host(82), GA(1438), NC(3944), OK(2872).

D. vernicosa. On charred substrates. **5:**E states(82), MO(82). **23:**OR(82,3202). **38:**ND(377). **65:**GA(1438). **77:**Range of host(82), NC(3944).

Dasyscyphella crystallina. **73:**CA(661).

Delacourea sp. **45:**GA(1438).

Dermateopsis tabacina. On dead limbs. **5:**NJ(82,3167). **12:**NJ(82,3167).

Diaporthe sp. **25:**GA(1438).

Diatrype albopruinosa. On dead limbs. **5:**IA(3629). **12:**MD(82), NJ(82). **22:**CO(1195), UT(82). **55:**KS(82). **60:**MD(82), NJ(82). **72:**NM(82). **74:**AR(82), IA(3629), MO(82), MS(82,2925). **77:**IA(3629), NC(3944).

D. americana f. *quercus*. **38:**ND(377).

D. bullata. **77:**OR(3202).

D. flavovirens. **77:**OR(3202).

D. prominens. **23:**OR(3202).

D. stigma. On dead limbs and trunks. **5:**Range of host(82), GA(1438), IA(3629). **8:**Range of host(82). **12:**GA(1438). **22:**CO(82), UT(82). **23:**OR(82,3202). **36:**Range of host(82). **38:**Range of host(82), IA(3629). **40:**GA(82). **45:**GA(82,1438). **53:**KS(82). **55:**GA(82,1438). **59:**GA(1438). **65:**Range of host(82). **74:**GA(1438), IA(3629). **77:**GA(1438), IA(3629), NC(1303,3944).

D. stigmaoides. **77:**OR(3202).

D. virescens. On dead limbs. **5:**Range of host(82). **8:**Range of host(82). **36:**Range of host(82). **38:**Range of host(82). **40:**OH(82). **65:**Range of host(82).

Diatrype sp. On dead limbs. **5:**Range of host(82). **60:**Range of host(82).

Diatrypella decorata. **77:**GA(1438).

D. eutypaeformis. **30:**IA(3629). **38:**IA(3629). **59:**IA(3629). **74:**IA(3629). **77:**IA(3629).

D. favacea. On limbs. **5:**IA(3629). **38:**IA(3629). **45:**GA(82,1438). **65:**GA(1438). **77:**IA(3629), NC(1303).

D. nitschkei. **5:**IA(3629). **59:**IA(3629). **77:**FL(2782), IA(3629).

D. pulvinata. **23:**WA(1248).

D. quercina. On dead limbs. **11:**CA(81). **45:**GA(82). **55:**GA(1438). **60:**ME(82). **72:**NM(82). **77:**NC(1303).

D. subfulva. **5:**GA(1438). **45:**GA(1438). **59:**GA(1438). **60:**GA(82).

Dicarpella bina. On living leaves. **2:**CA(1468).

D. quercifolia. On leaves. **12:**NJ(82,959), VA(82). **38:**NE(82).

Discohainesia oenotherae. On fallen leaves. **5:**VA(82). **20:**VA(82). **74:**VA(82).

Elsinoe quercicola. Spot anthracnose. **34:**FL(35). **77:**FL(35).

E. quercus-falcatae. Spot anthracnose, on leaves. **20:**GA(1438,1745,2402), NC(1303,2402). **74:**NC(1745).

Endothia gyrosa. Orange hobnail canker, on limbs and exposed roots. **2:**CA(82). **5:**E states(82), GA(1438), IN(82), LA(82). **10:**SC(82). **12:**SE states(82), AR(82), GA(1438), MI(82), NC(1303). **20:**SE states(82), AR(82), MI(82). **30:**IN(82). **36:**E states(82), IN(82), LA(82). **40:**IN(82), VA(2365). **45:**Gulf states(82), SE states(82), GA(1438), TN(82), TX(108). **52:**SE states(82), AL(2365), AR(82), GA(1438), MD(3495), MI(82), NC(1303), VA(2365,3493,3495). **53:**Gulf states(82), SE states(82), TN(82). **55:**IN(82). **58:**SC(82). **59:**GA(1438). **60:**SE states(82), AR(82), MI(82). **66:**SC(82), VA(2365). **74:**SE states(82), AR(82), GA(1438), MI(82). **75:**FL(35,82), GA(82,1438), MS(82,2720). **77:**FL(35), NC(3944), TX(108,652).

E. singularis. On dead limbs. **2:**CA(82). **22:**AZ(82), CO(82,2365), NM(82). **72:**CO(82).

Endothia sp. Canker. **52:**VA(3496).

Eriopezia microspora. **77:**MI(849).

Erysiphe polygoni. Powdery mildew. **13:**CA(1155). **16:**CA(1155). **19:**CA(1155). **62:**CA(1155). **71:**CA(1155).

Eutypa lata. **38:**ND(377). **77:**OR(3202).

E. limaeformis. **77:**MI(2925), MO(2925), NY(2925).

E. mela. On dead limbs and twigs. **5:**VA(82). **63:**NJ(2925). **77:**NJ(2925), PA(2925).

E. polycocca. **38:**ND(377).

E. spinosa. On wood. **5:**E states(82), MO(82), WI(82). **8:**E states(82), MO(82), WI(82). **36:**E states(82), MO(82), WI(82). **38:**E states(82), MO(82), WI(82). **65:**E states(82), MO(82), WI(82). **77:**FL(2782), GA(1438), IA(3629), NC(1303,1307).

Eutypella cerviculata. **74:**GA(1438). **77:**IA(3629).

E. leprosa. **38:**IA(3629). **77:**IA(3629).

Fenestella phaeospora. On dead limbs and twigs. **5:**GA(82,1438). **55:**GA(82,1438).

Galiella rufa. On bark. **74:**VA(638). **77:**GA(1438).

Glomerella cingulata. Leaf spot. **34:**FL(35).

Gloniopsis curvata. **65:**OK(660). **77:**ND(377).

Glonium abbreviatum. On decorticated wood. **5:**GA(1438). **77:**GA(1438), NC(1303).

G. lineare. **45:**GA(1438). **77:**GA(1438).

G. stellatum. **5:**GA(1438). **77:**GA(1438), NC(3944).

Glonium sp. **59:**GA(1438).

Gnomonia fasciculata. On dead leaves. **65:**OK(660).

G. setacea. On fallen leaves. **5:**GA(82,1438), MD(82). **23:**WA(3202). **33:**GA(1438). **38:**GA(82,1438). **39:**GA(1438). **40:**GA(82,1438). **50:**GA(1438). **55:**GA(82,1438), WV(82). **56:**GA(1438). **59:**GA(1438). **64:**GA(1438). **65:**GA(82,1438). **74:**GA(1438). **77:**GA(82).

Gnomonia sp. Anthracnose. **5:**VA(3493). **59:**WI(1345).

Godroniopsis quernea. Twig canker. **5:**NJ(82). **12:**AL(82), CT(82), NJ(82). **34:**FL(35). **40:**FL(82). **53:**FL(82), NC(82). **74:**NJ(82). **75:**AL(82), FL(35,82). **77:**FL(35).

Graphostroma platystoma. **77:**FL(634).

Hansenula polymorpha. In frass. **5:**SC(3596). **45:**LA(3596).

Helotium furfuraceum. On dead leaves. **2:**CA(2801).

Helvella lacunosa. On wood. **77:**NC(1303).

Hercospora taleola. On limbs. **38:**IA(1228). **77:**GA(2777), IA(1228).

Humaria hemisphaerica. **77:**NC(3944).

Hyalinia crystallina. On bark. **77:**NY(1958).

Hymenoscyphus epiphyllus. On leaves. **20:**NC(1303). **77:**NC(1307).

H. fructigenus. **77:**NY(2456).

Hypocrea rufa. **77:**LA(2675).

H. schweinitzii. **77:**GA(1438).

H. sulphurea. **45:**GA(1438). **77:**GA(1438).

Hypoderma ilicinum. On leaves. **40:**TX(82). **45:**AL(82), GA(82,1438). **75:**FL(35,82), GA(82,1438). **77:**MS(2720).

Hyponectria auripuncta. Leaf spot. **11:**CA(82). **22:**SW states(183). **76:**SW states(183), CA(82,1468).

Hypoxylon atropunctatum. Canker, dieback, sapwood rot. **5:**SE states(82), GA(1438), IA(1225), MI(82), NC(1303), VA(3493). **12:**GA(1438). **23:**OR(82,3202). **40:**OK(2872). **45:**GA(1438). **59:**GA(1438), IA(1225). **65:**SE states(82), MI(82), OK(660), TX(108). **74:**Unk.(3573), GA(1438), NC(1303). **75:**TX(108). **77:**Unk.(82), GA(1438,3617), NC(1303,3944), WV(3239).

H. cinereo-lilacinum. On wood. **77:**IA(1225).

H. cohaerens. **77:**OR(3202).

H. croceum. On wood. **77:**GA(2403).

H. crocopeplum. **45:**GA(1438). **77:**GA(1438).

H. fragiforme. **23:**OR(82,3202). **77:**Unk.(82).

H. fuscum. **2:**CA(81). **77:**SC(634).

H. howeianum. On wood. **5:**GA(82,1438), NY(82). **12:**GA(1438). **45:**GA(1438). **55:**E states(82), IN(82). **59:**GA(1438). **77:**GA(1438), IA(1225).

H. hypophlaeum. **77:**GA(1438).

H. illitum. On wood. **5:**GA(1438). **75:**GA(1438). **77:**GA(1438), IA(1225), VA(2403).

H. investiens. On dead limbs. **65:**OK(660). **77:**GA(1438), NC(1303).

H. jecorinum. On bark. **77:**GA(128).

H. maculum. **45:**GA(1438).

H. mammatum. **5:**NH(82), WI(82). **77:**Unk.(82).

H. mediterraneum. Coal canker, decline. **2:**CA(82), OR(82). **5:**GA(1438), IA(1225), VA(3493). **12:**GA(82), IL(82), IN(82), OH(82), VA(82). **14:**CA(82). **22:**UT(82). **23:**OR(82,3202), WA(82,3202). **25:**GA(1438). **45:**GA(82,1438). **55:**GA(82,1438). **58:**MD(82). **59:**GA(1438), IA(1225). **60:**GA(82), IL(82), IN(82), OH(82), VA(82). **74:**GA(82), IL(82), IN(82), OH(82), VA(82). **75:**GA(1438). **77:**CA(2439), IA(1225), NC(1303), OR(82), TX(108).

H. microplacum. **77:**TX(652).

H. multiforme. On wood. **5:**IA(82), IN(82), KS(82). **77:**IA(1225).
H. notatum. **12:**GA(1438). **45:**GA(1438).
H. nummularium var. *macrosporum*. **5:**IA(1225).
H. papillatum. **55:**GA(1438).
H. punctulatum. Decline. **5:**AL(82), GA(82,1438), MS(82), TN(82), VA(82,3493). **45:**GA(1438). **59:**GA(1438). **74:**GA(1438). **75:**GA(82,1438). **77:**Unk.(82), NC(1303,3944), WV(171,1811,3239).
H. regale. On wood. **77:**GA(1438,2403).
H. rubiginosum. On wood. **5:**IA(1225). **45:**GA(1438). **59:**GA(1438). **77:**Unk.(82), GA(1438), IA(1225), NC(3944), OR(3202).
H. rubrostromaticum. **77:**GA(1438).
H. serpens. On decaying wood. **5:**IA(1225). **55:**E states(82), GA(1438), IN(82). **65:**GA(1438). **77:**GA(1438), ID(2786).
H. serpens var. *macrospora*. Wood. **45:**GA(1438). **55:**NY(2786). **77:**CA(82,2786).
H. sphaeriostomum. On wood. **5:**GA(1438). **55:**GA(1438,2403). **74:**GA(1438). **77:**GA(1438).
H. stygium. **45:**GA(1438). **77:**GA(1438).
H. thouarsianum. On wood. **2:**CA(81,82). **75:**FL(128), GA(1438). **77:**FL(128), OR(3202).
H. tinctor. **45:**GA(1438). **77:**GA(1438).
H. truncatum. Sapwood rot. **5:**E states(82), Range of host(82), GA(82,1438), IA(1225), LA(82), MI(82), MO(82), NC(82,1303,1307), SC(1307), TN(82). **12:**GA(1438). **23:**OR(82,3202). **33:**FL(35). **34:**FL(35). **45:**GA(1438). **59:**GA(1438,1811), NC(1307), SC(1307). **60:**Unk.(82). **65:**E states(82), GA(1438), LA(82), MO(82). **74:**GA(1438), NC(1303,1307), SC(1307). **75:**FL(35,82), GA(82,1438).
H. vogesiacum. **77:**OR(3202).
H. vogesiacum var. *microspora*. On wood. **77:**CA(2403).
Hypoxylon sp. Sapwood rot. **2:**Unk.(82). **77:**MI(665).
Hysterium angustatum. **77:**FL(2782).
H. insidens. **77:**NC(1307).
H. magnosporum. **77:**GA(1438).
H. pulicare. **65:**OK(660). **77:**GA(1438).
Hysterographium flexuosum. On decorticated limbs. **12:**GA(1438). **20:**NC(1303). **45:**GA(1438). **74:**NC(1307). **77:**NC(3944).
H. mori. **77:**ND(377).
H. subrugosum. **12:**GA(1438). **33:**GA(1438).
Incrucipulum capitatum. On leaves. **5:**NY(3167).
I. ciliare. **5:**GA(1438). **45:**GA(1438). **53:**GA(1438). **59:**GA(1438). **64:**GA(1438). **74:**GA(1438).
Ionomidotis fulvotingens. **5:**GA(1438).
Irenopsis quercifolia. **45:**Unk.(1452).
Jattaea microtheca. On limbs. **77:**NJ(190).
Johansonia cupulata. On fallen leaves. **34:**FL(82).
Lachnum fuscescens. On leaves. **77:**NJ(849).
L. niveum. **77:**ND(377).
L. pollinarium. On leaves. **5:**GA(1438). **12:**GA(1438). **33:**GA(1438). **55:**GA(1438). **56:**GA(1438). **59:**GA(1438). **65:**GA(1438). **77:**NJ(849,3167).
Lachnum sp. **56:**GA(1438). **59:**GA(1438).
Lasiosphaeria foliicola. **77:**VA(1554).
L. hispida. Wood. **77:**GA(1438).
L. strigosa. **77:**GA(1438).
Lasiosphaeria sp. On dead limbs. **77:**NC(1303).
Lecanidion simile. **45:**GA(1438). **59:**GA(1438).
Lembosina liturae. Leaf spot. **34:**FL(35). **45:**FL(35), GA(82), TX(82).
L. quercina. Leaf blotch, black mildew, black leaf spot. **33:**GA(1438). **34:**GA(82). **40:**NC(82). **45:**FL(35,82). **53:**NC(82). **55:**GA(82,1438). **65:**GA(82), TX(82). **74:**GA(82,1438), NC(82,3944), TN(82), VA(82). **75:**SE states(82), FL(35), LA(82). **77:**FL(35,2782), GA(82), NC(82), OK(2872), TN(82), VA(82).
Leptophyma aurantiaca. On fallen leaves. **34:**FL(35,82).
Leptosphaeria dryophila. Leaf spot. **2:**CA(82). **74:**NC(82,3944).
Leptosphaeria sp. **45:**GA(1438).
Loculohypoxylon grandineum. On wood. **12:**NJ(182). **77:**GA(182,1438,2403).
Lopadostoma gastrinum. On bark. **5:**GA(82,1438). **45:**GA(1438). **55:**GA(82,1438). **77:**NC(1303).
L. sphinctrinum. On bark. **5:**IA(82), NJ(82), TX(82). **77:**CA(2439).
Lophiosphaera lophospora. Leaf spot. **45:**GA(82,1438). **77:**FL(35), OH(82).
L. velata. **55:**GA(1438). **59:**GA(1438). **77:**GA(1438).
Lophiostoma viridarium. **77:**IL(650).
Lophodermium maculare. On leaves. **77:**FL(634).
[*Melanconis chrysostroma*]. On dead limbs. **77:**NC(1303).
Microsphaera calocladophora. Powdery mildew. **5:**TN(82). **23:**WA(82). **34:**AL(82), FL(82), SC(82). **45:**AL(82), FL(82), SC(82).
M. extensa. Powdery mildew. **5:**GA(82), IA(82), IL(82), MA(82), NC(82,3944), NY(82). **17:**IA(82), IL(82), OH(82). **30:**MO(82). **45:**SC(82). **52:**IA(82), IL(82), OH(82). **54:**IA(82). **58:**KY(82). **59:**NC(3944), WI(1330). **60:**IA(82), IL(82), OH(82). **74:**IA(82), IL(82), OH(82).
M. penicillata. Powdery mildew. **2:**CA(82,1154). **5:**Range of host(82), GA(1307,1438), MS(2720), NC(1303,1307). **12:**MS(2720). **13:**CA(1154). **14:**CA(1154). **16:**Unk.(3976), CA(1154). **19:**CA(1154). **20:**NC(1303). **22:**CO(1195), NM(1195), UT(82). **23:**CA(81,82,1154), OR(82,3202), WA(82,3202). **24:**CA(1154). **32:**CA(81,1154). **34:**FL(35). **35:**CA(82,661,1154). **40:**Range of host(82). **43:**E states(82), IA(82), KS(82), MS(82), OK(2872). **45:**FL(35). **52:**NC(1303). **53:**NC(1303,3944). **54:**MS(2720). **55:**GA(1307), NC(1307). **58:**CA(1154), CT(82), IA(82), IL(82), IN(82), MD(82), NY(82). **59:**GA(1438), NC(1303,3944), OK(2872). **60:**IL(1052). **65:**NC(1303), OK(2872). **71:**CA(1154). **74:**GA(1307), MS(2720), NC(1303,1307). **75:**FL(35,82), MS(82,2720), TX(82). **77:**Range of host(82), FL(35), WI(1355).
M. quercina. **5:**GA(1438). **12:**GA(1438). **40:**GA(1438). **45:**GA(1438). **74:**GA(1438).
Microthyrium microscopicum. **12:**GA(1438). **59:**GA(1438). **65:**GA(1438).
Mitrula paludosa. On leaves in swamp. **45:**GA(1438).
Mollisia cinerea. **12:**GA(1438). **31:**GA(1438). **77:**CA(81).
M. melaleuca. **59:**GA(1438).
Mycosphaerella aquatica. On fallen leaves. **45:**GA(82,1438), SC(82). **75:**FL(35), GA(82,1438).
M. caespitosa. Leaf spot. **11:**CA(82). **62:**CA(81,638). **73:**CA(81). **75:**FL(35,82), TX(82).
M. catesbeyi. **33:**GA(1438,2399). **77:**GA(82), SC(82).
M. janus. Leaf spot. **62:**OR(3202). **75:**FL(82), TX(82). **77:**FL(35).
M. nigrita. **5:**NY(82). **36:**GA(82,1438).
M. operculata. **11:**CA(179), WA(3202).
M. psilospora. Leaf spot. **77:**IA(1229).
M. punctiformis. Leaf spot. **5:**E states(82), GA(1438), NY(82), WI(82). **12:**GA(82,1438). **22:**NM(82). **36:**GA(82). **40:**GA(82,1438). **43:**GA(82), NY(82), PA(82). **45:**GA(82,1438). **53:**GA(82,1438). **55:**GA(82,1438), NY(82), PA(82). **58:**MA(82), VA(82). **59:**GA(1438). **60:**GA(82), MA(82). **65:**VA(82). **74:**GA(82,1438), MA(82).
M. spleniata. Leaf spot. **5:**GA(82,1438), NY(82), WI(1361). **7:**GA(1438). **8:**NY(82,195), WI(1349). **22:**CO(82), UT(82). **38:**KS(82), NE(82), WI(82). **39:**GA(1438). **54:**NY(82). **55:**GA(82,1438), NY(82). **65:**GA(82,1438).
Mycosphaerella sp. Leaf spot. **23:**OR(82,3202). **60:**Unk.(82). **74:**Unk.(82). **77:**OK(2872).
Mycothyridium floridanum. **77:**FL(2785).
Myriangium duriaei. **45:**GA(1438). **77:**NC(3944).
Myriophacidium aphyophyllicum. On leaves. **62:**OR(3224).
Nectria cinnabarina. On wood. **5:**E states(82), MI(82). **12:**IN(82), PA(82), VA(82). **20:**AL(82). **33:**FL(35). **45:**AL(82), GA(82,1438), MS(82). **60:**IN(82), PA(82), VA(82). **77:**FL(35), GA(1438).
N. coccinea. **23:**OR(3202).
N. discophora. On dead limbs. **12:**MD(82). **74:**GA(82,1438).
N. episphaeria. Canker. **52:**NC(1533). **77:**CA(81).
N. flammea. **45:**GA(1438). **52:**GA(1438).
N. galligena. Trunk canker. **5:**NE states(82), WV(82). **8:**NE states(82), WV(82). **12:**CT(82), MA(82), NH(82), VT(82), WV(82). **23:**OR(82,3202). **55:**CT(82). **60:**CT(82), MA(82), NH(82), VT(82), WV(82).
N. peziza. On wood. **77:**CA(81), LA(2182), MI(3085), PA(3085).
N. punicea. On bark. **22:**OR(82). **23:**OR(3202).
Niesslia pulchriseta. **77:**GA(1438).
Nitschkia acanthostroma. On dead limbs. **77:**NC(1303).
N. confertula. **77:**GA(1438).
Nummularia broomeiana. On trunks and larger limbs. **77:**SE states(1810).

N. repanda f. *querceti*. **38**:ND(377).
N. signata. On trunks and larger limbs. **77**:SE states(1810).
Nummularia sp. **5**:GA(1438).
Ophiostoma leptographioides. Isolated from heartwood. **77**:OH(778), VA(128,778,3714).
O. megalobrunneum. On wood. **46**:MS(798). **77**:MS(3714).
O. piceae. Isolated from timber beetle galleries. **5**:WV(3916).
O. pluriannulatum. Sapwood stain. **17**:MN(484). **20**:MS(82,2720). **60**:IN(82), NC(3714). **77**:IN(1691), WV(3239).
O. rostrocylindricum. Isolated from heartwood. **77**:CT(128,778).
O. stenoceras. Isolated from heartwood of living trees. **52**:NY(82), VA(82). **77**:NJ(778), NY(82,778), VA(82,778).
Orbilia luteorubella. On bark of living trees. **20**:NC(1303).
Ostreichnion curtisii. **45**:GA(1438).
O. sassafras. **77**:GA(1438).
Ostreion sp. **5**:GA(1438).
Patellariopsis clavispora. **77**:GA(1438).
Penzigia frustulosa. **77**:NC(3944).
Perrotia flammea. **23**:OR(3202). **77**:OR(3202).
P. succina. On limbs. **77**:CA(849).
Pezicula cinnamomea. **77**:LA(2182).
Phaeosaccardinula martinii. On leaves. **34**:FL(82). **75**:MS(82,2720).
Phillipsiella atra. On living or fallen leaves. **34**:FL(82).
P. nigella. On dead leaves. **2**:CA(2801).
Phyllachora sp. Tar spot. **75**:FL(35). **77**:FL(35).
Phyllactinia guttata. Powdery mildew. **2**:CA(82,1154). **5**:NC(82,3944), NJ(82). **12**:SE states(82), CA(1154), IL(82), WI(82). **13**:CA(1155). **17**:WI(1352). **20**:NC(1303). **32**:CA(1154). **34**:FL(35). **35**:CA(1154,1155). **38**:IL(82), MS(82,2720), VA(82). **40**:NC(1303), OK(2872), TX(82). **45**:AL(82), FL(35,82), GA(82,1438), OH(82), SC(82). **52**:SE states(82), IL(82,1052), NC(1303), WI(82). **59**:WI(1330). **60**:SE states(82), CA(1154). **65**:AL(82), FL(82), GA(82,1438), MO(82), MS(82,2720), NJ(82). **74**:SE states(82), IN(82), NC(1303), VA(3493), WI(82). **76**:CA(1154,1155).
Physalospora sp. **77**:NC(1303).
Pichia sp. Isolated from timber beetle galleries. **5**:WV(3916).
Pleurostoma ootheca. On dead limbs. **45**:GA(82,1438). **77**:NC(82).
Plochmopeltis intricata. Black mildew, leaf spot. **53**:FL(82). **77**:FL(35), MS(2720).
Pocillum americanum. **77**:FL(2782).
Poculum petiolorum. On overwintered leaves. **5**:NY(1960).
P. sydowiana. **23**:OR(3202).
Proliferodiscus pulveraceus. **59**:GA(1438).
Propolidium glaucum. **77**:NJ(3225), PA(3225).
P. shearii. On wood. **77**:CA(527).
Propolis quercifolia. **5**:GA(1438). **45**:GA(1438). **74**:GA(1438).
Propolomyces versicolor. On branches and twigs. **45**:GA(1438). **59**:GA(1438). **77**:GA(1438), SC(634).
Protoventuria quercina. Leaf spot. **5**:NJ(82), NY(82). **12**:NJ(82), NY(82), VA(82). **38**:KS(82). **55**:E states(82), OH(82). **60**:NJ(82), NY(82), VA(82).
Pseudomassaria agrifolia. Leaf spot. **2**:CA(82,175,959).
P. polystigma. On fallen leaves. **5**:GA(82,1438). **12**:GA(82,1438), NJ(175), OH(82,175). **25**:GA(1438). **31**:GA(1438). **39**:GA(1438). **40**:GA(82,1438). **45**:GA(82,1438), NJ(175). **52**:GA(82,1438). **55**:GA(82,1438). **59**:GA(1438). **60**:GA(82), MA(175). **74**:GA(82,1438). **77**:GA(82).
Pseudovalsa lanciformis. **65**:NC(1307).
P. longipes. Twig canker. **2**:CA(82). **5**:IA(82), IL(82), MI(82), NJ(82), PA(82), VA(82), WI(82). **30**:IA(82), IL(82). **53**:E states(82), MO(82), WI(82). **55**:PA(82), WV(82). **74**:GA(1438). **75**:TX(82). **77**:E states(82), MO(82), WI(82).
Pyrenopeziza prinicola. On dead leaves. **55**:LA(965). **59**:GA(1438). **65**:GA(1438).
Requienella seminuda. On bark. **77**:NJ(320).
Rhynchostoma tinctum. **77**:OR(3202).
Rhytidhysteron rufulum. On dead limbs. **45**:AL(82), FL(82), GA(82,1438).
Rhytisma tostum. On leaves. **77**:FL(634).
Rosellinia aquila. On decaying wood. **5**:GA(1438). **68**:CA(2439). **77**:GA(1438), IA(1225).
R. corticium. On dead limbs. **5**:AR(82), GA(82), IN(82), WV(82).
R. glandiformis. **59**:GA(1438). **77**:GA(1438).
R. mammiformis. **23**:WA(3202). **77**:NY(2456).
R. subiculata. Wood. **75**:GA(1438). **77**:GA(1438), NC(3944).
Saccardia quercina. On leaves. **75**:GA(82).
Saccardoella transylvanica. **65**:OK(660).
Schizothyrium discoideum. Leaf spot. **34**:FL(82). **77**:FL(35).
Scutellinia scutellata. **77**:NC(3944).
Sphaerulina myriadea. On leaves. **5**:MD(82). **77**:SC(634).
Sphaerulina sp. On leaves. **17**:WI(1361).
Splanchnonema melanterum. On limbs. **22**:AZ(188). **72**:CO(188).
S. sporadicum. **77**:GA(1438).
Stictis dryophila. **77**:NJ(3225).
Taphrina caerulescens. Leaf blister, curl, gall. **2**:CA(82,2454). **5**:Range of host(82), GA(1438,2405), MA(2454), NC(1303,1533), NY(2454), OK(2872), VA(2454,3361). **9**:KS(2457,2458,3361). **12**:GA(1438,2454,3361), MI(2454), MS(2720), NC(1533), NJ(965), NY(2454). **13**:CA(82,2454). **17**:WI(2454). **20**:AL(2457), FL(35), MS(2720), VA(3493). **22**:CO(82,2454,2457), UT(82), WY(82,1195,2454). **23**:OR(3202). **25**:GA(1438,2454). **29**:GA(3361), NH(2454), NY(2454), VA(2454). **30**:MO(2454), NC(2454). **31**:NC(2454). **32**:CA(2454,3361). **33**:FL(35,2457), MS(2720). **34**:FL(35,2454,3361), GA(2405), MS(1525,2720). **35**:CA(82,2454). **38**:Range of host(82), IA(2454), KS(2454,2458), WI(2454). **40**:Range of host(82), GA(1438), KS(2454,2458,3361), MO(2454), MS(2720), NC(3944), OK(2872), VA(2454). **41**:NC(1303). **45**:AL(285,2454), FL(35,2454), GA(1438,1447), LA(2675), MA(2454), MS(1525,2720), NC(2454). **52**:GA(1438,2454), IL(1052), KS(3361), MS(2720), NC(1303), VA(3493). **53**:AL(2454), GA(1447), MS(1525,2720), NC(1303,2454). **54**:KS(2454). **55**:GA(82,1438), NC(2457). **60**:CT(2454), MA(656,2454), MI(2454), NH(2454), NY(3361), TN(2454). **65**:Range of host(82), IL(2454), NC(1303), SC(2454). **67**:TX(2457). **72**:CO(82,2454). **74**:AR(2454), CT(2454), GA(1438), KS(2458), MA(2454), MO(2454,3361), MS(2454), NC(1303), VA(3493), WI(2454), WV(2454). **75**:FL(35,82), GA(82,1438,2405), MS(1525,2720), NC(1303,1533,2454), SC(82), TX(82), VA(2454). **77**:Range of host(82), CT(3770,3780), FL(35), GA(2405), IN(663), KS(2457,2458), NC(1303,1533,3944), OK(2872).
Titanella macrospora. **77**:NJ(185).
T. pelorospora. **65**:OK(660).
Trabutia erythrospora. Black spot. **2**:CA(82,191). **6**:AZ(82). **11**:AZ(82), CA(82). **27**:AZ(191). **47**:AZ(82). **75**:FL(82), TX(82).
T. quercina. Tar spot. **2**:CA(81). **34**:SE states(191), AL(82), FL(35,82), MS(82,2720). **44**:MS(2720). **45**:SE states(191), MS(82,2720). **53**:AL(82). **75**:SE states(191), FL(35,82), GA(1438), MS(82,2720), SC(82). **77**:FL(35,82,2782).
T. rigida. On wood. **34**:FL(35).
Trematosphaeria confusa. On roots and at base of trees. **77**:FL(318), LA(318).
Trichosphaeria pilosa. **77**:GA(1438).
Tryblidaria sp. **59**:GA(1438).
Tubeufia pezizula. On bark. **20**:NC(1303,1307). **45**:GA(1438). **59**:GA(1438). **74**:GA(1438). **77**:FL(2782), GA(1438), NC(1303).
Typhulochaeta couchii. Powdery mildew. **6**:AZ(1195,3380).
Urnula craterium. On dead limbs. **20**:NC(1303). **77**:GA(1438), NC(3944), VA(780).
Ustulina deusta. On wood. **5**:GA(82), IL(82), MD(82), ME(82), PA(82), VA(82). **12**:IA(82), IN(82), MO(82). **46**:MS(2720). **59**:GA(1438). **60**:GA(82), VA(82). **77**:GA(1438).
Valsa ambiens. On dead leader of sapling. **12**:NJ(3415). **22**:NM(1003). **77**:IA(1226).
V. ceratosperma. Canker. **12**:MS(82), NJ(82,3415), TN(82). **25**:GA(1438). **31**:GA(1438). **38**:KS(3415), OK(2872). **49**:SC(634). **77**:IA(1226), PA(3415), SC(634).
V. coronata. **5**:GA(1438).
V. intermedia. On limbs and twigs. **5**:IL(82), NC(1303). **20**:NC(1303). **52**:IL(82). **60**:IL(82). **74**:IL(82).
Valsaria exasperans. On dead limbs. **77**:Range of host(82).
V. insitiva. **5**:GA(1438), IA(1226). **45**:GA(1438). **55**:GA(1438). **59**:IA(1226). **77**:IA(1226).
V. rubricosa. **5**:GA(1438). **45**:GA(1438). **53**:GA(1438). **59**:GA(1438).
Vertixore atronitidum. On leaves. **2**:CA(2434).
Vibrissea truncorum. **5**:GA(1438).

Winterella albofusca. On limbs. **5:**NJ(82). **65:**NJ(2948), SC(2948). **77:**NY(2948), WV(2948).
Xylaria acuta. At base of trees. **77:**CA(3017), PA(3017).
X. bulbosa. **77:**GA(1438).
X. corniformis. On dead limbs. **77:**NC(1303,1307), SC(1307).
X. cornu-damae. On wood. **77:**GA(1438), VA(3017).
X. cubensis. On wood. **77:**GA(1438).
X. digitata. On wood. **77:**IA(1225).
X. herculea. **77:**GA(1438).
X. hypoxylon. On wood. **5:**Unk.(82). **12:**GA(82,1438). **59:**NC(1303,1307), SC(1307). **74:**MO(82). **77:**GA(1438), NC(1307,3944), SC(1307).
X. longiana. **23:**OR(3202).
X. multiplex. On wood. **5:**Unk.(82). **31:**GA(1438).
X. poitei. On wood. **77:**FL(3016).
X. polymorpha. On wood. **5:**Unk.(82). **77:**NC(1307).

Basidiomycotina - Uredinales

Cronartium conigenum. Heteroecious: telial host. **6:**AZ(82,1203,3981). **18:**AZ(1195,1203,3981). **26:**AZ(1195,1203,3981). **47:**AZ(1203,3981). **51:**AZ(1203,3981). **61:**AZ(1203,3981).
C. quercuum. Heteroecious: telial host. **2:**CA(82). **5:**E states(82), GA(1439), MO(82), WI(82). **8:**GA(82), MD(82), MS(82,2720), WI(82). **11:**CA(81). **14:**CA(82,2439). **23:**CA(81,82). **30:**E states(82), AR(82), TN(82). **34:**FL(35). **38:**IA(82), KS(82), OK(82,2873), SD(82), WI(82). **40:**E states(82), AR(82), GA(1439), MS(2720), TN(82). **45:**E states(82), AR(82), FL(35), GA(1439), LA(658), TX(82). **52:**E states(82), Range of host(82), GA(1439), MN(82), TN(82). **53:**E states(82), LA(658). **54:**KS(82). **56:**GA(1439). **59:**GA(1307,1439), NC(1303,1307). **65:**GA(1439). **74:**E states(82), Range of host(82), GA(1307,1439), MN(82), NC(1307), TN(82). **75:**SE states(82), FL(35), TX(82). **77:**E states(82), Range of host(82), FL(35), LA(658), MN(82), NC(3944), OK(2872), TN(82).
C. quercuum f. sp. *fusiforme*. Heteroecious: telial host. **5:**MD(82). **8:**MD(82). **12:**NC(82,1303), VA(82), WV(82). **20:**NC(82,1303), VA(82), WV(82). **34:**AL(82), NC(82,1303). **45:**FL(739), NC(82). **52:**NC(82), VA(82), WV(82). **53:**NC(1303). **60:**NC(82), VA(82), WV(82). **65:**AL(82), FL(82), GA(82), MS(82,2720). **74:**NC(1303). **75:**FL(35,82), GA(82), MS(82,2720). **77:**FL(35).
C. strobilinum. Heteroecious: telial host. **5:**AR(82), IL(82), MS(82,2720). **34:**FL(82). **38:**IA(82), KS(82), MO(82). **45:**FL(82). **65:**MS(2720). **75:**FL(35).
Cronartium sp. **77:**OR(3202).

Basidiomycotina - Other

Abortiporus biennis. On wood. **23:**OR(3202). **52:**IL(82). **59:**NC(1303). **77:**AR(2675), LA(2675).
Acanthophysium bertii. On wood. **18:**AZ(1202). **22:**NM(2083). **27:**AZ(128,1202). **57:**AZ(1202).
A. cerussatum. On wood. **6:**AZ(1202). **22:**CO(2135). **77:**AZ(128), NM(2083), SD(2083).
A. diffissum. **77:**AZ(1202).
Aleurocystidiellum disciforme. On bark and wood. **6:**AZ(1202). **11:**CA(82,2083). **47:**AZ(1202). **77:**CA(2083), OR(2083,3202).
Aleurodiscus oakesii. On bark. **5:**AL(2083), IA(2083), MA(2083), MI(2083), NC(3944), PA(2083), TN(2083), VA(638,2083), WI(1360,2083). **6:**AZ(1202). **27:**AZ(128,1202). **38:**E states(82), MO(82), WI(82,1360). **57:**AZ(1202). **65:**NC(82,3944). **70:**AZ(1194). **77:**Unk.(82), AZ(128).
Amphinema byssoides. **22:**CO(2136), NM(1190).
Antrodia albida. Brown rot. **5:**GA(82), IL(82), LA(82), NJ(82). **18:**AZ(1202). **22:**AZ(1202). **27:**AZ(1202). **57:**AZ(1202). **77:**AZ(1202), OK(2872).
[*A. crassa*]. **23:**WA(3202).
A. heteromorpha. **70:**AZ(1202).
A. malicola. **77:**FL(128).
A. oleracea. Isolated from mine timbers. **77:**Unk.(1006).
A. radiculosa. On wood. **77:**NC(128,1303).
A. variiformis. **23:**OR(3202).
Antrodiella incrustans. **77:**NC(1307).
A. semisupina. **12:**NY(82). **74:**NY(82).
A. versicutis. On wood. **77:**SE states(2681), Unk.(1204), FL(2681), MS(2681).
Aporpium caryae. On wood. **77:**FL(2225).
Armillaria mellea. Spot anthracnose, butt, collar and shoestring root rot. **2:**CA(82,2898). **5:**NC states(82), NE states(82), IL(82), NC(1303), PA(3775), VT(59), WI(2877). **8:**NE states(82), IL(82), NC(82). **12:**PA(3458,3775), VA(3458). **14:**CA(82). **19:**CA(82). **23:**CO(2137), OR(82,3202). **27:**AZ(1202). **34:**FL(35,82). **35:**CA(82,2898). **36:**E states(82), NC states(82), IL(82). **38:**NC states(82), NE states(82), IL(82). **45:**LA(82). **55:**E states(82), WV(82,3775). **59:**PA(3458,3775), VA(3458), VT(59), WI(2877). **65:**NC states(82), NE states(82). **74:**Unk.(82), CT(59). **75:**CA(2897). **77:**CA(2898), CT(912), GA(1307), NC(1303,1307,3944).
A. putrida. On dead tree. **77:**CA(2439).
A. tabescens. Root rot, on wood. **5:**FL(82), MO(82). **20:**FL(35). **33:**FL(35). **34:**FL(35,82). **38:**OK(82,2872). **40:**OK(2872). **45:**FL(35,82), GA(2684,2963), MS(1055), OK(82). **53:**GA(2684). **54:**OK(82). **59:**NC(1303). **65:**GA(2684), OK(82). **75:**FL(35). **77:**FL(35,82), MO(82), NC(1303), OK(82), TX(82).
Asterostroma cervicolor. **6:**AZ(1202). **77:**LA(2675).
A. laxum. **77:**MD(128).
Athelia bombacina. On wood. **22:**CO(2135).
A. coprophila. **22:**CO(2136).
A. epiphylla. On wood. **27:**AZ(1178). **77:**MD(128).
A. salicum. On bark. **77:**WA(128).
A. teutoburgensis. On wood. **57:**AZ(1202).
Auricularia auricula. On wood. **5:**VA(914). **18:**AZ(1202). **22:**AZ(1202). **27:**AZ(1202). **45:**GA(914). **57:**AZ(1202).
Basidiodendron caesiocinereum. **18:**AZ(1202).
B. eyrei. On wood. **77:**IA(2185).
Bjerkandera adusta. On wood. **5:**Range of host(82). **22:**AZ(1202). **23:**OR(3202). **77:**AR(2675), LA(2675), MS(2675), NC(3944).
B. fumosa. On wood. **55:**GA(491,1439).
Bolbitius reticulatus. **77:**NC(3944).
Bondarzewia berkeleyi. White stringy butt rot, heart and root rot. **5:**E states(82), AR(2171), MO(82), NC(1303,3915,3941). **12:**NC(1303,1307). **23:**OR(3202). **59:**NC(1303). **65:**GA(491,1439). **74:**AR(2171). **77:**E states(82), AR(82), OR(4007), WI(82).
Botryobasidium botryosum. On wood. **22:**CO(2135).
Botryohypochnus chordulatus. On wood. **75:**FL(443).
B. isabellinus. **27:**AZ(128,1202).
B. verrucisporus. On dead limbs. **6:**AZ(438).
Byssocorticium atrovirens. On wood. **77:**IA(128).
Byssoporia terrestris. On roots. **77:**CA(128).
Calocera cornea. **77:**NY(2456), OR(3202).
Ceraceomyces serpens. On wood. **22:**CO(2137). **27:**AZ(1202).
Ceratobasidium cornigerum. On wood. **22:**CO(2137).
Cerinomyces pallidus. On wood. **77:**IA(128).
Ceriporia alachuana. **6:**AZ(1202). **18:**AZ(1202).
C. excelsa. On wood. **22:**CO(2135).
C. purpurea. **2:**CA(128). **18:**AZ(1202). **23:**OR(3202). **57:**AZ(1202). **74:**MO(82). **77:**AZ(1202).
C. reticulata. **22:**CO(2136), NM(1190).
C. spissa. On wood. **5:**IN(82), MI(82), VA(82). **11:**CA(81). **77:**CA(128).
C. tarda. **5:**IN(82), MD(82), PA(82), VA(82). **22:**CO(2136). **27:**AZ(128,1202). **30:**PA(82). **40:**PA(82). **57:**AZ(1202). **77:**AZ(1194,1202).
C. xylostromatoides. **77:**AZ(1202).
Ceriporiopsis mucida. **77:**OR(4007).
Cerocorticium confluens. **22:**CO(2136).
C. fuscum. On wood. **77:**PA(1239).
Cerrena unicolor. On wood. **5:**FL(82), NC(82,1303,3944), VA(82), WI(82). **23:**OR(3202). **74:**WI(82). **77:**GA(82), OK(2872).
Chionosphaera apobasidialis. On limbs. **38:**IL(700). **65:**IL(700).
Chondrostereum purpureum. On wood. **23:**OR(3906). **65:**OK(660). **77:**GA(491,1439), MA(543), OR(3202).
Claudopus affinis. **77:**OR(3202).
C. subargillaceus. **77:**OR(3202).
Clavicorona pyxidata. **5:**DE(884). **77:**NC(884), TN(884), VA(884).
Climacodon pulcherrimus. On wood. **5:**AR(82), IA(82), LA(82), MD(82). **34:**FL(128). **36:**AR(82), IA(82), LA(82), MD(82). **38:**AR(82), IA(82), LA(82), MD(82). **77:**AR(2675), FL(128), LA(2675).
C. septentrionalis. On wood. **12:**MI(82), NY(82). **45:**MS(2720). **53:**MS(2720).
Clitocybe eccentrica. On acorns and wood. **77:**TN(273).

C. setiseda. On dead portion of trunk of living tree. **74:**NC(1303).
Collybia dichrous. **5:**NC(3944).
Coniophora olivacea. **22:**NM(1190).
C. puteana. Brown "dry" rot. **23:**WA(3202). **60:**IA(82), IL(82), PA(82), VA(82). **77:**Unk.(1006), NY(128).
Coprinus disseminatus. **5:**NC(3944).
C. micaceus. Isolated from mine timbers. **77:**Unk.(1006), NC(1307).
Coriolopsis rigida. On wood. **34:**FL(82). **64:**FL(213). **77:**NC(1303).
Corticium maculare. Smooth patch. **5:**NC(82). **65:**NC(82).
C. roseum. **32:**CA(642).
Corticium sp. Limb blight. **77:**FL(35).
Crepidotus paxilloides. On rails. **77:**NC(1544).
C. smithii. On logs. **77:**TN(1543).
C. tennesseensis. On decayed log. **77:**TN(1543).
Crinipellis zonata. **77:**NC(594).
Crustoderma corneum. **27:**AZ(1202).
C. dryinum. **77:**AL(2555).
C. flavescens. On wood. **77:**MD(2559), WI(2559).
Cyathus striatus. **77:**NC(1307,3944).
Cyclomyces iodinus. On wood. **77:**SE states(2681).
Cylindrobasidium laeve. **77:**DC(128).
C. torrendii. On wood. **15:**GA(491,1439).
Cyphellopsis anomala. **57:**AZ(128,1202). **77:**GA(1439).
Cyptotrama asprata. **27:**AZ(1202).
Cystostereum murrayi. **77:**OR(3202,4007).
C. pini-canadense. Isolated from mine timbers. **77:**Unk.(1006).
Cytidia stereoides. On wood. **32:**CA(642).
Dacrymyces capitatus. On wood. **77:**AZ(1202), GA(2642).
D. minor. On wood. **18:**AZ(1202). **22:**CO(2135). **59:**NC(128). **77:**IA(128), NC(1303).
D. stillatus. On wood. **11:**CA(642). **32:**CA(642).
Dacrymyces sp. **77:**GA(1439).
Dacryobolus karstenii. **22:**NM(1190).
[*D. sudans*]. **27:**AZ(1202). **57:**AZ(1202).
Dacryopinax spathularia. On wood. **18:**AZ(1202). **55:**VA(128). **74:**NY(128). **77:**FL(128), GA(491,1439), MD(128), MO(128), PA(128), VA(128), WV(128).
Daedalea quercina. Brown heart rot. **5:**NE states(82), MN(82), MO(82). **8:**MD(82). **55:**PA(82). **58:**NY(82). **59:**NC(1303). **74:**VA(638). **77:**E states(82), GA(491,1439), MN(82), MO(82), NY(128), OR(3202), TX(652).
Daedaleopsis confragosa. **20:**AR(82), LA(82). **38:**E states(82), AR(82). **60:**NY(82). **77:**LA(2675), NC(3944).
Datronia mollis. On wood. **5:**NC states(82), NE states(82).
D. scutellata. **6:**AZ(128,1202), NM(1190). **27:**AZ(128,1202). **77:**AZ(128), NC(1307).
Dendrocorticium roseocarneum. On wood. **6:**AZ(1202). **18:**AZ(1202). **77:**GA(491,1439), NC(3944).
Dendrophora albobadia. On wood. **5:**VA(82). **6:**AZ(1194,1202). **22:**NM(128,1190). **45:**GA(1439). **77:**GA(543,1439).
D. versiformis. On wood. **5:**MD(82), PA(82,543), VA(82). **29:**NY(543). **77:**NJ(543).
Dendrothele acerina. On bark. **5:**MD(82,128), NY(82). **38:**IA(2084). **75:**TX(82).
D. alliacea. On bark. **38:**IA(2084).
D. candida. Smooth bark patch. **5:**CT(2084), GA(1439), MD(82), NC(82,1303,1307,3944), TN(82), VA(82). **11:**CA(82,638,2084). **20:**FL(35). **45:**NC(1303). **49:**SC(634). **55:**PA(2084). **59:**NC(128). **65:**NC(82,1307,3944). **74:**WV(2084).
D. griseo-cana. On bark. **5:**IA(82), PA(2084). **6:**AZ(1194,1202). **38:**IA(2084). **70:**AZ(128,1194,1202). **77:**FL(2084).
D. macrodens. On bark. **32:**OR(2084).
D. microspora. On bark. **5:**RI(2084). **77:**MA(2084), MD(2084), MO(2084), PA(2084).
D. strumosa. On bark. **77:**FL(2084).
Dichomitus campestris. **6:**NM(1190). **77:**AZ(128,1202).
Dichostereum effuscatum. **5:**CT(128).
D. pallescens. **77:**FL(128).
Ditiola radicata. **77:**AZ(1202).
Donkioporia expansa. **77:**OH(1204).
Ductifera sucina. On limbs. **77:**NC(2645).
Exidia cokeri. On limbs. **77:**NC(2640).
E. glandulosa. On wood. **22:**AZ(1202), CO(1195,2135). **74:**GA(491,1439). **77:**CA(2439), GA(491,1439), OR(3202).
E. nucleata. On limbs. **27:**AZ(1202). **68:**CA(2439). **77:**NC(1303).
E. recisa. On wood. **5:**GA(491,1439). **77:**GA(1439), NC(3944).
E. saccharina. **77:**AZ(1202).
E. tremelloides. On corticate limbs. **77:**LA(2643).
Exidiopsis grisea. **32:**CA(3853). **35:**CA(3853). **77:**CA(3853).
E. molybdea. On fallen logs. **22:**AZ(1202). **77:**IA(2318), NC(2640).
E. podlachica. On wood. **77:**GA(2642), IA(128).
Fibulobasidium inconspicuum. On limbs. **75:**LA(156).
Fibulomyces mutabilis. **22:**CO(2136).
Fistulina hepatica. Brown rot. **5:**NE states(82), NC(1303), TN(82). **12:**GA(1439), NJ(82), PA(82), VA(82), WV(82). **55:**VA(82). **60:**NJ(82), PA(82), VA(82), WV(82). **74:**NC(1303), NJ(82), PA(82), VA(82), WV(82). **77:**NC(1303,1307).
Flammulina velutipes. Sapwood rot. **60:**IN(82).
Fomes fasciatus. On wood. **5:**AL(82). **6:**AZ(1202). **34:**FL(35,82). **45:**FL(35,82), TX(82). **53:**MS(2720). **75:**FL(82). **77:**AZ(128), FL(35,82), GA(491,1439), LA(2675), TX(82).
F. fomentarius. White heart rot. **5:**AL(82). **20:**NC(82). **40:**VA(82).
Fomitella supina. **75:**AL(82), FL(82), LA(82). **77:**GA(1439), LA(2675).
Fomitopsis meliae. **77:**LA(2675).
F. spraguei. Brown cubical rot. **5:**NE states(82), LA(82), WI(82). **23:**OR(82,3202), WA(82,3202). **40:**IN(82). **55:**IN(82). **74:**NE states(82), NC(1303), WI(82). **77:**E states(2533), NE states(82), AR(2675), GA(491,1439), LA(2675), NC(1303), WI(82).
Fuscocerrena portoricensis. **20:**AR(82), LA(82). **77:**LA(2675).
Galerina castaneipes. On rotting logs. **77:**OR(3334).
G. marginata. On wood. **77:**KS(3193).
G. nigripes. On logs. **77:**OR(3334).
Ganoderma annularis. On wood. **77:**CA(1204).
G. applanatum. White rot. **2:**CA(82). **5:**C states(82), NE states(82), GA(491,1439), MO(82). **12:**GA(491,1439). **20:**SE states(82), IN(82), MN(82). **23:**OR(82,3202). **30:**MD(82). **34:**FL(35,82). **40:**IN(82). **55:**GA(491,1439). **75:**FL(35). **77:**E states(82), GA(1439), IN(82), ME(3242), MN(82), NC(3944).
G. australe. On wood. **34:**FL(35,82). **75:**FL(35,82).
G. brownii. On wood. **77:**CA(1204).
G. lobatum. On wood. **5:**AR(82), LA(82), MS(82), NC(82). **18:**AZ(1202). **40:**TX(82). **46:**MS(2720). **57:**AZ(1202). **60:**NY(82). **77:**GA(491,1439), IN(654), LA(2675), MO(82), MS(2675,2720), PA(82), SC(1307), TX(82).
G. lucidum. White root and butt rot. **2:**CA(82). **5:**AR(82), GA(82), MS(82,2720), NC(82), TX(82). **12:**IN(82), MD(82). **17:**LA(82). **18:**AZ(1202). **20:**SC(82), VA(82). **23:**OR(3202). **27:**AZ(17,18,1202). **35:**CA(82). **36:**MS(2720). **38:**IA(82). **45:**FL(35), GA(1439), LA(82), MS(82,2720), SC(82). **46:**MS(2720). **52:**MD(82). **53:**MS(2720), VA(82). **60:**IN(82). **74:**SC(1307). **77:**AR(2675), GA(491,1439), LA(2675), MS(2675), NC(1303,3944).
G. zonatum. Root rot. **65:**TX(2613). **77:**FL(35).
Gerronema subclavatum. On acorns and fallen twigs. **77:**TN(273).
Globifomes graveolens. White heart rot. **12:**E states(82), OH(82). **20:**NC(1303). **45:**AL(82), GA(491,1439). **53:**MS(2720). **59:**NC(1303). **60:**E states(82), OH(82). **74:**E states(82), NC(1303), OH(82). **77:**NC(3944).
Gloeocystidiellum fimbriatum. On wood. **45:**FL(436). **77:**FL(436).
G. marianum. On wood. **77:**MD(436).
G. porosum. White rot. **77:**NC(2553).
G. sinuosum. On wood. **65:**LA(436).
Gloeodontia discolor. On wood. **77:**GA(440), SC(440).
Gloeophyllum sepiarium. On wood. **22:**CO(2137).
G. trabeum. **22:**CO(2136).
Gloeoporus dichrous. **2:**CA(82). **5:**AR(82), IN(82), NJ(82), TN(82), VA(82), WI(82). **6:**AZ(1194). **18:**AZ(1202). **23:**OR(3202), WA(82,3202). **27:**NM(1190). **60:**ME(82). **77:**AR(2675), AZ(1202), GA(1439), LA(2675), OK(2872).
Grammothele lineata. On wood. **77:**FL(3079).
Grandinia arguta. **22:**CO(2136).
G. crustosa. On wood. **22:**CO(2135). **27:**AZ(1202).
G. pruni. **6:**AZ(128,1202).
G. spathulata. **77:**AZ(128).
G. subalutacea. On wood. **22:**CO(2135).
Grifola frondosa. White rot. **5:**NE states(82), AR(82,2171). **12:**GA(1439), NC(1303). **20:**VA(82). **45:**GA(491,1439). **55:**VA(82). **60:**MA(82). **77:**LA(2675), NC(1303,3944).
Gymnopilus luteoviridis. On stumps. **77:**TX(1542,3598).

G. rufosquamulosus. On wood. **77:**TX(1542).
G. sapineus. On wood. **22:**CO(2135).
G. subdryophilus. On logs. **77:**FL(1542).
G. subtropicus. On wood. **77:**MS(1542).
Hapalopilus croceus. White heart rot. **5:**E states(82), AR(82,2171), MN(82). **12:**E states(82), AR(82,2171), MN(82). **60:**E states(82), AR(82), MN(82). **67:**AR(2171). **74:**E states(82), AR(82,2171), MN(82). **77:**Unk.(1204,2681), GA(1439).
H. rutilans. White rot. **6:**AZ(1202). **18:**AZ(1202). **22:**AZ(1202,2136). **27:**AZ(1202). **65:**OK(660). **74:**NY(82). **77:**Unk.(2681), AZ(1202), NM(1190).
Helicobasidium candidum. **27:**AZ(1202).
Helicogloea contorta. On wood. **38:**IA(143).
H. lagerheimii. **27:**AZ(1202).
Henningsomyces candidus. On wood. **2:**CA(81).
Hericium americanum. On wood. **27:**AZ(1202). **59:**GA(491,1439). **77:**NC(1303).
H. coralloides. **2:**CA(82).
H. erinaceus. Spongy white rot. **5:**AL(82), AR(82), IA(82), NC(1303), NJ(1237), NY(82), PA(1237). **23:**OR(3202). **27:**AZ(1202). **34:**FL(35,82). **36:**MS(2720). **45:**FL(35,82), LA(82), MS(82,2720). **46:**MS(2720). **53:**MS(2720), NC(82). **55:**VA(82). **59:**GA(1439), VA(1237). **60:**WV(82). **74:**NC(82), PA(1237). **77:**GA(491,1439), MD(1237).
H. erinaceus subsp. *erinaceo-abietis*. On living tree. **77:**VA(442,1237).
Heterobasidion annosum. White root and butt rot. **22:**CO(82). **55:**PA(82).
Hexagonia hydnoides. On wood. **5:**TX(82). **53:**TX(82). **75:**FL(35,82), VA(82). **77:**CA(638).
Hohenbuehelia atrocaerulea. **77:**GA(1439).
H. mastrucata. On wood. **59:**NC(1303).
H. petalodes. **77:**AZ(82), NC(3944).
Hydnochaete olivacea. On wood. **45:**FL(35,82), GA(1439), LA(82), MS(82,2720). **74:**GA(1439), VA(638).
H. tabacina. On wood. **5:**AR(82), LA(82), MS(82,2720). **77:**LA(2675).
Hymenochaete arida. On wood. **22:**CO(2135), NM(1190).
H. badio-ferruginea. **23:**WA(3202).
H. cinnamomea. **27:**AZ(1202). **77:**AZ(128,1202).
H. corrugata. On wood. **45:**GA(491,1439). **77:**Unk.(82), NC(3944).
H. curtisii. On wood. **5:**GA(491,1439). **23:**OR(82,3202), WA(3202). **38:**IA(82), ND(82,378). **40:**LA(82,2675), MS(82,2720). **45:**GA(491,1439). **65:**MD(82), PA(82), VA(82). **77:**Unk.(82), GA(1439), NC(3944).
H. leonina. **6:**AZ(1194,1202).
H. rubiginosa. On wood. **2:**CA(82). **5:**IN(82), MD(82), MN(82), VA(82). **8:**OH(82). **23:**WA(82,3202). **38:**NY(82). **55:**VA(638). **65:**OK(660), TX(82). **77:**Unk.(82), CA(2439), NC(1303), NY(2456).
H. sallei. **77:**NC(3944).
H. spreta. **27:**AZ(1202).
H. tabacina. On wood. **5:**MD(82). **23:**OR(82). **65:**OK(660). **77:**Unk.(82), GA(491,1439).
Hyphoderma amoenum. **6:**AZ(1202). **27:**AZ(1202). **47:**AZ(1202).
H. anasaziense. On wood. **22:**CO(2137).
H. baculorubrense. On bark of living trees. **75:**LA(1184), MI(1184), TX(1184).
H. clavigerum. **27:**AZ(128,1202).
H. cristulatum. On wood. **32:**CA(642).
H. guttuliferum. **6:**AZ(1202). **77:**AZ(128).
H. medioburiense. On wood. **22:**CO(2137).
H. mutatum. **77:**GA(1439), NC(3944).
[*H. pilosum*]. **18:**AZ(1202). **27:**NM(1190).
H. praetermissum. On wood. **6:**AZ(1202). **22:**CO(2135). **27:**AZ(1202).
H. puberum. **77:**OR(3202).
H. roseocremeum. **18:**AZ(1202). **27:**AZ(1202). **57:**AZ(1202).
H. sambuci. **18:**AZ(1202).
H. setigerum. On wood. **5:**Range of host(82). **6:**AZ(1202). **70:**AZ(1194).
Hypholoma fasciculare. On dead wood. **6:**AZ(1202). **23:**OR(82,3202). **27:**AZ(1202). **77:**CA(2439).
H. sublateritium. On wood. **77:**Range of host(82).
Hypochniciellum subillaqueatum. **27:**AZ(1202).
Hypochnicium analogum. **18:**AZ(1202).
H. bombycinum. On wood. **6:**AZ(1194,1243). **18:**AZ(1202). **57:**AZ(1202).
Hypsizygus tessellatus. On wood. **5:**SE states(2421), MD(2421).
H. ulmarius. **77:**NC(3944).
Inonotus andersonii. White spongy heart rot. **2:**CA(82). **5:**E states(82), AR(82), WI(82). **12:**SC states(82), SE states(82), IL(82). **18:**AZ(1181,1194,1202), CA(128). **22:**AZ(82,1181,1202), CO(2135). **23:**OR(82,3202). **27:**AZ(1181,1202). **35:**CA(82). **45:**TX(82). **46:**MS(2720). **47:**AZ(1181,1202). **74:**C states(82), SE states(82). **77:**GA(491,1439).
I. cuticularis. **52:**MD(82). **77:**AR(2675), LA(2675).
I. dryadeus. White root and butt rot. **5:**NE states(82), AR(82), GA(490), MN(82), NC(1303), OK(2872). **20:**NC(1303). **23:**OR(3202). **30:**SE states(82), TX(82). **40:**OK(2872). **45:**E states(82), FL(35), GA(490,2963), TX(82). **53:**MS(2309). **55:**AR(82), MD(82), OK(82), TX(82), VA(82). **60:**SC states(82), SE states(82), OK(82).
I. dryophilus. White heart rot. **5:**NY(2533). **6:**AZ(1181), NM(1190). **18:**AZ(1181,1194,1202). **22:**AZ(1181,1202), CO(2135), NM(1190). **23:**OR(3202), WA(3202). **27:**AZ(1181,1202). **32:**OR(3202). **40:**OK(2872). **47:**AZ(1181,1202). **65:**GA(491,1439). **77:**LA(2675), MI(665).
I. glomeratus. On wood. **23:**OR(82,3202). **34:**FL(35,82). **77:**MT(3202).
I. hispidus. White heart rot. **5:**AR(82), CT(82), MS(82,2720), VA(82). **6:**AZ(1181,1202). **23:**OR(3202). **27:**AZ(1181,1202). **33:**FL(35). **34:**FL(35,82). **46:**MS(2720). **47:**AZ(1202). **53:**LA(82), MS(2312,2720), NC(82,3944). **55:**CT(82), VA(82). **59:**GA(1439), NC(1303). **60:**CT(82). **74:**NC(1303), NY(82). **77:**AR(2675), GA(491,1439), LA(2675), MS(2675), NC(1303), TX(108).
I. jamaicensis. On wood. **6:**AZ(1205).
I. ludovicianus. On wood. **2:**FL(82), LA(82). **11:**FL(82), LA(82). **34:**FL(35,82). **45:**FL(35), LA(82), MS(82,2720). **75:**FL(82), LA(82). **76:**FL(82), LA(82). **77:**AR(2675), FL(82), LA(82,2675).
I. munzii. **18:**AZ(1202).
I. patouillardii. Heart rot of living trees. **6:**AZ(1181,1194,1202). **47:**AZ(1181,1202).
I. porrectus. On wood. **75:**LA(1204).
I. quercustris. On living trees. **45:**LA(284).
I. rheades. White heart rot. **2:**CA(82), OR(82). **5:**E & C states(82), TX(82). **6:**AZ(82). **11:**CA(82), OR(82). **13:**CA(82). **22:**AZ(82), CO(82), NM(82), UT(82). **23:**CA(82), OR(82), WA(82). **35:**CA(82). **40:**LA(82), MO(82), OK(82), TX(82). **45:**TX(82). **47:**AZ(82), CA(82), NM(82). **53:**TX(82). **55:**E states(82), LA(82), OK(82). **60:**E states(82), AR(82). **72:**CO(82). **75:**TX(82). **76:**CA(82), OR(82). **77:**AZ(82), MS(2720), NM(82).
I. rickii. On wood. **77:**SE states(1204).
Irpex lacteus. On dead wood. **5:**Unk.(82). **6:**AZ(82). **22:**NM(1190). **77:**Unk.(82), NC(3944).
Ischnoderma resinosum. On wood. **5:**GA(491,1439), NC(1303), VA(82). **23:**OR(3202).
Junghuhnia nitida. **23:**OR(3202). **59:**VT(2168). **60:**MI(82). **77:**LA(2168,2675), MA(128,2168), MI(2168), PA(128,2168).
Kavinia himantia. On wood. **6:**AZ(1202). **22:**CO(2137). **23:**OR(3202), WA(3202). **77:**AZ(1202), NC(1303).
Laetiporus persicinus. **75:**FL(82).
L. sulphureus. Brown cubical rot. **2:**CA(82). **5:**E & C states(82), AR(82), NC(1303). **6:**AZ(1202). **11:**CA(82). **23:**OR(3202). **32:**CA(661), OR(3202). **34:**FL(35,82). **36:**MS(2720). **46:**MS(2309,2720). **53:**MS(2720). **59:**MN(1197). **75:**FL(35,82), SC(82), TX(82). **77:**Range of host(82), AZ(1202), GA(491,1439), LA(2675), NC(1303,1307,3944).
Laxitextum bicolor. On wood. **11:**TX(82). **45:**TX(82,543). **53:**TX(82). **59:**DC(543). **76:**TX(82). **77:**Unk.(82), AR(2675), FL(543), GA(491,1439), IA(543), LA(543,2675), NY(543), OH(543).
L. incrustatum. **77:**LA(1577).
Lazulinospora wakefieldii. On bark. **45:**FL(439).
Lentinellus cochleatus. On wood. **22:**CO(2135).
L. ursinus. **2:**CA(82). **27:**SW states(2423), AZ(1202). **77:**E states(2423).
Lentinus strigosus. Sapwood rot. **22:**CO(2135). **53:**NC(1303). **75:**FL(35,82). **77:**AR(2675), GA(1439), LA(2675), MS(2675),

NC(1303,3944).
L. tigrinus. Heart rot, sapwood rot, wound rot. **5**:LA(82), MS(82,2720). **45**:LA(82), MS(82,2720). **46**:MS(2720). **74**:NY(82).
L. torulosus. **77**:NC(3944).
L. velutinus. On wood. **5**:LA(82), MS(82,2720).
Lenzites betulina. White rot. **2**:CA(81,82). **5**:Range of host(82). **11**:CA(82), OR(3202). **23**:CA(82), OR(3202), WA(3202). **32**:OR(3202). **45**:GA(1439). **53**:MD(82). **77**:AR(2675), GA(1439), LA(2675), NC(3944).
Lenzites sp. On wood. **38**:OK(2872).
Leptosporomyces galzinii. On wood. **27**:AZ(1178). **77**:AZ(128,1202).
Lopharia cinerascens. **2**:CA(543). **77**:NY(543).
L. papyrina. **77**:FL(634).
Lycoperdon pyriforme. On wood. **77**:GA(1439), NC(1303).
Maireina cinerea. On bark. **77**:CA(2439).
M. texensis. **77**:TX(456).
Marasmiellus nigripes. On acorns and leaves. **77**:KS(3193).
M. ramealis var. *californicus*. **2**:CA(856).
Marasmius atlanticus. On leaves. **34**:FL(3288).
M. capillaris. On leaves. **20**:NC(1303,1307).
M. cohortalis var. *alachuanus*. **77**:FL(3288).
M. copelandii. **77**:CA(856).
M. copelandii var. *olidus*. On veins of fallen leaves. **77**:MI(1222).
M. corrugatus var. *aurantiacus*. On wood. **77**:FL(3288).
M. heliomyces. On leaves, logs and stumps. **75**:FL(3288).
M. insipidus. On fallen leaves. **77**:OH(1222).
M. plicatulus. On decayed leaves. **77**:CA(856,2439).
M. quercophilus. **77**:CA(856).
M. rotula. On bark. **77**:NC(1303,1307,3944).
M. spadiceus. On dead leaves. **77**:MI(1222,1223).
Marasmius sp. On acorns and leaves. **14**:CA(2439).
Megacollybia platyphylla. **22**:AZ(1202), CO(2137).
Megalocystidium lactescens. **23**:WA(3202). **77**:OR(3202).
M. leucoxanthum. **27**:AZ(1202).
Melanoporia nigra. On wood. **5**:GA(491,1439), IN(82), MO(82), VA(82). **12**:CT(82), IL(82), VA(82). **60**:CT(82), IL(82), VA(82). **74**:CT(82), IL(82), VA(82).
Meripilus sumstinei. White rot. **59**:GA(659). **74**:NC(1303). **77**:GA(491,1439), LA(2675), NC(1303).
Meruliopsis corium. **6**:AZ(1202). **22**:AZ(128,1202). **27**:AZ(1202). **57**:AZ(1202). **70**:AZ(1194). **75**:LA(128). **77**:CA(2439), FL(128), GA(1439).
Meruliporia incrassata. Dry rot of timber. **77**:Unk.(82).
Microporellus obovatus. **77**:AR(2675), LA(2675).
Microstroma album. Frosty mildew, leaf spot. **5**:FL(82), IL(82), WV(82). **23**:WA(3202). **30**:MO(82). **36**:IL(82), MS(2720). **38**:MS(2720). **77**:FL(35).
Mycena collybiiformis. On stumps. **77**:NY(3326).
M. galericulata. **77**:OR(3202).
M. meliigena. **77**:NC(3944).
M. mucor. On dead leaves. **77**:MI(3326).
M. subepipterygia. On wood. **77**:FL(3326).
M. subinclinata. On wood. **77**:FL(3326).
M. texensis. On wood. **77**:FL(3326), TX(3326).
Mycoacia fuscoatra. **6**:AZ(1202). **18**:AZ(1202).
M. meridionalis. **77**:SC(443).
Mycorrhaphium adustum. On dead limbs. **5**:MO(82), NC(82), VA(82). **77**:NC(128,1303,1307), PA(128), RI(128), VA(128).
Mylittopsis marmorata. On logs. **77**:FL(3012), LA(3012), SC(3012).
Neolentinus lepideus. **77**:OR(3202).
Odontia sp. On wood. **36**:MS(2720). **46**:MS(2720). **53**:MS(2720).
Omphalotus olearius. Root rot. **2**:CA(81). **5**:NC(1303). **77**:NC(1303).
O. olivascens. On stumps or buried wood. **2**:CA(274).
Oudemansiella canarii. On living and fallen trunks. **75**:FL(3284).
Oxyporus latemarginatus. On wood. **5**:AR(82). **23**:OR(3202). **46**:MS(2720). **77**:MD(2166), MS(2166), OR(4007), VA(2166).
O. populinus. Butt rot, heart rot. **5**:OH(82), VA(82). **60**:MA(82), VA(82).
Pachykytospora tuberculosa. On wood. **5**:PA(82). **6**:AZ(1202). **12**:IN(82), OH(82), VA(82). **27**:AZ(1202). **38**:NE(82). **55**:IN(82). **57**:AZ(1202). **60**:IN(82), OH(82), VA(82). **74**:IN(82), OH(82), VA(82). **77**:AZ(128).
Panellus pusillus. On wood. **77**:Unk.(2681), AR(2675), FL(441), GA(491,1439), LA(2675).
P. serotinus. **77**:OR(3202).
P. stipticus. Sapwood rot. **5**:Range of host(82). **45**:LA(82), MS(82). **59**:GA(1439). **77**:FL(35), LA(2675), MA(1394), NC(1303), OR(3202).
Paullicorticium niveocremeum. **27**:AZ(1202). **59**:MA(656).
Peniophora cinerea. On wood. **2**:CA(82). **5**:GA(1439). **23**:WA(3202). **32**:CA(642). **77**:Unk.(82), NC(1303).
P. decorticans. **23**:OR(3202), WA(3202).
P. fuscomarginata. **6**:AZ(128,1194,1202). **70**:AZ(1202).
P. incarnata. On wood. **23**:WA(3202). **55**:MD(128). **77**:OR(3202).
P. malenconii subsp. *americana*. On wood. **2**:CA(542).
P. nuda. On wood. **6**:AZ(1202). **22**:CO(2135). **27**:AZ(128,1202). **57**:AZ(1202). **77**:AZ(128,1202).
P. pseudoversicolor. On wood. **22**:CO(2135).
P. violaceolivida. On wood. **22**:AZ(1202), CO(2135). **57**:AZ(1202).
Peniophora sp. On rotten wood. **32**:CA(642).
Perenniporia compacta. On wood. **5**:NJ(82), PA(82). **12**:ME(82), VA(82). **36**:MS(2720). **60**:ME(82), VA(82). **65**:GA(491). **74**:ME(82), VA(82). **77**:Unk.(1006), GA(491,1439), ME(82), VA(82).
P. fraxinea. **77**:IL(82).
P. fraxinophila. On living tree. **38**:ND(82), SD(82). **70**:AZ(1194).
P. medulla-panis. On wood. **2**:CA(82). **5**:MD(82), SC(82), VA(82). **18**:AZ(1194,1202). **22**:AZ(1202), CO(2135). **23**:OR(3202). **27**:AZ(1202). **45**:IL(82). **55**:IN(82). **60**:IN(82), MI(82). **70**:AZ(1202). **74**:IN(82), MI(82). **75**:FL(82), SC(82). **77**:GA(1439), NY(128).
P. ohiensis. On wood. **6**:NM(1190). **27**:AZ(1202). **38**:ND(82,378). **47**:AZ(1202). **57**:AZ(1202). **65**:OK(660). **77**:NE states(82), IA(82), OK(2872).
P. phloiophila. On bark of living trees. **75**:FL(1184), LA(128,1184), MS(1184), SC(1184), TX(1184).
P. robiniophila. On trunks. **77**:Unk.(2681).
P. tenuis. On wood. **5**:MD(82), NY(82). **6**:AZ(1202). **18**:AZ(1202). **22**:AZ(1202), CO(2135). **55**:IN(82). **57**:AZ(1202).
Phaeolus schweinitzii. **23**:OR(3202). **77**:GA(1439).
Phaeomarasmius erinaceus. Stringy white rot. **22**:CO(2137).
Phanerochaete affinis. White rot. **23**:WA(3202). **59**:NY(128,433). **60**:CT(128). **77**:CT(128,433), VA(128).
P. arizonica. **22**:CO(2136).
P. avellanea. White rot. **77**:WI(433).
P. burtii. White rot. **6**:AZ(433,1202). **18**:AZ(433,1194). **22**:AZ(128), NM(128). **65**:OK(660). **77**:DC(433), MD(433), VA(433), WI(433).
P. carnosa. **22**:AZ(1202), CO(2136), NM(128,1190).
P. chrysorhiza. White rot. **6**:AZ(1202). **18**:AZ(1202). **77**:FL(433), MD(433), MS(433), SC(433).
P. crassa. White rot. **5**:AR(35,433), GA(491,1439). **6**:AZ(433,1202). **18**:AZ(1202). **77**:Unk.(82), MS(433).
P. ericina. White rot. **77**:GA(433).
P. filamentosa. White rot. **6**:AZ(1202). **22**:AZ(1202). **45**:AZ(433). **77**:AZ(1202), MD(128), MI(128,433), NC(433), NJ(128,433).
P. flavido-alba. White rot. **5**:AR(128). **77**:FL(128), LA(128), NC(128,433), OR(3202).
P. hiulca. White rot. **34**:FL(433). **77**:GA(433).
P. magnoliae. White rot. **77**:FL(433).
P. ravenelii. White rot. **2**:CA(433). **11**:CA(81). **75**:FL(433). **77**:AR(2675), LA(2675).
P. salmoneolutea. White rot. **6**:AZ(433).
P. sanguinea. White rot. **6**:AZ(433). **18**:AZ(433). **23**:CO(2137), OR(128,433). **27**:AZ(433). **45**:FL(433). **59**:MA(656), VA(433), WI(433). **77**:MS(433), NM(1190).
P. sordida. White rot. **22**:CO(2137). **77**:FL(433), NJ(128).
P. subceracea. White rot. **77**:MD(433).
P. tuberculata. On wood. **18**:AZ(1202). **22**:CO(2135). **47**:AZ(1202).
P. velutina. White rot. **18**:AZ(1202). **22**:AZ(1202), CO(2135), NM(128,433). **27**:AZ(433,1202). **57**:AZ(1202). **58**:NH(128). **59**:NH(128,433). **77**:Unk.(82), AZ(128), WI(433).
P. xerophila. **6**:AZ(433).
Phellinus bicuspidatus. White rot. **77**:LA(2169).
P. contiguus. **5**:CT(82), IN(82), PA(82), VA(82).
P. everhartii. White heart rot. **5**:E states(82), IA(82), MO(82), NC(1303). **6**:AZ(82,128,1202). **18**:AZ(1202). **22**:AZ(1202), CA(82), NM(82,128). **23**:OR(128). **30**:IN(82). **40**:GA(491,1439), NJ(82). **45**:GA(491). **47**:CA(82).

53:MS(2309), NC(3031). **55:**NC(82), OH(82), VA(82). **65:**NC(3944). **72:**CO(82), NM(82). **74:**NC(1307,3944), SC(1307). **77:**E states(82), MN(82), OK(2872).

P. ferreus. **23:**OR(82,3202), WA(82,3202). **32:**OR(3202). **35:**CA(82). **77:**MT(3202), OR(128), WA(128).

P. ferruginosus. On wood. **5:**Range of host(82). **18:**AZ(1202). **22:**CO(2135), NM(82,1190). **23:**OR(82,3202). **47:**AZ(1202). **59:**MN(128,1197).

P. gilvus. White rot. **2:**CA(82). **5:**Range of host(82). **6:**AZ(82,1194,1202). **18:**AZ(1202). **23:**OR(82). **32:**OR(3202). **34:**FL(82). **38:**ND(378). **45:**GA(1439). **47:**NM(82). **59:**MN(1197), NC(1303). **75:**FL(82), MS(2720), TX(82). **77:**AR(2675), AZ(82), GA(659,1439), LA(2675), MS(2675), NC(3944), OK(2872), PA(1044).

P. grenadensis. On wood. **75:**LA(1205).

P. igniarius. White heart rot. **5:**Range of host(82). **12:**NC states(82), NE states(82). **20:**NC(82). **22:**AZ(82), NM(82). **23:**OR(82,3202). **40:**TX(82). **46:**MS(2720). **55:**VA(82). **60:**NC states(82), NE states(82). **75:**LA(82). **77:**FL(35), NC(3944), OK(2872).

P. laevigatus. White laminated rot. **23:**OR(3202). **46:**MS(2720). **53:**MS(2720). **74:**VA(638). **77:**GA(491,1439), LA(2675).

P. linteus. On wood. **77:**FL(1205).

P. melleoporus. On wood. **77:**Gulf states(1205).

P. pini. On wood. **75:**FL(35).

P. punctatus. On wood. **5:**NE states(82), MO(82), WI(82). **45:**GA(491,1439). **77:**AR(2675), LA(2675).

P. robustus. White heart rot. **6:**AZ(1202), NM(1190). **20:**NC(1303). **34:**FL(35,82). **40:**GA(491,1439). **45:**LA(82). **46:**MS(2720). **53:**MS(2720). **60:**GA(491,1439), NY(82). **75:**FL(82), NC(82). **77:**FL(35), LA(82,2675), MS(82), NC(82), VA(82).

P. spiculosus. On wood. **20:**NC(82), VA(82,2169). **40:**NC(82). **45:**NC(82). **53:**DE(82), MS(2720), NC(82), VA(82). **77:**GA(1439).

P. viticola. White rot. **5:**SD(82), WI(82). **18:**AZ(1202). **23:**OR(82,3202), WA(3202). **27:**AZ(128,1202). **38:**SD(82), WI(82). **45:**IN(82). **65:**OK(660).

Phlebia albida. On wood. **22:**CO(2135). **27:**AZ(1202).

P. chrysocrea. On wood. **5:**Unk.(262), NC(82). **12:**Unk.(262). **36:**MS(2720). **52:**MD(82), NC(82), VA(82). **53:**MS(2720). **55:**Unk.(262). **59:**Unk.(262). **60:**MD(82), NC(82), VA(82). **74:**Unk.(262). **77:**FL(2165), PA(2165), TN(2165).

P. concentrica. White rot. **2:**OR(82,3202). **23:**OR(3202), WA(3202).

P. incarnata. On wood. **5:**NC(1303). **6:**AZ(128). **20:**FL(35). **55:**NC(1303,1307). **60:**Range of host(82). **77:**Range of host(82), FL(35), NC(3944).

P. lilascens. **22:**CO(2136).

P. ludoviciana. White rot. **27:**AZ(1202). **77:**AZ(128,1194,1202), FL(2556).

P. merismoides. Isolated from mine timbers. **22:**AZ(1202). **27:**AZ(1202). **77:**Unk.(1006), OR(3202,4007).

P. rufa. On wood. **2:**CA(128). **23:**OR(128). **27:**AZ(1202). **68:**CA(2439). **77:**AZ(128), CA(2439), NC(1303), OR(128,3202), VA(128).

P. subochracea. White rot. **6:**AZ(2556). **27:**AZ(2556). **77:**AZ(2556).

P. tremellosus. On wood. **5:**Range of host(82). **20:**FL(35). **23:**OR(128). **45:**LA(82), MS(82,2720). **46:**MS(2720). **74:**NY(128). **77:**FL(35), GA(1439), NC(3944), NY(128,2456).

P. umbrata. **27:**AZ(1202). **77:**AZ(1202).

Phlebiella subnitens. On wood. **77:**IA(2115).

P. tulasnelloidea. On wood. **5:**GA(491,1439).

Phleogena faginea. On wood. **77:**NC(1303).

Pholiota adiposa. Sprout butt rot, heart rot. **5:**Range of host(82). **12:**VA(82). **77:**FL(35).

Phyllotopsis nidulans. **77:**NC(3944).

Physisporinus sanguinolentus. **23:**OR(3202). **77:**OR(3202).

P. vitreus. On wood. **5:**AR(82), LA(82), MD(82). **77:**GA(491,1439).

Piptoporus soloniensis. On wood. **75:**FL(128). **77:**AL(2164), FL(128,2164).

Platygloea laplata. On wood. **22:**CO(2137).

P. sphaerospora. On wood. **20:**NJ(155).

Pleurotus cystidiosus. Isolated from white-rotted sapwood of fire-scarred trees. **46:**LA(2418). **74:**MO(2418).

P. dryinus. **77:**Unk.(82), AZ(1202).

P. minutus. **77:**NC(3944).

P. ostreatus. Sapwood rot, wound rot. **45:**LA(82), MS(82,2720). **46:**LA(63), MS(2720). **53:**MS(2720). **59:**NC(1307). **74:**NC(1303). **77:**Unk.(82), GA(1439), NC(1303,3944).

Plicaturopsis crispa. **77:**NC(3944).

Pluteus atricapillus. **77:**AZ(1202), NC(3944).

P. lilacinus. On wood. **77:**FL(3286), OH(3286).

P. petasatus. On wood. **34:**Fl(3285).

Polyporus arcularius. On wood. **18:**AZ(1202). **22:**AZ(1202), CO(2135). **23:**OR(3202). **77:**NC(1303,3944), NM(1190), OK(2872).

P. badius. **23:**OR(82,3202). **77:**NC(3944).

P. coronadensis. On wood. **27:**AZ(1201).

P. mori. White rot. **5:**Range of host(82). **6:**AZ(1194,1202). **27:**AZ(1202). **45:**GA(491,1439). **59:**NC(1303). **77:**Unk.(82), AZ(1202), NM(1190).

P. tenuiculus. On wood. **77:**SE states(2681).

P. virgatus. On wood. **77:**SE states(2681).

Porodisculus pendulus. **55:**Unk.(82). **74:**NY(82).

Porotheleum fimbriatum. On wood. **11:**CA(81). **27:**AZ(1202).

Postia caesia. **5:**E states(82). **23:**OR(3202). **77:**Unk.(82).

P. ceriflua. **23:**OR(3202).

P. guttulata. **23:**OR(3202).

P. placenta. **77:**NY(128).

P. stiptica. **77:**OR(3202).

P. undosa. **77:**NC(397).

Pseudoclitocybe cyathiformis. On wood. **77:**NC(273).

Pseudofistulina radicata. On wood. **5:**AL(82), NC(430).

Pseudohydnum gelatinosum. On decayed logs. **65:**NC(1303,1307). **77:**NY(2456).

Pulcherricium caeruleum. On wood. **5:**GA(491,1439), NC(82). **45:**GA(1439). **74:**GA(1439). **77:**NC(1303).

Punctularia atropurpurascens. On stumps. **77:**FL(3211).

Pycnoporus cinnabarinus. On wood. **5:**Range of host(82), GA(1439), SC(1307). **18:**AZ(1194,1202). **22:**CO(2136), NM(82). **34:**FL(2319). **45:**GA(1439). **53:**MS(2720). **77:**Unk.(82), AR(2675), FL(35), GA(1439), LA(2675), NC(3944).

P. sanguineus. **18:**AZ(1194,1202). **52:**LA(82). **77:**Unk.(82), LA(2675), NC(3944).

Ramaricium flavomarginatum. On bark and wood. **23:**WA(1233).

R. polyporoideum. On wood. **77:**FL(430), GA(128).

Rectipilus fasciculatus. **77:**PA(128), WV(128).

Resinomycena rhododendri. **77:**NC(3944).

Resupinatus applicatus. **18:**AZ(1202). **22:**CO(2136). **27:**AZ(1202). **57:**AZ(1202). **70:**AZ(1194,1202). **77:**TX(652).

Rigidoporus lineatus. On wood. **5:**AR(82), LA(82), MS(82,2720). **20:**MD(82). **34:**FL(35,82). **36:**AR(82), LA(82). **45:**LA(82), MS(82,2720). **46:**MS(2720). **53:**MS(2720). **55:**Unk.(82). **77:**AR(2675), LA(2675).

R. ulmarius. On wood. **20:**LA(82), MS(82,2720). **34:**FL(35,82). **45:**LA(82), MS(82). **46:**MS(2720). **77:**LA(82), MS(82,2720).

Saccoblastia farinacea. **6:**AZ(1202).

Schizophyllum commune. On wood. **2:**CA(82). **5:**Range of host(82). **23:**CA(82). **52:**MD(82), NY(82), VA(82). **60:**MD(82), NY(82), VA(82). **77:**Unk.(1006), GA(1439).

Schizopora apacheriensis. **18:**AZ(1193,1202).

S. paradoxa. **2:**CA(82). **5:**Range of host(82). **18:**AZ(128). **20:**MD(82). **23:**OR(82,3202), WA(3202). **65:**OK(660). **77:**AR(2675), LA(2675), NC(3944).

Scytinostroma galactinum. White root and butt rot. **5:**AR(3661). **52:**MD(82), VA(82). **60:**MD(82), VA(82). **74:**AR(82), MO(82), OK(82). **77:**AK(2090), MD(669,2090).

S. portentosum. **18:**AZ(1202). **22:**AZ(1202). **27:**AZ(1202). **77:**AZ(1202), CA(128).

Sebacina epigaea. On wood. **77:**GA(2642).

S. helvelloides. At base of living trees. **5:**IA(128). **77:**MA(128).

S. incrustans. At base of living trees. **38:**IA(128).

Septobasidium castaneum. Parasitic on scale insects. **34:**FL(35,82). **45:**FL(35,82), LA(82), SC(82). **53:**LA(82), NC(3944), SC(82). **75:**NC(1303). **77:**FL(35), MS(2720).

S. cokeri. Parasitic on scale insects. **20:**NC(82). **30:**TN(82). **34:**FL(35,82). **45:**FL(35), GA(1439). **53:**NC(82). **55:**NC(1303). **59:**NC(1303). **74:**NJ(82). **77:**FL(35,82).

S. pseudopedicellatum. Parasitic on scale insects. **12:**NJ(82). **34:**FL(35,82). **45:**AR(82), FL(35,82), NC(82), SC(82). **53:**NC(1303). **77:**FL(35).

Septobasidium sp. Parasitic on scale insects. **34:**FL(35). **45:**FL(35). **77:**FL(35).
Serpula himantioides. **77:**NM(1190).
Simocybe haustellaris. **6:**AZ(1194,1202).
Sistotrema brinkmannii. On wood. **22:**CO(2135). **45:**GA(491,1439).
S. coroniferum. **18:**AZ(128,1202).
S. hirschii. **23:**OR(3202).
Skeletocutis nivea. **18:**AZ(1202). **22:**CO(2136). **60:**WI(82). **77:**IN(654).
Spongipellis delectans. **23:**OR(3202). **77:**Unk.(82), OR(4007).
S. pachyodon. White rot. **5:**GA(3045), IA(82), MI(82), NJ(82), SC(3045), TX(82). **20:**GA(3045), SC(3045). **36:**MS(2720). **45:**GA(491,1439). **46:**MS(2720). **53:**MS(2309,2720). **55:**GA(491,1439,3045), SC(3045). **59:**GA(491,1439). **74:**GA(3045), SC(3045). **77:**FL(35), NC(3944).
S. unicolor. White rot. **5:**SE states(82), WI(82). **6:**AZ(82,1202). **18:**AZ(1202). **20:**AL(82). **22:**NM(82). **27:**AZ(1202), NM(1190). **34:**FL(35). **40:**NC(3944), OK(2872), SC(491). **46:**MS(2720). **57:**AZ(1202). **59:**GA(491,1439). **60:**MN(82). **65:**SE states(82), NC(3944), OK(660), WI(82). **74:**NC(82,3944). **77:**AZ(82,1202), LA(2675).
Steccherinum seriatum. On wood. **77:**FL(444).
Steccherinum ciliolatum. On logs. **77:**AZ(128), NC(1303).
S. fimbriatum. On wood. **22:**CO(2135).
S. laeticolor. On wood. **5:**AL(82), LA(82), TX(82). **6:**AZ(1194,1202). **18:**AZ(1202). **36:**AL(82), LA(82), TX(82). **57:**AZ(1202). **70:**AZ(1194,1202). **77:**AZ(128), FL(1175).
S. ochraceum. On wood. **2:**CA(82). **5:**Range of host(82). **12:**IN(82). **20:**MO(82). **43:**AL(82), AR(82). **45:**GA(1439). **55:**IN(82). **65:**OK(660). **68:**CA(2439). **77:**GA(1439), LA(2675), MS(2675).
Stereum gausapatum. White heart rot. **2:**CA(81). **5:**CT(543), IA(543), MA(543), NY(543), PA(543). **12:**DC(543). **18:**AZ(1194,1202). **23:**OR(543,3202). **34:**FL(543). **38:**ND(543). **40:**DC(543). **45:**NY(543). **46:**MS(2720). **52:**MD(128,543), NY(543). **55:**CT(543). **59:**MA(543), MD(543). **60:**CT(543), MA(543), VA(543). **74:**NY(543). **77:**Unk.(82,1006), AR(543,2675), AZ(543), DC(543), GA(491,1439), MD(543), NC(543,3944), NJ(543), NY(543), OH(543), OK(543), TX(543).
S. hirsutum. White rot. **2:**CA(82). **5:**GA(1439), MA(543). **6:**AZ(1202). **14:**CA(82,2439). **18:**AZ(1194,1202). **20:**MD(543). **22:**CO(2136), NM(543). **23:**OR(82,3202), WA(82,3202). **34:**FL(35). **45:**FL(35), GA(1439). **46:**MS(2720). **55:**WV(543). **59:**CT(543), GA(1439), NY(543). **60:**VT(543). **65:**NC(1303,1307). **77:**Unk.(82,1006), AR(2675), AZ(82,662), CA(2439), FL(35,634), GA(491), LA(2675), MA(543,656), NC(3944), NJ(543), NY(543,2456), OK(2872), TX(652).
S. ochraceo-flavum. Isolated from mine timbers. **5:**AR(543), MA(543). **6:**AZ(1202). **27:**AZ(1202). **29:**MA(543), NY(543). **43:**KS(543). **52:**NY(543). **60:**NH(543). **77:**Unk.(1006).
S. ostrea. White rot. **2:**CA(81,82). **5:**CT(543), GA(1439), MD(543), MN(543), NJ(543), PA(543). **27:**AZ(1202). **45:**GA(1439), TX(82,543). **52:**NY(543). **59:**CT(543), GA(1439), WI(543). **60:**CT(543). **65:**OK(660). **75:**FL(35,82). **77:**Unk.(82,1006), AZ(128,543,1202), FL(543), GA(1439), LA(2675), MS(2675), NC(543,3944), NJ(543), NY(543), OK(2872), OR(3202), PA(543).
S. striatum. On wood. **5:**NC(82), PA(82). **34:**FL(543). **77:**IN(543), NC(3944), NY(2456).
Stereum sp. On wood. **11:**CA(642). **32:**CA(642).
Stigmatolemma poriaeforme. **18:**AZ(1202). **27:**AZ(1202). **57:**AZ(1202).
Subulicystidium longisporum. On wood. **22:**CO(2137).
Thelephora albidobrunnea. On base of young trees. **5:**IN(82), NC(82,3944).
T. terrestris. On base of young trees. **5:**IN(82), NC(82).
Tinctoporellus epimiltinus. White rot. **77:**MS(2720).
Tomentella atrorubra. **77:**MD(2039), NY(2039).
T. botryoides. **77:**MD(2039), NC(2039), NY(2039), OH(2039), TN(2039).
T. bryophila. **27:**AZ(1202). **77:**AZ(2039), GA(2039), PA(2039).
T. chlorina. On fallen limbs. **27:**AZ(1202). **68:**CA(2439). **69:**CA(128). **77:**AZ(128,2039).
T. coerulea. **22:**NM(1190). **77:**AZ(2039), MD(2039), NM(2039).
T. crinalis. **77:**IN(2039), LA(2039).
T. ferruginea. On wood. **22:**CO(2135). **77:**AZ(2030,2039), NM(1190,2039), VA(2039), WI(128).
T. fuscoferruginosa. **77:**GA(2039), NC(2039), NY(2036,2039).
T. galzinii. **77:**AZ(2039).
T. griseoviolacea. **77:**TN(2039).
T. lateritia. On wood. **77:**AZ(2032,2039).
T. molybdaea. On wood. **22:**CO(2137).
T. muricata. On wood. **77:**NY(3746).
T. neobourdotii. **77:**AZ(1202,2039), WI(2039).
T. olivascens. **77:**MD(2039), NC(2039), NY(2039), OH(2039), PA(2039).
T. pilosa. On wood. **5:**WI(3746). **22:**CO(2136). **77:**AZ(1202,2039), MD(2039), NC(2039), NY(2039), WI(2039).
T. punicea. **77:**AZ(2039), FL(2039), LA(2039), NC(2039), TN(2039).
T. ramosissima. On wood. **22:**CO(2135). **77:**MI(2039).
T. rubiginosa. On wood. **5:**VA(3746). **27:**AZ(1202). **77:**AR(2039), IN(2039), LA(2039), MD(2039), OR(3202), PA(2039), VA(128).
T. sublilacina. **77:**NC(2039).
T. subtestacea. On wood. **22:**CO(2137).
T. terrestris. **77:**MI(2039).
T. umbrinospora. **77:**AZ(2039), FL(2030,2039).
T. violaceofusca. **77:**GA(2039), MD(2039), NC(2039), NY(2039), VT(2039).
Tomentellastrum brunneofirmum. On wood. **77:**AZ(128,2031).
Trametes cervina. **6:**AZ(1202). **18:**AZ(1202). **27:**AZ(1202).
T. cubensis. On wood. **77:**SE states(2681).
T. elegans. White rot. **5:**AR(82), MO(82), MS(82,2720), TX(82). **20:**FL(82). **75:**GA(82), LA(82). **77:**AR(2675), FL(82), LA(2675), MS(2675).
T. hirsuta. On wood. **5:**Range of host(82). **18:**AZ(1202). **22:**NM(1190). **23:**WA(82). **27:**AZ(1202). **38:**ND(378). **45:**GA(1439). **59:**NC(1303). **77:**Unk.(82), CA(2439), GA(1439), MS(2720), NC(1307,3944), OK(2872), SC(1307).
T. pubescens. On wood. **5:**NY(82).
T. subectypus. On wood. **77:**SE states(2681).
T. subnivosa. On wood. **77:**GA(491,1439).
T. trogii. **18:**AZ(1202).
T. versicolor. White rot. **2:**CA(82), OR(82,3202). **5:**Range of host(82). **6:**AZ(1202). **23:**OR(82,3202), WA(82). **27:**AZ(1202). **32:**OR(3202). **38:**ND(378), OK(2872). **45:**Unk.(82), GA(1439). **46:**MS(2720). **53:**MS(2720). **58:**NY(82). **77:**Unk.(82,1006), AZ(1202), CA(2439), GA(1439), NC(1303,3944), NY(2456), OK(2872).
Trechispora cohaerens. **77:**MN(1197).
T. farinacea. On wood. **22:**CO(2137). **77:**NM(1190).
T. fastidiosa. **27:**AZ(1202).
T. mollusca. **23:**OR(3202). **77:**OR(3202).
Tremella coalescens. On wood. **77:**LA(2643).
T. fimbriata. On wood. **5:**NC(1303). **6:**AZ(1202). **23:**OR(3202). **27:**AZ(1202). **77:**GA(2642), OR(3202).
T. fuciformis. **77:**NC(3944).
T. mesenterica. On wood. **32:**CA(642). **68:**CA(2439). **77:**AZ(1202), CA(2439), GA(1439), OR(3202).
Trichaptum biforme. White rot. **5:**Range of host(82). **23:**OR(82,3202), WA(82,3202). **32:**OR(3202). **59:**NC(1303). **60:**Unk.(82), IN(82). **75:**LA(82).
T. byssogenum. On wood. **77:**S states(2681).
T. sector. **34:**FL(82). **45:**FL(82). **52:**LA(82). **53:**FL(82). **75:**TX(82). **77:**AR(2675), LA(2675).
Tubulicrinis angustus. **77:**MA(3859).
T. calothrix. On wood. **22:**CO(2135).
Tulasnella albida. On wood. **77:**NC(2640).
T. allantospora. On wood. **27:**AZ(1202). **38:**IA(128). **77:**IA(128).
T. bifrons. **77:**IA(128).
T. calospora. On wood. **77:**IA(128).
T. eichleriana. On wood. **77:**PA(2673).
T. pinicola. On wood. **38:**IA(128). **77:**GA(2642).
T. violacea. On limbs. **77:**GA(2642), IA(128).
T. violea. **27:**AZ(1202). **77:**IL(128).
Tyromyces chioneus. **23:**OR(3202).
T. fissilis. On wood. **34:**FL(35,82). **36:**MS(2720). **45:**AR(82), LA(82), MS(82,2720). **46:**MS(2720). **53:**MS(2309,2720). **77:**GA(491,1439), LA(2675).
T. galactinus. On wood. **20:**MD(82). **77:**LA(2675), NC(1303).
T. humeanus. On wood. **77:**FL(2539).
T. pseudolacteus. On dead sticks. **77:**FL(2540).
Vararia investiens. **5:**CT(128). **59:**MA(128). **60:**MA(128).

77:MN(1197).
V. phyllophila. On leaves. **77:**FL(128).
Volvariella bombycina. Wound rot. **20:**GA(82).
V. lepiotospora. On dead roots. **75:**FL(3192,3292).
Wolfiporia cocos. Brown cubical butt and root rot. **5:**LA(82), MS(82,2720), NC(82), VA(82). **12:**VA(82), WV(82). **53:**FL(82). **60:**VA(82), WV(82). **74:**VA(82), WV(82). **77:**WV(1241).
W. dilatohypha. Brown heart rot. **45:**LA(82), MS(82). **55:**VA(82).
Wrightoporia cylindrospora. On wood. **77:**MD(1205).
Xenasma rimicola. **23:**OR(3202).
Xenosperma ludibundum. On bark. **77:**Unk.(2115).
Xeromphalina kauffmanii. On rotten logs and stumps. **77:**MI(3328), VA(3717), WV(3717).
Xylobolus frustulatus. White pocket rot. **5:**CT(543), IN(543), MN(543), PA(543). **36:**MS(2720). **38:**ND(378), SD(543). **43:**KS(543). **52:**NY(543). **53:**IN(543), MS(2720). **59:**CT(543), MN(128), NC(1303), VA(543). **60:**CT(543), VT(543). **74:**NY(543).
X. subpileatus. White pocket rot. **5:**DE(543), NC(1303). **23:**OR(3202,4007). **46:**MS(2720). **53:**MS(2720). **59:**GA(491). **65:**NC(1303,1307). **75:**FL(35).

Deuteromycotina - Hyphomycetes

Acrodictys atroapicula. On wood. **77:**NY(3771).
Alatospora acuminata. On submerged leaves. **77:**IL(3218).
Alternaria sp. On submerged leaves. **77:**IL(3218).
Anguillospora longissima. On submerged leaves. **77:**IL(3218).
Arthrographis cuboidea. Pink stain of lumber. **59:**Unk.(3135).
Articularia quercina. On dead leaves. **36:**MS(82,2720). **38:**MS(82,2720). **55:**MS(82,2720). **74:**IL(82), MS(82,2720).
A. quercina var. *minor*. Witches' broom. **22:**AZ(1499), CO(1499), NM(1499), UT(1499). **51:**AZ(2123).
Bacillispora aquatica. On submerged leaves. **77:**IL(3218).
Beltrania querna. On leaves. **2:**CA(1468).
B. rhombica. On leaf litter. **75:**FL(3548).
Berkleasmium concinnum. **77:**FL(2782).
B. conglobatum. **52:**DE(3543).
Cephalosporium sp. **77:**ME(3242), WV(3239).
Ceratophorum uncinatum. On leaves. **36:**IL(82), IN(82), MO(82). **38:**MS(2720). **65:**IL(82), IN(82), MO(82).
Cercospora polytricha. Leaf spot. **11:**CA(82). **75:**GA(82), SC(82).
C. querci. Leaf spot. **8:**LA(581). **34:**FL(35). **45:**FL(35). **75:**FL(35). **77:**FL(35).
Chaetoblastophorum ingramii. On dead leaves. **45:**AL(2488).
Chaetopsina auburnensis. On fallen leaves. **45:**AL(2494).
C. ramifera. On leaves. **75:**FL(3548).
Chaetopsis grisea. **77:**MI(862).
Chalara quercina. Wilt. **5:**IA(82), IL(82), MD(3793), MN(82), WI(82). **8:**IL(517). **12:**NC(1303). **17:**MN(1113). **20:**IA(82), IL(82), MN(82), WI(82). **30:**IA(82), MO(82). **33:**SC(3933). **38:**IA(82), IL(82), MN(82,1113), WI(82). **45:**NC(1303,1306), SC(3933). **52:**IL(82), MO(82). **53:**Unk.(1094), MO(1094), NC(1303). **55:**MD(3793), PA(1038). **59:**Unk.(1094), IL(513). **60:**IA(82), IL(82), MN(82,1113), WI(82). **65:**IA(82), KS(3237), MO(82). **74:**IA(82), IL(82), IN(82), KS(3238), MN(82), WI(82). **77:**AR(1094,1095,3924), IA(1094,1095), IL(1094,1095), IN(1094,1095), KS(1094,1095,2020), KY(1095), MD(1095,3793), MI(1095,3518,3519), MN(1094,1095), MO(1094), MS(1094,1095), NC(1095), NE(1094,1095,2020), NY(1095), OH(1094,1095), PA(1044,1094,1095,1406,2019), TN(1095), VA(1095), WI(1094,1335), WV(1095,3682).
Cladosporium brevipes. Leaf mold. **5:**NY(82).
C. fumagineum. **77:**WA(3202).
C. herbarum. **5:**IA(82).
Cladosporium sp. On leaves. **65:**NC(1303).
Clavariopsis aquatica. On submerged leaves. **77:**IL(3218).
Codinaea cylindrospora. On leaves. **45:**AL(2503).
C. simplex. On leaves. **45:**AL(2503).
C. unisetula. On leaves. **45:**AL(2503).
Confistulina hepatica. **12:**VA(3462). **74:**GA(3462).
Coniosporium olivaceum. **77:**NC(3944).
Conoplea fusca. **77:**NC(3944).
C. globosa. Trunk canker. **5:**NE states(82), CT(1672), MI(82), MO(82,1672), NY(1672), PA(1672). **23:**OR(82,1672,3202). **38:**PA(1037,1672). **45:**IL(82). **52:**CT(82). **55:**CT(82), PA(1672). **59:**NC(1303). **60:**CT(1672), MA(1672), MN(82), NH(1672), PA(82). **74:**MA(82). **77:**MA(1672), MI(1672), MN(1672), MO(1672), NC(1303), NJ(1672), NY(1672), PA(1672), RI(1672), TX(1672).
C. olivacea. **77:**CT(1672).
Corynespora pruni. **77:**DE(3543).
Cylindrocladium scoparium. On leaves of seedlings. **21:**MS(3360).
Cylindrocladium sp. Root rot. **77:**FL(35).
Dimorphospora foliicola. On submerged leaves. **77:**IL(3218).
Eversia subopaca. On wood. **77:**NJ(3149).
Exophiala jeanselmei. On wood. **77:**ME(3242).
Filosporella annelidica. On submerged leaves. **77:**IL(3218).
Flagellospora curvula. On submerged leaves. **77:**IL(3218).
Fumago vagans. Sooty mold. **23:**WA(3202). **77:**OK(2872).
Fusarium melanochlorum. **23:**OR(3202).
F. oxysporum. Dieback. **59:**VA(3493).
Fusarium sp. Canker. **77:**ME(3242), TX(108), WV(3239).
Gliocladium roseum. **77:**WV(3239).
Gliocladium sp. **77:**ME(3242).
Graphium rigidum. Sapwood stain. **12:**IN(82). **60:**IN(82). **77:**WV(3239).
G. rubrum. Sapwood stain. **5:**OH(82). **36:**AR(82), MS(82,2720), SC(82). **60:**IN(82).
Graphium sp. Sapwood stain. **5:**WV(3916). **17:**MN(484).
Haplotrichum conspersum. On wood. **77:**MA(2131).
H. curtisii. On bark. **77:**MO(2131).
H. ramosissimum. On bark, on railroad ties. **77:**MA(2131), NJ(2131), VA(2131).
Helicoma inflatum. On inner side of bark. **77:**MO(1271).
H. monilipes. On bark. **77:**MI(1271).
Helminthosporium macrocarpum var. *caudatum*. On twigs. **49:**SC(634).
H. variegatum. **77:**DE(3543).
Helminthosporium sp. Leaf spot. **77:**FL(35).
Idriella angustispora. On fallen leaves. **45:**AL(2493).
Lemonniera aquatica. On submerged leaves. **77:**IL(3218).
L. pseudofloscula. On submerged leaves. **77:**IL(3218).
L. terrestris. On submerged leaves. **77:**IL(3218).
Linodochium hyalinum. On leaves. **77:**NY(927).
Macrosporium subglobosum. On leaves. **77:**SC(634).
Nigrospora sphaerica. On twigs. **52:**IL(82). **65:**IL(82).
Oidium candicans. On wood. **22:**CO(2136). **77:**NY(2131).
Oramasia hirsuta var. *ramosa*. On dead needles. **75:**FL(3548).
Ovularia sp. Leaf spot. **55:**KS(82).
Passalora melioloides. Anthracnose, black leaf spot. **34:**FL(82). **75:**FL(35,82), SC(82).
Penicillium funiculosum. **77:**MD(651).
P. primulinum. On dead leaves. **77:**MD(651).
Penicillium sp. **77:**WV(3239).
Phialophora bubakii. Decline. **45:**TX(110). **65:**TX(110). **67:**TX(110). **75:**TX(110).
P. parasitica. Die-back, isolated from vascular tissue. **75:**TX(1494).
Phialophora sp. On wood. **77:**ME(3242).
Phymatotrichopsis omnivora. Root rot. **10:**TX(82). **38:**TX(82). **52:**TX(82). **53:**TX(82). **55:**TX(82). **75:**TX(82).
Polyscytalum cylindroides. On fallen leaves. **45:**LA(965). **75:**LA(965).
P. sericeum. **77:**DE(3543).
Ptychogaster cubensis. Brown heart rot, white heart rot. **45:**FL(35,82). **53:**LA(82). **75:**FL(35,82). **77:**FL(35,82).
Ramularia crypta. On fallen leaves. **20:**SC(82).
Ramularia sp. Leaf spot. **77:**FL(35).
Rhinocladiella sp. On wood. **77:**ME(3242).
Rhynchosporium secalis. Scald. **11:**NC states(82), NW states(82), AL(82), CA(82), PA(82), TX(82).
Sporidesmium sp. Leaf spot. **75:**FL(35).
Taeniolella alta. **77:**IN(1679).
Tetracladium marchalianum. On submerged leaves. **77:**IL(3218).
Trichoderma viride. On wood. **11:**CA(642). **77:**WV(3239).
Trichoderma sp. On wood. **77:**ME(3242).
Trichothecium roseum. **77:**WV(3239).
Tricladium angulatum. On submerged leaves. **77:**IL(3218).
Triscelophorus monosporus. On submerged leaves. **77:**IL(3218).
Tubercularia vulgaris. **77:**NY(2456).
Umbelopsis versiformis. Isolated from roots. **5:**WV(51). **12:**WV(51). **60:**WV(51). **74:**WV(51).
Vargamyces aquaticus. On submerged leaves. **77:**IL(3218).
Verticillium albo-atrum. Wilt. **52:**IL(1558). **59:**IL(1558).

Virgariella fusca. On twigs. **77**:SC(634).
Wiesneriomyces laurinus. On leaf litter. **75**:FL(3548).
Xenosporium berkeleyi. On bark. **77**:NC(1303).

Deuteromycotina - Coelomycetes

Aplosporella linearis. On twigs. **5**:NY(82).
Botryodiplodia gallae. Twig blight, canker, dieback, on galls and associated with scale injury. **2**:CA(82). **5**:Unk.(82), VA(349), WI(1346). **17**:MN(1172). **29**:MA(1390). **37**:CA(2439). **55**:VA(349). **58**:MD(82). **59**:WI(1346). **67**:TX(110). **77**:MI(728), SC(634).
B. ravenelii. Twig blight. **12**:OH(82).
Botryodiplodia sp. Leaf spot. **33**:FL(35). **75**:FL(35).
Chaetophoma quercifolia. On leaves. **2**:CA(82). **45**:NC(3549), SC(82).
C. setigera. On living leaves. **2**:CA(2434).
Cleistophoma dryina. On limbs. **77**:IL(82), PA(82).
Colletotrichum gloeosporioides. **32**:CA(81).
Coniothyrium microscopicum. On twigs. **52**:IL(82). **60**:IL(82). **74**:IL(82).
Coniothyrium sp. Leaf spot. **5**:IL(82). **8**:IL(82). **38**:IL(82). **75**:FL(35). **77**:FL(35).
Coryneum japonicum. **55**:WA(3547). **77**:DC(3549).
C. megaspermum. On dead limbs. **22**:CO(82,3547).
C. umbonatum. Branch and twig canker, dieback. **5**:PA(3547). **12**:PA(3547). **13**:CA(82). **17**:MN(1172). **43**:MD(82). **55**:PA(3547). **60**:MA(82), MI(82). **74**:AR(82). **77**:NY(3547), OK(2872).
Cryptocline cinerescens. Twig blight, leaf spot. **2**:CA(1508). **8**:Unk.(82). **35**:CA(1508). **76**:CA(1508).
Cryptospora leiphaemoides. On dead twigs. **5**:NY(82).
Cylindrosporium kelloggii. Leaf spot. **2**:CA(82). **23**:CA(82). **77**:CA(82).
C. microspilum. Leaf spot. **5**:IL(82), MO(82). **38**:IL(82), MO(82). **52**:IL(82).
C. quercinum. On twigs. **30**:IL(82).
Cylindrosporium sp. **23**:ID(3202).
Cylindroxyphium virginianum. On leaves. **75**:FL(3549).
Cytospora chrysosperma. Twig canker. **52**:OK(2872). **58**:NJ(82), RI(82).
C. pallida. On dead fallen limbs. **74**:NJ(970).
Cytospora sp. Canker. **5**:Unk.(82). **8**:Unk.(82). **17**:MN(1172). **36**:Unk.(82). **38**:Unk.(82). **55**:NJ(82). **60**:Unk.(82). **65**:Unk.(82). **77**:FL(35).
Dichaena strumosa. Branch and trunk canker. **12**:NJ(82). **45**:GA(82,1438), NC(82). **55**:WV(82). **74**:NJ(82), VA(82).
Didymosporium concinnum. **77**:VA(2780).
Diplodia longispora. Twig blight, canker. **5**:E states(82), IL(82), WI(82). **12**:NJ(82), VA(82). **38**:E states(82), IL(82), WI(82). **55**:E states(82). **60**:OH(82). **75**:FL(35), TX(82). **77**:FL(35).
D. quercina. Dieback, death. **2**:CA(1507).
D. suberina. **55**:NJ(82), PA(82).
Discella quercina. Leaf spot. **5**:WI(1343). **12**:NE states(82), NE(82), WI(82). **17**:NE states(82), NE(82), WI(82,1330). **30**:IL(82), MD(82), MO(82), NJ(82), OH(82). **38**:ND(378). **45**:AL(82), OK(82). **52**:IL(1052). **53**:MD(82). **60**:NE states(82), IL(1052), NE(82), WI(82). **74**:NE states(82), IL(1052), NE(82), WI(82).
Discosia artocreas. On fallen leaves. **12**:IL(82), ME(82). **60**:IL(82), ME(82). **77**:PA(3545).
Discosia sp. On submerged leaves. **77**:IL(3218).
Discula umbrinella. Leaf spot. **2**:CA(1468,1508). **5**:CT(82), IA(82), KY(82), MI(82), NY(82), WI(82,1355). **8**:CT(82), IA(82), KY(82), MI(82), NY(82), WI(82,1341). **23**:OR(82). **35**:CA(1508). **38**:CT(82), IA(82), KY(82), MI(82), NY(82), WI(82,1355). **52**:IL(1052). **59**:MS(2720). **65**:IL(2568). **76**:CA(1508). **77**:AK(526).
Discula sp. On leaves. **8**:WI(1355).
Dothiorella phomiformis. Leaf spot. **5**:NE states(82), MO(82), WI(82). **8**:WI(64,82). **17**:WI(1352). **38**:CT(82), IA(82). **43**:IL(82). **54**:NY(64). **60**:NY(82), VT(82). **65**:FL(82). **77**:FL(35).
Dothiorella sp. On twigs. **2**:Unk.(82). **77**:WV(3239).
Endothiella gyrosa. On exposed roots and dead limbs. **12**:NC(1303). **20**:NC(1303). **52**:NC(1303). **53**:NC(1303). **77**:LA(2675).
Fusicoccum ellisianum. Twig canker. **60**:IL(82), MA(82). **74**:NJ(82).
Gloeodes pomigena. On dead twigs. **5**:IN(82).
Gloeosporium quercuum. Leaf blight. **2**:CA(82).
Harknessia caudata. On leaves, petioles and twigs. **5**:SC(3549). **12**:SC(3549). **77**:MD(2783), NJ(2548), PA(2548).
Leptostroma querci. Leaf spot. **30**:IL(82).
Leptostroma sp. Leaf spot. **77**:FL(35).
Leptothyrium californicum. Leaf spot. **77**:CA(82).
L. castaneae var. *quercus*. On dead leaves. **55**:VA(82).
Leptothyrium sp. Leaf spot. **75**:FL(35).
Macrophoma nervicola. Leaf spot. **5**:OK(2872).
Macrophoma sp. Leaf spot. **75**:FL(35). **77**:FL(35), IL(1052).
Macrophomina phaseolina. Charcoal root rot. **77**:FL(35).
Marssonina martinii. Brown bordered leaf spot, on fruits. **2**:CA(81). **5**:E states(82), Range of host(82), MS(82,2720), ND(82), TX(82). **36**:E states(82), Range of host(82), MS(82), ND(82), TX(82). **38**:MS(2720), SD(2243). **43**:IN(82), KS(82), MS(82,2720), OK(82,2872), WI(82). **45**:AL(285), OK(2873). **54**:MS(2720). **55**:E states(82), MO(82), MS(82,2720). **58**:IA(82), IL(82), IN(82), VA(82). **65**:OK(2872).
Marssonina sp. Leaf spot. **77**:FL(35).
Mastigosporella hyalina. Leaf spot. **12**:NJ(2548). **55**:NJ(82).
Monochaetia concentrica. On leaves. **75**:FL(634).
M. hysteriiformis. On leaves. **45**:SC(634).
M. monochaeta. Leaf blister, leaf blotch, zonate leaf spot. **12**:NE states(82), KS(82), LA(965), NJ(965), TN(1509), TX(82). **20**:FL(35). **40**:AR(1509), IL(2568), OK(2872), TN(1509). **44**:FL(1509). **45**:GA(82), MS(82,2720), NC(82). **52**:NE states(82), KS(82), TX(82). **55**:GA(82,1509), NC(82,1509), TN(82,1509). **59**:FL(1509), NC(1509), TN(1509). **60**:NE states(82), GA(1509), KS(82), NC(1509), OH(1509), TN(1509), TX(82). **65**:E states(82), AR(1509), NC(1509), NJ(82,1509), TN(82,1509). **74**:NC(1509), OK(2873), TN(1509). **75**:FL(35,82,1509), LA(82), MS(82,2720), TN(82), TX(82). **77**:E states(82), SC states(82), AK(526), FL(35), MS(2720).
Myxosporium castaneum var. *quercus*. On dead limbs. **55**:NY(82).
M. stellatum. On twigs. **74**:NY(82).
Pestalotia breviseta. **74**:VA(82).
P. maura. On dead leaves. **20**:LA(82), NC(82). **52**:LA(966). **60**:IA(82). **75**:LA(966).
Pestalotiopsis clavispora. **20**:AL(82). **74**:IL(82).
P. montellica. **74**:NC(82), PA(82).
Phloeospora hansenii. Leaf spot. **77**:CA(82).
Phoma cava. Twig canker. **60**:IL(82).
P. dendritica. On fading leaves. **45**:SC(82).
P. discosiiformis. On leaves. **2**:CA(82).
P. quercina. On twigs. **5**:CT(82), MA(82), MD(82), NJ(82), NY(82), WI(82). **38**:CT(82), MA(82), MD(82), NJ(82), NY(82), WI(82). **65**:IL(82).
Phoma sp. Isolated from wood. **17**:MN(1172).
Phomopsis excelsior. On twigs. **74**:WI(1337).
P. glandicola. Dry rot of acorns. **5**:IN(82), MD(82), NY(82), OH(82). **23**:OR(3202,4007). **65**:Unk.(82). **74**:MD(82), NY(82). **75**:FL(82).
P. quercina. Trunk and twig canker, on acorns. **43**:OK(82,2873). **55**:VA(82).
Phomopsis sp. Canker, trunk gall. **5**:MD(82), TX(82), VA(82), WI(82). **34**:FL(35). **38**:MD(82), TX(82), VA(82), WI(82). **65**:MD(82), TX(82), VA(82), WI(82). **74**:MD(82), TX(82), VA(82), WI(82). **75**:FL(35).
Phyllosticta agrifolia. **2**:CA(82). **11**:CA(82).
P. livida. Leaf spot. **8**:WI(1344). **13**:CA(82). **17**:WI(1358). **38**:SD(2243), WI(64,82). **45**:FL(35,82). **65**:OK(2872). **75**:FL(35,82). **77**:FL(35).
P. ludoviciana. Leaf spot. **53**:LA(82). **77**:FL(35).
P. neuroterigallicola. **30**:IL(82).
P. quercea. On fading leaves or associated with distinct spots. **8**:WI(82).
P. quercus. On fading leaves or associated with distinct spots. **5**:WI(64). **38**:WI(64,82). **59**:WI(1342).
P. quercus-ilicis. **75**:MS(82).
P. quercus-prini. Leaf spot. **55**:WV(82). **77**:FL(35).
P. tumoricola. On fading leaves or associated with distinct spots. **5**:NY(82).
P. vesicatoria. On leaves. **77**:SC(82).
P. virens. Leaf spot. **75**:MS(2720).
P. wislizenii. **76**:CA(82).
Phyllosticta sp. Leaf spot. **8**:Unk.(82). **17**:WI(1360). **34**:FL(35). **77**:FL(35).

Pilidium concavum. On dead leaves. **3**:CA(81).
Polychaeton quercinum. On leaves. **49**:SC(634).
Polymorphum quercinum. On twigs and limbs. **2**:CA(82). **5**:AL(82), LA(82), NJ(82), SC(82), TX(82). **12**:NJ(82). **77**:FL(82,634).
Pyrenochaeta minuta. Twig canker. **5**:IL(82). **52**:IL(82).
Rhodosticta quercina. On dead twigs. **52**:IL(82).
[*Seimatosporium rhododendri*]. **55**:NC(1303,1307).
Septogloeum querceum. Leaf spot. **8**:WI(82).
Septoria dryina. Leaf spot. **12**:OK(2872). **20**:SC(82). **40**:OK(2872). **52**:OK(2872). **65**:OR(82,3202).
S. neglecta. Leaf spot. **53**:AL(82). **77**:MS(2720).
S. ocellata. Leaf spot. **23**:OR(82,3202).
S. querceti. Leaf spot. **60**:IA(82). **74**:SC(82). **77**:IA(1229).
S. quercicola. Leaf spot. **2**:CA(82). **11**:CA(81). **38**:WI(1355). **59**:NC(3944). **60**:MN(82), NC(82), PA(82). **75**:CA(82), FL(35). **77**:FL(35), OR(82).
S. quercicola var. *cinerea*. **77**:FL(82).
S. quercus. **5**:IL(82).
S. serpentaria. **34**:FL(82).
Septoria sp. Leaf spot. **53**:Unk.(82). **55**:PA(82). **77**:Unk.(82).
Sphaceloma sp. Spot anthracnose. **34**:FL(35). **45**:FL(35). **75**:FL(35). **77**:FL(35).
Sphaerellopsis filum. On leaves. **75**:GA(3549), MS(82,2720).
Sphaeropsis sp. Twig blight and canker. **33**:FL(35). **45**:FL(35). **55**:CT(82), NJ(82), NY(82), PA(82,1038). **58**:OH(82). **60**:IL(82). **75**:FL(35). **77**:CA(82), FL(35), MS(2720).
Stagonospora septorioides. On leaves. **30**:MS(82,2720).
S. virens. On leaves. **75**:FL(35,82).
Stigmella effigurata. On leaves. **13**:CA(82).
Tubakia dryina. Leaf spot. **1**:LA(1607). **5**:E states(82), IL(1052,1246), LA(1607), TX(82), VA(3493), WI(82). **8**:IL(1246). **12**:NC(1303). **20**:FL(35), IL(1246,2568), LA(1607), MS(2720). **30**:IL(1246). **33**:FL(35). **34**:FL(35). **38**:E states(82), IL(1052), OK(2873), OR(1246), TX(82), WI(82,1332). **40**:IL(1246,2568), OK(82,2873), WV(82). **43**:OK(2872). **45**:FL(35), GA(82), LA(1607,1805), MS(82,2720), OK(82,2873), TX(82). **52**:IL(1246), NC(1303), VA(3493). **53**:LA(658,1607), NC(82), SC(82). **58**:LA(1607). **59**:IL(1246), OK(2873). **64**:IL(1246). **65**:E states(82), IL(1246), MS(2720), OK(2873), TX(82), WI(82). **74**:IL(1052,1246), NC(1303), OK(2873). **75**:FL(35), LA(1607). **77**:E states(82), FL(35), IL(1246), NC(1303), OK(82), WI(82).

Deuteromycotina - Other
Rhizoctonia solani. Root and stem rot. **75**:FL(35). **77**:FL(35).
Rhizoctonia sp. Thread blight. **45**:IN(82). **74**:IN(82).

Flacourtiaceae

Azara Ruiz & Pav.
1. *A. microphylla* Hook. [aromo, chin-chin]

Deuteromycotina - Other
Sclerotium rolfsii. Stem rot. **1**:CA(82).

Dovyalis E. Mey. ex Arnott
1. *D. caffra* (J. D. Hook. & Harv.) Warb. (=*Aberia caffra* J. D. Hook. & Harv.) [kai-apple, kau-apple, kei-apple, umkokolo]

Deuteromycotina - Hyphomycetes
Cercospora sp. Leaf spot. **1**:FL(35).
Fusarium sp. Root rot. **1**:FL(35).

Deuteromycotina - Coelomycetes
Colletotrichum gloeosporioides. Anthracnose. **1**:FL(35).
Phyllosticta sp. Leaf spot. **1**:FL(35).

Flacourtia Commerson ex L'Hér.
1. *F. indica* (J. Burm.) Merr. [Batoko-plum, governor's-plum, Madagascar-plum, ramontchi]

Deuteromycotina - Coelomycetes
Colletotrichum gloeosporioides. Anthracnose. **1**:FL(35).
Phyllosticta sp. Leaf spot. **1**:FL(35).

Idesia Maxim.
1. *I. polycarpa* Maxim. [iigiri-tree]

Deuteromycotina - Hyphomycetes
Cercospora sp. Leaf spot. **1**:FL(35).

Oncoba Forssk.
1. *O. spinosa* Forssk.

Deuteromycotina - Hyphomycetes
Phymatotrichopsis omnivora. Root rot. **1**:TX(82).

Xylosma G. Forster
1. *X. congestum* (Lour.) Merr.

Ascomycotina
Botryosphaeria dothidea. Leaf spot. **1**:FL(35).

Basidiomycotina - Other
Armillaria mellea. Root rot. **1**:CA(2900).

Fouquieriaceae

Fouquieria Kunth in Humb., Bonpl. & Kunth
1. *F. spinosa* Torr. (=*Idria columnaria* Kellogg)
2. *F. splendens* Engelm. [candlewood, coach-whip, Jacob's-staff, ocotillo, vine cactus]
3. *Fouquieria* sp.

Ascomycotina
Lachnella alboviolascens. On wood. **2**:AZ(1202,2558).
Phyllactinia guttata. Powery mildew. **2**:CA(1218).
Xylaria hypoxylon. On wood. **2**:AZ(2558).

Basidiomycotina - Uredinales
Puccinia vexans. Heteroecious: aecial host. **1**:AZ(3981). **2**:AZ(82,747,1195,1203,3981). **3**:AZ(747).

Basidiomycotina - Other
Asterostroma muscicola. On wood. **2**:AZ(2558).
Athelia decipiens. On wood. **2**:AZ(2558).
Ceriporia purpurea. On wood. **2**:AZ(2558).
C. tarda. On wood. **2**:AZ(2558).
C. xylostromatoides. On wood. **2**:AZ(2558).
Cerocorticium confluens. **2**:AZ(1202).
Coprinus papagoensis. On wood. **2**:AZ(2558).
Cristinia sonorae. **2**:AZ(2558).
Dendrophora albobadia. On wood. **2**:AZ(2558).
Heliocybe sulcata. On wood. **2**:AZ(2558).
Hyphoderma fouquieriae. On wood. **2**:AZ(2558).
H. sambuci. On wood. **2**:AZ(2558).
Lentinus strigosus. On wood. **2**:AZ(2558).
Mycoacia austro-occidentale. On wood. **2**:AZ(2558).
Oxyporus corticola. **2**:AZ(1202).
Peniophora tamaricicola. On wood. **2**:AZ(1180,1202,2558).
Phanerochaete allantospora. On wood. **2**:AZ(2558).
P. arizonica. On wood. **2**:AZ(2558).
P. chrysorhiza. **2**:AZ(1202).
P. crassa. On wood. **2**:AZ(2558).
P. omnivora. On wood. **2**:AZ(433,2558).
P. tuberculata. On wood. **2**:AZ(1202,2558).
Phellinus texanus. On wood. **2**:AZ(2558).
Polyporus arcularius. On wood. **2**:AZ(1202,2558).
Pycnoporus cinnabarinus. On wood. **2**:AZ(2558).
Resupinatus applicatus. On wood. **2**:AZ(2558).
Schizophyllum commune. On wood. **2**:AZ(2558).
Theleporus ajovalliensis. Uniform white rot. **2**:AZ(1183).

Deuteromycotina - Hyphomycetes
Phymatotrichopsis omnivora. Root rot. **2**:TX(82).

Deuteromycotina - Coelomycetes
Colletogloeum fouquieriae. On leaves. **2**:AZ(2846,3549).

Fumariaceae

Dicentra Bernh.
1. *D. canadensis* (Goldie) Walp. [squirrel-corn]
2. *D. cucullaria* (L.) Bernh. [Dutchman's-breeches]
3. *D. formosa* (Andr.) Walp. [western bleeding-heart]
4. *D. spectabilis* (L.) Lem. [bleeding-heart]

Oomycetes
Peronospora corydalis. **3**:OR(3202), WA(3202).
P. dicentrae. Downy mildew. **1**:IN(82), MD(82), MI(82), MO(82), NY(82), VA(82), WI(82,1358). **2**:IN(82), MD(82), MI(82), MO(82), NY(82), VA(82), WI(82).

Ascomycotina
Sclerotinia sclerotiorum. Wilt. **4:**MN(82).
Basidiomycotina - Uredinales
Cerotelium dicentrae. Heteroecious: aecial host. **1:**WI(1358). **2:**NE states(82), KS(82), SD(82,2243).
Deuteromycotina - Hyphomycetes
Cercospora sp. Leaf spot. **4:**FL(35).
Fusarium sp. Wilt. **4:**NJ(82).
Stemphylium sp. Leaf spot. **4:**FL(35).
Deuteromycotina - Coelomycetes
Colletotrichum sp. Leaf spot. **4:**FL(35).
Deuteromycotina - Other
Sclerotium rolfsii. Stem rot. **4:**NY(82).

Garryaceae

Garrya Douglas ex Lindl.

1. *G. elliptica* Douglas ex Lindl.
2. *G. flavescens* S. Wats.
3. *G. fremontii* Torr.
4. *G. veatchii* Kellogg
5. *G. wrightii* Torr.
6. *Garrya* sp.

Ascomycotina
Aphanostigme lanugine. On leaves. **1:**CA(82,2434).
Dothidea sambuci. **3:**CA(81,329).
Harknessiella purpurea. On leaves. **1:**CA(82,2801).
Hyalopeziza tautilla. On living leaves. **1:**CA(82,849,2801,3167).
Hypoxylon rubiginosum. **1:**CA(81).
Lachnella alboviolascens. **5:**AZ(1202).
Lambertella garryae. On leaves. **1:**CA(81,527,909).
Meliola garryae. **6:**Unk.(1452).
Morenoina lucens. On living twigs. **1:**CA(82,1468).
Stictis radiata. **6:**CA(3225).
Basidiomycotina - Other
Hyphoderma amoenum. **5:**AZ(1202).
Perenniporia fraxinophila. **5:**AZ(1202).
Deuteromycotina - Hyphomycetes
Cercospora garryae. On leaves. **1:**CA(82,1468).
Janetia garryae. On leaves. **3:**CA(1682).
Phymatotrichopsis omnivora. Root rot. **2:**TX(82).
Podosporiella glomerata. On leaves. **1:**CA(82). **4:**CA(82).
Deuteromycotina - Coelomycetes
Ceuthospora foliicola. On dead leaves. **1:**CA(82).
Dothichiza garryae. Leaf spot. **1:**CA(81,82,325). **2:**CA(82,325).
Phyllosticta garryae. Leaf spot. **1:**CA(82), TX(82).

Gentianaceae

Centaurium J. Hill

1. *C. erythraea* Rafn (=*Erythraea centaurium* of authors, *C. minus* of authors, *C. umbellatum* Gilib.) [centaury]

Deuteromycotina - Hyphomycetes
Taeniolina centaurii. **1:**CA(81).

Eustoma Salisb.

1. *E. russellianum* (Hook.) G. Don (=*E. andrewsii* A. Nels.) [prairie-gentian]
2. *Eustoma* sp.

Deuteromycotina - Hyphomycetes
Alternaria sp. Stem blight. **1:**TX(82).
Cercospora eustomae. Leaf spot. **1:**CO(82,581), NE(82), TX(82,581). **2:**CA(82), TX(82).
Fusarium solani. Root rot. **1:**TX(82).
Deuteromycotina - Coelomycetes
Phyllosticta sp. Leaf spot. **1:**TX(82).
Sclerophoma eustomonis. Stem blight. **1:**TX(82).

Exacum L.

1. *E. affine* I. B. Balf. ex Regel [German violet, Persian violet]

Oomycetes
Pythium sp. Root rot. **1:**FL(35).
Ascomycotina
Nectria haematococca. Stem rot, wilt. **1:**CA(2806).
Deuteromycotina - Hyphomycetes
Botrytis cinerea. Foliage blight, stem canker. 'Botrytis blight.' **1:**FL(35), KS(82).
Fusarium sp. Stem rot. **1:**FL(35).
Deuteromycotina - Other
Rhizoctonia solani. Root rot. **1:**FL(35).
Sclerotium rolfsii. Basal stem rot. **1:**FL(35).

Frasera Walter

1. *F. albicaulis* Griseb.
2. *F. speciosa* Douglas ex Griseb. in Hook. (=*F. macrophylla* Greene, *Swertia radiatum* (Kellogg) Kuntze)
3. *F. thyrsiflora* Hook. (=*F. fastigiata* A. Heller)
4. *Frasera* sp.

Ascomycotina
Didymella delphinii. **3:**ID(3202).
Pleospora comata. **2:**MT(3202).
Basidiomycotina - Uredinales
Uromyces speciosus. Autoecious. **2:**CO(82), NM(82). **4:**CO(82).
Deuteromycotina - Hyphomycetes
Cercosporella fraserae. Leaf spot. **2:**CO(82), ID(1195,3378), WY(1195,3378). **4:**UT(82).
Deuteromycotina - Coelomycetes
Asteroma fraserae. Black mildew. **2:**CO(82), ID(82). **3:**ID(3202).
Cylindrosporium fraserae. Leaf spot. **2:**CO(666), WY(3812).
Didymochaeta fraserae. On stems. **1:**WA(3202). **2:**CO(81,82,3549). **3:**ID(3202), WA(3202). **4:**CO(82).
Marssonina fraserae. Leaf spot. **3:**WA(3202). **4:**ID(82), WA(82).
Phyllosticta fraserae. Leaf spot. **4:**CO(82).

Gentiana L.

1. *G. affinis* Griseb. (=*G. oregana* Engelm. ex A. Gray)
2. *G. alba* Muhl. (=*G. flavida* A. Gray)
3. *G. algida* Pall. (=*G. rumanzovii* Ledeb. ex Bunge)
4. *G. andrewsii* Griseb. [bottle gentian, closed gentian]
5. *G. bigelovii* A. Gray
6. *G. calycosa* Griseb.
7. *G. clausa* Raf. [blind gentian, bottle gentian, closed gentian]
8. *G. crinita* Froelich [fringed gentian]
9. *G. cruciata* L.
10. *G. glauca* Pall.
11. *G. lagodechiana* (Kusn.) Grossh.
12. *G. linearis* Froelich [closed gentian, narrow-leaved gentian]
13. *G. newberryi* A. Gray [alpine gentian]
14. *G. parryi* Engelm.
15. *G. procera* Holm
16. *G. puberula* Michx.
17. *G. purdomii* Marquand
18. *G. saponaria* L. [soapwort gentian]
19. *G. sceptrum* Griseb.
20. *G. setigera* A. Gray [Mendocino gentian]
21. *G. walujewii* Regel & Schmalh. (=*G. kesselringii* Regel)
22. *Gentiana* × sp. (unnamed hybrids)
23. *Gentiana* sp.

Chytridiomycetes
Synchytrium sp. On leaves. **1:**WA(1848,3202). **6:**WA(3202).
Ascomycotina
Mycosphaerella andrewsii. Leaf spot. **4:**ND(82), WI(82). **23:**DE(82).
M. gentianae. On leaves and stems. **23:**CO(82).
Wettsteinina gentianae. **6:**WY(3252).
Basidiomycotina - Uredinales
Puccinia gentianae. Autoecious. **1:**AZ(1203,3981), ID(3202), MT(3202), ND(378), NM(1195), OR(3202), SD(2243), WA(3202). **3:**AK(60,526), CO(1520). **4:**NC states(82), NE states(82), MN(82), NE(82), SD(2243). **5:**CO(1195), WY(1195,2924). **12:**NY(82). **14:**AZ(1203,3981), CO(1195), SD(2243). **16:**ND(378), SD(2243), WI(1347). **19:**CA(81), WA(3202). **23:**E & C states(82), W states(82), AK(82), WA(82).
P. haleniae. Autoecious. **6:**MT(3202), WY(1195). **23:**WY(82).
Uredo beringiana. **10:**AK(526,1575). **23:**AK(82).
Uromyces eugentianae. Autoecious. **23:**CO(82), IA(82), NC(82), NE(82), NM(82), NV(82), VT(82), WA(82), WY(82).

Deuteromycotina - Hyphomycetes
Botrytis cinerea. Stem canker. **4**:NY(82).
Cercospora gentianae. Leaf spot. **4**:ND(82). **12**:NY(82), VT(82).
C. gentianicola. Leaf spot. **8**:DE(82,3543). **15**:WI(82,581,1331). **23**:WI(82).
Fusarium solani. Root rot. **8**:MD(82).
Fusarium sp. **19**:WA(3202).
Deuteromycotina - Coelomycetes
Asteromella andrewsii. Leaf spot. **2**:WI(1333). **4**:IA(82), IL(82), NE(82), NJ(82), PA(82), WI(82,1330), WV(82). **6**:WA(3202). **7**:WI(1344,1347). **8**:DE(82). **9**:WI(1350). **11**:WI(1347). **13**:WI(1350). **14**:WI(1350). **16**:WI(1330). **17**:WI(1347). **18**:WI(1345). **20**:WI(1347). **21**:WI(1347). **22**: On *G. flavida* x *G. andersonii*. WI(1350). **23**:IA(82), WI(82).
Leptothyrium gentianaecola. **4**:ND(378).
Phyllosticta sp. On leaves. **4**:WI(1350).
Septoria gentianoides. On leaves. **23**:NY(82).
S. tosevi. **23**:CO(82).
Deuteromycotina - Other
Sclerotium sp. Stem rot. **4**:NY(82).

Gentianella Moench
1. *G. amarella* (L.) Boerner (=*Gentiana amarella* L.)
2. *G. propinqua* (Richardson) J. M. Gillett (=*Gentiana propinqua* Richardson)
3. *G. quinquefolia* (L.) Small

Ascomycotina
Mycosphaerella minor. On dead stems. **2**:AK(1938).
Basidiomycotina - Uredinales
Uromyces eugentianae. **1**:WA(3202). **3**:IA(2718).

Halenia Borkh.
1. *H. deflexa* (Sm.) Griseb.

Basidiomycotina - Uredinales
Puccinia haleniae. **1**:MN(2717).
Deuteromycotina - Hyphomycetes
Cercospora haleniae. **1**:WI(1349).

Sabatia Adans.
1. *S. angularis* (L.) Pursh [bitter-bloom, rose pink]
2. *S. campestris* Nutt.

Deuteromycotina - Hyphomycetes
Cercospora sabbatiae. Leaf spot. **1**:DE(82,581,3543), MS(82,2720), NC(82,3944), OK(82), TX(82). **2**:OK(2872).

Swertia L.
1. *S. perennis* L.

Basidiomycotina - Uredinales
Puccinia swertiae. Autoecious. **1**:AK(60,526), CO(1195), ID(3202).

Geraniaceae

Erodium L'Hér.
1. *E. cicutarium* (L.) L'Hér. [alfilaria, pin-clover, pin-grass, redstemmed filaree, wild musk]
2. *E. moschatum* (L.) L'Hér. ex Aiton [musk-clover, whitestem filaree]
3. *E. texanum* A. Gray
4. *Erodium* sp.

Chytridiomycetes
Synchytrium papillatum. Leaf gall. **1**:AZ(82), CA(81,82), OR(82,3202), TX(82). **2**:CA(82). **3**:TX(82). **4**:CA(3401).
Oomycetes
Peronospora erodii. Downy mildew. **1**:MT(82). **3**:TX(82).
Deuteromycotina - Hyphomycetes
Fusicladium sp. **1**:OR(3202).
Ramularia sp. **1**:WA(3202).
Deuteromycotina - Other
Rhizoctonia solani. Root rot. **1**:TX(82). **3**:TX(82).
Sclerotium rolfsii. Southern blight. **1**:TX(82). **3**:TX(82).

Geranium L.
1. *G. arboreum* A. Gray
2. *G. bicknellii* Britton
3. *G. caespitosum* James ex Torr.
4. *G. carolinianum* L. [Carolina geranium]
5. *G. dissectum* L. [cutleaf geranium]
6. *G. erianthum* DC.
7. *G. fremontii* Torr. ex A. Gray
8. *G. glabratum* Small
9. *G. maculatum* L. [alumroot, spotted cranesbill, spotted geranium, wild geranium]
10. *G. molle* L. [dove's-foot geranium]
11. *G. nervosum* Rydb.
12. *G. oreganum* Howell
13. *G. pusillum* J. Burm.
14. *G. richardsonii* Fisch. & Trautv.
15. *G. robertianum* L. [herb-robert, red-robin]
16. *G. sanguineum* L. [blood-red geranium]
17. *G. sibiricum* L.
18. *G. viscosissimum* Fisch. & C. A. Mey.
19. *Geranium* sp.

Chytridiomycetes
Synchytrium geranii. Leaf gall. **4**:FL(1841), LA(82,1841), MS(1841), OK(82,2872), TX(82,1841).
Oomycetes
Peronospora conglomerata. **13**:ID(3202), MT(3202), WA(3202).
Plasmopara geranii. Downy mildew. **4**:E states(82), FL(35), KS(82), MS(2720), NC(1303,3944), OK(2872), TX(82). **9**:NE states(82), IA(82), MA(656), WI(82). **12**:OR(3202). **14**:WY(1195,3378). **17**:WI(1349). **18**:MT(3202). **19**:C & W states(82), NE states(82), IA(82), MT(82,3202), UT(82), WI(82).
Pythium aphanidermatum. Root rot. **4**:FL(35).
P. splendens. Damping-off. **4**:FL(35).
Ascomycotina
Coleroa circinans. Leaf spot. **2**:WA(3202). **4**:CA(82), WA(3202). **5**:WA(3202). **6**:AK(526). **10**:NW states(3202). **19**:AK(82), OR(3202), WA(3202).
Diaporthe eres. On stems. **19**:CA(82).
Erysiphe polygoni. Powdery mildew. **9**:IL(82), IN(82), OH(82), PA(82), WI(82), WV(82). **19**:CO(82), WY(82).
Hormotheca robertiani. **15**:NY(2456). **16**:MA(82), NY(82), VT(82), WV(82).
Phaeosphaeria vagans. **18**:ID(3202).
Physalospora geranii. On stems. **19**:CA(82).
Seaverinia geranii. Rhizome rot. **9**:NY(82,1960,3167), OH(944), WI(82,3167).
Sphaerotheca fuliginea. Powdery mildew. **9**:MN(82), NE(82), PA(82), WI(82). **16**:ID(82), MT(82), WA(82).
S. macularis. Powdery mildew. **4**:FL(35). **7**:UT(1195), WY(1195). **11**:ID(1195). **12**:MT(3202), WA(3202). **14**:MT(3202). **16**:CA(82), MT(82), UT(82), WA(82), WY(82). **18**:ID(3202), WA(3202), WY(1195).
Venturia geranii. Black leaf speck. **4**:MD(82). **5**:CA(81).
Basidiomycotina - Uredinales
Puccinia callaquensis. **8**:HI(2903).
P. geranii-silvatici. **8**:HI(2903).
P. leveillei. Autoecious. **1**:HI(2903). **6**:AK(60,526). **7**:WY(1195). **12**:WA(3202). **18**:ID(638,3202), MT(3202), WA(3202). **19**:AK(82,526), CO(82), HI(82,2903), MT(82), UT(82), WA(82), WY(82).
P. polygoni-amphibii. Heteroecious: aecial host. **4**:KS(82), TX(82). **9**:E states(82), MN(82), MO(82), MS(82,2720), WI(1358). **12**:MT(3202). **18**:WA(3202). **19**:CT(82), MT(82).
Uromyces geranii. Autoecious. **6**:AK(60,526). **19**:AK(82), ME(82), WY(82).
Basidiomycotina - Other
Thanatephorus cucumeris. **4**:LA(3524).
Deuteromycotina - Hyphomycetes
Alternaria sp. Leaf spot. **4**:FL(35).
Botrytis cinerea. Gray mold, leaf spot. **19**:AK(82), FL(35), KS(82), MO(82).
Botrytis sp. **9**:WI(1361).
Cercospora brunkii. Leaf spot. **19**:FL(35).
C. geranii. Leaf spot. **4**:FL(35), KS(82). **9**:IA(82), MO(82), NY(82), TX(82), WI(82). **14**:WY(1195,3378). **18**:WY(1195,3378). **19**:CO(82), FL(35), KS(82), MT(82), TX(82), UT(82), WY(82).
C. ithacensis. On leaves. **9**:WI(1350). **18**:MT(3202), WA(3202), WY(1195,3378). **19**:WY(1195).
Cercosporella geranii. **18**:ID(3202), WA(3202).

Cylindrocladium sp. Collar rot. **19:**HI(2903).
Phymatotrichopsis omnivora. Root rot. **4:**TX(82).
Ramularia geranii. Leaf spot. **4:**FL(35), LA(82), WA(82), WY(82). **13:**ID(3202), OR(3202), WA(3202). **18:**OR(3202).

Deuteromycotina - Coelomycetes

Colletotrichum sp. Leaf spot. **9:**WI(1350).
Cylindrosporium geranii. Leaf spot. **3:**WY(2924). **4:**LA(82).
Lasiodiplodia theobromae. Leaf spot. **4:**FL(35).
Pestalozziella subsessilis. Leaf spot. **4:**FL(35,2549), MO(82,2549), MS(82,2549,2720), NJ(82,2549).
Phyllosticta geranii. Leaf spot. **4:**LA(82,963), TX(82).
Septoria expansa. Leaf spot. **4:**FL(35), KS(82), TX(82).

Pelargonium L'Hér.

1. *P.* × *domesticum* L. H. Bailey. A cultigen of complex hybrid origin. [fancy geranium, Lady Washington geranium, Martha Washington geranium, pansy-flowered geranium, regal geranium, show geranium, summer-azalea]
2. *P. graveolens* L'Hér. ex Aiton [rose geranium, sweet-scented geranium]
3. *P.* × *hortorum* L. H. Bailey. A cultigen of complex hybrid origin. [bedding geranium, fish geranium, horseshoe geranium, house geranium, zonal geranium]
4. *P. peltatum* (L.) L'Hér. ex Aiton [hanging geranium, ivy geranium]
5. *Pelargonium* × sp. (unnamed hybrids)
6. *Pelargonium* sp.

Oomycetes

Phytophthora sp. Root rot. **3:**FL(35).
Pythium aphanidermatum. Black leg, stem rot. **4:**FL(35). **3:**FL(35).
P. debaryanum. Black leg, root and stem rot. **2:**MD(82). **4:**FL(35). **1:**MD(82). **3:**FL(35), MD(82). **6:**NC(1303).
P. mamillatum. Black leg. **2:**NJ(82), NY(82). **1:**NJ(82), NY(82). **3:**NJ(82), NY(82).
P. splendens. Black leg, stem rot. **2:**MD(82). **4:**FL(35). **1:**MD(82). **3:**FL(35), MD(82).
P. ultimum. Black leg, stem rot. **2:**NJ(82), NY(82). **4:**FL(35). **1:**NJ(82), NY(82). **3:**FL(35), NJ(82), NY(82). **6:**WA(3202).
P. vexans. Black leg, stem rot. **2:**HI(82), MD(82), OK(82), PA(82). **1:**HI(82), MD(82), OK(82), PA(82). **3:**HI(82), MD(82), OK(82), PA(82). **6:**OK(2873).
Pythium sp. Black leg, root and stem rot. **4:**FL(35). **3:**Range of host(82), FL(35), HI(2903). **6:**NC(1303).

Ascomycotina

Diaporthe rudis. On stems. **2:**CA(82). **6:**CA(82).
Discohainesia oenotherae. On leaves. **2:**MD(82). **3:**NJ(82).
Erysiphe communis. **3:**AK(526).
Glomerella sp. Stem rot. **3:**FL(35).
Pleosphaerulina sp. **3:**PA(82). **6:**PA(82).

Basidiomycotina - Uredinales

Puccinia pelargonii-zonalis. Autoecious. **3:**AZ(1203,3981), CA(3806), FL(35,3806), GA(3806), HI(2299,2903), NY(3806). **5:** On *P. hortorum* x *P. zonale*. NC(1303). **6:**CA(882), HI(882), NY(882), PA(2601).

Basidiomycotina - Other

Armillaria mellea. Root rot. **4:**CA(2893,2898). **1:**CA(2893,2898).
Thanatephorus cucumeris. **6:**WA(3202).

Deuteromycotina - Hyphomycetes

Alternaria alternata. Leaf spot. **4:**FL(35). **3:**CA(2527), FL(35).
Alternaria sp. Secondary leaf spot. **6:**Range of host(82), NC(1303).
Aspergillus fischerianus. Stem rot. **6:**CA(82).
Bipolaris maydis. Leaf and petal spot. **3:**FL(35).
B. setariae. Leaf and petal spot. **3:**FL(35).
Botryosporium pulchrum. Leaf mold. **6:**Unk.(82).
Botrytis cinerea. Blossom blight, gray mold, cutting rot, leaf spot. **2:**Range of host(82). **4:**FL(35). **1:**WI(1359). **3:**FL(35). **6:**Range of host(82), HI(2903), NC(1303), WA(3202).
Cercospora brunkii. Leaf spot. **2:**SC states(82), SE states(82), MD(82), MD(82), NH(82), NH(82), OH(82), OH(82). **4:**SC states(82), SE states(82), FL(35), MD(82), NH(82), OH(82). **3:**SC states(82), SE states(82), FL(35), MD(82), NH(82), OH(82). **6:**MS(2720), NC(1303).
Cylindrocladium scoparium. Flower and leaf spot. **3:**FL(35).
Fusarium sp. Black rot, stem rot. **4:**FL(35). **3:**FL(35), IN(82), NY(82), OR(1961), WA(82). **6:**CT(3498), IN(82), NC(1303), NY(82), WA(82,3202).
Thielaviopsis basicola. Root rot. **6:**CT(82).
Verticillium albo-atrum. Wilt. **1:**CA(82). **3:**CA(82), OR(1961,3668). **6:**OR(3202).

Deuteromycotina - Coelomycetes

Ascochyta sp. Leaf spot. **6:**CT(82), NJ(82).
Colletotrichum sp. Leaf spot. **3:**FL(35).
Phyllosticta sp. Leaf spot. **3:**FL(35).

Deuteromycotina - Other

Rhizoctonia solani. Stem rot. **6:**NC(1303).
Rhizoctonia sp. Root rot. **6:**HI(2903).
Sclerotium sp. Stem rot. **3:**FL(35).

Gesneriaceae

Achimenes Pers.

1. *A. grandiflora* (Schiede) DC.
2. *Achimenes* sp.

Deuteromycotina - Hyphomycetes

Alternaria sp. Leaf necrosis. **2:**FL(35).
Corynespora sp. Leaf spot. **2:**FL(35).
Stemphylium sp. Leaf spot. **1:**FL(35).

Deuteromycotina - Other

Rhizoctonia sp. Root rot. **2:**FL(35).

Aeschynanthus W. Jack

1. *A. pulcher* (Blume) G. Don [climbing-beauty, lipstick-plant, pipe-plant, red-buglevine, royal-red-bugler, scarlet basketvine]
2. *Aeschynanthus* sp.

Oomycetes

Pythium splendens. Root rot. **1:**FL(35). **2:**FL(35).

Deuteromycotina - Hyphomycetes

Alternaria sp. Leaf spot. **1:**FL(35). **2:**FL(35).
Botrytis cinerea. Leaf spot. **2:**FL(35).
Cercospora sp. Leaf spot. **1:**FL(35). **2:**FL(35).
Corynespora cassiicola. 'Corynespora leaf spot.' **1:**FL(35,559).
Curvularia sp. Leaf spot. **1:**FL(35).
Fusarium sp. Root rot. **1:**FL(35).
Helminthosporium sp. Leaf spot. **2:**FL(35).
Myrothecium sp. Leaf blight. **1:**FL(35).
Stemphylium sp. Leaf spot. **1:**FL(35).

Deuteromycotina - Coelomycetes

Colletotrichum dematium. Leaf spot. **1:**FL(35).
Phomopsis sp. Leaf spot. **1:**FL(35).
Phyllosticta sp. Leaf spot. **1:**FL(35). **2:**FL(35).
Plenodomus sp. Stem canker. **2:**FL(35).

Deuteromycotina - Other

Rhizoctonia solani. Leaf blight. **1:**FL(35). **2:**FL(35).

Chrysothemis Decne.

1. *C. pulchella* (Donn ex Sims) Decne.

Oomycetes

Pythium sp. Root rot. **1:**HI(2903).

Columnea L.

1. *Columnea* sp.

Oomycetes

Pythium splendens. Root rot. **1:**FL(35).

Deuteromycotina - Hyphomycetes

Alternaria sp. Leaf spot. **1:**FL(35).
Corynespora cassiicola. Leaf spot. **1:**FL(35).
Fusarium sp. Root rot. **1:**FL(35).
Myrothecium sp. Leaf necrosis. **1:**FL(35).

Deuteromycotina - Coelomycetes

Colletotrichum sp. Leaf spot. **1:**FL(35).
Phyllosticta sp. Leaf spot. **1:**FL(35).

Deuteromycotina - Other

Rhizoctonia solani. Stem rot. **1:**FL(35).

Cyrtandra J. R. Forster & G. Forster

1. *C. cordifolia* Gaud.
2. *C. lessoniana* Gaud.
3. *Cyrtandra* sp.

Ascomycotina

Asteridiella cyrtandrae. Black mildew. **1:**HI(1272,1452,2903).

2:HI(1272,1452,2903).
Phragmocapnias smilacina. 3:HI(2784).

Episcia Mart.

1. *Episcia* sp.

Oomycetes
Phytophthora sp. Root rot. 1:FL(35).
Pythium sp. Root rot. 1:FL(35).
Deuteromycotina - Hyphomycetes
Alternaria sp. Leaf spot. 1:FL(35).
Botrytis cinerea. Leaf blight. 1:FL(35).
Deuteromycotina - Coelomycetes
Colletotrichum sp. Leaf spot. 1:FL(35).
Deuteromycotina - Other
Rhizoctonia solani. Leaf rot, basal stem rot. 1:FL(35).

Gloxinia L'Hér. in Aiton

1. *G. perennis* (L.) Fritsch (=*G. pallidiflora* Hook.)

Ascomycotina
Erysiphe cichoracearum. Powdery mildew. 1:CA(1155).

Kohleria Regel

1. *Kohleria* sp.

Oomycetes
Pythium sp. Root rot. 1:FL(35).

Nautilocalyx Linden

1. *N. lynchii* (J. D. Hook.) Sprague

Oomycetes
Pythium splendens. Root and basal stem rot. 1:FL(35).

Nematanthus Schrad.

1. *N. strigillosus* (Mart.) H. Moore (=*Hypocyrta glabra* Hook.)
2. *Nematanthus* sp. (=*Hypocyrta sp.*)

Oomycetes
Pythium sp. Root rot. 2:FL(35).
Deuteromycotina - Hyphomycetes
Corynespora cassiicola. Leaf spot. 2:FL(35).
Myrothecium sp. Stem rot, lower leaf rot. 1:FL(35).
Deuteromycotina - Coelomycetes
Colletotrichum sp. Leaf spot. 2:FL(35).
Deuteromycotina - Other
Rhizoctonia solani. Basal stem rot. 1:FL(35). 2:FL(35).
Rhizoctonia sp. Root rot. 2:FL(35).

Saintpaulia H. Wendl.

1. *S. ionantha* H. Wendl. [common African violet]
2. *Saintpaulia* sp.

Oomycetes
Phytophthora nicotianae. 1:NC(1303).
P. nicotianae var. *parasitica*. Crown and root rot. 'Phytophthora stem rot.' 1:FL(35,2835).
Phytophthora sp. Crown and root rot. 1:HI(2903), MD(82), NC(82).
Pythium splendens. Root rot. 1:FL(35).
P. ultimum. Crown and root rot. 1:CA(82), FL(35).
Pythium sp. Root rot. 2:NC(1303).
Deuteromycotina - Hyphomycetes
Alternaria sp. Leaf necrosis. 1:FL(35).
Botrytis cinerea. Soft rot. 'Gray mold.' 1:Range of host(82), FL(35), HI(2903), MI(82), NY(82), WA(3202).
Corynespora cassiicola. 'Corynespora leaf spot.' 1:FL(35).
Cylindrocarpon destructans. Root rot. 1:MD(82), NY(82).
Fusarium solani. Root rot. 1:MD(82), NY(82).
Fusarium sp. Root rot. 1:FL(35).
Oidium sp. Powdery mildew. 1:FL(35), MD(82), NJ(82), NY(82). 2:NC(1303), PA(1909).
Deuteromycotina - Other
Rhizoctonia solani. Leaf, root and stem rot. 1:FL(35), MD(82), NY(82).
Rhizoctonia sp. Root rot. 1:HI(2903).

Sinningia Nees

1. *S. speciosa* (Lodd.) Hiern [Brazilian gloxinia, violet-slipper gloxinia]

Oomycetes
Phytophthora cryptogea. Crown and root rot. 1:CA(82), FL(35).
P. nicotianae. Root rot. 1:NC(1303).
P. nicotianae var. *parasitica*. Crown, root and basal stem rot. 1:FL(35,2835).
Phytophthora sp. Crown rot. 1:HI(2903).
Pythium ultimum. Root rot. 1:CA(82).
Pythium sp. Damping-off, stem rot. 1:FL(35).
Ascomycotina
Sclerotinia sclerotiorum. Crown, flower and stem rot. 1:CA(82,2905), FL(35), NC(1303), WA(3202).
Deuteromycotina - Hyphomycetes
Botrytis cinerea. Blossom blight, gray mold, bud rot. 1:CA(82), FL(35), HI(2903), MO(82), WA(3202).
Cladosporium herbarum. Leaf rot. 1:NJ(82).
Myrothecium roridum. Crown rot. 1:FL(35).
Deuteromycotina - Coelomycetes
Colletotrichum sp. Anthracnose. 1:FL(35).
Plenodomus sp. Leaf spot. 1:FL(35).
Deuteromycotina - Other
Rhizoctonia solani. Crown and tuber rot. 1:FL(35).
Rhizoctonia sp. Root rot. 1:HI(2903).
Sclerotium rolfsii. Southern blight. 1:FL(35).

Streptocarpus Lindl.

1. *Streptocarpus* sp.

Oomycetes
Phytophthora sp. Leaf and stem rot. 1:FL(35).
Deuteromycotina - Hyphomycetes
Alternaria sp. Leaf spot. 1:FL(35).
Corynespora sp. Leaf spot. 1:FL(35).

Ginkgoaceae

Ginkgo L.

1. *G. biloba* L. [ginkgo, maidenhair-tree]

Ascomycotina
Glomerella cingulata. Leaf spot. 1:MD(82), TX(82).
Xylaria longiana. Seed rot. 1:PA(82).
Basidiomycotina - Other
Irpex lacteus. On wood. 1:MD(82).
Oxyporus populinus. 1:MD(82).
Trametes hirsuta. On wood. 1:MD(82).
T. versicolor. On wood. 1:MD(82).
Tyromyces chioneus. On wood. 1:MD(82).
Deuteromycotina - Hyphomycetes
Cercospora sp. Leaf spot. 1:FL(35).
Epicoccum nigrum. On leaves. 1:IL(2568).
Phymatotrichopsis omnivora. Root rot. 1:TX(82).
Deuteromycotina - Coelomycetes
Phyllosticta ginkgo. Leaf spot. 1:IN(82), MD(82), OH(82), PA(82), VA(82).
Phyllosticta sp. Leaf spot. 1:FL(35).

Goodeniaceae

Scaevola L.

1. *S. chamissoniana* Gaud.
2. *S. glabra* Hook. & Arnott
3. *S. mollis* Hook. & Arnott
4. *S. taccada* (Gaertn.) Roxb. (=*S. koenigii* Vahl, *S. sericea* Vahl)
5. *Scaevola* sp.

Oomycetes
Pythium sp. Root rot. 5:FL(35).
Ascomycotina
Amazonia psychotriae. Black mildew. 2:HI(2903). 5:HI(2903).
A. scaevolae. Black mildew. 2:HI(1268,1452,2903). 5:HI(1272,1452,2903).
Calloriopsis gelatinosa. 5:HI(2903).
Calothyriopeltis scaevolae. 5:HI(2903).
Meliola scaevolae. Black mildew. 2:HI(1272).
M. scaevolicola. Black mildew. 1:HI(1272,1452,2903). 2:HI(2903). 3:HI(1272,1452,2903).
Mycosphaerella scaevolae. Leaf spot of living leaves. 1:HI(2903,3480). 2:HI(2903,3480). 3:HI(2903,3480).

Pleospora scaevolae. **1**:HI(2903). **2**:HI(2903). **3**:HI(2903).
Sarcosoma godronioides. **5**:HI(2903).
Seuratia millardetii. **5**:HI(2784).
Deuteromycotina - Hyphomycetes
Alternaria sp. Leaf spot. **5**:FL(35).
Cercospora sp. Leaf spot. **4**:HI(2903). **5**:FL(35).
Corynespora sp. Leaf spot. **5**:FL(35).
Trichothallus hawaiiensis. **5**:HI(2903).
Verticillium sp. Wilt. **5**:FL(35).
Deuteromycotina - Coelomycetes
Asteromella scaevolae. On decaying leaves. **5**:HI(2784).
Dothiorella sp. Tip dieback. **5**:FL(35).
Phoma sp. Canker. **5**:FL(35).
Phyllosticta scaevolae. **1**:HI(2903).
Deuteromycotina - Other
Rhizoctonia solani. Root rot. **5**:FL(35).
Rhizoctonia sp. Root rot. **4**:HI(2903).

Grossulariaceae

Escallonia Mutis ex L. fil.

1. *Escallonia* sp.
Oomycetes
Pythium sp. Crown and root rot. **1**:FL(35).
Deuteromycotina - Coelomycetes
Colletotrichum sp. Leaf spot. **1**:FL(35).
Phyllosticta sp. Leaf spot. **1**:FL(35).

Itea L.

1. *I. virginica* L. [sweetspire, tassel-white, Virginia willow]
Ascomycotina
Botryosphaeria dothidea. **1**:GA(1438).
Deuteromycotina - Coelomycetes
Phyllosticta iteae. On living leaves. **1**:TN(837).
Phyllosticta sp. Leaf spot. **1**:FL(35).

Ribes L.

(= *Grossularia* Mill.)
1. *R. alpinum* L. [alpine currant, mountain currant]
2. *R. americanum* Mill. (=*R. floridum* L'Hér.) [American black currant, wild black currant]
3. *R. aureum* Pursh [buffalo currant, golden currant, Missouri currant]
4. *R. binominatum* A. Heller
5. *R. bracteosum* Douglas [California black currant, stink currant]
6. *R. cereum* Douglas [squaw currant, wax currant, white-flowered currant]
7. *R. cognatum* Greene
8. *R. coloradense* Crép.
9. *R. cruentum* Greene (=*R. roezli* Regel var. *cruentum* (Greene) Rehd.)
10. *R. curvatum* Small
11. *R. cynosbati* L. [dog-bramble, dogberry, pasture gooseberry, prickly gooseberry]
12. *R. diacantha* Pall.
13. *R. divaricatum* Douglas
14. *R. divericatum* Douglas var. *inerme* (Rydb.) McMinn (=*R. inerme* Rydb.)
15. *R. erythrocarpum* Crép. & Leiberg
16. *R. glandulosum* Grauer (=*R. prostratum* L'Hér.) [fetid currant, skunk currant]
17. *R. glutinosum* Benth.
18. *R.* × *gordonianum* Lem. *R. odoratum* × *R. sanguineum*.
19. *R. grossularis* L. [european gooseberry]
20. *R. hirtellum* Michx. [hairystem gooseberry]
21. *R. howellii* Greene
22. *R. hudsonianum* Richardson (=*R. petiolare* Douglas) [Hudson Bay currant, northern black currant]
23. *R. inebrians* Lindl.
24. *R. irriguum* Douglas
25. *R. japonicum* Maxim.
26. *R. klamathense* (Crép.) Fedde
27. *R. lacustre* (Pers.) Poir. [bristly black currant, swamp gooseberry]
28. *R. laxiflorum* Pursh
29. *R. leptanthum* A. Gray
30. *R. lobbii* A. Gray [gummy gooseberry]
31. *R. malvaceum* Sm. [Chaparral currant]
32. *R. marshallii* Greene
33. *R. menziesii* Pursh [canyon gooseberry]
34. *R. missouriense* Nutt. [Missouri gooseberry]
35. *R. montigenum* McClat.
36. *R. nevadense* Kellogg [Sierra currant]
37. *R. nigrum* L. [black currant, European black currant]
38. *R. niveum* Lindl.
39. *R. odoratum* H. Wendl. [buffalo currant, Missouri currant]
40. *R. oxyacanthoides* L. (=*R. saxosum* Hook.) [hawthorn-leaved gooseberry]
41. *R. pinetorum* Greene
42. *R. roezlii* Regel [Sierra gooseberry]
43. *R. roezlii* Regel var. *cruentum* (Greene) Rehd.
44. *R. rotundifolium* Michx.
45. *R. rubrum* L. (=*R. sativum* (Rchb.) Syme, *R. vulgare* Lam.) [northern red currant, white currant]
46. *R. sanguineum* Pursh (=*R. wolfii* Rothr.) [red-flowered currant]
47. *R. setosum* Lindl. [Missouri gooseberry]
48. *R. speciosum* Pursh [fuchsia-flowered gooseberry]
49. *R. triste* Pall. [swamp red currant]
50. *R. uva-crispa* L. (=*R. grossularia* L., *Grossularia reclinata* (L.) Mill.) [English gooseberry, European gooseberry]
51. *R. velutinum* Greene
52. *R. viscosissimum* Pursh
53. *R. watsonianum* Koehne
54. *Ribes* sp. (=*Grossularia sp.*)
55. *Ribes* × sp. (unnamed hybrids)
Oomycetes
Phytophthora sp. **20**:CA(1110). **45**:CA(1110).
Plasmopara ribicola. Downy mildew. **13**:WA(3202). **16**:NC states(82), NE states(82), NW states(82), WI(82). **33**:OR(3202). **49**:NC states(82), NE states(82), NW states(82), OR(3202), WI(82). **54**:MN(82), MT(3202), OR(82,3202), WI(82), WV(82).
Ascomycotina
Atropellis pinicola. **53**:OR(3202).
Botryosphaeria dothidea. **45**:CA(1110). **54**:IL(650).
B. obtusa. On dead canes. **54**:NE states(82), KS(82), MI(82).
B. ribis. Cane blight, canker, dieback, leaf spot. **54**:E states(82), CA(82), CT(82), IL(82), MD(82), NJ(82).
Capitotricha bicolor. On stems. **54**:WA(2219).
Chapeckia ribesia. On twigs. **28**:AK(526). **54**:AK(82).
Cucurbitaria obducens. **6**:ID(3202).
C. ribis. On dead canes. **48**:CO(82), MT(82), NM(82), VA(82). **54**:CO(82), MT(82,3202), NM(82), VA(82).
Dasyscyphus albolabra. **16**:CO(849).
Diapleella coniothyrium. **37**:AK(526). **54**:WA(3202).
Diaporthe eres. On dead canes and twigs. **54**:AK(82).
D. strumella. On dead canes, twigs and limbs. **5**:CA(81). **28**:AK(526). **54**:NC states(82), NE states(82), AK(82), CA(82).
Discohainesia oenotherae. On leaves. **54**:MD(82), ME(82).
Discostroma massarina. **44**:MT(3202). **54**:IL(650).
Dothiora ribesia. Twig knot. **10**:GA(1438). **22**:AK(526). **45**:AK(526). **54**:ID(3202), MS(2720), MT(3202), OR(3202), WA(3202).
Drepanopeziza ribis. Anthracnose, leaf skin and fruit spot. **1**:IL(2567). **5**:AK(526), OR(3202). **6**:WY(1195). **9**:OR(3202). **13**:OR(3202). **14**:WY(1195). **24**:MT(3202), OR(3202), WA(3202). **26**:OR(3202). **27**:WA(3202). **28**:AK(526). **30**:WA(3202). **32**:OR(3202). **36**:OR(3202). **37**:AK(526). **38**:WA(3202). **42**:OR(3202). **45**:AK(526). **49**:AK(526), WA(3202). **51**:OR(3202). **52**:OR(3202), WA(3202). **53**:OR(3202). **54**:Range of host(82), IA(82), MN(82), MT(82).
Encoeliopsis oricostata. On twigs. **46**:CO(521). **54**:CO(82), UT(82).
Eutypella leprosa. **34**:IA(3629). **54**:IA(3629).
Glomerella cingulata. Berry rot. **54**:CT(82), PA(82).
Gnomonia ribicola. **54**:OR(657).
Godronia cassandrae f. *ribicola*. **54**:CO(1382), NH(1382).
G. davidsonii. On stems. **46**:CO(128,1382). **55**: On *R. bracteosum* x *R. nigrum*. AK(526), OR(3202). **54**:AK(1382), OR(1382).
G. grossulariae. **53**:OR(1382,3202).
G. lobata. On dead twigs. **33**:CA(128,1382,3167). **54**:CA(82).
G. ribis. On twigs. **45**:CA(1110). **54**:CA(82).
G. tumoricola. On twig galls. **54**:CO(82).
G. uberiformis. On twigs. **2**:ND(128,377). **54**:CO(1382), ID(3202),

ND(1382).
G. urceolus. On twigs. **5:**AK(526). **54:**AK(82,526), CO(82), ND(82), NJ(82).
Herpotrichia herpotrichoides. On twigs. **54:**OR(189,657).
Hymenoscyphus calyculus. **28:**WA(3202).
Lophiostoma ribis. On stems. **16:**CO(128). **50:**AK(128,526).
Massarina microcarpa. On canes and twigs. **54:**MI(82).
Microsphaera grossulariae. Powdery mildew. **2:**MT(3202). **20:**CA(1110). **37:**MT(3202), WA(3202). **44:**MT(3202). **45:**CA(1110). **54:**CA(82), ID(82), MT(82), NE(82), NH(82).
Mycosphaerella ribis. Leaf spot. **5:**WA(3202). **6:**ID(3202). **13:**WA(3202). **14:**ID(3202), MT(3202), WA(3202). **22:**ID(3202). **27:**ID(3202), WA(3202). **45:**MS(2720). **47:**ID(3202), MT(3202). **54:**C states(82), E states(82), NE states(82), AR(82), OR(82), SD(82), WA(82,3202).
Nectria cinnabarina. Canker, dieback, coral spot. **5:**AK(526). **22:**AK(526). **37:**AK(526). **45:**AK(526). **46:**AK(526). **54:**MN(82).
N. ditissima. Canker. **54:**MN(82), NY(82).
Neokeissleria ribis. On canes. **5:**AK(526), OR(3202). **54:**AK(82), OR(82).
Nipterella parksii. **46:**CA(81).
Phyllactinia guttata. Powdery mildew. **13:**CA(81), WA(3202). **22:**WA(3202). **54:**NY(82).
Rhynchophoma raduloides. **5:**AK(526). **28:**AK(526). **54:**AK(82), CA(82).
Rosellinia parasitica. On dead canes. **54:**ID(82,3202), NM(82).
Sphaerotheca fuliginea. Powdery mildew. **17:**CA(1155). **31:**CA(1155). **46:**Unk.(3976).
S. macularis. Powdery mildew. **2:**MT(3202). **3:**AK(526). **5:**AK(526). **45:**AK(526). **55:** AK(526). **54:**AK(82), ID(82), MN(82), MT(82).
S. mors-uvae. American powdery mildew. **2:**ND(377). **5:**OR(3202). **6:**MT(3202). **19:**MS(2720). **22:**MT(3202). **24:**MT(3202). **27:**AK(526). **37:**WA(3202). **38:**ID(3202). **44:**MT(3202). **45:**AK(526), CA(1110). **50:**AK(526). **54:**C states(82), NE states(82), Range of host(82), AK(82), CA(82), CT(82), IA(82), MT(82), NE(82), WA(82).
Stictis radiata. **54:**CA(3225).
Tapesia ribicola. On twigs. **35:**CO(521).
Thyronectria berolinensis. Cane knot. **2:**ND(377), SD(2243). **45:**ND(377). **47:**MT(3202). **54:**C states(82), NE states(82), KS(82), MT(3202), OR(82,3202), SD(82), UT(82).
Valsa ambiens. On dead twigs and canes. **2:**IA(1226). **3:**KS(3415). **54:**CO(649), KS(82), MO(82), NJ(82).
V. ceratosperma. **54:**CA(3415).
V. ribesia. On dead twigs and canes. **54:**CA(82).
Xylaria hypoxylon. Black root rot. **54:**OR(82,3202).
Xylaria sp. **54:**IL(82).

Basidiomycotina - Uredinales

Coleosporium jonesii. Heteroecious: telial host. **2:**ID(3202), MT(3202). **3:**AZ(1203), ID(3202), MT(3202). **6:**AZ(1195,1203,3981), ID(3202), MT(3202), SD(2243). **14:**ID(3202), MT(3202), WY(1195). **23:**AZ(1195,1203,3981). **27:**MT(3202), WA(3202). **29:**AZ(1195,1203,3981). **40:**CO(1195). **41:**AZ(1195,1203,3981). **46:**AZ(3981). **47:**MT(3202). **51:**AZ(1195,1203,3981). **54:**C & W states(82), CA(82), WA(82).
Cronartium occidentale. Heteroecious: telial host. **3:**AZ(1203,3981). **37:**AZ(1203,3981). **39:**ID(3202), WA(3202). **48:**CA(82), ID(82), KS(82), MT(82), NE(82), NM(82), WA(82). **54:**SW states(82), CA(82), ID(82), MT(82), NE(82), WA(82).
C. ribicola. Heteroecious: telial host. **3:**NE states(82), CO(82), MN(82), WA(3202). **4:**OR(3202). **5:**CA(81), OR(3202), WA(3202). **6:**OR(3202). **8:**NW states(3202). **9:**OR(3202). **11:**IN(987), NC(1533), WI(1333). **12:**WI(1360). **13:**OR(3202), WA(3202). **14:**ID(3202), MT(3202), OR(3202), WA(3202). **15:**OR(3202). **16:**WI(1333). **20:**CA(1110), WI(1333). **22:**ID(3202), MT(3202), OR(3202), WA(3202). **24:**ID(3202). **26:**OR(3202). **27:**ID(3202), OR(3202), WA(3202). **28:**WA(3202). **30:**CA(81). **32:**OR(3202). **34:**WI(1354). **36:**OR(3202). **37:**ID(3202), OR(3202), WA(3202), WI(1333). **39:**NE states(82), CO(82), MN(82), WA(3202). **42:**OR(3202). **43:**CA(81). **44:**GA(1533), NC(1533). **46:**CA(81), OR(3202), WA(82,3202). **49:**NW states(3202), WI(1333). **50:**WI(1333). **51:**OR(3202). **52:**ID(638,3202), MT(3202), OR(3202), WA(3202). **54:**E states(82), W states(82), CA(82), IA(82), MN(82), NY(2456), OR(82), TN(82).
Melampsora epitea. Heteroecious: aecial host. **5:**WA(3202). **13:**WA(3202). **14:**UT(1195). **27:**OR(3202), WA(3202).
M. ribesii-purpureae. Heteroecious: aecial host. **3:**UT(82). **14:**WY(1195). **27:**WY(1195). **49:**AK(60,526). **54:**W states(82), AK(82), ID(82), WA(82).
Puccinia caricina. Heteroecious: aecial host. **1:**AK(526). **3:**MT(3202). **6:**MT(3202). **7:**WA(3202). **11:**IL(2567). **13:**CA(81), OR(3202), WA(3202). **14:**CA(81), MT(3202), SD(2768), WY(1195). **16:**AK(526). **20:**ID(3202). **21:**WA(3202). **22:**AK(60,526), ID(3202), WA(3202). **27:**AK(60,526), ID(3202), MT(3202), OR(3202), WA(3202), WI(1354). **28:**AK(60,526,3431), WA(3202). **34:**WI(1346). **39:**OK(2872). **40:**WI(1339), WY(1195,2924). **45:**AK(60,526). **46:**AK(60,526), CA(81), OR(3202), WA(3202). **47:**MT(3202), WY(1195,3379). **49:**AK(60,526). **50:**ND(378). **18:**AK(526). **54:**NC states(82), NE states(82), AK(82,526), CA(82), IA(82), MD(82), MS(2720), NY(82), SD(2243), VT(1224), WI(82).
P. micrantha. Heteroecious: aecial host. **3:**NE(82), WY(82). **54:**CO(82), MT(82), NE(82), SD(82), UT(82), WY(82).
P. parkerae. Autoecious. **27:**AK(60,526), ID(3202), MT(1195,3202), OR(3202), WA(3202). **46:**WA(82,3202). **54:**NW states(82), AK(82), MT(3202).
P. ribis. **45:**AK(526). **49:**AK(1904), ID(3202), WA(3202). **54:**NC states(82), NE states(82), NE(82), WA(82).

Basidiomycotina - Other

Aleurodiscus aurantius. On dead stems and twigs. **46:**CA(2083).
Armillaria mellea. Root rot. **45:**CA(1110). **54:**OR(3202), WA(3202).
Corticium sp. **53:**WA(3202).
Cylindrobasidium corrugum. On living plants. **54:**ID(128,3202).
Datronia stereoides. On dead canes. **54:**NC(82).
Dendrothele incrustans. **46:**CA(2084).
Hyphoderma setigerum. **13:**WA(3202).
Hyphodermella corrugata. On stems. **46:**CA(81).
Hypholoma sublateritium. Root rot. **54:**NY(82).
Pellicularia koleroga. Thread blight. **54:**FL(82).
Peniophora nuda. **13:**WA(3202).
Phylloporia ribis. White heart rot. **3:**ID(3202), NM(3485), WY(3485). **17:**CA(3485). **54:**C states(82), NE states(82), Unk.(1205), CA(2180), ID(2180,3202), MN(82), MT(2180,3202), NM(2180), OR(2180,3202,4007), UT(82), WY(2180).
Thelephora terrestris. Stem girdle. **54:**NH(82).
Trametes hirsuta. **54:**OR(3202).

Deuteromycotina - Hyphomycetes

Alternaria sp. Leaf spot. **34:**WI(1348). **54:**MI(82).
Botrytis cinerea. Dieback, fruit rot, leaf spot. **2:**AK(526). **20:**CA(1110). **37:**AK(526). **45:**AK(526). **46:**AK(526). **49:**AK(526). **54:**NE states(82), OR(82), WA(82).
Botrytis sp. **54:**AK(526).
Bullera alba. On rusted leaves. **54:**KS(2686).
Cercospora angulata. Leaf spot. **54:**KS(82).
C. ribicola. Leaf spot. **6:**OR(3202). **46:**OR(3202), WA(3202). **52:**ID(3202). **54:**CA(82), ID(82), OR(82), WA(82), WI(82), WY(82).
C. ribis. Leaf spot, on canes. **2:**WI(1335). **11:**WI(1345). **50:**IL(2567). **54:**AL(82), IA(82), IN(82), KS(82), WI(82).
C. saximontanensis. Leaf spot. **48:**CA(844). **52:**WY(844).
Cercospora sp. **20:**WI(1359).
Cladosporium herbarum. Leaf blight. **54:**AK(82), WA(82).
Cladosporium sp. Leaf blight. **22:**AK(526). **54:**AK(82), WA(82).
Conoplea olivacea. **54:**OH(1672).
Dematophora sp. Root rot. **54:**NY(82).
Dendrodochium berolinense. **54:**OR(3202).
Fumago vagans. **45:**WA(3202).
Fusarium avenaceum. On stems. **54:**OR(3202,4007).
F. lateritium. **54:**AK(526).
Heterosporium sp. **3:**WA(3202).
Phymatotrichopsis omnivora. Root rot. **54:**TX(82).
Ramularia coalescens. On leaves and dead stems. **2:**WI(1360). **5:**OR(648,3202). **16:**WI(1333). **54:**WA(82), WI(82).
Ramularia sp. Leaf spot. **5:**OR(3202), WA(3202). **22:**WA(3202). **27:**WA(3202). **46:**WA(3202). **54:**MI(82).
Verticillium dahliae. Wilt. **45:**CA(1110).

Verticillium sp. Wilt. **54:**NY(82).
Deuteromycotina - Coelomycetes
Ascochyta grossulariae. On canes. **54:**NY(82), OR(82,3202), VA(82).
A. philadelphi. **34:**WI(1357).
Asteroma ribicola. **2:**MT(3202). **54:**AK(526).
Clypeopycnis aeruginascens. On dead stems. **5:**AK(526). **34:**SD(3549). **54:**AK(82), SD(82).
Coniothyrium sp. On leaves. **11:**WI(1355). **34:**WI(1355).
Coryneum sp. **1:**AK(526). **49:**AK(526).
Cylindrosporium ribis. Leaf spot. **24:**ID(3202). **52:**ID(3202), WY(1195). **54:**SD(2243).
Cytospora grossulariae. On twigs. **54:**MD(82).
Discula sp. On leaves. **12:**WI(1355).
Dothiorella ribis. Canker. **25:**NY(3755).
Gloeosporidiella bartholomaei. Leaf spot. **5:**WA(3202). **54:**WA(82).
G. ribis. Leaf spot. **1:**IL(312). **6:**CO(649). **11:**WI(1350). **14:**WY(1195). **40:**WY(1195).
G. variabilis. **1:**WI(1355,1356). **34:**WI(1356).
Marssonina bracteosa. On leaves. **5:**AK(526), WA(3202). **54:**AK(82), WA(82).
M. grossulariae. Leaf spot. **16:**NY(2456). **54:**OH(82), WI(82).
M. ribicola. Leaf spot. **1:**IL(2568). **54:**CO(82).
Phoma terrestris. Isolated from rhizosphere. **30:**OR(2791).
Phyllosticta canescens. On leaves. **13:**ID(64,3202). **54:**ID(82).
P. ellisii. On leaves. **54:**VA(82), WV(82).
P. grossulariae. Fruit and leaf spot. **1:**WI(1347). **3:**OK(2872). **34:**WI(82,1335). **45:**OR(3202), WI(1357). **52:**ID(3202). **54:**C states(82), IN(82), KS(82), WI(82).
P. ribesicida. On leaves. **22:**ID(3202). **54:**AK(82,526).
P. succinosa. On living leaves. **2:**WI(1348). **34:**WI(1354).
Phyllosticta sp. On leaves. **11:**WI(1350).
Rhabdospora ribicola. On twigs. **54:**WI(82).
Septoria ribis. Leaf spot. **1:**WI(1355). **2:**ND(378). **5:**AK(526). **27:**ID(3202). **28:**AK(526). **34:**WI(1333). **37:**AK(526). **39:**OK(2872). **40:**WY(1195). **42:**CA(81). **45:**AK(526), ND(378). **50:**AK(526), ND(378). **52:**ID(3202). **54:**AK(526), OK(2872), SD(2243).
S. sanguinea. Leaf spot. **46:**OR(3202), WA(830,3202). **54:**WA(82).
S. siberica. **14:**WY(1195).
Sphaceloma ribis. Scab, on leaves. **54:**WA(82,3202).
Topospora uberiformis. On twigs. **54:**ND(3549).

Gunneraceae

Gunnera L.

1. *G. petaloidea* Gaud.
2. *Gunnera* sp.

Ascomycotina
Ceriospora sandwicensis. On living and withering petioles. **2:**HI(2781).
Mycosphaerella hawaiiensis. On living leaves. **1:**HI(2784,2903,3480).
Deuteromycotina - Coelomycetes
Harknessia gunnerae. Leaf spot. On dead leaf. On leaves. **1:**HI(2548,2903,3549).

Haemodoraceae

Lachnanthes Elliott

1. *L. caroliniana* (Lam.) Dandy [dyeroot, paintroot, redroot]

Ascomycotina
Phyllachora sp. Black spot of leaves and stems. **1:**FL(35).

Haloragaceae

Myriophyllum L.

1. *M. aquaticum* (Vell.) Verdc. (=*M. brasiliense* Cambess.) [parrot's-feather, waterfeather]
2. *M. spicatum* L. [Eurasian watermilfoil]

Oomycetes
Pythium carolinianum. Root and stem rot. **1:**CA(258).
Pythium sp. Leaf rot. **2:**FL(35).
Deuteromycotina - Hyphomycetes
Oidium sp. Powdery mildew. **1:**FL(35).

Haloragidaceae

Proserpinaca L.

1. *P. palustris* L.

Basidiomycotina - Uredinales
Puccinia proserpinacae. Autoecious. **1:**MA(1394).

Hamamelidaceae

Hamamelis L.

1. *H. virginiana* L. [witch-hazel]
2. *Hamamelis* sp.

Ascomycotina
Acrospermum foliicola. **1:**GA(1438).
Botryosphaeria obtusa. On limbs. **2:**Range of host(82).
Daldinia concentrica. On wood. **2:**Range of host(82).
Dermea hamamelidis. On bark. **1:**GA(1438), NY(2456). **2:**PA(82), WV(82).
Hainesia lythri. **1:**WI(1358).
Leptosphaeria hamamelidis. On limbs. **2:**NY(82).
Mycosphaerella punctiformis. **1:**GA(1438).
Mycosphaerella sp. Leaf spot. **1:**GA(1438). **2:**WV(82).
Phyllactinia guttata. Powdery mildew. **2:**MI(82), WI(82).
Podosphaera biuncinata. Powdery mildew. **1:**MS(2720). **2:**E states(82), SC states(82).
Schizoxylon insigne. **1:**GA(1438).
Valsa ambiens. On limbs. **2:**Range of host(82).
Basidiomycotina - Other
Datronia scutellata. Wood rot. **2:**Range of host(82).
Dendrophora versiformis. **1:**NY(543).
Junghuhnia nitida. **2:**VA(128,2168).
Perenniporia tenuis. Wood rot. **2:**Unk.(82).
Phellinus gilvus. Wood rot. **2:**Unk.(82).
Skeletocutis nivea. Wood rot. **2:**Unk.(82).
Stereum hirsutum. **1:**CT(543), ME(543).
Deuteromycotina - Hyphomycetes
Cercospora hamamelidis. Leaf spot. **1:**FL(35), NY(581), OK(2872). **2:**NE states(82), FL(82), OK(82), WI(82).
Conoplea olivacea. **2:**MI(1672).
Gonatobotryum apiculatum. Leaf spot. **1:**FL(35), NC(1307). **2:**NE states(82), FL(82), IN(82), NY(82), WI(82).
Graphium hamamelidis. Leaf spot. **1:**NC(1307). **2:**NC states(82), NE states(82), TN(82).
Deuteromycotina - Coelomycetes
Diplodia hamamelidis. On limbs. **2:**NY(82).
Discosia artocreas. Leaf spot. **1:**OK(2873). **2:**OK(82).
Hendersonia foliorum var. *hamamelidina*. On leaves. **2:**NY(82).
Monochaetia monochaeta. Zonate leaf spot. **1:**GA(1509), MA(1509), TN(1509), VA(1509). **2:**SE states(82), TN(82).
Phyllosticta hamamelidis. Leaf spot. **1:**NC(2). **2:**E states(82), Unk.(82), MS(82), TN(82), WI(82).
P. sphaeropsoidea. **2:**IL(82).
Septoria liquidambaris. **1:**WI(1360).

Liquidambar L.

1. *L. formosana* Hance [feng-hsiang-shu, Formosan gum]
2. *L. styraciflua* L. [American sweetgum, bilsted, red gum]
3. *Liquidambar* sp.

Chytridiomycetes
Synchytrium liquidambaris. Leaf gall. **2:**LA(627).
Ascomycotina
Anthostoma amygdalinum. On bark. **2:**AL(82), SC(82).
Botryosphaeria disrupta. Twig canker. **2:**FL(35).
B. dothidea. Canker, dieback. **2:**FL(35), NC(1303).
B. obtusa. On limbs. **2:**SE states(82).
B. quercuum. Twig canker. **2:**FL(35).
B. rhodina. On limbs. **2:**SE states(82).
B. ribis. Stem canker, bleeding necrosis, trunk lesion. **2:**Gulf states(3663), SE states(82), IL(2569), IN(2569), LA(82), MS(2720,3660).
Calonectria kyotensis. Root rot. **2:**NC(1303).
Ceratocystis coerulescens. Wood stain. **2:**Gulf States(82), SE states(779), FL(35).

C. moniliformis. Wood stain. **2:**Gulf States(82), FL(35).
Cryptodiaporthe liquidambaris. **2:**MD(2777).
Cryptosporella beltsvillensis. On decaying leaves. **2:**MD(2776).
Daldinia concentrica. On logs. **2:**NC(1303).
Diatrype asterostoma. On dead limbs. **2:**NC(1303).
Dicarpella georgiana. On leaves. **2:**GA(82,1438).
Didymella sp. **2:**GA(82,1438).
Didymosphaeria sp. On fruits. **2:**LA(658).
Endothia gyrosa. Canker, on dead limbs and exposed roots. **1:**GA(1438), MS(3362). **2:**Gulf states(82), SE states(82), GA(1438), NC(1303,3944).
Eutypella stellulata. On dead limbs. **2:**NC(1303).
Glonium lineare. On decorticated wood. **2:**NC(1303).
Gnomonia petiolorum. Leaf spot. **2:**GA(82,1438), MS(2720).
Hansenula polymorpha. In frass. **2:**LA(3596).
Hypocrea schweinitzii. **2:**GA(82,1438).
H. sulphurea. **3:**LA(2675).
Hypoxylon illitum. **2:**GA(82,1438).
H. sclerophaeum. **2:**GA(1438).
Melanconis xanthostroma. **2:**GA(82,1438).
Mycosphaerella punctiformis. **2:**GA(1438).
Mycosphaerella sp. **2:**GA(82,1438).
Myriangium duriaei. On scale insects infesting bark. **2:**Range of host(82), GA(1438).
Nectria sanguinea. **3:**LA(2675).
Nectria sp. Canker. **2:**GA(82).
Nitschkia acanthostroma. **2:**FL(2782).
N. broomeiana. **2:**FL(2782).
Ophiostoma piliferum. On wood or inner bark. **2:**AR(1364,1691,3714), MD(1364).
O. pluriannulatum. Wood stain. **2:**Gulf States(82), FL(35), MD(3714).
Peltella sp. Leaf spot. **2:**OK(2872).
Pezizella vulgaris var. *sanguinella*. **3:**SC(850).
Physalospora abdita. On limbs. **2:**SE states(82).
Physalospora sp. On limbs. **2:**SE states(82).
Pseudohelotium sacchariferum. **2:**GA(1438).
Tubeufia pezizula. Wood stain. **2:**FL(2782), IL(82).
Valsa ceratosperma. **2:**GA(82,1438).
V. liquidambaris. On limbs. **2:**GA(1438), NJ(82), SC(82).
Valsella laschii. **2:**GA(1438).
Velutarina rufo-olivacea. **2:**GA(1438).
Vibrissea truncorum. **2:**GA(1438).
Xylaria carpophila. On fallen fruits. **2:**NC(1303).
X. persicaria. On fruits. **2:**GA(82,539,1438,3014), LA(658), MD(651), TX(652). **3:**GA(659).

Basidiomycotina - Other

Aleurodiscus sp. **2:**TX(652).
Aporpium caryae. Wood rot. **2:**Unk.(82).
Armillaria mellea. Root rot. **2:**CA(2891,2898), FL(35).
A. tabescens. Root rot. **2:**FL(35,82).
Auricularia auricula. **3:**AR(2675), LA(2675).
A. cornea. **2:**LA(913).
Bjerkandera adusta. White rot. **2:**Range of host(82), FL(35), GA(1439), NC(1303). **3:**AR(2675), LA(2675), MS(2675).
Candelabrochaete langloisii. White rot. **2:**FL(432,443).
Ceriporia xylostromatoides. Wood rot. **2:**Unk.(82). **3:**LA(2675).
Cerrena unicolor. On wood. **2:**TX(82).
Clavicorona pyxidata. **2:**AL(884).
Climacodon pulcherrimus. Wood rot, sometimes on living trees. **2:**Gulf States(82), FL(35). **3:**AR(2675), LA(2675).
C. septentrionalis. On wood. **2:**MS(2720).
Coprinus disseminatus. **2:**NC(3944).
Coriolopsis rigida. White rot. **2:**S states(82).
Corticium stevensii. Thread blight. **2:**LA(82).
Cylindrobasidium laeve. On wood. **2:**GA(491,1439).
Dacryopinax spathularia. **3:**LA(2675).
Daedaleopsis confragosa. On wood. **2:**Range of host(82), FL(35), GA(1439), MS(2720), NC(1303,3944). **3:**LA(2675).
Dendrothele microspora. On bark. **2:**FL(2084), VA(2084).
D. strumosa. On bark. **2:**FL(2084).
Diplomitoporus lenis. **2:**Unk.(82).
Exidia recisa. On dead limbs. **2:**MD(128).
Fomes fasciatus. On wood. **2:**Unk.(82). **3:**LA(2675).
Fomitella supina. On wood. **2:**Unk.(82). **3:**LA(2675).
Fomitopsis spraguei. Wood rot chiefly of logs, sometimes of standing trees. **2:**Unk.(82). **3:**AR(2675), LA(2675).
Ganoderma applanatum. On wood. **2:**Unk.(82).
G. lucidum. White root and butt rot. **3:**LA(2675).
Gloeocystidiellum porosum. White rot. **2:**MD(2553).
G. turpe. On wood. **2:**FL(436).
Gloeodontia discolor. On wood. **2:**FL(440), MS(440), TX(440).
Gloeophyllum sepiarium. Wood rot of logs. **2:**Range of host(82).
G. trabeum. On wood. **2:**Range of host(82).
Gloeoporus dichrous. White rot. **2:**Unk.(82), NC(1303).
Gymnopilus armillatus. On living roots. **2:**FL(1542).
G. praefloccosus. On wood. **2:**FL(1542,2541).
Hericium erinaceus. Spongy white rot. **2:**Gulf States(82), FL(35).
Hexagonia hydnoides. On wood. **2:**Unk.(82). **3:**LA(2675).
Hymenochaete agglutinans. Stem canker. **2:**MD(82).
Hyphoderma mutatum. On wood. **2:**GA(1439), NC(1303,3944).
H. rimosum. White rot. **2:**FL(444).
H. setigerum. **3:**LA(2675).
Inonotus ludovicianus. Heart rot. **2:**SC states(82). **3:**AR(2675), LA(2675).
Junghuhnia nitida. **2:**MD(2168), NC(2168).
Laxitextum bicolor. On wood. **2:**MD(543), MS(2720), NC(543,1303,3944).
Lentinellus ursinus. Wood decay, heart rot. **2:**NC(1303).
Lentinus strigosus. On dead wood. **2:**NC(1303,3944).
L. tigrinus. Heart rot. **2:**LA(82), MS(82,2720).
Lenzites betulina. Wood rot of logs. **2:**Range of host(82). **3:**AR(2675), LA(2675).
Lenzites sp. Wood rot of logs. **2:**Range of host(82).
Meruliopsis corium. **3:**VA(128).
Neolentinus lepideus. Wood rot of timber. **2:**Range of host(82).
Nigroporus vinosus. **3:**LA(2675).
Oudemansiella canarii. On living and fallen trunks. **2:**FL(3284).
Oxyporus latemarginatus. Wood rot. **2:**Unk.(82). **3:**LA(2675).
Panellus pusillus. On wood. **2:**Unk.(82,2681), FL(441). **3:**AR(2675), LA(2675).
Perenniporia fraxinea. Wood rot of logs, timber, and sometimes living trees. **2:**Unk.(82).
Phaeolus schweinitzii. On wood. **2:**Unk.(82).
Phanerochaete burtii. White rot. **2:**FL(433), TN(433).
P. chrysorhiza. White rot. **2:**FL(433).
P. crassa. On wood. **2:**GA(491,1439).
P. flavido-alba. White rot. **2:**FL(433). **3:**FL(128).
P. velutina. **2:**NJ(128).
Phellinus conchatus. On wood. **2:**NC(1303).
P. ferruginosus. Wood rot. **2:**Unk.(82). **3:**LA(2675).
P. gilvus. White rot. **2:**Range of host(82), FL(35), NC(1303). **3:**AR(2675), LA(2675), MS(2675).
P. laevigatus. On wood. **2:**MS(2720).
P. melleoporus. On wood. **2:**Unk.(82). **3:**LA(2675).
P. punctatus. Wood rot. **2:**Unk.(82).
Phlebia chrysocrea. **2:**FL(2165).
P. ludoviciana. White rot. **2:**FL(2556).
Pholiota sp. **2:**MD(1017).
Pleurotus cystidiosus. Isolated from a fire-scarred tree. **2:**MS(2418).
P. dryinus. Sapwood rot. **2:**LA(82), MS(82,2720).
P. levis. Rot. **2:**AZ(82), LA(82). **3:**LA(2675).
P. ostreatus. On wood. **2:**LA(82), MS(82,2720).
Polyporus arcularius. On wood. **2:**NC(1303).
Pycnoporus cinnabarinus. On wood. **2:**Unk.(82), VA(2319).
P. sanguineus. On wood. **2:**Unk.(82). **3:**LA(2675).
Rigidoporus crocatus. Wood rot. **2:**Unk.(82).
R. lineatus. White pocket rot. **2:**C states(82), Unk.(82). **3:**AR(2675), LA(2675).
R. ulmarius. Wood rot of logs, timber, and sometimes living trees. **2:**Unk.(82). **3:**LA(2675), MS(2675).
Schizophyllum commune. Sapwood rot. **2:**Range of host(82), FL(35), GA(1439). **3:**AR(2675), LA(2675).
Schizopora paradoxa. Wood rot. **2:**Unk.(82).
Septobasidium pseudopedicellatum. Parasitic on scale insects infesting bark. **2:**S states(82), FL(35), NC(1303).
Septobasidium sp. Parasitic on scale insects. **2:**FL(35).
Simocybe atomacea. On rotten logs. **2:**FL(3287).
Spongipellis pachyodon. **3:**LA(2675).
Steccherinum ochraceum. Wood rot. **2:**Gulf States(82), FL(35). **3:**LA(2675), MS(2675).

Stereum hirsutum. On wood. **2**:CT(543), GA(1439), MS(2720), NC(1303,3944). **3**:TX(652).
S. ochraceo-flavum. **2**:MO(543).
S. ostrea. Wood rot. **2**:Unk.(82), GA(543,1439), IN(543).
S. striatum. On wood. **2**:AL(543), NC(1303,3944).
Tinctoporellus epimiltinus. On wood. **2**:GA(491,1439).
Tomentella botryoides. **3**:NJ(2039).
Trametes cubensis. On wood. **3**:SE states(2681).
T. hirsuta. On wood. **2**:Unk.(82), GA(1439), NC(1303).
T. subectypus. On wood. **2**:Unk.(82). **3**:LA(2675).
T. versicolor. White rot. **2**:Unk.(82), GA(1439), MS(2720), NC(1307), SC(1307). **3**:LA(2675).
Tremella mesenterica. On wood. **2**:NC(1303,3944).
Trichaptum biforme. On wood. **2**:Unk.(82), NC(1303). **3**:AR(2675), LA(2675).
T. sector. Butt rot, root rot. **2**:FL(35,82), MS(82,2720). **3**:AR(2675), LA(2675).
Tyromyces calkinsii. On wood. **3**:FL(2681).
T. fissilis. Heart rot. **2**:SC states(82), FL(35), MS(2720). **3**:LA(2675).
T. galactinus. Wood rot chiefly of logs, sometimes of standing trees. **2**:Unk.(82). **3**:LA(2675).
Volvariella bombycina. **3**:LA(2675).
Xylobolus subpileatus. Pocket rot of logs. **2**:Gulf States(82). **3**:AR(2675), LA(2675).

Deuteromycotina - Hyphomycetes

Beltrania rhombica. On fruits. **2**:GA(659).
Cercospora liquidambaris. Leaf spot. **2**:SE states(82), DE(3543), FL(35), MS(2720), NC(1303,3944), TX(82).
C. tuberculans. Leaf spot. **2**:FL(35,82), LA(82), MO(82), MS(82,2720).
Cladosporium sp. Leaf spot. **2**:OR(197,3202).
Cylindrocladium crotalariae. Root rot. **2**:NC(1303).
C. scoparium. Isolated from roots. **2**:GA(678,1694).
Fusarium moniliforme. Pink and purple wood stain. **2**:S states(82), FL(35).
F. solani. Pink and purple wood stain. **2**:S states(82), FL(35).
Gliocladium roseum. On fruits. **2**:GA(659).
Graphium rigidum. Wood stain. **2**:S states(82), FL(35).
G. rubrum. Wood stain. **2**:S states(82).
Helicosporium panacheum. On decorticated wood. **2**:NC(1303).
Helminthosporium asterinum. **3**:FL(634).
Leptographium microsporum. Wood stain. **2**:LA(777).
Oidium sp. Powdery mildew. **2**:NC(1303).
Penicillium sp. Green mold of timber. **2**:Range of host(82).
Phymatotrichopsis omnivora. Root rot. **2**:TX(82).
Stigmina liquidambaris. On living leaves. **2**:TX(82,2500,3597).
[*Torula ligniperda*]. Red wood stain. **2**:Range of host(82), FL(35).
Tubercularia vulgaris. Stem canker and dieback. **2**:MS(2720,3660).

Deuteromycotina - Coelomycetes

Aplosporella liquidambaris. Twig dieback. **2**:FL(35), NY(82).
Coniothyrium sp. Twig dieback. **2**:FL(35).
Discosia artocreas. Leaf spot. **2**:OK(82,2873).
Discula umbrinella. Leaf spot. **2**:FL(35), MS(2720).
Dothiorella berengeriana. Bleeding necrosis. **2**:NJ(82), NY(82).
Harknessia americana. On twigs. **3**:FL(3549), NJ(3549), SC(3549).
H. liquidambaris. On limbs. **2**:FL(2782), GA(2548), NJ(2548), SC(82).
Lasiodiplodia theobromae. Wood stain. **2**:Gulf States(82), FL(35).
Macrophoma sp. Leaf spot. **2**:FL(35).
Monochaetia sp. Leaf spot. **2**:FL(35).
Phoma sp. Twig dieback. **2**:FL(35).
Phomopsis sp. Twig dieback. **2**:FL(35).
Phyllosticta sp. Leaf spot. **2**:FL(35).
Pteridiospora spinosispora. Isolated from mycorrhizal roots. **2**:MS(1060).
Rhodosticta quercina. **3**:TX(652).
Seiridium sp. Stem canker. **2**:FL(35).
Septoria liquidambaris. Leaf spot. **2**:E states(82), FL(35), NC(1303), OK(2872), TX(82).
Tubakia dryina. Red leaf spot. **2**:FL(35), IL(82,1246), MD(82), NC(82,1303,3944), OK(2873).

Deuteromycotina - Other

Rhizoctonia solani. Root rot. **2**:FL(35).

Rhodoleia Champ. ex Hook.

1. *R. championii* Hook.

Deuteromycotina - Other

Rhizoctonia solani. Root and stem rot. **1**:FL(35).

Heliconiaceae

Heliconia L.

1. *H. psittacorum* L. fil. [parakeet-flower, parrot's-flower, parrot's-plantain]
2. *Heliconia* sp.

Oomycetes

Pythium sp. Root rot. **2**:FL(35).

Deuteromycotina - Hyphomycetes

Cercospora sp. Leaf spot. **2**:FL(35).
Curvularia sp. Leaf spot. **2**:FL(35).
Helminthosporium sp. Leaf spot. **2**:FL(35).

Deuteromycotina - Coelomycetes

Phomopsis sp. Leaf spot. **2**:FL(35).
Phyllosticta heliconiae. Leaf spot. **2**:HI(2903).
Septoria sp. Leaf spot. **1**:HI(2899,2903).

Deuteromycotina - Other

Rhizoctonia solani. Root rot. **2**:FL(35).

Hippocastanaceae

Aesculus L.

1. *A. ambigua* Hort.
2. *A. arnoldiana* Sarg.
3. *A.* × *bushii* C. K. Schneid. Probably a hybrid of *A. discolor* var. *mollis* and *A. glabra* var. *leucodermis*.
4. *A. californica* (Spach) Nutt. [California buckeye, California horsechestnut]
5. *A.* × *carnea* Hayne. *A. hippocastanum* × *A. pavia*. [red horsechestnut]
6. *A.* × *carnea* Hayne cv. 'briotii' [red horsechestnut]
7. *A.* × *carnea* Hayne var. *plantierensis* (André) Rehd.
8. *A. chinensis* Bunge
9. *A. discolor* Pursh var. *mollis* (Raf.) Sarg.
10. *A. dupontii* Sarg.
11. *A. glabra* Willd. (=*A. glabra* Willd. var. *leucodermis* Sarg., *A. glabra* Willd. var. *sargentii* Rehd.) [Ohio buckeye]
12. *A. glabra* Willd. var. *arguta* (Buckley) B. L. Robinson [Texas buckeye]
13. *A. glabra* Willd. var. *monticola* Sarg.
14. *A. glabra* Willd. f. *pallida* (Willd.) Scheele
15. *A. hippocastanum* L. [common horsechestnut, European horsechestnut]
16. *A. hippocastanum* L. cv. 'baumannii'
17. *A. hippocastanum* L. var. *laciniata* Jacq.
18. *A. hippocastanum* L. var. *luteo-variegata* Nestl.
19. *A. hippocastanum* L. var. *pyramidalis* Henry
20. *A. hippocastanum* L. var. *umbraculifera* Jacq.
21. *A.* × *hybrida* DC. Probably a hybrid of *A. octandra* and *A. pavia*.
22. *A.* × *mississippiensis* Sarg. Probably a hybrid of *A. glabra* and *A. pavia*.
23. *A.* × *mutabilis* (Spach) Scheele var. *harbisonii* (Sarg.) Rehd.
24. *A.* × *mutabilis* (Spach) Scheele var. *induta* Sarg.
25. *A.* × *mutabilis* (Spach) Scheele var. *penduliflora* Sarg.
26. *A. octandra* Marsh. [sweet buckeye, yellow buckeye]
27. *A. octandra* Marsh. var. *vestita* Sarg.
28. *A. parviflora* Walter (=*A. parviflora* Walter f. *serotina* Rehd.) [bottlebrush buckeye, dwarf horsechestnut]
29. *A. pavia* L. (=*A. discolor* Pursh, *A. splendens* Sarg.) [red buckeye]
30. *A. pavia* L. var. *atrosanguinea* Kirchn.
31. *A. pavia* L. var. *neglecta* Hort.
32. *A. pavia* L. var. *rosea* Hort.
33. *A. sylvatica* W. Bartram (=*A. glaucescens* Sarg., *A. neglecta* Lindl., *A. neglecta* Lindl. var. *georgiana* (Sarg.) Sarg., *A. georgiana* Sarg., *A. neglecta* Lindl. var. *lanceolata* (Sarg.) Sarg., *A. neglecta* Lindl. var. *pubescens* (Sarg.) Sarg., *A. neglecta* Lindl. var. *tomentosa* Sarg.)

34. *A. turbinata* Blume [Japanese horsechestnut]
35. *A. woerlitzensis* Koehne
36. *A. woerlitzensis* Koehne var. *ellwangeri* Rehd.
37. *Aesculus* sp.

Oomycetes

Phytophthora sp. Bleeding canker. **15**:RI(82).

Ascomycotina

Botryosphaeria dothidea. **26**:GA(1438). **29**:GA(1438).
B. obtusa. On limbs and twigs. **29**:GA(82). **37**:NJ(82), NY(82).
B. ribis. **26**:GA(82).
Cryptodiaporthe aesculi. On twigs. **4**:CA(82).
Diaporthe eres. Blight. **37**:Range of host(82), FL(35).
D. padi var. *patria*. **15**:GA(1438). **26**:GA(1438). **33**:GA(1438). **37**:GA(82).
Eutypella juglandicola. **26**:GA(1438). **29**:GA(1438). **33**:GA(1438). **37**:GA(82).
E. scoparia. **4**:CA(2925).
Fenestella erratica. **11**:OH(195).
Glomerella cingulata. Anthracnose, leaf blight, trunk canker. **15**:CT(82,3770), MD(82), MO(82), NJ(82), NY(82), TX(82).
Gnomoniella sp. **26**:GA(82,1438).
Guignardia aesculi. Leaf blotch. **1**:IL(2574). **2**:IL(2574). **4**:IL(2570). **8**:IL(2570). **9**:IL(2574). **10**:IL(2574). **11**:Range of host(82), IL(1052,2574), OK(2872). **14**:IL(2570). **13**:IL(2570). **15**:Range of host(82), CT(3770), IL(1052,2574), MA(1390), NY(3780), OK(2872). **16**:IL(2574). **17**:IL(2574). **18**:IL(2570). **19**:IL(2574). **20**:IL(2574). **26**:IA(82), IL(82,2570), OK(2872), TX(82). **27**:IL(2574). **28**:IL(2570). **29**:AL(82), FL(35), IL(2574), MS(82,2720). **30**:IL(2570). **31**:IL(2574). **32**:IL(2574). **34**:IL(2574). **35**:IL(2574). **36**:IL(2574). **3**:IL(2574). **5**:IL(2574). **6**:IL(2574). **7**:IL(2574). **21**:IL(2574). **22**:IL(2574). **23**:IL(2574). **24**:IL(2574). **25**:IL(2574). **37**:Range of host(82), NC(1303).
Hypoxylon rubiginosum. **33**:GA(1438). **37**:GA(82).
Massariovalsa sudans. **26**:GA(82,1438).
Mycosphaerella aesculi. Leaf spot. **4**:CA(82).
M. punctiformis. **29**:GA(82,1438).
Nectria cinnabarina. Twig blight. **37**:Range of host(82).
N. hippocastani. On pericarps. **15**:NY(82).
Ophiostoma piliferum. On inner bark or wood. **11**:NC(1364).
O. pluriannulatum. On wood. **11**:NC(3714).
Phyllactinia guttata. Powdery mildew. **4**:CA(81,82), TX(82).
Taphrina aesculi. Yellow leaf blister. **4**:CA(82,2454), TX(82).
Uncinuliella flexuosa. Powdery mildew. **11**:C states(82), NE states(82), MS(2720). **15**:Range of host(82), IN(663), NY(2456). **26**:Unk.(82). **29**:Range of host(82), AL(82), MS(82,2720). **5**:IN(82), PA(82). **37**:E & C states(82).
Valsaria insitiva. **26**:GA(82,1438).

Basidiomycotina - Uredinales

Puccinia andropogonis. **11**:IN(82), KS(82), MO(82), NE(82).

Basidiomycotina - Other

Armillaria mellea. Root rot. **4**:CA(2898). **37**:Range of host(82), FL(35).
Chondrostereum purpureum. **15**:OR(543).
Coriolopsis gallica. **15**:OR(3202).
Flammulina velutipes. Collar rot, white sapwood rot. **15**:RI(82). **37**:Unk.(82).
Ganoderma applanatum. On wood. **37**:Range of host(82), FL(35).
Gymnopilus fuscosquamulosus. On roots. **37**:NC(1542).
Peniophora malenconii subsp. *americana*. On wood. **4**:CA(542).
Pholiota sp. **15**:WA(3202).
Pleurotus sp. **15**:WA(3202).
Polyporus squamosus. White-mottled heart rot. **26**:NC(82).
Rhodotus palmatus. On wood. **15**:MI(2422).
Septobasidium pseudopedicellatum. Parasitic on scale insects. **29**:FL(35).
Septobasidium sp. Parasitic on scale insects. **37**:VA(1533).
Spongipellis spumeus. Wood rot. **37**:NY(82).
Stereum hirsutum. **4**:CA(543).
Trametes trogii. **4**:CA(82).

Deuteromycotina - Hyphomycetes

Cercospora aesculina. Leaf spot. **26**:WV(82). **29**:FL(35,82).
Cylindrocarpon album. On twigs. **4**:CA(82).
C. magnusianum. On twigs. **4**:CA(82).
Fusarium acuminatum. Dieback. **4**:CA(82).
Macrosporium baccatum. Leaf spot. **11**:KS(82). **12**:KS(82).
Phymatotrichopsis omnivora. Root rot. **15**:TX(82). **29**:TX(82). **37**:TX(82).
Verticillium albo-atrum. Wilt. **15**:PA(82).

Deuteromycotina - Coelomycetes

Aplosporella aesculi. On pericarps. **15**:NY(82).
Colletotrichum sp. Anthracnose. **37**:FL(35).
Diplodia aesculi. On twigs. **15**:NY(82).
Diplodina aesculi. On pericarps. **4**:CA(82).
Monochaetia monochaeta. Leaf spot. **37**:NC(82).
Phomopsis carposchiza. On nuts. **15**:NY(82).
Phyllosticta sphaeropsoidea. Leaf blotch and spot. **11**:MS(2720). **29**:FL(35,82), OK(2872), TX(82), WV(82). **37**:NC(1303).
Septoria glabra. Leaf spot. **11**:IN(82).
S. hippocastani. Leaf spot. **15**:PA(82), VT(82).

Hydrangeaceae

Broussaisia Gaud. in Freyc.

1. *B. pellucida* Gaud.
2. *Broussaisia* sp.

Oomycetes

Phytophthora cinnamomi. Root rot. **1**:HI(2903).

Deuteromycotina - Hyphomycetes

Trichothallus hawaiiensis. **2**:HI(2903).

Decumaria L.

1. *D. barbara* L. [climbing hydrangea, wood-vamp]

Ascomycotina

Diaporthe arctii. **1**:GA(1438).

Deuteromycotina - Hyphomycetes

Cercospora decumariae. Leaf spot. **1**:FL(35), MS(82,2720).

Deutzia Thunb.

1. *D. gracilis* Siebold & Zucc.
2. *D.* × *lemoinei* Hort. Lemoine ex Boiss. *D. gracilis* × *D. parviflora*.
3. *D. scabra* Thunb.
4. *Deutzia* sp.

Deuteromycotina - Hyphomycetes

Cercospora deutziae. Leaf spot. **1**:DE(82,581,3543), IA(82), TX(82). **3**:IL(313). **2**:OK(2872), WI(1354).

Deuteromycotina - Coelomycetes

Phyllosticta deutziae. On leaves. **4**:Unk.(966), AL(82), IA(82), NJ(82), TX(82).

Fendlera Engelm. & A. Gray

1. *F. rupicola* Engelm. & A. Gray
2. *Fendlera* sp.

Ascomycotina

Teichosporella dura. On twigs. **2**:CO(82).

Basidiomycotina - Uredinales

Gymnosporangium speciosum. Heteroecious: aecial host. **1**:AZ(1194,1195,1203,3981). **2**:AZ(82), CO(82), NM(82), UT(82).

Hydrangea L.

1. *H. anomala* D. Don subsp. *petiolaris* (Siebold & Zucc.) D. McClint. (=*H. petiolaris* Siebold & Zucc.) [climbing hydrangea]
2. *H. arborescens* L. [hills-of-snow, sevenbark, smooth hydrangea, wild hydrangea]
3. *H. arborescens* L. subsp. *discolor* (Ser.) McClint. (=*H. cinerea* Small)
4. *H. arborescens* L. subsp. *radiata* (Walter) McClint. (=*H. radiata* Walter)
5. *H. macrophylla* (Thunb.) Ser. [French hydrangea, hortensia, house hydrangea]
6. *H. paniculata* Siebold [panicle hydrangea]
7. *H. paniculata* Siebold cv. 'grandiflora' [peegee hydrangea]
8. *H. quercifolia* W. Bartram
9. *Hydrangea* sp.

Oomycetes

Peronospora calotheca. Downy mildew. **7**:IA(82), ND(82), WI(82).
Pythium sp. Root rot. **5**:FL(35). **9**:FL(35).

Ascomycotina
Diaporthe eres. On dead limbs. **2:**IL(650), WV(82).
Erysiphe polygoni. Powdery mildew. **1:**Range of host(82). **2:**MD(82), WV(82). **5:**FL(35), MN(82), OH(82). **6:**Range of host(82). **7:**Range of host(82). **9:**Range of host(82), FL(35), HI(2899).
Eutypella juglandicola. **2:**GA(1438).
Helotium sp. **2:**IN(647).
Massarina microcarpa. On limbs. **2:**OK(2872), WV(82).
Microsphaera friesii. **5:**WA(82).
M. penicillata. **5:**Unk.(3976).
Nectria cinnabarina. On limbs. **1:**VA(82), VA(82). **6:**VA(82). **7:**VA(82).
Nectriella pironii. Stem gall. **5:**FL(35). **9:**FL(35).
Stictis hydrangeae. On dead twigs. **2:**IN(654). **9:**IN(3225), NC(3225).
Venturia anthophila. On inflorescences. **1:**MD(82), MN(82). **6:**MD(82), MN(82). **7:**MD(82), MN(82). **9:**MD(82), MN(82).
Basidiomycotina - Uredinales
Pucciniastrum hydrangeae. Heteroecious: telial host. **2:**E states(82), AR(82), GA(1439), IL(82), IN(647,654), KY(1575), NC(1303), OH(1575), PA(1575), TN(1575), WV(1575). **3:**TN(1575). **4:**GA(1539), NC(1307), SC(1307). **6:**E states(82), IN(82), PA(1575), WV(82). **9:**E states(82), TN(82), WV(82).
Basidiomycotina - Other
Armillaria mellea. Root rot. **5:**CA(2900). **8:**CA(2897).
Trametes versicolor. Wound rot. **1:**CT(82), IA(82), MD(82). **6:**CT(82), IA(82), MD(82). **7:**CT(82), IA(82), MD(82). **9:**CT(82), IA(82), MD(82).
Deuteromycotina - Hyphomycetes
Alternaria sp. Leaf spot. **5:**FL(35). **9:**FL(35).
Botrytis cinerea. Flower blight, gray mold blight, leaf blight. **1:**Range of host(82). **2:**MS(2720), NC(1303). **5:**Range of host(82), FL(35), HI(2903). **6:**Range of host(82). **7:**Range of host(82). **9:**Range of host(82), FL(35), WA(3202).
Cercospora hydrangeae. Leaf spot. **2:**IL(82), OK(82,2872). **5:**FL(35), OK(82,2872). **9:**SE states(82), AL(82), FL(35), MS(2720), TX(82,3597).
Cercospora sp. Leaf spot. **2:**NC(1303). **5:**FL(82), MD(82).
Corynespora cassiicola. Leaf spot. **5:**FL(35,3367). **9:**FL(35).
Helminthosporium sp. Leaf spot. **5:**FL(35). **9:**FL(35).
Oidium sp. Powdery mildew. **5:**HI(2903), MD(82), PA(82), VA(82). **9:**HI(2903), WA(3202).
Phymatotrichopsis omnivora. Root rot. **1:**TX(82). **6:**TX(82). **7:**TX(82). **9:**TX(82).
Stachybotrys kampalensis. On dead twigs. **2:**AL(2486).
Deuteromycotina - Coelomycetes
Colletotrichum sp. Leaf spot. **5:**FL(35), MD(82), NJ(82). **9:**FL(35).
Cylindrosporium sp. Leaf spot. **9:**MS(2720).
Cytospora sacculus. On limbs. **1:**SC(82). **2:**WV(82). **6:**SC(82). **7:**SC(82). **9:**SC(82).
Hendersonia hydrangeae. On limbs. **1:**NY(82). **6:**NY(82). **7:**NY(82). **9:**NY(82).
Phoma exigua. Leaf spot. **1:**AK(82), NJ(82). **5:**NJ(82). **6:**AK(82), NJ(82). **7:**AK(82), NJ(82). **9:**AK(82,526), NJ(82).
Phomopsis sp. Canker. **5:**FL(35).
Phyllosticta hydrangeae. Leaf spot. **1:**Range of host(82). **2:**IN(82), NC(1303). **5:**DE(82), FL(35), NJ(82). **6:**Range of host(82). **7:**Range of host(82). **9:**Range of host(82), Unk.(966), FL(35).
Septoria hydrangeae. Leaf spot. **1:**CT(82), MS(82), OH(82). **2:**MS(2720), NC(1303). **6:**CT(82), MS(82,2720), OH(82). **7:**CT(82), MS(82), OH(82). **9:**CT(82), MS(82,2720), OH(82).
Deuteromycotina - Other
Rhizoctonia solani. Root and stem rot. **5:**FL(35), MD(82), OH(946).
Rhizoctonia sp. Crown and root rot. **5:**HI(2903).
Sclerotium rolfsii. Southern blight. **1:**FL(82), TX(82). **6:**FL(82), TX(82). **9:**FL(35,82), TX(82).

Jamesia Torr. & A. Gray
1. *J. americana* Torr. & A. Gray [cliffbush]
Deuteromycotina - Hyphomycetes
Ovularia edwiniae. **1:**CO(82,128).

Philadelphus L.
1. *P. coronarius* L. [sweet mock-orange]
2. *P. inodorus* L. var. *grandiflorus* (Willd.) A. Gray (=*P. grandiflorus* Willd.)
3. *P.* × *lemoinei* Hort. Lemoine. *P. coronarius* × *P. microphyllus*.
4. *P. lewisii* Pursh (=*P. gordonianus* Lindl.)
5. *P. microphyllus* A. Gray
6. *Philadelphus* sp.
Ascomycotina
Dasyscyphus patulus. **4:**ID(3202).
Leptosphaeria sp. **4:**WA(3202).
Lophiostoma submagnum. **4:**OR(3202).
Nectria cinnabarina. On twigs. **6:**ME(82), PA(82).
Nectria sp. **4:**OR(3202).
Phyllactinia guttata. Powdery mildew. **2:**GA(1440). **4:**MT(3202), WA(3202). **6:**MT(82), NC(1303), WA(82).
Strickeria kelseyi. **4:**MT(3202).
Basidiomycotina - Uredinales
Gymnosporangium speciosum. Heteroecious: aecial host. **5:**AZ(1194,1203,3981). **6:**CO(82), NM(82), TX(82), UT(82,2771).
[*G. tremelloides*]. **6:**AZ(1195).
Basidiomycotina - Other
Dendrophora albobadia. **6:**PA(543).
Exidiopsis calcea. **4:**WA(3202).
Trametes hirsuta. On dead limbs. **4:**ID(3202). **6:**OR(82).
T. versicolor. On dead limbs. **6:**OR(82).
Deuteromycotina - Hyphomycetes
Botrytis cinerea. Flower and shoot blight. **1:**FL(35). **6:**Range of host(82).
Cercospora angulata. Leaf spot. **1:**FL(35), OK(2872). **2:**OK(2872). **4:**ID(3202). **3:**OK(2872). **6:**IL(312), MO(82), TX(82).
Cercospora sp. Leaf spot. **1:**HI(2903).
Cladosporium herbarum. **4:**WA(3202).
Peyronelia sirodesmioides. **4:**WA(3202).
Phymatotrichopsis omnivora. Root rot. **6:**TX(82).
Ramularia philadelphi. Leaf spot. **4:**ID(638,3202), OR(3202), WA(3202). **6:**TX(82), WA(82).
[*Sarcinella heterospora*]. Sooty blotch. **6:**FL(82).
Deuteromycotina - Coelomycetes
Ascochyta philadelphi. Leaf spot. **6:**NY(82).
Colletotrichum sp. Anthracnose. **1:**FL(35).
Macrophoma sp. Stem canker. **1:**FL(35).
Microdiplodia microsporella. On twigs. **6:**ME(82).
Phoma philadelphi. On stems and twigs. **4:**OR(3202,4007). **6:**OR(82).
Phomopsis landeghemiae. On twigs. **6:**ME(82), VA(82).
Septoria philadelphi. Leaf spot. **4:**ID(3202), WA(3202). **6:**IA(82), ID(82), MT(82), WA(82).
Sphaeropsis sp. Stem gall. **1:**FL(35).

Whipplea Torr.
1. *W. modesta* Torr. [yerba-de-selva]
Oomycetes
Peronospora whippleae. Downy mildew. **1:**CA(82).
Ascomycotina
Erysiphe polygoni. Powdery mildew. **1:**CA(1155).

Hydrocharitaceae

Elodea Michx.
1. *E. canadensis* Michx. [elodea, waterweed]
Chytridiomycetes
Megachytrium westonii. **1:**NY(3407).

Hydrilla Rich.
1. *H. verticillata* (L. fil.) Royle [hydrilla]
Deuteromycotina - Other
Sclerotium rolfsii. Southern blight. **1:**FL(35).

Limnobium Rich.
1. *L. spongia* (Bosc) Rich. ex Steud.
Deuteromycotina - Hyphomycetes
Cercospora limnobii. Leaf spot. **1:**FL(35,614).

Thalassia Banks & Sol. ex K. D. Koenig
1. *T. testudinum* K. D. Koenig
Myxomycetes
Labyrinthula sp. **1**:FL(3566).

Hydrophyllaceae

Ellisia L.
1. *E. nyctelea* (L.) L.
2. *Ellisia* sp.
Oomycetes
Peronospora hydrophylli. **1**:WI(1351). **2**:MT(3202).
Basidiomycotina - Uredinales
Puccinia recondita. Heteroecious: aecial host. **1**:OK(2872), SD(2243).

Eriodictyon Benth.
1. *E. californicum* (Hook. & Arnott) Torr.
2. *E. crassifolium* Benth.
3. *Eriodictyon* sp.
Deuteromycotina - Hyphomycetes
Heterosporium californicum. **1**:Unk.(82).
Torula glutinosa. On leaves. **1**:CA(81). **2**:CA(81). **3**:CA(82).
Trimmatostroma eriodictyonis. Sooty blotch. **3**:CA(82).

Hesperochiron S. Wats.
1. *H. californicus* (Benth.) S. Wats.
2. *H. pumilus* (Griseb.) Porter
Chytridiomycetes
Synchytrium sp. On flower bracts, leaves, petioles and stems. **2**:WA(1847,3202).
Ascomycotina
Erysiphe cichoracearum. **1**:WA(3202).

Hydrolea L.
1. *H. ovata* Nutt. ex Choisy
Deuteromycotina - Hyphomycetes
Cercospora namae. Leaf spot. **1**:OK(2873).

Hydrophyllum L.
1. *H. canadense* L.
2. *H. capitatum* Douglas ex Benth. [cat's-breeches]
3. *H. fendleri* (A. Gray) A. Heller
4. *H. fendleri* (A. Gray) A. Heller var. *albifrons* (A. Heller) Macbr.
5. *H. macrophyllum* Nutt.
6. *H. occidentale* (S. Wats.) A. Gray
7. *H. tenuipes* A. Heller
8. *H. virginianum* L. [Indian-salad, John's cabbage, Shawnee-salad, Virginia waterleaf]
9. *Hydrophyllum* sp.
Oomycetes
Peronospora hydrophylli. Downy mildew. **2**:ID(3202), OR(3202), WA(82,3202). **3**:ID(3202), OR(3202), WA(3202). **6**:OR(3202). **7**:OR(3202), WA(3202). **8**:IA(82), IL(82), NY(82), WI(82). **9**:MD(82).
Ascomycotina
Erysiphe cichoracearum. Powdery mildew. **2**:OR(3202), WA(3202). **3**:ID(3202), WA(3202). **5**:NC(1303). **6**:WA(3202). **7**:CA(1155). **8**:ND(377). **9**:Unk.(82).
E. polygoni. Powdery mildew. **2**:ID(82). **9**:WA(82).
Sphaerotheca fuliginea. **8**:MN(82), ND(82,377).
S. macularis. Powdery mildew. **1**:OH(82). **3**:WA(3202). **8**:ND(82), SD(82,2243).
Basidiomycotina - Uredinales
Puccinia hydrophylli. Autoecious. **2**:CO(82), MT(3202), OR(3202), UT(82,1520), WA(3202). **3**:CO(1520), UT(1195). **6**:CA(82), UT(1520). **8**:E & C states(82), IA(1520), ND(82,1520), NE(1520), NY(1520), SD(1520,2243), VT(1520), WI(1520). **9**:NC(1303), OR(3202).
P. plumasensis. Autoecious. **2**:CO(1520). **6**:CA(1520).
P. recondita. Heteroecious: aecial host. **2**:W states(82), ID(3202), MT(3202), OR(3202), UT(82,1195), WA(82,3202), WY(1195,3379). **3**:CO(1195), ID(3202), OR(3202), WA(3202), WY(1195). **4**:CA(81). **6**:CA(81,82), WA(3202). **7**:CA(81), OR(3202), WA(3202). **8**:NC states(82), NE states(82), ND(378), WI(1330).
Basidiomycotina - Other
Ceratobasidium anceps. Stem necrosis. **1**:NY(82), WI(82). **8**:NY(82), WI(82,1342,1343).
Deuteromycotina - Hyphomycetes
Didymaria hydrophylli. Leaf spot. **2**:ID(3202), MT(3202), OR(3202), WA(82,970,3202), WY(1195).
Septocylindrium hydrophylli. Leaf spot. **1**:IL(82).
Deuteromycotina - Coelomycetes
Ascochyta hydrophylli-virginiani. On living leaves. **8**:WI(1339).
Gloeosporium hydrophylli. Leaf spot. **1**:NY(82). **7**:OR(3202,3421). **8**:NY(82). **9**:OR(82).
Septoria hydrophylli. Leaf spot. **1**:NY(82). **8**:NY(82).
Deuteromycotina - Other
Sclerotium deciduum. Stem necrosis. **8**:WI(82).

Nama L.
1. *N. hispidum* A. Gray
Chytridiomycetes
Synchytrium namae. On leaves. **1**:TX(1846).

Nemophila Nutt.
1. *N. breviflora* A. Gray
2. *N. kirtleyi* L. Henderson
3. *N. menziesii* Hook. & Arnott [baby-blue-eyes]
4. *N. menziesii* Hook. & Arnott var. *atomaria* (Fisch. & C. A. Mey.) H. P. Chandler
5. *N. parviflora* Douglas ex Benth.
6. *N. pedunculata* Douglas ex Benth. (=*N. sepulta* Parish)
7. *Nemophila* sp.
Oomycetes
Peronospora nemophilae. On living leaves. **1**:ID(3202), MT(3202), WA(3202). **2**:ID(3202), WA(3202). **5**:OR(3202), WA(3202). **6**:OR(3202), WA(3201).
Ascomycotina
Erysiphe cichoracearum. Powdery mildew. **3**:CA(82), TX(82). **4**:CA(81). **5**:CA(81), ID(3202), WA(3202). **7**:CA(82), NE(82), WA(82).
Sphaerotheca macularis. **5**:WA(3202).
Basidiomycotina - Uredinales
Puccinia recondita. **5**:WA(3202).
Deuteromycotina - Hyphomycetes
Fusarium sp. Root and stem rot. **3**:FL(35).

Phacelia Juss.
1. *P. bolanderi* A. Gray
2. *P. breweri* A. Gray
3. *P. crenulata* Torr. ex S. Wats.
4. *P. distans* Benth.
5. *P. eisenii* T. S. Brandegee
6. *P. hastata* Douglas ex Lehm. (=*P. leucophylla* Torr. in Frém.)
7. *P. heterophylla* Pursh
8. *P. humilis* Torr. & A. Gray
9. *P. idahoensis* L. Henderson
10. *P. linearis* (Pursh) Holz.
11. *P. magellanica* (Lam.) Coville
12. *P. mutabilis* Greene
13. *P. procera* A. Gray
14. *P. purpusii* T. S. Brandegee
15. *P. racemosa* (Kellogg) T. S. Brandegee
16. *P. ramosissima* Douglas ex Lehm.
17. *P. sericea* (R. J. Graham) A. Gray
18. *P. suaveolens* Greene
19. *Phacelia* sp.
Oomycetes
Peronospora hydrophylli. **6**:WA(3202). **7**:ID(3202), MT(3202), WA(3202). **10**:WA(3202).
Ascomycotina
Erysiphe cichoracearum. Powdery mildew. **1**:Unk.(3976). **6**:OR(3202), WA(3202). **7**:OR(3202). **10**:MT(3202), WA(3202). **12**:CA(642,661). **16**:WA(3202). **19**:CA(82), MT(82), NM(82), TX(82).
Leptosphaeria salsolae. **19**:WA(3202).
Pleospora comata. **17**:ID(3202).
Sphaerotheca fuliginea. Powdery mildew. **1**:CA(1155). **2**:CA(1155).

4:CA(1155). **5**:CA(1155). **14**:CA(1155). **15**:CA(1155). **18**:CA(1155).

Basidiomycotina - Uredinales

[*Puccinia aristidae*]. **3**:AZ(1203). **19**:Unk.(82).

P. phaceliae. Autoecious. **7**:WA(3202). **10**:ID(3202), WA(3202). **17**:WA(3202). **19**:WA(82).

P. recondita. Heteroecious: aecial host. **7**:MT(3202), OR(3202), WA(3202), WY(1195,2924). **8**:OR(3202). **11**:AZ(1203,3981). **19**:W states(82), CA(82), NM(82), OR(82).

P. subnitens. **3**:AZ(3981).

Uredo contraria. **19**:CA(82).

Deuteromycotina - Hyphomycetes

Phymatotrichopsis omnivora. Root rot. **19**:TX(82).

Ramularia phaceliae. On living leaves. **6**:ID(3202), OR(3202), WY(1195). **7**:ID(3202), OR(3202). **9**:MT(3202). **13**:CA(81), WA(3202).

Deuteromycotina - Coelomycetes

Cylindrosporium phaceliae. Leaf spot. **17**:MT(3202). **19**:MT(82), TX(82).

Pestalotia sp. On dead stems of overwintered herbage. **6**:CA(642).

Romanzoffia Cham. ex Nees

1. *R. sitchensis* Bong.

Basidiomycotina - Uredinales

Puccinia romanzoffiae. Autoecious. **1**:OR(82,3202).

Illiciaceae

Illicium L.

1. *I. floridanum* J. Ellis [purple anise]
2. *I. parviflorum* Venten.
3. *Illicium* sp.

Oomycetes

Pythium splendens. Root rot. **2**:FL(35). **3**:FL(35).

Ascomycotina

Asteridium illicii. On leaves. **1**:MS(82,2720).

Glomerella cingulata. Leaf spot. **2**:FL(35). **3**:FL(35).

Laestadia illiciicola. On leaves. **1**:MS(82,2720).

Lembosia illiciicola. Black mildew. **1**:AL(82), MS(82,2720).

Deuteromycotina - Hyphomycetes

Alternaria sp. Leaf spot. **2**:FL(35).

Cercospora sp. Leaf spot. **2**:FL(35).

Microxiphium fagi. Sooty mold. **1**:MS(82,2720).

Deuteromycotina - Coelomycetes

Phyllosticta sp. Leaf spot. **2**:FL(35). **3**:FL(35).

Sphaceloma sp. Leaf edema. **2**:FL(35).

Iridaceae

Belamcanda Adans.

1. *B. chinensis* (L.) DC. [blackberry-lily, leopard-flower]

Basidiomycotina - Uredinales

[*Puccinia iridis*]. **1**:FL(35,3368).

P. iridis f. sp. *australis*. **1**:FL(3368).

Deuteromycotina - Hyphomycetes

Alternaria sp. Leaf spot. **1**:KS(82), VA(82).

Cladosporium iridis. Leaf spot. **1**:CA(82), FL(35), IA(82), KS(82), NY(82), OK(82,2872), VA(82), VT(82), WI(1347).

Deuteromycotina - Coelomycetes

Colletotrichum sp. Anthracnose. **1**:FL(35).

Crocus L.

1. *Crocus* sp.

Ascomycotina

Stromatinia gladioli. Dry rot. **1**:NJ(82), NY(82), WA(3202).

Deuteromycotina - Hyphomycetes

Fusarium oxysporum. Corm rot. **1**:NY(82), PA(82).

Penicillium sp. Blue mold rot. **1**:Unk.(82).

Dietes Salisb. ex Klatt

1. *D. vegeta* N. E. Brown (=*Moraea iridioides* L.) [African iris]
2. *Dietes* sp.

Oomycetes

Phytophthora sp. Crown and root rot. **1**:FL(35).

Pythium splendens. Root rot. **1**:FL(35). **2**:FL(35).

Basidiomycotina - Uredinales

[*Puccinia iridis*]. **1**:FL(35,3368).

Deuteromycotina - Hyphomycetes

Alternaria sp. Leaf spot. **1**:FL(35).

Cercospora sp. Leaf spot. **2**:FL(35).

Fusarium sp. Crown and root rot. **1**:FL(35). **2**:FL(35).

Deuteromycotina - Other

Rhizoctonia solani. Root rot. **1**:FL(35).

Freesia Ecklon ex Klatt

1. *F.* × *hybrida* L. H. Bailey. A horticultural group of unknown origin.

Ascomycotina

Stromatinia gladioli. Dry rot. **1**:NJ(82), NY(82).

Deuteromycotina - Hyphomycetes

Cladosporium iridis. Leaf spot. **1**:CT(82).

Fusarium oxysporum. Root rot. **1**:Unk.(82).

F. oxysporum f. sp. *gladioli*. Root rot, present in commercial stocks. **1**:Unk.(82).

F. solani. **1**:TX(82).

Fusarium sp. Corm and root rot, wilt. **1**:CA(82), FL(35,82), TX(82).

Penicillium sp. Blue mold rot. **1**:Unk.(82).

Gladiolus L.

1. *G.* × *hortulanus* L. H. Bailey. Perhaps descended more or less directly from *G. natalensis*. [garden gladiola]
2. *Gladiolus* sp.

Oomycetes

Pythium sp. Root rot. **1**:FL(35).

Ascomycotina

Ophiostoma sp. On corms. **2**:OR(1666,3202).

Pleospora sp. Red leaf spot. **2**:CA(149).

Saccharomyces sp. Yeast. **2**:MS(2720).

Sclerotinia sp. Neck rot. **1**:NC(1303).

Stromatinia gladioli. Dry rot of corms, leaf, root and stalk rot. **1**:Range of host(82), FL(35), IL(312), NC(1303), WI(1361). **2**:CA(1666), FL(1666), IL(1666), MI(1666), NC(1666), NY(1666), OR(1666,3202), VA(1666), WA(3202), WI(1666).

Basidiomycotina - Ustilaginales

Urocystis gladiolicola. **2**:CA(1077).

Basidiomycotina - Other

Cellypha musaecola. On dead leaves. **2**:SC(456).

Thanatephorus cucumeris. **2**:OR(3202).

Deuteromycotina - Hyphomycetes

Alternaria alternata. Leaf spot. **1**:FL(35).

Alternaria sp. Leaf blight. **1**:Range of host(82), NC(1303). **2**:MI(1666).

Aspergillus sp. Storage rot. **2**:OK(2872).

Botrytis cinerea. Gray mold. **1**:FL(35). **2**:OR(3202), WA(3202).

B. elliptica. Leaf blight of seedlings. **1**:WA(82). **2**:WA(3202).

B. gladioli. **2**:WA(3202).

B. gladiolorum. Blight, corm and neck rot, leaf and flower spot. **1**:AK(82), CA(82), FL(35,82), IL(312), MA(82), MD(82), MI(82), NC(1303), NJ(82), NY(82), OR(82), WA(82), WI(82). **2**:FL(1666), MA(1666), MD(1666), MI(1666), NJ(1666), NY(1666), OR(3202), WA(3202).

Botrytis sp. Corm rot. **2**:MS(2720).

Cladosporium herbarum. Secondary leaf spot (or saprophytic). **1**:Range of host(82).

Curvularia gladioli. Leaf spot. **1**:FL(35).

C. lunata. Brown spot of foliage and flowers. **1**:FL(35,82), MD(82), MI(82), MS(82), NC(82), NY(82), SD(2243), VA(82), WI(82). **2**:CA(153), FL(1666), OR(3989), VA(1666).

C. trifolii. Leaf spot. **1**:FL(35). **2**:OR(3202).

C. trifolii f. sp. *gladioli*. Blight, brown spot of flowers and foliage, leaf spot. **1**:IL(312), NC(1303). **2**:NC(1303).

Curvularia sp. On corms. **2**:OR(54).

Fusarium oxysporum f. sp. *gladioli*. Brown rot, corm rot, wilt, yellows. **1**:E & C states(82), Range of host(82), CA(82), FL(35), NC(1303), SD(2243). **2**:CA(1666), DC(1666), FL(1666,3500), MD(1666), MI(1666), NC(1666), NJ(1666), NY(1666), OK(2872), TX(1666), VA(1666), WA(3202).

Fusarium sp. Corm rot, yellows. **1:**NC(1303). **2:**CA(1666), DC(1666), FL(1666), IL(1666), MS(2720), NY(1666), OR(1666,3202), TX(1666).
Heterosporium sp. Leaf spot. **1:**MD(82).
Papulaspora appendicularis. Saprophytic on decaying corms. **1:**NY(82).
P. coprophila. On corms. **1:**NY(82).
P. rubida. On corms. **1:**PA(82).
P. sepedonioides. On corms. **1:**NY(82).
Papulaspora sp. On corms. **1:**Unk.(82). **2:**WI(1666).
Penicillium funiculosum. Corm rot. **2:**FL(1712).
P. gladioli. Corm rot in storage. **1:**NC states(82), NE states(82), W states(82), CA(82), FL(35), TX(82), WA(82). **2:**FL(1666), IL(1666), MA(1666), MD(1666), MI(1666), MS(2720), NJ(1666), NY(1666), OK(2872), OR(1666,3202), WA(3202), WI(1666).
Penicillium sp. Associated with corm rot. **1:**NC(1303).
Stemphylium sp. Leaf and stem spot, red spot. **1:**FL(35,82), IL(312), MI(82), NC(1303), NJ(82), NY(82). **2:**CA(2589), FL(2589), MI(2589), NJ(2589), NY(2589).
Tritirachium dependens. On corms. **1:**TX(82). **2:**TX(1666).

Deuteromycotina - Coelomycetes

Macrophoma elongata. On stems. **1:**SC(82).
M. gladioli. On leaves. **1:**SC(82).
Phyllosticta gladioli. On leaves. **1:**AL(82), FL(35), SC(82).
Septoria gladioli. Blight, hard rot of corms, leaf spot. **1:**Range of host(82), FL(35), IL(312), NC(1303). **2:**ID(3202), MS(2720), NJ(1666), OK(2872), OR(1666,3202), WA(3202), WI(1348,1666).

Deuteromycotina - Other

Rhizoctonia solani. Collar, leaf and root rot. **1:**FL(35), IL(82), ND(82), NJ(82), TX(82). **2:**OR(1666).
Sclerotium rolfsii. Southern blight. **1:**FL(35,82).

Iris L.

1. *I. cristata* Aiton [dwarf-crested iris]
2. *I. douglasiana* Herb.
3. *I. fulva* Ker-Gawl. [copper iris, red iris]
4. *I.* × *germanica* L. [flag, fleur-de-lis]
5. *I. hexagona* Walter var. *savannarum* (Small) R. Foster
6. *I. kaempferi* Siebold ex Lem. [Japanese iris]
7. *I. laevigata* Fisch. ex Fisch. & C. A. Mey.
8. *I. macrosiphon* Torr.
9. *I. missouriensis* Nutt. (=*I. pelogonus* Goodd. in J. M. Coulter) [western blue flag]
10. *I. sibirica* L. var. *sanguinea* Ker-Gawl.
11. *I. tenax* Douglas ex Lindl.
12. *I. tingitana* Boiss. & Reut.
13. *I. versicolor* L. [larger blue flag, poison flag, wild iris]
14. *I. virginica* L.
15. *I. virginica* L. var. *schrevei* (Small) E. S. Anderson (=*I. schrevei* Small) [southern blue flag]
16. *I. xiphioides* Ehrh. [English iris]
17. *I. xiphium* L. [Spanish iris]
18. *Iris* × sp. (unnamed hybrids)
19. *Iris* sp.

Oomycetes

Phytophthora sp. Lower leaf rot. **7:**FL(35).
Pythium sp. Root rot. **7:**FL(35).

Ascomycotina

Botryotinia convoluta. Gray mold rot. **4:**MN(82), MT(3202), NJ(82), NY(82), OR(3202), WA(82,3202). **19:**MT(3202), WA(3202).
B. polyblastis. **19:**WA(3202).
Didymellina poecilospora. Leaf spot. **19:**OR(82,3202).
Didymosphaeria saccata. **9:**CO(82), ID(82,3202).
Glomerella cingulata. Flower spot. **17:**MD(82).
Mollisia iridis. On leaves. **13:**MS(2720).
Mycosphaerella macrospora. Leaf blight, leaf spot. **1:**NC(3944). **9:**ID(3202). **10:**KS(82). **13:**CA(82), IN(82), KS(82), MD(82), OK(82), OK(82), TX(82), TX(82). **17:**CA(82), NC(82), OR(82,3202), WA(82). **4:**Range of host(82), ID(3202), WA(638,3202). **19:**CA(82), GA(1438), ID(3202), IN(82), NC(82,1303), NE(82), OK(2872), OR(82,3202), TX(82), WA(82,3202).
Mycosphaerella sp. On leaves. **9:**NM(82).
Phaeosphaeria heterospora. On rhizomes and roots, probably saprophytic. **4:**MD(82), UT(318), VA(82). **19:**AR(318), DC(318), IA(318), IL(312), MA(318), MD(318).
Sclerotinia bulborum. **19:**WA(3202).
S. sclerotiorum. **19:**WA(3202).

Basidiomycotina - Uredinales

Puccinia iridis. Heteroecious: telial host. **2:**CA(81). **3:**IN(82), LA(82), NY(82), OH(82). **5:**FL(3368). **8:**CA(81). **9:**AZ(82,1195,1203,2709,3981), CA(82), CO(82,1195,2709), ID(82,638,3202), NM(82,2709), SD(2243), WA(82,3202), WY(1195). **11:**OR(82,3202). **13:**NE states(82), IA(82), LA(82). **16:**IN(82). **17:**CA(82,2709), IN(82), LA(82), NC(82). **4:**CA(81,82). **18:** On *I. xiphium* × *I. filifolia*. CA(82), IN(82), LA(82), NC(82). **19:**NE states(82), CA(2709), IA(82), IN(82), LA(82), TX(82).
P. iridis f. sp. *australis*. **3:**FL(3368).
P. iridis f. sp. *floridanum*. **2:**FL(3368). **19:**FL(35).
P. iridis f. sp. *septentrionalis*. **14:**FL(3368).
P. sessilis. Heteroecious: aecial host. **13:**NE states(82), IN(82), MN(82), NE(82). **15:**WI(1342).

Basidiomycotina - Other

Athelia rolfsii. **12:**WA(3202). **19:**OR(3202), WA(3202).
Scytinostroma galactinum. Root rot. **19:**MD(668).
Thanatephorus cucumeris. **12:**WA(3202). **17:**WA(3202). **19:**WA(3202).

Deuteromycotina - Hyphomycetes

Alternaria iridicola. Leaf spot. **9:**MT(82,3202), OR(3202), WA(82,3202). **4:**IL(82), WI(82). **19:**TX(82), WA(3202).
Alternaria sp. Leaf blight and spot. **9:**ID(3202). **19:**FL(35), NC(82), OR(82,3202).
Asperisporium punctulatum. On leaves. **19:**MS(82).
Aureobasidium microstictum. **4:**OR(3202).
Bipolaris iridis. Leaf blight. **17:**NC(82), OR(82), VA(82). **18:** On *I. xiphium* × *I. filifolia*. NC(82), OR(82), VA(82). **19:**OR(3202), WA(3202).
Botrytis cinerea. Blossom blight. **4:**MA(82), WA(3202).
B. convoluta. Firm, brown, odorless decay of rhizomes. **13:**OK(2872). **4:**OR(2218).
Botrytis sp. Gray mold blight. **19:**NC(1303), OR(82,3202).
Cercoseptoria iridis. **13:**IA(82), IN(82), NY(82), TX(82). **15:**WI(1347).
Cladosporium herbarum. Secondary leaf spot. **4:**Range of host(82), MI(82).
C. iridis. Leaf spot. **13:**MA(1390), OK(2872). **4:**WA(638). **19:**AL(285), FL(35), HI(2903), NC(1303), WY(1195).
Cladosporium sp. **19:**WA(3202).
Epicoccum nigrum. **19:**DE(3543).
Fusarium oxysporum. Bulb rot. **19:**FL(35), OR(3202), WA(3202).
Fusarium sp. Basal rot, on bulbs. **6:**NJ(82), NY(82). **17:**AZ(82), CA(82), NC(82), NY(82), OR(82), TX(82), VA(82), WA(82). **19:**AZ(82), CA(82,1666), NC(82,1303), NY(82,1666), OR(82,1666), TX(82), VA(82), WA(82,1666).
Helminthosporium sp. **17:**OR(3202).
Macrosporium iridis. **9:**ID(638). **19:**DE(3543).
Penicillium digitatum. Blue mold rot. **16:**Range of host(82). **17:**Range of host(82). **18:** On *I. xiphium* × *I. filifolia*. Range of host(82). **19:**WA(3202).
P. gladioli. Blue mold. **16:**Range of host(82). **17:**Range of host(82). **18:** On *I. xiphium* × *I. filifolia*. Range of host(82). **19:**WA(3202).
P. hirsutum. Blue mold rot. **16:**Range of host(82). **17:**Range of host(82). **18:** On *I. xiphium* × *I. filifolia*. Range of host(82). **19:**WA(3202).
Penicillium sp. Blue mold, on bulbs. **19:**NC(1303), OR(1666,3202), WA(1666).
Stemphylium sp. Leaf spot. **7:**FL(35).
Verticillium sp. On bulbs. **19:**OR(1666,3202).

Deuteromycotina - Coelomycetes

Ascochyta iridis. Leaf spot. **4:**MD(82), NY(82).
Asteroma venulosum. On leaves. **13:**TX(82), WI(82).
Colletotrichum dematium. Anthracnose. **1:**AL(82), WV(82). **9:**ID(3202). **19:**FL(35), OK(2872).
Colletotrichum sp. Anthracnose of flower stalks. **19:**IL(313).
Phoma iridina. **9:**MT(82,3202), WA(3202).
[*Phyllosticta convallariae*]. Leaf spot. **19:**IL(312).
P. iridis. Leaf spot. **13:**FL(82), MI(82), NY(82), OH(82), TX(82), WI(82). **15:**WI(1330). **19:**FL(35), WA(3202), WI(82).

Stictopatella iridis. Leaf spot. **15:**IL(82).
Deuteromycotina - Other
Rhizoctonia solani. Root rot. **7:**FL(35). **4:**Range of host(82). **18:** On *I. xiphium* × *I. filifolia*. NC(82), WA(82).
R. tuliparum. Bulb rot. **17:**NJ(82).
Sclerotium rolfsii. Southern blight, bulb, root and stem rot. **16:**CA(82), FL(82), MD(82), NC(82), OR(82), SC(82), TX(82). **17:**CA(82), FL(82), MD(82), NC(82), OR(82), SC(82), TX(82). **4:**E states(82), FL(35), TX(82). **18:** On *I. xiphium* × *I. filifolia*. CA(82), FL(82), MD(82), NC(82), OR(82), SC(82), TX(82). **19:**FL(35), NC(1303), WA(2886).
S. rolfsii var. *delphinii*. Rhizome rot. **1:**CT(82). **6:**CT(82). **10:**MA(82). **17:**CT(82).

Ixia L.
1. *Ixia* sp.

Deuteromycotina - Hyphomycetes
Fusarium oxysporum f. sp. *gladioli*. Corm rot. **1:**Unk.(82).
Deuteromycotina - Other
Sclerotium sp. Corm rot. **1:**Unk.(82).

Neomarica Sprague
1. *N. gracilis* (Herb.) Sprague

Deuteromycotina - Coelomycetes
Phyllosticta sp. Leaf spot. **1:**FL(35).
Deuteromycotina - Other
Sclerotium rolfsii. Southern blight. **1:**FL(35).

Sisyrinchium L.
1. *S. angustifolium* Mill. (=*S. gramineum* Lam., *S. graminoides* C. Bicknell)
2. *S. campestre* C. Bicknell
3. *S. douglasii* A. Dietr. [grass-widow, purple-eyed grass]
4. *S. douglasii* A. Dietr. var. *inflatum* (Suksd.) P. Holmgren (=*S. inflatum* (Suksd.) St. John)
5. *S. idahoense* C. Bicknell
6. *S. longipes* (C. Bicknell) Kearn. & Peebles
7. *S. mucronatum* Michx. var. *atlanticum* (C. Bicknell) W. Ahles (=*S. atlanticum* C. Bicknell)
8. *Sisyrinchium* sp.

Basidiomycotina - Uredinales
Aecidium residuum. **1:**OK(82,2872), TX(82).
Puccinia angulata. Autoecious. **6:**AZ(1203,3981).
Uromyces houstoniatus. Heteroecious: telial host. **1:**ME(82), WV(82). **7:**MA(1394).
U. probus. Autoecious. **1:**ID(3202), OR(3202), WA(3202). **3:**ID(82,3202), OR(82,3202), UT(82), WA(82,3202). **4:**WA(3202). **5:**WA(638). **8:**TX(82).
Deuteromycotina - Hyphomycetes
Cladosporium allii. **1:**WA(3202).
Didymaria puncta. **2:**WI(1349).
Deuteromycotina - Coelomycetes
Colletotrichum graminicola. Anthracnose. **8:**MS(2720).
Scolecosporiella sisyrinchii. leaf blight ? **1:**ND(82,378). **8:**CA(82), NM(82).

Tigridia Juss.
1. *T. pavonia* (L. fil.) Ker-Gawl. [tigerflower]

Deuteromycotina - Hyphomycetes
Fusarium oxysporum f. sp. *gladioli*. Internal rot. **1:**Unk.(82).
Penicillium gladioli. Storage rot. **1:**NY(82).

Tritonia Ker-Gawl.
1. *Tritonia* sp.

Deuteromycotina - Hyphomycetes
Alternaria sp. Leaf blight. **1:**NH(82).
Fusarium oxysporum f. sp. *gladioli*. Corm rot, yellows. **1:**CA(82).
Helminthosporium sp. Leaf spot. **1:**HI(2903).
Heterosporium sp. Leaf blight. **1:**OR(82,3202), WA(82).

Watsonia Mill.
1. *Watsonia* sp.

Basidiomycotina - Other
Armillaria mellea. Root rot. **1:**CA(82).

Juglandaceae

Carya Nutt.
1. *C. aquatica* (F. Michx.) Nutt. [bitter pecan, water hickory]
2. *C. ashei* (Sudw.) Kelsey & Dayton
3. *C. carolinae-septentrionalis* (Ashe) Engl. & Graebn.
4. *C. cordiformis* (Wangenh.) K. Koch (=*C. amara* Raf.) [bitternut hickory, pignut, swamp hickory]
5. *C. glabra* (Mill.) Sweet [broom hickory, pignut hickory, small-fruited hickory]
6. *C. illinoensis* (F. A. Wagenheim) K. Koch (=*C. oliviformis* (Michx.) Nutt., *C. pecan* (Marsh.) Engl. & Graebn.) [pecan]
7. *C. laciniosa* (F. Michx.) Loud. [kingnut, shellbark hickory]
8. *C. megacarpa* Sarg.
9. *C. ovalis* (Wangenh.) Sarg. (=*Hicoria microcarpa* (Wangenh.) Britton)
10. *C. ovata* (Mill.) K. Koch (=*C. alba* Nutt., *Hicorius alba* Raf., *Hicoria ovata* (Mill.) Britton). Cary alba (Mill.) K. Koch is a synonym of *Carya tomentosa* Nutt. while *Carya alba* Nutt. is a synonym of *Carya ovata* (Mill.) K. Koch. Reports in the literature of *Carya alba* which could not be clearly assigned to either *Carya ovata* or *Carya tomentosa* have been listed as *Carya ovata*. [shagbark hickory, shellbark hickory]
11. *C. pallida* (Ashe) Engl. & Graebn. [pale hickory, sand hickory]
12. *C. texana* Buckley [black hickory]
13. *C. tomentosa* (Poir.) Nutt. (=*Juglans alba* L.) [mockernut hickory, squarenut, white-heart hickory]
14. *Carya* sp.

Oomycetes
Phytophthora sp. Root rot. **6:**FL(35).
Pythium irregulare. **6:**GA(1516).
P. spinosum. **6:**GA(1516).
Pythium sp. Root rot. **6:**FL(35).
Zygomycotina
Absidia californica. Isolated from nuts. **6:**Unk.(1661).
A. corymbifera. Isolated from nuts. **6:**Unk.(1661).
Mucor hiemalis. Isolated from nuts. **6:**Unk.(1661).
M. plumbeus. Isolated from nuts. **6:**Unk.(1661).
M. racemosus. Isolated from nuts. **6:**Unk.(1661).
Rhizomucor pusillus. Isolated from nuts. **6:**Unk.(1661).
Rhizopus arrhizus. Isolated from nuts. **6:**Unk.(1661).
R. stolonifer. Isolated from nuts. **6:**Unk.(1661).
Syncephalastrum racemosum. Isolated from nuts. **6:**Unk.(1661).
Ascomycotina
Anthostoma juglandinum var. *caryae*. On twigs. **13:**IL(82).
Asteromassaria minor. On nuts. **6:**FL(35). **14:**Range of host(82), FL(35), NY(188).
A. seriata. On limbs. **14:**SC(82,188).
Biscogniauxia marginata. **10:**GA(1438). **13:**GA(82).
Botryosphaeria disrupta. Dieback. **6:**FL(35).
B. dothidea. Dieback. **4:**GA(1438). **6:**'Twig dieback.' SE states(82), AZ(82), FL(35,601), GA(601), LA(82). **10:**GA(1438). **11:**GA(1438). **14:**FL(35).
B. obtusa. On limbs. **6:**FL(82). **14:**S states(82), GA(82).
B. quercuum. Twig dieback. **6:**FL(35).
B. rhodina. On limbs. 'Physalospora canker.' **6:**FL(82).
B. ribis. On limbs. **6:**Range of host(82), S states(82).
Capnodium sp. Sooty mold. **6:**MS(2720).
Caryospora putaminum. On nuts. **5:**GA(1438). **10:**Range of host(82), GA(1438). **14:**Range of host(82), MS(2720).
Chaetomium spirale. Isolated from nuts. **6:**Unk.(1661).
Cryphonectria parasitica. On dead limbs. **10:**Unk.(82).
Cryptodiaporthe hicoriae. On limbs. **13:**AL(82).
Dactylospora stygia. On nuts. **14:**NY(82).
Diaporthe apocrypta. On limbs. **14:**NJ(82), OH(82).
D. hicoriae. On limbs. **5:**MI(82).
Diaporthe sp. Isolated from nuts. **6:**Unk.(1661).
Diatrype albopruinosa. **14:**IA(3629).
D. stigma. **14:**NC(1307), SC(1307).
Didymella exigua. Isolated from nuts. **6:**Unk.(1661).
D. nucis-hicoriae. On nuts. **14:**NY(82).
Discohainesia oenotherae. On twigs. **14:**VA(82).
Elsinoe randii. 'Nursery blight.' **6:**FL(35,82,1745), GA(82,1438,1745), LA(82), MS(82,2720), NC(82), OK(82,2872), SC(82), TX(82),

VA(82).
Emericella rugulosa. Isolated from nuts. **6:**Unk.(1661).
Eutypella juglandicola. **5:**GA(1438). **6:**GA(1438). **7:**GA(1438). **10:**GA(1438).
E. leprosa. **4:**IA(3629). **10:**IA(3629). **14:**IA(3629).
E. scoparia. On twigs. **6:**'Eutypa dieback.' OK(82,2873). **14:**LA(2925).
E. stellulata. On dead limbs and nuts. **14:**GA(659), NC(1303).
Gibberella cyanogena. On nuts. **14:**NY(82).
Glomerella cingulata. Anthracnose, pink rot, on nuts. **6:**Unk.(82), AL(285), FL(35), OK(2872). **14:**S states(82).
Gnomonia caryae. Anthracnose, leaf spot. **4:**WI(82). **5:**NY(82), OK(82,2872). **6:**FL(35). **10:**Range of host(82), GA(1438), IL(1052). **14:**E states(82), E states(82), Range of host(82), S states(82), AL(285), LA(82), VA(3493).
G. clavulata. On fallen leaves. **14:**NJ(82).
G. dispora. Leaf spot. **6:**'Gnomonia leaf spot.' FL(35,82), GA(82,1438). **14:**FL(35,82), GA(82).
G. nerviseda. Liver spot, vein spot disease, on leaves. **5:**NY(82). **6:**'Liver spot. Vein spot.' AL(2055), AR(82,600), GA(82,600), IL(3086), LA(82,600,3086,3087), MS(82,600,2720), TX(82,600). **14:**E states(82), N states(82).
G. setacea. Leaf spot. **6:**GA(1438), MS(2720). **7:**GA(1438). **10:**GA(1438). **14:**E states(82), S states(82), GA(82).
Gnomonia sp. **9:**GA(1438). **10:**GA(1438).
Herpotrichia herpotrichoides. On rachises. **14:**MA(189).
Hymenoscyphus fructigenus. On nuts. **14:**NY(2456), OH(653).
Hypocrea schweinitzii. **14:**GA(1438).
Hypocrea sp. **14:**Unk.(82).
Hypoxylon crocopeplum. **14:**GA(1438).
H. fuscum. On wood. **4:**IA(1225).
H. illitum. **14:**GA(1438).
H. morganii. On wood. **14:**OH(2403).
H. papillatum. On wood. **10:**GA(1438). **14:**OH(2403).
H. rubiginosum. On wood. **5:**GA(1438). **14:**GA(82,1438), IA(1225).
H. serpens. **14:**GA(1438).
Hysterographium nucicola. On old nuts. **14:**NJ(970).
Lambertella corni-maris. On hulls. **10:**IN(909), NY(909,1960).
Leptosphaeria cacuminispora. On nuts. **14:**NY(82).
Massariovalsa sudans. On limbs and twigs. **5:**GA(82), NJ(82). **6:**GA(1438). **10:**GA(1438). **14:**FL(2782), GA(82), NJ(82).
Melanconis juglandis var. *caryae*. On bark. **5:**GA(82,1438). **13:**GA(82).
M. pallida. On limbs. **4:**MA(82), MI(82). **10:**MA(82), MI(82). **14:**NE states(82).
Melanomma caryophagum. Stem canker, on nuts. **14:**Range of host(82).
Microascus trigonosporus. Isolated from nuts. **6:**Unk.(1661).
Microsphaera penicillata. Powdery mildew, leaf spot. **5:**FL(35). **10:**WI(1354). **13:**Range of host(82), MS(2720). **14:**Range of host(82), FL(35), TX(82), VA(3493).
Montagnula obtusa. Isolated from nuts. **6:**Unk.(1661).
Mycosphaerella caryigena. Leaf spot. **4:**WI(82). **6:**'Downy spot.' SE states(82), FL(35), GA(598,1438), OK(2872).
M. convexula. Leaf blotch and spot. **6:**AL(2918), FL(35,2918), GA(2918), OH(2918), SC(2918). **13:**PA(2918). **14:**FL(82).
M. dendroides. Leaf blotch. **1:**LA(82). **6:**'Leaf blotch.' SC states(82), SE states(82), FL(35), GA(598,1438), IN(82). **7:**GA(82,1438). **10:**GA(82). **13:**S states(82).
M. punctiformis. On leaves and nuts. **6:**GA(1438). **9:**GA(1438). **10:**GA(1438). **11:**GA(1438). **13:**NY(82). **14:**Unk.(3943), OK(660).
Myriangium duriaei. On scale insects. **6:**GA(82,1438).
M. tuberculans. Myriangium disease, black spot on bark. **5:**GA(1438). **6:**AL(285), FL(35), GA(82,1438,1745), MS(82,2720). **10:**GA(1438). **14:**GA(1438).
Nectria canadensis. **14:**PA(3178).
N. cinnabarina. On twigs and bark. **14:**Range of host(82).
N. coccinea. On limbs. **5:**GA(1438). **14:**GA(82,1438).
N. coccinea var. *faginata*. **14:**NE states(82).
N. galligena. Trunk canker. **4:**E states(82). **5:**CT(82), WV(82). **13:**E states(82). **14:**E states(82).
N. missouriensis. On bark. **14:**NC states(82), NE states(82).
N. ochroleuca. **10:**MO(3085).
N. peziza. On stumps. **14:**OH(3085).
N. pseudotrichia. **14:**Unk.(82).
Nematospora coryli. On nuts. **6:**S states(82), FL(35,82,204). **14:**FL(35).
Nummularia tinctor. **14:**FL(2782).
Ophiostoma piliferum. On inner bark or wood. **5:**TN(1364).
O. pluriannulatum. On wood. **5:**TN(3714).
Petriella guttulata. Isolated from nuts. **6:**Unk.(1661).
Phragmodiaporthe caryae. On branches. **10:**NY(184,195). **14:**IA(82), NY(82).
Phyllactinia guttata. Powdery mildew. **4:**MI(82). **6:**MS(2720), NC(1303). **14:**WI(82).
Physalospora abdita. On limbs. **6:**'Physalospora canker.' FL(82). **14:**S states(82), GA(82,1438).
Prosthecium ellipsosporum. On limbs. **14:**MI(82), NE(82), NJ(82).
Rosellinia aquila. On bark. **5:**NY(82).
R. caryae. Twig canker. **10:**MI(82).
Sarcoscypha occidentalis. On dead twigs. **14:**NC(1303).
Schizoxylon compositum. **14:**NY(3225).
S. insigne. **3:**GA(1438). **5:**GA(1438). **10:**GA(1438).
Scutellinia erinaceus. On wood. **14:**PA(848).
Splanchnonema hicoria. On limbs. **10:**MA(188).
Stictis radiata. On dead twigs. **14:**NC(1303).
Talaromyces flavus. Isolated from nuts. **6:**Unk.(1661).
T. trachyspermus. Isolated from nuts. **6:**Unk.(1661).
Valsa ambiens. **14:**NJ(3415).
V. ceratosperma. On dead limbs. **10:**IA(1226), WV(3415). **11:**GA(1438). **13:**NC(1303). **14:**GA(1438), IA(1226), NJ(3415).
Valsaria insitiva. **14:**OK(660).
V. rubricosa. **11:**GA(1438).
Valsella sp. **5:**GA(1438).
Xylaria bulbosa. On wood. **14:**FL(3015).
X. oxyacanthae. **10:**GA(1438).
Xylaria sp. Isolated from nuts. **6:**Unk.(1661).

Basidiomycotina - Other

Abortiporus biennis. **14:**SC states(82), AR(2675), LA(2675).
Aleurodiscus oakesii. On bark. **14:**MO(82), PA(82).
Antrodia albida. **14:**NY(82).
A. malicola. **14:**Unk.(82).
Armillaria mellea. Root rot. **6:**CA(1110). **14:**TX(82).
A. tabescens. Root rot. **6:**'Clitocybe root rot.' FL(35), GA(3574). **13:**OK(82,2872). **14:**FL(35).
Auricularia auricula. On wood. **6:**LA(914). **14:**GA(914,2642), KY(914).
Bjerkandera adusta. **10:**GA(1439). **14:**Unk.(82).
Ceriporia alachuana. **6:**FL(128).
C. xylostromatoides. **14:**LA(2675).
Cerrena unicolor. Wood rot. **14:**E states(82).
Chondrostereum purpureum. Wood rot. **4:**DE(543). **14:**IN(82), NY(82).
Clavicorona pyxidata. **14:**NC(884).
Climacodon septentrionalis. White spongy rot, occasionally on living trees. **14:**Unk.(82).
Coniophora puteana. Isolated from mine timbers. **14:**Unk.(1006).
Coriolopsis rigida. Wood rot. **14:**SC states(82), LA(2675).
Crepidotus maximus. On dead, standing trunk. **14:**TN(1544).
Crinipellis zonata. **10:**NC(594).
Cyathus striatus. **14:**GA(1439).
Cystostereum australe. White rot. **14:**FL(2554).
Daedalea quercina. Wood rot. **14:**E states(82).
Daedaleopsis confragosa. On wood. **14:**NC(1303).
Datronia mollis. **14:**NC(1307), NY(82).
Dendrophora albobadia. **14:**GA(1439).
D. versiformis. **14:**NY(543).
Dendrothele alliacea. On bark. **13:**AR(2084).
D. candida. On bark. **6:**NC(1303).
D. griseo-cana. On bark. **12:**LA(2084).
D. microspora. On bark. **5:**MA(2084). **10:**IA(2084). **14:**AL(2084), NY(2084).
D. pachysterigmata. On bark. **10:**NY(2084).
D. strumosa. **14:**FL(2084).
Eichleriella pulvinata. On dead twigs. **14:**NC(128).
Exidia cokeri. On limbs. **14:**GA(2642).
E. glandulosa. On dead limbs. **6:**NC(1303).
Fibulomyces mutabilis. **14:**MI(128).
Fomes fasciatus. On wood. **1:**MS(2720). **14:**SC states(82), GA(491,1439), LA(2675).

Fomitella supina. On wood. **6:**MS(2720). **14:**SC states(82), LA(2675).
Fomitopsis meliae. On stumps. **14:**MS(2320).
F. pinicola. Brown crumbly heart rot. **14:**WI(82).
Ganoderma applanatum. Heart rot. **14:**Range of host(82).
G. lucidum. On wood. **1:**MS(2720). **14:**SC states(82), AR(2675), LA(2675), MS(2675), NC(1303).
Gloeoporus dichrous. **14:**SC states(82), AR(2675), LA(2675).
Grandinia barba-jovis. **6:**FL(128).
Hapalopilus rutilans. White rot. **14:**Unk.(2681), LA(2675).
Helicobasidium brebissonii. 'Violet root rot.' **6:**TX(82).
Hericium americanum. On wood. **14:**PA(1237).
H. erinaceus. On wood. **1:**MS(2720).
Heterobasidion annosum. Root rot. **14:**NC(1303).
Hexagonia hydnoides. **14:**SC states(82), LA(2675).
Hohenbuehelia angustata. **14:**LA(2675).
Hydnum sp. White spongy rot. **14:**Unk.(82).
Hymenochaete borealis. On wood. **14:**NJ(128).
Hypochnicium analogum. Isolated from mine timbers. **14:**Unk.(1006).
Inonotus andersonii. On wood. **13:**GA(491,1439).
I. cuticularis. **6:**AZ(1181).
I. hispidus. On wood. **1:**MS(2720).
I. obliquus. On wood. **1:**MS(2720).
Irpex lacteus. **14:**IN(654).
Lopharia cinerascens. **6:**TX(543).
Merismodes ochraceus. Bark patch. **14:**MA(82).
Meruliopsis corium. On wood. **10:**GA(491,1439).
Microstroma juglandis. Catkin blight, leaf spot, witches' broom. **1:**LA(82). **4:**WI(82). **5:**Range of host(82), FL(35). **6:**SC states(82), SE states(82), GA(82), NC(1303), OK(608). **10:**Range of host(82), NJ(82). **12:**OK(2872). **13:**Range of host(82), MS(2720), NC(1303). **14:**Range of host(82), CT(82), NC(1303,3944).
M. juglandis var. *robustum*. Catkin blight. **6:**GA(1551), OK(608).
Mycena luteopallens. On nuts. **14:**IL(82).
Mycoacia meridionalis. **14:**FL(443).
Mycorrhaphium adustum. On dead limbs. **14:**NC(1303,1307).
Oudemansiella canarii. On living and fallen trunks. **8:**FL(3284).
Oxyporus latemarginatus. **14:**S states(82), FL(2166).
O. pellucida. **10:**IN(381).
O. populinus. On wood. **13:**GA(491,1439). **14:**NE states(82).
Pellicularia koleroga. 'Thread blight.' **6:**FL(82), LA(3640), NC(82).
Perenniporia fergusii. On wood. **10:**IL(1205).
P. medulla-panis. **14:**Unk.(82).
Phaeosolenia ravenelii. **14:**SC(456,634).
Phanerochaete ericina. White rot. **14:**GA(433).
P. ravenelii. **14:**AR(2675), LA(2675).
P. salmoneolutea. White rot. **14:**FL(433).
P. sordida. **6:**LA(128).
P. velutina. White rot. **14:**MD(433).
Phellinus conchatus. White rot. **6:**MS(2720). **14:**LA(82,2675), MS(82).
P. gilvus. On wood. **6:**GA(1439). **9:**NC(1303). **14:**NC(3944).
P. igniarius. White spongy heart rot. **5:**FL(35). **6:**FL(35). **14:**Range of host(82), FL(35).
P. spiculosus. White rot. **5:**NC(2169), PA(2169). **10:**PA(2169). **13:**GA(491,1439). **14:**E states(82), WV(82).
Phlebia tremellosus. **5:**NY(128).
Pholiota sp. **14:**IL(1017).
Pleurotus dryinus. Sapwood rot. **5:**FL(35). **14:**AR(82,2675).
P. minutus. **14:**NC(3944).
P. ostreatus. On wood. **1:**MS(2720). **4:**MN(63). **14:**NC(1307).
Polyporus mori. Heart rot. **13:**NY(82). **14:**VA(82), VT(82).
P. tenuiculus. On wood. **14:**SE states(2681).
Psathyrella incerta. On roots around stump. **10:**NY(3491).
Radulodon casearium. On wood. **14:**OH(1174).
Rigidoporus lineatus. On wood. **1:**MS(2720). **14:**SC states(82), AR(2675), LA(2675).
R. ulmarius. On wood. **1:**MS(2166). **6:**MS(2309).
Schizophyllum commune. On wood. **6:**'Heart rot.' OK(82,2872). **14:**Range of host(82), AR(2675), LA(2675), NC(1303), SC(1307).
Schizopora paradoxa. On wood. **14:**GA(491,1439).
Scytinostroma portentosum. **13:**VA(128).
Septobasidium castaneum. Felt fungus canker. **14:**SE states(82).
S. cokeri. Felt fungus canker. **1:**NC(82).
S. pseudopedicellatum. Parasitic on scale insects. **6:**FL(35). **14:**NC(1303).
Septobasidium sp. **14:**SE states(82).
Spongipellis delectans. On wood. **1:**MS(2720). **14:**NY(82).
Steccherinum ochraceum. **14:**IN(654).
Stereum hirsutum. Wood rot. **14:**CT(543), IN(82), NY(82).
S. ostrea. **1:**MS(543). **4:**DC(543). **5:**CT(543). **10:**GA(1439), IN(543). **14:**CT(543), NJ(543), NY(543).
Tomentella rubiginosa. **14:**NY(2039).
Trametes elegans. On wood. **14:**SC states(82), AR(2675), FL(35,82), LA(2675), MS(2675).
T. hirsuta. **14:**Unk.(82), GA(1439).
T. versicolor. Isolated from mine timbers. **10:**GA(1439). **14:**Unk.(82,1006), NC(1307), SC(1307).
Trichaptum biforme. **14:**SC states(82), Unk.(82), AR(2675), LA(2675).
T. sector. **14:**SC states(82), AR(2675), LA(2675).
Tulasnella pinicola. On wood. **14:**GA(2642).
T. violea. **10:**CT(128). **14:**PA(2673).
Typhula spathulata. **10:**NY(264).
Tyromyces fissilis. On wood. **1:**MS(2720). **6:**FL(35). **14:**FL(35,82).
Xylobolus frustulatus. **14:**NC(3944).

Deuteromycotina - Hyphomycetes

Acremonium strictum. Isolated from nuts. **6:**Unk.(1661).
Alternaria alternata. Isolated from nuts. **6:**Unk.(1661).
[*A. raphani*]. Isolated from nuts. **6:**Unk.(1661).
Alternaria sp. **6:**CA(1110).
[*Arthrinium phaeospermum*]. Isolated from nuts. **6:**Unk.(1661).
[*Articularia quercina*]. On leaves. **13:**MO(82).
A. quercina var. *minor*. On leaves. **6:**OK(608), TX(82,597).
Aspergillus candidus. Isolated from nuts. **6:**Unk.(1661).
A. duricaulis. Isolated from nuts. **6:**Unk.(1661).
A. equitis. Storage mold of nuts. **6:**S states(82).
A. ficuum. Isolated from nuts. **6:**Unk.(1661).
A. flavus. Isolated from nuts. **6:**Unk.(1661).
A. fumigatus. Isolated from nuts. **6:**Unk.(1661).
A. niger. Isolated from nuts. **6:**Unk.(1661).
A. ochraceus. Isolated from nuts. **6:**Unk.(1661).
A. oryzae. Isolated from nuts. **6:**Unk.(1661).
A. parasiticus. Isolated from nuts. **6:**Unk.(1661).
A. reptans. Isolated from nuts. **6:**Unk.(1661).
A. restrictus. Isolated from nuts. **6:**Unk.(1661).
A. rubrobrunneus. Isolated from nuts. **6:**Unk.(1661).
A. tamarii. Isolated from nuts. **6:**Unk.(1661).
A. terreus. Isolated from nuts. **6:**Unk.(1661).
A. ustus. Isolated from nuts. **6:**Unk.(1661).
A. versicolor. Isolated from nuts. **6:**Unk.(1661).
A. wentii. Isolated from nuts. **6:**Unk.(1661).
Aureobasidium pullulans. Isolated from nuts. **6:**Unk.(1661).
Cephalosporium sp. Isolated from nuts. **6:**Unk.(1661).
Ceratophorum epiphyllum. **4:**KS(3547). **9:**AR(3547).
C. uncinatum. On leaves. **4:**KS(82), MO(82). **13:**FL(82). **14:**AR(82), FL(35), KS(82), MO(82).
Cercospora halstedii. Leaf blotch. **5:**FL(35). **6:**FL(35), OK(2872). **14:**E states(82), S states(82).
Cladosporium caryigenum. Scab, leaf spot. **1:**FL(82). **4:**KS(82), WI(82). **5:**FL(35). **10:**WI(1341). **13:**IL(82), KS(82). **14:**Range of host(82), FL(35).
[*C. caryigenum* var. *carpinum*]. On leaves. **6:**GA(1281).
C. cladosporioides. Isolated from nuts. **6:**Unk.(1661).
C. sphaerospermum. Isolated from nuts. **6:**Unk.(1661).
Codinaeopsis gonytrichodes. **14:**IL(3215).
Coniosporium nucifoedum. On nuts. **14:**NY(82).
Conoplea geniculata. **14:**DE(3543).
C. globosa. Trunk canker. **14:**MO(1672), PA(82,1672).
C. olivacea. **14:**IL(1672), IN(1672), MI(1672), PA(1672), WI(1672).
Cristulariella moricola. 'Zonate leaf spot.' **6:**AL(2057), FL(35), GA(2057).
Curvularia senegalensis. Isolated from nuts. **6:**Unk.(1661).
Epicoccum nigrum. Isolated from nuts. **6:**Unk.(1661).
Eurotium chevalieri. Isolated from nuts. **6:**Unk.(1661).
E. repens. Isolated from nuts. **6:**Unk.(1661).
E. rubrum. Isolated from nuts. **6:**Unk.(1661).
Fusarium carpineum. Leaf spot. **4:**WI(82).
F. chlamydosporum. Isolated from nuts. **6:**Unk.(1661).
F. decemcellulare. Isolated from nuts. **6:**Unk.(2135).
F. lateritium. **14:**DE(3543).

F. merismoides. Isolated from nuts. **6:**Unk.(1661).
F. moniliforme. Isolated from nuts. **6:**Unk.(1661).
F. oxysporum. Isolated from nuts. **6:**Unk.(1661), GA(1516).
F. pallidoroseum. Isolated from nuts. **6:**Unk.(1661).
F. roseum. On nuts. **14:**NY(82).
Fusarium sp. Root rot. **6:**FL(35).
Fusicladium sp. **6:**CA(1110).
Geniculosporium serpens. Isolated from nuts. **6:**Unk.(1661).
Harpographium fasciculatum. On limbs. **14:**LA(82,2509).
Helicoma tenuifilum. On fruits. **14:**MO(1271).
Helicomyces roseus. On nuts. **14:**GA(659).
Itersonilia sp. Isolated from nuts. **6:**Unk.(1661).
Microdochium nivale. Isolated from nuts. **6:**Unk.(1661).
Nigrospora oryzae. Isolated from nuts. **6:**Unk.(1661).
N. sphaerica. Isolated from nuts. **6:**Unk.(1661).
Nodulisporium hinnuleum. Isolated from nuts. **6:**Unk.(1661).
Paecilomyces lilacinus. Isolated from nuts. **6:**Unk.(1661).
P. variotii. Isolated from nuts. **6:**Unk.(1661).
Penicillium aculeatum. Isolated from nuts. **6:**Unk.(1661).
P. aurantiogriseum. Isolated from nuts. **6:**Unk.(1661).
P. brevicompactum. Isolated from nuts. **6:**Unk.(1661).
P. camembertii. Isolated from nuts. **6:**Unk.(1661).
P. chrysogenum. Isolated from nuts. **6:**Unk.(1661).
P. citreonigrum. Isolated from nuts. **6:**Unk.(1661).
P. citrinum. Isolated from nuts. **6:**Unk.(1661).
P. corylophilum. Isolated from nuts. **6:**Unk.(1661).
P. crustosum. Isolated from nuts. **6:**Unk.(1661).
P. cyaneum. Isolated from nuts. **6:**Unk.(1661).
P. echinulatum. Isolated from nuts. **6:**Unk.(1661).
P. expansum. Isolated from nuts. **6:**Unk.(1661).
P. fellutanum. Isolated from nuts. **6:**Unk.(1661).
P. funiculosum. Isolated from nuts. **6:**Unk.(1661).
P. glabrum. Isolated from nuts. **6:**Unk.(1661).
P. griseofulvum. Isolated from nuts. **6:**Unk.(1661).
P. implicatum. Isolated from nuts. **6:**Unk.(1661).
P. islandicum. Isolated from nuts. **6:**Unk.(1661).
P. janczewskii. Isolated from nuts. **6:**Unk.(1661).
P. jensenii. Isolated from nuts. **6:**Unk.(1661).
P. puberulum. Isolated from nuts. **6:**Unk.(1661).
P. purpurogenum. Isolated from nuts. **6:**Unk.(1661).
P. resticulosum. Isolated from nuts. **6:**Unk.(1661).
P. roquefortii. Isolated from nuts. **6:**Unk.(1661).
P. thomii. Isolated from nuts. **6:**Unk.(1661).
P. variabile. Isolated from nuts. **6:**Unk.(1661).
Penicillium sp. **6:**GA(1516).
Periconia byssoides. On nuts. **14:**GA(659).
P. cookei. Isolated from nuts. **6:**Unk.(1661).
Phymatotrichopsis omnivora. Root rot. **6:**'Phymatotrichum root rot.' AZ(82), TX(82). **14:**TX(82).
Pseudocercosporella caryigena. Downy spot. **4:**Unk.(82). **6:**OK(2872).
Ramularia albomaculata. On leaves. **14:**MI(82), NY(82).
Rhinocladiella sp. Isolated from nuts. **6:**Unk.(1661).
Scopulariopsis koningii. Isolated from nuts. **6:**Unk.(1661).
Stilbella aterrima. **14:**SC(3178).
Thermomyces lanuginosus. Isolated from nuts. **6:**Unk.(1661).
Torula herbarum. Isolated from nuts. **6:**Unk.(1661).
Trichoderma koningii. Isolated from nuts. **6:**Unk.(1661).
Trichoderma sp. **6:**GA(1516).
Trichothecium roseum. Pink mold on nuts following scab. **6:**'Pink mold.' S states(82), Unk.(1661), FL(35), GA(598), NC(1303), OK(2873). **14:**Range of host(82), FL(35).
Tubercularia lateritia. On dead limbs. **6:**NC(3178).
Verticillium dahliae. **6:**CA(1110).
Verticillium sp. Wilt. **14:**VA(3493).
Volutella caryogena. On nuts. **14:**NY(82).
Wallemia sebi. Isolated from nuts. **6:**Unk.(1661).

Deuteromycotina - Coelomycetes

Aplosporella linearis. Leaf spot. **4:**OK(2872). **10:**OK(2872). **14:**MO(82), NC(1307), NY(82), WV(82).
Ascochyta caryae. On living leaves. **10:**WI(1358).
Ascochyta sp. Isolated from nuts. **6:**Unk.(1661).
Asteroma caryae. Nursery blight, leaf spot. **6:**Unk.(82), AL(285), FL(35), MS(2720). **10:**OK(2872). **13:**FL(82), NY(82), WI(82). **14:**Range of host(82), Unk.(82), FL(35), NC(3944), NJ(965), TX(82,3597).
Botryodiplodia sp. Canker, dieback. **6:**FL(35).
Chaetospermum carneum. On nuts. **14:**GA(659).
Colletotrichum sp. Leaf spot. **6:**FL(35). **10:**WI(1354).
Coniothyrium caryogenum. On nuts. **6:**S states(82).
C. fuckelii. Isolated from nuts. **6:**Unk.(1661).
C. pallidospermum. On living leaves. **4:**WI(1341).
Cytospora sacculus. On limbs. **14:**NC(1307), VA(82), WV(82).
Cytospora sp. Canker. **6:**AZ(1574).
Diplodia caryigena. Leaf spot. **4:**OK(2872).
Diplodia sp. Isolated from nuts. **6:**Unk.(1661).
Discosia artocreas. On nuts. **6:**FL(35). **14:**Range of host(82).
D. rugulosa. On leaves. **2:**FL(2782). **14:**FL(634,2782).
Dothiorella hicoriae. On twigs. **13:**NY(82).
D. nucis. On nuts. **14:**NY(82).
Gyrostroma missouriense. **13:**MO(3173), NY(3173). **14:**ME(3173).
Hendersonia davisii. Leaf spot. **10:**WI(966). **13:**WI(82).
Hendersonulina pustulata. On nuts. **14:**NJ(82), NY(82).
Lasiodiplodia theobromae. Isolated from nuts. **6:**Unk.(1661).
Leptothyrium nervisedum. Vein spot. **6:**OK(2872).
Libertella sp. Isolated from nuts. **6:**Unk.(1661).
Marssonina juglandis. Leaf spot. **14:**E states(82), NC states(82).
Melanconiopsis inquinans. **14:**Unk.(82).
Melanconium angustum. On limbs. **14:**Range of host(82).
M. bicolor. On limbs. **14:**Range of host(82).
M. gracile. On limbs. **14:**Range of host(82).
M. magnum. On limbs. **14:**Range of host(82).
M. pallidum. On limbs. **4:**OH(82), WV(82). **14:**Range of host(82).
Monochaetia monochaeta. Zonate leaf spot. **5:**MD(82,1509). **7:**NC(1509). **10:**TN(82,1509). **13:**TN(82,1509). **14:**SE states(82), TN(82).
Myxosporium luteum. On bark. **14:**WV(82).
Myxosporium sp. Leaf spot. **6:**OK(2872).
Pestalotia adusta. Isolated from nuts. **6:**Unk.(1661).
P. mangifolia. Isolated from nuts. **6:**Unk.(1661).
P. matildae. Isolated from nuts. **6:**Unk.(1661).
P. microspora. On nuts. **6:**Unk.(1661). **14:**NJ(82), NY(82).
P. neglecta. Isolated from nuts. **6:**Unk.(1661).
P. sphaerelloides. Leaf spot, isolated from nuts. **1:**LA(82). **6:**Unk.(1661).
Pestalotia sp. Isolated from nuts. **6:**Unk.(1661).
Pestalotiopsis sydowiana. Isolated from nuts. **6:**Unk.(1661).
P. uvicola. Leaf spot. **6:**FL(35,82), TX(82).
Phoma exocarpina. Nut decay. **6:**FL(35). **14:**MI(82), NY(82).
P. herbarum. Isolated from nuts. **6:**Unk.(1661).
P. macrostoma. Isolated from nuts. **6:**Unk.(1661).
Phoma sp. Blight. **6:**Unk.(1661), OK(2872).
Phomopsis sp. Dieback, twig gall. **6:**FL(35). **14:**E states(82).
Phyllosticta convexula. Leaf spot. **5:**VA(82). **6:**OK(82,2873). **13:**MO(64,82). **14:**Range of host(82).
P. subtilis. Leaf spot. **14:**Range of host(82), NY(64).
Phyllosticta sp. **4:**WI(1348). **10:**WI(1355).
Pyrenochaeta nucinata. On nuts. **14:**NY(82).
Septoria caryae. Leaf spot. **6:**'Septoria leaf spot.' TX(82). **10:**MI(82). **14:**DE(82).
S. hicoriae. Leaf spot. **4:**TX(82). **14:**TX(82,3597).
Sphaeropsis pallidula. On nuts. **14:**Range of host(82).
S. pericarpii. On nuts. **14:**Range of host(82).
Stagonospora linearis. On dead limbs. **14:**IL(82,2738).
Tubakia dryina. Leaf spot. **6:**LA(1607). **10:**IL(82,1246). **14:**NC(1303,1307).

Deuteromycotina - Other

Rhizoctonia solani. Thread blight, root rot. **6:**FL(35).

Juglans L.

1. *J. californica* S. Wats. [Southern California walnut]
2. *J. cinerea* L. [butternut, white walnut]
3. *J. cordiformis* Maxim. (=*J. ailantifolia* Carrière var. *cordiformis* (Maxim.) Rehd.) [heartnut, himegurumi]
4. *J. hindsii* (Jeps.) Jeps. ex R. E. Smith (=*J. californica* var. *hindsii* Jeps.) [Hind's walnut, Northern California black walnut]
5. *J. major* (Torr.) A. Heller [Arizona walnut, Nogal]
6. *J. microcarpa* Berland. [little walnut, nogal, river walnut, Texas walnut]
7. *J. nigra* L. [black walnut]
8. *J. regia* L. [English walnut, Madeira-nut, Persian walnut]

9. *J. sieboldiana* Maxim. (=*J. ailantifolia* Carrière) [Japanese walnut, onigurumi]
10. *Juglans* sp.
11. *Juglans* × sp. (unnamed hybrids)

Oomycetes

Phytophthora cactorum. Collar rot. **1:**CA(82). **4:**CA(82). **8:**'Phytophthora crown rot. Phytophthora root rot. Phytophthora trunk canker. Phytophthora scaffold branch canker.' CA(82,2446).
P. cinnamomi. Collar rot of seedlings. **7:**SE states(82,3624), LA(82). **8:**'Phytophthora crown rot. Phytophthora root rot.' SE states(3624), CA(2446).
P. citricola. Root rot of seedlings. **7:**IN(1326,1327). **8:**'Phytophthora crown rot. Phytophthora root rot. Phytophthora trunk canker. Phytophthora scaffold branch canker.' CA(2446).
P. citrophthora. **1:**Unk.(82). **8:**'Phytophthora trunk canker. Phytophthora scaffold branch canker. Phytophthora crown rot. Phytophthora root rot.' CA(2446).
P. megasperma. 'Phytophthora crown rot. Phytophthora root rot.' **8:**CA(2446).

Ascomycotina

Botryosphaeria dothidea. Twig dieback. **2:**GA(1438). **7:**CA(1110), GA(1438). **8:**CA(1110). **10:**FL(35).
B. obtusa. Black rot, leaf spot, isolated from cankers. **7:**IA(1883), IL(1883), IN(1883), MO(1883). **10:**Range of host(82), CA(1110).
B. quercuum. Dieback. **8:**FL(35). **10:**FL(35).
B. ribis. On limbs. **2:**GA(82).
Ceratocystiopsis alba. On injured bark tissue. **8:**CA(826).
Ceratocystis sp. On nuts. **7:**IN(663). **10:**IN(663).
Chaetomium sp. **10:**OR(3202).
Diaporthe spiculosa. On limbs. **10:**IA(82), MI(82), NY(82).
Diatrype albopruinosa. **7:**IA(3629).
Diatrypella eutypaeformis. **7:**IA(3629).
Eutypella dissepta. On dead limbs. **2:**IA(2925). **10:**IA(82).
E. juglandicola. **1:**CA(1110).
E. leprosa. **7:**IA(3629). **8:**IA(3629). **10:**IA(3629).
Gibberella baccata. **8:**CA(1110).
Gnomonia leptostyla. Anthracnose, on leaves, nuts, and stems, leaf spot. **2:**MS(2720), NC(1533). **7:**Range of host(82), Range of host(82), FL(35), IL(2571), IN(2571,3650), KY(2571), MO(2571), MS(2720), NC(1303), OH(2571), OK(2872), TN(2571). **8:**'Anthracnose.' Range of host(82), FL(35), OR(3202), WA(3202). **10:**Range of host(82), FL(35), MS(2720).
Grovesinia pyramidalis. Zonate leaf spot. **7:**IA(1886), MN(1886).
Hypoxylon quadratum. On wood. **10:**PA(2403).
H. rubiginosum. **7:**GA(1438). **10:**GA(82).
Lecanidion atratum. **5:**AZ(662).
Melanconis juglandis. Canker, dieback. **2:**Range of host(82), GA(82,1438), IA(82), VA(3493). **3:**Unk.(82), CT(82), NH(82), NY(82). **8:**CT(82), NJ(82). **9:**GA(1438). **10:**CT(82), MD(82), NY(82), PA(82).
Microsphaera penicillata. Powdery mildew. **7:**MS(2720). **10:**Range of host(82), CA(1110), FL(35).
Mycosphaerella juglandis. Leaf spot. **7:**IA(1885), IL(1884,1885), IN(1885), NC(1885). **8:**IL(1884).
Nectria cinnabarina. Canker. **4:**CA(1110). **7:**CA(1110), WA(3202). **8:**CA(1110), OK(2872), WA(3202). **10:**Range of host(82), FL(35).
N. coccinea. **10:**CT(82), OR(82), VA(82).
N. ditissima. **7:**E states(82), RI(82), WI(82). **10:**NE states(82), TN(82), WI(82).
N. galligena. Canker. **2:**E states(82). **7:**MI(3608), NC(1303). **8:**NC(1303). **10:**IN(82), NC(82), OH(82), PA(82), RI(82), VA(82), WV(82).
N. punicea. **8:**WA(3202). **10:**OR(3202).
Nectria sp. Canker. **2:**VA(1533). **8:**NY(82). **10:**OR(3202).
Phyllactinia guttata. Powdery mildew. **10:**IN(82), OH(82).
Sarcoscypha occidentalis. **2:**NC(1307).
Schizoxylon albo-atrum. **5:**AZ(662).
S. insigne. **7:**GA(1438).
Valsa ambiens. **7:**KS(3415).
V. ceratosperma. **7:**IA(1226).

Basidiomycotina - Uredinales

[*Gymnosporangium libocedri*]. **1:**CA(1110).

Basidiomycotina - Other

Acanthophysium diffissum. **5:**AZ(1202).
Aporpium caryae. On wood. **5:**AZ(1202).
Armillaria mellea. Root rot. **4:**CA(1110,2898), OR(3202). **8:**'Armillaria crown rot. Armillaria root rot.' CA(82,1110,2446), NC(82), NJ(82), OR(2424,2425,2426,2427,2429,2430,2431,3202), TX(82), WA(3202). **11:** Range of host(82), CA(2898,3202), FL(35), OR(2428).
A. tabescens. Root rot. **10:**FL(35).
Auricularia auricula. On wood. **2:**NC(1303,1307). **5:**AZ(1202).
Basidiodendron eyrei. **5:**AZ(1202).
Bjerkandera adusta. **7:**WA(3202). **10:**Unk.(82).
Ceriporia purpurea. **5:**AZ(1202). **7:**AZ(1194), WA(3202).
C. reticulata. **5:**AZ(128,1202). **7:**AZ(1194).
C. viridans. **5:**AZ(128,1202).
Cerocorticium confluens. **5:**AZ(128).
Coniophora arida. **5:**AZ(1202).
Coprinus micaceus. **2:**NC(1307).
Corticium canfieldii. **5:**AZ(2043).
C. roseum. **5:**AZ(128,1202).
Crepidotus nephrodes. **2:**NC(1307).
Cyphellopsis anomala. **5:**AZ(1202).
Daedalea quercina. Wood rot. **10:**Range of host(82).
Daedaleopsis confragosa. On wood. **10:**Range of host(82), FL(35).
Dendrophora albobadia. **7:**KS(543).
Dichostereum effuscatum. **2:**CT(128).
[*Fibricium rude*]. **5:**AZ(128).
Ganoderma applanatum. On wood. **2:**NC(1303).
Grandinia arguta. **5:**AZ(128).
G. spathulata. **5:**AZ(1202).
Hapalopilus rutilans. White rot. **10:**Unk.(2681).
Hymenochaete rubiginosa. **5:**AZ(1202).
Hyphoderma heterocystidia. **2:**PA(2321).
H. mutatum. **2:**PA(2321).
H. sambuci. **7:**WA(3202).
Hypochnicium geogenium. **5:**AZ(1202).
Inonotus hispidus. **5:**AZ(662,1181,1202). **7:**AZ(1194).
Laetiporus sulphureus. Brown cubical rot. **10:**Range of host(82), CA(1110), FL(35).
Lentinellus ursinus. Wood decay, heart rot. **2:**NC(1303,1307).
Marasmiellus candidus. **2:**NC(1307).
Megalocystidium lactescens. **5:**AZ(128).
Microstroma juglandis. White mold, downy leaf spot, witches' broom. **1:**CA(1110). **2:**Range of host(82), CA(1110), IA(82). **4:**CA(82,1110), NM(82). **6:**AZ(2123). **7:**Range of host(82), CA(1110), MS(2720). **8:**'Downy leaf spot.' E states(82), CA(82,1110), MS(82), OR(82,3202), WA(3202). **9:**CA(1110). **10:**Range of host(82), OR(2428,2429,2430), WA(3202).
Mycena excisa. **2:**NC(1307).
M. luteopallens. On buried nuts. **7:**VA(128,3717).
Oxyporus latemarginatus. **10:**Unk.(82).
Peniophora fuscomarginata. **5:**AZ(1194,1202).
P. nuda. **5:**AZ(1202).
P. tamaricicola. White rot. **5:**AZ(1180,1202).
Perenniporia medulla-panis. **5:**AZ(1202).
P. tenuis. **5:**AZ(1202). **10:**Unk.(82).
Phanerochaete allantospora. White rot. **5:**AZ(433,1202).
P. burtii. White rot. **10:**MI(128,433).
P. chrysorhiza. **5:**AZ(1202).
P. sordida. **2:**GA(1439).
P. tuberculata. **5:**AZ(1202).
Phellinus conchatus. Wood rot. **10:**NY(82).
P. everhartii. **10:**Range of host(82), SW states(82).
P. gilvus. **5:**AZ(1202). **10:**Unk.(82).
P. igniarius. White heart rot. **7:**OK(82,2873). **10:**Range of host(82), SW states(82).
P. punctatus. **10:**Unk.(82).
P. weirianus. On living trees. **5:**AZ(662,1202), NM(1190). **7:**AZ(1194).
Phlebia rufa. **2:**NE states(128).
Phlebiella vaga. **5:**AZ(128).
Pleurotus levis. **5:**AZ(1202).
P. ostreatus. **9:**GA(1439).
Polyporus admirabilis. **10:**Unk.(82).
Porodisculus pendulus. **2:**NY(2962).
Porotheleum fimbriatum. **2:**NC(1307).
Postia rancida. **5:**AZ(1202).

Pulcherricium caeruleum. **9**:GA(1439).
Pycnoporus cinnabarinus. **10**:Unk.(82).
P. sanguineus. **5**:AZ(1202).
Schizophyllum commune. On wood. **7**:FL(35). **8**:CA(82). **10**:Range of host(82), FL(35).
Schizopora apacheriensis. **5**:AZ(1193,1202).
S. paradoxa. **10**:MI(82).
Spongipellis delectans. **10**:NY(82).
S. spumeus. **10**:Unk.(82).
Steccherinum ochraceum. **5**:AZ(1202).
Stereum ostrea. On dead limbs. **7**:OK(82,2873).
Stereum sp. **7**:WA(3202).
Stigmatolemma poriaeforme. **5**:AZ(1202).
Tomentella chlorina. **5**:AZ(1202). **10**:AZ(2039).
T. ferruginea. **10**:AZ(2039), NM(2039).
T. sublilacina. **10**:AZ(2039).
T. viridescens. **10**:AZ(2039).
Trametes hirsuta. **7**:WA(3202). **8**:OR(3202). **10**:Unk.(82).
T. versicolor. **7**:WA(3202). **10**:Unk.(82).
Trechispora sphaerocystis. White rot. **5**:AZ(438).
Trichaptum biforme. **10**:NY(82).

Deuteromycotina - Hyphomycetes

Alternaria sp. On nuts. **8**:CA(82,1110).
Berkleasmium concinnum. On decorticated wood. **2**:NC(1303,1307), SC(1307).
B. opacum. On wood. **10**:NY(2468).
Cadophora sp. **10**:OR(3202).
Cercospora juglandis. Leaf spot. **2**:MA(82). **7**:KS(82).
Cercospora sp. Leaf spot. **7**:FL(35). **8**:FL(35).
Cladosporium pericarpium. On nuts. **7**:SC(82).
Cladosporium sp. **7**:MD(82), MN(82).
Conoplea globosa. **10**:PA(1672).
C. olivacea. **10**:MI(1672).
Cristulariella moricola. Zonate leaf spot. **2**:WV(815,816). **7**:IL(2575), WV(3680).
Cylindrocladiella parva. Seedling root rot. **7**:VA(3044).
Cylindrocladium scoparium. On dead limbs, isolated from rhizosphere and roots. **7**:KY(677), WV(1694). **10**:OH(82).
Cylindrocladium sp. **7**:TN(1694).
Dendrosporium lobatum. On submerged, decaying fruit. **7**:IL(720).
Fusarium lateritium. On nuts and twigs, isolated from cankers. **7**:IA(1883), IL(1883), IN(1883), MO(1883). **8**:CT(82), NY(82). **10**:Unk.(82).
F. oxysporum. Isolated from cankers. **7**:IA(1883), IL(1883), IN(1883), MO(1883).
F. semitectum var. *majus*. On nuts and twigs. **10**:Unk.(82).
F. solani. Stem canker. **7**:KS(3645).
Fusarium sp. **10**:OR(3202).
Helicoma morganii. On wood. **10**:OH(1271).
H. tenuifilum. On shell. **7**:VA(1271).
Helicomyces bellus. On wood. **10**:AL(1270).
Phymatotrichopsis omnivora. Root rot. **7**:TX(82). **8**:'Phymatotrichum root rot.' TX(82). **10**:TX(82).
Ramularia sp. **10**:OR(3202).
Trichoderma sp. **10**:OR(3202).
Trichothecium roseum. On nuts. **7**:IN(663).
Verticillium sp. **10**:OR(3202).

Deuteromycotina - Coelomycetes

Actinothecium juglandis. **2**:NY(82).
Ascochyta juglandis. 'Ascochyta ring spot.' **8**:OR(82,3202), WA(82).
Botryodiplodia congesta. On bark of grafted plants. **1**:OR(82,3202).
Colletotrichum gloeosporioides. Fruit drop. **7**:FL(35).
Cylindrosporium juglandis. Leaf blight and spot. **4**:CA(82,1110). **7**:AL(82), NC(1303), OK(82,2872), TN(82). **8**:AL(82), NC(82), OK(2872). **10**:SC states(82), SE states(82), CA(82,1110), MS(2720).
Cylindrosporium sp. Leaf spot. **2**:MN(82). **10**:FL(35).
Cytospora albiceps. On limbs. **7**:KS(82). **10**:KS(82).
C. leucosperma. On limbs. **7**:MS(2720). **10**:KS(82), MI(82).
Cytospora sp. Canker. **7**:AZ(82). **8**:CA(1110).
Dichomera juglandis. On dead limbs. **10**:OH(82).
Diplodia juglandis. Canker, dieback, on limbs. **1**:LA(82). **8**:CA(82,1110), FL(35). **10**:Range of host(82), CA(1110), FL(35).
Dothiorella gregaria. Dieback. **8**:CA(82).
Hendersonula toruloidea. On twigs. **1**:CA(1110). **8**:'Branch wilt.' CA(82,3382). **10**:CA(2633).
Marssonina californica. Leaf spot. **1**:CA(82,1110).
M. juglandis. Leaf spot. **1**:CA(1110). **2**:NC(3944). **7**:IL(1052), IN(663), OK(2872), SD(2243). **8**:OK(2872). **10**:Unk.(82).
Melanconium oblongum. Canker. **2**:NC(1303). **3**:Unk.(82). **7**:NC(1303).
Phloeospora multimaculans. Leaf spot. **7**:TX(82). **8**:OK(2872), TX(82). **10**:TX(82).
Phomopsis arnoldiae. Shoot dieback of seedlings. **7**:IN(1325).
P. juglandina. On limbs. **8**:CA(82). **10**:CA(1110).
Phyllosticta juglandina. On leaves. **8**:OR(82).
P. juglandis. Leaf spot. **8**:GA(82), IN(82), OR(82,3202), WA(82,3202). **10**:OR(2430).
Rhabdospora juglandis. On limbs. **7**:PA(82).
Sirococcus clavigignenti-juglandacearum. Canker. **2**:WI(2552,3646).
Sphaeropsis malorum. On nuts. **7**:PA(82).

Deuteromycotina - Other

Rhizoctonia sp. **10**:OR(3202).
Sclerotium rolfsii. Seedling blight. **7**:CA(1110), TX(82).

Pterocarya Kunth

1. *P. stenoptera* C. DC. [Chinese wingnut]

Basidiomycotina - Other

Armillaria mellea. Root rot. **1**:CA(2897).

Juncaceae

Juncus L.

1. *J. acuminatus* Michx.
2. *J. balticus* Willd. (=*J. balticus* Willd. var. *montanus* Engelm.) [Baltic rush, wire rush]
3. *J. balticus* Willd. var. *littoralis* Engelm.
4. *J. bolanderi* Engelm.
5. *J. brachycephalus* (Engelm.) Buchenau
6. *J. brevicaudatus* (Engelm.) Fernald
7. *J. bufonius* L. [toad rush]
8. *J. castaneus* Sm.
9. *J. confusus* Crép.
10. *J. covillei* Piper
11. *J. diffusissimus* Buckley
12. *J. drummondii* E. Mey.
13. *J. effusus* L. [bog rush, Japanese mat rush, soft rush]
14. *J. ensifolius* Wikstr. (=*J. saximontanus* A. Nels., *J. saximontanus* A. Nels. f. *brunnescens* F. J. Herm.)
15. *J. falcatus* E. Mey.
16. *J. greenei* Oakes & Tuck.
17. *J. howellii* F. J. Herm.
18. *J. lesueurii* Boland. [salt rush]
19. *J. longistylis* Torr.
20. *J. marginatus* Rostock
21. *J. maritimus* Lam.
22. *J. mertensianus* Bong.
23. *J. militaris* Bigel.
24. *J. nevadensis* S. Wats. (=*J. columbianus* Coville)
25. *J. orthophyllus* Crép.
26. *J. oxymeris* Engelm.
27. *J. parryi* Engelm.
28. *J. pelocarpus* E. Mey.
29. *J. phaeocephalus* Engelm.
30. *J. roemerianus* Scheele
31. *J. tenuis* Willd. (=*J. dudleyi* Wiegland, *J. interior* Wiegland) [path rush, slender rush]
32. *J. torreyi* Crép.
33. *J. vaseyi* Engelm.
34. *J. xiphioides* E. Mey.
35. *Juncus* sp.

Chytridiomycetes

Physoderma sp. On leaves. **28**:MI(3402).

Zygomycotina

Blakeslea trispora. On leaves. **30**:FL(1032).
Mucor sp. On leaves. **30**:FL(1032).
Syncephalastrum racemosum. On leaves. **30**:FL(1032).

Ascomycotina

Achaetomium sp. On leaves. **30**:FL(1032).

Aniptodera chesapeakensis. On dead leaves. **30:**NC(1945).
Belonopsis junciseda. **3:**WI(1333).
Bricookea sepalorum. On floral bracts. **27:**CA(638,642,661), CO(649), ID(3202), MT(3202), OR(81,3202), WA(638,3202), WY(2924). **35:**CA(187), OR(187), WA(187).
Ceratosphaeria sp. On leaves. **30:**FL(1032).
Chaetomium sp. On leaves. **30:**FL(1032).
Clathrospora bakeri. **35:**CO(3818).
C. elynae. On overwintered herbage. **22:**CA(642).
C. juncicola. On culms. **2:**CO(82).
Claviceps junci. Ergot. **35:**WI(82).
Crocicreas melanosporum. On overwintered stems, petioles and peduncles. **2:**CA(505,642).
Didymella juncina. On culms. **35:**SC(82).
Didymosamarospora euryhalina. On submerged culms. **30:**NC(1789).
Duplicaria acuminata. On culms and leaves. **12:**WA(3202,3815), WY(1195). **27:**CA(638,642,661), ID(3202), MT(3202), OR(3202), WA(3202), WY(2924). **35:**CA(82), CO(82).
Epichloe typhina. Cat-tail. **13:**NH(1901).
Glyphium elatum. **2:**CA(642).
Guignardia sp. On leaves. **30:**FL(1032).
Halosphaeria hamata. On leaves. **30:**FL(1032).
Keissleriella sp. On leaves. **30:**FL(1032).
Leptosphaeria australiensis. On leaves. **30:**FL(1032).
L. marina. On culms. **21:**NC(1793).
L. neomaritima. On culms. **21:**NC(1793).
Leptosphaeria sp. On dead stems. **1:**GA(1438). **2:**CA(82). **13:**GA(1438). **35:**GA(82), MS(82,2720).
Loramyces juncicola. **23:**MA(874), RI(874).
Massarina gloeospora. On dead stems. **13:**NJ(82). **35:**NY(82).
Massarina sp. On leaves. **30:**FL(1032).
Melanopsamma pomiformis. On leaves. **30:**FL(1032).
Melanospora sp. On leaves. **30:**FL(1032).
Mycosphaerella muhlenbergiae. **35:**WA(3202,3815).
M. tassiana. On dead sheaths and stems. **8:**AK(1938). **35:**WA(3202,3815).
M. tulasnei. **2:**CA(642).
Mycosphaerella sp. On leaves. **30:**FL(1032).
Myriosclerotinia curreyana. **35:**WA(3202).
Nectria sp. On leaves. **30:**FL(1032).
Nimbomollisia stictoidea. On culms. **35:**NJ(2563).
Niptera guestphalica. **13:**GA(1438).
Ophiobolus junci. On stems. **13:**GA(82,1438).
Otthia sp. On leaves. **30:**FL(1032).
Paraphaeosphaeria michotii. **25:**CA(642). **31:**OK(2873).
Phaeosphaeria caricinella. On overwintered herbage. **2:**CA(642). **25:**CA(642).
P. juncina. On leaves. **30:**FL(1032).
Phomatospora sp. On leaves. **30:**FL(1032).
Phyllachora junci. Stem spot, on culms. **13:**GA(82,1438), OH(82). **27:**CA(638,642), OR(3202), WA(3202). **31:**GA(1438), NY(82), WI(82). **33:**WY(2924). **35:**IA(82), OR(82).
Pleospora sp. On overwintered herbage. **22:**CA(642). **31:**GA(1438).
Pyrenophora trichostoma. **2:**CA(642).
Sphaerulina sp. On leaves. **30:**FL(1032).
Sporormia sp. On leaves. **30:**FL(1032).

Basidiomycotina - Uredinales

[*Puccinia caricina*]. **34:**AZ(1203).
Uromyces junci. Heteroecious: telial host. **2:**W states(82), CA(81,82,2439), ID(3202), MT(3202), ND(378), OR(3202), PA(82), WA(82,3202), WY(1195). **9:**MT(3202). **10:**CA(81). **13:**FL(35). **18:**CA(82), OR(82,3202). **22:**ID(3202). **24:**WA(3202). **27:**CA(638,642), WY(1195). **29:**CA(81). **32:**WA(3202). **34:**AZ(3981). **35:**NC(3944), SC(634).
U. junci-effusi. Telial host; aecial host unknown. **2:**WY(1195). **4:**OR(3202), WA(3202). **13:**E states(82), FL(35), GA(1439), MO(82), NC(1303,1307), WA(3202). **14:**AZ(1195,3379,3981), CA(81), ID(3202), MT(3202), OR(3202), WA(3202). **15:**WA(3202). **17:**CA(661). **18:**CA(81). **22:**OR(3202). **24:**WA(3202). **25:**OR(3202), WA(3202). **26:**OR(3202). **27:**WA(3202). **34:**AZ(3981), CA(81), MT(3202), WA(3202). **35:**NW states(82), AZ(1203), CA(81,82), WA(82).
U. silphii. Heteroecious: telial host. **1:**ID(3202). **3:**WI(1333). **13:**FL(35). **19:**ID(3202), MT(3202), WA(3202). **31:**E states(82), AZ(3981), CA(81), NC(82), ND(378), OK(2873), OR(3202), TX(82,742), WA(82,3202), WI(1333).

Basidiomycotina - Ustilaginales

[*Anthracoidea caricis*]. On plants. **27:**CA(638).
Cintractia junci. Inflorescence smut. **1:**MS(1077,2720), NY(1077). **7:**PA(1077). **11:**IN(1077). **13:**NJ(1077). **16:**WI(1348). **27:**WY(1077). **31:**NE states(82), CO(1077), CT(1077), DC(1077), DE(1077), IA(1077), IL(1077), IN(1077), MA(1077), MD(1077), NJ(1077), NY(1077), OH(1077), PA(1077), TX(82), VA(1077), VT(1077), WI(82,1077,1335). **35:**IA(82), MS(82,2720), NC(82), NY(82), PA(82), WI(82).
Tolyposporium junci. Inflorescence smut. **7:**OR(1077,3202), WA(3202). **35:**OR(82).
Urocystis junci. Stem smut. **2:**NV(82,1077), WY(82).

Basidiomycotina - Other

Stilbum sp. On leaves. **30:**FL(1032).

Deuteromycotina - Hyphomycetes

Acremonium sp. On leaves. **30:**FL(1032).
Alternaria alternata. On leaves. **30:**FL(1032).
[*A. longissima*]. On leaves. **30:**FL(1032).
Arthrinium cuspidatum. Sooty patches on dead culms. **2:**CA(82,638,639,642,661). **27:**WY(2924). **35:**OR(3202), WA(3202).
Arthrinium sp. On leaves. **30:**FL(1032).
Aspergillus niger. On leaves. **30:**FL(1032).
Aspergillus sp. On leaves. **30:**FL(1032).
Aureobasidium sp. On leaves. **30:**FL(1032).
Beltrania querna. On leaves. **30:**FL(1032).
Bipolaris hawaiiensis. On leaves. **30:**FL(1032).
B. palousensis. On living foliage. **14:**WA(3202,3440). **24:**OR(3202,3440).
Cercospora juncina. On leaves. **5:**WI(1348). **6:**WI(82,581,1348). **9:**WA(3202,3444). **14:**UT(3440). **16:**WI(1348). **31:**WI(1346,1348).
Circinotrichum maculiforme. On leaves. **30:**FL(1032).
Cirrenalia macrocephala. On leaves. **30:**FL(1032).
C. pseudomacrocephala. On leaves. **30:**FL(1032).
Cladobotryum sp. On leaves. **30:**FL(1032).
Cladosporium cladosporioides. On leaves. **30:**FL(1032).
C. sphaerospermum. On leaves. **30:**FL(1032).
Cladosporium sp. On leaves. **30:**FL(1032).
Cremasteria cymatilis. On leaves. **30:**FL(1032).
Crinula sp. On leaves. **30:**FL(1032).
Curvularia protuberata. On leaves. **30:**FL(1032).
C. tuberculata. On leaves. **30:**FL(1032).
Curvularia sp. On leaves. **30:**FL(1032).
Dendryphiella salina. On leaves. **30:**FL(1032).
Drechslera sp. On leaves. **30:**FL(1032).
Epicoccum nigrum. On leaves. **30:**FL(1032).
[*Exserohilum rostratum*]. On leaves. **30:**FL(1032).
Flagellospora sp. On leaves. **30:**FL(1032).
Fusariella obstipa. On leaves. **30:**FL(1032).
Fusarium sp. On leaves. **30:**FL(1032).
Geniculosporium sp. On leaves. **30:**FL(1032).
Gliocladium sp. On leaves. **30:**FL(1032).
Gliomastix sp. On leaves. **30:**FL(1032).
Haplobasidion lelebae. On leaves. **30:**FL(1032).
Heterosporium sp. **2:**WA(3202).
Humicola sp. On leaves. **30:**FL(1032).
Memnoniella echinata. On leaves. **30:**FL(1032).
Monilia sp. On leaves. **30:**FL(1032).
Monodictys austrina. On leaves. **30:**FL(1032).
Myrothecium leucotrichum. On leaves. **30:**FL(1032).
M. roridum. On leaves. **30:**FL(1032).
Nigrospora sphaerica. On leaves. **30:**FL(1032).
Paecilomyces sp. On leaves. **30:**FL(1032).
Papulaspora halima. On leaves. **30:**FL(1032).
Penicillium sp. On leaves. **30:**FL(1032).
Periconia cookei. On leaves. **30:**FL(1032).
P. digitata. On leaves. **30:**FL(1032).
P. echinochloae. On leaves. **30:**FL(1032).
P. igniaria. On leaves. **30:**FL(1032).
P. minutissima. On leaves. **30:**FL(1032).
Periconia sp. On leaves. **30:**FL(1032).
Pithomyces atro-olivaceus. On leaves. **30:**FL(1032).
P. chartarum. On leaves. **30:**FL(1032).

P. maydicus. On leaves. **30**:FL(1032).
Ramularia junci. Leaf spot. **20**:NY(82).
Rhinocladiella sp. On leaves. **30**:FL(1032).
Scolecobasidium humicola. On leaves. **30**:FL(1032).
Scopulariopsis sp. On leaves. **30**:FL(1032).
Septonema secedens. On leaves. **30**:FL(1032).
Spegazzinia tessarthra. On leaves. **30**:FL(1032).
Sporothrix sp. On leaves. **30**:FL(1032).
Stachybotrys chartarum. On leaves. **30**:FL(1032).
S. cylindrospora. **30**:FL(1032).
S. kampalensis. On leaves. **30**:FL(1032).
S. nephrospora. On leaves. **30**:FL(1032).
Stachybotrys sp. On leaves. **30**:FL(1032).
Stachylidium bicolor. On leaves. **30**:FL(1032).
Stemphylium lycopersici. On leaves. **30**:FL(1032).
S. vesicarium. On leaves. **30**:FL(1032).
Tetraploa aristata. On leaves. **30**:FL(1032).
Torula herbarum. On leaves. **30**:FL(1032).
Trichoderma viride. On leaves. **30**:FL(1032).
Trullula junci. On stems. **18**:CA(82).
Veronaea sp. On leaves. **30**:FL(1032).
Virgaria nigra. On leaves. **30**:FL(1032).
Zalerion varium. On leaves. **30**:FL(1032).
Zygosporium gibbum. On leaves. **30**:FL(1032).
Z. masonii. On leaves. **30**:FL(1032).
Zygosporium sp. On leaves. **30**:FL(1032).

Deuteromycotina - Coelomycetes

Aposphaeria mollis. On overwintered culms. **27**:CA(642).
Botryodiplodia sp. On leaves. **30**:FL(1032).
Coniothyrium junci. Dead scapes. **2**:OR(970,3202).
Coniothyrium sp. On leaves. **30**:FL(1032).
Crandallia juncicola. On dead leaves and living stems. **12**:CO(3549), WY(1195,3549,3550).
Cytosporina sp. On leaves. **30**:FL(1032).
Discella tenuispora. On stems. **35**:CA(82).
Hendersonia scirpicola. On dead stems. **18**:CA(82).
Hendersonia sp. On leaves and dead stems. **18**:CA(82). **30**:FL(1032).
Heteropatella umbilicata. **27**:WY(2924). **35**:CA(642).
Kellermania alpina. On stems. **35**:CA(82).
Leptostroma juncacearum. **35**:ID(3202), WA(3202).
Melanconium juncinum. On stems. **2**:CA(81), UT(82). **12**:AK(526). **18**:CA(82). **27**:CO(649), ID(3202), MT(3202), OR(81,3202), WA(3202). **35**:MT(3202), WA(3202).
Neottiospora sp. On leaves. **30**:FL(1032).
Pestalotia sp. On leaves. **30**:FL(1032).
Phoma junci. On stems. **35**:NC(82), VA(82).
Phoma sp. On leaves. **30**:FL(1032).
Phomopsis sp. On leaves. **30**:FL(1032).
Phyllosticta sp. **31**:WA(3202).
Placosphaeria junci. **27**:OR(3202). **31**:WA(3202).
Psammina sp. On leaves. **30**:FL(1032).
Pyrenochaeta sp. On leaves. **30**:FL(1032).
Rhabdospora junci. **35**:ID(3202).
Rhabdospora sp. Stem spot. **2**:MT(82,3202).
Selenophoma sp. On leaves. **30**:FL(1032).
Septoria sp. On leaves. **30**:FL(1032).
Sphaeronaema sp. On leaves. **30**:FL(1032).
Stagonospora californica. **35**:CA(2783).
S. caricinella. On dried basal leaves. **7**:AK(3444).
S. culmicola. On dead stems. **18**:CA(82).
Stagonospora sp. On leaves. **30**:FL(1032). **31**:WI(1347).
Stigmatomassaria sp. On leaves. **30**:FL(1032).
Zythia sp. On leaves. **30**:FL(1032).

Luzula DC. in Lam. & DC.

1. *L. arcuata* (Wahlenb.) Wahlenb.
2. *L. campestris* (L.) DC. [common woodrush, field woodrush]
3. *L. divaricata* S. Wats.
4. *L. echinata* (Small) F. J. Herm.
5. *L. glabrata* (Hoppe) Desv.
6. *L. multiflora* (Ehrh. ex Hoffm.) Lej.
7. *L. parviflora* (Ehrh.) Desv. [millet woodrush]
8. *L. piperi* (Coville) M. E. Jones
9. *L. spicata* (L.) DC.
10. *L. subcongesta* (S. Wats.) Jeps.
11. *L. wahlenbergii* Rupr.
12. *Luzula* sp.

Ascomycotina

[*Gaeumannomyces graminis*]. Take-all. **7**:WA(3202,3446).
Microthyrium sp. **11**:WA(3202).
Mycosphaerella alpina. **11**:WA(3202,3815).
Mycosphaerella sp. **12**:WA(3202).

Basidiomycotina - Uredinales

Puccinia obscura. Heteroecious: telial host. **2**:C states(82), NE states(82), CA(81), ID(82,3202), OR(3202), WA(3202). **5**:ID(3202), WA(3202). **6**:AK(60,526). **7**:CA(81), ID(82,3202), OR(82,3202), WA(82,3202). **8**:ID(3202,3440). **11**:AK(3444). **12**:IA(82), NY(82), WI(82).

Basidiomycotina - Ustilaginales

Ustilago luzulae. **2**:IN(1077). **4**:IL(312).
U. vuijckii. **5**:WA(1077,3202).

Deuteromycotina - Coelomycetes

Ascochyta luzulicola. On dried foliage. **7**:AK(3444).
A. paucisporula. On living leaves. **3**:ID(3202,3444). **8**:ID(3202).
[*A. zeicola*]. **3**:WA(3202,3816).
Phyllosticta sp. **6**:AK(3444).
Septoria chanousiana. **2**:ID(3202), OR(3202), WA(3202). **7**:AK(3444). **8**:WA(3202). **9**:OR(3202), WA(3202). **10**:OR(3202). **11**:WA(3202). **12**:MT(3202).
S. minuta. **2**:AK(3431). **6**:WI(1361). **10**:CA(642).
Stagonospora aquatica var. *luzulicola*. **1**:AK(526).

Juncaginaceae

Triglochin L.

1. *T. maritimum* L.

Basidiomycotina - Uredinales

[*Puccinia aristidae*]. **1**:WY(1195).

Krameriaceae

Krameria L. in Loefl.

1. *K. lanceolata* Torr.
2. *Krameria* sp.

Basidiomycotina - Uredinales

Uromyces krameriae. Autoecious. **1**:TX(82). **2**:FL(35).

Deuteromycotina - Hyphomycetes

Phymatotrichopsis omnivora. Root rot. **1**:TX(82).

Lamiaceae

Agastache Clayton ex Gronov.

1. *A. foeniculum* (Pursh) Kuntze [anise-hyssop, blue giant hyssop, fennel giant hyssop, fragrant giant hyssop]
2. *A. nepetoides* (L.) Kuntze [yellow giant hyssop]
3. *A. scrophulariifolia* (Willd.) Kuntze [purple giant hyssop]
4. *A. urticifolia* (Benth.) Kuntze [nettle-leaf giant hyssop]

Oomycetes

Peronospora lophanthi. Downy mildew. **2**:IA(82), IL(82). **3**:IA(82), WI(82).

Ascomycotina

Crocicreas cyathoideum var. *cacaliae*. On dead stems. **4**:UT(505).
C. nigrofuscum var. *allantosporum*. **4**:UT(505).
Leptosphaeria brightonensis. **4**:UT(3249).
L. darkeri. **4**:UT(3249).
Mycosphaerella tassiana. **4**:OR(3202).
Nodulosphaeria olivacea. **4**:WY(3810).
Pleospora helvetica. **4**:OR(3202).
P. herbarum var. *occidentalis*. **4**:WY(3810).
P. richtophensis. **4**:WY(3810).
P. scrophulariae var. *compositarum*. **4**:WY(3810).
Sphaerotheca fuliginea. Powdery mildew. **1**:MO(82), WI(82). **3**:WI(1348). **4**:UT(82), WA(82).
S. macularis. Powdery mildew. **4**:WA(82,3202).

Basidiomycotina - Uredinales

Puccinia hyssopi. Autoecious. **2**:NC states(82), NE states(82), IA(82). **3**:NC states(82), NE states(82), WI(82,1333).

Deuteromycotina - Hyphomycetes
Ramularia lophanthi. Leaf spot. **3:**CA(82). **4:**ID(1195,3202), MT(82,3202), WA(3202), WY(1195).
Deuteromycotina - Coelomycetes
Ascochyta lophanthi. Leaf spot. **1:**WI(82). **3:**WI(82).
A. lophanthi var. *osmophila*. **1:**WI(809).
Heteropatella umbilicata. **4:**OR(3202).
Kellermania alpina. On stems. **4:**OR(82).
Phoma exigua. **3:**WI(1361).
P. lophanthi. On stems. **2:**OH(82).
Septoria lophanthi. Leaf spot. **2:**IL(82), MO(82), OH(82). **3:**WI(82).

Ajuga L.
1. *A. reptans* L. [carpet bugleweed, creeping bugleweed]
2. *Ajuga* sp.
Oomycetes
Pythium sp. Root rot. **1:**FL(35).
Ascomycotina
Erysiphe cichoracearum. Powdery mildew. **1:**CA(1155).
Deuteromycotina - Hyphomycetes
Alternaria sp. Leaf spot. **1:**FL(35).
Cercospora sp. Leaf spot. **1:**FL(35). **2:**FL(35).
Corynespora cassiicola. Leaf spot. **1:**FL(35).
Fusarium oxysporum. Crown and root rot. **1:**FL(35).
Myrothecium roridum. Leaf spot. **1:**FL(35). **2:**FL(35).
Stemphylium sp. Leaf spot. **2:**FL(35).
Deuteromycotina - Coelomycetes
Colletotrichum sp. Leaf spot. **1:**FL(35).
Deuteromycotina - Other
Rhizoctonia solani. Foliage blight, damping-off, root rot. **1:**FL(35), NC(1303). **2:**FL(35).
Sclerotium rolfsii. Southern blight. Crown rot. **1:**CA(82), CT(82,3770), FL(35), HI(2299,2903), KS(82), NC(1303), NJ(82). **2:**FL(35), OH(946).
S. rolfsii var. *delphinii*. Crown rot. **2:**FL(35).

Blephilia Raf.
1. *B. ciliata* (L.) Benth.
2. *B. hirsuta* (Pursh) Benth. [wood mint]
Basidiomycotina - Uredinales
Puccinia menthae. Autoecious. **1:**IN(82), MD(82), MI(82). **2:**IA(82), IN(82), MO(82), TN(82), WI(82).
Deuteromycotina - Hyphomycetes
Cercoseptoria blephiliae. On leaves. **1:**WI(82,1331).
Cercospora blephiliae. On leaves. **1:**WI(1334).
Deuteromycotina - Coelomycetes
[*Septoria menthae*]. Leaf spot. **1:**WI(82,1330).

Clinopodium L.
1. *Clinopodium* sp.
Basidiomycotina - Uredinales
Puccinia menthae. Autoecious. **1:**NE states(82), CO(82), WI(82).
Deuteromycotina - Coelomycetes
Phoma exigua. Leaf spot. **1:**NY(82).

Collinsonia L.
1. *C. canadensis* L. [citronella, richweed, stoneroot]
Ascomycotina
Helotium rhizicola. On roots. **1:**NY(850).
Leptosphaeria collinsoniae. On stems. **1:**NY(82).
Phyllachora sp. Black stem spot. **1:**PA(82).
Deuteromycotina - Hyphomycetes
Dendryphiella vinosum. On stems. **1:**NY(82).
Stachylidium bicolor. On stems. **1:**NY(82).

Cunila D. Royen ex L.
1. *C. origanoides* (L.) Britton [American dittany, common dittany, stone mint, sweet horsemint]
Basidiomycotina - Uredinales
Puccinia menthae. Autoecious. **1:**E states(82), AR(82), IL(82), IN(987), NC(3944).
Deuteromycotina - Coelomycetes
Septoria cunillae. Leaf spot. **1:**IL(82).

Dracocephalum L.
1. *D. parviflorum* Nutt.
2. *D. virginianum* L. (=*Physostegia virginiana* (L.) Benth.)
3. *Dracocephalum* sp.
Oomycetes
Peronospora davisii. On living leaves. **1:**WI(1344,3201).
P. hedeomae. **1:**WI(82).
P. lophanthi var. *moldavicae*. **1:**WY(82).
Peronospora sp. Downy mildew. **1:**Unk.(82).
Ascomycotina
Mycosphaerella physostegiae. Leaf spot. **2:**GA(82), VA(82).
Sclerotinia sclerotiorum. Stem rot. **2:**ME(82).
Basidiomycotina - Uredinales
Puccinia physostegiae. Autoecious. **2:**IN(82,647), NH(82), NY(82).
Deuteromycotina - Coelomycetes
Phyllosticta dracocephali. Leaf spot. **1:**TX(82).
Rhabdospora physostegiae. **2:**NY(82).
Septoria dracocephali. Leaf spot. **1:**TX(82), WI(82).
S. physostegiae. On leaves. **2:**IL(82), WI(82,966).
Deuteromycotina - Other
Sclerotium rolfsii. Southern blight. **1:**TX(82). **2:**SC states(82), SE states(82), KS(82), OK(2872). **3:**IL(82).
S. rolfsii var. *delphinii*. **2:**CT(82), NJ(82), NY(82), WI(82).

Galeopsis L.
1. *G. tetrahit* L. [hemp-nettle]
Ascomycotina
Erysiphe cichoracearum. Powdery mildew. **1:**Range of host(82).
Deuteromycotina - Hyphomycetes
Cercospora sp. **1:**WI(1351).
Deuteromycotina - Coelomycetes
Phoma exigua. Leaf spot. **1:**WI(82,1333).
Septoria galeopsidis. Leaf spot. **1:**NY(82), VT(82).

Glechoma L.
1. *G. hederacea* L. (=*Nepeta hederacea* (L.) Trevir.) [alehoof, fieldbalm, gill-over-the-ground, groundivy, runaway-robin]
Basidiomycotina - Other
Sebacina incrustans. Encrusting living leaves and stems. **1:**DC(128).
Deuteromycotina - Hyphomycetes
Cercospora nepetae. **1:**WI(1352).
Deuteromycotina - Coelomycetes
Ascochyta leonuri. **1:**WI(1361).
Ascochyta sp. **1:**WI(1360).
Phoma exigua. Leaf spot. **1:**IL(82), IN(82), NJ(82), WI(64,82,1333).
Septoria alabamensis. Leaf spot. **1:**AL(82).

Hedeoma Pers.
1. *H. dentatum* Torr.
2. *H. hispidum* Pursh
3. *H. hyssopifolium* A. Gray
4. *H. longiflora* Rydb.
5. *H. nanum* (Torr.) Briq.
6. *H. palmeri* Hemsl.
7. *H. pulegioides* (L.) Pers. [American pennyroyal, mock pennyroyal, pudding-grass]
8. *Hedeoma* sp.
Oomycetes
Peronospora hedeomae. Downy mildew. **2:**WI(1334). **8:**KS(82), WI(82).
Basidiomycotina - Uredinales
Puccinia menthae. Autoecious. **1:**AZ(222,1203,3981). **3:**AZ(3981). **4:**AZ(3981). **5:**AZ(1203,3981). **6:**AZ(222). **7:**ME(82), PA(82), RI(82), WI(82). **8:**AZ(82), MT(82), NM(82), TX(82).
Deuteromycotina - Hyphomycetes
Verticillium albo-atrum. Wilt. **7:**FL(35).
Deuteromycotina - Coelomycetes
Rhabdospora hedeomina. On stems. **7:**NJ(82), NY(82). **8:**TX(82).
Septoria hedeomae. Leaf spot. **2:**WI(1338). **7:**NY(82).

Hyptis Jacq.
1. *H. alata* (Raf.) Shinners (=*H. radiata* Willd.)
2. *H. emoryi* Torr. [desert lavender]
3. *H. mutabilis* (A. Rich.) Briq.
4. *H. pectinata* (L.) Poit.

5. *Hyptis* sp.
Basidiomycotina - Uredinales
Puccinia distorta. Autoecious. **2**:AZ(221,747,1195,1203,3981), CA(81,221,747).
P. hyptidis. Autoecious. **1**:AL(221), FL(35,221), MS(221,2720), TX(221). **3**:FL(35). **5**:FL(35).
P. hyptidis-mutabilis. Autoecious. **3**:FL(221). **4**:FL(221).
Deuteromycotina - Other
Sclerotium rolfsii. Southern blight. **1**:FL(35).

Isanthus Michx.
1. *I. brachiatus* (L.) B. S. P.
Deuteromycotina - Hyphomycetes
Cercospora isanthi. **1**:WI(1333).

Lamium L.
1. *L. amplexicaule* L. [henbit]
Oomycetes
Peronospora lamii. Downy mildew. **1**:MS(82,2720), NC(82), NY(82), TX(82), WA(3202), WI(1359).
Ascomycotina
Erysiphe cichoracearum. Powdery mildew. **1**:CA(82), KS(82), OK(82,2872), SC(82).
E. galeopsidis. **1**:WA(3202).
Sclerotinia sclerotiorum. Crown and stem rot. **1**:NC(3855).
Deuteromycotina - Coelomycetes
Septoria lamiicola. Leaf spot. **1**:ID(3202), OK(82,2872), WA(3202).
Deuteromycotina - Other
Rhizoctonia solani. Root rot. **1**:TX(82).

Lavandula L.
1. *L. angustifolia* Mill. (=*L. officinalis* Chaix, *L. vera* DC.) [English lavender, lavender]
2. *Lavandula* sp.
Oomycetes
Pythium aphanidermatum. Root rot. **1**:HI(2903).
Basidiomycotina - Other
Armillaria mellea. Root rot. **1**:TX(82).
Deuteromycotina - Coelomycetes
Septoria lavandulae. Leaf spot. **1**:OH(82), OK(82). **2**:OK(2872).

Leonotis (Pers.) R. Br.
1. *L. nepetifolia* (L.) R. Br.
Basidiomycotina - Uredinales
Puccinia leonotidicola. Autoecious. **1**:FL(35,82).
Deuteromycotina - Hyphomycetes
Cercospora leonotidis. Leaf spot. **1**:LA(82,962).

Leonurus L.
1. *L. cardiaca* L. [motherwort]
2. *Leonurus* sp.
Ascomycotina
Leptosphaeria rubicunda. **1**:NY(3249).
Leptospora rubella. **2**:NY(3248).
Deuteromycotina - Hyphomycetes
Cercospora leonuri. **1**:WI(1352).
Deuteromycotina - Coelomycetes
Ascochyta leonuri. Leaf spot. **1**:LA(82), WI(1354,1360).
Ascochyta sp. **1**:WI(1348,1358).
Phoma exigua. Leaf spot. **1**:OH(82), TX(82), WI(82,1333).
Pseudolachnea hispidula. Leaf spot. **1**:TX(82).
Septoria lamii var. *brevior*. **1**:WI(1346).
S. lamiicola. Leaf spot. **1**:PA(82).

Lycopus L.
1. *L. americanus* Muhl. ex Barton
2. *L. rubellus* Moench
3. *L. uniflorus* Michx.
4. *L. virginicus* L.
5. *Lycopus* sp.
Chytridiomycetes
Physoderma lycopi. **1**:MI(3399).
Synchytrium cellulare. Leaf gall. **1**:WI(82). **5**:WI(82).
Basidiomycotina - Uredinales
Puccinia angustata. Heteroecious: aecial host. **1**:NE states(82), KS(82,3101), ND(82,378), NE(3101). **3**:NY(3101), WI(1343,3101). **4**:MN(3101). **5**:NE states(82), KS(82), ND(82).
Deuteromycotina - Hyphomycetes
Cercospora blephiliae. **3**:WI(1336).
Pseudocercospora lycopodis. Leaf spot. **2**:LA(82,581,962).
Deuteromycotina - Coelomycetes
Ascochyta lophanthi. Leaf spot. **1**:WI(82,1350). **5**:MA(82), WI(82).
A. lophanthi var. *lycopina*. On living leaves. **3**:MA(1391), WI(809).
Phoma exigua. Leaf spot. **1**:IA(82), OK(82,2872). **3**:WI(64,1333). **5**:OK(82), WI(82).
Phyllosticta monardae. **1**:WI(1344).
Septoria lycopi. Leaf spot. **1**:WI(82). **5**:ID(3202), WI(82).
Deuteromycotina - Other
Sclerotium rolfsii. **5**:CA(2886), KS(2886).

Marrubium L.
1. *M. vulgare* L. [common horehound, white horehound]
2. *Marrubium* sp.
Chytridiomycetes
Synchytrium marrubii. Leaf gall. **1**:TX(82).
Ascomycotina
Pleospora herbarum. On stems. **1**:CA(82).
Stictis lanuginicincta. On twigs. **1**:CA(2439). **2**:CA(3225).
Deuteromycotina - Hyphomycetes
Cercospora marrubii. On living leaves. **1**:OK(82,2872), SD(2243), TX(82,3597).
Deuteromycotina - Coelomycetes
Coniothyrium marrubii. Damping-off. **1**:CA(82,2439).
Diplodia herbarum. On stems. **1**:TX(82).
Phoma lanuginis. On stems. **1**:NY(82), TX(82).

Melissa L.
1. *M. officinalis* L. [bee balm, common balm, lemon balm, sweet balm]
Deuteromycotina - Hyphomycetes
Botrytis cinerea. Gray mold blight. **1**:AK(82).
Deuteromycotina - Coelomycetes
Phoma exigua. Leaf spot. **1**:NY(82).

Mentha L.
1. *M. aquatica* L. [water mint]
2. *M. arvensis* L. (=*M. penardii* (Briq.) Rydb.) [corn mint, field mint, Japanese mint]
3. *M. arvensis* L. var. *glabrata* (Benth.) Fernald
4. *M. arvensis* L. var. *villosa* (Benth.) S. R. Stewart (=*M. canadensis* L., *M. arvensis* L. var. *canadensis* (L.) Briq.)
5. *M. crispa* L.
6. *M.* × *gentilis* L. (=*M. cardiaca* (S. F. Gray) Baker). *M. arvensis* × *M. spicata*. [red mint, Scotch spearmint]
7. *M.* × *piperita* L. *M. aquatica* × *M. spicata*. [peppermint]
8. *M.* × *piperita* L. var. *citrata* (Ehrh.) Briq. (=*M. citrata* Ehrh.) [bergamot mint, lemon mint]
9. *M. piperita* L. var. *vulgaris* Sole
10. *M. spicata* L. (=*M. viridis* (L.) L.) [spearmint]
11. *Mentha* sp.
Chytridiomycetes
Physoderma menthae. On leaves and stems. **4**:MI(3400).
Oomycetes
Peronospora stigmaticola. On flowers. **4**:WI(1344,1351).
Pythium sp. Rhizome rot. **7**:OR(1633,3202).
Ascomycotina
Erysiphe cichoracearum. Powdery mildew. **2**:CO(82), IA(82), ID(3202), OR(3202), WA(82,3202), WY(1195). **4**:WI(1339). **10**:WA(3202). **7**:OR(3202), WA(2349,3202). **11**:UT(82).
E. galeopsidis. Powdery mildew. **2**:IA(82). **4**:WI(1330).
E. polygoni. Powdery mildew. **2**:TX(82).
Erysiphe sp. Powdery mildew. **7**:OR(1633).
Leptosphaeria substerilis. On leaves. **7**:NY(82,195).
Pleospora dichromatricha. On stems. **11**:CO(3817).
Sclerotinia sclerotiorum. Wilt. **7**:WA(3202,3309).
Sphaerotheca macularis. Powdery mildew. **2**:WA(82,3202).
Basidiomycotina - Uredinales
Puccinia angustata. Heteroecious: aecial host. **2**:MT(3202), WY(1195,3379). **3**:WI(3101). **10**:SD(2243).
P. menthae. Autoecious. **2**:E & C states(82), W states(82), AZ(1203,3981), CA(81), ID(3202), MA(1390), MT(3202),

OR(3202), SD(2243), WA(3202), WY(2924). **4:**AZ(3981), CA(81), ND(378), SD(2243), WY(1195). **10:**E & C states(82), W states(82), AZ(1195,1203,3981), ID(3202), OR(3202), WA(2349,3202). **7:**E & C states(82), W states(82), CA(81), GA(1439), OR(648,1633,3202), WI(1330). **8:**WI(1333). **11:**E & C states(82), W states(82), FL(35).

Basidiomycotina - Other

Thanatephorus cucumeris. **7:**OR(3202).

Typhula sp. Root rot. **7:**OR(1633,3476).

Deuteromycotina - Hyphomycetes

Alternaria sp. Stem canker, leaf spot. **1:**FL(35). **10:**MI(82). **7:**MI(82), WA(3202).

Cercospora menthicola. Leaf spot. **4:**IL(581).

Corynespora sp. Leaf spot. **11:**FL(35).

Fusarium roseum. Rhizome rot. **7:**OR(1633,3202).

Fusarium sp. Stem canker, stolon decay, rhizome rot. **10:**IN(1324), NC(1303). **6:**IN(1324). **7:**IN(1324), MI(82), OR(1633), WA(82).

Ramularia menthicola. Leaf spot. **2:**CA(81,82), ID(1195,3202,3378), ME(82), MT(82,3202), OR(82,3202), SD(2243), WA(82,3202), WI(82), WY(82,1195). **4:**WY(1195). **11:**SD(2243).

R. variata. **2:**WI(82). **10:**WI(1354).

Trichothecium sp. **7:**OR(3202).

Verticillium albo-atrum. Wilt. **10:**IN(82), MI(82), WA(3202). **7:**IN(82), MI(82), OR(82,1633,3202), WA(3202,3309).

V. albo-atrum var. *menthae*. Wilt. **10:**IN(1323), MI(1323), WI(1323). **6:**IN(1323), MI(1323), WI(1323). **7:**IN(1323), MI(1323), WI(1323).

V. dahliae. **7:**WA(3308).

V. nigrescens. **7:**Unk.(2346).

V. tricorpus. **7:**WA(3202,3308).

Deuteromycotina - Coelomycetes

Melanconium sp. Leaf spot. **1:**FL(35).

Phoma exigua. Leaf spot. **2:**NC states(82), NE states(82), IA(82), WI(82), WY(1195). **4:**WI(64,1333). **10:**NC states(82), NE states(82), IA(82), WI(82). **7:**NC states(82), NE states(82), IA(82), WI(82). **11:**NC states(82), NE states(82), IA(82), WI(82).

P. strasseri. Rhizome and stem rot. **7:**OR(1634,3202), WA(3202).

Phyllosticta monardae. **4:**WI(1340).

Phyllosticta sp. **4:**WI(1343).

Septoria menthae. Leaf spot. **2:**ID(3202), SD(2243), WI(82). **4:**WY(1195). **6:**IN(1321).

Sphaceloma menthae. Anthracnose, on leaves, runners and stems. **2:**IN(82,1745). **5:**IN(1745). **9:**IN(1745). **10:**IN(82,1745), MI(82). **7:**IN(82,1745), MD(82,1745), MI(82,1745). **11:**IN(82,1745), MD(1736).

Deuteromycotina - Other

Rhizoctonia solani. Leaf blight, rhizome rot. **1:**FL(35). **7:**OR(1633). **11:**FL(35).

Rhizoctonia sp. Stolon decay, root rot. **10:**IN(1324), NC(1303). **6:**IN(1324). **7:**IN(1324).

Moluccella L.

1. *M. laevis* L. [bells-of-Ireland, molucca balm, shellflower]

Ascomycotina

Sclerotinia sclerotiorum. Stem rot. **1:**FL(35).

Deuteromycotina - Hyphomycetes

Corynespora cassiicola. Flower and leaf blight. **1:**FL(35).

Helminthosporium sp. Leaf spot. **1:**FL(35).

Myrothecium sp. Crown necrosis of maturing plants. **1:**TX(3313).

Monarda L.

1. *M. citriodora* Cerv. ex Lag. (=*M. dispersa* Small) [lemon mint]
2. *M. clinopodia* L.
3. *M. clinopodioides* A. Gray
4. *M. didyma* L. [bee balm, Oswego tea]
5. *M. fistulosa* L. (=*M. media* Willd., *M. mollis* L.) [wild bergamot]
6. *M. fistulosa* L. var. *menthifolia* (R. C. Graham) Fernald (=*M. menthifolia* R. C. Graham)
7. *M. punctata* L. [dotted mint, horsemint, spotted bee balm]
8. *Monarda* sp.

Chytridiomycetes

Synchytrium holwayi. Leaf gall. **5:**IA(82), WI(82,1334). **8:**IA(82), WI(82).

Ascomycotina

Diaporthe arctii. **7:**GA(1438).

Diaporthopsis sp. **7:**GA(1438).

Erysiphe cichoracearum. On stems. **5:**WI(1335). **7:**WI(1361).

Leptosphaeria conoidea. **5:**NY(3249).

Leptosphaeria sp. On leaves. **5:**WI(1353). **7:**GA(1438).

Lophiostoma caulium. **7:**GA(1438).

Massarina gloeospora. **7:**GA(1438).

Mycosphaerella tassiana. **5:**ID(3202).

Ophiobolus anguillidus. **7:**GA(1438).

Basidiomycotina - Uredinales

Puccinia angustata. Heteroecious: aecial host. **5:**NE(82), WI(82,1332).

P. menthae. Autoecious. **4:**E states(82), ID(82), NC(1303,1307), PA(222), TX(82). **5:**E states(82), AK(60), AZ(3981), ID(82,3202), IN(647,987), MS(2720), MT(3202), ND(378), OK(2872), TX(82), WY(1195). **6:**AZ(1195,1203,3981), CO(1195), MT(1195), NM(1195), SD(2243). **7:**E states(82), FL(35), ID(82), TX(82). **8:**E states(82), ID(82), TX(82).

P. monardae. Autoecious. **2:**OH(222). **4:**WV(220). **5:**WV(220).

Basidiomycotina - Other

Ceratobasidium anceps. **5:**WI(1355).

Deuteromycotina - Hyphomycetes

Cercospora sp. Leaf spot. **3:**OK(2872). **8:**OK(82).

Cladosporium monardae. On living leaves. **7:**WI(1346).

Ramularia brevipes. Leaf spot. **2:**AL(82,969). **5:**TX(82). **7:**TX(82).

R. variata. **5:**WI(82).

Deuteromycotina - Coelomycetes

Ascochyta sp. On leaves. **7:**WI(1350).

Phoma exigua. Leaf spot. **1:**OK(2872). **5:**NE(82), TX(82), WI(1336). **7:**FL(35), OK(2872), TX(82), WI(64,1333). **8:**KS(82), OK(82).

Phyllosticta monardae. **5:**WI(1334). **7:**WI(1360). **8:**KS(82).

Septoria brunellae. **2:**NY(2456).

[*S. menthae*]. **5:**WI(1336).

Septoria sp. On leaves. **5:**WI(1351).

Deuteromycotina - Other

Sclerotium rolfsii. Southern blight. **5:**TX(82). **7:**TX(82). **8:**TX(82).

Monardella Benth.

1. *M. douglasii* Benth.
2. *M. odoratissima* Benth. (=*M. odoratissima* Benth. subsp. *pallida* (A. Heller) Epling) [mountain mint]
3. *M. undulata* Benth.
4. *M. villosa* Benth. [coyote mint]
5. *M. villosa* Benth. subsp. *sheltonii* (Torr.) Epling (=*M. villosa* Benth. var. *glabella* A. Gray)
6. *Monardella* sp.

Ascomycotina

Comoclathris permunda. On dead stems. **2:**CA(642).

Erysiphe cichoracearum. Powdery mildew. **1:**CA(1155).

Pleospora richtophensis. On stems. **2:**CA(661).

Basidiomycotina - Uredinales

Puccinia menthae. Autoecious. **2:**CA(81,82,638,642,661), NE(82), OR(82,3202), UT(82). **3:**CA(81). **4:**CA(81), OR(3202). **5:**CA(81). **6:**CA(82), NM(82), OR(82).

Deuteromycotina - Hyphomycetes

Stemphylium sp. On fruit heads. **2:**CA(642).

Deuteromycotina - Coelomycetes

Heteropatella umbilicata. **2:**CA(642).

Phyllosticta monardellae. On leaves. **2:**CA(82,638,642,661).

Siroplacodium shastense. On peduncles and stems. **2:**CA(82,638,642).

Nepeta L.

1. *N. cataria* L. [catmint, catnip]

Ascomycotina

Didymella catariae. On stems. **1:**NJ(82), OH(82).

Deuteromycotina - Hyphomycetes

Botrytis cinerea. Gray mold. **1:**AK(82).

Cercospora nepetae. Leaf spot. **1:**FL(35), IL(82,581), TX(82,581).

Fusarium sp. Wilt. **1:**GA(82).

Deuteromycotina - Coelomycetes

Ascochyta leonuri. Leaf spot. **1:**WI(82,1335).

Phoma exigua. Leaf spot. **1:**OH(82), WI(1333).

Septoria nepetae. **1:**TX(82), WI(82).

Deuteromycotina - Other

Rhizoctonia solani. Root rot. **1:**TX(82).

Sclerotium rolfsii. Southern blight. **1:**TX(82).

Ocimum L.
1. *O. basilicum* L. [common basil, sweet basil]
2. *O. gratissimum* L.
3. *O. sanctum* L.

Oomycetes
Pythium irregulare. Root rot. **3:**FL(35).
Pythium sp. Root rot. **1:**FL(35).
Deuteromycotina - Hyphomycetes
Alternaria sp. Leaf spot. **1:**FL(35).
Deuteromycotina - Coelomycetes
Colletotrichum sp. Leaf spot. **1:**FL(35).
Deuteromycotina - Other
Rhizoctonia solani. Basal stem rot. **1:**FL(35).
Rhizoctonia sp. Root rot. **2:**HI(2903).

Origanum L.
1. *O. majorana* L. [annual marjoram, sweet marjoram]
2. *O. vulgare* L. [marjoram, organy, origano, pot marjoram, wild marjoram]

Oomycetes
Pythium sp. Root rot. **1:**FL(35).
Deuteromycotina - Hyphomycetes
Alternaria sp. Leaf spot. **2:**FL(35).
Botrytis sp. Leaf spot. **2:**FL(35).
Helminthosporium sp. Leaf spot. **2:**FL(35).
Stemphylium sp. Leaf spot. **2:**FL(35).

Perilla L.
1. *P. frutescens* (L.) Britton [perilla]

Deuteromycotina - Hyphomycetes
Cristulariella moricola. On leaves. **1:**IL(2572).

Perovskia Karel.
1. *P. atriplicifolia* Benth.

Deuteromycotina - Coelomycetes
Cylindrosporium sp. Leaf spot. **1:**OK(2872).
Phoma sp. Stem canker. **1:**OK(2872).

Phlomis L.
1. *P. fruticosa* L. [Jerusalem sage]

Basidiomycotina - Other
Armillaria mellea. Root rot. **1:**CA(2900).

Phyllostegia Benth.
1. *P. floribunda* Benth.
2. *Phyllostegia* sp.

Ascomycotina
Asterina phyllostegiae. **2:**HI(2903).
Deuteromycotina - Hyphomycetes
Trichothallus hawaiiensis. **1:**HI(2903).

Physostegia Benth.
1. *P. parviflora* Nutt. ex A. Gray (=*Dracocephalum nuttallii* Britton)
2. *Physostegia* sp.

Oomycetes
Plasmopara cephalophora. Downy mildew. **1:**WI(82).
Ascomycotina
Erysiphe galeopsidis. **1:**WI(1341).
Rhynchosphaeria shearii. On dead leaves. **2:**HI(2781).
Basidiomycotina - Uredinales
Puccinia physostegiae. Autoecious. **1:**MT(3202), WA(3202). **2:**MT(82), NY(82).
Deuteromycotina - Hyphomycetes
Cercospora sp. Leaf spot. **2:**FL(35).
Deuteromycotina - Coelomycetes
Septoria physostegiae. **1:**WI(1339).
Deuteromycotina - Other
Sclerotium rolfsii. Leaf rot, basal stem rot. **2:**FL(35).

Plectranthus L'Hér.
1. *P. australis* R. Br.
2. *P. parviflorus* Willd.
3. *Plectranthus* sp.

Oomycetes
Phytophthora sp. Leaf rot. **1:**FL(35).
Pythium splendens. Root rot. **3:**FL(35).
Pythium sp. Root rot. **1:**FL(35). **3:**FL(35).
Deuteromycotina - Hyphomycetes
Alternaria alternata. Leaf spot. **3:**FL(35).
Alternaria sp. Leaf spot. **1:**FL(35). **3:**FL(35).
Botrytis cinerea. Leaf spot. **1:**FL(35).
Corynespora sp. Leaf spot. **1:**FL(35).
Fusarium oxysporum. Root rot. **3:**FL(35).
Fusarium sp. Root rot. **1:**FL(35).
Helminthosporium sp. Leaf and stem blight. **2:**HI(2903).
Myrothecium sp. Leaf necrosis. **1:**FL(35).
Deuteromycotina - Coelomycetes
Phomopsis sp. Dieback. **3:**FL(35).
Phyllosticta sp. Leaf spot. **1:**FL(35).
Deuteromycotina - Other
Rhizoctonia solani. Leaf, root and stem rot. **3:**FL(35).

Prunella L.
1. *P. vulgaris* L. (=*Brunella vulgaris* (L.) Greene) [carpenter-weed, heal-all, self-heal]
2. *P. vulgaris* L. var. *atropurpurea* Fernald (=*Brunella vulgaris* (L.) Greene var. *atropurpurea* (Fernald) Jeps.)
3. *Prunella* sp.

Oomycetes
Pythium palingenes. Root necrosis. **1:**VA(82).
P. polytylum. Root necrosis. **1:**VA(82).
Ascomycotina
Acrospermum compressum. **1:**GA(1438).
Erysiphe cichoracearum. Powdery mildew. **1:**PA(82).
Lophiostoma caulium. **1:**GA(1438).
Ophiobolus prunellae. Leaf spot. **1:**AK(82), CA(642), ID(82,3202), MT(3202), OR(3202,3248), WA(82,959,3202), WI(1357). **2:**CA(81). **3:**AK(526).
Rosenscheldia abundans. Tar spot. **1:**DC(128), ME(82), WA(82,3202), WI(1360).
Sphaerotheca fuliginea. Powdery mildew. **1:**Unk.(3976), IL(82), IN(82), MD(82), MS(82), WA(82), WI(82).
S. macularis. **1:**WA(3202).
Deuteromycotina - Hyphomycetes
Ramularia brunellae. Leaf spot. **1:**IL(82), IN(82,663), NY(82), OH(82), TX(82), VA(82), WI(82).
Deuteromycotina - Coelomycetes
Phyllosticta brunellae. Leaf spot. **1:**TX(82), WA(82,3202), WY(2924).
Septoria brunellae. Leaf spot. **1:**Range of host(82), ID(3202), MT(3202), WA(3202).
Deuteromycotina - Other
Rhizoctonia sp. Root rot. **1:**HI(2903).
Sclerotium rolfsii. Southern blight. **1:**TX(82).

Pycnanthemum Michx.
1. *P. flexuosum* (Walter) B. S. P.
2. *P. pilosum* Nutt.
3. *P. pycnanthemoides* (Leavenw.) Fernald
4. *P. virginianum* (L.) Th. Durand & B. D. Jacks. ex B. L. Robinson & Fernald
5. *Pycnanthemum* sp.

Chytridiomycetes
Synchytrium cellulare. Leaf gall. **1:**WI(82). **4:**WI(1329,1350).
Ascomycotina
Diaporthe arctii. **3:**GA(1438).
Basidiomycotina - Uredinales
Puccinia angustata. Heteroecious: aecial host. **4:**IN(82), WI(1352).
P. menthae. Autoecious. **1:**NE states(82), IA(82), TX(82). **2:**OK(2872). **3:**GA(1439). **4:**E states(82), GA(1439), MS(2720), ND(82). **5:**E states(82), CA(82), ND(82), OK(82).
Deuteromycotina - Hyphomycetes
Cercoseptoria blephiliae. Leaf spot. **4:**WI(82).
Cercosporella pycnanthemi. Leaf spot. **4:**AI(82).
Deuteromycotina - Coelomycetes
Phyllosticta monardae. **4:**WI(1342).

Rosmarinus L.
1. *R. officinalis* L. [rosemary]

Ascomycotina
Sphaerotheca fuliginea. Powdery mildew. **1:**NC(1303).

Basidiomycotina - Other
Armillaria tabescens. Root rot. **1**:FL(35).
Deuteromycotina - Other
Rhizoctonia sp. Root rot. **1**:HI(2903).

Salvia L.

1. *S. amissa* Epling
2. *S. apiana* Jeps. (=*Ramona polystachya* Greene) [greasewood, white sage]
3. *S. azurea* Lam. [blue sage]
4. *S. azurea* Lam. subsp. *pitcheri* (Torr. ex Benth.) Epling (=*S. pitcheri* Torr. ex Benth.)
5. *S. ballotiflora* Benth.
6. *S. coccinea* L. [scarlet sage, Texas sage]
7. *S. dorrii* (Kellogg) Abrams [gray-ball sage]
8. *S. farinacea* Benth. [mealy-cup sage]
9. *S. greggii* A. Gray [autumn sage]
10. *S. lemmonii* A. Gray
11. *S. leucophylla* Greene
12. *S. lycioides* A. Gray
13. *S. mellifera* Greene (=*Ramona stachyoides* Briq., *Audibertia stachyoides* Benth.) [black sage]
14. *S. officinalis* L. [common sage, garden sage]
15. *S. palmeri* (A. Gray) Greene (=*Audibertia palmeri* A. Gray)
16. *S. pinguifolia* (Fernald) Woot. & Standl.
17. *S. polystachya* Mart. & Galeotti
18. *S. reflexa* Hornem. (=*S. lanceolata* Brouss.)
19. *S. sonomensis* Greene (=*Audibertia humilis* Benth.) [creeping sage]
20. *S. spathacea* Greene [pitcher sage]
21. *S. splendens* Sellow ex Roem. & Schult. [scarlet sage]
22. *Salvia* sp.

Chytridiomycetes
Synchytrium salviae. On leaves, petioles and stems. **22**:CA(1847).
Oomycetes
Peronospora lamii. Downy mildew. **18**:KS(973). **22**:KS(82).
Phytophthora sp. Root rot. **21**:FL(35).
Pythium debaryanum. Damping-off. **21**:OH(82).
Pythium sp. Root rot. **21**:FL(35).
Ascomycotina
Didymella ramonae. On stems and twigs. **13**:CA(2439). **22**:CA(82).
Erysiphe cichoracearum. Powdery mildew. **20**:CA(1155). **22**:CA(82).
Mycosphaerella audibertiae. On stems. **17**:CA(2944). **22**:CA(82).
Myriangium catalinae. On stems. **13**:CA(2439).
Nectriella pironii. Stem gall. **21**:FL(35).
Ophiobolus claviger. On creeping stems. **19**:CA(1468).
Patellaria nigrocyanea. On dead stems. **13**:CA(2801).
Pleospora herbarum. On twigs. **13**:CA(2439).
Propolidium californicum. On wood. **11**:CA(527).
Basidiomycotina - Uredinales
Aecidium subsimulans. **10**:AZ(1203). **22**:AZ(82).
Puccinia ballotaeflorae. Telial host. **5**:TX(223).
P. biporula. Autoecious. **10**:AZ(1195,1203,3981). **12**:TX(742).
P. caulicola. Autoecious. **9**:TX(223). **18**:C states(82), CO(223), IA(223), KS(223), MO(223), NE(223), NM(82,223), OK(2872), SD(223,2243), TX(223), WI(1333). **22**:FL(35).
P. farinacea. Autoecious. **1**:AZ(3981). **3**:AL(82), KS(82), MO(82), MS(82,2720), NE(82), OK(82). **4**:OK(2872). **8**:TX(223). **22**:AZ(82), TX(82).
P. farinacea var. *azurea*. Autoecious. **3**:AL(223), FL(223), MS(223), NE(220), TX(223). **4**:KS(223), NE(223), OK(223). **18**:WI(220,1352). **22**:FL(223).
P. mellifera. Autoecious. **2**:CA(222). **7**:WA(3202). **13**:CA(82,220), NV(82). **19**:CA(220,638,642). **22**:CA(82).
P. salviicola. Autoecious. **6**:FL(82,223), TX(82,223). **22**:FL(35).
P. vertisepta. Autoecious. **16**:AZ(1195,1203,3981), NM(223). **22**:AZ(82), NM(82).
Deuteromycotina - Hyphomycetes
Botrytis sp. Blight. **21**:NC(1303).
Cercospora salviicola. Leaf spot. **3**:OK(82,2872). **8**:TX(3597). **14**:WV(82). **22**:FL(35), OK(82), WI(1335).
Phymatotrichopsis omnivora. Root rot. **9**:TX(82). **21**:TX(82).
Ramularia salviicola. Leaf spot. **8**:TX(3597). **22**:TX(82).
Stemphylium lycopersici. Leaf spot. **21**:FL(35).
Volutella flexuosa. On leaves. **22**:NJ(82).
Deuteromycotina - Coelomycetes
Camarosporium eriocryptum. On dead stems. **15**:CA(2439).
Macrophomina phaseolina. Charcoal rot. **22**:SC(82).
Microdiplodia ramonae. On twigs. **13**:CA(2439).
Septoria rhabdocarpa. On leaves. **2**:CA(2439).
S. salviae-pratensis. Leaf spot. **6**:HI(82,2903).
Sphaeropsis salviae. Stem rot. **22**:MS(82,2720).
Deuteromycotina - Other
Rhizoctonia solani. Root rot, sore shin, wire stem. **14**:CT(82). **21**:CT(82), FL(35), IL(82), NC(1303), NJ(82), OH(82).
Rhizoctonia sp. Root rot. **14**:HI(2903).
Sclerotium rolfsii. Southern blight. **14**:IL(82).

Satureja L.

1. *S. douglasii* (Benth.) Briq. (=*Micromeria chamissonis* (Benth.) Greene) [yerba-buena]
2. *S. vulgaris* (L.) Fritsch

Ascomycotina
Erysiphe cichoracearum. Powdery mildew. **1**:CA(1155).
Erysiphe sp. **1**:OR(3202).
Basidiomycotina - Uredinales
Puccinia menthae. Autoecious. **1**:CA(81,82), ID(82,3202), OR(82,3202), WA(82,3202).
Deuteromycotina - Hyphomycetes
Ramularia sp. On leaves. **2**:WI(1357).

Scutellaria L.

1. *S. angustifolia* Pursh
2. *S. galericulata* L.
3. *S. incana* Biehler
4. *S. integrifolia* L.
5. *S. lateriflora* L. [mad-dog skullcap]
6. *S. ovata* J. Hill var. *versicolor* (Nutt.) Fernald (=*S. versicolor* Nutt.)
7. *S. parvula* Michx.
8. *S. parvula* Michx. var. *ambigua* (Nutt.) Fernald
9. *Scutellaria* sp.

Ascomycotina
Erysiphe cichoracearum. **2**:WI(1352). **5**:WI(1336).
E. galeopsidis. Powdery mildew. **2**:MI(82), NY(82), WI(82). **5**:IA(82), IL(82), IN(82), KS(82), MI(82), NY(82), OH(82), WI(82,1351). **9**:MI(82), NY(82), WI(82).
Microsphaera sp. Powdery mildew. **5**:IL(82).
Deuteromycotina - Hyphomycetes
Botrytis cinerea. Stem rot. **2**:WA(3202). **9**:WA(82).
Cercospora scutellariae. Leaf spot. **3**:MS(2720). **9**:IL(82), MO(82), MS(82), TX(82).
Phymatotrichopsis omnivora. Root rot. **9**:TX(82).
Deuteromycotina - Coelomycetes
Phoma exigua. Leaf spot. **5**:WI(64,82,1333). **9**:TX(82).
Septoria scutellariae. Leaf spot. **1**:ID(3202), WA(3202). **2**:NC states(82), NE states(82), ID(1195,3378), WI(82). **4**:OK(2872). **5**:NC states(82), NE states(82), WA(3202), WI(82). **6**:MS(2720). **7**:OK(2872). **8**:WI(1334). **9**:NC states(82), NE states(82), CA(82), CO(82), MS(82), OK(82), WI(82).
Deuteromycotina - Other
Rhizoctonia solani. Root rot. **9**:TX(82).

Solenostemon Thonn.

(= *Coleus* Lour.)

1. *S. scutellarioides* (L.) Codd (=*Coleus blumei* Benth.)
2. *Solenostemon* sp. (=*Coleus sp.*)

Myxomycetes
Badhamia panicea. Slime mold. **1**:KS(82).
Oomycetes
Pythium debaryanum. **2**:Unk.(82).
P. splendens. Root rot. **1**:FL(35). **2**:Unk.(82).
P. vexans. **2**:Unk.(82).
Pythium sp. Damping-off. **1**:MS(2720). **2**:CA(82), MD(82).
Ascomycotina
Nectriella pironii. Stem gall. **1**:FL(35).
Sclerotinia sclerotiorum. Cottony blight. **1**:FL(35).
Deuteromycotina - Hyphomycetes
Alternaria sp. Leaf spot. **1**:NJ(82).
Botrytis cinerea. Leaf spot. **2**:AK(82).
Cercospora sp. Leaf spot. **1**:FL(35).

Stemphylium sp. Leaf spot. **1**:FL(35).
Verticillium sp. Wilt. **1**:CT(82).
Deuteromycotina - Coelomycetes
Phyllosticta sp. Leaf spot. **2**:NJ(82).
Deuteromycotina - Other
Rhizoctonia solani. Damping-off, root rot, stem rot. **1**:FL(35,82), IL(82), NY(82), TX(82). **2**:FL(82), IL(82), NY(82), TX(82).

Sphacele Benth.
1. *S. calycina* Benth. [pitcher-sage]
Basidiomycotina - Uredinales
Uredo sphacelicola. **1**:CA(81,82).
Deuteromycotina - Coelomycetes
Hendersonia varians. On stems. **1**:CA(82).

Stachys L.
1. *S. bullata* Benth.
2. *S. ciliata* Douglas ex Benth.
3. *S. coccinea* Jacq.
4. *S. floridana* Shuttleworth ex Benth.
5. *S. hispida* Pursh
6. *S. mexicana* Benth.
7. *S. palustris* L.
8. *S. tenuifolia* Willd.
9. *Stachys* sp.
Chytridiomycetes
Synchytrium stachydis. Leaf gall. **9**:LA(82).
Ascomycotina
Erysiphe cichoracearum. **1**:Unk.(3976).
E. galeopsidis. Powdery mildew. **2**:WA(82). **4**:FL(35). **6**:WA(3202). **7**:CT(82), MT(82,3202), ND(377), NJ(82), NY(82), OH(82), PA(82), SD(2243), WA(3202), WI(1330), WY(1195). **8**:NC states(82), NE states(82). **9**:NC states(82), NE states(82), NC(82), NM(82), TX(82).
Lophiostoma caulium. **8**:GA(1438).
Metasphaeria sp. **8**:GA(1438).
Sclerotinia sclerotiorum. Cottony blight. **4**:FL(35).
Sphaerotheca macularis. Powdery mildew. **1**:CA(81). **6**:WA(3202). **9**:CA(82).
Basidiomycotina - Uredinales
Puccinia pallidissima. Autoecious. **3**:AZ(1194,1195,1203,3379,3981). **9**:TX(82).
Basidiomycotina - Ustilaginales
Sporidiobolus pararoseus. On leaves. **9**:KS(2686).
Deuteromycotina - Hyphomycetes
Alternaria sp. Leaf spot. **4**:FL(35).
Cercospora stachydis. Leaf spot. **7**:IA(82), ME(82), WI(1339).
Mycovellosiella teucrii. **7**:WI(1341).
Ovularia bullata. Leaf spot. **1**:CA(525). **9**:CA(82).
O. stachydis-ciliatae. **2**:WA(82). **6**:WA(3202).
Ramularia stachydis. Leaf spot. **2**:OR(82). **6**:OR(3202).
Tilletiopsis washingtonensis. On leaves. **9**:KS(2686).
Deuteromycotina - Coelomycetes
Colletotrichum sp. Anthracnose. **4**:FL(35).
Cylindrosporium stachydis. Leaf spot. **7**:IL(82).
Phoma exigua. Leaf spot. **7**:MA(82), WI(64,82,1333). **8**:WI(64,82,1333).
Phyllosticta palustris. **5**:WI(1357). **7**:IL(82), OH(82), WI(1352).
Phyllosticta sp. Leaf spot. **7**:WI(1347).
Septoria stachydis. Leaf spot. **4**:FL(35). **7**:IL(82), MS(82,2720), WI(82). **8**:IA(82), NY(82). **9**:CA(82).

Teucrium L.
1. *T. botrys* L.
2. *T. canadense* L. [American germander, wood sage]
3. *T. occidentale* A. Gray
Oomycetes
Peronospora sp. Downy mildew. **2**:OK(82,2872).
Ascomycotina
Erysiphe cichoracearum. Powdery mildew. **1**:SD(2243). **2**:IL(82), PA(82), WI(82).
E. galeopsidis. **2**:WI(1348).
Basidiomycotina - Uredinales
Puccinia menthae. Autoecious. **2**:PA(82).
Deuteromycotina - Hyphomycetes
Coniosporium harknessioides. On leaves. **2**:IA(82).
Mycovellosiella teucrii. Leaf spot. **2**:E states(82), DE(3543), MS(82,2720), OK(2872), TX(82), WI(82). **3**:WI(1330,1342).
Deuteromycotina - Coelomycetes
Phoma exigua. Leaf spot. **2**:TX(82), WI(82).
Phyllosticta teucrii. **2**:IA(82).

Thymus L.
1. *T. serpyllum* L. [creeping thyme, lemon thyme, mother-of-thyme, wild thyme]
2. *T. vulgaris* L. [common thyme, garden thyme]
3. *Thymus* sp.
Ascomycotina
Mycosphaerella pachyasca. On leaves and stems. **1**:AK(82).
Deuteromycotina - Hyphomycetes
Botrytis cinerea. Gray mold. **2**:AK(82). **3**:AK(526).
Deuteromycotina - Other
Rhizoctonia solani. Root rot. **2**:MA(82).
Rhizoctonia sp. **2**:NC(1303).

Trichostema L.
1. *T. dichotomum* L. [bastard pennyroyal]
Deuteromycotina - Coelomycetes
Septoria trichostematis. Leaf spot. **1**:NY(82).

Westringia Sm.
1. *W. fruticosa* A. Druce
2. *W. rosmariniformis* Sm.
Deuteromycotina - Coelomycetes
Diplodia sp. Twig dieback. **2**:FL(35).
Deuteromycotina - Other
Rhizoctonia solani. Root and basal stem rot. **1**:FL(35). **2**:FL(35).

Lardizabalaceae

Akebia Decne.
1. *A. quinata* (Houtt.) Decne. [chocolate-vine, five-leaf akebia]
Deuteromycotina - Coelomycetes
Diplodia akebiae. On small limbs. **1**:NY(1012).
Phoma akebiae. On dead stems. **1**:NY(832).

Lauraceae

Cinnamomum Schaeff.
1. *C. camphora* (L.) J. Presl [camphor-tree]
2. *C. verum* J. Presl (=*C. zeylanicum* Blume) [Ceylon cinnamon, cinnamon]
Oomycetes
Phytophthora sp. Root rot. **1**:FL(35).
Pythium splendens. Root rot. **1**:HI(2903).
Pythium sp. Root rot. **1**:FL(35).
Ascomycotina
Botryosphaeria dothidea. Dieback. **1**:FL(35).
B. rhodina. On dead limbs. **1**:MS(82,2720).
B. ribis. On limbs. **1**:AL(82).
Echidnodella camphorae. Black mildew, leaf spot. **1**:FL(35,82).
Elsinoe cinnamomi. On leaves. **1**:MS(1745).
Elsinoe sp. Leaf spot. **1**:MS(82,2720).
Glomerella cingulata. Anthracnose. **1**:FL(35). **2**:FL(35).
Hypoxylon rubiginosum. **1**:GA(82,1438).
Microsphaera cinnamomi. Powdery mildew. **1**:LA(82).
Physalospora sp. Dieback. **1**:FL(35).
Schizothyrium pomi. On twigs. **1**:FL(35,82).
Basidiomycotina - Other
Armillaria mellea. Root rot. **1**:Range of host(82).
A. tabescens. Root rot. **1**:FL(35).
Pellicularia koleroga. **1**:LA(3640).
Deuteromycotina - Hyphomycetes
Alternaria sp. Leaf spot. **1**:FL(35).
Cercospora sp. Leaf spot. **1**:FL(35).
Cylindrocladium sp. Root and stem rot of cuttings. **2**:FL(35).
Fusarium sp. Dieback, root rot, wilt. **1**:FL(35).

Phymatotrichopsis omnivora. Root rot. **1**:TX(82).
Verticillium albo-atrum. **1**:CA(3896).
Deuteromycotina - Coelomycetes
Colletotrichum gloeosporioides. Anthracnose, canker, leaf spot. **2**:FL(82), TX(82,3597).
Colletotrichum sp. **1**:MS(82).
Coniothyrium sp. Canker, dieback. **1**:FL(35). **2**:FL(35).
Cytospora sp. Dieback. **1**:FL(35).
Diplodia camphorae. Canker. **1**:MS(2720).
Lasiodiplodia theobromae. Canker, dieback. **1**:Range of host(82), FL(35), TX(82).
Macrophoma sp. Canker, dieback. **1**:FL(35).
Monochaetia sp. On leaves. **1**:MS(82,2720).
Phoma sp. Stem dieback. **1**:FL(35), MS(82).
Phomopsis sp. Stem canker, root rot. **1**:FL(35,82).
Phyllosticta sp. Leaf spot. **1**:FL(35).
Rhabdospora sp. Leaf spot. **1**:FL(35).
Sphaeropsis tumefaciens. Stem gall. **1**:FL(35).
Deuteromycotina - Other
Rhizoctonia ramicola. Thread blight. **1**:FL(35).
R. solani. Root rot. **1**:FL(35). **2**:FL(35).

Cryptocarya R. Br.

1. *C. mannii* Hildebr.
2. *Cryptocarya* sp.

Ascomycotina
Asteridiella sheariana. Black mildew. **1**:HI(1272,1453,2903).
Meliola peleae. **2**:HI(2784).
M. praetervisa. Black mildew. **1**:HI(1272,2903).

Laurus L.

1. *L. nobilis* L. [bay, Grecian laurel, laurel, sweetbay]

Basidiomycotina - Other
Pellicularia koleroga. Thread blight. **1**:SC(82).

Lindera Thunb.

(= *Benzoin* Schaeff.)
1. *L. benzoin* (L.) Blume (=*Benzoin aestivale* (L.) Nees, *Laurus benzoin* L.) [Benjamin-bush, feverbush, spicebush, wild allspice]
2. *Lindera* sp. (=*Benzoin sp.*)

Ascomycotina
Botryosphaeria dothidea. **1**:GA(1438).
B. obtusa. On limbs. **2**:Unk.(82).
B. ribis. Canker. **2**:MD(82).
Diaporthe sociata. On limbs. **1**:GA(1438), MA(3807), MD(3807), NY(3807), WV(128). **2**:MA(82,3807), MD(82,3807), NY(82,3807), WV(82,3807).
Dothidea linderae. On limbs. **2**:NY(82), PA(82).
Eutypa rivulosa. **1**:PA(2925).
Eutypella goniostoma. On dead limbs. **1**:NY(195,2925). **2**:IN(82), NJ(82).
Glomerella cingulata. On fruits and twigs. **2**:NJ(82).
Hypoxylon crocopeplum. On limbs. **1**:VA(128).
H. rubiginosum. On limbs. **1**:GA(1438).
H. sassafras. On limbs. **1**:NY(195).
Massaria sp. **1**:GA(1438).
Microdothella ramularis. On limbs and twigs. **1**:MD(128). **2**:PA(82).
Nectria ochroleuca. On bark. **2**:PA(82).
N. pallidula. **1**:PA(3085), TN(3085). **2**:PA(3085).
Basidiomycotina - Uredinales
Puccinia coronopsora. **1**:FL(35).
Basidiomycotina - Other
Antrodia malicola. On wood. **2**:Unk.(82).
Cyphellopsis anomala. On limbs. **2**:TN(82).
Hymenochaete agglutinans. On shoots in contact with infected limbs. **1**:CT(1312).
Phlebiella tulasnelloidea. On wood. **1**:CT(128).
Deuteromycotina - Hyphomycetes
Cristulariella moricola. Leaf spot. **1**:WV(3680).
Isariopsis linderae. On leaves. **2**:AL(82), NJ(82), WV(82).
Phymatotrichopsis omnivora. Root rot. **2**:TX(82).
Deuteromycotina - Coelomycetes
Aplosporella linderae. On limbs and twigs. **1**:NY(128), WV(128). **2**:PA(128,2783).
Camarosporium linderae. On limbs. **1**:WV(128).
Diplodia linderae. On limbs. **2**:MI(82), NY(82).
Gloeodes pomigena. On limbs. **2**:IN(82).
Hendersonia linderae. On limbs. **2**:NY(82).
Monostichella falcata. On leaves. **2**:NY(82).
Phyllosticta linderae. Leaf spot. **1**:DE(82,959), IN(82), WV(82).
P. lindericola. Leaf spot. **1**:WV(959). **2**:WV(82).

Litsea Lam.

1. *L. aestivalis* (L.) Fernald [pondspice]

Deuteromycotina - Hyphomycetes
Cercospora seymouriana. Leaf spot. **1**:GA(82).

Nectandra Rolander & Rottb.

1. *N. coriacea* (Sw.) Griseb.

Basidiomycotina - Other
Oudemansiella canarii. On living and fallen trunks. **1**:FL(3284).
Phanerochaete chrysorhiza. White rot. **1**:FL(433).
Deuteromycotina - Coelomycetes
Phomopsis sp. Leaf spot. **1**:FL(35).

Ocotea Aubl.

1. *Ocotea* sp.

Ascomycotina
Phyllachora catesbyana. Anthracnose, black leaf spot. **1**:FL(35,82).
Deuteromycotina - Hyphomycetes
Fusarium sp. Stem lesions. **1**:FL(35).

Persea Mill.

1. *P. americana* Miller [alligator-pear, aquacate, avocado, palta]
2. *P. borbonia* (L.) Spreng. (=*P. pubescens* (Pursh) Sarg., *P. carolinensis* Nees) [Florida mahogany, laureltree, redbay, swamp redbay, sweet-bay, tisswood]
3. *P. palustris* Sarg.
4. *Persea* sp.

Oomycetes
Phytophthora cactorum. Collar rot. **1**:CA(1110). **2**:CA(82).
P. cinnamomi. Decline, root rot. **1**:CA(591,1110,3624,4009), FL(35), HI(2903). **2**:CA(82).
P. citricola. Trunk canker, fruit rot. **1**:CA(1110,1949,4009).
P. citrophthora. Fruit rot. **1**:CA(1110). **2**:CA(82).
P. nicotianae var. *parasitica*. Collar rot. **1**:FL(35). **2**:FL(82).
P. palmivora. Seedling blight, leaf spot. **1**:FL(35). **2**:FL(82).
Pythium ultimum. Associated with "decline" of trees. **1**:CA(1110). **2**:CA(82).
P. vexans. Secondary root rot. **1**:CA(1110). **2**:CA(82).
Pythium sp. Associated with "decline" of trees, root rot. **1**:CA(1110), FL(35), HI(2903). **2**:CA(82).
Zygomycotina
Rhizopus stolonifer. Fruit rot. **1**:CA(1110), FL(35). **2**:Range of host(82).
Ascomycotina
Asteridiella perseae. Black mildew. **1**:FL(35,82).
Asterina delitescens. Anthracnose, black mildew. **2**:SE states(82), GA(1438), MS(2720), TX(82). **4**:FL(35).
Botryosphaeria disrupta. Branch canker. **1**:FL(35).
B. dothidea. Twig dieback. **1**:CA(1110). **4**:FL(35).
B. obtusa. Canker. **1**:HI(2903). **2**:HI(82).
B. quercuum. On branches and fruit. **1**:FL(35).
B. rhodina. On limbs and fruit. **2**:FL(82).
B. ribis. Limb canker, dieback. **1**:CA(82). **2**:FL(82).
Calloriopsis gelatinosa. Leaf spot. **4**:FL(35,82).
Calycina maculosa. On living leaves. **3**:FL(850).
Cenangium magnoliae. On limbs. **4**:FL(634).
Coccoidella scutula. Tar spot on leaves. **2**:SE states(82), GA(1438), MS(2720). **3**:MS(2720). **4**:FL(35).
Daldinia concentrica. **4**:FL(634).
Echidnodella rugispora. Black mildew. **2**:MS(82,2720), NC(82). **3**:MS(2720).
Glomerella cingulata. Anthracnose. **1**:CA(1110), HI(2903).
Gloniopsis lineolata. On wood. **2**:FL(634).
Graphostroma platystoma. On bark. **4**:SC(634).
Hypoxylon hypophlaeum. **2**:GA(1438).
H. microplacum. **2**:GA(1438). **4**:GA(1438).
H. sassafras. **2**:GA(1438). **4**:GA(1438).
H. stygium. **2**:GA(1438).

Hypoxylon sp. **2:**GA(82).
Lembosia sp. Leaf spot. **4:**FL(35).
Meliola amphitricha. Black mildew. **2:**FL(82), GA(1438), MS(82,2720). **3:**MS(2720). **4:**FL(35,634,2782).
M. floridensis. **3:**FL(1452).
M. martiniana. Black mildew. **3:**FL(1452). **4:**Unk.(1452), FL(35).
Mycosphaerella exutans. On leaves. **2:**HI(82,179).
M. perseae. Leaf spot. **4:**FL(35).
Ostreichnion curtisii. On dead twigs. **3:**MS(2720).
Phyllachora perseae. On leaves. **2:**SC(1595).
Physalospora abdita. On limbs. **2:**FL(82).
Pseudomassaria carolinensis. On leaves. **2:**SC(1598).
Rhytidenglerula carnea. Anthracnose, black mildew. **2:**FL(82). **3:**MS(2720). **4:**FL(35).
Rosellinia necatrix. Root rot. **1:**CA(1110).
Sclerotinia sclerotiorum. Collar rot. **1:**CA(1110), FL(35). **2:**CA(82).

Basidiomycotina - Other

Armillaria mellea. Root rot. **1:**CA(82,591,1110).
A. tabescens. Root rot. **1:**FL(35,82). **4:**FL(35).
Cristinia gallica. **4:**FL(634).
Exidia glandulosa. On bark. **4:**FL(634).
Fomitella supina. On wood. **2:**FL(82). **4:**FL(35).
Ganoderma zonatum. Butt rot. **1:**FL(35).
Hyphoderma setigerum. **4:**FL(634).
Laetiporus sulphureus. On wood. **4:**FL(35).
Mycoacia meridionalis. **2:**MS(443).
Phellinus gilvus. Basal trunk rot. **4:**FL(35).
Pseudofavolus cucullatus. On wood. **4:**SE states(2681).
Rigidoporus ulmarius. On wood. **2:**FL(82). **4:**FL(35).
Septobasidium pseudopedicellatum. Parasitic on scale insects. **4:**FL(35).
Trametes hirsuta. On wood. **4:**FL(35).
T. versicolor. **1:**CA(1110).

Deuteromycotina - Hyphomycetes

Alternaria sp. Fruit rot. **1:**CA(82,1110).
Aureobasidium sp. **1:**CA(1110).
Cercospora sp. Cercospora blotch or spot of leaves and fruit. **1:**FL(82), GA(82), HI(82,2903), TX(82).
Chaetochalara aspera. On fallen leaves. **2:**SC(2818).
Circinotrichum fertile. On fallen leaves. **2:**SC(2818).
Cladosporium sp. **1:**CA(1110).
Cristulariella moricola. Leaf spot. **1:**WV(3680).
Cylindrocladium scoparium. Fruit rot. **1:**FL(35).
Cylindrocladium sp. Root rot. **1:**HI(2903).
Fusarium javanicum. Root rot. **4:**HI(2903).
F. solani. Root rot. **4:**HI(2903).
Fusarium sp. Bark, fruit and root rot. **1:**CA(82,1110), FL(35). **4:**FL(35).
Helminthosporium fumosum. Leaf spot. **2:**FL(82). **4:**FL(35).
Helminthosporium sp. Smudgy spot, on green twigs, rarely on leaves and fruit. **1:**CA(82,1110).
Menisporopsis profusa. On fallen leaves. **2:**SC(2818).
Oidium sp. Powdery mildew. **1:**FL(35). **2:**FL(82).
Penicillium expansum. Blue mold dry rot. **1:**CA(1110). **2:**CA(82).
Phymatotrichopsis omnivora. Root rot. **1:**TX(82).
Pseudocercospora purpurea. Leaf spot. **1:**FL(35). **2:**SE states(82), MS(82,2720). **3:**MS(2720).
Redbia elegans. On fallen leaves. **2:**SC(2818).
Scolecobasidium dendroides. On fallen leaves. **2:**SC(2818).
Sporidesmium velutinum. On wood. **4:**FL(634).
Thozetella cristata. On fallen leaves. **2:**SC(2818).
Uberispora simplex. On fallen leaves. **2:**SC(2818).
Verticillium albo-atrum. Wilt. **1:**FL(35,2252). **2:**CA(82).
V. dahliae. **1:**CA(591,1110).

Deuteromycotina - Coelomycetes

Colletotrichum gloeosporioides. Anthracnose, fruit and leaf spot. **1:**Range of host(82), FL(35), HI(2903), TX(82). **4:**FL(35).
Dothiorella gregaria. Postharvest decay of fruits. **1:**CA(939).
Dothiorella sp. Stem end rot. **1:**Unk.(82), FL(35), HI(2903).
Pestalotia maura. On dead leaves. **2:**LA(966).
Pestalotia sp. Fruit and leaf spot. **1:**CA(1110), FL(35). **2:**FL(82), TX(82). **4:**FL(35).
Phomopsis sp. Stem dieback, stem end rot. **1:**CA(1110), FL(35). **2:**FL(82), TX(82).
Phyllosticta micropuncta. Leaf spot. **1:**AL(285), FL(35). **2:**AL(82), MS(2720), NC(1303). **4:**FL(35).
Phyllosticta sp. Leaf spot. **2:**FL(82).
Sphaceloma perseae. Foliage and fruit scab. **1:**CA(1110), FL(35,82,1745), TX(82).

Deuteromycotina - Other

Rhizoctonia solani. Root and seed rot. **1:**CA(1110,2447), FL(35).
Rhizoctonia sp. Root rot. **1:**HI(2903).
Sclerotium rolfsii. Seedling blight. **1:**FL(35). **2:**FL(82).

Sassafras Nees & Eberm.

1. *S. albidum* (Nutt.) Nees (=*S. officinale* Nees & Eberm., *S. variifolium* (Salisb.) Kuntze) [white sassafras]
2. *Sassafras* sp.

Ascomycotina

Botryosphaeria obtusa. On limbs. **1:**E states(82), MS(82,2720).
Cryptodiaporthe corni. On dead twigs. **1:**PA(82).
Daldinia vernicosa. Wood rot. **1:**GA(82,1438), VA(82).
Diaporthe biglobosa. On limbs. **1:**NJ(82).
D. pardalota. **1:**NY(82).
Diatrypella sassafras. On limbs. **1:**NJ(82).
Discostroma corticola. On limbs. **1:**NJ(82).
Eutypella goniostoma. **2:**PA(2925).
Glomerella cingulata. **1:**TX(82).
Hypoxylon sassafras. Wood rot. **1:**E states(82), GA(1438), MO(82).
H. sclerophaeum. **1:**GA(1438).
Hysterographium flexuosum. On twigs. **1:**MA(1394).
Metasphaeria sassafrasicola. On leaves. **1:**IL(82).
Nectria verrucosa. On limbs. **1:**PA(82).
Nectria sp. Branch canker. **1:**NE states(82).
Phyllactinia guttata. Powdery mildew. **1:**MI(82).
Physalospora abdita. On limbs. **1:**AL(82).
Pleuroceras sassafras. On leaves. **1:**NJ(82), OH(82).
Schizothyrium pomi. Fly speck. **1:**NC(1303).
Stigmatophragmia sassafrasicola. On leaves. **1:**IL(82).
Valsa ambiens. On limbs. **1:**NJ(82), NY(3415). **2:**NJ(3415), WV(3415).
V. ceratosperma. On limbs. **1:**NJ(82).
V. subclypeata. On limbs. **1:**NY(82), VA(82).
Valsaria nigrofacta. **1:**NJ(82).
Xylaria polymorpha. On stumps. **2:**WV(3018).

Basidiomycotina - Other

Antrodia albida. Wood rot. **1:**IN(82).
Armillaria mellea. Root rot. **1:**PA(82).
Daedaleopsis confragosa. Trunk rot. **1:**IN(82), NY(82).
Dendrophora erumpens. **1:**NY(543).
Ganoderma sp. Root rot. **1:**NY(2604).
Hymenochaete agglutinans. **1:**VA(82).
Phellinus ferruginosus. Wood rot. **1:**Unk.(82).
P. gilvus. Wood rot. **1:**GA(1439), IN(82), LA(82), MD(82).
Phlebia chrysocrea. On wood. **1:**Unk.(262), GA(2165), TN(2165).
Phylloporia ribis. On wood. **1:**GA(491,1439).
Schizophyllum commune. Wood rot. **1:**PA(82), VA(82).
Tomentella subvinosa. **2:**NY(2039).
Trametes hirsuta. Wood rot. **1:**IN(82), NC(82,3944).
T. versicolor. Wood rot. **1:**IN(82), MD(82), VA(82).
Wolfiporia cocos. **1:**NC(1303).

Deuteromycotina - Hyphomycetes

Beltrania querna. On dead leaves. **1:**NC(1303).
Cristulariella moricola. Zonate leaf spot. **1:**AL(2056), GA(2056), NC(1302,1303), WV(815,816).
Haplotrichum ramosissimum. On stumps. **2:**NJ(2131).
Phymatotrichopsis omnivora. **1:**TX(82).
Trichoderma viride. Root rot. **1:**NY(2604).
Verticillium albo-atrum. Wilt. **1:**IL(1558).

Deuteromycotina - Coelomycetes

Cytospora leucosperma. On twigs. **1:**MI(82), NY(82), WV(82).
C. sassafrasicola. **1:**IL(82).
Diplodia officinalis. Canker. **1:**MI(82), NY(82).
D. sassafras. Canker. **1:**MS(82,2720).
D. subcuticularis. Canker. **1:**NY(82).
Diplodia sp. Stem canker. **1:**FL(35).
Diplopeltis sassafrasicola. On leaves. **1:**IL(82).
Discula sassafras. Leaf spot. **1:**Range of host(82), IL(82,2568), IN(647), MS(2720), OK(2872), TX(82).
Gloeodes pomigena. Sooty blotch. **1:**NC(1303).
Leptothyrium kellermanii. On leaves. **1:**OH(82).

Phyllosticta illinoensis. Leaf spot. **1**:IL(82), MA(82).
Pseudodictya sassafrasicola. On leaves. **1**:IL(82).
Septoria sp. Leaf spot. **1**:NY(82).

Umbellularia (Nees) Nutt.

1. *U. californica* (Hook. & Arnott) Nutt. [California bay, California laurel, California-olive, Oregon myrtle, pepperwood]
2. *Umbellularia* sp.

Ascomycotina
Allantoporthe decedens. **1**:OR(3202).
Asteromassaria pulchra. On dead limbs and stems. **1**:CA(188,1468).
Capnodium tuba. **1**:CA(2434).
Daldinia concentrica. **1**:CA(81).
Diatrype prominens. **1**:OR(3202).
Diatrypella decipiens. **1**:OR(3202).
[*Eutypella minuta*]. **1**:OR(3202).
Hypoxylon rubiginosum. **1**:OR(3202).
H. sassafras. **1**:CA(81).
Ionomidotis plicata. On wood. **1**:CA(2801).
Leptosphaeria odora. On limbs. **1**:CA(82).
Limacinula anomala. Black mildew. **1**:CA(82,2434).
Mycosphaerella arbuticola. On leaves. **1**:CA(81,82), OR(82).
Nectria cinnabarina. Canker. **1**:CA(82).
N. coccinea. Canker. **1**:CA(82).
N. peziza. **1**:CA(3085), CT(2836).
Phaeosaccardinula dematia. **1**:CA(81,2434), WA(3202).
Pleospora tragacanthae. **1**:ID(3202).
Sphaerella umbellulariae. **1**:ID(3202).
Stictis pupula. **1**:OR(3202).
S. radiata. **1**:OR(638,3202). **2**:CA(3225).
Thyridaria californica. On limbs. **1**:CA(82).
Valsa ceratosperma. On limbs. **1**:CA(82).
Vertixore atronitidum. On leaves. **1**:CA(81,2434).

Basidiomycotina - Other
Aleurodiscus aurantius. On dead stems and twigs. **1**:CA(2083).
Cyphellopsis anomala. **1**:OR(3202).
Dendrophora albobadia. Wood rot. **1**:CA(82).
Exidiopsis grisea. **1**:CA(3853).
Fomes fomentarius. White-mottled heart rot. **1**:CA(82), OR(3202).
Ganoderma annularis. On wood. **2**:CA(1204).
G. applanatum. White-mottled butt rot. **1**:CA(82), OR(3202).
G. brownii. On wood. **2**:CA(1204).
Lenzites betulina. **1**:CA(82).
Lopharia cinerascens. **1**:CA(543).
Marasmiellus candidus. **2**:CA(3287).
Oxyporus latemarginatus. Wood rot. **1**:CA(82).
Peniophora malenconii subsp. *americana*. On wood. **1**:CA(81,542).
Phellinus ferreus. **1**:OR(3202).
P. ferruginosus. Wood rot. **1**:CA(82).
P. igniarius. White-mottled heart rot. **1**:CA(82).
Schizophyllum commune. Wood rot. **1**:CA(82).
Steccherinum fimbriatum. **1**:OR(3202).
Stereum hirsutum. **1**:CA(81), OR(3202).
Trechispora mollusca. **1**:OR(3202).
Tremella mesenterica. On wood. **1**:CA(3949).

Deuteromycotina - Hyphomycetes
Aureobasidium umbellulariae. Anthracnose, leaf spot. **1**:CA(1486).

Deuteromycotina - Coelomycetes
Ceuthospora sp. On leaves. **1**:CA(82).
Chaetasbolisia falcata. **1**:CA(81,2434).
Colletotrichum gloeosporioides. On fruits and leaves. **1**:CA(81,82).
Cyclodomus umbellulariae. On leaves. **1**:CA(82,3549).
Diplodia umbellulariae. On dead limbs and twigs. **1**:CA(82,968).
Microdiplodia laurina. On leaves. **1**:CA(82).
Monochaetia coryneoidea. On dead twigs. **1**:CA(1468).

Lecythidaceae

Barringtonia J. R. Forster & G. Forster

1. *B. asiatica* (L.) Kurz
2. *B. racemosa* (L.) Roxb.

Deuteromycotina - Coelomycetes
Colletotrichum gloeosporioides. Anthracnose. **2**:HI(2903).
Dothiorella barringtoniae. On leaves. **2**:HI(2903).
Phyllosticta sp. Leaf spot. **1**:FL(35).

Bertholletia Bonpl.

1. *Bertholletia* sp.

Deuteromycotina - Hyphomycetes
Aspergillus flavus. Endosperm decay. **1**:Unk.(82).
Cephalosporium bertholletianum. White mold. **1**:Unk.(82).
Fusarium sp. Dry rot. **1**:Unk.(82).

Deuteromycotina - Coelomycetes
Myxosporium sp. Bitter rot. **1**:Unk.(82).
Pellionella macrospora. Black crust. **1**:Unk.(82).
Phomopsis bertholletianum. Endosperm decay. **1**:Unk.(82).

Leeaceae

Leea D. Royen ex L.

1. *L. coccinea* Planch. [West Indian holly]
2. *L. guineensis* G. Don

Oomycetes
Pythium aphanidermatum. Root rot. **1**:HI(2903).
P. splendens. Root rot. **1**:HI(2903). **2**:FL(35).
Pythium sp. Root rot. **1**:FL(35).

Ascomycotina
Calonectria crotalariae. Collar rot, leaf spot. **1**:HI(1937,2903).
Glomerella sp. Leaf spot. **1**:FL(35).
Nectriella pironii. Stem gall. **1**:FL(35).

Deuteromycotina - Hyphomycetes
Alternaria sp. Leaf spot. **2**:FL(35).
Cylindrocladium sp. Stem necrosis. **2**:FL(35).
Fusarium sp. Dieback, wilt. **1**:FL(35). **2**:FL(35).

Deuteromycotina - Coelomycetes
Colletotrichum sp. Anthracnose. **2**:FL(35).
Phyllosticta sp. Leaf spot. **1**:FL(35).

Deuteromycotina - Other
Rhizoctonia solani. Leaf necrosis, stem rot. **1**:FL(35). **2**:FL(35).
Rhizoctonia sp. Root rot. **1**:HI(2903).

Lemnaceae

Lemna L.

1. *L. minor* L. [common duckweed, lesser duckweed]

Hyphochytridiomycetes
Reessia amoeboides. In frond cells. **1**:MN(3745).

Spirodela Schleid.

1. *S. polyrhiza* (L.) Schleid. [giant duckweed, great duckweed, water-flaxseed]

Basidiomycotina - Ustilaginales
Tracya lemnae. **1**:CT(1077), MA(1077,3724), RI(1077), WI(1077).

Liliaceae

Agapanthus L'Hér.

1. *A. africanus* (L.) Hoffsgg. (=*A. umbellatus* L'Hér.) [African lily, blue agapanthus, lily, lily-of-the-Nile]
2. *Agapanthus* sp.

Oomycetes
Pythium sp. Root rot. **2**:FL(35).

Deuteromycotina - Hyphomycetes
Botrytis sp. Gray mold of leaves. **2**:FL(35).
Fusarium sp. Bulb and root rot. **2**:FL(35).
Myrothecium sp. Leaf spot. **2**:FL(35).

Deuteromycotina - Coelomycetes
Colletotrichum gloeosporioides. On dead plants. **2**:HI(2903).
Colletotrichum sp. Leaf spot. **2**:FL(35).
Phoma agapanthi. **1**:HI(2903).

Deuteromycotina - Other
Rhizoctonia solani. Bulb and root rot. **2**:FL(35).

Aletris L.

1. *A. aurea* Walter [yellow colicroot]
2. *A. farinosa* L. [aguerroot, crow-corn, unicornroot]
3. *Aletris* sp.

Basidiomycotina - Uredinales

Puccinia aletridis. Autoecious. **1:**FL(82), MS(82,2720), SC(82), TX(82). **2:**DE(82), FL(35), IL(82), IN(82), MA(82), NC(82), NJ(82), RI(82), TN(82), WI(82).

Deuteromycotina - Coelomycetes

Colletotrichum gloeosporioides. Leaf spot. **1:**MS(2720). **3:**MS(82).

Allium L.

1. *A. acuminatum* Hook.
2. *A. amplectens* Torr. (=*A. attenuifolium* Kellogg)
3. *A. ascalonicum* L. A confused name used by Linnaeus for what is probably a cultigen. [shallot]
4. *A. brandegei* S. Wats.
5. *A. brevistylum* S. Wats.
6. *A. campanulatum* S. Wats.
7. *A. canadense* L. (=*A. mutabile* Michx.) [meadow garlic, meadow leek, rose leek, wild garlic, wild onion]
8. *A. cepa* L. (=*A. cepa* L. var. *aggregatum* G. Don, *A. cepa* L. var. *proliferum* (Moench) Regel, *A. cepa* L. var. *solaninum* Alef., *A. cepa* L. var. *cepa*) [Egyptian onion, eschalot, multiplier onion, onion, potato onion, shallot, tree onion]
9. *A. cepa* L. var. *viviparum* (Metzg.) Alef. [top onion]
10. *A. cernuum* Roth [lady's leek, nodding onion, wild onion]
11. *A. douglasii* Hook.
12. *A. drummondii* Regel (=*A. nuttallii* S. Wats.)
13. *A. fistulosum* L. [bunching onion, ciboule, Japanese bunching onion, Spanish onion, spring onion, stone leek, two-bladed onion, Welsh onion]
14. *A. geyeri* S. Wats. (=*A. rubrum* Osterh.)
15. *A. gooddingii* Ownbey
16. *A. lemmonii* S. Wats.
17. *A. macropetalum* Rydb. (=*A. deserticola* (M. E. Jones) Woot. & Standl.)
18. *A. nevadense* S. Wats.
19. *A. plummerae* S. Wats.
20. *A. porrum* L. (=*A. ampeloprasum* var. *porrum* (L.) Gay) [leek, purret]
21. *A. punctum* L. Henderson
22. *A. sativum* L. [garlic]
23. *A. schoenoprasum* L. (=*A. sibiricum* L.) [chives, cive, ezo-negi, schnittlauch]
24. *A. stellatum* Nutt. ex Ker-Gawl. [prairie onion]
25. *A. textile* A. Nels. & Macbr. (=*A. reticulatum* Nutt. ex G. Don)
26. *A. tolmiei* C. Baker ex S. Wats.
27. *A. tricoccum* Aiton [ramp, wild leek]
28. *A. trinulatum* Wiegland
29. *A. tuberosum* Rottler ex Spreng. [Chinese chive, kui ts'ai, nira, oriental garlic]
30. *A. validum* S. Wats. [swamp onion]
31. *A. vineale* L. [crow garlic, field garlic, stag's garlic, wild garlic]
32. *A. watsonii* Howell
33. *Allium* sp.

Oomycetes

Peronospora destructor. Downy mildew. **3:**CA(82), LA(82). **8:**'Downy mildew.' Range of host(82), AZ(3381), CT(2974), ID(3202), LA(3635), MN(2874), MS(2720), NY(3354), OK(2872), OR(3202), TX(2414), WA(3202), WV(3907). **23:**CA(82).

Peronospora sp. Downy mildew. **8:**CO(49).

Phytophthora cinnamomi. Root rot. **8:**HI(2903).

P. drechsleri. Root rot. **8:**HI(82,2903).

P. nicotianae var. *parasitica*. **3:**HI(2903). **8:**HI(2903).

P. palmivora. **8:**HI(2903).

Pythium coloratum. Isolated from seeds. **8:**NY(413).

P. irregulare. Isolated from seeds. 'Pythium damping-off.' **8:**NY(413).

P. paroecandrum. Root tip rot, isolated from seeds. **8:**NY(413). **31:**VA(82).

P. rostratum. Isolated from seeds. **8:**NY(413).

P. spinosum. Damping-off. **8:**CA(82).

P. sylvaticum. Isolated from seeds. **8:**NY(413).

P. torulosum. Isolated from seeds. **8:**NY(413).

P. ultimum. Isolated from seeds. 'Pythium damping-off. Root rot.' **8:**CA(82), NY(413).

P. vexans. Root rot. **8:**HI(2903).

Pythium sp. Seedling blight, seed rot. **8:**CO(49), ID(82,3202), MA(82), NC(82), NY(82).

Zygomycotina

Rhizopus stolonifer. Bulb neck rot. 'Mushy rot.' **8:**IN(82), MA(82), NM(82,2107), TX(82).

Ascomycotina

Botryotinia squamosa. Leaf blight, small-sclerotial neck-rot. **8:**NY(254,1960).

Diaporthe phaseolorum. **8:**GA(1438). **22:**GA(1438).

Didymolepta mimuli. On dead stems. **30:**CA(642,661).

Leptosphaeria sp. **33:**MT(3202).

Mycosphaerella allicina. On leaves. **8:**WI(82,1335).

M. tassiana. On overwintered herbage. **30:**CA(642).

Pezizella helotioides. On overwintered leaves. **30:**CA(642).

Pleospora herbarum. Seedling root rot, on leaves. **8:**CA(82), NC(1303,3931), WA(3202).

Sclerotinia sclerotiorum. 'Sclerotinia rot.' **8:**CA(82), FL(82), HI(2903), ID(82), MI(82), OH(82), WA(3202), WV(82,3907).

Basidiomycotina - Uredinales

Puccinia allii. **3:**HI(2903). **8:**'Onion rust.' CA(82), CT(82). **9:**CT(82), NE(82). **13:**HI(2903). **22:**CA(82). **23:**CT(82), ID(3202), NY(82), WA(82,3202).

P. asparagi. Autoecious. **8:**'Asparagus rust.' CA(82), ID(3202,3277), MN(2874), OH(946), SD(2243), WI(1340). **9:**CT(82), IA(82), KS(82), MN(82,2874), NE(82), WI(82,1357,1359).

P. blasdalei. Autoecious. **1:**ID(82,3202), OR(82,3202), UT(82), WA(82,3202). **2:**CA(3097), OR(3202). **10:**CO(82,1195), WY(82). **14:**OR(3202). **16:**OR(3097,3202). **17:**AZ(747,1195,1203,3981), NM(3097). **21:**OR(3097,3202). **33:**OR(3202).

P. granulispora. Autoecious. **1:**WA(3202). **10:**AZ(3981), CO(3097), ID(82,3202), MT(82,1195,3202), SD(82,2243), WA(3202), WY(3097). **11:**OR(3202). **19:**AZ(3981).

P. mutabilis. Autoecious. **1:**AZ(1203,3981), UT(82). **6:**CA(661). **12:**CO(82). **14:**MT(1195,3202), OR(3202), WA(3202). **25:**MT(3202), WA(3202).

Uromyces aemulus. Autoecious. **1:**CO(82), ID(3202), UT(82), WA(3097,3202). **4:**ID(1195,3202). **5:**WY(1195,2924). **11:**OR(3097,3202). **15:**AZ(3981). **26:**ID(3097,3202). **30:**CA(82), NV(82), OR(82,3202). **33:**OR(3202).

U. aureus. Autoecious. **28:**WA(3202). **30:**CA(81,82,111,638,642).

U. bicolor. Autoecious. **7:**MA(82), MO(82), NY(82), OH(946), TN(1539), TX(82), WI(1338). **12:**MT(82). **30:**CA(81,82,642), ID(82,3202). **32:**CA(81).

U. sporoboli. Heteroecious: aecial host. **7:**IA(82), KS(753), MO(82,753), NE(82,753), TX(82). **24:**KS(753). **25:**KS(753).

Basidiomycotina - Ustilaginales

Urocystis colchici. **3:**NC(1303). **8:**'Smut.' AL(1077), CA(1077), CO(1077), CT(1077), IA(1077), ID(3202), IL(1077), IN(1077), KS(1077), KY(1077), MA(1077), MD(1077), ME(1077), MN(1077), MT(3202), NC(1303), ND(1077), NJ(1077), NY(1077), OH(1077), OR(1077,3202), PA(1077), SD(1077,2243), VT(1077), WA(1077,3202), WI(1077). **18:**NV(1077).

U. magica. **3:**MA(82). **8:**Range of host(82), MN(2874), OK(2872). **13:**MA(82), NY(82). **20:**MA(82), NY(82). **22:**MA(82), NY(82). **23:**MA(82).

Urocystis sp. **8:**CO(49).

Basidiomycotina - Other

Thanatephorus cucumeris. **8:**ID(3202), WA(3202).

Deuteromycotina - Hyphomycetes

Acrothecium melanopus. On stalks. **33:**PA(82).

Alternaria alternata. On leaves. **8:**Range of host(82).

A. porri. Blight, purple blotch, bulb rot. **3:**HI(2903), LA(82), TX(82). **8:**'Purple blotch.' E states(82), CO(49,82), HI(2899,2903), MN(82,2874), NC(3931), NM(2112), SD(2243), TX(82,2415), UT(82), WA(3202). **13:**MD(82). **20:**CT(82), HI(2903), LA(82), NJ(82), NY(82). **22:**ID(82), LA(82), TX(82). **33:**Unk.(82).

Alternaria sp. **8:**ID(3202).

Aspergillus alliaceus. **8:**TX(82). **22:**TX(82).

A. niger. Black mold, occasional in markets. **8:**'Blackmold.' Range of host(82), CA(82), GA(1438), ID(3202), MT(3202), NC(1303), OK(2872), OR(3202), TX(82), WA(3202). **22:**CA(82), OH(82), TX(82).

Botrytis aclada. Gray mold neck rot. **3:**LA(82). **8:**'Botrytis leaf spot. Botrytis brown stain. Flower blight. Gray mold neck.' E states(82), AL(285), CA(82), CO(82), ID(82,3202), MN(82,2874), NC(1303), NY(942,2917,3175), OK(2872), OR(82,3202), SD(2243), TX(82), UT(3620), WA(82,3202). **20:**NJ(82). **22:**CA(82), LA(82), TX(82).
B. byssoidea. 'Gray mold neck rot. Mycelial neck rot.' **8:**CT(82), ID(82), IL(82), IN(82), MA(82), NY(82), WI(82,1360).
B. cinerea. Gray mold blight on bulbs and foliage. **8:**'Botrytis leaf spot. Botrytis brown stain. Flower blight.' AK(2163), MS(2720), NY(584,942,2917,3175). **33:**CA(82), FL(82), LA(82), WA(82).
B. paeoniae. Blast, leaf spot. **8:**NY(3175).
B. squamosa. Small sclerotial neck rot. **8:**'Botrytis leaf blight. Flower blight. Gray mold neck rot.' HI(2903), NY(942,2917). **9:**FL(82), ID(82), IL(82), IN(82), LA(82), MA(82), ME(82), NY(82), OH(82), TX(82), WI(82).
B. tulipae. Blast, leaf spot. **8:**NY(3175).
Botrytis sp. Blast, leaf blight, neck rot. **3:**NC(1303). **8:**Unk.(82), AK(2162), CO(49), MT(3202), WV(3907). **9:**Unk.(82).
Cercospora sp. Leaf spot. **8:**TX(82).
Cladosporium allii. Leaf spot. **7:**MD(82), MS(2720), NY(82), PA(82), TX(82), WI(82). **8:**'Heterosporium blight.' CA(82), CO(49,82), WA(82). **10:**WY(2924). **20:**PA(82). **30:**CA(642), OR(638). **31:**DE(82), IL(82), NJ(82), TX(82).
C. iridis. Leaf spot. **7:**PA(82).
Curvularia lunata. On leaves. **8:**CA(82).
Cylindrocladiella parva. **3:**HI(2903).
Embellisia allii. Canker. **22:**LA(82), NM(82), OR(1423,3202), TX(82).
Fusarium acuminatum. **8:**MT(3202).
F. moniliforme. **8:**ID(3202).
F. oxysporum f. sp. *cepae*. Bulb rot. **8:**'Basal rot. Fusarium damping-off. Fusarium root rot.' AZ(2249), CO(49,1009), GA(3542), ID(3202), MN(2874), NC(1303), WA(3202). **22:**ID(82).
F. solani. Pink root, bulb rot. **3:**TX(82). **8:**FL(35), MN(2874), OK(2872), OR(3202). **22:**CT(82), TX(82).
Fusarium sp. Seedling blight, basal rot, bulb rot, wilt. **3:**NC(1303). **8:**E states(82), NC states(82), AK(2163), CA(82), CO(82), LA(3638), NC(1303), OR(49), WA(82). **20:**NJ(82). **22:**TX(82).
Oidiopsis sicula. 'Powdery mildew' **8:**CA(2010).
Olpitrichum tenellum. On rotten bulbs. **8:**SC(2131).
Penicillium hirsutum. Storage decay and field disease. **22:**CA(3316).
Penicillium sp. Blue mold rot, occasional in markets. **8:**ID(3202), OR(3202), WA(3202). **22:**CA(82), TX(82). **33:**Unk.(82).
Stemphylium botryosum. Leaf blight and mold, black stalk rot. **8:**'Blackstalk rot.' Range of host(82), HI(2903), LA(3635), MN(2874), MS(2720), NC(1303), OK(2872). **20:**CT(82), NJ(82), WA(82,3202). **22:**AR(82), CA(82), LA(82), NH(82), TX(82), WA(82,3202).
S. vesicarium. Purple blotch, water-soaked leaf spot. **8:**TX(2415). **33:**Unk.(82).
Stemphylium sp. **8:**ID(3202).
Thielaviopsis basicola. Secondary root rot. **8:**TX(82).
Trichoderma viride. Green mold rot. **22:**TX(82).

Deuteromycotina - Coelomycetes
Colletotrichum circinans. Smudge, leaf spot, on bulblets. **3:**HI(2903), IL(82), LA(82), WI(82). **8:**'Smudge.' E states(82), CO(49), MN(2874), MT(3202), NC(1303), WA(82,3202). **13:**HI(2903). **20:**IL(82), WI(82). **23:**FL(35). **27:**WI(1334).
Colletotrichum sp. **33:**Range of host(82).
Lasiodiplodia theobromae. Dry rot. **8:**'Diplodia stain.' TX(82). **22:**TX(82).
Macrophomina phaseolina. Charcoal rot, scale smudge on bulbs. **8:**'Charcoal rot.' CA(82), OK(82,2873), TX(82). **22:**TX(82).
Phoma allicola. Leaf spot. **7:**SD(2243), WI(82).
P. terrestris. Pink root. **3:**CO(82), LA(82,3636), TX(82). **7:**TX(82). **8:**'Pink rot.' CA(82), CO(49,82), HI(2903), IA(82), ID(82,3202), IN(82), LA(82,3638), MA(82), MI(82), NC(3931), NY(82), OR(82,2791,3202), TX(82), UT(82,3620), WA(3202), WI(82,1344). **10:**ND(82), SD(82). **20:**LA(82), NM(82), NY(82). **22:**CA(82), CO(82), CT(82), LA(82), NM(82), TX(82). **33:**HI(2903).
Phyllosticta allii. 'Phyllosticta leaf blight.' **8:**IL(82), NM(82,2107).
Septoria viridi-tingens. Leaf spot. **27:**MD(82), MN(2874), ND(82), NY(82), TN(82), WI(82).

Deuteromycotina - Other
Rhizoctonia solani. Seedling blight, damping-off, seed root rot, stem rot. **8:**'Rhizoctonia damping-off.' CO(49), HI(2899,2903), MS(2720), NC(1303). **33:**CO(82), ID(82), IN(82), OH(82), WA(82).
Rhizoctonia sp. Root rot. **3:**HI(2903). **29:**HI(2903).
Sclerotium cepivorum. White rot. **3:**LA(82), VA(82). **7:**LA(82). **8:**'White rot.' CA(82), HI(2903), KY(82), LA(82,3637), NC(1303), NJ(82), NY(2103), OH(82), OR(82,3202), PA(82), TX(82), VA(82), WA(3202). **13:**OR(82), VA(82). **20:**OR(82,3202), TX(82). **22:**CA(82), LA(82), OR(82,3202), TX(82). **23:**FL(35).
S. rolfsii. Southern blight, wilt. **3:**LA(82), NC(1303). **8:**'Southern blight.' AL(82), CA(82), FL(35), GA(82), NC(82,3931), OK(82,2872), SC(82), TX(82). **22:**CA(2886), GA(82), LA(82), TX(82). **33:**CA(2886).

Aloe L.

1. *A. barbadensis* Miller (=*A. vera* (L.) P. Webb & Benth.) [Barbados aloe, bitter aloe, Curaçao aloe, medicinal aloe, true aloe, unguentine-cactus]
2. *A. variegata* L. [kamniedood aloe, partridge-breast, pheasant's-wings, tiger aloe]
3. *Aloe* sp.

Oomycetes
Phytophthora nicotianae var. *parasitica*. Crown and root rot. **1:**FL(35).
Pythium ultimum. Stem rot. **1:**FL(35). **2:**CA(82). **3:**FL(35).

Ascomycotina
Phaeosphaeria nigrans. Leaf spot. **1:**FL(35). **3:**FL(35).
Physalospora sp. Leaf spot. **1:**FL(35).

Deuteromycotina - Hyphomycetes
Alternaria sp. Leaf spot. **1:**FL(35). **3:**FL(35).
Cercosporidium sp. Leaf spot. **1:**FL(35).
Corynespora sp. Leaf spot. **1:**FL(35).
[*Exserohilum rostratum*]. Leaf spot. **1:**FL(35).
Fusarium oxysporum. Crown and root rot. **1:**FL(35). **3:**FL(35).

Deuteromycotina - Coelomycetes
Botryodiplodia sp. Leaf spot. **1:**FL(35).
Macrophoma sp. Leaf spot. **1:**FL(35).
Microsphaeropsis concentrica. Leaf spot. **1:**FL(35). **3:**FL(35).
Phyllosticta sp. Leaf spot. **1:**FL(35).

Deuteromycotina - Other
Rhizoctonia solani. Root rot. **1:**FL(35). **3:**FL(35).
Sclerotium rolfsii. Southern blight. **1:**FL(35).

Amaryllis L.

1. *A. belladonna* L. [belladonna lily, cape belladonna, naked-lady lily]
2. *Amaryllis* sp.

Oomycetes
Pythium debaryanum. Seedling root rot. **2:**FL(82), TX(82).

Zygomycotina
Rhizopus stolonifer. Soft rot of bulbs. **2:**Unk.(82).

Ascomycotina
Botryotinia fuckeliana. On leaves. **2:**NC(1303).

Deuteromycotina - Hyphomycetes
Botrytis cinerea. Gray mold blight, bulb rot. **1:**Unk.(82). **2:**Unk.(82), NC(1303), OK(2872).
Cercospora amaryllidis. Leaf spot. **1:**FL(35), TX(82). **2:**AL(82), LA(82,962).
Cladosporium herbarum. Secondary leaf spot. **2:**MN(2874).
Epicoccum nigrum. Secondary leaf spot. **2:**CA(82).

Deuteromycotina - Coelomycetes
Colletotrichum dematium. Anthracnose. **2:**OK(2872).
Phyllosticta sp. **1:**Unk.(82).
Stagonospora curtisii. Red blotch, leaf scorch and spot. **1:**Range of host(82), FL(35), MS(2720). **2:**Range of host(82), HI(2903), NC(1303).

Deuteromycotina - Other
Rhizoctonia sp. Root rot. **2:**HI(2903).
Sclerotium rolfsii. Bulb rot. **1:**FL(82), TX(82). **2:**FL(82), TX(82).

Anthericum L.

1. *A. torreyi* Baker

Basidiomycotina - Uredinales
Uromyces eragrostidis. Heteroecious: aecial host. **1:**AZ(741,1203,3981).

Asparagus L.

1. *A. asparagoides* (L.) W. Wight [Smilax asparagus]
2. *A. densiflorus* (Kunth) Jessop (=*A. sprengeri* Regel, *A. myriocladus* Baker) [Sprenger asparagus]
3. *A. officinalis* L. [common asparagus, garden asparagus]
4. *A. officinalis* L. var. *altilis* L. [cultivated asparagus]
5. *A. plumosus* Baker [asparagus-fern]
6. *A. setaceus* (Kunth) Jessop [fern asparagus, lace-fern]
7. *A. verticillatus* L.
8. *Asparagus* sp.

Oomycetes

Phytophthora sp. Stem rot. **3:**CA(82).
Pythium spinosum. Root rot. **6:**FL(35).
P. splendens. Root rot. **6:**FL(35).
Pythium sp. Root rot. **2:**FL(35), HI(2903). **6:**HI(2903).

Ascomycotina

Didymella sp. **3:**GA(1438).
Didymosphaeria futilis. Blight, stem canker. **5:**FL(82,3938). **6:**FL(35).
Gibberella cyanogena. On stems. **3:**DE(82), IN(82), VA(82), WV(82).
Glomerella sp. Leaf spot. **2:**FL(35).
Leptosphaeria asparagina. On stems. **2:**FL(35). **5:**FL(82). **6:**FL(35).
Massarina gloeospora. On stems. **3:**GA(1438), NJ(82), NY(82). **8:**MA(269), NJ(269), NY(269).
Metasphaeria asparagi. **3:**IL(82).
Pleospora allii. Purple spot. **3:**WA(1764). **8:**MI(1008).
P. herbarum. On stems. **3:**Unk.(82), OK(615), OR(3202).
Sclerotinia sclerotiorum. Watery soft rot. **3:**S states(82).
Zopfia rhizophila. On roots. **3:**MA(185).

Basidiomycotina - Uredinales

Puccinia asparagi. Autoecious. **3:**Range of host(82), AZ(1203,3981), CA(81,1953), FL(35), GA(1439), ID(3202,3277), MS(2720), NC(1303), ND(378), OK(2872), OR(3202), SD(2243), WA(1763,3202), WI(1359). **6:**FL(35). **7:**MS(2720). **8:**WA(2349).

Basidiomycotina - Other

Armillaria mellea. Root rot. **3:**OR(82,3202).

Deuteromycotina - Hyphomycetes

Alternaria alternata. Leaf spot. **3:**OK(2872).
Alternaria sp. Branchlet blight, leaf spot. **2:**FL(35). **3:**IL(82), MA(82), NJ(82), NY(82), OK(82), SC(82), VA(82). **6:**FL(35).
Botrytis cinerea. Gray mold shoot blight, dieback. **1:**AK(82,526). **3:**CA(82), IL(82), MA(82), NH(2967), NY(82), WA(82,3202), WV(82).
Cercospora asparagi. Blight, branchlet and leaf spot. **2:**FL(35). **3:**SE states(82), CA(82), FL(35), HI(82), IL(312), NC(671,1303), NE(82), NM(82), OK(619,2872), TX(82). **4:**HI(2903). **5:**FL(82). **6:**FL(35).
Cladosporium herbarum. On branches, leaves and stems. **3:**Range of host(82), FL(35). **6:**FL(35).
Cladosporium sp. Associated with yellowing and dropping of leaflets. **2:**MS(2720). **5:**FL(82), MS(82,2720), TX(82).
Fusarium moniliforme. Crown, root and stem rot. **3:**MA(771,1200), NJ(1794), SC(82), WA(1701). **8:**MI(2008), WA(1375).
F. oxysporum. Leaf spot, isolated from internal crown tissue. **3:**MA(771). **6:**FL(35).
F. oxysporum f. sp. *asparagi*. Crown, root and stem rot, decline and replant problem, yellows. **3:**CA(82,1366), MA(1200), SC(82), WA(82,1701,3202). **4:**HI(2899,2903). **8:**MI(2008), WA(1375).
F. roseum. Root rot. **3:**IL(82), SC(82), WV(82).
F. solani. Isolated from internal crown tissue. **3:**MA(771).
F. subglutinans. Isolated from internal crown tissue. **3:**MA(771).
Fusarium sp. Root and stem rot, wilt. **1:**NJ(82). **2:**FL(35). **3:**Range of host(82). **4:**HI(2903). **5:**FL(82), WA(82). **6:**FL(35).
Helminthosporium sp. On branchlets. **5:**FL(82).
Penicillium aurantiogriseum. Crown rot. **3:**OR(3202), WA(2350,3202).
Penicillium sp. Blue mold rot. **3:**Unk.(82), CA(82), OR(82).
Stemphylium botryosum. Secondary branchlet blight. **3:**Range of host(82).
S. vesicarium. Purple spot on spears. **3:**CA(1769), MI(1769), WA(1768,1769).

Deuteromycotina - Coelomycetes

Ascochyta asparagina. Blight, stem canker. **5:**FL(3938), FL(82), TX(82). **6:**FL(35).
Ascochyta sp. Branchlet blight. **3:**DE(82), TX(82).
Colletotrichum dematium. On dead stems. **3:**Unk.(82).
C. gloeosporioides. Anthracnose, leaf blight. **2:**FL(35), HI(2899). **6:**FL(35).
Colletotrichum sp. On roots and stems. **3:**AL(82), CT(82), IL(82), OR(82,3202), SC(82), TX(82), VA(82). **5:**FL(82).
Confertopeltis asparagi. On dead stems. **3:**IL(82).
Diplodia asparagi. Root rot, on stems. **3:**NY(82), TX(82).
Diplodia sp. Leaf spot. **6:**FL(35).
Hendersonia asparagi. On branches and stems. **5:**FL(82). **6:**FL(35).
Macrophoma elongata. On stems. **3:**NJ(82).
M. smilacina var. *smilacis*. Leaf spot. **1:**WI(82).
Macrophoma sp. Leaf spot. **2:**FL(35).
Macrophomina phaseolina. Ashy stem blight. **3:**TX(82).
Phoma asparagina. **3:**IL(82).
P. media. On dead stems. **3:**IL(82), KS(82), NY(82), OH(82). **8:**NJ(966).
P. microspora. **3:**ND(82), SC(82).
Phoma sp. On stems. **3:**Unk.(82). **5:**FL(82).
Phomopsis asparagi. Canker. **2:**FL(35). **3:**AL(82), AR(82), FL(35), NJ(82), WA(3202). **6:**FL(35).
Phyllosticta sp. Leaf spot. **2:**FL(35).

Deuteromycotina - Other

Rhizoctonia solani. Stem canker, leaf, root and stem rot, red stain of stalks. **2:**FL(35), NY(82). **3:**Unk.(82), FL(35). **6:**FL(35).
Rhizoctonia sp. Blight, crown and root rot. **2:**HI(2903). **3:**NC(1303), WA(3202).
Sclerotium rolfsii. **3:**GA(2886).

Aspidistra Ker-Gawl.

1. *A. elatior* Blume [barroom-plant]
2. *Aspidistra* sp.

Oomycetes

Pythium sp. Rhizome rot. **1:**FL(35).

Deuteromycotina - Hyphomycetes

Cercospora sp. Leaf spot. **1:**FL(35).
Fusarium moniliforme. Leaf spot. **1:**FL(35).

Deuteromycotina - Coelomycetes

Ascochyta aspidistrae. Leaf spot. **1:**MN(82), NJ(82). **2:**FL(35).
Colletotrichum dematium. Anthracnose, leaf spot. **1:**CA(82), MO(82), NJ(82), OK(2872), PA(82), WV(82). **2:**FL(35).
Labrella aspidistrae. Leaf blight. **1:**IL(82), LA(82).

Brodiaea Sm.

1. *B. coronaria* (Salisb.) Engl. [harvest brodiaea]
2. *B. howellii* S. Wats.
3. *B. laxa* (Benth.) S. Wats. [grassnut, triplet-lily]
4. *B. palmeri* S. Wats. (=*Triteleiopsis palmeri* (S. Wats.) Hoover)
5. *Brodiaea* sp.

Ascomycotina

Mycosphaerella sp. **5:**WA(3202).

Basidiomycotina - Uredinales

Puccinia carnegiana. Autoecious. **5:**AZ(82).
P. dichelostemmae. Autoecious. **1:**OR(3202), WA(3202). **5:**CA(82).
P. nodosa. Autoecious. **4:**AZ(3981).
P. pattersoniana. Heteroecious: aecial host. **1:**WA(3202). **2:**WA(3202).
Uromyces brodiaeae. Autoecious. **1:**WA(3202). **3:**CA(972). **5:**WA(82).

Deuteromycotina - Coelomycetes

Stagonospora cylindrocarpa. On stems. **2:**WA(3202). **5:**WA(82).

Calochortus Pursh

1. *C. ambiguus* (M. E. Jones) Ownbey
2. *C. apiculatus* Baker
3. *C. elegans* Pursh
4. *C. gunnisonii* S. Wats.
5. *C. kennedyi* Porter [desert mariposa]
6. *C. longebarbatus* S. Wats.
7. *C. lyallii* Baker
8. *C. macrocarpus* Douglas [green-banded mariposa]
9. *C. nitidus* Douglas
10. *C. nudus* S. Wats. [Sierra star-tulip]
11. *C. nuttallii* Torr. & A. Gray [sego-lily]
12. *C. tolmiei* Hook. & Arnott [pussy-ears]
13. *Calochortus* sp.

Basidiomycotina - Uredinales
Puccinia calochorti. Autoecious. **1:**AZ(3981). **2:**ID(3202), MT(1195,3202). **3:**ID(3202), WA(3202). **4:**ID(3202), WY(1195,2924,3379). **5:**AZ(3981). **6:**WA(3202). **7:**WA(3202). **8:**OR(3202), WA(3202). **9:**ID(3202), OR(3202). **10:**CA(638,642). **11:**AZ(3981). **12:**OR(3202), WA(3202). **13:**W states(82), CA(82).
Deuteromycotina - Hyphomycetes
Epicoccum sp. **8:**WA(3202).

Camassia Lindl.
1. *C. leichtlinii* (Baker) S. Wats.
2. *C. quamash* (Pursh) Greene [camosh, common camas, quamash]
3. *C. scilloides* (Raf.) Cory [Eastern camass, indigo-squill, meadow-hyacinth, wild-hyacinth]
Basidiomycotina - Ustilaginales
Urocystis colchici. Leaf smut. **2:**IN(1077), OR(82,1077,3202). **3:**IN(82,1077).
Deuteromycotina - Hyphomycetes
Botrytis cinerea. Botrytis blight. **1:**OR(82,3202).
Phymatotrichopsis omnivora. Root rot. **3:**TX(82).
Deuteromycotina - Coelomycetes
Colletotrichum sp. On capsules and fruiting stalks. **3:**WI(1357).
Septoria chlorogali. Leaf spot. **1:**OR(82,3202).

Chlorogalum (Lindl.) Kunth
1. *C. pomeridianum* (DC.) Kunth [amole, soap-plant, wild-potato]
2. *Chlorogalum* sp.
Ascomycotina
Mycosphaerella chlorogali. On stems. **1:**CA(82,2439).
Pleospora chlorogalli. On stems. **1:**CA(82,2439).
Basidiomycotina - Uredinales
Uromyces aureus. Autoecious. **1:**CA(81,82,111).
Deuteromycotina - Hyphomycetes
Cladosporium iridis. Leaf spot. **2:**CA(82).
Deuteromycotina - Coelomycetes
Septoria chlorogali. On stems. **1:**CA(82).

Chlorophytum Ker-Gawl.
1. *C. capense* (L.) Kuntze
2. *C. comosum* Baker [ribbon-plant, spider-ivy, spider-plant, walking anthericum]
Oomycetes
Pythium splendens. Root rot. **2:**FL(35).
Deuteromycotina - Hyphomycetes
Alternaria sp. Leaf spot. **2:**FL(35).
Cercospora sp. Leaf spot. **2:**FL(35).
Fusarium sp. Leaf necrosis. **1:**FL(35). **2:**FL(35).
Deuteromycotina - Coelomycetes
Colletotrichum sp. Leaf spot. **2:**FL(35).
Phyllosticta sp. Leaf spot. **2:**FL(35).
Stagonospora curtisii. Leaf spot. **1:**HI(2903).
Deuteromycotina - Other
Rhizoctonia solani. Root rot. **2:**FL(35).
Sclerotium rolfsii. Southern blight. **1:**FL(35). **2:**FL(35).

Clintonia Raf.
1. *C. borealis* (Aiton) Raf. [bluebead lily, corn-lily, yellow clintonia]
2. *C. uniflora* (Menzies ex Schult. & J. H. Schult.) Kunth [bride's-bonnet, queencup]
3. *Clintonia* sp.
Chytridiomycetes
Synchytrium aureum. False rust. **1:**WI(82).
Basidiomycotina - Uredinales
Puccinia mesomajalis. Autoecious. **1:**MI(82,2862), MN(82), NH(82), NY(82), TN(82), VA(82), WI(82). **2:**AK(60,526), CA(81,82), ID(82,3202), MT(82,1195,3202), OR(82,3202), WA(82,638,3202). **3:**AK(526).
Basidiomycotina - Other
Ceratobasidium anceps. Leaf rot. **1:**ME(82), NY(82), WI(82).
Deuteromycotina - Hyphomycetes
Heterosporium sp. **2:**WA(3202).

Clivia Lindl.
1. *Clivia* sp.
Oomycetes
Pythium sp. Root rot. **1:**FL(35).
Deuteromycotina - Coelomycetes
Colletotrichum sp. Anthracnose. **1:**FL(35).
Phomopsis sp. Leaf necrosis. **1:**FL(35).

Colchicum L.
1. *C. autumnale* L. [autumn crocus, meadow-saffron, wonder-bulb, wysteria]
Basidiomycotina - Ustilaginales
Urocystis colchici. Leaf smut. **1:**DE(82), NY(82,1077), OH(82), PA(82), WA(82,3202).
Deuteromycotina - Hyphomycetes
Botrytis elliptica. Leaf spot. **1:**WA(82,3202).
Deuteromycotina - Coelomycetes
Colletotrichum dematium. On leaves. **1:**DE(82).

Convallaria L.
1. *C. majalis* L. [lily-of-the-valley]
Deuteromycotina - Hyphomycetes
Aureobasidium microstictum. Leaf spot. **1:**MD(82).
Botrytis paeoniae. Rhizome rot. **1:**IL(82), ME(82), PA(82).
Cercospora sp. On leaves. **1:**SC(82).
Deuteromycotina - Coelomycetes
Ascochyta majalis. Leaf blotch. **1:**PA(82).
Colletotrichum dematium. Anthracnose. **1:**IL(312).
Gloeosporium convallariae. Anthracnose, leaf spot. **1:**NC(1303), NY(82).
Phyllosticta vagans. On leaves. **1:**VA(82).
Phyllosticta sp. Leaf spot. **1:**NJ(82), NY(82).
Deuteromycotina - Other
Sclerotium rolfsii. Southern blight. **1:**MD(82), MS(2720).

Crinum L.
1. *C. americanum* L. [southern swamp crinum]
2. *C. longifolium* (L.) Thunb.
3. *Crinum* sp.
Deuteromycotina - Hyphomycetes
Cercospora pancratii. Leaf spot. **1:**FL(35), MS(82,2720). **2:**AL(82), FL(82). **3:**FL(35).
Deuteromycotina - Coelomycetes
Phyllosticta sp. Leaf spot. **1:**FL(35). **3:**FL(35).
Stagonospora curtisii. Leaf spot, red spot. **1:**CA(82), NY(82). **3:**FL(35), HI(2903).
Deuteromycotina - Other
Rhizoctonia solani. Root rot. **3:**FL(35).

Dianella Lam.
1. *D. nemorosa* Lam. (=*D. odorata* Blume)
2. *D. sandwicensis* Hook. & Arnott
3. *Dianella* sp.
Ascomycotina
Meliola gregoriana. Black mildew. **1:**HI(1272). **2:**HI(2903). **3:**HI(82,1452).
Mycosphaerella dianellae. Leaf spot. **1:**HI(3480). **2:**HI(2903).
Phaeosphaerella dianellae. **2:**HI(2903).
Deuteromycotina - Coelomycetes
Colletotrichum dianallae. On leaves. **2:**HI(2903).
Pirostoma dianellae. **2:**HI(2903).

Dichelostemma Kunth
1. *D. pulchellum* (Salisb.) A. Heller (=*Brodiaea pulchella* (Salisb.) Greene, *Brodiaea capitata* Benth.) [blue-dicks, wild hyacinth]
Basidiomycotina - Uredinales
Puccinia carnegiana. Autoecious. **1:**AZ(741).
P. dichelostemmae. **1:**OR(82,3202), WA(82,3202).
P. moreniana. **1:**CA(82).
P. nodosa. Autoecious. **1:**AZ(747,1195,1203,3981), CA(82,972).
P. subangulata. **1:**WA(82,3202).

P. tumamocensis. **1:**AZ(1203).

Disporum Salisb. ex D. Don

1. *D. hookeri* (Torr.) Nichols [fairy-bells]
2. *D. trachycarpum* (S. Wats.) Benth. & J. D. Hook.
3. *Disporum* sp.

Deuteromycotina - Hyphomycetes

Cercospora streptopi. **1:**WA(3202). **2:**MT(3202).
Cercosporella idahoensis. **1:**WA(3202).

Deuteromycotina - Coelomycetes

Septoria streptopidis. **1:**ID(3202), WA(3202). **2:**ID(3202). **3:**OR(3202), WA(3097).

Eremurus M. Bieb.

1. *E. robustus* Regel [desert-candle, king's-spear]

Deuteromycotina - Hyphomycetes

Myrothecium roridum. Leaf spot. **1:**OH(82).

Erythronium L.

1. *E. albidum* Nutt. [blonde-Lilian, white dogtooth-violet]
2. *E. americanum* Ker-Gawl. [amberbell, trout lily, yellow adder's-tongue]
3. *E. grandiflorum* Pursh (=*E. parviflorum* Goodd.) [avalanche lily]
4. *E. montanum* S. Wats.
5. *Erythronium* sp.

Chytridiomycetes

Synchytrium sp. On leaves. **2:**CA(1847).

Ascomycotina

Ciborinia erythronii. **1:**IL(208), IN(208), TX(82). **2:**MD(208), NY(82,208), PA(208).
C. gracilis. Leaf blight. **1:**IL(82), NE(82,208).

Basidiomycotina - Uredinales

Uromyces heterodermus. Autoecious. **1:**TX(82). **3:**CA(82), CO(82,649), ID(82,1195,3202), MT(82,1195,3202), OR(3202), UT(82), WA(82,638,3202), WY(82,1195,2924). **4:**OR(82), WA(82,3202). **5:**OR(657).

Basidiomycotina - Ustilaginales

Urocystis erythronii. Leaf smut. **1:**CT(82), NY(82). **2:**CT(82,1077), NY(82,1077), OH(1077).
Ustilago heufleri. Smut. **1:**MO(82,1077), TX(82), WI(1348). **2:**CT(1077), DC(1077), DE(82,1077), IL(1077), MA(1077), MD(82,1077), MI(82), MO(82,1077), NJ(82,1077), NY(82,1077), PA(82,1077). **5:**OH(1077).

Basidiomycotina - Other

Typhula sp. **3:**WA(3202).

Deuteromycotina - Hyphomycetes

Botrytis cinerea. **3:**CO(649).
B. elliptica. **3:**WA(82,3202).
Botrytis sp. Botrytis blight. **1:**IL(82), WA(82). **2:**NY(82), VT(82). **3:**WA(82). **4:**WA(82).
Epicoccum nigrum. Leaf spot. **2:**WI(1357).

Deuteromycotina - Coelomycetes

Asteroma tenerrimum. **3:**ID(3202), MT(3202), WA(3202), WY(1195).
A. tennerrimum var. *erythronii*. Black spot. **3:**ID(82), MT(82), WA(82).
Colletotrichum dematium. **3:**CO(649).
Phyllosticta erythronii. On leaves. **1:**WI(1352). **3:**WY(666,2924).
Phyllosticta sp. On leaves. **1:**WI(1350,1351).

Eucharis Planch. & Linden

1. *E. grandiflora* Planch. & Linden [Amazon lily, eucharist lily, lily-of-the-Amazon, madonna lily]
2. *Eucharis* sp.

Deuteromycotina - Hyphomycetes

Botrytis cinerea. Gray mold blight. **1:**FL(35,82).
Cercospora sp. Leaf spot. **1:**FL(35).
Fusarium sp. Bulb rot. **1:**FL(35).

Deuteromycotina - Coelomycetes

Phyllosticta sp. Leaf spot. **1:**FL(35).
Stagonospora curtisii. Red spot. **1:**CA(82), FL(35). **2:**FL(35).

Deuteromycotina - Other

Rhizoctonia solani. Root rot. **1:**FL(35).
Sclerotium rolfsii. Stem rot. **1:**FL(35).

Fritillaria L.

1. *F. atropurpurea* Nutt.
2. *F. camschatcensis* (L.) Ker-Gawl. [Kamchatka lily]
3. *F. lanceolata* Pursh [checker lily, narrow-leaved fritillary]

Basidiomycotina - Uredinales

Uromyces miurae. Autoecious. **2:**AK(60,82,526,3097). **3:**ID(3202), WA(82,638,3202).

Deuteromycotina - Coelomycetes

Asteroma tenerrimum. **3:**ID(3202).
Phyllosticta fritillariae. Leaf spot. **1:**CA(82,642). **3:**WA(3202).

Galanthus L.

1. *G. nivalis* L.

Deuteromycotina - Hyphomycetes

Botrytis galanthina. Blight. **1:**Unk.(82).

Deuteromycotina - Other

Sclerotium sp. **1:**Unk.(82).

Gasteria Duval.

1. *Gasteria* sp.

Oomycetes

Pythium sp. Root rot. **1:**FL(35).

Gloriosa L.

1. *G. rothschildiana* O'Brien [gloriosa lily]

Deuteromycotina - Hyphomycetes

Alternaria sp. Leaf spot. **1:**FL(35).
Fusarium sp. Tuber rot. **1:**FL(35).

Deuteromycotina - Coelomycetes

Colletotrichum dematium. Anthracnose. **1:**FL(35).
Phyllosticta sp. Leaf spot. **1:**FL(35).

Deuteromycotina - Other

Rhizoctonia solani. Bulb and root rot. **1:**FL(35).
Sclerotium rolfsii. Southern blight. **1:**FL(35).

Haemanthus L.

1. *H. coccineus* L. [African blood lily, blood lily]
2. *Haemanthus* sp.

Deuteromycotina - Hyphomycetes

Alternaria sp. Leaf spot. **2:**FL(35).

Deuteromycotina - Coelomycetes

Colletotrichum sp. Anthracnose. **1:**FL(35). **2:**FL(35).
Phyllosticta sp. Leaf spot. **1:**FL(35).

Haworthia Duval.

1. *Haworthia* sp.

Oomycetes

Pythium sp. Root rot. **1:**FL(35).

Deuteromycotina - Hyphomycetes

Cephalosporium sp. Leaf spot. **1:**FL(35).

Hemerocallis L.

1. *H. fulva* (L.) L. [fulvous daylily, orange daylily, tawny daylily]
2. *Hemerocallis* sp.

Oomycetes

Phytophthora sp. Crown and root rot. **1:**HI(2903). **2:**FL(35), HI(82,2903).
Pythium sp. Crown and root rot. **2:**FL(35).

Zygomycotina

Choanephora cucurbitarum. Flower blight. **2:**FL(35).

Ascomycotina

Ceratocystis paradoxa. **1:**HI(2903). **2:**HI(82,2903).

Deuteromycotina - Hyphomycetes

Alternaria sp. Leaf spot. **2:**FL(35).
Botrytis sp. Blight, leaf spot. **1:**WI(1355). **2:**FL(35), MD(82).
Cephalosporium sp. Leaf spot. **1:**NC(1303). **2:**FL(35).
Cercospora hemerocallidis. Leaf spot. **1:**IL(82). **2:**FL(35).
Collecephalus hemerocalli. Leaf streak. **2:**LA(1603), PA(2961).
C. hemerocalli var. *macrosporum*. On leaves and stems. **2:**MD(2190).
Fusarium sp. Crown, leaf and root spot. **2:**FL(35).
Gloeocephalus hemerocalli. Leaf streak. **2:**MS(3412).
Heterosporium sp. Leaf spot. **1:**NJ(82). **2:**MS(2720), TX(82).
Kabatiella sp. Leaf blight. **2:**MD(82).
Phymatotrichopsis omnivora. Root rot. **2:**TX(82).

Deuteromycotina - Coelomycetes
Colletotrichum dematium. Leaf scorch, on buds and stems. **1**:NC(1303). **2**:Range of host(82), AL(82), FL(35), MA(82), MD(82), NH(82), NY(82), OK(2872), PA(82).
C. lilii. Isolated from bulbs. **2**:LA(3373), PA(3373).
Phomopsis achilleae. On scapes and dead stems. **2**:HI(82,2903).
Phomopsis sp. Leaf spot. **2**:FL(35).
Phyllosticta sp. Leaf spot. **2**:FL(35).
Deuteromycotina - Other
Rhizoctonia solani. Root rot. **2**:FL(35).
Sclerotium rolfsii. Southern blight, southern stem rot. **1**:HI(2903), NC(1303). **2**:FL(35), HI(82,2903).
Sclerotium sp. Root rot. **2**:IN(82).

Hippeastrum Herb.
1. *H. vittatum* (L'Hér.) Herb. (=*Amaryllis vittata* L'Hér.)
2. *Hippeastrum* sp.

Oomycetes
Pythium debaryanum. Seedling root rot. **2**:FL(35).
Basidiomycotina - Other
Armillaria mellea. Bulb rot. **1**:CA(2893).
Deuteromycotina - Hyphomycetes
Alternaria sp. Leaf necrosis. **2**:FL(35).
Botrytis cinerea. **2**:AK(526).
Cephalosporium sp. Leaf spot. **2**:FL(35).
Cercospora amaryllidis. Leaf spot. **2**:FL(35).
Curvularia sp. Leaf spot. **2**:FL(35).
Fusarium sp. Bulb and root rot. **2**:FL(35).
Mycoleptodiscus indicus. Leaf spot. **2**:FL(35).
Stemphylium sp. Leaf necrosis. **2**:FL(35).
Deuteromycotina - Coelomycetes
Ascochyta sp. Leaf spot. **2**:FL(35).
Colletotrichum sp. Anthracnose. **2**:FL(35).
Pestalotia sp. Leaf spot. **2**:FL(35).
Phyllosticta sp. Leaf spot. **2**:FL(35).
Stagonospora curtisii. Red blotch, leaf scorch. **2**:FL(35).
Deuteromycotina - Other
Rhizoctonia solani. Root rot. **2**:FL(35).
Sclerotium rolfsii. Bulb and root rot. **2**:FL(35).

Hosta Tratt.
1. *H. plantaginea* (Lam.) Aschers. [fragrant plantain lily]
2. *Hosta* sp.

Deuteromycotina - Hyphomycetes
Alternaria sp. Leaf spot. **2**:NY(82).
Botrytis cinerea. Crown rot. **2**:AK(82), NJ(82).
Deuteromycotina - Coelomycetes
Colletotrichum dematium. On petioles and scapes. **1**:WI(1341). **2**:CT(82), NJ(82), VA(82).
Phyllosticta sp. Leaf spot. **2**:NJ(82), WA(3202).
Deuteromycotina - Other
Rhizoctonia solani. Root rot. **2**:NJ(82).
Sclerotium rolfsii. **2**:MD(82).
S. rolfsii var. *delphinii*. Crown rot. **2**:CT(82), MN(82), NJ(82), NY(82).

Hyacinthus L.
1. *H. orientalis* L. [common hyacinth, Dutch hyacinth, garden hyacinth]
2. *Hyacinthus* sp.

Oomycetes
Phytophthora sp. Root rot. **1**:NJ(82).
Ascomycotina
Sclerotinia bulborum. Black slime. **1**:Unk.(82).
Basidiomycotina - Ustilaginales
Ustilago vaillantii. **2**:LA(1077).
Deuteromycotina - Hyphomycetes
Aspergillus sp. Storage rot. **2**:OK(2872).
Botrytis hyacinthi. On leaves. **1**:WA(1289,3202).
Botrytis sp. Gray mold, storage rot, on leaves. **1**:NC(82), WI(1356). **2**:OK(2872).
Fusarium oxysporum. **1**:WA(3202).
Fusarium sp. Bulb rot. **1**:CO(82), MO(82), NJ(82), RI(82), TX(82), WA(82).
Penicillium hirsutum. Bulb rot. **1**:Range of host(82).
Penicillium sp. Blue mold, bud and storage rot. **1**:Range of host(82), NC(1303). **2**:OK(2872).

Hymenocallis Salisb.
1. *H. crassifolia* Herb. (=*Pancratium coronarium* Leconte)
2. *H. littoralis* (Jacq.) Salisb. (=*Pancratium littorale* Jacq.)
3. *H. narcissiflora* (Jacq.) Macbr. (=*H. calathina* (Ker-Gawl.) G. Nichols.) [basketflower, Peruvian daffodil]
4. *Hymenocallis* sp.

Oomycetes
Pythium sp. Root rot. **4**:FL(35).
Ascomycotina
Mycosphaerella aggregata. On leaves. **4**:TX(82).
Deuteromycotina - Hyphomycetes
Cercospora amaryllidis. Leaf spot. **4**:FL(35), TX(82).
C. pancratii. Leaf spot. **1**:FL(82), LA(82,962). **4**:FL(35), TX(82).
Deuteromycotina - Coelomycetes
Colletotrichum sp. Anthracnose. **4**:FL(35).
Gloeosporium hemerocallidis. Leaf spot. **4**:TX(82).
Phyllosticta hymenocallidis. **3**:Unk.(82).
Stagonospora curtisii. Leaf scorch, leaf spot, red spot. **2**:HI(2903). **3**:CA(82). **4**:MN(2874), TX(82).

Hypoxis L.
1. *H. hirsuta* (L.) Crép. [stargrass]
2. *Hypoxis* sp.

Myxomycetes
Mucilago crustacea. Smothering. **2**:OK(2872).
Basidiomycotina - Uredinales
Uromyces necopinus. Autoecious. **1**:NY(82).
U. primaverilis subsp. *nothoscordi*. Autoecious. **1**:FL(35), MO(82), MS(2720).
Basidiomycotina - Ustilaginales
Urocystis hypoxidis. Floral smut. **1**:CT(82,1077), MA(82,1077).
Deuteromycotina - Hyphomycetes
Alternaria sp. Leaf spot. **2**:OK(2872).
Deuteromycotina - Coelomycetes
Cylindrosporium guttatum. Leaf blotch and spot. **1**:OK(2872), WI(82).
Phyllosticta sp. Leaf spot. **2**:OK(2872).
Septoria hypoxis. Leaf spot. **1**:PA(82).

Ixiolirion Herb.
1. *I. tataricum* (Pall.) Herb. [Siberian lily, tartar lily]

Oomycetes
Phytophthora sp. Blight. **1**:MN(2874).

Kniphofia Moench
1. *K. schimperi* Baker (=*K. elegans* Engl.)
2. *K. uvaria* (L.) Hook. [common torch lily, flameflower, poker-plant]

Deuteromycotina - Hyphomycetes
Alternaria sp. Leaf spot. **2**:AL(82).
Deuteromycotina - Coelomycetes
Colletotrichum dematium. Anthracnose. **1**:OK(2872).

Leucocoryne Lindl.
1. *L. ixioides* (Sims) Lindl. (=*Brodiaea ixioides* Sims) [glory-of-the-sun]

Basidiomycotina - Uredinales
Uromyces brodiaeae. **1**:CA(82).

Leucocrinum Nutt. ex A. Gray
1. *L. montanum* Nutt. ex A. Gray

Basidiomycotina - Uredinales
Puccinia sporoboli var. *robusta*. Heteroecious: aecial host. **1**:CO(753).

Leucojum L.
1. *L. vernum* L. [spring-snowflake]

Deuteromycotina - Hyphomycetes
Botrytis sp. Scale speck. **1**:OR(82,3202).
Deuteromycotina - Coelomycetes
Stagonospora curtisii. Red spot. **1**:CA(82).

Lilium L.
1. *L. auratum* Lindl. [gold-banded lily, golden-rayed lily, mountain lily]
2. *L. canadense* L. [Canada lily, wild yellow lily, yellow-bell lily]
3. *L. candidum* L. [madonna lily]
4. *L. columbianum* Hanson ex Baker [Columbia lily, Oregon lily]
5. *L. croceum* Chaix (=*L. aurantium* Haw. in Loud.)

6. *L. davidii* Duchartre
7. *L. formosanum* Wallace
8. *L. hansonii* Leichtlin [Japanese turk's-cap lily]
9. *L. henryi* Baker
10. *L.* × *hollandicum* Bergmans ex Woodcock & Stearn. A group of lilies derived from crossing *L. bulbiferum* or *L. bulbiferum* var. *croceum* with forms of *L.* × *maculatum*. [candlestick lily]
11. *L. lancifolium* Thunb. (=*L. tigrinum* Ker-Gawl.) [tiger lily]
12. *L. leichtlinii* J. D. Hook.
13. *L. longiflorum* Thunb. [white-trumpet lily]
14. *L. longiflorum* Thunb. var. *eximium* G. Nichols. [Bermuda Easter lily, Easter lily]
15. *L. martagon* L. [Martagon lily, turban lily, turk's-cap lily]
16. *L. michiganense* Farw. [Michigan lily]
17. *L. monadelphum* M.-Bieb. [Caucasian lily]
18. *L. pardalinum* Kellogg [leopard lily, panther lily]
19. *L. pensylvanicum* Ker-Gawl. (=*L. dauricum* Ker-Gawl.) [candlestick lily]
20. *L. philadelphicum* L. [orange-cup lily, wild orange-red lily, wood lily]
21. *L. philadelphicum* L. var. *andinum* (Nutt.) Ker-Gawl. (=*L. umbellatum* Pursh). In horticulture the name *L. umbellatum* has been commonly applied to *L.* × *hollandicum*. [western orange-cup lily]
22. *L. philippinense* Baker
23. *L. regale* W. Wils. [regal lily, royal lily]
24. *L. sargentiae* W. Wils.
25. *L. speciosum* Thunb. [showy Japanese lily, showy lily]
26. *L. superbum* L. [American turk's-cap lily, lily-royal, swamp lily]
27. *L.* × *testaceum* Lindl. *L. candidum* × *L. chalcedonicum*. [Nankeen lily]
28. *L. washingtonianum* Kellogg [Washington lily]
29. *Lilium* sp.

Oomycetes

Phytophthora cactorum. Foot and stem rot. **3:**MD(82), MN(82). **13:**FL(35). **23:**IL(312), MD(82), MI(82), MN(82), NC(82), NJ(82). **25:**MD(82). **28:**MD(82). **29:**MD(82), MN(82), NY(82), OH(82), WA(82,3202), WI(82).
P. nicotianae var. *parasitica*. Stem, stump and top rot. **11:**FL(35). **13:**FL(35). **23:**TN(1027). **25:**MD(82), NJ(82), NY(82). **29:**IN(82).
Phytophthora sp. Stump rot. **13:**MI(3124), NJ(82), NY(82).
Pythium debaryanum. Damping-off. **23:**IN(82).
P. splendens. Root rot. **13:**FL(35,82).
P. ultimum. **13:**MI(3124).
Pythium sp. Root rot. **14:**NC(1303).

Zygomycotina

Mucor sp. **13:**MI(3124).
Rhizopus sp. Soft rot of bulbs. **1:**Range of host(82). **3:**Range of host(82). **11:**Range of host(82), FL(35). **13:**Range of host(82), TX(82). **19:**Range of host(82). **23:**Range of host(82). **25:**Range of host(82). **29:**Range of host(82), FL(35), OR(3202).

Ascomycotina

Botryotinia fuckeliana. Blight. **7:**NC(1303). **14:**NC(1303).
Chaetomium sp. **13:**MI(3124).
Mollisia atrata var. *liliaceara*. On dead stems. **18:**CA(2801).
Nectria ochroleuca. **5:**CA(3085).
N. radicicola. **13:**OR(3202).
Sclerotinia sclerotiorum. Stem rot. **3:**CT(82). **13:**CT(82), FL(35,82). **29:**TX(82), WA(82).

Basidiomycotina - Uredinales

Puccinia sporoboli. Heteroecious: aecial host. **16:**WI(1357). **21:**ND(82,378,753), NE(82,753).
Uromyces holwayi. Autoecious. **2:**NE states(82), NE(82), NJ(82), OR(3202). **3:**CT(82), MA(82), MI(82), NY(82), WA(82,3202). **4:**CA(82,642,3097), ID(82,3202), OR(82,3202), WA(82,3202). **16:**WI(3097). **18:**CA(82,3097), WA(82,3202). **20:**IL(82,3097), MA(82), NH(82), NY(82). **21:**ID(3202). **26:**NC states(82), NE states(82), MI(3097). **28:**CA(82), OR(82).
Uromyces sp. **29:**FL(35).

Basidiomycotina - Other

Athelia rolfsii. **1:**OR(3202). **3:**OR(3202). **23:**OR(3202). **25:**OR(3202). **10:**OR(3202).
Thanatephorus cucumeris. **13:**OR(3202), WA(3202).

Deuteromycotina - Hyphomycetes

Alternaria sp. Leaf spot. **13:**FL(35), MI(3124). **29:**FL(35).
Aureobasidium microstictum. On leaves. **1:**MD(82), NJ(82), NY(82), PA(82). **3:**MD(82), NJ(82), NY(82), PA(82). **29:**MD(82), NJ(82), NY(82), PA(82).
Botrytis cinerea. Botrytis blight, gray mold, leaf spot. **1:**Range of host(82). **2:**MD(82), NY(82), OR(82), VT(82), WA(82). **7:**NC(1303). **11:**FL(35). **13:**FL(35), HI(2903). **14:**NC(1303). **29:**FL(35), MS(2720).
B. elliptica. Blight, leaf spot. **1:**Range of host(82). **3:**WA(3202). **11:**FL(35). **13:**FL(35), WA(3202). **23:**WA(3202). **29:**OR(1666).
B. tulipae. **3:**WA(3202).
Botrytis sp. Blight. **13:**OR(2336). **16:**WI(1342). **23:**OK(2872).
Cercospora sp. Leaf spot. **11:**FL(35). **13:**FL(35,82).
Cercosporella inconspicua. Leaf spot. **2:**CT(82), NY(82,3097). **16:**WI(1341).
Cladosporium sp. Leaf mold. **1:**Range of host(82). **3:**Range of host(82). **11:**Range of host(82). **19:**Range of host(82). **23:**Range of host(82). **25:**Range of host(82). **29:**Range of host(82).
Cylindrocarpon destructans. Scale tip rot. **13:**NW states(82).
Fusarium oxysporum f. sp. *lilii*. Bulb, root and stem rot. **1:**E & C states(82), NW states(82), S states(82), OR(3202). **3:**E & C states(82), NW states(82), S states(82), OR(3202). **6:**OR(3202). **8:**OR(3202). **9:**OR(3202). **11:**FL(35), OR(3202). **12:**OR(3202). **13:**FL(35), OR(3202). **15:**OR(3202). **18:**NY(82), OR(3202). **23:**OR(3202). **24:**OR(3202). **25:**E & C states(82), NW states(82), S states(82), OR(3202). **26:**NY(82), OR(3202). **28:**VA(82). **10:**OR(3202). **27:**OR(3202). **29:**FL(35), OR(3202).
Fusarium sp. Root and stem rot. **13:**MI(3124), OR(82). **14:**NC(1303). **29:**NY(1666), OR(1666), WA(1666,3202).
Heterosporium sp. Leaf spot. **29:**MD(82).
Hormodendrum sp. **13:**MI(3124).
Penicillium aurantiogriseum. Blue mold of bulb scales. **1:**Range of host(82), OR(3202). **19:**Range of host(82). **25:**Range of host(82), OR(3202).
P. hirsutum. Blue mold of bulb scales. **1:**Range of host(82). **19:**Range of host(82). **23:**OR(3202). **25:**Range of host(82).
Penicillium sp. Blue mold of bulb scales. **1:**Range of host(82). **11:**FL(35). **13:**Range of host(82), FL(35), MI(3124). **19:**Range of host(82). **25:**Range of host(82). **29:**LA(1666), MI(1666), OH(1666), OR(1666,3202).
Ramularia sp. Leaf spot. **2:**WA(82). **29:**WA(3202).
Sporotrichum sp. Bud blight. **1:**VA(82).
Trichoderma sp. **13:**MI(3124).
Tritirachium dependens. **13:**LA(82). **29:**LA(1666).
Verticillium sp. **29:**NY(1666), OH(1666).

Deuteromycotina - Coelomycetes

Colletotrichum dematium. Anthracnose, black scale rot, on dead stems and isolated from bulbs. **11:**Range of host(82), IL(82), ME(82). **14:**NC(1303). **17:**OK(2872). **23:**Range of host(82), MD(82).
C. lilii. Isolated from bulbs. **7:**AL(3373). **13:**FL(35), LA(82,3373), OR(82,3202), WA(82,3202). **18:**OR(3202). **21:**MD(82), MS(2720).
Colletotrichum sp. Scale rot. **13:**OR(2336). **29:**LA(1666), OR(1666), WA(1666).
Macrophomina phaseolina. Charcoal rot of bulbs. **22:**CA(82).
Phoma sp. **13:**MI(3124).
Septoria sp. On leaves. **16:**WI(1351).

Deuteromycotina - Other

Rhizoctonia solani. Damping-off, yellow bulb discoloration, stem canker. **1:**Range of host(82), CT(82). **3:**Range of host(82), CT(82), MS(82,2720). **11:**FL(35). **13:**CA(82), FL(35), MI(3124), OR(82), WA(82). **25:**Range of host(82), MN(82), OR(82). **29:**Range of host(82), FL(35), MA(82), TX(82).
R. tuliparum. Stem rot. **13:**WA(82).
Rhizoctonia sp. Root rot. **14:**NC(1303).
Sclerotium rolfsii. Southern blight. **1:**Unk.(82). **3:**CT(82), IL(82), NY(82). **13:**FL(35,82), LA(82), OR(82). **23:**NJ(82), NY(82), OR(82), TN(82). **25:**Unk.(82). **29:**IN(82), MS(82,2720), NY(82), PA(82).

Liriope Lour.
1. *L. muscari* (Decne.) L. H. Bailey [big blue lilyturf]
2. *Liriope* sp.

Oomycetes

Phytophthora sp. Root rot. **1**:FL(35). **2**:FL(35).
Pythium splendens. Root rot. **1**:FL(35).
Pythium sp. Root rot. **2**:NC(1303).

Deuteromycotina - Hyphomycetes

Alternaria sp. Leaf spot. **1**:FL(35).
Cercospora liriopes. Leaf spot. **2**:FL(35).
Fusarium solani. Root rot. **1**:FL(35).
Fusarium sp. Dry rot. **1**:FL(35). **2**:FL(35).

Deuteromycotina - Coelomycetes

Colletotrichum sp. Anthracnose, leaf spot. **1**:FL(35). **2**:FL(35), NC(1303).
Leptothyrium sp. Leaf spot. **1**:FL(35).

Deuteromycotina - Other

Rhizoctonia solani. Root rot. **1**:FL(35). **2**:FL(35).
Sclerotium rolfsii. Southern blight. **1**:FL(35).

Lloydia Rchb.
1. *L. serotina* (L.) Salisb. ex Rchb.

Ascomycotina

Lewia scrophulariae. **1**:AK(526).
Pleospora helvetica. **1**:AK(526).

Basidiomycotina - Uredinales

Puccinia pagana. Autoecious. **1**:WY(749,3097).

Lycoris Herb.
1. *L. squamigera* Maxim. [magic lily, resurrection lily]

Deuteromycotina - Coelomycetes

Stagonospora curtisii. Red spot. **1**:CA(82).

Maianthemum Wigg.
1. *M. bifolium* DC. var. *kamtschaticum* (J. F. Gmelin) Jeps.
2. *M. canadense* Desf. [muguet, two-leaved Solomon's-seal, wild-lily-of-the-valley]
3. *M. dilatatum* (A. Wood) A. Nels. & Macbr.
4. *Maianthemum* sp.

Basidiomycotina - Uredinales

Puccinia sessilis. Heteroecious: aecial host. **2**:MA(82), MI(82), MN(82), NY(82), PA(82), WI(82). **3**:AK(526).
Uromyces acuminatus. Heteroecious: aecial host. **2**:ME(638), MI(82), NY(82), WI(82).

Deuteromycotina - Hyphomycetes

Botrytis sp. Botrytis blight. **2**:NY(82), WI(1342).
Ramularia subsanguinea. Leaf spot. **1**:CA(81). **2**:MI(82), NY(82), OH(82), PA(82), WI(82). **3**:W states(82), AK(82,526,3097), OR(3097,3202), WA(3097,3202). **4**:AK(526).

Deuteromycotina - Coelomycetes

Colletotrichum dematium. **2**:WI(1359).
Phoma majanthemi. **3**:AK(526).
Phyllosticta convallariae. Leaf spot. **3**:AK(82), WA(82,3202).
P. maianthemi. On leaves. **3**:WA(666).
Phyllosticta sp. **3**:AK(526).
Septoria sp. On leaves. **2**:WI(1341).

Medeola L.
1. *M. virginiana* L.

Ascomycotina

Medeolaria farlowii. Stem hypertrophy. **1**:MA(82), ME(82), NJ(82).

Deuteromycotina - Coelomycetes

Phyllosticta medeolae. Leaf spot. **1**:NY(64,82).

Muscari Mill.
1. *M. botryoides* (L.) Mill. [common grape-hyacinth]
2. *M. comosum* Mill. [tassel-hyacinth]
3. *Muscari* sp.

Ascomycotina

Sclerotinia bulborum. **3**:WA(3202).

Basidiomycotina - Ustilaginales

Ustilago vaillantii. Flower smut. **1**:WA(82,3202). **2**:MA(82,1077).

Deuteromycotina - Other

Sclerotium sp. Dry rot. **1**:MO(82), WA(82,3202).

Narcissus L.
1. *N. jonquilla* L. [jonquil]
2. *N. poeticus* L. [narcissus, pheasant's-eye, poet's narcissus]
3. *N. pseudonarcissus* L. [common daffodil, trumpet narcissus]
4. *N. tazetta* L. [polyanthus narcissus]
5. *Narcissus* sp.

Zygomycotina

Rhizopus stolonifer. Mushy rot, soft rot. **5**:Range of host(82), OR(3202), WA(3202).
Rhizopus sp. Soft rot. **2**:Range of Host(82).

Ascomycotina

Botryotinia narcissicola. Neck rot, smolder. **1**:NC(82), NY(82), OR(82), VA(82), WA(82). **2**:NC(82), NY(82), OR(82), VA(82), WA(82). **3**:NC(82), NY(82), OR(82), VA(82), WA(82). **5**:NC(82,1303), NY(82), OR(82,3202), VA(82), WA(82,3202).
B. polyblastis. Botrytis blight, flower and foliage fire. **3**:CA(82). **4**:WA(82). **5**:WA(569,3202).
Leptosphaeria sp. On leaves. **2**:WA(82). **5**:WA(3202).
Nectria radicicola. Secondary root rot. **5**:NC(1303), OR(3202), WA(3202).
Ophiostoma narcissi. On bulbs. **5**:OR(2125), WA(2125).
Ophiostoma sp. **5**:FL(1666), NY(1666), OR(1666,3202), WA(1666,3202).
Sclerotinia bulborum. **5**:WA(3202).
Stromatinia narcissi. On bulbs. **5**:Unk.(895), OR(3202), WA(3202).

Basidiomycotina - Other

Armillaria mellea. Bulb and root rot. **5**:CA(82), OR(82,3202), WA(82,3202).
Athelia rolfsii. **5**:OR(3202), WA(3202).

Deuteromycotina - Hyphomycetes

Aspergillus flavus. Black mold. **5**:Unk.(82).
A. niger. Black mold, secondary dry rot of bulb. **5**:Unk.(82), FL(35), VA(1666).
Aspergillus sp. Bulb rot. **5**:FL(1666), OK(2872).
Botrytis cinerea. Often on fading flowers and leaves. **5**:Unk.(82).
B. narcissicola. Sprout necrosis, neck rot, smolder. **5**:Unk.(82), NC(1303).
B. polyblastis. Leaf blight, fire. **5**:Unk.(82), WA(889).
Botrytis sp. Blight, gray mold. **5**:MS(2720), NC(1303), OK(2872).
Cladosporium herbarum. **5**:OR(3202), WA(3202).
Cylindrocarpon destructans. Secondary root rot. **5**:NC(82,1303), NY(82), OR(82), VA(82), WA(82).
Fusarium moniliforme. Bulb, root rot. **5**:Unk.(82), FL(35).
F. oxysporum f. sp. *narcissi*. Basal, bulb and root rot. **5**:Range of host(82), FL(35,1666), MS(2720), OR(1666,3202), VA(1666), WA(1666,3202).
Fusarium sp. Bulb rot. **5**:FL(1666), NC(1303), NY(1666), OR(1666).
Heterosporium sp. **5**:WA(3202).
Penicillium sp. Blue mold, bulb rot. **5**:Unk.(82), OK(2872), VA(1666).
Ramularia vallisumbrosae. Ramularia blight, white mold. **5**:OR(82,3202), WA(82,3202).
Trichoderma viride. Green mold. **5**:Range of host(82).
Verticillium sp. **5**:OR(1666,3202).

Deuteromycotina - Coelomycetes

Gloeosporium sp. Leaf and stem rot. **5**:LA(82), NC(82).
Septoria narcissi. Leaf spot. **5**:MS(2720).
Stagonospora curtisii. Leaf scorch. **5**:Range of host(82), FL(35), MS(2720), NC(1303), OR(3202), WA(3202).

Deuteromycotina - Other

Rhizoctonia solani. Bulb and root rot. **5**:FL(35).
Sclerotium rolfsii. Wet scale rot. **1**:CA(82), FL(82), NY(82), VA(82). **2**:CA(82), FL(82), NY(82), VA(82). **3**:CA(82), FL(82), NY(82), VA(82). **4**:FL(82). **5**:Fl(35).
S. rolfsii var. *delphinii*. **5**:OR(1666), WA(1666).
Sclerotium sp. Neck rot, large and small scale speck. **1**:MD(82), MO(82), NC(82), NY(82), OR(82), VA(82), WA(82). **2**:MD(82), MO(82), NC(82), NY(82), OR(82), VA(82), WA(82). **3**:MD(82), MO(82), NC(82), NY(82), OR(82), VA(82), WA(82). **4**:CA(82), FL(82), NC(82), NY(82), OR(82), TX(82), VA(82). **5**:CA(82), FL(35,82), NC(82,1303), NY(82), OR(82,3202), TX(82), VA(82,1666), WA(3202).

Nerine Herb.

1. *N. sarniensis* (L.) Herb. [Guernsey lily]
2. *Nerine* sp.

Deuteromycotina - Coelomycetes

Stagonospora curtisii. Leaf scorch, red spot. **1**:CA(82). **2**:CA(82).

Nothoscordum Kunth

1. *N. bivalve* (L.) Britton

Basidiomycotina - Uredinales

Uromyces hordeinus. Heteroecious: aecial host. **1**:KS(82), MS(2720), OK(82,2872), TX(82).
U. primaverilis. Autoecious. **1**:IL(3097), MO(82), OK(2872), TX(82,3097).

Deuteromycotina - Coelomycetes

Colletotrichum circinans. Anthracnose. **1**:OK(82,2873).
Colletotrichum sp. Anthracnose. **1**:OK(2872).

Ophiopogon Ker-Gawl.

1. *O. japonicus* (L. fil.) Ker-Gawl.
2. *Ophiopogon* sp.

Oomycetes

Phytophthora sp. Root rot. **1**:FL(35).
Pythium splendens. Root rot. **1**:FL(35).
Pythium sp. Root rot. **1**:HI(2903).

Deuteromycotina - Hyphomycetes

Alternaria sp. Leaf spot. **1**:FL(35).
Curvularia lunata. Leaf spot. **1**:HI(2903).
Fusarium sp. Damping-off. **1**:FL(35).

Deuteromycotina - Coelomycetes

Colletotrichum falcatum. Anthracnose. **1**:HI(2903).
Colletotrichum sp. Anthracnose, leaf spot. **1**:FL(35). **2**:NC(1303).
Vermicularia sp. Leaf spot. **1**:HI(2903).

Deuteromycotina - Other

Rhizoctonia solani. Root rot. **1**:FL(35).
Rhizoctonia sp. Root rot. **1**:HI(2903).

Ornithogalum L.

1. *O. umbellatum* L. [dove's-dung, nap-at-noon, star-of-Bethlehem, summer-snowflake]
2. *Ornithogalum* sp.

Ascomycotina

Mycosphaerella ornithogali. Leaf spot. **1**:IL(82), PA(82), WA(82,3202).

Basidiomycotina - Uredinales

Puccinia liliacearum. **1**:NY(128), PA(128).

Deuteromycotina - Coelomycetes

Septoria ornithogali. Leaf spot. **1**:CT(82), IL(312), MA(82).

Deuteromycotina - Other

Sclerotium rolfsii. Southern blight. **2**:CA(82).

Polygonatum Mill.

1. *P. biflorum* (Walter) Elliott (=*P. canaliculatum* (Muhl. in Willd.) Pursh)
2. *P. commutatum* (Schult. & J. H. Schult.) A. Dietr. (=*P. giganteum* A. Dietr.)
3. *Polygonatum* sp.

Ascomycotina

Guignardia polygonati. On leaves. **1**:PA(82).
Stromatinia rapulum. On dead root stocks and rhizomes. **1**:NY(82). **3**:NY(1960).

Basidiomycotina - Uredinales

Puccinia sessilis. Heteroecious: aecial host. **1**:AL(82), CT(82), IA(82), ID(82), MN(82), NY(82), OH(82), PA(82), WI(82), WY(82). **2**:ND(378), SD(2243).
Uromyces acuminatus. Heteroecious: aecial host. **1**:IA(82), IL(82), MN(82), ND(82), NE(82), SD(82).

Basidiomycotina - Ustilaginales

Urocystis colchici. Leaf smut. **1**:IA(82,1077).

Deuteromycotina - Hyphomycetes

Botrytis sp. **1**:WI(1342).

Deuteromycotina - Coelomycetes

Colletotrichum dematium. On dead stems. **1**:Range of host(82), IN(82), ME(82).
Discosia artocreas. **1**:NC(1307).
Macrophoma smilacina var. *smilacis*. **1**:NY(279), PA(279).
Phyllosticta convallariae. Leaf spot. **1**:Unk.(82), CT(82), IA(82), IN(82), NY(82), OH(82), VA(82), WI(82). **3**:WI(1341).
P. pallidior. **1**:Unk.(82), WI(1354).
Phyllosticta sp. **1**:WI(1354).

Rohdea Roth

1. *R. japonica* Roth

Deuteromycotina - Hyphomycetes

Cephalosporium sp. Leaf spot. **1**:FL(35).
Cercospora sp. Leaf spot. **1**:FL(35).

Scilla L.

1. *S. bifolia* L.
2. *S. hispanica* Mill. [blue-flowered squill]
3. *S. sibirica* Andr. [Siberian squill]
4. *S. violacea* Hutch.
5. *Scilla* sp.

Oomycetes

Pythium sp. Root rot. **4**:FL(35).

Basidiomycotina - Uredinales

Uromyces muscari. Autoecious. **2**:CA(111).

Basidiomycotina - Ustilaginales

Ustilago vaillantii. Flower smut. **1**:MA(1077). **5**:MA(82), WA(82).

Basidiomycotina - Other

Athelia rolfsii. **3**:WA(3202).

Deuteromycotina - Hyphomycetes

Penicillium gladioli. Blue mold rot. **5**:Unk.(82).

Deuteromycotina - Other

Sclerotium rolfsii var. *delphinii*. Bulb rot. **5**:WA(82).

Smilacina Desf.

1. *S. racemosa* (L.) Desf. (=*S. racemosa* (L.) Desf. var. *amplexicaulis* (Nutt.) S. Wats., *S. amplexicaulis* Nutt.) [false spikenard, fat Solomon, Solomon's-zigzag, treacleberry]
2. *S. stellata* (L.) Desf. (=*S. sessilifolia* (Baker) Nutt. ex S. Wats.) [lily-of-the-valley, starflower]
3. *S. trifolia* (L.) Desf.
4. *Smilacina* sp.

Ascomycotina

Discochora smilacinina. On leaves. **2**:NY(82).
Mycosphaerella vagnerae. On stems. **4**:NV(82).
Telimenella phacidioidea. On overwintered stalks. **1**:MA(183).

Basidiomycotina - Uredinales

[*Puccinia amphigena*]. **2**:WI(1352).
P. sessilis. Heteroecious: aecial host. **1**:IA(82), NY(82), OK(82,2872), PA(82). **2**:ID(82,3202), KS(82), MI(82), MT(82,3202), ND(82), NE(82), SD(82,2243), WA(3202). **4**:CA(82), ID(82), WA(82).
P. sporoboli var. *robusta*. Heteroecious: aecial host. **2**:IN(753).
Uromyces acuminatus. Heteroecious: aecial host. **1**:MN(82), WI(82). **2**:CO(82), IL(82), MT(82,3202), ND(82,378), NE(82), SD(82), WI(82), WY(1195).

Basidiomycotina - Ustilaginales

Tuburcinia clintoniae. **2**:WA(1077,3202), WI(1077).
Urocystis colchici. Leaf smut. **1**:MT(82,1077,3202). **2**:MT(82,1077,3202).

Deuteromycotina - Hyphomycetes

Cercosporella idahoensis. Leaf spot. **2**:ID(82,3202), WA(3202).
Heterosporium asperatum. On leaves. **2**:CA(642).
Heterosporium sp. **2**:ID(3202), OR(3202).
Ramularia smilacinae. Leaf spot. **2**:ID(3202), MT(3202), WY(82).
R. vagnerae. **1**:CA(82), CO(82).
Stemphylium arnyi. On living leaves. **2**:WI(2953).

Deuteromycotina - Coelomycetes

Colletotrichum dematium. On leaves and stems. **1**:IL(82), MD(82), NY(82), PA(82), VA(82), WV(82). **2**:MD(82), VA(82). **3**:NY(2456).
Colletotrichum sp. **3**:WI(1332).
Cylindrosporium smilacinae. Leaf spot. **1**:CO(82), MT(3202), OR(82,3202), WA(3202). **2**:CA(82,638,639,642), ID(3202), MT(3202), UT(82), WA(3202), WY(1195).
C. vagnerae. On living leaves. **1**:WI(1342).
Phloeospora vagnerae. Leaf spot. **2**:MT(3202). **4**:MT(82).
Phyllosticta pallidior. On leaves. **1**:CT(280), IN(280), MS(280), NC(3944), NJ(280), NY(280), PA(280), WI(64,280). **2**:MA(1394), MI(280), NY(64,280), WI(64,280). **4**:NY(280), WI(1341).

P. smilacinae. On yellowing leaves. **1:**CO(82,1195,3376).
P. smilacinae-trifoliae. **3:**WI(1359).
P. subeffusa. On leaves. **1:**VA(2), WI(2). **2:**ND(2), WI(2).
P. vagans. On leaves. **1:**CA(81), ID(3202), MD(82), ME(82), MO(82), MT(82,3202), NY(82), VT(82).
Septoria smilacinae. Leaf spot. **1:**Range of host(82), IN(647), MS(2720). **2:**IA(82).
Septoria sp. **1:**WI(1346).

Stenanthium (A. Gray) Kunth
1. *S. gramineum* (Ker-Gawl.) Morong [featherbells]
2. *S. occidentale* A. Gray

Basidiomycotina - Uredinales
Puccinia atropuncta. Heteroecious: telial host. **1:**GA(82).
P. grumosa. Autoecious. **2:**OR(82,2647,3202).

Sternbergia Waldst. & Kit.
1. *S. lutea* (L.) Ker-Gawl. [fall daffodil]

Deuteromycotina - Coelomycetes
Stagonospora curtisii. Red spot. **1:**CA(82).

Streptopus Michx.
1. *S. amplexifolius* (L.) DC. [white mandarin]
2. *S. curvipes* Vaill.
3. *S. roseus* Michx. [rose mandarin]
4. *Streptopus* sp.

Ascomycotina
Crocicreas cyathoideum. **4:**CO(505).
Basidiomycotina - Uredinales
Puccinia sessilis. Heteroecious: aecial host. **3:**NY(82).
Basidiomycotina - Ustilaginales
Tuburcinia clintoniae. Leaf smut. **2:**WA(1077). **3:**MT(3202), WA(3202), WI(82,1077).
Deuteromycotina - Hyphomycetes
Cercospora streptopi. Leaf spot. **1:**WA(82,3202).
Deuteromycotina - Coelomycetes
Colletotrichum dematium. On leaves. **3:**WI(82).
Septoria streptopidis. Leaf spot. **3:**MT(82).

Tofieldia Huds.
1. *T. glutinosa* (Michx.) Pers.
2. *T. palustris* Huds. (=*T. pusilla* (Michx.) Pers.)

Ascomycotina
Hypoderma commune. **1:**WA(3202).
Leptosphaeria sp. **1:**WA(3202).
Mycosphaerella minor. On dead stems. **2:**AK(1938).
M. tassiana. **2:**AK(1938).
Deuteromycotina - Coelomycetes
Septoria orchidearum. **2:**AK(526).

Trillium L.
1. *T. chloropetalum* (Torr.) Howell
2. *T. cuneatum* Raf.
3. *T. decumbens* Harb.
4. *T. erectum* L. [purple trillium, stinking-Benjamin]
5. *T. flexipes* Raf.
6. *T. gleasonii* Fernald (=*T. declinatum* (A. Gray) Gleason)
7. *T. grandiflorum* (Michx.) Salisb.
8. *T. maculatum* Raf.
9. *T. ovatum* Pursh
10. *T. petiolatum* Pursh
11. *T. recurvatum* L. C. Beck
12. *T. sessile* L.
13. *T. underwoodii* Small
14. *Trillium* sp.

Ascomycotina
Ciborinia trillii. **3:**AL(208). **4:**NY(208). **8:**AL(208), GA(208), NY(208).
Didymella sp. **9:**ID(3202).
Sclerotinia sp. Decay of petals and stems. **2:**NC(1303).
Basidiomycotina - Uredinales
Uromyces halstedii. Heteroecious: aecial host. **7:**WI(1346). **10:**WA(3202). **14:**IL(82), NY(82).
Basidiomycotina - Ustilaginales
Urocystis trillii. **1:**OR(1077,3202). **9:**ID(1077,3202), OR(1077,3202).
Deuteromycotina - Hyphomycetes
Heterosporium trillii. Leaf spot. **9:**ID(3202), MT(3202), WA(3202). **10:**WA(3202). **14:**WA(82).
Ramularia trillii. Leaf spot. **9:**ID(666,3202).
Deuteromycotina - Coelomycetes
Colletotrichum dematium. Leaf spot. **11:**WI(1341). **14:**E states(82), IL(82), MN(82).
C. gloeosporioides. **6:**WI(1329).
Cylindrogloeum trillii. Leaf spot. **9:**OR(3202), WA(3202). **10:**WA(3202). **12:**CA(82), OR(82), WA(82).
Gloeosporium brunneo-maculatum. On living leaves. **7:**WI(1338). **11:**WI(1349).
Phyllosticta trillii. **6:**WI(1334). **7:**WI(1329,1334). **10:**WA(3202). **14:**WI(1334).
Phyllosticta sp. **5:**WI(1359).
Septoria recurvata. **14:**IN(82).
S. trillii. Leaf spot. **11:**MS(2720), OK(2872). **13:**FL(35). **14:**E states(82), OK(82), WI(82).
Deuteromycotina - Other
Sclerotium rolfsii var. *delphinii*. Stem rot. **14:**NH(82), PA(82).
Sclerotium sp. On leaves. **14:**NY(82), VA(82).

Triteleia Douglas ex Lindl.
1. *T. grandiflora* Lindl. (=*Brodiaea douglasii* S. Wats.)
2. *T. hyacinthina* (Lindl.) Greene (=*Brodiaea hyacinthina* (Lindl.) Baker)
3. *T. ixioides* Greene [golden brodiaea, pretty-face]

Ascomycotina
Mycosphaerella macrospora. **1:**WA(3202).
Basidiomycotina - Uredinales
Puccinia dichelostemmae. **1:**WA(82,3202).
P. pattersoniana. **1:**ID(82,3202), MT(3202), UT(82), WA(82,3202).
P. subangulata. **1:**WA(3202).
Uromyces brodiaeae. **1:**OR(82,3202). **2:**WA(3202).
Deuteromycotina - Hyphomycetes
Cladosporium allii. On leaves. **3:**CA(642).
Cladosporium sp. **1:**ID(3202).
Heterosporium sp. **1:**ID(3202).
Deuteromycotina - Coelomycetes
Phyllosticta fritillariae. **1:**ID(3202), WA(3202).

Tulbaghia L.
1. *T. violacea* Harv. [society garlic]

Deuteromycotina - Coelomycetes
Colletotrichum sp. Flower spike lesions. **1:**FL(35).

Tulipa L.
1. *T. gesneriana* L.
2. *T. suaveolens* Roth in Usteri
3. *Tulipa* sp.

Oomycetes
Phytophthora cactorum. Stem rot. **1:**CA(82), IL(82), NJ(82), PA(82), SC(82), WA(82,3202).
Phytophthora sp. Bulb rot. **1:**MS(2720).
Pythium ultimum. Secondary bulb rot. **1:**WA(82,3202).
Zygomycotina
Rhizopus necans. **3:**WA(1666).
R. stolonifer. **1:**Range of host(82). **3:**WA(3202).
Rhizopus sp. **3:**WA(1666).
Ascomycotina
Nectria inventa. **3:**WA(3202).
Sclerotinia minor. Bulb rot. **1:**NY(82).
S. sclerotiorum. **1:**WA(3202).
Sclerotinia sp. White bulb and stem rot. **1:**FL(35), ME(82), OH(82), WA(82). **3:**HI(2903), WA(3202).
Basidiomycotina - Other
Athelia rolfsii. **1:**OR(3202), WA(3202).
Thanatephorus cucumeris. **3:**WA(3202).
Deuteromycotina - Hyphomycetes
Aspergillus niger. Black mold of bulbs. **1:**Range of host(82).
Aspergillus sp. Black mold of bulbs, bulb rot. **1:**FL(35). **3:**OK(2872).
Botrytis cinerea. Blight. **1:**NC states(82), NE states(82), NW states(82).
B. tulipae. Blight. **1:**NC states(82), NE states(82), NW states(82), CT(3770), FL(35), MS(2720), NC(1303), WA(3202), WI(1348). **2:**WI(1348). **3:**ID(3202), NJ(1666), OK(2872), OR(3202),

WA(3202), WV(3910).
Botrytis sp. Fire. **3:**CT(2974), HI(2903).
Fusarium oxysporum. Bulb rot. **1:**FL(35).
F. poae. On buds. **1:**NC(1303).
F. roseum. **3:**WA(3202).
Fusarium sp. Bulb and basal rot. **1:**NC(1303), OK(82), WA(82,3202). **3:**OK(2872), OR(1666,3202), WA(1666).
Penicillium aurantiogriseum. Blue mold rot of bulbs. **1:**Range of host(82).
P. hirsutum. Blue mold rot of bulbs. **1:**Range of host(82).
Penicillium sp. Blue mold rot of bulbs, bulb rot. **1:**FL(35), NC(1303). **3:**OK(2872), WA(1666,3202).
Verticillium tenerum. **3:**WA(1666).

Deuteromycotina - Other

Rhizoctonia solani. Bulb, root and stem rot. **1:**FL(35), MA(82), NY(82), WA(82).
R. tuliparum. Gray bulb rot. **1:**NE states(82), W states(82), OR(3202), WA(3202).
Sclerotium rolfsii. Southern blight, bulb rot. **1:**CA(82), CT(82), FL(35), GA(82), NY(82), OR(82).

Uvularia L.

(= *Oakesia* S. Wats.)
1. *U. grandiflora* Sm.
2. *U. perfoliata* L.
3. *U. sessilifolia* L. (=*Oakesia sessilifolia* S. Wats.) [wild-oats]
4. *Uvularia* sp. (=*Oakesia sp.*)

Ascomycotina

Guignardia sp. On leaves. **2:**GA(82,1438).

Basidiomycotina - Uredinales

Puccinia sessilis. Heteroecious: aecial host. **1:**IA(82), MN(82), MO(82), WI(82). **2:**NC(82). **3:**DE(82), MD(82), MN(82), MS(82,2720), NY(82), WI(82).
Uromyces acuminatus. Heteroecious: aecial host. **3:**MS(82), WI(82).

Deuteromycotina - Hyphomycetes

Botrytis sp. **1:**WI(1342).

Deuteromycotina - Coelomycetes

Colletotrichum dematium. On leaves and stems. **2:**VA(82). **3:**VA(82).
Macrophoma smilacina. **3:**NY(64).
M. smilacina var. *smilacis*. **3:**NY(279), WI(279). **4:**WI(1341).
Phyllosticta convallariae. Leaf spot. **1:**IA(82), IL(82), MO(82), VA(82), WI(82). **2:**IN(82). **3:**CT(82), NY(82), WI(82). **4:**WI(1341).
P. uvulariae. Leaf spot. **2:**PA(280), TN(280). **3:**WI(64). **4:**WI(1341).

Vallota Salisb. & Herb.

This genus is considered a synonym of *Cyrtanthus*, but the appropriate combination could not be located.
1. *V. speciosa* (L. fil.) E. M. Durand & Schinz. [Scarboro lily]

Deuteromycotina - Coelomycetes

Stagonospora curtisii. Red spot. **1:**LA(82).

Veratrum L.

1. *V. californicum* E. M. Durand
2. *V. californicum* E. M. Durand var. *caudatum* (A. Heller) C. Hitchc.
3. *V. eschscholtzii* (Roem. & Schult.) A. Gray
4. *V. fimbriatum* A. Gray
5. *V. parviflorum* Michx. (=*Melanthium parviflorum* (Michx.) S. Wats.)
6. *V. viride* Aiton [Indian poke, itchweed, white hellebore]
7. *V. woodii* J. W. Robbins ex A. Gray
8. *Veratrum* sp.

Ascomycotina

Crocicreas cyathoideum. **1:**CO(505).
C. nigrofuscum. **1:**UT(505).
Didymolepta mimuli. On overwintered herbage. **1:**CA(661).
Helotium sp. **6:**AK(3431).
Heterosphaeria patella. **1:**WA(3202).
Laetinaevia veratri. On overwintered herbage, on stems. **1:**CA(524,642), OR(3202).
Leptosphaeria veratri. On stems. **8:**CO(82).
Leptospora rubella. **6:**NY(3248).
Mycosphaerella tassiana. On overwintered stems. **1:**CA(642).
M. tassiana var. *arctica*. On dead stems. **1:**CA(661).
Nectria ochroleuca. **6:**NY(3085).
Nodulosphaeria octoseptata. On overwintered stems. **1:**CA(661).
Patinella aloysii-sabaudiae. **8:**AK(526).
Phyllachora melanoplaca. Tar spot. **6:**NY(82).
Pleospora sp. On overwintered herbage. **1:**CA(642).
Pyrenopeziza sp. On stems. **8:**CO(82).
Sclerotinia veratri. On stems. **8:**CO(82).
Trichopezizella nidulus var. *setigera*. **1:**WA(3202).
T. relicina var. *macrospora*. **2:**CA(661,1416). **8:**WA(1416).

Basidiomycotina - Uredinales

Puccinia atropuncta. Heteroecious: telial host. **5:**NC(82,1303,1307), TN(82), VA(82). **6:**TN(82), WV(82). **7:**OK(2872). **8:**MO(82), NC(82), OK(82), TN(82), VA(82).
P. veratri. Heteroecious: telial host. **1:**CA(81,82,642), CO(1195,1520,3379), ID(3202), NM(82), OR(82,3202), WA(82,3202). **3:**AK(526). **4:**CA(81). **6:**CA(81), CT(82), ID(3202), MS(2720), NH(82), NJ(82), NY(82), OR(3202), PA(82), VA(82), WA(3202), WV(82). **8:**CO(82), ID(82), MT(82).

Deuteromycotina - Hyphomycetes

Botrytis cinerea. On dead stems. **1:**CA(642,661).
Cercosporella veratri. On leaves. **6:**NY(82).

Deuteromycotina - Coelomycetes

Ascochyta veratrina. Leaf spot. **1:**WA(82,3202).
Colletotrichum dematium. On leaves and stems. **6:**DE(82), NJ(82).
Cylindrosporium veratrinum. Leaf spot. **1:**CA(81,638,639,642,661), ID(3202), UT(1195), WA(3202). **6:**NY(82), VA(82), WA(82,3202). **8:**CA(82), ID(3202), MT(3202), UT(82).
Diplodia veratri. On dead stems. **1:**CA(642,661), ID(3202). **8:**CO(82), NV(82).
Gloeosporium veratri. On leaves. **1:**CA(81), CO(82), WA(82,3202). **6:**WA(3202).
Heteropatella umbilicata. On overwintered stems. **1:**CA(661), WA(3202).
Phyllosticta melanoplaca. Leaf spot. **1:**CA(81,639,642,661), UT(1195), WA(3202). **6:**CA(82), CT(82). **8:**CA(82), CO(82), ID(82,3202), UT(82).
Septoria allardii. Leaf spot. **5:**VA(82).

Deuteromycotina - Other

Sclerotium durum. **8:**AK(526).
Sclerotium sp. On stems. **8:**AK(82).

Xerophyllum Michx.

1. *X. asphodeloides* (L.) Nutt. [turkey's-beard]
2. *X. tenax* (Pursh) Nutt.

Ascomycotina

Asteridium xerophylli. On old leaves. **1:**NJ(82).
Comoclathris permunda. On leaves. **2:**ID(82,3202), OR(3202).
Crocicreas hysterioides. **2:**OR(505).
Didymella sp. **2:**WA(3202).
Hypoderma commune. **2:**OR(3202).
Leptosphaeria xerophylli. On leaves. **1:**NJ(82,3249). **2:**MT(3202), OR(82,3202), WA(82,3202,3815).
Lophodermium sp. **2:**MT(3202), WA(3202).
Mycosphaerella tassiana. **2:**OR(657).
M. xerophylli. On dead stems. **2:**ID(82,3202).
Pleospora helvetica. **2:**OR(657).

Basidiomycotina - Uredinales

Puccinia atropuncta. Heteroecious: telial host. **1:**MS(82,2720).

Deuteromycotina - Coelomycetes

Creothyriella sideroxyfoliae. **2:**ID(3202).
Hendersonia xerophylli. On old leaves. **1:**NJ(82).
Leptostroma xerophylli. On leaves. **2:**ID(82,3202,3249), OR(3249), WA(3249).
Rhabdospora sp. **2:**WA(3202).

Zephyranthes Herb.

(= *Cooperia* Herb.)
1. *Z. atamasco* (L.) Herb. [atamasco lily]
2. *Z. brazosensis* Traub (=*Cooperia drummondii* Herb.)
3. *Z. candida* (Lindl.) Herb. [zephyr-lily]
4. *Zephyranthes* sp.

Ascomycotina

Stromatinia narcissi. On bulbs. **4:**Unk.(895).

Basidiomycotina - Uredinales

Puccinia cooperiae. **1:**AL(82), FL(35,82), NC(82,1303,3944). **2:**TX(82). **4:**FL(35,82), TX(82).

Deuteromycotina - Hyphomycetes
Cercospora amaryllidis. Leaf spot. **2**:TX(82).
Cercospora sp. Leaf spot. **4**:FL(35).
Deuteromycotina - Coelomycetes
Colletotrichum dematium. Leaf spot. **1**:NC(82,1303).
Colletotrichum sp. Anthracnose. **4**:FL(35).
Stagonospora curtisii. Red spot. **1**:CA(82). **3**:CA(82). **4**:FL(35).
Deuteromycotina - Other
Sclerotium rolfsii. Southern blight. **4**:FL(35).
Sclerotium sp. **3**:OR(82).

Zigadenus Michx.

1. *Z. elegans* Pursh (=*Z. coloradensis* Rydb.) [alkali-grass, white camas]
2. *Z. leimanthoides* A. Gray
3. *Z. nuttallii* (A. Gray) S. Wats.
4. *Z. paniculatus* (Nutt.) S. Wats.
5. *Z. venenosus* S. Wats. (=*Z. gramineus* Rydb.)
6. *Zigadenus* sp.

Ascomycotina
Mycosphaerella tassiana. On dead leaves. **1**:AK(1938).
Basidiomycotina - Uredinales
Puccinia atropuncta. Heteroecious: telial host. **1**:IA(82), MO(82), ND(82,378), WI(82). **6**:TX(82).
P. grumosa. Autoecious. **1**:AK(60,526,2647), CO(2647), MT(2647), WY(1195,2647). **5**:CO(82,1195,2647), MT(82,3202), WY(82,1195,2647). **6**:CO(82).
Uromyces zygadeni. Autoecious. **1**:CO(82), IA(82), ID(3202), MT(82,3202), WY(82,2924). **3**:OK(2872). **4**:ID(3202). **5**:ID(3202), MT(82,3202), SD(2243), TX(82), WA(82,3202), WY(82,1195). **6**:CA(82), CO(82), KS(82), NV(82), TX(82), UT(82), WA(3202).
Basidiomycotina - Ustilaginales
Urocystis flowersii. Leaf smut. **1**:UT(82,1077).
Deuteromycotina - Hyphomycetes
Fusarium oxysporum. **2**:DE(3543).
Stemphylium botryosum. **5**:OR(3202).
Deuteromycotina - Coelomycetes
Ascochyta veratri. **1**:WI(1345).
Septoria sp. On flowering stems, on blackened tips of leaves. **1**:WI(1345,1355).

Limnanthaceae

Floerkea Willd.

1. *F. proserpinacoides* Willd.

Oomycetes
Peronospora floerkeae. **1**:WY(1195,3202).
Basidiomycotina - Ustilaginales
Entyloma floerkeae. **1**:DE(1077), ID(3202), IL(1077), IN(1077), NY(1077), OH(1077), WA(3202), WI(1077,1338), WV(1077).

Limnanthes R. Br.

1. *L. douglasii* R. Br. (=*L. rosea* Hartw. ex Benth.)

Chytridiomycetes
Synchytrium sp. On leaves, petals, petioles, sepals and stems. **1**:CA(1847).

Linaceae

Linum L.

1. *L. digynum* A. Gray
2. *L. flavum* L.
3. *L. grandiflorum* Desf.
4. *L. micranthum* A. Gray
5. *L. perenne* L.
6. *L. perenne* L. subsp. *lewisii* (Pursh) Hultén (=*L. lewisii* Pursh) [perennial flax, prairie flax]
7. *L. rigidum* Pursh
8. *L. sulcatum* Riddell
9. *L. usitatissimum* L. [common flax]
10. *L. virginianum* L. [woodland flax]

Oomycetes
Aphanomyces cladogamus. On roots. **9**:WI(82).
Peronospora lini. Downy mildew. **8**:IA(82).
Pythium aphanidermatum. **9**:AZ(82).
P. debaryanum. **9**:IA(82).
Pythium sp. Root rot. **9**:AZ(82), IA(82), ND(82).
Ascomycotina
Comoclathris permunda. **5**:OR(3202). **6**:WY(3810).
Mycosphaerella linicola. **9**:AZ(82), CA(82), IA(82), KS(82), MI(82), MN(82), MT(3202), ND(82), SD(82), TX(82), WI(82).
Pleospora herbarum. On seeds. **9**:MN(82), ND(82).
P. stenospora. **6**:WY(3810).
Pyrenopeziza californica. On stems. **6**:CA(82).
Sclerotinia sclerotiorum. Stem rot. **2**:CA(82). **9**:AZ(82), CA(82), OR(82), TX(82).
Basidiomycotina - Uredinales
Melampsora lini. Autoecious. **1**:WA(3202). **4**:CA(638,642). **5**:ID(3202), MT(3202), OR(3202), WA(3202). **7**:MT(3202), ND(378). **8**:IA(82), ND(82,378), NE(82), SD(82,2243), WI(82), WY(82). **9**:Range of host(82), MT(3202), ND(378), OK(2872), OR(3202), SD(2243), WA(3202). **10**:NY(82).
Deuteromycotina - Hyphomycetes
Alternaria alternata. Leaf spot. **9**:IL(312).
Alternaria sp. Boll rot. **9**:MN(82), ND(82), TX(82), WA(82,3202).
Aureobasidium lini. On stems. **9**:AZ(82), IA(82), MI(82), MN(82), ND(82), OR(82).
A. pullulans. **9**:OR(3202).
Bipolaris sorokiniana. Seedling blight. **9**:MN(82).
Botrytis cinerea. Boll rot. **9**:CA(82).
Cercospora lini. Leaf spot. **6**:TX(82). **8**:TX(82). **10**:DE(82,962,3543).
Curvularia geniculata. Root rot, isolated from seeds. **9**:MI(82).
C. inaequalis. **9**:IL(82).
Fusarium acuminatum. Secondary root rot. **9**:ND(82).
F. equiseti. Secondary root rot. **9**:ND(82).
F. oxysporum f. sp. *lini*. Wilt. **9**:Range of host(82), CA(1643), MT(3202), ND(378), OK(2872), OR(3202), SD(2243), WA(3202).
F. solani. **9**:WA(3202).
Helminthosporium sp. **9**:MN(82).
Penicillium sp. Secondary root rot. **9**:WA(82,3202).
Phymatotrichopsis omnivora. Root rot. **6**:TX(82). **7**:TX(82). **8**:TX(82).
Thielaviopsis basicola. Root rot. **9**:MN(82), ND(82).
Deuteromycotina - Coelomycetes
Colletotrichum lini. Anthracnose. **9**:AZ(82), CA(82), IA(82), MI(82), MN(82), MT(82,3202), ND(82), SD(82), TX(82), WI(82).
C. linicola. **9**:CA(3163), IA(3163), MN(3163), ND(3163), SD(3163), WI(3163).
Phlyctema linicola. **9**:Unk.(82), WI(1340).
Phoma fusispora. On stems. **6**:WY(3812).
P. terrestris. Secondary root rot. **9**:MN(82), ND(82), SD(82).
Septoria linicola. Pasmo disease. **9**:OK(2872), SD(1048,2243).
Deuteromycotina - Other
Rhizoctonia solani. Damping-off, root rot. **3**:IL(82). **6**:TX(82). **8**:TX(82). **9**:CA(82), IA(82), MN(82), ND(82), OK(2872), TX(82).
Sclerotium rolfsii. **9**:TX(82).
S. rolfsii var. *delphinii*. Stem rot. **9**:CT(82).

Loasaceae

Mentzelia L.

1. *M. decapetala* (Pursh ex Sims) Urb. & Gilg
2. *M. floridana* Nutt.
3. *M. laevicaulis* (Douglas ex Hook.) Torr. & A. Gray
4. *M. multicaulis* (Osterh.) Darl. (=*Touterea multicaulis* Osterh.)
5. *M. nuda* (Pursh) Torr. & A. Gray
6. *Mentzelia* sp.

Ascomycotina
Pleospora dichromatricha. On stems. **4**:CO(3817).
Basidiomycotina - Uredinales
[*Puccinia aristidae*]. **6**:AZ(82), CO(82).
Uredo floridana. **2**:FL(35,82).
Deuteromycotina - Hyphomycetes
Cercospora sp. Leaf spot. **2**:FL(35).

Phymatotrichopsis omnivora. Root rot. **3**:OK(82), TX(82). **6**:OK(82), TX(82).
Deuteromycotina - Coelomycetes
Diplodina stevensii. **1**:ND(82).
Phyllosticta mentzeliae. Leaf spot. **5**:OK(2873). **6**:KS(82), TX(82).
Phyllosticta sp. Leaf spot. **2**:FL(35).
Septoria mentzeliae. Leaf spot. **1**:KS(82), SD(2243). **3**:WA(82,3202). **6**:KS(82), TX(82).
Deuteromycotina - Other
Rhizoctonia solani. Root rot. **1**:NJ(82). **6**:NJ(82).

Loganiaceae

Buddleja L.

1. *B. asiatica* Lour.
2. *B. davidii* Franch. [orange-eye buddleja, summer-lilac]
3. *B. diversifolia* (Ten.) Vahl (=*Nicodemia diversifolia* Ten.)
4. *B. japonica* Hemsl.
5. *B. officinalis* Maxim.
6. *Buddleja* sp.

Oomycetes
Pythium sp. Root rot. **6**:FL(35).
Ascomycotina
Gloniopsis praelonga. **6**:FL(2782).
Pleospora sp. **4**:GA(1438).
Basidiomycotina - Other
Armillaria tabescens. Root rot. **1**:FL(35).
Schizopora apacheriensis. **5**:AZ(1193,1202).
Deuteromycotina - Hyphomycetes
Cladosporium herbarum. **6**:WA(3202).
C. heugelinianum. Scab. **1**:VA(82).
Fusarium sp. Root rot. **6**:FL(35).
Phymatotrichopsis omnivora. Root rot. **2**:TX(82). **6**:TX(82).
Deuteromycotina - Coelomycetes
Lasiodiplodia theobromae. Dieback, stem gall. **6**:FL(35).
Phoma sp. Stem canker. **2**:AZ(82). **6**:AZ(82).
Phomopsis buddleiae. Stem canker. **6**:NC(1303).
Deuteromycotina - Other
Rhizoctonia solani. Basal stem rot. **3**:FL(35).
Sclerotium rolfsii. Southern blight. **6**:FL(35).

Gelsemium Juss.

1. *G. sempervirens* (L.) W. T. Aiton [Carolina jasmine, evening trumpet-flower, yellow jessamine]

Ascomycotina
Botryosphaeria dothidea. **1**:GA(1438).
B. obtusa. **1**:GA(82).
B. rhodina. On stems. **1**:SC(82).
[*Calothyrium stomatophorum*]. Black spot. **1**:FL(35,82).
Nectria rubicarpa. On stems. **1**:AL(82), SC(82).
Nectriella pironii. Stem gall. **1**:FL(35).
Saccardia sp. Leaf spot. **1**:FL(35).
Trichomerium grandisporum. Sooty mold. **1**:MS(82,2720), TX(82).
Deuteromycotina - Hyphomycetes
Cercospora sp. Leaf spot. **1**:FL(35).
Phymatotrichopsis omnivora. Root rot. **1**:Unk.(82).
Deuteromycotina - Coelomycetes
Phyllosticta gelsemii. Leaf spot. **1**:NJ(82), TX(82).
Phyllosticta sp. Leaf spot. **1**:FL(35).
Sphaceloma sp. **1**:FL(2525).
Deuteromycotina - Other
Rhizoctonia solani. Root rot. **1**:FL(35).

Labordia Gaud. in Freyc.

1. *Labordia* sp.

Ascomycotina
Amazonia psychotriae. **1**:HI(2903).
A. psychotriae var. *labordiae*. Black mildew. **1**:HI(1272,1452,2903).

Mitreola L.

1. *M. petiolata* Torr. & A. Gray

Deuteromycotina - Hyphomycetes
Cercospora torta. Leaf spot. **1**:MS(2720).

Spigelia L.

1. *S. anthelmia* L.
2. *S. marilandica* L. [Indian pink]

Deuteromycotina - Hyphomycetes
Cercospora sp. Leaf spot. **1**:FL(35).
Oidium sp. Powdery mildew. **2**:IL(312).

Lycopodiaceae

Lycopodium L.

1. *L. annotinum* L. [bristly club-moss, interrupted club-moss, stiff club-moss]
2. *L.* × *buttersii* Abbe
3. *L. clavatum* L. [buckhorn, common club-moss, coral-evergreen, running club-moss, staghorn-evergreen, wolf's-claws]
4. *L. complanatum* L. [Christmas-green, ground-cedar, running-evergreen, trailing-evergreen]
5. *L. complanatum* L. var. *flabelliforme* Fernald
6. *L. lucidulum* Michx. [shining club-moss]
7. *L. obscurum* L. [bunch-evergreen, flat-branch ground-pine, tree club-moss]
8. *L. obscurum* L. var. *dendroideum* (Michx.) D. C. Eaton [princess-pine, round branch ground-pine]
9. *L. selago* L.
10. *L. selago* L. var. *appressum* Desv.
11. *L. tristachyum* Pursh [ground-cedar, ground-pine]
12. *Lycopodium* sp.

Oomycetes
Phytophthora cinnamomi. **12**:HI(2903).
Ascomycotina
Mollisia lycopodii. On strobili. **3**:MA(656). **7**:MA(656).
Mycosphaerella lycopodii. On dead leaves and bracts of inflorescences, on scales of dead spikes, overwintering on sporophylls. **3**:NY(195). **8**:WI(1340). **10**:AK(1938). **12**:MI(179), NY(179).
M. polypodii. **4**:MD(651).
Phacidina gracilis. **7**:MA(656). **11**:MA(656).
Phaeosphaeria fuckelii. On strobili. **1**:MI(2862). **3**:NY(195). **5**:MI(2862).
P. lycopodina. On strobili. **1**:CO(649), MI(2862). **10**:AK(1938).
P. marcyensis. On sporophylls. **1**:NY(195). **9**:NY(195).
Pleospora heleocharidis var. *arctica*. On dead stems, leaves and bracts of inflorescences. **10**:AK(1938).
Pseudomassaria lycopodina. On overwintered leaves. **1**:MI(175). **3**:MA(656). **4**:MA(175,656). **6**:NY(175). **7**:MA(175,656), ME(175). **8**:NY(175). **2**:NY(175).
Basidiomycotina - Other
Tomentella sublilacina. **12**:NY(2039).
Deuteromycotina - Coelomycetes
Ceuthospora lycopodii. On dead leaves. **3**:MA(656). **4**:MA(656), MD(651). **7**:MD(651). **2**:NY(128).
Strasseria geniculata. **4**:MA(2546).
S. nigra. **3**:MA(656). **4**:MA(656).

Lythraceae

Ammannia L.

1. *A. coccinea* Rottb.

Deuteromycotina - Hyphomycetes
Cercospora ammanniae. On living leaves. **1**:TX(3597).

Cuphea P. Br.

1. *C. hyssopifolia* H. B. K. [elfin-herb, false heather]
2. *C. ignea* A. DC. (=*C. platycentra* Lem.) [cigar-flower, firecracker-plant, red-white-and-blue-flower]
3. *C. viscosissima* Jacq. (=*C. petiolata* (L.) Koehne) [blue waxweed, clammy cuphea, tarweed]
4. *Cuphea* sp.

Oomycetes
Phytophthora sp. Root rot. **1**:FL(35). **4**:FL(35).
Pythium sp. Cutting and root rot. **1**:FL(35).
Ascomycotina
Erysiphe polygoni. Powdery mildew. **3**:MD(82), VA(82).
Nectriella pironii. Stem gall. **1**:FL(35). **4**:FL(35).

Physalospora sp. Leaf spot. **1**:FL(35).
Basidiomycotina - Other
Armillaria tabescens. Root rot. **4**:FL(35).
Deuteromycotina - Hyphomycetes
Botrytis cinerea. Gray mold. **2**:Unk.(82).
Cristulariella moricola. On leaves. **3**:IL(2572).
Deuteromycotina - Coelomycetes
Phyllosticta sp. Leaf spot. **1**:FL(35).
Septoria maculifera. Leaf spot. **3**:NY(82), PA(82), VA(82), WV(82).
Deuteromycotina - Other
Rhizoctonia solani. Root and stem rot. **2**:IL(82). **4**:FL(35).

Decodon J. F. Gmelin

1. *D. verticillatus* (L.) Elliott (=*Nesaea verticillata* (L.) H. B. K.) [swamp loosestrife, water-oleander, water-willow]

Basidiomycotina - Uredinales
Puccinia minutissima. Heteroecious: aecial host. **1**:NE states(82), FL(35,82), IN(82), PA(1875), WI(82,1348).
Deuteromycotina - Hyphomycetes
Cercospora nesaeae. Leaf spot. **1**:DE(82,3543), FL(35), IL(82), NJ(82).
Deuteromycotina - Coelomycetes
Ascochyta sp. Leaf spot. **1**:WI(1361).
Phyllosticta nesaeae. Leaf spot. **1**:FL(35), NJ(82), NY(82), OH(82), VT(82).
Septoria sp. **1**:WI(1361).

Ginoria Jacq.

1. *G. glabra* Griseb.

Basidiomycotina - Uredinales
Uredo cupheae. **1**:FL(35).

Lagerstroemia L.

1. *L. indica* L. [crape-myrtle]
2. *L. parviflora* Roxb.
3. *Lagerstroemia* sp.

Oomycetes
Phytophthora sp. Crown and root rot. **1**:FL(35).
Pythium sp. Root rot. **1**:FL(35).
Ascomycotina
Botryosphaeria obtusa. On limbs. **1**:AL(82), FL(82).
B. quercuum. Dieback. **1**:FL(35).
B. ribis. On limbs. **1**:AL(82).
Botryosphaeria sp. Leaf spot. **3**:FL(35).
Erysiphe graminis f. sp. *avenae*. Powdery mildew. **1**:AR(3029).
E. lagerstroemiae. Powdery mildew. **1**:E states(82), FL(35), GA(1438,2405), MS(2720), NC(1303,1533), OK(82,2872), TX(82). **3**:FL(35).
Phyllactinia guttata. Powdery mildew. **1**:AL(82).
Uncinuliella australiana. Powdery mildew. **1**:GA(1438), HI(82), LA(82). **2**:GA(1438). **3**:AL(285).
Basidiomycotina - Other
Armillaria mellea. Root rot. **1**:CA(2900).
A. tabescens. Root rot. **1**:FL(35,82).
Dacryopinax spathularia. On stems. **1**:GA(2642).
Hyphoderma rimosum. White rot. **1**:MS(444).
Pellicularia koleroga. **1**:LA(3640).
Platygloea caroliniana. On dead limbs. **3**:NC(155).
P. lagerstroemiae. On dead limbs. **3**:NC(155).
Trametes versicolor. On wood. **1**:NC(1303).
Deuteromycotina - Hyphomycetes
Cercospora lythracearum. Leaf spot. **1**:FL(35), TX(82). **3**:FL(35).
Cercospora sp. Blotch, black spot. **1**:FL(35,82), TX(82).
Cylindrocladium scoparium. Leaf spot. **1**:FL(35).
Fusarium oxysporum. Wilt, root rot. **1**:FL(35).
F. solani. Dieback. **1**:FL(35).
Oidium sp. Powdery mildew. **1**:HI(2903).
Phymatotrichopsis omnivora. Root rot. **1**:TX(82).
Deuteromycotina - Coelomycetes
Pestalotia sp. Leaf tip blight. **1**:FL(35).
Phoma lagerstroemiae var. *foliicola*. On leaves. **1**:LA(965).
Phomopsis sp. Leaf spot. **1**:FL(35).
Phyllosticta lagerstroemiae. Leaf spot. **1**:FL(35), LA(82,965), OK(2872), TX(82). **3**:FL(35).
Sphaeropsis sp. Canker. **1**:FL(35).
[*Titaeospora equiseti*]. Leaf tip necrosis. **1**:FL(35).
Deuteromycotina - Other
Rhizoctonia ramicola. Thread blight. **1**:FL(35,82).
R. solani. Root rot. **1**:FL(35).

Lythrum L.

1. *L. alatum* Pursh [winged loosestrife]
2. *L. lineare* L.
3. *L. salicaria* L. [purple loosestrife, spiked loosestrife]
4. *Lythrum* sp.

Chytridiomycetes
Synchytrium lythrii. Leaf gall. **1**:LA(82).
Ascomycotina
Discohainesia oenotherae. Leaf spot. **3**:NY(82). **4**:MI(82).
Leptosphaeria lythri. On stems. **1**:KS(195).
Ophiobolus niesslii. **3**:Unk.(3248).
Deuteromycotina - Hyphomycetes
Stenella lythri. Leaf spot. **1**:WI(82,1334).
Deuteromycotina - Coelomycetes
Septoria lythrina. Leaf spot. **1**:KS(82), NY(82), WI(82,1334). **2**:FL(35). **3**:KS(82), NY(82), WI(1357).
Deuteromycotina - Other
Rhizoctonia solani. Root rot. **4**:IL(82).

Magnoliaceae

Liriodendron L.

1. *L. tulipifera* L. [tulip-poplar, tuliptree, whitewood, yellow-poplar]
2. *Liriodendron* sp.

Ascomycotina
Angelina rufescens. **1**:GA(82,1438).
Anthostomella sp. **1**:GA(82,1438).
Bertiella rhodospila. **1**:GA(1438).
Botryosphaeria dothidea. Dieback. **1**:FL(35), GA(1438).
B. obtusa. On limbs. **1**:SE states(82), TN(82).
B. ribis. On limbs. **1**:FL(82), GA(82), PA(82).
Calospora sp. **1**:GA(82,1438).
Camarops polysperma. **1**:GA(82,1438).
C. pugillus. **2**:VA(2561).
Capnodium elongatum. Sooty mold. **1**:Range of host(82), NC(1303).
Ceratocystis coerulescens. Wood stain. **1**:SE states(82,779), Unk.(82), NC(82), TN(1533).
Ceratostomella sp. Blue wood stain. **1**:SE states(82).
Chlorociboria aeruginosa. Green wood stain. **1**:Range of host(82).
Dactylospora stygia. **1**:GA(1438).
Daldinia vernicosa. Wood rot. **1**:Range of host(82).
Diaporthe eres. On dead limbs. **1**:MD(82).
Diaporthe sp. **1**:GA(1438).
Discohainesia oenotherae. **1**:GA(82,1438).
Erysiphe polygoni. Powdery mildew. **1**:Range of host(82), FL(35), MS(2720), NC(1303,3944).
Eutypa lata. **1**:GA(1438).
Eutypella goniostoma. **2**:SC(2925).
E. quaternata. **1**:GA(1438).
E. stellulata. On dead twigs. **1**:NC(1303).
Hypocrea patella. **1**:GA(1438).
Hypoxylon caries. On dead limbs. **1**:GA(2786), NC(1303).
H. croceum. On wood. **1**:GA(1438). **2**:GA(1438).
H. howeianum. On dead limbs. **1**:NC(1303).
H. hypophlaeum. **1**:GA(1438).
H. rubiginosum. **1**:NC(1307).
H. serpens. **1**:GA(1438).
H. tinctor. **1**:GA(1438).
Hypoxylon sp. **1**:GA(82).
Lasiosphaeria hispida. On decorticated branches. **1**:NC(1303).
L. ovina. On wood. **1**:GA(82,1438).
L. strigosa. **1**:GA(82,1438).
Lecanidion atratum. On scale insects. **1**:GA(82,1438).
Lophodermium punctiforme. **1**:GA(82,1438).
Lulworthia attenuata. On wood submerged in sea water. **1**:NC(1791).
Mycosphaerella infuscans. **1**:WV(179).
M. liriodendri. Leaf spot. **1**:FL(35), GA(82), MI(82), NY(82).
M. punctiformis. **1**:GA(1438).
M. tulipifera. Leaf spot. **1**:SE states(82), FL(35), GA(1438), MS(2720),

NC(1303).
Mycosphaerella sp. **1**:GA(1438).
Nectria magnoliae. Canker, on bark and limbs. **1**:E states(82), NC(1303,1533), OH(82), TN(82).
N. ochroleuca. On bark. **1**:GA(1438). **2**:SC(3085).
Nectria sp. Canker. **1**:SE states(82), TN(82).
Ophiostoma multiannulatum. Blue wood stain. **1**:SE states(82).
O. pluriannulatum. Blue stain of wood. **1**:SE states(82), DC(3714).
Patellariopsis clavispora. On scale insects. **1**:GA(82,1438).
Propolomyces versicolor. On logs. **1**:GA(1438), NC(1303,1307).
Rhytisma liriodendri. Anthracnose. **1**:CA(82), FL(35), TX(82), VA(3493).
Rosellinia obtusissima. **1**:GA(1438).
R. subiculata. **1**:GA(1438).
Stictis radiata. On dead twigs. **1**:NC(1303,1307).
Tubeufia pezizula. Wood stain. **1**:SC(82).
Ustulina deusta. On dead limbs. **1**:NC(1303).
Valsa ceratosperma. **1**:GA(1438).
Valsaria rubricosa. **1**:GA(1438).
Valsella laschii. **1**:GA(1438).
Venturia liriodendri. On living leaves. **1**:GA(1444).

Basidiomycotina - Other

Aporpium caryae. **1**:Lake States(82).
Armillaria mellea. Root rot, isolated from wood. **1**:CA(2900), NC(3944), NY(59), VA(82,3493), WV(1240).
Athelia scutellaris. On wood. **1**:GA(491,1439).
Bjerkandera adusta. White rot. **1**:Unk.(82,1006), GA(1439), NC(1303).
B. fumosa. On wood. **1**:GA(491,1439).
Candelabrochaete septocystidia. White rot. **1**:NC(432,1303).
Cerrena unicolor. Wood rot. **1**:Unk.(82).
Chondrostereum purpureum. **1**:MD(543), WV(543).
Coniophora puteana. Isolated from mine timbers. **1**:Unk.(1006).
Coprinus micaceus. Butt and root rot of living trees. **1**:VA(3717).
Corticium roseum. On wood. **1**:GA(491,1439).
Cylindrobasidium torrendii. On wood. **1**:Unk.(1006), GA(491,1439).
Cyphellopsis anomala. **1**:GA(1439).
Cystostereum pini-canadense. Isolated from mine timbers. **1**:Unk.(1006).
Dacryopinax spathularia. On wood. **1**:MD(128).
Daedaleopsis confragosa. On wood. **1**:GA(1439), NC(1303).
Dendrocorticium roseocarneum. On wood. **1**:GA(491,1439).
Dendrophora albobadia. **1**:DC(543), MD(543).
D. versiformis. **1**:MD(543).
Dendrothele griseo-cana. On bark. **1**:NY(2084).
Dentocorticium sulphurellum. White rot. **1**:NC(2043).
Ditiola pezizaeformis. On limbs. **1**:GA(2642).
Exidiopsis grisea. On limbs. **1**:GA(2642).
Flammulina velutipes. White spongy trunk rot. **1**:WV(82).
Ganoderma applanatum. White-mottled heart rot. **1**:Unk.(82).
Globifomes graveolens. Wood rot. **1**:Unk.(82).
Gloeophyllum sepiarium. Wood rot of logs and timber. **1**:Range of host(82).
Gloeoporus dichrous. White rot. **1**:Unk.(82), NC(1303,1307), SC(1307).
Hericium erinaceus. Usually on living trees. **1**:VA(82).
Hyphoderma mutatum. **1**:PA(2321).
Hyphodontia lanata. **1**:MS(443).
Hypholoma fasciculare. On decaying logs. **1**:NC(1303).
Hypholoma sp. Butt rot. **1**:VA(82).
Hypsizygus tessellatus. On wood. **1**:SE states(2421), MD(2421).
H. ulmarius. **1**:NC(3944).
Irpex lacteus. Wood rot. **1**:Unk.(82), GA(1439), NC(1303,3944).
Ischnoderma resinosum. Wood rot. **1**:Unk.(82).
Junghuhnia nitida. **1**:VA(128,2168).
Laetiporus sulphureus. Wood rot. **1**:Unk.(82).
Laxitextum bicolor. On wood. **1**:CT(543), GA(491,1439), MD(543), MS(543), NY(543), TN(543).
Lentinus strigosus. On wood. **1**:GA(1439), NC(435).
Lenzites betulina. Wood rot. **1**:Range of host(82), GA(1439), NC(1303,1307).
Leucogyrophana pinastri. On wood. **1**:NC(128,1303).
Mycena setulosa. On bark. **1**:TN(3326).
Peniophora cinerea. On wood. **1**:GA(491,1439).
Perenniporia robiniophila. On trunks. **2**:Unk.(2681).
P. subacida. On trunks. **1**:VA(638).
P. tenuis. Wood rot. **1**:Unk.(82).
Phanerochaete chrysosporium. **1**:MD(433,435).
P. crassa. Wood rot of logs. **1**:Unk.(82).
P. ericina. White rot. **1**:TN(433).
P. flavido-alba. **2**:GA(128).
P. sordida. White rot. **1**:TN(433).
P. velutina. White rot. **1**:TN(433).
Phellinus ferrugineo-velutinus. **1**:GA(1439).
P. igniarius. **1**:GA(1439).
P. punctatus. On wood. **1**:GA(491,1439).
Phlebia chrysocrea. **1**:SC(2165).
P. rufa. **1**:VA(128). **2**:VA(128).
Phyllotopsis nidulans. On logs and stumps. **1**:NC(1303).
Pleurotus ostreatus. Wood decay, sapwood rot. **1**:Unk.(82), GA(1439), NC(1303,1307,3944), OH(63).
Polyporus varius. **1**:NC(1307).
Postia caesia. Wood rot. **1**:Unk.(82).
Pulcherricium caeruleum. On wood. **1**:GA(491,1439), NC(1303,1307).
Pycnoporus cinnabarinus. On wood. **1**:Unk.(82), WV(2319).
Rhodotus palmatus. On wood. **1**:MD(2422).
Rigidoporus lineatus. Wood rot. **1**:Unk.(82).
Schizophyllum commune. Sapwood rot. **1**:Range of host(82).
Sebacina sublilacina. On limbs. **1**:GA(2642).
Sistotrema brinkmannii. On wood. **1**:GA(491,1439).
Steccherinum fimbriatum. On wood. **1**:GA(491,1439).
S. vagum. **1**:TN(443).
Stereum hirsutum. On wood. **1**:Unk.(82), GA(1439).
S. ostrea. **1**:GA(1439).
Subulicystidium longisporum. On wood. **1**:GA(491,1439).
Tomentella brunneorufa. On wood. **1**:MD(2039).
T. ferruginea. **2**:MD(2039).
T. griseoviolacea. **2**:MD(2039).
T. pilosa. **2**:NC(2039).
T. rubiginosa. **2**:MD(2039).
Trametes hirsuta. Wood rot. **1**:Unk.(82), GA(1439), NC(1307,3944), SC(1307).
T. pubescens. Wood rot. **1**:Unk.(82).
T. versicolor. Wood rot. **1**:Unk.(82), GA(1439), NC(1307,3944), SC(1307).
Tulasnella pinicola. On wood. **1**:GA(2642).
T. violea. On bark. **1**:CT(128).
Tyromyces chioneus. Wood rot. **1**:Unk.(82).
Wolfiporia cocos. Wood rot. **1**:Unk.(82), WV(1241).

Deuteromycotina - Hyphomycetes

Berkleasmium concinnum. On decorticated wood. **1**:NC(1303,1307), SC(1307).
Cladosporium cladosporioides. Canker. **1**:NC(1792).
Corynespora olivacea. **1**:PA(3547).
Cylindrocarpon liriodendri. Root rot. **1**:CA(2222).
Cylindrocladium scoparium. Collar rot of seedlings, black root rot. **1**:GA(3034), NC(676,1303), NJ(82), WV(1694).
Ectostroma liriodendri. Tar spot. **1**:Range of host(82), FL(35).
Fusarium solani. Canker. **1**:OH(67), SC(2154).
Fusarium sp. Damping-off, root rot. **1**:FL(35), NC(1533), WV(1240).
Graphium rubrum. Pink wood stain. **1**:SE states(82), MS(2720), SC(82).
Helminthosporium macrocarpum. **1**:DE(3543).
Isariopsis sp. Leaf spot. **1**:FL(35).
Margarinomyces sp. Isolated from wood. **1**:WV(1240).
Oidium sp. Powdery mildew. **1**:FL(35).
Olpitrichum macrosporum. **1**:KS(2131).
Oospora tulipiferae. **1**:NC(3944).
Periconia byssoides. On dead twigs. **1**:NC(1303).
Phaeoisariopsis liriodendri. Leaf spot. **1**:Unk.(82), AL(2496), FL(35), NC(1303,3944).
Phymatotrichopsis omnivora. Root rot. **1**:TX(82).
Ramularia liriodendri. Leaf spot. **1**:AL(82), DE(82).
Stigmina liriodendri. Leaf spot. **1**:MS(2720).
[*Torula ligniperda*]. Pink wood stain. **1**:Range of host(82).
Trichoderma viride. Isolated from wood. **1**:WV(1240).
Umbelopsis versiformis. Isolated from roots of living tree. **1**:WV(51).
Verticillium albo-atrum. Wilt. **1**:CT(1558), IL(992,1558), WV(1558).

Deuteromycotina - Coelomycetes

Asteroma liriodendri. On leaves. **1**:MD(82), SC(82).

Colletotrichum gloeosporioides. Anthracnose, leaf spot. **1**:CT(82), DE(963), NC(1303), NJ(82), NY(82), PA(82), TX(82).
Cylindrosporium cercosporoides. Leaf spot. **1**:DC(525,962), MD(82), WV(82).
Cytospora leucostoma. On limbs. **1**:MD(82), VA(82).
Diplodia liriodendri. On limbs. **1**:NY(82), WV(82).
Dothiorella minor. On limbs. **1**:MD(82), WV(82).
Dothiorella sp. Canker. **1**:PA(82).
Gloeodes pomigena. On limbs. **1**:IN(82), WV(82).
Leptothyrium liriodendri. On leaves. **1**:SC(82), WV(82).
Myxosporium coloratum. On twigs. **1**:NY(82).
M. liriodendri. On twigs. **1**:NY(82).
M. longisporum. On twigs. **1**:NY(82).
M. tulipiferae. On twigs. **1**:IA(82).
Myxosporium sp. Dieback. **1**:MS(3666), NC(1792), NY(82).
Pestalotia sp. Canker. **1**:NC(1792).
Phoma tulipiferae. On leaves. **1**:NY(82), PA(82).
Phyllosticta liriodendrica. Leaf spot. **1**:Range of host(82), FL(35), MS(2720), NC(3944), PA(82), VA(82), WV(82).
Septoria sp. **1**:NC(1303).

Deuteromycotina - Other

Rhizoctonia solani. Seedling blight. **1**:OH(82), VA(82).

Magnolia L.

1. *M. acuminata* (L.) L. [cucumbertree]
2. *M. cordata* Michx. [yellow cucumber-tree]
3. *M. fraseri* Walter [ear-leaved umbrella-tree, Fraser magnolia, mountain magnolia]
4. *M. grandiflora* L. [bull bay, southern magnolia]
5. *M. macrophylla* Michx. [great-leaved magnolia, large-leaved cucumber-tree]
6. *M. obovata* Thunb. (=*M. liliflora* Desr.)
7. *M.* × *soulangiana* (Buchoz) Dandy. *M. heptapeta* × *M. quinquepeta*. [Chinese magnolia, saucer magnolia]
8. *M. stellata* (Siebold & Zucc.) Maxim. [star magnolia]
9. *M. tripetala* (L.) L. [umbrella magnolia, umbrella-tree]
10. *M. virginiana* L. (=*M. glauca* L.) [sweet bay, sweetbay magnolia]
11. *Magnolia* sp.

Oomycetes

Phytophthora sp. Root rot. **4**:FL(35). **7**:FL(35). **11**:FL(35).
Pythium splendens. Root rot. **4**:FL(35).
Pythium sp. Root and stem rot. **10**:FL(35). **7**:FL(35).

Ascomycotina

Acanthonitschkea tristis. **4**:FL(2782).
Acanthostigma berenice. Leaf spot. **11**:SE states(82), FL(35), LA(82).
Anthostomella unguiculata. **11**:LA(1098).
[*Appendiculella araliae*]. Black mildew, sooty mold. **10**:MS(82,2720).
Biscogniauxia marginata. On limbs. **11**:Unk.(82).
Botryosphaeria disrupta. On limbs. **11**:FL(35).
B. dothidea. Stem necrosis. **4**:FL(35). **5**:GA(1438). **10**:GA(1438). **11**:FL(35).
B. obtusa. **11**:SE states(82).
Camarops polysperma. **11**:FL(2561).
Capnodium pelliculosum. **10**:GA(1438).
Ceratocystis coerulescens. Wood stain. **11**:SE states(779).
Ciborinia gracilipes. Petal rot. **10**:MD(82), NJ(202).
Coniochaeta pulveracea. **2**:GA(1438).
Cucurbitaria congesta. On dead limbs. **11**:NJ(82).
Daldinia concentrica. Wood rot. **11**:Range of host(82).
D. vernicosa. On wood. **11**:Range of host(82).
Dasyscyphus albopileatus. **10**:GA(1438), NJ(849).
D. hystriculus. On leaves. **4**:LA(849,965).
Diaporthe arctii. **9**:GA(1438).
D. binoculata. On limbs. **1**:NY(195). **10**:NC(82), NJ(82). **11**:NC(82), NJ(82), NY(82), WV(82).
D. eres. On limbs. **10**:GA(82,1438).
D. magnoliae. On limbs. **10**:FL(2782). **11**:AL(82), MD(82), NJ(82), NY(82).
Diaporthe sp. **11**:WA(82,3202).
Diatrype asterostoma. **10**:GA(1438).
D. disciformis. **11**:SC(634).
D. durieui. On limbs. **10**:SC(634).
D. stigma. **10**:GA(1438).
D. tremellophora. **10**:GA(1438).
Diatrypella favacea. **10**:GA(1438). **7**:GA(1438).
D. nitschkei. **10**:FL(2782).
D. quercina. **10**:GA(1438).
Didymosphaeria magnoliae. On dead leaves. **9**:GA(1438). **10**:GA(82,1438).
Dimerium piceum. **10**:GA(82,1438).
Discohainesia oenotherae. On limbs. **1**:NC(82).
Dothidasteromella magnoliae. Sooty mold. **4**:TX(82). **10**:MS(2720). **11**:TX(82).
Elsinoe magnoliae. Spot anthracnose. **4**:FL(35,2407,2524), GA(1438,2405,2407), LA(2407), MS(2407). **11**:NC(1303).
Eutypella goniostoma. **10**:SC(2925).
E. juglandicola. **4**:GA(1438).
E. leprosa. **1**:IA(3629).
E. scoparia. **1**:IA(3629).
Glomerella cingulata. Anthracnose, leaf spot. **4**:AL(82), FL(35,82), GA(82,2405), LA(82), MS(82), SC(82), TX(82). **11**:FL(35), LA(658), NC(1303).
Hypoderma commune. **9**:GA(1438).
Hyponectria magnoliae. Leaf spot. **4**:E states(183), GA(1438). **10**:E states(183), FL(35,82), GA(82,1438), NJ(82).
Hypoxylon hypophlaeum. **10**:GA(1438).
H. investiens var. *epiphaeum*. **10**:GA(1438).
H. microplacum. **10**:GA(1438).
H. rubiginosum. **10**:GA(1438).
[*H. sassafras*]. **10**:GA(1438).
H. serpens var. *effusum*. **11**:TX(652).
H. tinctor. **10**:GA(1438).
H. truncatum. **10**:GA(1438). **11**:FL(634).
Hypoxylon sp. Sapwood rot. **4**:FL(35).
Lophodermium maculare. Leaf spot. **11**:FL(35,82).
Massaria shearii. **11**:GA(2779).
Meliola amphitricha. Black mildew on leaves. **4**:Gulf States(82), FL(634), MS(2720). **10**:Gulf States(82), GA(1438), MS(2720). **11**:Gulf States(82).
M. magnoliae. Sooty mold. **4**:MS(2720). **10**:GA(82,1438). **11**:Gulf States(82).
M. nashii. Black mildew. **4**:AL(1452), FL(1452), LA(1452). **10**:FL(1452), MI(1452), MS(1452). **11**:FL(1452).
Micropeltis alabamensis. Black leaf spot. **10**:AL(82).
Microsphaera magnifica. **1**:PA(366).
M. penicillata. Powdery mildew. **1**:MD(82), PA(82). **4**:FL(35). **6**:Unk.(3976), AL(82). **7**:Unk.(82), FL(35), IL(1052). **11**:FL(35).
Microthyrium microscopicum. **4**:GA(1438).
Mycosphaerella annulata. Leaf spot. **10**:FL(35,82), LA(82), MS(82,2720). **11**:SC(634).
M. glauca. Leaf spot. **10**:FL(179), MS(82,2720), NJ(179), SC(82,634).
M. punctiformis. **3**:GA(1438). **9**:GA(1438).
Mycosphaerella sp. **2**:GA(1438). **10**:GA(82).
Naevia sp. **5**:GA(1438).
Nectria aureo-fulva. On bark. **11**:NJ(3085).
N. cinnabarina. On limbs. **11**:NY(82), WV(82).
N. magnoliae. Canker. **3**:TN(82), WV(82). **11**:NJ(82), WV(82).
N. ochroleuca. On old flower. **11**:OH(3085).
Nectria sp. Canker. **1**:WV(82).
Nummularia broomeiana. On trunks and larger limbs. **11**:SE states(1810).
Ophiodothella sp. **4**:GA(1438). **10**:GA(1438).
Phyllactinia guttata. Powdery mildew. **1**:OH(82), PA(82). **3**:Unk.(82).
Physalospora abdita. On limbs. **11**:FL(82).
Plagiostoma magnoliae. On fallen leaves. **10**:GA(82,1438), NJ(82).
Pleomassaria maxima. On limbs. **1**:PA(188), VA(188), WV(188). **6**:DC(188,3209), MD(188). **11**:LA(188), MA(188).
Rhytidhysteron rufulum. On twigs. **11**:Gulf States(82).
Rosellinia aquila. **10**:GA(1438).
Scirrhia concaviuscula. On dead limbs. **10**:NJ(82).
Splanchnonema clandestinum. On twigs. **1**:PA(188).
Thyridaria texensis. **11**:FL(2782).
Trichodothis comata. Black mildew. **4**:S states(82), GA(1438), MS(2720). **10**:GA(1438), MS(2720). **11**:Gulf States(82), FL(634,2782).
Urosporella magnoliae. On dead limbs. **3**:WV(176).
Valsa ambiens. On limbs. **10**:NY(82).
V. ceratosperma. **3**:WV(3415).
Valsaria insitiva. **10**:GA(1438).
V. rubricosa. **10**:GA(1438).

Venturia applanata. Leaf spot. **10**:FL(35,82).
Xylaria carpophila. On cones. **3**:NC(1307). **4**:NC(1303). **11**:LA(2182).
X. cubensis. On wood. **11**:NJ(3016).
X. ianthino-velutina. On cones. **4**:GA(1438). **11**:LA(2182), TX(652).
X. magnoliae. On fruits. **4**:FL(3014), GA(539), NC(3014). **11**:GA(659), LA(658,3014), WV(3014).

Basidiomycotina - Other

Abortiporus biennis. On wood. **11**:Unk.(82).
Antrodia sinuosa. Wood rot. **11**:Unk.(82).
Armillaria mellea. Root rot. **11**:CA(2898), FL(35).
A. tabescens. Root rot. **4**:FL(35), GA(2684).
Basidiodendron grandinioides. On wood. **11**:NJ(2185).
Campanella floridana. **11**:FL(2542).
Ceraceomyces serpens. **11**:FL(128).
Cerinomyces ceraceus. On wood. **4**:MS(1235,3176).
Crepidotus praelatifolius. **11**:FL(1544).
Cylindrobasidium laeve. **10**:NJ(128).
Dendrophora albobadia. **1**:TN(543). **11**:VA(543).
Exidia nucleata. **3**:WV(128).
E. tremelloides. On corticate limbs. **11**:LA(2643).
Fomes fasciatus. Heart rot of living trees. **4**:Unk.(82).
Fomitopsis pinicola. Wood rot. **11**:Unk.(82).
Ganoderma applanatum. Wood rot. **11**:MD(82).
Hydropus serifluus. On rotten wood. **11**:FL(3289,3326).
Hymenochaete corrugata. **11**:TX(652).
H. corticolor. On bark. **11**:FL(634).
Hyphoderma setigerum. **11**:FL(634).
Irpex lacteus. Wood rot. **11**:Unk.(82).
Laxitextum bicolor. **10**:DE(543).
Lazulinospora cinnamomea. **11**:FL(443).
Licrostroma subgiganteum. **11**:NJ(2084).
Marasmius magnoliae. On petioles and fallen leaves. **4**:FL(3284).
Meruliporia incrassata. Dry rot of timber. **11**:Unk.(82).
Oxyporus latemarginatus. On wood. **11**:Unk.(82).
Panellus pusillus. On wood. **11**:FL(441).
Pellicularia koleroga. Thread blight. **4**:SC(82).
Peniophora incarnata. On wood. **7**:MD(128).
Perenniporia ohiensis. Wood rot. **11**:VA(82).
Phanerochaete flavido-alba. **11**:NJ(128).
P. velutina. **11**:NJ(128).
Phellinus crocatus. Wood rot. **11**:Unk.(82).
Pleurotus ostreatus. Trunk rot. **4**:FL(35).
Polyporus squamosus. Wood rot. **11**:Unk.(82).
Pycnoporus cinnabarinus. Wood rot. **11**:Unk.(82).
Rigidoporus ulmarius. Butt and root rot. **4**:FL(35), LA(82), MS(2166), NY(82). **11**:SC states(82), FL(35,82), LA(2675), MS(2675), NJ(82).
R. vinctus. **11**:FL(3187).
Schizopora paradoxa. Wood rot. **11**:Unk.(82).
Septobasidium castaneum. Parasitic on scale insects. **10**:MS(2720).
S. pseudopedicellatum. Parasitic on scale insects. **10**:MS(2720).
Spongipellis spumeus. Wood rot. **11**:Unk.(82).
Stereum hirsutum. **11**:Unk.(82).
Strobilurus conigenoides. On cones and inflorescences. **3**:WV(128,959). **4**:GA(128). **11**:TX(652).
Tomentella muricata. On wood. **11**:FL(2039).
T. ruttneri. On wood. **11**:NJ(3746).
T. violaceofusca. **11**:MA(2039).
Trametes elegans. **11**:AL(82).
T. hirsuta. Wood rot. **11**:Unk.(82).
T. pubescens. On wood. **11**:Unk.(82).
Trechispora mollusca. Wood rot. **11**:Unk.(82).
Tyromyces calkinsii. On wood. **11**:FL(2681).
Volvariella bombycina var. *flaviceps*. On wood. **11**:FL(3192).
Wolfiporia cocos. Tuckahoe, on roots. **11**:FL(35,82).

Deuteromycotina - Hyphomycetes

Alternaria alternata. Leaf spot. **11**:TX(82).
Alternaria sp. Leaf spot. **4**:FL(35). **7**:FL(35). **11**:FL(35).
Botrytis cinerea. **11**:WA(3202).
Cercospora sp. Leaf spot. **7**:FL(35). **11**:NC(1303).
Cercosporidium magnoliae. Angular leaf spot. **4**:FL(35,82), LA(1600,2834), NC(1303), SC(1600), VA(82,1600). **7**:FL(35). **11**:FL(35).
Cladosporium herbarum. On leaves. **1**:OK(2872). **4**:SC states(82), SE states(82). **10**:MD(82), NJ(82), NY(82). **11**:Range of host(82).
Cristulariella moricola. Zonate leaf spot. **3**:NC(1303,1307), WV(815,816). **9**:WV(3680).
Cylindrocladium scoparium. Leaf spot. **4**:FL(32,35).
Dicyma pulvinata. On leaves. **11**:LA(658).
Epicoccum nigrum. Leaf spot. **4**:TX(82).
Eversia subopaca. On bark. **10**:NJ(3149).
Fusarium sp. Cutting rot. **7**:FL(35).
Helicoma phaeosporium. On wood. **11**:NJ(1271).
Helminthosporium macrocarpum. **10**:DE(3543).
Isariopsis sp. Angular leaf spot. **11**:NC(1303).
Johncouchia mangiferae. Parasitic on scale insects. **10**:LA(1683).
Oidium laevisporum. On bark and logs. **11**:FL(634,2131).
Periconia byssoides. On leaves. **11**:LA(658).
Phialocephala repens. On rotten wood. **11**:NJ(959).
Phymatotrichopsis omnivora. Root rot. **4**:TX(82).
Polyscytalum sericeum. On dead leaves. **11**:FL(634).
Spadicoides obovata. **11**:NJ(1675).
Sporidesmium ellipticum. On leaves. **10**:SC(82).
S. punctiphyllum. On leaves. **11**:SC(634).
S. rude. On leaves. **10**:FL(2782), NJ(82).
S. stictophyllum. On leaves. **4**:SC(82).
Stenella sp. On leaves. **11**:LA(658).
Stenellopsis magnoliae. Leaf spot. **4**:AL(2495), FL(35,82), NC(82), TX(82). **11**:FL(35).
Tubercularia sp. Twig blight. **7**:MI(82).
Verticillium albo-atrum. Wilt. **4**:CT(2296). **8**:IL(1558). **7**:IN(1558,3153). **11**:IL(992).
Xylocoremium flabelliforme. On wood. **11**:NJ(3016).
Zygosporium masonii. On leaves. **11**:LA(658).

Deuteromycotina - Coelomycetes

Ascochyta sp. Leaf spot. **4**:FL(35).
Colletotrichum gloeosporioides. Anthracnose. **4**:NC(1303).
Colletotrichum sp. Leaf spot, on twigs. **4**:FL(35,82), GA(82), SC(82). **10**:FL(35). **11**:E states(82), FL(35), NC(1303).
Coniothyrium sp. Leaf spot. **11**:NC(1303), TX(82).
Cytospora sacculus. On limbs. **3**:NC(1307), WV(82).
Diplodia punctipetiolae. On petioles. **4**:SC(634).
Diplodia sp. Twig dieback. **4**:FL(35). **7**:FL(35).
Exophoma magnoliae. Leaf spot. **4**:FL(35,82), TX(82). **11**:FL(35,82), TX(82).
Gloeodes pomigena. On limbs. **11**:IN(82).
Hendersonia magnoliae. Leaf spot. **4**:VA(82,3780).
Leptothyrella langloisii. On leaves. **4**:LA(82,966).
Macrophoma sp. Dieback. **4**:OK(82,2873). **7**:FL(35).
Melanconium magnoliae. On dead trunk. **10**:NJ(82).
Microsphaeropsis olivacea. Leaf spot. **4**:FL(35), TX(82).
Pestalotiopsis maculans. Leaf spot. **4**:Range of host(82), FL(35). **11**:S states(82), FL(35).
Phoma melaleuca. On limbs. **11**:Unk.(82).
P. pedunculi. On stems. **3**:WV(82).
Phoma sp. Leaf spot. **10**:FL(35).
Phomopsis magnoliicola. Leaf spot. **4**:FL(35). **10**:FL(35).
P. viridarii. **11**:Unk.(82).
Phomopsis sp. **10**:FL(82). **11**:NY(82), PA(82), VA(82).
Phyllosticta cookei. Leaf spot. **1**:WV(82). **4**:SE states(82), FL(35,82), MS(2720). **10**:FL(35,82), MS(82,2720), NJ(82), NY(82). **7**:FL(35). **11**:E states(82), AL(285), CA(82), FL(35), NC(1303).
P. glauca. Leaf spot. **10**:AL(82), FL(35,82,634), NC(82), SC(82). **11**:SE states(82), FL(35,82).
P. magnoliae. Large leaf spot. **3**:WV(82). **4**:FL(35). **10**:FL(35,82), NJ(82). **7**:NJ(82), NY(82), OH(82). **11**:E states(82), Gulf states(82).
P. terminalis. **11**:NJ(82).
Phyllosticta sp. Leaf spot. **11**:NC(1303).
Septoria magnoliae. Leaf spot. **4**:SE states(82), FL(35), TX(82). **11**:FL(634), OK(2872).
S. niphostoma. On leaves. **4**:FL(35,82,634), NC(82), SC(82).
Shearia fusa. On limbs. **6**:DC(3209). **11**:MS(3549), WA(3549).
Sphaceloma magnoliae. On leaves. **4**:GA(1745), LA(1745).
Sphaceloma sp. Anthracnose, leaf spot. **4**:FL(35), MS(2720). **10**:MS(2720). **11**:FL(35), GA(82), MS(82).
Spogotteria magnoliae. Leaf spot. **11**:C states(82), SE states(82), FL(35).

Deuteromycotina - Other
Rhizoctonia solani. Seedling blight, basal stem rot. **4**:FL(35). **10**:FL(35). **7**:FL(35). **11**:NJ(82).
Rhizoctonia sp. Damping-off. **11**:NC(1303).

Michelia L.

1. *M. figo* (Lour.) Spreng. (=*M. fuscata* (Andr.) Blume) [banana-shrub]

Oomycetes
Pythium sp. Root rot. **1**:FL(35).
Ascomycotina
Botryosphaeria sp. Leaf spot. **1**:FL(35).
Physalospora sp. Leaf spot. **1**:FL(35).
Basidiomycotina - Other
Armillaria tabescens. Root rot. **1**:FL(35).
Pellicularia koleroga. **1**:LA(3640).
Deuteromycotina - Hyphomycetes
Cylindrocladium sp. Root rot. **1**:FL(35).
Deuteromycotina - Other
Rhizoctonia solani. Root and stem rot. **1**:FL(35).

Malpighiaceae

Byrsonima Rich. ex Humb., Bonpl. & Kunth

1. *B. crassifolia* (L.) H. B. K. [nance]

Deuteromycotina - Hyphomycetes
Cercospora sp. Leaf spot. **1**:FL(35).

Galphimia Cav.

1. *G. glauca* Cav.

Deuteromycotina - Hyphomycetes
Alternaria sp. Leaf spot. **1**:FL(35).
Cylindrocladium scoparium. Cutting rot. **1**:FL(35).
Deuteromycotina - Other
Rhizoctonia solani. Basal stem rot. **1**:FL(35).

Malpighia L.

1. *M. coccigera* L. [dwarf holly, miniature holly, Singapore holly]
2. *M. glabra* L. (=*M. punicifolia* L.) [acerola, Barbados-cherry, huesito]

Oomycetes
Phytophthora sp. Root rot. **1**:FL(35).
Pythium sp. Root rot. **1**:FL(35).
Deuteromycotina - Hyphomycetes
Alternaria sp. Leaf spot. **1**:FL(35).
Cercospora bunchosiae. Leaf spot. **2**:FL(35,451), HI(1619,2903).
Cercospora sp. **2**:HI(2903).
Fusarium sp. Root rot. **1**:FL(35).
Helminthosporium sp. Leaf spot. **2**:FL(35).
Deuteromycotina - Coelomycetes
Colletotrichum gloeosporioides. Anthracnose. **2**:FL(35).
Fusicoccum sp. Dieback. **1**:FL(35).
Deuteromycotina - Other
Rhizoctonia solani. Root rot, basal stem rot. **1**:FL(35). **2**:FL(35).

Stigmaphyllon Adr. Juss. & St. Hil.

1. *S. ciliatum* (Lam.) Adr. Juss. [Amazon-vine, Brazilian golden-vine, butterfly-vine, golden-creeper, orchid-vine]

Deuteromycotina - Coelomycetes
Colletotrichum sp. Leaf spot. **1**:FL(35).
Phomopsis sp. Leaf spot. **1**:FL(35).

Malvaceae

Abelmoschus Medik.

1. *A. esculentus* (L.) Moench (=*Hibiscus esculentus* L.) [gumbo, lady's-finger, okra]

Oomycetes
Phytophthora nicotianae var. *parasitica*. Damping-off, root rot. **1**:FL(35).
Pythium sp. Damping-off, root rot, stem rot. **1**:NC(1303).
Zygomycotina
Choanephora cucurbitarum. Blossom blight, fruit rot. **1**:FL(35,82), GA(82), NC(1303), TX(82).
C. infundibulifera. On fallen flowers. **1**:GA(82).
Rhizopus sp. Secondary on pods. **1**:NC(1303).
Ascomycotina
Diaporthe arctii. On stems. **1**:GA(82,1438).
D. phaseolorum. On pods. **1**:GA(82,1438).
Erysiphe cichoracearum. Powdery mildew. **1**:CT(82), FL(35), NC(82), NJ(82), PA(82).
Hysteropatella prostii. **1**:GA(1438).
Leptosphaeria doliolum. **1**:GA(1438).
Leptospora rubella. **1**:GA(1438).
Ophiobolus consimilis. On dead stems. **1**:LA(82).
Sclerotinia sclerotiorum. Stem rot. **1**:FL(35), MA(82).
Deuteromycotina - Hyphomycetes
Alternaria brassicae. Leaf spot. **1**:FL(82), OH(82), PA(82), SC(82), UT(82), VA(82).
Alternaria sp. Leaf spot. **1**:FL(35,82), OH(82), PA(82), SC(82), UT(82), VA(82).
Botrytis sp. Pod rot. **1**:NY(82).
Cercospora malayensis. Leaf spot. **1**:SE states(82), FL(35), NJ(82), OK(82,2872), TX(82).
Cercospora sp. Leaf and pod spot. **1**:MS(2720).
Fusarium oxysporum. Wilt. **1**:E states(82), AZ(82), FL(35), MS(2720), TX(82).
F. oxysporum f. sp. *vasinfectum*. Wilt. **1**:AL(285), NC(1303).
Fusarium sp. Wilt, fruit rot. **1**:E states(82), AZ(82), FL(35), TX(82).
Oidium sp. Powdery mildew. **1**:HI(2903).
Phymatotrichopsis omnivora. Root rot. **1**:AZ(82), TX(82).
Pseudocercospora abelmoschi. Leaf spot. **1**:SE states(82), FL(35), MS(2720), TX(82).
Thielaviopsis basicola. Root rot. **1**:FL(35), NC(1303), NJ(82).
Verticillium albo-atrum. Wilt. **1**:E states(82), Range of host(82), S states(82), W states(82), FL(35), MS(2720), NC(1303), NM(2109,2112,2113,3455).
Verticillium sp. **1**:HI(2903).
Deuteromycotina - Coelomycetes
Ascochyta abelmoschi. Pod spot. **1**:FL(35), GA(82), MD(82), NJ(82), NY(82), PA(82).
Colletotrichum gloeosporioides. Anthracnose. **1**:FL(35,82).
Lasiodiplodia theobromae. On stems. **1**:AL(82).
Macrophomina phaseolina. Charcoal rot. **1**:TX(82).
Phoma exigua. Leaf spot. **1**:NC(1303).
P. okra. On stems. **1**:LA(82), SC(82,634).
Phyllosticta hibiscina. Leaf spot. **1**:AL(82), FL(35), IL(82), NC(82), NJ(82), OK(82,2872).
Deuteromycotina - Other
Rhizoctonia solani. Damping-off, root and stem rot, sore shin, target leaf spot. **1**:AL(82), FL(35,82), NC(1303), OK(82,2872), TX(82).
Sclerotium rolfsii. Southern blight, collar rot. **1**:AL(82), FL(35,82), HI(2903), NC(1303).

Abutilon Mill.

1. *A. californicum* Benth.
2. *A. grandifolium* (Willd.) Sweet
3. *A. hirtum* (Lam.) Sweet
4. *A. incanum* (Link) Sweet [Indian mallow]
5. *A. molle* Sweet
6. *A. parvulum* A. Gray
7. *A. pedunculare* H. B. K.
8. *A. pictum* (Gillies ex Hook. & Arnott) Walp. [flowering-maple]
9. *A. pringlei* Hochr.
10. *A. sonorae* A. Gray
11. *A. theophrastii* Medik. (=*A. avicennae* Gaertn.) [butterprint, China jute, Indian mallow, piemacker, velvetleaf]
12. *A. thurberi* A. Gray
13. *A. vitifolium* (Cav.) K. Presl
14. *Abutilon* sp.

Ascomycotina
Acarosporina hormospora. **2**:HI(2903).
Glomerella cingulata. **11**:IL(1526).
Orbilia abutilonis. **2**:HI(2903).
Stictis radiata. **2**:HI(2903).
Basidiomycotina - Uredinales
Puccinia heterospora. Autoecious. **1**:AZ(747,1195,1203,3981). **2**:HI(2903). **4**:AZ(82,747,1203,3981), HI(2903), OK(82,2873), TX(82). **6**:AZ(1203,3981). **7**:FL(35). **9**:AZ(1203,3981).

10:AZ(747,1203,3981). **11**:AZ(82), KS(82), MD(82), MO(82), NE(82), OK(82), TX(82). **12**:AZ(1203,3981). **14**:SC states(82), SE states(82), SW states(82), HI(82).
P. malvacearum. Autoecious. **2**:HI(2903). **5**:HI(82).
Basidiomycotina - Other
Armillaria mellea. Root rot. **8**:CA(2897). **13**:CA(209).
Deuteromycotina - Hyphomycetes
Alternaria abutilonis. **11**:WI(1335).
Alternaria sp. Leaf spot. **4**:IL(82), IN(82), NJ(82), NY(82), WI(82), WV(82). **11**:IL(82,1526), IN(82), NJ(82), NY(82), TX(82), WI(82), WV(82).
Cercospora abutilonis. Leaf spot. **11**:IL(82).
C. althaeina. **4**:TX(82). **11**:DE(3543), KS(82), TX(82).
C. avicennae. Leaf spot. **11**:KS(581), MO(581), VA(581), WI(1349).
C. brachypoda. Leaf spot. **3**:Unk.(82).
Cladosporium herbarum. Secondary leaf spot. **11**:NY(82).
Epicoccum sp. **11**:IL(1526).
Fusarium sp. On living leaves. **11**:IL(1526), WI(1349).
Gonatobotrys sp. **11**:IL(1526).
Macrosporium avicennae. **11**:DE(3543).
Nigrospora sp. **11**:IL(1526).
Phymatotrichopsis omnivora. Root rot. **4**:TX(82). **11**:TX(82).
Verticillium albo-atrum. Wilt. **14**:NJ(82).
V. dahliae. Wilt. **11**:WI(3265).
Deuteromycotina - Coelomycetes
Ascochyta sp. On leaves. **11**:WI(1348).
Colletotrichum coccodes. **11**:VT(1279).
C. gloeosporioides. **11**:IL(1526).
C. malvarum. Leaf spot. **11**:IA(82), WI(1358).
C. truncatum. Isolated from leaves and stems. **11**:IL(1482,1526).
Macrophomina phaseolina. Stem rot. **11**:IL(82,311,1526).
Phoma exigua. Leaf spot. **4**:TX(82). **11**:KS(82), TX(82), WV(82).
Phomopsis phaseoli. **11**:IL(1526).
Stagonospora sp. **11**:WI(1349).

Alcea L.

1. *A. ficifolia* (L.) Cav. (=*Althaea ficifolia* L.) [hollyhock]
2. *A. rosea* L. (=*Althaea rosea* (L.) Cav.) [hollyhock]
3. *Alcea* sp.

Oomycetes
Phytophthora megasperma. Crown rot. **2**:MD(82), VA(82).
Ascomycotina
Botryosphaeria obtusa. On dead stems. **2**:AL(82).
Erysiphe cichoracearum. Powdery mildew. **2**:MS(2720).
Sclerotinia sclerotiorum. Crown rot. **2**:MT(82,3202), NJ(82).
Basidiomycotina - Uredinales
Endophyllum tuberculatum. **2**:CO(82), KS(82), NE(82), OK(82,2872), WY(1195).
Kuehneola malvicola. Autoecious. **2**:FL(35). **3**:FL(35).
Puccinia heterospora. **2**:CA(82), KS(82), MS(2720), TX(82).
P. lobata. **2**:AZ(82,1203,3981), IL(394), NM(82), TX(82).
P. malvacearum. Autoecious. **1**:CA(81,82), CO(82), NY(82), OR(82,3202). **2**:Range of host(82), AZ(1203,3981), CA(81), FL(35), GA(1439), ID(3202), MS(2720), MT(3202), OK(2872), OR(3202), WA(3202), WY(1195).
P. schedonnardi. **2**:KS(82), MS(82,2720), ND(82), NE(82,753).
P. sherardiana. **2**:CA(82), WY(1195).
Deuteromycotina - Hyphomycetes
Alternaria sp. Secondary leaf spot. **2**:E states(82), FL(35), MN(82), NE(82).
Cercospora althaeina. Leaf spot. **2**:E & C states(82), FL(35), IA(82), IN(82), MN(82), MO(82), MS(2720), NJ(82), OH(82), OK(2872), SD(82), TX(82,3597). **3**:FL(35).
Corynespora sp. Leaf spot. **3**:FL(35).
Fusarium oxysporum. Stem rot. **2**:FL(35).
Myrothecium roridum. Leaf spot. **2**:MD(82).
Phymatotrichopsis omnivora. Root rot. **2**:AZ(82), TX(82).
Deuteromycotina - Coelomycetes
Colletotrichum malvarum. Anthracnose. **2**:E states(82), IA(82), MS(82,2720), TX(82).
Colletotrichum sp. Leaf spot. **3**:FL(35).
Phoma exigua. Wet weather blight, leaf spot. **2**:AL(82), CT(82), DE(82), IN(82), MD(82), NC(1303), NJ(82), NY(82), OH(82), PA(82), WV(82).
Phyllosticta sp. Leaf spot. **2**:FL(35).
Septoria fairmanii. On living leaves. **2**:NY(966).
S. malvicola. Leaf spot. **2**:OK(2872).
Deuteromycotina - Other
Rhizoctonia solani. Web blight. **2**:TX(82).
Sclerotium rolfsii. Southern blight. **2**:AR(82), FL(35).

Althaea L.

1. *A. cannabina* L. [hemp-leaved mallow]
2. *Althaea* sp.

Ascomycotina
Erysiphe cichoracearum. Powdery mildew. **1**:CA(82), MS(82).
E. polygoni. Powdery mildew. **1**:IA(82).
Nectria cinnabarina. Canker. **2**:OK(82,2872).
Basidiomycotina - Uredinales
Puccinia malvacearum. Autoecious. **2**:NC(1303).
Basidiomycotina - Ustilaginales
Sporidiobolus pararoseus. On healthy and rusted leaves. **2**:KS(2686).
S. salmonicolor. On rusted leaves. **2**:KS(2686).
Deuteromycotina - Hyphomycetes
Bullera alba. On healthy and rusted leaves. **2**:KS(2686).
Oidium sp. Powdery mildew. **2**:NC(1303).
Sporobolomyces albo-rubescens. On leaves. **2**:KS(2686).
Tilletiopsis washingtonensis. On rusted leaves. **2**:KS(2686).
Deuteromycotina - Coelomycetes
Colletotrichum malvarum. Anthracnose. **2**:NC(1303).
Phoma exigua. Leaf spot. **1**:NY(82).
Septoria malvicola. Leaf spot. **2**:MI(82), MN(82), NY(82), OH(82), OK(82), VT(82), WI(82).

Anoda Cav.

1. *A. cristata* (L.) Schlechtend. (=*A. hastata* Cav.) [spurred anoda, violettas]

Basidiomycotina - Uredinales
Puccinia sp. **1**:TX(82).
Deuteromycotina - Hyphomycetes
Alternaria macrospora. Stem lesions. **1**:AR(3594), MS(724,2638).
Fusarium lateritium. Stem lesions. **1**:MS(724).
Oidium erysiphoides. Powdery mildew. **1**:NM(82).

Callirhoe Nutt.

1. *C. alcaeoides* (Michx.) A. Gray
2. *C. digitata* Nutt.
3. *C. involucrata* (Nutt. ex Torr.) A. Gray
4. *C. papaver* (Cav.) A. Gray
5. *C. triangulata* (Leavenw.) A. Gray
6. *Callirhoe* sp.

Chytridiomycetes
Synchytrium australe. On leaves. **3**:KS(82,1840), TX(82).
S. callirrhoeae. On leaves. **3**:KS(82,1850), NE(1850), TX(82,1850).
Basidiomycotina - Uredinales
Aecidium avocense. **5**:WI(1348,1351).
Endophyllum tuberculatum. Autoecious. **1**:KS(82). **3**:IN(82), KS(82), NE(82), OK(82,2872), TX(82).
Puccinia interveniens. Heteroecious: aecial host. **1**:NE(82). **2**:TX(82). **5**:WI(1344). **6**:NY(82).
P. schedonnardi. Heteroecious: aecial host. **1**:KS(82,753), NE(82). **2**:OK(82,753,2872). **3**:CO(82), KS(82,753), NE(82,753), NY(753), OK(82,2872), TX(82), UT(82).
Deuteromycotina - Hyphomycetes
Cercospora althaeina. Leaf spot. **3**:KS(82), NE(82), TX(82).
C. praecincta. **5**:IL(82), WI(82).
Phymatotrichopsis omnivora. Root rot. **3**:TX(82).
Verticillium albo-atrum. Wilt. **4**:NY(82).
Deuteromycotina - Coelomycetes
Colletotrichum gloeosporioides. Leaf spot. **2**:TX(82). **3**:TX(82).
Phyllosticta sp. Leaf spot. **3**:OK(2872).

Gossypium L.

1. *G. arboreum* L. (=*G. nanking* Meyen) [Ceylon cotton, tree cotton]
2. *G. barbadense* L. (=*G. peruvianum* Cav., *G. vitifolium* Lam.) [American-Egyptian cotton, extra-long-staple cotton, Sea Island cotton, tree cotton]
3. *G. herbaceum* L. (=*G. obtusifolium* Roxb.) [Levant cotton]
4. *G. hirsutum* L. (=*G. mexicanum* Tod.) [upland cotton]
5. *G. sturtianum* J. H. Willis (=*G. sturtii* F. v. Muell.) [Sturt's

desert-rose]
6. *G. thurberi* Tod. [Arizona wild cotton]
7. *Gossypium* sp.

Oomycetes

Phytophthora nicotianae var. *parasitica*. Boll rot. **2:**FL(35). **4:**FL(35), LA(2811), MS(135).
Phytophthora sp. Boll rot. **4:**FL(2810), LA(2810), MS(135,2810).
Pythium aphanidermatum. Damping-off. **4:**FL(35).
P. debaryanum. Damping-off. **4:**AL(82), MS(82,2720), NC(82), TX(82).
P. ostracodes. Seedling stem rot, root rot. **4:**FL(35).
P. splendens. Root rot. **4:**CA(1436).
P. ultimum. Damping-off. **4:**NC(82), SC(82).
Pythium sp. Damping-off, root rot. **4:**Gulf States(82), AL(723), AR(723), AZ(723), CA(723), GA(723), LA(723), MO(723), MS(723,3060), NC(82,723,1303), NM(723), OK(723), SC(723), TN(82,723), TX(723).

Zygomycotina

Blakeslea trispora. Blossom-boll rot. **4:**FL(35).
Choanephora cucurbitarum. Boll rot. **4:**MS(135), NC(82), TX(82).
C. infundibulifera. On fallen flowers. **4:**GA(82).
Cunninghamella echinulata. Fiber deterioration. **4:**FL(82).
C. elegans. Fiber deterioration. **4:**FL(82).
Mucor sp. Fiber deterioration. **4:**Range of host(82).
Rhizopus stolonifer. Boll rot. **2:**FL(35). **4:**Gulf States(82), AZ(82), CA(82), FL(35), MS(135), OK(82,2872).
Rhizopus sp. Boll rot. **4:**AL(2108,2114,3341), AR(2108,2114,3341), AZ(2108,2114,3341), CA(2108,2114,3341), GA(2108,2114,3341), LA(2108,2114,3341), MO(2114,3341), MS(2108,2114,3341), NC(2108,2114,3341), NM(2114,3341), OK(2108,2114,3341), SC(2108,2114,3341), TN(2108,2114,3341), TX(2108,2114,3341). **7:**AR(324), AZ(324), CA(324), TX(324).

Ascomycotina

Ashbya gossypii. **4:**SC(204).
Botryosphaeria dothidea. Stem lesions, boll rot. **4:**FL(35), MS(135).
B. obtusa. **7:**HI(2903).
B. quercuum. Seedling blight. **4:**FL(35).
B. rhodina. Diplodia boll rot, seedling blight. **3:**SE states(82), IL(82), MD(82), OK(82), TX(82). **4:**SE states(82), IL(82), MD(82), MS(135,2720), OK(82,2873), TX(82).
B. ribis. Boll rot, on stems. **3:**AL(82), GA(82), LA(82), MS(82), SC(82), TX(82). **4:**AL(82), GA(82), LA(82), MS(82,135), SC(82), TX(82).
Chaetomium elatum. Fiber deterioration. **4:**Unk.(82).
C. funicola. Fiber deterioration. **4:**Unk.(82).
C. globosum. Fiber deterioration. **4:**Unk.(82).
Chaetomium sp. Fiber deterioration. **4:**Unk.(82).
Diaporthe sp. On limbs and isolated from seedlings. **4:**MS(3060). **7:**HI(2903).
Glomerella cingulata. Anthracnose, pink boll rot. **2:**FL(35). **4:**FL(35), NC(1303).
G. gossypii. Boll rot. 'Anthracnose.' **3:**Gulf states(82), SE states(82), HI(82), KY(82), MO(82), MS(2720), OK(82), TN(82). **4:**AL(285,2108,2114,3341), AR(2108,2114), GA(1438,2108,2114,2303,3341), LA(2108,2114,3341), MO(2114,3341), MS(135,2108,2114,2720), NC(2108,2114,3341), OK(2108,2114,2872), SC(2108,2114,3341), TN(2108,2114,3341), TX(2108,2114). **7:**HI(2903).
Hyponectria gossypii. On capsules. **4:**SC(82).
Leptosphaeria sp. Fiber deterioration. **2:**FL(35). **4:**FL(35,82).
Mycosphaerella areola. Frosty blight. 'Areolate mildew.' **2:**SE states(82), CA(82), FL(35), IL(82), OK(82), TX(82). **3:**MS(2720). **4:**SE states(82), CA(82), FL(35), GA(1438), IL(82), OK(82,2873), TX(82).
M. gossypina. 'Leaf spot.' **2:**FL(35). **3:**SE states(82), IL(82), MS(2720), OK(82), TX(82). **4:**FL(35), GA(1438), MS(2720).
Nectria cinnabarina. On stems. **4:**GA(82,1438), TX(82).
Nematospora coryli. Internal boll disease. **4:**CA(82).
Neocosmospora vasinfecta. On dead stems. **4:**S states(82), AL(82), AR(82), AZ(82), NC(82), SC(82), TX(82).
Physalospora abdita. Boll rot. **4:**FL(82), LA(82), TX(82).
Pleospora nigricantia. On leaves. **3:**Unk.(82). **4:**Unk.(82).
Talaromyces luteus. Boll rot. **4:**MS(135), TX(82).
Thielavia basicola. Isolated from seedlings. **4:**MS(3060).
Thielavia sp. Fiber deterioration. **4:**FL(82).
Valsa ceratosperma. On stems. **3:**GA(82), SC(82). **4:**GA(82,1438), SC(82). **7:**SC(3415).

Basidiomycotina - Uredinales

Phakopsora gossypii. Autoecious. 'Tropical cotton rust.' **1:**FL(82). **2:**FL(35,82).
Puccinia cacabata. Heteroecious: aecial host. 'Southwestern cotton rust.' **2:**AZ(82,3981), TX(82). **3:**AZ(82), TX(82). **4:**AZ(82,1195,1203,3981), NM(82,2113), OK(82,2873), TX(82). **6:**AZ(1203,3981). **7:**AZ(747).
P. schedonnardi. Heteroecious: aecial host. **4:**OK(2872).

Basidiomycotina - Other

Phanerochaete chrysorhiza. **4:**AZ(1202).
Schizophyllum commune. Root rot. **4:**CA(82), LA(82), TX(82).
Thanatephorus cucumeris. Seedling blight, boll rot. **4:**LA(2812,2813), MS(135,2720).

Deuteromycotina - Hyphomycetes

Acremonium strictum. Fiber deterioration. **2:**FL(35). **4:**FL(35,82).
Alternaria alternata. Boll rot, isolated from seedlings. 'Leaf spots.' **4:**GA(2303), MS(135,3060).
A. gossypii. Leaf spot. **4:**FL(35).
A. macrospora. 'Leaf spot.' **4:**LA(3164).
Alternaria sp. Blight, boll rot, leaf spot. **3:**SC states(82), SE states(82), SW states(82), MS(2720), OK(82). **4:**MO(478), MS(2720), NC(1303), NM(2112), OK(2872), SC(126).
Aspergillus flavus. Boll rot, seed contamination with aflatoxin. **4:**Unk.(82), AR(3276), AZ(3276), CA(3276), GA(3276), LA(3276), MS(135,3276), NC(3276), NM(3276), OK(3276), SC(3276), TN(3276), TX(3276). **7:**AR(324), AZ(324), CA(324), TX(324).
A. fumigatus. Boll rot. **4:**Unk.(82).
A. glaucus. Boll rot. **4:**Unk.(82).
A. nidulellus. Boll rot. **4:**Unk.(82), MS(135).
A. niger. Boll rot. **4:**Range of host(82), SW states(82), MS(135), OK(2872). **7:**AR(324), AZ(324), CA(324), TX(324).
A. niger var. *awamorii*. On roots. **4:**TX(82).
A. ochraceus. Boll rot. **4:**Unk.(82), MS(135).
A. penicillioides. Boll rot. **4:**Unk.(82).
A. reptans. Boll rot. **4:**Unk.(82), MS(135).
A. ustus. Boll rot. **4:**Unk.(82), MS(135).
A. versicolor. Boll rot. **4:**Unk.(82), MS(135).
Aspergillus sp. Boll rot. **4:**MS(135), OK(2872), SC(126).
Aureobasidium pullulans. Fiber deterioration, isolated from seedlings. **4:**Range of host(82), MS(3060).
Botrytis sp. Boll rot. **4:**SC(126).
Brachysporium sp. Boll rot. **4:**OK(82,2872), TX(82).
Cephalosporiopsis sp. Isolated from seedlings. **4:**MS(3060).
Cephalothecium sp. Boll rot. **4:**SC(126).
Cercospora althaeina. Leaf spot. **4:**IL(82), NC(82), SC(82), TX(82).
C. gossypina. Boll rot, leaf spot. **4:**FL(35), MS(135), OK(2872). **5:**OK(2872).
Cercospora sp. Leaf blight. **4:**MO(478).
Cladosporium herbarum. Boll rot. **4:**Range of host(82), MS(135,3060).
Curvularia eragrostidis. Fiber deterioration. **4:**FL(82), OK(82).
C. lunata. Fiber deterioration, isolated from seedlings. **4:**FL(82), MS(3060), OK(82).
Curvularia sp. Fiber deterioration, boll rot. **2:**FL(35). **4:**FL(35,82), MS(135), OK(82).
Epicoccum nigrum. Isolated from seedlings. **4:**MS(3060).
Fusarium acuminatum. Seedling disease. **4:**OK(2872).
F. chlamydosporum. Seedling disease. **4:**OK(2872).
F. equiseti. Seedling disease, isolated from bolls, seedlings and seeds. **3:**E states(82), S states(82).
F. moniliforme. Seedling disease, pink boll rot. **3:**E states(82), S states(82). **4:**Range of host(82), GA(2303), LA(82), MS(135,3060), NC(1303), OK(82,2872), SC(103), TX(82).
F. oxysporum. Boll rot, wilt. **4:**CA(1142), FL(35), GA(2303), MS(135,3060).
F. oxysporum f. sp. *vasinfectum*. Boll rot, wilt. **2:**FL(35). **3:**E states(82), Range of host(82), S states(82). **4:**AL(285,723,2108,2114,3341), AR(723,2108,2114,3341), AZ(295), CA(723), GA(723,2108,2114,3341), LA(723,2108,2114,3341), MO(723,2108,2114,3341), MS(723,2108,2114,2720,3341), NC(723,1303,2108,2114,3341), OK(723,2108,2114,2872,3341), SC(723,2108,2114,3341), TN(73,723,2108,2114,3341), TX(723,2108,2114,3341). **7:**TN(72).

F. pallidoroseum. Seedling disease. **4**:MS(3060), OK(2872).
F. roseum. Boll rot. **4**:GA(2303), MS(135), SC(103).
F. solani. Seedling disease, boll rot. **3**:E states(82), S states(82). **4**:E states(82), S states(82), GA(2303), MS(135,3060), OK(2872).
Fusarium sp. Damping-off, boll rot, wilt. **4**:AL(723), AR(723), AZ(723), CA(723), GA(723), LA(723), MO(723), MS(135,723), NC(723,1303), NM(723), OK(723), SC(103,126,723), TN(723), TX(723). **7**:TN(997).
Helminthosporium gossypii. Leaf spot, small reddish spots. **4**:TX(82). **7**:SC(104).
Helminthosporium sp. Boll rot. **4**:MS(135), OK(82,2872), TX(82).
Humicola fuscoatra. Fabric deterioration. **4**:MD(82).
Macrosporium nigricantium. Leaf spot. **4**:FL(35).
Memnoniella echinata. Fiber deterioration. **2**:FL(35). **4**:FL(35,82), MD(82), TX(82).
Myrothecium roridum. Boll rot. **4**:MS(135).
M. verrucaria. Fiber deterioration. **4**:FL(82), MD(82), MS(82,2720).
Nigrospora sphaerica. Seed mold, boll rot. **4**:MS(135).
Nigrospora sp. Boll rot. **4**:MS(135).
Oidiopsis sicula. Powdery mildew. **4**:CA(687).
Olpitrichum macrosporum. Secondary boll rot. **3**:Gulf states(82), SE states(82). **4**:Gulf states(82), SE states(82), AL(2131), FL(35), LA(2131).
O. tenellum. Secondary boll rot. **2**:FL(35). **4**:AL(82,2131), FL(82), LA(82), SC(82).
Ozonium auriconum. **4**:TX(82).
O. texanum. On dead roots and stems. **4**:TX(82).
Penicillium sp. Seedling disease, boll rot. **4**:Range of host(82), MS(135,3060), OK(2872), SC(126).
Phymatotrichopsis omnivora. 'Phymatotrichum root rot.' **3**:AR(82), AZ(82), CA(82), LA(82), NM(82), OK(82), TX(82). **4**:AR(723,2108,2114,3341), AZ(43,723,2108,2114,3341), CA(723,2108,2114,3341), LA(723,2108,2114,3341), NM(723,2108,2113,2114,3341), OK(723,2108,2114,2872,3341), TX(723,2108,2114,3075,3341).
Pithomyces chartarum. Isolated from seedlings. **4**:MS(3060).
Ramularia gossypii. Leaf spot. **4**:FL(35).
Stachybotrys chartarum. Fiber deterioration. **4**:MD(82).
Stemphylium solani. On leaves. **4**:AL(2503).
Stemphylium sp. Leaf spot. **4**:LA(3281).
Thermomyces lanuginosus. Isolated from seeds. **4**:AL(2484).
Thielaviopsis basicola. Internal collar and boll rot. 'Black root rot. Seedling disease complex.' **3**:AL(82), AZ(82), MS(82), TX(82). **4**:AL(82), AZ(82), MS(82,135,2720,3060), NC(1303), NM(3454), TX(82,2110). **7**:CA(2871).
Trichoderma viride. Boll rot. **4**:Range of host(82), MS(135,3060).
Trichothecium roseum. Boll rot. **4**:AL(82), CA(82), FL(35), MS(135), NM(82), TX(82).
Verticillium albo-atrum. Chlorosis, defoliaiton, hadromycosis, boll rot, wilt. **3**:Range of host(82). **4**:AL(723,2114,3338), AR(723,2108,2114,3341,3644), AZ(43,723,2108,2114,3341), CA(422,723,2108,2114,3141,3341), GA(723), LA(723,1613,2108,2114,2566,3341), MO(723,2114,3341), MS(135,723,2108,2114,2720,3341), NC(723,1303,2108,2114,3341), NM(723,2108,2109,2112,2113,2114,3341,3454), OK(723,2108,2114,2872,3341), SC(723), TN(73,723,2108,2114,3341), TX(723,2108,2114,3341). **7**:TN(72).
V. dahliae. 'Verticillium wilt.' **4**:AR(2881), CA(2881), MO(2881), MS(3060), NE(3142), TX(2881).
V. nigrescens. Weakly pathogenic, causing interveinal chlorosis, stunting and wilt. **4**:CA(422), GA(2303), MO(425).

Deuteromycotina - Coelomycetes

Ascochyta gossypii. Leaf and seedling blight, boll rot. 'Ascochyta blight.' **4**:SE states(82), AL(285,723,2114,3337,3341), AR(723,2114,3341), GA(723,2108,2114,3341), LA(723,2114,3341), MO(723,2114,3341), MS(135,723,2108,2720,3341), NC(723,2108,2114,3341), OK(82,723,2872), SC(126,723,2108,3341), SD(2114), TN(723,2108,2114,3341), TX(82,277,723,2108,2114,3341).
Colletotrichum capsici. Boll rot. **4**:LA(3364).
C. gossypii. Anthracnose, boll rot. 'Escobilla.' **4**:FL(35), SC(103,126).
Diplodia cowdellii. Fiber deterioration. **4**:FL(82).
Diplodia sp. Boll rot. **4**:AR(82), AZ(82), FL(82), GA(82), OK(2872), SC(82), TX(82).
Hendersonia sarmentorum. Fiber deterioration. **4**:FL(82).
Hendersonia sp. Fiber deterioration. **4**:FL(82).
Lasiodiplodia theobromae. Dieback, boll rot, isolated from seedlings. **6**:TX(82).
Pestalotia gracilis. Fiber deterioration. **4**:FL(82).
Pestalotia sp. Fiber deterioration. **4**:FL(35,82), MS(135).
Pestalozziella gossypina. Secondary boll rot or on stems. **4**:AL(82,2549), OK(82,2873), TX(82).
Phlyctema gossypii. On stems. **4**:AL(82), SC(82).
Phoma exigua. Wet weather blight. **4**:NC(1303).
P. gossypii. On stems. **4**:AL(82), AR(82), GA(82), MS(82,2720), SC(82), TX(82).
Phoma sp. Seedling blight, boll rot. **4**:MS(82,135,2720,3060), TX(82).
Phomopsis sp. Stem disease. **4**:TX(82).
Phyllosticta gossypina. Leaf spot. **3**:AL(82), LA(82), MS(82,2720), NC(82), SC(82), TX(82). **4**:AL(82), FL(35), LA(82), MS(82,2720), NC(82), SC(82), TX(82).
P. malkoffii. Leaf spot. **2**:TX(82). **4**:TX(82).
Septoria gossypina. On stems. **3**:MS(82), SC(82). **4**:MS(82,2720), SC(82).
Sphaeropsis malorum. **7**:HI(2903).

Deuteromycotina - Other

Rhizoctonia solani. Damping-off, sore shin, leaf spot. 'Seedling disease complex.' **2**:FL(35). **3**:Unk.(82). **4**:AL(723), AR(723), AZ(723), CA(12,723), FL(35), GA(723,2303), LA(723), MO(723), MS(723,3060), NC(723,1303), NM(723,2112), OK(723,2872), SC(723), TN(723), TX(723).
Rhizoctonia sp. Seedling blight, damping-off. **4**:AL(2108,2114,3341), AR(2108,2114,3341), AZ(2108,2114,3341), CA(2108,2114,3341), GA(2108,2114,3341), LA(2108,2114,3341), MO(2108,2114,3341), MS(2108,2114,3341), NC(2108,2114,3341), NM(2108,2113,2114,3341), OK(2108,2114,3341), SC(2108,2114), SD(3341), TN(2108,2114,3341), TX(2108,2114,3341).
Sclerotium rolfsii. Stem rot. 'Southern blight.' **2**:FL(35). **4**:Gulf States(82), AR(82), AZ(82), FL(35), NC(1303), OK(82,2872).

Herissantia Medik.

1. *H. crispa* (L.) Briz. (=*Bogenhardia crispa* (L.) Kearn.)

Basidiomycotina - Uredinales

Puccinia heterospora. Autoecious. **1**:AZ(747,1195,1203,3981).

Hibiscadelphus Rock

1. *H. hualalaiensis* Rock

Basidiomycotina - Other

Armillaria mellea. Root rot. **1**:HI(2011,2903).

Hibiscus L.

1. *H. cannabinus* L. [bastard jute, bimli jute, Deccan hemp, Indian hemp, kenaf]
2. *H. cardiophyllus* A. Gray
3. *H. grandiflorus* Michx. [great rose mallow]
4. *H. laevis* All. (=*H. militaris* Cav.) [Halberd-leaved mallow, soldier rose-mallow]
5. *H. lasiocarpus* Cav. [woolly rose mallow]
6. *H. moscheutos* L. [common rose mallow, hibiscus, mallow rose, swamp rose mallow, wild cotton]
7. *H. moscheutos* L. subsp. *palustris* (L.) R. T. Clausen (=*H. palustris* L.) [mallow-rose, marsh mallow, sea hollyhock, swamp rose mallow, swamp-rose]
8. *H. rosa-sinensis* L. [blackening-plant, China-rose, Chinese hibiscus, Hawaiian hibiscus, rose-of-China]
9. *H. sabdariffa* L. [Indian sorrel, Jamaican sorrel, roselle]
10. *H. syriacus* L. [rose-of-Sharon, shrub althea]
11. *H. tiliaceus* L. [mahoe, majagua, sea hibiscus]
12. *H. trionum* L. [flower-of-an-hour, Venice mallow]
13. *Hibiscus* sp.

Oomycetes

Phytophthora nicotianae var. *parasitica*. Leaf and stem blight, damping-off. **1**:FL(35). **8**:FL(35,3186), HI(2903). **9**:TX(82). **13**:FL(35), HI(2903).
Pythium sp. Root rot. **1**:FL(35). **8**:FL(35). **11**:HI(2903).

Zygomycotina

Choanephora cucurbitarum. Blossom blight. **8**:FL(35).
C. infundibulifera. Blossom blight. **8**:FL(35). **10**:FL(35). **13**:NC(82).

Ascomycotina

Botryosphaeria dothidea. **10**:GA(1438).

B. hibisci. **13:**LA(82), SC(82).
B. obtusa. **13:**FL(82).
B. ribis. Dieback. **7:**HI(82,2903). **9:**HI(2903). **10:**HI(2903), OK(2872). **11:**HI(2903). **13:**GA(82), HI(82).
Botryosphaeria sp. Leaf spot. **13:**FL(35).
Byssosphaeria schiedermayeriana. **11:**HI(2784).
Dactylospora haliotrepha. **13:**HI(2903).
Diaporthe eres. On twigs. **13:**NJ(82), SC(82).
Diaporthe sp. On limbs. **9:**HI(2903). **11:**HI(2903). **13:**HI(82).
Diatrypella favacea. **10:**GA(1438).
Dothidea puccinioides. **10:**GA(1438). **13:**GA(82).
Erysiphe cichoracearum. Powdery mildew. **1:**FL(35). **13:**FL(35).
Glomerella sp. Anthracnose. **8:**FL(35). **13:**FL(35).
Halosphaeria quadricornuta. On dead limbs, driftwood and pilings. **11:**HI(1942,2903).
Hydronectria tethys. On dead limbs. **11:**HI(1942,2903).
Hysteropatella prostii. **10:**GA(1438).
Lecanidion atratum. On stems. **11:**HI(2903).
Leptosphaeria agnita. Leaf spot. **8:**FL(35). **13:**FL(35).
L. australiensis. On submerged limbs. **11:**HI(1942,2903).
Leveillula taurica. Powdery mildew. **1:**FL(868,3501).
Lignincola laevis. On submerged limbs. **11:**HI(1942,2903).
Lulworthia sp. On submerged limbs. **11:**HI(1942,2903).
Microsphaera euphorbiae. Powdery mildew. **9:**AL(82), FL(82). **13:**FL(35), MS(2720).
Microthyriella hibisci. **13:**HI(2903).
Mollisia petiolorum. **11:**HI(2903).
Nectria cinnabarina. On limbs. **8:**FL(35). **10:**FL(35), GA(1438). **13:**GA(82), KY(82), MA(82), OH(82), SC(82).
N. gracilipes. On bark. **13:**SC(82).
Physalospora abdita. On limbs. **13:**FL(82), HI(82,2903).
Physalospora sp. Leaf spot. **8:**FL(35).
Rhytidhysteron rufulum. **10:**GA(1438).
Sclerotinia sclerotiorum. Stem rot. **1:**FL(35).
Valsaria insitiva. **10:**GA(1438).
Xylaria sp. Wilt. **13:**HI(82,2903).

Basidiomycotina - Uredinales

Kuehneola malvicola. Autoecious. **2:**TX(82). **8:**FL(35). **10:**Gulf States(82), AL(285), FL(35), GA(1439,2405), MS(2720). **13:**FL(35).
Puccinia heterospora. Autoecious. **2:**TX(82). **10:**FL(35).
P. malvacearum. Autoecious. **8:**FL(35).
P. schedonnardi. Heteroecious: aecial host. **4:**MD(753). **5:**IL(753). **6:**MO(3423), NC(1303), OH(3423). **7:**E states(82), NE(82), TX(82). **13:**E states(82), NE(82), TX(82).

Basidiomycotina - Other

Armillaria tabescens. Root rot. **8:**FL(35). **10:**FL(35). **13:**FL(35,82).
Auricularia cornea. **11:**HI(913,3948).
Campanella junghuhnii. **13:**HI(82,2903).
Crepidotus alabamensis. **11:**HI(2903,3706). **13:**HI(2903).
C. applanatus var. *globigera*. **11:**HI(3706).
C. citrinus. **11:**HI(3706).
C. mollis. **11:**HI(3706).
Dendrophora albobadia. **10:**PA(543).
Galerina reflexa. On limbs. **11:**HI(2903).
Pellicularia koleroga. Thread blight. **13:**FL(82).
Pleuroflammula flammea. **11:**HI(2903).
Scytinostroma galactinum. **10:**MD(2090).
Thanatephorus cucumeris. Web blight. **8:**FL(82).

Deuteromycotina - Hyphomycetes

Alternaria alternata. Leaf spot. **8:**FL(35). **13:**IN(82), NJ(82), PA(82).
Alternaria sp. Leaf spot. **8:**FL(35). **10:**FL(35), NC(1303). **11:**FL(35). **13:**FL(35).
Aureobasidium pullulans. **10:**OK(2872).
Botrytis cinerea. Gray mold blight. **8:**FL(35). **13:**FL(35), MD(82).
Botrytis sp. Stem canker. **1:**FL(2742), NJ(1408).
Cercospora althaeina. Leaf spot. **7:**IN(82). **12:**TX(82), WI(1333).
C. malayensis. Leaf spot. **8:**HI(2899,2903), OK(2872). **10:**FL(35), GA(82). **13:**HI(82,2903).
Cristulariella moricola. Zonate leaf spot. **1:**FL(35), MD(2848), NC(1302,1303).
Curvularia sp. Leaf spot. **8:**FL(35). **11:**FL(35).
Fusarium javanicum. Root rot. **9:**HI(82,2903). **13:**HI(2903).
F. lateritium. Dieback. **10:**OK(2872). **13:**FL(82), LA(82), NY(82).
F. oxysporum. Wilt. **8:**FL(35). **13:**FL(35).
Fusarium sp. Fruit rot, wilt. **1:**FL(3504). **8:**FL(35). **9:**FL(82). **13:**FL(35).
Myrothecium sp. Leaf spot. **1:**FL(35).
Oidiopsis macrospora. Powdery mildew. **1:**FL(35).
Periconia byssoides. **3:**TX(82).
Phymatotrichopsis omnivora. Root rot. **7:**TX(82). **9:**TX(82). **10:**TX(82). **12:**TX(82).
Pseudocercospora abelmoschi. Leaf spot. **8:**AL(82), FL(35), OK(82,2872), TX(82). **9:**TX(82). **10:**FL(35), OK(2872). **13:**FL(35).
Tubercularia vulgaris. **13:**MO(82).
Volutella sp. Canker. **8:**FL(35).
Xenosporium berkeleyi. On stems. **11:**HI(1269,2903).

Deuteromycotina - Coelomycetes

Aplosporella hibisci. On limbs. **13:**NY(82), SC(82).
Ascochyta abelmoschi. Leaf spot. **7:**NY(82).
A. gossypii. **13:**AR(82).
Ascochyta sp. Leaf blight. **1:**FL(3504).
Camarosporium lyndonvillae. On twigs. **13:**NY(82).
Colletotrichum gloeosporioides. Anthracnose, dieback. **1:**FL(35,3504). **8:**FL(35,82), TX(82). **9:**FL(82). **11:**FL(35). **13:**FL(35), TX(82).
C. truncatum. Isolated from leaves. **12:**IL(1482).
Coniothyrium hibisci. On twigs. **13:**SC(82).
Coniothyrium sp. Dieback. **8:**FL(35).
Cytosporina sp. **10:**OK(2872).
Dendrophoma sp. Leaf spot. **13:**FL(35).
Diplodia hibiscina. On limbs. **8:**HI(2903). **13:**NJ(82).
Diplodia sp. Leaf spot. **13:**FL(35).
Diplodina acerina. On twigs. **13:**PA(82).
Lasiodiplodia theobromae. **11:**HI(2903). **13:**HI(82,2903).
Macrophoma sp. Leaf spot. **8:**FL(35).
Macrophomina phaseolina. Stem blight, charcoal rot. **1:**FL(3504). **12:**IL(311).
Phoma macularis. **9:**HI(2903).
Phoma sp. Stem lesions, on submerged limb. **8:**FL(35). **11:**HI(2903).
Phomopsis hibisci. On twigs. **13:**NY(82).
P. lavaterae. **13:**MD(82).
P. ophites. **13:**PA(82).
Phomopsis sp. Limb dieback, leaf spot. **8:**FL(35). **13:**FL(35).
Phyllosticta hibiscina. Leaf spot. **6:**MA(1390). **7:**CT(82), FL(82), LA(82), MD(82), NJ(82), NY(82). **8:**FL(35), OK(2872). **10:**FL(35), OK(82,2873). **11:**FL(35). **12:**TX(82). **13:**FL(35), LA(82).
P. syriaca. **10:**NY(64,82).
Phyllosticta sp. Leaf spot. **10:**NC(1303).
Rhabdospora hibiscicola. On stems. **7:**Unk.(82).
Septoria sp. Leaf spot. **7:**NJ(82).
Sphaeropsis sp. Witches' broom. **8:**FL(35).

Deuteromycotina - Other

Rhizoctonia solani. Damping-off, root and stem rot, target leaf spot. **1:**FL(35). **8:**FL(35). **9:**TX(82). **12:**TX(82). **13:**FL(35).
Rhizoctonia sp. Leaf blight. **1:**FL(3504).
Sclerotium rolfsii. Southern blight, collar rot. **1:**FL(35). **9:**TX(82).

Iliamna Greene

1. *I. bakeri* (Jeps.) Wigg.
2. *I. rivularis* (Douglas ex Hook.) Greene (=*Sphaeralcea rivularis* (Douglas) Torr.)

Ascomycotina

Erysiphe polygoni. Powdery mildew. **2:**ID(82,3202).
Pleospora helvetica. **2:**MT(3202).
P. herbarum. On overwintered herbage. **1:**CA(642).
P. herbarum var. *occidentalis*. **2:**WY(3810).
Sclerotinia sclerotiorum. **2:**MT(3202).

Basidiomycotina - Uredinales

Puccinia interveniens. Heteroecious: aecial host. **2:**ID(82,3202), WA(82,3202).
P. sherardiana. Autoecious. **2:**WA(3202).

Deuteromycotina - Hyphomycetes

Kabatiella iliamnae. Leaf spot. **2:**MT(3202), OR(666,3202).
Ramularia sidalceae. **2:**ID(3202), OR(3202).

Deuteromycotina - Coelomycetes

Selenophoma linicola. On stems. **1:**CA(638,639,642).

Kokia Lewton
1. *K. drynarioides* (Seem.) Lewton (=*K. rockii* Lewton) [kokio]
2. *Kokia* sp.
Ascomycotina
Diaporthe kokiae. On decaying petioles. **1:**HI(2784).
Deuteromycotina - Coelomycetes
Libertella kokkiae. On decaying limbs. **2:**HI(2784).

Lavatera L.
1. *L. arborea* L. [tree mallow]
2. *Lavatera* sp.
Oomycetes
Pythium debaryanum. Root rot. **1:**FL(35).
Ascomycotina
Sclerotinia sclerotiorum. Stem rot. **1:**FL(35).
Valsa lavaterae. On stems. **2:**CA(82).
Basidiomycotina - Uredinales
Puccinia malvacearum. Autoecious. **2:**CA(82).
Deuteromycotina - Hyphomycetes
Phymatotrichopsis omnivora. Root rot. **2:**TX(82).
Deuteromycotina - Coelomycetes
Colletotrichum malvarum. Anthracnose. **2:**CA(82), TX(82).
Phoma sp. Twig dieback. **1:**FL(35).
Deuteromycotina - Other
Rhizoctonia solani. Damping-off, root rot. **1:**FL(35), IL(82).
Sclerotium rolfsii. Southern blight. **1:**FL(35).

Malachra L.
1. *M. capitata* L.
Deuteromycotina - Hyphomycetes
Cercospora malachrae. Leaf spot. **1:**TX(82).

Malacothamnus Greene
1. *M. fasciculatus* (Nutt. ex Torr. & A. Gray) Greene (=*Malvastrum fasciculatus* (Nutt. ex Torr. & A. Gray) Greene, *Malvastrum thurberi* A. Gray, *Sphaeralcea fasciculata* (Nutt. ex Torr. & A. Gray) Arth.)
Ascomycotina
Metasphaeria anisometra. On twigs. **1:**CA(2439).
Basidiomycotina - Uredinales
Puccinia interveniens. **1:**CA(81,82).
P. schedonnardi. **1:**CA(82,753).
P. sherardiana. **1:**CA(82,2439).

Malva L.
1. *M. neglecta* Wallr. [cheeses, common mallow]
2. *M. parviflora* L. [cheeseweed, little mallow]
3. *M. rotundifolia* L.
4. *Malva* sp.
Basidiomycotina - Uredinales
Puccinia malvacearum. Autoecious. **1:**ID(3202), OR(3202), WA(3202). **2:**CA(81,2439), HI(2903). **3:**CA(81), OK(2872), WY(1195). **4:**Range of host(82), CA(82), CO(82), FL(35), NY(2456), OR(82), VA(82), WV(82).
Basidiomycotina - Other
Thanatephorus cucumeris. **1:**WA(3202).
Deuteromycotina - Hyphomycetes
Alternaria sp. Leaf spot. **4:**IN(82), MI(82), NY(82).
Cercospora althaeina. Leaf spot. **3:**DE(3543), OK(2873). **4:**E states(82), KS(82), MS(82), TX(82).
C. beticola. **4:**IA(82).
C. malvarum. Leaf spot. **3:**SD(2243), WI(1332). **4:**IN(82), KS(82), NJ(82), WI(82).
C. malvicola. Leaf spot. **3:**PA(581).
Cladosporium sp. **4:**WA(82).
Heterosporium sp. **1:**WA(3202).
Phymatotrichopsis omnivora. Root rot. **4:**TX(82).
Deuteromycotina - Coelomycetes
Colletotrichum malvarum. Leaf spot. **2:**ND(378). **4:**ND(82).
Macrophomina phaseolina. Charcoal rot. **2:**CA(1171), HI(2903).
Phoma exigua. Leaf spot. **4:**MD(82).
Septoria malvicola. Leaf spot. **3:**ND(378). **4:**E states(82), SC states(82), ND(82).
Deuteromycotina - Other
Rhizoctonia solani. Root and stem rot. **4:**FL(35), WA(82).

Malvastrum A. Gray
1. *Malvastrum* sp.
Basidiomycotina - Uredinales
Puccinia heterospora. Autoecious. **1:**TX(82).
P. malvacearum. Autoecious. **1:**WI(82).
Deuteromycotina - Hyphomycetes
Phymatotrichopsis omnivora. Root rot. **1:**TX(82).

Malvaviscus Fabr.
1. *M. arboreus* Cav. [Turk's-cap, wax mallow]
2. *M. arboreus* Cav. var. *drummondii* (Torr. & A. Gray) Schery (=*M. drummondii* Torr. & A. Gray) [Drummond wax mallow, Texas mallow, Turk's-cap]
3. *Malvaviscus* sp.
Zygomycotina
Choanephora cucurbitarum. Blossom blight. **2:**FL(35).
Ascomycotina
Sclerotinia sclerotiorum. Twig blight. **3:**TX(82).
Basidiomycotina - Uredinales
Kuehneola malvicola. Autoecious. **2:**FL(35), TX(82).
Puccinia heterospora. Autoecious. **2:**TX(82).
Basidiomycotina - Other
Armillaria tabescens. Root rot. **1:**FL(82). **2:**FL(35).
Deuteromycotina - Hyphomycetes
Phymatotrichopsis omnivora. Root rot. **2:**TX(82).
Deuteromycotina - Coelomycetes
Colletotrichum sp. Leaf spot. **2:**FL(35).
Phyllosticta malvavisci. Leaf spot. **2:**FL(35), TX(82).
Deuteromycotina - Other
Rhizoctonia crocorum. Root rot. **2:**TX(82).

Malvella Jaub. & Spach
1. *M. lepidota* (A. Gray) Fryx.
2. *M. leprosa* (Ort.) Krapov. (=*Sida hederacea* (Douglas ex Hook.) Torr. ex A. Gray) [alkali mallow, alkali sida]
Basidiomycotina - Uredinales
Puccinia lobata. Autoecious. **1:**AZ(747,3981), CA(747). **2:**AZ(82,747,1203,3981), CA(82,747), NM(82), TX(82), UT(82).
P. malvacearum. **2:**CA(81).
P. schedonnardi. **2:**NM(82,753).
Deuteromycotina - Hyphomycetes
Phymatotrichopsis omnivora. Root rot. **2:**TX(82).

Modiola Moench
1. *M. caroliniana* (L.) G. Don
Chytridiomycetes
Synchytrium australe. On stems and leaves. **1:**LA(1840).
Deuteromycotina - Hyphomycetes
Cercospora althaeina. **1:**HI(2903).
C. modiolae. On living leaves. **1:**TX(3597).

Napaea L.
1. *N. dioica* L.
Ascomycotina
Erysiphe americana. Powdery mildew. **1:**WI(367).
Basidiomycotina - Uredinales
Puccinia schedonnardi. Heteroecious: aecial host. **1:**IL(753).
Deuteromycotina - Hyphomycetes
Cercospora althaeina. **1:**WI(1346).
Deuteromycotina - Coelomycetes
Ascochyta sp. **1:**WI(1347).
Septoria sp. On leaves. **1:**WI(1348).

Pavonia Cav.
1. *Pavonia* sp.
Deuteromycotina - Hyphomycetes
Phymatotrichopsis omnivora. Root rot. **1:**TX(82).

Sida L.
1. *S. acuta* J. Burm. [broomweed, southern sida]
2. *S. fallax* Walp.
3. *S. filicaulis* Torr. & A. Gray
4. *S. laurifolia* L.
5. *S. physocalyx* A. Gray (=*S. hastata* St. Hil.)
6. *S. procumbens* Sw.

7. *S. rhombifolia* L. [arrowleaf sida, Cuba jute]
8. *S. spinosa* L. (=*S. spinosa* L. var. *angustifolia* (Lam.) Trin. & Planch.) [prickly mallow, prickly sida]
9. *Sida* sp.

Ascomycotina

Diaporthe sp. **4:**GA(1438).
Dothidea puccinioides. On stems. **7:**GA(82,1438).
Nectria rubicarpa. **7:**GA(1438).
Ophioceras sp. **7:**GA(1438).

Basidiomycotina - Uredinales

Puccinia heterospora. Autoecious. **1:**FL(35). **2:**HI(2903). **3:**AZ(3981), TX(740). **5:**AZ(82,1203,3981), CA(82), HI(82), TX(82). **6:**AZ(1203). **8:**Gulf states(82), AZ(3981), IN(82), MO(82), MS(2720). **9:**HI(2903).
P. malvacearum. Autoecious. **2:**HI(2903). **9:**HI(82,2903).

Basidiomycotina - Other

Armillaria tabescens. Root rot. **7:**FL(35).

Deuteromycotina - Hyphomycetes

Alternaria sp. Leaf spot. **1:**FL(35).
Cercospora hyalospora. Leaf spot. **8:**IL(312).
C. sidaecola. Leaf spot. **7:**FL(35). **8:**LA(82,581), TX(82).
Cristulariella moricola. On leaves. **8:**IL(2572).
Fusarium lateritium. **8:**MS(3750).
Mycovellosiella sidarum. Leaf spot. **1:**FL(35,82). **9:**LA(82).
Oidium sp. Powdery mildew. **1:**FL(35).
Phymatotrichopsis omnivora. Root rot. **7:**TX(82). **8:**TX(82).
Ramularia gossypii. Leaf spot. **1:**FL(35,82). **7:**FL(35). **9:**FL(82).

Deuteromycotina - Coelomycetes

Colletotrichum malvarum. Leaf spot. **8:**KS(82), TX(82), UT(82). **9:**HI(2903).
Leptothyrium sidae. Leaf spot. **7:**HI(2903). **8:**HI(2903).
Macrophomina phaseolina. Charcoal rot. **8:**IL(311).
Phyllosticta spinosa. Leaf spot. **1:**FL(35). **8:**KS(82,973), TX(82).

Deuteromycotina - Other

Sclerotium rolfsii. Southern blight. **7:**FL(35,82).

Sidalcea A. Gray

1. *S. campestris* Greene
2. *S. candida* A. Gray
3. *S. glaucescens* Greene
4. *S. malviflora* (DC.) A. Gray ex Benth. (=*S. delphinifolia* Greene) [checkerbloom]
5. *S. malviflora* (DC.) A. Gray ex Benth. subsp. *asprella* (Greene) C. Hitchc. (=*S. asprella* Greene)
6. *S. neomexicana* A. Gray
7. *S. oregana* (Nutt. ex Torr. & A. Gray) A. Gray
8. *S. oregana* (Nutt. ex Torr. & A. Gray) A. Gray subsp. *spicata* (Regel) C. Hitchc. (=*S. spicata* Regel)
9. *S. virgata* Howell
10. *Sidalcea* sp.

Ascomycotina

Crocicreas nigrofuscum var. *allantosporum*. **10:**UT(505).
Leptosphaerulina sidalceae. On stems. **1:**OR(3202). **2:**OR(82).
Mycosphaerella sidaecola. **4:**CA(525).
Mycosphaerella sp. On leaves. **4:**CA(82). **7:**UT(82).

Basidiomycotina - Uredinales

Endophyllum tuberculatum. Autoecious. **2:**CO(82). **4:**CO(82), WY(82). **6:**AZ(1203,3981), CO(82,1195), WY(82).
Puccinia interveniens. Heteroecious: aecial host. **2:**UT(82). **4:**CA(81,82), MT(82,3202). **6:**CO(82), ID(82,3202), UT(82), WY(82). **7:**ID(3202), WA(638,3202). **10:**OR(3202), WA(3202).
P. schedonnardi. Heteroecious: aecial host. **2:**CO(82,753).
P. sherardiana. Autoecious. **4:**CA(81,82), NV(82). **5:**CA(81). **7:**CA(82), CO(82), ID(3202), OR(82,3202), WA(82,3202). **8:**CA(81). **9:**OR(3202).

Deuteromycotina - Hyphomycetes

Ramularia sidalceae. Leaf spot. **2:**CA(82), CO(82), WY(82). **3:**CA(81). **6:**CO(82). **7:**ID(3202), OR(3202), WA(3202).

Deuteromycotina - Coelomycetes

Phoma sidalceae. On stems. **6:**NM(82,2399).
Septoria destruens. **4:**CA(82).

Sphaeralcea St. Hil.

1. *S. ambigua* A. Gray [desert hollyhock, desert mallow]
2. *S. ambigua* A. Gray var. *rosacea* (Munz & I. M. Johnst.) Kearn.
3. *S. angustifolia* (Cav.) G. Don
4. *S. angustifolia* (Cav.) G. Don var. *cuspidata* (Britton) A. Gray
5. *S. coccinea* (Nutt.) Rydb. (=*Malvastrum coccineum* (Pursh) A. Gray) [prairie mallow, red false mallow]
6. *S. coulteri* (S. Wats.) A. Gray
7. *S. emoryi* Torr. ex A. Gray
8. *S. fendleri* A. Gray
9. *S. grossulariaefolia* Rydb.
10. *S. incana* Torr.
11. *S. laxa* Woot. & Standl.
12. *S. lobata* Woot.
13. *S. munroana* (Douglas ex Lindl.) Spach ex A. Gray
14. *S. orcuttii* J. Rose
15. *S. parvifolia* A. Nels. (=*S. marginata* Rydb.)
16. *S. pumila* Woot. & Standl.
17. *Sphaeralcea* sp.

Basidiomycotina - Uredinales

Puccinia malvacearum. Autoecious. **13:**WA(3202).
P. schedonnardi. Heteroecious: aecial host. **1:**AZ(82,753,1203). **2:**AZ(3981). **3:**NM(82), TX(82). **5:**CO(82,1195,3423), KS(82,3423), MT(82,3202), ND(82,378), NE(82), NM(82). **8:**AZ(741,753,1203,3981). **9:**AZ(753,1203,3981). **10:**AZ(1203,3981). **12:**NM(753). **13:**NM(82). **15:**AZ(753). **16:**AZ(753).
P. sherardiana. Autoecious. **1:**AZ(82,1203), CA(82), CO(82), NM(82), NV(82), TX(82), UT(82). **2:**AZ(3981). **3:**AZ(82,1195), CA(82), CO(82), NM(82), NV(82), TX(82,742), UT(82). **4:**AZ(1203,3981). **5:**CO(82), KS(82), MO(82), MT(82,3202), ND(82,378), NE(82), NM(82,1195), OK(82,2872), SD(82,2243), TX(82), UT(82), WA(3202), WY(82,1195). **6:**AZ(1195). **7:**AZ(747,1195,1203,3981), CA(747), NM(1195). **8:**AZ(82,1195,1203,3981), CA(82), CO(82), NM(82), NV(82), TX(82), UT(82). **11:**AZ(1195,1203,3981). **12:**NM(1195). **13:**ID(82,3202), WA(82,3202). **15:**AZ(1195,1203,3981). **17:**AZ(1203).
P. sphaeralceoides. Autoecious. **1:**CA(82). **14:**CA(81).

Deuteromycotina - Hyphomycetes

Phymatotrichopsis omnivora. Root rot. **3:**TX(82).
Ulocladium botrytis. **5:**ND(378).

Thespesia Sol. ex Corrêa

1. *T. grandiflora* DC. (=*Montezuma speciosissima* Sessé & Moc. ex DC.)
2. *Thespesia* sp.

Ascomycotina

Glomerella cingulata. Leaf blight. **1:**FL(35,82).
Lophodermium sp. On leaves. **2:**HI(82,2903).

Urena L.

1. *U. lobata* L. [aramina, bur mallow, cadillo, caesarweed, Congo jute]

Deuteromycotina - Hyphomycetes

Alternaria sp. Leaf spot. **1:**FL(35).
Cercospora malachrae. Leaf spot. **1:**FL(35).
Oidium sp. Powdery mildew. **1:**FL(35).

Deuteromycotina - Coelomycetes

Septoria sp. Leaf spot. **1:**FL(35).

Marantaceae

Calathea G. Mey.

1. *Calathea* sp.

Oomycetes

Pythium splendens. Root rot. **1:**FL(35).

Ascomycotina

Glomerella cingulata. Leaf spot. **1:**NJ(82).

Deuteromycotina - Hyphomycetes

Alternaria alternata. Leaf spot. **1:**FL(560).
Alternaria sp. Leaf spot. **1:**FL(35).
Bipolaris setariae. Leaf spot. **1:**FL(35,3275).
Botrytis cinerea. Flower blight. **1:**FL(35).
Cercospora calatheae. Leaf spot. **1:**FL(35).

Fusarium sp. Root rot. **1**:FL(35).
Deuteromycotina - Coelomycetes
Phyllosticta sp. Leaf spot. **1**:FL(35), NJ(82).
Deuteromycotina - Other
Rhizoctonia solani. Root rot, basal stem rot. **1**:FL(35).
Rhizoctonia sp. Root rot. **1**:HI(2903).

Ctenanthe Eichler
1. *C. oppenheimiana* (E. Morr.) K. Schum.
Oomycetes
Pythium sp. Root rot. **1**:FL(35).
Deuteromycotina - Other
Rhizoctonia solani. Root rot. **1**:FL(35).
Rhizoctonia sp. Root rot. **1**:HI(2903).

Maranta L.
1. *M. arundinacea* L. [arrowroot, obedience-plant]
2. *M. leuconeura* E. Morr. (=*M. leuconeura* E. Morr. var. *leuconeura*) [prayer-plant, ten-commandments]
3. *M. leuconeura* E. Morr. var. *erythroneura* Bunting
4. *M. leuconeura* E. Morr. var. *kerchoviana* E. Morr. [rabbit's-foot, rabbit's-tracks]
5. *M. ruiziana* Körn. (=*M. dichotoma* Poepp. ex Körn.)
6. *Maranta* sp.

Oomycetes
Phytophthora sp. Root rot. **4**:FL(35).
Pythium splendens. Root rot. **4**:FL(35).
Pythium sp. Root rot. **2**:FL(35). **3**:FL(35).
Ascomycotina
Glomerella cingulata. Leaf spot. **1**:MD(82), NJ(82). **6**:MD(82), NJ(82).
Glomerella sp. Leaf spot. **4**:FL(35).
Basidiomycotina - Uredinales
Puccinia thaliae. Autoecious. **1**:FL(35,82).
Deuteromycotina - Hyphomycetes
Alternaria sp. Leaf spot. **4**:FL(35).
Bipolaris setariae. Leaf spot. **1**:FL(35). **2**:'Helminthosporium leaf spot.' FL(35). **3**:FL(35). **4**:FL(35). **6**:FL(3275).
Corynespora sp. Leaf spot. **4**:FL(35).
Fusarium oxysporum. Root rot. **3**:FL(35).
Myrothecium sp. Leaf spot. **4**:FL(35).
Periconia digitata. **2**:WA(3202).
Deuteromycotina - Coelomycetes
Phyllosticta marantaceae. Leaf spot. **5**:HI(2903). **6**:HI(82).
Deuteromycotina - Other
Rhizoctonia solani. Crown and root rot. **2**:FL(35). **3**:FL(35). **4**:FL(35).

Stromanthe Sond.
1. *S. amabilis* (Linden) E. Morr.

Oomycetes
Phytophthora sp. Root rot. **1**:FL(35).
Deuteromycotina - Other
Rhizoctonia solani. Root rot. **1**:FL(35).

Thalia L.
1. *T. dealbata* Roscoe
2. *T. geniculata* L.

Basidiomycotina - Uredinales
Puccinia thaliae. Autoecious. **2**:FL(35,82).
Deuteromycotina - Hyphomycetes
Cercospora thaliae. Leaf spot. **1**:LA(82).

Melastomataceae

Clidemia D. Don
1. *C. hirta* (L.) D. Don [camasey, soapbush]

Deuteromycotina - Coelomycetes
Colletotrichum gloeosporioides. Possible biological control agent. **1**:HI(3686).
Microsphaeropsis clidemiae. On leaves. **1**:HI(2903).

Medinilla Gaud. in Freyc.
1. *M. magnifica* Lindl.

Oomycetes
Phytophthora sp. Root rot. **1**:FL(35).
Deuteromycotina - Other
Rhizoctonia sp. Basal stem rot. **1**:FL(35).

Melastoma L.
1. *M. malabathricum* L. [Indian rhododendron]

Oomycetes
Phytophthora cinnamomi. Root rot. **1**:HI(2903).

Miconia Ruiz & Pav.
1. *Miconia* sp.

Deuteromycotina - Hyphomycetes
Cercospora erythrogena. Leaf spot. **1**:AL(82), DE(82), MS(82), TN(82).
Deuteromycotina - Coelomycetes
Colletotrichum rhexiae. Leaf spot. **1**:DE(82).
Phyllosticta rhexiae. Leaf spot. **1**:FL(82).

Monolena Triana
1. *Monolena* sp.

Oomycetes
Pythium sp. Root rot. **1**:FL(35).

Rhexia L.
1. *R. ciliosa* Michx.
2. *R. mariana* L.
3. *Rhexia* sp.

Deuteromycotina - Hyphomycetes
Cercospora erythrogena. Leaf spot. **2**:MS(2720). **3**:AL(82), DE(82), FL(35), MS(82), TN(82).
Deuteromycotina - Coelomycetes
Colletotrichum rhexiae. Leaf spot. **3**:DE(82).
Phyllosticta rhexiae. Leaf spot. **1**:FL(64). **3**:FL(35,82).

Sonerila Roxb.
1. *S. margaritacea* Lindl.

Oomycetes
Pythium sp. Root rot. **1**:FL(35).
Deuteromycotina - Other
Rhizoctonia sp. Basal stem rot. **1**:FL(35).

Tetrazygia Rich. ex DC.
1. *T. bicolor* (Mill.) Cogn.
2. *T. elaeagnoides* (Sw.) DC. [cenizo]

Oomycetes
Phytophthora sp. Root rot. **1**:FL(35).
Basidiomycotina - Other
Armillaria tabescens. Root rot. **2**:FL(82).

Tibouchina Aubl.
1. *T. urvilleana* (DC.) Cogn. (=*T. semidecandra* Cogn.) [glory-bush, lasiandra, pleroma, princess-flower, purple glorytree]

Basidiomycotina - Other
Armillaria tabescens. Root rot. **1**:FL(35,82).
Deuteromycotina - Hyphomycetes
Alternaria sp. Leaf spot. **1**:FL(35).
Cercospora sp. Leaf spot. **1**:FL(35).
Fusarium sp. Root rot. **1**:FL(35).
Deuteromycotina - Coelomycetes
Colletotrichum sp. Leaf spot. **1**:FL(35).
Phomopsis sp. Dieback. **1**:FL(35).
Phyllosticta sp. Leaf spot. **1**:FL(35).

Meliaceae

Cedrela P. Br.
1. *C. sinensis* Juss. [cedrela]

Basidiomycotina - Other
Schizophyllum commune. Wound rot. **1**:NY(82).

Lansium Corrêa

1. *L. domesticum* Corrêa [langsat]

Ascomycotina

Nectriella pironii. Stem gall. **1**:FL(35).

Deuteromycotina - Coelomycetes

Phomopsis sp. Canker. **1**:FL(35).

Melia L.

1. *M. azedarach* L. [China-tree, chinaberry, Indian lilac, Japanese beadtree, paradise-tree, Persian lilac, pride-of-China, pride-of-India, Syrian beadtree]
2. *Melia* sp.

Oomycetes

Phytophthora nicotianae var. *parasitica*. Root rot. **1**:HI(2903).

Ascomycotina

Botryosphaeria dothidea. Dieback. **1**:FL(35), GA(1438).
B. obtusa. On limbs. **1**:AL(82), FL(82), GA(82).
B. quercuum. Twig dieback. **1**:FL(35).
B. rhodina. **1**:GA(82).
B. ribis. On limbs. **1**:AL(82), FL(82), GA(82), SC(82).
Diatrype azedarachtae. **1**:SC(2925).
Diatrype sp. **1**:GA(1438).
Didymosphaeria sp. **1**:GA(1438).
Dothidea puccinioides. **1**:GA(82,1438).
Eutypella juglandicola. **1**:GA(1438).
E. scoparia. **1**:GA(1438).
E. stellulata. Twig blight, on branches. **1**:OK(82,2873), TX(82).
Eutypella sp. **1**:GA(1438).
Glomerella cingulata. Anthracnose, on twigs and berries. **1**:LA(82), MS(2720).
Meliola sp. Black mildew. **1**:FL(35).
Melogramma meliae. **1**:GA(1438). **2**:SC(634).
Nectria cinnabarina. Canker, dieback, on twigs. **1**:Gulf States(82), CA(82), FL(35), GA(82,1438), OK(2872), SC(82).
N. coccinea. Canker. **1**:MS(82,2720), SC(82).
N. gracilipes. **2**:LA(3178).
N. oropensoides. **2**:LA(3085).
Nectriella pironii. Canker, stem gall. **1**:FL(35).
Nummularia sp. **1**:GA(1438).
Phyllactinia guttata. Powdery mildew. **1**:MS(82,2720).
Physalospora abdita. On limbs. **1**:GA(82).
Proliferodiscus pulveraceus. **1**:GA(1438).
Stictis radiata. **1**:GA(1438).
Valsaria rubricosa. **1**:GA(1438).

Basidiomycotina - Other

Abortiporus biennis. **1**:AZ(1202).
Amylosporus campbellii. At base of living tree. **1**:AZ(1199,1202).
Armillaria tabescens. Root rot. **1**:FL(35).
Fomitopsis meliae. Brown cubical rot. **1**:AL(82,128,3713).
Helicobasidium brebissonii. Root rot. **1**:TX(82).
Mycena meliigena. On bark. **2**:Unk.(634).
Oxyporus latemarginatus. **1**:FL(2166).
Pellicularia koleroga. Limb blight. **1**:FL(82), LA(3640).
Peniophora cinerea. **1**:GA(1439).
Phellinus punctatus. On wood. **1**:GA(491,1439).
Schizophyllum commune. Sapwood rot. **1**:OK(82,2873).
Trametes versicolor. **1**:GA(82,1439).

Deuteromycotina - Hyphomycetes

Cercospora meliae. Leaf spot. **1**:Gulf States(82), Unk.(962), FL(35), MS(2720).
Fusarium lateritium. Twig blight. **1**:SC(82), TX(82).
F. sambucinum. On berries. **1**:AL(82), LA(82).
Phymatotrichopsis omnivora. Root rot. **1**:AZ(82), TX(82).
Pseudocercospora subsessilis. Leaf spot. **1**:FL(35), TX(82).
Stilbella clavulata. **2**:LA(3178).

Deuteromycotina - Coelomycetes

Diplodia langloisii. On limbs. **1**:LA(82), OK(82,2873).
Diplodia sp. Dieback. **1**:FL(35).
Macrophoma subconica. On limbs. **1**:AL(82), TX(82).
Macrophoma sp. On limbs, stem gall. **1**:AL(82), FL(35), TX(82).
Phomopsis sp. Canker, dieback. **1**:FL(35).
Phyllosticta azedarachis. Leaf spot. **1**:AL(82).
P. meliae. **1**:LA(82), TX(82).
Phyllosticta sp. Leaf spot. **1**:FL(35).
Septoria sp. Leaf spot. **1**:FL(35).
Vermicularia rectispora. On petioles. **2**:SC(634).

Deuteromycotina - Other

Rhizoctonia solani. Root rot. **1**:FL(35).

Swietenia Jacq.

1. *S. mahagoni* (L.) Jacq. [Madeira redwood, Spanish mahogany, West Indian mahogany]

Oomycetes

Pythium splendens. Root rot. **1**:FL(35).

Ascomycotina

Calonectria pyrochroa. Leaf spot. **1**:FL(35).
Hypoxylon rubiginosum. Canker. **1**:FL(35).
Mycosphaerella sp. Leaf spot. **1**:FL(35).
Nectria swieteniae-mahagoni. Canker. **1**:FL(35).
Phyllachora swieteniae. On leaves. **1**:FL(35,82).

Deuteromycotina - Hyphomycetes

Alternaria sp. Leaf spot. **1**:FL(35).
Cylindrocarpon sp. Stem canker, gall. **1**:FL(35).
Fusarium sp. Root rot, leaf spot. **1**:FL(35).
Fusicladium sp. Leaf spot. **1**:FL(35).
Helminthosporium sp. Leaf spot. **1**:FL(35).
Oidium sp. Powdery mildew. **1**:FL(35).
Pseudocercospora subsessilis. Leaf spot. **1**:FL(35).

Deuteromycotina - Coelomycetes

Colletotrichum gloeosporioides. Anthracnose. **1**:FL(35).
Cytospora sp. Canker. **1**:FL(35).
Diplodia sp. Witches' broom. **1**:FL(35).
Pestalotia sorbi. Leaf spot. **1**:FL(35,82).
Phyllosticta sp. Leaf spot. **1**:FL(35).

Deuteromycotina - Other

Rhizoctonia sp. Root rot. **1**:FL(35).

Toona (Endl.) M. Roem.

1. *T. ciliata* M. Roem. var. *australis* F. v. Muell. [Australian-cedar]

Oomycetes

Phytophthora cinnamomi. Decline. **1**:HI(2903).

Menispermaceae

Calycocarpum (Nutt.) Spach

1. *C. lyonii* (Pursh) A. Gray

Deuteromycotina - Hyphomycetes

Phaeoisariopsis menispermi. Fruit and leaf spot. **1**:OK(2873).

Cocculus DC.

1. *C. carolinus* (L.) DC. [Carolina moonseed, coralbeads, red moonseed, snailseed]
2. *C. ferrandianus* Gaud.
3. *C. laurifolius* DC.
4. *C. trilobus* (Thunb.) DC.
5. *Cocculus* sp.

Chytridiomycetes

Synchytrium cocculi. Leaf gall. **1**:LA(627).

Ascomycotina

Botryosphaeria dothidea. **1**:GA(1438).
B. obtusa. **1**:GA(82). **4**:GA(82).
Lembosina cocculi. On leaves. **2**:HI(82,2903).

Basidiomycotina - Other

Phellinus texanus. **3**:AZ(1202).

Deuteromycotina - Hyphomycetes

Cercospora cocculicola. Leaf spot. **1**:OK(82,2872). **3**:FL(35). **4**:OK(82). **5**:FL(35).
Phaeoisariopsis menispermi. Leaf spot. **1**:MS(82), OK(2872), TX(82). **4**:MS(82), TX(82).
Phymatotrichopsis omnivora. Root rot. **1**:TX(82). **4**:TX(82).

Deuteromycotina - Coelomycetes

Pestalotiopsis maculans. Twig blight. **3**:FL(35).
Phyllosticta sp. Leaf spot. **1**:OK(2872). **3**:FL(35).

Menispermum L.

1. *M. canadense* L. [yellow parilla]
2. *Menispermum* sp.

Ascomycotina

Diaporthe beckhausii. On stems. **2**:NY(82).

Microsphaera penicillata. Powdery mildew. **2:**Range of host(82), IA(82), MI(82), NY(82), OH(82).
Valsa ambiens. On stems. **1:**IA(1226,3415). **2:**IA(82).
Basidiomycotina - Ustilaginales
Entyloma menispermi. Leaf smut. **1:**IA(1077), IL(1077), IN(1077), KS(1077), MN(1077), MO(1077), ND(378,1077), NE(1077), OH(1077), OK(1077), PA(1077), SD(378,1077,2243), VA(1077), WI(1077), WV(1077). **2:**NE states(82), KS(82), ND(82).
Deuteromycotina - Hyphomycetes
Botrytis sp. On leaves. **1:**WI(1350,1358).
Cristulariella moricola. Leaf spot. **1:**WV(3680).
Phaeoisariopsis menispermi. Leaf spot. **1:**MS(2720). **2:**NE states(82), KS(82), WI(82).
Tubercularia menispermi. **1:**DE(3543).
Deuteromycotina - Coelomycetes
Colletotrichum crassipes. Leaf spot. **1:**WI(1335). **2:**WI(82).
Diplodia fairmanii. On dead limbs. **1:**NY(971).
D. sarmentorum. On dead stems. **2:**NY(82).
Phomopsis menispermi. On stems. **2:**KS(82).
Phyllosticta menispermicola. Leaf spot. **1:**IL(3590), WI(1342).
Septoria abortiva. Leaf spot. **1:**WI(1346). **2:**IA(82), IL(82), KS(82), KY(82).
Sphaeropsis menispermi. On stems. **2:**NE states(82), Range of host(82), KS(82), MI(82).

Menyanthaceae

Menyanthes L.
1. *M. trifoliata* L. [bog-bean, buck-bean, marsh-trefoil]
Chytridiomycetes
Physoderma menyanthis. Leaf gall. **1:**AK(82,526), CO(1195), ID(3202), MI(82), NY(82), WA(82,3202), WI(82,1334), WY(1195).
Deuteromycotina - Hyphomycetes
Botrytis sp. Gray mold blight. **1:**NY(82).
Deuteromycotina - Coelomycetes
Ascochyta menyanthicola. Leaf spot. **1:**NY(82).
Phyllosticta sp. Leaf spot. **1:**WI(1360).
Septoria menyanthis. Leaf spot. **1:**WI(82).

Nephrophyllidium Gilg in Engl. & Prantl
1. *N. crista-galli* (Menzies ex Hook.) Gilg (=*Fauria crista-galli* (Menzies ex Hook.) Gilg)
Basidiomycotina - Uredinales
[*Puccinia areolata*]. **1:**AK(526).

Nymphoides J. Hill
1. *N. cordata* (Elliott) Fernald
2. *Nymphoides* sp.
Basidiomycotina - Uredinales
Puccinia scirpi. Heteroecious: aecial host. **2:**FL(35,82).
Basidiomycotina - Ustilaginales
Burrillia decipiens. **1:**NJ(1077), NY(1077). **2:**NJ(82).
Deuteromycotina - Hyphomycetes
Pseudocercospora nymphaeacea. Leaf spot. **2:**FL(35).

Molluginaceae

Mollugo L.
1. *M. verticillata* L. [carpetweed, kurumaba-zakuro-so]
Deuteromycotina - Hyphomycetes
Cercospora molluginis. Leaf spot. **1:**NJ(82), TX(82), WI(82).
Phymatotrichopsis omnivora. Root rot. **1:**TX(82).
Deuteromycotina - Coelomycetes
Phyllosticta molluginis. Leaf spot. **1:**NJ(82).
Deuteromycotina - Other
Rhizoctonia solani. Root rot. **1:**TX(82).
Sclerotium rolfsii. Southern blight. **1:**TX(82).

Monimiaceae

Peumus Molina
1. *P. boldus* Molina [boldo]
Basidiomycotina - Other
Armillaria mellea. Root rot. **1:**CA(2897,2898).

Monotropaceae

Monotropa L.
1. *M. uniflora* L. [convulsion-root, corpse-plant, fitsroot, Indian pipe, pine-sap]
Deuteromycotina - Coelomycetes
Colletotrichum dematium. **1:**MA(664).

Moraceae

Artocarpus J. R. Forster & G. Forster
1. *A. altilis* (S. Parkinson) Fosb. (=*A. incisus* L. fil., *A. communis* J. R. Forster) [breadfruit]
2. *A. heterophyllus* Lam. (=*A. integer* of authors, *A. integrifolius* of authors) [jackfruit]
3. *Artocarpus* sp.
Oomycetes
Pythium splendens. Root rot. **2:**FL(35).
Zygomycotina
Rhizopus sp. **1:**HI(2903).
Ascomycotina
Mycosphaerella artocarpi. On leaves. **1:**HI(82,2903,3480). **3:**HI(2784).
Orbilia epipora. On logs. **1:**HI(82,2903).
Deuteromycotina - Hyphomycetes
Botrytis cinerea. Gray mold of fruit. **2:**FL(35).
Deuteromycotina - Coelomycetes
Colletotrichum gloeosporioides. On living leaves. **1:**HI(82,2903).
Phoma sp. Twig dieback. **2:**FL(35).
Phyllosticta artocarpi. Leaf spot. **1:**HI(82,2903). **2:**FL(35).

Broussonetia L'Hér. ex Venten.
1. *B. papyrifera* (L.) Venten. [paper-mulberry, tapa-cloth-tree]
2. *Broussonetia* sp.
Ascomycotina
Acrospermum foliicola. **1:**NC(3944).
Botryosphaeria dothidea. **1:**GA(1438).
B. obtusa. On limbs. **1:**AL(82), MD(82), OK(82,2873).
Diatrypella favacea. **1:**GA(1438).
Eutypella scoparia. **1:**GA(1438).
Nectria cinnabarina. Dieback. **1:**AL(82), NY(82), OK(2872).
Basidiomycotina - Other
Armillaria mellea. Root rot. **1:**CA(2897).
Perenniporia robiniophila. On trunks. **2:**Unk.(2681).
Deuteromycotina - Hyphomycetes
Alternaria sp. Leaf spot. **1:**FL(35).
Cercospora broussonetiae. Leaf spot. **1:**FL(35).
Cercosporella mori. **1:**NC(82).
Corynespora cassiicola. Frog-eye spot, leaf spot, shot hole. **1:**AL(2845), FL(35).
Fusarium solani. Canker. **1:**OH(3152).
Phymatotrichopsis omnivora. Root rot. **1:**TX(82).

Ficus L.
1. *F. altissima* Blume [council-tree]
2. *F. aurea* Nutt. [Florida strangler fig, golden fig]
3. *F. benghalensis* L. [banyan fig, East Indian fig-tree, Indian banyan]
4. *F. benjamina* L. (=*F. nitida* Thunb.) [Benjamin-tree, Java fig, small-leaved rubberplant, tropic-laurel, weeping fig]
5. *F. carica* L. [common fig]
6. *F. citrifolia* Mill.
7. *F. deltoidea* W. Jack [mistletoe ficus, mistletoe rubberplant]
8. *F. elastica* Roxb. ex Hornem. [Assam rubber, India rubber fig, Indian rubbertree, rubberplant]
9. *F. elastica* Roxb. ex Hornem. cv. 'variegata'
10. *F. lyrata* Warb. [fiddleleaf ficus]
11. *F. microcarpa* L. fil. (=*F. retusa* of authors) [Indian laurel fig]

12. *F. pandurata* Hance. This name has been misapplied to *F. lyrata* and *F. willdemaniana*.
13. *F. pumila* L. (=*F. repens* Rottler) [climbing ficus, creeping ficus, creeping rubberplant]
14. *F. robusta* Corner
15. *Ficus* sp.

Oomycetes

Phytophthora cinnamomi. Root rot. **5**:FL(35).
P. palmivora. Fruit and root rot. **5**:HI(82,2903).
Phytophthora sp. Fruit and root rot. **4**:FL(35). **5**:FL(35). **7**:FL(35). **8**:FL(35). **13**:FL(35). **15**:FL(35).
Pythium splendens. Root rot. **4**:FL(35). **8**:FL(35). **10**:FL(35). **13**:FL(35).
Pythium sp. Root rot. **2**:FL(35). **3**:FL(35). **4**:FL(35). **7**:FL(35). **15**:FL(35).

Zygomycotina

Choanephora cucurbitarum. Fruit rot. **5**:TX(82).
Rhizopus stolonifer. Soft rot of fruit. **5**:Gulf States(82), CA(82,1110), FL(35).

Ascomycotina

Botryosphaeria disrupta. Twig dieback. **5**:FL(35).
B. dothidea. Dieback. **5**:FL(35).
B. ficus. **5**:NC(82).
B. obtusa. On limbs. **5**:AL(82), FL(82).
B. quercuum. Canker, dieback. **5**:FL(35). **8**:FL(35).
B. rhodina. Canker. **5**:AL(82), FL(82), TX(82). **8**:GA(82).
B. ribis. Dieback. **5**:FL(82), TX(82).
Botryosphaeria sp. **5**:NC(1303).
Capnodium sp. Sooty mold. **3**:HI(2903). **5**:TX(82).
Ceratocystis paradoxa. Leaf spot. **15**:FL(35).
Ceratostomella hystricina. **5**:SC(82).
Eutypa sp. Leaf spot. **5**:TX(82).
Eutypella scoparia. On dead twigs. **5**:LA(82), TX(82).
Gibberella baccata. **5**:CA(1110).
Glomerella cingulata. Anthracnose, fruit rot, leaf spot. **5**:SE states(82), FL(35), GA(1438), MS(2720), TX(82). **8**:Range of host(82), FL(35), MS(2720), NC(1303), SD(2243), WA(3202). **10**:WA(3202). **14**:HI(2903). **15**:FL(35).
Glomerella sp. Leaf spot. **4**:FL(35). **10**:FL(35). **12**:WA(1300).
Guignardia sp. Leaf spot. **8**:FL(35).
Hypoxylon sp. Canker. **3**:FL(35).
Mycosphaerella bolleana. Leaf spot. **5**:SC states(82), SE states(82), GA(1438), HI(82), MS(2720). **8**:GA(82).
Mycosphaerella sp. Leaf spot. **2**:FL(35). **5**:FL(35). **8**:FL(35).
Nectria cinnabarina. Canker. **5**:OK(2872), TX(82). **15**:OR(3202).
Nectria sp. On root galls. **8**:FL(35).
Nectriella pironii. Stem gall. **2**:FL(35). **5**:FL(31,35). **8**:FL(35).
Nitschkia broomeiana. Canker. **15**:NC(1303).
Ophiodothella fici. Anthracnose. **1**:FL(35). **2**:FL(35,82). **15**:FL(35).
O. floridana. Leaf spot. **15**:FL(35,82).
Phyllachora ficuum. Anthracnose, leaf spot. **8**:FL(35,82).
Physalospora abdita. On limbs. **5**:AL(82), FL(82).
Physalospora sp. Dieback. **2**:FL(35).
Rhytidhysteron rufulum. **5**:GA(1438).
Saccharomyces sp. Souring of fruit. **5**:CA(82), NC(82), TX(82), WA(82).
Sclerotinia sclerotiorum. Canker. **5**:CA(82,1110), TX(82).

Basidiomycotina - Uredinales

Cerotelium fici. Telial host. **2**:FL(35,82). **5**:SE states(82), AL(285), AR(763), FL(35), GA(1439), HI(2903), MS(2720), TX(82). **6**:FL(35). **8**:FL(35). **13**:FL(35). **15**:FL(35).

Basidiomycotina - Other

Armillaria mellea. Root rot. **5**:CA(82,1110).
A. tabescens. Root rot. **8**:FL(35). **15**:FL(82).
Athelia rolfsii. On leaves. **13**:FL(82).
Corticium microsclerotium. Web blight. **5**:LA(1010).
Cystostereum murrayi. On wood. **2**:FL(35).
Earliella scabrosa. Heart rot. **8**:HI(82,2903).
Erythricium salmonicolor. Limb blight. **5**:Gulf States(82), SE states(3639), FL(35), MS(2720). **15**:FL(35).
Inonotus cuticularis. **5**:AZ(1181).
Oudemansiella canarii. On living and fallen trunks. **2**:FL(3284). **8**:FL(3284).
Pellicularia koleroga. Thread blight. **5**:FL(82), LA(82,3640), MS(82,2720).
Pleurotus sp. **5**:GA(1439).
Schizophyllum commune. On wood. **2**:FL(35). **5**:CA(82), FL(35).
Septobasidium pseudopedicellatum. Felt fungus. **5**:MS(82).
Septobasidium sp. Felt fungus. **5**:MS(82).
Stilbum sp. Canker. **15**:FL(35).
Trametes subectypus. On wood. **15**:FL(35).

Deuteromycotina - Hyphomycetes

Acremonium strictum. **5**:LA(82).
A. zonatum. Leaf spot. **5**:LA(3641).
Alternaria alternata. Fruit spot in markets. **5**:Unk.(82), CA(1110).
Alternaria sp. Leaf spot. **2**:FL(35). **4**:FL(35). **5**:CA(82), FL(35), SC(82). **8**:FL(35), IN(82), OH(82), TX(82). **10**:FL(35). **15**:FL(35).
Aspergillus niger. Black mold. **5**:CA(82,1110), TX(82), WA(82,3202).
Botrytis cinerea. Leaf and stem blight, fruit rot. **2**:FL(35). **5**:CA(82,1110), MS(82,2720), WA(82,3202). **8**:'Gray mold.' FL(35). **10**:'Gray mold.' FL(35).
Cephalosporium sp. Wilt. **10**:FL(35).
Cercospora annulata. Leaf spot. **5**:FL(35).
C. elasticae. Leaf spot. **8**:FL(35).
C. fici. Blotch, leaf spot. **5**:SE states(82), FL(35), MS(2720), TX(82). **8**:FL(35). **13**:FL(35), SC(82).
C. ficina. On living leaves. **5**:TX(3597).
Cercospora sp. Leaf spot. **3**:FL(35). **4**:FL(35). **8**:FL(2250). **10**:FL(35). **13**:FL(35).
Chalara paradoxa. On leaves. **15**:FL(35,82).
Cladosporium herbarum. Fruit spot. **5**:Unk.(82), CA(1110).
Corynespora cassiicola. 'Corynespora leaf spot.' **4**:FL(561).
Corynespora sp. Leaf spot. **4**:FL(35).
Curvularia sp. Leaf spot. **4**:FL(35).
Cylindrocladium scoparium. Leaf necrosis of seedlings, leaf spot. **4**:FL(35). **8**:HI(2903).
Cylindrocladium sp. Leaf spot. **10**:FL(35).
Fumago vagans. Sooty mold. **5**:Gulf States(82).
Fusarium lateritium. Twig blight. **5**:CA(82).
F. moniliforme. Fruit and leaf spot. **2**:CA(82). **4**:FL(35). **5**:CA(1110).
Fusarium sp. Leaf necrosis, root and stem rot, wilt. **1**:FL(35). **4**:FL(35). **5**:AL(82), NC(82). **7**:FL(35). **8**:FL(35). **13**:FL(35). **15**:FL(35).
Fusicladium sp. Leaf spot. **8**:FL(35).
Gonatophragmium mori. Leaf spot. **5**:LA(82,3641).
Helminthosporium sp. Leaf spot. **2**:FL(35). **4**:FL(35). **8**:FL(35). **10**:FL(35). **15**:FL(35).
Hormodendrum sp. **5**:CA(1110).
Myrothecium roridum. Leaf spot. **8**:FL(35).
Olpitrichum macrosporum. Canker. **5**:MD(2092).
Oospora sp. Souring of fruit. **5**:TX(82).
Penicillium sp. **5**:CA(1110).
Phymatotrichopsis omnivora. Root rot. **5**:AZ(82), TX(82).
Pseudocercospora bolleana. Leaf spot. **5**:AL(285), HI(2903). **8**:Unk.(82).
Ramularia sp. Leaf spot. **2**:FL(35). **5**:FL(35). **15**:FL(35).
Stemphylium elasticae. Leaf spot. **8**:Unk.(82).
Trichothecium roseum. Fruit rot. **5**:Gulf States(82), FL(35).
Tubercularia fici. Leaf spot. **5**:GA(82), LA(82), MS(2720), TX(82), VA(82).
T. vulgaris. On twigs. **5**:AL(82), CA(82), TX(82).
Tubercularia sp. **5**:Unk.(82).
Verticillium sp. Wilt. **8**:FL(35).
Zygophiala jamaicensis. Fly speck. **5**:NC(1303).

Deuteromycotina - Coelomycetes

Ascochyta caricae. Leaf spot. **5**:OR(82,3202).
Ascochyta sp. Leaf spot. **8**:FL(35).
Botryodiplodia sp. Twig canker, leaf necrosis. **4**:FL(35). **8**:FL(35). **15**:FL(35).
Colletotrichum gloeosporioides. Anthracnose, fruit rot, leaf spot. **2**:FL(35,82). **15**:FL(35).
Colletotrichum sp. Anthracnose. **4**:FL(35). **10**:FL(35). **15**:FL(35).
Coryneum sp. Leaf spot. **3**:FL(35).
Cytospora sp. **5**:CA(1110).
Diplodia sycina. **5**:OR(3202).
D. sycina var. *syconophila*. Canker. **5**:NC(82), OR(82), TX(82).
Diplodia sp. Canker, dieback. **4**:FL(35). **5**:FL(35). **8**:FL(35). **15**:FL(35).
Dothiorella sp. Leaf spot. **15**:FL(35).

Gloeosporium sp. **8:**WA(1300). **9:**WA(1300). **12:**WA(1300).
Hendersonia findens. On bark. **15:**GA(82).
Hendersonula toruloidea. Branch wilt. **5:**CA(1110,2633).
Lasiodiplodia theobromae. Fruit rot. **5:**TX(82).
Leptostromella elastica. Leaf spot. **8:**NE states(82).
Leptostromella sp. Leaf spot. **4:**FL(35). **8:**FL(35). **10:**FL(35). **15:**FL(35).
Macrophoma fici. Canker. **5:**LA(82), MD(1729), MS(2720), NC(82), TX(82).
Macrophoma sp. Dieback, leaf spot. **4:**FL(35). **8:**FL(35). **10:**FL(35). **15:**FL(35).
Phlyctema ficuum. Leaf spot. **2:**FL(35,82). **15:**FL(35).
Phoma atrocincta. Dieback. **8:**FL(35,82).
Phomopsis cinerascens. Canker. **5:**CA(82,1110), MD(2092).
Phomopsis sp. Canker. **1:**FL(35). **4:**FL(35). **5:**FL(35). **6:**FL(35). **8:**FL(35). **10:**FL(35). **15:**FL(35).
Phyllosticta physopellae. Leaf spot. **2:**FL(35,82). **5:**FL(35). **8:**FL(35). **15:**FL(35).
P. robertii. Leaf spot. **2:**FL(35,82). **8:**Gulf States(82), FL(35), HI(82), MD(82), NY(82).
Phyllosticta sp. Leaf spot. **4:**FL(35). **7:**FL(35). **8:**Gulf States(82), HI(82), MD(82), NY(82). **10:**FL(35).
Plenodomus sp. Leaf spot. **5:**FL(35).
Sphaeropsis malorum. On limbs. **5:**MD(2092).
Sphaeropsis sp. Witches' broom. **5:**FL(35). **13:**FL(35). **15:**FL(35).

Deuteromycotina - Other

Rhizoctonia solani. Leaf and web blight, cutting and root rot. **4:**FL(35). **5:**Gulf states(82), FL(35), MS(2720), TX(82). **7:**FL(35). **8:**FL(35). **13:**FL(35). **15:**FL(35).
Rhizoctonia sp. Collar and root rot. **5:**TX(82). **11:**HI(2903).
Sclerotium rolfsii. Southern blight, on leaves. **4:**'Southern blight.' FL(35). **5:**FL(35,82). **13:**FL(82). **15:**FL(35).

Maclura Nutt.

1. *M. aurantiaca* Nutt.
2. *M. pomifera* (Raf.) C. K. Schneid. [bow-wood, osage-orange]
3. *Maclura* sp.

Oomycetes

Pythium ultimum. Damping-off. **2:**NE(82).

Ascomycotina

Botryosphaeria obtusa. On limbs. **2:**E states(82), KS(82), LA(82).
Daldinia concentrica. **2:**NC(82).
Diaporthe sp. **2:**GA(1438).
Diatrypella favacea. **2:**GA(1438).
Dothidea puccinioides. On twigs. **2:**IL(82), NJ(82), NY(82), PA(82).
Eutypella tetraploa. **1:**SC(634).
Leptosphaeria maclurae. On leaves. **1:**MO(967).
Mycosphaerella maclurae. On leaves. **2:**MO(82,967).
Nectria cinnabarina. On twigs. **2:**IN(82), NM(82), NY(82).
Nematospora coryli. **2:**KS(204).
Rhytidhysteron rufulum. On twigs. **2:**Gulf States(82), FL(35), KS(82).
Stictis radiata. **3:**CA(3225).
Valsa ceratosperma. On dead limbs. **1:**KS(3415), NJ(3415). **2:**KS(82), NJ(82).

Basidiomycotina - Uredinales

Cerotelium fici. Telial host. **2:**SE states(82), FL(35), TX(82).

Basidiomycotina - Other

Acanthophysium lapponicum. On wood. **2:**MO(2083).
Dendrophora albobadia. **2:**AZ(1202). **3:**KS(543).
Hyphoderma sambuci. **2:**AZ(1202).
H. setigerum. **2:**AZ(1202).
Phanerochaete omnivora. **2:**TX(128).
Phellinus ferruginosus. **2:**Unk.(82).
P. punctatus. On dead wood. **2:**Range of host(82).

Deuteromycotina - Hyphomycetes

Botrytis cinerea. On stems. **2:**OR(82,3202).
Cercospora maclurae. Leaf spot. **2:**AL(82).
Mycovellosiella maclurae. Leaf spot. **2:**AL(82), LA(82), MS(2720), TX(82).
Phymatotrichopsis omnivora. Root rot. **2:**OK(82,2873), TX(82).
Sporidesmium maclurae. Leaf blight. **2:**MO(82), SC(82), TX(82).
Verticillium sp. Wilt. **2:**CT(82,3780).

Deuteromycotina - Coelomycetes

Lasiodiplodia theobromae. Canker. **2:**OK(618).
Phyllosticta maclurae. Leaf spot. **1:**NJ(966). **2:**MO(82), NJ(82).
Septoria angustissima. Leaf spot. **2:**MS(82,2720).
Sphaeropsis malorum. Canker. **1:**SC(634). **2:**OK(618).

Deuteromycotina - Other

Rhizoctonia solani. Damping-off. **2:**NE(82).

Morus L.

1. *M. alba* L. [white mulberry]
2. *M. alba* L. var. *multicaulis* (Perr.) Loud. (=*M. multicaulis* Perr.) [silkworm mulberry]
3. *M. microphylla* Buckley
4. *M. nigra* L. [black mulberry]
5. *M. rubra* L. [American mulberry, red mulberry]
6. *Morus* sp.

Chytridiomycetes

Synchytrium sp. On leaves and petioles. **1:**LA(1847).

Ascomycotina

Acrospermoides subulata. On leaves. **5:**GA(82,1438).
Acrospermum compressum. **5:**NC(3944).
A. foliicola. **1:**GA(82). **5:**GA(1438).
Acrospermum sp. **5:**GA(1438).
Asteromassaria olivaceohirta. On limbs. **6:**AL(188), IN(188), KS(188), MA(188), MS(188), NJ(188), OK(188), PA(188), VA(188).
Botryosphaeria obtusa. On limbs. **1:**E states(82).
B. ribis. On limbs. **1:**GA(82), NJ(82).
Ciboria carunculoides. Popcorn disease. **1:**SE states(82), FL(35), TX(82), VA(3493). **5:**GA(1438), MS(2720). **6:**AL(285).
Dermea mori. On dead limbs. **1:**KS(82).
Diatrypella favacea. **5:**GA(1438).
D. nitschkei. **1:**FL(2782).
D. quercina. **5:**GA(82,1438).
Dothidea puccinioides. **5:**GA(1438).
Eutypa lata var. *aceri*. **5:**GA(1438).
Eutypella aequilinearis. **6:**SC(2925).
E. juglandicola. **5:**GA(1438).
E. leprosa. **1:**IA(3629). **6:**IA(3629).
E. scoparia. **5:**GA(1438).
Gibberella baccata. **1:**GA(82,1438). **5:**GA(1438). **6:**IA(82).
Glomerella sp. Leaf spot. **1:**FL(35).
Hysterographium mori. **5:**GA(82,1438).
Lecanidion atratum. **1:**WA(3202).
Lophiostoma quadrinucleatum. **5:**OK(660).
Massaria lantanae. **5:**GA(1438).
M. xanthoxyli. **5:**OK(660).
Mycosphaerella arachnoidea. False mildew. **5:**GA(82,1438), NC(82,3944).
M. mori. Leaf spot. **1:**Range of host(82), FL(35), GA(82,1438), IA(82), IL(2568), TX(82). **4:**FL(35). **5:**E states(82), FL(35), IL(312), NC(1303,3944), TX(82), WI(82), WV(82).
Nectria cinnabarina. Twig blight, canker. **1:**Range of host(82), FL(35), WI(1344). **5:**NY(82). **6:**SD(2243).
N. peziza. On rotten wood. **5:**DE(3085).
N. verrucosa. On limbs. **1:**NJ(82), PA(82), SC(82).
Nectria sp. Canker. **1:**Range of host(82).
Nitschkia broomeiana. **5:**GA(1438), OK(660).
Orbilia vinosa. **5:**GA(1438).
Patellaria sp. On decorticated wood. **5:**NC(1303).
Phyllactinia guttata. Powdery mildew. **5:**OH(82).
Rhytidhysteron rufulum. **5:**GA(1438). **6:**GA(1438).
Rosellinia aquila. Root rot, on dead limbs. **1:**IN(82).
Sclerotinia sp. Canker. **1:**TX(82).
Triblidium rufulum var. *microsporum*. On limbs. **1:**AL(82).
Tryblidiella nigrocinnabarina. On limbs. **1:**AL(82).
Uncinula geniculata. Powdery mildew. **5:**E states(82), AL(82), GA(1438), KS(82).
Valsa ambiens. **5:**GA(1438), IA(1226).
V. ceratosperma. Canker. **1:**GA(82,1438). **2:**SC(3415). **5:**OK(618).
Valsa sp. **5:**GA(1438).
Valsaria insitiva. On limbs. **1:**MD(82). **5:**GA(1438).
V. rubricosa. **5:**GA(1438).
Xylaria carpophila. On wood. **5:**OK(618).

Basidiomycotina - Uredinales

Cerotelium fici. Telial host. **5:**LA(82).

Basidiomycotina - Other

Armillaria mellea. Root rot. **1:**DE(82), NC(82), OK(82,2872).
A. tabescens. Root rot. **5:**OK(618).

Athelia scutellaris. On wood. **5:**GA(491,1439).
Coriolopsis gallica. On wood. **5:**OK(618).
Cylindrobasidium laeve. **1:**DC(128).
Dendrothele acerina. On bark. **5:**LA(2084), WI(2084).
Ganoderma applanatum. Heart rot. **5:**Unk.(82).
Ganoderma sp. Butt rot. **4:**FL(35).
Hapalopilus rutilans. **3:**AZ(1202).
Helicobasidium brebissonii. Root rot. **1:**TX(82).
Hymenochaete agglutinans. **5:**MD(82).
Hyphoderma sambuci. On decaying branch stub. **6:**VA(128).
Inonotus cuticularis. **1:**AZ(1181).
I. farlowii. Heart rot. **1:**AZ(82), CA(82), NM(82).
I. hispidus. Heart rot. **1:**CT(82).
I. munzii. **1:**AZ(1202).
Irpex sp. **1:**WA(3202).
Lopharia cinerascens. Wood rot. **1:**Unk.(82). **6:**PA(543).
Perenniporia robiniophila. On trunks. **6:**Unk.(2681), TN(128).
Phellinus gilvus. Wood rot. **5:**AL(82).
P. texanus. **3:**AZ(1202).
Pleurotus ostreatus. On wood. **5:**OK(618).
Polyporus arcularius. On wood. **5:**OK(618).
Schizophyllum commune. Wood rot. **1:**CA(82).

Deuteromycotina - Hyphomycetes

Cercospora missouriensis. Leaf spot. **5:**FL(35), KS(82), MO(82), MS(2720), NE(82), TX(82). **6:**TX(82).
C. moricola. Leaf spot. **1:**E states(82), FL(35), NE(82), OK(2872), TX(82). **5:**E states(82), Gulf states(82), FL(35), IL(312), KS(82), OK(82,2873).
Cercosporella arachnoidea. False mildew. **4:**FL(35). **5:**FL(35), NC(1303,3944).
C. mori. Leaf spot. **1:**FL(35), IL(312), NE(82), OK(82,2872), TX(82). **4:**FL(35). **5:**FL(35), NC(1303).
Coniosporium olivaceum. **1:**DE(3543).
Exosporium sp. Leaf spot. **5:**FL(35,82).
Fusarium lateritium. Twig blight, canker. **5:**MS(2720,3659).
F. lateritium f. sp. *mori*. Canker. **1:**Range of host(82), FL(35).
Periconia minutissima. **1:**DE(3543).
Phymatotrichopsis omnivora. Root rot. **1:**AZ(82), OK(82,2872), TX(82).
Stemphylium sp. **1:**WA(3202).

Deuteromycotina - Coelomycetes

Aplosporella mori. Canker. **1:**CT(82), KS(82), NY(82). **6:**OK(2872).
Botryodiplodia sp. Canker. **5:**OK(618).
Cylindrosporium ulmicola. Leaf spot. **1:**OK(2872). **5:**OK(2872).
Cylindrosporium sp. Leaf spot. **1:**FL(35). **4:**FL(35). **6:**NC(1303).
Cytospora sp. Twig canker, leaf spot. **1:**NJ(82), OK(82,2873), TX(82), VA(3493).
Dendrophoma longipes. On limbs. **1:**KS(82), MA(82), NY(82). **5:**Unk.(82), SC(82).
Dothiorella mori. Canker. **1:**NJ(82), TX(82).
Dothiorella sp. Canker. **1:**NJ(82), TX(82).
Libertella sp. Canker. **5:**OK(618).
Myxosporium diedickei. Twig blight. **1:**TX(82), WA(82,3202).
Phloeospora maculans. Leaf spot. **5:**NC(3944), OK(2872), WI(1329). **6:**IN(663).
P. moricola. Leaf spot. **5:**MD(82), SC(82).
Phomopsis moricola. **5:**WI(82,1333).
Phyllosticta moricola. Leaf spot. **5:**KS(82), SC(82).
Phyllosticta sp. Leaf spot. **1:**FL(35).
Sphaeropsis sp. On limbs. **1:**CT(82), KS(82), NY(82).
Stegonsporium mori. **1:**ID(3202), WA(3202).

Musaceae

Musa L.

1. *M. acuminata* Colla (=*M. cavendishii* Lamb. ex Paxt.) [dwarf banana, edible banana, plantain]
2. *M. nana* Lour. [dwarf banana]
3. *M.* × *paradisiaca* L. (=*M. sapientum* L.). *M. acuminata* × *M. balbisiana*. [plantain cooking banana]
4. *M. uranoscopos* Lour. (=*M. coccinea* Andr.)
5. *Musa* sp.

Oomycetes

Phytophthora sp. Leaf rot. **3:**FL(35).
Pythium aphanidermatum. Basal stem rot. **1:**HI(2903). **5:**HI(2351).
P. arrhenomanes. Root rot. **1:**HI(2903).
Pythium sp. Root rot, rootlet necrosis. **3:**FL(35), HI(82). **5:**HI(2351,2903).

Zygomycotina

Rhizopus stolonifer. Fruit rot. **5:**HI(2351,2903).

Ascomycotina

Ceratocystis paradoxa. Fruit rot. **1:**HI(2903). **5:**HI(2351).
Chaetothyrina musarum. Sooty blotch. **1:**HI(2903). **5:**HI(2351).
Glomerella cingulata. On fruit, leaves, and stems. **3:**Unk.(82).
Guignardia musae. **1:**HI(2903). **3:**HI(82). **5:**HI(2351,2903).
Mycosphaerella fijiensis. Leaf spot, black leaf streak. **1:**HI(2903). **5:**HI(2351,2353).
M. musae. Speckle. **5:**HI(2351).
M. musicola. Sigatoka disease, leaf spot. **1:**HI(2903). **2:**FL(35). **3:**FL(35). **5:**HI(2351).

Basidiomycotina - Other

Armillaria tabescens. Root rot. **2:**FL(35).
Glabrocyphella bananae. On dead leaves. **5:**FL(634).
Marasmiellus semiustus. Root, stem and trunk rot. **1:**HI(2903). **3:**HI(82). **5:**HI(2351).
Marasmius sp. Root rot. **3:**FL(35,82).

Deuteromycotina - Hyphomycetes

Aspergillus sp. **5:**HI(2351).
Chalara paradoxa. Basal stem rot. **3:**Range of host(82).
Cordana musae. Leaf spot. **1:**HI(2903). **5:**HI(2351).
Curvularia sp. Fine speckle, leaf spot. **1:**HI(2903). **5:**HI(2351).
Deightoniella torulosa. Anthracnose, speckle, eye spot, fruit and black spot, black tip. **1:**HI(2903). **2:**FL(35). **4:**HI(2903). **3:**Range of host(82), FL(35). **5:**HI(2351,2352).
Drechslera gigantea. Eye spot. **5:**HI(2351).
Fusarium lateritium. End rot of fruit. **3:**Unk.(82).
F. oxysporum. Panama disease, wilt. **3:**FL(35). **5:**HI(2903).
F. oxysporum f. sp. *cubense*. Panama disease, wilt. **1:**HI(2903). **2:**FL(35,82), HI(82). **3:**FL(35). **5:**HI(2351).
F. oxysporum f. sp. *vasinfectum*. **3:**HI(82).
F. subglutinans. Heart rot, end rot of fruit. **1:**HI(2903). **3:**Unk.(82), FL(35). **5:**HI(2351).
Fusarium sp. **5:**HI(2351).
Haplographium atrobrunneum. On leaves. **5:**FL(634).
Helminthosporium sp. Fine speckle, leaf spot. **1:**HI(2903). **5:**HI(2351).
Penicillium sp. **5:**HI(2351,2903).
Pseudocercospora musae. Sigatoka disease, leaf spot. **1:**HI(2903). **2:**FL(35). **3:**FL(35). **5:**HI(3685).
Verticillium theobromae. Cigar-end, fruit rot. **1:**HI(2903). **2:**FL(35). **3:**FL(35). **5:**HI(2351).
Zygosporium oscheoides. On leaves. **1:**HI(2903).

Deuteromycotina - Coelomycetes

Colletotrichum gloeosporioides. Anthracnose. **1:**HI(2903).
C. musae. Anthracnose, black end, fruit and stem rot, fruit and leaf spot. **2:**FL(35,82), HI(82), TX(82). **5:**HI(2351,2903).
Gloeosporium lagenarium var. *musarum*. On rind. **2:**NE(966).
Hendersonula toruloidea. Tip rot, leaf spot. **1:**HI(2903). **5:**HI(2351).
Lasiodiplodia theobromae. Fruit and stem rot. **5:**HI(2351).
Pestalotia leprogena. Ring spot. **1:**HI(2903). **5:**HI(2351).
Pestalotia sp. **3:**HI(82). **5:**HI(2903).
Phoma jolyana. **1:**HI(2903).
Phoma sp. Black finger disease. **3:**FL(35).
Phyllosticta musae. Leaf spot. **1:**HI(2903). **3:**HI(82). **5:**HI(2351,2903).
P. musarum. Freckle, black spot. **1:**HI(2903). **2:**HI(82). **3:**HI(82). **5:**HI(2351).
P. musicola. Leaf spot. **1:**HI(2903). **3:**HI(82). **5:**HI(2351,2903).

Deuteromycotina - Other

Rhizoctonia solani. Root and stem rot. **3:**FL(35).

Myoporaceae

Myoporum Banks & Sol. ex G. Forster

1. *M. laetum* G. Forster
2. *M. sandwicense* (A. DC.) A. Gray [bastard sandlewood, naio]

Ascomycotina

Peltosphaeria sandwicensis. On dead leaves. **2:**HI(2781).

Basidiomycotina - Other
Armillaria mellea. Root rot. **2**:HI(2011,2903).
Deuteromycotina - Other
Rhizoctonia sp. Root rot. **1**:HI(2903).

Myricaceae

Comptonia L'Hér. ex Aiton

1. *C. peregrina* (L.) J. M. Coulter [sweet-fern]
2. *C. peregrina* (L.) J. M. Coulter var. *asplenifolia* (L.) Fernald (=*C. asplenifolia* L., *Myrica asplenifolia* (L.) Aiton) [sweetfern]
3. *Comptonia* sp.

Oomycetes
Phytophthora cinnamomi. Root rot. **1**:NC(1303).
Ascomycotina
Cryptodiaporthe aubertii var. *comptoniae*. On stems. **1**:MA(82), NJ(82), NY(82).
Cucurbitaria comptoniae. On stems. **1**:NJ(82), NY(82).
Thyridaria comptoniae. On stems. **1**:NJ(82).
Trichobelonium rhabdocarpum. **3**:NJ(849).
Valsa ambiens. **3**:NJ(3415).
Basidiomycotina - Uredinales
Cronartium comptoniae. Heteroecious: telial host. **1**:NC states(82), NE states(82), NC(82,1303), OH(82). **2**:VA(1533), VT(1224).
Gymnosporangium ellisii. Heteroecious: aecial host. **1**:NJ(82).
Deuteromycotina - Hyphomycetes
Ramularia destructiva. **1**:WI(82).
Deuteromycotina - Coelomycetes
Uniseta flagellifera. On limbs and stems. **1**:Unk.(82). **2**:NJ(966,3549).

Myrica L.

1. *M. californica* Cham. [California bayberry, California wax-myrtle, Pacific wax-myrtle]
2. *M. cerifera* L. (=*M. caroliniensis* Mill.) [candleberry, wax myrtle, waxberry]
3. *M. gale* L. [bog myrtle, meadow-fern, sweetgale]
4. *M. pusilla* Raf.
5. *Myrica* sp.

Oomycetes
Phytophthora sp. Root rot. **2**:FL(35).
Pythium sp. Root rot. **2**:FL(35).
Ascomycotina
Appendiculella calostroma. Black mildew. **2**:Gulf States(82), FL(35).
Asterella myricae. On leaves. **2**:MS(82,2720).
Asteridiella manca. Black mildew. **2**:FL(35,82), MS(82,2720). **5**:FL(1452).
Botryosphaeria dothidea. Dieback. **2**:FL(35), GA(1438), NC(1303).
B. graphidea. **2**:NY(82).
B. obtusa. On limbs. **2**:FL(82), GA(82).
B. quercuum. Twig dieback. **2**:FL(35).
B. ribis. On roots and limbs. **2**:AL(82), FL(82).
Calycellina carolinensis. On leaves. **2**:NC(2181).
Ciboria acerina. On catkins. **3**:NY(82).
Cryptodiaporthe phomaspora. On dead stems and twigs. **3**:NY(82).
Dactylospora stygia. **2**:GA(82,1438).
Dasyscyphus sp. **2**:LA(658).
Eriopezia microspora. **5**:GA(849).
Glomerella sp. Leaf spot. **2**:FL(35).
Gnomonia myricae. On leaves. **1**:CA(81). **2**:GA(82,1438), MS(82,2720), NJ(82).
Hypoxylon rubiginosum. **2**:GA(1438).
H. stygium. **5**:GA(1438).
H. truncatum. **2**:GA(1438).
Leptosphaeria myricae. On twigs. **3**:NY(82).
Leptosphaeria sp. **2**:GA(82,1438).
Lophodermium foliicola. On leaves. **1**:CA(81,82).
Massarina myricae. On dead twigs. **3**:NY(82).
Mycosphaerella myricae. Leaf spot. **2**:GA(82,1438), MS(82,2720).
M. pardalota. On leaves. **2**:NJ(82).
Nectria erubescens. Leaf spot. **2**:FL(35,82).
Nitschkia broomeiana. **5**:GA(1438).
Propolis phacidioides. On leaves. **1**:CA(81,527).
Trichomerium grandisporum. Sooty mold. **2**:FL(82).
Trichopezizella barbata. **3**:NY(1416).
Valsa ceratosperma. **2**:GA(1438).
Zoellneria callochaetes. On fallen leaves. **2**:NJ(849,965).
Basidiomycotina - Uredinales
Cronartium comptoniae. Heteroecious: telial host. **1**:OR(82). **3**:NE states(82), AK(60,82,526), OR(3202), WA(82,3202), WI(1333).
Gymnosporangium ellisii. Heteroecious: aecial host. **2**:NE states(82), FL(35), VA(82). **3**:ME(82).
Basidiomycotina - Other
Aleurobotrys botryosus. On dead stems. **2**:FL(2083).
Armillaria tabescens. Root rot. **2**:FL(35,82).
Corticium floridensis. On wood. **5**:FL(2051).
Dendrophora versiformis. **2**:NY(543).
Exidia glandulosa. On bark. **5**:FL(634).
Hymenochaete rubiginosa. **2**:FL(634).
Inonotus rickii. On wood. **5**:SE states(1204).
Microporellus obovatus. On trunks. **2**:FL(634).
Pachykytospora alabamae. On wood. **2**:FL(634,2537).
Perenniporia tenuis. On dead wood. **2**:FL(82).
Phellinus ferreus. **1**:CA(81), WA(3202).
Pycnoporus sanguineus. On wood of saplings. **2**:FL(2319).
Septobasidium castaneum. Parasitic on scale insects. **2**:FL(35).
S. pseudopedicellatum. On scale insects infesting limbs. **2**:SE states(82).
Septobasidium sp. On scale insects infesting limbs. **2**:SE states(82).
Stereum hirsutum. **2**:CT(543).
S. ochraceo-flavum. On stems. **2**:FL(543), MA(543,1394). **4**:FL(543). **5**:MS(543).
S. ostrea. **2**:FL(543).
Trametes hirsuta. On dead limbs. **2**:MA(1394).
Deuteromycotina - Hyphomycetes
Cercospora sp. Leaf spot. **2**:FL(35).
Chaetochalara aspera. On dead leaves. **2**:NC(2550).
Fusarium oxysporum. Root rot, wilt. **2**:FL(35).
Fusarium sp. Stem lesions. **2**:FL(35).
Phymatotrichopsis omnivora. Root rot. **2**:TX(82).
Ptychogaster cubensis. Wound rot. **2**:FL(82).
Ramularia destructiva. On twigs. **3**:NE states(82), AK(82,526), WI(82).
R. monilioides. Leaf spot. **2**:OK(2872). **3**:NY(82).
Sporidesmium inflatum. On bark. **5**:SC(634).
Thozetella cristata. **2**:NC(2818).
Deuteromycotina - Coelomycetes
Aplosporella lathamii. On dead twigs. **2**:NY(82).
Diplodia sp. Dieback, twig blight. **2**:FL(35). **3**:NJ(82).
Fusicoccum sp. Stem canker. **2**:FL(35).
Macrophoma sp. Leaf spot. **2**:FL(35).
Pestalotia stellata. On leaves. **2**:FL(35,82).
Phyllosticta myricae. Leaf spot. **1**:CA(82). **2**:E states(82), FL(35), NY(64,82), SC(634), TX(82).
Septoria myricae. Leaf spot. **2**:FL(35), NC(1303), NJ(82).
S. myricata. Leaf spot. **3**:NY(82).
Sphaeropsis tumefaciens. Stem gall, witches' broom. **2**:FL(35).
Stagonospora myricae. On leaves. **2**:NJ(82,965).
Deuteromycotina - Other
Rhizoctonia ramicola. Leaf spot. **2**:FL(35).
R. solani. Seedling blight. **2**:NJ(82).

Myrsinaceae

Ardisia Sw.

1. *A. crenata* Sims [coralberry, spiceberry]
2. *A. crispa* (Thunb.) A. DC. [ardisia]
3. *A. escallonioides* Schlechtend. & Cham. [marlberry]
4. *Ardisia* sp.

Oomycetes
Phytophthora sp. Root rot. **1**:FL(35). **4**:FL(35).
Pythium splendens. Root rot. **1**:FL(35). **4**:FL(35).
Pythium sp. Root rot. **2**:FL(35).
Ascomycotina
Hysterostomella floridana. Leaf spot. **2**:FL(35). **3**:FL(82). **4**:FL(35).
Mycosphaerella sp. Leaf spot. **1**:FL(35).
Sclerotinia sclerotiorum. Stem rot. **1**:FL(35). **2**:FL(35).
Basidiomycotina - Other
Armillaria tabescens. Root rot. **1**:FL(35). **2**:FL(35).

Deuteromycotina - Hyphomycetes
Cercospora sp. Leaf spot. **1**:FL(35). **4**:FL(35).
Cylindrocarpon sp. Root rot. **1**:FL(35). **4**:FL(35).
Fusarium moniliforme. Stem necrosis. **1**:FL(35). **4**:FL(35).
F. oxysporum. Damping-off, stem rot. **2**:FL(35). **4**:FL(35).
Fusarium sp. Stem rot. **2**:FL(35).
Deuteromycotina - Coelomycetes
Phyllosticta sp. Leaf spot. **1**:FL(35). **4**:FL(35).
Deuteromycotina - Other
Rhizoctonia solani. Leaf rot. **1**:FL(35). **2**:FL(35). **4**:FL(35).
Sclerotium rolfsii. Southern blight. **1**:FL(35). **2**:FL(35).

Myrsine L.
1. *M. africana* L. [African boxwood, cape myrtle]
2. *M. kauaiensis* Hildebr.
3. *M. lanaiensis* Hildebr.
4. *M. lessertiana* A. DC. (=*Suttonia lessertiana* (A. DC.) Mez)
5. *M. sandwicensis* A. DC.
6. *Myrsine* sp.

Oomycetes
Phytophthora cinnamomi. Root rot, isolated from rootlet and soil samples. **4**:HI(232,2903).
Ascomycotina
Amazonia suttoniae. Black mildew. **2**:HI(2903). **4**:HI(1272,1452,2903).
Asterina suttoniae. **6**:HI(2903).
Beelia suttoniae. **3**:HI(2903).
Calothyrium suttoniae. **5**:HI(2903).
Meliola sp. Black mildew. **6**:HI(2903).
Oligostroma suttoniae. Leaf spot. **4**:HI(2903).
Stomiopeltis suttoniae. Anthracnose, black leaf spot. **4**:HI(2903). **6**:HI(82).
Trichothyrium reptans. **4**:HI(2903).
Basidiomycotina - Other
Armillaria mellea. Root rot. **4**:HI(2011,2903).
Deuteromycotina - Hyphomycetes
Cercospora maesae. Leaf spot. **1**:FL(35).
Deuteromycotina - Coelomycetes
Stagonospora nitida. On leaves. **6**:HI(82,2903).
Deuteromycotina - Other
Rhizoctonia solani. Root rot. **1**:FL(35).

Rapanea Aubl.
(= *Suttonia* A. Rich.)
1. *R. guianensis* Aubl. [Guiana rapanea]
2. *Rapanea* sp. (=*Suttonia sp.*)

Ascomycotina
Asterina suttoniae. **2**:HI(2784).
Seuratia millardetii. **2**:HI(2784).
Deuteromycotina - Hyphomycetes
Cercospora rapaneae. Leaf spot. **1**:FL(35).

Myrtaceae

Callistemon R. Br. in Flinders
1. *C. acuminatus* Cheel
2. *C. citrinus* (Curtis) Skeels (=*C. lanceolatus* (Sm.) DC.) [crimson bottlebrush]
3. *C. viminalis* (Sol. ex Gaertn.) Cheel [weeping bottlebrush]
4. *Callistemon* sp.

Oomycetes
Phytophthora nicotianae var. *parasitica*. Root rot. **2**:FL(35).
Pythium splendens. Root rot. **2**:FL(35). **4**:FL(35).
Ascomycotina
Botryosphaeria sp. Leaf spot. **2**:FL(35).
Leptosphaeria sp. Stem canker. **2**:FL(35).
Nectriella pironii. Stem gall. **4**:FL(35).
Physalospora sp. On stem galls. **2**:FL(35).
Basidiomycotina - Other
Armillaria mellea. Root rot. **1**:CA(2897).
A. tabescens. Root rot. **4**:FL(35).
Deuteromycotina - Hyphomycetes
Alternaria sp. Leaf spot. **2**:FL(35). **4**:FL(35).
Cercospora sp. Leaf spot. **2**:FL(35).
Cylindrocladium clavatum. Root rot. **2**:FL(35).
C. scoparium. Leaf spot. **2**:FL(35,1694). **4**:FL(35).
Fusarium oxysporum. Root rot. **2**:FL(35). **4**:FL(35).
Deuteromycotina - Coelomycetes
Colletotrichum sp. Leaf spot. **2**:FL(35). **4**:FL(35).
Macrophoma sp. Dieback, stem gall. **2**:FL(35). **4**:FL(35).
Phomopsis sp. Stem canker. **2**:FL(35). **4**:FL(35).
Phyllosticta sp. Leaf spot. **2**:FL(35). **4**:FL(35).
Selenophoma sp. Leaf spot. **2**:FL(35).
Sphaeropsis tumefaciens. Canker, stem gall. **2**:FL(35). **3**:FL(2258). **4**:FL(35).
Deuteromycotina - Other
Rhizoctonia solani. Basal stem rot. **2**:FL(35). **4**:FL(35).

Calothamnus Labill.
1. *C. sanguineus* Labill. [netbrush, one-sided bottlebrush]

Basidiomycotina - Other
Armillaria mellea. Root rot. **1**:CA(2897).

Calyptranthes Sw.
1. *C. pallens* (Poir.) Griseb.

Ascomycotina
Phyllachora urbaniana. Leaf spot. **1**:FL(35).
Phyllachora sp. Anthracnose. **1**:FL(35).
Deuteromycotina - Hyphomycetes
Fusarium sp. Witches' broom. **1**:FL(35).
Deuteromycotina - Coelomycetes
Phomopsis sp. Leaf spot. **1**:FL(35).
Sphaceloma sp. Spot anthracnose. **1**:FL(35).

Eucalyptus L'Hér.
1. *E. amplifolia* Naudin [cabbage gum]
2. *E. baxteri* (Benth.) Maiden & Blakely
3. *E. calophylla* Lindl. [marri, red gum]
4. *E. calycogona* Turcz. [gooseberry mallee]
5. *E. camaldulensis* Dehnh. (=*E. rostrata* Schlechtend., *E. longirostris* F. v. Muell.) [Murray red gum, Red River gum]
6. *E. campaspe* S. Moore
7. *E. cinerea* F. v. Muell. ex Benth. [mealy stringybark, silver-dollar-tree, spiral eucalyptus]
8. *E. citriodora* Hook. [lemon-scented gum]
9. *E. deglupta* Blume
10. *E. globulus* Labill. [Tasmanian blue gum]
11. *E. grandis* A. W. Hill ex Maiden [rose gum]
12. *E. lehmannii* (Schau) L. Preiss ex Benth. [bushy yate, Lehmann's gum]
13. *E. marginata* Sm. [Jarrah]
14. *E. microcarpa* Maiden
15. *E. niphophila* Maiden & Blakely [snow gum]
16. *E. obliqua* L'Hér. [messmate stringybark]
17. *E. odorata* Behr (=*E. porosa* Miq.)
18. *E. pauciflora* Sieber ex Spreng. [cabbage gum]
19. *E. pilularis* Sm. [blackbutt]
20. *E. platypus* Hook. [round-leaved moort]
21. *E. polyanthemos* Schauer [silver-dollar gum]
22. *E. populnea* F. v. Muell. [bimble-box]
23. *E. pulverulenta* Sims [money-tree, silver-leaved mountain gum]
24. *E. punctata* DC. [gray gum]
25. *E. robusta* Sm. [swamp-mahogany]
26. *E. saligna* Sm. [Sydney blue gum]
27. *E. sideroxylon* A. Cunn. ex Maiden (=*E. sideroxylon* A. Cunn. var. *rosea* Ingh.) [red ironbark]
28. *E. sieber* L. Johnson (=*E. sieberana* of authors, non F. v. Muell.) [black mountain-ash]
29. *E. tereticornis* Sm. [forest red gum]
30. *E. tetragona* (R. Br.) F. v. Muell. [white-leaved marlock]
31. *E. viminalis* Labill. [manna gum]
32. *E. viridis* R. Baker
33. *Eucalyptus* sp.

Myxomycetes
Reticularia lycoperdon. On decorticated log. **33**:CA(2439).
Oomycetes
Phytophthora cinnamomi. Decline, root rot. **2**:HI(2903). **3**:HI(2903). **8**:HI(2903). **13**:HI(2903). **16**:HI(2903). **19**:HI(2903). **25**:HI(2903). **26**:HI(2899,2903). **28**:HI(2903).

Phytophthora sp. Root rot. **7:**FL(35).
Pythium sp. Root rot. **7:**FL(35). **26:**HI(2903). **31:**FL(35).

Ascomycotina

Aulographina eucalypti. Leaf spot. **24:**HI(2903). **26:**HI(2903).
Bagnisiopsis eucalypti. On twigs. **33:**CA(82).
Botryosphaeria dothidea. Basal canker, twig dieback. **5:**FL(35). **10:**CA(82). **11:**FL(35,168). **25:**FL(35). **31:**FL(35). **33:**FL(35).
B. rhodina. On limbs. **10:**GA(82).
Botryotinia sp. Foliar necrosis. **31:**FL(35).
Calonectria crotalariae. Collar rot. **5:**HI(2903). **8:**HI(2903). **24:**HI(2903). **25:**HI(2903). **26:**HI(2903). **27:**HI(2903). **29:**HI(2903).
Chaetosphaeria dingleyae. On leaves. **26:**HI(2903).
C. ornata. Decaying bark. **33:**CA(1468).
Cryphonectria cubensis. Basal canker, stem canker. **5:**FL(35), GA(2592). **9:**HI(2903). **11:**FL(35,168,1599), HI(1599,2903). **26:**HI(1599,2366,2903). **33:**FL(35).
C. havanensis. **5:**FL(2366).
Dasyscyphus atropurpureus. **33:**CA(849).
Diaporthe eucalypti. On stems. **10:**CA(1468).
D. rudis. On leaves and twigs. **10:**CA(82).
Diaporthe sp. On limbs. **33:**HI(2903).
Didymosphaeria circinans. On leaves. **10:**CA(82,1468).
D. futilis. On limbs. **33:**CA(82).
Discohainesia oenotherae. On leaves. **10:**CA(82).
Emericella variecolor. On decaying leaves. **10:**CA(1468,2740).
Endothia sp. Canker. **11:**FL(35).
Erysiphe cichoracearum. Powdery mildew. **4:**CA(1155). **6:**CA(1155). **14:**CA(1155). **15:**CA(1155). **17:**CA(1155). **18:**CA(1155). **20:**CA(1155). **22:**CA(1155). **23:**CA(1155). **30:**CA(1155). **32:**CA(1155).
Glyphium elatum. On dead leaves. **33:**CA(1468).
Gnomoniella destruens. **10:**HI(193).
Hypocreopsis consimilis. On twigs. **33:**CA(82).
Hypoderma eucalypti. On leaves. **33:**CA(82).
Hysterium pulicare. On wood. **10:**CA(81).
Lembosia eucalypti. **33:**HI(2903).
Metasphaeria anisometra. On twigs. **33:**CA(2439).
Mollisia emergens. On dead leaves. **10:**CA(2801).
M. subcornea. On dead limbs. **10:**CA(2801).
Mycosphaerella molleriana. Leaf spot. **10:**CA(82).
Nectria eucalypti. On limbs. **33:**CA(82).
N. squamulosa. On wood. **33:**CA(3085).
Ophiostoma piliferum. **33:**CA(82).
Orbilia eucalypti. On wood. **10:**CA(2801).
Pezizella carneorosea. On twigs. **10:**CA(82). **33:**CA(850).
Phragmodothidea eucalypti. On bark. **33:**CA(82).
Phyllachora eucalypti. On leaves and twigs. **10:**CA(81,82), FL(35,82). **33:**CA(82), FL(35).
Physalospora eucalyptina. On leaves. **33:**CA(2439).
Pleosphaeria modesta. On limbs. **33:**CA(1468).
Propolis emarginata. On leaves. **5:**HI(2903). **10:**HI(2903). **11:**HI(2903). **25:**HI(2903). **26:**HI(2903).
Rosellinia spinosa. On decaying bark. **33:**CA(1468).
Schizoparme straminea. On leaves. **33:**FL(3210).
Scleroderris eucalypti. On limbs. **33:**CA(82).
Stomiopeltis sp. On leaves. **33:**LA(82).
Valsa eucalypti. On dead bark and twigs. **10:**CA(82). **33:**CA(2439), FL(2782).
Valsaria eucalypti. On twigs. **33:**CA(82).

Basidiomycotina - Other

Armillaria mellea. Root rot. **21:**CA(2897). **23:**CA(2897). **27:**CA(2897). **33:**CA(82).
A. tabescens. Root rot. **33:**FL(35,82).
Athelia epiphylla. On leaves. **33:**CA(82).
Basidioradulum radula. On wood. **33:**CA(2439).
Dacrymyces stillatus. On log. **33:**CA(2439).
Ganoderma applanatum. Heart rot. **10:**HI(82).
G. australe. **33:**HI(2903).
Laetiporus sulphureus. **10:**CA(82). **33:**AZ(1202).
Peniophora malenconii subsp. *americana*. On wood. **10:**CA(542). **25:**CA(542).
Phaeolus schweinitzii. **10:**CA(82).
Phellinus gilvus. **10:**CA(82).
P. robustus. **33:**CA(82).
Pleurotus ostreatus. **33:**CA(2439).
Schizopora paradoxa. **10:**CA(82).
Stereum hirsutum. White rot. **10:**CA(82,543).
S. ochraceo-flavum. **10:**CA(543).
Trametes hirsuta. White rot. **10:**CA(82), HI(82,2903). **33:**CA(2439).
T. versicolor. White rot. **10:**CA(82).
Wolfiporia cocos. On wood. **10:**FL(35,82).

Deuteromycotina - Hyphomycetes

Alternaria sp. Leaf spot. **1:**FL(35). **7:**FL(35). **11:**FL(35). **25:**FL(35). **31:**FL(35). **33:**FL(35).
Anungitea globosa. On leaves. **10:**HI(2903,3553). **26:**HI(2903,3553).
Beltrania querna. On leaves. **5:**HI(2903). **11:**HI(2903). **25:**HI(2903).
B. rhombica. On leaf litter. **33:**FL(3548).
Beltraniella portoricensis. On leaves. **26:**HI(2903).
Botrytis cinerea. On twigs. **33:**CA(82).
B. fusca. On twigs. **33:**CA(82).
Botrytis sp. Gray mold of seedlings. **31:**FL(35).
Camposporium antennatum. On bark. **10:**CA(1468,2740).
Cercospora epicoccoides. Leaf spot. **5:**FL(35). **7:**FL(35). **10:**FL(35). **11:**FL(35). **26:**FL(35). **33:**FL(35).
C. eucalypti. Leaf spot. **10:**FL(35). **33:**FL(35).
Chaetopsina fulva. On leaves. **26:**HI(2903).
C. splendida. On leaves. **25:**HI(2903).
Circinotrichum olivaceum. On leaf litter. **33:**FL(3548).
C. papakurae. On leaves. **26:**HI(2903).
Codinaea parva. On leaves. **25:**HI(2903).
C. simplex. On leaves. **10:**HI(2903). **25:**HI(2903). **26:**HI(2903).
Conoplea fusca. On leaves. **10:**HI(2903). **25:**HI(2903).
Corynespora sp. Leaf spot. **33:**FL(35).
Cryptophiale kakombensis. On leaves. **26:**HI(2903).
C. udagawae. On leaves. **11:**HI(2903). **25:**HI(2903). **26:**HI(2903).
Cylindrocladium pteridis. Leaf spot. **5:**FL(35). **7:**FL(35). **11:**FL(35). **25:**FL(35). **26:**FL(35). **29:**FL(35).
C. scoparium. Foliar blight, stem lesions, root rot, leaf spot. **1:**FL(35). **5:**FL(35). **7:**FL(35). **11:**FL(35,164). **25:**FL(35,164). **26:**FL(35). **29:**FL(35). **31:**FL(35). **33:**FL(35), NC(1303).
C. theae. **25:**HI(2903). **26:**HI(2903).
Fusarium oxysporum. Seedling blight, root and stem rot. **5:**CA(82). **7:**FL(35). **10:**CA(82), FL(35). **25:**CA(82). **33:**FL(35).
Gyrothrix circinata. On leaves. **26:**HI(2903).
Helicosporium pannosum. On bark. **33:**HI(1269).
Hemibeltrania navicularis. On leaf litter. **33:**FL(3548).
Heterosporium eucalypti. On leaves. **33:**CA(82).
Lobatopedis elegans. On fallen leaves. **26:**HI(1911).
Microdochium caespitosum. On leaves. **9:**HI(2903).
M. cylindricum. On leaves. **5:**HI(2903).
M. falcatum. On leaves. **10:**HI(2903). **25:**HI(2903). **26:**HI(2903).
M. griseum. On leaves. **33:**HI(2903).
Mycoleptodiscus brasiliensis. On leaves. **26:**HI(2903).
Olpitrichum macrosporum. On leaf litter. **33:**FL(3548).
Oramasia hirsuta. On leaves. **26:**HI(2903).
Penicillium expansum. On leaves. **33:**CA(82).
Phialocephala truncata. On leaves. **25:**HI(2903).
Phymatotrichopsis omnivora. Root rot. **5:**TX(82).
Polyscytalum truncatum. On leaves. **11:**HI(2903). **25:**HI(2903). **26:**HI(2903).
Pseudobeltrania chumrungensis. On leaves. **26:**HI(2903).
Ramichloridium indicum. On leaf litter. **33:**FL(3548).
Scolecobasidium echinophilum. On leaves. **25:**HI(2903).
Selenosporella verticillata. On leaves. **33:**HI(2903,3553).
Septosporium scyphophorum. On leaves. **10:**CA(82). **33:**CA(82).
Speiropsis pedatospora. On leaves. **26:**HI(2903).
Stilbella fimetaria. Stem gall. **33:**FL(35).
Taeniolella multiplex. On trunks. **10:**CA(82).
Troposporella fumosa. On leaves. **5:**HI(2903). **25:**HI(2903). **26:**HI(2903).
Verticillium tenuissimum. On leaves. **25:**HI(2903).
Volutella coronata. On twigs. **33:**CA(82).
Wiesneriomyces laurinus. On leaf litter. **25:**HI(2903). **26:**HI(2903). **33:**FL(3548).
Zygosporium gibbum. On leaves. **26:**HI(2903). **29:**GA(3548).

Deuteromycotina - Coelomycetes

Amerosporium sp. Leaf spot. **5:**FL(35).
Aplosporella novae-hollandiae. **33:**FL(2782).
Colletotrichum gloeosporioides. On leaves. **7:**CA(81). **10:**CA(81).

Colletotrichum sp. Leaf spot. **1:**FL(35).
Coniella castaneicola. On leaves. **11:**HI(2903). **25:**HI(2903). **26:**HI(2903).
Cryptosporium eucalypti. Leaf spot, on limbs. **10:**CA(82). **31:**FL(35).
Dendrophoma hispalensis. On leaves. **10:**CA(81).
Diplodia australiae. On bark. **10:**CA(82).
D. eucalypti. On limbs. **33:**CA(82).
D. tenuis. On limbs. **33:**CA(82).
Diplodia sp. Dieback. **11:**FL(35). **25:**FL(35). **33:**FL(35).
Discosia artocreas. On leaves. **26:**HI(2903).
Dothiorella eucalypti. Seed capsule abortion, twig dieback. **5:**FL(3798).
Eriosporella calami. On leaves. **11:**HI(2903).
Fairmaniella leprosa. On fruits and leaves. **5:**HI(2903). **33:**CA(2439), FL(2782), HI(2784,3549).
Fusicoccum sp. Stem canker. **33:**FL(35).
Gampsonema exile. On leaves. **26:**HI(2903). **33:**HI(3549).
Gloeosporium capsularum. On fruits. **33:**CA(82).
Harknessia eucalypti. On leaves and twigs. **33:**CA(2548), SC(3549).
H. globosa. On leaves. **10:**CA(2548).
H. hawaiiensis. On leaves. **10:**FL(2782). **25:**HI(2548,2903). **33:**HI(2548,2784,3549).
H. insueta. On leaves and twigs. **11:**FL(35).
H. uromycoides. On fruits, leaves and twigs. **10:**CA(81,82,2548), HI(2903). **17:**CA(1468). **33:**CA(82,2548).
H. ventricosa. On leaves. **26:**HI(2903).
Hendersonia corynoidea. On limbs. **33:**CA(82).
Hendersonia sp. Twig dieback. **33:**FL(35).
Macrophoma molleriana. Twig dieback, leaf spot. **7:**FL(35). **10:**CA(82). **33:**FL(35).
Melanconium globosum. On limbs. **33:**CA(82).
Melanconium sp. **33:**HI(2903).
Microdiplodia microsporella. On leaves. **33:**CA(82).
Microsphaeropsis callista. On fallen leaves. **12:**CA(81).
Monochaetia monochaeta. On leaves. **10:**CA(82).
Mycotribulus mirabilis. On leaves. **11:**HI(2903). **25:**HI(2903). **26:**HI(2903).
Pestalotia sp. **10:**HI(2903).
[*Pestalotiopsis funerea*]. Leaf spot. **33:**FL(35).
Phaeoseptoria eucalypti. **33:**HI(2903).
Phomopsis sp. Leaf spot. **7:**FL(35). **25:**FL(35).
Phyllosticta extensa. Leaf spot. **7:**FL(35). **10:**CA(81,82).
Phyllosticta sp. Leaf spot. **1:**FL(35). **11:**FL(35). **25:**FL(35). **33:**FL(35).
Pilidium acerinum. On leaves. **10:**HI(2903). **25:**HI(2903). **26:**HI(2903).
Pleurophomopsis eucalypti. On decorticated wood. **33:**HI(2784).
Satchmopsis brasiliensis. On leaves. **5:**HI(2903). **25:**HI(2903). **33:**HI(3549).
Seimatosporium lichenicola. On leaves. **33:**CA(82).
Septoria ceuthosporioides. On leaves. **33:**CA(82).
S. mortolensis. On leaves. **33:**CA(82).
Sonderhenia eucalypticola. On leaves. **10:**CA(82). **33:**CA(82).
Sphaeronaema eucalypti. On limbs. **33:**CA(82).
Sphaeropsis macrospermum. On leaves. **33:**CA(82).
S. tumefaciens. Dieback, stem gall. **7:**FL(35).
Sphaeropsis sp. Dieback, stem gall. **25:**FL(35). **31:**FL(35). **33:**FL(35).
Truncatella angustata. On leaves. **10:**CA(82).
Waydora typica. On leaves. **5:**HI(2903). **25:**HI(2903).

Deuteromycotina - Other

Rhizoctonia solani. Damping-off, root and basal stem rot. **7:**FL(35). **11:**FL(35). **31:**FL(35). **33:**FL(35).

Eugenia L.

1. *E. sandwicensis* A. Gray
2. *E. uniflora* L. (=*E. michelii* Lam.) [Brazil-cherry, pitanga, Surinam-cherry]
3. *Eugenia* sp.

Oomycetes

Phytophthora sp. Root rot. **2:**FL(35).
Pythium sp. Root rot. **2:**FL(35). **3:**FL(35).

Ascomycotina

Asterinella puiggarii. Black mildew. **2:**FL(35). **3:**FL(35,82).
Discohainesia oenotherae. Leaf spot. **3:**NY(82).
Meliola hawaiiensis. Black mildew. **1:**HI(1272,1452,2903). **3:**HI(82,1272,1452).
Mycosphaerella eugeniae. **1:**HI(2903,3480). **3:**HI(2903,3480).

Basidiomycotina - Other

Armillaria tabescens. Root rot. **2:**FL(35). **3:**FL(35,82).
Corticium stevensii. Thread blight. **3:**FL(35).

Deuteromycotina - Hyphomycetes

Cercospora eugeniae. Leaf spot. **2:**FL(35). **3:**FL(35).
Curvularia sp. Leaf spot. **3:**FL(35).
Fusarium sp. Root rot. **2:**FL(35). **3:**FL(35).
Helminthosporium sp. Leaf spot. **3:**FL(35).

Deuteromycotina - Coelomycetes

Colletotrichum gloeosporioides. Anthracnose, on leaves. **2:**FL(35). **3:**FL(35), MD(82).
Coniothyrium sp. Stem canker. **3:**FL(35).
Lasiodiplodia theobromae. Dieback. **2:**FL(35).
Macrophoma sp. Dieback. **2:**FL(35).
Phomopsis sp. Witches' broom. **2:**FL(35). **3:**FL(35).
Phyllosticta eugeniae. Leaf spot. **2:**FL(35). **3:**FL(35).
Sphaeropsis tumefaciens. Stem gall. **2:**FL(35). **3:**FL(35).

Deuteromycotina - Other

Rhizoctonia solani. Twig dieback, root rot. **3:**FL(35).

Feijoa O. Berg

1. *F. sellowiana* (O. Berg) O. Berg [feijoa, pineapple guava]

Oomycetes

Peronospora sparsa. **1:**CA(1110).
Phytophthora sp. Root rot. **1:**CA(1110), FL(35).
Pythium sp. Root rot. **1:**CA(1110), FL(35).

Ascomycotina

Glomerella cingulata. **1:**CA(1110).

Basidiomycotina - Other

Armillaria mellea. **1:**CA(1110).
A. tabescens. Root rot. **1:**FL(35).
Corticium stevensii. Thread blight. **1:**FL(35).

Deuteromycotina - Hyphomycetes

Botrytis cinerea. Fruit rot. **1:**CA(82,1110).
Cercospora sp. Leaf spot. **1:**FL(35).
Cylindrocladium scoparium. Leaf spot. **1:**FL(35).
Penicillium expansum. Fruit rot. **1:**CA(82,1110).
Phymatotrichopsis omnivora. Root rot. **1:**TX(82).
Ramularia sp. Leaf spot. **1:**FL(35).

Deuteromycotina - Coelomycetes

Entomosporium mespili. Leaf spot. **1:**FL(35).
Lasiodiplodia theobromae. Dieback. **1:**FL(35).
Phyllosticta sp. Leaf spot. **1:**FL(35).
Pyrenochaeta sp. Leaf spot. **1:**FL(35).
Sphaceloma psidii. On leaves. **1:**FL(1745,2524).

Deuteromycotina - Other

Rhizoctonia ramicola. Thread blight. **1:**FL(35,82).
R. solani. Root rot. **1:**FL(35).

Leptospermum J. R. Forster & G. Forster

1. *L. scoparium* J. R. Forster [manuka, New Zealand tea-tree]

Basidiomycotina - Other

Armillaria mellea. Root rot. **1:**CA(2897).

Melaleuca L.

1. *M. leucadendra* (L.) L. [cajeput, river tea-tree, weeping tea-tree]
2. *M. quinquenervia* (Cav.) S. T. Blake (=*M. leucadendron* of authors) [broad-leaved tea-tree, cajeput-tree, paperbark-tree, punk-tree, swamp tea-tree]
3. *M. radula* Lindl.
4. *Melaleuca* sp.

Oomycetes

Pythium sp. Damping-off, root rot. **1:**HI(2903). **2:**FL(35).

Ascomycotina

Botryosphaeria sp. Dieback. **2:**FL(35).
Physalospora sp. Dieback. **2:**FL(35).

Basidiomycotina - Uredinales

Puccinia psidii. Autoecious. **2:**FL(35).

Basidiomycotina - Other

Armillaria mellea. Root rot. **3:**CA(2900).
A. tabescens. Root rot. **2:**FL(35). **4:**FL(35).

Deuteromycotina - Hyphomycetes

Alternaria sp. Leaf spot. **2:**FL(35).

Cercospora sp. Leaf spot. **2**:FL(35). **4**:FL(35).
Cylindrocladium pteridis. Leaf spot. **2**:FL(35).
C. scoparium. Leaf spot. **2**:FL(35).
Fusarium sp. Root rot. **2**:FL(35).
Deuteromycotina - Coelomycetes
Dothiorella sp. Leaf spot. **2**:FL(35).
Macrophoma sp. Dieback. **2**:FL(35). **4**:FL(35).
Phoma sp. Twig dieback. **2**:FL(35).
Phomopsis sp. Canker, dieback. **2**:FL(35).
Phyllosticta sp. Leaf spot. **2**:FL(35). **4**:FL(35).
Sphaeropsis sp. Stem gall. **2**:FL(35).
Deuteromycotina - Other
Rhizoctonia solani. Damping-off. **2**:FL(35).

Metrosideros Banks & Gaertn.

1. *M. collina* (J. R. Forster) A. Gray
2. *M. macropus* Hook. & Arnott
3. *M. polymorpha* Gaud. (=*M. collina* (J. R. Forster) A. Gray subsp. *polymorpha* (Gaud.) Rock) ['ohi'a lehua]
4. *M. tremuloides* (A. Heller) P. Knuth
5. *Metrosideros* sp.

Oomycetes
Phytophthora cinnamomi. Decline, root rot. **3**:HI(232,1918,1920,2903).
Pythium vexans. Rootlet necrosis, root rot. **3**:HI(1919,2903).
Zygomycotina
Mortierella isabellina. On leaf surface. **1**:HI(144). **3**:HI(2903).
Mortierella sp. On leaves. **1**:HI(144). **3**:HI(2903).
Ascomycotina
Asteridiella ohiana. Black mildew. **1**:HI(2903). **3**:HI(1272). **5**:HI(82).
Asterina rickii. Black leaf patch. **5**:HI(82,2903).
Calonectria theae. Shoot blight, leaf spot. **1**:HI(2903). **3**:HI(2609).
Calothyriopeltis metrosideri. **5**:HI(2903).
Capronia mansonii. On leaf surface. **1**:HI(144).
Chaetomium globosum. On leaves. **1**:HI(144). **3**:HI(2903).
Coniochaeta sp. On leaves. **1**:HI(144). **3**:HI(2903).
Dasyscyphus citrinoalbus. **1**:HI(2903).
Diatrype princeps. **1**:HI(2903).
Endothia metrosideri. On bark. **3**:HI(2995).
Khuskia oryzae. On leaf surface. **1**:HI(144).
Leptosphaeria sp. On leaves. **1**:HI(144). **3**:HI(2903).
Meliola sydowiana. Black mildew. **5**:HI(1272).
Meliola sp. Sooty mold. **3**:HI(2903).
Meliolina haplochaeta. Black mildew. **3**:HI(2903). **5**:HI(82,2784).
M. mollis. Black mildew. **1**:HI(1272). **3**:HI(2903).
M. novae-zealandiae. Black mildew. **1**:HI(1272). **3**:HI(2903).
M. sydowiana. Black mildew. **2**:HI(1272,2903). **3**:HI(2903). **5**:HI(82,2784,2903).
Metasphaeria hawaiiensis. Leaf spot. **3**:HI(2903). **5**:HI(82).
Mycosphaerella metrosideri. Leaf spot. **1**:HI(144). **3**:HI(82,2903,3480).
Neurospora sp. **3**:HI(2903).
Nummularia mauritanica. **3**:HI(2903).
Ostropella leve. **3**:HI(2903).
O. verrucosa. **3**:HI(2903).
Preussia vulgare. On leaves. **1**:HI(144). **3**:HI(2903).
Rhizodiscina lignyota. **4**:HI(2784).
Seuratia maunauluana. On leaves. **3**:HI(2337).
S. millardetii. On leaves. **3**:HI(2337).
Trichothyrium reptans. **3**:HI(2903).
Xylaria anisopleura. **3**:HI(2903). **5**:HI(2903).
Basidiomycotina - Other
Acanthophysium mirabile. **5**:HI(2083).
Antrodia malicola. **5**:HI(82,2903).
Armillaria mellea. Root rot. **1**:HI(2903). **3**:HI(2011,2899).
Auricularia cornea. **3**:HI(3948).
Gymnopilus pacificus. On fence post. **3**:HI(1542).
Phellinus inermis. Heart rot. **3**:HI(2903).
Polyporus arcularius. On wood. **5**:HI(82,2903).
Schizophyllum sp. On leaves. **1**:HI(144). **3**:HI(2903).
Deuteromycotina - Hyphomycetes
Acremonium curvulum. On leaves. **1**:HI(144). **3**:HI(2903).
A. strictum. On leaves. **1**:HI(144). **3**:HI(2903).
Acremonium sp. On leaves. **1**:HI(144). **3**:HI(2903).
Alternaria alternata. On leaves. **1**:HI(144). **3**:HI(2903).
Arthrinium sporophleum. On leaves. **1**:HI(144). **3**:HI(2903).
Aspergillus panamensis. On leaves. **1**:HI(144). **3**:HI(2903).
A. ustus. On leaves. **1**:HI(144). **3**:HI(2903).
Atichia solaridiscoidea. On leaves. **3**:HI(2337).
Aureobasidium pullulans. On leaves. **1**:HI(144). **3**:HI(2903).
Beauveria sp. On leaves. **1**:HI(144). **3**:HI(2903).
Botrytis sp. On leaves. **1**:HI(144). **3**:HI(2903).
Calcarisporium sp. On leaves. **1**:HI(144). **3**:HI(2903).
Cercospora sp. On leaves. **1**:HI(144). **3**:HI(2903).
Chalara sp. On leaves. **1**:HI(144). **3**:HI(2903).
Cladosporium carpophilum. On leaves. **3**:HI(2903).
C. cladosporioides. On leaves. **1**:HI(144,2903).
Cladosporium sp. On leaves. **1**:HI(144). **3**:HI(2903).
Coremiella sp. On leaves. **1**:HI(144). **3**:HI(2903).
Cylindrocarpon sp. On leaves. **1**:HI(144). **3**:HI(2903).
Cylindrocephalum sp. On leaves. **1**:HI(2903).
Drechslera biseptata. On leaves. **1**:HI(144). **3**:HI(2903).
Drechslera sp. On leaves. **1**:HI(144). **3**:HI(2903).
Epicoccum nigrum. On leaves. **1**:HI(144). **3**:HI(2903).
Exophiala mansonii. On leaves. **3**:HI(2903).
Fusarium episphaeria. On leaves. **1**:HI(144). **3**:HI(2903).
F. lateritium. On leaves. **1**:HI(144). **3**:HI(2903).
F. roseum. On leaves. **1**:HI(144). **3**:HI(2903).
Fusarium sp. On leaves. **3**:HI(2903).
Fusidium sp. On leaves. **1**:HI(144). **3**:HI(2903).
Gilmaniella sp. On leaves. **1**:HI(144). **3**:HI(2903).
Gliocladium roseum. On leaves. **1**:HI(144). **3**:HI(2903).
G. viride. On leaves. **1**:HI(144). **3**:HI(2903).
Gliomastix murorum var. *felina*. On leaves. **1**:HI(144). **3**:HI(2903).
Humicola fuscoatra. On leaves. **1**:HI(144). **3**:HI(2903).
Malbranchea sp. On leaves. **1**:HI(144). **3**:HI(2903).
Microdochium nivale. On leaves. **1**:HI(144). **3**:HI(2903).
Monilia sp. On leaves. **1**:HI(144). **3**:HI(2903).
Monocillium sp. On leaves. **1**:HI(144). **3**:HI(2903).
Monodictys glauca. On leaves. **1**:HI(144). **3**:HI(2903).
Nigrospora oryzae. On leaves. **3**:HI(2903).
Nigrospora sp. On leaves. **1**:HI(144). **3**:HI(2903).
Nodulisporium gregarium. On leaves. **1**:HI(144). **3**:HI(2903).
Oidiodendron sp. On leaves. **1**:HI(144). **3**:HI(2903).
Paecilomyces sp. On leaves. **1**:HI(144). **3**:HI(2903).
Penicillium camembertii. On leaves. **1**:HI(144). **3**:HI(2903).
P. canescens. On leaves. **1**:HI(144). **3**:HI(2903).
P. capsulatum. On leaves. **3**:HI(2903).
P. citreonigrum. On leaves. **1**:HI(144). **3**:HI(2903).
P. citrinum. On leaves. **1**:HI(144). **3**:HI(2903).
P. cyaneum. On leaves. **1**:HI(144). **3**:HI(2903).
P. diversum. On leaves. **1**:HI(144). **3**:HI(2903).
P. funiculosum. On leaves. **1**:HI(144). **3**:HI(2903).
P. glabrum. On leaves. **1**:HI(144). **3**:HI(2903).
P. herquei. On leaves. **1**:HI(144). **3**:HI(2903).
P. janthinellum. On leaves. **1**:HI(144). **3**:HI(2903).
P. miczynskii. On leaves. **1**:HI(144). **3**:HI(2903).
P. simplicissimum. On leaves. **1**:HI(144). **3**:HI(2903).
P. spinulosum. On leaves. **1**:HI(144). **3**:HI(2903).
Penicillium sp. On leaves. **1**:HI(144). **3**:HI(2903).
Periconia sp. On leaves. **1**:HI(144). **3**:HI(2903).
Periconiella angusiana. On leaves. **3**:HI(2903).
Phialophora sp. On leaves. **1**:HI(144). **3**:HI(2903).
Piricauda sp. On leaves. **1**:HI(144). **3**:HI(2903).
Pithomyces chartarum. On leaves. **1**:HI(144). **3**:HI(2903).
Pyricularia sp. On leaves. **1**:HI(144). **3**:HI(2903).
Rhinocladiella sp. On leaves. **1**:HI(144). **3**:HI(2903).
Rhodotorula sp. On leaves. **1**:HI(144). **3**:HI(2903).
Scolecobasidium humicola. On leaves. **1**:HI(144). **3**:HI(2903).
Sporidesmium sp. On leaves. **1**:HI(144). **3**:HI(2903).
Stachybotrys chartarum. On leaves. **1**:HI(144). **3**:HI(2903).
Stemphylium sp. On leaves. **3**:HI(2903).
Sympodiella sp. On leaves. **1**:HI(144). **3**:HI(2903).
Thermomyces lanuginosus. On leaves. **1**:HI(144). **3**:HI(2903).
Thielaviopsis sp. On leaves. **1**:HI(144). **3**:HI(2903).
Torula sp. On leaves. **1**:HI(144). **3**:HI(2903).
Trichoderma viride. On leaves. **1**:HI(144). **3**:HI(2903).
Trichothallus hawaiiensis. **3**:HI(2903).
Tritirachium sp. On leaves. **1**:HI(144). **3**:HI(2903).
Verticillium sp. On leaves. **1**:HI(144). **3**:HI(2903).
Zasmidium cellare. On leaves. **1**:HI(144). **3**:HI(2903).
Zygosporium gibbum. On leaves. **1**:HI(144). **3**:HI(2903).

Deuteromycotina - Coelomycetes
Asteromella sp. On leaves. **1**:HI(144). **3**:HI(2903).
Botryodiplodia sp. On leaves. **1**:HI(144). **3**:HI(2903).
Chaetophoma sp. On leaves. **1**:HI(144). **3**:HI(2903).
Coniothyrium sp. On leaves. **1**:HI(144). **3**:HI(2903).
Cylindrosporium sp. On leaves. **1**:HI(2903).
Dinemasporium sp. On leaves. **1**:HI(144,2903). **3**:HI(2903).
Diplodia sp. On leaves. **3**:HI(2903).
Diplodina sp. On leaves. **1**:HI(144,2903).
Harknessia shearii. On dead leaves. **5**:HI(2548,2784,3549).
Marssonina sp. On leaves. **1**:HI(144). **3**:HI(2903).
Melanconium sp. On leaves. **1**:HI(144). **3**:HI(2903).
Monochaetia sp. On leaves. **1**:HI(144). **3**:HI(2903).
Pestalotia stevensonii. On leaves. **1**:HI(144). **3**:HI(2903).
Pestalotia sp. On leaves. **1**:HI(144). **3**:HI(2903).
Phoma glomerata. On leaves. **3**:HI(2903).
Phoma sp. On leaves. **3**:HI(2903).
Phyllosticta sp. On leaves. **1**:HI(144). **3**:HI(2903).
Sphaceloma sp. On leaves. **1**:HI(144). **3**:HI(2903).
Sphaeropsis sp. On leaves. **1**:HI(144). **3**:HI(2903).
Sporonema sp. On leaves. **1**:HI(144). **3**:HI(2903).
Stagonospora sp. On leaves. **1**:HI(144). **3**:HI(2903).

Myrciaria O. Berg
1. *M. cauliflora* (DC.) Bergius [jaboticaba]
Ascomycotina
Botryosphaeria quercuum. Dieback. **1**:FL(35).

Myrtus L.
1. *M. communis* L. [Greek myrtle, Indian buchu, myrtle, Swedish myrtle]
2. *M. compacta* Ridl.
Oomycetes
Phytophthora cinnamomi. Root rot. **2**:CA(3624,4010).
Basidiomycotina - Other
Armillaria tabescens. Root rot. **1**:FL(35).
Ceriporia xylostromatoides. **1**:AZ(1202).
Phlebia concentrica. **1**:OR(1174).
Deuteromycotina - Hyphomycetes
Alternaria sp. Leaf spot. **1**:FL(35).
Cylindrocladium sp. Leaf spot. **1**:FL(35).
Pseudocercospora myrticola. Leaf spot. **1**:FL(35).
Deuteromycotina - Coelomycetes
Pestalotia unicolor. Leaf spot. **1**:LA(82).
Deuteromycotina - Other
Sclerotium rolfsii. Southern blight, root and stem rot. **1**:FL(35,82).

Pimenta Lindl.
1. *P. dioica* (L.) Merr. (=*P. officinalis* Lindl.) [allspice, pimento]
Basidiomycotina - Uredinales
Puccinia psidii. Autoecious. **1**:Unk.(82), FL(35,2253).
Deuteromycotina - Hyphomycetes
Cylindrocladium scoparium. Root rot. **1**:FL(35).

Pseudanamomis Kausel
1. *P. umbellulifera* (Kunth) Kausel
Oomycetes
Pythium sp. Root rot. **1**:FL(35).

Psidium L.
1. *P. guajava* L. [apple guava, common guava, yellow guava]
2. *P. littorale* Raddi var. *longipes* (O. Berg) Fosb. (=*P. cattleianum* Sabine) [cattley guava, purple strawberry guava]
3. *Psidium* sp.
Oomycetes
Phytophthora cinnamomi. Root rot. **2**:HI(2903).
P. citricola. Fruit rot. **1**:HI(1936).
Pythium sp. Root rot. **1**:FL(35). **2**:FL(35).
Zygomycotina
Mucor hiemalis. Soft rot. **1**:HI(2903).
Rhizopus stolonifer. Blossom-end rot, fruit rot. **1**:HI(2648,2903).
Ascomycotina
Botryosphaeria dothidea. Twig dieback, brown rot of fruit. **1**:FL(35), HI(2001).
B. ribis. On limbs. **1**:FL(82), HI(82,2903).
Diaporthe sp. On limbs. **1**:HI(2903).
Glomerella cingulata. Anthracnose. **1**:FL(35,82), HI(82,2903), TX(82). **2**:FL(35).
Guignardia sp. Blossom-end rot. **1**:HI(2903).
Limaciniella psidii. **1**:HI(2903).
Orbilia leucostigma. **1**:HI(2903).
Seuratia globifera. On leaves. **2**:HI(2337).
Trichothyrium reptans. **1**:HI(2903).
Basidiomycotina - Other
Armillaria tabescens. Root rot. **1**:FL(35,82). **2**:FL(35).
Auricularia auricula. **1**:HI(2903).
Crepidotus alabamensis. **1**:HI(3706).
C. nephrodes. **1**:HI(2903,3706).
Cyptotrama asprata. On branches. **3**:HI(2903).
Earliella scabrosa. **1**:HI(2903).
Pellicularia koleroga. Thread blight. **1**:FL(82).
Trametes versicolor. Wound rot. **1**:CA(82). **3**:CA(1110).
Deuteromycotina - Hyphomycetes
Alternaria citri. Fruit decay. **1**:CA(82). **3**:CA(1110).
Alternaria sp. Leaf spot. **1**:FL(35).
Atichia solaridiscoidea. On leaves. **2**:HI(2337).
Cercospora psidii. Leaf spot. **1**:FL(35,82). **2**:FL(35).
Corynespora cassiicola. Leaf spot. **1**:FL(35).
Fusarium sp. Root rot. **1**:FL(35).
Gliocephalotrichum bulbilium. On decaying wood. **1**:HI(2903).
Phymatotrichopsis omnivora. Root rot. **1**:Unk.(82).
Triscelophorus monosporus. On leaf litter. **1**:HI(2903). **2**:HI(2903).
Deuteromycotina - Coelomycetes
Colletotrichum gloeosporioides. Anthracnose. **1**:Unk.(82), FL(35). **2**:FL(35).
[*Coryneum umbonatum*]. Stem canker. **1**:FL(35).
Phoma sp. Blossom-end rot, fruit rot. **1**:FL(35), HI(2903).
Phomopsis psidii. Fruit spot. **1**:HI(2903).
Phomopsis sp. Leaf spot. **1**:FL(35).
Phyllosticta psidiicola. On fruits. **1**:HI(2784).
Sphaceloma psidii. Spot anthracnose. **1**:FL(35).
Deuteromycotina - Other
Rhizoctonia solani. Thread blight. **1**:FL(35).
Rhizoctonia sp. Root rot. **1**:HI(2903).

Syzygium Gaertn.
(= *Jambosa* Adans.)
1. *S. cumini* (L.) Skeels [blackplum, jambolan, jambolan-plum, jambool, jambu, Java-plum]
2. *S. jambos* (L.) Alston (=*Eugenia jambos* L.) [Malabar-plum, roseapple]
3. *S. malaccense* (L.) Merr. & L. M. Perry (=*Eugenia malaccensis* L.) [large-fruited roseapple, Malay-apple, pomerac jambos, roseapple]
4. *Syzygium* sp. (=*Jambosa sp.*)
Ascomycotina
Asterinella puiggarii. Anthracnose, leaf spot. **1**:FL(35). **2**:FL(35).
Glomerella cingulata. Anthracnose. **1**:FL(35). **2**:FL(35).
Seuratia millardetii. **4**:HI(2784).
Basidiomycotina - Other
Armillaria tabescens. Root rot. **1**:FL(35). **2**:FL(35).
Crepidotus alabamensis. **2**:HI(3706). **3**:HI(2903).
C. amygdalosporus. **3**:HI(3706).
C. mollis. **2**:HI(3706). **3**:HI(3706).
Deuteromycotina - Hyphomycetes
Cercospora sp. Leaf spot. **1**:FL(35).
Fusarium sp. Root rot. **2**:FL(35).
Deuteromycotina - Coelomycetes
Colletotrichum gloeosporioides. Anthracnose. **1**:FL(35). **2**:FL(35).
Pestalotia sp. **3**:HI(2903).
Phyllosticta eugeniae. Leaf spot. **1**:FL(35). **2**:FL(35).
P. jambosicola. On dead limbs. **4**:HI(2784).

Tristania R. Br. in Aiton
1. *T. conferta* R. Br. [brisbane-box]
Basidiomycotina - Other
Pseudomerulius curtisii. **1**:HI(2903).
Deuteromycotina - Coelomycetes
Macrophomina phaseolina. Charcoal rot. **1**:HI(2903).

Nelumboaceae

Nelumbo Adans.

1. *N. lutea* (Willd.) Pers. (=*N. pentapetala* (Walter) Fernald) [American lotus, pondnuts, water chinquapin, wonkapin, yanquapin]
2. *N. nucifera* Gaertn. [East Indian lotus, Hindu lotus, oriental lotus, sacred lotus]

Deuteromycotina - Hyphomycetes
Alternaria nelumbii. Leaf spot. **1:**OK(82,2872). **2:**MD(82), NJ(82), NY(82), PA(82).
Cercospora sp. Leaf spot. **1:**FL(35).
Pseudocercospora nymphaeacea. Leaf spot. **1:**IL(312), IN(82), TX(82,3597).
Deuteromycotina - Coelomycetes
Dothiorella nelumbii. On flower parts. **1:**DE(82), MD(82).

Nepenthaceae

Nepenthes L.

1. *Nepenthes* sp.

Deuteromycotina - Hyphomycetes
Cercospora sp. Leaf spot. **1:**FL(35).
Deuteromycotina - Coelomycetes
Colletotrichum sp. Leaf spot. **1:**FL(35).

Nyctaginaceae

Abronia Juss.

1. *A. latifolia* Eschsch. [yellow sandverbena]
2. *A. mellifera* Douglas ex Hook.
3. *Abronia* sp.

Oomycetes
Peronospora oxybaphi. Downy mildew. **3:**TX(82).
Basidiomycotina - Uredinales
[*Puccinia aristidae*]. **3:**AZ(82), CA(82), CO(82), NM(82).
Deuteromycotina - Hyphomycetes
Heterosporium abroniae. Leaf spot. **1:**CA(1468). **3:**CA(82), TX(82).
Deuteromycotina - Coelomycetes
Ascochyta oxybaphi. On living leaves. **2:**OR(3202,3421).

Allionia L.

1. *A. incarnata* L. [trailing four-o'clock]
2. *Allionia* sp.

Oomycetes
Albugo platensis. White-rust. **1:**AZ(82), NM(82), TN(82), TX(82).
Basidiomycotina - Uredinales
[*Puccinia aristidae*]. **2:**AZ(82), NM(82), TX(82).
Deuteromycotina - Hyphomycetes
Cercospora oxybaphi. Leaf spot. **2:**IA(82), KS(82), WI(82).
Heterosporium oxybaphi. On stems. **2:**NE(82), TX(82).
Phymatotrichopsis omnivora. Root rot. **2:**TX(82).

Boerhavia L.

1. *B. coulteri* (J. D. Hook.) S. Wats.
2. *B. erecta* L.
3. *B. repens* L. (=*B. coccinea* Mill.)
4. *Boerhavia* sp.

Oomycetes
Albugo platensis. White-rust. **2:**TX(82). **3:**HI(82,2903). **4:**AZ(82,1195), FL(35,82), NM(82), TX(82).
Basidiomycotina - Uredinales
[*Puccinia aristidae*]. **2:**AZ(741,1203).
P. aristidae var. *chaetariae.* **4:**AZ(3981).
Deuteromycotina - Hyphomycetes
Cercospora furfurella. Leaf spot. **3:**FL(35). **4:**TX(82).
Phymatotrichopsis omnivora. Root rot. **2:**TX(82). **4:**TX(82).
Deuteromycotina - Coelomycetes
Ascochyta oxybaphi. Leaf spot. **2:**TX(82,3597).
Microdiplodia bonnayae. **1:**CA(2780).
Deuteromycotina - Other
Rhizoctonia solani. Web blight. **3:**FL(35).

Bougainvillea Commerson & Juss.

1. *B. glabra* Choisy ex DC. [paperflower]
2. *B. spectabilis* Willd.
3. *B. umbellifera* (J. R. Forster & G. Forster) Seem. [kepau, papala]
4. *Bougainvillea* sp.

Oomycetes
Phytophthora nicotianae var. *parasitica.* Flower and leaf blight. **1:**FL(35). **4:**FL(29,35).
Pythium splendens. Root rot. **3:**HI(2903).
Ascomycotina
Glomerella sp. Leaf spot. **4:**FL(35).
Nectria cinnabarina. Dieback. **2:**FL(35).
Basidiomycotina - Other
Corticium sp. Tip dieback. **4:**FL(35).
Scytinostroma portentosum. **2:**FL(128).
Deuteromycotina - Hyphomycetes
Alternaria sp. Leaf spot. **4:**FL(35).
Cercosporidium bougainvilleae. Leaf spot. **1:**FL(35), HI(2903). **2:**FL(35,3372). **4:**FL(35).
Cladosporium arthrinioides. On leaves. **4:**TX(82).
Fusarium sp. Root rot. **2:**FL(35). **4:**FL(35).
Deuteromycotina - Coelomycetes
Colletotrichum sp. Anthracnose. **1:**FL(35). **2:**FL(35). **4:**FL(35).
Coniothyrium sp. Stem rot of cuttings. **4:**FL(35).
Phoma sp. Leaf spot. **1:**FL(35).
Phomopsis sp. Leaf spot. **1:**FL(35).
Phyllosticta sp. Leaf spot. **4:**FL(35).
Deuteromycotina - Other
Rhizoctonia solani. Leaf, root and stem rot. **4:**FL(35).
Rhizoctonia sp. **3:**HI(2903).

Mirabilis L.

1. *M. comatus* (Standl. in Sm.) Standl. (=*Oxybaphus comatus* (Standl. in Sm.) Weath.)
2. *M. jalapa* L. [beauty-of-the-night, common four-o'clock, marvel-of-Peru]
3. *M. longiflora* L.
4. *M. multiflora* (Torr.) A. Gray [Colorado four-o'clock]
5. *M. nyctaginea* (Michx.) MacMill. (=*Oxybaphus nyctagineus* (Michx.) Sweet)
6. *M. wrightiana* A. Gray (=*M. longiflora* L. var. *wrightiana* (A. Gray) Kearn. & Peebles)
7. *Mirabilis* sp.

Oomycetes
Albugo platensis. White rust. **2:**FL(35), TX(82). **5:**AL(82).
Peronospora oxybaphi. Downy mildew. **5:**KS(82), SD(82).
Basidiomycotina - Uredinales
[*Puccinia aristidae*]. **1:**AZ(1203). **4:**AZ(82,1203).
P. esclavensis. Heteroecious: aecial host. **3:**AZ(1195,1203). **4:**AZ(82), NM(82). **6:**AZ(1203,3981). **7:**AZ(1203).
P. subnitens. **1:**AZ(3981). **4:**AZ(3981).
Deuteromycotina - Hyphomycetes
Cercospora mirabilis. On living leaves. **2:**FL(35), TX(82,3597).
C. oxybaphi. Leaf spot. **5:**IA(82), IL(82), KS(82), NE(82), OH(82), TX(82), WI(1332).
Phymatotrichopsis omnivora. Root rot. **2:**TX(82). **5:**TX(82).
Deuteromycotina - Coelomycetes
Ascochyta oxybaphi. Leaf spot. **5:**IA(82), WI(82,1338).
Phyllosticta sp. Leaf spot. **5:**WI(1360).
Deuteromycotina - Other
Rhizoctonia solani. Root and stem rot. **2:**FL(35).

Pisonia L.

1. *P. aculeata* L.
2. *P. sandwicensis* Hildebr.
3. *P. umbellifera* (J. Presl & G. Forster) Seem. [bird-catcher-tree, para-para]

Oomycetes
Phytophthora sp. Basal stem rot. **3:**FL(35).
Ascomycotina
Echidnodes pisoniae. **2:**HI(2903). **3:**HI(2903).
Stigmochora variegata. On leaves. **2:**HI(193).
Deuteromycotina - Hyphomycetes
Fusarium sp. Root rot. **1:**FL(35).

Deuteromycotina - Coelomycetes
Colletotrichum sp. Anthracnose. **1**:FL(35).
Phyllosticta sp. Leaf spot. **1**:FL(35).

Rockia Heimerl
1. *R. sandwicensis* Heimerl
Ascomycotina
Echidnodes pisoniae. **1**:HI(2903).

Nymphaeaceae

Nuphar Sm. in Sibth. & Sm.
1. *N. advena* (Sol. in Aiton) W. T. Aiton (=*Nymphaea advena* Sol. in Aiton) [common spatterdock]
2. *N. luteum* (L.) Sibth. & Sm. [yellow water-lily]
3. *N. luteum* (L.) Sibth. & Sm. subsp. *variegatum* (Engelm. ex G. W. Clinton) E. Beal (=*N. variegatum* Engelm. ex G. W. Clinton)
4. *N. polysepalus* Engelm.

Ascomycotina
Mycosphaerella pontederiae. Leaf spot. **1**:ME(82), MI(82), NY(82), VA(82), WI(82,1330).
Basidiomycotina - Ustilaginales
Rhamphospora nymphaeae. White smut. **1**:CT(82,1077), IL(82,1077), MA(82,1077), NY(1077), WI(82,1077).
Deuteromycotina - Hyphomycetes
Botryosporium madrasense. **2**:FL(1684).
Dichotomophthoropsis nymphaearum. Isolated from leaves. **3**:MN(1767).
Pseudocercospora nymphaeacea. **1**:DE(3543).
Deuteromycotina - Coelomycetes
Phyllosticta fatiscens. Leaf spot. **1**:IA(82), NJ(82), NY(82), WI(82). **4**:WA(3202).
P. nymphaeacea. **1**:TX(82).
P. nymphaeicola. On leaves. **1**:IL(82,3590).
Deuteromycotina - Other
Sclerotium hydrophilum. **3**:MN(2888).
Sclerotium sp. Isolated from leaves. **3**:MN(1771).

Nymphaea L.
1. *N. alba* L. [European white water-lily]
2. *N. odorata* Aiton (=*Castalia odorata* Woodv. & W. Wood) [American water-lily, fragrant water-lily, pond lily]
3. *N. tuberosa* Paine [magnolia water-lily, tuberous water-lily]
4. *Nymphaea* sp.

Oomycetes
Pythium marsipium. Leaf and stem rot. **2**:MA(82), NY(82), WI(82). **3**:MA(82), NY(82), WI(82).
Ascomycotina
Mycosphaerella pontederiae. Leaf spot. **2**:VA(82). **4**:MI(82).
Basidiomycotina - Ustilaginales
Rhamphospora nymphaeae. White smut. **1**:DC(1077). **2**:NE states(82), CT(1077), MA(1077), NY(1077), OH(1077), OK(82,1077,2872), WI(82,1077). **3**:NE states(82), IA(1077), IL(1077), NY(1077), OK(82), WI(82,1077). **4**:IN(1077), MA(82), MD(82), NJ(82), NY(82), TX(82), WI(82).
Deuteromycotina - Hyphomycetes
Alternaria sp. Leaf spot. **4**:TX(82).
Dichotomophthoropsis nymphaearum. Leaf spot. **2**:MN(1767). **3**:MN(1767). **4**:MD(82), NJ(82), NY(82).
Pseudocercospora nymphaeacea. Leaf spot. **3**:IL(82), MI(82), MO(82), WI(82). **4**:IL(82), MS(82,2720), TX(82), VA(82).
Ramularia nymphaearum. Leaf spot. **2**:WA(82,3202). **4**:MD(82), NY(82), WA(82).
Deuteromycotina - Coelomycetes
Phyllosticta fatiscens. Leaf spot. **2**:VT(82).
P. nymphaeacea. **4**:AL(82), TX(82).
Deuteromycotina - Other
Sclerotium hydrophilum. **2**:MN(2888).
S. rolfsii. Southern blight. **2**:FL(35).
Sclerotium sp. Isolated from leaves. **2**:MN(1771).

Victoria Lindl.
1. *V. amazonica* (Poepp.) Sowerby [Amazon water-lily, Amazon water-platter, royal water-lily, water maize]

Oomycetes
Pythium sp. Leaf spot. **1**:FL(35).
Deuteromycotina - Coelomycetes
Septoria sp. Leaf spot. **1**:FL(35).

Nyssaceae

Camptotheca Decne.
1. *C. acuminata* Decne.

Basidiomycotina - Uredinales
Puccinia sp. **1**:FL(35).
Deuteromycotina - Hyphomycetes
Cercospora camptothecae. Leaf spot. **1**:FL(35).

Nyssa L.
1. *N. aquatica* L. [cotton-gum, large tupelo, water tupelo, wild olive]
2. *N. ogeche* J. Bartram ex Marsh. [Ogeechee lime tupelo]
3. *N. sylvatica* Marsh. (=*N. multiflora* Wangenh.) [black gum, pepperidge, sour gum, upland tupelo]
4. *N. sylvatica* Marsh. var. *biflora* (Walter) Sarg. (=*N. biflora* Walter) [swamp tupelo]
5. *Nyssa* sp.

Oomycetes
Phytophthora cactorum. Seedling top wilt. **3**:MO(82).
Ascomycotina
Botryosphaeria ribis. On limbs. **3**:PA(82).
Byssosphaeria diffusa. On dead wood. **3**:GA(82). **5**:GA(82,1438).
Ceratocystis coerulescens. Wood stain. **1**:SE states(82,779), FL(35). **3**:FL(35). **5**:S states(82).
C. moniliformis. Blue stain of wood. **1**:AR(82), FL(35), LA(82). **3**:AR(82), FL(35), LA(82). **5**:S states(82).
Ceratostomella nyssicola. **3**:PA(82).
Ceratostomella sp. Blue stain of wood. **5**:Range of host(82).
Ciboria nyssaegena. On fallen fruits. **3**:NJ(82,850).
Daldinia vernicosa. On wood. **3**:MA(82), VA(82).
Dermatella cucurbitaria. On twigs. **3**:NJ(82).
Diaporthe paulula. On limbs and twigs. **3**:AL(82), GA(82,1438), NY(82). **5**:MD(2783), NJ(82).
Diatrype asterostoma. **3**:GA(1438). **4**:GA(1438).
Diatrypella favacea. On dead limbs. **3**:GA(1438), NC(1303,1307).
D. subfulva. **3**:GA(1438).
Dicarpella georgiana. On leaves. **2**:GA(82,1438). **3**:GA(82,1438). **4**:GA(1438).
Didymella segna. On twigs. **3**:NJ(82).
Discohainesia oenotherae. On leaves and limbs. **3**:VA(82). **5**:VA(82).
Endoxyla cirrhosa. **3**:NJ(82).
E. rostrata. **3**:NJ(82).
Eutypella juglandicola. **3**:GA(1438).
E. leprosa. On dead twigs. **3**:NC(1303).
Hansenula polymorpha. In frass. **3**:LA(3596).
Hypoxylon fuscum. **5**:Range of host(82).
H. rubiginosum. On dead branches. **3**:AL(82), GA(1438). **5**:Range of host(82).
H. tinctor. **5**:GA(82,1438).
H. weldenii. **5**:LA(658).
Isothea nyssae. On leaves. **1**:Unk.(82).
Massariovalsa sudans. On dead limbs. **4**:GA(1438). **5**:NJ(82), NY(82), VA(82).
Mollisia sp. **3**:GA(1438).
Mycosphaerella nyssicola. Leaf spot. **1**:SE states(82), AL(82), FL(35), GA(82). **2**:GA(82,1438). **3**:E states(82), FL(82), GA(82,1438), MI(82), NC(82,3940), NY(82), PA(82), SC(82), VA(82), WV(82). **5**:SE states(82), FL(634), GA(82).
Myriangium duriaei. Parasitic on scale insects. **1**:SE states(82), LA(82). **2**:SE states(82), GA(1438), LA(82). **3**:E states(82), GA(1438,1745), LA(82), NC(3944). **4**:GA(1438). **5**:Gulf states(82), SE states(82).
Nectria galligena. Trunk canker. **3**:E states(82), CT(82).
Nectria sp. Canker. **3**:CT(82).
Requienella seminuda. On bark. **3**:NJ(82). **5**:NJ(320).
Rhytidhysteron rufulum. On dead limbs. **1**:FL(82), GA(82). **2**:FL(82),

GA(82). **3**:GA(1438).
Rosellinia corticium. On dead limbs. **5**:AL(82).
Rosellinia sp. On dead limbs. **3**:NC(1303).
Rutstroemia renispora. On fallen leaves. **3**:PA(82), SC(82), VA(82).
Tubeufia pezizula. Grayish-olive stain of wood. **5**:LA(82).
Ustulina deusta. **3**:GA(1438). **5**:Range of host(82).
Valsa ambiens. On dead limbs. **3**:NY(3415). **5**:NE states(82).
V. ceratosperma. On limbs and twigs. **3**:GA(1438), NJ(3415), NY(82), SC(82), WV(82). **5**:NE states(82), SC(634,3415).

Basidiomycotina - Uredinales

Aplopsora nyssae. Telial host. **1**:KY(82), MS(82,2720), TX(82), VA(82). **2**:MS(82,2720). **3**:AL(82), MD(82), SC(82), VA(82).

Basidiomycotina - Other

Abortiporus biennis. **3**:IN(82).
Armillaria mellea. Butt rot, root and shoestring rot. **3**:FL(35), NC(1303). **5**:FL(35).
Bjerkandera adusta. Wood rot. **5**:MD(82), MS(82).
Ceriporia tarda. Wood rot. **3**:IN(82). **5**:Unk.(82).
Cerrena unicolor. Wood rot. **3**:VA(82). **5**:Unk.(82).
Chondrostereum purpureum. Wood rot. **3**:VA(543). **5**:IN(82).
Coriolopsis rigida. On wood. **1**:FL(35). **4**:FL(213).
Corticium appalachiensis. **3**:TN(2043).
C. roseum. On dead limbs. **3**:MD(128).
Crepidotus nyssicola. On wood. **4**:FL(1544).
Daedaleopsis confragosa. On wood. **3**:IN(82). **5**:Unk.(82), IN(82).
Dendrocorticium roseocarneum. On wood. **3**:GA(491,1439).
Dendrothele macrodens. On bark. **3**:LA(2084).
D. microspora. On bark. **3**:PA(2084).
Fomitella supina. Wood rot. **5**:TX(82).
Ganoderma applanatum. White rot. **3**:MD(82), VA(82). **5**:Range of host(82).
Hericium erinaceus. Spongy white rot. **3**:NC(82). **5**:NC(82).
Hymenochaete agglutinans. **3**:NC(1307).
Ischnoderma resinosum. Wood rot. **3**:IN(82). **5**:IN(82).
Laxitextum bicolor. Wood rot. **1**:LA(82,543). **5**:LA(82).
Lentinellus ursinus. On wood. **3**:NC(1303,1307).
Lentinus crinitus. Wood rot. **1**:AL(82).
L. tigrinus. On wood. **1**:LA(82), MS(82,2720). **3**:LA(82), MS(82,2720).
Lenzites betulina. Wood rot. **1**:LA(82). **3**:IN(82), MD(82), TX(82). **5**:Range of host(82).
Meruliopsis corium. **5**:DC(128), SC(634).
Microporellus obovatus. Wood rot. **1**:AR(82), LA(82). **3**:AR(82), LA(82). **5**:AR(82,2675), LA(82,2675).
Oxyporus latemarginatus. White rot. **1**:FL(35). **2**:FL(82). **5**:FL(82).
O. populinus. White spongy rot. **3**:MA(82), PA(82). **5**:CT(82).
Panellus pusillus. Wood rot. **1**:AL(82). **5**:AR(82).
Pellicularia koleroga. Thread blight. **1**:LA(82). **3**:LA(82).
Perenniporia fraxinea. **3**:AR(82), LA(82).
Phanerochaete rimosa. White rot. **5**:FL(433).
P. sordida. White rot. **5**:FL(433).
Phellinus punctatus. Wood rot. **2**:FL(82). **3**:FL(35). **5**:Gulf States(82).
P. robustus. **3**:GA(82). **5**:GA(82).
Phlebia rufa. **3**:DC(128).
Pleurotus ostreatus. Invades sapwood and sometimes heartwood of living trees after fire damage. **1**:LA(82), MS(82,2720). **3**:LA(82), MS(82).
Pseudoclitocybe cyathiformis. **3**:NC(1303,1307).
Pycnoporus cinnabarinus. Wood rot. **5**:Unk.(82).
P. sanguineus. **1**:LA(82).
Rigidoporus lineatus. White pocket rot. **1**:AR(82), LA(82), MS(82,2720). **2**:AR(82), LA(82), MS(82,2720). **5**:AR(82), MS(82).
R. ulmarius. On wood. **3**:FL(35,82). **5**:FL(82).
Schizophyllum commune. On dead limbs. **5**:Range of host(82).
Schizopora paradoxa. Wood rot. **3**:IN(82), MI(82), PA(82). **5**:Unk.(82).
Sebacina incrustans. On saplings. **5**:NC(128).
Septobasidium fumigatum. Parasitic on scale insects. **5**:E states(82), Gulf states(82), MS(82).
S. pseudopedicellatum. Parasitic on scale insects. **5**:E states(82), Gulf states(82), AL(82), FL(82).
Septobasidium sp. Parasitic on scale insects. **3**:FL(35).
Steccherinum ochraceum. **5**:LA(82,2675), MS(2675).
Stereum hirsutum. White rot. **3**:LA(82), MS(82,2720), PA(543), VA(543). **5**:IN(82), LA(82), MS(82), NY(82), OH(82), VA(82).
S. ostrea. White rot. **1**:MS(82). **3**:MS(2720), NC(1303). **5**:MS(82).
S. striatum. **3**:CT(543), NJ(543). **5**:FL(543), NJ(543).
Trametes hirsuta. Wood rot. **5**:Range of host(82), LA(2675).
T. subectypus. On wood.*T. subectypus*. On wood. **5**:SE states(2681).
T. versicolor. Wood rot. **5**:Range of host(82).
Trichaptum biforme. Wood rot. **5**:IN(82).
Tulasnella violea. **5**:PA(2673).
Tyromyces galactinus. Wood rot. **3**:TN(82). **5**:TN(82).

Deuteromycotina - Hyphomycetes

Cercospora nyssae. Leaf spot. **3**:FL(35), TX(82,3597). **5**:FL(35).
C. nyssae-sylvaticae. On living leaves. **3**:WI(1354).
Conoplea globosa. Canker. **3**:NE states(82).
Morrisographium hispidulum. On limbs. **3**:NJ(82), NY(82), PA(82), VA(82).
Phymatotrichopsis omnivora. Root rot. **5**:TX(82).
Verticillium albo-atrum. Wilt. **3**:IL(992,1558).

Deuteromycotina - Coelomycetes

Macrophomina phaseolina. Charcoal root rot. **3**:FL(35).
Mastigosporella nyssae. On leaves. **3**:GA(2548). **4**:GA(2548).
Phoma nyssicarpa. On fruits and limbs. **1**:SC(82). **3**:WV(82).
Phoma sp. Dieback. **1**:FL(35).
Phyllosticta nyssae. Leaf spot. **3**:NC(1303,3940). **4**:NC(3940). **5**:SC states(82), SE states(82).
Sphaceloma sp. Spot anthracnose. **5**:FL(35).
Stagonospora nyssaecola. On twigs. **3**:NY(82).
Tubakia dryina. Leaf spot. **3**:AL(82), IL(82), OK(82,2873), VA(82,1246).

Deuteromycotina - Other

Rhizoctonia solani. Damping-off, basal stem rot. **1**:FL(35).

Oleaceae

Chionanthus L.

1. *C. virginicus* L. [fringetree, old-man's-beard]
2. *Chionanthus* sp.

Ascomycotina

Botryosphaeria pyriospora. On twigs. **1**:NJ(82), WV(82).
Dermea chionanthi. On limbs. **1**:MD(82), WV(82).
Didymosphaeria chionanthi. On dead leaves. **1**:GA(82,1438).
Guignardia sp. Leaf spot. **1**:FL(35).
Hypoxylon argillaceum. On limbs. **2**:VA(2403).
Hysterographium mori. On limbs. **1**:MD(82), VA(82).
Mycosphaerella punctiformis. **1**:GA(1438).
Nectria pseudotrichia. On bark. **1**:NC(82), NJ(82).
Phyllactinia guttata. Powdery mildew. **1**:MD(82).
Valsa cypri. On dead limbs. **1**:WV(82).

Basidiomycotina - Other

Daedaleopsis confragosa. On wood. **1**:MD(82).

Deuteromycotina - Hyphomycetes

Alternaria sp. Leaf spot. **1**:FL(35).
Cercospora chionanthi. Leaf spot. **1**:E states(82), FL(35), NC(3944), OK(2872), WV(82).
Phymatotrichopsis omnivora. Root rot. **1**:TX(82).

Deuteromycotina - Coelomycetes

Coniothyrium chionanthi. On limbs. **1**:NY(1012).
Dothiorella diatrypea. On dead limbs. **1**:NJ(82).
Macrophoma sp. Leaf spot. **1**:FL(35).
Phomopsis sp. Leaf spot. **1**:FL(35).
Phyllosticta chionanthi. Leaf spot. **1**:FL(35), NJ(82), WV(82).
Septoria chionanthi. Leaf spot. **1**:SC(82,634), TX(82).
S. eleospora. Leaf spot. **1**:SC(82), TX(82).
Septoria sp. Leaf spot. **1**:FL(35).

Deuteromycotina - Other

Rhizoctonia solani. Root rot. **1**:FL(35).

Fontanesia Labill.

1. *F. fortunei* Carrière

Basidiomycotina - Other

Armillaria mellea. Root rot. **1**:CA(2900).

Forestiera Poir. in Lam.

1. *F. acuminata* (Michx.) Poir. [swamp privet]
2. *F. neomexicana* A. Gray [desert olive]
3. *Forestiera* sp.

Ascomycotina

Microsphaera neomexicana. **2:**AZ(366).
M. penicillata. Powdery mildew. **2:**NM(2112). **3:**IL(82), TX(82).

Basidiomycotina - Uredinales

Coleosporium minutum. Heteroecious: telial host. **1:**FL(35,82), TX(82).
Puccinia sparganioides. **1:**FL(35,82), TX(82).

Deuteromycotina - Hyphomycetes

Phymatotrichopsis omnivora. Root rot. **1:**TX(82).

Deuteromycotina - Coelomycetes

Diplodia sp. Stem gall. **1:**FL(35).
Phyllosticta sp. On leaves. **1:**FL(82).

Forsythia Vahl

1. *F. × intermedia* Zab. *F. suspensa × F. viridissima*.
2. *F. suspensa* (Thunb.) Vahl
3. *F. viridissima* Lindl.
4. *Forsythia* sp.

Oomycetes

Phytophthora cinnamomi. Root rot. **4:**NC(1303).

Ascomycotina

Botryosphaeria ribis. Cane blight. **2:**MD(2262). **1:**MD(2262).
Nectriella pironii. Stem gall. **4:**FL(35).
Sclerotinia sclerotiorum. Twig blight. **4:**NC(82).

Basidiomycotina - Other

Dendrophora albobadia. On wood. **4:**MD(543), OK(2872).
Trametes hirsuta. On wood. **4:**OK(2872).

Deuteromycotina - Hyphomycetes

Alternaria forsythiae. **2:**DC(1481).
Alternaria sp. Leaf spot. **1:**OK(2872). **4:**IA(82).
Botrytis sp. **4:**CT(2974).
Phymatotrichopsis omnivora. Root rot. **4:**TX(82).

Deuteromycotina - Coelomycetes

Colletotrichum sp. Anthracnose. **4:**FL(35).
[*Coryneum umbonatum*]. Blight. **4:**MS(2720).
Diplodia sp. On stems. **4:**FL(35).
Phomopsis sp. Stem galls. **4:**KY(82), WA(3202).
Phyllosticta discincola. Leaf spot. **4:**MD(82).
P. forsythiae. **4:**CT(82).
P. terminalis. Leaf spot. **3:**OK(2872). **4:**FL(35,82), TX(82).

Deuteromycotina - Other

Sclerotium rolfsii. Southern blight. **4:**GA(82).

Fraxinus L.

1. *F. americana* L. [white ash]
2. *F. americana* L. var. *biltmoreana* (Beadle) W. Wright ex Fernald (=*F. biltmoreana* Beadle)
3. *F. caroliniana* Mill. (=*F. platycarpa* Michx.) [Carolina ash, pop ash, water ash]
4. *F. excelsior* L. [European ash]
5. *F. latifolia* Benth. (=*F. oregona* Nutt.) [Oregon ash]
6. *F. macropetala* Eastw.
7. *F. nigra* Marsh. [black ash]
8. *F. pennsylvanica* Marsh. (=*F. pennsylvanica* Marsh. var. *pennsylvanica* , *F. pennsylvanica* Marsh. var. *lanceolata* (Borkh.) Sarg., *F. pennsylvanica* Marsh. var. *subintegerrima* (Vahl) Fernald, *F. viridis* F. Michx., *F. lanceolata* Borkh., *F. tomentosa* Michx., *F. profunda* (Bush) Bush) [green ash, red ash]
9. *F. quadrangulata* Michx. [blue ash]
10. *F. velutina* Torr. [velvet ash]
11. *F. velutina* Torr. var. *toumeyi* Rehd.
12. *Fraxinus* sp.

Chytridiomycetes

Synchytrium fraxini. Leaf gall. **12:**LA(627).

Oomycetes

Pythium sp. Root rot. **12:**FL(35).

Ascomycotina

Acrospermum foliicola. **8:**GA(1438).
Anthostomella picaceum. **8:**ND(377).
Biscogniauxia nummularia var. *rumpens*. On wood. **12:**SC(2403).
Botryosphaeria dothidea. **8:**GA(1438).
B. obtusa. On limbs. **1:**E states(82). **12:**E states(82).
B. ribis. On limbs. **8:**GA(82).
Ceratocystis fraxinopennsylvanica. Brown stain, associated with *Leperisinus californicus* (Coleoptera). **8:**ND(1565).
Chlorociboria aeruginosa. Green wood stain. **12:**Range of host(82).
Cryptosphaeria eunomia. On dead limbs. **8:**ND(82,377). **12:**IA(3629).
C. eunomia var. *fraxini*. **8:**ND(377).
Cucurbitaria fraxini. On limbs. **1:**WV(82). **8:**ND(82,377).
Dactylospora stygia. **12:**ND(377).
Daldinia concentrica. On wood. **12:**Range of host(82), IA(1225).
D. vernicosa. **12:**ND(377).
Dasyscyphus puberulus. On leaves. **12:**SC(849).
Diatrype albopruinosa. **1:**IA(3629).
Diatrypella favacea. **1:**GA(1438). **8:**GA(1438).
D. fraxini. **8:**GA(1438).
D. quercina. **1:**GA(1438).
Discostroma corticola. **8:**GA(1438).
Dothidea puccinioides. **8:**Unk.(82), GA(1438).
Durandiella fraxini. **1:**GA(128), MA(128), NH(128), VT(128). **2:**GA(1438). **7:**NH(128), VT(128). **12:**CT(1381), GA(1381), MA(1381), MI(1381), NH(1381), NY(128,1381), VT(1381).
Erysiphe cichoracearum. Powdery mildew. **1:**CA(3975).
Eutypa lata. **8:**ND(377).
E. scabrosa. **8:**ND(377).
Eutypa sp. **1:**GA(1438).
Eutypella angulosa f. *fraxini*. **8:**ND(377).
E. juglandicola. **1:**GA(1438). **8:**GA(1438). **12:**MD(2925).
E. leprosa. **8:**IA(3629). **12:**IA(3629).
E. scoparia. **12:**LA(2925).
Eutypella sp. **8:**GA(1438).
Glonium sp. **8:**GA(1438).
Herpotrichia macrotricha. On leaves. **12:**ME(189).
Hypoderma commune. **1:**GA(1438).
Hypomyces rosellus. **8:**GA(1438).
Hypoxylon albocinctum. On wood. **12:**IA(1225).
H. argillaceum. **2:**GA(1438). **8:**GA(1438).
H. fuscum. **2:**GA(1438).
H. hypophlaeum. **8:**GA(1438).
H. illitum. On wood. **12:**VA(2403).
H. investiens. **8:**GA(1438).
H. regale. On wood. **12:**IA(1225), OH(2403).
H. rubiginosum. **8:**GA(1438). **12:**LA(2675).
H. rubrostromaticum. **8:**GA(1438).
H. stygium. **12:**GA(1438).
H. tinctor. **12:**GA(82,1438).
Hypoxylon sp. **12:**GA(82).
Hysterographium fraxini. On dead limbs. **5:**CA(81), OR(82,3202). **8:**ND(377), SD(2243). **10:**AZ(662). **12:**ME(82), MI(82), ND(82), OR(82).
Laestadia fraxinicola. On leaves. **1:**NY(82).
Lanzia longipes. On petioles. **1:**NY(505,1960).
L. luteovirescens. **12:**ND(377).
Leptosphaeria fraxini. On leaves. **1:**MO(82).
Lophiostoma triseptatum f. *fraxini*. **8:**ND(377).
Lophodermium punctiforme. On leaves. **12:**NY(82).
Melanomma pulvis-pyrius f. *fraxini*. **8:**ND(377).
Microsphaera penicillata. **12:**IL(82).
M. syringae. On leaves. **1:**PA(363).
Microthyrium microscopicum. On leaves. **12:**ME(82).
Mycosphaerella effigurata. Leaf spot. **1:**E states(82), GA(1438), IA(82), IN(3943), MS(82), NC(3943), NY(2456), OK(82), VA(3943). **2:**FL(35), GA(82,1438), NC(3943). **5:**OR(82,3202), WA(82,3202). **7:**NC(82), WI(82). **8:**C states(82), NE states(82), GA(1438), ND(3943), VA(3943). **9:**MO(967). **12:**IN(82), NC(82), ND(82), NY(82), TN(82), VA(82).
M. fraxini. On fallen leaves. **12:**VA(82).
M. fraxinicola. Leaf spot. **1:**E states(82), FL(35), KS(82), MN(82), NC(1303). **2:**FL(35), GA(82,1438). **5:**OR(3202), WA(3202). **6:**NM(1195). **8:**E states(82), GA(82,1438), IL(82). **9:**KS(82), MO(82). **10:**ID(1195), OK(82,2872,2873). **12:**E & C states(82).
M. punctiformis. On leaves. **12:**ME(82).
Mycosphaerella sp. **3:**GA(82,1438).
Nectria cinnabarina. On limbs. **12:**LA(82), ME(82).
N. coccinea. Branch canker. **7:**NY(82). **12:**NE states(82).
N. flammea. **8:**GA(1438).

N. galligena. Canker. **7:**NY(82).
N. pseudotrichia. **8:**GA(1438).
Nitschkia confertula. **1:**NC(3944). **8:**GA(1438). **9:**GA(82).
Nummularia broomeiana. On wood. **7:**IA(1225). **8:**GA(1438). **12:**SE states(1810).
Phyllactinia guttata. Powdery mildew. **1:**NC states(82). **5:**W states(82), CA(81), WA(3202). **6:**NM(1195). **7:**C states(82). **8:**C states(82), NC(1303), ND(82,377), SD(2243). **9:**IN(82), MI(82). **10:**UT(1195). **12:**C states(82), NE states(82), W states(82).
Pleosphaeria echinata. **8:**GA(1438).
Pulveria porrecta. On wood. **12:**NY(2232).
Requienella seminuda. On bark. **12:**MI(320), NY(320).
Rosellinia aquila. **12:**GA(1438).
Schiffnerula pulchra. Sooty mold. **12:**Unk.(82), IA(1681), MS(2720).
Schizoxylon berkeleyanum. **12:**ND(377).
S. compositum. **12:**KS(3225).
S. insigne. **8:**GA(1438).
Sphaeronaemella carnea. On bark. **12:**NY(966).
Stictis schizoxyloides. **12:**CO(3225).
Teichospora gregaria. **8:**ND(377).
Thyronectria austroamericana. On bark. **8:**NE(82).
Trematosphaeria pertusa. On dead limbs. **8:**KS(318).
Valsa ambiens. On limbs. **12:**Range of host(82), IA(1226).
V. ambiens subsp. *leucostomoides*. On limbs. **8:**IA(1226). **12:**Range of host(82), IA(1226).
V. cypri. **12:**Range of host(82).
V. leucopsis. On limbs. **12:**Range of host(82).
Valsa sp. On limbs. **12:**Range of host(82).
Valsaria insitiva. **8:**IA(1226).
Venturia fraxini. Root rot. **12:**FL(35,82).
Xylaria carpophila. On wood. **8:**OK(618).

Basidiomycotina - Uredinales

Puccinia sparganioides. Heteroecious: aecial host. **1:**E states(82), CT(3770), FL(35), IL(2568), MS(2720). **7:**NE states(82), MN(82). **9:**IL(82), PA(82). **10:**KS(82), TX(82). **12:**E & C states(82), FL(35), MA(2127), OK(2872), PA(3780).

Basidiomycotina - Other

Acanthophysium cerussatum. **10:**AZ(1202).
Aleurodiscus oakesii. On bark. **12:**MI(2083), NY(2083).
Aleurodiscus sp. **5:**WA(3202).
Antrodia albida. **10:**AZ(1202).
A. heteromorpha. **10:**AZ(1202).
A. malicola. Brown spongy rot. **12:**NY(82).
A. oleracea. **10:**AZ(1202).
A. radiculosa. **12:**AZ(2167).
Armillaria mellea. Root rot. **10:**CA(2900). **12:**VT(59).
Basidiodendron caesiocinereum. **10:**AZ(1202).
Bjerkandera adusta. On wood. **1:**ID(3202). **8:**OK(618). **12:**Unk.(82).
Bourdotia galzinii. **10:**AZ(1202).
Ceriporia tarda. Wood rot of logs. **12:**Unk.(82).
C. viridans. **7:**MN(128).
C. xylostromatoides. **10:**AZ(1202).
Clavicorona pyxidata. **1:**NH(884).
Coniophora arida var. *suffocata*. **5:**WA(3202).
C. puteana. **5:**WA(3202).
Coriolopsis gallica. **10:**AZ(1202).
C. rigida. On wood. **12:**AR(82), LA(82,2675).
Corticium spretum. **5:**WA(2118,3202).
Crepidotus subfibrillosus. On logs. **12:**MI(1544).
Cylindrobasidium torrendii. **5:**WA(3202).
Cyphellopsis anomala. **7:**MN(1197).
Cystostereum murrayi. **7:**MN(1197).
Cytidia patelliformis. **5:**WA(3202).
Dacrymyces minor. On wood. **12:**IA(128).
Daedaleopsis confragosa. On wood. **12:**Unk.(82).
Datronia mollis. On wood. **8:**OK(618).
Dendrophora albobadia. On wood. **10:**AZ(542,543).
D. versiformis. On wood. **12:**Unk.(82).
Dendrothele acerina. On bark. **12:**E states(82).
D. candida. On bark. **1:**NC(3944). **12:**E states(82), MO(2084).
D. griseo-cana. **1:**OH(653).
D. macrodens. On bark. **1:**CT(2084), GA(2084), WV(2084). **7:**CT(2084), OH(2084). **12:**GA(2084), MA(2084), MI(2084), NC(1303,2084), NH(2084), NY(2084), PA(2084).
D. microspora. On bark. **12:**LA(2084), NY(2084).
D. seriata. On bark. **8:**FL(2084). **12:**FL(2084), LA(2084).
D. strumosa. On bark. **8:**LA(2084).
Dichostereum effuscatum. **7:**MN(128,1197).
Exidiopsis glaira. On limbs. **12:**IA(128).
E. grisea. On bark. **12:**IA(2318).
Fomitella supina. **12:**LA(2675).
Fomitopsis meliae. Brown cubical rot. **12:**E states(82), S states(82), AR(2675), LA(2320,2675), MS(2320), VA(2320).
F. pinicola. On wood. **12:**E states(82), S states(82).
F. spraguei. On wood. **12:**Unk.(82), AR(2675), LA(2675).
Ganoderma applanatum. On wood. **12:**E states(82), S states(82), FL(35).
G. australe. **12:**FL(82).
G. lucidum. Trunk rot. **10:**AZ(1202). **12:**LA(82), MS(82,2720).
Gloeoporus dichrous. On wood. **8:**ND(378). **12:**Unk.(82), AR(2675), LA(2675).
Grandinia pruni. **10:**AZ(1202).
G. spathulata. **1:**AZ(128). **10:**AZ(1194,1202).
Helicobasidium brebissonii. Collar rot. **12:**TX(82).
Helicogloea lagerheimii. **10:**AZ(1202).
Heliocybe sulcata. **10:**AZ(1202).
Henningsomyces candidus. **10:**AZ(1202).
Hericium erinaceus. Usually on living trees. **12:**Unk.(82).
Hexagonia hydnoides. Wood rot. **12:**AR(82), LA(82), TX(82).
Hymenochaete agglutinans. **7:**NY(128).
H. leonina. **10:**AZ(1202).
Hyphoderma roseocremeum. **10:**AZ(1202).
Hyphodermopsis polonensis. **7:**MN(128,1197).
Hyphodontia stipata. **10:**AZ(1202).
Hypochnicium analogum. **10:**AZ(128,1194,1202).
H. lundellii. **10:**AZ(1202).
Inonotus hispidus. White heart rot. **7:**MI(82), MN(3488). **8:**MS(2720).
Intextomyces contiguus. **5:**WA(3202).
Irpex lacteus. On dead wood. **8:**ND(378). **12:**Unk.(82).
Irpex sp. **5:**OR(3202).
Ischnoderma resinosum. **5:**OR(3202).
Junghuhnia nitida. Wood rot of logs. **12:**Unk.(82).
Laetiporus sulphureus. On wood. **10:**AZ(1202). **12:**Unk.(82).
Laxitextum bicolor. **1:**IN(543). **12:**Unk.(82), AR(2675), LA(2675), NY(543).
Lentinellus ursinus. **10:**AZ(1194,1202).
Lentinus tigrinus. Sapwood rot. **8:**MS(2720). **10:**AZ(1194,1202). **12:**MS(82).
Oxyporus latemarginatus. **12:**S states(82).
Paullicorticium pearsonii. **10:**AZ(1202).
Peniophora cinerea. **5:**WA(3202).
P. fuscomarginata. **10:**AZ(1202).
P. nuda. **10:**AZ(128,1202).
P. tamaricicola. White rot. **10:**AZ(1180,1202).
Perenniporia fraxinea. On wood. **12:**E states(82), S states(82), LA(2675).
P. fraxinophila. White rot. **1:**NC states(82), NE states(82), IA(82), MS(82), OH(653), OK(82,2872). **5:**WA(3202). **8:**W states(82), ND(3761), NE(2982,2984), NM(82), SD(378,2243). **10:**AZ(82,1202), NM(82). **12:**E & C states(82), MT(3202,3832).
P. ohiensis. **1:**MN(1197).
P. tenuis. On wood. **10:**AZ(1194,1202). **12:**Unk.(82).
Phanerochaete allantospora. White rot. **10:**AZ(433,1202).
P. chrysorhiza. White rot. **10:**AZ(1202). **12:**NC(433).
P. rimosa. **7:**MN(128,1197).
Phellinus conchatus. On wood. **7:**MN(128,1197). **8:**MS(2720). **12:**E states(82), S states(82), LA(2675), VT(1224).
P. ferreus. **5:**OR(3202).
P. ferruginosus. **12:**E states(82), S states(82).
P. gilvus. On wood. **10:**AZ(1202). **12:**Unk.(82).
P. igniarius. On wood. **8:**MS(2720). **12:**E states(82), S states(82).
P. johnsonianus. White rot. **12:**LA(801,2169), VA(801,2169).
P. punctatus. White rot. **8:**ND(3759). **12:**Unk.(82), AR(2675), LA(2675), MS(2720).
P. robustus. On wood. **8:**MS(2720).
[*P. torulosus*]. **12:**AR(2675).
Phlebia ludoviciana. White rot. **12:**IL(2556).
P. rufa. **7:**MN(128,1197).

Physisporinus vitreus. Wood rot of logs. **12:**Unk.(82), LA(2675).
Pleurotus ostreatus. **1:**Unk.(82). **12:**Unk.(82).
Polyporus admirabilis. On wood. **12:**Unk.(82).
P. mori. On dead wood. **12:**Range of host(82), MT(3202).
P. squamosus. On wood. **12:**Unk.(82).
Porodisculus pendulus. On wood. **12:**NC(1303).
Postia rancida. **10:**AZ(1202).
P. stiptica. **12:**NY(82).
Pycnoporus cinnabarinus. On wood. **12:**Unk.(82).
Resupinatus applicatus. **10:**AZ(1202).
Rigidoporus ulmarius. **12:**Unk.(82).
Schizophyllum commune. On wood. **10:**AZ(1202). **12:**Range of host(82), FL(35).
Sebacina epigaea. On wood. **7:**WI(128).
S. helvelloides. At base of living trees. **7:**CT(128).
Septobasidium pseudopedicellatum. Felt fungus, parasitic on scale insects. **1:**NC(3944). **8:**GA(1439). **12:**SE states(82), FL(35), NC(1303).
Septobasidium sp. Felt fungus. **12:**SE states(82).
Sphaerobolus stellatus. **12:**ID(3202).
Spongipellis spumeus. On wood. **12:**Unk.(82).
Stereum ostrea. **12:**PA(543).
Subulicystidium longisporum. **7:**MN(128,1197).
Thelephora albidobrunnea. **1:**NC(3944).
Tomentella chlorina. **10:**AZ(1202).
T. crinalis. **7:**ME(128). **12:**ME(2039).
T. punicea. **12:**MN(2039).
T. rubiginosa. **12:**FL(2039).
T. violaceofusca. **12:**ME(2039).
Trametes ectypus. On wood. **8:**MS(2720).
T. hirsuta. On wood. **5:**OR(3202). **12:**Unk.(82).
T. versicolor. White rot. **5:**OR(3202), WA(3202). **8:**NC(1303). **12:**Unk.(82), OK(2872).
T. villosa. On wood. **12:**Unk.(82), AR(2675), LA(2675).
Trechispora farinacea. **10:**AZ(1202).
T. variseptata. White rot. **10:**AZ(438).
Tremella mesenterica. **2:**GA(1439).
Trichaptum biforme. On wood. **12:**Unk.(82).
Tulasnella violea. **10:**AZ(1202).
Tyromyces chioneus. **12:**NY(82).
T. fissilis. **12:**LA(82).
Wolfiporia cocos. **12:**VT(82).
Xeromphalina fraxinophila. On dead limbs and trunks. **12:**MI(3328).

Deuteromycotina - Hyphomycetes

Alternaria inconspicua. On living leaves. **8:**WI(1348).
Cercospora fraxinites. Leaf spot. **1:**FL(35), TX(82). **8:**AL(82). **12:**FL(35,82), LA(82,962), TX(82).
[*C. gymnocladi*]. **12:**IA(82).
C. lumbricoides. **12:**Unk.(82).
C. texensis. Leaf spot. **8:**TX(82,581).
Cercosporella trichophila. On leaves. **8:**WI(82). **12:**FL(35,82), WI(82).
Conoplea olivacea. **12:**OH(1672).
Corynespora olivacea. **1:**NY(3547).
Cristulariella moricola. Zonate leaf spot. **1:**WV(3680). **12:**FL(35).
Helicoma muelleri. **12:**MA(1271).
Helicomyces bellus. On wood. **12:**MI(1270).
Heterosporium fraxini. **12:**WA(3202).
Phymatotrichopsis omnivora. Root rot. **12:**AZ(82), TX(82).
Ramularia fraxinea. On leaves. **8:**WI(808). **12:**FL(35,82).
Sarcinella heterospora. **12:**IA(1681).
S. quadrata. Leaf spot. **3:**FL(1681). **12:**FL(35), SC(1681), TX(1681).
Sphaerosporium lignatile. **1:**DE(3543).
Spilocaea nebulosa. On leaves. **1:**WI(82).
Tubercularia nigricans. **1:**DE(3543).
Verticillium albo-atrum. Wilt. **1:**IL(1558). **4:**IL(1558). **7:**IL(1558). **8:**CO(82,1558,3611). **9:**IL(1558).
Verticillium sp. Wilt. **1:**PA(1090). **8:**PA(1090).
Xenosporium berkeleyi. **1:**NC(3944).

Deuteromycotina - Coelomycetes

Coniothyrium orni. **5:**CA(81).
Cornicurariella spina. On dead limbs. **1:**NC(3944). **12:**Range of host(82), CT(861), GA(861), MA(861), MI(861), NH(861), NY(861,2456,3549), VT(861).
Cylindrosporium fraxini. Leaf spot. **1:**NC states(82), NE states(82), W states(82), FL(35), IA(82), IN(82), KS(82), MO(82), MS(82,2720), NY(82), WI(82). **5:**W states(82), OR(3202), WA(3202). **7:**MI(82), OH(82), OK(2872). **8:**C states(82), MN(82), MO(82), ND(82,378), TX(82), WI(82,1348). **9:**MO(82). **10:**AZ(1194), OK(82,2872). **11:**OK(2872). **12:**NC states(82), NE states(82), W states(82), FL(35,82,634), IA(82), IN(82), KS(82), LA(82), MN(82), MO(82), OK(2872), TX(82).
Cytospora fugax. On limbs. **12:**ME(82).
C. minuta. **8:**NY(82). **12:**WV(82).
C. pruinosa. Twig canker, dieback. **1:**NC states(82). **7:**IA(82), OK(2872). **8:**IA(82), IN(82), OK(618). **12:**NC states(82).
C. sacculus. On limbs. **8:**WV(82).
Cytospora sp. Leaf spot. **1:**VA(3493). **12:**WA(3202).
Dendrophoma pruinosa. On dead limbs. **12:**NC states(82), NE states(82), WV(82).
Discula fraxinea. Anthracnose, leaf spot. **1:**E states(82), IA(82), MS(2720), NC(3944), VA(3493), WI(82,962). **7:**WI(82). **8:**IA(82), MI(82), NY(82), OK(618), WI(82), WV(82). **9:**IL(82). **10:**CA(82). **12:**E & C states(82), NC states(82), NE states(82), KY(1777), NC(1303).
Dothiorella concaviuscula. On limbs. **12:**Unk.(82).
D. everhartii. **12:**Unk.(82).
D. fraxinicola. Branch canker. **1:**IA(82). **8:**IA(82). **12:**IA(82), KS(82), NE(82).
Fusicoccum sp. Canker, dieback. **1:**NY(3032), WV(3910).
Gloeodes pomigena. On limbs. **12:**IN(82).
Hendersonulina fraxini. On dead limbs. **8:**KS(82).
Libertella sp. Twig dieback. **8:**OK(618).
Macrophoma hyalina. **12:**Unk.(82).
Marssonina gloeodes. On living leaves. **1:**WI(1336,1338).
M. punctiformis. On leaves. **1:**DE(82,962). **12:**DE(82), PA(82).
Micropera fraxini. **12:**OH(653).
Phloeosporina fraxini. Leaf spot. **5:**CA(81,961,1468). **8:**KS(973), LA(2463).
Phoma exigua. **8:**WI(1359).
P. infossa. On dead limbs. **8:**Unk.(82). **12:**NY(965).
P. pezizoides. On decorticated wood. **12:**WV(307).
Phyllosticta fraxinicola. Leaf spot. **1:**IL(1052). **8:**NY(2456), SD(2243). **10:**OK(2872).
P. innumera. Leaf spot. **5:**W states(82), CA(81). **12:**CA(82).
P. viridis. Leaf spot. **1:**NC(3944), OK(2872). **8:**ND(378), WI(1344). **12:**FL(35).
Phyllosticta sp. Leaf spot. **8:**WI(1350). **12:**FL(35), SD(2243).
Piggotia fraxini. Leaf spot. **1:**E states(82), IA(82), KS(963), MS(82,2720), OK(82,2872), WI(1329,1334). **5:**OR(82), WA(82). **7:**NC(82), WI(82). **8:**C states(82), NE states(82), MS(2720), OK(618), SD(2243), WI(1341). **12:**E states(82), WA(82), WI(82).
Pteridiospora spinosispora. Isolated from roots. **8:**MS(1060).
Pyrenochaeta fraxinina. On petioles. **12:**NY(1012).
Schizothyrella fraxini. On fallen leaves. **8:**KS(82).
Septogloeum sp. On leaves. **12:**NE(82).
Septoria submaculata. **1:**MO(82).
Sphaeropsis hyalina. Canker, twig dieback. **1:**NE states(82). **8:**KS(82), OK(618), SD(2243). **12:**NE states(82), Range of host(82).
S. malorum. Canker, on limbs. **8:**Unk.(82). **12:**Range of host(82).
S. profundae. Canker. **12:**Range of host(82), FL(35).
Sphaeropsis sp. Canker. **8:**Unk.(82).
Sporonema fraxini. On leaves. **5:**OR(3202), WA(2128).
Stagonospora sp. **8:**WI(1355).
Tubakia dryina. **1:**IL(82,1246). **7:**IL(1246). **8:**IL(82,1246).

Deuteromycotina - Other

Rhizoctonia crocorum. **12:**Unk.(82).
R. solani. Seedling blight. **8:**OK(2873). **12:**OK(82,2872).

Jasminum L.

1. *J. floridum* Bunge
2. *J. heterophyllum* Roxb.
3. *J. humile* L.
4. *J. multiflorum* (J. Burm.) Andr. [star jasmine]
5. *J. nitidum* Skan [angel-wing jasmine, Confederate jasmine, star jasmine, windmill jasmine]
6. *J. nudiflorum* Lindl. [winter jasmine]
7. *J. sambac* (L.) Aiton [Arabian jasmine]
8. *J. volubile* Jacq.
9. *Jasminum* sp.

Oomycetes
Phytophthora nicotianae var. *parasitica*. Root rot. **4**:HI(2903).
Phytophthora sp. Root rot. **9**:FL(35).
Pythium splendens. Root rot. **9**:FL(35), HI(2903).
Pythium sp. Root rot. **7**:HI(2903).
Zygomycotina
Choanephora infundibulifera. Blossom blight. **9**:FL(35,82).
Ascomycotina
Botryosphaeria obtusa. On twigs. **9**:AL(82).
Elsinoe jasmini. **2**:FL(2525). **3**:FL(2525).
Leptosphaeria sp. Leaf spot. **9**:FL(35).
Nectria pseudotrichia. **9**:AL(82).
Nectriella pironii. Stem canker. **9**:FL(35).
Schizothyrium rufulum. **9**:FL(2782).
Sclerotinia sp. Stem rot. **9**:FL(35).
Basidiomycotina - Uredinales
Gymnosporangium sp. **9**:FL(35).
Basidiomycotina - Other
Armillaria tabescens. Root rot. **9**:FL(35,82).
Erythricium salmonicolor. Limb blight. **8**:FL(35). **9**:FL(35).
Marasmiellus ramealis. On dead roots. **9**:TX(82).
Scytinostroma galactinum. White root rot. **9**:MD(82).
Deuteromycotina - Hyphomycetes
Alternaria sp. Leaf spot. **5**:FL(35). **8**:FL(35). **9**:FL(35).
Corynespora cassiicola. Leaf spot. **9**:FL(35).
Fusarium sp. **4**:HI(2903).
Pseudocercospora jasminicola. Leaf spot. **9**:FL(35).
Stemphylium sp. Leaf spot. **9**:FL(35).
Deuteromycotina - Coelomycetes
Actinopelte sp. Leaf spot. **9**:FL(35).
Aplosporella jasmini. On twigs. **9**:AL(82).
Colletotrichum gloeosporioides. Anthracnose, leaf spot. **9**:FL(35,82), TX(82).
Colletotrichum sp. Flower blight. **7**:HI(2299,2903).
Coniothyrium sp. Stem canker, dieback. **1**:OK(2872). **3**:OK(2872). **6**:OK(2872). **9**:FL(35).
Diplodia sp. Stem gall. **8**:FL(35).
Phoma sp. **6**:TX(1555).
Phomopsis sp. Stem gall. **9**:FL(35,82), TX(82).
Phyllosticta sp. Leaf spot. **9**:FL(35).
Rhabdospora jasmini. On dead stems. **9**:PA(82).
Deuteromycotina - Other
Rhizoctonia solani. Seedling blight, root and stem rot. **3**:OK(2872). **9**:FL(35).
Rhizoctonia sp. Root rot. **7**:HI(2903).
Sclerotium rolfsii. Southern blight and wilt. **4**:HI(2903). **9**:FL(35), Fl(82).

Ligustrum L.

1. *L. amurense* Carrière [Amur privet]
2. *L.* × *ibolium* E. F. Coe ex Rehd. *L. obtusifolium* × *L. ovalifolium*.
3. *L. indicum* (Lour.) Merr.
4. *L. japonicum* Thunb. [Japanese privet, wax-leaf privet]
5. *L. lucidum* Aiton [Chinese privet, glossy privet, Nepal privet, wax-leaf privet, white wax-tree]
6. *L. obtusifolium* Siebold & Zucc. (=*L. ibota* Siebold) [California privet]
7. *L. ovalifolium* Hassk. [California privet]
8. *L. sinense* Lour. [Chinese privet]
9. *L. vulgare* L. [common privet]
10. *Ligustrum* sp.

Oomycetes
Phytophthora sp. Root rot. **4**:FL(35). **5**:FL(35). **8**:FL(35). **10**:CA(82).
Pythium irregulare. Root rot. **4**:FL(35).
P. myriotylum. **5**:GA(1208).
P. splendens. Root rot. **4**:FL(35). **5**:FL(35).
Pythium sp. Root rot. **5**:GA(1208). **8**:FL(35). **10**:HI(2903), NC(1303).
Ascomycotina
Botryosphaeria dothidea. **9**:GA(1438).
B. obtusa. On limbs. **10**:AL(82), VA(82).
Diaporthe eres. On dead twigs. **10**:NY(82), VA(82).
D. ligustrina. **10**:Unk.(82).
Discohainesia oenotherae. On leaves and stems. **10**:Unk.(82).
Glomerella cingulata. Anthracnose, canker, dieback. **3**:MS(2720). **5**:FL(35). **7**:MS(2720). **8**:FL(35). **9**:OK(2872). **10**:E states(82), Range of host(82), NE(82), TX(82).
Hypoxylon rubiginosum. **9**:GA(1438).
Microsphaera penicillata. Powdery mildew. **1**:OK(2872). **9**:WI(1330). **10**:IN(82), MS(2720), NC(1303), NJ(82), OH(82).
Microsphaera sp. Powdery mildew. **4**:LA(1604).
Nectria cinnabarina. Canker. **9**:OK(2872).
Nectriella pironii. Stem gall. **4**:FL(35). **5**:FL(35). **8**:FL(35).
Rosellinia necatrix. Root rot. **10**:CA(82).
Schizoxylon ligustri. **10**:FL(3225), PA(3225).
Stictis radiata. **9**:GA(1438).
Togninia minima. On limbs. **9**:MA(190).
Basidiomycotina - Other
Armillaria mellea. Root rot. **9**:OK(2872). **10**:AR(82), CA(82), MS(82,2720), TX(82).
A. tabescens. Root rot. **4**:FL(35). **5**:FL(35). **10**:AR(82), FL(82).
Dendrophora albobadia. On wood. **1**:GA(491,1439). **9**:MD(543).
Dendrothele griseo-cana. On bark. **10**:MI(2084).
Erythricium salmonicolor. Limb blight. **5**:FL(35).
Ganoderma applanatum. Collar rot. **10**:NY(82).
Hyphodontia lanata. **10**:MS(443).
Meruliopsis corium. On wood. **1**:GA(491,1439). **9**:MD(128).
Pellicularia koleroga. Thread blight. **10**:FL(82).
Peniophora cinerea. On wood. **1**:GA(491,1439).
Phanerochaete ravenelii. On wood. **1**:GA(491,1439).
Stereum hirsutum. White rot. **5**:FL(35). **9**:CT(543). **10**:Range of host(82), PA(543).
Trametes versicolor. Wound rot. **10**:Range of host(82).
T. villosa. **10**:TX(82).
Deuteromycotina - Hyphomycetes
Alternaria alternata. Leaf spot. **4**:FL(35). **5**:FL(35).
Cercospora adusta. Leaf spot. **1**:OK(2872). **4**:FL(35). **5**:FL(35). **7**:OK(2872), TX(581). **2**:OK(2872). **10**:SE states(82), TX(82).
C. ligustricola. Leaf spot. **5**:FL(35). **8**:FL(35).
C. lilacis. Leaf spot. **4**:FL(35). **5**:FL(35). **8**:FL(35).
Corynespora cassiicola. Leaf abscission, leaf spot. **4**:FL(35). **5**:FL(35). **8**:FL(35,2412).
Cylindrocladium sp. Cutting rot. **8**:FL(35).
Fusarium oxysporum. Root rot. **8**:FL(35).
Fusarium sp. Root rot. **4**:FL(35). **5**:FL(35). **9**:OK(2872).
Helminthosporium sp. Leaf spot. **4**:FL(35). **5**:FL(35).
Oidium sp. Powdery mildew. **4**:FL(35).
Phymatotrichopsis omnivora. Root rot. **10**:AZ(82), OK(82,2872), TX(82).
Pseudocercospora ligustri. Cane blight, leaf spot. **4**:FL(35), GA(841), MS(2720). **5**:FL(35), TX(841). **7**:TX(841). **8**:FL(35). **9**:MS(2720). **10**:Gulf States(82), AL(285), MS(2720), NC(1303).
Ramularia sp. Leaf blight. **4**:FL(35). **5**:FL(35). **10**:NJ(82), WA(82,3202).
Thedgonia ligustrina. **10**:WA(3202).
Verticillium albo-atrum. Wilt. **1**:IL(1558).
Deuteromycotina - Coelomycetes
Colletotrichum sp. Anthracnose. **4**:FL(35). **5**:FL(35).
Coniothyrium sp. **9**:OK(2872).
Cytospora sp. Dieback. **7**:OK(2873).
Cytosporella sp. **9**:OK(2872).
Discogloeum concentricum. Leaf spot. **10**:TX(82).
Lasiodiplodia theobromae. Dieback. **4**:FL(35). **8**:FL(35).
Macrophoma sp. Leaf spot. **4**:FL(35).
Macrophomina phaseolina. Charcoal root rot. **4**:FL(35). **5**:FL(35).
Phoma ligustrina. On limbs. **6**:NY(3080).
Phomopsis ligustri-vulgaris. Phomopsis blight. **10**:PA(82).
Phomopsis sp. Stem gall, cutting rot, leaf spot. **4**:FL(35). **5**:FL(35). **8**:FL(35). **10**:TN(82).
Phyllosticta ovalifolii. Leaf spot. **10**:MD(82), MS(82,2720), TX(82).
Phyllosticta sp. Leaf spot. **4**:FL(35). **5**:FL(35). **8**:FL(35). **10**:TX(82), VA(82).
Sphaceloma sp. Scab. **5**:FL(35). **8**:FL(35).
Sphaeropsis sp. Twig dieback. **4**:FL(35).
Deuteromycotina - Other
Rhizoctonia ramicola. Thread blight. **5**:FL(35). **8**:FL(35).
R. solani. Root and stem rot. **4**:FL(35). **5**:FL(35). **8**:FL(35).
Sclerotium rolfsii. Southern blight. **4**:FL(35).

Noronhia J. F. Stadtmann ex Thouars

1. *N. emarginata* (Lam.) Thouars
2. *Noronhia* sp.

Oomycetes

Phytophthora sp. Root rot. **1**:FL(35).

Ascomycotina

Leptosphaeria sp. Leaf spot. **2**:FL(35).

Deuteromycotina - Hyphomycetes

Helminthosporium sp. Leaf spot. **1**:FL(35).

Deuteromycotina - Coelomycetes

Colletotrichum gloeosporioides. Leaf spot. **1**:FL(35).
Phyllosticta sp. Leaf spot. **1**:FL(35).

Olea L.

1. *O. europaea* L. [common olive]

Oomycetes

Phytophthora sp. **1**:CA(1110).
Pythium sp. Crown and root rot. **1**:FL(35), HI(2903).

Ascomycotina

Glomerella cingulata. **1**:CA(1110).
Zukalia purpurea. Black mildew, leaf spot. **1**:FL(35,82), GA(82).

Basidiomycotina - Other

Abortiporus biennis. Butt rot, root rot. **1**:CA(328,1110).
Armillaria mellea. Root rot. **1**:CA(82,1110), TX(82).
Ganoderma lucidum. **1**:AZ(18,1202).
Omphalotus olearius. **1**:CA(1110).

Deuteromycotina - Hyphomycetes

Mycocentrospora cladosporioides. Fruit spot. **1**:CA(82,1110).
Phymatotrichopsis omnivora. Root rot. **1**:TX(82).
Spilocaea oleaginea. Peacock spot. **1**:CA(81,82,1110).
Verticillium albo-atrum. Wilt. **1**:AZ(359), CA(3366).
V. dahliae. **1**:CA(1110,2881).

Deuteromycotina - Coelomycetes

Coleophoma oleae. **1**:CA(81,1110).
Colletotrichum gloeosporioides. Anthracnose. **1**:CA(82).
Phyllosticta oleae. Leaf spot. **1**:FL(35).
Septoria serpentaria. Leaf spot. **1**:FL(35).

Osmanthus Lour.

1. *O. americanus* (L.) Benth. & J. D. Hook. [American olive, devilwood, wild olive]
2. *O. × fortunei* Carrière. Probably a hybrid between *O. fragrans* and *O. heterophyllus*.
3. *O. fragrans* (Thunb.) Lour. [fragrant olive, sweet olive, tea olive]
4. *O. heterophyllus* (G. Don) P. Green (=*Olea aquifolium* Siebold & Zucc., *Olea ilicifolia* Hassk., *Ilex heterophylla* G. Don, *O. ilicifolius* (Hassk.) Hort. ex Carrière) [Chinese olive, false holly, holly olive, holly osmanthus]
5. *O. sandwicensis* (A. Gray) Knobl.
6. *Osmanthus* sp.

Ascomycotina

Asteridiella americana. **1**:Unk.(1452).
Asterina asterophora. Anthracnose. **1**:FL(35,82), GA(82).
Asterina sp. Anthracnose. **3**:MS(82,2720).
Asterinella intensa. Anthracnose, black leaf spot. **5**:HI(2903). **6**:HI(82).
Aulacostroma osmanthi. On leaves. **5**:HI(2784,2903). **6**:HI(82).
Calothyriella osmanthi. On leaves. **5**:HI(2903). **6**:HI(82).
Calothyrium osmanthi. On leaves. **5**:HI(2903). **6**:HI(82).
Capnodium elongatum. Sooty mold. **4**:TX(82).
Lembosia oleae. Black mildew. **1**:MS(82,2720).
Meliola amphitricha. Black mildew. **1**:S states(82), FL(35), MS(2720).
M. osmanthi. Black mildew. **5**:HI(1272,2903). **6**:HI(82,2784).
M. osmanthi var. *hawaiiensis*. Black mildew. **5**:HI(1272,1452,2903).
M. osmanthicola. Black mildew. **5**:HI(1272,1452).
M. osmanthina. Black mildew. **5**:HI(1272,1452).
Mycosphaerella oleina. On leaves. **1**:GA(82,1438).
Nectria erubescens. Leaf spot. **1**:FL(35,82).
Nectriella pironii. Stem gall. **3**:FL(35).
Rosellinia necatrix. Root rot. **4**:CA(82).
Schizothyrium discoideum. Anthracnose. **1**:FL(35,82), GA(82).
Venturia formosa. On leaves. **1**:FL(35,82).
Zukalia purpurea. Anthracnose, black leaf spot. **1**:FL(35). **6**:FL(82).

Basidiomycotina - Other

Armillaria mellea. Root rot. **3**:CA(82). **5**:HI(2011,2903).
A. tabescens. Dieback, root rot. **3**:LA(2815).
Fomes hawaiiensis. On wood. **5**:HI(2903). **6**:HI(82).

Deuteromycotina - Hyphomycetes

Alternaria sp. Leaf spot. **3**:FL(35).
Cercospora sp. Leaf spot. **3**:FL(35). **6**:FL(35).
Fumago salicina. Sooty mold. **4**:TX(82).
Isariopsis carnea. On leaves. **1**:FL(35,82).
Phymatotrichopsis omnivora. Root rot. **4**:TX(82).
Verticillium albo-atrum. Wilt. **4**:VA(1386).

Deuteromycotina - Coelomycetes

Colletotrichum gloeosporioides. Leaf spot. **3**:FL(35), MD(82).
Colletotrichum sp. Anthracnose. **3**:FL(35).
Phyllosticta oleae. Leaf spot. **1**:FL(35,82), NC(82,1303). **4**:TX(82).
P. osmanthi. On leaves. **2**:LA(82).
P. sinuosa. Leaf spot. **1**:FL(35,82), MS(82,2720). **4**:TX(82).
P. terminalis. Leaf spot. **1**:FL(35,82).
Septoria sp. On leaves. **1**:MS(82,2720).

Syringa L.

1. *S. amurensis* Rupr. [Amur lilac]
2. *S. × chinensis* Willd. *S. × persica × S. vulgaris*. [Chinese lilac]
3. *S. josikaea* J. Jacq. [Hungarian lilac]
4. *S. × persica* L. *S. afghanica × S. laciniata*. [Persian lilac]
5. *S. vulgaris* L. [common lilac]
6. *Syringa* sp.

Oomycetes

Phytophthora cactorum. Blossom blight, dieback. **5**:IA(82), MA(82), MD(82), MN(82), NJ(82), SD(2243).
P. syringae. Shoot blight. **5**:MD(82), NY(82).

Ascomycotina

Botryosphaeria obtusa. On limbs and twigs. **5**:NE states(82), OH(82).
Diaporthe rudis. On limbs. **5**:ME(82).
Erysiphe graminis f. sp. *avenae*. Powdery mildew. **5**:AR(3029).
Microsphaera penicillata. Powdery mildew. **1**:WI(1351). **5**:Range of host(82), Unk.(3976), FL(35), GA(1438,2405), MA(1390), MS(2720), NC(1303), OK(2872), SD(2243), WA(3202), WI(1348). **4**:IA(82), IL(82), MN(82), NE(82), OK(2872), WI(1339). **6**:FL(35), ND(377).
Pleospora herbarum. Secondary leaf spot. **5**:MD(82).
Sclerotinia sclerotiorum. Shoot blight. **5**:WA(82).
Venturia syringae. **5**:MA(177).
Xylaria sp. **5**:WA(3202).

Basidiomycotina - Other

Armillaria mellea. Root rot. **5**:CA(82), MS(82,2720).
Ceriporia xylostromatoides. **4**:AZ(1202).
Chondrostereum purpureum. On wood. **5**:OK(82,2872).
Dendrophora albobadia. **5**:DC(128,543).
Ganoderma applanatum. **5**:WA(3202).
Hymenochaete agglutinans. Stem girdle. **5**:CT(82,128).
Oxyporus latemarginatus. **5**:DC(2166).
Pellicularia koleroga. Thread blight. **5**:FL(82), MS(82,2720), NC(82).
Phellinus gilvus. Wood rot. **5**:MD(82).
P. punctatus. On living trees. **5**:ND(3759).
Trametes versicolor. **5**:MO(82), NC(82), NY(82), WA(3202).

Deuteromycotina - Hyphomycetes

Alternaria alternata. Leaf spot. **5**:Range of host(82).
A. brassicae. Leaf spot. **5**:Range of host(82).
Botrytis cinerea. Gray mold blight. **5**:NE states(82), NW states(82), WA(3202).
Botrytis sp. **6**:CT(2974).
Cercospora lilacis. Leaf spot. **5**:Range of host(82), AL(285), TX(82). **4**:MS(82,2720). **6**:FL(35).
C. macromaculans. Leaf spot. **5**:IA(82), OK(82,2872), TX(82), WI(82,1335).
Cercospora sp. On leaves. **6**:WI(1361).
Cladosporium herbarum. Leaf blotch. **5**:OK(2872).
Cladosporium sp. Leaf blotch. **5**:Range of host(82).
Cylindrocladium scoparium. Cutting rot. **5**:FL(35).
Heterosporium syringae. Blight, leaf blotch. **5**:NC(1303), NJ(82). **6**:MS(2720).
Macrosporium sp. **5**:Unk.(82).
Phymatotrichopsis omnivora. Root rot. **1**:TX(82). **5**:AZ(82), TX(82). **2**:Unk.(82).
Thielaviopsis basicola. Root rot. **5**:CT(82).

Deuteromycotina - Coelomycetes
Ascochyta syringae. Blight. **3**:WI(1347). **5**:NC(1303), OR(82), WA(3202), WI(82,1335).
Colletotrichum gloeosporioides. **5**:CT(82), MA(82).
Dendrophoma albomaculans. On twigs. **5**:NY(82).
D. syringae. On twigs. **5**:NY(82,832).
Dothiorella berengeriana. On limbs. **5**:OH(82).
Macrophoma halstedii. Leaf spot. **5**:CT(82), NJ(82), NY(82).
Myxosporium depressum. On twigs. **5**:NY(82,3080), OH(82), VA(82).
Phomopsis depressa. On twigs and capsules. **5**:NY(82), OH(82).
Phyllosticta halstedii. On living leaves. **5**:NJ(82,966).
P. porteri. Leaf spot. **5**:IL(82,3590).
P. syringae. Leaf spot. **5**:NY(82), OK(2872), WA(82), WI(64,82).
Phyllosticta sp. Leaf spot. **5**:MA(82).
Sphaeropsis syringae. On leaves and twigs, probably secondary. **5**:NJ(82), OK(2872), PA(82), SD(2243).
Deuteromycotina - Other
Rhizoctonia solani. Thread blight. **5**:FL(35).
Sclerotium rolfsii. Southern blight. **5**:MS(2720).

Onagraceae

Boisduvalia Spach
1. *B. densiflora* (Lindl.) S. Wats. [spike primrose]
2. *B. glabella* (Nutt.) Walp.
3. *B. stricta* (A. Gray) Greene
4. *Boisduvalia* sp.

Chytridiomycetes
Synchytrium sp. **1**:CA(1843,3401), OR(1843), OR(3202). **2**:ID(3202), UT(1843).
Basidiomycotina - Uredinales
Puccinia glabella. Autoecious. **2**:OR(3202). **4**:NV(82), OR(82), UT(82).
P. oenotherae. Autoecious. **1**:CA(81,82), ID(82,3202), NV(82), OR(82,3202), WA(82,3202). **3**:ID(3202), OR(3202), WA(3202). **4**:CA(82), ID(82), NV(82), OR(82), WA(82).
P. pulverulenta. Autoecious. **2**:ID(3202). **3**:OR(3202). **4**:ID(82), NV(82), OR(82), UT(82).
Deuteromycotina - Hyphomycetes
Ramularia montana. **2**:CA(81).
Ramularia sp. **1**:ID(3202), OR(3202), WA(3202). **2**:ID(3202), OR(3202), WA(3202). **3**:WA(3202).

Calylophus Spach
1. *C. serrulata* (Nutt.) Raven (=*Oenothera serrulata* Nutt., *Meriolix serrulata* Raf.)

Chytridiomycetes
Synchytrium sp. **1**:NE(1843).
Basidiomycotina - Uredinales
Puccinia dioicae. **1**:ND(378).
Deuteromycotina - Hyphomycetes
Cercospora oenotherae. **1**:WI(1357).
Deuteromycotina - Coelomycetes
Cryptostictis oenotherae. Leaf spot. **1**:OK(2872).
Septoria oenotherae. Leaf spot. **1**:OK(2872).

Camissonia Link
1. *C. andina* (Nutt. in Torr. & A. Gray) Raven (=*Oenothera andina* Nutt. in Torr. & A. Gray)
2. *C. bistorta* (Nutt. ex Torr. & A. Gray) Raven (=*Oenothera bistorta* Nutt. ex Torr. & A. Gray) [suncup]
3. *C. boothii* (Douglas in Hook.) Raven (=*Oenothera boothii* Douglas in Hook.)
4. *C. boothii* (Douglas) Raven subsp. *alyssoides* (Hook. & Arnott) Raven (=*Oenothera alyssoides* Hook. & Arnott)
5. *C. breviflora* (Torr. & A. Gray) Raven (=*Oenothera breviflora* Torr. & A. Gray)
6. *C. brevipes* (A. Gray) Raven (=*Oenothera brevipes* A. Gray)
7. *C. californica* (Nutt. ex Torr. & A. Gray) Raven (=*Oenothera leptocarpa* Greene)
8. *C. campestris* (Greene) Raven (=*Oenothera dentata* Cav.)
9. *C. chamaenerioides* (A. Gray) Raven (=*Oenothera chamaenerioides* A. Gray)
10. *C. cheiranthifolia* (Hornem. ex Spreng.) Raim. (=*Oenothera cheiranthifolia* Hornem. ex Spreng.)
11. *C. claviformis* (Torr. & Frém.) Raven
12. *C. contorta* (Douglas) Kearn. (=*Oenothera contorta* Douglas)
13. *C. graciliflora* (Hook. & Arnott) Raven (=*Oenothera graciliflora* Hook. & Arnott)
14. *C. minor* (A. Nels.) Raven (=*Oenothera minor* (A. Nels.) Munz)
15. *C. ovata* (Nutt. ex Torr. & A. Gray) Raven (=*Oenothera ovata* Nutt. ex Torr. & A. Gray)
16. *C. scapoidea* (Torr. & A. Gray) Raven (=*Oenothera scapoidea* Torr. & A. Gray)
17. *C. subacaulis* (Pursh) Raven (=*Oenothera heterantha* Nutt.) [suncup]

Chytridiomycetes
Synchytrium fulgens. **2**:CA(1843).
Synchytrium sp. **5**:MT(3202).
Oomycetes
Peronospora arthurii. On leaves. **9**:AZ(1195,3378). **14**:ID(3202), WA(3202).
Basidiomycotina - Uredinales
[*Puccinia aristidae*]. **9**:AZ(1203).
P. oenotherae. Autoecious. **1**:ID(3202), WA(3202). **3**:OR(3202), WA(3202). **4**:ID(3202), WA(3202). **5**:ID(3202), MT(3202). **6**:AZ(747,3981), CA(81). **7**:AZ(747,1195,1203,3379,3981). **8**:OR(3202). **10**:CA(81). **11**:AZ(747,3981). **12**:ID(3202), WA(3202). **13**:OR(3202). **14**:WA(3202). **15**:CA(82). **16**:CA(81). **17**:ID(3202), MT(3202), WA(3202), WY(1195).
P. pulverulenta. **16**:ID(3202).
P. subnitens. **9**:AZ(3981).
Uromyces plumbarius. **15**:CA(82).
Deuteromycotina - Coelomycetes
Septoria oenotherae. Leaf spot. **15**:CA(82).

Circaea L.
1. *C. alpina* L. (=*C. alpina* L. var. *alpina*)
2. *C. alpina* L. var. *pacifica* (Aschers. & Magnus) M. E. Jones (=*C. pacifica* Aschers. & Magnus)
3. *C. quadrisulcata* (Maxim.) Franch. & Sav. (=*C. latifolia* J. Hill)
4. *C. quadrisulcata* (Maxim.) Franch. & Sav. var. *canadensis* (L.) H. Hara
5. *Circaea* sp.

Basidiomycotina - Uredinales
Puccinia circaeae. Autoecious. **1**:AK(60,526,3098,3431), ID(3202), MT(3202), OR(3202), SD(2768), WA(3202). **2**:CA(81,642). **3**:IN(987). **4**:MA(3098), MD(3098), MI(3098), MS(2720), NJ(3098), NY(2456,3098), VA(3098). **5**:AK(526).
Basidiomycotina - Ustilaginales
Entyloma arnicale. **1**:OR(1077,3202).
Deuteromycotina - Coelomycetes
Phoma exigua. **3**:WI(1352).
Septoria sp. On leaves. **3**:WI(1354).
Stagonospora sp. **3**:WI(1355).

Clarkia Pursh
(= *Godetia* Spach.)
1. *C. amoena* (Lehm.) A. Nels. (=*Godetia amoena* (Lehm.) G. Don) [satinflower]
2. *C. gracilis* (Piper) A. Nels. & Macbr.
3. *C. pulchella* Pursh
4. *C. purpurea* (Curtis) A. Nels. & Macbr.
5. *C. rhomboidea* Douglas ex Hook.
6. *C. unguiculata* Lindl. (=*C. elegans* Douglas)
7. *Clarkia* sp. (=*Godetia sp.*)

Chytridiomycetes
Synchytrium echinulatum. **1**:CA(1849).
S. fulgens. Leaf gall. **6**:CA(82,1843).
Oomycetes
Peronospora arthurii. Downy mildew. **1**:CA(82). **6**:CA(82).
Phytophthora cactorum. Stem rot. **6**:NY(82).
Pythium debaryanum. Damping-off. **6**:CA(82).
Ascomycotina
Erysiphe cichoracearum. Powdery mildew. **1**:CA(1155).
Phaeosphaeria agminalis. **7**:OR(3202).
Basidiomycotina - Uredinales
Puccinia oenotherae. Autoecious. **1**:CA(81,82), OR(82,3202), WA(3202). **2**:OR(3202). **3**:ID(3202), OR(3202), WA(3202).

4:WA(3202). **5:**CA(81), ID(3202), WA(3202). **6:**CA(82), WA(82). **7:**CA(82), ID(82), NV(82), OR(82), WA(82).
P. pulverulenta. **5:**CA(642). **7:**CA(82).
Pucciniastrum pustulatum. Heteroecious: telial host. **1:**AK(526). **6:**AK(60,82,526), NY(82).

Deuteromycotina - Hyphomycetes

Alternaria alternata. Leaf spot. **6:**CA(82).
Botrytis cinerea. Gray mold leaf spot. **1:**AK(82). **6:**CA(82), NY(82).
Fusarium sp. Stem rot. **6:**CA(82).
Verticillium albo-atrum. Wilt. **6:**CA(82).

Deuteromycotina - Coelomycetes

Colletotrichum sp. Anthracnose. **6:**PA(82).

Deuteromycotina - Other

Rhizoctonia solani. Damping-off, root rot. **6:**CT(82). **7:**IL(82).

Epilobium L.

1. *E. alpinum* L. (=*E. anagallidifolium* Lam.)
2. *E. americanum* Hausskn.
3. *E. angustifolium* L. (=*Chamaenerion angustifolium* (L.) Scop.) [fireweed, great willowherb, wickup]
4. *E. boreale* Hausskn.
5. *E. brevistylum* Barbey (=*E. pringleanum* Hausskn., *E. brevistylum* Barbey var. *pringleanum* (Hausskn.) Jeps.)
6. *E. californicum* Hausskn.
7. *E. ciliatum* Raf. (=*E. adenocaulon* Hausskn., *E. glandulosum* Lehm. var. *adenocaulon* (Hausskn.) Fernald)
8. *E. ciliatum* Raf. subsp. *glandulosum* (Lehm.) P. Hoch & Raven (=*E. glandulosum* Lehm.)
9. *E. clavatum* Trel.
10. *E. coloratum* Biehler
11. *E. exaltatum* E. R. Drew
12. *E. glaberrimum* Barbey
13. *E. halleanum* Hausskn. (=*E. drummondii* Hausskn.)
14. *E. hirsutum* L.
15. *E. holosericeum* Trel.
16. *E. hornemannii* Rchb.
17. *E. hornemannii* Rchb. subsp. *behringianum* (Hausskn.) P. Hoch & Raven (=*E. behringianum* Hausskn.)
18. *E. junceum* G. Forster
19. *E. lactiflorum* Hausskn.
20. *E. latifolium* L. [river-beauty]
21. *E. leptocarpum* Hausskn. (=*E. leptocarpum* Hausskn. var. *macounii* Trel.)
22. *E. luteum* Pursh
23. *E. minutum* Lindl. ex Lehm.
24. *E. novomexicanum* Hausskn.
25. *E. occidentale* (Trel.) Rydb.
26. *E. palustre* L.
27. *E. paniculatum* Nutt. ex Torr. & A. Gray
28. *E. perplexans* Trel. ex A. Nels.
29. *E. platyphyllum* Rydb.
30. *E. rigidum* Hausskn.
31. *E. saximontanum* Hausskn. (=*E. stramineum* Rydb.)
32. *E. suffruticosum* Nutt.
33. *E. watsonii* Barbey (=*E. franciscanum* Barbey)
34. *Epilobium* sp.

Chytridiomycetes

Synchytrium epilobii. Stunting and curling of leaves. **1:**WA(3202). **19:**WY(1195). **33:**WA(1841,3202).

Oomycetes

Peronospora arthurii. On leaves. **23:**AZ(1195,3381).
Plasmopara epilobii. Downy mildew. **20:**AK(82,526,3431). **26:**AK(3098). **34:**IL(82), NY(82).
P. latifolii. **20:**AK(3098).
Pythium debaryanum. **20:**AK(3431).

Ascomycotina

Crocicreas cyathoideum. **3:**CO(505). **20:**AK(3431).
C. cyathoideum var. *cacaliae*. On dead stems. **3:**UT(505).
Diapleella coniothyrium. **34:**WA(3202).
Discohainesia oenotherae. On leaves. **3:**ME(82).
Ditopellopsis racemula. On stems. **3:**NY(82,195).
Dothidella adusta. On leaves. **3:**AK(82).
Erysiphe polygoni. Powdery mildew. **8:**WA(3202). **34:**WA(82).
Herpotrichia herpotrichoides. On stalks. **3:**OR(657). **34:**OR(189).
Hymenoscyphus scutula. **20:**AK(3431).
Leptosphaeria cylindrospora. On dead stems. **3:**CO(82), ID(82,3202), WA(82,3202,3249). **34:**AK(3249).
Massariosphaeria rubicunda. **3:**OR(657).
Microsphaera sp. Powdery mildew. **34:**IL(82).
Mycosphaerella hermione. **34:**WY(2924).
M. minor. On dead stem. **20:**AK(1938).
M. tassiana. On dead stem. **20:**AK(1938).
Nectria cinnabarina. On stems. **3:**AK(82,526).
Phaeosphaerella sp. **17:**AK(3431).
Phomatospora sp. **3:**AK(526).
Phyllactinia guttata. **33:**WA(3202).
Pleospora herbarum. On dead stem. **20:**AK(1938).
Sphaerotheca epilobii. **19:**CA(638).
S. fuliginea. Powdery mildew. **18:**CA(1155).
S. macularis. Powdery mildew. **1:**AK(3431). **5:**CA(81). **8:**AK(526). **10:**WA(3202). **16:**AK(526). **17:**AK(3431). **20:**AK(3431). **27:**CA(81), ID(3202), WA(3202). **33:**WA(3202). **34:**AK(82), CA(82), IA(82), KS(82), MI(82), NJ(82), OR(82,3202), SD(82), WA(82,3202), WI(82), WY(82).
Sydowiella fenestrans. On overwintered stems. **3:**AK(82,526), ID(82,3202), OR(657). **34:**AK(3431), WA(3202).
Venturia adusta. **3:**AK(526). **4:**AK(526). **16:**AK(526). **34:**AK(526).
V. asteromorpha. On stems. **3:**NY(82).

Basidiomycotina - Uredinales

Endophyllum alaskanum. **1:**AK(3098).
Puccinia dioicae. **3:**CO(82).
P. epilobii. Autoecious. **27:**ND(378). **34:**MI(82), WY(82).
P. gigantea. Autoecious. **3:**AK(60,82,526,3098), ID(82,3202), MT(82,3202), OR(3202), TX(82), WA(82,3202), WY(82). **34:**AK(526).
P. oenotherae. Autoecious. **5:**CA(81). **23:**OR(3202), WA(3202). **30:**OR(3202). **34:**CA(82).
P. pulverulenta. Autoecious. **7:**WY(2924). **8:**AK(60,526). **10:**WA(3202). **16:**AK(526). **20:**AK(526). **23:**CA(81), OR(3202), WA(3202). **27:**CA(81), ID(3202), MT(3202), OR(3202), SD(2243), WA(638,3202), WY(1195). **28:**WY(2924). **30:**OR(3202). **33:**ID(3202), MT(3202), WA(3202). **34:**W states(82), AK(82,526), CA(82,642), MT(3202).
P. scandica. Autoecious. **1:**AK(3098), WA(3202). **9:**CA(638,642,3098), WA(3098). **34:**UT(82), WA(82), WY(82).
P. veratri. Heteroecious: aecial host. **1:**AK(60,526), ID(3202), MT(1195,3202), WA(3202). **7:**AK(526). **16:**AK(60). **20:**AK(60,526,3097,3098). **27:**ID(3202), MT(3202), WA(3202). **34:**W states(82), AK(82,526,3098), NH(82), OR(3202), WA(82).
Pucciniastrum epilobii. Heteroecious: telial host. **1:**AK(1575), CO(1575), ID(3202), MT(1575,3202), UT(1575), WA(3202). **2:**UT(1575). **3:**AK(1575,3098), CA(1575), CO(1575), ID(3098,3202), ME(1575), MI(1575), MT(1575,3098,3202), ND(1575), NH(1575), NJ(1575), NM(1575), NY(1575), OR(3202), SD(1575), UT(1575), VT(1575), WA(3098,3202), WI(1575), WY(1195). **5:**CA(1575), OR(1575), UT(1575). **6:**AZ(3981), CA(1575). **7:**AK(1575), AZ(1203,3981), CA(1575), CO(1575), ID(1575), ME(1575), MI(1575), MT(1575), ND(1575), NE(1575), NY(1575), OR(1575), PA(1575), SD(1575), UT(1575), VA(1575), WA(1575), WI(1575), WV(1575), WY(1195,3379). **8:**AK(1575), MT(3202), OR(3202), WA(3202), WY(1195). **9:**WA(1575). **10:**CA(1575), DE(1575), IA(1575), MN(1575), MT(3202), NE(1575), NY(1575), PA(1575), TN(1575). **11:**AK(1575). **12:**ID(3202), MT(3202), WA(3202). **13:**AZ(1203,3981), UT(1575). **14:**MA(1575), NY(1575). **15:**CA(1575). **16:**ID(1195), WY(1195). **24:**NM(1575). **25:**CA(1575), WY(1575). **26:**AK(1575). **27:**MT(3202), WA(3202). **29:**MT(1575). **31:**UT(1575). **33:**CA(1575), ID(3202), MT(3202), OR(3202), WA(3202). **34:**OR(648), WY(1195).
P. pustulatum. Heteroecious: telial host. **1:**AK(526). **3:**NC states(82), NE states(82), NW states(82), AK(60,526,1904), CA(81,82,638,642), CO(82), ND(378), NM(82), OR(638). **5:**CA(81,3098). **6:**CA(3098). **7:**AK(60,526,3098), CA(638,642), ND(378), NV(3098), NY(3098), OR(3098), SD(2243), WY(1195,3098). **8:**AK(60,526). **11:**CA(81). **20:**AK(60,82,526). **26:**AK(60,526,3098). **31:**CO(649). **33:**CA(81,3098). **34:**NC states(82), NE states(82), NW states(82), AK(526), CA(82), CO(82), NM(82), SD(2243).

Basidiomycotina - Ustilaginales
Doassansia epilobii. Leaf smut. **1:**NH(1077), OR(3202), WA(3202). **8:**OR(3202). **22:**WA(3202). **34:**CO(82,1077), NH(82).
Basidiomycotina - Other
Typhula uncialis. On dead stems, fallen leaves. **3:**AK(3431), NY(459).
Deuteromycotina - Hyphomycetes
Botrytis cinerea. Gray mold blight. **3:**AK(82). **34:**AK(82).
Cercospora sp. **7:**WI(1354).
Didymaria sp. **3:**OR(3202).
Epicoccum nigrum. Leaf spot. **34:**FL(82).
Heterosporium sp. **27:**WA(3202).
Phymatotrichopsis omnivora. Root rot. **3:**TX(82).
Ramularia montana. Leaf spot. **1:**OR(3202), WA(3202). **7:**NY(3098), WY(1195). **8:**ID(3202). **10:**ND(378). **16:**AK(3098,3431). **17:**AK(3431). **20:**MT(3202), WA(3202). **23:**WA(3202). **26:**AK(3098). **27:**ID(3202), MT(3202), WA(3202). **32:**ID(3202), MT(3202). **33:**ID(3202), MT(3202), OR(3202). **34:**AK(82,526), ND(82), NH(82), NM(82), OR(82), WI(82), WY(82).
R. punctiformis. **34:**AK(526).
Ramularia sp. **3:**SD(2243).
Deuteromycotina - Coelomycetes
Leptothyrium protuberans. On capsules. **30:**CA(81).
Phyllosticta chamaenerii. Leaf spot. **3:**OR(82,3202).
P. wyomingensis. **3:**WY(82,1195).
Seimatosporium kriegerianum. Leaf spot. **3:**NY(82), WI(82,1361). **34:**AK(526).
Septoria epilobii. Leaf spot. **21:**AK(3431). **34:**AK(82,526), CA(82), DE(82), IL(82), VT(82), WA(3202), WI(82).
Deuteromycotina - Other
Sclerotium rolfsii. Southern blight. **3:**TX(82).

Fuchsia L.

1. *F.* × *hybrida* Hort. ex Vilm. A name applied to commonly cultivated fuchsia hybrids, said to be derived from *F. fulgens* and *F. magellanica*.
2. *F. speciosa* Hort. A listed name of no botanical standing often used for *F. hybrida*.
3. *Fuchsia* sp.

Oomycetes
Phytophthora nicotianae var. *parasitica*. Root and stem rot. **1:**OR(3202). **3:**NC(1303).
Pythium sp. Root and stem rot. **3:**FL(35).
Ascomycotina
Didymella fuchsiae. On stems. **3:**CA(82).
Basidiomycotina - Uredinales
Pucciniastrum epilobii. **1:**HI(1143,2903), OR(3202), WA(3202). **3:**FL(35), NC(1303,3515), OH(82,1575), OR(2176,3202).
P. pustulatum. Heteroecious: telial host. **2:**AK(60,526).
Basidiomycotina - Other
Armillaria mellea. Root rot. **1:**CA(2893,2898).
Deuteromycotina - Hyphomycetes
Botrytis cinerea. Gray mold blight. **1:**WA(3202). **3:**AK(82), FL(35), WV(82).
Cercospora fuchsiae. Leaf spot. **3:**FL(35).
Thielaviopsis sp. **3:**OR(3202).
Verticillium albo-atrum. Wilt. **3:**CA(82).
Deuteromycotina - Coelomycetes
Diplodia fuchsiae. On stems. **3:**CA(82).
Phomopsis sp. **3:**VA(82).
Septoria sp. Leaf spot. **3:**FL(35), OK(82,2872).
Deuteromycotina - Other
Rhizoctonia sp. Root and stem rot. **3:**FL(35), NC(1303).

Gaura L.

1. *G. angustifolia* Michx.
2. *G. biennis* L.
3. *G. coccinea* Nutt. ex Pursh
4. *G. filiformis* Small (=*G. filiformis* Small var. *filiformis*)
5. *G. parviflora* Douglas ex Lehm.
6. *G. sinuata* Nutt. ex Ser.
7. *G. suffulta* Engelm. ex A. Gray
8. *Gaura* sp.

Chytridiomycetes
Synchytrium fulgens. Leaf gall. **8:**TX(82).
Oomycetes
Peronospora arthurii. Downy mildew. **8:**KS(82), NE(82).
Ascomycotina
Discohainesia oenotherae. On leaves. **2:**VA(82).
Erysiphe polygoni. Powdery mildew. **8:**CO(82), TX(82).
Mycosphaerella sp. On stems. **2:**KS(82).
Basidiomycotina - Uredinales
Puccinia dioicae. Heteroecious: aecial host. **3:**NE(82). **6:**TX(82).
[*P. plumbaria*]. **3:**MT(3202).
Uromyces plumbarius. Autoecious. **1:**MS(2720). **2:**E states(82), OK(2872), TX(82), WI(82). **3:**SC states(82), W states(82), MS(2720), MT(3202), ND(378), NM(82), OK(2872), SD(2243). **6:**OK(82,2872), TX(82). **8:**Range of host(82), FL(35).
Deuteromycotina - Hyphomycetes
Cercospora flagellifera. Leaf spot. **5:**OK(2873). **8:**OK(82).
C. gaurae. Leaf spot. **2:**NY(82), OK(82,2872), TX(82). **3:**OK(2873). **6:**OK(82,2872), TX(82).
Phymatotrichopsis omnivora. Root rot. **2:**TX(82). **3:**TX(82). **8:**TX(82).
Deuteromycotina - Coelomycetes
Pestalotiopsis uvicola. **5:**KS(1393).
Phyllosticta sp. Leaf spot. **8:**FL(35).
Septoria gaurina. Leaf spot. **2:**IL(82), KS(82), OK(82,2872), WI(1341). **3:**MT(3202), ND(82,378), OK(82,2872). **4:**OK(2873). **7:**OK(2873). **8:**KS(82), NE(82), NM(82), TX(82).
S. oenotherae. **2:**WI(1340). **8:**TX(82).
Septoria sp. Leaf spot. **8:**FL(35).

Gayophytum Adr. Juss.

1. *G. diffusum* Torr. & A. Gray
2. *G. diffusum* Torr. & A. Gray subsp. *parviflorum* Harms & Szweyk.
3. *G. humile* Adr. Juss.
4. *G. intermedium* Rydb.
5. *G. lasiospermum* Greene
6. *G. racemosum* Torr. & A. Gray (=*G. caesium* Torr. & A. Gray, *G. nuttallii* Torr. & A. Gray)
7. *G. ramosissimum* Torr. & A. Gray
8. *Gayophytum* sp.

Basidiomycotina - Uredinales
Puccinia pulverulenta. Autoecious. **2:**AZ(3981). **3:**CA(81), ID(3202). **5:**CA(638). **6:**AZ(1195,1203,3981), CO(1195), ID(3202), OR(3202), WA(3202), WY(1195). **7:**AZ(1195,1203,3981), CA(638), ID(3202), MT(1195,3202), OR(3202), WA(3202). **8:**W states(82), AZ(1203), CA(82,642,661), WA(82).
Basidiomycotina - Ustilaginales
Ustilago gayophyti. Seed smut. **4:**UT(1077). **5:**OR(1077). **6:**NV(1077), OR(1077,3202), UT(1077). **7:**CA(1077,1468), OR(1077,3202), UT(1077). **8:**CA(82), NV(82), OR(82), UT(82).
Deuteromycotina - Hyphomycetes
Phaeoramularia gayophyti. Leaf spot. **1:**WY(1195). **7:**CA(81). **8:**CA(82).
Ramularia sp. **3:**WA(3202). **6:**ID(3202), MT(3202), OR(3202). **7:**ID(3202).

Ludwigia L.

(= *Jussiaea* L.)

1. *L. alternifolia* L. [rattlebox, seedbox]
2. *L. decurrens* Walter (=*Jussiaea decurrens* (Walter) DC.) [winged waterprimrose]
3. *L. hirtella* Raf.
4. *L. octovalvis* (Jacq.) Raven (=*Jussiaea scabra* Willd.) [primrose-willow]
5. *L. palustris* (L.) Elliott [water-purslane]
6. *L. peploides* (H. B. K.) Raven (=*Jussiaea californica* (S. Wats.) Jeps., *Jussiaea diffusa* Forssk.)
7. *L. peruviana* (L.) H. Hara
8. *L. pilosa* (H. B. K.) Walter (=*Jussiaea pilosa* H. B. K.)
9. *L. polycarpa* Short & Peter
10. *Ludwigia* sp. (=*Jussiaea sp.*)

Ascomycotina
Botryosphaeria dothidea. **1:**GA(1438).
B. obtusa. On stems. **1:**GA(82), SC(82).
Diaporthe austro-americana. **4:**FL(2782).
Discohainesia oenotherae. Leaf spot. **1:**VA(82).
Guignardia jussiaeae. **4:**HI(2903).

Hainesia lythri. **9**:WI(1352).
Nectriella pironii. Stem gall. **7**:FL(35).
Basidiomycotina - Uredinales
Aecidium bethelii. **6**:CA(82).
Puccinia jussiaeae. Autoecious. **1**:FL(82), MD(82). **2**:MS(82,2720). **3**:MS(2720). **5**:FL(35,82), LA(82), OH(82), TN(1539), TX(82). **9**:WI(1333). **10**:E states(82), CA(82), FL(35), MS(82), WI(82).
Uredo guaynabensis. **5**:FL(35). **10**:FL(35,82).
Deuteromycotina - Hyphomycetes
Alternaria sp. Leaf spot. **2**:OK(82,2872).
Cercospora ludwigiae. Leaf spot. **1**:AL(82), OK(2872). **7**:FL(35). **9**:WI(1333).
Pseudocercospora jussiaeae. Leaf spot. **2**:AL(82), OK(82), TX(82). **6**:OK(2872). **8**:AL(581). **10**:FL(35), OK(82), TX(82).
Deuteromycotina - Coelomycetes
Colletotrichum gloeosporioides. Leaf spot. **2**:AL(82), AR(3594), TX(82).
Phyllosticta ludwigiae. Leaf spot. **5**:NY(82). **10**:IA(82), WI(82).
Septoria jussiaeae. Leaf spot. **5**:FL(35). **10**:AL(82), FL(35,82), LA(82), TX(82).
S. ludwigiae. Leaf spot. **1**:IL(82), MD(82), MS(82,2720), PA(82). **5**:CA(82), DE(82), FL(35,82), IL(82), LA(658), NY(82), OR(82,3202), SC(82), WI(82). **9**:WI(1356).
S. oenotherae. Leaf spot. **10**:FL(35).

Oenothera L.

(= *Anogra* Spach, *Onagra* Mill.)
1. *O. albicaulis* Pursh
2. *O. biennis* L. [common evening-primrose, German rampion]
3. *O. biennis* L. var. *hirsutissima* A. Gray (=*Onagra strigosa* (Rydb.) Mackenz. & Bush, *O. strigosa* Rydb.)
4. *O. caespitosa* Nutt.
5. *O. deltoides* Torr. & Frém. [desert evening-primrose]
6. *O. elata* Kunth (=*O. hookeri* Torr. & A. Gray)
7. *O. flava* (A. Nels.) Garrett
8. *O. fruticosa* L. [sundrops]
9. *O. laciniata* J. Hill
10. *O. macrocarpa* Nutt. (=*O. missouriensis* Sims)
11. *O. macrocarpa* Nutt. subsp. *fremontii* (S. Wats.) W. Wagner (=*O. fremontii* S. Wats.)
12. *O. marginata* Nutt. ex Hook. & Arnott (=*Pachylophus marginatus* Rydb.)
13. *O. missouriensis* Sims var. *incana* A. Gray
14. *O. missouriensis* Sims var. *oklahomensis* (Norton) Munz (=*O. oklahomensis* (Norton) Norton ex F. Gates)
15. *O. nuttallii* Sweet (=*O. tanacetifolia* Torr. & A. Gray)
16. *O. pallida* Lindl. (=*Anogra pallida* (Lindl.) Britton, *O. californica* S. Wats.)
17. *O. parviflora* L.
18. *O. pilosella* Raf.
19. *O. primiveris* A. Gray
20. *O. rhombipetala* Nutt. ex Torr. & A. Gray
21. *O. sinuata* L.
22. *O. speciosa* Nutt. [white evening primrose]
23. *O. whitneyi* A. Gray (=*Godetia grandiflora* Lindl.)
24. *Oenothera* sp.

Chytridiomycetes
Synchytrium fulgens. Leaf gall. **2**:DE(1843), IA(82), IN(1838), KS(82), NJ(1843), NY(82,1843), SD(2243). **8**:MS(82,2720). **9**:AL(82,1843), LA(82,1838,1843), MS(82,2720), NC(82,3944), OK(82,2872), TX(82,1838), VA(1838). **20**:NE(1843), WI(1355). **24**:W states(82), AL(82), IA(82), LA(82), MS(82), NC(82), OK(82), TX(82).
Oomycetes
Peronospora arthurii. Downy mildew. **2**:E states(82), FL(35), ID(3202), MS(82,2720), MT(3202), NE(82), OR(3202), SD(82), WA(3202). **3**:ND(377), WI(1361). **7**:MT(3202). **9**:KS(82), OK(82,2872). **16**:ID(3202), WA(3202). **19**:AZ(1195,3381). **24**:MT(82), ND(82).
Pythium ultimum. Root rot. **23**:MO(82).
Ascomycotina
Botryosphaeria dothidea. **9**:GA(1438).
B. obtusa. On stems. **2**:WV(82). **9**:GA(82).
B. rhodina. **2**:NJ(82).
Diaporthe arctii. On stems. **2**:GA(82,1438).
Discohainesia oenotherae. **2**:GA(1438). **24**:NJ(850).
Erysiphe polygoni. Powdery mildew. **1**:Unk.(82). **2**:Unk.(82), FL(35), GA(1438), MS(2720), NC(1303,3944), WA(3202). **8**:Unk.(82). **9**:Unk.(82), GA(1438), NC(1303,3944). **15**:MT(3202). **24**:Unk.(82).
Hainesia lythri. **4**:WI(1356). **18**:WI(1349). **20**:WI(1354).
Leptosphaeria ellisiana. On dead stems. **2**:NJ(3249). **3**:ND(377,3249). **24**:MI(3249), ND(82), NJ(82).
Mycosphaerella oenotherae. **2**:NJ(959,967).
Mycosphaerella sp. On seed pods and stems. **2**:NJ(82).
Ophiobolus vermisporus. **2**:GA(1438), NJ(3248). **24**:MO(3248).
Basidiomycotina - Uredinales
[*Puccinia aristidae*]. **24**:AZ(82), NV(82).
P. dioicae. Heteroecious: aecial host. **2**:E states(82), CO(82), ID(3202), MS(2720), NM(82), WI(1353). **3**:ND(378), WI(1357). **4**:CO(1195). **6**:AZ(1203,3981). **8**:E states(82), CO(82), MS(2720), NC(3944), NM(82). **9**:E states(82), AZ(1203,3981), CO(82), NC(3944), NM(82). **18**:WI(1349). **24**:E states(82), AZ(1203), CA(82), CO(82), NM(82), OK(2872), TX(82).
P. oenotherae. Autoecious. **2**:MT(3202), WA(3202). **4**:OR(3202). **15**:OR(3202). **16**:AZ(3981). **23**:CA(82). **24**:W states(82), CA(82), ID(82), WA(82).
P. pulverulenta. Autoecious. **2**:MT(3202), WA(3202). **9**:GA(1439).
P. redfieldiae. Heteroecious: aecial host. **1**:NE(82). **15**:CO(1195).
Pucciniastrum pustulatum. **23**:AK(60,82).
Uromyces plumbarius. Autoecious. **2**:Range of host(82), FL(35), OK(2872). **4**:ID(3202), MT(3202), OR(3202). **9**:Range of host(82), FL(35), GA(1439), MS(2720), NC(1303,3944), OK(2872). **12**:ID(1195). **15**:OR(3202). **24**:E states(82), W states(82), FL(35).
Basidiomycotina - Other
Glabrocyphella ellisiana. **24**:NJ(646).
Deuteromycotina - Hyphomycetes
Alternaria alternata. Leaf spot. **19**:AZ(1195,3378). **24**:NJ(82).
Alternaria sp. **16**:OR(3202).
Aspergillus niger. **15**:OR(3202).
Botrytis cinerea. Gray-mold blight. **24**:AK(82).
Cercospora didymospora. On leaves and seed pods. **1**:OR(82). **2**:KS(82,581). **11**:KS(82,581).
C. oenotherae. Leaf spot. **1**:KS(82). **2**:KS(82). **15**:OR(3202). **22**:KS(82).
C. oenotherae-sinuatae. Leaf spot. **9**:AL(82,581), NC(82,581).
Cercospora sp. Leaf spot. **2**:WI(1351). **24**:FL(35).
Didymaria sp. **16**:OR(3202).
Phymatotrichopsis omnivora. Root rot. **9**:TX(82). **24**:TX(82).
Ramularia sp. **5**:OR(3202).
Deuteromycotina - Coelomycetes
Cryptostictis oenotherae. Leaf spot. **2**:OH(82). **24**:OK(82).
Macrophoma oenotherae-biennis. On seed pods and stems. **2**:NY(82).
Macrophomina phaseolina. Charcoal rot. **2**:IL(311).
Phomopsis oenotherae. On seed pods. **2**:NY(82).
Septoria gaurina. **2**:WI(1359).
S. oenotherae. Leaf spot. **1**:ND(82). **2**:E states(82), FL(35), ID(3202), MA(1390), MS(2720), MT(3202), NC(3944), OK(82,2872), SD(82), WA(3202). **3**:ND(378), WI(1356). **8**:E states(82), OK(82), SD(82). **9**:E states(82), MS(2720), NC(3944), OK(82,2872), SD(82). **10**:OK(2872). **13**:OK(2872). **14**:OK(2872). **16**:ND(378), OK(2872). **17**:WI(1349). **20**:OK(2872). **21**:SC(634). **22**:OK(2872). **24**:E states(82), FL(35), OK(82), SD(82).
Deuteromycotina - Other
Rhizoctonia solani. Root rot. **9**:TX(82).

Stenosiphon Spach

1. *S. linifolius* (Nutt. ex E. James) Heynh.

Deuteromycotina - Hyphomycetes
Cercospora sp. Leaf spot. **1**:OK(2872).

Zauschneria K. Presl

1. *Z. californica* K. Presl [California fuchsia, fire-chalice]
2. *Z. garrettii* A. Nels.

Ascomycotina
Erysiphe cichoracearum. Powdery mildew. **1**:CA(1155).
Basidiomycotina - Uredinales
Puccinia oenotherae. Autoecious. **1**:CA(81,82). **2**:UT(82).

Deuteromycotina - Coelomycetes
Septoria gaurina. **1**:AZ(1195).

Orchidaceae

Records with host identified to family only.
1. *Orchidaceae (exotic taxa)*
2. *Orchidaceae (native taxa)*
3. *Orchidaceae* sp.

Oomycetes
Phytophthora cactorum. Black rot. **3**:FL(35).
Pythium splendens. Root rot. **3**:FL(35).
P. ultimum. Black rot. **3**:FL(35).
Ascomycotina
Botryosphaeria vanillae. Leaf and pod spot. **3**:FL(35).
Glomerella cingulata. Anthracnose. **1**:Unk.(82). **3**:FL(35).
Leptosphaeria corallorhizae. On stems. **2**:NY(82).
Schizothyrium perexiguum. Fly speck. **3**:FL(35).
Basidiomycotina - Uredinales
Coleosporium bletiae. Heteroecious: telial host. **3**:FL(35).
Sphenospora kevorkianii. Autoecious. **3**:FL(35).
S. mera. Autoecious. **3**:FL(35).
Uredo behnickiana. **3**:FL(35).
U. nigropuncta. **3**:FL(35).
Deuteromycotina - Hyphomycetes
Botrytis cinerea. Gray mold and flower spot in greenhouses. **1**:Unk.(82). **3**:FL(35).
Cercospora angraeci. Leaf spot. **3**:FL(35).
C. cypripedii. Leaf spot. **3**:FL(35).
C. epipactidis. Leaf spot. **3**:FL(35).
C. odontoglossi. Leaf spot. **3**:FL(35).
C. peristeriae. Leaf spot. **3**:FL(35).
Cercospora sp. Leaf spot. **3**:FL(35).
Fusarium oxysporum f. sp. *cattleyae*. Bulb and root rot. **3**:FL(35).
F. oxysporum f. sp. *vanillae*. Root rot. **3**:FL(35).
Myrothecium sp. Leaf necrosis. **3**:FL(35).
Volutella vanillae. Leaf and pod spot. **3**:FL(35).
Deuteromycotina - Coelomycetes
Colletotrichum gloeosporioides. Anthracnose. **1**:Range of host(82), Unk.(82). **3**:FL(35).
Gloeodes pomigena. Sooty blotch. **3**:FL(35).
Lasiodiplodia theobromae. Leaf spot. **3**:FL(35).
Phoma corallorhizae. On stems. **2**:NY(82).
Phyllosticta capitalensis. Leaf spot. **3**:FL(35).
Phyllosticta sp. Leaf spot. **3**:FL(35).
Deuteromycotina - Other
Rhizoctonia monilioides. On roots. **2**:WI(82).
R. repens. On roots. **2**:NH(82), WI(82).
R. solani. Root and stem rot. **3**:FL(35).
Sclerotium rolfsii. Basal rot. **3**:FL(35).

Aplectrum Torr.
1. *A. hyemale* (Muhl. ex Willd.) Torr. [Adam-and-Eve, puttyroot]
2. *Aplectrum* sp.

Deuteromycotina - Hyphomycetes
Fusicladium aplectri. Leaf spot. **2**:DE(82).
Deuteromycotina - Coelomycetes
Phyllosticta aplectri. Leaf spot. **1**:DE(2). **2**:DE(82).
Deuteromycotina - Other
Rhizoctonia neottiae. On roots. **2**:WI(82).

×*Brassolaeliocattleya* Hort.
A name of no botanical standing for hybrids of *Cattelya* × *Laelia* × *Rhyncholaelia*.
1. *Brassolaeliocattleya* sp.

Oomycetes
Phytophthora cactorum. Black rot. **1**:FL(35).
Pythium ultimum. Black rot. **1**:FL(35).
Ascomycotina
Schizothyrium perexiguum. Fly speck. **1**:FL(35).
Deuteromycotina - Hyphomycetes
Botrytis cinerea. Gray mold. **1**:FL(35).
Cercospora odontoglossi. Leaf spot. **1**:FL(35).
Cercospora sp. Leaf spot. **1**:FL(35).
Fusarium oxysporum f. sp. *cattleyae*. Bulb and root rot. **1**:FL(35).
Deuteromycotina - Coelomycetes
Colletotrichum gloeosporioides. Anthracnose. **1**:FL(35).
Gloeodes pomigena. Sooty blotch. **1**:FL(35).
Phyllosticta capitalensis. Leaf spot. **1**:FL(35).
Phyllosticta sp. Leaf spot. **1**:FL(35).
Deuteromycotina - Other
Rhizoctonia solani. Root and stem rot. **1**:FL(35).

Calopogon R. Br.
1. *Calopogon* sp.

Basidiomycotina - Uredinales
Puccinia cypripedii. **1**:IN(82), NJ(82).
Deuteromycotina - Coelomycetes
Septoria calypsonis. Leaf spot. **1**:MI(82).
Deuteromycotina - Other
Rhizoctonia lanuginosa. On roots. **1**:WI(82).
R. repens. On roots. **1**:NH(82), WI(82).

Cattleya Lindl.
1. *Cattleya* sp.

Oomycetes
Phytophthora cactorum. Black rot. **1**:FL(35).
P. palmivora. Root rot. **1**:HI(2903).
Pythium splendens. Leaf rot. **1**:Unk.(82).
P. ultimum. Black rot. **1**:FL(35).
Ascomycotina
Glomerella cingulata. Anthracnose, leaf and stem spot of imported plants. **1**:Unk.(82), FL(35).
Schizothyrium perexiguum. Fly speck. **1**:FL(35).
Basidiomycotina - Uredinales
Sphenospora kevorkianii. Autoecious. **1**:FL(35).
Uredo behnickiana. **1**:FL(35), NJ(82), NY(82).
U. nigropuncta. **1**:Unk.(82).
Deuteromycotina - Hyphomycetes
Alternaria sp. Leaf spot. **1**:FL(35), HI(2903).
Botrytis cinerea. Gray mold. **1**:FL(35).
Cercospora angraeci. Leaf spot. **1**:FL(35).
C. odontoglossi. Leaf spot. **1**:FL(35).
Cercospora sp. Leaf spot. **1**:FL(35).
Fusarium oxysporum f. sp. *cattleyae*. Bulb and root rot. **1**:FL(35).
Fusarium sp. **1**:HI(2903).
Stephanoma phaeospora. Isolated from roots. **1**:CA(468).
Deuteromycotina - Coelomycetes
Botryodiplodia oncidii. On leaves and stems of imported plants. **1**:Unk.(82).
Chaetodiplodia sp. Leaf spot. **1**:Unk.(82).
Colletotrichum sp. Anthracnose. **1**:HI(2903).
Coniothyrium sp. Leaf spot. **1**:FL(35).
Diplodia laelio-cattleyae. Leaf spot. **1**:FL(35).
D. paraphysaria. On old leaves and stems, on imported plants. **1**:MD(82), MO(82).
Gloeodes pomigena. Sooty blotch. **1**:FL(35).
Gloeosporium cattleyae. Anthracnose on imported plants. **1**:VA(82).
Lasiodiplodia theobromae. Leaf spot. **1**:Unk.(82), FL(35).
Phoma oncidii-sphacelati. Leaf spot. **1**:FL(35).
Phyllosticta capitalensis. Leaf spot. **1**:FL(35).
Septoria selenophomoides. Leaf spot. **1**:FL(35).
Deuteromycotina - Other
Rhizoctonia solani. Root and stem rot. **1**:FL(35).
Sclerotium rolfsii. Basal rot. **1**:Unk.(82), FL(35).

Corallorrhiza Gagnebin
1. *C. maculata* Raf. (=*C. multiflora* Nutt.) [large coralroot, spotted coralroot]

Ascomycotina
Leptosphaeria corallorhizae. On stems. **1**:NY(195).
Deuteromycotina - Coelomycetes
Phoma corallorhizae. **1**:MI(525).

Cymbidium Sw.
1. *Cymbidium* sp.

Oomycetes
Pythium ultimum. Black rot. **1**:FL(35).

Deuteromycotina - Hyphomycetes
Cercospora sp. Leaf spot. **1:**FL(35).
Fusarium oxysporum f. sp. *cattleyae*. Bulb and root rot. **1:**FL(35).
Deuteromycotina - Coelomycetes
Colletotrichum gloeosporioides. Anthracnose. **1:**FL(35).
Phyllosticta capitalensis. Leaf spot. **1:**FL(35).
Selenophoma sp. Leaf spot of imported plants. **1:**Unk.(82).
Septoria selenophomoides. Leaf spot. **1:**FL(35).
Deuteromycotina - Other
Rhizoctonia solani. Root and stem rot. **1:**FL(35).

Cypripedium L.
1. *C. calceolus* L. var. *pubescens* (Willd.) D. S. Correll (=*C. pubescens* Willd.) [American valerian, golden-slipper, large lady's-slipper, nerveroot, Noah's-ark, umbilroot, Venus's-shoe, whippoorwill-shoe, yellow Indian-shoe]
2. *C. candidum* Muhl. ex Willd. [small white lady's-slipper]
3. *C. reginae* Walter [showy lady's-slipper]
4. *Cypripedium* sp.

Ascomycotina
Mycosphaerella cypripedii. Leaf spot. **4:**NY(82).
Basidiomycotina - Uredinales
Puccinia cypripedii. **4:**IA(82), MI(82), VA(82), WI(82).
Deuteromycotina - Hyphomycetes
Cercospora cypripedii. Leaf spot. **1:**WI(1330). **2:**WI(1332). **3:**WI(1330). **4:**NY(82), WI(82).
Deuteromycotina - Coelomycetes
Ascochyta cypripedii. On living leaves. **2:**WI(1343).
Deuteromycotina - Other
Rhizoctonia robusta. On roots. **4:**WI(82).
R. sclerotica. On roots. **4:**WI(82).
R. subtilis. On roots. **4:**WI(82).

Dendrobium Sw.
1. *D. nobile* Lindl.
2. *Dendrobium* × sp. (unnamed hybrids)
3. *Dendrobium* sp.

Oomycetes
Phytophthora cactorum. Black rot. **3:**FL(35).
P. palmivora. Black rot. **3:**HI(2903).
Pythium ultimum. Black rot. **3:**FL(35).
Ascomycotina
Glomerella cingulata. Leaf and stem spot of imported plants. **3:**Unk.(82).
Schizothyrium perexiguum. Fly speck. **3:**FL(35).
Deuteromycotina - Hyphomycetes
Alternaria alternata. Flecks on flowers. **3:**HI(2903).
Alternaria sp. Leaf spot. **3:**FL(35).
Bipolaris setariae. Flecks on flowers. **3:**HI(2903).
B. sorokiniana. Flecks on flowers. **3:**HI(2903).
B. urochloae. Flecks on flowers. **3:**HI(2903).
Botrytis cinerea. Blossom blight, flecks on flowers, gray mold. **3:**FL(35), HI(1704,2903).
Botrytis sp. Blossom blight. **1:**HI(2903). **2:** HI(1704).
Cercospora dendrobii. Leaf spot. **3:**FL(35).
C. odontoglossi. Leaf spot. **3:**FL(35).
Exserohilum rostratum. Flecks on flowers. **3:**HI(2903).
Fusarium oxysporum f. sp. *cattleyae*. Bulb and root rot. **3:**FL(35).
Fusarium sp. **3:**HI(2903).
Stemphylium sp. Flecks on flowers. **3:**HI(2903).
Deuteromycotina - Coelomycetes
Botryodiplodia oncidii. On leaves and stems of imported plants. **3:**Unk.(82).
Colletotrichum gloeosporioides. Anthracnose. **1:**WI(1330). **3:**Unk.(82), FL(35).
Gloeodes pomigena. Sooty blotch. **3:**FL(35).
Macrophoma sp. Leaf spot. **3:**FL(35).
Phyllosticta capitalensis. Leaf spot. **3:**FL(35).
Phyllosticta sp. Leaf spot. **3:**FL(35).
Selenophoma sp. Leaf spot. **3:**NJ(82), NY(82).
Septoria selenophomoides. Leaf spot. **3:**FL(35).
Deuteromycotina - Other
Rhizoctonia solani. Root and stem rot. **3:**FL(35).

Epidendrum L.
1. *Epidendrum* sp.

Oomycetes
Phytophthora cactorum. Black rot. **1:**FL(35).
P. palmivora. **1:**HI(2903).
Pythium ultimum. Black rot. **1:**FL(35).
Ascomycotina
Glomerella cingulata. Leaf and stem spot of imported plants. **1:**Unk.(82).
Schizothyrium perexiguum. Fly speck. **1:**FL(35).
Basidiomycotina - Uredinales
Sphenospora kevorkianii. Autoecious. **1:**FL(35).
Uredo behnickiana. **1:**NJ(82).
U. epidendri. **1:**Unk.(82).
U. nigropuncta. **1:**FL(35,82).
Deuteromycotina - Hyphomycetes
Alternaria sp. Leaf spot. **1:**FL(35).
Botrytis cinerea. Gray mold. **1:**FL(35).
Cercospora epidendronis. Leaf spot. **1:**FL(35).
C. odontoglossi. Leaf spot. **1:**FL(35).
Cercospora sp. Leaf spot. **1:**FL(35).
Fusarium oxysporum f. sp. *cattleyae*. Bulb and root rot. **1:**FL(35).
Deuteromycotina - Coelomycetes
Gloeodes pomigena. Sooty blotch. **1:**FL(35).
Lasiodiplodia theobromae. Leaf spot. **1:**FL(35).
Macrophoma sp. Leaf spot. **1:**FL(35).
Phyllosticta capitalensis. Leaf spot. **1:**FL(35).
Deuteromycotina - Other
Rhizoctonia solani. Root and stem rot. **1:**FL(35), OK(82,2872).

Goodyera R. Br. in Aiton & W. T. Aiton
1. *G. oblongifolia* Raf. (=*Peramium decipiens* (Hook.) Piper)
2. *Goodyera* sp.

Basidiomycotina - Uredinales
Uredo goodyerae. **1:**CA(81), OR(3202). **2:**CA(82), CO(82), NM(82), OR(82), WA(82).
Deuteromycotina - Other
Rhizoctonia borealis. On roots. **2:**WI(82).
R. lanuginosa. On roots. **2:**WI(82).
R. repens. On roots. **2:**NH(82), WI(82).

Habenaria Willd.
1. *H. dilatata* (Pursh) Hook. [bog-candle, leaf white orchid, scented bottle, tall white bog orchid]
2. *H. flava* (L.) R. Br. (=*H. herbiola* R. Br.)
3. *H. hyperborea* (L.) R. Br. in Aiton [northern green orchid]
4. *H. leucophaea* A. Gray
5. *H. leucostachys* (Lindl.) S. Wats. [slender bog orchid]
6. *H. saccata* Greene
7. *H. viridis* (L.) R. Br. var. *bracteata* (Muhl. ex Willd.) A. Gray (=*Coeloglossum viride* (L.) Hartm. var. *bracteatum* (Muhl. ex Willd.) Hultén) [bracted green-onion]
8. *Habenaria* sp.

Basidiomycotina - Uredinales
Aecidium graebnerianum. **1:**AK(60). **3:**AK(60). **5:**CA(642). **6:**AK(60). **7:**AK(60). **8:**AK(82), CA(82), MT(82), OR(82), WA(82).
Puccinia praegracilis. Heteroecious: aecial host. **1:**AK(526), MT(3202), OR(3202), WA(3202). **3:**AK(526). **6:**AK(526), WA(3202). **7:**AK(526). **8:**AK(526), CA(661).
Deuteromycotina - Coelomycetes
Phyllosticta sp. **4:**WI(1343).
Septoria sp. Leaf spot. **4:**WI(1341).
Stagonospora sp. On leaves. **2:**WI(1348).
Deuteromycotina - Other
Rhizoctonia monilioides. On roots. **8:**WI(82).
R. repens. On roots. **8:**NH(82), WI(82).
R. robusta. On roots. **8:**WI(82).
R. sclerotica. On roots. **8:**WI(82).
R. sphacelati. On roots. **8:**WI(82).
R. stahlii. On roots. **8:**WI(82).
R. subtilis. On roots. **8:**WI(82).

Laelia Lindl.
1. *Laelia* sp.
Oomycetes
Pythium ultimum. Black rot. **1:**FL(35).
Basidiomycotina - Uredinales
Uredo behnickiana. **1:**FL(35).
Deuteromycotina - Hyphomycetes
Botrytis cinerea. Gray mold. **1:**FL(35).
Cercospora angraeci. Leaf spot. **1:**Unk.(82).
C. odontoglossi. Leaf spot. **1:**Unk.(82), FL(35).
Cercospora sp. Leaf spot. **1:**FL(35).
Fusarium oxysporum f. sp. *cattleyae*. Bulb and root rot. **1:**FL(35).
Deuteromycotina - Coelomycetes
Botryodiplodia oncidii. On leaves and stems of imported plants. **1:**Unk.(82).
Phyllosticta capitalensis. Leaf spot. **1:**FL(35).
P. laeliae. Leaf spot. **1:**Unk.(82).
Phyllosticta sp. Leaf spot. **1:**PA(82).
Selenophoma sp. Leaf spot. **1:**Unk.(82).

×*Laeliocattleya* Hort.
A name of no botanical standing for hybrids of *Cattleya* × *Laelia*.
1. × *Laeliocattleya* sp. (unnamed hybrids)
Oomycetes
Phytophthora cactorum. Black rot, root rot. **1:** FL(35), HI(2903).
Pythium ultimum. Black rot. **1:** FL(35).
Deuteromycotina - Hyphomycetes
Botrytis cinerea. Gray mold. **1:** FL(35).
Cercospora odontoglossi. Leaf spot. **1:** FL(35).
Cercospora sp. Leaf spot. **1:** FL(35).
Fusarium oxysporum f. sp. *cattleyae*. Bulb and root rot. **1:** FL(35).
Deuteromycotina - Coelomycetes
Colletotrichum gloeosporioides. Anthracnose. **1:** FL(35).
Gloeodes pomigena. Sooty blotch. **1:** FL(35).
Phyllosticta capitalensis. Leaf spot. **1:** FL(35).
Deuteromycotina - Other
Rhizoctonia solani. Root and stem rot. **1:** FL(35).

Odontoglossum Kunth
1. *Odontoglossum* sp.
Ascomycotina
Glomerella cingulata. Anthracnose. **1:**FL(35), OR(3202).
Basidiomycotina - Uredinales
Uredo nigropuncta. **1:**Unk.(82), FL(35).
Deuteromycotina - Hyphomycetes
Cercospora angraeci. Leaf spot. **1:**Unk.(82).
C. odontoglossi. Leaf spot. **1:**Unk.(82).
Cercospora sp. Leaf spot. **1:**FL(35).
Deuteromycotina - Coelomycetes
Colletotrichum gloeosporioides. Anthracnose. **1:**FL(35).
Phyllosticta capitalensis. Leaf spot. **1:**FL(35).
Selenophoma sp. Leaf spot. **1:**Unk.(82).
Septoria selenophomoides. Leaf spot. **1:**FL(35).

Oncidium Sw.
1. *O. sphacelatum* Lindl.
2. *Oncidium* sp.
Oomycetes
Phytophthora cactorum. Black rot. **2:**FL(35).
Pythium ultimum. Black rot. **2:**FL(35).
Ascomycotina
Botryosphaeria vanillae. Leaf and pod spot. **2:**FL(35).
Glomerella cingulata. Anthracnose, leaf and stem spot of imported plants. **2:**Unk.(82), FL(35).
Basidiomycotina - Uredinales
Sphenospora saphena. Autoecious. **2:**FL(35).
Uredo behnickiana. **2:**Unk.(82).
U. epidendri. **2:**FL(35).
U. oncidii. **2:**HI(2903).
Deuteromycotina - Hyphomycetes
Alternaria sp. Leaf spot. **2:**FL(35).
Botrytis cinerea. Gray mold. **2:**FL(35).
Cercospora angraeci. Leaf spot. **2:**Unk.(82).
C. odontoglossi. Leaf spot. **2:**Unk.(82).
Cercospora sp. Leaf spot. **2:**FL(35).
Fusarium oxysporum f. sp. *cattleyae*. Bulb and root rot. **2:**FL(35).
Fusarium sp. **2:**HI(2903).
Deuteromycotina - Coelomycetes
Ascochyta sp. Leaf spot. **2:**FL(35).
Botryodiplodia oncidii. On leaves and stems of imported plants. **2:**Unk.(82).
Colletotrichum gloeosporioides. Anthracnose. **1:**WI(1330).
Coniothyrium sp. Leaf spot. **2:**FL(35).
Phyllosticta capitalensis. Leaf spot. **2:**FL(35).
P. nigromaculans. Leaf spot. **2:**Unk.(82).
Selenophoma sp. Leaf spot of imported plants. **2:**Unk.(82).
Septoria selenophomoides. Leaf spot. **2:**FL(35).
Deuteromycotina - Other
Rhizoctonia solani. Root and stem rot. **2:**FL(35).

Orchis L.
1. *O. aristata* Fisch. ex Lindl.
2. *Orchis* sp.
Basidiomycotina - Uredinales
Aecidium graebnerianum. **1:**AK(60).
Puccinia praegracilis. Heteroecious: aecial host. **1:**AK(526). **2:**AK(526).

Phalaenopsis Blume
1. *Phalaenopsis* sp.
Ascomycotina
Glomerella cingulata. Leaf and stem spot of imported plants. **1:**Unk.(82).
Nectria bulbicola. Dry rot of imported plants. **1:**Unk.(82).
Deuteromycotina - Hyphomycetes
Alternaria sp. Leaf spot. **1:**FL(35).
Botrytis cinerea. Gray mold. **1:**FL(35).
Curvularia lunata. Leaf spot. **1:**FL(35).
Fusarium oxysporum f. sp. *cattleyae*. Bulb and root rot. **1:**FL(35).
Fusarium sp. **1:**HI(2903).
Volutella albo-pila. Leaf spot. **1:**Unk.(82).
Deuteromycotina - Coelomycetes
Botryodiplodia oncidii. On leaves and stems of imported plants. **1:**Unk.(82).
Phyllosticta capitalensis. Leaf spot. **1:**FL(35).
Septoria selenophomoides. Leaf spot. **1:**FL(35).
Deuteromycotina - Other
Rhizoctonia solani. Root and stem rot. **1:**FL(35).
Sclerotium rolfsii. Basal rot. **1:**FL(35).

Pogonia Juss.
1. *Pogonia* sp.
Deuteromycotina - Other
Rhizoctonia monilioides. On roots. **1:**WI(82).
R. repens. On roots. **1:**NH(82), WI(82).

Spathoglottis Blume
1. *Spathoglottis* sp.
Oomycetes
Pythium splendens. Root rot. **1:**HI(2903).

Spiranthes Rich.
1. *Spiranthes* sp.
Deuteromycotina - Other
Rhizoctonia borealis. On roots. **1:**WI(82).
R. monilioides. On roots. **1:**WI(82).
R. repens. On roots. **1:**NH(82), WI(82).
R. sclerotica. On roots. **1:**WI(82).
R. stahlii. On roots. **1:**WI(82).
R. subtilis var. *nigra*. On roots. **1:**WI(82).

Vanda Jones & R. Br.
1. *V. tricolor* Lindl.
2. *Vanda* sp.
Oomycetes
Phytophthora cactorum. Black rot. **2:**FL(35).
P. palmivora. **2:**HI(2903).
Phytophthora sp. **2:**HI(2903).
Pythium ultimum. Black rot. **2:**FL(35).

Ascomycotina
Glomerella cingulata. Anthracnose. **2:**FL(35).
Glomerella sp. Black spot. **1:**HI(98).
Schizothyrium perexiguum. Fly speck. **2:**FL(35).
Deuteromycotina - Hyphomycetes
Alternaria sp. Leaf spot. **2:**FL(35), HI(2903).
Botrytis cinerea. Blossom blight, gray mold. **2:**FL(35), HI(2903).
Botrytis sp. **2:**HI(2903).
Cercospora sp. Leaf spot. **2:**FL(35).
Fusarium oxysporum f. sp. *cattleyae*. Bulb and root rot. **2:**FL(35).
Fusarium sp. **2:**HI(2903).
Deuteromycotina - Coelomycetes
Colletotrichum gloeosporioides. Anthracnose. **2:**FL(35).
Colletotrichum sp. Anthracnose. **2:**HI(2903).
Gloeodes pomigena. Sooty blotch. **2:**FL(35).
Phyllosticta capitalensis. Leaf spot. **2:**FL(35).
Deuteromycotina - Other
Rhizoctonia solani. Root and stem rot. **2:**FL(35).
Rhizoctonia sp. Crown and root rot. **2:**HI(2903).
Sclerotium rolfsii. Basal rot. **2:**FL(35).

Vanilla Mill.
1. *V. planifolia* Andr. (=*V. fragrans* (Salisb.) Ames) [vanilla]
2. *Vanilla* sp.

Ascomycotina
Botryosphaeria vanillae. Leaf spot. **1:**FL(82).
Glomerella cingulata. Leaf spot. **1:**HI(82).
Lembosia rolfsii. Black mildew. **1:**FL(82).
Deuteromycotina - Hyphomycetes
Volutella vanillae. Leaf spot. **1:**FL(82).
Deuteromycotina - Coelomycetes
Colletotrichum gloeosporioides. **2:**HI(2784).

Orobanchaceae

Boschniakia C. A. Mey. ex Bong.
1. *B. rossica* (Cham. & Schlechtend.) Fedch.

Ascomycotina
Crocicreas cyathoideum. **1:**AK(3431).
C. cyathoideum var. *cacaliae*. **1:**AK(505).

Orobanche L.
1. *O. fasciculata* Nutt.

Deuteromycotina - Hyphomycetes
Botrytis cinerea. **1:**WA(3202).

Oxalidaceae

Averrhoa L.
1. *A. carambola* L. [blimbing, caramba, carambola, country gooseberry, starfruit]
2. *Averrhoa* sp.

Deuteromycotina - Hyphomycetes
Cercospora averrhoae. Leaf spot. **1:**FL(35).
Deuteromycotina - Coelomycetes
Colletotrichum gloeosporioides. Anthracnose. **1:**FL(35). **2:**FL(35).
Phomopsis sp. Leaf spot. **1:**FL(35).
Phyllosticta sp. Leaf spot. **1:**FL(35).

Oxalis L.
1. *O. acetosella* L. [European wood sorrel, Irish shamrock]
2. *O. amplifolia* (Trel.) R. Knuth
3. *O. braziliensis* Lodd. ex Westc. & G. Knowles
4. *O. corniculata* L. (=*O. europea* Jord. in F. W. Schultz) [creeping lady's-sorrel, creeping woodsorrel]
5. *O. filipes* Small in Britton & A. Brown
6. *O. oregana* Nutt. [red woodsorrel]
7. *O. rubra* St. Hil.
8. *O. stricta* L.
9. *O. suksdorfii* Trel.
10. *O. violacea* L. [violet woodsorrel]
11. *Oxalis* sp.

Oomycetes
Pythium sp. Root rot. **11:**FL(35), NC(1303).
Ascomycotina
Microsphaera russellii. Powdery mildew. **4:**E states(82), KS(82), MN(82), SD(2243). **8:**E states(82), KS(82), MN(82), WA(82). **9:**OR(3202), WA(3202). **11:**OH(82), SD(2243).
Basidiomycotina - Uredinales
Puccinia andropogonis. Heteroecious: aecial host. **2:**AZ(741,1203,3981). **4:**OK(82,2872), TX(82).
P. oxalidis. Heteroecious: telial host. **7:**CA(81). **10:**GA(1439), LA(82,658), MS(82,2720), NM(82), TX(82). **11:**FL(35,82), GA(82), NC(1303), PA(1875), SC(82), TN(82), TX(82).
P. sorghi. Heteroecious: aecial host. **3:**NC(1303). **4:**IA(82), IN(82), KS(82), MI(82), ND(82), SD(82,2243), TX(82), WI(1344). **5:**OK(2872). **8:**GA(1439), IA(82), IN(82), KS(82), MI(82), ND(82,378), SD(82), TX(82). **10:**IA(82), KS(82), MS(82,2720), NC(1303), NE(82), OK(2872). **11:**FL(35), HI(2903), NC(1303), OK(82), TX(82).
Basidiomycotina - Ustilaginales
Melanotaenium oxalidis. On petioles and leaf midribs. **6:**WA(3202). **11:**WA(873).
Ustilago oxalidis. Seed smut. **4:**IL(1077), MO(82), NY(1077), PA(82,1077), TX(82), WV(82,1077). **8:**E states(82), CT(1077), IA(1077), IL(1077), IN(1077), KS(1077), MO(1077), MS(82,1077,2720), NC(1077), NE(1077), NJ(1077), NY(1077), PA(1077), SC(1077), SD(1077,2243), TX(82,1077), WI(82,1077,1341). **11:**OH(82), TX(82).
Deuteromycotina - Hyphomycetes
Botrytis cinerea. Gray mold blight. **11:**AK(82), FL(35).
Cercospora oxalidiphila. Leaf spot. **4:**WI(1360). **8:**WI(82).
C. oxalidis. Leaf spot. **11:**FL(35).
Cercospora sp. **8:**WI(1332).
Ramularia oxalidis. Leaf spot. **1:**NH(82), PA(82), TN(82), VT(82). **4:**VT(82). **6:**OR(648,3202), WA(3202). **10:**NE(82), NH(82), NM(82). **11:**FL(35), OR(82).
Thielaviopsis basicola. Root rot. **8:**CT(82).
Deuteromycotina - Coelomycetes
Ascochyta sp. On dead portions of leaves. **4:**WI(1352).
Colletotrichum sp. Anthracnose. **11:**FL(35).
[*Phomopsis oxalina*]. Tar spot. **4:**DE(82). **8:**ME(82), VT(82).
Phyllosticta guttulatae. Leaf spot. **4:**NE states(82), IN(82), WI(82). **8:**NE states(82), IN(82), WI(82,1331). **9:**WA(3202). **10:**WI(1341).
P. oxalidis. **4:**TX(82). **8:**TX(82), WI(64,82). **11:**TX(82).
Septoria acetosella. Leaf spot. **1:**NY(82).
S. oxalidis. **4:**WI(1344). **6:**OR(648,3202). **11:**OR(82).
S. violacea. On living leaves. **10:**WI(1341).
Deuteromycotina - Other
Rhizoctonia solani. Root rot. **11:**FL(35).
Sclerotium rolfsii. Southern blight. **11:**FL(35).

Paeoniaceae

Paeonia L.
1. *P. brownii* Douglas ex Hook.
2. *P. lactiflora* Pall. (=*P. albiflora* Pall.) [Chinese peony, common garden peony]
3. *P. officinalis* L. [piney]
4. *P. suffruticosa* Andr. [tree peony]
5. *Paeonia* sp.

Oomycetes
Phytophthora cactorum. Tip blight. **2:**CT(82), IL(82), IN(82), KS(82), MN(82), NJ(82), NY(82), OH(82). **3:**CT(82), IL(82), IN(82), KS(82), MN(82), NJ(82), NY(82), OH(82), SD(2243).
Ascomycotina
Discohainesia oenotherae. Leaf spot. **2:**MD(82), PA(82).
Erysiphe polygoni. Powdery mildew. **2:**TX(82).
Omphalospora stellariae. **1:**WA(3202).
Sclerotinia sclerotiorum. Stem rot, wilt. **2:**IL(82), MD(82), ME(82), MN(82), NJ(82), NY(82), OH(82), OK(82). **3:**NC(1303).
Basidiomycotina - Other
Armillaria mellea. Root rot. **2:**CA(82), IA(82), MI(82), OR(82).
Scytinostroma galactinum. Root rot. **3:**MD(668).
Thanatephorus cucumeris. **5:**WA(3202).

Deuteromycotina - Hyphomycetes
Alternaria sp. Leaf spot. **2:**C states(82), NE states(82). **3:**C states(82), NE states(82). **5:**OK(2872).
Botrytis cinerea. **5:**WA(3202).
B. paeoniae. Blight. **3:**MA(1390). **5:**AL(285), ID(3202), MS(2720), OK(2872), OR(3202,4007), WA(3202).
Botrytis sp. Blight. **2:**Range of host(82). **3:**Range of host(82), CT(3770), NC(1303), WI(1350). **5:**CT(2974).
Cercospora paeoniae. Leaf spot. **3:**IL(82), SD(2243). **5:**OK(2872).
C. variicolor. Leaf spot. **2:**E states(82), NE(82), WI(82,1330). **3:**WI(1336).
Cladosporium herbarum. Leaf blotch. **2:**Unk.(82).
C. paeoniae. Leaf blotch. **2:**Range of host(82). **3:**Range of host(82). **4:**MA(82). **5:**AK(526), AL(285), FL(35), MS(2720), NC(1303), OK(2872), WA(3202).
Fusarium sp. Root and stem rot. **2:**CO(82), IN(82), MO(82), NE(82), NJ(82), OK(82).
Phymatotrichopsis omnivora. Root rot. **2:**AZ(82), TX(82).
Thielaviopsis basicola. Root rot. **2:**CT(82).
Verticillium albo-atrum. Wilt. **2:**IL(82), KS(82), MD(82), NY(82), OH(82). **3:**IL(82), KS(82), MD(82), NY(82), OH(82). **5:**KS(3022).
Deuteromycotina - Coelomycetes
Coniothyrium sp. Stem canker. **4:**CA(82).
Cryptostictis paeoniae. Leaf spot. **3:**IL(82).
Discosia artocreas. Leaf spot. **5:**OK(2872).
Monochaetia paeoniae. On stems. **4:**MD(82), NJ(82).
Phyllosticta baldensis. **2:**AK(82). **3:**PA(82). **5:**AK(526).
P. commonsii. Leaf spot. **2:**NE states(82), IN(82), OK(82). **3:**NE states(82), IN(82), OK(82). **5:**DE(966), OK(2872).
Phyllosticta sp. Leaf spot. **2:**NJ(82), PA(82), VA(82). **3:**NC(1303).
Septoria paeoniae. Blight, leaf spot. **2:**ME(82), MI(82), MN(82), NJ(82), NY(82), OR(82,3202), RI(82), WA(82,3202), WI(82). **5:**MS(2720).
Deuteromycotina - Other
Rhizoctonia solani. Root rot. **2:**CT(82), IL(82), MD(82), MN(82), NY(82), PA(82), VA(82).
Rhizoctonia sp. Root rot. **4:**HI(2903).
Sclerotium rolfsii. Southern blight. **2:**MS(82,2720), TX(82).

Pandanaceae

Freycinetia Gaud.
1. *F. arborea* Gaud.
2. *F. arnottii* Gaud.
3. *Freycinetia* sp.

Oomycetes
Phytophthora cinnamomi. Root rot. **1:**HI(2903).
Ascomycotina
Clypeosphaeria stevensii. **2:**HI(2903).
Eutypella sp. **1:**HI(2903).
Gibberella lagerheimii. **1:**HI(2903).
Melanomma clypeatum. **3:**HI(2903).
Meliola juddiana. **3:**HI(2903).
Meliola sp. **1:**HI(1268,2903).
Mollisia petiolorum. **3:**HI(2903).
Mycosphaerella freycinetiae. Leaf spot. **1:**HI(2903). **2:**HI(3480).
Myiocopron freycinetiae. **1:**HI(2903). **2:**HI(2784). **3:**HI(2784).
Phyllachora freycinetiae. **1:**HI(2903). **2:**HI(2903).
Deuteromycotina - Hyphomycetes
Trichothallus hawaiiensis. **1:**HI(2903). **2:**HI(2903).

Pandanus S. Parkinson
1. *P. odoratissimus* L. fil. (=*P. odorifer* Kuntze) [breadfruit, pandang]
2. *P. tectorius* S. Parkinson ex Du Roi [pandanus-palm, thatch screw-pine]
3. *P. utilis* Bory [common screw-pine]
4. *P. veitchii* J. H. Veitch ex Mast. & T. Moore [Veitch screw-pine]
5. *Pandanus* sp.

Myxomycetes
Physarella oblonga. On bark. **2:**HI(2903).
Oomycetes
Phytophthora cinnamomi. Root rot. **5:**HI(2903).
Pythium sp. Root rot. **3:**FL(35).
Ascomycotina
Botryosphaeria ribis. On stems. **1:**HI(2903).
Botryosphaeria sp. Crown rot. **3:**FL(35).
Diaporthe sp. On limbs. **1:**HI(2903).
Echidnodes sandwicensis. On dead leaves. **1:**HI(2781). **2:**HI(2781).
Glomerella cingulata. On leaves. **2:**HI(2903).
Linocarpon pandani. **2:**HI(2784). **5:**HI(2784).
Oxydothis pandanicola. On decaying leaves. **1:**HI(2903). **5:**HI(82,2781).
Phomatospora cylindrotheca. On decaying leaves. **5:**HI(2784).
P. pandani. On leaves. **5:**TN(82).
Physalospora sp. Leaf spot. **3:**FL(35).
Rosellinia sublimbata. **5:**HI(2784).
Basidiomycotina - Other
Earliella scabrosa. On wood. **1:**HI(2903). **5:**HI(82).
Pleurotus ostreatus. On decaying trunk. **2:**HI(2903).
Deuteromycotina - Hyphomycetes
Alternaria alternata. Leaf spot. **5:**FL(35).
[*Cladosporium iridis*]. Leaf spot. **5:**IA(82).
Fusarium sp. Bud rot. **3:**FL(35).
Hobsonia mirabilis. On bark. **2:**HI(2903).
Volutella mellea. On leaves. **5:**NY(82).
Zygosporium oscheoides. On leaves. **2:**HI(2903).
Deuteromycotina - Coelomycetes
Colletotrichum gloeosporioides. Anthracnose. **3:**FL(35). **5:**Range of host(82), FL(35).
Coniothyrium pandanicola. On fruits. **5:**HI(2784).
Lasiodiplodia theobromae. Leaf spot. **3:**FL(35). **5:**Range of host(82).
Macrophoma pandani. Leaf spot. **3:**FL(35). **5:**CA(82), FL(35,82).
Melanconium pandani. Leaf spot, on fruits. **1:**HI(2903). **2:**HI(2903). **3:**FL(35). **5:**FL(35,82), HI(82), MD(82).
Pestalotiopsis palmarum. On leaves. **5:**FL(35,82).
Phoma coryphae. Leaf spot. **3:**FL(35).
Phomopsis sp. Leaf spot. **3:**FL(35). **5:**NJ(82).
Phyllosticta pandanicola. Leaf spot. **4:**NC(1303).
Phyllosticta sp. Leaf spot. **3:**FL(35).
Sphaeropsis sp. Dieback. **3:**FL(35).
Deuteromycotina - Other
Rhizoctonia solani. Basal stem rot. **3:**FL(35).
Rhizoctonia sp. Root rot. **5:**HI(2903).

Papaveraceae

Argemone L.
1. *A. mexicana* L. (=*A. alba* Lestib.) [Mexican pricklypoppy]
2. *A. platyceras* Link & Otto [crested pricklepoppy]
3. *A. polyanthemos* (Fedde) Ownbey (=*A. intermedia* Eastw.)
4. *Argemone* sp.

Oomycetes
Peronospora arborescens. Downy mildew. **1:**TX(82). **3:**WY(1195,3378).
Basidiomycotina - Uredinales
Aecidium plenum. **3:**TX(82).
Deuteromycotina - Hyphomycetes
Alternaria sp. Leaf spot. **1:**FL(35).
Cercosporidium guanicense. Leaf spot. **1:**IL(2844).
Deightoniella argemonensis. Leaf spot. **1:**FL(2844).
Phymatotrichopsis omnivora. Root rot. **1:**TX(82). **2:**TX(82). **4:**TX(82).
Stemphylium lancipes. Leaf spot. **1:**TX(82). **2:**KS(82), TX(82). **4:**TX(82).
Deuteromycotina - Coelomycetes
Gloeosporium argemonis. Leaf spot. **2:**KS(82,963), TX(82,3597). **3:**OK(2873). **4:**NE(82), OK(82), TX(82).

Chelidonium L.
1. *C. majus* L. [greater celandine]

Deuteromycotina - Hyphomycetes
Phymatotrichopsis omnivora. Root rot. **1:**TX(82).
Deuteromycotina - Coelomycetes
Ascochyta sp. On leaves. **1:**WI(1357).
Septoria chelidonii. Leaf spot. **1:**TX(82).

Corydalis Venten.
1. *C. aurea* Willd.
2. *C. brandegei* S. Wats.
3. *C. scouleri* Hook.
4. *C. sempervirens* (L.) Pers. (=*C. glauca* Pursh) [rock-harlequin, Roman wormwood]
5. *Corydalis* sp.

Oomycetes
Peronospora corydalis. Downy mildew. **1**:AZ(1195,3378), IN(82), MT(3202). **4**:MA(82), MD(82), WI(1333). **5**:KS(82), MD(82), NE(82), OK(82,2872), TX(82).
Ascomycotina
Crocicreas cyathoideum. On stems. **2**:CO(505).
Basidiomycotina - Uredinales
[*Puccinia aristidae*]. **1**:CO(82), KS(82), NE(82). **5**:CO(82), KS(82), NE(82).
P. brandegei. Autoecious. **3**:WA(3202). **5**:CO(82), WA(82).
Deuteromycotina - Coelomycetes
Phyllosticta corydalis. **4**:WI(1348).
Septoria corydalis. Leaf spot. **4**:WI(82,970). **5**:TX(82), WI(82).

Dendromecon Benth.
1. *D. rigida* Benth. [bushpoppy, treepoppy]

Ascomycotina
Massaria cleistotheca. On dead stems. **1**:CA(1468).
Pleosphaerulina californica. On limbs. **1**:CA(82).
Pleospora argyrospora. On dead limbs. **1**:CA(1468).
Pocosphaeria dendromeconis. On limbs. **1**:CA(82).
Basidiomycotina - Ustilaginales
Entyloma eschscholtziae. Leaf smut. **1**:CA(82,1077).

Eschscholzia Cham. in Nees
1. *E. californica* Cham. [California poppy]

Ascomycotina
Crocicreas eschscholtziae. On dead stems. **1**:CA(505,850,2801).
Erysiphe polygoni. Powdery mildew. **1**:CA(82).
Basidiomycotina - Ustilaginales
Entyloma eschscholtziae. On living leaves. **1**:CA(82,1077,1468,3092).
Deuteromycotina - Hyphomycetes
Acroconidiella eschscholtziae. On capsules, leaflets, petioles and stems. **1**:CA(82,813,1468).
Alternaria sp. Collar rot. **1**:TX(82).
Botrytis cinerea. Gray mold blight. **1**:AK(82,526), FL(35).
Fusarium sp. Wilt. **1**:FL(35).
Verticillium albo-atrum. Wilt. **1**:CA(82).
Deuteromycotina - Other
Sclerotium rolfsii. Southern blight. **1**:FL(35).

Meconopsis Vig.
1. *M. betonicifolia* Franch. [blue poppy]

Ascomycotina
Erysiphe cichoracearum. Powdery mildew. **1**:CA(1155).

Papaver L.
1. *P. alboroseum* Hultén
2. *P. dubium* L.
3. *P. macounii* Greene
4. *P. nudicaule* L. [arctic poppy, Iceland poppy]
5. *P. orientale* L. [oriental poppy]
6. *P. rhoeas* L. [corn poppy, field poppy, Flanders poppy]
7. *P. somniferum* L. [opium poppy]
8. *Papaver* sp.

Ascomycotina
Clathrospora pentamera. On dead stem. **8**:AK(1938).
Erysiphe polygoni. Powdery mildew. **4**:CA(2892). **7**:OR(82,3202). **8**:FL(35).
Mycosphaerella tassiana. On dead stem. **1**:AK(1938). **3**:AK(1938).
Pleospora helvetica. On dead stem. **3**:AK(1938).
P. herbarum. **4**:AK(526).
P. paucitricha. **4**:AK(526).
P. penicillus. **4**:AK(82).
Sclerotinia sclerotiorum. **7**:OR(3202).
Basidiomycotina - Ustilaginales
Entyloma fuscum. Leaf smut. **2**:LA(1077). **4**:IA(82), ME(82). **7**:NY(1077). **8**:IA(1077), ME(1077), TX(82).
Basidiomycotina - Other
Thanatephorus cucumeris. **7**:ID(3202).
Deuteromycotina - Hyphomycetes
Botrytis cinerea. Gray mold blight. **4**:AK(82). **5**:MD(82), WI(82). **6**:AK(82). **8**:AK(526).
Cercospora papavericola. Leaf spot. **8**:AL(82), FL(35), NC(1303), TX(82).
Verticillium albo-atrum. Wilt. **5**:NY(82). **8**:NY(82).
Deuteromycotina - Coelomycetes
Septoria sp. Leaf spot. **8**:IA(82).
Deuteromycotina - Other
Rhizoctonia solani. Damping-off, root and stem rot. **4**:ME(82), NY(82). **5**:IN(82), NJ(82). **6**:KS(82), NJ(82). **7**:ID(82).

Sanguinaria L.
1. *S. canadensis* L. [bloodroot]

Oomycetes
Pythium paroecandrum. Root necrosis. **1**:VA(82).
Deuteromycotina - Hyphomycetes
Aureobasidium sanguinariae. Leaf spot. **1**:OH(82), TX(82), WV(82).
Botrytis sp. Gray mold blight. **1**:NY(82), WI(1345).
Cercospora sanguinariae. Leaf spot. **1**:DE(3543), MD(82), MO(82), NY(82), PA(82), TX(82), WI(82,1334).
Deuteromycotina - Coelomycetes
Ascochyta sp. Leaf spot. **1**:WI(1359,1361).
Colletotrichum sp. **1**:WI(1329).
Cylindrosporium circinans. Leaf spot. **1**:MD(82), MO(82), WI(82).
Phyllosticta sanguinariae. Leaf spot. **1**:MO(82), TX(82), WV(82).
Phyllosticta sp. **1**:WI(1352).

Passifloraceae

Passiflora L.
1. *P. caerulea* L. [blue passionflower]
2. *P. edulis* Sims [passionfruit, purple granadilla]
3. *P. edulis* Sims f. *flavicarpa* Degener
4. *P. foetida* L. [love-in-a-mist, running-pop, wild-water-lemon]
5. *P. incarnata* L. [apricotvine, maypop passionflower, wild passionflower]
6. *P. laurifolia* L. [belle-apple, Jamaica honeysuckle, pomme-de-liane, vinegar-pear, yellow granadilla]
7. *P. quadrangularis* L. [giant granadilla]
8. *P. sp. (edible species)*
9. *P. sp. (native species)*
10. *P. sp. (ornamental species)*
11. *P. suberosa* L.
12. *Passiflora* sp.

Oomycetes
Phytophthora sp. Root rot. **3**:HI(2903).
Pythium aphanidermatum. Root rot. **3**:HI(2903).
P. splendens. Root rot. **3**:HI(2903). **12**:FL(35).
Ascomycotina
Erysiphe polygoni. Powdery mildew. **1**:CA(1155).
Leptosphaeria doliolum. **5**:GA(1438).
Sclerotinia sp. Collar rot. **12**:CA(82).
Seuratia millardetii. **12**:HI(2784).
Deuteromycotina - Hyphomycetes
Alternaria passiflorae. Leaf and pod spot. **2**:HI(2903). **3**:HI(2903). **4**:HI(2903). **7**:HI(2903). **11**:HI(2903).
A. tenuissima. **2**:HI(2903). **3**:HI(2903).
Alternaria sp. **2**:HI(2903). **3**:HI(2903). **4**:HI(2903). **7**:HI(2903).
Botrytis sp. **12**:HI(2903). **8**:HI(82).
Cercospora biformis. Leaf spot. **5**:OK(2872). **12**:FL(35). **9**:AR(82), NC(82). **10**:AR(82), OK(82).
C. fusco-virens. Leaf spot. **5**:NC(3944). **9**:E & C states(82), TX(82).
C. passiflorae. Leaf spot. **12**:FL(35).
C. regalis. On living leaves. **12**:TX(3597). **9**:TX(82).
C. truncatella. **5**:NC(3944). **9**:AL(82), NC(82), TX(82).
Corynespora cassiicola. Leaf spot. **12**:FL(35).
Fusarium sp. Root rot. **2**:FL(35).
Phymatotrichopsis omnivora. Root rot. **9**:TX(82).
Stemphylium sp. **4**:HI(2903).
Deuteromycotina - Coelomycetes
Botryodiplodia sp. Stem rot. **2**:FL(35).

Colletotrichum gloeosporioides. Leaf and stem spot, on fruits and leaves. **2**:FL(35), HI(2903). **3**:HI(2903). **6**:HI(2903). **12**:FL(35). **8**:FL(82), HI(82). **9**:LA(82).
Phomopsis sp. **3**:HI(2903).
Phyllosticta sp. Leaf spot. **2**:FL(35). **10**:NJ(82).
Septoria fructigena. On fruits. **10**:SC(82).
Deuteromycotina - Other
Rhizoctonia solani. Root rot. **12**:FL(35).
Rhizoctonia sp. Root rot. **3**:HI(2903).
Sclerotium rolfsii. Southern blight. **12**:FL(35). **9**:FL(82).

Pedaliaceae

Proboscidea Schmidel

1. *P. louisianica* (Mill.) Thell. (=*P. jussieui* of authors, *P. louisiana* (Mill.) Woot. & Standl., *Martynia louisiana* Mill., *Martynia proboscidea* Gloxin) [common unicorn-flower, devil's-claw, proboscis-flower, rams-horn]

Ascomycotina
Sclerotinia sclerotiorum. Stem rot. **1**:MA(82).
Deuteromycotina - Hyphomycetes
Cercospora beticola. Leaf spot. **1**:IA(82), OK(82,2873), TX(82), WI(82,1330).
Phymatotrichopsis omnivora. Root rot. **1**:TX(82).
Deuteromycotina - Other
Sclerotium rolfsii. Southern blight. **1**:TX(82).

Sesamum L.

1. *S. indicum* L. (=*S. orientale* L.) [sesame]
2. *Sesamum* sp.

Deuteromycotina - Hyphomycetes
Alternaria sesami. Blight, leaf spot. **1**:FL(35), MS(737).
Cercoseptoria sesami. Leaf spot. **1**:FL(35,2652), SC(2652).
Cercospora sesami. Leaf spot. **1**:FL(35,82), GA(82), MS(2720), NC(1303), SC(82).
Corynespora cassiicola. Blight. **1**:MS(737,2720).
Fusarium sp. Wilt. **2**:GA(101).
Helminthosporium sesami. Aerial stem rot. **1**:TX(2855).
Thielaviopsis basicola. Root and stem rot. **1**:MD(3605).
Verticillium albo-atrum. Wilt. **1**:NM(577).
Deuteromycotina - Coelomycetes
Macrophomina phaseolina. Charcoal rot. **1**:CA(82), TX(82).
Deuteromycotina - Other
Rhizoctonia sp. Damping-off. **1**:NC(1303).

Phytolaccaceae

Agdestis Moc. & Sessé

1. *A. clematidea* Moc. & Sessé ex DC.

Deuteromycotina - Coelomycetes
Colletotrichum sp. Anthracnose. **1**:FL(35).

Phytolacca L.

1. *P. americana* L. (=*P. decandra* L.) [garget, pigeon-berry, pocan, pokeberry, pokeweed, scoke, Virginian poke]
2. *P. brachystachys* Moq. in DC.
3. *P. dioica* L.
4. *Phytolacca* sp.

Ascomycotina
Calycellina pulviscula. On stems. **4**:NY(850,2181,3167).
Crocicreas cyathoideum var. *cacaliae*. On stems. **1**:NY(505).
Cryptosphaerella shearii. On dead stems. **2**:HI(2781).
Diaporthe arctii. On dead stems. **1**:GA(1438), OK(2873). **4**:E states(82), KY(82).
Leptosphaeria clavigera. **4**:NJ(967).
L. doliolum. **1**:GA(1438).
[*L. fulgida*]. **1**:GA(1438).
Stictis stellata. **4**:NJ(3225).
Deuteromycotina - Hyphomycetes
Alternaria sp. Leaf spot. **1**:FL(35), WI(1332). **4**:NY(82), TX(82), WI(82).
Cercospora flagellaris. Leaf spot. **1**:E states(82), DE(3543), FL(35), IL(82), NC(3944), OK(2872), TX(82). **4**:FL(35).
Cristulariella moricola. On leaves. **1**:IL(2572).
Dendryphiella vinosum. Leaf spot. **1**:DE(3543), SC(82), TX(82). **4**:SC(634).
Fusarium roseum. **1**:DE(3543).
Helminthosporium sp. Leaf spot. **1**:FL(35).
Phymatotrichopsis omnivora. Root rot. **1**:TX(82).
Deuteromycotina - Coelomycetes
Asteroma sp. On stems. **4**:NC(82), PA(82).
Colletotrichum gloeosporioides. Dieback. **3**:CA(81).
Colletotrichum sp. Anthracnose. **1**:OK(2872).
Phoma apocrypta. On stems. **1**:TX(82).
Phomopsis phytolaccae. On stems. **4**:E states(82), KY(82).
Phyllosticta phytolaccae. Leaf spot. **4**:AL(82), NJ(82), SC(82).
Septoria phlyctanoides. Leaf spot. **1**:NJ(82), NY(82), OH(82), TX(82).
Deuteromycotina - Other
Rhizoctonia crocorum. Root rot. **4**:TX(82).
R. solani. Root rot. **1**:FL(35,82), TX(82).

Rivina L.

1. *R. humilis* L. [baby-pepper, bloodberry, rougeplant]

Basidiomycotina - Uredinales
Puccinia raunkaerii. Autoecious. **1**:FL(35,82), TX(82).
Deuteromycotina - Hyphomycetes
[*Cercospora flagellaris*]. Leaf spot. **1**:FL(35).
Phymatotrichopsis omnivora. Root rot. **1**:TX(82).
Deuteromycotina - Coelomycetes
Septoria riviniae. Leaf spot. **1**:TX(82).
Deuteromycotina - Other
Rhizoctonia crocorum. Root rot. **1**:TX(82).

Pinaceae

Abies Mill.

1. *A. alba* Mill. [silver fir]
2. *A. amabilis* Douglas ex Forbes [Cascade fir, Pacific silverfir, white fir]
3. *A. balsamea* (L.) Mill. [balsam fir, fir-balsam]
4. *A. bracteata* (D. Don) D. Don ex Poit. [bristlecone fir, Santa Lucia fir]
5. *A. cephalonica* Loud. [Greek fir]
6. *A. concolor* (Gordon & Glend.) Lindl. ex Hildebr. (=*Pinus concolor* Engelm. ex Gordon) [white fir]
7. *A. fraseri* (Pursh) Poir.
8. *A. grandis* (Douglas ex D. Don) Lindl. [grand fir, lowland white fir]
9. *A. lasiocarpa* (Hook.) Nutt. (=*A. lasiocarpa* (Hook.) Nutt. var. *lasiocarpa*) [alpine fir, subalpine fir]
10. *A. lasiocarpa* (Hook.) Nutt. var. *arizonica* (Merriam) Lemmon (=*A. arizonica* Merriam) [cork fir]
11. *A. magnifica* Andr. Murray (=*A. magnifica* Andr. Murray var. *shastensis* Lemmon) [California red fir, red fir]
12. *A. nordmanniana* (Steven) Spach
13. *A. procera* Rehd. (=*A. nobilis* (Douglas ex D. Don) Lindl.) [noble fir]
14. *A. religiosa* Lindl.
15. *A. sibirica* Ledeb.
16. *Abies* sp.

Oomycetes
Phytophthora cactorum. Isolated from roots. **2**:OR(1430), WA(1430). **11**:WA(1429). **13**:OR(1430), WA(1430).
P. cinnamomi. Root rot. **1**:SE states(3624). **3**:NC(1303,1873). **6**:NC(1303). **7**:NC(1301,1303,1994), WV(3907). **12**:SE states(3624). **13**:NC(1303). **15**:SE states(3624).
P. citricola. Shoot blight, canker, root rot. **6**:CA(2298). **7**:NC(1303,3235). **11**:CA(2298).
P. cryptogea. Isolated from roots. **6**:OR(1430), WA(1430). **11**:OR(1430), WA(1430). **13**:OR(1430), WA(1430).
P. drechsleri. Root rot. **7**:NC(1303).
Pythium sp. Root decay. **7**:NC(1303).
Ascomycotina
Asterina sp. On needles. **3**:NY(3247). **8**:ID(3202).
Belonium inconspicuum. On twigs. **16**:CO(521).
Bertia moriformis. On bark and limbs. **11**:CA(638,642).
B. moriformis var. *latispora*. On wood. **3**:NH(684).
Bombardia solaris. **11**:CA(642).

Botryosphaeria abietina. On needles. **6:**CA(661). **8:**ID(3202).
Cenangium ferruginosum. Twig blight. **3:**MI(82), PA(82).
Cenangium sp. On dead limbs. **11:**CA(642).
Ceratocystis brunnea. Associated with bark beetle galleries. **9:**CO(784).
C. leucocarpa. On dead logs. **9:**CO(785).
Chaetomium murorum. **11:**CA(642).
Chaetomium sp. On decaying cuttings. **7:**NC(1303).
Chlorociboria aeruginosa. **8:**WA(3202). **9:**OR(3202).
Ciboria rufo-fusca. On cone scales. **9:**OR(3202). **11:**CA(642). **16:**WA(3202).
Clithris graphis. **16:**WA(3202).
Coccomyces pseudotsugae. On bark. **8:**CA(3229).
Colpoma crispum. On wood. **9:**ID(3202), OR(3202), WA(638,3202). **11:**CA(638,642,661).
Coniochaeta ligniaria. **8:**ID(3202).
Darkera abietis. Needle blight. **6:**OR(3884).
Dasyscyphus serinellus. On needles. **11:**CA(638).
Delphinella abietis. **3:**Unk.(82). **5:**Unk.(82). **6:**Unk.(82). **7:**Unk.(82). **13:**Unk.(82).
D. balsameae. Needle blight. **3:**ME(82), NH(82), VT(82). **5:**RI(82). **6:**MA(82), ME(82), NH(82), NY(82), RI(82). **7:**MA(82). **13:**MA(82).
Dermea balsamea. On dead limbs and logs. **7:**NC(1303,1307).
D. piceae. **8:**ID(3202).
Diaporthe conorum. On dead limbs and trunks. **16:**Unk.(82).
Dimerium balsamicola. On needles. **3:**NY(82,3081,3247). **8:**ID(3202), WA(3202).
Discina ancilis. On litter. **11:**CA(638).
Durella sp. On wood. **11:**CA(642).
Euantennaria alaskensis. **2:**WA(3202).
Exarmidium diaphanum. **16:**WA(3202).
E. hysteriiforme. **16:**ID(3202).
E. morthieri. **8:**ID(3202).
Gibbera sp. On limbs. **11:**CA(642).
Gloniella abietina. **8:**ID(3202).
Grovesiella abieticola. Canker. **2:**OR(82,3202). **6:**CA(81,2390), WA(3202). **8:**ID(3202), OR(82,3202). **9:**CO(3459).
Herpotrichia juniperi. Brown felt blight, snow mold. **2:**OR(3202), WA(3202). **6:**CA(642,661), OR(3202). **8:**ID(3202), MT(3202), WA(3202). **9:**CO(1195,3274), ID(3202), MT(3202), OR(3202), WA(3202), WY(2924). **11:**CA(642,661). **13:**OR(3202). **16:**W states(82), OR(657).
H. parasitica. Needle blight. **16:**Unk.(82).
Hyaloscypha stevensonii. On logs. **6:**CA(661). **11:**CA(642).
Hypoderma sp. On bark. **11:**CA(642).
Hypomyces aurantius. **11:**CA(661).
Hysterium angustatum. On wood. **11:**CA(639,642).
Incrucipulum sulphurellum. **7:**NC(1307).
Isthmiella abietis. Needle cast. **9:**CO(82,3457), ID(82,3202), NM(82), UT(82). **16:**CO(49).
I. faullii. Needle cast. **3:**ME(82), MI(82), NH(82), NY(82), VT(1224), WI(1336).
Kirschsteiniothelia thujina. On wood. **3:**ME(2849). **7:**NC(1303).
Lachnellula agassizii. On wood. **3:**ME(82), MI(82,2862), VT(1224). **7:**NC(1303,1307). **8:**WA(3202). **9:**ID(3202), MT(3202), OR(3202), WA(3202), WY(2924). **11:**CA(642,661). **13:**WA(3202). **16:**NW states(82).
L. arida. On wood. **3:**MI(82). **8:**OR(638,3202). **9:**ID(3202), MT(3202), OR(3202), UT(860), WA(3202), WY(1195,2924,3378). **11:**CA(81,638,642,661,860). **13:**WA(3202).
L. calyciformis. **8:**ID(3202).
L. resinaria. Canker, on resin of dead limbs. **3:**MN(82,2874). **7:**NC(1303,1307).
L. subtilissima. On bark. **3:**ME(82).
L. suecica. On bark. **9:**OR(3202), WY(1195,3378). **11:**CA(642,661).
Lachnum niveum. **6:**AZ(662).
Leucostoma kunzei. On twigs and stems. **3:**MI(1878).
Lirula abietis-concoloris. Foliage disease, needle blight and cast. **2:**OR(3202). **6:**CA(81,82,642,661,3121), CO(3457), OR(3202). **7:**NC(82). **8:**ID(3202), MT(3202), OR(3202), WA(3202). **9:**CO(3457), WY(2924). **13:**OR(3202). **16:**Range of host(82), CO(49).
L. mirabilis. Needle cast. **3:**MI(82), WI(1343).
L. nervata. Needle cast. **3:**ME(82), NH(82), VT(82,1224). **7:**NC(1303).
L. nervisequia var. *conspicua*. Needle cast. **4:**CA(3123).
L. punctata. Needle cast. **2:**OR(82,3202). **6:**CA(81), OR(82,3202). **8:**ID(82,3202). **11:**WA(3202). **13:**OR(82,3202).
Lophodermium consociatum. Needle cast. **2:**ID(3202), OR(82,3202), WA(82,3202). **6:**OR(82), WA(82). **8:**OR(82), WA(82). **9:**OR(82), WA(82). **11:**OR(82), WA(82). **13:**OR(82), WA(82). **16:**OR(82), WA(82).
L. decorum. Needle blight and cast. **6:**CA(81,661), CO(1195,3457). **8:**ID(3202), MT(82,3202), OR(82,3202), WA(3202). **9:**WA(3202). **13:**OR(3228). **16:**CO(49).
L. lacerum. Secondary needle cast. **3:**NH(82), NY(82), PA(82), VT(82).
L. piceae. Tar spot needle blight. **6:**OR(3202). **9:**OR(3202). **11:**CA(642). **16:**Range of host(82).
L. uncinatum. Needle blight and cast. **2:**OR(82,3202), WA(82,3202). **13:**OR(3228).
Lophodermium sp. Needle cast. **4:**CA(2394).
Lophomerum autumnale. Needle cast. **2:**OR(82,2669,3202). **3:**MI(82,2669). **6:**CA(81,82,642,2669), OR(82,2669,3202). **9:**CO(2669,3457), ID(82,2669,3202), UT(82,2669), WA(3202), WY(82,2669). **11:**CA(82,2669). **13:**OR(82,2669,3202). **16:**CO(49).
Maurodothina farrae. Black mildew on living needles. **3:**NY(2819).
Melittosporium hysterinum. **16:**WA(3202).
Melomastia shastensis. **11:**CA(642).
Mollisia cinerea. On wood. **11:**CA(661).
M. conigena. **9:**OR(3202).
M. piceae. On dead needles. **6:**CA(81,661).
Mytilinidion gemmigenum. **9:**WA(3202).
M. mytilinellum. **7:**NC(1307). **8:**ID(3202). **9:**ID(3202), WA(3202).
Nectria fuckeliana. On bark. **2:**WA(3202). **3:**ME(82), MI(82). **9:**CO(649), MT(3202).
N. pinea. On dead twigs. **7:**NC(1303).
Niesslia coloradensis. On wood. **9:**CO(528). **11:**CA(638,642).
N. exilis. On dead limbs. **11:**CA(81).
Nothophacidium abietinellum. Needle blight, snow blight. **3:**NH(2952), NY(82,2952).
Odontotrema majusculum. On wood. **11:**CA(81,3232).
O. minus. **9:**WA(3202).
Ophiostoma distortum. On wood, associated with bark beetles. **3:**NY(786,3714). **6:**AZ(786,3714).
O. minus. Isolated from bark. **9:**CO(784).
O. nigrocarpum. On wood, associated with wood borers. **16:**ID(785,824,3714).
O. nigrum. On wood. **6:**WY(1364). **9:**CO(784,1364,3714).
O. olivaceum. On wood. **9:**CO(3714).
O. penicillatum. Isolated from bark. **9:**CO(784).
O. piceae. Lumber and sapwood stain. **10:**AZ(3714). **13:**OR(782,3202), WA(782,3202).
O. stenoceras. On decaying cuttings. **7:**NC(1303).
O. subannulatum. **8:**ID(2156).
O. wageneri. **2:**OR(1067). **6:**OR(1258). **8:**OR(1067,1258). **13:**OR(1067).
Ostreola formosa. On twigs. **11:**CA(638,642).
Ostreola sp. **9:**MT(3202).
Patinella abietina. On dead limbs. **11:**CA(642).
Pezicula livida. **8:**ID(3202).
Phacidium abietis. Needle blight, snow blight. **6:**ID(82,3202), MT(3202), OR(82,3202), WA(3202). **8:**ID(82,3202), MT(82,3202), OR(82,3202), WA(3202). **9:**ID(82), MT(82), OR(82), WA(82).
P. balsamicola. On limbs and twigs. **8:**ID(3202), MT(3202). **16:**ID(864).
P. infestans. Snow blight. **3:**NE states(82). **6:**OR(3202). **9:**ID(3202), MT(3202), WA(3202). **16:**Unk.(82).
Phaeocryptopus nudus. Needle cast, black mildew. **3:**ME(82), NC(82), NY(82,3247), WI(82). **9:**CO(3457), WY(1195). **16:**CO(49).
Phaeophacidium abietinum. **8:**ID(3202), MT(3202), WA(3202).
Phaeosaccardinula dematia. **2:**CA(2434).
[*Pithya cupressina*]. **8:**ID(3202). **13:**WA(3202).
Pleospora sp. Twig dieback. **6:**OR(2177,3202). **8:**OR(2177).
Rasutoria abietis. Black mildew. **2:**OR(3202,3247), WA(82,3202,3247). **8:**CA(81,2434,3247), ID(82,3202), OR(82,3202,3247), WA(82,3202,3247). **16:**OR(3247).
R. tsugae. **8:**WA(3202).

Rebentischia massalongii. On limbs. **3:**ME(186).
Rhizina undulata. Root rot of seedlings. **8:**ID(82,3202), MT(82).
Rhizocalyx abietis. **9:**ID(3202).
Rosellinia diathrausta. On limbs. **6:**CA(638,2403). **9:**CO(649,2403), ID(3202), OR(1811,3202), WA(3202).
Sarcosphaera coronaria. **6:**CA(661).
Sarcotrochila balsameae. Needle blight and cast. **3:**NE states(82), WI(327). **6:**CA(81,327,661), MT(82), OR(82,327,3202). **9:**CO(82), WA(82,3202). **16:**CO(49,2951), ME(2951), NH(2951), OR(2951).
Scoleconectria cucurbitula. Canker. **3:**Range of host(82), MI(3041), MN(2874), NH(3041). **6:**WA(3202). **7:**NC(1303,1307). **8:**OR(3202). **16:**MI(3041).
Scutellinia scutellata. On wood. **7:**NC(1307). **8:**ID(3202). **11:**CA(638).
Strigopodia resinae. **3:**NH(1674).
Sydowia polyspora. Tip dieback. **6:**CA(2393).
Therrya sp. On dead limbs. **7:**NC(1303).
Thyronectria balsamea. Bark canker. **3:**MI(82,2862), MN(82,2874), NH(3041,3173), NY(82,3041,3173), PA(82), VT(3173). **9:**CO(82). **16:**CO(49), MN(3041).
Tryblidiopsis pinastri. **3:**NE states(82).
Tympanis laricina. **3:**ME(1380), NH(1380), NY(1380). **6:**CA(661). **8:**ID(3202). **16:**ID(1380), ME(2670), NY(2670), OR(3202), UT(2670), WY(1380,2670).
T. truncatula. On limbs. **3:**MI(1380), NH(1380), NY(2670). **9:**ID(3202). **16:**ID(1380,3202).
Tympanis sp. Trunk canker. **3:**MN(2874).
Valsa abietis. Branch canker. **3:**ME(82), PA(82). **6:**CA(82), OR(82,3202). **8:**ID(3202), WA(3202). **9:**ID(3202). **11:**CA(82).
V. pini. **8:**WA(3202).
Virgella robusta. Needle cast. **2:**OR(774,3202), WA(3202). **6:**CA(638,642,774), CO(3457), OR(774,3202), WA(82). **8:**CA(82,774), ID(82,774,3202), OR(82,638,3202), WA(82,3202). **11:**CA(774). **13:**OR(774,3202). **16:**CO(49).
Xenomeris abietis. **8:**ID(3202).
Xylopezia hemisphaerica. On decorticated wood. **6:**CA(3233).

Basidiomycotina - Uredinales

Hyalopsora aspidiotus. Heteroecious: aecial host. **2:**OR(1979). **3:**Unk.(82). **8:**ID(1979), OR(1979).
Melampsora abieti-capraearum. Heteroecious: aecial host. **3:**Range of host(82). **6:**CO(82), UT(82). **8:**ID(82), MT(82), OR(82), WA(82). **9:**ID(82), MT(82), WA(82), WY(1195). **16:**CO(49).
M. epitea. Heteroecious: aecial host. **8:**ID(3202), MT(3202), OR(3202), WA(3202). **9:**WA(3202).
M. medusae. **9:**OR(82).
Melampsorella caryophyllacearum. Heteroecious: aecial host. **2:**WA(82,3202). **3:**Range of host(82), MA(1575), ME(1575), MI(1575), MN(2874), NH(3866), NY(1575), VT(1224,1575), WI(1575). **5:**CO(1575), ID(1575), OR(1575), UT(1575). **6:**AZ(1203,2769,3981), CO(82,2769), ID(82,3202), NM(2769), NV(2769), OR(82,3202), UT(82,1575). **8:**ID(82,3202), MT(82,3202), OR(82,3202), WA(82,3202), WY(82). **9:**Range of host(82), AZ(1203,2769), CO(1195), ID(1195,2769,3202), MT(3202), NM(1575), OR(3202), UT(1195,2769,3620), WA(3202), WY(1195,2769). **10:**AZ(3981). **11:**CA(82,661,2769), NV(2769). **13:**Unk.(82). **16:**Range of host(3866), CO(49), UT(1575).
Milesina fructuosa. Heteroecious: aecial host. **3:**ME(82,1575), NH(82,1575), NY(82,1575).
M. marginalis. Heteroecious: aecial host. **3:**MA(82), NH(82), NY(82).
M. pycnograndis. Heteroecious: aecial host. **3:**ME(82), NH(82), NY(82), VT(82).
Peridermium holwayi. **9:**WY(1195).
P. ornamentale. **6:**NM(82), NV(82), OR(82). **9:**ID(82), MT(82), OR(82), WA(82), WY(82). **13:**OR(82), WA(82).
P. rugosum. **2:**WA(82,3202). **8:**CA(82), OR(82,3202), WA(82,3202).
Peridermium sp. **2:**OR(82).
Pucciniastrum epilobii. Heteroecious: aecial host. **2:**WA(3202). **3:**MI(1575), NY(1575), VT(1575), WI(1575). **6:**AZ(3981), OR(3202). **8:**ID(3202), MT(3202), WA(3202). **9:**AK(526), ID(3202), MT(3098,3202), OR(3202), WA(3202), WY(1195). **10:**CO(1575). **13:**OR(3202).
P. goeppertianum. Heteroecious: aecial host. **2:**OR(82), WA(82,936,3202). **3:**ME(82), PA(82), WI(82). **6:**CA(81,82,638,936), CO(82), NM(82), OR(82,3202). **8:**ID(936,3202), MT(82), OR(82,3202), WA(82,936,3202). **9:**Range of host(82), AK(60), CO(1575), ID(1195,1575,3202), MT(1575,3202), OR(936,1575,3202), WA(1575,3202), WY(1195,1575). **11:**CA(81,82,661,936), OR(82,3202). **13:**OR(82), WA(82). **16:**CO(49).
P. pustulatum. Heteroecious: aecial host. **2:**WA(82). **3:**MI(82), MN(2874), NY(82), WI(82,1345). **6:**CO(82), OR(82). **8:**ID(82,3098), MT(82), OR(82), WA(82,3098). **9:**Range of host(82), AK(60,526), CO(82). **13:**OR(82). **16:**CO(49).
Uredinopsis americana. Heteroecious: aecial host. **3:**Range of host(82), CT(1575), MA(1575), ME(1575), MI(1575), MN(1575,2874), NY(1575), VT(8,1575), WI(1575), WV(1575). **7:**NC(1303,1307), TN(82), WV(82,1575,3907).
U. longimucronata. Heteroecious: aecial host. **8:**ID(3202), OR(3202), WA(3202). **9:**MT(2769,3202), WA(3202).
U. osmundae. Heteroecious: aecial host. **3:**Range of host(82).
U. phegopteridis. Heteroecious: aecial host. **3:**WI(82).
U. pteridis. Heteroecious: aecial host. **2:**OR(82,3202), WA(82,3202). **6:**CA(81). **8:**CA(81,82), ID(82,3202), MT(82,1195,3202), OR(82,648,3202), WA(82,3202). **9:**WA(82). **13:**OR(82,3202).
U. struthiopteridis. Heteroecious: aecial host. **3:**MI(82). **8:**ID(82), OR(82), WA(82). **9:**ID(82), MT(1195), OR(82), WA(82). **13:**OR(82), WA(82).

Basidiomycotina - Other

Acanthophysium cerussatum. On wood. **6:**AZ(128,1202). **9:**ID(2083,3202), WA(128).
A. lividocaeruleum. On decorticated wood. **6:**AZ(128,1202), NM(128,1190). **8:**ID(2083,3202).
A. weirii. On bark. **6:**AZ(1202). **8:**ID(2083,3202).
Aleurocystidiellum subcruentatum. **8:**OR(3202).
Aleurodiscus abietis. On living and dead limbs. **3:**NY(2083), VT(2083). **8:**ID(2083,3202).
A. amorphus. On suppressed, injured, or recently dead trees. **1:**AZ(128). **2:**OR(1449,3202), WA(1449,2083,3202). **3:**CT(1449), ME(1449), MI(1449,2083,2862), MN(1197,1236,2083), NH(1449,2083), NY(1449,2083), VT(1449,2083). **6:**Unk.(82), AZ(128,1202,1236), CA(638,642,1449,2083), WA(2083,3202). **7:**MA(1449), TN(2083). **8:**CA(2083), ID(1449), MT(128), OR(1449), WA(638,1449,2083). **9:**ID(1449,3202), MT(1449,3202), OR(1449,2083,3202), WA(3202). **10:**AZ(1202). **11:**Unk.(82), CA(642). **13:**OR(1449). **16:**ID(1449), MA(1449), MI(2862), MO(1449), NH(1449), NY(1449), OR(1449), WA(1449), WI(1449).
A. farlowii. **8:**ID(3202).
A. grantii. On limbs and trunks. **2:**WA(1236). **6:**CA(661,1236), WA(1236). **8:**CA(1236), ID(1236), MT(1236), WA(1236). **9:**ID(1236), MT(1236), OR(1236), WA(1236). **11:**CA(1236).
Amphinema byssoides. **6:**AZ(128,1202), NM(1190).
Amylocystis lapponica. **8:**ID(3202), MT(3202).
Amylostereum chailletii. White rot. **2:**WA(3202). **3:**NH(543), NY(543). **6:**AZ(1202), NM(543), OR(543). **8:**MT(543). **10:**AZ(1202). **16:**Unk.(82), CO(543), ID(543).
Anomoporia albolutescens. On wood. **8:**ID(3202). **9:**CO(2167). **11:**CA(642).
A. bombycina. **8:**ID(3202).
Antrodia albobrunnea. **8:**ID(128,3202), MT(3202).
A. carbonica. **8:**ID(128,3202). **10:**AZ(1202).
A. heteromorpha. Brown cubical rot. **3:**NE states(82), NY(82). **8:**ID(638,3202), MT(3202,3832), OR(3202), WA(3202). **9:**MT(3202).
A. serialis. **8:**ID(3202), MT(3202), WA(3202). **10:**AZ(1202).
A. sinuosa. Brown cubical rot. **3:**NE states(82). **8:**ID(3202), MT(3202,3832). **11:**CA(642).
A. vaillantii. **8:**ID(3202). **16:**OR(4007).
A. xantha. Brown cubical rot. **3:**MN(128,2874). **8:**ID(3202). **16:**NW states(82).
Armillaria mellea. Root, stem and shoestring rot. **2:**OR(1067), WA(3202). **3:**NE states(82), MN(2874), NH(3643), NY(82), VT(59). **6:**AZ(1202), CO(1724), NM(82), OR(1062,1068,1258), WA(1068). **8:**ID(59,538,1583,2387,3202), OR(1067,1068,1258), WA(59,1068,1258,3202). **9:**AZ(82), CO(82,1724), ID(538,1583), MT(3202), NM(82), UT(82). **11:**OR(1062). **13:**OR(1067,1258), WA(1258). **16:**NE states(82), CO(49).

Asterostroma cervicolor. **6:**AZ(128,1202).
A. laxum. **8:**ID(3202).
A. ochroleucum. **6:**NM(1190).
Athelia decipiens. **6:**AZ(1202), NM(1190). **10:**AZ(1202).
Auricularia auricula. On wood. **2:**WA(2219,3202). **3:**MN(914). **6:**AZ(1202), CA(638,642,914), ID(3202). **8:**ID(914,3202), MT(914,3202), OR(638,914,3202). **9:**ID(914), OR(914,3202), WA(914,3202). **10:**AZ(1202).
Bjerkandera adusta. **6:**AZ(1202).
Bondarzewia berkeleyi. **8:**ID(3202), MT(3832), OR(3202), WA(3202).
B. montana. Root rot. **6:**CA(661). **8:**ID(538,1583,2387).
Botryobasidium ansosum. **11:**CA(128).
B. botryosum. **6:**NM(1190).
B. flavescens. On wood. **11:**CA(642).
B. subcoronatum. On wood. **6:**CA(661). **11:**CA(128).
B. vagum. On wood. **8:**WA(3202). **10:**AZ(1202). **11:**CA(642).
Botryohypochnus isabellinus. **6:**AZ(128,1202).
Ceraceomyces serpens. On wood. **6:**NM(1190). **11:**CA(642).
C. sublaevis. On wood. **8:**ID(128).
Ceriporia tarda. **10:**AZ(1202).
Ceriporiopsis rivulosa. **8:**ID(128,3202).
C. subvermispora. **6:**AZ(128,1202).
Cerocorticium cremoricolor. On wood. **11:**CA(642).
C. sulfureo-isabellinum. **3:**AZ(128), MN(1197), NH(1715).
Cerrena unicolor. **9:**ID(3202). **16:**WA(3202).
Chaetoderma luna. **9:**ID(128).
Chondrostereum purpureum. On wood. **2:**WA(3202). **8:**ID(543). **16:**Unk.(82).
Clavicorona divaricata. On wood. **16:**ID(2073).
Climacocystis borealis. On wood. **6:**AZ(1202), CA(642). **7:**NC(1303,1307). **8:**MT(3832), MT(3202). **9:**ID(3202).
Climacodon pulcherrimus. **6:**NM(1190).
Collybia acervata. On debris, logs and stumps. **7:**NC(3717), TN(3717), VA(3717).
C. bakerensis. On bark. **16:**CA(854).
Columnocystis abietina. On logs. **2:**OR(3202). **3:**VT(543). **6:**CA(642,661). **8:**ID(3202). **10:**AZ(1202). **11:**CA(543,638,642,661), OR(638,3202). **13:**OR(3202).
C. ambigua. On wood. **7:**NC(430). **9:**ID(3202). **16:**NC(543).
Confertobasidium montanum. **9:**OR(1707).
C. olivaceo-album. On wood. **6:**NM(1190). **10:**AZ(1178,1202).
Coniophora arida. On wood. **6:**AZ(1202). **7:**NC(1303,1307). **10:**AZ(1202).
C. arida var. *suffocata*. **8:**ID(128).
C. olivacea. On wood. **8:**WA(3202). **11:**CA(642,661).
C. puteana. Brown rot. **3:**NE states(82), NY(82). **6:**NM(128,1190).
Coniophora sp. **8:**ID(128).
Corticium roseum. **11:**CA(642).
Crepidotus fusisporus var. *abietinus*. On dead twigs. **3:**MI(1544).
C. versutus. **9:**WA(3202).
Crepidotus sp. On limbs. **11:**CA(642).
Cryptoporus volvatus. On wood. **6:**CA(642,661), OR(3202). **8:**ID(1469,3202), MT(3202,3832), OR(3202), WA(3202). **9:**ID(3202), OR(3202). **11:**CA(638,642,661). **13:**OR(3202).
Cyathus striatus. **16:**OR(4007).
Cylindrobasidium corrugum. On wood. **8:**WA(3202). **10:**AZ(1178,1202). **11:**CA(642,661). **16:**CO(49).
C. torrendii. **7:**NC(1307).
Cyphellopsis anomala. **3:**NH(128).
Cyptotrama asprata. **6:**AZ(1202). **10:**AZ(1202).
Cystostereum pini-canadense. **3:**MN(128,1197).
Dacrymyces chrysospermus. On wood. **6:**AZ(1202). **7:**NC(1303,1307). **8:**ID(3202), MT(3176), WA(3202). **9:**MT(3202).
D. stillatus. On limbs, logs, twigs and hand hewn boards. **8:**ID(3202), WA(3202). **11:**CA(642).
Dacryobolus karstenii. **8:**CA(81).
Dacryopinax spathularia. On wood. **8:**ID(128).
Daedaleopsis confragosa. On wood. **6:**AZ(1202), NM(1190). **8:**ID(3202), WA(82,3202).
Dichomitus squalens. White pocket heart rot. **8:**MT(82,3202), WA(3202).
Diplomitoporus crustulinus. **9:**AZ(128). **10:**AZ(1202).
D. lenis. On logs. **6:**AZ(1202). **8:**ID(128,3202), MT(128,3202). **11:**CA(638,642). **13:**OR(3202).
D. linbladii. **2:**WA(3202). **8:**ID(3202). **9:**ID(3202), MT(3202), WA(3202).
Echinodontium tinctorium. Heart rot of living trees. **2:**OR(3202), WA(1070). **6:**AZ(662,1202), CA(642,661), NM(1190), OR(22,3202), WA(3202). **8:**ID(1583,3202), MT(3202), OR(3202), WA(3202). **9:**ID(538,1583,3202), MT(3202), WA(3202). **10:**AZ(1202). **11:**CA(642). **13:**WA(3202). **16:**Range of host(82), AK(2180), AZ(2180), CA(2180), CO(49,2180), ID(2180), MT(2180), NM(2180), OR(657,2180), WA(2180).
Exidiopsis calcea. On roots of living trees. **9:**AK(128).
Fibricium rude. **6:**AZ(1202).
Fibulomyces mutabilis. **9:**OR(3202).
Flammulina velutipes. **8:**ID(3202).
Fomitopsis cajanderi. On wood. **6:**AZ(1202), NM(1190). **8:**ID(3202), MT(3202), OR(3202), WA(3202). **11:**CA(642).
F. officinalis. Brown trunk rot. **3:**MN(2874). **6:**CA(81,661). **8:**ID(538,2387,3202), WA(3202). **11:**CA(642). **13:**OR(3202). **16:**Unk.(82), MT(3202).
F. pinicola. Brown cubical rot. **2:**OR(3202), WA(3202). **3:**MN(2874). **6:**AZ(1202), CA(661), ID(3202), NM(1190), OR(22). **7:**NC(1303,1307). **8:**ID(3202), MT(3202,3832), OR(3202), WA(3202). **9:**ID(3202), MT(2045,2046,3202,3832), WA(3202). **11:**CA(642,661). **13:**WA(3202). **16:**Range of host(82).
F. rosea. Brown cubical rot. **6:**AZ(1202). **8:**ID(3202), OR(3202). **16:**Range of host(82).
Galerina autumnalis. Inside rotting logs. **6:**CA(661).
G. megalocystis. On logs. **16:**CA(3333).
G. subbadia. On logs. **16:**OR(3333).
Ganoderma applanatum. On wood. **2:**WA(3202). **3:**NY(2962). **6:**CA(642,661), OR(3202). **8:**ID(3202), MT(3202,3832), OR(3202). **9:**OR(3202). **16:**Range of host(82).
G. oregonense. White rot. **2:**OR(82), WA(82,3202). **6:**CA(661). **8:**ID(538,1583,3202), OR(3202). **9:**ID(538), OR(3202). **11:**CA(642,661).
G. tsugae. **6:**AZ(17,18,1202). **16:**CA(17).
Gloeophyllum carbonarium. **6:**NM(1190).
G. odoratum. **6:**NM(1190). **8:**ID(3202). **9:**ID(3202).
G. sepiarium. Brown rot. **3:**NE states(82), MN(2874), NY(82). **6:**AZ(1202). **8:**ID(1583,3202), OR(3202), WA(3202). **9:**ID(3202), MT(3202), OR(3202), WA(3202). **11:**CA(642), OR(3202). **16:**WA(82).
Gloeoporus dichrous. **8:**ID(3202), WA(3202). **9:**ID(3202). **16:**MN(655).
Gloiodon strigosus. **10:**AZ(1202).
Grandinia abieticola. **16:**WA(3202).
G. arguta. **6:**AZ(1202), NM(1190). **10:**AZ(1202).
G. barba-jovis. **8:**MT(128). **16:**MT(128).
G. breviseta. **3:**MN(1197).
G. hastata. **8:**OR(3202).
G. pallidula. On wood. **6:**AZ(128,1202), CA(661). **11:**CA(642).
Guepiniopsis alpina. On wood. **9:**OR(3202), WA(3202). **10:**AZ(1202). **11:**CA(638,642,661). **16:**OR(3202).
G. buccina. **8:**ID(3202).
Gymnopilus bellulus. On wood. **7:**VA(3717).
G. oregonensis. **16:**OR(3202).
G. penetrans. **11:**CA(642).
G. sapineus. **10:**AZ(1202).
Hapalopilus rutilans. **6:**NM(1190).
Helicobasidium corticioides. On wood. **9:**CO(154). **16:**CO(154).
Henningsomyces candidus. On wood. **6:**AZ(128,662,1202), CA(642), NM(1190). **11:**CA(638,642).
Hericium abietis. Long pocket heart rot of living trees. **2:**WA(1070). **6:**OR(22). **8:**ID(82,2235,3202), MT(82), WA(82). **9:**ID(82,3202), MT(3202), OR(3202), WA(82,3202).
H. americanum. White rot. **6:**ID(82), WA(82). **8:**ID(3202), WA(3202).
Heterobasidion annosum. White root and butt rot. **2:**OR(1067,3202), WA(1070,3202). **4:**CA(3349). **6:**W states(3584), AZ(128,1202), CA(82,661,1051), CO(568,1723,1724), NM(128,1190), NV(3055), OR(568,1068), UT(3055), WA(1068). **7:**NC(675,1303). **8:**ID(538,1583,2387,3202), OR(1067,1068,1258,3202), WA(1068). **9:**W states(3584), CO(1724), ID(538,1583,3055,3202), UT(3055). **11:**CA(82,642,661). **13:**OR(1067,1258), WA(1258). **16:**Range of host(82), CO(49).
Hohenbuehelia petalodes. On dead log. **6:**AZ(1202), CA(661).

Hydnum sp. Stem rot. **8:**ID(538,1583). **9:**ID(538,1583).
Hydropus marginellus. On dead logs and stumps. **7:**NC(3717), VA(3717).
Hymenochaete corrugata. **10:**AZ(1202).
H. tabacina. On wood. **3:**NE states(82), NY(82). **8:**ID(3202), OR(3202). **9:**ID(3202). **16:**WA(542).
Hyphoderma argillaceum. **6:**AZ(1202).
H. pilosum. **3:**MN(1197).
H. praetermissum. **6:**AZ(1202).
H. setigerum. **3:**MN(128,1197). **9:**MT(3202).
Hyphodermopsis polonensis. **3:**MN(1197).
Hyphodontia stipata. **3:**MN(1197).
Hypholoma capnoides. On wood. **7:**NC(1303,1307).
Hypochnopsis mustialaensis. **6:**AZ(1202).
Inonotus circinatus. White root and butt rot. **3:**NE states(82), MN(82,2874), NY(82). **9:**CO(82). **16:**CO(49).
I. dryadeus. White root and butt rot. **6:**CA(82,661), NM(82,1181,1190), OR(3202). **8:**ID(82,3202), MT(82,3202), OR(82,3202), WA(82,3202). **14:**NM(1181).
[*I. dryophilus*]. **6:**AZ(128).
I. hispidus. **8:**OR(3202).
I. tomentosus. Root rot. **8:**ID(538,1583,2387,3202). **9:**ID(538,1583,3202).
Irpex lacteus. **8:**ID(3202).
Ischnoderma resinosum. White rot. **6:**CA(661). **8:**ID(2387,3202), MT(3202), OR(3202,4007), WA(3202). **11:**CA(661).
Jahnoporus hirtus. On roots. **3:**MI(82). **6:**AZ(1202).
Junghuhnia collabens. **8:**ID(128,2168,3202).
J. luteoalba. **8:**ID(2168,3202), MT(2168,3202). **13:**OR(2168,3202).
J. zonata. **8:**ID(3202). **9:**ID(3202).
Kavinia alboviridis. **6:**NM(128,1190). **8:**ID(3202).
K. himantia. **6:**AZ(128,1202).
Kuehneromyces lignicola. **16:**ID(3202).
K. mutabilis. **16:**ID(3202).
Laetiporus sulphureus. Brown cubical rot. **2:**WA(1070). **6:**AZ(1202). **7:**NC(1303,1307). **8:**ID(2387,3202), MT(3202), OR(3202). **9:**ID(3202). **11:**CA(642,661). **16:**Range of host(82), WA(3202).
Laurilia sulcata. On wood. **2:**OR(3202). **9:**ID(538,1583).
Lentinellus micheneri. **9:**Unk.(2423), ID(3202), MT(3202), OR(3202), WA(3202).
L. montanus. On wood. **9:**W states(2417,2423). **11:**W states(2417,2423). **16:**ID(3202), MT(3202), OR(3202), WA(3202).
L. ursinus. **6:**W states(2423), AZ(1202), ID(3202), MT(3202), OR(3202), WA(3202). **8:**W states(2423), ID(3202), MT(3202), OR(3202), WA(3202). **9:**OR(3202).
Lentinus sp. On logs. **9:**OR(3202), WA(3202). **11:**CA(642), OR(3202).
Leptoporus mollis. **8:**MT(3202,3832).
Leptosporomyces galzinii. **6:**AZ(1202).
L. mundus. **11:**CA(128).
L. raunkiaerii. On wood. **11:**CA(661).
Leucogyrophana mollusca. On wood. **3:**ME(128). **6:**AZ(1202). **11:**CA(642). **16:**NY(128).
Marasmiellus filopes. **3:**NY(2930).
Megacollybia platyphylla. **10:**AZ(1202).
Megalocystidium luridum. On rotten wood. **11:**CA(642).
Meruliopsis albostramineus. On wood. **6:**AZ(128), CA(128). **8:**ID(128,3202). **9:**AZ(128). **10:**AZ(128). **11:**CA(642,661).
M. bellus. On wood. **6:**CA(642). **11:**CA(638,642).
M. corium. **6:**OR(128), WA(128).
M. hirtellus. On fallen twigs and limbs. **6:**AZ(1202). **10:**AZ(1178,1202).
M. taxicola. On wood. **6:**CA(642). **9:**WA(3202). **11:**CA(638,642).
Merulius sp. **9:**OR(3202).
Mucronella aggregata. **6:**AZ(128).
Mycena borealis. On rotten logs and stumps. **7:**NC(3717), TN(3717), VA(3717).
M. elegantula var. *inflata*. On wood. **16:**CO(2450).
M. griseoviridis var. *cascadensis*. On wood. **16:**WA(3326).
M. overholtsii. On rotten logs and stumps. **11:**CA(661).
M. tenax. On dead needles. **3:**NY(3326).
Neolentinus lepideus. **8:**ID(3202). **9:**ID(3202).
Nivatogastrium nubigenum. On duff, litter and logs. **11:**CA(638), OR(3202).
Odonticium romellii. **8:**ID(128,3202).
Omphalotus olearius. **9:**OR(3202).
Ossicaulis lignatilis. **6:**CA(2937).
Osteina obducta. **8:**ID(3202).
Oxyporus corticola. **8:**ID(3202), WA(3202).
O. nobilissimus. On wood. **13:**OR(82,3202), WA(82,3202). **16:**OR(2180), WA(2180).
Panellus serotinus. **8:**WA(3202).
P. violaceofulvus. **16:**MA(3623).
Parmastomyces transmutans. **6:**AZ(1202), NM(1190).
Pellicularia sp. On wood. **6:**CA(661).
Peniophora aurantiaca. **8:**WA(3202).
P. cinerea. **6:**NM(1190).
P. exima. On wood. **11:**CA(642,1716).
P. piceae. **3:**MN(128,1197).
P. pithya. **6:**AZ(128,1202).
P. violaceolivida. On wood. **11:**CA(661).
Perenniporia medulla-panis. **8:**ID(3202). **9:**MT(3202).
P. subacida. White rot. **3:**Lake states(82), NE states(82), MN(128,2874). **7:**NC(1303). **8:**ID(82,538,1583,2387,3202), MT(82,3202,3832), OR(3202). **9:**ID(538,1583,3202), MT(3202). **16:**W states(82), ID(3202).
Phaeolus schweinitzii. Brown cubical butt and root rot. **2:**WA(3202). **3:**NE states(82), NY(82). **6:**AZ(1202), CA(81,661). **7:**NC(675,1303,1307). **8:**ID(1583,2387,3202), OR(3202), WA(3202). **9:**ID(1583,3202). **16:**Range of host(82).
Phanerochaete carnosa. White rot. **3:**MI(433), MN(1197). **6:**AZ(128), NM(128,1190). **8:**ID(128). **9:**ID(3202). **11:**CA(642). **16:**NY(128).
P. filamentosa. White rot. **6:**AZ(433,1202).
P. gigantea. White rot. **3:**MN(433). **6:**NM(1190). **10:**AZ(433). **16:**NM(128).
P. sanguinea. **8:**ID(3202).
P. viticola. White rot. **3:**NH(128). **7:**NC(433).
Phellinus ferreus. White rot. **11:**CA(642,661). **16:**OR(3202).
P. ferrugineofuscus. White laminated rot. **8:**ID(3202), WA(3202). **9:**ID(3202), MT(3202). **11:**CA(642,661). **13:**WA(3202). **16:**W states(82).
P. ferruginosus. **9:**WA(3202). **10:**AZ(1202).
P. nigrolimitatus. Root and stem rot. **8:**ID(538,1583,2387,3202), WA(3202). **9:**AZ(128), CO(128), ID(128,3202), MT(128), OR(3202). **10:**AZ(1202). **11:**CA(642), WA(3202). **16:**WA(3202).
P. pini. White pocket heart rot. **2:**OR(3202). **3:**MN(292,2874). **6:**OR(292,2049,3202). **7:**NC(675). **9:**CO(82), ID(538,3202), MT(3202,3832). **11:**OR(3202). **16:**Range of host(82), CO(49).
P. repandus. **8:**ID(128,3202).
[*P. robustus*]. **3:**MN(1197). **9:**ID(3202). **13:**OR(3202). **16:**Unk.(82).
P. viticola. **3:**MN(128,1197). **8:**ID(3202).
P. weirii. Butt rot, root rot. **2:**OR(1067). **6:**OR(1068,1258), WA(1068). **8:**ID(538,1583,2387), OR(1067,1068,1072,1258,3202), WA(1068,1072,1258,3202). **9:**ID(538,1583). **11:**OR(1072), WA(1072). **13:**OR(1067,1258), WA(1258).
Phlebia albida. On logs. **6:**AZ(128,1202), CA(642,661), NM(1190). **9:**AZ(128), WA(3202). **10:**AZ(1202). **11:**CA(642,661), OR(3202).
P. incarnata. **6:**NM(1190).
P. merismoides. On logs. **11:**CA(638). **16:**OR(3202,4007).
P. subalbida. On wood. **11:**CA(643).
P. subserialis. White rot. **6:**AZ(1202,2556).
P. tremellosus. On wood. **6:**CA(661).
Phlebiella filicina. On wood. **11:**CA(661).
P. vaga. On wood. **3:**NY(128). **6:**CA(661), NM(1190). **8:**ID(3202), WA(3202).
Pholiota abieticola. On wood. **7:**NC(1542).
P. adiposa. Yellow stringy rot. **6:**AZ(1202). **8:**ID(1583,2235,3202).
P. alnicola. **6:**AZ(1202).
P. bridgii. On rotten logs. **6:**CA(1542). **11:**CA(661).
P. decorata. **9:**AZ(1202).
P. flammans. Heart rot. **6:**Range of host(82). **8:**Range of host(82).
P. graveolens. On rotten wood. **11:**CA(642).
P. limonella. On wood. **6:**AZ(1202), CA(661). **16:**WA(1542).
P. malicola var. *macropoda*. On wood. **7:**NC(1303,1307).
P. squarrosoides. **6:**AZ(1202).
Pholiota sp. Isolated from wood. **6:**OR(22).

Phyllotopsis nidulans. **9:**AZ(1202).
Physisporinus sanguinolentus. **6:**AZ(128,1202), NM(128,1190). **8:**ID(3202).
P. vitreus. **8:**MT(3832).
Piloderma bicolor. **6:**NM(1190).
Pleurocybella porrigens. On logs and stumps. **7:**NC(1541), TN(1541).
Pleurotus ostreatus. **6:**AZ(1202). **10:**AZ(1202).
Pleurotus sp. **8:**OR(3202).
Plicatura nivea. **8:**MT(3202,3832).
Pluteus atricapillus. **6:**AZ(1202). **10:**AZ(1202).
Polyporus badius. White rot. **8:**ID(3202). **9:**AZ(1202), MT(3202). **10:**AZ(1202).
P. melanopus. **8:**ID(3202).
P. varius. **6:**AZ(1202). **8:**ID(3202), WA(3202).
Porotheleum fimbriatum. **6:**AZ(128,1202).
Postia amara. **8:**ID(82,3202).
P. balsamea. Brown cubical butt rot of living trees. **3:**E states(82), MN(82,2874), TN(82). **9:**UT(82). **16:**OR(3202), WA(3202).
P. caesia. On logs. **3:**MN(2874). **6:**NM(1190). **8:**ID(3202), MT(3202), OR(3202), WA(3202). **11:**CA(661). **13:**OR(3202).
P. fragilis. Brown cubical rot. **3:**MI(82). **8:**ID(3202), OR(3202), WA(3202).
P. guttulata. Brown cubical rot. **3:**Lake States(82), NE states(82), MN(1054,2874). **8:**ID(3202).
P. hibernica. **6:**AZ(1202). **10:**AZ(1202).
P. leucospongia. Brown cubical rot. **6:**AZ(1202), CA(81,642). **9:**ID(3202), MT(3832), MT(3202), OR(3202), WA(3202). **10:**AZ(1202). **13:**WA(3202).
P. perdelicata. **8:**ID(3202), WA(3202).
P. placenta. **8:**ID(3202), MT(128). **16:**MT(128,3202).
P. rancida. **6:**NM(1190).
P. sericeomollis. Brown cubical rot. **8:**WA(3202). **9:**ID(538). **11:**CA(642).
P. stiptica. **8:**ID(3202).
P. undosa. **2:**WA(397,3202). **3:**NH(397). **7:**TN(397). **8:**MT(3832).
Pseudohydnum gelatinosum. On logs. **6:**AZ(1202), CA(661). **11:**CA(642,661).
Pseudotomentella tristis. **6:**AZ(1202), NM(1190).
Pycnoporellus alboluteus. Brown cubical rot. **6:**AZ(1202), NM(1190), WA(3434). **8:**ID(3202), MT(3202), OR(3202), WA(3202). **9:**ID(3202), OR(3202), WA(3202).
P. fulgens. On wind-thrown trees. **2:**WA(3202). **6:**OR(3202). **8:**ID(3202), MT(3202,3832). **9:**MT(3202). **16:**NW states(82).
Pycnoporus cinnabarinus. **8:**ID(3202).
Repetobasidium mirificum. **6:**NM(1190).
Resinicium bicolor. On wood. **8:**ID(2235,3202).
R. furfuraceum. **8:**ID(2116,3202).
Rigidoporus crocatus. **6:**AZ(128,1202). **8:**ID(128), MT(3202).
Scytinostroma galactinum. **3:**ME(128), MN(1197), NY(128). **8:**OR(3202).
Serpula himantioides. On logs. **6:**AZ(128,1202), CA(642), NM(128,1190). **11:**CA(638,642).
S. lacrimans. On logs. **11:**CA(661).
Sistotrema brinkmannii. On wood. **6:**NM(1190). **11:**CA(642).
S. subtrigonospermum. **6:**AZ(1202).
Skeletocutis amorpha. **8:**ID(3202), MT(3202).
S. nivea. **8:**ID(3202).
S. stellae. **6:**AZ(1202). **11:**CA(128).
S. subincarnata. On wood. **6:**AZ(1202). **11:**CA(642).
Sparassis crispa. **6:**AZ(1202,2266).
Sphaerobasidium minutum. On wood. **16:**OR(128).
Steccherinum ciliolatum. On wood. **6:**NM(1190). **9:**CO(1175). **10:**AZ(1175).
S. fimbriatum. **6:**NM(1190).
Stereum sanguinolentum. Heart rot. **3:**NE states(82), ME(543), MI(543), NH(543), NY(82,543), VT(543). **6:**AZ(128,543,1202), NM(1190). **7:**TN(82,543). **8:**ID(82). **9:**CO(82), ID(82,538,1583). **10:**AZ(1202). **16:**CO(543), ME(543), NY(543).
Subulicystidium longisporum. **3:**MN(1197).
Tapinella panuoides. On logs. **6:**AZ(1202). **11:**CA(642).
Thelephora terrestris. Smothering seedlings. **8:**ID(82,3202).
Tomentella avellanea. **16:**WA(2039).
T. bresadolae. **6:**NM(1190). **16:**NM(2039).
T. bryophila. **6:**NM(1190). **10:**AZ(1202). **16:**AZ(2039), NM(2039).
T. chlorina. **6:**NM(1190). **16:**AZ(2039), NM(2039).
T. cinerascens. **6:**AZ(1202).
T. crinalis. **6:**AZ(128,1202). **10:**AZ(1202). **16:**AZ(2039).
T. ferruginea. **6:**NM(1190). **16:**AZ(2039), NM(2039).
T. griseoviolacea. **16:**AZ(2039), NY(2039).
T. molybdaea. **6:**NM(1190). **16:**NM(2039).
T. neobourdotii. **16:**AZ(2039).
T. ramosissima. **6:**AZ(1202). **16:**AZ(2039).
T. ruttneri. On wood. **16:**NY(128,2035).
T. sublilacina. **16:**AZ(2039), MI(2039), NY(2039).
T. terrestris. **16:**AZ(2039).
T. viridescens. **16:**AZ(2039), NM(2039).
Tomentellopsis echinospora. **8:**ID(128).
Trametes hirsuta. **6:**CA(642), NM(1190). **8:**ID(3202).
T. versicolor. On wood. **2:**WA(3202). **3:**NY(2962). **8:**ID(3202). **9:**WA(3202). **16:**Range of host(82).
Trechispora cohaerens. **3:**MN(128). **6:**NM(1190).
T. mollusca. On wood. **10:**AZ(1202). **11:**CA(642,661). **16:**WA(128).
Trechispora sp. **8:**WA(3202).
Trichaptum abietinum. White rot. **6:**AZ(662,1202), CA(661), NM(1190), OR(3202). **8:**CA(81), ID(82,3202), MT(82,3202), OR(3202), WA(82,3202). **9:**ID(3202), MT(3202), OR(3202), WA(3202). **11:**CA(81,642,661), OR(3202). **13:**WA(3202). **16:**Range of host(82).
T. fuscoviolaceum. On wood. **16:**Unk.(1205).
T. laricinum. **6:**CA(661). **11:**CA(661). **16:**MN(655).
Tricholomopsis bella. **6:**AZ(1202).
T. decora. On wood. **7:**NC(1303,1307,3717), TN(3717), VA(3717).
T. rutilans. **6:**AZ(1202).
Tubulicrinis propinquus. **6:**NM(128,1190).
T. subulatus. **6:**AZ(128,1202). **11:**CA(638).
Tulasnella fuscoviolacea. On bark. **3:**NH(2673). **11:**CA(128,642).
Tulasnella sp. On rotten wood. **11:**CA(642).
Tyromyces chioneus. **8:**ID(3202), MT(3202,3832), WA(3202).
Vararia boreale. **6:**AZ(1202). **8:**ID(128,3202), MT(128,3202).
V. investiens. **3:**ME(128). **8:**ID(128). **9:**ID(128).
V. racemosa. **8:**ID(3202), WA(3202). **9:**CO(128). **10:**AZ(1202).
Vararia sp. **9:**CO(128).
Veluticeps berkeleyi. Brown cubical trunk rot. **6:**NM(1198,2265).
Vesiculomyces citrinus. On wood. **2:**WA(3202). **6:**NM(128). **9:**AZ(128), CO(797). **10:**AZ(1202). **11:**CA(642).
Wolfiporia cocos. Brown cubical butt and root rot. **3:**MN(793,1241), PA(1241). **16:**CA(793).
Wrightoporia subrutilans. White rot. **3:**ME(2682).
Xeromphalina campanella. On logs, on stumps. **6:**AZ(1202). **7:**NC(1303,1307). **8:**ID(3202).

Deuteromycotina - Hyphomycetes

Atichia glomerulosa. **8:**ID(3202).
Aureobasidium pullulans. **8:**ID(3202).
Berkleasmium corticola. **7:**NC(1307).
Botrytis cinerea. Seedling twig blight, twig canker. **7:**NC(1303). **8:**ID(3202). **9:**ID(3202), MT(82).
Botrytis sp. Twig blight of seedlings. **8:**ID(82,3202), WA(3202).
Cephalosporium sp. Canker. **3:**MN(82,2874), WI(82).
Chalara acuaria. On needles. **16:**NJ(2550).
Chloridium sp. Obtained from rotting wood. **11:**CA(642).
Conoplea abietina. **8:**ID(3202), WA(3202).
C. geniculata. **3:**NY(1672).
[*C. juniperi*]. **8:**ID(3202). **9:**WA(3202).
C. juniperi var. *robusta*. **16:**ID(1672), WA(1672).
Costantinella terrestris. On wood. **6:**CA(661).
Cylindrocarpon cylindroides. Canker, dieback. **2:**OR(1069). **13:**OR(1069).
Cylindrocarpon sp. Isolated from rotting roots. **7:**NC(1303).
Cylindrocladium scoparium. Root rot. **7:**NC(1303), WV(3907).
Cylindrocolla sp. On wood, on boards. **11:**CA(642).
Eustilbum aureum. On resin. **3:**NY(3180).
Fusarium oxysporum. Hypocotyl rot. **6:**CA(411). **11:**CA(411).
F. solani. Root rot. **7:**PA(2359).
Fusarium sp. Root rot. **7:**PA(3858).
Gliocladium catenulatum. **8:**ID(3202).
Haplotrichum magnisporum. On rotten wood. **11:**CA(642,2131).
Helminthosporium sp. **8:**ID(3202).
Hyphosoma lumbricoideum. On twigs and older needles. **16:**WA(666,3202).

Kmetia corticola. On bark. **11**:CA(661,666).
Leptographium penicillatum. Isolated from roots. **8**:ID(2372).
L. wageneri. Root disease. **6**:W states(1257).
Oidiodendron truncatum. On wood. **11**:CA(661).
Septonema fasciculare. On bark. **11**:CA(639,642).
Sphaeridium luteum. On rotten wood. **11**:CA(642).
Sporothrix subannulata. **8**:ID(2156).
Toxosporium camptospermum. **8**:OR(3202).
Trichoderma viride. On wood. **11**:CA(642).
Trichosporum symbioticum. Tracheomycosis, sapwood stain. **6**:CA(82).

Deuteromycotina - Coelomycetes

Bothrodiscus berenice. On branches. **3**:MI(3549), MN(2874).
Cryptocline abietina. On living needles. **6**:OR(2481).
C. effusa. On living needles. **6**:OR(2481).
Cryptosporium candidum. **8**:ID(3202).
C. macrospermum. Canker. **3**:NE states(82), NY(82).
Cytospora abietis. Canker, dieback. **2**:OR(1069). **6**:CA(3120,3964), NV(3964). **8**:OR(1066). **11**:CA(3120,3964). **13**:OR(1069).
C. kunzei. Canker. **3**:ME(3781), VT(3781).
C. pinastri. Twig and branch canker, on needles. **3**:ME(82), MN(2874), WI(82). **7**:VA(82).
Cytospora sp. Cytospora canker. **16**:CO(49).
Foveostroma abietinum. **3**:Unk.(861), ME(82).
Furcaspora pinicola. On needles. **6**:CA(81,2390,3549).
Gelatinosporium pinicola. Canker, dieback. **2**:OR(1069). **13**:OR(1069).
Macrophomina phaseolina. Root rot of seedlings. **6**:CA(3348). **11**:CA(3348).
Melanconium sp. On needles. **6**:CA(661).
Phacidiopycnis boycei. Canker. **8**:ID(82), MT(82).
P. pseudotsugae. **3**:NE states(1413).
Phoma dura. On dead needles. **6**:CA(81).
P. eupyrena. Needle cast, blight in nursery. **11**:CA(1917).
Phomopsis abietina. Twig canker. **6**:ID(82), MT(82), WA(82). **8**:ID(3202), MT(3202), WA(3202). **9**:ID(82), MT(82), WA(82).
P. montanensis. Limb canker. **8**:MT(3202). **9**:ID(82,3202), MT(82,3202).
P. occulta. A weak parasite on needles of nursery seedlings. **16**:Unk.(82).
Phomopsis sp. Dieback. **13**:NC(1303).
Rhabdogloeopsis balsameae. Needle blight. **3**:WI(82,810). **7**:NC(82). **16**:NC(3549), WI(3549).
Rhizosphaera pini. On needles. **7**:NC(82,1303,1307).
Rhizothyrium abietis. Needle discoloration. **9**:ID(3204).
Sphaeropsis abietis. On needles. **3**:MI(82,2862).
Sphaeropsis sp. Dieback of cultivated trees. **6**:MD(82), NY(82). **7**:NC(1303), PA(82).
Stegonsporiopsis cenangioides. On dead limbs. **3**:PA(960).
Tiarosporella abietis. Needle blight. **6**:OR(3884). **8**:OR(3884).
T. parca. Needle blight and cast. **8**:CO(82), ID(82,3202), OR(82,3202), WA(82,3202). **9**:MT(82,3202). **16**:CO(49).

Deuteromycotina - Other

Apiosporium abietis. On needles. **8**:ID(3202), OR(3202).

Cedrus Trew

1. *C. atlantica* (Endl.) G. Manetti ex Carrière [atlas cedar]
2. *C. deodara* (D. Don) G. Don [deodar]
3. *Cedrus* sp.

Oomycetes

Phytophthora cinnamomi. Root rot. **1**:SE states(3624). **2**:SE states(3624), CA(3624,4010), NC(1303).
Phytophthora sp. Root rot. **2**:FL(35), NC(1303). **3**:FL(35).
Pythium sp. Root rot. **2**:FL(35). **3**:FL(35).

Basidiomycotina - Uredinales

[*Gymnosporangium juniperi-virginianae*]. **3**:FL(35).

Basidiomycotina - Other

Armillaria mellea. Root rot. **2**:CA(2898). **3**:MS(82,2720).
A. tabescens. Root rot. **2**:FL(35). **3**:FL(35,82).
Phellinus pini. Root rot. **3**:Unk.(82).
Tomentella atrorubra. **3**:NJ(2039).

Deuteromycotina - Hyphomycetes

Cercospora sp. Needle blight. **3**:FL(35).
Cylindrocladium sp. Root rot. **2**:NC(1303).
Fusarium sp. Root rot. **2**:FL(35).
Phymatotrichopsis omnivora. Root rot. **3**:TX(82).

Deuteromycotina - Coelomycetes

Diplodia sp. Needle necrosis. **2**:FL(35).
Macrophoma sp. Needle blight. **2**:FL(35).
Phomopsis juniperovora. Twig blight. **2**:FL(35). **3**:FL(35).
Phyllosticta sp. Needle necrosis. **2**:FL(35).
Sphaeropsis sapinea. Canker. **3**:AL(82).
Sphaeropsis sp. Dieback. **2**:FL(35). **3**:FL(35).
Stagonospora sp. Twig dieback. **3**:FL(35).

Deuteromycotina - Other

Rhizoctonia solani. Root rot. **2**:FL(35). **3**:FL(35).

Larix Mill.

1. *L. decidua* Mill. (=*L. europaea* DC.) [European larch]
2. *L.* × *eurolepis* Hort.
3. *L. kaempferi* (Lamb.) Carrière (=*L. leptolepis* (Siebold & Zucc.) Gordon) [Japanese larch]
4. *L. laricina* (Du Roi) K. Koch [American larch, black larch, hackmatack, tamarack]
5. *L. lyallii* Parl.
6. *L. occidentalis* Nutt. [western larch]
7. *Larix* sp.

Myxomycetes

Licea operculata. Slime mold. **7**:MN(82).

Oomycetes

Phytophthora cinnamomi. Seedling root rot. **1**:SE states(3624), MD(82). **3**:SE states(3624), MD(82).

Ascomycotina

Botryosphaeria obtusa. On limbs. **7**:Unk.(82).
Coccodinium corticola. **6**:ID(3202).
C. laricis. On bark. **6**:ID(641).
Colpoma crispum. **6**:WA(3202).
Coniochaeta ligniaria. **6**:ID(3202).
Dasyscyphus ellisianus. On twigs. **1**:NY(82), RI(82). **3**:MA(82). **7**:MA(82), NY(82), RI(82).
D. oblongosporus. On dead limbs. **1**:MA(82), ME(82), MI(82), NY(82), PA(82). **3**:MA(82). **4**:NE states(82), MI(82). **7**:NC states(82), NE states(82).
Durella sp. **4**:AK(1904).
Encoeliopsis laricina. **6**:ID(3202).
Hypodermella laricis. Needle blight. **6**:NW states(82), ID(3202), MT(3202), OR(3202), WA(3202). **7**:NC states(82), NW states(82).
Lachnellula arida. **6**:MT(82,3202), WA(3202).
L. calyciformis. **6**:ID(82,3202).
L. laricis. On dead limbs. **4**:VT(860).
L. occidentalis. On dead limbs. **1**:MA(82), NY(82). **3**:MA(82). **4**:AK(526), NY(82), PA(82), VT(82). **6**:NW states(82), ID(3202), MT(3202), OR(3202), WA(3202). **7**:NE states(82), NW states(82).
L. subtilissima. On dead limbs. **1**:MA(82). **3**:MA(82). **4**:CT(82), VT(82). **7**:MA(82).
L. willkommii. Canker. **1**:MA(82). **3**:MA(82). **4**:ME(2436). **7**:MA(82).
Lophodermium pinastri. Needle blight. **5**:ID(3202). **6**:NW states(82), ID(82).
Lophodermium sp. Leaf cast. **7**:WA(82).
Mycosphaerella laricina. Needle cast. **1**:IA(2692,2728), MI(2692,2728), VT(251), WI(2692,2728). **6**:IA(2667), WI(2667). **2**:IA(2728), MI(2728), WI(2728).
Mytilinidion gemmigenum. **6**:ID(3202), MT(3202), WA(3202).
Ophiostoma ips. Blue stain of wood. **6**:OR(3068).
O. minus. Wood stain. **6**:OR(82,3202), WA(82,3068,3202).
Pezoloma laricina. On fallen needles. **7**:MI(850).
Phialea acicularum. On fallen needles. **4**:MI(82).
Plectania melastoma. **6**:ID(3202).
Propolis leonis. **6**:WA(3202).
Rhizina undulata. Seedling blight. **6**:ID(82,3202).
Scoleconectria cucurbitula. On limbs. **1**:MA(3041).
Strigopodia batistae. **6**:ID(3202).
Teichospora sp. **6**:ID(3202).
Tympanis laricina. **6**:ID(3202). **7**:ID(3202).
Valsa abietis. On dead limbs. **4**:NY(82). **6**:OR(3202), WA(3202). **7**:NY(82), OH(82), PA(82).
V. ambiens. **4**:MI(82).

Basidiomycotina - Uredinales

Melampsora epitea. Heteroecious: aecial host. **6**:ID(3202), MT(3202), OR(3202).

M. medusae. Heteroecious: aecial host. **4**:NC states(82), NE states(82), IN(82). **6**:ID(3202), MT(82,3202), OR(82). **7**:NE states(82), IA(3891), MN(3891), WI(3891).

M. paradoxa. Heteroecious: aecial host. **1**:IN(82), NY(82). **4**:Lake states(82), NE states(82), AK(60,82,526). **6**:NW states(82). **7**:NC states(82), NE states(82), W states(82), OR(82).

Melampsoridium betulinum. Heteroecious: aecial host. **4**:CT(82), WI(82). **7**:CT(82), WI(82).

Basidiomycotina - Other

Acanthophysium lividocaeruleum. On wood. **6**:ID(2083), MT(3202). **7**:MT(128).

[*A. mirabile*]. On bark. **6**:ID(82,3202), MT(82,3202). **7**:NW states(82).

A. weirii. On bark. **4**:MI(2083). **6**:ID(82,2083,3202), MT(128). **7**:NW states(82), MT(128).

Aleurodiscus amorphus. On suppressed, injured, or recently dead trees. **4**:MN(1449). **6**:ID(82,1449,3202), MT(82). **7**:NW states(82), MN(82).

Amylocorticium subsulphureum. **6**:ID(3202).

Amylocystis lapponica. **6**:ID(3202).

Amylostereum chailletii. White rot. **1**:WI(543). **7**:Range of host(82).

Anomoporia albolutescens. **7**:ID(128).

A. bombycina. **6**:ID(3202).

Antrodia albobrunnea. **6**:ID(128).

A. alpina. On logs. **6**:ID(2167,3202), MT(3202). **7**:ID(2167), OR(3202).

A. carbonica. **6**:WA(128).

A. heteromorpha. **6**:ID(3202), MT(3202), WA(3202). **7**:Unk.(82).

A. serialis. **6**:ID(3202), MT(128), OR(3202), WA(3202). **7**:MT(128).

A. sinuosa. Brown cubical rot. **6**:ID(82,3202), MT(128,3202), WA(82,3202). **7**:ID(128).

A. vaillantii. **6**:ID(3202).

A. variiformis. **6**:MT(3202,3832). **7**:Unk.(82).

A. xantha. **6**:Unk.(82), ID(3202), MT(3202), WA(3202). **7**:ID(128), MT(128).

Antrodiella romellii. **7**:ID(128).

Armillaria mellea. Root and stem rot. **6**:ID(82,538,1583,3202), MT(82,3202). **7**:Range of host(82).

Athelia arachnoidea. **6**:ID(128,3202).

Bjerkandera adusta. Wood rot. **7**:Range of host(82).

Bondarzewia berkeleyi. Butt rot. **6**:ID(3202), MT(3202,3832). **7**:ID(82), MT(82).

Byssoporia terrestris. **7**:ID(128).

Ceraceomyces serpens. **6**:MT(128).

Ceriporiopsis rivulosa. **7**:ID(128).

Chaetoderma luna. **6**:ID(128).

Climacocystis borealis. **6**:ID(3202).

Columnocystis abietina. Brown cubical rot. **6**:ID(543). **7**:Range of host(82).

Coniophora puteana. **6**:ID(128,1583).

Cryptoporus volvatus. **6**:ID(3202), MT(3202), OR(3202), WA(3202).

Cystostereum pini-canadense. White rot. **6**:MT(541).

Dacrymyces chrysospermus. **6**:WA(3202).

D. stillatus. **6**:WA(3202).

Dacryobolus sudans. **6**:MT(128).

Dichomitus squalens. **6**:ID(3202), MT(3202,3832), OR(3202), WA(3202).

Diplomitoporus lenis. **6**:ID(3202), MT(3202).

D. linbladii. **6**:ID(128,3202), MT(3202).

Echinodontium tinctorium. **6**:ID(1583).

Fomitopsis cajanderi. Brown pocket heart rot. **4**:Range of host(82). **5**:MT(3202). **6**:Range of host(82), ID(3202), MT(3202), OR(3202). **7**:Range of host(82).

F. meliae. Isolated from utility pole. **7**:IL(2320).

F. officinalis. Brown trunk rot. **4**:WI(82). **6**:NW states(82), ID(538,3202), MT(3202,3832), OR(3202), WA(3202). **7**:W states(82).

F. pinicola. Brown crumbly heart rot. **4**:NY(82), WI(82). **5**:MT(3202,3832). **6**:Range of host(82), ID(3202), MT(3202,3832), WA(3202). **7**:Range of host(82).

Gloeophyllum carbonarium. **6**:ID(3202).

G. odoratum. **6**:MT(3202), OR(3202), WA(3202).

G. sepiarium. Brown rot. **4**:AK(1904). **6**:ID(3202), MT(3202). **7**:IN(82).

Grandinia arguta. **6**:MT(128).

G. floccosa. **6**:MT(3202).

Heterobasidion annosum. White root and butt rot. **6**:NW states(82), ID(3202), MT(3202,3832). **7**:Range of host(82).

Hymenochaete tabacina. **6**:ID(3202), OR(3202).

Hymenochaete sp. **6**:WA(3202).

Hyphoderma setigerum. **5**:MT(3202).

Inonotus circinatus. White root and butt rot. **6**:ID(82), MT(82). **7**:Range of host(82).

I. tomentosus. Stem rot. **6**:ID(538,3202), WA(3202).

Ischnoderma resinosum. **6**:WA(3202).

Junghuhnia collabens. **6**:MT(2168,3202).

J. nitida. **6**:MT(3832).

J. zonata. **6**:ID(3202).

Laetiporus sulphureus. Brown cubical rot. **6**:ID(82,3202), MT(82,3202,3832), OR(82,3202), WA(82,3202). **7**:Range of host(82).

Laurilia sulcata. White pocket rot. **5**:MT(3202). **6**:ID(128,543,3202), MT(543,3202). **7**:Range of host(82).

Leptoporus mollis. **6**:MT(3202,3832).

Leucogyrophana mollusca. **6**:ID(128).

Meruliopsis ambiguus. **6**:MT(128).

M. corium. **6**:MT(3202,3832).

Neolentinus lepideus. Brown cubical rot. **6**:ID(3202). **7**:Range of host(82).

Odonticium romellii. **6**:MT(128).

Osteina obducta. **6**:ID(3202), WA(3202).

Perenniporia medulla-panis. **6**:MT(3202).

P. subacida. Wood rot. **6**:ID(3202), MT(3202,3832), OR(3202), WA(3202). **7**:NE states(82), MT(128).

Phaeolus schweinitzii. Root rot. **4**:Range of host(82). **5**:ID(3202). **6**:Range of host(82), ID(3202), MT(3202,3832), OR(3202). **7**:Range of host(82).

Phanerochaete carnosa. White rot. **6**:MT(128,433), WA(128).

P. gigantea. White rot. **6**:MT(433).

P. sordida. **6**:MT(128).

Phellinus ferrugineofuscus. **5**:MT(3202). **6**:ID(128,3202), MT(128,3202). **7**:MT(128).

P. nigrolimitatus. **5**:MT(3202).

P. pini. White pocket heart rot. **4**:NC states(82), NE states(82). **5**:MT(3202). **6**:Range of host(82), ID(538,1583,3202), MT(3202,3832), OR(3202), WA(3202). **7**:Range of host(82).

P. repandus. **6**:ID(3202), MT(3202).

P. viticola. On wood. **4**:AK(1904). **6**:ID(3202), MT(3202,3832), WA(3202).

P. weirii. Root rot. **6**:ID(1583), OR(1072), WA(1072).

Phlebiella filicina. On wood. **6**:MT(128).

Pleurotus ostreatus. **6**:WA(3202).

Postia balsamea. **6**:ID(3202).

P. caesia. **6**:ID(3202), MT(3202), WA(3202).

P. floriformis. Butt rot. **7**:MI(82).

P. fragilis. **6**:ID(3202), WA(3202).

P. guttulata. **6**:ID(3202).

P. leucomallela. On wood. **7**:W states(1205).

P. leucospongia. **5**:MT(3202,3832). **6**:OR(3202), WA(3202).

P. perdelicata. **6**:ID(3202).

P. placenta. **6**:WA(3202).

P. sericeomollis. **6**:ID(128), MT(3202). **7**:ID(3202).

Pseudomerulius aureus. **6**:MT(3202,3832).

Pycnoporellus alboluteus. Wood rot. **5**:MT(3202,3832). **6**:ID(3202), MT(3202,3832). **7**:Range of host(82).

P. fulgens. Wood rot. **6**:ID(3202). **7**:Range of host(82), IN(82), MN(82).

Resinicium bicolor. **6**:MT(128).

R. furfuraceum. **6**:ID(128,2116,3202).

Schizopora paradoxa. **6**:Unk.(82).

Scytinostroma galactinum. **4**:NY(128). **6**:ID(128).

Skeletocutis amorpha. **6**:ID(3202), MT(3202).

S. nivea. **6**:OR(3202), WA(3202).

Sparassis crispa. Root rot. **6**:NW states(82,3202), WA(82).

Stereum hirsutum. **3**:MA(543).

S. sanguinolentum. Heart rot. **1**:CT(543), ID(82). **3**:MA(543). **6**:ID(82,638,3202), MT(543), WA(3202). **7**:Range of host(82), ID(543), NY(543).

Thelephora caryophyllea. Girdle of seedlings. **6:**ID(82,3202), MT(82).
T. terrestris. **6:**MT(3202).
Tomentella chlorina. **6:**ID(128,3202). **7:**ID(2039).
T. neobourdotii. On wood. **7:**MT(2033,2039,3202).
T. ochracea. **6:**ID(2036,3202). **7:**ID(2039).
T. ramosissima. **6:**MT(3202). **7:**MT(2039).
T. violaceofusca. **6:**MT(3202).
Tomentellina fibrosa. **6:**ID(3202). **7:**ID(128).
Trametes versicolor. **6:**ID(3202). **7:**Range of host(82), ME(82).
Trechispora mollusca. **7:**ID(128).
Trichaptum abietinum. On wood. **4:**AK(1904). **5:**MT(3202). **6:**ID(3202), MT(3202), OR(3202).
Tyromyces chioneus. **6:**ID(3202).
Vararia boreale. **6:**ID(3202), MT(128,3202). **7:**MT(128).
V. investiens. **6:**MT(3202). **7:**MT(128).
V. racemosa. **6:**ID(3202), WA(3202).

Deuteromycotina - Hyphomycetes

Botrytis douglasii. Gray mold of seedlings. **6:**NW states(82), ID(3202).
Cladosporium sp. Needle cast. **1:**ME(481).
Cylindrocladium scoparium. Seedling root rot. **3:**NJ(82).
Heterosporium laricis. **6:**WA(3202).
Hyphosoma resinacea. **6:**ID(3202).
Leptographium penicillatum. Isolated from roots. **6:**ID(2372).
Meria laricis. Needle blight, seedling blight. **6:**ID(82,3202), MT(3202), WA(670).
Phymatotrichopsis omnivora. Root rot. **4:**TX(82).

Deuteromycotina - Coelomycetes

Cytospora sp. On twigs. **1:**MA(82).
Phacidiopycnis pseudotsugae. Canker. **1:**NE states(1413). **3:**NE states(1413). **6:**ID(3202,3888), WA(3202,3888).
Phoma sp. **6:**ID(3202), WA(3202).
Phomopsis occulta. On dead limbs. **3:**Range of host(82). **7:**Range of host(82).

Deuteromycotina - Other

Rhizoctonia solani. Damping-off. **7:**Range of host(82).

Picea A. Dietr.

1. *P. abies* (L.) H. Karst. (=*P. excelsa* (Lam.) Link) [Norway spruce]
2. *P. breweriana* S. Wats.
3. *P. engelmannii* Parry ex Engelm. [Engelmann spruce]
4. *P. glauca* (Moench) Voss (=*Pinus alba* (Aiton) Link, *P. canadensis* (Mill.) B. S. P.) [cat spruce, white spruce]
5. *P. glauca* (Moench) Voss var. *albertiana* (S. Brown) Sarg. [Alberta spruce]
6. *P. mariana* (Mill.) B. S. P. (=*P. nigra* (Aiton) Link) [black spruce, bog spruce, double spruce]
7. *P. orientalis* (L.) Link
8. *P. pungens* Engelm. [Colorado blue spruce, Colorado spruce]
9. *P. rubens* Sarg. (=*P. rubra* Sarg.) [he balsam, red spruce]
10. *P. sitchensis* (Bong.) Carrière [Sitka spruce]
11. *Picea* sp.

Oomycetes

Phytophthora cactorum. Seedling blight. **11:**NY(82), VA(82).
P. cinnamomi. Seedling blight, root rot. **1:**SE states(3624), MD(82), NC(1303,1873), VA(82). **8:**SE states(3624), MD(82), NC(1303). **11:**NY(82), VA(82).
Pythium debaryanum. Damping-off. **3:**TX(82).
P. ultimum. Damping-off. **11:**Range of host(82).
Pythium sp. Root rot. **1:**NC(1303). **4:**NC(1303). **5:**NC(1303). **8:**NC(1303).

Ascomycotina

Bertia moriformis. On wood. **11:**CO(684).
Bombardia solaris. **3:**WY(2924).
Botryosphaeria ribis. On twigs. **1:**IL(82).
Botryotinia fuckeliana. **1:**NC(82).
Camarops ohiensis. **11:**AK(526).
Cenangium ferruginosum. On cones and twigs. **11:**CO(82), ME(82), VA(82).
Cenangium sp. **3:**ID(3202).
Ceratocystiopsis alba. On wood. **3:**CO(3714).
Ceratocystis coerulescens. Wood stain associated with bark beetles. **3:**CO(783). **4:**AK(3492).
C. piceaperda. On wood. **4:**AK(3714).
C. tenella. On wood, associated with bark beetles. **3:**CO(128,1364,3714).
Cistella parksii. On bark. **10:**CA(527).
Claussenomyces olivaceus. On resin. **3:**CO(1385).
Coccomyces bipartitus. On decorticated wood. **11:**MI(3229).
C. irretitus. On bark. **11:**ME(3228,3229).
Dasyscyphus ellisianus. On twigs. **3:**MA(82). **8:**MA(82). **11:**MA(82).
D. oblongosporus. On dead twigs. **8:**MA(82), PA(82).
Dasyscyphus sp. On cones. **10:**AK(3431).
Dimerium balsamicola. **3:**WA(3202).
Discocainia treleasei. On bark. **10:**AK(82,3167), OR(3228).
Eriosphaeria vermicularia. **3:**ID(3202).
Eutypella leprosa. **11:**IA(3629).
Glonium strobilarum. On dead cone scales. **10:**CA(81), OR(81,648,3202), WA(3202).
Hansenula capsulata. **3:**WY(2924).
Herpotrichia juniperi. Brown felt blight. **1:**CO(82), OR(82), WY(82). **3:**NW states(82), W states(82), CO(649,3274,3457), ID(3202), MT(3202), OR(3202), WA(3202), WY(1195). **8:**MT(82,3202). **10:**AK(3431), OR(3202). **11:**NW states(82).
Heyderia abietis. **3:**MT(3202).
Hypomyces papyraceus. **3:**WY(2924).
Hypoxylon fragiforme. **10:**AK(3431).
Isthmiella crepidiformis. On needles. **3:**MT(3202,3461).
Lachnellula abietis. On twigs. **9:**NY(82), VT(82).
L. agassizii. On twigs. **3:**CO(82), WA(82), WY(2924). **6:**MI(82), NY(82). **9:**NY(82), VT(82).
L. arida. On bark. **11:**CO(82).
L. subtilissima. On twigs. **6:**MI(82), NY(82).
L. suecica. **3:**WA(3202). **10:**AK(526).
Leucostoma kunzei. Canker. **3:**CO(1498).
Lirula macrospora. Needle blight and cast, tar spot. **3:**CO(82). **4:**AK(526,1904), ND(3760). **6:**AK(526,1904). **8:**ND(3760). **9:**NY(82), VT(1224). **10:**AK(1904), OR(1692,3202). **11:**AK(82,3780), CO(82), NY(82).
Lophodermium piceae. Needle cast, tar spot. **1:**MA(82), MI(82). **3:**CO(3457), MT(3202), OR(82,3202), WA(82). **4:**MI(82). **6:**ME(82). **8:**MI(82). **9:**NY(82), VT(1224). **10:**AK(82,526,1904), CA(82), OR(82,3202). **11:**Lake states(82), NE states(82), NW states(82), NY(82).
Lophodermium sp. **10:**AK(526).
Lophomerum septatum. Tar spot. **10:**OR(82,2669,3202).
Lophophacidium hyperboreum. **11:**ME(2951).
Lulworthia conica. On wood submerged in sea water. **11:**NC(1791).
L. submersa. On panels submerged in sea water. **11:**NC(1791).
Mytilinidion gemmigenum. **3:**CO(82), ID(82,3202), MT(3202), OR(3202), WA(3202), WY(2924). **11:**OR(82).
Neopeckia coulteri. Brown felt blight. **3:**WY(1195). **11:**CO(82).
Ophiostoma abiocarpum. On wood. **3:**AZ(785), CO(785,3714), NM(785). **4:**AK(3714).
O. bicolor. Wood stain associated with bark beetles. **3:**Unk.(784), CO(783,1364,3714).
O. distortum. On wood, associated with bark beetles. **3:**AZ(786,3714). **10:**AK(786).
O. ips. On wood, associated with bark beetles. **4:**AK(3714).
O. minus. On bark. **11:**CO(784).
O. olivaceum. On wood. **3:**CO(1364), OR(3714). **4:**AK(3714). **11:**AK(3714).
O. piceae. Lumber and sapwood stain. **10:**NW states(3202), OR(782), WA(782).
O. piliferum. Sapwood stain. **11:**MN(484).
O. sparsum. On wood, associated with bark beetles. **4:**AK(786,3714).
Pezizella aurantiaca. On bark. **11:**CO(521).
Phacidium infestans. Snow blight. **4:**NE states(82), ME(82). **6:**NE states(82). **9:**NE states(82). **11:**NE states(82), ME(82,3780).
Rhizina undulata. **9:**VT(3619).
Rosellinia aquila. **10:**AK(526).
R. herpotrichioides. On seedlings. **10:**WA(3202,3207).
R. thelena. On limbs. **10:**AK(82).
R. weiriana. On twigs. **3:**WA(82,3202).
Sarcotrochila piniperda. On needles and twigs. **6:**WI(82). **11:**NH(2951), VT(2951), WI(2951).
Scoleconectria cucurbitula. On bark. **3:**MT(3202), WA(3202). **6:**NY(3041). **11:**NE states(82).
Stictis mollis. **11:**CO(3225).
S. radiata. **11:**OH(3225).
Strigopodia resinae. **9:**ME(1674).

Tryblidiopsis pinastri. On twigs. **3:**ID(3202). **4:**NH(3228).
Unguicularia sp. **10:**OR(648).

Basidiomycotina - Uredinales

Chrysomyxa arctostaphyli. Heteroecious: aecial host. **1:**MT(3202), WA(3202). **3:**AZ(1203,3981), CO(649), ID(3202), MT(3202), OR(3202), WA(3202), WY(1195). **4:**WA(3202), WY(2767). **8:**AZ(1203,3981), UT(2767). **10:**WA(3202). **11:**AZ(1203).
C. empetri. Heteroecious: aecial host. **9:**ME(82). **10:**AK(526).
C. ledi. Heteroecious: aecial host. **4:**SD(2243). **6:**Lake states(82), NE states(82), WI(1344). **8:**MN(82). **9:**ME(82). **11:**Lake states(82), NE states(82).
C. ledi var. *cassandrae*. Heteroecious: aecial host. **3:**WI(82). **4:**MN(82), WI(82). **6:**NC states(82), NE states(82), Unk.(82), AK(60,526), CT(82), MI(2862). **8:**MN(82), NH(82), WI(82,1335). **9:**MI(82), NY(82). **11:**Lake states(82), NE states(82).
C. ledicola. Heteroecious: aecial host. **3:**CO(82), ID(82,3202). **4:**AK(60,82,526,1904), MN(82). **6:**Lake states(82), NE states(82), AK(60,526,1904). **8:**ME(82), MN(82), NY(82), WI(82,1335). **9:**NH(82), NY(82). **10:**AK(60,82,526,1904), ID(3202), WA(82,936,3202). **11:**N states(82), AK(82,2294).
C. monesis. **11:**W states(2699).
C. piperiana. Heteroecious: aecial host. **10:**CA(81,82), OR(82,3202). **11:**OR(81,3202).
C. pirolata. Heteroecious: aecial host. **3:**CO(82), MT(82,3202), OR(82,3202), WA(3202). **4:**AK(1904), ME(82), MI(82). **6:**AK(82,1904), MA(82), NH(82), PA(82). **9:**ME(82), NY(82), PA(82), VT(82). **10:**AK(60,526,1904). **11:**NC states(82), NE states(82), NW states(82).
C. roanensis. Heteroecious: aecial host. **9:**TN(82).
C. weirii. Autoecious. **3:**NW states(82), ID(3202), MT(3202), OR(638,3202), WA(82,3202). **4:**SD(2767). **8:**VT(8,252). **9:**TN(82), VT(1224), WV(82).
Chrysomyxa sp. **6:**MN(82), WI(82).
[*Melampsorella caryophyllacearum*]. **1:**ID(82), MT(82), WA(82), WY(82). **3:**W states(82), AZ(1195), CO(1575), ID(1575), MT(1575), OR(1575), UT(1195), WA(1575), WY(1195,1575,2924). **4:**NC states(82), NE states(82), NW states(82), AK(60,526,1575). **6:**AK(60,82,526), ME(82), MI(82), WI(82). **8:**CO(1575), NM(1575), SD(1575), WY(1195). **9:**ME(82), NY(82). **10:**W states(82), AK(60,82,526,3431), MT(82). **11:**Range of host(82), OK(2872).
Peridermium coloradense. **4:**AK(526,1904). **6:**AK(526,1904). **10:**AK(526).
Pucciniastrum americanum. Heteroecious: aecial host. **4:**Unk.(82).
P. arcticum. Heteroecious: aecial host. **4:**NC states(82), NE states(82), SD(1575), VT(1575), WI(1575).

Basidiomycotina - Other

Acanthophysium fennicum. On limbs and twigs. **3:**AZ(1202). **9:**NH(2083,2267), NY(2267), VT(2083).
A. lividocaeruleum. On decorticated wood, associated with a white rot. **3:**AZ(1202,2267), CO(2267), UT(2267). **4:**AK(2267). **11:**MT(128,2267).
A. weirii. White rot. **10:**AK(2083,2267). **11:**MT(128).
Aleurocystidiellum subcruentatum. On limbs and trunks, sometimes of living trees. **9:**NH(2267), NY(2267). **10:**AK(2267), CA(82,2083,2267), OR(82,2083,2267,3202), WA(2083,2267,3202).
Aleurodiscus amorphus. On suppressed, injured, or recently dead trees. **3:**ID(1449,2083,3202). **6:**NY(1449). **8:**CO(2267), ID(2267). **10:**CA(81,2083,2267), OR(1236,1449,2267,3202). **11:**Range of host(82), CO(1449,2267), MA(1449), NY(1449).
A. grantii. On limbs and trunks. **10:**CA(1236). **11:**WA(1236).
A. penicillatus. White rot. **9:**NH(2083,2267). **10:**OR(2083,2267,3202).
A. piceinus. On dead stems and twigs. **9:**NH(2083,2267), VT(2083,2267).
Amphinema byssoides. White rot. **3:**AZ(1202,2267), CO(2267). **4:**AK(2267). **8:**CO(2267).
Amyloathelia amylacea. **11:**ID(3202).
Amylocorticium subincarnatum. White rot. **3:**AZ(1202,2267). **4:**MN(128,1197,2267).
A. subsulphureum. White rot. **3:**MT(2267).
Amylocystis lapponica. Brown cubical rot. **3:**AZ(1202,2269), MT(2269,3202), UT(2269), WY(2924). **4:**NH(2269). **9:**NH(82). **11:**AZ(2269), CO(82,2269), WY(82,2269).
Amylostereum chailletii. White rot. **3:**AK(543), CO(2270), OR(2270), WA(2270). **4:**AK(2270). **9:**NY(543). **11:**NC states(82), NE states(82), NW states(82), ID(543), NY(2270), WA(2270).
Anomoporia albolutescens. **3:**CO(2167). **11:**CO(128).
A. bombycina. **3:**CO(2167), MT(3202). **11:**MT(128).
A. myceliosa. White rot. **3:**ID(3202). **8:**CO(2269). **11:**CO(128).
Antrodia albida. Brown rot. **4:**AK(2269). **9:**CT(2269), ME(2269), NH(2269), NY(2269). **10:**OR(3202).
A. albobrunnea. Brown cubical rot. **4:**AK(2269). **10:**AK(1904), OR(2269,3202).
A. alpina. Brown cubical rot. **3:**MT(2269).
A. carbonica. Brown cubical rot. **3:**AZ(1202,2269), ID(2269,2682,3202).
A. heteromorpha. Brown cubical rot. **3:**CO(82,2269), ID(2269,3202). **9:**CT(2269), ME(2269), NH(2269), NY(2269). **11:**NC states(82), NE states(82), NW states(82).
A. odora. Brown rot. **4:**AK(2269).
A. serialis. Brown cubical rot. **3:**AZ(128,1202,2269), CO(82,2269), ID(2269,3202), MT(2269,3202), UT(2269), WA(3202). **4:**AK(214,526,1904,2269). **6:**MN(1197,2269). **8:**CO(2269). **9:**NE states(82), CT(2269), ME(2269), NH(2269), NY(2269), PA(2269). **10:**NW states(82), AK(214,526,1904,2269), OR(2269,3202), WA(2269,3202).
A. sinuosa. Brown cubical rot. **3:**AZ(2269), CO(2167,2269), MT(128,2269). **4:**AK(2269). **6:**Lake States(82). **9:**NE states(82). **10:**AK(82), WA(3202). **11:**Range of host(82).
A. sitchensis. Brown cubical rot. **3:**MT(2269). **10:**AK(82,526,1904,2269), CA(82,2269), OR(82,2269,3202). **11:**AK(82).
A. sordida. Brown cubical rot. **6:**MN(2269). **9:**NH(128). **11:**MN(128,2269).
A. vaillantii. Brown cubical rot. **10:**WA(2269,3202).
A. variiformis. Brown rot. **3:**ID(2269,3202), MT(2269,3202), UT(2269). **9:**NE states(82), ME(2269), NH(2269), NY(2269), VT(2269). **10:**AK(82,214,526,2269), WA(2269,3202). **11:**AK(82,214,526), CO(128), MT(82), NY(82), PA(82).
A. xantha. Brown cubical rot. **3:**MT(128,2269,3202). **4:**AK(526,2269). **9:**ME(82,2269). **11:**NW states(82), AK(82,526), ID(128).
[*Antrodiella americana*]. **3:**OR(3202).
A. overholtsii. White rot. **4:**MN(2269). **11:**ID(1204), MN(1204), MT(1204), OR(1204), WA(1204).
A. romellii. White rot. **3:**CO(2269), MT(128).
Armillaria mellea. Butt rot, white root rot. **3:**AZ(82), CO(82,1724), ID(538,1583), NM(82), UT(82). **4:**VT(59). **8:**PA(2172). **9:**CT(2270), MA(498), ME(2270), NH(498,2270,3643), NY(498,2270), VT(498). **10:**OR(82,2270), WA(82,2270,3202). **11:**AZ(82,2270), CA(2898), CO(82,2270), NM(82,2270), WA(82,2270).
Asterodon ferruginosum. White rot. **3:**MT(128). **4:**AK(2270). **9:**CT(2270), ME(2270), NH(2270), NY(2270). **11:**MI(2862).
Athelia decipiens. White rot. **3:**CO(2267), UT(2267).
A. neuhoffii. White rot. **4:**AK(2267).
Auricularia auricula. White rot. **3:**ID(2270), UT(2270). **4:**MN(2270). **9:**CT(2270), ME(2270), NH(2270), NY(2270). **11:**CO(649,2270), MN(2270).
Basidiodendron eyrei. **8:**CO(2270).
Basidioradulum radula. **11:**NY(128).
Bjerkandera adusta. White rot. **3:**MT(2269,3202). **4:**MN(1197,2269). **10:**OR(2269,3202).
Bondarzewia berkeleyi. Root rot. **10:**CA(82,2269).
Boreostereum radiatum. On wood. **3:**ID(543). **4:**AZ(128), MN(543,2270). **6:**MN(543,2270), NY(543). **9:**ME(543). **11:**ID(82,2270), MT(82,2270).
Botryobasidium ansosum. White rot. **3:**AZ(2267). **10:**WA(3202).
B. botryosum. **10:**AK(2267).
B. pruinatum. White rot. **4:**MN(1197,2267). **10:**AK(2267).
B. subcoronatum. White rot. **3:**MT(2267). **9:**NY(2267).
B. vagum. White rot. **3:**AZ(1202,2267), MT(2267). **10:**AK(2267). **11:**AK(82), SD(82).
Byssoporia terrestris. **3:**AZ(1202,2269). **11:**AZ(128).
Ceraceomyces borealis. **10:**WA(128).
C. serpens. White rot. **3:**AZ(128,1202,2267).
C. tessulatus. On wood. **11:**ME(128).
Ceriporia purpurea. On wood. **4:**AK(526). **11:**Unk.(82).
C. reticulata. White rot. **3:**CO(2269). **10:**WA(82,2269,3202).
Ceriporiopsis subvermispora. White rot. **3:**CO(2269).

Cerocorticium hiemale. **10**:AK(3431).
Chaetoderma luna. Brown rot. **3**:CO(2267), MT(2267), UT(2267). **4**:AK(2267). **6**:AK(2267).
Clavariadelphus ligula. **3**:AZ(2270), MT(2270,3202).
Climacocystis borealis. Butt rot, root rot. **3**:AZ(1202,2269), CO(2269), ID(2269,3202), MN(2269), MT(2269,3202,3832), OR(2269,3202). **6**:MN(82,2269), NY(82), TN(82). **9**:NY(2269), TN(2269), VT(2269). **10**:AK(82,526,1904,2269), OR(2269,3202). **11**:NC states(82), NE states(82), NW states(82).
Collybia acervata. On debris, logs and stumps. **9**:NC(3717), TN(3717), VA(3717), WV(3717). **10**:AK(128). **11**:NC(1539).
Collybia sp. **3**:ID(3202).
Columnocystis abietina. Brown cubical rot. **3**:AK(2270), CO(2270), MT(543), UT(2270), WY(1195,2270,2924). **4**:AK(526,2270). **6**:AK(2270), NY(82,2270). **9**:NH(543). **10**:AK(2270), OR(543). **11**:CO(543), ID(543), NH(543), NY(82,543,2270), OR(2270,3202,4007).
C. ambigua. Brown cubical rot. **9**:ME(430,543,2270), NC(430,1303,2270), NH(543), TN(543). **11**:MT(82), NC(2270), NY(82,430,543,2270), TN(82,543,2270).
Confertobasidium olivaceo-album. White rot. **3**:AZ(2267).
Coniophora arida. Brown cubical rot. **3**:AZ(128,1202), ID(3202). **11**:Range of host(82).
C. olivacea. Brown rot. **3**:CO(2270), MT(2270). **11**:CO(2270), MN(1197,2270).
C. puteana. Brown cubical rot. **3**:AR(128), AZ(1202,2270), CO(82), ID(1583), MT(2270). **4**:AK(1904,2270). **11**:CO(128,2270).
Crepidotus ramosus. On wood. **11**:CO(1544,2270).
C. sphaerosporus. **8**:AZ(1202).
Crinipellis piceae. **11**:CA(3291).
Crustoderma dryinum. **3**:AZ(2267). **4**:AK(526,2267). **11**:CO(2555).
Cryptoporus volvatus. On wood. **3**:CO(2269), MT(2269,3202,3832), OR(2269,3202). **6**:Unk.(82). **10**:OR(2269,3202), WA(82,2269,3202). **11**:Range of host(82).
Cylindrobasidium corrugum. On wood. **3**:AZ(2270), CO(82), ID(82), UT(2270), WA(3202), WY(82,638,2924). **8**:AK(2270), AZ(1202), CO(128). **11**:AZ(128,1178).
Cyptotrama asprata. White rot. **3**:AZ(1202,2270). **9**:NC(1513), TN(1513).
Cystostereum murrayi. **9**:NH(543).
C. pini-canadense. White rot. **1**:NY(541). **9**:NH(541,2270).
Dacrymyces chrysospermus. Witch's butter. **3**:AZ(1202,2270).
D. stillatus. **3**:CO(2270).
D. tortus. On wood. **3**:MT(2270). **4**:NY(128).
Dacryobolus karstenii. Brown rot. **3**:ID(2267,3202), MT(2267). **4**:AK(526,2267). **10**:AK(526,2267).
Dacryopinax spathularia. On wood. **11**:MI(2862).
Dendrothele incrustans. **10**:CA(2084).
Dichomitus squalens. White pocket rot. **3**:AZ(2269), CO(82,2269), MT(2269), NM(82,2269). **4**:AK(526,2269), MN(1197,2269). **6**:MN(1197,2269), NY(82,2269). **9**:CT(2269), ME(2269), NH(2269), NY(2269), VT(2269). **10**:AK(82,2269). **11**:NC states(82), NE states(82), NW states(82).
Diplomitoporus crustulinus. On wood. **3**:AZ(128,1202,2269), CO(2269), ID(2269,3202), MT(2269,3202), WA(2269,3202), WY(2269). **4**:AK(526,1904,2269). **6**:AK(2269). **10**:AK(82,2269), MT(82,2269,3202), WA(82,2269,3202). **11**:NW states(82), AK(82), NY(82).
D. lenis. White rot. **3**:ID(2269,3202), MT(2269,3202). **4**:Unk.(82), AK(526,2269), MN(128). **10**:AK(82,2269), CA(82,2269), OR(82,2269,3202). **11**:MN(128,2269).
D. linbladii. White rot. **3**:AZ(1202,2269), MT(2269), OR(2269,3202). **4**:AK(526,2269). **9**:CT(2269), ME(2269), NH(2269), NY(2269). **10**:AK(2269). **11**:MN(2269).
Echinodontium tinctorium. Heart rot of living trees. **3**:AZ(1202,2270), ID(82,1583,2270,3202), MT(82,2270,3202), NM(82,2270), UT(82).
Eichleriella macrospora. On fallen logs. **3**:CO(128).
Exidiopsis calcea. White rot. **3**:AZ(1202,2270), MT(2270). **4**:AK(526,2270). **6**:MN(128,2270). **11**:NY(128).
Fibricium rude. **3**:AZ(1202).
Fibulomyces septentrionalis. On wood. **3**:AZ(1202,2267). **4**:AK(2267). **9**:AK(2267).
Flammulina velutipes. White rot. **3**:CO(2270).
Fomitopsis cajanderi. Brown cubical rot. **1**:NC(1303). **3**:Unk.(82), AZ(1202,2269), ID(2269,3202), MT(2269,3202), TX(82). **4**:Range of host(82), MN(1197), NH(2269). **9**:NE states(82), ME(2591), NY(2269). **10**:NW states(82), OR(2269,3202), WA(2269,3202).
F. officinalis. Brown trunk rot. **3**:Range of host(82), ID(2269,3202). **4**:MN(82), SD(82). **6**:MN(82,2269). **10**:AK(82,526,2269), OR(82,2269), WA(82,2269,3202). **11**:NC states(82), W states(82), CO(128), MI(2862).
F. pinicola. Brown cubical rot. **3**:Range of host(82), AZ(1202,2269), CO(1195,2269), ID(2269,3202), MT(2269,3202,3832), OR(2269,3202), UT(2269), WY(2269). **4**:Range of host(82), AK(526,1904,2269). **6**:Range of host(82), MN(128,1197). **8**:CO(82,2269). **9**:NE states(82), CT(2269), ME(2269), NH(2269), NY(2269), TN(82,2269). **10**:AK(82,526,1904,2269,3431), OR(82,2269,3202), WA(2269,3202). **11**:Range of host(82).
F. rosea. Brown cubical rot. **3**:Unk.(82), AZ(1202,2269), CO(2269), ID(2269), MT(2269,3832), TX(82), UT(2269). **4**:Range of host(82), AK(526,1904,2269). **6**:MN(1197,2269). **8**:CO(2269). **9**:NE states(82), CT(2269), ME(2269), NH(2269), NY(2269). **10**:NW states(82), WA(2269,3202). **11**:NE states(82), NW states(82), MN(1197,2269), NY(2269), PA(2269), TX(82), WA(128).
Ganoderma applanatum. White rot. **3**:ID(2269,3202). **10**:AK(526,1904,2269), CA(2269), OR(82), WA(82,2269,3202). **11**:NW states(82).
G. lucidum. White spongy sapwood rot. **1**:PA(82).
G. oregonense. White rot. **10**:AK(1904,2269), OR(82,2269), WA(82,2269,3202).
G. tsugae. **1**:PA(2962). **11**:MN(17,18).
Gerronema chrysophyllum. On logs. **11**:NC(273).
Gloeocystidiellum ochraceum. **10**:AK(2267).
G. porosum. White rot. **3**:MT(2267). **8**:CO(2267). **9**:NY(2267).
Gloeophyllum carbonarium. Brown rot. **3**:AZ(1202,2269).
G. odoratum. Brown rot. **3**:MT(3202). **4**:AK(526,2269). **11**:Range of host(82), MT(128).
G. sepiarium. Brown rot. **3**:AZ(1202,2269), CO(82,1195,2269), ID(2269,3202), MT(2269,3202), NM(82,2269), VT(2269), WA(82,2269,3202), WY(82,2269,2924). **4**:AK(82,526,1904,2269), ME(82,2269), MT(2269), NM(2269), NY(82,2269), SD(82,2243), VT(82,2269), WA(2269), WY(2269). **6**:MN(1197), NY(82). **8**:AZ(1202,2269). **9**:NC(1307). **10**:NW states(82), AK(526,1904,2269), OR(2269,3202), WA(2269,3202). **11**:Range of host(82), MI(2862).
G. trabeum. Brown rot. **6**:NY(82). **9**:CT(2269), ME(2269), NH(2269), NY(2269). **11**:NY(82,2269), PA(2269).
Gloeoporus dichrous. **3**:MT(2269,3202). **9**:NY(2269).
Grandinia alutaria. White rot. **4**:MN(128,1197,2267).
G. arguta. **3**:MT(128).
G. barba-jovis. **3**:MT(128).
G. breviseta. White rot. **3**:MT(2267). **6**:MN(128,1197,2267). **11**:MN(1197).
G. crustosa. White rot. **3**:CO(2267).
G. granulosa. White rot. **3**:AZ(1202,2267), MT(2267). **4**:MN(1197,2267).
G. pallidula. White rot. **3**:AZ(1202,2267), CO(2267).
G. stenospora. White rot. **3**:AZ(1202,2267). **4**:AK(2267), MN(2267). **10**:WA(2267,3202).
G. subalutacea. White rot. **3**:UT(2267).
Guepiniopsis alpina. **3**:CO(2270), ID(2270,3202), MT(2270,3202), OR(2270,3202), UT(2270), WY(2270,2924).
Hapalopilus rutilans. White rot. **3**:AZ(1202,2269).
Helicobasidium corticioides. Brown cubical pocket rot. **3**:CO(2270). **11**:CO(154).
Hericium abietis. White rot. **3**:ID(2270), MT(2270), WA(2270). **10**:AK(2270).
H. americanum. White pocket heart rot. **3**:ID(82,3202), MT(82,3202), WA(82,3202).
Heterobasidion annosum. White root and butt rot. **3**:W states(82,3584), ID(1583,2269,3202), NM(82,2269), OR(2269,3202), UT(294), WA(2269). **4**:MN(1197,2269). **9**:ME(82,2269), VT(82,2269). **10**:AK(82,526,2269,3431), CA(82,2269), OR(82,2269,3202), WA(82,2269,3202). **11**:NE states(82), NW states(82), AK(568), WA(128).
Hydropus marginellus. On dead logs and stumps. **9**:NC(3717), TN(3717), VA(3717), WV(3717).

Hymenochaete tabacina. White rot. **3:**ID(3202). **9:**CT(2270), ME(2270), NH(2270), NY(2270).
H. tenuis. White rot. **4:**AK(2270).
Hyphoderma amoenum. **11:**CO(128).
H. argillaceum. White rot. **3:**CO(2267).
H. medioburiense. White rot. **3:**CO(2267), MT(2267).
H. praetermissum. White rot. **3:**AZ(1202,2267), CO(2267), MT(2267). **8:**CO(2267).
H. sambuci. White rot. **8:**CO(2267).
Hypochnicium bombycinum. White rot. **3:**AZ(128), ID(2267).
Inonotus circinatus. White root and butt rot. **3:**AZ(1202), CO(82,2270), ID(82,2270), NM(82), OR(2270), UT(82), WA(2270). **4:**MN(82,2270). **6:**MN(82,2270). **11:**NC states(82), NE states(82), NW states(82), ID(2270), NY(2270).
I. dryadeus. White root and butt rot. **3:**NW states(82), CO(2270), ID(2270,3202), MT(2270,3202), OR(2270,3202), WA(2270,3202). **10:**CA(2270), WA(2270,3202). **11:**W states(82).
I. tomentosus. White pocket heart rot. **1:**WV(3911). **3:**AZ(1181,1202,2270), CO(2270), ID(538,1583,2270,3202), OR(2270), WA(2270,3202). **4:**AK(1904,2270). **10:**OR(2270,3202). **11:**Unk.(82).
Irpex lacteus. **11:**NW states(3202).
Ischnoderma resinosum. White rot. **9:**CT(2269), ME(2269), NH(2269), NY(2269). **10:**WA(2269,3202). **11:**NC states(82), NE states(82), W states(82), MI(2862), NY(2269).
Jahnoporus hirtus. White rot. **3:**AZ(2269). **11:**AZ(2269).
Junghuhnia collabens. White laminated rot. **3:**MT(2168,2269,3202). **4:**SD(2168,2269). **6:**MN(128,2269). **9:**NH(2168,2269), NY(2168,2269). **11:**NY(2168,2269).
J. luteoalba. White laminated or pitted rot. **9:**NH(2168,2269).
J. nitida. White rot. **9:**NY(2168,2269).
J. subfimbriata. White rot. **9:**ME(82,2269).
Laetiporus sulphureus. Brown cubical rot. **3:**ID(2269,3202). **4:**ME(2269). **9:**ME(82). **10:**AK(82,526,1904,2269), OR(2269,3202), WA(2269,3202). **11:**NW states(82), ME(82).
Laurilia sulcata. On wood. **3:**CO(82,2270), ID(1583,2270,3202), MT(2270,3202). **4:**AK(526,2270). **8:**CO(2270). **10:**WA(543,3202). **11:**E & C states(82), NW states(82), ID(543,2270), NY(543).
Lentinellus micheneri. **3:**NW states(3202), Unk.(2423).
L. montanus. On wood. **3:**W states(2417,2423), CO(1566,2270). **11:**NW states(3202).
Leptoporus mollis. Brown rot. **3:**AZ(2269).
Leptosporomyces galzinii. White rot. **3:**CO(2267), MT(2267).
L. mundus. White rot. **3:**AZ(2267), WY(1716).
Leucogyrophana mollusca. Brown rot. **3:**MT(2267). **4:**AK(526). **10:**AK(2267).
L. olivascens. **8:**CO(2267).
L. pinastri. Brown rot. **3:**AZ(1202,2267). **10:**AK(2267).
Marasmius pallidocephalus. On fallen needles. **11:**E & C states(1222).
M. plicatulus. **11:**CA(856).
Megalocystidium leucoxanthum. White rot. **3:**CO(2267).
M. luridum. White rot. **3:**AZ(1202,2267).
Meruliopsis corium. White rot. **6:**MN(1197,2267).
M. hirtellus. White rot. **3:**AZ(1178,1202,2267), CO(2267), UT(2267). **9:**NY(2267).
M. taxicola. On wood. **3:**CO(2167,2269), ID(2269,3202), WA(128,2269,3202). **9:**NH(128). **11:**Lake States(82), Unk.(82), AK(82), MI(2862), NY(128).
Mucronella aggregata. White rot. **3:**AZ(1202,2270), CO(2270).
Mycena austinii. On wood. **11:**NY(3326).
M. borealis. On wood. **9:**VA(3717), WV(3717).
M. elegantula var. *inflata*. On wood. **11:**CO(2450).
M. longiseta. On dead needles. **9:**VA(3717).
M. plicosa. On dead needles. **11:**MI(3326).
M. plumbea. On dead needles. **11:**CO(3326), MI(3326), NY(3326).
M. sanguinolenta. On dead needles and twigs. **9:**WV(3717).
M. subplicosa. On dead needles. **11:**CA(3326), MI(3326), NY(3326), OR(3326), WA(3326).
M. tenax. On dead needles. **9:**NY(3326).
Neolentinus kauffmanii. Brown cubical pocket rot. **10:**AK(82,526,2270), CA(82,2270), OR(82,2270,3202), WA(82,2270,3202).
N. lepideus. Brown cubical rot. **3:**AZ(82,2270). **9:**CT(2270), ME(2270), NH(2270), NY(2270). **11:**Range of host(82).
Osteina obducta. Brown rot. **3:**CO(2269).
Panellus stipticus. White rot. **9:**CT(2270), ME(2270), NH(2270), NY(2270).
Peniophora cinerea. **3:**CO(2267). **11:**CO(2267).
P. pithya. **3:**AZ(1202,2267), UT(2267).
Perenniporia subacida. White rot. **3:**ID(2269,3202), MT(2269,3202,3832), OR(2269), WA(2269). **4:**ME(82,2269). **9:**CT(2269), ME(2269), NH(2269), NY(2269), VT(82,2269). **10:**AK(82,526,2269), OR(2269,3202), WA(82,2269,3202). **11:**AK(526,2269), MI(2862), WA(128).
P. tenuis. On wood. **11:**NY(82), PA(82).
Phaeolus schweinitzii. Brown cubical butt and root rot. **3:**W states(82), AZ(1202,2269), ID(2269,3202), MT(2269,3202), OR(2269,3202). **4:**Range of host(82). **6:**NE states(82), MN(82,2269). **9:**MA(2269), ME(82,2269), NY(82,2269), VT(82,2269). **10:**AK(82,526,1904,2269), CA(82,2269), OR(82,2269), WA(82,2269,3202). **11:**Range of host(82), MI(2862).
Phanerochaete carnosa. White rot. **3:**AZ(2267), ID(128). **6:**MN(128). **9:**MA(128), ME(128), NH(128), VT(128).
P. gigantea. White rot. **3:**ID(2267,3202), MT(2267), OR(2267,3202). **4:**AK(526). **11:**Range of host(82).
P. sordida. White rot. **11:**CO(2267).
P. velutina. White rot. **3:**CO(128,433). **11:**NY(2267).
P. viticola. White rot. **11:**NY(128,433).
Phellinus chrysoloma. White pocket rot. **3:**AZ(2270), UT(2270).
P. ferrugineofuscus. White laminated rot. **11:**CO(2047), ID(2047).
P. ferruginosus. **11:**MN(128,1197).
P. laevigatus. **3:**MT(3202,3832).
P. nigrolimitatus. White pocket rot. **10:**AK(526,1904,2270), OR(2270,3202). **11:**NW states(82), AK(526), CO(128).
P. pini. White pocket heart rot. **2:**OR(2270,3202). **3:**AZ(1202), CO(82,1195), ID(538,1583,3202), MT(3202,3832), VT(2270), WA(3202), WY(82,1195). **4:**Range of host(82), AK(526,1904). **6:**Range of host(82), AK(526,2270), MN(128,1197). **8:**CO(82,2270). **9:**NE states(82), CT(2270), ME(2270), NC(82,1303,2270), NH(2270), NY(2270), WV(3911). **10:**NW states(82), AK(82,526,1904,2270), CA(82,2270), OR(2270,3202), WA(2270,3202). **11:**Range of host(82), AK(526,2270), CO(128), MI(2862), MN(1197).
P. punctatus. On wood. **11:**Unk.(82), ID(3202), OR(3202).
P. repandus. White pocket rot. **3:**MT(2270,3202).
P. viticola. White rot. **1:**NW states(82), AK(82). **6:**AK(82). **11:**MT(128).
P. weirii. **10:**AK(128).
Phlebia albida. White rot. **3:**AZ(1202,2267), MT(2267).
P. livida. White rot. **3:**AZ(128,1202,2165,2267), ID(3202). **11:**OR(2165).
P. ludoviciana. White rot. **3:**AZ(128), CO(2267), MT(2267).
P. subserialis. White rot. **3:**CO(2267). **8:**CO(2267).
Phlebiella filicina. White rot. **11:**CO(2267).
P. ralla. **10:**NW states(3202).
P. vaga. White rot. **3:**CO(2267). **8:**CO(2267). **10:**AK(3431).
Pholiota adiposa. White rot. **3:**CO(2270).
P. bakerensis. White rot. **3:**AZ(1202,2270).
P. decorata. White rot. **3:**AZ(1202,2270).
Physisporinus sanguinolentus. **11:**WA(128).
Piloderma bicolor. **4:**AK(2267). **11:**CO(2267).
P. byssinum. On wood. **3:**AZ(1202,2267), MT(2267). **4:**MN(1197,2267).
P. reticulatum. White rot. **3:**AZ(2267).
Pleurotus ostreatus. White rot. **11:**CO(2270).
Pluteus atricapillus. White rot. **3:**AZ(1202,2270).
Polyporus badius. White rot. **3:**MT(3202), WA(3202). **10:**AK(2269).
P. melanopus. White rot. **3:**MT(2269).
P. umbellatus. **3:**MT(3832), MT(2269).
P. varius. White rot. **3:**ID(2269,3202).
Postia balsamea. Brown rot, root rot. **3:**CO(82,2269), ID(2269,3202). **10:**AK(82,2269), WA(2269,3202).
P. caesia. Brown rot. **3:**AZ(1202,2269). **4:**CA(2269). **6:**AK(2269). **9:**CT(2269), ME(2269), NH(2269), NY(2269). **10:**AK(2269), CA(82,2269). **11:**CA(2269).
P. fragilis. Brown cubical rot. **3:**AZ(1202,2269), ID(2269,3202). **6:**MN(82,2269). **10:**AK(3431), WA(2269,3202). **11:**NC states(82), NE states(82), MN(2269), NY(2269), OR(2269,3202).
P. guttulata. Brown cubical rot. **3:**MN(2269). **4:**Range of host(82).

6:MN(82,2269). 9:NY(2269), PA(2269), WA(2269). 10:WA(2269,3202). 11:NC states(82), NE states(82), NW states(82), OR(3202).
P. hibernica. Brown rot. 11:AZ(2269).
P. leucomallela. On wood. 11:W states(1205).
P. leucospongia. Brown cubical rot. 3:CO(1195,2269), ID(2269,3202), UT(2269), WY(2269,2924). 11:W states(82), CA(2269).
P. mappa. White rot. 6:AK(2269).
P. perdelicata. Brown rot. 3:AZ(1202,2269), MT(2269).
P. placenta. Brown cubical rot. 3:CO(2269), ID(2269,3202). 10:AK(128), OR(2269,3202).
P. sericeomollis. Brown cubical rot. 3:MT(2269,3202). 11:ID(82), MT(82).
P. stiptica. Brown cubical rot. 3:ID(2269,3202). 6:ME(82). 11:AZ(2269), MT(2269), OR(2269).
P. tephroleuca. Brown rot. 3:AZ(2269).
P. undosa. Brown rot. 3:AZ(397,1202,2269), MT(2269,3202). 9:ME(397,2269), NH(397,2269), NY(397,2269). 10:WA(397,2269,3202).
Pseudohydnum gelatinosum. 3:CO(2270).
Pseudomerulius aureus. Brown rot. 3:AZ(1202), MT(2267). 10:WA(128).
Pseudotomentella mucidula. 11:WA(2270).
P. nigra. On wood. 11:CO(2032,2270).
P. tristis. A base of living tree. 9:NY(2270). 11:NY(128).
Pycnoporellus alboluteus. Brown cubical rot. 3:W states(82), AZ(1202,2269), CO(1195,2269), ID(2269,3202), MT(2269,3202,3832), OR(2269,3202), UT(2269), WA(82,2269,3202), WY(2269,2924). 4:AK(526,2269). 6:MI(82,2269). 8:AZ(1202,2269). 9:NY(82,2269). 10:AK(82,526,2269), WA(82,2269,3202). 11:NC states(82), W states(82), AK(82).
P. fulgens. Brown cubical rot. 4:AK(1904). 6:MN(82,2269). 9:NC(82,2269), NY(82,2269). 10:NW states(82), AK(526,1904,2269), OR(2269,3202), WA(2269,3202). 11:E states(82), MI(2862), MT(82), NC(2269), WA(82).
Pycnoporus cinnabarinus. White rot. 3:CT(2269), ME(2269), NH(2269), NY(2269). 9:NY(2962). 11:NC(82,2269), NY(82,2269).
Resinicium bicolor. 3:MT(128). 11:NY(128).
R. chiricahuaensis. White rot. 3:AZ(1202,2267).
R. furfuraceum. White rot. 3:MT(2267).
Schizophyllum commune. White rot. 9:CT(2270), ME(2270), NH(2270), NY(2270). 11:PA(2270).
Scytinostroma arachnoideum. White rot. 3:AZ(1202,2270). 9:NH(2270). 10:WA(3202). 11:WA(128,2270).
S. galactinum. White root and butt rot. 3:ID(128), OR(2270), WA(2270). 4:MN(128,1197). 9:CT(2270), ME(128,2270), NH(128,2270), NY(128). 11:NY(128,2270).
Scytinostromella nannfeldtii. On wood. 4:AK(2267).
Sebacina helvelloides. On roots. 11:MA(128).
Serpula himantioides. Brown rot. 3:AZ(2270), CO(2270), MT(3202,3832).
Sistotrema coroniferum. 3:CO(2267).
S. diademiferum. On wood. 10:OR(128).
S. raduloides. White rot. 4:AK(2267).
Skeletocutis alutacea. 11:ID(128).
S. amorpha. White rot. 3:ID(2269,3202), MT(2269,3202,3832). 10:AK(82,2269). 11:NC states(82), NE states(82), NW states(82), AK(82,2269), PA(2269).
S. stellae. White rot. 3:CO(2269), MT(128,2269,3202).
S. subincarnata. White rot. 3:AZ(1202,2269). 4:AK(2269).
Sparassis crispa. Butt rot, brown cubical root rot. 3:NW states(3202), ID(82,2270), MT(82,2270), OR(82,2270), WA(82,2270).
Sphaerobasidium minutum. 3:AZ(1202,2267), MT(2267).
Steccherinum ciliolatum. 3:AZ(1202).
S. ochraceum. 8:AZ(128,1202).
Stereum hirsutum. White rot. 10:AK(2270).
S. sanguinolentum. Heart rot. 1:ID(82), NY(2985), VT(543). 3:AZ(1202,2270), CO(82,2270), ID(538,1583), OR(2270), WA(2270), WY(2924). 4:AK(526,1904,2270). 9:CT(2270), ME(543,2270), NH(543,2270), NY(543,2270), VT(543). 10:AK(526,1904,2270). 11:Range of host(82), AK(526), NY(543).
Stereum sp. 11:WA(3202).
Strobilurus occidentalis. On cones. 4:AK(3854). 10:AK(3854).
Tapinella panuoides. On wood. 10:AK(526,1904).
Thanatephorus cucumeris. 4:AK(526).
Thelephora terrestris. Smothering of seedlings. 3:ID(3202). 6:MN(82). 11:ID(82), OR(82).
Tomentella atrorubra. On wood. 11:NY(2030,2039,2270).
T. avellanea. On wood. 11:WA(2039,2270).
T. bresadolae. On wood. 11:NY(2039,2270).
T. bryophila. On wood. 3:AZ(2270). 11:AZ(2039).
T. chlorina. On wood. 3:AZ(128,2270). 11:AZ(2039).
T. cinerascens. On wood. 11:MT(2039,2270).
T. lateritia. On wood. 11:CO(2032,2039).
T. nitellina. On wood. 10:WA(2032,2270,3202). 11:WA(2039).
T. ochracea. On wood. 3:AZ(1202,2270). 11:AZ(2039,2270), NJ(2039,2270), NY(2036,2039,2270).
T. olivascens. On wood. 11:NY(2039,2270).
T. ramosissima. On wood. 3:MT(2270). 11:MT(2039).
T. rubiginosa. 9:NH(128).
T. ruttneri. On wood. 11:MI(2270), NY(2035,2270).
T. sublilacina. On wood. 11:MI(2039,2270), NY(2039,2270).
T. subvinosa. On wood. 3:AZ(2270). 11:AZ(2039).
T. terrestris. On wood. 3:MT(2270). 11:MT(2039).
T. violaceofusca. On wood. 4:MN(2270).
Tomentellina fibrosa. 4:AK(82).
Trametes hirsuta. 10:WA(3202).
T. pubescens. White rot. 3:MT(2269,3202).
T. versicolor. White rot. 3:ID(2269,3202), OR(2269), WA(2269,3202). 9:Range of host(82), CT(2269), ME(2269), NH(2269), NY(2269). 10:NW states(82), AK(2269,3431), CA(2269), OR(2269,3202), WA(2269,3202). 11:Range of host(82), PA(2269).
Trechispora cohaerens. White rot. 3:CO(2267), MN(2267). 6:AK(2267), MN(1197).
T. farinacea. White rot. 8:CO(2267).
T. microspora. White rot. 8:CO(2267).
T. mollusca. On wood. 11:Unk.(82), NY(82).
Trichaptum abietinum. White rot. 3:AZ(2269), CO(2269), ID(2269,3202), MT(2269,3202), OR(2269,3202), VT(2269), WA(2269,3202), WY(82,2269). 4:AK(82,526,1904,2269), ME(82,2269). 6:Unk.(82). 9:CT(2269), ME(2269), NH(2269), NY(2269), PA(2269). 10:NW states(82), AK(526,1904,2269), WA(2269,3202). 11:Range of host(82).
T. biforme. On wood. 4:AK(526,1904).
T. laricinum. White rot. 9:NY(2269).
Tricholomopsis rutilans. White rot. 10:AK(2270). 11:NH(2270).
Tubulicrinis glebulosus. White rot. 11:MT(2267).
T. propinquus. 4:MA(656).
T. subulatus. On wood. 8:CO(2267).
Tulasnella fuscoviolacea. 9:NH(128). 11:NY(128).
T. pruinosa. 3:MT(2270).
T. violea. 8:CO(2270).
Tyromyces destructor. 10:WA(82).
Vararia athabascensis. White rot. 3:AZ(128,1202,2270).
V. boreale. On wood. 3:ID(128), MT(128). 11:MT(128).
V. investiens. White rot. 8:CO(2270). 9:NH(128).
V. racemosa. White rot. 3:AZ(1202,2270), CO(2270), MT(128,2270,3202), UT(2270). 11:CO(128).
Vesiculomyces citrinus. White rot. 3:AZ(1202,2267), CO(2267).
Xeromphalina campanella. On rotten logs and stumps. 3:CO(2270). 9:NC(3717), TN(3717), VA(3717), WV(3717).

Deuteromycotina - Hyphomycetes

Ambrosiella ferruginea. 3:WA(3202).
Botrytis cinerea. Gray mold blight of seedlings. 1:NJ(82). 9:NC(1303). 11:Range of host(82).
Botrytis sp. Damping-off, root rot. 1:DE(702).
Cladosporium sp. Needle cast. 1:ME(481). 4:ME(481).
Conoplea abietina. 11:NY(1672).
Cylindrocladium scoparium. Damping-off. 1:NJ(82). 4:MN(1694). 6:MN(1694).
Dictyocatenulata alba. On bark. 10:AK(526). 11:ME(3181).
Fusarium acuminatum. Snow mold of nursery plants. 3:CO(82).
F. avenaceum. 3:ID(3202), MT(3202).
F. oxysporum. 1:WA(3202).
F. roseum. 1:WA(3202). 3:ID(3202), MT(3202).
Fusarium sp. Damping-off, root rot. 1:DE(702). 4:PA(3858). 8:PA(3858).
Geniculodendron pyriforme. Seed infection, resulting in germination

failure. **3:**AZ(3890), CO(3890), ID(3890), UT(3890), WY(3890).
Leptographium engelmanii. Wood stain associated with bark beetles. **3:**CO(783).
Leptographium sp. Sapwood stain. **11:**MN(484).
Phialophora sp. Isolated from wood chips. **4:**WI(1005).
Phymatotrichopsis omnivora. Root rot. **3:**TX(82). **11:**TX(82).
Trichoderma sp. Damping-off, root rot. **1:**DE(702).
Tubercularia atomospora. **3:**MT(3202), WA(3202).

Deuteromycotina - Coelomycetes

Cytospora chrysosperma. Canker. **1:**NE states(82), IL(82).
C. kunzei. Twig blight, canker. **1:**CT(3781), MA(3781). **4:**CT(3781), MA(82), NJ(82). **7:**CT(3781), PA(3781). **8:**NC states(82), NE states(82), CT(3781), MN(82), NJ(3781), NY(3781), SD(2243). **9:**ME(3781), NH(3781), NY(82). **11:**NE states(82), MA(3780), NY(82,3780).
C. leucostoma. Canker. **1:**NE states(82), IL(82).
Cytospora sp. Canker. **1:**VA(3493).
Diedickea piceae. Needle cast. **10:**CA(81,82,325).
Leptostroma sp. **3:**WA(3202).
Leptostromella conigena. **3:**WY(2924).
Macrophomina phaseolina. Damping-off. **1:**NC(82).
Pestalotia stevensonii. On cone scales. **1:**PA(82).
P. thujicola. On shoots. **1:**MA(2215).
Pestalotia sp. **3:**TX(82).
Pestalotiopsis funerea. On cones and twigs. **1:**MA(82).
Phoma lineolata. **3:**MT(3202).
P. piceina. On twigs. **1:**MD(82), NY(82), OH(82). **9:**NY(82).
Phoma sp. Damping-off, root rot. **1:**DE(702).
Phomopsis conorum. On snow-blighted seedlings. **3:**WA(3202). **11:**ME(82).
P. occulta. Tip blight, needle necrosis. **8:**NJ(82), WI(3088). **11:**NE states(82).
Phomopsis sp. Dieback. **8:**NC(1303).
Phragmotrichum chailletii. On cones. **9:**NC(82), NY(82).
Rhizosphaera kalkhoffii. Swiss needle cast. **1:**PA(2356). **3:**CO(1501). **4:**PA(2356), WI(1350). **6:**PA(2172,2356). **8:**AZ(1500), CT(82), IN(2597), MN(2597), NC(1303,3263), NY(82), PA(2356), VA(82), WI(1360,2597).
Rhizosphaera sp. Leaf cast. **8:**VA(3780).
Sclerophoma pythiophila. On cone scales. **1:**ME(82), OH(82).
Sirococcus conigenus. Seedling twig blight, tip blight. **1:**NC(82). **4:**NC(82). **8:**KS(3194), ME(82), NC(1303). **9:**NC(82). **10:**AK(526). **11:**ME(82,3780), NC(82).
Sphaeropsis sapinea. Twig blight, collar rot. **1:**NJ(82). **10:**NY(82).
Truncatella angustata. On cones and twigs. **1:**NY(82).

Deuteromycotina - Other

Racodium resinae. **10:**AK(526), WA(3202).
Rhizoctonia solani. Damping-off. **3:**TX(82). **11:**Range of host(82).
Rhizoctonia sp. Damping-off, root rot. **1:**DE(702).

Pinus L.

(= *Strobus* (Endl.) Opiz.)
1. *P. albicaulis* Engelm. [white-barked pine]
2. *P. aristata* Engelm. [bristlecone pine, hickory pine]
3. *P. aristata* Engelm. var. *longaeva* (D. Bailey) Little (=*P. longaeva* D. Bailey)
4. *P. attenuata* Lemmon [knobcone pine]
5. *P.* × *attenuradiata* Stockw. & Righter
6. *P. ayacahuite* Ehrenb. ex Schlechtend. (=*P. strobiformis* Engelm., *P. durangensis* Roezl ex Gordon). *P. attenuata* × *P. radiata*.
7. *P. balfouriana* Grev. & Balf. [foxtail pine]
8. *P. banksiana* Lamb. [gray pine, Jack pine, scrub pine]
9. *P. canariensis* Chr. Sweet ex Spreng. [Canary Island pine]
10. *P. caribaea* P. M. Morelet [Caribbean pine, Cuban pine]
11. *P. cembra* L. [Russian cedar, Swiss stone pine]
12. *P. cembroides* Zucc. [Mexican stone pine]
13. *P. clausa* (Chapm. ex Engelm.) Vasey ex Sarg.
14. *P. clausa* (Chapm. ex Engelm.) Vasey ex Sarg. var. *clausa*
15. *P. clausa* (Chapm. ex Engelm.) Vasey ex Sarg. var. *immuginata* Warder
16. *P. contorta* Douglas ex Loud. [beach pine, lodgepole pine, shore pine]
17. *P. contorta* Douglas & Loud. var. *latifolia* Engelm. ex S. Wats. (=*P. murrayana* Grev. & Balf., *P. contorta* Douglas ex Loud. var. *murrayana* (Grev. & Balf.) Engelm.) [lodgepole pine]
18. *P. coulteri* D. Don [big-cone pine]
19. *P. densiflora* Siebold & Zucc. [Japanese red pine]
20. *P. echinata* Mill. [long-tag pine, shortleaf pine, yellow pine]
21. *P. edulis* Engelm. [pinyon pine, two-leaved nut pine]
22. *P. eldarica* Medw.
23. *P. elliottii* Engelm. (=*P. elliottii* Engelm. var. *elliottii*) [slash pine]
24. *P. elliottii* Engelm. var. *densa* Little & Dorman
25. *P. engelmannii* Carrière [Apache pine]
26. *P. flexilis* E. James [limber pine]
27. *P. glabra* Walter [cedar pine, spruce pine]
28. *P. halepensis* Mill. [Aleppo pine, Jerusalem pine]
29. *P. inops* Sol. in Aiton
30. *P. lambertiana* Douglas [giant pine, sugar pine]
31. *P. latifolia* Mayr
32. *P. leiophylla* Schlechtend. & Cham.
33. *P. leiophylla* Schlechtend. & Cham. var. *chihuahuana* (Engelm.) Shaw (=*P. chihuahuana* Engelm.)
34. *P. monophylla* Torr. & Frém. (=*P. cembroides* Zucc. var. *monophylla* (Torr. & Frém.) Voss) [nut pine, single-leaf pinyon pine, stone pine]
35. *P. montezumae* Lamb. [rough-barked Mexican pine]
36. *P. monticola* Douglas ex D. Don [western white pine]
37. *P. mugo* Turra (=*P. montana* Mill.) [Swiss mountain pine]
38. *P. muricata* D. Don [bishop pine]
39. *P. nigra* Arnold [Austrian pine, black pine, Corsican pine]
40. *P. nigra* Arnold var. *austriaca* (Hoess) H. Badoux (=*P. austriaca* (Hoess) Link, *P. laricio* Poir., *P. laricio* Poir. var. *austriaca* (Hoess) Loud.)
41. *P. occidentalis* Sw.
42. *P. palustris* Mill. (=*P. australis* F. Michx.) [Georgia pine, longleaf pine, southern yellow pine]
43. *P. patula* Schlechtend. & Cham. [Mexican yellow pine]
44. *P. pinaster* Aiton [cluster pine, maritime pine]
45. *P. pinea* L. [Italian stone pine, umbrella pine]
46. *P. ponderosa* Douglas ex P. Laws. & C. Laws. [ponderosa pine, western yellow pine]
47. *P. ponderosa* Douglas ex P. Laws. & C. Laws. var. *arizonica* (Engelm.) Shaw
48. *P. ponderosa* Douglas ex P. Laws. & C. Laws. var. *jeffreyi* (E. Murray) S. Wats. (=*P. jeffreyi* E. Murray) [Jeffrey pine]
49. *P. ponderosa* Douglas ex P. Laws. & C. Laws. var. *scopulorum* Engelm. [Rocky Mountain yellow pine]
50. *P. pungens* Lamb. [prickly pine, table mountain pine]
51. *P. quadrifolia* Parl. ex Sudw. [Parry pinyon pine]
52. *P. radiata* D. Don (=*P. insignis* Douglas ex Loud.) [Monterey pine]
53. *P. resinosa* Aiton [Norway pine, red pine]
54. *P. rigida* Mill. [pitch pine]
55. *P. sabiniana* Douglas ex D. Don [digger pine]
56. *P. serotina* Michx. (=*P. rigida* Mill. var. *serotina* (Michx.) Loud.) [pond pine]
57. *P. strobus* L. [eastern white pine]
58. *P. sylvestris* L. [Scotch pine, Scots pine]
59. *P. taeda* L. [frankincense pine, loblolly pine, old-field pine]
60. *P. thunbergiana* Franco (=*P. thunbergii* Parl.) [Japanese black pine]
61. *P. torreyana* Parry ex Carrière [soledad pine, Torrey pine]
62. *P. virginiana* Mill. [Jersey pine, poverty pine, scrub pine, spruce pine, Virginia pine]
63. *P. wallichiana* A. B. Jackson (=*P. excelsa* Wall ex D. Don, *P. griffithii* M'Clell.) [Bhutan pine, blue pine, Himalayan white pine]
64. *Pinus* sp. (=*Strobus sp.*)
65. *Pinus* × sp. (unnamed hybrids)

Myxomycetes

Arcyria occidentalis. **64:**NC(3944).
Licea operculata. Slime mold, on bark. **57:**IA(82).
Physarum citrinum. **64:**NC(3944).
P. pulcherripes. **64:**NC(3944).
Stemonitis splendens. **64:**NC(3944).

Chytridiomycetes

Phlyctochytrium lackeyi. Parasitic on pollen in water after pollen shower. **64:**FL(3403).
P. macrosporum. Parasitic on pollen in water. **64:**FL(3405).
Rhizophydium annulatum. Parasitic on pollen in water. **64:**FL(3404).
Sporophlyctidium neustonicum. Parasitic on pollen in water. **64:**FL(3405).

Oomycetes

Phytophthora cactorum. Seedling blight, root rot. **53:**MN(82). **58:**MN(82). **64:**Lake states(82), NE states(82).

P. cinnamomi. Decay of feeder roots, little leaf disease, root rot. **9:**CA(3624,4010). **13:**FL(166). **20:**SE states(3624), NC(1303,1722), VA(3493). **52:**CA(3624,4010). **53:**SE states(3624), DE(82), MD(82). **57:**MD(82), NC(1303). **58:**SE states(3624), MD(82), VA(82). **59:**SE states(3624), LA(2178), VA(3493). **64:**MD(82).

P. drechsleri. **52:**CA(1580).

P. nicotianae. Root rot. **52:**CA(2528).

P. nicotianae var. *parasitica*. Isolated from roots. **13:**FL(166).

Phytophthora sp. Little leaf, root rot. **4:**HI(2899,2903). **42:**MS(2720).

Pythium aphanidermatum. Damping-off. **64:**Range of host(82).

P. debaryanum. Damping-off. **64:**Range of host(82).

P. dimorphum. Isolated from dead feeder roots of declining trees. **59:**LA(1517).

P. irregulare. **53:**WI(82).

P. splendens. Root rot. **28:**HI(2903).

P. ultimum. Damping-off. **53:**Range of host(82). **57:**Range of host(82). **64:**Range of host(82).

Pythium sp. Damping-off, root rot. **60:**HI(2903). **64:**Range of host(82).

Zygomycotina

Mortierella sp. Isolated from treated southern yellow pine utility poles. **64:**E states(3996), E states(3996).

Rhizopus sp. Isolated from treated southern yellow pine utility poles. **64:**E states(3996), E states(3996).

Ascomycotina

Acanthostigma decastylum. On decorticated wood. **59:**NC(1303).

Anthostomella sabiniana. On needles. **55:**CA(1098).

Appendiculella pinicola. Black mildew. **20:**Unk.(1452), NC(82).

Ascocalyx abietina. Canker. **8:**MI(679,1119), MN(3302), NY(3188), WI(1119). **46:**NY(3305). **53:**ME(2622), MI(679,1119), MN(3302), NH(246), NY(1119,3188), WI(1119). **58:**NY(3188), VT(3576).

Ascocoryne sarcoides. On decorticated wood. **59:**NC(1303).

Atropellis apiculata. Twig canker. **10:**NC(82), VA(82).

A. pinicola. Branch, limb and trunk canker. **1:**OR(82). **16:**NW states(82), ID(3202), OR(82,3202), WA(3202). **17:**OR(82). **30:**OR(82,3202). **36:**NW states(82), ID(82,3202), MT(3202), OR(3202), WA(82,3202). **57:**OR(82,3202).

A. piniphila. Branch canker. **1:**NW States(82), AL(82), AZ(82), NM(82), OR(3202), TN(82). **8:**SD(82). **16:**NW states(82), AL(82), AZ(82), ID(3202), MT(3202), NM(82), TN(82), WA(3202). **17:**ID(82), MT(82), WA(82). **20:**NC(1303). **36:**ID(3202). **46:**NW States(82), AL(82), AZ(82), ID(3202), MT(82,3202), NM(82), TN(82). **59:**AL(82).

A. tingens. Limb and twig canker. **8:**NC(82), PA(82). **10:**GA(82). **19:**VA(82). **20:**GA(1438), NC(1303). **23:**NC(1303). **39:**VA(82). **50:**NC(82). **53:**MA(82), NH(82). **54:**NC(82), PA(82), VA(82). **57:**VA(82). **58:**OH(82). **59:**GA(82,1438), NC(1303), VA(82). **62:**GA(82,1438). **64:**E states(82), AR(82), OH(82), OK(82), TX(82).

Atropellis sp. Canker. **58:**MA(82), NH(82), PA(82).

Bertia moriformis var. *latispora*. On wood. **57:**NH(684). **64:**VA(684).

Bifusella linearis. Needle cast, tar spot. **1:**CA(2388), VA(3351). **7:**CA(82). **17:**CO(82). **26:**CA(2388,3351), CO(82,1195,3457), MT(82,3202), NM(82). **36:**ID(82,3202), OR(82,3202), WA(82,3202). **57:**E states(82), MN(82), NC(1533), VA(1533), WV(3910). **64:**E states(82), W states(82), MN(82), WA(82).

B. pini. Needle cast. **26:**ID(3202). **34:**CA(82), NV(82,1195).

B. saccata. Needle cast. **1:**CA(3351). **12:**CO(82), NM(82). **21:**CO(82), NM(82). **26:**CO(82,1195,3457).

Bisporella citrina. **36:**ID(3202). **64:**ID(3202).

Botryosphaeria dothidea. Death, wilt. **23:**HI(1594). **57:**GA(1438). **59:**HI(1594). **62:**GA(1438).

B. obtusa. Cone rot, on limbs. **39:**CT(82), IN(82). **57:**NY(82). **62:**GA(82).

B. ribis. On limbs. **57:**GA(82).

Caliciopsis pinea. Canker, gall. **54:**NC(82), NJ(82). **57:**NC(1303,1533). **62:**NC(82), VA(3493). **64:**MA(82), NJ(82), NY(82), PA(82), VT(82).

Capnodium pini. Black mildew on needles. **57:**Range of host(82), MI(2862), NC(1303).

Capnodium sp. **57:**WI(1334).

Cenangium acicola. On needles. **53:**OH(1960). **57:**WI(1354). **59:**GA(1438).

C. alpinum. **17:**CO(82).

C. atropurpureum. On dead twigs. **10:**GA(82), SC(82). **37:**GA(1438), MD(82,528). **39:**MD(82,528). **46:**MA(1960). **50:**MD(82,528). **54:**MD(82,528). **58:**PA(82,528). **59:**GA(82,528,1438). **62:**MD(82,528).

C. ferruginosum. Twig blight, canker, dieback. **16:**ID(3202), MT(3202), WA(3202). **17:**NW states(82), WA(82). **19:**MI(82), NY(3283). **20:**GA(82,1438). **26:**AZ(82). **36:**CO(82), ID(82,3202). **37:**NY(82). **39:**NY(82,3283). **46:**Range of host(82), MT(3202), OR(3202). **48:**CA(81,82), OR(82,3202). **53:**NY(82), PA(82). **55:**CA(82). **57:**NE states(82), NY(1960). **58:**NY(82), PA(82). **59:**AL(82), GA(82,1438), NC(82). **60:**NY(3283). **62:**MD(82), VA(82). **64:**NW states(82), ME(82), NJ(82), OH(82), WA(82).

C. purpureum. **16:**ID(3202). **17:**ID(82).

Ceratocystiopsis minuta. On wood. **16:**CO(1364).

C. minuta-bicolor. On wood, associated with bark beetles. **16:**AZ(3714), CO(1364,3714). **46:**CA(785), CO(785).

C. ranaculosus. **59:**LA(385).

Ceratocystis coerulescens. Sapwood stain. **8:**MN(484). **53:**MN(484,3492). **64:**S states(82), LA(3732).

C. denticulata. Associated with bark beetle attack. **46:**CO(790).

C. hyalothecium. Isolated from bark beetle infested trunk. **16:**WY(787,3714).

C. leucocarpa. In bark beetle infested trees. **16:**AZ(785), CO(785), NM(785). **30:**OR(785). **46:**CO(785).

C. moniliformis. Wood stain. **64:**S states(82), LA(3732).

C. olivaceapinii. On wood, associated with bark beetles. **46:**AZ(786,3714), NM(786,3714).

C. sagmatospora. Isolated from wood chips. **64:**TX(1005,3714).

Ceratostoma marylandicum. **42:**FL(2779).

Ceratostomella sp. Wood stain. **64:**LA(3732).

Ceriosporopsis halima. On driftwood in ocean. **59:**NC(1941).

Chaetomium erectum. **36:**ID(3202).

C. globosum. Isolated from treated southern yellow pine utility poles. **64:**E states(3996), E states(3996).

Cistella parksii. On bark. **38:**CA(527).

Clithris graphis. **64:**WA(3202).

Coccomyces parvulus. On wood. **64:**WA(3229).

C. strobi. Canker. **57:**CT(2950), MA(1391,2950,3228), MN(2950), NC(1303), NH(2950), NJ(2950), NY(2950), WI(2950).

Coccomycetella richardsonii. On wood. **64:**OR(3232).

Coccophacidium sp. **1:**ID(3202). **36:**ID(3202).

Colpoma crispum. **36:**ID(3202).

Coniochaeta ligniaria. **16:**ID(3202). **46:**ID(3202), OR(3202), WA(3202).

C. malacotricha. On end of cut logs. **46:**ID(3202). **59:**NC(1303).

Coniochaeta sp. **16:**ID(3202). **64:**TX(652).

Crumenulopsis lacrimiformia. On dead tips. **42:**FL(1131).

C. sororia. **46:**NM(128).

Cucurbidothis pithyophila. Twig canker. **26:**ID(781,3202). **36:**CO(781), WA(781,3202). **57:**MI(82), WA(82).

Cyclaneusma minus. Needle cast. **39:**PA(2361). **58:**MA(1099).

C. niveum. Needle cast. **16:**MT(3202), WA(3202). **17:**W states(82), WA(82). **34:**CA(81). **39:**ID(3202), OR(82,3202). **44:**HI(2903). **46:**CA(81,82), ID(3202), WA(3202). **48:**CA(82). **52:**CA(81,82). **55:**CA(81). **58:**GA(82,1438), MA(82), MI(70,82), PA(2355), WI(1334).

Dasyscyphus ellisianus. Twig canker. **10:**GA(1438). **11:**CT(82). **20:**GA(82,1438), NC(82). **26:**MA(82). **36:**MA(82). **39:**CT(82), OH(82). **42:**GA(1438). **46:**MA(82). **48:**GA(1438). **50:**NC(82). **53:**NE states(82). **54:**E states(82), WV(82). **57:**E states(82), GA(1438). **58:**MA(82), NJ(82), PA(82), RI(82). **59:**DE(82), GA(82,1438), LA(82), NC(1303), NJ(82). **62:**MD(82), PA(82), VA(82). **64:**NW states(82), Unk.(82), MD(651), NC(3944).

D. oblongosporus. Commonly saprophytic on limbs and twigs. **16:**WY(1195,3378). **50:**PA(82). **64:**NE states(82).

D. serinellus. **64:**OR(82).

Dasyscyphus sp. **57:**GA(1438). **58:**ID(3202). **59:**GA(1438).

Davisomycella ampla. Tar spot. **8:**Lake States(82). **16:**CA(81).

D. lacrimiformis. Needle cast. **4:**CA(82), OR(82,3202).

D. limitata. Needle cast. **52:**CA(82).

D. medusa. Needle cast. **16:**CA(82), CO(82,3457), WA(3202).

17:CO(82). **46:**CA(82,642), CO(82,3457), WA(3202). **48:**CA(81,82,661). **49:**CA(82).
D. montana. Needle cast. **16:**W states(82), ID(3202), OR(3202), WA(3202). **17:**W states(82), CA(81).
D. ponderosae. On needles. **46:**CO(3456).
Didymopeltis pini. On needles. **9:**CA(81).
Didymosphaeria futilis. **1:**ID(3202).
Dimerosporiopsis macrospora. **36:**ID(3202).
Discina ancilis. **16:**WY(1195).
Discohainesia oenotherae. On twigs. **62:**VA(82).
Dothiora slippii. **1:**ID(179).
Elytroderma deformans. Needle blight and cast, witches' broom. **4:**CA(469). **12:**AZ(82), CO(82), NM(82). **16:**CA(82), ID(3202), MT(82,3202), OR(82,3202), WA(3202). **17:**NW states(82). **18:**CA(3738). **20:**GA(82). **21:**AZ(82), CO(82,3457), NM(82). **46:**Unk.(3048), W states(82), CA(82,642,3738), CO(3457), ID(3202), MT(3202), OR(3202), WA(82,3202). **48:**CA(81,82,642,661,3738). **49:**SD(82). **64:**SE states(82), W states(82).
Endoxyla operculata. **46:**OR(3202).
E. parallela. **46:**OR(3202).
Eriosphaeria vermicularia. **1:**ID(3202).
Eutypella leprosa. **57:**IA(3629).
Gorgoniceps aridula. **16:**ID(3202). **64:**NJ(849).
Grovesiella abieticola. **36:**ID(3202).
Gyromitra infula. On rotting logs. **17:**CA(661).
Haligena viscidula. **64:**NC(3215).
Halosphaeria cucullata. On driftwood in ocean. **59:**NC(1941).
H. mediosetigera. On driftwood in ocean. **59:**NC(1941).
Hansenula canadensis. On dead tree. **17:**WY(2924).
H. capsulata. **17:**WY(2924).
Helotium sp. **46:**ID(3202).
Hemiphacidium convexum. On bark, needles and wood. **54:**NC(82,2951).
H. planum. Needle blight, on bark and wood. **26:**CO(82,3457). **46:**CO(1195,3457). **57:**WI(82). **64:**AZ(2951), CO(2951), WI(2951).
Herpotrichia juniperi. Brown felt blight. **1:**NW states(82), ID(82). **36:**NW states(82), ID(82). **64:**W states(82).
Hyaloscypha stevensonii. **57:**NY(2456).
Hymenoscyphus lutescens. **46:**WA(3202).
Hypocrea chionea. **64:**LA(658).
H. schweinitzii. **10:**GA(1438). **64:**NC(3944).
Hypoderma sp. **51:**CA(82).
Hypodermella sp. Needle cast, on wood. **1:**CA(642). **4:**CA(470), OR(470). **36:**MT(82), WA(82). **46:**CA(470), OR(470). **62:**OH(82).
Hysterium angustatum. On boards. **1:**WA(3202). **46:**CA(639,642).
H. macrosporum. **64:**GA(1438).
Hysterium sp. **52:**HI(2903).
Hysterographium flexuosum. On living bark. **20:**GA(1438). **62:**NJ(82).
H. mori. **1:**CA(661), ID(3202), OR(3202), WA(3202). **36:**ID(3202). **64:**GA(1438).
Kalmusia pinicola. On needles. **34:**CA(81). **55:**CA(81).
Karschia sp. **20:**GA(82,1438).
Lachnella villosa. **64:**GA(1439).
Lachnellula agassizii. On blister rust cankers. **1:**CA(642), ID(82,3202), WA(3202). **4:**OR(3202). **16:**CA(81). **17:**CA(661). **26:**ID(3202). **36:**ID(82,3202), OR(82,3202). **46:**ID(3202). **57:**NC states(82), NE states(82). **64:**Range of host(82), CA(638).
L. arida. On limbs and twigs. **1:**CA(82,642,661), ID(82,3202), WA(3202). **16:**CA(81), ID(3202), OR(3202), WA(3202), WY(1195). **17:**CA(661). **26:**MT(82,3202). **30:**CA(642). **36:**ID(3202). **64:**W states(82), OR(3202).
L. calyciformis. **54:**MI(82).
L. fuscosanguinea. Branch and twig canker. **16:**WY(1195,3378). **36:**NW states(82), ID(3202). **57:**MI(82).
L. pini. **36:**ID(3202), MT(3202), WA(3202).
L. subtilissima. On twigs. **36:**ID(3202). **46:**WA(3202). **58:**MA(82).
L. suecica. **1:**CA(661). **16:**ID(3202).
Lachnellula sp. **16:**OR(3202).
Lecanidiella contortae. On bark. **64:**OR(3231).
Lembosia acicola. Black mildew. **55:**CA(82,1468).
Leucostoma kunzei. On limbs and twigs. **64:**MI(82).
Ligníncola laevis. On driftwood in ocean. **59:**NC(1941).
Lophodermella arcuata. Defoliation, needle cast. **26:**CO(1195,3457), MT(3202). **30:**CA(81,82,449), OR(82,3202). **36:**ID(3202,3203).
L. cerina. Needle blight and cast. **16:**CA(82). **23:**AL(760), FL(760), GA(760), LA(760), MS(760), TX(760). **46:**AZ(1195), CA(82). **59:**AL(760), FL(760), GA(760), LA(760), MS(760), TX(760).
L. concolor. Needle cast. **16:**W states(82), CO(3457), ID(3202), MT(3202), OR(3202), WA(3202). **17:**W states(82), OR(82).
L. montivaga. Needle cast. **16:**W states(82), CO(1195,3457), ID(3202), MT(3202), OR(3202), WA(3202), WY(1195). **17:**W states(82). **26:**MT(3202). **36:**WA(3202). **52:**CA(82).
L. morbida. Needle cast. **4:**CA(3460). **46:**OR(1485,3202,3460), WA(3202,3460).
Lophodermium australe. Needle cast. **23:**HI(2903). **42:**FL(833), MS(2720). **44:**HI(2903). **52:**HI(2903). **59:**FL(833).
L. baculiferum. On needles. **48:**CA(2442,2443).
L. durilabrum. Needle cast. **1:**OR(82), WA(3815). **7:**CA(2391). **36:**OR(3202), WA(3202,3815). **57:**WI(1340). **64:**OR(2443).
L. molitoris. On needles. **64:**GA(2443).
L. nitens. Needle cast. **1:**CA(642). **16:**CA(81). **26:**MT(3202). **30:**CA(81,82), OR(82,3202). **36:**NW states(82), ID(3202), MT(3202), OR(3202), WA(3202). **52:**CA(82). **57:**E states(82), GA(1438), MI(82), NC(1303,1307), WA(82,3202), WI(1340). **63:**GA(82,1438). **64:**NE states(82), NW states(82), NW states(82), CA(638,2443), ID(82,2443,3202), MA(2443), ME(2443), MI(2443), MT(3202), NH(2443), OR(2443,3202), PA(2443), WA(2443,3202).
L. pinastri. Needle cast. **1:**NW states(82), MA(82), OH(82), OR(3202), WA(3202). **4:**OR(82,3202). **7:**MT(82). **8:**Lake States(82), ID(3202). **10:**FL(82), GA(82,1438), MS(82,2720). **16:**NW states(82), AK(82), CA(82), ID(3202), MT(3202), OR(3202), WA(3202). **17:**W states(82), AK(526). **20:**S states(82), GA(1438), NC(3944). **26:**ID(3202), MT(3202). **27:**S states(82). **30:**CA(642). **36:**NW states(82), CA(661), MA(82), OH(82). **37:**GA(82,1438), IL(82), MA(82), NC(1303). **39:**CT(82), OR(82,3202). **42:**GA(1438), SC(634). **46:**Range of host(82), CA(81,642), GA(1438), ID(3202), MT(3202), OR(3202), WA(2219,3202). **48:**CA(81,82,661), GA(1438), WA(82,3202). **49:**SD(2243). **50:**GA(1438). **52:**CA(82). **53:**Range of host(82), MA(3228), MI(2594), MN(2594), WI(2594), WV(3911). **54:**E states(82), GA(82,1438), MA(1390), NC(1303,1307), PA(2508), WI(1340), WV(82). **55:**CA(81,82). **57:**Range of host(82), GA(82,1438), ME(82), MT(3202), NC(1303), OH(82). **58:**NE states(82), GA(82), IL(82), MI(3304), PA(2355,2357), WA(3202). **59:**SE states(82), GA(1438), MS(82,2720), NC(1303,3944). **60:**MD(82). **62:**GA(82,1438), MD(82), PA(82), VA(82). **64:**Range of host(82), AK(526), ME(2443), MI(2443), OR(2443), WA(2443).
L. ravenelii. On needles. **64:**SC(2443), TX(2443).
L. seditiosum. **6:**IN(2443), MI(2443), OR(2443), WA(2443). **16:**IN(2443), MI(2443), OR(2443), WA(2443). **17:**IN(2443), MI(2443), OR(2443), WA(2443). **19:**IN(2443), MI(2443), OR(2443), WA(2443). **35:**IN(2443), MI(2443), OR(2443), WA(2443). **42:**IN(2443), MI(2443), OR(2443), WA(2443). **53:**IN(2443), MI(2443), OR(2443), WA(2443). **62:**IN(2443), MI(2443), OR(2443), WA(2443).
L. staleyi. On needles. **58:**OR(2443).
Lophodermium sp. Needle cast. **20:**NC(1303). **46:**CO(3457), WA(2219). **52:**HI(235). **61:**CA(82).
Lulworthia sp. On driftwood in ocean. **59:**NC(1941).
Meloderma desmazieresii. Needle cast. **39:**NJ(82), NY(82). **53:**NY(82). **54:**NE states(82), NY(82). **57:**E states(82), GA(82,1438), NC(1303), WI(82). **64:**E states(82), NY(2456).
Microthyrium microscopicum. On needles. **30:**CA(81).
Mitrula paludosa. **64:**GA(1438).
Mollisia cinerea. **36:**WA(3202).
Mollisia sp. On dead needles. **57:**WI(1344).
Monascostroma euryasca. **16:**MT(3202).
Monascus floridanus. Isolated from roots. **13:**FL(167). **23:**FL(167). **42:**FL(167).
Mycosphaerella acicola. On needles. **46:**CA(81). **55:**CA(81). **64:**CA(82).
M. dearnessii. Brown spot needle blight, black spot. **4:**FL(82), LA(82), OR(82). **10:**FL(82), GA(82), LA(82), MS(82,2720), SC(82), TX(82). **17:**FL(82), LA(82), OH(82). **20:**LA(82). **27:**FL(82). **28:**FL(82). **37:**NY(3283). **39:**KS(82), KY(82), MO(82),

NC(1303), OH(82), VA(1533). **42:**AL(1827), FL(82,1827), GA(82,1438,1827), LA(838,1827,3747), MS(1827,2720), NC(82), SC(82), TX(82), VA(3493). **46:**ID(82), MO(82,3477), OH(82), TN(82). **48:**MO(82), OH(82), TN(82). **49:**FL(82), LA(82), TN(82). **53:**WI(2596). **54:**LA(82), NC(82), TN(82). **55:**FL(82), LA(82). **56:**LA(82). **57:**NC(82,354). **58:**IA(2598), KS(2598), KY(2598), MN(2598), MO(2598), OH(82), WI(2598,2875). **59:**FL(82), GA(82), MS(2720), NC(82), SC(82), TN(82), TX(82). **60:**FL(82). **62:**GA(82), NC(82). **64:**NW states(82), S states(82).

M. hypodermellae. Needle spot. **17:**WY(82,3811).

M. pini. **4:**OR(3202). **16:**ID(3202), OR(3202), WA(3202). **44:**OR(3202). **46:**ID(3202), MT(3202). **48:**OR(3202). **52:**OR(3202). **60:**OR(3202).

Mytilinidion gemmigenum. **1:**ID(3202). **16:**ID(3202), MT(3202), OR(3202), WA(3202). **36:**ID(3202). **46:**ID(3202).

M. mytilinellum. On twigs. **16:**ID(3202). **36:**ID(3202). **57:**MI(82).

M. rhenanum. **20:**GA(1438).

M. tortile. **1:**WA(3202). **16:**OR(3202). **46:**ID(3202).

Naemacyclus fimbriatus. On cone scales. **37:**GA(1438). **53:**MA(3228). **59:**GA(1438).

N. stictiformis. **42:**FL(2777).

Nectria episphaeria. **1:**ID(3202). **36:**ID(3202).

N. fuckeliana. **36:**ID(3202).

N. pithoides. **36:**WA(3202).

Nectria sp. Trunk canker. **57:**MI(82).

Neopeckia coulteri. Brown felt blight. **1:**W states(82), CA(638,642,661), ID(3202), OR(3202), WA(3202). **7:**CA(82). **16:**CA(81,82), CO(3457), ID(3202), MT(3202), OR(3202), WA(3202), WY(1195). **17:**NW states(82), CA(661), CO(82), WA(82). **26:**CA(82), ID(3202), MT(82,3202). **36:**W states(82), CA(81,661), ID(3202), OR(3202). **46:**W states(82), OR(3202), WA(2219,3202). **48:**CA(661). **49:**UT(82). **57:**MT(82,3202). **64:**Range of host(82).

Niesslia coloradensis. On boards. **46:**CA(642).

N. exilis. On needles. **38:**CA(81).

Nitschkia broomeiana. **57:**GA(82,1438). **59:**GA(82,1438).

Odontotrema minus. **1:**WA(3202).

O. oregonense. On wood. **46:**OR(3232).

Ophiostoma adjuncti. Wood stain associated with bark beetles. **46:**NM(788).

O. aureum. Blue stain of wood. **17:**WY(3008). **64:**W states(3714).

O. bicolor. On wood. **16:**CO(1364).

O. clavigerum. Blue stain of wood. **16:**OR(3202), WA(3202), WY(3008). **36:**OR(3008), WA(3008). **46:**CO(3008,3714).

O. huntii. Sapwood stain. **36:**WA(804,3202). **46:**AZ(804,3714), CO(804), OR(804,3202). **57:**NY(804). **64:**NY(3714).

O. ips. Blue stain of wood, associated with bark beetles. **8:**MN(484). **16:**CO(1364), ID(3202), MT(3516). **17:**ID(3069), UT(3069), WY(3069). **20:**NC(3067). **26:**WY(3069). **36:**WA(3069,3202,3714). **42:**FL(3067). **46:**AZ(3714), CO(3069,3714), ID(3068,3714), NM(82,3714), OR(82,3068,3202,3714), SD(3069), UT(3069), WA(3068), WY(3069). **53:**MN(82,484), PA(82,3067). **54:**NJ(3067). **57:**NY(3714). **58:**MA(2921). **60:**MA(2921).

O. longirostellatum. On wood. **64:**CA(3714).

O. minus. Blue stain, tracheomycosis. **8:**NC(82), VA(82). **10:**NC(82), VA(82). **16:**CO(784). **20:**S states(82), GA(1438). **46:**CA(1364), ID(82,3202), OR(82,3202), SD(82), WA(1364,3202). **50:**NC(82), VA(82). **53:**NC(82), VA(82). **54:**NC(82), VA(82). **56:**NC(82), VA(82). **59:**GA(82,1438).

O. multiannulatum. Wood stain. **64:**S states(82), FL(777), LA(777,3732), MS(777), NC(3714).

O. nigrocarpum. On wood, associated with bark beetles. **46:**CA(785,3714).

O. obscurum. Wood stain. **64:**S states(82,1691), FL(777), LA(777,3732).

O. olivaceum. On wood. **36:**OR(3714).

O. piliferum. Blue-black stain of wood. **8:**MN(484). **16:**CO(3714). **20:**LA(3714), MO(3714). **36:**Range of host(82), CO(3714), ID(82,3202). **46:**CA(82), CO(3714), ID(3202), WY(82,1364). **53:**HI(2903). **54:**PA(82). **57:**MN(484).

O. pluriannulatum. Wood stain. **64:**S states(82), LA(3732).

O. populinum. Blue stain. **46:**CO(1565).

O. robustum. Blue stain of wood. **46:**CO(3008), ID(3202,3714), OR(3008,3202).

O. stenoceras. Isolated from wood infested with bark beetles. **16:**CO(786). **26:**CO(786). **36:**WA(786,3202). **46:**AZ(786), CA(786), CO(786), NM(786,3714), OR(786,3202). **50:**NM(786). **57:**NH(786), NY(786).

O. wageneri. Black stain rot. **16:**OR(1067). **36:**OR(1067). **46:**CA(589,1256,3902), OR(1067).

Orbilia sp. **36:**ID(3202).

Ostreola formosa. On bark. **64:**CA(642).

Ostreola sp. **1:**ID(3202), WA(3202). **26:**WA(3202).

Patinella atroviridis. On limbs. **1:**CA(638,642), ID(3202). **16:**ID(3202), OR(3202).

Patinella sp. **36:**ID(3202).

Pezicula livida. On limbs. **57:**GA(1438), IA(82), MA(664), NC(82).

Pezizella minuta. On needles. **20:**NC(82). **54:**NC(82). **59:**GA(82,1438), NC(82). **62:**NC(82).

Phacidium coniferarum. On trunks and limbs. **57:**ME(1413).

P. infestans. Snow blight and mold. **57:**NE states(82). **64:**NE states(82).

Phaeocryptopus pinastri. On needles following brown spot. **4:**Unk.(82). **54:**GA(1438), NJ(82).

Plectania nannfeldtii. **1:**ID(3202).

Pleospora sp. **26:**ID(3202).

Ploioderma hedgcockii. Needle blight and cast. **10:**FL(82), NC(82). **13:**FL(82). **20:**MS(82,2720), NC(82), VA(82). **23:**AL(760), FL(760), GA(760), LA(760), MS(760), TX(760). **42:**FL(82). **54:**NC(82). **59:**AL(760), FL(760), GA(760), LA(760), MS(760), TX(760). **62:**SE states(82), MS(82,2720), TN(82). **64:**SE states(82).

P. lethale. Gray needle blight, needle cast. **10:**FL(82,1533), MS(82,2720), SC(1533). **13:**FL(82). **20:**SE states(82), GA(1438), LA(82), MS(2720). **23:**AL(760), FL(760), GA(760,1970), LA(760), MS(760), TX(760). **39:**MO(82), NY(82), PA(82). **53:**NY(82). **54:**E states(82), MS(82,2720), PA(2508), WV(3910). **56:**FL(82), NC(1303), SC(82). **59:**SE states(82), AL(760), FL(760), GA(760,1533,1970), LA(760), MS(760), NC(1303), SC(1533), TX(760). **62:**SE states(82), TN(82), VA(3493). **64:**E states(82), Gulf states(82).

P. pedatum. Needle cast. **52:**CA(81,82).

Pragmopora pini. **36:**ID(3202).

P. pithya. Isolated from basal cankers. **57:**NY(1644). **64:**ID(3202).

Propolis decidua. **64:**NJ(3225).

P. leonis. On bark and cones. **10:**GA(1438). **16:**CA(81), ID(3202), WA(3202). **17:**WA(638). **20:**GA(1438). **36:**ID(3202). **46:**ID(638,3202), WA(3202). **58:**GA(1438).

P. rhodoleuca. **16:**ID(3202). **20:**GA(1438). **57:**GA(1438).

P. strobilina. **37:**GA(1438).

Propolomyces versicolor. **16:**MT(3202). **64:**MD(651).

Quasiconcha reticulata. On roots. **28:**AZ(283).

Rhizina undulata. Seedling blight. **8:**MN(82). **16:**MT(3202), OR(3202). **17:**W states(82). **36:**ID(82,3202), MT(82,3202). **46:**MT(82,3202). **52:**CA(82). **53:**MD(82).

Rhizodiscina lignyota. **20:**GA(82,1438).

Rosellinia aquila. **36:**ID(3202).

R. diathrausta. **1:**ID(3202). **16:**ID(3202), MT(3202).

R. limoniispora. **36:**ID(3202).

R. obliquata. **46:**MT(3202).

R. thelena. On roots. **36:**ID(82,3202).

R. trichota. **59:**GA(1438).

Rosellinia sp. On freshly cut stump. **57:**NC(1303).

Sarea resinae. On resin. **8:**MI(1495). **16:**OR(1495). **36:**ID(3202).

Schizothyrium acuum. On dead needles. **46:**CA(81).

Schizoxylon sp. On scales of fallen cones. **20:**NC(1303).

Scoleconectria cucurbitula. Burn blight, canker, parasitic on moribund limbs. **8:**WI(82,1335,1338). **16:**ID(3202). **52:**CA(3041). **53:**WI(82,1338,1345). **54:**NJ(82). **58:**IA(82). **59:**GA(82,1438). **62:**MD(82), VA(82).

Scorias spongiosa. Snow mold. **57:**IN(82).

Soleella striiformis. Needle cast. **18:**CA(81,82). **55:**CA(82). **61:**CA(82).

Sphaerella sp. On cone scales. **1:**CA(642).

Stictis radiata. **64:**NC(3944).

Stictis sp. **42:**GA(1438).

Stomiopeltis cassiae. **36:**ID(3202).

Tapesia lividofusca. On scales of fallen cones. **30:**CA(81).

Therrya fuckelii. On dead limbs. **53:**MI(2950), NH(3228).

T. pini. On limbs and twigs. **57:**E states(82), GA(1438), MI(82). **63:**GA(1438).
Thyridaria rubronotatum. **1:**ID(3202).
Thyronectria balsamea. **57:**NY(3041).
Trichoscyphella tenuipilosa. On twigs. **46:**WA(641,3202).
Tryblidiopsis pinastri. Limb canker. **1:**CA(82), ID(82,3202). **36:**CA(82), ID(82,3202). **57:**NE states(82), PA(82).
Tubeufia pezizula. Wood stain. **64:**S states(82).
T. scopula. **64:**GA(1438).
Tympanis confusa. Canker. **1:**ID(3202). **16:**ID(3202). **57:**CT(82), NY(82). **62:**VA(3493). **64:**ID(3202), OR(3202).
T. hypopodia. On blister rust cankers. **36:**ID(82,3202). **57:**CT(82), MA(82). **64:**CT(2670), ID(2670,3202), MA(2670).
T. laricina. **64:**NY(2670).
T. neopithya. **64:**MI(2670), NC(2670).
T. truncatula. On bark. **1:**CA(638,642), WA(3202). **36:**ID(3202). **64:**ID(3202).
Tympanis sp. Trunk canker. **19:**CT(82), PA(82). **46:**ID(3202). **53:**NC states(82), NE states(82), OH(82). **58:**CT(82). **62:**VA(82).
Unguiculella uncinata. **64:**CA(849).
Valsa abietis. On limbs and twigs. **46:**WA(3202). **64:**NE states(82).
V. collicula. On limbs and twigs. **57:**NY(82), PA(82). **58:**PA(82). **64:**NE states(82).
V. pini. On limbs and twigs. **16:**OR(3202). **36:**ID(3202), WA(3202). **53:**IA(1226). **57:**Range of host(82), GA(1438), IA(1226), MA(656). **58:**IA(1226). **59:**GA(1438). **62:**VA(82). **64:**MI(82).
V. superficialis. Branch canker. **8:**MI(82). **64:**MI(82).
Valsaria gemmata. **59:**GA(1438).
Xerotrema megalosporum. On wood. **16:**OR(3232).
Xylaria cornu-damae. On wood. **57:**NC(1303).
X. hypoxylon. On logs. **59:**NC(1303).
Xylopezia hemisphaerica. On cone scales and wood. **30:**CA(3233). **46:**OR(3233).

Basidiomycotina - Uredinales

Coleosporium apocynaceum. Heteroecious: aecial host. **10:**S states(82), AL(82), FL(82), TX(82). **42:**S states(82), FL(82), GA(82). **59:**GA(82).
C. asterum. Heteroecious: aecial host. **8:**NC states(82), NE states(82). **16:**NC(1303). **17:**W states(82), WA(82). **20:**E states(82), AR(82), GA(1439), MS(2720), SC(1533). **37:**NJ(82). **39:**CO(82), NJ(82), PA(82). **46:**MT(3202). **49:**MT(82). **53:**NE states(82), IL(1052), MN(82). **54:**E states(82), MA(1390), MD(651), NC(1303), OH(82), PA(2508). **58:**NJ(82). **59:**E states(82), GA(1439), MS(2720), NC(1303,1533,3944), SC(1533). **60:**MD(82). **62:**GA(1439), PA(82), TN(82), VA(82,638). **64:**E states(82).
C. crowellii. Autoecious. **12:**AZ(82), CO(82), NM(82). **21:**AZ(82,1195,1203,3981), CO(82), NM(82,1195), UT(2772). **26:**AZ(82), CO(82), NM(82). **34:**CA(2772).
C. delicatulum. Heteroecious: aecial host. **10:**E & C states(82), W states(82). **20:**E & C states(82), W states(82), AL(285). **39:**PA(82). **42:**E & C states(82), W states(82), AL(285), MS(2720). **50:**E states(82). **53:**E & C states(82), W states(82), VT(82). **54:**E & C states(82), W states(82), NC(1303). **56:**E & C states(82), W states(82). **59:**E & C states(82), W states(82), AL(285), NC(1303).
C. helianthi. Heteroecious: aecial host. **8:**NC states(82), NE states(82). **20:**E states(82), SE states(82), NC(1303), OH(82). **42:**SC(82). **54:**NC(82,1303), SC(82). **56:**SE states(82). **59:**SE states(82), AL(82), SC(82). **62:**E states(82), SE states(82), NC(1303), OH(82), SC(82).
C. inconspicuum. Heteroecious: aecial host. **20:**SE states(82), AL(285), NC(1303), OH(82), TN(82). **42:**SE states(82), OH(82), TN(82). **62:**SE states(82), NC(1303), OH(82), TN(82).
C. ipomoeae. Heteroecious: aecial host. **10:**E states(82), AZ(82), IL(82). **20:**E states(82), AZ(82), IL(82), NC(1303), OK(2872). **33:**AZ(82,1203,3981). **42:**E states(82), AL(285), AZ(82), IL(82), MS(2720). **54:**E states(82), AL(82). **56:**E states(82), IL(82). **59:**E states(82), AZ(82), GA(1439), IL(82), NC(1303).
C. jonesii. Heteroecious: aecial host. **12:**AZ(82), CO(82), ID(82), NM(82), UT(82). **21:**AZ(82,1195,1203,3981), CO(82,1195), ID(82,3202), NM(82), UT(82).
C. laciniariae. Heteroecious: aecial host. **42:**E states(82). **54:**E states(82), MD(82). **59:**E states(82), VA(2989).
C. minutum. Heteroecious: aecial host. **27:**FL(82), FL(82).
C. pacificum. Heteroecious: aecial host. **18:**OR(3202). **48:**OR(82,3202). **52:**CA(82).
C. pinicola. Autoecious. **54:**MA(82). **59:**MD(82). **62:**E states(82), IN(82), NC(1303), TN(82), VA(3493).
C. tussilaginis. Heteroecious: aecial host. **8:**NC states(82), NE states(82). **53:**MI(82), NH(82), NY(82). **54:**E states(82), OH(82). **58:**NY(82), WI(82). **62:**E & C states(82), NC(82). **64:**Unk.(82).
C. vernoniae. Heteroecious: aecial host. **10:**E states(82), SE states(82), GA(1439), IL(82), KY(82), LA(82), MS(82,2720), NJ(82), TX(82). **13:**FL(82). **17:**KS(82). **20:**E states(82), SE states(82), GA(1439), IL(82), KS(82), NC(1303,3944), NJ(82), TX(82). **27:**FL(82). **37:**OH(82). **39:**IN(82), KY(82), OH(82). **42:**E states(82), IL(82), MS(2720), NJ(82), TX(82). **46:**Unk.(82). **54:**E states(82), SE states(82), IL(82), KY(82), NC(1303), TN(82). **56:**E states(82), IL(82), TX(82). **58:**OH(82). **59:**E states(82), SE states(82), AR(3918), GA(1439), IL(82), KS(82), MS(2720), NC(1303,3944), NJ(82), SC(1533), TX(82). **64:**E states(82), KS(82), TX(82).
Coleosporium sp. **17:**ID(82), MT(82), WA(82). **59:**AR(3918), VA(3493). **64:**Unk.(82).
Cronartium appalachianum. Heteroecious: aecial host. **62:**NC(1531,2770), TN(1531), VA(1531,3493), WV(1531).
C. arizonicum. **46:**AZ(3981).
C. coleosporioides. Heteroecious: aecial host. **4:**CA(111). **8:**MI(57), MN(55,57). **16:**W states(82), AK(60,82,526,3431), ID(3202), MT(3202), OR(3202), SD(2765), WA(3202), WY(1195). **17:**W states(82), AK(526), CA(661), WA(638). **18:**CA(82). **28:**CA(81). **33:**AZ(82). **46:**W states(82), AK(82), ID(3202), MT(2066,3202), NE(82), OR(3202), SD(82,2243,2766), WA(3202). **48:**W states(82), AK(82), CA(82), NE(82), NV(82), OR(82,3202). **49:**W states(82), AZ(1496), NE(82), SD(82). **52:**CA(111). **55:**CA(81,82). **58:**SD(2765).
C. comandrae. Heteroecious: aecial host. **4:**OR(82,3202). **8:**E states(82), MI(82), MN(82,248), WI(82). **16:**W states(82,1977), CO(1772), ID(3202), MT(3202), WA(3202), WY(77,1160,1195). **17:**W states(82). **20:**AR(885,2866,3946), MO(263), MS(2866). **22:**AZ(3981). **23:**TN(88). **27:**TN(88). **46:**W states(82,1977), AZ(3981), CA(81,642), MT(3202), OR(1063,3202), WA(3202). **47:**W states(82). **48:**OR(82,3202). **49:**CO(82). **50:**E states(82), PA(82). **53:**VT(82). **54:**CT(82), NJ(82). **56:**TN(674). **58:**E states(82). **59:**KY(2288), TN(674,2865,3946).
C. comptoniae. Heteroecious: aecial host. **8:**NE states(82), MN(82,291), WI(1116). **16:**CA(82), WA(3202). **17:**NE states(82), CA(82,642), WA(82). **18:**NH(82). **19:**PA(82). **37:**NE states(82), WA(3202). **39:**NE states(82), OH(82), WI(82). **46:**CA(82), MT(82,3202), VT(1224). **53:**NE states(82), MN(82). **54:**NE states(82). **58:**C states(82), NE states(82), WI(82,1352). **59:**E states(82), MI(82). **62:**MD(82), NC(1533), NJ(82), PA(82). **64:**C states(82), NE states(82).
C. conigenum. Heteroecious: aecial host. **33:**AZ(82,1195,1203,3981).
C. occidentale. Heteroecious: aecial host. **12:**SW states(82), NM(82). **21:**W states(82), NM(82). **34:**SW states(82), AZ(1203,3981), CA(82), NV(82).
C. quercuum. Heteroecious: aecial host. **8:**CT(82), MI(82,1478), MN(55,82), WI(82). **10:**MS(2720), TX(82). **13:**FL(82). **20:**SE states(82), AR(82), SC(1307). **28:**CA(82). **33:**AZ(82). **42:**AL(285), TX(82). **46:**AZ(82), CA(82), CT(82), ID(82), NE(82), PA(82), SD(82), WA(82). **49:**NE(82), SD(2243), WY(82). **50:**NC(1533), TN(82). **52:**CA(81,82). **53:**MN(82), NJ(82). **54:**NE states(82), KY(82), TN(82). **58:**MI(1478), NC(1303). **59:**SE states(82), AR(82), GA(1439), MS(2720). **62:**Range of host(82), NC(1533), VA(3493). **64:**E states(82), NC states(82), NW states(82), NC(3944).
C. quercuum f. sp. *banksianae*. Heteroecious: aecial host. **8:**MN(872), WI(1116).
C. quercuum f. sp. *echinatae*. Heteroecious: aecial host. **20:**NC(1303).
C. quercuum f. sp. *fusiforme*. Heteroecious: aecial host. **10:**S states(82), FL(1533), GA(1533), MS(2720), SC(1533). **20:**NC(1722). **42:**SE states(82), AL(82), MS(82,2720), NC(1303), SC(1969), TX(82). **54:**NC(1303). **56:**NC(1303,1722). **59:**S states(82), SE states(3139), AL(1622,2492,2796,3968), AR(2796), FL(3137), GA(35,2796,3137,3968), LA(2796), MS(2720,2796,3968), NC(1303,1722,2796), SC(1533,2796,3968), TX(3769), VA(3493). **65:** On *P. taeda* x *P. sondereggeri*. MS(1828).

C. quercuum f. sp. *virginianae*. Heteroecious: aecial host. **62:**NC(1303).
C. ribicola. Heteroecious: aecial host. **1:**ID(82,3202), MT(3202), OR(82,3202), WA(82,3202). **6:**NY(82). **7:**CA(2389). **26:**IA(82), ID(1974,3202), MA(82), MN(82), MT(398,3202), WY(398,399,1974). **30:**CA(81,82), NY(82), OR(82,3202). **36:**ID(82,3202), MA(82), MT(3202), NC(1533), NY(82), OR(82,3202), WA(82,3202). **53:**MN(2157). **57:**E states(82), Lake states(82), NC(82,1303,1533), TN(82), VA(3493), WA(82,3202).
C. strobilinum. Heteroecious: aecial host. **10:**FL(82), GA(3997), MS(82,2720). **23:**FL(2236), GA(2236). **42:**FL(82), MS(82,2720).
Cronartium sp. **17:**OR(657).
Endocronartium harknessii. Autoecious. **4:**CA(82). **8:**ID(3202), MI(1478), MN(872). **16:**W states(82), ID(3202), MT(3202), WA(3202). **17:**W states(82). **18:**CA(82,3737). **37:**MN(247). **38:**CA(3737). **46:**W states(82), AZ(3981), ID(3202), MI(3609), MO(3477), MT(3202), OR(3202), PA(2360), WA(3202). **48:**W states(82), CA(82), MI(2354), NV(82), OR(82), WA(82). **49:**W states(82). **52:**CA(82,3737), OR(3202), WA(3202). **55:**W states(82), CA(82), WA(82). **58:**NE states(82), ID(3202), MA(2920), MD(2354), MI(82,1478,2354), ND(3762), NY(2354), PA(2172,2354), VA(3300), WA(3202).
Peridermium cerebrum. **54:**NY(2770).
P. comptoniae. **58:**CT(2770).
P. filamentosum. **46:**AZ(1195,1203,2770), NM(1195).
P. fusiforme. **59:**AL(2770).
P. guatemalense. **42:**FL(82).
P. harknessii. **16:**SD(2766), WA(2219). **46:**AZ(1203), CA(1468,2770). **48:**MI(1978). **52:**CA(1468). **55:**CA(1468). **58:**SD(2766).
P. occidentale. **21:**CO(2770).
P. stalactiforme. **8:**MN(61). **16:**ID(3202), WA(2770,3202). **19:**WA(3202).
P. strobilinum. **42:**FL(2770).
P. weirii. **16:**ID(3202). **17:**ID(82).
Peridermium sp. **62:**NC(1529).

Basidiomycotina - Other

Acanthophysium canadense. **46:**SW states(1179), AZ(1202).
A. lividocaeruleum. On wood. **3:**UT(2144). **16:**ID(2083,3202), MT(128). **46:**SW states(1179), NM(128,1190).
Aleurodiscus amorphus. On suppressed, injured, or recently dead trees. **38:**CA(2083). **57:**Range of host(82).
A. grantii. On limbs and trunks. **38:**CA(1236).
A. mesaverdensis. On bark. **21:**CO(2138).
A. penicillatus. **46:**Unk.(1179).
A. utahensis. White rot. **3:**UT(128,2144).
Amphinema byssoides. White rot. **3:**UT(2144). **25:**AZ(1202). **46:**SW states(1179), AZ(1202), NM(1190). **58:**OH(128).
Amylocorticium subincarnatum. **16:**ID(3202). **36:**WA(82,3202).
A. subsulphureum. **46:**SW states(1179), AZ(128,1202).
Amylocystis lapponica. **16:**ID(3202), OR(3202). **36:**ID(3202), MT(3202). **46:**Unk.(1179), MT(3202). **64:**NC(3944).
Amylostereum chailletii. **58:**CT(543).
Anomoporia bombycina. **16:**CO(128,2167), ID(128,3202). **36:**ID(3202). **46:**SW states(1179), AZ(128,1202), NM(1190). **64:**GA(659).
Antrodia albida. **21:**AZ(1202). **36:**ID(3202).
A. albobrunnea. Brown cubical rot. **16:**ID(3202), MT(128,3202), OR(2167,3202). **36:**ID(82,128,3202). **46:**SW states(1179), AZ(128,1202). **64:**ID(82).
A. carbonica. Brown cubical rot. **46:**SW states(1179), AZ(1202), MT(2682,3202), WA(3202).
A. crassa. **36:**WA(3202). **46:**SW states(1179), AZ(128,1202), NM(1190), OR(3202).
[*A. ferox*]. **21:**AZ(1202).
A. heteromorpha. **16:**ID(3202), WA(3202). **36:**ID(3202), MT(3202,3832). **46:**Unk.(1179), ID(3202), MT(3202), OR(3202).
A. odora. **36:**ID(3202). **46:**SW states(1179), AZ(128,1202), NM(1190).
A. radiculosa. **25:**AZ(1202,2167). **42:**FL(2167). **46:**SW states(1179), AZ(1202). **59:**TX(652). **64:**AZ(128).
A. serialis. Brown cubical rot. **16:**ID(3202), MT(3202,3832), WA(3202). **17:**CO(82), MT(82). **26:**MT(3202). **30:**OR(3202). **36:**ID(128,3202), MT(82,3202,3832), WA(3202). **53:**MN(82,128). **59:**LA(82), NC(1303). **62:**MD(82), PA(82), VA(82). **64:**N states(82), Range of host(82), Unk.(1006).
A. sinuosa. Brown cubical rot. **16:**ID(3202), MT(3202). **30:**CA(82), OR(82,3202). **36:**ID(2167,3202), MT(3202), OR(3202), WA(3202). **46:**SW states(1179), AZ(128,1202), ID(3202), NM(128,1190), OR(3202). **64:**Range of host(82).
A. sitchensis. **1:**ID(128). **6:**AZ(1202). **33:**AZ(1202). **46:**SW states(1179), AZ(128,1202), NM(128,1190).
A. sordida. **6:**AZ(1202). **46:**AZ(1202).
A. vaillantii. **36:**ID(128,3202). **46:**Unk.(1179), AZ(1202), ID(3202), MT(3202). **64:**AZ(128), WA(3202).
A. variiformis. Brown rot. **16:**AK(214,526). **36:**ID(82,3202), MT(128,3202). **64:**N states(82), Range of host(82), MI(2862).
A. xantha. **16:**AK(526), ID(3202), MT(3202). **36:**ID(3202), MT(3202,3832). **46:**SW states(1179), AZ(1202,2167), MT(3202), NM(128,1190), OR(3202). **52:**CA(128). **64:**Range of host(82).
Antrodiella romellii. **46:**SW states(1179), AZ(128,1202).
A. semisupina. **46:**OR(3202).
Armillaria mellea. Butt rot, root, stem and white rot. **8:**MI(82), MN(82). **12:**AZ(82), CO(82), NM(82), UT(82). **16:**ID(538,1583), OR(1062,1067), WA(59). **17:**ID(82). **20:**HI(2011,2903,2907). **21:**AZ(82), CO(82), NM(82), UT(82). **26:**MT(82,3202). **28:**HI(2011,2903,2907). **30:**CA(82,2898). **35:**HI(2011,2903,2907). **36:**ID(82,538,1583,1998,3202), MT(3202), OR(82,1062,1067,3202). **39:**MN(2157). **41:**HI(2011,2903,2907). **43:**CA(2897). **44:**HI(2011,2903). **45:**HI(2011,2903). **46:**NW states(82), SW states(1179), AZ(1202), CA(2898), ID(3202), MT(3202), NM(3835), OR(1062,1067,3202), WA(59,1071,3202). **48:**CA(82). **53:**MN(82), WI(1345,2876). **54:**PA(82). **57:**Range of host(82), VT(59). **64:**Range of host(82).
A. tabescens. Decay of roots and stumps, root rot. **10:**FL(82). **13:**FL(82,166). **59:**NC(1303). **64:**NC(1303).
Arrhytidia involuta. On wood. **16:**MT(128). **36:**ID(128). **46:**ID(128). **54:**NY(128). **64:**NC(3944), NJ(128).
Asterodon ferruginosum. **36:**WA(128).
Asterostroma cervicolor. **53:**NY(128).
A. medium. **53:**NY(128).
Athelia arachnoidea. **16:**WA(3202). **46:**WA(3202).
A. decipiens. On wood. **46:**SW states(1179), AZ(1178,1202), NM(1190).
A. epiphylla. **46:**AZ(1178,1202).
A. scutellaris. **46:**SW states(1179), NM(1190).
A. teutoburgensis. On wood. **46:**AZ(1202).
Auriporia aurea. **46:**SW states(1179).
Auriscalpium vulgare. On cones. **46:**ID(3202), WA(638,3202). **54:**MA(1394). **58:**KS(3193).
Baeospora myosura. On cones. **57:**MN(3134).
Basidiodendron caesiocinereum. On bark and wood. **46:**SW states(1179), AZ(1202). **57:**IA(2185).
B. cinereum. White rot. **3:**UT(2144). **16:**OR(128). **46:**SW states(1179), AZ(1202).
B. eyrei. White rot. **3:**UT(2144). **16:**OR(2185,3202). **64:**MS(2185).
Bjerkandera adusta. Isolated from treated utility poles. **6:**AZ(1202). **64:**E states(3996), NC(3944).
Bondarzewia berkeleyi. Butt rot. **57:**MN(82).
B. montana. **36:**ID(3202).
Boreostereum radiatum. On logs. **64:**MI(2862).
Botryobasidium ansosum. **46:**SW states(1179), AZ(128,1202).
B. candicans. **46:**AZ(1202).
B. laeve. **53:**MA(656).
B. medium. **46:**SW states(1179), AZ(1202).
B. pruinatum. **46:**SW states(1179), AZ(1202).
B. subcoronatum. On rotten wood. **6:**AZ(1202). **32:**AZ(128). **33:**AZ(1202). **46:**SW states(1179), AZ(128,1202), CA(642), ID(3202), NM(1190), WA(3202). **57:**MA(656).
B. vagum. On rotten wood. **12:**AZ(1202). **25:**AZ(1202). **46:**SW states(1179), AZ(128,1202), CA(642), ID(3202), NM(1190), WA(3202). **57:**NY(128). **64:**NY(2456).
Botryohypochnus isabellinus. **46:**SW states(1179), AZ(128,1202), NM(1190), WA(3202).
Bulbillomyces farinosa. White rot. **3:**UT(2144).
Byssocorticium atrovirens. On logs. **54:**NY(128). **64:**MA(128).
B. californicum. On log. **38:**CA(1707).
B. neomexicanum. On wood. **46:**SW states(1179), NM(1190), NM(128,1188).
Byssoporia terrestris. **16:**CO(128). **46:**SW states(1179), AZ(128,1202).
Callistosporium luteo-olivaceum. On wood. **59:**VA(3717).

Calocera cornea. **6:**AZ(1202). **16:**WA(3202). **46:**Unk.(1179), ID(3202).
Candelabrochaete langloisii. **46:**SW states(1179), AZ(1202).
C. septocystidia. White rot. **64:**NJ(128,432).
Ceraceomyces borealis. **64:**MN(1197).
C. serpens. **46:**SW states(1179), AZ(128,1202). **64:**NJ(128).
C. sulphurinus. **46:**SW states(1179), AZ(1202), NM(1190). **58:**NY(128).
C. tessulatus. On bark. **46:**SW states(1179), AZ(1202). **64:**SC(634).
Ceriporia purpurea. On wood. **16:**ID(3202). **46:**SW states(1179), AZ(1202), CA(128), OR(3202). **64:**Range of host(82).
C. spissa. Isolated from treated utility poles. **25:**AZ(128,1202). **36:**MT(3202). **46:**Unk.(1179), AZ(1202), ID(3202), OR(128,3202), WA(3202). **64:**E states(3996), E states(3996).
C. tarda. **46:**SW states(1179), NM(1190).
C. viridans. **21:**AZ(1202).
Ceriporiopsis gilvescens. **36:**ID(3202).
C. pannocincta. Isolated from treated utility poles. **64:**E states(3996), E states(3996).
C. subrufa. **36:**ID(3202).
C. subvermispora. **25:**AZ(128,1202). **32:**AZ(128). **33:**AZ(1202). **46:**SW states(1179), AZ(1202), NM(1190).
Cerrena unicolor. On wood. **62:**VA(638).
Chaetoderma luna. Brown rot. **1:**ID(128). **3:**UT(2144). **16:**CO(797), ID(3202), WA(3202). **36:**ID(3202). **46:**SW states(1179), AZ(1202), NM(1190).
Chaetoporellus latitans. **32:**AZ(128). **33:**AZ(1202). **46:**SW states(1179), NM(1190). **64:**NC(128), TN(128).
Clavicorona piperata. **36:**ID(884,3202).
Climacocystis borealis. Rot. **36:**ID(3202). **64:**Range of host(82).
Climacodon pulcherrimus. On wood. **10:**FL(128). **33:**AZ(1202). **46:**SW states(1179), AZ(1202). **64:**FL(128).
Clitocybe martiorum. On dead needles. **57:**MI(275). **58:**MI(275).
Coltriciella dependens. On wood. **20:**GA(491,1439). **54:**NJ(3717).
Columnocystis abietina. Brown cubical rot. **17:**CA(661), MT(543), WA(82). **36:**CA(661). **46:**Unk.(1179), WA(3202). **64:**NE states(82), W states(82).
C. ambigua. Brown cubical rot. **64:**NE states(82), VT(543).
C. pimeriensis. **46:**SW states(1179), AZ(128,1202).
Confertobasidium olivaceo-album. On wood. **46:**SW states(1179), AZ(1178,1202), OR(3202). **53:**MA(656). **57:**MA(128).
Coniophora arida. Brown cubical rot. **6:**AZ(1202). **12:**AZ(1202). **21:**AZ(1202). **46:**SW states(1179), AZ(1202), NM(1190), WA(3202). **62:**MD(128). **64:**Range of host(82), Unk.(1006), NJ(128).
C. arida var. *suffocata*. **4:**CA(128). **46:**NM(1190). **54:**PA(128). **57:**MA(128).
C. fusispora. **54:**MA(128).
C. olivacea. **33:**AZ(1202). **36:**ID(3202). **46:**SW states(1179), AZ(128,1202), NM(128,1190), WA(3202). **64:**NJ(128), NY(128).
C. puteana. Brown cubical rot. **25:**AZ(1202). **36:**ID(128). **38:**CA(81). **46:**SW states(1179), AZ(1202), NM(1190). **64:**Range of host(82), Unk.(1006).
Coriolopsis gallica. On wood. **59:**NC(1303).
Corticium minnsiae. **16:**OR(128). **46:**Unk.(1179).
C. pini. **57:**NY(2456).
C. rubicundum. **64:**WA(3202).
Crepidotus payettensis. On decorticated logs. **17:**ID(1544).
Crustoderma corneum. Brown rot. **16:**MT(2555). **25:**AZ(1202). **46:**SW states(1179), AZ(128,1202), MT(2555).
C. dryinum. **36:**ID(2555). **46:**SW states(1179), AZ(128,1202), NM(128,1190), OR(3202).
C. marianum. Brown rot. **64:**MD(2555).
C. resinosum. On wood. **16:**MT(2559).
Cryptoporus volvatus. On wood. **16:**ID(3202), MT(3202,3832). **17:**CA(661), CO(3209). **21:**AZ(1202). **36:**ID(1469,3202), WA(3202). **46:**Range of host(82), SW states(1179), AZ(1202), CA(642), ID(3202), MT(3202,3832), OR(3202), WA(2219,3202). **50:**VA(638). **53:**MI(82), MN(82). **54:**MA(82,1394), NY(82), VT(82). **59:**AL(82). **62:**DC(3209), MD(82), VA(82). **64:**Range of host(82), CA(81).
Cylindrobasidium corrugum. On wood. **3:**UT(2144). **17:**CA(661). **26:**CO(128,1195). **46:**SW states(1179), AZ(82,1202), CO(82), WY(82).
C. laeve. **46:**CT(128).
C. torrendii. **46:**WA(3202).
Cyptotrama asprata. **33:**AZ(1202). **46:**SW states(1179), AZ(1202).
Cystobasidium proliferans. On insect excreta on inner surface of bark of dead tree. **64:**NJ(2644).
Cystostereum pini-canadense. Isolated from mine timbers. **64:**Unk.(1006).
Dacrymyces chrysospermus. On logs and wood. **6:**AZ(1202). **16:**ID(3202), WA(3202). **25:**AZ(1202). **46:**SW states(1179), AZ(1202), CA(642), ID(3202), WA(3202). **59:**NC(1303).
D. cokeri. On dead limbs. **59:**NC(1303).
D. dictyosporus. On wood. **25:**AZ(1202,3176). **46:**SW states(1179), AZ(1202).
D. minor. On wood. **46:**AZ(128). **57:**NY(2456). **64:**GA(2642).
D. pedunculatus. On wood. **64:**NC(128).
D. stillatus. On wood. **4:**CA(642). **16:**OR(3202). **46:**Unk.(1179), CA(642), ID(3202), WA(3202). **59:**NC(1303), TX(652). **64:**GA(2642), SC(634).
D. tortus. **46:**SW states(1179), AZ(1202), ID(3202), WA(3202).
D. variisporus. On wood. **46:**WA(2329).
Dacrymyces sp. Sapwood stain. **20:**NC(82). **64:**Unk.(82), GA(1439).
Dacryobolus karstenii. On wood. **33:**AZ(1202). **46:**SW states(1179), AZ(128,1202), CA(642), ND(378), NM(128,1190), OR(3202).
D. sudans. Brown cubical rot. **3:**UT(2144). **21:**NM(1190). **46:**SW states(1179), AZ(1202).
Dacryopinax spathularia. On wood. **23:**FL(128). **42:**FL(128). **64:**FL(128), NJ(128).
Daedaleopsis confragosa. **64:**NC(3944).
Dichomitus squalens. White pocket rot. **6:**AZ(1202). **8:**MI(82). **12:**AZ(1202). **16:**ID(3202), MT(3202,3832), OR(3202), WA(3202). **17:**MT(82), OR(82). **20:**GA(491,1439). **25:**AZ(1202). **33:**AZ(1202). **36:**ID(82,3202), MT(3202), WA(3202). **53:**MN(82,294,1197), NH(82). **57:**MI(82), VT(82). **59:**MS(82,2675,2720).
Dichostereum pallescens. **42:**TX(128).
Diplomitoporus lenis. **16:**ID(3202), MT(3202). **36:**ID(3202). **46:**SW states(1179), AZ(128,1202), ID(3202), MT(3202), NM(128,1190), OR(3202). **64:**MI(2862).
D. linbladii. **1:**ID(128). **16:**ID(3202), MT(128,3202), OR(3202), WA(3202). **25:**AZ(1202). **31:**AZ(128). **33:**AZ(1202). **36:**ID(3202), MT(3202). **46:**SW states(1179), AZ(1202), ID(3202), MT(3202), NM(128,1190), OR(3202), WA(3202). **53:**MA(656). **64:**NC(3944).
Ditiola radicata. On limbs. **64:**GA(2642), MD(651), NC(3944).
Entomocorticium dendroctoni. In bark beetle galleries. **17:**MT(3883).
Exidia glandulosa. **64:**NC(3944).
E. saccharina. **16:**WA(3202). **46:**WA(3202).
Exidiopsis calcea. On bark. **54:**NH(128).
E. fugacissima. **25:**AZ(1202).
E. glaira. **32:**AZ(1202).
E. podlachica. On fallen logs. **57:**MA(128).
E. sordida. On limbs. **64:**NC(2640).
Fibricium rude. **46:**SW states(1179), AZ(1202).
Fomitopsis cajanderi. Brown cubical rot. **16:**ID(3202), WI(2591). **17:**CO(82), ID(82). **36:**MT(82,3202), WA(3202). **46:**Range of host(82), SW states(1179), AZ(1202,2591), MT(3202,3832), ND(378), OR(3202), WA(3202). **47:**Range of host(82). **54:**Unk.(82). **62:**MD(82), VA(82,638). **64:**Range of host(82).
F. officinalis. Brown trunk rot. **17:**Range of host(82), CA(661), CO(82). **30:**NW states(82), CA(82). **36:**ID(82,3202). **46:**Range of host(82), SW states(1179), AZ(1202), CO(82), MT(3202), OR(3202), SD(378), WA(3202). **57:**MI(82). **64:**NE states(82), W states(82).
F. palustris. Wood rot. **10:**GA(491,1439). **20:**GA(491,1439). **42:**GA(491,1439). **59:**GA(82). **64:**S states(82), LA(82,2675), MS(2675).
F. pinicola. Brown cubical rot. **1:**Range of host(82), ID(3202), MT(3202,3832), WA(3202). **2:**CO(82). **6:**AZ(82,1202). **8:**Range of host(82). **16:**ID(3202), MT(3202,3832). **17:**ID(82), MT(82). **26:**MT(82,3202,3832). **30:**CA(82). **31:**Range of host(82). **33:**Range of host(82). **34:**Range of host(82). **36:**Range of host(82), ID(3202), MT(3202,3832), WA(3202). **46:**SW states(1179), AZ(1202), ID(3202), MT(3202,3832), OR(3202), SD(378), WA(3202). **47:**Range of host(82). **48:**Range of host(82), CA(661). **53:**Range of host(82). **54:**Range of host(82). **57:**Range of host(82). **62:**E states(82), TN(82).

64: Range of host(82).
F. rosea. Brown cubical rot. **6:**AZ(82). **8:**Range of host(82). **16:**MT(3202,3832). **17:**MT(82). **30:**Range of host(82). **36:**MT(3202), WA(82,3202). **46:**Range of host(82). **47:**Range of host(82). **57:**Range of host(82). **62:**MD(82), VA(82). **64:**ID(3202).
Galerina autumnalis. On dead limbs. **62:**VA(3717).
Galzinia cymosa. On wood. **53:**MA(128). **54:**MA(3011).
G. occidentalis. **16:**OR(3202). **33:**AZ(1202).
G. pedicellata. On wood. **16:**OR(128). **50:**OR(3202).
Ganoderma applanatum. White rot. **36:**ID(3202). **48:**CA(661). **57:**NY(82).
G. lucidum. White root and butt rot. **48:**CA(661). **64:**ME(82), MI(82,2862).
G. oregonense. **16:**OR(3202).
G. tsugae. **54:**PA(2962).
Gloeocystidiellum ochraceum. **53:**MA(656).
G. porosum. White rot. **53:**WI(2553).
G. sinuosum. On wood. **64:**FL(436).
Gloeophyllum carbonarium. **6:**AZ(1202). **16:**MT(3202,3832). **36:**ID(3202). **46:**SW states(1179), AZ(128,1202), ID(3202), NM(128,1190). **64:**MI(2862).
G. mexicanum. On logs. **64:**FL(634).
G. odoratum. Wood rot. **16:**CO(128), ID(3202), MT(3202), OR(3202), WA(3202). **36:**ID(3202), MT(3202), WA(3202). **46:**Range of host(82), SW states(1179), AZ(1202), ID(3202), MT(3202), WA(3202). **62:**MD(82), PA(82), VA(82). **64:**N states(82), Range of host(82).
G. protractum. **16:**MT(3832). **36:**MT(3832).
G. sepiarium. Brown rot. **1:**WA(82,3202). **16:**ID(3202), MT(3202), OR(3202), WA(3202). **17:**OR(657). **33:**AZ(1202). **36:**ID(3202), MT(3202), WA(3202). **46:**SW states(1179), AZ(1202), ID(3202), ND(378), NM(1190), OR(3202), SD(2243), WA(3202). **57:**Range of host(82), MA(656). **59:**NC(1303). **64:**Range of host(82), Unk.(1006), AR(2675), LA(2675), MI(2862), MS(2675), NC(3944), OK(2872), TX(652).
G. trabeum. Brown cubical rot. **16:**MT(3202,3832). **46:**Unk.(1179), ID(3202), OR(3202). **64:**Range of host(82), ND(378).
Gloeoporus dichrous. **16:**OR(3202), WY(1195,3378). **26:**MT(3202). **36:**ID(3202). **46:**SW states(1179), ID(3202), NM(1190), WA(3202). **53:**MN(1197).
Grandinia abieticola. **46:**SW states(1179), AZ(128,1202).
G. alutaria. **46:**SW states(1179), AZ(1202), NM(1190).
G. arguta. On wood. **46:**SW states(1179), NM(1190). **59:**GA(491,1439). **64:**NY(128).
G. barba-jovis. **36:**ID(128).
G. breviseta. **46:**SW states(1179), AZ(1202), NM(1190).
G. crustosa. **46:**Unk.(1179), ID(3202).
G. floccosa. **46:**SW states(1179), AZ(128,1202), ID(3202). **53:**NY(128).
G. granulosa. **46:**AZ(1202).
G. hastata. **46:**SW states(1179), AZ(1202), NM(128,1190), WA(3202).
G. pallidula. **32:**AZ(128). **33:**AZ(1202). **46:**SW states(1179), AZ(128,1202), NM(1190). **53:**MN(1197).
G. spathulata. On wood. **42:**SC(1175). **46:**NM(1190). **53:**MN(128,1197), NY(128). **64:**MN(1197), NC(128).
G. stenospora. **16:**ID(128,3202), MT(128,3202). **36:**ID(128,3202). **46:**SW states(1179), AZ(128,1202), MT(3202). **53:**MN(128,1197), NY(128). **57:**MA(128). **64:**SC(128).
G. subalutacea. White rot. **3:**UT(2144). **16:**WA(3202). **20:**GA(491,1439). **46:**SW states(1179), AZ(128,1202), OR(128), WA(3202). **53:**MN(128,1197). **57:**NY(128).
Grifola frondosa. **46:**MT(3202,3832).
Guepiniopsis alpina. **1:**ID(3202). **46:**SW states(1179), AZ(1202).
G. buccina. **17:**CA(661). **46:**ID(3202), WA(3202).
G. chrysocomus. On limbs. **64:**GA(2642).
Gymnopilus amarissimus. On wood. **64:**FL(1542,2541).
G. croceoluteus. On fallen cones. **59:**NC(1303).
G. decurrens. **64:**CA(1542).
G. liquiritiae. **46:**WA(3202).
G. melleus. On stumps. **64:**AL(1542).
G. penetrans. On stumps. **64:**MA(1394), OR(3202).
G. sapineus. On wood. **46:**SW states(1179), AZ(1202). **64:**GA(491).
G. underwoodii. On wood. **64:**AL(1542), FL(1542), VA(1542).
Hapalopilus rutilans. White rot. **46:**SW states(1179), AZ(1202). **64:**AZ(1204).
Helicobasidium corticioides. Brown cubical pocket rot. **3:**UT(2144).
Henningsomyces candidus. **46:**Unk.(1179), ID(3202). **57:**NY(2456).
Heterobasidion annosum. White root and butt rot. **1:**ID(82,3202), MT(82). **8:**Range of host(82), IA(404). **13:**FL(166), GA(3036). **16:**W states(3584), ID(3055), MT(3202,3832), OR(1067,3599). **17:**CA(82,661). **20:**AR(3919), GA(491,902,1439), MO(1809), NC(1303), OH(1808). **21:**W states(3584), UT(3055). **23:**AL(901,1122), AR(3919), FL(901,2864), GA(901,903,904,1122,2864), LA(1122), MS(901,1122), SC(489,2864), TX(1122). **25:**ID(3055). **30:**CA(82). **36:**ID(82,538,3202), MT(82,3202,3832), OR(1067). **42:**Range of host(82). **43:**CA(3349). **46:**SW states(1179), W states(3584), CA(82), ID(3055), MT(568,3202,3832,3909), NM(82), OR(568,1067,3599), UT(82,3055), WA(3202). **48:**W states(3584), CA(82), NV(3055), OR(3599). **53:**NE states(82), ME(3642), MI(294), MN(82), NH(3243), NY(3837), VT(567). **54:**NE states(82), MA(1394), NC(82,3046). **57:**NC states(82), NE states(82), IA(69), NC(82,1303,1307). **58:**CT(82), MI(3521). **59:**AL(1122), AR(3919), GA(902,1122,1442,3035), LA(1122), MS(1122,2122), NC(1303), OK(26), TN(87), TX(1122), VA(360,3493). **62:**MD(82), NC(1303), OH(1808), VA(82,3493). **64:**NE states(82), NW states(82), FL(82), MS(2720), NC(3944).
Hohenbuehelia unguicularis. On twigs. **59:**NC(2421).
Hymenochaete agglutinans. **57:**PA(128).
H. cinnamomea. White rot. **3:**UT(2144).
H. pinnatifida. On wood. **59:**GA(491,1439).
H. tabacina. **16:**WA(3202). **46:**ID(3202).
H. tenuis. **46:**SW states(1179).
Hyphoderma argillaceum. **36:**ID(3202). **46:**SW states(1179), AZ(128,1202), NM(1190).
H. cristulatum. **46:**Unk.(1179), WA(3202).
H. definitum. **25:**AZ(1202). **46:**SW states(1179), AZ(1202).
H. medioburiense. **46:**AZ(1202).
H. pallidum. **46:**SW states(1179), AZ(1202).
H. pilosum. **46:**SW states(1179), AZ(128,1202).
H. praetermissum. Isolated from treated utility poles. **6:**AZ(1202). **46:**SW states(1179), AZ(128,1202), NM(1190), WA(3202). **53:**MA(656). **64:**E states(3996), E states(3996).
H. puberum. Isolated from treated utility poles. **46:**SW states(1179), NM(128,1190). **64:**E states(3996), E states(3996).
H. sambuci. On wood. **46:**WA(3202). **59:**GA(491,1439).
H. setigerum. On wood. **16:**WA(3202). **20:**GA(491,1439). **46:**SW states(1179), AZ(1202).
Hyphodontia stipata. **4:**OR(128).
Hypholoma fasciculare. **57:**NC(1307).
H. sublateritium. **46:**OR(3202).
Hypochniciellum molle. **46:**OR(3202), WA(3202).
H. subillaqueatum. **46:**SW states(1179), AZ(1202).
Hypochnicium bombycinum. **25:**AZ(1202). **46:**SW states(1179), AZ(1202), NM(1190).
H. geogenium. **46:**SW states(1179), AZ(128,1202), NM(1190).
H. lundellii. **46:**AZ(1202), NM(128).
H. punctulatum. On wood chips. **46:**SW states(1179), AZ(1202). **64:**SC(634).
H. sphaerosporum. **46:**SW states(1179), AZ(128,1202).
Hypochnopsis mustialaensis. **46:**SW states(1179), NM(1190).
Inonotus circinatus. White root and butt rot. **8:**Range of host(82). **13:**FL(166). **17:**CO(82). **23:**FL(3033), GA(355,3033), SC(3033). **36:**ID(82), MT(3832). **46:**SW states(1179), AZ(1181,1202). **53:**MN(82). **54:**PA(82). **57:**MN(82), NC(82), NY(82). **64:**NC states(82), NE states(82), NW states(82), FL(82).
I. dryadeus. **46:**CA(661).
I. tomentosus. White pocket heart rot. **16:**ID(538,1583,3202), MT(3202), OR(3202). **36:**ID(538,1583,3202), MT(3202). **46:**SW states(1179), AZ(1181,1202), OR(3202). **57:**NC(1303,1307).
Irpex lacteus. Isolated from treated utility poles and mine timbers. **46:**SW states(1179), NM(1190). **64:**E states(3996), Unk.(1006).
Ischnoderma resinosum. Stem rot. **16:**ID(538,1583,3202). **36:**ID(3202). **46:**Unk.(1179), ID(3202), OR(3202), WA(3202). **57:**NY(2962).
Junghuhnia collabens. **16:**ID(2168,3202), MT(2168,3202). **36:**ID(2168,3202), MT(128,2168,3202). **46:**SW states(1179), AZ(128,1202,2168), SD(2168). **64:**CO(2168), ID(2168), LA(2168).

J. luteoalba. **16:**ID(2168,3202). **36:**MT(3202). **46:**SW states(1179), AZ(128,1202,2168), OR(2168,3202). **54:**PA(2168). **57:**MA(128,2168), NY(2168). **58:**NY(2168). **59:**FL(2168). **64:**GA(2168), MD(2168), NC(128,2168), TN(2168), VA(2168).
J. nitida. **16:**MT(3832). **36:**MT(3832).
J. subfimbriata. On stumps. **64:**SC(634).
Kavinia alboviridis. **46:**SW states(1179), NM(128,1190).
K. himantia. **46:**SW states(1179), AZ(1202), NM(1190).
Laetiporus sulphureus. Brown cubical rot. **16:**ID(3202). **36:**ID(3202). **46:**Unk.(1179), MT(82,3202,3832). **53:**MN(82). **57:**WI(82). **64:**Range of host(82).
Laurilia sulcata. On wood. **16:**ID(538,1583). **36:**ID(3202). **46:**ID(3202). **64:**MI(2862).
Laxitextum bicolor. **59:**FL(543).
Lentinellus cochleatus. **64:**NC(3944).
L. montanus. **16:**NW states(3202), W states(2423), CO(1566).
L. ursinus. **46:**NW states(3202), SW states(1179), W states(2423), AZ(1202).
Lentinus strigosus. **64:**AR(2675), LA(2675), MS(2675).
Lenzites sp. On wood. **64:**Range of host(82).
Leptoporus mollis. **16:**MT(3202,3832). **46:**SW states(1179), AZ(1202), NM(1190).
Leptosporomyces galzinii. **46:**SW states(1179), AZ(128,1202), NM(1190).
Leucogyrophana arizonica. On wood. **46:**AZ(1232). **62:**Unk.(1232). **64:**Unk.(1232).
L. mollusca. **6:**AZ(1202). **16:**MT(3202,3832). **25:**AZ(1202). **36:**ID(128), MT(128,3202,3832). **46:**SW states(1179), AZ(1202), NM(128,1190). **64:**VT(128).
L. montana. On wood. **16:**Unk.(1232). **36:**Unk.(1232), ID(3202). **57:**Unk.(1232).
L. olivascens. **46:**SW states(1179), AZ(128,1202), NM(1190). **64:**NC(3944).
L. pinastri. **21:**NM(1190). **46:**SW states(1179), AZ(1202).
L. sororia. **64:**MD(128).
Lindtneria leucobryophila. **53:**MN(1197).
Marasmius androsaceus. On needles. **52:**CA(81). **64:**WA(3202).
M. plicatulus. **64:**WA(3202).
M. rotula. **46:**AZ(1202).
M. thiersii. **64:**CA(856).
Marasmius sp. On needles. **59:**TX(652).
Megacollybia platyphylla. **46:**SW states(1179), AZ(1202).
Melanotus textilis. On limb. **38:**CA(2938).
Meripilus sumstinei. On wood. **59:**NC(1303).
Meruliopsis albostramineus. **36:**ID(128). **46:**AZ(128).
M. ambiguus. On wood. **16:**ID(128), MT(128). **20:**GA(491,1439). **25:**AZ(1202). **59:**GA(491). **64:**FL(634), NC(3944).
M. bellus. On rotten wood. **4:**CA(642). **16:**WA(3202).
M. corium. **38:**CA(128). **46:**SW states(1179), AZ(1202), NM(1190).
M. hirtellus. On wood. **46:**SW states(1179), AZ(1178,1202).
M. taxicola. On wood. **4:**OR(3202). **16:**MT(3202,3832). **36:**ID(3202), MT(3202,3832). **46:**SW states(1179), AZ(1202), CA(128,642), ID(3202), WA(3202). **59:**GA(491,1439), TX(128).
Meruliporia incrassata. Decay of structural timber. **46:**ID(3202). **64:**Range of host(82), NC(1303).
Merulius sp. Wood rot. **46:**ID(3202). **64:**Range of host(82).
Micromphale perforans. **46:**WA(3202).
Microporellus obovatus. **64:**GA(1439).
Mucronella aggregata. **46:**SW states(1179), AZ(128,1202), WA(3202).
Mycena citrinomarginata. **46:**WA(3202).
M. clavicularis. On dead needles. **62:**VA(3717).
M. intertexta. On dead wood and needles. **64:**SC(3326).
M. plicosa. On dead needles. **64:**MI(3326).
M. subincarnata. On dead needles. **57:**NY(3326).
Mycoacia meridionalis. **42:**MS(443). **59:**FL(443), MS(443). **64:**FL(443), GA(443), MS(443).
Mycorrhaphium adustum. On dead limbs. **20:**NC(1303).
Neolentinus lepideus. Brown cubical sapwood rot. **1:**CA(82,642). **6:**AZ(1202). **8:**MN(82). **16:**ID(3202), MT(3202). **17:**AZ(82). **20:**NC(1303). **36:**ID(82,3202). **46:**Range of host(82), SW states(1179), AZ(1202), CA(642), ID(3202), OR(3202). **48:**OR(82,3202). **57:**Range of host(82). **64:**Range of host(82), GA(491).
N. ponderosus. On wood. **17:**CA(661). **36:**CA(661). **46:**SW states(1179), AZ(1202), ID(2417,3202).
Nigroporus vinosus. On wood. **20:**GA(491,1439). **59:**NC(1303).
Odonticium romellii. **16:**ID(128,3202). **36:**ID(128,3202). **46:**ID(128,3202).
Oliveonia subviolacea. On bark. **57:**MA(128).
Osteina obducta. Brown butt rot. **17:**CO(82). **36:**ID(3202).
Oxyporus corticola. **46:**SW states(1179), AZ(128,1202).
Panellus pusillus. **57:**SC(1307).
Parmastomyces transmutans. **46:**SW states(1179), AZ(1202), NM(1190).
Paullicorticium pearsonii. **46:**SW states(1179), AZ(1202).
Paxillus atrotomentosus. Brown root rot. **8:**WI(803).
Peniophora cinerea. **36:**ID(3202). **46:**Unk.(1179), ID(3202).
P. pini subsp. *duplex*. On wood. **40:**NY(3860). **46:**MA(3860). **54:**CT(3860), MA(3860), NY(3860), PA(3860). **57:**MA(3860), PA(3860). **58:**MA(3860), PA(3860). **59:**GA(491,1439,3860). **62:**MD(3860), PA(3860). **64:**FL(3860), NJ(3860).
P. pseudopini. **46:**MA(3860). **53:**NY(3860). **57:**MA(3860), PA(3860). **64:**MA(3860).
Peniophora sp. Wood rot. **64:**Range of host(82).
Perenniporia medulla-panis. **16:**ID(3202). **36:**ID(3202), MT(3202,3832), OR(3202). **46:**MT(3202).
P. subacida. White rot. **16:**CA(82), ID(538,1583,3202), WA(3202). **36:**ID(82,538,3202). **46:**SW states(1179), AZ(128,1202), ID(3202), MT(3832), NM(128,1190). **53:**Lake States(82), MN(128). **57:**NE states(82). **59:**NC(1303). **62:**MD(82), VA(82,638). **64:**Range of host(82), GA(491,1439), MI(2862).
P. tenuis. **46:**Unk.(1179), AZ(128,1202), MT(3202,3832).
Phaeolus schweinitzii. Brown cubical butt rot, root rot. **1:**MT(82,3202). **2:**CO(82). **6:**AZ(1202). **7:**CO(82). **8:**Range of host(82). **13:**FL(166). **16:**ID(1583,3202), MT(3202), OR(3202). **17:**Range of host(82), CA(661), MT(3832). **19:**Range of host(82). **23:**FL(287). **26:**Range of host(82), ID(3202). **30:**Range of host(82). **31:**Range of host(82). **33:**Range of host(82). **34:**Range of host(82). **36:**ID(82,1583,1998,3202), MT(82,3202,3832), WA(3202). **42:**Unk.(82). **46:**SW states(1179), AZ(1202,2268), CA(82,661), ID(82,3202), MT(82,3202,3832), OR(82,3202), SD(82). **47:**Range of host(82). **53:**Range of host(82). **57:**Range of host(82), NY(3790). **58:**NY(82). **59:**GA(1439), NC(1303). **60:**Range of host(82). **62:**MD(82), VA(82,3493). **64:**Range of host(82), GA(491), LA(2675), MS(2720), NC(3944).
Phanerochaete affinis. White rot. **16:**WA(3202). **46:**Unk.(1179), MT(128,433,3202), WA(3202).
P. allantospora. **46:**AZ(1202).
P. burtii. White rot. **42:**FL(433). **46:**AZ(128), MT(128). **59:**FL(433).
P. cacaina. White rot. **46:**SW states(1179), AZ(433).
P. carnosa. White rot. **36:**ID(128,424), MT(128), OR(128). **46:**SW states(1179), AZ(128,1202), MT(128), NM(1190), SD(128). **47:**AZ(433). **54:**NH(128). **57:**MA(128,433), MN(128,433).
P. chrysorhiza. White rot. **59:**MS(433).
P. flavido-alba. Isolated from treated utility poles. **46:**AZ(128). **64:**E states(3996), E states(3996).
P. gigantea. White rot. **8:**MI(433). **16:**CA(81), ID(3202). **20:**GA(491,1439). **23:**GA(904). **32:**AZ(128). **33:**AZ(1202). **36:**ID(3202). **46:**SW states(1179), AZ(128,1202), NM(128,1190), OR(3202). **52:**CA(81). **53:**WI(433). **57:**Range of host(82). **64:**Range of host(82), GA(1439), LA(2675), NC(433,3944), VA(433).
P. sanguinea. White rot. **8:**MI(433). **46:**SW states(1179), AZ(128,1202), CA(642), ID(3202), NM(1190), WA(3202). **53:**MA(656), MN(128,1197). **54:**NY(128). **64:**FL(433), NY(128).
P. sordida. White rot, isolated from treated utility poles. **16:**FL(433), ID(128), WA(3202). **46:**SW states(1179), AZ(1202), ID(3202), WA(3202). **59:**FL(433). **64:**E states(3996), NY(128).
P. velutina. White rot. **12:**AZ(1202). **25:**AZ(1202). **46:**SW states(1179), AZ(128,433,1202), NM(1190). **53:**WI(433). **57:**MI(128).
Phellinus ferrugineofuscus. **16:**ID(3202), MT(128,3202,3832), WA(3202). **36:**ID(3202). **46:**Unk.(1179), AZ(1202), ID(3202), MT(3202), WA(3202).
P. ferruginosus. **16:**ID(3202). **64:**NC(3944).
P. gilvus. Wood rot. **64:**Range of host(82), AR(2675), LA(2675), MS(2675).
P. nigrolimitatus. White pocket rot. **16:**ID(3202). **17:**WY(2924).

36:ID(82,538,1583,3202). **46:**NW states(82).
P. pini. White pocket heart rot. **1:**Range of host(82), ID(82,3202), MT(82,3202,3832), WA(3202). **2:**AZ(2144). **6:**AZ(82,1202). **8:**Range of host(82), MN(291). **10:**MS(2720). **12:**Range of host(82). **13:**FL(82). **16:**ID(538,1583,3202), MT(3202,3832), WA(3202). **17:**CO(82), ID(82), MT(82), OR(82), WY(2924). **20:**GA(491,1439), MS(2720), NC(1303). **21:**Range of host(82). **26:**MT(82,3202,3832). **30:**OR(3202). **36:**Range of host(82), ID(82,128,538,1583,1998,3202), MT(82,3202,3832), WA(3202). **42:**MS(2720). **46:**SW states(1179), AZ(1202), ID(3202), MT(3202,3832), NM(1190), OK(618), OR(3202), SD(378), WA(3202). **53:**Range of host(82). **54:**Range of host(82), GA(491,1439). **57:**Range of host(82), NC(1303). **58:**Range of host(82). **59:**GA(491,1439), MS(2720), NC(1303). **62:**E states(82), TN(82). **64:**Range of host(82), NC(3944).
P. repandus. On wood. **36:**ID(2047,3202), WA(3202).
[*P. robustus*]. **10:**LA(82). **64:**WA(3202).
P. torulosus. Butt rot, wood rot. **6:**AZ(1189,1202). **46:**SW states(1179), AZ(1189,1202).
P. viticola. White rot. **1:**MT(82,3202). **4:**OR(3202). **16:**ID(3202), MT(128,3202), WA(3202). **17:**Range of host(82), ID(82), MT(82). **20:**GA(1439). **36:**ID(3202), MT(82,3202,3832). **46:**Range of host(82), SW states(1179), MT(3202,3832), ND(378), NM(1190). **64:**N states(82), Range of host(82), MT(128).
P. weirii. Butt rot, root rot. **16:**OR(1067,1072), WA(1072). **36:**ID(1583), OR(1067), WA(3202). **46:**Unk.(1179), OR(1067).
Phlebia albida. **25:**AZ(1202). **46:**SW states(1179), AZ(128,1202), WA(3202).
P. brevispora. On wood. **24:**FL(2556,2557). **64:**E states(3996), E states(3996).
P. incarnata. **46:**SW states(1179), AZ(1202).
P. livida. On wood. **16:**MT(2165). **46:**SW states(1179), AZ(128,1202), NM(1190). **57:**NC(1303,2165). **64:**NC(1303,2165).
P. merismoides. Isolated from treated utility poles. **16:**ID(3202). **36:**ID(3202). **64:**E states(3996), E states(3996).
P. murrillii. On limbs. **42:**FL(643).
P. rufa. Isolated from treated utility poles. **64:**E states(3996), E states(3996).
P. subserialis. On wood. **16:**MT(2556). **20:**FL(2556). **25:**AZ(1202,2556). **33:**AZ(1202). **64:**E states(3996), E states(3996).
P. tremellosus. Wood rot. **36:**WA(3202). **46:**MT(3202,3832). **64:**Range of host(82), NC(128), OH(128).
Phlebia sp. **46:**AZ(662).
Phlebiella tulasnelloidea. **57:**NY(2456).
P. vaga. White rot. **3:**UT(2144). **25:**AZ(1202). **33:**AZ(1202). **36:**ID(3202). **46:**SW states(1179), AZ(128,1202), NM(1190). **57:**NY(128). **59:**AL(128).
Pholiota adiposa. Heart rot, sometimes of living trees. **1:**ID(82), MT(82).
P. bakerensis. **46:**SW states(1179), AZ(1202).
P. decorata. **46:**SW states(1179), AZ(1202).
P. spumosa. On wood. **59:**NC(1303).
Phyllotopsis nidulans. On wood. **20:**GA(1439). **46:**OR(3202). **59:**NC(1303). **64:**NC(3944).
Physisporinus sanguinolentus. On living trees. **16:**ID(3202). **36:**ID(3202), MT(3202,3832).
P. vitreus. Isolated from treated utility poles. **16:**ID(128,3202), MT(3202,3832). **64:**E states(3996), E states(3996).
Piloderma bicolor. **16:**ID(3202). **36:**WA(3202). **46:**SW states(1179), AZ(1202), NM(128,1190).
P. byssinum. White rot. **3:**UT(2144). **46:**SW states(1179), AZ(128,1202).
Pleurotellus hypnophilus. **1:**ID(3202).
Pluteus atricapillus. **46:**SW states(1179), AZ(1202). **64:**WA(3202).
P. fuliginosus. On stumps. **57:**NY(3336).
Polyporus badius. **16:**ID(3202). **46:**Unk.(1179), ID(3202).
P. melanopus. **16:**ID(3202), MT(3202).
P. varius. **16:**ID(3202), MT(3202), OR(3202).
Porotheleum fimbriatum. On wood. **64:**MI(2945).
Postia balsamea. **16:**ID(3202). **36:**ID(3202), MT(3202), WA(3202). **46:**SW states(1179), AZ(1202), MT(3202), NM(1190).
P. caesia. On wood. **4:**CA(642). **16:**ID(3202), WA(3202). **20:**GA(491,1439). **36:**ID(3202). **46:**SW states(1179), ID(3202), NM(1190), OR(3202), WA(3202). **59:**NC(1303). **64:**NC(3944).
P. ceriflua. On wood. **46:**ID(3202). **64:**Unk.(1205).
P. floriformis. **36:**ID(3202).
P. fragilis. On wood. **16:**ID(3202), WA(3202). **36:**ID(3202), MT(3202,3832). **46:**SW states(1179), AZ(1202), ID(3202), MT(3202), NM(1190), OR(3202), WA(3202). **57:**NC(1303,1307).
P. guttulata. Brown cubical rot. **16:**ID(3202). **36:**ID(3202). **46:**Unk.(1179), OR(3202). **59:**NC(1303). **64:**Range of host(82).
P. leucomallela. On wood. **64:**W states(1205).
P. leucospongia. Brown cubical rot. **1:**CA(642), ID(3202), MT(3202), OR(3202). **3:**UT(2144). **16:**ID(3202), MT(3202,3832), OR(3202), WA(3202). **17:**CA(81,661). **26:**ID(3202). **36:**CA(661). **46:**AZ(1202).
P. perdelicata. **16:**ID(3202), WA(3202). **36:**ID(3202). **46:**Unk.(1179), WA(3202). **64:**ID(3202).
P. placenta. Brown cubical rot. **16:**MT(128), WA(128,3202). **36:**ID(128,3202), MT(3202), WA(3202). **46:**AZ(1202), NM(1190). **64:**E states(3996), Unk.(82).
P. salmonicolor. **36:**ID(128,3202). **46:**SW states(1179), AZ(1202). **64:**FL(128).
P. sericeomollis. Stem rot. **36:**ID(538,1583). **46:**Unk.(1179), ID(3202), MT(3202), OR(3202). **64:**ID(3202).
P. stiptica. **36:**MT(3202,3832).
P. tephroleuca. **46:**SW states(1179), AZ(1202).
P. undosa. **16:**ID(3202), MT(397,3202). **36:**ID(397,3202). **46:**SW states(1179), AZ(1202), ID(3202). **54:**CT(397). **57:**CT(397), NY(397).
Pseudohydnum gelatinosum. **64:**ID(3202), NC(3944).
Pseudomerulius aureus. Wood rot. **16:**ID(128,3202). **25:**AZ(1202). **36:**ID(128), MT(128,3202,3832). **46:**SW states(1179), AZ(128,1202), ID(128), NM(1190), WA(3202). **64:**Range of host(82), MD(651).
P. curtisii. On wood. **59:**NC(1303).
Pseudotomentella atrofusca. On wood. **46:**AZ(1202,2037).
P. griseopergamacea. On wood. **53:**NY(128,2037). **57:**NY(2037).
P. mucidula. **46:**SW states(1179), AZ(1202).
P. nigra. On wood. **57:**NY(2032).
P. tristis. **16:**MT(3202). **46:**SW states(1179), AZ(1202), MT(128).
Pycnoporellus alboluteus. **16:**WA(3202). **26:**MT(3202). **36:**ID(3202), MT(3202), WA(3202). **46:**SW states(1179), AZ(1202), OR(3202). **64:**ID(3202), MT(3202).
P. fulgens. Wood rot. **36:**ID(3202). **64:**Range of host(82).
Pycnoporus cinnabarinus. Wood rot. **21:**AZ(1202). **33:**AZ(1202). **46:**SW states(1179), AZ(1202). **64:**Range of host(82), NC(3944).
P. sanguineus. On wood. **23:**FL(2319). **64:**GA(491,1439).
Ramaricium polyporoideum. On wood. **58:**NY(128).
Rectipilus fasciculatus. **64:**VA(128).
R. idahoensis. **36:**ID(3202).
Resinicium bicolor. On wood. **16:**ID(538,1583), MT(128). **36:**ID(538,1583,1998). **46:**SW states(1179), AZ(128,1202). **53:**MA(656). **57:**NY(128). **64:**NC(128).
R. chiricahuaensis. White rot. **46:**SW states(1179), AZ(128,1188,1202), NM(128,1188,1190).
R. furfuraceum. **36:**ID(2116,3202), WA(3202). **46:**SW states(1179), AZ(1202), WA(3202). **53:**MA(656). **57:**MA(656), MN(1197). **64:**NC(2116).
Resupinatus applicatus. **46:**SW states(1179), AZ(1202). **64:**NC(3944).
Rigidoporus crocatus. **46:**SW states(1179), NM(1190).
[*R. vinctus*]. **20:**FL(3187).
Schizophyllum commune. On wood. **20:**NC(1303). **59:**NC(1303). **64:**Unk.(82).
Schizopora paradoxa. **16:**ID(128), MT(128), OR(3202).
Scytinostroma arachnoideum. **46:**SW states(1179), AZ(128), NM(128).
S. fallax. **59:**FL(443), MS(443). **64:**MS(443).
S. galactinum. Root rot. **46:**SW states(1179), AZ(1202), MT(128), NM(128,1190), SD(128). **53:**MN(1197). **57:**CT(2090). **59:**LA(2675). **64:**ID(128), MN(1197), TX(128).
S. ochroleucum. **46:**AZ(1202).
Scytinostromella heterogenea. **46:**SW states(1179), NM(128,1190).
Serpula himantioides. On logs. **6:**AZ(1202). **36:**MT(3202,3832). **46:**SW states(1179), AZ(128,1202), CA(642), NM(128,1190), WA(638,3202). **64:**MD(1698), NY(128), VA(128).
S. lacrimans. Dry rot of structural timber, on leaves and wood. **16:**MT(3202,3832). **46:**CA(642). **64:**Range of host(82), SC(634).
Sistotrema brinkmannii. On wood. **46:**SW states(1179), AZ(128,1202). **57:**GA(491,1439).

S. confluens. On fallen needles. **16:**MT(128). **64:**WI(128).
S. subtrigonospermum. **46:**AZ(1202).
Sistotrema sp. Isolated from treated utility poles. **64:**E states(3996), E states(3996).
Skeletocutis amorpha. White rot. **4:**OR(82,3202). **16:**AK(82), ID(82,3202), MT(82,3202,3832). **20:**GA(491,1439), MS(82,2720). **33:**AZ(1202). **36:**ID(3202), MT(3202). **46:**SW states(1179), W states(82), AZ(1202), ID(3202), MT(3202,3832), OR(3202), SD(378). **50:**Unk.(82). **54:**PA(82). **59:**SC(82). **64:**Range of host(82).
S. nivea. **59:**TX(652).
S. stellae. **16:**WY(128). **46:**SW states(1179), AZ(128).
S. subincarnata. **16:**ID(128,3202), MT(3202), OR(3202). **36:**ID(3202). **46:**SW states(1179), AZ(128,1202), NM(1190), OR(3202).
Sparassis crispa. Yellow-mottled root rot. **6:**AZ(1202). **36:**NW states(82,3202), WA(82). **46:**SW states(1179), AZ(1202). **64:**NC(3944).
Sphaerobasidium minutum. On wood. **16:**OR(128). **46:**SW states(1179), AZ(1202), OR(128).
Steccherinum fimbriatum. **46:**SW states(1179), AZ(1202). **53:**NY(128).
S. laeticolor. Isolated from treated utility poles. **64:**E states(3996), E states(3996).
S. subcrinale. **57:**MA(128).
Stella sp. Black mildew on needles. **57:**NC(1303).
Stereum sanguinolentum. Heart rot. **2:**CO(2144). **8:**CT(543). **16:**CO(543), ID(3202), WA(3202). **36:**ID(82,1998,3202). **46:**SW states(1179), AZ(1202), ID(3202), MA(543), WA(3202). **53:**CT(543), MA(543), NH(543), NY(543). **54:**MA(543). **57:**Range of host(82), CT(543), ID(82,3202), MA(543), ME(543), NH(543), VT(543). **64:**NC states(82), NE states(82), NW states(82), NY(543).
Tapinella panuoides. **16:**OR(3202). **33:**AZ(1202). **46:**SW states(1179), AZ(1202).
Thanatephorus cucumeris. On bark. **64:**SC(634).
Thelephora terrestris. Seedling blight. **8:**MI(82). **16:**MT(3202). **17:**MT(82), WY(82). **36:**ID(3202), WA(82,3202). **46:**SW states(1179), AZ(1202), ID(3202), OR(3202), SD(2243). **57:**ME(82), NH(82), OH(82). **59:**NC(1303). **64:**ND(378).
Tomentella atrorubra. On logs. **57:**NY(2030). **64:**MA(2030,2039), MD(2039), NC(2030,2039), NY(2039).
T. avellanea. **64:**NY(2039).
T. botryoides. **46:**SW states(1179), AZ(1202). **64:**AZ(2039), MD(2039), NY(2039).
T. bresadolae. On wood. **46:**NM(1190). **57:**TN(2035). **64:**NM(2039), TN(2039).
T. brunneorufa. On wood. **64:**MD(2039).
T. bryophila. **46:**SW states(1179), NM(1190). **64:**NM(2039).
T. carbonaria. On wood. **16:**NM(2040).
T. chlorina. **6:**AZ(1202). **21:**AZ(1202). **46:**SW states(1179), AZ(1202), NM(1190). **64:**AZ(2039), NM(2039).
T. clavigera. On wood. **64:**NJ(2039).
T. coerulea. **46:**MT(3202). **64:**MT(2039).
T. crinalis. **16:**MT(128,3202). **64:**MT(2039).
T. ferruginea. **46:**SW states(1179), AZ(1202), NM(1190). **59:**SC(128). **64:**NM(2039), NY(2030,2039).
T. fuscoferruginosa. **64:**TN(2039).
T. griseo-umbrina. On wood. **64:**AZ(2032,2039).
T. lateritia. On wood. **25:**AZ(1202). **46:**SW states(1179), AZ(1202), NM(1190). **64:**AZ(2032,2039), NM(2039), NY(2039), TN(2039).
T. muricata. On wood. **64:**NJ(2039,3746).
T. neobourdotii. On wood. **64:**AZ(2039), MI(2039), NY(2033,2039).
T. ochracea. **46:**AZ(1202).
T. olivascens. **57:**MA(128). **64:**NC(2039), NY(2039), PA(2039), SC(2039), TN(2039).
T. pilosa. **46:**SW states(1179), AZ(1202). **64:**MT(2039), NY(2039).
T. punicea. **64:**NY(128), TN(2039).
T. ramosissima. On wood. **46:**SW states(1179), AZ(1202), NM(1190). **64:**AZ(2039), MT(2039), NM(2039), NY(2039), SC(2039,3746).
T. rubiginosa. **64:**NY(2039).
T. ruttneri. On wood. **64:**AR(2035), NY(2035).
T. sublilacina. **46:**NM(1190). **64:**MA(2039), MD(2039), NM(2039), NY(2039), TN(2039).
T. terrestris. **46:**SW states(1179), AZ(1202), NM(1190). **64:**AZ(2039), MT(2039), NM(2039), NY(2039).
T. umbrinospora. **64:**AL(2030,2039).
T. violaceofusca. On bark. **46:**SW states(1179). **64:**MD(2039), NJ(965).
T. viridescens. **64:**AZ(2039), NM(2039).
T. viridula. **64:**NM(2039).
Tomentella sp. **36:**ID(3202).
Tomentellastrum badium. **46:**SW states(1179), AZ(1202).
Tomentellina fibrosa. **3:**UT(2144). **46:**SW states(1179), AZ(1202). **64:**AL(128).
Tomentellopsis zygodesmoides. On wood. **64:**Unk.(458).
Trametes hirsuta. **36:**ID(3202). **46:**Unk.(1179), MT(3202).
T. versicolor. Wood rot. **1:**ID(3202). **16:**OR(3202). **36:**ID(3202), MT(82,3202). **46:**ID(3202). **64:**E states(3996), E states(3996), Range of host(82).
Trechispora amianthina. **46:**SW states(1179), NM(1190).
T. cohaerens. **53:**MN(1197).
T. farinacea. White rot. **3:**UT(2144). **12:**AZ(1202). **25:**AZ(1202). **46:**SW states(1179), AZ(128,1202), MT(128), NM(1190). **53:**NY(128). **58:**NY(128).
T. lunata. **53:**MA(656).
T. microspora. **46:**SW states(1179), AZ(1202).
T. mollusca. On wood. **36:**ID(3202). **46:**SW states(1179), AZ(1202), ID(3202), NM(1190). **48:**CA(661).
T. pallidoaurantiaca. On wood. **46:**SW states(1179), NM(1188,1190). **58:**NY(1188).
T. praefocata. **46:**SW states(1179), AZ(1202).
Trichaptum abietinum. White rot. **1:**ID(3202), MT(3202). **16:**ID(3202), MT(3202), OR(3202), WA(3202). **20:**GA(1439). **21:**AZ(1202), NM(1190). **26:**MT(3202). **30:**OR(3202). **33:**AZ(1202). **36:**ID(3202), MT(3202), WA(3202). **38:**CA(81). **46:**SW states(1179), AZ(1202), ID(3202), MT(3202), OR(3202), WA(3202). **53:**MI(665). **59:**LA(658), TX(652).
T. biforme. Isolated from treated utility poles. **64:**E states(3996), NC(3944).
T. byssogenum. On wood. **64:**S states(2681).
T. fuscoviolaceum. On wood. **64:**Unk.(1205).
T. laricinum. **46:**Unk.(1179). **48:**CA(661).
Tricholomopsis rutilans. On wood. **59:**NC(1303).
T. thompsoniana. **46:**SW states(1179).
Tubulicrinis accedens. **57:**MA(3859).
T. angustus. **16:**OR(3202,3859). **57:**RI(3859).
T. calothrix. **12:**AZ(1202). **32:**AZ(128). **33:**AZ(1202). **46:**SW states(1179), AZ(1202). **57:**MA(3859).
T. chaetophorus. **36:**ID(3202). **46:**SW states(1179), AZ(1202).
T. glebulosus. On wood. **46:**SW states(1179), AZ(1202), CA(642), NM(128,1190), WA(3202).
T. inornatus. On wood. **16:**OR(1714,3202).
T. prominens. **36:**ID(1716,3202). **46:**SW states(1179), AZ(1202).
T. subulatus. **25:**AZ(1202). **46:**SW states(1179), AZ(128,1202), NM(128,1190). **53:**MA(656). **57:**MA(656).
Tulasnella allantospora. **46:**SW states(1179), AZ(1202).
T. calospora. On wood. **17:**OR(128).
T. fuscoviolacea. **1:**ID(3202).
T. pinicola. On wood. **64:**MA(128).
T. violea. **16:**ID(3202). **46:**SW states(1179), AZ(1202), WA(3202).
Tyromyces chioneus. On wood. **16:**ID(3202). **20:**GA(491,1439). **36:**ID(3202). **59:**NC(1303).
Uthatobasidium fusisporum. **54:**PA(128).
Vararia athabascensis. White rot. **3:**UT(2144).
V. boreale. **1:**MT(3202). **16:**MT(128), WA(3202). **36:**ID(128,3202), MT(128,3202). **46:**SW states(1179), AZ(128,1202), ID(3202), MT(128,3202), NM(128,1190), SD(128). **53:**MN(128,1197).
V. investiens. **1:**ID(128,3202). **36:**ID(3202). **57:**NY(128).
V. phyllophila. **36:**ID(3202).
V. racemosa. **16:**MT(128).
Vararia sp. **59:**LA(128).
Veluticeps berkeleyi. Brown rot. **25:**AZ(1198,1202,2265). **46:**SW states(1179), AZ(128,1198,1202), NM(1190,1198), SD(1198), WA(1198,3202).
Vesiculomyces citrinus. **46:**SW states(1179), AZ(128,1202), NM(1190).
Wolfiporia cocos. Brown cubical butt rot, root rot. **20:**GA(491,1439), SC(1241). **36:**ID(793,3202). **46:**Unk.(1179). **64:**SE states(82), NC(3944).
Wrightoporia lenta. **46:**AZ(1202).

Xeromphalina campanella. **46:**SW states(1179), AZ(1202).
X. cauticinalis. On cone debris. **64:**MN(3795).
[*Xerula radicata*]. Stem rot. **36:**ID(538).
Xylobolus frustulatus. On stumps. **57:**NC(1303,1307), SC(1307).

Deuteromycotina - Hyphomycetes

Acremonium sp. Isolated from treated southern yellow pine utility poles. **64:**E states(3996), E states(3996).
Acrodontium intermissum. Isolated from treated southern yellow pine utility poles. **64:**E states(3996), E states(3996).
Alternaria alternata. Wood stain. **16:**CO(2247). **64:**E states(3996), LA(3732).
Alternaria sp. Isolated from basal cankers. **57:**NY(1644).
Ambrosiella ferruginea. **36:**OR(3202).
Arthrographis cuboidea. Isolated from treated southern yellow pine utility poles. **64:**E states(3996), E states(3996).
Aspergillus sp. Isolated from treated southern yellow pine utility poles. **64:**E states(3996), E states(3996).
Atichia glomerulosa. **46:**WA(3202).
Aureobasidium pullulans. Needle blight following insect damage, gray stain of wood, isolated from stored pulpwood and treated southern yellow pine utility poles. **16:**CO(2247). **53:**NE states(82), WI(1005). **64:**E states(3996), S states(82).
Aureobasidium sp. Isolated from wood chips. **8:**WI(1005).
Berkleasmium concinnum. On wood. **64:**NC(1303).
Botryophialophora sp. **36:**ID(3202).
Botrytis cinerea. Seedling blight, gray mold. **28:**OR(82,3202). **46:**WA(3202). **52:**OR(3202). **64:**Range of host(82).
Botrytis sp. Damping-off, root rot. **16:**CO(2247). **57:**DE(702).
Cephalosporium sp. **16:**CO(2247).
Chalara acuaria. On needles. **64:**Unk.(2550).
Cirrenalia macrocephala. On driftwood in ocean. **59:**NC(1941).
Cladosporium cladosporioides. Isolated from treated southern yellow pine utility poles. **64:**E states(3996), E states(3996).
C. herbarum. **16:**CO(2247). **26:**WY(2924).
Cladosporium sp. Isolated from basal cankers. **36:**ID(3202). **57:**NY(1644).
Confistulina hepatica. **62:**MD(3462).
Conoplea elegantula. On fallen cones. **59:**NC(1303).
Costantinella micheneri. **16:**ID(3202).
Curvularia lunata. Isolated from treated southern yellow pine utility poles. **64:**E states(3996), E states(3996).
Cylindrocarpon gillii. **16:**CO(2247).
Cylindrocladium scoparium. Seedling blight, root decay and rot. **46:**MT(3202). **53:**MN(1694). **54:**NJ(82), PA(82). **57:**NC(1303), NJ(82), WV(1694). **58:**MD(82), NJ(82), PA(82), WA(82,3202).
Doratomyces stemonitis. **46:**ID(3202).
Epicoccum nigrum. Isolated from treated southern yellow pine utility poles. **16:**CO(2247). **64:**E states(3996), E states(3996).
Epicoccum sp. Isolated from basal cankers. **57:**NY(1644).
Eversia subopaca. On rotten wood. **29:**WV(3149).
Exophiala jeanselmei var. *heteromorpha*. Isolated from treated southern yellow pine utility poles. **64:**E states(3996), E states(3996).
E. mansonii. Isolated from treated southern yellow pine utility poles. **64:**E states(3996), E states(3996).
Fumago sp. Sooty mold. **58:**NC(1303). **59:**NC(1303).
Fusarium arthrosporioides. Seedling blight. **64:**Range of host(82).
F. avenaceum. Preemergence seedling blight. **53:**PA(82). **64:**Range of host(82).
F. lateritium. Isolated from cankers. **57:**VA(112).
F. lateritium f. sp. *pini*. Pitch canker. **10:**FL(1533), GA(1533), NC(82), SC(1533), VA(82). **20:**NC(1533), SC(1533). **23:**FL(3138). **42:**FL(1533), SC(1533). **54:**NC(1533). **62:**FL(1533), GA(1533), NC(1533), SC(1533), VA(1533,3493).
F. moniliforme. Seedling blight. **64:**Range of host(82).
F. oxysporum. Seedling blight, damping-off, hypocotyl and root rot. **30:**CA(411). **48:**CA(82). **53:**NY(3283), PA(82). **59:**NC(1303). **64:**Range of host(82).
F. oxysporum f. sp. *vasinfectum*. Seedling blight. **64:**Range of host(82).
F. roseum. **46:**WA(3202).
F. sambucinum. Seedling blight. **64:**Range of host(82).
F. solani. Seedling blight. **64:**Range of host(82).
F. sporotrichioides. Seedling blight. **64:**Range of host(82).
F. subglutinans. Shoot dieback, seedling pitch canker. **13:**FL(290). **15:**FL(290). **20:**NC(1303). **23:**FL(165,288,289), SC(1969). **42:**NC(1303), SC(1969). **46:**NC(1303). **56:**NC(1303,1993). **58:**NC(1303). **59:**NC(1303,1993), SC(196).
Fusarium sp. Preemergence seedling blight, damping-off, black root rot, seedling root rot. **10:**GA(82), SC(82). **23:**FL(3317). **42:**GA(82). **46:**Range of host(82). **53:**PA(82). **57:**Range of host(82), DE(702), NC(1533), NY(1644), PA(3858). **58:**PA(3858).
Geotrichum sp. Pink wood stain. **20:**S states(82).
Gliocladium roseum. Isolated from treated southern yellow pine utility poles. **46:**ID(3202). **64:**E states(3996), E states(3996).
G. virens. Isolated from treated southern yellow pine utility poles. **64:**E states(3996), E states(3996).
Gonatobotrys sp. Red-mottled heart rot. **16:**CO(2247).
Graphium rigidum. Gray stain of wood, on scales of cones. **48:**CA(81). **64:**S states(82).
G. rubrum. Gray wood stain. **64:**S states(82).
Graphium sp. Sapwood stain. **8:**MN(484). **53:**MN(484). **57:**MN(484).
Haplotrichum curtisii. On wood. **64:**PA(2131).
H. ramosissimum. On railroad ties. **64:**NJ(2131).
Helicosporium aureum. On fallen cones. **59:**NC(1303).
Heterosporium sp. **46:**ID(3202).
Hormiscium gelatinosum. Blue stain of wood. **64:**Unk.(82).
Hormoconis resinae. Isolated from treated southern yellow pine utility poles. **64:**E states(3996), E states(3996).
Hyalotrochophora lignitalis. On wood. **57:**MA(2510).
Illosporium coccinellum. On palings. **64:**SC(634).
I. conicola. On cones. **62:**DC(3209).
Leptodontium elatius. Isolated from treated southern yellow pine utility poles. **64:**E states(3996), E states(3996).
Leptographium abietinum. **36:**ID(1998).
L. antibioticum. **36:**ID(1998).
L. engelmanii. Wood stain associated with bark beetles. **16:**CO(783).
L. lundbergii. **36:**ID(1665,3202).
L. penicillatum. Isolated from roots. **16:**ID(2372). **36:**ID(1998). **46:**ID(2372).
L. procerum. Stem and root canker, root rot, wilt. **13:**FL(166). **39:**MN(2157), VA(2006). **53:**MN(2157), NY(3283). **57:**MN(2157), NY(1644,3283), PA(3670), VA(1635,2006,2007). **58:**MA(2921), VA(1635,2006,2007). **60:**MA(2921).
L. pyrinum. Wood stain associated with bark beetles. **46:**NM(788).
L. terebrantis. Blue stain, associated with *Dendroctonus terebrans* (Coleoptera). **58:**MA(1553,2921). **60:**MA(1553,2921).
L. wageneri. Black stain root disease. **21:**W states(1257), CO(1471,2022), NM(1471,3767). **30:**W states(1257). **34:**CA(1471,2396), NV(1471). **36:**ID(1998). **46:**W states(1257), CA(1256,2396), OR(1259). **48:**W states(1257), CA(2396).
L. wageneri var. *ponderosum*. Black stain root disease. **16:**OR(1471). **46:**CA(1471), ID(1471). **48:**CA(1471). **57:**MT(1471).
Leptographium sp. Sapwood stain. **8:**MN(484). **16:**OR(1005). **36:**ID(3202), MT(3202), WA(3202). **53:**MN(484). **57:**MN(484).
Lylea catenulata. On twigs. **59:**AL(1191,2483,2497).
Moniliella sp. Isolated from treated southern yellow pine utility poles. **64:**E states(3996), E states(3996).
Oidiodendron griseum. Isolated from treated southern yellow pine utility poles. **64:**E states(3996), E states(3996).
Paecilomyces farinosus. **46:**ID(3202).
P. marquandii. **46:**WA(3202).
P. variotii. Isolated from treated southern yellow pine utility poles. **64:**E states(3996), E states(3996).
Penicillium chrysogenum. **46:**WA(3202).
P. decumbens. **46:**WA(3202).
P. purpurogenum. Red lumber stain. **46:**Unk.(791).
Penicillium sp. Isolated from treated southern yellow pine utility poles. **16:**CO(2247). **64:**E states(3996), E states(3996).
Phaeococcus sp. Isolated from treated southern yellow pine utility poles. **64:**E states(3996), E states(3996).
Phialocephala dimorphospora. Isolated from treated southern yellow pine utility poles. **64:**E states(3996), E states(3996).
Phialophora fastigiata. Isolated from treated southern yellow pine utility poles. **64:**E states(3996), E states(3996).
P. olivacea. Isolated from treated southern yellow pine utility poles. **64:**E states(3996), E states(3996).
P. repens. Wood stain. **64:**FL(82,777), LA(82).
P. richardsiae. Wood stain. **64:**E states(3996), FL(82,777), LA(82,777).

Phialophora sp. Isolated from treated southern yellow pine utility poles. **64:**E states(3996), E states(3996).
Phymatotrichopsis omnivora. Root rot. **20:**TX(82).
Ptychogaster rubescens. Isolated from treated southern yellow pine utility poles. **64:**E states(3996), E states(3996).
Pullularia sp. Isolated from basal cankers. **57:**NY(1644).
Rhinocladiella atrovirens. Isolated from treated southern yellow pine utility poles. **64:**E states(3996), E states(3996).
Scopularia sp. **36:**ID(82), MT(82), WA(82).
Scoriomyces andersonii. **46:**MT(3202).
Scytalidium lignicola. Isolated from treated southern yellow pine utility poles. **64:**E states(3996), E states(3996).
Septonema sp. Isolated from treated southern yellow pine utility poles. **64:**E states(3996), E states(3996).
Spiniger meineckellum. Root rot. **46:**CA(82). **48:**CA(82).
Sporothrix sp. Isolated from treated southern yellow pine utility poles. **64:**E states(3996), E states(3996).
Sporotrichum bombacinum. **64:**DE(3543).
Stigmina sp. **46:**WA(3202).
Taeniolella rudis. **64:**NC(1680).
[*T. stilbospora*]. **16:**MT(3202).
Tetraploa aristata. On twigs. **64:**AL(2507).
Tilachlidium brachiatum. **46:**OR(3202).
Torula herbarum. **16:**CO(2247).
T. marginata. Seedling root rot. **42:**MS(2720).
Toxosporium camptospermum. Twig blight, on needles. **42:**FL(82,3547). **57:**NC(353).
Trichoderma harzianum. Isolated from treated southern yellow pine utility poles. **64:**E states(3996), E states(3996).
T. viride. **46:**WA(3202).
Trichoderma sp. Damping-off, root rot. **46:**WA(1071). **57:**DE(702). **64:**E states(3996), E states(3996).
Trichosporum pinicola. **36:**ID(3202).
Trichothecium roseum. **1:**CA(642).
Tuberculina maxima. **57:**MI(3991).
Verticicladiella sp. Root stain. **21:**AZ(3743), CO(3743). **34:**CA(3743). **36:**ID(1998). **46:**CA(3743). **48:**CA(3743). **57:**MT(2067,3202).
Volutella aciculosa. On leaves. **46:**CA(972).

Deuteromycotina - Coelomycetes

Aplosporella pini. **57:**NY(82).
Aplosporella sp. **7:**MT(82,3202).
Asteromella sp. Isolated from treated southern yellow pine utility poles. **64:**E states(3996), E states(3996).
Camaropycnis libocedri. On limbs. **34:**CA(3549).
Camarosporium pini. On twigs. **57:**PA(82).
Colletotrichum dematium. **46:**ID(3202).
Cryptocoryneum simmonsii. **46:**WA(3202).
Cryptosporium lunasporum. On living blister rust cankers. **36:**ID(82,3202).
Cylindrosporium acicola. **4:**CA(81).
Cytospora kunzei. Canker. **57:**CT(3781), ME(3781), NC(1303), NY(3781), VA(3781). **63:**MD(3781).
C. pinastri. Needle blight. **58:**ME(82), NJ(82), PA(82). **64:**ME(82), NJ(82).
Cytospora sp. Brown sapwood stain. **46:**WA(3020).
Diplodia megalospora. Blue-gray stain. **57:**NY(82). **59:**Gulf states(82). **62:**VA(82). **64:**E states(82), IL(82).
D. pinastri. Blue-gray stain. **64:**Range of host(82).
Discosia pini. On seedlings. **46:**NE(82).
Dothiorella pinastri. On live blister rust-cankered branch. **36:**ID(82,3202).
Dothiorella sp. On tree. **1:**CA(661).
Dothistroma septospora. Needle blight, needle cast. **20:**OK(2872). **36:**OH(82). **39:**IA(82), IL(82), KY(1780), MD(82), MN(2157,2595), NE(2753,2756), OH(82), OK(82), PA(3673). **40:**OK(2872), WI(1361). **46:**NE(2753,2756). **52:**CA(81). **53:**OH(82), OK(2873). **58:**VA(3299). **64:**IA(82), IL(82), OH(82), OK(82).
D. septospora var. *lineare.* Red band disease of needles. **46:**ID(3625).
Furcaspora pinicola. On needles. **34:**CA(2390,3549).
Fusicoccum sp. **46:**OR(3202).
Gelatinosporium pinicola. Associated with blister rust cankers. **36:**ID(82,3202).
Gloeocoryneum cinereum. On needles. **4:**OR(3202). **16:**CA(81), CO(3457), OR(3202,3547), WA(3202). **17:**CO(82). **30:**CA(81). **34:**CA(81,3547,3549), OR(3549). **46:**CA(81), CO(3457), ID(3202). **48:**CA(81,3547). **61:**CA(81).
Hendersonia pinicola. On living needles. **16:**ID(3202), MT(3202). **17:**WY(82,3812).
Hendersonia sp. **16:**ID(3202).
Hendersonula pinicola. **57:**NC(82), TN(82).
[*Kirschsteiniella applanata*]. **36:**ID(3202).
Lasiodiplodia theobromae. Wood stain. **64:**S states(82), LA(3732).
Lecanosticta acicola. Brown spot needle blight. **20:**NC(3944). **42:**NC(3944). **58:**WI(1357,2317). **59:**NC(3944).
Lecanosticta sp. Needle blight. **36:**ID(3202,3203). **57:**ID(3202), MT(3202).
Leptostroma decipiens. On needles. **16:**ID(3202), WA(3202). **46:**CA(2442), ID(82,2442,3202), MT(82,3202), OR(3202), WA(3202). **48:**CA(2442).
L. durissimum. On needles. **59:**NC(2442), SC(2442). **64:**SC(2442).
L. hedgcockii. On green needles. **10:**FL(350).
L. pinorum var. *graupianum.* **64:**MA(2443), MD(2443), MI(2443), NJ(2443).
L. rostrupii. On needles. **58:**WA(2442).
Leptostroma sp. On needles. **30:**CA(2442). **36:**ID(2442). **57:**NH(2442). **58:**OR(2442).
Leptothyrium stenosporum. On needles. **57:**GA(82).
Macrophoma magnifructa. Twig blight. **28:**TX(82). **64:**TX(82).
Macrophomina phaseolina. Black root rot, charcoal rot, seedling root rot. **4:**HI(2903). **13:**FL(166). **18:**CA(3348). **23:**FL(3317). **30:**CA(3348). **42:**MS(2720). **46:**CA(3348). **48:**CA(3348). **58:**CA(3348). **5:**CA(3348). **65:** CA(3348).
Microdiplodia conigena. On cones. **64:**CA(2439).
Neofuckelia pinicola. On limbs. **17:**OR(82). **26:**ID(82), OR(82), WA(82). **64:**NW states(82).
Pestalotia foedans. On needles and twigs. **39:**NJ(82).
Pestalotia sp. On dying needles. **37:**NJ(82). **57:**NC(1303).
Pestalotiopsis funerea. Seedling blight and root rot, on needles. **46:**NE(82). **53:**MD(82), TX(82). **57:**Range of host(82), WI(1337). **58:**Range of host(82), PA(82). **64:**Range of host(82), MD(82).
Phacidiopycnis pseudotsugae. Canker, dieback. **38:**CA(81). **57:**NE states(1413,1414), NY(1414,1644).
Phoma acuum. On twigs and cones. **64:**NJ(959), NY(82).
P. bacteriophila. **57:**NY(82).
P. harknessii. On twigs and cones. **1:**CA(642). **30:**CA(82). **57:**NE states(82). **64:**CA(82), MA(82).
P. terrestris. Isolated from rhizosphere. **46:**OR(2791).
Phoma sp. Damping-off, root rot. **16:**CO(2247). **57:**DE(702).
Phomopsis conorum. On twigs. **57:**IA(82). **58:**IA(82).
P. strobi. On limbs, associated with rust galls. **57:**ME(82).
Phomopsis sp. **39:**WA(3202).
Phragmotrichum pini. On twigs. **36:**WA(666,3202,3549).
Pragmopycnis pithya. **64:**NY(3552).
Pseudocenangium succineum. On needles. **54:**NJ(927). **64:**NJ(927), OH(927).
Rhabdospora mirabilissima. Stem canker of seedlings. **57:**NY(82).
R. pinea. On limbs. **58:**WI(1352).
R. pini. On limbs. **64:**Unk.(82).
Rhizosphaera kalkhoffii. **36:**WA(3202). **46:**WA(3202).
Sclerophoma pityella. **57:**NY(82), PA(82).
S. pythiophila. On twigs, cones and needles. **8:**WI(1334). **16:**OR(3202). **17:**CA(81). **26:**CO(82), WI(1352). **34:**CA(81). **39:**NJ(82). **46:**CA(81), OR(3202). **53:**WI(1334). **55:**CA(81). **57:**WI(1334). **58:**WI(1334). **64:**MI(82), OR(82,3202).
Sclerophoma sp. On needles. **21:**NM(1003).
Seimatosporium foliicola. On needles. **57:**DC(82), MA(82).
Septopatella septata. On needles. **46:**CA(927). **64:**CA(3549).
Septoria pinicola. Needle blight, lesions on needles. **59:**NC(1303). **62:**VA(82).
S. spadicea. Needle blight. **57:**NH(82), NY(82), VT(82).
Septoria sp. Black spot. **39:**KS(82), VA(82,3780). **64:**KS(3780).
Sirococcus conigenus. Shoot blight, tip dieback. **18:**CA(3352). **30:**CA(3352). **46:**CA(3352). **48:**CA(3352). **53:**MI(2621), MN(2621), WI(2621).
Sirothyriella pini-austriacae. On needles. **53:**WI(1334).
Sphaeronaema pithyum. On limbs and trunks. **57:**NY(82).
Sphaeropsis malorum. On dead areas of limbs and trunk. **57:**Unk.(82).
S. sapinea. Seedling, shoot and tip blight, bleeding canker, dieback,

collar rot. **10:**Unk.(82). **12:**KS(82). **21:**KS(82). **23:**HI(2903). **34:**NV(81). **37:**CT(82), IA(82,3780), IL(2568), NJ(82), NY(82), VA(82). **38:**CA(1693). **40:**OK(2873), WI(1329). **44:**HI(235,2903). **46:**ND(3758), NE(2757), NY(82), SD(1775). **52:**CA(1693), HI(2903), WA(3202). **53:**NC states(2691), MD(82), NJ(82), PA(82), WI(82,2690,2691). **54:**MA(82). **55:**CA(1693). **57:**NJ(82), NY(82), WI(82). **59:**NC(1303). **60:**NE states(133). **62:**PA(82). **63:**MD(82). **64:**E states(82), CT(3770), KS(82), OK(82), SD(82).
Sphaeropsis sp. Twig blight. **57:**NC(1533), VA(1533).
Stilbospora pinicola. Needle blight. **16:**OR(3202). **17:**OR(82). **42:**MS(82,2720).
Thyrinula sp. On fallen needles. **4:**CA(642).
Tuberculariella ips. Blue stain. **53:**MN(82).

Deuteromycotina - Other

Rhizoctonia solani. Damping-off. **4:**Range of host(82). **16:**Range of host(82). **17:**Range of host(82). **18:**Range of host(82). **37:**NH(82). **42:**NC(12). **46:**CA(12). **53:**Range of host(82). **57:**Range of host(82). **58:**Range of host(82). **64:**Range of host(82).
Rhizoctonia sp. Bud and needle blight, damping-off, root rot. **42:**NC(1303). **57:**DE(702). **64:**HI(2903).

Pseudolarix Gordon

1. *P. kaempferi* Gordon (=*P. amabilis* (J. Nels.) Rehd., *Chrysolarix amabilis* (J. Nels.) H. Moore)

Ascomycotina

Lachnellula willkommii. Canker. **1:**MA(82).

Pseudotsuga Carrière

1. *P. macrocarpa* (Vasey) Mayr [big-cone spruce]
2. *P. menziesii* (Mirb.) Franco (=*P. menziesii* (Mirb.) Franco var. *menziesii* , *P. mucronata* (Raf.) Sudw., *P. taxifolia* (Lamb.) Britton) [douglas fir]
3. *P. menziesii* (Mirb.) Franco var. *glauca* (Beissn.) Franco (=*P. taxifolia* (Lamb.) Britton var. *glauca* (Beissn.) Sudw.)
4. *Pseudotsuga* sp.

Chytridiomycetes

Rhizophydium pollinis-pini. **2:**MT(3202).

Oomycetes

Phytophthora cactorum. Isolated from roots. **2:**OR(1430), WA(1429,1430).
P. cinnamomi. Root rot. **2:**NC(1303), OR(2870,3049,3202), WA(2870).
P. cryptogea. Root rot. **2:**OR(1430,1580,2870), WA(1430,2870).
P. drechsleri. Root rot. **2:**OR(1580,2870), WA(2870).
P. megasperma. **2:**OR(1428), WA(1428).
P. pseudotsugae. **2:**OR(1429), WA(1429).
Phytophthora sp. Damping-off, root rot. **2:**HI(2899,2903).
Pythium ultimum. Damping-off, root rot. **2:**CO(82), HI(2899,2903).

Zygomycotina

Mortierella ramanniana var. *angulispora*. On leaf surface. **2:**OR(3042).

Ascomycotina

Antennatula sp. On twigs. **2:**OR(3230).
Bisporella citrina. **2:**ID(3202).
Botryosphaeria dothidea. Twig blight. **3:**KS(3195).
Caliciopsis pseudotsugae. On twigs. **2:**ID(3202), OR(3202,3230).
Cenangium leptospermum. **2:**OR(3202).
Ceratocystiopsis minuta. On wood. **2:**WA(1364,3202).
Ceratocystis angusticollis. On bark and wood. **2:**WA(1364,3202).
C. coerulescens. Sapwood stain. **2:**W states(782), CA(81,82).
C. piceaperda. Sapwood stain. **2:**WA(804,3202,3714).
Chaetomium elatum. On leaf surface. **2:**OR(3042).
C. funicola. On leaf surface. **2:**OR(3042).
Chlorociboria aeruginosa. **2:**ID(3202).
Claussenomyces pseudotsugae. **2:**OR(3202).
Coccomyces parvulus. On decorticated wood. **2:**OR(3229).
Colpoma morbida. **2:**OR(3202).
Coniochaeta sp. **2:**OR(3202), WA(3202).
Dasyscyphus ellisianus. Twig canker. **2:**MA(82), NC(82), RI(82).
D. oblongosporus. Saprophytic on limbs and twigs. **2:**NE states(82).
Dermea pseudotsugae. Branch canker, dieback. **2:**CA(2392), ID(3202), WA(3202).
Diaporthe eres. On twigs. **2:**OR(3230).
D. lokoyae. **2:**OR(3202).
Diatrype diffidens. **2:**OR(3202).
Didymella douglasii. **2:**MT(3202).
Durandiella pseudotsugae. **2:**ID(3202), OR(3202).
Glonium tryblidioides. **2:**OR(3202).
Herpotrichia juniperi. Brown felt blight. **2:**NW states(82), W states(82), ID(3202).
Heyderia abietis. **2:**WA(3202).
Hyalopeziza ciliata. **2:**ID(3202), OR(3202), WA(3202).
Hyaloscypha stevensonii. On old cones. **2:**CA(81).
Khuskia oryzae. On leaf surface. **2:**OR(3042).
Lachnellula agassizii. Saprophytic on limbs and twigs. **2:**W states(82), WA(3202).
L. arida. Saprophytic on limbs and twigs. **2:**W states(82).
L. ciliata. Saprophytic on limbs and twigs. **2:**W states(82), ID(3202), OR(860,3202), WA(3202).
L. pseudotsugae. Branch canker. **2:**W states(82).
L. subtilissima. Saprophytic on limbs and twigs. **2:**MA(82), RI(82).
Leucostoma kunzei. **2:**ID(3202), WA(3202).
Limacinia lithocarpi. On twigs. **2:**OR(3230).
Lophium sp. On twigs. **2:**OR(3230).
Metacapnodium sp. On twigs. **2:**OR(3230).
Mytilinidion tortile. **2:**ID(3202), WA(3202).
Ophiostoma minus. Blue-black stain of wood. **2:**OR(82,3068,3202), WA(82,1364,3068,3202).
O. nigrocarpum. On wood, associated with wood borers. **2:**ID(785,824,3714).
O. piceae. Lumber and sapwood stain. **2:**OR(782,3202), WA(782,3202).
O. wageneri. Black stain root. **2:**CA(589), OR(1067,1258,1546), WA(1258).
Patellaria signata. **2:**WA(3202).
Patellea californica. **2:**OR(3202).
Phacidium abietis. **2:**ID(3202), OR(3202), WA(3202).
P. infestans. Snow blight. **2:**ID(82).
Phaeocryptopus gaeumannii. Needle cast. **2:**W states(82), AZ(2515), CA(2515), CT(2515), MA(2515), ME(2515), MI(2515), MN(249), NH(2515), NM(2515), NY(2515), OR(2367,3230), PA(2358), RI(2515), VT(2515), WA(2367,2515), WI(2515).
Pithya cupressina. **2:**OR(3202).
P. vulgaris. On limbs and twigs. **2:**OR(638).
Propolis leonis. On fallen cones, on wood. **2:**CA(81), WA(3202).
Propolomyces versicolor. **2:**OR(3202).
Pseudographis elatina. On bark. **2:**OR(3228).
Pseudographis sp. On bark. **2:**OR(3228).
Rasutoria pseudotsugae. On twigs and needles. **2:**CA(81,2434,3247), OR(3230).
Rhabdocline parkeri. On needles. **2:**OR(3234).
R. pseudotsugae. Needle cast. **1:**CA(2395). **2:**NW states(82), Range of host(82), AZ(82,2696), CA(81,82,2696), CO(82,2696,3457), ID(1195,2395,3202), MA(82), MT(2395,2696,3202), NM(2696), NY(82), OR(2696,3202), PA(82,2313,3463), RI(82), WA(2696,3202), WI(3303).
R. pseudotsugae subsp. *epiphylla*. On foliage. **2:**ID(2696), OR(2696), WA(2696).
R. weirii. Needle cast. **2:**CA(2696), ID(2696,3202), MI(2514), MT(2696,3202), OR(2696,3202), WA(3202).
R. weirii subsp. *oblonga*. Needle cast. **2:**MA(2696), NM(2696), PA(2313), RI(2696).
R. weirii subsp. *obovata*. On foliage. **2:**CA(2696), MT(2696), OR(2696,3228), WA(2696).
Rhizina undulata. Seedling blight, root rot. **2:**NW states(82), WA(2479,3202).
Rosellinia diathrausta. **2:**OR(3202).
R. herpotrichioides. Needle blight. **2:**CA(3350), WA(3202).
Sarcotrochila balsameae. **1:**CA(327).
Sarea difformis. On resin. **4:**OR(1495).
Schizothyrium sp. On needles. **2:**OR(3230).
Scutellinia scutellata. **2:**ID(3202).
Sordaria fimicola. On leaf surface. **2:**OR(3042).
S. minima. On leaf surface. **2:**OR(3042).
Stictis radiata. **4:**CA(3225).
Strigopodia batistae. On twigs. **2:**ID(3202), OR(657,3230), WA(3202).
S. spongiosa. **2:**WA(3202).
Teichosporina sp. On twigs. **2:**OR(3230).
Tryblidiopsis pinastri. **2:**WA(3202).

Tympanis truncatula. **2:**OR(3202).
Tympanis sp. **2:**ID(3202).
Valsa abietis. On limbs and twigs. **2:**CO(82), ID(82), MT(3202), OR(82,3202), WA(82,3202).
Winteria caerulea. On twigs. **2:**OR(3230).
Xenomeris abietis. Dieback, stromata on leaf scars. **2:**ID(1132).
Xerotrema megalosporum. On wood. **2:**OR(3232).
Xylaria digitata. **2:**ID(3202), WA(3202). **3:**WA(638).

Basidiomycotina - Uredinales

Melampsora medusae. Heteroecious: aecial host. **1:**CA(82). **3:**AZ(3981), WA(638).
M. occidentalis. Heteroecious: aecial host. **2:**ID(3202), MT(1195,3202), OR(3202), WA(3202).

Basidiomycotina - Other

Acanthophysium lividocaeruleum. On wood. **2:**ID(2083,3202), OR(3202).
A. spiniger. On bark. **2:**AZ(128,1202), ID(2083,3202), OR(2083,3202).
A. weirii. On bark. **2:**MT(128,2083,3202), OR(128).
Aleurocystidiellum subcruentatum. On bark, sometimes weakly parasitic. **2:**CA(82), OR(82,3202).
Aleurodiscus amorphus. On suppressed, injured, or recently dead trees. **2:**CA(2083), ID(1449,3202), MA(1449), MT(3202), OR(1449,3202), WA(2083,3202).
A. grantii. On limbs and trunks. **2:**CA(1236), WA(1236).
A. penicillatus. On dead limbs and twigs. **2:**CA(2083), OR(82,2083,3202), WA(3202).
Amylocystis lapponica. **2:**ID(3202), OR(3202).
Amylostereum chailletii. White rot. **2:**NW states(82), AZ(128,543,1202).
Anomoporia albolutescens. **2:**ID(3202), WA(3202).
A. bombycina. **2:**WA(3202).
Antrodia albobrunnea. Wood rot. **2:**Unk.(82), ID(3202), MT(128,3202).
A. carbonica. Brown cubical rot. **2:**NE states(3995), AZ(1202), ID(82), NM(1190), OR(82,2682,3202), WA(82,2167,2682,3202).
A. crassa. **2:**WA(3202).
[*A. ferox*]. **2:**WA(3202).
A. gossypina. **2:**WA(3202).
A. radiculosa. **2:**AZ(1202).
A. sitchensis. **2:**OR(3202).
A. vaillantii. Isolated from treated utility poles. **2:**NE states(3995), NE states(3995), OR(3202), WA(3202).
A. xantha. **2:**Unk.(82), AZ(128,1202), ID(3202), OR(3202), WA(3202).
Antrodiella romellii. **2:**AZ(1202).
Armillaria mellea. Shoestring root rot. **2:**Range of host(82), AZ(1202), CA(2898), ID(538,1583,3202), MT(3202), OR(1067,1258,1259,3202,3600), PA(2172), WA(1065,1258,3202). **3:**CO(1724).
Asterostroma muscicola. **2:**AZ(1202).
Auricularia auricula. **2:**ID(3202).
Auriscalpium vulgare. **2:**ID(3202), OR(3202), WA(3202).
Basidioradulum radula. **2:**ID(3202).
Bondarzewia berkeleyi. **2:**ID(3202), WA(3202).
Botryobasidium subcoronatum. **2:**AZ(1202).
Byssoporia terrestris. **2:**OR(3202).
Calocera cornea. **2:**ID(3202), OR(128), WA(3202).
Caulorhiza umbonata. **2:**OR(3202).
Ceraceomyces tessulatus. **2:**AZ(1202).
Ceriporia purpurea. Wood rot. **2:**Unk.(82).
C. spissa. Isolated from treated utility poles. **2:**NE states(3995), NE states(3995).
Ceriporiopsis rivulosa. **2:**OR(3202).
C. subvermispora. **2:**AZ(1202).
Chaetoderma luna. **2:**WA(3202).
Clavicorona cristata. On wood. **2:**OR(884).
Climacocystis borealis. **2:**AZ(662,1202), ID(3202).
Clitocybe isabellina. On logs. **2:**WA(3335).
Columnocystis abietina. **2:**ID(3202).
Confertobasidium olivaceo-album. On wood. **2:**AZ(1178,1202).
Coniophora arida. **2:**AZ(1202).
C. olivacea. **2:**AZ(1202), MT(82,3202).
C. puteana. Brown rot. **2:**ID(82,3202), OR(3202).
Coriolopsis gallica. Wood rot. **2:**NW states(82), WA(3202).
Corticium minnsiae. **2:**OR(128,3202).
Crucibulum laeve. **2:**ID(3202).
Crustoderma dryinum. **2:**AZ(128,1202).
Cryptoporus volvatus. On wood. **2:**Range of host(82), CA(642), ID(1469,3202), MT(3202), OR(3202), WA(3202).
Cyathus olla. **2:**WA(3202).
Cylindrobasidium corrugum. **2:**WA(128).
C. laeve. **4:**WA(128).
Dacrymyces chrysospermus. **2:**AZ(1202), ID(3202), OR(3202), WA(3202).
Dacryobolus karstenii. **2:**NM(128,1190).
Daedaleopsis confragosa. **2:**AZ(1202).
Dendrothele incrustans. **2:**CA(2084).
Dichomitus squalens. Red ray rot. **2:**ID(82,3202), MT(3202), OR(82,3202), WA(82,3202).
Dichostereum pallescens. **2:**OR(3202).
Diplomitoporus lenis. Wood rot. **2:**Unk.(82), AZ(1202), ID(3202), OR(3202), WA(3202).
D. linbladii. Isolated from treated utility poles. **2:**NE states(3995), Unk.(82), ID(3202), MT(3202), OR(3202), WA(3202).
Echinodontium tinctorium. Heart rot of living trees. **2:**W states(82), ID(3202), MT(3202), OR(82,3202), WA(82).
Exidia glandulosa. **2:**AZ(1202).
Exidia sp. **2:**ID(3202).
Exidiopsis calcea. **2:**AZ(1202).
Fibricium rude. **2:**AZ(1202), NM(128), WA(3202).
Fibulomyces mutabilis. **2:**AZ(128).
Fomes fomentarius. **2:**WA(3202).
Fomitopsis officinalis. Brown trunk rot. **2:**W states(82), ID(538,3202), MT(3202,3832), OR(82,3202), WA(82,128,3202).
F. palustris. **2:**AZ(1202).
F. pinicola. Brown cubical rot. **2:**Range of host(82), AZ(1202,1859), ID(3202), MT(2045,3202,3832), NM(1190), OR(3202), WA(2219,3202).
F. rosea. Brown trunk rot. **2:**Range of host(82), AZ(1202), CA(81), MT(3202,3832), OR(3202), WA(3202).
Galerina laticeps. On logs. **2:**OR(3329).
Galzinia occidentalis. On wood. **2:**OR(128,3202).
Ganoderma lucidum. **2:**MT(3832).
G. oregonense. White pocket rot. **2:**ID(3202), MT(82,3202), OR(3202), WA(82,3202).
G. tsugae. **2:**AZ(18,1202).
Gloeocystidiellum karstenii. **2:**AZ(1202).
Gloeomucro nodosus. **2:**WA(3202).
Gloeophyllum trabeum. Brown sapwood rot. **2:**Range of host(82).
Gloeoporus dichrous. **2:**ID(3202), MT(3202).
Grandinia abieticola. **2:**AZ(1202).
G. breviseta. **2:**NM(128).
G. pallidula. **2:**AZ(1202).
Grifola frondosa. **2:**MT(3202,3832).
Guepiniopsis buccina. **2:**ID(3202).
Gymnopilus spectabilis. **2:**OR(3202), WA(3202).
Hapalopilus rutilans. White rot. **4:**AZ(1204).
Helicogloea longispora. On wood. **2:**OR(143,3202).
Henningsomyces candidus. **2:**AZ(1202), WA(3202).
Hericium abietis. **2:**WA(3202).
H. americanum. White pocket rot, sometimes of living trees. **2:**NW states(82), OR(3202), WA(3202).
H. erinaceus. **2:**OR(3202).
Hydnochaete olivacea. **2:**ID(3202).
Hymenochaete spreta. **2:**OR(3202).
H. tabacina. **2:**ID(3202), WA(3202).
H. tenuis. **2:**AZ(1202).
Hyphoderma praetermissum. **2:**AZ(1202).
Hypholoma fasciculare. **2:**AZ(1202).
Hypochnicium geogenium. **2:**NM(1190).
H. punctulatum. **2:**NM(128), WA(3202).
Inonotus circinatus. White root and butt rot. **2:**ID(82).
I. tomentosus. **2:**ID(1583,3202), WA(3202).
Irpex lacteus. Isolated from treated utility poles. **2:**NE states(3995), NE states(3995).
Irpex sp. **2:**WA(3202).
Jahnoporus hirtus. **2:**AZ(1202), OR(3202), WA(3202).
Junghuhnia collabens. **2:**OR(3202).
J. luteoalba. **2:**OR(3202).
J. zonata. **2:**ID(3202), WA(128).

Kavinia alboviridis. **2**:NM(1190).
Laetiporus sulphureus. Brown cubical heart rot. **2**:Range of host(82), ID(3202), MT(3202,3832), OR(3202), WA(3202).
Laurilia sulcata. White pocket rot of logs and slash. **2**:NW states(82).
Lentinellus cochleatus. **2**:AZ(1202).
L. ursinus. **2**:AZ(1202).
Leptosporomyces galzinii. **2**:AZ(1202).
Leucogyrophana arizonica. On wood. **2**:Unk.(1232).
Marasmius albogriseus. **2**:OR(3202).
M. plicatulus. **4**:CA(856).
Megacollybia platyphylla. **2**:AZ(1202).
Meripilus sumstinei. **2**:ID(3202).
Meruliporia incrassata. Dry rot. **2**:NW states(82), ID(3202), OR(3202).
Mucronella aggregata. **2**:AZ(1202).
Mycena fuliginella. On wood. **2**:WA(3326).
M. lilacifolia. **2**:ID(3202).
M. subcana. **2**:WA(3202).
M. subvitrea. On needles. **2**:OR(3326).
Mycoacia uda. On wood. **2**:OR(638,3202).
Neolentinus lepideus. Brown cubical rot. **2**:NW states(82,3202), AZ(1202), ID(3202).
Odonticium romellii. **2**:ID(3202).
Osteina obducta. **2**:AZ(1202), ID(3202), NM(1190), OR(3202).
Oxyporus corticola. **2**:AZ(1202), ID(3202).
Panellus mitis. **2**:ID(3202), OR(3202).
Parmastomyces transmutans. **2**:NM(1190).
Peniophora perexigua. **2**:AZ(1202).
Perenniporia medulla-panis. **2**:MT(3202,3832), OR(3202).
P. subacida. White rot. **2**:Unk.(82), ID(538,1583,3202), MT(3832), OR(128,3202), WA(3202).
P. tenuis. Isolated from treated utility poles. **2**:NE states(3995).
Phaeolus schweinitzii. Brown cubical butt and root rot. **2**:Range of host(82), AZ(1202,2268), ID(1583,3202), MT(3202), NM(1190), OR(3202), WA(3202).
Phanerochaete burtii. **2**:AZ(128).
P. carnosa. White rot. **2**:AZ(433,1202), ID(128,433), MT(128).
P. sanguinea. **2**:ID(3202), OR(3202).
P. velutina. White rot. **2**:AZ(433,1202).
Phellinus ferreus. **2**:OR(3202).
P. ferrugineofuscus. **2**:MT(3202), OR(3202), WA(3202).
P. ferruginosus. **2**:NM(128,1190), WA(3202).
P. igniarius. **2**:ID(3202).
P. nigrolimitatus. White pocket rot. **2**:ID(538,3202), MT(82,128,3202), OR(3202), WA(82,128,3202).
P. pini. White pocket heart rot. **2**:Range of host(82), AZ(1202), CA(81), ID(538,1583,3202), MT(3202), OR(3202), WA(3202). **4**:AZ(128).
P. repandus. On wood. **2**:MT(2047,3202), OR(3202), WA(128,3202).
[*P. robustus*]. **2**:WA(3202).
[*P. torulosus*]. **2**:AZ(1189,1202).
P. viticola. On wood. **2**:NW states(82), ID(3202), MT(3202), OR(3202), WA(3202).
P. weirii. Butt rot, root rot. **2**:ID(538,3202), OR(82,1067,1072,1258,1450,2579,3202,3600,3601), WA(82,1065,1072,1258,3202).
Phlebia merismoides. Isolated from treated utility poles. **2**:NE states(3995), NE states(3995).
P. subserialis. **2**:AZ(1202).
Phlebiella filicina. **2**:ID(3202).
P. subnitens. On wood. **2**:OR(2115).
P. vaga. **2**:WA(3202).
Pholiota adiposa. **2**:AZ(1202).
P. bakerensis. **2**:AZ(1202).
P. decorata. **2**:AZ(1202).
P. limonella. **2**:AZ(1202).
P. olympiana. On wood. **2**:WA(3327).
P. spumosa. **2**:AZ(1202).
Physisporinus sanguinolentus. **2**:OR(128), WA(3202).
Pleurocybella porrigens. **2**:WA(3202).
Pleurotus albolanatus. On logs. **4**:MI(3330).
P. ostreatus. **2**:AZ(1202).
Pluteus atricapillus. **2**:AZ(1202), ID(3202).
Polyporus badius. **2**:AZ(1202), ID(3202).
P. varius. **2**:AZ(1202), OR(3202).
Postia angulopora. On wood. **2**:OR(2048).
P. balsamea. **2**:AZ(1202), ID(3202), MT(3202), NM(1190), OR(3202), WA(3202).
P. caesia. **2**:ID(3202), OR(3202), WA(3202).
P. ceriflua. **2**:WA(3202).
P. fragilis. **2**:ID(3202), OR(3202), WA(3202).
P. guttulata. **2**:Range of host(82), OR(3202), WA(3202).
P. leucospongia. Wood rot. **2**:Range of host(82), ID(3202), OR(3202).
P. sericeomollis. **2**:WA(3202).
P. stiptica. **2**:MT(3832).
P. tephroleuca. **2**:AZ(1202).
P. undosa. **2**:ID(3202), MT(3202), WA(3202).
Psathyrella piluliformis. **2**:ID(3202).
Pseudohydnum gelatinosum. **2**:AZ(1202).
Pseudomerulius aureus. **2**:NM(128,1190).
Pseudotomentella fumosa. On wood. **2**:OR(2041).
P. griseoveneta. **2**:OR(2038).
P. kaniksuensis. On wood. **2**:ID(2041).
P. molybdea. On wood. **2**:MT(2041).
Pycnoporellus alboluteus. Wood rot. **2**:Range of host(82), AZ(1202), ID(3202), WA(3202).
P. fulgens. Wood rot. **2**:Range of host(82), ID(3202), MT(3202,3832), OR(3202), WA(3202).
Pycnoporus cinnabarinus. **2**:ID(3202).
Resinicium bicolor. **2**:ID(3202).
R. chiricahuaensis. White rot. **2**:AZ(1202), NM(128,1188).
R. furfuraceum. **2**:AZ(128,1202).
Schizophyllum commune. On cones in storage. **2**:NW states(3208), OR(3202), WA(3202).
Schizopora paradoxa. **2**:OR(3202).
Scytinostroma arachnoideum. **2**:WA(128,3202).
S. galactinum. **2**:MT(128).
S. ochroleucum. On wood. **2**:CA(81,128).
S. portentosum. **2**:ID(128).
Serpula himantioides. **2**:OR(3202,4007).
S. lacrimans. **2**:MT(3202,3832).
Skeletocutis amorpha. Wood rot. **2**:Range of host(82), ID(3202), OR(3202), WA(3202).
S. nivea. **2**:MT(3202).
S. subincarnata. **2**:AZ(1202), ID(128,3202).
Solenia sp. **2**:ID(3202).
Sparassis crispa. Brown rot. **2**:NW states(82), AZ(1202,2268), MT(3202), NM(128,1190), OR(3202).
Sphaerobolus stellatus. **2**:WA(3202).
Steccherinum ciliolatum. On wood. **2**:AZ(1202), MT(1175,3202).
S. ochraceum. **2**:OR(3202).
[*Stereum gausapatum*]. Isolated from treated utility poles. **2**:NE states(3995), NE states(3995).
S. hirsutum. **2**:OR(543).
S. sanguinolentum. White pocket rot. **2**:NW states(82), AZ(1202), ID(3202), MA(543), OR(3202), WA(3202). **4**:CA(543), OR(4007).
Stereum sp. Isolated from treated utility poles. **2**:NE states(3995), NE states(3995).
Strobilurus trullisatus. On cones. **2**:OR(3202,3854), WA(3854).
Stypella minor. On wood. **2**:OR(128).
Thelephora terrestris. Smothering of seedlings. **2**:NW states(82), ID(3202), WA(3202).
Tomentella avellanea. **2**:WA(3202).
T. ferruginea. **2**:ID(3202), MT(128).
T. lateritia. **2**:AZ(1202), NM(1190). **4**:AZ(2039), MT(2039), NM(2039).
T. pilosa. **2**:NM(1190). **4**:NM(2039).
T. ramosissima. **4**:NM(2039).
Tomentellina fibrosa. **2**:AZ(128,1202), MT(3202).
Trametes hirsuta. Wood rot. **2**:Range of host(82), WA(3202).
T. versicolor. White rot. **2**:NE states(3995), Range of host(82), ID(3202), MT(3202), OR(3202), WA(3202).
Trechispora farinacea. **2**:WA(3202).
T. lunata. **2**:AZ(1202).
T. mollusca. **2**:OR(3202).
Trichaptum biforme. Isolated from treated utility poles. **2**:NE states(3995), NE states(3995).
T. laricinum. On logs. **4**:CO(2226).
T. sector. **2**:AZ(1202).
Tricholomopsis bella. **2**:AZ(1202).

T. radicata. **2**:ID(3202).
T. rutilans. On wood. **2**:AZ(1202), OR(4007).
Tubulicrinis accedens. **2**:OR(3202,3859).
T. calothrix. **2**:AZ(1202).
T. glebulosus. **2**:ID(3202).
T. hamatus. **2**:AZ(1202).
T. inornatus. On wood. **2**:OR(1714,3202).
Tyromyces chioneus. **2**:OR(3202).
T. galactinus. **2**:OR(3202).
Vararia boreale. **2**:ID(128,3202), MT(3202), OR(3202), WA(128,3202).
V. racemosa. **2**:ID(3202), WA(3202).
Xeromphalina campanella. **2**:ID(3202), WA(3202).
X. fulvipes. **2**:ID(3202).
Xylobolus frustulatus. Isolated from treated utility poles. **2**:NE states(3995), NE states(3995).

Deuteromycotina - Hyphomycetes

Alternaria alternata. On leaf surface. **2**:OR(3042).
Alternaria sp. **2**:WA(3202).
Aspergillus clavatus. On leaf surface. **2**:OR(3042).
A. niger. On leaf surface. **2**:OR(3042).
Aspergillus sp. Isolated from planks. **2**:OR(2386), WA(3202).
Atichia glomerulosa. **2**:CA(81), WA(3202).
Atichia sp. On twigs and needles. **2**:OR(3230).
Aureobasidium pullulans. On twigs and needles. **2**:OR(3042,3230), WA(3202).
Botrytis cinerea. Gray mold twig blight. **2**:Range of host(82), ID(3202), MT(3202), OR(3042).
Botrytis sp. Damping-off, root rot. **2**:DE(702), OR(1419,3202).
Capnocybe spongiosa. On limbs and twigs. **2**:CA(638,2434).
Cephalosporium sp. On twigs. **2**:OR(3230).
Chalara ungeri. On wood. **2**:WA(2550).
Cladobotryum verticillatum. On leaf surface. **2**:OR(3042).
Cladosporium cladosporioides. On leaf surface. **2**:OR(3042).
C. herbarum. On twigs and needles. **2**:OR(3042,3230).
C. macrocarpum. On leaf surface. **2**:OR(3042).
C. oxysporum. On twigs and needles. **2**:OR(3042,3230).
C. sphaerospermum. On leaf surface. **2**:OR(3042).
Cladosporium sp. **2**:WA(3202).
Conoplea fusca. **4**:CO(1672).
Costantinella micheneri. **2**:OR(3202).
Cylindrocarpon obtusisporum. On leaf surface. **2**:OR(3042).
Epicoccum nigrum. On twigs. **2**:OR(3230).
Epicoccum sp. **2**:WA(3202).
Fusarium avenaceum. Seed rot. **2**:Range of host(82).
F. moniliforme. Seed rot. **2**:Range of host(82).
F. oxysporum. Seedling root rot. **2**:Range of host(82), NY(3283), OK(3282), OR(1431), PA(2359).
F. roseum. Stem rot. **2**:OR(2478).
F. solani. Root and seed rot. **2**:Range of host(82), PA(2359).
Fusarium sp. Damping-off, root rot. **2**:DE(702), MT(3202), PA(3858).
Gliocladium virens. Isolated from roots. **2**:OR(2578).
Graphium sp. Isolated from plants. **2**:OR(2386).
Heteroconium chaetospira. On leaf surface. **2**:OR(3042).
Humicola fuscoatra. On leaf surface. **2**:OR(3042).
Hyalodendron sp. On twigs. **2**:OR(3230).
Hyphosoma resinacea. **2**:ID(3202), WA(3202). **3**:WA(641).
Leptographium penicillatum. Isolated from roots. **2**:ID(2372).
L. wageneri. Black stain root disease. **2**:W states(1257), CA(661,1471,1473,2396), CO(1471), ID(1471), MT(1471), NM(1471), OR(1259,1450,1471,3935), WA(1259,1471).
L. wageneri var. *pseudotsugae*. **2**:CA(1472), MT(1472), WA(1472).
Leptographium sp. Isolated from planks. **2**:OR(2386).
Memnoniella echinata. On leaf surface. **2**:OR(3042).
Meria parkeri. On needles. **2**:OR(3234).
Microdochium tabacinum. On leaf surface. **2**:OR(3042).
Monodictys castaneae. On leaf surface. **2**:OR(3042).
Nigrospora sp. On twigs. **2**:OR(3230).
Nodulisporium sp. On leaf surface. **2**:OR(3042).
Pachnocybe ferruginea. Isolated from utility poles in service. **2**:OR(1985).
Penicillium sp. On leaf surface. **2**:OR(2386,3042), WA(3202).
Periconiella verrucosa. On marine waterfront pilings and utility poles. **2**:OR(3489).
Phialophora hoffmannii. On leaf surface. **2**:OR(3042).
Phymatotrichopsis omnivora. Root rot. **2**:TX(82).
Scopulariopsis asperula. **2**:WA(3202).
S. brevicaulis. On leaf surface. **2**:OR(3042).
Scytalidium lignicola. On leaf surface. **2**:OR(3042).
Scytalidium sp. **2**:OR(3202).
Sporotrichum sp. **2**:WA(3202).
Stephanosporium cereale. On leaf surface. **2**:OR(3042).
Torula conglutinata. **2**:OR(3202).
Trichoderma citrinoviride. Isolated from roots. **2**:OR(2578).
T. hamatum. Isolated from roots. **2**:OR(2578).
T. harzianum. Isolated from roots. **2**:OR(2578).
T. longibrachiatum. Isolated from roots. **2**:OR(2578).
T. polysporum. Isolated from roots. **2**:OR(2578).
T. saturnisporum. Isolated from roots. **2**:OR(2578).
T. viride. Isolated from roots. **2**:OR(2578).
Trichoderma sp. Damping-off, root rot, isolated from planks. **2**:DE(702), OR(2386), WA(3202).
Trichothecium roseum. On leaf surface. **2**:OR(3042).
Ulocladium botrytis. On twigs and needles. **2**:OR(3042).
U. chartarum. On leaf surface. **2**:OR(3042).
Varicosporium sp. On leaf surface. **2**:OR(3042).
Verticicladiella sp. **3**:CO(1724).

Deuteromycotina - Coelomycetes

Corniculariella pseudotsugae. Bark canker. **2**:OR(82,861).
Cryptocline hypophyllum. On needles. **2**:AZ(82), NM(82,2481).
Cytospora kunzei. Canker. **2**:CO(3957), MA(3781), NH(3781), NM(1003), PA(3781), VT(3781), WA(3781).
Cytospora sp. Twig canker. **2**:CO(82), NJ(82), OR(82).
Dinemasporium sp. On leaf surface. **2**:OR(3042).
Foveostroma abietinum. **2**:ID(3202).
F. boycei. Bark canker. **2**:WA(82).
Gelatinosporium sp. On leaf surface. **2**:OR(3042).
Leptothyrium pseudotsugae. Fly speck on needles. **2**:CO(82), ID(3202).
Macrophomina phaseolina. Root rot of seedlings. **2**:CA(3348).
Melanconium sp. On leaf surface. **2**:OR(3042).
Monochaetia sp. On leaf surface. **2**:OR(3042).
Pestalotia vaccinii. On leaf surface. **2**:OR(3042).
Pestalotiopsis clavispora. On leaf surface. **2**:OR(3042).
Phacidiopycnis pseudotsugae. **3**:NE states(1413).
Phoma douglasii. On twigs. **2**:CO(82).
P. eupyrena. Needle blight and cast. **2**:CA(1917).
P. glomerata. On leaf surface. **2**:OR(3042).
P. herbarum. On leaf surface. **2**:OR(3042).
Phoma sp. Damping-off, root rot. **2**:DE(702), OR(3202).
Phomopsis lokoyae. Stem canker. **2**:W states(82), CA(233).
P. occulta. On dead twigs. **2**:Range of host(82).
Phomopsis sp. **2**:WA(3202).
Phyllosticta sp. On needles. **2**:OR(82,3202).
Pragmopycnis pithya. On twigs. **2**:NY(3549).
Rhabdogloeum pseudotsugae. **2**:WY(1195).
Sclerophoma pythiophila. On bracts and cone scales. **2**:CA(81).
Selenophoma sp. On leaf surface. **2**:OR(3042).
Sphaeropsis sapinea. Dieback. **1**:CA(82), KS(82), NJ(82), NY(82). **2**:CA(82), KS(82), NJ(82), NY(82).

Deuteromycotina - Other

Rhizoctonia solani. Damping-off, root rot. **2**:Range of host(82), HI(2899).
Rhizoctonia sp. Damping-off, root rot. **2**:DE(702), HI(2903).

Tsuga Carrière

1. *T. canadensis* (L.) Carrière [Canada hemlock]
2. *T. caroliniana* Engelm. [Carolina hemlock]
3. *T. diversifolia* (Maxim.) Mast. [northern Japanese hemlock]
4. *T. heterophylla* (Raf.) Sarg. [western hemlock]
5. *T. mertensiana* (Bong.) Carrière [mountain hemlock]
6. *Tsuga* sp.

Oomycetes

Phytophthora cinnamomi. Root decay. **1**:NC(1303).

Ascomycotina

Amphisphaeria juniperi. On wood. **6**:WA(3202,3815).
Anthostomella brachystoma. **5**:WA(3202).
Atropellis sp. **4**:WA(3202).
Bertia moriformis. **6**:WA(3202).
B. moriformis var. *latispora*. On wood. **1**:NH(684), NY(684).
Caliciopsis pinea. Canker. **1**:PA(82).

Ceratocystis seticollis. On wood, associated with bark beetles. **1:**NY(785,3714).
Ceratostomella sp. Blue stain. **4:**OR(82), WA(82).
Ciboria sp. **6:**AK(526).
Cistella parksii. On bark. **4:**CA(527).
Clithris graphis. **4:**ID(3202).
Coccomyces heterophyllae. On twigs. **4:**OR(3229), WA(3202). **5:**OR(3229).
C. mertensianae. On dead corticate twigs. **5:**OR(3231).
Coccophacidium sp. **4:**WA(3202).
Colpoma crispum. On wood. **4:**ID(3202). **5:**CA(661).
Coniochaeta ligniaria. **4:**ID(3202).
Dasyscyphus serinellus. On needles. **5:**CA(661), OR(638,3202). **6:**OR(3202).
Delphinella peckii. On cone scales. **1:**NY(82).
Dermea balsamea. Canker. **1:**GA(82,1438), MA(664).
Desmazierella acicola. **6:**WA(3202).
Diatrypella rimosa. **5:**CA(661).
Discocainia treleasei. **4:**AK(128).
Fabrella tsugae. Needle blight, needle cast. **1:**MA(82), NC(1303), NH(82), NY(3228), PA(1912), WI(82). **2:**NC(1303,1307). **4:**AK(82). **6:**AK(526).
Glonium stellatum. **1:**NC(1307).
Glonium sp. **4:**WA(3202).
Herpotrichia juniperi. Snow blight. **5:**NW states(82), AK(82,526), CA(642,661), ID(3202), MT(3202), OR(3202), WA(3202).
Heyderia abietis. **4:**WA(3202).
Hyaloscypha stevensonii. **5:**CA(661).
Hysterium angustatum. On bark. **5:**CA(639,642).
Hysterium sp. **4:**WA(3202).
Korfia tsugae. Yellowing of needles. **1:**NC(82,528,1303).
Lachnellula agassizii. On wood. **1:**NY(82). **5:**CA(642,661), ID(3202), OR(3202).
L. arida. **5:**CA(642,661), ID(3202), OR(3202).
L. subtilissima. **5:**CA(661).
L. suecica. **5:**ID(3202), MT(3202).
Lasiosphaeria stuppea. **5:**WA(3202).
Lophodermium sp. **4:**AK(526).
Microthyrium harrimanii. **4:**AK(526).
Mollisia fumigata. **1:**NC(1307).
Odontotrema minus. **5:**OR(3202), WA(3202).
Ophiostoma nigrocarpum. On logs. **1:**NH(785).
O. piceae. Lumber and sapwood stain. **4:**OR(782,3202), WA(782,3202).
O. wageneri. **5:**OR(1258).
Orbilia xanthostigma. **6:**NY(2456).
Patellaria signata. **5:**WA(3202).
Phaeocryptopus nudus. On needles. **1:**NC(1303,1307), NY(82).
Plectania nannfeldtii. **6:**WA(3202).
Propolis leonis. **1:**GA(1438). **4:**ID(3202).
Rasutoria tsugae. On needles. **1:**CT(3247), NC(3247), TN(3247), WV(3247). **3:**GA(82,1438), NC(82), NC(82), TN(82), VA(82). **4:**AK(82,526), ID(3202,3247), OR(82,3202,3247), WA(82,3202,3247). **5:**OR(3202,3247).
Rhizina undulata. Seedling and root rot. **1:**NY(82). **4:**ID(82,3202), WA(2480).
Rosellinia diathrausta. On limbs. **5:**CA(2403).
R. herpotrichioides. Needle and twig blight. **1:**NC(82,1303,1532).
Scutellinia scutellata. On wood. **1:**NC(1303,1307), NY(848). **4:**WA(3202).
Tympanis hypopodia. **1:**CT(2670).
Valsa abietis. On limbs. **1:**VA(82), WV(82). **4:**OR(648,3202).
Xylaria curta. On wood. **1:**ME(3015).
X. hypoxylon. On wood. **1:**NC(1303,1307), SC(1307).
X. longipes. On logs. **1:**NC(1303).
Xyloschizon weirianum. **5:**ID(3202).

Basidiomycotina - Uredinales

Melampsora abietis-canadensis. Heteroecious: aecial host. **1:**NC states(82), NE states(82), NC(82,1303), VA(3493), WI(1355). **2:**CT(82).
M. epitea. Heteroecious: aecial host. **4:**ID(3202), MT(3202), OR(3202), WA(3202). **5:**ID(3202), OR(3202), WA(3202).
M. epitea f. sp. *tsugae*. **4:**ID(82), MT(82), OR(82), WA(82).
M. farlowii. Autoecious. **1:**E states(82), NC(1303,1533), VA(3493), WI(82). **2:**NC(82,1303).
Pucciniastrum hydrangeae. Heteroecious: aecial host. **1:**IN(82), MD(82), NC(82,1303), PA(1575), TN(82), VA(82), WV(82). **2:**NC(111).
P. vaccinii. Heteroecious: aecial host. **1:**E states(82), GA(1439), IN(82), WI(82). **2:**NC(82). **4:**AK(60,82,526,1904), ID(3202), MT(3202), OR(82), WA(82). **5:**ID(82), MT(82).

Basidiomycotina - Other

Acanthophysium lividocaeruleum. On wood. **4:**ID(2083,3202).
A. weirii. On bark. **4:**AK(2083), ID(3202). **6:**AK(526).
Aleurodiscus abietis. **5:**OR(2083,3202).
A. amorphus. On suppressed, injured, or recently dead trees. **1:**NE states(82), NH(1449), NY(1449).
A. farlowii. On suppressed or recently dead trees. **1:**NE states(82), ME(2083), NH(2083), NY(2083).
A. grantii. On limbs and trunks. **4:**ID(2083,3202), WA(1236).
A. penicillatus. On dead limbs and twigs. **1:**VT(2083). **4:**ID(2083,3202), OR(2083,3202). **6:**OR(4007).
Aleurodiscus sp. **5:**OR(3202).
Amphinema byssoides. **4:**ID(3202). **6:**NY(2456).
Amylocorticium canadense. **6:**NY(128).
Amylocystis lapponica. Brown cubical rot. **4:**ID(82,3202), OR(3202).
Amylostereum chailletii. White pocket rot. **1:**NY(543), VT(543). **4:**NW states(82). **6:**NC(543).
Anomoporia albolutescens. **4:**OR(3202).
A. bombycina. **4:**WA(3202).
A. myceliosa. **4:**WA(3202). **6:**WA(128).
Antrodia albida. Wood rot. **1:**NY(82,2962). **4:**AK(3431).
A. albobrunnea. **4:**AK(526), ID(128,3202), MT(3202).
A. crassa. On wood. **5:**AK(526,1904).
A. heteromorpha. On wood. **1:**ME(82), NY(82), VA(82). **4:**AK(214,526,1904), ID(3202), MT(3202,3832), WA(3202). **5:**AK(214,526). **6:**AK(214,526).
A. radiculosa. **6:**NY(128).
A. serialis. On wood. **1:**ME(82), NY(82), PA(82). **4:**AK(214,526), ID(82,3202), OR(3202), WA(82,3202). **5:**AK(214,526).
A. sinuosa. **1:**NY(2167). **4:**ID(3202), MT(128). **5:**CA(642). **6:**MT(128).
A. sitchensis. **4:**AK(526).
A. vaillantii. On wood. **1:**Range of host(82).
A. variiformis. **4:**AK(214,526), WA(3202). **6:**AK(214).
A. xantha. **4:**WA(3202).
Armillaria mellea. Wood decay, butt and root rot. **1:**NE states(82), GA(1307), MI(82,2862), NC(1303,1307). **4:**ID(538,1583,3202), OR(82,1258), WA(82,1258,3202). **6:**MI(2862).
Arrhytidia involuta. On wood. **1:**WV(128). **6:**VA(128).
Asterodon ferruginosum. **4:**ID(128). **6:**ID(128).
Athelia laxa. On wood. **1:**CT(128), PA(128), VA(128).
Auriporia aurea. **6:**NY(128).
Basidioradulum radula. **1:**NY(2611).
Bjerkandera adusta. **1:**SC(1307). **4:**ID(3202).
Bondarzewia montana. **4:**ID(3202).
Boreostereum radiatum. White pocket rot. **1:**NH(543), NY(543), PA(82,543), VA(82,543), WI(82,543).
Botryobasidium subcoronatum. **1:**WI(2116).
B. vagum. **4:**ID(3202). **6:**NY(2456).
Calocera cornea. **1:**MA(1781).
Cerinomyces pallidus. On wood. **1:**WI(2116).
Ceriporiopsis rivulosa. **4:**ID(3202), WA(3202). **6:**WA(128).
Cerrena unicolor. **4:**WA(3202).
Chondrostereum purpureum. **4:**WA(3202).
Clavicorona piperata. On logs. **6:**ID(884,2073,3202).
C. pyxidata. **1:**NC(884), VA(884). **6:**NC(884), NY(884).
Climacocystis borealis. White mottled rot. **1:**MA(82,1781), NC(1307), NY(82), TN(82), VA(82). **4:**MT(3832), MT(82,3202).
Clitocybula abundans. On wood. **1:**VT(128).
Collybia bakerensis. **6:**CO(854), ID(854), WA(854).
Columnocystis abietina. On fallen logs. **1:**MI(543). **4:**AK(526), WA(543,3202). **5:**AK(543), CA(642,661), OR(3202), WA(3202). **6:**OR(3202).
C. ambigua. **6:**TN(128).
Coniophora arida. **6:**WA(128).
C. arida var. *suffocata*. **6:**PA(128).
C. puteana. Brown rot. **1:**Range of host(82). **4:**ID(82,1583), OR(3202), WA(82,3202).
Corticium minnsiae. On wood. **1:**ME(128), NC(1303), NH(128),

NY(128), VT(128).
C. roseum. **5**:CA(661).
Cryptoporus volvatus. Sapwood rot. **4**:WA(82,3202). **6**:OR(3202).
Cylindrobasidium corrugum. On bark and wood. **4**:ID(128). **5**:CA(642,661).
Cystostereum pini-canadense. White rot. **1**:MI(541).
Dacrymyces minutus. On wood. **1**:NC(2645).
D. stillatus. **5**:OR(3202).
Dacryobolus karstenii. **4**:AK(526). **6**:AK(526).
Dacryopinax spathularia. On wood. **1**:NY(128).
Dendrothele incrustans. On bark and wood. **4**:CA(2084).
Dichomitus squalens. Red ray rot. **1**:MA(82). **4**:ID(3202).
Dichostereum effuscatum. **1**:CT(128).
Diplomitoporus crustulinus. **4**:ID(3202).
D. lenis. Inside rotten logs. **4**:ID(3202), WA(3202). **5**:CA(642).
D. linbladii. **4**:ID(3202), WA(3202). **5**:ID(3202).
Echinodontium tinctorium. Heart rot of living trees. **4**:Range of host(82), AK(526), ID(1583,3202), MT(2045,3202), OR(3202), WA(3202). **5**:NW states(82), AK(526), ID(3202), OR(3202). **6**:AK(526).
Exidiopsis calcea. On bark. **1**:NY(128).
E. fuliginea. On wood. **4**:ID(128).
Fomitopsis cajanderi. On wood. **1**:NC(1303), NY(82), PA(82), VA(82). **4**:NW states(82), ID(3202), MT(3202,3832), OR(3202), WA(3202).
F. officinalis. Brown trunk rot. **4**:ID(3202), OR(82,3202), WA(82,3202). **5**:CA(661), ID(82,3202), OR(3202). **6**:MI(2862).
F. pinicola. Brown cubical rot. **1**:Range of host(82). **4**:Range of host(82), AK(526,1904,3431), ID(3202), MT(2045,3202,3832), OR(3202), WA(3202). **5**:Range of host(82), AK(526,1904), CA(642,661), MT(3202), OR(3202), WA(3202). **6**:WA(128).
F. rosea. Brown cubical trunk rot. **1**:E states(82).
F. spraguei. Sapwood rot. **1**:NC(1303,1307), NY(82).
Galerina tsugae. On logs. **1**:MI(3333).
Ganoderma applanatum. White-mottled butt rot. **1**:Unk.(82), MI(2862), NY(2962), PA(2962). **4**:AK(526,1904), ID(3202), MT(3202), OR(82,3202), WA(82,3202). **6**:AK(526), MI(2862).
G. lucidum. On wood. **1**:NC states(82), NE states(82), WI(82). **6**:MI(2862).
G. oregonense. White rot. **4**:AK(526), ID(3202), OR(82,3202), WA(3202). **5**:CA(661), ID(82,3202), OR(3202), WA(3202).
G. tsugae. On wood. **1**:Range of host(2533), CT(17,18), GA(491,1439), MA(656), MN(294), NY(2456), WI(293).
Gloeophyllum carbonarium. Wood rot. **1**:NY(82), PA(82).
G. odoratum. Wood rot. **1**:ME(82), WI(82). **4**:ID(3202), WA(3202).
G. sepiarium. Brown cubical rot. **1**:Range of host(82), NC(1303,1307). **4**:AK(1904), ID(82,3202), MT(3202), OR(3202), WA(82,3202). **5**:WA(3202).
G. trabeum. Brown cubical rot. **1**:Range of host(82). **4**:WA(3202).
Gloeoporus dichrous. On wood. **1**:NY(82). **4**:ID(3202), MT(3202), WA(3202).
Grandinia pallidula. **1**:WI(2116). **4**:ID(3202).
G. spathulata. **1**:NY(128). **6**:NY(128).
Guepiniopsis alpina. **4**:WA(3202). **5**:CA(661), ID(3202), OR(3202), WA(3202).
G. chrysocomus. **4**:WA(3202).
Gymnopilus naucorioides. On stumps. **6**:TN(1542).
G. pulchrifolius. On wood. **6**:NY(1542).
Hapalopilus rutilans. White rot. **4**:WA(3202). **6**:Unk.(2681).
Henningsomyces candidus. On logs. **1**:TN(1539).
Hericium abietis. Long pocket rot. **4**:ID(3202), OR(82,3202), WA(82,3202).
H. americanum. On wood. **1**:NY(1237).
Heterobasidion annosum. White root and butt rot. **1**:CT(82). **4**:AK(526,1904,3431), ID(82,3202), OR(82,128,906,1258,3202), WA(82,906,1258,3202). **5**:CA(642), OR(1258). **6**:WA(128).
Hydnum sp. **4**:ID(1583).
Hymenochaete agglutinans. Canker. **1**:PA(82).
H. fusca. **4**:WA(3202).
Hyphoderma argillaceum. **1**:WI(2116).
H. tsugae. On wood. **1**:NH(460).
Hypholoma capnoides. On wood. **1**:NC(1303,1307). **4**:ID(3202).
H. fasciculare. **4**:ID(3202).
Hypsizygus tessellatus. On wood. **1**:NC(1303).
Inonotus circinatus. White root and butt rot. **1**:NY(82), WI(82). **4**:ID(82).
I. dryadeus. White root rot. **4**:NW states(82), ID(3202), MT(3202), OR(3202), WA(82,3202). **5**:NW states(82), ID(3202), OR(3202), WA(82,3202).
I. tomentosus. **4**:ID(3202), WA(3202).
Irpex lacteus. Sapwood rot. **1**:VT(82). **4**:NW states(3202).
Ischnoderma resinosum. White rot. **1**:ME(82), NC(1303), TN(82), VT(82). **4**:AK(1904), ID(3202), MT(3202,3832), WA(82,3202). **5**:AK(526,1904), ID(3202), MT(3202), OR(3202).
Junghuhnia collabens. **1**:NY(2168). **4**:ID(2168,3202), MT(3202). **5**:ID(3202).
J. luteoalba. **1**:PA(2168). **4**:ID(3202).
J. zonata. **1**:ID(128). **4**:ID(3202).
Kuehneromyces mutabilis. **6**:OR(3202).
Laetiporus sulphureus. Brown cubical heart rot. **1**:Unk.(82), NC(1307). **4**:AK(526,1904), ID(3202), MT(3202), OR(82), WA(82,3202).
Laurilia sulcata. On wood. **1**:NH(543), NY(543), PA(82), VT(82,543), WI(82). **4**:NW states(82), ID(538,1583,3202), OR(3202), WA(3202). **5**:ID(82,3202), MT(82), WA(82,3202). **6**:ID(543).
Lentinellus montanus. On wood. **5**:CA(661).
L. ursinus. **4**:NW states(3202), W states(2423).
Lentinus sp. On logs. **5**:CA(642), WA(3202).
Lenzites betulina. White sapwood rot. **1**:NY(82).
Leptosporomyces galzinii. On wood. **5**:CA(661).
L. mundus. On bark. **5**:CA(661).
Lindtneria thujatsugina. On duff. **4**:ID(2042).
Lycoperdon pyriforme. **1**:NC(1307).
Marasmiellus filopes. **6**:AK(526).
Marasmius pallidocephalus. On fallen needles. **6**:E & C states(1222).
Meruliopsis albostramineus. On wood. **1**:NC(1303).
M. bellus. On rotten wood. **5**:CA(642).
Meruliporia incrassata. Dry rot. **1**:Range of host(82).
Merulius sp. **6**:ID(3202).
Micromphale perforans. On fallen needles and twigs. **6**:AK(526).
Mycena fusco-ocula. On dead needles. **6**:WA(3326).
M. metata. **5**:WA(3202).
M. overholtsii. On rotten logs and stumps. **5**:CA(661).
M. plicosa. On dead needles. **6**:NC(3326).
Neolentinus adhaerens. On fallen trunks, and stumps. **1**:ME(266), NH(266).
N. kauffmanii. **4**:OR(3202), WA(3202).
N. lepideus. **4**:OR(3202), WA(3202).
Nivatogastrium nubigenum. **6**:OR(3202).
Oligoporus subpendulus. On wood. **1**:NY(82,1205,2681).
Oliveonia subviolacea. On bark. **1**:NC(128), NY(128).
Omphalina epichysium. On stumps. **1**:TN(273).
Osteina obducta. **4**:ID(3202).
Oxyporus corticola. **4**:WA(3202).
O. cuneatus. On logs. **4**:WA(638,3202).
O. nobilissimus. Butt rot. **4**:WA(82,3202).
O. similis. **4**:OR(3202), WA(3202).
Panellus serotinus. **4**:ID(3202).
Paullicorticium pearsonii. **1**:WI(2116).
Paxillus atrotomentosus. On decaying log. **1**:GA(1307), NC(1303,1307), SC(1307).
Peniophora pusilla. On bark. **1**:NH(1714).
Peniophora sp. **4**:WA(3202). **5**:OR(3202).
Perenniporia medulla-panis. **4**:MT(3202).
P. subacida. White rot. **1**:E states(82). **4**:AK(526,1904), ID(82,128,538,1583,3202), OR(82,3202), WA(82,3202). **5**:ID(3202).
P. tenuis. On wood. **1**:NC(1303).
Phaeolus schweinitzii. Brown cubical butt rot, root rot. **4**:AK(1904), ID(82,1583,3202), OR(82,3202), WA(82,3202). **5**:MT(82), OR(82).
Phanerochaete gigantea. White rot. **1**:MI(433). **4**:ID(3202).
P. sanguinea. White rot. **1**:WI(433).
Phellinus ferreus. White rot. **5**:CA(642,661).
P. ferrugineofuscus. **4**:AK(526), ID(3202).
P. ferruginosus. **6**:MT(128).
P. gilvus. Sapwood rot. **1**:PA(82,2962).
P. igniarius. **4**:WA(3202).
P. merrillii. On wood. **6**:TN(1205).
P. nigrolimitatus. White pocket rot. **4**:AK(526), ID(128,3202), OR(3202), WA(3202). **5**:CA(642,661), OR(82,3202).

6:OR(2047).
P. pini. White pocket heart rot. **1:**E states(82). **4:**Range of host(82), AK(1904), ID(3202), MT(3202,3832), OR(3202), WA(3202). **5:**AK(1904), ID(82,3202), OR(82,3202). **6:**MT(128).
P. punctatus. White rot. **1:**E states(82), GA(491,1439), OH(653). **4:**ID(82), WA(82). **5:**CA(638), OR(3202).
P. repandus. **4:**ID(128,3202), WA(3202).
[*P. robustus*]. On wood. **1:**Range of host(82), MA(656). **4:**AK(526). **5:**AK(526,1904). **6:**WA(128).
P. viticola. White pocket rot. **1:**NY(82). **4:**AK(82), ID(82,3202), MT(3202), WA(3202).
P. weirii. Root rot. **4:**ID(3202), OR(3202), WA(82,3202). **5:**OR(1072,1258), WA(1072).
Phlebia albida. On logs. **4:**ID(3202). **5:**CA(642,661), OR(3202).
P. livida. **4:**ID(3202). **6:**WA(2165).
P. merismoides. **4:**OR(4007).
P. rufa. **6:**OR(128).
P. tremellosus. **4:**WA(128). **6:**NY(128).
Phlebiella vaga. **1:**MA(128).
Pholiota adiposa. Brown-mottled butt rot. **4:**AK(1904), ID(82,3202).
P. alnicola. **1:**NC(1307).
P. flammans. On wood. **1:**NC(1303,1307).
P. squarrosoides. **1:**NC(1307).
Pholiota sp. **6:**ID(1017).
Physisporinus sanguinolentus. **4:**WA(3202). **6:**ID(128).
Piloderma bicolor. **1:**WI(2116).
Platygloea acanthophysa. On wood. **1:**WI(434).
Polyporus badius. Sapwood rot. **1:**MI(82). **4:**AK(526), WA(3202).
P. melanopus. Heart rot, sap rot. **4:**ID(82,3202).
P. varius. On wood. **4:**AK(1904), ID(3202).
Postia balsamea. **4:**WA(3202).
P. caesia. Brown rot. **1:**ME(82), NC(1303). **4:**ID(82).
P. ceriflua. On wood. **6:**Unk.(1205).
P. fragilis. On wood. **1:**NC(1303). **4:**ID(3202), WA(3202).
P. guttulata. Brown cubical rot. **1:**NY(82). **4:**ID(82,3202), MT(3202), WA(3202).
P. johnstonii. On wood. **5:**CA(661).
P. leucospongia. On logs. **5:**CA(642,661), ID(3202), OR(3202), WA(3202).
P. perdelicata. **4:**ID(3202), OR(3202), WA(3202).
P. placenta. **4:**ID(3202).
P. sericeomollis. **4:**ID(1583).
P. stiptica. Sapwood rot. **1:**NY(82). **4:**ID(3202).
P. undosa. Sapwood rot. **1:**NY(82,397). **4:**ID(397,3202).
Protodontia piceicola. **1:**WI(2116).
Psathyrella atomatoides. **1:**NC(1307).
Pseudohydnum gelatinosum. **1:**NC(1307).
Pseudomerulius aureus. **6:**NY(128).
P. curtisii. **1:**NC(1307).
Pseudotomentella griseopergamacea. On wood. **1:**NY(2037).
P. griseoveneta. **6:**OR(2038).
P. tristis. **6:**WA(3202).
Pycnoporellus alboluteus. Brown cubical rot. **4:**ID(82,3202), MT(3202), WA(3202). **5:**CA(642,661).
P. fulgens. Brown cubical rot. **1:**NY(82), TN(82). **4:**AK(1904), ID(82,3202), OR(3202), WA(3202).
Pycnoporus cinnabarinus. Sapwood rot. **1:**PA(82,2962), VT(82). **4:**ID(3202).
Resinicium bicolor. **4:**ID(1583).
R. furfuraceum. **1:**NH(128), WI(2116). **4:**ID(3202), WI(2116).
Schizophyllum commune. On wood. **1:**NC(1303). **4:**CA(82), WA(3202).
Scytinostroma arachnoideum. **4:**AK(128).
S. galactinum. Basal heart rot. **1:**MD(2090). **4:**OR(3202).
S. ochroleucum. **1:**NH(128).
Serpula himantioides. On wood. **5:**CA(642,661).
S. lacrimans. On logs. **5:**CA(661).
Sistotrema brinkmannii. On wood. **1:**CT(128).
S. raduloides. **4:**OR(3202), WA(3202).
Skeletocutis alutacea. **4:**WA(128).
S. amorpha. On wood. **1:**PA(82). **4:**ID(82,3202), OR(82,3202).
Sphaerobasidium minutum. **1:**WI(2116).
Spongipellis spumeus. Sapwood rot. **1:**ME(82).
Steccherinum subcrinale. **1:**NY(128). **6:**WA(3202).
Stereum sanguinolentum. Heart rot. **1:**Range of host(82), CT(543), NH(543), NY(543), PA(543). **4:**NW states(82), AK(1904), ID(538,1583,3202). **5:**ID(82,3202), MT(82), WA(82). **6:**MA(543), ME(543), NY(2456).
Suillosporium cystidiatum. On wood. **1:**CT(128).
Tomentella atrorubra. On bark. **1:**MA(2030). **6:**MA(2039), NY(2039).
T. botryoides. **6:**NC(2039).
T. bryophila. **6:**MI(2039), MT(2039), NY(2039), VT(2039).
T. coerulea. **6:**MI(2039).
T. crinalis. **1:**NY(128). **6:**MT(3202), NY(2039).
T. griseo-umbrina. On wood. **1:**NY(2032,2039).
T. lateritia. On wood. **1:**NY(2032). **6:**MI(2039), NY(2039).
T. neobourdotii. **6:**MI(2039), TN(2039).
T. olivascens. **6:**MA(2039), NY(2039).
T. punicea. **6:**MI(2039).
T. ruttneri. On wood. **6:**NY(2035).
T. sublilacina. **6:**MA(2039), NY(2039), WA(2039).
T. terrestris. **6:**NY(2039).
Trametes hirsuta. Sapwood rot. **1:**NY(82,2962). **4:**ID(3202).
T. pubescens. **4:**ID(3202).
T. versicolor. Sapwood rot. **1:**Range of host(82), NC(1307), NY(2962), PA(2962), SC(1307). **4:**Range of host(82), ID(3202), MT(3202), OR(3202), WA(3202).
Trechispora farinacea. On wood. **1:**WI(2116). **5:**CA(661).
T. fastidiosa. **5:**CA(661).
T. mollusca. **4:**ID(3202). **6:**WA(3202).
Trichaptum abietinum. White sapwood rot. **1:**Range of host(82). **4:**Range of host(82), AK(526,1904), ID(3202), MT(3202), OR(3202), WA(3202). **5:**CA(661), OR(3202), WA(3202).
T. biforme. Sapwood rot. **1:**GA(1439), NY(82), PA(82).
T. fuscoviolaceum. On wood. **6:**Unk.(1205).
Tricholomopsis decora. **1:**NC(1303,1307).
T. sulfureoides. On wood. **1:**NY(1853).
Tubulicrinis accedens. **1:**RI(3859).
T. angustus. **1:**WI(2116).
T. calothrix. **1:**WI(2116).
T. chaetophorus. On rotten wood. **5:**CA(642).
T. glebulosus. On wood. **1:**WI(2116).
T. sceptriferus. On wood. **1:**WI(2116).
Tulasnella fuscoviolacea. **4:**WA(3202).
Vararia investiens. **1:**CT(128).
Wolfiporia cocos. Brown cubical butt rot, root rot. **1:**WI(793). **4:**OR(793,3202). **5:**CA(661).
Wrightoporia lenta. Brown rot. **1:**NY(2682).
W. subrutilans. White rot. **1:**VT(2682).
Xenasma pulverulentum. **1:**WI(2116).
Xeromphalina campanella. **4:**ID(3202).

Deuteromycotina - Hyphomycetes

Berkleasmium concinnum. **1:**NC(1307), SC(1307).
Botrytis cinerea. Seedling twig blight. **4:**ID(82).
Botrytis sp. Twig blight. **1:**NJ(82).
Cladosporium herbarum. **4:**WA(3202).
Conoplea sp. **4:**WA(3202).
Corynespora olivacea. On limbs. **1:**WV(128,3547).
Cylindrocladium scoparium. Seedling root rot. **1:**NJ(82).
Fusarium sp. Root rot. **1:**NC(1303), PA(3858).
Haplotrichum ramosissimum. On roots. **1:**VT(2131).
Helminthosporium sp. **4:**ID(3202), WA(3202).
Leptographium wageneri. Black stain root disease. **5:**OR(1259).
L. wageneri var. *ponderosum*. Black stain root disease. **4:**OR(1471). **5:**OR(1471).
Phymatotrichopsis omnivora. Root rot. **3:**TX(82).
Tubercularia sp. On dead twigs. **1:**NC(1303).

Deuteromycotina - Coelomycetes

Ascochyta conicola. On cone scales. **1:**NY(82).
Coryneum stromatoideum. On dead limbs. **1:**NY(832,3547,3549).
Cytospora kunzei. Basal stem rot. **1:**CT(3781), NC(1303), NH(3781).
Cytospora sp. Twig and stem canker. **1:**MD(82), VA(82,3493).
Dothiorella sp. On exposed roots. **1:**NC(1303).
Foveostroma abietinum. On stems. **1:**Unk.(861), NC(1303,1307).
[*Kirschsteiniella applanata*]. **4:**ID(3202).
Phacidiopycnis pseudotsugae. Isolated from bleeding cankers. **1:**PA(1370).
Phomopsis occulta. On blighted twigs. **1:**MA(82), NJ(82), NY(82).
Sirococcus conigenus. Shoot-blight. **4:**AK(1130). **5:**AK(526).

Deuteromycotina - Other
Rhizoctonia solani. Damping-off. **1:**Range of host(82).

Piperaceae

Peperomia Ruiz & Pav.

1. *P. bicolor* Sodiro
2. *P. caperata* Yunck. [emerald-ripple peperomia, green-ripple peperomia, little-fantasy peperomia]
3. *P. metallica* (L.) Linden & Rodigas
4. *P. obtusifolia* (L.) A. Dietr. [American radiator-plant, baby-rubberplant, oval-leaf peperomia, pepper-face]
5. *P. obtusifolia* (L.) A. Dietr. cv. 'variegata' [oval-leaf peperomia]
6. *P. reflexa* A. Dietr.
7. *P. serpens* (Sw.) Loud.
8. *Peperomia* sp.

Oomycetes
Phytophthora nicotianae var. *parasitica*. Leaf rot. 'Phytophthora stem rot. Phytophthora root rot.' **1:**FL(35). **2:**FL(35). **4:**FL(35). **8:**FL(35), OH(3293).
P. palmivora. Root rot. **4:**HI(2903). **8:**CA(96).
Phytophthora sp. Cutting rot. **5:**CA(2529).
Pythium aphanidermatum. Leaf, root and stem rot. **4:**FL(35).
P. myriotylum. Root and stem rot. **4:**FL(35).
P. splendens. Cutting and stem rot, stunting. 'Pythium root rot.' **2:**FL(35). **4:**FL(35). **5:**CA(563). **7:**FL(35). **8:**FL(35), HI(2903).
Pythium sp. Cutting, leaf, root and stem rot. **1:**FL(35). **2:**HI(2903). **5:**CA(2529). **6:**HI(2903).

Ascomycotina
Glomerella cingulata. **4:**WA(3202).
Glomerella sp. Anthracnose. **4:**FL(35). **7:**FL(35).
Trichothyrium reptans. **8:**HI(2903).

Deuteromycotina - Hyphomycetes
Alternaria sp. Leaf spot. **4:**FL(35).
Botrytis cinerea. Gray mold. **4:**FL(35). **8:**FL(35).
Cephalosporium sp. Foliage rot, stem rot. **4:**FL(35).
Cercospora sp. Edema. **4:**FL(35). **7:**FL(35). **8:**FL(35).
Corynespora cassiicola. Leaf spot. **4:**FL(35).
Fusarium sp. Leaf and stem rot. **4:**FL(35).
Gliocladium sp. Stem rot. **4:**FL(35).
Myrothecium roridum. 'Myrothecium leaf spot.' **8:**NC(1303).
Myrothecium sp. Leaf and stem rot. **1:**FL(35). **2:**FL(35). **4:**FL(35). **8:**FL(35).

Deuteromycotina - Coelomycetes
Ascochyta sp. Leaf spot. **4:**FL(35).
Colletotrichum sp. **4:**Unk.(1300).
Gloeosporium sp. **4:**Unk.(1300).
Phyllosticta sp. Leaf spot. **4:**FL(35).

Deuteromycotina - Other
Rhizoctonia solani. Leaf and stem blight, cutting rot. 'Rhizoctonia leaf spot.' **1:**FL(35). **2:**FL(35). **4:**FL(35). **5:**CA(2529). **7:**FL(35). **8:**FL(35).
Rhizoctonia sp. Root rot. **3:**HI(2903). **4:**HI(2903).
Sclerotium rolfsii. 'Southern blight.' **4:**FL(35). **8:**FL(35).

Piper L.

1. *P. nigrum* L. [black pepper, pepperplant, white pepper]
2. *P. ornatum* N. E. Brown [Celebes pepper]
3. *Piper* sp.

Oomycetes
Pythium splendens. Root rot. **2:**HI(2903).

Deuteromycotina - Hyphomycetes
Cercospora sp. Leaf spot. **3:**FL(35).

Deuteromycotina - Coelomycetes
Colletotrichum sp. Leaf spot. **3:**FL(35).

Deuteromycotina - Other
Sclerotium rolfsii. Southern blight and wilt. **1:**FL(35,82).

Pittosporaceae

Pittosporum Banks ex Sol. in Gaertn.

1. *P. crassifolium* Sol. ex Putterl. [karo]
2. *P. tobira* (Thunb.) W. T. Aiton [Australian-laurel, house blooming mock-orange, Japanese pittosporum]
3. *Pittosporum* sp.

Oomycetes
Phytophthora sp. Root rot. **2:**FL(35).
Pythium irregulare. Root rot. **2:**FL(35).
P. splendens. Root rot. **2:**FL(35). **3:**FL(35).
Pythium sp. Root rot. **1:**HI(2903). **2:**HI(2903).

Ascomycotina
Botryosphaeria quercuum. Dieback. **2:**FL(35).
B. rhodina. On limbs. **3:**FL(82).
Capnodium elongatum. On leaves and twigs. **3:**CA(2434).
Diaporthe sp. Stem canker. **2:**FL(35).
Hypoxylon sp. Dieback. **2:**FL(35).
Leptosphaeria sp. Stem gall. **2:**FL(35).
Mycosphaerella pittospori. On leaves. **3:**CA(82), SC(82).
Nectriella pironii. Stem gall. **2:**FL(35).
Stictis monilifera. **3:**CA(3225).

Basidiomycotina - Other
Armillaria mellea. **2:**CA(2898).
A. tabescens. Root rot. **2:**FL(35).
Ceratobasidium ramicola. On leaves. **2:**FL(3689).
Erythricium salmonicolor. Limb blight. **2:**FL(35).
Pellicularia koleroga. **2:**LA(3640).

Deuteromycotina - Hyphomycetes
Alternaria tenuissima. Leaf spot. **2:**FL(35). **3:**FL(35).
Cercospora pittospori. Angular leaf spot. **2:**SC states(82), SE states(82,581), FL(35), HI(2903), MS(2720). **3:**FL(35).
Corynespora sp. Leaf spot. **2:**FL(35).
Fusarium oxysporum. Stem canker, leaf spot. **2:**FL(35).
Fusarium sp. Root rot. **2:**FL(35).
Phymatotrichopsis omnivora. Root rot. **3:**TX(82).
Verticillium albo-atrum. Wilt. **3:**CA(82).

Deuteromycotina - Coelomycetes
Colletotrichum sp. Anthracnose. **2:**FL(35). **3:**FL(35).
Coniothyrium sp. Twig dieback. **2:**FL(35).
Diplodia mutila. On twigs. **3:**CA(82).
Diplodia sp. Foot rot. **2:**FL(35). **3:**FL(82).
Fusicoccum sp. Dieback. **2:**FL(35).
Phomopsis archeri. Stem canker. **2:**FL(35).
P. pittospori. On twigs. **3:**CA(82,637).
Phyllosticta sp. Leaf spot. **2:**FL(35). **3:**AL(82), MS(2720), NC(1303).
Pyrenochaeta sp. Basal stem rot. **2:**FL(35).
Rhabdospora pittospori. **3:**HI(2903).
Sphaeropsis tumefaciens. Dieback, stem gall. **2:**FL(35).
Stagonospora sp. Stem knot. **2:**FL(35).

Deuteromycotina - Other
Rhizoctonia ramicola. Thread blight. **2:**FL(35). **3:**FL(82).
R. solani. Root rot. **2:**FL(35). **3:**FL(35).
Rhizoctonia sp. Root rot. **2:**HI(2903).
Sclerotium rolfsii. Southern blight and wilt. **2:**FL(35). **3:**FL(35,82), TX(82).

Plantaginaceae

Plantago L.

1. *P. aristata* Michx. [bracted plantain]
2. *P. eriopoda* Torr.
3. *P. heterophylla* Nutt.
4. *P. hirtella* H. B. K.
5. *P. lanceolata* L. [buckhorn plantain, English plantain, narrow-leaved plantain, ribgrass, ripplegrass]
6. *P. macrocarpa* Cham. & Schlechtend.
7. *P. major* L. (=*P. asiatica* L.) [broadleaf plantain, cart-track-plant, common plantain, great plantain, white-man's-foot]
8. *P. maritima* L. (=*P. alpina* Sieber)
9. *P. ovata* Forssk.
10. *P. purshii* Roem. & Schult.
11. *P. rhodosperma* Decne.
12. *P. rugelii* Decne. [blackseed plantain, Rugel's plantain]

13. *P. tweedyi* A. Gray
14. *P. virginica* L. [paleseed plantain]
15. *Plantago* sp.

Myxomycetes

Physarum cinereum. Slime mold. **5**:E & C states(82). **7**:E & C states(82).

Chytridiomycetes

Synchytrium plantagineum. Leaf gall. **5**:MS(82,2720). **14**:MS(82,2720), PA(82), TX(82).
S. texanum. **11**:TX(1851).
Synchytrium sp. **13**:WY(2924).

Oomycetes

Peronospora alta. Downy mildew. **1**:SE states(82), TX(82), WI(82,1335). **5**:FL(35). **7**:NE states(82), ID(3202), MO(82), MS(2720), OR(3202), SD(2243), WA(82,3202). **10**:WI(82,1331). **12**:C states(82), NE states(82), WI(82). **15**:AK(82,526), MT(82,3202), TX(82).
P. plantaginis. On leaves. **1**:AL(3713), NC(3944), WI(1340). **10**:WI(1340).

Ascomycotina

Elsinoe sp. Scab. **5**:MD(1736). **12**:MD(1736).
Erysiphe cichoracearum. Powdery mildew. **1**:TX(82). **5**:NC(82,1303,3944), TX(82). **7**:E states(82), CA(81), CO(82), MA(1390), MS(2720), MT(3202), NC(1303,3944), ND(377), OK(2872), SD(2243), WA(82,3202), WY(1195). **8**:CA(82). **12**:E states(82), GA(1438), MT(82), NC(1303,3944), NE(82). **15**:NC(82), TX(82).
Eudimeriolum plantaginis. On leaves. **14**:IL(82,3590).
Hormotheca plantaginis. On leaves. **7**:CA(82), IL(82), MD(82), NJ(82), PA(82), VA(82). **12**:VT(82), WI(82).
Leptosphaeria sp. **5**:ID(3202). **8**:AK(3431).
Mycosphaerella columbi. Leaf spot. **12**:OH(82).
M. pachyasca. **7**:ME(82), WV(82). **15**:ND(82).
M. plantaginicola. **5**:PA(82). **7**:NY(82).
Mycosphaerella sp. On leaves. **12**:WI(1360).
Nodulosphaeria winteri. **8**:CA(3245).
Phyllactinia guttata. **7**:WA(3202).
[*Sclerotinia trifoliorum*]. Root rot. **5**:KY(82), OH(82).
Sphaerotheca macularis. Powdery mildew. **2**:WY(82,1195). **7**:WI(82). **15**:NJ(82).

Basidiomycotina - Uredinales

[*Puccinia aristidae*]. **1**:TX(82). **2**:CO(82). **15**:TX(82).
P. pacifica. Autoecious. **8**:CA(81,82).
P. subnitens. Heteroecious: aecial host. **2**:CO(1195), ND(82), NE(82), SD(82), WA(82). **5**:WY(1195).
[*Uromyces peckianus*]. **1**:OK(2872). **2**:MT(3202), WA(3202). **3**:OK(2872). **10**:OK(2872), WI(1333). **14**:MS(2720), OK(2872).
U. seditiosus. Heteroecious: aecial host. **1**:C states(82). **10**:NE(82), OK(82), TX(82). **12**:MO(82). **14**:SE states(82), IL(82).

Basidiomycotina - Ustilaginales

Sporidiobolus pararoseus. On mildewed leaves. **15**:KS(2686).

Basidiomycotina - Other

Herpobasidium filicinum. Frond spot. **12**:NY(82).
Thanatephorus cucumeris. On leafstalks. **5**:WI(1355). **7**:IL(82), WA(3202).

Deuteromycotina - Hyphomycetes

Cercospora plantaginis. Leaf spot. **5**:E states(82), FL(35), KS(82), NC(82), WI(1338). **7**:E states(82), KS(82). **12**:IL(82), IN(82), WI(82,1340). **15**:HI(82,2903), WI(1333).
Fusarium oxysporum. Wilt. **9**:AZ(3077).
Phymatotrichopsis omnivora. Root rot. **1**:TX(82). **5**:TX(82). **15**:TX(82).
Ramularia lanceolata. Leaf spot. **5**:CA(82), MA(82,1394), NY(82).
R. rhabdospora. Leaf spot. **1**:TX(82). **4**:CA(81). **5**:NY(82), OR(82,3202), TX(82), WA(82,3202), WI(1352). **7**:E states(82), DE(3543), ID(3202), MN(82), MS(2720), NE(82), WA(3202). **12**:IN(82), NY(82), WI(82,1360). **14**:NE(82). **15**:AK(526), ID(3202), TX(82), WA(3202).
Ramularia sp. **7**:AK(526).

Deuteromycotina - Coelomycetes

Ascochyta sodalis. Leaf spot. **1**:OK(82,2872). **7**:OK(82,2872). **12**:IL(82).
Phoma exigua. **12**:WI(1352).
P. polygramma var. *plantaginis*. Stem blight. **5**:NJ(82). **12**:WI(1360).
Phomopsis subordinaria. On stems. **1**:IA(82). **5**:NY(82), PA(82), WA(3202).
Phyllosticta atomata. Leaf spot. **12**:IL(82).
P. plantaginella. Leaf spot. **7**:NY(82).
P. plantaginicola. Leaf spot. **14**:IL(82,3590).
P. plantaginis. Leaf spot. **6**:AK(526). **15**:AK(82).
Phyllosticta sp. Leaf spot. **7**:Unk.(82). **12**:Unk.(82), WI(1347,1354). **14**:Unk.(82). **15**:Unk.(82).
Rhabdospora continua. On flower stalks. **7**:MI(82), PA(82).
Septoria inconspicua. Leaf spot. **1**:MO(82), TX(82). **5**:ID(3202), NJ(82), TX(82), WA(3202), WI(1338). **7**:NY(82), WI(1340). **10**:OK(82,2872).
S. plantaginis. Leaf spot. **1**:WI(1335). **7**:DE(82), VA(82), WI(82). **10**:WI(82).
S. plantaginis-majoris. **10**:WI(1329).
Septoria sp. Leaf spot. **7**:OK(2872).
Sphaceloma plantaginis. Spot anthracnose, scab. **5**:MD(1745). **7**:MA(82,1745), ME(82), OK(82,2873). **12**:NE states(82), DC(1745), IL(82), IN(1745), MD(1745), NY(1745,2456), OH(1745), OK(82,2873), PA(1745), VA(1745), WI(82,1332,1745). **15**:IA(82,3749), IL(82,3749), IN(82,3749), KS(82,3749), MA(1745), MD(1736), MO(82,3749), NJ(1745), OH(82,3749), WI(82,3749).

Deuteromycotina - Other

Rhizoctonia solani. Damping-off. **1**:IL(82), TX(82). **5**:IL(82), TX(82). **7**:IL(82). **15**:TX(82).
Sclerotium rolfsii. Southern blight. **5**:FL(35). **12**:MD(82).

Platanaceae

Platanus L.

1. *P.* × *acerifolia* (Aiton) Willd. *P. occidentalis* × *P. orientalis*. [London planetree]
2. *P. occidentalis* L. [American sycamore, buttonwood, eastern sycamore, planetree]
3. *P. orientalis* L. [oriental planetree]
4. *P. racemosa* Nutt.
5. *P. wrightii* S. Wats.
6. *Platanus* sp.

Oomycetes

Phytophthora cinnamomi. Seedling root rot. **3**:SE states(3624), MD(82).
Phytophthora sp. Root rot. **2**:FL(35).
Pythium sp. Root rot. **2**:FL(35).

Ascomycotina

Apiognomonia errabunda. Anthracnose. **2**:NC(1303).
A. veneta. Anthracnose, twig blight. **2**:Range of host(82), CT(3770), FL(35), IL(1052), MA(3780), MD(3780), MO(3780), MS(2720), NC(1533,3939,3944), OK(618,2872), PA(3780). **3**:Range of host(82). **4**:CA(82). **1**:CA(82), NJ(82), NY(82). **6**:FL(35), WI(1342).
Botryosphaeria dothidea. Canker. **2**:FL(35), GA(1438).
B. obtusa. On limbs. **2**:SE states(82), GA(82).
B. rhodina. On limbs. **2**:NC(82).
B. ribis. On limbs. **2**:FL(82), PA(82).
Ceratocystis fimbriata f. *platani*. Blight, canker. **2**:AR(2310), GA(3037), MS(82,2720), PA(82), SC(3037), VA(82), WV(82). **1**:DE(82), KY(82), MD(82), MO(82), MS(82), NC(82), NJ(82), NY(82), OH(82), PA(82), TN(82), VA(82), WV(82).
Chaetomium globosum. Isolated from achenes. **2**:SE states(1014).
Chaetosphaeria pannicola. **2**:GA(82,1438).
Coniochaeta pulveracea. **2**:GA(1438).
Daldinia eschscholzii. On roots of living tree. **5**:AZ(1194).
Diaporthe arctii. **2**:GA(82,1438).
Diatrypella prominens. **2**:GA(1438), IA(3629), NC(3944).
Eutypella leprosa. **2**:IA(3629).
E. platani. On trunks and limbs. **2**:GA(1438), MS(2720).
Eutypella sp. **2**:GA(82).
Gnomonia sp. Anthracnose. **2**:VA(3493).
Hapalocystis berkeleyi. On dead limbs. **2**:NY(959).
H. corticale. On twigs. **2**:NC(3809), NY(3809).
Hypoxylon illitum. **2**:GA(1438).
H. rubiginosum. **2**:GA(1438).
H. tinctor. Canker. **2**:GA(1438,1811,2292).
Hypoxylon sp. Dieback. **2**:GA(82), OK(2872).

Lachnella alboviolascens. **5**:AZ(1202).
Leptosphaeria platanicola. **2**:GA(1438), MD(2783). **6**:NY(967).
Lophiostoma caulium. **2**:GA(1438).
Microsphaera extensa. Powdery mildew. **2**:NC(82,1303).
M. penicillata. Powdery mildew. **2**:Range of host(82), FL(35), GA(1438), MS(2720), NC(3944), OK(2872). **3**:Range of host(82). **4**:CA(82). **1**:Unk.(3976), CA(81), NJ(82), NY(82). **6**:FL(35).
Mycosphaerella platanifolia. Leaf spot. **2**:SC states(82), SE states(82), FL(35), GA(1438), IA(82), NC(3939,3944).
M. punctiformis. On leaves. **6**:Unk.(3943).
M. stigmina-platani. Leaf blight and spot. **2**:NC(3939,3944), OK(2872).
Nectria ochroleuca. **2**:DE(3085).
N. peziza. On wood. **2**:GA(1438).
Nectria sp. Canker. **2**:OK(618).
Ophiodothella leucospila. On fallen leaves. **2**:GA(82,1438).
Ophiostoma piliferum. On wood. **2**:DC(3714).
Phyllactinia guttata. Powdery mildew. **2**:IN(82).
Physalospora abdita. On limbs. **2**:NC(82).
Rhytidhysteron rufulum. **2**:GA(1438).
Rosellinia aquila. On decaying wood. **6**:IA(1225).
Schizoxylon insigne. **2**:GA(1438).
S. melleum. **6**:VT(3225).
Sordaria fimicola. Isolated from achenes. **2**:SE states(1014).
Sphaerella platani. **2**:KS(82).
Splanchnonema platani. Branch canker. **2**:FL(35), GA(82,1438), IA(82), IN(82), KS(82). **4**:CA(82). **1**:NJ(82). **6**:CA(188), DC(188), IA(188), MA(188), MD(188), SC(188), VA(188).
S. semitectum. On limbs. **6**:VA(188).
S. sporadicum. On limbs. **6**:DC(188).
Valsella sp. Canker. **2**:OK(618).
Xylaria sp. Isolated from achenes. **2**:SE states(1014).

Basidiomycotina - Other

Antrodia albida. **5**:AZ(1202).
A. heteromorpha. **5**:AZ(1202).
A. malicola. **5**:AZ(1202).
A. radiculosa. **5**:AZ(1202).
Armillaria mellea. Root rot. **2**:MD(82), OK(2872), TX(82), WV(82). **4**:CA(82).
Basidiodendron caesiocinereum. **5**:AZ(1202).
B. cinereum. **5**:AZ(1202).
Bjerkandera adusta. On wood. **2**:Unk.(82), NC(1303,3944).
Ceraceomyces serpens. **5**:AZ(1202).
Ceriporia purpurea. **5**:AZ(128,1202).
C. spissa. **5**:AZ(1202).
C. viridans. **5**:AZ(128,1202).
C. xylostromatoides. **5**:AZ(1202).
[*Ceriporiopsis rivulosa*]. **5**:AZ(1202). **6**:AZ(128).
Chondrostereum purpureum. Wood rot. **2**:Range of host(82).
Climacodon pulcherrimus. **5**:AZ(1202).
Coniophora arida. **5**:AZ(1202).
C. arida var. *suffocata*. **5**:AZ(128).
C. olivacea. **5**:AZ(1202).
Coriolopsis gallica. **5**:AZ(1202).
Dacrymyces capitatus. **2**:NY(2456).
[*Dacryobolus sudans*]. **5**:AZ(128,1202).
Dacryopinax spathularia. On wood. **2**:VA(128). **5**:AZ(1202).
Daedaleopsis confragosa. Wood rot. **2**:Range of host(82).
Datronia scutellata. **2**:MD(82).
Dendrophora albobadia. **2**:GA(543). **5**:AZ(1202).
Dendrothele griseo-cana. On bark. **2**:NY(2084).
D. microspora. On bark. **2**:MA(2084), NY(2084).
Diplomitoporus overholtsii. **5**:AZ(128,1202). **6**:AZ(128).
Fomitopsis pinicola. **2**:Range of host(82).
Ganoderma applanatum. White-mottled heart rot. **2**:CT(82).
G. lucidum. Heart and root rot. **2**:NC(3944), OK(618). **5**:AZ(1202).
Gloeoporus dichrous. **5**:AZ(1202).
Grandinia pruni. **5**:AZ(128,1202).
G. spathulata. **5**:AZ(1202).
Hapalopilus rutilans. **5**:AZ(1202).
Heliocybe sulcata. **5**:AZ(1202).
Hericium americanum. On wood. **6**:PA(1237).
H. erinaceus. Usually on living trees. **2**:NC(82).
Hyphoderma guttuliferum. **5**:AZ(128,1202).
H. praetermissum. **5**:AZ(128,1202).
H. roseocremeum. **5**:AZ(1202).
H. setigerum. **5**:AZ(1202).
Hypsizygus ulmarius. On butt of living tree. **2**:WV().
Inonotus arizonicus. Heart rot of living trees. **4**:AZ(1181), CA(1176). **5**:AZ(128,1176,1181,1194,1202,1260).
I. munzii. **5**:AZ(1202).
Laxitextum bicolor. Wood rot. **2**:Range of host(82), IN(543).
Lentinus tigrinus. **5**:AZ(1202).
Leucogyrophana olivascens. **2**:CT(128).
Lopharia cinerascens. **6**:OH(543).
Oxyporus corticola. **5**:AZ(1194,1202).
O. populinus. **5**:NM(1190).
Peniophora fuscomarginata. **5**:AZ(1202).
P. incarnata. On wood. **2**:GA(491,1439).
P. malenconii subsp. *americana*. On wood. **6**:CA(542).
P. nuda. **5**:AZ(1194,1202).
Phanerochaete allantospora. **5**:AZ(1202).
P. arizonica. White rot. **5**:AZ(433).
P. avellanea. **5**:AZ(1202).
P. burtii. **5**:AZ(1202).
P. carnosa. **5**:NM(128).
P. chrysorhiza. **5**:AZ(1202).
P. chrysosporium. **5**:AZ(433,435).
P. crassa. White rot. **5**:AZ(433,1202).
P. filamentosa. White rot. **5**:AZ(433).
P. omnivora. White rot. **5**:AZ(433).
P. sordida. **5**:AZ(1202).
P. tuberculata. White rot. **5**:AZ(433,1194,1202).
P. xerophila. **5**:AZ(433).
Phellinus gilvus. On wood. **2**:Unk.(82), NC(1303,3944). **5**:AZ(1194,1202).
Phlebia ludoviciana. **5**:AZ(128,1202).
P. merismoides. **5**:AZ(1202).
P. subochracea. White rot. **5**:AZ(2556).
Plicaturopsis crispa. **2**:NC(3944).
Pluteus longistriatus. **5**:AZ(1202).
Polyporus arcularius. **5**:AZ(1202).
Postia stiptica. **5**:AZ(1202).
P. tephroleuca. Wood rot. **2**:Unk.(82).
Pycnoporus sanguineus. Wood rot. **2**:Unk.(82).
Resupinatus applicatus. **5**:AZ(1202).
Rigidoporus lineatus. On wood. **2**:Unk.(82).
Steccherinum laeticolor. On wood. **6**:MI(1175).
Stereum hirsutum. **2**:DC(543).
S. ostrea. On wood. **2**:OK(82,2873).
Stigmatolemma poriaeforme. **2**:VA(128). **5**:AZ(128,1202).
Tomentella chlorina. **6**:AZ(2039).
T. ferruginea. **5**:AZ(2030). **6**:AZ(2039).
T. lateritia. On wood. **6**:AZ(2032,2039).
T. pilosa. **5**:AZ(1202). **6**:AZ(2039).
T. viridescens. **6**:AZ(2039).
T. viridula. **6**:AZ(2039).
Trametes cervina. **5**:AZ(1202).
T. elegans. Wood rot. **2**:Range of host(82).
T. hirsuta. Wood rot. **2**:Unk.(82).
T. versicolor. Wound rot. **2**:MD(82), NC(3944). **5**:AZ(1202).
Trechispora farinacea. **5**:AZ(128,1202).
Tyromyces galactinus. Wood rot. **2**:Unk.(82).
Xylobolus subpileatus. On wood. **2**:GA(491,1439).

Deuteromycotina - Hyphomycetes

Alternaria sp. Isolated from achenes. **2**:SE states(1014).
Aspergillus hollandicus. Isolated from achenes. **2**:SE states(1014).
A. niger. Isolated from achenes. **2**:SE states(1014).
A. ochraceus. Isolated from achenes. **2**:SE states(1014).
A. reptans. Isolated from achenes. **2**:SE states(1014).
A. rubrobrunneus. Isolated from achenes. **2**:SE states(1014).
Aureobasidium pullulans. Isolated from achenes. **2**:SE states(1014).
Bipolaris spicifera. Isolated from achenes. **2**:SE states(1014).
Bipolaris sp. Isolated from achenes. **2**:SE states(1014).
Calcarisporium sp. Isolated from achenes. **2**:SE states(1014).
Cercospora platanicola. Leaf spot. **2**:Unk.(82), FL(35), NC(1303,3939,3944). **6**:FL(35).
Cladosporium cladosporioides. Isolated from achenes. **2**:SE states(1014).
Cristulariella moricola. Leaf spot. **2**:FL(35), IL(2572).

Curvularia intermedia. Isolated from achenes. **2**:SE states(1014).
Dendrodochium rubellum. **2**:DE(3543).
Epicoccum nigrum. Isolated from achenes. **2**:SE states(1014).
Exosporium platani. Leaf spot. **2**:MS(2720).
Filosporella annelidica. On submerged leaves. **2**:IL(3216).
Fusarium merismoides. Stem canker. **2**:FL(35).
F. moniliforme. Isolated from achenes. **2**:SE states(1014).
F. roseum. Isolated from achenes. **2**:SE states(1014).
Fusicladium sp. Leaf spot. **6**:FL(35).
Harpographium fasciculatum. On bark and wood. **2**:IN(2509). **6**:OH(2509).
Helicoma perelegans. On bark and wood. **6**:MA(1271).
Helicomyces ambiguus. On wood. **6**:OH(1270).
H. roseus. On bark. **6**:IN(1270).
Monodictys antiqua. **2**:DE(3543).
Nigrospora sp. Isolated from achenes. **2**:SE states(1014).
Nodulisporium hinnuleum. Isolated from achenes. **2**:SE states(1014).
Oidium obductum. Powdery mildew. **3**:PA(82), VA(82), WV(82).
Penicillium brevicompactum. Isolated from achenes. **2**:SE states(1014).
Phymatotrichopsis omnivora. Root rot. **2**:TX(82). **5**:AZ(82).
Polyschema bicellularis. On twigs. **2**:IL(3214).
Ramularia sp. Leaf spot. **2**:FL(35).
Stemphylium sp. Isolated from achenes. **2**:SE states(1014).
Stigmina platani. Leaf spot. **2**:Unk.(82), FL(35), MS(2720), NC(3939,3944). **3**:CA(82).
S. platani-racemosae. Leaf spot. **4**:CA(82).
S. tharpii. Leaf spot. **2**:FL(35), TX(82,2500,3597). **6**:FL(35).
Trichoderma koningii. **2**:SE states(1014).
Trichothecium roseum. **2**:DE(3543).
Xenosporium berkeleyi. **2**:NC(3944).
Deuteromycotina - Coelomycetes
Botryodiplodia sp. Dieback. **2**:FL(35).
Coniothyrium sp. Isolated from achenes. **2**:SE states(1014).
Cytospora platani. Canker. **2**:E states(82), IA(82), OK(2872), TX(82).
Cytospora sp. Isolated from achenes. **2**:SE states(1014).
Diplodia sp. Stem canker. **2**:SE states(1014), FL(35).
Diplodina sp. Isolated from achenes. **2**:SE states(1014).
Discosia artocreas. **2**:NC(3944).
Discula platani. On petioles. **6**:SC(634).
Dothiorella sp. Branch canker. **2**:FL(35). **1**:NY(82).
Gloeodes pomigena. On limbs. **2**:IN(82).
Lasiodiplodia theobromae. Basal canker, twig canker. **2**:AR(2310), GA(3037), MS(1059,3662), SC(3037).
Libertella platani. **2**:OH(653).
Libertella sp. On limbs. **2**:LA(82).
Macrodiplodiopsis desmazieresii. On dead limbs. **6**:CA(3549).
Pestalotia sp. Isolated from achenes. **2**:SE states(1014).
Peyronellaea sp. Isolated from achenes. **2**:SE states(1014).
Phloeospora multimaculans. Leaf blight. **2**:IN(82), TX(82).
Phoma pezizoides. On decorticated wood. **2**:WV(307).
Phoma sp. Isolated from achenes. **2**:SE states(1014).
Phomopsis scabra. On twigs. **2**:OH(82), OK(82,2873), WV(82).
Phomopsis sp. Stem canker. **2**:SE states(1014), FL(35).
Phyllosticta platani. Leaf spot. **2**:AL(82), KS(82), MA(82), MD(82), VA(82). **6**:FL(35).
Septoria platanifolia. Leaf spot. **2**:GA(82), IA(82), MD(82), SC(82), TX(82), WV(82). **6**:SC(634).
Sphaceloma sp. Spot anthracnose. **2**:FL(35,2525).
Stilbospora sp. **6**:AZ(662).
Deuteromycotina - Other
Rhizoctonia solani. Seedling root rot. **2**:FL(35).

Plumbaginaceae

Armeria (DC.) Willd.

1. *A. maritima* (Mill.) Willd. (=*A. vulgaris* Willd.)
2. *A. maritima* (Mill.) Willd. var. *californica* (Boiss.) G. Lawr.

Basidiomycotina - Uredinales
Uromyces armeriae subsp. *pacificus*. Autoecious. **2**:OR(3114).

Limonium Mill.

1. *L. bonduellii* (Lestib.) Kuntze
2. *L. californicum* (Boiss. in DC.) A. Heller (=*L. commune* S. F. Gray var. *californicum* (Boiss. in DC.) Greene)
3. *L. carolinianum* (Walter) Britton
4. *L. latifolium* (Sm.) Kuntze
5. *L. limbatum* Small
6. *L. sinuatum* (L.) Mill.
7. *L. vulgare* Mill. (=*Statice limonium* L.)
8. *Limonium* sp.

Oomycetes
Phytophthora sp. Root rot. **8**:FL(35).
Pythium irregulare. Root rot. **8**:FL(35).
Basidiomycotina - Uredinales
Uromyces limonii. Autoecious. **2**:AZ(747,3981), CA(81,82,3114). **3**:MA(1390), MS(2720). **4**:CT(82,3834). **5**:NM(3114), TX(3114). **8**:FL(35).
U. limonii-caroliniani. Autoecious. **3**:E states(82), MS(3114), TX(82).
Deuteromycotina - Hyphomycetes
Alternaria sp. Flower blast, leaf and stem spot. **4**:CT(82). **6**:FL(35,709,710). **8**:FL(35).
Botrytis cinerea. Blossom, leaf and stem blight, damping-off. **1**:FL(1709). **6**:NC(3511). **8**:FL(35), HI(2903).
Cercospora insulana. Leaf spot. **1**:FL(35,1710). **6**:FL(35,709,1710). **8**:FL(35).
Cercospora sp. Leaf and stem spot. **3**:TX(82). **6**:HI(2899,2903). **8**:HI(2903).
Fusarium sp. Stem lesions. **8**:FL(35).
Fusicladium staticis. Leaf spot. **7**:DE(3543). **8**:TX(82).
Phymatotrichopsis omnivora. Root rot. **8**:TX(82).
Stemphylium sp. Flower blight, stem lesions. **8**:FL(35).
Deuteromycotina - Coelomycetes
Ascochyta plumbaginicola. Leaf spot. **4**:IA(82). **8**:FL(35).
Colletotrichum gloeosporioides. Flower, leaf and stem lesions, crown rot. **1**:FL(35,3371). **6**:FL(35).
Phyllosticta armeriae. **8**:WI(1355).
P. staticis. **3**:NY(82).
Phyllosticta sp. Leaf spot. **4**:CT(82).
Deuteromycotina - Other
Rhizoctonia solani. Crown and root rot. **1**:FL(1709). **6**:FL(35,1709). **8**:FL(35).
Rhizoctonia sp. Root rot. **6**:HI(2903).
Sclerotium rolfsii. Southern blight, crown rot. **1**:FL(1709). **6**:FL(35,1709). **8**:FL(711).

Plumbago L.

1. *P. auriculata* Lam. [Cape leadwort]
2. *Plumbago* sp.

Oomycetes
Phytophthora sp. Root rot. **1**:FL(35).
Pythium sp. Root rot. **1**:FL(35).
Ascomycotina
Nectriella pironii. Stem gall. **1**:FL(35).
Basidiomycotina - Other
Armillaria tabescens. Root rot. **1**:FL(35).
Deuteromycotina - Hyphomycetes
Alternaria sp. Leaf spot. **1**:FL(35).
Cercospora sp. Leaf spot. **1**:FL(35). **2**:FL(35).
Fusarium sp. Root rot. **1**:FL(35).
Oidium sp. Powdery mildew. **2**:FL(35).
Deuteromycotina - Coelomycetes
Phyllosticta sp. Leaf spot. **1**:FL(35).

Poaceae

Records with host identified to family only.

1. *Poaceae (Bambuseae)*

Ascomycotina
Apiospora sp. On dead culms. **1**:FL(82), GA(82).
Daldinia concentrica. On dead culms. **1**:MD(82).
Guignardia bambusae. On culms. **1**:MD(82).
Mycosphaerella sp. Leaf spot. **1**:CA(82).
Placostroma bambusae. On dead culms. **1**:FL(82), GA(82).
Scirrhia lophodermioides. On dead culms. **1**:HI(82).
Basidiomycotina - Uredinales
[*Puccinia melanocephala*]. **1**:FL(82), GA(82), MS(82), TX(82).
Uredo ignava. **1**:GA(82).

Basidiomycotina - Ustilaginales
Ustilago shiraiana. **1:**CA(82), FL(82), LA(82), MD(82), MS(82), TX(82).
Basidiomycotina - Other
Schizophyllum commune. On dead culms. **1:**CA(82), TX(82).
Deuteromycotina - Hyphomycetes
Arthrinium phaeospermum. On culms. **1:**Range of host(82).
Charomyces amphimelas. On debris. **1:**HI(3179).
Cladosporium herbarum. On leaves. **1:**SC(82).
Helminthosporium sp. Leaf spot. **1:**FL(82).
Papularia vinosa. On dead or living culms. **1:**TX(82).
Deuteromycotina - Coelomycetes
Cylindrosporium bambusae. On culms. **1:**GA(82).
Diplodia bambusae. On culms. **1:**LA(82), TX(82).
Hendersonia sp. On culms. **1:**LA(82).
Pseudoseptoria donacis. Culm spot. **1:**CA(82).

Aegilops L.

1. *A. cylindrica* Host [jointed goatgrass]
2. *A. ovata* L.
3. *A. triuncialis* L. (=*Triticum persicum* (Boiss.) Aitch. & Hemsl., *A. squarrosa* L.) [barb goat grass]
4. *Aegilops* sp. [goat grass]

Oomycetes
Pythium arrhenomanes. Root browning. **1:**ND(82).
P. debaryanum. Damping-off, root necrosis, seed rot. **1:**WA(82,3423). **3:**ND(3423).
P. graminicola. **3:**ND(3423).
Basidiomycotina - Uredinales
[*Puccinia graminis*]. **1:**KS(3021,3023), OK(2872). **4:**KS(82), MI(82), MN(82).
P. recondita. Heteroecious: telial host. **1:**KS(82), MI(82), OR(82), WA(82). **4:**KS(82), MI(82), OR(82), WA(82).
P. striiformis. **1:**WA(82).
Uromyces graminicola. **1:**OK(2872).
Basidiomycotina - Ustilaginales
Tilletia controversa. Dwarf bunt. **1:**UT(859).
Deuteromycotina - Hyphomycetes
Fusarium acuminatum. Pink mold, secondary root rot. **1:**OR(3423), WA(82,3423). **2:**OR(3423). **3:**OR(3423).
Fusarium sp. Head mold. **4:**MI(82).
Pseudocercosporella herpotrichoides. Foot rot, strawbreaker. **1:**OR(82,3423). **2:**OR(3423). **3:**OR(3423). **4:**OR(82).
Deuteromycotina - Coelomycetes
Ascochyta sp. Leaf spot. **1:**OK(2872).

×*Agrohordeum* E. G. Camus ex A. Camus

Agropyron × *Hordeum.*
1. × *A. macounii* (Vasey) Lepage (=*Elymus macounii* Vasey). *Agropyron trachycaulum* × *Hordeum jubatum.*

Oomycetes
Pythium debaryanum. Damping-off, root necrosis, seed rot. **1:**ND(3423).
P. graminicola. **1:**ND(3423).
Sclerophthora macrospora. **1:**SD(3184).
Ascomycotina
Claviceps purpurea. Ergot. **1:**MT(3423), ND(3423), SD(2242), WA(3423).
Phyllachora graminis. **1:**SD(2242).
Basidiomycotina - Uredinales
Puccinia graminis. **1:**SD(2242).
P. montanensis. **1:**CO(754).
Basidiomycotina - Ustilaginales
Urocystis agropyri. **1:**CO(1077), WA(1077).
Ustilago bullata. **1:**MT(2342), ND(1077), UT(1077), WA(1077), WY(1077).
U. striiformis. **1:**ID(1077), NE(1077), NM(1077), SD(1077,2242), WA(1077).
U. trebouxii. **1:**NV(1077).
Basidiomycotina - Other
Thanatephorus cucumeris. Brown patch, root rot, sharp eye spot. **1:**ND(3423).
Deuteromycotina - Hyphomycetes
Bipolaris sorokiniana. Seedling blight, spot blotch, black point, root rot. **1:**ND(3423).
Cercosporidium graminis. Leaf streak. **1:**CO(3423), ND(3423), WY(3423).
Fusarium acuminatum. Secondary root rot. **1:**ND(3423), WY(3423).
Microdochium bolleyi. Root necrosis. **1:**ND(3423).
M. nivale. Snow mold. **1:**WA(3423).
Deuteromycotina - Coelomycetes
Ascochyta brachypodii. Leaf spot. **1:**ND(3423).
A. sorghi. Leaf spot. **1:**MT(3423), ND(3423).
Septoria agropyrina. Leaf blotch, leaf spot. **1:**ND(3423), SD(2242).
Stagonospora avenae f. sp. *triticea.* On dead leaf tips. **1:**MT(3433).

×*Agropogon* P. Fourn.

Agrostis × *Polypogon.*
1. × *A. littoralis* (Sm.) Hubb. (=*Polypogon littoralis* Sm.)

Basidiomycotina - Uredinales
Puccinia coronata. **1:**CA(81).

Agropyron Gaertn.

1. *A. albicans* Lams.-Scribn. & J. G. Sm. (=*A. griffithsii* Lams.-Scribn. & J. G. Sm. ex Piper)
2. *A. arizonicum* Lams.-Scribn. & J. G. Sm.
3. *A. bakeri* E. E. Nels. [Baker wheatgrass]
4. *A. boreale* (Turcz.) Drobov var. *alaskanum* (Lams.-Scribn. & Merr.) Welsh (=*A. alaskanum* Lams.-Scribn. & Merr. var. *arcticum* Hultén)
5. *A. cristatum* (L.) Gaertn. [fairway crested wheatgrass]
6. *A. dasystachyum* (Hook.) Lams.-Scribn. (=*A. riparium* Lams.-Scribn. & J. G. Sm.) [thick-spike wheatgrass]
7. *A. desertorum* (Fisch. ex Link) Schult. [standard crested wheatgrass]
8. *A. elmeri* Lams.-Scribn.
9. *A. elongatum* (Host) P. Beauv. (=*A. rigidum* (Schrad.) P. Beauv.) [tall wheatgrass]
10. *A. fragile* (Roth) Candargy subsp. *sibiricum* Willd. (=*A. sibiricum* (Willd.) P. Beauv.) [Siberian wheatgrass]
11. *A. gmelinii* Lams.-Scribn. & Sm.
12. *A. intermedium* (Host) P. Beauv. (=*A. trichophorum* (Link) K. Richt.) [intermediate wheatgrass, pubescent wheatgrass]
13. *A. michnoi* Roshevitz
14. *A. mongolicum* Keng
15. *A.* × *pseudorepens* Lams.-Scribn. & Sm. *Agropyron trachycaulum* × *Agropyron smithii.*
16. *A. repens* (L.) P. Beauv. (=*Triticum repens* L., *Elytrigia repens* (L.) Nevski) [couchgrass, quackgrass]
17. *A. scabraglume* (Hack.) Parodi
18. *A. spicatum* (Pursh) Lams.-Scribn. & J. G. Sm. (=*A. inerme* (Lams.-Scribn. & J. G. Sm.) Rydb.) [beardless wheatgrass, bluebunch wheatgrass]
19. *A.* × *subsecundum* (Link) Hitchc. var. *andinum* (Lams.-Scribn. & Sm.) Hitchc. *Agropyron scribneri* × *Agropyron violaceum.* [bearded wheatgrass]
20. *A. trachycaulum* (Link) Malte (=*A. caninum* (L.) P. Beauv., *A. pauciflorum* (Schwein.) Hitchc., *A. richardsonii* Schrad., *A. subsecundum* (Link) Hitchc., *A. tenerum* Vasey) [bearded couch, slender wheatgrass]
21. *A. trachycaulum* (Link) Malte ex H. Lewis var. *glaucum* (Pease & A. H. Moore) Malte [slender wheatgrass]
22. *Agropyron* sp. [cheatgrass]

Chytridiomycetes
Physoderma graminis. **16:**WI(1338).
Physoderma sp. Brown streak. **16:**IL(82), IN(82), MI(82,3408), WI(82).
Synchytrium graminicola. **20:**WA(3430).
Oomycetes
Phytophthora sp. Leaf rot. **10:**ND(82).
Pythium aristosporum. Root browning. **5:**MT(82). **20:**ND(82).
P. arrhenomanes. Root browning. **5:**ND(82), SD(82). **6:**ND(82), SD(82). **7:**ND(82), SD(82). **10:**ND(82), SD(82). **16:**MN(82), ND(82), SD(82). **18:**ND(82), SD(82). **20:**ND(82), SD(82). **22:**ND(82), SD(82).
P. debaryanum. Damping-off, root necrosis, seed rot. **5:**ID(3423), MN(3423), MT(3423), ND(82,3423), NE(3423), OR(3423), SD(2242,3423), WA(3423), WY(3423). **6:**ND(82,3423), SD(82), WA(3423), WY(82). **7:**ND(82,3423), SD(2242,3423). **10:**MN(3976), ND(82,3976), NE(3976), OR(3976), SD(3976), WA(3976), WY(3976). **12:**ND(3423), SD(2242,3423), WA(3423).

13:ND(3423), SD(3423). **14:**ND(3423). **16:**ID(3423), MN(82,3423), ND(82,3423), NE(3423), OR(3423), SD(2242,3423), WA(3423), WY(3423). **18:**OR(3423), WA(3423). **20:**MT(3976), ND(82,3423), NE(3976), SD(82), WA(82,3976), WY(82,3423). **22:**ND(82), SD(82), WY(82).

P. graminicola. Root browning. **5:**MN(82), MT(3423), ND(82,3423), NE(3423), SD(82,2242,3423), WA(3423). **6:**ND(3423), WA(3423). **7:**ND(82,3423), SD(2242,3423). **9:**ND(3423), NE(3423). **10:**MN(3423), ND(82,3423), NE(3423), SD(3423). **12:**ND(3423), SD(2242,3423), WA(3423). **13:**ND(3423), SD(3423), WA(3423). **14:**ND(3423). **16:**MN(3423), ND(3423), NE(3423), SD(2242,3423), WA(3423). **18:**ND(3423), WA(3423). **20:**ND(3423), WA(3423). **22:**ND(82,3423).

P. irregulare. Root necrosis. **5:**ND(3423). **12:**ND(3423), SD(2242,3423), WA(3423). **20:**ND(3423).

P. periilum. Root necrosis. **16:**MN(3423).

P. ultimum. Root necrosis. **5:**ND(82,3423), WA(3423). **7:**ND(82,3423). **10:**ND(82). **12:**ND(3423), SD(2242,3423). **16:**MN(82,3423), ND(82), WA(3423). **20:**ND(3423). **22:**ND(82).

Sclerophthora macrospora. **5:**SD(2242). **16:**SD(2242). **20:**SD(2242).

Ascomycotina

Ascospora graminis. On leaves. **6:**OR(82).

Botryosphaeria festucae. On dead leaves. **20:**ND(82,377).

Claviceps purpurea. Ergot. **1:**ND(3423). **5:**Range of host(82), IN(3423), MI(3423), MN(3423), MT(3423), ND(3423), OR(3423), SD(2242,3423), WA(638,3423), WY(3423). **6:**NC states(82), NW states(82), ID(3423), MT(3423), ND(3423), SD(2242), WA(3423), WY(1195,3423). **7:**MI(3423), ND(3423), SD(2242), WA(3423), WY(1195). **10:**Range of host(82), MI(3423), MN(3423), MT(3423), ND(3423). **11:**ND(3423). **12:**ND(3423), SD(2242,3423), WA(3423). **13:**ND(3423). **16:**Range of host(82), IA(3423), KS(3423), MA(3423), MI(3423), MN(3423), MO(3423), MT(3423), ND(377,3423), NE(3423), NY(3423), OH(3423), OK(2872,3423), OR(3423), SD(2242,3423), WA(3423), WI(3423), WV(952,3423). **18:**NC states(82), NGP(82), NW states(82), NW states(82), IA(3423), ID(3423), MT(3423), ND(3423), OR(3423), UT(3423), WA(3423), WY(3423). **21:**WI(1338). **22:**Range of host(82), WY(819).

Epichloe typhina. Choke. **18:**CO(1944), ID(3423), MT(3423), WA(3423). **20:**MT(3423), ND(3423), WA(3423).

Erysiphe graminis. Powdery mildew. **5:**Range of host(82), WA(3423). **6:**Range of host(82), CO(1195), ID(3423), OR(3423), WY(1195). **7:**Range of host(82). **10:**WA(3423). **12:**ND(3423), UT(3423). **16:**E states(3423), Range of host(82), CA(3423), CO(1195), IA(3423), ID(3423), KS(3021), MI(3423), MN(3423), MT(3423), ND(3423), NE(3423), NH(1890), OR(3423), SD(2242,3423), UT(3423), WA(3423), WI(1331,1352,3423), WV(953). **18:**Range of host(82), CO(3423), MT(3423), NV(3423), OR(3423), WA(3423). **20:**Range of host(82), CA(3423), CO(3423), MN(3423), MT(3423), ND(3423), NV(3430), OR(3423), SD(2242,3423), UT(3423), WA(3423), WI(1339,1352,3423), WY(1195,2924,3423). **22:**Range of host(82).

Gaeumannomyces graminis. Take-all. **16:**NY(82,3423). **20:**AK(3431).

Herpotrichia macrotricha. On culms. **16:**NY(82).

Leptosphaerulina trifolii. **16:**AL(759).

Lophiostoma arundinis. **22:**ND(377).

Lophodermium arundinaceum. **20:**AK(3431).

Mollisia sp. **20:**AK(3431).

Mycosphaerella tassiana. On culms. **5:**OR(82). **18:**AK(526).

M. tulasnei. **16:**AK(3431). **20:**AK(3431).

Ophiosphaerella herpotricha. On culms. **16:**MD(82), NY(82), VA(82), WI(82).

Phaeosphaeria culmorum. **20:**ND(377).

P. herpotrichoides. **22:**WA(3814).

Phyllachora graminis. Tar spot. **5:**MI(82). **16:**NE states(82), IA(82,3423), MA(3423), ME(3423), MN(82,3423), NH(1895,3423), NY(3423), PA(3423), RI(3423), SD(2242), VT(3423), WI(3423), WV(3423). **20:**NY(82,3423), WI(1339,3423).

Pleospora lutea. On dead stem and sheath. **4:**AK(1938).

Pyrenopeziza karstenii. **20:**AK(3431).

Sclerotinia trifoliorum. Crown and stem rot. **16:**NC(48).

Basidiomycotina - Uredinales

Puccinia coronata. Heteroecious: telial host. **10:**MI(82). **12:**SD(2242). **16:**MA(82), NH(82), SD(2242), VT(82), WI(1336). **20:**SD(2242), WI(1350), WY(749,1195).

P. graminis. Heteroecious: telial host. **12:**SD(2242). **15:**ND(378).

P. montanensis. Heteroecious: telial host. **2:**TX(742,754). **5:**ND(82,754), SD(2242). **6:**C states(82), W states(82), ID(754), ND(82), SD(2242), WA(82). **9:**ND(754). **10:**ND(82). **12:**ND(754), SD(2242). **16:**Range of host(82), MT(82), ND(82), SD(82,2242). **18:**C & W states(82), Range of host(82), ND(82), WA(82), WY(749,754,1195). **20:**C & W states(82), AK(754), IA(754), MI(82), ND(82,378), NM(82), SD(2242), UT(754), WA(82), WY(754,1195). **15:**CO(754), NM(3425). **22:**W states(82), ND(82).

P. pattersoniana. Heteroecious: telial host. **18:**ID(82), MT(82,1195), OR(82), UT(82), WA(82), WY(749).

P. recondita. Heteroecious: telial host. **1:**WY(1195). **2:**AZ(1195,1203,3981), TX(742). **5:**MI(82), ND(82). **6:**Unk.(82), MT(1195), WI(1357). **10:**MI(82), ND(82). **16:**SD(2242), WV(953). **20:**Unk.(82), AK(3431), AZ(1195,1203,3981), CO(1195), SD(2242), UT(1195), WY(1195). **21:**WI(1338). **15:**CO(1195). **22:**Unk.(82), WY(2924).

P. striiformis. **5:**CA(82), ID(82), OR(82), WA(82). **6:**W states(82), OR(82), WA(82). **7:**WA(82). **10:**WA(82). **12:**WA(2341). **14:**WA(2341). **16:**ID(82), MT(82), WA(82). **17:**WA(2341). **18:**W states(82), OR(82), WA(82). **20:**W states(82), OR(82), WA(82). **22:**WA(82).

Puccinia sp. **16:**ND(3162), NH(1890). **22:**ND(3162).

Basidiomycotina - Ustilaginales

Tilletia caries. **5:**WA(82,1077). **9:**OR(1459). **12:**ID(2999). **18:**WA(82). **20:**ID(82,1077,2999), WA(82).

T. controversa. Dwarf bunt. **6:**OR(2343). **16:**OR(2343). **18:**OR(2344). **20:**ID(1077,2344), OR(1465).

T. earlei. **16:**ND(82,378,1077).

T. laevis. **5:**WA(82).

Tilletia sp. **5:**ID(1464), OR(1464). **7:**ID(1464), OR(1464). **9:**OR(1464). **10:**ID(1464). **12:**ID(1464), OR(1464).

Urocystis agropyri. Flag smut. **3:**WA(2342). **16:**NC states(82), NE states(82), CT(1077), IN(1077), MA(1077), MN(1077), NH(1077), NY(1077), PA(1077), SD(2242), VT(1077), WA(1077), WI(82,1077,1332). **18:**MT(2342), OR(82,1077), UT(82,1077), WA(82,1077). **20:**AK(526), CO(2342), OR(1077), UT(82,1077), WA(1077), WY(82,1077,2924). **19:**CO(1077), WY(1077). **22:**MO(1077).

Ustilago bullata. **5:**ND(1077), OR(1077), SD(2242), WA(82,1077). **6:**W states(82), ID(1077), MT(1077), OR(1077), UT(1077), WA(1077). **18:**W states(82), UT(1077), WA(1077). **20:**W states(82,1077), W states(82), ID(1077), MI(82,1077), MN(1077), OR(82,1077), SD(2242), UT(1077), WA(82,1077), WY(1077). **22:**NE(82).

U. hordei. Covered kernel smut. **5:**MT(82,1077), ND(82,1077), WA(82,1077). **20:**WA(1077).

U. hypodytes. Stem smut. **5:**OR(82,1077), WA(82,1077). **9:**WA(1077). **10:**WA(82,1077). **16:**CA(2342), ID(82,1077), MI(82,1077), ND(82,1077), NY(82,1077), OR(82,1077), SD(82,1077,2242), WA(82,638,1077). **20:**AZ(82), CO(82,1077), ID(1077), OR(1077), WA(82,1077). **22:**WA(82).

U. serpens. **5:**WA(82,1077). **16:**IA(1077), MA(1077), MN(1077), MO(1077), WI(1077,1338). **18:**OR(1077), UT(1077), WA(1077), WY(1077). **20:**ID(1077), NV(1077), UT(82,1077).

U. striiformis. Stripe smut. **5:**UT(1077), WA(1077). **6:**ID(1077). **16:**IA(1077), MA(1077,1890), NY(82,1077), PA(82,1077), TX(82,1077), WI(82,1077). **18:**OR(82,1077), UT(82,1077), WA(82,1077). **20:**AZ(82), CO(82,1077), ID(1077), MO(1077), MT(82,1077), NM(1077), OR(82,1077), SD(82,1077,2242), UT(82,1077), WA(82,1077), WY(1077). **22:**NY(82), PA(82), TX(82), WI(82).

U. tritici. Loose kernel smut. **10:**WA(82,1077). **20:**AK(1077).

Basidiomycotina - Other

Marasmius sp. On dead culms. **16:**IL(82), MN(82), WI(82,1335).

Thanatephorus cucumeris. Brown patch, root rot, sharp eye spot. **5:**MT(3423), ND(3423), NE(3423), OR(3423), SD(3423), WA(3423), WY(3423). **6:**ND(3423), OR(3423). **7:**ND(3423), OR(3423). **9:**ND(3423), NE(3423). **10:**MT(3423), ND(3423), OR(3423), SD(3423), WA(3423). **12:**ND(3423), SD(3423). **13:**ND(3423), SD(3423). **14:**ND(3423). **16:**ID(3423), MN(3423), ND(3423), NE(3423), OR(3423), SD(3423), WA(3423). **20:**ID(3423), MT(3423), ND(3423), NE(3423), SD(3423),

WY(3423).
Typhula idahoensis. Snow mold. **5**:ID(82,2955,3423), WA(3423). **12**:WA(3423). **18**:WA(3423).
T. incarnata. Snow mold. **18**:WA(3423).

Deuteromycotina - Hyphomycetes

Acremoniella sp. **20**:AK(3431).
Alternaria alternata. In commercial seed lots. **5**:Unk.(82). **16**:AK(3431).
Arthrinium phaeospermum. On stolons and culms. **16**:NY(82).
Bipolaris sorokiniana. Seedling blight, spot blotch, black point, root and seed rot. **5**:NC states(82), MD(3423), MN(3423), MT(3423), ND(3423), NE(3423), OR(3423), SD(2242,3423), WA(82,3423), WY(3423). **6**:MN(82), MT(82), ND(82,3423), NE(82), SD(82), WY(82). **7**:ND(3423), SD(2242,3423). **9**:ND(3423), NE(3423). **10**:MN(3423), ND(82,3423), NE(3423), OR(3423), SD(3423), WY(3423). **12**:ND(3423), SD(2242,3423). **13**:ND(3423), SD(3423). **14**:ND(3423). **16**:NC states(82), IA(3423), MN(3423), ND(3423), NE(3423), NY(82), SD(2242,3423), WA(82), WI(1334,3423). **18**:MN(82), MT(82), ND(82), NE(82), SD(82), WY(82). **20**:MN(82), MT(82,3423), ND(82,3423), NE(82,3423), SD(82,2242,3423), WY(82,3423). **22**:MN(82), MT(82), ND(82), NE(82), SD(82), WY(82).
B. spicifera. **5**:SC(82).
B. victoriae. Blight. **5**:IA(3423).
Cercospora zebrina. **16**:AL(759).
Cercosporidium graminis. Leaf streak, brown stripe. **1**:SD(3423). **5**:ND(82,3423). **6**:NC states(82), NW states(82), Range of host(82,3423), ND(3423), OR(3423), WA(3423), WI(1357), WY(3423). **12**:WA(2341). **16**:AK(3431), ID(3423), MN(3423), ND(82), NE(3423), SD(82,2242), UT(3423), WA(82,3423), WI(3423). **18**:NC states(82), NW states(82), Range of host(82), CO(3423), ID(3423), MT(3423), NV(3423), OR(3423), UT(3423), WA(3423), WY(2924,3423). **22**:NC states(82,3423), NW states(82), Range of host(82), ID(3423), OR(3423), UT(3423), WA(3423), WY(3423).
Cladosporium herbarum. Leaf mold. **16**:IA(82). **20**:AK(3431).
C. phlei. **13**:AK(526). **20**:AK(526).
Curvularia geniculata. Secondary root rot. **5**:ND(82), SD(82,2242). **10**:ND(82), SD(82). **20**:SD(2242). **22**:ND(82).
C. trifolii. **16**:AL(759).
Drechslera gigantea. Zonate leaf spot. **16**:MD(82,3423), MO(3423), TX(82,3423), VA(82,3423).
D. tritici-repentis. Leaf blight. **5**:SD(82). **7**:SD(2242). **12**:ND(3423). **16**:E states(82), MI(3423), ND(3423), NH(1895), NY(3423), OR(82,3423), SD(82,2242), WA(82), WI(1352,3423), WV(953). **18**:ID(3424). **20**:AK(3428), ID(3438), ND(82,3423).
Epicoccum nigrum. Head mold. **5**:OR(82).
Exserohilum rostratum. Mold. **16**:ND(3423).
Fusarium acuminatum. Root and stem rot. **1**:MT(3423). **5**:NGP(82), ID(3423), MN(3423), MT(3423), ND(3423), NE(3423), OR(3423), SD(2242,3423), WA(3423), WY(3423). **6**:ND(3423), WA(3423), WY(3423). **7**:MT(3423), ND(3423), NE(3423), OR(3423), SD(2242,3423), WY(3423). **9**:ND(3423), NE(3423). **10**:MN(3423), MT(3423), ND(82,3423), NE(82,3423), OR(3423), SD(3423), WA(3423), WY(3423). **12**:MN(3423), ND(3423), NE(3423), SD(2242,3423), WA(3423). **13**:ID(3423), ND(3423), SD(3423), WA(3423). **14**:ND(3423). **16**:MN(82,3423), MT(3423), ND(82,3423), NE(3423), OR(3423), SC(82), SD(2242,3423), WY(3423). **18**:ID(3423), NV(3423), OR(3423), WA(3423). **22**:ND(82,3423), NE(82).
F. culmorum. Head mold, culm and root rot, scab. **5**:OR(82,3423), WA(82,3423). **7**:OR(3423). **9**:WA(3423). **10**:ND(3423), OR(3423). **18**:ID(3423), OR(3423), WA(3423). **20**:WA(82).
F. equiseti. Culm and root rot. **5**:ND(82), NE(82), SD(82). **6**:MT(82), ND(82), SD(82), WY(82). **16**:ND(82), SD(82).
F. graminearum. Seedling blight, culm and root rot, scab. **5**:ND(82,3423). **12**:SD(2242). **20**:ND(82,3423).
F. moniliforme. Secondary root rot. **10**:ND(3423).
F. oxysporum. Culm and root rot. **5**:ND(82), OR(82), SD(82). **6**:ND(82). **20**:ND(82).
F. sporotrichioides. Culm and root rot. **20**:MT(82), ND(82), SD(82), WY(82).
Fusarium sp. Stem rot. **10**:Unk.(82). **22**:Unk.(82).
Helminthosporium sp. On necrotic roots. **5**:SD(3423). **18**:MI(82). **20**:MN(82).
Heterosporium avenae. **20**:WA(3430).
Hymenula cerealis. Stripe. **16**:WA(415).
Microdochium bolleyi. Root necrosis, rot. **1**:MT(3423), ND(35). **5**:MT(3423), ND(82,3423), SD(82,2242,3423), WY(3423). **6**:WY(3423). **7**:ND(3423), SD(2242,3423). **9**:ND(3423). **10**:MT(3423), ND(3423), SD(3423). **12**:ND(3423). **13**:ND(3423). **14**:ND(3423). **16**:MN(3423), ND(3423), SD(2242,3423). **18**:WY(3423). **20**:MT(3423), ND(3423), NE(3423), SD(2242,3423), WY(3423).
M. nivale. Snow mold, leaf spot. **5**:ID(3423), ND(3429), WA(3423). **6**:WA(3423), WY(3423). **7**:WA(3423). **9**:WA(3423). **10**:ID(3423), MT(3423), OR(3423), WA(3423). **12**:ID(3423), WA(3423). **13**:WA(3423). **16**:ID(3423), WA(3423). **18**:OR(3423), WA(3423). **20**:ID(3423), WA(3423).
Nigrospora sphaerica. Black mold of culms. **20**:CO(82).
Penicillium expansum. Blue mold. **5**:ND(3423), OR(3423), SD(3423), WA(3423).
Periconia circinata. Root rot. **5**:MT(3423), ND(3423). **12**:ND(3423). **13**:ND(3423). **18**:WA(3423). **20**:ND(3423).
Pseudocercosporella herpotrichoides. Foot rot, strawbreaker. **5**:WA(82,3423). **6**:ID(82), OR(82), WA(82,3423). **18**:ID(82), OR(82), WA(82,3423). **22**:ID(82), OR(82), WA(82).
Ramularia pusilla. Leaf spot. **5**:CO(82,3423,3429). **13**:AK(526). **16**:AK(3428,3431), WI(82). **20**:AK(526), MT(3429), OR(3433), WI(82,3423,3429).
Rhynchosporium orthosporum. Leaf spot, scald, white leaf stripe. **20**:AK(3426,3431), ID(3423,3424), UT(3423,3424), WY(2924). **19**:CO(3424).
R. secalis. Scald, leaf spot. **6**:OR(82,3423), WA(2341,3423). **7**:WA(2341). **8**:WA(3423). **12**:ID(3429). **16**:CA(82,3423), CO(3423), ID(3423), MN(82,3423), MT(3429), ND(82,3423), OR(82,3423), SD(2242), WA(82,3423), WI(82,3423), WV(955). **17**:WA(2341). **20**:AK(526,3426), CO(3423), ID(82,3423,3424), WA(82,3423), WY(3423). **22**:WA(82).
Sporotrichum columbiense. White stipple. **18**:MT(3423), WA(3423). **20**:ID(3423).
Stemphylium botryosum. On leaves. **5**:SD(82).
Stemphylium sp. **16**:AL(759).

Deuteromycotina - Coelomycetes

Ascochyta agropyri-repentis. On leaves. **16**:WI(1342,3423).
A. agropyrina. Leaf rot and spot. **3**:NM(3423). **5**:MT(82,3423,3449), ND(82,3423,3449), WA(82,3423,3449). **10**:MI(82), ND(3423,3449). **12**:ND(3423,3449). **13**:ND(3423,3449). **18**:MT(3423,3449), ND(82), WA(3423). **22**:ND(82), NM(82).
A. graminicola. On leaves. **16**:MA(82,1391).
A. sorghi. Leaf spot. **7**:ND(3423,3449), SD(2242). **16**:MA(3423). **18**:AK(526), MT(3423,3449). **20**:ID(3423), WY(2924).
A. utahensis. Black spot, leaf spot. **16**:AK(3441). **18**:CO(1195), ID(3442), OR(3433), UT(3423,3449). **20**:CO(3423,3449), ID(3425), WY(3423,3449).
Cheilaria agrostis. Blotch spot, char spot. **16**:ND(82,3423). **18**:WY(82,3423). **20**:AK(3431), MT(82,3423,3442), ND(82,3423), SD(82,3423).
Colletotrichum caudatum. Leaf spot. **20**:OK(82,2872,3423).
C. destructivum. **16**:AL(759).
[*C. graminicola*]. Anthracnose. **7**:MN(82,3423), ND(3423). **9**:OK(2872,3423). **12**:OK(2872,3423). **16**:AL(759), MN(82,3423), ND(82,3423), NE(3423), PA(82,3423), TX(82,3423). **20**:MN(82), ND(82,3423), NE(2927), OR(82).
Dilophospora alopecuri. Twist disease, leaf spot. **18**:MT(3423), UT(3423), WA(82,3423). **20**:ND(82,3423), UT(3435), WY(82,3423).
Dinemasporium graminum var. *strigosulum*. **20**:ND(378).
Macrophoma sorghicola. On dead culms. **5**:MT(3442).
Phaeoseptoria festucae var. *muhlenbergiae*. **16**:WI(1352).
Phloeospora graminearum. Leaf spot. **16**:MI(82,3423).
Phoma terrestris. Pink root. **5**:MT(3423), ND(82,3423), SD(3423), WY(3423). **7**:MT(3423), ND(3423), SD(2242,3423), WY(3423). **9**:ND(3423). **10**:ND(3423), WY(3423). **12**:ND(3423). **13**:ND(3423), SD(3423). **14**:ND(3423). **16**:MN(3423), ND(82,3423), NE(3423), SD(3423). **20**:MN(82), ND(82,3423). **22**:OR(2791).
Phyllosticta sp. On leaves. **16**:WI(1357).
Pseudoseptoria donacis. Eye spot. **1**:ND(3423,3450). **5**:MT(3433), WY(3423,3450). **6**:AZ(1194), ND(3423,3450), OR(3423,3450),

WY(3423,3450). **7:**MT(3433). **12:**WA(3442). **16:**MT(3433), ND(3423,3450), WA(3423,3450). **18:**CO(1195,3423,3450), ID(3450), ND(3423,3450), NV(3423,3450), UT(3423,3450), WA(3423,3450), WY(3423,3450). **20:**ID(3430), MT(3423,3450), ND(638,3423,3450), UT(3430,3450), WY(3423,3450). **22:**KS(3021).

P. everhartii. **16:**WI(1343).

P. obtusa. Stem speckle, eye spot. **18:**OR(3438), WA(82,2695,3423,3450). **20:**WY(3450). **19:**CO(3423,3450).

P. stomaticola. Brown cubical butt rot, stem speckle. **1:**WY(2924). **5:**WY(82). **6:**NC states(82), NW states(82), Range of host(82). **16:**ND(82), WA(82). **18:**NC states(82), NW states(82), Range of host(82). **20:**NC states(82), NW states(82), Range of host(82), AK(526). **22:**NC states(82), NW states(82), Range of host(82).

Selenophoma donacis var. *linearis*. On fallen leaves. **18:**WA(3450).

Septoria agropyrina. Leaf blotch, leaf spot. **16:**MN(3423), ND(3423), SD(82,2242,3423), WA(3423), WI(1335,3423). **18:**WA(3423). **20:**MN(3430), MT(82,3423), ND(82,3423), NE(82,3423), SD(2242,3423), WA(3423), WY(3423). **21:**WI(1342). **22:**ND(82).

S. elymi. Speckled leaf blotch, leaf spot. **16:**NC states(82), IA(3423), KY(3423), MI(3423), MN(3423), MT(3423), ND(3423), NE(3423), OH(3423), OR(82,3423), PA(3423), SD(2242,3423), WA(82,3423), WI(1345,3423). **18:**OR(82,3423), WA(82,3423), WY(82,3423). **20:**W states(3442), AZ(3435,3438), IA(82), MN(82), MT(82,3423), ND(82,3423), NE(3423), SD(2242,3423), WY(3423).

S. elymi f. *elymina*. **18:**WY(1195).

S. infuscans. Brown leaf blotch, leaf spot. **18:**AZ(1194), OR(82), WA(82,3429).

Septoria sp. Leaf spot. **5:**OK(2872). **20:**OK(2872).

Stagonospora arenaria. Purple-brown blotch. **5:**ND(3423). **7:**ND(3423), SD(2242). **12:**KS(3021), ND(3423). **16:**OR(82,3423). **18:**ND(3423). **20:**MI(82,3423), ND(82,3423), SD(82). **22:**ND(82), SD(82).

S. avenae. **5:**SD(2242).

S. avenae f. sp. *triticea*. Blotch. **5:**ND(82). **16:**MN(3442), SD(2242). **20:**ND(3423).

S. culmicola. Leaf rot. **16:**ND(3423). **18:**OR(3423). **20:**WA(3423).

S. nodorum. Glume blotch, leaf spot. **16:**ND(3423), SD(2242), WA(3432), WI(1357), WV(953). **18:**WA(3423,3424). **20:**ND(3423). **15:**NM(3425).

S. simplicior. Leaf spot. **20:**AK(3428,3431).

Stagonospora sp. **16:**AL(759).

Wojnowicia hirta. Secondary foot rot, rot of mature straw. **5:**ND(82,3423), OR(82,3423), WA(82,3423). **6:**WA(3423). **12:**ND(3423). **16:**ND(82,3423). **18:**KS(82), MT(82), ND(82), WA(82,3423). **20:**KS(82), MT(82), ND(82,3423), WA(82). **22:**KS(82), MT(82), ND(82), WA(82).

Deuteromycotina - Other

Rhizoctonia solani. Secondary root rot. **5:**ND(82), SD(82,2242), WA(82). **10:**MN(82), ND(82). **12:**SD(2242). **16:**AK(3441), MN(82), ND(82), SD(2242). **20:**NGP(82), Unk.(82), AK(3441), SD(2242).

Sclerotium rhizodes. On leaves. **20:**WI(3423).

S. rolfsii. **16:**AL(759).

×*Agrositanion* Bowden

Agrositanion × *Sitanion*.

1. × *A. saxicola* (Lams.-Scribn. & J. G. Sm.) Bowden (=*Agropyron saxicola* (Lams.-Scribn. & J. G. Sm.) Piper). *Sitanion hystrix* × *Agropyron spicatum*.

Deuteromycotina - Hyphomycetes

Cercosporidium graminis. **1:**UT(3430).

Agrostis L.

1. *A. aequivalvis* (Trin.) Trin.
2. *A. altissima* (Walter) Tuck.
3. *A. borealis* Hartm.
4. *A. canina* L. [brown bentgrass, velvet bentgrass]
5. *A. capillaris* L. (=*A. tenuis* Sibth., *A. tenuis* Sibth. cv. 'astoria') [browntop, colonial bentgrass, Rhode Island bentgrass]
6. *A. castellana* Boiss. & Reut.
7. *A. diegoensis* Vasey [thin grass]
8. *A. exarata* Trin. (=*A. exarata* Trin. var. *monolepis* (Torr.) A. Hitchc.) [spike redtop]
9. *A. exarata* Trin. var. *ampla* (Hitchc.) Hitchc.
10. *A. gigantea* Roth [redtop]
11. *A. hallii* Vasey
12. *A. humilis* Vasey
13. *A. hyemalis* (Walter) B. S. P.
14. *A. idahoensis* Nash [Idaho redtop]
15. *A. longiligula* Hitchc.
16. *A. microphylla* Steud.
17. *A. oregonensis* Vasey [Oregon bentgrass]
18. *A. pallens* Trin. [dune bentgrass]
19. *A. perennans* (Walter) Tuck. [autumn bentgrass, brown bentgrass, upland bentgrass]
20. *A. rossae* Vasey [Ross redtop]
21. *A. scabra* Willd. (=*A. geminata* Trin., *A. scabra* Willd. var. *geminata* (Trin.) Swall.) [rough bentgrass]
22. *A. semiverticillata* (Forssk.) C. Chr. (=*A. verticillata* Vill.)
23. *A. stolonifera* L. (=*A. alba* of American authors, *A. palustris* Huds.) [creeping bentgrass]
24. *A. thurberiana* Hitchc. [Thurber redtop]
25. *A. variabilis* Rydb.
26. *Agrostis* sp.

Myxomycetes

Fuligo septica. Slime mold. **4:**Unk.(82). **5:**Unk.(82). **23:**Unk.(82). **26:**Unk.(82).

Physarum cinereum. Slime mold. **4:**Unk.(82). **5:**Unk.(82). **23:**Unk.(82). **26:**Unk.(82), NC(1303).

Plasmodiophoromycetes

Polymyxa graminis. On roots. 'Polymyxa root rot.' **23:**IL(393).

Chytridiomycetes

Olpidium brassicae. On roots. **23:**IL(393).

Physoderma graminis. Damping-off. **23:**CT(82,3423).

Synchytrium graminicola. Purple leaf speckle. **4:**OR(3423). **5:**OR(3423), WA(3423). **8:**OR(3423). **11:**OR(3423). **12:**WA(3430). **23:**OR(3423).

Synchytrium sp. Purple leaf speckle. **4:**OR(82), WA(82). **5:**OR(82), WA(82). **8:**OR(82). **23:**OR(82), WA(82). **26:**OR(82).

Oomycetes

[*Pythium aphanidermatum*]. Spot blight, stalk rot of corn, grease spot, turf spot. **4:**'Pythium blight. Pythium root rot.' E states(82), NC states(82), IL(82), IN(82). **5:**E states(82), NC states(82). **23:**'Pythium blight. Pythium root rot.' E states(82), NC states(82), DC(3423), IL(3423), MO(3423), PA(3423), VA(3423). **26:**IL(82), IN(82).

P. aristosporum. Pathogens of secondary roots. **23:**IA(1589).

P. arrhenomanes. Pathogens of secondary roots. **23:**IA(1589).

P. debaryanum. Damping-off, root necrosis, seed rot. **13:**ND(82). **19:**ND(82). **21:**ND(3423). **23:**CT(82), ND(82,3976), NE(3976).

P. graminicola. 'Pythium blight. Pythium root rot.' **23:**WY(3423).

P. ultimum. Damping-off, root necrosis. 'Pythium blight.' **23:**CT(82), ND(82,3423).

P. vanterpoolii. Associated with roots. 'Pythium blight.' **23:**IA(1589).

Pythium sp. Blight, damping-off, root rot. **23:**CA(988). **26:**FL(35), NC(1303).

Sclerophthora macrospora. Yellow tuft. **23:**'Downy mildew.' SD(3184). **26:**RI(2522).

Ascomycotina

Acrospermum compressum. Leaf spot. **19:**MS(2720).

Balansia epichloe. **23:**GA(1438).

Claviceps purpurea. Ergot. **5:**OR(82,3423), WA(3423). **8:**AK(526,3423), OR(82,3423). **13:**ND(82), OK(82), WI(82). **21:**ND(3423), OK(2872,3423), WI(3423). **23:**NC states(82), NE states(82), IA(3423), IL(312,3423), MI(82,2862), MN(3423), ND(82,3423), OH(82), OR(82,3423).

Crocicreas melanosporum. On overwintered stems, petioles and peduncles. **24:**CA(642).

Epichloe typhina. Choke. **13:**WI(1342). **19:**VA(82). **21:**WI(3423).

Erysiphe graminis. Powdery mildew. **7:**CA(82,3423). **8:**AK(82,526,3423), CA(82,638,642,3423), ID(82,3423), MT(82,3423), NV(3430), OR(82,3423), WA(82,3423), WY(3423). **20:**NV(3430). **22:**OR(3423). **23:**IA(82,3423), ID(3423), MN(82,3423), WA(3430).

Euryachora aristidae. Tar spot. **26:**OR(82).

Gaeumannomyces graminis. Stunting and black basal decay, foot rot, take-all. **8:**AK(3428,3441), ID(3423). **23:**ID(3423), NY(82).

G. graminis var. *avenae*. Patch disease, crown rot. **23:**MD(852), NC(1303). **26:**NC(1303).
Gloeotinia granigena. **8:**OR(1075).
Herpotrichia macrotricha. On culms. **23:**NY(82).
Leptosphaeria asparagina. **26:**WA(3814).
L. leersiana. **26:**AK(526).
Leptosphaeria sp. On lower leaves. **8:**AK(3431). **19:**WI(1357). **21:**AK(3431). **24:**OR(3430).
Lophodermium arundinaceum. On dead culms and leaves. **8:**AK(3431). **17:**WA(3423). **24:**OR(3430).
Mycosphaerella ignobilis. **21:**AK(526).
M. tulasnei. **8:**AK(3431).
Mycosphaerella sp. On culms. **13:**AK(82).
Ophiobolus sp. **5:**WA(1290).
Phaeosphaeria caricinella. On overwintered herbage. **26:**CA(642).
P. herpotrichoides. **26:**WA(3814).
P. microscopica. **24:**CA(642).
P. typharum. **26:**WA(3814).
Phyllachora graminis. Tar spot. **23:**CA(82,3423), NY(82,3423), VA(82,3423).
Pleospora herbarum. On overwintered herbage. **24:**CA(642).
P. rainierensis. **26:**WA(3813,3814).
Pyrenopeziza karstenii. **8:**AK(3431).
Sclerotinia homoeocarpa. Turf rot, dollar spot. **4:**'Dollar spot.' MA(3423), RI(3423), VA(3423). **5:**DC(3423), MA(3423), PA(3423), RI(3423), VA(3423). **23:**'Dollar spot.' CA(988), DC(3423), HI(2299,2903), MA(3423), PA(3423), RI(3423), VA(3423). **26:**NC(1303).
Sphaerella californica. **26:**AK(526).
Trichothyrina alpestris. Leaf spot. **21:**AK(3426,3431).

Basidiomycotina - Uredinales

Puccinia brachypodii var. *poae-nemoralis*. 'Leaf rust.' **23:**WY(3377).
P. coronata. Heteroecious: telial host. **4:**'Leaf rust.' OR(82). **5:**MA(82), ME(82), NH(82), OR(82), VT(82). **7:**OR(82). **8:**CA(81). **13:**CA(82), OR(82), WA(82). **15:**CA(81). **21:**AK(3441). **23:**'Leaf rust.' CA(81), FL(35,82), GA(82), KY(82), LA(82), MA(82), ME(82), MT(1195), NH(82), OR(82), UT(1195), VT(82), WI(1339), WV(953), WY(1195). **26:**CA(82), WA(82).
P. graminis. Heteroecious: telial host. **4:**'Stem rust.' OR(82). **5:**NE states(82), IA(82), MA(664), OR(82). **7:**CA(81,82). **8:**AK(3431), CA(82), ID(82), MT(82), NE(82), OR(82). **10:**OK(2872). **13:**NE states(82), IA(82), KS(3023), ND(82). **19:**NE states(82). **23:**'Stem rust.' C & W states(82), NE states(82), CA(81,82), MA(1390), ME(82), MS(2720), PA(82), SD(2242), VT(82), WV(953). **26:**AK(526), CA(82).
P. poarum. Heteroecious: telial host. **8:**CO(82). **13:**CO(82), NM(82). **21:**AZ(1195,1203,3981).
P. pygmaea. Heteroecious: telial host. **8:**CA(754).
P. recondita. Heteroecious: telial host. **13:**NE states(82), FL(35). **14:**MT(3430). **19:**NE states(82), IN(987), TX(82), WI(82), WV(953). **21:**MA(664). **23:**'Leaf rust.' GA(1439), KS(3021), MI(82), NY(82), TN(82), VA(82), WI(82).
P. striiformis. 'Stripe rust.' **23:**MT(82).
Uromyces fragilipes. Heteroecious: aecial host. **7:**CA(82). **8:**OR(82). **23:**OR(82). **26:**CA(82), OR(82).

Basidiomycotina - Ustilaginales

Entyloma dactylidis. Leaf smut. **5:**WI(82,1077). **22:**TX(1077). **23:**WI(82,1077,1357). **26:**CA(1077), WI(1077).
Tilletia pallida. **2:**TN(1077). **4:**'Covered floral smut.' NJ(82,1077), OH(82,1077), RI(82,1077). **23:**'Covered floral smut.' OR(82,1077).
T. sphaerococca. **5:**OR(2342). **23:**'Covered floral smut.' OR(2342).
Urocystis agropyri. Flag smut. **5:**VT(1077). **7:**CA(81). **23:**'Flag floral smut.' VT(1077), WI(82,1077).
Ustilago striiformis. Stripe smut. **5:**NC states(82), NE states(82), CT(1077), IA(82,1077), IL(1077), IN(1077), MA(1077), MD(1077), NY(1077), OR(82,1077), PA(1077), WA(82,1077), WI(1077), WV(1077). **6:**WA(1077). **8:**OR(82,1077), UT(82,1077). **12:**WY(1077). **13:**WI(1342). **16:**OR(1077). **19:**PA(82,1077), WI(1357), WV(82,1077). **20:**WY(1077). **21:**WY(1077). **23:**'Stripe floral smut.' NC states(82), NE states(82), NW states(82), AZ(1077), CA(1077), CO(2342), CT(1077), IA(82,1077,1586,1962), ID(1077), IL(82,1077), IN(1077), KY(1077), MA(1077,1962), MD(1077), MO(1077,1962), MT(2342), NV(1077), OR(82,1077,1962), UT(1077), WA(82,1077,1962), WI(1077), WY(1077,2924). **26:**OR(82), WA(82).

Basidiomycotina - Other

Laetisaria fuciformis. Pink patch. **4:**RI(3423). **5:**RI(3423). **23:**'Red thread.' RI(3423), WA(3423).
Marasmius graminum. On base of culms. **23:**IL(82).
Thanatephorus cucumeris. Slime mold, brown patch, root rot, sharp eye spot. **4:**DC(3423), OR(3423), PA(3423), RI(3423), VA(3423). **5:**CT(3423), DC(3423), OR(3423), PA(3423), RI(3423), VA(3423), WA(3423). **8:**OR(3423), WA(3423). **11:**OR(3423). **23:**CA(3423), CT(3423), DC(3423), DE(3423), GA(1439), IL(3423), MA(3423), MD(3423), MN(3423), MO(3423), ND(3423), NE(3423), NJ(3423), NY(3423), OH(3423), OR(3423), PA(3423), RI(3423), VA(3423), WA(3423), WI(3423), WV(3423).
Typhula incarnata. Snow mold, snow scald. **4:**'Typhula blight.' MA(82), MN(82), NJ(82), NY(82), PA(82). **5:**MA(82), MN(82), NJ(82), NY(82), PA(82,3423). **23:**'Typhula blight.' ID(3423), MA(82,3423), MN(82), NJ(82), NY(82,3423), PA(82).
Typhula sp. Leaf spot. **21:**AK(3426,3431).

Deuteromycotina - Hyphomycetes

Acremoniella alascensis. On leaves. **8:**AK(3441).
Acremonium typhinum. **13:**TX(3872).
Bipolaris sorokiniana. Seedling blight, spot blotch, black point, root rot. 'Helminthosporium leaf spot.' **23:**CA(988), HI(2299,2903), WY(3423).
Cercospora agrostidis. Leaf spot. **19:**AL(3423), OK(2872). **21:**ID(3423). **26:**AL(3423).
Cercosporidium graminis. Leaf streak, brown stripe. **5:**IN(82), MA(82), ME(82), NH(82), VT(82). **6:**OR(3423). **8:**CA(82), ID(82,3423), MT(3423), NV(3430), OR(82,3423), SD(82), WA(3429,3430). **11:**OR(3423). **13:**CA(82), ID(82), OR(82), SD(82). **17:**ID(3423), MT(3423). **20:**WA(3423). **21:**CA(3423), CO(3423), ID(3442), ND(3423), OR(3423), SD(2242,3423), UT(3423), WA(3429), WY(3423). **23:**'Brown stripe.' NC states(82), NE states(82), CO(3423), ID(3423), MA(82), ME(82), MN(3423), MT(3423), ND(3423), NE(3423), NH(82), OR(3423), SD(2242,3423), UT(3423), VT(82), WA(3423), WI(3423), WY(3423). **25:**OR(3433). **26:**CA(82), ID(82), OR(82), SD(82).
Cladosporium herbarum. **21:**AK(3431).
Curvularia lunata. Leaf mold. 'Curvularia blight.' **23:**VA(82,2519).
Curvularia sp. **26:**NC(1303).
Drechslera catenaria. Leaf blight, crown rot. **5:**PA(376). **23:**'Drechslera leaf blight. Drechslera crown rot.' Unk.(3416), OH(2052).
D. dematioidea. Leaf mold. **19:**CT(82,3423), NY(82,3423). **23:**CT(82,3423), NY(82,3423).
D. erythrospila. Leaf spot. **4:**'Red leaf spot.' VA(3423). **5:**VA(3423). **19:**CT(82), NY(82,3423). **23:**'Red leaf spot.' NE states(82), CT(3423), DC(3423), IN(82,3423), MD(376,3423), MI(3423), NY(3423), OH(3423), PA(376), VA(82,3423), WI(82,3423), WV(955). **26:**AK(526).
D. fugax. Leaf mold. **5:**CT(3423), OR(82,3423). **12:**OR(3432). **17:**ID(3423). **21:**CO(3423), ID(3423), WA(3433). **23:**CT(82,3423), DC(3423), IA(82), MD(82), MN(3423), MT(3423), NC(82,3423), WV(952). **25:**WA(3435).
D. gigantea. Zonate eye spot, patch. **4:**'Zonate eyespot.' MD(82), RI(1719). **5:**RI(1719). **23:**'Zonate eyespot.' CT(82), MD(82), MI(82), NY(82), TX(82), VA(82).
D. triseptata. On leaves. **8:**OR(82,3423). **23:**IL(312), MD(82), NY(82,3423), VA(82), WI(1344).
Endoconidium temulentum. Blind seed disease. **8:**OR(82).
Fusarium acuminatum. Secondary root rot. **4:**OR(3423). **21:**SD(3423). **23:**MN(3423), ND(3423), NE(3423), OR(82,3423), WA(3423), WY(3423).
F. avenaceum. Root rot. **23:**CA(988).
F. culmorum. Head mold, root rot, scab. **5:**MN(3423), OR(82,3423), WA(3423). **23:**'Fusarium blight.' MN(3423), OR(82,3423), WA(3423).
F. graminearum. Scab, seedling blight. **26:**ND(3423).
Fusarium sp. Root rot. **23:**OK(2872). **26:**NC(1303).
Gloeocercospora sorghi. Copper spot, zonate leaf spot. **4:**'Copper spot.' PA(3423), RI(3423). **5:**PA(3423), RI(3423). **23:**'Copper spot.' PA(3423), RI(3423).

Helminthosporium sp. Melting-out disease, leaf spot, on necrotic roots. **4:**MI(82). **23:**VA(3423), WI(1356). **26:**MI(75).
Mastigosporium rubricosum. Red eye spot, leaf spot. **1:**AK(3428). **4:**OR(82,3423), WA(82). **5:**AK(3441), OR(82,3423), WA(82,3423). **8:**AK(3441), OR(82,3423). **12:**MT(3430), WA(3430). **22:**OR(3423). **23:**AK(3428), OR(82,3423), WA(82,3423), WY(82,3423). **26:**OR(82), WA(82).
Microdochium bolleyi. Root necrosis. **21:**ND(3423), SD(3423). **23:**'Anthracnose.' ND(3423), SD(3423), WY(3423).
M. nivale. Snow mold, patch. **4:**'Fusarium patch.' MI(3423), MN(3423), OR(3423). **5:**MI(3423), MN(3423), OR(3423), WA(3423). **8:**OR(3423). **23:**'Fusarium patch.' MI(3423), MN(3423), OR(82,3423), WA(3423). **26:**WA(1286).
Ovularia sp. Leaf spot. **23:**PA(82).
Periconia circinata. Root rot. **26:**ND(3423).
Periconia sp. **1:**AK(3431).
Pyricularia grisea. Blast, leaf blight, gray spot. **23:**TX(82,3423).
Ramularia graminicola. On leaves. **23:**WI(82).
R. pusilla. Leaf blight and spot. **1:**AK(3429). **5:**UT(82,3423,3429). **8:**AK(3429,3431). **12:**WY(3423,3424,3429). **17:**MT(3423,3429). **23:**OR(82,3423,3429), UT(82,3423,3429), WA(3429), WI(1341,3423,3429).
[*Ramulispora sorghi*]. Copper spot. **4:**RI(82). **5:**RI(82). **23:**NJ(82), PA(82), RI(82).
Rhynchosporium orthosporum. Scald, white leaf stripe. **8:**AK(3426). **23:**OR(82,3423).
R. secalis. Scald. **23:**OR(82,3423).
Spermospora subulata. Blast. **23:**WA(3423).

Deuteromycotina - Coelomycetes

Ascochyta graminicola. On leaves. **4:**MA(82). **23:**NY(82).
A. sorghi. Leaf spot. **4:**'Ascochyta leaf spot.' MA(3423). **12:**OR(3432). **21:**AK(3428,3431), ID(3449), OR(3433). **23:**CA(81), ID(3423,3449), NY(3423).
Ascochyta sp. On dead lower leaves. **23:**WI(1357).
Cheilaria agrostis. Blotch spot, char spot. **1:**AK(3426). **8:**ID(3442), WA(3438), WY(3423). **11:**OR(3423). **13:**OR(82). **18:**WA(3430). **21:**OR(3423), WA(3433). **23:**CO(3423), UT(3423), VA(82), WA(3423), WY(82,2924,3423). **24:**OR(3432), WY(2924,3425). **26:**OR(82), WA(82).
Colletotrichum graminicola. Anthracnose. **8:**WA(3423). **17:**AK(3430). **21:**CO(3423). **23:**'Anthracnose.' NE states(82), IA(82,3423), ID(3423), MN(3423), OH(3423), OR(82), WA(82,3423), WI(1349,3423), WV(952).
Colletotrichum sp. **26:**NC(1303).
Cylindrosporium glyceriae. Leaf spot. **23:**NY(82).
Macrophoma phlei. On leaves. **23:**OR(82,3423).
M. spartiicola. On culms. **23:**MI(82,2862).
Phaeoseptoria airae. On dead leaves. **21:**NM(3425).
P. calamagrostidis. Leaf mold. **23:**OR(82).
Phoma terrestris. Isolated from thatch samples of turf. 'Pyrenochaeta root rot.' **23:**IL(2364).
Pseudorobillarda agrostidis. Leaf rot. **5:**OR(82).
Pseudoseptoria donacis. **21:**SD(2242).
P. everhartii. Stem speckle, eye spot. **7:**CA(82,3423,3450). **13:**ND(82). **21:**ND(3423,3450), SD(2242,3423,3450).
Septoria agrosticola. **5:**OR(3438). **9:**OR(3438). **14:**AK(3441). **23:**OR(3438).
S. calamagrostidis. Leaf spot. **1:**AK(3428). **3:**AK(3430). **7:**OR(82,3423). **8:**AK(526,3423,3431), OR(3423). **13:**AK(82), OR(82), WI(1342). **20:**WY(3423). **21:**AK(3423), WI(3423). **23:**OR(82,3423), WA(3423).
S. graminum. **8:**AK(526).
S. gryllii. **8:**AK(526). **21:**AK(526).
S. secalis var. *stipae*. Leaf spot. **11:**OR(3423). **21:**ND(3423).
S. triseti. Leaf mottle, gray leaf speckle. **5:**OR(82,3423), WA(82,3423). **6:**OR(3423). **8:**OR(82). **9:**OR(3423). **23:**OR(82,3423), WA(3423). **26:**OR(82), WA(82).
Septoria sp. On dead leaves. **23:**WI(1355).
Stagonospora avenae. Speckled blotch, leaf spot. **3:**AK(3426). **8:**AK(3431), ID(3424), OR(3423). **23:**'Septoria leaf spot.' WA(3430).
S. bellunensis. On culms. **3:**AK(530,3441).
S. culmicola. Leaf mold. **8:**OR(82). **26:**OR(82).
S. foliicola. Culm and leaf rot. **23:**WI(1332,3423,3429).
S. mariae. Leaf spot. **8:**AK(3441). **17:**Unk.(3433), WA(3425). **23:**UT(3435).

Deuteromycotina - Other

Rhizoctonia cerealis. Cool weather brown patch. 'Yellow patch.' **23:**NC(1303,2271).
R. solani. Leaf tip blight, brown patch, root rot. **4:**'Rhizoctonia blight.' Range of host(82). **5:**Range of host(82). **13:**WI(82). **23:**'Rhizoctonia blight.' Range of host(82). **26:**Range of host(82), FL(35), NC(1303), OK(2872), OR(82).
R. zeae. Foliar blight, brown patch. 'Rhizoctonia leaf spot. Rhizoctonia sheath spot.' **23:**NC(1303,2271).
Rhizoctonia sp. Root rot. **23:**HI(2903).
Sclerotium rhizodes. Tip blight, leaf rot. **21:**WI(3423). **23:**CT(82,3423), MA(82), PA(82,3423), UT(3423), WI(82), WY(82,3423).
S. rolfsii. Southern blight. 'Sclerotium blight.' **23:**CA(2887), NC(2183).
Sclerotium sp. **4:**Unk.(82). **5:**Unk.(82). **23:**NC states(82), NE states(82), Unk.(82). **26:**Unk.(82).

Aira L.

1. *A. caryophyllea* L.

Ascomycotina

Cochliobolus sativus. Blind seed disease. **1:**OR(3423).
Gloeotinia granigena. **1:**OR(1075).

Alopecurus L.

1. *A. aequalis* Sobol. [short-awn foxtail]
2. *A. alpinus* Sm. [alpine foxtail]
3. *A. arundinaceus* Poir. [creeping foxtail, reed foxtail]
4. *A. carolinianus* Walter
5. *A. geniculatus* L. [water foxtail]
6. *A. pratensis* L. [meadow foxtail]
7. *Alopecurus* sp.

Oomycetes

Sclerospora graminicola. Downy mildew. **6:**MS(82,2720).

Ascomycotina

Claviceps purpurea. Ergot. **5:**ND(82,3423). **6:**IN(82), MI(82), MT(82,3423), ND(3423), OR(82,3423), SD(2242).
Cochliobolus sativus. Blind seed disease. **5:**OR(3423).
Epichloe typhina. Choke. **1:**MI(82).
Gloeotinia granigena. **5:**OR(1075).
Mycosphaerella tassiana. On dead leaves and sheaths. **2:**AK(1938).
Pleospora chlamydospora. On dead sheaths and stems. **2:**AK(1938).
P. lutea. On dead sheath and stem. **2:**AK(1938).

Basidiomycotina - Uredinales

Puccinia brachypodii var. *poae-nemoralis*. **1:**IN(82), NE(82), UT(82), WI(1357). **2:**CO(3424). **3:**WA(2341). **4:**WI(1357). **5:**IN(82), NE(82). **7:**CO(82).
P. coronata. Heteroecious: telial host. **1:**IN(82). **5:**LA(82), MN(82). **6:**OR(82).
P. graminis. Heteroecious: telial host. **1:**CA(81,82), IN(82), MN(82), ND(82), WI(82). **5:**MN(82), ND(82), WI(82). **6:**GA(1439), IA(82), MN(82), OR(82), SD(82,2242), WA(82). **7:**AZ(82), OR(82).
P. recondita. Heteroecious: telial host. **6:**MA(82).
Uromyces dactylidis. Heteroecious: telial host. **1:**AZ(1203,3981). **5:**AZ(1203,3981).
U. dactylidis var. *poae*. Heteroecious: telial host. **1:**CO(82), IA(82), MN(82), NE(82), WY(82).

Basidiomycotina - Ustilaginales

Entyloma dactylidis. Leaf smut. **5:**TX(82,1077).
Tilletia youngii. **4:**AR(1077,3594).
Urocystis agropyri. **2:**ID(2342).
Ustilago striiformis. **2:**ID(2342).

Basidiomycotina - Other

Thanatephorus cucumeris. **4:**LA(3524).

Deuteromycotina - Hyphomycetes

Cercosporidium graminis. Leaf streak, brown stripe. **1:**E states(82), AK(526), MS(82), OR(82,3423), SD(2242), TX(82), WA(3423). **2:**ID(638). **4:**LA(3423), OK(2872,3423), WI(1332). **5:**E states(82), IA(3423), ID(3423), MS(82,2720), OK(2872,3423), OR(82,3423), TX(82), WA(3423), WI(3423). **6:**E states(82), MS(82), ND(3423), OR(82,3423), TX(82), WI(1349). **7:**E states(82), OR(82,2720), TX(82).
Endoconidium temulentum. Blind seed disease. **5:**OR(82).
Fusicladium alopecuri. Mold. **5:**MO(3423).

[*F. fasciculatum*]. Leaf mold. **1**:MS(2720). **5**:MO(82).
Helminthosporium sp. Leaf spot. **6**:ND(82).
Mastigosporium album. Leaf spot. **6**:NY(3038).
Ramularia pusilla. On leaves. **2**:AK(526), MT(3423,3424,3429).
Rhynchosporium orthosporum. Scald, leaf spot, white leaf stripe. **3**:WA(2341). **6**:OR(82,3423), WA(3425).
Deuteromycotina - Coelomycetes
Ascochyta sorghi. Leaf spot. **1**:NM(3425).
Cheilaria agrostis. Leaf spot. **2**:ID(638,3425).
Colletotrichum graminicola. Anthracnose. **1**:MI(82), MN(82,3423), MT(82), OR(82). **6**:MI(82), MN(82,3423), MT(82), OR(82).
Pseudoseptoria everhartii. **6**:AK(3441).
Septoria oudemansii. On leaves. **6**:AK(3441).
Septoria sp. On leaves. **6**:WI(1357).
Stagonospora nodorum. **1**:WI(1355).

Ammophila Host

1. *A. arenaria* (L.) Link [European beachgrass, marram grass]
2. *A. breviligulata* Fernald [American beachgrass]

Oomycetes
Pythium arrhenomanes. Root browning. **1**:ND(82).
P. graminicola. **1**:ND(3423).
Ascomycotina
Claviceps purpurea. Ergot. **1**:WI(3423). **2**:IN(82), MI(82), NC(1303), OH(82).
Lachnella villosa. **2**:MA(1394).
Lophodermium arundinaceum. **1**:OR(3228).
Mollisina ammophilae. On stems. **1**:CA(527).
Phyllachora ammophilae. Tar spot. **1**:MA(82,3423), NJ(82,3423), NY(82,3423). **2**:MA(82).
P. graminis. On leaves and stems. **2**:MA(1394).
Phyllachora sp. Leaf spot. **2**:NC(1303).
Telimena sp. Tar spot. **1**:MA(82).
Basidiomycotina - Uredinales
[*Puccinia amphigena*]. **2**:IN(82).
P. coronata. Heteroecious: telial host. **1**:MA(82), NH(82).
P. graminis. Heteroecious: telial host. **1**:CA(81,82).
P. pygmaea. Heteroecious: telial host. **2**:MI(754).
P. pygmaea var. *ammophilina*. Telial host. **1**:OR(82,739,754). **2**:MI(82).
Basidiomycotina - Ustilaginales
Ustilago striiformis. Stripe smut. **1**:CT(82,1077), MA(82,1077), ME(82,1077). **2**:CT(82), MA(82,1077), ME(82).
Basidiomycotina - Other
Marasmiellus mesosporus. Blight. **2**:NC(1303,3290).
Marasmius sp. Blight, root and basal stem rot. **2**:NC(2184,3779).
Deuteromycotina - Hyphomycetes
Alternaria sp. Leaf spot. **2**:NC(1303).
Bipolaris sorokiniana. Spot blotch, root rot, seedling blight, black point. **1**:ND(82,3423).
Cercosporidium graminis. **1**:CA(81).
[*Fusicladium fasciculatum*]. Leaf mold. **1**:CA(82).
Tetraploa aristata. On stems. **1**:CA(81).
Deuteromycotina - Coelomycetes
Amarenographium metableticum. On culms. **2**:MA(82,1391), NY(82).
Colletotrichum graminicola. Anthracnose. **2**:WI(1343,3423).
Tiarospora perforans. On leaves. **1**:OR(82).
Deuteromycotina - Other
Rhizoctonia solani. Stem lesions. **2**:NC(1303).

Andropogon L.

1. *A. brachystachys* Chapm.
2. *A. elliottii* Chapm. [Elliott's beardgrass]
3. *A. floridanus* Lams.-Scribn.
4. *A. gerardii* Vitm. (=*A. furcatus* Mühlenb., *A. provincialis* Lams.-Scribn., *A. ternarius* Michx.) [big bluestem, turkeyfoot]
5. *A. glomeratus* (Walter) B. S. P. [bluestem, bushy beardstem]
6. *A. hallii* Hack. [sand bluestem, turkeyfoot]
7. *A. longiberbis* Hack.
8. *A. tenarius* Michx.
9. *A. virginicus* L. [broom-sedge]
10. *A. virginicus* L. var. *hirsutior* (Hack.) Hitchc.
11. *Andropogon* sp.

Oomycetes
Pythium arrhenomanes. Root browning. **4**:ND(82). **11**:ND(82).
P. debaryanum. Damping-off, root necrosis, seed rot. **4**:ND(82,3423), NE(82,3423), SD(2242,3423), WY(3423). **6**:ND(3423), SD(3423). **11**:ND(82), NE(82).
P. graminicola. **4**:ND(3423). **6**:ND(3423).
P. irregulare. Root necrosis. **6**:NE(3423).
P. ultimum. Root necrosis, seed rot. **4**:ND(82), NE(82,3423). **11**:ND(82), NE(82).
Pythium sp. Root rot. **11**:FL(35).
Ascomycotina
Anthostomella chionostoma. On stems. **9**:NJ(1098).
Apiospora montagnei. On culms. **5**:GA(1438). **9**:GA(82,1438). **11**:FL(2782), GA(82).
Balansia epichloe. On leaves. **9**:GA(3423). **11**:MO(3423).
B. henningsiana. Black choke, black ring. **9**:AL(82), FL(82), MD(82), NC(82), NY(82), TN(82). **11**:FL(35,82), GA(82), MS(82,2720).
B. strangulans f. *discoidea*. **11**:KS(82).
Clathrospora pentamera. **9**:GA(1438).
Claviceps purpurea. Ergot. **4**:GA(82), KS(82), NE(2927), OK(2872,3423), TX(82,3423), WI(3423). **6**:NE(2927), OK(2872,3423), TX(3423). **9**:FL(82), MS(82), OK(82). **11**:FL(35,82), MS(82), OK(82).
Claviceps sp. Ergot. **11**:MS(2720).
Clypeoporthe andropogonis. On dead culms and leaves. **9**:MS(82). **11**:NJ(82).
Crocicreas culmicola. On stems. **11**:NJ(505).
Dasyscyphella sp. **4**:GA(1438).
Dasyscyphus distinctus. **9**:GA(1438).
Dasyscyphus sp. **9**:GA(1438).
Entosordaria sp. **5**:GA(1438).
Epichloe typhina. Choke. **4**:WI(82,3423).
Erysiphe graminis. Powdery mildew. **11**:OK(2872).
Graphyllium hysterioides. On culms. **4**:KS(82). **9**:GA(1438). **11**:GA(82).
Lachnum pygmaeum. On exposed dead roots. **11**:NJ(965).
Leptosphaeria muricata. **9**:GA(82,1438).
Leptosphaeria sp. **9**:TX(652).
Meliola panici. Black mildew. **11**:CA(82).
Metasphaeria infuscans. On dead culms. **9**:AL(82), GA(82,1438), MS(82).
Mollisia atriella. **5**:GA(1438). **9**:GA(1438).
M. culmina. **9**:GA(1438).
Mollisia sp. **5**:GA(1438).
Myriogenospora atramentosa. **9**:GA(1438).
Naemacyclus culmigenus. **5**:GA(1438). **9**:GA(1438). **11**:FL(522).
Naemacyclus sp. **4**:GA(1438).
Nimbomollisia aberrans. On culms and panicle branches. **2**:GA(1438). **3**:GA(1438). **4**:GA(1438). **5**:GA(1438). **9**:GA(1438).
Paraphaeosphaeria michotii. **9**:TX(652).
Phaeosphaeria clavispora. On culms. **4**:GA(82). **9**:GA(1438).
P. culmorum. On culms. **4**:ND(82,377). **5**:GA(1438). **9**:GA(82,1438).
Phyllachora americana. Leaf spot, tar spot. **2**:FL(3438), GA(1438), SC(3423). **4**:AL(3423), GA(1438,3423). **5**:FL(3423), GA(1438,3423), TX(3423). **7**:AL(3423), FL(3423). **9**:GA(1438). **11**:LA(3423), MS(3423), PA(3423), VA(3423).
P. graminis. **9**:MA(1394).
P. luteo-maculata. Black spot, tar spot. **4**:C states(82), E states(82), ID(3423), IL(3423), MA(3423), MI(3423), NE(3423), NY(3423), OH(638), OK(2872,3423), SD(82,2242,3423), WI(3423). **5**:FL(35). **6**:NE(2927), NM(3423). **9**:E states(82), AL(3423), GA(1438), LA(82,3423), NC(1303,3423), NJ(3423), VA(3423). **11**:NM(82), SC(3423), TX(82).
P. quadraspora. Tar spot. **11**:FL(82,3423).
Stictis arundinacea. On culms and leaves. **11**:MS(3225,3226), SC(3225,3226).
S. maydis. **9**:GA(1438).
Telimena sp. Tar spot. **4**:IL(82). **11**:FL(35,82).
Basidiomycotina - Uredinales
Puccinia andropogonis. Heteroecious: telial host. **4**:E states(82), KS(3021,3023), MS(2720), ND(82,378), OK(2872), SD(2242), TX(82), WI(1346). **5**:FL(35), OK(2872). **6**:CO(1195), KS(3021), ND(378), NM(1195). **8**:NC(1303). **9**:E states(82), GA(1439), KS(82), NC(1303), OK(2872), TX(82). **11**:S states(82), FL(35), ND(82), NV(82), TX(82).
P. ellisiana. Heteroecious: telial host. **6**:NE(2927). **9**:MD(82), NC(1303), NJ(82), TN(82), TX(652), VA(82). **11**:FL(35,82),

NC(82), NJ(82).
P. posadensis. Telial host; aecial host unknown. **11**:CA(82), FL(35,82), TX(740).
P. tripsaci. Heteroecious: telial host. **4**:C states(82), CO(82), KS(3021,3023), OK(2872), SD(2242). **11**:KS(82), ND(82), NE(82).
Uromyces andropogonis. Heteroecious: telial host. **4**:FL(1519), TN(1519). **5**:AL(1519), DE(1519), FL(1519), MD(1519), MS(1519,2720), TN(1519). **9**:E states(82), AL(1519), AR(1519), DC(1519), DE(1519), FL(1519), GA(1439), IL(1519), IN(1519), KY(1519), LA(82), MD(651,1519), MO(82,1519), MS(1519,2720), NC(1519), NY(1519), OH(638,1519), SC(1519), TN(1519), TX(652), VA(1519), WV(953,1519). **11**:E states(82), FL(35), GA(1439), LA(82), MO(82), NC(82), NJ(82).

Basidiomycotina - Ustilaginales

Sorosporium ellisii. Head smut. **2**:VA(1077). **4**:OK(2872). **5**:LA(1077), MS(2720), NJ(1077). **9**:AL(1077), AR(82,1077), DE(82,1077), NC(3944), NJ(82,1077), OH(638,1077). **11**:LA(82), MD(82,1077), MS(82), NJ(82), VA(82).
S. everhartii. Seed smut. **1**:FL(1077). **3**:FL(1077). **4**:FL(1077), GA(1077), MO(82,1077), NC(1077,1303), NE(1077). **5**:FL(1077), GA(1077). **8**:GA(1439). **9**:SE states(82), AL(1077), FL(1077), GA(1077,1439), LA(1077), LA(82), MS(1077,2720), NC(1077,1303), NJ(1077), NY(1077), PA(1077), TN(1077,1539). **10**:FL(1077). **11**:FL(35,82), VA(82).
S. provinciale. Head smut. **4**:C states(82), IA(1077), IN(1077), KS(3021), MO(1077), NE(1077), TX(1077). **6**:KS(3021). **11**:TX(82).
Sphacelotheca andropogonis-hirtifolii. **6**:AZ(1077). **11**:TX(1077).
S. monilifera. **5**:VA(1077).
S. occidentalis. Seed smut. **5**:CA(1077), TX(1077). **6**:IL(1077), KS(1077,3021), ND(378,1077), NE(1077), SD(1077). **9**:NJ(1077), WV(955). **11**:CA(82), FL(35), KS(82), ND(82), NE(82), SD(82).
S. seymouriana. **4**:GA(1077), NC(1077,1303), OH(946), VA(1077). **8**:GA(1439). **9**:AL(82,1077), GA(1077), NC(82,1077). **11**:GA(82).
Sporisorium andropogonis. Head smut. **4**:KS(82,1077), SD(82,1077,2242), TN(1077), TX(82,1077). **6**:AZ(1077), CA(1077), KS(1077).
S. cenchri. **9**:GA(1439).
Tolyposporella brunkii. **4**:TX(1077).
Ustilago striiformis. Stripe smut. **9**:OH(82).

Deuteromycotina - Hyphomycetes

Arthrinium phaeospermum. On culms. **9**:NJ(82). **11**:FL(35).
Bipolaris sorokiniana. Seedling blight, spot blotch, black point, root rot. **4**:ND(82). **6**:ND(3423). **11**:ND(82).
Cerebella andropogonis. Blackhead. **4**:GA(82), KS(82), MS(2720), OK(2872,3423), TX(82). **11**:KS(82), MS(82,2720), TX(82).
Fusarium acuminatum. Secondary root rot. **4**:ND(3423), NE(3423), SD(2242,3423), WY(3423). **6**:ND(3423), SD(3423).
F. equiseti. Secondary root rot. **4**:ND(82), NE(82). **11**:ND(82).
Helminthosporium sp. Leaf spot. **4**:MN(82).
Microdochium bolleyi. Root necrosis. **4**:ND(3423), NE(3423), SD(3423). **6**:NE(3423), SD(3423).
Papularia vinosa. **9**:FL(82).
Periconia circinata. Root rot. **6**:ND(3423).
Pyricularia grisea. Leaf spot. **11**:TX(82).
Tetraploa ellisii. On leaves. **11**:AL(2486).

Deuteromycotina - Coelomycetes

Ascochyta brachypodii. Leaf spot. **4**:ND(3423,3449), NY(4001), PA(4001), SD(2242).
A. graminicola. Leaf spot. **4**:ND(82).
Asteroma graminis. On culms. **4**:KS(82), MO(82).
Colletotrichum caudatum. Leaf spot. **6**:MS(2544), OK(2872,3423). **11**:GA(2544), MS(82,2720), WI(82).
C. dematium. **4**:ND(378).
C. graminicola. Anthracnose. **4**:CO(3423), KY(82), ND(82,3423), NE(2927,3423), OR(82), WI(82). **6**:MS(3423). **11**:KY(82).
Colletotrichum sp. **4**:WI(1332).
Cylindrosporium andropogoni. On living leaves. **4**:KS(3432).
Davisiella elymina. Tar spot. **4**:WI(3423).
Leptostromella andropogonis. On culms. **9**:AL(82).
Neottiosporina paspali. **4**:WI(82).
Phaeoseptoria festucae var. *andropogonis*. On living leaves. **4**:WI(1339).
Phoma terrestris. Pink rot. **4**:ND(82,3423). **6**:ND(3423). **11**:ND(82).
Phyllosticta andropogonivora. On living leaves. **4**:KS(3451), MN(3442).
Phyllosticta sp. Leaf spot. **4**:ND(82), WI(1343).
Septoria andropogonis. Leaf spot. **4**:KS(82,3023,3423), ND(3442), NE(3423), OK(2872,3423), SD(2242,3423), WI(82,1332,3423).
Stagonospora andropogonis. On leaves. **4**:OK(530).
S. ischaemi. Leaf spot. **4**:AL(82), OK(2872,3423). **6**:OK(2872,3423).

Deuteromycotina - Other

Rhizoctonia solani. Summer blight. **9**:WV(955).

Anthaenantia P. Beauv.

1. *A. rufa* (Elliott) Schutt
2. *Anthaenantia* sp.

Ascomycotina

Claviceps sp. Ergot. **1**:MS(2720). **2**:MS(82).

Deuteromycotina - Hyphomycetes

Cerebella andropogonis. Blackhead. **1**:MS(2720,3423). **2**:MS(82).

Deuteromycotina - Coelomycetes

Colletotrichum graminicola. Anthracnose. **1**:FL(3423). **2**:FL(82).

Anthoxanthum L.

1. *A. aristatum* Boiss.
2. *A. odoratum* L. [sweet vernalgrass]

Chytridiomycetes

Synchytrium graminicola. Purple leaf speckle. **2**:OR(3423).
Synchytrium sp. Purple leaf speck. **2**:OR(82).

Basidiomycotina - Uredinales

Puccinia brachypodii var. *poae-nemoralis*. **2**:HI(2903), ME(82).
P. graminis. Heteroecious: telial host. **1**:MS(82,2720). **2**:DE(82), ME(82), NJ(82), NY(82), WV(955).

Basidiomycotina - Ustilaginales

Tilletia holci. Seed smut. **2**:CT(82,1077), PA(82,1077).

Deuteromycotina - Hyphomycetes

Cladosporium herbarum. Leaf mold. **2**:OR(82).
Drechslera dematioidea. Mold, secondary leaf spot. **2**:DC(3423), MD(82), NY(82,3423), OR(82,3423), WA(3423), WV(955).

Deuteromycotina - Coelomycetes

Ascochyta sorghi. Leaf spot. **2**:WV(955).
Colletotrichum graminicola. Anthracnose. **2**:KY(82), OR(82,3423).
Phyllosticta anthoxella. On leaves. **2**:OR(82,3423).
Septoria anthoxanthina. Leaf spot. **2**:OR(3432).

Deuteromycotina - Other

Rhizoctonia solani. Summer blight. **2**:WV(955).

Apera Adans.

1. *A. interrupta* (L.) P. Beauv. (=*Agrostis interrupta* L.)

Deuteromycotina - Hyphomycetes

Drechslera fugax. Leaf spot. **1**:ID(3424).

Arctagrostis Griseb. in Ledeb.

1. *A. latifolia* (R. Br.) Griseb. (=*A. latifolia* (R. Br.) Griseb. var. *arundinacea* (Trin.) Griseb., *A. arundinacea* (Trin.) Beal)

Ascomycotina

Apiospora parallela. **1**:AK(526).
Claviceps purpurea. **1**:AK(526).
Mycosphaerella minor. On dead sheaths. **1**:AK(1938).
M. recutita. **1**:AK(526).
Phaeosphaeria insignis. On dead sheaths and stems. **1**:AK(1938).

Basidiomycotina - Ustilaginales

Ustilago striiformis. **1**:AK(526).

Deuteromycotina - Hyphomycetes

Cercosporidium graminis. **1**:AK(526).
Cladosporium phlei. **1**:AK(526).
Ramularia pusilla. Leaf spot. **1**:AK(3429).

Deuteromycotina - Coelomycetes

Sphaerellopsis filum. **1**:AK(526).

Arctophila Rupr. ex Anderss.

1. *A. fulva* (Trin.) Anderss. (=*Colpodium fulvum* (Trin.) Griseb.)

Ascomycotina

Claviceps purpurea. Ergot. **1**:AK(526,3423).
Mycosphaerella pusilla. **1**:AK(526).
Pyrenophora schroeteri. **1**:AK(526).

Aristida L.

1. *A. adoensis* Hochst. in A. Rich.
2. *A. adscensionis* L. (=*A. oligantha* Michx.) [prairie three-awn, six-weeks three-awn]
3. *A. arizonica* Vasey [Arizona three-awn grass]
4. *A. basiramea* Engelm. ex Vasey
5. *A. curtissii* (A. Gray) Nash
6. *A. desmantha* Trin. & Rupr.
7. *A. dichotoma* Michx.
8. *A. divaricata* Humb. & Bonpl. ex Willd. [poverty three-awn]
9. *A. hamulosa* Henry
10. *A. lanosa* Muhl. ex Elliott
11. *A. longespica* Poir. (=*A. intermedia* Lams.-Scribn. & J. Ball) [cana-brava]
12. *A. murina* Cav.
13. *A. orcuttiana* Vasey [beggartick grass]
14. *A. pansa* Woot. & Standl. [Wooton three-awn]
15. *A. purpurascens* Poir. (=*A. virgata* Trin.) [arrow-feather]
16. *A. purpurea* Nutt. (=*A. glauca* (Nees) Walp., *A. longiseta* Steud., *A. longiseta* Steud. var. *robusta* Merr.) [purple three-awn, red three-awn, Reverchon three-awn]
17. *A. purpurea* Nutt. var. *laxiflora* Merr.
18. *A. ramosissima* Engelm. ex A. Gray
19. *A. roemeriana* Scheele
20. *A. spiciformis* Elliott
21. *A. stricta* Michx. [pineland three-awn]
22. *A. ternipes* Cav. [spidergrass]
23. *A. tuberculosa* Nutt.
24. *Aristida* sp.

Oomycetes

Pythium graminicola. **14:**AZ(3437). **16:**MN(3423).
Pythium sp. Root rot. **24:**FL(35).

Ascomycotina

Anthostomella sp. **21:**GA(1438).
Apiospora montagnei. **10:**GA(1438). **15:**GA(1438).
Balansia aristidae. Black choke and ring, on leaves. **2:**MS(2720). **7:**MS(2720). **15:**AL(3423), GA(1438). **21:**GA(1438,3423). **24:**AL(82), FL(35,82), GA(82), MS(82,2720), TX(82), VA(82).
B. hemicrypta. **24:**NM(82), TX(82).
Dothidella aristidae. Char spot, tar spot. **2:**KS(3021). **7:**NC(1303,3423). **15:**MS(3423).
Euryachora aristidae. Tar spot. **15:**FL(82), GA(82), MD(82), MS(82,2720), NC(82), NJ(82), PA(82). **24:**FL(35).
Graphyllium chloes. On culms. **24:**ND(82), NE(82).
G. hysterioides. **15:**GA(1438). **24:**GA(82).
Nimbomollisia aberrans. **10:**GA(1438).

Basidiomycotina - Uredinales

Puccinia aristidae. **2:**AZ(741,1203), CA(82), NM(82), TX(742). **23:**MN(2874). **24:**TX(82).
P. aristidae var. *chaetariae*. Heteroecious: telial host. **1:**AZ(755,3981), CA(755), NM(755), TX(755). **2:**AZ(747).
P. sonorica. Telial host; aecial host unknown. **9:**AZ(747,3981). **22:**AZ(747,755,1194,1203,3981).
P. unica. Telial host; aecial host unknown. **3:**AZ(1194,3981). **9:**TX(742,755). **13:**AZ(755,1203,3981), TX(742,755). **22:**AZ(3981), TX(742,755).
Uromyces aristidae. Telial host; aecial host unknown. **3:**NM(111,755). [*U. peckianus*]. **2:**OK(2872). **4:**OK(2872), WI(1357). **5:**OK(2872). **6:**OK(2872). **11:**OK(2872). **23:**WI(1337). **24:**FL(35).
U. seditiosus. Heteroecious: telial host. **2:**AL(82), LA(755), NE(82), NY(82), TN(755), TX(82,755), WI(82). **4:**KS(755), NE(755). **5:**OK(755). **6:**OK(755). **7:**AL(755), AR(755), KY(755). **11:**IL(755), IN(755), NJ(755). **15:**VA(755). **16:**AL(82), CO(755), MO(755), NE(82), NY(82), TX(82), WI(82). **18:**IN(755). **19:**TX(755). **23:**NY(755), WI(755).

Basidiomycotina - Ustilaginales

Sorosporium confusum. **7:**CA(1077), DC(1077), DE(1077), OH(1077), PA(1077). **11:**IN(1077). **15:**AL(1077). **16:**CO(1077). **20:**FL(1077). **21:**FL(35,1077). **24:**CO(1077), FL(35), TX(1077).
Sporisorium consanguineum. Seed smut. **2:**TX(82,1077). **3:**AZ(1077). **4:**SD(1077,2242). **8:**NM(1202). **12:**FL(1077). **13:**AZ(1077,1195). **15:**CO(1077). **17:**AZ(1077), NM(1077). **22:**AZ(1077). **24:**AZ(82), CA(82), CO(82), KS(82), NM(82), SD(82), TX(82).
Tilletia oklahomae. **11:**OK(1077,2872). **24:**OK(82).

Deuteromycotina - Hyphomycetes

Curvularia sp. Leaf spot. **24:**FL(35).
Fusarium acuminatum. Secondary root rot. **16:**NE(3423), WY(3423).
Helminthosporium sp. Leaf spot. **24:**FL(35).
Microdochium bolleyi. Root necrosis. **16:**MT(3423), NE(3423).

Deuteromycotina - Coelomycetes

Colletotrichum caudatum. Leaf spot. **5:**OK(2872,3423).
C. graminicola. Anthracnose. **7:**OK(2872,3423).
Hendersonia effusa. On dead culms. **15:**OK(2872). **24:**AL(82), MS(82), NC(82).
Phyllosticta sp. **11:**WI(1357).
Pseudoseptoria everhartii. Eye spot, leaf spot. **2:**IA(82,3423). **16:**IA(3450), ND(82,3423,3450).
P. stomaticola. **23:**WI(1346).
Tiarosporella paludosa. **15:**WI(1358).

Deuteromycotina - Other

Rhizoctonia solani. Summer blight. **2:**WV(953).
Rhizoctonia sp. Root rot. **24:**FL(35).

Arrhenatherum P. Beauv.

1. *A. elatius* (L.) J. Presl & K. Presl (=*A. elatius* (L.) J. Presl & K. Presl var. *bulbosum* (Willd.) Saint Amans) [tall oatgrass, tuber oatgrass]
2. *Arrhenatherum* sp.

Oomycetes

Pythium aristosporum. Root browning. **1:**MT(82).
P. arrhenomanes. **1:**ND(82).
P. debaryanum. Damping-off, root necrosis, seed rot. **1:**ND(82,3423), WA(3423).
P. graminicola. **1:**ND(3423), WA(3423).
P. irregulare. Root necrosis. **1:**WA(3423).

Ascomycotina

Claviceps purpurea. Ergot. **1:**IN(82), KY(82), MO(82,3423), MT(82,3423), OH(82,3423), OR(82,3423), WI(82,3423), WV(953).
Erysiphe graminis f. sp. *avenae*. Powdery mildew. **2:**MS(3051).

Basidiomycotina - Uredinales

Puccinia coronata. Heteroecious: telial host. **1:**LA(82), MN(82), WV(952).
P. graminis. Heteroecious: telial host. **1:**IN(82), LA(82), MI(82), MN(82), NY(82), OR(82), WA(82), WI(1348), WV(82).
P. montanensis. Heteroecious: telial host. **1:**MI(82).

Basidiomycotina - Ustilaginales

Tilletia caries. **1:**OR(1459).
T. controversa. Dwarf bunt. **1:**ID(1465), OR(1462,1463,2343).
Ustilago avenae. **1:**E states(82), CT(1077), GA(1077,1439), IA(1077), ID(1077), IL(1077), IN(1077), LA(1077), MD(1077), MI(1077), MN(82,1077), MO(82,1077), MS(1077,2720), NC(1077,1303,3944), NJ(1077), NY(1077), OH(1077), OR(82,1077), PA(1077), SD(2242), VA(1077), VT(1077), WA(82,1077), WI(1077), WV(1077).

Basidiomycotina - Other

Athelia rolfsii. Seedling blight, stalk rot, basal sheath. **1:**LA(3423).
Thanatephorus cucumeris. Brown patch, root rot, sharp eye spot. **1:**CA(3423), OR(3423), WA(3423).

Deuteromycotina - Hyphomycetes

Cercosporidium graminis. Leaf streak. **1:**CA(3423), ID(3423), KY(82,3423), MD(82,3423), MI(82), MO(3423), MS(82,2720,3423), MT(82,3423), NY(82,3423), OR(82,3423), PA(82,3423), WA(82,3423), WV(952).
Cladosporium herbarum. Leaf mold. **1:**WA(82).
Drechslera avenae. Leaf blotch. **1:**CA(82,3423).
Ramularia pusilla. Leaf spot. **1:**OR(82,3423,3429).
Spermospora avenae. Red leather-leaf. **1:**ID(3423), WA(82,3423), WV(952,953).

Deuteromycotina - Coelomycetes

Ascochyta graminicola. Purple leaf speck. **1:**OR(82).
A. sorghi. Leaf spot. **1:**OR(3423,3449).
Cheilaria agrostis. Blotch spot, char spot. **1:**OR(82,3423), WA(3423).
Colletotrichum graminicola. Anthracnose. **1:**MN(82,3423), ND(82), OH(82,3423), OR(82), PA(3423), VA(82), WV(952).
Macrophomina phaseolina. Charcoal rot. **1:**ND(82).
Pseudoseptoria stomaticola. Eye spot, leaf spot. **1:**ID(3423,3450), MT(82,3423,3450).
Stagonospora avenae. Leaf blotch, speckled blotch. **1:**PA(82,3423), WA(82,3423), WV(952).

Deuteromycotina - Other
Sclerotium rolfsii. Southern blight. **1**:LA(82).

Arthraxon P. Beauv.

1. *A. hispidus* (Thunb.) Makino
2. *A. hispidus* (Thunb.) Makino var. *cryptatherus* (Hack.) Honda

Deuteromycotina - Hyphomycetes
Bipolaris sorokiniana. Seedling blight, spot blotch, black point, root rot. **1**:ND(3423).
Periconia circinata. Root rot. **2**:ND(3423).
Deuteromycotina - Coelomycetes
Phoma terrestris. Pink root. **2**:ND(3423).

Arundinaria Michx.

1. *A. gigantea* (Walter) Muhl. (=*A. macrosperma* Michx.) [canebreak bamboo, large cane, southern cane]
2. *A. gigantea* (Walter) Muhl. subsp. *tecta* (Walter) Maclure (=*A. tecta* (Walter) Muhl.) [giant cane, switch cane]
3. *Arundinaria* sp.

Ascomycotina
Apiospora montagnei. On culms and dead sheaths. **1**:AL(82), LA(82), MS(82). **2**:AL(82), GA(1438), LA(82), MS(82). **3**:GA(659).
Aulographum arundinariae. On culms. **1**:GA(82). **2**:GA(1438).
Botryosphaeria arundinariae. On culms. **1**:AL(82), LA(82).
B. dothidea. **2**:GA(1438).
B. festucae. On culms. **1**:LA(82).
Caryospora langloisii. **3**:LA(185).
Diatrype arundinariae. On dead culms. **3**:LA(2925).
Didymella eumorpha. **1**:OK(2872).
Didymosphaeria arundinariae. On culms. **2**:GA(82,1438).
Discocurtisia arundinariae. On culms and sheaths. **1**:AL(82), MS(82,2720), NC(82), SC(82,2563). **2**:AL(82), GA(1438), MS(82,2720), NC(82), SC(82).
Dothidella minima. On leaves. **1**:LA(82,3423).
Eupropolella arundinariae. **2**:GA(1438).
Eutypa consobrina. On culms. **1**:LA(82), MS(82). **2**:LA(82), MS(82).
Herpotrichia macrotricha. **3**:GA(189).
Hypocrella tuberiformis. On culms. **2**:AL(82).
Hypoderma scirpinum. On culms. **1**:MS(82,2720).
Hypoxylon culmorum. On culms. **1**:FL(35,82), GA(82), LA(82). **2**:FL(82), GA(82,1438), LA(82). **3**:FL(35).
H. rubiginosum. **1**:AL(82), GA(82). **2**:AL(82), GA(82,1438).
Konenia bambusae. **3**:NC(2780).
Lachnum arundinariae. On leaves. **2**:GA(849,1438,3167).
Leptosphaeria eumorpha. **2**:IN(82), SC(82).
Lophodermium arundinaceum. On culms. **1**:AL(82), GA(82), SC(82). **2**:AL(82), GA(82,1438), SC(82).
Massarina arundinacea. On dead culms. **1**:SC(82).
Meliola tenuis. Black mildew. **1**:AL(82), GA(82,1438), MS(82,2720). **2**:Unk.(1452), AL(82), GA(82,1438), MS(82). **3**:Unk.(1452).
Metasphaeria rimularum. On culms. **1**:FL(35,82), GA(82).
Mycosphaerella arundinariae. Leaf spot. **2**:AL(82).
Ophiobolus stictisporus. On culms. **2**:GA(82,1438).
Phaeosphaeria nigrans. **2**:GA(82,1438).
Phyllachora shiraiana. Black leaf speck, tar spot. **1**:NC(1303). **2**:AL(82,3423), GA(82,1438,3423), MS(82,2720), SC(82,3423).
Phyllachora sp. **2**:GA(1438).
Rosellinia sublimbata. **3**:NC(2783), SC(2783).
Stictis pustulata. **3**:MS(3225).
S. stellata. **3**:LA(3225).
Stigmatea sclerotidea. On leaves. **2**:GA(82,1438), SC(82).
Stomiopeltis polyloculata. **2**:GA(1438).
Basidiomycotina - Uredinales
Puccinia arundinariae. Heteroecious: telial host. **1**:SE states(82), FL(35), OK(82,2872), TX(82). **2**:SE states(82), GA(1439), MS(2720), NC(1303), OK(82), TX(82).
Basidiomycotina - Other
Athelia scutellaris. On wood. **2**:GA(491,1439).
Cyphellathelia pezizoidea. **3**:LA(1707), NJ(1707).
Dicellomyces gloeosporus. Leaf parasite, leaf spot. **2**:AL(82), GA(82,3423), NC(82,2641,3423). **3**:LA(2643).
Deuteromycotina - Hyphomycetes
Alternaria sp. **1**:GA(659).
Arthrinium arundinis. **3**:GA(659).
A. phaeospermum. On dead culms and leaves. **1**:Gulf states(3423), Range of host(82), NC(1303). **2**:Range of host(82), MS(2720).
Cercosporidium compactum. Leaf spot. **2**:AL(3423).
C. graminis. Leaf streak, brown stripe. **1**:AL(3423). **2**:AL(82).
Cladosporium sp. **1**:GA(659).
Coniosporium gramineum. On dead culms. **3**:Unk.(82), LA(965).
Papularia vinosa. Leaf spot. **1**:FL(35,82), LA(82). **2**:FL(82), LA(82). **3**:FL(35).
Periconia minutissima. **3**:GA(659).
Volutella tecticola. Leaf spot. **2**:AL(82).
Deuteromycotina - Coelomycetes
Colletotrichum graminicola. Anthracnose. **1**:MS(82,2720), SC(82).
Leptothyrium cylindricum. Leaf spot. **1**:AL(3423). **2**:AL(82).
Stagonospora simplicior. **3**:GA(659).
Deuteromycotina - Other
Sclerotium sacidioides. On leaves. **2**:VA(82).

Arundo L.

1. *A. donax* L. [cana-brava, carrizo, giant reed]
2. *Arundo* sp.

Ascomycotina
Anthostomella sp. **2**:HI(2903).
Apiospora montagnei. **1**:GA(1438).
Basidiomycotina - Uredinales
Puccinia coronata. Heteroecious: telial host. **1**:CA(82).
Basidiomycotina - Other
Armillaria mellea. Root rot. **1**:MD(82).
Deuteromycotina - Hyphomycetes
Arthrinium phaeospermum. **1**:Range of host(82), CA(3423).
Papularia vinosa. Anthracnose-like lesions, dieback, leaf spot. **1**:TX(82).
Scolicotrichum maculicola. Leaf spot. **1**:OK(2872,3423).
Deuteromycotina - Coelomycetes
Leptostroma donacis. On culms. **1**:NC(82).
Pseudoseptoria donacis. Stem speckle, eye spot. **1**:CA(82,3423,3450).

Avena L.

1. *A. barbata* Brot. [slender oat]
2. *A. brevis* Roth
3. *A. fatua* L. (=*A. orientalis* Schreb.) [drake, flaver, potato oats, red oats, Tartarian oat, wild oats]
4. *A. nuda* L. [naked oat]
5. *A. sativa* L. (=*A. byzantina* K. Koch) [common oat]
6. *A. sterilis* L. [animated oat]
7. *A. strigosa* Schreb. [bristle oat, sand oat, small oat]
8. *Avena* sp.

Plasmodiophoromycetes
Polymyxa graminis. In roots. **5**:Unk.(2145), SC(82).
Oomycetes
Aphanomyces camptostylus. Root necrosis. **5**:MN(82,3423), MT(3423), OR(3423), WI(82,3423).
Aphanomyces sp. **5**:MT(82), OR(82).
Phytophthora cactorum. Rot. **3**:CA(82,3423), OR(82,3423).
Phytophthora sp. **5**:CA(82).
Pythium aphanidermatum. Seedling blight, damping-off. **5**:FL(35,3169).
P. aristosporum. Root browning. **5**:OR(82).
P. arrhenomanes. Root browning. **3**:ND(82). **5**:MN(82), ND(82), SD(82).
P. debaryanum. Damping-off, root necrosis, seed rot. **3**:ND(3423), OR(3423), WA(3423), WY(3423). **4**:ND(3423). **5**:'Root rot.' NC states(82), NW states(82), CO(3423), IA(3423), MN(3423), MT(3423), ND(3423), NE(3423), OR(3423), SD(2242,3423), WA(3423).
P. graminicola. Root browning. **3**:ND(3423), WY(3423). **4**:ND(3423). **5**:IA(3423), MN(3423), MO(3423), ND(3423), OR(3423), SD(2242,3423), VA(3423), WY(3423). **8**:CA(82), IA(82).
P. hypogynum. Root necrosis. **5**:NE(3423).
P. irregulare. Preemergence damping-off, root necrosis. **4**:ND(3423). **5**:'Root rot.' NC states(82), NW states(82), IA(3423), MN(1489,3423), ND(3423), OR(3423), WA(82).
P. rostratum. Root necrosis. **5**:CA(82), WA(82,3423).
P. tardicrescens. Root browning, root rot. **3**:WA(82,3423).
P. ultimum. Preemergence damping-off. 'Root rot.' **5**:NC states(82), NW states(82), OR(3423), WA(82,3423).
Pythium sp. Seedling blight, preemergence damping-off, root rot.

5:NC(1303).
Sclerophthora macrospora. Downy mildew. 3:SD(2242,3184). 5:'Downy mildew.' AR(3537,3538), CO(82), FL(930,931,3877), GA(2406,3537,3538), ID(82,638,3423), IL(312), IN(82,3423), LA(82,3423,3537), MI(3423), MN(1280), MS(82,2720,3050,3423,3537,3538,3539), NC(1303), SD(2242), TN(3537), VA(2989).

Zygomycotina

Rhizopus sp. Isolated from basal parts and seeds. 5:MN(1489).

Ascomycotina

Claviceps purpurea. 'Ergot.' 5:E states(3423), CO(82), IA(82,3423), IL(312), KS(82), MI(82), MN(82), ND(82,3423), NY(82), OH(946), OK(82), SD(2242), WA(82,3423).
Erysiphe graminis. Powdery mildew. 3:CA(3423), OR(82,3423). 5:'Powdery mildew.' E states(82), N states(3423), AZ(82), CA(82,3423), FL(35), GA(3423), ID(82), IN(3117), MI(82), MO(82,3423), NC(2516), OR(82,3423), TX(3423), UT(82,3423), WA(82,3423). 6:OR(82).
E. graminis f. sp. *avenae*. Powdery mildew. 5:AR(3029), FL(35,929,931), GA(1023,1275), TN(72,73), VA(2993). 8:MS(3051).
Gaeumannomyces graminis. Foot rot, take-all. 5:'Take-all.' CA(3423), OR(82,3423). 6:CA(82), OR(82).
Gibberella zeae. Head mold, scab. 5:NE states(82), ND(82), OK(2872), TX(82).
Lewia scrophulariae. On stubble and straw. 5:WI(3878).
Mycosphaerella holci. Leaf spot. 5:IL(312), MS(3539).
M. tassiana. 5:AK(526).
Phaeosphaeria avenaria. Black stem. 5:E & C states(82), IA(3468), IL(3468), MN(3468), MS(3539), ND(3468), SC(932), SD(3468), WI(3468).
Phyllachora graminis. Leaf spot. 5:MS(3539).
Pyrenophora avenae. Leaf blotch. 3:Range of host(82). 5:Range of host(82), AR(3026,3027,3028), TX(1134,1284). 6:Range of host(82).
P. teres. Node canker. 5:IL(312).

Basidiomycotina - Uredinales

Puccinia coronata. Heteroecious: telial host. 1:W states(82). 2:Range of host(82), S states(82). 3:C & W states(82), CA(81), SD(2242). 4:Range of host(82), S states(82). 5:'Crown rust.' AL(2213), AR(3026,3027,3028), CA(81), FL(35), GA(1439,2404), HI(2903), IL(309), KY(1815), MA(1394), MS(2720), NC(2516), NE(2155), OK(2872), SD(2242), TX(119). 6:W states(82). 7:Range of host(82), S states(82).
P. coronata var. *avenae*. Heteroecious: telial host. 5:Range of host(82), AL(285), FL(929,931), GA(1023,1024,1275,1439,2406), MS(1706,2720,3539), NC(1303), ND(378), NJ(496), SC(932,933), TN(72,73), TX(1134), VA(2993). 8:TX(1135).
P. graminis. Heteroecious: telial host. 1:CA(81,82), OR(82). 2:Range of host(82). 3:C & W states(82), CO(1195), ND(378), SD(2242). 4:Range of host(82). 5:'Stem rust.' Range of host(82), AR(3026,3027,3028), CA(81), FL(35,929,931), GA(1439,2406), MN(3009), MS(2720,3539), NC(1303,2516), ND(378,3009), OK(2872), SC(932), SD(2242,3009), TX(119,1134,1135), VA(2993), WI(1348), WV(3911). 6:CA(82), OR(82). 7:Range of host(82).
P. recondita. Heteroecious: telial host. 1:CA(82,2439). 3:CA(82). 5:AL(82), CA(82), IN(82), NC(82), NH(1895), OK(82), TX(82).

Basidiomycotina - Ustilaginales

Ustilago avenae. 1:CA(1077). 3:CA(1077), CO(1077), MD(1077), WA(1077). 4:MO(82,1077). 5:'Loose smut.' Range of host(82,1077), AK(526,2078), AL(285,2213), AR(3026,3027,3028), CA(81), FL(35,929,931), GA(1023,1024,1439,2406), HI(2903), IA(1077), MA(1390), MN(1077), MS(1706,2720,3539), NC(1303,2516), NJ(496), NM(1077), OK(2872), OR(1077), SC(932), SD(2242), TX(1134), VA(2993), WA(1077), WY(819). 6:CA(82), MD(82), OR(82,1077), TX(82,1077), WA(82,1077).
U. hordei. 3:CO(1077), KS(1077), OR(1077), TX(1077), WA(1077). 5:Range of host(1077), IA(1077), MN(1077), MS(2720), NC(1303), ND(378), OR(1077), SD(2242), WA(1077), WY(1195).
U. segetum. Covered smut. 3:KS(82), OR(82), TX(82), WA(82). 5:'Covered smut.' Range of host(82), AK(526), AR(3027,3028), FL(35,929,931), MS(3539), NC(2516), NJ(496), OK(2872), SC(932), TX(1134), VA(2993). 6:KS(82), TX(82), WA(82).
Ustilago sp. 5:AK(2078), AR(3026), TN(72,73), TX(119).

Basidiomycotina - Other

Agrocybe molesta. Attached to or associated with crowns. 5:OR(3422).
A. praecox. Attached to or associated with crowns and roots. 5:OR(82,3422).
Athelia rolfsii. Seedling blight, basal sheath and stalk rot. 5:AR(3423), MS(3423).
Marasmius graminum. On culms. 5:IL(82).
Thanatephorus cucumeris. Node and sheath blight, brown patch, root rot, sharp eye spot. 3:ND(3423), OR(3423), WA(3423). 5:IA(3423), IL(312), MN(3423), ND(3423), OR(3423), SD(3423), WA(3423), WY(3423).
Typhula incarnata. 'Speckled or gray snow mold.' 5:PA(2264).
Typhula sp. Scald. 5:MS(3539).

Deuteromycotina - Hyphomycetes

Alternaria alternata. Spikelet blight. 5:IL(312).
Alternaria sp. Glume spot, leaf spot. 5:FL(35), ID(82), IN(82), MN(1489), MO(82), NY(82), VT(82).
Aspergillus flavus. On stored seeds. 5:Unk.(3089).
A. glaucus. On stored seeds. 5:Unk.(3089).
A. restrictus. On stored seeds. 5:Unk.(3089).
Aspergillus sp. Isolated from basal parts and seeds. 5:MN(1489).
Bipolaris sorokiniana. Seedling blight, spot blotch, black point, crown, culm or root rot. 3:MN(3423), ND(82,3423), SD(2242,3423). 5:'Head blight.' E & C states(82), AR(3026,3027,3028), CA(3116), CO(3423), FL(931), GA(1023,2404,2406,3423), HI(1272), IA(3423), MN(3423), MS(2720), NC(2516), ND(82,3423), OK(2872,3423), OR(82,3423), SC(932,933), SD(2242,3423), TX(119). 6:OR(82).
B. victoriae. Victoria blight, victoria blotch, leaf spot. 5:'Victoria blight.' AL(929,2213), AR(3026,3423), FL(35,929,931,2587,3423), GA(1023,1024,1275,2404,2406,2587,3423,3644), IA(2587,3423), ID(3423), KY(1815,3423), LA(124,2587), MN(3423), MO(3423), MS(1706,2587,2720,3539), NC(1303,2516,2587), ND(3423), NE(2155,3423), NY(3423), SC(929,932,933), SD(2242,3423), TX(119,1134,1135,2587,3423), VA(2993), WI(2587), WV(954), WY(2587). 8:E & C states(82), FL(82), MT(82), TX(82).
Botrytis cinerea. Gray mold, head mold. 5:MN(82), WA(3423).
Cercospora secalis. Leaf spot. 5:IL(312).
Cercosporidium graminis. Brown stripe, leaf streak. 5:IA(82), IL(82,3423), KS(82), NY(82,3423).
Cladosporium herbarum. On leaves. 5:AL(82), IA(82), MN(82), TX(82), VA(82), WV(82). 6:OR(82), WA(82).
Drechslera avenacea. Leaf blotch. 3:MN(3442). 5:'Head blight.' FL(35), IL(312), ND(378).
D. avenae. Leaf blotch, spikelet drop, leaf spot, black stem. 1:CA(3423). 3:MT(3433), OR(3423). 5:AK(526,2078,3423), AL(2213), AR(3423), CA(3423), CT(3423), FL(929,931,2188,3423), GA(1023,1024,1275,2404,2406,3423), IA(3423), ID(3423), IL(3423), IN(3423), KY(3423), LA(3423), ME(3423), MN(1280,3423), MO(3423), MS(1705,1706,2720,3539), MT(3423), NC(1303,2516), ND(3423), NE(3423), NY(3423), OH(3423), OK(3423), OR(3423), PA(3423), SC(932,933), SD(2242,3423), TX(1135), VA(2993), WA(3423), WI(1337,3423), WV(954), WY(3423).
Epicoccum nigrum. Leaf mold, on seeds. 3:OR(82). 5:AK(526), FL(35), NY(732).
Fusarium avenaceum. Head mold, stem rot. 5:CA(3423), MT(3200), OH(3423). 8:CA(82), OH(82).
F. culmorum. Seedling blight, red leaf, head mold, foot and root rot, scab. 3:Unk.(82), ID(3423), OR(3423), WA(3423). 5:'Fusarium foot rot. Seedling blight.' CA(3116,3423), ID(3423), IL(3423), IN(3423), MI(3423), MT(3423), OH(3423), OR(3423), WA(3423), WI(3423). 6:Unk.(82). 8:W states(82), IA(82), MO(82), MS(82), UT(82).
F. equiseti. Secondary root rot. 1:ND(82). 2:ND(82). 5:ND(82).
F. graminearum. Seedling blight, root rot, scab. 'Head blight.' 5:CA(3116), HI(2903), IA(3423), IL(3423), IN(3423), MI(3423), ND(3423), OH(3423), OK(3423), TX(3423), WI(3423).
F. moniliforme. Seedling blight, secondary root rot. 5:MN(82), ND(82,3423).
F. oxysporum. Secondary root rot. 5:ND(82), SD(2242). 6:Unk.(82).
F. poae. Head blight and mold, secondary root rot, silver top. 5:OR(82,3423), SD(2242). 6:OR(82).

Fusarium sp. Seedling blight, foot rot. **4:**Unk.(82). **5:**FL(35), MN(1489), NY(732).
Fusicladium destruens. On leaves. **5:**NY(82,3423).
Helminthosporium sp. Leaf spot, root rot. **5:**MN(1489), MS(1706,2720), NH(1895).
Heterosporium avenae. Spikelet blight, on leaves. **5:**IL(312), OR(3423), WA(3423). **8:**OR(82).
Microdochium bolleyi. Root necrosis and rot. **3:**ND(3423), WA(3423), WY(3423). **5:**CA(3116), MN(3423), ND(3423), NE(3423), SD(3423), WY(3423).
M. nivale. 'Pink snow mold.' **5:**NJ(3423), WA(3441).
Nigrospora sphaerica. Node, canker, sheath, spikelet blight. **5:**IL(312).
Penicillium expansum. Blue mold. **5:**CO(3423), IA(3423), ID(3423), MN(3423), MT(3423), ND(3423), NE(3423), OR(3423), SD(3423), WA(3423), WY(3423).
Periconia circinata. Root rot. **3:**WA(3423). **4:**ND(3423). **5:**MN(3423), ND(3423).
Pseudocercosporella herpotrichoides. Foot rot, strawbreaker. **3:**OR(82), WA(82). **5:**OR(3423), WA(3423). **8:**OR(82), WA(82).
Spermospora avenae. Red leather-leaf. **5:**OR(82,3423), WA(82,638,3423). **6:**OR(82), WA(82).
Trichoderma sp. Isolated from basal parts. **5:**MN(1489).
Deuteromycotina - Coelomycetes
Ascochyta avenae. White leaf spot. **3:**WA(82,3423,3449). **5:**OR(3423,3449). **6:**OR(82).
A. brachypodii. Leaf spot. **5:**ND(3423,3449).
A. graminicola. Leaf spot. **5:**GA(82).
A. sorghi. Leaf spot. **5:**GA(3423,3449).
Colletotrichum graminicola. Anthracnose. **3:**ND(3423). **5:**'Anthracnose.' Range of host(82), AL(3550), AR(3026,3027,3028,3423), FL(35), GA(2406), IA(3423), IL(3423), KS(3423), KY(1815,3423), MN(1280), MO(3423), MS(2720,3539), ND(3423), NJ(3423), NY(3423), OH(3423), OK(2872,3423), OR(3423), PA(3423), SD(2242), WI(1355).
Leptothyrium avenae. On glumes and pedicels. **5:**IL(3423).
Macrophoma secalina. Leaf blight, spikelet blight. **5:**IL(312).
Macrophomina phaseolina. Charcoal rot. **5:**IL(311).
Pestalotia avenae. **5:**WA(3442).
Phoma terrestris. Pink root rot. **5:**CA(3423), MN(1489,3423), MT(3423), ND(82,3423), NE(3423), SD(2242,3423), WY(3423).
Phyllosticta avenophila. Leaf spot. **5:**IL(82,3423), MS(82,2720,3423).
Pseudoseptoria donacis. Eye spot. **5:**ND(3423,3450), SD(2242).
Septoria tritici f. *avenae*. Speckled leaf blotch, leaf spot. **3:**OR(82,3423), WA(82,3423). **5:**OR(82,3423), WA(82,3423). **6:**OR(82), WA(82).
Septoria sp. **5:**TX(1134).
Sphaerographium avenaceum. On dead leaves. **1:**CA(2439).
Stagonospora avenae. Speckled leaf blotch, stem rot. **3:**CA(3430). **5:**'Septoria blight.' FL(931), IA(3423), KY(3423), MA(3423), MI(76), MN(1280,3423), MS(3539), ND(3423), OH(946), WI(3423).
Wojnowicia hirta. Secondary foot rot, rot of mature straw. **5:**OR(82,3423), WA(82,3423).
Deuteromycotina - Other
Rhizoctonia oryzae. Root rot, sheath spot. **5:**ND(3423).
R. solani. Root rot. **3:**OR(82). **5:**'Seedling blight.' MI(82), MN(82,1489), ND(82), OR(82), SD(2242), WA(82). **6:**OR(82).
R. zeae. Root rot. **5:**NC(1303).
Sclerotium rolfsii. Seedling blight. **5:**AR(82), FL(35), MS(82,2720).

Axonopus P. Beauv.

1. *A. affinis* Chase [carpetgrass]
2. *A. compressus* (Sw.) P. Beauv. (=*Paspalum platycaule* Willd.) [tropical carpetgrass]
3. *A. furcatus* (Flüggé) Hitchc.
4. *Axonopus* sp.

Myxomycetes
Physarum cinereum. Slime mold. **2:**MS(2720).
Chytridiomycetes
Physoderma sp. **1:**HI(2903).
Ascomycotina
Balansia strangulans. Black choke, black ring. **1:**FL(35,82,3423). **2:**FL(82). **4:**FL(35).
Claviceps sp. Ergot. **1:**FL(35,82), GA(82).
Clypeoporthe iliau. On culms. **3:**FL(82). **4:**FL(35).
Basidiomycotina - Uredinales
Physopella compressa. Autoecious. **2:**LA(82).
Basidiomycotina - Ustilaginales
Sorosporium paranensis. **1:**FL(35).
Basidiomycotina - Other
Thanatephorus cucumeris. Brown patch, root rot, sharp eye spot. **1:**S states(3423), GA(1439), HI(3423), LA(3524).
Deuteromycotina - Hyphomycetes
Bipolaris ravenelii. False smut. **1:**MS(82,2720).
Cerebella andropogonis. Blackhead. **1:**Gulf states(82), SE states(82), FL(3423), GA(3423), LA(3423), MD(3423), NC(1303,3423), SC(3423), VA(3423). **2:**FL(82), LA(3423), MS(2720). **4:**FL(35).
Curvularia sp. Mold, leaf spot. **1:**GA(82), NC(1303).
Fusarium heterosporum. Head mold. **1:**GA(82), MS(82,2720), TX(82).
Helminthosporium sp. Leaf blight and spot. **1:**TX(82,3423). **4:**FL(35).
Deuteromycotina - Coelomycetes
Ascochyta sp. On seed. **1:**MD(82).
Massariothea paspali. **2:**LA(3547).
Stagonospora sp. **2:**OK(2872).
Deuteromycotina - Other
Rhizoctonia solani. Brown patch, root rot. **1:**FL(35,82), HI(82,2903), LA(82), NC(1303). **2:**FL(82), HI(82,2903), LA(82). **4:**FL(35).

Bambusa Schreb.

1. *B. multiplex* (Lour.) Raeusch. [hedge bamboo]
2. *B. vulgaris* Schrad. ex Wendl. [common bamboo]
3. *Bambusa* sp.

Ascomycotina
Apiospora sp. On dead culms. **3:**FL(82), GA(82).
Guignardia bambusae. On culms. **3:**MD(82).
Laestadia sp. On culms. **3:**FL(35).
Lageniformia bambusae. **3:**HI(2903).
Mycosphaerella sp. Leaf spot. **3:**CA(82).
Placostroma bambusae. On culms. **3:**FL(35,82,3423), GA(82,3423).
Scirrhia lophodermioides. On dead culms. **3:**FL(35), HI(82,2903).
Basidiomycotina - Uredinales
[*Puccinia melanocephala*]. **2:**FL(82), GA(82), MS(82,2720), NC(1303), TX(82). **3:**FL(35,82).
Puccinia sp. **3:**NC(1303).
Uredo ignava. **3:**FL(2305), GA(82).
Basidiomycotina - Ustilaginales
Ustilago shiraiana. **3:**FL(35).
Basidiomycotina - Other
Schizophyllum commune. On dead culms. **2:**CA(82), TX(82). **3:**TX(82).
Deuteromycotina - Hyphomycetes
Alternaria sp. Leaf spot. **3:**FL(35).
Arthrinium phaeospermum. On dead or dying culms. **1:**Range of host(82). **2:**Range of host(82). **3:**Range of host(82), CA(3423), FL(3423), GA(3423), TX(3423).
Cladosporium herbarum. On leaves. **3:**SC(82).
C. oxysporum. On leaves. **3:**AL(2486).
Corynespora foveolata. On culms. **3:**AL(2503).
Fusarium merismoides. On cut canes. **3:**GA(82).
Helicomyces roseus. **3:**HI(1270).
Helminthosporium sp. Leaf spot. **3:**FL(35,82).
Idriella bambusae. On dead stems. **3:**AL(2493).
Memnoniella echinata. On leaves. **3:**AL(2486).
Papularia vinosa. Dieback, on culms. **2:**FL(82). **3:**FL(35,82).
Phaeoisariopsis bambusicola. On living leaves. **3:**AL(2491).
Rutola graminis. **3:**SC(634).
Stachybotrys elegans. On leaves. **3:**AL(2486).
Torula herbarum. On dead stems. **3:**CA(81).
Deuteromycotina - Coelomycetes
Cylindrosporium bambusae. On culms. **3:**GA(128,3423).
Diplodia bambusae. On shoots and culms. **3:**LA(82,3423), TX(82).
Hendersonia sp. On culms. **3:**LA(82).
Phyllostictina divergens. **3:**FL(2782).
Pseudoseptoria donacis. Culm spot. **3:**CA(82).

Beckmannia Host

1. *B. erucaeformis* (L.) Host
2. *B. syzigachne* (Steud.) Fernald [American sloughgrass]
3. *Beckmannia* sp.

Chytridiomycetes
Physoderma beckmanniae. On living leaves. **2**:NV(3430).
Oomycetes
Pythium debaryanum. Damping-off, root necrosis, seed rot. **2**:ND(3423).
Ascomycotina
Erysiphe graminis. Powdery mildew. **2**:ID(3423), MN(3423), MO(82,3423), MT(82,3423), SD(82,2242,3423). **3**:AK(526).
Basidiomycotina - Uredinales
Puccinia coronata. **2**:SD(2242).
P. graminis. Heteroecious: telial host. **1**:MI(82), ND(378). **2**:ND(82), OR(82), UT(82).
Uromyces beckmanniae. Telial host; aecial host unknown. **1**:OR(82). **2**:OR(82).
Basidiomycotina - Ustilaginales
Ustilago striiformis. Stripe smut. **2**:CA(1077), CO(1077), OR(82,1077), WA(1077).
Deuteromycotina - Hyphomycetes
Cercosporidium graminis. Leaf streak, brown stripe. **2**:IA(3430), ID(3423), MT(3423), ND(82,3423), NE(3423), OR(82,3423), SD(2242,3423), WA(82,3423), WI(1357), WY(3423).
Drechslera catenaria. **2**:ND(82).
Helminthosporium sp. Leaf spot. **1**:MI(82). **2**:ND(3423).
Deuteromycotina - Coelomycetes
Ascochyta sorghi. Leaf spot. **2**:ND(3423,3449).
Colletotrichum graminicola. Anthracnose. **2**:MN(82,3423), ND(82,3423), NE(3423).
Septoria spartinae. Leaf spot. **2**:ND(82,3423).

Blepharoneuron Nash

1. *B. tricholepis* (Torr.) Nash [hairy dropseed]

Ascomycotina
Phyllachora vulgata. Tar spot. **1**:NM(82,3423).

Bothriochloa Kuntze

1. *B. barbinodis* (Lag.) Herter (=*Andropogon barbinodis* Lag.) [cane bluestem]
2. *B. barbinodis* (Lag.) Herter var. *perforata* (Trin. ex E. Fourn.) Gould (=*Andropogon perforatus* Trin. ex E. Fourn.)
3. *B. ischaemum* (L.) Keng (=*Andropogon ischaemum* L.) [King Ranch bluestem, Turkestan bluestem, yellow bluestem]
4. *B. saccharoides* (Sw.) Rydb. (=*Andropogon saccharoides* Sw.) [silver beardgrass, silver bluestem]

Ascomycotina
Claviceps purpurea. Ergot. **4**:TX(82).
Phyllachora luteo-maculata. Black spot, tar spot. **2**:TX(3423). **4**:OK(82,2872,3423), TX(82,3423).
Basidiomycotina - Uredinales
Puccinia cesatii. Autoecious. **3**:TX(742).
Basidiomycotina - Ustilaginales
Sphacelotheca andropogonis-hirtifolii. **4**:AZ(82,1077).
Sporisorium andropogonis. Head smut. **1**:CA(1077). **4**:AZ(82,1077), CA(1077), TX(82,1077).
Tolyposporella brunkii. Leaf smut. **4**:TX(82,1077).
Ustilago sp. **4**:OK(2872).
Deuteromycotina - Hyphomycetes
Cerebella andropogonis. Blackhead. **3**:OK(2872,3423). **4**:OK(82,2872,3423), TX(82).
Helminthosporium sp. Leaf spot. **4**:OK(2872).

Bouteloua Lag.

1. *B. aristidoides* (H. B. K.) Griseb. [needle grama]
2. *B. barbata* Lag. [six-weeks grama]
3. *B. breviseta* Vasey
4. *B. chondrosioides* (H. B. K.) Benth. ex S. Wats.
5. *B. curtipendula* (Michx.) A. Gray (=*Antheropogon curtipendulus* (Michx.) E. Fourn.) [side-oats grama]
6. *B. eriopoda* (Torr.) Torr. [black grama]
7. *B. gracilis* (H. B. K.) Lag. ex Steud. (=*B. oligostachya* (Nutt.) Torr. ex A. Gray) [blue grama]
8. *B. hirsuta* Lag. [hairy grama]
9. *B. radicosa* (E. Fourn.) D. Griffiths [purple grama]
10. *B. rigidiseta* (Steud.) Hitchc.
11. *B. rothrockii* Vasey [Rothrock grama]
12. *B. simplex* Lag. [mat grama]
13. *Bouteloua* sp.

Oomycetes
Pythium arrhenomanes. Root browning. **5**:ND(82). **7**:ND(82).
P. debaryanum. Damping-off, root necrosis, seed rot. **5**:ND(82,3423). **7**:MT(3423), ND(82,3423), SD(2242,3423).
P. graminicola. **5**:ND(3423). **7**:ND(3423), WA(3423).
P. irregulare. Root necrosis. **7**:SD(2242,3423).
Ascomycotina
Balansia strangulans. Black ring. **5**:KS(3423).
B. strangulans f. *discoidea*. Black ring. **5**:KS(82,3023), NE(2927).
Claviceps purpurea. Ergot. **5**:NE(82,3423), OK(3423). **7**:OK(2872,3423). **13**:OK(82).
Cochliobolus bouteloue. On leaves. **7**:KS(3023), NM(3425).
Epichloe typhina. Choke. **7**:IA(3423), KS(82,3423), NE(82,3423).
Graphyllium chloes. On dead culms. **7**:NE(82).
Lophodermium arundinaceum. On dead leaves. **13**:Unk.(82).
Phyllachora cynodontis. Tar spot. **5**:AZ(1194), IA(82,3423), IL(82,3423), KS(82,3023), NE(2927), OK(2872,3423), SD(2242), TX(82,3423), WI(82,1332,1334,1338,3423). **7**:ND(82,3423), NE(82,3423), NM(82,3423), SD(82,2242,3423). **8**:WI(1330,3423), WY(82).
Pleospora pellita var. *oligostachyae*. On dead leaves. **7**:KS(82).
Basidiomycotina - Uredinales
Puccinia boutelouae. Telial host. **5**:AZ(82,1203,3981), KS(3023), OK(2872), TX(82,1521).
P. cacabata. Heteroecious: telial host. **1**:AZ(747,1195,1203,3981), NM(3356). **2**:AZ(747,1195,1203,3981), NM(2113,3356), TX(742). **6**:AZ(1203,3981). **8**:AZ(1203,3981). **11**:AZ(1195,1203). **13**:AZ(82), TX(82).
P. chloridis. Heteroecious: telial host. **5**:C states(82), AZ(82,1203,3981), KS(3023), MD(82), MN(2720), OK(2872), SD(82,2242), TX(742). **7**:W states(82), AZ(82,1203,3981), ND(378), SD(2242). **8**:AZ(1203,3981), NE(82).
P. diplachnis. **7**:TX(742).
P. exasperans. **5**:NE(1521), TX(1521).
P. graminis. Heteroecious: telial host. **5**:Unk.(82). **7**:Unk.(82).
P. vexans. Heteroecious: telial host. **5**:C states(82), AZ(82,747,1194,1195,1203,3981), IA(1521), KS(3021,3023), ND(82,378), OK(2872), SD(2242), TX(742). **7**:NE(82), SD(2242).
Basidiomycotina - Ustilaginales
Sporisorium neglectum. Head smut. **1**:AZ(1195,3378). **7**:KS(82,1077,3021), LA(82,1077), NE(82,1077), OK(82,1077,2872), TX(82,1077). **8**:OK(2872), TX(82,1077). **12**:AZ(1077), SD(2242). **13**:AZ(82).
Ustilago buchloes. Leaf smut. **1**:AZ(1077). **2**:AZ(1077). **3**:AZ(1077), NM(1077). **4**:AZ(1077). **5**:C states(82), AZ(1077), CA(1077), CO(1077), IN(1077), KS(1077,3023), ND(82,378,1077), TX(1077), UT(82,1077). **6**:AZ(82,1077), NM(82,1077). **7**:W states(82), AZ(1077), CO(1077), KS(1077), MT(82,1077), ND(1077), NM(1077), TX(1077), UT(82,1077). **9**:AZ(1077). **13**:AZ(82,922), CO(82), NM(82).
U. calcara. **3**:NM(1077).
U. minor. **1**:AZ(1077). **6**:AZ(1077). **8**:KS(82,1077), NE(82,1077).
Basidiomycotina - Other
Thanatephorus cucumeris. Brown patch, root rot, sharp eye spot. **5**:ND(3423), SD(3423). **7**:MT(3423), ND(3423), SD(3423).
Deuteromycotina - Hyphomycetes
Bipolaris buchloes. Mold, leaf spot. **5**:KS(82,3423), ND(82,3423), NE(82,3423), SD(2242,3423), WI(82,3423), WY(3423). **7**:CO(3425), KS(82,3423), MT(3423), ND(82,3423), NE(82,3423), SD(2242,3423), WY(3423). **8**:TX(3430), WI(82,3423).
B. sorokiniana. Seedling blight, spot blotch, black point, root rot. **5**:MN(82), ND(82,3423), SD(3423). **7**:MN(82), ND(82,3423), SD(2242,3423).
Cercospora boutelouae. Leaf spot. **5**:OK(82,2873,3423), SD(2242), WI(82,1331,3423). **7**:CO(3423,3424), SD(2242). **8**:IL(3430), OK(3430), WI(1359).
Cercospora sp. Leaf spot. **13**:KY(82).
Cercosporella boutelouae. On living leaves. **5**:KS(3451).
Curvularia geniculata. Secondary root rot. **5**:SD(2242). **7**:MT(82), ND(82), SD(2242).
Exserohilum turcicum. **5**:WI(1334). **8**:WI(1334).
Fusarium acuminatum. Culm and root rot. **5**:ND(3423), OK(3430), SD(2242,3423), WY(3423). **7**:MT(3423), ND(82,3423),

NE(82,3423), SD(2242,3423), WA(3423), WY(3423). **13**:Unk.(82).
F. equiseti. Secondary root rot. **7**:MT(3423), ND(3423), NE(3423), SD(3423). **13**:Unk.(82).
F. oxysporum. Secondary root rot. **5**:ND(3423), SD(2242,3423), WY(3423). **7**:MT(3423), ND(3423), NE(3423), SD(2242,3423), WY(3423). **13**:Unk.(82).
F. poae. Secondary root rot, silver top. **5**:ND(3423), SD(2242,3423). **7**:ND(82,3423), NE(82,3423), SD(2242,3423). **13**:Unk.(82).
F. sporotrichioides. Culm and root rot. **5**:ND(82), NE(82). **7**:ND(82), NE(82).
Fusarium sp. Culm and root rot. **5**:ND(82), NE(82). **7**:ND(82), NE(82).
Helminthosporium sp. Leaf spot. **5**:WI(1352). **7**:KS(3023). **8**:WI(1352).
Heterosporium avenae. On dead leaves. **5**:TX(3430).
Microdochium bolleyi. Root necrosis. **7**:MT(3423), ND(3423), NE(3423), SD(2242,3423), WY(3423).
Rhynchosporium secalis. Scald. **7**:NE(3423). **8**:NE(82).
Deuteromycotina - Coelomycetes
Ascochyta boutelouae. On leaves. **7**:ND(82), NM(82).
[*A. graminea*]. On leaves. **5**:KS(3023), OK(3430), SD(2242), TX(3430). **7**:KS(3430), ND(3423,3449), NM(3423,3425), SD(2242).
Ascochyta sp. Leaf spot. **5**:OK(2872). **7**:OK(2872). **10**:OK(2872).
Colletotrichum graminicola. Anthracnose. **5**:KS(3023). **8**:WI(1341).
Davisiella elymina. Tar spot. **5**:WI(3423).
Macrophoma sporoboli. **2**:AZ(3438).
Macrophoma sp. On lower leaves. **5**:WI(1357).
Macrophomina phaseolina. Charcoal rot. **5**:OK(2872,3423).
Phaeodiscula texensis. On dead leaves. **7**:TX(3430).
Phoma terrestris. Pink root. **5**:ND(82,3423), SD(3423). **7**:MT(3423), ND(82,3423), SD(2242,3423).
Pseudoseptoria donacis. On culms. **6**:AZ(3435).
P. everhartii. Eye spot, leaf spot. **5**:AZ(1194), CO(3430), ND(82,3430,3450), WI(82,1338,3450).
Scaphidium boutelouae. Leaf lesions, narrow brown spot. **5**:AZ(3423,3424), NM(1195,3378,3423,3424,3549,3550). **8**:NE(82,3423).
Septoria andropogonis. On faded leaves. **5**:KS(3023), SD(2242,3423).
Septoria sp. Leaf spot. **5**:OK(2872). **7**:OK(2872).
Stagonospora arenaria. Purple-brown blotch, leaf spot. **5**:OK(2872,3423).
S. subchlorinoflavida. On living leaves. **5**:KS(3451).
Wojnowicia hirta. Secondary foot rot, rot of mature straw. **7**:ND(3423).
Deuteromycotina - Other
Rhizoctonia solani. Root rot. **5**:ND(82). **7**:SD(2242).

Brachiaria (Trin.) Griseb.

1. *B. ciliatissima* (Buckley) Chase
2. *B. mutica* (Forssk.) Stapf
3. *B. platyphylla* (Griseb.) Nash (=*B. extensa* Chase) [broadleaf signalgrass]
4. *Brachiaria* sp.

Ascomycotina
Claviceps purpurea. Ergot. **3**:FL(3423).
Basidiomycotina - Ustilaginales
Sorosporium ovarium. **4**:FL(35).
Deuteromycotina - Hyphomycetes
Bipolaris setariae. Leaf spot. **2**:HI(2903).
Bipolaris sp. Leaf spot. **3**:NC(1303).
Cerebella andropogonis. Blackhead. **3**:MD(3423).
Curvularia lunata. On roots. **4**:FL(35,82).
Helminthosporium sp. Leaf spot. **4**:FL(35,82).
Deuteromycotina - Coelomycetes
Phoma sorghina. **1**:TX(3430).

Brachyelytrum P. Beauv.

1. *B. erectum* (Schreb.) P. Beauv.

Ascomycotina
Leptosphaeria sp. On living leaves. **1**:WI(1345).
Mycosphaerella sp. **1**:WI(1344).
Phyllachora graminis. Tar spot. **1**:NE(3423), VT(3423).
Basidiomycotina - Uredinales
Uromyces halstedii. Heteroecious: telial host. **1**:NY(82).
Basidiomycotina - Other
Kriegeria eriophori. **1**:WI(1344).
Deuteromycotina - Hyphomycetes
Cladosporium brachyelytri. On living leaves. **1**:WI(1359).
Deuteromycotina - Coelomycetes
Stagonospora brachyelytri. On leaves. **1**:GA(3430), NY(82), WI(82,1337,1345,3423), WV(953).

Brachypodium P. Beauv.

1. *B. mucronatum* Willk.
2. *B. pinnatum* (L.) P. Beauv. [Japanese false bromegrass]
3. *B. sylvaticum* (Huds.) P. Beauv. [slender false bromegrass]

Deuteromycotina - Hyphomycetes
Curvularia geniculata. Secondary root rot. **3**:ND(82).
Fusarium acuminatum. Secondary root rot. **3**:ND(82).
Helminthosporium sp. **2**:MD(3423).
Deuteromycotina - Coelomycetes
Colletotrichum graminicola. Anthracnose. **1**:MD(3423). **3**:MD(82).

Briza L.

1. *B. media* L. [perennial quaking-grass]
2. *B. minor* L. [little quaking-grass]
3. *Briza* sp.

Basidiomycotina - Uredinales
Puccinia graminis. Heteroecious: telial host. **3**:MN(82), SD(82), WI(82).
Basidiomycotina - Other
Laetisaria fuciformis. Pink patch. **1**:RI(3423).
Deuteromycotina - Hyphomycetes
Helminthosporium sp. **2**:OR(3430).

Bromus L.

1. *B. anomalus* Rupr. ex E. Fourn. (=*B. porteri* (J. M. Coulter) Nash) [nodding brome]
2. *B. arcticus* (Shear ex Lams.-Scribn.) Merr.
3. *B. arenarius* Labill. [Australian chess, sand brome]
4. *B. arvensis* L. [field brome]
5. *B. brevis* Nees ex Steud.
6. *B. brizaeformis* Fisch. & C. A. Mey. [rattlesnake brome, rattlesnake chess]
7. *B. carinatus* Hook. & Arnott (=*B. aleutensis* Rydb., *B. breviaristatus* Buckley, *B. marginatus* Nees ex Steud., *B. polyanthus* Lams.-Scribn. ex Shear) [California brome, mountain brome]
8. *B. catharticus* Vahl [rescue grass]
9. *B. ciliatus* L. [fringe brome]
10. *B. commutatus* Schrad. (=*B. commutatus* Schrad. var. *aparicorum* Simonk.) [hairy chess]
11. *B. erectus* Huds. [meadow brome, upright brome]
12. *B. frondosus* (Shear) Woot. & Standl.
13. *B. inermis* Leyss. (=*B. pumpellianus* Lams.-Scribn.) [awnless bromegrass, smooth brome]
14. *B. japonicus* Thunb. ex J. A. Murray [Japanese brome, Japanese chess]
15. *B. kalmii* A. Gray
16. *B. laevipes* Shear
17. *B. latiglumis* (Shear) Hitchc.
18. *B. macrostachys* Desf.
19. *B. madritensis* L. [compact brome, Madrid brome, Spanish brome, Spanish bromegrass]
20. *B. mollis* L. (=*B. racemosus* L., *B. hordeaceus* L.) [soft brome, soft chess]
21. *B. orcuttianus* Vasey
22. *B. pacificus* Shear
23. *B. purgans* L. [Canada brome]
24. *B. rigidus* Roth (=*B. villosus* Forssk., *B. gussonii* Parl.) [ripgut brome, ripgut bromegrass, ripgut grass]
25. *B. rubens* L. [foxtail brome, foxtail bromegrass, foxtail chess, red brome]
26. *B. secalinus* L. (=*B. secalinus* L. var. *velutinus* K. Koch) [cheat, cheatgrass, chess, rye brome]
27. *B. sibiricus* Drobov
28. *B. sitchensis* Trin.
29. *B. squarrosus* L.
30. *B. sterilis* L. [barren brome, poverty brome, sterile brome]
31. *B. suksdorfii* Vasey

32. *B. tectorum* L. [cheatgrass, downy brome, downy chess, drooping brome]
33. *B. texensis* (Shear) Hitchc.
34. *B. tomentellus* Boiss.
35. *B. triniana* (Schult.) Holub (=*B. tomentosus* Trin. ex Ledeb.)
36. *B. trinii* Desv. [Chilean chess]
37. *B. vulgaris* (Hook.) Shear (=*B. vulgaris* (Hook.) Shear var. *eximius* Shear)
38. *Bromus* sp.

Chytridiomycetes

Synchytrium graminicola. Purple leaf speckle. **7:**OR(3423).
Synchytrium sp. Purple leaf speckle. **7:**OR(82).

Oomycetes

Pythium aristosporum. Root browning. **13:**ND(82).
P. arrhenomanes. Root browning. **1:**ND(82). **7:**ND(82). **11:**ND(82). **13:**NC states(82), ND(82).
P. debaryanum. Damping-off, root necrosis, seed rot. **4:**ND(3423). **7:**ND(3423), WA(3423), WY(3423). **11:**WA(3423). **13:**NC states(82), MN(3423), ND(3423), NE(3423), OR(3423), SD(2242,3423), WA(3423), WY(3423). **14:**WA(3423). **19:**WA(3423). **32:**ID(3423), NE(3423), WY(3423). **38:**ND(82).
P. graminicola. **4:**ND(3423). **6:**WA(3423). **7:**ND(3423), WA(3423), WY(3423). **13:**IA(3423), MN(82,3423), ND(3423), NE(3423), SD(2242,3423), WA(3423), WY(3423). **14:**SD(2242,3423). **19:**WA(3423). **25:**NV(3437). **32:**NE(3423), SD(2242,3423), WA(3423).
P. irregulare. Root necrosis. **7:**WA(3423). **13:**MN(3423), ND(82,3423), NE(3423). **32:**ND(82), NE(82,3423).
P. middletonii. Root browning and necrosis. **13:**NE(82). **32:**NE(82).
P. perniciosum. **32:**WA(82).
P. ultimum. Root necrosis. **4:**ND(3423). **7:**WA(3423). **11:**WA(3423). **13:**MN(82), ND(82,3423), NE(3423), WA(3423), WY(3423).
Pythium sp. Root browning. **13:**Unk.(82).
Sclerophthora macrospora. Powdery mildew. **10:**KY(3423), TN(3423). **13:**SD(2242). **14:**SD(2242,3184).

Ascomycotina

Claviceps purpurea. Ergot. **7:**W states(82), ID(3423), IN(3423), MT(3423), OR(3423), UT(3423), WA(82,3423), WY(3423). **9:**MT(82,3423), WI(3423). **11:**MN(82,3423), MT(82,3423), NY(82). **13:**Range of host(82), AZ(3423), CO(3423), DC(3423), IA(3423), ID(638,3423), IL(312,3423), KS(3021), MD(3423), MI(3423), MN(3423), MO(3423), MT(3423), ND(377,3423), NE(3423), SD(2242,3423), WA(3423), WI(3423), WV(952), WY(3423). **26:**IN(82,3423), KY(82), WA(3423), WV(952). **38:**AK(526), NH(1890), WY(819).
Cochliobolus sativus. Blind seed disease. **13:**MN(1988), MT(1988), ND(1988), SD(1988). **20:**OR(3423).
Crocicreas melanosporum. **38:**UT(505).
Epichloe typhina. **38:**MO(1944).
Erysiphe graminis. Powdery mildew. **7:**CO(3423), ID(82,3423), MT(82,3423), OR(82,3423), WA(82,3423), WY(82,1195,3423). **8:**FL(35), GA(82,2198,3423), MS(2720), OK(2872,3423), SD(82,3423), TX(82,3423). **10:**WA(3423). **11:**MN(82), WA(82,3423). **13:**MN(82,3423), WA(82,3423). **20:**OR(3423), WA(3423). **24:**OR(82), WA(82,3423). **26:**OR(82), WA(82). **30:**OR(3423). **31:**WA(3423). **32:**NE(82,3423), WA(82,3423). **37:**ID(3423). **38:**CT(1890), GA(1438), MT(82), NH(1890), NY(1890), OR(82), RI(1890), WA(82), WY(82).
Gaeumannomyces graminis. Foot rot, take-all. **7:**ID(3423). **9:**AK(3441). **13:**ND(82). **14:**ID(3423). **21:**OR(3423). **26:**AR(82,3423), NY(82,3423). **28:**AK(3441). **32:**ID(3423). **37:**WA(3423,3446,3754). **38:**OR(82).
G. graminis var. *tritici*. **37:**WA(3754).
Gibberella zeae. Leaf blight. **26:**MI(82), WI(82).
Gloeotinia granigena. **20:**OR(1075).
Leptosphaeria sp. **13:**WI(1345).
Lewia infectoria. On dead culms. **9:**CA(82).
Mycosphaerella clallamensis. On living leaves. **16:**WA(3433).
M. longissima. **26:**AK(526).
M. recutita. **13:**AK(526).
Mycosphaerella sp. Leaf spot. **26:**AK(82).
Ophiobolus sp. **13:**WI(1344).
Phaeosphaeria nodorum. On leaves. **13:**MN(1988), MT(1988), ND(1988), SD(1988).
Phyllachora graminis. Common tar spot. **1:**WY(82,1195). **9:**IN(82,3423), MT(82), WI(3423). **13:**ND(82). **23:**IA(82), PA(82,3423). **36:**CA(3423).
Pleospora sp. Leaf spot. **32:**OK(2872).
Pyrenophora bromi. Black spot, leaf spot, on roots. **7:**KY(82), MI(82), WA(82). **13:**NE states(82), IA(3628), KY(82), ND(82), SD(2242). **38:**NE states(1890), NH(1897), NJ(494), NY(1890,2998,2999,3004,3005).
P. trichostoma. Leaf spot. **13:**KS(3021).
P. tritici-repentis. On leaves. **13:**MN(1988), MT(1988), ND(1988), SD(1988).
Trichopezizella horridula. **13:**NY(1416).

Basidiomycotina - Uredinales

Puccinia brachypodii-phoenicoidis var. *chisosana*. Autoecious. **1:**TX(742,754).
P. coronata. Heteroecious: telial host. **1:**TX(742), WY(1195). **2:**AK(526). **6:**MI(82). **7:**TX(742). **9:**WY(1195,2924). **13:**AK(526), SD(82). **26:**GA(82,1439), OR(82). **38:**CO(82), MI(82), NY(82).
P. graminis. Heteroecious: telial host. **7:**CA(82), IN(82), MN(82), ND(82), OR(82). **9:**CA(82), IN(82), MN(82), ND(82), OR(82). **10:**OK(2872). **14:**CO(82), KS(82). **23:**CA(82), IN(82), MN(82), ND(82), OR(82). **24:**CA(82). **26:**C states(82), NE states(82), OK(2872), WI(82).
P. graminis subsp. *graminicola*. Heteroecious: telial host. **32:**CA(82), CO(82), WA(82).
P. recondita. Heteroecious: telial host. **1:**AZ(1203,3981), CO(82), NM(82), UT(82), WY(1195). **7:**C states(82), CA(81,82), CO(1195), WA(82), WY(1195). **8:**OR(82), WA(82). **10:**OK(2872). **12:**AZ(1195,1203,3981). **13:**CO(82), PA(4005). **14:**IN(82). **17:**WI(1332). **20:**CA(81). **24:**W states(82), CA(81). **26:**Range of host(82), NC(1303). **32:**W states(82). **37:**CA(81). **38:**CA(82), NJ(494), OR(82).
P. striiformis. **1:**W states(82), CA(82), WA(82). **7:**W states(82), CA(82,642), WA(82). **8:**CA(82). **9:**W states(82), CA(82), WA(82). **11:**ID(82). **23:**W states(82), CA(82), WA(82). **26:**WA(82).
P. tomipara. Heteroecious: telial host. **1:**MI(82), WI(82). **9:**MI(82), WI(82,1351). **23:**ND(378).

Basidiomycotina - Ustilaginales

Tilletia bromi. **6:**ID(82,1077), OR(2342), WA(82,1077). **7:**ID(1077), MT(1077). **10:**OR(2342). **14:**ID(1077), MT(2342), UT(2342), WA(82,1077). **20:**ID(1077), OR(1077,2342), WA(1077). **26:**ID(82), OR(82), WA(82). **30:**WA(2342). **32:**WY(2342). **38:**CA(1077).
T. caries. Rough-spored bunt. **7:**ID(82), MT(82). **9:**WY(82). **24:**WA(82,1077). **32:**CO(82,1077), ID(82,1077), MT(82,1077), NV(1077), UT(82,1077), WA(82,1077).
T. controversa. Dwarf bunt. **7:**ID(1465), OR(1465,2344).
Urocystis agropyri. Flag smut. **7:**WA(82,1077), WI(1077). **9:**CO(1077), IA(82,1077), KS(82,1077). **13:**WY(2342). **38:**MN(1077).
Ustilago bullata. Head and loose smut. **1:**W states(82), IA(82), ID(1077), KY(82), MD(82), MI(82), MT(1077), WA(1077). **4:**CO(1077), NY(1077). **5:**WA(1077). **6:**ID(1077), MN(1077), OR(1077), WA(1077). **8:**NGP(82), NW states(82), AR(82,1077), CO(1077), GA(1439,2198), HI(2903), ID(1077), LA(1077), MI(1077), MS(2720), ND(1077), NE(1077), NY(82,1077), OK(1077,2872), OR(1077), TN(1077), TX(1077), WA(1077). **9:**W states(82), CO(1077), DC(1077), IA(82,1077), KY(82), MD(82), MI(82), MT(1077), NM(1077), UT(1077), WA(1077), WY(1077). **10:**ID(1077), OR(1077), SD(2242), WA(1077). **11:**ID(82,1077), ND(82,1077), OR(1077), SD(2242), WA(82,1077). **13:**NGP(82), W states(82), DC(1077), IA(82,1077), KY(82), MD(82), MI(82,1077), MN(1077), MT(1077), ND(1077), OK(2872), OR(1077), WA(1077), WI(1340). **14:**W states(82), CO(1077), ID(1077), KS(82,1077,3021,3023), LA(82), MT(1077), ND(82,1077), SD(1077,2242), WA(1077), WY(1077). **15:**MT(1077), UT(1077). **18:**WA(1077). **20:**CA(81,1077), CO(1077), ID(1077), MT(1077), OR(1077), UT(1077), WA(1077). **23:**W states(82), MI(1077), WA(1077), WY(1077). **24:**CA(2439), OR(1077), WA(82,1077). **25:**AZ(1077,1195), CA(1077), NV(1077,1195), OR(1077), UT(1077). **26:**W states(82), CA(1077), KS(82), LA(82), MT(1077), ND(82), OR(1077), UT(1077), WA(1077). **27:**WA(1077). **29:**WA(1077). **30:**WA(1077). **32:**W states(82), AZ(1077,1195), CA(638,642,1077), CO(1077,1195), ID(1077),

MT(1077), ND(1077), NE(1077), NV(1077), OR(1077), SD(1077,2242), UT(1077), WA(1077), WY(1077,1195). **34:**WA(1077). **36:**CA(1077). **37:**CA(1077), MT(1077), OR(1077), WA(1077). **38:**CA(81), NC(3944), OR(657).

U. bullata var. *macrospora*. **9:**CO(1077), IA(1077).

U. serpens. **13:**IA(1077).

U. striiformis. Stripe smut. **9:**CO(1077), WY(1077). **13:**MN(1077), SD(2242), WY(2342). **38:**ND(82).

Basidiomycotina - Other

Thanatephorus cucumeris. Brown patch, root rot, sharp eye spot. **4:**ND(3423). **7:**MT(3423), ND(3423), OR(3423), WA(3423), WY(3423). **8:**GA(1439), LA(3524), OR(3423). **11:**ND(3423), OR(3423). **14:**MN(3423), MT(3423), ND(3423), SD(3423), WY(3423). **19:**OR(3423). **20:**OR(3423), WA(3423). **24:**OR(3423), WA(3423). **26:**ID(3423), LA(3524), OR(3423), WA(3423). **32:**ID(3423), MT(3423), ND(3423), NE(3423), OR(3423), SD(3423), WA(3423), WY(3423).

Typhula idahoensis. Snow mold. **7:**ID(3423), WA(3423). **13:**WA(3423). **32:**WA(3423).

T. incarnata. Snow mold. **32:**WA(3423).

Typhula sp. Snow blight. **26:**WA(82). **32:**WA(82).

Deuteromycotina - Hyphomycetes

Acremonium starrii. **1:**TX(3873).

Alternaria alternata. On seeds. **13:**MN(82), MT(82), OR(82).

Alternaria sp. Inflorescence mold. **7:**OR(82).

Bipolaris leersiae. Leaf spot. **13:**OK(2872).

B. ravenelii. **9:**WI(1342).

B. sorokiniana. Seedling blight, spot blotch, black point, root rot. **4:**ND(3423). **5:**NY(373). **7:**ND(82,3423), OR(82), WA(82,3423). **8:**GA(2198), ND(3423). **9:**WI(1352). **11:**ND(3423). **12:**NY(373). **13:**NC states(82), GA(2198), IA(3423), MN(3423), MT(82), ND(3423), NE(3423), SD(2242,3423), VA(82), WY(3423). **14:**CA(82,3423). **19:**ND(3423). **32:**ND(82,3423), NE(3423), SD(2242,3423). **35:**NY(373). **38:**ND(82).

B. spicifera. **13:**GA(2198).

Botrytis cinerea. Gray mold. **20:**WA(3423).

Cercospora festucae. Leaf spot. **13:**GA(2198), TX(82).

Cercospora sp. Leaf spot. **23:**OK(2872).

Cercosporidium graminis. Leaf streak, brown stripe. **1:**WY(1195). **3:**OR(3423). **7:**W states(82), CA(3423), CO(3423), ID(3423), MT(3423), NY(3423), OR(3423), SD(3423), UT(3423), WA(82,3423), WY(1195,3423). **8:**KS(3023), NY(82,3423), WA(82,3423). **9:**CO(3423), NE(3430), WY(3423). **12:**CO(3423). **13:**AZ(82,3423), ID(3423), OR(82,3423). **16:**MT(3423). **20:**WA(3423). **21:**OR(3423). **22:**OR(3423). **23:**CO(3423). **26:**OR(82), WA(82). **28:**WA(3423). **37:**ID(3423). **38:**CO(82), OR(82), WA(82).

Cladosporium herbarum. Leaf mold. **23:**IA(82). **38:**CA(661).

C. phlei. **13:**AK(526).

Cladosporium sp. On leaves. **23:**WI(1355).

Curvularia geniculata. Secondary root rot. **13:**SD(82,2242).

Drechslera bromi. Leaf blight, leaf spot. **7:**ID(3423,3442), KY(3423), ND(3423), WA(3423). **9:**ID(3442), WI(3423), WY(3423). **13:**AK(2078), IL(3423), IN(3423), MD(3423), MN(3423), ND(3423), NY(3423), OH(3423), PA(3423), WI(1332,3423), WV(952). **14:**KS(3023), MT(82,3423), SD(2242,3423), WI(1340,3423), WY(82,3423). **24:**OR(82,3423). **26:**CA(3423). **32:**WA(82,3423), WI(1341,3423). **37:**ID(3442), WA(3438).

D. gigantea. Leaf spot. **13:**MD(82).

Fusarium acuminatum. Secondary root rot. **4:**ND(3423). **7:**MT(3423), ND(3423), OR(3423), WA(3423), WY(3423). **10:**OR(3423). **11:**MT(82), ND(82,3423), SD(82), WA(3423). **13:**ID(3423), MN(3423), MT(82), ND(82,3423), SD(82,2242,3423), WY(3423). **14:**ID(3423), MN(3423), MT(3423), ND(82,3423), SD(82,2242,3423), WA(3423), WY(3423). **20:**ID(3423), OR(3423), WA(3423). **25:**AZ(3437), NV(3437). **32:**ID(3423), MT(3423), ND(82,3423), NE(3423), OR(3423), SD(82,2242,3423), WA(3423), WY(3423). **38:**ND(82), SD(82).

F. avenaceum. Stem rot. **13:**WY(3423).

F. culmorum. Head mold, root rot, scab. **7:**ID(3423). **13:**ID(3423). **20:**ID(3423). **32:**ID(3423), OR(82,3423), WA(82,3423).

F. equiseti. Secondary root rot. **13:**ND(82), NE(82), SD(82). **14:**ND(82), SD(82). **32:**ND(82), SD(82). **38:**ND(82), SD(82).

F. graminearum. Scab, seedling blight. **23:**MN(3438). **32:**SD(2242,3423).

F. oxysporum. Secondary root rot. **32:**SD(2242,3423). **38:**ND(82).

F. poae. Secondary root rot, silver top. **13:**MN(3423), ND(82,3423), NE(3423), SD(82,2242,3423). **25:**AZ(3437).

F. sporotrichioides. Secondary root rot. **13:**ND(82).

Helminthosporium sp. Leaf spot. **8:**FL(35,82,3423), GA(82,3423), MS(2720). **13:**ND(3423), SD(3423).

Heterosporium sp. On leaves. **7:**OR(82).

Hymenula cerealis. Stripe. **7:**WA(415).

Mastigosporium cylindricum. Black spot, leaf spot. **37:**OR(3423). **38:**OR(82).

Microdochium bolleyi. Root necrosis. **1:**MT(3423). **4:**ND(3423). **7:**ND(3423), WY(3423). **11:**ND(3423). **13:**MN(3423), ND(3423), SD(2242,3423), WY(3423). **14:**MT(3423), SD(2242,3423), WA(3423), WY(3423). **32:**MT(3423), ND(3423), NE(3423), SD(3423), WY(3423).

M. nivale. Snow mold. **7:**ID(3423), OR(82), WA(82,3423). **10:**OR(3423), WA(3423). **11:**WA(3423). **13:**ID(3423), WA(3423). **14:**WA(3423). **20:**ID(3423), WA(3423). **23:**WA(3423). **24:**OR(82,3423), WA(82,3423). **26:**OR(82), WA(82,3423). **32:**ID(3423), OR(82,3423), WA(82,3423).

Periconia circinata. Root rot. **13:**ND(3423).

Pseudocercosporella herpotrichoides. Foot rot, strawbreaker. **7:**ID(82,3423). **13:**ID(82,3423). **14:**ID(3423). **30:**ID(3423). **32:**ID(3423), OR(82,3423), WA(82,3423).

Ramularia collo-cygni. **13:**AK(526). **38:**AK(526).

R. pusilla. Gray leaf spot. **7:**ID(3423,3429), MT(3433), WA(82,3423,3429,3433). **9:**ID(3442), WI(1358). **13:**ID(3423,3429), SD(2242), WA(3429), WI(3423,3429). **37:**MT(3433), WA(3432).

Ramulispora bromi. Blotch, leaf spot. **24:**OR(82,3423), WA(82). **37:**OR(82,3423).

Rhynchosporium secalis. Leaf scald. **7:**ID(3423), OR(3442), WA(82,3423,3430). **9:**AK(526), ID(3442). **12:**WY(3423). **13:**AK(526), CA(82,3423), IA(3628), KS(3021), NH(1895), OR(3423), SD(2242), WA(82), WI(82,1332,3423), WY(3423). **26:**CA(82,3423). **37:**WA(3435). **38:**NY(1890,2998,2999,3005).

Spermospora subulata. Blast. **7:**WA(2341). **37:**MT(3423).

Deuteromycotina - Coelomycetes

Ascochyta brachypodii. **7:**WA(3442).

A. graminicola. **14:**WI(1338). **32:**WI(1344).

A. hordei. Leaf spot. **7:**UT(3423,3449), WA(3423,3449).

A. leptospora. Leaf spot, on roots. **7:**OR(82). **13:**KY(82), MN(82).

A. macrospora. On necrotic leaves. **1:**NV(3430).

A. skagwayensis. On leaves. **13:**AK(3441).

A. sorghi. Leaf spot. **7:**OR(3423,3449), UT(3423,3449), WA(3423,3449). **10:**WI(3423). **13:**IA(3628), KY(3423), MN(3423,3449), MO(3449), MT(3449), ND(3423,3449). **14:**WI(3423). **20:**WA(3438). **26:**WA(3432), WV(955). **32:**MN(3423,3449). **37:**ID(3442).

Ascochyta sp. Leaf spot. **9:**NM(3425). **10:**OK(2872). **13:**OK(2872). **23:**OK(2872).

Cheilaria agrostis. Blotch spot, char spot. **7:**CO(3423). **9:**CO(82,3423).

Colletotrichum destructivum. Leaf spot. **13:**IA(3628).

C. graminicola. Anthracnose. **7:**OK(82,2872,3423). **8:**TX(82,3423). **13:**NE(82,3423). **15:**MN(3438). **17:**ND(3423). **23:**IA(82,3423), KS(3023), MN(82), ND(82), WI(82). **26:**MS(2720), NY(82), OH(82,3423), PA(82,3423), TX(82,3423). **38:**IA(82), MN(82), ND(82), WI(82).

Hendersonia culmicola var. *minor*. Associated with Rhizoctonia blight lesions. **13:**GA(2198).

H. stipae-pennatae. **1:**NV(3430).

Neophoma graminella. On leaves. **26:**MS(82,2720).

Phaeoseptoria airae. On leaves. **28:**AK(3441).

Phoma terrestris. Pink root. **4:**ND(3423). **7:**ND(82,3423). **11:**ND(3423). **13:**ND(82,3423), NE(3423), SD(2242,3423).

Phomopsis sp. **10:**OK(2872).

Phyllosticta bromivora. Leaf spot. **9:**NM(3425).

Phyllosticta sp. Leaf spot. **7:**WY(1195). **9:**NV(3438). **13:**OK(2872). **26:**GA(82).

Pseudoseptoria bromigena. Leaf lesions, leaf spot. **6:**CO(3450). **7:**CA(3423,3450), MT(3433), ND(82,3423,3450), WA(82,2695,3423,3432,3450), WY(82,2695,3423,3450). **9:**AZ(3435), CO(82,2695,3423,3430,3450), NV(3430). **12:**CO(3450). **13:**NC states(82), W states(3450), AK(526,2078),

AZ(82,3423,3450), IA(3628), ID(3423,3450), KS(3021), MN(1988,3423,3450), MT(82,1988,3423,3450), ND(378,1988,3081,3423,3450,3549), NE(3423,3450), SD(1988,2242,3423,3450), UT(3423,3450), WA(82,2695,3423,3450), WI(1332,3423), WY(3423,3450). **14:**SD(2242). **15:**WI(1344). **17:**WI(1356). **23:**CO(3423), WI(3442). **38:**MT(3433).

Septoria bromi. Leaf blotch, leaf spot. **7:**OK(2872,3423). **9:**ND(3423), NE(3430). **10:**ID(3423), OK(2872,3423), OR(3423), WI(3423). **13:**IA(3423,3628), ID(3423), MT(82,3423), ND(82,3423), OR(82,3423), WA(82,3423), WI(82,3423). **14:**ID(3423), KS(3021,3023), NE(3423), UT(3423), WA(82,3423), WI(1338), WV(3423), WY(82,3423). **17:**WI(3423). **20:**CA(3442), OR(3423), WA(3423). **23:**OK(82,2872,3423). **26:**NW states(82), SC states(82), SE states(82), IA(3423), ID(3423), IL(3423), IN(3423), KS(3021), KY(3423), MO(3423), NC(3423), OH(3423), OK(2872,3423), OR(3423), TN(3423), TX(3423), WI(82,3423). **33:**OK(2872,3423). **37:**WA(3432). **38:**OR(82), WI(82).

S. jacucella. Western brome speckle, leaf spot. **7:**CO(3423), ID(82,3423), OR(82,1501,3423), WA(82,3423). **9:**AZ(82,3423), MT(3423), OR(638). **16:**CA(3423), OR(638). **22:**OR(3438). **24:**OR(82,638,3423), WA(82,3423). **28:**AK(3426). **32:**OR(638), WA(82,3423). **38:**CA(82).

Stagonospora brevispora. Leaf spot. **9:**WI(82,3423).

S. bromi. Leaf blotch, purple brown blotch. **7:**ID(3423), KY(82), MI(82,3423), ND(82,3423). **8:**KY(82,3423). **9:**CO(3423), WI(82,1337,3423). **12:**CO(3423). **13:**AK(526,3441), IA(3628), KS(3021,3023), KY(82,3423), SD(2242), WI(82,1332,3423), WV(955). **14:**WI(1357). **15:**WI(1337,3423). **17:**WI(1344). **23:**WI(82,1337,3423). **37:**ID(3423), MN(3423). **38:**MN(82).

S. culmicola. Leaf rot. **37:**OR(3423).

S. mariae. **7:**WA(3433). **28:**WA(3433).

S. nodorum. Glume blotch, leaf spot. **7:**WA(3433). **9:**NM(3425). **38:**OK(2872).

Wojnowicia hirta. Secondary foot rot, rot of mature straw. **13:**IA(3628), ND(3423). **32:**ID(3423), OR(82,3423), WA(3423).

Deuteromycotina - Other

Rhizoctonia oryzae. Root rot, sheath spot. **32:**ID(3423).

R. solani. Blight, leaf decay, leaf and root rot. **7:**ND(82), OR(82). **8:**GA(2198). **13:**GA(2198), ND(82), SD(82,2242), WV(953). **14:**SD(2242). **32:**MN(82), SD(2242). **37:**ID(3442).

Rhizoctonia sp. Leaf spot. **32:**ID(82), WA(82).

Sclerotium rhizodes. On leaves. **9:**WI(82,3423).

Buchloe Engelm.

1. *B. dactyloides* (Nutt.) Engelm. [buffalograss]

Ascomycotina

Erysiphe graminis. Powdery mildew. **1:**OK(2872,3423).

Phyllachora cynodontis. Black spot, tar spot. **1:**AR(82,3423), KS(82,3423), OK(2872,3423), TX(82,3423).

Basidiomycotina - Uredinales

Puccinia graminis. Heteroecious: telial host. **1:**Unk.(82).

P. kansensis. Heteroecious: telial host. **1:**KS(82,1521,3023), OK(2872), TX(82).

Basidiomycotina - Ustilaginales

Sporisorium neglectum. **1:**KS(1077).

Tilletia buchloeana. Kernel smut. **1:**KS(82,1077,3023), NE(1077), OK(1669), TX(82,1077).

Ustilago buchloes. Leaf smut. **1:**CO(1077), NE(82,1077), NM(82,1077).

Basidiomycotina - Other

Thanatephorus cucumeris. Brown patch, root rot, sharp eye spot. **1:**ND(3423), NE(3423).

Deuteromycotina - Hyphomycetes

Alternaria sp. Glume spot. **1:**TX(82).

Bipolaris buchloes. Leaf spot. **1:**KS(82,3021,3423), MT(3423), ND(3423), NE(82,3423), NM(3425), OK(82,3423), SD(2242,3423), TX(82,3423), WI(82,3423).

B. sorokiniana. Melting out. **1:**KS(3023).

Cercospora seminalis. Seed blight, glume and leaf spot. **1:**W states(82), CO(82,3423), KS(3021,3023,3423), ND(3423), NE(3423), OK(2872,3423), SD(2242,3423), TX(3423), WI(1334,3423).

Exserohilum turcicum. Leaf blight. **1:**OK(2872), WI(1334).

Fusarium acuminatum. Secondary root rot. **1:**ND(3423), NE(3423).

F. equiseti. Secondary root rot. **1:**ND(82).

Microdochium bolleyi. Root necrosis. **1:**ND(3423).

Deuteromycotina - Coelomycetes

[*Ascochyta graminea*]. On dying leaves. **1:**KS(3023), SD(2242).

A. sorghi. **1:**SD(2242).

Colletotrichum graminicola. Anthracnose. **1:**NE(82,3423).

Calamagrostis Adans.

1. *C. breweri* Thurb.
2. *C. canadensis* (Michx.) P. Beauv. [bluejoint]
3. *C. canadensis* (Michx.) P. Beauv. var. *scabra* (J. Presl) Hitchc.
4. *C. epigeios* (L.) Roth
5. *C. koelerioides* Vasey
6. *C. montanensis* (Lams.-Scribn.) Lams.-Scribn. [plains reedgrass]
7. *C. neglecta* (Ehrh.) Gaertn., B. Mey. & Scherb. (=*C. inexpansa* A. Gray, *C. inexpansa* A. Gray var. *brevior* (Vasey) Stebb., *C. hyperborea* Lange) [northern reedgrass]
8. *C. nutkaensis* (J. Presl) Steud. (=*C. aleutica* Trin.) [Pacific reedgrass]
9. *C. pickeringii* A. Gray
10. *C. purpurascens* R. Br. [purple reedgrass]
11. *C. rubescens* Buckley (=*C. rubescens* Buckley f. *ciliata*) [pinegrass]
12. *C. scribneri* Beal [Scribner reedgrass]
13. *Calamagrostis* sp.

Ascomycotina

Acrospermum compressum. On dead culms. **2:**NY(82).

Balansia epichloe. Black ring, on leaves. **2:**MN(3423), ND(82), WI(3423). **7:**ND(82,3423).

Claviceps purpurea. Ergot. **2:**NC states(82), NE states(82), IA(3423), ME(3423), MI(3423), MN(3423), OR(3423), SD(2242), WA(82,3423), WI(3423). **3:**AK(526,3423). **8:**AK(526,3423), OR(3423). **13:**AK(82), MN(82), MT(82), ND(82), OR(82).

Epichloe typhina. Choke. **2:**MN(82,3423), WI(82,3423).

Erysiphe graminis. Powdery mildew. **2:**OH(82,3423). **11:**MT(82,3423), WA(82).

Herpotrichia graminea. Leaf mold. **2:**NY(82).

H. purpurea. **13:**WA(82).

Lachnum carneolum. **2:**AK(526).

Leptosphaeria asparagina. **2:**WA(3814).

L. elongata. **2:**WA(3814).

Lophodermium apiculatum. On culms. **2:**CO(82).

L. arundinaceum. On dead culms and leaves. **1:**KS(3423). **2:**AK(526). **3:**AK(3431), CA(3423). **7:**NM(82). **13:**CA(82).

Mycosphaerella greenei. On culms and leaves. **2:**WI(82,1335,3423). **13:**OR(82).

M. tassiana. On dead culms. **13:**CO(82), OR(82).

M. tulasnei. Leaf mold. **7:**AK(3431). **13:**CA(82).

Phaeosphaeria herpotrichoides. On dead culms. **2:**NY(82). **8:**AK(526). **13:**AK(82), WA(3814).

Phyllachora graminis. Common tar spot. **2:**MI(82,3423), MN(82), NE(82,3423), NY(82,3423), WI(82,3423). **3:**AK(526). **4:**SD(2242). **7:**NE(82), SD(2242), WI(1349). **13:**AK(82).

Trichopeziza agrostina. On dead stems. **2:**NY(3167).

Basidiomycotina - Uredinales

Puccinia brachypodii var. *poae-nemoralis*. Heteroecious: telial host. **11:**Unk.(754).

P. coronata. Heteroecious: telial host. **2:**Range of host(82), AK(60,526), MI(2862), MT(1195), WY(1195). **3:**AK(3431). **5:**WY(2924). **7:**MI(82), ND(82,378), OR(82), WY(1195). **8:**CA(81). **11:**MT(1195), WY(82,1195). **13:**CA(82), CO(82), MT(82,1195), OR(82), UT(82), WA(82), WY(82).

P. graminis. Heteroecious: telial host. **2:**CT(82), OH(82), WI(82). **7:**CA(82). **13:**CA(82).

P. pygmaea. Heteroecious: telial host. **2:**AK(754), MI(82), WI(82,754). **7:**OR(82,754). **8:**AK(60,526,754), CA(754). **13:**AK(82), OR(82).

Basidiomycotina - Ustilaginales

Urocystis agropyri. Flag smut. **2:**MT(2342), OR(82,1077).

Ustilago calamagrostidis. **2:**AK(526), OH(82), VT(82), WI(82), WY(82). **13:**NY(82).

U. striiformis. **2:**AK(526,2078), CO(1077), ID(1077), IN(1077), MT(1077), OH(1077), VT(1077), WI(1077), WY(1077). **6:**CO(1077). **7:**MT(2342). **9:**NY(1077). **12:**ID(1077), PA(1077), TN(1077), VA(1077), WV(1077).

Basidiomycotina - Other

Coprinus psychromorbidus. Snow mold. **2:**AK(2074,3676).

3:AK(3441).
Thanatephorus cucumeris. Brown patch, root rot, sharp eye spot. **6**:SD(3423).
Typhula incarnata. On leaves. **3**:AK(3441).

Deuteromycotina - Hyphomycetes

Acremoniella sp. Leaf spot. **3**:AK(3428).
Bipolaris sorokiniana. **13**:VA(82).
Cercosporidium graminis. Leaf spot, leaf streak, brown stripe. **2**:CO(82), ID(3423), MT(3423), ND(82,3423), SD(2242,3423), WA(82,3423), WI(3423), WY(2924). **3**:AK(3426), WA(3423). **7**:WA(3423).
Cladosporium herbarum. Leaf mold. **2**:AK(526). **11**:OR(82), WA(82).
C. phlei. **2**:AK(526).
Fusarium acuminatum. Secondary root rot. **6**:SD(3423).
Hadrotrichum lineare. Mold. **2**:NY(82,3423). **3**:AK(526). **13**:AK(82).
Helminthosporium sp. Leaf spot. **2**:MN(82,3423).
Heterosporium avenae. Leaf mold. **11**:MT(82).
Macrosporium pinguedinis. **2**:MI(2862).
Mastigosporium rubricosum. Red eye spot, leaf spot. **2**:AK(82,526,3426), ID(3423), MT(82,3423), OR(82), UT(82,3423), WA(82,3423), WI(82,1346,3423), WY(82,3423). **3**:AK(526,3423,3431), MT(3423). **8**:AK(3429). **11**:MT(3433). **13**:MT(82), OR(82).
Microdochium bolleyi. Root necrosis. **4**:ND(3423). **6**:SD(3423).
M. nivale. Snow mold. **1**:CA(3423). **2**:MN(3423). **13**:CA(82).
Ramularia pusilla. Leaf spot. **2**:MI(3429), MT(82,3423,3429). **3**:AK(3428). **11**:OR(3433).
Rhynchosporium orthosporum. Scald, white leaf stripe. **2**:SD(82,2242,3423), UT(3435), WY(82,3423). **3**:AK(3426).
Spermospora subulata. Leaf blast. **11**:WA(82,3423). **12**:WY(3423). **13**:WY(82).

Deuteromycotina - Coelomycetes

Ascochyta graminicola. On leaves. **2**:WI(82).
A. minor. On living leaves. **2**:WI(1338,3423).
A. sorghi. Leaf spot. **2**:WI(3423).
Ascochyta sp. On leaves. **2**:WI(1352).
Cheilaria agrostis. Blotch spot, char spot. **2**:AK(526), MT(3423), UT(3423), WY(1195). **7**:ID(82), ND(82,3423). **11**:ID(82,3423). **13**:WY(1195).
Colletotrichum graminicola. Anthracnose. **2**:AK(526), ID(82,3423), MN(82,3423), MT(82,3423), ND(82,3423), OH(82,3423), WI(82,3423). **7**:MN(82), MT(82,3423), ND(82,3423), SD(2242), WA(3423). **8**:AK(3441). **10**:AK(526), MT(3423). **11**:MT(82,3423), OR(82,3423), WA(3423). **13**:MT(82).
Coniothyrium psammae. Leaf spot. **3**:AK(3428). **8**:OR(3423). **13**:OR(82).
Cylindrosporium calamagrostidis. Leaf spot. **2**:WI(82,3423). **7**:WI(1357).
Dilophospora alopecuri. Twist disease. **2**:MN(3442), WI(82,1350,3423). **11**:MT(3433).
Hendersonia calamovilfae. Leaf mold. **2**:SD(82), WY(1195).
H. culmicola var. *minor*. **2**:AK(526).
Hyalothyridium calamagrostidis. Leaf spot. **2**:IA(1228), WI(82,1336,3423).
Phaeoseptoria airae. **11**:WA(82).
P. calamagrostidis. On leaves. **2**:WI(82). **13**:OR(82).
Phyllosticta sp. On leaves. **2**:WI(1353).
Pseudoseptoria everhartii. Speckle, eye spot. **2**:AK(3423), WI(82,3423). **5**:AZ(1194), WY(3450). **6**:ND(82,3423,3450). **7**:WA(3433). **11**:WA(82,2695,3423,3450). **12**:WY(3423,3450). **13**:AK(82), WY(82).
Septoria arctica. Leaf spot. **8**:AK(526,3423), OR(3423). **13**:AK(82), OR(82).
S. calamagrostidis. **7**:SD(2242).
S. graminum. **13**:AK(526).
Septoria sp. On leaves. **2**:WI(1355).
Stagonospora arenaria. Leaf mold. **7**:MN(82).
S. avenae. Speckled blotch. **2**:MT(82,3423,3424). **3**:AK(3428). **7**:SD(2242).
S. culmicola. **7**:MT(82).
S. foliicola. **2**:WY(1195).
S. simplicior. **3**:AK(3441).
Wojnowicia hirta. Secondary foot rot, rot of mature straw, leaf spot. **3**:AK(3423,3428), AK(526). **11**:MT(3423).

Deuteromycotina - Other

Rhizoctonia solani. Root rot. **6**:SD(82).
Sclerotium rhizodes. Leaf spot. **2**:AK(526), ID(3423), MA(82,3423), NH(82,3423), NJ(82,3423), NY(82), WA(3423), WI(82,3423). **3**:WY(3433). **7**:WI(3423).

Calamovilfa (A. Gray) Hack. ex Lams.-Scribn. & E. A. Southworth

1. *C. gigantea* (Nutt.) Lams.-Scribn. & Merr.
2. *C. longifolia* (Hook.) Lams.-Scribn. (=*Ammophila longifolia* Benth. ex Vasey) [prairie sandreed]
3. *Calamovilfa* sp.

Oomycetes

Pythium arrhenomanes. Root rot. **2**:ND(82).
P. debaryanum. Damping-off, root necrosis, seed rot. **2**:NE(3423), SD(3423).
P. graminicola. **2**:ND(3423), WA(3423).

Ascomycotina

Epichloe typhina. Choke. **3**:Unk.(82).
Mollisia clavigera. On dead leaves. **2**:MT(965).
Mycosphaerella tassiana. On culms. **2**:CO(82).
Pyrenopeziza ellisii. On culms. **2**:MT(82).

Basidiomycotina - Uredinales

Puccinia amphigena. Heteroecious: telial host. **1**:CO(82), MT(82), ND(82), OK(82,2872). **2**:C states(82), CO(753), IL(753), IN(753), KS(753), MI(753), MS(2720), MT(82), ND(378,753), OK(2872), SD(2242), WI(1332), WY(1195).
P. graminis. Heteroecious: telial host. **2**:MT(82), ND(82), NE(82).
P. sporoboli var. *robusta*. Heteroecious: telial host. **1**:OK(753). **2**:CO(753,1195), IN(753), MI(82,665), MT(753), NE(753), SD(753), WI(753,1353), WY(753). **3**:WI(1358).

Deuteromycotina - Hyphomycetes

Bipolaris sorokiniana. Seedling blight, spot blotch, black point, root rot. **2**:ND(82,3423).
Cladosporium herbarum. Leaf mold. **2**:ND(82).
Fusarium acuminatum. Secondary root rot. **2**:ND(82,3423), SD(3423).
F. moniliforme. Secondary root rot. **2**:NE(3423).
Microdochium bolleyi. Root necrosis. **1**:NE(3423), SD(3423). **2**:MT(3423), ND(3423), NE(3423), SD(3423), WY(3423).
Periconia circinata. Root rot. **2**:MT(3423).

Deuteromycotina - Coelomycetes

Ascochyta sp. Anthracnose. **1**:OK(2872).
Colletotrichum graminicola. Anthracnose. **1**:OK(2873,3423). **2**:NE(2927), OK(82), WI(82,3423).
Hendersonia calamovilfae. Leaf mold. **1**:ND(82), WA(82). **2**:ND(82), SD(2242), WI(82,1332).
Phoma terrestris. Pink root. **2**:ND(3423).
Pseudoseptoria stomaticola. Leaf spot. **2**:ND(82,3423,3450), SD(82,3423,3450).
Septoria calamovilfae. Leaf spot. **2**:MT(3423), ND(82,3423), SD(82,2242,3423), WY(82,3423).
Stagonospora simplicior. Leaf mold. **2**:ND(82).
Tiarosporella paludosa. **2**:WI(1358).
Wojnowicia hirta. Secondary foot rot, rot of mature straw. **2**:ND(3423).

Catabrosa P. Beauv.

1. *C. aquatica* (L.) P. Beauv.

Ascomycotina

Erysiphe graminis. Powdery mildew. **1**:MT(82,3423), SD(82,2242,3423).

Basidiomycotina - Uredinales

Puccinia brachypodii var. *poae-nemoralis*. Heteroecious: telial host. **1**:Unk.(754), ID(82), IN(82), MT(82), NE(82), SD(82,2242), UT(82), WY(82,1195,3377).
P. graminis. Heteroecious: telial host. **1**:NE(82), WI(82).

Basidiomycotina - Ustilaginales

Entyloma dactylidis. Leaf smut. **1**:UT(82,1077).

Deuteromycotina - Coelomycetes

Colletotrichum aquatilis. On leaves. **1**:ID(3442), OR(3433).
C. graminicola. Anthracnose on floating leaves. **1**:OR(3425).
Phaeoseptoria airae. On leaves. **1**:WA(3442).

Cenchrus L.

1. *C. biflorus* Roxb.
2. *C. echinatus* L. [hedgehog grass, southern sandbur]

3. *C. echinatus* L. var. *hillebrandianus* (Hitchc.) R. Br.
4. *C. incertus* M. A. Curtis (=*C. pauciflorus* Benth., *C. carolinianus* Walter) [burgrass, coast sandbur, field sandbur]
5. *C. myosuroides* H. B. K.
6. *C. tribuloides* L. [dune sandbur]
7. *Cenchrus* sp.

Oomycetes

Pythium graminicola. **4**:MN(3423), ND(3423), NE(3423).

Ascomycotina

Balansia claviceps. Inflorescence blight. **2**:FL(3423), VA(3423).
B. obtecta. Black ring, sterility disease. **2**:FL(35,82). **7**:FL(35,82).
Gibberella zeae. On culms. **2**:FL(35,82).
Ophiobolus sp. Leaf spot. **7**:FL(35).
Phyllachora sphaerosperma. Tar spot. **2**:CA(82,3423), FL(35,82), MS(82,2720). **4**:MS(3423). **5**:CA(82), FL(82), MS(82,2720). **6**:CA(82), FL(82), MS(82,2720). **7**:CA(82), FL(35,82), MS(82).

Basidiomycotina - Uredinales

Puccinia cenchri. Autoecious. **2**:Gulf states(82), FL(35), HI(82), OK(82). **3**:HI(2903). **4**:Gulf states(82), HI(82), OK(82,2872). **7**:Gulf states(82), FL(35), HI(82), OK(82).

Basidiomycotina - Ustilaginales

Sporisorium cenchri. Inflorescence smut. **2**:FL(35), TX(82,1077), WI(1077). **4**:Range of host(82), AZ(1194), CO(1077), FL(1077), IA(1077), ID(1077), IN(1077), KS(3021,3023), MN(1077), ND(1077), NE(1077), OK(1077,2872), OR(1077), PA(1077), SD(1077,2242), WA(1077), WI(1077). **6**:CO(1077), CT(1077), FL(1077), IA(1077), IL(1077), IN(1077), KS(1077), MI(1077), MN(1077), ND(378,1077), NE(1077), NJ(1077), NY(1077), OR(1077), PA(1077), SD(1077), TX(1077), WI(1077). **7**:AR(1077), CO(1077), FL(35,82,1077), KS(1077), NM(1077), TX(1077).

Basidiomycotina - Other

Thanatephorus cucumeris. Brown patch, root rot, sharp eye spot. **4**:ND(3423).

Deuteromycotina - Hyphomycetes

Bipolaris sorokiniana. Seedling blight, spot blotch, black point, root rot. **4**:ND(3423).
Cladosporium spongiosum. On burs. **7**:FL(35,82).
Curvularia sp. On leaves. **7**:FL(35,82).
Fusarium acuminatum. Secondary root rot. **4**:ND(3423), SD(3423).
F. graminearum. Seedling blight, scab. **2**:FL(3423). **4**:CA(82).
Helminthosporium sp. Leaf spot. **1**:FL(3423). **7**:FL(35,82).
Microdochium bolleyi. Root necrosis. **4**:ND(3423), NE(3423), SD(3423).

Deuteromycotina - Coelomycetes

Amerosporium sp. **7**:FL(82).
Colletotrichum graminicola. Anthracnose. **4**:DE(82), NE(3423), WI(1361). **6**:DE(3423).
Phoma terrestris. Pink root. **4**:ND(3423).
Phyllosticta sp. On leaves. **4**:WI(1338).
Septoria cenchrina. Leaf spot. **4**:MN(82,3423), ND(82,3423), WI(82,3423). **5**:TX(82,3423).

Chasmanthium Link

1. *C. latifolium* (Michx.) Yates (=*Uniola latifolia* Michx.) [wild oats]
2. *C. sessiliflorum* (Poir.) Yates (=*Uniola sessiliflora* Poir.)
3. *Chasmanthium* sp.

Oomycetes

Pythium graminicola. **1**:MI(3423).

Ascomycotina

Apiospora montagnei. **1**:GA(1438), OK(660).
Epipolaeum sp. **2**:TX(652). **3**:TX(652).
Keissleriella culmifida. **1**:OK(660).
Mollisia culmina. **1**:GA(1438).
Montagnula spinosella. **1**:OK(660).
Mycosphaerella recutita. **1**:OK(660).
Phaeosphaeria herpotrichoides. **1**:OK(660).
P. nigrans. On dead culms. **1**:GA(82,1438).
Pleospora sp. **1**:GA(1438).

Basidiomycotina - Ustilaginales

Ustilago uniolae. **2**:LA(1077).

Deuteromycotina - Hyphomycetes

Helminthosporium sp. Leaf spot. **1**:OK(2872).

Deuteromycotina - Coelomycetes

Colletotrichum graminicola. **1**:OK(660).
Phoma glumarum. Leaf spot, on glumes. **1**:MS(82,2720,3423).

Chloris Sw.

1. *C. andropogonoides* E. Fourn.
2. *C. chloridea* (J. Presl) Hitchc.
3. *C. ciliata* Sw.
4. *C. gayana* Kunth [Rhodes grass]
5. *C. petraea* Sw.
6. *C. verticillata* Nutt. [tumble windmill-grass, windmill-grass]
7. *C. virgata* Sw. [feather fingergrass]
8. *Chloris* sp.

Oomycetes

Sclerospora farlowii. Downy mildew. **7**:AZ(82,3423).

Ascomycotina

Balansia epichloe. Black ring, sterility disease. **5**:FL(3423). **7**:FL(3423). **8**:FL(35,82).
B. henningsiana. Black ring. **8**:FL(35,82).
Myriogenospora atramentosa. Black choke. **4**:FL(35).
Phyllachora cynodontis. Tar spot. **2**:TX(3423). **4**:FL(35). **7**:CA(3423), NM(82). **8**:TX(82).

Basidiomycotina - Uredinales

Puccinia cacabata. Heteroecious: telial host. **3**:TX(740).
P. chloridis. Heteroecious: telial host. **4**:NY(82). **6**:KS(3021,3023), OK(2872). **7**:AZ(1203,3981). **8**:KS(82), NM(82), OK(82), TX(82).
P. dietelii. Heteroecious: telial host. **1**:TX(740). **7**:AZ(741,747,1194,1203,3981).
P. striiformis. **8**:WA(82).
Uromyces archerianus. Autoecious. **7**:ND(82), NM(1521).

Basidiomycotina - Ustilaginales

Ustilago chloridicola. Inflorescence smut. **6**:CA(81,1077). **8**:CA(82).
U. elegans. **7**:AZ(82,1077).

Deuteromycotina - Hyphomycetes

Bipolaris hawaiiensis. Leaf spot, culm and leaf spot. **4**:FL(35,3388).
B. sorokiniana. Seedling blight, spot blotch, black point, root rot. **6**:KS(3023), MN(3423). **8**:MN(82).
B. victoriae. Blight. **4**:IA(3423).
Cercospora caespitosa. Leaf spot. **4**:FL(35). **5**:MS(2720), WI(3423).
Curvularia sp. On leaves. **8**:FL(35,82).
Drechslera graminea. Leaf spot, leaf stripe. **4**:FL(35).
Fusarium acuminatum. Secondary root rot. **7**:ND(82,3423).
Helminthosporium sp. Leaf spot. **4**:FL(82,3423).

Deuteromycotina - Coelomycetes

Ascochyta brachypodii. Leaf spot. **6**:KS(3023). **7**:CO(3423), ND(3449).
Colletotrichum graminicola. Anthracnose. **6**:OK(2872,3423).
Cylindrosporium campicola. On living leaves. **6**:KS(3451).
Phoma sorghina. **7**:AZ(3435).
Polynema tricristatum. Leaf spot. **4**:FL(35).
Pyrenochaeta graminis. Leaf blight. **8**:KS(82), TX(82).

Cinna L.

1. *C. arundinacea* L. [stout woodreed]
2. *C. latifolia* (Trevir. ex Goepp.) Griseb. (=*Agrostis latifolia* Trevir. ex Goepp.) [drooping woodreed]
3. *Cinna* sp.

Ascomycotina

Acrospermum compressum. On culms. **1**:PA(82).
Apiospora montagnei. **1**:GA(1438).
Clathrospora pentamera. On dead sheath. **2**:AK(1938).
Claviceps purpurea. Ergot. **1**:WI(3423).
Epichloe typhina. Choke. **1**:AR(82,3423), MD(82,3423), VA(82,3423).
Erysiphe graminis. Powdery mildew. **1**:IA(82,3423), WI(1353). **2**:ID(3442), OR(3433).
Gaeumannomyces graminis. Culms and sheaths. **2**:AK(3441).
Heptameria tenera. **2**:WA(3814).
Mycosphaerella cinnafolia. On leaves. **2**:OR(3435).
M. tulasnei. **2**:AK(3431).
Phaeosphaeria herpotrichoides. **2**:WA(3814).
Phyllachora graminis. Tar spot. **1**:IL(82,3423), KS(82,3423), MD(82,3423), VA(82,3423), WI(82,1344).
Pleospora sp. On culms and leaves. **2**:WI(82,1332).

Basidiomycotina - Uredinales

Puccinia coronata. Heteroecious: telial host. **1**:GA(1439), IA(82), IN(82), NY(82), WI(82). **2**:ID(82), MI(82,2862), MT(82),

WA(82), WI(1357).
P. graminis. Heteroecious: telial host. **1:**CT(82), IN(82), MI(82), MO(82), NE(82), NY(82), OH(82), SD(2242), VA(82), WI(82). **2:**CT(82), NY(82), WI(82).
P. graminis subsp. *graminicola*. Heteroecious: telial host. **3:**WI(82).
P. recondita. Heteroecious: telial host. **1:**GA(1439), MD(82), MO(82), NY(82), WI(82). **2:**WI(82).
P. striiformis. **3:**ID(82).
Deuteromycotina - Hyphomycetes
Cercosporidium graminis. Leaf streak, brown stripe. **1:**WI(3423). **2:**ID(3423), WI(82).
Drechslera catenaria. Leaf blight. **1:**NY(82,3423), PA(376).
Exserohilum turcicum. On leaves. **1:**AL(82).
Microdochium nivale. Leaf spot. **2:**WA(3432).
Deuteromycotina - Coelomycetes
Hendersonia sp. On living culms. **2:**WI(1332).
Stagonospora arenaria. Leaf blotch, purple brown blotch. **2:**MN(82,3423), OR(82,3423).
S. foliicola. Culm and leaf rot. **1:**WI(82,3423). **2:**AK(3441), WI(82,3423), WV(953).
S. nodorum. Glume blotch. **2:**MN(3423).
Stagonospora sp. **2:**WI(1358).
Wojnowicia hirta. **2:**AK(3431).

Coix L.
1. *C. lacryma-jobi* L. (=*C. lacryma* L.) [Job's-tears]
Ascomycotina
Melanomma trochus. **1:**HI(2784).
Basidiomycotina - Ustilaginales
Ustilago coicis. Head smut. **1:**Unk.(82), DC(1077).
Deuteromycotina - Coelomycetes
Phyllosticta sp. Leaf spot. **1:**HI(82,2903).

Cortaderia Stapf
1. *C. selloana* (Schult. & J. H. Schult.) Aschers. & Graebn. [pampasgrass]
2. *Cortaderia* sp.
Oomycetes
Pythium sp. Root rot. **1:**FL(35).
Ascomycotina
Apiospora montagnei. On dead leaves. **1:**HI(2903), SC(82).
Deuteromycotina - Hyphomycetes
Arthrinium arundinis. On culms. **1:**CA(82).
A. phaeospermum. Mold. **1:**CA(3423).
Chaetostroma aterrima. On leaves. **1:**SC(82).
Curvularia sp. Leaf spot. **1:**FL(35).
Fusarium graminearum. Seedling blight, scab. **1:**CA(3423).
Fusarium sp. Leaf spot. **2:**FL(35).
Helminthosporium parvulum. On leaves. **1:**SC(82).
Helminthosporium sp. Leaf spot. **1:**FL(35), GA(82,3423). **2:**FL(35).
Mystrosporium consors. On dead leaves. **1:**SC(82).
Pyricularia sp. Leaf spot. **1:**FL(35).
Deuteromycotina - Coelomycetes
Hendersonia culmiseda. Leaf mold. **1:**OR(82).
Phoma gyneriicola. On leaves. **1:**CA(82), FL(35).
P. linearis. On culms. **1:**CA(82).
Phyllosticta sp. Leaf spot. **1:**KY(82).
Deuteromycotina - Other
Rhizoctonia solani. Root rot. **1:**FL(35).

Ctenium Panz.
1. *C. aromaticum* (Walter) A. Wood [toothache grass]
Ascomycotina
Balansia epichloe. On leaves. **1:**NC(1303,3423).

Cymbopogon Spreng.
1. *C. citratus* (DC. ex Nees) Stapf (=*Andropogon citratus* DC. ex Nees) [fevergrass, lemongrass, West Indian lemongrass]
Ascomycotina
Myriogenospora atramentosa. Dwarfing, fasciation and tangle top. **1:**FL(35,82,3423).
Basidiomycotina - Uredinales
Puccinia nakanishikii. **1:**HI(1149,2903).
Deuteromycotina - Hyphomycetes
Bipolaris sacchari. Eye spot. **1:**FL(35,82,3423).

Cynodon Rich.
1. *C. dactylon* (L.) Pers. (=*C. dactylon* (L.) Pers. var. *dactylon*) [bermudagrass, devilgrass, scutch grass]
2. *Cynodon* sp.
Myxomycetes
Mucilago crustacea. Smothering. **1:**OK(2872).
Physarum cinereum. Slime mold, smothering, on inflorescences and leaves. **1:**FL(82), KS(3023), MS(2720), NC(1303), OK(82,2872), TX(82).
Physarum sp. Slime mold. **1:**NC(1303).
Plasmodiophoromycetes
Polymyxa graminis. 'Polymyxa root rot.' **1:**FL(35).
Oomycetes
Phytophthora sp. Leaf spot. **1:**HI(2299,2903).
Pythium aphanidermatum. 'Pythium root rot. Pythium blight.' **1:**FL(35), HI(2903).
P. debaryanum. Damping-off, root necrosis, seed rot. **1:**ND(3423).
P. ultimum. Root rot. 'Pythium blight.' **1:**FL(35).
Pythium sp. Rootlet necrosis, greasy spot. **1:**NC(1303), OK(82).
Sclerospora farlowii. Downy mildew. **1:**OK(82,2872,3423).
Ascomycotina
Apiospora montagnei. On dead leaves. **1:**GA(82,1438).
Cochliobolus spicifer. On hay. **1:**AL(2507).
Erysiphe graminis. 'Powdery mildew.' **1:**CA(82,3423).
Lasiostemma erysipheoides. On dead leaves. **1:**LA(82,1018), TX(1018).
Leptosphaeria korrae. 'Spring dead spot.' **1:**CA(989).
Phyllachora sp. Tar spot. **1:**FL(35).
Sclerotinia homoeocarpa. 'Dollar spot.' **1:**FL(35), HI(2299,2903), NC(1303).
Basidiomycotina - Uredinales
Puccinia cynodontis. Heteroecious: telial host. **1:**Gulf states(82), SW states(82), AZ(747,1203,3981), CA(81,747), FL(35), HI(2903), MS(2720), NM(2112), OK(2872), TX(742). **2:**HI(2903).
P. graminis. Heteroecious: telial host. **1:**Unk.(82).
Basidiomycotina - Ustilaginales
Sporisorium cenchri. Inflorescence smut. **1:**CA(82), TX(82).
Ustilago cynodontis. Inflorescence smut. **1:**AR(1077), AZ(82,1077), CA(81,82,1077), HI(2903), KS(3021), MO(82,1077), MS(1077,2720), NC(1303), OK(82,1077,2872), TX(82,1077).
Basidiomycotina - Other
Amylosporus campbellii. **1:**AZ(1199).
Athelia rolfsii. Seedling blight, stalk rot, basal sheath. **1:**FL(3423).
Laetisaria fuciformis. Pink patch. 'Red thread.' **1:**FL(35), MS(1058).
Lepiota sp. Fairy ring. **1:**FL(35).
Marasmius sp. Fairy ring. **1:**FL(35).
Thanatephorus cucumeris. Banded sclerotial disease, brown patch, root rot, sharp eye spot. **1:**AL(82), AZ(3423), GA(1439), HI(3423), LA(82,3423,3524), MS(82), TX(3423).
Deuteromycotina - Hyphomycetes
Arthrinium arundinis. On culms and stolons. **1:**LA(82), MS(82,2720).
A. phaeospermum. Mold. **1:**MS(3423).
Bipolaris cynodontis. Leaf blotch. 'Helminthosporium leaf blotch.' **1:**AL(2484), AZ(3435), CA(988,3423), FL(35,82,3423), GA(82,3423), HI(2903), KS(3021), NC(82,1303,3423), OK(2872,3423), TX(3423), VA(82,3423).
B. sorokiniana. On crowns, leaves, roots and stems. **1:**CA(988).
B. spicifera. Seed disorder and mold, leaf spot. 'Helminthosporium crown necrosis. Helminthosporium stem necrosis.' **1:**AZ(1107), FL(35).
B. stenospila. Leaf spot. **1:**FL(35,1104).
Cercospora seminalis. Leaf spot. **1:**TX(82).
Coniosporium rhizophilum. **1:**Unk.(82), OK(2872).
Curvularia intermedia. **1:**AL(2484).
C. lunata. Leaf spot. **1:**FL(35), HI(2903).
Curvularia sp. Fading out. **1:**NC(1303).
Drechslera gigantea. Zonate leaf spot. 'Zonate eyespot.' **1:**FL(35,82,3423), IL(312), MD(82), MO(3423), NC(48,82,3423,3944), OK(2872,3423), TX(82,3423).
Ephelis mexicana. Black rings. **1:**FL(35).
Exserohilum rostratum. Leaf spot. **1:**AL(2507), FL(35), MS(2720), NC(1303).
Fusarium sp. **1:**NC(1303).
Helminthosporium sp. Leaf blight, spring dead spot. **1:**CA(82), FL(82,3423), GA(82,3423), HI(2903), OK(82,2872,3742), TX(82).
Microdochium bolleyi. Root necrosis. 'Anthracnose.' **1:**ND(3423).

M. nivale. Snow mold. 'Fusarium patch.' **1**:WA(82).
Nigrospora sphaerica. On hay. **1**:AL(2507).
Periconia lateralis. On moldy hay. **1**:AL(2482).
P. minutissima. On moldy hay. **1**:AL(2482).
Phymatotrichopsis omnivora. On roots. **1**:LA(82), MS(82).
Pithomyces chartarum. On hay. **1**:AL(2507).
Pyricularia grisea. Blast. 'Gray leaf spot.' **1**:FL(35), OK(2872,3423).
Deuteromycotina - Coelomycetes
Ascochyta graminea. On culms and leaves. **1**:CA(3423,3449), KS(3021), SC(82,3423).
Colletotrichum graminicola. 'Anthracnose.' **1**:HI(2903), OK(2872,3423).
Colletotrichum sp. **1**:NC(1303).
Cylindrosporium campicola. On living leaves. **1**:KS(3451).
Leptostromella cynodontis. Decline. **1**:AL(3423).
Macrophoma sp. On culms. **1**:SC(82).
Phoma terrestris. 'Pyrenochaeta root rot.' **1**:ND(3423).
Septoria cynodontis. Leaf spot. **1**:FL(35), HI(2299,2903), TX(82,3423).
Septoria sp. Leaf spot. **1**:OK(2872).
Deuteromycotina - Other
Rhizoctonia solani. Foliar blight, brown patch. **1**:'Rhizoctonia blight.' FL(35), HI(82), LA(82), NC(1303), OK(82), TX(82). **2**:FL(3392), NC(48).
Rhizoctonia sp. Root rot. **1**:AZ(82), HI(2903), OK(82,2872), TX(82).
Sclerotium rolfsii. Southern blight. 'Sclerotium blight.' **1**:FL(35,82).

Cynosurus L.

1. *C. cristatus* L. [crested dogtail]
2. *C. echinatus* L.

Ascomycotina
Cochliobolus sativus. Blind seed disease. **2**:OR(3423).
Gloeotinia granigena. Blind seed disease. **1**:OR(82). **2**:OR(1075).
Deuteromycotina - Hyphomycetes
Cercosporidium graminis. Leaf streak, brown stripe. **1**:NY(3423). **2**:NY(82).
Deuteromycotina - Coelomycetes
Ascochyta subalpina. **2**:CA(3442).

Dactylis L.

1. *D. glomerata* L. [cock's-foot, orchardgrass]

Myxomycetes
Physarum sp. Slime mold. **1**:NC(1303).
Chytridiomycetes
Synchytrium graminicola. Purple leaf speckle. **1**:OR(3423).
Synchytrium sp. Leaf speckle. **1**:OR(82).
Oomycetes
Pythium aristosporum. Root browning. **1**:MT(82).
P. debaryanum. Damping-off, root necrosis, seed rot. **1**:SD(2242,3423), WA(3423).
P. graminicola. **1**:ND(3423), SD(2242), WA(82,3423).
P. ultimum. Root necrosis. **1**:WA(3423).
Ascomycotina
Acrospermum graminum. On culms and leaves. **1**:OR(3228).
Claviceps purpurea. Ergot. **1**:E states(3423), NE states(82), AK(526,3423), AZ(3423), CT(3423), KY(3423), MD(1729), ME(3423), MI(3423), MN(3423), MT(3423), NH(1890), OK(82,2873,3423), OR(3423), TN(3423), WA(82,3423), WI(3423).
Epichloe typhina. Choke. **1**:NJ(82,3423).
Erysiphe graminis. Powdery mildew. **1**:C states(82), E states(82), IA(3423), ID(3423), MI(3423), MN(3423), MO(3423), MT(3423), NY(1890,2998,3005,3423), OH(3423), OR(82,3423), PA(3423), UT(3423), WA(82,3423), WI(3423).
E. graminis f. sp. *avenae*. Powdery mildew. **1**:AR(3029).
Phaeosphaeria culmorum. On dead culms. **1**:CA(82), NY(82).
Phyllachora sp. Black spot. **1**:OK(2872).
Pleospora herbarum. **1**:CA(81).
Pleospora sp. **1**:PA(1292).
Pyrenophora phaeocomes. Foliar disease. **1**:PA(1294).
[*Sclerotinia trifoliorum*]. Anthracnose, crown rot. **1**:KY(3719), MA(1890).
Basidiomycotina - Uredinales
Puccinia coronata. Heteroecious: telial host. **1**:IA(82), TN(82), WI(82).
P. graminis. Heteroecious: telial host. **1**:Range of host(82), HI(2903), IN(663), KS(3023), MS(2720), NC(3944), NY(2998), OK(2872), SD(2242), WV(952).
P. graminis subsp. *graminicola*. Heteroecious: telial host. **1**:Range of host(82), NY(2998,3003).
P. recondita. Heteroecious: telial host. **1**:AL(759), MT(82), NY(2999).
Uromyces dactylidis. Heteroecious: telial host. **1**:E states(82), CA(81), KY(82), NC(48), NC(1303), OK(2872), PA(1875).
Basidiomycotina - Ustilaginales
Tilletia sp. **1**:OR(1464).
Urocystis agropyri. **1**:NH(1077).
Ustilago striiformis. On leaves. **1**:E states(82), KS(3021,3023), KY(82,1077), MA(1077), NJ(494), NY(1077), OH(946), OR(82,1077), PA(1077), UT(82,1077), VA(638,1077), WA(82,1077), WI(1341), WV(952).
Basidiomycotina - Other
Thanatephorus cucumeris. Brown patch, root rot, sharp eye spot. **1**:MT(3423), OR(3423).
Typhula incarnata. Snow mold. **1**:CT(82,3423), ID(82,2955,3423), NY(82).
T. phacorrhiza. **1**:PA(82).
Deuteromycotina - Hyphomycetes
Bipolaris sorokiniana. Seedling blight, spot blotch, black point, root rot. **1**:MN(3423), ND(82,3423), SD(2242,3423), WA(3423).
B. victoriae. Blight. **1**:IA(3423).
Cercosporidium graminis. Leaf blotch, leaf streak, brown stripe. **1**:E states(3423), Range of host(82), CA(81), CO(3423), DE(3543), IA(3423), ID(3423), KS(3021,3023), KY(3423), LA(3423), MD(3423), MO(3423), MS(2720), NC(45,48,1303), NH(1895), NY(2998,2999,3004,3005,3423), OK(2872,3423), OR(3423), SD(2242), UT(3423), VA(3423), VT(3423), WA(3423), WI(3423), WV(952).
Curvularia inaequalis. Mold. **1**:IL(3423).
Curvularia sp. **1**:AL(759).
Drechslera biseptata. From seed. **1**:OR(2065).
D. dactylidis. Leaf blotch, leaf spot. **1**:PA(4000).
D. triseptata. Leaf mold. **1**:NY(82,3423).
Drechslera sp. Leaf spot. **1**:NC(1303).
Fusarium acuminatum. Secondary root rot. **1**:MN(3423), MT(3423), ND(82,3423), SD(3423), WA(3423).
F. avenaceum. Canker, stem rot. **1**:OR(82,3423).
F. culmorum. Head mold, root rot, scab. **1**:WA(3423), WV(82).
F. poae. Secondary root rot, silver top. **1**:NY(82,3423), PA(82,3423).
Fusarium sp. **1**:PA(1292).
Helminthosporium sp. Leaf spot. **1**:AL(759), MA(82), MD(82), MN(82), PA(82), WA(82,3423).
Hymenula cerealis. Stripe. **1**:WA(415).
Mastigosporium rubricosum. Leaf fleck, eye spot. **1**:NY(2998,3005), OR(82,3423), WA(82,3423).
Microdochium bolleyi. Root necrosis. **1**:MD(3423), MT(3423).
M. nivale. Snow mold. **1**:ID(3423), WA(3423).
Periconia circinata. Root rot. **1**:ND(3423).
Pithomyces chartarum. On seeds. **1**:OR(2064).
Rhynchosporium orthosporum. Scald, white leaf stripe. **1**:CA(3423), ID(82,3423), MT(3423), NJ(494), NY(2998,2999,3005), OH(946), OR(82), PA(82), UT(82,3423), WA(3423), WI(82,3423), WV(955), WY(3423).
R. secalis. **1**:CA(82), MT(82), OR(82), WI(82).
Deuteromycotina - Coelomycetes
Ascochyta graminicola. Leaf spot. **1**:DE(82), PA(82), VT(82).
A. sorghi. Leaf spot. **1**:DE(3423).
Colletotrichum cereale. Anthracnose. **1**:KY(3719).
C. graminicola. Anthracnose. **1**:E states(82), KS(3021,3023), KY(3423), MD(3423), MN(82,3423), MS(3423), NC(1303,3423), NJ(3423), OH(3423), OK(82,2872,3423), OR(3423), VA(3423), WV(952).
Dilophospora alopecuri. Twist. **1**:VA(3903).
Phoma terrestris. Pink root. **1**:ND(82,3423).
Phoma sp. **1**:PA(1292).
Phyllosticta owensii. Gray leaf streak, leaf spot. **1**:OR(82,3423,3424).
Pseudoseptoria donacis. **1**:WA(638).
P. stomaticola. Stem speckle, eye spot. **1**:ID(82,3423,3450), OR(82,3423,3450), WA(82,3423,3450).
Septoria sp. Leaf spot. **1**:Unk.(82), NC(1303), OK(2872).
Stagonospora arenaria. Leaf blotch, purple brown blotch. **1**:IA(82), KY(82,3423), MD(3423), MI(82,3423), OH(82,3423),

OR(82,3423), WA(82,3423), WI(1357), WV(953).
S. maculata. Purple leaf spot. **1:**MD(82,3423), NC(48), NH(82), NJ(494), NY(2998,2999,3004,3005), PA(82,3423), VT(82), WV(952).
Wojnowicia hirta. Secondary foot rot, rot of mature straw. **1:**ND(82,3423), WA(3423), WY(82,3423).
Deuteromycotina - Other
Rhizoctonia solani. Foliage blight. **1:**NC(45,48), VA(892), WA(3432), WV(953).
Rhizoctonia sp. **1:**PA(1292).

Dactyloctenium Willd.

1. *D. aegyptium* (L.) Willd. [crowfoot grass]

Ascomycotina
Gloeotinia granigena. Blind seed disease. **1:**OR(82).
Setosphaeria holmii. **1:**GA(1438).
Basidiomycotina - Uredinales
Puccinia graminis. Heteroecious: telial host. **1:**FL(35).
Basidiomycotina - Ustilaginales
Ustilago sparsa. Head smut. **1:**AL(82,1077,3713), FL(1077), GA(1077), LA(82,1077), MS(2720), SC(82,1077).
Deuteromycotina - Hyphomycetes
Cercospora tessellata. Leaf spot. **1:**AL(82,3423).
Exserohilum holmii. Leaf blight. **1:**GA(2203), NC(1303).

Danthonia DC. in Lam. & DC.

1. *D. californica* Boland. (=*D. americana* Lams.-Scribn., *D. californica* Boland. var. *americana* (Lams.-Scribn.) Hitchc.) [California oatgrass]
2. *D. compressa* Aust.
3. *D. intermedia* Vasey [timber oatgrass]
4. *D. parryi* Lams.-Scribn. [Parry oatgrass]
5. *D. pilosa* R. Br.
6. *D. sericea* Nutt. [downy oatgrass]
7. *D. spicata* (L.) P. Beauv. ex Roem. & Schult. [poverty oatgrass]
8. *D. unispicata* (Thurb.) Munro ex Macoun [one-spike oatgrass]
9. *Danthonia* sp.

Ascomycotina
Atkinsonella hypoxylon. Black band, black choke. **2:**E states(3423), NC(82), OH(82,3423), TN(82), VA(82), WV(82). **6:**GA(1438,3423). **7:**E states(82,3423), GA(1438), IA(3423), MD(638), MO(3423), NC(1303,3423), OH(82,3423), TN(82,3423), WI(1344). **9:**GA(82), MA(82).
Claviceps purpurea. Ergot. **4:**CO(3423). **7:**MI(82,2862,3423). **9:**CO(82).
Cochliobolus sativus. Blind seed disease. **1:**OR(3423).
Epichloe typhina. Choke. **7:**MD(3423), VA(82,3423).
Gloeotinia granigena. **1:**OR(1075).
Lewia infectoria. **9:**NM(82).
Lophodermium arundinaceum. On dead culms, leaves and sheaths. **3:**CA(3423), CO(3423). **9:**CA(82), NM(82).
L. danthoniae. **7:**IL(82), WI(82).
Ophiobolus stictisporus. **6:**GA(1438). **9:**GA(82).
Phaeosphaeria caricinella. **3:**WA(3814).
Phyllachora graminis. Tar spot. **9:**ME(82).
Pleospora sp. Leaf rot. **9:**OR(82).
Basidiomycotina - Uredinales
Puccinia graminis. Heteroecious: telial host. **7:**WV(82).
Basidiomycotina - Ustilaginales
Ustilago residua. Panicle smut. **1:**CA(81,1077), ID(1077), KY(1077), MT(1077), OR(1077), UT(1077), WA(1077), WY(1077). **2:**IA(1077), IN(1077), KY(1077), NC(1077,1303), NY(1077), PA(1077), TN(1077), VA(1077), VT(1077). **3:**CA(1077), CO(1077), MT(1077), WY(1077). **4:**CO(1077), NM(1077). **7:**IA(1077), ID(1077), IN(1077), NH(1077), NY(1077), OH(1077), PA(1077), VA(1077), WA(1077), WV(1077), WY(1077). **8:**CA(1077), WY(1195). **9:**Range of host(82), CO(922).
U. striiformis. **3:**WY(1077,2924).
Basidiomycotina - Other
Thanatephorus cucumeris. Brown patch, root rot, sharp eye spot. **1:**OR(3423).
Deuteromycotina - Hyphomycetes
Cercosporidium graminis. Leaf streak, brown stripe. **1:**CA(82), OR(82,3423). **3:**WY(2924).
Drechslera campanulata. Leaf spot. **1:**OR(3423). **7:**CA(82), MA(82,3423), ME(82,3423), NY(2456), OR(82), WV(955).
Ephelis borealis. **1:**Unk.(82).
Fusarium acuminatum. Secondary root rot. **5:**ND(3423).
Helminthosporium sp. Leaf spot. **7:**NY(82).
Heterosporium avenae. Leaf mold. **4:**CA(3423). **9:**CA(82).
Microdochium nivale. Leaf spot. **1:**ID(3442).
Rhynchosporium secalis. Scald. **9:**OR(82).
Deuteromycotina - Coelomycetes
Colletotrichum graminicola. Anthracnose. **1:**ID(82,3423), WA(82,3423).
Macrophoma phlei. **2:**WV(953).
Neottiospora sp. **7:**NC(1307).
Phaeoseptoria airae. On leaves. **9:**CO(82), OR(82).
Pseudoseptoria everhartii. **7:**WI(1350).
P. stomaticola. Eye spot, leaf spot. **1:**CA(3442), OR(82,3423,3450). **3:**MT(3433). **7:**NY(3423,3450).
Stagonospora subseriata. Leaf mold and rot. **3:**CO(3423). **9:**CO(82).

Deschampsia P. Beauv.

1. *D. atropurpurea* (Wahlenb.) Scheele [mountain hairgrass]
2. *D. caespitosa* (L.) P. Beauv. [tufted hairgrass]
3. *D. caespitosa* (L.) P. Beauv. subsp. *beringensis* (Hultén) W. E. Lawr. (=*D. beringensis* Hultén)
4. *D. danthonioides* (Trin.) Munro ex Benth. [annual hairgrass]
5. *D. elongata* (Hook.) Munro ex Benth. [slender hairgrass]
6. *Deschampsia* sp.

Ascomycotina
Claviceps purpurea. Ergot. **2:**OR(82,3423).
Cochliobolus sativus. Blind seed disease. **2:**OR(3423).
Didymosphaeria arenaria var. *macrospora*. **2:**AK(526).
Erysiphe graminis. Powdery mildew. **4:**OR(82).
Gaeumannomyces graminis. Root rot, take-all. **4:**ID(82,3423), WA(3446,3754).
Gloeotinia granigena. **2:**OR(1075).
Mycosphaerella deschampsiae. On living leaves. **1:**AK(3426,3428,3431).
Trichothyrina alpestris. **1:**AK(3431).
Basidiomycotina - Uredinales
[*Puccinia brachypodii*]. **2:**WY(1195).
P. brachypodii var. *arrhenatheri*. Heteroecious: telial host. **2:**Unk.(754), AK(749), CO(749,1195), WY(749).
P. brachypodii var. *poae-nemoralis*. **2:**AK(526), CO(82). **6:**WA(82).
P. coronata. Heteroecious: telial host. **2:**AK(60), OR(82), WA(82), WY(82,1195).
P. graminis. Heteroecious: telial host. **1:**WA(82). **2:**MI(82), WA(82). **4:**CA(82), WA(82).
Uromyces fragilipes. Heteroecious: aecial host. **2:**CA(82), WA(82). **4:**CA(82), OR(82), WA(82). **5:**CA(81). **6:**ID(82), MI(82), OR(82), WA(82).
Basidiomycotina - Ustilaginales
Entyloma dactylidis. **2:**UT(1077).
Tilletia cerebrina. Bunt. **2:**AK(526,1077), ID(1077), OR(1077), WA(82,1077), WY(82,1077,2924). **4:**CA(82,1077), ID(2342), OR(82,1077), WA(82,1077,3264), WY(82). **5:**ID(1077,3264), OR(1077), WA(1077). **6:**ID(82), OR(82).
Ustilago striiformis. **1:**MT(1077), WY(1077). **2:**CO(1077), OR(1077), WY(1077).
Basidiomycotina - Other
Coprinus psychromorbidus. **1:**AK(3441).
Thanatephorus cucumeris. Brown patch, root rot, sharp eye spot. **1:**WA(3423).
Typhula idahoensis. Snow mold. **5:**ID(3423). **6:**ID(82).
Deuteromycotina - Hyphomycetes
Cercosporidium graminis. Leaf streak, brown stripe. **1:**WA(82,3423). **2:**OR(82,3423), WA(82,3423). **4:**ID(82,3423), OR(82,3423), WA(82). **5:**ID(3423), MT(3423), OR(3423), WA(3423). **6:**ID(82), MT(82), OR(82), WA(82).
Fusarium acuminatum. Secondary root rot. **5:**WY(3423).
Helminthosporium sp. **4:**ID(3423). **5:**MT(3423).
Mastigosporium rubricosum. On living leaves. **1:**WA(3425).
Ramularia pusilla. Leaf spot. **1:**AK(3429,3431).
Spermospora subulata. Blast. **2:**ID(82,3423), MT(82,3423), WY(82,3423).
Deuteromycotina - Coelomycetes
Ascochyta graminicola. Leaf spot. **6:**OR(82).

A. sorghi. Leaf spot. **5:**OR(3423).
A. subalpina. Leaf spot. **5:**WA(3449).
Dilophospora alopecuri. Twist disease. **2:**OR(82,3423).
Hendersonia culmicola var. *minor*. **2:**OR(82).
Hendersonia sp. Leaf mold. **2:**OR(82).
Phaeoseptoria airae. Leaf mold. **2:**OR(82).
Pseudoseptoria everhartii. Speckle, eye spot. **1:**MT(82,3423), WA(82,3423,3450), WY(82,2924,3423,3450). **2:**MT(82,3423,3450), ND(3442), UT(3438), WY(82,3423,3450). **4:**ID(3423,3450), WA(82,3423,3450). **5:**WA(3423,3450). **6:**WA(82).
P. stomaticola. Leaf spot. **4:**ID(82). **5:**WA(3450). **6:**WA(82).
Septoria arundinacea. Leaf spot. **2:**ID(3424).
S. poliomela. Leaf spot. **2:**OR(82,3423), WA(82), WY(3433). **4:**OR(3423), WA(3423). **5:**ID(3435), WA(3432).
Stagonospora culmicola. Leaf rot. **2:**OR(3423).
S. graminum. Rot. **1:**AK(82,526,3423).
S. nodorum. Glume blotch, leaf spot. **1:**WA(82,3423). **4:**WA(82). **5:**ID(3442), WA(3423). **6:**WA(82).
S. subseriata. On leaves. **2:**OR(82). **6:**CO(82).
S. vexatula. Leaf spot. **2:**ID(82,3423,3425). **3:**AK(3428). **5:**UT(3435).

Deuteromycotina - Other

Rhizoctonia solani. Root rot. **1:**WA(82).

Diarrhena P. Beauv.

1. *D. americana* P. Beauv.

Deuteromycotina - Coelomycetes

Colletotrichum graminicola. Anthracnose. **1:**OK(2872,3423), WV(955).

Dichanthelium (Hitchc. & Chase) Gould

1. *D. lanuginosum* (Elliott) Gould (=*Panicum lanuginosum* Elliott, *Panicum huachucae* Ashe, *Panicum huachucae* Ashe var. *fasciculatum* (Torr.) Hubb., *Panicum pacificum* Hitchc. & Chase, *Panicum tennesseense* Ashe)
2. *D. linearifolium* (Lams.-Scribn.) Gould (=*Panicum linearifolium* Lams.-Scribn., *Panicum perlongum* Nash)
3. *D. oligosanthes* (Schult.) Gould (=*Panicum scribnerianum* Nash)

Ascomycotina

Balansia strangulans. Black ring. **3:**NE(82,2927,3423).
Phyllachora punctum. Black spot, tar spot. **2:**TX(3423), WI(1343). **3:**AL(3423), NE(3423), OK(2873,3423), WI(3423).

Basidiomycotina - Other

Thanatephorus cucumeris. Brown patch, root rot, sharp eye spot. **1:**NE(3423).

Deuteromycotina - Hyphomycetes

Bipolaris sorokiniana. Seedling blight, spot blotch, black point, root rot. **2:**NE(3423).
Fusarium acuminatum. Secondary root rot. **2:**NE(3423).
F. moniliforme. Secondary root rot. **2:**NE(3423).
Microdochium bolleyi. Root necrosis. **1:**NE(3423).
Phaeoramularia fusimaculans. Leaf spot. **1:**OR(3423). **2:**WI(1332,3423). **3:**WI(82,1332,3423).

Deuteromycotina - Coelomycetes

Colletotrichum graminicola. Anthracnose. **1:**NE(3423). **3:**NE(2927).
Phyllosticta healdii. Leaf spot. **1:**NE(3423).
Phyllosticta sp. **3:**WI(1352).
Pseudoseptoria everhartii. Leaf spot. **3:**KS(3021).
Septoria tandilensis. Leaf spot. **3:**KS(3023), NE(2927,3429), WI(1333,3423).
Septoria sp. Leaf spot. **3:**NE(82), WI(82,1332).
Stagonospora simplicior. **3:**WI(1343).

Dichanthium Willemet

1. *D. annulatum* (Forssk.) Stapf (=*Andropogon annulatus* Forssk.) [brahman grass, Diaz bluestem, Kleberg grass, ringed beardgrass]

Deuteromycotina - Hyphomycetes

Curvularia robusta. On leaves. **1:**TX(1900).

Deuteromycotina - Coelomycetes

Colletotrichum caudatum. Leaf spot. **1:**GA(3423).

Digitaria Haller

(= *Leptoloma* Chase, *Trichachne* Nees.)

1. *D. californica* (Benth.) Henry (=*Trichachne californica* (Benth.) Chase) [Arizona cottontop]
2. *D. cognata* (Schult.) Pilg. in Engl. & Prantl (=*Leptoloma cognatum* (Schult.) Chase)
3. *D. eriantha* Steud. subsp. *pentzii* (Stent) Kok (=*D. decumbens* Stent, *D. pentzii* Stent, *D. valida* Stent) [pangola grass, woolly fingergrass]
4. *D. filiformis* (L.) Koeler
5. *D. hitchcockii* (Chase) Stuck. (=*Trichachne hitchcockii* (Chase) Chase & Hitchc.)
6. *D. ischaemum* (Schreb.) Schreb. ex Muhl. [smooth crabgrass]
7. *D. patens* (Swall.) Henr. (=*Trichachne patens* Swall.)
8. *D. sanguinalis* (L.) Scop. (=*Panicum sanguinale* L.) [hairy crabgrass, large crabgrass]
9. *D. serotina* (Walter) Michx.
10. *D. setigera* Roem. & F. W. Schultz (=*D. pruriens* (Fisch. ex Trin.) Buse)
11. *D. villosa* (Walter) Pers. (=*D. filiformis* (L.) Koeler var. *villosa* (Walter) Fernald)
12. *D. violascens* Link
13. *Digitaria* sp.

Oomycetes

Sclerophthora macrospora. Downy mildew, crazy top. **6:**SD(2242). **8:**IN(3708), SD(2242). **13:**MO(3877).

Ascomycotina

Apiospora montagnei. **2:**GA(1438).
Erysiphe graminis. Powdery mildew. **8:**TX(82,3423).
Mycosphaerella sp. On culms, leaves. **2:**OK(82,2873), WI(1345).
Phyllachora graminis. **8:**NC(3944).
P. paspalicola. Tar spot. **7:**TX(3423).
P. punctum. Tar spot. **2:**TX(82,3423), WI(82,1332,3423).
Phyllachora sp. Tar spot. **8:**NC(82).

Basidiomycotina - Uredinales

Puccinia esclavensis. **1:**AZ(1203,3981), TX(742). **5:**TX(742).
P. esclavensis var. *panicophila*. **1:**AZ(82), NM(82), TX(82). **2:**TX(82).
P. oahuensis. Autoecious. **3:**HI(2903). **6:**LA(3523). **8:**FL(35,82), GA(82), LA(3523), TX(82). **9:**FL(82). **10:**HI(2903). **12:**HI(82,2903). **13:**FL(35), HI(2903).
P. substriata. Heteroecious: telial host. **1:**AZ(1203,3981). **13:**FL(35), NC(1303).
P. substriata var. *imposita*. Heteroecious: telial host. **2:**AL(82), KS(82,3023), NM(82), OK(82,2872), TX(82).

Basidiomycotina - Ustilaginales

Sphacelotheca cordobensis. Inflorescence smut. **1:**AZ(82,1077).
S. digitariae. **8:**CA(1077).
S. diplospora. Head smut. **8:**CA(1077), FL(1077), IA(82,1077), IL(82,1077), MS(82,1077,2720). **11:**FL(1077).
Sporisorium cenchri. **2:**IA(1077), IN(1077).
Tilletia barclayana. **6:**MS(1077,2720). **8:**MS(82,1077,2720).
Ustilago syntherismae. Loose smut. **4:**IN(1077), MO(1077), NJ(1077), NY(1077), VA(1077). **6:**IA(1077), MI(1077), MO(82), NJ(82), SD(2242). **8:**AL(1077), AR(1077), CA(1077), CT(1077), DC(1077), DE(1077), FL(35), GA(1077,1439), IA(1077), IL(1077), IN(1077), KS(1077,3021), KY(1077), LA(1077), MA(1077), MD(1077), ME(1077), MN(1077), MO(82,1077), MS(1077,2720), NC(1077,1303,3944), NE(1077), NH(1077), NJ(82,1077), NY(1077), OH(1077), OK(2872), PA(1077), SC(1077), SD(2242), TN(1077), TX(1077), VA(1077), WI(1077), WV(953,1077). **13:**FL(35), OK(1077).

Basidiomycotina - Other

Thanatephorus cucumeris. Brown patch, root rot, sharp eye spot. **8:**OH(3423). **13:**LA(3524).

Deuteromycotina - Hyphomycetes

Alternaria sp. Leaf spot. **8:**FL(35).
Bipolaris sorokiniana. Seedling blight, spot blotch, black point, root rot. **8:**MN(3423), ND(82,3423).
Cercospora sp. On leaves. **6:**WI(1348).
Cercosporidium graminis. Brown stripe, leaf streak. **8:**VA(82,3423).
Curvularia geniculata. Leaf mold and spot. **6:**WA(3429). **8:**ND(82). **13:**FL(35).
C. lunata. **3:**HI(2903).
C. trifolii. **8:**ND(82).
Curvularia sp. **13:**FL(82).
Fusarium acuminatum. Secondary root rot. **8:**ND(82,3423).
F. heterosporum. Head mold. **13:**TX(82).
Helminthosporium sp. Leaf spot. **8:**FL(35), MA(82), MN(82,3423), MS(2720).
Phaeoramularia fusimaculans. Leaf spot. **2:**KS(3021,3023), WI(82,1332,1334,3423).

Pyricularia grisea. Blast, gray spot. **3:**HI(2903). **6:**Range of host(82), WV(955). **8:**Range of host(82,3423), AL(2497), DE(3543), KS(3023), NC(1303,3944), NE(2927), OK(2872). **9:**FL(82,3423). **13:**FL(35), NC(1303).

Deuteromycotina - Coelomycetes

Colletotrichum graminicola. Anthracnose. **2:**OK(82,2873,3423). **4:**OK(2872,3423). **6:**WI(1348). **8:**KS(3021), MD(82), MI(82), ND(82,3423), NJ(3423), OK(82,2873,3423), WV(955). **13:**NC(1303).

Phyllosticta digitariae. On sheaths. **6:**UT(3432).

P. rogleri. White leaf spot. **8:**IA(82,3423).

Phyllosticta sp. Leaf spot. **8:**KY(82).

Septoria digitarivora. Leaf spot. **8:**WV(82,3423).

S. tandilensis. Leaf spot. **8:**NJ(3429).

Septoria sp. Leaf spot. **8:**AL(82), NJ(82).

Deuteromycotina - Other

Rhizoctonia solani. Summer blight, brown patch, root rot. **6:**WV(955). **8:**FL(35), OH(82).

Distichlis Raf.

1. *D. spicata* (L.) Greene (=*D. maritima* Raf.) [seashore saltgrass]
2. *D. stricta* (Torr.) Rydb. [desert saltgrass]

Oomycetes

Pythium debaryanum. Damping-off, root necrosis, seed rot. **2:**SD(3423).

Ascomycotina

Claviceps purpurea. Ergot. **1:**MD(3423), MS(2720), WA(3423). **2:**MS(82), WA(82).

Dothidella aristidae. Char spot. **1:**CA(3423), OR(3423). **2:**MT(3433), ND(3423), OR(3423), SD(3423), TX(3423), UT(3423), WA(3423).

Euryachora aristidae. Char spot. **1:**CA(82), OR(82). **2:**ND(82), NE(82), OR(82), SD(82), TX(82), UT(82), WA(82).

Phyllachora cynodontis. On culms. **1:**CA(81,82), MS(82,2720).

P. sylvatica. Tar spot. **2:**CA(82,3423), CO(82,3423), KS(3021,3423), ND(82,3423), NM(3423), OK(82,2873,3423), SD(2242,3423), TX(82,3423).

Basidiomycotina - Uredinales

[*Puccinia aristidae*]. **1:**W states(82), CA(81), DE(82), ME(82), NC(82), OK(2872), TX(82), WA(638), WY(1195). **2:**C states(82), CA(82), CO(2657), KS(3023), MT(1195), OK(2872), SD(2242), WA(82).

P. graminis. Heteroecious: telial host. **2:**MT(82).

P. subnitens. Heteroecious: telial host. **1:**AZ(3981), CA(81), ND(378), WY(1195). **2:**AZ(747,3981).

Uromyces peckianus. Heteroecious: telial host. **1:**CA(81,82), CT(82), FL(35), MA(82,743), NY(82), WA(82).

U. peckianus subsp. *rugulosus*. **1:**MA(3110), ME(3110), WA(3110).

Basidiomycotina - Ustilaginales

Ustilago halophila. Stem smut. **1:**AZ(82,1077), CA(82,1077), CO(1077), CT(1077), ID(82,1077), NM(82,1077), NV(82,1077), OR(1077), TX(1077), UT(1077), WA(1077). **2:**AZ(1077), CA(1077), CO(1077), ID(1077), NM(1077), NV(1077), OR(82,1077), UT(82,1077), WA(82,1077), WY(1077).

U. halophiloides. **2:**OR(1077).

U. trebouxii. **2:**WA(1077).

Deuteromycotina - Hyphomycetes

Cerebella andropogonis. On heads following ergot. **1:**MS(82,2720).

Exserohilum rostratum. Leaf blotch, mold. **1:**NY(82,3423).

Fusarium acuminatum. Secondary root rot. **1:**ND(82). **2:**MT(3423), ND(3423), SD(3423), WY(3423).

Microdochium bolleyi. Root necrosis. **2:**ND(3423), SD(3423).

Deuteromycotina - Coelomycetes

Ascochyta subalpina. Leaf spot. **2:**MT(3449).

Phoma terrestris. Pink root. **1:**ND(82). **2:**SD(3423).

Pyrenochaeta graminis. **1:**Unk.(82). **2:**Unk.(82).

Tracylla spartinae. On culms. **2:**TX(82).

Wojnowicia hirta. Secondary foot rot, rot of mature straw. **2:**ND(82,3423).

Dupontia R. Br.

1. *D. fischeri* R. Br.

Ascomycotina

Clathrospora pentamera. **1:**AK(526).

Mycosphaerella recutita. On dead sheath. **1:**AK(1938).

Pleospora heleocharidis var. *arctica*. On dead sheaths and stems. **1:**AK(1938).

Echinochloa P. Beauv.

1. *E. colonum* (L.) Link [jungle ricegrass]
2. *E. crusgalli* (L.) P. Beauv. (=*Panicum crusgalli* L., *E. pungens* (Poir.) Rydb.) [barnyard grass, jungle ricegrass]
3. *E. crusgalli* (L.) P. Beauv. var. *frumentacea* (Link) Wight [billion-dollar grass, Japanese millet, Sanwa millet]
4. *E. crusgalli* (L.) P. Beauv. var. *zelayensis* (H. B. K.) A. Hitchc.
5. *E. walteri* (Pursh) A. Heller
6. *Echinochloa* sp.

Chytridiomycetes

Physoderma sp. Brown stripe. **2:**AR(82).

Oomycetes

Phytophthora cactorum. Root rot. **2:**SD(82,2242,3423).

Pythium aristosporum. **2:**MT(82).

P. arrhenomanes. **2:**Unk.(82).

P. debaryanum. Damping-off, root necrosis, seed rot. **2:**MN(3423), ND(82,3423), WY(3423).

P. graminicola. **2:**MN(3423), ND(3423), SD(2242,3423), WY(3423).

P. tardicrescens. Root browning and rot. **2:**MN(82,3423).

P. ultimum. Root necrosis. **2:**ND(3423).

Pythium sp. Root browning. **2:**Unk.(82).

Ascomycotina

Leptosphaeria occidentalis. On dead culms. **2:**KS(82).

Magnaporthe salvinii. Stem rot. **1:**AR(3423).

Mycosphaerella crus-galli. On leaves. **2:**KS(82,967,973).

Basidiomycotina - Uredinales

Puccinia abnormis. **2:**NC states(82), FL(35), NE(82), NM(82), OK(2872).

[*P. graminis*]. **2:**IA(82).

Basidiomycotina - Ustilaginales

Moesziomyces bullatus. Seed smut. **2:**E states(82), CA(82,1077), CO(1077), CT(1077), FL(35,1077), IA(1077), IL(1077), IN(1077), LA(1077), MA(1077), NC(1077,1303), NE(82,1077), PA(1077), SD(82,1077,2242), WI(1077). **3:**CT(1077), IN(1077), MD(1077), SD(1077). **5:**IN(82,1077), MD(82), NY(1077).

Sphacelotheca diplospora. Covered smut. **2:**IL(82,1077), TX(82,1077).

Ustilago trichophora. Loose smut, node smut. **1:**LA(82,1077), TX(82,1077). **2:**Range of host(82), AR(1077), CA(82,1077), CO(82,1077), CT(1077), DC(1077), DE(1077), FL(35,1077), IA(1077), ID(922,1077), IL(1077), IN(1077), KS(3021), MA(1077), MD(1077), MN(1077), MO(1077), NC(1077,1303), NE(1077), NH(1077,1895), NM(1077), NV(1077), NY(1077), OH(1077), OK(1077,2872), OR(1077), PA(1077), SD(1077,2242), UT(1077), WA(1077), WI(82,1077,1338), WY(1077,1195). **3:**CT(82). **4:**NM(1077). **5:**CT(1077), DE(82,1077), FL(1077), IN(82,1077), NY(82,1077).

Basidiomycotina - Other

Corticium microsclerotium. **1:**FL(82).

Thanatephorus cucumeris. Brown patch, root rot, sharp eye spot. **1:**FL(3423). **2:**MN(3423), ND(3423), OR(3423).

Deuteromycotina - Hyphomycetes

Alternaria sp. Basal culm rot. **2:**MN(82), ND(82), SD(82).

Bipolaris sorokiniana. Seedling blight, spot blotch, black point, root rot. **1:**MN(82), ND(82). **2:**MN(3423), ND(3423), WY(3423).

Cercospora echinochloae. Leaf spot. **2:**KS(3021), WI(82,581).

C. sorghi. Leaf spot. **2:**ND(82,3423), SD(2242).

Cladosporium herbarum. Basal culm rot. **1:**MN(82), ND(82).

Curvularia trifolii. Mold, root rot. **2:**MN(3423), ND(82,3423), SD(3423).

Drechslera gigantea. Zonate leaf spot. **2:**MO(3423).

Epicoccum nigrum. **2:**DE(3543).

Fusarium acuminatum. Secondary root rot. **2:**NGP(82), MN(3423), ND(3423), SD(2242,3423), WY(3423). **3:**ND(3423).

F. avenaceum. Head mold. **2:**WI(82).

F. culmorum. Head mold, root rot, scab. **2:**MN(82,3423), ND(82,3423), SD(2242,3423). **3:**ND(3423).

F. equiseti. Root rot. **2:**NGP(82).

F. graminearum. Seedling blight, root rot, scab. **2:**NGP(82), MN(3423), ND(3423), SD(2242,3423).

F. oxysporum. Root rot. **2:**NGP(82).

Microdochium bolleyi. Root necrosis. **2:**WY(3423). **3:**ND(3423).

Nakataea sigmoidea. Stem rot. **1:**NC(1303).

Phaeoramularia fusimaculans. Leaf spot. **2:**IA(82,3423), WI(82,3423).
Pyricularia grisea. Blast, gray spot. **1:**OK(82,2873,3423), TX(82,3423). **2:**WI(82).

Deuteromycotina - Coelomycetes

Colletotrichum graminicola. Anthracnose. **2:**E & C states(82), FL(35), GA(3423), KS(3021,3023,3423), LA(3423), MO(3423), MS(2720,3423), NC(1303), OH(3423), OK(2872,3423), SD(2242), WI(1341,3423), WV(3423). **6:**FL(35).
Phoma terrestris. Pink root. **2:**ND(82,3423), SD(3423). **3:**ND(3423).

Deuteromycotina - Other

Rhizoctonia solani. Web blight, root rot. **2:**MN(82), ND(82). **6:**FL(35).
Sclerotium hydrophilum. Leaf sheath rot. **1:**AR(82,3423). **2:**AR(82,3423).
S. oryzae. Culm rot. **1:**AR(82). **2:**AR(82).

Eleusine Gaertn.

1. *E. coracana* (L.) Gaertn. [African millet, finger millet, korakan, ragi]
2. *E. indica* (L.) Gaertn. [goosegrass, wiregrass, yardgrass]

Oomycetes

Sclerophthora macrospora. Downy mildew, crazy top. **2:**AR(3594), IL(312), IN(3708).

Ascomycotina

Cochliobolus nodulosus. **2:**GA(1438).

Basidiomycotina - Other

Corticium microsclerotium. Web blight. **2:**FL(82).
Thanatephorus cucumeris. Brown patch, root rot, sharp eye spot. **2:**FL(3423), LA(3524).

Deuteromycotina - Hyphomycetes

Alternaria sp. Leaf spot. **1:**NC(1303).
Bipolaris cynodontis. Mold, leaf spot. **2:**FL(35,82,3423).
B. hadrotrichoides. Leaf spot, mold. **2:**AL(3423), DC(3423), KS(3423), KY(3423), LA(3423), MD(3423), WI(3423), WV(955).
Bipolaris sp. Leaf spot. **1:**NC(1303).
Drechslera fugax. On leaves. **2:**MD(82,3423).
D. gigantea. Eye spot, zonate leaf spot. **2:**FL(35,82,3423), MD(82,3423), MO(3423), NC(1303).
Helminthosporium sp. Leaf spot. **2:**HI(2903).

Deuteromycotina - Coelomycetes

Colletotrichum graminicola. Anthracnose. **2:**FL(35,82,2872), OK(3423).

Deuteromycotina - Other

Rhizoctonia solani. Web blight, root rot. **2:**FL(35).
Sclerotium rolfsii. Southern blight. **2:**FL(35).

Elymus L.

(= *Elytrigia* Desv.)

1. *E. alaskanus* K. Presl subsp. *latiglumis* (Lams.-Scribn. & J. G. Sm.) A. Löve (=*Agropyron latiglume* (Lams.-Scribn. & J. G. Sm.) Rydb.)
2. *E. ambiguus* Vasey & Lams.-Scribn.
3. *E. antarcticus* J. D. Hook.
4. *E. aristatus* Merr.
5. *E. californicus* (Boland.) Gould (=*Hystrix californica* (Boland.) Kuntze)
6. *E. canadensis* L. (=*E. wiegandii* Fisch.) [Canada wild-rye]
7. *E. canadensis* L. var. *robustus* (Lams.-Scribn. & Sm.) Mackenz. & Bush
8. *E. ciliaris* (Trin.) Tzvelev (=*Agropyron ciliare* (Trin.) Franch.)
9. *E. ciliaris* (Trin.) Tzvelev subsp. *amurensis* Drobov (=*Agropyron amurense* Drobov in Trav.)
10. *E. cinereus* Lams.-Scribn. & Merr. (=*E. cinereus* Lams.-Scribn. & Merr. var. *pubens* (Piper) C. Hitchc., *E. condensatus* K. Presl) [basin wild-rye, giant wild-rye]
11. *E. dahuricus* Turcz. ex Griseb.
12. *E. dahuricus* Turcz. ex Griseb. subsp. *excelsus* (Turcz. ex Griseb.) Tzvelev (=*E. excelsus* Turcz. ex Griseb.)
13. *E. dasystachys* Trin.
14. *E. flavescens* Lams.-Scribn. & J. G. Sm. (=*E. arenicolus* Lams.-Scribn. & J. G. Sm.)
15. *E. glaucus* Buckley (=*E. glaucus* Buckley var. *jepsonii* J. B. Davy) [blue wild-rye]
16. *E. innovatus* Beal
17. *E. interruptus* Buckley
18. *E. junceus* Fisch. [Russian wild rice]
19. *E. macrourus* (Turcz.) Tzvelev (=*Agropyron sericeum* Hitchc.)
20. *E. mollis* Trin. in Spreng. [American dunegrass]
21. *E. mutabilis* (Drobov) Tzvelev (=*Agropyron angustiglume* Lams.-Scribn. & J. G. Sm.)
22. *E. panormitanus* (Parl.) Tzvelev (=*Agropyron panormitanum* Parl.)
23. *E. pendulinus* (Nevski) Tzvelev (=*Agropyron pendulinum* (Nevski) Popov)
24. *E. riparius* Wiegland
25. *E. salina* M. E. Jones [Salina wild-rye]
26. *E. scabrus* (R. Br.) A. Löve (=*Agropyron scabrum* (R. Br.) P. Beauv.)
27. *E. semicostatus* Nees ex Steud. (=*Agropyron semicostatum* (Nees ex Steud.) Boiss.)
28. *E. semicostatus* (Nees ex Steud.) Meld. subsp. *striatus* (Nees ex Steud.) A. Löve (=*Agropyron striatum* (Nees ex Steud.) Hook.)
29. *E. sibiricus* L. [Siberian wild-rye]
30. *E. smithii* (Rydb.) Gould (=*Elytrigia smithii* (Rydb.) A. Löve, *Agropyron occidentale* (Lams.-Scribn.) Lams.-Scribn., *Pascopyrum smithii* (Rydb.) A. Löve, *Agropyron smithii* Rydb., *Agropyron smithii* Rydb. var. *molle* (Lams.-Scribn. & J. G. Smith) M. E. Jones, *Agropyron spicatum* (Pursh) Lams.-Scribn. & J. G. Sm. var. *molle* Lams.-Scribn. & J. G. Smith) [western wheatgrass]
31. *E. tauri* (Boiss. & Balb.) Meld. subsp. *pertenuis* (C. A. Mey.) Meld. (=*Agropyron pertenuis* (C. A. Mey.) Nevski)
32. *E. triticoides* Buckley [beardless wild-rye]
33. *E. villosus* Muhl. ex Willd. (=*E. hirsutus* Schreb. ex Roem. & Schult.) [hairy wild-rye]
34. *E. virescens* Piper
35. *E. virginicus* L. (=*E. striatus* Willd.) [Virginia wild-rye]
36. *E. virginicus* L. var. *australis* (Lams.-Scribn. & C. Ball) Hitchc. in Deam (=*E. australis* Lams.-Scribn. & C. Ball)
37. *E. virginicus* L. var. *intermedius* (Vasey) Bush (=*E. hirsutiglumis* Lams.-Scribn.)
38. *Elymus* sp.

Chytridiomycetes

Physoderma graminis. **15:**MT(3433).
Synchytrium graminicola. Purple leaf speckle. **15:**OR(3423).
Synchytrium sp. Purple leaf speckle. **15:**OR(82).

Oomycetes

Pythium arrhenomanes. Root browning. **15:**ND(82). **30:**ND(82), SD(82). **38:**ND(82).
P. debaryanum. Damping-off, root necrosis, seed rot. **6:**ND(3423), SD(2242,3423), WA(3423), WY(3423). **8:**ND(3423). **11:**ND(3423), WA(3423). **12:**ND(3423). **15:**WA(3423). **17:**ND(3423). **18:**ND(3423). **27:**WA(3423). **29:**ND(3423), WA(3423). **30:**MT(3423), ND(82,3423), NE(3423), SD(82), WY(82). **35:**ND(3423). **38:**ND(82).
P. graminicola. **6:**ND(3423), WA(82,3423), WY(3423). **8:**ND(3423). **10:**ND(3423). **11:**ND(3423). **12:**ND(3423). **15:**ND(3423). **17:**ND(3423). **18:**ND(3423), NE(3423), SD(2242,3423). **21:**ND(3423). **23:**ND(3423). **27:**ND(3423). **29:**ND(3423). **30:**MT(3423), ND(3423), WA(3423). **35:**ND(3423).
P. irregulare. Root necrosis. **18:**ND(3423). **38:**ND(82).
P. ultimum. Root necrosis. **6:**WA(3423). **11:**WA(3423). **27:**WA(3423). **30:**ND(82,3423). **35:**WA(3423).
Sclerophthora macrospora. **30:**SD(2242,3184).

Ascomycotina

Apiospora montagnei. On dead culms. **35:**GA(82). **36:**GA(1438).
Claviceps purpurea. Ergot. **2:**MT(3423). **6:**E states(3423), IA(3423), ID(3423), KS(3021), MN(3423), MO(3423), ND(3423), NE(3423), NM(3423), OK(2872,3423), OR(3423), SD(2242,3423), TX(3423), UT(3423), WA(3423), WI(3423), WY(3423). **7:**IA(3423). **10:**CA(3435), CO(3423), ID(3423), MT(3423), ND(3423), OR(3423), UT(3423), WA(3423), WY(1195,3423). **11:**ND(3423), WA(3423). **14:**OR(3423). **15:**CA(3423), ID(3423), OR(3423), WA(3423), WY(1195,3423). **18:**MT(3423), ND(3423), WA(3423). **20:**AK(526,3423,3431), OR(3423). **27:**MI(3423), ND(3423), WA(3423). **30:**Range of host(82), AZ(3423), CO(3423), IA(3423), MI(3423), MN(3423), MO(3423), MT(3423), ND(377,3423), NE(3423), NM(3423), OK(2872,3423), SD(2242,3423), TX(3423), WA(3423), WI(1330,3423), WY(1195,3423). **32:**CA(3423), NV(3423), OR(3423), WA(3423). **35:**E states(3423), IA(3423), KS(3021,3423), MI(3423), MO(3423), ND(377,3423), NE(3423), OK(2872,3423), SD(2242), TX(3423), WI(1344). **38:**Range of host(82), MT(82), ND(82),

WA(82).
Cochliobolus sativus. Leaf spot. **18:**ND(1990).
Crocicreas cyathoideum var. *cacaliae*. On dead stems. **10:**CO(505).
C. melanosporum. **38:**UT(505).
Didymosphaeria arenaria var. *macrospora*. **15:**WA(3815).
Epichloe typhina. Cat-tail, choke. **6:**MN(82), ND(82), OK(82), SD(82), WA(82,3423), WY(82). **11:**WA(82,3423). **15:**CA(1944). **30:**AZ(3423), KS(3023), MT(3423), ND(3423), SD(2242). **35:**IA(3423), KS(3021,3023), MN(82,3423), ND(82,1944,3423), NE(2927), OK(82,2873,3423), SD(82,1944,2242,3423), WA(82,3423), WY(82). **38:**MN(82), ND(82), OK(82), SD(82), WA(82), WY(82).
Erysiphe graminis. Powdery mildew. **1:**AK(526). **6:**Range of host(82), IA(3423), MI(3423), MN(3423), MO(3423), MT(3423), NE(3423), PA(3423), WA(3423), WI(3423). **10:**Range of host(82), CA(3423), ID(3423), NV(3423), OR(3423), WA(3423), WY(3423). **11:**MI(82,3423). **15:**Range of host(82), MT(3423), ND(3423), WA(3423). **18:**AK(526), MT(3423), ND(3423), WA(3423). **19:**AK(526). **28:**WA(3423). **29:**ND(82,3423). **30:**Range of host(82), CO(3423), MT(3423), ND(3423), NE(3423), SD(2242,3423), WY(3423). **32:**Range of host(82), MI(3423). **33:**WI(1341,3423). **35:**Range of host(82), MN(3423), MO(3423). **38:**Range of host(82), WA(82).
Euryachora aristidae. Char spot. **10:**MT(82), WY(82).
Gaeumannomyces graminis. Take-all. **15:**ID(82,3423). **33:**OK(3423).
Glomerella sp. Leaf spot. **6:**OK(2872).
Graphyllium planispora. On culms. **38:**UT(959,967).
Leptosphaeria asparagina. **15:**WA(3814). **34:**WA(3814).
L. elongata. **15:**WA(3814).
L. elymi. **6:**ND(82), WI(1360). **33:**WI(1346). **35:**AL(82).
L. spectabilis. **38:**TX(82).
Leptosphaeria sp. **20:**AK(3431).
Lophodermium arundinaceum. On culms. **20:**AK(82,526,3431).
Mycosphaerella tassiana. On dead sheaths and stems. **15:**WA(3815). **20:**AK(1938).
M. tulasnei. On leaves. **30:**ID(3423), ND(3423).
Mycosphaerella sp. On leaves. **10:**ND(82), WA(82). **32:**ND(82), WA(82).
Ophiobolus cariceti. **33:**OK(2872).
Phaeosphaeria culmorum. **6:**ND(377).
P. herpotrichoides. On dead culms. **6:**NM(82). **15:**CA(82), WA(3814). **35:**ND(82). **37:**ND(377).
Phomatospora luzulae. **15:**WA(3815).
Phyllachora cynodontis. Tar spot. **32:**CA(82,3423).
P. graminis. Common tar spot. **6:**Range of host(82), DC(3423), IA(3423), IL(3423), IN(3423), KS(3423), MD(3423), MI(3423), MN(3423), MS(2720,3423), ND(3423), NE(3423), NY(3423), OH(3423), PA(3423), SD(2242,3423), VA(3423), VT(3423), WI(1355,1360,3423), WV(3423). **10:**CA(82,3423). **15:**CA(82,3423), MT(82,3423), NY(82,3423). **24:**IN(654), NY(2456,3423), PA(3423), WI(1338,3423), WV(3423). **30:**NE(82,3423), NY(82), WI(1357). **32:**CA(82,3423). **35:**Range of host(82), AR(3423), DC(3423), GA(1438,3423), IA(3423), IL(3423), IN(3423), KS(3423), KY(3423), MA(3423), MI(3423), MN(3423), MO(3423), MT(3423), NC(3423), ND(3423), NE(3423), NY(3423), OH(3423), PA(3423), SD(2242,3423), VA(3423), WI(3423), WV(3423). **36:**GA(1438). **38:**NY(82), PA(82), WV(82).
Phyllactinia guttata. Powdery mildew. **15:**OR(82).
Pleospora longispora. On dead sheaths and stems. **20:**AK(1938).
Pleospora sp. **20:**AK(3431).
Pyrenophora trichostoma. Leaf spot. **10:**ND(1990). **18:**ND(1987,1990). **30:**ND(1987).
Sphaerulina divergens. On dead culms. **6:**ND(82,377).
Telimena sp. Tar spot. **32:**WA(82).

Basidiomycotina - Uredinales

Phragmidium sp. **30:**CA(2774).
Puccinia coronata. Heteroecious: telial host. **6:**MI(82), WI(82,1333). **11:**MI(82). **16:**SD(2242). **32:**MI(82). **35:**MI(82). **38:**SD(82).
P. graminis. Heteroecious: telial host. **15:**C & W states(82), CA(81), OK(2872), WA(82). **18:**OK(2872), SD(2242). **32:**CA(81). **35:**C & W states(82), KS(3021), OK(2872), SD(2242), TX(742), WA(82). **38:**C & W states(82), ND(82), NY(82), TX(82), WA(82).
P. graminis subsp. *graminicola*. Heteroecious: telial host. **35:**C & W states(82), WA(82).
P. montanensis. Heteroecious: telial host. **6:**C states(82), AZ(82), CA(82), CO(754), IN(82,754), KS(754), KY(82), MN(754), MO(754), ND(378), OH(82), SD(2242), WA(82), WI(754). **8:**ND(754). **15:**C states(82), AZ(82), CA(82), CO(754), WA(82). **22:**SD(2242). **29:**MI(82), ND(82). **30:**C & W states(82), Unk.(754), AZ(1203,3981), ND(82,378), SD(2242), WA(82), WY(1195). **32:**C states(82), AZ(82), CA(82), WA(82). **35:**C states(82), AZ(82), CA(82), KY(82,754), NM(754), SD(2242), WA(82), WV(754). **38:**C states(82), AZ(82), CA(82), WA(82).
P. pattersoniana. Heteroecious: telial host. **10:**UT(82). **32:**NM(82), UT(82).
P. procera. Heteroecious: telial host. **10:**CA(82), OR(82), WA(82), WY(82). **11:**AK(82,526). **20:**CA(82). **34:**AK(526).
P. recondita. Heteroecious: telial host. **10:**Unk.(82), CA(81,638,2439), CO(1195), ID(1195), MT(1195), WY(1195). **12:**WA(82). **20:**Unk.(82), AK(60,526), CA(638). **29:**ND(82), OR(82). **32:**Unk.(82), AZ(1203,3981). **33:**AK(526). **34:**AK(60,526). **36:**GA(1439).
P. striiformis. **5:**CA(82). **6:**W states(82), WA(82). **9:**WA(2341). **10:**W states(82), WA(82). **11:**WA(2341). **12:**WA(82). **15:**W states(82), CA(638,642), WA(82). **29:**WA(82). **30:**W states(82), OR(82), WA(82). **35:**W states(82), WA(82). **38:**W states(82), WA(82).

Basidiomycotina - Ustilaginales

Tilletia controversa. Dwarf bunt. **6:**ID(1465), OR(1465). **29:**OR(1465). **32:**OR(2343).
T. earlei. **30:**ND(1077), SD(82,1077,2242).
T. elymi. **6:**CO(1077), MT(82,1077). **15:**CA(1077), CO(82,1077), ID(1077), MT(82,1077), OR(82,1077), WA(82,1077), WY(82,1077,1195,2924).
T. laevis. **6:**WA(2342).
Urocystis agropyri. Flag and leaf smut. **4:**MO(1077). **6:**Range of host(82), IA(1077), IL(1077), IN(1077), KS(1077,3021,3023), MN(1077), MO(1077), MS(1077,2720), NE(1077), NM(1077), SD(1077,2242), UT(1077), WA(1077), WI(1077). **7:**IA(1077). **10:**CA(82,1077), ID(1077), UT(82,1077), WA(82,1077), WY(82,1077). **15:**CA(638,642,1077), CO(1077), ID(1077), MT(2342), OR(1077), UT(82,1077), WA(82,1077), WY(82,1077,1195). **20:**WA(1077). **24:**OH(946). **26:**DC(1077). **30:**AZ(82,1077), CO(1077), ID(2342), KS(3023), MT(1077), ND(82,1077), NM(82,1077), SD(82,1077,2242), UT(82,1077). **32:**CA(1077), NV(82,1077), OR(1077), UT(82,1077), WA(82,1077), WY(82,1077). **35:**IL(82,1077), IN(1077), KS(82,3021,3023), MO(82,1077), OK(1077,2872), WA(1077), WI(82,1077). **38:**AR(1077), NY(1077), PA(1077), TN(1077), WV(1077).
Ustilago bullata. **6:**ID(1077), MI(82,1077), MT(82,1077), ND(82,1077), OR(1077), WA(82,1077). **12:**WA(1077). **15:**MT(82,1077), ND(82,1077), WA(82,1077). **18:**ID(1077), MT(1077), ND(1077), WA(1077). **27:**NE(1077). **29:**ID(82,1077), WA(82,1077). **38:**MT(82), ND(82), WA(82).
U. hordei. Head smut. **6:**WA(82,1077). **12:**WA(82,1077). **15:**WA(82,1077). **27:**WA(1077).
U. hypodytes. Stem smut. **6:**Unk.(82), IA(1077), MT(82,1077), UT(82,1077), WA(82,1077). **9:**WA(1077). **10:**Unk.(82), CA(1077), MT(82,1077), NV(1077), OR(1077), UT(82,1077), WA(82,1077). **11:**WA(82,1077). **12:**WA(82,1077). **13:**WA(1077). **18:**WA(2342). **23:**WA(1077). **27:**WA(1077). **29:**WA(82,1077).
U. serpens. Stripe smut. **6:**CO(1077), ID(82,1077), MT(82,1077), ND(82,1077), NE(1077), NM(1077), SD(1077,2242), UT(82,1077), WA(82,1077), WI(1358). **7:**UT(1077). **10:**MT(82,1077), ND(82,1077), OR(82,1077), UT(82,1077), WA(82,1077). **15:**CO(1077), MT(2342), WY(1077). **32:**NV(82,1077). **35:**KS(3023).
U. striiformis. Stripe smut. **6:**C states(82), CO(1077), KS(3023), KY(1077), MN(1077), MO(1077), NE(1077), SD(1077,2242), TX(1077), UT(82,1077), WA(82,1077), WI(1077,1342). **10:**UT(1077). **15:**CA(642), OR(82,1077), UT(82,1077), WA(82,1077). **28:**WA(1077). **29:**WA(82,1077). **30:**OR(82), WA(82). **32:**NV(1077). **35:**IA(82,1077), IL(82,1077), IN(82,1077), KS(82,1077), KY(1077), OK(2872), WI(1077). **38:**NM(82).
U. trebouxii. Loose smut. **6:**CA(1077). **10:**CA(1077). **25:**CO(1077).

32:CA(1077), WA(82,1077).

Basidiomycotina - Other

Thanatephorus cucumeris. Brown patch, root rot, sharp eye spot. **3:**ND(3423). **6:**ID(3423), MN(3423), ND(3423), NE(3423), SD(3423), WY(3423). **8:**ND(3423). **11:**ND(3423). **15:**OR(3423), WA(3423). **17:**ND(3423). **18:**ND(3423), SD(3423). **21:**ND(3423). **29:**ND(3423). **30:**MT(3423), ND(3423), NE(3423), OR(3423), SD(3423), WY(3423).

Typhula idahoensis. Snow mold. **30:**WA(3423).

T. incarnata. Snow mold. **15:**ID(82,3423).

Deuteromycotina - Hyphomycetes

Acremonium typhinum. **6:**TX(3872,3873).

Bipolaris sorokiniana. Seedling blight, spot blotch, black point, root rot. **6:**MN(3423), MN(82), ND(82,3423), NE(3423), SD(82,2242,3423), WI(1338,3423). **7:**ND(3423). **8:**ND(3423). **11:**ND(82,3423). **12:**ND(82). **13:**ND(3423). **15:**MN(82), ND(82), SD(82). **17:**ND(3423). **18:**ND(3423), SD(2242,3423). **23:**ND(3423). **27:**ND(3423), WA(3423). **29:**ND(82,3423). **30:**CO(3423), MN(82,3423), MT(82,3423), ND(82,3423), NE(82,3423), SD(82,2242,3423), WY(82,3423). **35:**MN(82,3423), ND(82), SD(82). **38:**MN(82), ND(82), SD(82).

Cercosporidium graminis. Leaf streak, brown stripe. **1:**AK(526,3431). **3:**ND(3423). **4:**OR(3423). **6:**IA(3423), ID(3423), KS(3021,3023,3423), MN(3423), MT(3423), ND(3423), NE(3423), NY(3423), SD(2242,3423), UT(3423), WA(3423), WI(3423), WY(3423). **10:**CO(3423), ID(3423), MT(3423), UT(3423), WA(3423). **14:**MT(3423), WA(3423). **15:**CA(3423), CO(3423), ID(3423), KS(3423), MT(3442), OR(3423), WA(3423), WY(2924,3423). **19:**AK(526). **20:**AK(526,3423). **29:**WA(3423). **30:**NC states(82), NW states(82), Range of host(3423), ND(3423), NE(3423), OR(3423), SD(2242,3423), WY(3423). **32:**CA(3423), OR(3423). **35:**KS(3021,3023,3423). **38:**Range of host(82).

Cladosporium herbarum. On leaves. **1:**AK(3431). **30:**ND(82), SD(82). **38:**IA(82), ND(82), OR(82).

C. phlei. **18:**AK(526).

Curvularia geniculata. Secondary root rot. **30:**SD(2242). **38:**ND(82).

Drechslera gigantea. Zonate leaf spot. **35:**MD(82,3423), PA(82,3423), WA(82).

D. tritici-repentis. Leaf blight and blotch, leaf spot. **15:**ID(3442), OR(3432), WA(82,3423). **16:**AK(526). **18:**ND(3423). **30:**MN(82,3423), MT(82,3423), ND(82,3423), NY(82), SD(82,2242,3423), WY(82,3423). **33:**OR(3423). **38:**WA(82).

Fusarium acuminatum. Secondary root rot. **2:**ND(3423). **6:**MN(3423), MT(3423), ND(3423), NE(3423), SD(2242,3423), WA(3423), WY(3423). **7:**ND(3423). **8:**ND(3423). **11:**ND(82,3423). **12:**ND(3423). **13:**ND(3423). **15:**ND(82,3423). **17:**ND(3423). **18:**ND(3423), NE(3423). **21:**ND(3423). **29:**ND(3423). **30:**CO(3423), MN(3423), MT(3423), ND(3423), NE(3423), SD(2242,3423), WY(3423). **32:**ND(82,3423). **38:**ND(82).

F. avenaceum. Stem rot. **15:**WA(3423).

F. culmorum. Head mold, root rot, scab. **10:**ID(82,3423). **30:**MN(82,3423).

F. equiseti. Culm and root rot. **11:**ND(82). **29:**ND(82). **30:**MN(82), MT(82), ND(82), NE(82), SD(82), WY(82).

F. heterosporum. Head mold. **35:**MO(82).

F. moniliforme. Root and culm rot. **30:**NE(82,3423).

F. oxysporum. Culm and root rot. **11:**ND(82). **29:**ND(82). **30:**ND(82).

F. poae. Secondary root rot, silver top. **18:**ND(3423).

F. sporotrichioides. Culm and root rot. **30:**SD(82). **38:**SD(82).

Fusarium sp. Secondary root rot. **11:**Unk.(82).

Helminthosporium sp. Leaf spot. **6:**MN(82). **10:**MI(82).

Heterosporium avenae. Leaf spot. **10:**ID(3423,3433), OR(82,3423), WA(82,3423), WY(82,3423).

Hymenula cerealis. Stripe. **15:**WA(415).

Microdochium bolleyi. Root necrosis, seedling root rot. **2:**ND(3423). **3:**ND(3423). **6:**ND(82,3423), NE(3423), SD(2242,3423), WY(3423). **10:**ND(3423), WY(3423). **11:**ND(3423). **12:**ND(3423). **13:**ND(3423). **15:**ND(3423). **17:**ND(3423). **18:**ND(3423). **21:**ND(3423). **27:**ND(3423). **29:**ND(3423). **30:**MT(3423), ND(3423), NE(3423), SD(2242,3423), WY(3423).

M. nivale. Snow mold, secondary root rot. **6:**WA(82,3423). **15:**WA(82,3423). **27:**WA(3423). **29:**WA(82,3423). **30:**WA(82,3423).

Muirella alascensis. Leaf spot. **15:**WA(3432,3435).

Periconia circinata. Root rot. **11:**ND(3423). **18:**ND(3423). **27:**ND(3423).

Pyricularia parasitica. Tar spot. **35:**WI(3423).

Ramularia pusilla. Leaf spot. **15:**MT(3429), WA(3429,3432).

Rhynchosporium orthosporum. Leaf spot, scald, white leaf stripe. **15:**ID(3442), OR(82,3423,3424). **33:**AK(3441).

R. secalis. Leaf spot, scald. **6:**WA(82,3423), WI(3423). **10:**ID(82), OR(82), WA(82). **15:**ID(82,3423), MT(3433), OR(82,3423), WA(82,3423), WY(82,3423). **16:**SD(2242,3423). **18:**WA(3423). **27:**WA(3423). **32:**WA(82,3423). **35:**IN(3423). **38:**SD(82), WA(82).

Deuteromycotina - Coelomycetes

[*Ascochyta agropyrina*]. Leaf rot and spot. **6:**ND(82,3423,3449). **12:**ND(82), WA(82). **27:**ND(3423,3449), WA(3423,3449). **30:**MT(82,3423,3449), UT(82). **35:**ND(82,3423,3449), WV(953). **38:**ND(82).

A. graminicola. **10:**WA(82). **35:**TX(82), WI(1343). **38:**MT(82), ND(82).

A. leptospora. Leaf spot. **6:**ND(82), OK(2872). **35:**ND(82). **38:**ND(82).

A. sorghi. Leaf spot. **1:**AK(526). **6:**MO(3423,3449), MT(3449), ND(3423,3449), OK(3423), WI(3423). **10:**ID(3423), MT(3423), WA(3423,3449). **15:**ID(82,3423). **30:**CO(3423,3449). **35:**IL(3423), TX(3423).

A. utahensis. Leaf spot. **10:**OR(3425). **32:**CO(3429).

Ascochyta sp. On dead lower leaves. **30:**WI(1357).

Cheilaria agrostis. Blotch spot, char spot. **10:**CA(82,3423), CO(3423), ID(82,3423), MT(82,3423), OR(82,3423), UT(3423), WA(82,3423), WY(82,3423). **15:**ID(82,3423), MT(3442), UT(82,3423), WA(82,3423), WY(82,3423). **32:**CA(82).

Colletotrichum graminicola. Anthracnose. **6:**WI(82,3423). **18:**OK(2872,3423). **20:**AK(3431). **30:**NE(2927), OK(82,2872,3423), WI(1354). **33:**WI(1356). **35:**MS(82,2720), OK(2872,3423). **38:**WI(1330).

Davisiella elymina. Tar spot. **6:**WI(3423). **33:**WI(3423). **35:**WI(3423).

Dilophospora alopecuri. Twist disease. **15:**WY(82,3423). **30:**ID(3423), UT(82). **32:**NV(82,3423).

Hendersonia agropyri-repentis. **15:**WA(82).

Hendersonia sp. Leaf mold. **15:**OR(82).

Leptothyrium petrakii. **10:**WY(1195).

Macrophoma phlei. Leaf spot. **10:**NV(3432). **32:**ND(82,3423).

Macrophomina phaseolina. Root rot. **38:**ND(82).

Phaeoseptoria elymi. Leaf spot. **35:**NE(82).

P. festucae var. *muhlenbergiae*. **6:**WI(1345). **35:**WI(82,1335).

Phloeospora graminearum. Leaf spot. **6:**MI(3423).

Phoma terrestris. Pink root. **6:**ND(3423). **8:**ND(3423). **11:**ND(3423). **17:**ND(3423). **18:**ND(3423). **23:**ND(3423). **27:**ND(3423). **29:**ND(3423). **30:**MT(3423), ND(3423), NE(3423), SD(3423), WY(3423). **38:**ND(82), NE(82), SD(82).

Phoma sp. On culms. **30:**ND(82), UT(82).

Phyllosticta sp. Leaf spot. **10:**OR(82).

Pseudoseptoria donacis. Stem speckle, eye spot. **4:**WA(3423,3450). **6:**ND(3423,3450). **10:**CA(82,3423,3450), MT(82,3423,3450), ND(82,3423,3450), NV(3423,3450), OR(82,3423,3450), WA(82,3423,3450). **14:**WA(3423,3450). **15:**CA(642), OR(3450), WA(2695,3450), WY(82,2924,3450). **18:**ND(3423,3450). **20:**AK(3441). **23:**ND(3423,3450). **30:**AZ(3423,3450), MT(3423), ND(3423,3450), NE(3423,3450), SD(2242,3423,3450), WY(3423,3450). **33:**ND(3423,3450). **35:**ND(3423). **38:**WA(82).

P. obtusa. Eye spot, leaf spot. **10:**MT(3423), ND(3423,3450), OR(3423,3450), WA(82,3423,3450). **15:**CA(82,2695,3423,3450), MT(3433), OR(82,3423,3450), UT(3435), WA(82,3423,3450), WY(3423,3450).

P. stomaticola. Stem speckle, leaf spot. **6:**ND(82). **10:**NV(82). **15:**ID(3432), OR(82), WA(82). **30:**W states(82), AZ(82). **35:**ND(82), WI(1348). **38:**ND(82), OR(82), WA(82).

Pyrenochaeta elymi. On dead leaf tips. **6:**UT(82).

Septoria agropyrina. Blotch, leaf spot. **6:**IA(82), MI(82), ND(82,3423), SD(82,2242,3423). **10:**ND(3423). **15:**OR(3423), WA(82,3423). **18:**ND(3423). **20:**AK(3426). **27:**ND(3423). **30:**MT(82,3423), ND(82,3423), NE(82,3423), SD(2242,3423), WY(3423). **31:**ND(3423). **32:**CA(3423). **35:**MS(3423), ND(3423), WI(1345). **38:**ID(82), ND(82).

S. elymi. Speckled leaf blotch, leaf spot. **6:**IA(82), MI(82), MN(82,3423), MS(2720), ND(82,3423), OK(2872,3423), SD(82,2242,3423), WI(82,1332,3423). **7:**IA(3423), MI(3423), WI(3423). **15:**MI(82,3423), OR(82,3423), WA(82,3423). **30:**NC states(82), NW states(82), MI(3423), ND(3423), NE(3423), SD(2242,3423), UT(3423), WI(3423). **32:**CA(82,3423). **33:**WA(3433). **35:**MS(82,2720), WI(82,3423).
S. elymi f. *elymina*. **30:**WI(1341).
S. infuscans. Brown leaf blotch, leaf spot. **6:**NM(3425,3435). **15:**W states(82), AZ(1194). **20:**W states(82). **32:**W states(82), AZ(1194), CA(638).
S. pacifica. Leaf blotch, leaf spot. **10:**CA(82,3423). **15:**ND(82). **20:**AK(3426,3431), OR(82,3423).
S. passerinii. **6:**WI(1357). **33:**WI(1358).
S. spraguei. Leaf spot. **18:**ND(1990), SD(3704). **32:**ND(1990).
Stagonospora arenaria. Purple-brown blotch. **6:**C states(82), KY(3423), MI(3423), ND(82,3423). **8:**ND(3423). **11:**ND(82,3423). **15:**OR(82,3423), WA(82,3423). **18:**AK(526). **20:**AK(82,526,3423). **29:**ND(82). **30:**MI(82), ND(82,3423), SD(82). **32:**CA(82,3423). **35:**MI(3423). **38:**ND(82).
S. avenae. Blotch. **18:**ND(3423).
S. avenae f. sp. *triticea*. Blotch. **3:**WA(3423). **12:**ND(3423). **15:**AZ(1194), ID(82). **29:**ND(82,3423). **30:**MT(3423), ND(3423).
S. culmicola. Leaf rot. **6:**PA(82), TX(82). **15:**OR(3423), WA(3423). **27:**ND(3423).
S. nodorum. Glume blotch, leaf spot. **6:**MI(3423). **12:**ND(82,3423). **15:**WA(82,3423), WY(2924,3425). **18:**ND(3423). **29:**ND(82,3423). **30:**ND(3423), SD(2242). **35:**WV(953).
S. simplicior. **1:**AK(3431).
S. subseriata. Leaf rot. **20:**OR(82,3423), WA(3424).
Stigmopeltis graminicola. **15:**WA(3816).
Tiarospora perforans. Leaf speckle. **20:**OR(82).
Wojnowicia hirta. Leaf mold, secondary foot rot, rot of mature straw. **2:**ND(3423). **30:**KS(82), MT(82,3423), ND(82,3423), WA(82), WY(3423). **38:**ND(82).

Deuteromycotina - Other

Rhizoctonia solani. Secondary root rot, on culms and basal leaf sheaths. **6:**ID(82), ND(82), NE(82), SD(2242). **15:**OR(82). **18:**SD(2242). **20:**AK(3441). **30:**NGP(82), Unk.(82), SD(2242). **38:**ND(82).

×*Elysitanion* Bowden

Elymus × *Sitanion*.

1. × *E. hansenii* (Lams.-Scribn.) Bowden (=*Elymus hansenii* Lams.-Scribn., *Sitanion hansenii* (Lams.-Scribn.) J. G. Sm.). *Elymus glaucus* × *Sitanion* spp.

Ascomycotina

Erysiphe graminis. Powdery mildew. **1:**CA(3423).

Basidiomycotina - Uredinales

Puccinia pattersoniana. **1:**CA(642).
P. striiformis. **1:**CA(638,642).

Basidiomycotina - Ustilaginales

Ustilago trebouxii. **1:**OR(1077), WA(1077).

Deuteromycotina - Hyphomycetes

Cercosporidium graminis. Leaf streak. **1:**CA(3423), WA(3423).

Deuteromycotina - Coelomycetes

Pseudoseptoria donacis. On leaves. **1:**CA(642).
P. obtusa. Eye spot. **1:**CA(3423,3450), OR(3450), WA(3423,3450).
P. stomaticola. **1:**WA(3450).

Enneapogon Desv. ex P. Beauv.

1. *E. desvauxii* P. Beauv.

Basidiomycotina - Ustilaginales

Ustilago enneapogonis. **1:**AZ(1077).

Eragrostis Wolf

1. *E. capillaris* (L.) Nees [lacegrass]
2. *E. cilianensis* (All.) Link ex Vign. (=*E. major* Host) [stinkgrass]
3. *E. curvula* (Schrad.) Nees [Boer lovegrass, weeping lovegrass]
4. *E. glomerata* (Walter) L. Dewey
5. *E. hirsuta* (Michx.) Nees
6. *E. hypnoides* (Lam.) B. S. P. (=*E. reptans* (Michx.) Nees)
7. *E. intermedia* Hitchc. [plains lovegrass]
8. *E. lehmanniana* Nees [Lehmann lovegrass]
9. *E. lugens* Nees
10. *E. mexicana* (Hornem.) Link [Mexican lovegrass]
11. *E. neomexicana* Vasey
12. *E. palmeri* S. Wats.
13. *E. pectinacea* (Michx.) Nees (=*E. diffusa* Buckley) [tufted lovegrass]
14. *E. pilosa* (L.) P. Beauv. [India lovegrass]
15. *E. poaeoides* P. Beauv. ex Roem. & Schult. [little lovegrass]
16. *E. refracta* (Muhl.) Lams.-Scribn.
17. *E. secundiflora* J. Presl & K. Presl
18. *E. spectabilis* (Pursh) Steud. [purple lovegrass]
19. *E. tenella* (L.) P. Beauv. ex Roem. & Schult.
20. *E. tremula* Hochst. (=*E. rachitricha* Hochst. ex Miq.)
21. *E. trichodes* (Nutt.) A. Wood [sand lovegrass]
22. *E. variabilis* (Gaud.) Steud.
23. *Eragrostis* sp.

Oomycetes

Pythium graminicola. **2:**ND(3423). **3:**ND(3423).
P. ultimum. **10:**AZ(3437).
Sclerophthora macrospora. Crazy top. **2:**IN(3708), SD(2242). **13:**IN(3708), SD(2242).

Ascomycotina

Balansia epichloe. Black choke, sterility disease, black ring. **1:**AL(82), FL(82), GA(82), SC(82,3423), VA(82). **5:**AL(82), FL(82), GA(82,1438,3423), SC(82), VA(82). **9:**GA(1438). **16:**FL(82,3423), GA(82,1438,3423), LA(82), SC(82), VA(82,3423). **23:**FL(35,3423), SC(3423).
Phyllachora eragrostidis. Tar spot. **1:**AL(82,3423), GA(82), NE(82,3423), NM(82), OK(82), TX(82,3423). **5:**AL(82), GA(82,1438,3423), NE(82), NM(82), OK(82), TX(82). **9:**OK(2873,3423). **12:**TX(3423). **21:**OK(660). **22:**HI(3423). **23:**NM(3423).
P. graminis. Leaf spot. **22:**HI(2903).
Phyllachora sp. **5:**GA(1438).
Venturia nebulosa. On dead leaves. **23:**AL(82).

Basidiomycotina - Uredinales

Uromyces eragrostidis. Heteroecious: telial host. **1:**E states(82), AZ(82), NE(82). **5:**GA(1439). **7:**AZ(741,747,1195,1203,3981), TX(742). **8:**AZ(1203,3981). **9:**OK(2873). **13:**E states(82), AZ(82), MS(2720), NE(82), OK(2872). **18:**OK(2872).

Basidiomycotina - Ustilaginales

Sporisorium montaniensis. Inflorescence smut. **10:**AZ(82,1077). **11:**AZ(1077). **13:**AZ(82,1077).
Tilletia eragrostidis. **4:**MS(82,1077,2720).
Ustilago spermophora. Seed smut. **2:**E states(82), CT(1077), DE(1077), IA(1077), IL(1077), IN(1077), KS(1077), LA(82,1077), MA(1077), MD(1077), MS(1077,2720), NC(1077,1303), ND(82,1077), NE(1077), NY(1077), SC(1077), SD(1077), WI(1077). **6:**DC(1077), SD(1077,2242), WI(1077). **14:**IA(1077). **15:**IA(82,1077), IL(82,1077). **23:**MD(82), SD(82), WI(82).

Basidiomycotina - Other

Thanatephorus cucumeris. Brown patch, root rot, sharp eye spot. **3:**ND(3423). **21:**ND(3423).

Deuteromycotina - Hyphomycetes

Alternaria sp. Leaf spot. **23:**FL(35).
Bipolaris hadrotrichoides. On roots and leaves. **2:**DE(82,128,3423,3543), KS(3023), MD(3423), MT(82,3423), ND(82,3423), VA(3423), WI(1352). **13:**WI(1352).
B. kusanoi. **14:**KS(3023).
B. nodulosa. Leaf spot. **2:**DE(82), MD(82), VA(82).
B. ravenelii. On glumes. **14:**MS(82,2720).
B. sorokiniana. Seedling blight, spot blotch, black point, root rot. **2:**ND(3423). **3:**ND(82,3423). **14:**ND(82,3423).
Cercospora clemensiae. **19:**HI(2903).
Cercosporidium graminis. Leaf streak, brown stripe. **17:**OK(2873,3423). **23:**OK(82).
Curvularia geniculata. Leaf mold and spot. **18:**WI(1352). **20:**ID(3423), MN(3423), MT(3423), ND(3423), NE(3423), SD(3423), WA(3423), WY(3423). **23:**IA(82), MS(82,2720).
C. lunata. **10:**AZ(3437).
Drechslera gigantea. Zonate leaf spot. **3:**MD(82,3423), VA(3423). **13:**MD(3423), VA(3423). **23:**MD(82), TX(82).
Exserohilum rostratum. Leaf spot. **2:**DC(3423), IA(82), KY(82), MD(82), WI(82,3423). **13:**WI(1335,3423). **18:**WI(1339,3423). **23:**WI(82).

Fusarium acuminatum. Secondary root rot. **3**:ND(82,3423). **14**:ND(82,3423). **21**:ND(3423).
Helminthosporium sp. Leaf spot. **23**:FL(35).
Microdochium bolleyi. Root necrosis. **3**:ND(3423).
Periconia circinata. Root rot. **2**:ND(3423).
Pyricularia grisea. Blast, gray leaf spot. **9**:OK(2873,3423). **23**:OK(82).
Deuteromycotina - Coelomycetes
Colletotrichum caudatum. Leaf spot. **17**:OK(2872,3423). **18**:OK(2872,3423).
C. graminicola. Anthracnose. **3**:OK(82,2873,3423). **21**:OK(2872,3423).
Gloeosporium eragrostidis. On living leaves. **18**:WI(1349).
Phoma terrestris. Pink root. **14**:ND(82).
Deuteromycotina - Other
Rhizoctonia solani. Secondary root rot. **3**:ND(82). **10**:AZ(3437).

Eremochloa Buse in Miq.

1. *E. ophiuroides* (Munro) Hack. [centipede grass, lazy-man's grass]

Myxomycetes
Physarum cinereum. Slime mold. **1**:NC(1303).
Oomycetes
Pythium debaryanum. Blight. **1**:FL(35).
Pythium sp. Root rot, greasy spot. **1**:HI(2903), NC(1303).
Ascomycotina
Myriogenospora atramentosa. On leaves. **1**:LA(1608).
Sclerotinia homoeocarpa. 'Dollar spot.' **1**:FL(35).
Stictis pustulata. On dead culms. **1**:AL(82).
Basidiomycotina - Other
Lepiota sp. Fairy ring. **1**:FL(35).
Marasmius sp. Fairy ring. **1**:FL(35).
Deuteromycotina - Hyphomycetes
Alternaria sp. Leaf spot. **1**:FL(35).
Curvularia geniculata. **1**:OK(2872).
C. lunata. Leaf spot, fading out. **1**:FL(35), HI(2903).
Curvularia sp. Leaf mold, fading out. **1**:LA(82), NC(1303).
Fusarium roseum. Blight. **1**:AL(3532).
Pleurophragmium flumeanum. Leaf blight. **1**:FL(35).
Pyricularia grisea. Gray leaf spot. **1**:FL(35), NC(1303).
P. oryzae. **1**:FL(1790).
Stachybotrys sp. Leaf mold. **1**:MD(82).
Deuteromycotina - Coelomycetes
Colletotrichum falcatum. Leaf spot. **1**:HI(2903).
C. graminicola. 'Anthracnose.' **1**:FL(35,82,3423), NC(1303), OK(2872).
Deuteromycotina - Other
Rhizoctonia solani. Brown patch. **1**:FL(35), NC(1303).
Rhizoctonia sp. Root rot. **1**:HI(2903).

Eremopyrum (Ledeb.) Jaub. & Spach

1. *E. buonapartis* (Spreng.) Nevski (=*Agropyron buonapartis* (Spreng.) Th. Durand & Schinz.)
2. *E. distans* (K. Koch) Nevski (=*Agropyron orientale* (L.) Roem. & Schult. var. *lasianthum* (Boiss.) Boiss.)
3. *E. orientale* (L.) Jaub. & Spach (=*Agropyron orientale* (L.) Roem. & Schult.)

Oomycetes
Pythium graminicola. **2**:WA(3423).
Basidiomycotina - Uredinales
Puccinia montanensis. **3**:SD(2242).
Basidiomycotina - Ustilaginales
Ustilago turcomanica. **2**:WA(1077,1079).
Deuteromycotina - Hyphomycetes
Bipolaris sorokiniana. Seedling blight, spot blotch, black point, root rot. **1**:ND(3423).

Erianthus Michx.

This genus is considered a synonym of *Saccharum*, but the appropriate combinations could not be located.

1. *E. brevibarbis* Michx. [brown plumegrass]
2. *E. contortus* Baldw. ex Elliott [bent-awn plumegrass]
3. *E. giganteus* (Walter) Muhl. [sugarcane plumegrass]
4. *E. hostii* Griseb.

Ascomycotina
Acrospermum graminum. **1**:GA(1438).
Apiospora montagnei. **1**:GA(1438).
Dasyscyphus distinctus. **1**:GA(1438).
Glomerella tucumanensis. Anthracnose, red rot, leaf spot. **3**:LA(3423), MS(3423).
Lasiosphaeria sp. **1**:GA(1438).
Leptosphaeria orthogramma. **1**:GA(1438).
Nimbomollisia aberrans. **1**:GA(1438).
Phaeosphaeria culmorum. **1**:GA(1438).
Phyllachora dimeriae. Tar spot. **1**:FL(3423), GA(1438,3423). **2**:GA(1438,3423). **3**:AL(82,3423), FL(82,3423), GA(82), SC(82).
Deuteromycotina - Hyphomycetes
Cerebella andropogonis. **2**:MS(2720).
Helminthosporium sp. **4**:FL(3423).
Deuteromycotina - Coelomycetes
Colletotrichum falcatum. Anthracnose. **3**:LA(82).

Eriochloa Kunth in Humb., Bonpl. & Kunth

1. *E. lemmonii* Vasey & Lams.-Scribn. var. *gracilis* (E. Fourn.) Gould (=*E. gracilis* (E. Fourn.) Hitchc.)
2. *E. michauxii* (Poir.) Hitchc.
3. *E. polystachya* H. B. K. (=*E. acuminata* (K. Presl) Kunth) [caribgrass, malojilla]
4. *E. punctata* (L.) Desv.
5. *Eriochloa* sp.

Basidiomycotina - Uredinales
Uromyces setariae-italicae. Heteroecious: telial host. **1**:AZ(747). **2**:Unk.(2910). **3**:Unk.(82).
Basidiomycotina - Ustilaginales
Sorosporium mixtum. Seed smut. **1**:AZ(1077), CA(1077). **3**:AZ(1077). **4**:AZ(1077). **5**:AZ(82,922), CA(82).
Deuteromycotina - Hyphomycetes
Helminthosporium sp. Leaf spot. **3**:FL(35,82,3423).

Erioneuron Nash in Small

1. *E. pulchellum* (H. B. K.) Tateoka (=*Triodia pulchella* H. B. K., *Tridens pulchellus* (H. B. K.) Hitchc.)

Basidiomycotina - Ustilaginales
Ustilago buchloes. Leaf smut. **1**:AZ(82,1077), NM(82,1077).
Deuteromycotina - Hyphomycetes
Curvularia lunata. **1**:AZ(3437).
Fusarium acuminatum. **1**:AZ(3437).

Festuca L.

1. *F. altaica* Trin.
2. *F. arizonica* Vasey [Arizona fescue]
3. *F. arundinacea* Schreb. (=*F. elatior* L. var. *arundinacea* (Schreb.) Wimm.) [tall fescue]
4. *F. capillata* Lam. [hair fescue, slender fescue]
5. *F. idahoensis* Elmer [bluebunch fescue, Idaho fescue]
6. *F. mairei* Saint-Yves
7. *F. mollis* Kunth (=*Bromus lanatus* H. B. K.)
8. *F. obtusa* Biehler [nodding fescue]
9. *F. occidentalis* Hook. [western fescue]
10. *F. ovina* L. (=*F. brachyphylla* Schult., *F. ovina* L. var. *brachyphylla* (Schult.) Hitchc.) [sheep fescue]
11. *F. ovina* L. var. *duriuscula* (L.) W. Koch [hard fescue]
12. *F. paradoxa* Desv. (=*F. nutans* Spreng.)
13. *F. pratensis* Huds. (=*F. elatior* L.) [English bluegrass, meadow fescue]
14. *F. rubra* L. (=*F. rubra* L. subsp. *rubra* , *F. rubra* L. var. *commutata* Gaudin, *F. rubra* L. var. *lanuginosa* Mert. & W. Koch, *F. rubra* L. var. *prolifera* Piper) [chewing festuca, red fescue]
15. *F. rubra* L. var. *heterophylla* (Lam.) Mutel [shade fescue]
16. *F. scabrella* Torr. ex Hook. [rough fescue]
17. *F. sororia* Piper
18. *F. subulata* Trin.
19. *F. subuliflora* Lams.-Scribn.
20. *F. thurberi* Vasey [Thurber fescue]
21. *F. versuta* Beal
22. *F. viridula* Vasey [green-leaf fescue]
23. *Festuca* sp.

Myxomycetes
Fuligo septica. Slime mold. **13**:NC(1303).
Physarum cinereum. Slime mold. **13**:NC(1303).
Chytridiomycetes
Synchytrium graminicola. Purple leaf speckle. **14**:OR(3423).

Synchytrium sp. Leaf speckle. **14:**OR(82).

Oomycetes

Aphanomyces camptostylus. Root necrosis. **13:**ND(3423). **14:**ND(3423).

Aphanomyces sp. Secondary root rot. **13:**ND(82). **14:**ND(82).

Pythium arrhenomanes. Root browning. **13:**ND(82). **14:**ND(82).

P. debaryanum. Damping-off, root browning and necrosis, seed rot. **3:**MT(3423), ND(3423), NE(3423). **10:**WA(3423). **13:**ND(82,3423), WA(3423), WY(3423). **14:**ND(82,3423).

P. graminicola. Root necrosis. **3:**'Pythium blight. Pythium root rot.' MT(3423), ND(3423). **5:**WA(82,3423). **13:**ND(3423), WY(3423).

P. irregulare. Root browning, necrosis. **3:**MT(3423). **13:**ND(82,3423).

Pythium sp. Root browning, damping-off. **13:**Unk.(82), NC(1303).

Sclerophthora macrospora. Crazy top. **3:**'Downy mildew.' IN(1158). **13:**AR(3594).

Ascomycotina

Chaetomium sp. **13:**GA(1438).

Claviceps purpurea. Ergot. **3:**KY(3423), ND(3423), OR(3423), WA(3423). **5:**ID(82,3423), MT(3423), ND(3423), OR(82,3423). **8:**WI(1346). **10:**MD(82), MO(82,3423), WI(1343,3423). **13:**E states(3423), Range of host(82), FL(35), NC(1303), ND(3423), OK(2873,3423), OR(3423), WA(3423), WI(1347). **14:**E states(82), AK(526), KY(3423), MN(82,3423), ND(3423), NY(3423), OR(3423), SD(2242), WA(3423). **23:**CO(82), UT(82), WY(82).

Cochliobolus sativus. Blind seed disease. **13:**OR(3423).

Didymella festucae. On culms. **5:**OR(681).

Epichloe typhina. Choke. **3:**MO(3078). **12:**IN(1944). **14:**PA(3423).

Erysiphe graminis. Powdery mildew. **5:**WA(82,3423). **10:**WA(82,3423). **13:**NC(1303). **14:**MI(82,3423), OR(82,3423). **15:**MI(82), OR(82).

Gaeumannomyces graminis. Root rot, take-all. **13:**AR(82,3423). **18:**WA(3446,3754). **23:**AR(82), OR(82).

G. graminis var. *avenae*. Patch. **13:**NC(1303).

Gloeotinia granigena. **13:**OR(1075).

Leptosphaeria asparagina. **18:**WA(3814).

L. ophiopogonis f. *graminum*. **14:**AK(526).

Lewia scrophulariae. On dead stem. **10:**AK(1938).

Lophodermium alpinum. On leaves. **5:**WY(203,1195). **23:**CO(82).

L. arundinaceum. On dead culms and leaves. **2:**AZ(82,3423). **10:**AK(3431), CA(82,3423), CO(3423), MT(3423), WA(3423), WY(3423). **13:**CA(82).

Mycosphaerella muhlenbergiae. **23:**WA(3815).

M. tassiana. On leaves. **23:**WY(82).

M. tulasnei. Leaf mold. **5:**W states(3423), W states(3423), WA(82). **13:**OR(82). **14:**AK(3431), OR(82). **23:**OR(82).

Mycosphaerella sp. On leaves. **23:**OR(82).

Ophiobolus sp. **14:**WA(1290).

Phaeosphaeria caricinella. **14:**WA(3814).

P. herpotrichoides. On dead culms. **14:**WA(3814). **22:**WA(3814). **23:**OR(82).

P. typharum. **22:**WA(3814). **23:**WA(3814).

Phyllachora cynodontis. **19:**CA(81).

P. graminis. Tar spot. **13:**ME(82,3423), RI(82).

P. sylvatica. Tar spot. **5:**CA(81,82,3423). **9:**OR(3423). **14:**CA(82,3423), OR(82,3423). **23:**OR(82).

Pleospora rainierensis. **14:**WA(3814).

Pyrenopeziza karstenii. **10:**AK(3431).

Sclerotinia homoeocarpa. Turf rot, dollar spot, leaf spot. **10:**MA(3423). **11:**IA(1588). **13:**NC(1303). **14:**IA(1588), MA(3423).

Trichothyrina alpestris. **10:**AK(3431).

Basidiomycotina - Uredinales

Puccinia brachypodii var. *arrhenatheri*. Heteroecious: telial host. **10:**Unk.(754).

P. brachypodii var. *poae-nemoralis*. Heteroecious: telial host. **10:**AZ(1195). **13:**Unk.(754), OR(82).

P. cockerelliana. Heteroecious: telial host. **14:**AK(526). **23:**CO(82).

P. coronata. Heteroecious: telial host. **10:**NY(82). **13:**Range of host(82), FL(35), GA(1439). **18:**OR(82).

P. crandallii. Heteroecious: telial host. **5:**W states(82), CA(82), ID(638), MT(1195), WA(82). **10:**CO(82), MT(1195), ND(82,378), NV(82), OR(82), WA(82), WY(82). **14:**CO(82), ND(82), NV(82), OR(82), WA(82), WY(82). **16:**MT(1195). **22:**WA(82). **23:**W states(82), CA(82), WA(82).

P. festucae. Heteroecious: telial host. **1:**AK(526).

P. graminis. Heteroecious: telial host. **2:**CO(82). **5:**MI(82), OR(82), WA(82). **13:**Range of host(82), FL(35). **14:**CA(82), WA(82). **23:**CA(81,82), CO(82), MI(82), MN(82), OR(82), WA(82), WY(82).

P. graminis subsp. *graminicola*. Heteroecious: telial host. **2:**CO(82). **5:**MI(82), OR(82), WA(82). **13:**Range of host(82). **14:**Range of host(82). **23:**CA(82), CO(82), WY(82).

P. piperi. **13:**UT(82). **23:**CA(82).

P. poarum. Heteroecious: telial host. **2:**AZ(741,1203,3981). **17:**AZ(1203,3981).

P. pygmaea. Heteroecious: telial host. **5:**OR(754). **14:**NY(2790). **19:**CA(754).

P. recondita. Heteroecious: telial host. **3:**'Leaf rust.' AL(759). **13:**ND(82), VT(82). **20:**CO(1195). **23:**ND(82).

Puccinia sp. **13:**NC(1303). **14:**NC(1303).

Uromyces dactylidis. Heteroecious: telial host. **10:**CO(1195).

Basidiomycotina - Ustilaginales

Tilletia bromi. Seed smut. **23:**CA(82), WA(82).

T. caries. **14:**OR(1077).

T. controversa. Dwarf bunt. **5:**OR(2343). **13:**OR(1465). **14:**OR(1462).

T. texana. Seed smut. **23:**OR(82).

Urocystis agropyri. **5:**MT(2342). **14:**CA(1077).

Ustilago bullata. **5:**WA(82,1077). **22:**CO(1077).

U. mulfordiana. Head smut. **23:**CA(1077), ID(82), OR(82), WA(82).

U. striiformis. Stripe smut. **5:**CO(1077), MT(1077), OR(1077). **8:**DC(1077), IA(82), MD(82,1077), VA(1077). **9:**MT(2342), WA(1077). **10:**CO(1077), ND(82,1077), WY(1077,2342). **13:**IL(1077), NC(1303). **20:**NM(1077).

Ustilago sp. Seed smut. **22:**CO(82).

Basidiomycotina - Other

Laetisaria fuciformis. Pink patch, red thread. **11:**RI(3423). **23:**OR(82).

Marasmius sp. Collar rot. **4:**Unk.(82). **13:**NC(1303).

Thanatephorus cucumeris. Brown patch, root rot, sharp eye spot. **3:**GA(1439), LA(3524), MD(3423). **5:**OR(3423), WA(3423). **10:**ID(3423), MA(3423), MD(3423), ND(3423), OH(3423), RI(3423), WA(3423). **13:**MT(3423), OR(3423). **14:**CT(3423), DC(3423), ID(3423), IN(3423), KY(3423), MA(3423), MD(3423), MI(3423), MO(3423), ND(3423), NY(3423), OH(3423), OR(3423), PA(3423), RI(3423), VA(3423), WA(3423), WI(3423).

Thelephora sp. Fruiting structure; nonparasitic. **13:**NC(1303).

Deuteromycotina - Hyphomycetes

Acremonium coenophialum. Endophyte; associated with toxicity syndrome in cattle. **3:**AL(3869), KY(3856), OR(3856), TN(2170), TX(3869).

A. huerfanum. **2:**CO(3873).

A. starrii. **2:**AZ(3873), TX(3873). **8:**TX(3873). **18:**WA(3873).

Alternaria sp. Net blotch. **3:**GA(2196).

Aspergillus reptans. On stored kernels. **13:**NC(1303).

Bipolaris sorokiniana. Seedling blight, spot blotch, black point, root rot. **3:**'Helminthosporium leaf spot.' GA(2196). **13:**IA(82,3423), ND(3423). **14:**CA(988), ND(82,3423), VA(82,3423).

B. spicifera. On dead culms and roots. **5:**WA(82).

Cercospora festucae. Leaf spot. **3:**'Cercospora leaf spot.' GA(2196), KY(3423), MS(2720), OR(3423). **13:**KY(82), OR(82), TX(82).

Cercospora sp. Leaf spot. **13:**OK(2872).

Cercosporidium graminis. Leaf streak, brown stripe. **13:**KS(82,3423). **14:**SD(2242), WA(82), WI(82,1332,3423). **18:**WA(3442).

Cladosporium herbarum. On dead leaves. **10:**CA(82).

C. phlei. **1:**AK(526). **14:**AK(526).

Curvularia inaequalis. Seed mold. **13:**IL(3423), OR(82). **14:**IL(3423).

Curvularia sp. Fading out. **13:**NC(1303). **14:**NC(1303).

Drechslera dictyoides. Net blotch, leaf spot. **3:**'Brown blight. Helminthosporium blight.' GA(2196), KS(3021), MS(2720), NC(45), NY(376), PA(376), WA(3423), WV(376,952). **9:**WA(3423). **13:**E states(82), CT(3423), DC(3423), IA(82,3423), ID(82,3423), LA(82), MA(3423), MD(3423), ME(3423), MI(3423), MT(3433), NC(376,1303), NY(376,3423), PA(376), VA(3423), WA(82,3423), WI(1339,3423), WV(376). **14:**OR(82,3423). **23:**LA(120), NC(48), WA(82).

Drechslera sp. Melting out. **14:**NC(1303).

Endoconidium temulentum. Blind seed disease. **13:**OR(82). **23:**Unk.(82).

Epicoccum sp. Secondary leaf spot. **13:**KY(82).
Fusarium acuminatum. Secondary root rot. **3:**ND(82,3423), OR(3423). **10:**ND(82,3423). **13:**ND(82,3423), OR(3423), WY(3423). **23:**ND(82).
F. avenaceum. Stem rot. **14:**OR(3423).
F. culmorum. Head mold, root rot, scab. **14:**WA(82,3423).
F. heterosporum. Head-scab. **3:**MO(1093).
F. moniliforme. Secondary root rot. **13:**WA(3423).
F. poae. Secondary root rot, silver top. **14:**ND(3423), PA(82,3423).
Fusarium sp. Blight, dying out. **13:**NC(1303). **14:**NC(1303).
Helminthosporium sp. **3:**AL(759). **10:**MD(3423). **14:**VA(3423).
Microdochium bolleyi. Root necrosis. **3:**'Anthracnose.' MT(3423), WY(3423). **10:**WY(3423). **13:**ND(3423). **14:**ND(3423).
M. nivale. Pink snow mold, leaf spot. **3:**'Fusarium patch.' ID(3423). **5:**ID(82,3423), MT(3429), OR(82,3423), WA(82,3423). **10:**ID(82,3423). **13:**ID(82,3423), WA(3423). **14:**AK(3441), ID(82,3423), MI(82,3423), MN(82,3423), ND(82), NY(82,3423), OR(3423), WA(82,3423), WI(82,3423). **22:**WA(3435). **23:**AK(2293), WA(82,1286).
Periconia circinata. Root rot. **13:**ND(3423).
Pseudocercosporella herpotrichoides. Foot rot. **5:**ID(82), OR(82), WA(82).
Pyricularia sp. Leaf spot. **13:**NC(1303).
Ramularia collo-cygni. Leaf spot. **23:**UT(82).
R. pusilla. Tan leaf spot. **2:**AZ(3435). **5:**ID(82,3423,3429). **13:**WI(1358), WV(953). **23:**OR(82).
Rhynchosporium orthosporum. **18:**WA(3442).
Spermospora ciliata. Leaf spot. **13:**WA(3432).
S. subulata. Blast, eye spot. **14:**AK(3431), OR(82,3423), WA(82,3423).
Stemphylium botryosum. Seed mold. **3:**Unk.(82). **10:**Unk.(82). **13:**Unk.(82). **14:**Unk.(82). **15:**Unk.(82). **23:**Unk.(82).

Deuteromycotina - Coelomycetes

Ascochyta brachypodii. Leaf spot. **10:**CO(3425). **14:**ND(3423), OR(3423).
A. graminicola. Leaf spot. **10:**WI(1342). **14:**ND(82), OR(82).
A. hordei. Leaf spot. **13:**UT(82,3423,3449).
A. sorghi. Leaf spot. **10:**WI(3423). **13:**WY(3429). **14:**ND(3423,3449), OR(3423,3449).
Cheilaria agrostis. Leaf spot. **5:**ID(3425), MT(3425), WY(3425). **9:**ID(3425), MT(3425). **13:**AK(3425), ID(3425). **14:**AK(3425).
Colletotrichum destructivum. **3:**AL(759).
C. graminicola. Anthracnose, net blotch. **3:**'Anthracnose.' AL(759), GA(2196), WV(952). **10:**NE states(82), MN(82,3423), OK(82), SD(2242), WI(1346). **13:**NE states(82), KY(3423), MN(82), NC(1303), NY(3423), OK(82,2872,3423), PA(3423). **14:**E states(82), MN(3423), OK(82), PA(3423), VA(3423), WA(82,3423). **18:**ID(3423). **21:**OK(2872,3423). **23:**E states(82), ID(82), MN(82), OK(82), WA(82).
Dilophospora alopecuri. Twist disease. **10:**WI(1341,3423).
Discella quercina. On living leaves. **23:**WI(1345).
Hendersonia culmicola var. *minor*. On dying leaves, sheaths, or culms. **5:**CA(82).
Hendersonia sp. Leaf mold. **10:**WY(82).
Leptothyrella graminis. **22:**WA(3816).
Phaeoseptoria airae. Leaf spot. **1:**AK(3428). **10:**CA(82). **14:**OR(82).
P. festucae var. *andropogonis*. On leaves. **18:**ID(3442).
Phloeospora idahoensis. Leaf and pedicel blight, stem eye spot. **5:**ID(82,3423), OR(3342). **9:**OR(3342). **10:**ID(3438). **14:**OR(3342). **19:**OR(3342).
Phoma sclerotioides. On roots. **14:**AK(307).
P. terrestris. Pink root. **7:**OR(2791). **14:**ND(3423).
Phoma sp. Black stem. **5:**ID(82,3423), MT(3423). **9:**MT(3423). **13:**AK(526,3423), ID(82,3423). **14:**AK(3423), AK(526). **23:**MT(82).
Phyllosticta sp. Leaf spot. **10:**WI(1350). **23:**ID(82).
Pseudoseptoria donacis. **10:**SD(2242).
P. everhartii. Eye spot. **10:**CA(82,3423,3450), CO(3450), MT(82,3423,3450), WA(82), WY(82,3423,3450). **14:**AK(3441), MT(82,3423,3450), WA(3423,3450). **16:**MT(3438). **23:**MT(82), UT(82).
P. stomaticola. Leaf speckle, eye spot. **5:**WA(82,3423,3450), WY(82). **10:**Unk.(3450), WA(82,3423). **13:**CA(82,3423,3450), CO(3450).
Septogloeum sp. Leaf spot. **5:**ID(82).
Septoria tenella. Leaf blotch, leaf spot. **5:**MT(3429), OR(82,3423), WA(82,3423), WY(82,3423). **6:**OR(3423). **9:**OR(3433), WA(3432). **10:**AK(3426), WA(82,3423), WY(82,3423). **13:**AK(3441), IL(82), WA(82,3423), WI(1359). **23:**OR(82).
Septoria sp. Leaf spot. **13:**NC(1303).
Stagonospora avenae. Leaf spot. **13:**UT(82,3423).
S. culmicola. Leaf rot and spot. **1:**AK(3428). **5:**CA(3423), OR(3423). **10:**CA(82), MT(3423). **14:**OR(82,3423).
S. nodorum. Brown blight, glume blotch, leaf spot. **8:**WI(1341,3423). **18:**ID(3442), WA(3423,3425). **23:**WA(82).
S. simplicior. On leaves. **10:**CO(82,3424).
Vermicularia sp. **14:**AK(3431).
Wojnowicia hirta. Secondary foot rot, rot of mature straw, leaf spot. **5:**OR(82,3423). **13:**WY(3429). **14:**OR(3423).

Deuteromycotina - Other

Rhizoctonia cerealis. Brown and yellow patch. **3:**'Yellow patch.' NC(2271). **13:**NC(1303).
R. solani. Summer blight, leaf blotch, brown patch, culm, root and stem rot. **3:**'Rhizoctonia blight.' GA(2196), MS(2720), NC(45,2271), VA(892), WV(955). **10:**Range of host(82), MD(82), NC(82). **13:**FL(35), KS(3023), MD(82), NC(82,1303). **14:**Range of host(82), NC(1303,2271). **23:**NC(48), ND(82), OR(82), WA(82).
R. zeae. Foliar blight, brown patch. **3:**'Rhizoctonia leaf spot. Rhizoctonia sheath spot.' NC(2271). **13:**NC(1303).
Rhizoctonia sp. Dollar spot. **10:**NC states(82), NE states(82). **14:**NC states(82), NE states(82).
Sclerotium rolfsii. Southern blight, white mold. **3:**GA(2196). **13:**FL(35).

Gastridium P. Beauv.

1. *G. ventricosum* (Gouan) Schinz. & Thell.

Basidiomycotina - Uredinales

Puccinia graminis. Heteroecious: telial host. **1:**AZ(82).

Deuteromycotina - Hyphomycetes

Drechslera brizae. **1:**CA(3438,3442).
Helminthosporium sp. Leaf spot. **1:**OR(82,3423).

Glyceria R. Br.

1. *G. borealis* (Nash) Batsch [northern mannagrass]
2. *G. canadensis* (Michx.) Trin. (=*G. canadensis* (Michx.) Trin. var. *laxa* (Lams.-Scribn.) Hitchc.) [rattlesnake mannagrass]
3. *G. elata* (Nash) M. E. Jones (=*Panicularia elata* Nash in Rydb.) [tall mannagrass]
4. *G. erecta* Hitchc. in Jeps.
5. *G. fluitans* (L.) R. Br. [water mannagrass]
6. *G. grandis* S. Wats. ex A. Gray (=*Panicularia americana* (Torr.) MacMill.) [American mannagrass]
7. *G. leptostachya* Buckley
8. *G. obtusa* (Muhl.) Trin.
9. *G. occidentalis* (Piper) J. C. Nelson
10. *G. pallida* (Torr.) Trin.
11. *G. septentrionalis* Hitchc. [eastern mannagrass]
12. *G. striata* (Lam.) A. Hitchc. (=*G. nervata* (Willd.) Trin.) [fowl mannagrass]
13. *Glyceria* sp.

Chytridiomycetes

Physoderma gerhardtii. Leaf smut. **6:**CO(1077), MT(1077), UT(1077), WY(1077). **13:**WI(82).

Ascomycotina

Claviceps purpurea. Ergot. **1:**ID(3423), ME(3423), MI(3423), WI(3423). **2:**ME(3423). **5:**WI(82). **6:**IA(3423), ND(3423), WI(1344). **11:**IA(82), IN(82,3423), WI(82,3423). **12:**OK(82,2873,3423), WA(82,3423), WI(82,3423). **13:**ID(82), MA(82), ME(82), MI(82), WI(82).
Cochliobolus sativus. Blind seed disease. **1:**OR(3423).
Epichloe typhina. Cat-tail, choke. **2:**MI(3423). **5:**MI(82). **11:**MI(82), NY(82,3423), WV(1944). **12:**DE(1944), GA(82,1438,3423), IA(82,3423), IL(82,3423), IN(1944), MD(82,3423), MI(82,1944,3423), MN(1944), MO(1944), MS(82,2720,3423), NY(1944), OH(946), WI(82,3423). **13:**MI(82).
Erysiphe graminis. Powdery mildew. **12:**MT(82,3423). **13:**PA(82).
Gloeotinia granigena. **1:**OR(1075).
Mollisia sp. **12:**GA(1438).
Phaeosphaeria herpotrichoides. **13:**WA(3814).

Basidiomycotina - Uredinales

Puccinia brachypodii var. *poae-nemoralis*. Heteroecious: telial host.

12:Unk.(754).
P. coronata. Heteroecious: telial host. **3**:CA(82), OR(82). **6**:ND(82). **12**:ID(82).
P. graminis. Heteroecious: telial host. **1**:WI(1350). **6**:MO(82), WI(82). **12**:CA(82).
P. recondita. Heteroecious: telial host. **6**:Unk.(82), IA(82), IN(82), MN(82), SD(82,2242).
Uromyces amphidymus. Telial host; aecial host unknown. **1**:WI(1349). **5**:WI(82). **11**:IL(82), IN(82), NJ(82), WI(82). **13**:RI(82).
Basidiomycotina - Ustilaginales
Entyloma dactylidis. **6**:MT(1077). **10**:WI(1077).
Urocystis agropyri. **12**:NY(1077), WY(2342).
Ustilago arthurii. Leaf smut. **6**:IA(82,1077).
U. davisii. Leaf smut. **5**:IL(82,1077), ME(82,1077), MI(82,1077), MS(1077), SD(1077,2242), WI(1077). **11**:IL(82), ME(82), MI(82), MS(82,2720), WI(1077). **13**:ME(82), WI(82).
U. longissima. **1**:OR(1077). **3**:OR(82,1077). **5**:IL(1077). **6**:Range of host(82), CO(1077), CT(1077), IA(1077), ID(1077), MA(1077), MI(1077), MN(1077), MT(1077), ND(378,1077), NE(1077), NY(1077), OR(1077), PA(1077), SD(2242), UT(1077), VT(1077), WA(1077), WI(1077). **8**:PA(1077). **12**:ME(82,1077), MN(1077), PA(82,1077), UT(82,1077). **13**:NH(82,1077), OR(82), PA(82).
Deuteromycotina - Hyphomycetes
Cercosporidium graminis. Leaf streak, brown stripe. **1**:WI(3423). **2**:WI(1343). **3**:CA(3423), WA(82,3423). **5**:MI(82,3423), NY(82,3423), OR(82,3423), WI(3423). **6**:ID(82,3423), MT(3423), ND(82,3423), OR(82,3423), SD(2242), WA(82), WI(1338,3423). **7**:WA(3423,3424). **11**:NY(3423), OR(82,3423), WI(82,3423). **12**:OR(3423). **13**:WI(82).
Endoconidium temulentum. Blind seed disease. **13**:OR(82).
Fusarium heterosporum. Head mold following ergot. **5**:WI(82). **11**:WI(82).
Ramularia collo-cygni. Leaf spot. **6**:WI(1341,3423).
R. pusilla. Leaf spot. **3**:CA(661), WA(82,3423,3429). **4**:CA(3423,3429). **6**:WI(3429). **13**:CA(82).
Deuteromycotina - Coelomycetes
[*Ascochyta graminea*]. On culms. **12**:TX(82).
Cheilaria agrostis. Blotch spot, char spot. **3**:WA(3423).
Colletotrichum graminicola. Anthracnose. **6**:SD(2242). **12**:WI(1342,3423).
Cylindrosporium glyceriae. Leaf spot. **2**:WI(3423). **12**:WI(82,3423). **13**:WI(82).
Phyllosticta helenae. Spot. **9**:ID(3423).
Phyllosticta sp. **6**:WI(1359).
Pyrenochaeta sp. **12**:OK(2872).
Septoria glycericola. Buff leaf spot. **2**:MA(3423). **3**:OR(82,3423). **6**:CO(3423), MT(3423,3424), ND(82,3423,3424). **12**:KY(82,3423), OR(82,3423), WA(82,3423). **13**:MA(82).
Stagonospora avenae. Speckled blotch, leaf spot. **3**:ID(3435), OR(3435), WA(638), WY(82,3423). **6**:UT(3438), WY(82,2924,3423). **9**:ID(3423). **12**:CO(1195), WY(3423). **13**:MT(3433).
S. foliicola. Tawny blotch, leaf spot. **3**:CO(82,3423).
S. glycericola. Leaf rot and spot. **6**:ND(82,3423), WI(1347). **12**:MN(82,3423).
S. nodorum. Glume blotch. **12**:OR(3423), WA(3423).
Deuteromycotina - Other
Sclerotium globuliferum. On culms and leaves. **6**:WI(82). **11**:WI(82).
S. hydrophilum. **5**:WI(2888).
S. oryzae. **5**:WI(2888). **11**:WI(2888).
S. rhizodes. On leaves. **6**:ID(3423). **12**:WI(82,3423).

Gymnopogon P. Beauv.
1. *G. ambiguus* (Michx.) B. S. P.
Ascomycotina
Apiospora montagnei. **1**:GA(1438).
Balansia epichloe. Black ring. **1**:GA(82,3423), NC(82,3423,3944), VA(82,3423).
B. henningsiana. Black ring. **1**:MS(2720).

Hackelochloa Kuntze
1. *H. granularis* (L.) Kuntze
Basidiomycotina - Ustilaginales
Sporisorium holci-sorghi. **1**:AZ(1077).

Helictotrichon Besser ex Schult. & J. H. Schult.
1. *H. hookeri* (Lams.-Scribn.) Henr. (=*Avena hookeri* Lams.-Scribn. in Hack.)
2. *H. mortonianum* (Lams.-Scribn. in J.M. Coulter) Henr.
Basidiomycotina - Uredinales
Puccinia coronata. **1**:MT(82).
P. graminis. **1**:Unk.(82).
Deuteromycotina - Coelomycetes
Ascochyta avenae. **2**:CO(3430).
Colletotrichum graminicola. Anthracnose. **1**:ND(82,3423).

Hemarthria R. Br.
1. *H. altissima* (Poir.) Stapf & Hubb. [limpo grass]
Basidiomycotina - Uredinales
Uromyces sp. **1**:FL(35).

Hesperochloa (Piper) Rydb.
1. *Hesperochloa* sp.
Ascomycotina
Pleospora longispora. On dead stems. **1**:AK(1938).

Heteropogon Pers.
1. *H. contortus* (L.) P. Beauv. ex Roem. & Schult. [tanglehead]
2. *H. melanocarpus* (Elliott) Benth. [sweet tanglehead]
Ascomycotina
Phyllachora graminis. Tar spot. **2**:FL(35).
Basidiomycotina - Uredinales
Puccinia versicolor. Heteroecious: telial host. **1**:AZ(1195,1203,3981), HI(82,2903), TX(740,742).
Basidiomycotina - Ustilaginales
Sorosporium contortum. Loose smut. **1**:AZ(82,1077), NM(82,1077).
S. ellisii. **1**:AZ(1077), NM(1077).
Sphacelotheca monilifera. **1**:AZ(82,1077), HI(82,2903).
S. nealii. **2**:FL(35,82,1077).
S. occidentalis. **1**:AZ(1077).
Sporisorium andropogonis. Covered smut. **1**:AZ(82,1077).
Deuteromycotina - Hyphomycetes
Cerebella andropogonis. On inflorescences associated with ergot, blackhead. **1**:AL(82,3423).
Helminthosporium sp. Leaf spot. **2**:FL(35,82).

Hierochloe R. Br.
1. *H. alpina* (Sw.) Roem. & Schutt
2. *H. occidentalis* Buckley (=*H. macrophylla* Thurb. ex Boland.) [California sweetgrass]
3. *H. odorata* (L.) P. Beauv. [sweetgrass]
4. *H. pauciflora* R. Br.
Ascomycotina
Claviceps purpurea. Ergot. **3**:ND(3423).
Mycosphaerella tassiana. **4**:AK(526).
Mycosphaerella sp. **1**:AK(3431).
Pleospora heleocharidis var. *arctica*. **4**:AK(526).
Pyrenopeziza stictoidea. **1**:AK(526). **3**:AK(526).
Pyrenophora trichostoma. **1**:AK(526,3431).
Basidiomycotina - Uredinales
Puccinia coronata. Heteroecious: telial host. **3**:ND(82), NY(82).
P. graminis. Heteroecious: telial host. **2**:CA(81,82), OR(82). **3**:CA(82), WI(1349).
P. striiformis. **2**:CA(81).
Basidiomycotina - Other
Thanatephorus cucumeris. Brown patch, root rot, sharp eye spot. **3**:WY(3423).
Deuteromycotina - Hyphomycetes
Helminthosporium sp. Leaf blotch and rot. **3**:CA(82,3423).
Ramularia pusilla. **3**:WI(1349).
Deuteromycotina - Coelomycetes
Ascochyta graminicola. Leaf spot. **3**:ND(82), SD(82), WI(82,1330).
A. sorghi. Leaf spot. **3**:Unk.(3449), ND(3423,3449), SD(2242,3423,3449), WI(3423).
Colletotrichum graminicola. Anthracnose. **3**:WI(1338,3423).
Septoria oudemansii. On leaves and sheaths. **3**:WI(1350,3432).
Septoria sp. Leaf spot. **3**:MN(82).
Deuteromycotina - Other
Rhizoctonia solani. On basal leaf sheaths and culms. **3**:AK(3441).

Hilaria Kunth in Humb., Bonpl. & Kunth

1. *H. belangeri* (Steud.) Nash (=*H. cenchroides* H. B. K.) [curly mesquitegrass]
2. *H. jamesii* (Torr.) Benth. [galleta grass]
3. *H. mutica* (Buckley) Benth. [tobosa grass]
4. *H. rigida* (Thurb.) Benth. ex Lams.-Scribn. [big galleta]
5. *Hilaria* sp.

Ascomycotina

Claviceps cinerea. Ergot. **1:**AZ(82,3423). **2:**OK(82). **3:**AZ(82,3423), NM(82,3423), OK(82,2873,3423), TX(82).

Basidiomycotina - Uredinales

[*Puccinia aristidae*]. **2:**AZ(82,1203), NM(82). **3:**NM(82).
P. aristidae var. *chaetariae*. **2:**AZ(3981).

Basidiomycotina - Ustilaginales

Ustilago aegopogonis. Head smut. **1:**AZ(82,922,1077), TX(82,1077).
U. affinis var. *hilariae*. **1:**NM(82,1077), TX(82,1077). **3:**AZ(1077), NM(1077).
U. hilariicola. **2:**AZ(1077), NM(1077), NV(1077), UT(1077).
U. hypodytes. Stem smut. **2:**UT(82).
U. vilfae. Smut. **1:**TX(1077). **2:**AZ(1077), CO(82,1077), NM(82,1077,1195), UT(1077). **3:**AZ(1077), NM(82,1077). **5:**AZ(922).

Deuteromycotina - Hyphomycetes

Fusarium acuminatum. Secondary root rot. **2:**ND(82,3423). **3:**ND(82,3423). **4:**AZ(3437).
F. oxysporum. **4:**AZ(3437).
Helminthosporium sp. Leaf spot. **1:**TX(82).
Microdochium bolleyi. Root necrosis. **3:**ND(3423).

Deuteromycotina - Coelomycetes

[*Ascochyta graminea*]. Leaf spot. **2:**NM(3425).

Holcus L.

1. *H. lanatus* L. (=*Notholcus lanatus* (L.) Nash ex Hitchc.) [velvetgrass, Yorkshire-fog]
2. *Holcus* sp.

Ascomycotina

Claviceps purpurea. Ergot. **1:**OR(82,3423), WA(82).
Cochliobolus sativus. Blind seed disease. **1:**OR(3423), WA(3423).
Gaeumannomyces graminis. Take-all. **1:**OR(82,3423).
Gloeotinia granigena. **1:**OR(1075).
Mycosphaerella tulasnei. Leaf mold. **1:**WA(82).

Basidiomycotina - Uredinales

Puccinia coronata. Heteroecious: telial host. **1:**E states(82), AZ(1195,1203,3981), CA(81), HI(82,2903), MA(1394), MS(82,2720), NC(1303), OR(82), WA(82), WV(952).
P. graminis. Heteroecious: telial host. **1:**Unk.(82), IA(82), OR(82).
P. hordei. **1:**Unk.(82), CA(81).

Basidiomycotina - Ustilaginales

Tilletia holci. **1:**MA(82,1077), OR(82,1077), PA(82,1077), WA(82,1077).
Ustilago striiformis. Stripe smut. **1:**CA(1077), OR(82,1077), WA(82,1077).

Basidiomycotina - Other

Thanatephorus cucumeris. Brown patch, root rot, sharp eye spot. **1:**OR(3423), WA(3423).

Deuteromycotina - Hyphomycetes

Cercosporella holci. Buff scald, tan leaf spot. **1:**OR(82,3423), WA(82,3423).
Drechslera triseptata. On leaves. **1:**MI(82,3423), NY(82,3423), OR(82,3423), WA(82,3423).
Endoconidium temulentum. Blind seed disease. **1:**OR(82).
Helminthosporium sp. Leaf spot. **1:**WV(952).
Microdochium nivale. Snow mold. **1:**MI(82,3423), MN(82,3423), OR(3423), WA(82,3423), WI(82,3423).
Ramularia pusilla. Leaf spot. **1:**OR(82,3423,3429).

Deuteromycotina - Coelomycetes

Ascochyta graminicola. Leaf spot. **2:**FL(35).
A. graminicola var. *holci*. On leaves. **1:**WA(82).
A. hordei. Leaf spot. **1:**WA(3423,3449).
Colletotrichum graminicola. Anthracnose. **1:**IN(82,3423), MD(82,3423), PA(82,3423), WA(82,3423), WV(952). **2:**FL(35).
Dilophospora alopecuri. Twist disease. **1:**CA(82,3423), OR(82,3421,3423), WA(82,3423).
Phoma terrestris. Isolated from rhizosphere. **1:**OR(2791).
Septoria tritici f. *holci*. Speckled leaf blotch. **1:**OR(82,3423), WA(82,3423).

Deuteromycotina - Other

Rhizoctonia solani. Brown patch. **1:**OR(82).

Hordeum L.

1. *H. brachyantherum* Nevski (=*H. boreale* Lams.-Scribn. & J. G. Sm., *H. nodosum* L. var. *boreale* (Lams.-Scribn. & J. G. Sm.) Hitchc., *H. nodosum* L.)
2. *H. brevisubulatum* (Trin.) Link [short-awned barley]
3. *H. bulbosum* L. [bulbous barley]
4. *H. geniculatum* All. (=*H. gussoneanum* Parl., *H. hystrix* Roth)
5. *H. glaucum* Steud. (=*H. stebbinsii* Covas)
6. *H. jubatum* L. (=*H. jubatum* L. var. *caespitosum* (Lams.-Scribn.) Hitchc.) [foxtail barley, squirrel-tail barley]
7. *H. leporinum* Link
8. *H. murinum* L. [mouse barley, wall barley]
9. *H. pusillum* Nutt. [little barley]
10. *H. vulgare* L. (=*H. distichon* L., *H. hexastichon* L., *H. irregulare* Aberg & Wiebe, *H. sativum* Pers., *H. aegiceras* Royle ex Walp., *H. vulgare* L. var. *trifurcatum* (Schlechtend.) Alef.) [common barley, Nepal barley]
11. *Hordeum* sp.

Oomycetes

Aphanomyces euteiches. In roots. **10:**WI(82).
Phytophthora cactorum. Leaf rot. **6:**ND(82).
Pythium aphanidermatum. Damping-off. **10:**FL(3169).
P. aristosporum. Root rot. **10:**ND(82).
P. arrhenomanes. Root browning. **6:**ND(82). **10:**'Pythium root rot.' SD(414). **11:**ND(82).
P. debaryanum. Damping-off, root necrosis, root and seed rot. **2:**ND(3423). **3:**ND(3423), WA(3423). **6:**MN(3423), ND(3423), SD(2242,3423). **10:**CO(82,3423), IA(82,3423), MN(3423), ND(3423), NE(3423), OR(3423), SD(2242,3423), WA(82,3423), WY(3423). **11:**ND(82).
P. graminicola. Root necrosis, root rot. **2:**ND(3423), WA(3423). **6:**MN(3423), ND(3423), SD(2242,3423). **8:**WA(82,3423). **10:**'Pythium root rot.' CA(82,3423), CO(3423), IA(82,1581,3423), MN(3423), ND(3423), NE(3423), SD(2242,3423), WA(3423), WY(3423). **11:**WA(82).
P. hypogynum. Root rot. **10:**MO(82,3423).
P. irregulare. Root necrosis. **3:**WA(3423). **10:**ND(3423).
P. monospermum. Root necrosis. **10:**MO(3423).
P. rostratum. Root rot. **10:**CA(82).
P. ultimum. Root necrosis. **10:**ND(3423).
Pythium sp. Seedling blight, root rot. **10:**FL(35).
Sclerophthora macrospora. Downy mildew. **6:**SD(2242,3184). **9:**MS(3050). **10:**'Downy mildew.' AZ(3681), CA(82,3423), IL(312), MO(82,3877), ND(1813), OK(2872,3423), SD(2242). **11:**MO(1908), MS(3839), VA(2989).

Ascomycotina

Claviceps purpurea. Ergot. **1:**AK(82,526,3423), CO(3423), MN(82,3423), WA(3423). **6:**IA(82,3423), MT(82,3423), ND(3423), NE(2927), OK(82,2873,3423), OR(82,3423), WA(82,3423), WI(82,3423). **8:**WA(82). **10:**'Ergot.' E states(3423), Range of host(82), CO(3423), ID(3423), MN(3423), MO(3423), MT(3423), ND(3423), OK(2872,3423), OR(3423), SD(2242), WA(3423), WY(3423). **11:**WY(819).
Cochliobolus sativus. Spot blotch, blind seed disease, black spot, isolated from leaf lesions. **4:**OR(3423). **10:**'Spot blotch. Kernel blight.' IN(1697), UT(1888). **11:**VA(2993).
Erysiphe graminis. Powdery mildew. **1:**AK(526), CO(3423), OR(3423). **2:**ND(3423). **6:**CO(82,3423), IA(3423), MN(82,3423), MO(82,3423), MS(82,2720,3423), MT(82,3423), ND(3423), NE(82), TX(82), WI(82,1340,3423). **8:**CO(82), OR(82). **9:**GA(82,1438,3423), KS(82,3023,3423), MS(2720), OK(2872,3423). **11:**GA(2404,2406), NC(2516), SC(932), TN(72,73).
E. graminis f. sp. *avenae*. Powdery mildew. **11:**AR(3029), TX(1135).
E. graminis f. sp. *hordei*. Powdery mildew. **10:**'Powdery mildew.' TX(1134). **11:**FL(929), GA(1023,1275), KY(1815), MS(3539), NJ(496), VA(2993), WI(1355), WV(3244).
Gaeumannomyces graminis. Foot rot, take-all. **9:**AR(82,3423). **10:**'Take-all.' CA(82,3423), ID(3423), KS(3423), MO(3423), OK(2872,3423), OR(82,3423), WA(3423).
Gibberella cyanogena. Scab. **11:**SC(933).

G. zeae. Head and seedling blight, patch, foot rot, scab. **1:**SD(82). **11:**NC(2516), NJ(496).
Gloeotinia granigena. On old seeds. **4:**OR(1075).
Lophodermium arundinaceum. **1:**AK(3431).
Monographella nivalis. **10:**WA(629).
Mycosphaerella holci. Leaf spot. **10:**IL(312).
M. tulasnei. **1:**AK(3431).
Phaeosphaeria avenaria. Leaf spot. **10:**IL(312).
Phialea campanula. **1:**AK(3431).
Phyllachora graminis. Tar spot. **6:**IL(82), WI(82,1332,3423). **11:**CA(82,3423).
Pyrenophora teres. Net blotch. **8:**CA(82). **10:**'Net blotch.' E states(82), NC states(82), Range of host(82), CA(82), ID(82), IN(1697), OR(82), PA(846), TX(1134), WI(3265). **11:**KY(1815), NM(2109).
P. tritici-repentis. Leaf spot. **9:**IL(312).

Basidiomycotina - Uredinales

Puccinia brachypodii var. *arrhenatheri*. Heteroecious: telial host. **1:**AK(754).
P. coronata. Heteroecious: telial host. **1:**WA(82). **10:**OH(949). **11:**WA(82).
P. graminis. Heteroecious: telial host. **1:**CA(82), ID(82), MI(82), OR(82), WA(82), WY(82). **3:**OK(2872). **6:**CA(82), KS(3021), KY(82), MS(2720), ND(378), PA(82), SD(2242), WA(82). **8:**CA(82), OR(82), WA(82). **9:**KS(82), OK(2872), SD(2242). **11:**CA(82), MD(82), MI(82), MS(3539), NC(2516), ND(82).
P. hordei. Heteroecious: telial host. **1:**CA(81,82), OR(82), WA(82). **8:**CA(81,82), OR(82), WA(82). **9:**IL(312). **10:**'Leaf rust.' Range of host(82), MS(2720), NC(1303), OK(2872), SD(2242), WV(954).
P. montanensis. Heteroecious: telial host. **5:**CA(754). **6:**CO(82), MT(82), UT(82,754). **11:**ND(82).
P. pygmaea. Heteroecious: telial host. **1:**AK(82,526).
P. recondita. Heteroecious: telial host. **1:**CO(82), ID(82), OR(82), UT(82), WA(82). **6:**NC states(82), CO(82), ND(378), NY(82), OR(82), SD(2242), VA(82), WA(82). **9:**KS(82), TX(82).
P. striiformis. **1:**CA(81,82), CO(82), ID(82), OR(82), WA(82). **6:**W states(82), CA(82), SD(2242), WA(82). **8:**CA(82), WA(82). **9:**OK(2872), TX(82), UT(82). **11:**AZ(82), CA(82), CO(82), ID(82), OR(82), WA(82).
Uromyces fragilipes. Heteroecious: aecial host. **1:**CA(81,82), CO(82), OR(82), UT(82), WA(82). **6:**CO(82), ID(82), NM(82), OR(82), UT(82).
U. hordeinus. Heteroecious: telial host. **1:**SC states(82), CA(82), MS(2720), WA(82). **6:**MS(82,2720). **9:**SE states(82), GA(1439), IL(312), KS(3023), MS(2720), NE(82), OK(2872), TX(82).

Basidiomycotina - Ustilaginales

Tilletia controversa. 'Dwarf bunt.' **10:**UT(858).
T. laevis. Bunt. **1:**Unk.(82).
T. texana. Bunt. **1:**MO(1077), TX(82,1077). **9:**LA(1077), MO(82,1077), OK(1077,2872), TX(1077).
Urocystis agropyri. **1:**CO(1077), OR(82,1077), WY(1077). **6:**OR(82,1077), SD(2242), WI(82,1077,1332).
Ustilago avenae. **9:**TX(1077). **10:**'Semi-loose smut.' Range of host(82), AR(82,1077), CO(82,1077), DE(82,1077), GA(1077), IA(82,1077), ID(1077), IL(82,1077), IN(1077), KS(82,1077), KY(1077), MD(82,1077), MI(1077), MN(1077), MO(1077), MS(1077,2720), MT(1077), NC(1077,1303), ND(1077), NE(1077), NY(82,1077), OH(1077), OK(1077,2872), OR(1077), PA(82,1077), SC(1077), SD(1077,2242), TN(1077), TX(1077), VA(1077), WA(82,1077), WI(82,1077), WV(1077), WY(1077). **11:**MN(1053).
U. bullata. **1:**CA(81,82,1077), CO(1077), ID(1077), OR(82,1077), UT(82,1077), WA(82,1077), WY(1077). **4:**ID(1077). **6:**NC states(82), W states(82), AK(526), CO(1077), IA(1077), ID(1077), MN(1077), MT(1077), ND(378,1077), NV(1077), OR(1077), SD(1077,2242), UT(1077), WA(1077), WI(1077,1332), WY(1077,1195). **7:**CA(1077), OR(1077), WA(1077). **8:**CA(82), OR(82), WA(82). **9:**CA(82,1077), NE(2927), UT(82,1077), WY(1077). **11:**ID(82), KS(82).
U. hordei. **3:**WA(1077). **11:**AK(2078), AL(1077), AR(1077), AZ(1077), CA(81,82,1077), CO(1077), CT(1077), DE(1077), GA(1023,1077,1275), IA(1077), ID(1077), IL(1077), IN(1077), KS(1077), KY(1077,1815), MD(1077), MN(1077), MO(1077), MT(1077), NC(1077), ND(1077), NE(1077), NH(1077), NJ(1077), NM(1077), NV(1077), NY(1077), OH(1077), OK(1077), OR(1077), PA(1077), SD(1077), TN(1077), TX(1077), UT(1077), VA(1077), VT(1077), WA(1077), WI(1077), WY(1077).
U. hypodytes. **1:**WA(82,1077). **2:**WA(1077). **6:**UT(1077).
U. striiformis. **6:**SD(1077,2242).
U. trebouxii. **1:**WA(1077). **6:**UT(1077), WA(82,1077).
U. tritici. Deep loose smut. **11:**GA(1023,1024,1275,2404,2406), KY(1815), MN(1053), MS(3539), NC(2516), NJ(496), SC(932,933), TN(72,73), VA(2993), WY(819).
U. williamsii. **6:**UT(82).
Ustilago sp. **11:**IL(309).

Basidiomycotina - Other

Agrocybe molesta. Attached to or associated with crowns. **10:**OR(3422).
A. praecox. Attached to or associated with crowns. **10:**OR(3422).
Athelia rolfsii. Seedling blight, stalk rot, basal sheath. **10:**CA(3423).
Coprinus psychromorbidus. **1:**AK(3441).
Marasmius graminum. Root rot. **10:**IL(82).
Naucoria cerealis. Collar rot and possibly root rot. **10:**IL(82,308).
Thanatephorus cucumeris. Brown patch, root rot, sharp eye spot. **1:**OR(3423). **2:**ND(3423). **4:**OR(3423). **6:**MN(3423), ND(3423), OR(3423), SD(3423). **8:**OR(3423). **9:**MT(3423), NE(3423), SD(3423). **10:**MN(3423), MT(3423), ND(3423), NE(3423), OR(3423), SD(3423), WA(3423), WY(3423). **11:**VA(2993).
Typhula idahoensis. Snow mold. **1:**WA(3423).
T. incarnata. 'Speckled or gray snow mold.' **10:**ID(82,3423).

Deuteromycotina - Hyphomycetes

Alternaria alternata. Black point. **10:**IL(312).
Alternaria sp. Black point. **10:**Range of host(82).
Aspergillus flavus. Storage decay. **10:**Unk.(99).
A. hollandicus. Storage decay. **10:**Unk.(99).
A. niger. Storage decay. **10:**Unk.(99).
Bipolaris sorokiniana. Seedling blight, leaf and stem blotch, rusty blotch, spot blotch, black point, root rot. **2:**ND(3423). **3:**ND(3423). **6:**MN(82,3423), ND(82,3423), SD(2242,3423). **8:**CA(82), OR(3423), WA(82,3423). **11:**CA(3116), FL(929), GA(1023,1275,2406), KY(1815), MS(3539), NC(2516), ND(82), SC(932,933), TN(72,73).
B. victoriae. Blight. **10:**GA(3125), IA(3423), MA(3125), MD(3125), MT(3125), WI(3125).
Cercospora secalis. Leaf spot. **10:**IL(312).
Cercosporidium graminis. Brown spot, leaf spot, leaf streak, brown stripe. **1:**AK(526), ID(82,3423), KS(82,3423), MT(82,3423), OR(82,3423), UT(3438), WA(82,3423), WY(1195,3423). **6:**NC states(82), NE states(82), AK(526), CO(3423), ID(3423), KS(82,3023), MN(3423), MT(3423), ND(3423), NE(3423), NV(3423), OR(3423), SD(2242,3423), UT(3423), WA(82,3423), WI(3423), WY(1195,3423). **9:**KS(82,3021,3023,3423), MO(82,3423), MT(82,3423), OK(2872,3423), SD(82,2242,3423). **10:**AK(526), IA(82,3423), KS(82), OH(82,3423), OK(2872,3423), UT(82,3423).
Cladosporium herbarum. Secondary on leaves, seed mold. **1:**AK(3431). **10:**Range of host(82).
Curvularia geniculata. **1:**OR(3437). **6:**SD(2242).
Drechslera avenae. Leaf spot. **10:**OH(949).
D. graminea. Stripe blight, leaf stripe. **11:**FL(929), GA(1023,1024,1275,2404,2406), IL(309), MS(3539), NC(2516), SC(932,933).
D. teres. Net blotch, leaf spot. **1:**AK(3426,3431). **8:**CA(3423). **11:**GA(1023,1024,1275,2406), KY(1815), MS(3539), NJ(496), TX(1135), VA(2993).
D. tritici-repentis. Leaf blight. **6:**IA(3423), SD(2242).
Fusarium acuminatum. Secondary root rot. **1:**AK(3431), OR(3437). **2:**ND(3423). **3:**ND(3423). **6:**MN(3423), ND(82,3423), OR(3423), SD(2242,3423), WY(3423). **9:**MT(3423), SD(2242,3423). **11:**ND(82).
F. arthrosporioides. Head mold. **10:**WI(82).
F. avenaceum. Head mold, stem rot. **1:**AK(3431). **10:**WI(82,3423).
F. culmorum. Head blight and mold, culm and root rot, scab. **1:**WA(82). **6:**WA(82,3423). **11:**CA(3116).
F. equiseti. Secondary root rot. **1:**AK(3431). **6:**ND(82). **10:**CA(82), ND(82), WI(82). **11:**ND(82).
F. graminearum. Seedling blight, root rot, scab. **1:**SD(3423). **10:**'Scab.' CA(3423), IA(3423), IL(3423), IN(3423), KS(3423), KY(3423),

MD(3423), MI(3423), MN(3423), MO(3423), NC(1303,3423), ND(3423), NE(3423), NY(3423), OH(3423), OK(3423), PA(3423), SC(3423), SD(2242,3423), VA(3423), WV(3423). **11:**CA(3116).
F. moniliforme. Secondary root rot, storage decay. **10:**Unk.(99), MN(3423), ND(3423).
F. oxysporum. Secondary root rot. **10:**ND(82). **11:**ND(82).
F. poae. Secondary root rot, silver top. **6:**ND(3423). **10:**ND(3423).
F. sporotrichioides. Head mold, secondary root rot. **6:**ND(82).
F. subglutinans. Sheath spot. **1:**WA(3432).
Fusarium sp. Seedling blight, root rot, scab. **10:**FL(35). **11:**NC(2516).
Helminthosporium sp. Leaf spot. **10:**MS(2720). **11:**TX(119).
Heterosporium avenae. Spikelet blight on leaves. **10:**IL(312), MT(82,3423), WA(3423).
H. hordei. **11:**AK(526).
Hormodendrum hordei. Leaf mold. **10:**MT(82).
Microdochium bolleyi. Root necrosis, root and stem rot. **1:**MT(3423), SD(3423). **2:**ND(3423). **6:**MT(3423), ND(3423), SD(2242,3423). **9:**ND(3423), SD(2242,3423), WY(3423). **10:**MN(3423), MT(3423), ND(82,3423), NE(3423), SD(3423), WA(82,3423), WY(3423). **11:**CA(3116).
M. nivale. Snow mold. **1:**OR(82,3423), WA(82,3423). **2:**WA(3423). **3:**WA(3423). **6:**AK(3441), WA(82,3423). **8:**OR(82,3423), WA(82,3423). **10:**'Pink snow mold.' MI(82), MN(82,3423), OR(3423), WA(82,3423), WI(82). **11:**WA(82).
Penicillium aurantiogriseum. Storage decay. **10:**Unk.(99).
P. chrysogenum. Storage decay. **10:**Unk.(99).
P. expansum. Blue mold. **10:**IA(3423), ID(3423), IL(3423), MN(3423), MO(3423), ND(3423), NE(3423), OR(3423), SD(3423), WA(3423), WI(3423), WY(3423).
Periconia circinata. Root rot. **2:**ND(3423). **10:**ND(3423).
Pseudocercosporella herpotrichoides. Foot rot, strawbreaker. 'Eyespot.' **10:**ID(82,3423), OR(82,3423), WA(82,3423).
Ramularia pusilla. Leaf spot. **1:**AK(3429,3431).
Rhynchosporium secalis. Scald, leaf scorch. **1:**AK(3441), UT(3438). **6:**OR(82,3423), SD(3423), WI(3423). **8:**CA(3423), OR(82,3423). **11:**GA(1023,1024,1275,2404,2406), KY(1815), MS(3539), NC(2516), NJ(496), PA(1910), SC(932,933), TN(72,73,3644), VA(2993).
Trichoderma viride. Seed rot. **10:**MN(82).
Verticillium dahliae. Leaf stripe. **10:**ID(2280).

Deuteromycotina - Coelomycetes

[*Ascochyta graminea*]. **10:**VA(2992).
A. graminicola. Leaf spot. **1:**AK(526). **6:**WI(1344). **8:**ID(82).
A. hordei. Leaf spot. **8:**ID(82,3423,3449). **10:**WA(3423,3449), WV(954). **11:**OH(947).
A. hordei var. *americana*. Leaf spot. **10:**NY(1670).
A. sorghi. Leaf spot. **1:**AK(3423,3449). **6:**IA(3423), SD(2242). **8:**ID(3423,3449). **10:**IL(3423).
Ascochyta sp. Leaf spot. **10:**IL(82), WA(82).
Cheilaria agrostis. Leaf spot. **1:**AK(3428,3431).
Colletotrichum graminicola. Anthracnose. **3:**OK(2872,3423). **6:**PA(82). **9:**IL(312), KS(82,3423). **10:**'Anthracnose.' E states(82), AL(3423), GA(3423), IA(3423), IN(3423), MD(3423), MN(3423), NC(1303,3423), ND(82), NJ(3423), NY(3423), OK(2872,3423), OR(82), PA(3423), SC(3423), TX(82,3423), VA(3423), WA(82).
Dilophospora alopecuri. Twist disease. **1:**OR(82,3423). **6:**UT(82).
Macrophoma secalina. Leaf blight. **10:**IL(312).
Phaeoseptoria airae. **1:**AK(3431).
Phoma terrestris. Pink root, secondary root rot. **2:**ND(3423). **10:**NC states(82), IA(3423), MN(3423), MT(82,3423), ND(3423), NE(3423), SD(3423), WY(3423).
Pseudoseptoria donacis. Speckle. 'Halo spot.' **10:**OR(82).
P. stomaticola. Eye spot, speckle. **2:**ND(3423,3450). **10:**AZ(1194), OR(3450). **11:**ND(82).
Septoria passerinii. Leaf speckle, leaf spot. **1:**AK(3431), IA(82), MT(82,3423), OR(82), WY(82). **2:**ND(3423). **6:**IA(82,3423), MN(82,3423), ND(82,3423), SD(82,2242,3423), WI(82,3423). **10:**NC states(82), AR(3423), IA(3423), ID(82,3423), IL(3423), IN(3423), MI(3423), MN(3423), NC(1303,3423), ND(3423), OH(3423), OK(2872,3423), OR(82,3423), PA(3423), SC(3423), SD(2242,3423), VA(3423), WI(3423), WV(3423), WY(3423). **11:**ND(82).
Stagonospora avenae. Leaf spot. **8:**OR(82). **10:**OH(949).
S. avenae f. sp. *triticea*. Blotch. **8:**OR(3423).
S. nodorum. Glume blotch, leaf spot. **1:**AK(3431), ID(82,3423). **6:**ND(82,3423). **10:**GA(757).
Wojnowicia hirta. Secondary foot rot, rot of mature straw. **6:**ND(3423). **10:**CA(82,3423), CO(3423), ID(3423), ND(3423), OR(82,3423), WA(82,3423).

Deuteromycotina - Other

Rhizoctonia oryzae. Root rot, sheath spot. **10:**ND(3423).
R. solani. On culms and basal leaf sheaths. **1:**AK(3441). **6:**ND(82), SD(2242). **9:**MT(82), NE(82), SD(82,2242). **10:**'Sharp eyespot. Rhizoctonia root rot.' NW states(3838), NC(1303), OR(82), SD(2242), WA(82). **11:**ND(82), OR(82).
Rhizoctonia sp. Seedling blight, root rot. **10:**FL(35).
Sclerotium rhizodes. On leaves. **6:**WI(82).
S. rolfsii. Root rot. **10:**CA(82).

Hyparrhenia Anderss. ex E. Fourn.

1. *H. hirta* (L.) Stapf

Basidiomycotina - Uredinales

Puccinia andropogonis-hirti. Telial host. **1:**FL(35).

Hystrix Moench

1. *H. patula* Moench (=*Elymus hystrix* L., *H. patula* Moench. var. *bigeloviana* (Fernald) Deam, *Asperella hystrix* (L.) Humb.) [bottlebrush]

Ascomycotina

Claviceps purpurea. Ergot. **1:**IA(82,1272), IN(82,987,1272), ND(1272).
Erysiphe graminis. Powdery mildew. **1:**WI(3423).
Mycosphaerella sp. On basal leaves. **1:**WI(1341).
Phyllachora graminis. Tar spot. **1:**Range of host(82), IA(3423), IL(3423), IN(647,987,3423), KY(3423), MI(3423), NJ(3423), NY(3423), OH(3423), PA(3423), SD(2242), VA(3423), VT(3423), WI(3423), WV(3423).

Basidiomycotina - Uredinales

Puccinia coronata. Heteroecious: telial host. **1:**WI(1341).
P. montanensis. Heteroecious: telial host. **1:**Unk.(754), IA(82).
P. recondita. Heteroecious: telial host. **1:**IA(82), IN(82,987), KY(82), MI(82), MN(82), MO(82), NY(82), VA(82), WI(82).
P. striiformis. **1:**ID(82).

Basidiomycotina - Ustilaginales

Ustilago striiformis. Stripe smut. **1:**KS(82,1077).

Deuteromycotina - Hyphomycetes

Bipolaris sorokiniana. Seedling blight, spot blotch, black point, root rot. **1:**MN(82,3423).
Cercospora asprellae. Leaf spot. **1:**IL(3423).
Cercospora sp. Leaf spot. **1:**IL(82).
Cercosporidium graminis. Leaf streak. **1:**WI(3423).
Microdochium nivale. Snow mold. **1:**WA(82,3423).
Pyricularia grisea. Blast, gray spot. **1:**WI(3423).
P. parasitica. Leaf spot, tar spot. **1:**WI(82,3423).

Deuteromycotina - Coelomycetes

Ascochyta graminicola. **1:**WI(1341).
A. sorghi. Leaf spot. **1:**WI(3423).
Davisiella elymina. Tar spot. **1:**WI(3423).
Pseudoseptoria everhartii. **1:**WI(1345).
Septoria bromi. Leaf spot. **1:**MS(2720).
S. microspora. Leaf blotch. **1:**IN(82).
S. passerinii. Leaf blotch and speckle. **1:**IN(3423), MN(82,3423), WI(1350).
Stagonospora nodorum. Glume blotch. **1:**MN(82,3423).

Imperata Cirillo

1. *I. brevifolia* Vasey (=*I. hookeri* (Rupr. ex Anderss.) Rupr. ex Hack.)

Basidiomycotina - Uredinales

Puccinia microspora. Telial host; aecial host unknown. **1:**AZ(82).
P. posadensis. Telial host; aecial host unknown. **1:**AZ(1203,3981).

Koeleria Pers.

1. *K. nitida* Nutt. (=*K. cristata* Pers.) [June grass]
2. *K. phleoides* (Vill.) Pers.

Oomycetes

Pythium debaryanum. Damping-off, root necrosis, seed rot. **1:**WY(3423).

Ascomycotina

Claviceps purpurea. Ergot. **1:**AZ(82,3423), IA(82,3423), ID(82,3423),

KS(82,3423), MO(3423), MT(3423), ND(3423), OK(82,2873,3423), WI(82,3423).
Epichloe typhina. Cat-tail, choke. **1**:MN(82), ND(82,377,1944,3423), NE(82), PA(82,3423), SD(82,1944,2242,3423), WI(82,3423).
Erysiphe graminis. Powdery mildew. **1**:WA(82,3423), WI(1338,3423).
Basidiomycotina - Uredinales
Puccinia brachypodii var. *arrhenatheri*. Heteroecious: telial host. **1**:Unk.(754), W states(82), CA(82), ND(378), SD(2242), WA(82).
P. coronata. Heteroecious: telial host. **1**:WY(82,1195).
P. crandallii. **1**:AZ(3981).
P. graminis subsp. *graminicola*. Heteroecious: telial host. **1**:CA(82), KS(82), MI(82), ND(82), NE(82), OR(82), SD(82), WA(82), WI(82).
P. monoica. Heteroecious: telial host. **1**:CA(82), CO(82,1195), MT(1195), NM(82), OR(82), WA(82), WY(82,2924).
P. pygmaea. Heteroecious: telial host. **1**:AZ(1195,1203,3981), CO(1195), WI(1353).
[*P. stipae*]. **1**:CO(82).
P. striiformis. **1**:WA(82).
Basidiomycotina - Ustilaginales
Tilletia controversa. Dwarf bunt. **1**:OR(2343).
Urocystis agropyri. Flag smut. **1**:AZ(82,1077), ID(1077), OR(82,1077).
Ustilago striiformis. Stripe smut. **1**:CO(1077), ID(1077), OR(82,1077).
Basidiomycotina - Other
Thanatephorus cucumeris. Brown patch, root rot, sharp eye spot. **1**:MN(3423), ND(3423), OR(3423).
Deuteromycotina - Hyphomycetes
Bipolaris sorokiniana. Seedling blight, spot blotch, black point, root rot. **1**:ND(3423).
Cercosporidium graminis. Leaf streak, brown stripe. **1**:MT(82,3423), ND(82,3423), NE(3423), OR(82), WY(82,3423).
Curvularia geniculata. Secondary root rot. **1**:MT(82), WA(82).
Drechslera avenae. Leaf blotch. **1**:SD(82,2242,3423), WA(82,3423).
D. poae. **1**:WI(1352).
Fusarium acuminatum. Secondary root rot. **1**:MT(3423), ND(82,3423), OR(3423), SD(3423), WA(3423), WY(3423).
F. equiseti. Secondary root rot. **2**:ND(82).
Helminthosporium sp. On leaves. **1**:WI(1344).
Heterosporium avenae. On leaves. **1**:WA(82,3423).
Microdochium bolleyi. Root necrosis. **1**:SD(3423), WY(3423).
M. nivale. Pink snow mold. **1**:ID(82,3423), OR(82,3423), WA(82,3423).
Periconia circinata. Root rot. **1**:ND(3423).
Pseudocercosporella herpotrichoides. Foot rot, strawbreaker. **1**:WA(82,3423).
Deuteromycotina - Coelomycetes
Colletotrichum caudatum. Leaf spot. **1**:OK(2872,3423), WI(1355).
C. graminicola. **1**:WI(1344).
Phoma terrestris. Pink root. **1**:MN(82), ND(82,3423), NE(82), SD(82).
Pseudoseptoria stomaticola. Leaf speckle, eye spot. **1**:NW states(82), ID(3423,3450), MT(3423), ND(3423,3450), SD(2242,3423), WA(82,3423,3450), WY(3423,3450).
Septoria calamagrostidis f. sp. *koeleriae*. Leaf blotch, leaf spot. **1**:NW states(82), AZ(82,3423), CO(3423), ID(3423), ND(3423), OR(3423), WA(82,3423), WY(3423).
S. quinqueseptata. Leaf spot. **1**:ND(82,3423), WA(82,3423), WI(1347).
Stagonospora culmicola. Leaf rot. **1**:OR(3423).
Wojnowicia hirta. Secondary foot rot, rot of mature straw. **1**:ID(3423), WA(3423).
Deuteromycotina - Other
Rhizoctonia solani. Secondary root rot. **1**:MN(82), ND(82).

Lagurus L.
1. *L. ovatus* L. [hare's-tail, rabbit's-tail grass]
Basidiomycotina - Uredinales
Puccinia coronata. Heteroecious: telial host. **1**:Unk.(82).

Lamarckia Moench
1. *L. aurea* (L.) Moench [goldentop]
Basidiomycotina - Uredinales
Puccinia coronata. Heteroecious: telial host. **1**:CA(82).
P. graminis. Heteroecious: telial host. **1**:CA(82).

Lasiacis (Griseb.) Hitchc.
1. *L. divaricata* (L.) Hitchc. [tibisee]
2. *Lasiacis* sp.
Ascomycotina
Phyllachora graminis. Tar spot. **2**:MO(82).
Basidiomycotina - Uredinales
Uromyces costaricensis. Telial host; aecial host unknown. **1**:FL(35,2284).

Leersia Sw.
1. *L. hexandra* Sw.
2. *L. lenticularis* Michx. [catchfly grass]
3. *L. oryzoides* (L.) Sw. [rice cutgrass]
4. *L. virginica* Willd.
Ascomycotina
Epichloe typhina. Cat-tail, choke. **3**:NY(82,3423).
Mollisia culmina. **4**:GA(1438).
Mycosphaerella sp. On leaves. **3**:WI(82), WV(953).
Ophiobolus cariceti. **4**:GA(1438).
Phaeosphaeria nigrans. On leaves. **3**:NY(82). **4**:WI(1360).
Basidiomycotina - Uredinales
Uromyces halstedii. Heteroecious: telial host. **3**:DE(82), IL(82), IN(82), MD(82), MI(82), OH(82), PA(82), SD(82,2242), TX(82). **4**:AL(82), IA(82), LA(82), MD(82), MS(2720), WI(82).
Basidiomycotina - Ustilaginales
Moesziomyces bullatus. Smut. **1**:AL(82,1077), TX(82,1077).
Tilletia corona. **1**:AL(82), TX(82). **2**:MS(2720), MS(1077), TX(1077), WI(1077). **3**:DC(1077), MD(82,1077), MO(82,1077), VA(82,1077), WI(82,1077). **4**:DC(1077), GA(1439), IA(82,1077), IL(82,1077), IN(82,1077), MD(82,1077), MS(82,1077,2720), OH(82,1077), WI(82,1077).
Deuteromycotina - Hyphomycetes
Bipolaris leersiae. Leaf spot. **4**:AL(82,3423), CT(82,3423), DC(3423), IN(3438), KS(3021), MD(82,3423).
Cercosporidium graminis. Leaf streak, brown stripe. **3**:NY(82,3423).
Cladotrichum leersiae. Leaf blotch. **3**:WI(82). **4**:AL(82).
Dactylaria graminum. Leaf spot. **3**:NH(82,128), PA(3423).
Drechslera gigantea. Zonate leaf spot. **3**:WI(1349). **4**:MD(82,3423).
Pyricularia grisea. Blast, gray spot. **3**:IN(3438), NE(3423), WI(82,3423). **4**:NE(3423), WI(82,3423).
Ramularia sp. Leaf spot. **1**:LA(82).
Deuteromycotina - Coelomycetes
Dilophospora alopecuri. Twist disease. **3**:WI(1338,3423).

Leptochloa P. Beauv.
1. *L. dubia* (H. B. K.) Nees [green sprangletop]
2. *L. filiformis* (Lam.) P. Beauv. [red sprangletop]
3. *L. scabra* Nees
4. *L. uninervia* (J. Presl) Hitchc. & Chase
5. *L. virgata* (L.) P. Beauv.
6. *L. viscida* (Lams.-Scribn.) Beal
Ascomycotina
Glomerella tucumanensis. Anthracnose, red rot, leaf spot. **1**:LA(3423). **2**:LA(82,3423).
Basidiomycotina - Uredinales
Puccinia chloridis. Heteroecious: telial host. **1**:AZ(1203), TX(82). **2**:OK(82,2872).
P. diplachnis. Heteroecious: telial host. **1**:AZ(747,1194,1203,3981), TX(742).
P. leptochloae. Heteroecious: telial host. **2**:AZ(741,1203,3981), TX(82).
Basidiomycotina - Ustilaginales
Ustilago heterogena. Stem smut. **2**:AZ(82,1077), NM(82,1077). **3**:AZ(1077), LA(1077). **5**:AZ(1077). **6**:AZ(1077).
U. ornata. Seed smut. **2**:LA(1077), MS(82,1077,2720). **4**:MS(1077,2720).

Leucopoa Griseb. in Ledeb.
1. *L. kingii* (S. Wats.) W. A. Weber (=*Festuca kingii* (S. Wats.) Lams.-Scribn., *Hesperochloa kingii* (S. Wats..) Rydb.)
Oomycetes
Pythium graminicola. **1**:WY(3423).
Ascomycotina
Claviceps purpurea. Ergot. **1**:CO(3423), WY(3423).

Basidiomycotina - Uredinales
Puccinia crandallii. **1**:CO(1195), WY(1195).
Basidiomycotina - Ustilaginales
Urocystis agropyri. **1**:UT(1077).
Ustilago hypodytes. **1**:WY(1077).
U. striiformis. **1**:CO(2342).
Deuteromycotina - Hyphomycetes
Cercosporidium graminis. **1**:WY(2924).
Fusarium avenaceum. On blighted heads. **1**:UT(3438).
Ramularia collo-cygni. Leaf spot. **1**:CO(3423), UT(3423).
R. pusilla. Leaf spot. **1**:CO(3429), UT(3429).
Deuteromycotina - Coelomycetes
Dilophospora alopecuri. **1**:UT(3438).
Pseudoseptoria everhartii. **1**:MT(3450), WY(3450).
P. stomaticola. **1**:MT(3423), UT(3423,3450), WY(3423,3450).

Leymus Hochst.

1. *L. angustus* (Trin.) Pilg. (=*Elymus angustus* Trin.) [Altai wild-rye]
2. *L. arenarius* (L.) Hochst. (=*Elymus arenarius* L.) [European dunegrass, lyme grass]
3. *L. chinensis* (Trin.) Tzvelev (=*Elymus chinensis* (Trin.) Keng, *Elymus pseudoagropyrum* (Trin. ex Griseb.) Turcz.) [Chinese wild-rice, false wheatgrass]
4. *L. racemosus* (Lam.) Tzvelev (=*Elymus giganteus* Vahl) [mammoth wild-rye]
5. *L. secalinus* (Georgi) Tzvelev (=*Elymus salsuginosus* Turcz. ex Ledeb.)

Oomycetes
Pythium graminicola. **4**:ND(3423).
Ascomycotina
Claviceps purpurea. Ergot. **3**:MT(3423). **4**:ND(3423), WA(3423). **5**:ND(3423).
Pyrenophora trichostoma. Leaf spot. **1**:ND(1987). **2**:ND(1990). **4**:ND(1987).
Basidiomycotina - Uredinales
Puccinia striiformis. **3**:WA(2341).
Basidiomycotina - Ustilaginales
Ustilago hypodytes. **1**:WA(1077).
U. striiformis. **3**:WA(2342).
Basidiomycotina - Other
Thanatephorus cucumeris. Brown patch, root rot, sharp eye spot. **4**:ND(3423).
Deuteromycotina - Hyphomycetes
Fusarium acuminatum. Secondary root rot. **4**:ND(3423).
Microdochium bolleyi. Root necrosis. **3**:ND(3423). **4**:ND(3423). **5**:ND(3423).
Periconia circinata. Root rot. **5**:ND(3423).
Rhynchosporium secalis. Scald. **1**:WA(2341). **3**:WA(2341).
Deuteromycotina - Coelomycetes
Pseudoseptoria donacis. Eye spot. **4**:OR(3423), WA(3423,3450).
Septoria agropyrina. Leaf blotch, leaf spot. **4**:ND(3423).
S. spraguei. Leaf spot. **1**:ND(1990). **2**:ND(1990).
Stagonospora arenaria. Purple brown blotch. **3**:ND(3423).

Limnodea L. Dewey in J. M. Coulter

1. *L. arkansana* (Nutt.) L. Dewey

Basidiomycotina - Uredinales
Puccinia coronata. Heteroecious: telial host. **1**:OK(82,2873).
P. graminis. Heteroecious: telial host. **1**:OK(82,2873).
P. limnodeae. Autoecious. **1**:TX(743).

Lolium L.

1. *L. multiflorum* Lam. [annual ryegrass, Australian ryegrass, Italian ryegrass]
2. *L. multiflorum* Lam. cv. 'gulf' [Australian ryegrass, Italian ryegrass]
3. *L. perenne* L. (=*L. marschallii* Steven) [English ryegrass, lyme ryegrass, perennial ryegrass, strand-wheat, Terrell grass]
4. *L. remotum* Schrank
5. *L. subulatum* Vis.
6. *L. temulentum* L. [darnel]
7. *Lolium* × sp. (unnamed hybrids)
8. *Lolium* sp.

Oomycetes
Phytophthora cactorum. Leaf rot. **3**:OR(82).
Pythium aphanidermatum. 'Pythium blight. Pythium root rot.' **1**:FL(35).
P. debaryanum. Damping-off, root necrosis, seed rot. **4**:WA(3423).
P. graminicola. 'Pythium root rot. Pythium blight.' **1**:AZ(3437).
P. myriotylum. 'Pythium root rot. Pythium blight.' **1**:FL(35).
Pythium sp. Root rot. **1**:CA(988), NC(1303). **3**:CA(988).
Sclerophthora macrospora. Downy mildew. **1**:MS(3050).
Ascomycotina
Acrospermum graminum. **8**:OR(3225).
Claviceps purpurea. Ergot. **1**:CA(82,3423), KY(82,3423), OK(82,2872,3423), OR(82,3423), WA(82,3423). **3**:CA(82,3423), ID(82,3423), KS(82,3423), MD(82,3423), MI(82,3423), MN(82,3423), ND(3423), NE(82,3423), OK(82,2872,3423), OR(82,3423), SD(82,3423), VA(82,3423), WA(82), WI(1341,3423), WV(82,3423). **5**:CA(3438).
Cochliobolus sativus. Blind seed disease. **3**:OR(3423). **6**:OR(3423).
Erysiphe graminis. Powdery mildew. **1**:NC(1303).
Gaeumannomyces cylindrosporus. On roots. **3**:RI(1720).
Gibberella sp. Root rot. **3**:PA(82).
Gloeotinia granigena. Blind seed disease. **3**:OR(82,1075,1458,1461). **6**:OR(82,1075).
Leptosphaeria sp. **1**:OR(82). **3**:OR(82), PA(82), TX(82).
Sclerotinia homoeocarpa. 'Dollar spot.' **1**:FL(35), NC(1303).
Basidiomycotina - Uredinales
Puccinia coronata. Heteroecious: telial host. **1**:'Leaf rust.' CA(81,82), FL(35,82), KY(82), LA(82), MS(2720), NC(1303), OK(82,2872), OR(82), TX(82), WA(82), WI(1348). **3**:'Leaf rust.' AL(82), AZ(82,1203,3981), CA(81,82), GA(1439), HI(2532,2903), KS(3021,3023), KY(82), LA(658), OK(2872), OR(82), PA(82), SC(82), UT(82), WA(82), WI(1341), WV(953). **6**:GA(1439). **8**:FL(35), NC(1303).
P. graminis. Heteroecious: telial host. 'Stem rust.' **1**:MS(2720), NY(82), OR(82). **3**:AZ(3981), MI(82), MN(82), NY(82), PA(82).
P. graminis subsp. *graminicola*. Heteroecious: telial host. **1**:NY(82). **3**:NY(82).
P. montanensis. Heteroecious: telial host. **1**:MI(82).
P. recondita. 'Leaf rust.' **3**:OK(2872).
P. striiformis. Telial host. 'Stripe rust.' **3**:HI(2532), OR(82), WA(82).
Basidiomycotina - Ustilaginales
Tilletia controversa. Dwarf bunt. **1**:ID(1465), NY(857). **3**:NY(857).
Ustilago compacta. **1**:OR(1077).
U. striiformis. 'Stripe floral smut.' **3**:OR(82,1077).
U. tritici. **6**:AL(1077).
Ustilago sp. Head smut. **1**:OR(82). **6**:AL(82).
Basidiomycotina - Other
Thanatephorus cucumeris. Brown patch, root rot, sharp eye spot. **1**:GA(1439), ND(3423), OR(3423), WA(3423). **3**:OR(3423), WA(3423).
Deuteromycotina - Hyphomycetes
Bipolaris sorokiniana. Seedling blight, spot blotch, black point, crown and root rot, leaf spot. **3**:'Helminthosporium leaf spot.' CA(988), MN(82,3423), OR(82). **8**:FL(35).
Cercosporidium graminis. Leaf streak, brown stripe. 'Brown stripe.' **1**:OR(3423). **3**:OR(82).
Cladosporium herbarum. Leaf mold. **1**:OR(82). **3**:OR(82).
Curvularia lunata. **1**:AZ(3437).
Curvularia sp. Leaf spot. **1**:NC(1303).
Drechslera dictyoides. **1**:'Brown blight.' NY(376), PA(376). **3**:'Brown blight.' NY(376), PA(376), WV(376). **7**: On *Lolium perenne* x *Festuca elatior*. PA(376).
D. nobleae. Leaf blight. **1**:AL(2511).
D. poae. Leaf spot. 'Melting-out.' **1**:FL(35).
D. siccans. Leaf blight, leaf spot. 'Brown blight.' **1**:CA(3423), DC(3423), FL(35), KY(82,3423), MD(82,3423), NJ(82), OR(82), VA(82,3423), WA(3423), WV(952). **3**:MD(82,376,3423), NJ(82), OH(82,3423), OR(82,3423), VA(82,3423), WA(3423), WV(376,953).
D. teres. Blight, leaf spot. **1**:NC(1303). **3**:MD(82).
Fusarium acuminatum. Secondary root rot. **1**:AZ(3437), ND(3423), OR(3423), WA(3423). **4**:WA(3423).
F. culmorum. Head mold, root rot, scab. 'Fusarium blight.' **1**:Unk.(82), OR(82,3423), WA(3423). **3**:OR(3423).
F. graminearum. Seedling blight, scab. **3**:PA(3423).
Helminthosporium sp. Leaf spot. **1**:MS(2720). **3**:HI(2903).
Heterosporium sp. Secondary leaf spot. **1**:OR(82). **3**:OR(82).

Microdochium nivale. Snow mold. 'Fusarium patch.' **1:**OR(3423), WA(3423). **3:**OR(3423), WA(82,3423).
Ovularia lolii. Leaf spot. **3:**OR(82).
Pyricularia sp. Leaf spot. **2:**LA(520).
Ramularia collo-cygni. Leaf spot. **3:**OR(3423).
R. pusilla. Leaf spot. **1:**OR(3429). **3:**OR(3423,3429).
Rhynchosporium orthosporum. Scald, white leaf stripe. **1:**OR(82,3423). **3:**OR(82,3423). **4:**WA(2341).
R. secalis. 'Scald.' **1:**OR(82,3423). **3:**OR(82,3423).
Titaea sp. Leaf mold. **1:**OR(82). **3:**OR(82).
Deuteromycotina - Coelomycetes
Ascochyta brachypodii. White leaf spot. **3:**OR(82,3423,3449).
A. desmazieresii. 'Ascochyta leaf spot.' **1:**CA(82,3423), KY(82), OR(82,3423,3449), WA(82,3423,3442,3449). **3:**CA(82,3423,3449), OR(82,3423,3449), WA(82,3423).
Colletotrichum graminicola. Anthracnose. **1:**FL(35), OR(82,3423), WA(82,3423). **3:**'Anthracnose.' OR(82,3423), PA(82,3423), WV(953). **6:**TX(82).
Phyllosticta sp. Leaf spot. **1:**FL(35).
Septoria elymi. **1:**WI(1357).
S. loligena. Leaf spot. **1:**CA(82,3423).
S. tritici var. *lolicola*. Speckled leaf blotch. **1:**OR(82,3423), WA(3442). **3:**OR(82,3423).
Stagonospora arenaria. **1:**WI(1355).
Deuteromycotina - Other
Rhizoctonia solani. Brown patch, crown and root rot. 'Rhizoctonia blight.' **1:**Range of host(82), AZ(3437), FL(35), NC(1303). **3:**Range of host(82), NC(2271).
Sclerotium rolfsii. Stalk rot, basal sheath. **1:**GA(3842).

Lycurus Humb., Bonpl. & Kunth

1. *L. phleoides* H. B. K.

Basidiomycotina - Uredinales
Puccinia schedonnardi. Heteroecious: telial host. **1:**AZ(747,753,1194,1195,1203,3981), TX(742).
Basidiomycotina - Ustilaginales
Ustilago lycuroides. Seed smut. **1:**AZ(82,1077).

Melica L.

1. *M. bulbosa* Geyer ex Porter & J. M. Coulter [oniongrass]
2. *M. californica* Lams.-Scribn.
3. *M. geyeri* Munro [Geyer oniongrass]
4. *M. harfordii* Boland. [Harford melic]
5. *M. imperfecta* Trin. [California melic]
6. *M. mutica* Walter [two-flower melic]
7. *M. nitens* (Lams.-Scribn.) Nutt. ex Piper [three-flower melic]
8. *M. porteri* Lams.-Scribn. [Porter melic]
9. *M. scabrosa* Trin.
10. *M. smithii* (Porter ex A. Gray) Vasey [Smith melic]
11. *M. spectabilis* Lams.-Scribn. [purple oniongrass]
12. *M. subulata* (Griseb.) Lams.-Scribn. [Alaska oniongrass]
13. *M. virgata* Turcz. ex Trin.
14. *Melica* sp.

Oomycetes
Pythium graminicola. **9:**ND(3423).
Ascomycotina
Claviceps cinerea. Ergot. **14:**AZ(82).
Epichloe typhina. Choke. **1:**WA(82,3423).
Erysiphe graminis. Powdery mildew. **2:**CA(3423). **3:**CA(81). **14:**CA(82).
Basidiomycotina - Uredinales
Puccinia brachypodii var. *poae-nemoralis*. Heteroecious: telial host. **5:**Unk.(754). **12:**Unk.(754).
P. graminis. Heteroecious: telial host. **1:**CA(82). **14:**CA(82).
P. montanensis. Heteroecious: telial host. **5:**CA(82).
P. paradoxica. Autoecious. **14:**MI(82).
P. pygmaea. Heteroecious: telial host. **12:**CA(82).
P. recondita. Heteroecious: telial host. **14:**CA(82).
P. schedonnardi. Heteroecious: telial host. **6:**MS(2720). **8:**CO(82). **14:**IN(82).
Basidiomycotina - Ustilaginales
Urocystis agropyri. Flag smut. **5:**AZ(82,1077), CA(82,1077).
Ustilago hypodytes. Stem smut. **1:**CA(82,1077). **4:**WA(1077). **14:**WA(82).
U. striiformis. **11:**WY(1077).
Deuteromycotina - Hyphomycetes
Cercosporidium graminis. Leaf streak, brown stripe. **1:**OR(82,3423), UT(82). **3:**WY(3423). **10:**ID(3423). **11:**WY(3423). **12:**ID(82,3423), WA(82). **14:**CA(82), WY(82).
Fusarium acuminatum. Secondary root rot. **13:**ND(3423).
F. oxysporum. Secondary root rot. **9:**ND(3423).
Microdochium bolleyi. Root necrosis. **9:**ND(3423). **13:**ND(3423).
Ramularia pusilla. Leaf spot. **1:**MT(3425,3429). **12:**WA(3432).
Spermospora subulata. Leaf blast. **1:**WY(82,3423). **10:**ID(3423). **11:**WY(3423). **12:**OR(82). **14:**WY(82).
Deuteromycotina - Coelomycetes
Ascochyta brachypodii. Leaf spot. **13:**ND(3423,3449).
A. sorghi. Leaf spot. **1:**WY(3423,3449).
Colletotrichum graminicola. Anthracnose, leaf spot. **7:**OK(2872,3423). **12:**MT(3429).
Dilophospora alopecuri. Twist disease. **1:**WY(82,3423).
Phyllosticta sp. On leaves. **12:**WA(3425).
Pseudoseptoria stomaticola. Leaf speckle, eye spot. **4:**CA(3423,3450). **14:**CA(82).
Septoria melicae. Leaf blotch, leaf spot. **9:**ND(3423). **14:**ND(82).
Stagonospora culmicola. Leaf rot, leaf spot. **5:**CA(3432). **12:**ID(3423).
S. nodorum. Glume blotch, leaf spot. **1:**CO(3424), UT(3423). **2:**CA(638,642,3423). **5:**CA(3432). **10:**MT(3429). **12:**WA(3425). **14:**ND(82).

Melinis P. Beauv.

1. *M. minutiflora* P. Beauv. [molasses grass]

Deuteromycotina - Hyphomycetes
Bipolaris urochloae. Leaf spot. **1:**HI(2903).

Milium L.

1. *M. effusum* L.

Ascomycotina
Erysiphe graminis. Powdery mildew. **1:**MI(82,3423).
Basidiomycotina - Uredinales
Puccinia graminis. Heteroecious: telial host. **1:**MI(82), WI(82).
P. pygmaea. Heteroecious: telial host. **1:**WI(82).
Puccinia sp. **1:**NC(1303).
Deuteromycotina - Hyphomycetes
Bipolaris sp. Leaf spot. **1:**NC(1303).
Cercosporidium graminis. Leaf streak, brown stripe. **1:**NY(82,3423), WI(3423).
Deuteromycotina - Coelomycetes
Colletotrichum graminicola. Anthracnose. **1:**MI(82,3423).

Molinia Schrank

1. *M. caerulea* (L.) Moench [moor grass]

Basidiomycotina - Uredinales
Puccinia graminis. Heteroecious: telial host. **1:**Unk.(82).

Monanthochloe Engelm.

1. *M. littoralis* Engelm.

Ascomycotina
Phyllachora monanthochloes. **1:**CA(2785).
Basidiomycotina - Uredinales
Puccinia schedonnardi. Heteroecious: telial host. **1:**TX(82).

Muhlenbergia Schreb.

1. *M. andina* (Nutt.) Hitchc. [foxtail muhly]
2. *M. arenacea* (Buckley) Hitchc.
3. *M. arizonica* Lams.-Scribn.
4. *M. asperifolia* (Nees & C. A. Mey.) Parodi (=*Sporobolus asperifolius* Nees & C. A. Mey.) [alkali muhly, scratchgrass]
5. *M. californica* Vasey
6. *M. capillaris* (Lam.) Trin.
7. *M. cuspidata* (Torr. ex Hook.) Rydb. (=*Sporobolus cuspidatus* A. Wood) [plains muhly]
8. *M. emersleyi* Vasey [bullgrass]
9. *M. filiculmis* Vasey [slimstem muhly]
10. *M. filiformis* (Thurb. ex S. Wats.) Rydb. [pull-up muhly]
11. *M. foliosa* (Roem. & Schult.) Trin.
12. *M. frondosa* (Poir.) Fernald
13. *M. glauca* (Nees) Mez
14. *M. glomerata* (Willd.) Trin.
15. *M. japonica* Steud.

16. *M. longiligula* Hitchc.
17. *M. metcalfei* (M. E. Jones) Woot. & Standl.
18. *M. mexicana* (L.) Trin. [Mexican muhly]
19. *M. minutissima* (Steud.) Swall.
20. *M. montana* (Nutt.) Hitchc. [mountain muhly]
21. *M. parviglumis* Vasey
22. *M. pauciflora* Buckley [New Mexican muhly]
23. *M. porteri* Lams.-Scribn. ex Beal [bush muhly]
24. *M. pungens* Thurb.
25. *M. racemosa* (Michx.) B. S. P.
26. *M. repens* (J. Presl) Hitchc. [creeping muhly]
27. *M. reverchonii* Vasey & Lams.-Scribn.
28. *M. richardsonii* (Trin.) Rydb. (=*M. squarrosa* (Trin.) Rydb.)
29. *M. rigens* (Benth.) Hitchc. [deergrass]
30. *M. schreberi* J. F. Gmelin [nimblewill]
31. *M. sobolifera* (Muhl. ex Willd.) Trin.
32. *M. sylvatica* (Torr.) Torr.
33. *M. tenuiflora* (Willd.) B. S. P.
34. *M. tenuifolia* (H. B. K.) Trin.
35. *M. texana* Buckley
36. *M. torreyi* (Kunth) Hitchc. [ring grass]
37. *M. uniflora* (Muhl.) Fernald
38. *M. utilis* (Torr.) Hitchc. [Aparejo grass]
39. *M. wrightii* Vasey [spike muhly]
40. *Muhlenbergia* sp.

Oomycetes

Pythium arrhenomanes. Root browning. **25:**ND(82).
P. graminicola. **25:**MN(3423), ND(3423), SD(2242,3423).
Sclerophthora macrospora. **4:**SD(2242).

Ascomycotina

Clathrospora multiseptata. **10:**WA(3814).
Comoclathris permunda. On dead culms. **40:**NM(82).
Didymella culmigena. On dead culms. **18:**MS(82,2720).
Epichloe typhina. Cat-tail. **40:**IA(82).
Erysiphe sp. Powdery mildew. **40:**MD(82).
Leptosphaeria georgius-fischeri. On living leaves. **10:**OR(3435).
L. muhlenbergiae. On culms. **25:**ND(82). **31:**OK(2872).
L. sporoboli. **40:**MT(82).
Leptosphaeria sp. On living leaves. **33:**WI(1354).
Lophodermium arundinaceum. On dead culms and leaves. **40:**CA(3423).
Mollisia atriella. **6:**GA(1438).
Mycosphaerella muhlenbergiae. On leaves. **10:**WA(3815). **40:**NJ(967).
M. tassiana. **10:**WA(3815).
Mycosphaerella sp. On leaves. **18:**NJ(82).
Nimbomollisia aberrans. **6:**GA(1438).
Phaeosphaeria herpotrichoides. **10:**WA(3814).
P. typharum. **10:**WA(3814).
Phyllachora coloradensis. Tar spot. **20:**CO(82).
P. epicampis. Tar spot. **8:**AZ(1195,3423). **16:**AZ(1195,3423). **17:**NM(1195,3378). **29:**AZ(82,1194,1195,3378,3423), CA(82,3423).
P. paspalicola. Tar spot. **20:**CO(3423).
P. texensis. Tar spot. **27:**Unk.(82), TX(3423).
P. vulgata. Black spot, tar spot. **4:**Unk.(82), CO(3423), MT(3423), ND(3423), NE(3423), NM(3423), OR(3423), SD(2242), WA(3423). **5:**CA(3423). **6:**OK(2872,3423). **7:**IA(3423), IN(3423), MN(3423), ND(3423), NE(2927,3423), SD(2242,3423), WI(1330,1335), WY(3423). **11:**PA(3423), WI(1332,3423). **13:**AZ(1195,3423). **14:**WI(1345). **18:**Unk.(82), FL(35), IA(3423), IL(3423), IN(3423), KS(3021), MD(3423), MI(3423), NE(3423), NY(3423), OH(3423), PA(3423), SD(2242,3423), WI(1330,1335), WV(3423). **20:**C states(82), AZ(82,3423), CO(3423), ND(82), NM(82), WY(82). **22:**CO(82), NM(82,3423). **23:**AZ(82,3423), NM(82). **25:**Unk.(82), CO(3423), GA(3423), IL(3423), KS(3423), MI(3423), ND(3423), NE(3423), NY(3423), SD(2242,3423), WI(3423). **27:**MD(3423). **28:**Unk.(82), CO(3423), MT(3423), ND(3423), NM(3423), WY(3423). **30:**Unk.(82), GA(1438,3423), IA(3423), IL(3423), LA(3423), MD(3423), MS(2720,3423), ND(3423), NY(3423), PA(3423), TN(3423), WV(3423). **31:**IA(3423), OK(660), PA(3423). **32:**Unk.(82), KS(3423), ME(3423), MO(3423), WI(3423). **34:**WI(3423). **37:**VT(3423). **40:**Unk.(82), AL(3423), MS(2720), RI(3423).
Placostroma sporoboli. On dead culms. **4:**OK(82,2873,3423).
Pleospora rainierensis. **10:**WA(3813,3814).
Scirrhia pluristriae. On leaves. **4:**WA(3438).

Basidiomycotina - Uredinales

[*Puccinia cryptandri* var. *luxurians*]. **4:**MT(82).
P. dochmia. Autoecious. **22:**TX(742).
P. graminis. Heteroecious: telial host. **4:**IA(82), MO(82). **7:**IA(82), NE(82). **18:**IA(82), MO(82).
P. schedonnardi. Heteroecious: telial host. **2:**TX(753). **4:**Unk.(82), AZ(3981), CA(753), CO(753), KS(753), MT(753), ND(378,753), NE(753), NM(753,1195), SD(2242), UT(753), WA(753), WY(753). **7:**IA(82), ND(82,378,753,2703), NE(82,753), SD(82,753,2242). **8:**AZ(1203,3981), TX(742). **9:**CO(753). **10:**NM(82,753). **17:**TX(742). **18:**Unk.(82), FL(35), IA(3423), KS(3423), MS(2720), NE(3423), NY(3423), OH(3423), OK(2872), TN(3423), WI(2703,3423), WV(3423). **20:**AZ(741,1203,3981), CO(82,753,1195), NM(82), UT(82). **21:**TX(753). **23:**AZ(82,753,1203,3981). **25:**Unk.(82), KS(753,3021), ND(378,753), NE(753), OK(2873), SD(753,2242), WI(2703). **26:**AZ(747,1203,3981), TX(742). **28:**Unk.(82), AZ(753). **29:**AZ(3981), NM(753). **30:**Unk.(82), AR(753), GA(1439), IN(987), KS(753), KY(753), MO(753), NC(1303), NJ(753), OH(753), OK(753,2872), SC(753), TN(753), WI(1353). **31:**VA(753). **32:**Unk.(82), WI(753). **33:**IN(753). **36:**CO(82,753). **38:**AZ(753,3981). **40:**CA(82), NM(82).
Uromyces epicampis. Autoecious. **29:**AZ(82,753,1195,1203,3981), CA(753), TX(742).
U. major. Telial host; aecial host unknown. **27:**TX(753).
U. minimus. Telial host; aecial host unknown. **1:**OR(753). **25:**MI(753), OR(82,753), WA(82). **32:**WI(82,753). **40:**OR(82).
U. tenuicutis. Autoecious. **40:**TX(82).

Basidiomycotina - Ustilaginales

Entyloma dactylidis. Leaf smut. **4:**UT(82,1077). **18:**IN(1077).
Sporisorium montaniensis. Inflorescence smut. **4:**MT(82,1077), ND(82,1077). **7:**CO(1077). **14:**MT(1077), ND(378,1077). **25:**MT(1077), ND(1077).
Tilletia asperifolia. Seed smut. **2:**NM(1077). **4:**W states(82), AZ(1077), CA(1077), CO(1077), ID(1077), MT(1077), NM(1077), NV(82,1077), OR(1077), UT(1077), WA(82,1077), WY(1077,1195). **10:**MT(1077), WY(2342). **19:**CO(1077).
T. asperifolioides. **10:**CO(1077), OR(1077), WY(922,1077).
T. montana. Seed smut. **10:**MT(82,1077).
Ustilago bethelii. Leaf smut. **20:**AZ(82,1077), CO(82,922,1077), NM(1077).
U. buchloes. Leaf smut. **28:**CO(82,1077). **36:**CO(82,1077).
U. hyalino-bipolaris. Inflorescence smut. **23:**AZ(82,1077), NM(82,1077). **35:**AZ(82,1077). **40:**AZ(82).
U. mexicana. **8:**CA(81).
U. minor. **2:**CA(1077).
U. muhlenbergiae. **22:**AZ(1077), NM(82,1077). **23:**NM(1077). **35:**AZ(1077).
U. striiformis. Stripe smut. **40:**CA(82).

Basidiomycotina - Other

Thanatephorus cucumeris. Brown patch, root rot, sharp eye spot. **25:**MN(3423), SD(3423). **28:**ND(3423). **30:**LA(3524).

Deuteromycotina - Hyphomycetes

Bipolaris arizonica. **39:**AZ(3438).
B. cynodontis. Leaf blotch and mold. **18:**FL(35,82), NY(3423), VA(3423).
B. sorokiniana. Seedling blight, spot blotch, black point, root rot. **11:**WI(1336,3423). **18:**WI(82). **20:**ND(82). **25:**ND(3423).
Cercospora muhlenbergiae. Leaf spot, stripe, on nodes and sheath collars. **14:**WI(1357). **18:**AL(82,3423), WI(1346). **25:**ND(82,3423), WI(82,1334,3423). **30:**AL(82,3423). **32:**WI(82,3423). **40:**IA(3423), KS(3451), NE(82).
Cercosporidium graminis. Leaf streak, brown stripe. **10:**ID(82,3423). **18:**IA(82,3423), NJ(3543). **25:**SD(2242), WI(82,1334,3423). **32:**WI(3423).
Cladosporium sp. On leaves. **12:**WI(1354).
Curvularia geniculata. Leaf mold, secondary root rot. **20:**ND(82). **40:**ND(82).
Drechslera gigantea. Zonate leaf spot. **18:**MD(82,3423). **30:**MD(82,3423). **32:**WI(1357). **37:**WI(1357). **40:**MD(82).
Exserohilum rostratum. Mold, secondary root rot. **15:**ND(3423). **18:**ND(82).
Fusarium acuminatum. Secondary root rot. **24:**NE(3423).

25:MN(3423), ND(82,3423). **28:**ND(3423). **40:**NE(82).
F. avenaceum. Leaf spot. **25:**AZ(3435).
F. equiseti. Secondary root rot. **20:**ND(82). **40:**ND(82).
F. oxysporum. **20:**ND(82).
F. poae. Secondary root rot, silver top. **25:**MN(82,3423).
Helminthosporium sp. Leaf blotch. **4:**MN(82), ND(3423). **40:**VA(82).
Microdochium bolleyi. Root necrosis. **15:**ND(3423). **25:**MN(3423), ND(3423).
Pyricularia grisea. Blast, gray spot. **25:**NE(3423). **40:**KS(82).
P. parasitica. Tar spot. **11:**WI(3423).
Ramularia pusilla. Leaf spot. **10:**CA(3438).

Deuteromycotina - Coelomycetes

Amerosporium atrum. **18:**OK(2873).
[*Ascochyta graminea*]. Leaf spot. **25:**NM(3425). **36:**NM(3425).
A. graminicola. Leaf spot. **11:**WI(1336). **25:**WI(1355). **30:**WI(1346). **40:**WI(82).
A. sorghi. Leaf spot. **11:**WI(3423).
Cheilaria agrostis. Blotch spot, char spot. **4:**ND(82,3423).
Clypeostagonospora muhlenbergiae. Anthracnose. **18:**MI(2883).
Colletotrichum georgius-fischeri. On living leaves. **10:**Unk.(3435).
C. graminicola. Anthracnose. **6:**OK(2872,3423). **18:**OK(2872,3423). **25:**NE(2927), OK(2872,3423). **30:**OK(82,2873,3423).
Cylindrosporium calamagrostidis. On leaves. **10:**UT(82,3423).
Davisiella elymina. Tar spot. **25:**WI(3423,3549). **30:**WV(955).
Hendersonia sp. Leaf spot. **36:**NM(3425).
Leptothyrella graminis. **10:**WA(3816).
Macrophomina phaseolina. Charcoal stem rot. **18:**IN(82,3423). **30:**VA(82,3423).
Phaeoseptoria airae. Leaf spot. **18:**MI(82). **25:**MI(82).
P. festucae var. *muhlenbergiae*. **33:**WI(1361).
Phloeospora muhlenbergiae. Leaf spot. **3:**AZ(1195,3376,3423). **4:**AZ(82). **10:**MT(3429).
Phoma terrestris. Pink root. **28:**ND(3423).
Phyllosticta minutaspora. Yellow leaf. **10:**ID(82,3423).
Pseudorobillarda agrostidis. On leaves. **10:**CA(3425).
Pseudoseptoria everhartii. Leaf speckle, eye spot. **7:**ND(82,3423,3450). **25:**ND(82,3423,3450).
Septoria mississippiensis. Leaf spot. **4:**ND(82,3423). **18:**MN(82,3423), WI(1356). **28:**MN(3442). **33:**WI(1354).
Stagonospora montagnei. On leaves. **40:**NM(82).
Wojnowicia hirta. On basal sheaths. **9:**WA(3438).

Deuteromycotina - Other

Rhizoctonia solani. Summer blight, root rot. **25:**MN(82), ND(82), SD(2242). **30:**WV(955).

Munroa Torr.

1. *M. squarrosa* (Nutt.) Torr. [false buffalograss]

Deuteromycotina - Hyphomycetes

Cercospora sp. Leaf spot. **1:**CO(82), KS(82).

Deuteromycotina - Coelomycetes

Septoria munroae. Leaf spot. **1:**CO(3423), KS(3423), OK(2872,3423).

Oplismenus P. Beauv.

1. *O. hirtellus* (L.) P. Beauv. [basket grass]
2. *O. setarius* (Lam.) Roem. & Schult.

Ascomycotina

Phyllachora punctum. Tar spot. **1:**FL(82), LA(82), SC(82,3423). **2:**FL(35,82,3423), LA(82,3423), SC(82).

Oryza L.

1. *O. sativa* L. [rice, upland rice]

Oomycetes

Pythium aphanidermatum. Root rot. **1:**HI(2903).
P. graminicola. **1:**HI(82,3423).
P. nagaii. Leaf rot of seedlings. **1:**AR(82).
P. rostratum. **1:**CA(82).
Pythium sp. Root-tip rot. **1:**AR(82), HI(2903), LA(82).
Sclerophthora macrospora. Downy mildew. **1:**AR(3594,3595).

Ascomycotina

Balansia oryzae-sativae. Black ring. **1:**LA(82).
Chaetomium sp. Kernel discoloration. **1:**Unk.(82).
Cochliobolus miyabeanus. Leaf blight, black spot. **1:**FL(1328,3500).
Gaeumannomyces graminis. Black sheath rot. **1:**AR(3423,3754), LA(3423), TX(3423).
Leptosphaerulina trifolii. On discolored glumes. **1:**VA(82).
Magnaporthe salvinii. Stem rot. **1:**AR(82,1971,3423), CA(82,3423), LA(82,3423), NC(3423), TX(82,3423).
Monascus purpureus. Kernel discoloration. **1:**Unk.(82).
Mycosphaerella oryzae. On sheaths. **1:**TX(82).
Ophiosphaerella herpotricha. On dead culms. **1:**LA(82).
Wolkia decolorans. Kernel discoloration. **1:**AR(82), CA(82).

Basidiomycotina - Ustilaginales

Entyloma dactylidis. **1:**LA(1077), TX(1077).
E. oryzae. Leaf smut. **1:**AR(82), LA(82), TX(82).

Basidiomycotina - Other

Athelia rolfsii. Seedling blight, basal sheath and stalk rot. **1:**FL(1328), LA(3423).
Thanatephorus cucumeris. Brown patch, root rot, sharp eye spot, sheath spot. **1:**AR(82,3423), CA(3423), LA(82,3423,3524), TX(82,3423).

Deuteromycotina - Hyphomycetes

Alternaria sp. Kernel discoloration, leaf spot. **1:**AR(82), CA(82), FL(35), LA(82).
Aspergillus sp. Kernel discoloration. **1:**TX(82).
Bipolaris hawaiiensis. On grains. **1:**HI(982,2903).
B. oryzae. Grain discoloration, brown spot. **1:**AR(82,3423), CA(82,3423), FL(35,82,3423), LA(82,3423), MS(82,2720,3423), TX(82,3157,3423).
B. victoriae. **1:**TX(3125).
Cercospora oryzae. Narrow brown spot, glume spot. **1:**AL(82,3423), AR(82,3423), FL(1328), LA(82,3423), TX(82,3423).
Cladosporium herbarum. Kernel discoloration. **1:**Unk.(82).
Curvularia brachyspora. **1:**TX(82).
C. eragrostidis. **1:**AR(82), LA(82), TX(82).
C. lunata. **1:**AR(82), LA(82), TX(82).
Curvularia sp. Glume mold. **1:**FL(35).
Epicoccum nigrum. Kernel discoloration. **1:**Range of host(82), MS(2720).
Exserohilum rostratum. Isolated from spotted leaves. **1:**SC(123).
Fusarium moniliforme. Bakanae disease, kernel discoloration, secondary root rot. **1:**AR(82), LA(82), TX(122,3423).
F. pallidoroseum. Dry rot of grain. **1:**TX(82).
Fusarium sp. Bakanae disease, kernel discoloration. **1:**AR(82), LA(82), TX(122).
Helicoceras oryzae. Kernel discoloration, reddish brown sheath rot. **1:**AR(82,3423), LA(82,3423), TX(82,3423).
Helminthosporium sp. Black spot. **1:**HI(2903).
Heterosporium avenae. Leaf mold. **1:**CA(82), TX(82).
Microdochium oryzae. Leaf scald. **1:**LA(3076).
Nakataea sigmoidea var. *irregulare*. Stem rot. **1:**AR(3423), LA(3423), TX(3423).
Nigrospora oryzae. Kernel discoloration. **1:**AR(82), LA(82), TX(82).
Penicillium sp. Blue mold of grains. **1:**AR(82).
Pyricularia grisea. Blast. **1:**HI(2903).
P. oryzae. Blast, seedling blight, rotten neck. **1:**Unk.(82), AR(1690,2018,3423), FL(35,1690), HI(1272,3423), LA(121,1690,2018,3423), MS(2720), TX(121,1690,2018,3423).
Sarocladium oryzae. Sheath rot. **1:**LA(3196).
Trichoconis caudata. Kernel discoloration. **1:**AR(82), LA(82), TX(82).
Trichoderma viride. Secondary sheath spot. **1:**AR(82), LA(82), TX(82).
Ustilaginoidea virens. False smut. **1:**CA(82,3423), GA(82,3423), LA(82,3423), SC(82,3423), TX(82).

Deuteromycotina - Coelomycetes

Ascochyta oryzae. On glumes, discolored culms, and discolored leaves. **1:**VA(82,3423).
Phoma glumarum. Glume spot, brown-bordered leaf spot. **1:**AR(82,3423), LA(82,3423), MI(306), MS(82,2720,3423), TX(82,3423).
P. necatrix. **1:**AR(82), CA(82), TX(82).
P. sorghina. On culms. **1:**FL(35), LA(82).
Phoma sp. Kernel discoloration. **1:**AR(82), LA(82), TX(82).
Phyllosticta japonica. **1:**SC(82).
P. miyakei. On rice leaves and glumes. **1:**SC(3423), TX(3423).
Pyrenochaeta oryzae. On sheaths and glumes. **1:**AR(82).
Septoria oryzae. On glumes. **1:**FL(35,82,3423).

Deuteromycotina - Other

Rhizoctonia oryzae. Root rot, sheath spot. **1:**AR(12,3423), CA(3423), FL(1328), LA(3423), TX(3423).
R. oryzae-sativae. Aggregate sheath spot. **1:**CA(12,1403).

R. solani. Sheath and seedling blight, sheath spot, stem rot. **1**:AR(2076), FL(35), LA(12).
Rhizoctonia sp. Sheath spot. **1**:Range of host(82), AR(82), CA(82), LA(82), TX(82).
Sclerotium hydrophilum. Sheath rot, stem rot. **1**:LA(2888,3197).
S. oryzae. Stem rot. **1**:Unk.(82), CA(3805), LA(3197).
S. oryzae var. *irregulare*. Stem rot, sheath blight. **1**:LA(3197).
S. rolfsii. Seedling blight, sheath and stem rot. **1**:LA(82,3197), TX(82).

Oryzopsis Michx.

1. *O. asperifolia* Michx.
2. *O. exigua* Thurb. ex Torr. [little ricegrass]
3. *O. hymenoides* (Roem. & Schult.) Ricker [Indian mullet, Indian rice, silk grass]
4. *O. micrantha* (Trin. & Rupr.) Thurb. [littleseed ricegrass]
5. *O. miliacea* (L.) Benth. & J. D. Hook. ex Aschers. & Schweinf. [smilograss]
6. *O. pungens* (Torr. ex Spreng.) Hitchc.
7. *O. racemosa* (Sm.) Ricker ex Hitchc.
8. *Oryzopsis* sp.

Oomycetes
Pythium arrhenomanes. Root browning. **3**:ND(82).
P. debaryanum. Damping-off, root necrosis, seed rot. **3**:ND(82,3423), SD(2242,3423), WA(3423). **5**:ND(3423). **8**:ND(82).
P. graminicola. **3**:MT(3423), ND(3423), SD(3423), WA(82,3423).

Ascomycotina
Asterina graminicola. On leaves. **1**:MI(82).
Balansia epichloe. Black ring. **1**:PA(82). **3**:PA(3423).
Claviceps sp. Ergot. **8**:WI(82).
Coleroa sporoboli. **1**:WI(1359,1361).
Lophodermium arundinaceum. On dead culms and leaves. **2**:WA(3423), WY(2924). **8**:WA(82).
Mycosphaerella sp. On leaves. **1**:MI(82).
Phyllachora oryzopsidis. Tar spot. **1**:MA(82,3423), MI(82,3423), MN(82,3423), NY(82,3423), SD(2242,3423), VT(82,3423), WI(82).
Physalospora sp. On leaves. **1**:WI(1344).

Basidiomycotina - Uredinales
Puccinia brachypodii-phoenicoidis var. *davisii*. Autoecious. **1**:WI(754,1361).
P. burnettii. Heteroecious: telial host. **3**:CA(82), CO(82), MT(82), NE(82), UT(82), WY(82).
P. graminis. Heteroecious: telial host. **6**:WI(1357).
P. micrantha. Heteroecious: telial host. **4**:CO(82,1195), MT(82), NE(82), SD(82,2242).
P. monoica. Heteroecious: telial host. **3**:AZ(1195,1203,3981).
P. pygmaea. Heteroecious: telial host. **1**:MI(82), SD(2242), WI(82).
[*P. stipae*]. **3**:AZ(82,1203,3981), CA(82), CO(82), WA(82).
P. substerilis. **3**:W states(82), CA(82), WA(82).
P. substerilis var. *oryzopsidis*. Autoecious. **3**:W states(1362), AZ(1203,3981), CO(1195), WY(1195).

Basidiomycotina - Ustilaginales
Ustilago hypodytes. Stem smut. **3**:AZ(82,1077,1195), CA(82,1077), CO(82,1077), ID(82,1077), MT(82,1077), NM(1077), NV(82,1077), OK(82,1077,2873), OR(82,1077), UT(82,1077), WA(82,1077), WY(82,1077).
U. minima. **3**:AZ(82,1077), CA(82,1077), CO(82,1077), NM(82,1077).
U. williamsii. **3**:CA(1077), MT(82,1077), NM(1077), NV(1077), UT(82,1077), WY(82,1077). **8**:WA(82).

Basidiomycotina - Other
Thanatephorus cucumeris. Brown patch, root rot, sharp eye spot. **3**:ND(3423), SD(3423), WY(3423).

Deuteromycotina - Hyphomycetes
Bipolaris sorokiniana. Seedling blight, spot blotch, black point, root rot. **3**:ND(82,3423), SD(2242,3423), WA(82,3423).
Cercosporidium graminis. Brown stripe, leaf streak. **3**:MT(3423), UT(82,3423), WA(82,3423), WY(82,3423).
Curvularia geniculata. **3**:SD(2242).
Fusarium acuminatum. Secondary root rot. **3**:ND(82,3423), NV(3437), SD(2242,3423), WA(82,3423), WY(3423).
F. equiseti. Secondary root rot. **3**:ND(82).
F. moniliforme. Secondary root rot. **3**:ND(3423). **4**:ND(3423).
F. oxysporum. Secondary root rot. **3**:AZ(3437), ND(82,3423). **4**:ND(3423).
F. sporotrichioides. Secondary root rot. **3**:ND(82).
Fusarium sp. Secondary root rot. **3**:ND(82).
Microdochium bolleyi. Root necrosis. **3**:ND(3423).
M. nivale. Snow mold. **3**:CA(3423).

Deuteromycotina - Coelomycetes
Ascochyta graminicola. **1**:WI(1355).
A. subalpina. Leaf spot. **1**:SD(3449).
Colletotrichum graminicola. Anthracnose. **4**:ND(82,3423).
Davisiella elymina. Tar spot. **1**:WI(3423).
Phoma terrestris. Pink root. **1**:ND(82). **3**:ND(82,3423), SD(2242,3423).
Pseudoseptoria donacis. **3**:SD(2242,3450).
P. stomaticola. Stem speckle, eye spot. **3**:CA(82,3423,3450), MT(82,3423,3450), ND(82,3423,3450), OR(82,3423,3450), SD(3450), UT(82,3423,3450), WY(3423,3450).
Stagonospora arenaria. Purple brown blotch. **3**:MT(82). **7**:WI(1338,1351,3423).
S. avenae. Speckled blotch. **3**:MT(3423).
S. culmicola. Leaf rot. On culms. **1**:ND(82,3423), SD(2242).
Wojnowicia hirta. Leaf mold. **3**:ND(82).

Deuteromycotina - Other
Rhizoctonia solani. Secondary root rot. **3**:ND(82), SD(2242).

Panicum L.

1. *P. aciculare* Desv. in Poir.
2. *P. albemarlense* Ashe (=*P. meridionale* Ashe var. *albemarlense* Fernald)
3. *P. altum* Hitchc. & Chase
4. *P. amarulum* Hitchc. & Chase
5. *P. amarum* Elliott
6. *P. anceps* Michx. [panicgrass]
7. *P. angustifolium* Elliott (=*P. curtisii* Steud.)
8. *P. boreale* Nash
9. *P. boscii* Poir.
10. *P. boscii* Poir. var. *molle* (Vasey) Hitchc. & Chase
11. *P. bulbosum* H. B. K. (=*P. plenum* Hitchc. & Chase) [bulb panicgrass]
12. *P. capillare* L. (=*P. capillare* L. var. *occidentale* Rydb.) [witch grass]
13. *P. capillarioides* Vasey
14. *P. clandestinum* L. [deertongue]
15. *P. columbianum* Lams.-Scribn.
16. *P. depauperatum* (Muhl.) Gould
17. *P. dichotomiflorum* Michx. [fall panicgrass]
18. *P. dichotomum* (L.) Gould (=*P. commutatum* Roem. & Schult., *P. consanguineum* Kunth, *P. lancearum* Trin., *P. laxiflorum* Lam.)
19. *P. dictyoneura* Fig. & De Not. (=*Brachiaria dictyoneura* (Fig. & De Not.) Stapf)
20. *P. euchlamydeum* Shinners
21. *P. fasciculatum* Sw. [browntop millet]
22. *P. firmulum* (Hitchc. & Chase) Pilg.
23. *P. flavovirens* Nash
24. *P. geminatum* Forssk.
25. *P. helleri* Nash
26. *P. hemitomon* Schult. [maidencane]
27. *P. hians* Elliott
28. *P. hirticaule* J. Presl
29. *P. implicatum* Lams.-Scribn.
30. *P. joori* Vasey
31. *P. latifolium* L.
32. *P. leibergii* (Vasey) Lams.-Scribn.
33. *P. lindheimeri* Nash
34. *P. longifolium* Torr.
35. *P. maximum* Jacq. [Guinea grass]
36. *P. meridionale* Ashe
37. *P. microcarpon* Muhl. (=*P. sphaerocarpon* Elliott)
38. *P. miliaceum* L. [broom millet, broomcorn, common millet, hog millet, proso millet]
39. *P. nephelophilum* Gaud.
40. *P. obtusum* H. B. K. [vine mesquitegrass]
41. *P. pedicellatum* Vasey
42. *P. praecocius* Hitchc. & Chase
43. *P. purpurascens* Raddi [Pará grass]
44. *P. repens* L. [torpedograss]
45. *P. rigidulum* Bosc (=*P. agrostoides* Spreng.)
46. *P. scabriusculum* Elliott
47. *P. scoparium* Lams.-Scribn.

48. *P. subvillosum* Ashe
49. *P. texanum* Buckley [Texas millet]
50. *P. tsugetorum* Nash
51. *P. virgatum* L. [switchgrass]
52. *P. wilcoxianum* Vasey
53. *P. wrightianum* Lams.-Scribn.
54. *P. xalepense* H. B. K.
55. *Panicum* sp.

Plasmodiophoromycetes

Ligniera junci. On roots. **17:**NC(1303).

Chytridiomycetes

Rhizophydium graminis. Root and rootlet necrosis. **55:**MA(3423).

Oomycetes

Phytophthora cactorum. Rot. **12:**MT(3423).

Pythium arrhenomanes. Root necrosis, root rot. **21:**HI(82). **35:**HI(2903). **38:**ND(82). **43:**HI(82,2903). **51:**ND(82).

P. debaryanum. Damping-off, root necrosis, seed rot. **12:**MN(3423), ND(3423). **38:**MN(3423), ND(3423), OR(3423), SD(2242,3423), WA(3423). **51:**ND(3423), NE(82,3423).

P. graminicola. Root necrosis. **12:**MN(3423), ND(3423). **38:**IA(82), MN(3423), ND(3423), SD(2242,3423). **43:**HI(3423). **48:**MN(3423). **51:**ND(3423).

P. hydnosporum. Rot. **43:**HI(2903,3423).

P. monospermum. Root necrosis. **38:**ND(82,3423).

P. rostratum. Root rot. **21:**HI(82). **35:**HI(2903). **43:**HI(82,2903).

P. ultimum. Root necrosis. **38:**ND(3423).

Pythium sp. Root necrosis. **38:**Unk.(82).

Sclerophthora macrospora. **12:**SD(2242). **38:**SD(2242). **51:**SD(2242,3184).

Sclerospora graminicola. Downy mildew. **17:**TX(82). **38:**IA(82,3423), MN(3423), ND(3423), SD(2242,3423). **43:**HI(2899,2903).

Ascomycotina

Acrospermum graminum. **55:**GA(1438).

Apiospora montagnei. **1:**GA(1438). **14:**GA(1438).

A. spinulosa. On dry culms. **55:**HI(2784).

Balansia epichloe. Black ring. **36:**NY(3423). **45:**MO(82,3423), MS(3423). **51:**KY(3423). **55:**MS(3423).

B. henningsiana. Black ring. **6:**MO(82), MS(82,2720). **45:**MS(82,2720). **51:**KY(82), VA(82). **55:**VA(82).

B. strangulans. Black ring. **1:**GA(1438,3423). **14:**NY(82,3423), PA(82,3423), VA(82), WV(955). **18:**FL(35,82,3423). **26:**FL(35,82,3423). **31:**WV(955). **47:**GA(1438,3423). **55:**E states(82), GA(1438), KS(82), MS(82), OH(653), VA(3423).

Botryosphaeria festucae. On dead culms. **26:**LA(82), MS(82,2720).

Claviceps sp. Ergot. **6:**AR(764). **51:**MS(82,2720).

Crocicreas culmicola. **1:**GA(1438).

Elsinoe panici. On culms, leaves, panicles and sheaths. **51:**IA(1745,3630), WI(1357).

Exarmidium fusariisporum. On dead culms. **51:**KS(82).

Gaeumannomyces graminis. **7:**LA(3754).

Graphyllium hysterioides. On culms. **51:**KS(82).

Griphosphaeria stevensonii. Leaf spot. **26:**FL(35,82).

Lasiostemma erysipheoides. **55:**TX(652).

Leptosphaerulina trifolii. **21:**GA(1297,1438).

Massarina gloeospora. **6:**GA(1438).

Metasphaeria panicorum. On leaves. **18:**GA(82,1438), SC(82). **55:**NH(82), SC(634).

M. punctulata. On dead culms. **14:**GA(82,1438). **18:**GA(82,1438). **26:**LA(82). **55:**TX(652).

M. stenotheca. On dead culms. **26:**LA(82,963).

M. subseriata. On culms. **51:**KS(82,970).

Mollisia sp. **45:**GA(1438).

Mycosphaerella sp. **29:**WI(1333).

Myriogenospora atramentosa. Black crust, black streak. **6:**VA(82). **43:**FL(35,82).

M. bresadoleana. Black crust. **26:**FL(35,82).

Naemacyclus culmigenus. On culms. **55:**FL(35,82,522).

[*Ophiobolus acuminatus*]. On dead culms. **14:**GA(82,1438).

O. brachyascus. **55:**NJ(82).

Perisporium zeae. **43:**FL(82).

Phaeosphaeria nigrans. **45:**GA(1438).

Phyllachora bonariensis. Tar spot. **27:**AL(3423). **34:**NJ(82,3423). **55:**AL(82).

P. graminis. Black spot, tar spot. **5:**MS(2720). **12:**NH(82), TX(82). **18:**MS(2720). **38:**ND(82). **49:**TX(82). **51:**IA(3423), KS(3021,3423), KY(3423), MD(3423), MS(2720), NC(1303,3423), NE(3423), NY(3423), OK(2872,3423), RI(3423), SD(2242), WI(1333). **55:**FL(35), IN(987), LA(2675).

P. minutissima. Tar spot. **6:**GA(82,1438,3423). **34:**AL(82,3423). **45:**AL(82,3423), GA(82,1438), IL(82,3423).

P. paspalicola. Black spot, tar spot. **24:**FL(35,82,3423). **40:**NM(82,3423), OK(82,2872,3423).

P. punctum. Black spot, tar spot. **8:**NH(3423), WI(1341,3423). **9:**FL(3423), GA(1438,3423), IL(3423), KY(3423), MI(3423), MO(3423), NY(3423), WI(3423). **14:**DC(3423), DE(3423), GA(1438,3423), IA(3423), IN(3423), MD(3423), NJ(3423), NY(3423), OH(3423), PA(3423), RI(3423), VA(3423), WV(3423). **16:**VA(3423). **20:**WI(1333,3423). **23:**FL(3423). **25:**TX(3423). **29:**WI(1332,1335,3423). **30:**FL(3423), LA(3423). **31:**DE(3423), IA(3423), IL(3423), IN(3423), KY(3423), MI(3423), MO(3423), NY(3423), OH(638,3423), WI(1332,3423), WV(3423). **33:**IN(3423), LA(3423), WI(3423). **35:**FL(35). **37:**AL(3423), GA(1438,3423), NC(3423), PA(3423), VA(3423), WV(3423). **41:**TX(3423). **45:**GA(1438). **46:**AL(3423), GA(3423). **47:**AL(3423). **48:**WI(1347). **52:**WI(1354). **53:**NY(3423). **54:**GA(1438,3423), MS(3423). **55:**E states(82), AZ(82), CA(82), GA(1438), TX(82), WI(82,1333).

Phyllachora sp. **14:**GA(1438). **18:**GA(1438). **37:**GA(1438). **54:**GA(1438).

Venturia erysipheoides. On culms. **26:**LA(82,963).

Basidiomycotina - Uredinales

Physopella cameliae. Telial host. **6:**TX(740). **22:**TX(740).

Puccinia coronata. Heteroecious: telial host. **45:**MS(82).

P. dolosa var. *circumdata*. Telial host; aecial host unknown. **55:**TX(740).

P. emaculata. Heteroecious: telial host. **12:**E states(82), MS(82,2720), ND(82), OK(82,2872), SD(2242). **18:**TX(82). **38:**IN(82). **51:**E states(82), MS(2720), OK(2872), SD(82,2242), TX(82). **55:**IN(82), NC(82), NH(82), VA(82).

P. esclavensis. Heteroecious: telial host. **11:**AZ(1194,1203,3981), TX(742). **39:**HI(2903).

P. esclavensis var. *panicophila*. Telial host. **11:**AZ(82,1195), NM(82,1195).

P. graminis. Heteroecious: telial host. **51:**KS(82).

P. substriata. Heteroecious: telial host. **49:**TX(82).

Uromyces graminicola. Heteroecious: telial host. **3:**FL(2910). **4:**VA(2910). **5:**FL(2910). **6:**MS(2910). **26:**FL(35). **35:**AL(82), FL(35,82), HI(2903). **51:**E states(82), CT(2910), KS(2910,3021,3023), MS(2720), NC(1303), NE(2910), NJ(638), NM(82,2910), OK(2872), SD(82,2242), TX(740), WI(2910). **55:**AZ(1203,3981), FL(35,82), VA(82).

U. setariae-italicae. Heteroecious: telial host. **43:**Unk.(2910), FL(35,82), HI(2903), TX(82). **49:**Unk.(2910), TX(82).

U. vossiae. Autoecious. **16:**Unk.(2910).

Basidiomycotina - Ustilaginales

Entyloma dactylidis. Leaf smut. **4:**NC(1303). **17:**IL(82,1077), IN(82,1077), LA(82,1077), MD(82,1077). **55:**NC(1303).

Moesziomyces bullatus. **17:**IA(1077).

Sporisorium cenchri. Inflorescence, loose smut, seed smut. **12:**NC states(82), NE states(82), CO(1077), CT(1077), IA(1077), ID(82,1077), IN(1077), KS(82,1077), MA(1077), MN(1077), MO(1077), NC(1077), ND(1077), NE(1077), SD(1077,2242), WI(1077). **14:**IA(1077). **17:**E states(82), CT(1202), DC(1202), DE(1202), IA(1202), ID(1202), IL(1202), IN(1202), KS(1202,3021), MA(1202), MD(1202), MO(1202), MS(82,1202,2720), NC(1202,1303), NE(82,1202), NJ(1202), NY(1202), OH(1202), PA(82,1202), VA(1202), WI(82,1202), WV(953,1202). **28:**AZ(1077). **35:**FL(35). **38:**ID(1077), MN(1077), MO(1077), ND(1077), NE(1077), OR(1077), SD(1077), WA(1077), WI(1077). **45:**MO(82,1077). **51:**WI(82,1077). **55:**AZ(82), NC(1303), SC(634), WY(1077).

S. destruens. Head smut. **12:**NE(1077). **17:**KS(3021). **35:**FL(35). **38:**C states(82), NE states(82), Range of host(82), CA(82,1077), CO(1077), MN(1077), MO(1077), ND(378,1077), NE(1077), OR(1077), SD(1077,2242), WA(82,1077), WI(1077), WY(1077). **55:**FL(35).

Tilletia barclayana. Seed smut. **12:**PA(82,1077). **40:**AZ(82,1077). **51:**AZ(82,1077), IA(82,1077), IL(82,1077), KS(82,1077), MS(82,1077), NE(82,1077), OK(82,1077,2872).

T. maclaganii. Seed smut. **51:**CT(82,1077), IA(82,1077), IL(82,1077),

KS(82), NE(82,1077), NY(82,1077).
Ustilago heterogena. Seed smut. **17:**DC(1077), IA(82,1077), IL(82,1077), IN(82,1077), KS(82,1077), LA(1077), MD(82,1077), MS(82,1077,2720), NE(82,1077), NJ(82,1077), NY(82,1077). **51:**TX(82,1077).
U. panici-proliferi. **17:**KY(1077). **44:**HI(2299,2903).
U. trebouxii. Leaf smut. **11:**AZ(1194). **51:**NY(82), WI(1360).
Basidiomycotina - Other
Thanatephorus cucumeris. Brown patch, root rot, sharp eye spot. **12:**ND(3423). **51:**SD(3423). **55:**LA(3524).
Deuteromycotina - Hyphomycetes
Alternaria flagelloideum. Leaf spot. **14:**AL(82).
Alternaria sp. On leaves. **51:**WI(1349).
Beniowskia sphaeroidea. On leaves. **51:**KS(1445), NE(1445,2927), WI(1356,1361,3442).
Bipolaris sacchari. Leaf spot. **43:**HI(2903).
B. setariae. Leaf spot. **14:**HI(2903). **21:**GA(2208).
B. sorokiniana. Seedling blight, spot blotch, black point, root rot. **12:**MN(3423), ND(82,3423), SD(2242,3423), WI(82,1332,3423), WY(3423). **29:**WI(1343,3423). **38:**MN(3423), ND(3423). **51:**NE(82,3423), PA(4002).
B. yamadai. **12:**WI(1352). **29:**WI(1352).
Cerebella andropogonis. Blackhead, ergot. **51:**MS(82,2720,3423).
Cladosporium piricularioides. Leaf spot. **55:**NY(82).
Curvularia geniculata. Seedling blight, secondary leaf spot. **38:**ND(82). **51:**KS(82), NE(82).
C. protuberata. **49:**TX(652).
Curvularia sp. **55:**MD(651).
Drechslera gigantea. Zonate eye spot, zonate leaf spot. **6:**MD(82). **12:**WI(1345). **14:**MD(82,3423). **17:**MD(82), MO(3423). **55:**MD(82).
Exserohilum monoceras. Leaf blotch, leaf spot. **38:**ND(82,3423).
E. rostratum. Mold, secondary root rot. **38:**MN(82), ND(82,3423), SD(82). **49:**TX(652).
E. turcicum. Secondary root rot. **38:**ND(82).
Fusarium acuminatum. Secondary root rot. **12:**MT(3423), ND(82,3423), SD(82,2242,3423). **17:**NE(82,3423). **38:**ND(82,3423), NE(3423), SD(82,2242,3423), WA(3423). **48:**MN(3423). **51:**ND(82,3423), NE(82,3423), SD(2242,3423), WY(3423).
F. equiseti. Secondary root rot. **12:**Unk.(82). **38:**Unk.(82). **51:**ND(82), NE(82).
F. moniliforme. Secondary root rot. **38:**ND(3423), WA(3423).
F. oxysporum. Secondary root rot. **12:**Unk.(82). **38:**Unk.(82), CA(82), MN(3423), ND(3423), SD(2242,3423).
F. poae. Secondary root rot, silver top. **12:**ND(82,3423). **38:**ND(82,3423).
F. sporotrichioides. Secondary root rot. **12:**ND(82).
Helminthosporium sp. Leaf spot. **12:**IA(82,3423), ID(82,3423), MN(82,3423), ND(82,3423). **19:**FL(3423). **26:**FL(35). **35:**FL(35,82,3423). **38:**ND(82), NJ(82), TX(82). **43:**FL(35,82,3423), HI(2903). **55:**ND(82).
Microdochium bolleyi. Root necrosis. **12:**ND(3423). **38:**ND(3423), SD(2242,3423), WA(3423). **51:**ND(3423).
Nigrospora oryzae. Leaf mold. **26:**FL(35,82).
Ovularia sp. Leaf streak. **26:**FL(35,82).
Penicillium expansum. Blue mold. **38:**MN(3423), MT(3423), ND(3423), SD(3423).
Periconia circinata. Root rot. **38:**ND(3423).
Phaeoramularia fusimaculans. Leaf spot. **15:**WI(1357). **17:**IA(82,3423), NC(1303). **18:**AL(3423). **31:**WI(82,1333,3423). **32:**WI(1332,3423). **35:**FL(35). **42:**WI(1332,3423). **51:**ND(82,3423), WI(82,1332,3423). **52:**WI(1355). **54:**OK(2873,3423). **55:**AL(82), FL(35), OK(82), OR(82), TX(652), WI(82).
Pyricularia grisea. Blast gray spot. **38:**DE(82,3423), IL(82,3423), MN(82,3423), OH(82,3423). **49:**TX(82,3423).
Tetraploa divergens. Leaf spot. **9:**NC(82,3944). **31:**NC(82). **45:**MS(82,2720).
Deuteromycotina - Coelomycetes
Amerosporium atrum. **49:**OK(2872).
Ascochyta graminicola. Leaf spot. **12:**WI(82,1340).
A. missouriensis. Leaf rot and spot. **17:**ND(3423,3449). **51:**ND(3423,3449).
A. sorghi. Leaf spot. **12:**WI(3423,3449).
Ascochyta sp. Leaf rot, leaf spot. **17:**ND(82). **51:**ND(82).
Colletotrichum caudatum. On leaves. **55:**NM(82).
C. graminicola. Anthracnose. **10:**OK(2872,3423). **12:**MN(3423). **26:**FL(35,82,3423), LA(82). **29:**WI(3423). **38:**IA(82), MN(82). **51:**AL(82,3423), KS(82,3023,3423), LA(82,3423), MN(82,3423), NE(2927), SD(2242), WI(82,1332,3423). **55:**IA(3423), WI(82).
Colletotrichum sp. **29:**WI(1332).
Hendersonia panicicola. On culms. **51:**ND(82).
Leptostromella panici. On leaves. **27:**SC(3423). **55:**SC(82).
Phloeospora panici. On living leaves. **2:**WI(1357). **42:**WI(1345).
Phoma sorghina. Leaf spot. **17:**FL(35,82,3423). **51:**KS(3021).
P. terrestris. Pink root. **12:**MT(3423), ND(3423). **38:**ND(3423), SD(2242,3423). **51:**ND(3423).
Phyllosticta owensii. Gray leaf streak. **12:**WA(3423,3424).
P. panici. Leaf spot. **35:**FL(82,3423). **51:**OK(82,2873,3423).
Phyllosticta sp. Leaf spot. **40:**OK(2872).
Pseudoseptoria donacis. Eye spot, leaf spot. **51:**KS(82,3021,3423,3450), ND(82,3423,3450), SD(2242), WI(82,1334,3423,3450). **52:**NE(3423,3450).
Septoria arechavaletae. Leaf spot. **12:**VA(82,3423).
S. graminum. Leaf spot. **14:**IL(312).
S. sigmoidea. **51:**IA(82), WI(82,1346).
S. tandilensis. Black spot, leaf spot. **2:**WI(1357). **8:**WI(3423). **12:**ND(3423). **14:**WV(953,3429). **18:**MS(3429). **20:**WI(1333). **29:**WI(3423). **31:**WV(952,953). **42:**WI(1332,3423). **48:**WI(1342). **50:**WI(1333). **55:**FL(3423), GA(3423), NC(1303,3423), OK(3423), SC(3423), TX(3423), WI(82,1340).
Septoria sp. Leaf spot. **9:**MI(82). **14:**MD(82), OK(82,2872). **51:**OK(2872).
Stagonospora curvula. Leaf spot. **13:**TX(3423). **55:**TX(82).
S. simplicior. **29:**WI(1343). **31:**WI(1343), WV(953). **32:**WI(1343).
Wojnowicia hirta. On leaves. **51:**WI(1361).
Deuteromycotina - Other
Rhizoctonia solani. Secondary root rot. **12:**ND(82). **51:**SD(2242).

Pappophorum Schreb.

1. *P. mucronulatum* Nees

Ascomycotina
Phyllachora graminis. Common tar spot. **1:**TX(3423).

Paspalum L.

1. *P. almum* Chase [Combs paspalum]
2. *P. boscianum* Flüggé [bull paspalum]
3. *P. caespitosum* Flüggé
4. *P. ciliatifolium* Michx. [fringeleaf paspalum]
5. *P. conjugatum* Bergius [sour paspalum, sourgrass]
6. *P. dilatatum* Poir. [Dallis grass]
7. *P. distichum* L. [knot grass]
8. *P. floridanum* Michx.
9. *P. floridanum* Michx. var. *glabratum* Engelm. ex Vasey
10. *P. giganteum* Baldw.
11. *P. intermedium* Munro ex Morong & Britton
12. *P. laeve* Michx. [field paspalum]
13. *P. longepedunculatum* Leconte
14. *P. longipilum* Nash
15. *P. monostachyum* Vasey
16. *P. notatum* Flüggé [Bahia grass]
17. *P. paniculatum* L.
18. *P. plicatulum* Michx. [brownseed paspalum]
19. *P. pubescens* Muhl. (=*P. muhlenbergii* Nash)
20. *P. pubiflorum* Rupr. ex E. Fourn.
21. *P. pubiflorum* Rupr. ex E. Fourn. var. *glabrum* Vasey ex Lams.-Scribn.
22. *P. quadrifarium* Lam.
23. *P. scrobiculatum* L. (=*P. orbiculare* G. Forster)
24. *P. setaceum* Michx.
25. *P. stramineum* Nash
26. *P. supinum* Bosc
27. *P. urvillei* Steud. [Vasey grass]
28. *P. vaginatum* Sw.
29. *P. virgatum* L.
30. *Paspalum* sp.

Oomycetes
Pythium sp. Root rot. **16:**FL(35).

Ascomycotina
Acrospermum graminum. **30:**GA(1438).
Apiospora montagnei. **10:**GA(1438).
Balansia epichloe. On leaves. **5:**VA(3423).
Balansiopsis pilulaeformis. Black choke, black crust. **4:**FL(82,3423), GA(82), VA(82,3423). **19:**GA(1438,3423). **30:**FL(35,82), GA(82), VA(82).
Claviceps paspali. Ergot. **3:**FL(3423). **4:**Range of host(82), FL(3423), GA(3423), VA(3423). **6:**SE states(2205), SE states(82), AL(3423), AR(3423), CA(82), FL(35,3423), GA(1438,3423), HI(82,2903,3423), LA(3423), MD(3423), MS(2720,3423), NC(1303,3423,3944), OK(82,2873,3423), SC(3423), TX(82,3423), VA(3423). **7:**Range of host(82), MS(3423), OK(2873,3423), SC(3423), VA(3423). **8:**FL(3423), GA(3423), NC(1303,3423,3944), VA(3423). **10:**FL(3423). **11:**VA(3423). **12:**Range of host(82), AR(764,3423), MD(3423), NC(1303,3423,3944), OK(2873,3423), WV(953). **14:**VA(3423). **15:**TX(3423). **16:**Range of host(82), AZ(3423), FL(35,3423), GA(3423), LA(3423). **18:**Range of host(82), FL(3423), LA(3423), TX(3423). **19:**VA(3423). **20:**Range of host(82), GA(3423), TX(3423), VA(3423). **22:**VA(3423). **23:**HI(2903,3423). **26:**VA(3423). **27:**SE states(82), FL(3423), GA(1438), MD(3423), NC(1303,3423), OK(2873,3423), TX(3423). **28:**Range of host(82). **30:**Range of host(82), FL(35), GA(1438), MS(2720).
C. purpurea. **6:**GA(1438).
Leptosphaeria proteispora. **5:**HI(2903).
Myriogenospora atramentosa. Dwarfing, fasciation, black streak. **4:**VA(82). **12:**AL(82,3423). **16:**FL(35). **30:**FL(35).
Naemacyclus culmigenus. **30:**GA(1438).
Phaeosphaerella paspali. Leaf spot. **30:**OK(2872).
Phyllachora graminis. **30:**TX(652).
P. minutissima. Tar spot. **7:**FL(82,3423). **12:**AL(82), FL(82,3423). **16:**FL(35). **29:**AK(82), FL(82), TX(82). **30:**FL(35), TX(82).
P. paspalicola. Black spot, tar spot. **4:**AL(82,3423), FL(82,3423), GA(82,1438,3423). **19:**AL(3423), IL(3423), MO(3423), NC(3423), TX(82), VA(3423). **21:**TX(3423). **24:**AL(3423), DC(3423), NC(1303), NJ(3423), SC(3423). **25:**KS(3023), NE(82,3423), OK(82,2872,3423), TX(82,3423), WI(82,1333,3423). **26:**MO(3423). **30:**SE states(82), FL(35,82,3423), LA(3423), MO(82), MS(3423), NM(82).
P. punctum. Leaf spot, tar spot. **4:**FL(82,3423). **12:**LA(82,3423). **13:**FL(3438). **30:**FL(35).
Phyllachora sp. Black spot, tar spot. **4:**GA(1438). **6:**MS(82,2720).
Sclerotinia homoeocarpa. Blight, dollar spot. **16:**AL(1399), FL(35).
Basidiomycotina - Uredinales
Physopella compressa. Autoecious. **5:**FL(82). **30:**FL(35), TX(82).
Puccinia chaetochloae. Autoecious. **30:**FL(35).
P. coronata. Heteroecious: telial host. **24:**TX(82).
P. emaculata. Heteroecious: telial host. **25:**OK(82,2872).
P. esclavensis var. *panicophila*. Telial host. **30:**Unk.(82).
[*P. huberi*]. **6:**HI(2903).
P. levis. Telial host; aecial host unknown. **23:**HI(2903). **27:**FL(82), GA(82), LA(82), MS(82,2720), TX(82). **30:**FL(35).
P. substriata. Heteroecious: telial host. **4:**FL(82), TX(82). **5:**FL(82), TX(82). **8:**OK(2873). **16:**FL(35). **20:**TX(740). **25:**KS(82), NE(82), OK(82,2872), VA(82). **30:**FL(35), TX(652).
Basidiomycotina - Ustilaginales
Sorosporium paspali-thunbergii. Head smut. **5:**HI(2903). **6:**HI(82,2903). **23:**HI(2903). **30:**HI(82,2903).
Sphacelotheca cordobensis. Head smut. **30:**FL(35,82,1077).
Tilletia rugispora. Seed smut. **18:**LA(82,1077), TX(82,1077).
Ustilago schroeteriana. Seed smut. **7:**FL(82). **30:**CA(82), FL(35), HI(2903).
Basidiomycotina - Other
Lepiota sp. Fairy ring. **16:**FL(35).
Marasmius sp. Fairy ring. **16:**FL(35).
Thanatephorus cucumeris. **30:**LA(3524).
Deuteromycotina - Hyphomycetes
Alternaria sp. Leaf spot. **18:**TX(82).
Bipolaris micropus. Seedling blight, leaf spot. **2:**FL(82,3423). **30:**FL(35).
B. victoriae. Blight. **16:**IA(3423).
Cercospora setariae. Leaf spot. **25:**OK(82,2872,3423). **30:**OK(82).
Cerebella andropogonis. Blackhead. **6:**AL(82,3423), LA(82,3423). **8:**FL(3423), GA(3423). **27:**AL(82), LA(82).
Cladosporium sp. Head mold. **6:**LA(82), TX(82). **27:**LA(82), TX(82).
Curvularia sp. Leaf spot. **5:**HI(2299,2903). **16:**FL(35,82). **30:**FL(35,82).
Exserohilum rostratum. Leaf spot. **2:**GA(82,3423).
Fusarium heterosporum. Head mold. **6:**FL(35), MS(2720). **7:**FL(82), LA(82), MS(2720). **8:**MS(2720). **16:**FL(35). **30:**FL(35,82), LA(82).
Fusarium sp. Head mold, root rot. **6:**Range of host(82). **18:**TX(82). **27:**Range of host(82).
Helminthosporium sp. Leaf blight, leaf spot. **1:**TX(3423). **6:**FL(35). **16:**FL(35,82,3423), MS(2720). **17:**FL(3423). **27:**FL(82,3423). **30:**FL(35,82).
Pyricularia grisea. Blast, gray leaf spot. **18:**AL(82,3423), FL(82), KS(82). **25:**OK(2873,3423). **30:**AL(82), FL(35,82), KS(82).
Deuteromycotina - Coelomycetes
Ascochyta paspali. On leaves and sheaths. **30:**SE states(421).
Chaetospermum chaetosporum. On dead culms. **4:**FL(82). **30:**FL(35).
Colletotrichum graminicola. Anthracnose. **6:**FL(35,82,3423), GA(82,3423). **25:**OK(82,2873,3423), WI(82,1332,1356,3423).
Melanconium sp. On culms. **16:**FL(35).
Neottiosporina paspali. Leaf spot. **6:**GA(82,3423), NC(82,3423), TX(3423). **8:**NC(1303). **9:**NC(3423). **12:**AL(82,3423,3549). **16:**FL(35). **30:**FL(35), NC(82).
Phyllosticta sp. Leaf spot. **6:**SC(82).
Deuteromycotina - Other
Rhizoctonia solani. Summer blight, brown patch. **16:**FL(35). **19:**WV(955).
Rhizoctonia sp. Root rot. **5:**HI(2903).

Pennisetum Rich. ex Pers.

1. *P. alopecuroides* (L.) Spreng. [Chinese pennisetum]
2. *P. clandestinum* Hochst. ex Chiov. [kikuyu grass]
3. *P. glaucum* (L.) R. Br. (=*P. americanum* (L.) K. Schum., *P. spicatum* (L.) Körn., *P. typhoides* (N. L. Burm.) Stapf & Hubb., *P. typhoideum* Rich.) [bulrush millet, pearl millet]
4. *P. nervosum* (Nees) Trin.
5. *P. orientale* Rich.
6. *P. purpureum* Schumach. [elephant grass, napier grass]

Oomycetes
Pythium sp. Root rot. **3:**FL(35).
Ascomycotina
[*Leptosphaeria sacchari*]. **6:**HI(2903).
Basidiomycotina - Uredinales
Phakopsora apoda. **2:**HI(1147).
Phakopsora sp. **2:**HI(2903).
Puccinia stenotaphricola. **6:**FL(35).
P. substriata. **3:**GA(2201).
P. substriata var. *indica*. Heteroecious: telial host. **3:**AL(3843), FL(3843), GA(3843), TX(3843).
Puccinia sp. **2:**HI(2903).
Basidiomycotina - Ustilaginales
Moesziomyces bullatus. **3:**GA(1439,3844).
Sorosporium texanum. **4:**TX(1077).
Sphacelotheca pamparum. **1:**GA(1077), VA(1077).
Ustilago penniseti. Seed smut. **1:**VA(82).
Basidiomycotina - Other
Athelia rolfsii. Southern blight. **3:**FL(3802).
Thanatephorus cucumeris. Banded disease, root rot. **3:**FL(3802).
Deuteromycotina - Hyphomycetes
Bipolaris sacchari. Eye spot, leaf spot. **3:**GA(2201), HI(82,2903,3423). **6:**FL(35,82,3423), GA(82,3423), HI(82,2903,3423).
B. stenospila. Leaf spot. **3:**GA(2201).
Cercospora penniseti. Leaf spot. **3:**GA(2201).
Curvularia eragrostidis. On leaves. **3:**GA(2201).
C. geniculata. On leaves. **3:**GA(2201).
C. lunata. Head mold. **3:**GA(2201).
Curvularia sp. Leaf mold. **2:**HI(2903). **3:**GA(82). **6:**GA(82).
Drechslera gigantea. Zonate leaf spot. **1:**MD(82,3423).
D. graminea. Leaf spot. **6:**FL(35).
Exserohilum rostratum. Head mold, zonate eye spot. **3:**GA(82,2201,3423).
E. turcicum. Leaf spot. **2:**HI(2903).
Fusarium moniliforme. Head mold. **3:**GA(2201,3841).
Gloeocercospora sorghi. Zonate leaf spot. **3:**GA(82,2201).

Helminthosporium sp. **5:**GA(3423).
Olpitrichum tenellum. Head mold. **3:**GA(2201).

Deuteromycotina - Coelomycetes

Phyllosticta sp. Leaf spot. **6:**HI(2903).

Deuteromycotina - Other

Rhizoctonia solani. **3:**GA(2201).
R. zeae. **3:**GA(2201).
Sclerotium rolfsii. Southern blight. **3:**GA(2201).

Phalaris L.

1. *P. angusta* Nees ex Trin.
2. *P. aquatica* L. (=*P. tuberosa* L.) [bulbous canarygrass]
3. *P. arundinacea* L. (=*P. arundinacea* L. var. *picta* L.) [reed canarygrass]
4. *P. californica* Hook. & Arnott
5. *P. canariensis* L. [birdseed grass, canarygrass]
6. *P. caroliniana* Walter [May grass]
7. *P. minor* Retz. [littleseed canarygrass]
8. *P. stenoptera* Hack. (=*P. tuberosa* L. var. *stenoptera* (Hack.) Hitchc.) [Harding grass]
9. *Phalaris* sp.

Oomycetes

Pythium debaryanum. Damping-off, root necrosis, root and seed rot. **3:**ND(82,3423).
Sclerophthora macrospora. **3:**SD(2242).

Ascomycotina

Claviceps purpurea. Ergot. **3:**E states(1272), NE states(82), IL(1272), MN(1272), MT(82,1272), ND(1272), NY(82,1272), OK(82,1272,2873), SD(2242), WI(1272). **5:**MI(82,2862,3423), MN(82,3423).
Crocicreas culmicola. On culms. **9:**NY(505).
Diaporthe phaseolorum. **3:**IA(1228).
Erysiphe graminis. Powdery mildew. **3:**WA(82,3423), WI(82,3423).
Leptosphaeria sp. Leaf spot. **3:**WA(3432), WI(1347).
Phaeosphaeria sp. On living leaves. **3:**WI(1352).
Phyllachora phalaridis. Tar spot. **3:**MA(3423), MS(82,2720).
Phyllachora sp. **3:**WI(1360).

Basidiomycotina - Uredinales

Puccinia coronata. Heteroecious: telial host. **3:**WA(82), WY(1195). **5:**Unk.(82). **6:**Unk.(82), OK(2872).
P. graminis. Heteroecious: telial host. **1:**OK(2872). **3:**CA(82), NY(82). **4:**CA(82). **6:**OK(82,2873), TX(82). **9:**OK(82).
P. recondita. Heteroecious: telial host. **3:**AL(759).
P. sessilis. Heteroecious: telial host. **1:**OK(2872). **2:**OR(82). **3:**NC states(82), NE states(82), CO(82), OR(82), SD(2242). **4:**CA(82). **5:**CA(81). **6:**CA(82), KS(3021). **9:**OK(82).
P. sessilis var. *minor*. Heteroecious: telial host. **6:**TX(743).
P. striiformis. Telial host. **3:**AZ(1203). **9:**CA(82).

Basidiomycotina - Ustilaginales

Ustilago echinata. Leaf smut. **3:**MI(82), ND(82), NE(82), SD(82), WA(82).
U. hypodytes. Stem smut. **3:**NE(1077), WA(82,1077).
U. serpens. **3:**MI(1077), MN(1077), ND(1077), NE(1077), NY(1077), SD(1077,2242), WA(1077), WY(2342).
U. striiformis. Stripe smut. **3:**ME(82,1077), ND(1077), NE(1077), SD(1077), WA(1077).

Basidiomycotina - Other

Thanatephorus cucumeris. Brown patch, root rot, sharp eye spot. **2:**OR(3423). **3:**LA(3524), MN(3423), OR(3423), SD(3423). **5:**LA(3524). **6:**LA(3524). **7:**LA(3524).

Deuteromycotina - Hyphomycetes

Bipolaris sorokiniana. Seedling blight, spot blotch, black point, root rot. **2:**OR(3423). **3:**MD(82,3423), WI(82,1335,3423).
B. victoriae. **3:**CA(3125).
Cercospora sp. Leaf spot. **8:**OK(2872).
Cladosporium herbarum. Leaf mold. **2:**OR(82).
C. velutinum. On leaves. **5:**MS(82,2720).
Curvularia sp. **3:**AL(759).
Drechslera catenaria. Leaf spot. **3:**PA(3998).
D. gigantea. Zonate eye spot, zonate leaf spot. **3:**MD(82,3423), WI(1349).
Fusarium acuminatum. Secondary root rot. **2:**ND(82,3423). **3:**MN(3423).
F. equiseti. Root rot. **2:**ND(82).
Helminthosporium sp. Leaf spot. **3:**AL(759), MN(82), NC(82,3423).
Heterosporium avenae. On leaves. **3:**NE(82,3423).
Microdochium bolleyi. Root necrosis. **3:**MN(3423), ND(3423).
M. nivale. Leaf spot. **3:**AK(3428).
Ramularia collo-cygni. Leaf spot. **3:**ID(3423), MT(3423), ND(82,3423), SD(82,2242,3423), WI(82,1330,1340,3423), WY(82,3423).
R. pusilla. Leaf spot. **3:**ID(3429), MT(3429), ND(3429), WI(3429), WY(3429).
Rhynchosporium secalis. Scald. **3:**ID(3423), MT(82,3423), ND(82,3423), SD(82,2242,3423), WY(82,3423).

Deuteromycotina - Coelomycetes

Ascochyta brachypodii. Leaf spot. **3:**ND(82,3423,3449).
A. sorghi. **3:**WV(953).
Ascochyta sp. **3:**AL(759).
Colletotrichum destructivum. **3:**AL(759).
C. graminicola. Anthracnose, leaf spot. **3:**AK(3428), WA(82). **8:**OK(2872,3423).
Dilophospora alopecuri. Twist disease. **3:**WI(1340,3423).
Phaeoseptoria airae. **3:**AK(3441).
P. phalaridis. Leaf mold. **2:**OR(82). **3:**OR(82).
Pseudoseptoria donacis. Eye spot, leaf spot. **3:**ND(82,3423,3450), SD(2242), WA(82,3423,3450).
Septoria bromi var. *phalaricola*. Buff spot, leaf spot. **3:**OR(82,3423), WA(82).
Septoria sp. On leaves. **3:**PA(3999).
Stagonospora avenae f. sp. *triticea*. **3:**MT(3433).
S. foliicola. Tawny blotch, buff spot. **3:**CO(3423), IA(82,3423), ID(3423), KS(3021), KY(82,3423), MN(82,3423), MT(3423), ND(82,3423), SD(82,2242,3423), WI(82,1341,3423), WV(955). **4:**CA(82,3423).

Deuteromycotina - Other

Rhizoctonia solani. Secondary root rot, leaf spot. **2:**OR(82). **3:**SD(2242), WV(953).
Sclerotium rhizodes. Leaf rot. **3:**WI(82,3423).

Phleum L.

1. *P. alpinum* L. [alpine cat's timothy, alpine timothy]
2. *P. graecum* Boiss. ex Heldr.
3. *P. paniculatum* Huds.
4. *P. phleoides* (L.) Karsten
5. *P. pratense* L. [common timothy, herd's grass, mountain timothy]
6. *Phleum* sp.

Oomycetes

Pythium debaryanum. Damping-off, root necrosis, seed rot. **5:**MN(3423), WA(3423).
P. graminicola. Root necrosis. **5:**MN(3423), ND(3423), WA(82,3423).
Sclerophthora macrospora. **5:**SD(2242).

Ascomycotina

Balansia strangulans. **5:**OR(3435).
Botryosphaeria festucae. On dead culms and leaves. **1:**NJ(82). **5:**IL(82,3590). **6:**NJ(82).
Claviceps purpurea. Ergot. **1:**E states(82), AK(82), MN(82), OR(82), WA(82). **5:**E states(3423), AK(526,3423), CT(3423), IA(3423), MI(2862), MN(3423), ND(3423), OR(3423), SD(2242), WI(3423).
Claviceps sp. Ergot. **5:**MS(2720).
Cochliobolus sativus. Blind seed disease. **5:**OR(3423).
Crocicreas gramineum. **1:**AK(3431).
Diaporthe arctii. On dead culms. **5:**NJ(82).
Epichloe typhina. Choke. **5:**IA(82), ME(82,3423), NE(82,3423), PA(82,3423).
Erysiphe graminis. Powdery mildew. **5:**MN(82,3423), NY(1890), PA(82,3423), WI(3423).
Gaeumannomyces graminis. Take-all, on culms and sheaths. **5:**AK(3441), ID(82,3423), NY(82,3423).
Gloeotinia granigena. **5:**OR(1075).
Leptosphaeria asparagina. **5:**WA(3814).
Leptosphaerulina argentinensis. **5:**PA(1297).
Lophodermium arundinaceum. **1:**AK(3431).
L. arundinaceum var. *gramineum*. On leaves. **5:**NJ(82), OH(82), VA(82).
L. phlei. **5:**NJ(82).
Mycosphaerella tassiana. On overwintered herbage. **1:**CA(642), WA(3815).
M. tulasnei. **1:**AK(3431).
Phaeosphaeria caricinella. On dead culms and overwintered herbage.

1:CA(82,642), WA(3814).
P. herpotrichoides. On culms. 5:NJ(82), WA(3814).
P. typharum. On overwintered culms and leaves. 1:CA(642). 6:WA(3814).
Phyllachora graminis. Tar spot. 5:ME(82,3423), NH(1897).
Pleospora herbarum. 1:WA(3814).
Trichothyrina alpestris. Leaf spot. 1:AK(3426,3431).

Basidiomycotina - Uredinales

[*Puccinia brachypodii*]. 5:WY(1195).
P. brachypodii var. *poae-nemoralis*. Heteroecious: telial host. 1:Unk.(754), AK(526,3431), CO(82). 5:Unk.(754), NY(2999), WY(1195,3377).
P. coronata. Heteroecious: telial host. 1:IA(82), WV(82). 5:IA(82), WV(82). 6:IA(82), WV(82).
P. graminis. Heteroecious: telial host. 2:WA(82). 3:OR(82). 4:CA(82), WA(82). 5:AZ(3981), MA(1390), MT(1195), NH(1890), OR(648), SD(2242), WV(952).
P. graminis subsp. *graminicola*. Heteroecious: telial host. 5:Range of host(82), GA(1439), ID(638), NH(1897), NY(2998), OK(2872). 6:CA(82), WA(82).

Basidiomycotina - Ustilaginales

Entyloma dactylidis. Leaf smut. 5:IA(82,1077), IL(82,1077), WI(82,1077).
Urocystis agropyri. Flag smut. 1:CO(1077), UT(1077), WY(1077,2924). 5:MO(82,1077), NH(82), PA(82), VT(82,1077), WI(1341).
Ustilago hypodytes. Stem smut. 4:WA(1077). 5:WA(82,1077). 6:WA(82).
U. striiformis. Stripe smut. 1:MT(1077), WY(1077). 5:Range of host(82), CA(1077), CO(1077), CT(1077), IA(1077), ID(1077), IL(1077), IN(1077), KS(1077), KY(1077), MA(1077), MD(1077), MI(1077), MN(1077), MO(1077), MT(1077), ND(1077), NJ(494,1077), NY(1077,3005), OH(1077), OR(1077), PA(1077), UT(1077), WA(1077), WI(1077), WV(952,1077), WY(1195).

Basidiomycotina - Other

Coprinus psychromorbidus. 1:AK(3441).
Thanatephorus cucumeris. Brown patch, root rot, sharp eye spot. 5:MN(3423), MT(3423), ND(3423), OR(3423), SD(3423), WY(3423).

Deuteromycotina - Hyphomycetes

Bipolaris sorokiniana. Seedling blight, spot blotch, black point, root rot, leaf spot. 5:ND(3423), WA(82,3423).
B. victoriae. Blight, on seed. 5:IA(82,3423), ND(3423).
Cercosporidium graminis. Leaf spot, leaf streak, brown stripe. 1:AK(526,3431), CO(3423), WY(2924,3423). 5:E states(3423), Range of host(82), AK(526), CO(3423), DE(3543), IA(3423), ID(3423), KS(3021), KY(3423), MD(3423), MN(3423), MO(3423), MS(2720), MT(3423), ND(3423), NY(2999,3005), OR(3423), SD(2242,3423), UT(3423), VT(3423), WA(3423), WI(3423), WV(952), WY(2924). 6:NY(2998,3004).
Cladosporium herbarum. Leaf mold. 1:AK(3431), CA(82). 5:MN(82), NY(82).
C. phlei. Eye spot, leaf spot, purple spot. 1:AK(3428,3431), CA(3423), CO(3423), WA(3438), WY(2924). 4:WA(82,3423). 5:E & C states(82), AZ(82,3423), CA(82,3423), CO(82), CT(3423), ID(82,3423), IL(3423), IN(3423), KY(3423), MI(3423), MN(3423), MO(3423), MT(3433), NJ(494), NV(3423), NY(2999,3005,3423), OH(3423), OR(82,3423), PA(3423), WA(82,3423), WI(3423), WV(952), WY(82,3423). 6:NY(2998,3004).
Curvularia sp. Leaf mold. 5:MO(82).
Drechslera gigantea. Zonate leaf spot. 5:MD(82,3423), NH(1895).
D. phlei. Leaf streak. 5:NE states(1293), NY(376), PA(376,1293).
D. triseptata. Leaf spot. 5:MD(82), VA(82).
Endoconidium temulentum. Blind seed disease. 5:OR(82).
Fusarium acuminatum. Root rot. 1:ND(3423). 4:ND(82,3423). 5:Unk.(82), MN(3423), MT(3423), ND(3423), SD(3423), WA(3423), WY(3423).
F. avenaceum. 1:AK(3431).
F. equiseti. Secondary root rot. 5:ND(82).
F. heterosporum. Head mold, probably following ergot. 5:MN(82), WI(82), WV(82).
F. oxysporum. Secondary root rot. 6:Unk.(82).
F. poae. Secondary root rot, silver top. 5:IA(82,3423), IN(82,3423), MN(82,3423), NY(82,3423).
Fusarium sp. Seedling blight. 1:Unk.(82).
Helminthosporium sp. Leaf blight, leaf spot. 5:CT(82), MN(82,3423), MO(82), NH(1895), NY(2999), WV(952).
Mastigosporium rubricosum. Eye spot, leaf spot. 1:AK(3428). 5:ID(3423), WA(3423).
Microdochium bolleyi. Root necrosis. 1:ND(3423). 4:ND(3423). 5:MN(3423), ND(3423), NE(3423), SD(3423), WY(3423).
M. nivale. Pink snow mold. 4:WA(82). 5:WA(82,3423).
Periconia circinata. Root rot. 5:MT(3423), ND(3423).
Ramularia pusilla. Leaf spot. 1:AK(3431,3432).
Rhynchosporium secalis. Scald. 5:NY(2998,3005).
Stemphylium botryosum. Seed mold. 5:NY(82).

Deuteromycotina - Coelomycetes

Ascochyta hordei. Leaf spot. 4:WA(3423,3449).
A. phleina. Leaf spot. 5:IL(312), MN(82,3423,3449).
A. sorghi. 5:WV(953).
Colletotrichum graminicola. Anthracnose. 5:NE states(82), IL(3423), IN(3423), MN(82,3423), MO(82,3423), NJ(3423), OH(3423), OK(2872,3423), PA(3423), WV(952).
Hendersonia sp. Leaf mold. 5:MN(82).
Leptothyrella graminis. 6:WA(3816).
Macrophoma phlei. On leaves. 5:IL(82,3423).
Phaeoseptoria airae. Leaf spot. 1:AK(3431,3432).
P. poae. On dead leaves. 5:WA(3425).
Phoma terrestris. Pink root. 1:ND(3423). 5:MN(82), ND(82,3423), SD(3423).
Pseudoseptoria donacis. Leaf speckle. 5:AK(82), ND(82).
P. stomaticola. Leaf speckle, eye spot. 1:MT(3450). 5:AK(526,3423,3450), CO(82,3423,3450), ND(3423,3450), WA(3423,3450).
Rhynchosporina meinersii. Leaf spot. 1:WA(3425). 5:MT(3433), WA(82,3423).
Septoria sp. Leaf spot. 5:IL(82), PA(82).
Sirococcus phlei. On culms. 5:IL(82,3590).
Sphaeropsis malorum. On leaves. 5:NJ(82), VA(82).
Stagonospora subseriata. Leaf mold. 5:MN(82).
Wojnowicia hirta. Secondary foot rot, rot of mature straw. 5:MN(3423), ND(3423), WY(3423).

Deuteromycotina - Other

Rhizoctonia solani. Summer blight, root rot. 5:ND(82), WV(955).
Sclerotium rhizodes. On leaves. 5:WI(3423).
Sclerotium sp. Leaf rot. 5:MA(82), WI(82).

Phragmites Adans.

1. *P. australis* (Cav.) Trin. ex Steud. (=*P. communis* Trin.) [carrizo, common reed]
2. *Phragmites* sp.

Oomycetes

Pythium debaryanum. Damping-off, root necrosis, seed rot. 1:Unk.(82), ND(3423).
P. graminicola. 1:ND(3423).
P. ultimum. Root rot and necrosis. 1:ND(82,3423).

Ascomycotina

Claviceps purpurea. Ergot. 1:IA(82,3423), ND(3423), NJ(82,3423), OK(82,2873,3423).
Graphyllium dakotense. 1:Unk.(82).
G. hysterioides. On dead culms. 1:ND(82,377).
Lophiostoma arundinis. On dead culms. 1:ND(82,377).
Mycosphaerella sp. On leaves. 1:IN(82).
Phaeosphaeria clavispora. On dead culms. 1:IA(82), ND(82,377), NY(82), SD(82), WI(82).
P. nigrans. On dead culms. 1:IA(82).
Pseudographis phragmitis. On dead culms. 1:NY(82).

Basidiomycotina - Uredinales

Puccinia magnusiana. Heteroecious: telial host. 1:IA(82), ND(82,378), NE(82), SD(82,2242), WI(82).
P. phragmitis. Heteroecious: telial host. 1:NE states(82), CO(82), ND(378), WA(82), WY(1195).

Basidiomycotina - Ustilaginales

Neovossia iowensis. Seed smut. 1:CT(82,1077), IA(82,1077), IL(1077), MD(82,1077).
Ustilago grandis. 1:CA(1077).

Deuteromycotina - Hyphomycetes

Arthrinium arundinis. 1:Unk.(82).
A. phaeospermum. Black mold of culms and leaves. 1:Range of host(82).

Cercosporidium graminis. Leaf streak, brown stripe. **1**:OR(82), WI(82,3423).
Deightoniella arundinacea. Black leaf mold, leaf spot. **1**:MN(82,3423), ND(82,3423), NE(82,3423), SD(2242,3423), WI(82).
Fusarium acuminatum. Secondary root rot. **1**:ND(3423).
Helminthosporium sp. Leaf spot. **1**:FL(35).
Scolicotrichum maculicola. On living leaves. **1**:IA(82,3423), KS(82,973,3423), ND(3423), NE(82,3423), OK(2872,3423), OR(82,3423), SD(2242), WA(82), WI(3423).

Deuteromycotina - Coelomycetes

Hendersonia arundinacea. On culms and leaf sheaths. **1**:ND(82,378), TX(82).
H. grantii. On leaf sheaths. **1**:WA(82). **2**:WA(832).
Phoma sorghina. Eye spot. **1**:MN(3442).
Phragmopeltis phragmitis. On dead culms. **1**:WA(82).
Pseudoseptoria donacis. Eye spot, leaf spot. **1**:ND(82,3423,3450), SD(2242).
Stagonospora montagnei. **1**:ND(82,378).
S. phragmitis. **1**:IN(82).
S. simplicior. **1**:WI(82).
S. vexatula. Leaf spot. **1**:KS(3021), NJ(3423).

Phyllostachys Siebold & Zucc.

1. *P. aurea* Rivière & C. Rivière [fishpole bamboo, golden bamboo]
2. *P. bambusoides* Siebold & Zucc. (=*P. quilioi* Rivière & C. Rivière) [hardy timber bamboo, Japenese timber bamboo, madake]
3. *P. nigra* (Lodd.) Munro (=*P. henonis* Mitf.) [black bamboo]
4. *Phyllostachys* sp.

Ascomycotina

Apiospora sp. On dead culms. **2**:FL(35,82), GA(82).
Laestadia sp. On culms. **2**:FL(82).
Mycosphaerella sp. Leaf spot. **2**:CA(82). **3**:CA(82).
Scirrhia lophodermioides. On dead culms. **2**:FL(35,82).

Basidiomycotina - Uredinales

[*Puccinia melanocephala*]. **1**:FL(35,82), GA(82), MS(82), TX(82). **2**:FL(35,82), GA(82,1439), MS(82), TX(82).
P. phyllostachydis. Telial host; aecial host unknown. **1**:MS(2720). **2**:MS(2720). **4**:MS(2720), TX(740).

Basidiomycotina - Ustilaginales

Ustilago shiraiana. Stem smut. **1**:FL(35). **2**:CA(82,1077), FL(35,82,1077), LA(82,1077), MD(82,1077), MS(82,1077,2720), TX(82,1077). **3**:CA(82,1077), FL(82,1077), LA(82,1077), MD(82,1077), MS(82,1077,2720), TX(82,1077). **4**:MS(2720), NC(1303).

Basidiomycotina - Other

Schizophyllum commune. On dead culms. **1**:CA(82), TX(82). **2**:CA(82), TX(82).

Deuteromycotina - Hyphomycetes

Arthrinium phaeospermum. On dead or dying culms. **1**:Range of host(82). **2**:Range of host(82). **3**:Range of host(82).
Cladosporium herbarum. On leaves. **1**:OR(82).
Nigrospora oryzae. On senescent leaves. **2**:FL(35,82).
Papularia vinosa. On culms. **1**:FL(35). **2**:GA(82).

Piptochaetium J. Presl in K. Presl

1. *P. fimbriatum* (H. B. K.) Hitchc. [pinyon ricegrass]

Basidiomycotina - Uredinales

Puccinia chisosensis. Autoecious. **1**:TX(742).
P. durangensis. **1**:AZ(3981).
P. neocoronata. **1**:AZ(1195,1203,3981), TX(742).

Pleuropogon R. Br.

1. *P. refractus* (A. Gray) Benth. ex Vasey [nodding semaphore-grass]

Deuteromycotina - Coelomycetes

Ascochyta brachypodii. Leaf spot. **1**:OR(3423,3449).

Poa L.

1. *P. alpina* L. [alpine bluegrass]
2. *P. alsodes* A. Gray
3. *P. ampla* Merr. [big bluegrass]
4. *P. annua* L. [annual bluegrass, dwarf meadow grass, low speargrass, six-weeks grass]
5. *P. arachnifera* Torr. [Texas bluegrass]
6. *P. arctica* R. Br. [arctic bluegrass]
7. *P. arida* Vasey [plains bluegrass]
8. *P. autumnalis* Muhl. ex Elliott
9. *P. bigelovii* Vasey & Lams.-Scribn. [Bigelow bluegrass]
10. *P. bolanderi* Vasey
11. *P. bulbosa* L. [bulbous bluegrass]
12. *P. canbyi* (Lams.-Scribn.) Howell [Canby bluegrass]
13. *P. chapmaniana* Lams.-Scribn.
14. *P. compressa* L. [Canada bluegrass, wiregrass]
15. *P. curtifolia* Lams.-Scribn.
16. *P. cusickii* Vasey [Cusick bluegrass]
17. *P. eminens* K. Presl
18. *P. epilis* Lams.-Scribn. [skyline bluegrass]
19. *P. fendleriana* (Steud.) Vasey (=*Aristida fendleriana* Steud.) [mutton bluegrass]
20. *P. glauca* Vahl
21. *P. glaucifolia* Lams.-Scribn. & B. S. Williams
22. *P. gracillima* Vasey [Pacific bluegrass, slender bluegrass]
23. *P. interior* Rydb. [inland bluegrass]
24. *P. involuta* Hitchc.
25. *P. juncifolia* Lams.-Scribn. [alkali bluegrass]
26. *P. kelloggii* Vasey
27. *P. laxiflora* Buckley
28. *P. leibergii* Lams.-Scribn. [Leiberg bluegrass]
29. *P. leptocoma* Trin. [bog bluegrass]
30. *P. longifolia* Trin.
31. *P. longiligula* Lams.-Scribn. & Willd. [longtongue mutton grass]
32. *P. macrantha* Vasey
33. *P. macrocalyx* Trautv. ex C. A. Mey. (=*P. hispidula* Vasey)
34. *P. marcida* Hitchc.
35. *P. nemoralis* L. [wood bluegrass]
36. *P. nervosa* (Hook.) Vasey (=*P. wheeleri* Vasey) [Wheeler bluegrass]
37. *P. nevadensis* Vasey ex Lams.-Scribn. [Nevada bluegrass]
38. *P. palustris* L. (=*P. serotina* Ehrh.) [fowl bluegrass, roughstalk bluegrass]
39. *P. paucispicula* Lams.-Scribn. & Merr.
40. *P. pratensis* L. [Kentucky bluegrass, speargrass]
41. *P. pratensis* L. subsp. *irrigata* (Lindm.) H. Lindb. (=*P. irrigata* Lindm.)
42. *P. reflexa* Vasey & Lams.-Scribn. ex Vasey
43. *P. rupicola* Nash [timberline bluegrass]
44. *P. sandbergii* Vasey [Sandberg bluegrass]
45. *P. scabrella* (Thurb.) Benth. ex Vasey [pine bluegrass]
46. *P. secunda* J. Presl [Sandberg bluegrass]
47. *P. stenantha* Trin.
48. *P. suksdorfii* (Beal) Vasey ex Piper
49. *P. sylvestris* A. Gray
50. *P. trivialis* L. [roughstalk bluegrass]
51. *P. vaseyochloa* Lams.-Scribn.
52. *Poa* sp.

Myxomycetes

Physarum cinereum. Slime mold. **14**:Unk.(82). **40**:Unk.(82), KS(3021).
Physarum sp. Slime mold. **40**:NC(1303).

Plasmodiophoromycetes

Sorosphaera radicalis. On roots. **4**:CA(82).

Chytridiomycetes

Physoderma alpina. On living leaves. **22**:UT(3438). **42**:UT(3438).
P. gerhardtii. 'Spot floral spot.' **40**:ND(82,1077).
Synchytrium graminicola. Purple leaf speckle. **4**:OR(3423). **25**:OR(3423). **40**:OR(3423), WA(3423). **42**:UT(3435). **46**:WA(3423).
Synchytrium sp. Purple speckle. **4**:OR(82), WA(82). **25**:OR(82), WA(82). **40**:OR(82), WA(82). **44**:OR(82), WA(82).

Oomycetes

Phytophthora cactorum. Leaf rot. **52**:OR(82).
Pythium debaryanum. Damping-off, root necrosis, seed rot. **3**:ND(3423), WA(3423), WY(3423). **11**:ND(82,3423). **37**:WA(3423). **38**:MN(3423), ND(82,3423), SD(2242,3423). **40**:MN(3423), ND(3423), SD(2242,3423), WA(3423). **44**:ND(82), WA(82). **46**:ND(3423), WA(3423). **52**:ND(82).
P. graminicola. **25**:ND(3423). **37**:ND(3423). **38**:MT(3423), SD(2242). **40**:'Pythium root rot. Pythium blight.' ND(3423). **46**:WY(3423).
P. ultimum. Root necrosis. **3**:ND(3423), WA(3423), WY(3423). **12**:ND(3423), WY(3423). **44**:ND(82). **46**:ND(3423).

52:ND(82).
Pythium sp. Cottony blight, root rot. **4:**CA(988). **40:**CA(988), NC(1303). **50:**HI(2903).
Sclerophthora macrospora. Yellow tuft. **40:**'Downy mildew.' SD(2242). **52:**RI(2522).

Ascomycotina

Clathrospora pentamera. On dead sheaths and stems. **52:**AK(1938).
Claviceps purpurea. Ergot. **3:**ID(3423), MT(3423), ND(3423), WA(3423). **4:**MO(3423), OK(82,2873,3423). **7:**ND(82,3423). **12:**ID(3423), MT(3423), ND(3423), WA(3423), WY(3423). **14:**NC states(82), NE states(82), IL(312), KY(82), MN(3423), MO(3423), MT(82,3423), ND(3423), OK(82,2873,3423), WA(82,3423), WI(3423). **19:**W states(82), UT(3423), WA(82). **25:**OR(3423). **30:**UT(3423). **35:**WA(3423). **36:**WY(3423). **37:**W states(82), MT(3423), ND(3423), OR(3423), WA(82,3423). **38:**MN(3423), ND(82,3423), SD(2242,3423). **40:**NC states(82), NE states(82), AK(3441), IA(3423), KY(82,3423), MN(3423), MT(82,3423), ND(3423), NY(3423), OH(3423), OK(82,2873,3423), PA(3423), UT(3423), WA(3423), WI(3423), WY(3423). **44:**W states(82), WA(82). **45:**W states(82), UT(3423), WA(82). **46:**ID(3423), WA(3423). **47:**WA(3423). **52:**W states(82), CA(82), MO(3423), WA(82), WY(819).
Crocicreas gramineum. On culms and leaves. **1:**AK(3431). **6:**AK(1384). **14:**AK(3431).
C. melanosporum. Brown patch. **36:**UT(505). **52:**UT(505).
Dothidea atramentaria. On leaves. **52:**SC(634).
Epichloe typhina. Cat-tail, choke. **14:**ND(82,3423). **38:**DE(82,3423), NY(82,3423). **40:**DE(82,1944,3423), IA(3423), ND(82), NE(3423). **44:**ND(82). **45:**WA(82). **46:**ND(3423). **47:**WA(3423). **52:**GA(1438).
Erysiphe graminis. Powdery mildew. **1:**WY(2924,3423). **2:**SD(2242). **3:**MT(3423), OR(3423), WA(3423), WY(1195,3423). **4:**WI(1357). **5:**IA(3423). **7:**MT(3423), ND(3423). **12:**CO(3423), MT(3423), ND(3423), OR(3423), SD(2242,3423), UT(3423), WA(3423), WY(1195,3423). **14:**AK(3431), ID(3429), KS(3021), MN(82,3423), MT(82,3423), OK(2872,3423), WI(3423), WV(82,955,3423). **16:**MT(3423), WY(3423). **18:**WA(3423). **21:**ND(3423). **22:**OR(3423), WA(3423), WY(3423). **23:**ME(82,3423), MI(82), MN(82,3423), MT(3423), ND(82), OR(82), SD(82,2242,3423), WA(82,3423), WY(82,3423). **25:**C states(82), CA(82), OR(3423), WA(82). **29:**OR(3423). **30:**WA(3423). **35:**MT(3423), SD(3423), WA(3423), WI(1356), WY(3423). **36:**VT(3423), WY(82,3423). **37:**ND(3423), WA(3423). **38:**CO(82,3423), IA(3423), ME(82), MI(82,3423), MN(82), MT(3423), ND(82,3423), OR(82,3423), SD(82,2242,3423), UT(82,3423), WA(82,3423), WI(3423), WY(82,3423). **40:**'Powdery mildew.' E states(3423), Range of host(82), AK(526,3423,3431), AR(3029), CA(638,642,3423), CO(3423), DE(3423), IA(3423), ID(3423), KS(3021,3023), MI(3423), MN(3423), MO(3423), MS(2720), MT(3423), NC(1303), ND(377,3423), NE(3423), NM(3423), OR(3423), SD(2242,3423), UT(3423), WA(3423), WI(1357,3423), WY(1195,3423). **44:**C states(82), CA(82), WA(82). **45:**C states(82), CA(82), MT(3423), OR(3423), WA(82,3423). **46:**CA(3423), CO(3423), MT(3423), OR(3423), WA(3423), WY(2924). **49:**WA(3423). **50:**OR(3432). **51:**OR(3423), WA(3423). **52:**C states(82), CA(82), WA(82).
Gaeumannomyces cylindrosporus. On roots. **4:**RI(1720).
G. graminis. Take-all, on culms and sheaths. **12:**WA(3423). **14:**NY(82,3423). **40:**'Take-all patch.' ID(3423,3754), WA(3446). **41:**AK(3441). **52:**WA(82).
Gibberella zeae. Culm rot. **4:**NY(82), WI(82). **40:**NY(82).
Leptosphaeria korrae. Patch disease. 'Necrotic ring spot.' **40:**NY(3324), WA(1433), WI(3954).
L. muirensis. On living leaves. **4:**AK(3428,3431).
Leptosphaeria sp. **1:**AK(3431).
Leptosphaerulina trifolii. On dead leaves. **4:**GA(1297,1438). **38:**NY(82). **50:**NY(82).
Lewia infectoria. **20:**AK(526).
Lophodermium arundinaceum. On dead culms and leaves. **1:**AK(3431), CO(3423), WY(2924,3423). **6:**WA(3423). **14:**CA(82,3423). **18:**CA(3423). **19:**AZ(82,3423), CO(82), UT(82,3423), WY(2924,3423). **25:**AZ(82,3423). **28:**CA(3423). **31:**AZ(3423). **40:**CA(82), ID(82,3423). **43:**CA(3423). **44:**AZ(82). **46:**AZ(3423), WY(2924). **47:**AK(3431). **52:**AZ(82), CA(82).
L. culmigenum. On culms. **14:**NJ(82).
L. gramineum. **52:**WA(3815).
Magnaporthe sp. Summer patch disease. **4:**MD(2024), MI(2024), NE(2024), NY(2024), RI(2024). **40:**MD(2024), MI(2024), NE(2024), NY(2024), RI(2024).
Mycosphaerella muhlenbergiae. **52:**WA(3815).
M. tassiana. On dead sheath. **25:**OR(82). **47:**AK(1938).
M. tulasnei. **1:**AK(3431). **47:**AK(3431). **52:**OR(82).
Mycosphaerella sp. On leaves. **17:**AK(3441). **37:**CO(82). **44:**WA(82). **52:**AK(82), WA(82).
Phaeosphaeria caricinella. **18:**WA(3814).
P. herpotrichoides. **18:**WA(3814). **52:**WA(3814).
P. heterospora. On overwintered herbage. **18:**CA(642).
P. typharum. **18:**WA(3814). **52:**WA(3814).
P. vagans. Leaf mold. **38:**ND(82,377).
Pleospora heleocharidis var. *arctica*. On leaves. **45:**OR(82).
P. magnusiana. **6:**AK(526).
P. rainierensis. **52:**WA(3814).
Pyrenopeziza karstenii. **1:**AK(3431). **40:**AK(3431). **47:**AK(3431).
Pyrenophora trichostoma. **40:**AK(526).
Sclerotinia homoeocarpa. Turf rot. 'Dollar spot.' **4:**CA(988), MA(3423). **40:**NE states(3423), NE states(3423), MA(3423), PA(3423), RI(3423).
Scopinella solani. On dead culms. **37:**WY(82).
Sphaerella graminum. **47:**AK(526).
Trichothyrina alpestris. On sheaths. **6:**AK(3431). **47:**AK(3426,3431).
Wettsteinina niesslii. On dead stems. **36:**CA(81).
Xylaria multipartita. On dead roots. **40:**NC(2408).

Basidiomycotina - Uredinales

[*Puccinia brachypodii*]. **1:**WY(1195). **23:**CO(1195), WY(1195). **29:**WY(1195). **42:**WY(1195).
P. brachypodii var. *arrhenatheri*. Heteroecious: telial host. **10:**Unk.(754).
P. brachypodii var. *poae-nemoralis*. Heteroecious: telial host. **1:**AK(3426,3428,3431), WY(2924,3377). **4:**'Leaf rust.' Unk.(754), CA(82), DE(82), NY(82), OR(82). **5:**Range of host(82). **7:**Range of host(82), Unk.(754), WY(3377). **9:**Unk.(754), NM(82). **11:**OR(82), WA(82). **12:**Unk.(754). **14:**Range of host(82), WV(955). **18:**WY(3377). **19:**Range of host(82), Unk.(754). **20:**AK(526). **25:**Range of host(82). **29:**Unk.(754), CO(1195), WY(3377). **32:**Unk.(754). **33:**AK(3441). **36:**Range of host(82), Unk.(754), WY(2924,3377). **37:**Range of host(82), Unk.(754). **38:**Unk.(754), CO(82), ID(82), MT(82), NY(82), OR(82), SD(2242), WA(82), WI(82), WV(82), WY(1195,2924,3377). **39:**AK(3431). **42:**Unk.(754), CO(1195). **44:**Range of host(82). **45:**Range of host(82). **47:**Unk.(754), AK(526,3431). **50:**Unk.(754), OR(82), WA(82). **51:**Unk.(754). **52:**Range of host(82), WY(2924,3377).
P. coronata. Heteroecious: telial host. **5:**AL(82). **14:**KS(82), MO(82), PA(82). **24:**TX(742).
P. crandallii. Heteroecious: telial host. **19:**AZ(1195,1203,3379,3981), CO(82,1195). **31:**AZ(1195,1203,3981).
P. graminis. Heteroecious: telial host. **4:**'Leaf rust.' KS(3023), WI(82). **5:**ND(82). **7:**KS(3021,3023), MI(82), NE(82), TX(82). **11:**ND(82). **13:**Unk.(82). **14:**E states(82), IL(82), WI(82), WI(1332), WV(955). **23:**WA(82). **25:**Unk.(82). **37:**Unk.(82), WA(82). **38:**CA(82), IN(82). **40:**'Stem rust.' E states(82), IL(82), KS(3021,3023,3024), NC(3944), WI(82), WV(953), WY(819). **45:**Unk.(82). **50:**TN(82), WI(82). **52:**CA(82).
P. monoica. Heteroecious: telial host. **46:**WY(749,1195).
P. poarum. Heteroecious: telial host. **40:**CA(81), ND(378).
P. recondita. Heteroecious: telial host. **7:**MT(82), NE(82). **19:**W states(82), CA(82), CO(82), WA(82). **23:**CO(82). **25:**W states(82), CA(82), CO(82), WA(82). **36:**MT(82), UT(82). **37:**W states(82), CA(82), CO(82), WA(82). **40:**'Leaf rust.' E states(82), CA(82), HI(2903), NC(1303), NE(2926), WA(82). **44:**W states(82), CA(82), CO(82), WA(82). **45:**W states(82), CA(81), CO(82), WA(82). **46:**SD(2242). **52:**W states(82), CO(82), WA(82).
P. striiformis. Telial host. 'Stripe rust.' **40:**HI(2532).
Puccinia sp. **4:**HI(2903).
Uromyces dactylidis. Heteroecious: telial host. **12:**WY(1195). **40:**IN(82), MI(82), SD(82,2242), VT(82). **44:**ND(82). **45:**OR(82). **46:**SD(2242). **52:**UT(82), WA(82).

Basidiomycotina - Ustilaginales

Entyloma dactylidis. Dark leaf smut, pale leaf smut. **4:**'Blister.' AK(3431), CA(81,1077), OR(82,1077), WA(82,1077). **12:**WY(1077). **14:**OR(1077), WA(82,1077). **25:**OR(82,1077), WA(82). **37:**WY(1077). **46:**SD(1077,2242). **51:**WA(1077). **52:**WA(82).

E. lineatum. **40:**WA(1076).

Sporisorium consanguineum. **19:**UT(1077).

Tilletia bromi. **42:**WY(1077).

T. sterilis. **46:**ID(1077), OR(2342).

Urocystis agropyri. Flag smut. **3:**WA(1077), WY(1077). **12:**WY(1077). **36:**CO(1077), WY(1077,2924). **40:**'Flag floral smut.' CA(82,1077), IA(1077), KS(3023), MI(82,1077), NH(82,1077), OH(2606), PA(82,1077), SD(2242), UT(1077), WA(82,1077), WI(82,1077,1332). **44:**ID(82), WA(82). **46:**ID(1077), OR(2342), WA(1077), WY(1077). **52:**AK(526), WY(82).

Ustilago hypodytes. Stem smut. **3:**WA(1077). **12:**WA(1077). **25:**WA(82,1077). **37:**WA(82,1077). **38:**SD(1077), WA(82,1077). **45:**NE(82), NV(1077). **46:**WA(1077). **50:**SD(1077,2242). **52:**NY(1077), WA(82).

U. striiformis. Stripe smut. **1:**MT(2342), WY(1077). **4:**'Stripe floral smut.' MA(82,1077). **6:**AK(1077). **12:**ND(1077), WA(1077), WY(1077). **13:**KY(82). **14:**CA(1077), ID(1077), IN(82,1077), KS(3021), PA(82,1077), WI(1341), WV(82,953,1077). **15:**WY(1077). **25:**OR(1077), WA(1077). **30:**WA(1077). **37:**NV(2342). **38:**SD(82,1077,2242), WA(82,1077). **40:**'Stripe floral smut.' NE states(82), DE(1077), GA(1962), IA(1077), ID(82,1077), IL(1077), IN(1077), KS(3021), KY(1077), MA(1077), ME(1077), MN(82,1077), MO(82,1077), NC(1303), NH(1077), NM(82,1077), NY(1077), OH(1077,2606), OK(2872), OR(1077), PA(1077), SD(2242), UT(1077), VA(1077), WA(82,1077), WI(1077), WV(952,1077). **42:**WY(1077). **44:**WY(82). **46:**WY(1077,2342,2924). **50:**CT(82,1077), WV(82,1077). **52:**ND(82), OR(82), WA(82).

U. trebouxii. **40:**'Loose floral smut.' NC(1303), OR(1077). **46:**OR(1077).

Basidiomycotina - Other

Coprinus psychromorbidus. 'Coprinus snow mold.' **40:**AK(2074,3676).

Laetisaria fuciformis. Pink patch, red thread. **4:**RI(3423), WA(82,3423). **40:**'Red thread.' NC(1303), RI(3423).

Marasmius sp. Blight. **40:**NC(1303).

Sebacina incrustans. Encrusting living plants. **40:**MD(128).

Thanatephorus cucumeris. Brown patch, root rot, sharp eye spot. **3:**MT(3423), ND(3423), WY(3423). **4:**ID(3423), OR(3423), WA(3423). **11:**ND(3423), OR(3423), WA(3423). **14:**MN(3423), ND(3423), OR(3423), WA(3423). **38:**MN(3423), ND(3423). **40:**CA(3423), CT(3423), DC(3423), IA(3423), ID(3423), IL(3423), KY(3423), MD(3423), MN(3423), MO(3423), MT(3423), ND(3423), NE(3423), OH(3423), OR(3423), PA(3423), SD(3423), VA(3423), WA(3423), WY(3423). **46:**WA(3423). **50:**DC(3423), OR(3423), VA(3423).

Trechispora alnicola. 'Yellow ring.' **40:**IA(3900), IL(3900), IN(3900), NJ(3900), NY(3900), OH(3900), PA(3900), WI(3900).

Typhula idahoensis. Snow mold. **44:**WA(82).

T. incarnata. Snow scald, snow mold. 'Typhula blight.' **40:**ID(3423), MN(82,3423), PA(82), WA(3423).

Deuteromycotina - Hyphomycetes

Acremonium coenophialum. **8:**TX(3873).

Bipolaris sorokiniana. Seedling blight, spot blotch, black point, foot and root rot, leaf spot. **4:**'Helminthosporium leaf spot.' CA(988). **11:**ND(82,3423). **14:**ND(3423), OR(3423), SD(3423), WA(3423). **25:**ND(3423). **38:**MN(3423), ND(82,3423). **40:**'Helminthosporium leaf spot.' CA(988), IA(1587), KS(3021,3023,3024), MN(228), ND(82,3423), NY(82), SD(2242,3423), TX(82). **44:**ND(82). **46:**ND(3423), OR(3423).

B. spicifera. Leaf mold. **40:**NY(82).

Botrytis cinerea. Gray mold. **46:**WA(3423).

Cercosporella poagena. Eye spot, leaf spot. **40:**AK(3426), OR(82,3423), WA(3432).

Cercosporidium graminis. Leaf streak, brown stripe, on culms. **1:**WY(2924,3423). **3:**MT(3423), WA(3423), WY(3423). **7:**MT(82,3423), ND(82,3423). **12:**CO(3423), MT(3423), ND(3423), SD(2242,3423), WA(3423), WY(3423). **14:**E states(3423), Range of host(82), AK(3431), CO(3423), IA(3423), ID(3423), KS(3021,3023), MD(3423), MN(3423), MO(3423), MT(3423), ND(3423), NE(3423), OR(3423), PA(3423), SD(2242,3423), WA(3423), WI(3423), WV(953), WY(3423). **18:**WA(3432), WY(3423). **22:**ID(3423). **23:**ID(3423), IN(82), MN(82), NY(82), WA(82), WY(82). **25:**ID(3423), OR(). **31:**WY(3423). **36:**OR(3423), WY(3423). **37:**W states(82), AZ(82), CO(3423), MI(82,3423), ND(82,3423), OR(3423), WA(82,3423), WY(3423). **38:**IN(82), MN(82,3423), NY(82), OR(82,3423), WA(82), WI(1340,3423). **40:**'Brown stripe.' Range of host(82), AK(526,3423), IA(3423), ID(3442), MN(3423), MT(3423), NY(3423), WI(3423). **44:**NW states(82), AZ(82), MI(82), WA(82). **46:**AZ(3423), ND(3423), WY(2924). **47:**MI(3423). **50:**IN(82), MN(82), NY(82,3423), WA(82). **52:**NW states(82), AZ(82), MI(82), WA(82).

Cladosporium herbarum. Leaf mold. **14:**Unk.(82). **38:**SD(82). **40:**Unk.(82). **45:**MT(82). **47:**AK(3431).

C. phlei. **40:**AK(526).

Curvularia geniculata. Fading out. **14:**SD(2242). **40:**'Curvularia blight.' KS(3021,3023).

Curvularia sp. Leaf spot. **4:**NY(82), PA(82). **14:**ND(82), NY(82), VA(82). **40:**NC(1303), ND(82), NY(82), VA(82).

Cylindrocarpon magnusianum. **1:**AK(3428,3431).

Dendryphion sp. On overwintered herbage. **40:**CA(642).

Drechslera avenacea. Basal culm decay, leaf rot. **46:**WA(3442).

D. campanulata. Leaf spot. **40:**WA(3423).

D. dictyoides. Leaf spot. 'Helminthosporium blight.' **40:**MN(228).

D. gigantea. Zonate leaf spot. **14:**MD(82,3423), VA(82,3423). **40:**'Zonate eyespot.' MD(82,3423), VA(82).

D. poae. Leaf mold, foot and sheath rot, purple spot. **7:**ND(82,3423). **11:**WA(82,3423). **14:**NC states(82), NE states(82), NW states(82), AK(3426), ID(3423), ND(3423), OR(3423), WA(3423). **38:**WA(3432). **40:**'Melting-out.' NC states(82), NE states(82), NW states(82), AK(526,3426), AZ(82,3423), CA(638,642,988,3423), CT(3423,3770), DC(3423), ID(3423), IL(3423), IN(3423), MA(3423), MD(3423), ME(3423), MI(3423), MN(3423), MT(3423), ND(3423), NE(3423), NJ(3423), NY(3423), OH(3423), OR(3423), PA(3423), SD(2242,3423), UT(82,3423), VA(3423), WA(3423), WI(1332,3423), WV(952), WY(3423). **44:**ID(82), OR(82), WA(82). **46:**ID(3423), OR(3423), WA(3423,3425). **47:**AK(3426). **50:**OR(82,3423), WA(3432).

D. siccans. Leaf spot. **40:**NC(1303).

D. triseptata. Leaf spot. **40:**MD(2661).

Fusarium acuminatum. Secondary root rot. **1:**MT(3423), WY(3423). **3:**MT(3423), ND(3423), WA(3423). **4:**ND(82,3423), OR(3423), WA(3423). **7:**MN(82), ND(82,3423), NE(82), SD(82). **11:**ID(82,3423), ND(3423), WA(82,3423). **12:**ND(3423), SD(3423), WY(3423). **14:**MN(82,3423), ND(82,3423), NE(82), OR(3423), SD(82), WY(3423). **16:**WY(3423). **21:**ND(3423). **23:**WY(3423). **37:**ND(3423). **38:**MN(3423), MT(3423), ND(82,3423), SD(3423). **40:**CA(3423), ID(3423), MN(82,3423), ND(82,3423), NE(82,3423), OR(3423), SD(82,2242,3423), WA(3423), WY(3423). **44:**WA(82). **45:**ND(3423), OR(3423), WA(3423). **46:**MT(3423), ND(3423), OR(3423), SD(3423), WA(3423), WY(3423).

F. avenaceum. Root rot. **4:**CA(988). **40:**CA(988).

F. culmorum. Head mold, culm and root rot, scab. **4:**'Fusarium blight.' OR(82,3423), WA(3423). **25:**OR(3423). **40:**'Fusarium blight.' WA(82,3423).

F. equiseti. Secondary root rot. **11:**WA(82). **14:**MN(82). **38:**ND(82). **44:**ND(82).

F. graminearum. Seedling blight, scab. **4:**WI(3423). **40:**WI(3423), WY(3423).

F. heterosporum. Head mold, often following ergot. **14:**IL(82). **40:**IL(82).

F. poae. Head mold, secondary root rot, silver top. **38:**MN(3423), SD(3423), WY(3423). **40:**'Fusarium blight.' IA(82,3423), MN(82,3423), ND(3423), NE(3423), NY(82,3423), PA(82), WA(3423), WI(82,1332,3423). **50:**MN(82).

F. roseum. **40:**Unk.(3631).

Fusarium sp. Leaf blight, snow mold. **4:**OR(82), WA(82). **11:**ID(82), WA(82). **38:**WA(82). **40:**NC(1303), WA(82). **44:**OR(82), WA(82). **50:**WA(82). **52:**ID(82), OR(82), WA(82).

Helminthosporium sp. Leaf spot. **37:**MI(82). **52:**MI(82).

Heterosporium avenae. Leaf mold. **22:**UT(3423). **37:**OR(82,3423). **47:**AK(3431). **52:**UT(82).

Microdochium bolleyi. Root necrosis, seedling root rot. **11:**ND(3423). **12:**ND(3423). **14:**MN(3423), ND(3423), SD(3423). **21:**ND(3423). **23:**WY(3423). **37:**ND(3423). **38:**MN(3423), MT(3423), ND(3423). **40:**'Anthracnose.' MN(3423), ND(3423), NE(3423), WA(82,3423), WY(3423). **46:**ND(3423).
M. nivale. Snow mold. **1:**AK(3431). **3:**ID(3423), OR(3423), WA(3423). **4:**'Fusarium patch.' ID(3423), OR(3423), WA(3423). **6:**WY(3423). **11:**ID(3423), WA(3423). **14:**AK(3441). **16:**OR(3423). **29:**OR(3423). **38:**OR(3423), WA(3423). **40:**'Fusarium patch.' AK(2293), ID(3423), MI(3423), MN(3423), OR(3423), WA(3423). **46:**OR(3423), WA(3423). **47:**AK(3431). **52:**WA(1286).
Periconia circinata. Root rot. **25:**ND(3423).
Phialophora graminicola. 'Summer patch.' **40:**NY(3324,3325).
Phialophora sp. Summer patch disease. **4:**MD(2024), MI(2024), NE(2024), NY(2024), RI(2024). **40:**MD(2024), MI(2024), NE(2024), NY(2024), RI(2024).
Pseudocercosporella herpotrichoides. Culm and foot rot, strawbreaker. **44:**WA(82). **46:**WA(3423).
Pyricularia grisea. Blast, gray spot. **40:**NE(3423).
Ramularia graminicola. Leaf spot. **38:**NY(82,3423).
R. pusilla. Leaf spot. **3:**ID(3423,3429), WY(3423,3429). **27:**AK(3429,3431). **30:**WY(3423). **31:**WY(3429). **40:**AK(3429,3431,3441). **52:**ID(82), WY(82).
Rhynchosporium orthosporum. Leaf spot. **27:**AK(3428).
R. secalis. **17:**AK(526).
Spermospora ciliata. Leaf spot. **50:**AK(3428).
Viscomacula aenea. On living leaves. **3:**CO(3424).

Deuteromycotina - Coelomycetes

Ascochyta brachypodii. Leaf spot. **16:**OR(3423,3449). **52:**OR(82).
A. graminicola. Leaf spot. **38:**WI(1356). **40:**AK(526), IL(312), NC(82), ND(82), OH(82), TX(82), WA(82), WI(1357).
A. macrospora. On leaves. **3:**CO(3424).
A. sorghi. Leaf spot. **11:**WA(3423,3449). **22:**OR(3423,3449). **29:**AK(3441). **40:**'Ascochyta leaf spot.' CO(3423), IA(3423,3449), KS(3021), NC(1303,3423), ND(3423,3449), OH(3423), SD(2242), TX(3423), WA(3423), WV(953), WY(3423,3449). **46:**ND(3423,3449), WA(3423,3449). **47:**AK(3431).
Cheilaria agrostis. **17:**AK(526).
Colletotrichum georgius-fischeri. On basal leaves. **10:**CA(3435).
C. graminicola. Anthracnose, leaf spot. **3:**OK(2872,3423). **12:**OK(2872,3423), SD(2242,3423). **14:**NE states(82), KY(82), MN(3423), ND(82), OK(82,2873,3423), WV(955). **38:**MN(3423), ND(82), SD(2242). **40:**'Anthracnose.' E states(82), KS(3021), MN(3423), NC(1303), ND(82,3423), NE(2927), NJ(3423), NY(3423), OH(3423), OK(82), PA(3423), SD(3423), TX(3423), WV(952,3423). **47:**AK(3431,3432). **52:**SD(82).
Coniothyrium poavora. On living leaves. **40:**WI(1357).
Coniothyrium sp. On leaves. **40:**WI(1354).
Dilophospora alopecuri. Twist disease. **44:**UT(82). **46:**UT(3423).
Dinemasporium strigosum. On fading leaves. **14:**KS(82), MD(82), NC(82). **40:**KS(82), MD(82), NC(82).
Discula graminum. Anthracnose. **40:**OR(82,3423).
Hendersonia calospora. Leaf mold. **14:**ND(82).
H. culmicola var. *minor*. Leaf mold. **44:**UT(82). **52:**ID(82), UT(82).
Macrophoma oblongata. On culms and leaves. **40:**IL(82,3423).
Phaeoseptoria airae. Leaf mold, leaf spot. **37:**ND(82). **40:**AK(3428).
P. poae. On dead leaves. **3:**WY(2924). **40:**WA(3425). **52:**MT(82).
Phoma graminis. **6:**AK(526).
P. sclerotioides. On roots. **40:**AK(307).
P. terrestris. Pink root. **14:**ND(3423). **40:**'Pyrenochaeta root rot.' ND(82,3423). **44:**ND(82). **46:**SD(3423).
Pseudoseptoria donacis. Eye spot, leaf spot. **1:**CO(3423,3450), MT(3423,3450). **3:**ID(3423), MT(3423,3450), OR(3423,3450), WA(2695,3423,3450), WY(3423). **12:**OR(3438), WA(3450). **14:**CO(3423,3450), ND(3423,3450). **16:**OR(3423,3450). **18:**CA(3423,3450), WY(3423,3450). **19:**UT(3450). **22:**OR(3423,3450), UT(3423,3450). **25:**OR(3423,3450), UT(3423,3450), WA(3423,3450). **37:**OR(3423,3450), WA(3423,3450). **38:**CO(82,3423,3450), MT(3433). **46:**WA(3450). **52:**CO(82), MT(82), WY(82).
P. everhartii. **6:**MT(3450). **27:**AK(3442). **36:**MT(3450), WY(3450).
P. obtusa. Eye spot. **7:**ND(82,3423,3450). **44:**OR(82). **46:**OR(3423,3450).
P. stomaticola. Stem speckle, eye spot, leaf spot. **14:**CO(82,3450), MT(82), ND(82), WY(82). **19:**W states(82), CA(82), ND(82), WA(82). **23:**MT(82,3450), ND(82), WY(82). **25:**W states(82), CA(82), ND(82), WA(82). **36:**MT(82,3423), WY(82,3423). **37:**W states(82), CA(82), ND(82), WA(82). **38:**AK(3441). **40:**AK(3441), CO(3423), MT(82,3423,3450), ND(82,3423,3450), WA(82,3423), WY(82,3423). **44:**NW states(82), CA(82), WA(82). **46:**MT(3423,3450), ND(3423,3450), NV(3423,3450), OR(3423,3450), SD(2242,3423,3450), WA(3423,3450), WY(3423,3450). **47:**AK(3426,3431). **48:**WA(3429). **52:**W states(82), CA(82), WA(82).
Rhynchosporina meinersii. Leaf spot. **1:**CO(3423). **52:**CO(82).
Septoria graminum. **52:**Unk.(82).
S. macropoda. Leaf blotch, leaf spot. **4:**AK(3426), HI(2903), OR(82,3423), WA(82,3423), WI(82,1338,3423). **7:**SD(2242). **26:**OR(3423). **45:**OR(82), WA(82). **52:**OR(82), WA(82).
S. macropoda f. *sitadakaensis*. On leaves. **39:**AK(3426,3431).
S. macropoda var. *grandis*. Leaf blotch, leaf spot. **3:**WA(3423). **7:**SD(82,3423). **12:**MT(3423), SD(2242,3423), WA(3423). **14:**MT(82). **16:**MT(3423), OR(3423). **22:**WA(3423). **36:**MT(3423). **37:**NW states(82), CA(82,3423), OR(3423), WA(82). **40:**'Septoria leaf spot.' WY(82,3423). **44:**NW states(82), WA(82). **45:**NW states(82), CA(82), WA(82,3423). **46:**ID(3423), MT(3423), ND(3423), OR(3423), SD(2242,3423), WA(3423), WY(3423). **51:**WA(3423). **52:**NW states(82), WA(82).
S. macropoda var. *septulata*. Leaf blotch. **14:**ID(3423), ND(3423), OR(82,3423), SD(82,2242,3423). **36:**OR(3433), WA(82,3423). **40:**'Septoria leaf spot.' NE states(82), NW states(82), IA(3423), ID(3423), KS(3021,3023), KY(3423), MI(3423), MN(3423), MT(3423), ND(3423), NE(3423), OH(3423), OR(3423), PA(3423), SD(2242,3423), TN(3423), VA(82,3423), WA(3423), WI(1340,3423), WV(955), WY(3423).
S. oudemansii. Eye spot, tan leaf spot. **1:**AK(3431). **3:**MT(3423), ND(3423). **5:**ND(82,3423). **6:**AK(3426). **12:**MT(3423), ND(3423). **14:**AK(3431), ND(3423), OR(82,3423), WA(82), WV(953). **23:**AK(3428,3431). **25:**ND(82,3423), SD(82). **31:**AZ(3438). **37:**ND(82), SD(82). **38:**CO(82,3423), ND(82,3423). **40:**'Septoria leaf spot.' C states(82), AK(3431), AZ(3435), CA(82,642,3423), CO(3423), IA(3423), ID(3423), KY(3423), MI(3423), MN(3423), MT(3423), ND(3423), NE(82,3423), OR(3423), UT(3423), WA(82,3423), WV(953), WY(3423). **42:**CO(3423). **44:**ID(82), MT(82), ND(82), OR(82), SD(82), WA(82). **46:**ID(3423), ND(3423), WA(3423), WY(3423). **47:**AK(3431). **52:**MT(82), ND(82), SD(82).
Sphaerellopsis filum. **40:**WV(955).
Stagonospora culmicola. Leaf mold, leaf rot. **12:**ND(3423), SD(3423). **14:**CA(3423). **22:**CA(3423), OR(3423), UT(3423). **29:**AK(3441). **38:**SD(3423). **40:**MN(3423), ND(82,3423). **44:**WA(82). **46:**WA(3423).
S. foliicola. Culm and leaf rot, leaf spot. **5:**MS(82,2720,3423).
S. nodorum. Glume blotch, glume spot. **36:**WA(82). **38:**SD(82,2242,3423). **40:**'Septoria leaf spot.' AR(82), MD(82), ND(82), OH(82), OR(82,3423), SD(82), WI(82).
S. subseriata. Leaf rot. **40:**MN(82,3423).
S. vexatula. Leaf spot. **34:**WA(3432).
Wojnowicia hirta. Leaf mold, secondary foot rot, rot of mature straw, leaf spot. **3:**WA(3423). **4:**AK(3428). **11:**WA(3423). **12:**ND(3423), SD(2242,3423). **23:**WY(3423). **40:**MT(82), ND(82), WY(82,3423). **46:**MT(3423), ND(3423), WA(3423).

Deuteromycotina - Other

Rhizoctonia cerealis. 'Yellow patch.' **40:**NC(2271).
R. solani. Brown patch, secondary root rot. **4:**'Rhizoctonia blight.' Unk.(82). **11:**Unk.(82). **14:**Unk.(82). **17:**AK(3441). **23:**Unk.(82). **38:**Unk.(82), SD(2242). **40:**'Rhizoctonia blight.' Unk.(82), AK(3441), KS(3021,3023), NC(1303), OH(1816), VA(892), WV(955). **44:**WA(82). **50:**Unk.(82). **52:**ND(82), WA(82), WY(82).
Rhizoctonia sp. Brown patch. **50:**NC(1303). **52:**CA(988).
Sclerotium rhizodes. 'Frost scorch.' **40:**PA(82,3423), WI(82,3423).
S. rolfsii. Blight. **4:**CA(2887). **40:**'Sclerotium blight.' MD(2886), NC(1303).
Sclerotium sp. Leaf rot. **4:**MA(82). **5:**Unk.(82). **7:**Unk.(82). **14:**Unk.(82). **36:**Unk.(82). **40:**Unk.(82). **52:**Unk.(82).

Polypogon Desf.

1. *P. maritimus* Willd.
2. *P. monspeliensis* (L.) Desf. [annual beardgrass, rabbitfoot grass]
3. *Polypogon* sp.

Ascomycotina

Erysiphe graminis. Powdery mildew. **2:**CA(82,3423), NM(3423), OK(2872,3423), OR(82,3423), UT(82), WA(82,3423).

Basidiomycotina - Uredinales

Puccinia coronata. Heteroecious: telial host. **2:**CA(81,82), OK(2872). **3:**CA(82).

P. graminis. Heteroecious: telial host. **2:**NE(82).

Deuteromycotina - Coelomycetes

Colletotrichum graminicola. Anthracnose, leaf spot. **1:**OR(3425). **2:**FL(35).

Puccinellia Parl.

1. *P. borealis* Swall.
2. *P. distans* (L.) Parl. [alkaligrass, weeping alkaligrass]
3. *P. langeana* (A. Berlin) T. Sorensen ex Hultén (=*P. paupercula* Fernald & Weath. var. *alaskana* (Lams.-Scribn. & Merr.) Fernald & Weath.)
4. *P. nutkaensis* (J. Presl) Fernald & Weath.
5. *P. nuttalliana* (Schult.) Hitchc. (=*P. airoides* (Nutt.) S. Wats. & J. M. Coulter) [Nuttall alkaligrass]
6. *P. pauciflora* (K. Presl) Munz (=*Glyceria pauciflora* K. Presl)
7. *P. tenella* (Lange) Holmb. ex Porsild
8. *Puccinellia* sp.

Ascomycotina

Clathrospora pentamera. **7:**AK(526).

Claviceps purpurea. Ergot. **2:**WA(82,3423). **6:**AK(526,3423).

Erysiphe graminis. Powdery mildew, leaf spot. **1:**AK(526). **2:**WA(82,3423,3425). **5:**NV(3425), WY(2924).

Lophodermium arundinaceum. On dead culms. **4:**AK(3431). **8:**Unk.(82).

Mycosphaerella tassiana. **7:**AK(526).

Pyrenophora trichostoma. **1:**AK(526).

Basidiomycotina - Uredinales

Puccinia brachypodii var. *poae-nemoralis*. **5:**WA(82).

P. coronata. **6:**CA(81,82), OR(82), WA(82).

P. graminis. Heteroecious: telial host. **2:**MI(82), WA(82). **4:**WA(82). **5:**CA(82), WA(82). **6:**CA(82), ID(82), MT(82).

P. recondita. Heteroecious: telial host. **2:**OR(82), WA(82). **4:**AK(3431), OR(82), WA(82). **5:**CA(82), CO(82), ND(82,378), OR(82), UT(82), WA(82), WY(82). **8:**AK(526).

P. striiformis. **4:**WA(82). **5:**WA(82). **8:**WA(82).

Basidiomycotina - Ustilaginales

Ustilago hypodytes. Stem smut. **2:**OR(1077), WA(82,1077).

U. longissima. **6:**WA(1077).

U. striiformis. Stripe smut. **5:**OR(82,1077).

U. trebouxii. **2:**OR(1077). **5:**CA(1077).

Deuteromycotina - Hyphomycetes

Cercosporidium graminis. Leaf streak, brown stripe. **6:**CA(3423), WY(82,2924,3423). **8:**AK(3431).

Cladosporium herbarum. Leaf mold. **2:**OR(82).

Ramularia pusilla. Leaf spot. **6:**AK(3426,3429).

Deuteromycotina - Coelomycetes

Cheilaria agrostis. **6:**ID(3442).

Colletotrichum aquatilis. **6:**AK(3441).

C. caudatum. Leaf spot. **6:**WY(3423).

C. graminicola. On leaves. **6:**ID(3438), WY(3432).

Phaeoseptoria airae. Leaf spot. **6:**ND(82).

Pseudoseptoria everhartii. Eye spot. **5:**ND(3423,3450), SD(3450).

P. stomaticola. Eye spot, leaf spot. **3:**AK(3428,3431). **5:**ND(82,3423), SD(2242,3423).

Septoria glycericola. Leaf spot. **6:**CA(3423), CO(82,3423), WA(82,3423), WY(82).

Stagonospora avenae. Speckled blotch, leaf spot. **6:**ID(3423), MT(3442), WA(82,3423).

Redfieldia Vasey

1. *R. flexuosa* (Thunb.) Vasey

Oomycetes

Pythium graminicola. **1:**NE(3423), SD(2242,3423).

Ascomycotina

Physalospora sp. On awns. **1:**ND(82).

Basidiomycotina - Uredinales

Puccinia redfieldiae. Heteroecious: telial host. **1:**CO(82), KS(82), ND(82), NE(82).

Basidiomycotina - Ustilaginales

Tilletia redfieldiae. Seed smut. **1:**NE(82,1077).

Basidiomycotina - Other

Thanatephorus cucumeris. Brown patch, root rot, sharp eye spot. **1:**NE(3423), SD(3423).

Deuteromycotina - Hyphomycetes

Helminthosporium sp. **1:**SD(3423).

Deuteromycotina - Other

Rhizoctonia solani. **1:**SD(2242).

Rhynchelytrum Nees

1. *R. repens* (Willd.) Hubb. (=*Tricholaena rosea* Nees, *R. roseum* (Nees) Stapf & Hubb. ex Bews, *Tricholaena repens* (Willd.) Hitchc.) [Natal grass]

Basidiomycotina - Uredinales

Puccinia levis. Telial host; aecial host unknown. **1:**FL(35,82).

Deuteromycotina - Hyphomycetes

Cerebella andropogonis. Blackhead, head mold. **1:**Unk.(82), FL(3423).

Rottboellia L. fil.

(= *Manisuris* L.)

1. *R. cochinchinensis* (Lour.) W. Clayton (=*R. exaltata* L. fil.) [itchgrass]
2. *R. cylindrica* (Michx.) Torr. (=*Manisuris cylindrica* (Michx.) Kuntze)
3. *R. rugosa* Nutt. (=*Manisuris rugosa* (Nutt.) Kuntze)
4. *Rottboellia* sp. (=*Manisuris sp.*)

Ascomycotina

Phyllachora quadraspora. Tar spot. **3:**MS(82,2720,3423).

Deuteromycotina - Hyphomycetes

Curvularia cymbopogonis. **1:**LA(3753).

Helminthosporium sp. Leaf spot. **1:**FL(35).

Deuteromycotina - Coelomycetes

Colletotrichum caudatum. Leaf spot. **2:**OK(2544,2872,3423).

Septoria sp. Leaf spot. **2:**OK(2872).

Stagonospora maritima. Leaf spot. **4:**TX(82,3423).

Saccharum L.

(= *Erianthus* Michx.)

1. *S. alopecuroideum* (L.) Nutt. (=*Erianthus alopecuroides* (L.) Elliott) [silver plumegrass]
2. *S. barberi* Jesw.
3. *S. officinarum* L. [sugarcane]
4. *S. robustum* Brandes & Jesw. ex Grassl
5. *S. sinense* Roxb.
6. *S. spontaneum* L.
7. *Saccharum* × sp. (unnamed hybrids)
8. *Saccharum* sp. (=*Erianthus sp.*)

Chytridiomycetes

Chytridium sp. On roots. **3:**HI(82,2903).

Oomycetes

Phytophthora erythroseptica. Seed piece rot. **3:**LA(82,4028).

P. megasperma. 'Phytophthora rot of cuttings.' **3:**LA(4028).

Pythium aphanidermatum. Root rot. **3:**FL(82), HI(82,2903), LA(82,3423).

P. arrhenomanes. 'Root rot.' **3:**S states(82), HI(2903).

P. debaryanum. Damping-off, root necrosis, seed rot. **3:**GA(82), HI(82,2903,3423), LA(82).

P. dissotocum. **3:**LA(82,3423).

P. graminicola. 'Root rot.' **2:**LA(3423). **3:**HI(82,2903,3423), LA(3423). **6:**LA(3423).

P. helicoides. **3:**FL(82).

P. hydnosporum. Rot. **3:**HI(82,2903,3423), LA(82).

P. inflatum. **3:**LA(2821).

P. irregulare. Root necrosis. **3:**LA(82,3423).

P. mamillatum. **3:**LA(82).

P. monospermum. Root necrosis. **3:**LA(82,3423).

P. periilum. Root necrosis. **3:**LA(82,3423).

P. rostratum. Root rot. **3:**HI(82,2903).

P. splendens. Root rot. **3:**HI(2903,3423).

P. ultimum. **3:**LA(82).

P. vexans. **3:**LA(82).

Pythium sp. Root rot. **3:**FL(35), HI(2903).

Zygomycotina

Rhizopus sp. **3:**HI(2903).

Ascomycotina

Apiospora montagnei. **1:**GA(1438).

Capnodium sp. **3:**HI(2903).

Ceratocystis adiposa. 'Black rot.' **3:**HI(82,2903), LA(82,3423). **8:**HI(472).

C. paradoxa. 'Pineapple disease.' **3:**Gulf States(82), FL(35), HI(82,2903,3423), LA(3423).

Claviceps purpurea. Ergot. **1:**AL(3423), OK(82,2873,3423). **8:**AL(82).

Clypeoporthe iliau. Stem rot, leaf sheath spot. 'Iliau.' **3:**HI(82,2903,3423), LA(82,3423), MS(82,2720).

Cochliobolus lunatus. Seedling blight. **8:**HI(473).

C. stenospilus. Brown stripe. **3:**HI(2903).

Dasyscyphella sp. **1:**GA(1438).

Dimeriella sacchari. 'Red leaf spot.' **3:**HI(82,2903).

Elsinoe sacchari. Spot anthracnose, white speck. 'White rash.' **3:**Unk.(3298), FL(1745,3648).

Elsinoe sp. **3:**FL(35).

Gibberella fujikuroi. Top rot. **3:**Gulf States(82), HI(82,2903).

G. subglutinans. 'Fusarium set. Fusarium stem rot. Fusarium knife cut. Fusarium pokkah boeng.' **3:**HI(82,2903), LA(82).

Glomerella tucumanensis. Anthracnose, leaf spot. 'Red rot.' **2:**LA(3423). **3:**FL(35,3423), HI(2903,3423), LA(3423), MS(3423). **4:**LA(3423). **5:**LA(3423). **6:**LA(3423).

Leptosphaeria orthogramma. On dead culms. **1:**AL(82). **8:**AL(82).

L. sacchari. On leaves. 'Ring spot.' **3:**Range of host(82), FL(35), HI(2784,2903).

Lophodermium sacchari. On leaves. **3:**HI(82,2903).

Massarina gloeospora. **1:**GA(1438).

Mycosphaerella striatiformans. 'Downy mildew, leaf splitting form.' **3:**HI(82,2903,3298).

Myriogenospora aciculispora. Tangle top. 'Myriogenospora leaf binding.' **3:**LA(82,3423).

Naemacyclus culmigenus. **1:**GA(1438).

Naemacyclus sp. **1:**GA(1438).

Nectria sp. On dead tissues. **3:**HI(2903).

Neocosmospora vasinfecta. **3:**GA(1438).

Neurospora sitophila. On post-harvest debris. **3:**HI(82,2903), LA(82).

Nimbomollisia aberrans. On dead culms. **1:**GA(1438). **8:**NJ(82).

Peziza sp. **3:**HI(2903).

Phyllachora dimeriae. Tar spot. **1:**AL(82), FL(35,82), GA(82), SC(82,3423). **8:**AL(82), FL(35,82), GA(82), SC(82).

P. graminis. Tar spot. **1:**FL(35).

Saccharomyces zopfii. Sugar deterioration. **3:**LA(82).

Basidiomycotina - Uredinales

Puccinia melanocephala. 'Common rust.' **3:**FL(35), HI(2903), MS(4023). **8:**FL(2890), HI(603), LA(1947).

P. polysora. **1:**FL(35,82).

P. virgata. **8:**Unk.(82).

Basidiomycotina - Ustilaginales

Ustilago scitaminea. 'Culmicolous smut.' **3:**FL(35), HI(474,475,2903), LA(1948), TX(3736). **7:** LA(1657).

Basidiomycotina - Other

Athelia rolfsii. Seedling blight, stalk rot. 'Red rot of leaf sheath. Sprout rot.' **3:**FL(3423), HI(3423), LA(3423).

Dictyophora sp. **3:**HI(2903).

Marasmiellus stenophyllus. Root rot. 'Marasmius sheath blight. Marasmius shoot blight.' **3:**Gulf States(82), FL(35).

Marasmius sacchari. 'Root rot.' **3:**HI(82,2903,3423), LA(3423).

Odontia saccharicola. Leaf sheath adhesion and rot. **3:**HI(82,2903).

Oxyporus latemarginatus. On dead culms. **3:**FL(82).

Psathyrella atomatoides. Discoloration and deterioration of bagasse. **3:**LA(82).

Schizophyllum commune. Wound rot of canes. 'Schizophyllum rot.' **3:**FL(82), HI(82,2903), LA(82).

Thanatephorus cucumeris. Brown patch, root rot, sharp eye spot. 'Banded sclerotial.' **3:**LA(82,3423).

Deuteromycotina - Hyphomycetes

Allantospora radicicola. On roots and other parts. **3:**HI(82,2903).

Alternaria graminum. **3:**HI(2903).

Alternaria sp. Leaf spot. **3:**Range of host(82), FL(35).

Arthrinium phaeospermum. On dead culms. **3:**Range of host(82), HI(2903), LA(3423).

Aspergillus flavus. Mold on canes, sugar, and byproducts. **3:**Range of host(82).

A. fumigatus. Mold on canes, sugar, and byproducts. **3:**Range of host(82).

A. glaucus. Mold on canes, sugar, and byproducts. **3:**Range of host(82).

A. nidulellus. Mold on canes, sugar, and byproducts. **3:**Range of host(82).

A. niger. Mold on canes, sugar, and byproducts. **3:**Range of host(82).

A. parasiticus. Mold on canes, sugar, and byproducts. **3:**HI(82).

A. penicillioides. Mold on canes, sugar, and byproducts. **3:**Range of host(82).

A. reptans. Mold on canes, sugar, and byproducts. **3:**Range of host(82).

A. sydowii. Mold on canes, sugar, and byproducts. **3:**Range of host(82).

A. terreus. Mold on canes, sugar, and byproducts. **3:**Range of host(82).

Aspergillus sp. Mold. **3:**HI(2903).

Bipolaris hawaiiensis. Seedling blight. **8:**HI(473).

B. sacchari. 'Eye spot.' **3:**Gulf States(82), AL(285,3423), FL(35,3423), HI(82,2903,3423), LA(3423).

B. stenospila. 'Brown stripe.' **3:**FL(35,82,3423), GA(82,3423), HI(82,2903,3423), LA(82,3423).

Cercospora longipes. 'Brown spot.' **3:**AL(3423), FL(35,3423), HI(2903), LA(3423).

Cercospora sp. Red spot of leaf-sheath spot. **3:**HI(2903).

Cerebella andropogonis. On ergot-infected inflorescences. **8:**AL(82).

Chalara paradoxa. **3:**HI(2903).

Chrysonilia sitophilia. **3:**HI(2903).

Cladosporium erianthi. Leaf mold. **1:**SC(82).

C. herbarum. Gray-green mold of cut leaves and cane. **3:**Range of host(82), HI(2903).

Curvularia lunata. Secondary leaf spot. **3:**HI(82,2903).

C. senegalensis. Seedling blight. **8:**HI(473).

Curvularia sp. Secondary root rot. **3:**LA(82). **8:**FL(35,82).

Cylindrocarpon destructans. **3:**HI(2903).

Exserohilum rostratum. Seedling blight, leaf spot, red spot. **3:**FL(35), MS(4024). **8:**HI(473).

Fusarium moniliforme. Root and top rot. **3:**Unk.(82), AL(3423), FL(35), HI(2903,3423), LA(3423), MS(3423).

F. oxysporum. **3:**HI(2903).

Fusarium sp. Cane, leaf, root, seed and sugar rot. **3:**Gulf States(82), HI(82,2903).

Gloeocercospora sorghi. Copper spot. 'Zonate leaf spot.' **3:**LA(82), MS(2720,3423).

Graphium sp. On leaves. **3:**HI(2903).

Helminthosporium sp. Target blotch. **3:**FL(3649). **8:**FL(35,82).

Hormiactella sacchari. On leaves. **3:**FL(82), HI(2903).

Mycovellosiella koepkei. Brown spot, on leaves. 'Yellow spot.' **3:**AL(82), FL(35,82,128), LA(82).

M. vaginae. 'Red spot of leaf sheath.' **3:**FL(35,82,3423), HI(82,3423), LA(82,3423), TX(82).

Nigrospora oryzae. Dry rot, black mold, on canes and leaves. **3:**Range of host(3423), Range of host(82), HI(2903).

Papularia vinosa. Leaf spot. **3:**LA(82).

Penicillium expansum. Blue mold. **3:**HI(2903,3423), LA(3423).

Penicillium sp. **3:**HI(2903).

Periconia sacchari. On dying leaves. **3:**LA(82).

Strumella sacchari. **3:**HI(2903).

Trichoderma viride. Green mold of canes. **3:**Range of host(82), HI(2903).

Deuteromycotina - Coelomycetes

Colletotrichum falcatum. Red rot. **3:**Range of host(82), FL(35), HI(2903), MS(2720).

C. graminicola. On dead tissues. **3:**HI(2903).

Cytospora sacchari. Stem canker. 'Cytospora sheath rot.' **3:**GA(82), LA(82,3423).

Hendersonia donacis. On dead culms. **8:**AL(82).

H. erianthi. On dead culms. **8:**AL(82).

Macrophoma sacchari. On culms. **3:**GA(82).

Melanconium sp. **3:**HI(2903).

Pestalotia sp. **3:**HI(2903).

Phaeocytostroma sacchari. 'Rind disease.' **8:**HI(2784).

Phoma sorghina. Leaf spot. **3:**Gulf States(82), FL(3423), HI(82,3423), LA(3423).

P. terrestris. Pink root. **3:**LA(3423).

Phyllosticta hawaiiensis. 'Phyllosticta leaf spot.' **3:**FL(35), HI(2903,3298).

Stagonospora sacchari. 'Leaf scorch.' **3:**HI(2903).

Deuteromycotina - Other
Himantia stellifera. Web blight. **3:**HI(82,2903).
Rhizoctonia solani. Secondary root rot. 'Rhizoctonia sheath rot. Rhizoctonia shoot rot.' **3:**FL(35), LA(82).
Rhizoctonia sp. Root, sheath and stem rot. **3:**HI(2903), NC(1303).
Sclerotium rolfsii. Southern blight, leaf sheath rot, basal stalk rot. **3:**Gulf States(82), FL(35), HI(82,2903), MS(2720).
Sclerotium sp. **3:**HI(2903).

Schedonnardus Steud.

1. *S. paniculatus* (Nutt.) Trel. [tumblegrass]

Oomycetes
Pythium debaryanum. Damping-off, root necrosis, seed rot. **1:**MT(3423), ND(3423), WY(3423).
P. graminicola. **1:**MT(3423), ND(3423), SD(2242,3423).
Sclerophthora macrospora. **1:**SD(2242,3184).
Basidiomycotina - Uredinales
Puccinia schedonnardi. Heteroecious: telial host. **1:**CO(82), KS(82,753,1521,3023), ND(82), NE(82,753), SD(82,2242).
Basidiomycotina - Other
Thanatephorus cucumeris. Brown patch, root rot, sharp eye spot. **1:**SD(3423).
Deuteromycotina - Hyphomycetes
Exserohilum rostratum. Leaf mold. **1:**MN(82), ND(82,3423), SD(2242).
Fusarium acuminatum. Secondary root rot. **1:**MT(3423), ND(82,3423), NE(3423), SD(3423), WY(3423).
Helminthosporium sp. Isolated from roots. **1:**NE(3423).
Microdochium bolleyi. Root necrosis. **1:**NE(3423), SD(3423).
Deuteromycotina - Coelomycetes
Ascochyta brachypodii. **1:**NM(3425).
Phoma terrestris. Pink root. **1:**MN(82), MT(3423), ND(82,3423).

Schizachne Hack.

1. *S. purpurascens* (Torr.) Swall. [false melic]

Ascomycotina
Leptosphaeria sp. **1:**WI(1356).
Phyllachora graminis. Tar spot. **1:**NY(82), WI(82,3423).
Basidiomycotina - Uredinales
Puccinia coronata. Heteroecious: telial host. **1:**WI(82,1357).
Deuteromycotina - Coelomycetes
Ascochyta graminicola. Leaf spot. **1:**WI(82,1333).
A. sorghi. Leaf spot. **1:**WI(3423).
Colletotrichum graminicola. Anthracnose. **1:**MN(82,3423), ND(82), WI(82,3423).
Hendersonia sp. Leaf rot. **1:**MN(82).
Phyllosticta sp. Leaf spot. **1:**MN(82).
Septoria melicae. Leaf spot. **1:**Unk.(82).
Stagonospora avenae. Speckled blotch. **1:**NE(3423).
S. culmicola. Leaf rot. **1:**MN(3423).

Schizachyrium Nees

1. *S. sanguineum* (Retz.) Alston (=*Andropogon hirtiflorus* (Nees) Kunth)
2. *S. scoparium* (Michx.) Nash (=*Andropogon scoparius* Michx.) [bluestem, broom beardgrass, broom wiregrass, bunchgrass, little blue-stem, prairie beardgrass]
3. *S. scoparium* (Michx.) Nash var. *littorale* (Nash) Gould (=*Andropogon littoralis* Nash)
4. *S. tenerum* Nees (=*Andropogon tener* (Nees) Kunth)

Oomycetes
Pythium debaryanum. Damping-off, root necrosis, root and seed rot. **2:**ND(82,3423), SD(2242,3423).
Ascomycotina
Apiospora montagnei. **2:**GA(1438).
Balansia epichloe. Black choke. **2:**NC(82), VA(3423).
B. henningsiana. Black choke. **2:**FL(35,82), KS(82), LA(82), MD(82), MO(82), VA(82).
Claviceps purpurea. Ergot. **2:**FL(35), OK(2872,3423).
Clypeoporthe andropogonis. On culms and leaves. **2:**NY(82).
Coleroa sporoboli. **2:**WI(1357,1359).
Didymosphaeria futilis. On culms. **2:**ND(82,377).
Epichloe typhina. Choke. **2:**WI(3423).
Gibberella zeae. On dead culms and sheaths. **2:**KS(3021).
Graphyllium dakotense. On culms. **2:**ND(82,377).
G. hysterioides. **2:**GA(1438).
Mollisia culmina. **2:**GA(1438).
Mycosphaerella sp. On leaves. **2:**WI(1354).
Myriogenospora atramentosa. Dwarfing, fasciation, tangle top. **2:**LA(82,3423), TX(82).
Naemacyclus culmigenus. **2:**GA(522,1438).
Nimbomollisia aberrans. **2:**GA(1438).
Phyllachora luteo-maculata. Black spot, tar spot. **2:**GA(82,1438,3423), MA(82,3423), MN(82,3423), OK(2872,3423), PA(82,3423), VA(82,3423).
Venturia erysipheoides. **2:**WI(1361).
Basidiomycotina - Uredinales
Puccinia andropogonis. **1:**AZ(1194,1203,3981). **2:**Range of host(82), AZ(1203,3981), FL(35), GA(1439), KS(3023), MS(2720), NC(1303), ND(378), NM(1195), OK(2872), SD(2242), TX(742), WI(1350). **3:**NJ(1252).
P. ellisiana. **2:**Range of host(82), FL(35), MS(2720), ND(378), OK(2872), SD(2242).
P. posadensis. **2:**FL(35).
Uromyces andropogonis. **2:**NC(82,1303,1519), NJ(82,1519).
Basidiomycotina - Ustilaginales
Sorosporium ellisii. Head smut. **2:**CT(82,1077), FL(35,82,1077), GA(82,1077), IL(82,1077), IN(1077), KS(82,1077), ME(1077), MS(82,1077,2720), OH(946), SD(82,1077,2242).
S. everhartii. Kernel smut. **2:**E states(82), AL(1077), CT(1077), GA(1077,1439), ME(1077), MO(1077), MS(82,1077,2720), NC(3944), NH(1077), NJ(1077), NY(1077), PA(1077), VA(1077), VT(1077), WI(1354).
Sphacelotheca occidentalis. Seed smut. **2:**KS(3021), OH(946), OK(331,2872), TX(1077), VA(82,1077).
S. seymouriana. **2:**VA(82,1077).
Sporisorium andropogonis. Head smut. **2:**IL(82,1077), KS(82,1077), MS(82,1077,2720), NC(82,1077,1303), NJ(82,1077), VA(82,1077).
Deuteromycotina - Hyphomycetes
Cercoseptoria andropogonis. On living leaves. **2:**WI(1354).
Cerebella andropogonis. Blackhead. **2:**FL(35), KS(82), OK(2872,3423). **4:**MS(2720).
Fusarium acuminatum. Secondary root rot. **2:**MT(3423), NE(3423), SD(3423).
Microdochium bolleyi. Root necrosis. **2:**MN(3423), MT(3423), NE(3423).
Periconia circinata. Root rot. **2:**MN(3423), SD(3423).
Deuteromycotina - Coelomycetes
Ascochyta brachypodii. Leaf spot. **2:**AZ(3423,3424), NY(4001), PA(4001).
A. graminicola. On leaves. **2:**MA(82,1394).
Colletotrichum caudatum. Leaf spot. **2:**KS(82), OK(82,2872,3423), WI(1338,1352,2544,3423).
C. graminicola. Anthracnose. **2:**CT(82), KY(82), OK(2872,3423), SD(2242).
Dinemasporium strigosum. On leaves. **2:**MA(1394).
Leptostromella andropogonis. On culms. **2:**AL(82).
Pirostoma circinans. **2:**WI(1361).
Pseudoseptoria donacis. **2:**SD(2242).
Septoria andropogonis. Leaf spot. **2:**NE(3423), SD(2242), WI(1347).
Stagonospora arenaria. **2:**SD(2242).
S. simplicior. **2:**WI(1354).
Deuteromycotina - Other
Rhizoctonia oryzae. **2:**SD(2242).

Sclerochloa P. Beauv.

1. *S. dura* P. Beauv.

Deuteromycotina - Coelomycetes
Ascochyta sorghi. Leaf spot. **1:**ID(3423,3449).

Scleropogon Phil.

1. *S. brevifolius* Phil.

Basidiomycotina - Uredinales
[*Puccinia aristidae*]. **1:**NM(82).
P. scleropogonis. Heteroecious: telial host. **1:**TX(742).

Scolochloa Link

1. *S. festucacea* (Willd.) Link (=*Fluminia festucacea* (Willd.) Hitchc.)

Oomycetes
Pythium graminicola. **1:**ND(3423).

Basidiomycotina - Uredinales

Puccinia coronata. Heteroecious: telial host. **1:**MT(82), ND(82,378), NY(82), SD(82,2242).

P. recondita. **1:**Unk.(82), MT(82), ND(82).

Basidiomycotina - Ustilaginales

Ustilago arthurii. Leaf smut. **1:**IL(1077), NE(1077), OR(82,1077), SD(1077,2242).

Deuteromycotina - Hyphomycetes

Pyricularia grisea. Blast, gray leaf spot. **1:**ND(82,3423).

Deuteromycotina - Coelomycetes

Stagonospora arenaria. Purple leaf spot. **1:**ND(82).

S. avenae. Speckled blotch, leaf spot. **1:**ND(3423), SD(82,2242,3423).

Secale L.

1. *S. cereale* L. [common rye]
2. *S. montanum* Guss. [mountain rye]
3. *Secale* × sp. (unnamed hybrids)

Oomycetes

Pythium aphanidermatum. Root rot on early plantings. **1:**FL(35,3169).

P. arrhenomanes. **1:**NGP(3726), MO(82), ND(82).

P. debaryanum. Damping-off, root necrosis, seed rot. **1:**MT(3423), ND(82,3423), NE(3423), SD(2242,3423), WA(3423). **2:**ND(82), WA(3423). **3:** WA(3423).

P. graminicola. Root browning. **1:**MN(2874,3423), ND(3423), SD(2242,3423).

P. ultimum. Root necrosis. **1:**ND(3423).

Sclerophthora macrospora. Downy mildew. **1:**Range of host(2620), AR(769), MS(3050).

Ascomycotina

Clathrospora pentamera. Leaf spot. **1:**NC states(2620).

Claviceps purpurea. Ergot. **1:**Range of host(82), Unk.(3423), FL(35), MD(1729), MN(2874), NC(1303), ND(377), OK(2872), SD(2242), WI(3191), WV(3907), WY(819). **2:**Range of host(82). **3:** MT(3423), ND(3423), SD(3423), WA(3423).

Erysiphe graminis. Powdery mildew. **1:**Range of host(82), AZ(3423), CT(3423), FL(35), IA(3423), MI(3423), MN(2874,3423), OH(950), OK(2872,3423), OR(3423), WA(3423), WI(1352,3423), WV(3907).

E. graminis f. sp. *secalis*. Powdery mildew. **1:**GA(1023), SC(933).

Gaeumannomyces graminis. Take-all. **1:**IL(82), NY(82,3423), OK(82,2872,3423).

Gibberella zeae. Seedling blight, foot rot, scab. **1:**E & C states(82), CA(82), IL(309), MN(2874), OK(2872).

Lophodermium culmigenum. On dead culms. **1:**NY(82).

Massarina sp. **1:**GA(1438).

Monographella nivalis. Pink snow mold. **1:**OR(3442), WA(629).

Mycosphaerella zeae. Leaf spot. **1:**IL(312).

Myriosclerotinia borealis. Snow mold. **1:**AK(2074).

Nectria secalina. On dead culms. **1:**AL(82).

Phaeosphaeria herpotrichoides. Node canker, secondary root rot. **1:**IL(312), ND(82,3423), WA(82).

P. microscopica. **1:**WA(3439).

Pyrenophora japonica. Leaf spot. **1:**WI(1343,3878).

P. trichostoma. On dead leaves and culms. **1:**GA(82,1438), PA(82).

P. tritici-repentis. On leaves. **1:**WV(3907).

Basidiomycotina - Uredinales

Puccinia graminis. Heteroecious: telial host. **1:**Range of host(82), FL(35,929,931), GA(1023,1275,2406), MN(2874), MS(3539), OK(2872), SC(932), SD(2242), WV(3907).

P. recondita. Heteroecious: telial host. **1:**Range of host(82), AZ(3981), FL(35,929,931), GA(931,1023,1024,1275,1439,2406), IL(309), IN(2230), KS(2230), MN(2874), MS(2720,3539), NC(1303), NE(2230), OK(2872), SC(932,933), SD(2242), WA(2230), WI(2230), WV(3907). **2:**Range of host(82).

P. striiformis. **1:**ID(82), OR(82), WA(82). **2:**WA(82).

Basidiomycotina - Ustilaginales

Tilletia caries. **1:**CA(82,921,1077,3953), NY(82,921,1077), OR(82,921,1077), UT(3953), WA(82,921,1077,1140).

T. controversa. **1:**ID(921,1077,1465), MT(921,1465), NY(921,1078), OR(921), UT(921).

T. laevis. **1:**CA(82,921,1077), KS(82,921,1077), MI(82,921,1077), WA(82,921,1077).

Tilletia sp. **1:**OR(1464), UT(1464).

Urocystis occulta. **1:**AL(82,1077), CO(1077), CT(82,1077), GA(1439), IA(1077), ID(1077), IL(1077), IN(1077), MA(1077), MI(1077), MN(1077,2874), MO(1077), NC(1303), ND(1077), NJ(1077), NY(1077), OH(1077), OK(2872), PA(1077), RI(1077), SD(1077), TN(1077), TX(1077), VA(1077), WI(1077), WV(3907).

Ustilago hordei. **1:**WA(82,1077).

U. hypodytes. **2:**WA(82,1077). **3:** On *S. cereale* x *S. montanum*.. WA(1077).

U. tritici. **1:**AK(526,1077), IL(1077), IN(1077), KY(1077), MN(1077,2874), MO(1077), ND(1077), NY(1077), OK(1077,2872), TN(1077), VA(1077), WA(1077), WV(1077).

Basidiomycotina - Other

Marasmius graminum. On culms, roots and stalks. **1:**IL(82), SC(634).

Naucoria cerealis. On culms and roots. **1:**IL(82,308).

Thanatephorus cucumeris. Brown patch, root rot, sharp eye spot. **1:**MN(2874,3423), MS(3423), ND(3423), SD(3423).

Typhula idahoensis. Speckled snow mold. **1:**NW states(2620).

T. incarnata. Speckled snow mold. **1:**NW states(2620), WA(82,3423).

Deuteromycotina - Hyphomycetes

Alternaria alternata. Leaf spot, on glumes. **1:**DE(82), MD(82), ND(82), NJ(82).

Alternaria sp. Black-chaff disease. **1:**FL(35).

Bipolaris sorokiniana. Seedling blight, leaf blotch, crown necrosis, black point, root rot. **1:**E & C states(82), CA(3423), CO(3423), FL(35), ID(3423), IL(3423), IN(3423), MI(3423), MN(2874,3423), MT(3423), NC(1303), ND(3423), NE(3423), OH(3423), OR(3423), PA(3423), SD(2242,3423), UT(3423), WA(3423), WI(3423), WY(3423).

Cercospora secalis. Leaf spot. **1:**IL(312).

Cercosporidium graminis. Leaf streak, brown stripe. **1:**OK(82,2873), OR(82,3423), SC(82,3423), WA(82). **2:**WA(82,3423). **3:** WA(3423).

Cladosporium herbarum. Associated with, or simulating black-chaff disease. **1:**MN(82), OH(82), TN(82).

Curvularia inaequalis. Seed mold. **1:**IL(3423), MN(35,2874), ND(82).

Drechslera campanulata. Seedling mortality. **1:**UT(1980).

D. tritici-repentis. Blight, scorch, yellow leaf spot. **1:**FL(931), GA(2406), IL(312), MN(2874,3423), SC(932,933), WV(82,954).

D. tuberosa. On dead culms. **1:**AL(82), DE(82), WI(1352).

Fusarium acuminatum. Secondary root rot. **1:**CA(3423), CO(3423), IA(3423), ID(3423), MN(2874,3423), MT(3423), ND(82,3423), NE(3423), OR(82,3423), SD(2242,3423), WA(3423).

F. avenaceum. Leaf mold, root and stem rot. **1:**IN(82,3423), OH(82,3423).

F. culmorum. Seedling blight, head and snow mold, foot and root rot, scab. **1:**E & C states(82), AR(3423), CA(82,3423), CT(3423), MD(3423), MN(2874), NJ(3423), OR(82,3423), PA(3423), VA(3423), WA(82,3423).

F. equiseti. Secondary root rot. **1:**ND(82).

F. graminearum. Seedling blight, scab. **1:**CA(3423), IA(3423), MD(3423), MN(2874,3423), MO(3423), NC(3423), NY(3423), OK(3423), PA(3423), VA(3423).

F. heterosporum. Head mold, often following ergot. **1:**IL(82).

Fusarium sp. Root rot, on seeds. **1:**MN(2874), NY(732).

Hymenula cerealis. Stripe. **1:**Unk.(2620), WA(415).

Microdochium bolleyi. Root necrosis. **1:**MN(2874,3423), ND(3423), NE(3423), SD(2242,3423). **3:** ND(3423).

M. nivale. Snow mold. **1:**AK(2293), MN(2874,3423), ND(82,3423), WA(82,3423). **3:** WA(3423).

Nigrospora sphaerica. Node canker. **1:**IL(312).

Pseudocercosporella herpotrichoides. Eye spot, foot rot, strawbreaker. **1:**ID(82,3423), OR(82,3423), WA(82,3423).

Rhynchosporium secalis. Scald, leaf spot. **1:**CA(3423), IL(309,3423), KY(1815), MS(3539), OK(82), OR(82), SC(82), UT(3423), VA(2993), WA(82,3423), WI(3423). **2:**WA(82,3423).

Stemphylium botryosum. Seed mold. **1:**Range of host(82), FL(35).

Deuteromycotina - Coelomycetes

Ascochyta graminicola. Leaf spot. **1:**IL(82).

A. hordei. Leaf spot. **1:**WA(3439).

A. sorghi. Leaf spot. **1:**IL(3423).

Colletotrichum graminicola. Anthracnose. **1:**C states(), Range of host(2620), SE states(82), AL(285), AR(3423), FL(35,2187,3170), IA(3423), IL(3423), IN(3423), KY(1815,3423), MD(3423), MI(3423), MN(2874,3423), MO(3423), MS(2720,3423,3539), NC(1303,3423), ND(3423), NJ(3423), NY(3423), OH(3423), OK(2872,3423), PA(3423), SC(3423), SD(2242), TN(3423), TX(3423), VA(3423).

Dilophospora alopecuri. Twist disease. **1**:OR(82,3423).
Dinemasporium strigosum. On leaves. **1**:Range of host(82).
Macrophoma secalina. On culms and leaves. **1**:IL(82,3423,3585).
Phaeoseptoria airae. **1**:WA(3439).
Phyllosticta sp. **1**:WA(3439).
Pseudoseptoria donacis. Eye spot. **1**:MN(2874,3423,3450), ND(3423,3450).
P. stomaticola. Leaf spot. **1**:MN(82), ND(82).
Septoria secalis. Leaf blotch, leaf spot. **1**:AL(3423), DC(3423), IA(3423), IL(3423), MN(2874,3423), NC(1303,3423), ND(3423), SD(2242,3423), TN(3423), TX(3423), VA(3423), WI(3423,3799), WV(3423).
S. tritici. Speckled leaf and sheath blotch. **1**:NC(1303), WA(3423,3424).
Stagonospora nodorum. Glume blotch. **1**:AR(82,3423), IN(82,3423), MD(82,3423), OH(82,3423), WI(82,3423).
Wojnowicia hirta. Secondary foot rot, rot of mature straw. **1**:ID(3423), MT(82,3423), OR(82,3423), WA(82,3423).

Deuteromycotina - Other

Rhizoctonia solani. Brown patch, root and stem rot. **1**:Range of host(2620), FL(35), MS(82,2720), SD(2242).
Rhizoctonia sp. Isolated from seeds. **1**:Unk.(731).

Setaria P. Beauv.

1. *S. faberi* Herrm. [giant foxtail, mutton bluegrass]
2. *S. geniculata* (Lam.) P. Beauv. [knotroot bristlegrass]
3. *S. glauca* (L.) P. Beauv. (=*Chaetochloa glauca* (L.) Lams.-Scribn.) [yellow foxtail]
4. *S. grisebachii* E. Fourn. [Grisebach bristlegrass]
5. *S. italica* (L.) P. Beauv. [Bengal grass, foxtail millet, German millet, Hungarian millet, Italian millet, Japanese millet]
6. *S. macrostachya* H. B. K. [plains bristlegrass]
7. *S. magna* Griseb.
8. *S. pumila* (Poir.) Roem. & Schult. (=*S. lutescens* (Weigel) Hubb.) [yellow foxtail]
9. *S. verticillata* (L.) P. Beauv. [bristly foxtail, bur bristlegrass]
10. *S. viridis* (L.) P. Beauv. (=*Chaetochloa viridis* (L.) Lams.-Scribn.) [green foxtail]
11. *Setaria* sp.

Oomycetes

Aphanomyces camptostylus. Root necrosis. **10**:MT(3423).
Aphanomyces sp. Root necrosis. **10**:MT(82).
Pythium arrhenomanes. Root browning and necrosis. **5**:MN(82), ND(82), SD(82). **10**:MN(82), ND(82), SD(82).
P. debaryanum. Damping-off, root necrosis, seed rot. **5**:MN(82), ND(82,3423), SD(82,2242). **8**:ND(3423), SD(2242,3423). **10**:MN(82,3423), ND(82,3423), SD(82,2242,3423).
P. graminicola. **5**:MN(3423), ND(3423), SD(2242,3423). **8**:ND(3423), SD(2242,3423). **10**:MN(3423), MT(3423), ND(3423), NE(3423), SD(2242,3423).
Sclerophthora macrospora. Crazy top. **8**:SD(2242). **10**:IN(3708), SD(2242).
Sclerospora graminicola. Downy mildew, powdery mildew. **5**:C states(82), HI(82,2903), IA(3423), MI(3423), MN(3423), MT(82), NY(3423), SD(2242), WI(3423). **7**:FL(3423). **8**:FL(35), IA(82,3423), NE(3423), NY(82,3423), SD(2242), WI(82,3423). **10**:E & C states(82), W states(82), IA(3423), MN(3423), MT(82,3423), ND(377,3423), NE(3423), NM(3423), NY(3423), SD(2242,3423), WA(3423), WI(3423). **11**:FL(82), MS(2720).

Ascomycotina

Claviceps purpurea. Ergot. **6**:NM(82,3423). **10**:NM(82,3423).
C. ranunculoides. Ergot. **6**:AZ(82,3423), NM(3423), TX(82).
Gaeumannomyces graminis. Take-all. **2**:AR(82,3423).
Gibberella zeae. Root rot. **8**:WI(82).
Mycosphaerella tassiana. **11**:OR(657).
[*Ophiobolus acuminatus*]. On dead culms. **3**:GA(1438). **8**:GA(82).
Phaeosphaeria clavispora. On dead culms. **3**:GA(1438). **8**:GA(82).
P. herpotrichoides. **11**:OR(657).
Sclerotinia sclerotiorum. **10**:CO(447).

Basidiomycotina - Uredinales

Physopella cameliae. **6**:TX(82). **11**:TX(82).
Puccinia chaetochloae. Autoecious. **11**:FL(35,82).
P. esclavensis. Heteroecious: telial host. **4**:AZ(1203,3981).
P. esclavensis var. *panicophila*. Telial host. **4**:AZ(1195).
P. setariae. Telial host; aecial host unknown. **2**:LA(82). **8**:FL(35). **11**:FL(35).
P. substriata. Heteroecious: telial host. **8**:FL(35), TX(82). **9**:HI(2299,2903).
Uredo sp. **9**:HI(2903). **11**:HI(2299).
Uromyces setariae-italicae. Heteroecious: telial host. **2**:Unk.(2910), FL(82). **11**:FL(35).

Basidiomycotina - Ustilaginales

Sphacelotheca magnusiana. Seed smut. **8**:WI(82).
Sporisorium neglectum. Head smut. **3**:ND(378). **8**:Range of host(82), CA(1077), CO(1077), CT(1077), FL(35), IA(1077), IL(1077), IN(1077), KS(1077), MA(1077), ME(1077), MI(1077), MN(1077), MO(1077), MS(1077,2720), ND(1077), NE(1077), NH(1077), NJ(1077), NY(1077), OH(1077), PA(1077), SD(1077,2242), UT(1077), VT(1077), WA(1077), WI(1077). **10**:Unk.(82), IA(1077), MN(1077), NE(1077), UT(1077).
Ustilago crameri. Head smut. **5**:Range of host(82), CO(1077), CT(1077), DC(1077), IA(1077), IL(1077), IN(1077), ME(1077), MI(1077), MN(1077), MO(1077), MT(1077), ND(1077), OH(1077), PA(1077), SD(1077,2242). **10**:SD(1077,2242).

Basidiomycotina - Other

Thanatephorus cucumeris. Brown patch, root rot, sharp eye spot. **8**:MN(3423), ND(3423). **10**:MN(3423), ND(3423), NE(3423), SD(3423).

Deuteromycotina - Hyphomycetes

Beniowskia sphaeroidea. Leaf blight, blister and spot. **2**:GA(401,1445), TX(1445,3569). **8**:IL(82,3423), KY(82), MO(3423), OK(82,2873,3423). **10**:MO(3423). **11**:GA(1445).
Bipolaris setariae. Spot blotch, leaf spot, secondary root rot. **5**:MD(3423), MN(82), ND(82), NJ(82). **8**:IA(3423), IL(312). **10**:IA(3423), MN(82), ND(82,3423), NE(3423), SD(2242,3423).
B. sorokiniana. Seedling blight, spot blotch, black point, root rot. **5**:MN(82,3423), ND(82,3423), SD(2242). **8**:KY(82), MN(82), ND(82,3423). **10**:IA(3423), KY(82), MN(82,3423), ND(82,3423), SD(2242,3423).
B. victoriae. Blight. **10**:IA(3423).
Cercospora setariae. Leaf spot. **2**:OK(82,2873,3423), WV(952). **8**:AL(82,3423), FL(35), IA(82,3423), IL(82,3423), KS(3023), KY(82,3423), MI(82,3423), ND(82,3423), NY(82,3423), OK(2872,3423), TX(82), WI(82,3423).
Cladosporium herbarum. Leaf mold. **5**:Range of host(82).
Curvularia geniculata. Secondary root rot. **10**:MN(82), ND(82), SD(2242).
Epicoccum nigrum. Leaf mold. **5**:OR(82).
Exserohilum monoceras. Blotch, leaf spot. **10**:ND(82,3423).
E. rostratum. Mold, secondary root rot. **5**:MN(82), ND(82,3423), SD(82,2242). **10**:MN(82,3423), ND(82,3423), SD(82,2242,3423).
E. turcicum. Leaf stripe. **5**:OK(2872). **10**:SD(2242).
Fusarium acuminatum. Secondary root rot. **5**:IA(3423), ND(82,3423). **8**:Unk.(82), ND(3423), SD(2242,3423). **9**:Unk.(82). **10**:Unk.(82), MN(3423), MT(3423), ND(3423), NE(3423), SD(2242,3423), WY(3423). **11**:Unk.(82).
F. equiseti. **5**:ND(82).
F. graminearum. Scab, seedling blight. **8**:IA(3423), IL(3423), IN(3423), WI(3423).
F. moniliforme. Secondary root rot. **5**:ND(82). **8**:Unk.(82), MN(3423), ND(3423). **9**:Unk.(82). **10**:Unk.(82), ND(3423). **11**:Unk.(82).
F. oxysporum. Secondary root rot. **5**:ND(82), SD(2242). **8**:Unk.(82), ND(3423). **9**:Unk.(82). **10**:Unk.(82), IA(3423), MN(3423), MT(3423), ND(3423), NE(3423), SD(2242,3423), WY(3423). **11**:Unk.(82).
F. poae. Secondary root rot, silver top. **8**:ND(3423). **10**:ND(3423).
F. solani. **5**:ND(82). **10**:ND(82).
Fusarium sp. Leaf spot. **11**:FL(35).
Helminthosporium sp. Leaf spot. **2**:TX(82). **8**:MN(82), TX(82). **10**:SD(82).
Microdochium bolleyi. Root necrosis, rot. **5**:MN(3423), ND(3423). **8**:ND(3423), SD(2242,3423). **10**:ND(82,3423), NE(3423), SD(2242,3423).
M. nivale. Snow mold. **9**:CA(82).
Nigrospora sphaerica. Node blight, sheath blight. **8**:IL(312).
Periconia circinata. Root rot. **5**:ND(3423). **10**:ND(3423).
Pyricularia grisea. Blast, head blight, gray leaf spot. **1**:KS(3023), NC(1303). **5**:E states(82), AL(3423), CT(3423), DE(3423), FL(3423), GA(3423), IA(3423), MA(3423), MD(3423),

MN(3423), MO(3423), MS(2720), NC(1303,3423), ND(3423), NE(3423), NJ(3423), RI(3423), SC(3423), SD(2242), TX(82,3423), VA(3423), WI(82,3423). **8:**NE states(82), CT(3423), DE(3423), IN(3423), KS(3021,3023), KY(3423), MA(3423), MD(3423), MI(3423), NE(2927,3423), NJ(3423), NY(3423), OH(3423), OK(2872,3423), PA(3423), RI(3423), SD(2242), TX(82), VA(3423), VT(3423), WI(82,3423), WV(3423). **10:**NE states(82), IA(3423), IL(312,3423), IN(3423), KS(3023,3423), MN(3423), MO(3423), NC(1303), ND(3423), NE(3423), OK(2873,3423), TX(82), WI(82,3423), WV(3423).
Stemphylium botryosum. Seed mold. **10:**ND(82), SD(82).
Ustilaginoidea setariae. False smut. **11:**LA(82).

Deuteromycotina - Coelomycetes
Ascochyta graminicola. **10:**WI(1361).
A. sorghi. Leaf spot. **8:**MO(3423). **10:**KS(3021).
Colletotrichum caudatum. Leaf spot. **8:**OK(2872,3423).
Macrophomina phaseolina. Charcoal rot. **8:**IL(311).
Melasmia setariae. Leaf spot. **8:**AL(82), SC(82). **10:**AL(3423).
Phoma sorghina. Leaf spot. **9:**CO(3423). **10:**KS(3021), ND(82,3423).
P. terrestris. Pink root. **5:**IA(82,3423), MN(82), ND(82,3423). **8:**MN(82), ND(82,3423), SD(82). **10:**MN(82), ND(82,3423), SD(82).
Pyrenochaeta setariae. On living leaves. **8:**WI(1359).
Septoria graminum. Leaf spot. **10:**WI(82).
S. oudemansii. Leaf spot. **10:**WA(3429).

Deuteromycotina - Other
Rhizoctonia solani. Summer blight, secondary root rot. **8:**WV(955). **10:**ND(82), NE(82), SD(82,2242).

Sitanion Raf.

1. *S. hystrix* (Nutt.) J. G. Sm. (=*Elymus elymoides* (Raf.) Swezey) [squirreltail]
2. *S. jubatum* J. G. Sm.
3. *Sitanion* sp.

Oomycetes
Pythium debaryanum. Damping-off, root necrosis, seed rot. **1:**WY(3423).

Ascomycotina
Claviceps purpurea. Ergot. **1:**OR(82,3423).
Crocicreas melanosporum. **1:**CA(505).
Didymosphaeria massarioides. **1:**WA(3815).
Erysiphe graminis. Powdery mildew. **1:**AZ(82), CA(82,638,642,3423), NV(82,3423). **2:**UT(82,3423).
Lophodermium arundinaceum. On dead culms and leaves. **1:**AZ(3423), CA(82,3423), WA(3423).
Mycosphaerella tassiana. **1:**WA(3815).
Phaeosphaeria caricinella. On overwintered herbage. **2:**CA(81,661).
Pleospora herbarum. **1:**WA(3814).

Basidiomycotina - Uredinales
Puccinia graminis. Heteroecious: telial host. **1:**CO(82), MI(82), NM(82), SD(2242), UT(82), WA(82), WY(82). **2:**ID(82), MI(82).
P. montanensis. Heteroecious: telial host. **1:**AZ(754,1195,1203,3981), CO(82,754), MT(82), UT(82).
P. pattersoniana. Heteroecious: telial host. **1:**ID(82), OR(82), UT(82). **2:**CA(81,82), WA(82). **3:**CA(82).
P. recondita. Heteroecious: telial host. **1:**AZ(82,1195,1203,3981), CA(82), CO(82,1195), ID(82), MI(82), NM(82), OR(82), UT(82), WA(82), WY(82). **2:**MI(82), OR(82), UT(82), WA(82).
P. striiformis. **1:**CA(81,82,642), CO(82), OR(82), WA(82), WY(82). **2:**CA(82), OR(82), WA(82).

Basidiomycotina - Ustilaginales
Urocystis agropyri. Flag smut. **2:**WA(82,1077).
Ustilago bullata. **1:**UT(82,1077), WY(1077). **2:**CA(1077).
U. hypodytes. **2:**WA(1077). **3:**CA(82).
U. striiformis. Stripe smut. **1:**CA(82,661,1077), ID(82,1077). **2:**CA(1077), OR(82,1077), WA(1077). **3:**UT(82,1077).
U. trebouxii. Loose smut. **1:**CA(1077), NM(1077), OR(82,1077), UT(1077), WA(82,1077), WY(1077). **2:**OR(82,1077), WA(82,1077).

Basidiomycotina - Other
Thanatephorus cucumeris. Brown patch, root rot, sharp eye spot. **1:**WY(3423).

Deuteromycotina - Hyphomycetes
Cercosporidium graminis. Leaf streak, brown stripe. **1:**AZ(82,3423), CA(661), MT(82,3423), NE(82), NM(1195,3423), NV(3423), OR(82,3423), SD(2242), WY(82,3423). **2:**CA(82), OR(3423), UT(82,3423), WA(3423). **3:**OR(82), WA(82).
Fusarium acuminatum. Secondary root rot. **1:**ID(3423), MT(3423), OR(3423), WA(3423), WY(3423).
F. poae. Leaf rot. **1:**AZ(3438).
Fusarium sp. Snow mold. **1:**CA(82). **2:**WA(82).
Microdochium bolleyi. Root necrosis. **1:**WY(3423).
M. nivale. Snow mold. **1:**CA(3423), WA(3423). **2:**WA(3423).
Pseudocercosporella herpotrichoides. Foot rot, strawbreaker. **1:**ID(82,3423).

Deuteromycotina - Coelomycetes
Ascochyta graminicola var. *holci*. Leaf spot. **1:**OR(82). **2:**WA(82).
A. sorghi. Leaf spot. **1:**OR(3423,3449). **2:**WA(3423,3449).
Dilophospora alopecuri. Twist disease. **1:**NV(3423), WA(82). **2:**OR(82), WA(82,3423).
Leptostromella hysterioides var. *graminicola*. On dead leaves. **1:**CA(642).
Phoma terrestris. Isolated from rhizosphere. **1:**OR(2791).
Pseudoseptoria donacis. On leaves. **1:**CA(642).
P. obtusa. Eye spot. **1:**CA(82,638,3423,3450), ID(82), MT(3423,3450), UT(82,3423,3450), WA(3450). **3:**CA(82).
P. stomaticola. Stem speckle, eye spot. **1:**MT(82), NM(82,3423,3450), OR(82,3423,3450), WA(82,3423).
Septoria passerinii. Leaf blotch, speckle. **1:**AZ(3435), OR(82,3423).
Stagonospora avenae. **1:**WA(3816).

Deuteromycotina - Other
Rhizoctonia solani. Secondary root rot. **1:**WY(82).

Sorghastrum Nash in Britton

1. *S. elliottii* (D. Mohr) Nash
2. *S. nutans* (L.) Nash [Indian grass]

Oomycetes
Pythium arrhenomanes. Root browning. **2:**ND(82).
P. graminicola. **2:**ND(3423).

Ascomycotina
Apiospora montagnei. **2:**GA(1438).
Claviceps purpurea. Ergot. **2:**KS(82,3423), OK(82,3423), TX(82,3423).
Claviceps sp. Ergot. **2:**OK(2872).
Massarina gloeospora. **2:**GA(1438).
Myriogenospora atramentosa. **2:**SD(2242).
Naemacyclus culmigenus. **1:**GA(1438). **2:**GA(1438).
Phyllachora luteo-maculata. Tar spot. **2:**IA(82,3423), SD(2242), TX(82).

Basidiomycotina - Uredinales
Puccinia virgata. Autoecious. **2:**E states(82), FL(35), KS(3021), MS(2720), NC(3944), NE(82), OK(2872), SD(2242), TX(82), WI(82,1334).

Basidiomycotina - Ustilaginales
Sphacelotheca andropogonis-hirtifolii. Head smut. **1:**GA(82,1077), TN(1077), VA(82,1077).
Sporisorium cruentum. **2:**MD(1077).
Tolyposporella chrysopogonis. Leaf smut. **2:**AL(82,1077), KS(82,1077), NE(82,1077), TX(82,1077,3602).

Deuteromycotina - Hyphomycetes
Cerebella andropogonis. Blackhead, head mold. **2:**AL(82,3423), AR(82,3423), KS(82,3423), MS(82,2720,3423), OK(2872,3423), TX(82,3423).
Cladosporium herbarum. Leaf mold. **2:**AL(82).
Helminthosporium sp. Leaf spot. **2:**KS(82).
Microdochium bolleyi. Root necrosis. **2:**ND(3423).
Periconia circinata. Root rot. **2:**ND(3423).

Deuteromycotina - Coelomycetes
Ascochyta brachypodii. Leaf spot. **2:**ND(3423,3449), NY(4001), PA(4001).
Colletotrichum caudatum. Leaf spot. **2:**NJ(82,3423), OK(2872,3423), PA(4004), SD(2242), WI(82,1338,1352,2544).
C. graminicola. Anthracnose. **2:**KS(82,3423), OK(82,3423), WI(82,3423), WV(953).
Colletotrichum sp. **2:**WI(1332).
Phoma sorghina. Leaf spot. **2:**WV(953).
Phyllosticta sp. Leaf spot. **2:**OK(2872), WI(1361).
Pirostoma circinans. On basal leaves. **2:**WI(1355).
Septogloeum bartholomaei. On leaves. **2:**KS(82,3423).
Septoria andropogonis. **2:**WI(1332).
S. andropogonis var. *sorghastri*. On leaves. **2:**NE(82,3423),

WI(82,1331,3423).
Stagonospora simplicior. Leaf mold, leaf spot. **2**:KS(3023), ND(82), WI(82,1340,1343,3423).
Stagonospora sp. Leaf spot. **2**:OK(2872).
Wojnowicia hirta. Secondary foot rot, rot of mature straw. **2**:ND(3423).

Sorghum Moench

1. *S. bicolor* (L.) Moench (=*S. vulgare* Pers., *S. saccharatum* (L.) Moench, *S. vulgare* Pers. var. *caffrorum* (Retz.) Hubb. & Rehd., *S. vulgare* Pers. var. *saccharatum* (L.) Boerl., *S. vulgare* Pers. var. *subglabrescens* Schwein. & Aschers., *S. vulgare* Pers. var. *technicum* (Körn.) Jáv., *Holcus sorghum* L., *S. caudatum* (Hack.) Stapf, *S. caffrorum* (Retz.) P. Beauv., *S. dochna* (Forssk.) Snowden, *S. durra* (Forssk.) Stapf, *S. guineense* Stapf) [broomcorn, durra, grain sorghum, kaffir, kaffir-corn, milo, sorghum, sweet sorghum]
2. *S.* × *drummondii* (Nees ex Steud.) Millsp. & Chase (=*S. sudanense* (Piper) Stapf, *S. vulgare* Pers. var. *sudanense* (Piper) A. Hitchc.). *S. bicolor* × *S. arundinaceum*. [Sudan grass]
3. *S. halepense* (L.) Pers. (=*S. miliaceum* (Roxb.) Snowden, *Holcus halepensis* L.) [aleppo grass, Egyptian millet, Johnson grass, means-grass]
4. *Sorghum* sp.
5. *Sorghum* × sp. (unnamed hybrids)

Oomycetes
Peronosclerospora sorghi. Downy mildew, powdery mildew. **3**:AL(1136), GA(1136), KS(1136), LA(1136), MS(1136), NM(1136), OK(1136), TN(1136), TX(1136). **2**:KY(3904). **4**:TX(2959).
Pythium arrhenomanes. Root browning, seedling disease, root rot. **1**:AR(82), CA(82), CO(82), KS(82), ND(82), NE(82), NM(82), OK(82,2872), TX(82,1086). **2**:ND(82).
P. debaryanum. Damping-off, root necrosis, seed rot. **1**:MN(3423), ND(82,3423), NE(3423), SD(2242,3423). **2**:ND(82,3423), NE(3423), SD(2242,3423).
P. graminicola. **1**:'Pythium root rot.' CA(3423), CO(3423), IA(82,3423), KS(3423), MN(3423), ND(3423), NE(3423), NM(3423), OK(3423), SD(2242), TX(3423). **2**:CA(3423), IA(3423), ND(3423), NE(3423), SD(2242,3423).
P. ultimum. Root necrosis. **1**:CA(82,3423), MO(82).
Pythium sp. Root rot. **1**:FL(35), HI(2903), NC(1303). **2**:GA(82).
Sclerophthora macrospora. Crazy top. **1**:'Crazy top downy mildew.' AZ(3681), SD(2242), TX(3877). **2**:IN(3710), SD(2242). **4**:IN(3710).
Sclerospora graminicola. Downy mildew. **1**:IA(82).
Sclerospora sp. Downy mildew, witches' broom. **3**:LA(1022), MS(3539).

Zygomycotina
Blakeslea trispora. On leaves. **1**:NC(1303).

Ascomycotina
Acanthophiobolus helicosporus. On culms. **3**:LA(82,3754).
Apiospora montagnei. **3**:GA(1438).
Botryosphaeria rhodina. **1**:TX(3423).
Gibberella fujikuroi. Seed rot. **1**:CA(82), GA(82), KS(82), ND(82), OK(82), TX(82).
G. zeae. Head mold, on culms. **1**:GA(82,2194). **3**:GA(82,1438).
Glomerella tucumanensis. Anthracnose, red rot, leaf spot. **1**:AL(3423), FL(3423), GA(3423), LA(3423), MN(3423), MS(3423), SC(3423), TX(3423), WV(3423). **3**:AL(3423), FL(3423), GA(3423), LA(3423), MS(3423), NC(3423), SC(3423), TX(3423). **2**:MS(3423), TX(3423).
Leptosphaeria sp. **3**:GA(1438).
Leptosphaerulina argentinensis. **3**:GA(1438).
Leptosphora implexa. On dead stems and adventitious roots. **3**:LA(3754).
Metasphaeria panicorum. **3**:GA(1438).
Monascus purpureus. Mold. **3**:OK(2872).
Mycosphaerella holci. Leaf spot. **1**:IL(82,3423).
Ophiobolus acuminatus. **3**:GA(1438).
Sclerotinia sclerotiorum. Culm rot. **1**:CA(82).
Setosphaeria turcica. **3**:GA(1438). **2**:GA(1438).
Stictis radiata. **3**:GA(1438).

Basidiomycotina - Uredinales
Puccinia purpurea. Heteroecious: telial host. **3**:SE states(82), AZ(1203,3981), CA(82), FL(35), GA(1439), HI(82,2903), IN(3710), MS(2720), OK(82,2872), TX(82,742). **2**:HI(82,2903), IN(3710), LA(82), OK(82), TX(82), WI(82,1330). **5**: On S. *vulgare* x *S. vulgare* var. *sudanensis*. IN(3710), OK(2873).
[*Uromyces andropogonis*]. **3**:TN(82).

Basidiomycotina - Ustilaginales
Sporisorium cruentum. Kernel smut. **3**:AR(82,1077), AZ(1077), CA(81,1077), GA(1077), KS(1077,3021,3023), MS(2720), NM(1077), OK(82,1077,2872), OR(1077), TN(82,1077), TX(82,652,1077), UT(1077). **2**:C states(1077), S states(1077), MS(2720), TX(82). **4**:IL(1077).
S. holci-sorghi. Downy mildew, head smut. **3**:NC(1303). **2**:CA(81,82,1077,1421), TX(82), WA(82,1077), WI(82,1358,1455). **4**:HI(2899), IN(3710), TX(2959).
S. sorghi. Covered kernel smut. **1**:'Covered kernel smut.' Range of host(82), AL(1077), AZ(1077), CA(82,1077), CO(82,1077), CT(1077), DC(1077), FL(35,1077), GA(82,1077,1439), HI(2903), IA(82,1077), IL(1077), IN(1077), KS(82,1077), MD(1077), MI(1077), MN(82,1077), MO(82,1077), MS(1077,2720), MT(1077), NC(1077,1303), ND(1077), NE(1077), NJ(1077), NM(1077), OH(1077), OK(1077,2872), OR(1077), PA(82,1077), SD(1077,2242), TN(1077), TX(82,1077), UT(82,1077), VA(1077), WA(82,1077), WI(82,1077), WY(1077). **3**:AR(1077), AZ(82,1077,1195), CA(81,82,1077), GA(1439), IN(82,1077), KS(82,1077,3023), LA(1077), NM(82,1077), OK(2872), SD(82,1077), TX(82,1077), UT(1077), WA(82,1077). **2**:AR(1077), AZ(82,1077), CA(82,1077), IA(82,1077), KS(82), NC(82), ND(1077), NE(82,1077), OR(82,1077), TX(82,1077), UT(1077), WA(82,1077), WI(82,1330).

Basidiomycotina - Other
Athelia rolfsii. Seedling blight, basal sheath and stalk rot. **1**:FL(35,3802), GA(3423), LA(3423).
Schizophyllum commune. In stored grain, possibly producing a mycotoxin. **1**:MO(1092).
Thanatephorus cucumeris. Banded disease, brown patch, root rot, sharp eye spot. **3**:GA(1439), LA(3524). **2**:ND(3423).

Deuteromycotina - Hyphomycetes
Acremonium strictum. Secondary root rot. 'Acremonium wilt.' **1**:IL(82), TX(2565).
Alternaria sp. Secondary leaf spot, stalk rot. **1**:AR(82), FL(35), OK(2872), TX(82), UT(82). **4**:NM(1660).
Aspergillus niger. **1**:OK(2872).
Aspergillus sp. Head mold. **1**:Unk.(82), FL(35).
Bipolaris cookei. Leaf spot, target spot, on culms. **1**:'Target leaf spot.' FL(35), NC(82). **4**:SC(634).
B. maydis. Leaf blight. **1**:FL(35).
B. sorghicola. Leaf spot, target spot. **3**:GA(82,3423), NC(1303), TX(82,3423). **2**:FL(82,3423), GA(82,3423). **5**: GA(3423).
B. sorokiniana. Seedling blight, spot blotch, black point, root rot. **1**:MN(82,3423), ND(82,3423), OK(82,2873). **2**:ND(3423).
B. victoriae. Blight. **1**:IA(3423).
B. zeicola. Leaf spot. **1**:NC(1303).
Cephalothecium sp. Leaf spot. **1**:OK(2872).
Cercospora sorghi. Gray leaf spot. **1**:'Gray leaf spot.' SE states(82), AL(82,3423), FL(35,3423), GA(2194,3423), IA(82), LA(3423), MO(82), MS(2720,3423), NC(1303,3423), NE(2630), SC(3423), SD(2242,3423), TX(82,3757), VA(3423). **3**:Gulf States(82), AL(3423), FL(35), IN(3710), KS(82), LA(962,3423), MS(2720,3423), NC(1303), NE(82,3423), OK(2872,3423), SD(82,3423), TN(82,3423), TX(3423). **2**:Gulf States(82).
Cladosporium herbarum. Leaf mold. **1**:Range of host(82), FL(35).
[*C. variabile*]. Leaf mold. **1**:OR(82).
Corethropsis sp. Head mold. **1**:GA(82,2194).
Curvularia geniculata. **1**:AL(2497).
Curvularia sp. Head mold, leaf spot. **1**:GA(82,2194), HI(2903), MD(651). **3**:IL(312).
Drechslera catenaria. Leaf blight. **2**:VA(3423).
Exserohilum rostratum. Seedling blight, leaf spot, red spot. **2**:FL(3423).
E. turcicum. Seedling blight, root rot, leaf spot. **3**:E states(82), AL(3423), FL(3423), GA(3423), LA(3423), MS(3423), NC(1303,3423), OH(3423), OK(2872), SC(3423), TX(82,3423).
Fusarium acuminatum. Head mold, secondary root rot. **1**:Range of host(82), IA(3423), MN(3423), ND(3423), NE(3423), SD(2242,3423). **2**:Unk.(82), KS(3423), ND(3423).
F. culmorum. Head mold. **1**:Range of host(82).
F. equiseti. Head mold. **1**:Range of host(82). **2**:Unk.(82).
F. graminearum. Scab, seedling blight. **3**:GA(3423).

F. moniliforme. Seedling blight, head mold, secondary root rot, pink rot, stalk rot. **1**:'Seedling blight and seed rot.' CA(3423), GA(2194), HI(2903), KS(3423), ND(3423), NE(3423), OK(2872,2873), SD(2242), TX(3423). **2**:ND(3423). **4**:NM(1660).
F. oxysporum. Head mold, secondary root rot. **1**:Range of host(82), ND(3423), SD(2242,3423). **2**:Unk.(82), ND(3423).
F. tricinctum. Head blight. **1**:WI(1358,1455).
Fusarium sp. Seedling blight, head mold, red root rot, stalk rot. **1**:FL(35), GA(2194), NC(1303), OK(2872). **2**:OK(2872). **4**:NM(1660).
Gloeocercospora sorghi. Copper spot, zonate leaf spot. **3**:IN(3710), LA(82,3423), MS(82,2720,3423). **2**:LA(82,3423), TX(82), VA(82,3423). **4**:KS(2632), NE(2632).
Helminthosporium sorghi. Leaf spot. **1**:PA(82).
Helminthosporium sp. Root and stalk rot. **1**:HI(2903), OK(2872). **2**:OK(2872). **4**:NM(1660).
Microdochium bolleyi. Root necrosis. **1**:ND(3423), SD(2242). **3**:ND(3423). **2**:ND(3423).
Nigrospora oryzae. Secondary root rot. **1**:IL(82).
N. sphaerica. Dry rot, kernel rot and secondary stalk rot. **1**:CA(82), OK(2872,3423).
Penicillium expansum. Blue mold. **1**:CO(3423), IA(3423), IL(3423), IN(3423), KS(3423), MO(3423), MT(3423), ND(3423), NE(3423), SD(3423), WA(3423), WI(3423).
P. oxalicum. Seed rot. **1**:MD(82).
Penicillium sp. Head, seed and stalk rot. **1**:AR(82), GA(2194), TX(82). **4**:NM(1660).
Periconia circinata. Milo disease, root rot. **1**:'Periconia root rot.' CA(82), KS(82), MD(3423), MN(3423), MT(3423), ND(3423), NM(82), OK(82), SD(3423), TX(82), WY(3423). **3**:ND(3423). **2**:ND(3423).
P. cookei. On leaves. **3**:AL(2503).
P. macrospinosa. On roots. **1**:KS(82).
Pyricularia grisea. Blast, gray spot. **2**:ND(82,3423).
Ramulispora sorghi. Sooty blotch, leaf spot, sooty stripe. **3**:AL(82,3423), IN(3710), LA(82), MS(82,2720,3423), NC(82,3423), NE(2631), OK(2872). **2**:FL(82), GA(82), TX(82), VA(82). **4**:AL(3423), AR(3423), FL(3423), GA(3423), LA(3423), MS(3423), OK(3423), TX(3423), VA(3423).
Trichothecium roseum. Head mold. **1**:KS(82).

Deuteromycotina - Coelomycetes

Ascochyta sorghi. Leaf spot. **1**:FL(35). **2**:SD(82,3423,3449).
A. sorghina. Rough leaf spot. **1**:'Rough leaf spot.' SE states(82), AL(3423), FL(35,82,3423), GA(2194,3423), KS(3021), LA(82), MS(82,2720,3423), NC(1303,3423), OK(82,2872,3423), SC(3423). **3**:GA(82,3423), IN(3710), LA(82,3423), MS(2720), NC(82,1303,3423). **2**:GA(82,3423), SD(82,2242).
Colletotrichum falcatum. Anthracnose, red rot, leaf spot. **3**:MS(2720), NC(1303).
C. graminicola. Anthracnose, head mold, red rot, red leaf spot, stalk rot. **3**:KS(3021), NC(1303), OK(2872). **2**:Unk.(82), GA(3423), MO(3423), ND(3423), NE(3423), VA(3423), WI(1332,3423). **4**:HI(2903), NM(1660).
C. lineola var. *halepense*. **3**:NC(3944).
Colletotrichum sp. Anthracnose, stalk rot, red leaf spot. **1**:C states(82), N states(82), Range of host(82), S states(82), FL(35). **3**:FL(35).
Lasiodiplodia theobromae. Seedling blight. **1**:TX(82).
Macrophomina phaseolina. Dry root rot, charcoal stalk rot. **4**:NM(1660).
Microdiplodia sp. Leaf spot. **1**:LA(3423).
Phoma sorghina. Leaf spot. **3**:FL(3423), MS(82,2720), TX(82,3423). **2**:MD(82,3423), VA(82,3423).
P. terrestris. Pink root. **1**:IA(3423), ND(3423), NE(3423), SD(3423). **2**:ND(3423).
Phoma sp. Seedling blight, root rot. **1**:AR(82), OK(82,2872).
Phyllosticta sp. Leaf spot. **1**:IL(82), MO(82).
Septoria pertusa. Leaf blight, leaf spot. **3**:FL(35), TX(82,3423).
Septoria sp. Leaf blight. **1**:IL(82).

Deuteromycotina - Other

Rhizoctonia solani. Leaf blight, sheath blight, root and stalk rot. **1**:GA(238,2194), IA(82), LA(2623), NC(1303), OK(82,2873), SD(2242), TX(82). **4**:NM(1660).
Rhizoctonia sp. Root and stalk rot. **1**:FL(35), HI(2903).
Sclerotium rolfsii. Southern sclerotial rot. **1**:GA(82,2194), LA(82), TX(82).

Spartina Schreb.

1. *S. alterniflora* Loisel. [salt marsh cordgrass, smooth cordgrass]
2. *S. alterniflora* Loisel. var. *glabra* (Muhl. ex Bigel.) Fernald
3. *S. alterniflora* Loisel. var. *pilosa* (Merr.) Fernald
4. *S. cynosuroides* (L.) Roth [big cordgrass]
5. *S. foliosa* Trin. (=*S. leiantha* Benth.)
6. *S. gracilis* Trin. [alkali cordgrass]
7. *S. maritima* (Curtis) Fernald (=*S. stricta* Roth)
8. *S. patens* (Aiton) Muhl. [salt meadow cordgrass]
9. *S. pectinata* Link (=*S. michauxiana* Hitchc.) [freshwater cordgrass, prairie cordgrass, slough grass]
10. *Spartina* sp.

Ascomycotina

Aniptodera chesapeakensis. On dead leaves. **1**:NC(1945).
Anthostomella sp. On dead plants. **1**:CT(1169), FL(1169), NC(1169,1303).
Buergenerula spartinae. **1**:CT(1169), FL(1169), GA(2592), MA(1170), ME(1169), NC(1169,1303), NJ(1169), RI(1166), VA(1169). **10**:SC(1169).
Ceriosporopsis halima. On seeds. **1**:NC(1303).
Chaetomium funicola. **1**:RI(1168).
Claviceps purpurea. Ergot. **1**:DE(82,3423), LA(82,3423), MA(82,3423), ME(82,3423), MS(82,940,2720,3423), NC(1169,1303), NJ(82,3423), NY(82,3423), RI(1166,1168). **3**:MA(1391,1394). **4**:FL(82), NY(82,3423). **5**:CA(81). **8**:MD(82,3423), NC(1303). **9**:IA(3423), MT(82,3423), NE(3423), OK(82,3423), SD(82,2242,3423). **10**:AL(941), DE(941), FL(35,941), GA(941), LA(941), MA(941), MD(941), MS(941), NC(941), NH(941), NJ(941), RI(941), SC(941), TX(941), VA(941).
Corollospora maritima. On seeds. **1**:NC(1303), RI(1168).
Didymella sp. On culms. **4**:LA(82).
Discocurtisia arundinariae. **9**:MA(1394).
Gnomonia salina. **1**:CT(1169).
Haligena spartinae. **1**:RI(1168). **8**:RI(1169).
Halosphaeria hamata. On seeds. **1**:FL(1169), ME(1169), NC(1303), RI(1166,1168). **8**:RI(1169).
Halosphaeria sp. **1**:MA(1170), NC(1303).
Leptosphaeria albopunctata. On culms and sheaths. **1**:CT(1169), FL(1169), GA(2592), MA(1170), NC(1303,1793), NJ(1169), RI(1166,1168). **10**:SC(1169).
L. halima. On culms and stems. **1**:GA(2592), MA(1170), NC(1303,1793).
L. marina. On culms. **1**:CT(1169), GA(2592), ME(1169), NC(1303,1793), NJ(1169), RI(1166,1169).
L. neomaritima. **1**:NC(1303).
L. obiones. On culms and stems. **1**:CT(1169), FL(1169), GA(2592), MA(1170), ME(1169), NC(1303,1793), NJ(1169), RI(1166,1168). **8**:RI(1169).
L. oraemaris. On culms. **1**:NC(1303,1793), RI(1169).
L. pelagica. **1**:CT(1169), RI(1166,1168). **8**:RI(1169).
Lignincola laevis. **8**:NC(1303).
Lulworthia floridana. **1**:NC(1303), RI(1168).
L. medusa. On culms. **10**:NJ(967,3248).
Lulworthia sp. On dead plants. **1**:NC(1169), RI(1166,1169). **8**:NC(1303). **10**:ME(1169), NC(1169).
Massarina arundinacea. **10**:NC(1303).
M. lacustris. **10**:NC(1303).
Massariosphaeria scirpina. **10**:NC(1303).
Microthecium levitum. **1**:RI(1168).
Mycosphaerella spartinae. On dead leaves and sheaths. **4**:NE(965,967).
Mycosphaerella sp. On leaves and sheaths. **9**:NE(82), WI(1356).
Nais inornata. **1**:RI(1168).
Phaeosphaeria caricinella. **8**:NC(1303).
P. herpotrichoides. **8**:NC(1303).
P. typharum. On leaves. **1**:CT(1169), FL(1169), GA(2592), MA(1170), NC(1169,1303), NJ(1169), RI(1166,1168), VA(1169). **8**:RI(1169). **10**:ME(1169).
P. vagans. **1**:MA(1170), RI(1166,1169).
Phaeosphaeria sp. On seeds. **1**:NC(1303).
Phyllachora cynodontis. Tar spot. **5**:CA(82,3423). **7**:CA(970).
P. graminis. **1**:MS(82). **2**:MA(1394).
P. spartinae. Tar spot. **1**:FL(35,82,3423), GA(82,3423), MD(82,3423).

P. sylvatica. Tar spot. **8:**SC(82,3423).
Phyllachora sp. Tar spot. **9:**MN(82).
Pleospora herbarum. **1:**RI(1168).
P. pelagica. On culms. **1:**FL(1169), MA(1170), NC(1303,1793), RI(1166,1169). **10:**SC(1169).
Pleospora sp. On seeds. **1:**NC(1303), RI(1168).
Rosellinia sp. **1:**RI(1168).
Sphaerulina pedicellata. **1:**RI(1168).
Xylaria psamathos. On buried culms. **8:**NC(316,1303).
Basidiomycotina - Uredinales
Puccinia distichlidis. Heteroecious: telial host. **6:**MT(82,1521), ND(378), NE(82), SD(82,2242), WY(82). **9:**NE states(82), SC states(82), CO(82), MT(82), ND(378), OK(2873), SD(2242). **10:**FL(35).
P. seymouriana. Heteroecious: telial host. **1:**MS(82,2720), NY(82). **4:**NY(82). **9:**E states(82), KS(3023), MS(2720), MT(82), OK(82,2872), WI(1521).
P. sparganioides. Heteroecious: telial host. **1:**CT(1169), DE(82), FL(35,82), MA(82), ME(1169), MS(82,2720), NC(82,1303), NH(82), RI(1166), VA(1169), VT(82). **2:**MA(1394). **4:**DE(82), FL(82), LA(82), MD(82), MS(2720), NJ(82), SC(82). **6:**KS(82), ND(82,378), NE(82), OK(82,2872), SD(82,2242). **8:**CT(82), MD(82), NC(1303), NJ(82), NY(82). **10:**FL(35).
Uromyces acuminatus. Heteroecious: telial host. **1:**CT(82), DE(82), FL(35,82), MA(82), MD(82), ME(82), NH(82), NJ(82), NY(82). **3:**MA(1394). **4:**FL(82). **6:**MT(82), ND(82), WY(82). **8:**CT(82), DE(82), FL(82), MA(82), MD(82), ME(82), NH(82), NJ(82), NY(82). **9:**NC states(82), NE states(82), CO(82), IA(1521), MT(82), ND(378), NE(2927), SD(2242). **10:**FL(35).
U. argutus. Telial host; aecial host unknown. **1:**FL(35,82,743,1521).
Uromyces sp. **1:**MA(1170).
Basidiomycotina - Other
Nia vibrissa. **1:**NC(1303,1943).
Deuteromycotina - Hyphomycetes
Alternaria alternata. On seeds. **1:**CT(1169), FL(1169), ME(1169), NC(1303), RI(1166), VA(1169).
A. maritima. **1:**NC(1303).
Alternaria sp. **1:**RI(1167,1168).
Arthrinium phaeospermum. **1:**RI(1168). **8:**RI(1169).
Aspergillus niger. **1:**RI(1167).
A. terreus. **1:**RI(1167).
Cephalosporium sp. **1:**RI(1168).
Cercosporidium graminis. Leaf streak, brown stripe. **6:**OR(82,3423).
Cirrenalia macrocephala. On seeds. **1:**NC(1303), RI(1168).
Cladosporium sp. On seeds. **1:**NC(1303). **8:**RI(1169).
Curvularia sp. On seeds. **1:**NC(1303).
Dendryphiella arenaria. **1:**RI(1168).
D. salina. **1:**RI(1168).
Dictyosporium pelagicum. **1:**RI(1168).
Drechslera sp. On seeds. **1:**NC(1303).
Epicoccum nigrum. On seeds. **1:**CT(1169), FL(1169), NC(1303), RI(1166,1168), VA(1169).
Exserohilum rostratum. On dead plants and seeds. **1:**FL(1169), NC(1169,1303), RI(1166,1169).
Fusarium acuminatum. Secondary root rot. **9:**ND(3423).
Fusarium sp. On seeds. **1:**NC(1303).
Monodictys pelagica. **1:**RI(1168).
Penicillium sp. On seeds. **1:**NC(1303).
Periconia sp. On seeds. **1:**NC(1303).
Rhinocladiella sp. On seeds. **1:**NC(1303).
Deuteromycotina - Coelomycetes
Cylindrosporium interstitialis. On living leaves. **9:**WI(1347).
Hendersonia spartinae. On living leaves. **9:**WI(1341).
Hendersonia sp. Leaf spot. **9:**WI(1360).
Pestalotia planimi. **1:**RI(1168).
Phoma sp. **1:**CT(1169), FL(1169), MA(1170), ME(1169), NC(1169,1303), NJ(1169), RI(1166,1168), VA(1169). **8:**RI(1169). **10:**ME(1169), SC(1169).
Phyllosticta sp. On leaves. **4:**MD(82).
Septogloeum spartinae. On leaves. **5:**CA(82,3423). **7:**CA(970).
Septoria spartinae. Leaf spot. **6:**UT(82,3423), WY(3433). **9:**IA(82), SD(82,2242,3423), WI(82,1340,1341,3423).
Septoria sp. On leaves. **1:**MA(1170), NC(1303), RI(1168). **4:**MD(82).
Stagonospora spartinicola. Streak spot on leaves. **9:**ND(3423,3424), SD(2242), WI(1344,3423,3424).
Stagonospora sp. On culms. **1:**CT(1169), FL(1169), MA(1170), ME(1169), NC(1169,1303), NJ(1169), RI(1166), VA(1169). **4:**MD(82). **8:**RI(1169).
Tracylla spartinae. Leaf spot. **8:**MS(82). **10:**FL(35).
Wojnowicia hirta. Secondary foot rot, rot of mature straw. **9:**ND(3423).

Sphenopholis Lams.-Scribn.

1. *S. nitida* (Biehler) Lams.-Scribn.
2. *S. obtusata* (Michx.) Lams.-Scribn. (=*Eatonia obtusata* A. Gray, *S. intermedia* Rydb.) [prairie wedge grass]
3. *S. obtusata* (Michx.) Lams.-Scribn. var. *major* (Torr.) Erdman (=*S. pallens* Lams.-Scribn.)
4. *Sphenopholis* sp.

Oomycetes
Pythium graminicola. **2:**ND(3423).
Ascomycotina
Epichloe typhina. Cat-tail, choke. **1:**MD(3423), PA(1944). **2:**GA(82,1944,3423), MS(82,2720,3423), NC(82,3423,3944). **3:**OH(1944,3423). **4:**MD(82), NC(82), OH(82).
Erysiphe graminis. Powdery mildew. **2:**IA(82,3423).
Basidiomycotina - Uredinales
Puccinia eatoniae. Heteroecious: telial host. **1:**KS(3021), MS(2720). **2:**E states(82), FL(35), MS(2720), ND(82,378), OK(82,2872), SD(2242), WV(82). **3:**ND(378). **4:**NM(82).
P. graminis. Heteroecious: telial host. **2:**CA(82), KS(82), LA(82), ND(82).
Deuteromycotina - Hyphomycetes
Acremonium typhinum. **2:**GA(3873).
Cercospora agrostidis. Leaf spot. **2:**ND(3423).
Cercospora sp. **2:**WI(1342).
Deuteromycotina - Coelomycetes
Ascochyta graminicola. Leaf spot. **2:**OK(2872).
A. sorghi. Leaf spot. **2:**OK(3423).
Colletotrichum graminicola. Anthracnose. **2:**ND(82,3423), SD(2242).
Pseudoseptoria everhartii. Speckle, eye spot. **2:**ND(82,3423,3450).
Septoria andropogonis. Leaf spot. **2:**ND(82,3423).
S. quinqueseptata. Leaf spot. **2:**ND(82,3423).
Deuteromycotina - Other
Sclerotium rhizodes. Leaf rot. **2:**WI(82). **3:**WI(3423).

Sporobolus R. Br.

1. *S. africanus* (Poir.) Robyns & Tourn.
2. *S. airoides* (Torr.) Torr. (=*S. wrightii* Munro ex Lams.-Scribn.) [alkali sacaton]
3. *S. argutus* (Nees) Kunth
4. *S. asper* (Michx.) Kunth
5. *S. asper* (Michx.) Kunth var. *hookeri* (Trin.) Vasey [meadow dropseed]
6. *S. buckleyi* Vasey
7. *S. clandestinus* (Biehler) Hitchc.
8. *S. contractus* Hitchc. [spike dropseed]
9. *S. cryptandrus* (Torr.) A. Gray [sand dropseed]
10. *S. elongatus* R. Br.
11. *S. giganteus* Nash [giant dropseed]
12. *S. heterolepis* (A. Gray) A. Gray [prairie dropseed]
13. *S. indicus* (L.) R. Br. (=*S. poiretii* (Roem. & Schult.) Hitchc., *S. berteroanus* (Trin.) Hitchc. & Chase)
14. *S. junceus* (Michx.) Kunth (=*S. gracilis* (Trin.) Merr.) [pineywood's dropseed]
15. *S. neglectus* Nash [annual dropseed]
16. *S. pyramidatus* (Lam.) Hitchc.
17. *S. vaginiflorus* (Torr.) A. Wood ex A. Gray (=*S. vaginiflorus* (Torr.) Wood ex A. Gray var. *inaequalis* Fernald) [poverty dropseed]
18. *Sporobolus* sp.

Oomycetes
Pythium debaryanum. Damping-off, root necrosis, seed rot. **4:**MT(3423).
P. graminicola. **9:**SD(2242), WY(3423).
P. irregulare. Root necrosis. **9:**NE(3423).
P. monospermum. Root necrosis. **9:**SD(82,2242,3423).
Sclerophthora macrospora. **15:**SD(2242).
Ascomycotina
Apiospora montagnei. **7:**GA(1438).
Balansia epichloe. Sterility disease, black ring. **12:**NE(2927). **18:**FL(35), MS(2720).

Coleroa sporoboli. On living leaves. **9:**WI(1344,1361). **12:**WI(1349,1361).
Erysiphe graminis. Powdery mildew. **11:**AZ(82,3423).
Graphyllium hysterioides. On dead culms. **4:**KS(82).
Myriogenospora atramentosa. **13:**LA(2675).
Nimbomollisia aberrans. On dead culms. **7:**GA(1438), KS(82).
Phyllachora graminis. **9:**WI(1330).
P. vulgata. Black spot, tar spot. **2:**CO(82,3423). **3:**TX(3423). **4:**IL(82,3423), NE(2927). **9:**KS(82,3021,3023,3423), NE(2927), OK(2872,3423), TX(82,3423), WI(82). **13:**KS(82). **18:**SD(3423), WI(82).
Placostroma sporoboli. On culms. **4:**AL(82,3423).
Rosellinia sp. **2:**TX(82).

Basidiomycotina - Uredinales

Puccinia cryptandri. **8:**AZ(753), CO(753), KS(753), MT(753), NE(753), NM(753), SD(753), TX(753), WI(753), WY(753). **9:**C states(82), AZ(82,1203,3981), ID(82), KS(753,3023), OK(2872), SD(2242), TX(742), WI(1349). **11:**AZ(3981). **18:**CO(82), NM(82).
P. graminis. Heteroecious: telial host. **4:**AL(82), IL(82). **9:**CO(82), KS(82,3021), ND(82), SD(82,2242), TX(82).
P. schedonnardi. Heteroecious: telial host. **2:**ND(82), SD(82), WY(82). **4:**WI(1348). **18:**TX(82).
P. sporoboli. Heteroecious: telial host. **4:**IA(82), ND(82), NE(82,2927), WI(753,1349). **12:**IA(82,753), ND(82,378,753), NE(82), WI(753,1339).
P. sporoboli var. *robusta*. Heteroecious: telial host. **4:**WI(753). **12:**ND(753).
P. vilfae. Heteroecious: telial host. **2:**TX(753). **4:**C states(82), SE states(82), AL(753), IN(753), KS(753,3023), MS(2720), OK(82,2872), SD(82,753,2242), TN(753), WI(753). **6:**TX(753). **12:**NE(2927). **15:**KS(753), NE(82,753), OK(2872).
Uromyces sporoboli. Heteroecious: telial host. **4:**IN(753), KS(82,3023), SD(82,2242), TX(82), WI(753,1349). **9:**IA(82), SD(82,753,2242). **12:**WI(1360). **13:**IA(82), KS(82), NE(82), SD(82). **15:**IA(82), KS(82,753), NE(82), SD(82,2242). **17:**KS(753), SD(2242).
U. sporobolicola. Telial host; aecial host unknown. **3:**TX(753). **16:**TX(742,753). **18:**TX(82).

Basidiomycotina - Ustilaginales

Sphacelotheca utahensis. Seed smut. **2:**UT(82,1077).
Ustilago hypodytes. Seed and stem smut. **9:**CO(82). **14:**MO(1077), MS(82,2720).
U. striiformis. **18:**KS(1077).
U. vilfae. Seed smut. **15:**IA(82,1077), IN(1077), KS(82,1077), NE(82,1077), PA(82,1077), SD(1077,2242), WI(82,1077,1330). **17:**IA(82,1077), IN(1077), KS(82,1077), NE(82,1077), PA(82,1077), WI(82,1077).

Basidiomycotina - Other

Thanatephorus cucumeris. Brown patch, root rot, sharp eye spot. **9:**NE(3423), SD(3423), WY(3423). **15:**ND(3423).

Deuteromycotina - Hyphomycetes

Bipolaris ravenelii. Head mold, sooty spike. **1:**HI(2903). **9:**WI(1342,3423). **10:**HI(3423). **15:**WI(82,1332).
Cercospora seriata. Leaf spot. **4:**AL(82,3423). **7:**OK(2872,3423). **9:**WI(1342).
C. setariae. Leaf spot. **9:**WI(3423).
Curvularia geniculata. Leaf mold. **13:**FL(35). **15:**ND(82).
Drechslera gigantea. **15:**KS(3023).
Exserohilum rostratum. On leaves. **9:**KS(3021).
Fusarium acuminatum. Secondary root rot. **4:**MT(3423). **9:**ID(3423), ND(82,3423), WY(3423).
F. oxysporum. **9:**ND(82).
Microdochium bolleyi. Root necrosis. **9:**ND(3423). **15:**ND(3423).
Volutella bartholomaei. On leaves. **4:**KS(82). **7:**KS(82). **12:**KS(82). **14:**KS(82).

Deuteromycotina - Coelomycetes

Amerosporium atrum. **4:**OK(2872). **17:**OK(2872).
Ascochyta stipae. Leaf spot. **2:**WA(82). **18:**CA(82).
Colletotrichum caudatum. Leaf spot. **4:**WI(1349). **12:**WI(1338,1339,1352,3423).
C. graminicola. Anthracnose. **4:**OK(2872,3423). **7:**OK(82,2873,3423). **12:**ND(82,3423), OK(2872,3423).
Macrophoma sporoboli. On dead leaves. **11:**AZ(82). **18:**NM(82).
Pseudoseptoria donacis. Speckle, eye spot, leaf spot. **4:**KS(3023). **7:**AZ(1194), OK(2873). **9:**AZ(3435), SD(2242).
P. stomaticola. Speckle, eye spot. **4:**AZ(1194), IA(82,3450), MT(82,3450), ND(82,3450), WI(82,1344). **7:**OK(82). **9:**ID(82,3423,3450), ND(82,3423,3450), SD(3423,3450), WY(3423,3450). **11:**AZ(82,3423,3450). **12:**NE(2927), WI(1340,3423,3450).
Selenophoma sp. **5:**OK(2872). **17:**OK(2872).
Septoria andropogonis. Leaf spot. **12:**AZ(1194).
S. andropogonis f. *sporobolicola*. Leaf blotch. **4:**WI(1349). **12:**ND(82,3423), NE(2927), WI(1338,3423).
Septoria sp. Leaf spot. **4:**WI(1354). **7:**OK(2872).
Stagonospora culmicola. Leaf mold, leaf rot. **2:**WA(82,3423). **4:**WA(82,3423).
S. subseriata. Leaf mold. **12:**ND(82).
Tiarosporella paludosa. **17:**WI(1358).
Wojnowicia hirta. On leaves. **4:**WI(1352,1361). **12:**WI(1338).

Deuteromycotina - Other

Rhizoctonia solani. Summer blight. **9:**SD(2242). **17:**WV(955).

Stenotaphrum Trin.

1. *S. secundatum* (Walter) Kuntze [St.-Augustine grass]

Myxomycetes

Physarum cinereum. Slime mold. **1:**FL(82), MS(2720), TX(82).

Oomycetes

Pythium sp. Root rot. **1:**FL(35).
Sclerophthora macrospora. 'Downy mildew.' **1:**FL(35,420), MS(419), TX(420).
Sclerophthora sp. Downy mildew. **1:**FL(1798), LA(1606), TX(1798).

Ascomycotina

Physalospora sp. **1:**AL(3182).
Sclerotinia homoeocarpa. 'Dollar spot.' **1:**FL(35).

Basidiomycotina - Uredinales

Puccinia stenotaphricola. **1:**FL(35,3647), HI(2903), LA(1606), TX(2077).
[*P. windsoriae*]. **1:**NC(1303).
Uromyces tenuicutis. **1:**HI(2903).

Basidiomycotina - Ustilaginales

Ustilago affinis. Purple top. 'Loose floral smut.' **1:**FL(35,1077), NC(1303).

Basidiomycotina - Other

Lepiota sp. Fairy ring. **1:**FL(35).
Marasmius sp. Fairy ring. **1:**FL(35).
Thanatephorus cucumeris. Brown patch, root rot, sharp eye spot. **1:**TX(3423).

Deuteromycotina - Hyphomycetes

Curvularia lunata. Leaf spot. **1:**HI(2903).
Curvularia sp. Fading out. **1:**FL(35), NC(1303).
Drechslera tritici-repentis. Leaf spot. **1:**FL(35).
Exserohilum rostratum. Leaf spot. **1:**FL(35).
Fusarium sp. Crown and root rot. **1:**FL(35).
Helminthosporium sp. Leaf blight, foot rot. **1:**HI(2903), LA(82,3423), TX(82).
Nigrospora sphaerica. Dry rot, leaf spot. 'Nigrospora blight.' **1:**FL(35,82,3423).
Phaeoramularia fusimaculans. 'Cercospora leaf spot.' **1:**FL(35,1105).
Pyricularia grisea. Blast, gray leaf spot. 'Gray leaf spot.' **1:**FL(35,82,2231,3423), HI(2299,2903), MS(2720), NC(1303), TX(82,3423).

Deuteromycotina - Coelomycetes

Ascochyta graminicola. Leaf spot. **1:**FL(35).
Colletotrichum sp. Anthracnose. **1:**FL(35), NC(1303).

Deuteromycotina - Other

Rhizoctonia oryzae. Root rot, leaf and sheath spot. **1:**FL(35,82,3423).
R. solani. Brown patch. 'Rhizoctonia blight.' **1:**FL(35), NC(1303), TX(82,1696).
Rhizoctonia sp. Brown patch. **1:**HI(2903), MS(2720).

Stipa L.

1. *S. arida* M. E. Jones
2. *S. avenacea* L. [black oatgrass, blackseed needlegrass]
3. *S. baicalensis* Roshevitz
4. *S. californica* Merr. & J. B. Davy ex H. Hall
5. *S. cernua* Stebb. & D. Löve
6. *S. columbiana* Macoun (=*S. williamsii* Lams.-Scribn.) [Columbia needlegrass]
7. *S. columbiana* Macoun var. *nelsonii* (Lams.-Scribn.) St. John

8. *S. comata* Trin. & Rupr. [needle-and-thread]
9. *S. coronata* Thurb.
10. *S. eminens* Cav.
11. *S. lemmonii* (Vasey) Lams.-Scribn. [Lemmon needlegrass]
12. *S. lepida* Hitchc. [foothill needlegrass]
13. *S. lepida* Hitchc. var. *andersonii* (Vasey) Hitchc.
14. *S. lettermanii* Vasey [Letterman needlegrass]
15. *S. leucotricha* Trin. & Rupr. [Texas needlegrass]
16. *S. mucronata* Kunth (=*S. setigera* J. Presl)
17. *S. neomexicana* (Thurb.) Lams.-Scribn. [New Mexican feathergrass]
18. *S. occidentalis* Thurb. (=*S. elmeri* Lams.-Scribn.). *S. elmeri* is a herbarium name published without description by Lams.-Scribn. [western needlegrass]
19. *S. pinetorum* M. E. Jones
20. *S. pringlei* Lams.-Scribn. [Pringle needlegrass]
21. *S. pulchra* Hitchc. [California needlegrass, purple needlegrass]
22. *S. richardsonii* Link [Richardson needlegrass]
23. *S. robusta* (Vasey) Lams.-Scribn. [sleepy grass]
24. *S. scribneri* Vasey [Scribner needlegrass]
25. *S. sibirica* (L.) Lam. [needlegrass]
26. *S. spartea* Trin. [porcupine grass]
27. *S. speciosa* Trin. & Rupr. [desert needlegrass]
28. *S. thurberiana* Piper [Thurber needlegrass]
29. *S. viridula* Trin. [feather bunchgrass, green needlegrass]
30. *Stipa* sp.

Oomycetes

Aphanomyces camptostylus. Root necrosis. **29:**ND(3423).
Aphanomyces sp. Root necrosis. **29:**ND(82).
Pythium arrhenomanes. Root browning. **8:**ND(82). **26:**ND(82). **29:**ND(82).
P. debaryanum. Damping-off, root necrosis, seed rot. **8:**WY(3423). **25:**ND(3423). **26:**SD(3423). **29:**MN(3423), MT(3423), ND(82,3423), SD(3423).
P. graminicola. **3:**ND(3423). **6:**ND(3423). **8:**MT(3423), ND(3423). **25:**ND(3423). **26:**ND(3423). **29:**MT(3423), ND(3423).
P. irregulare. Root necrosis. **8:**WY(3423).

Ascomycotina

Atkinsonella hypoxylon. Black choke. **15:**TX(82,3423). **21:**CA(3423). **29:**TX(82,3423).
Clathrospora multiseptata. On overwintered culms and leaves. **4:**CA(642), WA(3814).
Claviceps purpurea. Ergot. **6:**ID(3423), IN(82), UT(82,3423), WA(82). **7:**WA(3423). **8:**ND(3423). **14:**WA(82,3423). **23:**MT(3423). **26:**ND(3423), WI(1344). **29:**MN(82,3423), ND(82,3423), SD(2242). **30:**MT(82).
Crocicreas gramineum. On dead culms and leaves. **14:**UT(505).
Erysiphe graminis. Powdery mildew. **4:**CA(3423).
Phaeosphaeria heterospora. On overwintered herbage. **18:**CA(642).

Basidiomycotina - Uredinales

Puccinia avocensis. Heteroecious: telial host. **26:**WI(1348,1351,1362).
P. burnettii. Heteroecious: telial host. **8:**WY(1362).
P. durangensis. Telial host; aecial host unknown. **20:**AZ(1195,3981), CO(3379).
P. graminella. Autoecious. **12:**CA(1362). **30:**CA(82).
P. graminis. Heteroecious: telial host. **8:**WA(82). **21:**CA(82). **30:**CA(82).
P. interveniens. Heteroecious: telial host. **6:**CO(82), UT(82). **8:**CO(82), MT(82), WY(82). **11:**CA(81). **14:**CO(82). **21:**CA(1362). **27:**AZ(1203,3981). **30:**AZ(82), CA(82).
P. mexicensis. **30:**NM(743).
P. monoica. Heteroecious: telial host. **18:**CA(82). **30:**NE(82).
P. neocoronata. **20:**AZ(1203,3981).
P. stipae. Heteroecious: telial host. **6:**CA(82). **8:**W states(82), AZ(1195,1203,3981), CO(1195), MT(1195), SD(2242), WA(82), WY(1195,2924). **16:**CA(81). **20:**AZ(1195,1203). **26:**CA(82), IA(82,1362), KS(82), ND(82,378), NE(82), SD(2242), WI(82). **30:**CA(82).
P. substerilis. Autoecious. **6:**CO(82), ID(82), MN(82), WY(82,1195). **8:**MT(82). **14:**AZ(1195,1203,3981), CO(82,1195), NM(82), UT(82). **18:**WY(82). **23:**CO(1195,3424), SD(2242). **24:**NM(1195). **29:**C & W states(82), CO(1195,1362), MT(82), ND(378), SD(2242). **30:**ID(82), OR(82).
P. substerilis var. *oryzopsidis*. Autoecious. **1:**W states(1362).
P. substerilis var. *scribneri*. Autoecious. **24:**CO(1362), NM(1362).
[*P. windsoriae*]. **8:**VA(82).

Basidiomycotina - Ustilaginales

Tilletia hyalospora. Seed smut. **12:**CA(82,1077). **13:**CA(1077).
Urocystis fraseri. Culm, inflorescence and stem smut. **6:**ID(1077), WY(2342). **8:**CO(82,1077), ID(1077), MT(82,1077), ND(1077), NE(82,1077), OR(1077), SD(2242), UT(1077), WY(82,922,1077). **14:**NM(1077). **26:**NE(1077). **29:**CO(82,1077). **30:**OR(82).
U. granulosa. Inflorescence smut. **8:**ID(82,1077).
Ustilago hypodytes. Brown cubical rot, stem smut. **2:**IL(82). **4:**CA(642,661,1077). **6:**CO(82,1077), OR(82,1077). **9:**CA(1077). **10:**CA(1077). **11:**CA(1077). **15:**TX(82,1077). **16:**CA(1077), TX(1077). **21:**CA(81,82,1077). **23:**NM(1077). **24:**CO(1077). **26:**IA(82,1077), IL(82,1077), MN(1077), ND(378,1077), SD(1077,2242), WI(82,1077,1339). **27:**CA(1077), NV(1077). **28:**MT(1077), WA(1077). **29:**CO(1077), IL(1077), MN(1077), MT(82,1077), ND(82,1077), SD(82,1077,2242).
U. jacksonii. **6:**WY(2342). **14:**WY(2924).
U. minima. Stem smut. **8:**CA(82,1077), MT(82,1077), NE(1077), OK(1077,2872). **10:**CA(1077). **17:**AZ(1077). **24:**CO(1077). **26:**IA(82,1077), IL(82,1077), NE(1077), SD(82,1077,2242). **30:**AZ(82), CO(82).
U. trebouxii. **9:**CA(1077).
U. williamsii. **4:**CA(638,1077). **5:**CA(1077). **6:**ID(1077), UT(1077), WA(1077). **8:**CA(82,1077), MT(82,1077), OR(82,1077), UT(1077), WA(82,1077). **9:**CA(1077). **14:**CO(1077), ID(1077), NM(1077), OR(1077), UT(1077), WY(82,1077). **18:**CA(82,642,1077), MT(1077), OR(82,1077), WA(1077). **21:**CA(82,1077). **22:**MT(1077), WY(1077). **26:**SD(2242). **27:**CA(1077), NV(1077). **28:**MT(1077), OR(1077), WA(1077). **29:**MT(82,1077). **30:**CA(82), MT(82), OR(82), WA(82).

Basidiomycotina - Other

Thanatephorus cucumeris. Brown patch, root rot, sharp eye spot. **8:**MT(3423), ND(3423), SD(3423), WY(3423). **25:**ND(3423). **26:**NE(3423), SD(3423). **29:**MT(3423), ND(3423), NE(3423), SD(3423), WY(3423).
Typhula idahoensis. Snow mold. **7:**WA(3423). **8:**WA(3423).
T. incarnata. Snow mold. **7:**WA(3423).

Deuteromycotina - Hyphomycetes

Acremonium chisosum. Isolated from culms, leaf sheaths, and seeds. **10:**TX(3870).
Bipolaris sorokiniana. Seedling blight, spot blotch, black point, root rot. **3:**ND(3423). **8:**MN(82), ND(82,3423), SD(82). **21:**ND(3423). **25:**ND(3423). **26:**MN(82), ND(82,3423), SD(82), WI(1349). **29:**MN(82), ND(82,3423), SD(82,2242,3423). **30:**MN(82), ND(82), SD(82).
Cercosporidium graminis. Leaf streak, brown stripe. **4:**OR(3435). **6:**NW states(82), ID(3423), MT(3423), OR(3423), UT(3423), WA(82,3423), WY(3423). **7:**CA(3423), MT(3423), WA(3423). **8:**NW states(82), CO(3423), MT(3423), NE(3423), OR(3423), SD(2242,3423), UT(3423), WA(82), WY(3423). **9:**CA(3423). **11:**ID(3423), OR(3423), WA(3423). **14:**NW states(82), OR(3423), UT(3423), WA(82,3423), WY(3423). **18:**NW states(82), CA(3423), OR(3423), WA(82,3423). **19:**NV(3425). **27:**CA(3423). **28:**NV(3423), OR(3423), WA(3423), WY(3423). **29:**NW states(82), ID(3423), MT(3423), UT(3423), WA(82). **30:**NW states(82), WA(82).
Cladosporium herbarum. Leaf mold. **29:**ND(82).
C. stipae. On living leaves. **26:**WI(1345).
Curvularia geniculata. Secondary root rot. **8:**ND(82). **29:**ND(82), SD(2242).
Epicoccum nigrum. Leaf mold. **8:**NE(82).
Fusarium acuminatum. Secondary root rot. **6:**ND(3423). **7:**WA(3423). **8:**ND(82,3423), SD(3423), WA(3423), WY(3423). **25:**ND(3423). **29:**ND(82,3423), SD(2242,3423), WY(3423).
F. equiseti. Secondary root rot. **8:**ND(82). **26:**ND(82). **29:**ND(82).
Fusarium sp. Secondary root rot. **6:**Unk.(82). **26:**Unk.(82). **29:**Unk.(82).
Helminthosporium sp. Leaf spot. **6:**WA(3423). **26:**MN(82). **30:**KY(82), WA(82).
Microdochium bolleyi. Root necrosis. **3:**ND(3423). **6:**ND(3423). **8:**ND(3423), SD(3423), WY(3423). **26:**NE(3423). **29:**ND(3423), SD(2242,3423), WY(3423).
M. nivale. Pink snow mold. **6:**CA(82,3423), WA(82). **7:**WA(3423). **8:**CA(82,3423), WA(3423). **11:**WA(3423). **30:**WA(82).

Deuteromycotina - Coelomycetes

Ascochyta brachypodii. Leaf spot. **6:**OR(3432). **8:**MT(82,3423,3449),

OR(82). **29:**MN(82,3423,3449), ND(82).
A. hordei. Leaf spot. **14:**ID(3425). **23:**CO(3423,3449). **30:**CO(82).
A. sorghi. Leaf spot. **18:**OR(3423). **29:**SD(3423,3449).
A. stipae. **30:**CA(82), OR(82).
Colletotrichum graminicola. **29:**SD(2242).
Phoma terrestris. Pink root. **6:**ND(3423). **8:**MN(82), MT(3423), ND(82), SD(82,3423). **26:**MN(82), ND(82), SD(82). **29:**MN(82), ND(82,3423), SD(82).
Pseudoseptoria donacis. Speckle, eye spot. **6:**WA(3423,3450). **8:**MT(82,3423,3450), ND(82,638,3423,3450), NE(82,3423,3450), SD(82,2242,3423,3450), WI(3450), WY(82,3423,3450). **11:**WA(3423,3450). **18:**OR(3450). **22:**WY(3423,3450). **26:**MT(82), ND(82,3423,3450), SD(82). **29:**MT(82,3423,3450), ND(82,638,3423,3450), SD(82,2242,3423). **30:**WA(82), WY(82).
P. obtusa. Eye spot. **6:**WA(82). **9:**CA(3423,3450). **22:**MT(3423,3450), WY(3450). **30:**CA(82), MT(82), WY(82).
P. stomaticola. Eye spot. **11:**WA(3423).
Septoria andropogonis. Leaf spot. **8:**SD(2242). **23:**CO(3424), NM(3425). **29:**SD(2242).
S. andropogonis f. *sporobolicola*. Brown blotch. **8:**MT(82,3423), ND(82,3423,3424). **29:**ND(638,3423), SD(82,3423).
S. secalis. **29:**SD(2242).
S. secalis var. *stipae*. Leaf spot, white spot. **6:**ND(3423), WA(82). **7:**WA(3423). **29:**MN(82), ND(82,3423), SD(82,3423). **30:**MN(82), ND(82), SD(82).
S. stipina. Leaf spot. **6:**OR(82). **7:**OR(3423).
Stagonospora bellunensis. Leaf blotch. **23:**CO(3424). **29:**ND(82,638).
S. nodorum. Glume blotch. **29:**CO(3424), ND(82,3423), SD(2242).
S. simplicior. Leaf spot. **6:**MT(3433), UT(82), WY(3423,3424). **7:**CA(642,3423,3424), UT(3423,3424). **8:**NV(3438). **11:**OR(3423,3424). **14:**CO(3423,3424), OR(3432), WA(3442). **18:**ND(3424), OR(3423). **29:**ND(3423).
S. spiculisperma. On living leaves. **26:**WI(1343).
Urohendersonia stipae. On living leaves. **26:**WI(1347).
Wojnowicia hirta. Secondary culm and foot rot, rot of mature straw. **11:**OR(3423). **14:**ID(3425). **29:**MT(3423), ND(3423). **30:**MN(82), OR(82).

Deuteromycotina - Other
Rhizoctonia solani. Secondary root rot. **8:**ND(82). **26:**NE(82). **29:**ND(82), SD(82). **30:**ND(82).

×*Stiporyzopsis* B. Johnson & Rogler
Oryzopsis × *Stipa*.
1. × *S. bloomeri* (Boland.) B. Johnson (=*Oryzopsis bloomeri* (Boland.) Ricker ex Piper). *Oryzopsis hymenoides* × *Stipa* spp.

Basidiomycotina - Ustilaginales
Ustilago williamsii. **1:**WA(1077).

Deuteromycotina - Hyphomycetes
Microdochium nivale. Snow mold. **1:**CA(3423).

Deuteromycotina - Coelomycetes
Stagonospora simplicior. Leaf spot. **1:**CA(3423,3424).

Taeniatherum Nevski
1. *T. caput-medusae* (L.) Nevski (=*Elymus caput-medusae* L.)

Deuteromycotina - Coelomycetes
Pseudoseptoria stomaticola. Eye spot. **1:**WA(3423,3450).

Tridens Roem. & Schult.
1. *T. albescens* (Vasey) Woot. & Standl. (=*Triodia albescens* Vasey)
2. *T. elongata* (Buckley) Nash in Small (=*Triodia elongata* Buckley)
3. *T. flavus* (L.) Hitchc. (=*Triodia flava* (L.) Smyth) [purpletop]
4. *T. strictus* (Nutt.) Nash in Small (=*Triodia stricta* (Nutt.) Benth. & Vasey)

Ascomycotina
Apiospora montagnei. **3:**GA(1438).
Balansia epichloe. On leaves. **3:**KS(3021,3023).
Mycosphaerella crus-galli. On dead leaves. **3:**NY(82).
Myriogenospora atramentosa. **3:**OK(2872).
Phaeosphaeria culmorum. **3:**GA(1438).
Phyllachora eragrostidis. Tar spot. **1:**NM(82), TX(82,3423). **3:**GA(82,1438,3423), TX(82,3423).
Phyllachora sp. **3:**GA(1438).

Basidiomycotina - Uredinales
Puccinia graminis. **3:**AL(82).
P. windsoriae. Heteroecious: telial host. **3:**E states(82), GA(1439), IN(647), KS(3021,3023), MD(651), MI(665), MS(2720), NC(3944), ND(82), OK(660,2872), TX(82).

Basidiomycotina - Ustilaginales
Ustilago tricuspidis. Seed smut. **3:**MO(82,1077), OK(1077,2872), TN(1077), WV(82,1077).

Deuteromycotina - Hyphomycetes
Curvularia geniculata. On leaves. **3:**KS(3023), OK(660).
Nigrospora sphaerica. **3:**MD(651).
Periconia lateralis. On culms. **3:**NE(82).

Deuteromycotina - Coelomycetes
Colletotrichum graminicola. Anthracnose. **1:**OK(2872,3423). **2:**OK(2872,3423). **3:**AL(82,3423).
Neottiosporina paspali. Leaf spot. **3:**IN(654), VA(82,3423).
Rhynchosporina tridentis. On living leaves. **3:**IN(987), KS(3451), OH(653).
Selenophoma sp. Leaf spot. **3:**OK(2872). **4:**OK(2872).
Stagonospora tridentis. On living leaves. **3:**KS(3451).

Deuteromycotina - Other
Rhizoctonia solani. Summer blight. **3:**WV(955).

Triodia R. Br.
1. *Triodia* sp.

Basidiomycotina - Uredinales
[*Puccinia cryptandri*]. **1:**TX(82).
[*P. windsoriae*]. **1:**TX(82).

Triplasis P. Beauv.
1. *T. americana* P. Beauv.
2. *T. purpurea* (Walter) Chapm.

Basidiomycotina - Uredinales
Puccinia schedonnardi. Heteroecious: telial host. **2:**KS(82,753), OK(82,2872), TX(82,753).

Basidiomycotina - Ustilaginales
Ustilago sieglingiae. **2:**FL(35,82,1077), GA(1443), IA(82,1077), IL(82,1077), NC(82,1077).
U. triplasidis. Inflorescence smut. **1:**FL(35,82,1077), MS(82,1077,2720), SC(82,1077).

Deuteromycotina - Hyphomycetes
Bipolaris spicifera. **2:**WI(1354).

Tripsacum L.
1. *T. dactyloides* (L.) L. [eastern gamagrass]
2. *T. floridanum* Porter ex Vasey [Florida gamagrass]
3. *Tripsacum* sp.

Ascomycotina
Apiospora montagnei. **1:**GA(1438).
Claviceps purpurea. Ergot. **1:**MS(82,2720), OK(82,2873,3423).
C. tripsaci. Ergot. **1:**GA(82,1438,3423), MD(82,3423), MS(82,2720,3423), NC(82,1303,3423,3944), VA(82,3423).
Dasyscyphella sp. **1:**GA(1438).
Mollisia sp. **1:**GA(1438).
Stictis maydis. **1:**GA(1438).

Basidiomycotina - Uredinales
Physopella pallescens. Telial host. **3:**FL(35,3190).
Puccinia polysora. Telial host; aecial host unknown. **1:**AL(82,3713), FL(35,82), IA(82), NC(82,3944), NJ(82), TX(82,742), VA(82). **2:**FL(35), NC(1303). **3:**FL(35), VA(82).
P. tripsaci. Heteroecious: telial host. **1:**MS(2720), NC(1303,3944). **3:**IN(82), KS(82), MD(82), MO(82), MS(82), NC(82), NE(82), TX(82).

Deuteromycotina - Hyphomycetes
Alternaria sp. Leaf spot. **3:**FL(35).
Bipolaris zeicola. Leaf spot. **1:**NC(1303).
Epicoccum nigrum. On leaves. **1:**MS(82,2720).
Fusarium sp. Leaf spot. **1:**FL(35). **3:**FL(35).
Helminthosporium sp. Leaf spot. **1:**FL(35). **3:**FL(35).
Ramularia carniformis. Head mold. **1:**MS(2720). **3:**MS(82).

Deuteromycotina - Coelomycetes
Colletotrichum graminicola. Anthracnose. **1:**FL(35), NY(82,3423), OK(2872,3423), SC(82,3423), VA(82,3423). **2:**FL(35). **3:**FL(35).
Phyllosticta sp. Leaf spot. **1:**FL(35). **3:**FL(35).

Trisetum Pers.

1. *T. canescens* Buckley [tall trisetum]
2. *T. cernuum* Trin. [nodding trisetum]
3. *T. flavescens* (L.) P. Beauv.
4. *T. glomerata* (Kunth) Trin. (=*Koeleria glomerata* Kunth)
5. *T. melicoides* (Michx.) Lams.-Scribn.
6. *T. montanum* Vasey
7. *T. spicatum* (L.) Richter [spike trisetum]
8. *T. wolfii* Vasey [Wolf's trisetum]

Ascomycotina

Claviceps purpurea. Ergot. **1:**ID(82,3423).
Erysiphe graminis. **7:**CA(661).
Gaeumannomyces graminis. Take-all. **2:**AK(3441), WA(3446,3754).
Leptosphaeria sp. **7:**AK(3431).
Lophodermium arundinaceum. On dead culms and leaves. **7:**AK(3431), CA(82,3423), CO(3423), MT(3423), WY(3423).
Mycosphaerella tulasnei. **7:**AK(3431), CA(642).
Phaeosphaeria caricinella. On overwintered herbage. **7:**CA(642).
P. eustoma. On dead sheath. **7:**AK(1938).
P. heterospora. On overwintered herbage. **7:**CA(642).
P. microscopica. **7:**CA(642).
Pleospora eocoronis. On dead stem. **7:**AK(1938).
P. penicillus. On dead stem. **7:**AK(1938).
Pleospora sp. **7:**AK(3431).

Basidiomycotina - Uredinales

Puccinia brachypodii var. *arrhenatheri*. Heteroecious: telial host. **7:**Unk.(754).
P. brachypodii var. *poae-nemoralis*. Heteroecious: telial host. **6:**Unk.(754). **7:**AK(526,3431), NV(82), OR(3438), WA(82), WY(2924,3377).
P. coronata. Heteroecious: telial host. **1:**OR(82). **2:**CA(81,82), OR(82), WA(82). **7:**CO(82).
P. graminis. Heteroecious: telial host. **1:**CA(82), ID(82), MT(82), OR(82). **5:**WI(1349).
P. monoica. Heteroecious: telial host. **7:**CO(82,1195), UT(82,1362), WA(82), WY(2924).
P. striiformis. **2:**WA(82).

Basidiomycotina - Ustilaginales

Urocystis agropyri. **7:**WY(1077,2924).
Ustilago striiformis. **7:**ID(2342), WY(1077).

Basidiomycotina - Other

Coprinus psychromorbidus. **7:**AK(3441).
Thanatephorus cucumeris. Brown patch, root rot, sharp eye spot. **1:**WA(3423).

Deuteromycotina - Hyphomycetes

Cercosporidium graminis. Leaf streak, brown stripe. **1:**ID(82,3423), MT(82,3423). **3:**NY(3423). **6:**MT(3423). **7:**NY(82), WA(3442).
Cladosporium herbarum. Leaf mold. **1:**WA(82).
Drechslera avenacea. Leaf tip blight. **7:**UT(3438).
D. campanulata. On dead leaves. **7:**AK(3428,3431).
Fusarium acuminatum. Secondary root rot. **2:**WA(82,3423).
Hadrotrichum triseti. **2:**ID(3442).
Mastigosporium rubricosum. Leaf spot. **2:**AK(3426,3441), OR(82,3423).
Muirella alascensis. On living leaves. **7:**AK(3426,3431,3435).
Ramularia pusilla. Leaf spot. **7:**WY(82,3423,3425,3429).
Spermospora subulata. Blast, leaf spot. **7:**MT(3423,3424).

Deuteromycotina - Coelomycetes

Ascochyta sorghi. Leaf spot. **1:**ID(82,3423,3449).
Cheilaria agrostis. Leaf spot. **7:**MT(3433).
Colletotrichum graminicola. Anthracnose. **1:**OR(82), WA(3423). **2:**MI(82), OR(82,3423). **3:**MI(3423). **7:**OR(82).
Hendersonia mouiensis. **4:**HI(2775).
H. rockiana. **4:**HI(2775).
Pseudoseptoria everhartii. Leaf speckle, eye spot. **7:**CO(82,3423,3450), MT(82,3423,3450), WY(82,3423). **8:**WY(3423,3450).
P. stomaticola. **7:**WY(3450).
Septoria calamagrostidis. Leaf spot. **1:**ID(3423), OR(82,3423). **2:**OR(82,3423). **6:**WY(2924,3425). **7:**AK(3431), CA(82), ID(3442), UT(3425), WA(82).
Stagonospora avenae. Leaf spot. **6:**NM(3425).
S. culmicola. Leaf mold, rot. **7:**MT(82,3423).
S. mariae. **2:**WA(3433).

Triticum L.

1. *T. aestivum* L. (=*T. sativum* Lam., *T. vulgare* Vill., *T. timopheevii* (Zhuk.) Zhuk.) [bread wheat, common wheat]
2. *T. dicoccoides* (Kornh. ex Aschers. & Graebn.) Aarons. [wild emmer]
3. *T. dicoccon* Schrank (=*T. dicoccum* Schrank) [emmer]
4. *T. dicoccum* Schrank var. *farrum* Bayle-Bar.
5. *T. durum* Desf. (=*T. pyramidale* Percival) [durum wheat]
6. *T. monococcum* L. [einkorn]
7. *T. polonicum* L. [Polish wheat]
8. *T. spelta* L. [spelt]
9. *T. turgidum* L. [Alaska wheat, cone wheat, English wheat, Mediterranean wheat, poulard wheat, rivet wheat]
10. *Triticum* sp.

Plasmodiophoromycetes

Polymyxa graminis. **1:**Unk.(2145). **8:**MD(2145).

Chytridiomycetes

Olpidium brassicae. **1:**SD(2242).
Rhizophydium graminis. Rootlet necrosis. **1:**MA(3423).

Oomycetes

Phytophthora sp. Leaf rot. **1:**ND(82), WA(82).
Pythium aphanidermatum. Damping-off. 'Pythium root rot.' **1:**FL(35,3169).
P. aristosporum. Isolated from dead plants. **1:**MT(82), ND(82), WA(2151).
P. arrhenomanes. 'Pythium root rot.' **1:**CA(82), HI(2903), IA(82), MO(82), MT(82), NE(82), SD(2242), WA(82). **3:**ND(82).
P. debaryanum. Damping-off, root necrosis, seed rot. **1:**NC states(82), NW states(82), AZ(3423), CA(3423), CO(3423), IA(3423), KS(3423), MN(3423), MO(3423), MT(3423), ND(3423), NE(3423), NM(3423), OR(3423), SD(2242,3423), UT(3423), WA(3423), WY(3423). **3:**ND(3423). **5:**MN(3423), ND(3423), SD(2242,3423), WA(3423).
P. graminicola. 'Pythium root rot.' **1:**CA(3423), CO(3423), HI(3423), IA(82,3423), MN(3423), MT(3423), ND(3423), NE(3423), OK(3423), SD(82,2242,3423), WA(3423). **3:**ND(3423). **5:**MN(3423), ND(3423), SD(2242).
P. heterothallicum. **1:**Unk.(3893).
P. hypogynum. Root necrosis. **1:**ND(82,3423).
P. irregulare. Root necrosis. **1:**MN(82,3423), ND(82,3423), NE(82,3423), SD(82,2242,3423).
P. iwayamai. Isolated from dead plants. **1:**WA(2151).
P. myriotylum. 'Pythium root rot.' **1:**Unk.(3893).
P. okanoganense. Isolated from decaying leaves. **1:**WA(2149).
P. oligandrum. **1:**TX(82,3423).
P. ostracodes. Roots. **1:**TX(82,3423).
P. splendens. Root necrosis, rot. **1:**HI(2903,3423). **5:**ND(3423).
P. sylvaticum. **1:**Unk.(3893).
P. torulosum. **1:**Unk.(3893).
P. ultimum. Damping-off, root necrosis, seed rot. **1:**NC states(82), NW states(82), MN(3423), MT(3423), ND(3423), NE(3423), SD(2242,3423), WA(2151,3423). **5:**ND(3423).
P. ultimum var. *sporangiiferum*. **1:**Unk.(3893).
P. volutum. 'Pythium root rot.' **1:**Unk.(3893).
Pythium sp. Root rot. **1:**OK(2872).
Sclerophthora macrospora. 'Downy mildew.' **1:**AZ(3681), CA(82), DE(3423), IL(312), KY(82,3423,3877), MS(3423,3839), NY(82,3423), TN(82,3423). **5:**ND(1813). **10:**MS(3050), VA(2989).

Zygomycotina

Rhizopus stolonifer. Black mold seed rot. **1:**Range of host(82).

Ascomycotina

Botryosphaeria zeae. Node canker. **1:**IL(312).
Chaetomium aterrimum. On damaged plants. **10:**KS(3306).
Clathrospora pentamera. On dead culms, rachises and glumes. 'Platyspora leaf spot.' **1:**ND(1639).
Claviceps purpurea. 'Ergot.' **1:**AZ(3423), IA(3423), KS(3423), MI(3833), MN(3423), MO(3423), ND(3423), NY(3423), OH(3423), OK(2872,3423), OR(3423), PA(3423), WI(3423). **3:**ND(3423), WI(82). **5:**ND(3423). **10:**WY(819).
Cochliobolus sativus. On leaves. **1:**NY(2212).
Endomycopsis burtonii. Isolated from wheat samples, and from flour. **1:**SE states(2003), OK(2003), TX(2003).
Erysiphe graminis. 'Powdery mildew.' **1:**Range of host(82), CA(3423), CO(3423), DE(3423), GA(1438,3423), IA(3423), ID(3423), IL(3423), MD(3423), MN(3423), MO(3423), MT(3423),

NC(1303), NE(3423), NJ(3423), OH(949,3423), OK(2872,3423), OR(3423), SD(2242), UT(3423), WA(3423), WY(3423). **10:**GA(2404), IN(3117), NC(2516), PA(1910), SC(932).
E. graminis f. sp. *avenae*. Powdery mildew. **10:**AR(3029).
E. graminis f. sp. *tritici*. Powdery mildew. **1:**FL(35), GA(2406), NY(2560), PA(3062). **10:**FL(929,931), GA(1023,1275), IL(309), MS(3539), NJ(496), SC(933), TX(1134,1135), VA(2993).
Gaeumannomyces graminis. Foot rot, white-heads. 'Take-all.' **1:**E states(82), AR(3423), CA(82,3423), ID(82,630,3423), IL(1761,3423), IN(3423), KS(3423), MD(3423), MI(3833), MO(3423), MS(3539), MT(2282), NC(1303,3423), NE(82,3423), NY(3423), OK(82,3423), OR(82,630,3423), PA(3423), SD(2242), TN(3423), TX(3423), VA(3423), WA(82,630,3423), WI(3423). **3:**NY(82). **8:**NY(82). **10:**GA(1475), IL(1761), MS(3539), MT(3200), NE(2155), VA(2993).
G. graminis var. *tritici*. 'Take-all.' **1:**GA(3052), NC(3754), TX(2263). **10:**IN(1662).
Gibberella avenacea. **1:**WA(628).
G. cyanogena. **1:**GA(1438).
G. fujikuroi. Seedling root rot. **1:**IN(82), MN(82), MO(82), TN(82).
G. zeae. Seedling blight, scab. 'Crown rot.' **1:**E & C states(82), SW states(82), HI(2903), MS(2720), NC(1303), OH(949), OK(2872). **3:**OH(82), SD(82). **8:**IL(82), MN(82), WI(82). **10:**NC(2516), NJ(496), WA(1702).
Gibberella sp. Scab. **10:**TN(72,73).
Hansenula anomala. Isolated from flour. **1:**KS(2003), MT(2003), ND(2003), NE(2003), OK(2003), TX(2003).
Lophodermium culmigenum. On old leaves. **1:**KS(82), MI(82), NY(82), OH(82), PA(82).
Microascus trigonosporus. On seeds. **1:**WI(82).
Monographella nivalis. **1:**WA(629).
Mycosphaerella tulasnei. **1:**Unk.(82).
Myriosclerotinia borealis. 'Snow scald.' **1:**WA(3442).
Phaeosphaeria avenaria f. sp. *triticea*. On leaves. **1:**NY(2211,2212).
P. herpotrichoides. On culms, secondary root rot. 'Herpotrichoides leaf spot.' **1:**MN(1640), ND(1640), WA(82,3423).
P. nodorum. On leaves. **1:**NY(2212).
P. tritici. Leaf and sheath spot. **1:**IA(82), IL(312), NE(82), NY(82).
Phyllachora graminis. 'Tar spot.' **1:**IA(82,3423), MI(82).
Pichia farinosa. Isolated from flour. **1:**MT(2003), ND(2003).
P. fermentans. Isolated from wheat samples. **1:**SE states(2003).
Pleospora herbarum. Glume spot. **1:**CA(82), NY(82), OK(82), TX(82), UT(82).
Pyrenophora trichostoma. Tan spot. **1:**GA(82,1438), MI(82), ND(1987), WA(82). **10:**NE(3785).
P. tritici-repentis. 'Tan spot. Pink grain.' **1:**IL(312), IN(1664), NY(2212).
Zygosaccharomyces microellipsodes. Isolated from flour. **1:**KS(2003), NE(2003).

Basidiomycotina - Uredinales

Puccinia graminis. Heteroecious: telial host. 'Stem rust.' **2:**MN(82), TX(82). **3:**CA(82), MO(82), OK(82,2872), TX(82), WA(82). **5:**SD(2242). **6:**MO(82). **7:**IA(82), MN(82), MO(82). **8:**KS(82), MN(82), MO(82), ND(82,378), OK(2872), OR(82), SD(82), WA(82), WY(82). **9:**MN(82), MO(82), PA(82). **10:**FL(929,931), GA(1024,2404), IL(309), KY(1815), MD(2850), MS(3539), ND(380), NE(2155), SC(932,933), SD(2551), TN(72), TX(119,1134,1135), VA(2993), WY(819).
P. recondita. Heteroecious: telial host. 'Leaf rust.' **2:**TX(82). **3:**GA(82,1439), MO(82), OR(82), SD(82), TX(82). **5:**SD(2242). **6:**MO(82). **7:**KS(82). **8:**IL(82), KS(82), MD(82), MN(82), MO(82), OH(949), OK(82,2872), OR(82), TX(82), WI(82). **9:**KS(82). **10:**FL(929,931), GA(1023,1024,1275,2404), IL(309), KY(1815), MS(3539), NC(2516), NE(2155), NJ(496), NM(2109,2112), PA(1910), SC(932), TN(72,73), TX(119,1134,1135), VA(2993), WY(819).
P. striiformis. Telial host. 'Stripe rust.' **3:**ID(82), OR(82). **7:**WA(82). **8:**ID(82). **10:**ID(2857), KS(2688), MN(2397), MT(2857), ND(1285), NE(770), OK(3987), OR(2857), SD(770,1523), TX(1133,1134,1135,1137), UT(2857), WA(2857), WY(387,770,819).
P. triticina. Heteroecious: telial host. **10:**ID(3343), PA(3118), SC(933).

Basidiomycotina - Ustilaginales

Tilletia caries. Rough-spored bunt. **1:**Range of host(82), AK(1077), AR(1077), AZ(1077), CA(1077), CO(82,1077), CT(1077), DE(1077), GA(1077,1439), IA(1077), ID(82,1077), IN(1077), KS(1077), MA(1077), MD(1077), MI(1077), MS(2720), MT(82,1077), NC(1077,1303), ND(378,1077), NE(1077), NJ(1077), NM(1077), NY(82,1077), OH(1077), OK(2872,3905), OR(1077), SD(1077), TN(1077), TX(1077), UT(82,1077), VA(1077), WA(82,1077), WI(1077), WV(1077), WY(82,1077). **3:**CA(1077), IN(1077), MN(82,1077), ND(82), SD(82,1077), WA(82,1077). **5:**CA(1077), ND(1077), WA(1077). **8:**ID(1077), IN(1077), ND(82,378,1077), SD(82,1077), UT(1077). **10:**AK(526), WY(819).
T. controversa. 'Dwarf bunt.' **1:**NW states(1363). **10:**CA(1078), ID(1078), IN(1078), NY(1078), OR(1078), UT(1078).
T. laevis. 'Common bunt.' **1:**Range of host(82), AR(1077), AZ(1077), CA(1077), CO(1077), CT(1077), DE(1077), GA(1077), IA(1077), IL(1077), IN(1077), KY(1077), MA(1077), MI(1077), MO(1077), MS(1077,2720), MT(1077), NC(1077,1303), ND(378,1077), NE(1077), NJ(1077), NM(1077), NV(1077), NY(1077), OH(949,1077), OK(1077,2872), OR(1077), PA(1077), SC(1077), SD(1077,2242), TN(1077), TX(1077), UT(1077), VA(1077), WA(1077), WI(1077), WY(1077). **3:**MO(1077). **5:**ND(1077). **8:**MN(82). **10:**AK(526), VA(2993), WY(819).
Tilletia sp. Bunt. **10:**AK(2078), TX(119,1134).
Urocystis agropyri. 'Flag smut.' **1:**IL(1077), KS(1077), MD(1077), MO(1077), OR(1077), WA(1077).
U. tritici. Flag smut. **1:**IL(82), KS(82), MO(82), WA(82).
Ustilago tritici. 'Loose smut.' **5:**MN(1077), ND(1077). **8:**IL(82,1077), NY(82,1077), OR(82,1077). **10:**FL(929,931), GA(1023,1024,2746), KY(1815), MS(3539), NC(2516), NE(2155), NJ(496), SC(932,933), TN(72,73), TX(119), VA(2993), WY(819).

Basidiomycotina - Other

Agrocybe molesta. Attached to or associated with crowns. **1:**OR(3422).
A. praecox. Attached to or associated with crowns. **1:**OR(3422).
Athelia rolfsii. **1:**LA(3423).
Marasmius graminum. On base of culms. **1:**IL(82).
Naucoria cerealis. On roots and tillers. **1:**IL(82,308).
Thanatephorus cucumeris. Brown patch, root rot, sharp eye spot. **1:**CO(3423), ID(3423), MT(3423), ND(3423), NE(3423), OR(3423), SD(3423), UT(3423), WA(3423), WY(3423). **3:**ND(3423), OR(3423), WA(3423). **5:**MN(3423), ND(3423), NE(3423), SD(3423), WA(3423). **6:**WA(3423). **8:**OR(3423).
Typhula idahoensis. 'Speckled snow mold.' **1:**ID(416,3423), MT(416,2955,3423), UT(416,3423), WA(416,638,3423). **10:**WA(1617).
T. incarnata. 'Speckled snow mold.' **1:**ID(3423), WA(3423). **10:**WA(1617).
T. ishikariensis. 'Speckled snow mold.' **1:**ID(416), MT(416), WA(416).
Typhula sp. Snow mold. **10:**OK(620).

Deuteromycotina - Hyphomycetes

Acremoniella sp. On seeds. **10:**NY(732).
Acremonium strictum. Seedling root rot. **1:**MN(82).
Alternaria alternata. Black point. **1:**Range of host(82), IL(312).
A. brassicae. Black point (kernel discoloration). **1:**NY(82).
A. peglionii. Black point (kernel discoloration). **1:**ND(82).
Alternaria sp. Head disease, mold. **1:**MI(3833), NC(1303), NM(1658). **8:**OK(2872).
Aspergillus candidus. Grain mold. **1:**Range of host(82).
A. flavus. Grain mold, on stored seeds. **1:**Range of host(82), Unk.(3089).
A. glaucus. Grain mold. **1:**Range of host(82), Unk.(3089).
A. niger. Grain mold. **1:**Range of host(82).
A. ochraceus. On stored seed. **1:**Unk.(3893).
A. restrictus. On stored seeds. **1:**Unk.(3089).
Aureobasidium pullulans. Isolated from wheat samples, and from flour. **1:**SE states(2003), IN(2003), MI(2003), MT(2003), ND(2003), OK(2003), TX(2003).
Bipolaris sorokiniana. Seedling blight, spot blotch, black point. 'Common root rot.' **1:**Unk.(3423), FL(35), MI(3833), MT(3119), ND(3423,3453), NM(1658), NY(2212), OK(2872), SD(2242). **3:**MN(3423), ND(82,3423). **5:**MN(3423), ND(3423), SD(2242,3423). **6:**MN(3423). **7:**MN(3423). **8:**MN(3423), ND(3423), OR(82,3423). **9:**MN(3423). **10:**CA(3116), CO(1557), FL(929), GA(1024), NM(2112), PA(395), WY(1557).
Candida sp. Isolated from flour. **1:**OK(2003), TX(2003).
Cercospora secalis. Leaf spot. **1:**IL(312).
Cladosporium herbarum. Glume and leaf mold. 'Black head mold.'

1:Range of host(82), AK(526). 7:OR(82). 10:KY(1815).
Cladosporium sp. Head disease, mold. 1:MI(3833), OK(2872). 10:NY(732).
Cryptococcus albidus. Isolated from wheat samples. 1:NW states(2003), SE states(), IN(2003), KS(2003), MI(2003), MT(2003), ND(2003), NE(2003), OK(2003), TX(2003).
C. laurentii. Isolated from wheat samples. 1:IN(2003), MI(2003).
Curvularia geniculata. Leaf mold. 1:ND(82).
C. inaequalis. On discolored seeds. 1:MN(82).
Curvularia sp. 1:NM(1658).
Dendrodochium pezizoides. 1:DE(3543).
Drechslera campanulata. Leaf spot. 1:UT(82), WA(3423).
D. tritici-repentis. Leaf blight, secondary root rot, yellow leaf spot. 3:ND(3423). 10:MI(79).
Embellisia chlamydospora. On roots. 1:WA(1601). 10:ID(3270).
Epicoccum nigrum. Glume spot. 1:DE(82), IL(82), OH(82), PA(82).
Epicoccum sp. On ripening glumes and straw. 1:MI(3833).
Exserohilum pedicellatum. Root rot. 1:MN(82).
Fusarium acuminatum. Root rot. 1:Unk.(3423), ND(3423), SD(2242). 3:ND(82,3423), OR(3423), WA(3423). 5:ND(3423), OR(3423), WA(3423). 6:OR(3423), WA(3423). 8:ND(3423), OR(3423), WA(3423).
F. arthrosporioides. Head mold. 1:NY(82), OR(82), WI(82), WV(82).
F. avenaceum. Head mold, stem rot. 'Scab.' 1:CA(3423), NY(82), OR(82), WA(3509), WI(82), WV(82).
F. culmorum. Seedling blight, head mold. 'Foot rot. Scab.' 1:E & C states(82), CA(82,3423), CT(3423), ID(3423), MN(3423), MS(2720), MT(3423), ND(82,3423), NE(3423), OR(82,3423), SD(2242,3423), WA(3423,3509). 3:OR(82,3423), WA(82,3423). 5:OR(3423), WA(3423). 6:WA(82,3423). 8:OR(82,3423), WA(82,3423). 10:CA(3116).
F. equiseti. Root rot. 1:CA(3423), MT(82), ND(82), OR(82), WI(82). 3:ND(82).
F. graminearum. Seedling blight, root rot. 'Scab.' 1:AK(526,3423), CA(3423), GA(3423), IA(3423), IL(3423), IN(3423), KS(3423), KY(3423), MD(3423), MI(3423), MN(3423), MO(3423), NC(1303,3423), ND(3423), NE(3423), NY(3423), OH(3423), OK(3423), PA(3423), SC(3423), SD(2242,3423), TN(3423), VA(3423), WA(3509), WI(3423). 3:SD(3423). 5:MN(3423), ND(3423), SD(2242,3423). 8:MN(3423). 10:CA(3116).
F. moniliforme. Secondary root rot. 1:FL(35), NY(3423). 5:ND(3423).
F. oxysporum. Root rot. 1:ND(82), OR(82). 3:ND(82).
F. poae. Head mold, secondary root rot, silver top. 1:ND(3423), SD(2242,3423), WA(82,3423). 5:ND(3423).
F. roseum. Root rot. 10:CO(1557), WY(1557).
F. sporotrichioides. Head mold. 1:MN(82,3727), ND(82), WY(82).
Fusarium sp. Seedling blight, foot and root rot, scab. 1:FL(35), MI(3833), NM(1658), OK(2872). 10:CO(1557), NC(2516), NE(2155), NY(732), TN(72,73), WY(1557).
Helminthosporium sp. Leaf spot. 1:MI(3833), OK(2872). 10:NJ(496).
Heterosporium avenae. Leaf mold, spikelet and leaf blight. 1:IL(312), WA(82,3423), WY(82,1195,3423).
Heterosporium sp. Glume blotch. 1:OK(2872).
Hymenula cerealis. Leaf stripe. 'Cephalosporium stripe.' 1:ID(105), KS(3423,3914), MI(3833), MT(2852), NY(3699), VA(2994), WA(415,418), WY(1049). 10:KS(305), MT(3200).
Microdochium bolleyi. Root necrosis, rot. 'Aureobasidium decay.' 1:NC states(82), NW states(82), CA(3423), MN(3423), MT(3423), ND(3423), NE(3423), OR(3423), SD(2242,3423), WA(3423), WY(3423). 3:ND(3423). 5:ND(3423), SD(3423). 8:ND(3423). 10:CA(3116).
M. nivale. 'Pink snow mold. Scab.' 1:AK(2293), CA(82,3423), ID(82,3423), OR(3423), WA(82,3423,3509). 10:WA(1617).
Nigrospora sphaerica. Node canker, sheath. 1:IL(312).
Penicillium expansum. Blue mold. 1:IA(3423), ID(3423), IL(3423), IN(3423), MN(3423), MO(3423), MT(3423), ND(3423), NY(3423), OH(3423), OR(3423), SD(3423), WA(3423), WI(3423). 5:ND(3423), SD(3423). 8:ND(3423), OR(3423), WA(3423).
Penicillium sp. Blue mold seed rot. 1:Range of host(82).
Periconia circinata. Root rot. 1:ND(3423). 3:ND(3423). 5:ND(3423).
P. macrospinosa. On roots. 1:NC(1303).
Phialophora graminicola. 1:Unk.(3893).
Pseudocercosporella herpotrichoides. 'Eyespot and foot rot (strawbreaker).' 1:ID(82,3423), MI(3833), OR(82,3423), WA(82,417,3423). 3:OR(82,3423), WA(82,3423). 5:OR(3423), WA(3423). 6:OR(82,3423), WA(82,3423). 8:OR(82,3423), WA(82,3423). 10:MT(3200), NY(3261).
Rhodotorula glutinis. Isolated from wheat samples. 1:IN(2003), KS(2003), MI(2003), NE(2003).
Sporobolomyces sp. 1:Unk.(3893).
Stemphylium tritici. 1:OK(2872).
S. vesicarium. Black point. 1:MN(82), TX(82), UT(82), WA(82).
Stemphylium sp. On seeds. 10:NY(732).
Trichoderma viride. Seed mold. 1:MN(82).
Trichothecium roseum. Pink mold of heads. 1:KS(82), MD(82).
Ulocladium consortiale. Mold. 1:WA(3423).

Deuteromycotina - Coelomycetes

Ascochyta graminicola. Leaf spot. 1:OK(82,2873).
A. sorghi. Leaf spot. 1:OK(3423), VA(2992), WA(3442).
A. tritici. 'Ascochyta leaf spot.' 1:MI(3833), MT(3119).
Colletotrichum graminicola. 'Anthracnose.' 1:E & C states(82,285), OK(2872). 3:NJ(3423), OH(82). 8:IL(82).
Dilophospora alopecuri. 'Dilophospora leaf spot. Dilophospora twist.' 1:GA(2406), NC(1303), SC(82), VA(3423). 10:NC(2516).
Diplodia sp. On seeds. 10:NY(732).
Macrophoma secalina. Leaf blight. 1:IL(312).
Phoma americana. On leaves. 1:AL(2504).
P. harknessii. 6:CA(82).
P. terrestris. Pink root. 1:CA(82), CO(82), IA(82), MN(82), MT(3423), ND(82,3423), TX(82). 3:ND(82,3423). 5:ND(3423). 8:ND(82,3423). 10:OR(2791).
Pseudoseptoria donacis. 'Halo spot.' 1:ID(82,2695,3423,3450), WA(82,2695,3423,3450).
Septoria tritici. Glume blotch, speckled leaf blotch, crown rot. 'Speckled leaf blotch.' 1:C states(82), NW states(82), CA(82,3423), DC(3423), FL(35), GA(2406,3423,3833), IA(3423), ID(3423), IL(3423), IN(3423), KS(3423), KY(3423), MI(1480,3423,3833), MN(3423), MO(3423), MS(2720), MT(3119,3423), NC(1303,3423), NE(3423), NY(3423), OH(949,3423), OK(2872,3423), OR(3423), PA(3423), SD(2242), TN(3423), UT(82), WA(3423), WI(3423), WV(3423). 3:OR(82). 4:OR(3423). 8:IL(82), OR(82,3423). 9:OR(82,3423). 10:FL(931), GA(1023,1024,1275,2404), KY(1815), MI(76), NC(2516), SC(932,933), TX(119,1135), WY(819).
Stagonospora avenae. 'Speckled leaf blotch.' 5:ND(3423).
S. avenae f. sp. *triticea*. Blotch. 1:MN(1490), ND(82,3423), NY(2212), WA(82,3423). 3:ND(3423).
S. nodorum. 'Glume blotch.' 1:E states(82), NC states(82), AL(285), AR(3423), DE(3423), FL(35,934), GA(756,2406,2581,3423), IA(3423), IL(3423), IN(3423), KS(3423), KY(3423), MD(3423), MI(1480,3833), MN(1490,3423), MO(3423), MS(2720), MT(82,3119,3423), NC(1303,3423), ND(3423), NE(3423), NJ(3423), NY(2212,3423), OH(949,3423), OK(2872,3423), PA(1420,3423), SD(2242,3423), TN(3423), TX(82), VA(3423), WA(82,3423), WI(3423), WV(954,3423). 3:MD(82), ND(3423). 5:SD(2242). 8:MD(82), MN(82,3423), OR(82,3423). 10:FL(934), GA(1023,1024,1275,2404), MI(76), MS(3539), NC(130,2516), SC(932,933), TN(72,73), TX(119,1135), VA(2993).
Wojnowicia hirta. Secondary foot rot, rot of mature straw. 1:NW states(82), AR(82), CA(82,3423), CO(82,3423), ID(3423), KS(82,3423), MT(3423), NY(82), OR(3423), WA(3423), WY(3423). 3:OR(82,3423), WA(82,3423). 5:OR(3423), WA(3423). 6:OR(82,3423), WA(82,3423). 8:ND(3423), OR(82,3423), WA(82,3423).

Deuteromycotina - Other

Rhizoctonia cerealis. 'Sharp eyespot. Rhizoctonia root rot.' 1:OH(2152).
R. oryzae. Root rot, sheath spot. 1:ND(3423).
R. solani. White foot rot. 'Sharp eyespot. Rhizoctonia root rot.' 1:NM(1658), OR(82), SD(2242), WA(82). 5:SD(2242). 8:OR(82). 10:NW states(3838), AR(3479), NJ(496).
R. zeae. Secondary root. 1:ND(82).
Rhizoctonia sp. Seedling blight, root rot. 1:FL(35), OK(2872). 10:NM(2112), NY(732).
Sclerotium rolfsii. 'Southern blight.' 1:AL(82), AR(2886), GA(82), MS(2720), TX(82).

Uniola L.

1. *U. laxa* B. S. P.
2. *U. paniculata* L. [sea-oats]

Oomycetes

Pythium sp. Damping-off. **2:**NC(1303).

Ascomycotina

Apiospora montagnei. **1:**GA(1438).
Asterina sp. Black mildew. **2:**NC(1303).
Balansiopsis pilulaeformis. Black ring. **1:**MS(2720).
Lasiostemma erysipheoides. **1:**TX(652).
Mollisia culmina. **1:**GA(1438).
Phyllachora graminis. Tar spot. **1:**GA(82,1438,3423).
Phyllachora sp. **1:**GA(1438).

Basidiomycotina - Ustilaginales

Ustilago uniolae. Seed smut. **1:**AL(1077), DE(1077), GA(1077), MS(1077,2720), NC(1077,1303), SC(1077), TX(1077), VA(1077). **2:**E states(82), TX(82).

Deuteromycotina - Hyphomycetes

Cercospora sp. Blight, leaf spot. **2:**NC(1303).
Cercosporidium sp. Leaf spot. **2:**FL(35).
Exserohilum rostratum. Leaf spot. **2:**FL(35), NC(1303).

Deuteromycotina - Coelomycetes

Phoma glumarum. Leaf spot. **2:**MS(82,2720,3423).
Tunicago uniolae. On inflorescences and leaves. **2:**FL(3549,3557).

Deuteromycotina - Other

Rhizoctonia sp. Damping-off. **2:**NC(1303).

Vetiveria Bory in Lem.

1. *V. zizanioides* (L.) Nash (=*Andropogon muricatus* Retz.) [khus-khus, vetiver]

Ascomycotina

Didymella andropogonis. On leaves. **1:**LA(82), MS(82,2720).
Didymosphaeria andropogonis. On dead culms. **1:**LA(82).
Lulworthia medusa. On culms. **1:**LA(967).
Ophiosphaerella herpotricha. **1:**LA(3248,3754).

Vulpia C. C. Gmelin

1. *V. bromoides* (L.) A. Gray (=*Festuca dertonensis* (All.) Aschers. & Graebn.) [squirreltail fescue]
2. *V. microstachys* (Nutt.) Benth. (=*Festuca microstachys* Nutt., *Festuca pacifica* Piper, *Festuca reflexa* Buckley)
3. *V. myuros* (L.) C. C. Gmelin (=*V. megalura* (Nutt.) Rydb., *Festuca megalura* Nutt., *Festuca myuros* L.) [Foxtail fescue, rattail fescue]
4. *V. octoflora* (Walter) Rydb. (=*Festuca octoflora* Walter, *Festuca octoflora* Walter var. *tenella* Fernald)

Oomycetes

Pythium arrhenomanes. Root browning. **4:**ND(82).
P. debaryanum. Damping-off, root necrosis, seed rot. **2:**WA(3423). **3:**OR(3423). **4:**AZ(3437), ND(3423), NE(3423), WA(3423).
P. graminicola. **4:**ND(3423).

Ascomycotina

Claviceps purpurea. Ergot. **4:**MI(82,3423).
Cochliobolus sativus. Blind seed disease. **3:**OR(3423).
Gaeumannomyces graminis. Take-all. **1:**OR(3423), WA(3423). **3:**OR(3423), WA(3423). **4:**AR(3423).
Gloeotinia granigena. **3:**OR(1075).
Phyllachora sylvatica. Tar spot. **1:**OR(3423). **3:**OR(3423).

Basidiomycotina - Uredinales

Puccinia graminis. **2:**OR(82). **4:**KS(82), SD(2242).
P. piperi. **2:**CA(82), OR(82).
P. striiformis. **4:**WA(82).
Uromyces fragilipes. **2:**CA(81,82).
U. hordeinus. **4:**KS(3023), MS(82,2720), TX(82).

Basidiomycotina - Ustilaginales

Tilletia bromi. Seed smut. **1:**WA(1077). **3:**CA(1077). **4:**ID(82,1077), MT(82,1077), NE(1077), NV(82,1077), OR(82,1077), WA(82,1077), WY(82,1077).
T. texana. Seed smut. **4:**MO(82,1077).
Ustilago mulfordiana. Head smut. **1:**WA(1077). **4:**AZ(82,1077), CO(82,1077), ID(82,1077), MT(82,1077), OR(82,1077), UT(82,1077), WA(82,1077), WY(82,1077).

Basidiomycotina - Other

Laetisaria fuciformis. Pink patch. **3:**OR(3423).
Thanatephorus cucumeris. Brown patch, root rot, sharp eye spot. **1:**OR(3423). **3:**OR(3423), WA(3423). **4:**ND(3423), OR(3423), SD(3423), WA(3423).

Deuteromycotina - Hyphomycetes

Bipolaris sorokiniana. Seedling blight, spot blotch, black point. **4:**ND(82,3423).
Drechslera dictyoides. Net blotch, leaf spot. **2:**WA(3423,3424).
Fusarium acuminatum. Secondary root rot. **1:**OR(3423), WA(3423). **3:**OR(3423), WA(3423). **4:**AZ(3437), ID(3423), ND(3423), SD(2242,3423), WA(3423), WY(3423).
F. poae. Head mold, secondary root rot, silver top. **4:**WA(82,3423).
Microdochium bolleyi. Root necrosis. **2:**WA(3423). **4:**ND(3423), NE(3423), SD(2242,3423).
M. nivale. Snow mold. **1:**OR(3423), WA(3423). **3:**OR(3423), WA(3423).
Periconia circinata. Root rot. **4:**SD(3423).
Ramularia pusilla. Leaf spot. **3:**OR(3423,3429).

Deuteromycotina - Coelomycetes

Colletotrichum graminicola. Anthracnose. **3:**WA(3423). **4:**ND(82,3423).
Phoma terrestris. Pink root. **4:**ND(3423).
Pseudoseptoria everhartii. Eye spot. **3:**ID(2695). **4:**SD(2242,3423,3450).
P. stomaticola. Eye spot. **4:**NE(3423,3450).
Selenophoma donacis var. *linearis*. Leaf spot. **2:**ID(3432).
Septoria tenella. Leaf blotch, leaf spot. **1:**OR(3423). **2:**CA(81). **4:**W states(82), CO(82,3423), ID(3423), IL(3423), KS(3423), MT(3423), ND(3423), NE(3423), NJ(82,3423), SD(2242,3423), WA(82), WY(3423).
Stagonospora culmicola. Leaf rot. **2:**OR(3423).
Wojnowicia hirta. Secondary foot rot, rot of mature straw. **2:**WA(3423).

Deuteromycotina - Other

Rhizoctonia solani. **4:**SD(2242).

Zea L.

(= *Euchlaena* Schrad.)

1. *Z. mays* L. (=*Z. mays* L. subsp. *mays* , *Z. mays* L. var. *saccharata* (Sturtev.) L. H. Bailey, *Z. mays* L. var. *rugosa* Bonaf.) [Dent corn, field corn, flint corn, maize, pod corn, popcorn, sweet corn]
2. *Z. mays* L. var. *everta* (Sturtev.) L. H. Bailey
3. *Z. mays* L. subsp. *mexicana* (Schrad.) Iltis (=*Z. mexicana* (Schrad.) Kuntze, *Euchlaena mexicana* Schrad.) [teosinte]
4. *Z. perennis* (Hitchc.) Reeves & Mangelsd. (=*Euchlaena perennis* Hitchc.) [perennial teosinte]
5. *Zea* sp. (=*Euchlaena sp.*)

Myxomycetes

Didymium iridis. **1:**NJ(1446).
Perichaena vermicularis. On dead stalks. **1:**NY(1446).
Physarum pusillum. **1:**OH(1446).

Plasmodiophoromycetes

Ligniera junci. On roots. **1:**NC(1303,1310).

Chytridiomycetes

Physoderma maydis. Black spot, brown spot. **3:**FL(35,82), LA(82), SC(82).

Hyphochytridiomycetes

Hyphochytrium catenoides. On leaves and roots. **1:**NY(82,1446,3423).

Oomycetes

Peronosclerospora sorghi. 'Sorghum downy mildew.' **1:**AK(1446), AL(1446), AR(1446), GA(1446), IL(3867), IN(1446,3778), KS(1446), KY(1446,3904), LA(1446), MI(1446), MO(1446), MS(1446), NE(2726), NM(1446), OK(1446), TN(1446), TX(1102,1446).
Phytophthora cactorum. Root rot. **1:**CT(1446), IA(1446), SD(1446).
P. drechsleri. Root rot. **1:**HI(1446,2903).
P. nicotianae var. *parasitica*. Crown, root and stalk rot. **1:**IN(82,1446,3423).
Phytophthora sp. Root and stalk rot, soft rot. **1:**AZ(1446), HI(1446), SD(1446).
Pythium acanthicum. Seedling blight. **1:**IA(1446).
P. adhaerens. **1:**Unk.(1446).
P. angustatum. **1:**Unk.(1446).
P. aphanidermatum. Seedling blight, stalk rot of corn, root rot, turf spot. 'Pythium stalk rot.' **1:**FL(35), HI(2903), IA(82,1446), IL(1446), IN(82,1446,3707), KY(82,1446), MO(82,1446), SD(1446), TX(82,1446,3423), VA(82,1446,2989,3423).
P. arrhenomanes. Seedling blight. 'Pythium root rot.' **1:**HI(1446,2903),

IA(1446), IL(82,1446), IN(82,1446), KY(82,1446), MD(1446), MN(82), MO(82,1446), OH(82,1446), PA(1446), SD(2242), TN(1446), WI(82,1446).
P. debaryanum. Damping-off, seedling blight, root and seed rot. **1:**'Pythium root rot.' CT(82,1446), FL(35,82,1446), HI(1446,2903,3423), IA(82,1446,3423), MN(1446,3423), MO(82,1446), MT(1446,3423), ND(82,1446,3423), OK(1446,2872,3423), SD(1446,2242,3423), WI(82,1446), WY(1446,3423). **2:**IA(82).
P. graminicola. Seedling blight. **1:**'Pythium root rot.' CA(1446), CO(1446,3423), DC(1446,3423), HI(1446,3423), IA(82,1446,3423), IL(1446,3423), KY(82,1446,3423), MN(1446,3423), MO(3423), ND(1446,3423), OH(1446), SD(1446,2242,3423), WI(3423). **2:**IA(82).
P. irregulare. Seedling blight, root rot. **1:**IA(1446), WI(82,1446).
P. paroecandrum. Seedling blight, root rot. **1:**IA(1446), WI(82,1446).
P. pulchrum. Seedling blight. **1:**WI(1446).
P. rostratum. Seedling blight, root rot. **1:**HI(1446,2903), IA(1446), WI(82,1446).
P. splendens. Seedling blight, root rot. **1:**HI(2903), IA(1446).
P. ultimum. Seedling blight, root rot. **1:**MA(1446), ND(1446,3423), SD(1446,2242,3423), WI(82,1446).
P. vexans. Seedling blight, root rot. **1:**WI(82,1446).
Pythium sp. Seedling blight, root and stalk rot. **1:**AR(1446), CA(1446), CO(1446), FL(1446), HI(1446,2903), IA(82,1446), IL(82,1446), IN(1446), KY(1446), LA(1446), MA(1446), MO(1446), NC(1303,1446), ND(1446), SC(1446), TN(1446), TX(1446), WI(82,1446).
Sclerophthora macrospora. Crazy top. 'Crazy top downy mildew.' **1:**IA(1446), IL(1446), IN(1446,3707,3710,3712), KY(1446), MO(1446,3877), MS(412,1446), NC(1303,1446), ND(1812), NY(1446), PA(1446), SD(1446,2242), TN(1446), TX(1446,3876,3877), VA(1446,2988).
Sclerospora graminicola. Downy mildew. **1:**'Green ear downy mildew.' IA(82,1446,3423), OK(1446,2872,3423), WI(82,1343,1446,3423). **3:**IA(82). **2:**IA(82).
Sclerospora sp. Crazy top. **1:**KY(1780).

Zygomycotina

Absidia corymbifera. On seeds. **1:**GA(1446), IA(1446), WI(1446).
A. repens. Isolated from corn samples. **1:**IA(2121).
Absidia sp. Isolated from corn samples. **1:**IA(1446), IL(1446), MD(1446), MN(1446), MO(1446,1547), NE(1446), OH(1446), PA(1446), VA(1446).
Actinomucor elegans. Isolated from corn samples. **1:**MO(1547).
Blakeslea trispora. On leaves. **1:**NC(1303,1446).
Circinella muscae. Isolated from corn samples. **1:**IA(2121).
Circinella sp. Isolated from corn samples. **1:**MO(1547).
Cunninghamella sp. Ear rot. **1:**MO(1547), VA(82,1446,2987).
Mortierella sp. On roots and stalks. **1:**FL(3990).
Mucor circinelloides. On seeds. **1:**WI(1446).
M. fragilis. On seeds. **1:**WI(1446).
M. hiemalis. On seeds. **1:**GA(1446).
M. mucedo. Isolated from corn samples. **1:**IA(2121).
M. plumbeus. On seeds. **1:**GA(1446).
M. racemosus. On seeds. **1:**GA(1446), IA(1446).
Mucor sp. Root and stalk rot. **1:**C states(82), Range of host(82), SW states(82), Unk.(1446), AL(1446), AR(1446), FL(1446), IA(1446,2121), IL(1446), IN(1446,3693), KS(1446), KY(1446), LA(1446), MD(1446), ME(1446), MI(1446), MN(825,1446), MO(1446,1547), MS(1446), ND(1446), NE(1446), NJ(1446), NY(1446), OH(1446), OK(1446), PA(1446), SD(1446), TN(1446), TX(1446), VA(1446), WI(1446).
Phycomyces nitens. **1:**IL(1446).
Phycomyces sp. On seeds. **1:**WI(1446).
Rhizomucor pusillus. From moldy seed. **1:**WI(1446).
Rhizopus arrhizus. Root and scutellum rot. **1:**Unk.(82,1446), GA(1446), IA(1446), MO(1446).
R. microsporus. Rot. **1:**IL(1446), WI(1446).
R. microsporus var. *rhizopodiformis*. On seeds. **1:**WI(1446).
R. stolonifer. Ear and scutellum rot. **1:**Unk.(1446), AL(1446), AR(1446), FL(1446), IA(1401,1446,2121), ID(1446), IL(82,1446), IN(82,1446), KS(1446,2345), KY(1446), LA(1446), MD(1446), ME(1446), MI(1446), MN(1446), MO(1446), MS(1446), NE(1446), NY(1446), OH(1446), OK(1446,2872), PA(82,1446), SD(1446), TN(1446), TX(1446), VA(1446), WI(1446).
Rhizopus sp. On ears, roots, seedlings, seeds and stalks. **1:**Range of host(82), Unk.(1446), FL(35,1446,3990), GA(1446), IA(1446), IL(1446), IN(1446,3694), MD(1446), MN(825,1446), MO(1446,1547), NE(1446), OH(1446), PA(1446), VA(1446), WI(1446).
Syncephalastrum racemosum. On seeds. **1:**IA(1446,2121), MO(1547).
Syncephalastrum sp. On seeds and meal. **1:**Unk.(1446), MO(1446).
Thamnidium elegans. Isolated from corn samples. **1:**IA(2121).

Ascomycotina

Botryosphaeria disrupta. On leaves and stalks. **1:**FL(35,1446).
B. festucae. Stalk rot. 'Physalospora ear rot.' **1:**FL(35,82,1446,3423), LA(82,1446,3423), NJ(1446), OK(82,1446,3423).
B. quercuum. On ears and stalks. **1:**AL(1446), FL(35,1446), TX(1446).
B. rhodina. **1:**FL(82), TX(3423).
B. zeae. 'Gray ear rot.' **1:**IL(82,1446,3423), IN(82,1446,3423,3707).
Byssochlamys nivea. Isolated from seeds. **1:**IA(1228,2121).
Chaetomium bostrychodes. Isolated from corn samples. **1:**IA(2121).
C. brasiliense. On seeds. **1:**KS(1446,2345).
C. dolichotrichum. Isolated from corn samples. **1:**IA(2121).
C. funicola. Isolated from corn samples. **1:**IA(2121).
C. indicum. On seeds. **1:**IA(2121), KS(2345), LA(82), NJ(82), NY(82).
C. murorum. Isolated from corn samples. **1:**IA(2121).
C. torulosum. On husks. **1:**NH(3306).
Chaetomium sp. On roots, seeds and stalks. **1:**DE(1446), GA(1446), IA(1446), IL(1446), IN(1446), MN(825,1446), MO(1446), NE(1446), OH(1446), SC(1446), TN(1446).
Cochliobolus carbonum. 'Helminthosporium ear rot.' **1:**NC(2071).
C. heterostrophus. Seedling blight. 'Southern corn leaf blight.' **1:**NE states(82), AL(2470), FL(461,1446,2470,3129,3423), GA(1446,2470,3423), IL(1446,2470), IN(1446,2470), KY(1446,2470), LA(2470), MD(1446,3423), MN(2470), MS(1446,2470,2720,3423), NC(1446,2470,3423), OH(1446), OK(2470), PA(1446,3423), SC(1446,2470,3423), TN(1446,3423), TX(82,1446,2470,3423), VA(1446,2987,3423), WI(2470), WV(1446).
C. sativus. Leaf spot. **1:**NY(1446).
Corynascus sepedonium. Isolated from corn samples. **1:**IA(2121).
Crocicreas cyathoideum. **1:**NY(505,1446).
Dactylospora imperfecta. **1:**NJ(1446).
Diaporthe phaseolorum. On dead stalks. **1:**KY(82,1446), LA(82,1446), NJ(82,1446).
Dictyochora gambellii. On dead stalks. **1:**NJ(1446).
Didymosphaeria graminicola. **1:**GA(1438).
Didymosphaeria sp. **1:**GA(1438).
Eidamella sp. Isolated from corn samples. **1:**IA(2121).
Gaeumannomyces graminis. Sheath spot. **1:**IL(312).
Gibberella cyanogena. Root and stalk rot, on seeds. **1:**AR(1446), CT(1446), DE(471), IN(1446), KY(1446), MA(1446), MD(1446), MN(1446), NC(1446), NJ(1446), NY(1446), OH(1446), PA(1446), WI(1446).
G. pulicaris. Ear rot, stalk rot. **1:**ND(377), NH(1446), NY(1446).
G. subglutinans. Stalk rot on seeds. **1:**CO(1446), IA(1446), IL(1446), IN(1446), KS(1446), MD(1446), MN(1446), MO(1446), NE(1446), NJ(82,1446), OH(82,1446), TN(1446), WI(1446).
G. zeae. Seedling blight, ear and stalk rot. **2:**IA(82).
Gibberella sp. Stalk rot on seeds. **1:**CA(1446), DE(1446), FL(1446), IA(1446), IL(1446), IN(1446), MN(1446), MT(1446), NE(1446), NJ(1446), OH(1446), PA(1446), WI(1446).
Glomerella graminicola. Anthracnose. **1:**KY(1446).
G. tucumanensis. Anthracnose. **1:**KY(1446).
Hansenula anomala. On seeds. **1:**IA(1446).
Lecanidion atratum. On dead stalks. **1:**MS(1446).
Leptosphaeria macrospora. **1:**AL(3249).
L. maydis. On leaves. **1:**IL(82,1446).
L. variiseptata. On leaves and stalks. **1:**IL(82,1446).
Leptosphaeria sp. On leaves and stalks. **1:**SC(1446).
Leptosphaerulina trifolii. On leaves. **1:**IL(82,1446).
Lophiosphaera zeicola. **1:**KS(1446).
Lophiostoma arundinis. **1:**SC(1446).
Massarina arundinacea. On stalks. **1:**AL(82,1446), KS(82,1446), NJ(1446), VA(82,1446), WV(82,1446).
Melanospora zamiae. On seeds. **1:**NE(1446).
Melanospora sp. Isolated from corn samples. **1:**IA(2121).
Meliola bidentata. **1:**HI(2903).
Metasphaeria sp. **1:**GA(1438).

Microascus cinereus. On seeds. **1**:IA(1446,2121).
M. cirrosus. Isolated from corn samples. **1**:IA(2121).
M. desmosporus. On seeds. **1**:IA(1446).
M. longirostris. Isolated from corn samples. **1**:IA(1446,2121).
Monascus purpureus. Silage mold. **1**:CT(82,1446), IA(82,1446), KS(82,1446).
Monascus sp. On seeds. **1**:IL(1446).
Mycosphaerella zeae. On leaves. **1**:IL(82,1446), IN(1446).
M. zeae-maydis. Leaf blight. **1**:NY(1446).
Mycosphaerella sp. **1**:GA(1438), HI(2903).
Odontotrema sp. **1**:NC(1446), VA(1446).
[*Ophiobolus acuminatus*]. **1**:WV(1446).
Ophiosphaerella herpotricha. Leaf blight. **1**:FL(1446,3130).
Paraphaeosphaeria michotii. Leaf spot. **1**:IL(1446,3251).
Perisporium zeae. **1**:AL(1446), SC(1446).
Phaeosphaeria eustoma. On leaves and stalks. **1**:MI(82,1446).
Physalospora abdita. **1**:FL(82).
Pithoascus intermedius. On seeds. **1**:IA(1446,2121).
P. schumacheri. On seeds. **1**:IA(1446,2121).
Pleospora straminis. On overwintered stalks. **1**:MA(270).
Podospora minor. **1**:KS(1446).
Pyronema omphalodes. Isolated from corn samples. **1**:IA(2121).
Rhopographus zeae. Stalk rot. **1**:FL(82,1446), GA(82,1438,1446), MN(1446), WI(82,1446).
Setosphaeria prolata. Isolated from living leaves. **1**:MS(2097).
S. turcica. Leaf blight. 'Northern corn leaf blight.' **1**:GA(1438), HI(1446,2903), NC(2096).
Sordaria fimicola. Isolated from corn samples. **1**:IA(2121).
Sphaerella paulula. On sheaths. **1**:FL(634).
Stictis radiata. **1**:AL(1446), GA(1438,1446).
S. stellata. **5**:OH(3225).
Talaromyces luteus. On seeds and meal. **1**:Unk.(1446), IN(1446,2449).
T. stipitatus. Isolated from corn samples. **1**:IA(2121).
Tubeufia cylindrothecia. **1**:OH(1446).
T. paludosa. On old stalks. **1**:OH(3041).

Basidiomycotina - Uredinales

Physopella pallescens. **3**:FL(35). **4**:FL(3190).
P. zeae. **4**:FL(3190).
Puccinia polysora. Telial host; aecial host unknown. **1**:'Southern corn rust.' S states(1612), AR(763,1446), FL(1446), GA(82,1446), HI(2903), IL(1446,1623), IN(1446,3707,3710), KS(3268), LA(1446), MA(1446), MD(1446), MO(1446), MS(82,1446,2720), NC(1303,1446,2095), OK(1446), SC(1446), TX(82,1446), VA(1446), WI(1355,1446,2729). **3**:FL(35).
P. sorghi. Heteroecious: telial host. **3**:CT(82), IA(82), NE(82). **2**:IA(82).

Basidiomycotina - Ustilaginales

Sporisorium holci-sorghi. Head smut. **1**:'Head smut.' W states(82), CA(82,1077,1446), ID(1077,1446), IN(82,1077,1446), KS(1077,1446), MN(2289,2290,3517), NE(1077,1446), NM(1077), NY(82,1446), OH(1077,1446), OK(2872), OR(1077,1446), TX(1101), UT(1077), WA(82,1101,1446). **2**:WA(82).
Ustilago zeae. Smut. **3**:E states(82), AL(1077), CA(1077), CT(1077), DC(1077), IA(1077), IL(1077), IN(1077), KS(82,1077), NY(1077), WI(82,1077). **2**:IA(82), IL(82), IN(82), RI(82), VT(82). **5**:OK(2872).

Basidiomycotina - Other

Athelia rolfsii. Seedling blight, basal sheath and stalk rot. **1**:AL(3423), FL(3423), TX(3423), VA(3423).
Coniophora sp. **1**:WV(1446).
Corticium sp. **1**:SC(1446).
Glabrocyphella ellisiana. **1**:NJ(646).
Thanatephorus cucumeris. Damping-off, brown patch, root rot, sharp eye spot. **1**:AZ(1446,3423), GA(1446), IA(1446,3423), IL(1446,3423), MA(1446,3423), MN(1446,3423), MS(1446), MT(1446,3423), NC(1446), ND(1446,3423), NY(1446), OH(1446,3423), OK(1446,3423), OR(1446,3423), SC(1446), SD(1446,3423), WA(1446,3423).
Tomentella punicea. **1**:NJ(2039).
Typhula phacorrhiza. **1**:NY(1446).
T. variabilis. On stalks and leaves. **1**:NY(1446,2955).
Typhula sp. On dead stalks. **1**:NY(1446).
Wolfiporia cocos. Root rot. **1**:FL(35,82,1446), NC(82,1446).

Deuteromycotina - Hyphomycetes

Acremoniella sp. On seeds. **1**:GA(1446), IA(1446,2121), IN(3694).
Acremonium strictum. Black bundle disease, leaf lesions, stalk rot. 'Black bundle disease. Cephalosporium kernel rot.' **1**:Unk.(82), AL(1446), AR(1446,3423), CA(1446,3423), CT(1446,3423), DE(1446,3423), FL(35), IA(1446,3423), IL(1446,1625,3423), IN(1446,3423,3693,3694), KS(1446,3423), KY(1446,3423), MA(1446), MD(1446,3423), ME(1446), MI(1446,3423), MN(1446,3423), MO(1446,3423), MS(1446,3423), MT(82,1446), NC(1303,1446,3423), ND(1446,3423), NE(1446,3423), NJ(1446,3423), NY(1446,3423), OH(1446,3423), PA(3423), SC(1446,3423), SD(3423), TX(3423), VA(2987), WI(3423).
Acrodictys erecta. On stalks. **1**:IA(980), LA(980).
Alternaria alternata. Ear and seed mold, root rot, leaf spot. **1**:Unk.(82), CA(1446), DE(3543), IA(1401,2121), KS(1446,2345), KY(1446), LA(1446), MO(1446), NC(1303), NJ(1446), WI(1446), WV(1446).
A. brassicae. **1**:CA(1446).
Alternaria sp. Ear and seed mold, root and stalk rot, leaf spot. **1**:Range of host(82), AL(1446), AR(1446), CT(1446), FL(1446), GA(1446), HI(2903), IA(1446,2121), ID(1446), IL(1446), IN(1446,3693,3694), KS(1446), KY(1446), LA(1446), MA(1446), MD(1446), ME(1446), MI(1446), MN(825,1446), MO(1446), MS(1446), ND(1446), NE(1446), NJ(1446), NY(1446), OH(1446), OK(1446,2872), PA(1446), RI(1446), SC(1446), SD(1446), TN(1446), TX(1446), VA(1446), WI(1446).
Arthrinium phaeospermum. Isolated from corn samples. **1**:IA(2121).
Arthrobotrys sp. Isolated from corn samples. **1**:IA(2121).
Aspergillus alliaceus. Isolated from corn samples. **1**:IA(2121).
A. caespitosus. **1**:TX(1446).
A. candidus. On seeds and meal. **1**:Unk.(82), FL(1446), IA(1401,1446,2121), IL(1446), KY(1446), MD(1446), MN(1446), MO(1446), NE(1446), OH(1446), PA(1446), TX(1446), VA(1446), WI(1446).
A. carbonarius. On seeds. **1**:KS(1446,2345).
A. chevalieri var. *intermedius*. Isolated from corn samples. **1**:IA(2121).
A. clavatus. On seeds and meal. **1**:IA(1401), KS(1446,2345), WI(1446).
A. echinulatus. Isolated from corn samples. **1**:IA(2121).
A. elegans. Isolated from corn samples. **1**:IA(2121).
A. equitis. On seeds. **1**:IA(1446,2121), IL(1446).
A. flavipes. Isolated from corn samples. **1**:IA(2121).
A. flavus. On seeds. 'Aspergillus ear rot. Aspergillus storage rot.' **1**:Unk.(3089), AL(1446), FL(35,82,1446), GA(1446), HI(2650,2903), IA(82,1401,1446,2121), IL(82,1446), IN(1446,3693,3694), KS(1446,2345), KY(1446), LA(1446), MD(1446), MN(1446), MO(1446,1547), MS(1446), NC(1303,1446), NE(1446), NY(1446), OH(1446), OR(1446), PA(1446), SC(1446), TN(1446), TX(82,1446), VA(1446), WI(1446).
A. fumigatus. On seeds and meal. **1**:IA(1401,2121), IL(1446), MD(1446), MO(1547), WI(1446).
A. glaucus. Isolated from corn samples. **1**:Unk.(3089), AL(1446), AR(1446), IA(1401,1446,2121), IL(82,1446), IN(1446,3693,3694), KS(1446), KY(1446), LA(1446), MD(82,1446), ME(1446), MI(1446), MN(1446), MO(1446), MS(1446), NC(1446), NE(1446), NY(1446), OH(1446), OK(1446), OR(1446), PA(1446), SD(1446), TN(1446), TX(1446), VA(1446), WI(1446).
A. hollandicus. On seeds. **1**:IA(1446,2121), IL(1446).
A. mangini. Isolated from corn samples. **1**:IA(1446,2121).
A. nidulellus. On seeds. **1**:IA(1446).
A. niger. On seeds. 'Aspergillus kernel rot.' **1**:AL(1446), AR(1446), CA(1446), IA(1401,1446,2121), IL(1446), IN(1446,3693), KS(1446,2345), KY(1446), LA(1446), MD(1446), ME(1446), MI(1446), MN(1446), MO(1446,1547), MS(1446), NC(1303), NE(1446), NY(1446), OH(1446), OK(1446,2872), PA(1446), SD(1446), TN(1446), TX(1446), VA(1446,2987), WI(1446).
A. ochraceus. On seeds and meal. **1**:Unk.(82), IA(1446,2121), IL(1446), IN(3693), MD(1446), MN(1446), MO(1446), NE(1446), OH(1446), PA(1446), VA(1446).
A. parasiticus. On ears and grain samples. **1**:HI(2650,2903), IA(1446), IN(1446), VA(1446).
A. reptans. On seeds and meal. **1**:GA(1446), IA(1446,2121), MN(1446), NC(1303,1446).
A. restrictus. Blue-eye, on stored seeds. **1**:Unk.(3089), IA(1446,2121), IL(1446), MD(1446), MN(1446), MO(1446), NE(1446), OH(1446), PA(1446), VA(1446).
A. rubrobrunneus. On stored kernels. **1**:IA(1446,2121), NC(1303,1446).
A. stellifer. Isolated from corn samples. **1**:IA(2121).

A. sulphureus. Isolated from corn samples. **1**:IA(2121).
A. sydowii. Isolated from corn samples. **1**:IA(2121).
A. tamarii. On seeds and meal. **1**:Unk.(82), AL(1446), AR(1446), IA(1446), IL(1446), IN(1446), KS(1446), KY(1446), LA(1446), MD(1446), ME(1446), MI(1446), MN(1446), MO(1446), MS(1446), NE(1446), NY(1446), OH(1446), OK(1446), PA(1446), SD(1446), TN(1446), TX(1446), VA(1446), WI(1446).
A. terreus. On seeds. **1**:IA(1446,2121), KS(1446,2345), MN(1446), MO(82,1446,1547).
A. unguis. Isolated from corn samples. **1**:IA(2121).
A. ustus. Isolated from corn samples. **1**:IA(2121).
A. versicolor. On seeds. **1**:IA(1446,2121), IL(82,1446), IN(3693), WI(1446).
A. wentii. On seeds. **1**:IA(1446,2121), IL(1446).
Aspergillus sp. On leaves, roots, seeds and stalks. **1**:AL(1446), AR(1446), CA(1446), CO(1446), CT(1446), DE(1446), FL(1446), GA(1446), HI(2903), IA(1446), IL(1446), IN(1446), KS(1446), KY(1446), LA(1446), MA(1446), MD(1446), MN(825,1446), MO(1446), MS(1446), NE(1446), OH(1446), OK(1446), PA(1446), SC(1446), SD(1446), TX(1446), VA(1446).
Aureobasidium pullulans. On seeds. **1**:MO(1446).
A. zeae. 'Eyespot.' **1**:IA(1446,3711), IL(107,1446,3711), IN(1446,3711), MI(1446), MN(1446,3711), NY(1446), SD(512,1446,3711), WI(1446,3711).
Basidiobotrys pallida. **1**:OH(1446).
Bipolaris maydis. Southern leaf blight, blackening of silks, stalk rot, leaf spot. **3**:NC(1303).
B. sacchari. **1**:HI(1446).
B. sorghicola. Flecks. **1**:NE(1446).
B. sorokiniana. Seedling blight, spot blotch, black point, root rot. **1**:IA(82,1446,3423), IL(1446), MN(1446,3423), NC(1446), ND(1446,3423), NJ(471,1446), SD(82,1446,3423).
B. victoriae. Leaf spot. **1**:NY(1446).
B. zeicola. Seedling blight, charred ear mold, stalk rot, leaf spot. **3**:NC(1303).
Botrytis cinerea. Ear rot. **1**:IA(2121), ID(1446), VA(82,1446,2987), WA(82,1446).
Botrytis sp. Isolated from roots and stalks. **1**:MN(825,1446), NY(1446).
Candida albicans. On seeds. **1**:IA(1446), IL(1446), IN(1446), OH(1446).
C. guilliermondii. On seeds. **1**:NC(1303).
C. intermedia. **1**:IA(1446).
C. krusei. **1**:IA(1446).
C. parapsilosis. **1**:IA(1446).
C. pseudotropicalis. On seeds. **1**:IA(1446,2121), MN(1446).
Candida sp. On seeds. **1**:GA(1446).
Cephalosporium sp. On roots, seeds and stalks. **1**:FL(1446), IA(1446,2121), IL(1446), MN(825,1446), MO(1446), NC(1446), NE(1446), NJ(1446), OH(1446), WI(1446).
Ceratosporella bicornis. On dead stalks. **1**:OH(1446).
Cercospora sorghi. Crown, root and stalk rot, leaf spot. 'Gray leaf spot.' **1**:AL(82,1446,3423), FL(35,82,1446,3423), LA(82,1446,3423), MS(82,1446,2720,3423), NC(82,1303,1446), SC(82,1446,3423), TN(82,1446,3423), TX(82,1446), VA(82,1446,3423), WV(82,1446,3423).
C. zeae-maydis. Leaf spot. **1**:'Gray leaf spot.' DE(140,1492), IL(82,1446,3423), KY(82,1446,3070,3423), MD(140), NC(140,1303,1307,1446,2095), OH(2150), PA(140), SC(140,1446,1906), TN(82,140,1446,3423), VA(82,140,1446,2987,2990), WV(3911). **2**:VA(82).
Cercospora sp. Leaf spot. **1**:FL(1446), NC(1446), SC(1446). **3**:NC(1303).
Chalara sp. **1**:AL(1446), AR(1446), IA(1446), IL(1446), IN(1446), KS(1446), KY(1446), LA(1446), MD(1446), ME(1446), MI(1446), MN(1446), MO(1446), MS(1446), NE(1446), NY(1446), OH(1446), OK(1446), PA(1446), SD(1446), TN(1446), TX(1446), VA(1446), WI(1446).
Chrysonilia sitophilia. **1**:VA(1446).
Cladosporium cladosporioides. On seeds. 'Hormodendrum ear rot.' **1**:NC(1303), NY(1446), SD(1446).
C. herbarum. Black mold of ears, seeds and stalks. 'Hormodendrum ear rot.' **1**:Range of host(82), AL(1446), DC(1446), FL(35,1446), HI(2903), IA(1401,1446), NJ(1446), NY(1446), PA(1446), SC(1446), UT(1446), WI(1446).
C. tenuissimum. Black mold on sheaths. **1**:Unk.(82), IN(1446), SC(634,1446).
C. zeae. Black mold. **1**:Unk.(82), CT(1446).
Cladosporium sp. Ear, root and stalk rot, on seeds. **1**:IA(1446,2121), IL(1446), IN(1446,3693), MD(1446), MN(825,1446), MO(1446), NE(1446), PA(1446), VA(1446), WI(1446).
Coniosporium sp. Cob rot. **1**:OH(1446).
Cryptococcus laurentii. On seeds. **1**:IL(1446).
Curvularia eragrostidis. Leaf spot. 'Curvularia leaf spot.' **1**:AL(2497), GA(1446,2586), NC(1446,2586).
C. geniculata. **1**:AL(2497).
C. gudauskasii. On leaves. **1**:AL(2486,2498).
C. inaequalis. **1**:IL(1446,3423).
C. intermedia. On dead leaves. **1**:NC(1303,1446).
C. lunata. On seeds. 'Curvularia leaf spot.' **1**:AL(2484), NC(1303).
C. pallescens. Ear rot. **1**:IA(1401).
C. protuberata. On kernels. **1**:NC(1303,1446).
Curvularia sp. Root, seedling and stalk rot. **1**:FL(3990), IA(1446), NJ(471,1446).
Dematium sp. On seeds. **1**:IA(1446).
Dictyochaeta fertilis. Root rot. **1**:NC(1303).
Doratomyces stemonitis. Ear rot. **1**:IA(1401).
Epicoccum nigrum. Kernel discoloration, red kernel disease, ear mold, ear, leaf and seed rot, leaf spot. **1**:DE(3543), FL(1446,3959), IA(1401,1446), IL(1446,3959), MI(82,1446), NJ(82,1446), NY(1446), OR(82,1446), VA(82,1446), WA(82,1446), WV(1446).
Epicoccum sp. Isolated from corn samples. **1**:GA(1446), IA(2121).
Eurotium rubrum. On stored kernels. **1**:GA(1446), NC(1303).
Exserohilum pedicellatum. Seedling rot. 'Helminthosporium root rot.' **1**:CA(1446), HI(2903), IA(1446), NY(1446), OH(1446).
E. prolatum. Leaf spot. **1**:MS(1446,2097).
E. rostratum. Ear, leaf and stalk rot, leaf spot, on seeds. **1**:AL(2497), FL(35,1446,1992,3990), GA(82,1446,3423), MS(82,1446,3423), NC(1303,1446,2588), TX(1446).
E. turcicum. White blast, northern leaf blight, seedling blight, crown and stalk rot, stripe. **3**:TX(82).
Fusarium acuminatum. Root rot. **1**:Unk.(82), IA(1446,3423), ID(1446,3423), MN(1446,3423), MT(1446,3423), ND(1446,3423), NE(1446,3423), OR(1446,3423), SD(1446,2242,3423), WA(1446,3423), WY(1446,3423).
F. avenaceum. Seedling root rot, stalk rot. **1**:MN(1446), PA(1446), WI(82,1446).
F. culmorum. Head mold, ear and seedling root rot, scab. **1**:CA(1446,3423), IA(1401), MN(1446), NJ(1446), OR(1446,3423), TN(1446), WV(1446).
F. episphaeria. Stalk rot. **1**:MN(1446), PA(127,1446).
F. equiseti. Root rot, on seeds. **1**:IA(1401), KS(1446,2345), ND(82,1446).
F. graminearum. Seedling blight, ear and stalk rot, isolated from kernels. **1**:AR(3423), CO(1187), CT(3423), DE(3423), GA(1446), IA(1401,3423), IL(3423), IN(5,3423), KY(3423), MA(3423), MD(3423), MN(825,1446,3423), MO(3423), NC(1303,1446,3423), ND(3423), NE(3423), NJ(3423), NY(3423), OH(3423), OK(1446,3423), PA(127,3423), SD(2242,3423), WI(3423).
F. merismoides. Stalk rot. **1**:DE(3543), NY(82,1446), OR(82,1446,4007).
F. moniliforme. Seedling blight, bud, ear, root and stalk rot, pink ear rot. **1**:'Seed rot-seedling blight. Fusarium kernel rot. Fusarium stalk rot.' AL(1446,3423), AR(1446,3423), AZ(82,1446,3423), CA(3423), CO(1187,3423), CT(1446,3423), DE(471,3423), FL(35,3423,3990), GA(1446,3423), HI(2903), IA(82,1401,1446,3423), ID(1446), IL(53,1446,1625,1940,3423), IN(1446,2603,3423,3693,3694), KS(1446,2345,3423), KY(1446,3423), LA(1446,3423), MA(1446,3423), MD(82,1446,3423), ME(1446), MI(1446,3423), MN(1446,1951,1984,3423), MO(477,1446,3423), MS(1446,3423), NC(1303,1446,2098,2588,3423), ND(1446,3423), NE(1446,3423), NJ(471,1446), NY(82,1446,3423), OH(82,1446,3423), OK(1446,2872,3423), PA(82,127,471,1446), SC(1446,3423), SD(1446,3423), TN(1446,3423), TX(1446,3423), VA(1446,2987,3423), WA(1446,3423), WI(1446,3423), WV(1446,3423), WY(3423). **2**:IA(82).
F. oxysporum. Root and stalk rot. **1**:IA(1446,3423), KS(1446,2345), MN(1446,3423), MT(1446,3423), ND(82,1446,3423), NE(1446,3423), OR(1446,3423), PA(127,1446), SD(1446,2242,3423), WY(1446,3423).

F. pallidoroseum. Root rot. **1**:Unk.(1446).
F. poae. Root and stalk rot, on seeds. **1**:Unk.(1446), IN(1446), KS(1446,2345), MN(1446), PA(1446), WY(1446,3423).
F. proliferatum. On roots and stalks. **1**:MN(1950).
F. roseum. Root and stalk rot, on seeds. **1**:FL(35), MA(1446), MN(1446,1951,1984), NY(1446), PA(127,1446), WI(1446).
F. sacchari. Isolated from seeds. **1**:DE(471).
F. solani. Stalk rot, on roots. **1**:FL(3990), MN(1446), PA(127,1446).
F. subglutinans. Ear and stalk rot, isolated from kernels. **1**:CO(1187), IA(1401), PA(127,471,1446).
F. tricinctum. Stalk rot. **1**:PA(127).
Fusidium griseum. On seeds. **1**:IA(1446).
Fusidium sp. Isolated from corn samples. **1**:IA(2121).
Fusisporium cerealis. On sheaths. **1**:FL(634).
Geotrichum sp. On seeds. **1**:IA(2121), IL(1446).
Gloeocercospora sorghi. 'Zonate leaf spot.' **1**:KS(1446), LA(82,1446,3423), MS(82,1446,2720,3423), NC(1303).
Gonatobotrys simplex. Ear rot. **1**:IA(1401).
G. zeae. Kernel rot. **1**:MS(1446).
Gonatobotrys sp. Stalk rot. **1**:IA(2121), NJ(1446).
Graphium penicillioides. Leaf spot. **1**:LA(1446).
Harzia acremonioides. On seeds. **1**:IL(1446), WI(1446).
Helicosporium panacheum. On decaying stems. **1**:IA(2464).
Helminthosporium ahmadii. On dead stalks. **1**:NY(1446).
Helminthosporium sp. Blight, root and stalk rot, leaf spot, on seeds. **1**:AR(1446), CA(1446), CT(1446), DE(1446), FL(1446), GA(1446), HI(2903), IA(1446,2121), IL(1446,1624,1625,1940), IN(1446,1624), KY(1446), MI(1446,1624), MN(825,1446), MO(1446), MS(1446), NC(1446), NJ(1446), OH(1446), OK(2872), SD(1446), TN(72,73,1446), VA(1446), WI(1446,1624), WV(1446). **3**:FL(35).
Hormodendrum sp. Ear and kernel rot, on seeds. **1**:IA(1446,2121), IL(1446), IN(1446), ND(1446), NY(1446), WI(1446).
Illosporium pallidum. **1**:SC(634).
Isariopsis subulata. **1**:KS(1446), LA(1446).
Macrosporium maculatum. **1**:DE(3543).
Mariannaea elegans. Stalk rot. **1**:TX(82,1446).
Microdochium bolleyi. Root necrosis. **1**:CA(1446,3423), MN(1446,3423), ND(1446,3423), SD(1446,3423).
M. nivale. **1**:OH(1446).
Monilia sp. Isolated from corn samples. **1**:IA(2121), IL(1446).
Myrothecium cinctum. **1**:FL(1446).
M. gramineum. Shuck rot. **1**:IL(312).
M. verrucaria. Isolated from corn samples. **1**:IA(2121).
Nigrospora sphaerica. Dry ear rot, cob, kernel and stalk rot, on seeds. **1**:C states(1446), E & C states(3423), E states(1446), Unk.(82), DE(1446), IA(1446), IL(1446), IN(1446), KS(1446), MD(1446), MN(1446), MO(1446), NE(1446), OH(1446), OK(1446), SC(1446), SD(2242).
Nigrospora sp. Dry ear rot, cob and stalk rot, on seeds. **1**:C states(1446), E states(1446), Unk.(82), HI(2903), IA(1446), IL(1446), IN(1446), MI(78), MN(825,1446), MO(1446), NE(1446), NJ(1446), WI(1446).
Oedemium sp. **1**:IA(1446).
Olpitrichum sp. On seeds. **1**:IL(1446), MO(1446).
Oospora sp. On seeds. **1**:IA(1446,2121).
Paecilomyces sp. On seeds. **1**:IL(1446), MO(1446).
Papulaspora sp. Isolated from corn samples. **1**:IA(2121).
Penicillium brevicompactum. On seeds. **1**:IN(1446,2449).
P. canescens. On seeds. **1**:FL(1446).
P. chrysogenum. Blue-eye of kernels, on seeds. **1**:IA(82,1401,1446,2121), IL(1446), IN(1446,2449), KS(1446,2345), MN(1446), OH(1446), TN(1446), WI(1446).
P. citrinum. On seeds. **1**:GA(1446), IA(2121), IL(1446), IN(1446,2449), MO(1547).
P. claviforme. On moldy seed. **1**:WI(1446).
P. crustosum. **1**:IL(1446).
P. cyaneum. On seeds. **1**:IA(1446).
P. dangeardii. Isolated from corn samples. **1**:IA(2121).
P. dierckxii. On seeds. **1**:IL(1446).
P. digitatum. On seeds. **1**:IN(1446,2449).
P. expansum. Blue mold. **1**:Range of host(3423), Unk.(1446), IA(2121), IN(1446,2449), KY(1446).
P. fellutanum. On seeds. **1**:IN(1446,2449).
P. funiculosum. On seeds. **1**:IA(2121), IL(1446), IN(1446,2449), KS(1446,2345), MD(1446), NC(1303).
P. glabrum. On seeds. **1**:IA(2121), IN(1446,2449), KS(1446,2345), MO(1446).
P. granulatum. On seeds. **1**:GA(1446), IL(1446), IN(369,1446).
P. griseofulvum. On seeds. **1**:IA(2121), IN(1446,2449).
P. herquei. Isolated from corn samples. **1**:IA(2121).
P. implicatum. On seeds. **1**:IN(1446,2449).
P. janthinellum. On seeds. **1**:IN(1446,2449).
P. oxalicum. On seeds. 'Penicillium ear rot.' **1**:Unk.(1446), CT(1446), DC(1446,3423), IA(82,1401,1446,2121,3423), IL(82,1446,3423), IN(1446,2449,3423,3707), KS(1446,2345), KY(82,1446,3423), MD(1446), MN(1446), NC(1303), NY(1446), OH(1446,3423), VA(1446,3423), WI(82,1446,3423).
P. purpurogenum. On seeds and meal. **1**:Unk.(1446), AL(1446), FL(1446), GA(1446), IA(2121), IN(1446,2449), WI(1446).
P. roquefortii. On seeds. **1**:GA(1446), IL(1446).
P. rugulosum. On seeds. **1**:IA(82,1446,2121).
P. sclerotiorum. On seeds. **1**:IA(2121), IN(1446,2449).
P. thomii. On seeds. **1**:IA(1446).
P. variabile. On seeds. **1**:GA(1446), IA(1446,2121), IL(1446), IN(1446,2449), NC(1303).
P. verrucosum. On seeds. **1**:IA(1446).
P. waksmanii. On seeds. **1**:IA(1446).
Penicillium sp. On kernels, roots and stalks. **1**:Unk.(1446), AL(1446), AR(1446), CA(1446), CO(1446), DE(1446), FL(3990), GA(1446), IA(1446), IL(1446), IN(1446,3693,3694), KS(1446), KY(1446), LA(1446), MA(1446), MD(1446), ME(1446), MI(1446), MN(825,1446), MO(1446), MS(1446), NC(1303,1446), NE(1446), NJ(1446), NY(1446), OH(1446), OK(1446,2872), PA(1446), SC(1446), SD(1446), TN(1446), TX(1446), VA(1446,2987), WI(1446).
Periconia circinata. Root rot. **1**:ND(1446,3423).
Phymatotrichopsis omnivora. On roots. **1**:OK(82), TX(82,1446).
Pithomyces maydicus. Ear rot. **1**:IA(1401).
Polyschema olivacea. On dead stalks. **1**:NJ(82,1446).
Rhinotrichum sp. On seeds. **1**:S states(1446), S states(82), IA(1446).
Rhodotorula sp. **1**:IA(1446).
Scopulariopsis brevicaulis. Isolated from corn samples. **1**:IA(2121), IL(1446).
S. brumptii. Ear rot. **1**:IA(1401).
Scopulariopsis sp. On seeds. **1**:MN(1446).
Sepedonium sp. Stalk rot on seeds. **1**:IL(1446), MO(1446), NJ(1446).
Spegazzinia tessarthra. On stalks. **1**:LA(1446), VA(1446).
Spicaria sp. Leaf, root, seed and stalk rot. **1**:DE(1446), MN(825,1446), SC(82,1446).
Sporendonema sp. On seeds. **1**:MN(1446).
Sporidesmium folliculatum. **1**:WV(1446).
Sporotrichum sp. Isolated from corn samples. **1**:IA(2121).
Stachybotrys zeae. On leaves. **1**:AL(2499).
Stemphylium sp. **1**:IN(1446).
Thielaviopsis sp. **1**:HI(2903).
Torula sulphurea. Isolated from corn samples. **1**:IA(2121).
Torula sp. On seeds. **1**:DE(1446).
Trichoderma koningii. Isolated from corn samples. **1**:IA(2121), OR(1446).
T. viride. Root, seed and stalk rot. 'Trichoderma ear rot.' **1**:Unk.(82), FL(3990), IA(1401,1446,2121), IL(1446), IN(1446,3694), KS(1446,2345), MD(1446), MN(825,1446), NC(1303), NE(1446), NY(1446), SC(1446), WI(1446).
Trichoderma sp. Root, seed and stalk rot. **1**:Unk.(1446), IA(82,1446), IL(1446), MD(82,1446), MN(82,1446), MO(864,1446), NC(1446), NE(82,1446), NJ(471,1446), NY(1446), OH(1446), SC(82).
Trichothecium roseum. Pink mold, on seeds and meal. **1**:AL(1446), AR(1446), CT(1446), IA(1446,2121), IL(1446), IN(1446), KS(1446), KY(1446), LA(1446), MA(1446), MD(1446), ME(1446), MI(1446), MN(1446), MO(1446), MS(1446), NE(1446), NY(1446), OH(1446), OK(1446), PA(1446), SD(1446), TN(1446), TX(1446), VA(1446), WI(1446).
Trichothecium sp. On seeds. **1**:MN(1446).
Tritirachium oryzae. Isolated from corn samples. **1**:IA(2121).
Ulocladium lanuginosum. Isolated from corn samples. **1**:IA(2121).
Ustilaginoidea virens. False smut. **1**:LA(82,1446,3423).
Verticillium tenerum. **1**:NE(1446).
Verticillium sp. **1**:IA(1446).

Deuteromycotina - Coelomycetes
Ascochyta ischaemi. Yellow leaf blight. **1**:IL(82,1446,3423), OH(1446).
A. maydis. Leaf spot. **1**:IL(82,1446,3423).
A. tritici. Leaf spot. **1**:MD(1446).
A. zeicola. On dead stalks. **1**:NJ(82,1446).
Colletotrichum cereale. On seeds. **1**:DE(1446), IN(1446), MN(1446), NE(1446).
C. graminicola. 'Anthracnose stalk rot. Anthracnose leaf blight.' **1**:E states(82), S states(82), Unk.(3059), AR(765,1446), CA(1446), CT(1446,3423), DE(1446), GA(1446), IL(53,1446,1625,2743,3423), IN(1446,3423,3777), KS(1446), KY(1446), LA(1446,3423), MD(1446,2476,3423), MS(1446,2720), NC(1303,1307,1446,2095,2098,3423), NJ(1446,3423), NY(1446), OH(1446,3423), PA(1446,3423), SC(1446,3423), VA(1446,2987,3423), WV(1446).
Colletotrichum sp. Anthracnose. **1**:CT(1446), DE(1446), NM(2112), SC(1446).
Coniothyrium scirpi. On leaves. **1**:IL(82,1446,3423).
Dendrophoma zeae. On leaves. **1**:IL(82,1446).
Diplodia frumenti. Ear rot. 'Diplodia stalk rot.' **1**:AL(1446), GA(1446), OK(2872), SC(1446), TN(1446), TX(1446).
D. maydis. 'Diplodia ear rot. Diplodia stalk rot. Seed rot-seedling blight.' **1**:C states(1446), E states(1446), AL(1446), AR(1446), CA(1446), CT(1446), DC(1446), DE(1446), FL(1446,3990), GA(1446), IA(1446), IL(1446,1625), IN(1446,3693), KS(1446), KY(1446), LA(1446), MA(1446), MD(1446), ME(1446), MI(1446), MN(1446), MO(1446), MS(1446), NC(1446), ND(1446), NE(1446), NJ(1446), NY(1446), OH(1446), OK(1446), PA(1446), RI(1446), SC(1446), SD(1446), TN(1446), TX(1446), VA(1446), WA(1446), WI(1446), WV(1446).
Diplodia sp. Seedling blight, ear, root and stalk rot. **1**:AR(1446), DE(1446), FL(1446), GA(1446), IA(1446), IL(1446), IN(1446), KS(1446), KY(1446), LA(1446), MD(1446), MI(1446), MN(1446), NE(1446), NJ(1446), OH(1446), OK(1446), PA(1446), SD(1446), TN(72,73,1446), TX(1446), VA(1446).
Leptothyrium zeae. Leaf spot. **1**:IL(82,1446,3251,3423).
Macrophoma zeae. **1**:IL(3590).
Macrophomina phaseolina. Seedling root rot, stalk rot, on seeds. 'Charcoal rot.' **1**:AR(82,1446,3423), CA(82,1446,3423), CO(82,1187,1446,3423), FL(35,3990), GA(82,1446,3423), IA(82,1446,3423), IL(53,82,311,1446,1625,1940,3423), IN(1446), KS(82,1446,3423), KY(1446), MO(82,1446,3423), NC(1303,1446,2098), NE(82,1446,3423), OK(82,1446,2872,3423), SC(82,1446,3423), SD(1446,2242), TX(82,1446,3423).
Phaeocytostroma ambiguum. Root and stalk rot. **1**:Unk.(1446), IL(82,1446,1940), IN(1446).
Phoma americana. On roots. **1**:GA(2504).
P. terrestris. Pink root, secondary root rot, stalk rot. **1**:CO(82,1446,3423), FL(3990), IA(82,1446), IL(1446), LA(1446), MN(82,1446), ND(82,1446,3423), SD(1446,3423), WI(82,1446,3423).
P. zeicola. On kernels and leaves. **1**:AL(82,1446), IA(1401), LA(82,1446), TX(82,1446), WV(82,1446).
Phoma sp. Leaf spot. **1**:HI(2903), IA(1446,2121,3185).
Phomopsis sp. Seed rot. **1**:Unk.(1446).
Phyllosticta maydis. 'Yellow leaf blight.' **1**:IA(1446), IL(1446), MI(1446), WI(1446).
P. zeae. Blight, yellow leaf spot. **1**:CT(1446), IA(1446), IL(82,1446), OH(1446), PA(1446,3126).
Phyllosticta sp. Yellow leaf blight. **1**:CT(1446), FL(35,82,1446), HI(2903), IA(1446), IL(1446), IN(1446), MI(1446), MN(1446), MO(1446), NC(1303,1446), NE(1446), NY(1446), TX(82,1446), WI(107).
Pyrenochaeta sp. Stalk rot. **1**:NE(1446).
Septoria zeae. On leaves. **1**:IL(82,1446).
S. zeicola. On leaves. **1**:IL(82,1446).
S. zeina. On leaves. **1**:IL(82,1446).
Septoria sp. Leaf spot. **1**:NC(1303,1446).
Sphaeropsis sp. On stalks. **1**:CA(82,1446).
Stauronema cruciferum. **1**:SC(634).
Stenocarpella macrospora. Dry rot of ears and stalks. 'Diplodia stalk rot. Southern leaf spot. Diplodia ear rot.' **1**:SE states(82), AL(1446,3423), CT(1446,3423), DE(1446,3423), FL(35,1446,3423), GA(1446,3423), LA(82,1446,3423), MD(1446,3423), MS(82,1446,2720,3423), NC(1303,1446,2731,3423), OK(1446,3423), SC(1446,3423), TN(1446,3423), TX(82).
S. maydis. White ear rot, root and stalk rot, isolated from seeds. **2**:IA(82).
Stigmella sp. Isolated from corn samples. **1**:IA(2121).
Deuteromycotina - Other
Rhizoctonia solani. Damping-off, stalk rot. **1**:AZ(82), FL(35,3990), GA(3541), IA(82), IL(82), MA(82), MN(82), MS(2720), NC(1303), OH(82), OK(2872), SC(82), SD(2242), WA(82).
R. zeae. Ear rot, isolated from roots. 'Rhizoctonia leaf spot.' **1**:FL(35,82,1446), GA(3541), IN(1446), MS(1446).
Rhizoctonia sp. Root and stalk rot. **1**:AZ(1446), GA(3541), HI(2903), IN(1446), MA(1446), MN(825,1446), NC(1446), OH(1446), TN(1446), WA(1446).
Sclerotium rolfsii. Southern blight, root rot, basal sheath and stalk rot. **1**:AL(82,1446), FL(35,82,1446,3990), HI(2903), IL(1446), NY(1446), TX(82,1446), VA(82,1446). **2**:FL(82).

Zizania L.
1. *Z. aquatica* L. [annual wild rice, eastern wild rice, Indian rice]
2. *Z. aquatica* L. var. *angustifolia* Hitchc.
3. *Z. latifolia* (Griseb.) Turcz. ex Stapf (=*Z. caduciflora* (Trin.) Hand.-Mazz.) [Manchurian wild rice]
4. *Z. palustris* L. [cultivated northern wild rice]

Ascomycotina
Claviceps purpurea. Ergot. **1**:IA(3423), MD(3423), ME(3423), MN(3423), ND(3423), NE(3423), WI(3423).
C. zizaniae. Ergot. **1**:MN(1880).
Claviceps sp. Ergot. **1**:IA(82), MD(82), ME(82), MN(82), MO(82), ND(82), NE(82), WI(82).
Gaeumannomyces graminis. Culm rot. **1**:VA(82,3754).
Mycosphaerella zizaniae. On leaves. **1**:DE(82,967), MI(82).
Basidiomycotina - Ustilaginales
Doassansia zizaniae. Stem smut. **1**:WI(82).
Entyloma dactylidis. **1**:IA(1077).
E. lineatum. Leaf smut. **1**:Range of host(82), CT(1077), DC(1077), DE(1077), GA(1077), IA(1077), ID(344), IL(1077), IN(1077), MA(1394), MI(1077), MN(1077,1880), NE(1077), NJ(1077), SD(1077,2242), WI(1077). **2**:FL(1077), ME(1077), NJ(1077), WI(1077).
Ustilago esculenta. **1**:DC(1077). **3**:DC(1077).
Deuteromycotina - Hyphomycetes
Bipolaris oryzae. Blight, black spot, leaf spot. **1**:MD(82), MN(82,227,1880).
B. sorokiniana. Blight. **1**:ID(344), MN(1880).
Drechslera gigantea. Zonate leaf spot. **1**:MN(1836).
Helminthosporium sp. Leaf spot. **1**:MN(3423).
Nakataea sigmoidea. Stem rot. **1**:MN(1880,2512).
Deuteromycotina - Coelomycetes
Colletotrichum graminicola. Anthracnose. **1**:MN(346).
Diplodia oryzae. On dead culms. **1**:LA(82).
Deuteromycotina - Other
Sclerotium hydrophilum. **1**:ID(344), MN(2888), WI(2888).
S. oryzae. **1**:DC(2888), ID(344), MN(2888), WI(2888). **4**:WI(2888).
S. zizaniae. On culms. **1**:MD(82), WI(82).
Sclerotium sp. Stem rot. **1**:MN(1880,2512).

Zizaniopsis Döll & Aschers. ex Döll in Mart.
1. *Z. miliacea* (Michx.) Döll & Aschers. [giant cutgrass]

Ascomycotina
Didymella eumorpha. On culms. **1**:LA(82).
Magnaporthe salvinii. Stem rot. **1**:AR(3423), LA(3423), TX(3423).
Basidiomycotina - Uredinales
Uromyces coronatus. **1**:FL(35).
Deuteromycotina - Hyphomycetes
Nakataea sigmoidea. Leaf spot. **1**:AR(82), LA(82), TX(82).
Deuteromycotina - Other
Sclerotium hydrophilum. On dead culms. **1**:LA(82).

Zoysia Willd.
1. *Z. japonica* Steud. [Japanese lawngrass, Korean grass, zoysia]
2. *Z. matrella* (L.) Merr. [flawn, Japanese carpet grass, Manila grass, zoysia grass]
3. *Z. tenuifolia* Willd. ex Trin. (=*Osterdamia tenuifolia* (Trin.) Kuntze) [Mascarene grass]

4. *Zoysia* sp.
5. *Zoysia* sp. (unnamed hybrids)

Oomycetes

Pythium aphanidermatum. Root rot. **4:**HI(2903).
Pythium sp. Root rot. **1:**FL(35).

Ascomycotina

Claviceps yanagawaensis. Ergot. **1:**Unk.(82), DC(3423), FL(35). **4:**FL(35).
Sclerotinia homoeocarpa. 'Dollar spot.' **1:**FL(35). **4:**FL(35).

Basidiomycotina - Uredinales

Puccinia zoysiae. Heteroecious: telial host. **1:**AL(1398,1819), AR(1819), FL(35,1106,1819), GA(1819), HI(2299,2903), LA(1819), MD(1819), MO(1819,1981), MS(1819), NE(1819), TX(1819). **2:**AL(1398), FL(1106), GA(1963), NC(1303). **3:**FL(1106), HI(2903). **5:** On *S. japonica* x *S. tenuifolia*. AL(1283,1398), FL(35,1106,1283), GA(1283), LA(1283), MD(1283), MO(1283), MS(139,1283), TX(1283).

Deuteromycotina - Hyphomycetes

Curvularia lunata. Leaf spot. **1:**FL(35). **3:**HI(2903). **4:**FL(35).
Curvularia sp. Blight disease, fading-out, leaf spot. **1:**AR(766), NC(1303). **4:**NC(1303).
Drechslera dictyoides. Blight. **4:**NC(1303).
Exserohilum sp. Leaf spot. **3:**HI(2903).
Fusarium sp. Associated with blight disease. **1:**AR(766).
Helminthosporium sp. Leaf spot. **1:**FL(35). **2:**HI(2903). **4:**FL(35), NC(1303).

Deuteromycotina - Coelomycetes

Colletotrichum falcatum. Anthracnose. **1:**HI(2903). **3:**HI(2903).

Deuteromycotina - Other

Rhizoctonia cerealis. Yellow patch. **5:** NC(2271).
R. solani. Dieback, brown patch, root and stem rot, leaf spot. 'Rhizoctonia blight.' **1:**FL(35), KS(3023), WV(953). **4:**FL(35), NC(1303).
Rhizoctonia sp. Frog-eye, root rot. **1:**AR(767), HI(2903). **3:**HI(2903).

Podocarpaceae

Podocarpus L'Hér. ex Pers.

1. *P. macrophyllus* (Thunb.) D. Don [yew podocarpus]
2. *Podocarpus* sp.

Oomycetes

Phytophthora sp. Root rot. **1:**FL(35). **2:**FL(35).
Pythium irregulare. Root rot. **1:**FL(35).
P. splendens. Root rot. **1:**FL(35). **2:**FL(35).
Pythium sp. Root rot. **1:**HI(2903).

Ascomycotina

Botryosphaeria sp. Dieback. **1:**FL(35). **2:**FL(35).

Basidiomycotina - Other

Armillaria tabescens. Root rot. **1:**FL(35).

Deuteromycotina - Hyphomycetes

Alternaria sp. Leaf spot. **1:**FL(35). **2:**FL(35).
Beltrania rhombica. On leaf litter. **2:**FL(3548).
Cercospora sp. Leaf spot. **1:**FL(35). **2:**FL(35).
Cylindrocladium sp. Root rot. **2:**FL(35).
Flosculomyces floridaensis. On dead leaves. **2:**FL(3548).
Fusarium oxysporum. Root rot, wilt. **1:**FL(35). **2:**FL(35).
F. solani. Basal stem rot. **1:**FL(35).
Hemibeltrania navicularis. On leaf litter. **2:**FL(3548).
Heterosporium sp. Leaf spot. **1:**FL(35).
Mycoleptodiscus sp. Leaf spot. **1:**FL(35).
Zygosporium gibbum. On leaf litter. **2:**FL(3548).
Z. oscheoides. On leaf litter. **2:**FL(3548).

Deuteromycotina - Coelomycetes

Botryodiplodia sp. Dieback. **2:**FL(35).
Colletotrichum gloeosporioides. Anthracnose. **1:**FL(35). **2:**FL(35).
Coniothyrium sp. Dieback. **2:**FL(35).
Diplodia sp. Dieback. **1:**FL(35). **2:**FL(35).
Endophragmiella boewei. On leaf litter. **2:**FL(3548).
Macrophoma sp. Tip dieback, leaf spot. **1:**FL(35). **2:**FL(35).
Phoma sp. Dieback. **1:**FL(35). **2:**FL(35).
Phomopsis sp. Dieback, stem gall. **1:**FL(35). **2:**FL(35).
Phyllosticta sp. Leaf spot. **1:**FL(35). **2:**FL(35).
Sphaeropsis sp. Dieback. **2:**FL(35).

Deuteromycotina - Other

Rhizoctonia solani. Root and stem rot. **1:**FL(35). **2:**FL(35).
Rhizoctonia sp. Root rot. **1:**HI(2903).

Polemoniaceae

Collomia Nutt.

1. *C. grandiflora* Douglas ex Lindl.
2. *C. heterophylla* Hook.
3. *C. linearis* Nutt. (=*Gilia linearis* (Nutt.) A. Gray)
4. *C. tinctoria* Kellogg

Oomycetes

Peronospora phlogina. **1:**ID(3202), OR(3202), WA(3202). **2:**ID(3202), OR(3202), WA(3202). **3:**ID(3202), MT(3202), OR(3202), WA(3202).

Ascomycotina

Sphaerotheca collomiae. On leaves and stems. **2:**CA(370).
S. fuliginea. Powdery mildew. **3:**CO(82), ID(82), MT(82), ND(82,377), WA(82), WY(82).
S. macularis. Powdery mildew. **1:**CA(81,82), WA(3202). **2:**CA(81), WA(3202). **3:**CA(81), CO(82), ID(82,3202), MT(82,3202), ND(82), OR(3202), WA(82,3202), WI(1341), WY(82,1195). **4:**WA(3202).

Basidiomycotina - Uredinales

Puccinia giliae. Autoecious. **1:**CA(82,2708), WA(82,2708,3202).
P. plumbaria. Autoecious. **3:**NV(82).
Uromyces acuminatus. Heteroecious: aecial host. **3:**CO(82), ND(82,378), NE(82), SD(2243).

Deuteromycotina - Coelomycetes

Phyllosticta sp. **2:**WA(3202).
Septoria sp. **1:**ID(3202), WA(3202). **3:**ID(3202), OR(3202), WA(3202).

Eriastrum Woot. & Standl.

1. *E. diffusum* (A. Gray) H. L. Mason (=*Gilia filifolia* Nutt.)

Basidiomycotina - Uredinales

Puccinia giliae. Autoecious. **1:**AZ(747,1203,2708,3981).

Gilia Ruiz & Pav.

1. *G. aggregata* (Pursh) Spreng.
2. *G. capitata* Sims
3. *G. gilioides* (Benth.) Greene (=*Allophyllum gilioides* (Benth.) A. D. Grant & V. Grant)
4. *G. inconspicua* (Sm.) Sweet var. *sinuata* (Douglas ex Benth.) A. Gray (=*G. sinuata* Douglas ex Benth.)
5. *G. pinnatifida* Nutt. ex A. Gray
6. *G. rubra* (L.) A. Heller (=*Ipomopsis rubra* (L.) Wherry)
7. *G. tenerrima* A. Gray
8. *G. tricolor* Benth. [bird's-eyes]
9. *Gilia* sp.

Oomycetes

Peronospora phlogina. Downy mildew. **9:**TX(82).

Ascomycotina

Pleospora helvetica. **9:**OR(657).
Sphaerotheca macularis. Powdery mildew. **3:**AZ(1195). **7:**ID(3202), WA(3202). **8:**CA(81). **9:**CA(82), TX(82), WA(82).

Basidiomycotina - Uredinales

[*Puccinia aristidae*]. **4:**AZ(1203). **9:**AZ(82), CO(82).
P. giliae. Autoecious. **2:**CA(81,2708), OR(3202). **3:**CA(2708). **5:**CO(2708). **9:**AZ(82), CA(82), CO(82), NV(82), OR(82), WA(82).
P. plumbaria. Autoecious. **2:**WA(3202). **9:**CA(82), CO(82), NV(82), UT(82), WY(82).
P. subnitens. **4:**AZ(3981).
P. yosemitana. Autoecious. **9:**CA(82), CO(82).

Basidiomycotina - Ustilaginales

Urocystis giliae. **9:**WY(1077).

Deuteromycotina - Hyphomycetes

Botrytis cinerea. **9:**WA(3202).
Heterosporium sp. **1:**WA(3202).
Phymatotrichopsis omnivora. Root rot. **9:**TX(82).

Deuteromycotina - Coelomycetes

Diplodia leptodactyli. On stems. **9:**TX(82).
Heteropatella conglutinata. On dead stems. **1:**CA(642).

Septoria sp. **9:**WA(3202).
Deuteromycotina - Other
Sclerotium rolfsii. Southern blight, root rot. **6:**FL(35).

Leptodactylon Hook. & Arnott

1. *L. californicum* Hook. & Arnott [prickly-phlox]
2. *L. pungens* (Torr.) Rydb. (=*L. pungens* (Torr.) Rydb. subsp. *hookeri* (Douglas ex Hook.) Wherry) [granite gilia]
3. *L. watsonii* (A. Gray) Rydb. (=*Gilia watsonii* A. Gray)

Ascomycotina
Pleospora pleosphaerioides. **3:**WY(3810).
Basidiomycotina - Uredinales
Puccinia giliicola. Autoecious. **1:**CA(2708).
P. yosemitana. Autoecious. **2:**CA(81,2708), CO(2708).
Deuteromycotina - Coelomycetes
Phoma linearispora. On stems. **3:**WY(3812).

Linanthus Benth.

1. *L. bicolor* (Nutt.) Greene (=*Gilia bicolor* (Nutt.) Piper)
2. *L. bigelovii* (A. Gray) Greene
3. *L. ciliatus* (Benth.) Greene (=*Gilia ciliata* Benth.)
4. *L. nuttallii* (S. F. Gray) Greene (=*Linanthastrum nuttallii* (S. F. Gray) Ewan)
5. *L. pharnaceoides* Greene
6. *L. septentrionalis* H. L. Mason

Ascomycotina
Sphaerotheca macularis. **1:**WA(3202).
Basidiomycotina - Uredinales
[*Puccinia aristidae*]. **2:**AZ(1203).
P. fragilis. **4:**CA(2708), CO(2708), UT(2708), WY(2708).
P. giliae. Autoecious. **2:**AZ(1203,2708). **3:**CA(1468,2708). **5:**WA(3202).
P. subnitens. **2:**AZ(3981).
Deuteromycotina - Hyphomycetes
Ramularia giliae. Leaf spot. **1:**OR(82,3202,3421). **6:**ID(3202), MT(3202), WA(3202).

Microsteris Greene

1. *M. gracilis* (Douglas ex Hook.) Greene (=*Gilia gracilis* Douglas ex Hook., *Phlox gracilis* (Douglas ex Hook.) Greene, *M. micrantha* Greene)

Chytridiomycetes
Synchytrium polemonii. **1:**WA(3202).
Synchytrium sp. On leaves. **1:**WA(1847).
Oomycetes
Peronospora phlogina. Downy mildew. **1:**ID(82,3202), MT(3202), OR(3202), WA(3202).
Ascomycotina
Erysiphe cichoracearum. **1:**WA(3202).
Sphaerotheca fuliginea. **1:**MT(82).
Deuteromycotina - Hyphomycetes
Alternaria sp. **1:**ID(3202).
Epicoccum nigrum. **1:**ID(3202).
Ramularia giliae. **1:**ID(3202).
Deuteromycotina - Coelomycetes
Septoria sp. **1:**OR(3202).

Navarretia Ruiz & Pav.

1. *N. atractyloides* (Benth.) Hook. & Arnott
2. *N. divaricata* Greene
3. *N. intertexta* (Benth.) Hook.
4. *N. squarrosa* (Eschsch.) Hook. & Arnott
5. *N. tagetina* Greene

Oomycetes
Peronospora phlogina. **3:**ID(3202), OR(3202), WA(3202).
Ascomycotina
Erysiphe cichoracearum. Powdery mildew. **4:**CA(1155).
Sphaerotheca macularis. **2:**CA(81). **3:**WA(3202).
Basidiomycotina - Uredinales
Puccinia giliae. Autoecious. **1:**CA(2708). **3:**OR(2708,3202). **5:**OR(3202).
Deuteromycotina - Hyphomycetes
Alternaria sp. **3:**WA(3202).
Deuteromycotina - Coelomycetes
Diplodia sp. **3:**ID(3202).
Septoria sp. **3:**ID(3202).

Phlox L.

1. *P. albomarginata* M. E. Jones
2. *P. alyssifolia* Greene
3. *P. amoena* Sims
4. *P. austromontana* Crép.
5. *P. caespitosa* Nutt.
6. *P. carolina* L. [thick-leaf phlox]
7. *P. diffusa* Benth.
8. *P. divaricata* L. [blue phlox, wild sweet-William]
9. *P. douglasii* Hook.
10. *P. drummondii* Hook. [annual phlox, Drummond phlox]
11. *P. glaberrima* L. [smooth phlox]
12. *P. hoodii* Richardson (=*P. glabrata* (E. E. Nels.) A. Brand)
13. *P. kelseyi* Britton
14. *P. longifolia* Nutt. (=*P. stansburyi* A. Heller)
15. *P. maculata* L. [meadow phlox, wild sweet-William]
16. *P. multiflora* A. Nels. (=*P. depressa* (E. E. Nels.) Rydb.)
17. *P. nana* Nutt.
18. *P. paniculata* L. (=*P. decussata* Lyon ex Pursh) [fall phlox, perennial phlox, summer perennial phlox]
19. *P. pilosa* L.
20. *P. speciosa* Pursh
21. *P. subulata* L. [ground pink, moss pink]
22. *Phlox* × sp. (unnamed hybrids)
23. *Phlox* sp.

Oomycetes
Peronospora phlogina. Downy mildew. **8:**IA(82), WI(82).
Pythium aphanidermatum. Root rot. **23:**HI(2903).
Ascomycotina
Dasyscyphus phlogis. On living and dead stems. **9:**CA(524,642).
Erysiphe cichoracearum. Powdery mildew. **7:**WA(3202). **8:**IN(82), NY(82), OH(82), WI(82). **10:**Range of host(82), FL(35). **15:**MI(82). **18:**Range of host(82), MA(1390), MS(2720), OR(81,3202), WA(3202). **23:**FL(35), NC(1303), OR(3202), WA(3202).
Leptosphaeria phlogis. **8:**Unk.(82).
Lophodermium phloxii. On leaves. **9:**CA(638,642). **23:**CA(82).
Mollisia hysteroides. On dead limbs. **9:**CA(524,642).
Mycosphaerella phlogina. On dead leaves. **14:**MT(967,3202).
M. tassiana. **5:**MT(3202).
Mycosphaerella sp. On leaves. **14:**ID(3202). **23:**MT(82), NY(82).
Pleospora comata. **5:**WA(3202). **14:**ID(3202).
P. magnifica. On stems. **23:**UT(195).
P. tragacanthae. On leaves. **7:**CA(661). **9:**CA(642).
Sphaerotheca macularis. Powdery mildew. **14:**WY(1195). **18:**KS(82), NH(82), NY(82), OH(82), WA(82).
Basidiomycotina - Uredinales
Puccinia douglasii. Autoecious. **1:**MT(2708,3202). **2:**MT(3202). **4:**WY(2708). **5:**UT(2708), WA(2708,3202), WY(2708). **7:**OR(2708,3202), UT(2708), WA(2708,3202), WY(2708). **9:**CA(638,642). **12:**MT(3202), WY(1195). **14:**NM(2708). **16:**CO(2708), WY(2708). **17:**NM(2708). **23:**CO(82), ID(2708), MT(82), NE(82), NM(82), OR(82), UT(82), WA(82), WY(82,2708).
P. douglasii var. *hoodii*. Autoecious. **12:**MT(2708), WY(2708).
P. fragilis. **12:**CO(2708). **14:**AZ(2708), CA(2708), CO(2708), ID(2708), NM(2708), NV(2708), UT(2708), WY(2708). **16:**CO(2708), WY(2708). **19:**TX(2708). **20:**CA(2708), OR(2708).
P. fragilis var. *divaricatae*. Autoecious. **8:**IA(2708), IL(2708), WI(2708).
P. plumbaria. Autoecious. **5:**ID(3202), WA(3202). **8:**IA(82), IL(82), MO(82), WI(1354,1359). **14:**ID(638,1195,3202), WA(3202), WY(1195). **16:**WY(1195). **18:**IA(82). **19:**IA(82), SD(2243), TX(82). **20:**OR(3202), WA(3202). **23:**W states(82).
Uromyces acuminatus. Heteroecious: aecial host. **8:**IA(82), MN(82). **11:**WI(1348). **18:**IA(82), MS(82,2720), TX(82). **19:**IA(82), IL(82), SD(82), WI(82).
Deuteromycotina - Hyphomycetes
Alternaria sp. Leaf spot. **23:**FL(35).
Botrytis cinerea. Gray mold. **18:**AK(82).
Cercospora omphacodes. Leaf spot. **6:**IA(82). **8:**NC states(82), NE states(82), NJ(1818), WI(82). **11:**WI(1344). **18:**NC states(82),

NE states(82), MN(82), NY(82), VA(82), WI(82,1346). **19:**OK(82,2872), WI(82,1332). **23:**AL(82), MS(82,2720), WV(82).
Fusarium sp. Isolated from decayed stems. **18:**NC(1303).
Heterosporium sp. **14:**ID(3202), MT(3202).
Oidium sp. Powdery mildew. **10:**FL(35).
Phymatotrichopsis omnivora. Root rot. **18:**NJ(82), PA(82), TX(82).
Ramularia giliae. **13:**MT(3202). **20:**WA(3202).
Ramularia sp. Leaf spot. **18:**WA(82).
Stemphylium botryosum. Leaf spot. **10:**NJ(1817,2760).
Stemphylium sp. Leaf spot. **10:**FL(35).
Thielaviopsis basicola. Crown, root and basal stem rot. **18:**NJ(2759).
Verticillium albo-atrum. Wilt. **6:**MN(82). **10:**NY(82). **18:**NY(82).
Volutella phlogina. Leaf spot. **18:**LA(82,128).
Deuteromycotina - Coelomycetes
Ascochyta phlogis. Leaf spot. **10:**NY(82).
Coleophoma cylindrospora. Leaf spot. **23:**CA(82).
Colletotrichum sp. Stem canker, leaf necrosis. **10:**FL(35). **23:**FL(82), NC(1303).
Macrophomina phaseolina. Charcoal rot. **18:**IL(311).
Phyllosticta sp. Leaf spot. **18:**WA(82). **23:**WA(3202).
Pyrenochaeta phlogis. Stem blight. **18:**NY(82).
Septoria phlogis. Leaf spot. **3:**WI(1342). **8:**IA(82), IL(82), IN(82), NJ(82), NY(82,966), WI(82,1340). **10:**AK(82,526), CT(82), FL(35,82), MI(82), NY(82), PA(82). **11:**WI(1344). **18:**E states(82), AK(82), CT(82), FL(82), MI(82), MS(82,2720), PA(82), TX(82), WI(82,1332). **22:** AL(285), MS(2720), OK(2872), WI(1330).
Septoria sp. Leaf spot. **23:**HI(2903).
Vermicularia phlogina. On leaves and stems. **8:**CT(82), WI(1338). **18:**CT(82), IL(312), MD(82), NY(82). **21:**NY(82). **23:**MS(82,2720), VA(82).
Deuteromycotina - Other
Rhizoctonia solani. Stem rot. **18:**MN(82), NJ(82), NY(82).
Rhizoctonia sp. Crown rot. **10:**FL(35).
Sclerotium rolfsii. Southern blight, stem rot. **8:**CT(82). **18:**MD(82), NJ(82), NY(82), OH(82), TX(82), VA(82). **21:**CT(82), FL(82), IL(82). **23:**FL(35), NC(1303).

Polemonium L.
1. *P. acutiflorum* Willd. ex Roem. & Schutt
2. *P. boreale* Adans. (=*P. humile* Willd. ex Roem. & Schult.)
3. *P. caeruleum* L. subsp. *villosum* (J. H. Rudolph ex Georgi) A. Brand [charity, Greek valerian, Jacob's-ladder]
4. *P. californicum* Eastw.
5. *P. intermedium* (A. Brand) Rydb.
6. *P. occidentale* Greene
7. *P. pulcherrimum* Hook. (=*P. delicatum* Rydb.)
8. *P. reptans* L.
9. *Polemonium* sp.

Chytridiomycetes
Synchytrium polemonii. On leaves. **6:**WA(1846).
Ascomycotina
Erysiphe cichoracearum. Powdery mildew. **8:**WI(1350). **9:**UT(82).
Mycosphaerella tassiana. On dead stem. **2:**AK(1938). **3:**AK(1938).
Phyllachora polemonii. **7:**CO(649).
Pleospora anthyllidis var. *abbreviata*. On dead stems and leaves. **2:**AK(526). **7:**AK(82).
P. comata. **9:**AK(1938).
Sphaerotheca macularis. Powdery mildew. **9:**CO(649), WA(82).
Basidiomycotina - Uredinales
Puccinia gulosa. Autoecious. **7:**CA(82,2708).
P. polemonii. Autoecious. **1:**AK(60,526). **2:**AK(526,2708). **3:**AK(1938). **5:**ID(2708). **6:**CA(2708), MT(2708). **8:**ID(82), IN(82). **9:**AK(82,2708), CA(82), ID(82).
Uromyces acuminatus. Heteroecious: aecial host. **8:**IA(82), IL(82), IN(82), WI(82).
Basidiomycotina - Ustilaginales
Entyloma lapponicum. **4:**WA(1077).
Deuteromycotina - Hyphomycetes
Cercospora omphacodes. Leaf spot. **8:**PA(82).
Fusarium sp. Wilt. **9:**NJ(82).
Ramularia polemoniae. Leaf spot. **4:**WA(666).
Verticillium albo-atrum. Wilt. **9:**NY(82).
Deuteromycotina - Coelomycetes
Ascochyta sp. **8:**WI(1360).
Leptostroma herbarum. **3:**AK(526).
Phoma minuta. On stems. **6:**WY(3812).
Septoria polemonii. Leaf spot. **6:**WY(1195). **8:**CT(82), IN(82), MO(82), WI(82).

Polygalaceae

Polygala L.
1. *P. cruciata* L.
2. *P. lutea* L. [candyweed, yellow bachelor's-button, yellow milkwort]
3. *P. nana* (Michx.) DC.
4. *P. paucifolia* Willd. [bird-on-the-wing, flowering-wintergreen, fringed polygala]
5. *P. polygama* Walter
6. *P. sanguinea* L.
7. *P. senega* L. [Seneca snakeroot]
8. *P. verticillata* L.
9. *Polygala* sp.

Ascomycotina
Phomatospora berkeleyi. On stems. **9:**CA(82).
Phyllachora polygalae. Leaf spot. **9:**FL(35).
Basidiomycotina - Uredinales
Aecidium renatum. **9:**NM(82).
Puccinia andropogonis. Heteroecious: aecial host. **5:**WI(1331). **9:**IA(82), MI(82), WI(82).
P. pyrolae. Autoecious. **4:**CT(82), ME(82), MI(82), MN(82), NH(82), NY(82), WI(82).
Deuteromycotina - Hyphomycetes
Cercospora grisea. Leaf spot. **1:**NJ(82). **2:**MS(82,2720). **3:**MS(2720). **5:**WI(1338). **6:**MS(82,2720), WI(1334). **9:**VA(82).
Cladosporium sp. **8:**WI(1345).
Curvularia lunata. **8:**WI(1343,1351).
Deuteromycotina - Coelomycetes
Gloeosporidiella ramosa. Anthracnose. **5:**NJ(966). **6:**WI(1330). **9:**IN(82), NJ(82), WI(82).
Phoma polygalae. On stems. **9:**CA(82).
Septoria consocia. Leaf spot. **7:**WI(1348). **9:**IN(82), MI(82).
S. polygalae. **4:**NY(82).

Polygonaceae

Antigonon Endl.
1. *A. leptopus* Hook. & Arnott

Basidiomycotina - Other
Armillaria tabescens. Root rot. **1:**FL(35).
Deuteromycotina - Hyphomycetes
Cercospora sp. Leaf spot. **1:**FL(35).
Phaeoramularia antigoni. On leaves. **1:**TX(843).
Deuteromycotina - Coelomycetes
Colletotrichum sp. Leaf spot. **1:**FL(35).
Deuteromycotina - Other
Rhizoctonia sp. Root rot. **1:**HI(2903).

Bistorta (L.) Adans.
This genus is considered a synonym of *Polygonum*, but the appropriate combination could not be located.
1. *B. hirtifolia* Greene

Basidiomycotina - Ustilaginales
Ustilago hydropiperis. **1:**CO(1077).

Brunnichia Banks ex Gaertn.
1. *B. cirrhosa* Gaertn. [buckwheat-vine, lady's-eardrops]

Ascomycotina
Physalospora abdita. On stems. **1:**LA(82).
Deuteromycotina - Coelomycetes
Leptothyrium brunnichiae. **1:**IL(82).

Chorizanthe R. Br. ex Benth.
1. *C. pungens* Benth.
2. *C. watsonii* Torr. & A. Gray

Basidiomycotina - Uredinales
Uromyces intricatus. Autoecious. **1:**CA(81,972).
Deuteromycotina - Coelomycetes
Phyllosticta sp. **2:**WA(3202).

Coccoloba P. Br.

1. *C. diversifolia* Jacq. (=*C. floridana* Meisn.) [pigeon-plum, snailseed]
2. *C. pubescens* L. (=*C. grandifolia* Jacq.)
3. *C. uvifera* (L.) L. [kino, platterleaf, sea-grape]
4. *Coccoloba* sp.

Oomycetes
Pythium splendens. Root rot. **3:**FL(35).
Ascomycotina
Cryphonectria coccolobii. **3:**FL(2365,2366).
Endothia sp. On roots. **3:**FL(35).
Glomerella cingulata. Anthracnose. **3:**FL(35).
Phyllachora simplex. Anthracnose, tar spot. **1:**FL(2287). **4:**FL(35).
Rhytisma sp. Tar spot. **4:**FL(35).
Basidiomycotina - Uredinales
Uredo coccolobae. **3:**FL(35,82).
Basidiomycotina - Other
Armillaria tabescens. Root rot. **3:**FL(35).
Oudemansiella canarii. On living and fallen trunks. **1:**FL(3284). **2:**FL(82).
Septobasidium fumigatum. Parasitic on scale insects. **3:**FL(35).
Deuteromycotina - Hyphomycetes
Alternaria sp. Leaf spot. **3:**FL(35). **4:**FL(35).
Cercospora bicolor. Leaf spot. **3:**FL(35). **4:**FL(35).
Fusarium solani. Dieback. **3:**FL(35).
Helminthosporium sp. Leaf spot. **3:**FL(35).
Verticicladium effusum. On leaves. **3:**FL(35,82).
Deuteromycotina - Coelomycetes
Botryodiplodia sp. Leaf spot. **3:**FL(35).
Colletotrichum sp. Anthracnose. **3:**FL(35). **4:**FL(35).
Dendrophoma sp. Stem canker. **3:**FL(35).
Diplodia sp. Leaf spot. **3:**FL(35).
Diplodina sp. Leaf spot. **3:**FL(35).
Pestalotia versicolor. Leaf spot. **3:**FL(35,82). **4:**FL(35).
Phomopsis sp. Dieback, leaf spot. **3:**FL(35). **4:**FL(35).
Phyllosticta coccolobae. Leaf spot. **3:**FL(35).
Septoria sp. Leaf spot. **3:**FL(35).
Sphaeropsis sp. Stem canker. **3:**FL(35).
Deuteromycotina - Other
Rhizoctonia solani. Damping-off, basal stem rot. **3:**FL(35).

Eriogonum Michx.

1. *E. alatum* Torr.
2. *E. arborescens* Greene
3. *E. blissianum* H. L. Mason
4. *E. brevicaule* Nutt. (=*E. campanulatum* Nutt.)
5. *E. chrysops* Rydb.
6. *E. compositum* Douglas ex Benth.
7. *E. deflexum* Torr.
8. *E. effusum* Nutt.
9. *E. elatum* Douglas ex Benth.
10. *E. elongatum* Benth.
11. *E. fasciculatum* Benth. (=*E. polifolium* Benth. in DC.) [California buckwheat]
12. *E. flavum* Nutt.
13. *E. giganteum* S. Wats. [St.-Catharine's-lace]
14. *E. heracleoides* Nutt.
15. *E. jamesii* Benth.
16. *E. latifolium* Sm.
17. *E. marifolium* Torr. & A. Gray
18. *E. microthecum* Nutt.
19. *E. molle* Greene
20. *E. multiceps* Nees
21. *E. niveum* Douglas ex Benth.
22. *E. nudum* Douglas ex Benth.
23. *E. ochroleucum* Small ex Rydb.
24. *E. ovalifolium* Nutt.
25. *E. pauciflorum* Pursh
26. *E. piperi* Greene
27. *E. pyrolifolium* Hook.
28. *E. racemosum* Nutt.
29. *E. sphaerocephalum* Douglas ex Benth.
30. *E. strictum* Benth.
31. *E. subalpinum* Greene
32. *E. subreniforme* S. Wats.
33. *E. thymoides* Benth.
34. *E. tomentosum* Michx.
35. *E. trichopes* Torr.
36. *E. umbellatum* Torr. (=*E. stellatum* Benth.) [sulphur-flower]
37. *E. umbellatum* Torr. subsp. *polyanthum* S. Stokes
38. *E. vimineum* Douglas ex Benth.
39. *E. virgatum* Benth. in DC.
40. *E. wrightii* Torr. ex Benth.
41. *Eriogonum* sp.

Oomycetes
Albugo sp. **9:**WA(3202).
Peronospora eriogoni. On leaves. **7:**AZ(3381).
Peronospora sp. Downy mildew. **3:**CA(82).
Ascomycotina
Cystotheca lanestris. Powdery mildew. **2:**CA(1155).
Erysiphe californica. Powdery mildew. **3:**CA(369). **13:**CA(369). **17:**CA(369).
E. cichoracearum. Powdery mildew. **3:**CA(81). **9:**WA(82,3202). **13:**CA(81).
E. polygoni. Powdery mildew. **17:**CA(82,642). **41:**TX(82).
Mollisia shastensis. On dead roots. **17:**CA(642). **27:**CA(524,642).
Mycosphaerella tassiana. On dead stems. **16:**CA(81).
Phialea campanula. **17:**CA(638,642).
Pleospora herbarum var. *occidentalis*. On stems. **41:**CO(82).
Sclerotinia sp. **14:**WA(3202).
Basidiomycotina - Uredinales
[*Puccinia aristidae*]. **32:**AZ(1203). **35:**AZ(82,1203).
P. subnitens. **32:**AZ(3981). **35:**AZ(3981).
Uromyces intricatus. Autoecious. **4:**CO(1195), WY(1195,3379). **5:**ID(3202). **6:**ID(3202), OR(3202), WA(3202). **7:**AZ(747,3981). **8:**ID(3202), WA(3202), WY(1195). **10:**CA(81). **11:**AZ(82,747,1203,3981), CA(82). **12:**ID(3202), MT(1195,3202), ND(378), WA(3202). **14:**ID(3202), UT(1195), WA(3202). **15:**WY(1195). **16:**CA(81,638,642,661). **17:**CA(638,642), OR(3202). **18:**MT(3202), OR(3202), WA(3202). **20:**MT(3202), ND(378). **21:**WA(3202). **22:**CA(81,661,2439), OR(3202). **23:**MT(3202). **24:**ID(3202), MT(1195,3202), OR(3202). **25:**MT(3202). **26:**ID(3202), MT(3202), OR(3202), WA(3202). **28:**AZ(1195,1203,3981). **29:**OR(3202), WA(3202). **30:**OR(3202). **31:**ID(3202), MT(3202), WA(3202). **33:**OR(3202). **36:**CA(661), ID(3202), MT(3202), OR(3202), WA(3202), WY(2924). **37:**CA(638,642). **38:**CA(81), ID(3202), OR(3202), WA(3202). **39:**CA(972), OR(3202). **40:**AZ(747,1203,3981). **41:**W states(82), CA(82), WA(82).
Deuteromycotina - Hyphomycetes
Cercospora eriogoni. On leaves. **19:**CA(82).
C. rubella. Leaf spot. **34:**SC(82,581). **41:**TX(82).
Heterosporium sphaeriiforme. On stems. **41:**CO(82).
Phymatotrichopsis omnivora. Root rot. **41:**TX(82).
Ramularia sp. **9:**WA(3202). **14:**WA(3202).
Deuteromycotina - Coelomycetes
Cytospora eriogoni. On dead underground stems. **27:**CA(642,666).
Hendersonia eriogoni. On stems. **1:**NM(2399).
Leptostromella eriogoni. On stems. **41:**NV(82).
Phoma eriogoni. On dead stems and root stocks. **17:**CA(642,666).
Phyllosticta eriogoni. **9:**WA(666,3202).
Sarcophoma eriogoni. Leaf spot. **11:**CO(82). **12:**ID(3202), MT(3202). **14:**WA(3202).
Seimatosporium anomalum. On twigs. **16:**CA(81). **41:**CA(3549).
Septoria sp. **9:**WA(3202).

Fagopyrum Mill.

1. *F. esculentum* Moench (=*F. sagittatum* Gilib., *F. vulgare* T. Nees) [common buckwheat, notch-seeded buckwheat]

Ascomycotina
Erysiphe polygoni. Powdery mildew. **1:**CA(82).
Sclerotinia sclerotiorum. Stem rot. **1:**DE(82).
Basidiomycotina - Other
Thanatephorus cucumeris. **1:**WA(3202).
Typhula variabilis. On straw. **1:**NY(2955).

Deuteromycotina - Hyphomycetes
Alternaria alternata. Seed mold, leaf spot. **1**:IA(82), IL(312), MN(82), VT(82).
Cercospora fagopyri. Leaf spot. **1**:IL(312).
Ramularia anomala. Leaf spot. **1**:CT(82), IN(82), WI(82).
Deuteromycotina - Coelomycetes
Ascochyta bresadolae. Leaf spot. **1**:CT(82), NY(82), PA(82), VT(82), WI(82).
Macrophomina phaseolina. Charcoal rot. **1**:IL(311).
Phomopsis fagopyri. On stems. **1**:DE(82).
Phyllosticta polygonorum. Leaf spot. **1**:DE(82), IL(312), NJ(82), PA(82), WV(82).
Deuteromycotina - Other
Rhizoctonia solani. Damping-off. **1**:MN(82), NC(82), WA(82).

Homalocladium (F. v. Muell.) L. H. Bailey

1. *H. platycladum* (F. v. Muell.) L. H. Bailey (=*Muehlenbeckia platyclada* (F. v. Muell.) Meisn.) [centipede-plant, ribbon-bush, tapeworm-plant]

Ascomycotina
Erysiphe polygoni. Powdery mildew. **1**:FL(35), NY(82), PA(82), WI(82,1330).
Deuteromycotina - Coelomycetes
Ascochyta sp. Leaf spot. **1**:FL(35).

Oxyria J. Hill

1. *O. digyna* (L.) J. Hill [mountain-sorrel]

Ascomycotina
Pleospora cerastii. **1**:AK(526).
P. comata. **1**:AK(526).
Basidiomycotina - Uredinales
Puccinia oxyriae. Autoecious. **1**:AK(60,82,526,1938), CA(21,82,638,642), CO(82,1520), ID(21,82,3202), OR(21,82,3202), UT(21,82).
Basidiomycotina - Ustilaginales
Ustilago vinosa. Floral smut. **1**:AK(82,526,1077), CA(82,1077), CO(82,1077), WA(82,1077,3202), WY(82,1077,1195).

Polygonella Michx.

1. *P. articulata* (L.) Meisn.

Ascomycotina
Jattaea curvicolla. On stalks. **1**:NY(190).
Deuteromycotina - Hyphomycetes
Cercospora sp. On stems. **1**:WI(1357).

Polygonum L.

(= *Bistorta* (L.) Adans.)

1. *P. achoreum* S. F. Blake
2. *P. alaskanum* W. Wight ex Harshb.
3. *P. alpinum* All. (=*P. polymorphum* Ledeb.)
4. *P. amphibium* L. (=*P. natans* (Michx.) D. C. Eaton f. *hartwrightii* (A. Gray) E. E. Stanford) [water smartweed]
5. *P. amphibium* L. var. *stipulaceum* N. Coleman (=*P. hartwrightii* A. Gray)
6. *P. arifolium* L.
7. *P. atlanticum* C. Bicknell
8. *P. aviculare* L. (=*P. buxiforme* Small) [prostrate knotweed]
9. *P. bistortoides* Pursh (=*Bistorta bistortoides* (Pursh) Small, *P. bistorta* L., *P. plumosum* Small, *P. bistorta* L. subsp. *plumosum* (Small) Hultén) [bistort, snakeweed]
10. *P. capitatum* F. Ham. ex D. Don
11. *P. careyi* Olney
12. *P. cilinode* Michx.
13. *P. coccineum* Muhl. ex Willd. (=*P. muhlenbergii* (Meisn.) S. Wats., *P. emersum* Britton) [swamp smartweed]
14. *P. coccineum* Muhl. var. *pratincola* (Greene) E. E. Stanford
15. *P. confertiflorum* Willd.
16. *P. convolvulus* L. [black bindweed, corn bindweed, wild buckwheat]
17. *P. cuspidatum* Siebold & Zucc. (=*Reynoutria japonica* Houtt.) [fleece-flower, Japanese knotweed, Mexican-bamboo]
18. *P. davisiae* W. H. Brewer ex A. Gray
19. *P. douglasii* Greene
20. *P. dumetorum* L.
21. *P. erectum* L. [erect knotweed]
22. *P. glabrum* Willd.
23. *P. hydropiper* L. [common smartweed, marshpepper smartweed]
24. *P. hydropiperoides* Michx. [mild water-pepper]
25. *P. hydropiperoides* Michx. var. *psilostachyum* St. John
26. *P. kelloggii* Greene
27. *P. lapathifolium* L.
28. *P. newberryi* Small
29. *P. orientale* L. [kiss-me-over-the-garden-gate, prince's-feather, princess-feather]
30. *P. pensylvanicum* L. [Pennsylvania smartweed]
31. *P. persicaria* L. [hearts-ease, lady's-thumb]
32. *P. phytolaccaefolium* Meisn. ex Small
33. *P. punctatum* Elliott (=*P. acre* H. B. K.) [dotted smartweed]
34. *P. ramosissimum* Michx. (=*P. exsertum* Small, *P. prolificum* (Small) B. L. Robinson) [bushy knotweed]
35. *P. sagittatum* L. [arrow-leaved tearthumb]
36. *P. scandens* L. [climbing false buckwheat]
37. *P. scandens* L. var. *cristatum* (Engelm. & A. Gray) Gleason (=*P. cristatum* Engelm. & A. Gray)
38. *P. shastense* W. H. Brewer
39. *P. spergulariaeforme* Meisn. & Small
40. *P. tenue* Michx.
41. *P. virginianum* L. (=*Tovara virginiana* (L.) Raf.)
42. *P. viviparum* L. [alpine bistort, serpent-grass]
43. *Polygonum* sp.

Plasmodiophoromycetes
Ligniera junci. On roots. **27**:NC(1303).
Chytridiomycetes
Synchytrium polygoni. Leaf gall. **43**:LA(627).
Synchytrium sp. **9**:WY(2924).
Oomycetes
Peronospora americana. Downy mildew. **15**:MT(3202). **34**:IL(82), NE(82).
P. polygoni. Downy mildew. **8**:IA(82). **16**:IA(82), IN(82), ND(82,377), OH(82), WA(3202). **19**:ID(3202), OR(3202), WA(3202). **20**:IA(82), IL(82), IN(82), KS(82), KY(82), NE(82), NY(82), OH(82), WI(82). **36**:IA(82), IL(82), IN(82), KS(82), KY(82), NE(82), NY(82), OH(82), WI(82).
Ascomycotina
Botryosphaeria obtusa. On stems. **43**:AL(82), NJ(82).
Comoclathris permunda. On dead stems. **3**:CA(1468).
Diaporthe arctii. On dead stems. **30**:GA(1438). **35**:GA(1438). **43**:GA(82), LA(82).
D. dakotensis. On stems. **13**:ND(3807).
Erysiphe polygoni. Powdery mildew. **1**:WI(1351). **7**:WI(1333). **8**:Range of host(82), Unk.(3976), FL(35), GA(1438), ND(377), OK(2872), SD(2243), WI(1333), WY(1195). **16**:IN(82). **21**:Range of host(82), ID(3202), MS(2720), ND(377), SD(2243), WA(3202). **23**:PA(82). **30**:GA(1438). **31**:MT(82,3202), NY(82). **39**:OR(3202). **43**:NM(82), UT(82), WY(82).
Leptosphaeria complanata. **30**:GA(1438).
L. fusispora. On stems. **17**:NY(82).
Massarina gloeospora. **35**:GA(1438).
Mycosphaerella tassiana. On dead stem. **42**:AK(1938).
Mycosphaerella sp. Leaf spot. **4**:IL(82). **14**:WI(1358). **35**:WI(1341). **41**:WI(1358). **43**:NY(82).
Nitschkia polygoni. On stems. **43**:IL(82,3590).
Phyllactinia guttata. Powdery mildew. **41**:MI(82).
Plagiostoma devexum. On leaves. **4**:NY(82).
Pseudorhytisma bistortae. Leaf spot. **42**:AK(82,526).
Pyrenopeziza polygoni. **30**:GA(1438).
Stigmatea polygonorum. On leaves. **42**:CA(82).
Trichopezizella nidulus var. *setigera*. **3**:OR(1416).
Wettsteinina eucarpa. On dead leaves. **9**:AK(1938).
Basidiomycotina - Uredinales
Puccinia amphispilusa. **28**:CA(661).
[*P. aristidae*]. **8**:CO(82), DE(82), ND(82), NE(82), SD(82). **21**:CO(82). **34**:CO(82).
P. bistortae. Heteroecious: telial host. **15**:OR(3202). **26**:OR(3202). **28**:OR(3202), WA(3202). **32**:OR(3202). **42**:AK(60,82,526), CO(82), NH(82), WY(82,1195,2924).
P. parca. Autoecious. **36**:NC(82), NJ(82), NY(82). **43**:AK(526).
P. polygoni-alpini. **2**:AK(60,526). **3**:CA(81). **32**:ID(3202), OR(657,3202), WA(3202).
P. polygoni-amphibii. Heteroecious: telial host. **4**:E & C states(82), Range of host(82), W states(82), CA(81). **5**:CA(81), OK(2872),

SD(2243). **16:**E & C states(82), ND(82), SD(2243), TX(82). **20:**E & C states(82), ND(82), OK(2872), TX(82). **22:**HI(2903). **23:**E & C states(82), Range of host(82), W states(82), NC(1303). **24:**MS(2720), WA(3202). **27:**E & C states(82), Range of host(82), W states(82), OK(2872), TX(742). **31:**E & C states(82), Range of host(82), W states(82), MS(2720), NC(1303), OK(2873), OR(3202). **33:**Range of host(82), GA(1439), MS(2720), OK(2872). **36:**E & C states(82), IN(647), MS(2720), ND(82), OK(660,2872), TX(82). **37:**MS(2720). **41:**E states(82), IN(647), MI(665), TX(82), WI(82).

P. septentrionalis. Heteroecious: telial host. **42:**AK(60,82,526).

Uromyces polygoni-avicularae. Autoecious. **13:**WA(3202). **21:**Range of host(82), MS(2720), ND(378), SD(2243). **24:**MS(2720). **34:**Range of host(82), MT(3202). **39:**CA(81). **43:**CA(82), CO(82), FL(35), LA(82), MA(82), ME(82).

U. rickerianus. Autoecious. **3:**ID(1195). **32:**ID(3202).

Basidiomycotina - Ustilaginales

Melanopsichium austro-americanum. **8:**CA(82). **27:**CA(82), OK(2872), TX(82). **30:**OK(2872). **31:**KS(2458).

M. pennsylvanicum. **8:**CA(1077), NY(82). **23:**DC(1077), IL(1077), MD(82). **30:**IL(1077,3132), IN(1077), KS(1077), MO(1077), NJ(1077), NY(1077). **41:**MO(1077). **43:**DC(1077), DE(82), IA(82), IL(82), IN(82), KS(82), KY(82), MN(1077), MO(82), NJ(82), NY(82), OH(82), OK(82), TX(1077), VA(82), WY(82).

Ustilago anomala. Floral smut. **12:**CT(1077), MA(1077), ME(1077), MN(1077), NH(1077), NJ(1077), NY(1077), PA(1077), WI(1350). **16:**NE states(82), IL(82,1077), IN(1077), MO(82), SD(1077). **20:**NE states(82), CT(1077), IL(82), IN(1077), MO(82), PA(1077), WV(1077). **36:**NE states(82), CT(1077), DE(1077), IL(82), IN(1077), MO(82,1077), PA(1077), VT(1077), WV(1077). **43:**NE states(82), IL(82), MO(82).

U. bistortarum. Inflorescence and leaf smut. **9:**CO(82), NV(82), UT(82), WA(82,1077), WY(82). **42:**AK(82,526,1077), CO(82,1077), WY(82,1077,1195). **43:**CA(82), OR(82), UT(1077).

U. bosniaca. **3:**CA(1077), OR(1077). **18:**CA(81,1077), OR(1077,3202). **28:**WA(1077,3202). **32:**OR(3202), WA(3202). **33:**OK(1077).

U. hydropiperis. Covered seed smut. **9:**CA(661), CO(1077), WA(1077), WY(1077). **23:**NE states(82), ME(1077), NE(82,1077), RI(1077), WA(82,1077,3202). **24:**ME(1077), OR(1077,3202), WA(1077,3202). **25:**WA(1077). **31:**NE states(82), NE(82), NY(1077), WA(82,1077,3202). **33:**CT(1077), DE(1077), IL(1077), MA(1077), PA(1077), WV(1077). **35:**E states(82), CT(1077), DE(1077), IL(82,1077), MA(1077), ME(1077), MN(1077), NC(1077), NE(1077), NH(1077), NJ(1077), NY(1077), OH(1077), PA(1077), RI(1077), TN(1077), VA(1077), VT(1077), WI(82,1077), WV(1077). **42:**AK(1077), CO(1077), WY(1077). **43:**E states(82), CA(1077), CO(922), MI(1077), MO(1077), NE(82), WA(82).

U. piperi. **18:**CA(1077). **27:**ID(3202). **32:**ID(1077,3202). **43:**ID(1077), NV(1077).

U. shastensis. **38:**CA(638,642,661,922,1077,4026).

U. tenuispora. **33:**FL(82,1077). **43:**FL(35).

U. utriculosa. Inflorescence smut, loose smut, seed smut. **4:**E states(82), CA(81,1077), IN(1077), PA(1077), TX(82), WA(82). **5:**IN(1077). **8:**CA(82,1077), IA(1077). **11:**ME(1077), WI(1077). **13:**FL(1077). **21:**MS(82,2720). **22:**HI(2903). **23:**E states(82), AL(1077), CT(1077), IA(1077), TX(82), WA(82). **24:**CT(1077), IN(1077), MS(1077,2720), VT(1077), WA(1077,3202), WI(1077). **27:**E states(82), CA(81), CT(1077), DC(1077), IA(1077), IL(1077), IN(1077), MD(1077), MI(1077), MO(1077), ND(1077), NY(1077), PA(1077), TX(82), WA(82), WI(1077). **30:**AL(1077), CT(1077), DE(1077), FL(1077), IA(1077), IL(1077), IN(1077), KS(1077), LA(1077), MA(1077), MD(1077), MO(1077), MS(1077,2720), NE(1077), NH(1077), NJ(1077), NY(1077), OH(1077), PA(1077), RI(1077), VA(1077), VT(1077), WI(1077), WV(1077). **31:**CA(1077), IA(1077), IN(1077), MN(1077), OR(1077,3202). **33:**DE(1077), HI(2899,2903), IL(1077), KS(1077), MI(1077), MS(1077,2720), PA(1045). **35:**ME(1077), NY(82,1077). **43:**FL(35), NC(1077,1303), OR(1077), TX(1077), WA(1077).

Basidiomycotina - Other

Thanatephorus cucumeris. Web blight. **16:**WA(3202). **43:**FL(82).

Deuteromycotina - Hyphomycetes

Bostrichonema polygonia. On leaves. **9:**CO(82), WA(82). **42:**AK(82,526), WY(82).

Botrytis sp. **36:**WI(1343).

Cercospora avicularis. Leaf spot. **1:**WI(1357). **8:**E states(82), DE(3543), ND(82,378), SD(2243), TX(82). **21:**IA(82), MN(82), SD(82), WI(82). **27:**AL(82), TX(82). **30:**OK(2872). **34:**IA(82), KS(82), NE(82), WA(3202). **43:**OK(82), TX(82).

C. beticola. Leaf spot. **16:**IA(82).

C. effusa. Leaf spot. **8:**DE(3543). **23:**E states(82), E states(82), S states(82), MO(82), MT(82,3202). **27:**C states(82), E states(82), E states(82), S states(82), MO(82), MT(82,3202). **30:**NC(3944), OK(2872). **31:**C states(82), E states(82), E states(82), S states(82), MO(82), MT(82). **33:**SC(634), WI(1343). **35:**DE(3543).

C. polygonacea. Leaf spot. **16:**IA(82), IL(82), MS(82,2720), NJ(82), OH(82), TX(82), WI(1334). **18:**OR(3202). **20:**DE(3543), MD(82), NJ(82). **35:**AL(82). **36:**AL(82), TX(82), WI(1343). **40:**WI(1330). **43:**FL(35).

C. polygonacea var. *avicularis*. **8:**DE(3543).

Cercospora sp. **19:**ID(3202).

Cristulariella moricola. On leaves. **30:**IL(2572). **36:**IL(2572).

Ovularia avicularis. On leaves. **8:**NE(2739).

Phymatotrichopsis omnivora. Root rot. **8:**TX(82). **27:**TX(82). **31:**TX(82). **43:**TX(82).

Ramularia anomala. Leaf spot. **16:**IN(82), NE(82), WI(82). **31:**WI(1340). **33:**WI(1340). **36:**IN(82), NE(82), WI(82).

R. bistortae. Leaf spot. **9:**WY(1195).

R. cilinodis. Leaf spot. **43:**PA(82), WI(82).

R. rigidula. Leaf spot. **8:**IN(82), NE(82), WI(82). **21:**WI(82).

R. rufomaculans. Leaf spot. **4:**IN(82), KS(82), MD(82), ND(82), NE(82), NY(82), WI(82,1339). **8:**IA(82). **13:**ND(378). **20:**IA(82). **23:**IN(82), KS(82), MD(82), ND(82), NE(82), NY(82), WI(82). **31:**IN(82), KS(82), MD(82), ND(82), NE(82), NY(82), WI(82). **32:**ID(3202). **43:**IN(82), KS(82), MD(82), ND(82), NE(82), NY(82), WI(82).

Deuteromycotina - Coelomycetes

Ascochyta biguttulata. Leaf spot. **16:**IL(82). **31:**OK(82,2873).

A. volubilis. Leaf spot. **6:**WI(1344). **31:**MA(82,1394). **35:**WI(1359). **43:**NY(82).

Ascochyta sp. **23:**WI(1358).

Chaetoconis polygoni. On stems. **9:**CA(82).

Colletotrichum truncatum. Isolated from leaves and stems. **30:**IL(1482).

Diplodia polygonicola. On stems. **27:**KS(82).

Macrophoma pulchrispora. **43:**NY(82).

Macrophoma sp. On stems. **43:**MD(82).

Macrophomina phaseolina. Charcoal rot. **27:**IL(311).

Phlyctema complanata. On stems. **41:**NY(82), PA(82).

Phyllosticta polygonorum. Leaf spot. **4:**WI(82,1335). **43:**IL(82).

Phyllosticta sp. **16:**WI(1343). **35:**WI(1352).

Rhabdospora polygoni. On stems. **41:**NY(82).

Septoria polygonorum. Leaf spot. **8:**MN(82), SD(2243), WI(82). **14:**WI(1335). **16:**WI(82). **21:**MN(82), WI(82). **23:**E states(82), KS(82), MI(82), MS(82,2720), WA(82,3202). **27:**E states(82), KS(82), MS(82,2720), MT(3202), ND(378), WA(82,3202), WI(1333). **29:**E states(82), KS(82), MS(82,2720), WA(82). **30:**MS(2720), MT(3202), OK(2872). **31:**E states(82), KS(82), MA(1390), MS(82,2720), WA(82,3202), WI(1334), WY(1195). **36:**VT(82).

Deuteromycotina - Other

Rhizoctonia solani. Web blight, root rot. **16:**WA(82). **43:**FL(35).

Rhizoctonia sp. Root rot. **10:**HI(2903).

Rheum L.

1. *R. officinale* Baill. [Chinese rhubarb]
2. *R. rhabarbarum* L. [pieplant, rhubarb, wineplant]
3. *R. rhaponticum* L. A name misapplied to rhubarb.
4. *Rheum* sp.

Oomycetes

Peronospora rumicis. Downy mildew. **3:**CA(82).

Phytophthora cactorum. Crown and root rot. **3:**CA(82), MO(82), OK(82,2872), PA(82).

P. nicotianae var. *parasitica*. Root rot. **3:**IL(82), KS(82), LA(82), MD(82), MO(82), NY(82), TX(82), VA(82).

Phytophthora sp. Crown rot. **3:**NC(1303).
Pythium anandrum. Damping-off. **3:**CA(82), MD(82).
P. irregulare. Damping-off. **1:**CA(82).
P. oligandrum. Damping-off. **3:**CA(82).
P. splendens. Damping-off. **1:**CA(82). **3:**CA(82).
P. ultimum. Damping-off. **3:**CA(82).
Pythium sp. Damping-off, crown rot. **1:**Unk.(82). **3:**NC(1303).

Ascomycotina

Hyponectria depressa. On overwintered stems. **4:**NE states(183).

Basidiomycotina - Uredinales

Puccinia phragmitis. Heteroecious: aecial host. **3:**CA(82), MN(82), NE(82), SD(2243).

Basidiomycotina - Other

Armillaria mellea. Root rot. **3:**CA(82), TX(82).
Thanatephorus cucumeris. Stem rot. **3:**OK(2872), WA(3202).

Deuteromycotina - Hyphomycetes

Alternaria sp. Secondary leaf and stalk spot, storage rot. **3:**CA(82), MN(82), NE(82), NJ(82), OK(2872), PA(82).
Botrytis cinerea. Gray mold rot, leaf rot. **1:**AK(82,526). **2:**MI(2531). **3:**ID(3202), MI(82), MO(82), VA(82), WA(82,3202).
Cercospora rhapontici. **3:**IL(82).
Cercospora sp. Leaf spot. **1:**DE(82), MD(82), NE(82).
Cladosporium sp. Leaf spot. **3:**CA(82), WA(82,3202).
Fusarium sp. Root rot. **3:**NC(82,1303), OK(82,2872), WA(82,3202).
Phymatotrichopsis omnivora. Root rot. **3:**AZ(82), TX(82).
Ramularia rhei. Leaf spot. **3:**CA(82), WA(3202).
Verticillium albo-atrum. **3:**CA(3896).

Deuteromycotina - Coelomycetes

Ascochyta rhei. Leaf spot. **1:**NJ(82,966). **3:**E & C states(82), Unk.(82), HI(2903), IL(312), KS(82), MS(2720), WA(82,3202).
Ascochyta sp. Leaf spot. **2:**NY(580).
Colletotrichum dematium. Anthracnose. **3:**IA(82), IL(82), MO(82), OK(82,2873), PA(82), WI(82), WV(82).
Macrophoma straminella. Leaf, stalk and stem spot. **2:**FL(35). **3:**Range of host(82), IL(64), IN(64), OK(2873).
Septoria rhapontici. Leaf spot. **3:**IA(82).
Vermicularia polygoni-virginica. **3:**SD(2243).

Deuteromycotina - Other

Rhizoctonia solani. Crown and root rot, sore shin. **3:**CA(82), CT(82), HI(2899), IL(82), MN(82), NC(1303), NY(82), OK(82), TX(82), WA(82).
Rhizoctonia sp. **3:**HI(2903).
Sclerotium rolfsii. Southern blight. **2:**FL(35). **3:**FL(82), MS(82,2720), NC(1303), TX(82), VA(82).

Rumex L.

1. *R. acetosa* L. [garden sorrel, sour dock]
2. *R. acetosella* L. [common sorrel, red sorrel, sheep sorrel]
3. *R. alpinus* L. (=*R. pauciflorus* Campd.) [monk's rhubarb, mountain rhubarb]
4. *R. altissimus* A. Wood
5. *R. britannica* L.
6. *R. crispus* L. [curly dock, yellow dock]
7. *R. hastatulus* Baldw.
8. *R. hymenosepalus* Torr. [canaigre, tanner's dock, wild-rhubarb]
9. *R. maritimus* L. (=*R. persicarioides* L.)
10. *R. mexicanus* Meisn.
11. *R. obtusifolius* L. [bitter dock, blunt-leaved dock, broadleaf dock, red-veined dock]
12. *R. occidentalis* S. Wats.
13. *R. patientia* L. [herb-patience, monk's rhubarb, patience dock, spinach dock]
14. *R. paucifolius* Nutt. ex S. Wats.
15. *R. salicifolius* Weinm.
16. *R. venosus* Pursh [sour-greens, wild begonia, wild hydrangea]
17. *R. verticillatus* L.
18. *Rumex* sp.

Chytridiomycetes

Physoderma majus. On leaves. **5:**ND(377). **18:**ND(82), WI(82).
Synchytrium anomalum. On leaves. **1:**IA(82).

Oomycetes

Pythium helicandrum. **2:**MD(2821).

Ascomycotina

Apiospora sp. **6:**AK(3431).
Botryosphaeria dothidea. **6:**GA(1438). **11:**GA(1438).
B. obtusa. On stems. **6:**GA(82). **11:**GA(82).
Diaporthe arctii. On stems. **6:**GA(82,1438).
Erysiphe polygoni. **15:**ID(3202).
Leptosphaeria macrospora. **11:**ME(272).
Lophiosphaera lophospora. On dead stems. **2:**NJ(681).
Lophiostoma caulium. **6:**GA(1438).
[*Ophiobolus acuminatus*]. **17:**GA(1438).
Phyllactinia guttata. Powdery mildew. **12:**WA(82,3202).
Stictis radiata. **6:**GA(1438).
Venturia rumicis. Leaf spot. **6:**WA(82,3202).

Basidiomycotina - Uredinales

Aecidium minutulum. **18:**TX(82).
Puccinia acetosae. Autoecious. **1:**FL(82), NY(82), PA(82). **2:**E states(82), AL(82), CA(81,82), GA(1439), IN(82), NC(3944), OR(82,3202), WA(3202), WI(1339). **7:**FL(35,82), MA(82), SC(82).
[*P. aristidae*]. **8:**AZ(82,1203).
P. lapathicola. **7:**FL(35). **8:**CA(82). **9:**CA(81). **12:**CA(82). **15:**CA(81). **18:**CA(82), KS(82), TX(82), WI(82).
P. ornata. Autoecious. **12:**GA(1438), ND(82,378). **18:**ME(82), MN(82), NH(82), NY(82), TX(82), WI(82).
P. phragmitis. Heteroecious: aecial host. **6:**NE(82). **15:**ND(378). **16:**OK(2872), SD(2243). **18:**IA(82), MN(82), ND(82), NE(82), OK(82), WI(82).
P. subnitens. **8:**AZ(3981).
Uromyces rickerianus. Autoecious. **3:**CO(1195). **14:**ID(3202), OR(3202). **18:**ID(82), UT(82), WY(82).

Basidiomycotina - Ustilaginales

Ustilago kuehneana. Inflorescence smut. **2:**SC(82). **7:**AL(82), FL(35,82).
U. parlatorei. Inflorescence smut. **2:**SC(1077). **4:**IL(1077). **5:**MO(1077), NE(1077). **7:**AL(1077), FL(1077), NH(1077). **18:**IL(82), MO(82), NE(82).

Deuteromycotina - Hyphomycetes

Botrytis cinerea. Blight. **18:**NC(1303).
Botrytis sp. **2:**WI(1361).
Cercospora acetosellae. Leaf spot. **1:**TX(82). **2:**LA(82), NC(82,3944), NJ(82), TX(82), WI(1334). **6:**SD(2243). **7:**FL(35).
C. peckiana. Leaf spot. **6:**NY(82).
C. rumicis. Leaf spot. **6:**IA(82). **11:**LA(82), TX(82).
Cercospora sp. Leaf spot. **2:**HI(2903).
Cristulariella moricola. On leaves. **6:**IL(2572).
Heterosporium caulicola. On dead stems. **12:**CO(82).
Phymatotrichopsis omnivora. Root rot. **6:**TX(82). **18:**TX(82).
Ramularia decipiens. **8:**Unk.(82). **16:**WY(1195). **18:**WY(1195).
R. macrospora. **11:**DE(3543). **13:**SD(82,2243).
R. occidentalis. Leaf spot. **2:**WI(1333). **5:**KS(973). **6:**MS(2720), WI(1342). **18:**IA(82), IL(82), KS(82), LA(82), MS(82), NE(82), NH(82), WI(82).
R. pratensis. **2:**WI(1338). **6:**DE(3543). **9:**WI(82). **11:**DE(3543), WI(1342). **12:**AK(82,526). **18:**DE(82,3543), MD(82), WI(82).
R. rubella. Leaf and stem spot. **2:**ID(3202), OR(3202), WA(3202). **7:**OK(2872). **8:**AZ(407,1195,2882,3378), NM(82). **9:**WA(3202). **12:**WA(3202). **15:**ID(3202), WA(3202). **16:**OK(2872), WA(3202).
R. rumicis. **18:**AK(526).
Ramularia sp. **6:**Unk.(82). **11:**Unk.(82). **18:**Unk.(82).

Deuteromycotina - Coelomycetes

Aplosporella rumicicola. On dead stems. **18:**NY(82), WA(3202).
Colletotrichum dematium. Anthracnose, stem spot. **6:**OK(2872). **13:**Unk.(82).
Cylindrosporium pulchrum. Leaf spot. **9:**MO(82), SD(82). **11:**LA(82). **18:**MO(962), WA(3202).
Gloeosporium rumicis. Leaf spot. **1:**NY(82), TX(82). **2:**TX(82). **11:**TX(82).
Phoma herbarum. **6:**WI(1358).
P. terrestris. Isolated from rhizosphere. **6:**OR(2791).
Phyllosticta acetosellae. Leaf spot. **2:**WI(82,1335).
P. circuligerens. Leaf spot. **4:**IL(3590), OK(2872).
Phyllosticta sp. Leaf spot. **1:**NY(82). **11:**WI(1354).
Septoria pleosporoides. Leaf spot. **1:**TX(82). **2:**TX(82).
S. rumicis. Leaf spot. **4:**OK(2872), WI(1340). **6:**ID(3202), MN(82), MT(3202), WA(3202). **10:**SD(2243). **15:**ND(378). **16:**WY(1195). **18:**IL(82), KS(82), ND(82), NE(82), NY(82), WA(3202), WI(82,1341), WY(82).

Sphaeropsis sphaerelloides. **18:**OH(82).
Deuteromycotina - Other
Rhizoctonia solani. Root rot. **1:**TX(82). **2:**TX(82).

Pontederiaceae

Eichhornia Kunth

1. *E. crassipes* (Mart.) Solms. [water hyacinth]
2. *E. speciosa* Kunth (=*Piaropus crassipes* Raf.)

Oomycetes
Pythium sp. Root rot. **1:**FL(35,554).
Zygomycotina
Mucor sp. Associated with insect damage. **1:**FL(554).
Rhizopus sp. Associated with insect damage. **1:**FL(554).
Ascomycotina
Chaetomium sp. Associated with insect damage. **1:**FL(554).
Didymella exigua. **1:**FL(616).
Leptosphaerulina sp. **1:**FL(616).
Melanospora zamiae. **1:**FL(616).
Mycosphaerella tassiana. **1:**FL(616).
Mycosphaerella sp. Associated with insect damage. **1:**FL(554).
Deuteromycotina - Hyphomycetes
Acremonium zonatum. Zonate leaf spot. **1:**FL(35,616).
Acremonium sp. Associated with insect damage. **1:**FL(554).
Alternaria sp. Leaf spot, associated with insect damage. **1:**FL(35,554,616).
Aspergillus flavus. **1:**FL(616).
A. niger. **1:**FL(616).
Aspergillus sp. Associated with insect damage. **1:**FL(554).
Beltrania sp. Associated with insect damage. **1:**FL(554).
Bipolaris sp. **1:**FL(616).
Brachysporium sp. Associated with insect damage. **1:**FL(554).
Cephalotrichum sp. **1:**FL(616).
Cercospora piaropi. Leaf spot. **1:**FL(35,1108). **2:**TX(3597).
C. rodmanii. Leaf spot. **1:**FL(35,613), TX(612).
Cercospora sp. Associated with insect damage. **1:**FL(554,616).
Cercosporella sp. Associated with insect damage. **1:**FL(554).
Cladosporium sp. Associated with insect damage. **1:**FL(554,616).
Curvularia affinis. **1:**FL(616).
C. penniseti. **1:**FL(616).
Curvularia sp. Associated with insect damage. **1:**FL(554).
Dendryphiella infuscans. **1:**FL(616).
Doratomyces eichhornius. On laminae. **1:**FL(617).
Epicoccum nigrum. **1:**FL(616).
Exserohilum prolatum. **1:**FL(616).
Fusarium sp. Associated with insect damage. **1:**FL(554).
Gilmaniella sp. Associated with insect damage. **1:**FL(554).
Helminthosporium sp. Associated with insect damage. **1:**FL(554).
Memnoniella subsimplex. **1:**FL(616).
Mycoleptodiscus terrestris. Leaf spot. **1:**FL(35,553,616).
Myrothecium cinctum. **1:**FL(616).
Nigrospora oryzae. Associated with insect damage. **1:**FL(554,616).
N. sphaerica. Associated with insect damage. **1:**FL(554,616).
Ovularia sp. Associated with insect damage. **1:**FL(554).
Passalora sp. Associated with insect damage. **1:**FL(554).
Penicillium sp. Associated with insect damage. **1:**FL(554).
Periconia byssoides. **1:**FL(616).
Periconia sp. Associated with insect damage. **1:**FL(554).
Pithomyces chartarum. **1:**FL(616).
Scolecobasidium humicola. **1:**FL(616).
Spadicoides sp. Associated with insect damage. **1:**FL(554).
Sporobolomyces sp. **1:**FL(616).
Stemphylium vesicarium. **1:**FL(616).
Thysanophora longispora. **1:**FL(616).
Torula sp. Associated with insect damage. **1:**FL(554).
Trichoderma sp. Associated with insect damage. **1:**FL(554).
Deuteromycotina - Coelomycetes
Botryodiplodia sp. Associated with insect damage. **1:**FL(554,616).
Chaetomella sp. Associated with insect damage. **1:**FL(554).
Coniothyrium sp. Associated with insect damage **1:**FL(554).
Macrophoma sp. Associated with insect damage. **1:**FL(554).
Monochaetia sp. Associated with insect damage. **1:**FL(554).
Pestalotia sp. Associated with insect damage. **1:**FL(554).
Phlyctema sp. Associated with insect damage. **1:**FL(554).
Phoma sp. **1:**FL(616).
Sphaeropsis sp. Associated with insect damage. **1:**FL(554).
Deuteromycotina - Other
Rhizoctonia sp. Foliar blight. **1:**LA(1109).

Heteranthera Ruiz & Pav.

1. *H. dubia* (Jacq.) MacMill. [water stargrass]

Plasmodiophoromycetes
*Membranosorus heteranthera*e. Root hypertrophy. **1:**NY(82), VT(82).

Pontederia L.

1. *P. cordata* L. [pickerelweed]
2. *P. cordata* L. var. *lancifolia* (Muhl.) Torr. (=*P. lanceolata* Nutt.) [lanceleaf pickerelweed]

Ascomycotina
Mycosphaerella pontederiae. Leaf spot. **1:**NY(82), TX(82), VA(82).
Basidiomycotina - Uredinales
Uromyces pontederiae. Autoecious. **1:**DE(82), FL(35,82,552), GA(82), MO(82), NC(82), NY(82), PA(82), TX(82), VA(82). **2:**FL(552).
Deuteromycotina - Hyphomycetes
Cercospora sp. Leaf spot. **1:**FL(35).
Cercosporella pontederiae. Leaf spot. **1:**TX(82), WI(82).
Deuteromycotina - Coelomycetes
Phyllosticta sp. Leaf spot. **1:**FL(35).

Portulacaceae

Calandrinia Kunth in Humb., Bonpl. & Kunth

1. *C. breweri* S. Wats.
2. *C. ciliata* (Ruiz & Pav.) DC. (=*C. caulescens* H. B. K.) [redmaids]
3. *C. ciliata* (Ruiz & Pav.) DC. var. *menziesii* (Hook.) Macbr. (=*C. menziesii* (Hook.) Torr. & A. Gray)

Oomycetes
Peronospora claytoniae. **2:**CA(81), OR(3202), WA(3202).
Basidiomycotina - Ustilaginales
Ustilago calandriniae. **1:**CA(1077,3115). **2:**CA(1077). **3:**CA(1077).

Calyptridium Nutt. ex Torr. & A. Gray

1. *C. roseum* S. Wats.
2. *C. umbellatum* (Torr.) Greene (=*Spraguea umbellata* Torr.) [pussy-paws]
3. *C. umbellatum* (Torr.) Greene var. *caudiciferum* (A. Gray) Jeps. (=*Spraguea umbellata* Torr. var. *caudicifera* A. Gray)

Basidiomycotina - Uredinales
Uromyces unitus. Autoecious. **1:**OR(3202). **2:**MT(3202), OR(3202), WA(3202).
U. unitus subsp. *pacificus*. **2:**CA(3115), OR(3115).
U. unitus subsp. *spragueae*. Autoecious. **1:**OR(3115). **3:**CA(638,642,3115), OR(3115), WY(3115).

Claytonia L.

1. *C. acutifolia* Pall. ex Schult. subsp. *graminifolia* Hultén
2. *C. arctica* M. F. Adams
3. *C. caroliniana* Michx.
4. *C. lanceolata* Pursh
5. *C. megarrhiza* (A. Gray) Parry ex S. Wats. [spring-beauty]
6. *C. virginica* L. [spring-beauty]
7. *Claytonia* sp.

Chytridiomycetes
Physoderma australasica. **6:**MI(3115).
P. claytoniana. Leaf gall. **5:**MI(82), WI(82). **6:**WI(1331,1335,1342).
Physoderma sp. On flower parts, petioles and stems. **5:**UT(1844).
Synchytrium sp. On flowers and leaves. **5:**UT(1848).
Oomycetes
Peronospora claytoniae. Downy mildew. **6:**ID(82), KY(3115), MD(82), TX(82). **7:**CA(82), WA(82).
Ascomycotina
Pleospora helvetica. On dead stem. **1:**AK(1938).
Basidiomycotina - Uredinales
Puccinia agnita. Autoecious. **5:**CO(82,111,3115).
P. claytoniicola. Autoecious. **4:**WY(738,1195).
P. mariae-wilsoniae. **2:**AK(526). **3:**NH(82,111), NY(82,111), VT(82,111), WI(1357). **4:**OR(3202). **5:**CO(82,111). **6:**NE states(82), DC(111), DE(111), IA(111), IL(111), IN(111),

KY(111), MD(111), MI(111), MO(82,111), NJ(111), NY(111), OH(111), PA(111), TN(111), VA(111), VT(111), WI(82,111), WV(111). **7:**AK(82), UT(82), WA(82).
Uromyces claytoniae. Autoecious. **3:**NY(82,111,3115).
Basidiomycotina - Ustilaginales
Entyloma sp. **4:**WA(3202).
Deuteromycotina - Hyphomycetes
Phymatotrichopsis omnivora. Root rot. **3:**TX(82).
Ramularia claytoniae. Leaf spot. **7:**CA(82).

Lewisia Pursh
1. *L. columbiana* (Howell) B. L. Robinson
2. *L. nevadensis* (A. Gray) B. L. Robinson
3. *L. rediviva* Pursh [bitteroot]
4. *Lewisia* sp.
Ascomycotina
Comoclathris permunda. **1:**OR(3202), WA(3202).
Basidiomycotina - Uredinales
Uromyces unitus. Autoecious. **1:**MT(1195,3202), WA(111,3115,3202). **3:**CA(82), MT(82,111,3202), WA(3202). **4:**CA(82), WA(82).
U. unitus subsp. *montanensis*. Autoecious. **3:**MT(3115).
U. unitus subsp. *spragueae*. **1:**MT(3115).
Basidiomycotina - Ustilaginales
Ustilago calandriniae. **2:**CA(1077).

Montia L.
1. *M. cordifolia* (S. Wats.) Pax & O. Hoffm.
2. *M. fontana* L.
3. *M. linearis* (Douglas ex Hook.) Greene (=*Claytonia linearis* Douglas ex Hook.) [Indian lettuce]
4. *M. perfoliata* (Donn ex Willd.) Howell (=*Claytonia perfoliata* Donn ex Willd.) [miner's-lettuce, winter purslane]
5. *M. sibirica* (L.) J. T. Howell (=*Claytonia sibirica* L.) [Siberian purslane]
Oomycetes
Peronospora claytoniae. **4:**ID(3202), OR(3202), WA(3202).
Basidiomycotina - Uredinales
Puccinia mariae-wilsoniae var. *montiae*. Autoecious. **1:**CA(3115). **2:**ID(3202). **5:**WA(3115,3202).
Basidiomycotina - Ustilaginales
Tolyposporium montiae. **3:**ID(1077,3202), OR(1077,3202), WA(1077,3202).
Ustilago claytoniae. **3:**WA(82,1077,3115,3202). **4:**OR(3202).
Deuteromycotina - Coelomycetes
Phyllosticta sp. **1:**WA(3202).

Portulaca L.
1. *P. grandiflora* Hook. [eleven-o'clock, rose-moss, sunplant]
2. *P. lutea* Sol. ex G. Forster
3. *P. oleracea* L. [common purslane]
Oomycetes
Albugo portulacae. White rust. **3:**E states(82), FL(35), MS(2720), MT(82,3202), OK(2872), SD(2243), TX(82).
Pythium sp. Seedling damping-off. **1:**FL(35). **3:**FL(35).
Deuteromycotina - Hyphomycetes
Alternaria sp. Stem tip blight. **3:**FL(35).
Dichotomophthora lutea. Stem blight. **3:**VA(211).
D. portulacae. Black stem rot, on leaves. **1:**HI(2903). **2:**HI(2903). **3:**CA(1926), HI(982,2338,2903), TX(2452).
Dichotomophthora sp. Leaf necrosis. **3:**FL(35).
Drechslera portulacae. Damping-off, on stems and leaves. **1:**NC(3514). **3:**NY(2909,3115).
Fusarium sp. Stem tip rot. **3:**FL(35).
Helminthosporium sp. Leaf and stem blight. **3:**HI(2903).
Phymatotrichopsis omnivora. Root rot. **1:**TX(82).
Verticillium albo-atrum. Wilt. **1:**FL(35). **3:**FL(35).
Deuteromycotina - Coelomycetes
Pyrenochaeta sp. Basal stem rot. **3:**FL(35).
Deuteromycotina - Other
Rhizoctonia solani. Damping-off, basal stem rot. **1:**FL(35), IL(82), TX(82). **3:**FL(35).
Rhizoctonia sp. Root rot. **3:**HI(2903).

Portulacaria Jacq.
1. *P. afra* (L.) Jacq. [elephant-bush]
Oomycetes
Albugo portulacae. **1:**FL(35).
Deuteromycotina - Other
Rhizoctonia sp. Root rot. **1:**HI(2903).

Spraguea Torr.
1. *Spraguea* sp.
Basidiomycotina - Uredinales
Uromyces unitus subsp. *spragueae*. Autoecious. **1:**CA(82), OR(82), WY(82).

Talinum Adans.
1. *T. paniculatum* (Jacq.) Gaertn. [fameflower, jewels-of-Opar]
2. *T. rugospermum* Holz.
Basidiomycotina - Uredinales
Puccinia leptochloae. Heteroecious: aecial host. **1:**AZ(741,1203,3981).
Deuteromycotina - Coelomycetes
Phoma sp. On stems. **2:**WI(1357).

Potamogetonaceae

Potamogeton L.
1. *P. crispus* L. [curlyleaf pondweed]
2. *P. epihydrus* Raf.
3. *P. gramineus* L. (=*P. heterophyllus* Schreb.)
4. *P. natans* L.
5. *P. nodosus* Poir. (=*P. americanus* Cham. & Schlechtend.) [American pondweed]
6. *P. pectinatus* L. [sago pondweed]
7. *P. perfoliatus* L.
8. *P. spirillus* Tuck.
9. *Potamogeton* sp.
Oomycetes
Lagenidium muenscheri. Root hypertrophy. **8:**NY(82).
Pythium carolinianum. On tubers and turions. **1:**CA(260).
Basidiomycotina - Ustilaginales
Doassansiopsis hydrophila. **3:**MI(1077), WA(3202), WI(1357). **4:**MA(1077), ME(1077), NC(3944), WA(3202). **5:**NY(1077). **9:**E states(82), CT(1077), IL(82,1077), NY(1077), WA(82,1077), WI(82,1077).
D. occulta. **2:**CT(82,1077), MD(82,1077), NY(82,1077), VT(1077). **3:**WI(1077,1334). **5:**NY(1077). **9:**NE states(82), CA(82), IL(82,1077), KS(82,1077).
Deuteromycotina - Hyphomycetes
Botrytis cinerea. **5:**CA(260).
Cylindrocarpon sp. **5:**CA(260).
Fusarium crookwellense. **1:**CA(260). **5:**CA(260).
Geotrichum sp. **1:**CA(260).
Heliscus lugdunensis. **5:**CA(260).
Mycocentrospora acerina. **5:**CA(260).
Papulaspora aspera. **1:**CA(259,260). **5:**CA(259,260).
Ramularia aquatilis. Leaf spot. **5:**KS(82), NY(82).
Deuteromycotina - Other
Rhizoctonia solani. Stem rot. **6:**NC(82), VA(82). **7:**NC(82), VA(82).
Rhizoctonia sp. **5:**CA(12,260).

Primulaceae

Anagallis L.
1. *A. arvensis* L. [poorman's-weatherglass, scarlet pimpernel, shepherd's-clock]
Deuteromycotina - Coelomycetes
Septoria anagallidis. Leaf spot. **1:**NJ(82), TX(82).

Androsace L.
1. *A. chamaejasme* Jacq. ex Willd.
2. *A. chamaejasme* Jacq. ex Willd. var. *lehmanniana* (Spreng.) Hultén
3. *A. occidentalis* Pursh
4. *Androsace* sp.
Chytridiomycetes
Synchytrium shawii. On leaves, petioles, and stems. **3:**SD(1848).

Oomycetes
Peronospora candida. Downy mildew. **3**:KS(82).
Ascomycotina
Mycosphaerella primulae. Leaf spot. **4**:NM(82).
M. tassiana. **1**:AK(526).
Pleospora penicillus. **1**:AK(526).
Basidiomycotina - Uredinales
Puccinia volkartiana. Autoecious. **1**:AK(526). **2**:AK(60,526). **4**:AK(82).

Cyclamen L.

1. *C. persicum* Mill. [florist's cyclamen]
2. *Cyclamen* sp.

Ascomycotina
Glomerella cingulata. Bud and leaf blight, leaf spot. **1**:FL(35), IN(82), MA(82), MO(82), NC(1303), NJ(82), OH(82), PA(82), TX(82), VA(82).
Deuteromycotina - Hyphomycetes
Botrytis cinerea. Blight, gray mold. **1**:Range of host(82), CA(82), FL(35), NC(1303).
Cylindrocladiella peruviana. Root rot, basal stem rot. **1**:FL(35).
Fusarium oxysporum f. sp. *cyclaminis*. Wilt. **1**:CA(3653).
Fusarium sp. Wilt. **1**:NJ(82).
Ramularia cyclamanicola. Stunt. **1**:CA(82), IL(82), NJ(82), OH(82), PA(82). **2**:OK(2872).
R. cyclaminicola. Leaf spot. **1**:CA(82), IL(82), MN(82), OH(82).
Thielaviopsis basicola. Root rot. **1**:CT(82), MD(1867).
Deuteromycotina - Coelomycetes
Phyllosticta cyclaminicola. **1**:IL(82), OH(82), TX(82).
P. cyclaminis. **1**:VA(82).
Phyllosticta sp. Leaf spot. **1**:NJ(82).
Deuteromycotina - Other
Rhizoctonia solani. Leaf rot, on stems and petioles. **1**:FL(35).

Dodecatheon L.

1. *D. alpinum* (A. Gray) Greene [alpine shooting-star]
2. *D. clevelandii* Greene subsp. *sanctarum* (Greene) Abrams (=*D. macrocarpum* (A. Gray) R. Knuth)
3. *D. conjugens* Greene [American cowslip, shooting-star]
4. *D. frigidum* Cham. & Schlechtend.
5. *D. hendersonii* A. Gray [mosquito-bills, sailor-caps]
6. *D. meadia* L. [American cowslip]
7. *D. pulchellum* (Raf.) Merr. (=*D. amethystinum* (Fassett) Fassett, *D. pauciflorum* Greene, *D. thornense* Lunell)
8. *D. viviparum* Greene
9. *Dodecatheon* sp.

Basidiomycotina - Uredinales
Puccinia melanconioides. Autoecious. **5**:CA(81), OR(3202). **7**:ID(2705), OR(3202), WA(3202). **9**:CA(82,642), OR(82).
[*P. nodosa*]. **6**:CA(972).
P. ortonii. Autoecious. **1**:WA(3202). **2**:AK(60,526). **3**:OR(3202). **4**:AK(60,526). **5**:CA(2705). **7**:AK(526), ID(3202). **8**:AK(60,526). **9**:AK(82), CA(82), OR(82), SD(82), UT(82), WA(82).
P. solheimii. **7**:WY(82).
Uromyces acuminatus. Heteroecious: aecial host. **6**:NE(82). **7**:ND(378). **9**:ND(82).
Deuteromycotina - Hyphomycetes
Botrytis sp. **9**:OR(3202).
Heterosporium sp. Leaf spot. **9**:AK(82).
Deuteromycotina - Coelomycetes
Coniothyrium sp. **6**:OK(2872).
Phyllosticta dodecathei. Leaf spot. **6**:TX(82), WI(82).
Septoria dodecatheonis. **7**:WI(1356).

Glaux L.

1. *G. maritima* L. [sea-milkwort]

Basidiomycotina - Uredinales
Puccinia distichlidis. Heteroecious: aecial host. **1**:MT(82,3202), WY(82).
Deuteromycotina - Coelomycetes
Ascochyta sp. Leaf spot. **1**:AK(82).

Lysimachia L.

(= *Steironema* Raf.)

1. *L. ciliata* L. (=*Steironema ciliatum* (L.) Raf.)
2. *L. lanceolata* Walter (=*Steironema lanceolatum* (Walter) A. Gray)
3. *L. lunellii* (Greene) Hand.-Mazz. (=*Steironema lunellii* Greene)
4. *L. nummularia* L. [creeping-Charlie, creeping-Jennie, moneywort]
5. *L. quadrifolia* L.
6. *L. terrestris* (L.) B. S. P. [swamp loosestrife]
7. *L. thyrsiflora* L. [tufted loosestrife]
8. *L. vulgaris* L. [garden loosestrife]
9. *Lysimachia* sp. (=*Steironema sp.*)

Chytridiomycetes
Synchytrium aureum. Leaf gall. **6**:WI(82).
Ascomycotina
Botryosphaeria dothidea. **1**:GA(1438).
Discohainesia oenotherae. On leaves and stems. **1**:VA(82), WI(82).
Hainesia lythri. **2**:WI(1357). **4**:WI(1359).
Hymenoscyphus dearnessii. **1**:GA(1438).
H. scutula. **1**:GA(1438).
Kalmusia clivensis. **1**:GA(1438).
Leptosphaerulina americana. On leaves. **4**:VA(82).
Metasphaeria sp. **1**:GA(1438).
Mycosphaerella sp. Leaf spot. **1**:NY(82). **4**:FL(35).
Basidiomycotina - Uredinales
[*Coleosporium tussilaginis*]. **9**:TN(82).
Puccinia dayi. Autoecious. **1**:E states(82), GA(1439), IL(82), MT(82,3202). **9**:MI(82), WI(82).
P. distichlidis. **1**:CO(82), ND(82,378), TX(82).
P. limosae. Heteroecious: aecial host. **5**:MA(2704), NC(1303). **6**:E states(82), MI(82). **7**:IL(82), MA(82), NE(82), NY(82), WI(82). **9**:Unk.(82).
Uromyces acuminatus. **1**:NC states(82), NE states(82), W states(82), SD(82,2243). **3**:ND(378). **9**:ND(82).
Basidiomycotina - Other
Ceratobasidium anceps. On leaves. **1**:WI(1343,1350).
Deuteromycotina - Hyphomycetes
Cladosporium lysimachiae. Leaf spot, on leaves and stems. **6**:WI(1336). **8**:MA(82,1391).
Phymatotrichopsis omnivora. Root rot. **1**:TX(82). **9**:TX(82).
Pseudocercospora lysimachiae. Leaf spot. **6**:NJ(82).
Ramularia lysimachiae. Leaf spot. **1**:NC(82,3944), WI(82). **4**:TX(82). **6**:WI(82). **7**:WI(82). **9**:WI(82).
Deuteromycotina - Coelomycetes
Cylindrosporium steironematis. Leaf spot. **1**:NY(82).
Phoma exigua. **1**:TX(82), WI(64,1333).
Phyllosticta lysimachiae. Leaf spot. **1**:NY(82).
P. steironematis. **1**:MA(664), NY(64,82), PA(82), WI(82).
Phyllosticta sp. **4**:WI(1359).
Septoria conspicua. Leaf spot. **1**:E states(82), CO(82), IN(654), MS(82,2720), WA(3202), WI(82). **2**:ND(378). **4**:TX(82). **9**:IA(82), IL(82), ND(82), NY(82), VT(82), WI(82).
S. lysimachiae. On leaves. **1**:MI(82), NY(82), SD(2243). **4**:IA(1340). **6**:WI(1334). **7**:WI(1340).
Deuteromycotina - Other
Rhizoctonia solani. Leaf necrosis. **4**:FL(35).
Sclerotium deciduum. Stem necrosis. **1**:WI(82).
S. rolfsii. Southern blight, root and stem rot. **4**:FL(35).
S. rolfsii var. *delphinii*. Stem rot. **4**:KS(82).

Primula L.

1. *P. borealis* Duby
2. *P. bulleyana* Forrest
3. *P. elatior* (L.) J. Hill [oxslip]
4. *P. malacoides* Franch. [baby primrose, fairy primrose]
5. *P. mistassinica* Michx. var. *noveboracensis* Fernald
6. *P. obconica* Hance [German primrose]
7. *P. obconica* Hance cv. 'gigantea' [German primrose, poison primrose]
8. *P.* × *polyantha* Hort. A hybrid group probably having the parentage of *P. veris*, *P. elatior* and *P. vulgaris*. [polyanthus]
9. *P. sinensis* Sabine ex Lindl. [Chinese primrose]
10. *P. suffrutescens* A. Gray
11. *P. veris* L. [cowslip]
12. *P. vulgaris* Huds. [English primrose]
13. *Primula* sp.

Oomycetes
Pythium irregulare. Root rot. **6:**CA(82).
Ascomycotina
Cyclaneusma niveum. Needle cast. **3:**Unk.(82).
Erysiphe polygoni. Powdery mildew. **12:**CT(82). **8:**NJ(82), VA(82).
Mycosphaerella sp. Leaf spot. **13:**AZ(82).
Pleospora herbarum. **1:**AK(526).
Basidiomycotina - Uredinales
[*Puccinia aristidae*]. **3:**ME(82). **4:**ME(82). **6:**ME(82). **9:**ME(82). **12:**ME(82). **8:**ME(82). **13:**ME(82).
Uromyces apiosporus. Autoecious. **10:**CA(1468). **13:**CA(82), NV(82).
Deuteromycotina - Hyphomycetes
Alternaria sp. Stem rot. **13:**CT(82).
Botrytis cinerea. Gray mold blight. **3:**Unk.(82). **6:**Unk.(82). **7:**NC(1303). **9:**Unk.(82). **8:**Unk.(82). **13:**Unk.(82), FL(35).
Cercospora sp. **5:**WI(1356).
Phymatotrichopsis omnivora. Root rot. **6:**TX(82).
Ramularia primulae. Leaf spot. **3:**DE(82), WA(82). **9:**CA(82). **11:**WA(3202). **8:**CT(82), NY(82).
Deuteromycotina - Coelomycetes
Asteroma garrettianum. Black leaf spot. **13:**CO(82), UT(82).
Colletotrichum gloeosporioides. Anthracnose, leaf spot. **2:**NC(1303). **8:**FL(82). **13:**FL(35).
Phyllosticta primulicola. **6:**WI(1361).
Seimatosporium caudatum. On leaves. **13:**NM(82).
Septoria primulae. Leaf spot. **13:**FL(35).
Deuteromycotina - Other
Rhizoctonia solani. Root rot. **4:**IL(82). **13:**FL(35,82), TX(82).

Samolus L.
1. *S. ebracteatus* H. B. K.

Deuteromycotina - Coelomycetes
Septoria sp. Leaf spot. **1:**FL(35).

Trientalis L.
1. *T. arctica* Fisch. ex Hook. (=*T. europaea* L. var. *arctica* (Fisch. ex Hook.) Lehm.)
2. *T. borealis* Raf. (=*T. americana* Pursh) [starflower]
3. *T. europaea* L.
4. *T. latifolia* Hook. (=*T. europaea* L. var. *latifolia* (Hook.) Torr.)
5. *Trientalis* sp.

Chytridiomycetes
Synchytrium aureum. Leaf gall. **2:**PA(82).
Ascomycotina
Sphaerella sp. **4:**WA(3202).
Basidiomycotina - Uredinales
Puccinia karelica. Heteroecious: aecial host. **1:**AK(60,526). **5:**AK(82).
P. karelica subsp. *laurentina*. Autoecious. **2:**NY(3091), WI(3091).
Basidiomycotina - Ustilaginales
Urocystis trientalis. Leaf smut. **1:**AK(526). **3:**AK(1077). **4:**CA(81), OR(1077,3202), WA(1077,3202). **5:**AK(82), OR(82), WA(82).
Basidiomycotina - Other
Ceratobasidium anceps. Leaf rot. **5:**WI(82).
Deuteromycotina - Hyphomycetes
Cylindrocarpon magnusianum. Leaf spot. **2:**MA(82), MI(82), NY(82), WI(82,1351). **5:**CA(82).
Ramularia sp. **4:**WA(3202).
Deuteromycotina - Coelomycetes
Septoria trientalis. Leaf spot. **1:**ID(3202). **2:**ME(82), MI(82), NY(82), VT(82), WI(82). **4:**CA(81). **5:**CA(82).
Deuteromycotina - Other
Sclerotium deciduum. **2:**WI(1342).

Proteaceae

Banksia L. fil.
1. *B. media* R. Br.
2. *B. menziesii* R. Br.
3. *B. speciosa* R. Br.

Oomycetes
Phytophthora cinnamomi. Root rot. **2:**HI(2903). **3:**HI(579).
Pythium sp. Root rot. **1:**FL(35).

Grevillea R. Br. ex J. Knight
1. *G. robusta* A. Cunn. ex R. Br. [silk-oak]
2. *Grevillea* sp.

Oomycetes
Phytophthora sp. Root rot. **1:**FL(35).
Pythium sp. Root rot. **1:**FL(35).
Ascomycotina
Botryosphaeria quercuum. Dieback, gumming. **1:**FL(35).
B. ribis. On limbs. **2:**HI(82,2903).
Nectria pseudotrichia. **2:**FL(2782).
Rhytidhysteron rufulum. **1:**FL(2782).
Seuratia globifera. Anthracnose, black leaf spot. **1:**HI(82,2903). **2:**HI(82).
Basidiomycotina - Other
Armillaria mellea. **1:**HI(2011).
A. tabescens. Root rot. **1:**FL(35).
Ganoderma zonatum. Root and basal stem rot. **1:**FL(35).
Deuteromycotina - Hyphomycetes
Fusarium sp. Root rot. **1:**FL(35).
Phymatotrichopsis omnivora. Root rot. **1:**AZ(82). **2:**AZ(82).
Ramularia sp. Leaf spot. **1:**FL(35).
Deuteromycotina - Coelomycetes
Diplodia sp. Dieback, gumming. **1:**FL(35,82). **2:**FL(82).
Deuteromycotina - Other
Rhizoctonia solani. Basal stem rot. **1:**FL(35).

Hakea Schrad.
1. *H. petiolaris* Meisn.
2. *Hakea* sp.

Oomycetes
Pythium sp. Seedling dieback, root rot. **1:**FL(35).
Basidiomycotina - Other
Ganoderma brownii. On wood. **2:**CA(1204).

Leucospermum R. Br.
1. *L. cordifolium* (J. Knight) Fourc.
2. *L. nutans* R. Br. (=*Protea nutans* (R. Br.) Poir. ex Steud.)

Oomycetes
Phytophthora cinnamomi. Root rot. **1:**HI(2903).
Pythium sp. Root rot. **1:**HI(2903).
Deuteromycotina - Hyphomycetes
Botrytis cinerea. Flower and shoot blight. **1:**HI(578,2903). **2:**HI(2903).
Deuteromycotina - Other
Rhizoctonia sp. Root rot. **1:**HI(2903).

Macadamia F. v. Muell.
1. *M. integrifolia* Maiden & Betcke [Australian-nut, macadamia nut, Queensland-nut]
2. *M. tetraphylla* L. Johnson [small-fruited Queensland-nut]

Oomycetes
Phytophthora capsici. Leaf and shoot blight, raceme. **1:**HI(95,2000,2903).
P. cinnamomi. Trunk canker, root rot. **1:**CA(1110,4008), HI(1571,2903). **2:**CA(1110).
P. nicotianae var. *parasitica*. Nut and raceme blight. **1:**HI(1695,2903).
P. palmivora. Leaf and shoot blight. **1:**HI(95,1695,2903).
Pythium sp. Root rot. **1:**FL(35).
Ascomycotina
Kretzschmaria clavus. Root decay, decline. **1:**HI(1935,2903).
Xylaria sp. Root rot. **1:**HI(2903).
Basidiomycotina - Other
Ganoderma lucidum. On wood. **1:**HI(2903).
Pycnoporus sp. On wood. **1:**HI(2903).
Septobasidium pseudopedicellatum. Parasitic on scale insects. **1:**HI(2903).
Deuteromycotina - Hyphomycetes
Botrytis cinerea. Raceme blight. **1:**HI(1618,2903).
Fusarium sp. Root rot. **1:**FL(35).
Deuteromycotina - Coelomycetes
Colletotrichum sp. Anthracnose. **1:**FL(35).
Diplodia sp. Leaf spot. **1:**FL(35).
Macrophoma sp. Leaf spot. **1:**FL(35).
Pestalotia sp. Internal rot. **1:**FL(35).
Phomopsis sp. Twig dieback. **1:**FL(35).

Deuteromycotina - Other
Rhizoctonia solani. Damping-off, root rot. **1**:FL(35).
Rhizoctonia sp. Root rot. **1**:HI(2903).

Oreocallis R. Br.
1. *O. pinnata* (C. T. White) Sleum.
Oomycetes
Phytophthora sp. Root rot. **1**:FL(35).
Deuteromycotina - Other
Rhizoctonia solani. Root rot. **1**:FL(35).

Protea L.
1. *P. cynaroides* L. [giant protea, king protea]
2. *Protea* sp.
Oomycetes
Phytophthora cinnamomi. Root rot. **1**:HI(2903). **2**:HI(2903).
Pythium sp. Root rot. **2**:HI(2903).
Deuteromycotina - Coelomycetes
Phoma sp. Dieback. **2**:HI(2903).
Deuteromycotina - Other
Rhizoctonia sp. Root rot. **2**:HI(2903).

Stenocarpus R. Br.
1. *S. sinuatus* Endl. [firewheel-tree, wheel-of-fire]
Oomycetes
Pythium sp. Root rot. **1**:FL(35).
Deuteromycotina - Hyphomycetes
Fusarium sp. Root rot. **1**:FL(35).

Psilotaceae

Psilotum Sw.
1. *P. nudum* (L.) P. Beauv. [whisk-fern]
Deuteromycotina - Hyphomycetes
Gliocladium roseum. On stems. **1**:Unk.(1700).

Pteridophyta

Acrostichum L.
1. *A. aureum* L. [leather fern]
2. *Acrostichum* sp.
Deuteromycotina - Hyphomycetes
Cercospora sp. Leaf spot. **1**:FL(35).
Deuteromycotina - Coelomycetes
Pestalotia sp. **2**:HI(2903).

Adiantum L.
1. *A. pedatum* L. [American maidenhair fern, five-finger fern, northern maidenhair fern]
2. *A. raddianum* K. Presl cv. 'decorum' (=*A. decorum* T. Moore) [delta maidenhair fern]
3. *A. tenerum* Sw. [brittle maidenhair fern, fan maidenhair fern]
4. *Adiantum* sp.
Oomycetes
Pythium splendens. Root rot. **4**:FL(35).
Pythium sp. Root rot. **2**:HI(2903).
Ascomycotina
Glomerella sp. Leaf spot. **3**:FL(35). **4**:FL(35).
Mycosphaerella pteridis. Leaf spot. **3**:FL(35).
Mycosphaerella sp. Leaf spot. **1**:IA(82). **4**:FL(82).
Physalospora sp. Leaf spot. **4**:FL(35).
Psilachnum chrysostigmum. On rhizomes. **1**:MI(82).
Basidiomycotina - Other
Ceratobasidium anceps. **1**:WA(3202), WI(1343).
Sebacina incrustans. Encrusting living rachises. **1**:MD(128).
Deuteromycotina - Hyphomycetes
Corynespora cassiicola. Leaf spot. **3**:FL(35).
Deuteromycotina - Coelomycetes
Phyllosticta sp. Leaf spot. **1**:WI(1351). **4**:FL(35).
Deuteromycotina - Other
Rhizoctonia solani. Stem rot. **4**:FL(35).
Sclerotium deciduum. On rachises and pinnae. **1**:WI(1342). **4**:WI(82).

Aglaomorpha Schott
1. *A. meyeniana* Schott [bear's-paw fern]
Oomycetes
Pythium sp. Root rot. **1**:FL(35).
Deuteromycotina - Hyphomycetes
Cladosporium sp. Leaf necrosis. **1**:FL(35).

Alsophila R. Br.
1. *A. australis* R. Br. [Australian tree fern]
2. *A. quadripinnata* C. Chr.
3. *Alsophila* sp.
Oomycetes
Pythium splendens. Root rot. **1**:FL(35).
Pythium sp. Root rot. **3**:FL(35).
Ascomycotina
Cyanodermella candidum. On leaves. **2**:CA(81). **3**:AZ(82).
Basidiomycotina - Other
Armillaria mellea. Root rot. **1**:CA(82).
Deuteromycotina - Hyphomycetes
Botrytis cinerea. Gray mold of fronds. **1**:FL(35).
Cephalosporium sp. Dieback, wilt. **1**:FL(35).
Cercospora sp. Leaf spot. **3**:FL(35).
Fusarium sp. On fronds. **1**:FL(35).
Deuteromycotina - Other
Rhizoctonia solani. Foliage blight, stem rot. **1**:FL(35).

Angiopteris Hoffm.
1. *Angiopteris* sp.
Deuteromycotina - Hyphomycetes
Cercospora phyllitidis. Leaf spot. **1**:FL(35).

Asplenium L.
1. *A. nidus* L. [bird's-nest fern]
2. *Asplenium* sp.
Oomycetes
Pythium sp. Root rot. **1**:FL(35). **2**:FL(35).
Deuteromycotina - Hyphomycetes
Alternaria sp. Leaf spot. **1**:FL(35).
Cercospora asplenii. Leaf spot, brown streak. **1**:FL(35). **2**:FL(35).
Cercospora sp. Leaf spot. **2**:FL(82).
Fusarium sp. Leaf necrosis, root rot. **1**:FL(35). **2**:FL(35).
Myrothecium sp. Leaf spot. **1**:FL(35).
Deuteromycotina - Coelomycetes
Colletotrichum sp. Anthracnose. **1**:FL(35).
Phyllosticta sp. Leaf spot. **1**:FL(35).
Deuteromycotina - Other
Rhizoctonia solani. Leaf necrosis, root rot. **1**:FL(35). **2**:FL(35).
Sclerotium rolfsii. Southern blight. **1**:FL(35).

Athyrium Roth
1. *A. alpestre* (Hoppe) Rylands (=*A. americanum* (F. K. Butters) Maxon)
2. *A. felix-femina* (L.) Roth [ladyfern]
3. *A. felix-femina* (L.) Roth var. *michauxii* (Spreng.) Farw. (=*A. angustum* (Willd.) J. Presl) [northern ladyfern]
4. *A. felix-femina* (L.) Roth var. *sitchense* Rupr. (=*A. felix-femina* (L.) Roth var. *californica* F. K. Butters, *A. cyclosorum* Rupr.)
5. *A. thelypteroides* (Michx.) Desv. (=*A. acrostichoides* (Sw.) Diels) [silvery spleenwort]
6. *Athyrium* sp.
Ascomycotina
Hysterium magnosporum. On fronds, stalks and overwintered stems. **1**:CA(638,642).
Basidiomycotina - Uredinales
Uredinopsis acrostichoides. Telial host; aecial host unknown. **5**:NH(82,1575), NY(82,1575), WI(82,1575).
U. copelandii. Telial host; aecial host unknown. **2**:CA(82). **4**:CA(638,642).
U. longimucronata. Heteroecious: telial host. **2**:NE states(82), ID(3202), MT(3202), OR(3202), WA(3202), WI(82). **3**:CT(1575), MA(1575), ME(1575), NH(1575), NY(1575), PA(1575), RI(1575), VT(1575), WI(1575). **4**:AK(82,1575), CA(82,1575), ID(82,1575), MT(82,1575), OR(82,1575), WA(82,1575). **6**:OR(648).
U. struthiopteridis. Heteroecious: telial host. **2**:AK(60,526).

4:AK(526), CA(81).
Basidiomycotina - Other
Mycena lohwagii. On subterranean rhizomes. **2:**OR(2933), WA(2933).
M. pterigena. On fallen, senescent fronds. **2:**OR(2933), WA(2933).
Typhula athyrii. On overwintered stems. **3:**NY(2955).
Deuteromycotina - Hyphomycetes
Botrytis cinerea. Gray mold. **6:**AK(82).
Deuteromycotina - Coelomycetes
Phomopsis filicina. On living leaves. **3:**WI(1357).
Phomopsis sp. On dead, blackened portions of fronds. **3:**WI(1346).
Pleurothyrium longissimum. On dead leaf stalks. **4:**CA(81).
Septoria asplenii. Leaf spot. **6:**MI(82).

Blechnum L.

1. *B. gibbum* (Labill.) Mett.

Ascomycotina
Mycosphaerella filicum. On leaves. **1:**FL(82).

Botrychium Sw.

1. *B. multifidum* (S. G. Gmelin) Rupr. subsp. *silaifolium* (K. Presl) R. T. Clausen (=*B. multifidum* (S. G. Gmelin) Rupr. var. *intermedium* (D. C. Eaton) Farw.)
2. *B. virginianum* (L.) Sw. [rattlesnake fern]

Ascomycotina
Mycosphaerella botrychii. On fronds. **1:**AK(3095).
Mycosphaerella sp. On overwintered leaves. **2:**WI(1343).
Deuteromycotina - Coelomycetes
Phyllosticta sp. On fronds. **2:**WI(1344).

Camptosorus Link

1. *C. rhizophyllus* (L.) Link [walking fern]

Deuteromycotina - Hyphomycetes
Cercospora camptosori. Leaf spot. **1:**WI(82,1361).

Ceratopteris Brongn.

1. *C. thalictroides* (L.) Brongn. [floating water fern, water fern]

Oomycetes
Pythium sp. Leaf rot. **1:**FL(35).
Deuteromycotina - Hyphomycetes
Alternaria sp. Leaf rot. **1:**FL(35).

Cheilanthes Sw.

1. *C. pringlei* Davenp.

Basidiomycotina - Uredinales
Hyalopsora cheilanthis. Telial host. **1:**AZ(1203,1575,3981).

Cibotium Kaulf.

1. *C. chamissoi* Kaulf. (=*C. menziesii* Hook.) [hapuu-ii, Hawaiian tree fern]
2. *C. glaucum* (Sm.) Hook. & Arnott
3. *C. splendens* (Gaud.) Kraj. [blond tree fern, hapuu, Hawaiian tree fern, man tree fern]
4. *Cibotium* sp.

Oomycetes
Phytophthora cinnamomi. Root rot. **1:**HI(2903). **2:**HI(2903). **3:**HI(2903). **4:**HI(232).
Pythium sp. Root rot. **4:**FL(35).
Ascomycotina
Asterina sp. **4:**HI(2903).
Calloria cremea. **4:**HI(2903).
Coccodiella nuda. **1:**HI(2903).
C. polymorpha. Black spot. **1:**HI(2903). **4:**Unk.(82).
Dasyscyphus oncospermatis. **1:**HI(2903).
Lachnum flavidulum. **4:**HI(2903).
Phyllachora cibotii. Stipe spot. **2:**HI(1592,2903).
Psilachnum chrysostigmum. On leaf stalks. **4:**HI(82,2903).
Sphaerulina cibotii. Leaf spot. **1:**HI(2903). **4:**HI(82,2784,2903).
Stictis radiata. **4:**HI(2903).
Yoshinagella polymorpha var. *pauciseta*. **1:**HI(2903).
Basidiomycotina - Other
Athelia arachnoidea. On dead stems. **4:**HI(2903).
Epithele hydnoides. On petioles and dead stems. **1:**HI(341). **4:**HI(128,2903).
Deuteromycotina - Hyphomycetes
Chalara cibotti. **1:**HI(2550,2903).
Helminthosporium cibotii. Leaf spot. **4:**HI(82,2903).
Trichothallus hawaiiensis. Fly speck on dead leaves. **4:**HI(82,2903).
Deuteromycotina - Coelomycetes
Pestalotia cibotii. Leaf spot. **4:**NJ(82).

Cryptogramma R. Br. in Franklin

1. *C. acrostichoides* R. Br.
2. *C. crispa* (L.) R. Br. [European parsley fern, mountain parsley fern]
3. *C. stelleri* (S. G. Gmelin) Prantl (=*Pellaea gracilis* Hook.) [fragile cliffbrake, slender cliffbrake]
4. *Cryptogramma* sp.

Basidiomycotina - Uredinales
Hyalopsora cheilanthis. Telial host. **3:**IA(1575), MI(1575), MT(1575,3202), WI(1575). **4:**IA(82), MI(82), MT(82), WI(82).
Milesina darkeri. Heteroecious: telial host. **1:**WY(2924). **2:**OR(3202). **4:**CA(82), OR(82).
Deuteromycotina - Hyphomycetes
Chromosporium sp. On dead leaves. **1:**CA(642).

Cyathea Sm.

1. *C. arborea* (L.) Sm. [West Indian tree fern]
2. *C. cooperi* (F. v. Muell.) Domin (=*Sphaeropteris cooperi* (F. v. Muell.) R. Tryon) [Australian tree fern]
3. *Cyathea* sp.

Ascomycotina
Nectria filicina. On leaf stalks. **3:**CA(82).
Deuteromycotina - Coelomycetes
Stagonospora sp. On fading leaves. **1:**CA(82).
Deuteromycotina - Other
Rhizoctonia sp. Basal frond necrosis. **2:**FL(35).

Cystopteris Bernh.

1. *C. bulbifera* (L.) Bernh. [berry bladderfern, bulblet bladderfern]
2. *C. fragilis* (L.) Bernh. [brittle fern, fragile fern]
3. *C. fragilis* (L.) Bernh. var. *protrusa* Weath.

Ascomycotina
Pleospora helvetica. On dead leaves and stems. **2:**AK(1938).
Taphrina cystopteridis. Leaf gall. **2:**IN(82,2454), KS(82,2454,2458,3361), WI(82,1334,2454). **3:**KS(2457).
Basidiomycotina - Uredinales
Hyalopsora polypodii. Telial host. **2:**N states(82), Range of host(82), W states(82), AK(60,526), AZ(1195,1203,3981), CA(81,638,642), CO(649), IA(2315), IN(987), MS(2720), OH(638), SD(2769), WY(1195,2924).
Uredinopsis ceratophora. Telial host; aecial host unknown. **1:**IN(82), NY(82), WI(82). **2:**SD(2769).
U. glabra. Telial host; aecial host unknown. **2:**AZ(1195,1203), CO(2769), NM(82).
Deuteromycotina - Coelomycetes
Phyllosticta cystopteridis. On living leaves. **2:**WI(1360).
Phyllosticta sp. **1:**WI(1343).

Davallia Sm.

1. *D. trichomanoides* Blume [ball fern, squirrel-foot fern]
2. *D. trichomanoides* Blume f. *barbata* (Alderwerelt) Backer & Posth.
3. *Davallia* sp.

Oomycetes
Pythium sp. Root rot. **2:**FL(35).
Deuteromycotina - Hyphomycetes
Alternaria sp. Leaf spot. **2:**FL(35).
Botrytis cinerea. Gray mold. **2:**FL(35). **3:**FL(35).
Cercospora phyllitidis. Leaf spot. **2:**FL(35).
Curvularia sp. Petiole lesions. **1:**FL(35).
Fusarium sp. Rhizome and root rot, leaf spot. **2:**FL(35).
Deuteromycotina - Other
Rhizoctonia solani. Leaf blight. **3:**FL(35).
Sclerotium rolfsii. Southern blight. **3:**FL(35).

Dennstaedtia Bernh.

1. *D. punctilobula* (Michx.) T. Moore [boulder fern, hay-scented fern]

Ascomycotina
Hymenoscyphus scutula. **1:**VA(82).
Pezizella pteridina. On leaf stalks. **1:**PA(82).

Dicranopteris Bernh.
1. *D. linearis* (J. Burm.) Underw.
2. *Dicranopteris* sp.
Oomycetes
Phytophthora cinnamomi. Root rot. **1**:HI(2903).
Ascomycotina
Calloria cremea. **2**:HI(2903).
Dasyscyphus ulei. **2**:HI(2903).
Lachnum flavidulum. On rachis. **2**:HI(1417,2903).
Deuteromycotina - Hyphomycetes
Helminthosporium gleicheniae. **1**:HI(2903).
Deuteromycotina - Coelomycetes
Leptothyrium gleicheniae. **2**:HI(2903).

Doryopteris John Sm.
1. *D. pedata* (L.) Fée [hand fern]
2. *D. pedata* (L.) Fée var. *palmata* (Willd.) Hicken
Oomycetes
Pythium sp. Root rot. **2**:FL(35).
Deuteromycotina - Coelomycetes
Phyllosticta sp. Leaf spot. **1**:FL(35).
Deuteromycotina - Other
Rhizoctonia solani. On fronds. **1**:FL(35).

Dryopteris Adans.
1. *D. arguta* (Kaulf.) D. Watt [coastal woodfern]
2. *D. austriaca* (Jacq.) Woyn.
3. *D. austriaca* (Jacq.) Woyn. var. *spinulosa* (O. Müll.) Fiori (=*Aspidium spinulosum* (O. Müll.) A. Gray, *D. spinulosa* (O. Müll.) D. Watt) [fancy fern, florist's fern, spinulose wood fern]
4. *D. dilatata* (Hoffm.) A. Gray
5. *D. marginalis* (L.) A. Gray (=*Aspidium marginale* (L.) Sw.) [leather woodfern, marginal shieldfern]
6. *Dryopteris* sp.
Ascomycotina
Cryptomycina pteridis. Tar spot. **5**:WI(82). **6**:FL(35).
Incrupila aspidii. **4**:AK(3431).
Mollisia sp. **4**:AK(3431).
Taphrina californica. Leaf gall. **1**:CA(82,2454,2457), OR(82,2454,2457,3202).
T. filicina. According to Mix this in known only on (5) and its varieties from the vicinity of Ithaca, NY. **1**:CA(81). **3**:NY(82,2454), PA(82,2454).
T. fusca. **3**:NH(82,2454), VT(82,2454), WV(82,2454).
T. gracilis. **5**:NY(82,2454).
T. lutescens. Leaf gall. **6**:FL(35).
Basidiomycotina - Uredinales
Milesina dilatata. Heteroecious: telial host. **2**:OR(3202). **6**:OR(82).
M. fructuosa. Heteroecious: telial host. **3**:MA(82), ME(82), NH(82), NY(82), VT(82).
M. marginalis. Heteroecious: telial host. **5**:MA(82,1575), NH(82,1575), NY(82,1575), WI(1343).
Uredinopsis phegopteridis. Heteroecious: telial host. **2**:MT(3202).
Basidiomycotina - Other
Helicobasidium holospirum. **3**:NY(2456). **5**:NY(2456).
Deuteromycotina - Hyphomycetes
Cercospora camptosori. Leaf spot. **6**:FL(35).
Cylindrocladium pteridis. Leaf spot. **6**:FL(35,82).
Deuteromycotina - Coelomycetes
Colletotrichum sp. Leaf spot. **6**:FL(35).
Leptostromella filicina. On leaf stalks. **3**:WV(82).
Phoma sp. Leaf spot. **6**:FL(35).
Deuteromycotina - Other
Sclerotium rolfsii. Southern blight. **6**:FL(35).

Elaphoglossum Schott ex Sm.
1. *Elaphoglossum* sp.
Deuteromycotina - Hyphomycetes
Trichothallus hawaiiensis. **1**:HI(2903).
Deuteromycotina - Coelomycetes
Pestalotia sp. **1**:HI(2903).

Gleichenia Sm.
1. *G. glauca* (Thunb.) Hook.
2. *G. linearis* (N. L. Burm.) C. B. Clarke (=*G. dichotoma* Hook.) [savannah fern]
3. *G. longissima* Blume
4. *Gleichenia* sp.
Oomycetes
Phytophthora cinnamomi. **4**:HI(2903).
Ascomycotina
Calloria cremea. On stipes. **4**:HI(2903).
Dasyscyphus ulei. **1**:HI(2903). **4**:HI(2903).
Lachnum flavidulum. On stipes. **4**:HI(1417,2903).
Deuteromycotina - Hyphomycetes
Helminthosporium gleicheniae. **2**:HI(2903).
Deuteromycotina - Coelomycetes
Leptothyrium gleicheniae. On living fronds. **3**:HI(2903).

Gymnocarpium Newm.
1. *G. dryopteris* (L.) Newm. (=*Phegopteris dryopteris* (L.) Fée, *Dryopteris disjuncta* (Ledeb.) C. Mort., *Dryopteris linnaeana* C. Chr.) [oak fern]
Basidiomycotina - Uredinales
Uredinopsis phegopteridis. **1**:ME(82), NH(82), WI(82).
Basidiomycotina - Other
Herpobasidium filicinum. **1**:VT(1224), WI(1343).

Hicriopteris K. Presl
1. *H. pinnata* (Kunze) Ching
Ascomycotina
Dasyscyphus ulei. **1**:HI(2903).
Deuteromycotina - Coelomycetes
Leptothyrium gleicheniae. **1**:HI(2903).

Lygodium Sw.
1. *L. japonicum* (Thunb.) Sw. [Japanese climbing fern]
Basidiomycotina - Uredinales
Uredinopsis sp. **1**:FL(35).

Marsilea L.
1. *M. vestita* Hook. & Grev.
Chytridiomycetes
Physoderma marsiliae. Galls of leaflets and petioles. **1**:KS(384).
Basidiomycotina - Ustilaginales
Entyloma sp. **1**:WA(3202).

Matteuccia Tod.
1. *M. pensylvanica* (Willd.) Raym. (=*Pteretis nodulosa* (Willd.) Fernald, *Pteretis pensylvanica* (Willd.) Fernald)
2. *M. struthiopteris* (L.) Tod. (=*Onoclea struthiopteris* (L.) Hoffm.) [ostrich fern]
Ascomycotina
Taphrina hiratsukae. Leaf blister. **1**:WI(82,1348,2454).
Basidiomycotina - Uredinales
Uredinopsis struthiopteridis. **1**:NY(82), VT(82), WI(82). **2**:NY(1575), VT(1575), WI(1575).
Basidiomycotina - Other
Ceratobasidium anceps. On fronds. **1**:WI(82,1343).
Deuteromycotina - Other
Sclerotium deciduum. **2**:WI(1342).

Microlepia K. Presl
1. *Microlepia* sp.
Deuteromycotina - Hyphomycetes
Ramularia microlepiae. On fronds. **1**:HI(2903).

Microsorium Link
1. *M. punctatum* (L.) H. Copel. [crested fern]
Deuteromycotina - Other
Rhizoctonia sp. Root rot. **1**:HI(2903).

Nephrolepis Schott
1. *N. biserrata* (Sw.) Schott [sword fern]
2. *N. exaltata* (L.) Schott ex C. Chr.
3. *Nephrolepis* sp.

Oomycetes
Phytophthora sp. Root rot. **2:**FL(35).
Pythium splendens. Root rot. **2:**FL(35).
Pythium sp. Root rot. **1:**HI(2903). **2:**HI(2903). **3:**FL(35).
Ascomycotina
Glomerella nephrolepis. Anthracnose. **2:**NY(82), OH(82).
Basidiomycotina - Uredinales
Desmella aneimiae. **2:**FL(35).
Deuteromycotina - Hyphomycetes
Alternaria sp. On fronds. **2:**FL(35). **3:**FL(35).
Botrytis cinerea. On fronds. **2:**'Gray mold.' AK(82), FL(35). **3:**FL(35).
Cercospora phyllitidis. Leaf spot. **2:**FL(35). **3:**FL(35).
Cercospora sp. Leaf spot. **2:**IN(82).
Corynespora sp. Lower frond blight. **2:**FL(35).
Cylindrocladium pteridis. Brown leaf spot. **2:**FL(35,82). **3:**FL(35).
Fusarium sp. On fronds. **2:**FL(35).
Myrothecium sp. Frond dieback. **2:**FL(35).
Ramularia nephrolepis. Leaf spot. **2:**HI(2903).
Deuteromycotina - Coelomycetes
Phyllosticta sp. Leaf spot. **2:**FL(35). **3:**FL(35).
Deuteromycotina - Other
Rhizoctonia solani. On fronds. **2:**'Rhizoctonia aerial blight.' FL(35). **3:**FL(35).
Rhizoctonia sp. Foliage blight, damping-off, root rot. **2:**FL(82,1929). **3:**HI(2903).

Notholaena R. Br.
1. *N. sinuata* (Lag. ex Sw.) Kaulf. [waxy cloak fern]
Basidiomycotina - Uredinales
Hyalopsora cheilanthis. Telial host. **1:**AZ(1203,3981), NM(1575).

Onoclea L.
1. *O. sensibilis* L. [sensitive fern]
Ascomycotina
Dasyscyphus pteridis. **1:**MA(656).
Taphrina filicina. Leaf blister. **1:**Unk.(82).
T. hiratsukae. **1:**NY(82,2454), PA(82,2454).
Trichopezizella nidulus var. *onocleae*. On rachises. **1:**NY(1416).
Basidiomycotina - Uredinales
Uredinopsis americana. Heteroecious: telial host. **1:**NE states(82), CT(1575), DE(1575), IL(1575), IN(1575), MA(1575,1781), MD(1575), ME(1575), MI(1575), MN(82,1575), NC(1575), NE(82,1575), NH(1575), NJ(1575), NY(1575,2456), PA(1575), VA(1575), VT(1575), WA(1575), WI(1575), WV(1575).
Basidiomycotina - Other
Ceratobasidium anceps. On fronds. **1:**NY(82), WI(82,1342,1343).
Mycena pterigena. On fallen, senescent fronds. **1:**MI(2933), NY(2933).
Sebacina incrustans. Encrusting living rachises. **1:**CT(128).
Deuteromycotina - Coelomycetes
Phyllosticta sp. **1:**WI(1358).

Ophioglossum L.
1. *O. vulgatum* L.
Deuteromycotina - Hyphomycetes
Curvularia crepinii. Blight. **1:**OH(945).

Osmunda L.
1. *O. cinnamomea* L. [buckhorn, cinnamon fern, fiddleheads]
2. *O. claytoniana* L. [interrupted fern]
3. *O. regalis* L. [flowering fern, royal fern]
4. *Osmunda* sp.
Ascomycotina
Calycellina osmundae. On fronds. **4:**NJ(849,959,2181).
Cryptomycina filicina. On dead leaf stalks. **1:**ME(82), PA(82).
Dasyscyphus pteridis. **4:**NJ(849).
Psilachnum chrysostigmum. **1:**MA(656). **2:**MA(656).
Scirrhia osmundae. On dead leaf stalks. **3:**NY(82).
Stictis filicina. **4:**NY(3225).
Basidiomycotina - Uredinales
Uredinopsis osmundae. Heteroecious: telial host. **1:**E states(82), Lake states(82), AL(82), PA(1575). **2:**Lake states(82), NE states(82), PA(1575). **3:**E states(82), Lake states(82), AL(82), FL(35), MI(2862). **4:**FL(35), MA(1781).
Basidiomycotina - Ustilaginales
Mycosyrinx osmundae. Inflorescence smut. **1:**NY(82). **3:**MI(82), NY(82), WI(82).
Basidiomycotina - Other
Phlebiella filicina. **1:**NY(2456).
Typhula filicina. On dead leaf stalks. **1:**NY(82,264), VA(82).
Deuteromycotina - Hyphomycetes
Cercospora phyllitidis. Leaf spot. **3:**FL(35).
Deuteromycotina - Coelomycetes
Ascochyta osmundae. On living leaves. **2:**WI(1357).
Cylindrosporium sp. Leaf necrosis. **3:**FL(35).
Leptostromella filicina. On dead leaf stalks. **1:**MS(82), OH(82), PA(82), TN(82), VA(82). **3:**NY(82).
Leptothyrium litigiosum. On dead leaf stalks. **1:**GA(82), SC(82).
Phyllosticta sp. Leaf spot. **4:**FL(35).

Pellaea Link
1. *P. andromedifolia* (Kaulf.) Fée [coffee fern]
2. *P. glabella* Mett. ex Kühn [smooth cliffbrake]
3. *P. ovata* (Desv.) Weath. (=*P. flexuosa* Kaulf. ex Schlechtend. & Cham.)
4. *P. rotundifolia* (G. Forster) Hook. [button fern, New Zealand cliffbrake]
5. *Pellaea* sp.
Oomycetes
Pythium sp. Root rot. **4:**FL(35).
Ascomycotina
Hexagonella peleae. **4:**HI(2903).
Meliola juddiana. Black mildew. **4:**HI(1272,2903).
M. peleae. Black mildew. **4:**HI(1272,1452,2903).
Basidiomycotina - Uredinales
Hyalopsora cheilanthis. Telial host. **1:**CA(1575). **2:**CA(1575). **3:**TX(1575). **5:**CA(82), TX(82).
Deuteromycotina - Hyphomycetes
Fusarium sp. On fronds. **4:**FL(35).
Myrothecium sp. On fronds. **4:**FL(35).
Deuteromycotina - Other
Rhizoctonia solani. Basal frond rot, root rot. **4:**FL(35).

Phanerophlebia K. Presl
(= *Cyrtomium* K. Presl.)
1. *P. falcata* (L. fil.) H. Copel. (=*Cyrtomium falcatum* (L. fil.) K. Presl) [house hollyfern, Japanese hollyfern]
2. *Phanerophlebia* sp. (=*Cyrtomium sp.*)
Oomycetes
Pythium sp. Root rot. **1:**FL(35). **2:**FL(35).
Zygomycotina
Completoria complens. Damping-off of prothallia. **1:**NY(82).
Basidiomycotina - Uredinales
Milesina laeviuscula. **1:**CA(81).
Deuteromycotina - Hyphomycetes
Alternaria sp. Leaf spot. **1:**FL(35).
Botrytis cinerea. Gray mold. **1:**AK(82).
Cercospora sp. Leaf spot. **1:**FL(35).
Fusarium oxysporum. On fronds. **1:**FL(35).
Deuteromycotina - Coelomycetes
Phyllosticta sp. Leaf spot. **1:**FL(35).
Deuteromycotina - Other
Rhizoctonia solani. Root rot. **1:**FL(35). **2:**FL(35).

Pityrogramma Link
1. *P. calomelanos* (L.) Link [silver fern]
2. *P. sulphurea* (Sw.) Maxon [Jamaica gold fern]
3. *P. triangularis* (Kaulf.) Maxon [California gold-bark fern]
Basidiomycotina - Uredinales
Hyalopsora cheilanthis. **3:**CA(1575).
Deuteromycotina - Hyphomycetes
Cercospora sp. Leaf spot. **1:**FL(35).
Deuteromycotina - Other
Rhizoctonia sp. Basal necrosis. **1:**FL(35). **2:**FL(35).

Platycerium Desv.
1. *P. bifurcatum* (Cav.) C. Chr. [common staghorn fern]
2. *P. grande* Sm. ex Hook. & Baker
3. *Platycerium* sp.
Oomycetes
Phytophthora sp. Root rot. **1:**FL(35).

Pythium sp. Root rot. **1**:FL(35). **3**:FL(35).
Ascomycotina
Botryosphaeria sp. Leaf spot. **1**:FL(35).
Botryotinia sp. On leaves. **3**:FL(35).
Physalospora sp. Leaf spot. **1**:FL(35).
Deuteromycotina - Hyphomycetes
Alternaria sp. Leaf spot. **1**:FL(35). **3**:FL(35).
Botrytis sp. Gray mold. **3**:FL(35).
Cercospora platycerii. Leaf spot. **1**:FL(35). **3**:FL(35).
Cercospora sp. On fronds. **2**:HI(2903).
Fusarium sp. Leaf spot. **1**:FL(35).
Deuteromycotina - Coelomycetes
Actinopelte sp. Leaf spot. **3**:FL(35).
Colletotrichum sp. Leaf spot. **1**:FL(35). **3**:FL(35).
Macrophoma sp. Leaf spot. **3**:FL(35).
Phyllosticta sp. Leaf spot. **1**:FL(35).
Deuteromycotina - Other
Rhizoctonia solani. Leaf and root rot. 'Rhizoctonia aerial blight.' **1**:FL(35). **3**:FL(35).
Sclerotium rolfsii. Southern blight. **1**:FL(35).

Polypodium L.

1. *P. californicum* Kaulf. [California polypody]
2. *P. glycyrrhiza* D. C. Eaton (=*P. vulgare* L. var. *occidentale* Hook.) [licorice fern]
3. *P. hesperium* Maxon [western polypody]
4. *P. phyllitidis* L. [ribbon fern, strap fern]
5. *P. polypodioides* (L.) D. Watt [resurrection fern]
6. *P. virginianum* L. [American wall fern, rock polypody]
7. *Polypodium* sp.

Oomycetes
Pythium splendens. Root rot. **7**:FL(35).
Ascomycotina
Glomerella sp. Stem rot. **7**:FL(35).
Leptosphaeria sp. Leaf spot. **7**:FL(35).
Mycosphaerella sp. On fronds. **6**:WI(1345).
Sclerotinia sp. Leaf blight. **7**:FL(35).
Basidiomycotina - Uredinales
Milesina laeviuscula. Heteroecious: telial host. **1**:CA(81,1575). **3**:WA(3202). **7**:CA(82).
M. pycnograndis. Heteroecious: telial host. **6**:CT(82), MA(82), ME(82), NH(82), NY(82), PA(82), TN(82), VT(82,1224).
Deuteromycotina - Hyphomycetes
Cercospora phyllitidis. Leaf spot. **4**:FL(82). **5**:FL(35).
Cercospora sp. Leaf spot. **4**:FL(35). **7**:FL(35).
Fusarium sp. Rhizome rot. **7**:FL(35).
Deuteromycotina - Coelomycetes
Colletotrichum sp. Leaf spot. **7**:FL(35).
Phomopsis sp. Leaf spot. **7**:FL(35).
Phyllosticta sp. Leaf spot. **6**:VA(82). **7**:FL(35).
Deuteromycotina - Other
Rhizoctonia solani. Blight, leaf spot. **7**:FL(35).

Polystichum Roth

1. *P. acrostichoides* (Michx.) Schott [cankerbrake, Christmas fern, daggerfern]
2. *P. lonchitis* (L.) Roth [mountain hollyfern, northern hollyfern]
3. *P. munitum* (Kaulf.) K. Presl [giant hollyfern, Pacific Christmas fern, western swordfern]
4. *P. tsus-simense* (Hook.) Small

Oomycetes
Phytophthora cinnamomi. Root rot. **3**:CA(3627).
Ascomycotina
Helotium polystichi. On stems. **3**:CA(527).
Pezizella pteridina. On leaf stalks. **3**:CA(82).
Phyllachora filicina. Tar spot. **2**:AK(82,526,3095).
Taphrina faulliana. Leaf blister. **3**:ID(3202), OR(82,2012,2454,3202), WA(3202).
T. polystichi. On leaves. **1**:CT(82), GA(1438), MA(82), MD(82), ME(82), NC(82,1303,2454,3944), NY(82,2454,3361), OH(82,638,2454), PA(82,2454), TN(82,2454,3361), VA(82), WV(82,2454).
Basidiomycotina - Uredinales
Milesina vogesiaca. Heteroecious: telial host. **3**:CA(2769), OR(82,3202).
M. winelandii. Heteroecious: telial host. **3**:CA(81,1575), ID(82,3202), MT(82,3202), OR(82,3202), WA(82,3202).
Basidiomycotina - Other
Herpobasidium abnorme. On croziers. **3**:ID(2628).
Mycena fragillima. On debris and stems. **3**:CA(3326), OR(3326), WA(3326).
Sebacina incrustans. **1**:NY(2456).
Deuteromycotina - Coelomycetes
Chaetasbolisia falcata. **3**:CA(81,2434).
Phomopsis sp. On fronds. **3**:OR(2012,3202).
Phyllosticta sp. On fronds. **3**:OR(2012,3202).
Deuteromycotina - Other
Rhizoctonia sp. Root rot. **4**:FL(35).

Pteridium Gled. ex Scop.

1. *P. aquilinum* (L.) Kühn (=*Pteris aquilina* L.) [bracken fern, pasture brake]
2. *P. aquilinum* (L.) Kühn var. *decompositum* (Gaud.) R. Tryon
3. *P. aquilinum* (L.) Kühn var. *latiusculum* (Desv.) Underw. ex A. Heller (=*P. latiusculum* (Desv.) Hieron. ex Fr.)
4. *P. aquilinum* (L.) Kühn var. *pseudocaudatum* (Clute) A. Heller
5. *P. aquilinum* (L.) Kühn var. *pubescens* Underw. (=*P. aquilinum* (L.) Kühn var. *lanuginosum* (Bong.) Fernald, *Pteris aquilina* L. var. *lanuginosa* (Bong.) Kühn, *Pteris aquilina* L. var. *pubescens* (Underw.) Clute)
6. *P. caudatum* (L.) Maxon
7. *Pteridium* sp.

Ascomycotina
Albotricha washingtonensis. **1**:WA(3202). **7**:WA(849).
Cryptomycina pteridis. Leaf roll, tar spot. **1**:ID(3202), ME(3819), MT(3202), OR(3202), WA(3202). **3**:E states(82), IA(82), NC(3944), NY(131), WI(82). **4**:AL(82), GA(82), MS(82,2720). **5**:AK(82,526), CA(82,642,661), ID(82,638), MT(82), OR(82), WA(82).
Lachnaster miniatus. **1**:WA(1834).
Leptopeltis litigiosa. **1**:WA(3202).
Mycosphaerella pteridicola. On old leaves or stalks. **3**:NY(82).
M. pteridis. On old leaves and stalks. **1**:NY(195). **3**:MA(82), NY(82,195). **5**:CA(82).
Mycosphaerella sp. Leaf spot. **1**:FL(35).
Peltella sp. **1**:ID(3202).
Phyllachora flabella. Black mildew. **1**:GA(1438), WA(3202). **3**:MD(82), NJ(82), PA(82), TN(82), WI(82,1340), WV(82). **4**:GA(82), SC(82).
Rhopographus filicinus. On leaf stalks. **3**:ME(82), NJ(82), NY(82). **5**:CA(82).
Stictis radiata. **7**:CA(3225).
Basidiomycotina - Uredinales
Uredinopsis hashiokai. Telial host; aecial host unknown. **1**:HI(2903), OR(3202). **2**:CA(1575), HI(1575). **5**:CA(82), HI(82).
U. pteridis. Heteroecious: telial host. **1**:AZ(1195,1203), FL(35), GA(1439), ID(3202), MT(1195,3202), NC(1303), NM(2769), OR(648,3202), WA(3202). **3**:NC(3944), NJ(638). **4**:AK(1575), AL(82), FL(82), GA(82,1575), KY(1575), MD(1575), MS(82,1575,2720), NC(1575), NJ(1575), NY(1575), SC(1575), TN(1575), TX(1575), VA(1575). **5**:AZ(3379,3981), CA(81,638,642). **6**:FL(82), GA(82).
U. virginiana. Telial host; aecial host unknown. **3**:E states(82), TN(82). **4**:SC states(82), SE states(82).
Basidiomycotina - Other
Ceratobasidium anceps. **1**:OR(3202), WA(3202). **3**:WI(1343).
Nidula candida. **1**:OR(3202).
Deuteromycotina - Hyphomycetes
Cercospora pteridis. **3**:WI(1354).
Cladosporium grumosum. **3**:NY(2456).
Deuteromycotina - Coelomycetes
Camarosporium pteridis. On living leaves. **3**:WI(1359).
Cryptomycella pteridis. Leaf roll. **5**:CA(81,638,642), OR(657).
Leptostroma pteridis. On leaves. **3**:ME(82).
Leptostromella filicina. On leaf stalks. **7**:MI(82).
Leptothyrium litigiosum. On leaf stalks. **3**:NY(82). **5**:CA(82), WA(82).
Melasmia imitans. **1**:WY(1195).
Phyllosticta pteridis. Leaf spot. **7**:ME(82), NJ(82).
Phyllosticta sp. Leaf spot. **3**:WI(1360).
Septoria pteridis. **5**:Unk.(82).

Deuteromycotina - Other
Rhizoctonia sp. Scurf, on leaf stalks. **5:**OR(82).
Sclerotium deciduum. **1:**WI(1342).

Pteris L.
1. *P. cretica* L. [Cretan brake]
2. *P. ensiformis* J. Burm. [sword brake]
3. *P. excelsa* Gaud.
4. *Pteris* sp.

Oomycetes
Phytophthora cinnamomi. Root rot. **2:**FL(35).
Pythium intermedium. Damping-off. **1:**NY(82). **4:**NY(82).
P. splendens. Root rot. **2:**FL(35). **4:**FL(35).
Pythium sp. Root rot. **4:**HI(2903).
Zygomycotina
Completoria complens. Damping-off, on prothallia. **1:**NY(82). **4:**NY(82).
Ascomycotina
Dothidea flabella. **4:**SC(634).
Mycosphaerella olindensis. On rhizoids. **3:**HI(2784).
Rhopographus anceps. On decaying fronds. **4:**HI(2784).
Basidiomycotina - Uredinales
[*Uredinopsis pteridis*]. **4:**FL(35).
Deuteromycotina - Hyphomycetes
Alternaria sp. Leaf spot. **4:**FL(35).
Botrytis cinerea. 'Gray mold.' **2:**FL(35).
Cercospora phyllitidis. Leaf spot. **2:**FL(35).
Fusarium sp. Leaf necrosis. **2:**FL(35). **4:**FL(35).
Trichothecium roseum. Damping-off, on prothallia. **4:**IN(82).
Deuteromycotina - Coelomycetes
Phyllosticta pteridis. Tip blight, leaf spot. **1:**MS(82,2720), NJ(82).
Deuteromycotina - Other
Rhizoctonia solani. 'Rhizoctonia aerial blight.' **2:**FL(35). **4:**FL(35).

Rumohra Raddi
1. *R. adiantiformis* (G. Forster) Ching (=*Polystichum adiantiforme* (G. Forster) Small) [iron fern, leather fern]

Oomycetes
Pythium sp. Root and stem rot. **1:**FL(35), HI(2903).
Ascomycotina
Sclerotinia sclerotiorum. Root and stem rot. **1:**FL(35).
Basidiomycotina - Other
Armillaria tabescens. Root rot. **1:**FL(35).
Deuteromycotina - Hyphomycetes
Cercospora phyllitidis. Leaf spot. **1:**FL(35).
Cylindrocladium heptaseptatum. Decay, leaf spot. **1:**FL(35,2261).
C. pteridis. Decay, leaf spot. **1:**FL(35,82,2261,3936), HI(2903).
C. scoparium. Leaf spot. **1:**FL(35).
Deuteromycotina - Coelomycetes
Ascochyta necans. Leaf spot. **1:**FL(35).
Deuteromycotina - Other
Rhizoctonia solani. Root and stem rot. **1:**FL(35).
Rhizoctonia sp. Decay, root rot. **1:**FL(2261), HI(2903).

Sadleria Kaulf.
1. *Sadleria* sp.

Ascomycotina
Calycellina sadleriae. **1:**HI(849,2181,2903).
Psilachnum chrysostigmum. **1:**HI(2903).
Stictis radiata. **1:**HI(2903).
Basidiomycotina - Other
Lycoperdon sp. **1:**HI(2903).

Sphenomeris Maxon
1. *S. chusana* (L.) H. Copel. [lace fern]

Deuteromycotina - Hyphomycetes
Cercospora sp. Blight. **1:**HI(2903).

Stangeria T. Moore
1. *S. eriopus* (Kuntze) Nash

Oomycetes
Pythium splendens. Root rot. **1:**FL(35).

Struthiopteris Scop.
1. *Struthiopteris* sp.

Ascomycotina
Urceolella amphipila. On stalks. **1:**VT(1688).

Tectaria Cav.
1. *T. heracleifolia* (Willd.) Underw. [halberd fern]

Deuteromycotina - Hyphomycetes
Cercospora sp. Leaf spot. **1:**FL(35).

Thelypteris Schmidel
1. *T. noveboracensis* (L.) Nieuwl. [New York fern]
2. *T. palustris* Schott (=*Dryopteris thelypteris* (L.) A. Gray, *T. thelypteris* (L.) Nieuwl.) [marsh fern, meadow fern, snuffbox fern]
3. *T. palustris* Schott var. *pubescens* (G. Laws.) Fernald (=*Dryopteris thelypteris* (L.) A. Gray var. *pubescens* (G. Laws.) Nakai)
4. *T. tetragona* (Sw.) Small

Ascomycotina
Taphrina lutescens. **2:**ME(82,2454), MN(82), NY(82,2454), WI(82,1331,2454).
Basidiomycotina - Uredinales
Uredinopsis atkinsonii. **3:**E states(82), IN(1575), MI(1575), MS(82), MT(1575), ND(82), NE(82), WI(1575).
Basidiomycotina - Other
Ceratobasidium anceps. **2:**NH(82).
Sebacina incrustans. On leaves and stems. **1:**NC(1307).
Deuteromycotina - Hyphomycetes
Cercospora phyllitidis. Leaf spot. **4:**FL(35).
Deuteromycotina - Coelomycetes
Phyllosticta dryopteridis. On living fronds. **2:**WI(1359).

Woodsia R. Br.
1. *W. glabella* R. Br. [smooth woodsia]
2. *W. oregana* D. C. Eaton [Oregon woodsia]
3. *W. scopulina* D. C. Eaton [Rocky Mountain woodsia]

Basidiomycotina - Uredinales
Hyalopsora polypodii. Telial host. **1:**MI(82). **2:**ID(3202), WA(3202). **3:**ID(82).

Woodwardia Sm.
1. *W. areolata* (L.) T. Moore [nettle chain fern]
2. *W. fimbriata* Sm. [giant chain fern]
3. *W. radicans* (L.) Sm. [European chain fern]
4. *W. virginica* (L.) Sm. [Virginia chain fern]

Basidiomycotina - Uredinales
[*Uredinopsis americana*]. **1:**GA(1439).
U. arthurii. **4:**E states(82), IN(82), MI(82).
U. maculata. Telial host; aecial host unknown. **1:**E states(82), AL(1575), FL(35). **4:**E states(82).
Uredo acuta. Heteroecious: telial host. **2:**CA(81,1575).
Deuteromycotina - Coelomycetes
Chaetasbolisia falcata. **3:**CA(2434).

Punicaceae

Punica L.
1. *P. granatum* L. [pomegranate]
2. *P. granatum* L. cv. 'nana' (=*P. granatum* L. var. *nana* (L.) Pers.) [dwarf pomegranate]

Oomycetes
Phytophthora sp. Root rot. **1:**FL(35).
Pythium splendens. Root rot. **1:**FL(35).
Ascomycotina
Botryosphaeria dothidea. Canker. **1:**NC(1303).
Mycosphaerella lythracearum. Leaf blotch, fruit spot. **1:**SE states(82), AL(3937), FL(35,3937), MS(2720), TX(82,3937).
Nematospora coryli. Fruit rot. **1:**CA(82,1110).
Basidiomycotina - Other
Armillaria mellea. **1:**CA(1110).
A. tabescens. Root rot. **1:**FL(35,82).
Corticium stevensii. Thread blight. **1:**OK(2873).
Pellicularia koleroga. Thread blight. **1:**FL(82), LA(3640), OK(82).
Deuteromycotina - Hyphomycetes
Alternaria sp. Fruit rot. **1:**CA(82,1110).

Aspergillus niger. Internal rot. **1:**AZ(82), CA(82,1110), TX(82).
Botrytis cinerea. Gray mold rot. **1:**Range of host(82), CA(1110), FL(35).
Cercospora lythracearum. Leaf spot. **1:**FL(35).
Fusarium solani. **1:**CA(1110).
Penicillium expansum. Blue mold. **1:**Range of host(82), CA(1110), FL(35).
Phymatotrichopsis omnivora. Root rot. **1:**TX(82).
Pseudocercospora punicae. Leaf spot. **1:**FL(35), HI(2903).
Deuteromycotina - Coelomycetes
Colletotrichum gloeosporioides. Anthracnose, fruit spot. **1:**FL(35).
Colletotrichum sp. Anthracnose. **1:**FL(82).
Coniella granati. Fruit rot. **1:**NC(1303,1506).
Macrophoma sp. Dieback. **1:**FL(35).
Phomopsis sp. On twigs. **1:**OK(82,2873).
Phyllosticta sp. Leaf spot. **1:**FL(35).
Sphaceloma punicae. Scab. **1:**FL(35,2525).
Sphaceloma sp. **1:**LA(82).
Deuteromycotina - Other
Rhizoctonia solani. Foliage blight. **1:**CA(1110), FL(35).
Rhizoctonia sp. **2:**HI(2903).

Pyrolaceae

Chimaphila Pursh
1. *C. maculata* (L.) Pursh [spotted wintergreen]
2. *C. umbellata* (L.) Barton (=*C. corymbosa* Pursh)
3. *C. umbellata* (L.) Barton var. *cisatlantica* S. F. Blake [pipsissewa, prince's-pipe]
4. *Chimaphila* sp.

Ascomycotina
Mycosphaerella chimaphilae. Leaf spot. **2:**DE(82), ID(3202), MT(3202), NY(82), OR(82,3202), PA(82), WA(82,3202). **3:**MI(2699), NY(2699).
M. tassiana. **2:**WA(3202).
Stigmatea pyrolae. On leaves. **2:**NY(82).
Basidiomycotina - Uredinales
Pucciniastrum pyrolae. Telial host. **1:**E states(82), NC(3944), OR(1575), PA(1575), TN(82), WI(82). **2:**CA(82,638,642,936), GA(1439), MT(82,1195,3202), NY(82), OR(82,936,3202), PA(82), WA(82,936,3202), WI(82,1575). **4:**CA(936).
Basidiomycotina - Other
Thanatephorus cucumeris. On leaves. **1:**MA(82).
Deuteromycotina - Hyphomycetes
Ramularia chimaphilae. On living leaves. **2:**WI(1339,1353,2699).
Deuteromycotina - Coelomycetes
Septoria ellisiana. Leaf spot. **1:**DE(82). **2:**DE(2699).

Moneses Salisb. ex A. Gray
1. *M. uniflora* (L.) A. Gray (=*Pyrola uniflora* L.) [one-flowered pyrola, one-flowered shinleaf, woodnymph]
2. *M. uniflora* (L.) A. Gray var. *reticulata* (Nutt.) S. F. Blake

Basidiomycotina - Uredinales
Chrysomyxa monesis. Heteroecious: telial host. **1:**AK(4012), WA(3202,4012). **2:**AK(4012).
C. pirolata. **1:**AK(60,82,526), CO(82), ID(3202), ME(82), MT(82,3202), NM(82), WA(82,3202), WY(82). **2:**AK(526).

Pyrola L.
1. *P. asarifolia* Michx. [pink pyrola, pink wintergreen]
2. *P. asarifolia* Michx. var. *bracteata* (Hook.) Jeps. (=*P. bracteata* Hook.)
3. *P. asarifolia* Michx. var. *incarnata* (DC.) Fernald
4. *P. asarifolia* Michx. var. *purpurea* (Bunge) Fernald
5. *P. chlorantha* Sw.
6. *P. dentata* Sm. in Rees
7. *P. dentata* Sm. in Rees var. *integra* A. Gray
8. *P. elliptica* Nutt. [shinleaf, wild-lily-of-the-valley]
9. *P. glandiflora* Radius [arctic pyrola, arctic wintergreen]
10. *P. minor* L.
11. *P. pallida* Greene
12. *P. picta* Sm. [white-veined shinleaf]
13. *P. rotundifolia* L. [wild-lily-of-the-valley]
14. *P. rotundifolia* L. var. *americana* (Sweet) Fernald (=*P. americana* Sweet)
15. *P. secunda* L. (=*Ramischia secunda* (L.) Garcke) [one-sided pyrola, one-sided wintergreen]
16. *P. secunda* L. var. *obtusata* Turcz.
17. *P. virens* Schweigg. & Koerte
18. *Pyrola* sp.

Ascomycotina
Lophodermium exaridum. **1:**WA(936,3202).
L. maculare. **15:**AK(936,3431). **18:**AK(526).
L. pyrolae. **1:**AK(2699).
Mycosphaerella chimaphilae. Leaf spot. **15:**MI(82).
M. isariphora. **15:**CO(82).
M. minor. On dead stems. **9:**AK(1938).
Sphaerulina marginata. **15:**AK(526).
Basidiomycotina - Uredinales
Chrysomyxa pirolata. Heteroecious: telial host. **1:**NE states(82), Range of host(82), W states(82), AK(82,526), CA(82), IA(82), ID(3202), MN(82), MT(3202), OR(3202), WA(3202), WY(1195,3379). **2:**AK(526). **3:**AK(60,526). **5:**AK(60,526), WY(1195,3379). **8:**NE states(82), W states(82), AK(82), CA(82), IA(82), MN(82), WI(1355). **10:**AK(60,526), SD(2767). **14:**NE states(82), W states(82), AK(82), CA(82), IA(82), MN(82). **15:**NE states(82), W states(82), AK(60,82,526), CA(82), IA(82), ID(3202), MT(3202), OR(3202), WA(3202), WY(2924). **17:**AZ(1203,3981), MT(3202), WA(3202). **18:**Range of host(82), W states(82), AK(82,526), CA(82), IA(82), MN(82), WY(2924).
Pucciniastrum pyrolae. Telial host. **1:**NC states(82), NE states(82), W states(82), AK(82,526,3431), CA(82,936), MT(3202), WA(3202). **3:**AK(60,526). **5:**MT(1575), WI(1343), WY(1575). **6:**CA(1575). **7:**CA(638,642). **8:**NC states(82), NE states(82), W states(82), AK(82), CA(82). **10:**AK(60,526,936), CO(1575), ID(3202), MT(3202), OR(3202), WA(1575). **11:**NV(1575). **12:**AZ(1203,3981), CA(642), ID(936,3202). **13:**CA(81). **14:**NC states(82), NE states(82), W states(82), AK(82), CA(82), ND(1575), PA(1575), SD(1575,2243), WI(1575). **15:**AK(60,526,936,3431), CA(642,1575), CO(1575), ID(936,3202), MT(3202), OR(3202), WA(936,3202), WI(1575), WY(1195,2924). **16:**AK(1938). **17:**MT(3202), WY(1195,3379). **18:**NC states(82), NE states(82), W states(82), AK(82), CA(82).
Basidiomycotina - Other
Tomentella pirolae. On wood. **18:**NJ(2039).
Deuteromycotina - Hyphomycetes
Botrytis cinerea. Leaf blight. **14:**MD(82).
Deuteromycotina - Coelomycetes
Colletotrichum pyrolae. Leaf spot. **8:**WI(82,1341,2699). **13:**WI(2699). **14:**DE(82), MT(82), WI(82).
Phyllosticta pyrolae. Leaf spot. **1:**MT(3202). **4:**AK(2699). **8:**WI(82). **13:**DE(2,966). **14:**DE(82), MT(82).
Septoria chlorantha. On leaves. **5:**NY(2699).
S. ellisiana. Leaf spot. **15:**DE(2699), MI(82,974).
Strasseria geniculata. On leaves. **15:**NY(2546).
S. nigra. **15:**NY(2699). **17:**NY(2699). **18:**NY(2710).

Ranunculaceae

Aconitum L.
1. *A. carmichaelii* Debeaux [azure monkshood]
2. *A. columbianum* Nutt.
3. *A. delphinifolium* DC.
4. *A. maximum* Pall. ex DC.
5. *A. napellus* L. [aconite, bear's-foot, friar's-cap, garden monkshood, garden wolfbane, helmut-flower, soldier's-cap, Turk's-cap]
6. *A. noveboracense* A. Gray
7. *Aconitum* sp.

Oomycetes
Plasmopara pygmaea. Downy mildew. **2:**WY(1195). **7:**AK(82).
Ascomycotina
Erysiphe polygoni. Powdery mildew. **1:**NY(82), WV(82). **6:**WI(1358). **7:**TX(82).
Leptosphaeria ogilviensis. On dead stems. **3:**AK(1938).
Mycosphaerella tassiana. On dead stem. **3:**AK(1938).
Sclerotinia sclerotiorum. Stem rot. **1:**CO(82).
Trichopezizella relicina var. *macrospora*. **7:**CO(1416).

Basidiomycotina - Uredinales
Puccinia recondita. Heteroecious: aecial host. **3:**AK(60,526). **4:**AK(60,526). **7:**AK(82), CO(82).
Uromyces lycoctoni. Autoecious. **2:**ID(3202), WY(1195). **7:**CA(82), CO(82), TX(82), UT(82), WY(82).
Basidiomycotina - Ustilaginales
Urocystis anemones. **2:**UT(1077).
U. carcinodes. Leaf smut. **2:**CO(1077), UT(1077). **7:**UT(82).
U. sorosporioides. Leaf smut. **2:**UT(1077). **7:**UT(82).
Deuteromycotina - Hyphomycetes
Phymatotrichopsis omnivora. Root rot. **7:**TX(82).
Verticillium albo-atrum. Wilt. **1:**MA(82), NJ(82), NY(82). **5:**Unk.(82), MA(82).
Deuteromycotina - Coelomycetes
Septoria sp. **2:**ID(3202), WA(3202).
Deuteromycotina - Other
Rhizoctonia solani. Root rot. **5:**CT(82), NJ(82).
Sclerotium rolfsii var. *delphinii*. Stem rot. **5:**CT(82), DE(82), ME(82), MN(82), NJ(82), NY(82).

Actaea L.

1. *A. alba* (L.) Mill. [doll's-eyes, white baneberry, white cohosh]
2. *A. rubra* (Aiton) Willd. (=*A. arguta* Nutt. ex Torr. & A. Gray, *A. spicata* L. var. *arguta* (Nutt. ex Torr. & A. Gray) E. Murray, *A. eburnea* Rydb.) [red baneberry, snakeberry]
3. *Actaea* sp.

Ascomycotina
Crocicreas cyathoideum var. *cacaliae*. On stems. **2:**UT(505).
Nodulosphaeria modesta. **2:**WY(3250). **3:**AZ(662).
Basidiomycotina - Ustilaginales
Urocystis carcinodes. Leaf smut. **1:**PA(82,1077), WV(82,1077). **2:**AK(526), ID(82,1077,3202), OR(1077,3202), UT(82,1077). **3:**UT(82).
Deuteromycotina - Hyphomycetes
Ramularia actaeae. Leaf spot. **1:**IA(82), VT(82). **2:**AK(3431), CO(82), ID(3202), MT(3202), NM(82), WA(3202), WI(82), WY(1195). **3:**WI(82).
Deuteromycotina - Coelomycetes
Ascochyta actaeae. Leaf spot. **1:**WI(1342). **2:**ID(3202), WI(82).
Phyllosticta sp. **2:**ID(3202), MT(3202).

Anemone L.

1. *A. acutiloba* G. Laws. (=*Hepatica acutiloba* DC.)
2. *A. baldensis* L.
3. *A. canadensis* L. [meadow anemone]
4. *A. caroliniana* Walter
5. *A. coronaria* L. [poppy anemone]
6. *A. cylindrica* A. Gray [candle anemone, long-headed anemone, thimbleweed]
7. *A. deltoidea* Hook.
8. *A. drummondii* S. Wats.
9. *A. hupehensis* Hort. Lemoine var. *japonica* (Thunb.) Bowles & Stearn (=*A. japonica* (Thunb.) Siebold & Zucc.) [Japanese anemone]
10. *A. multifida* Poir. (=*A. globosa* Nutt.)
11. *A. narcissiflora* L.
12. *A. narcissiflora* L. var. *villosissima* (DC.) Welsh
13. *A. nemorosa* L. [European wood anemone]
14. *A. nuttalliana* DC. (=*A. patens* L. var. *wolfgangiana* (Besser) K. Koch, *Pulsatilla hirsutissima* Britton) [hartshorn-plant, lion's-beard, pasque-flower, prairie-smoke, wild crocus]
15. *A. occidentalis* S. Wats.
16. *A. oregana* A. Gray
17. *A. parviflora* Michx.
18. *A. patens* L. [American pasque-flower, prairie-smoke, wild crocus]
19. *A. patens* (L.) Miller subsp. *multifida* (Pritz.) Zam.
20. *A. piperi* Britton
21. *A. quinquefolia* L. [wood anemone]
22. *A. quinquefolia* L. subsp. *grayi* (Behr & Kellogg) E. Murray
23. *A. tuberosa* Rydb.
24. *A. virginiana* L. (=*A. riparia* Fernald) [thimbleweed]
25. *A. zephyra* A. Nels.
26. *Anemone* sp.

Chytridiomycetes
Synchytrium anemones. Leaf gall. **6:**IA(82), MN(82). **16:**WA(3202). **20:**WA(3202). **21:**DE(82), IA(82), MA(82), MI(82), VT(82), WI(82). **24:**VT(82).
Oomycetes
Plasmopara pygmaea. Downy mildew. **2:**OR(3202). **3:**NC states(82), NE states(82), ND(82,377). **4:**IA(82). **7:**OR(3202). **10:**ID(3202), MT(3202). **15:**WA(3202). **20:**ID(3202), WA(3202). **21:**NC states(82), NE states(82), WI(82). **24:**IL(82), NY(82), WI(82). **26:**MT(82), SD(82).
Ascomycotina
Comoclathris permunda. On dead stems. **15:**CA(642).
Crocicreas melanosporum. On overwintered peduncles, petioles and stems. **15:**CA(505,642).
Erysiphe communis. **3:**ND(377).
E. polygoni. Powdery mildew. **3:**IA(82), IL(82), ND(82), NJ(82), WI(82,1342). **24:**IA(82), MI(82), MN(82), SD(2243). **26:**IL(82), IN(82).
Mazzantia borealis. **26:**AK(526).
Mycosphaerella ranunculi. On dead stems. **8:**AK(1938). **17:**AK(526).
Mycosphaerella sp. On leaves. **3:**WI(1350).
Pleospora anthyllidis var. *abbreviata*. On leaves and stems. **26:**UT(82).
P. comata. **8:**AK(526).
P. penicillus. **15:**WA(3202).
Sclerotinia sclerotiorum. Wilt. **5:**FL(35).
Sphaerella sp. **16:**WA(3202).
Wettsteinina duplex. **26:**WA(3252).
W. macrospora. **15:**WA(3202).
Basidiomycotina - Uredinales
Aecidium ranunculacearum. **6:**ND(378).
Puccinia anemones-virginianae. Autoecious. **3:**IA(82), IN(647), MI(82), NY(82). **6:**NC states(82), NE states(82), WI(1330). **24:**E states(82), MS(82,2720), ND(82,378), OK(82,2873), WI(1333,1344). **26:**MS(2720).
P. gigantispora. Autoecious. **6:**CO(82), IL(82), ND(82,378), SD(2243), WI(82). **10:**AK(60,526), AZ(3379), CO(1195), ID(3202), MT(3202), WA(3202), WY(1195). **11:**AK(526). **26:**AK(82), CO(82), ID(82), MT(82), OR(3202), WY(82).
P. magnusiana. Heteroecious: aecial host. **3:**C states(82), NE states(82), ND(82,378), WI(1357).
P. pulsatillae. Autoecious. **2:**OR(3202). **11:**AK(60,526). **14:**AK(526), ND(378). **17:**AK(60,526). **18:**AK(82), CO(82), IA(82), ND(82), SD(82). **19:**AK(60). **26:**AK(82,1938), OR(82), WA(3202).
P. recondita. Heteroecious: aecial host. **3:**C states(82), NE states(82), CO(82), MT(82), SD(2243). **4:**E & C states(82), CO(82), MT(82). **6:**Unk.(82), CO(1195), MT(3202), SD(2243), WI(1342,1347). **10:**MT(3202), OR(3202), WA(3202). **20:**ID(3202), WA(3202). **21:**E & C states(82), CO(82), MT(82). **24:**E & C states(82), CO(82), MT(82), WI(1356). **26:**E & C states(82), CO(82), MT(82).
P. retecta. Autoecious. **11:**AK(60,526). **25:**CO(1195), WY(1195,1520). **26:**AK(82), CO(82).
P. vesiculosa. **12:**AK(526). **26:**AK(82).
Tranzschelia anemones. Autoecious. **10:**MT(3202). **14:**AK(526), ND(378), WI(1349). **16:**OR(3202). **17:**ID(3202). **18:**AK(82), CO(82), IA(82), MT(82,3202), ND(82), SD(82,2243). **19:**AK(60,526). **20:**ID(3202), WA(3202). **21:**NE states(82), CA(82), IL(82), MN(82), OR(82), VT(1224). **22:**CA(81). **26:**ID(82), OR(82), UT(82), WA(82), WY(82).
T. cohaesa. Autoecious. **9:**TX(82). **23:**AZ(744,747,1195,1203,3379,3981). **26:**AZ(82), TX(82).
T. discolor. Heteroecious: aecial host. **5:**CA(82), OR(82,3202).
T. pruni-spinosae. Heteroecious: aecial host. **1:**OH(638), WI(1333,1342). **4:**GA(1439), KS(82), LA(82), NE(82), TX(82). **21:**MA(1390).
Basidiomycotina - Ustilaginales
Entyloma ficariae. White smut. **21:**WI(82,1077).
Urocystis anemones. Leaf and stem smut. **1:**IA(1077), IL(1077), IN(1077), ME(1077), MN(1077), NY(1077), OH(638,1077), PA(1077), WI(1077). **3:**MN(82), NY(82,1077), WI(82,1077). **4:**KS(82,1077), SD(1077), TX(82,1077). **6:**CO(1077). **11:**AK(526). **13:**CT(1077), IA(1077), MA(1077), ME(1077), MI(1077), NY(1077), WI(1077). **14:**ND(378), WI(1349). **18:**CO(82,1077), ND(82,1077), NM(82,1077), SD(1077,2243). **20:**ID(3202). **21:**NE states(82), DE(1077), IA(82), MA(1077), MN(1077), NY(1077), OH(946), WI(82,1077). **24:**IA(82,1077), IN(1077), NY(1077), TX(1077). **26:**AK(82), CO(82), KS(82),

TX(82).
U. sorosporioides. On leaves, petioles and stems. **11:**AK(526,1077). **15:**CA(638,642,1077). **26:**AK(82).
Deuteromycotina - Hyphomycetes
Alternaria sp. On heads. **6:**WI(1344). **24:**WI(1351).
Botrytis cinerea. Collar rot. **5:**NJ(82), PA(82). **26:**NJ(82), PA(82).
Cercospora filiformis. Leaf spot. **14:**WI(1329,1350,1355). **18:**CO(82), WI(82).
Ramularia didyma. Leaf spot. **3:**IA(82), MI(82), NY(82), SD(2243), WI(82). **6:**IA(82), MI(82), WI(82). **24:**IA(82), MI(82), WI(82).
R. ovularioides. On leaves. **6:**WI(1337).
Deuteromycotina - Coelomycetes
Ascochyta vodakii. **1:**WI(1337).
Ascochyta sp. Leaf spot. **3:**WI(1361). **6:**WI(1354).
Phoma anemone. On stems. **5:**FL(35). **24:**FL(82).
P. regina. On stems. **9:**NY(82).
Phyllosticta anemonicola. Leaf spot. **3:**IL(82), MI(82), WI(82,1359). **6:**IL(82), NE(82), WI(82,1332). **9:**MD(82). **26:**MD(82), MI(82), VA(82).
P. ellisiana. **24:**VT(82).
Phyllosticta sp. **24:**WI(1358).
Rhabdospora camptospora. **11:**AK(526).
Selenophoma sp. **15:**WA(3202).
Septoria anemones. Leaf spot. **3:**IA(82), IL(82), WI(1345). **4:**MS(82,2720), TX(82,3597). **6:**WI(82,1331). **21:**WI(82,1330). **24:**IA(82), IL(82), MO(82), MS(82,2720), VT(82), WI(82,1330). **26:**MS(2720).
S. cylindrica. **6:**MT(82,3202). **10:**MT(3202). **24:**VA(82).
S. punicea. Leaf speck, leaf spot. **6:**IA(82), NE(82), WI(1344,1355). **21:**WI(1343). **24:**MA(82), MI(82), MS(82,2720), WI(82,1361). **26:**KS(82,973).
Stagonospora canadensis. On living leaves. **3:**WI(1341).
S. pulsatillae. On stems. **26:**AK(82,526).
Stagonospora sp. **2:**OR(3202).
Deuteromycotina - Other
Rhizoctonia sp. Damping-off, crown rot. **26:**NC(1303).
Sclerotium rolfsii. Southern blight. **5:**CA(82). **26:**MA(82), MD(82).

Anemonella Spach

1. *A. thalictroides* (L.) Spach (=*Anemone thalictroides* L.) [rue anemone]

Ascomycotina
Erysiphe polygoni. Powdery mildew. **1:**IA(82), WI(1339).
Basidiomycotina - Uredinales
Puccinia recondita. Heteroecious: aecial host. **1:**IA(82), IN(82), MO(82), WI(1338).
Basidiomycotina - Ustilaginales
Urocystis anemones. Leaf smut. **1:**IA(1077), MN(1077), NY(82,1077).
U. sorosporioides. **1:**IA(1077).
Deuteromycotina - Hyphomycetes
Cercospora caulophylli. Leaf spot. **1:**MO(82).
Deuteromycotina - Coelomycetes
Colletotrichum sp. On leaves. **1:**WI(1359).

Aquilegia L.

1. *A. buergerana* Siebold & Zucc.
2. *A. caerulea* E. James
3. *A. canadensis* L. [honeysuckle, meeting-houses, wild columbine]
4. *A. chrysantha* A. Gray
5. *A. flavescens* S. Wats.
6. *A. formosa* Fisch. ex DC.
7. *A. formosa* Fisch. ex DC. var. *truncata* (Fisch. & C. A. Mey.) Baker (=*A. truncata* Fisch. & C. A. Mey.)
8. *A. glandulosa* Fisch. ex Link
9. *A. jonesii* Parry
10. *A. leptocera* Fisch. & C. A. Mey.
11. *A. pubescens* Crép.
12. *A. vulgaris* L. [European columbine, European crowfoot, garden crowfoot]
13. *Aquilegia* sp.

Oomycetes
Pythium sp. Root rot. **12:**HI(2903).
Ascomycotina
Crocicreas cyathoideum var. *cacaliae*. On stems. **2:**CO(505), UT(505). **5:**UT(505). **13:**UT(505).
Erysiphe communis. **7:**CA(81).
E. polygoni. Powdery mildew. **2:**UT(82), WY(82,1195). **3:**IA(82), IL(82), IN(82), NY(82), OH(82), PA(82), WI(82,1330). **5:**MT(3202). **6:**AK(526), WA(3202). **7:**Unk.(3976). **8:**CA(1155). **9:**MT(3202). **11:**CA(1155). **12:**NJ(82), SD(2243). **13:**AK(82), TX(82).
Hyponectria sceptri. On dead stems. **2:**UT(3084).
Leptosphaeria erigerontis. **2:**WY(3810).
L. oreophila. **2:**WY(3810).
Lewia scrophulariae. **2:**WY(3810).
Mycosphaerella aquilegiae-jonesii. On leaves and stems. **13:**MT(82).
M. coerulea. On stems. **2:**CO(82,525,959).
M. tassiana. **9:**MT(3202).
Nodulosphaeria olivacea. **2:**WY(3250).
[*Phaeosphaeria eustoma*]. **2:**WY(3810).
Sclerotinia sclerotiorum. Stem rot. **12:**DE(82), OH(82), PA(82), TX(82).
Urceolella tuberculiformis. On stems. **2:**CO(849).
Basidiomycotina - Uredinales
Puccinia recondita. Heteroecious: aecial host. **5:**ID(3202), MT(3202). **7:**CA(81). **13:**CA(82), CO(82), ID(82), MT(82), OR(82), WA(82), WY(82).
Basidiomycotina - Ustilaginales
Urocystis sorosporioides. Leaf smut. **2:**UT(82,1077). **5:**UT(1077). **10:**UT(1077). **13:**UT(82).
Deuteromycotina - Hyphomycetes
Botrytis cinerea. Gray-mold blight. **12:**VA(82).
Cercospora aquilegiae. Leaf spot. **3:**KS(82), OR(3202). **6:**OR(3202). **12:**KS(82). **13:**OR(82).
Haplobasidion thalictri. Leaf spot. **12:**AK(82). **13:**AK(526).
Heterosporium sp. **6:**WA(3202).
Phymatotrichopsis omnivora. Root rot. **3:**TX(82).
Stemphylium lancipes. **13:**KS(1195,3379).
Stemphylium sp. Leaf spot. **2:**CA(149).
Deuteromycotina - Coelomycetes
Ascochyta aquilegiae. Leaf spot. **1:**WI(1355). **3:**IL(82), WI(82,1354). **5:**MT(3202). **6:**ID(3202). **12:**AK(82), CT(82), IA(82), MD(82), NJ(82), NY(82), PA(82), TX(82), WI(82). **13:**AK(82,526), CT(82), IA(82), MD(82), NJ(82), NY(82), PA(82), TX(82), WI(82,1343).
Kellermania alpina. On stems. **2:**CO(82).
Phoma sp. Stem rot. **12:**PA(82).
Phomopsis sp. On capsules and stems. **3:**WI(1361).
Phyllosticta aquilegicola. Stem rot. **12:**WA(82). **13:**WA(3202).
Phyllosticta sp. Leaf spot. **3:**WI(1361).
Rhabdospora clarkeana. On stems. **3:**NY(82,3081).
Septoria aquilegiae. Leaf spot. **3:**IN(82), MI(82), NY(82), OH(82), VT(82), WI(82). **4:**AZ(662). **12:**MI(82), OH(82), VA(82), WI(82). **13:**CT(82).
S. longispora. **3:**PA(82).
Sirexcipula wyomingensis. On stems. **2:**WY(3812).
Deuteromycotina - Other
Rhizoctonia solani. Root rot. **12:**IL(82). **13:**TX(82).
Rhizoctonia sp. Crown rot. **13:**NC(1303).
Sclerotium rolfsii. Southern wilt. **13:**OK(2872).

Caltha L.

1. *C. leptosepala* DC. (=*C. rotundifolia* Greene)
2. *C. palustris* L. (=*C. biflora* DC., *C. howellii* Greene) [cowslip, king's-cup, marsh marigold, May-blob, meadow-bright]
3. *C. palustris* L. var. *asarifolia* (DC.) Hultén
4. *Caltha* sp.

Chytridiomycetes
Physoderma johnsii. On leaves. **2:**MI(3400).
Synchytrium aureum. Leaf gall. **2:**WI(82).
Ascomycotina
Botryotinia calthae. On petioles, on leaf veins. **2:**NY(1518,1960).
Erysiphe polygoni. Powdery mildew. **2:**MI(82), OH(82), WI(82).
Pseudopeziza calthae. Leaf spot. **2:**CA(661), WI(82). **4:**CA(82).
Verpatinia calthicola. On overwintered petioles. **2:**NY(1960).
Basidiomycotina - Uredinales
Puccinia areolata. Autoecious. **1:**AK(60,526,2715). **2:**AK(526), WA(3202). **3:**AK(60,526). **4:**AK(82), CA(82), CO(82), WA(82).
P. calthae. Autoecious. **2:**NE states(82), IA(82), ND(82), SD(2243), WI(1346). **4:**IA(2716), ND(2716), NJ(2716), NY(2716).

P. calthicola. Autoecious. **2:**NC states(82), NE states(82), MN(82).
P. gemella. Autoecious. **1:**AK(60,526), ID(3202), WA(3202). **2:**CA(81), ID(3202), OR(3202), WA(3202). **4:**AK(82), CA(82), ID(82), MT(82), OR(82), WA(82).
P. treleasiana. Autoecious. **1:**CO(1195), MT(1520,3202), WA(3202), WY(1195,2924). **2:**WA(3202). **4:**CO(82), NV(82), UT(82), WA(82), WY(82).
Deuteromycotina - Hyphomycetes
Botrytis sp. On leaves. **2:**WI(1358).
Ramularia didyma. Leaf spot. **1:**WA(3202). **2:**NY(82), WI(82).
Deuteromycotina - Coelomycetes
Cylindrosporium sp. Leaf spot. **2:**NY(82).

Cimicifuga Wernisch.

1. *C. racemosa* (L.) Nutt. [black snakeroot]

Ascomycotina
Leptosphaeria clavigera. **1:**GA(1438).
Ophiobolus nigro-clypeata. **1:**GA(1438).
Basidiomycotina - Uredinales
Puccinia recondita. Heteroecious: aecial host. **1:**NC(3944).
Basidiomycotina - Ustilaginales
Urocystis carcinodes. **1:**NC(1077,1303), NY(1077), OH(1077), PA(1077), TN(1077), VA(1077).
Deuteromycotina - Hyphomycetes
Ectostroma afflatum. Leaf spot. **1:**VA(82).
Deuteromycotina - Coelomycetes
Ascochyta actaeae. Leaf spot. **1:**CT(82), NY(82).

Clematis L.

1. *C. columbiana* (Nutt.) Torr. & A. Gray (=*C. pseudoalpina* (Kuntze) A. Nels.)
2. *C. drummondii* Torr. & A. Gray
3. *C. hirsutissima* Pursh (=*C. douglasii* Hook.)
4. *C. integrifolia* L.
5. *C.* × *jackmanii* T. Moore. *C. lanuginosa* × *C. viticella*.
6. *C. lasiantha* Nutt.
7. *C. ligusticifolia* Nutt.
8. *C. occidentalis* (Hornem.) DC. (=*Atragene occidentalis* Hornem., *C. verticillaris* DC.)
9. *C. pitcheri* Torr. & A. Gray
10. *C. versicolor* Small ex Rydb. [leatherflower]
11. *C. viorna* L. [leather-flower, vase-vine]
12. *C. virginiana* L. [devil's-darning-needle, leatherflower, virgin's-bower, woodbine]
13. *Clematis* sp.

Chytridiomycetes
Synchytrium clematidis. Leaf gall. **13:**LA(627).
Oomycetes
Pythium sp. **13:**AL(82).
Ascomycotina
Botryosphaeria obtusa. On stems. **13:**NY(82).
Broomella montaniensis. On stems. **7:**MT(3202). **13:**CO(82), MT(82,3202), NM(82).
Clathrospora pentamera. **3:**WY(3810).
Comoclathris permunda. **3:**WY(3810). **7:**MT(3202).
Crocicreas cyathoideum var. *cacaliae*. On stems. **7:**UT(505).
Didymosphaeria futilis. **7:**WY(2924).
Erysiphe communis. **12:**GA(1438).
E. polygoni. Powdery mildew. **7:**CA(81), MT(3202), NM(1195), WA(3202). **11:**E states(82), CA(82), CO(82), WA(82). **12:**NY(2456). **13:**E states(82), CA(82), CO(82), FL(35), WA(82).
Glomerella cingulata. Leaf spot. **13:**FL(82).
Heptameria tenera. **3:**WY(3810).
Kalmusia clivensis. **12:**NY(3249).
Massarina microcarpa. On dead stems. **13:**NY(82).
Mycosphaerella applanata. Leaf spot. **7:**MT(965,3202). **13:**MT(82), TX(82).
M. dolichospora. On stems. **3:**WY(3811).
Mycosphaerella sp. On tendrils. **8:**WI(1333).
Otthia fruticola. On stems. **7:**MT(965,3202), WA(3202). **13:**CO(82), MT(82), NM(82).
Platystomum compressum var. *pseudomacrostomum*. **7:**WA(3202).
Pleospora vitalbae. **12:**ME(272).
Scirrhia insculpta. On stems. **13:**MT(82), NM(82).
Basidiomycotina - Uredinales
Puccinia pulsatillae. Autoecious. **6:**CA(82).
P. recondita. Heteroecious: aecial host. **2:**AZ(1203,3981). **3:**CO(1195), ID(3202), MT(3202), WA(3202), WY(1195). **7:**AZ(1203,3981), CA(638,642), CO(1195,3379), ID(3202), MT(1195,3202), OR(3202), SD(2243), WA(3202), WY(1195). **9:**IL(312). **11:**NC(3944). **12:**GA(1439), MI(2862), NC(1303), WI(1359). **13:**Unk.(82), NY(2456), OK(2872).
P. stromatica. Autoecious. **13:**AL(82).
Tranzschelia viornae. Autoecious. **13:**TX(82).
Basidiomycotina - Ustilaginales
Urocystis carcinodes. **1:**UT(82). **8:**UT(1077).
Deuteromycotina - Hyphomycetes
Botrytis elliptica. **3:**ID(3202).
Cercospora ligusticicola. **13:**WA(82).
C. squalidula. Leaf spot. **2:**OK(2873). **3:**ID(3202), MT(3202), WY(1195). **4:**WI(1330). **7:**CA(81,2439), MT(3202), OK(2873), OR(3202), WA(3202), WY(1195). **9:**OK(2872). **10:**OK(2873). **11:**NC(3944). **12:**DE(3543), OK(2872). **13:**AL(82), CO(82), CT(82), IA(82), LA(82), MA(82), MS(82), MT(82), NC(82,1303), ND(82), NE(82), NY(82), OK(82), OR(82), UT(82), WA(82), WI(82), WY(82).
Cladosporium herbarum. **7:**WA(3202).
Cladosporium sp. Leaf spot. **12:**OK(2872).
Didymaria clematidis. On leaves. **7:**CO(1195), MT(3202), WY(3378). **13:**CA(82), MT(82), UT(82).
Macrosporium clematidis. On leaves. **13:**KS(82).
Phymatotrichopsis omnivora. Root rot. **13:**TX(82).
Ramularia clematidis. Leaf spot. **1:**OR(3202), WA(3202). **7:**ID(3202), MT(3202). **13:**MT(82), TX(82).
R. saximontanensis. **1:**WY(1195). **13:**WY(82).
Deuteromycotina - Coelomycetes
Aplosporella clematidis. **13:**NY(82).
Ascochyta aquilegiae. Stem canker. **13:**NC(1303).
Cylindrosporium clematidis. Leaf spot. **12:**DE(962), MS(2720). **5:**NY(966). **13:**AL(82,285), CT(82), DE(82), IN(82), MI(82), MS(82), NY(82), WI(82).
Diplodia hortensis. On stems. **13:**MI(82), NY(82).
D. thalictricola. On stems. **7:**MI(82), NY(82). **13:**MI(82), NY(82).
Diplodia sp. On stems. **13:**MI(82), NY(82).
Dothiorella phomopsis. On stems. **13:**NM(82).
Hendersonia hortilecta. On stems. **13:**NY(82).
Pestalotia pestalozzioides. **7:**NM(3259).
Phloeospora adusta. Leaf blight. **13:**TX(82).
Phoma clematidina. Stem canker and rot, leaf spot. **7:**OR(3202). **12:**MS(2720), SD(2243). **13:**MS(82), NC(1303), NJ(82), NY(82), OH(82), OR(82), WI(82), WV(82).
Phyllosticta clematidis. Leaf spot. **7:**WA(3202). **13:**MT(82), NC(1303), UT(82), VA(82).
P. tetonensis. **3:**WY(1195).
Septogloeum sp. **13:**LA(82).
Septoria clematidis. Leaf spot. **3:**MT(3202). **7:**OR(3202), WA(3202). **12:**OH(653). **13:**NY(82), WA(82), WI(82).

Coptis Salisb.

1. *C. aspleniifolia* Salisb.
2. *C. groenlandica* (Oed.) Fernald [cankerroot]
3. *C. occidentalis* (Nutt.) Torr. & A. Gray
4. *C. trifolia* (L.) Salisb.
5. *Coptis* sp.

Ascomycotina
Mycosphaerella coptis. Leaf spot. **2:**ME(82), NY(82), VT(82).
Deuteromycotina - Hyphomycetes
Eurotium herbariorum. **4:**AK(526).
Deuteromycotina - Coelomycetes
Colletotrichum dematium. **4:**MA(656).
Phyllosticta helleboricola var. *coptidis*. Leaf spot. **4:**AK(64,526). **5:**AK(82).
Septoria coptis. Leaf spot. **1:**AK(526). **2:**MI(82), NY(82), VT(82), WI(82). **3:**ID(638,3202), WA(3202). **4:**WI(1333). **5:**AK(82), ID(82), WA(82).
Vermicularia coptina. Leaf spot. **2:**NY(82). **4:**WI(1329).

Delphinium L.

1. *D. ajacis* L. [rocket larkspur]
2. *D. barbeyi* (Huth.) Huth.
3. *D.* × *belladonna* Hort. ex Bergmans. *D. elatum* × *D. grandiflorum*.
4. *D. bicolor* Nutt.
5. *D. brownii* Rydb.
6. *D. burkei* Greene
7. *D. cardinale* Hook. [scarlet larkspur]
8. *D. consolida* L. [field larkspur]
9. *D.* × *cultorum* Voss. A name used for various garden plants of hybrid origin.
10. *D. decorum* Fisch. & C. A. Mey.
11. *D. depauperatum* Nutt.
12. *D. divaricatum* Ledeb.
13. *D. elatum* L. [candle larkspur]
14. *D. geyeri* Greene
15. *D. glareosum* Greene
16. *D. glaucum* S. Wats.
17. *D. grandiflorum* L. [bouquet larkspur]
18. *D. menziesii* DC.
19. *D. nuttallianum* Pritz. ex Walp. (=*D. pauciflorum* Nutt. ex Torr. & A. Gray)
20. *D. occidentale* (S. Wats.) S. Wats.
21. *D. reticulatum* Rydb.
22. *D. semibarbatum* Bien. ex Boiss. (=*D. zalil* Aitch. & Hemsl.)
23. *D. stachydeum* (A. Gray) A. Nels. & Macbr.
24. *D. tricorne* Michx. [dwarf larkspur]
25. *D. trolliifolium* A. Gray
26. *D. virescens* Nutt.
27. *Delphinium* × sp. (unnamed hybrids)
28. *Delphinium* sp.

Chytridiomycetes

Synchytrium aureum. Leaf gall. **28:**IA(82).

Oomycetes

Peronospora ficariae. **11:**WA(3202).
Phytophthora sp. Stem rot. **13:**MN(82).
Pythium aphanidermatum. Root rot. **1:**VA(82).
P. ultimum. Crown and root rot. **1:**CA(82), NY(82). **7:**CA(82), NY(82). **9:**CA(82), NY(82). **28:**MS(2720).
P. vexans. Root rot. **1:**CA(82).
Pythium sp. Root rot. **28:**FL(35).

Ascomycotina

Belonidium elegantulum. **28:**WA(3202).
Clathrospora pentamera. **5:**WY(3810).
Comoclathris permunda. **5:**WY(3810).
Crocicreas cyathoideum. On stems. **28:**WA(505).
C. cyathoideum var. *cacaliae*. On dead stems. **28:**UT(505).
C. nigrofuscum var. *allantosporum*. On stems. **20:**UT(505).
Diaporthe arctii. Stem canker. **1:**MD(82), NC(82), NY(82), OH(82), PA(82). **12:**IL(312). **9:**IL(312), NY(82).
Diaporthe sp. Stem canker. **28:**FL(35).
Didymella delphinii. On stems. **28:**NV(82).
Erysiphe cichoracearum. Powdery mildew. **9:**MA(82), MN(82), NY(82), SD(2243), WA(82). **28:**WA(3202).
E. polygoni. Powdery mildew. **1:**CA(82), FL(35). **12:**IL(312). **13:**ME(82), PA(82). **17:**NY(82). **26:**WI(1351). **3:**CA(81). **9:**Range of host(82), IL(312), WI(1334). **28:**ID(3202), OK(2872), OR(3202), WA(3202).
Hyalopeziza ciliata. **28:**WA(3202).
Leptosphaeria erigerontis. **5:**WY(3810). **21:**WY(2924).
L. oreophila. **5:**WY(3810).
Mycosphaerella delphiniicola. On stems. **28:**CO(82).
Sclerotinia sclerotiorum. Stem rot, wilt. **1:**MO(82), TX(82). **13:**DE(82), IL(82), MA(82), MI(82), OH(82), WA(82). **17:**DE(82), IL(82), MA(82), MI(82), OH(82), WA(82). **9:**DE(82), IL(82), MA(82), MI(82), OH(82), WA(82). **28:**FL(35), NC(1303), TX(82), WA(3202).
Sphaerotheca fuliginea. Powdery mildew. **1:**CA(82). **9:**CA(82). **28:**CA(82).
S. macularis. **8:**CA(81). **25:**CA(81).
Trichopezizella relicina. **23:**CO(1416). **28:**WA(1416).

Basidiomycotina - Uredinales

Puccinia delphinii. Autoecious. **4:**SD(2243). **28:**CA(82).
P. recondita. Heteroecious: aecial host. **4:**ID(3202), MT(3202), OR(3202), WA(3202). **11:**OR(3202). **14:**CO(1195), WY(1195). **15:**WA(3202). **18:**OR(3202), WA(3202). **19:**ID(3202), WA(3202). **20:**ID(1195,3202), WY(1195). **28:**SC states(82), CA(82), CO(1195,3379), NM(82), WA(82).

Basidiomycotina - Ustilaginales

Entyloma winteri. Leaf spot smut, white smut. **2:**WY(1077,1195). **19:**OR(3202). **20:**WY(1195,3378). **21:**WY(1077). **28:**CA(82), WY(82).
Urocystis sorosporioides. Leaf smut. **10:**CA(1077). **19:**OR(3202). **24:**KY(1077). **28:**AZ(1077), CA(82), KY(82), VA(82).

Deuteromycotina - Hyphomycetes

Botrytis cinerea. Gray mold blight, bud rot. **1:**NY(82). **9:**CT(82), MA(82), MS(82), NY(82), WI(82), WV(82). **28:**FL(35).
Botrytis sp. Crown and stem rot. **28:**NC(1303).
Cercospora delphinii. Leaf spot. **28:**CO(82), MO(82), TX(82).
Fusarium oxysporum f. sp. *delphinii*. Stem canker, wilt. **13:**NY(82). **17:**NY(82). **9:**NY(82). **28:**FL(35), WA(3202).
Fusarium sp. Root rot. **28:**OK(2872).
Heterosporium sp. **19:**WA(3202).
Ovularia delphinii. Leaf spot. **2:**WY(1195). **4:**WY(1195). **6:**WY(1195). **19:**ID(3202). **28:**WY(82).
Phymatotrichopsis omnivora. Root rot. **1:**TX(82). **28:**TX(82).
Pseudocercosporella herpotrichoides. On stems. **28:**ID(82), OR(82), WA(82).
Ramularia delphinii. Leaf spot. **4:**MT(3202). **11:**ID(3202), MT(3202), OR(3202), WA(3202). **19:**CA(642), ID(3202), OR(3202). **20:**ID(3202), MT(3202), OR(3202). **28:**CA(82), CO(82), UT(82).
R. sheldonii. On leaves. **20:**ID(3378), WY(1195). **28:**CO(82).
Sphaeromyces delphinii. On stems. **28:**UT(82).
Verticillium albo-atrum. Wilt. **1:**NY(82). **28:**WA(82,3202).
Volutella sp. Stem canker. **9:**MD(82).

Deuteromycotina - Coelomycetes

Ascochyta aquilegiae. Collar rot, leaf spot. **1:**CA(82). **13:**CT(82), NJ(82,2053), NY(82,2053). **22:**CT(2053). **9:**CT(82), NY(82). **27:** CT(2053). **28:**WI(1358).
Ascochyta sp. Leaf spot. **28:**FL(35).
Phoma delphiniicola. On stems. **28:**CO(82), TX(82).
Phoma sp. Stem canker. **9:**CT(82), NJ(82), NY(82).
Phyllosticta delphinii. **6:**WY(1195). **16:**WA(3202). **19:**ID(3202), OR(3202), WA(3202). **20:**MT(3202). **28:**CO(82).
Phyllosticta sp. Leaf spot. **9:**NY(82).
Septoria delphinella. Leaf spot. **1:**IL(82), SD(2243). **4:**SD(2243). **28:**KS(82), WI(82).

Deuteromycotina - Other

Rhizoctonia solani. Damping-off, root rot, sore shin. **1:**FL(35). **13:**Range of host(82). **17:**Range of host(82). **9:**Range of host(82). **28:**Range of host(82), NC(1303).
Sclerotium rolfsii. Southern blight, stem rot. **1:**FL(35), TX(82). **17:**FL(82). **9:**E states(82), CA(82), HI(82), NC(1303), TX(82). **28:**AL(285), HI(2903), MS(2720), NC(1303), TX(82).
S. rolfsii var. *delphinii*. Collar rot. **1:**KS(82), NC(82).

Helleborus L.

1. *H. niger* L. [Christmas-rose]

Deuteromycotina - Hyphomycetes

Botrytis cinerea. Flower spot. **1:**NJ(82).

Deuteromycotina - Coelomycetes

Coniothyrium hellebori. Black leaf spot. **1:**CA(2126), MD(82), NJ(2126), NY(82), OK(2126), OR(82,3202,4007).

Hepatica Mill.

1. *H. americana* (DC.) Ker-Gawl. (=*H. triloba* of authors)
2. *Hepatica* sp.

Oomycetes

Plasmopara pygmaea. Downy mildew. **2:**IL(82), NY(82), PA(82), WI(82).
P. pygmaea var. *fusca*. **1:**WI(1329). **2:**NY(82), WI(82).

Basidiomycotina - Uredinales

Tranzschelia pruni-spinosae. Heteroecious: aecial host. **1:**WI(1332). **2:**NE states(82), MN(82), TN(82).

Basidiomycotina - Ustilaginales

Urocystis anemones. Leaf smut. **1:**MN(1077), MO(1077), NY(1077), OH(1077), VA(1077). **2:**C states(82), NE states(82), MI(82), MO(82), WI(82).

Deuteromycotina - Coelomycetes
Colletotrichum hepaticae. On leaves. **1:**WI(1332). **2:**NY(82), WI(82).
Discosia artocreas. Secondary leaf spot. **2:**IA(82).
Septoria hepaticicola. Leaf spot. **2:**MI(82), NC(82).

Hydrastis L.

1. *H. canadensis* L. [goldenseal, orangeroot, tumeric]
2. *Hydrastis* sp.

Basidiomycotina - Other
Thanatephorus cucumeris. **2:**WA(3202).
Deuteromycotina - Hyphomycetes
Alternaria sp. Leaf blight. **1:**MI(82), NY(82), OH(82).
Botrytis sp. Leaf blight. **1:**E states(82), IN(82), WA(82), WI(82). **2:**WA(3202).
Fusarium sp. Wilt. **1:**IL(82), NY(82), OH(82), WA(82). **2:**WA(3202).
Phymatotrichopsis omnivora. Root rot. **1:**TX(82).
Streptobotrys streptothrix. **1:**DE(3543).
Deuteromycotina - Other
Rhizoctonia solani. Root rot. **1:**NC(82).

Isopyrum L.

1. *I. biternatum* Torr. & A. Gray
2. *I. hallii* A. Gray

Oomycetes
Plasmopara pygmaea. **2:**WA(3202).
Basidiomycotina - Uredinales
Puccinia recondita. Heteroecious: aecial host. **1:**IL(312), WI(1358).
Deuteromycotina - Hyphomycetes
Cercospora sp. On leaves. **1:**WI(1354).

Ranunculus L.

1. *R. abortivus* L. [kidneyleaf, small-flowered buttercup, small-flowered crowfoot]
2. *R. acriformis* A. Gray
3. *R. acris* L. [meadow buttercup, tall buttercup]
4. *R. adoneus* A. Gray
5. *R. alismifolius* Geyer ex Benth.
6. *R. andersonii* A. Gray (=*Beckwithia andersonii* (A. Gray) Jeps.)
7. *R. asiaticus* L. [Persian buttercup, turban buttercup]
8. *R. bulbosus* L. [bulbous buttercup]
9. *R. cardiophyllus* Hook.
10. *R. cymbalaria* Pursh (=*Oxygraphis cymbalaria* (Pursh) Prantl)
11. *R. eschscholtzii* Schlechtend.
12. *R. fascicularis* Muhl. ex Bigel. [early buttercup]
13. *R. ficaria* L. (=*R. calthaefolius* Jord.) [lesser celandine, pilewort, small celandine]
14. *R. flabellaris* Raf. (=*R. delphinifolius* Torr., *R. multifidus* Pursh) [yellow water buttercup, yellow water crowfoot]
15. *R. flammula* L.
16. *R. glaberrimus* Hook.
17. *R. hebecarpus* Hook. & Arnott
18. *R. inamoenus* Greene
19. *R. jovis* A. Nels.
20. *R. lobbii* (Hiern) A. Gray
21. *R. macounii* Britton
22. *R. macranthus* Scheele
23. *R. nivalis* L. (=*R. sulphureus* Sol. in Phipps)
24. *R. occidentalis* Nutt. (=*R. bongardii* Greene)
25. *R. orthorhynchus* Hook.
26. *R. parviflorus* L.
27. *R. pedatifidus* Sm.
28. *R. pensylvanicus* L. fil.
29. *R. pusillus* Poir.
30. *R. ranunculinus* (Nutt.) Rydb.
31. *R. recurvatus* Poir.
32. *R. repens* L. [butter-daisy, creeping buttercup, yellow-gowan]
33. *R. rhomboideus* Goldie
34. *R. sabinei* R. Br.
35. *R. sceleratus* L. (=*R. apiophyllus* St. Lag. in Cariot) [celery-leaved buttercup, cursed crowfoot]
36. *R. sceleratus* L. var. *multifidus* Nutt. in Torr. & A. Gray (=*R. eremogenes* Greene)
37. *R. septentrionalis* Poir. [swamp buttercup]
38. *R. uncinatus* D. Don ex G. Don
39. *Ranunculus* sp.

Chytridiomycetes
Physoderma sp. On leaves and stems. **3:**MI(3400). **20:**CA(3401). **37:**MI(3410).
Synchytrium anomalum. Leaf gall. **39:**IA(82).
S. aureum. **32:**IL(82). **39:**CA(82), WI(82).
S. cinnamomeum. **37:**WI(82). **39:**OR(3202), WI(82).
S. ranunculi. Leaf gall. **29:**LA(1839).
S. variabilum. Leaf and petiole galls. **24:**CA(1845).
Oomycetes
Peronospora ficariae. Downy mildew. **2:**WY(1195,3378). **3:**NE states(82), IA(82), MN(82), WI(1361). **8:**NE states(82), IA(82), MN(82). **32:**NE states(82), CA(82), IA(82), MN(82), WA(3202). **37:**NE states(82), IA(82), MN(82). **38:**ID(3202), WA(3202). **39:**NE states(82), AK(526), IA(82), MN(82), OR(3202).
P. hiemalis. Downy mildew. **3:**MA(82), NY(82).
P. pennsylvanica. Downy mildew. **39:**NY(82).
P. ranunculi. Downy mildew. **8:**MA(82). **32:**MI(82).
Pythium debaryanum. Root rot. **7:**CA(82).
P. irregulare. Root rot. **7:**CA(82).
P. ultimum. Root rot. **7:**CA(82).
Ascomycotina
Botryotinia ranunculi. On living leaves, petioles, and stems. **37:**NY(1518).
Crocicreas starbaeckii. **3:**NY(505).
Erysiphe polygoni. Powdery mildew. **1:**MS(2720), ND(377), NY(2456), SD(2243). **3:**E & C states(82), MT(3202). **10:**ND(377), WA(3202). **14:**Unk.(82). **15:**WA(3202). **22:**MT(3202). **25:**ID(3202), WA(3202). **26:**MS(2720). **28:**WI(1342). **32:**Unk.(82), MT(3202). **35:**MT(3202). **37:**WI(1351). **38:**OR(3202). **39:**E & C states(82), AK(526), MS(82), MT(82), NV(82), WI(1352), WY(82).
Leptotrochila ranunculi. Leaf spot. **3:**NY(82). **24:**AK(526). **37:**WI(82). **39:**AK(82), CA(82), NE(82), NY(82), WI(82).
Metasphaeria annae. **23:**AK(526).
Mycosphaerella ranunculi. **23:**AK(526).
Pleospora comata. **23:**AK(526).
P. herbarum. **23:**AK(526).
P. richtophensis. **39:**OR(3202), WA(3202).
P. richtophensis var. *pallida*. **39:**WA(3814).
Pseudopeziza singularis. On living leaves. **28:**IN(3255), WI(3255).
Sclerotinia sclerotiorum. Stem rot. **39:**AZ(82), CA(82).
Sphaerotheca macularis. Powdery mildew. **39:**CO(82).
Stigmatea ranunculi. On leaves. **39:**CA(82).
Basidiomycotina - Uredinales
Aecidium ranunculacearum. **35:**ND(378).
Puccinia andina. Autoecious. **37:**IL(82), IN(82).
P. blyttiana. Autoecious. **9:**AZ(1203). **11:**AK(60,526), WA(3202). **27:**WY(1520). **39:**AK(82), AZ(82), CO(82), UT(82), WA(82), WY(82).
P. eatoniae. Heteroecious: aecial host. **1:**GA(1439), MS(2720), NC(1303), ND(378), SD(2243). **12:**MS(2720). **39:**E states(82), CA(82), CO(82), MS(82), ND(82).
P. gibberulosa. **9:**AZ(3981).
P. recondita. Heteroecious: aecial host. **3:**C states(82), AK(82), CA(82), TX(82), WA(82). **10:**AK(60,526), MT(3202), ND(378), WY(1195). **16:**ID(638,3202), MT(3202), WA(3202). **17:**ID(3202), WA(3202). **18:**ID(3202). **24:**AK(60,526), ID(3202), WA(3202). **32:**C states(82), AK(82), CA(82), TX(82), WA(82). **37:**C states(82), AK(82), CA(82), TX(82), WA(82). **39:**C states(82), AK(82), CA(82), WA(82).
P. ustalis. Autoecious. **34:**AK(2714).
Uromyces dactylidis. Heteroecious: aecial host. **32:**MA(82).
U. dactylidis var. *poae*. Heteroecious: aecial host. **39:**CO(82), TX(82).
U. jonesii. Autoecious. **5:**CO(1195), MT(3202), WY(1195,3379). **13:**WY(1195,2924). **39:**CA(82), CO(82), MT(82), WY(82).
Basidiomycotina - Ustilaginales
Doassansia ranunculina. Leaf smut. **14:**IN(82,1077), MD(1077), WI(82,1077).
Entyloma ficariae. **14:**IN(82,1077), WI(1077). **24:**CO(1077), OR(1077), WA(1077,3202). **28:**WI(1077). **32:**VA(1077). **35:**MT(3202), OR(3202), WA(3202). **36:**OR(1077), WA(1077). **37:**ME(1077). **38:**ID(3202), MT(3202), OR(3202), WA(3202). **39:**MT(1077), OR(82,3202), WA(82), WI(82).
E. microsporum. White smut. **12:**WI(1077). **32:**IL(82,1077), KY(82), VA(82), WA(3202). **37:**IA(82,1077), IL(82,1077), IN(82,1077),

ME(1077), NY(1077), WI(82,1077). **39:**Unk.(82), IA(1077), OR(648,3202).
Urocystis anemones. leaf smut **4:**UT(1077). **11:**UT(1077). **12:**IL(1077). **19:**UT(1077). **30:**WY(1077). **35:**WY(1077). **37:**MO(1077). **39:**IL(82), UT(82), WY(82).
U. sorosporioides. **6:**NV(1077).
Basidiomycotina - Other
Ceratobasidium anceps. Leaf rot. **1:**WI(1356). **37:**WI(82).
Deuteromycotina - Hyphomycetes
Botrytis cinerea. Gray mold. **7:**CA(82), WA(82). **39:**NY(82), WI(82).
Botrytis sp. On leaves. **1:**WI(1329,1361). **31:**WI(1358). **33:**WI(1333).
Cercospora ranunculi. Leaf spot. **32:**IA(82), WI(82). **37:**IA(82), WI(82).
Ovularia decipiens. Leaf spot. **3:**NY(82). **21:**MT(3202), WY(1195). **25:**WA(3202). **32:**TX(82). **39:**NY(82), WY(82).
Phymatotrichopsis omnivora. Root rot. **32:**TX(82). **39:**TX(82).
Ramularia aequivoca. Leaf spot. **28:**WI(1336). **32:**IA(82), IL(82), WI(82). **35:**WI(1356). **37:**IA(82), IL(82), WI(82). **38:**OR(3202), WA(3202). **39:**AK(82,526), IA(82), IL(82), OR(82), WA(3202), WI(82).
R. didyma. Leaf spot. **3:**ME(82), NY(82), VT(82), WI(82). **12:**WI(1344). **21:**AK(526). **25:**WA(3202). **32:**IL(82), MA(82), WI(82). **37:**IA(82), IN(82), MI(82), MS(82,2720), NY(82), WI(82). **39:**Unk.(82), DE(3543), IA(82), IN(82), MD(82), NY(82), VT(82).
R. ranunculi-lyallii. **38:**WA(3202). **39:**WA(82).
Ramularia sp. **10:**ID(3202). **18:**MT(3202). **21:**WA(3202). **38:**ID(3202), MT(3202), OR(3202), WA(3202).
Sclerotiomyces colchicus. **37:**WI(1356).
Septocylindrium ranunculi. Leaf spot. **3:**NY(82). **38:**ID(3202). **39:**IL(82), NY(82), WA(82), WI(82).
Deuteromycotina - Coelomycetes
Ascochyta infuscans. Leaf spot. **39:**WI(82).
Cylindrosporium ranunculi. **21:**WY(2924).
Gloeosporium ficariae. Leaf spot. **39:**WA(82).
Heteropatella umbilicata. **23:**AK(526).
Phyllosticta sp. On leaves. **3:**WI(1352). **28:**WI(1346).
Septoria ficarioides. Leaf spot. **39:**NE(82).
S. polaris. Leaf spot. **37:**WI(82). **39:**LA(82), MS(82,2720), WI(82).
S. ranunculacearum. Leaf spot. **26:**MS(2720). **39:**LA(82), MS(82).
Septoria sp. Leaf spot. **3:**PA(82). **39:**NE(82).
Deuteromycotina - Other
Rhizoctonia solani. Basal stem rot. **39:**FL(35).
Sclerotium deciduum. **37:**WI(1342).
S. rolfsii. Stem rot. **39:**CA(82).

Thalictrum L.

1. *T. alpinum* L. [alpine meadowrue]
2. *T. aquilegifolium* L.
3. *T. dasycarpum* Fisch. & Avé-Lall. [purple meadowrue]
4. *T. dioicum* L. [early meadowrue, quick-silverweed]
5. *T. fendleri* Engelm. ex A. Gray
6. *T. glaucum* Desf.
7. *T. hultenii* Boivin
8. *T. macrostylum* Small & A. Heller
9. *T. occidentale* A. Gray
10. *T. polycarpum* (Torr.) S. Wats.
11. *T. polygamum* Muhl. [king-of-the-meadow, muskrat-weed, tall meadowrue]
12. *T. purpurascens* L.
13. *T. revolutum* DC.
14. *T. sparsiflorum* Turcz. ex Fisch. & C. A. Mey.
15. *T. thyrsoideum* Greene
16. *T. vegetum* Greene
17. *T. venulosum* Trel. (=*T. lunellii* Greene)
18. *Thalictrum* sp.

Chytridiomycetes
Synchytrium sp. On leaves. **9:**WA(1847,3202).
Oomycetes
Phytophthora infestans f. sp. *thalictri*. Downy mildew. **3:**WI(82). **11:**CT(82), NY(82), WI(82). **18:**WI(82).
Ascomycotina
Didymosphaeria thalictri. **8:**GA(1438).
Erysiphe polygoni. Powdery mildew. **2:**WI(1330). **3:**NE states(82), IL(82), ND(82). **6:**WI(1330). **10:**CA(1155). **11:**NE states(82), IL(82), ND(82). **12:**ND(377), SD(2243). **13:**WI(1329). **18:**NE states(82), IL(82), ND(82).
Leptosphaeria conoidea. **11:**NY(3249).
L. thalictri. **3:**ND(377).
Mycosphaerella fendleri. On stems. **18:**CO(82).
M. punctata. On stems. **11:**NY(82).
M. thalictri. Leaf spot. **3:**IA(82), WI(82,1329). **4:**IA(82), NJ(82), NY(82), VT(82), WI(1344). **11:**IA(82), NY(82), WI(82). **18:**IA(82), PA(82).
[*Phaeosphaeria eustoma*]. **9:**WY(3810).
Pyrenopeziza thalictri. On stems. **18:**NY(82).
Basidiomycotina - Uredinales
Puccinia cockerelliana. Heteroecious: aecial host. **18:**CO(82).
[*P. montanensis*]. **3:**ND(378).
P. recondita. Heteroecious: aecial host. **1:**CO(82). **2:**MI(82). **3:**NC states(82), NE states(82), W states(82), MT(3202), SD(2243). **4:**NC states(82), NE states(82), W states(82), MN(2102), MT(82,3202). **5:**AZ(1195,1203,3981), CA(81), CO(1195), ID(1195,3202), MT(3202), UT(1195). **7:**AK(60,526). **9:**CO(1195), ID(3202), MT(1195,3202), OR(3202), WA(3202), WY(1195). **10:**CA(81), OR(3202). **11:**NC states(82), NE states(82), W states(82), MT(82). **17:**MT(3202), WA(3202). **18:**NC states(82), NE states(82), W states(82), VT(1224), WA(82), WY(2924).
P. rhytismoides. **1:**AK(1938).
P. septentrionalis. Heteroecious: aecial host. **1:**AK(60,82,526). **12:**ND(378).
Tranzschelia pruni-spinosae. Heteroecious: aecial host. **3:**CO(82), IA(82), KS(82), ND(82), NE(82), SD(82). **4:**IA(82), IN(82), PA(82). **11:**OH(82). **12:**SD(2243).
T. thalictri. Autoecious. **3:**E & C states(82), MS(2720), ND(378). **4:**E & C states(82), MS(2720). **5:**AZ(82,1203,3981), CO(1195), UT(1195). **9:**ID(3202), MT(3202), OR(3202). **10:**CA(81). **11:**E & C states(82), MS(2720). **15:**ND(378). **16:**ND(378). **17:**MT(3202), ND(378). **18:**AK(526), CA(82), ID(82), NM(82).
Basidiomycotina - Ustilaginales
Entyloma ficariae. **3:**IL(1077), WI(1077).
E. thalictri. White smut. **3:**IL(82,1077), WI(82,1077). **4:**IN(82,1077), WI(82,1077). **11:**CT(82,1077), NH(82,1077). **18:**IL(82), NY(1077), WI(82).
Urocystis sorosporioides. Leaf and stem smut. **3:**MA(1077). **5:**UT(1077). **11:**NY(82,1077). **14:**UT(1077). **18:**AZ(82,1077), MA(82,1077), NY(82,1077), UT(82).
Deuteromycotina - Hyphomycetes
Botrytis sp. On leaves. **3:**WI(1348).
Cercospora filiformis. Leaf spot. **3:**WI(82,1329).
C. fingens. Leaf spot. **3:**WI(82). **4:**WI(82). **18:**IL(82).
Cercosporella sp. **9:**MT(3202).
Haplobasidion thalictri. **9:**WA(3202).
Deuteromycotina - Coelomycetes
Aplosporella thalictri. On stems. **18:**NY(82).
Ascochyta clematidina var. *thalictri*. Leaf spot. **3:**WI(82). **4:**WI(82).
Cylindrosporium thalictri. Leaf spot. **3:**IN(82), KS(82), SD(2243), WI(82,1334). **4:**WI(82). **9:**ID(3202), OR(3202). **12:**KS(963).
Diplodia thalictri. On leaves and stems. **3:**WI(1340). **11:**NY(82).
Gloeosporium thalictri. Leaf spot. **3:**WI(82).
Phoma nebulosa. On stems. **18:**CA(82).
P. spermoides. On stems. **3:**NY(82), PA(82). **18:**NY(832).
Phomopsis thalictri. On living leaflets. **3:**WI(1356).
P. thalictrina. On stems. **3:**NM(82). **18:**NM(82).
Phyllosticta sp. On leaflets. **3:**WI(1354).
Septoria thalictri. Leaf spot. **3:**KS(82).
Stagonospora sp. **18:**WI(1343).

Trautvetteria Fisch. & C. A. Mey.

1. *T. carolinensis* (Walter) Britton ex Vail (=*T. grandis* Nutt. ex Torr. & A. Gray) [tassel-rue]

Oomycetes
Peronospora ficariae. Downy mildew. **1:**TN(82).
Basidiomycotina - Uredinales
Puccinia pulsatillae. Autoecious. **1:**ID(3202), OR(3202), WA(3202).
P. trautvetteriae. Autoecious. **1:**ID(82,111), OR(82,111), WA(82,111).
Basidiomycotina - Ustilaginales
Urccystis anemones. Leaf smut. **1:**UT(82,1077).

Deuteromycotina - Hyphomycetes
Ramularia trautvetteriae. **1:**ID(3202), OR(3202), WA(3202).
Deuteromycotina - Coelomycetes
Septoria trautvetteriae. Leaf spot. **1:**WV(82).

Trollius L.

1. *T. laxus* Salisb. (=*T. albiflorus* (A. Gray) Rydb.) [spreading globeflower]
2. *Trollius* sp.

Ascomycotina
Mycosphaerella hypsicola. On stems. **1:**CO(82).
Basidiomycotina - Ustilaginales
Urocystis anemones. Leaf smut. **1:**NY(82). **2:**MD(82), NY(1077).
Deuteromycotina - Hyphomycetes
Cercosporella trollii. **2:**CO(649).
Deuteromycotina - Coelomycetes
Ascochyta sp. Leaf spot. **2:**NY(82).
Cylindrosporium montenegrinum. Leaf spot. **1:**WY(82,1195,2924).
Phomopsis trollii. On stems. **2:**NY(82).
Phyllosticta trollii. Leaf spot. **1:**CO(649), WY(82,1195,2924).

Xanthorhiza Marsh.

1. *X. simplicissima* Marsh. [shrub yellowroot]

Deuteromycotina - Coelomycetes
Phyllosticta xanthorrhizae. Leaf spot. **1:**NC(82), WV(82,959).

Resedaceae

Reseda L.

1. *R. odorata* L. [common mignonette]

Deuteromycotina - Hyphomycetes
Cercospora resedae. Leaf spot. **1:**E states(82), IA(82), MO(82), MS(2720).
Verticillium albo-atrum. Wilt. **1:**NY(82).
Deuteromycotina - Other
Rhizoctonia solani. Damping-off. **1:**CT(82), TX(82).

Rhamnaceae

Alphitonia Reissek ex Endl.

1. *A. excelsa* Reissek ex Endl.
2. *A. ponderosa* Hildebr.
3. *Alphitonia* sp.

Ascomycotina
Appendiculella splendens. Black mildew. **1:**HI(1272,2903). **3:**HI(1452).
Uleomyces sandicensis. **2:**HI(2903).

Berchemia Neck. ex DC.

1. *B. scandens* (J. Hill) Trel. (=*B. volubilis* (L. fil.) DC.) [Alabama supplejack, rattan-vine, supplejack]
2. *Berchemia* sp.

Ascomycotina
Biscogniauxia nummularia var. *rumpens*. On wood. **2:**LA(2403).
Botryosphaeria dothidea. **1:**GA(1438).
Diatrypella favacea. **1:**GA(1438).
D. quercina. **1:**GA(1438).
Eutypella goniostoma. **1:**GA(1438).
Gloniopsis praelonga. **1:**GA(1438).
Massarina berchemiae. **1:**MD(2780).
Ostreichnion curtisii. On dead limbs. **1:**SC states(82), SE states(82), LA(1438).
Rosellinia aquila. **1:**GA(1438).
Basidiomycotina - Uredinales
Puccinia coronata. Heteroecious: aecial host. **1:**SC states(82), SE states(82), MS(2720).
Deuteromycotina - Hyphomycetes
Paratomenticola lanceolatus. On wood. **1:**SC(128).

Ceanothus L.

1. *C. americanus* L. [Jersey tea ceanothus, mountain-sweet, New Jersey tea, wild snowball]
2. *C. arboreus* Greene [Catalina ceanothus, Catalina mountain-lilac, felt-leaf ceanothus]
3. *C. cordulatus* Kellogg [snowbush]
4. *C. cuneatus* (Hook.) Nutt. ex Torr. & A. Gray [buckbrush]
5. *C. dentatus* Torr. & A. Gray
6. *C. divergens* Parry
7. *C. fendleri* A. Gray
8. *C. gloriosus* J. T. Howell [Point Reyes ceanothus, Point Reyes creeper]
9. *C. greggii* A. Gray
10. *C. griseus* (Trel.) McMinn (=*C. thyrsiflorus* Eschsch. var. *griseus* Trel.) [carmel ceanothus]
11. *C. griseus* (Trel.) McMinn var. *horizontalis* McMinn [carmel creeper, Yankee point ceanothus]
12. *C. integerrimus* Hook. & Arnott [deerbrush, deerbush]
13. *C. integerrimus* Hook. & Arnott var. *macrothyrsus* (Torr.) Benson
14. *C. ovatus* Desf.
15. *C. rigidus* Nutt. [Monterey ceanothus]
16. *C. sanguineus* Pursh [wild lilac]
17. *C. thyrsiflorus* Eschsch. [blueblossum, bluebrush]
18. *C. velutinus* Douglas ex Hook.
19. *Ceanothus* sp.

Ascomycotina
Botryosphaeria dothidea. **1:**GA(1438).
B. obtusa. On twigs. **19:**Unk.(82).
B. ribis. On dead stems. **1:**Range of host(82).
Capitotricha bicolor. On dead limbs and twigs. **18:**CA(642,661).
Cenangium aureum. On dead limbs and stems. **18:**CA(661). **19:**CO(82).
Coniochaeta ligniaria. **18:**ID(3202).
Cucurbitaria ceanothi. On dead stems. **1:**NY(82).
Dasyscyphus cerinus. **18:**ID(3202).
Dasyscyphus sp. **16:**ID(3202). **18:**WA(3202).
Diapleella coniothyrium. **18:**WA(3202).
Diaporthe ceanothi. On stems. **1:**Unk.(82), NY(82). **19:**KS(82).
Diatrype flavovirens. **4:**OR(3202).
D. stigma. **17:**CA(81). **19:**CA(81).
Didymella wehmeyeri. On dead stems. **1:**NY(82).
Didymosphaeria ceanothi. On twigs. **19:**CA(82).
D. housei. On dead stems. **1:**NY(82,830).
Erysiphe polygoni. Powdery mildew. **5:**CA(1155). **6:**CA(1155). **8:**CA(1155). **13:**CA(1155). **15:**CA(1155).
Eutypa armeniacae. Canker, limb dieback. **19:**CA(2460).
Glyphium corrugatum. **18:**WA(3202).
Hymenopleella sp. **18:**CA(661).
Hypoxylon mediterraneum. On dead stems. **18:**CA(661).
H. serpens. On dead limbs. **18:**CA(638,642).
Jattaea ceanothina. On limbs and stems. **1:**GA(82), NY(82,190).
Lachnella alboviolascens. **12:**AZ(1202).
Leptosphaeria ceanothi. On twigs. **19:**CA(82).
Microsphaera penicillata. Powdery mildew. **1:**Range of host(82), GA(82,1438), WI(1343). **8:**Unk.(3976). **16:**ID(3202), WA(3202). **18:**ID(3202), WA(3202).
M. sydowiana. **18:**WA(365).
Mollisia cinerea. **3:**CA(661).
Mycothyridium ceanothi. On dead twigs. **1:**NY(82).
Nectria coccinea. **18:**ID(3202).
N. sanguinea. On dead stems. **3:**CA(81).
Perrotia succina. On twigs. **18:**CA(81,642,661). **19:**CA(82).
Stictis chrysopsis. On stems. **12:**WA(3202,3226). **19:**WA(3225).
S. radiata. **19:**CA(3225).
Tapesia fusca. On twigs. **4:**CA(81). **18:**CA(642).
Tapesia sp. **18:**ID(3202).
Thyridium tuberculatum. On dead limbs. **19:**CA(82).
Vertixore atronitidum. On living leaves. **10:**CA(81,2434).
Xylaria hypoxylon. **18:**WA(3202).
Xylopezia hemisphaerica. On wood. **18:**OR(3233).
Basidiomycotina - Uredinales
Puccinia tripsaci. Heteroecious: aecial host. **19:**CO(82), KS(82), NE(82), WI(82).
Basidiomycotina - Other
Acanthophysium diffissum. **9:**AZ(1202).
A. fruticetorum. On twigs and stems. **18:**CA(642,2083).
Antrodia albida. **12:**AZ(1202).
Armillaria mellea. Root rot. **2:**CA(2898). **11:**CA(2900). **18:**ID(3202), MT(3202), OR(3202), WA(3202). **19:**CA(82).

Ceriporia viridans. **12:**AZ(1202).
Cytidia stereoides. On wood. **18:**CA(640,642).
Dendrothele incrustans. **17:**CA(2084).
Hymenochaete arida. On dead limbs. **7:**AZ(1202). **18:**CA(661).
H. cinnamomea. **19:**CA(81).
H. spreta. On wood. **18:**CA(642).
H. tabacina. On limbs. **18:**CA(638,642), ID(3202).
Hypochnicium punctulatum. **12:**WA(3202).
Polyporus varius. **18:**CA(642).
Schizophyllum commune. **19:**CA(82).
Scytinostroma galactinum. **17:**CA(128).
Stereum sp. On wood. **18:**CA(642).
Deuteromycotina - Hyphomycetes
Cercospora ceanothi. Leaf spot. **1:**C states(82). **19:**CA(2439), KS(82), WI(82).
C. macclatchieana. Leaf spot. **2:**CA(581). **14:**WI(1343).
Septonema secedens. **18:**ID(3202).
Verticillium albo-atrum. Wilt. **12:**CA(1470).
Deuteromycotina - Coelomycetes
Camarosporium ceanothi. On dead stems. **1:**MD(82).
Cryptocline ceanothi. On living leaves. **1:**WI(1344).
Cytospora ceanothi. On twigs. **1:**PA(82).
Harknessia aggregata. On leaves. **19:**UT(82).
Microdiplodia ceanothi. On dead stems. **1:**NY(82).
Myxofusicoccum peckianum. On dead stems. **1:**NY(82).
Phloeosporella ceanothi. Leaf spot. **16:**OR(3202), WA(3202). **18:**ID(3202), OR(3202), WA(3202). **19:**W states(82), CA(3549), CO(3549), IA(3549), OR(3549), WA(3549).
Phyllosticta ceanothi. Leaf spot. **1:**MS(82,2720).
Phyllosticta sp. On leaves. **1:**WI(1354).
Septoria ceanothi. Leaf spot. **16:**ID(3202), WA(3202). **18:**ID(3202). **19:**ID(82).

Colubrina Rich. ex Brongn.

1. *C. arborescens* (Mill.) Sarg. [wild coffee]
2. *C. elliptica* (Sw.) Briz. & F. C. Stern
3. *Colubrina* sp.

Basidiomycotina - Uredinales
Phakopsora sp. **1:**FL(35). **2:**FL(35).
Deuteromycotina - Hyphomycetes
Cercospora sp. Leaf spot. **3:**FL(35).

Condalia Cav.

1. *C. mexicana* Schlechtend. [Mexican bluewood, Mexican condalia]
2. *C. spathulata* A. Gray [squawbush]

Basidiomycotina - Other
Antrodia heteromorpha. **1:**AZ(1202).
Dendrophora albobadia. **1:**AZ(1202).
Peniophora tamaricicola. White rot. **1:**AZ(128,1180,1202). **2:**AZ(1202).
Phanerochaete tuberculata. **1:**AZ(1202).

Gouania Jacq.

1. *Gouania* sp.

Basidiomycotina - Uredinales
Puccinia invaginata. Autoecious. **1:**FL(35,82).

Hovenia Thunb.

1. *H. dulcis* Thunb. [Japanese raisin-tree]

Basidiomycotina - Other
Armillaria mellea. Root rot. **1:**CA(2897).

Krugiodendron Urb.

1. *K. ferreum* (Vahl) Urb.

Ascomycotina
Botryosphaeria sp. Anthracnose. **1:**FL(35).
Meliola krugiodendri. **1:**FL(3483).

Rhamnus L.

1. *R. alnifolia* L'Hér. [alder-leaved buckthorn]
2. *R. betulaefolia* Greene
3. *R. californica* Eschsch. [California buckthorn, coffeeberry]
4. *R. caroliniana* Walter [Carolina buckthorn, Indian cherry, yellow buckthorn]
5. *R. cathartica* L. [common buckthorn]
6. *R. crocea* Nutt. [redberry]
7. *R. crocea* Nutt. subsp. *ilicifolia* (Kellogg) C. Wolf [hollyleaf buckthorn, red-berried buckthorn]
8. *R. crocea* Nutt. subsp. *insula* (Kellogg) C. Wolf (=*R. insularis* Greene, *R. insula* Kellogg)
9. *R. crocea* Nutt. subsp. *pyrifolia* (Greene) C. Wolf
10. *R. davurica* Pall.
11. *R. frangula* L. (=*Frangula alnus* Mill.) [alder buckthorn, glossy buckthorn]
12. *R. lanceolata* Pursh [lame-leaved buckthorn]
13. *R. purshiana* DC. [bearberry, cascara buckthorn, cascara sagrada]
14. *R. smithii* Greene
15. *Rhamnus* sp.

Ascomycotina
Botryosphaeria dothidea. **4:**GA(1438).
B. obtusa. On limbs. **4:**GA(82).
B. ribis. On limbs. **13:**WA(82,3202).
Capnodium rhamni. **3:**CA(2434).
Capnodium sp. Sooty mold. **3:**CA(82).
Dermatea sp. **13:**WA(82).
Dermea sp. **13:**ID(3202), WA(3202).
Diaporthe syngenesia. On limbs. **1:**NE(82). **13:**WA(3202).
Hypoxylon atropurpureum. On dead log. **15:**CA(2439).
H. rubiginosum. **8:**CA(2439).
Hysterographium sp. **13:**ID(3202), WA(3202).
Hysteropatella sp. **13:**ID(3202).
Lophiostoma quadrinucleatum. **13:**ID(3202).
L. quadrinucleatum var. *triseptatum*. Stem end rot, tuber rot, wilt. **12:**NE(82).
Microsphaera penicillata. Powdery mildew. **1:**WI(82).
Pezicula frangulae. On limbs. **3:**CA(82). **13:**CA(82), ID(82), WA(3202).
Phaeangium sphaeroides. On bleached wood. **8:**CA(2439).
Platystomum compressum var. *pseudomacrostomum*. **3:**CA(2783).
Pleospora herbarum. On leaves. **3:**CA(82).
Stictis radiata. **13:**CA(81). **15:**CA(3225).
Venturia rhamni. Leaf spot. **3:**CA(82,325).
Basidiomycotina - Uredinales
Puccinia coronata. Heteroecious: aecial host. **1:**NC states(82), NE states(82), ID(3202), MO(82), MT(3202), WA(82,3202), WY(1195). **4:**IN(82), TN(82), TX(82). **5:**NE states(82), MN(2191), MO(82), MT(82), SD(2243). **10:**SD(2243). **12:**IA(82), IN(82), MO(82), NE(82), WI(82). **13:**W states(82), ID(3202), MI(82), MT(82,3202), OR(3202), WA(82,3202). **14:**CO(82). **15:**OH(949).
P. mesnieriana. Autoecious. **3:**CA(82). **6:**CA(81,972). **7:**CA(81,82). **8:**CA(82). **9:**CA(82).
Basidiomycotina - Other
Cerrena unicolor. Heart rot. **13:**WA(82,3202).
Coniophora olivacea. **13:**ID(3202).
Exidia tremelloides. On corticate limbs. **15:**LA(2643).
Hymenochaete rubiginosa. **8:**CA(2439).
Inonotus hispidus. **3:**AZ(1181,1202).
Lopharia cinerascens. **11:**NY(543).
Meruliopsis corium. **13:**CA(81).
Phellinus igniarius. White heart rot. **13:**ID(82,3202).
P. punctatus. On living trees. **5:**ND(3759).
Deuteromycotina - Hyphomycetes
Cercospora bacilligera. Leaf spot. **4:**FL(35). **11:**Unk.(82), WI(1332). **12:**MO(82), NE(82), SC(82). **15:**SC(634).
C. rhamni. Leaf spot. **1:**WI(82). **4:**LA(82), NE(82), TX(82). **5:**NJ(82), NY(82), WI(82).
Fusarium sp. Leaf spot. **4:**FL(35).
Ovularia rhamnigena. On leaves. **3:**OR(82,3202).
Phymatotrichopsis omnivora. Root rot. **4:**TX(82).
Tubercularia ulmea. Basal stem canker. **11:**IL(3148).
Deuteromycotina - Coelomycetes
Ascochyta paliuri. **13:**ID(666,3202), WA(3202).
Ceuthospora foliicola. On leaves. **7:**CA(82).
Chaetasbolisia falcata. **13:**CA(2434).
Coniothyrium rhamnigenum. Leaf spot. **13:**OR(3202), WA(82,3202).
Cylindrosporium rhamni. Leaf spot. **13:**ID(82,3202). **15:**CA(642).
Diplodia frangulae. On twigs. **3:**CA(82).
Marssonina rhamni. Leaf spot. **13:**WA(82,3202).
Phoma rhamnicola. On twigs. **3:**CA(82,637).

Phomopsis communis. On twigs. **3:**CA(82).
Phyllosticta cinerea. **2:**AZ(1195).
P. rhamni. **5:**WI(82,1335).
Phyllosticta sp. Leaf spot. **5:**WI(1347).
Septoria blasdalei. Leaf spot. **13:**CA(82,642), ID(82,3202), OR(82,3202), TX(82), WA(3202).
S. rhamni-catharticae. On leaves. **15:**CA(82).
Sphaerographium niveum. On limbs. **1:**NY(82).
Sphaeropsis rhamni. Leaf spot. **5:**OK(2872).

Sageretia Brongn.
1. *S. wrightii* S. Wats.
2. *Sageretia* sp.

Ascomycotina
Stictis sagaretiae. **2:**FL(3225).
Basidiomycotina - Other
Hyphoderma amoenum. **1:**AZ(128).
Hypochnicium punctulatum. **1:**AZ(1194,1202).

Ziziphus Mill.
1. *Z. jujuba* Mill. [Chinese date, Chinese jujube, common jujube]
2. *Z. mauritiana* Lam. [cottony jujube, Indian jujube]

Ascomycotina
Botryosphaeria quercuum. Stem canker. **1:**FL(35).
Basidiomycotina - Uredinales
Phakopsora zizyphi-vulgaris. **1:**FL(35,82).
Deuteromycotina - Hyphomycetes
Alternaria sp. Leaf spot. **1:**OK(2872).
Cercospora jujubae. Leaf spot. **1:**FL(35,1261).
Cladosporium ziziphi. Leaf spot. **1:**FL(35).
Pseudocercospora vitis. Leaf spot. **1:**FL(35).
Deuteromycotina - Coelomycetes
Coniothyrium sp. **1:**OK(2873).
Deuteromycotina - Other
Rhizoctonia sp. Root rot. **2:**HI(2903).

Rhizophoraceae

Rhizophora L.
1. *R. mangle* L. [American mangrove, mangrove, red mangrove]

Oomycetes
Phytophthora epistomium. **1:**FL(1033).
P. spinosa. **1:**FL(1033).
P. vesicula. **1:**FL(1033).
Pythium grandisporangium. **1:**FL(1033).
Pythium sp. Root rot. **1:**FL(35).
Ascomycotina
Anthostomella rhizomorphae. Leaf spot. **1:**FL(35).
Botryosphaeria dothidea. Leaf spot. **1:**FL(35).
Dactylospora haliotrepha. On wood core of prop roots. **1:**FL(1943).
Didymosphaeria enalia. **1:**HI(2903).
Glomerella sp. Leaf spot. **1:**FL(35).
Helicascus kanaloanus. On dead prop roots. **1:**HI(1942,1945,2903).
Hydronectria tethys. On dead roots. **1:**FL(1942), HI(2903).
Keissleriella blepharospora. On bark. **1:**NC(1943).
Leptosphaeria australiensis. On prop roots. **1:**HI(1942,2903).
Lignincola laevis. On seedlings. **1:**HI(2903).
Lulworthia sp. On bark. **1:**HI(2903).
Mycosphaerella sp. Leaf spot. **1:**FL(35).
Basidiomycotina - Other
Halocyphina villosa. On roots. **1:**HI(1942).
Deuteromycotina - Hyphomycetes
Alternaria sp. Leaf spot. **1:**FL(35).
Cercospora rhizophorae. On leaves. **1:**FL(35,725).
Cladosporium sp. Leaf spot, isolated from macerated roots. **1:**FL(35), HI(2075).
Cylindrocarpon didymum. Stem root gall, prop root gall. **1:**FL(35,2639).
Cylindrocladiella parva. Isolated from macerated roots. **1:**HI(2075).
Fusarium oxysporum. Isolated from macerated roots. **1:**HI(2075).
F. roseum. Isolated from macerated roots. **1:**HI(2075).
F. solani. Isolated from macerated roots. **1:**HI(2075).
Geotrichum sp. Isolated from macerated roots. **1:**HI(2075).
Gliocladium roseum. Isolated from macerated roots. **1:**HI(2075).
Penicillium citrinum. Isolated from macerated roots. **1:**HI(2075).
Scopulariopsis sp. Isolated from macerated roots. **1:**HI(2075).
Trichoderma viride. Isolated from macerated roots. **1:**HI(2075).
Deuteromycotina - Coelomycetes
Cytospora rhizophorae. On submerged tips of aerial roots, on prop roots and green seedlings. **1:**FL(1943), HI(1943).
Cytospora sp. On prop roots. **1:**HI(1942,2903).
Pestalotiopsis disseminata. Leaf spot. **1:**FL(35).
Phoma eupyrena. Leaf spot. **1:**FL(35).
Phoma sp. Isolated from macerated roots. **1:**HI(2075).
Robillarda rhizophorae. On roots and seedlings. **1:**FL(1945), HI(1942,1945,2903).
Selenophoma sp. Leaf spot. **1:**FL(35).

Rosaceae

Acaena Mutis ex L.
1. *A. californica* Bitter
2. *A. fissistipula* Bitter
3. *A. novae-zelandiae* T. Kirk.
4. *A. tridactyla* J. Presl

Ascomycotina
Erysiphe cichoracearum. Powdery mildew. **2:**CA(3975). **3:**CA(3975).
Basidiomycotina - Uredinales
Phragmidium ivesiae. **1:**CA(3111). **4:**CA(81).

Adenostoma J. D. Hook. & Arnott
1. *A. fasciculatum* J. D. Hook. & Arnott [chamise, greasewood]

Ascomycotina
Eutypella domicalis. **1:**CA(2439).
Hypoxylon sp. **1:**CA(2439).
Hysterographium flexuosum. **1:**CA(2439).
Schizoxylon insigne. **1:**CA(2439).
Basidiomycotina - Other
Exidia recisa. On bark. **1:**CA(2439).
Peniophora malenconii subsp. *americana*. On wood. **1:**CA(542).
Stereum hirsutum. On bark. **1:**CA(2439).
S. ochraceo-flavum. **1:**CA(543).

Agrimonia L.
1. *A. eupatoria* L. [agrimony]
2. *A. gryposepala* Wallr.
3. *A. hirsuta* Bong.
4. *A. incisa* Torr. & A. Gray
5. *A. microcarpa* Wallr.
6. *A. parviflora* Sol. in Aiton
7. *A. pubescens* Wallr.
8. *A. rostellata* Wallr.
9. *A. striata* Michx.
10. *Agrimonia* sp.

Oomycetes
Peronospora potentillae. Downy mildew. **1:**NY(82), WI(82). **7:**IA(82), WI(82). **9:**WI(82). **10:**IA(82), NC(82), TX(82), WI(82).
Ascomycotina
Sphaerotheca macularis. Powdery mildew. **1:**E states(82), NE(82), WI(82). **3:**ND(377). **7:**E states(82), NE(82), WI(82,1330). **9:**E states(82), NE(82), NY(2456), WI(82). **10:**E states(82), NE(82), WI(82).
Basidiomycotina - Uredinales
Pucciniastrum agrimoniae. Telial host; aecial host unknown. **1:**MS(2720). **2:**AZ(1203,3981), DE(1575), IA(1575), IL(1575), IN(987), MD(1575), MN(1575), ND(1575), NE(1575), NY(1575), PA(1575), SD(1575), VT(1575), WI(1340), WV(1575). **3:**ND(378). **4:**NC(1303). **5:**FL(1575), IN(1575), KS(1575), LA(1575), MI(1575), MO(1575). **6:**AL(1575), AR(1575), DE(1575), GA(1439), IL(1575), IN(1575), KY(1575), MI(1575), MS(2720), NC(1303), NJ(1575), NY(1575), OH(1575), PA(1575), TN(1575), VA(1575), WV(1575). **7:**E states(82), Range of host(82), CO(82), IN(654), ND(82), NM(82). **8:**IN(1575), NY(1575). **9:**E states(82), Range of host(82), CO(82), ND(82), NE(1575), NM(82), PA(1575), SD(1575), VA(1575), WI(1575), WV(1575). **10:**E states(82), Range of host(82), CO(82), FL(35), MD(82), NM(82), NY(2456).

Deuteromycotina - Hyphomycetes
Cercospora sp. Leaf spot. **1:**IL(82).
Phymatotrichopsis omnivora. Root rot. **7:**TX(82). **10:**TX(82).
Deuteromycotina - Coelomycetes
Phoma exigua. Leaf spot. **2:**WI(1333). **7:**TX(82), WI(82). **9:**NC(82,3944), WI(82,1333). **10:**TX(82), WI(82).
P. herbarum. Stem blight. **7:**TX(82). **10:**TX(82).
Phyllosticta sp. On leaves. **2:**WI(1353).
Septoria agrimoniae. Leaf spot. **1:**MO(82). **7:**IA(82), TX(82). **9:**WI(82). **10:**IL(82), IN(82), NC(82), NY(82), TX(82), WI(82).
S. agrimoniae-eupatoriae. **7:**IA(82), WI(1333).

Amelanchier Medik.

1. *A. alnifolia* (Nutt.) Nutt. (=*A. florida* Lindl.) [Saskatoon serviceberry, western serviceberry]
2. *A. alnifolia* (Nutt.) Nutt. var. *cusickii* (Fernald) C. Hitchc. (=*A. cusickii* Fernald) [Cusick serviceberry]
3. *A.* × *amabilis* Wiegland
4. *A. arborea* (F. Michx.) Fernald [downy serviceberry]
5. *A. canadensis* (L.) Medik. (=*A. intermedia* Spach, *A. oblongifolia* M. Roem.) [downy serviceberry]
6. *A. goldmanii* Woot. & Standl.
7. *A. interior* P. Nielsen
8. *A. laevis* Wiegland [Allegheny serviceberry]
9. *A. ovalis* Medik. (=*A. rotundifolia* Dum. Cours.)
10. *A. pallida* Greene
11. *A. pumila* Nutt. (=*A. polycarpa* Greene) [cluster serviceberry]
12. *A. sanguinea* (Pursh) DC.
13. *A. spicata* (Lam.) K. Koch (=*A. stolonifera* Wiegland, *A. humilis* Wiegland)
14. *A. utahensis* Koehne (=*A. bakeri* Greene, *A. elliptica* A. Nels., *A. mormonica* C. K. Schneid., *A. oreophila* A. Nels., *A. prunifolia* Greene)
15. *A. wiegandii* P. Nielsen
16. *Amelanchier* sp.
17. *Amelanchier* × sp. (unnamed hybrids)

Ascomycotina
Apiosporina collinsii. Black mildew, witches' broom. **5:**Range of host(82), MA(1391), WI(1340). **10:**OR(657). **11:**NM(82). **16:**Range of host(82), ID(3202), VA(3493).
Biscogniauxia marginata. Blister canker. **5:**GA(82,1438), IA(82,1225), KS(82), MA(82), OH(82).
Cenangium sp. **16:**ID(3202).
Coccomyces tumidus. Leaf spot. **1:**MO(82), MT(82).
Coniochaeta pulveracea. **16:**ID(3202).
Dermea bicolor. On limbs. **1:**MT(82,3202), ND(82). **16:**ND(82).
Diaporthe tuberculosa. On limbs and twigs. **5:**MI(3807), NJ(3807), NY(3807). **8:**GA(1438).
Diatrype albopruinosa. **5:**IA(3629).
D. flavovirens. **16:**ID(3202).
Diatrypella eutypaeformis. **16:**IA(3629).
Durella atrocyanea. **16:**ID(3202).
Erysiphe polygoni. Powdery mildew. **1:**ID(82,3202), MT(82,3202). **16:**Unk.(82).
Eutypella cerviculata. **16:**IA(3629).
E. juglandicola. **5:**GA(82,1438).
E. leprosa. **16:**IA(3629).
Glyphium corrugatum. **1:**ID(3202), MT(3202), WA(3202).
Hypoxylon rubiginosum. **8:**GA(1438). **16:**OR(3202).
H. vogesiacum. **1:**OR(3202).
Hysterographium sp. **1:**ID(3202), OR(3202), WA(3202).
Lophiostoma hysterioides. **1:**OR(3202).
L. quadrinucleatum. **1:**ID(3202).
Lophodermium aucupariae. **1:**ID(3202), MT(3202), OR(3202), WA(3202).
L. foliicola. On overwintered leaves. **1:**ID(638).
Massaria conspurcata. On dead limbs. **5:**NY(82).
M. inquinans. **5:**GA(1438).
Massaria sp. **5:**GA(1438).
Monilinia amelanchieris. Mummified fruits, fruit rot. **16:**NY(82,1621).
M. fructicola. Brown rot. **16:**MI(82), OH(82).
M. gregaria. Leaf blight. **1:**CO(82), WA(3202). **2:**WA(82). **16:**IA(82), WA(82).
Mycosphaerella punctiformis. On fallen leaves. **5:**GA(82,1438), NY(82). **8:**GA(1438).
Nectria cinnabarina. Dieback. **5:**GA(82,1438). **16:**Unk.(82), AK(526), ID(3202), OR(3202).
N. galligena. **8:**NY(82).
N. pallidula. **8:**NC(3085).
Pezicula pruinosa. On limbs. **1:**ID(3202), MT(3202). **16:**Range of host(82).
Phyllactinia guttata. Powdery mildew. **1:**WA(3202). **14:**CO(1195). **16:**OR(82), WA(82).
Pleospora amelanchieris. **14:**WY(3810).
Podosphaera clandestina. Powdery mildew. **1:**ID(3202). **8:**WI(1358). **16:**Unk.(82).
Propolomyces versicolor. **8:**GA(1438).
Schizoxylon insigne. **16:**ID(3202).
Taphrina amelanchieri. Witches' broom. **1:**CA(82,2454).
T. japonica. **16:**CA(81).
Triblidium fatiscens. **16:**OR(3202).
Tympanis amelanchieris. **1:**ID(3202). **16:**ID(3202).
Valsa ceratosperma. **5:**GA(82,1438).
Valsella laschii. **5:**GA(82,1438).
Basidiomycotina - Uredinales
Gymnosporangium biseptatum. Heteroecious: aecial host. **5:**E states(82), NE states(82). **15:**ME(82).
G. clavariiforme. Heteroecious: aecial host. **1:**W states(82), CO(1195), MT(3202), UT(2771), WY(1195). **11:**W states(82), W states(82). **14:**UT(2771).
G. clavipes. Heteroecious: aecial host. **1:**MT(82,3202), SD(2768), WY(82). **4:**NC(1303). **12:**C states(82), E states(82), ME(82). **13:**E & C states(82), ME(82). **16:**WI(1335).
G. cunninghamianum. Heteroecious: aecial host. **10:**CA(2764,2773).
G. globosum. Heteroecious: aecial host. **17:** On *A. laevis* x *A. humilis*. WI(1333).
G. harknessianum. Heteroecious: aecial host. **1:**CA(82), OR(82,3202). **10:**CA(2771), OR(2771). **14:**NV(2771). **16:**OR(3202).
G. inconspicuum. Heteroecious: aecial host. **1:**AZ(82), CO(82,1195), NM(82), UT(82). **6:**AZ(1203). **16:**AZ(1203).
G. kernianum. Heteroecious: aecial host. **1:**AZ(82), CO(82), OR(82,3202). **14:**AZ(82,1203,3981), CO(82), NV(2771), OR(82).
G. libocedri. Heteroecious: aecial host. **10:**CA(81,642), OR(657). **16:**NV(2771).
G. nelsonii. Heteroecious: aecial host. **1:**W states(82), AK(60), CA(2771), CO(1195), ID(3202), MT(1195,3202), NM(82), OR(3202), SD(2768), WY(1195). **6:**AZ(1203). **9:**SD(2243). **11:**W states(82), NM(82), WY(1195). **12:**C states(82), E states(82), ME(82).
G. nidus-avis. Heteroecious: aecial host. **7:**WI(1355). **8:**E & C states(82), ME(82). **11:**W states(82), W states(82). **13:**IL(82), ME(82), WI(82). **14:**W states(82), AZ(1203,3981), WY(82). **3:**WI(1356). **16:**ID(2771), ME(82), UT(2771), WY(2771).
G. nootkatense. Heteroecious: aecial host. **1:**WA(3202).
Gymnosporangium sp. **16:**NC(1303).
Basidiomycotina - Other
Cerrena unicolor. Wood rot. **1:**MT(3202). **16:**Range of host(82).
Chondrostereum purpureum. **5:**IN(543).
Crepidotus variabilis. **16:**ID(3202).
Daedaleopsis confragosa. **1:**ID(3202).
Datronia scutellata. **1:**MT(3202,3832).
Dendrophora versiformis. **13:**NY(543).
Exidia recisa. **1:**ID(3202).
E. umbrinella. **16:**OR(3202).
Hymenochaete tabacina. **1:**ID(3202), ND(378).
Peniophora hydnoidea. **10:**CA(661).
Phanerochaete velutina. White rot. **16:**NJ(128,433).
Phellinus ferreus. **1:**ID(3202).
P. igniarius. **1:**OR(1524).
Polyporus varius. **1:**ID(3202), MT(3202,3832).
Steccherinum fimbriatum. **1:**WA(3202).
Stereum hirsutum. **1:**WA(3202).
S. ochraceo-flavum. **16:**MA(543).
S. sanguinolentum. **2:**ID(543).
Thelephora albidobrunnea. On crowns and stems of young trees. **5:**NC(82,3944). **16:**NC(82).
Tremella mesenterica. **1:**ID(3202).
Trichaptum biforme. **1:**ID(3202).
Deuteromycotina - Hyphomycetes
Alternaria sp. Leaf spot. **16:**FL(35).

Botrytis cinerea. **1**:AK(526).
Cristulariella moricola. Zonate leaf spot. **4**:NC(1302,1303).
Diplocarpon mespili. Leaf blight. **1**:Range of host(82), AK(526), ID(3202), MT(3202), OR(82,3202), WA(3202). **5**:GA(82,1438), IL(82), IN(82). **13**:IA(82). **14**:CO(1195). **16**:Range of host(82).
Fusicladium sp. Witches'-broom. **1**:WA(82). **2**:WA(82).
Oidium sp. Powdery mildew. **4**:NC(1303).
Phymatotrichopsis omnivora. Root rot. **1**:TX(82).
Septonema sp. **1**:ID(3202).
Deuteromycotina - Coelomycetes
Ascochyta sp. **1**:WA(3202).
Coryneum longestipitatum. On limbs. **16**:AK(82).
C. longistipitatum. **16**:AK(526).
Cryptosporiopsis curvispora. Canker. **1**:OR(82).
C. pruinosa. On twigs. **16**:ID(3549), MI(2862), ND(3549), NY(3549).
Cylindrosporium aroniae. On leaves. **1**:MT(82,3202), WA(3202).
Diplodia amelanchieris. On limbs. **1**:CO(82).
Entomosporium mespili. Leaf spot, on blades and petioles. **1**:CA(81). **5**:MA(1390). **3**:WI(1359). **16**:FL(35).
Phyllosticta innumerabilis. Leaf spot. **1**:IA(82), MT(3202), ND(82), NE(82). **12**:MN(82). **16**:NE(64,2738).
P. paupercula. Leaf spot. **1**:KS(64,82), MT(82).
P. virginiana. Leaf spot. **5**:NY(82), PA(82), WI(82).
Sphaeronaema acerinum. **16**:MI(2862).
Sphaeropsis amelanchieris. On limbs. **5**:NY(82,832).

Aphanes L.
1. *Aphanes* sp.
Basidiomycotina - Other
Ganoderma zonatum. Bud rot. **1**:FL(35).
Deuteromycotina - Hyphomycetes
Cercospora sp. Leaf spot. **1**:FL(35).
Deuteromycotina - Coelomycetes
Pestalotia sp. Leaf spot. **1**:FL(35).
Phyllosticta sp. Leaf spot. **1**:FL(35).

Armeniaca Scop.
This genus is considered a synonym of *Prunus*, but the appropriate combination could not be located.
1. *A. vulgaris* Trin. & Rupr.
Basidiomycotina - Uredinales
[*Tranzschelia pruni-spinosae*]. **1**:MS(2720).

Aronia Medik.
1. *A. arbutifolia* (L.) Pers. (=*Pyrus arbutifolia* (L.) L. fil.) [red chokeberry]
2. *A. melanocarpa* (Michx.) Elliott [black chokeberry]
3. *Aronia* sp.
Ascomycotina
Botryosphaeria dothidea. **1**:GA(1438).
B. obtusa. **1**:GA(82).
Calosphaeria pulchella. On limbs. **1**:NY(82).
Mycosphaerella arbutifoliae. Leaf spot. **1**:GA(82,1438), NY(82). **2**:NY(82,195).
Mycosphaerella sp. Leaf spot. **1**:FL(35).
Sclerotinia sp. On young shoots. **1**:NY(82).
Valsa ceratosperma. **1**:NJ(3415).
Basidiomycotina - Uredinales
Gymnosporangium clavariiforme. **1**:MA(82).
G. clavipes. Heteroecious: aecial host. **1**:MA(82), ME(82), TX(82). **2**:CT(82), ME(82).
G. davisii. Heteroecious: aecial host. **1**:ME(82). **2**:ME(82), MI(82), NH(82), WI(82).
G. fraternum. Heteroecious: aecial host. **1**:DE(82), FL(35), MA(82), ME(82), NJ(82), PA(82). **2**:ME(82).
Basidiomycotina - Other
Stereum ochraceo-flavum. **1**:MD(543).
Deuteromycotina - Hyphomycetes
Cercospora pyri. Leaf spot. **1**:FL(35), MI(82), NH(82), WI(82). **2**:FL(35), MI(82), WI(82).
Cercosporella pyrina. On leaves. **2**:MI(82).
Phymatotrichopsis omnivora. Root rot. **2**:TX(82).
Pseudocercospora mali. Leaf spot. **1**:AL(82), TX(82).
Deuteromycotina - Coelomycetes
Phoma exigua. Leaf spot. **1**:WI(82).
Phyllosticta arbutifoliae. Leaf spot. **1**:NJ(82). **3**:OK(2872).
Sphaeropsis viscosa. On leaves. **1**:MA(1390).

Aruncus L.
1. *A. sylvester* Kostel. [goatsbeard]
Ascomycotina
Brunnipila clandestina. On dead stems. **1**:CA(81).
Leptosphaeria arunci. On stems. **1**:OR(82,3202).
Sphaerotheca fuliginea. Powdery mildew. **1**:CA(1155).
Deuteromycotina - Hyphomycetes
Cercospora sp. **1**:OR(82).
Ramularia ulmariae. Leaf spot. **1**:AK(82).
Deuteromycotina - Coelomycetes
Cylindrosporium filipendulae. **1**:OR(3202).
Phyllosticta sp. **1**:WI(1338).

Cercocarpus H. B. K.
1. *C. betuloides* Nutt. ex Torr. & A. Gray [birchlead mountain-mahogany]
2. *C. betuloides* Nutt. ex Torr. & A. Gray var. *multiflorus* Jeps.
3. *C. betuloides* Nutt. ex Torr. & A. Gray var. *traskiae* (Eastw.) Dunkle
4. *C. hypoleucus* Rydb.
5. *C. ledifolius* Nutt. ex Torr. & A. Gray
6. *C. montanus* Raf. (=*C. betulaefolius* Nutt. ex Hook.) [mountain-mahogany]
7. *C. montanus* Raf. var. *paucidentatus* (S. Wats.) F. L. Martin (=*C. breviflorus* A. Gray)
8. *Cercocarpus* sp.
Ascomycotina
Coniochaeta ligniaria. **5**:WA(3202).
Diatrype standleyi. On dead limbs. **6**:NM(82).
Gibbera cercocarpi. On leaves. **8**:CA(82).
Glonium abbreviatum. **6**:CA(2439).
Hypoxylon rubiginosum. **6**:CA(2439).
Hysterographium fraxini. **5**:WA(3202).
H. mori. On bark and wood. **6**:CA(2439).
H. subrugosum. **5**:OR(3202).
Hysterographium sp. **5**:ID(3202).
Leptosphaeria cercocarpi. On leaves. **5**:ID(3202), UT(3249,3565).
Perrotia flammea. On dead limbs. **5**:CA(642).
Sphaerotheca macularis. Powdery mildew. **2**:CA(1155). **3**:CA(1155).
Teichospora cercocarpi. On dead limbs. **6**:CO(82), NM(82).
Basidiomycotina - Other
Acanthophysium bertii. **1**:AZ(1202). **7**:AZ(1202).
A. cerussatum. **7**:AZ(128,1202).
Cytidia stereoides. On wood. **5**:CA(640,642).
Dacrymyces stillatus. On wood. **5**:CA(642).
Exidiopsis calcea. **7**:AZ(1202).
Heterobasidion annosum. Heart rot. **8**:W states(3584), NV(3055).
Meruliopsis bellus. On rotten wood. **5**:CA(642).
Peniophora cinerea. **5**:CA(642).
P. nuda. **1**:AZ(1202).
Peniophora sp. On rotten wood. **5**:CA(642).
Perenniporia medulla-panis. **7**:AZ(1202).
Phellinus arctostaphyli. **1**:AZ(1202).
P. robustus. On trunk. **8**:CA(2439).
P. texanus. **1**:AZ(1202). **7**:AZ(1202).
Stereum hirsutum. White rot. **1**:OR(3202). **8**:OR(82).
Stereum sp. **5**:WA(3202).
Trametes hirsuta. **5**:ID(3202).
T. versicolor. On wood. **5**:CA(642).
Deuteromycotina - Coelomycetes
Camarosporellum cercocarpi. On leaves. **4**:WA(3555). **5**:ID(3555), MT(3555), NV(3555), UT(3555).
Coryneum sp. **5**:WA(3202).
Gloeosporidina cercocarpi. Lesions on leaves. **1**:CA(82,3549,3556).
Phloeospora cercocarpi. Leaf spot. **1**:CA(82,3549,3555).
Sphaceloma cercocarpi. Anthracnose. **2**:CA(82,3556).
Stegonsporium sp. **5**:WA(3202).

Chaenomeles Lindl.
1. *C. japonica* (Thunb.) Lindl. ex Spach (=*Cydonia japonica* (Thunb.) Pers.) [flowering quince]
2. *C. lagenaria* (Loisel.) Koidz. [Japanese quince]
3. *C. sinensis* (Thouin) Koehne (=*Pyrus sinensis* Lindl., *Cydonia*

sinensis Thouin) [Chinese quince]
4. *Chaenomeles* sp.
Ascomycotina
Botryosphaeria obtusa. Canker. **4:**E states(82), TX(82).
Monilinia fructicola. Brown rot. **2:**CA(2635). **4:**MI(82).
M. laxa. Brown rot. **1:**OR(3202). **2:**CA(2635). **4:**CA(82).
Mycosphaerella pomi. Leaf blotch, fruit spot. **1:**IL(312). **3:**MD(82).
M. pyrina. **3:**GA(1438).
Nectria cinnabarina. On limbs. **2:**WA(3202). **4:**CA(82).
Basidiomycotina - Uredinales
Gymnosporangium clavipes. Heteroecious: aecial host. **1:**FL(35), MS(2720). **4:**CT(82), NC(1303), NJ(82).
G. libocedri. Heteroecious: aecial host. **1:**OR(3202). **4:**OR(82).
Basidiomycotina - Other
Pellicularia koleroga. **2:**LA(3640).
Septobasidium pseudopedicellatum. Parasitic on scale insects. **1:**FL(35).
Deuteromycotina - Hyphomycetes
Cercospora cydoniae. Leaf spot. **1:**FL(35). **4:**AL(82), GA(82), MS(2720).
Diplocarpon mespili. Leaf blight. **3:**GA(1438). **4:**AL(82), CT(82), NY(82).
Fusicladium pyrorum. Scab. **2:**OK(2872).
Phymatotrichopsis omnivora. Root rot. **4:**TX(82).
Stigmina pallida. On limbs. **1:**NY(3547). **4:**NY(82).
Deuteromycotina - Coelomycetes
Aplosporella cydoniae. **1:**VA(2783).
Phoma sp. Canker. **4:**MD(82), TX(82).
Phyllosticta sp. On leaves. **4:**MO(82), NE(82).

Chamaebatia Benth.
1. *C. foliolosa* Benth. [mountain-misery]
Ascomycotina
Sphaerotheca macularis. Powdery mildew. **1:**CA(1155).

Chamaebatiaria (Porter) Maxim.
1. *C. millefolium* (Torr.) Maxim.
Ascomycotina
Cucurbitaria obducens. **1:**ID(3202).

Cotoneaster Medik.
1. *C. franchetii* Boiss.
2. *Cotoneaster* sp.
Oomycetes
Phytophthora cinnamomi. Root rot. **2:**NC(1303).
Phytophthora sp. Collar and root rot. **2:**CA(82), NC(1303).
Ascomycotina
Botryosphaeria dothidea. **1:**GA(1438).
B. obtusa. Canker. **2:**NE states(82), TX(82).
Botryosphaeria sp. Stem canker. **2:**FL(35).
Gibberella baccata. On twigs. **2:**CA(82).
Nectria cinnabarina. On twigs. **2:**CT(82), NC(1303), SD(2243).
Podosphaera sp. Powdery mildew. **2:**CA(82).
Venturia sp. Scab. **2:**WA(82,3202).
Basidiomycotina - Other
Armillaria mellea. Root rot. **2:**CA(82,2898).
A. tabescens. Root rot. **2:**FL(35,82).
Deuteromycotina - Hyphomycetes
Cercospora sp. Leaf spot. **2:**FL(35).
Diplocarpon mespili. Leaf spot. **2:**CA(82), IA(82).
Phymatotrichopsis omnivora. Root rot. **2:**AZ(82), TX(82).
Deuteromycotina - Coelomycetes
Coniothyrium sp. **2:**WI(1360).
Diplodia sp. Twig blight. **2:**FL(35), TX(82).
Myriellina cydoniae. Leaf spot. **2:**FL(35), MS(82,2720).
Phyllosticta cotoneastri. Leaf spot. **2:**MD(82).
Deuteromycotina - Other
Rhizoctonia solani. Web blight. **2:**FL(35), NC(1303).

Cowania D. Don
1. *C. mexicana* D. Don (=*C. stansburiana* Torr.)
2. *Cowania* sp.
Basidiomycotina - Uredinales
[*Phragmidium andersonii*]. **1:**ID(3202). **2:**ID(82).
Basidiomycotina - Other
Grandinia floccosa. **1:**AZ(1202).
G. pruni. **1:**AZ(1202).
Phellinus texanus. **1:**AZ(1202).
Polyporus arcularius. **1:**AZ(1202).
Schizopora apacheriensis. **1:**AZ(1193,1202).

Crataegus L.
1. *C. aestivalis* (Walter) Torr. & A. Gray [hawthorn]
2. *C. brevispina* Douglas ex Steud.
3. *C. calpodendron* (Ehrh.) Medik. [black-thorn, pear-thorn]
4. *C. coccinea* L. (=*C. holmesiana* Ashe)
5. *C. columbiana* Howell
6. *C. crus-galli* L. [cockspur, cockspur-thorn]
7. *C. douglasii* Lindl. [black hawthorn]
8. *C. erythropoda* Ashe
9. *C. flava* Aiton (=*C. michauxii* Pers.) [summer-haw, yellow-fruited-thorn]
10. *C. laevigata* (Poir.) DC. (=*C. oxycantha* of authors, non L.) [English hawthorn, pearthorn, quick-set-thorn]
11. *C.* × *lavallei* Hérincq. *C. crus-galli* × *C. pubescens*.
12. *C. macrantha* Hort. A name of no botanical standing, perhaps meant for *Crataegus macracantha* which is a synonym of *Crataegus succulenta* var. *macrancantha*.
13. *C. marshallii* Eggl. (=*C. apiifolia* (Marsh.) Michx.)
14. *C. mollis* (Torr. & A. Gray) Scheele [downy hawthorn]
15. *C. monogyna* Jacq. [English hawthorn]
16. *C. oxyacantha* L.
17. *C. phaenopyrum* (L. fil.) Medik. [Washington thorn]
18. *C. piperi* Britton
19. *C. praecoqua* Sarg. (=*C. praecox* Sarg.)
20. *C. punctata* Jacq. (=*C. collina* Chapm.)
21. *C. ravenelii* Sarg.
22. *C. rivularis* Nutt. ex Torr. & A. Gray
23. *C. rotundifolia* Moench (=*C. margaretta* Ashe)
24. *C. spathulata* Michx.
25. *C. succulenta* Schrad. ex Link
26. *C. tomentosa* L.
27. *C. uniflora* Münchh.
28. *C. venusta* Beadle
29. *C. viridis* L. [green hawthorn]
30. *C. watsonii* Murrill
31. *Crataegus* sp.
Ascomycotina
Acrospermum foliicola. **9:**GA(1438). **27:**GA(1438).
Arxiomyces vitis. On bark. **14:**ND(377).
Biscogniauxia nummularia var. *pseudopachyloma*. **24:**GA(1438).
B. nummularia var. *rumpens*. **24:**GA(1438). **29:**GA(1438). **31:**GA(82).
Bisporella citrina. **31:**ID(3202).
Botryosphaeria dothidea. **13:**GA(1438). **31:**GA(1438).
B. melathroa. On dead limbs. **31:**MD(82), PA(82).
B. obtusa. On limbs. **31:**C states(82), E states(82), GA(82), IA(82).
B. quercuum. Branch canker. **31:**FL(35).
Cenangium crataegi. On limbs. **31:**PA(82).
Coniochaeta ligniaria. **31:**ID(3202).
Cucurbitaria conglobata var. *crataegi*. **14:**ND(377).
C. crataegi. On limbs. **14:**ND(377). **31:**MI(82), ND(82), PA(82).
C. obducens. **31:**ID(3202), OR(3202).
Diaporthe aliena. **20:**GA(1438).
Diatrype stigma. **23:**IA(3629). **31:**ID(3202), WA(3202).
Diatrypella favacea. **29:**GA(1438).
D. nitschkei. **14:**IA(3629). **31:**IA(3629).
D. quercina. **24:**GA(1438). **29:**GA(1438).
D. subfulva. **6:**GA(1438).
Durandiella lenticellicola. **31:**NY(128).
Eutypa spinosa. **31:**IA(3629).
Eutypella goniostoma. On limbs. **31:**NY(195).
E. leprosa. **31:**IA(3629).
Fabraea thuemenii. **16:**WI(1350).
Fenestella phaeospora. **7:**ID(3202).
Glomerella cingulata. On fruit. **31:**MD(82).
Haematomyxa vinosa. On decorticated wood. **31:**NC(1303,1307).
Hypoxylon albocinctum. On bark. **31:**OH(967).
H. cinereo-lilacinum. On wood. **31:**IA(1225).
H. rubiginosum. On wood. **24:**GA(1438). **31:**GA(82), IA(1225).
Hysterographium sp. **7:**OR(3202). **31:**NC(1307).

Leucostoma cincta. **31:**IA(1226).
Monilinia johnsonii. Leaf blight. **31:**NC states(82), NE states(82), IA(82).
Mycosphaerella punctiformis. **9:**GA(1438). **13:**GA(1438).
Mycosphaerella sp. Leaf spot. **9:**GA(1438). **21:**GA(1438). **31:**GA(82).
Myriangium duriaei. **31:**GA(1438).
M. superficialis. Parasitic on scale insects. **6:**GA(1438). **9:**GA(1438,1745). **20:**GA(1438). **24:**GA(1438). **27:**GA(1438). **29:**GA(1438). **31:**E states(82).
M. tuberculans. Black spot. **14:**MS(2720).
Nectria galligena. **31:**WA(3202).
Otthia spiraeae. On limbs. **14:**ND(377). **31:**ND(82).
Patellariopsis clavispora. **29:**GA(1438). **31:**GA(82).
Pezicula sepium. On dead limbs. **9:**GA(1438). **24:**GA(1438). **31:**GA(82), IA(82).
Phyllactinia guttata. Powdery mildew. **5:**WA(3202). **7:**MT(3202), WA(3202). **12:**WI(1359). **31:**Range of host(82), FL(35).
Pleosphaeria echinata. **24:**GA(1438).
Pleosphaerulina corticola f. *crataegi*. **14:**ND(377).
Podosphaera clandestina. Powdery mildew. **4:**MS(2720). **6:**MS(2720). **14:**ND(377). **16:**Unk.(3976). **31:**E & C states(82), FL(35).
Schizoxylon insigne. **14:**ND(377).
Valsa ambiens. On twigs. **31:**Range of host(82).
V. ambiens f. *crataegi*. **14:**ND(377).
V. ceratosperma. **9:**GA(1438). **24:**GA(1438). **28:**GA(1438).
Venturia crataegi. Scab. **31:**FL(35,82).
V. inaequalis. Scab. **31:**Range of host(82), WA(3202).
Xylaria carpophila. On decaying fruit and leaves. **31:**IA(1225).
X. digitata. Grayish-white root rot. **7:**ID(3202). **31:**Unk.(82).
X. hypoxylon. Root rot. **7:**ID(3202). **31:**IN(82), OH(82).
X. oxyacanthae. On fruits. **31:**MI(3506).
Xyloschizon atratum. **7:**ID(3202).

Basidiomycotina - Uredinales

Gymnosporangium bethelii. Heteroecious: aecial host. **5:**MT(3202). **7:**ID(3202), MT(3202), OR(3202), WA(3202), WY(1195). **22:**UT(1195). **25:**CO(1195), ND(2700), SD(2243). **31:**NW states(82), CO(649), ND(82,378), SD(2700), WA(82).
G. clavariiforme. Heteroecious: aecial host. **5:**OR(2771). **31:**GA(1439).
G. clavipes. Heteroecious: aecial host. **1:**GA(1439). **4:**MS(2720). **6:**MA(1390), MS(2720). **7:**MT(3202), WA(3202). **9:**GA(1439). **16:**MA(1390). **24:**GA(1439). **31:**E & C states(82), FL(35), MS(2720), NC(1303), WI(1335).
G. confusum. Heteroecious: aecial host. **16:**CA(160). **31:**NC(1303,1307).
G. exiguum. Heteroecious: aecial host. **31:**TX(82).
G. floriforme. Heteroecious: aecial host. **13:**TX(652). **24:**MS(2720), TX(652). **31:**SE states(82), FL(35), OK(82), TX(82).
G. globosum. Heteroecious: aecial host. **3:**WI(1333). **4:**MS(2720), WI(1333). **6:**OK(2873), WI(1330). **14:**WI(1334). **15:**WI(1334). **20:**WI(1333). **25:**SD(2243), WI(1333). **29:**OK(2872). **11:**WI(1353). **31:**NC states(82), NE states(82), FL(35,82), GA(82), IL(1052), MS(2720), NC(1303,3944), NY(2456), OK(2872), TX(82).
G. hyalinum. Heteroecious: aecial host. **31:**SE states(82), FL(35), SC(634).
[*G. juniperi-virginianae*]. **31:**GA(1439), IA(82), IL(1052).
G. libocedri. Heteroecious: aecial host. **7:**CA(81,638,642), OR(3202). **31:**CA(82), OR(82).
G. nelsonii. **16:**AZ(3981). **31:**WY(82).
G. trachysorum. Heteroecious: aecial host. **13:**MS(2720). **31:**E states(82), Gulf states(82), FL(35), LA(82), MS(2720), TX(82).
Gymnosporangium sp. **16:**AZ(1856). **31:**OK(2872).

Basidiomycotina - Ustilaginales

Sporidiobolus pararoseus. On rusted leaves. **31:**KS(2686).

Basidiomycotina - Other

Armillaria mellea. Root rot. **17:**CA(2897). **31:**OK(82,2872).
A. tabescens. Root rot. **31:**FL(35).
Coniophora puteana. **7:**WA(3202).
Crepidotus nephrodes. **31:**WA(3202).
Cytidia patelliformis. **7:**WA(3202).
Daedaleopsis confragosa. On wood. **31:**Range of host(82).
Dendrocorticium roseocarneum. **31:**PA(128).
Dendrophora versiformis. **24:**AL(543).
Eichleriella leveilliana. **31:**LA(2675).
Ganoderma applanatum. White heart rot. **31:**CA(82), MI(82).
Melzericium udicola. **7:**WA(3202).
Peniophora incarnata. **7:**WA(3202).
Phaeomarasmius distans. On bark. **26:**NY(1544).
P. rufolateritius. On bark. **31:**MO(1544).
Phanerochaete ravenelii. **31:**AR(2675), LA(2675).
P. sordida. **31:**LA(2675).
Phellinus conchatus. Wood rot. **31:**MI(82), ND(82).
P. gilvus. Wood rot. **31:**SC states(82), AR(2675), LA(2675), MS(2675).
P. occidentalis. **5:**OR(3202), WA(128,3202). **7:**ID(128,3202), MT(128,3202), UT(128). **18:**OR(128). **31:**ID(128), MT(128).
P. pini. Wood rot. **7:**ID(3202). **31:**CO(82), MT(82).
P. pomaceus. White pocket rot. **7:**MT(3202). **31:**ID(3202), MI(82), WY(128).
Phylloporia ribis. **31:**LA(82).
Sarcodontia setosa. On dead trunks. **31:**C states(82), E states(82).
Septobasidium castaneum. Parasitic on scale insects. **31:**FL(35,82).
S. pseudopedicellatum. Felt fungus. **31:**Gulf states(82), SE states(82), FL(35), LA(2675).
Septobasidium sp. Parasitic on scale insects. **31:**Gulf states(82), SE states(82), FL(35).
Trametes versicolor. Wood rot. **7:**ID(3202), OR(3202), WA(3202). **31:**Range of host(82).
Vararia boreale. **2:**ID(128).

Deuteromycotina - Hyphomycetes

Alternaria sp. Leaf spot. **31:**FL(35).
Botrytis cinerea. Fruit rot. **31:**MA(82).
Cercoseptoria crataegi. **14:**WI(1354). **31:**Unk.(82), OR(4007), WV(82).
Cercospora confluens. **31:**TX(82).
C. crataegi. Leaf spot. **13:**TX(3597). **31:**FL(35), TX(82).
Cercosporella mirabilis. Leaf spot. **31:**CO(82), NY(82), WI(82).
Cladosporium sp. On fruits. **31:**IA(82).
Coniosporium olivaceum. On dead limbs. **31:**NC(1303).
Conoplea olivacea. **31:**MI(1672).
Diplocarpon mespili. Leaf spot. **7:**WA(3202). **10:**CT(3770). **16:**WI(1330). **31:**SC states(82), SE states(82), IL(1052), MS(2720), NY(3780), PA(3780), WA(3202).
Hirudinaria macrospora. On leaves. **6:**MS(2720). **31:**IL(82), SC(82).
Monilia crataegi. **14:**WI(1332).
Oidium pirinum. **31:**WI(1349).
Phymatotrichopsis omnivora. Root rot. **31:**TX(82).
Pithomyces quadrata. On leaves. **31:**AL(82), NJ(82), SC(82).

Deuteromycotina - Coelomycetes

Coniothyrium pyrinum. On leaves. **14:**IL(2568).
Cornicular iella harpographoidea. On limbs, on or around lenticels. **31:**MI(82,861), NY(861).
Cylindrosporium brevispina. Leaf spot. **7:**MT(832,3202), WA(3202). **31:**W states(82), OR(3202), WA(82).
Cytospora microspora. On twigs. **31:**MI(82), NJ(82).
Discosia artocreas. Leaf spot. **14:**MS(2720).
Entomosporium mespili. Leaf blight and spot. **10:**MI(3520), WI(3505). **15:**IL(2567). **31:**FL(35).
Gloeodes pomigena. On twigs. **31:**IN(82).
Hendersonia crataegicola. Leaf spot. **31:**AL(82), TX(82).
Monochaetia turgida. Leaf spot. **27:**FL(82). **30:**FL(82). **31:**FL(35).
Phoma leprosa. On fruit. **31:**NY(82).
Phyllosticta crataegi. **4:**NY(64). **8:**NM(64). **31:**NC states(82), NE states(82), NM(82), SC(82).
P. crataegicola. **14:**ND(64,378). **31:**CO(82), ND(82).
P. grisea. **19:**NY(64). **31:**NY(82).
P. solitaria. **31:**IL(82), NJ(82).
Phyllosticta sp. Leaf spot. **7:**MT(3202). **31:**OK(2872).
Seimatosporium caudatum. **31:**TX(82).
S. lichenicola. **14:**ND(378). **31:**ND(82).
Septoria oxyacanthae. Leaf spot. **12:**WI(1359). **14:**ND(378), WI(1354). **20:**OK(2873). **31:**NC states(82), OK(82), WI(1352).
Sphaceloma crataegi. On living leaves. **2:**WA(82). **7:**WA(3202,3556).
Sphaeronaema longirostre. On limbs. **31:**NE(82).
Sphaeropsis foliicola. On leaves. **31:**NY(82), WI(1331).
Strasseria geniculata. On fruits. **20:**DC(2546).

Deuteromycotina - Other

Sclerotium rolfsii. Seedling blight. **31:**FL(35,82).

Cydonia Mill.

1. *C. oblonga* Mill. (=*Pyrus cydonia* L., *C. vulgaris* Pers.) [common quince]
2. *Cydonia* sp.

Ascomycotina

Asteromassaria seriata. **1**:GA(1438).
Botryosphaeria obtusa. Black rot of fruit. **1**:E & C states(82), MS(2720).
B. quercuum. Canker, black rot of fruit. **1**:FL(35), NC(1303).
B. ribis. Dothiorella twig blight. **1**:TX(82).
Elsinoe piri. Anthracnose. **1**:WA(82,3202).
Glomerella cingulata. Bitter rot. **1**:C states(82), E states(82), MS(2720), TX(82).
Leucostoma persoonii. Canker. **1**:Unk.(82).
Monilinia fructicola. Brown rot. **1**:E states(82), MS(82), NC(1303), OR(82,3202), TX(82).
M. laxa. Brown rot. **1**:NW states(3202), W states(82).
Mycosphaerella pomi. Fruit spot. **1**:NE states(82).
Nectria cinnabarina. On limbs. **1**:NH(82), WA(82).
N. galligena. Trunk canker. **1**:OR(82,3202).
Pezicula malicorticis. Northwestern anthracnose, perennial canker. **1**:OR(82,3202), WA(82,3202).
Phorcys seriata. **1**:GA(1438).
Phyllactinia guttata. Powdery mildew. **1**:VA(82).
Podosphaera clandestina. **1**:NC states(82), NE states(82), NY(82).
P. leucotricha. Powdery mildew. **1**:CA(82,1110), WA(82,3202).
Schizothyrium pomi. Fruit spot. **1**:MO(82).
Venturia pyrina. Scab. **1**:CT(82).

Basidiomycotina - Uredinales

Gymnosporangium clavariiforme. **1**:CT(82), ME(82), NH(82).
G. clavipes. Heteroecious: aecial host. **1**:E & C states(82), CA(1110), MA(1390), MS(2720), NC(1303), OK(2872), TX(82).
G. libocedri. Heteroecious: aecial host. **1**:OR(82,3202).
G. nelsonii. **1**:AZ(82,1203,3981), CO(82).
G. nidus-avis. Heteroecious: aecial host. **1**:CT(82), NY(82).
Gymnosporangium sp. **2**:AZ(1856).

Basidiomycotina - Other

Chondrostereum purpureum. **1**:WA(3202).
Pellicularia koleroga. Thread blight. **1**:Gulf States(82), LA(3640), NC(82).

Deuteromycotina - Hyphomycetes

Alternaria mali. Fruit rot. **1**:IN(82).
Botrytis cinerea. Gray mold fruit rot. **1**:Unk.(82).
Exosporium sp. **1**:GA(82).
Macrosporium cydoniae. **1**:DE(3543).
Penicillium expansum. Blue mold rot. **1**:Range of host(82), FL(35).
Phymatotrichopsis omnivora. Root rot. **1**:TX(82).
Trichothecium roseum. Pink mold fruit rot. **1**:Unk.(82).

Deuteromycotina - Coelomycetes

Coniothyrium pyrinum. On leaves. **1**:WV(82).
Discosia artocreas. Leaf spot. **1**:MS(2720).
Entomosporium mespili. Leaf blight, leaf spot, on fruit. **1**:Unk.(82), FL(35), MA(1390), NC(1303), OK(2872).
Hendersonia cydoniae. On leaves. **1**:CT(82), NJ(82).
Phoma cydoniae. Pale rot of fruit. **1**:IL(82), MI(82).
P. macrostoma. Fruit rot. **1**:Unk.(82), IL(82).
Phyllosticta sp. Leaf spot. **1**:DE(82).

Deuteromycotina - Other

Rhizoctonia solani. Thread blight. **1**:FL(35).

Dalibarda L.

This genus is considered a synonym of *Rubus*, but the appropriate combination could not be located.

1. *D. repens* L. [dewdrop, robin-run-away]

Ascomycotina

Phyllachora dalibardae. On leaves. **1**:NY(82,195).

Deuteromycotina - Coelomycetes

Septoria dalibardae. Leaf spot. **1**:ME(82), MI(82), NH(82), NY(82), VT(82).

Dryas L.

1. *D. drummondii* Richardson ex Hook.
2. *D. integrifolia* Vahl
3. *D. octopetala* L. [mountain avens]
4. *D. octopetala* L. subsp. *alaskensis* (Porsild) Hultén (=*D. alaskensis* Porsild) [mountain avens]
5. *Dryas* sp.

Ascomycotina

Hymenoscyphus scutula. **1**:AK(3431).
Melanomma dryadis. On flowers. **1**:AK(3431).
Mycosphaerella ootheca. On leaves. **3**:AK(526). **5**:AK(82).
Wettsteinina dryadis. On leaves. **2**:AK(526,3431). **5**:AK(82).

Basidiomycotina - Other

Marasmius epidryas. On dead limbs. **1**:AK(2416). **2**:AK(2416). **3**:AK(2416). **4**:AK(2416). **5**:AK(2939).

Duchesnea Sm.

1. *D. indica* (Andr.) Focke [false strawberry, Indian strawberry, mock-strawberry]

Chytridiomycetes

Synchytrium duchesneae. Leaf gall. **1**:LA(627).
S. globosum. Leaf gall. **1**:SC(82).

Oomycetes

Peronospora potentillae. Downy mildew. **1**:NC(82).

Ascomycotina

Discohainesia oenotherae. Leaf spot. **1**:VA(82).

Basidiomycotina - Uredinales

Frommeella duchesneae. Autoecious. **1**:SC states(82), SE states(82), SW states(82), AZ(3981), FL(35), GA(1439), IN(3066), KY(82), NC(1303), PA(1875).
[*Phragmidium potentillae*]. **1**:MI(3490), MN(3490), MS(2720).

Eriobotrya Lindl.

1. *E. deflexa* (Hemsl.) Nakai
2. *E. japonica* (Thunb.) Lindl. [Japanese medlar, Japanese plum, loquat]

Oomycetes

Phytophthora cactorum. Collar rot. **2**:CA(82,1110).
Pythium sp. Root rot. **2**:FL(35).

Ascomycotina

Botryosphaeria obtusa. On limbs. **2**:CA(1110), FL(82).
B. quercuum. Limb canker. **2**:FL(35).
B. ribis. **2**:HI(2903).
Capnodium caespitosum. On leaves. **2**:CA(2434).
Glomerella cingulata. Anthracnose, flower blight, on fruit and twigs. **2**:Gulf States(82), FL(35).
Hypoxylon sp. **2**:CA(1110).
Nectria cinnabarina. **2**:CA(1110).
Venturia inaequalis. **2**:CA(1110).

Basidiomycotina - Other

Armillaria mellea. Root rot. **2**:CA(82,1110,2898).
A. tabescens. Root rot. **2**:FL(35,82).
Ganoderma lucidum. **2**:AZ(1202).

Deuteromycotina - Hyphomycetes

Alternaria sp. Leaf spot. **2**:FL(35).
Cercospora eriobotryae. Leaf spot. **2**:FL(35).
Corynespora sp. Leaf spot. **2**:FL(35).
Diplocarpon mespili. **1**:CA(1110). **2**:CA(1110).
Fusarium sp. Root rot. **2**:FL(35).
Phymatotrichopsis omnivora. Root rot. **2**:TX(82).
Spilocaea pomi. Leaf scab. **2**:FL(35).
S. pyracanthae. Leaf scab, on fruits, leaves and stems. **2**:Range of host(82), CA(1110), FL(35), WA(3202).

Deuteromycotina - Coelomycetes

Botryodiplodia sp. Dieback. **2**:FL(35).
Colletotrichum gloeosporioides. Anthracnose, flower blight. **2**:FL(35,82), TX(82).
Diplodia eriobotryae. Twig dieback. **2**:FL(35).
Entomosporium mespili. Leaf spot. **2**:FL(35,82), LA(2832,3471), MS(2720), NC(1303).
Monochaetia sp. Leaf spot. **2**:FL(35).
Pestalotia longiaristata. On leaves. **2**:FL(35,82).
Pestalotia sp. On leaves. **2**:FL(82).
Phoma sp. Dieback. **2**:FL(35).
Phomopsis sp. **2**:CA(1110).
Phyllosticta eriobotryae. Leaf spot. **2**:FL(35,82), NC(1303).
Protostroma sp. Leaf spot. **2**:FL(35).
Septoria eriobotryae. Leaf spot. **2**:FL(2306).
Sphaeropsis sp. Leaf spot. **2**:FL(35).

Deuteromycotina - Other
Rhizoctonia solani. Root rot. **2:**FL(35).

Exochorda Lindl.

1. *E. racemosa* (Lindl.) Rehd. [common pearlbush]

Basidiomycotina - Other
Armillaria mellea. Root rot. **1:**CA(2900).
Scytinostroma galactinum. Root rot. **1:**MD(668).

Filipendula Mill.

(= *Ulmaria* Mill.)
1. *F. purpurea* Maxim.
2. *F. rubra* (J. Hill) B. L. Robinson (=*Spiraea rubra* Britton) [queen-of-the-prairie]
3. *F. ulmaria* (L.) Maxim. [queen-of-the-meadow]
4. *F. vulgaris* Moench (=*Spiraea filipendula* L.) [dropwort]
5. *Filipendula* sp. (=*Ulmaria sp.*)

Ascomycotina
Psilachnum lanceolato-paraphysatum. On dead stems. **4:**NY(850).
Sphaerotheca macularis. Powdery mildew. **2:**IN(82), NY(82), VT(82). **3:**VT(82).

Basidiomycotina - Uredinales
Triphragmium ulmariae. Autoecious. **2:**IN(111,2461). **3:**IN(82). **5:**IN(2461).

Deuteromycotina - Coelomycetes
Cylindrosporium sp. Leaf spot. **1:**CT(82). **3:**CT(82).
Septoria ulmariae. Leaf spot. **1:**WI(82). **3:**CT(82), WI(1346). **5:**WI(1335).

Fragaria L.

1. *F.* × *ananassa* Duchesne (=*F. chiloensis* (L.) Duchesne var. *ananassa* (Duchesne) Hort. ex L. H. Bailey). *F. chiloensis* × *F. virginiana*. [cultivated strawberry, garden strawberry]
2. *F.* × *ananassa* Duchesne var. *cuneifolia* (Nutt. ex Howell) Staudt (=*F. cuneifolia* Nutt. ex Howell)
3. *F. californica* Cham. & Schlechtend.
4. *F. chiloensis* (L.) Duchesne [beach strawberry, Chilean strawberry, chiloe strawberry]
5. *F. ovalis* (Lehm.) Rydb.
6. *F. vesca* L. [alpine strawberry, European strawberry, sow-teat strawberry, woodland strawberry]
7. *F. vesca* L. subsp. *americana* (Porter) Staudt (=*F. americana* (Porter) Britton, *F. vesca* L. var. *americana* Porter) [woodland strawberry]
8. *F. vesca* L. subsp. *bracteata* (A. Heller) Staudt (=*F. bracteata* A. Heller)
9. *F. virginiana* Duchesne [Virginia strawberry, wild strawberry]
10. *F. virginiana* Duchesne subsp. *platypetala* (Rydb.) Staudt (=*F. platypetala* Rydb., *F. virginiana* Duchesne var. *platypetala* (Rydb.) H. Hall)
11. *Fragaria* sp.

Myxomycetes
Diachea leucopodia. On leaves and stolons. **4:**MS(2720). **1:**CA(82,1110), IL(82), KS(82), LA(82), MO(82), MS(82), TN(82), TX(82). **11:**MI(1129).
Fuligo septica. On leaves and stolons. **1:**IA(82), KS(82), MN(82), NE(82), NJ(82), TX(82), WA(82).
Mucilago crustacea. Smothering, on leaves and stolons. **1:**KS(82), MI(82), MO(82), NE(82), OK(82). **11:**OK(2872).
Physarum cinereum. On leaves and stolons. **1:**IL(82), NE(82), NJ(82).
P. leucophaeum. Slime mold. **11:**MI(1129).
Stemonitis herbatica. Slime mold. **1:**NC(1303).

Chytridiomycetes
Olpidium brassicae. Black root. **1:**UT(82).
Synchytrium aureum. Leaf gall. **9:**WI(82).
S. fragariae. Root gall. **4:**WA(3202). **1:**WA(82).
Synchytrium sp. On leaves, petioles and stems. **4:**WA(3202). **2:**WA(1848).

Oomycetes
Peronospora fragariae. Downy mildew. **1:**IA(82).
Phytophthora cactorum. Blight, postharvest decay of fruits, leather rot of fruits. **1:**'Leather rot.' S states(82), AZ(82), CA(1110), MD(82), MI(82), NC(1303), OH(978,1372,1373), OK(82), VA(82). **11:**CA(939), IA(3958), IL(3958), OK(2872), TN(1031).
P. fragariae. Red stele disease. **4:**CA(609), CT(3770), OR(609,3202), WA(609,3202). **1:**'Red stele.' NE states(82), W states(82), CA(1110), CO(82), FL(35), IA(82), MD(611), MI(82), NC(1303), NJ(611), OK(82), WI(82). **11:**FL(35), KY(1776), MD(1729), MI(1129), OK(2872), OR(3202).
P. nicotianae var. *parasitica*. Root rot. **1:**HI(2903), TN(1027).
Phytophthora sp. Bud, flower and leaf blight, root rot. **11:**TN(1030).
Pythium acanthicum. **1:**IL(2590).
P. dissotocum. **1:**IL(2590).
P. hypogynum. **1:**IL(2590).
P. irregulare. **1:**IL(2590).
P. perniciosum. **1:**IL(2590).
P. rostratum. **1:**IL(2590).
P. sylvaticum. **1:**IL(2590).
P. ultimum. Root rot. 'Black root.' **1:**CA(82,1110), IL(2590).
Pythium sp. Root rot. **4:**ID(3202). **1:**CA(1110), FL(35,82), ID(82), LA(82), NC(1303), NJ(82), OR(82), TN(82), UT(82), WA(82). **11:**FL(35).

Zygomycotina
Mucor recurvus. Isolated from roots. **11:**MN(464).
Rhizopus stolonifer. Leak of fruits, black rot, soft rot. **4:**ID(3202), MS(2720), OR(3202), WA(3202). **1:**'Rhizopus rot.' Range of host(82), FL(35), NC(1303). **11:**OK(2872).
Rhizopus sp. Black rot of fruit. **1:**Range of Host(82), CA(1110).

Ascomycotina
Chaetomium sp. **4:**OR(3202). **11:**CA(3894).
Diapleella coniothyrium. Black root. **4:**OR(3202). **1:**OR(82).
Discohainesia oenotherae. Tan rot of fruit, secondary rot of leaves and stolons, leaf spot. **4:**OR(3202). **8:**MD(82), ME(82), NJ(82), NY(82), WI(82). **1:**S states(82), AK(82), FL(35), IA(82), IL(82), ME(82), MI(82), OH(82), OK(82), OR(82), TN(82), WI(82). **11:**MS(2720), OK(2873).
Erysiphe graminis f. sp. *avenae*. Powdery mildew. **11:**AR(3029).
Gnomonia comari. Leaf blotch. **1:**'Leaf blotch.' FL(35,1648), NC(1303). **11:**CA(3894), ME(1083).
G. fragariae. On calyces, fruits, leaves and pedicles. 'Leaf blotch.' **1:**CA(1110), MI(28), NJ(82).
Hainesia lythri. 'Tan-brown rot.' **9:**WI(1357). **1:**Unk.(82).
Hainesia sp. Leaf spot. **1:**NC(1303).
Leptotrochila repanda. On leaves. **7:**MT(82).
Mycosphaerella fragariae. Leaf spot. 'Common leaf spot and black seed disease.' **4:**AK(82,526), ID(3202), MS(2720), OR(3202), WA(82,3202). **5:**ID(1195). **6:**ID(3202), MT(3202), OR(3202), WA(3202). **7:**Range of host(82). **8:**CO(82), UT(82). **9:**Range of host(82), GA(1438), ID(3202), MS(2720), NC(1303), OK(2872). **1:**Range of host(82), CA(1110), FL(35), NC(1303), SD(2243). **11:**AK(526), AL(285), AR(768), FL(35,3500), HI(2903), KY(1776), ND(377), OK(2872).
M. louisianae. Purple leaf spot. **4:**MS(2720). **1:**LA(82), MS(82), NC(82).
Phyllactinia guttata. Powdery mildew. **6:**WA(3202). **8:**WA(82).
Pithoascus intermedius. On roots. **1:**NC(82). **11:**NC(685).
Rosellinia necatrix. Root rot. 'Dematophora crown rot. Dematophora root rot.' **1:**CA(1110,3897).
Sclerotinia sclerotiorum. Crown rot, on fruit. **1:**'Sclerotinia crown rot. Sclerotinia fruit rot.' SE states(82), CA(82,1110), FL(35), IA(82), IL(82), MN(82), NC(1303), TX(82). **11:**WA(2324).
Scorias spongiosa. Sooty mold on leaves. **1:**ME(82).
Sphaerotheca fuliginea. **1:**WA(82).
S. macularis. Powdery mildew. **4:**Unk.(3976), AK(82,526), FL(453), ID(3202), OR(3202), WA(3202). **8:**PA(82). **9:**MA(82), MT(3202). **10:**AK(526). **1:**'Powdery mildew.' C states(82), N states(82), AL(82), AR(82), CA(1110), FL(35), NC(1303), OK(82), TX(82). **11:**AK(526), FL(1653), OK(2872).

Basidiomycotina - Uredinales
[*Phragmidium potentillae*]. **1:**FL(35). **11:**FL(35).

Basidiomycotina - Other
Armillaria mellea. Root rot. **4:**OR(3202), WA(3202). **1:**'Shoestring root rot.' CA(82,1110), NC(1303), OR(82), WA(82).
A. tabescens. Root rot. **1:**NC(1303).
Ceratobasidium anceps. Leaf rot. **9:**WI(82).
Ceratobasidium sp. **1:**CA(1110).
Thanatephorus cucumeris. **4:**ID(3202), OR(3202), WA(3202).

Deuteromycotina - Hyphomycetes
Alternaria tenuissima. Fruit rot. 'Alternaria fruit rot.' **1:**FL(35,1652).
Alternaria sp. Associated with black-root and root rot. **4:**OR(3202).

1:UT(82).
Aspergillus sp. 4:OR(3202).
Botrytis cinerea. Blight, gray mold fruit rot. 4:ID(3202), OR(3202), WA(3202). 1:'Gray mold.' Range of host(82), CA(1110), FL(35), NC(1303). 11:AK(526), CA(939), MS(2720), TN(72,73).
Botrytis sp. Fruit rot, gray mold rot. 1:HI(2899,2903). 11:ME(1083), OK(2872).
Cercospora vexans. Leaf spot. 7:WI(82). 9:NY(82), WI(82). 1:FL(35). 11:FL(35).
Cercospora sp. Leaf spot. 1:HI(2903), LA(82).
Cladosporium herbarum. Black root, secondary leaf and fruit rot. 4:OR(3202). 1:OR(82).
Cylindrocarpon destructans. 11:CA(3894).
Cylindrocladium scoparium. Root rot. 1:TN(82).
Dicoccum sp. 4:OR(3202).
Diplocarpon earlianum. Leaf scorch. 4:AK(3431), CA(82), ID(3202), MS(2720), MT(3202), OR(82,3202), WA(3202). 6:ID(3202), MT(3202), OR(3202), WA(3202). 7:MO(82), NE(82). 8:NM(82). 9:GA(82,1438), MS(82,2720), NC(82), NY(82), WA(3202). 1:'Leaf scorch.' Range of host(82), CA(1110), FL(35), SD(2243). 11:FL(35), KY(1776), NC(1303), OK(2872).
Fusarium oxysporum. 1:CA(1110), MD(82), UT(82). 11:CA(3894).
F. roseum. 11:CA(3894).
F. solani. 1:UT(82).
Fusarium sp. Root rot. 4:ID(3202), OR(3202). 1:E & C states(82), NW states(82), SE states(82), CA(1110), FL(35), NC(1303). 11:FL(35).
Gliocladium roseum. 11:CA(3894).
Hormodendrum sp. 4:OR(3202).
Idriella lunata. Root rot. 'Black root.' 1:CA(2582,2584), MD(2214).
Isaria sp. 4:OR(3202).
Mycogone sp. 4:OR(3202).
Oidium sp. Powdery mildew. 1:HI(2903).
Penicillium sp. Secondary fruit rot. 4:ID(3202). 1:CA(1110), ID(82), NY(82), OR(82), UT(82), WA(82).
Phialophora cinerescens. 4:OR(3202).
Ramularia brunnea. Black seed disease, common leaf spot. 3:CA(638,642). 5:SD(2243). 7:WI(1339), WY(2924). 9:DE(3543), MA(1390), NC(3944), NY(2456). 1:HI(2903), NC(1303).
R. fragariae. Leaf spot. 1:FL(35), HI(2899,2903). 11:FL(35).
Ramularia sp. Root rot. 1:ID(82), OR(82), WA(82).
Stemphylium sp. Black root. 4:OR(3202). 1:OR(82).
Trichothecium sp. 1:CA(1110).
Verticillium albo-atrum. Wilt. 4:OR(3202). 1:'Verticillium wilt.' CA(82), FL(35,1654), NY(82), OR(82).
V. dahliae. 'Verticillium wilt.' 1:CA(1110).
Verticillium sp. Black root. 4:ID(3202). 1:ID(82). 11:OR(3202).

Deuteromycotina - Coelomycetes

Ascochyta sp. Leaf spot. 1:NC(1303).
Colletotrichum acutatum. Crown, fruit and petiole lesions. 'Anthracnose. Anthracnose fruit rot.' 1:AR(3339), CA(1110,3339), FL(3339), MO(3339), MS(3339).
C. dematium. 'Anthracnose. Anthracnose fruit rot.' 1:IL(245).
C. gloeosporioides. Anthracnose, ripe fruit rot. 9:NC(1303). 1:'Anthracnose. Anthracnose fruit rot.' FL(35,82,1647,1650,1655,1656), LA(1650), NC(1303), OH(976). 11:FL(35,162).
Colletotrichum sp. Anthracnose. 1:MD(2216).
Coniella castaneicola. Isolated from fruits. 1:MD(2217).
Coniothyrium fuckelii. Black root. 1:MI(82), NC(82), OK(82), UT(82).
Coniothyrium sp. Black root, black rot. 1:MI(82), NC(82), OK(82), UT(82). 11:OK(2873).
Diplodia sp. Root rot. 1:FL(35,82).
Gloeosporium sp. Fruit rot. 11:LA(3962).
Kabatia fragariae. On living leaves. 5:WY(1195,3376).
Macrophomina phaseolina. 11:CA(3894), IL(3698).
Marssonina fragariae. Leaf scorch. 10:CA(81). 1:Unk.(82), FL(35).
M. potentillae var. *tormentillae*. 7:WI(1337).
Melanconium sp. 4:NW states(3202).
Oothecium sp. 4:OR(3202).
Pestalotia longisetula. Cortical decay of petioles and stolons, fruit rot. 1:FL(1651), IL(3063).
Pestalotia sp. Necrotic leaf spot. 4:OR(3202). 1:NC(1303).
Phoma terrestris. 11:CA(3894).
Phoma sp. Leaf blight, on leaves, roots and stolons. 4:OR(3202). 1:CT(82), FL(35,82), IA(82), ID(82), IL(82), IN(82), NC(82), NJ(82), OH(82). 11:FL(35).
Phomopsis obscurans. Leaf blight and spot, stem end rot of fruit, tan rot of fruit. 4:OR(3202), WA(3202). 9:CT(82), IN(82), MD(82), MS(2720), NC(82). 1:'Phomopsis leaf blight. Phomopsis soft rot.' E states(82), FL(35,82,1649), IL(82), LA(82), MI(82), MN(82), NE(82), OR(82), SD(2243), TX(82). 11:FL(35), OK(2872).
Phomopsis sp. On decaying roots and stolons. 1:NC(1303).
Phyllosticta fragaricola. Leaf spot. 7:NC(82). 9:WI(1357). 1:CA(1110), CT(82), FL(35,82), IA(82), IN(82), NC(82), NJ(82). 11:FL(35).
Phyllosticta sp. 9:WI(1354).
Rhabdospora fragariae. On leaves. 7:NY(82). 1:MD(82), NY(82).
Septoria aciculosa. Leaf spot. 9:WI(1333). 1:IL(82), MD(82), NY(82), VA(82), WV(82). 11:FL(35).
S. fragariae. 4:CA(81).
Septoria sp. 4:OR(3202).
Sphaeropsis sp. Fruit rot. 1:CO(82), IL(82).
Zythia fragariae. Leaf blotch. 1:NC(1303). 11:CA(3894), ME(1083).

Deuteromycotina - Other

Rhizoctonia fragariae. Black root rot. 11:WV(2965).
R. solani. Blossom, foliage and web blight, bud and root rot. 4:MS(2720), WA(82). 1:'Hard brown rot.' Range of host(82), CA(1110), FL(35), NC(1303). 11:CA(3894), FL(35), OK(2872).
Rhizoctonia sp. Leaf blight, crown and root rot. 1:FL(82), HI(2903). 11:MD(2477).
Sclerotium deciduum. 9:WI(1342).
S. rolfsii. Southern blight. 7:FL(82), NC(1303). 9:FL(82). 1:'Southern blight.' AL(82), FL(35,82), NC(82), TX(82). 11:FL(35).

Geum L.

1. *G. aleppicum* Jacq. (=*G. aleppicum* Jacq. var. *strictum* (Aiton) Fernald, *G. strictum* Aiton) [avens]
2. *G. canadense* Jacq.
3. *G. chiloense* Balb. ex Ser. in DC. A name of no botanical standing for *G. quellyon*.
4. *G. glaciale* Adams
5. *G. macrophyllum* Willd.
6. *G. radiatum* Michx.
7. *G. rivale* L. [chocolate-root, Indian chocolate, purple avens, water avens]
8. *G. rossii* (R. Br.) Ser.
9. *G. triflorum* Pursh (=*Sieversia ciliata* G. Don)
10. *G. triflorum* Pursh var. *ciliatum* (Pursh) Fassett
11. *G. turbinatum* Rydb. (=*Sieversia turbinata* (Rydb.) Greene)
12. *G. virginianum* L.
13. *Geum* sp.

Chytridiomycetes

Synchytrium aureum. Leaf gall. 1:WI(82). 2:WI(82). 12:WI(82).

Oomycetes

Peronospora potentillae. Downy mildew. 2:IA(82), IL(82), IN(82), KS(82), MS(2720), NE(82), WI(82). 3:CA(82). 5:WI(82). 7:WI(82). 9:WI(82). 13:WA(82).

Ascomycotina

Crocicreas cyathoideum var. *cacaliae*. On dead stems. 5:UT(505).
Erysiphe polygoni. Powdery mildew. 5:AK(82,526), WA(3202).
Hainesia lythri. 9:WI(1342,1361).
Lophiostoma sieversiae. On stems. 11:UT(195).
Mycosphaerella tassiana. On dead stems. 4:AK(1938). 9:OR(3202).
Pleospora potentillae. 9:WY(3810).
Sphaerotheca macularis. Powdery mildew. 1:PA(82), WA(82,3202), WI(82). 2:IA(82), IN(82), MD(82), ND(82,377), NE(82), OH(82), WI(82). 5:CO(1195), WA(82,3202), WI(82). 7:MI(82). 12:NY(82), WI(82).

Basidiomycotina - Uredinales

Puccinia sieversiae. Autoecious. 8:AK(1938), CO(82). 11:AK(526), CO(111), UT(111,1520).

Basidiomycotina - Ustilaginales

Ustacystis waldsteiniae. Leaf smut. 9:MT(3202), WA(3202). 10:MT(1077), WA(1077). 12:MT(82), WA(82).

Deuteromycotina - Hyphomycetes

Eurotium herbariorum. 13:AK(526).
Phymatotrichopsis omnivora. Root rot. 1:TX(82). 2:TX(82).

Ramularia gei. Leaf spot. **1**:WI(82,1332,1357). **5**:ID(3202), MT(3202), WA(3202), WY(1195). **9**:MT(3202), WA(3202), WI(82,1330,1333). **13**:MO(82).
Sclerotiomyces colchicus. **2**:WI(1356).
Deuteromycotina - Coelomycetes
Cylindrosporium gei. Leaf spot. **6**:WI(82).
C. triflori. On leaves. **9**:WI(82,1331).
Kabatia fragariae. On leaves. **13**:WY(82).
Marssonia adunca. Leaf spot. **1**:ID(82). **5**:MT(82), WA(82).
Marssonina fragariae. **1**:ID(3202), WA(3202). **5**:MT(3202), WA(3202).
Phoma exigua. **1**:WI(1336). **2**:TX(82), WI(64,82,1333).
Phyllosticta sp. Leaf spot. **2**:WI(82,1361), WV(82).
Seimatosporium discosioides. On leaves. **9**:WI(82,1330).
Septoria gei. Leaf spot. **1**:MI(82), WI(82). **2**:IL(82), MI(82), NE(82), OH(82), WI(82). **5**:ID(3202), MT(3202), VT(82), WI(82), WY(82). **12**:DE(82), MA(82), NY(82), WI(82). **13**:AK(82,526), IL(82).

Gillenia Moench

1. *G. stipulata* (Muhl.) Baill. [American ipecac]
2. *G. trifoliata* (L.) Moench [bowman's-root, Indian-physic]

Basidiomycotina - Uredinales
Gymnosporangium exterum. Heteroecious: aecial host. **1**:IN(82), KY(82), MO(82), NC(82,3944). **2**:TN(82), VA(82), WV(82).

Heteromeles M. Roem.

1. *H. arbutifolia* M. Roem. (=*Photinia arbutifolia* Lindl.) [Christmasberry]
2. *Heteromeles* sp.

Oomycetes
Phytophthora cactorum. Crown rot. **1**:CA(1865).
Ascomycotina
Dasyscyphus cerinus. **2**:CA(2439).
Eutypella ceranata. On dead wood. **1**:CA(2439).
Gloniopsis praelonga. On dead wood. **2**:CA(2439).
Glonium abbreviatum. On dead dry wood. **1**:CA(2439).
Helotiella microspora. On bark. **1**:CA(2439).
Hypoxylon rubiginosum. **2**:CA(2439).
H. truncatum. On dead log. **1**:CA(2439).
Hysterographium bakeri. On dead log. **1**:CA(2439).
H. flexuosum. **1**:CA(2439).
Limacinula anomala. On leaves and limbs. **1**:CA(81,2434,2439).
Lophodermium heteromelis. Leaf spot. **1**:CA(82).
Myriangium duriaei. On scale insects infesting bark. **1**:CA(82).
Phacidium heteromelis. **1**:CA(864).
Physalospora heteromelina. On leaves. **1**:CA(2439).
Pyrenopeziza heteromeles. On leaves. **1**:CA(82).
Stictis radiata. On dead bark. **2**:CA(2439).
Vertixore atronitidum. On leaves. **1**:CA(81,2434).
Basidiomycotina - Uredinales
Gymnosporangium exiguum. Heteroecious: aecial host. **1**:CA(81).
G. japonicum. **1**:CA(82).
Basidiomycotina - Other
Athelia arachnoidea. On dead leaves and twigs. **2**:CA(2439).
Ceriporia viridans. **2**:CA(2439).
Grandinia sp. On bark. **1**:CA(2439).
Hyphoderma mutatum. On wood. **2**:CA(2439).
Phanerochaete chrysosporium. **1**:CA(128).
P. sordida. White rot. **1**:CA(433).
Pleurotus septicus. On bark. **2**:CA(2439).
Resupinatus applicatus. On bark. **1**:CA(2439).
Stereum hirsutum. **2**:CA(543,2439).
Tremella mesenterica. On dead limbs and wood. **1**:CA(3949). **2**:CA(2439).
Deuteromycotina - Hyphomycetes
Cercospora sp. **1**:GA(82).
Diplocarpon mespili. Leaf spot. **1**:CA(82).
Hadrotrichum heteromelis. On leaves. **1**:CA(82).
Phymatotrichopsis omnivora. Root rot. **1**:TX(82).
Spilocaea photinicola. Scab, on leaves and berries. **1**:CA(81,82), TX(82).
Stigmina heteromelis. Leaf spot. **1**:CA(82,581,1468), TX(82).
Deuteromycotina - Coelomycetes
Ceuthospora brevispora. On leaves. **1**:CA(82).
Chaetophoma setigera. On leaves. **1**:CA(2434).
Diplodia heteromelina. On dead twigs. **2**:CA(2439).
Discosia poikilimera. On dead leaves. **1**:CA(2439).
Entomosporium mespili. Leaf spot. **1**:CA(81), FL(35).
Phyllosticta heteromeles. On leaves. **1**:CA(82,864,2439), TX(82).
Septoria photiniae. Leaf spot. **1**:CA(82).
S. rhabdocarpa. On dead leaves. **1**:CA(2439).

Holodiscus (K. Koch) Maxim.

1. *H. discolor* (Pursh) Maxim. (=*H. ariaefolius* (Sm.) Greene) [creambush, oceanspray, rock spirea]
2. *H. discolor* (Pursh) Maxim. var. *glabrescens* A. Heller
3. *H. dumosus* (Nutt. ex Hook.) A. Heller [mountain-glory, rock spiraea]
4. *H. franciscana* (Rydb.) Rehd. (=*Sericotheca franciscana* Rydb.)
5. *H. microphyllus* Rydb.
6. *Holodiscus* sp.

Ascomycotina
Biscogniauxia nummularia. **1**:CA(81).
Coniochaeta sp. **5**:CA(661).
Dasyscyphus patulus. **1**:ID(3202).
Glyphium corrugatum. **1**:WA(3202).
Helotium sp. **2**:CA(642).
Leptosphaeria sambuci. On limbs. **1**:WA(3202). **6**:WA(82).
Metasphaeria anisometra. On dead twigs. **1**:CA(2439).
Mycomicrothelia macularis. **5**:CA(661).
Mycosphaerella tassiana. **1**:ID(3202).
Nectria cinnabarina. Coral spot. **1**:WA(3202). **6**:WA(82).
Phyllactinia guttata. Powdery mildew. **1**:WA(3202). **6**:WA(82).
Podosphaera clandestina. Powdery mildew. **6**:ID(82,3202).
Propolomyces versicolor. **1**:ID(3202).
Schizoxylon insigne. **4**:CA(2439).
Valsa ceratosperma. On limbs. **1**:WA(3202,3415). **6**:Unk.(82).
V. holodiscina. On dead twigs. **1**:CA(2439).
Xylaria hypoxylon. **1**:WA(3202).
Basidiomycotina - Other
Dendrothele incrustans. **1**:CA(2084).
Exidia glandulosa. **1**:ID(3202).
Hymenochaete tabacina. **1**:ID(3202), WA(3202).
Hyphoderma amoenum. **1**:ID(3202).
Schizopora paradoxa. **1**:ID(3202), WA(3202).
Trametes hirsuta. **1**:ID(3202).
Deuteromycotina - Hyphomycetes
Peyronelia sirodesmioides. On dead limbs and twigs. **2**:CA(639,642).
Rhopalidium cercosporelloides. Leaf spot. **6**:ID(82).
Deuteromycotina - Coelomycetes
Camarosporium sp. **1**:ID(3202).
Cylindrosporium filipendulae. **1**:CA(81), ID(3202), WA(3202). **6**:WA(82).
Cylindrosporium sp. **3**:ID(3202).
Diplodia holodisci. On limbs. **6**:WA(82,3202).
Phloeosporella ariaefoliae. Leaf spot. **1**:ID(3202), OR(3202), WA(3202). **6**:OR(82).
Septogloeum schizonoti. Leaf spot. **1**:WA(3202). **6**:WA(82).

Horkelia Cham. & Schlechtend.

1. *H. congesta* Douglas ex Hook.
2. *H. fusca* Lindl.
3. *H. tridentata* Torr.

Basidiomycotina - Uredinales
Phragmidium ivesiae. **2**:UT(3111). **3**:CA(3111).
Deuteromycotina - Hyphomycetes
Ramularia arvensis. **1**:OR(3202).

Ivesia Torr. & A. Gray

1. *I. baileyi* S. Wats.
2. *I. gordonii* (Hook.) Torr. & A. Gray (=*Horkelia gordonii* Hook.)
3. *I. sericoleuca* Rydb.

Ascomycotina
Crocicreas cyathoideum var. *cacaliae*. On dead stems. **2**:UT(505).
Pleospora herbarum var. *occidentalis*. **2**:WY(3810).
Basidiomycotina - Uredinales
Phragmidium horkeliae. Autoecious. **2**:UT(82,3111).
P. ivesiae. Autoecious. **3**:CA(82,3111).
P. jonesii. Autoecious. **1**:NV(82), OR(82,3202). **2**:UT(82).

Deuteromycotina - Coelomycetes
Septoria sp. **2:**MT(3202).

Kerria DC.

1. *K. japonica* (L.) DC. [globeflower, Japanese-rose, kerria]

Ascomycotina
Blumeriella kerriae. Leaf and twig blight. **1:**E & C states(82), Range of host(82), GA(82,1438), MS(2720), NC(1303), TX(82).
Nectria cinnabarina. Coral spot. **1:**NY(82), OR(82,3202), WA(82,3202).
Deuteromycotina - Hyphomycetes
Phymatotrichopsis omnivora. Root rot. **1:**TX(82).
Deuteromycotina - Coelomycetes
Cylindrosporium kerriae. Leaf and twig blight. **1:**NC(1303).
Phomopsis japonica. Twig blight, canker. **1:**NC(1303), NJ(82), OH(82), TX(82).
Septoria sp. Leaf spot. **1:**MD(82), NJ(82).

Laurocerasus Duhamel

This genus is considered a synonym of *Prunus*, but the appropriate combination could not be located.

1. *L. lyonii* Britton

Ascomycotina
Propolomyces versicolor. **1:**CA(2439).
Schizoxylon insigne. **1:**CA(2439).
Basidiomycotina - Other
Peniophora malenconii subsp. *americana*. On wood. **1:**CA(542).
Pleurotus ostreatus. On bole of living tree. **1:**CA(2439).
Deuteromycotina - Coelomycetes
Discosia poikilimera. On dead leaves. **1:**CA(2439).
Phyllosticta laurocerasi. On leaves. **1:**CA(2439).

Lyonothamnus A. Gray

1. *L. floribundus* A. Gray

Ascomycotina
Hysterographium flexuosum. On dead wood. **1:**CA(2439).
H. mori. On bare wood. **1:**CA(2439).
Strickeria catalinae. On bare wood. **1:**CA(2439).
Basidiomycotina - Other
Antrodiella incrustans. On bark. **1:**CA(2439).
Hymenochaete tabacina. On wood. **1:**CA(2439).
[*Stereum gausapatum*]. On bark. **1:**CA(2439).
S. hirsutum. On wood. **1:**CA(2439).

Malus Mill.

1. *M. angustifolia* (Aiton) Michx. (=*Pyrus angustifolia* Aiton) [American crabapple, southern wild crabapple]
2. *M.* × *arnoldiana* (Rehd.) Rehd. (=*Pyrus arnoldiana* (Rehd.) Bean). *M. baccata* × *M. floribunda*.
3. *M. baccata* (L.) Borkh. (=*Pyrus baccata* L.) [Siberian crabapple]
4. *M. betulifolia* Wenz.
5. *M. coronaria* (L.) Mill. (=*Pyrus coronaria* L.) [American crabapple, garland crabapple, sweet-scented crabapple, wild sweet crabapple]
6. *M. domestica* Borkh. (=*Pyrus malus* L.) [common apple]
7. *M. floribunda* Siebold ex Van Houtte [Japanese flowering crabapple, purple chokeberry, showy crabapple]
8. *M. fusca* (Raf.) C. K. Schneid. (=*M. diversifolia* (Bong.) M. Roem., *Pyrus fusca* Raf., *Pyrus rivularis* Douglas ex Hook.) [Oregon crabapple]
9. *M. ioensis* (A. Wood) Britton (=*Pyrus ioensis* (A. Wood) L. H. Bailey) [prairie crabapple, wild crabapple]
10. *M. prunifolia* (Willd.) Borkh. (=*Pyrus prunifolia* Willd.) [Chinese apple, plum-leaved apple]
11. *M. pumila* (L.) Mill. [paradise apple]
12. *M.* × *purpurea* (Barbier) Rehd. *M.* × *atrosanguinea* × *M. pumila* cv. 'Niedzwetzkyana'.
13. *M.* × *robusta* (Carrière) Rehd. *M. baccata* × *M. prunifolia*.
14. *M. rockii* Rehd.
15. *M. sieboldii* (Regel) Rehd. [Toringo crabapple]
16. *M. spectabilis* (Aiton) Borkh. [Chinese flowering crabapple]
17. *M. sylvestris* Mill. [apple, crabapple, paradise apple, wild apple]
18. *Malus* sp.

Oomycetes
Phytophthora cactorum. Crown and fruit rot. **17:**'Collar rot. Sprinkler rot.' Range of host(82), AR(82), CA(1110), CT(3770), FL(35), ID(3202), ME(908), NC(1303), NY(24), PA(1073), WA(3202). **18:**MO(1429), NC(1429), OR(1429), WA(1429), WI(1429).
P. cambivora. Collar and crown rot. **17:**CA(1110), NC(1303).
P. cryptogea. **17:**CA(1110).
P. drechsleri. **17:**CA(1110).
P. megasperma. **17:**CA(1110).
Phytophthora sp. Collar and root rot. **17:**NC(1303,3559).
Pythium splendens. Root rot. **17:**FL(35).
Pythium sp. **17:**CA(1110).
Zygomycotina
Mucor mucedo. **17:**WA(3202).
M. piriformis. Fruit rot. **17:**NW states(3419), WA(82,3202).
M. racemosus. **17:**WA(3202).
Rhizopus stolonifer. Fruit rot, on moldy carpels. **17:**Range of host(82), CA(1110), FL(35), MT(3202), NY(536), OR(3202), WA(3202), WV(16).
Ascomycotina
Biscogniauxia marginata. Blister canker. **5:**WV(82). **6:**IA(1225), NC(3944). **11:**GA(1438). **17:**'Blister canker.' E & C states(82), MS(2720), OK(2872).
Botryosphaeria dothidea. Dieback, isolated from stem lesions. **11:**GA(1438). **17:**'Botryosphaeria canker. Fruit rot.' CA(1110), FL(35), NC(1303). **18:**GA(400).
B. obtusa. Trunk canker, black rot of fruit, frog-eye leaf spot. **3:**E & C states(82). **5:**IL(82), NY(82). **6:**GA(237). **7:**AL(82), NJ(82). **9:**IL(82). **17:**'Black rot. Frogeye leaf spot. Canker.' E & C states(82), W states(82), CA(82,1110), CT(3770), GA(3581), ID(3202), MS(2720), OK(2872), OR(3202), WA(3202). **18:**CT(2974,3770), MI(2023).
B. quercuum. Limb and twig canker, black rot of fruit, frog-eye leaf spot. **1:**FL(35). **17:**AL(285), FL(35), NC(1303).
B. rhodina. On roots. **17:**KY(82).
B. ribis. Canker, dieback. 'Botryosphaeria canker. Botryosphaeria fruit rot.' **17:**E & C states(82), E states(82), S states(82), GA(3581), OK(2872), WA(695,3202), WV(16).
B. stevensii. Black rot. **17:**CA(82,1110), MT(82,3202), OR(82,3202), WA(82,3202).
Botryosphaeria sp. Black fruit rot. **17:**HI(2903).
Cenangium tuberculiforme. On dead limbs. **17:**NC(1303). **18:**NC(1307).
Chaetomium funicola. On fruits. **17:**WV(16).
C. globosum. On fruits. **17:**WV(16).
C. trilaterale. On fruits. **17:**WV(16).
Daldinia concentrica. Wood rot. **17:**Range of host(82), WA(3202).
D. occidentalis. **17:**ID(3202).
D. vernicosa. On wood. **6:**WA(879).
Diapleella coniothyrium. 'Leptosphaeria canker. Fruit rot.' **17:**C states(82), E states(82), NY(536), OK(82,2873), WA(82,3202).
Diaporthe perniciosa. Leaf spot. **17:**AR(82).
Discohainesia oenotherae. On leaves and twigs. **17:**MD(82).
Elsinoe piri. Anthracnose, on fruit and leaves. **17:**OR(82,3202), WA(82,1745,3202).
Endomycopsis mali. Fruit rot. **17:**ME(82).
Erysiphe graminis f. sp. *avenae*. Powdery mildew. **17:**AR(3029).
Eutypella leprosa. **6:**IA(3629).
E. prunastri. **17:**OR(3202).
E. scoparia. **6:**IA(3629).
Gibberella baccata. Twig canker. **17:**OR(82,3202), WA(3202).
Glomerella cingulata. Branch, limb, trunk and twig canker, fruit rot, withertip. **1:**FL(35). **6:**SE states(3561). **11:**GA(1438). **17:**'Bitter rot.' Range of host(82), SE states(82), W states(82), AL(285), CA(1110), FL(35), IL(82), MS(2720), NC(1303,3199), OK(2872), OR(3202), TX(82), VA(3494), WA(3202). **18:**MD(162).
Gloniopsis praelonga. **8:**CA(81).
Glyphium corrugatum. **17:**WA(3202).
Hypoxylon atropunctatum. Sapwood rot. **17:**Unk.(82).
H. howeianum. Sapwood rot. **17:**Unk.(82).
H. mammatum. **17:**MT(3202).
H. mediterraneum. On fruits. **17:**WV(16).
H. rubiginosum. Sapwood rot. **17:**Unk.(82).
H. serpens. Sapwood rot. **17:**Unk.(82).
Hysterium pulicare. **17:**AK(526).
Hysteropatella sp. **17:**WA(3202).
Jattaea microtheca. On limbs. **18:**NJ(190).

Leptosphaeria sp. Stem canker. **1:**FL(35). **17:**FL(35).
Leucostoma persoonii. Canker dieback. **3:**WV(82). **5:**WV(82). **6:**IA(1226). **9:**IA(1226). **17:**NC states(82), NE states(82), W states(82), ID(3202), OK(2872), WA(3202). **18:**AK(526).
Massaria pyri. **17:**ME(2783), NY(2783), TN(2783), VA(2783).
Melanopsamma improvisa. **17:**ND(377).
M. pomiformis. **17:**OR(3202).
Mollisia caespiticia. **18:**WA(3202).
Monilinia fructicola. American brown rot of fruit. 'American brown rot.' **17:**Unk.(82), OK(2872), SD(2243), TX(82), WA(3202), WV(16).
M. laxa. 'Brown rot blossom blight.' **17:**OR(82,3202), WA(82,3202).
Mycosphaerella pomi. Fruit spot. **6:**NC(3558). **17:**'Phoma fruit spot.' E states(82), AR(82), IA(82), MO(82).
M. pyri. Leaf spot. **17:**IL(82), NJ(82), PA(82).
M. tulasnei. Fruit rot. **17:**OR(3202), WA(82,3202).
Mycosphaerella sp. On leaves. **6:**WI(1345).
Myriangium superficialis. On scale insects infesting bark. **1:**GA(1438). **11:**GA(1438). **17:**GA(82).
Nectria cinnabarina. Twig blight, branch canker, coral spot. **3:**AK(82,526). **8:**AK(82,526). **17:**'Nectria twig blight.' N states(82), AK(526), NC(1303), OR(3202), SD(2243), WA(3202). **18:**AK(526), NY(1796).
N. coccinea. **17:**OR(3202).
N. ditissima. **17:**WA(3202).
N. galligena. 'Nectria canker.' **17:**E & C states(82), W states(82), CA(907,1110,2304,2599), MS(2720), NC(1303), OR(3202), WA(3202).
N. sanguinea. **17:**OR(3202).
N. vulpina. On dead wood. **18:**NJ(3085).
Nectria sp. Twig canker. **1:**FL(35).
Otthia amica. **17:**WA(3202).
Pezicula malicorticis. Anthracnose, black spot canker, fruit rot. **8:**OR(82,3202). **17:**'Anthracnose. Bull's eye rot. Perennial canker.' CA(82,1110), ID(82,3202), IL(82), MA(82), ME(82), MT(82,3202), OK(82,2872), OR(82,3202), WA(82,3202).
P. pruinosa. **17:**OR(3202), WA(3202).
Peziza repanda. **17:**WA(3202).
Pezizella regalis. On bark. **18:**NJ(850).
Phyllactinia guttata. Powdery mildew. **3:**WA(82,3202). **8:**WA(82,3202).
Pleospora herbarum. Fruit rot. **17:**W states(82), CA(1110), OR(3202), WA(82,3202).
P. mali. On twigs. **17:**NY(82).
Podosphaera clandestina. Powdery mildew. **3:**WA(82). **5:**WI(82). **17:**Range of host(82), CA(1110), FL(35), ID(3202), MS(2720), SD(2243), WA(3202).
P. leucotricha. Powdery mildew. **1:**FL(35). **3:**WA(3202). **7:**IL(82). **11:**GA(1438). **15:**IA(82). **17:**'Powdery mildew.' Range of host(82), CA(1110), FL(35), GA(3581), ID(3202), NC(1303), NM(2112), OK(2872), OR(3202), WA(3202). **18:**AK(526), MI(3560), NC(1303).
Podosphaera sp. **9:**CA(1110). **17:**AR(3311).
Roesleria pallida. Root rot of seedlings. **17:**NY(82).
Rosellinia aquila. **6:**GA(1438).
R. necatrix. 'Dematophora root rot.' **17:**CA(82,1110).
Schizothyrium perexiguum. Fly speck of fruit. **17:**CA(925,1110), FL(35).
S. pomi. Fly speck of fruit. **16:**FL(82). **17:**'Flyspeck.' Range of host(82), FL(35), MS(2720), NC(402,1303), OK(2872), WA(3202). **18:**FL(35), NC(1303).
Sclerotinia sclerotiorum. Calyx end rot. **6:**WA(696).
Sclerotinia sp. Blossom end rot. **17:**NC(1303).
Sphaerotheca fuliginea. Powdery mildew. **14:**CA(1155).
S. macularis. **6:**GA(1438).
Sphaerotheca sp. Powdery mildew. **17:**FL(35).
Sporormia sp. **17:**WA(3202).
Taphrina bullata. **17:**WA(3202).
Tympanis alnea. **18:**MD(2670), ME(2670), NJ(2670), NY(2670), WV(2670).
T. conspersa. On twigs. **17:**NH(82), NY(82), VA(82).
Valsa ambiens. Canker, dieback. **3:**Unk.(82). **5:**IA(1226). **6:**IA(1226). **17:**NC states(82), NE states(82), W states(82), OK(2872), OR(3202).
V. ambiens subsp. *leucostomoides*. **6:**IA(1226).
V. ceratosperma. **11:**GA(1438). **17:**GA(82).
V. papyriferae. On winter-injured bark. **17:**OR(82,3202).
Valsa sp. On twigs. **17:**NC states(82), NE states(82), W states(82), ID(3202), MT(3202), WA(3202).
Valsaria insitiva. **6:**IA(1226).
Valsella melastoma. On limbs. **6:**IA(1226). **17:**WA(82,3202).
Venturia inaequalis. On fruits and leaves, scab. **1:**FL(35), IL(1052), VA(82). **3:**Range of host(82), AK(526). **5:**Range of host(82). **6:**MA(1394). **9:**Range of host(82), CA(1110). **11:**GA(1438). **15:**Range of host(82). **17:**'Scab.' Range of host(82), AK(526), AL(285), CA(1110), CT(3770), FL(35), ID(3202), MS(2720), MT(3202), NC(1303), OK(2872), OR(3202), SD(2243), TN(72,73), WA(303,3202). **18:**CT(2974,3770), MA(2602), NC(1303), NH(1139), OH(2602), PA(2602), RI(2602), SD(2243).
Xenotypa aterrima. On bark. **17:**MN(82).
Xylaria curta. On roots. **18:**IN(3015).
X. longiana. Root rot. **17:**VA(82).
X. mali. Black root rot. **11:**GA(1438). **17:**'Black root rot.' SE states(82), AL(82), AR(82), IL(82), NC(1303,3559), OK(2872), PA(82). **18:**MD(3017), TN(3017), VA(3017).
X. polymorpha. Black root rot, wood decay. **17:**E states(82), GA(2400), NC(1303), OR(82,3202), VA(3018).
Xylaria sp. Black root-rot. **17:**AL(285).

Basidiomycotina - Uredinales

Gymnosporangium clavipes. Heteroecious: aecial host. **1:**FL(35). **7:**NJ(82). **16:**CT(82). **17:**'Quince rust.' E states(82), AR(82), MI(82), MS(2720). **18:**FL(35), NC(1303).
G. globosum. Heteroecious: aecial host. **1:**AL(82), GA(2405), IN(82), SC(82), VA(82). **3:**KS(82), OK(2872). **7:**NJ(82). **17:**'Hawthorn rust.' E states(82), AL(82,285), GA(1439), KS(82), MS(2720), NC(1303), NE(82).
G. juniperi-virginianae. Heteroecious: aecial host. **1:**E states(82), AL(82), FL(35), GA(1439,2405), IL(1052), MS(2720), NE(82). **3:**NE states(82), MO(82), NE(82). **5:**Range of host(82), NC(3944). **6:**IN(987). **7:**Range of host(82). **8:**Unk.(82). **9:**Range of host(82), MS(2720), ND(378), OK(2872). **11:**GA(1439). **15:**TN(82). **16:**CT(82). **17:**'Cedar apple rust.' E states(82), Range of host(82), AL(285), GA(3581), MS(2720), OK(2872), SD(2243). **18:**CT(2974), FL(35), NC(1303,1307).
G. libocedri. Heteroecious: aecial host. **3:**OR(82,3202). **7:**OR(82). **8:**AK(82), OR(82,3202). **9:**OR(3202). **17:**CA(82,1110), OR(82,3202).
G. nelsonii. **8:**AK(60,82,526), WA(82,3202).
G. nidus-avis. Heteroecious: aecial host. **17:**IN(82), MD(82), MS(82,2720), NJ(82).
G. nootkatense. Heteroecious: aecial host. **4:**AK(526). **8:**AK(60,82,526), WA(82,3202).
Gymnosporangium sp. **3:**FL(82). **17:**TN(72,73).

Basidiomycotina - Ustilaginales

Sporidiobolus pararoseus. On rusted leaves. **18:**KS(2686).

Basidiomycotina - Other

Abortiporus biennis. **17:**OR(3202). **18:**OR(4007).
Antrodia malicola. On wood. **17:**Range of host(82), WA(3202).
Armillaria mellea. Root rot. **8:**WA(82,3202). **17:**'Shoestring root rot.' Range of host(82), CA(1110,2897,2898), HI(2011,2903), ID(3202), MS(2720), NC(1303,3559), OK(2872), OR(3202), WA(3202). **12:**CA(2897). **13:**CA(2897). **18:**NC(1303).
A. tabescens. Root rot. **17:**'Clitocybe root rot.' AR(82), FL(35,82), NC(1303,3559), OK(82,2872), TX(82), VA(82). **18:**FL(35).
Athelia bombacina. Isolated from leaves. **11:**WI(1549).
Bjerkandera adusta. White rot. **6:**WA(879). **11:**MN(250). **17:**NE states(82), NY(82).
[*Boreostereum radiatum*]. **17:**NY(543).
Butlerelfia eustacei. 'Fish-eye rot.' **17:**E states(82), NW states(82), ID(82), IL(82), NY(82), OR(82), VA(82), WA(82,3202).
Ceriporia spissa. Wood rot. **6:**NY(2456), OH(638). **17:**MI(82).
Ceriporiopsis pannocincta. Wood rot. **17:**NY(82).
Cerrena unicolor. Wood rot. **11:**MN(250). **17:**Range of host(82).
Chondrostereum purpureum. Silver leaf, heart rot. **17:**'Silver leaf.' NW states(82), CA(1110), ID(3202), KS(82), ME(82,543), MN(82), NY(82), OR(3202), WA(543,3202), WI(3189). **18:**MN(543), WA(3447,3448).
Climacodon septentrionalis. Sapwood rot. **17:**Range of host(82).
Collybia sp. **17:**WA(3202).
Coniophora puteana. **17:**Range of host(82).
Coriolopsis gallica. White rot. **6:**WA(879). **17:**CO(82), MT(82,3202),

WA(82,3202), WI(82).
Corticium stevensii. Hypochnose. **11:**GA(491). **17:**AL(285), GA(1439).
Cristinia gallica. On wood. **17:**Unk.(82).
Cystostereum murrayi. White rot. **17:**VT(82,543). **18:**NY(543).
Dacrymyces minor. On wood. **17:**IA(128).
Dacryopinax spathularia. On wood. **11:**AZ(1202). **17:**IA(128).
Daedaleopsis confragosa. On wood. **17:**Range of host(82).
Dendrocorticium roseocarneum. **6:**NC(3944).
Dendrophora albobadia. White rot. **17:**TN(82).
D. erumpens. **17:**OR(543,3202). **18:**OR(4007).
D. versiformis. **11:**AZ(1202). **17:**MA(543).
Erythricium salmonicolor. Branch and limb blight. **17:**FL(35,82), LA(82), MS(2720). **18:**SE states(3639).
Flammulina velutipes. On wood. **6:**WA(879).
Fomes fomentarius. White spongy heart rot. **17:**NE states(82), ID(3202), MT(3202), NY(82), OR(82).
Fomitopsis pinicola. Brown cubical rot. **6:**WA(879). **17:**ID(82,3202), OR(82), WA(3202).
Ganoderma applanatum. White rot. **6:**WA(879). **8:**Unk.(82), WA(3202). **17:**C states(82), E states(82), ID(3202), MT(3202), NC(1303), WA(3202).
G. lucidum. White root and butt rot. **17:**GA(1439), NC(1303), NY(82).
Gloeophyllum sepiarium. Sapwood rot. **17:**Range of host(82), MT(3202), OR(3202).
G. trabeum. On wood. **17:**Range of host(82), OR(3202).
Grandinia spathulata. **11:**AZ(128,1202).
Heterobasidion annosum. Root rot. **1:**GA(2405).
Hymenochaete agglutinans. On limbs adhering to infected limbs. **17:**CT(82,1312).
Hyphoderma litschaueri. On bark. **17:**ND(82), OR(82,3202).
Hypholoma fasciculare. On wood. **6:**WA(879).
H. sublateritium. Root rot. **17:**NY(82).
Hypsizygus ulmarius. Wound rot. **17:**Unk.(82).
Irpex lacteus. On wood. **6:**NC(3944). **11:**MN(250). **17:**MD(82), MN(82), ND(82), NY(82), VT(82).
Ischnoderma resinosum. Sapwood rot. **17:**MN(82).
Laetiporus sulphureus. Sapwood rot. **17:**Unk.(82).
Laxitextum bicolor. **18:**IA(543).
Lenzites betulina. Sapwood rot. **11:**MN(250). **17:**Range of host(82).
Maireina marginata. On twigs. **17:**OR(82,3202), WA(3202).
Marasmius minutus. On twigs. **17:**IL(82).
Merulius sp. On wood. **6:**WA(879). **11:**MN(250).
Omphalotus olearius. Root rot. **18:**NC(1303).
Oxyporus latemarginatus. White rot. **17:**CA(82,1110), MS(2720), TN(82).
Panellus serotinus. **17:**OR(3202).
P. stipticus. **11:**MN(250).
Pellicularia koleroga. Thread blight. **1:**LA(82). **17:**'Hypochnus leaf blight.' Gulf states(82), SE states(82), IN(82), LA(3640), MS(2720). **18:**LA(3640).
Peniophora cinerea. Twig rot. **11:**MN(250). **17:**ME(82), NY(82).
Perenniporia tenuis. **11:**AZ(128,1202).
Phanerochaete flavido-alba. **6:**OH(128).
Phellinus ferruginosus. On wood. **6:**WA(879).
P. gilvus. Sapwood rot. **17:**IN(82).
P. igniarius. On wood. **6:**WA(879). **17:**CA(1110), OR(3202).
P. pomaceus. White heart rot. **6:**NC(3944). **17:**OH(82).
Phlebia merismoides. **6:**CT(128).
P. rufa. **6:**NH(128).
P. tremellosus. **6:**CT(128).
Pholiota adiposa. Brown cubical heart rot. **6:**WA(879). **17:**MN(82).
P. squarrosa. On wood. **6:**WA(879).
Pleurotus dryinus. **17:**GA(1439).
P. ostreatus. **11:**MN(250). **17:**GA(1439).
Pleurotus sp. On wood. **6:**WA(879).
Polyporus admirabilis. Sapwood rot. **17:**NE states(82), NY(82).
Postia caesia. Sapwood rot. **17:**NY(82).
P. stiptica. Sapwood rot. **17:**NY(82).
P. tephroleuca. **17:**NW states(82), IN(82), NC(82), VA(82).
Psathyrella incerta. On roots around stump. **6:**NY(3491).
Pulcherricium caeruleum. On wood. **17:**NC(1303).
Sarcodontia setosa. **6:**NY(2456). **17:**C states(82), E states(82).
Schizophyllum commune. On wood. **11:**MN(250). **17:**Unk.(82), CA(1110), GA(1439), OK(82,2873), OR(3202), SD(2243), WA(3202). **18:**AL(2054), GA(491).
Scytinostroma galactinum. Root rot. **17:**'Eastern white root rot.' AK(82), AL(2090), AR(2090), DE(82,669), IL(82,2090), IN(669), KY(2090), MD(82,669), MO(2090), NC(1303,3559), OK(2090), TN(669), VA(82,669), WV(2090). **18:**VA(667).
Septobasidium pseudopedicellatum. Parasitic on scale insects. **1:**FL(35). **17:**E states(82), FL(35), TX(82), WV(82).
Septobasidium sp. Parasitic on scale insects. **1:**FL(35). **17:**FL(35).
Spongipellis spumeus. Sapwood rot. **17:**CT(82).
Steccherinum ciliolatum. On wood. **18:**NY(1175).
S. ochraceum. Sapwood rot. **17:**Range of host(82).
[*Stereum gausapatum*]. White rot. **17:**OH(82).
S. hirsutum. **11:**MN(250). **17:**ID(3202), MT(82,3202).
Trametes hirsuta. **11:**MN(250). **17:**Range of host(82), MT(3202), WA(3202).
T. pubescens. White rot. **6:**WA(879). **17:**MN(82).
T. versicolor. White rot. **6:**WA(879,880). **11:**MN(250). **17:**Range of host(82), CA(1110), ID(3202), OK(2872), OR(3202), WA(697,3202).
Trechispora farinacea. **6:**OH(128).
Tyromyces chioneus. Sapwood rot. **17:**NW states(82), ID(3202), IN(82), NC(82), OR(3202), VA(82), WA(3202).
T. fissilis. Sapwood rot. **17:**NE states(82), NY(82).
T. galactinus. Sapwood rot. **17:**NE states(82), NY(82).
T. subgiganteus. Sapwood rot. **17:**NY(82).

Deuteromycotina - Hyphomycetes

Alternaria alternata. 'Moldy core.' **17:**CA(1110), MT(3202), NY(536), WA(3202).
A. mali. Cork rot. **17:**'Alternaria leaf spot. Alternaria storage rot.' Range of host(82), FL(35), WA(3202). **18:**IL(3697).
A. pomicola. **17:**WA(3202).
Alternaria sp. Fruit rot, leaf spot. **17:**NE states(2004), Range of host(82), ID(3202), NC(1303), NJ(2004), NY(2004), OR(3202), WV(16). **18:**FL(35).
Aspergillus clavatus. On fruits. **17:**WV(16).
A. elegans. On fruits. **17:**WV(16).
A. fumigatus. On fruits. **17:**WV(16).
A. niger. Fruit rot. **17:**Range of host(82), WV(16).
A. sclerotiorum. Fruit rot. **17:**Range of host(82), WA(3202).
A. wentii. On fruits. **17:**WV(16).
Aspergillus sp. Fruit rot. **17:**Range of host(82), CA(1110).
Aureobasidium pullulans. **17:**OR(3202), WA(3202).
Berkleasmium moriforme. On wood. **18:**NY(2468).
Botrytis cinerea. Rot. 'Gray mold.' **17:**NW states(3419), Range of host(82), CA(1110), GA(3581), NY(2693), WA(3202), WV(16).
B. mali. Fruit rot. **17:**WA(82,3202).
Botrytis sp. Gray mold rot. **17:**Range of host(82), NC(585), OR(3202).
Cephalosporium carpogenum. Fruit rot. **17:**PA(1073), WA(82,3202).
Cercospora pyri. **17:**IA(82).
Cercosporella pyrina. Leaf spot. **5:**IL(82), WI(82). **9:**WI(82).
Cladosporium cladosporioides. Fruit rot. **17:**CA(1110), WA(82,3202), WV(16).
C. elatum. On fruits. **17:**WV(16).
C. extorre. **5:**WA(3202).
C. herbarum. Fruit rot. **17:**Unk.(82), CA(1110), WA(3202).
C. malorum. Fruit rot. **17:**Unk.(82), WA(3202).
Cladosporium sp. Fruit rot. **17:**Unk.(82), OR(3202).
Clasterosporium sp. Papular scurf, on bark and twigs. **17:**WV(82).
Coniothecium sp. Blister. **17:**NM(82), WA(82,3202).
Cristulariella moricola. Leaf spot. **17:**NC(1303). **18:**FL(35).
Cylindrocarpon angustum. On bark. **17:**OR(82,3202).
C. magnusianum. Fruit rot. **17:**NW states(82), WA(3202).
Cylindrocarpon sp. **17:**WA(3202).
Cylindrocladium scoparium. Root rot. **17:**FL(35), NC(1303).
Diplocarpon mespili. Leaf spot. **3:**MD(82). **8:**AK(82,526), WA(3202). **17:**IA(82).
Epicoccum nigrum. Fruit rot. **17:**WA(82,3202), WV(16).
Fumago vagans. Sooty mold. **17:**Unk.(82).
Fusarium acuminatum. Canker. **17:**NW states(82).
F. avenaceum. Bull's-eye rot of dropped or stored fruit. **17:**E states(82), NW states(82), OR(3202), WA(3202).
F. equiseti. **17:**OR(3202), WA(3202).
F. heterosporum. **17:**OR(3202).
F. lateritium. Bull's-eye rot of dropped or stored fruit. **17:**E states(82),

NW states(82).
F. oxysporum. Root rot. **6:**ID(1514). **17:**NW states(82), OR(3202).
F. roseum. **17:**WA(3202).
F. solani. Root rot. **6:**ID(1514). **17:**CA(1110).
Fusarium sp. Canker, moldy carpels and core, bull's-eye rot of dropped or stored fruits. **17:**E states(82), NW states(82), CA(1110), MT(3202), NC(1303), NY(536), WV(16).
Fusicladium alopecuri. **17:**MT(3202).
Gliocladium viride. Fruit rot. **17:**IL(82).
Haplotrichum conspersum. In hollow stump. **18:**MA(2131).
Helminthosporium papulosum. Black pox, leaf spot. **17:**'Black pox.' GA(3580), IN(82), KY(1776), MA(82), MS(82,2720), NC(1303), NJ(82), OH(82), OR(3202), PA(82), WV(16,82). **18:**GA(3580,3582).
Helminthosporium sp. On fruits. **17:**WV(16).
Heterosporium maculatum. **17:**WA(3202).
Hormiscium sp. Fruit rot. **17:**MI(1797).
Illosporium malifoliorum. Leaf spot. **5:**WV(82). **17:**E states(82), IN(82).
Monilia sp. On fruits. **17:**WV(16).
Nematogonum ferrugineum. **17:**ND(378).
Nigrospora sphaerica. On fruits. **17:**WV(16).
Oidium pirinum. Powdery mildew. **5:**WI(82). **18:**FL(35).
Oospora mali. **17:**WA(3202).
O. otophila. **17:**WA(3202).
Oospora sp. Fruit rot. **17:**Unk.(82).
Penicillium aurantiogriseum. Blue-mold rot. **17:**Range of host(82), WA(3202).
P. expansum. 'Blue mold.' **17:**NW states(3419), Range of host(82), CA(1110), ID(3202), OR(3202), WA(3202), WV(16).
P. puberulum. Blue-mold rot. **17:**Range of host(82), WA(3202).
P. verrucosum. Blue-mold rot. **17:**Range of host(82), WA(3202).
P. viridicatum. Blue mold rot. **17:**Range of host(82), WA(3202).
Penicillium sp. Blue mold fruit rot. **17:**CA(1110), FL(35), NC(1303), OK(2872).
Phialophora malorum. Fruit rot and spot. 'Side rot.' **17:**CA(1110), IN(82), OR(3202), PA(3577), VA(82), WA(82,3202), WV(16).
Phymatotrichopsis omnivora. Root rot. **3:**TX(82). **17:**'Phymatotrichum root rot.' AR(82), AZ(82), NM(82), NV(82), TX(82).
Pseudocercospora mali. Leaf spot. **1:**FL(35). **17:**Gulf States(82), FL(35), MS(2720), VT(82). **18:**FL(35).
Ramularia macrospora. Fruit rot. **17:**MA(82), MD(82), WA(3202).
Spilocaea pomi. Scab. **17:**DE(3543), NC(1303), ND(378), OK(2872). **2:**WI(1355).
Stemphylium congestum. Fruit spot. **17:**CA(1110), WA(82).
S. congestum var. *minor*. Fruit spot. **17:**WA(82).
S. graminis. **17:**WA(3202).
Trichoderma sp. **17:**WA(3202).
Trichosporum sp. **17:**WA(3202).
Trichothecium roseum. Pink mold rot. 'Pink rot.' **17:**Range of host(82), OK(2872), WA(3202), WV(16).
Tripospermum myrti. On scab lesions. **17:**NC(1303).
Ulocladium consortiale. **17:**WA(3202).
Verticillium sp. **17:**WA(3202).
Zygophiala jamaicensis. Fly speck. **18:**CA(150).

Deuteromycotina - Coelomycetes

Aposphaeria fuscomaculans. Canker. **17:**CA(82,1110), MI(82).
Ascochyta mali. On limbs. **17:**IN(82), MI(82).
Ascochyta sp. Leaf spot. **17:**NC(1303).
Chaetomella sp. Fruit rot. **17:**WA(82,3202).
Coleophoma empetri. Fruit rot. **17:**MD(82).
Colletotrichum dematium. Spongy dry rot of fruit, on leaves and twigs. **17:**E states(82), IL(82).
C. gloeosporioides. Anthracnose, fruit rot. **1:**FL(35). **17:**Range of host(82), Unk.(3059), FL(35), NC(1303).
Coniothyrium convolutum. **17:**WA(3202).
C. cydoniae. **17:**WA(3202).
C. fuckelii. Bark canker, on fruits. **17:**NC(1303), WV(16).
C. pyrinum. Fruit and leaf spot, on twigs. **1:**FL(35). **5:**WI(82). **17:**E & C States(82), FL(35), WA(3202). **18:**WA(3202).
Coniothyrium sp. Fruit rot. **17:**WA(82).
Coryneum longestipitatum. On limbs. **17:**AK(82).
C. longistipitatum. **17:**AK(526).
Cryptosporiopsis corticola. 'Superficial bark canker. Fruit rot.' **17:**NE states(82), IL(82), MI(82), NC(1303), OK(82), OR(82,3202), SD(82), WA(3202).
C. curvispora. Canker. **18:**MI(2741).
C. pruinosa. Canker. **17:**OR(82).
Cylindrosporium sp. On fruits. **17:**WV(16).
Cytospora chrysosperma. 'Cytospora canker. Dieback' **17:**Range of host(82).
C. leucosperma. 'Cytospora canker. Dieback' **17:**Range of host(82).
C. leucosticta. 'Cytospora canker. Dieback' **17:**Range of host(82).
C. leucostoma. Canker. **17:**'Cytospora canker. Dieback' Range of host(82), OK(2872). **18:**NM(2111).
C. microspora. 'Cytospora canker. Dieback' **17:**Range of host(82).
C. personata. 'Cytospora canker. Dieback' **17:**Range of host(82).
C. rubescens. 'Cytospora canker. Dieback' **17:**Range of host(82).
Cytospora sp. Canker, dieback. **17:**Range of host(82), CA(1110), FL(35), NM(82), WV(16). **18:**FL(35).
Dendrophoma sp. **17:**WA(3202).
Diplodia malorum. On limbs. **17:**NC(82), NY(82), SC(82).
D. mutila. Canker, dieback, black rot, leaf spot. **17:**OR(82).
D. sarmentorum. On limbs. **17:**OR(82,3202).
Diplodia sp. On limbs. **17:**Unk.(82), FL(35).
Dothichiza sp. **17:**WA(3202).
Dothiorella mali var. *fructus*. On decaying fruit. **18:**TN(837).
Dothiorella sp. Canker. **17:**NC(1303).
Entomosporium mespili. Leaf spot. **17:**FL(35).
Fusicoccum pyrorum. Dieback. **17:**OR(3202). **18:**OR(4007).
Fusicoccum sp. **17:**WA(3202).
Gloeodes pomigena. Sooty blotch of fruit. **5:**IN(82). **17:**'Sooty blotch.' E & C states(82), AL(285), FL(35), MS(2720), NC(402,1303), OK(2872), WA(3202). **18:**NC(1303).
Gloeosporium sp. Fruit rot. **18:**MI(2741).
Greeneria uvicola. On fruits. **17:**WV(16).
Hendersonia cydoniae. On leaves. **17:**IL(82), NY(82).
H. mali. On living leaves. **17:**WI(1348).
Macrophoma sp. Dieback. **17:**FL(35).
Marssonia coronariae. **9:**WI(1332).
Marssonina coronaria. Leaf spot. **5:**IA(82), IN(82), WI(82). **9:**MO(82), WI(82).
Marssonina sp. Leaf spot. **1:**FL(35).
Microdiplodia sp. Wound rot of fruit. **17:**WA(82,3202).
Monochaetia concentrica. Leaf spot. **17:**SE states(82), ID(82,3202), IN(82).
Pestalotia hartigii. Leaf spot. **17:**WA(82,3202).
Pestalotia sp. Leaf spot. **17:**Unk.(82).
Phlyctema vagabunda. **17:**WA(3202).
Phoma fuliginea. **17:**WA(3202).
P. macrostoma. Fruit spot, on leaves, petioles and twigs. **6:**WI(1361). **17:**NE states(82), AR(82), FL(35), IN(82), OK(2872), WA(3202), WV(16,82).
P. pomorum. Scurfy bark, twig blight, fruit spot. **17:**SE states(82), KS(82), OH(82), WA(3202).
Phoma sp. Twig blight, dieback. **17:**Range of host(82), FL(35).
Phomopsis ambigua. Twig blight. **17:**Range of host(82).
P. mali. 'Rough bark.' **17:**C states(82), W states(82), CA(729,1110), NC(1303), OH(82), OR(3202), VA(82).
Phomopsis sp. Stem canker, on twigs. **3:**AK(82,526). **17:**FL(35).
Phyllosticta clypeata. On leaves, petioles, and twigs. **6:**OR(64,970). **17:**OR(82,3202).
P. solitaria. Fruit blotch, twig canker, leaf spot. **1:**KS(82). **3:**OH(82), TX(82). **5:**Unk.(2), IA(82), IN(82), WV(82). **6:**C states(64). **10:**WI(1340). **17:**'Blotch. Leaf spot. Twig canker.' C & W states(82), AL(285), FL(35), MS(2720), NC(1303), OK(2872), TX(82), WA(3202). **18:**KS(2), NC(1303).
P. zonata. Leaf spot. **9:**IA(82).
Phyllosticta sp. Leaf spot. **1:**FL(35). **6:**WI(1352). **7:**IA(82). **9:**WI(1361).
Pyrenochaeta mali. Fruit rot. **18:**IL(3345), WA(3345).
Scolicosporium pedicellatum. On bark. **17:**PA(82).
Seimatosporium lichenicola. Twig canker, on leaves. **17:**Range of host(82), AK(526), WA(3202).
Seiridium unicorne. Canker, on leaves. 'Twig canker.' **17:**E states(82), IL(2863), MO(82), NJ(970), WV(82).
Septoria pyricola. Leaf spot. **5:**IA(82). **17:**IA(82).
Sphaeropsis malorum. Canker. **1:**FL(35). **6:**MA(1390). **10:**NC(3944). **17:**S states(82), FL(35), NC(1303), WV(16).

Trichoseptoria fructigena. Fruit rot. **17:**NY(82).
Deuteromycotina - Other
Rhizoctonia solani. Root rot. **17:**CA(1110), FL(35), NY(455), TX(82), WA(82).
Rhizoctonia sp. Thread blight. **17:**NC(1303).
Sclerotium rolfsii. Seedling and southern blight, crown and southern stem rot. **17:**CA(1110), FL(35), KY(82), MD(82), NC(1303), TX(82), VA(82). **18:**CA(2886), IN(3206).

Mespilus L.
1. *M. germanica* L. [medlar]
Basidiomycotina - Uredinales
Gymnosporangium clavipes. **1:**IN(3756).
Deuteromycotina - Hyphomycetes
Diplocarpon mespili. Leaf spot. **1:**CA(82,1110).

Neillia D. Don
1. *N. opulifolia* Benth. & J. D. Hook.
Deuteromycotina - Coelomycetes
Marssonina neilliae. On living leaves. **1:**CA(1468).

Oemleria Rchb.
1. *O. cerasiformis* (Hook. & Arnott) J. W. Landon (=*Nuttallia cerasiformis* Hook. & Arnott, *Osmaronia cerasiformis* (Hook. & Arnott) Greene) [osoberry]
Ascomycotina
Nectria flammea. On limbs. **1:**CA(82).
Phyllactinia guttata. Powdery mildew. **1:**OR(82,3202).
Deuteromycotina - Coelomycetes
Ascochyta idaei. Leaf spot. **1:**WA(82,3202).
Cylindrosporium nuttallii. Leaf spot. **1:**CA(81,82,1468), OR(82,3202,4007), WA(82,3202).

Osteomeles Lindl.
1. *O. anthyllidifolia* (Sm.) Lindl.
2. *Osteomeles* sp.
Ascomycotina
Appendiculella calostroma. Black mildew. **2:**HI(82,2903).
Botryosphaeria obtusa. On limbs. **1:**HI(2903). **2:**HI(82).
Diaporthe sp. On limbs. **1:**HI(2903).

Peraphyllum Nutt. ex Torr. & A. Gray
1. *P. ramosissimum* Nutt. ex Torr. & A. Gray [squaw-apple]
Basidiomycotina - Uredinales
Gymnosporangium inconspicuum. Heteroecious: aecial host. **1:**CO(82).
G. nelsonii. **1:**CA(2771), CO(82), UT(82).
Deuteromycotina - Coelomycetes
Septoria peraphylli. Leaf spot. **1:**UT(82).

Photinia Lindl.
(= *Stranvaesia* Lindl.)
1. *P. davidiana* (Decne.) Cardot (=*Stranvaesia davidiana* Decne.)
2. *P. glabra* (Thunb.) Franch. & Sav. [Japanese photinia]
3. *P. serratifolia* (Desf.) Kalkman (=*P. serrulata* Lindl.)
4. *P. villosa* (Thunb.) DC.
5. *Photinia* sp. (=*Stranvaesia sp.*)
6. *Photinia* × sp. (unnamed hybrids)
Oomycetes
Phytophthora sp. Root rot. **2:**FL(35). **3:**FL(35).
Pythium sp. Root rot. **2:**FL(35). **5:**FL(35).
Ascomycotina
Botryosphaeria dothidea. Canker, dieback. **6:** NC(1303).
Glomerella sp. Leaf spot. **2:**FL(35).
Mycosphaerella sp. Leaf spot. **2:**FL(35).
Podosphaera leucotricha. Powdery mildew. **2:**CA(82), NC(1303). **3:**NC(1303).
Sphaerotheca pannosa. Powdery mildew. **3:**FL(35), NC(1303), OR(82,3202).
Basidiomycotina - Uredinales
Gymnosporangium clavipes. Heteroecious: aecial host. **4:**Unk.(82).
Basidiomycotina - Other
Armillaria mellea. Root rot. **6:** NC(1303). **5:**NC(1303).
A. tabescens. Dieback, root rot. **1:**FL(82). **2:**LA(2815). **5:**FL(35).
Trametes versicolor. On wood. **2:**NC(1303).
Deuteromycotina - Hyphomycetes
Alternaria sp. Leaf spot. **3:**FL(35). **5:**FL(35).
Cercospora photiniae-serrulatae. Leaf spot. **2:**FL(35). **3:**FL(35), LA(85). **5:**FL(35).
Cercospora sp. Leaf spot. **2:**NC(1303). **3:**GA(82), MS(2720), NC(1303).
Cylindrocladium scoparium. Leaf spot. **2:**FL(35).
Diplocarpon mespili. Leaf spot. **3:**MS(2720). **5:**NC(1303).
Fusarium sp. Root rot. **5:**FL(35).
Phymatotrichopsis omnivora. Root rot. **3:**TX(82).
Spilocaea photinicola. On berries and leaves, scab. **4:**OK(82,2873).
Stigmina heteromelis. Leaf spot. **3:**FL(35). **5:**FL(35), OK(2872).
Deuteromycotina - Coelomycetes
Cytospora sp. Dieback. **4:**OK(82,2873).
Discosia baarnensis. **5:**CA(3723).
Entomosporium mespili. Defoliation, leaf spot. **2:**FL(35), LA(2832,3471). **3:**FL(35), LA(3471). **5:**FL(35), NC(1303).
Entomosporium sp. Leaf spot. **2:**NC(1303).
Pestalotia sp. Leaf spot. **2:**FL(35). **3:**FL(35), NJ(82). **5:**FL(35).
Phaeoseptoria sp. Leaf spot. **2:**FL(35).
Phyllosticta sp. Leaf spot. **2:**FL(35). **5:**FL(35).
Septoria photiniae. Leaf spot. **3:**PA(82). **5:**NC(1303).
Sphaeropsis sp. Leaf spot. **6:** NC(1303).
Deuteromycotina - Other
Rhizoctonia solani. Root and stem rot. **2:**FL(35). **3:**FL(35). **5:**FL(35).

Physocarpus Maxim.
1. *P. alternans* (M. E. Jones) Howell
2. *P. bracteatus* (Rydb.) Rehd.
3. *P. capitatus* (Pursh) Kuntze
4. *P. malvaceus* (Greene) Kuntze
5. *P. opulifolius* (L.) Maxim. (=*Spiraea opulifolia* L.) [ninebark]
6. *Physocarpus* sp.
Ascomycotina
Botryosphaeria dothidea. **5:**GA(1438).
B. obtusa. On limbs. **6:**VA(82).
Coniochaeta ligniaria. **4:**ID(3202).
C. pulveracea. **6:**CO(649).
Cucurbitaria obducens. **4:**ID(3202).
Dasyscyphus corticalis. On dead twigs. **3:**CA(81).
D. patulus. **4:**ID(3202).
Dermea spiraeae. On limbs. **6:**NC(82), PA(82).
Diaporthe eres. On limbs. **6:**NY(82), WV(82).
Eutypella leprosa. **5:**IA(3629).
Haematomyxa tetraspora. On twigs. **3:**CA(527).
Leptosphaeria conoidea. **5:**NY(3249).
Lophiostoma spiraeae. On dead limbs. **6:**MI(82), NY(82), PA(82).
Mollisia cinerea. **3:**CA(81).
Phyllactinia guttata. Powdery mildew. **3:**CA(1155).
Sphaerotheca aphanis. Powdery mildew. **1:**CO(368). **5:**WI(368).
S. macularis. Powdery mildew, witches' broom. **5:**WI(1350). **6:**NC states(82), NE states(82), NY(2456).
Stictis radiata. **6:**CA(3225).
Teichospora sp. **3:**OR(3202). **4:**ID(3202), OR(3202).
Valsa ambiens. On dead limbs. **6:**NY(82), WV(82).
V. opulifoliae. On dead limbs. **6:**NY(82), WV(82).
Basidiomycotina - Other
Acanthophysium cerussatum. On wood. **4:**ID(3202). **6:**ID(2083).
Hymenochaete tabacina. **4:**ID(3202).
Phellinus conchatus. Wood rot. **6:**NY(82).
Stereum hirsutum. **3:**ID(3202).
Deuteromycotina - Hyphomycetes
Cercospora physocarpi. Leaf spot. **2:**OK(2872). **5:**OK(2872).
C. spiraeae. Leaf spot. **6:**IN(82).
Chaetotrichum macrosporum. **3:**CA(81).
Phymatotrichopsis omnivora. Root rot. **6:**TX(82).
Ramularia spiraeae. Leaf spot. **5:**NY(82). **6:**MI(82), NY(82), WI(82).
Stysanus sp. On living leaves. **5:**WI(1348).
Deuteromycotina - Coelomycetes
Cylindrosporium sp. **4:**MT(3202).
Cytospora globifera. **4:**ID(3202).
Kabatia lonicerae. Leaf spot. **6:**OR(82).
Marssonina neilliae. Leaf spot. **3:**ID(3202), OR(3202). **5:**CA(82), TX(82), WI(82). **6:**CA(82), TX(82), WI(82).

Monochaetia ellisiana. On dead twigs. **6:**WV(82).
Phyllosticta opulasteris. Leaf spot. **4:**ID(3202). **6:**ID(82).
P. physocarpi. On living leaves. **5:**WI(1340).
Seimatosporium parasiticum. On leaves. **6:**NY(82), WV(82).
Sphaeronaema physocarpi. On limbs. **6:**VA(82).
Sphaeropsis physocarpi. On limbs. **6:**WV(82).
Stagonospora physocarpi. On twigs. **5:**WV(82).

Potentilla L.

1. *P. amadorensis* Rydb.
2. *P. anserina* L. [goose-grass, goose-tansy, silverweed cinquefoil]
3. *P. argentea* L. [hoary cinquefoil, silvery cinquefoil]
4. *P. arguta* Pursh [tall cinquefoil]
5. *P. biennis* Greene
6. *P. biflora* Willd. ex Schlechtend.
7. *P. bipinnatifida* Douglas ex Hook.
8. *P. breweri* S. Wats.
9. *P. canadensis* L. (=*P. pumila* Poir.) [common cinquefoil]
10. *P. concinna* Richardson
11. *P. diversifolia* Lehm. (=*P. glaucophylla* Lehm.)
12. *P. drummondii* Lehm.
13. *P. fissa* Nutt.
14. *P. flabellifolia* Hook. ex Torr. & A. Gray
15. *P. fruticosa* L. [bush cinquefoil, golden hardhack, shrubby cinquefoil, widdy]
16. *P. fruticosa* L. var. *tenuifolia* Lehm. (=*Dasiphora fruticosa* (L.) Rydb. var. *tenuifolia* (Willd.) Rydb.)
17. *P. glandulosa* Lindl.
18. *P. glandulosa* Lindl. subsp. *nevadensis* (S. Wats.) D. Keck
19. *P. gracilis* Douglas ex Hook. (=*P. gracilis* Douglas ex Hook. var. *gracilis* , *P. concinnea* A. Gray var. *concinnea* , *P. flabelliformis* Lehm.)
20. *P. gracilis* Douglas ex Hook. var. *brunnescens* (Rydb.) C. Hitchc.
21. *P. gracilis* Douglas ex Hook. var. *elmeri* (Rydb.) Jeps.
22. *P. gracilis* Douglas ex Hook. var. *flabelliformis* (Lehm.) Nutt. ex Torr. & A. Gray
23. *P. gracilis* Douglas ex Hook. var. *glabrata* (Lehm.) C. Hitchc.
24. *P. gracilis* Douglas ex Hook. subsp. *nuttallii* (Lehm.) D. Keck (=*P. nuttallii* Lehm., *P. blaschkeana* Turcz. ex Lehm.)
25. *P. gracilis* Douglas ex Hook. var. *permollis* (Rydb.) C. Hitchc.
26. *P. gracilis* Douglas ex Hook. var. *pulcherrima* (Lehm.) Fernald (=*P. pulcherrima* Lehm., *P. filipes* Rydb.)
27. *P. grandiflora* L.
28. *P. hippiana* Lehm.
29. *P. hookeriana* Lehm.
30. *P. hyparctica* Malte
31. *P. millefolia* Rydb.
32. *P. nepalensis* Hook.
33. *P. newberryi* A. Gray
34. *P. norvegica* L. [rough cinquefoil]
35. *P. norvegica* L. var. *hirsuta* (Michx.) Lehm.
36. *P. palustris* (L.) Scop. [marsh firefinger]
37. *P. paradoxa* Nutt.
38. *P. pectinisecta* Rydb. (=*P. bakeri* Rydb.)
39. *P. pensylvanica* L. (=*P. strigosa* Pall. ex Pursh, *P. pensylvanica* L. var. *strigosa* (Pall. ex Pursh) Pursh)
40. *P. pentandra* Engelm.
41. *P. pulchella* R. Br.
42. *P. recta* L. [sulphur cinquefoil, upright cinquefoil]
43. *P. rivalis* Nutt.
44. *P. simplex* Michx. [old-field cinquefoil]
45. *P. subjuga* Rydb.
46. *P. tridentata* Aiton [three-toothed cinquefoil]
47. *P. trina* A. Nels.
48. *P. villosa* Pall. ex Pursh
49. *Potentilla* sp.

Chytridiomycetes
Physoderma comari. On leaves. **36:**MI(3400).
Synchytrium aureum. Leaf gall. **9:**PA(82). **34:**TX(82).
S. pilificum. On leaves, fruits and runners. **44:**NJ(1852).
S. potentillae. **11:**MT(3202), WA(3202). **17:**ID(3202), OR(3202).

Oomycetes
Peronospora potentillae. Downy mildew. **3:**IA(82), TX(82). **19:**ID(3202), OR(3202), WA(3202), WY(1195). **24:**WY(82,1195). **26:**WY(1195). **27:**NJ(82). **32:**NJ(82). **34:**NE states(82), TX(82), WI(82). **40:**ND(377). **49:**AK(526), ND(82).

Ascomycotina
Discohainesia oenotherae. Leaf blight. **9:**VA(82).
Erysiphe polygoni. Powdery mildew. **15:**CO(82).
Hainesia lythri. **3:**WI(1357). **4:**WI(1347). **44:**WI(1349).
Leptotrochila repanda. Leaf spot. **3:**IA(82), NY(82), VT(82). **34:**E states(82), MO(82), MS(2720), ND(82,377), SD(2243). **35:**WI(1358). **49:**MT(3202).
Massariosphaeria rubelloides. **14:**WA(3814).
M. rubicunda. **14:**WA(3202).
Mycosphaerella ranunculi. On leaves and dead stems. **6:**AK(1938). **36:**WI(82). **49:**AK(1938).
Mycosphaerella sp. **9:**CO(82). **34:**MN(82).
Physalospora potentillae. Leaf blight. **49:**TX(82).
Pleospora coloradensis. **14:**WA(3202).
P. herbarum. **41:**AK(526).
P. longispora. **14:**WA(3814). **49:**WA(3813).
Pyrenopeziza coloradensis. On stems. **9:**CO(82).
P. potentillae. On dead leaves. **49:**AK(1938).
Sphaerotheca macularis. Powdery mildew. **4:**WA(3202). **15:**CO(82), WI(82). **17:**WA(3202). **19:**OR(3202), WA(3202). **26:**OR(82), WA(82). **34:**IA(82), MI(82), TX(82). **36:**MI(82), WI(82). **49:**CA(82), CO(82).
Taphrina potentillae. Leaf blister. **4:**CT(2454), MI(2454). **9:**CT(82,2454), IN(82), MA(82,664,2454), MS(82,2720), NC(3944), NJ(82,2454), NY(82), OH(82,2454), TN(82,2454), VA(82), VT(2454), WI(82). **14:**CA(642,2454). **17:**CA(2454), WA(3202). **18:**CA(638,642). **34:**TX(82). **44:**MA(656), NY(2457,3361). **49:**CA(2454), CO(82), CT(82), MI(82), NC(82), WI(82).
Venturia palustris. **36:**WI(1341).
V. potentillae. On leaves. **9:**MA(82), NC(3944). **49:**AK(526), CA(82), NC(82).

Basidiomycotina - Uredinales
Phragmidium andersonii. **15:**Range of host(82), AK(60,526), CO(1195), ID(3202), MT(3202,3209), SD(2243), WY(1195,2924,3209). **16:**ND(378). **49:**FL(35).
P. biloculare. Autoecious. **14:**CA(3111), ID(3111), WA(3111,3202). **19:**WA(3202). **24:**WA(82,111). **47:**ID(3202). **49:**CA(82), ID(82), WA(82).
P. boreale. Autoecious. **11:**CO(1520), MT(1520,3111), WY(3111).
P. ivesiae. Autoecious. **1:**CA(81). **8:**CA(3111). **10:**MT(3202), WY(1195). **11:**W states(82), CA(82), MT(3202), WA(82), WY(1195,2768). **19:**CA(3111), ID(3202), MT(1195,3111,3202), OR(3111,3202), WA(3111,3202), WY(1195,2924). **24:**Range of host(82), W states(82), CA(81,82,638,642), WA(82,638), WY(1195,3379). **20:**WY(3111). **21:**CA(3111), CO(3111), ID(3111), MT(3111), UT(3111), WA(3111). **22:**CA(3111), ID(3111), MT(3111), OR(3111), WA(3111), WY(3111). **23:**CA(3111), CO(3111), MT(3111), WY(3111). **25:**CA(3111), MT(3111). **26:**W states(82), CA(82), CO(1195,3111), MT(3111), NE(3111), UT(3111), WA(82), WY(1195,3111). **28:**CO(3111), MT(3202), WY(3111). **34:**CO(82,3111), MT(82,3202). **35:**WI(1335). **37:**IA(3111), NE(3111), SD(3111). **38:**ID(3202). **42:**IL(312), IN(3111), MD(3111), MI(3111), MN(3111), NJ(3111), NY(3111), PA(3111), WI(1330,3111). **45:**WY(2924). **48:**WA(3202). **49:**Range of host(82), W states(82), CA(82,661), WA(82).
P. potentillae. Autoecious. **5:**WA(3202). **7:**ND(378). **9:**GA(1439), MA(664), MS(2720), NC(1303,3944). **28:**OR(3202). **29:**AK(3102). **34:**CT(82), ME(82), PA(82), TX(82). **39:**AK(60,526), CO(1195), MT(3202), ND(378), OR(3202), SD(2243), WY(1195). **44:**IN(987). **49:**E states(82), NC states(82), AK(526), AR(82), ME(82), MN(82), NC(82), NH(82), OH(653), OR(82,3202).
Pucciniastrum potentillae. Telial host; aecial host unknown. **34:**FL(82). **46:**E states(82), FL(35), MI(1575), MN(82), TN(1575), WI(1575).

Basidiomycotina - Other
Marasmius epiphyllus. **49:**WA(3202).
Sebacina incrustans. Encrusting living stems. **49:**WI(128).

Deuteromycotina - Hyphomycetes
Cercospora comari. Leaf spot. **36:**NY(82).
C. potentillae. On leaves. **35:**WI(1334). **42:**WI(1334).
Phymatotrichopsis omnivora. Root rot. **49:**TX(82).

Ramularia arvensis. Leaf spot. **2:**AK(82,526), WI(82). **4:**CO(1195), WI(1338), WY(1195). **9:**DE(3543), WI(82,1334). **11:**ID(3202), WA(3202). **12:**OR(3202), WA(3202). **14:**WA(3202). **17:**ID(3202), OR(3202), WA(3202). **19:**ID(3202), MT(3202), OR(3202), WA(3202), WY(1195). **26:**WY(1195). **31:**OR(3202). **33:**OR(3202). **34:**Range of host(82), AK(526), SD(2243), WA(3202), WY(1195,3378). **36:**WA(3202). **40:**ND(378). **49:**AK(82,3431), FL(35), NY(82), WY(82).
R. punctiformis. **2:**AK(526). **34:**AK(526).

Deuteromycotina - Coelomycetes

Colletotrichum gloeosporioides. Anthracnose. **9:**NC(1303).
Cylindrosporium triflori. **19:**WY(1195).
Diplodina lyngei. **6:**AK(526).
Discosia potentillae. Leaf spot. **9:**IL(82).
Leptothyrium arcticum. **30:**AK(2059).
Marssonia potentillae var. *helleri*. **18:**CA(638).
Marssonina fragariae. Leaf spot. **2:**CA(81,82), CO(82), MA(82), MI(82), NM(82), NY(82), WA(3202), WI(82). **3:**NM(82), WA(3202). **4:**MT(3202), OR(3202), WA(3202). **9:**WI(82). **13:**WY(1195,3378). **15:**WA(3202). **17:**CA(81), ID(3202), MT(3202), OR(3202), WA(3202). **18:**CA(642). **34:**MI(82), TX(82), WI(82). **36:**ID(3202), MT(3202), WI(82). **37:**SD(2243). **43:**ID(3202). **49:**CA(82), WA(82), WI(82).
Phoma potentillica. Stem blight. **34:**TX(82).
Phyllosticta anserinae. Leaf spot. **2:**IL(82).
P. fragaricola. **4:**WI(1340). **35:**WI(1355). **42:**WI(1345).
P. numerospora. On dead leaves. **3:**WI(1346,1357).
Phyllosticta sp. Leaf spot. **42:**WI(1360).
Septoria fragariae. Leaf spot. **36:**ID(3202), WI(82).
S. purpurascens. Leaf spot. **2:**MT(3202). **19:**ID(3202), MT(3202), WY(1195). **34:**NY(82). **36:**WA(3202). **42:**WI(82). **49:**WI(82).
S. rectae. On leaves. **42:**WI(1335).
Stagonospora sp. **15:**ID(3202).

Deuteromycotina - Other

Rhizoctonia solani. Damping-off. **34:**TX(82).
Sclerotium rolfsii var. *delphinii*. Crown rot. **9:**CT(82).

Prunus L.

(= *Armeniaca* Scop., *Laurocerasus* Duhamel.)
1. *P. alabamensis* C. Mohr
2. *P. alleghaniensis* Porter [Alleghany plum, sloe]
3. *P. americana* Marsh. [American plum, August plum, goose plum, hog plum, sloe, wild plum]
4. *P. americana* Marsh. var. *lanata* Sudw. (=*P. lanata* (Sudw.) Mackenz. & Bush)
5. *P. andersonii* A. Gray [desert peach]
6. *P. angustifolia* Marsh. [chickasaw plum, sand plum]
7. *P. armeniaca* L. [apricot]
8. *P. avium* (L.) L. [bird cherry, gean, mazzard cherry, sweet cherry]
9. *P. besseyi* L. H. Bailey [Bessey cherry, western sand cherry]
10. *P. caroliniana* (Mill.) Aiton (=*Laurocerasus caroliniana* (Mill.) M. Roem.) [Caroline cherry-laurel, mock-orange, wild-orange]
11. *P. cerasifera* Ehrh. [cherry plum, Myrobalan plum]
12. *P. cerasus* L. [pie cherry, sour cherry]
13. *P. cerasus* L. var. *austera* L. [morello cherry]
14. *P. chicasa* Michx.
15. *P. davidiana* (Carrière) Franch. [Chinese wild peach]
16. *P. domestica* L. (=*P. domestica* L. subsp. *domestica*) [common plum, European plum, garden plum, prune plum]
17. *P. domestica* L. subsp. *insititia* (L.) C. K. Schneid. [bullace plum, Damson plum]
18. *P. dulcis* (Mill.) D. Webb (=*P. amygdalus* Batsch, *P. communis* (L.) Arcang., non Huds., *Amygdalus communis* L., *Amygdalus dulcis* Mill.) [almond]
19. *P. emarginata* (Douglas ex Hook.) Walp. [bitter cherry, Oregon cherry]
20. *P. fenzliana* Fritsch (=*Amygdalus fenzliana* (Fritsch) Lipsky)
21. *P. fruticosa* Pall. [European dwarf cherry, European ground cherry]
22. *P. glandulosa* Thunb. ex J. A. Murray
23. *P. hortulana* L. H. Bailey [hortulan plum, wild-goose plum]
24. *P. ilicifolia* (Nutt. ex Hook. & Arnott) Walp. [evergreen cherry, hollyleaf cherry, islay, mountain holly, wild cherry]
25. *P. laurocerasus* L. (=*Laurocerasus officinalis* M. Roem., *Cerasus laurocerasus* (L.) Loisel.) [English cherry-laurel]
26. *P. lusitanica* L. [Portuguese cherry-laurel]
27. *P. lyonii* (Eastw.) Sarg. (=*P. ilicifolia* (Nutt.) Walp. var. *occidentalis* (Nutt.) T. S. Brandegee) [Catalina Island cherry]
28. *P. mahaleb* L. (=*Cerasus mahaleb* (L.) Mill.) [mahaleb cherry, perfumed cherry, St.-Lucie cherry]
29. *P. maritima* Marsh. [beach plum, shore plum]
30. *P. mexicana* S. Wats. [Mexican plum]
31. *P. mitis* Beadle
32. *P. munsoniana* W. Wight & Hedr. [wild-goose plum]
33. *P. nigra* Aiton [Canada plum]
34. *P. orthosepala* Koehne
35. *P. padus* L. [European bird cherry, hagberry]
36. *P. pensylvanica* L. fil. [bird cherry, fire cherry, pin cherry, wild red cherry]
37. *P. persica* (L.) Batsch (=*Amygdalus persica* L., *Persica vulgaris* Mill.) [peach]
38. *P. persica* (L.) Batsch var. *nucipersica* (Suckow) C. K. Schneid. (=*P. persica* (L.) Batsch var. *nectarina* (Aiton) Maxim.) [nectarine]
39. *P. pumila* L. (=*P. depressa* Pursh) [dwarf cherry, sand cherry]
40. *P. pumila* L. var. *susquehanae* (Willd.) Jaeger
41. *P. salicina* Lindl. (=*P. triflora* Roxb.) [Japanese plum]
42. *P. serotina* Ehrh. [rum cherry, wild black cherry]
43. *P. serotina* Ehrh. subsp. *virens* (Woot. & Standl.) McVaugh [southwestern chokecherry]
44. *P. serrulata* Lindl. [Japanese flowering cherry, oriental cherry]
45. *P. simonii* Carrière [apricot plum, Simon plum]
46. *P. spachiana* (Lavalle ex H. Otto) Kitam. (=*P. subhirtella* Miq. var. *pendula* (Siebold ex Maxim.) Tanaka)
47. *P. spinosa* L. [blackthorn, sloe]
48. *P. subcordata* Benth. [Klamath plum, Pacific plum, Sierra plum]
49. *P. tangutica* (Batalin) Koehne (=*Amygdalus tangutica* (Batalin) Korshinsky)
50. *P. tarda* Sarg. in J. M. Coulter
51. *P. tenella* Batsch [dwarf Russian almond]
52. *P. triloba* Lindl. [flowering almond]
53. *P. umbellata* Elliott [flatwoods plum, hog plum, sloe plum]
54. *P. virens* Shreve ex Sarg.
55. *P. virginiana* L. [chokecherry]
56. *P. virginiana* L. var. *demissa* (Nutt.) Torr. (=*P. demissa* (Nutt.) Walp.)
57. *P. virginiana* L. var. *melanocarpa* (A. Nels.) Sarg. (=*P. melanocarpa* (A. Nels.) Rydb.) [black chokecherry]
58. *Prunus* sp.

Oomycetes

Phytophthora cactorum. Bark, collar, stem and trunk canker, collar, crown and root rot. **3:**CA(1110). **7:**CA(82,1110), WA(3202). **8:**CA(82,1110). **11:**CA(1110). **12:**CA(1110). **16:**CA(82,1110), IN(82). **18:**CA(82,1110). **28:**CA(1110). **37:**'Phytophthora root rot.' AR(82), CA(82,1110), MS(1502), WA(3202). **38:**CA(1110). **41:**CA(1110). **58:**CA(82), IN(82).
P. cambivora. Crown and root rot. **3:**CA(1110). **7:**CA(1110). **8:**CA(1110). **11:**CA(1110). **12:**CA(1110). **16:**CA(1110). **18:**CA(1110). **28:**CA(1110). **37:**CA(1110). **38:**CA(1110). **41:**CA(1110). **58:**CA(2445).
P. cinnamomi. Stem canker, crown and root rot. **3:**CA(1110). **7:**CA(1110). **37:**MS(1502).
P. citricola. Crown and root rot. **7:**CA(1110). **18:**CA(1110). **37:**CA(1110).
P. citrophthora. Collar, crown and trunk canker, dieback, crown and root rot. **7:**CA(82,1110). **8:**CA(82,1110). **18:**CA(82,1110). **37:**CA(82,1110). **58:**CA(82).
P. cryptogea. Crown and root rot. **7:**CA(1110). **8:**CA(1110). **28:**CA(1110). **37:**CA(1110).
P. drechsleri. Crown and root rot. **7:**CA(1110). **8:**CA(1110). **37:**CA(1110). **58:**CA(2445).
P. megasperma. Crown and root rot. **8:**CA(1110). **16:**CA(1110). **18:**CA(1110). **37:**CA(1110). **58:**CA(2445).
P. nicotianae var. *parasitica*. Stem canker, root rot. **37:**MS(1502).
P. syringae. Canker, brown rot, crown and root rot. **7:**CA(1110). **8:**CA(1110). **18:**CA(343,1110). **37:**CA(1110).
Phytophthora sp. Root rot. **37:**FL(35).
Pythium ultimum. On roots. **37:**CA(1570).
Pythium sp. Tree decline, root rot. **3:**CA(1110), FL(35). **6:**FL(35). **7:**FL(35). **8:**CA(1110). **10:**FL(35). **16:**FL(35). **18:**CA(1110), FL(35), HI(2903). **25:**FL(35). **32:**FL(35). **42:**FL(35).

Zygomycotina

Gilbertella persicaria. Fruit rot. **7:**CA(1110). **37:**CA(1110), NY(82).
Mucor piriformis. Postharvest decay of fruits. **37:**CA(1110), MD(3358). **38:**MD(3358).
Mucor sp. Fruit rot. **7:**CA(1110). **8:**OR(3202), WA(3202). **16:**CA(1110,2369).
Rhizopus arrhizus. Hull rot. **18:**CA(2448).
R. circinans. Hull rot. **18:**CA(2448).
R. stolonifer. Postharvest decay of fruits. **7:**'Rhizopus fruit rot.' WA(3202). **8:**'Rhizopus rot.' Unk.(82), ID(3202), OR(3202), WA(3202). **16:**'Rhizopus rot.' ID(3202). **18:**'Hull rot.' CA(2448,2798). **37:**'Rhizopus rot.' Range of host(82), CA(939), FL(35), ID(3202), OK(2872), OR(3202).
Rhizopus sp. Fruit rot. **3:**CA(1110). **7:**CA(1110). **8:**CA(1110). **11:**CA(1110). **12:**CA(1110). **16:**CA(1110). **18:**CA(1110,2889). **28:**CA(1110). **37:**CA(1110), NC(1303). **38:**CA(1110), NC(1303). **41:**CA(1110).

Ascomycotina

Acrospermum sp. **6:**GA(1438). **53:**GA(1438).
Apiosporina morbosa. Dieback, black knot. 'Black knot.' **3:**Range of host(82), AL(285), FL(35), MS(2720), NC(1303), ND(377), OK(82,2873), TX(82). **6:**FL(35), GA(1438). **7:**CO(82), FL(35), IA(82), NY(82). **8:**NC states(82), NE states(82), NC(82,3944), TX(82). **9:**ND(377). **12:**E states(82), GA(1438), NC(1303). **16:**Range of host(82), FL(35), KY(1777), MS(2720). **17:**NC(1303). **19:**ID(3202), MT(3202), OR(3202), WA(3202). **29:**AL(82), MA(82), NY(82), OR(82). **32:**AL(82), FL(35), GA(82), MS(82,2720). **36:**Range of host(82), NC(1303). **39:**MI(638), ND(82), WI(82). **42:**Range of host(82), GA(82,659,1438), MS(2720), NC(1303), VA(3493), WI(1358). **48:**OR(3202). **53:**GA(1438). **55:**Range of host(82), CO(1195), ID(3202), MS(2720), MT(3202), NC(1303), SC(1307), SD(2243), WA(3202). **56:**CA(638,642). **57:**SD(2243). **58:**Range of host(82), GA(1438), MA(1781), MD(651), NC(1303), OK(2872), SD(2243), VT(1224), WI(1342).
Auerswaldiella puccinioides. Leaf blister. **10:**FL(35), GA(82,1438). **25:**FL(35). **42:**FL(35).
Bisporella citrina. **19:**ID(3202).
Blumeriella jaapii. Shot hole. 'Cherry leaf spot.' **3:**FL(35), IA(82), IN(82), MS(2720). **4:**OK(2873). **6:**FL(35). **7:**CA(1110), FL(35), OK(2872). **8:**Range of host(82), CA(1110), GA(1438), MT(3202), OR(3202), WA(3202). **10:**FL(35). **12:**Range of host(82), GA(1438), MS(2720), MT(3202), NC(1303), OK(2872), OR(3202), WA(3202). **19:**ID(3202), MT(82,3202), OR(3202), WA(82,3202). **25:**FL(35), MS(82,2720). **36:**Range of host(82), NC(1303). **37:**FL(35). **39:**WI(82). **41:**Range of host(82). **44:**CT(3770), MA(82), NJ(82). **55:**Range of host(82), ID(3202), IN(987), MS(2720), MT(3202), NE(1703), OR(3202), SD(2243), WA(3202). **58:**Range of host(82), NC(1303), OK(2872).
Botryosphaeria dothidea. Canker, gummosis, fruit rot. **18:**'Band canker.' CA(996,1110). **37:**AL(2954,2986), FL(35,2954), GA(391,392,1438,2954,2986,3791), LA(2954), TN(2954), TX(2954). **42:**GA(1438). **44:**GA(1438).
B. obtusa. Branch canker, gummosis, fruit rot. **32:**AL(82). **37:**E states(82), S states(82), AL(2986), GA(391,392,2986). **42:**GA(82). **55:**KS(82), NY(82). **58:**Range of host(82).
B. rhodina. Gummosis, fruit rot. **37:**AL(2986), GA(391,392,2986). **58:**SC(82).
B. ribis. Canker, dieback, fruit rot. **37:**FL(82), GA(82), IL(312). **44:**GA(82).
Botryosphaeria sp. On limbs. **46:**NC(1303).
Calosphaeria pulchella. On limbs and trunks. **3:**ND(377). **19:**WA(3202). **37:**NC(1303).
Capnodium pelliculosum. On leaves. **14:**SC(634).
Caryospora minima. On endocarps on ground in orchard. **37:**MD(185).
C. putaminum. On endocarps. **3:**NC(3944). **37:**Range of host(82), GA(1438), MD(185), MS(2720), NC(1303,3944).
Cenangium populneum var. *prunicola*. **57:**ND(377).
Ceratocystiopsis alba. On injured bark tissue. **7:**CA(826). **18:**CA(826,3714).
Ceratocystis fimbriata. Canker. **7:**'Ceratocystis canker.' CA(1110). **16:**CA(1110). **18:**'Ceratocystis canker.' CA(827,1110). **37:**CA(1110).
Coccomyces sp. Leaf spot. **7:**CA(82), IA(82), MA(82), TX(82), VT(82).
Cucurbitaria interstitialis. **58:**NY(195).
Daldinia sp. **58:**WA(3202).
Dermea cerasi. On limbs and twigs. **8:**AK(82,526), OR(82,3202). **12:**AK(526), NJ(82), NY(82). **16:**NY(82). **19:**WA(3202). **36:**Range of host(82), MA(128), ME(128), NC(128), NH(128), NY(128), PA(128), VT(128), WV(128). **42:**NY(82), VT(1224). **55:**MA(82), ND(82). **58:**Range of host(82), ID(3202).
D. padi. On twigs. **58:**NY(1379).
D. prunastri. **58:**ID(3202), WA(3202).
Diapleella coniothyrium. **58:**WA(3202).
Diaporthe decorticans. Canker. **12:**Unk.(82). **32:**IA(82).
D. eres. On limbs. **58:**Unk.(82).
D. pennsylvanica. On limbs. **36:**Unk.(82). **42:**NY(82). **55:**Unk.(82).
D. pruni. Twig canker. **6:**IA(82). **32:**IA(82). **42:**Unk.(82), IA(82), MA(82), MI(82), NY(82). **55:**Unk.(82).
D. prunicola. Twig canker. **3:**IA(82), NE(82). **36:**Unk.(82), NY(195). **42:**Unk.(82).
D. tuberculosa var. *pruni*. On limbs. **42:**NY(82,3807).
Diatrype albopruinosa. **58:**IA(3629).
D. asterostoma. **57:**ND(377).
D. tumida f. *pruni*. **3:**ND(377).
Diatrypella favacea. **6:**GA(1438). **58:**IA(3629).
D. nitschkei. **42:**IA(3629).
Discohainesia oenotherae. On twigs. **42:**VA(82).
Dothidea puccinioides. **37:**GA(1438). **42:**GA(1438).
Eutypa armeniacae. Canker. **7:**'Eutypa dieback.' CA(1249,3716), WA(1247). **56:**CA(995).
E. lata. **3:**ND(377). **7:**CA(1110).
Eutypella cerviculata. **42:**GA(1438).
E. leprosa. **42:**IA(3629). **55:**IA(3629).
E. prunastri. **58:**IA(3629).
Eutypella sp. Dieback. **37:**FL(35).
Gibberella baccata. Twig blight, bud rot. **37:**CA(1110).
Glomerella cingulata. Twig blight, fruit rot. **6:**GA(2749). **12:**IN(82). **37:**'Anthracnose.' AL(2986), AR(82), FL(35,82), GA(82,2915,2986), MD(82), MI(3344), NC(1303), NJ(82), OH(82), SC(2915), TN(82), TX(82), WV(3359).
Glonium stellatum. On logs. **42:**VA(638).
Godronia confertus. **36:**NH(128), VT(128). **58:**NH(1382), NJ(1382).
Hypoxylon fuscum. **42:**GA(1438).
H. rubiginosum. On wood. **42:**GA(1438). **53:**GA(1438). **55:**IA(1225).
H. serpens var. *macrospora*. **53:**GA(1438).
Hysterographium flexuosum. **58:**FL(2782).
Lambertella jasmini. Fruit rot. **8:**OR(82,909,3202).
L. pruni. Fruit rot. **7:**CA(82,1110). **8:**OR(82,909,3202). **16:**CA(82,909,1110).
Lambertella sp. Fruit rot. **8:**OR(82).
Lecanidion atratum. **57:**ND(377).
Leucostoma cincta. Canker, dieback. **35:**IA(1226). **37:**E & C states(82), Range of host(82), NC(1303), OK(82), TX(82). **55:**KS(82), MT(82), ND(82). **57:**ND(377). **58:**Range of host(82).
L. persoonii. Canker, dieback. **3:**FL(35), SD(2243). **6:**FL(35). **7:**FL(35), MO(82), TX(82). **8:**OR(3202), WA(3202). **10:**GA(1438). **15:**GA(82). **16:**Range of host(82), FL(35), ID(3202), WA(3202). **32:**FL(35). **35:**AK(526). **37:**'Cytospora canker.' E & C states(82), Range of host(82), GA(1438), ID(3202), MS(2720), NC(1303), OK(82,2872), OR(3202), TX(82), WA(3202). **38:**DC(82). **42:**Range of host(82), GA(82,1438), IA(1226), NC(1303). **55:**KS(82), MT(82,3202), ND(82), WA(3202). **58:**Range of host(82), IA(1226), OK(2872).
Lophiostoma quadrinucleatum var. *triseptatum*. **3:**ND(377). **57:**ND(377).
L. starbackii. **57:**ND(377).
Massaria conspurcata. On dead limbs. **3:**ND(377). **58:**Range of host(82), SD(2243), WV(2783).
Mollisia melaleuca. **37:**GA(1438). **42:**GA(1438).
Mollisia sp. **42:**GA(1438).
Monilinia demissa. **55:**ID(3202), WA(3202).
M. fructicola. Blossom and twig blight, gummosis, brown rot, fruit and hull rot. **3:**'Brown rot. Blossom blight. Fruit rot.' Range of host(82), AL(285), CA(1110), FL(35), MS(2720), NC(1303,3944), SD(2243), WI(1361). **6:**'Brown rot. Blossom blight. Fruit rot.' FL(35), GA(1438), NC(3944). **7:**'Brown rot. Blossom and twig blight. Fruit rot.' Range of host(82), CA(1110,2368,2635), FL(35), OK(2872), WA(3202). **8:**'Brown rot. Blossom blight. Fruit rot.'

Range of host(82), CA(1110,2635), GA(1438), NC(1303), OR(3202), WA(3202). **9:**'Brown rot. Blossom blight. Fruit rot.' SD(2243). **10:**'Brown rot. Blossom blight. Fruit rot.' FL(35). **11:**'Brown rot. Blossom blight. Fruit rot.' CA(1110). **12:**'Brown rot. Blossom blight. Fruit rot.' Range of host(82), CA(1110), GA(82,1438), MS(2720), WA(3202), WI(1361). **16:**'Brown rot. Blossom blight. Fruit rot.' Range of host(82), CA(1110,2368,2369,2635), FL(35), GA(1438), ID(3202), OR(3202), WA(3202). **17:**'Brown rot. Blossom blight. Fruit rot.' NC(1303). **18:**'Brown rot. Blossom blight. Hull rot.' CA(1110,2448,2635,2798). **25:**'Brown rot. Blossom blight. Fruit rot.' CA(82), FL(35). **28:**'Brown rot. Blossom blight. Fruit rot.' CA(1110). **29:**'Brown rot. Blossom blight. Fruit rot.' CA(82), FL(82), MA(1390). **32:**'Brown rot. Blossom blight. Fruit rot.' Range of host(82), FL(35). **33:**'Brown rot. Blossom blight. Fruit rot.' WI(1360). **38:**'Brown rot. Blossom blight. Fruit rot.' CA(1110,2635), CT(82), NC(1303), NY(82), TX(82). **39:**'Brown rot. Blossom blight. Fruit rot.' Range of host(82), PA(82), SD(82), WA(82), WI(1344). **41:**'Brown rot. Blossom blight. Fruit rot.' CA(1110). **42:**'Brown rot. Blossom blight. Fruit rot.' Range of host(82), FL(35), MS(2720). **52:**'Brown rot. Blossom blight. Fruit rot.' CT(82), KS(82). **53:**'Brown rot. Blossom blight. Fruit rot.' GA(1438). **55:**'Brown rot. Blossom blight. Fruit rot.' SD(2243). **58:**'Brown rot. Blossom blight. Fruit rot.' Range of host(82), NC(1303), OK(2872), SD(2243).

M. fructigena. Fruit rot. **37:**FL(35).

M. laxa. Blossom and twig blight, gummosis, brown rot, hull and fruit rot. **3:**'Brown rot. Blossom blight. Fruit rot.' CA(1110). **7:**'Brown rot. Blossom and twig blight. Fruit rot.' W states(82), CA(1110,2368,2635), OR(3202), WA(3202). **8:**'Brown rot. Blossom blight. Fruit rot.' CA(1110,2635), ID(82,3202), OR(82,3202), WA(82,3202). **11:**'Brown rot. Blossom blight. Fruit rot.' CA(1110). **12:**'Brown rot. Blossom blight. Fruit rot.' CA(1110), MI(82), OR(82,3202), WA(82,3202), WI(82,1334). **13:**'Brown rot. Blossom blight. Fruit rot.' NY(1824). **16:**'Brown rot. Blossom blight. Fruit rot.' CA(82,1110,2368,2369,2635), HI(2899,2903), OR(82,3202), WA(82,3202). **25:**'Brown rot. Blossom blight. Fruit rot.' CA(82). **28:**'Brown rot. Blossom blight. Fruit rot.' CA(1110). **37:**'Brown rot blossom blight.' W states(82), CA(1110,2635), HI(2899,2903), OR(3202), WA(3202). **38:**'Brown rot. Blossom blight. Fruit rot.' CA(82), WA(82). **41:**'Brown rot. Blossom blight. Fruit rot.' CA(1110). **58:**'Brown rot. Blossom blight. Fruit rot.' W states(82), WI(82).

M. padi. On fruits, leaves and twigs. **55:**NY(3167).

M. seaveri. Leaf and twig blight, mummified fruits, fruit rot. **12:**AR(82). **37:**NC(1303). **42:**E & C states(82), GA(1438), IA(82), MS(2720), NC(1303), NY(1960), WI(1337).

[*M. vaccinii-corymbosi*]. Mummified fruits. **36:**MI(1129).

Monilinia sp. **11:**HI(2903).

Mycosphaerella cerasella. Leaf spot. **3:**FL(35). **6:**FL(35). **7:**FL(35). **8:**SE states(82), MS(82,2720). **10:**FL(35). **12:**SE states(82), MS(82). **16:**FL(35). **25:**FL(35). **28:**MS(2720). **32:**FL(35). **42:**FL(35), NY(82). **55:**KS(82). **58:**E states(82), SC states(82).

M. nigerristigma. Leaf spot. **19:**MS(2720). **36:**GA(82), MS(82,2720), WI(1352).

M. pruni-persicae. Frosty mildew. **37:**Range of host(82), FL(35), GA(1438), MS(2720), NC(3944).

M. punctiformis. **42:**GA(1438).

Mycosphaerella sp. **42:**GA(1438).

Nectria cinnabarina. Twig blight, dieback, coral spot. **7:**CA(1110), IN(82), WA(82,3202). **8:**WA(3202). **10:**FL(35). **12:**AK(526). **16:**AK(526). **19:**ID(3202), MT(3202), WA(3202). **25:**FL(35). **35:**AK(526). **36:**NC(82), TN(82), WA(82). **37:**AL(82), WA(3202). **42:**FL(35). **55:**Range of host(82), ID(3202), MT(3202), WA(3202). **57:**ND(377). **58:**Range of host(82), AK(526), ID(3202).

Nectria sp. Canker. **36:**NE states(82). **42:**NE states(82), NY(82).

Nitschkia broomeiana. **58:**ID(3202).

N. crustacea. **57:**ND(377).

Ophiostoma longirostellatum. On injured bark tissue. **16:**CA(824,826).

Peroneutypa corniculata. **57:**ND(377).

Perrotia succina. **19:**CA(642).

Phyllachora beaumontii. Leaf spot. **25:**AL(82).

Phyllactinia guttata. Powdery mildew. **7:**WA(3202). **55:**ID(3202), WA(82,3202).

Physalospora abdita. On limbs. **37:**S states(82). **58:**SC(82).

Physalospora sp. Twig dieback. **10:**FL(35). **25:**FL(35). **42:**FL(35).

Pleospora vulgaris f. *astericola*. **57:**ND(377).

Pleospora sp. **8:**OR(3202), WA(3202).

Podosphaera clandestina. 'Powdery mildew.' **3:**Range of host(82), CA(1110), FL(35), ND(377). **6:**FL(35). **7:**CA(82), FL(35), IA(82). **8:**Unk.(82), IA(82), NC(1303), PA(82), SD(2243), WA(82,1374,3202). **9:**ND(377), SD(2243). **12:**Range of host(82), CA(1110), ID(3202), MS(2720), MT(3202), NM(2112), OK(2872), SD(2243), WA(3202). **16:**Unk.(82), CA(1110), FL(35), GA(82,1438). **18:**ID(3202). **19:**ID(3202), WA(3202). **27:**Unk.(3976). **29:**FL(82), MD(82). **32:**FL(35), IA(82), MD(82), NE(82). **36:**IL(82), PA(82), WA(82), WI(82). **37:**Range of host(82), Unk.(1861), CA(1110), FL(35), ID(3202), WA(3202). **38:**ID(82), WA(82). **39:**NC states(82). **41:**CA(1110). **42:**NC states(82), NE states(82), NE(82). **52:**IA(82), MS(2720). **53:**GA(1438). **55:**MT(3202), OR(3202), WA(3202). **57:**ND(377), SD(2243), WY(1195). **58:**Range of host(82), IL(1052), NC(1303), OK(2872), SD(2243).

P. leucotricha. Powdery mildew. **18:**CA(3975). **37:**CA(1110), IL(2979).

P. tridactyla. Powdery mildew on leaves and twigs. **7:**'Powdery mildew.' CA(81,1110). **8:**CA(1110). **16:**Unk.(82), GA(82). **18:**'Powdery mildew.' CA(1110), WA(3822). **36:**IL(82), PA(82), WA(82), WI(82). **58:**NW states(82).

Proliferodiscus pulveraceus. **53:**GA(1438).

Propolomyces versicolor. **37:**GA(1438).

Pyrenopeziza laurocerasi. On fallen leaves. **25:**CA(81).

Romellia tympanoides. On limbs. **36:**MA(190).

Rosellinia aquila. **55:**WA(3202).

R. necatrix. Root rot. **7:**CA(1110).

Rosellinia sp. **41:**GA(1438). **42:**GA(1438).

Schizothyrium pomi. Fly speck on fruits and twigs. **3:**IL(82). **32:**MO(82). **37:**IN(82), WA(82,3202). **58:**C states(82), E states(82).

Schizoxylon compositum. On bark and wood. **57:**ND(377).

Sclerotinia sclerotiorum. Blossom blight, green fruit rot. **7:**'Green fruit rot.' CA(82,939,1110). **8:**CA(1110). **18:**CA(82,1110). **37:**CA(82,1110). **58:**CA(82).

Sphaerotheca pannosa. Powdery mildew, rusty spot. **7:**'Powdery mildew.' CA(1110). **18:**'Powdery mildew.' CA(1110), WA(3202). **37:**'Powdery mildew.' CA(1110), ID(3202), NJ(761), OR(3202), WA(3202). **38:**CA(1110), ID(82), NY(82), OR(82). **41:**CA(1110). **56:**OR(81).

S. pannosa var. *persicae*. Powdery mildew. **7:**NY(82). **37:**Range of host(82), FL(35).

Sydowia pruni. **58:**WA(179), WY(179).

Taphrina armeniacae. **37:**CA(2454), IL(2454), KS(2454), MI(2454), NC(2454), VA(2454).

T. communis. Leaf blister and curl, bud, fruit and shoot hypertrophy, bladder plum, plum pocket. **3:**AL(285), AR(82,2454), FL(35), IA(82,2454), IL(82), KS(82,2454,3361), MA(82,2454), MI(82,2454), MS(82,2720), MT(3202), NC(82,3944), NY(82,2454), OK(2872), SD(2243), TX(82), WI(82). **4:**AR(2454). **6:**AL(2454), AR(2454,3361), FL(35,2454), GA(1438,2454), IA(2454), KS(2454), MS(2454), NC(3944), OK(2872), TN(2454), VA(2454). **7:**FL(35). **9:**NE(82), SD(2243). **10:**FL(35). **16:**AL(82), FL(35). **23:**AR(2454), IA(2454), WI(2454). **25:**FL(35). **29:**E states(82), MA(2454), WA(3202). **33:**MI(2454), NY(2454). **39:**C states(82), NE states(82), ND(82). **41:**AL(2454), CT(2454). **42:**FL(35). **50:**TX(2457). **53:**GA(1438,2454).

T. confusa. On fruits, leaves and shoots, witches' broom. **1:**GA(1438). **55:**NC states(82), NE states(82), W states(82), FL(2454), ID(3202), MA(2454), MI(2454), MN(2454), MT(3202), NE(2454), NY(2454), OR(3202), WA(3202,3361), WY(1195). **56:**CA(642,2454), CO(2454), ID(2454), OR(2454), WA(2454). **57:**CO(2454), MT(2454), SD(2243), WA(638).

T. deformans. Blight, leaf curl. **3:**FL(35), ND(377). **6:**FL(35). **7:**FL(35). **9:**ND(377). **10:**FL(35). **16:**FL(35). **25:**FL(35). **32:**FL(35). **37:**'Leaf curl.' Range of host(82), AK(526), AL(285), CA(1110,3361), CT(3770), FL(35), GA(1438,3692), HI(2903), ID(3202), KS(1967,3361), KY(1777), MS(2720), MT(3202), NC(1303,3944), OK(2872), OR(3202,3361), WA(82). **38:**CA(82,1110), NC(1303), OR(82), TX(82), WA(82).

42:FL(35). 57:ND(377). 58:GA(2405).
T. farlowii. Blister, leaf curl and spot, fruit, leaf and shoot hypertrophy, pocket. **10**:FL(35). **25**:FL(35). **39**:NY(82). **42**:E states(82), AL(2454), AR(82,2454,3361), FL(35,2454), GA(1438,2454,3361), KS(82), MA(1390), MS(2720), NC(1303,2454,3944), TX(82), VA(3493), VT(2454). **55**:NC(1303,3944).
T. flavorubra. Shoot atrophy, pocket. **3**:KS(3361). **9**:KS(2454), NE(2454), OR(2454,3202). **39**:KS(2454), ME(2454), MI(2454), NY(2457,3361), SD(2454). **40**:NY(2454), WA(82), WI(1348,2454).
T. flectans. Witches' brooms. **19**:CA(81,638,642,661,2454,3361), ID(2454,3202), OR(2454,3202).
T. jenkinsoniana. **5**:NV(2457).
T. pruni. Leaf blister, plum pocket, witches' broom. **3**:CA(1110), FL(35), ND(377). **6**:FL(35). **7**:FL(35). **9**:ND(377). **16**:AL(285), CA(1110), FL(35), NJ(82,2454), NY(82,2454), WA(3202). **32**:FL(35). **36**:NC states(82), NE states(82), CO(82). **41**:CA(1110). **57**:ND(377). **58**:NC(1303).
T. pruni-subcordatae. Witches' brooms. **16**:OR(1512,3202). **48**:CA(638,642,2454,2457), CO(2454), OR(1512,3202).
T. thomasii. Witches' brooms. **24**:CA(82,2454).
T. wiesneri. Leaf curl, witches' broom. 'Witches' broom.' **8**:NW states(82), AK(526), CA(81,1110,2454), ID(3202), NY(82,2454), OR(2454,3202), PA(82), TX(82), WA(2454,3202). **12**:NE states(82), CA(1110), MN(82), OR(3202), WA(3202). **36**:CO(2454,3361), ME(2454), MN(2454), NC(2454), NH(82,2454), NY(2454,3361), WI(1348). **44**:MD(82), NJ(82).
Taphrina sp. Leaf curl, plum pocket. **7**:SC(82). **16**:KY(1777). **36**:NC(1303), ND(82), VT(82), WA(82). **58**:SC states(82), SE states(82), KS(82), MN(82), SD(82).
Teichospora helenae. **55**:MT(3202).
T. pruni-americanae. **3**:ND(377).
Tubeufia pezizula. **42**:GA(1438).
Valsa auerswaldii. **55**:WA(3202).
V. ceratosperma. **3**:ND(377). **37**:GA(82,1438).
Valsa sp. **8**:ID(3202), OR(3202), WA(3202). **55**:ID(3202), MT(3202).
Valsaria exasperans. **42**:GA(82,1438).
Venturia sp. Fruit spot. **7**:KY(82).
Xylaria longiana. Root rot. **8**:VA(82).
X. mali. Root rot. **8**:VA(82).
Xylaria sp. Black root rot. **12**:TN(82). **58**:OK(2872).

Basidiomycotina - Uredinales

Tranzschelia arthurii. Autoecious. **42**:IA(82), MI(82). **58**:IA(82), MI(82).
T. discolor. Heteroecious: telial host. 'Leaf rust.' **3**:CA(322,1110), FL(35). **6**:FL(35). **7**:CA(82,322,1110), FL(35), MS(82,2720), NM(82), TX(82). **8**:CA(322,1110). **11**:CA(1110). **12**:CA(1110). **15**:GA(82), TX(82). **18**:CA(82,322,1110,2634). **23**:C states(82), S states(82). **28**:CA(1110). **30**:C states(82), S states(82). **32**:FL(35). **38**:CA(322,1110). **39**:IA(82), KS(82), MN(82), SD(82), WI(82). **55**:MT(3202). **58**:Unk.(82).
T. discolor f. sp. *dulcis*. **18**:CA(1110).
T. discolor f. sp. *persica*. **37**:CA(1110).
T. pruni-spinosae. Heteroecious: telial host. 'Leaf rust.' **3**:Range of host(82), S states(82), FL(35), MS(2720), ND(378), OK(2872). **6**:FL(35), GA(1439), OK(2872). **7**:CA(82), FL(35), MS(82,2720), NM(82), TX(82). **8**:NY(82), OK(82). **10**:FL(35). **12**:MS(2720), NC(82,3944). **16**:Range of host(82), S states(82), W states(82), FL(35), GA(1439). **17**:NC(1303). **18**:CA(81). **20**:CA(81). **25**:FL(35). **29**:WI(1354). **32**:FL(35), MS(2720), OK(2872). **34**:OK(2872). **36**:IA(82), IL(82), WI(82). **37**:CA(81), FL(35), GA(1439), NC(1303,3944). **39**:CA(81), IA(82), KS(82), MN(82), SD(82), WI(82). **42**:E states(82), FL(35), GA(1439), MA(1390), NC(1303,3944), NE(82), TX(82). **43**:AZ(3981). **45**:MS(2720). **47**:MS(2720). **48**:CA(82). **49**:CA(81). **53**:FL(82), GA(82). **54**:AZ(1194,1195,1203). **55**:NC states(82), NE states(82), AZ(1203,3981), FL(82), WI(82). **58**:Range of host(82), OK(2872).
Tranzschelia sp. **58**:Unk.(82).

Basidiomycotina - Other

Acanthophysium cerussatum. **43**:AZ(1202).
Antrodia heteromorpha. **43**:AZ(1202). **55**:ID(3202).
A. serialis. Isolated from wood. **42**:PA(792).
A. sinuosa. **42**:E states(82), NC states(82).
A. xantha. Isolated from wood. **42**:PA(792).
Armillaria mellea. Crown and root rot. **2**:'Shoestring root rot.' CA(2897). **3**:'Shoestring root rot.' CA(1110), FL(35). **6**:'Shoestring root rot.' FL(35). **7**:'Shoestring crown rot. Root rot.' CA(82,1110), FL(35), TX(82). **8**:'Shoestring root rot.' CA(1110), OK(82), OR(82). **10**:'Shoestring root rot.' CA(2897). **11**:'Shoestring root rot.' CA(1110). **12**:'Shoestring root rot.' CA(1110). **15**:'Shoestring root rot.' CA(82). **16**:'Shoestring root rot.' Range of host(82), CA(1110,2898), FL(35), HI(2011,2903), OR(3202), WA(3202). **18**:'Armillaria crown rot. Armillaria root rot.' CA(82,1110,2898), NC(82). **19**:'Shoestring root rot.' CA(2900), WA(3202). **28**:'Shoestring root rot.' CA(1110,2898). **32**:'Shoestring root rot.' FL(35). **36**:'Shoestring root rot.' WA(82). **37**:'Shoestring root rot.' Range of host(82), CA(1110,2898), FL(35), HI(2011,2903), ID(3202), MS(2720), NC(1303), OK(2872), OR(3202), SC(2747), WA(3202). **38**:'Shoestring root rot.' CA(1110). **41**:'Shoestring root rot.' CA(1110). **42**:'Shoestring root rot.' NH(82), NY(59). **52**:'Shoestring root rot.' MS(82). **56**:'Shoestring root rot.' CA(2900). **58**:'Shoestring root rot.' Range of host(82), NC(1303), OK(2872).
A. tabescens. Root rot. **3**:FL(35). **6**:FL(35). **7**:FL(35,82). **8**:NC(1303). **10**:FL(35). **16**:FL(35). **25**:FL(35,82). **32**:FL(35). **42**:FL(35). **58**:FL(82), OK(2872).
Basidiodendron eyrei. On wood. **3**:IA(2185).
Basidioradulum radula. **12**:NY(128).
Bjerkandera adusta. **37**:SC(2748). **55**:MT(3202).
Bondarzewia berkeleyi. Butt rot. **42**:PA(792). **44**:PA(82).
Cerrena unicolor. Wood rot. **19**:WA(3202). **58**:Unk.(82), OR(3202).
Chondrostereum purpureum. Silver leaf, heart rot. 'Silver leaf.' **7**:CA(82,1110). **8**:MT(3202). **12**:NY(82), OR(3906). **16**:CA(1110), WA(82,3202). **37**:Unk.(82), CA(545,1110), OK(2872). **58**:W states(82), CA(1110), NY(82,543).
Clitocybe sp. Root rot. **58**:OK(2872).
Coniophora arida. **42**:CT(128).
C. puteana. Brown rot. **42**:PA(82,792). **58**:Unk.(1006).
Coriolopsis gallica. White rot. **7**:CO(82). **58**:CO(82), OR(82), WY(82).
Dacryopinax spathularia. On wood. **42**:KS(128).
Daedaleopsis confragosa. On wood. **58**:Unk.(82).
Dendrocorticium roseocarneum. **36**:VT(128).
Dendrophora albobadia. On wood. **37**:SC(2747). **42**:GA(491,1439).
D. erumpens. **8**:NY(543). **58**:OR(543), WA(3202).
D. versiformis. **42**:NY(543). **58**:MA(543).
Dendrothele maculata. On stems. **42**:CT(2084). **55**:ME(2084).
Exidiopsis molybdea. On limbs. **42**:GA(2642).
E. podlachica. On wood. **42**:GA(2642).
Fomes fasciatus. **58**:SC(82).
F. fomentarius. Trunk rot. **16**:OR(128,3202). **42**:NE states(82). **55**:WA(3202). **58**:NE states(82).
Fomitella supina. **58**:S states(82).
Fomitopsis cajanderi. Brown cubical rot. **8**:WA(3202). **16**:OR(3202), WA(3202). **37**:OR(82,3202), SC(2747). **58**:W states(82), WI(2591).
F. meliae. On wood. **37**:GA(2963), SC(2320,2963).
F. palustris. **37**:SC(2748).
F. pinicola. Brown cubical rot. **8**:WA(3202). **16**:NW states(82), OR(3202). **37**:NC(82), OR(82,3202). **42**:NE states(82), PA(82,792), VA(82). **58**:Range of host(82).
F. rosea. Brown pocket rot. **37**:OR(82), WV(82). **58**:OR(82).
F. spraguei. Brown cubical rot. **42**:PA(792). **44**:PA(82).
Ganoderma applanatum. White rot. **16**:OR(82,3202). **37**:CT(82), NJ(82), SC(2747). **58**:NY(2456), OR(82), WA(3202).
G. lobatum. White, soft decay. **37**:SC(2747).
G. lucidum. White root and butt rot. **3**:FL(35). **6**:FL(35). **7**:FL(35). **16**:FL(35). **32**:FL(35). **37**:FL(35,2963), NC(82), SC(2748), VA(82).
G. zonatum. Basal trunk rot. **10**:FL(35). **25**:FL(35). **42**:FL(35).
Gloeocystidiellum porosum. White rot. **58**:MI(2553), NH(2553).
Gloeophyllum sepiarium. Brown rot. **8**:PA(2962), WA(3202). **16**:NW states(82), OR(3202). **37**:OR(82,3202), SC(2747). **58**:ID(82), OR(82), WA(82).
G. trabeum. Brown rot. **37**:SC(2747). **55**:MT(3202). **56**:MT(3832). **58**:ID(82), OR(82), WA(82).
Gloeoporus dichrous. **37**:SC(2748).
Grandinia pruni. **56**:KS(128).

G. spathulata. Brown rot. **37:**SC(2747).
Grifola frondosa. Isolated from wood. **42:**PA(792).
Hapalopilus mutans. Trunk rot. **42:**PA(82,792).
H. rutilans. White rot. **43:**AZ(1202). **58:**Unk.(2681).
Heterobasidion annosum. Trunk rot. **7:**CA(1110). **58:**NE states(82).
Hydnochaete olivacea. **42:**MA(656).
Hydnum sp. Isolated from wood. **42:**PA(792).
Hymenochaete agglutinans. **58:**MA(128).
H. pinnatifida. On wood. **42:**GA(491,1439).
H. tabacina. **19:**ID(3202), WA(3202). **55:**ID(3202). **58:**NY(128).
Hyphoderma amoenum. **43:**AZ(1202). **54:**AZ(1194).
H. roseocremeum. **43:**AZ(1202).
H. setigerum. **43:**AZ(1202).
Hyphodermella corrugata. **19:**WA(3202).
Hypochnicium bombycinum. **43:**AZ(1202).
Inonotus dryophilus. Sapwood. **42:**Range of host(82). **58:**ME(82), MI(82).
Irpex lacteus. On wood. **3:**ND(378). **8:**CA(1110), OR(3202). **12:**OR(3202). **19:**ID(3202), WA(3202). **37:**GA(2963), NC(1303), SC(2748,2963). **55:**MN(1197), MT(3202). **58:**NE states(82).
Irpex sp. **8:**WA(3202).
Junghuhnia nitida. **36:**NH(128,2168).
Laetiporus sulphureus. Brown cubical rot. **3:**CA(1110). **10:**FL(35). **25:**FL(35). **37:**SC(2748). **42:**Range of host(82), FL(35), PA(82,792). **58:**NE states(82).
Laxitextum bicolor. White rot. **36:**VT(543). **37:**SC(543,2747).
Lentinus strigosus. On limbs. **37:**SC(2747).
Lenzites betulina. On limbs. **37:**SC(2747). **58:**MI(82), NY(82).
Lopharia cinerascens. **42:**MD(543). **55:**MT(3202).
Maireina marginata. Twig blight. **18:**OR(82,3202). **37:**OR(82,3202).
Meruliopsis corium. On limbs and trunks. **37:**SC(2747). **55:**MT(3202). **56:**MT(3832).
Microstroma sp. **37:**SC(867).
Mycena incarnatifolia. On wood. **42:**MI(3326).
Oxyporus corticola. On wood. **37:**SC(2747).
O. latemarginatus. White rot. **8:**CA(82,828,1110). **18:**CA(1110). **28:**CA(828). **37:**SC(2747). **58:**CA(82).
O. populinus. **37:**CT(82).
Panellus stipticus. Pruning cuts. **37:**SC(2747).
Parmastomyces transmutans. Isolated from wood. **42:**PA(792,2680,2681).
Pellicularia koleroga. Thread blight. **6:**LA(3640). **10:**LA(3640). **16:**LA(82). **22:**LA(3640). **32:**FL(82). **58:**Fl(82).
Peniophora nuda. On wood. **42:**GA(491,1439).
Perenniporia fraxinophila. **58:**OR(82).
P. medulla-panis. **43:**AZ(1202). **54:**AZ(128).
P. tenuis. **42:**E states(82), NC states(82).
Phaeolus schweinitzii. Isolated from wood. **42:**PA(792).
Phanerochaete affinis. **36:**NH(128). **37:**SC(2747). **55:**NY(128).
P. arizonica. White rot. **37:**CA(433).
P. burtii. **42:**WV(128).
P. crassa. White rot. **37:**SC(2747).
P. filamentosa. **42:**MA(656).
P. sanguinea. **55:**ID(3202).
Phellinus ferreus. **19:**OR(3202).
P. ferruginosus. **19:**OR(3202).
P. gilvus. White rot. **3:**ND(378). **37:**FL(35), GA(2963), SC(2748,2963). **43:**AZ(1202). **58:**Range of host(82).
P. igniarius. White heart rot. **3:**MO(82). **19:**MT(3202,3832), WA(3202). **55:**ID(3202), MT(3202), WA(3202). **56:**MT(3832). **58:**ID(82), MT(82).
P. laevigatus. **58:**MT(3202).
P. pomaceus. White heart rot. **3:**Range of host(82), NC(3944), ND(378), SD(378). **6:**GA(491,1439). **16:**Range of host(82), MT(3202). **19:**MT(3202,3832), WA(3202). **32:**IA(82), VA(82). **36:**TN(82). **37:**MT(3202). **53:**GA(491,1439). **55:**MT(3202,3832), ND(82), SD(82), WA(3202). **57:**ND(378), SD(378). **58:**Range of host(82), ID(3202), OR(3202).
P. prunicola. White rot. **42:**E states(82), NC states(82), ME(82), NY(82), PA(82,792). **58:**Unk.(82), MI(2862).
P. punctatus. White rot. **3:**ND(378,3759). **55:**ND(3759).
P. texanus. **43:**AZ(1202).
P. viticola. **43:**AZ(1202). **58:**NY(2456).
Phlebia chrysocrea. On wood. **42:**PA(262,792). **58:**TN(2165).
P. concentrica. **58:**WA(128).
P. rufa. **36:**NY(128).
Plicaturopsis crispa. **16:**OR(3202).
Polyporus arcularius. **37:**SC(2748).
P. varius. **19:**CA(642). **55:**ID(3202), WA(3202). **58:**ID(3202).
Porodisculus pendulus. On dead limbs. **42:**NC(1303).
Postia balsamea. Isolated from wood. **42:**PA(792).
P. sericeomollis. Brown heart rot. **42:**MI(82), PA(82,792).
P. tephroleuca. Brown cubical rot. **8:**OR(82). **37:**NC(82). **58:**Range of host(82).
Pycnoporellus fulgens. Isolated from wood. **42:**PA(792).
Pycnoporus cinnabarinus. On wood. **37:**GA(2963), NC(1303), SC(2963). **42:**NC(1303,3944). **55:**MT(3202). **58:**Range of host(82).
P. sanguineus. **37:**SC(2748).
Sarcodontia setosa. **58:**MA(1175).
Schizophyllum commune. On wood. **7:**CA(1110), TX(82), WA(82,3202). **8:**CA(1110), WA(82). **16:**WA(82,3202). **18:**'Wood rot.' CA(82,1110). **37:**Range of host(82), CA(1110), SC(2747). **58:**Range of host(82), GA(491,1439), MI(2862), OK(2872), OR(3202), WA(3202).
Schizopora paradoxa. **42:**E states(82), NC states(82), GA(1439).
Scytinostroma galactinum. Root rot. **22:**MD(668). **52:**MD(82,668).
Sebacina helvelloides. **58:**IA(128).
Septobasidium pseudopedicellatum. On scale insects infesting bark. **7:**MS(82).
Steccherinum laeticolor. **43:**AZ(1202). **54:**AZ(128,1194).
S. ochraceum. Sapwood rot. **36:**NY(128). **37:**SC(2747).
Stereum hirsutum. White rot. **3:**NC(3944). **12:**NY(543), VA(543). **16:**OR(3202). **37:**OR(82,3202), SC(2747). **42:**MA(543), NY(82), PA(82,792). **58:**Range of host(82), NY(82,543), OR(543), PA(82), VT(543).
S. ochraceo-flavum. On wood. **7:**CA(543). **36:**NH(543), NY(543). **37:**SC(2747). **58:**CT(543).
S. striatum. On wood. **36:**WV(543). **42:**NC(3944). **58:**NC(82), WV(82).
Stereum sp. **19:**WA(3202). **37:**GA(2963), SC(2963).
Tomentella lateritia. **42:**MA(656).
T. olivascens. **58:**MA(2039), NY(2039).
T. rubiginosa. **58:**PA(2039).
Trametes hirsuta. White rot. **3:**ND(378). **7:**CA(1110). **8:**OR(82), WA(3202). **16:**Range of host(82), ID(3202), OR(3202). **19:**ID(3202), OR(3202), WA(3202). **25:**OR(3202). **37:**Range of host(82), AZ(1202), FL(35), GA(2963), ID(3202), NC(1303), OR(3202), SC(2748,2963), WA(3202). **42:**Range of host(82). **55:**ID(3202), MT(3202), WA(3202). **58:**Range of host(82).
T. pubescens. **8:**OR(3202).
T. versicolor. White rot. **3:**ND(378). **7:**CA(1110). **10:**FL(35). **18:**'Wood rot.' CA(1110), OR(82). **19:**WA(3202). **25:**FL(35). **26:**OR(3202). **42:**Range of host(82), FL(35), PA(792). **43:**AZ(1202). **55:**WA(3202). **58:**Range of host(82), Unk.(1006), MA(1390).
Tremella mesenterica. On dead limbs. **42:**NC(1303).
Trichaptum biforme. White rot. **37:**SC(2748). **58:**Range of host(82).
Typhula gyrans. On leaves and petioles. **42:**NY(2955).
Tyromyces chioneus. White rot. **37:**NC(82), SC(2748). **42:**Range of host(82). **58:**Range of host(82), ID(3202), OR(3202), WA(3202).
T. galactinus. **16:**OR(3202). **58:**NY(82), OR(82).
Wolfiporia cocos. Isolated from wood. **42:**PA(792).
W. dilatohypha. Trunk rot. **42:**PA(82,792).
Xeromphalina campanella. Isolated from wood. **42:**PA(792).

Deuteromycotina - Hyphomycetes

Alternaria alternata. Isolated from fruits and twigs. **7:**'Alternaria spot. Alternaria fruit rot.' WA(2028). **8:**CA(1110), ID(3202). **16:**CA(1110,2369). **37:**Unk.(3064).
A. citri. Blossom end rot. **7:**CA(82).
A. mali. **8:**OR(3202), WA(3202).
A. tenuissima. Leaf spot. **8:**CA(82).
Alternaria sp. Blossom end rot, green mold of fruit, fruit rot, leaf spot. **7:**CA(82). **8:**CA(82), ID(82,3202), MI(82), WA(3202). **12:**MI(82). **16:**ID(82), MI(82), OR(82). **18:**CA(2889). **25:**TX(82). **37:**IL(1141).
Aspergillus ficuum. Isolated from fruits. **18:**CA(2798).
A. flavus. Isolated from fruits. **18:**CA(2798,2889).
A. glaucus. Isolated from fruits. **18:**CA(2889).

A. niger. 'Black mold rot.' **3:**CA(1110). **7:**CA(1110). **8:**CA(1110), WA(3202). **11:**CA(1110). **12:**CA(1110). **16:**CA(1110). **18:**CA(1110,2889). **28:**CA(1110). **37:**Range of host(82), CA(1110). **38:**CA(1110). **41:**CA(1110).
A. ochraceus. Isolated from fruits. **18:**CA(2889).
A. wentii. Isolated from fruits. **18:**CA(2889).
Aureobasidium prunicola. Leaf spot, on fruits. **8:**MA(2089). **55:**NY(82), WI(82,963).
A. pullulans. **8:**OR(3202), WA(3202).
Bipolaris spicifera. Isolated from fruits. **18:**CA(2798).
Botrytis cinerea. Blossom blight, gray-mold rot, on fruits. **3:**'Gray mold rot.' CA(1110). **7:**'Green fruit rot.' CA(82,1110). **8:**'Gray mold rot.' NW states(82), CA(82,1110), ID(3202), WA(3202). **11:**'Gray mold rot.' CA(1110). **12:**'Gray mold rot.' Unk.(82), CA(1110). **16:**'Gray mold rot.' CA(82,1110,2369), WA(82,3202). **18:**'Green fruit rot.' CA(1110). **28:**'Gray mold rot.' CA(1110). **37:**'Gray mold rot.' Range of host(82), CA(1110), FL(35), NC(1303). **38:**'Gray mold rot.' CA(1110). **41:**'Gray mold rot.' CA(1110). **52:**'Gray mold rot.' NY(82). **58:**'Gray mold rot.' Range of host(82), AK(526).
Candida albicans. **37:**CA(1110).
Candida sp. Isolated from cankers. **37:**IL(1141).
Cercoseptoria prunicola. Leaf spot. **3:**AL(82), LA(82).
Cercospora circumscissa. Shot hole, leaf spot. **3:**FL(35), IL(82), ND(82,378), NE(82), WI(1358). **6:**FL(35), KS(82). **7:**FL(35), TX(82). **8:**NJ(82), OH(82), PA(82), TX(82). **10:**FL(35), NC(1303). **12:**DE(3543), NC(3944), NJ(82), TX(82). **16:**CA(82,1110), FL(35,82), MA(82), MS(2720), TX(82), WA(82,3202). **18:**CA(82,1110), OR(82,3202). **19:**ID(3202). **25:**CA(82), FL(35). **32:**FL(35), KS(82), MS(2720). **36:**WI(82). **37:**Range of host(82), FL(35), IA(82), MS(2720), TX(82). **39:**DE(82). **42:**FL(35). **51:**KS(82). **55:**NE states(82), W states(82), IA(82), MT(82,3202), WI(82). **57:**ND(378). **58:**Range of host(82), Unk.(82), SD(2243), WA(3202).
C. graphioides. Leaf spot. **10:**FL(35). **25:**FL(35). **42:**FL(35), IA(82), WI(82).
C. rubrotincta. Leaf spot. **37:**IL(82), LA(82,962).
Cercospora sp. Leaf spot. **10:**OK(2872). **58:**IL(1052).
Chrysosporium sp. **16:**CA(1110).
Cladosporium americanum. On living leaves. **3:**WI(1339).
C. carpophilum. Freckle, scab, black spot. **3:**Range of host(82), FL(35). **6:**FL(35). **7:**'Scab.' Range of host(82), CA(1110), FL(35), OK(82,2873), TX(82). **10:**FL(35). **12:**IA(82), NJ(82), NY(82). **16:**Range of host(82), CA(82), FL(35), IA(82). **18:**'Scab.' CA(82,1110), CT(82), OR(82,3202), WA(3202). **25:**FL(35). **32:**FL(35), IA(82). **35:**AK(526). **38:**CA(1110), CT(82), DE(82), IL(82), NC(1303), NY(82), PA(82), TX(82). **42:**FL(35). **44:**MS(82,2720). **58:**Range of host(82), OK(2872).
C. cladosporioides. Isolated from fruits. **18:**CA(2798).
C. herbarum. 'Cladosporium rot.' **8:**NW states(82), ID(82,3202), OR(3202), WA(3202). **16:**CA(2369).
Cladosporium sp. Green mold of fruit, scab. **16:**CA(1110), ID(82,3202), OR(82,3202), WA(3202). **18:**CA(2636). **28:**AK(526). **36:**NC(1303). **37:**IL(1141).
Conoplea olivacea. **58:**IA(1672), WI(1672).
Corynespora pruni. On bark. **42:**SC(256).
Cristulariella moricola. Leaf spot. **37:**FL(35).
Cylindrocladium scoparium. Decline, root rot. **37:**FL(35,3370,3375), GA(1694).
Diplocarpon mespili. Leaf blight. **37:**CA(82,1110), NJ(82).
Epicoccum nigrum. Isolated from twigs. **37:**Unk.(3064).
Epicoccum sp. Isolated from cankers. **37:**IL(1141).
Eurotium sp. On dead leaves. **27:**CA(2439).
Fumago vagans. Sooty mold. **16:**WA(3202). **37:**Range of host(82), NC(1303).
Fusarium avenaceum. Fruit rot. **37:**CA(82,1110), IL(82), NC(82), TX(82).
F. lateritium. Bud rot. **37:**CA(82), GA(82), KS(82).
F. oxysporum. Fruit rot, basal stem rot. **37:**CA(82,1110), FL(35), IL(82), NC(82), TX(82).
F. roseum. Isolated from fruits. **18:**CA(2798). **37:**CA(1110).
F. solani. Fruit rot. **37:**CA(82,1110), IL(82), NC(82), TX(82).
Fusarium sp. On roots. **3:**CA(1110). **7:**CA(1110). **8:**CA(1110), WA(3202). **11:**CA(1110). **12:**CA(1110). **16:**CA(1110). **18:**CA(1110). **28:**CA(1110). **37:**CA(1110,1570). **38:**CA(1110). **41:**CA(1110).
Fusicladium cerasi. Scab. **12:**NC states(82), NE states(82), WA(3202).
Geotrichum candidum. Sour rot. **37:**GA(462,3851), NC(462), NJ(462), PA(462).
Geotrichum sp. Isolated from cankers. **37:**IL(1141).
Helicoma olivaceum. On bark. **58:**NY(1271).
Miuraea persica. Frosty mildew. **37:**NC(3944), OK(2872).
Monilia angustior. Shoot blight. **55:**E & C states(82), ND(82). **57:**ND(378).
M. implicata. Sour rot market disease. **37:**GA(3851).
M. laxa. **42:**NC(3944).
M. seaveri. Leaf blight. **42:**NC(1303).
Monilia sp. Brown rot. **19:**ID(3202), WA(3202). **58:**Unk.(82).
Morrisographium persicae. On limbs and bark. **37:**C states(82), S states(82). **58:**NC(82), NJ(82), NY(82), VA(82), VT(82).
Mycocentrospora cladosporioides. Leaf spot. **25:**LA(82), TX(82).
Nigrospora sp. Isolated from cankers. **37:**IL(1141).
Oidium sp. **37:**VA(894).
Penicillium expansum. **7:**'Penicillium fruit rot.' CA(1110). **8:**'Blue mold rot.' NW states(82), CA(1110), OR(3202), WA(3202). **16:**'Blue mold rot.' ID(3202), OR(3202), WA(3202).
P. funiculosum. Isolated from fruits. **18:**CA(2798).
P. italicum. Fruit rot. **8:**CA(1110).
Penicillium sp. Seedling tip necrosis, blue mold rot, fruit and seed rot. **16:**CA(1110,2369). **18:**CA(1110,2889), WA(3202). **37:**CA(1110), TX(1422). **58:**NW states(82).
Phymatotrichopsis omnivora. Root rot. **7:**'Phymatotrichum root rot.' AZ(82), OK(82,2872), TX(82). **12:**TX(82). **16:**TX(82). **18:**AZ(82), TX(82). **25:**TX(82). **32:**TX(82). **37:**'Phymatotrichum root rot.' AZ(82), NM(82), TX(82). **42:**TX(82). **58:**AZ(82), OK(82), TX(82).
Stemphylium sp. **8:**OR(3202), WA(3202).
Stigmina carpophila. Coryneum blight, bud, leaf and shot blight, shot hole, pustular spot. 'Shothole.' **3:**CA(1110), FL(35), OR(82). **6:**FL(35). **7:**Range of host(82), CA(1110), FL(35), ID(3202), WA(3202). **8:**CA(1110), ID(82,3202), OR(82,3202), WA(82,3202). **11:**CA(1110). **12:**CA(82,1110), ID(82,3202), WA(82,3202). **15:**CA(82). **16:**NW states(82), CA(82,1110), FL(35), ID(3202), WA(3202). **19:**ID(3202). **25:**ID(3202). **27:**CA(2894). **28:**CA(1110). **32:**FL(35). **36:**ID(82). **38:**CA(82,1110), OR(82), WA(82). **39:**WA(82). **41:**CA(1110). **58:**Range of host(82), W states(82).
Taeniolella muricata. **55:**MT(3202).
Trichothecium roseum. Pink mold fruit rot. **37:**Range of host(82), CA(1110), OK(2872).
Trimmatostroma americana. **58:**WA(3202).
Ulocladium atrum. Isolated from fruits. **18:**CA(2798).
U. chartarum. Isolated from fruits. **18:**CA(2798).
Verticillium albo-atrum. Blackheart. 'Verticillium wilt.' **7:**CA(82), UT(82), WA(82,3202). **8:**CA(3895), WA(3202). **16:**CA(82), WA(3202). **18:**CA(82), IL(82,1558), WA(3202). **25:**CA(82). **37:**NE states(82), Unk.(82), CA(82), OR(82,3202), WA(3202).
V. dahliae. 'Verticillium wilt.' **3:**CA(1110). **7:**CA(1110). **8:**CA(1110), WA(3308). **11:**CA(1110). **12:**CA(1110). **16:**CA(1110). **18:**CA(1110). **28:**CA(1110). **37:**CA(1110). **38:**CA(1110). **41:**CA(1110).
V. nigrescens. **8:**WA(3308).

Deuteromycotina - Coelomycetes

Aplosporella amygdalina. **37:**KS(82).
Catenophora pruni. On twigs. **8:**CA(1110). **44:**NC(82,3549).
Colletotrichum gloeosporioides. Anthracnose, seedling tip blight, kernel rot. **3:**FL(35). **6:**FL(35). **7:**FL(35). **16:**FL(35). **18:**CA(82,1110). **21:**NY(1230). **32:**FL(35). **37:**FL(35).
C. prunicola. On living leaves. **42:**WI(1342).
Coniothyrium fuckelii. Branch and stem canker. **3:**FL(35). **6:**FL(35). **7:**FL(35). **16:**FL(35). **32:**FL(35). **37:**WV(82).
Coniothyrium sp. Twig blight, stem canker, leaf spot. **10:**FL(35). **25:**FL(35). **37:**FL(35), TX(82). **42:**FL(35). **55:**WI(1354). **58:**AK(526).
Cylindrosporium nuttallii. Leaf spot. **55:**ID(3202), OR(3202). **56:**OR(82).
Cylindrosporium sp. Leaf spot. **29:**FL(82). **33:**WI(1340). **42:**WI(1346). **55:**IN(987). **58:**Unk.(82).
Cytospora chrysosperma. On twigs. **55:**MT(82).
C. cincta. Canker, dieback. **16:**ID(3202). **37:**Unk.(3064), ID(3202),

NC(1303). **58:**Unk.(82), SD(2243).
C. leucosperma. Twig canker, dieback. **3:**FL(35). **6:**FL(35). **7:**FL(35). **16:**FL(35). **31:**FL(82). **32:**FL(35). **37:**FL(35).
C. leucostoma. Dieback. 'Cytospora canker.' **7:**AZ(82), NY(82). **16:**CA(1110). **37:**CA(1110), CO(2186), FL(35), NY(1435), OK(2872). **38:**CA(1110). **42:**PA(3672). **58:**Unk.(82).
C. persicae. On limbs and twigs. **37:**E states(82), AZ(82), MS(82).
Cytospora sp. Canker, dieback. **7:**AZ(82), NY(82). **8:**NY(1823). **25:**NC(1303). **37:**IL(1141).
Cytosporina ludibunda. **12:**AK(526).
Dendrophoma sp. **37:**OR(3202).
Diplodia persicina. On limbs. **37:**LA(82), PA(82).
D. pruni. Twig blight, gumming disease. **3:**FL(35). **6:**FL(35). **7:**FL(35). **16:**FL(35,82), NJ(82). **32:**FL(35,82).
D. vulgaris. Twig blight. **16:**FL(82), NJ(82).
Diplodia sp. Twig blight, fruit rot. **7:**CA(82,1110). **16:**FL(82), NJ(82). **37:**GA(3346), NC(1303).
Diplodina persicae. Fruit rot. **37:**LA(1629). **58:**SC(3961).
Diplodina sp. Brown fruit rot. **37:**WI(1628).
Dothichiza serotina. On leaves. **42:**AL(82), WV(82).
Foveostroma drupacearum. On limbs. **58:**MD(861), ME(861), MI(861), NH(861), NJ(861), NY(861), PA(861), WA(3549).
Fusicoccum amygdali. Canker, leaf spot. 'Fusicoccum canker.' **37:**MA(82,1392), NC(1303), NJ(82), NY(82).
F. dakotense. **57:**ND(378).
F. persicae. **37:**GA(82), LA(82), PA(82).
Gloeodes pomigena. On limbs. **37:**IN(82).
Hendersonia jonesiana. **5:**NV(2785).
Hendersonula toruloidea. Canker. **7:**CA(1110). **18:**CA(1110,2805). **37:**CA(1110).
Lasiodiplodia theobromae. Gumming disease. **3:**FL(35). **6:**FL(35). **7:**FL(35). **10:**FL(35). **16:**FL(35). **25:**FL(35). **32:**FL(35). **37:**AL(82), FL(35,82), TX(82). **42:**FL(35).
Macrophoma persicina. On limbs. **37:**MO(82), PA(82).
Macrophoma sp. Dieback. **3:**FL(35). **6:**FL(35). **7:**FL(35). **16:**FL(35). **32:**FL(35). **37:**FL(35).
Macrophomina phaseolina. Charcoal root rot. **37:**FL(35).
Melanconium cerasinum. On limbs. **55:**KS(82), ND(82). **57:**ND(378).
Microgloeum pruni. Shot hole. **58:**ID(3202,3549).
Micropera spuria. **58:**MA(664).
Microsphaeropsis olivacea. Isolated from twigs. **37:**Unk.(3064).
Myxosporium pruni. On limbs. **55:**NY(82).
Pestalotia adusta. On leaves. **12:**NY(82). **42:**MA(1390), NY(82), VA(82).
P. laurocerasi. Limb and trunk gall. **7:**CA(82,1110).
Phloeosporella padi. Shot hole, leaf spot. **3:**FL(35). **6:**FL(35). **7:**FL(35). **10:**FL(35). **12:**NC(3944). **16:**FL(35). **19:**CA(81). **25:**FL(35). **32:**FL(35). **37:**FL(35). **42:**FL(35), NC(1303,3944). **56:**CA(81). **58:**Unk.(82), CA(3549), NC(1303), NY(3549), OR(3549).
Phoma persicae. Stem canker. **37:**E states(82), S states(82), IL(1141).
P. pomorum. Scurfy bark, leaf spot. **3:**IA(82). **8:**OR(82,3202). **12:**MS(2720). **33:**WI(64,1343). **36:**IA(82). **42:**IA(82).
Phomopsis mali. Bark canker, fruit rot. **37:**NY(3030). **58:**VA(82).
P. padina. Canker. **8:**WA(82,3202). **12:**NY(82), PA(82).
P. prunorum. **16:**WA(3202).
Phomopsis sp. Dieback, constriction disease, fruit rot. **12:**AK(526). **37:**MD(1729,3792), NJ(762). **58:**DC(3780).
Phyllosticta circumscissa. Leaf spot. Leaf spot. **3:**CA(82,1110), IL(82), WI(1346). **7:**CA(82,1110). **10:**FL(35). **12:**NJ(82). **16:**IA(82), WA(82,3202). **25:**FL(35). **37:**Range of host(82), FL(35), HI(2903), NY(64). **42:**FL(35). **55:**KS(82), WA(82,3202). **58:**Range of host(82).
P. congesta. Leaf blotch and spot, on fruit, leaves and twigs. **16:**GA(82), TX(82). **32:**TX(82). **36:**WI(82). **58:**IL(82), TX(64,82), WI(82).
P. destruens. On leaves. **55:**IA(82), NY(82). **57:**ND(64,378).
P. laurocerasi. Leaf spot. **3:**FL(35). **6:**FL(35). **7:**FL(35). **10:**FL(35). **16:**FL(35). **25:**CA(81,82), FL(35,82), NC(1303), WA(3202). **32:**FL(35). **37:**FL(35). **42:**FL(35).
P. pennsylvanica. On living leaves. **36:**WI(1343).
P. persicae. Leaf spot. **37:**MD(82), NE(82), OH(82).
P. pruni-virginianae. On living leaves. **55:**WI(1358).
P. serotina. Leaf spot. **10:**FL(35). **25:**FL(35). **42:**Range of host(82), FL(35), MS(2720). **55:**NE(82), PA(82), SC(82). **58:**E states(82), WI(82).
P. virginiana. Leaf blotch and spot. **3:**IA(82), KS(82). **8:**WA(82). **28:**WA(3202). **36:**ME(82). **39:**WI(1361). **42:**IA(82), KS(82). **51:**MT(82,3202). **54:**AZ(1195). **55:**E & C states(82), IA(82), MT(82,3202). **58:**C states(82), MT(82), NM(1195).
P. vulgaris. **55:**IA(82).
Phyllosticta sp. Leaf spot. **55:**WI(1354,1361). **58:**NC(1303).
Rhodosticta quercina. Canker. **16:**CA(1110).
Seimatosporium lichenicola. Leaf blight. **18:**CA(1110).
Septoria pruni. Leaf spot. **3:**IA(82), KS(82), KY(82), ND(82,378), NE(82). **16:**TX(2093). **55:**MI(82). **56:**CA(638,642).
S. purpureo-cincta. Leaf spot. **3:**KS(82).
S. ravenelii. Leaf spot. **25:**SC(82).
Septoria sp. Leaf spot. **10:**FL(35). **16:**TX(2093). **25:**FL(35). **42:**FL(35).
Sphaceloma sp. Anthracnose. **24:**CA(82).
Sphaeropsis malorum. On twigs. **58:**NY(82).
S. peckii. On twigs. **3:**ND(378). **58:**KS(82), ND(82).
Sphaeropsis sp. Canker. **10:**FL(35). **25:**FL(35). **37:**FL(35), IL(1141). **42:**FL(35). **58:**OK(2872).

Deuteromycotina - Other

Rhizoctonia solani. Thread blight, damping-off, root rot. **3:**FL(35). **6:**FL(35). **7:**FL(35). **8:**CA(1110). **10:**FL(35). **16:**CA(1110), FL(35). **25:**FL(35). **32:**FL(35). **37:**AR(82), CA(1110), CT(82), FL(35). **42:**FL(35). **58:**OK(2872).
Rhizoctonia sp. On roots. **37:**CA(1570).
Sclerotium rolfsii. Seedling stem rot. **37:**S states(82), FL(35).

Purshia DC. ex Poir. in Lam.

1. *P. tridentata* (Pursh) DC.

Ascomycotina

Odontotrema oregonense. On wood. **1:**OR(3232).
Orbilia coccinella. **1:**CA(642).
Thyridaria californica. **1:**CA(642).

Basidiomycotina - Other

Cytidia patelliformis. **1:**WA(3202).
C. stereoides. On wood. **1:**CA(640,642).
Dacrymyces stillatus. On dead sticks. **1:**CA(642).
Heterobasidion annosum. **1:**W states(3584).
Hyphoderma assimile. On limbs. **1:**CA(642).
Peniophora cinerea. **1:**WA(3202).

Deuteromycotina - Coelomycetes

Phoma terrestris. Isolated from rhizosphere. **1:**OR(2791).

Pyracantha M. Roem.

1. *P. angustifolia* (Franch.) C. K. Schneid. [narrowleaf fire-thorn]
2. *P. coccinea* M. Roem. [scarlet firethorn]
3. *P. coccinea* Roem. cv. 'lalandei'
4. *P. crenato-serrata* (Hance) Rehd.
5. *Pyracantha* sp.

Oomycetes

Phytophthora cactorum. Root rot, fruit spot. **2:**FL(35).
P. cinnamomi. Root rot. **5:**NC(1303).
Pythium sp. Root rot. **2:**FL(35), GA(1208).

Ascomycotina

Botryosphaeria dothidea. Canker. **2:**FL(35). **5:**NC(1303).
B. obtusa. On limbs. **5:**AL(82).
B. ribis. Cane blight. **2:**MS(2720).
Physalospora sp. Canker, dieback. **2:**FL(35).
Podosphaera clandestina. Powdery mildew. **5:**NC(1303).
Venturia inaequalis. Scab. **2:**OR(3202), WA(3202). **3:**GA(1438), OK(2872).

Basidiomycotina - Other

Armillaria mellea. Root rot. **2:**FL(35). **5:**CA(82).
A. tabescens. Root rot. **2:**FL(35).
Septobasidium cokeri. On scale insects infesting bark. **5:**NJ(82).
Trametes versicolor. **2:**OR(3202).

Deuteromycotina - Hyphomycetes

Botrytis cinerea. Gray mold of berries. **2:**FL(35).
Cercospora pyracanthae. Leaf spot. **2:**FL(35). **5:**FL(35).
Clasterosporium sp. Leaf spot. **2:**FL(35).
Fusarium oxysporum. Wilt. **1:**Unk.(2332). **2:**FL(35). **5:**FL(2333).
Fusarium sp. Stem lesions, root rot. **2:**FL(35).
Phymatotrichopsis omnivora. Root rot. **5:**TX(82).
Spilocaea pyracanthae. Fruit and leaf scab. **2:**FL(35), GA(2405),

NC(1303). **3**:DE(703). **4**:IL(312). **5**:Range of host(82), FL(35), MO(2437), NC(1303).

Deuteromycotina - Coelomycetes

Colletotrichum gloeosporioides. Fruit rot. **1**:NC(1303).
Colletotrichum sp. Fruit rot. **5**:NC(1303).
Coniothyrium sp. Limb canker, dieback. **2**:FL(35).
Diplodia crataegi. Twig blight. **2**:FL(35). **5**:PA(82).
Fusicoccum sp. Canker. **2**:OK(2872).
Macrophoma sp. Dieback, fruit rot. **2**:FL(35).
Macrophomina phaseolina. Charcoal root rot. **2**:FL(35).
Phoma sp. Twig dieback. **2**:FL(35).
Phomopsis sp. Canker, dieback. **2**:FL(35). **5**:FL(35), NC(1303).
Phyllosticta sp. Leaf spot. **2**:FL(35). **5**:FL(35).
Sphaeropsis tumefaciens. Stem gall. **2**:FL(35).

Deuteromycotina - Other

Rhizoctonia solani. Leaf and stem blight. **2**:FL(35). **5**:FL(35).

Pyrus L.

1. *P. calleryana* Decne. [Bradford pear, Callery pear]
2. *P. communis* L. [common pear]
3. *P. pyrifolia* (N. L. Burm.) Nakai (=*P. serotina* Rehd.) [Asian pear, Chinese pear, Japanese pear, oriental pear, sand pear]
4. *Pyrus* × sp. (unnamed hybrids)
5. *Pyrus* sp.

Oomycetes

Phytophthora cactorum. Collar and fruit rot. 'Sprinkler rot. Phytophthora crown rot. Phytophthora root rot.' **2**:Range of host(82), AL(285), CA(1110), FL(35), ID(3202), OR(3202), WA(3202).
P. cambivora. **2**:CA(1110).
P. cinnamomi. **2**:OR(3202).
P. citrophthora. Collar rot. **2**:CA(82,1110).
P. megasperma. Crown and root rot. **2**:CA(1110).
P. nicotianae var. *parasitica*. **2**:OR(3202).
P. syringae. **2**:OR(3202).
Phytophthora sp. **2**:WA(3434).
Pythium debaryanum. **2**:WA(3202,3434).
P. dictyosporum. **2**:WA(3434).
P. megalacanthum. **2**:WA(3434).
P. pulchrum. **2**:WA(3434).
P. splendens. **2**:WA(3434).
P. ultimum. **2**:WA(3434).
P. vexans. **2**:WA(3434).
Pythium sp. Root rot. **2**:FL(35), WA(3434).

Zygomycotina

Mortierella sp. **2**:WA(3434).
Mucor piriformis. Fruit rot. **2**:NW states(3418), NW states(3419), OR(2370,3417), WA(3202).
Mucor sp. **2**:WA(3434).
Rhizopus stolonifer. Black mold rot. 'Rhizopus rot.' **2**:Range of host(82), FL(35), ID(3202), OR(3202), WA(3202).

Ascomycotina

Biscogniauxia marginata. Blister canker. **2**:DE(82), IA(82,1225).
Botryosphaeria dothidea. Canker, black rot of fruit, white rot of fruit. **1**:AL(2523). **2**:NC(1303). **3**:FL(35).
B. obtusa. Black rot. **2**:'Black rot. Leaf spot. Canker.' Range of host(82), MS(2720), OK(2872), OR(3202). **3**:MS(82,2720).
B. quercuum. Canker, black rot, leaf spot. **2**:FL(35), NC(1303). **3**:FL(35).
B. rhodina. On limbs. **2**:FL(82).
B. ribis. Black rot of fruit. **2**:MS(2720), VA(82).
Chaetomium globosum. **2**:OR(3202).
Diaporthe eres. Twig blight. **2**:FL(35), WA(3202).
Diatrypella nitschkei. **2**:IA(3629).
Dothiora pyrenophora. On limbs and twigs. **2**:NJ(82).
Elsinoe piri. Anthracnose. 'Elsinoe leaf spot. Elsinoe fruit spot.' **2**:OR(82,3202), WA(82,3202).
Endomyces sp. **2**:WA(3434).
Gibberella baccata. 'Gibberella canker.' **2**:CA(1110).
Glomerella cingulata. Bitter rot, fruit rot. **2**:'Bitter rot.' Range of host(82), AL(285), CA(1110), GA(1438), NC(1303), TX(82). **3**:MS(82,2720).
Hypoxylon rubiginosum. **2**:GA(1438).
Leptosphaeria sp. Leaf spot. **3**:FL(35,82).
Leucostoma persoonii. Twig blight. **2**:WA(82,3202).
Monilinia fructicola. Brown rot. **2**:'Brown rot.' E states(82), AL(285), MS(2720), TX(82), WA(82,3202). **3**:VA(82).
M. fructigena. Brown fruit rot. **5**:MD(206).
M. laxa. 'Brown rot.' **2**:W states(82), OR(3202), WA(3202).
Mycosphaerella pyri. Fruit and leaf spot. **2**:'Mycosphaerella leaf spot.' E states(82), Range of host(82), FL(35), GA(1438), IA(82), MS(2720), OK(2872), TX(82), WA(3202). **3**:KS(82), MS(82,2720), TX(82).
M. pyrina. On leaves. **2**:GA(1438), NJ(967). **3**:GA(82).
M. tulasnei. **2**:OR(82,3202), WA(82,3202).
Nectria cinnabarina. Coral spot. 'Nectria canker.' **2**:Range of host(82), FL(35), OR(3202), WA(3202).
N. coccinea. **2**:OR(3202).
N. galligena. 'European canker.' **2**:CA(1110), OR(82,3202), WA(82,3202).
Neurospora sitophila. Ripe rot. **2**:NC(82).
Pezicula pruinosa. **2**:OR(3202).
Physalospora abdita. **2**:NJ(82), VA(82).
Pleospora herbarum. Storage rot. **2**:WA(82,3202).
Podosphaera clandestina. **2**:NJ(82).
P. leucotricha. 'Powdery mildew.' **2**:CO(82), OR(82,3202), WA(82,3202).
Rhytidhysteron rufulum. On dead branches. **3**:FL(82).
Rosellinia aquila. **2**:WA(3202).
R. necatrix. 'Dematophora root rot.' **2**:CA(1110).
Schizothyrium pomi. 'Fly speck.' **2**:E states(82), FL(35), NC(1303), TX(82).
Schizoxylon albo-atrum. **5**:OR(3225).
Sclerotinia sclerotiorum. Fruit rot. **2**:CA(82,1110), WA(82,3202).
Sphaerotheca macularis. **5**:GA(1438).
Sphaerulina potebniae. **2**:ID(3202).
Stictis radiata. **5**:CA(3225).
Taphrina bullata. On leaves. **2**:WA(82,2457,3202).
Valsa ambiens. **5**:IA(1226).
V. ambiens subsp. *leucostomoides*. **5**:IA(1226).
V. ceratosperma. **5**:IA(1226).
Valsa sp. **2**:OR(3202), WA(3202).
Venturia inaequalis. **4**: On *P. niedzwetzkyana* x *P. baccata*. WI(1342).
V. pyrina. Scab. **2**:'Scab.' Range of host(82), AL(285), CA(1110), FL(35), GA(1438), ID(3202), MT(3202), NC(1303), OK(2872), OR(3202), WA(3202). **3**:CA(1110).
Xylaria mali. Root rot. **2**:VA(82). **5**:VA(82).
Xylaria sp. Root rot. **2**:ID(82,3202), IN(82).

Basidiomycotina - Uredinales

Gymnosporangium asiaticum. Heteroecious: aecial host. **3**:CA(82,1110), OR(82,3202), WA(3202).
G. clavariiforme. **2**:FL(35), SC(82). **3**:FL(35). **5**:FL(82).
G. clavipes. Heteroecious: aecial host. **2**:AZ(1203,3981), LA(82), MS(2720), NC(1303), TX(82). **3**:AR(82). **5**:MS(2675).
G. fuscum. Heteroecious: aecial host. 'Pear trellis rust.' **2**:CA(1110,2295).
G. globosum. Heteroecious: aecial host. **2**:E states(82), NC states(82).
[*G. hyalinum*]. **2**:FL(35,82).
G. libocedri. Heteroecious: aecial host. 'Pear rust.' **2**:CA(82,1110), OR(3202), WA(3202).
G. nelsonii. **2**:AZ(82,1203,3981), CO(82).
Gymnosporangium sp. **2**:AZ(1203,1856,1858).

Basidiomycotina - Other

Armillaria mellea. 'Shoestring root rot.' **2**:Range of host(82), CA(1110,2898), MS(2720), OR(3202), TX(82), WA(3202).
A. tabescens. Root rot. **2**:'Clitocybe root rot.' FL(35,82), LA(82). **3**:FL(35).
Chondrostereum purpureum. Heart rot. 'Silver leaf.' **2**:NY(82), OR(82,3202).
Coprinus sp. Storage rot. **2**:OR(3420).
Corticium stevensii. **2**:GA(1439).
Erythricium salmonicolor. Branch and twig blight. **2**:SE states(3639), FL(35,82), LA(82), MS(2720).
Fomitopsis pinicola. **2**:Unk.(82).
Hapalopilus rutilans. White rot. **5**:Unk.(2681).
Laetiporus sulphureus. **2**:Range of host(82).
Pellicularia koleroga. Thread blight. **2**:LA(3640). **3**:MS(82,2720).
Phellinus gilvus. **2**:GA(1439), IN(82), NY(82).
P. igniarius. White heart rot. **2**:Unk.(82), OR(3202).
Polyporus squamosus. **2**:Range of host(82).

Postia tephroleuca. **2**:Unk.(82).
Schizophyllum commune. Trunk rot. **2**:Range of host(82).
Septobasidium pseudopedicellatum. Parasitic on scale insects. **2**:SE states(82), FL(35), TX(82). **3**:FL(35).
Septobasidium sp. Parasitic on scale insects. **2**:FL(35). **3**:FL(35).
Stereum hirsutum. Trunk rot. **2**:ID(82).
Trametes hirsuta. **2**:OR(82), WA(3202).
T. versicolor. 'Wood rot.' **2**:Range of host(82), CA(1110), OR(3202).
Trichaptum biforme. **2**:OR(3202).

Deuteromycotina - Hyphomycetes

Alternaria alternata. **2**:WA(3434).
A. mali. **2**:WA(3202).
Alternaria sp. Black-mold rot, leaf spot. **2**:W states(82), CA(1110), FL(35), OR(3202), WA(3434).
Aspergillus flavus. Storage rot. **2**:Range of host(82), WA(3202).
A. niger. **2**:WA(3202).
A. niger var. *phoenicis*. Storage rot. **2**:Range of host(82).
Aspergillus sp. Storage rot. **2**:Range of host(82), FL(35), WA(3434).
Aureobasidium pullulans. **2**:WA(3202,3434).
Botrytis cinerea. 'Botrytis spur. Botrytis blossom blight. Gray mold rot.' **2**:NW states(3418), NW states(3419), Range of host(82), CA(1110,3384), FL(35), ID(3202), OR(3202,3417), WA(3202).
Cephalosporium carpogenum. Storage rot. **2**:OR(82,3202), WA(82).
Cephalosporium sp. **2**:CA(1110).
Cercospora pyri. Leaf spot. **2**:MI(82). **3**:FL(35).
Cladosporium sp. Green mold rot. **2**:Unk.(82), OR(3202), WA(3202,3434).
Clonostachys araucaria. Isolated from dead bark. **2**:OR(3202). **5**:OR(4007).
Corynespora cassiicola. Leaf spot. **2**:FL(35).
Curvularia geniculata. **2**:WA(3434).
C. inaequalis. **2**:WA(3434).
Cylindrocarpon destructans. Limb canker. **2**:Unk.(82), WA(3434).
C. magnusianum. Canker. **2**:Unk.(82), OR(3202).
C. obtusisporum. Canker. **2**:Unk.(82), OR(3202).
Dendrosporium lobatum. Isolated from bark. **3**:LA(720).
Diplocarpon mespili. Leaf blight, fruit spot. **2**:'Fabraea leaf spot. Fabraea fruit spot.' Range of host(82), CA(1110), GA(1438), ID(3202), MS(2720), NC(1303), OK(2872), WA(3202). **3**:CA(1110), IA(82), LA(82), MD(82), MS(82,2720), NY(82), WA(3202). **4**: GA(1438), VA(82). **5**:VA(82).
Fusarium acuminatum. Twig blight. **2**:Unk.(82), WA(3434).
F. avenaceum. Twig blight. **2**:Unk.(82), OR(3202).
F. equiseti. **2**:OR(3202), WA(3434).
F. lateritium. Twig blight. **2**:Unk.(82).
F. merismoides. **2**:WA(3434).
F. oxysporum. **2**:WA(3434).
F. redolens. **2**:WA(3434).
F. sambucinum. Twig blight. **2**:Unk.(82), OR(3202).
F. solani. **2**:WA(3434).
Fusarium sp. Canker, twig dieback. **2**:CA(1110), WA(3202). **3**:FL(35).
Fusicladium pyrorum. Scab, black spot. **2**:FL(35).
Helminthosporium papulosum. Bark canker. **2**:MS(82,2720).
H. papulosum f. *pyri*. Blister canker. **2**:MA(9).
Helminthosporium sp. **2**:PA(82), WA(3202).
Hormodendrum pyri. **2**:WA(3202).
Penicillium aurantiogriseum. Blue mold. **2**:Unk.(82), WA(3202).
P. chrysogenum. **2**:WA(3434).
P. crustosum. Blue mold. **2**:Unk.(82), WA(3202).
P. expansum. Storage decay. 'Blue mold rot.' **2**:NW states(3418), NW states(3419), Unk.(82), CA(939), ID(3202), OR(3202,3417), WA(3202,3434).
P. puberulum. Blue mold. **2**:Unk.(82), WA(3202).
P. roquefortii. Blue mold. **2**:Unk.(82), WA(3202).
P. rugulosum. Blue mold. **2**:Unk.(82), WA(3202).
Penicillium sp. **2**:OR(3202), WA(3434).
Phialophora malorum. Storage rot. 'Side rot.' **2**:NW states(3418), OR(82,265,3202,3417), WA(82,3202).
Phymatotrichopsis omnivora. Root rot. **2**:SW states(82), OK(2872).
Ramularia sp. **2**:WA(3434).
Stemphylium botryosum. **2**:WA(3434).
Stemphylium sp. **2**:WA(3202).
Trichoderma koningii. **2**:WA(3434).
Trichoderma sp. **2**:WA(3434).
Trichothecium roseum. 'Pink mold rot.' **2**:Unk.(82), WA(3202).
Trichothecium sp. **2**:WA(3202).
Ulocladium consortiale. **2**:WA(3434).
Verticillium dahliae. Wilt. **2**:CA(1110).
Verticillium sp. **2**:WA(3434).

Deuteromycotina - Coelomycetes

Coniothyrium pyrinum. Leaf spot. **2**:E states(82), IA(82), TX(82).
Coniothyrium sp. Canker, leaf spot. **1**:DC(1913). **2**:FL(35).
Coryneum longestipitatum. On twigs. **2**:FL(82), NJ(82), WA(3202).
Cryptosporiopsis corticola. Bark canker. **2**:NC states(82), NE states(82), MS(82,2720), OR(82,3202).
C. pruinosa. Bark canker. **2**:OR(82).
Cylindrosporium sp. **2**:Unk.(82).
Cytospora sp. Canker. **2**:OR(82), VA(82), WA(82,3202).
Diplodia malorum. On limbs and twigs. **2**:AL(82), KS(82), NJ(82), OR(3202). **5**:OR(4007).
D. sarmentorum. On limbs and twigs. **2**:OR(82), WA(3202).
Diplodia sp. Stem canker, leaf spot. **2**:FL(35).
Entomosporium mespili. Leaf blight, fruit spot. **2**:FL(35), NC(1303), OK(2872). **3**:FL(35). **5**:HI(2903), LA(3471).
Gloeodes pomigena. 'Sooty blotch.' **2**:E states(82), SC states(82), NC(1303), OK(2872).
Gloeosporium sp. Fruit rot. **2**:Range of host(82).
Hendersonia cydoniae. Leaf spot. **2**:NY(82).
[*H. pinicola*]. Leaf spot. **2**:FL(35).
Hendersonia sp. **2**:WA(3202).
Lasiodiplodia sp. Twig dieback. **2**:FL(35).
Leptothyrium carpophilum. Fly speck. **2**:AL(285).
Macrophoma sp. Leaf spot. **2**:FL(35).
Melanconium sp. **2**:WA(3434).
Pestalotia hartigii. **2**:OR(3202). **5**:OR(4007).
Pestalotia sp. **2**:WA(3434).
Pestalotiopsis maculans. Leaf spot. **2**:AL(285).
Phoma exigua. Storage rot. **2**:WA(82).
P. macrostoma. Fruit rot. **2**:NJ(82), WA(82,3202).
P. pomorum. Leaf spot. **2**:AL(285).
Phoma sp. Twig dieback. **2**:FL(35), WA(3434).
Phomopsis ambigua. Twig blight. **2**:Range of host(82), WA(3202).
P. mali. Bark canker. **2**:CA(82,1110).
Phomopsis sp. Twig dieback. **2**:FL(35).
Phyllosticta pyrorum. Leaf spot. **2**:FL(35), IL(82), MS(82,2720), SC(82).
P. solitaria. Blotch. **2**:MD(82), OK(2872).
Phyllosticta sp. **2**:WA(3202).
Seimatosporium caudatum. **2**:TX(82).
S. discosioides. On twigs. **2**:FL(35).
S. lichenicola. Leaf spot. **2**:FL(82), IN(82), NJ(82).
Septoria pyricola. Leaf spot. **2**:OK(2872).
Sphaeropsis malorum. Stem canker, twig dieback. **2**:FL(35).

Deuteromycotina - Other

Rhizoctonia solani. **2**:CA(1110), WA(3434).
Rhizoctonia sp. Root rot. **2**:HI(2903).

Rhaphiolepis Lindl.

1. *R. indica* (L.) Lindl. [Indian hawthorn]
2. *R. umbellata* (Thunb.) Makino [Yedda-hawthorn]
3. *R. umbellata* (Thunb.) Makino var. *integerrima* Rehd.

Oomycetes

Pythium sp. Root rot. **1**:FL(35).

Ascomycotina

Leptosphaeria sp. Rough-bark. **2**:FL(35).
Metasphaeria sp. Rough-bark. **1**:FL(35).

Basidiomycotina - Other

Armillaria mellea. Root rot. **1**:CA(2895).
A. tabescens. Root rot. **2**:FL(35).

Deuteromycotina - Hyphomycetes

Alternaria sp. Leaf spot. **1**:FL(35).
Cercospora ariae. Leaf spot. **1**:FL(35). **2**:FL(35).
C. violamaculans. Leaf spot. **1**:FL(35). **2**:FL(35).
Verticillium albo-atrum. **1**:CA(3896). **3**:CA(3896).

Deuteromycotina - Coelomycetes

Colletotrichum gloeosporioides. Anthracnose. **1**:FL(35). **2**:FL(35).
Colletotrichum sp. Anthracnose. **1**:NC(1303).
Dothiorella sp. Leaf spot. **1**:FL(35).
Entomosporium mespili. Leaf blight and spot. **1**:CA(2904), FL(35),

HI(2903). **2:**FL(35).
Entomosporium sp. Leaf spot. **1:**NC(1303), OR(3202).
Macrophoma sp. Twig dieback. **1:**FL(35).
Phoma sp. Rough-bark. **2:**FL(35).
Phomopsis sp. Leaf spot. **1:**FL(35).
Phyllosticta sp. Leaf spot. **1:**FL(35).
Deuteromycotina - Other
Rhizoctonia ramicola. Thread blight. **1:**FL(35). **2:**FL(35).
R. solani. Web blight. **1:**FL(35).
Sclerotium rolfsii. Southern blight. **1:**FL(35,82). **2:**FL(35).

Rhodotypos Siebold & Zucc.

1. *R. scandens* (Thunb.) Makino [black jetbead]

Ascomycotina
Nectria cinnabarina. Twig blight, coral spot. **1:**MA(82).
Deuteromycotina - Coelomycetes
Ascochyta rhodotypi. Leaf spot. **1:**IL(82).

Rosa L.

1. *R. acicularis* Lindl. (=*R. engelmannii* S. Wats.) [prickly rose]
2. *R. arizonica* Rydb. var. *granulifera* (Rydb.) Kearn. & Peebles
3. *R. arkansana* Porter (=*R. praticola* Greene) [Arkansas rose, Prairie rose]
4. *R. blanda* Aiton
5. *R. bracteata* Wendl. [Macartney rose]
6. *R. californica* Cham. & Schlechtend. [California wild rose]
7. *R. canina* L. [dog rose, dogbrier]
8. *R. carolina* L. (=*R. humilis* Marsh.) [pasture rose]
9. *R. centifolia* L. [cabbage rose]
10. *R. chinensis* Jacq. [Bengal rose, China rose]
11. *R. corymbosa* Moench
12. *R. dilecta* Rehd.
13. *R. eglanteria* L. (=*R. rubiginosa* L.) [eglantine, sweetbrier]
14. *R. foetida* Herrm. [Austrian brier]
15. *R. foliolosa* Nutt. ex Torr. & A. Gray
16. *R. foliosa* Léman
17. *R. gallica* L. [French rose]
18. *R. gymnocarpa* Nutt. [wood rose]
19. *R. hemisphaerica* Herrm.
20. *R. indica* L.
21. *R. laevigata* Michx. [Cherokee rose]
22. *R. manca* Greene
23. *R. multiflora* Thunb. ex J. A. Murray [baby rose, multiflora rose]
24. *R.* × *noisettiana* Thory cv. 'manettii' (=*R. manettii* Bals.-Criv. ex T. Rivers). *R. chinensis* × *R. moschata*. [Manettii rose]
25. *R. nutkana* K. Presl (=*R. aciculata* Cockerell) [nootka rose, nutka rose]
26. *R. nuttallii* J. Presl
27. *R. odorata* (Andr.) Sweet [tea rose]
28. *R. palustris* Marsh. [swamp rose]
29. *R. pisocarpa* A. Gray [cluster rose]
30. *R. rubus* H. Lév. & Vaniot
31. *R. rugosa* Thunb. [Japanese rose, rugosa rose, Turkestan rose]
32. *R. setigera* Michx. [climbing rose, prairie rose]
33. *R. spaldingii* Crép.
34. *R. spinosissima* L. [Burnet rose, Scotch rose]
35. *R. subnuda* Lunell
36. *R. suffulta* Greene (=*R. heliophila* Greene)
37. *R. virginiana* Mill. (=*R. lucida* Ehrh.)
38. *R. wichuraiana* Crép. [memorial rose]
39. *R. woodsii* Lindl. (=*R. fendleri* Crép., *R. macounii* Greene, *R. neomexicana* Cockerell) [Wood's rose]
40. *R. xanthina* Lindl.
41. *R. xanthina* Lindl. f. *hugonis* (Hemsl.) A. V. Roberts (=*R. hugonis* Hemsl.)
42. *Rosa* sp.

Oomycetes
Peronospora sparsa. 'Downy mildew.' **6:**Unk.(1636), CA(145). **7:**Unk.(1636). **9:**Unk.(1636). **13:**Unk.(1636). **20:**Unk.(1636). **42:**E states(82), CA(82,3978), GA(1210), IA(82), IN(145).
Phytophthora megasperma. **42:**Unk.(1636).
Phytophthora sp. Root rot. **42:**FL(35).
Pythium aphanidermatum. Root rot. **42:**FL(35).
P. splendens. Root rot. **42:**FL(35).
Pythium sp. Root rot. **27:**HI(2903).
Zygomycotina
Pilobolus sp. On leaves of greenhouse roses; superficial but unsightly, the sporangia discharged from coprophilous fungi. **42:**Unk.(82).
Ascomycotina
Botryosphaeria disrupta. Cane blight. **42:**FL(35).
B. dothidea. Dieback. 'Cane blight canker.' **42:**FL(35), NC(1303).
B. obtusa. Cane blight. **42:**E states(82), KS(82), MD(82), ND(82), TX(82), VA(82).
B. quercuum. Cane blight. **42:**FL(35).
B. rhodina. On canes. **42:**FL(82), NC(82).
B. ribis. 'Cane blight canker.' **42:**Unk.(1636).
Cryptosporella umbrina. Brown and crown canker, on buds and flowers. 'Brown canker.' **42:**E states(82), AL(285), CA(82), CT(1739), DC(1739), FL(35), GA(1438,1739), ID(82,3202), IL(312), MD(82), MI(82), MS(2720), NC(1303), NE(82), NJ(82), NY(82), OK(2872), TX(82), VA(82,1739), WV(82,1739).
Cucurbitaria rosae. On canes. **42:**NM(82), NY(82).
Dermea bicolor. On twigs. **42:**ND(82,377).
Diapleella coniothyrium. Canker, dieback, leaf spot. 'Common stem canker.' **42:**Range of host(82), FL(35), MS(2720), NC(1303), VA(82), WA(3202).
Diaporthe eres. On buds, canes and twigs. 'Fungal canker.' **42:**CA(82), GA(82), LA(82), MD(82), MO(82), NM(82).
Didymella nigrificans. On canes. **42:**NM(82).
Discohainesia oenotherae. Cane and leaf spot. **42:**E states(82), FL(35), MI(82), TX(82), WA(3202).
Discostroma corticola. Canker. **42:**E states(82), NW states(82), ND(82), WA(3202).
Durandiella rosae. **11:**PA(1381).
Elsinoe rosarum. On small living stems. 'Anthracnose.' **25:**OR(3202). **42:**OR(1747,3202), WA(3202).
Elsinoe sp. Anthracnose, scab. **23:**MD(1736). **42:**MD(1736).
Erysiphe graminis f. sp. *avenae*. Powdery mildew. **42:**AR(3029).
Eutypella juglandicola. **42:**GA(1438).
Glomerella cingulata. Canker, dieback. **42:**MD(82), NC(1303), NJ(82), VA(82).
Gnomonia rauii. On dead limbs. **42:**PA(967).
Guignardia spinicola. On thorns. **42:**PA(82).
Hainesia lythri. **4:**WI(1347). **42:**WI(1343).
Hymenoscyphus calyculus. **25:**ID(3202), WA(3202).
Mycosphaerella rosicola. 'Mycosphaerella leaf spot.' **16:**OK(82). **18:**WA(3202). **21:**OK(82), TX(82). **25:**WA(3202). **32:**OK(82). **42:**SC states(82), SE states(82), WA(82).
M. rosigena. Leaf spot. **42:**HI(2903), LA(82), MD(82), NC(82), NJ(82), NY(82), OH(82), TN(82).
Nectria cinnabarina. Coral spot of canes and twigs. 'Fungal canker.' **25:**AK(526). **31:**AK(526). **42:**AK(82), NY(82).
Niptera stictella. **25:**ID(3202).
Phomatospora rosae. On dead stems. **42:**ND(82,377).
Phyllachora rosae. Tar spot. **8:**MS(2720).
Phyllactinia guttata. Powdery mildew. **18:**WA(3202). **42:**WA(82).
Physalospora abdita. On canes. **42:**FL(82), GA(82).
Pleosphaerulina corticola f. *rosae*. **42:**ND(377).
Pleospora herbarum. **25:**WA(3202).
Propolomyces versicolor. **25:**WA(3202).
Pseudomassaria sepincolaeformis. Dieback. **42:**MD(82).
Rosellinia medullaris. **25:**ID(3202).
R. rosarum. **42:**ND(82,377), NM(82).
Schizoparme straminea. On dead leaves. **31:**VA(3210).
Sphaeronaemella rosae. On canes, old calyx tubes and twigs. **37:**NJ(966). **42:**MD(82), NJ(82), VA(82).
Sphaerotheca fuliginea. Powdery mildew. **40:**CA(1155).
S. macularis. Powdery mildew. Powdery mildew. **8:**WI(1343). **25:**WA(3202). **30:**Unk.(3976). **37:**WI(1353). **42:**NE states(82), CA(82), FL(35), NC(1303), OR(3202), TX(82), WA(82).
S. pannosa. 'Powdery mildew.' **1:**AK(526). **8:**NC(1303), ND(377). **18:**ID(3202). **37:**NC(3944). **42:**Range of host(82), AK(526), AL(285), GA(1438,2405), HI(2903), ID(3202), MS(2720), MT(3202), NC(1303), OK(2872), WA(3202).
S. pannosa var. *rosae*. Powdery mildew. **3:**SD(2243). **23:**OR(240). **31:**OR(240,241). **42:**FL(35), OR(240).
Sphaerotheca sp. Powdery mildew. **38:**WA(3202). **42:**OK(2872).
Stictis chrysopsis. **42:**OR(3225).
S. sphaeroboloidea. **42:**IL(3225).
Tapesia rosae. **42:**NY(82), PA(82).

Valsa ambiens. Twig dieback. **42:**FL(35), IA(1226).
V. ambiens f. *rosae*. **36:**ND(377).
V. ceratosperma. **42:**PA(3415).
Valseutypella tristicha. On dead stems. **35:**ND(377). **42:**CA(82), ND(82).
Velutarina rufo-olivacea. **42:**PA(849).

Basidiomycotina - Uredinales

Phragmidium americanum. Autoecious. **4:**E states(82), ND(82), ND(378), TX(82). **32:**E states(82), GA(1439), ND(82), TX(82), WI(1347). **36:**WI(1354). **37:**E states(82), MA(1390), NC(3944), ND(82), TX(82). **42:**E & C states(82), W states(82), AK(526), MT(3202), SD(2243), TX(82).
P. fusiforme. Autoecious. **1:**AK(60,526,1904), CO(649,1195), ID(3202), MT(1195,3202,3379), WY(1195,3379). **18:**ID(3202), MT(3202), WA(3202). **25:**AK(60,526), ID(3202), MT(3202), OR(3202), WA(3202). **29:**CA(81), ID(3202). **31:**AK(60,526). **33:**WA(638). **36:**ND(378). **39:**MT(1195,3202), WA(3202), WY(1195). **42:**NC states(82), W states(82), CA(82), CO(1195), OR(657), WY(82,2924).
P. fusiforme var. *novi-boreale*. **25:**AK(3104).
P. montivagum. Autoecious. **1:**CO(1195), ID(3202), MT(3202), WY(1195). **13:**WA(3202). **18:**MT(3202), OR(3202), WA(3202). **22:**AZ(1195,1203). **25:**ID(3202), WA(3202). **26:**ID(3202). **29:**OR(3202). **39:**AK(60,526), AZ(1195,1203,3981), CO(1195), ID(1195,3202), MT(1195,3202), UT(1195), WA(3202), WY(1195,3379). **42:**W states(82), AK(526), AZ(82), ND(378), WA(82), WY(1195).
P. mucronatum. Autoecious. 'Rust.' **3:**SD(2243). **14:**NC(1303). **17:**ID(3202). **19:**AK(526). **25:**WA(3202). **27:**HI(2903). **31:**AK(60,526). **39:**SD(2243). **42:**E & C states(82), NW states(82), CA(81), CO(82), FL(35), HI(82,2903), IL(82), KS(82), ME(82), MN(82), MT(82,3202), OR(3202), PA(1875), SD(2243), WI(82).
P. rosae-arkansanae. Autoecious. **3:**SD(2243). **39:**MT(3202). **42:**WY(1195,2924).
P. rosae-californicae. Autoecious. **1:**AK(60,526). **2:**AZ(1203,3981). **6:**CA(81,2439), OR(3202). **18:**CA(81,638,642), OR(648,3202), WA(3202). **25:**AK(60,526), CA(81), OR(3202), WA(3202). **29:**ID(3202), OR(3202), WA(3202). **39:**ID(3202), WA(3202). **42:**AK(526), AZ(82), CA(82), MT(82), OR(82).
P. rosae-pimpinellifoliae. Autoecious. **14:**WY(1195). **25:**WA(3202). **29:**WA(3202). **34:**Unk.(1636). **39:**WA(3202). **42:**ID(3202).
P. rosicola. Autoecious. **42:**CO(82), MT(82), NE(82).
P. speciosum. Autoecious. **3:**ND(378). **4:**E & C states(82), W states(82), ND(378). **8:**E & C states(82), W states(82), GA(1439), MS(2720), NC(1303). **18:**OR(3202). **25:**ID(3202), SD(2243), WA(3202). **27:**HI(2903). **28:**FL(82). **29:**ID(3202), OR(3202), WA(3202). **31:**SD(2243). **33:**WA(638). **36:**ND(378). **37:**E & C states(82), W states(82), MA(1394), MS(2720). **39:**E & C states(82), W states(82), MT(3202), WA(3202). **42:**E & C states(82), W states(82), FL(35), GA(1439), ID(3202), SD(2243), WA(3202).
Phragmidium sp. **32:**ID(3202). **39:**OR(3202). **42:**LA(2824), OK(2872).

Basidiomycotina - Other

Armillaria mellea. Root rot. **42:**CA(82,2898), MS(82,2720), OK(2872), OR(82,3202), TX(82), WA(82,3202).
A. tabescens. Dieback, root rot. **42:**FL(35,82), LA(2815).
Athelia arachnoidea. **25:**WA(3202).
Ceriporia xylostromatoides. **42:**AZ(128,1202).
Cytidia patelliformis. **25:**WA(3202).
Flagelloscypha libertiana. **25:**ID(3202).
Hapalopilus rutilans. **25:**WA(3202).
Hemimycena lactea. **42:**WA(3202).
Hymenochaete tabacina. **25:**ID(3202).
Lentinellus micheneri. **42:**WA(3202).
Melzericium udicola. **25:**ID(3202), WA(3202).
Mycena albidula. **42:**WA(3202).
M. capillaris. **42:**ID(3202).
Pellicularia koleroga. Thread blight. **42:**FL(82), LA(82,3640).
Peniophora malenconii subsp. *americana*. On wood. **42:**UT(542).
Stereum hirsutum. **25:**WA(3202).
Tubaria pellucida. **42:**WA(3202).

Deuteromycotina - Hyphomycetes

[*Alternaria brassicicola*]. 'Alternaria leaf spot.' **42:**TX(82).
Alternaria sp. Leaf spot. **42:**SE states(82), FL(35), TX(82).
Bipolaris setariae. Flower petal spot. **42:**FL(35).
Botrytis cinerea. Blossom, bud and twig blight, storage mold, blossom rot. 'Botrytis bud blight. Botrytis twig blight.' **27:**HI(2903). **42:**Range of host(82), AK(526), FL(35), NC(1303), OK(2872), TX(82), WA(3202).
Botrytis sp. **42:**MT(3202), OR(3202), WA(3202).
Cercospora hyalina. Leaf spot. **42:**FL(35), OK(2872).
C. rosicola. Leaf spot. **3:**SD(2243). **15:**OK(2873). **21:**OK(2873). **32:**OK(2873). **36:**OK(2873). **37:**NC(3944), OK(2873). **39:**OK(2873). **42:**CO(1195,3378), FL(35), HI(2903), TX(82,3597).
Cercospora sp. Leaf spot. **42:**MI(82).
Chalara thielavioides. "Manetti mold", root rot. 'Black mold.' **23:**Unk.(1636). **27:**Unk.(1636). **24:**Unk.(1636). **42:**CA(146), IL(82), NY(82), OR(82,3202), PA(82), WA(82).
Cladosporium herbarum. Leaf spot. **34:**WA(3202). **42:**AK(82), CA(82), MD(82), MN(82), MS(82,2720), OK(82), TX(82), WA(3202).
Cladosporium sp. Bud blight, on leaves. **42:**AK(82), CA(82), MD(82), MN(82), MS(82), OK(82,2873), TX(82).
Coniosporium limoniiforme. On leaves. **42:**UT(82).
Cylindrocladium scoparium. Basal canker, crown canker. 'Crown canker.' **23:**DE(704). **42:**E states(82), DE(704), FL(35), IL(82), TX(82).
Diplocarpon rosae. 'Black spot.' **5:**NC(82). **8:**OK(2873). **10:**OK(2873). **17:**OK(2873). **25:**ID(3202), OR(3202), WA(3202). **27:**HI(2903). **29:**ID(3202). **31:**AK(526). **37:**MA(1390). **39:**ID(3202), OK(2873), WA(3202). **42:**Range of host(82), AK(526), AL(285), FL(35), GA(1438,2405), HI(2903), ID(3202), MS(2720), MT(3202), NC(1303), OK(2872), OR(3202), SD(2243).
Fusarium oxysporum. Root and stem rot. **42:**FL(35).
Fusarium sp. Root and stem rot. **42:**S states(82), OK(2872).
Helminthosporium sp. Flower blight. **42:**FL(35).
Oidium sp. Powdery mildew. **27:**HI(2903). **42:**HI(2903).
Phymatotrichopsis omnivora. Root rot. **42:**AZ(82), TX(82).
Pseudocercospora puderi. Leaf spot. 'Cercospora leaf spot.' **42:**FL(35,82), GA(82).
P. rubi. **42:**IL(82).
Ramularia macrospora. Root rot. **42:**MD(82).
Tilletiopsis washingtonensis. On leaves. **42:**KS(2686).
Trimmatostroma brencklei. **36:**ND(378,3081).
Verticillium albo-atrum. 'Verticillium wilt.' **23:**IL(992,1558). **42:**Unk.(1636), CA(82,1407), IL(82), NJ(82,1407).
V. dahliae. 'Verticillium wilt.' **42:**Unk.(1636).
Verticillium sp. Wilt. **41:**NY(82). **42:**AR(82), NY(82).

Deuteromycotina - Coelomycetes

Botryodiplodia sp. Graft canker. **42:**FL(35).
Colletotrichum capsici. **42:**Unk.(1636).
Coniothyrium fuckelii. Graft and stem canker. **42:**FL(35), NC(1303), OK(2872).
C. rosarum. 'Graft canker.' **42:**CA(82), IA(82), MA(82), MN(82), NJ(82), PA(82), TX(82).
C. wernsdorffiae. 'Brand canker.' **42:**CO(82), FL(35), IN(82), MN(82), MS(82,2720), NC(1303), NY(82), PA(82), TX(82), WA(3202).
Coniothyrium sp. Stem canker. **27:**HI(2903).
Cryptosporium minimum. Canker. 'Fungal canker.' **42:**OR(82,3202), PA(82).
Cryptosporium sp. Canker. **42:**FL(35).
Cytospora sp. Dieback. **17:**SD(2243). **42:**FL(35), KY(82), PA(82), VA(82), WA(82).
Diplodia rosarum. Canker, dieback. **42:**CA(81), TX(82).
D. sarmentorum. On limbs. **42:**CA(82).
Diplodia sp. Canker, dieback. **42:**E states(82), FL(35), NC(1303), NE(82), TX(82).
Discosia artocreas. On buds, flowers and stems, probably saprophytic. **42:**LA(82), MO(82), MS(82,2720), OH(82), TX(82).
Dothiorella sp. Blossom blight, canker, on canes and twigs. **42:**LA(82), MS(2720), VA(82).
Gloeosporium rosae. Anthracnose, twig blight. **42:**AL(285), OK(2872).
Hendersonia hypocarpa. On flower stems and pedicels. **14:**NY(1012). **42:**NY(82).
Lasiodiplodia theobromae. **42:**TX(82).
Leptothyrium rosarum. Cane and leaf spot. **42:**FL(35).

Macrophoma sp. Dieback. **42:**FL(35), TX(82), VA(82).
Marssonina rosae. Black spot. **4:**WI(1354). **12:**NC(1303). **36:**ND(378). **42:**NC(1303), SC(634).
Monochaetia sp. Cane dieback. **42:**FL(35).
Myxosporium rosae. **42:**Unk.(82).
Pestalotia longisetula. On stems. **42:**MD(128).
P. oleandri. **42:**SC(1393).
P. pezizoides. Leaf spot. **42:**FL(35,82).
P. rosae. **42:**E states(82), TX(82).
P. suffocata. Leaf spot. **42:**FL(35), PA(82).
Pestalotia sp. Leaf and stem blight. **42:**MS(2720), WA(3202).
Phoma rhodocarpa. On fruit. **42:**NY(82).
P. sepincola. On twigs. On living stems. **42:**NE(82), WI(1337).
Phoma sp. Stem dieback. **42:**Range of host(82), FL(35).
Phomopsis rosae. **25:**CA(81).
Phomopsis sp. On bark, buds and thorns. **42:**Range of host(82), FL(35), WI(1360).
Phyllosticta erratica. **42:**AL(82).
P. indianensis. **32:**IN(82).
P. rosae. Leaf spot. **32:**WI(1349). **42:**E states(82), CO(82), IN(82).
P. rosae-setigerae. On leaves. **32:**IN(82).
P. rosicola. **42:**CO(1195), NJ(82).
Phyllosticta sp. Leaf spot. **42:**HI(2903), WI(1359).
Seimatosporium caninum. On stems. **42:**NY(82).
S. discosioides. On leaves. **31:**WI(1360). **42:**DE(82,3549), MI(82), NY(82), WI(3549).
S. lichenicola. Canker. **42:**AK(526), OK(2872).
Septoria rosae. Leaf spot. 'Septoria leaf spot.' **42:**MS(82,2720), NJ(82), SC(82).
Sphaceloma rosarum. Anthracnose. **23:**MD(1737), NY(1745). **31:**WI(1360). **42:**NC states(82), Unk.(1636), CA(82), NC(82), OR(82), RI(1741), TN(82).
Sphaeropsis malorum. Stem gall. **4:**ND(378). **42:**FL(35).
S. rhodocarpa. On fruits. **14:**NY(1012).
Vermicularia sp. Petal spot. **42:**FL(35).

Deuteromycotina - Other

Rhizoctonia solani. Root rot. **42:**FL(35).
Rhizoctonia sp. Thread blight. **42:**NC(1303).
Sclerotium rolfsii. Southern blight. **42:**FL(35,82), KS(82), MS(2720), TX(82).
Sclerotium sp. **31:**AK(526).

Rubus L.

1. *R. alaskensis* L. H. Bailey
2. *R. allegheniensis* Porter (=*R. nigrobaccus* L. H. Bailey) [Allegheny blackberry, common blackberry, sow-teat blackberry]
3. *R. arcticus* L. [arctic blackberry, crimson berry, crimson blackberry]
4. *R. argutus* Link (=*R. floridus* Tratt.)
5. *R. arizonensis* Focke
6. *R. bartonianus* Peck
7. *R. betulifolius* Small
8. *R. canadensis* L. (=*R. procumbens* Muhl.)
9. *R. carolinianus* Rydb.
10. *R. chamaemorus* L. [baked-apple berry, cloudberry, malka, salmonberry, yellowberry]
11. *R. cuneifolius* Pursh [sand blackberry]
12. *R. deliciosus* Torr. [Rocky Mountain raspberry]
13. *R. flagellaris* Willd. [American dewberry, northern dewberry]
14. *R. hawaiiensis* A. Gray
15. *R. hispidus* L. [running blackberry, swamp blackberry, swamp dewberry]
16. *R. idaeus* L. (=*R. idaeus* L. var. *idaeus*) [European red raspberry, framboise, red raspberry]
17. *R. idaeus* L. var. *sachalinensis* (Lév.) Focke (=*R. melanolasius* Focke)
18. *R. laciniatus* Willd. [cutleaf blackberry, evergreen blackberry, parsley-leaved blackberry]
19. *R. lasiococcus* A. Gray
20. *R. leucodermis* Douglas ex Torr. & A. Gray
21. *R. loganobaccus* L. H. Bailey (=*R. ursinus* var. *loganobaccus* (L. H. Bailey) L. H. Bailey) [boysenberry, loganberry]
22. *R. neglectus* Peck
23. *R. neomexicanus* A. Gray [New Mexico raspberry]
24. *R. occidentalis* L. [black raspberry, blackcap, thimbleberry]
25. *R. odoratus* L. [purple flowering raspberry, thimbleberry]
26. *R. ostryifolius* Rydb.
27. *R. parviflorus* Nutt. (=*R. nutkanus* Moc. ex Ser.) [salmonberry, thimbleberry]
28. *R. pedatus* Sm.
29. *R. procerus* P. J. Müll. [Himalaya berry]
30. *R. pubescens* Raf. [dwarf red raspberry]
31. *R. rosifolius* Sm. (=*R. rosaefolius* Sm.) [Mauritius raspberry]
32. *R. spectabilis* Pursh [salmonberry]
33. *R. stellatus* Sm. [knesheneka, nagoonberry]
34. *R. strigosus* Michx. (=*R. idaeus* L. var. *strigosus* (Michx.) Maxim.) [American red raspberry]
35. *R. strigosus* Michx. var. *arizonicus* (Greene) Kearn. & Peebles
36. *R. triflorus* Richardson
37. *R. trivialis* Michx. [southern dewberry]
38. *R. ursinus* Cham. & Schlechtend. [California blackberry, Pacific blackberry, Pacific dewberry]
39. *R. villosus* Aiton
40. *R. vitifolius* Cham. & Schlechtend. (=*R. ursinus* Cham. & Schlechtend. var. *vitifolius* (Cham. & Schlechtend.) Focke)
41. *Rubus* sp.

Chytridiomycetes

Synchytrium aureum. Leaf gall. **30:**WI(82).
S. davisii. **15:**WI(1355,1849). **36:**WI(1355,1849).
Synchytrium sp. On leaves and petioles. **37:**LA(1847). **41:**LA(1847).

Oomycetes

Peronospora potentillae. 'Downy mildew.' **41:**CT(82), IL(82), LA(82).
P. rubi. 'Downy mildew.' **18:**OR(3202). **20:**WA(3202). **41:**FL(82), MD(82), WA(82), WI(82).
P. sparsa. Downy mildew. **16:**CA(1110). **21:**CA(1110). **40:**CA(1110).
Phytophthora cinnamomi. Root rot. **31:**HI(2903).
P. erythroseptica. 'Wet soil root rot.' **16:**WA(610,3202).
Pythium sp. Root rot. **41:**FL(35).

Zygomycotina

Rhizopus stolonifer. Black mold of fruit. **41:**Range of host(82), FL(35).
Rhizopus sp. Fruit rot. **41:**ID(3202), NC(1303), OR(3202).

Ascomycotina

Acanthostigma decastylum. On dead canes. **2:**NC(1303,1307).
Anomothallus erraticus. **14:**HI(2903).
Appendiculella calostroma. Black mildew. **14:**HI(1272,2903). **37:**Unk.(1452). **39:**Unk.(1452). **41:**AL(82), HI(82,1452), LA(82).
Ascospora ruborum. Cane spot. **16:**OR(3202), WA(3202). **18:**WA(3202). **24:**OR(3202). **41:**AL(82), CA(82), MA(82), OR(82), WA(82).
Belonidium sulphureum. **41:**WA(3202).
Belonopsis sp. **41:**WA(3202).
Botryosphaeria dothidea. Canker. **4:**GA(1438). **41:**GA(1438), MD(2220).
B. obtusa. Cane canker. **41:**IA(82), MD(82), MI(82), MO(82), ND(82), NY(82), OH(979), PA(82).
B. quercuum. On stems. **41:**NC(1303).
B. rhodina. On canes. **41:**AL(82).
Brunnipila clandestina. **27:**ID(3202), WA(3202). **41:**WA(3202).
Calycella subochracea. On stems. **25:**NY(849). **27:**WA(3202).
Capitotricha bicolor. **27:**CA(81), ID(3202), WA(3202). **41:**OR(3202).
Ceratostoma subrufum. **34:**ND(377).
Coleroa chaetomium. On fallen leaves. **20:**WA(3202). **24:**CA(1110). **27:**OR(3202). **28:**AK(526). **41:**AK(526), CA(82), CO(1195).
Crocicreas cyathoideum. **41:**PA(505).
Cryptodiaporthe vepris. On twigs and stems. **4:**GA(1438). **41:**NC(82), VA(82), WV(82).
Dasyscyphus bicolor var. *indicus*. **27:**CA(849).
D. corticalis. **41:**WA(3202).
D. scabrovillosus. **27:**ID(3202), WA(3202).
Diapleella coniothyrium. Cane blight, canker, leaf spot. 'Cane blight.' **16:**GA(1438), OK(2872), OR(3202). **21:**CA(1110), WA(3202). **22:**GA(1438). **24:**OR(3202). **34:**ND(377). **38:**WA(3202). **41:**AL(285), CA(1110), FL(35), GA(1438), HI(82,2903), ID(3202), MA(82), MI(82), MS(2720), NC(1303), NJ(82), NY(82), OH(979), OK(2872), WA(3202).
Didymella applanata. Spur blight, on dead canes. 'Spur blight.' **4:**NC(1303). **16:**CA(1110), OR(3202), WA(3202). **21:**OR(3202), WA(3202). **38:**ID(3202). **41:**Range of host(82), KS(82), OR(82), PA(82), WI(82).
Discohainesia oenotherae. Canker, cane and leaf spot. 'Leaf spot.' **41:**MD(82), OH(82), VA(82).

Elsinoe veneta. Anthracnose, on canes and leaves. 'Anthranose.' **2:**IL(1745), MS(2720), NY(1745), PA(1745), WI(1348,1745). **16:**GA(1438), OR(3730), WA(3202,3730). **21:**WA(3202). **24:**GA(1438), MD(1745), OR(3202,3730), SD(2243), WA(3202,3730), WI(1329,1348). **27:**WI(1355). **34:**IA(1745), IL(1745), MS(2720), NY(1745), SD(2243), WI(1349). **37:**LA(1745), MS(2720). **38:**WA(3202). **40:**OR(1745), WA(1745). **41:**Range of host(82), AL(285), OK(3525), OR(82).
Fabraea cincta. **41:**AK(526).
Glomerella cingulata. Anthracnose, white bud, on canes. **8:**MS(2720). **24:**DC(970). **41:**FL(35), ID(3202), IL(82), NJ(82), WA(3202), WV(82).
Gnomonia rostellata. On canes and stems. **41:**NC(82), OR(82).
G. rubi. Cane canker, dieback. **16:**OR(3202). **21:**OR(3202). **24:**OR(3202). **41:**OR(82).
G. setacea. **21:**OR(3202).
Gnomonia sp. Twig tip blight. **31:**HI(2903).
Godronia sp. **27:**ID(3202).
Hainesia lythri. **2:**WI(1355). **12:**WI(1360). **30:**WI(1358).
Hormotheca rubicola. Black spot of canes and leaves. **8:**WI(1361). **15:**WI(1361). **16:**MT(3202). **20:**CA(81). **24:**NW states(3202). **30:**NY(2456). **34:**AZ(1195,3378), WI(1355).
Hyaloscypha hyalina. **41:**WA(3202).
Hymenoscyphus scutula. **41:**WA(3202).
Hypoderma rubi. On dead canes and stems. **16:**OR(3202). **41:**NY(82,2456), WV(82).
Lachnum crystalligerum. On dead limbs. **27:**ID(3202), OR(3202), WA(3202). **32:**MT(638,3202), WA(3202).
L. niveum. **41:**WA(3202).
Lasiosphaeria spermoides. **4:**GA(1438).
Leptosphaeria thomasiana. Cane blight. **16:**OR(3202). **21:**OR(3202), WA(3202). **41:**OR(82), WA(82).
Lophodermium rubiicola. On canes. **4:**GA(1438). **41:**AL(82), GA(81,82,1438).
Microthyrium rubicola. On still green, early-season fruiting canes. **2:**WI(1356).
Monilinia fructicola. Brown rot of berries. **41:**IL(82).
Mycosphaerella confusa. Blotch, leaf spot. 'Blotch.' **4:**GA(1438). **13:**GA(1438). **16:**GA(1438). **22:**GA(1438). **37:**MS(2720). **41:**SE states(82), GA(82), IL(82), MS(82), TX(82).
M. fructicum. On leaves. **41:**AK(82,526).
M. ligea. On leaves. **18:**OR(82,3202). **21:**OR(82). **29:**OR(82,3202). **41:**OR(82).
M. rubi. Cane and leaf spot. **4:**GA(1438). **13:**GA(1438). **16:**CA(1110). **21:**CA(1110). **34:**WI(82). **41:**MD(82).
Mycosphaerella sp. Leaf spot. **38:**WA(3202). **41:**OK(2872).
Nectria cinnabarina. Coral spot. **27:**AK(526). **32:**AK(526). **34:**AK(526). **41:**AK(82).
Ostropa cinerea. **41:**WA(3202).
Otthia rubicola. On dead canes. **41:**HI(2784).
Pezicula rubi. On canes. **2:**NC(1303,1307). **4:**GA(1438), NC(1303). **32:**CA(81). **41:**ME(82), NY(82), VA(82).
Phyllactinia guttata. Powdery mildew. **41:**MI(82).
Physalospora minima. On dead canes. **34:**AL(970).
Pleospora herbarum. **41:**OR(3202).
Pringsheimia sepincola. On canes. **4:**FL(2782). **16:**OR(3202), WA(3202). **18:**OR(3202). **20:**OR(3202), WA(3202). **41:**OR(82).
Pyrenopeziza lacerata. On dead canes. **41:**NY(82), WV(82).
P. rubi. On canes. **16:**CA(1110). **20:**OR(3202). **21:**CA(1110). **30:**CA(82). **32:**WA(3202). **41:**CA(82), ND(82), OR(82,3202).
Rebentischia massalongii. On limbs. **41:**MA(186).
Rosellinia necatrix. Root rot. **21:**CA(1110).
Schizothyrium pomi. Fly speck on canes and leaves. **16:**OK(2872). **41:**NC(82,1303).
Sclerotinia sp. Cane blight. **41:**NY(82).
Sphaerotheca macularis. 'Powdery mildew.' **16:**ID(3202), OR(3202), WA(3202,3730). **18:**OR(3202). **21:**OR(3730). **24:**CA(1110), OR(3202), WA(3202,3730). **27:**Unk.(3976), AK(526), CA(81), WA(3202), WI(1356). **30:**WI(1356). **32:**AK(526), CA(81), OR(3202), WA(3202). **38:**ID(3202), OR(3202), WA(3202).
Sphaerulina rubi. 'Leaf spot.' **16:**CA(1110). **32:**ID(3202), MT(3202), WA(3202). **41:**CA(1110), MD(82), MO(82), NC(82).
Stictis hawaiiensis. On living stems. **31:**HI(2903). **41:**HI(82,3225).
S. radiata. On dead canes. **27:**OR(3228). **41:**CA(3225).
Stomiopeltis versicolor. On canes. **41:**VA(82).
Sydowiella depressula. On dead canes. **41:**AK(82,526).
Trematosphaeria sp. **27:**ID(3202).
Valsa ambiens. On dead canes. **34:**ND(377). **41:**ND(82), NY(82).
V. ceratosperma. On canes. **41:**NY(82,3415), PA(82), VA(82).
Velutarina rufo-olivacea. **27:**ID(3202). **41:**OR(3202).
Venturia kunzei var. *ramicola*. **33:**AK(526).
Xylaria sp. Root rot. **41:**WA(82,3202).

Basidiomycotina - Uredinales

Arthuriomyces peckianus. Autoecious. 'Orange rust.' **2:**IN(987), NC(1303). **3:**AK(60,526). **4:**GA(1439), NC(1303). **8:**MS(2720), NC(1303). **11:**GA(1439). **13:**GA(1439), MA(1390). **16:**CA(1110). **21:**AZ(1203,3981), CA(1110). **24:**MS(2720), NC(1303). **30:**AK(60), ME(82), NH(82), NY(2456), WI(82). **33:**AK(526). **37:**NC(3944), TX(82). **41:**E states(82), NC states(82), W states(82), AL(285), IL(82), MN(82), NC(3944).
Gerwasia epiphylla. Autoecious. **37:**TX(82).
Gymnoconia nitens. Autoecious. **2:**NC(1303). **4:**NC(1303). **8:**MS(2720). **13:**NC(1303). **15:**OK(2872). **21:**CA(1110). **26:**IL(2568). **27:**CA(82). **30:**MT(1195,3202), VA(82). **37:**NC(1303). **40:**CA(81). **41:**CA(1110).
Kuehneola uredinis. Autoecious. 'Cane rust. Leaf rust.' **2:**GA(1439), IN(987), MN(2706), MS(2720), NC(1303,1307), WI(1357). **4:**GA(1439), NC(1303,1307,3944). **7:**NC(1307). **8:**MS(2720). **11:**GA(1439). **15:**MA(656), NC(1303). **16:**CA(1110). **18:**W states(82), ID(3202), OR(648,3202,3730), WA(3202,3730). **21:**CA(1110). **30:**MA(82), MI(82), NY(82). **37:**MS(2720). **38:**WA(3202). **39:**HI(2903). **40:**CA(2439). **41:**C states(82), E states(82), HI(82), IL(82), OR(82), PA(82), WI(82).
Phragmidium alaskanum. Autoecious. **33:**AK(60,526).
P. occidentale. Autoecious. **25:**E states(82), MI(82), NC(1303). **27:**W states(82), AK(60,82,526), AZ(1203,3981), CA(81,638,642), ID(3202), MT(1195,3202), OR(648,3202), WA(3202), WY(1195). **41:**ID(3108), MI(3108), NC(1303), NY(3108), WA(3108), WY(3108).
P. peckianum. Autoecious. **6:**ID(3202). **12:**W states(82), CO(649,1195), NM(82), WY(1195). **23:**AZ(1195,1203,3981). **41:**ID(3202).
P. rubi-idaei. Autoecious. 'Yellow rust.' **3:**AK(60,526). **16:**AK(1904), CA(1110), ID(3202), MT(1195,3202), OR(3202,3730), WA(3202,3730), WY(1195). **17:**WY(3379). **20:**CA(81,642), MT(1195), OR(3202), WA(3202). **22:**OR(3202). **24:**CA(1110). **27:**CA(642). **28:**AK(3431). **32:**OR(82), WA(82). **33:**AK(60,526). **34:**AK(60,526), AZ(1195,1203), CA(81), ND(378), WY(1195,2924). **35:**AZ(3981). **41:**NC states(82), NE states(82), W states(82), AK(526), CA(81,82), CO(1195), MA(82), MS(82,2720), OR(82), WI(82).
Phragmidium sp. **16:**MT(3202). **38:**WA(3202).
Pucciniastrum americanum. Heteroecious: telial host. 'Late leaf rust.' **9:**TN(1575). **16:**ID(3202), MT(3202), OH(977). **17:**ID(1575), MT(1575). **20:**MD(1575). **22:**MA(1575). **34:**CT(1575), IA(1575), IL(1575), MA(1575), MD(1575), ME(1575), MI(1575), MT(1575), ND(378,1575), NH(1575), NJ(1575), NY(1575), OH(1575), PA(1575), VT(1575), WI(1575), WV(1575). **41:**E states(82), ID(82), IL(82), WI(1356).
P. arcticum. Heteroecious: telial host. **1:**AK(60,526). **3:**AK(60,526). **10:**AK(60,526). **30:**NC states(82), NE states(82), AK(82). **33:**AK(60,526). **36:**CT(1575), MI(1575), MN(1575), NE(1575), NH(1575), VT(1575), WI(1575).

Basidiomycotina - Ustilaginales

Sporidiobolus johnsonii. On leaves. **16:**WA(2618,3202).

Basidiomycotina - Other

[*Aleurobotrys botryosus*]. On dead stems. **4:**GA(1439,2083). **24:**MD(2083). **32:**CA(2083). **41:**LA(2083), MD(128,2083).
Aleurodiscus aurantius. On dead stems and twigs. **38:**CA(2083).
Armillaria mellea. 'Shoestring root rot.' **16:**CA(1110), OR(3202), WA(3202). **21:**WA(3202). **41:**OK(2872), OR(82), TX(82), WA(82,3202).
Ceratobasidium anceps. Leaf spot. **2:**WI(82).
Collybia dryophila. 'Root rot.' **41:**NC(82).
Dichostereum pallescens. On dead limbs. **32:**CA(81).
Grandinia pruni. **5:**AZ(1202).
Helicobasidium brebissonii. 'Root rot.' **41:**NC(82), TX(82).
Kuehneromyces mutabilis. **27:**WA(3202).
Mycena cinerella. **41:**WA(3202,3326).

M. fallax. On canes. **41**:OR(3335).
Nidula candida. On dead stems. **32**:AK(1904).
Pellicularia koleroga. Thread blight. **37**:LA(3640). **41**:FL(82), LA(82).
Peniophora fuscomarginata. **5**:AZ(1202).
P. isabellina. **41**:VA(128,2117).
Phellinus ferreus. **41**:WA(3202).
Polyporus varius. On wood. **32**:AK(1904). **41**:OR(3202).
Psathyrella rubicola. On decaying canes and roots. **41**:WA(3335).
Scytinostroma galactinum. White root and butt rot. **41**:AR(82,2090), MD(82), VA(82).
Thanatephorus cucumeris. **21**:WA(3202). **24**:WA(3202).
Typhula gyrans. **41**:NY(264,2955).
T. sphaeroidea. On stems. **2**:NC(1303,1307). **41**:NH(2955), NY(264).

Deuteromycotina - Hyphomycetes

Acremonium sp. **2**:WI(1357).
Alternaria sp. Fruit rot, leaf spot. **21**:CA(1110). **41**:CA(1110), FL(35), MA(82), NY(82), OH(979).
Aspergillus sp. **34**:AK(526).
Botrytis cinerea. Gray mold of flowers and fruits. 'Gray mold.' **16**:AK(526), CA(1110). **21**:CA(1110). **27**:AK(526). **32**:AK(526). **34**:AK(526). **38**:WA(3202). **41**:NC(82).
Botrytis sp. On fruits and leaves. **16**:WA(3202). **18**:WA(3202). **30**:WI(1348). **34**:WI(1354).
Cercosporella rubi. Double blossom, rosette, leaf spot. 'Rosette.' **41**:E states(82), AL(285), IL(82), LA(82), MS(82), TX(82).
Cladosporium herbarum. On canes. **24**:OR(3202). **41**:OR(4007).
Cladosporium sp. Fruit mold. **16**:CA(1110). **41**:WA(82,3202).
Codinaeopsis gonytrichodes. **41**:AL(2485).
Epicoccum nigrum. Isolated from cankers. **41**:OH(979).
Eurotium herbariorum. On leaves and stems. **41**:MA(1394).
Fusarium oxysporum. **16**:CA(1110).
Fusarium sp. Root rot, isolated from cankers. **16**:CA(1110). **41**:FL(35), OH(979), WA(3202).
Graphium gracile. On leaves. **41**:IA(82), IN(82), NY(82), WI(82).
Oidium sp. Powdery mildew. **41**:FL(35).
Penicillium sp. **41**:ID(3202).
Phymatotrichopsis omnivora. 'Root rot.' **41**:TX(82).
Stilbella flavipes. **41**:NY(3178).
Trichothallus hawaiiensis. **14**:HI(2903).
Verticillium albo-atrum. Hadromycosis, wilt. 'Verticillium wilt.' **2**:CA(3366). **16**:OR(3202). **21**:WA(3202). **24**:IL(312), OR(3202), WA(3202). **38**:CA(3899), OR(3202). **41**:NE states(82), CA(82), ID(3202), NY(82), OH(82), OR(82), WA(82,3202).
V. dahliae. 'Verticillium wilt.' **16**:CA(1110). **21**:CA(1110).

Deuteromycotina - Coelomycetes

Aposphaeria major. On stems. **27**:UT(3565).
Colletotrichum gloeosporioides. **39**:WI(1344). **41**:WI(1338).
Coniothyrium fuckelii. Cane blight, canker. **2**:WI(1356). **16**:MA(1390). **24**:WI(1359). **30**:WI(1341,1356). **34**:WI(1360). **41**:NC(1303), OK(2872), OR(82,4007).
Coniothyrium sp. Cane blight, canker. **24**:WI(1356). **41**:NC(1303), OK(2872).
Cryptostictis libertiana. **41**:AK(526).
Cylindrosporium rubi. Leaf spot. **16**:E & C states(82). **24**:MS(2720), OK(2872). **25**:WI(1360). **34**:E & C states(82). **41**:NC(1303,3944), OK(2872), WI(1356).
Cytospora rubi. On dead stems. **41**:ID(82), NJ(82), OR(82), PA(82).
Cytospora sp. Dieback, isolated from cankers. **16**:ID(3202). **41**:FL(35), ID(82), NJ(82), OH(979), OR(82), PA(82).
Diplodia rubi. On canes. **41**:VA(82), WA(3202).
Discosia artocreas. On leaves. **41**:AL(82), NC(3944).
Gloeodes pomigena. Sooty blotch. **41**:SE states(82), IN(82), NC(82), PA(82), TX(82).
Gloeosporium papulatum. On leaves. **41**:TN(82).
Hapalosphaeria deformans. Dry berry, stamen blight. 'Stamen blight.' **18**:OR(3202). **21**:OR(3202). **38**:OR(3202). **41**:OR(82), WA(82).
Leptothyrium clypeosphaerioides. On dead stems. **10**:AK(526). **41**:AK(82).
L. vulgare. **33**:AK(526).
Leptothyrium sp. Fly speck on cane. **41**:FL(35).
Macrophoma rubi. Stem canker. **41**:IL(82).
Marssonina potentillae var. *tormentillae*. Leaf spot. **41**:WI(82).
Microsphaeropsis olivacea. **34**:AK(526).
Monochaetia monochaeta. On stems and leaves. **41**:AK(82), IA(82), ME(82), NC(82), TN(82), VA(82).
Monochaetia sp. Leaf spot. **41**:FL(35).
Myxosporium sp. **41**:OR(3202).
Neottiospora sp. **41**:ID(3202).
Pestalotia laurocerasi. **34**:AK(526).
Pestalotiopsis sp. Isolated from cankers. **41**:OH(979).
Phlyctema vagabunda. Anthracnose. **16**:OR(3202), WA(3202). **20**:OR(3202), WA(3202). **41**:OR(82), WA(82).
Phoma rubicola. Cane and spur blight. **16**:OK(2872). **41**:MN(82), NY(82).
Phoma sp. **21**:OR(3202).
Phomopsis rubiseda. On canes. **41**:NY(82).
Phomopsis sp. Canker, on canes. **18**:WA(3202). **41**:IA(82).
Phyllosticta allegheniensis. Leaf spot. **41**:FL(82), IL(82), NC(82), NH(82), WV(82).
P. argillacea. **34**:WI(1360).
P. carpogena. 'Black rot.' **41**:MD(82).
P. dearnessii. On leaves. **2**:WI(1359). **27**:WI(1354). **29**:OR(82,3202). **30**:NY(82), WI(1358). **34**:WI(1354). **41**:WI(1355).
P. rubicola. On leaves. **41**:IN(82).
P. ruborum. Leaf spot. **41**:FL(82), IL(82), NC(82), NH(82), WV(82).
P. variabilis. Leaf spot. **25**:NY(64). **41**:FL(82), NY(82), VT(82), WA(82).
P. villosa. Leaf spot. **41**:FL(82), IL(82), NC(82), NH(82), WV(82).
Phyllosticta sp. Leaf spot. **2**:WI(1356). **16**:WA(3202). **24**:WI(1355). **34**:WI(1353). **41**:FL(35,82), NY(82), WA(82).
Seimatosporium caudatum. On stems. **41**:VA(82), VT(82).
S. lichenicola. Cane spot. **21**:CA(1110). **41**:IL(82), ME(82).
Septocyta ruborum. On canes, on fruiting laterals. **20**:OR(4007), WA(3202). **34**:AK(526). **41**:AK(82), AL(82), DE(82), IL(82), NJ(82), PA(82), WA(82).
Septoria darrowii. Leaf spot. **15**:NY(82). **41**:MD(82), NC(82).
S. rubi. Canker, dieback, cane and leaf spot. 'Blackberry leaf spot. Blackberry cane spot.' **8**:MS(2720). **13**:MA(1390). **16**:MA(1390). **18**:WA(3202). **19**:OR(3202), WA(3202). **20**:CA(81). **21**:ID(3202), OR(3202), WA(3202). **24**:SD(2243). **27**:CA(81), ID(3202), OR(3202), WA(3202). **32**:AK(526), CA(81), WA(3202). **34**:SD(2243). **37**:MS(2720). **38**:ID(3202), OR(3202), WA(3202). **40**:CA(2439). **41**:E states(82), Range of host(82), AR(82), FL(35), GA(82), IA(82), LA(82), NC(82,3944), NE(82), OK(2872), OR(82,3730), SC(82), TX(82), WA(3730).
S. rubi var. *pallida*. **41**:LA(2675).
Sphaceloma necator. Anthracnose, on living canes. **41**:FL(35), IL(963).
Sphaeropsis malorum. **34**:ND(378).
Truncatella angustata. On dead canes. **41**:OR(3202,4007).

Deuteromycotina - Other

Rhizoctonia solani. Root, collar and web blight. **16**:CA(1110). **21**:CA(1110). **41**:CO(82), FL(35,82), ID(82), NC(1303), WA(82).
Rhizoctonia sp. Root rot. **41**:HI(2903).
Sclerotium deciduum. **2**:WI(1342).

Sanguisorba L.

1. *S. annua* (Nutt. ex Hook.) Nutt. ex Torr. & A. Gray (=*S. occidentalis* Nutt. ex Torr. & A. Gray)
2. *S. canadensis* L. [American burnet, Canadian burnet]
3. *S. canadensis* L. subsp. *latifolia* (Hook.) Calder & N. Tayl.
4. *S. officinalis* L. (=*Poterium officinale* (L.) A. Gray) [burnet bloodwort, great burnet]
5. *S. officinalis* L. subsp. *microcephala* (J. Presl) Calder & N. Tayl. (=*S. microcephala* J. Presl)
6. *S. sitchensis* (C. A. Mey.) Koidz.
7. *S. stipulata* Raf. [Sitka burnet]
8. *Sanguisorba* sp.

Ascomycotina

Leptotrochila sanguisorbae. **8**:AK(526).
Mycosphaerella sp. **1**:ID(3202).
Podosphaera clandestina. Powdery mildew. **2**:IA(82).
Sphaerotheca macularis. Powdery mildew. **2**:MA(82), NY(82), PA(82). **4**:AK(526). **8**:AK(82).

Basidiomycotina - Uredinales

Phragmidium jonesii. Autoecious. **1**:WA(3202).
Xenodochus carbonarius. Autoecious. **4**:AK(60,526). **8**:AK(82).
X. minor. Autoecious. **3**:AK(2707). **6**:AK(60,526). **7**:AK(82).

Deuteromycotina - Hyphomycetes
Graphium sessile. Leaf spot. **2:**NY(82).
Heterosporium sp. **4:**ID(3202).
Isariopsis bulbigera. Leaf spot. **2:**IL(82). **6:**AK(526), ID(1195,3202), WA(3202). **7:**AK(82).
Ramularia sp. **1:**ID(3202), WA(3202).
Deuteromycotina - Coelomycetes
Marssonina sennenis. **5:**AK(3431).

Sibbaldia L.

1. *S. procumbens* L.

Chytridiomycetes
Synchytrium sp. **1:**WY(2924).
Ascomycotina
Mycosphaerella tassiana. **1:**WA(3202).

Sorbaria (Ser. ex DC.) A. Braun in Aschers.

1. *S. sorbifolia* (L.) A. Braun

Deuteromycotina - Coelomycetes
Aplosporella neilliae. **1:**NY(2783).
Phyllosticta sorbariae. **1:**WI(1346).

Sorbus L.

1. *S. americana* Marsh. [American mountain-ash, dogberry, missey-moosey, roundwood]
2. *S. aucuparia* L. [European mountain-ash, quickbeam, rowan]
3. *S. californica* Greene
4. *S. domestica* L. [service tree]
5. *S. dumosa* Greene
6. *S. hybrida* L.
7. *S. occidentalis* (S. Wats.) Greene
8. *S. scopulina* Greene
9. *S. sitchensis* M. Roem. [Pacific mountain-ash]
10. *Sorbus* sp.

Ascomycotina
Biscogniauxia marginata. Blister canker. **1:**IA(82), MA(82), MN(82), WV(82). **2:**IA(82,1225). **10:**IA(1225).
B. repanda. On limbs. **1:**FL(2782), PA(82).
Bisporella citrina. **9:**ID(3202).
Botryosphaeria obtusa. Black rot. **1:**NE states(82), MI(82), TX(82). **2:**IN(82), OH(82).
Cucurbitaria sorbi. **10:**ME(2779).
Dermea ariae. On twigs. **1:**MI(82), NY(82), PA(82).
Diaporthe impulsa. On limbs. **1:**MA(82), MI(82), NH(82), NY(82). **9:**CA(82).
Eutypella sorbi. On limbs. **1:**MT(82,3202). **9:**CO(82), ID(3202), MT(82). **10:**AK(526), OR(3202).
Fenestella subcaespitosa. On limbs. **1:**ME(180), NY(180). **10:**ME(180).
Glomerella cingulata. Branch canker. **1:**IN(82).
Glyphium corrugatum. **8:**ID(3202).
Hysteropatella sp. **8:**ID(3202).
Leptosphaeria suffulta. **10:**ID(3202).
Leucostoma massariana. Canker, dieback. **1:**ID(82,3202).
L. persoonii. Twig blight. **1:**MT(82), NY(82). **2:**OH(82), WV(82). **9:**ID(3202).
Myriangium superficialis. On living bark. **1:**NY(195).
Nectria cinnabarina. Coral spot. **1:**NC(82). **9:**AK(82,526). **10:**AK(526), WA(3202).
Nitschkia callista. **1:**GA(1438).
Phyllactinia guttata. **10:**ID(3202).
Podosphaera tridactyla. Powdery mildew. **2:**WA(82).
Pseudopeziza pyri. On leaves. **1:**NY(82).
Thyridaria sorbi. On limbs. **1:**NY(82).
Valsa sordida. On limbs and twigs, sometimes causing cankers. **1:**Range of host(82). **8:**MT(3202).
Valsa sp. **10:**MT(3202).
Valsaria rubricosa. **9:**WA(3202).
Venturia inaequalis. Scab. **1:**IL(82), MN(82), NY(82), WA(82). **2:**CT(3770), IL(82). **10:**WA(3202).
Basidiomycotina - Uredinales
Gymnosporangium cornutum. Heteroecious: aecial host. **1:**NE states(82), WI(82). **2:**CT(82), MI(82), NJ(82). **8:**AK(60,526), MT(3202), WY(1195). **9:**W states(82), AK(60,82,526), CO(82), WA(3202).
G. globosum. Heteroecious: aecial host. **1:**NE states(82), WI(1355).
G. libocedri. Heteroecious: aecial host. **2:**OR(82,3202). **6:**OR(3202).
G. nelsonii. **8:**MT(3202). **9:**MT(3202), WA(3202), WY(82).
G. nootkatense. Heteroecious: aecial host. **7:**AK(526). **9:**AK(60,82,526), OR(82,3202), WA(82,3202).
G. tremelloides. Heteroecious: aecial host. **5:**AZ(3981). **8:**CO(1195), MT(3202), WA(3202). **9:**OR(3202), WA(3202), WY(1195).
Gymnosporangium sp. **1:**ID(3202). **8:**ID(3202). **9:**ID(3202), MT(3202), WA(3202).
Basidiomycotina - Other
Armillaria mellea. Root rot. **1:**NJ(82).
Corticium sp. **8:**WA(3202).
Hapalopilus rutilans. White rot. **10:**Unk.(2681).
Hymenochaete tabacina. **8:**WA(3202).
Panellus serotinus. **10:**WA(3202).
Phlebia tremellosus. **10:**MA(128).
Pholiota adiposa. **1:**ID(1017).
Trametes hirsuta. White heart rot. **1:**MI(82), WA(82). **9:**OR(3202).
T. pubescens. **10:**WA(3202).
T. versicolor. Wound rot. **1:**WI(82). **10:**WA(3202).
Deuteromycotina - Hyphomycetes
Alternaria sp. Leaf spot. **1:**IA(82). **10:**FL(35).
Cercospora sp. Leaf spot. **2:**WI(1352). **10:**FL(35).
Diplocarpon mespili. Leaf blight. **8:**ID(3202). **9:**AK(82,526), WA(3202).
Graphium sorbi. Leaf spot. **1:**NY(82), WI(82).
Phymatotrichopsis omnivora. Root rot. **1:**TX(82).
Ramularia destruens. On leaves. **1:**NY(82).
Deuteromycotina - Coelomycetes
Cytospora chrysosperma. Canker. **1:**MN(82), MT(82), NE(82), NJ(82), WA(82). **9:**CO(82), MT(82), NE(82).
C. leucostoma. Canker. **1:**MT(82).
C. microspora. Canker. **1:**MT(82).
C. rubescens. Trunk canker. **2:**IA(82).
Cytospora sp. Canker. **2:**VA(3493). **10:**SD(2243), WA(3202).
Dothichiza sorbi. On limbs. **1:**KS(82), RI(82). **10:**AK(526).
Dothiorella scopulina. On limbs. **9:**WY(82).
Entomosporium mespili. Leaf spot. **1:**WI(82). **2:**WI(82).
Fusicoccum sp. Canker. **2:**IL(3146).
Gloeodes pomigena. On twigs. **1:**IN(82).
Leucocytospora massariana. Canker, dieback. **1:**ID(82).
Phomopsis sp. Twig blight. **1:**MA(82).
Phyllosticta globigera. Leaf spot. **4:**ID(3202). **8:**WA(3202). **9:**ID(82,3202), WA(82,3202).
P. sorbi. Leaf spot. **1:**IA(82), ME(82), MO(82), OK(82,2872), TX(82). **2:**IA(82).
Rhabdospora inaequalis. On twigs. **9:**AK(82,526).
Seimatosporium cassiopes. On twigs. **3:**CA(3547). **9:**CA(82).
Septoria sitchensis. Leaf spot. **8:**ID(3202), WA(3202). **9:**ID(82,3202), OR(3202).
S. sorbi. Leaf spot. **2:**IA(82).
Sphaeronaema pallidum. On limbs. **1:**NY(82), WI(82).

Spiraea L.

1. *S. alba* Du Roi [meadowsweet]
2. *S. aruncus* L. (=*Aruncus vulgaris* Raf.)
3. *S. beauverdiana* C. K. Schneid.
4. *S. betulifolia* Pall.
5. *S.* × *bumalda* F. Burvenich. *S. albiflora* × *S. japonica*.
6. *S. densiflora* Nutt. ex Rydb.
7. *S. densiflora* Nutt. ex Rydb. subsp. *splendens* (E. Baumann ex K. Koch) Abrams (=*S. splendens* E. Baumann ex K. Koch)
8. *S. discolor* Pursh
9. *S. douglasii* Hook.
10. *S. japonica* L. fil. [Japanese spirea]
11. *S. latifolia* (Aiton) Borkh. [meadowsweet]
12. *S. lucida* Douglas ex Hook.
13. *S. menziesii* Hook.
14. *S. prunifolia* Siebold & Zucc. [bridal-wreath]
15. *S.* × *pyramidata* Greene. *S. lucida* × *S. menziesii*.
16. *S. salicifolia* L. (=*S. lanceolata* Borkh.)
17. *S. thunbergii* Siebold ex Blume
18. *S. tomentosa* L. [hardhack, steeplebush]
19. *S.* × *vanhouttei* (Briot) Zab. *S. cantoniensis* × *S. trilobata*. [bridal-wreath]

20. *Spiraea* sp.
Ascomycotina
Botryosphaeria dothidea. **14:**GA(1438). **17:**GA(1438).
B. obtusa. On twigs. **20:**GA(82), NY(82).
Cryptodiaporthe macounii. On twigs. **20:**NY(82).
Diaporthe viburni var. *spiraeicola*. On twigs. **18:**MA(3807). **20:**MA(82), NY(82).
Dothidea sambuci. **9:**OR(3202).
Godronia spiraeae. On dead twigs. **4:**MT(3202). **20:**NH(1382), NY(82,1382), PA(1382), WA(1382,3202).
Herpotrichia sp. **4:**MT(3202).
Microsphaera penicillata. Powdery mildew. **20:**CT(82), FL(35).
Mycosphaerella minor. On dead leaves and stem. **3:**AK(1938).
Mycosphaerella sp. On leaves. **20:**AK(82).
Nectria cinnabarina. **20:**WA(3202).
Podosphaera clandestina. Powdery mildew. **4:**WA(3202). **9:**CA(81), WA(3202). **12:**WY(1195). **16:**NY(128). **20:**Range of host(82).
P. tridactyla. **20:**WA(82).
Schizoxylon spiraeae. On dead stems. **16:**NH(3227).
Sphaerognomonia papillostoma. On leaves. **20:**NY(82).
Sphaerotheca fuliginea. Powdery mildew. **20:**PA(82).
S. macularis. Powdery mildew. **20:**CT(82), MI(82), NY(82), PA(82), TX(82).
Stictis radiata. **20:**CA(3225).
Tapesia sp. **4:**ID(3202).
Trichopezizella barbata. **11:**NY(1416).
Basidiomycotina - Other
Armillaria tabescens. Root rot. **20:**FL(35).
Dendrothele incrustans. On bark and wood. **9:**CA(2084).
Ganoderma applanatum. **20:**WA(3202).
Scytinostroma galactinum. Root rot. **17:**MD(668).
Thelephora terrestris. Seedling stem girdle. **20:**ID(82).
Deuteromycotina - Hyphomycetes
Alternaria sp. Leaf spot. **20:**FL(35).
Cercospora physocarpi. Leaf spot. **10:**OK(2872). **5:**IL(312).
C. rubigo. Leaf spot. **20:**CA(82), FL(35), KS(82), OR(82), WI(82).
C. spiraeicola. Leaf spot. **20:**FL(35).
Heterosporium spiraeae. On leaves. **20:**AK(82,526).
Phymatotrichopsis omnivora. Root rot. **20:**TX(82).
Ramularia ulmariae. **2:**AK(526).
Rhopalidium cercosporelloides. **9:**ID(3202).
Sporidesmium spiraeacola. On leaves. **20:**SC(82,634).
Deuteromycotina - Coelomycetes
Coniothyrium sp. Leaf spot. **20:**FL(35).
Cylindrosporium fairmanianum. Leaf spot. **16:**NY(3080).
C. filipendulae. Leaf spot. **4:**ID(3202), MT(3202), WA(3202). **6:**ID(3202). **7:**WY(1195). **9:**ID(3202), MT(3202), WA(3202). **13:**WI(1359). **15:**ID(3202), MT(3202), WA(3202). **19:**KS(1041), WI(1352).
Microdiplodia spiraeicola. On twigs. **20:**NY(82).
Phloeospora spiraeicola. Leaf spot. **4:**ID(3202), WA(3202). **9:**WI(1359). **12:**WY(1195). **20:**ID(82).
Phloeosporella ariaefoliae. On leaves. **8:**CA(3549).
Phoma sp. **1:**WI(1347).
Phomopsis spiraeae. On twigs. **20:**MD(82).
Phyllosticta sp. Leaf spot. **20:**FL(35).
Septoria salicifolia. Leaf spot. **20:**FL(35), TX(82).

Vauquelinia Corrêa ex Humb. & Bonpl.
1. *V. californica* (Torr.) Sarg.
Ascomycotina
Erysiphe cichoracearum. Powdery mildew. **1:**CA(1155).
Basidiomycotina - Uredinales
Gymnosporangium vauqueliniae. **1:**AZ(1203,3981).
Basidiomycotina - Other
Phanerochaete tuberculata. **1:**AZ(1202).
Phellinus texanus. **1:**AZ(1202).

Waldsteinia Willd.
1. *W. fragarioides* (Michx.) Tratt. [barren strawberry]
2. *W. idahoensis* Piper
Ascomycotina
Gnomonia waldsteiniae. On leaves and petioles. **1:**NY(82).
Basidiomycotina - Uredinales
Puccinia waldsteiniae. Autoecious. **1:**MI(82), NY(82), VT(82), WI(82). **2:**ID(3202).
Basidiomycotina - Ustilaginales
Ustacystis waldsteiniae. **1:**MI(1077), MN(1077), NY(82,1077), WI(82,1077).
Deuteromycotina - Hyphomycetes
Ramularia waldsteiniae. Leaf spot. **1:**WI(82).
Deuteromycotina - Coelomycetes
Septoria waldsteiniae. Leaf spot. **1:**MI(82), NY(82), VT(82), WI(1340).

Rubiaceae

Borreria G. Mey.
1. *B. micrantha* Torr. & A. Gray
2. *B. terminalis* Small
3. *Borreria* sp.
Ascomycotina
Trichothyrina tracyi. On leaves. **1:**MS(82).
Basidiomycotina - Uredinales
Puccinia lateritia. Autoecious. **2:**FL(35). **3:**FL(82).
Deuteromycotina - Hyphomycetes
Pseudocercospora borreriae. Leaf spot. **3:**MS(82).

Bouvardia Salisb.
1. *B. glaberrima* Engelm.
2. *B. ternifolia* (Cav.) Schlechtend.
3. *Bouvardia* sp.
Basidiomycotina - Uredinales
Aecidium bouvardiae. **1:**AZ(749).
[*Puccinia anisacanthi*]. **1:**AZ(1203). **2:**AZ(1203). **3:**AZ(82).
P. chloridis. **1:**AZ(1203).
P. diplachnis. Heteroecious: aecial host. **1:**AZ(741,747,1203,3981).

Cephalanthus L.
1. *C. occidentalis* L. [buttonbush, common buttonbush]
2. *Cephalanthus* sp.
Ascomycotina
Botryosphaeria dothidea. Twig dieback. **1:**FL(35), GA(1438).
B. obtusa. On dead stems. **1:**GA(82), VA(82).
B. ribis. On dead stems. **1:**FL(82).
Diatrypella cephalanthi. On dead limbs. **1:**IA(3629), NY(82), OK(82,2873).
D. favacea. **1:**GA(1438).
Godronia cephalanthi. On bark. **1:**AL(82), GA(1438), NY(82).
Lambertella cephalanthi. On leaves. **1:**MA(909), NY(909).
Leptosphaerulina sp. **1:**GA(1438).
Lophiostoma prominens. On limbs. **1:**NY(195).
Massaria lantanae. On dead shoots. **1:**NY(82,3257).
Microsphaera penicillata. Powdery mildew. **1:**Range of host(82), FL(35).
Phyllactinia guttata. Powdery mildew. **1:**IN(82).
Tympanis cephalanthi. On dead twigs. **1:**NY(82).
Xylaria arbuscula. **1:**GA(82,1438).
Basidiomycotina - Uredinales
Puccinia seymouriana. **1:**NE states(82), FL(35).
Uredo cephalanthi. **1:**FL(35,82).
Basidiomycotina - Other
Armillaria mellea. Root rot. **1:**CA(2900).
Corticium stevensii. Thread blight. **1:**FL(35).
Gloeoporus dichrous. On dead stems. **1:**SC states(82). **2:**AR(2675), LA(2675).
Heterochaete spinulosa. On wood. **2:**AL(1175,2946).
Inonotus radiatus. On dead stems. **1:**SC states(82), IL(82), LA(2675), MI(82).
Pellicularia koleroga. Thread blight. **1:**FL(82).
Phellinus gilvus. On dead stems. **1:**SC states(82). **2:**AR(2675), LA(2675), MS(2675).
Stereum hirsutum. White rot. **1:**SC states(82), OK(82,2873). **2:**AR(2675), LA(2675).
Deuteromycotina - Hyphomycetes
Cercospora cephalanthi. Leaf blight, spot. **1:**AL(82), DE(3543), FL(35), KS(82), LA(82), MS(2720), NY(82), OK(2872), TX(82,581), WI(82).
Deuteromycotina - Coelomycetes
Ascochyta cephalanthi. Leaf spot. **1:**LA(82), TX(82).

Coniothyrium cephalanthi. Leaf spot. **1**:LA(82). **2**:LA(966).
Dendrophoma cephalanthi. On dead limbs. **1**:NY(82).
Pestalotia sp. **1**:TX(82).
Phoma cephalanthi. On dead stems. **1**:NY(82).
Phyllosticta cephalanthi. On living leaves. **1**:FL(35), TX(64,3597).
Rhabdospora verruciformis. On dead stems. **1**:AL(82).
Septoria cephalanthi. Leaf spot. **1**:KS(82), WI(82,1333).
Deuteromycotina - Other
Rhizoctonia solani. Thread blight. **1**:FL(35).

Chiococca P. Br.

1. *C. alba* (L.) Hitchc. [snowberry]

Ascomycotina
Meliola psychotriae var. *chiococcae*. **1**:FL(3483).
Phyllachora sp. Leaf spot. **1**:FL(35).
Deuteromycotina - Hyphomycetes
Cercospora sp. Leaf spot. **1**:FL(35).

Cinchona L.

1. *C. officinalis* L. [quinine]
2. *Cinchona* sp.

Deuteromycotina - Hyphomycetes
Cercospora cinchonae. Leaf spot. **1**:LA(82). **2**:LA(82).

Coccocypselum P. Br.

1. *C. lanceolatum* Pers. (=*Condalia obovata* Ruiz & Pav.)

Ascomycotina
Meliola condaliae. **1**:TX(1452).

Coffea L.

1. *C. arabica* L. [arabica coffee, common coffee]
2. *C. canephora* Pierre ex Froehner (=*C. robusta* L. Linden) [wild robusta ocffee]
3. *Coffea* sp.

Oomycetes
Pythium sp. Root rot. **1**:FL(35).
Ascomycotina
Botryotinia fuckeliana. Gray mold. **1**:HI(2903).
Capnodium sp. Sooty mold. **1**:Range of host(82). **3**:HI(2903).
Glomerella cingulata. Dieback, leaf spot. **1**:Range of host(82), FL(35).
Massarina coffeicola. **3**:HI(2784).
Seuratia millardetii. On leaves. **3**:HI(2337).
Deuteromycotina - Hyphomycetes
Cercospora coffeicola. Eye spot, leaf spot, brown spot of leaves and berries. **1**:Range of host(82), FL(35), HI(2903). **2**:HI(2903).
Fusarium sp. Seedling blight, neck rot. **1**:FL(35).
Myrothecium sp. Leaf spot. **1**:FL(35).
Trichothecium roseum. Pink mold. **1**:FL(35).
Deuteromycotina - Coelomycetes
Colletotrichum coffeanum. Anthracnose, brown blight. **1**:Range of host(82), FL(35).
Phomopsis sp. Leaf spot. **1**:FL(35).
Phyllosticta sp. Leaf spot. **1**:FL(35).
Deuteromycotina - Other
Rhizoctonia solani. Damping-off. **1**:Range of host(82), FL(35).

Coprosma J. R. Forster & G. Forster

1. *C. ernodeoides* A. Gray
2. *C. menziesii* A. Gray
3. *C. repens* A. Rich. [looking-glass-plant, mirror-plant]
4. *Coprosma* sp.

Oomycetes
Phytophthora cinnamomi. Root rot. **1**:HI(2903). **2**:HI(2903).
Ascomycotina
Amazonia psychotriae. Black mildew. **4**:HI(1272,2903).
Asteridiella coprosmae. **4**:HI(1452).
Venturia coprosmae. **4**:HI(2903).
Basidiomycotina - Other
Armillaria mellea. Root rot. **4**:HI(2011,2903).
Deuteromycotina - Other
Rhizoctonia sp. Root rot. **3**:FL(35).

Diodia L.

1. *D. teres* Walter
2. *D. virginiana* L.
3. *Diodia* sp.

Oomycetes
Peronospora sp. **2**:NC(3944).
Basidiomycotina - Uredinales
Puccinia lateritia. Autoecious. **1**:FL(35).
Uromyces spermacoces. Autoecious. **1**:FL(35), GA(1439), IN(987), MS(2720), OK(2872), TX(652). **2**:NC(3944). **3**:SC(634).
Deuteromycotina - Hyphomycetes
Cercospora diodiae. Leaf spot. **1**:DE(3543), FL(35), MS(2720), OK(2872), TX(652), WI(1330).
C. diodiae-virginianae. Leaf spot. **1**:NC(3944). **2**:AL(581).
Deuteromycotina - Other
Rhizoctonia solani. Dry rot. **1**:FL(35).
Sclerotium rolfsii. Southern blight. **1**:FL(35).

Ernodea Sw.

1. *E. littoralis* Sw.
2. *Ernodea* sp.

Basidiomycotina - Uredinales
Puccinia lateritia. Autoecious. **1**:FL(35). **2**:FL(82).
Deuteromycotina - Other
Sclerotium rolfsii. Southern blight. **1**:FL(35).

Galium L.

1. *G. angustifolium* Nutt. ex A. Gray
2. *G. aparine* L. [catchweed bedstraw, cleavers]
3. *G. bifolium* S. Wats.
4. *G. boreale* L. [northern bedstraw]
5. *G. brevipes* Fernald & Wiegland
6. *G. circaezans* Michx.
7. *G. concinnum* Torr. & A. Gray
8. *G. grayanum* Ehrend.
9. *G. labradoricum* (Wiegland) Wiegland
10. *G. lanceolatum* Torr.
11. *G. linearifolium* Hoehne
12. *G. mexicanum* H. B. K. subsp. *asperrimum* (A. Gray) Dempster (=*G. asperrimum* A. Gray)
13. *G. multiflorum* Kellogg
14. *G. obtusum* Bigel.
15. *G. oreganum* Britton
16. *G. pilosum* Aiton
17. *G. tinctorium* L.
18. *G. trifidum* L.
19. *G. triflorum* Michx. [sweet-scented bedstraw]
20. *Galium* sp.

Oomycetes
Peronospora calotheca. Downy mildew. **2**:SE states(82), FL(35), KS(82), WA(3202), WI(82). **3**:ID(3202). **4**:IA(82), ID(3202), MT(3202), ND(82,377), OR(3202), WI(82). **17**:WI(1333). **18**:MT(3202). **19**:WA(3202), WI(82). **20**:WI(82).
Ascomycotina
Erysiphe cichoracearum. Powdery mildew. **2**:Range of host(82), ID(3202), IL(312), MT(3202), WA(3202), WI(1334). **4**:WA(3202). **18**:WA(3202). **19**:CA(81,82), ID(3202), MT(82,3202), OR(82,3202), WA(3202). **20**:CA(82), PA(82), WA(82).
E. polygoni. Powdery mildew. **2**:KS(82).
Hainesia borealis. On leaves. **3**:WA(3202). **4**:ID(638,3202), IN(82), MT(3202), OR(3202), WA(82,3202), WY(2924). **19**:OR(3202).
Leptosphaeria galiorum. On overwintered stems. **2**:WI(1348).
Leptotrochila repanda. Leaf spot. **2**:OR(82,3202), WA(3202). **4**:IA(82), IL(82), ND(82), OR(82,3202), WI(82), WY(1195). **9**:WI(1358). **14**:WI(1358). **16**:CT(82), GA(82). **17**:NY(82), WI(82). **18**:WI(1358). **19**:CA(82), OR(82,3202). **20**:PA(82), WI(82).
Massarina gloeospora. **16**:GA(1438).
Mazzantia galii. On overwintered stalks. **4**:NM(82), WI(128,1346). **19**:OR(82,3202).
Metasphaeria anisometra. On stems. **1**:CA(2439).
Mollisia nigritella. On dead stems. **4**:CA(2801).
Mycosphaerella sp. **7**:WI(1357).
Nectria galii. **19**:CA(2836).

Pyrenopeziza sp. **16:**GA(1438).

Basidiomycotina - Uredinales

Puccinia difformis. Autoecious. **2:**C & W states(82), W states(82), ID(3202), MT(82,3202), ND(378), OR(3202), SD(2243), WA(3202), WI(1352). **19:**ID(3202).

P. punctata. **2:**CA(82), ID(3202), IL(82), MI(82), MS(82,2720), OR(82,3202), WA(82,3202), WI(82), WY(82). **7:**IL(650), IN(647), WI(1357). **12:**AZ(1203,3981), ID(3202), OR(3202), WA(3202). **16:**AZ(3981), NC(82). **19:**W states(82), AZ(1203,3981), CA(81), ID(3202), MS(2720), MT(1195,3202), OR(3202), SD(2243), WA(3202). **20:**E states(82), FL(35), MS(82).

P. punctata var. *troglodytes*. Autoecious. **19:**NC states(82), NE states(82), IN(987), SD(82), WA(82), WI(1354).

P. rubefaciens. Autoecious. **4:**C states(82), NC states(82), AK(60,82,526), CO(1195), ID(638,3202), MT(1195,3202), ND(378), OR(3202), WA(82,3202), WY(1195). **11:**ND(378). **13:**WA(3202). **20:**CA(82).

Pucciniastrum guttatum. Telial host; aecial host unknown. **19:**NC states(82), NE states(82), AZ(3981), CA(81,82,1575), CO(82,1575), ID(3202), NY(1575), OR(82,3202), PA(1575), UT(1575), WA(3202), WI(1575).

Uromyces galii-californici. Autoecious. **20:**CA(82).

Basidiomycotina - Other

Ceratobasidium anceps. **19:**WI(1354).

Deuteromycotina - Hyphomycetes

Cercospora tenuis. **14:**WI(1359). **16:**NY(82).

Cercosporidium galii. Leaf spot. **2:**IA(82), ID(3202), OR(82,3202), WI(82). **15:**OR(3202). **16:**AL(82,2484), DE(3543). **17:**NY(82). **19:**AK(82,526), OR(82,3202), WA(3202), WI(82). **20:**FL(35), IA(82), NH(82), WI(82).

Phymatotrichopsis omnivora. Root rot. **20:**TX(82).

Deuteromycotina - Coelomycetes

Marssonia galii. **12:**AZ(1195).

Melasmia galii. Leaf spot. **4:**IA(82).

Phyllosticta sp. Leaf spot. **2:**WI(1357). **6:**WI(1357). **19:**WI(1357).

Placosphaeria galii. **4:**ND(378).

P. punctiformis. **2:**Unk.(82). **4:**Unk.(82). **16:**Unk.(82). **17:**Unk.(82). **19:**Unk.(82). **20:**Unk.(82).

Rhabdospora galiorum. On stems. **2:**NJ(82). **20:**NJ(82), NY(82).

Septoria cruciatae. Leaf spot on leaves and stems. **2:**KS(82), WI(82). **4:**ID(3202), WA(3202). **16:**NJ(82). **18:**WA(3202). **20:**IN(82), MI(82), NY(82), WI(82), WV(82).

S. psilostega. **5:**WI(1358). **10:**WI(1357). **18:**WI(1349).

Sphaeropsis nebelina. On stems. **1:**CA(2439).

Sporonema punctiforme. **4:**WY(1195). **8:**CA(81).

Gardenia J. Ellis

1. *G. jasminoides* J. Ellis (=*G. florida* L.) [Cape jasmine, common gardenia]
2. *Gardenia* sp.

Oomycetes

Phytophthora sp. Root rot. **1:**FL(35). **2:**FL(35).

Pythium spinosum. On roots. **1:**FL(35).

P. splendens. Root rot. **1:**FL(35). **2:**FL(35).

Pythium sp. Root rot. **2:**HI(2903).

Ascomycotina

Botryotinia sp. Foliar blight. **1:**FL(35).

Capnodium citri. **1:**GA(1438).

Capnodium sp. Sooty mold. **1:**Gulf States(82), GA(2405). **2:**MS(2720).

Erysiphe polygoni. Powdery mildew. **1:**TX(82).

Meliola sp. Sooty mold, following white flies. **1:**NC(1303).

Mycosphaerella gardeniae. On leaves. **1:**FL(35), GA(82), SC(82).

Nectria sp. Stem gall. **1:**FL(35).

Physalospora sp. Stem dieback. **2:**FL(35).

Sclerotinia sp. Foliar blight. **1:**FL(35).

Seuratia millardetii. **2:**HI(2784).

Basidiomycotina - Other

Armillaria mellea. Root rot. **1:**NC(1303).

A. tabescens. Root rot. **2:**FL(35).

Erythricium salmonicolor. Limb blight. **1:**FL(35). **2:**FL(35).

Deuteromycotina - Hyphomycetes

Alternaria sp. Leaf spot. **1:**FL(35). **2:**FL(35).

Botrytis cinerea. Petal blight, bud rot. **1:**Unk.(82), FL(35), HI(2903).

Botrytis sp. **2:**WA(3202).

Cercospora sp. Leaf spot. **1:**FL(35).

Cylindrocladium scoparium. Root rot. **2:**FL(35).

Fusarium lateritium. Bud rot. **1:**FL(35).

F. oxysporum. Root rot. **2:**FL(35).

F. roseum. Bud rot. **1:**FL(35).

Myrothecium roridum. Leaf spot in greenhouses. **1:**FL(35). **2:**PA(1040).

Phymatotrichopsis omnivora. Root rot. **1:**TX(82).

Deuteromycotina - Coelomycetes

Colletotrichum gloeosporioides. Anthracnose. **1:**FL(35).

Conidioxyphium gardeniorum. On leaves. **1:**LA(3549,3550).

Diplodia sp. Tip dieback. **1:**FL(35).

Leptothyrium sp. Tip necrosis. **1:**FL(35).

Pestalotia langloisii. **1:**AL(3260), FL(3260).

Phoma sp. Stem canker. **1:**FL(35). **2:**FL(35).

Phomopsis gardeniae. Canker. **1:**CA(82), FL(35). **2:**FL(35), OK(2872), WA(3202).

Phyllosticta sp. Leaf spot. **1:**FL(35), MS(82), NC(1303), NJ(82), TX(82). **2:**MS(2720).

Sphaeropsis sp. Leaf spot. **1:**FL(35).

Deuteromycotina - Other

Rhizoctonia solani. Thread blight, root rot. **1:**FL(35). **2:**FL(35).

Rhizoctonia sp. Crown and root rot, leaf spot. **1:**NJ(82). **2:**HI(2903).

Sclerotium rolfsii. Leaf blight. **1:**FL(35).

Genipa L.

1. *G. americana* L. [genipap, marmelade, marmelade-box]
2. *G. clusiifolia* (Jacq.) Griseb.

Deuteromycotina - Hyphomycetes

Cercospora sp. Leaf spot. **1:**FL(35). **2:**FL(35).

Deuteromycotina - Coelomycetes

Septoria cephalanthi. Fruit and leaf spot. **1:**FL(35).

Gouldia A. Gray

1. *G. coriacea* (Hook. & Arnott) Hillebr.
2. *G. elongata* A. Heller
3. *G. lanceolata* A. Heller
4. *G. terminalis* (Hook. & Arnott) Hildebr.
5. *G. terminalis* (Hook. & Arnott) Hildebr. var. *macrocarpa* (Hildebr.) Fosb. (=*G. macrocarpa* Hillebr.)
6. *Gouldia* sp.

Oomycetes

Phytophthora cinnamomi. Decline. **5:**HI(2903).

Ascomycotina

Amazonia asterinoides. Black mildew. **6:**HI(82,2903).

Asterina gouldiae. Black leaf spot. **1:**HI(2903). **6:**HI(82).

Dothidella rubiacearum. Anthracnose, black leaf spot. **1:**HI(2903). **6:**HI(82).

Meliola gouldiae. Black mildew. **1:**HI(1272,1452,2903).

M. kaduae. Black mildew. **3:**HI(1272,2903). **4:**HI(1272). **5:**HI(2903). **6:**HI(1452).

M. sandwicensis. Black mildew. **1:**HI(1272,1452,2903). **2:**HI(2903). **3:**HI(1272,1452). **5:**HI(1272,2903). **6:**HI(82,1272,2784,2903).

M. sandwicensis var. *gouldiae*. Black mildew. **1:**HI(1272,1452,2903). **2:**HI(1272,2903). **3:**HI(1272,1452,2903). **4:**HI(1272,1452). **5:**HI(1272). **6:**HI(1272,1452).

M. sandwicensis var. *major*. Black mildew. **1:**HI(1272,1452,1453,2903). **3:**HI(2903). **5:**HI(1272,1452). **6:**HI(1272,2903).

Trichothyrium reptans. On leaves. **6:**HI(82,2903).

Basidiomycotina - Other

Armillaria mellea. **6:**HI(2011,2903).

Deuteromycotina - Hyphomycetes

Cercospora sp. Leaf spot. **6:**HI(82,2903).

Deuteromycotina - Coelomycetes

Dendrophoma gouldiae. On leaves. **1:**HI(2903). **6:**HI(82).

Enthallopycnidium gouldiae. On leaves. **6:**HI(82,2903).

Septoria gouldiae. Leaf spot. **3:**HI(2903). **6:**HI(82).

S. hawaiiensis. Leaf spot. **6:**HI(82,2903).

Sphaeropsis gouldiae. On leaves. **6:**HI(82,2903).

Guettarda L.

1. *G. scabra* (L.) Venten.

Ascomycotina

Meliola psychotriae. **1:**FL(3483).

Hamelia Jacq.
1. *H. patens* Jacq. (=*H. erecta* Jacq.) [firebush, scarletbush]
2. *Hamelia* sp.
Basidiomycotina - Uredinales
Uredo hameliae. **1**:FL(35,2286).
Basidiomycotina - Other
Armillaria tabescens. Root rot. **1**:FL(35). **2**:FL(82).
Deuteromycotina - Hyphomycetes
Cercospora sp. Leaf spot. **1**:FL(35).

Hedyotis L.
(= *Houstonia* L.)
1. *H. caerulea* (L.) Hook. (=*Houstonia caerulea* L.)
2. *H. purpurea* (L.) Torr. & A. Gray (=*Houstonia purpurea* L.)
3. *H. purpurea* (L.) Torr. & A. Gray var. *longifolia* (Gaertn.) Fosb. (=*Houstonia longifolia* (Gaertn.) Hook.)
4. *Hedyotis* sp. (=*Houstonia sp.*)
Oomycetes
Peronospora calotheca. Downy mildew. **1**:IL(82).
P. seymourii. Downy mildew. **1**:MS(82,2720). **4**:AL(82,3713), AR(82), IA(82), IL(82,3713), IN(82), TX(82).
Basidiomycotina - Uredinales
Puccinia lateritia. **4**:TX(82).
Uromyces houstoniatus. **1**:E states(82), IL(82), MA(1394), MO(82), MS(82,2720). **2**:MS(82,2720), TN(82), WV(82). **3**:NY(82), PA(82), WI(82), WV(82). **4**:KS(82), NC(82), TX(82).
U. peckianus. **2**:MS(82,2720). **4**:AL(82), MS(82), TX(82).
Deuteromycotina - Hyphomycetes
Cercospora houstoniae. Leaf spot. **1**:DE(82,3543). **3**:WI(1338).
Phymatotrichopsis omnivora. Root rot. **4**:TX(82).
Deuteromycotina - Coelomycetes
Septoria sp. Leaf spot. **4**:TX(82).

Houstonia L.
This genus is considered a synonym of *Hedyotis*, but appropriate combinations could not be located.
1. *H. pusilla* Schöpf (=*H. patens* Elliott)
Basidiomycotina - Uredinales
Uromyces houstoniatus. Heteroecious: aecial host. **1**:NC(3944).

Hydnophytum W. Jack
1. *H. andamanense* Becc. (=*H. formicarum* Kurz)
Oomycetes
Phytophthora sp. Basal stem rot. **1**:FL(35).
Pythium sp. Root rot. **1**:FL(35).
Deuteromycotina - Hyphomycetes
Fusarium sp. Basal stem rot. **1**:FL(35).

Ixora L.
1. *I. coccinea* L. [flame-of-the-woods, scarlet ixora]
2. *Ixora* sp.
Oomycetes
Phytophthora sp. Root rot. **2**:FL(35).
Pythium splendens. Root rot. **1**:FL(35). **2**:FL(35).
Ascomycotina
Physalospora sp. Leaf spot. **1**:FL(35). **2**:FL(35).
Basidiomycotina - Other
Armillaria tabescens. Root rot. **1**:FL(35). **2**:FL(35,82).
Deuteromycotina - Hyphomycetes
Alternaria sp. Leaf spot. **1**:FL(35). **2**:FL(35).
Fusarium sp. Root rot. **1**:FL(35).
Helminthosporium sp. Leaf spot. **2**:FL(35).
Pseudocercospora ixorae. Leaf spot. **1**:FL(35), HI(2903). **2**:FL(35).
Deuteromycotina - Coelomycetes
Colletotrichum sp. Leaf spot. **1**:FL(35).
Pestalotia sp. Leaf spot. **1**:FL(35). **2**:FL(35).
Phomopsis sp. Leaf spot. **1**:FL(35). **2**:FL(35).
Phyllosticta ixorae. Leaf spot. **1**:FL(35). **2**:FL(35).
Sphaeropsis tumefaciens. Twig canker, stem gall. **2**:FL(35).
Deuteromycotina - Other
Rhizoctonia solani. Root rot. **1**:FL(35). **2**:FL(35).

Kadua Cham. & Schlechtend.
1. *K. glomerata* Hook. & Arnott
2. *K. grandis* A. Gray
3. *K. knudsenii* Hillebr.
4. *Kadua* sp.
Ascomycotina
Dothidella kaduae. **1**:HI(2903).
Meliola kaduae. Black mildew. **4**:HI(82,1272,1452,2903).
M. kauaiensis. Black mildew. **3**:HI(1452,2903). **4**:HI(1272,1452,2903).
M. sandwicensis. Black mildew. **3**:HI(2903). **4**:HI(1272,1452,2903).
M. sandwicensis var. *gouldiae*. Black mildew. **3**:HI(1452). **4**:HI(1272,1452).
Mycosphaerella kaduae. On living leaves. **2**:HI(2903,3480). **4**:HI(82,2903,3480).
Phragmocapnias smilacina. **4**:HI(2784).
Trichothyrium reptans. **1**:HI(2903).

Kelloggia Torr. ex Benth. in Benth. & J. D. Hook.
1. *K. galioides* Torr.
Deuteromycotina - Coelomycetes
Phyllosticta crustosa. On leaves. **1**:CA(638,639,642).

Mitchella L.
1. *M. repens* L. [partridgeberry, squawberry]
Chytridiomycetes
Synchytrium mitchellae. Leaf gall. **1**:LA(627).
Ascomycotina
Meliola mitchellae. Black mildew, on leaves. **1**:Unk.(1452), AL(82), FL(35,82,634,959,1452), MS(82,2720), PA(82).
Microthyrium mitchellae. On leaves. **1**:FL(35,82).
Deuteromycotina - Other
Rhizoctonia sp. Leaf blight, root and stem rot. **1**:FL(35).
Sclerotium rolfsii. Southern blight, stem rot. **1**:MD(82), NC(1303).

Morinda L.
1. *M. citrifolia* L. [Indian mulberry]
2. *M. royoc* L.
Ascomycotina
Chaetothyrium hawaiiense. **1**:HI(2903).
Limacinula theae. **1**:HI(2903).
Deuteromycotina - Hyphomycetes
Cercospora morindicola. Leaf spot. **2**:FL(35,82).
Deuteromycotina - Coelomycetes
Sphaceloma morindae. Scab, on leaves and stems. **2**:FL(35,82,1745).

Paederia L.
1. *P. foetida* L.
Basidiomycotina - Uredinales
Coleosporium paederiae. Heteroecious: telial host. **1**:HI(2903).

Pentas Benth. in Hook.
1. *P. lanceolata* (Forssk.) Defl. [Egyptian star-cluster, star-cluster]
2. *Pentas* sp.
Oomycetes
Pythium sp. Root rot. **1**:FL(35).
Deuteromycotina - Hyphomycetes
Alternaria sp. Leaf spot. **2**:FL(35).
Cercospora sp. Leaf spot. **1**:FL(35).
Corynespora sp. Leaf spot. **1**:FL(35).
Fusarium sp. Root rot. **2**:FL(35).
Oidium sp. Powdery mildew. **1**:FL(35). **2**:FL(35).
Deuteromycotina - Coelomycetes
Phoma sp. Stem gall. **2**:FL(35).
Phyllosticta sp. Leaf spot. **1**:FL(35). **2**:FL(35).
Deuteromycotina - Other
Rhizoctonia solani. Root and stem rot. **1**:FL(35). **2**:FL(35).
Sclerotium rolfsii. Southern blight. **1**:HI(2903).

Pinckneya Michx.
1. *P. pubens* Michx. [fevertree]
Ascomycotina
Diaporthe sp. **1**:GA(1438).

Portlandia P. Br.
1. *P. grandiflora* L.

Deuteromycotina - Coelomycetes

Sphaeropsis tumefaciens. Stem gall. **1**:FL(35).

Psychotria L.
1. *P. hawaiiensis* (A. Gray) Fosb. var. *hillebrandii* (Rock) Fosb.
2. *P. kuduana* (Cham. & Schlechtend.) Fosb. [kopiko-tea]
3. *P. mariana* Müll. Arg. in Mart.
4. *P. nervosa* Sw.
5. *Psychotria* sp.

Oomycetes

Phytophthora cinnamomi. Root rot. **1**:HI(2903).

Ascomycotina

Amazonia polypoda. Black mildew. **1**:HI(2903). **2**:HI(2903). **3**:HI(2903). **5**:HI(2903).
A. psychotriae. **5**:HI(1268).
Calothyrium ryanae. Black patch. **4**:FL(35).
Chaetothyrium straussiae. **3**:HI(2903).
Lyonella neurophila. **5**:HI(2903).
Meliola bidentata. Black mildew. **1**:HI(2903).
M. kaduae. **2**:HI(2903).
M. straussiae. **5**:HI(2903).
Metasphaeria bifoveolata. Leaf spot. **4**:FL(35).
Nectriella pironii. Canker, stem gall. **4**:FL(35).
Trichothyrium reptans. **5**:HI(2903).

Basidiomycotina - Other

Armillaria mellea. Root rot. **3**:HI(2903).
Septobasidium sp. On scale insects. **3**:HI(2903).

Deuteromycotina - Hyphomycetes

Cercospora psychotriae. Leaf spot. **4**:FL(35).
Trichothallus hawaiiensis. **1**:HI(2903).

Deuteromycotina - Coelomycetes

Macrophoma sp. Leaf spot. **4**:FL(35).
Phyllosticta sp. Leaf spot. **4**:FL(35).

Randia L.
1. *R. aculeata* L. [inkberry]
2. *R. formosa* (Jacq.) K. Schum. (=*R. mussaenda* (L. fil.) DC.)

Oomycetes

Pythium sp. Root rot. **2**:FL(35).

Deuteromycotina - Coelomycetes

Diplodia sp. Stem gall. **1**:FL(35).

Richardia L.
1. *R. brasiliensis* Gomes
2. *R. scabra* L. (=*Richardsonia scabra* (L.) St. Hil.) [Mexican-clover]

Oomycetes

Peronospora sp. Downy mildew. **2**:FL(35,82).
Phytophthora cinnamomi. Root rot. **1**:HI(2903). **2**:HI(82).
P. nicotianae var. *parasitica*. Root rot. **1**:HI(2903). **2**:HI(82).
P. palmivora. Root rot. **1**:HI(2903).

Deuteromycotina - Hyphomycetes

Alternaria multirostrata. **2**:GA(1713).
Alternaria sp. Leaf spot. **1**:HI(2899,2903).
Cercospora carveriana. Spot. **2**:AL(82).
Cercospora sp. Leaf spot. **2**:FL(35).
Curvularia richardiae. Leaf spot. **2**:NC(1303).
Fusarium oxysporum f. sp. *batatas*. On roots and stems. **2**:SC(82).

Deuteromycotina - Coelomycetes

Phyllosticta richardsoniae. Leaf spot. **2**:AL(64,82).

Deuteromycotina - Other

Rhizoctonia solani. Root rot. **2**:FL(35,82).
Sclerotium rolfsii. Southern blight. **2**:FL(35,82).

Spermacoce L.
1. *S. ocymoides* N. L. Burm.
2. *S. parvifolia* (J. D. Hook.) Kuntze

Ascomycotina

Trichothyrina tracyi. On leaves. **2**:MS(2720).

Deuteromycotina - Hyphomycetes

Pseudocercospora borreriae. Leaf spot. **1**:MS(2720).

Straussia (DC.) A. Gray
This genus is considered a synonym of *Psychotria*, but appropriate combinations could not be located.
1. *S. hawaiensis* A. Gray
2. *S. kaduana* A. Gray
3. *S. mariniana* A. Gray
4. *Straussia* sp.

Ascomycotina

Amazonia polypoda. Black mildew. **1**:HI(1272,1452). **2**:HI(1272,1452). **3**:HI(1272,1452). **4**:HI(1272,1452).
Meliola kaduae. **4**:HI(1452).
M. straussiae. Black mildew. **4**:HI(1272,1452).
Seuratia millardetii. **4**:HI(2784).

Basidiomycotina - Other

Armillaria mellea. **4**:HI(2011).

Rutaceae

Casimiroa Llave & Lex.
1. *C. edulis* Llave & Lex. [Mexican apple, white sapote, zapote-blanco]

Deuteromycotina - Hyphomycetes

Cercospora coleroides. Leaf spot. **1**:FL(35).

Deuteromycotina - Coelomycetes

Colletotrichum gloeosporioides. Anthracnose. **1**:FL(35).
Phyllosticta casimiroae. **1**:HI(2903).
Phyllosticta sp. Leaf spot. **1**:FL(35).

Choisya Kunth in H. B. K.
1. *C. mollis* Standl.

Deuteromycotina - Coelomycetes

Sphaceloma choisyae. Spot anthracnose. **1**:AZ(3788).

×*Citrofortunella* J. Ingram & H. Moore
Citrus × *Fortunella*.
1. × *C. mitis* (Blanco) J. Ingram & H. Moore. *Citrus reticulata* × *Fortunella* sp. [calamondin, Panama orange]

Oomycetes

Phytophthora nicotianae var. *parasitica*. Collar and foot rot. **1**:FL(35).
Pythium ultimum. Root rot. **1**:FL(35).

Ascomycotina

Diaporthe citri. Stem end rot. **1**:FL(35).
Elsinoe fawcettii. Scab. **1**:FL(35).
Glomerella cingulata. Anthracnose. **1**:FL(35).
Mycosphaerella citri. Greasy spot. **1**:FL(35).
M. horii. Leaf spot. **1**:FL(35).

Basidiomycotina - Other

Ganoderma lucidum. Heart rot. **1**:FL(35).
Septobasidium pseudopedicellatum. Parasitic on scale insects. **1**:FL(35).

Deuteromycotina - Hyphomycetes

Alternaria citri. Center rot, leaf spot. **1**:FL(35).
Fusarium solani. Collar and root rot. **1**:FL(35).
Penicillium digitatum. Green mold. **1**:FL(35).

Deuteromycotina - Coelomycetes

Colletotrichum gloeosporioides. Anthracnose. **1**:FL(35).
Lasiodiplodia theobromae. Dieback, stem-end rot. **1**:FL(35).
Phyllosticta hesperidearum. Leaf spot. **1**:FL(35).
Sphaeropsis tumefaciens. Limb knot. **1**:FL(35).

Deuteromycotina - Other

Rhizoctonia solani. Root rot. **1**:FL(35).

Citrus L.
1. *C. aurantifolia* (L.) Swingle [lime]
2. *C. aurantium* L. [bigarade, bitter orange, Seville orange, sour orange]
3. *C. jambhiri* Lush. [rough lemon]
4. *C. latifolia* Tanaka [Persian lime]
5. *C. limon* (L.) N. L. Burm. [lemon]
6. *C.* × *limonia* Osbeck. *C. limon* × *C. reticulata*. [lemandarin, mandarin lime, Rangpur lime]
7. *C. macrophylla* P. J. Wester [alimau]
8. *C. maxima* (J. Burm.) Merr. (=*C. decumana* (L.) L., *C. grandis* (L.) Osbeck) [pummelo, shaddock]
9. *C. medica* L. [citron]
10. *C. meyerii* Y. Tanaka
11. *C.* × *paradisi* Macfady. *C. maxima* × *C. sinensis*.

[grapefruit]
12. *C. reticulata* Blanco (=*C. reticulata* Blanco var. *deliciosa* Swingle, *C. nobilis* Andr., *C. nobilis* Andr. var. *deliciosa* Swingle) [king orange, Mandarin orange, Satsuma orange, tangerine]
13. *C. sinensis* (L.) Osbeck (=*C. aurantium* var. *sinensis* L.) [sweet orange]
14. *C.* × *tangelo* J. Ingram & H. Moore. *C. paradisi* × *C. reticulata*. [tangelo]
15. *C. temple* Hort. ex Y. Tanaka
16. *C. unshiu* Markovich (=*C. nobilis* Andr. var. *unshiu* Swingle) [Satsuma orange]
17. *Citrus* sp.

Oomycetes

Phytophthora cactorum. Gummosis, brown rot. **13:**CA(1110).

P. cinnamomi. Root rot. **17:**CA(82,1110).

P. citrophthora. Limb blight. 'Brown rot. Phytophthora foot rot. Phytophthora gummosis. Phytophthora root rot.' **1:**CA(1110), HI(82,2903). **2:**CA(82), HI(2903). **4:**CA(1110). **5:**CA(82,1110). **8:**CA(1110). **10:**CA(1110). **12:**CA(82,1110). **13:**Gulf States(82), CA(82,1025,1110,1425), FL(35), HI(2899,2903). **11:**AZ(82), CA(82,1025,1110), FL(35,82). **14:**CA(82,1110). **17:**AZ(82), CA(82,286,1025,1110), FL(82), HI(82), LA(82).

P. drechsleri. **17:**CA(1580).

P. hibernalis. Damping-off. 'Brown rot (fruit); Phytophthora foot rot, gummosis, and root rot.' **17:**CA(1110).

P. megasperma. Root rot. **17:**CA(82,1110).

P. nicotianae var. *parasitica*. Shoot blight, damping-off, collar rot. 'Brown rot. Phytophthora foot rot. Phytophthora gummosis. Phytophthora root rot.' **1:**CA(1110), FL(35,82). **2:**FL(35,82). **3:**FL(35). **4:**CA(1110). **5:**CA(82,1110), FL(35,82). **8:**CA(1110), FL(35). **9:**FL(35). **10:**CA(1110). **12:**CA(82,1110), FL(35,82). **13:**CA(82,1110,2210), FL(35,82,3861), TX(82). **6:**FL(35). **11:**CA(82,1110), FL(35,82,3861). **14:**CA(1110). **17:**Gulf States(82), CA(82,286,1110), FL(82).

P. syringae. Leaf blight. 'Phytophthora foot rot. Phytophthora gummosis. Phytophthora root rot. Brown rot.' **1:**CA(1110). **4:**CA(1110). **5:**CA(1110). **8:**CA(1110). **10:**CA(1110). **12:**CA(82,1110). **13:**CA(82,1110). **16:**CA(82). **11:**CA(1110). **14:**CA(1110). **17:**CA(82,1110).

Phytophthora sp. **1:**HI(2903).

Pythium aphanidermatum. Damping-off. **1:**CA(1110). **4:**CA(1110). **5:**CA(829,1110). **8:**CA(1110). **10:**CA(1110). **12:**CA(1110). **13:**CA(1110). **11:**CA(1110). **14:**CA(1110). **17:**CA(1110).

P. debaryanum. Damping-off, soft brown rot of fruit, secondary root rot. **12:**CA(82). **17:**CA(82).

P. irregulare. Secondary root rot. **12:**CA(82).

P. rostratum. Damping-off. Rootlet rot. **17:**CA(82,1110).

P. ultimum. Damping-off, soft brown rot of fruit, root and rootlet rot. **3:**FL(35). **12:**CA(82). **13:**CA(82,1110). **17:**CA(82,1110).

P. vexans. Damping-off, secondary root rot. **12:**CA(82). **17:**Unk.(82), CA(1110).

Pythium sp. Damping-off, root rot. **1:**FL(35). **2:**FL(35). **3:**FL(35). **5:**FL(35). **8:**FL(35). **9:**FL(35). **12:**CA(82), FL(35), HI(2903). **13:**FL(35). **16:**CA(82). **6:**FL(35). **11:**FL(35).

Zygomycotina

Mucor paronychia. Fruit rot. **17:**CA(82,1110).

M. racemosus. Fruit rot. **17:**CA(82,1110).

Rhizopus stolonifer. Black rot of fruit. **17:**Unk.(82).

Ascomycotina

Ashbya gossypii. Dry rot of fruit, associated with insect punctures. **13:**Range of host(82), FL(204). **17:**Range of host(82).

Botryosphaeria dothidea. Stem end rot. 'Dothiorella gummosis.' **1:**CA(1110), FL(35). **4:**CA(1110). **5:**CA(1110), FL(35). **8:**CA(1110), FL(35). **10:**CA(1110). **12:**CA(1110). **13:**CA(1110), FL(35). **11:**CA(1110), FL(35). **14:**CA(1110).

B. obtusa. On limbs and twigs. **12:**Unk.(82). **17:**Gulf States(82), CA(82).

B. quercuum. Dieback, stem end rot. **1:**FL(35). **2:**FL(35). **3:**FL(35). **5:**FL(35). **8:**FL(35). **12:**FL(35). **13:**FL(35). **6:**FL(35). **11:**FL(35).

B. rhodina. 'Diplodia gummosis. Diplodia stem-end rot.' **1:**CA(1110). **2:**Unk.(82). **4:**CA(1110). **5:**CA(1110). **8:**CA(1110). **10:**CA(1110). **12:**CA(1110). **13:**CA(1110). **11:**Unk.(82), CA(1110). **14:**CA(1110).

B. ribis. Trunk canker, gummosis, stem end rot. **5:**CA(82). **13:**CA(82). **11:**CA(82), FL(82), TX(82).

Camarops polysperma. **17:**FL(2561).

Capnodium citri. Sooty mold, on leaves. 'Sooty mold.' **1:**CA(1110). **2:**CA(2434), MS(2720). **4:**CA(1110). **5:**CA(1110). **8:**CA(1110). **10:**CA(1110). **12:**CA(1110). **13:**CA(1110). **16:**MS(2720). **11:**CA(1110). **14:**CA(1110). **17:**Gulf States(82).

C. citricola. Sooty mold. **17:**Gulf States(82).

Capnodium sp. Sooty mold. **12:**HI(2903). **13:**HI(2903). **17:**Gulf States(82).

Chaetothyrium hawaiiense. On leaves. **8:**FL(35). **11:**FL(35,82).

Daldinia concentrica. Wood rot. **13:**FL(35). **11:**FL(35). **17:**Unk.(82).

Daldinia sp. Wood rot. **17:**Unk.(82).

Diaporthe citri. Shell-bark, decorticosis, gummosis, melanose, stem end rot. **1:**Range of host(82), FL(35), HI(2903). **2:**Gulf States(82), FL(35). **4:**FL(2257). **5:**CA(82), FL(35,82). **8:**FL(35). **12:**Range of host(82), FL(35), HI(2899,2903). **13:**Range of host(82), FL(35), HI(2899,2903). **6:**FL(35), HI(2899). **11:**Range of host(82), FL(35), HI(2903). **17:**Range of host(82).

D. rudis. 'Phomopsis stem-end rot. Melanose.' **1:**CA(1110). **4:**CA(1110). **5:**CA(1110). **8:**CA(1110). **10:**CA(1110). **12:**CA(1110). **13:**CA(1110). **11:**CA(1110). **14:**CA(1110).

Elsinoe fawcettii. Verrucosis, on fruits, leaves, stems and twigs. 'Scab.' **1:**FL(35,82,1745), HI(82,2903). **2:**AL(82), FL(35,82,1745), HI(82,2903), LA(82), MS(82,2720), TX(82). **3:**FL(35,3881). **5:**Gulf States(82), FL(35,1745), MS(2720), TX(1745). **8:**FL(35). **9:**FL(35). **12:**Gulf States(82), AL(1745), FL(35), HI(2899,2903), MS(2720), TX(1745). **13:**FL(35,82,3881), HI(2899,2903), MS(2720). **15:**FL(3881). **16:**GA(1438), MS(2720). **6:**FL(35), HI(2899,2903). **11:**Gulf states(82), FL(35,1745,3881), HI(2903), MS(2720). **17:**Gulf States(82).

Eutypella citricola. On dead twigs. **17:**FL(82).

Gibberella baccata. On twigs. **17:**CA(82,1110).

Glomerella cingulata. Withertip. 'Anthracnose. Postbloom fruit drop disease.' **1:**Range of host(82), CA(1110), FL(35). **2:**Range of host(82), FL(35), MS(2720). **3:**FL(35). **4:**CA(1110). **5:**Range of host(82), CA(1110), FL(35). **8:**CA(1110), FL(35). **10:**CA(1110). **12:**CA(1110), FL(35), MS(2720), TX(82). **13:**Range of host(82), CA(1110), FL(35). **16:**MS(2720). **6:**FL(35). **11:**Range of host(82), CA(1110), FL(35), MS(2720), OK(2872). **14:**CA(1110). **17:**Range of host(82), CA(1110).

Mycosphaerella citri. 'Greasy spot. Rind blotch.' **1:**FL(35). **2:**FL(35). **3:**FL(35). **4:**FL(2257). **5:**FL(35). **8:**FL(35). **12:**FL(35). **13:**FL(35). **6:**FL(35). **11:**FL(35). **17:**FL(3879).

M. horii. Leaf spot. **13:**FL(35). **11:**FL(35).

M. lageniformis. Leaf spot. **2:**CA(82). **17:**CA(82,1110).

Mycosphaerella sp. Greasy spot. **17:**CA(1110).

Myriangium citri. Parasitic on scale insects, also causes a fruit speck. Fruit speck. **17:**Gulf States(82), CA(82,1110).

M. duriaei. **17:**FL(2782).

Nectria cinnabarina. On limbs. **5:**CA(1110).

N. diploa. Parasitic on scale insects. **17:**Gulf States(82), FL(82).

N. flammea. Parasitic on scale insects. **17:**FL(82).

N. mauritiicola. On roots, associated with red root disease but not proved pathogenic. **17:**Unk.(82).

N. ochroleuca. **13:**LA(3085).

N. pseudotrichia. On bark. **17:**LA(3178).

Nematospora coryli. Dry rot of fruit, associated with insect punctures. **1:**CA(1110). **4:**CA(1110). **5:**CA(1110). **8:**CA(1110), FL(35). **10:**CA(1110). **12:**CA(82,1110), FL(35,82,3800). **13:**Range of host(82), CA(1110), FL(35,204). **11:**CA(82,1110), FL(35,82,204,3800). **14:**CA(1110). **17:**Range of host(82).

Nummularia hypophloea. **17:**FL(2779).

Physalospora abdita. On limbs and twigs. **17:**Gulf States(82), CA(82).

Pleospora herbarum. **17:**CA(82,1110).

Podonectria coccicola. **17:**FL(82,3041).

Rhytidhysteron rufulum. Twig blight. **5:**TX(82).

Rosellinia subiculata. **17:**FL(2782).

Schizothyrium pomi. Fly speck. **2:**FL(35). **12:**FL(35). **13:**FL(35). **11:**FL(35,82). **17:**Gulf States(82).

Sclerotinia sclerotiorum. Bark, cottony fruit rot. 'Sclerotinia twig blight. Sclerotinia fruit rot. Sclerotinia root rot.' **1:**CA(1110). **4:**CA(1110). **5:**CA(82,1110), TX(82). **8:**CA(1110). **10:**CA(1110). **12:**CA(1110). **13:**CA(82,1110). **11:**CA(82,1110). **14:**CA(1110). **17:**CA(82,1110), TX(82).

Ustulina deusta. Following foot rot. **17:**FL(82).

Xylaria polymorpha. On living and dead roots. **17**:FL(3018).

Basidiomycotina - Other

Amylosporus campbellii. **17**:AZ(1199,1202).
Antrodia sinuosa. Brown cubical rot. **5**:CA(82,1110). **17**:CA(82,1110).
Armillaria mellea. 'Shoestring root rot.' **1**:CA(1110). **2**:CA(82). **4**:CA(1110). **5**:CA(82,1110), TX(82). **8**:CA(1110). **10**:CA(1110). **12**:CA(1110). **13**:CA(82,1110). **11**:CA(1110). **14**:CA(1110). **17**:CA(1110).
A. tabescens. 'Clitocybe root rot.' **3**:FL(35). **5**:FL(35). **8**:FL(35). **12**:FL(35,82). **13**:FL(35,82). **11**:FL(35,82). **17**:FL(82).
Bjerkandera adusta. **17**:Unk.(82).
Ceriporia xylostromatoides. **5**:AZ(1202).
Coprinus atramentarius. Root rot. **17**:CA(82).
Coriolopsis gallica. Wood rot. **17**:CA(82,1110).
Corticium stevensii. Thread blight. **13**:FL(35). **11**:FL(35).
Dacryopinax spathularia. On wood. **17**:FL(128).
Ganoderma applanatum. On wood. 'Heart rot.' **1**:FL(35). **12**:FL(35). **13**:CA(1110), FL(35). **11**:FL(35,574).
G. brownii. **1**:CA(1110). **4**:CA(1110). **5**:CA(1110). **8**:CA(1110). **10**:CA(1110). **12**:CA(1110). **13**:CA(1110). **11**:CA(1110). **14**:CA(1110).
G. lucidum. White root and butt rot. 'Heart rot.' **1**:FL(35). **2**:FL(35). **3**:FL(35). **5**:FL(35). **8**:FL(35). **9**:FL(35). **12**:FL(35). **13**:FL(35). **6**:FL(35). **11**:FL(35). **17**:FL(82).
Hexagonia hydnoides. Wood rot. **17**:CA(82,1110).
Meruliopsis corium. **17**:FL(128).
Oxyporus latemarginatus. **17**:CA(82).
Pellicularia koleroga. Thread blight. **2**:FL(82). **12**:LA(3640). **17**:Unk.(82).
Peniophora malenconii subsp. *americana*. On wood. **5**:CA(542). **8**:CA(542). **13**:CA(542).
Phanerochaete chrysorhiza. **17**:AZ(1202).
P. flavido-alba. **13**:FL(128).
Phellinus gilvus. **17**:Unk.(82).
Phylloporia frutica. On wood. **17**:SE states(2681).
Rigidoporus microporus. **17**:Unk.(82).
Schizophyllum commune. Wood rot. **1**:CA(1110). **4**:CA(1110). **5**:CA(1110). **8**:CA(1110). **10**:CA(1110). **12**:CA(1110). **13**:CA(1110). **11**:CA(1110). **14**:CA(1110). **17**:Range of host(82), CA(1110).
Septobasidium pseudopedicellatum. Felt fungus, parasitic on scale insects. **2**:FL(35). **5**:FL(35). **13**:FL(35). **11**:FL(35). **17**:Gulf States(82).
Septobasidium sp. **17**:Gulf States(82).
Trametes hirsuta. **17**:Unk.(82).
T. versicolor. Wound rot. **17**:Unk.(82).
Wolfiporia cocos. On roots. **17**:FL(82).

Deuteromycotina - Hyphomycetes

Alternaria citri. Leaf rot, fruit rot. 'Alternaria stem-end rot. Black rot. Alternaria leaf spot. Alternaria fruit spot.' **1**:CA(1110), FL(35). **2**:Range of host(82), FL(35). **4**:CA(1110). **5**:CA(82,1110), FL(35,82). **8**:CA(1110). **10**:CA(1110). **12**:Range of host(82), CA(1110), FL(35,3880). **13**:Range of host(82), CA(1110), FL(35), TX(82). **11**:AZ(82), CA(1110), FL(35). **14**:CA(1110). **17**:Range of host(82).
Aspergillus alliaceus. Fruit rot. **17**:Unk.(82).
A. flavus. Fruit and root rot. 'Albinism.' **1**:CA(1110). **5**:CA(1110). **8**:CA(1110). **10**:CA(1110). **12**:CA(1110). **13**:CA(1110). **11**:CA(1110). **14**:CA(1110). **17**:Unk.(82), CA(82,2898).
A. foetidus. Dieback. **5**:CA(503). **7**:CA(503).
A. niger. Fruit rot. **17**:Unk.(82).
Botrytis cinerea. Gummosis 'Gray mold. Botrytis blossom blight. Botrytis twig blight.' **1**:CA(1110), FL(35). **2**:FL(35). **3**:FL(35). **4**:CA(1110). **5**:CA(82,1110), FL(35). **8**:CA(1110), FL(35). **9**:FL(35). **10**:CA(1110). **12**:CA(1110), FL(35). **13**:CA(82,1110), FL(35). **6**:FL(35). **11**:CA(82,1110), FL(35). **14**:CA(1110). **17**:Unk.(82).
Candida krusei. Fruit rot. **5**:CA(1457).
Cercospora gigantea. Tar spot. **12**:FL(35). **13**:FL(35). **11**:FL(35).
C. penzigii. Leaf spot. **12**:MS(82). **13**:FL(35), TX(82). **11**:TX(82). **17**:FL(82), TX(82).
Cercospora sp. Pink pitting. **11**:FL(3536). **17**:FL(82), TX(82).
Cladosporium citri. Scab. **17**:AL(285).
C. herbarum. Sooty mold associated with scale insect infestation. **5**:CA(2362). **8**:CA(2362). **11**:CA(2362). **17**:CA(2362).
C. herbarum var. *citricola*. Fruit blemish, leprosis. **2**:FL(35). **12**:FL(35). **13**:FL(35). **17**:Gulf States(82).
C. oxysporum. Leaf spot. **1**:FL(35,1080). **5**:FL(35). **12**:FL(35,1080). **13**:FL(35,1080). **11**:FL(35,1080).
Cladosporium sp. **13**:CA(1110).
Cylindrocladium citri. Fruit rot. **12**:FL(35). **13**:CA(82,1110), FL(35,82,1694).
Fusarium avenaceum. Fruit rot. **17**:CA(82,1110), TX(82).
F. coccophilum. Parasitic on scale insects. **17**:Gulf States(82).
F. equiseti. Parasitic on scale insects. **17**:Gulf States(82).
F. juruanum. Parasitic on scale insects. **17**:Gulf States(82).
F. lateritium. Stylar end decay, gummosis, collar and foot rot. **13**:FL(35). **17**:CA(82), FL(82).
F. oxysporum. Dieback, fruit rot. 'Fusarium wilt.' **1**:FL(3633).
F. oxysporum f. sp. *citri*. Dieback, wilt. **1**:FL(35,3632).
F. solani. Gummosis, collar and fruit rot. **1**:CA(1110), FL(35). **2**:FL(35). **4**:CA(1110). **5**:CA(1110), FL(35). **8**:CA(1110). **9**:FL(35). **10**:CA(1110). **12**:CA(1110), FL(35). **13**:CA(1110), FL(35). **11**:CA(1110), FL(35). **14**:CA(1110). **17**:CA(82), FL(82,3222).
Fusarium sp. Fruit and root rot. **13**:Unk.(82). **11**:AZ(82). **17**:AZ(82), CA(82), FL(82).
Geotrichum candidum. 'Sour rot.' **12**:CA(1110), FL(35). **13**:CA(1110), FL(35). **17**:CA(939).
G. citri-aurantii. Sour rot. **5**:FL(35). **12**:FL(35). **13**:Unk.(82), FL(35). **11**:Unk.(82), FL(35). **17**:Range of host(82).
Gliocladium roseum. Pink mold, fruit rot. **1**:FL(35). **2**:FL(35). **3**:FL(35). **5**:Range of host(82), FL(35). **8**:FL(35). **9**:FL(35). **12**:FL(35). **13**:FL(35). **6**:FL(35). **11**:FL(35). **17**:Range of host(82).
Oidium tingitaninum. Powdery mildew. **12**:CA(82,1110). **17**:CA(82).
Oidium sp. Powdery mildew. **17**:CA(82).
Oospora sp. Fruit rot. **12**:LA(1630). **13**:LA(1630).
Penicillium brevicompactum. On all citrus fruits. **17**:Range of host(82).
P. digitatum. 'Green mold.' **1**:Range of host(82), CA(1110), FL(35). **2**:FL(35). **3**:FL(35). **4**:CA(1110). **5**:Range of host(82), CA(1110), FL(35). **8**:CA(1110), FL(35). **9**:FL(35). **10**:CA(1110). **12**:CA(1110), FL(35). **13**:CA(1110), FL(35), HI(2903). **6**:FL(35). **11**:Range of host(82), CA(1110), FL(35), OK(2872). **14**:CA(1110). **17**:Range of host(82), CA(939).
P. italicum. 'Blue mold.' **1**:Range of host(82), CA(1110), FL(35), HI(2903). **2**:FL(35), HI(2903). **3**:FL(35). **4**:CA(1110). **5**:Range of host(82), CA(1110), FL(35). **8**:CA(1110), FL(35). **9**:FL(35). **10**:CA(1110). **12**:CA(1110), FL(35), HI(2903). **13**:CA(1110), FL(35), HI(2903). **6**:FL(35). **11**:Range of host(82), CA(1110), FL(35). **14**:CA(1110). **17**:Range of host(82), CA(939,1110).
Penicillium sp. Blue mold, fruit rot. **2**:Range of host(82). **12**:HI(2899). **17**:Range of host(82).
Phymatotrichopsis omnivora. Root rot. **1**:TX(82). **2**:TX(82). **5**:TX(82). **12**:TX(82). **13**:TX(82). **16**:TX(82). **11**:TX(82). **17**:TX(82).
Stemphylium citri. End rot of fruit. **17**:AZ(82).
Thielaviopsis basicola. 'Black root rot.' **1**:CA(1110). **4**:CA(1110). **5**:CA(1110), FL(35). **8**:CA(1110). **9**:FL(35). **10**:CA(1110). **12**:CA(1110). **13**:CA(1110). **11**:CA(1110). **14**:CA(1110). **17**:CA(1110,3688).
Trichoderma viride. Fruit rot. **17**:Unk.(82), HI(2784).
Verticillium lecanii. Parasitic on scale insects. **17**:FL(82).

Deuteromycotina - Coelomycetes

Ascochyta sp. Gummosis. **17**:CA(82,1110).
Botryodiplodia lecanidion. **17**:FL(2782).
Botryodiplodia sp. Dieback. **13**:FL(35).
Colletotrichum gloeosporioides. Anthracnose, dieback, fruit rot, withertip, on leaves. 'Lime anthracnose.' **1**:CA(82), FL(35,82), HI(82,2903). **2**:FL(35), HI(2903). **3**:FL(35). **6**:HI(2903). **17**:AL(285), FL(82), HI(2903).
Colletotrichum sp. **6**:HI(2903).
Dothiorella sp. Stem-end rot. **17**:Range of host(82).
Gloeodes pomigena. 'Sooty blotch.' **12**:FL(35). **13**:FL(35). **11**:FL(35,82). **17**:Gulf States(82).
Hendersonula toruloidea. Canker, dieback, gummosis, wilt. 'Hendersonula branch wilt.' **1**:CA(1110). **4**:CA(1110). **5**:CA(82,1110). **8**:CA(1110). **10**:CA(1110). **12**:CA(1110). **13**:CA(1110). **11**:CA(82,476,1110). **14**:CA(1110). **17**:CA(82).

Lasiodiplodia theobromae. Dieback, trunk lesions, melanose, charcoal rot, Diplodia rot, fruit and stem end rot. **1:**Range of host(82), FL(35). **2:**Range of host(82), FL(35), MS(2720). **3:**FL(35). **5:**Range of host(82), FL(35). **8:**FL(35). **12:**Range of host(82), FL(35). **13:**Range of host(82), FL(35,163). **6:**FL(35). **11:**Range of host(82), FL(35), Tx(814). **17:**Range of host(82), FL(403).
Macrophomina phaseolina. Charcoal root rot. **1:**CA(1110), FL(35). **3:**FL(35). **4:**CA(1110). **5:**AZ(82), CA(82,1110). **8:**CA(1110), FL(35). **10:**CA(1110). **12:**CA(1110). **13:**CA(1110). **11:**CA(1110), FL(35). **14:**CA(1110). **17:**AZ(82), CA(82).
Pestalotiopsis maculans. On twigs. **12:**MS(2720). **17:**Gulf States(82).
Phoma socia. **17:**Gulf States(82).
Phomopsis citri. Postharvest decay of fruits, melanose. **17:**AL(285), CA(939), FL(403).
Phyllosticta aurantiicola. On leaves. **17:**Gulf States(82).
P. citricarpa. On decaying leaves. **17:**FL(2782).
P. erratica. On leaves. **17:**Gulf States(82).
P. hesperidearum. Leaf spot. **5:**FL(82). **11:**FL(35). **17:**Gulf States(82).
P. longispora. On leaves. **17:**Gulf States(82).
Phyllosticta sp. On leaves. **17:**Gulf States(82).
Septoria citri. Tear-stain of fruit. 'Septoria spot.' **5:**CA(82,1110). **13:**CA(1110). **11:**CA(82,1110). **17:**CA(82), TX(82).
S. limonum. Fruit spot. **5:**CA(82).
Sirodothis citri. On leaves. **17:**FL(201).
Sphaceloma fawcettii. **17:**HI(2903).
Sphaeropsis tumefaciens. Branch knot, limb knot. **1:**FL(35,82). **2:**FL(35,82). **3:**FL(35). **8:**FL(35). **13:**FL(35,82). **11:**FL(35,82). **17:**FL(82), HI(2903).
Deuteromycotina - Other
Rhizoctonia solani. Root rot. 'Damping-off.' **1:**CA(1110), FL(35). **2:**FL(35). **3:**FL(35). **4:**CA(1110). **5:**CA(1110), FL(35). **8:**CA(1110), FL(35). **9:**FL(35). **10:**CA(1110). **12:**CA(1110), FL(35). **13:**CA(1110), FL(35). **6:**FL(35). **11:**CA(1110), FL(35). **14:**CA(1110). **17:**Range of host(82), CA(1110).
Sclerotium rolfsii. Seedling blight, fruit rot. **8:**FL(35). **11:**FL(35).

Clausena N. L. Burm.
1. *C. lansium* (Lour.) Skeels [wampi]
Ascomycotina
Elsinoe fawcettii. Scab. **1:**FL(35,82).
Glomerella cingulata. Leaf spot. **1:**MD(82).

Dictamnus L.
1. *D. albus* L. (=*D. fraxinellus* Pers.) [burning-bush, dittany, fraxinella, gas-plant]
Deuteromycotina - Coelomycetes
Phyllosticta dictamni. **1:**NY(64).

Flindersia R. Br. in Flinders
1. *F. brayleana* F. v. Muell.
Oomycetes
Phytophthora cinnamomi. Decline. **1:**HI(2903).
Ascomycotina
Calonectria sp. Collar rot. **1:**HI(2903).
Deuteromycotina - Other
Rhizoctonia solani. Wilt. **1:**HI(1934).
Rhizoctonia sp. Seedling blight, root rot. **1:**HI(2903).

Fortunella Swingle
1. *F. hindsii* (Champ. ex Benth.) Swingle [Hong Kong kumquat]
2. *F. margarita* (Lour.) Swingle [oval kumquat]
3. *Fortunella* sp.
Oomycetes
Phytophthora cactorum. **2:**CA(1110).
P. nicotianae var. *parasitica*. Foot rot. **2:**FL(35).
Phytophthora sp. **1:**CA(1110).
Pythium sp. Root rot. **2:**FL(35).
Ascomycotina
Botryosphaeria quercuum. Gummosis, stem end rot. **3:**FL(35).
B. rhodina. On branch. **2:**FL(82). **3:**FL(82).
Diaporthe citri. Melanose, stem end rot. **2:**CA(82), FL(35). **3:**FL(35), MS(2720).
D. rudis. **2:**CA(1110).
Elsinoe fawcettii. Scab. **2:**FL(35).
Mycosphaerella citri. Greasy spot. **2:**FL(35).
Basidiomycotina - Other
Dendrophora albobadia. On wood. **3:**MS(2720).
Deuteromycotina - Hyphomycetes
Alternaria citri. Black rot. **2:**CA(82,1110). **3:**CA(82).
Fusarium sp. Root rot. **2:**CA(1110), FL(35).
Oospora sp. Fruit rot. **3:**LA(1630).
Deuteromycotina - Coelomycetes
Colletotrichum gloeosporioides. Anthracnose. **2:**FL(35). **3:**FL(35).
Lasiodiplodia theobromae. Trunk rot. **2:**FL(35).
Phoma socia. On twigs. **2:**AL(82). **3:**AL(82).
Phyllosticta citricola. Leaf spot. **1:**MS(2720). **2:**MS(82,2720).
Sphaceloma fawcettii. Scab, verrucosis. **2:**FL(35).

Limonia L.
1. *L. acidissima* L. (=*Feronia limonia* (L.) Swingle) [elephant-apple, Indian woodapple, woodapple]
Basidiomycotina - Other
Ganoderma lucidum. On wood. **1:**FL(35).

Murraya J. König ex L.
1. *M. paniculata* (L.) W. Jack [China-box, cosmetic-bark-tree, orange jessamine, orange-jasmine, satinwood]
2. *Murraya* sp.
Oomycetes
Phytophthora nicotianae var. *parasitica*. Root and stem rot. **1:**FL(35).
Pythium splendens. Root rot. **1:**FL(35). **2:**FL(35).
Pythium sp. Root rot. **1:**HI(2903).
Ascomycotina
Mycosphaerella citri. Greasy spot. **1:**FL(35). **2:**FL(35).
Deuteromycotina - Hyphomycetes
Alternaria citri. Leaf spot. **1:**FL(35).
Fusarium solani. Root rot. **1:**FL(35).
Deuteromycotina - Coelomycetes
Colletotrichum sp. Anthracnose. **1:**FL(35). **2:**FL(35).
Phomopsis sp. Leaf spot. **1:**FL(35).
Phyllosticta sp. Leaf spot. **1:**FL(35).
Deuteromycotina - Other
Rhizoctonia solani. Damping-off of seedlings, root rot. **1:**FL(35). **2:**FL(35).

Pelea A. Gray
1. *P. anisata* H. Mann
2. *P. barbigera* (A. Gray) Hildebr.
3. *P. cinerea* (A. Gray) Hildebr.
4. *P. clusiifolia* A. Gray
5. *P. elliptica* (A. Gray) Hildebr.
6. *P. hawaiiensis* Wawra
7. *P. kauaiensis* St. John
8. *P. parvifolia* Hildebr.
9. *P. sandwicensis* (Hook. & Arnott) A. Gray
10. *Pelea* sp.
Ascomycotina
Meliola juddiana. Black mildew. **3:**HI(1272,2903). **4:**HI(1272,2903). **5:**HI(2903). **6:**HI(1272,2903). **8:**HI(1272,2903). **9:**HI(1272,2903). **10:**HI(1268,1272,1452,2903).
M. kisubiensis var. *peleicola*. Black mildew. **10:**HI(1272,1452,2903).
M. peleae. Black mildew. **2:**HI(1272,1452,2903). **3:**HI(1272,1452,2903). **5:**HI(1452,2903). **8:**HI(1452). **9:**HI(1272,1452,2903). **10:**HI(1268,1272,1452,2903).
M. recurvipoda. Black mildew. **6:**HI(1272,1452,2903).
Phragmocapnias smilacina. **10:**HI(2903).
Trichothyrium reptans. **7:**HI(2903). **10:**HI(2903).
Basidiomycotina - Uredinales
Kernella lauricola. **1:**HI(2903).
Deuteromycotina - Hyphomycetes
Trichothallus hawaiiensis. **10:**HI(2903).
Trichothecium roseum. Pink mold. **10:**HI(2903).
Deuteromycotina - Coelomycetes
Colletotrichum gloeosporioides. On galls. **10:**HI(2903).

Phellodendron Rupr.
1. *P. amurense* Rupr.
Deuteromycotina - Hyphomycetes
Cercospora sp. Leaf spot. **1:**OK(2872).

Verticillium albo-atrum. Wilt. **1**:IL(1558).

Platydesma H. Mann
1. *P. spathulata* (A. Gray) Stone
Deuteromycotina - Coelomycetes
Clypeoseptoria rockii. **1**:HI(2903).

Poncirus Raf.
1. *P. trifoliata* (L.) Raf. (=*Citrus trifoliata* L.) [hardy orange, trifoliate orange]
Oomycetes
Phytophthora citrophthora. Brown rot of fruit, foot rot. **1**:FL(35).
P. nicotianae var. *parasitica*. Collar and foot rot. **1**:FL(35).
Ascomycotina
Botryosphaeria disrupta. On twigs. **1**:FL(35).
B. dothidea. Twig blight. **1**:FL(35), GA(1438).
B. obtusa. On twigs. **1**:Gulf States(82), MS(2720).
B. quercuum. Twig dieback. **1**:FL(35).
Diaporthe citri. Melanose, stem end rot. **1**:Gulf States(82), FL(35), MS(2720).
Elsinoe fawcettii. Scab. **1**:FL(35), MS(82,2720).
Eutypella citricola. Twig dieback. **1**:FL(35).
Glomerella cingulata. Anthracnose, withertip. **1**:Gulf States(82), FL(35), MS(2720).
Mycosphaerella citri. Greasy spot. **1**:FL(35).
Rhytidhysteron rufulum. Twig blight. **1**:FL(35).
Schizothyrium pomi. Fly speck. **1**:FL(35).
Ustulina deusta. On wood. **1**:FL(35).
Xylaria polymorpha. Root rot. **1**:LA(82).
Basidiomycotina - Other
Armillaria tabescens. Root rot. **1**:FL(35).
Ganoderma lucidum. Trunk rot. **1**:FL(35).
Schizophyllum commune. On wood. **1**:FL(35).
Septobasidium pseudopedicellatum. Parasitic on scale insects. **1**:FL(35).
Wolfiporia cocos. Root rot. **1**:FL(35).
Deuteromycotina - Hyphomycetes
Cercospora penzigii. Leaf spot. **1**:FL(35).
Cladosporium herbarum var. *citricola*. Fruit blemish, leprosis. **1**:FL(35).
Fusarium solani. Root rot, wilt. **1**:FL(35).
Geotrichum citri-aurantii. Sour rot. **1**:FL(35).
Gliocladium roseum. Pink mold. **1**:FL(35).
Oidium tingitaninum. Powdery mildew. **1**:FL(35).
Penicillium digitatum. Green mold. **1**:FL(35).
P. italicum. Blue mold. **1**:FL(35).
Phymatotrichopsis omnivora. Root rot. **1**:TX(82).
Deuteromycotina - Coelomycetes
Botryodiplodia diplocarpa. Twig blight. **1**:FL(35).
Colletotrichum gloeosporioides. Anthracnose. **1**:FL(35).
Gloeodes pomigena. Sooty blotch. **1**:FL(35).
Lasiodiplodia theobromae. Dieback. **1**:Range of host(82), FL(35), MS(2720), OK(2872).
Pestalotiopsis maculans. On twigs. **1**:FL(35).
Phyllosticta hesperidearum. Leaf spot. **1**:FL(35).
Sphaeropsis tumefaciens. Limb knot. **1**:FL(35).
Deuteromycotina - Other
Rhizoctonia solani. Damping-off, root rot. **1**:FL(35).
Sclerotium rolfsii. Seedling blight, fruit rot. **1**:FL(35).

Ptelea L.
1. *P. angustifolia* Benth.
2. *P. trifoliata* L. [stinking-ash, water-ash]
Ascomycotina
Karstenula sp. **2**:GA(82,1438).
Mycosphaerella sp. On leaves. **2**:TX(82).
Basidiomycotina - Uredinales
[*Puccinia windsoriae*]. **2**:E states(82), GA(1439), KS(82), OK(2872), TX(82).
Basidiomycotina - Other
Dendrophora albobadia. **1**:AZ(1202).
Deuteromycotina - Hyphomycetes
Alternaria alternata. Leaf spot. **2**:OK(2872).
Cercospora pteleae. Leaf spot. **2**:IA(82), IL(1052,2567), IN(82), MO(82), OK(2872), TX(82), WI(1350).
Phymatotrichopsis omnivora. Root rot. **2**:TX(82).
Deuteromycotina - Coelomycetes
Phloeospora pteleae. Leaf spot. **2**:TX(82,3597).
Phyllosticta pteleicola. Leaf spot. **2**:IL(82).
Septoria pteleae. Leaf spot. **2**:IN(82), OK(2872), WI(82,1340).

Severinia Ten. & Endl.
1. *S. buxifolia* (Poir.) Ten. [Chinese box orange]
Oomycetes
Pythium sp. Root rot. **1**:FL(35).
Basidiomycotina - Other
Erythricium salmonicolor. Limb blight. **1**:FL(35).
Deuteromycotina - Hyphomycetes
Alternaria sp. Leaf spot. **1**:FL(35).
Fusarium sp. Root rot. **1**:FL(35).
Volutella sp. Stem canker. **1**:FL(35).
Deuteromycotina - Coelomycetes
Phyllosticta sp. Leaf spot. **1**:FL(35).

Triphasia Lour.
1. *T. trifolia* (J. Burm.) P. Wils. [limeberry]
Oomycetes
Pythium sp. Root rot. **1**:HI(2903).
Deuteromycotina - Coelomycetes
Phomopsis sp. Twig dieback, leaf spot. **1**:FL(35).
Phyllosticta sp. Leaf spot. **1**:FL(35).

Zanthoxylum L.
1. *Z. americanum* Mill. [northern pricklyash, toothache tree]
2. *Z. clava-herculis* L. [Hercules-club, pepperwood, sea-ash, southern pricklyash]
3. *Z. dipetalum* H. Mann
4. *Z. fagara* (L.) Sarg. [lime pricklyash, wild lime]
5. *Zanthoxylum* sp.
Ascomycotina
Dermea xanthoxyli. On dead limbs. **1**:NY(82).
Eutypella leprosa. **1**:IA(3629).
Lachnum sclerotii. On dead stems. **4**:FL(522).
Massaria xanthoxyli. **1**:ND(3258), NY(3258).
Phyllactinia guttata. Powdery mildew. **1**:Range of host(82), MS(2720). **4**:FL(35). **5**:FL(35).
Schizothyrium pomi. Fly speck. **4**:FL(35).
Thyronectria xanthoxyli. On stems. **1**:MI(82), NY(82,3173).
Basidiomycotina - Uredinales
Puccinia andropogonis. Heteroecious: aecial host. **1**:IA(82), IL(312), KS(82), MO(82), NE(82), OK(2872), TX(82), WI(82). **2**:FL(82), OK(2872), TX(82). **5**:FL(35).
Basidiomycotina - Other
Armillaria mellea. Root rot. **3**:HI(2011,2903).
Phellinus igniarius. White heart rot. **1**:Unk.(82).
Deuteromycotina - Hyphomycetes
Cercospora xanthoxyli. Leaf spot. **1**:IN(82), TX(82). **2**:FL(82), GA(82), TX(82). **4**:FL(35). **5**:FL(35).
Ovulariopsis farinosa. Powdery mildew. **4**:FL(35,82).
Periconia lateralis. On leaves. **2**:TX(82).
Sclerotiomyces colchicus. **1**:WI(1355).
Deuteromycotina - Coelomycetes
Discula xanthoxyli. On dead limbs and stems. **1**:LA(82). **5**:LA(966).
Lasiodiplodia theobromae. Stem canker. **1**:TX(82).
Macrophoma xanthoxyli. On limbs. **1**:LA(82). **5**:LA(965).
Phyllosticta zanthoxyli. On living leaves. **1**:WI(1339).
Phyllosticta sp. Leaf spot. **4**:FL(35). **5**:FL(35).
Septoria pachyspora. Leaf spot. **1**:IA(82), NE(82), SD(2243), WI(1358). **2**:TX(82).

Salicaceae

Populus L.
1. *P.* × *acuminata* Rydb. *P. angustifolia* × *P. deltoides* subsp. *monilifera*.
2. *P.* × *alba* L. [abele, silver-leaved poplar, white poplar]
3. *P. alba* L. cv. 'pyramidalis' (=*P. alba* L. var. *pyramidalis* Bunge)
4. *P. angustifolia* E. James ex Torr.
5. *P. balsamifera* L. (=*P. tacamahaca* Mill., *P. candicans* Aiton) [balsam poplar, hackmatack, tacamahac]

6. *P.* × *berolinensis* Dippel. *P. laurifolia* × *P. nigra* cv. 'Italica'.
7. *P.* × *canadensis* Moench. *Populus deltoides* × *P. nigra*.
8. *P. deltoides* J. Bartram ex Marsh. (=*P. angulata* Aiton, *P. deltoides* J. Bartram ex Marsh. var. *virginiana* (Foug.) Sudw.) [eastern cottonwood, necklace poplar, southern cottonwood]
9. *P. deltoides* J. Bartram ex Marsh. subsp. *monilifera* (Aiton) J. E. Eckenwalder (=*P. monilifera* Aiton, *P. sargentii* Dode, *P. occidentalis* (Rydb.) Britton) [plains cottonwood]
10. *P.* × *euramericana* (Dode) Guinier
11. *P. fremontii* S. Wats. [Fremont cottonwood]
12. *P. grandidentata* Michx. [bigtooth aspen, large-toothed aspen]
13. *P. heterophylla* L. [black cottonwood, downy poplar, swamp cottonwood]
14. *P. nigra* L. [black poplar]
15. *P. nigra* L. cv. 'italica' (=*P. nigra* L. var. *italica* Münchh.) [black poplar, Lombardy poplar]
16. *P. nigra* L. var. *betulifolia* (Pursh) Torr.
17. *P. petrowskiana* (Regel) C. K. Schneid. *P. deltoides* × *P. laurifolia*.
18. *P. simonii* Carrière [Simon poplar]
19. *P. sp. (aspen)* . Entries under this name apply to taxa listed in Lindsey & Gilbertson (Basidiomycetes that decay aspen in North America. 1978). In this work, the term "aspen" applies to *P. tremuloides* and *P. grandidentata* and their hybrids.
20. *P. tomentosa* Carrière [Chinese white poplar]
21. *P. tremuloides* Michx. [quaking aspen, quiverleaf, trembling aspen]
22. *P. tremuloides* Michx. var. *aurea* (Tidestr.) Dandy
23. *P. trichocarpa* Torr. & A. Gray [black cottonwood, Western balsam poplar]
24. *Populus* × sp. (unnamed hybrids)
25. *Populus* sp.

Chytridiomycetes

Blastocladia ramosa. **23:**MT(3202).
Synchytrium sp. On leaves and petioles. **8:**LA(1847).

Oomycetes

Phytophthora sp. Root rot. **8:**FL(35).
Pythium sp. Root rot. **25:**HI(2903).

Ascomycotina

Apioplagiostoma populi. On overwintered leaves. **24:** MA(529).
Ascocoryne sp. Isolated from roots. **21:**WY(3040).
Belonium aggregatum. On wood. **21:**CO(521).
Belonium sp. **21:**OR(3202).
Berlesiella sp. **23:**CA(661).
Bertia moriformis. On wood. **21:**ID(3202). **22:**ID(684). **23:**WA(3202,3815). **25:**VT(684).
Bertiella rhodospila. On wood. **25:**MI(195).
Bisporella citrina. On rotting wood. **21:**CA(661), WA(3202). **23:**AK(3431), ID(3202).
Botryosphaeria dothidea. Canker. **8:**GA(1438). **14:**GA(1438). **15:**NC(1303).
B. obtusa. On limbs. **8:**GA(82), KS(82). **2:**NY(82), VA(82).
B. ribis. On limbs. **14:**GA(82,2405). **7:**AR(82).
Botryosphaeria sp. Blight. **25:**NC(1303).
Byssosphaeria rhodomphala. On decaying log. **25:**LA(3085).
Caliciopsis calicioides. **23:**ID(3202), MT(3202), WA(3202). **25:**ID(3202).
Capitotricha bicolor. On dead twigs. **21:**CA(661), OR(3202), WY(2924).
Capnodium salicinum. **21:**MT(3202).
Ceratocystiopsis alba. Canker. **21:**AK(1567), CO(3714), MN(1562).
C. retusi. In wood, associated with *Typodendronis retuso* (Coleoptera). **21:**CO(1564,3714).
Ceratocystis cana. Canker. **21:**MN(2240).
C. fimbriata. Canker. **21:**W states(1568), AK(1567), CO(1822,3714), MN(2240,3951), NM(3714), NY(2241), PA(3950).
C. moniliformis. Sapwood stain. **21:**CO(3714), MN(484).
[*C. serpens*]. Canker. **21:**MN(2240).
Ceratocystis sp. On wood. **21:**CA(661).
Chaetomium murorum. **25:**MT(3202).
Chaetoplea oblongispora. On dead twigs. **4:**NM(1003).
Chlorociboria aeruginosa. **25:**AK(526), ID(3202).
Chlorociboria sp. On wood. **21:**CA(661).
Ciboria caucus. **25:**OR(3202).
Ciborinia davidsoniana. On overwintered leaves. **21:**CO(1383).
C. pseudobifrons. On fallen leaves and twigs. **21:**NY(1383).
C. seaveri. Ink spot, black leaf spot. **5:**CO(82). **8:**CO(82). **21:**W states(82), CO(1383), ID(82,3202), MT(3202), OR(82), WA(82), WY(1195,3378). **25:**CO(82).
C. whetzelii. Anthracnose, ink spot. **14:**MA(82). **21:**MI(82), MN(82), NY(82), VT(82), WI(82), WY(2924). **25:**E & C states(82).
[*Colpoma crispum*]. On wood. **23:**CA(661).
Coniochaeta ligniaria. **23:**ID(3202).
C. pulveracea. **23:**CA(661).
Cryptodiaporthe salicella. Ink spot, on limbs. **8:**NY(82). **25:**MI(82), NY(82).
Cryptodiscus foveolaris. On wood. **25:**NY(3225).
Cryptosphaeria lignyota. Canker. **8:**IA(3629), ND(377), SD(2243). **12:**IA(3629). **15:**IA(3629). **21:**AK(1567), CO(1561,1822), IA(3629), MT(3202), UT(1561), WA(3202). **2:**IA(3629). **25:**Range of host(82), IA(3629), OR(3202).
C. pullmanensis. On fallen limbs. **23:**WA(1244).
Cucurbitaria staphula. Rough bark. **21:**AK(1567).
Dactylospora stygia. **23:**WA(3202).
Daldinia occidentalis. On lower trunk of living trees. **21:**AZ(1003). **23:**ID(3202), MT(3202).
D. vernicosa. **25:**ND(377).
Dasyscyphus populicolus. On stems. **21:**CO(1418), UT(1418), WA(1418).
Dermatella populina. **21:**ID(3202).
Dermea sp. **21:**MT(3202).
Diatrype albopruinosa. **25:**IA(3629).
D. bullata. **23:**AK(526).
D. flavovirens. **23:**OR(3202).
D. tumida f. *populi*. **21:**ND(377).
D. virescens. **25:**MN(1248).
Diatrypella favacea. **21:**IA(3629).
D. nitschkei. **8:**IA(3629).
D. populi. **8:**IA(3629). **14:**IA(3629). **25:**IA(3629).
D. quercina. **8:**GA(1438).
Dimerosporium populi. **21:**MT(3202). **25:**NJ(82).
Diplonaevia melaleuca. **25:**MT(3202).
Discohainesia oenotherae. On leaves. **14:**VA(82).
Drepanopeziza populi-albae. On leaves. **2:**CT(82), GA(82,1438), IA(82), MI(82), NY(82). **25:**IA(82).
D. populorum. **5:**ID(3202). **21:**ID(3202), OR(3202), WA(3202). **23:**WA(3202). **2:**OR(3202), WA(3202).
Encoelia fascicularis. On wood. **5:**AK(1567). **21:**CA(81), NY(1960). **25:**Range of host(82).
E. pruinosa. Sooty bark canker. **5:**AK(1567). **23:**ID(3202). **25:**NY(1960).
Erysiphe cichoracearum. Powdery mildew. **21:**UT(82).
E. polygoni. Powdery mildew. **11:**CA(1155).
Eutypa lata. **21:**IA(3629). **25:**IA(3629).
Eutypella leprosa. **8:**IA(3629). **25:**IA(3629).
E. populi. **23:**CA(2439).
E. stellulata. On bark. **4:**NM(1003).
Glyphium corrugatum. **21:**ID(3202), OR(3202), WA(3202). **25:**ID(3202), MT(3202), WA(3202).
G. elatum. On trees. **23:**CA(661).
G. schizosporum. **21:**MT(3202).
Hansenula polymorpha. In frass. **21:**MN(3596).
Herpotrichiella sp. **23:**CA(661).
Humaria lusatiae. **25:**WA(3202).
Hymenoscyphus conscriptus. **23:**WA(3202).
H. salicellus. **23:**WA(3202).
Hyponectria populi. On overwintered leaves. **5:**NE states(183). **21:**NE states(183).
Hypoxylon howeianum. **25:**NY(82).
H. mammatum. Limb and trunk canker. **5:**NE states(82), MI(82), WI(1350). **12:**ME(82), MI(82), MN(82), NY(82), WI(82,1350). **21:**Lake states(82), NE states(82), AZ(2980), CO(1560,1822,2980), IA(82), MI(58,2862), MN(56,58), ND(58), NM(2980), NY(544,2241), WI(58,68,1335,1350), WY(1195,2980). **25:**Lake states(82), NE states(82), IA(82,1225).
H. mediterraneum. **23:**MT(3202), WA(3202).
H. rubiginosum. **14:**GA(82,1438). **25:**Unk.(82).
H. serpens. **21:**OR(3202). **23:**ID(82,3202).
H. serpens var. *macrospora*. **5:**ID(2786).
Hypoxylon sp. Canker. **8:**GA(82), OK(618). **21:**CO(796).

Hysterium angustatum. **21:**OR(3202).
Hysterographium sp. **21:**ID(3202). **25:**WA(3202).
Jattaea microtheca. On limbs. **25:**ME(190).
Karschia melaspileoides. **23:**WA(3202).
Kirschsteiniothelia aethiops. **21:**CA(661).
Lachnum virgineum. On twigs. **21:**CA(661).
Lahmia kunzei. Decline. **24:** On *P. tremuloides* × *P. tremula*. WI(2666).
Lasiosphaeria canescens. **23:**WA(3202,3815).
L. caudata. On wood. **21:**CA(661).
Leptosphaeria sp. On leaves. **8:**WI(1344). **2:**WI(1352).
Leucostoma nivea. Canker on limbs and twigs. **5:**Range of host(82). **8:**Range of host(82), FL(35). **12:**Unk.(82), IA(1226). **14:**Range of host(82), FL(35). **21:**Range of host(82), CO(1195,3378), IA(1226), MT(3202). **2:**Range of host(82). **25:**Unk.(82), FL(35), IA(1226), WA(3202).
L. persoonii. On dead limbs. **21:**MI(2862). **25:**Range of host(82), Unk.(82).
L. translucens. On bark. **21:**NM(1003). **25:**IA(1226).
Linospora tetraspora. Leaf blight. **5:**AK(82), WI(82). **23:**AK(526). **25:**AK(526).
Linospora sp. **21:**ID(3202).
Lojkania separans. On limbs. **4:**CO(1195).
Lophiostoma alpigenum. **25:**MT(3202).
L. caulium. **25:**MT(3202).
L. macrostomoides. **25:**ND(377).
L. quadrinucleatum. **25:**ID(3202).
Macrodiaporthe occulta. On limbs. **8:**MI(82). **21:**MI(82), NY(82).
Melanconis apocrypta. On limbs. **25:**IA(82).
Melanomma pulvis-pyrius. **25:**OR(3202).
Mollisia cinerea. **21:**CA(661). **23:**ID(3202).
Mollisia sp. **21:**WA(3202).
Mollisina acerina. On fallen leaves. **21:**CO(2181).
Mycosphaerella populi. **5:**AK(526). **23:**AK(526).
M. populicola. Leaf spot. **5:**NC states(82), NE states(82), ID(82), NE(82). **23:**NC states(82), NE states(82), ID(82), NE(82). **25:**NC states(82), NE states(82), ID(82,3202), NE(82).
M. populifolia. On leaves. **5:**NY(82). **8:**SC(82). **23:**ID(3202).
M. populnea. On leaves. **5:**NY(3081). **25:**Unk.(82).
M. populorum. Limb and twig canker, leaf spot. **4:**MT(3202), OR(3202). **5:**Range of host(82), ID(3202). **8:**E states(82), GA(1438), TX(82), WI(82). **9:**CA(82), UT(82). **12:**IN(82), NY(82), WI(82). **14:**GA(82), NY(82). **21:**NJ(82), NY(82), OH(82), WA(3202), WI(82). **23:**ID(3202), MT(3202), OR(3202), WA(3202). **7:**NY(82). **10:**IA(2610). **25:**E & C states(82), W states(82), IA(2189), MT(3202).
M. punctiformis. On fallen leaves. **14:**GA(82,1438). **20:**GA(1438). **2:**GA(82,1438).
Nectria cinnabarina. Canker. **8:**OK(2872). **23:**ID(3202). **25:**Unk.(82).
N. coccinea. On dead limbs. **23:**CA(81).
N. coryli. On dead limbs. **25:**NY(891).
N. galligena. Canker. Canker. **12:**NE states(82), MN(2240). **21:**NE states(82), MI(82), MN(82,2240), NH(82), NY(2241). **25:**Lake states(82), NE states(82).
N. peziza. **25:**CA(3085), NH(3085).
N. sulphurata. **21:**MT(3202).
N. verrucosa. On limbs and twigs. **8:**ND(377). **25:**Unk.(82).
Nectria sp. Canker. **8:**MN(82), ND(82).
Neobulgaria pura. **23:**WA(3202).
Odontotrema majusculum. On wood. **9:**MT(3232).
Ophiostoma brevicolla. **21:**CO(3714). **19:**CO(784).
O. crassivaginatum. Canker. **21:**AK(1567), MN(1562).
O. piceae. Sapwood stain. **21:**MN(484).
O. piliferum. Canker. **21:**MN(2240).
O. pluriannulatum. Sapwood stain. **21:**MN(484). **25:**WI(1005).
O. populinum. Branch and stem canker. **21:**CO(1563,3714), MN(1562).
O. tremulo-aureum. Canker. **21:**CO(798,3714).
Perrotia populina. On wood. **21:**CO(849,1418), MT(3202), OR(3202), UT(1418).
Pezicula ocellata. On limbs. **21:**NY(82).
Phaeotrype sp. **25:**OK(2872).
Phyllactinia guttata. **4:**NM(1195).
Pleuroceras populi. **12:**WI(1349). **21:**WI(1349).
Propolis faginea var. *atra*. On weathered wood. **21:**CO(521).
Pycnopeziza sympodialis. On buds and catkins. **21:**NY(35).
Rhytidiella baranyayi. Rough bark. **21:**AK(1567), CA(661).
Rosellinia ovalis. **23:**CA(661).
R. riminicola. **21:**WA(3202). **23:**WA(3202).
R. subcompressa. **8:**SD(2243).
Schizoxylon berkeleyanum. **25:**ND(377).
S. compositum. **25:**KS(3225), NY(3225).
Sclerotinia sp. **21:**WA(3202).
Scutellinia erinaceus. **23:**WA(3202).
S. hirta. **23:**ID(3202).
Septotinia populiperda. Leaf blotch. **11:**MD(82,3782). **23:**MD(3782). **7:**CT(82), MA(82), MD(82), ME(82), NY(82), PA(82), VT(82). **24:** CT(3782), MA(3782), MD(3782), ME(3782), NY(3782), PA(3782), VT(3782).
Stictis bicolor. **25:**CO(3225).
S. radiata. **23:**WA(3202).
S. sphaeroboloidea. **25:**NY(3225).
Sydowia dothideoides. Canker. **21:**AK(1569), CO(82,1569), NM(1569).
Talaromyces flavus. Isolated from roots. **21:**WY(3040).
Taphrina johansonii. Catkin deformity. **8:**VT(82). **9:**Unk.(82). **12:**MA(82,2454), NY(82), WI(82). **21:**IA(82), MA(82,2454), MI(82,2454), NY(82,2454,3361), WA(2457,3202), WI(82,2454). **2:**NY(82). **25:**Range of host(82).
T. populi-salicis. Yellow leaf spot. **11:**Unk.(82), CA(1967,2454). **23:**AK(3431), CA(81,82,1967,2454), ID(3202), MT(3202), OR(82,657,2454,3202), WA(3202).
T. populina. Yellow leaf blister. **8:**AL(2457), LA(2457), SC(82,2457). **14:**IA(82,2454), OR(3202), WA(82,2454,3202), WV(82,2454). **15:**IA(82,2454), OR(82,2454), VA(2989), VT(82,2454). **16:**IA(2457). **17:**IA(2454). **23:**CA(81). **1:**OR(2457). **6:**IA(2454). **25:**Lake states(82), NE states(82), W states(82).
T. rhizophora. **2:**NY(2457), WI(2457).
Teichospora fulgurata. **8:**ND(377).
T. helenae. **25:**ND(377).
T. nautica. **25:**MT(3202).
T. patellarioides. **21:**WA(3202).
T. populi. On dead limbs. **4:**NM(1003).
Trematosphaeria pertusa. On roots and wood. **8:**KS(318). **25:**KS(318).
Trichopezizella relicina var. *macrospora*. On dead twigs. **21:**CA(661).
Tryblidaria fenestrata. **21:**ND(377).
Tryblidiella clavispora. **25:**ND(377).
Tympanis spermatiospora. On dead limbs. **21:**CA(661), ID(3202). **25:**ID(3202).
Uncinula adunca. Powdery mildew. **4:**MT(3202). **5:**Range of host(82), CO(1195), MT(3202), WA(3202), WY(1195). **8:**FL(35), IL(82), MO(82), SC(82). **12:**Range of host(82). **13:**IA(82), TX(82). **21:**Range of host(82), ID(3202), MS(2720), MT(3202), SD(2243), WA(3202), WY(1195). **23:**AK(526), CA(81), ID(3202), OR(3202), WA(3202). **25:**Range of host(82), FL(35).
Valsa ambiens. On dead limbs. **21:**WA(3202). **2:**GA(82,1438). **25:**Range of host(82).
V. ceratosperma. **9:**MO(3415).
V. sordida. Canker, dieback. **4:**MT(3202), WA(3202). **5:**Range of host(82), AK(526), ID(3202). **8:**Range of host(82), FL(35), MT(3202). **9:**Range of host(82). **12:**Unk.(82). **13:**TX(82). **14:**Range of host(82), FL(35), WA(3202). **21:**IA(1226), ID(3202), MT(3202), WA(3202). **23:**ID(3202), MT(3202), WA(3202). **1:**MT(3202). **2:**Range of host(82), WA(3202). **7:**Range of host(82). **25:**Range of host(82), FL(35), IA(1226,3415), OR(3202).
Valsa sp. Canker. **25:**ID(3202), OK(2872).
Venturia macularis. Leaf, shoot and twig blight, parasitic or occurring on fallen leaves. **8:**IA(82), KS(82), OK(618). **12:**NY(82). **21:**ID(3202), MN(62), MT(3202), NY(82), WA(3202).
V. populina. Shoot blight, leaf spot. **5:**AK(1567), WA(3202). **12:**WI(82). **15:**KY(82), MN(82). **21:**NC states(82), NE states(82), OR(82), WA(82). **23:**OR(3202). **2:**ME(82). **25:**NC states(82), NE states(82).
Xylaria digitata. Root rot. **5:**ID(82). **23:**ID(3202).
Xylaria sp. Isolated from wood. **8:**GA(1007).

Basidiomycotina - Uredinales

Melampsora abietis-canadensis. Heteroecious: telial host. **5:**NC states(82), NE states(82). **8:**IN(82), ND(82). **9:**NC states(82), W states(82), OK(2872). **12:**NC states(82), NE states(82),

VA(3493), WI(82). **14:**Unk.(82). **21:**NC states(82), NE states(82). **2:**PA(82). **25:**NC states(82), NE states(82), W states(82).
M. cumminsii. Telial host. **25:**NH(136).
M. medusae. Heteroecious: telial host. **4:**CO(1195), OK(2872), OR(3202), WY(1195). **8:**Range of host(82), FL(35), LA(2675), MS(82,2720), MT(3202), NC(1303), ND(378), OK(2872), SD(2243), WI(3133). **9:**C states(82), W states(82), CO(82), MT(82), ND(378), SD(2243). **12:**WI(82). **14:**MA(82), MO(82), PA(82). **18:**WI(1358). **21:**Range of host(82), W states(82), AK(60,82,526,1904), AZ(1195,1203,3379,3981), CA(82,642), CO(1195), ID(3202), MT(3202), ND(378), OR(3202), WA(3202), WI(82), WY(1195,2924). **23:**CA(81). **10:**WI(3133). **24:** On *P. balsamifera* x *P. tristis* and *P. deltoides* x *P. balsamifera*. WI(3133). **25:**Range of host(82), W states(82), AK(82), CA(82), IA(3891,3892), MN(3891,3892), OK(2872), WI(3891,3892), WY(1195).
M. occidentalis. Heteroecious: telial host. **4:**MT(1195,3202), WA(3202). **5:**NC states(82), W states(82), ID(3202), MT(3202), WA(82,3202). **8:**ID(82). **9:**CA(82). **11:**MT(3202). **14:**CA(82). **23:**AK(3431), CA(81), ID(3202), MT(3202), OR(3202), WA(3202). **1:**MT(3202). **2:**CA(81,82). **25:**W states(82), WA(82).
M. populnea. Telial host. **2:**W states(82), CA(111), CO(82,111), OR(111), RI(82,111), WA(111).
Melampsora sp. **25:**MS(2720).

Basidiomycotina - Other

Acanthophysium lapponicum. **19:**Unk.(2142).
[*A. lividocaeruleum*]. **25:**CO(128).
Amphinema byssoides. **21:**AZ(1202), NM(1190), NY(128). **25:**CO(128). **19:**Unk.(2142).
[*Anomoporia albolutescens*]. **23:**ID(3202).
[*A. bombycina*]. **19:**Unk.(2142).
Antrodia heteromorpha. **1:**MT(3202).
A. malicola. **21:**MT(3202,3832). **19:**Unk.(2142).
A. serialis. **21:**AZ(1202). **25:**AZ(128). **19:**Unk.(2142).
A. variiformis. On wood. **23:**AK(526,1904).
A. xantha. **21:**MN(128). **23:**MT(3202). **19:**Unk.(2142).
Antrodiella romellii. **19:**Unk.(2142).
Aporpium caryae. On wood. **25:**ID(3202), NH(3592). **19:**Unk.(2142).
Armillaria mellea. Root rot. **21:**AK(1567), CO(1195), WA(3202), WY(3040). **23:**WA(3202). **25:**Unk.(82), CA(82). **19:**Unk.(2142).
Athelia bombacina. On wood. **21:**CA(661).
A. coprophila. **21:**MN(1197).
A. epiphylla. **21:**AZ(1202). **19:**Unk.(2142).
A. fibulata. **19:**Unk.(2142).
A. neuhoffii. **21:**AZ(1202).
Athelia sp. On wood. **21:**CA(661).
Basidiodendron eyrei. On wood. **11:**AZ(1202). **23:**AK(128). **25:**NY(2185). **19:**Unk.(2142).
B. grandinioides. **21:**AZ(1202).
Bjerkandera adusta. White rot. **4:**SD(378). **5:**AK(526). **8:**Range of host(82), GA(1007), ND(378). **12:**Lake states(82), NE states(82), MI(2862). **14:**ID(3202), WA(3202). **23:**CA(661), ID(3202), MT(3202,3832), OR(3202), WA(3202). **1:**ID(3202), OR(3202). **25:**Range of host(82). **19:**Unk.(2142).
B. fumosa. Wood rot. **21:**MT(3202,3832). **23:**ID(3202), MT(3202,3832), OR(3202), WA(3202). **1:**ID(3202). **25:**Range of host(82).
Botryobasidium flavescens. **19:**Unk.(2142).
B. pruinatum. **21:**MN(128,1197).
B. vagum. **21:**MN(128,1197,2142).
Botryohypochnus isabellinus. **21:**AZ(1202). **19:**Unk.(2142).
Bourdotia galzinii. **21:**AZ(1202). **19:**Unk.(2142).
Byssoporia terrestris. **19:**NY(128).
Calathella eruciformis. On fallen limbs. **21:**MI(2945), UT(2945). **23:**ID(2945,3202).
Calocera cornea. On wood. **21:**AZ(1202). **25:**MI(2862). **19:**Unk.(2142).
Candelabrochaete septocystidia. White rot. **5:**MN(432,1197). **25:**MN(432).
Ceraceomyces borealis. **21:**MN(128,1197). **19:**Unk.(2142).
C. serpens. **21:**AZ(128,1202). **19:**Unk.(2142).
Ceriporia purpurea. **11:**AZ(1202). **21:**NM(128,1190). **19:**Unk.(2142).
C. reticulata. **21:**AZ(128,1202,2167), CO(2167), NM(1190). **23:**WA(2167,3202). **19:**Unk.(2142).
C. spissa. **21:**MN(128). **19:**Unk.(2142).
C. tarda. **21:**MN(128). **19:**Unk.(2142).
C. viridans. **21:**AZ(1202), NM(1190). **25:**AZ(128), NM(128), WA(128). **19:**Unk.(2142).
C. xylostromatoides. **11:**AZ(128,1202).
Ceriporiopsis aneirina. **5:**AK(526). **21:**AZ(128,1202), CO(2167), ID(3202), OR(3202). **23:**ID(128,3202), WA(3202). **25:**CO(128), MT(128). **19:**Unk.(2142).
C. pannocincta. **21:**MN(128), NM(128,1190), WA(3202). **23:**ID(3202). **25:**CO(128), MT(128,3202), NY(128). **19:**Unk.(2142).
Cerocorticium confluens. On wood. **21:**AZ(1202), NY(544).
Cerrena unicolor. **21:**ID(3202). **25:**Unk.(82), MT(128). **19:**Unk.(2142).
Chondrostereum purpureum. Wood rot. **5:**AK(526), ID(3202). **8:**GA(1007). **12:**CT(543), MA(543), ME(543), NH(543). **14:**WA(3202). **21:**ME(543), MT(543). **23:**AK(1904), OR(543). **25:**Range of host(82), ME(543), NH(543), NY(543). **19:**Unk.(2142).
Clavicorona divaricata. On logs. **8:**ID(2073). **25:**ID(3202).
C. pyxidata. **21:**MN(1197), NM(884). **25:**ME(884), MI(884), WY(884). **19:**Unk.(2142).
Climacodon pulcherrimus. Trunk rot. **25:**WI(82).
C. septentrionalis. **19:**Unk.(2142).
Clitocybe sp. **25:**WA(3202).
Confertobasidium olivaceo-album. On wood. **21:**AZ(1178,1202), MN(1197). **19:**Unk.(2142).
Coniophora arida. **21:**AZ(1202). **19:**Unk.(2142).
C. arida var. *suffocata*. **21:**AZ(128).
C. olivacea. **21:**AZ(1202), NM(128,1190). **19:**Unk.(2142).
C. puteana. Brown rot. **21:**Range of host(82). **19:**Unk.(2142).
Conohypha terricola. **21:**NM(1190). **19:**Unk.(2142).
Coprinus atramentarius. At base of living trees, isolated from roots. **21:**CO(2450), WY(3040). **19:**Unk.(2142).
C. lagopus var. *sphaerosporus*. Isolated from wood. **8:**GA(1007).
C. micaceus. **19:**Unk.(2142).
C. radians. Isolated from wood. **8:**GA(1007).
Coprinus sp. **8:**OK(618).
Coriolopsis byrsina. On wood. **25:**SE states(2681).
C. gallica. White rot. **4:**MT(3202), ND(378). **5:**Range of host(82). **8:**Unk.(82), MT(3202), ND(378). **11:**AZ(1202), CA(81). **13:**AL(82). **14:**ID(3202), WA(3202). **21:**Range of host(82), CO(1195), OR(3202). **23:**ID(3202), MT(3202), OR(3202), WA(3202). **1:**MT(3202). **25:**Range of host(82), W states(82), ID(128), MT(3832). **19:**Unk.(2142).
Corticium lombardiae. On wood. **25:**WI(2044). **19:**Unk.(2142).
C. roseum. On dead, attached limbs. **21:**AZ(128,1202), NM(1190), NY(544), OR(128), WA(3202). **19:**Unk.(2142).
Crepidotus cinnabarinus. On wood. **25:**MI(2192).
C. coloradensis. On wood. **25:**CO(1544), MI(1544), NM(1544).
C. contortus. Decaying twigs. **25:**WA(1544).
C. flexuosus. On wood. **25:**MI(1544).
C. mollis. **21:**AZ(1202). **19:**Unk.(2142).
C. nephrodes. **25:**WA(3202).
C. occidentalis. **19:**Unk.(2142).
C. subverrucisporus var. *megalosporus*. On wood. **25:**CO(1544).
C. versutus. **25:**WA(3202).
Cristinia helvetica. **19:**Unk.(2142).
Crustoderma dryinum. **21:**NM(1190). **19:**Unk.(2142).
Cylindrobasidium corrugum. On wood. **21:**AZ(1178,1202). **19:**Unk.(2142).
C. laeve. On dead, attached limbs. **12:**NH(128). **21:**AZ(2141), NY(544). **19:**Unk.(2142).
C. torrendii. **23:**WA(3202).
Cyphellopsis anomala. On dead, attached limbs. **21:**AZ(128,1202), MN(128,1197), NM(1190), NY(544). **23:**CA(661). **25:**CO(128). **19:**Unk.(2142).
Cystostereum murrayi. **25:**ME(543). **19:**Unk.(2142).
C. pini-canadense. White rot. **5:**MN(1197). **21:**MN(128,541,1197). **25:**MN(1197), WI(541). **19:**Unk.(2142).
Cytidia salicina. On dead, attached limbs. **21:**NY(544).
Dacrymyces stillatus. **19:**Unk.(2142).
[*Dacryobolus sudans*]. **19:**Unk.(2142).

Daedaleopsis confragosa. White rot. **21:**MT(3202,3832), NM(1190), NY(544), WA(3202). **23:**ID(3202). **25:**Unk.(82). **19:**Unk.(2142).
Datronia mollis. **21:**ID(3202). **23:**MT(3202,3832). **25:**MT(128). **19:**Unk.(2142).
D. stereoides. On wood. **21:**AZ(1202). **19:**Unk.(2142).
Dendrophora versiformis. **5:**NJ(543). **12:**NH(543).
Dendrothele microspora. On bark. **5:**VT(2084).
Dentipellis fragilis. **5:**AK(526). **23:**WA(128).
Dichostereum effuscatum. **25:**NY(128).
Diplomitoporus crustulinus. **21:**MT(3202). **19:**Unk.(2142).
D. linbladii. **23:**WA(3202).
Eichleriella deglubens. On wood. **5:**AK(1904). **19:**Unk.(2142).
Erythricium laetum. **19:**Unk.(2142).
Exidia glandulosa. On dead, attached limbs. **21:**MI(2862), NY(544).
E. nucleata. On dead, attached limbs. **21:**NY(544).
Exidiopsis calcea. **19:**Unk.(2142).
E. fuliginea. On wood. **8:**WA(128). **21:**WA(128). **23:**ID(3202). **25:**ID(2318).
E. grisea. On wood. **23:**WA(2318,3202,3853).
E. molybdea. On bark. **12:**IA(128). **25:**IA(2318).
[*Fibricium rude*]. **19:**Unk.(2142).
Flagelloscypha minutissima. On bark. **25:**NH(456).
Flammulina velutipes. Root rot, on wood. **21:**AK(1567), AZ(1202). **25:**W states(82). **19:**Unk.(2142).
Fomes fomentarius. White heart rot. **21:**Unk.(82), MT(3202,3832). **23:**MT(3202,3832), WA(3202). **2:**ID(3202). **25:**NC states(82), W states(82), MI(2862). **19:**Unk.(2142).
Fomitopsis cajanderi. **21:**AZ(1202). **19:**Unk.(2142).
F. meliae. **25:**LA(2675).
F. pinicola. Brown cubical rot. **12:**NY(82), VT(82). **21:**Unk.(82), AZ(1202), CA(661), MN(1197), NM(1190). **23:**ID(3202), WA(3202). **25:**NC states(82), W states(82), WA(3202). **19:**Unk.(2142).
F. rosea. **21:**AZ(128,1202). **19:**Unk.(2142).
Galerina olympiana. On wood. **23:**WA(3329).
Ganoderma applanatum. White rot. **4:**MT(3202,3832). **5:**Range of host(82), AK(526), MT(3202,3832). **8:**Range of host(82), FL(35), LA(2675), MS(2720), MT(3832), SD(378,2243). **9:**W states(82). **12:**Lake states(82), NE states(82). **13:**TX(82). **23:**AK(526,1904), ID(3202), MT(3202,3832), OR(3202), WA(3202). **1:**ID(3202). **2:**CT(82), UT(82). **7:**NY(82). **25:**Range of host(82). **19:**Unk.(2142).
G. lucidum. **11:**AZ(1202).
Glabrocyphella ellisiana. **25:**NJ(646).
Gloeocystidiellum clavuligerum. White rot. **5:**MN(2553). **21:**MA(2553), MN(2553). **25:**MI(2553), WI(2553).
G. convolvens. **21:**AZ(128,1202), NM(1190). **19:**Unk.(2142).
G. heimii. **21:**MN(1197).
G. karstenii. **5:**AK(526). **19:**Unk.(2142).
G. ochraceum. **21:**MN(1197).
G. porosum. **5:**MN(1197). **21:**AZ(1202), MN(1197), NM(128,1190), WA(3202). **23:**WA(3202). **19:**Unk.(2142).
Gloeocystidiellum sp. **21:**SD(128).
Gloeodontia columbiensis. On wood. **21:**ID(440), MN(440), MT(440), OR(440). **23:**ID(440). **19:**Unk.(2142).
Gloeophyllum sepiarium. Sapwood rot. **21:**AZ(1202), MT(3202), WA(3202). **23:**MT(3202). **25:**Range of host(82). **19:**Unk.(2142).
G. trabeum. Sapwood rot. **23:**ID(3202), OR(3202). **25:**Range of host(82), MT(128,3202).
Gloeoporus dichrous. Wood rot. **21:**MT(3202,3832), ND(378). **23:**MT(3202,3832). **25:**Range of host(82). **19:**Unk.(2142).
Grandinia alutaria. **19:**Unk.(2142).
G. arguta. **5:**OR(128). **21:**AZ(1202). **19:**Unk.(2142).
G. breviseta. **21:**AZ(1202). **19:**Unk.(2142).
G. crustosa. **21:**AZ(1202). **19:**Unk.(2142).
G. granulosa. **21:**AZ(128,1202), MN(1197). **19:**Unk.(2142).
G. pallidula. **21:**AZ(128,1202). **19:**Unk.(2142).
G. pruni. **21:**AZ(128,1202). **19:**Unk.(2142).
G. spathulata. **19:**Unk.(2142).
G. stenospora. **19:**Unk.(2142).
Gymnopilus spectabilis. **19:**Unk.(2142).
Hapalopilus rutilans. White rot. **25:**Unk.(2681).
Heliocybe sulcata. On wood. **21:**MT(3202). **25:**MI(3328).
Henningsomyces candidus. **21:**AZ(1202), WA(3202). **19:**Unk.(2142).
Hericium americanum. **5:**MN(1197). **23:**ID(3202). **25:**NW states(82), WA(3202). **19:**Unk.(2142).
H. coralloides. On wood. **23:**AK(526,1904), WA(3202). **19:**Unk.(2142).
Heterobasidion annosum. White root and butt rot. **5:**MT(82). **23:**MT(3202,3832). **25:**NC states(82), W states(82).
Hydnochaete olivacea. On wood. **25:**MI(430).
Hymenochaete tabacina. **21:**ID(3202). **19:**Unk.(2142).
Hymenochaete sp. **23:**WA(3202).
Hyphoderma argillaceum. **21:**MN(128,1197). **19:**Unk.(2142).
H. budingtonii. On wood. **21:**AZ(2141). **19:**Unk.(2142).
H. guttuliferum. **21:**AZ(1202). **19:**Unk.(2142).
H. karstenii. **21:**MN(128,1197). **19:**Unk.(2142).
H. medioburiense. **21:**MN(1197). **19:**Unk.(2142).
H. mutatum. **4:**ND(378). **12:**PA(2321). **25:**MI(2321). **19:**Unk.(2142).
H. pallidum. **23:**ID(3202).
[*H. pilosum*]. **21:**AZ(1202). **19:**Unk.(2142).
H. populneum. **21:**AZ(1202), NM(128,1190). **25:**NY(3315). **19:**Unk.(2142).
H. praetermissum. **21:**AZ(1202), MN(1197), NM(1190). **19:**Unk.(2142).
H. puberum. **21:**MN(128,1197). **19:**Unk.(2142).
H. sambuci. **21:**MN(1197), NM(1190). **19:**Unk.(2142).
H. setigerum. **19:**Unk.(2142).
Hyphodermella corrugata. **19:**Unk.(2142).
Hyphodontia albicans. **19:**Unk.(2142).
H. magnacystidiata. **21:**NY(128,2141). **19:**Unk.(2142).
H. stipata. **21:**ID(128). **23:**ID(128).
Hypochnicium analogum. **19:**Unk.(2142).
H. bombycinum. **19:**Unk.(2142).
H. lundellii. **19:**Unk.(2142).
H. punctulatum. **11:**AZ(1202).
H. vellereum. **21:**SD(378). **19:**Unk.(2142).
Hypsizygus tessellatus. On wood. **23:**ID(2421).
Inonotus cuticularis. **11:**AZ(1181).
I. dryophilus. White heart rot. **8:**Range of host(82). **12:**Lake states(82), NE states(82). **21:**Range of host(82), ID(3202), MT(3202), WA(3202), WY(82). **23:**ID(3202). **25:**Range of host(82).
I. farlowii. White heart rot. **14:**AZ(82), NM(82). **25:**Range of host(82).
I. glomeratus. **5:**MN(1197). **23:**ID(3202), MT(3202).
I. munzii. **11:**AZ(1202).
I. rheades. On living tree. **12:**MI(2862). **21:**MT(3832). **25:**ID(3202), MN(2513), MT(3202), WA(3202). **19:**Unk.(2142).
Irpex lacteus. On dead, attached limbs. **8:**ND(378). **21:**ID(3202), NY(544). **19:**Unk.(2142).
Ischnoderma resinosum. Wood rot. **21:**MN(1197). **25:**Range of host(82). **19:**Unk.(2142).
Junghuhnia fimbriatella. **25:**WA(128,3202). **19:**Unk.(2142).
J. nitida. **12:**OH(128,2168). **21:**MI(2168), NY(128,2168), VT(128,2168), WI(128,2168). **23:**ID(2168,3202), WA(2168,3202). **19:**Unk.(2142).
J. subfimbriata. **21:**AZ(1202), MN(128). **19:**Unk.(2142).
Kavinia alboviridis. **21:**AZ(128,1202). **19:**Unk.(2142).
Kuehneromyces mutabilis. **23:**WA(3202).
Laeticorticium simplicibasidium. On wood. **21:**AZ(2141).
Laetiporus sulphureus. **8:**AK(526). **11:**AZ(1202). **21:**MN(1197). **19:**Unk.(2142).
Laxitextum bicolor. **21:**WI(543). **23:**ID(543). **25:**ME(543).
Lentinellus cochleatus. **21:**AZ(1202). **19:**Unk.(2142).
L. flabelliformis. **25:**Unk.(2423).
L. ursinus. On wood. **21:**AZ(1202). **25:**NC(1303).
L. vulpinus. **25:**Unk.(2423).
Lentinus strigosus. **21:**AZ(1202).
Lenzites betulina. White rot. **21:**AK(1904). **25:**Range of host(82).
Leptosporomyces raunkiaerii. **19:**Unk.(2142).
Leucogyrophana mollusca. **21:**WA(128). **23:**MT(128).
L. olivascens. **8:**LA(2675).
L. pinastri. **21:**NM(128,1190). **19:**Unk.(2142).
Lopharia cinerascens. **25:**NY(543).
Megacollybia platyphylla. **21:**AZ(1202).
Megalocystidium lactescens. **21:**AZ(1202). **19:**Unk.(2142).

M. leucoxanthum. **19**:Unk.(2142).
Meruliopsis albostramineus. **21**:AZ(128).
M. corium. **21**:AZ(128,1202). **19**:Unk.(2142).
M. hirtellus. **19**:Unk.(2142).
Mucronella aggregata. **5**:MN(128,1197).
Mycena haematopoda var. *cuspidata*. On rotten wood. **21**:CO(2450).
Mycoacia aurea. **21**:MN(1197). **19**:Unk.(2142).
M. fuscoatra. **21**:ID(128), WA(128). **19**:Unk.(2142).
M. uda. **19**:Unk.(2142).
Oxyporus corticola. **11**:AZ(1202). **21**:ID(3202), MN(128), MT(3202,3832), WA(3202). **23**:ID(3202), MT(3202,3832), WA(3202). **1**:MT(3202). **25**:MT(128). **19**:Unk.(2142).
O. latemarginatus. Isolated from wood. **8**:GA(1007), LA(2166). **11**:AZ(128,1202). **23**:WA(3202).
O. populinus. Butt and heart rot. **5**:NE states(82). **21**:NE states(82). **25**:NE states(82). **19**:Unk.(2142).
O. similis. **21**:MT(3202), WA(128,3202). **23**:ID(128,3202), MT(128,3202), OR(3202). **1**:MT(128). **25**:MT(128). **19**:Unk.(2142).
Panellus serotinus. Sapwood rot. **25**:Range of host(82).
Parmastomyces transmutans. **21**:AZ(1202). **19**:Unk.(2142).
Peniophora cinerea. **25**:WA(3202). **19**:Unk.(2142).
P. incarnata. On branch stubs and dead trunks. **21**:MI(2862), NY(544).
P. polygonia. On wood. **21**:AK(1567,1904), AZ(1202), CO(797), ID(3202), MN(1197), NM(128,1190), WA(3202). **23**:ID(3202), WA(3202). **25**:ID(3202). **19**:Unk.(2142).
P. rufa. On wood. **12**:NY(2456). **23**:OR(3202). **25**:Range of host(82), MT(3202). **19**:Unk.(2142).
P. violaceolivida. On dead, attached limbs. **21**:AZ(128,1202), NY(544). **19**:Unk.(2142).
Perenniporia medulla-panis. **21**:NM(128). **19**:Unk.(2142).
P. narymica. **19**:Unk.(2142).
P. subacida. **21**:MN(128). **23**:WA(3202). **25**:NY(128). **19**:Unk.(2142).
P. tenuis. **23**:MT(3202). **25**:CO(128). **19**:Unk.(2142).
Phanerochaete affinis. **12**:NY(128). **21**:WI(128). **23**:WA(3202). **19**:Unk.(2142).
P. arizonica. White rot. **11**:AZ(433). **21**:MN(1190).
P. burtii. White rot. **11**:AZ(433).
P. carnosa. **12**:NH(128). **21**:AZ(1202), NM(128), SD(128). **19**:Unk.(2142).
P. chrysorhiza. White rot. **12**:NY(433). **21**:WI(433). **19**:Unk.(2142).
P. filamentosa. **19**:Unk.(2142).
P. flavido-alba. Isolated from wood. **8**:GA(1007).
P. gigantea. **19**:Unk.(2142).
P. magnoliae. **25**:MN(1197).
P. rimosa. **21**:AZ(1202), MN(128,1197). **19**:Unk.(2142).
P. sanguinea. **19**:Unk.(2142).
P. sordida. White rot. **23**:ID(128), WA(128). **25**:AK(128,433), AZ(433). **19**:Unk.(2142).
P. tuberculata. **11**:AZ(1202).
P. velutina. White rot. **21**:AZ(128,1202), CO(128,433), CT(128), MN(1197). **19**:Unk.(2142).
Phellinus conchatus. **23**:MT(3202,3832). **19**:Unk.(2142).
P. crocatus. On wood. **25**:FL(2681), LA(2681).
P. everhartii. On living trunks. **23**:MT(3202,3832).
P. ferreus. On twigs. **25**:MI(2862), MT(3202).
P. ferruginosus. **5**:MN(1197). **21**:AZ(128,1202). **19**:Unk.(2142).
P. gilvus. **23**:OR(3202).
P. igniarius. White heart rot. **4**:MT(3202,3832). **5**:Range of host(82). **8**:Range of host(82), FL(35), TX(82). **23**:CA(661), ID(3202), MT(3202,3832), OR(1524,3202), WA(3202), WY(2924). **25**:Range of host(82).
P. laevigatus. **21**:MN(128,1197). **19**:Unk.(2142).
P. tremulae. White rot. **12**:Lake states(82), NE states(82). **21**:Range of host(82), AK(526,1567,1904), AZ(1202), CA(661), CO(1822), ID(3202), MN(1197), MT(128,3202,3832), NY(2241), OR(1524,3202), SD(378), UT(1524), WA(638,3202), WI(1524), WY(82,1195,3378). **22**:AZ(1855). **19**:Unk.(2142).
P. vaninii. On wood. **21**:MN(1205).
P. viticola. On wood. **21**:AK(128,1904).
Phlebia albida. **21**:MT(3202). **19**:Unk.(2142).
P. ludoviciana. **21**:NM(1190).
P. merismoides. **25**:MA(128). **19**:Unk.(2142).
P. rufa. **19**:Unk.(2142).
P. subochracea. White rot. **21**:MT(2556). **23**:ID(2118,3202). **25**:MT(2556).
P. subserialis. **19**:Unk.(2142).
P. tremellosus. **23**:ID(3202). **19**:Unk.(2142).
Phlebiella vaga. **21**:AZ(1202), MN(1197). **19**:Unk.(2142).
Pholiota adiposa. Canker, brown stem rot, wound rot. **8**:MS(2720). **21**:AZ(1202), MN(2240). **25**:NW states(82), AK(82).
P. decorata. **21**:AZ(1202).
P. limonella. **21**:WA(3202).
P. populnea. Wound rot. **23**:ID(3202), WA(3202). **25**:NW states(82), AK(82), ID(3202).
P. squarrosa. Root rot, wood rot. **21**:AK(1567), AZ(1202), WY(3040). **25**:NY(1017).
Pholiota sp. **25**:NY(1017).
Phyllotopsis nidulans. **21**:AZ(1202).
Physisporinus sanguinolentus. **23**:ID(128,3202).
Piloderma bicolor. **21**:NM(1190). **19**:Unk.(2142).
P. byssinum. **21**:AZ(1202). **19**:Unk.(2142).
Pleurotus cystidiosus. On wood. **8**:GA(1007,2418).
P. ostreatus. Sapwood rot. **8**:CA(2439), MS(2720), ND(82,378), WY(82). **11**:AZ(1202). **15**:GA(1439). **21**:AK(526,1904), ID(3202), MN(63), NY(63), WY(3040). **23**:ID(3202). **25**:Range of host(82), ID(3202).
Plicatura nivea. **25**:OR(3202).
Plicaturopsis crispa. On dead trunks. **21**:NY(544).
Pluteus atricapillus. **21**:AZ(1202).
Polyporus arcularius. Wood rot. **21**:NM(1190). **25**:Range of host(82).
P. badius. White rot. **21**:CA(661), ID(3202), MT(3832). **23**:AK(1904), ID(3202), MT(3202,3832), OR(3202), WA(3202). **25**:Range of host(82). **19**:Unk.(2142).
P. brumalis. **23**:MT(3202,3832).
P. melanopus. On wood. **21**:CA(661).
P. mori. Wound rot. **21**:NM(1190). **25**:Unk.(82), MT(3202).
P. squamosus. White heart rot. **21**:AZ(1202), CA(661). **23**:MT(3202), WA(3202). **25**:Range of host(82). **19**:Unk.(2142).
P. varius. Wood rot. **21**:AZ(1202), CA(661), MT(3202,3832), NM(1190). **23**:AK(3431). **25**:Range of host(82). **19**:Unk.(2142).
Porotheleum fimbriatum. **21**:AZ(128,1202), MN(1197), WA(3202). **23**:ID(3202), MT(3202). **25**:AZ(128), MN(128,1197). **19**:Unk.(2142).
Postia balsamea. **23**:MT(3832).
P. caesia. Wood rot. **5**:AK(526). **21**:AZ(1202), NM(1190). **23**:MT(3202,3832). **25**:Range of host(82). **19**:Unk.(2142).
P. ceriflua. **23**:ID(3202), MT(3832).
P. fragilis. **23**:AK(3431), MT(3202).
P. guttulata. **4**:ID(3202).
P. leucospongia. **19**:Unk.(2142).
P. perdelicata. **19**:Unk.(2142).
P. placenta. **23**:ID(3202).
P. undosa. **21**:CO(397). **19**:Unk.(2142).
Psathyrella sublateritia var. *intermedia*. At base of living trees. **8**:CO(2450).
Pseudotomentella tristis. **21**:AZ(1202).
Punctularia strigoso-zonata. On wood. **21**:ND(378), NY(544). **25**:MI(2862). **19**:Unk.(2142).
Pycnoporellus alboluteus. **21**:AZ(1202), NM(1190). **19**:Unk.(2142).
Pycnoporus cinnabarinus. Wood rot. **4**:NM(1190). **11**:AZ(1202). **21**:MT(3202,3832). **23**:MT(3202,3832). **25**:Range of host(82). **19**:Unk.(2142).
Radulodon americanus. **21**:ID(128), MN(128,1197). **23**:MT(128). **25**:AZ(128), MN(1197). **19**:Unk.(2142).
Resinicium bicolor. **19**:Unk.(2142).
Rigidoporus lineatus. **25**:AR(2675), LA(2675).
R. ulmarius. **11**:AZ(128,1202). **25**:LA(2675), MS(2675).
Schizophyllum commune. On wood. **8**:GA(1007), OK(618). **2**:OK(2873). **25**:Range of host(82), WA(3202).
Schizopora paradoxa. **23**:WA(3202).
Scytinostroma galactinum. **5**:MN(1197). **21**:FL(128), MN(1197). **19**:Unk.(2142).
Scytinostroma sp. On wood. **21**:CA(661).
Sebacina epigaea. **19**:Unk.(2142).
Serpula himantioides. **21**:NM(1190). **19**:Unk.(2142).
Simocybe haustellaris. **25**:WA(3202).
Sistotrema brinkmannii. On dead, attached limbs. **21**:AZ(1202), NM(1190), NY(544). **19**:Unk.(2142).

S. confluens. **19**:Unk.(2142).
S. coroniferum. **21**:AZ(1202). **19**:Unk.(2142).
S. diademiferum. On fallen logs. **25**:CO(128).
S. muscicola. **21**:AZ(1202). **19**:Unk.(2142).
S. raduloides. **5**:MN(128). **21**:AZ(1202), MN(128,1197). **25**:MT(128). **19**:Unk.(2142).
S. subtrigonospermum. **21**:AZ(1202). **19**:Unk.(2142).
Sistotremastrum suecicum. **19**:Unk.(2142).
Skeletocutis nivea. **19**:Unk.(2142).
Spongipellis spumeus. **4**:ND(378). **21**:MT(3202), WA(3202). **23**:ID(3202), MT(3202,3832), WA(3202). **1**:MT(3202). **25**:CO(82).
Steccherinum ciliolatum. On wood. **5**:MN(128). **21**:AZ(1202), CO(1175), NM(1190), NY(1175). **19**:Unk.(2142).
S. fimbriatum. **21**:AZ(128,1202), NM(1190). **25**:NY(128). **19**:Unk.(2142).
S. ochraceum. **5**:AK(526). **8**:IA(82). **25**:ID(3202). **19**:Unk.(2142).
S. oreophilum. On wood. **21**:AZ(128,1234,2139), UT(128). **19**:Unk.(2142).
Stereum hirsutum. White rot. **8**:GA(1007). **12**:CT(543), NH(543). **21**:AK(1904), CA(661), SD(543), WY(2924). **23**:CA(661). **25**:Range of host(82), ID(3202).
S. ostrea. **21**:ND(378). **23**:WA(3202). **19**:Unk.(2142).
Stigmatolemma poriaeforme. **21**:AZ(1202). **19**:Unk.(2142).
Subulicystidium longisporum. **21**:AZ(128,1202), MN(128,1197). **25**:MN(1197). **19**:Unk.(2142).
Thanatephorus cucumeris. **25**:ID(3202).
Tomentella asperula. **19**:Unk.(2142).
T. atrorubra. On wood. **25**:MA(2039).
T. botryoides. **25**:NC(2039), NY(2039). **19**:Unk.(2142).
T. bryophila. **21**:AZ(1202), NM(1190). **25**:AZ(2039), CO(2039), MI(2039), MT(2039), NM(2039), NY(128,2039). **19**:Unk.(2142).
T. calcicola. **21**:WA(128,3202). **23**:ID(128,3202), MT(3202). **25**:IA(2039), ID(2039), MT(2039,3202). **19**:Unk.(2142).
T. chlorina. **21**:AZ(1202). **25**:AZ(2039), NM(2039). **19**:Unk.(2142).
T. cinerascens. **25**:MI(2039), MT(2039), NY(2039), SD(2039). **19**:Unk.(2142).
T. coerulea. **21**:AZ(2036). **25**:AZ(2039), MT(2039). **19**:Unk.(2142).
T. crinalis. **21**:AZ(1202), CO(128), MN(128), SD(128). **25**:AZ(2039), CO(128,2039), ID(2039,3202), MN(128), MT(2039,3202), NY(2039), SD(2039). **19**:Unk.(2142).
T. ferruginea. **21**:NM(1190). **25**:CO(128), ID(2039), MI(2039), MT(2039,3202), NM(2039), NY(2030,2039). **19**:Unk.(2142).
T. ferruginella. On wood. **21**:NM(1190). **25**:NM(2039).
T. fuscoferruginosa. **19**:Unk.(2142).
T. griseo-umbrina. **25**:AZ(2039).
T. griseoviolacea. **25**:NY(2036,2039).
T. lateritia. **21**:AZ(1202), NM(1190). **25**:AZ(2039), MT(2039), NM(2039), NY(2039). **19**:Unk.(2142).
T. molybdaea. **25**:MT(2036,2039,3202), NY(2039).
T. neobourdotii. On wood. **21**:NM(1190). **25**:AZ(2039), MI(2039), MT(2033,2039,3202), NM(2039), NY(2033,2039). **19**:Unk.(2142).
T. olivascens. **21**:CT(128). **25**:NY(2039).
T. pilatii. **21**:NM(1190).
T. pilosa. **21**:MT(3202), NM(1190). **25**:MI(2039), MT(2039), NM(2039), NY(2039). **19**:Unk.(2142).
T. punicea. **25**:NY(2030). **19**:Unk.(2142), NY(2039).
T. ramosissima. **25**:MS(2039), MT(2039,3202), NY(2039). **19**:Unk.(2142).
T. rubiginosa. **25**:NY(2039). **19**:Unk.(2142).
T. ruttneri. On wood. **25**:NY(2035).
T. sublilacina. **25**:NY(2039).
T. subvinosa. **25**:NY(2039).
T. terrestris. **25**:AZ(2032), MT(2039). **19**:Unk.(2142).
T. umbrinospora. **19**:Unk.(2142).
T. violaceofusca. **25**:NY(2039).
T. viridescens. **25**:AZ(2039). **19**:Unk.(2142).
T. viridula. **25**:AZ(2039).
Tomentellastrum badium. On wood. **25**:NY(2032). **19**:Unk.(2142).
T. montanensis. **21**:NM(1190). **25**:MT(2034,3202). **19**:Unk.(2142).
Tomentellina fibrosa. **21**:AZ(1202). **19**:Unk.(2142).
Trametes cervina. **21**:AZ(1202). **19**:Unk.(2142).
T. hirsuta. White rot. **8**:Range of host(82). **21**:AK(526,1904), AZ(1202), ID(3202), OR(3202), WA(3202). **23**:ID(3202), MT(3202), OR(3202), WA(3202). **2**:TN(82). **25**:Range of host(82), LA(2675). **19**:Unk.(2142).
T. pubescens. On wood. **21**:AK(1904), AZ(1202), ID(3202), MT(3202), NM(1190), WA(3202). **23**:WA(3202). **19**:Unk.(2142).
T. suaveolens. White rot. **5**:Range of host(82), AK(1904). **21**:AK(1904), AZ(1202), ID(3202), MT(3202,3832), ND(378). **23**:MT(3202,3832). **25**:Range of host(82). **19**:Unk.(2142).
T. trogii. Wound rot. **21**:MN(1197). **23**:ID(128,3202). **25**:Range of host(82), NY(128). **19**:Unk.(2142).
T. versicolor. White rot. **12**:NY(2456). **21**:AZ(1202), ID(3202), MT(3202), OR(3202). **23**:AK(526,1904), ID(3202), MT(3202), OR(3202), WA(3202). **25**:Range of host(82), WA(3202). **19**:Unk.(2142).
Trechispora cohaerens. **21**:MN(1197). **19**:Unk.(2142).
T. farinacea. **21**:AZ(1202), MN(1197). **19**:Unk.(2142).
T. fastidiosa. **19**:Unk.(2142).
T. mollusca. **21**:MN(1197). **19**:Unk.(2142).
T. pallidoaurantiaca. **19**:Unk.(2142).
Tremella mesenterica. **19**:Unk.(2142).
Trichaptum biforme. White rot. **5**:AK(1904). **21**:AK(1904), ID(3202), MN(1197,2240), ND(378), NY(544). **1**:ID(3202). **25**:ID(3202), MI(2862), WI(82). **19**:Unk.(2142).
T. subchartaceum. Wood rot. **21**:AZ(1202), NM(1190). **25**:Range of host(82). **19**:Unk.(2142).
Tricholomopsis decora. **23**:WA(3202).
Tubulicrinis chaetophorus. On wood. **21**:CA(661).
Tulasnella eichleriana. On wood. **25**:PA(2673).
T. hyalina. On wood. **25**:IA(128).
T. traumatica. On wood. **19**:IA(128).
Typhula variabilis. On leaves. **25**:NY(2955).
Tyromyces chioneus. On wood. **21**:AK(1904), MT(3202). **23**:MT(3202). **25**:ID(3202). **19**:Unk.(2142).
T. galactinus. Wood rot. **23**:MT(3202). **25**:Range of host(82).
T. subgiganteus. **19**:Unk.(2142).
Vararia athabascensis. **21**:AZ(128,1177,1202). **19**:Unk.(2142).
V. investiens. **21**:MN(128,1197). **25**:CO(128). **19**:Unk.(2142).
V. phyllophila. **21**:NM(1190). **19**:Unk.(2142).
V. racemosa. **19**:Unk.(2142).
Vesiculomyces citrinus. **19**:Unk.(2142).
Volvariella bombycina. **11**:AZ(1202).
Wolfiporia cocos. **25**:MN(128). **19**:Unk.(2142).
Xerula americana. On rotten wood. **21**:CO(2450), ID(2450), NM(2450), UT(2450), WY(2450).

Deuteromycotina - Hyphomycetes

Acrodictys sp. On branches. **21**:CA(661).
Alternaria alternata. **8**:ND(378).
Alternaria sp. Leaf spot. **25**:FL(35).
Botrytis vulgaris var. *interrupta*. **8**:ND(378).
Cercospora populicola. On leaves. **8**:FL(35), TX(82,581,3597).
Cercospora sp. Leaf spot. **8**:OK(618).
Cladosporium letiferum. Leaf blight. **21**:NY(82).
C. rambosum. **21**:IL(82).
Cladosporium sp. **14**:WA(3202).
Coniosporium olivaceum. **25**:OR(3202).
C. subgranulosum. **25**:MT(3202).
Curvularia fallax. **15**:AL(2482).
Fusariella kansensis. On wood. **25**:KS(525).
Fusarium solani. **8**:IA(883).
Fusarium sp. Cutting rot. **8**:FL(35).
Fusicladium tremulae. **23**:AK(3431).
Graphium rubrum. Pink wood stain. **8**:C states(82), S states(82).
Helicoma morganii. On wood. **25**:IA(1271).
H. taenia. On wood. **25**:IA(1271,2465).
Helicomyces roseus. **25**:IA(1270).
H. scandens. On bark. **25**:MA(1270).
Helicosporium lumbricoides. **25**:NY(2456).
Heterosporium sp. **8**:WI(1354).
Phaeoramularia maculicola. Leaf spot. **21**:CO(82), UT(82), WI(82). **25**:AK(526).
Phaeostalagmus arbusculus. On rotten wood. **25**:NY(3771).
Phialophora sp. Isolated from roots. **21**:WY(3040).
Phymatotrichopsis omnivora. Root rot. **8**:OK(82,2873), TX(82). **14**:TX(82). **2**:TX(82). **7**:TX(82). **25**:OK(2872).
Pollaccia radiosa. Shepherds crook, leaf spot. **3**:SD(2243). **21**:AK(526,1567), MT(82), NY(82), SD(82).

Pseudocercospora salicina. Leaf spot. **8:**AL(82), FL(35), LA(82,962), TX(82). **14:**AL(82), FL(35,82). **2:**LA(962). **25:**FL(35).
Rhinocladium dingleyae. On bark. **25:**NY(3771).
Septonema griseofulvum. **21:**MT(3202).
Sporidesmium tabacinum. **21:**MT(3202).
Stigmina sp. **8:**IL(82).
Toxosporium effusum. On wood. **9:**KS(128,3547).
Tubercularia sp. Branck dieback. **8:**OK(618), SD(2243).

Deuteromycotina - Coelomycetes

Ascochyta sp. On juvenile leaves. **21:**WI(1359).
Colletotrichum gloeosporioides. Anthracnose, leaf and shoot blight, shoot dieback. **21:**WI(2248). **2:**NC(1303), WI(2248). **25:**NC(1303).
C. lucidae. **21:**WI(1354).
Corniculariella populi. **25:**MI(861).
Cylindrosporium oculatum. Leaf spot. **4:**NM(1195). **5:**ID(82). **8:**MS(2720). **23:**AK(3431). **25:**FL(35).
C. populinum. Leaf spot. **12:**NJ(82), NY(82), PA(82), WI(82). **14:**MS(2720). **25:**Range of host(82).
C. saximontanense. Leaf spot. **4:**CO(82,1195,3376), NM(1195), WY(1195).
Cytospora chrysosperma. Canker, dieback. **8:**MS(1057), NE(3956), OK(618,2872), SD(2243). **15:**SD(2243). **21:**AK(1567), CO(1560,1822), WY(1195,3040). **23:**CA(2439). **2:**NC(1303). **25:**CA(3549), KS(3549), OK(2872), OR(4007), WY(1195,3549).
C. leucosperma. On bark. **4:**NM(1003).
C. translucens. On bark. **21:**NM(1003).
Cytospora sp. **25:**CO(649).
Diplodia populina. Twig blight. **12:**PA(82).
D. tumefaciens. Bark canker. **12:**MN(82). **21:**MN(82). **23:**ID(3202), MT(3202).
Discella populina. On twigs. **14:**IA(82). **2:**IA(82).
Discosporium populeum. Canker, dieback. **3:**SD(2243). **8:**Range of host(82), FL(35). **14:**C states(82), E states(82), IA(82). **15:**C states(82), E states(82), NC(1303), OK(2872). **21:**WI(82). **2:**CT(82), IA(82), IN(82), MA(82). **7:**CT(82), OH(82). **25:**Range of host(82), OK(2872).
D. tremuloides. On leaves. **21:**WI(2481,3549).
Dothiorella sp. **18:**IL(2567).
Hendersonula toruloidea. **11:**CA(2633).
Libertella sp. Canker. **21:**CO(1561), WY(3040).
Marssonina carpogena. On capsules. **8:**CA(636).
M. castagnei. Anthracnose, leaf spot. **5:**Range of host(82). **8:**SD(2243). **12:**NY(82), WI(82). **21:**Range of host(82), CA(661). **2:**E & C states(82), W states(82), NC(1303), SD(2243). **25:**Range of host(82), NC(1303).
M. piriformis. Leaf spot. **2:**OR(4007). **25:**Range of host(82).
M. populi. Leaf spot. **4:**WY(1195). **5:**Range of host(82). **8:**C states(82), E states(82). **9:**CA(82). **21:**Range of host(82), AK(1567), MS(2720), OR(81), WY(2924). **23:**AK(3431). **2:**OK(2872). **25:**Range of host(82), WY(1195).
M. stenospora. Leaf spot. **25:**Range of host(82).
M. tremulae. Leaf spot. **5:**Range of host(82), NJ(966). **8:**C states(1795), C states(1795). **21:**Range of host(82), AZ(1195,3378), WY(1195). **7:**IA(82), NJ(82). **25:**Range of host(82).
Marssonina sp. Leaf spot. **5:**Range of host(82). **14:**IN(82), NC(82). **21:**Range of host(82). **7:**IA(82), MD(82).
Melanconium atrum. **4:**WA(3202).
Myrioconium comitatum. On leaves. **21:**WI(82).
Myxosporium alboluteum. On bark. **9:**KS(82,832).
Phoma terrestris. Isolated from rhizosphere. **21:**OR(2791).
P. urens. On dead limbs. **25:**MA(965).
Phomopsis macrospora. Basal canker. **8:**MS(1057).
Phyllosticta alcides. Leaf spot. **23:**ID(3202). **2:**OH(82). **7:**AL(82). **25:**Unk.(82), AK(526).
P. brunnea. Leaf spot. **3:**OK(2872). **4:**CO(64,1195), NM(1195), OK(2872). **5:**CO(82), WI(82). **11:**AZ(1195). **25:**Unk.(82), MT(3202).
P. intermixta. Leaf spot. **5:**NC(82). **8:**KS(82), MO(82), NE(82), WI(82,1358). **23:**WA(3202). **25:**Unk.(82), AK(526).
P. maculans. Leaf spot. **8:**KS(82). **12:**NY(82). **21:**NY(82). **23:**CA(2439). **25:**Unk.(82).
P. osteospora. **23:**AK(526).
Phyllosticta sp. Leaf spot. **12:**WI(1354). **25:**FL(35).
Septogloeum rhopaloideum. Leaf spot. **21:**WY(82,1195,3378).
Septoria musiva. Canker, leaf spot. **8:**MS(1061), OK(2872), SD(2243). **18:**WI(1357). **21:**CA(81). **23:**AK(3431). **2:**WI(1352).
S. populi. Leaf spot. **5:**WY(1195). **8:**AR(82), OK(2872). **14:**WI(1330). **23:**OR(4007). **25:**Unk.(82), WI(1334).
S. populicola. **8:**TX(82).
S. rhabdocarpa. On dead leaves. **9:**CA(2439).
Sirodothis populnea. On limbs. **21:**CO(3552). **25:**CO(3549).

Deuteromycotina - Other

Rhizoctonia solani. Root rot. **8:**FL(35). **25:**FL(35).

Salix L.

1. *S. alaxensis* (Anderss.) Coville [feltleaf willow]
2. *S. alba* L. [white willow]
3. *S. alba* L. var. *vitellina* (L.) Stokes [golden willow]
4. *S. amygdaloides* Anderss. [peach-leaved willow]
5. *S. arbusculoides* Anderss. (=*S. arbusculoides* Anderss. var. *glabra* Anderss.) [littletree willow]
6. *S. arctica* Pall. (=*S. petrophila* Rydb., *S. anglorum* Cham.) [arctic willow]
7. *S. babylonica* L. [weeping willow]
8. *S. barclayi* Anderss. [Barkley willow]
9. *S. bebbiana* Sarg. (=*S. bebbiana* Sarg. var. *perrostrata* (Rydb.) C. K. Schneid., *S. perrostrata* Rydb., *S. rostrata* Richardson) [Bebb willow]
10. *S. bonplandiana* H. B. K.
11. *S. brachycarpa* Nutt. subsp. *niphoclada* (Rydb.) Argus (=*S. niphoclada* Rydb.) [barren-ground willow]
12. *S. breweri* Bebb
13. *S. candida* Flüggé ex Willd. [silver willow]
14. *S. caroliniana* Michx. (=*S. longipes* Shuttleworth ex Anderss.)
15. *S. chamissonis* Anderss. [Chamisso willow]
16. *S. commutata* Bebb [undergreen willow]
17. *S. cordata* Michx. (=*S. adenophylla* Hook.) [heart-leaved willow]
18. *S. curtiflora* Anderss.
19. *S. delnortensis* C. K. Schneid.
20. *S. discolor* Muhl. [large pussy willow]
21. *S. drummondiana* J. Barratt ex Hook. (=*S. subcoerulea* Piper)
22. *S. eriocephala* Michx. (=*S. rigida* Muhl.)
23. *S. exigua* Nutt. (=*S. interior* Rowlee, *S. longifolia* Muhl.) [ditchbank willow, sandbar willow]
24. *S. fluviatilis* Nutt.
25. *S. fragilis* L. [brittle willow, crack willow]
26. *S. fuscescens* Anderss. [Alaska bog willow]
27. *S. geyeriana* Anderss.
28. *S. glauca* L. (=*S. glauca* L. var. *acutifolia* (Hook.) C. K. Schneid., *S. glauca* L. var. *aliceae* C. Ball, *S. glauca* L. var. *glabrescens* C. K. Schneid., *S. glaucops* Anderss., *S. seemanii* Rydb.) [grayleaf willow]
29. *S. gooddingii* C. Ball
30. *S. hookeriana* J. Barratt [Hooker willow]
31. *S. humilis* Marsh. [gray willow, prairie willow, small pussy willow]
32. *S. humilis* Marsh. var. *microphylla* (Anderss.) Fernald (=*S. tristis* Aiton)
33. *S. irrorata* Anderss.
34. *S. jepsonii* C. K. Schneid.
35. *S. laevigata* Bebb [polished willow, red willow]
36. *S. lanata* L. subsp. *richardsonii* (Hook.) Skvortzov (=*S. richardsonii* Hook.) [Richardson willow]
37. *S. lasiandra* Benth. [Pacific willow]
38. *S. lasiandra* Benth. var. *abramsii* C. Ball
39. *S. lasiolepis* Benth.
40. *S. lemmonii* Bebb
41. *S. ligulifolia* (C. Ball) C. Ball ex C. K. Schneid.
42. *S. longifolia* Muhl. var. *argyrophylla* Anderss.
43. *S. lucida* Muhl.
44. *S. lutea* Nutt.
45. *S. melanopsis* Nutt.
46. *S. melanopsis* Nutt. var. *gracilipes* C. Ball
47. *S. melanopsis* Nutt. var. *tenerrima* C. Ball
48. *S. monticola* Bebb (=*S. padophylla* Rydb.) [park willow]
49. *S. myricoides* Muhl. (=*S. glaucophylloides* Fernald)
50. *S. myrtillifolia* Anderss. [low blueberry willow]
51. *S. nigra* Marsh. (=*S. nigra* Marsh. var. *falcata* (Pursh) Torr.) [black willow]
52. *S. nigra* Marsh. var. *aurea* Sw.

53. *S. novaeangliae* Anderss. (=*S. pseudocordata* Anderss.) [tall blueberry willow]
54. *S. ovalifolia* Trautv. (=*S. ovalifolia* Trautv. var. *camdensis* C. K. Schneid.) [oval-leaf willow]
55. *S. pedicellaris* Pursh
56. *S. pentandra* L. [bay willow, bay-leaved willow, laurel willow]
57. *S. petiolaris* Sm. [skeleton-leaf willow]
58. *S. phlebophylla* Anderss.
59. *S. phylicifolia* L.
60. *S. piperi* Bebb
61. *S. planifolia* Pursh (=*S. nelsonii* C. Ball) [diamond-leaf willow]
62. *S. polaris* Wahlenb.
63. *S. pulchra* Cham.
64. *S. purpurea* L. [basket willow, purple osier, purple willow]
65. *S. pyrifolia* Anderss.
66. *S. reticulata* L. (=*S. reticulata* L. var. *gigantifolia* C. Ball, *S. nivalis* Hook.) [netleaf willow]
67. *S. rotundifolia* Trautv. [least willow]
68. *S. scouleriana* J. Barratt ex Hook. (=*S. flavescens* Nutt., *S. nuttallii* Sarg., *S. brachystachys* Benth.) [mountain willow, Scouler willow]
69. *S. sericea* Marsh. [silky willow]
70. *S. serissima* (L. H. Bailey) Fernald
71. *S. sitchensis* Sanson ex Bong. (=*S. coulteri* Anderss.) [Sitka willow]
72. *S. stolonifera* Crêp. [oval-leaf willow]
73. *S. taxifolia* H. B. K. (=*S. microphylla* Cham. & Schlechtend.)
74. *S. tracyi* C. Ball
75. *S. viminalis* L. [basket willow, common osier]
76. *Salix* sp.
77. *Salix* × sp. (unnamed hybrids)

Myxomycetes

Craterium leucocephalum. On leaves. **39:**CA(2439).
Trichia decipiens. On dead leaves, on twigs. **39:**CA(2439).

Oomycetes

Pythium sp. Root rot. **7:**FL(35). **76:**FL(35).

Ascomycotina

Allantoporthe tessella. On limbs. **76:**IA(82), LA(82), NY(82).
Bertia moriformis. On wood. **68:**CA(684).
Bisporella citrina. On logs. **51:**GA(1438). **76:**ID(3202), WA(3202).
Botryosphaeria dothidea. Limb and trunk canker. **2:**GA(1438). **7:**FL(35), GA(1438), NC(1303). **20:**NC(1303). **31:**GA(1438). **51:**GA(1438). **52:**GA(1438). **76:**FL(35).
B. obtusa. On limbs. **7:**GA(82). **76:**SE states(82), TN(82).
B. quercuum. Twig dieback. **76:**FL(35).
B. rhodina. On limbs. **76:**SC(82).
B. ribis. Branch and trunk canker. **2:**GA(82), MO(82). **7:**MD(82). **20:**MS(2720). **51:**CA(82), WV(82). **76:**SE states(82), AR(82), WV(82).
Brunnipila calyculiformis. On wood. **71:**AK(3431). **76:**CO(521).
B. clandestina. **76:**WA(3202).
Calosphaeria arctica. On dead limbs. **76:**AK(1938).
Calyptronectria ohiensis. **51:**GA(1438).
Capitotricha bicolor. On dead twigs. **76:**CA(661), OR(3202).
Capnodium salicinum. Sooty mold. **76:**CA(82), WI(82).
Cenangium tryblidioides. **76:**WA(3202).
Chaetosphaeria pannicola. **51:**GA(1438).
Chlorociboria aeruginosa. Green wood stain. **76:**Range of host(82), ID(3202), WA(3202).
Ciboria acerina. On staminate catkins. **76:**NY(82).
C. caucus. **76:**NY(82), OR(82,3202).
Ciborinia foliicola. On leaves. **31:**CA(202), CO(202). **76:**CO(82).
C. wisconsinensis. On leaves. **57:**WI(202).
Crocicreas amenti. On catkins. **76:**AK(505), OR(505).
Cryptodiaporthe salicella. Branch canker. **2:**MD(82), NY(82). **9:**VA(82). **17:**SD(2243). **23:**MS(82), SD(82). **51:**GA(82,1438). **76:**NE states(82), AK(82,526), CA(82), KS(82), OK(82), SD(82), WA(82).
Cryptomyces maximus. Bark blister, blister canker. **69:**Unk.(82). **75:**Unk.(82). **76:**NM(82), UT(82).
Cryptosphaeria lignyota. On dead limbs. **75:**PA(82).
C. subcutanea. **20:**IA(3629), NY(2925). **51:**IA(3629). **76:**IA(3629), NY(2925).
Daldinia concentrica. Wood rot. **76:**Range of host(82).
D. grandis. **76:**W states(82), MT(3202).
D. occidentalis. **76:**W states(82), ID(3202), MT(3202).
Diatrype bullata. On wood. **68:**WA(3202). **76:**AK(526), ID(3202), OR(3202), WA(3019).
D. dakotensis. **76:**ND(377).
D. disciformis. **30:**CA(81).
D. linearis f. *salicis*. **76:**ND(377).
D. stigma. **76:**ND(377), OR(3202).
Diatrypella decorata. **76:**ID(3202).
D. quercina. **51:**GA(1438).
Didymosphaeria futilis. **76:**ND(377).
Discohainesia oenotherae. On leaves. **76:**VA(82).
Drepanopeziza salicis. Leaf spot. **76:**MN(82), WA(3202), WI(1357).
Encoelia fascicularis. On dead limbs. **76:**CO(82).
Eutypa laevata. **76:**ND(377).
E. scabrosa. **76:**ND(377).
Eutypella canodisca. **76:**ND(377).
E. juglandicola. **51:**GA(1438).
E. leprosa. **76:**IA(3629).
E. quaternata. **76:**IA(3629).
Gibberella acervalis. On dead stems. **23:**ID(82).
Glomerella cingulata. Black canker, dieback, on leaves and twigs. **2:**CT(82). **7:**WV(82). **76:**NE states(82), FL(35), NY(82).
Glyphium corrugatum. On decorticated stems. **76:**ID(3202), MT(3202), NY(2456), OR(3202), WA(3202), WY(1195,3376).
G. elatum. On dead twigs. **1:**AK(526). **76:**CO(82), ID(3202), WA(3202).
G. schizosporum. **76:**ID(3202).
Gnomonia sp. **36:**AK(526).
Godronia fuliginosa. On limbs. **36:**AK(526). **76:**AK(526), CO(1382), NY(82,1382).
Hyaloscypha hyalina. **76:**WA(3202).
Hymenoscyphus calyculus. **53:**CA(661).
H. conscriptus. On base of dead stems. **76:**AK(1938), CA(661), WA(3202).
H. salicellus. **76:**WA(3202).
Hypocrea gelatinosa. **76:**WA(3202).
Hypoxylon cinereo-lilacinum. On wood. **76:**IA(1225).
H. fuscum. **76:**ID(3202).
H. mammatum. Branch canker. **1:**AK(526). **23:**IA(82,1225). **76:**Range of host(82).
H. mediterraneum. On limbs. **76:**MD(82).
H. rubiginosum. On wood. **2:**GA(1438), IA(1225). **51:**AL(82). **76:**Range of host(82), IA(1225), WA(3202).
H. vogesiacum var. *macrospora*. On wood. **76:**CO(2403).
Hysterium magnosporum. **51:**GA(1438).
Hysterium sp. **76:**ID(3202), WA(3202).
Hysterographium flexuosum. On stems. **7:**GA(1438). **39:**CA(81,2801). **76:**OR(3202).
H. fraxini. **76:**ID(3202).
H. mori. **76:**ND(377).
Hysterographium sp. **76:**ID(3202), WA(3202).
Hysteropatella prostii. **76:**ID(3202).
Lachnum virgineum. **1:**AK(3431).
Lasiobelonium boreale. On stems. **76:**WA(849,3202).
Lasiosphaeria sp. **51:**GA(1438).
Leptosphaeria borealis. **76:**MT(3202), ND(377).
Leucostoma nivea. Twig blight and canker. **4:**UT(82). **23:**KS(82). **76:**W states(82), CO(649).
L. translucens. Twig canker. **2:**NY(82). **23:**ID(82), ND(82,377). **25:**SD(82,2243). **51:**IN(82). **68:**WA(3202). **76:**ID(3202), MT(3202).
Lewia scrophulariae. **1:**AK(526).
Linospora capreae. **76:**ID(3202).
Lophidium trifidum. **76:**WA(3202).
Lophiosphaera fluviatilis. **76:**MT(3202).
L. gloniospora. **76:**MT(3202).
Lophiostoma insidiosum f. *salicis*. **76:**ND(377).
L. macrostomoides. **76:**ND(377).
L. quadrinucleatum. **76:**ID(3202).
L. vagabundum. **51:**GA(1438).
Lophodermium maculare. **76:**AK(526).
L. versicolor. On leaves. **43:**PA(82). **76:**AK(82), CO(82), PA(82).
Melanomma cupulata. **76:**WA(3202).
M. minutum. **76:**ND(377).
M. pulvis-pyrius. **76:**AK(526).
Melanopsamma salicaria. **76:**ND(377).
M. salicaria subsp. *fallax*. **76:**ND(377).

Melittosporiella macrospora. On wood. **76:**OR(3231).
Mollisia melaleuca. **51:**GA(1438).
M. sublividula. **1:**AK(526).
Mollisia sp. **51:**GA(1438).
Mycosphaerella minor. On dead leaves. **6:**AK(1938). **67:**AK(1938).
M. salicina. On twigs. **17:**KS(82). **76:**KS(82).
Mycosphaerella sp. Leaf spot. **7:**GA(1438). **76:**FL(35).
Nectria cinnabarina. On twigs, associated with dieback. **17:**NM(82). **23:**NM(82). **64:**AK(526). **68:**WA(82,3202). **75:**AK(82,526). **76:**Range of host(82), AK(526).
N. coccinea. Canker. **76:**OR(82).
N. coryli. Canker. **76:**AK(526), OR(82,3202).
N. galligena. Canker. **76:**OR(82,3202).
N. peziza. On rotten wood. **39:**CA(3085). **76:**NY(3085).
N. ramulariae. **76:**Unk.(82).
N. sanguinea. **76:**CA(81).
Nectriella pironii. Stem gall. **76:**FL(35).
Nitschkia broomeiana. **51:**GA(1438).
Orbilia cruenta. **76:**AK(526).
O. eucalypti. **76:**WA(3202).
Orbiliopsis sp. **76:**MT(3202).
Otthia diminuta. **76:**AK(526).
Patellea californica. On dead limbs. **68:**CA(642).
Penzigia frustulosa. On wood. **51:**GA(1438). **76:**IA(1225).
Perrotia flammea. **76:**CA(661).
Pezicula ocellata. Branch canker. **17:**ND(82). **69:**GA(1438). **76:**CO(82), NC(82), ND(82), OR(3202), PA(82), WA(3202).
Pezizella vulgaris. **76:**WA(3202).
Phyllactinia guttata. Powdery mildew. **9:**WA(82,3202). **17:**WA(82). **22:**WA(3202). **76:**WA(82).
Physalospora abdita. On limbs. **76:**OR(82,3202), SC(82).
P. gregaria. Twig canker. **76:**WV(82).
Pleospora cerastii. **67:**AK(526).
P. herbarum. On twigs. **76:**CA(81).
P. moravica. On decorticated wood. **76:**AK(1938).
Pleospora sp. On twigs. **51:**WA(82).
Pleuroceras pleurostylum. **76:**OR(657).
Propolomyces versicolor. **37:**WA(3202). **76:**ID(3202).
Rhytisma salicinum. Tar spot. **1:**AK(3431). **4:**SD(82,2243). **8:**AK(526,3431). **9:**W states(82), WA(82). **16:**AK(3431). **17:**Range of host(82), MS(2720), SD(2243). **22:**ID(3202), WA(3202). **23:**CO(82), MT(82), ND(82), SD(82), WY(82,1195). **24:**WY(1195). **30:**CA(81). **32:**GA(1438). **35:**WA(82). **36:**AK(526). **37:**ID(3202), WA(3202). **43:**IA(82), ME(82), MI(82), NY(82). **48:**WY(1195). **51:**MS(82,2720), TX(82). **57:**WI(1350). **60:**CA(81). **63:**AK(526). **66:**AK(3431). **68:**AK(526,3431), MT(3202), WA(3202), WY(1195). **71:**AK(526,3431), WA(3202). **72:**AK(3431). **76:**Range of host(82), AK(526), ID(1195,3202), MI(2862), MT(3202), ND(377), OR(3202), SD(2243), WA(3202).
Rhytisma sp. **67:**AK(526).
Rosellinia albolanata. **76:**MT(3202).
R. aquila. **2:**GA(1438). **76:**WA(3202).
Schizothyrium perexiguum. On twigs. **76:**IN(82).
Schizoxylon albo-atrum. **76:**NY(3225).
S. compositum. **76:**KS(3225).
S. insigne. **1:**AK(526). **2:**GA(1438).
Scutellinia scutellata. On wood. **76:**CA(661).
Sphaerella grossulariae var. *salicella*. **76:**AK(526).
Sphaerulina salicina. On limbs. **23:**ND(82,377). **76:**ND(82).
Stictis chrysopsis. **76:**CA(3225).
S. radiata. **8:**WA(3202). **76:**CA(3225).
S. schizoxyloides. On dead limbs. **76:**CO(3225), ID(3202), MT(82,3202,3225).
S. serpentaria. On bark. **39:**CA(81,1277). **76:**CO(1277), OR(1277,3202), WA(1277,3202,3225).
Strickeria inclusa. **76:**ID(3202).
Sydowia dothideoides. Twig canker. **76:**CO(82).
Tapesia fusca. **71:**AK(3431). **76:**ID(3202), WA(3202).
T. mollisioides. **76:**CA(661).
Taphrina populi-salicis. Yellow leaf blister. **35:**CA(82,661,2454).
Teichospora aberrans. **76:**ND(377).
T. helenae. **76:**MT(3202), ND(377), OR(3202).
T. megastega. **76:**MT(3202).
Teichospora sp. **76:**AK(526).
Teichosporella dura. **76:**MT(3202).
Trematosphaeria confusa. **51:**LA(318).
Tryblidiella clavispora. On dead limbs. **51:**Unk.(2543). **76:**ND(377).
Tubeufia pezizula. **76:**ID(3202).
Uncinula adunca. Powdery mildew. **1:**AK(526). **2:**CT(82), NY(82). **4:**Range of host(82), MT(3202), WY(1195). **7:**NC(1303). **8:**AK(526). **17:**Range of host(82), ND(377), SD(2243), WI(1354,1358). **22:**MT(3202), WA(3202). **23:**IA(82), MI(82), MT(82,3202), NE(82). **25:**IA(82), MA(82), NY(82). **28:**MT(3202). **30:**CA(81). **31:**GA(1438). **35:**CA(82). **37:**WA(3202). **39:**CA(82). **44:**SD(2243). **48:**WY(1195). **50:**AK(526). **51:**Range of host(82), MS(2720), NC(1303,3944). **53:**WY(1195). **60:**CA(81). **68:**CA(81), MT(82,3202), OR(82,3202), WA(82,3202), WY(82,1195). **69:**IN(82). **71:**AK(526), WA(3202). **76:**Range of host(82), AK(1904), ID(3202), MI(2862), MS(2720), NC(1303), WA(638,3202).
Valsa ambiens. Twig blight. **51:**GA(1438). **76:**Range of host(82), CO(649,3415).
V. boreella. Twig blight, canker. **1:**AK(526). **9:**ID(82,3202), MT(82), WA(82,3202). **17:**KS(82). **23:**KS(82). **76:**ID(82), KS(82), MI(82).
V. ceratosperma. **51:**GA(1438). **68:**WA(3202).
V. germanica. **76:**MT(3202).
V. pallida. **51:**GA(1438).
V. salicina. Twig blight and canker. **7:**MD(82). **9:**ID(82), MT(82), WA(82). **11:**AK(526). **17:**KS(3415), MT(82), SD(82). **22:**MT(3202). **44:**MT(3415). **51:**AL(82), GA(1438), NE(82), WV(82). **68:**WA(3202). **75:**CA(82), IA(82). **76:**Range of host(82), IA(1226), ID(3202), ND(377), OR(3202), WA(3202).
V. sordida. Canker, dieback. **2:**NJ(82), NY(82), OK(82), WI(82). **4:**CO(82), SD(82). **7:**IA(82), ID(3202), MD(82), NJ(82), NM(82), TN(82), WA(3202). **9:**ID(82), MT(82), WA(82). **20:**MS(2720). **23:**ND(82,377). **25:**MA(82). **35:**ID(82). **37:**ID(3202). **39:**CA(82). **51:**NV(82). **76:**Range of host(82), AK(526), CO(3415), IA(1226), ID(3202), MT(3202), NY(3415), WA(3202).
Valsa sp. **2:**ID(3202). **76:**ID(3202).
Valsella nigroannulata. **23:**ND(377).
Venturia chlorospora. On leaves. **76:**CO(82), OR(657).
V. saliciperda. **76:**WA(3202).
V. subcutanea. On leaves. **66:**AK(526,682). **76:**AK(82).
Xylaria hypoxylon. **51:**GA(1438).
Xylaria sp. Isolated from wood. **51:**GA(1007).

Basidiomycotina - Uredinales

Melampsora abieti-capraearum. Heteroecious: telial host. **2:**CT(82), NY(82), PA(82), WA(82). **4:**WY(1195). **7:**MO(82), WV(82). **9:**NC states(82), NE states(82), W states(82), AZ(1195), WA(82), WI(1354), WY(1195). **14:**OK(2872). **17:**E states(82), KS(82), MT(82), WI(1360). **20:**WI(1344). **21:**WY(1195). **25:**IA(82), IL(82), NY(82), PA(82). **29:**CA(81). **43:**NC states(82), NE states(82), WI(82). **49:**WI(1355). **51:**Range of host(82), MS(2720), NC(3944), OK(2872). **55:**WI(1344). **57:**WI(1350). **69:**CT(82), GA(1439), IN(82), NY(82), VA(82). **75:**NY(82), PA(82). **76:**Range of host(82), FL(35), IN(647), LA(2675), MI(2862), NC(1307), WY(1195).
M. arctica. Heteroecious: telial host. **1:**AK(3431). **5:**AK(526). **6:**AK(60,526,3431), WY(749,1195). **8:**AK(3431). **15:**AK(60,526). **16:**AK(3431). **26:**AK(60,526). **28:**AK(526). **54:**AK(60,526). **62:**AK(60,526). **63:**AK(60,526). **66:**WY(1520). **71:**AK(60,526), CA(638,642). **72:**AK(60,526). **76:**AK(82,526), CO(82), NH(82).
M. epitea. Heteroecious: telial host. **2:**WA(3202). **4:**ID(3202), MT(3202), WA(3202), WY(1195). **9:**AZ(1203,3981), CO(649), ID(3202), MT(3202), OR(3202), WA(3202). **10:**AZ(1195,1203,3981). **16:**WA(3202). **18:**OR(3202). **21:**ID(3202), MT(3202), UT(1195), WA(3202). **22:**ID(3202), MT(3202), OR(3202), WA(3202). **23:**AZ(3981), ID(3202), MT(3202), OR(3202), WA(3202). **24:**MT(3202), OR(3202). **27:**ID(3202), OR(3202), WA(3202), WY(1195,3379). **28:**ID(3202). **29:**AZ(3981), TX(742). **30:**WA(3202). **33:**ID(3202). **35:**AZ(1203). **37:**ID(3202), MT(3202), OR(3202), WA(3202). **39:**OR(3202), WA(3202). **48:**MT(3202). **50:**ID(3202), MT(3202), OR(3202), WA(3202). **55:**WA(3202). **59:**WA(3202). **60:**OR(3202), WA(3202). **68:**AZ(1195,1203,3379,3981), ID(3202), MT(3202), OR(3202), WA(3202). **71:**OR(3202), WA(3202). **73:**AZ(1194,3981).

76:AZ(1203), CA(661), CO(649), WY(1195,3379).

M. paradoxa. Heteroecious: telial host. **1:**AK(60,526,3431). **2:**CT(82), NE(82), WA(82), WI(82). **4:**Range of host(82), IL(312), ND(378), SD(2243), WY(1195). **5:**AK(526). **6:**AK(60,526,3431). **7:**FL(35), WI(1354). **8:**AK(60,526,3431). **9:**WI(1350), WY(1195,3379). **10:**AZ(1203). **12:**CA(81). **13:**ND(378). **16:**AK(3431). **17:**W states(82). **23:**W states(82), CA(81), ND(378), WA(82), WY(1195). **24:**W states(82), WA(82), WY(1195). **26:**AK(60). **27:**CA(81), WY(1195). **28:**AK(60,526). **29:**AZ(747,1203), CA(81). **34:**CA(81). **35:**NC states(82), NW states(82), AZ(1203), CA(81,82), WA(82). **38:**CA(81). **39:**CA(81,82). **40:**CA(81). **42:**CA(81). **44:**W states(82), WY(1195,3379). **45:**CA(81). **46:**CA(81). **47:**CA(81). **50:**WY(1195). **51:**GA(1439), MS(2720), NC(3944). **53:**WY(1195). **54:**AK(60,526). **58:**AK(60,526). **61:**WY(1195). **62:**AK(60). **63:**AK(60,526). **66:**AK(60,526,1938). **67:**AK(60,526). **68:**W states(82), CA(81), SD(2243), WA(82), WY(1195). **70:**WI(1359). **71:**AK(60,526,3431), CA(81). **72:**AK(526,3431). **74:**CA(81). **76:**E states(82), NC states(82), NW states(82), AK(526,1904), FL(35), IA(82), NC(1303), NM(82), OK(2872), SD(2243).

M. ribesii-purpureae. Heteroecious: telial host. **1:**AK(526,1904). **4:**WY(1195). **5:**AK(526). **9:**WY(1195). **16:**AK(3431). **19:**CA(81). **21:**CO(1195), WY(1195). **26:**AK(1904). **27:**WY(1195). **28:**AK(526). **35:**AZ(1195), NV(82). **39:**AZ(82), CA(82), OR(82), WA(82). **41:**ID(1195). **44:**W states(82), CA(82), OR(82), UT(1195). **48:**AK(1904). **53:**WY(1195). **61:**WY(2924). **66:**AK(526). **71:**CA(81,638). **74:**CA(81). **76:**W states(82), AK(82,526), CA(82), CO(82,1195), ID(1195), NM(1195), UT(1195), WY(1195).

Uredo mckinleyensis. **66:**AK(526).

U. phoradendri. **35:**CA(81).

Basidiomycotina - Other

Aleurodiscus oakesii. On bark. **51:**NJ(2083). **76:**E states(82).

Antrodia albida. Wood rot. **76:**NY(82).

A. heteromorpha. **76:**AK(214,526).

A. malicola. Wood rot. **76:**WI(82).

A. sinuosa. **76:**CA(2439).

Armillaria mellea. Root rot. **7:**NC(1303). **76:**CA(82), WA(82,3202).

A. tabescens. Root rot. **76:**FL(35).

Athelia arachnoidea. **76:**WA(3202).

Basidiodendron cinereum. On wood. **76:**IA(2185).

Bjerkandera adusta. White rot. **37:**WA(3202). **51:**GA(1007), NC(3944). **68:**MT(3202). **76:**LA(82), NC(82), OR(3202), WI(82).

B. fumosa. On wood. **76:**CA(82).

Bondarzewia montana. **71:**AK(3431).

Calathella ellisii. **76:**OR(3202).

C. eruciformis. On twigs. **76:**CA(661).

Ceratobasidium anceps. **57:**WI(1352).

Ceriporia reticulata. **76:**AK(82,526).

Cerrena unicolor. **76:**NE states(82), ID(3202), KS(82), OR(3202), SD(82).

Chondrostereum purpureum. Wood rot. **7:**WA(3202). **51:**GA(491,1007,1439). **76:**WA(82,3202).

Clavicorona pyxidata. **76:**MO(884).

Coniophora arida var. *suffocata*. **76:**IA(128).

Coriolopsis gallica. Trunk rot. **10:**AZ(1194,1202). **29:**AZ(1202). **51:**AR(82), LA(82). **76:**Range of host(82), ID(3202), MT(3202,3832), OR(3202).

C. rigida. Wood rot. **76:**LA(82).

Corticium argentatum. **76:**KS(2118).

Crepidotus kauffmanii. **76:**ID(3202).

Cylindrobasidium laeve. **76:**OR(4007).

C. torrendii. Isolated from wood. **51:**GA(1007). **76:**WA(3202).

Cyphellopsis anomala. On wood. **1:**AK(526). **37:**CA(661). **68:**CA(642). **76:**WA(3202), WY(2924).

C. mellea. On wood. **51:**LA(456).

Cytidia flocculenta. On limbs and trunks. **76:**AK(82,526), MT(82), WY(82).

C. salicina. On dead limbs. **1:**AK(526). **68:**ID(638,3202). **76:**Range of host(82), AK(526), ID(3202), OR(3202), WA(3202), WY(2924).

Daedaleopsis confragosa. White rot. **1:**AK(1904). **4:**IN(82), MA(82). **7:**PA(82). **9:**MT(82,3202). **22:**MT(3202). **23:**MO(82), SD(82). **25:**MA(82). **51:**Range of host(82), GA(1439), NC(3944). **76:**Range of host(82), AK(526), ID(3202), MA(1391), MT(3832), NC(1303), ND(378), OR(3202), SD(2243), WA(3202).

Datronia mollis. Wood rot. **76:**ME(82).

D. stereoides. On wood. **76:**CO(82).

Dendrophora erumpens. **68:**WA(543,3202).

D. versiformis. **76:**MI(543).

Dendrothele macrodens. On bark. **76:**E states(82), E states(82).

D. microspora. On bark. **76:**LA(2084), PA(2084).

Exidia glandulosa. **1:**AK(526). **76:**ID(3202), WA(3202).

E. repanda. **68:**ID(3202). **71:**WA(3202).

E. umbrinella. **76:**OR(3202).

Exidia sp. **9:**ID(3202). **76:**MT(3202), OR(3202).

Exidiopsis fugacissima. On wood. **76:**IA(128).

E. molybdea. On wood. **76:**IA(128).

Flammulina velutipes. Commonly on stumps and logs, sometimes in wounds of living trunks. **76:**ID(3202), ND(82), NY(82).

Fomes fasciatus. On wood. **51:**MS(2720).

F. fomentarius. Wood rot. **76:**AL(82), ID(3202).

Fomitella supina. Isolated from wood. **51:**GA(1007).

Ganoderma applanatum. White rot. **2:**CT(82). **51:**Unk.(82), MS(2720). **76:**CT(82), NY(82), SD(2243), WA(3202), WI(82).

G. lucidum. White root and butt rot. **7:**MD(82). **51:**MS(2720). **76:**CA(82), VA(82,638).

Gloeocystidiellum porosum. **1:**AK(3431).

G. sinuosum. On wood. **76:**AR(436).

Gloeodontia discolor. **29:**AZ(128,440,1202).

Gloeophyllum sepiarium. Brown cubical rot. **76:**Range of host(82).

G. trabeum. Brown cubical rot. **76:**Range of host(82).

Gloeoporus dichrous. On wood. **76:**NY(82).

Grandinia arguta. **25:**NY(128).

Gymnopilus spectabilis. Sapwood rot. **76:**N states(82).

Hapalopilus rutilans. White rot. **76:**Unk.(2681).

Helicobasidium brebissonii. Root rot. **51:**TX(82). **76:**TX(82).

Hericium erinaceus. Isolated from wood. **51:**GA(1007).

Hydnochaete olivacea. Isolated from wood. **51:**GA(1007).

Hymenochaete agglutinans. **76:**AK(526).

H. cinnamomea. **76:**WA(3202).

H. tabacina. On wood. **1:**AK(1904). **76:**AK(526), ID(3202), OR(3202), WA(3202).

Hypochnicium analogum. **29:**AZ(1202).

H. bombycinum. On bark. **76:**NE states(82), TX(82), WA(82,3202).

Hypsizygus ulmarius. Sapwood rot, wound rot. **51:**NC(82,3944). **76:**NC(82), SD(2243).

Inonotus andersonii. On wood. **51:**MD(82). **76:**MD(82).

I. cuticularis. **10:**AZ(1181). **29:**AZ(1181).

I. dryophilus. On wood. **76:**OH(82).

I. farlowii. On wood. **76:**SW states(82), AZ(1181).

I. munzii. **10:**AZ(1202). **29:**AZ(1194,1202).

I. radiatus. On wood. **51:**MS(2720).

Irpex lacteus. On wood. **76:**NE states(82), OR(3202).

Junghuhnia nitida. **76:**NY(128,2168), WA(3202).

Laetiporus sulphureus. On wood. **76:**Range of host(82).

Laxitextum bicolor. **76:**LA(543).

Lentinellus ursinus. **29:**AZ(1202).

Lentinus tigrinus. **29:**AZ(1202).

Lenzites betulina. On wood. **76:**Range of host(82).

Lopharia cinerascens. **4:**MT(3202). **76:**CO(649), IA(543).

Merismodes fasciculatus. On limbs and twigs. **76:**NY(2945).

M. ochraceus. On bark, sometimes of living trees. **76:**N states(82).

Meruliopsis corium. **1:**AK(3431).

Merulius sp. **76:**WA(3202).

Mycena alcalina. **51:**NC(3944).

Oxyporus corticola. **76:**WA(3202).

O. latemarginatus. White rot. **51:**GA(1007), MD(82), TX(82). **76:**AR(2166), TX(82), VA(2166).

O. populinus. **51:**Unk.(82). **76:**NY(82).

Panellus serotinus. Sapwood rot. **76:**MI(82).

Peniophora aurantiaca. **76:**ID(128,3202).

P. cinerea. Isolated from wood. **2:**GA(1439). **51:**GA(1007).

P. incarnata. **76:**WA(3202).

P. rufa. On wood. **76:**CO(82), ID(82).

Perenniporia fraxinophila. Wood rot. **76:**NY(82).

P. medulla-panis. On wood. **51:**MI(82), NY(82). **76:**Range of host(82), ID(3202).

Phanerochaete affinis. **76:**OR(3202), WA(3202).
P. flavido-alba. Isolated from wood. **51:**GA(1007), LA(128).
P. magnoliae. White rot. **51:**NY(128,433).
Phellinus conchatus. White rot. **9:**MT(82,3202,3832), WA(82). **17:**SD(82). **23:**ND(82,378). **37:**WA(3202). **76:**MT(82), ND(82), NY(82), SD(82).
P. crocatus. Wood rot. **51:**LA(82).
P. ferreus. **76:**OR(3202), WA(3202).
P. ferruginosus. Wood rot. **1:**AK(526). **76:**AK(82), MT(3202).
P. gilvus. On wood. **29:**AZ(1202). **37:**MT(3202). **76:**NE(82).
P. igniarius. White heart rot. **1:**AK(1904,3431). **7:**WA(3202). **9:**MT(82), WA(82,3202). **35:**WA(82). **37:**ID(3202), WA(3202). **39:**CA(2439). **51:**Unk.(82). **60:**OR(3202). **68:**ID(3202), MT(3202), OR(1524), WA(3202). **71:**OR(3202). **76:**CA(661), ID(3202).
P. nigricans. White spongy heart rot. **1:**AK(526). **76:**AK(82,526), ID(82), IN(82), NY(82), WA(82), WY(82).
P. punctatus. White rot. **2:**ND(3759). **3:**ND(3759). **9:**ND(3759). **22:**ND(3759). **76:**MI(82), MN(82).
P. robustus. **76:**MN(128,1197).
P. viticola. Wood rot. **76:**NC(82), PA(82).
Phlebia albida. **68:**MT(3202).
P. concentrica. **76:**ID(3202), WA(3202).
P. merismoides. **76:**ID(3202).
P. subochracea. White rot. **76:**AR(2556), MS(2556).
Pholiota adiposa. Sapwood rot. **7:**WA(3202). **76:**N states(82), AK(526).
P. aurivelloides. Sapwood rot. **76:**CO(82), NM(82), WY(82).
P. limonella. **7:**WA(3202).
P. oregonensis. Sapwood rot. **76:**OR(82,3202).
P. squarrosa. **7:**WA(3202).
Pleurotus cystidiosus. Isolated from wood. **51:**AR(2418).
P. levis. **29:**AZ(1202).
P. ostreatus. Sapwood rot. **7:**WA(3202). **23:**WA(3202). **29:**AZ(1202). **51:**GA(1007), MS(2720).
Plicaturopsis crispa. **51:**NC(3944).
Polyporus admirabilis. On wood. **76:**NY(82).
P. arcularius. Isolated from wood. **51:**GA(1007).
P. badius. **76:**ID(3202), OR(3202), WA(3202).
P. brumalis. **76:**ID(3202), MT(3202).
P. mori. **76:**AK(526).
P. squamosus. White heart rot. **7:**WA(3202). **51:**MA(82). **76:**NE states(82), OR(3202), WA(3202).
P. varius. Wood rot. **1:**AK(526,3431). **9:**WA(3202). **24:**ID(3202). **71:**AK(3431). **76:**AK(82,2419), CA(661), ID(82,3202), OR(3202), WA(82), WY(82).
Postia caesia. On wood. **76:**ID(3202), MT(3832), MT(82,3202), WA(3202).
Pycnoporus cinnabarinus. On wood. **4:**WA(3202). **76:**NY(82), WA(82).
P. sanguineus. On wood. **29:**AZ(1202). **76:**MO(82).
Schizophyllum commune. On wood. **29:**AZ(1202). **51:**GA(1007). **76:**Range of host(82), CA(661), ID(3202), WA(3202).
Sistotrema brinkmannii. On wood. **76:**GA(491,1439).
S. hirschii. **29:**AZ(1202).
Sphaerobolus stellatus. **76:**ID(3202).
Spongipellis pachyodon. Isolated from wood. **51:**GA(1007).
S. unicolor. **68:**MT(3202).
Steccherinum fimbriatum. **76:**WA(3202).
S. laeticolor. **76:**ID(3202).
S. ochraceum. On dead wood. **76:**Range of host(82), AK(526).
Stereum hirsutum. White rot. **37:**WA(3202). **51:**GA(1007). **64:**PA(543). **76:**Range of host(82), W states(82), CA(661), ID(3202), MA(543), OR(3202).
S. ochraceo-flavum. **76:**CT(543), FL(543).
S. rugosum. White rot. **76:**MN(82).
Stereum sp. On wood. **68:**CA(642).
Thelephora caryophyllea. **71:**AK(3431).
Trametes cervina. **29:**AZ(1202).
T. elegans. Wood rot. **51:**GA(82). **76:**NC(82), VA(82).
T. hirsuta. White rot. **51:**NC(82,3944). **68:**MT(3202). **76:**Range of host(82), ID(3202), WA(3202).
T. pavonia. On wood. **76:**FL(2681), LA(2681).
T. pubescens. On wood. **9:**ID(3202). **37:**WA(3202). **76:**NW states(82), ID(3202), MT(3202,3832), NY(82), WA(3202).
T. suaveolens. On wood. **1:**AK(526). **2:**CT(82), ME(82). **76:**NC states(82), NE states(82), NW states(82), AK(82), MA(1394), MT(3202,3832).
T. trogii. Wood rot. **29:**AZ(1202). **76:**MI(82).
T. versicolor. White rot. **1:**AK(3431). **37:**WA(3202). **39:**CA(2439). **51:**GA(1007), LA(82). **76:**Range of host(82), OR(3202), WA(3202).
Trechispora fastidiosa. **29:**AZ(1202).
Tremella mesenterica. On wood. **76:**CA(3949).
Trichaptum biforme. On wood. **37:**MT(3202,3832). **76:**Range of host(82).
Tubaria pellucida. **76:**WA(3202).
Tulasnella eichleriana. **76:**IA(128).
T. pinicola. **76:**IA(128).
T. pruinosa. **76:**WA(3202).
Tyromyces chioneus. **76:**WA(3202).
T. fumidiceps. On wood. **76:**MO(82).
T. galactinus. On wood. **76:**NY(82).
Vesiculomyces citrinus. Isolated from wood. **51:**GA(1007).
Volvariella bombycina. **29:**AZ(1202).

Deuteromycotina - Hyphomycetes

Alternaria sp. Leaf spot. **76:**FL(35).
Bahusakala cookei. On branches. **76:**CA(661).
Bispora betulina. **76:**DE(3543).
Cercospora salicis. On leaves. **2:**WI(581,1341). **23:**WI(1350).
Cercospora sp. Leaf spot. **7:**NC(1303). **17:**WI(1356).
Cladosporium herbarum. **76:**AK(526).
C. salicis-sitchensis. **71:**WA(3202).
Costantinella terrestris. **1:**AK(3431).
Dendrodochium densipes. **76:**DE(3543).
Dendryphiopsis atra. On twigs. **76:**CA(661).
Fumago salicina. On leaves. **76:**SC(634).
Helicomyces colligatus. On wood. **76:**IA(1270).
Helminthosporium sp. **31:**DE(3543).
Kabatiella boreale. Leaf spot. **17:**VT(82). **28:**WY(1195). **48:**WY(1195). **76:**AZ(1195), VT(82), WY(82).
Ozonium auriconum. **1:**AK(526).
Phymatotrichopsis omnivora. Root rot. **51:**TX(82). **76:**TX(82).
Pollaccia saliciperda. Shoot and twig blight, scab. **2:**CT(82), MA(82), ME(82), NH(82), NY(82). **7:**MA(82). **17:**CT(82), MA(82), NY(82), PA(82). **25:**MA(82), NY(82). **43:**CT(82), ME(82). **51:**CT(82), MA(82), NY(82). **69:**CT(82), NY(82). **76:**NE states(82), NC(82,1303), PA(82).
Pseudocercospora salicina. Leaf spot. **2:**LA(82). **7:**FL(35), MD(82). **23:**LA(82). **51:**LA(82,962), TX(82). **76:**FL(35), IL(82), LA(82), MD(82), SD(2243), TX(82).
Ramularia lucidae. Leaf spot. **43:**WI(82).
R. rosea. Leaf spot. **5:**AK(526). **9:**AK(526), WI(82). **76:**CO(82), MT(82,3202), WA(3202), WI(82).
Taeniolella stilbospora. **63:**AK(526).
Torula fuckelii. **25:**DE(3543).
Trimmatostroma americana. On dead twigs. **68:**ID(3202). **76:**Range of host(82), DE(3543).
T. salicis. **76:**ND(378).
Tubercularia vulgaris. **76:**NM(1003).
Verticicladium pulvereum. **68:**WA(3202).

Deuteromycotina - Coelomycetes

Aplosporella salicicola. On dead twigs. **17:**KS(82). **76:**KS(82), MD(82), PA(82), TX(82).
Ascochyta babylonica. On leaves. **77:** WI(1359).
A. salicis. Leaf spot. **35:**CA(82).
Asteroma capreae. On leaves. **2:**CT(82). **76:**CT(82).
Colletotrichum lucidae. On living leaves. **43:**WI(1350). **65:**WI(1359).
Coniothyrium fuligineum. **54:**AK(526). **76:**AK(82).
C. spokanense. **76:**WA(82,3202).
Coniothyrium sp. On twigs. **20:**WI(1360). **76:**CT(82).
Coryneum salicinum. On limbs and twigs. **23:**ND(82,378,3547). **76:**AK(526), ND(82).
Coryneum sp. Leaf spot. **76:**OK(2872).
Cryptosporium sp. Branch canker. **39:**CA(82).
Cylindrosporella conspicua. On living leaves. **49:**WI(1354).
Cylindrosporium nuttallii. On leaves. **37:**WY(1195,3378).
C. salicinum. Leaf spot. **9:**WI(82). **17:**NY(82). **25:**WI(82,1330). **43:**NC states(82), NE states(82), IA(82). **51:**MS(2720). **76:**C & W states(82), NE states(82), MS(82), NC(1303), OK(2872),

OR(3202,4007), WA(3202), WI(82,1350).
Cytospora chrysosperma. Canker, dieback. **3:**OK(2873). **7:**WY(1195). **20:**NC(1303), SD(2243). **76:**OK(2872).
C. fugax. On dead twigs. **39:**CA(2439).
Cytospora sp. Canker. **7:**NC(1303).
Dendrophoma caespitosa. **51:**WA(82,3202).
Diplodia salicina. Twig dieback. **7:**FL(35). **51:**WV(82). **76:**FL(35), TX(82), WV(82).
Diplodia sp. On limbs. **76:**CA(82), TX(82), WV(82).
Diplodina sp. Twig blight. **76:**MA(82), MD(82), WV(82).
Discella microsperma. Twig blight and canker. **2:**CT(82). **7:**MA(82), NM(82). **17:**SD(82). **76:**NE states(82), MS(2720), OK(2873).
Discula brenckleana. On limbs and twigs. **76:**MT(82).
D. microsperma. On twigs. **76:**CA(82), MS(2720).
Dothiorella gregaria. Black canker. **39:**CA(82).
D. pyrenophora var. *salicis*. On twigs. **76:**AK(82,526).
D. ribis. Branch canker. **20:**OK(2872).
Dothiorella sp. Branch canker. **76:**AR(82), ND(82).
Fusamen fructigenum. On catkins. **76:**CO(82), ID(3202), MT(82), WA(3202).
Gloeodes pomigena. On limbs and twigs. **51:**IN(82). **76:**IN(82).
Gloeosporium sp. Twig blight, leaf spot. **76:**CT(82), DE(82), MA(82).
Hendersonia lineolans. On limbs. **76:**PA(82).
Leptothyrium pulchrum. **63:**AK(526).
L. salicicola. On living leaves. **57:**WI(1355).
Macrophoma sp. Branch and limb canker. **7:**FL(35), TX(82). **20:**MS(2720), OK(2872). **76:**AR(82), KY(82), MS(82), NC(82), SC(82), TX(82).
Marssonia sp. Twig canker. **7:**NY(82).
Marssonina apicalis. Leaf spot, on twigs. **35:**CA(82). **39:**CA(82). **43:**WI(82). **72:**AK(3431). **76:**CA(82), WI(82), WY(82).
M. kriegeriana. Leaf spot, on twigs. **4:**IL(312). **7:**WA(3202). **8:**AK(3431). **16:**AK(3431). **17:**WI(82). **21:**WY(1195). **28:**WY(1195). **48:**WY(1195). **70:**WI(1356). **76:**AZ(1195), CA(82), SD(2243), WI(82), WY(82).
M. nigricans. Leaf spot, on twigs. **39:**CA(82). **76:**CA(82).
M. populi. Leaf spot. **76:**OK(2872).
M. rubiginosa. Leaf spot, on twigs. **76:**ID(82,3202), WI(82).
M. salicicola. Leaf spot. **7:**NC(1303). **76:**MD(82), NY(82).
M. salicina. Leaf spot, on twigs. **51:**IL(82). **76:**IL(82).
M. salicis. Leaf spot. **76:**MA(82).
Marssonina sp. Twig blight, leaf spot. **9:**WA(82). **25:**GA(82), MA(82). **35:**OR(82). **68:**WA(3202). **76:**MT(3202), WA(3202).
Melanconium salicinum. On twigs. **51:**LA(82).
Melanconium sp. Twig canker. **51:**CO(82).
Monostichella salicis. Twig blight, leaf spot. **2:**SD(2243), WI(82). **23:**NY(82), WI(82). **25:**WI(82). **43:**IL(82), WI(82). **76:**NE states(82), MS(82,2720), SD(2243), WI(82).
Myrioconium comitatum var. *salicarium*. Leaf spot. **23:**WI(82). **57:**WI(1360).
Myxofusicoccum salicis. On limbs. **37:**ID(3202). **68:**WA(82,3202). **76:**WA(82).
Myxosporium cytosporeum. On limbs. **76:**WA(82,3202).
M. salicinum. **76:**MA(82).
Phoma pezizoides. On decorticated wood. **51:**WV(307).
P. platysperma. On limbs and trunks. **51:**IL(82,2738), WV(82). **76:**IL(82).
P. terrestris. Isolated from rhizosphere. **76:**OR(2791).
Phoma sp. Twig blight and dieback. **7:**FL(35). **51:**IL(82), WV(82). **76:**AK(526), AR(82), NC(82).
Phomopsis pallida. **7:**WA(3202).
P. salicina. Twig canker. **76:**AK(526), IA(82), MA(82), VA(82), WA(3202).
Phomopsis sp. Canker, twig dieback. **7:**FL(35), NC(1303).
Phyllosticta apicalis. Leaf spot. **2:**KS(82). **43:**WI(64,82). **76:**KS(82), WI(82).
P. eminens. On living leaves. **20:**WI(1361). **76:**WI(1354).
P. salicicola. **76:**CT(82), ID(82,3202), WI(1336).
P. salicis. **76:**KS(82).
Phyllosticta sp. Leaf spot. **7:**FL(35). **17:**WI(1359). **76:**FL(35).
Sclerophoma salicis. Twig blight. **76:**CA(82).
Seimatosporium pezizoides. On bark and limbs. **76:**CO(82,959,3547,3549), ID(82).
Septogloeum maculans. Leaf spot. **39:**CA(82,1468). **76:**CA(82), MT(3202).
S. salicinum. Leaf spot. **9:**WI(82,1352). **57:**WI(1352). **69:**NY(82). **71:**AK(526). **76:**AK(82), NY(82), WI(82).
S. salicis-fendlerianae. Leaf spot. **35:**ID(82), MT(82), WY(82). **37:**ID(3202), MT(3202), WA(3202), WY(1195). **76:**ID(82).
Septomyxa grisea. On twigs. **69:**NY(82).
Septoria didyma. Leaf spot. **23:**WI(1350). **25:**WI(82). **51:**WI(1350). **76:**WI(82,1350).
S. didyma f. *santonensis*. **56:**WI(1357).
S. rhabdocarpa. On leaves. **39:**CA(82,2439).
S. salicicola. **76:**AK(82), MT(3202), OR(82,3202).
S. salicinia. **70:**WI(1356).
S. salicis. Leaf spot. **17:**KS(82).
Septoria sp. Leaf spot. **7:**NC(1303). **76:**ID(3202), OR(3202), WA(3202).
Sphaceloma murrayae. Gray scab. **3:**WI(1354). **20:**WI(1350). **23:**WI(1350). **35:**OR(82), WA(82). **37:**OR(3202), WA(1745,3202). **39:**CA(82). **43:**WI(1350). **51:**WI(1350). **57:**WI(1356). **76:**WA(3202).
Sphaeropsis sp. Twig dieback. **76:**FL(35).
Stagonospora samarorum. **52:**AK(526).
Truncatella angustata. **64:**AK(526).
Deuteromycotina - Other
Rhizoctonia solani. Basal stem rot. **7:**FL(35).
Rhizoctonia sp. Root rot. **7:**HI(2903).

Santalaceae

Buckleya Torr.

1. *B. distichophylla* (Nutt.) Torr.

Basidiomycotina - Uredinales
Cronartium appalachianum. Heteroecious: telial host. **1:**NC(1530), TN(82,1530), VA(1530).

Comandra Nutt.

1. *C. pallida* A. DC.
2. *C. richardsiana* Fernald
3. *C. umbellata* (L.) Nutt. (=*C. umbellata* (L.) Nutt. subsp. *umbellata*)
4. *Comandra* sp.

Zygomycotina
Absidia spinosa. **1:**WY(1195).
Basidiomycotina - Uredinales
Cronartium comandrae. Heteroecious: telial host. **1:**AZ(1203,3981), ID(3202), MT(1195,3202), ND(378), OR(3202), SD(2243), WA(3202), WY(1195,2924). **2:**WI(1333). **3:**Range of host(82), W states(1977), AR(2866), MN(248), MS(2720,2866), MT(3202), OR(3202), TN(674), WA(3202).
Puccinia andropogonis. Heteroecious: aecial host. **1:**ND(378). **3:**C states(82), NE states(82), WI(82). **4:**C & W states(82), ND(82).
P. comandrae. Autoecious. **1:**ID(3202), MT(3202), OR(3202), WA(3202). **3:**CA(642), OR(3202), WA(3202).
Deuteromycotina - Hyphomycetes
Cercospora comandrae. **2:**WI(1333).
Cladosporium echinulatum. **3:**WA(3202).

Geocaulon Fernald

1. *G. lividum* (Richardson) Fernald (=*Comandra livida* Richardson)

Basidiomycotina - Uredinales
Cronartium comandrae. **1:**WA(3202).

Pyrularia Michx.

1. *P. pubera* Michx.

Ascomycotina
Melogramma sp. **1:**GA(1438).

Santalum L.

1. *Santalum* sp.

Ascomycotina
Asterina congesta. **1:**HI(2784).

Sapindaceae

Blighia K. D. Koenig
1. *B. sapida* K. D. Koenig [akee]
Deuteromycotina - Coelomycetes
Colletotrichum sp. Anthracnose. **1**:FL(35).
Phoma sp. Fruit rot. **1**:FL(35).
Phyllosticta sp. Leaf spot. **1**:FL(35).

Cardiospermum L.
1. *C. halicacabum* L. [balloonvine]
Chytridiomycetes
Synchytrium cardiospermi. Leaf gall. **1**:LA(627).

Cupaniopsis Radlk.
1. *C. anacardiopsis* (A. Rich.) Radlk.
Deuteromycotina - Hyphomycetes
Corynespora sp. Leaf spot. **1**:FL(35).
Deuteromycotina - Coelomycetes
Phomopsis sp. Stem canker. **1**:FL(35).
Phyllosticta sp. Leaf spot. **1**:FL(35).

Dodonaea Mill.
1. *D. viscosa* (L.) Jacq.
2. *D. viscosa* (L.) Jacq. var. *arborescens* (J. D. Hook.) Jacq.
3. *Dodonaea* sp.
Oomycetes
Pythium sp. Root rot. **1**:FL(35).
Ascomycotina
Episphaerella dodonaeae. On leaves. **1**:FL(1021).
Meliola lyoni. Black mildew. **1**:HI(82,1272,1452,2903). **3**:HI(82,1452,2784).
Basidiomycotina - Other
Campanella junghuhnii. **1**:HI(82). **2**:HI(2903). **3**:HI(82).
Phellinus robustus. **1**:AZ(1202).
P. sonorae. At base of dead or dying plants; possibly a root rot pathogen. **1**:AZ(1182).

Exothea Macfady.
1. *E. paniculata* Walp. ex Radlk. [butterbough]
Basidiomycotina - Other
Phellinus extensus. Butt rot, heart rot. **1**:FL(35,82).

Filicium Thwaites ex J. D. Hook. in Benth. & J. D. Hook.
1. *F. decipiens* (Wight & Arnott) Thwaites
Deuteromycotina - Other
Rhizoctonia sp. Root rot. **1**:HI(2903).

Harpullia Roxb.
1. *H. arborea* (Blanco) Radlk.
2. *Harpullia* sp.
Oomycetes
Pythium sp. Root rot. **1**:FL(35).
Deuteromycotina - Hyphomycetes
Cercospora pteleae. Leaf spot. **2**:FL(35).
Deuteromycotina - Coelomycetes
Colletotrichum sp. Leaf spot. **1**:FL(35).

Koelreuteria Laxm.
1. *K. elegans* (Seem.) A. C. Sm. [flamegold]
2. *K. paniculata* Laxm. (=*K. apiculata* Rehd. & E. Wils.) [varnish tree]
3. *Koelreuteria* sp.
Oomycetes
Pythium sp. Root rot. **1**:FL(35).
Ascomycotina
Botryosphaeria obtusa. On limbs. **3**:MD(82).
Nectria cinnabarina. Canker. **2**:OK(2872). **3**:CA(82), CT(82).
Basidiomycotina - Other
Armillaria tabescens. Root rot. **1**:FL(35). **2**:FL(35).
Deuteromycotina - Hyphomycetes
Cercospora sp. Leaf spot. **1**:FL(35). **2**:FL(35). **3**:FL(82).
Corynespora sp. Leaf spot. **1**:FL(35).
Helminthosporium sp. Leaf spot. **2**:FL(35).
Phymatotrichopsis omnivora. Root rot. **3**:TX(82).
Verticillium albo-atrum. Wilt. **2**:IL(992,1558).
Verticillium sp. Wilt. **3**:NJ(82).
Deuteromycotina - Coelomycetes
Botryodiplodia sp. Twig dieback. **2**:FL(35).
Cytosporina sp. **2**:OK(2872).
Diplodia koelreuteriae. On limbs. **3**:NJ(82).
Macrophomina phaseolina. Charcoal rot. **1**:FL(35).
Monochaetia sp. Stem canker. **2**:FL(35).
Phoma sp. Stem canker. **2**:FL(35).
Phyllosticta sp. Leaf spot. **1**:FL(35). **2**:FL(35), OK(2872).
Deuteromycotina - Other
Rhizoctonia sp. Root rot. **1**:FL(35).

Litchi Sonn.
1. *L. chinensis* Sonn. (=*Nephelium litchi* Cambess.) [leechee, litchi, lychee]
Oomycetes
Pythium sp. Root rot. **1**:FL(35).
Ascomycotina
Glomerella sp. **1**:HI(2903).
Leptosphaeria sp. Twig dieback. **1**:FL(35).
Basidiomycotina - Other
Armillaria tabescens. Root rot. **1**:FL(35,82,592).
Deuteromycotina - Coelomycetes
Colletotrichum sp. Fruit rot. **1**:FL(35).
Diplodia sp. Dieback. **1**:FL(35).
Phomopsis sp. Dieback, leaf spot. **1**:FL(35).
Sphaceloma sp. **1**:FL(2525).
Deuteromycotina - Other
Rhizoctonia solani. Root and stem rot. **1**:FL(35).

Melicoccus P. Br.
1. *M. bijugatus* Jacq. [genipe, honeyberry, mamoncillo, Spanish lime]
Basidiomycotina - Other
Armillaria tabescens. Root rot. **1**:FL(35).
Deuteromycotina - Hyphomycetes
Fusarium sp. Leaf spot. **1**:FL(35).
Deuteromycotina - Coelomycetes
Phyllosticta sp. Leaf spot. **1**:FL(35).

Nephelium L.
1. *N. lappaceum* L. [rambutan]
Deuteromycotina - Coelomycetes
Phyllosticta sp. Leaf spot. **1**:FL(35).

Sapindus L.
1. *S. oahuensis* Hildebr. ex Radlk. [Hawaiian soap-tree]
2. *S. saponaria* L. [false dogwood, jaboncillo, southern soapberry, wingleaf soapberry]
3. *S. saponaria* L. var. *drummondii* (Hook. & Arnott) L. Benson (=*S. drummondii* Hook. & Arnott) [jaboncillo, western soapberry]
Ascomycotina
Brasiliomyces setosus. Powdery mildew. **1**:HI(1597).
Dothivalsaria megalospora. **3**:OK(660).
Glomerella cingulata. Anthracnose, leaf spot. **2**:FL(35,82). **3**:TX(82).
[*Mycosphaerella effigurata*]. Leaf spot. **3**:MO(82).
Teichospora trabicola. **3**:OK(660).
Uncinula circinata. Powdery mildew. **3**:TX(82).
Basidiomycotina - Other
Helicobasidium brebissonii. Root rot. **3**:TX(82).
Scytinostroma portentosum. **3**:AZ(1202).
Deuteromycotina - Hyphomycetes
Phymatotrichopsis omnivora. Root rot. **3**:OK(82,2872), TX(82).
Deuteromycotina - Coelomycetes
Cylindrosporium griseum. Leaf blight. **3**:OK(82,2872), TX(82).
Cytosporella sp. Canker. **3**:OK(618).
Phomopsis sp. On limbs and twigs. **3**:OK(82,2873).
Phyllosticta sapindii. Leaf spot. **2**:FL(35,82).

Serjania Mill.
1. *S. brachycarpa* A. Gray
2. *Serjania* sp.
Ascomycotina
Meliola serjaniae. **1**:TX(3483).
Basidiomycotina - Uredinales
Puccinia arechavaletae. Autoecious. **2**:TX(82).

Ungnadia Endl.
1. *U. speciosa* Endl. [Mexican-buckeye]
Ascomycotina
Dothivalsaria megalospora. **1:**TX(185).
Deuteromycotina - Hyphomycetes
Phymatotrichopsis omnivora. Root rot. **1:**TX(82).

Sapotaceae

Bumelia Sw.
1. *B. microcarpa* Small
2. *B. parviflora* (Lundell) Lundell
3. *Bumelia* sp.
Ascomycotina
Meliola bumeliae. **1:**FL(3483). **2:**FL(1452). **3:**FL(3483).
Mycosphaerella hermione. **3:**GA(82,1438).

Chrysophyllum L.
1. *Chrysophyllum* sp.
Deuteromycotina - Coelomycetes
Phomopsis sp. Leaf spot. **1:**FL(35).
Phyllosticta sp. Leaf spot. **1:**FL(35).

Manilkara Adans.
1. *M. jaimiqui* (W. Wright ex Griseb.) Dubard subsp. *emarginata* (L.) Cronq. (=*M. bahamensis* (Baker) H. J. Lam & A. D. J. Meeuse) [wild dilly, wild sapodilla]
2. *M. zapota* (L.) D. Royen (=*Achras zapota* L.) [chicle, chicozapote, naseberry, nispero, sapodilla, sapotilla]
Ascomycotina
Elsinoe lepagei. Scab, on leaves. **2:**FL(35,593).
Guignardia sp. Leaf spot. **1:**FL(35).
Basidiomycotina - Uredinales
Maravalia sapotae. **2:**FL(35,82).
Basidiomycotina - Other
Armillaria tabescens. Root rot. **2:**FL(35).
Ganoderma sp. Butt rot. **2:**FL(35).
Deuteromycotina - Hyphomycetes
Phymatotrichopsis omnivora. Root rot. **2:**TX(82).
Deuteromycotina - Coelomycetes
Pestalotia scirrofaciens. Limb gall. **2:**FL(35,82), TX(82).
Phyllosticta sp. Leaf spot. **2:**FL(35,82).
Septoria sp. Leaf spot. **2:**FL(35,82).

Mimusops L.
1. *M. elengi* L. [medlar, Spanish cherry]
2. *Mimusops* sp.
Ascomycotina
Mycosphaerella sp. Leaf spot. **2:**FL(35).
Deuteromycotina - Coelomycetes
Pestalotiopsis glandicola. Leaf spot. **2:**FL(35).
Phyllosticta sp. Leaf spot. **1:**FL(35). **2:**FL(35).

Pouteria Aubl.
(= *Lucuma* Molina.)
1. *P. campechiana* (Kunth) Baehni (=*Lucuma nervosa* A. DC., *Lucuma salicifolia* Kunth) [canistel, egg-fruit-tree, sapote borracho, sapote-amarillo, ti-es]
2. *P. sapota* (Jacq.) H. Moore & Stearn (=*Lucuma mammosa* of authors, *Calocarpum sapota* (Jacq.) Merr.) [mamey-colorado, mammee sapote, marmalade plum, marmalade-fruit, sapota, sapote]
3. *Pouteria* sp. (=*Lucuma sp.*)
Oomycetes
Pythium sp. Root rot. **2:**FL(35).
Ascomycotina
Botryosphaeria obtusa. On limbs. **3:**SC(82).
Elsinoe lepagei. Scab, leaf spot. **1:**FL(35,593).
Phyllachora sp. Anthracnose. **2:**FL(35).
Basidiomycotina - Uredinales
Achrotelium lucumae. **1:**FL(35,739). **3:**FL(82).
Deuteromycotina - Coelomycetes
Colletotrichum gloeosporioides. Fruit spot. **1:**FL(35). **2:**FL(35). **3:**FL(82).
Phyllosticta sp. Leaf spot. **2:**FL(35).

Sideroxylon L.
1. *S. celastrinum* (H. B. K.) Pennell (=*Bumelia angustifolia* Nutt.)
2. *S. foetidissimum* Jacq. (=*Mastichodendron foetidissimum* (Jacq.) Cronq.) [false mastic]
3. *S. lanuginosa* Michx. (=*Bumelia lanuginosa* (Michx.) Pers.) [chittamwood, false buckthorn, gum bumelia, shittimwood]
4. *S. lycioides* L. (=*Bumelia lycioides* (L.) Pers.) [ironwood, mock-orange, southern buckthorn]
5. *S. reclinatum* Michx. (=*Bumelia reclinata* (Michx.) Venten.)
6. *S. rhynchospermum* Rock
7. *S. salicifolium* (L.) Lam. (=*Dipholis salicifolia* (L.) A. DC.)
8. *S. sandwicense* (A. Gray) Benth. & J. D. Hook. ex Drake
9. *S. tenax* L. (=*Bumelia tenax* (L.) Willd.) [tough buckthorn]
10. *Sideroxylon* sp.
Ascomycotina
Botryosphaeria sp. Leaf spot. **3:**FL(35).
Guignardia bumeliae. On fallen leaves. **4:**GA(82,1438).
Lachnum sp. **4:**GA(1438).
Meliola bumeliae. **1:**FL(3483). **3:**FL(3483). **5:**FL(3483). **9:**FL(3483).
M. sideroxyli. Black mildew. **8:**HI(1272,2903). **10:**HI(82,1452).
M. sideroxylicola. On living leaves. **2:**FL(3483).
Mycosphaerella hermione. **4:**GA(1438).
Pauahia sideroxyli. **6:**HI(2903).
Phyllachora halstediana. Anthracnose, black leaf spot. **2:**FL(35). **10:**FL(82).
Polystigma bumeliae. On leaves. **4:**AR(82).
Basidiomycotina - Other
Gloeocystidiellum tropicale. On wood. **7:**FL(436).
Helicobasidium brebissonii. Root rot. **3:**TX(82).
Hexagonia hydnoides. On wood. **2:**FL(35).
Trametes hirsuta. White rot. **2:**FL(35). **10:**FL(82).
Deuteromycotina - Hyphomycetes
Cercospora lanuginosa. Leaf spot. **3:**TX(82,581).
Phymatotrichopsis omnivora. Root rot. **3:**TX(82).
Deuteromycotina - Coelomycetes
Phoma bumeliae. On stems. **3:**AL(82), MS(82,2720).
Phyllosticta bumeliifolia. Leaf spot. **3:**AL(82), OK(2872), TX(64,82). **4:**AL(82), MS(82,2720).
P. curtisii. Leaf spot. **3:**FL(35,82), MO(82), TX(82).
Septoria bumeliae. Leaf spot. **3:**TX(82). **4:**MS(82,2720).

Synsepalum (A. DC.) Baill.
1. *S. dulcificum* (Schumach. & Thonn.) W. F. Daniell (=*Pouteria dulcifica* (Schumach. & Thonn.) Baehni) [miraculous berry]
Deuteromycotina - Hyphomycetes
Cylindrocladium pteridis. Collar rot. **1:**HI(2903).
Deuteromycotina - Coelomycetes
Phyllosticta sp. Leaf spot. **1:**FL(35).

Sarraceniaceae

Darlingtonia Torr.
1. *D. californica* Torr. [California pitcherplant]
Ascomycotina
Mycosphaerella sarraceniae. Leaf spot. **1:**CA(82).
Deuteromycotina - Coelomycetes
Septoria darlingtoniae. Leaf spot. **1:**OR(82,3202).

Sarracenia L.
1. *S. alata* (A. Wood) A. Wood [yellow trumpets]
2. *S. flava* L. [huntsman's-horn, trumpet-leaf, trumpets, umbrella-trumpets, watches, yellow pitcherplant]
3. *S. purpurea* L. [common pitcherplant, huntsman's-cup, Indian cup-plant, side-saddle-flower, southern pitcherplant, sweet pitcherplant]
4. *Sarracenia* sp.
Oomycetes
Pythium graminicola. Root rot. **4:**NC(82).
Ascomycotina
Mollisia atrata. On leaves. **2:**GA(1438). **3:**FL(35), MN(82). **4:**FL(82), GA(82).

Mollisia sp. **2:**GA(1438).
Mycosphaerella pontederiae. **3:**GA(1438).
M. sarraceniae. Leaf spot. **1:**MS(2720). **2:**GA(1438). **3:**GA(1438), ME(82), MI(82), MN(82), NY(82), PA(82), WI(1361). **4:**GA(82), MS(82), SC(82).
Deuteromycotina - Hyphomycetes
Helminthosporium sarraceniae. Secondary leaf spot. **3:**MN(82).
Deuteromycotina - Coelomycetes
Colletotrichum gloeosporioides. Leaf spot. **3:**NJ(82). **4:**TX(82).
Drudeola sarraceniae. On leaves. **3:**MI(82), NY(82).
Pestalotiopsis aquatica. Secondary leaf spot. **3:**MN(82). **4:**MD(82), TX(82).
Deuteromycotina - Other
Rhizoctonia solani. Root rot. **4:**TX(82).
Sclerotium rolfsii. Southern blight. **4:**TX(82).

Saururaceae

Anemopsis Hook. & Arnott
1. *A. californica* (Nutt.) Hook. & Arnott
Basidiomycotina - Uredinales
[*Puccinia aristidae*]. **1:**NV(3378).
Uredinopsis sp. **1:**AZ(1203).

Saururus L.
1. *S. cernuus* L. [lizard-tail, swamp-lily, water-dragon]
Chytridiomycetes
Physoderma sp. Leaf gall. **1:**VA(82).
Deuteromycotina - Hyphomycetes
Alternaria sp. Leaf spot. **1:**FL(35).
Cercospora saururi. Leaf spot. **1:**AL(82), FL(35,82), IL(82), IN(82), LA(82,962), MS(2720), NY(82), OK(82,2873), TX(82,3597).
Deuteromycotina - Coelomycetes
Gloeosporium saururi. Anthracnose, leaf spot. **1:**FL(35).
Deuteromycotina - Other
Rhizoctonia solani. Root rot. **1:**FL(35).

Saxifragaceae

Astilbe F. Ham. ex D. Don
1. *A. simplicifolia* Makino
2. *Astilbe* sp.
Ascomycotina
Erysiphe polygoni. Powdery mildew. **2:**MA(82).
Deuteromycotina - Hyphomycetes
Cercospora sp. Leaf spot. **1:**FL(35).
Fusarium sp. Wilt. **2:**WA(82).

Bensoniella C. Mort.
1. *B. oregana* (Abrams & Bacigal.) C. Mort.
Ascomycotina
Sphaerotheca fuliginea. Powdery mildew. **1:**CA(1155).

Bergenia Moench
1. *B. cordifolia* (Howell) Sternb.
Basidiomycotina - Other
Armillaria mellea. Root rot. **1:**CA(2897,2898).

Boykinia Nutt.
1. *B. elata* (Nutt.) Greene
2. *B. major* A. Gray
3. *B. richardsonii* (Hook.) A. Gray
Ascomycotina
Sphaerotheca fuliginea. Powdery mildew. **1:**CA(1155).
Basidiomycotina - Ustilaginales
Urocystis alaskana. **3:**AK(526,1077,4026).
Deuteromycotina - Hyphomycetes
Ramularia sp. **2:**ID(3202).

Elmera Rydb.
1. *E. racemosa* (S. Wats.) Rydb.
Basidiomycotina - Uredinales
Puccinia heucherae. Autoecious. **1:**WA(3202).

Heuchera L.
1. *H. americana* L. [black-geranium]
2. *H. bracteata* (Torr.) Ser.
3. *H. chlorantha* Piper
4. *H. cylindrica* Douglas ex Hook. (=*H. ovalifolia* Nutt. ex Torr. & A. Gray)
5. *H. glabra* Willd. ex Roem. & Schult.
6. *H. grossulariifolia* Rydb.
7. *H. hallii* A. Gray
8. *H. micrantha* Douglas ex Lindl.
9. *H. parvifolia* Nutt. ex Torr. & A. Gray
10. *H. parvifolia* Nutt. ex Torr. & A. Gray var. *major* Rosend., F. K. Butters & Lakela
11. *H. richardsonii* R. Br.
12. *H. richardsonii* R. Br. var. *affinis* Rosend., F. K. Butters & Lakela (=*H. hispida* Pursh)
13. *H. rubescens* Torr.
14. *H. sanguinea* Engelm. [coralbells]
15. *H. sanguinea* Engelm. var. *pulchra* (Rydb.) Rosend.
16. *H. villosa* Michx.
17. *Heuchera* sp.
Ascomycotina
Erysiphe cichoracearum. Powdery mildew. **3:**CA(1155). **17:**CA(82).
Mycosphaerella heucherae. On leaves. **4:**ID(3202), WA(3202). **6:**ID(3202). **17:**MT(3202), WA(82).
M. tassiana. **4:**ID(3202).
Phyllactinia guttata. Powdery mildew. **17:**MT(82,3202).
Sphaerotheca alpina. **9:**NM(128). **17:**CO(128).
S. fuliginea. Powdery mildew. **14:**CA(2896).
S. macularis. Powdery mildew. **4:**OR(3202). **17:**MT(82,3202), NM(82).
S. mors-uvae. **8:**CA(81).
Basidiomycotina - Uredinales
Puccinia heucherae. Autoecious. **1:**E states(82), NC(3094), SD(82). **2:**CO(649,1195). **4:**ID(3202), MT(3202), OR(3202), WA(3202). **5:**AK(60,526,3094,3431). **7:**CO(3094). **8:**CA(81,972,3094), OR(3094,3202), WA(3202). **9:**MT(3202). **11:**WI(1361). **12:**ND(378), SD(2243), WI(1332). **13:**AZ(1203,3981). **14:**CA(82). **15:**AZ(3981). **16:**ID(3094), NC(3094), VA(3094). **17:**ID(3094).
P. heucherae var. *saxifragae*. **4:**WA(3094). **10:**CO(3094). **11:**NE(3094), SD(3094), WI(3094).
Basidiomycotina - Ustilaginales
Urocystis heucherae. **9:**UT(1077).
U. lithophragmae. Leaf smut. **17:**UT(82).
Deuteromycotina - Hyphomycetes
Botrytis cinerea. **14:**WA(3202).
Cercospora heucherae. Leaf spot. **1:**DE(3543), IN(82), OH(82), PA(82), VA(82), WV(82). **17:**IA(82), IL(82), IN(82), WI(82).
Ramularia mitellae. **5:**WA(3202).
R. mitellae var. *heucherae*. Leaf spot. **5:**WA(2817). **12:**SD(2243). **17:**WA(82).
Deuteromycotina - Coelomycetes
Colletotrichum gloeosporioides. **17:**IL(82).
Leptostroma herbarum. **5:**AK(526).
Phyllosticta excavata. Leaf spot. **5:**AK(526). **17:**AK(82), ID(82).
P. heucherae. Leaf spot. **4:**ID(525).
Septoria heucherae. Leaf spot. **1:**IN(82).
S. saxifragae. **1:**WV(82). **4:**WA(3202). **14:**WA(82).
Septoria sp. **11:**WI(1342).

Leptarrhena R. Br.
1. *L. pyrolifolia* (D. Don) Ser.
Ascomycotina
Laestadia saxifragae. **1:**AK(3431).
Physalospora sp. **1:**AK(3431).

Lithophragma (Nutt.) Torr. & A. Gray
1. *L. affine* A. Gray
2. *L. glabrum* Nutt. (=*L. bulbiferum* Rydb.)
3. *L. parviflorum* (Hook.) Nutt. ex Torr. & A. Gray (=*Tellima parviflora* Hook.)
4. *Lithophragma* sp.
Chytridiomycetes
Synchytrium aureum. **2:**WA(3202).

S. lithophragmae. On leaves. **2:**ID(1847,3202), WA(1847,3202). **3:**OR(1847,3202).

Oomycetes

Peronospora saxifragae. **2:**WA(3202). **3:**WA(3202).

Ascomycotina

Mycosphaerella sp. **2:**ID(3202).

Basidiomycotina - Uredinales

Puccinia heucherae. Autoecious. **3:**ID(3202), WA(3202). **4:**CA(82), UT(82), WA(82).

P. heucherae var. *lithophragmae*. **1:**CA(3094). **3:**UT(3094).

Basidiomycotina - Ustilaginales

Entyloma lithophragmae. **3:**WA(1077,3202).

Urocystis lithophragmae. Leaf smut. **2:**UT(1077). **4:**UT(82).

Deuteromycotina - Coelomycetes

Phyllosticta sp. **2:**ID(3202).

Septoria tellimae. On leaves. **3:**WA(666,3202).

Mitella L.

1. *M. breweri* A. Gray
2. *M. caulescens* Nutt.
3. *M. diphylla* L. [coolwort]
4. *M. diversifolia* (Greene) Rydb.
5. *M. nuda* L.
6. *M. ovalis* Greene
7. *M. pentandra* Hook.
8. *M. stauropetala* Piper
9. *M. trifida* R. J. Graham

Ascomycotina

Sphaerotheca fuliginea. Powdery mildew. **6:**CA(1155). **8:**CA(1155). **9:**CA(1155).

S. macularis. Powdery mildew. **3:**WI(82).

Basidiomycotina - Uredinales

Puccinia heucherae. Autoecious. **1:**OR(3202), WA(3202). **2:**WA(3202). **3:**E states(82), IA(82), NY(3094), WI(82). **4:**WA(3202). **5:**MT(3202). **6:**OR(3202). **7:**AK(60,526), CA(638,642,661), CO(1195), ID(3202), MT(3202), WA(3202), WY(1195). **8:**ID(3202), OR(3202), WA(3202), WY(1195). **9:**ID(3202).

P. heucherae var. *austroberingiana*. **1:**WA(3094). **7:**AK(3094), CA(3094), UT(3094), WA(3094). **9:**WA(3094).

Deuteromycotina - Hyphomycetes

Cercospora mitellae. Leaf spot. **3:**MI(82).

Ramularia mitellae. Leaf spot. **1:**WA(3202). **3:**IL(82), IN(82), MI(82), NY(82). **5:**WI(1357).

Deuteromycotina - Coelomycetes

Ascochyta sp. On leaves. **3:**WI(1358,1361).

Phyllosticta mitellae. Leaf spot. **3:**NY(82), WI(82).

Phyllosticta sp. **3:**WI(1354).

Septoria mitellae. Leaf spot. **3:**MI(82), WI(82,1334).

Deuteromycotina - Other

Sclerotium deciduum. Leaf rot. **3:**WI(82,1342).

Parnassia L.

1. *P. caroliniana* Michx.
2. *P. fimbriata* K. D. Koenig
3. *P. kotzebuei* Cham. ex Spreng.
4. *P. palustris* L.
5. *Parnassia* sp.

Chytridiomycetes

Synchytrium aureum. **2:**WA(3202).

Synchytrium sp. On leaves. **2:**MT(1849), WA(1848).

Ascomycotina

Erysiphe polygoni. Powdery mildew. **1:**NY(82).

Mycosphaerella tassiana. On dead stem. **3:**AK(1938).

Basidiomycotina - Uredinales

Puccinia caricina var. *uliginosa*. **4:**AK(60,526). **5:**AK(82).

P. parnassiae. Autoecious. **2:**MT(1195,3202), UT(82).

Penthorum L.

1. *P. sedoides* L.

Ascomycotina

Hainesia lythri. **1:**WI(1342).

Deuteromycotina - Hyphomycetes

Cercospora sedoides. Leaf spot. **1:**IA(581), KS(581), WI(581).

Saxifraga L.

1. *S. arguta* D. Don
2. *S. bracteata* D. Don
3. *S. bronchialis* L.
4. *S. bronchialis* L. var. *austromontana* (Wiegland) Piper ex G. N. Jones
5. *S. caespitosa* L.
6. *S. californica* Greene (=*S. parvifolia* Greene)
7. *S. cernua* L.
8. *S. ferruginea* R. J. Graham
9. *S. flagellaris* Willd. ex Sternb.
10. *S. hirculus* L.
11. *S. hirculus* L. var. *propinqua* (R. Br.) Simmons
12. *S. integrifolia* Hook.
13. *S. jamesii* Torr. (=*Boykinia jamesii* Engl. in Engl. & Prantl)
14. *S. lyallii* Engl.
15. *S. marshallii* Greene
16. *S. mertensiana* Bong.
17. *S. nelsoniana* D. Don (=*S. punctata* L. var. *nelsoniana* (D. Don) Hultén)
18. *S. nidifica* Greene
19. *S. nidifica* Greene var. *claytoniifolia* (Canby ex Small) P. Elvander (=*S. fragosa* Suksd.)
20. *S. nudicaulis* D. Don
21. *S. occidentalis* S. Wats.
22. *S. odontoloma* Piper
23. *S. oregana* Howell
24. *S. pensylvanica* L. [swamp saxifraga, wild beet]
25. *S. punctata* L.
26. *S. ranunculifolia* Hook. (=*Boykinia ranunculifolia* Greene)
27. *S. rhomboidea* Greene
28. *S. rivularis* L. [alpine-brook saxifrage]
29. *S. stolonifera* Meerb. [beefsteak-geranium, creeping-sailor, mother-of-thousands, strawberry-begonia, strawberry-geranium]
30. *S. tolmiei* Torr. & A. Gray
31. *S. tricuspidata* Rottb.
32. *S. virginiensis* Michx. [early saxifrage]
33. *Saxifraga* sp.

Chytridiomycetes

Synchytrium aureum. **1:**WA(3202). **22:**CO(649).

Synchytrium sp. On leaves and petioles. **1:**WA(1847), WY(1847,2924). **6:**CA(1847). **12:**ID(3202), WA(3202). **18:**CA(1847). **19:**ID(1847), WA(1847). **26:**CA(1847). **33:**ID(1847).

Oomycetes

Phytophthora sp. Root rot. **29:**FL(35).

Pythium sp. Root rot. **29:**FL(35).

Ascomycotina

Dothidella sphaerelloides. **10:**AK(526).

Laestadia saxifragae. On leaves. **6:**AK(526). **33:**AK(82).

Leptosphaeria oreophila. **13:**WY(3810).

Mycosphaerella minor. On dead leaves and stems. **33:**AK(1938).

M. tassiana. On dead stems. **11:**AK(1938). **33:**AK(1938).

Mycosphaerella sp. On leaves. **33:**AK(82).

Pleospora helvetica. On dead stem. **33:**AK(1938).

P. herbarum. **5:**WA(3202).

P. penicillus. On dead stems. **7:**AK(1938). **33:**AK(1938), WA(3202).

Sphaerotheca fuliginea. **1:**WY(82). **22:**CO(649). **32:**PA(82). **33:**CO(82), WY(82).

S. macularis. Powdery mildew. **1:**WY(1195). **17:**AK(526). **33:**AK(82).

Basidiomycotina - Uredinales

Melampsora arctica. Heteroecious: aecial host. **2:**AK(60,526). **9:**CO(82). **33:**AK(82).

Puccinia heucherae. Autoecious. **1:**W states(82), ID(3202), MT(3202), OR(3202), WA(82,3202), WY(1195,2924). **6:**OR(3202). **12:**MT(3202), WA(3202). **14:**AK(60,526), MT(1195). **15:**OR(3202). **16:**ID(82,3202), WA(82,3202), WY(82). **17:**AK(60,526). **21:**WA(3202). **22:**CO(649). **23:**WA(3202). **24:**NC states(82), NE states(82). **25:**CO(3094), ID(3202), OR(3202), UT(3094), WA(3094,3202). **27:**CO(1520), MT(1195,3202). **32:**NE states(82), OH(82). **33:**W states(82), AK(82,526), NY(2456).

P. heucherae var. *austroberingiana*. Autoecious. **1:**WY(3379). **14:**AK(3094). **16:**WA(3094). **25:**AK(3094), ID(3094).

P. heucherae var. *saxifragae*. **6:**CA(3094). **7:**AK(1938). **19:**WA(3094). **28:**AK(3094).

P. laurentiana. **20:**AK(60,526,3094).
P. pazschkei. Autoecious. **16:**ID(82,3202), MT(82,3202), WA(82,3202). **31:**AK(526).
P. pazschkei var. *heterisiae*. **16:**MT(3094), WA(3094).
P. pazschkei var. *tricuspidatae*. Autoecious. **3:**CO(3094), UT(3094). **4:**CO(649). **31:**AK(3094). **33:**CO(82), UT(82).
Deuteromycotina - Hyphomycetes
Botrytis cinerea. Gray mold. **33:**AK(82).
Cercospora sp. Leaf spot. **29:**FL(35). **33:**WI(1343).
Cercosporella saxifragae. Leaf spot. **24:**WI(82,1337,1341).
Ramularia sp. Leaf spot. **16:**WA(3202). **33:**AK(82).
Deuteromycotina - Coelomycetes
Asteroma saxifragae. On leaves. **33:**AK(82).
Marssonina saxifragae. **16:**WA(3202).
Phyllosticta saxifragarum. Leaf spot. **1:**WY(82,1195).
Septoria albicans. **8:**MT(3202). **24:**WI(82).
Sirexcipula wyomingensis. **30:**CA(642).
Deuteromycotina - Other
Rhizoctonia solani. Lower leaf rot, root rot. **29:**FL(35).

Sullivantia Torr. & A. Gray
1. *S. renifolia* Rosend.
Deuteromycotina - Coelomycetes
Ascochyta sp. **1:**WI(1341).

Tellima R. Br. in Franklin
1. *T. grandiflora* (Pursh) Douglas ex Lindl. [fingercups]
Ascomycotina
Sphaerotheca alpina. **1:**AK(128), CA(128).
S. fuliginea. Powdery mildew. **1:**Unk.(3976), AK(82,526).
Basidiomycotina - Uredinales
Puccinia heucherae. Autoecious. **1:**AK(60,82,526), CA(81,82), OR(82,3202), WA(82,3202).
P. heucherae var. *austroberingiana*. **1:**WA(3094).

Tiarella L.
1. *T. cordifolia* L. [foamflower]
2. *T. trifoliata* L.
3. *T. unifoliata* Hook. [sugar-scoop]
4. *Tiarella* sp.
Chytridiomycetes
Synchytrium sp. **3:**WA(3202).
Ascomycotina
Sphaerotheca fuliginea. Powdery mildew. **2:**AK(526). **4:**AK(82).
Basidiomycotina - Uredinales
Puccinia heucherae. Autoecious. **1:**E states(82), CO(82), MI(82), NY(3094), TN(82), VT(1224,3094), WI(82). **2:**AK(60,526), WA(3094,3202). **3:**AK(60,526), CA(81), ID(638,3202), MT(3202), OR(3202), WA(3094,3202). **4:**AK(82), CA(82), ID(82), MT(82), OR(82), WA(82).
Deuteromycotina - Hyphomycetes
Ramularia sp. **3:**MT(3202).
Deuteromycotina - Coelomycetes
Septoria sp. **3:**ID(3202), MT(3202).

Tolmiea Torr. & A. Gray
1. *T. menziesii* (Pursh) Torr. & A. Gray [pickaback-plant, piggyback-plant, thousand-mothers, youth-on-age]
Oomycetes
Pythium sp. Root rot. **1:**FL(35).
Ascomycotina
Sphaerotheca mors-uvae. **1:**OR(3202).
Sphaerotheca sp. Powdery mildew. **1:**OR(2175).
Basidiomycotina - Uredinales
Puccinia heucherae. Autoecious. **1:**AK(60,526), CA(81,3094), OR(3094,3202), WA(3094,3202).
Deuteromycotina - Hyphomycetes
Alternaria sp. Leaf spot. **1:**FL(35).
Cercospora sp. Leaf spot. **1:**FL(35).
Corynespora cassiicola. Leaf spot. **1:**FL(35).
Fusarium sp. Root rot. **1:**FL(35).
Deuteromycotina - Coelomycetes
Colletotrichum sp. Leaf spot. **1:**FL(35).
Phomopsis sp. Dieback. **1:**FL(35).
Phyllosticta sp. Leaf spot. **1:**FL(35).

Scheuchzeriaceae

Scheuchzeria L.
1. *S. palustris* L. var. *americana* Fernald
Ascomycotina
Mycosphaerella bacillifera. **1:**WI(1357).

Scrophulariaceae

Agalinis Raf.
1. *A. fasciculata* (Elliott) Raf.
2. *A. purpurea* (L.) Pennell (=*Gerardia paupercula* (A. Gray) Britton)
Deuteromycotina - Hyphomycetes
Cercospora sp. Leaf spot. **1:**FL(35).
Deuteromycotina - Coelomycetes
Septoria scrophulariae var. *agalinis*. On living leaves. **2:**WI(1339).

Angelonia Bonpl.
1. *A. salicariaefolia* Humb. & Bonpl.
Ascomycotina
Erysiphe cichoracearum. Powdery mildew. **1:**HI(2899).
Deuteromycotina - Hyphomycetes
Oidium sp. Powdery mildew. **1:**HI(2903).

Antirrhinum L.
1. *A. majus* L. [common snapdragon, garden snapdragon]
2. *Antirrhinum* sp.
Oomycetes
Peronospora antirrhini. Downy mildew. **1:**CA(82), FL(35), MD(1727), NC(1303), NY(82), OK(82,2872), OR(82,3202), PA(82), WA(3202).
Phytophthora cactorum. Stem rot. **1:**CA(82), FL(35), IL(82), MN(82), NJ(82), NY(82).
P. cryptogea. Damping-off. **1:**CA(82), OK(82,2872).
P. nicotianae var. *parasitica*. Branch, root and stem rot, wilt. **1:**FL(35), GA(1206), HI(82,2903), NC(1303).
Pythium debaryanum. **1:**ND(82), TX(82).
P. irregulare. **1:**CA(82).
P. mamillatum. **1:**CA(82).
P. myriotylum. **1:**GA(1208).
P. pulchrum. **1:**CA(82).
P. rostratum. **1:**CA(82).
P. splendens. Damping-off. **1:**FL(35).
P. ultimum. **1:**CA(82).
P. vexans. **1:**CA(82).
Pythium sp. Damping-off, root rot. **1:**Range of host(82), GA(1208), NC(1303).
Ascomycotina
Erysiphe cichoracearum. **1:**WA(3202).
E. graminis f. sp. *avenae*. Powdery mildew. **1:**AR(3029).
Sclerotinia minor. **1:**CT(82).
S. sclerotiorum. Stem rot, wilt. **1:**FL(35), IL(312), OH(946).
Thielavia basicola. On dead roots. **1:**CT(82).
Basidiomycotina - Uredinales
Puccinia antirrhini. Autoecious. **1:**Range of host(82), AL(285), AZ(1203,3981), FL(35), GA(1439), HI(2903), ID(3202), MA(1390), MS(2720), NC(1303), OK(2872), OR(3202), SD(2243), WA(3202). **2:**CA(81).
Deuteromycotina - Hyphomycetes
Alternaria alternata. Blossom blight. **1:**FL(35).
Arthrobotrys oligospora. On decaying leaves and stems. **1:**NC(1303).
Bipolaris maydis. Petal spot. **1:**FL(35).
B. setariae. Flower petal spot. **1:**FL(35).
Botrytis cinerea. Gray mold blight on blossoms, leaves and stems. **1:**Range of host(82), FL(35), OR(3202).
Botrytis sp. Blight. **1:**OK(2872).
Cercospora antirrhini. Blight, leaf spot. **1:**FL(35,321,1708), HI(2899), IL(312), NC(1303).
Cercospora sp. Leaf spot. **1:**NC(1303).
Fusarium sp. Root and stem rot, wilt. **1:**CT(82), FL(35,82), GA(82), MS(2720), NC(1303), OK(82,2872), TN(82), WA(82).
Myrothecium roridum. Crown and stem canker. **1:**IL(312).
Oidium sp. Powdery mildew. **1:**CA(151,2224), FL(35), MA(82),

NC(1303), NY(82), OH(946), PA(82), WA(151).
Phymatotrichopsis omnivora. Root rot. **1**:TX(82).
Stemphylium lycopersici. Leaf spot. **1**:FL(35).
Thielaviopsis basicola. Root rot. **1**:CT(82), NC(1303), NJ(82).
Verticillium albo-atrum. Wilt. **1**:CA(82), CT(82), FL(35), MA(82), ME(82), MN(82), NC(1303), NJ(82), NY(82), OK(2872), PA(82).
Deuteromycotina - Coelomycetes
Colletotrichum gloeosporioides. Anthracnose. **1**:E states(82), SC states(82), SW states(82), FL(35), MS(2720).
Macrophomina phaseolina. Charcoal stem rot. **1**:OK(82,2873).
Phoma sp. Stem rot. **1**:CT(82), IN(82), MA(82), MI(82), NY(82), OH(82), PA(82), VA(82).
Phyllosticta antirrhini. Canker, stem rot, leaf spot. **1**:E states(82), NC states(82), FL(35), MA(64), NC(1303), TX(82), WA(82,3202).
Deuteromycotina - Other
Rhizoctonia solani. Root and stem blight, damping-off. **1**:Range of host(82), FL(35), NC(1303), OK(2872).
Sclerotium rolfsii. Southern blight, southern stem rot. **1**:CA(82), FL(35,82), MS(82,2720), NC(1303), NJ(82), NY(82), TX(82).
S. rolfsii var. *delphinii*. Southern blight. **1**:CA(82), FL(82), MS(82), NJ(82), NY(82), TX(82).

Aureolaria Raf.

1. *A. flava* (L.) Fée (=*Gerardia flava* L.)
2. *A. grandiflora* (Benth. in Hook.) Pennell (=*Gerardia grandiflora* Benth. in Hook.)
3. *A. pedicularia* (L.) Raf. (=*Gerardia pedicularia* L.) [false foxglove]
4. *A. virginica* (L.) Pennell (=*Gerardia quercifolia* Pursh, *Gerardia virginica* (L.) B. S. P.) [downy false foxglove]

Ascomycotina
Diaporthe arctii. **1**:GA(1438).
Erysiphe galeopsidis. **4**:GA(1438).
Leptosphaeria doliolum. **4**:NY(3249).
Massarina gloeospora. **4**:GA(1438).
Mycosphaerella sp. **4**:GA(1438).
Sphaerotheca fuliginea. **3**:WI(1344).
Basidiomycotina - Uredinales
Puccinia andropogonis. Heteroecious: aecial host. **1**:GA(1439), NC(1303,3944). **4**:NC(1303).
Deuteromycotina - Hyphomycetes
Cercospora gerardiae. Leaf spot. **1**:NC(3944). **4**:MS(2720).
C. madisonensis. On leaves. **2**:WI(1332).
Deuteromycotina - Coelomycetes
Phyllosticta sp. Leaf spot, on pods. **3**:WI(1358,1360).

Besseya Rydb.

1. *B. bullii* (Eaton) Rydb. (=*Wulfenia bullii* (Eaton) Barnhart)
2. *B. cinerea* (Raf.) Pennell (=*Synthyris wyomingensis* (A. Nels.) A. Heller)
3. *B. rubra* (Douglas ex Hook.) Rydb.

Ascomycotina
Sphaerotheca macularis. **2**:WY(1195). **3**:WA(3202).
Basidiomycotina - Uredinales
Puccinia acrophila. Autoecious. **2**:WY(1520). **3**:MT(3202).
P. wulfeniae. Autoecious. **3**:WA(3202).
Deuteromycotina - Coelomycetes
Leptothyrium sp. On capsules and flower stalks. **1**:WI(1347).
Phomopsis sp. On leaves. **1**:WI(1350).
Phyllosticta wulfeniae. **1**:WI(1343).
Septoria digitalidis. **1**:WI(1341).

Buchnera L.

1. *B. floridana* Gand.

Basidiomycotina - Uredinales
Puccinia buchnerae. **1**:FL(35).

Calceolaria L.

1. *C. crenatiflora* Cav.
2. *Calceolaria* sp.

Oomycetes
Phytophthora sp. Root and basal stem rot. **1**:FL(35).
Pythium mastophorum. Root rot. **2**:CA(82).
P. ultimum. Root rot. **2**:CA(82).
Pythium sp. Root rot. **1**:FL(35).
Ascomycotina
Sclerotinia sclerotiorum. Stem rot. **2**:NY(82), WA(82,3202).
Deuteromycotina - Hyphomycetes
Botrytis cinerea. Gray mold. **2**:AK(82).
Myrothecium sp. Stem rot. **1**:FL(35).
Verticillium albo-atrum. Wilt. **2**:NY(82), WA(82,3202).

Castilleja Mutis ex L. fil.

1. *C. affinis* Hook. & Arnott (=*C. californica* Abrams) [Indian paintbrush]
2. *C. angustifolia* (Nutt.) G. Don
3. *C. applegatei* Fernald var. *pinetorum* (Fernald) N. Holmgren (=*C. pinetorum* Fernald)
4. *C. arachnoidea* Greenm. var. *shastensis* Pennell
5. *C. coccinea* (L.) Spreng. [scarlet paintbrush]
6. *C. flava* S. Wats.
7. *C. foliolosa* Hook. & Arnott [woolly painted-cup]
8. *C. gracillima* Rydb.
9. *C. hispida* Benth.
10. *C. integra* A. Gray
11. *C. latifolia* Hook. & Arnott [seaside painted-cup]
12. *C. laxa* A. Gray
13. *C. linariifolia* Benth.
14. *C. lutescens* (Greenm.) Rydb.
15. *C. miniata* Douglas ex Hook.
16. *C. minor* (A. Gray) A. Gray
17. *C. pallescens* (A. Gray) Greenm.
18. *C. pallida* (L.) Spreng.
19. *C. parviflora* Bong.
20. *C. parviflora* Bong. var. *oreopola* (Greenm.) Ownbey (=*C. oreopola* Greenm. in J. M. Coulter)
21. *C. patriotica* Fernald
22. *C. rhexifolia* Rydb.
23. *C. septentrionalis* Lindl.
24. *C. sessiliflora* Pursh
25. *C. suksdorfii* A. Gray
26. *C. sulphurea* Rydb.
27. *C. thompsonii* Pennell
28. *Castilleja* sp.

Ascomycotina
Comoclathris permunda. On dead stems. **3**:CA(642). **15**:CA(642), ID(3202), WA(3202).
Crocicreas cyathoideum var. *cacaliae*. On stems. **18**:UT(505). **22**:UT(505). **28**:CO(505), UT(505).
Didymella castillejae. On stems. **15**:WY(681).
Erysiphe polygoni. Powdery mildew. **28**:CO(82), WA(82).
Leptosphaeria agnita. **15**:WY(3810).
L. castillejae. **15**:MT(3249).
L. concinna. **15**:WA(3202).
L. oreophila. **13**:WY(3810).
L. salsolae. **8**:ID(3202).
Lewia scrophulariae. **6**:WY(3810). **15**:WA(3202). **19**:WA(3202). **20**:WA(3814). **28**:ID(3202).
Mycosphaerella sp. On stems. **15**:ID(3202), WA(3202). **28**:CO(82).
Nectriella sp. On overwintered herbage. **4**:CA(642).
Nodulosphaeria modesta. **15**:WA(3250). **18**:AK(526). **28**:UT(3250).
Ophiobolus castillejae. **15**:WA(3248).
Pleospora helvetica. **28**:ID(3202).
P. njegusensis. **15**:WA(3202).
P. richtophensis. **13**:WY(3810).
P. scrophulariae var. *compositarum*. **13**:WA(3810).
P. tomentosa. **13**:WA(3810).
P. tragacanthae. **3**:CA(642).
Pleospora sp. On overwintered herbage. **4**:CA(642).
Sphaerotheca fuliginea. Powdery mildew. **2**:ID(3202). **15**:WA(3202). **19**:WA(3202). **23**:CO(649). **24**:WI(1330). **25**:WA(3202). **27**:WA(3202). **28**:CO(82), WA(82), WI(82).
S. macularis. **15**:WY(1195).
Basidiomycotina - Uredinales
Cronartium arizonicum. Heteroecious: telial host. **10**:AZ(3981). **12**:AZ(3981). **13**:AZ(3981). **16**:AZ(3981). **21**:AZ(748,3981).
C. coleosporioides. Heteroecious: telial host. **1**:CA(82). **10**:AZ(1203). **11**:CA(81). **12**:AZ(1203). **13**:AZ(1195,1203), CO(82), NM(82), UT(82), WY(82). **15**:ID(3202), MT(1195,3202), OR(3202), WA(3202). **16**:AZ(1203). **21**:AZ(1203), WY(1195).

24:SD(2765,2766). **25:**WA(3202). **26:**SD(2765,2766). **28:**W states(82), AZ(1203), CA(82), ID(3202), SD(2243), WA(82,3202).
Puccinia andropogonis. Heteroecious: aecial host. **5:**IA(82). **24:**WI(1330). **28:**NC states(82), W states(82), CA(82), MT(82,3202), SD(2243).
P. castillejae. Telial host. **7:**CA(81). **28:**CA(82), UT(82).
Basidiomycotina - Ustilaginales
Entyloma sp. **15:**WA(3202).
Deuteromycotina - Hyphomycetes
Botrytis cinerea. **15:**ID(3202).
Cercospora sp. Leaf spot. **18:**AK(526). **28:**AK(82).
Fusarium avenaceum. **18:**AK(3431).
F. pallidoroseum. **18:**AK(3431).
Ramularia castillejae. On leaves. **9:**WA(3202). **14:**MT(3202). **15:**ID(3202), MT(3202), OR(3202), WA(3202), WY(1195). **17:**MT(3202). **22:**MT(3202). **28:**WY(2924).
Deuteromycotina - Coelomycetes
Heteropatella umbilicata. **4:**CA(642). **15:**CA(642). **19:**WA(3202). **28:**MT(3202).
Leptostroma sp. **22:**OR(3202).
Phoma herbarum. On overwintering remains. **4:**CA(642).
Phoma sp. **4:**CA(642).
Phyllosticta sp. On capsules. **24:**WI(1347,1350).
Septoria sp. Leaf spot. **5:**WI(1356). **24:**WI(1359).
Sirexcipula wyomingensis. **22:**MT(3202).
Sphaeropsis microscopica. On overwintered remains. **4:**CA(642).

Chaenorrhinum (DC. ex Duby) Rchb.

1. *C. minus* (L.) Lange

Deuteromycotina - Coelomycetes
Septoria chaenorrhini. On leaves and inflorescence. **1:**WI(1356).

Chelone L.

1. *C. glabra* L. [balmony, snakehead, turtlehead]
2. *C. lyonii* Pursh
3. *Chelone* sp.

Ascomycotina
Diaporthe arctii. **1:**GA(1438).
Erysiphe cichoracearum. **1:**IN(82), MA(82), MT(3202).
E. galeopsidis. Powdery mildew. **1:**DE(82), MI(82), MS(82,2720), NY(82), OH(82), PA(82), WI(82). **3:**NC(82).
E. polygoni. Powdery mildew. **1:**MA(82), MI(82), NY(82), PA(82). **2:**NC(1303,1307).
Hypoderma commune. **1:**GA(1438).
Leptosphaeria conoidea. **1:**NY(3249).
L. doliolum. **1:**GA(1438).
Mollisia atrata. **1:**GA(1438).
Ophioceras sp. **1:**GA(1438).
Phyllactinia guttata. **1:**OH(82).
Basidiomycotina - Uredinales
Puccinia andropogonis. **1:**CT(82), MA(82), NJ(82), NY(82), PA(82). **3:**TN(82).
P. chelonis. Autoecious. **3:**OR(82), WA(82).
Deuteromycotina - Hyphomycetes
Cercospora sp. On leaves. **1:**WI(1357).
Deuteromycotina - Coelomycetes
Septoria mariae-wilsonii. Leaf spot. **1:**NE states(82), NC(1303,1307), OH(82), WI(82).

Collinsia Nutt.

1. *C. bartsiaefolia* Benth. in DC.
2. *C. grandiflora* Douglas ex Lindl. [bluelips]
3. *C. parviflora* Lindl.
4. *C. rattanii* A. Gray
5. *C. verna* Nutt. [blue-eyed-Mary, innocence]
6. *Collinsia* sp.

Chytridiomycetes
Synchytrium sp. **3:**WY(1195).
Oomycetes
[*Peronospora grisea*]. **3:**MT(3202), OR(3202), SD(2243), WA(3202).
Pythium mamillatum. Root rot. **6:**CA(82).
Ascomycotina
Sphaerotheca fuliginea. **3:**ID(3202), OR(3202).
S. macularis. **3:**WY(1195).
Basidiomycotina - Uredinales
Aecidium insulsum. **3:**UT(82).
Puccinia collinsiae. Autoecious. **2:**CA(81). **3:**OR(3202), WA(3202). **4:**WA(3202). **6:**CA(82), OR(82), UT(82), WA(82).
Basidiomycotina - Ustilaginales
Entyloma linariae. White smut. **1:**CA(1077,1468). **2:**ID(3202), OR(82,1077,3202). **3:**OR(1077,3202). **6:**CA(82).
Deuteromycotina - Hyphomycetes
Ramularia sp. **3:**OR(3202).
Deuteromycotina - Coelomycetes
Septoria collinsiae. Leaf spot. **5:**IL(82).

Cymbalaria J. Hill

1. *C. muralis* P. Gaertn., B. Mey. & Scherb. (=*Linaria cymbalaria* (L.) Mill.) [coliseum ivy, Kenilworth ivy, pennywort]

Deuteromycotina - Hyphomycetes
Botrytis cinerea. **1:**AK(526).
Cercospora sp. Leaf spot. **1:**FL(35).

Digitalis L.

1. *D. grandiflora* Mill. (=*D. ambigua* J. A. Murray)
2. *D. lanata* Ehrh. [digitalis, Grecian foxglove]
3. *D. lutea* L. [straw foxglove]
4. *D. purpurea* L. (=*D. nevadensis* Kunze) [common foxglove, foxglove]
5. *Digitalis* sp.

Ascomycotina
Sclerotinia sclerotiorum. Stem rot. **4:**NY(82).
Deuteromycotina - Hyphomycetes
Cladosporium sp. Leaf spot. **4:**NJ(82).
Fusarium sp. Root rot, wilt. **4:**CA(82). **5:**NH(82).
Ramularia variabilis. Leaf spot. **4:**HI(2899,2903), OR(82,3202), WA(3202).
Verticillium albo-atrum. Wilt. **4:**NY(82).
Deuteromycotina - Coelomycetes
Colletotrichum fuscum. Anthracnose. **1:**SD(2243). **2:**SD(2243). **3:**SD(2243). **4:**MD(3603), OR(3202), PA(3603), SD(2243). **5:**CT(82), PA(82).
Phyllosticta digitalis. Leaf spot. **4:**AK(82,526), TX(82). **5:**AK(82), NY(64,82).
Deuteromycotina - Other
Rhizoctonia solani. Root and stem rot. **4:**NJ(82).
Sclerotium rolfsii. Stem rot. **4:**IN(82), NJ(82), TX(82). **5:**FL(35), IN(82), NJ(82), TX(82).

Euphrasia L.

1. *E. subarctica* Raup

Deuteromycotina - Hyphomycetes
Cylindrocarpon magnusianum. **1:**AK(3431).

Gratiola L.

1. *G. virginiana* L.

Basidiomycotina - Ustilaginales
Entyloma gratiolae. **1:**WI(1077).

Hebe Commerson ex Juss.

1. *H.* × *andersonii* (Lindl. & Paxt.) Cockayne. *H. salicifolia* × *H. speciosa*.
2. *H. odora* (J. D. Hook.) Cockayne (=*H. buxifolia* (Benth.) Cockayne & Allan)
3. *H. speciosa* (R. Cunn. ex A. Cunn.) Cockayne & Allan [showy speedwell]
4. *H. subalpina* (Cockayne) Cockayne & Allan (=*Veronica subalpina* Cockayne)
5. *Hebe* sp.

Ascomycotina
Erysiphe cichoracearum. Powdery mildew. **1:**CA(1155).
Basidiomycotina - Uredinales
Puccinia albulensis. **4:**WY(2924).
Basidiomycotina - Other
Armillaria mellea. Root rot. **2:**CA(2897,2898).
Deuteromycotina - Coelomycetes
Septoria exotica. Leaf spot. **3:**HI(2899,2903). **5:**CA(82).

Keckiella Straw
1. *K. antirrhinoides* (Benth.) Straw subsp. *microphylla* (A. Gray) Straw
2. *K. breviflorus* (Lindl.) Straw (=*Penstemon breviflorus* Lindl.)
Ascomycotina
Triblidium turgidulum. On dead stems. **2:**CA(2801).
Basidiomycotina - Uredinales
Puccinia confraga. **1:**AZ(747,3981).

Lagotis Gaertn.
1. *L. glauca* Gaertn.
Basidiomycotina - Uredinales
Puccinia gymnandrae. **1:**AK(60,526).

Leucophyllum Bonpl. in Humb. & Bonpl.
1. *L. frutescens* (Berland.) I. M. Johnst. [barometer-bush, cenizo]
Oomycetes
Pythium sp. Root rot. **1:**FL(35).
Ascomycotina
Nectria sp. Canker. **1:**TX(82).
Nectriella pironii. Canker, stem gall. **1:**FL(35).
Deuteromycotina - Hyphomycetes
Alternaria sp. Leaf spot. **1:**FL(35).
Fusarium lateritium. Leaf spot. **1:**FL(35).
F. oxysporum. Basal stem rot. **1:**FL(35).
Myrothecium roridum. Leaf necrosis. **1:**FL(35).
Oidium sp. Powdery mildew. **1:**FL(35).
Phymatotrichopsis omnivora. Root rot. **1:**TX(82).
Volutella sp. **1:**TX(82).
Deuteromycotina - Coelomycetes
Phomopsis sp. Canker. **1:**FL(35).
Deuteromycotina - Other
Rhizoctonia solani. Leaf and stem blight. **1:**FL(35).

Limosella L.
1. *L. aquatica* L.
Basidiomycotina - Ustilaginales
Doassansia limosellae. **1:**MT(3202), OR(3202).

Linaria Mill.
1. *L. canadensis* (L.) Dum. Cours. [blue toadflax, old-field toadflax]
2. *L. vulgaris* Mill. [butter-and-eggs, wild snapdragon, yellow toadflax]
3. *Linaria* sp.
Chytridiomycetes
Synchytrium linariae. On leaves and stems. **1:**FL(1849).
Oomycetes
Peronospora linariae. Downy mildew. **1:**FL(35), OK(2872). **3:**FL(82), MA(82), OK(82), WI(82).
Ascomycotina
Comoclathris permunda. **2:**WY(3810).
Erysiphe cichoracearum. Powdery mildew. **3:**CA(82).
Heterosphaeria patella. On stems. **2:**NY(82).
Sclerotinia sclerotiorum. Stem rot. **1:**FL(35). **3:**AZ(82).
Basidiomycotina - Uredinales
Aecidium sp. **1:**WI(82).
[*Puccinia antirrhini*]. **3:**CA(82).
Basidiomycotina - Ustilaginales
Entyloma linariae. White smut. **1:**FL(35). **2:**CT(82,1077), IL(1077), NJ(82,1077), NY(1077), PA(82,1077), WI(1077,1348).
Deuteromycotina - Hyphomycetes
Alternaria sp. Leaf spot. **2:**MI(82).
Botrytis cinerea. Gray mold. **3:**AK(82).
B. viciae. **2:**WI(1352).
Phymatotrichopsis omnivora. Root rot. **2:**TX(82). **3:**TX(82).
Thielaviopsis basicola. Root rot. **1:**CT(82).
Deuteromycotina - Coelomycetes
Colletotrichum gloeosporioides. Anthracnose, on lower leaves and stems. **2:**MA(82), NJ(82), NY(82), TX(82), WI(82,1330).
Phoma linariae. On stems. **2:**NY(82), TX(82).
Phoma sp. **2:**WI(1352).
Septoria linariae. Leaf spot. **1:**WI(82,1332,1334).
Septoria sp. **1:**WI(1329).
Deuteromycotina - Other
Rhizoctonia solani. Root rot. **3:**IL(82).
Sclerotium rolfsii. Southern blight. **1:**FL(35). **2:**TX(82).

Maurandya Ort.
1. *M. antirrhiniflora* Humb. & Bonpl. ex Willd.
Deuteromycotina - Coelomycetes
Septoria antirrhinorum. Leaf spot. **1:**TX(82,3597).

Micranthemum Michx.
(= *Globifera* J. F. Gmelin.)
1. *Micranthemum* sp. (=*Globifera sp.*)
Ascomycotina
Seuratia globifera. On leaves. **1:**HI(2337).

Mimulus L.
1. *M. alatus* Aiton
2. *M. cardinalis* Douglas ex Benth. [scarlet monkey-flower]
3. *M. dentatus* Nutt.
4. *M. floribundus* Lindl.
5. *M. guttatus* DC. [common monkey-flower]
6. *M. lewisii* Pursh
7. *M.* × *linearis* Benth. (=*Diplacus linearis* (Benth.) Greene)
8. *M. luteus* L.
9. *M. moschatus* Douglas ex Lindl. [muskflower, muskplant]
10. *M. pulsiferae* A. Gray
11. *M. ringens* L. [Allegheny monkey-flower]
12. *M. tilingii* Regel
13. *Mimulus* sp.
Chytridiomycetes
Physoderma sp. On leaves and stems. **11:**MI(3400).
Oomycetes
Peronospora jacksonii. On leaves. **5:**OR(3201,3202), WA(3202). **10:**WA(3202).
Ascomycotina
Didymolepta mimuli. On leaves. **6:**WY(3811).
Erysiphe brunneopunctata. Powdery mildew. **5:**CA(369).
E. cichoracearum. Powdery mildew. **5:**CA(81,661). **8:**UT(82). **13:**Range of host(82), CA(82), PA(82).
Melanomma dothideasporum. On dead stems. **13:**CA(82).
Metasphaeria anisometra. On dead stems. **13:**CA(82).
Mycomicrothelia dothideaspora. On dead stems. **7:**CA(2439).
Basidiomycotina - Uredinales
Puccinia andropogonis. **13:**MO(82), WI(82).
Uredo sp. **11:**WI(82).
Basidiomycotina - Ustilaginales
Entyloma clintonianum. **4:**WA(1077,3202). **5:**OR(3202). **12:**WA(3202).
Deuteromycotina - Hyphomycetes
Botrytis cinerea. Gray mold. **8:**AK(82,526).
Cercospora mimuli. On leaves. **1:**MO(82,962).
Ramularia mimuli. Leaf spot. **3:**WA(3202). **5:**MT(3202), OR(3202), WA(3202). **6:**ID(3202), OR(3202), WA(3202). **8:**CA(82). **11:**NY(82). **12:**CA(638,639,642), OR(3202). **13:**OH(82), WY(82).
Deuteromycotina - Coelomycetes
Macrophoma mimuli. On stems. **6:**WY(3812).
Microdiplodia mimuli. On leaves. **2:**CA(2439).
Rhabdospora kellermanii. On leaves. **11:**OH(82).
Septoria mimuli. Leaf spot. **1:**IL(82), IN(82), MS(82,2720). **9:**ID(3202), MT(3202), OR(3202), WA(3202). **11:**NC states(82), NE states(82), ND(82), NE(82). **12:**WA(3202). **13:**SE states(82), KS(82), MS(82), ND(82).

Nothochelone (A. Gray) Straw
1. *N. nemorosa* (Douglas ex Lindl.) Straw
Ascomycotina
Erysiphe polygoni. **1:**WA(3202).
Basidiomycotina - Uredinales
Puccinia chelonis. Autoecious. **1:**OR(3202), WA(3202).

Orthocarpus Nutt.
1. *O. hispidus* Benth.
2. *O. luteus* Nutt.
3. *O. pusillus* Benth.
4. *O. tenuifolius* (Pursh) Benth.
5. *Orthocarpus* sp.
Oomycetes
Peronospora sordida. **1:**OR(3202).

Basidiomycotina - Uredinales
Cronartium coleosporioides. Heteroecious: telial host. **2**:SD(2765,2766). **5**:CO(82), ID(82), UT(82).
Puccinia andropogonis. **2**:SD(2768).
Deuteromycotina - Coelomycetes
Ascochyta garrettiana. Leaf spot. **1**:ID(3202), OR(3202). **3**:OR(3202), WA(3202). **4**:MT(3202), WA(3202). **5**:OR(82), UT(82).

Paulownia Siebold & Zucc.

1. *P. tomentosa* (Thunb.) Steud. [kiri-tree, princess-tree, royal paulownia]
2. *Paulownia* sp.

Ascomycotina
Botryosphaeria dothidea. **1**:GA(1438).
B. obtusa. On limbs. **1**:MD(82).
Hypoxylon mediterraneum. **1**:GA(1438).
H. rubiginosum. **1**:GA(1438).
Mollisia cinerea. **1**:GA(1438).
Valsaria rubricosa. **1**:GA(1438).
Basidiomycotina - Other
Exidiopsis calcea. On wood. **1**:GA(491,1439).
Fomitopsis spraguei. Wood rot. **1**:AL(82).
Perenniporia robiniophila. On trunks. **2**:Unk.(2681).
Deuteromycotina - Hyphomycetes
Phymatotrichopsis omnivora. Root rot. **1**:TX(82).
Deuteromycotina - Coelomycetes
Ascochyta paulowniae. Leaf spot. **1**:MD(82).
Phomopsis imperialis. On twigs. **1**:NY(82).
Phyllosticta paulowniae. Leaf spot. **1**:AL(82,285), MD(82), NY(82), OK(82,2873).

Pedicularis L.

1. *P. bracteosa* Benth. in Hook. (=*P. bracteosa* Benth. in Hook. var. *paysoniana* (Pennell) Cronq., *P. paysoniana* Pennell)
2. *P. bracteosa* Benth. in Hook. var. *latifolia* (Pennell) Cronq. (=*P. latifolia* Pennell)
3. *P. canadensis* L. [common lousewort, woodbetony]
4. *P. canadensis* L. var. *fluviatilis* (A. Heller) Macbr. (=*P. fluviatilis* A. Heller)
5. *P. capitata* M. F. Adams
6. *P. centranthera* A. Gray
7. *P. contorta* Benth. ex Hook.
8. *P. densiflora* Benth. ex Hook. [Indian warrior]
9. *P. groenlandica* Retz. [elephant-heads]
10. *P. lanata* Cham. & Schlechtend.
11. *P. lanceolata* Michx.
12. *P. racemosa* Douglas ex Hook. (=*P. racemosa* Douglas ex Hook. subsp. *alba* Pennell)
13. *P. semibarbata* A. Gray
14. *Pedicularis* sp.

Chytridiomycetes
Synchytrium sp. **14**:WA(3202).
Ascomycotina
Apiosporella alpina. On leaves. **1**:WA(3202), WY(1195,3811). **2**:WA(3815). **7**:WA(3202,3815). **14**:WA(3815).
Crocicreas cyathoideum var. *cacaliae*. On dead stems. **12**:UT(505).
Entodesmium niessleanum. **7**:OR(3202). **12**:ID(3202), OR(3202), WA(3202).
Hyponectria sceptri. On dead stems. **12**:UT(3084). **14**:CO(3084).
Leptosphaeria bupleuri. **7**:WA(3202,3814).
L. doliolum. **11**:NY(3249).
L. erigerontis. **1**:WA(3202). **2**:WA(3814).
L. jacksonensis. **1**:WY(3249).
L. oreophila. **1**:WY(3810).
[*L. tetonensis*]. **1**:WY(3810).
Leptosphaeria sp. **7**:WA(3202).
Lewia scrophulariae. **9**:WY(3810).
Massariosphaeria rubelloides. **2**:WA(3814).
M. rubicunda. **1**:WA(3202).
Mycosphaerella tassiana. **7**:WA(3202).
Nodulosphaeria modesta. **12**:ID(3250), OR(3250).
Pleospora cerastii. **10**:AK(526).
P. helvetica. On dead leaves and stems. **7**:WA(3202,3814). **14**:AK(1938).
P. herbarum. **12**:ID(3202).
Sphaerotheca fuliginea. On leaves. **1**:CO(649,1195,3378), MT(3202), OR(3202), WA(3202). **3**:WI(1338). **12**:CO(649), WY(2924).
S. macularis. **1**:WY(1195). **14**:CO(649).
Basidiomycotina - Uredinales
Cronartium coleosporioides. Heteroecious: telial host. **9**:ID(3202), WA(3202).
Puccinia clintonii. Autoecious. **1**:ID(3202), WA(3202), WY(1195). **3**:MI(3100), NY(3100), WI(3100). **4**:NM(3100). **9**:CO(1195), ID(3202), OR(3202), WA(3202). **12**:OR(3202), WA(3202).
P. clintonii var. *albae*. **12**:WY(3100).
P. clintonii var. *bracteosae*. **1**:CO(3100), WY(3100).
P. clintonii var. *groenlandicae*. **9**:CO(3100), ID(3100), OR(35,2534), WA(3100).
P. clintonii var. *racemosae*. **12**:OR(3100).
P. helicalis. **5**:AK(3100).
P. rufescens. Autoecious. **6**:AZ(1195,1203,3981). **8**:CA(638,642). **13**:CA(81).
Deuteromycotina - Hyphomycetes
Eriomycopsis sp. On living leaves. **11**:WI(1351).
Ramularia obducens. On leaves. **1**:WY(1195). **8**:CA(638,642).
Deuteromycotina - Coelomycetes
Diplodina pedicularidis. **9**:CO(649).
Heteropatella conglutinata. **12**:WA(3202).
H. umbilicata. **1**:WA(3202). **2**:WA(3816). **7**:WA(3202,3816). **14**:WA(3816).
Leptostroma lupini. **12**:WA(3816).
Leptothyrella graminis. **7**:WA(3816). **14**:WA(3816).
Phoma coloradoensis. On stems. **12**:CO(128).
P. herbicola. On stems. **1**:WY(3812).
P. minuta. **7**:WA(3202).
P. pedicularis. On stems. **1**:WY(3812). **7**:WA(3202). **12**:WY(2924,3812).
Phyllosticta pedicularidis. **1**:WY(1195).

Penstemon Schmidel

1. *P. albertinus* Greene
2. *P. albidus* Nutt.
3. *P. antirrhinoides* Benth. ssp. *microphyllus* (A. Gray) D. Keck (=*P. microphyllus* A. Gray)
4. *P. aridus* Rydb.
5. *P. attenuatus* Douglas ex Lindl.
6. *P. australis* Small
7. *P. azureus* Benth.
8. *P. barbatus* (Cav.) Roth
9. *P. canescens* (Britton) Britton
10. *P. centranthifolius* Benth. [scarlet bugler]
11. *P. cobaea* Nutt.
12. *P. confertus* Douglas ex Lindl. (=*P. tolmiei* Hook., *P. procerus* Douglas ex Graham)
13. *P. cordifolius* Benth.
14. *P. corymbosus* Benth.
15. *P. davidsonii* Greene (=*P. menziesii* Hook. subsp. *davidsonii* (Greene) Piper)
16. *P. deustus* Douglas ex Lindl.
17. *P. ellipticus* J. M. Coulter & E. M. Fisher
18. *P. fruticosus* (Pursh) Greene
19. *P. gairdneri* Hook.
20. *P. glaber* Pursh (=*P. alpinus* Torr.)
21. *P. globosus* (Piper) Pennell & D. Keck
22. *P. gloxinioides* Hort. A listed name of no botanical standing.
23. *P. gracilentus* A. Gray
24. *P. gracilis* Nutt.
25. *P. gracilis* Nutt. var. *wisconsinensis* (Pennell) Fassett
26. *P. grandiflorus* Nutt.
27. *P. hirsutus* (L.) Willd.
28. *P. humilis* Nutt. ex A. Gray
29. *P. jamesii* Benth.
30. *P. laetus* A. Gray
31. *P. laevigatus* (L.) Aiton (=*P. digitalis* Nutt.)
32. *P. linarioides* A. Gray
33. *P. montanus* Greene
34. *P. newberryi* A. Gray [mountain-pride]
35. *P. nitidus* Douglas ex Benth.
36. *P. ovatus* Douglas
37. *P. pallidus* Small

38. *P. payettensis* A. Nels. & Macbr.
39. *P. pinifolius* Greene
40. *P. pruinosus* Douglas ex Lindl.
41. *P. pseudospectabilis* M. E. Jones (=*P. connatifolius* A. Nels.)
42. *P. radicosus* A. Nels.
43. *P. rattanii* A. Gray
44. *P. rattanii* A. Gray var. *kleei* (Greene) D. Keck
45. *P. richardsonii* Douglas ex Lindl.
46. *P. rostriflorus* Kellogg (=*P. bridgesii* A. Gray)
47. *P. rupicola* (Piper) Howell
48. *P.* × *rupicola* (Piper) Howell var. *edithae* C. English. *P. rupicola* × *P. barrettae*.
49. *P. rydbergii* A. Nels.
50. *P. rydbergii* A. Nels. var. *oreocharis* (Greene) N. Holmgren (=*P. oreocharis* Greene)
51. *P. serrulatus* Menzies ex Sm.
52. *P. shastensis* D. Keck
53. *P. speciosus* Douglas ex Lindl.
54. *P. spectabilis* Thurb. ex Torr. & A. Gray
55. *P. tenuiflorus* Pennell
56. *P. triphyllus* Douglas ex Lindl.
57. *P. tubiflorus* Nutt.
58. *P. unilateralis* Rydb.
59. *P. venustus* Douglas ex Lindl.
60. *P. virgatus* A. Gray
61. *P. watsonii* A. Gray
62. *P. whippleanus* A. Gray (=*P. stenosepalus* (A. Gray) Howell)
63. *P. willcoxii* Rydb.
64. *Penstemon* sp.

Chytridiomycetes

Synchytrium rugulosum. **21**:ID(3202).

Synchytrium sp. On leaves and petioles. **12**:ID(3202). **20**:CO(1847). **21**:ID(1847). **36**:CA(1847). **62**:CO(1849). **64**:WY(1847,2924).

Ascomycotina

Acerbia bacillata. On dead stems. **13**:CA(2439).

Antennulariella alpina. Black crust, on foliage. **15**:CA(638,639,642), WA(3202). **64**:CA(82).

Comoclathris permunda. On dead stems. **20**:WY(3810). **23**:CA(638,642). **40**:WA(3202). **62**:WY(3810). **64**:WY(3810).

Crocicreas melanosporum. On overwintered peduncles, petioles and stems. **23**:CA(505,642).

Erysiphe cichoracearum. Powdery mildew. **7**:CA(1155). **8**:WA(3202). **10**:CA(1155). **12**:ID(3202), WA(82,3202). **14**:CA(1155). **22**:Unk.(3976), CA(1155). **27**:WA(3202). **30**:CA(1155). **44**:CA(1155). **50**:CA(1155). **58**:WA(3202).

Graphyllium planispora. **20**:WY(3810).

Guignardia caryophyllea. **12**:WA(3202,3815).

Leptosphaeria lethalis. **12**:ID(3202), WA(3202).

Lewia scrophulariae. **18**:WA(3202). **20**:WY(3810). **42**:ID(3202).

Mycosphaerella pentstemonis. On leaves and stems. **64**:CO(82).

Mycosphaerella sp. On overwintered basal leaves. **25**:WI(1357).

Niesslia exilis. **36**:ID(3202).

Pleospora herbarum. **4**:ID(3202). **19**:WA(3202). **40**:WA(3202).

P. herbarum var. *occidentalis*. **20**:WY(3810).

P. scrophulariae var. *compositarum*. **20**:WY(3810). **49**:WY(3810). **62**:WY(3810).

Robergea obliqua. On dead stems. **13**:CA(3225,3226).

Basidiomycotina - Uredinales

Puccinia andropogonis. Heteroecious: aecial host. **2**:SD(2243). **6**:MS(2720), NC(1303). **9**:NC(3944). **11**:KS(82). **18**:MT(3202). **20**:NM(1195). **24**:IA(82), MO(82), SD(82,2243), WI(82). **26**:IA(82), KS(82), ND(82), NE(82), SD(82,2243), WI(1333). **27**:AL(82), GA(82,1439), IL(82), IN(82), MS(82,2720), NY(82), OK(82), PA(82), WI(82). **29**:OK(2872). **31**:NC(3944), WI(1350,1359). **35**:MT(3202). **36**:ID(3202). **37**:WI(1361). **55**:OK(2872). **57**:MS(2720), WI(1338). **60**:AZ(1203,3981). **64**:Range of host(82), ID(3202), MT(3202), WA(3202).

P. confraga. **3**:AZ(1203). **64**:AZ(82).

P. palmeri. Autoecious. **1**:MT(3202). **5**:ID(3202), MT(3202), WA(3202). **12**:AK(60), AZ(82), CA(82), ID(82,3202), MT(1195,3202), WA(82,3202), WY(1195). **15**:CA(642), OR(3202), WA(3202). **16**:ID(3202), MT(3202). **17**:MT(1195). **18**:WA(3202). **21**:OR(3202). **28**:MT(3202). **34**:CA(638,642). **49**:MT(3202). **59**:OR(3202), WA(3202). **62**:WY(2924). **63**:ID(3202), OR(3202), WA(3202). **64**:W states(82), CA(82), ID(1195), WA(82).

P. pentstemonis. Autoecious. **12**:UT(82), WA(3202). **16**:CA(81,638,642), ID(3202), OR(3202), WA(3202), WY(1195). **26**:SD(2243). **32**:AZ(1203,3981). **39**:AZ(1195,3379,3981). **41**:AZ(1195,1203,3981). **45**:OR(3202). **46**:AZ(1195,1203,3981). **51**:OR(3202). **54**:AZ(82), CA(82), OR(82), WY(82). **56**:ID(3202).

Deuteromycotina - Hyphomycetes

Cercospora pentstemonis. Leaf spot. **11**:KS(82), OK(82,2872), TX(82). **24**:MT(82), WI(1344). **26**:KS(82), ND(82,378), NE(82), SD(82,2243), WI(82). **27**:AL(82). **31**:WI(1335). **37**:OK(2872).

Phymatotrichopsis omnivora. Root rot. **11**:TX(82). **64**:TX(82).

Ramularia nivosa. Leaf spot. **1**:ID(3202). **2**:MT(3202). **12**:ID(3202), MT(3202), OR(3202), WA(3202). **16**:ID(3202), WA(3202). **18**:ID(3202). **21**:ID(3202). **28**:MT(3202). **33**:ID(3202). **35**:MT(3202). **36**:ID(3202). **38**:ID(3202). **40**:WA(3202). **42**:ID(3202). **43**:OR(3202). **49**:OR(3202). **51**:OR(3202), WA(3202). **52**:CA(642). **53**:ID(3202), WA(3202). **59**:ID(3202), OR(3202), WA(3202). **61**:ID(3202). **63**:ID(3202), MT(3202), OR(3202). **64**:CO(82), ID(82,1195), OH(82), WA(82).

R. pentstemonis. Leaf spot. **52**:CA(639). **64**:CA(82).

Deuteromycotina - Coelomycetes

Ascochyta garrettiana. **12**:WA(3202,3816).

Ascochytella pentstemonii. Stem spot. **54**:CA(82).

Coniothyrium penstemonis. On leaves and stems. **10**:CA(81). **15**:CA(661). **64**:CO(82).

Gloeosporium pentstemonicola. On lower leaves. **27**:WI(1344).

Heteropatella umbilicata. **12**:WA(3202,3816). **23**:CA(642). **47**:WA(3202). **64**:ID(3202).

Kellermania alpina. On stems. **64**:NV(82).

Phoma jejuna. On stems. **62**:WY(3812).

P. wyomingensis. On stems. **20**:WY(3812).

Phyllosticta antirrhini. Leaf spot. **27**:IL(82).

Phyllosticta sp. Leaf spot. **25**:WI(1360). **57**:WI(1352).

Septoria pentstemonicola. On leaves. **24**:SD(82,970,2243), WI(82). **37**:WI(1361).

S. pentstemonis. Leaf spot. **6**:MS(2720). **11**:TX(82). **12**:WA(3202). **26**:WI(1342). **27**:IL(82). **31**:MS(2720), WI(1333,1359). **37**:OK(2873). **57**:WI(1338). **64**:CA(82), ME(82), MI(82), MO(82), MS(82), NY(82), OK(82), TX(82), WI(82).

Sphaceloma penstemonis. Spot anthracnose of leaves. **48**:NY(3788).

Sphaeropsis sp. **16**:OR(3202).

Deuteromycotina - Other

Sclerotium rolfsii. Stem rot. **8**:CT(82), NJ(82). **11**:TX(82). **27**:IL(82). **64**:MA(82), NJ(82), TX(82).

Rhinanthus L.

1. *R. crista-galli* L. (=*R. kyrollae* Chabrey) [rattlebox]
2. *Rhinanthus* sp.

Basidiomycotina - Uredinales

Cronartium coleosporioides. Heteroecious: telial host. **1**:MT(1195,3202).

Deuteromycotina - Hyphomycetes

Botrytis cinerea. **2**:AK(526).

Russelia Jacq.

1. *R. equisetiformis* Schlechtend. & Cham. [coralplant, fountainbush, fountainplant]

Oomycetes

Pythium sp. Root rot. **1**:FL(35).

Deuteromycotina - Hyphomycetes

Corynespora cassiicola. Leaf spot. **1**:FL(35).

Deuteromycotina - Other

Rhizoctonia solani. Root and stem rot. **1**:FL(35).

Scoparia L.

1. *S. dulcis* L.

Deuteromycotina - Hyphomycetes

Oidium sp. Powdery mildew. **1**:FL(35).

Deuteromycotina - Coelomycetes

Phomopsis sp. Stem canker. **1**:FL(35).

Scrophularia L.

1. *S. californica* Cham. & Schlechtend.
2. *S. lanceolata* Pursh (=*S. leporella* C. Bickenll)
3. *S. marilandica* L. (=*S. neglecta* Rydb.) [carpenter's-square]
4. *S. nodosa* L.
5. *S. parviflora* Woot. & Standl.
6. *Scrophularia* sp.

Oomycetes

Peronospora sordida. Downy mildew. **1**:CA(82). **2**:IA(82), IL(82), NY(82), OR(3202), WI(82). **3**:SC states(82), SE states(82), NE(82), WI(82).

Ascomycotina

Crocicreas cyathoideum var. *cacaliae*. On dead stems. **2**:UT(505).
Erysiphe cichoracearum. Powdery mildew. **1**:CA(3975).
Leptosphaeria consessa. On stems. **1**:CA(81,82), OR(82). **2**:WA(3202).
Mycosphaerella decidua. Leaf spot. **4**:KS(82,967).
Nodulosphaeria erythrospora. **2**:NY(3248).
N. modesta. **5**:AZ(662).
Pezizellaster herbarum. On dead stems. **1**:CA(527).
Stictis ostropoides. **6**:CA(3225).

Deuteromycotina - Coelomycetes

Coniothyrium scrophulariae. On stems. **1**:CA(82).
Cylindrosporium scrophulariae. Leaf spot. **3**:IL(82), OK(82,2872), PA(82).
Phyllosticta sp. Leaf spot. **3**:WI(1358,1360).
Septoria scrophulariae. Leaf spot. **1**:CA(81,82), OR(82,3202). **2**:E states(82), CO(82), ID(3202), MS(82), ND(378), WA(82,3202). **3**:E states(82,2720), CO(82), MS(82,2873), OK(2873), WA(82).

Striga Lour.

1. *S. asiatica* (L.) Kuntze (=*S. lutea* Lour.) [witchweed]
2. *S. gesnerioides* (Willd.) Vatke [tobacco witchweed]

Deuteromycotina - Hyphomycetes

Curvularia geniculata. **1**:NC(1303).
Fusarium roseum. Root and stem rot. **1**:NC(1303).
F. solani. Root and stem rot. **1**:NC(1303).

Deuteromycotina - Other

Rhizoctonia solani. Root and stem rot. **1**:NC(1303).
Rhizoctonia sp. Root and stem rot. **2**:FL(35).
Sclerotium rolfsii. Stem rot. **1**:NC(1303).

Synthyris Benth. in A. DC.

1. *S. dissecta* Rydb.
2. *S. missurica* (Raf.) Pennell
3. *S. pinnatifida* S. Wats. (=*S. pinnatifida* S. Wats. var. *pinnatifida*)
4. *S. platycarpa* Gail & Pennell
5. *S. reniformis* (Douglas ex Benth.) Benth.
6. *Synthyris* sp.

Ascomycotina

Comoclathris permunda. **1**:WY(3810).
Sphaerotheca fuliginea. **2**:ID(3202).

Basidiomycotina - Uredinales

Puccinia acrophila. Autoecious. **2**:WA(3202). **3**:WY(749,1195,3379). **6**:W states(82).
P. wulfeniae. Autoecious. **2**:ID(3202), WA(3202). **4**:ID(3202). **5**:CA(82), ID(82,3202), OR(82,3202), WA(82,3202). **6**:ID(82), WA(82), WY(82).

Deuteromycotina - Hyphomycetes

Ramularia synthyridis. Leaf spot. **2**:ID(3202). **5**:OR(3202), WA(666,3202).
Ramularia sp. Leaf spot. **6**:OR(82).

Deuteromycotina - Coelomycetes

Phoma herbicola. On stems. **1**:WY(3812).

Tonella Nutt. ex A. Gray

1. *T. floribunda* A. Gray
2. *T. tenella* (Benth.) A. Heller

Basidiomycotina - Uredinales

Puccinia collinsiae. Autoecious. **2**:WA(3202).

Basidiomycotina - Ustilaginales

Entyloma linariae. **2**:OR(3202).

Deuteromycotina - Hyphomycetes

Cercosporella sp. **1**:ID(3202).

Torenia L.

1. *Torenia* sp.

Deuteromycotina - Coelomycetes

Phyllosticta sp. Leaf spot. **1**:FL(35).

Verbascum L.

1. *V. blattaria* L. [moth mullein]
2. *V. blattaria* L. var. *albiflora* (D. Don) House [white moth mullein]
3. *V. densiflorum* Bertol. (=*V. thapsiforme* Schrad.)
4. *V. nigrum* L. (=*V. urticifolium* Lam.)
5. *V. simplex* Labill.
6. *V. thapsus* L. [common mullein, flannelplant, mullein, velvetplant]
7. *Verbascum* sp.

Oomycetes

Peronospora sordida. Downy mildew. **7**:NJ(82).

Ascomycotina

Crocicreas nigrescens. **7**:NJ(505).
Erysiphe cichoracearum. Powdery mildew. **4**:MS(2720). **5**:MS(2720). **6**:PA(82). **7**:MS(2720).
Lophiostoma caulium. **6**:GA(1438).
Mycosphaerella verbascicola. On leaves. **6**:ME(82), NC(82), NY(82), PA(82).
Pleospora sp. **6**:WA(3202).

Deuteromycotina - Hyphomycetes

Cercospora septatissima. Leaf spot. **7**:MS(2720).
C. verbascicola. Leaf spot. **1**:TX(82). **3**:WI(1339). **6**:AL(82), DE(82,3543), IA(82), IL(312), MD(82), WI(82,1335).
Oidium sp. Powdery mildew. **1**:NJ(82).
Phymatotrichopsis omnivora. Root rot. **1**:TX(82). **6**:TX(82).
Ramularia variabilis. Leaf spot. **6**:E states(82), ID(3202), MS(82,2720), MT(3202), TX(82), WA(82,3202). **7**:CT(82).

Deuteromycotina - Coelomycetes

Phoma terrestris. Isolated from rhizosphere. **6**:OR(2791).
P. thapsi. On stems. **1**:TX(82). **6**:NJ(82), TX(82).
P. verbascicarpa. On stems. **1**:NY(82).
P. verbascicola. **6**:Unk.(82).
Phoma sp. On stems. **1**:Unk.(82).
Phyllosticta verbascicola. Leaf spot. **1**:TX(82), WI(1357). **3**:WI(1335). **6**:IN(82), KS(82), NJ(82), NY(82), TX(82), WI(82). **7**:NY(82), WI(82).
Septoria verbascicola. Leaf spot. **1**:E states(82), ID(3202), MO(82), NC(3944), OK(2872), OR(3202), TX(82). **2**:MS(2720).

Veronica L.

1. *V. alpina* L.
2. *V. americana* Schwein. ex Benth. [American brooklime]
3. *V. arvensis* L. [corn speedwell]
4. *V. cusickii* A. Gray
5. *V. longifolia* L. (=*V. maritima* L.) [beach speedwell]
6. *V. officinalis* L. [common speedwell, gypsyweed]
7. *V. peregrina* L. [neckweed, purslane speedwell]
8. *V. peregrina* L. var. *xalapensis* (H. B. K.) Pennell
9. *V. persica* Poir.
10. *V. scutellata* L. [marsh speedwell]
11. *V. serpyllifolia* L. (=*V. humifusa* Dickson) [thyme-leaved speedwell]
12. *V. spicata* L. [spike speedwell]
13. *V. wormskjoldii* Roem. & Schult. (=*V. alpina* L. var. *unalaschcensis* Cham. & Schlechtend.)
14. *Veronica* sp.

Plasmodiophoromycetes

Ligniera verrucosa. On roots. **9**:OH(2385).
Sorosphaera veronicae. Leaf and stem galls. **3**:MS(82,2720). **14**:CO(82).

Chytridiomycetes

Synchytrium globosum. Leaf gall. **7**:LA(82). **11**:WA(3202). **13**:WA(3202).
Synchytrium sp. On leaves, petioles and stems. **1**:WY(1848,2924). **8**:CA(1848). **11**:WA(1848).

Oomycetes

Peronospora grisea. Downy mildew. **3**:ID(3202), IL(82), IN(82), NY(82), WA(3202), WI(82). **7**:GA(82), ID(3202), IN(82), KS(82), OR(3202), WA(3202), WI(82). **11**:ID(3202), WA(3202). **13**:CA(642). **14**:CA(82), MO(82), TX(82), WI(82).

Ascomycotina
Sphaerotheca fuliginea. **12:**WI(1343).
S. macularis. Powdery mildew. **5:**CT(82), OR(82). **11:**WI(82).
Basidiomycotina - Uredinales
Puccinia albulensis. Autoecious. **4:**WA(3202). **10:**WA(3202). **13:**MT(3202), OR(3202), WA(3202). **14:**CO(82), MT(82), OR(82), UT(82), WA(82), WY(82).
P. cascadensis. Autoecious. **4:**ID(3202), OR(3202), WA(3110,3202). **13:**WY(1520). **14:**WA(82).
[*P. gentianae*]. **11:**NM(82).
P. rhaetica subsp. *americana*. **4:**WA(3110). **13:**WY(1520).
P. veronicarum. Autoecious. **11:**IA(82). **14:**IA(82), WI(82).
P. wulfeniae. Autoecious. **13:**WY(749,1195).
Basidiomycotina - Ustilaginales
Entyloma linariae. Leaf smut. **2:**CO(1077), ID(1077,3202), MT(3202), NY(1077), OR(3202), WA(3202), WY(1195). **3:**TX(82). **6:**TX(82). **7:**CT(82,1077), IA(82,1077), ID(3202), IL(82,1077), IN(1077), KS(82), MD(1077), MO(82,1077), MS(82,1077,2720), TX(82), WA(3202), WI(82,1077). **11:**ID(3202), WA(3202). **14:**CO(82), NY(82), TX(82).
Basidiomycotina - Other
Thanatephorus cucumeris. **11:**WI(1357).
Deuteromycotina - Hyphomycetes
Cercospora tortipes. Leaf spot. **10:**NY(581), OR(3202), WI(82,581).
Fusarium sp. Root rot. **5:**NJ(82). **12:**NJ(82).
Phymatotrichopsis omnivora. Root rot. **3:**TX(82). **6:**TX(82). **7:**TX(82). **14:**TX(82).
Ramularia anagallidis. **14:**WI(1343).
R. veronicae. Leaf spot. **3:**TX(82). **6:**TX(82). **7:**OK(82,2872), TX(82), WI(82). **11:**WI(82). **14:**TX(82).
Thallospora aspera. **7:**WI(1341).
Deuteromycotina - Coelomycetes
Discogloeum veronicae. Leaf spot. **6:**NY(82). **9:**OR(3202).
Septoria exotica. **12:**WI(1340).
S. veronicae. Leaf spot. **3:**MI(82), OR(3202), WA(3202), WI(82). **14:**CA(82), FL(35,82), IA(82), OH(82), WI(82).
Deuteromycotina - Other
Rhizoctonia solani. Root rot. **11:**IL(82), MD(82).
Sclerotium rolfsii. Stem rot. **6:**NJ(82). **14:**CT(82), OH(82).

Veronicastrum Heist. ex Fabr.

1. *V. virginicum* (L.) Fée (=*Veronica virginica* L.) [blackroot, bowman's-root, Culver's-root]

Ascomycotina
Erysiphe cichoracearum. Powdery mildew. **1:**IL(82).
Sphaerotheca fuliginea. Powdery mildew. **1:**CT(82), IA(82), IL(82), MD(82), MI(82), MO(82), WI(82).
Basidiomycotina - Uredinales
Puccinia veronicarum. Autoecious. **1:**IA(82), TX(82), WI(82).
Basidiomycotina - Other
Ceratobasidium anceps. **1:**WI(1355).
Deuteromycotina - Hyphomycetes
Cercospora leptandrae. Leaf spot. **1:**WI(82,1329,1358).
Phymatotrichopsis omnivora. Root rot. **1:**TX(82).
Ramularia veronicae. Leaf spot. **1:**TX(82).
Deuteromycotina - Coelomycetes
Ascochyta sp. On leaves. **1:**WI(1360).
Phoma exigua. Leaf spot. **1:**TX(82), WI(82,1333).
Septoria veronicae. Leaf spot. **1:**WI(82).
Deuteromycotina - Other
Rhizoctonia solani. Root rot. **1:**TX(82).

Selaginellaceae

Selaginella P. Beauv.

1. *S. apoda* (L.) Fernald
2. *S. arenicola* Underw.
3. *S. arenicola* Underw. subsp. *acanthonota* (Underw.) R. Tryon
4. *S. arizonica* Maxon
5. *S. asprella* Maxon
6. *S. bigelovii* Underw.
7. *S. densa* Rydb. var. *scopulorum* (Maxon) R. Tryon
8. *S. hansenii* Hieron.
9. *S. lepidophylla* (Hook. & Grev.) Spring
10. *S. mutica* D. C. Eaton ex Underw.
11. *S. mutica* D. C. Eaton ex Underw. var. *limitanea* Weath.
12. *S. oregana* D. C. Eaton
13. *S. pilifera* A. Brown var. *pringlei* (Baker) Morton
14. *S. rupestris* (L.) Spring [dwarf lycopod, rock spike-moss]
15. *S. underwoodii* Hieron.
16. *S. utahensis* Flowers
17. *S. viridissima* Weath.
18. *S. wallacei* Hieron.
19. *S. watsonii* Underw.
20. *S. weatherbiana* R. Tryon
21. *S. wrightii* Hieron.
22. *Selaginella* sp.

Oomycetes
Pythium sp. Root rot. **22:**FL(35).
Ascomycotina
Acrospermum urceolatum. **4:**AZ(1019), TX(1019). **8:**CA(1019). **11:**NM(1019). **14:**MN(1019), VT(1019). **22:**WA(3202).
Clathrospora platysporioides. **8:**CA(1019). **9:**TX(1019). **10:**AZ(1019), TX(1019). **11:**NM(1019). **12:**CA(1019), OR(1019). **13:**TX(1019). **14:**WV(1019). **15:**CO(1019). **16:**UT(1019). **17:**TX(1019). **20:**CO(1019). **21:**TX(1019).
Gibbera selaginellae. **12:**OR(1019), WA(1019).
Leptosphaeria helvetica. **3:**FL(1019). **5:**CA(1019). **10:**TX(1019). **11:**NM(1019). **14:**IN(1019), SD(1019). **18:**OR(1019), WA(1019).
L. mellispora. **15:**AZ(1019), CO(1019), NM(1019).
L. rostrata. **3:**FL(1019). **14:**VA(1019).
Morenoina selaginellae. **11:**TX(1019). **13:**TX(1019). **21:**TX(1019).
Mycosphaerella sp. **22:**AZ(662).
Plectosphaera selaginellae. **14:**NC(1019).
Pleospora farlowiana. **3:**FL(1019). **4:**AZ(1019), TX(1019). **6:**CA(1019). **8:**CA(1019). **9:**TX(1019). **10:**AZ(1019). **11:**NM(1019), TX(1019). **13:**TX(1019). **14:**IN(1019), MA(1019), NY(1019), SD(1019), VT(1019), WI(1019). **15:**NM(1019). **17:**TX(1019). **18:**CA(1019).
P. nivalis. **5:**CA(1019). **6:**CA(1019). **7:**CO(1019), OR(1019), WY(1019). **19:**UT(1019).
Pleospora sp. **5:**CA(1019). **10:**TX(1019). **15:**AZ(1019). **17:**TX(1019).
Talaromyces flavus. **12:**OR(1019).
Basidiomycotina - Other
Amphinema byssoides. **1:**TX(1019).
Deuteromycotina - Hyphomycetes
Acrotheca sp. **12:**WA(1019). **13:**TX(1019).
Alternaria sp. **14:**VT(1019).
Botryosporium pulchrum. **22:**DC(1019).
Cercospora selaginellarum. **1:**MI(1019).
Cercospora sp. Leaf spot. **22:**FL(35).
Chalara selaginellae. On leaves. **2:**FL(2550). **3:**FL(1019). **14:**NC(1019,2550).
Cladosporium sp. **3:**FL(1019).
Heterosporium selaginellarum. **3:**FL(1019).
Papulospora selaginellae. **7:**WY(1019).
Papulospora sp. **1:**TX(1019). **3:**FL(1019). **5:**CA(1019).
Phaeodochium myrothecioides. **9:**TX(1019).
Deuteromycotina - Coelomycetes
Belaina asclepiadis. **14:**NC(1019).
Coniothyrium rupestris. **14:**VA(1019).
Hendersonia fuscispora. **5:**CA(1019). **16:**UT(1019).
H. pallidispora. **5:**CA(1019).
Hendersonia sp. **5:**CA(1019). **15:**AZ(1019), CO(1019).
Phyllosticta sp. Leaf spot. **22:**FL(35).
Deuteromycotina - Other
Rhizoctonia solani. Foliage blight, root and stem rot. **22:**FL(35).

Simaroubaceae

Ailanthus Desf.

1. *A. altissima* (Mill.) Swingle (=*A. glandulosa* Desf.) [ailanthus, copa-tree, tree-of-heaven, varnish-tree]
2. *Ailanthus* sp.

Ascomycotina
Botryosphaeria dothidea. **1:**GA(1438).

B. obtusa. On limbs and twigs. **1:**KS(82), MI(82), NY(82).
B. ribis. On twigs. **1:**GA(82), MD(82).
Daldinia concentrica. **1:**ID(3202).
Diaporthe rudis. On twigs. **1:**DE(82), NJ(82), NY(82), OH(82).
Eutypella glandulosa. On limbs. **1:**Range of host(82), GA(82).
E. juglandicola. **2:**NJ(2925).
E. leprosa. **1:**IA(3629).
E. scoparia. **1:**GA(1438), IA(3629). **2:**NY(2925).
Guignardia ailanthi. On twigs. **1:**SC(82).
Hypoxylon rubiginosum. On dead limbs. **1:**NC(1303).
Jattaea microtheca. On limbs. **2:**NJ(190).
Mycosphaerella ailanthi. On leaves and twigs. **1:**NY(82). **2:**SC(634).
Nectria cinnabarina. Dieback. **1:**KS(82), NJ(82,891), OK(2872), SC(82), WA(3202).
N. coccinea. Trunk canker. **1:**Unk.(82).
Thyridium vestitum. **1:**IA(1226).
Basidiomycotina - Other
Antrodia malicola. **1:**Unk.(82).
Armillaria mellea. Root rot. **1:**NY(82).
Cerrena unicolor. Butt rot. **1:**Unk.(82).
Glabrocyphella ailanthi. On bark. **1:**NJ(646).
Phellinus punctatus. On trees. **1:**OH(638).
Schizophyllum commune. Wound rot. **1:**ID(3202), KS(82), MO(82), NY(82).
Stereum ochraceo-flavum. **1:**TN(543).
Trametes hirsuta. **1:**GA(1439).
T. versicolor. White rot. **1:**Unk.(82), NC(1303).
Tyromyces chioneus. White rot. **1:**Unk.(82).
Deuteromycotina - Hyphomycetes
Cercospora glandulosa. Leaf spot. **1:**Range of host(82), OK(2872), TX(82).
Cristulariella moricola. Leaf spot. **2:**FL(35).
Fusarium lateritium. Twig blight. **1:**VA(82).
Phymatotrichopsis omnivora. Root rot. **1:**AZ(82), TX(82).
Tubercularia vulgaris. **1:**Unk.(82), SC(3178).
Verticillium albo-atrum. Wilt. **1:**NY(82), PA(82), VA(82).
Deuteromycotina - Coelomycetes
Aplosporella ailanthi. On dry twigs. **1:**KS(82), NY(82,966).
Botryodiplodia ailanthi. Twig blight. **1:**AL(82), SC(82,634), TX(82).
Camarosporium berkeleyanum. On twigs. **1:**NY(82).
Colletotrichum gloeosporioides. Leaf spot. **1:**LA(82), SC(82), TX(82), WV(82).
Cytoplea insitiva. On twigs. **1:**Range of host(82).
Cytospora sacculus. On twigs. **1:**KS(82), MS(2720).
Dothiorella glandulosa. On bark. **1:**SC(634).
Lasiodiplodia theobromae. Twig blight. **1:**TX(82).
Leptothyrium petiolorum. On petioles. **1:**NJ(82), OH(82).
Phoma glandulosum. On petioles. **1:**SC(634).
Phomopsis ailanthi. Stem blight. **1:**TX(82).
Phyllosticta ailanthi. Leaf spot. **1:**VA(82).

Castela Turpin
1. *C. tortuosa* Liebm.
Ascomycotina
Meliola castelae. On living leaves. **1:**TX(3483).

Picramnia Sw.
1. *P. pentandra* Sw. [bitterbush]
Ascomycotina
Phyllachora domingensis. Tar spot. **1:**FL(35,2287).

Simarouba Aubl.
This genus is considered by some to be a synonym of *Quassia*, but following Wm. W. Thomas (pers. comm.) we retain *Simarouba* as a separate genus.
1. *S. glauca* DC. [aceituna, bitter damson, bitterwood, paradise-tree]
Deuteromycotina - Coelomycetes
Phomopsis sp. Twig dieback. **1:**FL(35).

Simmondsiaceae

Simmondsia Nutt.
1. *S. chinensis* (Link) C. K. Schneid. (=*S. californica* Nutt.) [deernut, goatnut, jojoba]
Basidiomycotina - Other
Dendrophora albobadia. **1:**AZ(1202).
Hypochnicium punctulatum. **1:**AZ(1202).
Meruliopsis corium. **1:**AZ(1202).
Deuteromycotina - Hyphomycetes
Strumella simmondsiae. On flowers and leaves. **1:**CA(325).
Deuteromycotina - Coelomycetes
Coniothyrium sp. On leaves and stems. **1:**AZ(3984).

Smilacaceae

Smilax L.
1. *S. auriculata* Walter
2. *S. balbisiana* Kunth (=*S. pseudo-china* Bertero ex Kunth)
3. *S. bona-nox* L.
4. *S. bona-nox* L. var. *hederaefolia* (Beyrich) Fernald
5. *S. ecirrhata* (Engelm.) S. Wats.
6. *S. glauca* Walter [cat greenbrier, catbrier, sawbrier]
7. *S. glauca* Walter var. *genuina* S. F. Blake
8. *S. herbacea* L. [carrionflower, Jacob's ladder]
9. *S. hispida* Muhl. (=*S. tamnoides* L. var. *hispida* Muhl.) [bristly greenbrier, hagbrier, hellfetter]
10. *S. lanceolata* L. [Jackson brier]
11. *S. lasioneuron* Hook. (=*S. herbacea* L. var. *lasioneuron* (Hook.) A. DC.)
12. *S. laurifolia* L. [bamboo-vine, blaspheme-vine, laurel-leaved greenbrier]
13. *S. pulverulenta* Michx. (=*S. herbacea* L. var. *pulverulenta* (Michx.) A. Gray)
14. *S. rotundifolia* L. [bullbrier, catbrier, common greenbrier, horsebrier]
15. *S. sandwicensis* Kunth
16. *S. tamnoides* L. [China-root, hellfetter]
17. *Smilax* sp.
Myxomycetes
Diachea leucopodia. **17:**NC(3944).
Chytridiomycetes
Synchytrium smilacis. Galls of leaf blades, petioles and stems. **17:**LA(326).
Ascomycotina
Acrospermum sp. **6:**GA(1438).
Anthostomella ludoviciana. On dead stems. **17:**LA(82).
A. sepelibilis. On dead stems. **3:**GA(1438). **12:**GA(1438). **17:**SE states(82), GA(1438), TN(82).
Anthostomella sp. On dead stems. **12:**GA(1438). **17:**Unk.(82).
Apiospora montagnei. **8:**GA(1438).
Botryosphaeria disrupta. On stems. **14:**MS(2720). **17:**AL(82), MS(82), SC(82).
B. muriculata. On dead stems. **17:**AL(82).
B. obtusa. On stems. **17:**VA(82).
B. ribis. **12:**AL(2505).
Diatrypella favacea. **12:**GA(1438).
Didymosphaeria polysticta. **17:**MD(2783).
Didymosphaeria sp. **12:**GA(1438).
Dimerosporium nimbosum. On stems. **17:**FL(82), SC(82).
Discohainesia oenotherae. On leaves. **17:**VA(82).
Eutypa sp. **12:**GA(1438).
Eutypella juglandicola. **6:**GA(1438).
E. leprosa. **14:**IA(3629). **17:**IA(3629).
E. quadrifida. **17:**NC(3944), TX(2925).
E. scoparia. **6:**GA(1438). **12:**GA(1438). **14:**GA(1438). **17:**GA(1438), SC(2925).
Glomerella cingulata. On leaves. **17:**MD(82).
Gloniopsis curvata. On stems. **6:**GA(1438), NC(3944). **12:**GA(1438). **14:**GA(1438). **17:**E states(82), FL(35), LA(82).
Hypoderma smilacis. On stems. **17:**E states(82), LA(82), NC(1303).
Hypoxylon smilacicola. On dead stems. **14:**VA(2403). **17:**NC(2403), NY(2403), PA(2403).
[*Hysterium macrosporum*]. **17:**FL(2782).

Lophodermium sp. **12:**GA(1438).
Mycosphaerella smilacicola. Leaf spot. **6:**GA(1438). **12:**GA(1438). **14:**GA(1438). **17:**FL(2782), SC(82,634).
M. smilacina. On stems. **17:**NJ(82,967).
Mycosphaerella sp. Leaf spot. **6:**GA(82).
Myiocopron smilacis. On stems. **3:**GA(1438). **17:**E states(82), GA(1438), SC(634), TX(82).
Ophiobolus nigro-clypeata. **12:**GA(1438).
Ostreichnion curtisii. **12:**GA(1438).
Pezizella sp. **12:**GA(1438).
Phragmocapnias smilacina. **17:**HI(2903).
Phyllactinia guttata. Powdery mildew. **17:**MI(82).
Pyrenopeziza smilacicola. On stems. **17:**NJ(82), NY(82).
Schizothyrium pomi. Fly speck. **17:**NC(1303).
Seuratia millardetii. **17:**HI(2784).
Stictis radiata. **6:**GA(1438).
Trichothyrium reptans. **15:**HI(2903).

Basidiomycotina - Uredinales

Puccinia amphigena. Heteroecious: aecial host. **3:**OK(82,753,2872). **8:**IN(753), KS(82), ND(82,378), OK(2872). **9:**IN(753), KS(82), MI(82,753), NE(82,753). **14:**WI(753).
P. arundinariae. Heteroecious: aecial host. **17:**NC(82), SC(82).
P. eumacrospora. Heteroecious: aecial host. **6:**NJ(82). **9:**KS(82), NJ(82). **14:**DE(82), NJ(82), NY(82).
P. smilacis. Heteroecious: telial host. **3:**GA(1439), MS(2720), OH(638), OK(2872). **6:**GA(1439), MS(2720), NC(1303), OK(2872), SC(634), VA(541). **9:**NC(1307,3944). **14:**GA(1439), NC(1303), OK(2872). **16:**IN(987). **17:**E states(82), LA(2675), MS(2720), NE(82), OK(2872), TX(82).

Basidiomycotina - Other

Crinipellis foliicola. **17:**FL(3292).
Helicobasidium brebissonii. Root rot. **17:**TX(82).
Septobasidium pseudopedicellatum. Parasitic on scale insects. **1:**FL(82). **17:**FL(35).

Deuteromycotina - Hyphomycetes

Cercospora mississippiensis. Leaf spot. **6:**NC(3944). **9:**WI(1351). **17:**FL(35), NC(1303).
C. petersii. Leaf spot. **6:**NY(581), SC(581,634). **17:**FL(35).
C. smilacis. Leaf spot. **2:**OK(2872). **3:**OK(2872). **6:**DE(3543), MS(2720), NC(3944). **12:**MS(2720). **13:**WI(1329). **14:**DE(3543), MS(2720), OK(2873). **16:**MS(2720). **17:**E states(82), FL(35), MN(82), NE(82), OK(82), TX(82).
Corynespora aterrima. On dead stems. **14:**FL(82), LA(82), PA(82).
Cristulariella moricola. Zonate leaf spot. **13:**NC(1302,1303).
Endocalyx melanoxanthus. **17:**FL(2782).
Fusarium avenaceum. **9:**DE(3543).
Heterosporium asperatum. On leaves. **17:**Unk.(82).
Phaeoisariopsis pubescens. On twigs. **17:**NJ(525).
Ramularia subrufa. Leaf spot. **8:**WI(1330). **13:**WI(1329). **17:**IA(82), MS(82,2720), NE(82), WI(82).
Sporidesmium fragilissimum. On limbs and stems. **17:**FL(634), SC(634).
Taeniolella subsessilis. On dead stems. **9:**KS(963).
Trichothallus hawaiiensis. **17:**HI(2903).

Deuteromycotina - Coelomycetes

Aplosporella smilacis. On stems. **9:**NE(966). **17:**NY(82).
Ascochyta confusa. Leaf spot. **9:**NY(971), WV(971). **17:**NY(82), WI(82).
Colletotrichum dematium. Anthracnose, leaf spot. **17:**IL(82), OK(2872).
C. gloeosporioides. Leaf spot. **17:**OK(2872).
Colletotrichum sp. **5:**WI(1340,1354).
Coniothyrium fuckelii. Canker, on stems. **3:**MS(2720). **17:**MO(82), MS(82,2720).
Cylindrosporium smilacis. Leaf spot. **17:**AL(82), FL(35).
Cytospora sacculus. On exposed roots. **17:**GA(82).
Diplodia smilacina. On leaves and stems. **17:**LA(82), MA(82), MO(82), MS(82,2720), RI(82), TX(82).
Diplodina ramulorum. On bleached stems. **17:**NJ(966).
D. smilacis. On stems. **17:**NJ(82).
Discosia maculicola. Leaf spot. **16:**MS(2720).
D. smilacina. Leaf spot. **9:**MS(2720).
Gloeodes pomigena. On stems. **17:**IN(82).
Leptostroma smilacis. On stems. **12:**SC(634). **17:**NY(82), SC(82).
Leptothyrium smilacis. On stems. **17:**NY(82).
Macrophoma smilacina. Leaf spot. **3:**OK(2872). **6:**OK(2872). **8:**ND(378), OK(2873). **9:**OK(2873). **14:**NY(64), OK(2872). **15:**HI(2903). **17:**E states(82), FL(35), LA(82), ND(82), TX(82).
M. smilacina var. *smilacis*. Leaf spot. **3:**NY(279). **6:**FL(279), NY(279). **7:**IN(279). **8:**ME(279), NE(279), NY(279), WI(279). **9:**MI(279), NY(279,970). **11:**MN(279). **13:**ND(378). **14:**ME(279), NY(279), PA(279). **17:**NE states(82), CT(279), MD(279), ND(82), NY(82), OH(279), TX(82), VA(279).
Pestalotia clavata. Leaf spot. **12:**MS(2720). **17:**E states(82), FL(35,82), LA(82), MS(2720), OK(82,2873).
Pestalotiopsis smilacis. **14:**PA(3547).
Phlyctema smilacis. On stems. **17:**FL(35,82), TX(82).
Phoma smilacis. On stems. **17:**NY(82).
Phoma sp. Canker. **17:**MS(82,2720).
Phyllosticta convallariae. Leaf spot. **8:**IA(82). **9:**IA(82). **14:**IA(82).
P. crypta. On leaves. **6:**NY(280). **12:**GA(280). **14:**WV(280). **17:**MD(280).
P. cumminsii. On leaves. **3:**OK(280), TN(280), TX(280). **4:**IN(280). **9:**WI(280). **14:**AR(280), IN(280), LA(280), VA(280). **17:**KS(280).
P. hispida. **5:**WI(1360). **9:**WI(1334).
P. subeffusa. Leaf spot. **3:**NY(280). **14:**NY(280), WV(280). **17:**DC(280), IL(82), KS(82), NY(280), WV(82).
Phyllosticta sp. Leaf spot. **8:**WI(1354). **9:**OK(2872). **10:**OK(2872). **14:**NC(1303). **17:**Unk.(82), FL(35), NC(1303).
Septogloeum subnudum. Leaf spot. **17:**IL(82), WI(82).
Septoria smilacinae. On leaves. **17:**LA(82).
S. smilacis. **17:**WV(82,971).
Septoria sp. Leaf spot. **17:**FL(35).
Stilbospora macrosperma. Leaf spot. **17:**FL(35).

Solanaceae

Atropa L.

1. *A. belladonna* L. [belladonna, deadly nightshade]

Oomycetes

Peronospora sp. Downy mildew. **1:**CA(82).
Phytophthora nicotianae var. *parasitica*. Root rot. **1:**CA(82).
Pythium debaryanum. Damping-off. **1:**CA(82).
P. irregulare. Damping-off. **1:**CA(82).
P. ultimum. Damping-off. **1:**CA(82).
Pythium sp. Damping-off. **1:**CA(82).

Ascomycotina

Mycosphaerella sp. Leaf spot. **1:**CA(82).

Deuteromycotina - Hyphomycetes

Alternaria sp. Leaf spot. **1:**CA(82).
Cercospora atropae. Leaf spot. **1:**VA(82).
Fusarium sp. Root rot. **1:**CA(82).
Thielaviopsis basicola. Root rot. **1:**CA(82).
Verticillium albo-atrum. Wilt. **1:**CA(82).

Deuteromycotina - Other

Rhizoctonia solani. Damping-off. **1:**CA(82), FL(35).

Browallia L.

1. *Browallia* sp.

Deuteromycotina - Hyphomycetes

Fusarium sp. Wilt. **1:**DE(82).

Brunfelsia L.

1. *B. americana* L. [lady-of-the-night]

Deuteromycotina - Hyphomycetes

Cercospora sp. Leaf spot. **1:**FL(35).
Fusarium sp. Stem rot. **1:**FL(35).

Deuteromycotina - Other

Rhizoctonia solani. Thread blight. **1:**FL(35).

Capsicum L.

1. *C. annuum* L. (=*C. annuum* L. var. *annuum*) [bell pepper, Cayenne pepper, chili pepper, common garden pepper, green pepper, mango pepper, paprika pepper, pimento]
2. *C. frutescens* L. [tabasco pepper]
3. *C. frutescens* L. cv. 'grossum' [bell pepper, sweet pepper]
4. *Capsicum* sp.

Oomycetes
Aphanomyces sp. Seedling root rot. **2:**NJ(82).
Peronospora tabacina. Downy mildew. **1:**FL(35). **2:**FL(35,82), GA(82), NC(82,1303), SC(82), TX(82).
Phytophthora capsici. Blight, crown, fruit and root rot, wilt. **1:**CA(2838), CT(2975), FL(35), NC(2838,3931), NM(2113,2838), WV(2838). **2:**CA(82), CO(82), FL(35,82), HI(2903), LA(82), MO(82), NC(1303), NM(82), NY(82), OH(82), TX(82), VA(82).
P. nicotianae var. *parasitica*. Fruit rot. **1:**FL(35). **2:**FL(35), IL(82), IN(82).
Phytophthora sp. Gray mold, stem blight. **2:**LA(3280), MS(2720).
Pythium aphanidermatum. Damping-off, root rot. **1:**FL(35), NC(86). **2:**CA(82), HI(2903).
P. debaryanum. Damping-off, fruit rot. **2:**CT(82), LA(82), MA(82), NC(82), NJ(82), NY(82).
P. irregulare. Damping-off, root rot. **1:**FL(35).
P. ultimum. **2:**CA(82).
P. vexans. Root rot. **2:**MD(82), VA(82).
Pythium sp. Damping-off, root rot. **1:**FL(35). **2:**Range of host(82), FL(35), MS(2720).
Zygomycotina
Choanephora cucurbitarum. Blossom rot. **1:**FL(35). **2:**FL(35,82), NC(82).
Rhizopus stolonifer. **2:**WA(3202).
Rhizopus sp. Blue mold, fruit rot. **1:**FL(35). **2:**Unk.(82), FL(82), TX(82), WA(82).
Ascomycotina
Chaetomium aureum. Isolated from seeds. **1:**NJ(3306).
C. bostrychodes. Isolated from seeds. **1:**MI(3306).
Diaporthe phaseolorum. Fruit rot. **2:**Unk.(82), GA(1438), MO(82), MS(82,2720).
Erysiphe cichoracearum. Powdery mildew. **4:**FL(2469).
Glomerella cingulata. Anthracnose, internal mold of fruit. **1:**FL(35). **2:**E states(82), KS(82), MS(2720), NC(1303), OK(2872), TX(82), WI(1349). **4:**GA(1438).
Leveillula taurica. Powdery mildew. **1:**UT(3622).
Nematospora coryli. Pod spot, associated with injury by plant bugs. **2:**FL(82,3800).
Neocosmospora vasinfecta. **4:**GA(1438).
Sclerotinia sclerotiorum. Pink joint, fruit and stem rot. **1:**FL(35). **2:**CA(82), CT(82), FL(35,82), HI(2903), MA(82), TX(82).
Basidiomycotina - Other
Thanatephorus cucumeris. **2:**WA(3202).
Deuteromycotina - Hyphomycetes
Alternaria solani. Early blight, fruit rot. **1:**FL(35). **2:**FL(35), OK(2872).
Alternaria sp. Fruit rot, leaf spot. **2:**Range of host(82), NC(1303). **3:**FL(533), NY(533).
Botrytis cinerea. Stem canker, gray mold, fruit rot, on blossoms, on flower parts adhering to fruit. **1:**AK(2163), FL(35). **2:**CA(82), CT(82), NC(1303,3932), NJ(82), PA(82), WA(82,3202).
Cercospora capsici. Fruit and leaf spot. **1:**NC(3931). **2:**E states(82,1505), CA(82), HI(2903), MO(82), MS(2720), NC(1303), OK(2872), TX(82).
Cladosporium herbarum. Velvet mold, fruit rot. **1:**CA(2913). **2:**CA(82), GA(82), TX(82), WA(82,3202).
Corynespora cassiicola. Leaf spot. **1:**FL(35).
Curvularia lunata. Pod rot. **1:**FL(35). **2:**FL(35,82), MI(82).
Fusarium annuum. Wilt. **2:**AZ(82), CO(82), LA(82), MD(82), MS(82), NJ(82), NM(82,2112), OK(82), TX(82).
F. oxysporum. Fruit and stem rot, wilt. **1:**FL(35). **2:**FL(35).
Fusarium sp. Damping-off, fruit and stem rot, wilt. **2:**HI(2903), NC(1303), OK(2872).
Mycoleptodiscus terrestris. Leaf spot. **1:**FL(35).
Oidiopsis sp. Powdery mildew. **1:**FL(35,298). **2:**FL(35).
Penicillium sp. Internal mold of fruit, on leaves. **2:**GA(82).
Phaeoramularia capsicicola. Frog-eye leaf spot. **2:**FL(35).
Phymatotrichopsis omnivora. Root rot. **2:**AZ(82), TX(82).
Stemphylium botryosum. Seed mold, leaf spot. **1:**FL(35). **2:**CT(82), FL(35,82).
S. lycopersici. Leaf spot. **1:**FL(35).
S. solani. Leaf spot. **1:**FL(35). **2:**FL(35,82,296), HI(2903).
Stemphylium sp. Leaf spot. **2:**LA(3280).
Verticillium albo-atrum. Wilt. **1:**FL(35). **2:**CA(82), CO(82), CT(82), FL(35), NY(82), TX(82).
Deuteromycotina - Coelomycetes
Colletotrichum capsici. Anthracnose. **1:**NC(3931). **2:**GA(82), TX(82).
C. gloeosporioides. Anthracnose, fruit rot, fruit, leaf and stem spot. **1:**FL(35), NC(3931), WI(1341).
Colletotrichum sp. Anthracnose. **2:**NC(1303).
Macrophomina phaseolina. Charcoal rot. **1:**FL(35). **2:**CA(82), GA(82), IL(311), KS(82), NJ(82), TX(82).
Phoma destructiva. Fruit rot. **1:**AL(285), FL(35). **2:**AL(82), DE(82), FL(35,82), GA(82), MS(82,2720), NC(1303), NY(82), PA(82), SC(82).
P. exigua. Leaf spot. **1:**FL(35), NC(3931). **2:**FL(35), NC(1303), WA(82,3202).
Phyllosticta sp. Leaf spot. **2:**AL(82), DE(82), GA(82), HI(2903), NC(1303), NY(82), VA(82), WV(82).
Deuteromycotina - Other
Rhizoctonia solani. Damping-off, root and stem rot, soil rot of fruits. **1:**FL(35). **2:**Range of host(82), FL(35), NC(1303).
Rhizoctonia sp. Root rot. **2:**HI(2903).
Sclerotium rolfsii. Southern blight. **1:**AL(285), FL(35), NC(3931). **2:**SE states(82), FL(35), HI(2903), MS(2720), NC(1303), OK(2873), TX(82).

Cestrum L.

1. *C. diurnum* L. [day jessamine, day-blooming cestrum]
2. *C. nocturnum* L. [night jessamine]
3. *Cestrum* sp.

Ascomycotina
Erysiphe sp. Powdery mildew. **1:**FL(35).
Basidiomycotina - Uredinales
Chrysocyclus cestri. **1:**FL(35).
Uromyces cestri. Autoecious. **1:**FL(35,2841). **2:**FL(35). **3:**FL(82).
Deuteromycotina - Hyphomycetes
Cercospora uromycestri. Leaf spot. **1:**FL(35).
Cladosporium sp. **3:**TX(82).
Fusarium sp. Root rot. **1:**FL(35).
Deuteromycotina - Coelomycetes
Phomopsis sp. Dieback. **2:**FL(35).
Phyllosticta sp. Leaf spot. **2:**FL(35).
Sphaeropsis sp. Stem gall. **2:**FL(35).

Cyphomandra Mart. ex Sendtn.

1. *C. betacea* (Cav.) Sendtn. (=*C. crassifolia* (Ort.) Kuntze) [tomato-tree, tree tomato]

Ascomycotina
Nectriella pironii. Stem canker. **1:**FL(35).
Deuteromycotina - Hyphomycetes
Oidium sp. Powdery mildew. **1:**MD(82).
Deuteromycotina - Coelomycetes
Phomopsis sp. Stem gall. **1:**FL(35).

Datura L.

1. *D. inoxia* Mill. (=*D. meteloides* DC. ex Dunal) [angel's-trumpet, downy thornapple, Indian apple, sacred datura, tolguacha]
2. *D. metel* L. [downy thornapple, Hindu datura, horn-of-plenty, metel]
3. *D. stramonium* L. [common thornapple, Jamestown-weed, jimsonweed, Stramonium thornapple]
4. *Datura* sp.

Ascomycotina
Botryosphaeria rhodina. On stems. **2:**AL(82).
Leptosphaeria doliolum. **3:**GA(1438).
Basidiomycotina - Uredinales
[*Puccinia aristidae*]. **1:**AZ(82,1203). **4:**FL(35).
P. subnitens. **1:**AZ(3981).
Deuteromycotina - Hyphomycetes
Alternaria crassa. Leaf spot. **1:**OK(82,2872), TX(82). **2:**FL(35,82), WI(82). **3:**NC states(82), NE states(82), DE(3543), IA(82), IL(82), OK(2872), SD(2243), TX(82), WI(82). **4:**FL(35).
A. solani. Leaf spot. **3:**IL(82), MS(2720), NC(1303,3944). **4:**FL(35).
Cercospora daturicola. Leaf spot. **2:**FL(35). **3:**FL(35), OK(82,2872). **4:**FL(35).
Curvularia sp. Leaf spot. **2:**FL(35).
Phymatotrichopsis omnivora. Root rot. **1:**TX(82). **3:**TX(82).
Pseudocercospora jamaicensis. Leaf spot. **4:**FL(35).
Thielaviopsis basicola. Root rot. **2:**WI(82). **3:**WI(82).

Deuteromycotina - Coelomycetes
Colletotrichum truncatum. Isolated from stems. **3:**IL(1482).
Diplodia atrocoerulea. On leaves and stems. **3:**AL(82), TX(82).
Dothiorella stramonii. On rotting stems. **3:**HI(2784).
Macrophomina phaseolina. Charcoal rot. **3:**IL(311).
Phoma exigua. Leaf spot. **3:**WV(82).
Phomopsis venenosa. On stems. **3:**NY(82), PA(82), SC(82).
[*Septoria lycopersici*]. Leaf spot. **3:**MD(82), TX(82), VA(82).
Deuteromycotina - Other
Rhizoctonia solani. Root rot. **2:**NY(82), VA(82), WI(82).
Sclerotium rolfsii. Southern blight. **2:**FL(35,82). **3:**FL(35,82).

Duboisia R. Br.

1. *D. leichhardtii* (F. v. Muell.) F. v. Muell.

Ascomycotina
Erysiphe polygoni. Powdery mildew. **1:**CA(1155).

Fabiana Ruiz & Pav.

1. *F. imbricata* Ruiz & Pav. [pichi, pichi-pichi]

Basidiomycotina - Other
Armillaria mellea. Root rot. **1:**CA(2897).

Hyoscyamus L.

1. *H. niger* L. [black henbane, henbane, stinking nightshade]

Deuteromycotina - Hyphomycetes
Alternaria solani. Leaf spot. **1:**WI(82).

Leucophysalis Rydb.

1. *L. nana* (A. Gray) J. E. Averett (=*Chamaesaracha nana* A. Gray)

Basidiomycotina - Uredinales
Puccinia chamaesarachae. **1:**CA(81,111,638,642), NV(111).

Lycium L.

1. *L. andersonii* A. Gray
2. *L. barbarum* L. (=*L. halimifolium* Mill., *L. vulgare* Dunal) [barbary matrimony-vine]
3. *L. berlandieri* Dunal
4. *L. berlandieri* Dunal var. *parviflorum* (A. Gray) A. Terrac.
5. *L. brevipes* Benth.
6. *L. californicum* Nutt.
7. *L. carolinianum* Walter [Christmas-berry]
8. *L. chinense* Mill. [Chinese matrimony-vine, gow, gow-kee]
9. *L. exsertum* A. Gray
10. *L. fremontii* (A. Gray) McClint. & Epling
11. *L. pallidum* Miers [desert-thorn]
12. *L. parishii* A. Gray
13. *L. torreyi* A. Gray
14. *Lycium* sp.

Ascomycotina
Erysiphe polygoni. Powdery mildew. **2:**CT(82), DE(82), MD(82), NJ(82), PA(82), WA(3202). **8:**MN(82).
Lachnella alboviolascens. **14:**AZ(1202).
Microsphaera diffusa. Powdery mildew. **2:**OH(82), PA(82), UT(82), WA(3202).
Sphaerotheca pannosa. **2:**ID(3202), WA(3202).
Basidiomycotina - Uredinales
Aecidium lycii. **11:**AZ(82,1203,3981).
Puccinia globosipes. Autoecious. **1:**AZ(82), CA(82), NM(82), NV(82), UT(82). **3:**AZ(747,1195), CA(747). **4:**AZ(1203,3981). **5:**AZ(747), CA(747). **6:**AZ(82,1195,1203,3981), CA(82), NM(82), NV(82), UT(82). **9:**AZ(1195,1203,3981). **10:**AZ(1195,1203,3981). **12:**AZ(3981). **13:**AZ(1203). **14:**AZ(1203).
P. tumidipes. Autoecious. **2:**E states(82), MS(2720), SD(82,2243), TX(82), WI(1333,1359). **4:**AZ(1203,3981). **7:**TX(82,740). **8:**MN(82), SD(82,2243), WI(1359). **11:**AZ(82,1195,1203,3981), CO(82,1195), NM(82), OK(82,2872), TX(82), UT(82).
Puccinia sp. **8:**FL(35).
Basidiomycotina - Other
Acanthophysium diffissum. **1:**AZ(1202).
Dendrophora albobadia. **14:**AZ(1202).
Exidiopsis calcea. **14:**AZ(1202).
Grandinia pruni. **14:**AZ(1202).
Phanerochaete chrysorhiza. **14:**AZ(1202).
P. tuberculata. **14:**AZ(1202).
Phlebia ludoviciana. **14:**AZ(1202).
Deuteromycotina - Hyphomycetes
Alternaria sp. Leaf spot. **2:**IA(82).
Cercospora lycii. Leaf spot. **2:**IA(82), WI(1337). **11:**OK(82,2872).
Phymatotrichopsis omnivora. Root rot. **7:**TX(82).
Deuteromycotina - Coelomycetes
Diplodina ramulorum. On bleached stems. **14:**NJ(966).
Phyllosticta lycii. Leaf spot. **2:**NY(82), OH(82).
Septoria sp. On leaves. **2:**WI(1340).
Sphaeropsis lycii. On stems. **2:**MO(82).
Stagonospora sp. On leaves. **2:**WI(1340).

Lycopersicon Mill.

1. *L. esculentum* Miller [goldapple, loveapple, tomato]
2. *L. esculentum* Mill. var. *cerasiforme* (Dunal) A. Gray (=*L. cerasiforme* Dunal) [cherry tomato]
3. *L. esculentum* Mill. var. *pyriforme* (Dunal) L. H. Bailey (=*L. lycopersicum* (L.) Karsten var. *pyriforme* (Dunal) Alef., *L. esculentum* Mill. f. *pyriforme* (Dunal) C. Mull., *L. pyriforme* Dunal) [pear tomato]
4. *L. pimpinellifolium* (L.) Mill. [currant tomato]
5. *Lycopersicon* sp.

Plasmodiophoromycetes
Spongospora subterranea. Powder scab, on roots. **1:**PA(82).
Chytridiomycetes
Olpidium brassicae. On roots. **1:**WI(82).
Synchytrium endobioticum. Wart. **1:**PA(82).
Oomycetes
Aphanomyces cladogamus. Rootlet necrosis. **1:**VA(82).
Peronospora tabacina. Downy mildew. **1:**GA(82), NC(82,1303), SC(82).
Phytophthora cactorum. Stem rot. **1:**NY(82), PA(82), WI(82).
P. capsici. 'Phytophthora root rot.' **1:**CA(82,2838), CO(82).
P. cinnamomi. Root rot. **1:**HI(2903).
P. cryptogea. **1:**IL(82), NH(82).
P. drechsleri. **1:**CA(82).
P. infestans. 'Late blight.' **1:**Range of host(82), CA(3728,3729), FL(35,3500), GA(2302), HI(2903), KS(1905), KY(1776,1777,1778), MA(356), MS(2720), NC(1303,3931), OR(3202), SD(2243), TX(3988), VA(2991), WA(3202).
P. nicotianae. Fruit rot. 'Buckeye rot. Phytophthora damping-off.' **1:**OK(2872).
P. nicotianae var. *parasitica*. Buckeye rot of fruit, stem rot. 'Phytophthora root rot.' **1:**E states(82), AZ(82), CA(82), FL(35), HI(2903), IL(82), MS(2720), NC(1303), TN(72,73), TX(82).
P. palmivora. Root rot. **1:**HI(2903).
Phytophthora sp. Fruit rot. **1:**LA(3280), NC(1303).
Pythium aphanidermatum. Damping-off, collar, fruit, root and soil rot. **1:**CA(2735), FL(35,3386), HI(2903), NC(1303), OK(82,2872).
P. arrhenomanes. Stunt and root rot of seedlings. **1:**FL(3385).
P. debaryanum. Damping-off, collar rot, soil rot. **1:**NC states(82), NE states(82), FL(35), NC(1303). **2:**NC states(82), NE states(82). **3:**NC states(82), NE states(82).
P. megalacanthum. Root rot. **1:**CA(82).
P. myriotylum. Stunt and rootlet rot of seedlings. **1:**FL(3385), GA(1208), NC(1303), SC(82).
P. oligandrum. Damping-off, collar rot, soil rot. **1:**FL(35,82), VA(82).
P. salpingophorum. **1:**MD(82).
P. ultimum. Damping-off, collar, fruit and soil rot. **1:**'Water mold.' CA(82,2735), CO(82), FL(35), MD(82), NC(1303), NY(82), OH(82). **2:**CA(82), CO(82), MD(82), NY(82), OH(82). **3:**CA(82), CO(82), MD(82), NY(82), OH(82).
P. vexans. Root rot. **1:**MD(82), VA(82).
Pythium sp. Damping-off, collar rot, aerial leaf rot, soil rot. **1:**FL(35), GA(1208), NC(1303).
Zygomycotina
Mucor circinelloides. Postharvest decay of fruits. **1:**VA(3358).
M. hiemalis. Fruit rot. **1:**CA(465).
Mucor sp. Fruit rot. **1:**CA(465).
Rhizopus stolonifer. Stem blight. 'Ripe rot.' **1:**Range of host(82), CA(465), FL(35), IL(312), NC(1303).
Ascomycotina
Chaetomium bostrychodes. On leaves. **1:**TX(82).
C. globosum. Isolated from seeds. **1:**MI(3306).
Diaporthe phaseolorum. Fruit rot. **1:**GA(1438), MS(82,2720), TX(82).

Erysiphe polygoni. Powdery mildew on seedlings in greenhouse. **1**:NC(82).
Farrowia seminuda. Isolated from seeds. **1**:PA(3306).
Gibberella zeae. Fruit rot. **1**:NY(736).
Glomerella cingulata. Fruit rot. **1**:FL(35,82), LA(82), ME(82), MI(82), NC(1303), NJ(82), NY(82), OH(3562), WI(1349).
Leptospora rubella. **5**:DE(3248).
Leveillula taurica. 'Powdery mildew.' **1**:AZ(2374), CA(688), UT(3622).
Melanospora interna. On stems. **1**:IL(82).
Mycosphaerella sp. On leaves. **1**:SC(82).
Nectria sp. **1**:GA(1438).
Nematospora coryli. Cloudy spot of fruit. **1**:CA(82), FL(35,82,3800), GA(82).
N. lycopersici. Isolated from fruits. **1**:SC(204).
Ophioceras sp. **1**:GA(1438).
Pleospora herbarum. 'Black mold rot.' **1**:CA(82).
Sclerotinia minor. Fruit rot. **1**:TX(82).
S. sclerotiorum. Lesions of leaves, stems and twigs. 'Sclerotinia fruit rot. Sclerotinia stem rot.' **1**:Unk.(82), CA(161), FL(35), HI(1272,2903), LA(2832), NC(1303), WA(3202).
Sclerotinia sp. Blossom blight. **1**:CA(82), FL(82), NY(82), OH(82).

Basidiomycotina - Ustilaginales

Entyloma australe. **5**:CT(1077).

Basidiomycotina - Other

Thanatephorus cucumeris. Damping-off, stem rot. **1**:Range of host(82), AL(285), MS(2720), OK(2872), WA(3202).

Deuteromycotina - Hyphomycetes

Alternaria solani. Early blight, collar and fruit rot, on fruits and leaves. **1**:'Early blight.' Range of host(82), FL(35), GA(2302), HI(2903), KY(1778), MA(1391), MS(2720), NC(1303,3931), OK(2872), OR(3202), SD(2243), TX(1255), WA(3202), WI(1341). **2**:Range of host(82). **3**:Range of host(82). **4**:Range of host(82).
A. tenuissima. On fruits and stems. 'Nailhead spot.' **1**:E states(82), AL(285), FL(35), HI(2903), NC(1303,3931), ND(82,378), OK(2872), TX(82,1255).
Alternaria sp. Fruit rot. **1**:Range of host(82), LA(3280), NC(1303), NY(3158).
Aspergillus flavus. **1**:TX(82).
A. luteo-niger. **1**:TX(82).
A. nidulellus. **1**:TX(82).
A. ochraceus. **1**:TX(82).
A. tamarii. **1**:TX(82).
A. terreus. **1**:TX(82).
Aspergillus sp. Fruit rot. **1**:CA(465), TX(82).
Aureobasidium pullulans. Fruit rot and spot. **1**:CA(465), WV(82).
Botryosporium pulchrum. On leaves. **1**:PA(82), TX(82).
Botrytis cinerea. Gray mold, on fruits, leaves and stems. **1**:'Gray mold and ghost spot.' Unk.(82), AK(526), CA(465), FL(35,773,3500,3502), KY(1776), NC(1303,3931). **2**:Unk.(82), AK(526).
Botrytis sp. Stem rot, leaf spot. **1**:AK(2162).
Cercospora canescens. 'Cercospora leaf spot.' **1**:AL(82), IA(82), KS(82), MO(82), NJ(82), SC(82).
C. diffusa. Leaf spot. **1**:Unk.(82).
C. nicotianae. **1**:TX(82).
Cercospora sp. Leaf spot. **1**:NC(1303).
Cladosporium herbarum. Fruit rot. **1**:Range of host(82), CA(465), WA(3202). **2**:Range of host(82). **3**:Range of host(82).
Cladosporium sp. On leaves. **1**:AK(2078).
Corynespora cassiicola. Leaf blight. 'Target spot.' **1**:FL(35,297,299).
Cristulariella moricola. **1**:WV(3680).
Fulvia fulva. On leaves. **1**:'Leaf mold.' Range of host(82), AK(2162), CT(2968), FL(35), HI(2903), ID(3202), KY(1776,1778), LA(3279), MS(2720), NC(1303), OK(2872), OR(3202), TX(1255), WA(3202). **2**:Range of host(82). **3**:Range of host(82). **5**:DE(3543).
Fusarium moniliforme. In transit and markets. **1**:Unk.(82), MA(2245). **2**:Unk.(82).
F. oxysporum. Fruit and root rot. 'Crown rot. Fusarium wilt.' **1**:CA(465), FL(3389), MA(2245), NH(2617).
F. oxysporum f. sp. *lycopersici*. Wilt. 'Crown rot.' **1**:Range of host(82), AL(285), FL(35), HI(2903), ID(3202), KY(1778,1780), MS(2720), MT(3202), NC(1303,3931), OK(2872), OR(3202), TN(72,73), WA(3202).
F. oxysporum f. sp. *radicis-lycopersici*. Crown, fruit and root rot. **1**:CA(1726), FL(35,1726), NC(1303), OH(1726), PA(2977), TX(170).
F. retusum. **1**:IN(82).
F. roseum. Fruit rot. **1**:NW states(3202), CA(465), MA(2245).
F. solani. In transit and markets. **1**:Unk.(82), MA(2245).
Fusarium sp. Fruit and root rot. **1**:CA(465), FL(35), HI(2903), LA(3280), NY(3158), OK(2872).
Fusidium sp. On dead stems. **1**:NC(1303).
Geotrichum candidum. Watery fruit rot. 'Sour rot.' **1**:CA(465), FL(35), IN(82), MD(82), MO(82), MS(82,2720), NC(1303), NJ(82,2856), TX(82), VA(82).
Gliocladium roseum. Fruit rot. **1**:VA(82).
Helminthosporium tomato. Fruit rot. **1**:KS(82), TX(82).
Helminthosporium sp. Stem canker. **1**:TX(82).
Myrothecium roridum. On fruits 'Ring rot.' **1**:OH(82), TX(82), VA(82), WI(82).
Myrothecium sp. Stem canker. **1**:TX(82).
Nigrospora oryzae. Fruit rot. **1**:CA(82).
Oidiopsis sicula. Powdery mildew. **1**:CA(487,1957).
Penicillium sp. Fruit rot. **1**:CA(465).
Phymatotrichopsis omnivora. Root rot. **1**:AZ(82), OK(82,2872), TX(82).
Pseudocercospora cruenta. Leaf spot. **1**:AL(82), IA(82), KS(82), MO(82), NJ(82), SC(82).
P. fuligena. Defoliation, leaf spot. **1**:FL(35,301).
Sporotrichum sp. Fruit rot. **1**:IN(82), TX(82).
Stachybotrys chartarum. On stems. **1**:LA(965).
Stemphylium botryosum. Fruit rot. **1**:CA(465,467).
S. lycopersici. Leaf spot. 'Gray leaf spot.' **1**:FL(35,1448).
S. solani. On leaves. **1**:'Gray leaf spot.' FL(35,82,3500), GA(82), HI(2903), IN(82), LA(82), MS(2720), NC(82,3931), NJ(82), SC(82), TN(82), TX(82,1255), VA(82). **5**:MD(1729).
Stemphylium sp. Leaf spot. **1**:CA(82), DE(82), IN(82), SC(82).
Thielaviopsis basicola. Root rot. **1**:TX(82).
Trichothecium roseum. Fruit rot. **1**:MD(82), NC(1303,3836), OH(82).
Ulocladium consortiale. Fruit rot. **1**:CA(465).
Verticillium albo-atrum. Fruit rot, wilt. **1**:'Verticillium wilt.' Range of host(82), CA(465,3366), DC(3796), FL(35), HI(2903), ID(3202), KY(1778,1780), MS(2720), NC(1303), NM(2112), NY(2878), OR(3202), WA(3202). **2**:Range of host(82). **3**:Range of host(82).
V. dahliae. 'Verticillium wilt.' **1**:CA(82,2881), MS(82,2720), OH(82).
V. lycopersici. Collar rot in seedbeds. **1**:CO(82), DE(82), IL(82), MD(82), NJ(82), SC(82).
Verticillium sp. **1**:HI(2903).

Deuteromycotina - Coelomycetes

Colletotrichum coccodes. Fruit anthracnose, basal stem canker, corky root, wilt. 'Black dot root rot.' **1**:CA(488,3145), IN(3486), MA(2244), MD(162), NC(1303), PA(82), VA(82).
C. dematium. Fruit anthracnose and rot. **1**:IN(3486), SC(162), TX(82).
C. gloeosporioides. On fruits and leaves. 'Anthracnose.' **1**:Range of host(82), FL(35), HI(2903), MD(162), MS(2720), NC(3931), OK(2872), SD(2243), WI(1340).
Colletotrichum sp. Fruit rot. **1**:NY(3158).
Lasiodiplodia theobromae. Fruit rot. **1**:AL(82).
Macrophomina phaseolina. 'Charcoal rot.' **1**:CA(82), HI(2903), IL(311), MS(2720), TX(82).
Pestalotia sp. Fruit rot. **1**:TX(82).
Phoma destructiva. Leaf and stem blight, fruit rot, black spot. 'Black spot.' **1**:E states(82), SC states(82), Unk.(82), W states(82), FL(35), HI(2629,2903), OK(2872), WA(3202), WI(1340).
P. exigua. Leaf spot. **1**:NC(1303,3931), NY(82).
P. lycopersici. Leaf spot. **1**:Unk.(82), DE(82), FL(35,82), NC(82), NJ(82), OR(82,3202), VA(82), WI(82).
P. terrestris. Secondary root rot. **1**:IA(82), IL(82), NJ(82).
Phomopsis sp. Stem canker, soft rot. **1**:FL(35), HI(2903), NY(535).
Phyllosticta sp. Leaf spot. **1**:NC(1303).
Pyrenochaeta lycopersici. Brown root rot. 'Corky root.' **1**:CA(488,1371), FL(3739), MA(2246).
Pyrenochaeta sp. Root rot. **1**:HI(2903).
Septoria lycopersici. Leaf spot. **1**:'Septoria leaf spot.' AL(285), FL(35), HI(2903), LA(3279), MS(2720), NC(1303), OK(2872), SD(2243), WA(3202). **2**:E & C states(82), SW states(). **4**:Range of host(82).

Deuteromycotina - Other

Rhizoctonia solani. Fruit and stem rot. **1:**'Rhizoctonia damping-off. Rhizoctonia stem canker. Rhizoctonia soil rot.' CA(465), FL(35), MA(2244), MI(3522), NC(1303,3931), OK(2872). **2:**Unk.(82). **3:**Unk.(82).

Rhizoctonia sp. Collar and fruit rot. **1:**HI(2903), NC(1303), NY(3158), TN(1028).

Sclerotium rolfsii. 'Southern blight.' **1:**SE states(82), AL(285,900), CA(82,1203), FL(35,1203), GA(2302), HI(2903), IN(82), KS(82), MD(82), MS(2720), NC(1303,3931), NJ(82), OK(2872), TN(72,73), TX(82).

Nicandra Adans.

1. *N. physalodes* (L.) Gaertn. [apple-of-Peru, shoo-fly-plant]

Ascomycotina

Sclerotinia sclerotiorum. Rot. **1:**HI(2903).

Deuteromycotina - Hyphomycetes

Phymatotrichopsis omnivora. Root rot. **1:**TX(82).

Deuteromycotina - Coelomycetes

Macrophomina phaseolina. Charcoal rot. **1:**HI(2903).

Nicotiana L.

1. *N. alata* Link & Otto [flowering tobacco, jasmine tobacco, winged tobacco]
2. *N. attenuata* Torr. ex S. Wats.
3. *N. glauca* Graham [mustard-tree, tree tobacco]
4. *N. glutinosa* L.
5. *N. quadrivalis* Pursh (=*N. bigelovii* (Torr.) S. Wats.)
6. *N. repanda* Willd. ex Lehm. [fiddle-leaf tobacco, tabacco cimarron, wild tobacco]
7. *N. sylvestris* Speg.
8. *N. tabacum* L. (=*Tabacum nicotianum* Bercht. & Opiz) [tobacco]
9. *N. trigonophylla* Dunal [desert tobacco]
10. *Nicotiana* sp.

Plasmodiophoromycetes

Plasmodiophora brassicae. Root rot. **8:**NC(82).

Chytridiomycetes

Olpidium brassicae. 'Olpidium seeding blight.' **8:**NC(1303), WI(82).

Oomycetes

Peronospora hyoscyami. On leaves. **3:**CA(2439).

P. tabacina. Downy mildew. **2:**WA(3202). **3:**CA(82), TX(82). **5:**NV(82). **6:**TX(82). **7:**Unk.(82). **8:**'Blue mold.' E states(82), CT(3322,3744), FL(35,3322), GA(1751,3322), IN(82), KY(1776,1777,1778,1779,1780,3321,3322,3720), MA(3322), MD(3322), NC(1303,1755,3322), SC(3322), TN(72,73,1510), TX(82), VA(1755,3322), WI(82,1340,3322). **9:**AZ(1195,3381).

Phytophthora nicotianae. Black shank. **8:**AL(82), CT(3770), FL(35,82), GA(82,1082,1751), KY(82,1082,1776,1777,1778,1779,1780,3320,3321,3644), MD(82), NC(82,1082,1755), PA(212,3266), TN(72,73,82,1082,1510), TX(82), VA(82,1755,3644). **10:**MD(1729).

P. nicotianae var. *parasitica*. 'Black shank.' **8:**NC(1303).

Pythium aphanidermatum. Seedling stem burn, root rot. 'Pythium damping-off.' **8:**CT(82), FL(35), MA(82).

P. debaryanum. 'Pythium damping-off.' **8:**CT(82), FL(35), KY(82), OH(82), PA(82), WI(82).

P. irregulare. Damping-off. **8:**KY(82).

P. ultimum. Damping-off. **5:**CA(82). **8:**'Pythium damping-off. Stem rot of transplants.' FL(35), MO(82), OH(82). **9:**CA(82).

Pythium sp. Damping-off, soft rot, stem rot. **8:**Unk.(82), FL(35), GA(1751), KY(82,1776,1777,1778,1779,1780,3320,3321,3720), NC(1303,1755), VA(1755,1758).

Zygomycotina

Mucor mucedo. Mold. **8:**Range of host(82).

M. racemosus. Mold. **8:**Range of host(82).

Mucor sp. Storage rot. **8:**Range of host(82), NC(1303).

Rhizopus sp. Storage and barn rot. **8:**GA(1751), NC(1303).

Syncephalastrum sp. Storage rot. **8:**NC(1303).

Ascomycotina

Botryosphaeria sp. On bark. **3:**CA(2439).

Diporotheca sp. **8:**CT(3578).

Erysiphe cichoracearum. 'Powdery mildew.' **8:**PA(82).

Glonium vestigiale. On twigs. **3:**CA(2439).

Metasphaeria anisometra. On twigs. **3:**CA(2439).

Physalospora erratica. On limbs. **3:**CA(2439).

Sclerotinia minor. Seedling blight. **8:**CT(82).

S. sclerotiorum. Seedling blight. **8:**CT(82), KY(82), NC(82).

Sclerotinia sp. Collar rot. **8:**GA(1751).

Thielavia basicola. Root rot. **8:**CT(82), NC(82), OK(2872), WI(82).

Basidiomycotina - Uredinales

[*Puccinia aristidae*]. **9:**AZ(82,1203).

P. subnitens. **9:**AZ(3981).

Uredo nicotianae. **5:**CA(82), NV(82).

Basidiomycotina - Other

Athelia rolfsii. Southern stem rot. **8:**GA(1751).

Stereum ochraceo-flavum. On bark. **3:**CA(2439).

Thanatephorus cucumeris. Sore shin. **8:**GA(1751).

Deuteromycotina - Hyphomycetes

Alternaria alternata. Leaf spot. 'Brown spot.' **8:**CT(82), FL(35), GA(82), WI(82), WV(3911).

A. longipes. Black spot. **3:**TX(82). **8:**FL(35,82), GA(82,1751), KY(3323), MD(82), MO(82), NC(82,1303), PA(82), SC(82), TN(1510), VA(82).

A. tenuissima. Leaf and white spot. **8:**CT(82), MA(82), MO(82), NC(82), OH(82), PA(82).

Aspergillus candidus. Storage rot. **8:**Range of host(82), NC(1303).

A. flavus. Storage rot. **8:**NC(1303).

A. glaucus. Mold of manufactured tobacco. **8:**Range of host(82).

A. hollandicus. Storage rot. **8:**NC(1303).

A. niger. Storage rot. **8:**Range of host(82), FL(35), NC(1303).

A. oryzae. Stem rot. **8:**FL(35,82).

A. reptans. Storage rot. **8:**NC(1303).

A. restrictus. Storage rot. **8:**NC(1303).

A. rubrobrunneus. Storage rot. **8:**NC(1303).

A. tamarii. Storage rot. **8:**NC(1303).

A. versicolor. Storage rot. **8:**NC(1303).

Aspergillus sp. Shed burn, pole rot. **8:**Range of host(82).

Botryosporium pulchrum. White mold. **8:**CT(82).

Botrytis cinerea. Seedling blight, pole rot of cured tobacco. 'Collar rot. Gray mold.' **8:**AK(526), CT(82), KY(82), NC(82,1303), VA(82).

Botrytis sp. On stems. **8:**WI(1360).

Cercospora apii. Frog eye. **8:**KY(1777,1779).

C. nicotianae. Leaf spot. **6:**TX(82). **8:**'Barn spot. Frogeye leaf spot.' E states(82), FL(35), GA(1751), HI(2903), KY(1778), LA(82), MS(2720), NC(1303), TN(72,73,1510), WI(82).

Cladosporium herbarum. Blossom spot. **8:**CT(82), PA(82).

Cladosporium sp. Storage rot. **8:**NC(1303).

Epicoccum sp. Storage rot. **8:**NC(1303).

Fumago vagans. Sooty mold. **8:**GA(1751), NC(1303).

Fusarium oxysporum f. sp. *nicotianae*. Wilt. **8:**E states(82), FL(35), GA(1751), MD(82), MS(82,2720), NC(82), SC(82), TN(72,73), VA(82,1757), WI(82).

Nigrospora sp. Storage rot. **8:**NC(1303).

Oidium sp. Powdery mildew. **1:**KY(82). **8:**KY(547). **9:**NM(82).

Oospora nicotianae. Mold. **8:**KY(82), OH(82).

Penicillium aurantiogriseum. Storage rot. **8:**NC(1303).

Penicillium sp. **8:**Range of host(82).

Phymatotrichopsis omnivora. Root rot. **3:**TX(82). **8:**TX(82).

Thielaviopsis basicola. Black root rot. **4:**VA(82), WI(82). **8:**'Black root rot.' Range of host(82), FL(35), KY(1776,1778,1779,1780,3320,3321), NC(1303), VA(3411).

Verticillium albo-atrum. 'Verticillium wilt.' **8:**TN(82).

Deuteromycotina - Coelomycetes

Ascochyta sp. Leaf spot. **8:**SC(1610).

Colletotrichum destructivum. 'Anthracnose.' **8:**GA(1751), KY(82,1776,1777,1778,1779,1780,3320,3321,3720), NC(1303), TN(162).

Colletotrichum sp. Anthracnose. **8:**GA(2745), MD(82), NC(1755,2616), PA(82), TN(72,73), VA(82,1755).

Phomopsis nicotianae. On bark and stems. **3:**CA(82,2439).

Phyllosticta nicotiana. Leaf spot of seedlings. 'Phyllosticta leaf spot.' **8:**AL(82), FL(35,82), GA(82), MA(82), NC(82), SC(82), VA(82).

P. tabaci. Leaf spot. **8:**FL(35).

Phyllosticta sp. Leaf spot. **8:**FL(3474).

Septomyxa affinis. Scab, blotch. **8:**CT(82), FL(35,82), KY(82), OH(82), WI(82).

Deuteromycotina - Other

Rhizoctonia solani. Root and stem rot, leaf spot. 'Sore shin and damping-off.' **8:**Range of host(82), FL(35), KY(1776,3320,3321), NC(1303,3236).

Rhizoctonia sp. Soft rot of stems, on seedlings. **8:**KY(3720), NC(1755), VA(1755).
Sclerotium rolfsii. Stem rot. 'Southern blight.' **8:**FL(35,82), GA(82), NC(82,2886), SC(82), VA(82).

Petunia Juss.

1. *P. axillaris* (Lam.) B. S. P. [large white petunia, whitemoon petunia]
2. *P.* × *hybrida* Hort. Vilm.-Andr. A cultigen of complex hybrid origin. [common garden petunia]
3. *P. parviflora* Juss. [seaside petunia, wild petunia]
4. *Petunia* sp.

Oomycetes
Phytophthora nicotianae var. *parasitica*. Root and crown rot. **2:**FL(35), NC(1303). **4:**NC(1303).
Pythium sp. Root, collar, and stem rot, wilt. **2:**FL(35), HI(2903), IL(312).
Zygomycotina
Choanephora infundibulifera. On old flowers. **2:**GA(82).
Choanephora sp. Flower blight. **2:**FL(35).
Ascomycotina
Sclerotinia sclerotiorum. Sclerotinia wilt. **1:**WA(82). **2:**FL(35). **4:**WA(3202).
Basidiomycotina - Uredinales
[*Puccinia aristidae*]. **3:**AZ(82,1203).
P. subnitens. **3:**AZ(3981).
Basidiomycotina - Other
Thanatephorus cucumeris. **4:**OR(3202).
Deuteromycotina - Hyphomycetes
Alternaria alternata. Leaf spot. **2:**FL(35).
A. solani. Leaf spot. **2:**FL(35).
Botrytis cinerea. Leaf and flower blight. **2:**FL(35), HI(2903), IL(312). **4:**NC(1303).
Cercospora petuniae. Leaf blotch. **2:**FL(35,82), OK(82,2872), WI(1341). **4:**NC(1303).
Corynespora sp. Leaf spot. **2:**FL(35).
Fusarium sp. Wilt, stem rot. **2:**FL(35), WA(82,3202). **4:**NC(1303).
Helminthosporium sp. Leaf spot. **2:**FL(35).
Heterosporium petuniae. Leaf spot. **2:**WA(3202,3445).
Oidium sp. Powdery mildew. **1:**MN(82), NY(82), WV(82). **2:**FL(35).
Stemphylium botryosum. Secondary black stem. **2:**TX(82).
Deuteromycotina - Coelomycetes
Ascochyta petuniae. Leaf spot. **2:**TX(82).
Colletotrichum sp. Stem decay. **4:**NC(1303).
Deuteromycotina - Other
Rhizoctonia solani. Damping-off, stem and root rot. **2:**FL(35,82), NC(82), NJ(82), NY(82), OK(82), OR(197), PA(82), TX(82). **4:**NC(1303).
Rhizoctonia sp. **2:**HI(2903).
Sclerotium rolfsii. Basal stem rot. **2:**FL(35). **4:**NC(1303).

Physalis L.

1. *P. acutifolia* (Miers) Sandw. (=*P. wrightii* A. Gray)
2. *P. alkekengi* L. [alkekengi, Chinese lanternplant, Japanese-lantern, strawberry groundcherry, winter-cherry]
3. *P. angulata* L.
4. *P. ciliosa* Rydb.
5. *P. heterophylla* Nees [clammy groundcherry]
6. *P. lanceolata* Michx.
7. *P. lobata* Torr. (=*Quincula lobata* (Torr.) Raf.) [purple groundcherry]
8. *P. longifolia* Nutt. (=*P. subglabrata* Mackenz. & Bush)
9. *P. minima* L.
10. *P. mollis* Nutt.
11. *P. neomexicana* Rydb.
12. *P. peruviana* L. (=*P. edulis* Sims) [Barbados gooseberry, Cape gooseberry, cherry tomato, gooseberry tomato, Peruvian ground-cherry, poha, strawberry tomato, winter-cherry]
13. *P. pruinosa* L. [dwarf cape gooseberry, hairy groundcherry, strawberry tomato]
14. *P. pubescens* L. [downy groundcherry, strawberry tomato]
15. *P. pumila* Nutt.
16. *P. virginiana* Mill. (=*P. virginica* A. Gray)
17. *P. viscosa* L. (=*P. maritima* M. A. Curtis)
18. *Physalis* sp.

Plasmodiophoromycetes
Ligniera junci. On roots. **3:**NC(1303).
Ascomycotina
Asteridiella acervata. Black mildew. **12:**HI(1272,1452,2903).
A. inermis. Black mildew. **12:**HI(82).
Leptosphaeria physalidis. Leaf spot. **14:**KY(82).
Meliola sp. Black mildew. **12:**HI(2903).
Sphaerotheca fuliginea. Powdery mildew. **2:**CA(1155).
Basidiomycotina - Uredinales
Aecidium physalidis. **1:**AZ(1203). **5:**C states(82), CO(82), NM(82), WI(1340). **6:**C states(82), CO(82), MS(2720), NM(82). **8:**C states(82), CO(82), NM(82). **15:**OK(2872). **16:**C states(82), CO(82), NM(82), OK(2872), WI(1340).
[*Puccinia aristidae*]. **7:**CO(82).
P. physalidis. Autoecious. **5:**CO(82), IA(82), MN(82), NE(82), WI(82,1332,1343,1344). **6:**CO(82), IA(82), MN(82), NE(82), WI(82). **16:**CO(82), IA(82), MN(82), NE(82), WI(82).
P. substriata. **1:**AZ(3981).
Basidiomycotina - Ustilaginales
Entyloma australe. White smut. **2:**CT(82), NY(82). **3:**MS(1077,2720), OK(2872). **4:**FL(1077). **5:**IA(1077), PA(1077), WI(1077). **6:**FL(1077), IL(1077), KS(1077), WY(1077). **7:**CO(1077). **8:**CO(1077), IL(1077), IN(1077), KS(1077), NE(1077), NY(1077), PA(1077), WI(1349). **9:**MO(1077). **10:**FL(1077), OK(1077,2872). **11:**NM(1077). **13:**NY(1077). **14:**E states(82), CT(1077), FL(1077), IA(1077), IL(1077), IN(1077), KS(1077), MN(1077), NM(82), TX(82), VA(1077). **16:**CT(1077), FL(1077), IA(1077), IL(1077), MA(1077), MI(1077), ND(1077), NJ(1077), PA(1077), WI(1077). **17:**FL(1077), IL(1077), MA(1077), TX(1077). **18:**E states(82), AR(1077,2675), FL(35), KY(1077), LA(1077), NM(82), OK(2872), TX(82), WA(1077,3202), WV(1077).
Basidiomycotina - Other
Thanatephorus cucumeris. On stem bases. **5:**WI(82).
Deuteromycotina - Hyphomycetes
Alternaria solani. Leaf spot. **18:**Unk.(82).
Alternaria sp. Leaf spot. **18:**FL(35).
Cercospora diffusa. Leaf spot. **8:**WI(1343). **12:**HI(2903). **18:**IL(82), KS(82), WI(82).
C. physalidis. Leaf spot. **7:**KS(82). **8:**WI(1342). **10:**OK(2872). **14:**GA(82), IN(82), WI(82). **16:**KS(581). **18:**E states(82), FL(35), IA(82), IN(82), KS(82), KY(82), MI(82), NE(82), TX(82), WI(82).
Curvularia sp. Leaf spot. **18:**FL(35).
Oidium sp. Powdery mildew. **18:**FL(35).
Phymatotrichopsis omnivora. Root rot. **18:**TX(82).
Stemphylium solani. Gray leaf spot. **14:**FL(82). **18:**FL(35).
Verticillium albo-atrum. Wilt. **2:**NY(82).
Deuteromycotina - Coelomycetes
Phyllosticta physaleos. Leaf spot. **2:**CT(82).
Phyllosticta sp. Leaf spot. **10:**OK(2873). **18:**OK(82).
Septoria sp. Leaf spot. **18:**NE(82).
Deuteromycotina - Other
Sclerotium rolfsii. Southern blight. **12:**HI(2903). **18:**FL(35).

Salpiglossis Ruiz & Pav.

1. *S. sinuata* Ruiz & Pav. [painted-tongue]

Deuteromycotina - Hyphomycetes
Fusarium sp. Wilt. **1:**WA(82).
Verticillium albo-atrum. Wilt. **1:**NY(82).

Schizanthus Ruiz & Pav.

1. *Schizanthus* sp.

Oomycetes
Pythium ultimum. Damping-off. **1:**MO(82).
Ascomycotina
Sclerotinia sclerotiorum. Stem rot. **1:**MO(82), MS(82,2720).
Deuteromycotina - Coelomycetes
Colletotrichum schizanthi. Anthracnose. **1:**NY(82).
Deuteromycotina - Other
Rhizoctonia solani. Damping-off. **1:**NY(82).

Solanum L.

1. *S. capsicastrum* Link ex Schauer [false Jerusalem-cherry]
2. *S. carolinense* L. [ball-nightshade, Carolina nettle, horsenettle]
3. *S. chacoense* Bitter
4. *S. dulcamara* L. [bittersweet nightshade, deadly nightshade, poisonous nightshade]
5. *S. elaeagnifolium* Cav. [bull-nettle, silver horse-nettle, silver-leaved nettle, silver-leaved nightshade, trompillo, white horse-nettle]
6. *S. integrifolium* Poir. in Lam. [fruited eggplant, scarlet eggplant, tomato fruited eggplant]
7. *S. melongena* L. [aubergine, eggplant]
8. *S. melongena* L. var. *esculentum* Nees [aubergine, eggplant, madapple, melongene]
9. *S. nigrum* L. (=*S. ptycanthum* Dunal in DC.) [black nightshade, common nightshade, poisonberry]
10. *S. pseudocapsicum* L. [Jerusalem-cherry]
11. *S. rostratum* Dunal [buffalobur]
12. *S. sarachioides* Rusby
13. *S. torvum* Sw. [terongan, turkeyberry]
14. *S. triflorum* Nutt.
15. *S. tuberosum* L. [potato, white Irish potato]
16. *S. villosum* Mill. (=*S. nigrum* L. var. *villosum* L.)
17. *S. wallacei* (A. Gray) Parish
18. *Solanum* sp.

Plasmodiophoromycetes

Spongospora subterranea. 'Powdery scab.' **15:**NE states(82), AL(82,285), FL(35,82), HI(2903), ID(3202), MN(82), MS(82,2720), MT(3202), OK(82,2872), OR(3202), SC(82), WA(82,3202), WY(82).

Chytridiomycetes

Synchytrium endobioticum. 'Wart.' **15:**MD(82,1730), PA(82,1699), WV(82).

Oomycetes

Peronospora tabacina. Downy mildew. **8:**SC(82).
Phytophthora cryptogea. 'Pink rot.' **15:**OH(3058).
P. drechsleri. Tuber rot. **15:**ID(82,3202), NE(82).
P. erythroseptica. Watery rot, wilt. 'Pink rot.' **15:**DE(510), ID(82,3202), LA(82), MA(82), ME(82), NE(82,339), NY(82), OH(3058), OK(82), OR(3202), RI(82).
P. infestans. Late blight. **4:**MD(82), NY(82). **8:**FL(35,82), NY(82). **9:**OR(82,3202). **12:**CA(82). **15:**'Late blight.' E states(82), Gulf States(82), NC states(82), Range of host(82), W states(82), AK(526), AL(285), FL(35,3500,3826), HI(82,2903), ID(3202), ME(332), MS(2720), MT(19,3202), NC(1303), OR(3202), SD(2243), WA(3202).
P. nicotianae var. *parasitica*. Fruit, root and tuber rot. **7:**HI(2903), NC(1303). **8:**FL(35,82), IN(82). **10:**NC(1303).
Phytophthora sp. Pink rot, root rot. **10:**FL(35). **15:**OK(2872).
Pythium acanthicum. Cottony leak. **8:**CA(82).
P. aphanidermatum. Cottony leak, seed piece rot. **7:**NC(1303). **8:**CA(82), FL(35,82), TN(82). **15:**CA(82), FL(35).
P. arrhenomanes. Watery rot. **15:**HI(82,2903).
P. debaryanum. Damping-off, watery rot. **7:**MS(2720). **8:**CT(82), LA(82), NJ(82), NY(82). **15:**'Leak. Pythium seed rot. Pythium damping-off.' CA(82), CO(82), CT(82), FL(35), HI(2903), ID(82,3202), LA(82), MN(82), MS(82,2720), MT(82), ND(82), NE(82), NJ(82), NY(82), OR(3202), WA(82,3202), WY(82).
P. irregulare. Root rot. **8:**FL(35).
P. myriotylum. Cottony leak, tuber rot. **8:**FL(35,82).
P. rostratum. Root rot, watery rot. **15:**HI(82,2903).
P. ultimum. Watery rot. 'Leak.' **15:**CA(82), NC(1303).
Pythium sp. Root rot, watery rot. **7:**HI(2903). **9:**FL(35). **15:**Unk.(82).

Zygomycotina

Choanephora cucurbitarum. Flower blight. **8:**FL(35).
Rhizopus stolonifer. Leak, fruit rot. **8:**CA(82), FL(35), IN(82), TX(82). **15:**'Rhizopus soft rot.' CA(82), ID(82,3202), OH(82), TX(82), WA(82,3202).

Ascomycotina

Crocicreas cyathoideum. **15:**NY(505).
Diaporthe arctii. On stems. **7:**GA(1438). **8:**GA(82).
Diporotheca rhizophila. Parasitic on roots. **7:**WA(1273,3202). **9:**WA(1273,3202). **15:**WA(1273,3202).
Erysiphe cichoracearum. Powdery mildew. **2:**AL(82), IA(82), MD(82), NC(82,1303), PA(82), TX(82). **9:**CA(82), WA(82,3202). **15:**'Powdery mildew.' KY(82), NJ(82), OH(3054), UT(3678), WA(82,2348,3202). **18:**OK(2873).
Gibberella pulicaris. On dead fruits. **18:**HI(2903).
Glomerella cingulata. Anthracnose. **8:**IA(82).
Gnomonia sp. On dead stems. **8:**LA(82).
Metasphaeria anisometra. On dead stems. **17:**CA(2439).
Metasphaeria sp. **7:**GA(1438).
Microsphaera sp. Powdery mildew. **2:**IL(82).
Mycosphaerella solani. Leaf spot, on stems and tubers. **4:**OH(82). **15:**ME(82), NY(82).
Nectria brassicae. **15:**NJ(82).
N. cinnabarina. On dead upper portions of still living stems. **4:**WI(1356).
N. inventa. **15:**ID(3202), OR(3202), WA(3202).
N. ipomoeae. On tubers. **15:**MD(82), NJ(82).
Nectria sp. On stems and tubers. **15:**LA(82), MD(82), NJ(82), VA(82).
Neocosmospora vasinfecta. On tubers. **15:**AL(82).
Sclerotinia minor. **15:**CA(82).
S. sclerotiorum. Stem blight and rot. **3:**HI(2899). **5:**TX(82). **7:**HI(2903). **8:**CA(82), FL(35,82), IN(82), LA(82), OR(82), TX(82). **9:**FL(35). **13:**FL(35). **15:**'White mold.' CO(447), FL(35,82), MA(82), MT(82,3202), NY(82), OR(82), TX(82), WA(82,3202).
Sclerotinia sp. Stem rot. **15:**FL(82), ME(82), SC(82).
Xylaria apiculata. Tuber rot. **15:**FL(35,82).

Basidiomycotina - Uredinales

Puccinia substriata. Heteroecious: aecial host. **5:**TX(82,742). **9:**TX(82). **16:**TX(82).

Basidiomycotina - Ustilaginales

Entyloma australe. White smut. **9:**FL(35), IA(82,1077), KS(82,1077). **14:**CO(1077), MT(82,3202), ND(82,378,1077), NM(1077), WY(1077,1195). **18:**FL(1077).

Basidiomycotina - Other

Armillaria mellea. Root rot. **15:**CA(82), FL(35,82), MI(82), OR(82,3202), WA(82,3202), WI(82), WY(82).
Helicobasidium brebissonii. **15:**MT(3202), OK(2872), OR(3202), WA(3202).
Hohenbuehelia cyphelliformis. On dead stems. **4:**MI(2421).
Thanatephorus cucumeris. Canker, stem rot, black scurf. **2:**LA(3524). **4:**WI(1347). **7:**WA(3202). **15:**Range of host(82), AK(499,526,2078,2162), AL(285), ID(3202), MT(3202), OK(2872), OR(3202), WA(3202).

Deuteromycotina - Hyphomycetes

Alternaria solani. Early blight, tuber rot, leaf spot. **1:**Unk.(82). **7:**HI(2903), NC(1303), WI(1340). **8:**FL(35). **9:**FL(35), ID(3202), OR(3202). **10:**FL(35). **11:**KS(82), TX(82). **15:**'Early blight.' AK(2162), AL(285), AR(82), AZ(82), CO(82), FL(35,82,3500), HI(2903), ID(890,1404,3202), MA(82,1391), MI(82), MN(82), MS(2720), MT(3202), NC(82,1303), ND(82), OK(2872), SD(2243), WA(890,3202), WI(82,2880). **18:**FL(35).
Alternaria sp. **7:**WA(3202).
Aspergillus niger. Secondary tuber rot. **15:**Unk.(82).
Botrytis cinerea. Shoot blight, stem girdle, storage rot. **8:**CA(82), CT(82), FL(35), MA(82), NJ(82), VA(82), WA(82). **15:**'Gray mold.' AK(82,526), CT(82), FL(35), ID(82,3202), ME(82), NC(1303), OH(82).
Botrytis sp. Flower blight, leaf spot. **7:**WA(3202). **15:**AK(2162).
Cercospora carolinensis. On leaves. **2:**TX(82,3597).
C. dulcamarae. On leaves. **4:**MI(82), NY(82), WI(82,1329,1336).
C. melongenae. Leaf spot. **8:**CA(82), HI(82,2903).
C. solani. **9:**LA(82), TX(82). **15:**AL(82), TX(82).
C. solanicola. Leaf spot. **15:**AL(581).
Cercospora sp. Leaf spot. **8:**FL(35). **9:**FL(35). **15:**FL(35). **18:**FL(35).
Clonostachys araucaria var. *rosea*. Tuber rot. **15:**Unk.(82).
Corynespora sp. Leaf spot. **8:**FL(35).
Cristulariella moricola. **15:**WV(3680).
Cylindrocarpon destructans. On necrotic tubers and roots. **15:**OR(82).
C. magnusianum. On necrotic tubers and roots. **15:**ID(82), NY(82).
Doratomyces stemonitis. Speck rot, secondary tuber rot. **7:**VA(2989). **15:**ID(82), IN(82), MA(82), MI(82), NE(82), NY(82), OR(82,3202,4007), WA(3202).
Fusarium acuminatum. Tuber and seedpiece rots, vascular discoloration. **15:**Range of host(82).
F. arthrosporioides. Tuber and seedpiece rots, vascular discoloration. **15:**MT(82), NY(82).

F. avenaceum. Tuber and seedpiece rot, vascular discoloration, stem end rot, wilt. **15:**ME(82), MN(82), MT(82,3202), NY(82), WI(82).
F. culmorum. Tuber and seedpiece rot, vascular discoloration, dry rot. **15:**Range of host(82), FL(35).
F. equiseti. Tuber and seedpiece rots, vascular discoloration. **15:**Range of host(82), IA(82), MT(82), NY(82).
F. flocciferum. Tuber and seedpiece rots, vascular discoloration. **15:**ID(82), MT(82), NY(82).
F. javanicum. Black rot of tubers, seedpiece rot, stem end rot, wilt. **15:**N states(82), W states(82), FL(35), HI(2903), ID(3202), NE(82), OR(3202), WA(3202).
F. moniliforme. Tuber and seedpiece rots, vascular discoloration. **15:**CA(82).
F. oxysporum. Tuber and seedpiece rots, vascular discoloration, stem end rot, wilt. **8:**FL(35). **15:**Range of host(82), AK(526), AL(82,285), CA(82), FL(35), HI(2903), ID(3202), MT(3202), NC(1303), NY(82), OK(2872), OR(3202), WA(3202).
F. oxysporum f. sp. *batatas*. Tuber and seedpiece rots, vascular discoloration. **15:**MS(82,2720).
[*F. oxysporum* f. sp. *conglutinans*]. Tuber and seedpiece rots, vascular discoloration. **15:**ME(82).
F. poae. Tuber and seedpiece rots, vascular discoloration. **15:**ME(82), NY(82).
F. redolens. Tuber and seedpiece rots, vascular discoloration. **15:**NY(82).
F. roseum. Seedpiece decay **15:**ID(3202), MT(3202), NC(1303), OR(3202), WA(3202).
F. sambucinum. Tuber and seedpiece rots, vascular discoloration. **15:**C states(82), N states(82), LA(82), MT(82), NY(82).
F. solani. Tuber and seedpiece rots, vascular discoloration. **15:**Range of host(82), CO(82), HI(2903), NC(1303).
F. solani f. sp. *eumartii*. Dry rot, stem end rot, wilt. **15:**NE states(82), CO(82), ID(82,1405,3202), MT(82,3202), NE(82), OK(82,2872), OR(3202), WI(82), WY(82).
F. solani var. *coeruleum*. Dry, powdery rot in storage and markets. **15:**NE states(82), HI(2903), ID(3202), MO(82), MT(3202), OR(3202), WA(82,3202).
F. sporotrichioides. Tuber and seedpiece rots, vascular discoloration. **15:**ME(82), NY(82).
F. trichothecioides. Dry rot in storage and markets, field rot, wilt. **15:**C & W states(82), ID(3202), OR(3202), WA(3202).
F. ventricosum. Tuber and seedpiece rots, vascular discoloration. **15:**AL(82), LA(82), MD(82), ME(82), NJ(82), NY(82).
Fusarium sp. Dry rot, tuber rot, wilt. **7:**HI(2903), MT(3202), OR(3202). **8:**E states(82). **15:**NC(1303).
Gliocladium sp. Secondary tuber rot. **15:**TX(82).
Helminthosporium solani. 'Silver scurf.' **15:**C states(82), E states(82), NW states(82), Range of host(82), FL(35), HI(2903), ID(3202), OR(3202), WA(3202).
Microdochium dimerum. Tuber and seedpiece rots, vascular discoloration. **15:**MN(82), NY(82).
Mycovellosiella concors. 'Leaf blotch.' **15:**GA(82), IA(82), IN(82), MI(82), MS(2720), NY(82), VT(82), WA(3202), WV(82).
Oidium sp. Powdery mildew. **8:**FL(35). **15:**MD(82), NJ(82), PA(82).
Papulaspora coprophila. On tubers following powdery scab. **15:**ME(82), PA(82).
Penicillium sp. Blue mold rot. **7:**WA(3202). **15:**ID(82,3202), IN(82), NY(82), VA(82).
Phymatotrichopsis omnivora. Root rot. **2:**TX(82). **5:**TX(82). **8:**TX(82). **9:**TX(82). **11:**TX(82). **15:**TX(82).
Polyscytalum pustulans. Skin spot of stored tubers. **15:**AK(2163), ME(82,1085), PA(82), WA(82,3202).
Pseudocercospora atromarginalis. Leaf spot. **2:**TX(82). **5:**TX(82). **9:**AL(82), FL(35,82), GA(82), HI(2903), TX(82,3597), WI(1345).
Ramularia solani. On necrotic tubers. **15:**NY(82).
Stemphylium lycopersici. Leaf spot. **10:**FL(35).
S. solani. Gray leaf spot. **8:**FL(35,82). **10:**FL(35,82).
Trichothecium roseum. Pink-mold rot. **15:**ID(82), MN(82).
Ulocladium consortiale. Tuber blight. **15:**ID(890,3202), WA(890).
Verticillium albo-atrum. Early dying disease, hadromycosis, wilt. **2:**MD(82), NJ(82), NY(82). **5:**NY(82). **6:**NY(82). **7:**FL(706,3500,3503), NC(1303), OR(3202), WA(3202). **8:**Range of host(82), W states(82), FL(35). **9:**CA(82,3366). **11:**NY(82). **15:**'Verticillium wilt.' NE states(82), W states(82), AR(82), CT(1887), FL(35,82,3827), HI(2903), ID(3202), IN(1322), LA(82), NM(3455), OR(3202), VA(82), WA(3202).
V. dahliae. Wilt. **8:**MA(82), NY(82), WA(82). **15:**'Verticillium wilt.' ID(812,2881), OH(3057), TX(689).
V. lycopersici. Wilt. **8:**UT(82).
V. tenerum. On tubers. **15:**Range of host(82).
V. tricorpus. **15:**WA(3308).
Verticillium sp. Wilt. **7:**HI(2903). **15:**AK(2078), HI(2903), MA(356).

Deuteromycotina - Coelomycetes

Ascochyta petuniae. **8:**NY(82).
Botryodiplodia sp. **3:**HI(2903).
Colletotrichum coccodes. Anthracnose, black dot disease, corky root, on stems and stolons. **5:**TX(82). **7:**OK(2872). **9:**CA(488), MN(71), TX(82). **11:**TX(82). **15:**'Black dot.' FL(35), IN(3487), NJ(82), NY(82), OH(82), OK(82,2872), OR(82,3202), PA(82), TX(82), UT(82), VA(82), WV(82).
C. gloeosporioides. Anthracnose, black dot, on stems and stolons. **7:**NC(1303,3931). **8:**E states(82), FL(35), IA(82), TX(82). **15:**NC(1303).
C. truncatum. Fruit rot, on leaves and stems. **2:**IL(1482). **7:**MS(2720). **8:**MS(82). **9:**IL(1482).
Colletotrichum sp. Anthracnose, black dot of tubers. **8:**E states(82), FL(35), OK(82), TX(82), UT(82). **10:**FL(35). **15:**ME(1084).
Lasiodiplodia theobromae. Fruit spot. **8:**FL(35,82).
Macrophoma subconica. Leaf spot. **9:**LA(82,966), TX(82).
Macrophomina phaseolina. Charcoal rot. **2:**IL(311). **7:**HI(2903). **8:**NJ(82). **15:**'Charcoal rot.' CA(82), GA(82), IL(82,311), MD(82), OK(82,2872), TN(1026), TX(82).
Phacidiopycnis tuberivora. Stem-end hard rot. **15:**WA(82,3202).
Phlyctema arcuata. On dead stems. **17:**CA(2439).
Phoma destructiva. Leaf spot. **15:**HI(2629,2903).
P. eupyrena. On dead stems. **15:**NY(82). **17:**CA(2439).
P. exigua. Button and tuber rot, leaf spot. **2:**WI(1338). **4:**WI(1336). **7:**HI(2903), NC(1303), WI(64). **15:**'Gangrene' AK(2163), MD(82), ME(82), NC(1303), PA(82), WV(82).
P. lycopersici. Leaf spot. **2:**DE(82). **4:**NY(82), OH(82). **7:**OR(3202). **8:**DE(82), IL(82), IN(82), NY(82). **15:**AK(82,526), OR(82,3202).
P. terrestris. Stem canker. **2:**OK(82,2872).
Phoma sp. Fruit and stem rot. **7:**HI(2903). **8:**FL(35). **15:**NE states(82), CA(82), MI(82), WA(3202).
Phomopsis dulcamarina. **15:**OR(82).
P. vexans. Blight, fruit rot, leaf spot, tip over. **7:**AL(285), FL(3500), HI(2903), MS(2720), NC(1303,3931), OK(2872). **8:**Range of host(82), FL(35).
Phomopsis sp. On stems. **15:**TX(82).
Phyllosticta dulcamarae. Leaf spot. **2:**IN(82), MO(82), TX(82). **4:**WI(1345). **9:**FL(35). **10:**FL(35). **18:**FL(35).
P. pseudocapsici. Leaf spot. **10:**LA(82).
P. solani. Leaf spot. **2:**KY(82), OH(82). **7:**MS(2720). **8:**FL(35), LA(82).
Phyllosticta sp. **15:**AK(526).
Pyrenochaeta lycopersici. **9:**CA(488).
Septomyxa affinis. On necrotic stems and tubers. **15:**ME(82), NY(82).
[*Septoria lycopersici*]. Leaf spot. **2:**DE(82), MD(82), TX(82), VA(82). **7:**HI(2903). **8:**FL(35), HI(82), IN(82), MD(82), NC(82), VA(82).
S. solanina. **9:**CA(82).

Deuteromycotina - Other

Rhizoctonia crocorum. Violet root rot. **5:**TX(82). **15:**AK(2163), MA(82), MT(82), ND(82), NE(82), NY(82), OK(82), OR(82), TX(82), WA(82).
R. solani. Damping-off, fruit, root, stem and tuber rot, black scurf. **5:**TX(82). **7:**NC(1303). **8:**Range of host(82), FL(35). **9:**FL(35), TX(82). **10:**FL(35). **11:**TX(82). **15:**'Rhizoctonia canker.' Range of host(82), AK(499,2162), CA(12,82), FL(35), HI(2903), ID(82), MS(2720), NC(1303), OK(2872). **18:**FL(35).
Rhizoctonia sp. Root rot. **7:**HI(2903). **9:**HI(2903). **10:**NC(1303). **15:**HI(2903).
Sclerotium rolfsii. Southern blight. **7:**HI(2903), MS(2720), NC(1303,3931). **8:**SE states(82), FL(35), TX(82). **9:**FL(35,82), TX(82). **10:**FL(35). **15:**'Southern blight.' SE states(82), AL(285), AZ(82), FL(35), HI(82,2903), MS(2720), NC(1303), OK(82,2872).

Sparganiaceae

Sparganium L.

1. *S. chlorocarpum* Rydb.
2. *S. eurycarpum* Engelm. ex A. Gray

Basidiomycotina - Uredinales
Uromyces sparganii. Heteroecious: telial host. **2:**E states(82), NE(82), WI(82).
Basidiomycotina - Ustilaginales
Nannfeldtiomyces anomalus. **1:**CT(3724).
Deuteromycotina - Coelomycetes
Cylindrosporium sparganii. Leaf spot. **2:**WI(82).
Phyllosticta sp. On leaves. **1:**WI(1357). **2:**WI(1357).
Stagonospora sparganii. Leaf spot. **2:**WI(82,1340).

Staphyleaceae

Staphylea L.

1. *S. rufa* Masson
2. *S. trifolia* L. [American bladdernut]

Ascomycotina
Diaporthe staphylina. On limbs. **2:**MI(82).
Discostroma corticola. On limbs. **2:**NY(82).
Dothiora staphylina. On limbs and twigs. **2:**NY(82).
Eutypella juglandicola. On limbs. **2:**GA(82,1438).
E. staphyleae. **2:**NY(82).
Leptosphaeria rubrotincta. On limbs. **2:**PA(82).
Massarina staphyleae. **1:**FL(2779).
Mycosphaerella staphylina. Leaf spot. **2:**GA(82,1438), KS(82,963).
Nectria atrofusca. On dead limbs. **2:**GA(82,1438), MD(82), PA(82), VA(82).
N. ipomoeae. Twig blight. **2:**CT(818), IL(818), MA(82,818), NY(818), OH(818), VT(818).
Deuteromycotina - Hyphomycetes
Botrytis cinerea. **2:**DE(3543).
Cercospora staphyleae. Leaf spot. **2:**OK(2872).
Helminthosporium velutinum. **2:**IN(663).
Ovularia isarioides. Leaf spot. **2:**C states(82), NE states(82), IA(82), NC(3944), WI(1330).
Deuteromycotina - Coelomycetes
Gloeodes pomigena. On limbs. **2:**IN(82).
Hendersonia staphyleae. On twigs. **2:**PA(82).
Phyllosticta staphyleae. On capsules and leaves. **2:**NY(82), WI(1340).
Phyllosticta sp. Leaf spot. **2:**WI(1360).
Seimatosporium vaccinii. Twig blight. **2:**MA(82,3547).
Septoria cirrhosa. Leaf spot. **2:**MO(82).
Sphaeropsis staphyleae. On dead twigs. **2:**NJ(82), NY(82).

Sterculiaceae

Brachychiton Schott & Endl.

1. *B. acerifolius* A. Cunn. ex F. v. Muell. [Australian flame-tree, flame bottle-tree]

Deuteromycotina - Coelomycetes
Diplodia sp. Stem canker. **1:**FL(35).

Dombeya Cav.

1. *Dombeya* sp.

Ascomycotina
Nectriella pironii. Stem gall. **1:**FL(35).
Basidiomycotina - Other
Armillaria tabescens. Root rot. **1:**FL(35).

Firmiana Marsili

1. *F. simplex* (L.) W. Wight (=*F. platanifolia* (L. fil.) Marsili, *Sterculia platanifolia* L. fil.) [Chinese parasol-tree, Japanese varnish-tree, Phoenix-tree]

Ascomycotina
Botryosphaeria ribis. On limbs. **1:**AL(82).
Diaporthe arctii. **1:**GA(82,1438).
Mycosphaerella punctiformis. **1:**GA(1438).
Nectria cinnabarina. Dieback, coral spot. **1:**OK(82,2873).
Ophiobolus nigro-clypeata. **1:**GA(1438).
Deuteromycotina - Hyphomycetes
Phymatotrichopsis omnivora. Root rot. **1:**TX(82).
Deuteromycotina - Coelomycetes
Phomopsis sp. On branches. **1:**OK(82,2873).
Deuteromycotina - Other
Rhizoctonia solani. Web blight. **1:**FL(35,82).

Fremontodendron Coville

1. *F. californica* (Torr.) J. M. Coulter (=*Fremontia californica* Torr.) [California fremontia, flannel-bush]

Oomycetes
Phytophthora nicotianae. Basal stem canker, crown and root rot. **1:**CA(4011).
Phytophthora sp. Collar rot. **1:**CA(82).
Deuteromycotina - Hyphomycetes
Verticillium albo-atrum. Wilt. **1:**CA(82).
Deuteromycotina - Coelomycetes
Ascochyta fremontiae. Leaf spot. **1:**CA(82).

Pterygota Schott & Endl.

1. *P. alata* (Roxb.) R. Br.

Basidiomycotina - Other
Armillaria tabescens. Root rot. **1:**FL(35).

Theobroma L.

1. *T. cacao* L. [cacao]

Oomycetes
Pythium sp. Root rot. **1:**FL(35).
Ascomycotina
Nectria ochroleuca. **1:**Unk.(82).
Deuteromycotina - Hyphomycetes
Fusarium decemcellulare. **1:**Unk.(82).
Deuteromycotina - Coelomycetes
Lasiodiplodia theobromae. Dieback, brown rot of pods. **1:**FL(35,82), HI(2903).

Strelitziaceae

Ravenala Adans.

1. *R. madagascariensis* Sonn. [traveler's-palm]

Deuteromycotina - Hyphomycetes
Alternaria sp. Leaf spot. **1:**FL(35).
Curvularia lunata. Leaf spot. **1:**FL(35).
Deightoniella torulosa. Black spot of midrib and petiole. **1:**FL(35).
Fusarium solani. Root rot. **1:**FL(35).

Strelitzia Banks in Aiton

1. *S. reginae* Banks ex Aiton [bird-of-paradise, crane-lily, craneflower, queen's bird-of-paradise]
2. *Strelitzia* sp.

Oomycetes
Pythium splendens. Root rot. **1:**FL(35).
Pythium sp. Root rot. **1:**HI(2903).
Ascomycotina
Leptosphaeria sp. Leaf spot. **1:**FL(35).
Basidiomycotina - Other
Thanatephorus cucumeris. Stem rot. **1:**FL(35).
Deuteromycotina - Hyphomycetes
Alternaria sp. Flower and leaf spot. **1:**FL(35).
Botrytis sp. Flower spot. **1:**FL(35), HI(2903).
Cercospora sp. Leaf spot. **1:**FL(35).
Curvularia sp. Flower spot, leaf spot. **1:**FL(35).
Cylindrocladium pteridis. Leaf spot. **1:**FL(35).
Fusarium sp. Leaf spot, root rot. **1:**FL(35), HI(2903).
Helminthosporium sp. Leaf spot. **1:**FL(35).
Nigrospora sp. Petal spot. **1:**FL(35).
Phaeotrichoconis crotalariae. Leaf spot. **1:**FL(35).
Stemphylium sp. Flower spot. **1:**FL(35).
Deuteromycotina - Coelomycetes
Colletotrichum gloeosporioides. Leaf and petal spot. **1:**FL(35). **2:**FL(35).
C. truncatum. Petal spot. **1:**FL(35).
Phomopsis sp. Leaf spot. **1:**FL(35).
Phyllosticta sp. Leaf spot. **1:**FL(35). **2:**FL(35).

Deuteromycotina - Other
Rhizoctonia solani. Root rot. **1**:FL(35).

Styracaceae

Halesia J. Ellis ex L.

1. *H. carolina* L. (=*H. tetraptera* L.) [oppossumwood, shittimwood, wild-olive]
2. *H. diptera* J. Ellis
3. *H. monticola* Sarg.
4. *Halesia* sp.

Ascomycotina
Diaporthe beckhausii. On limbs. **4**:WV(82).
D. tetraptera. On limbs. **1**:GA(1438), WV(3807). **4**:GA(82).
Hypoxylon punctulatum. **1**:GA(1438). **4**:GA(82).
Lachnum halesiae. On leaves. **1**:GA(849,1438).
Lophiotrema halesiae. On bark. **4**:NY(82).
Mycosphaerella punctiformis. **1**:GA(1438).
Basidiomycotina - Other
Bjerkandera adusta. Wood rot. **4**:Unk.(82).
Deuteromycotina - Hyphomycetes
Cercospora halesiae. Leaf spot. **3**:OK(2872). **4**:TN(82).
Deuteromycotina - Coelomycetes
Cytospora halesiae. On limbs. **4**:WV(82).
Phoma halesiae. On twigs. **4**:NY(82).
Phomopsis sp. Leaf spot. **2**:FL(35).
Sirococcus halesiae. On limbs. **4**:WV(82).

Styrax L.

1. *S. americana* Lam. [American snowbell, mock-orange]
2. *S. grandifolia* Aiton [bigleaf snowbell]

Ascomycotina
Leptosphaeria subcutanea. **1**:GA(1438).
Mycosphaerella punctiformis. **2**:GA(82,1438).
Deuteromycotina - Hyphomycetes
Cercospora styracis. Leaf spot. **1**:MS(2720).

Symplocaceae

Symplocos Jacq.

1. *S. tinctoria* (L.) L'Hér. [common sweetleaf, horse-sugar, sapphire-berry sweetleaf, yellow-wood]

Ascomycotina
Mycosphaerella sp. On leaves. **1**:GA(82,1438).
Basidiomycotina - Other
Exobasidium symploci. Bud gall. **1**:SE states(82), AL(457), FL(35,457), IN(457), MS(2720), NC(1303).
Deuteromycotina - Hyphomycetes
Phymatotrichopsis omnivora. Root rot. **1**:TX(82).
Deuteromycotina - Coelomycetes
Leptothyrium symploci. On leaves. **1**:SC(82).
Septoria stigma. Leaf spot. **1**:AL(82).
S. symploci. Leaf spot. **1**:FL(35,82), MS(82,2720), NC(82), OK(82,2873), TX(82).
S. tinctoria. **1**:AR(82).
Septoria sp. **1**:NC(82).

Taccaceae

Tacca J. R. Forster & G. Forster

1. *Tacca* sp.

Deuteromycotina - Hyphomycetes
Botrytis sp. Leaf necrosis. **1**:FL(35).
Cercospora sp. Leaf spot. **1**:FL(35).

Tamaricaceae

Tamarix L.

1. *T. aphylla* (L.) Karsten [athel tamarisk]
2. *T. hispida* Willd. [Kashgar tamarix]
3. *Tamarix* sp.

Ascomycotina
Botryosphaeria tamaricis. On limbs. **3**:SC(82).
Corollospora pulchella. **1**:CA(3215).
Leptosphaeria tamaricis. On limbs. **2**:OK(2873). **3**:OK(82), PA(82,967).
Mollisia ligni. On dead stems. **3**:NY(82).
Sphaerotheca macularis. Powdery mildew. **3**:IN(82).
Basidiomycotina - Other
Laetiporus sulphureus. Wood rot. **3**:MD(82).
Deuteromycotina - Hyphomycetes
Clavariopsis bulbosa. On wood of tree submerged in salt lake. **1**:CA(52).
Phymatotrichopsis omnivora. Root rot. **3**:CA(82), TX(82).
Deuteromycotina - Coelomycetes
Diplodia tamariscina. On limbs. **3**:NY(82).

Taxaceae

Taxus L.

1. *T. baccata* L. [English yew]
2. *T. baccata* L. cv. 'fastigiata'
3. *T. brevifolia* Nutt. [Pacific yew, western yew]
4. *T. canadensis* Marsh. [American yew, Canada yew, ground-hemlock]
5. *T. cuspidata* Siebold & Zucc. [Japanese yew]
6. *T. cuspidata* Siebold & Zucc. cv. 'capitata'
7. *T. cuspidata* Siebold & Zucc. cv. 'nana'
8. *T.* × *media* Rehd. *T. baccata* × *T. cuspidata*.
9. *Taxus* sp.

Oomycetes
Phytophthora cinnamomi. Seedling blight, dieback, root rot. **1**:SE states(3624), OR(3202). **2**:OR(3624). **5**:SE states(3624), OR(3202). **8**:SE states(3624). **9**:IN(3154), MD(82,1729,3624), VA(82).
Ascomycotina
Asteridiella pitya. Black mildew. **4**:Unk.(1452), NY(2819).
Botryosphaeria ribis. On twigs. **9**:PA(82).
Dothiora taxicola. Leaf and needle blight. **1**:OR(3202). **3**:CA(81,82), ID(82,3202), OR(82), WA(82,3202). **4**:NY(82), PA(82). **9**:OR(82).
Herpotrichia juniperi. Needle blight. **3**:ID(82,3202).
Mycosphaerella taxi. **3**:ID(3202), OR(3202).
Phacidium taxi. On needles. **4**:WI(82).
P. taxicola. On leaves and wood. **3**:ID(3202). **4**:NJ(864), NY(82,864). **9**:ID(2951), NY(2951).
Physalospora gregaria. **3**:ID(3202), MT(3202), WA(3202).
Basidiomycotina - Other
Amylocystis lapponica. **3**:ID(3202).
Armillaria mellea. Root rot. **3**:ID(82,3202).
Basidioradulum radula. **3**:ID(3202).
Fomitopsis rosea. On wood. **3**:ID(82).
Hymenochaete tabacina. **3**:ID(3202).
Phaeolus schweinitzii. Butt rot. **3**:ID(82,3202).
Phellinus hartigii. On wood. **3**:OR(82).
P. nigrolimitatus. **3**:OR(3202).
P. pini. **3**:OR(3202).
[*P. robustus*]. **3**:OR(3202).
Postia fragilis. **3**:ID(3202).
Deuteromycotina - Hyphomycetes
Alternaria sp. Leaf drop. **5**:RI(3526). **6**:RI(3526). **7**:RI(3526). **8**:RI(3526).
Conoplea olivacea. **9**:IA(1672).
Cylindrocladium sp. Cutting rot. **9**:FL(35).
Deuteromycotina - Coelomycetes
Apostrasseria lunata. On leaves and twigs. **4**:NY(2546).
A. robusta. On leaves and twigs. **3**:ID(2546).
Coleophoma taxi. On needles. **3**:ID(82,3202).
Cryptocline taxicola. On needles. **1**:VT(2481).
Diplodia taxi. On leaves and needles. **3**:OR(82,3202). **4**:OR(82). **9**:WI(1337).
Endophragmiella taxi. On dead leaves. **9**:WV(981).
Leptothyrium sp. On needles. **4**:NY(82).
Macrophoma taxi. **3**:WA(3202).
Pestalotia sp. On twig cankers. **9**:PA(82).
Pestalotiopsis funerea. Twig blight. **9**:MA(82).

Phoma macrostoma. On needles. **9**:VA(82).
Phomopsis juniperovora. **1**:ID(3202).
P. occulta. On twigs. **9**:NE states(82).
Phomopsis sp. Twig blight. **9**:FL(35).
Phyllosticta concentrica. Twig blight. **3**:ID(82,3202), MT(82), WA(82,3202).
Sphaeropsis sp. Twig blight. **9**:NJ(82).
Deuteromycotina - Other
Rhizoctonia solani. Damping-off. **4**:CT(82). **9**:CT(82).
Rhizoctonia sp. Root rot. **9**:FL(35).

Torreya Arnott
1. *T. californica* Torr. [California-nutmeg]
2. *T. taxifolia* Arnott [stinking-cedar]
3. *Torreya* sp.
Oomycetes
Phytophthora sp. Root rot. **2**:FL(35).
Pythium sp. Root rot. **2**:FL(35).
Ascomycotina
Botryosphaeria sp. Needle spot. **2**:FL(35).
Physalospora sp. Needle, stem and twig blight. **2**:FL(35,37).
Basidiomycotina - Uredinales
Caeoma torreyae. **1**:CA(81,326).
Deuteromycotina - Hyphomycetes
Alternaria sp. Needle spot. **2**:FL(35).
Fusarium lateritium. Isolated from spots on needles. **2**:FL(937).
Fusarium sp. Root rot. **2**:FL(35).
Janetia bonarii. On leaves. **1**:CA(326,1682).
Sporidesmium fragilissimum. On needles. **1**:CA(81).
Deuteromycotina - Coelomycetes
Lasiodiplodia theobromae. Twig dieback. **2**:FL(35).
Macrophoma sp. Needle and stem blight. **2**:FL(35,37).
Phomopsis sp. **3**:OR(3202).
Phyllosticta sp. Needle spot. **2**:FL(35).
Sphaeropsis sp. Needle blight. **2**:FL(35).
Deuteromycotina - Other
Rhizoctonia solani. Root rot. **2**:FL(35).
Sclerotium rolfsii. Southern blight. **2**:FL(35).

Taxodiaceae

Athrotaxis D. Don
1. *A. selaginoides* D. Don
Oomycetes
Pythium sp. Root rot. **1**:FL(35).

Cryptomeria D. Don
1. *C. japonica* (L. fil.) D. Don [common cryptomeria, Japanese cedar]
Oomycetes
Phytophthora cinnamomi. Root rot. **1**:NC(1303).
Phytophthora sp. Root rot. **1**:FL(35).
Pythium sp. Root rot. **1**:FL(35).
Ascomycotina
Diaporthe eres. On twigs. **1**:NY(82).
Deuteromycotina - Hyphomycetes
Chalara thielavioides. Black mold. **1**:NJ(1545).
Cylindrocladium sp. Root rot. **1**:FL(35).
Deuteromycotina - Coelomycetes
Coniothyrium ovispermum. **1**:WA(3202).
Pestalotia cryptomeriae. On needles. **1**:SC(82).
Pestalotiopsis funerea. On fallen leaves. **1**:CA(81), NJ(82).
Phyllosticta cryptomeriae. Needle blight. **1**:DC(390), VA(390).
Phyllosticta sp. Needle blight. **1**:FL(35).

Cunninghamia R. Br. in Rich.
1. *C. lanceolata* (Lamb.) Hook.
2. *Cunninghamia* sp.
Oomycetes
Pythium sp. Root rot. **1**:FL(35).
Ascomycotina
Botryosphaeria sp. Leaf necrosis. **1**:FL(35).
Deuteromycotina - Hyphomycetes
Cercospora sp. Leaf spot. **1**:OK(2872).
Helminthosporium sp. Leaf and stem lesions. **1**:FL(35).
Deuteromycotina - Coelomycetes
Phomopsis sp. **2**:OR(3202).
Phyllosticta sp. **1**:NC(1303).

Metasequoia Miki ex Hu & W. C. Cheng
1. *M. glyptostroboides* Hu & W. C. Cheng [dawn redwood]
Ascomycotina
Botryosphaeria dothidea. Canker. **1**:VA(3493).
Botryosphaeria sp. Limb dieback. **1**:FL(35).

Sciadopitys Siebold & Zucc.
1. *S. verticillata* (Thunb.) Siebold & Zucc. [Japanese umbrella-pine]
Oomycetes
Phytophthora cinnamomi. Root rot, wilt. **1**:VA(2017).
Basidiomycotina - Other
Mycena meliigena. On bark. **1**:PA(3326).
Deuteromycotina - Coelomycetes
Cytospora sp. Stem canker, leaf spot. **1**:OR(197,3202).
Pestalotia versicolor. **1**:GA(659).
Phyllosticta sp. Leaf spot. **1**:RI(82).
Sphaeropsis sp. Twig blight. **1**:NJ(82).
Deuteromycotina - Other
Rhizoctonia solani. Damping-off. **1**:CT(82).

Sequoia Endl.
1. *S. sempervirens* (D. Don) Endl. [coast redwood, redwood]
2. *Sequoia* sp.
Oomycetes
Pythium sp. Root rot. **1**:FL(35), HI(2903).
Ascomycotina
Botryosphaeria dothidea. Stem canker. **1**:CA(3955).
Chloroscypha chloromela. **1**:CA(81,82).
Clithris sequoiae. On twigs. **1**:CA(82,325).
Hymenoscyphus sphaerophoroides. On dead leaves. **1**:CA(2801).
Hypoxylon thouarsianum. On wood. **1**:CA(128).
Mycosphaerella sequoiae. Needle blight. **1**:CA(81,82,325).
Pezicula livida. Bark canker. **1**:CA(81,82).
Physalospora sp. Twig dieback. **1**:FL(35).
Pyronema omphalodes. **1**:CA(81).
Stictis radiata. **2**:CA(3225).
Basidiomycotina - Other
Amylostereum chailletii. **1**:CA(543).
Armillaria mellea. Root rot. **1**:CA(82,2898).
Caulorhiza umbonata. On roots. **1**:CA(81).
Ceriporiopsis rivulosa. Heart rot. **1**:CA(1903).
Fomitopsis meliae. Isolated from cooling tower. **2**:OH(2320).
Galerina filiformis. On logs. **1**:CA(3334).
G. perangusta. On trunks. **1**:CA(3334).
Ganoderma oregonense. Trunk rot. **1**:CA(82).
Gloeophyllum carbonarium. Brown cubical rot. **1**:CA(82,1903).
G. odoratum. Brown rot. **1**:CA(82).
Heterobasidion annosum. White root and butt rot. **1**:CA(82,1903).
Hymenochaete tabacina. Wood rot. **1**:CA(82).
Laetiporus sulphureus. Wood rot. **1**:CA(82).
Meruliporia incrassata. Dry rot. **1**:CA(82).
Micromphale sequoiae. On needles and stems. **1**:CA(853).
Mycena albissima. On wood. **2**:CA(3326).
M. clavata. On wood. **2**:CA(3326).
M. fusco-ocula. On dead needles. **2**:CA(3326).
M. paucilamellata. On twigs. **1**:CA(3326).
Phanerochaete burtii. **1**:CA(128).
P. sanguinea. White rot. **1**:CA(433).
Phellinus viticola. Wood rot. **1**:CA(82).
Postia sequoiae. Brown cubical heart rot. **1**:CA(82,1903).
Schizophyllum commune. Wood rot. **1**:CA(82).
Schizopora paradoxa. Heart rot. **1**:CA(1903).
Serpula hexagonoides. On wood. **1**:CA(82,128,644).
Skeletocutis amorpha. Wood rot. **1**:CA(82).
Stereum hirsutum. Wood rot. **1**:CA(82).
S. ostrea. Wood rot. **1**:CA(82).
Trametes versicolor. Wood rot. **1**:CA(82).
Trichaptum abietinum. Heart rot. **1**:CA(1903).
Xeromphalina campanella. On debris. **1**:CA(81).
X. orickiana. On logs. **1**:MI(3328).

Deuteromycotina - Hyphomycetes
Aureobasidium pullulans. Isolated from discolored roof shingles. **1**:OR(3136).
Botrytis cinerea. Twig blight. **1**:MD(82).
B. douglasii. Seedling blight. **1**:CA(82).
Brachysporiella turbinata. On leaves and decorticated wood. **1**:CA(81). **2**:CA(637).
Cercospora sequoiae. Needle blight. **1**:FL(35).
Phymatotrichopsis omnivora. Root rot. **1**:TX(82).
Deuteromycotina - Coelomycetes
Chaetasbolisia falcata. On leaves. **1**:CA(81,2434).
Cytospora pinastri. On twigs. **1**:CA(82).
Leptostroma sequoiae. On twigs. **1**:CA(82).
Macrophoma sp. Twig dieback. **1**:FL(35), MD(82).
Oncospora abietina. On wood. **1**:CA(81).
Phomopsis occulta. On dead twigs. **1**:MD(82).
Phyllosticta sp. Needle blight. **1**:FL(35).

Sequoiadendron Buchholz
1. *S. giganteum* (Lindl.) Buchholz (=*Sequoia gigantea* (Lindl.) Buchholz) [bigtree, giant redwood, giant sequoia]

Oomycetes
Phytophthora sp. Damping-off, root rot. **1**:HI(2899,2903).
Pythium ultimum. Damping-off, root rot. **1**:HI(2899,2903).
Ascomycotina
Acanthostigma sequoiae. On needles. **1**:CA(82).
Botryosphaeria dothidea. Limb and twig dieback. **1**:CA(3955).
Ceratophacidium aristosporum. On leaves and twigs. **1**:CA(81,325).
Haematomyxa sequoiae. On bark. **1**:CA(81).
Murangium sequoiae. On bark. **1**:CA(3167).
Basidiomycotina - Other
Armillaria mellea. Root rot. **1**:CA(2897,2898).
Gloeophyllum sepiarium. Wood rot. **1**:CA(82).
Heterobasidion annosum. Trunk rot. **1**:CA(82,2808).
Meruliporia incrassata. Butt rot, root rot. **1**:CA(2809).
Deuteromycotina - Hyphomycetes
Botrytis cinerea. Gray mold of nursery seedlings. **1**:CA(3353).
B. douglasii. Seedling blight. **1**:CA(82), OH(82), PA(82).
Cercospora sequoiae. Needle blight of seedlings. **1**:MD(82), PA(82,962,1591).
Deuteromycotina - Coelomycetes
Hendersonula toruloidea. Dieback of twigs. **1**:CA(81).
Macrophomina phaseolina. Root rot of seedlings. **1**:CA(3348).
Microdiplodia cupressina. On dead twigs. **1**:CA(81).
Pestalotiopsis funerea. Needle blight. **1**:TX(82).
Phomopsis juniperovora. Twig blight. **1**:CA(82).
P. occulta. On dead twigs. **1**:CA(82).

Taxodium Rich.
1. *T. distichum* (L.) Rich. [bald-cypress]
2. *T. distichum* (L.) Rich. var. *imbricarium* (Nutt.) H. B. Croom (=*T. ascendens* Brongn., *T. distichum* (L.) Rich. var. *nutans* of authors) [pond cypress]
3. *Taxodium* sp.

Oomycetes
Pythium sp. Root rot. **1**:FL(35). **3**:FL(35).
Ascomycotina
Hypoxylon rubiginosum. **3**:LA(2675).
Lachnum virgineum. **1**:GA(1438).
Metasphaeria cavernosa. **1**:GA(1438).
Mycosphaerella taxodii. On needles. **3**:SC(82).
Basidiomycotina - Other
Antrodia juniperina. On wood. **3**:AR(82).
Arrhytidia involuta. **3**:LA(2675).
Ceriporia xylostromatoides. On wood. **3**:IL(82), LA(2675).
Coniophora arida. On plank. **1**:LA(128).
Dacryopinax spathularia. **3**:LA(2675).
Dendrocorticium piceinum. **1**:SC(2043).
[*Dendrothele microspora*]. **1**:LA(2084).
Fomitopsis cajanderi. **3**:FL(82).
Ganoderma applanatum. Butt rot. **1**:FL(35). **3**:FL(35,82).
Gloeocystidiellum wakullum. On wood. **1**:FL(436).
Gloeophyllum sepiarium. **3**:FL(82).
G. striatum. **3**:FL(82).
Henningsomyces candidus. **3**:LA(2675).
Hydropus taxodii. On dead standing trunks. **1**:FL(3326).
Hyphoderma praetermissum. **1**:TX(2117).
Laurilia taxodii. Pecky cypress. **1**:FL(800,805), GA(805), LA(800), MS(800,805), NC(805), SC(128), TX(800).
Meruliporia incrassata. On wood. **3**:S states(82).
Phellinus extensus. White pocket heart rot. **1**:FL(35). **3**:FL(35,82).
P. gilvus. On wood. **1**:FL(35). **3**:AR(2675), FL(35,82), LA(82,2675), MS(2675).
[*P. robustus*]. **3**:GA(82).
Pseudofavolus cucullatus. On wood. **3**:SE states(2681).
Rigidoporus ulmarius. Root and butt rot. **3**:SC states(82), SE states(82).
Rogersiomyces okefenokeensis. On submerged, decayed leaves. **2**:GA(722).
Septobasidium pseudopedicellatum. Felt fungus. **3**:LA(82).
Trametes drummondii. On wood. **3**:FL(1205,2681), LA(1205,2681).
T. hirsuta. On wood. **3**:SC(82).
T. membranacea. On wood. **3**:SE states(2681), LA(2675).
T. versicolor. On wood. **3**:FL(82), LA(82).
Trichaptum abietinum. On wood. **3**:SC(82).
T. byssogenum. On wood. **3**:S states(2681), GA(82), LA(82,2675).
Tricholomopsis decora. **3**:LA(2675).
Wolfiporia cocos. On wood. **3**:TN(82).
Xylobolus subpileatus. **1**:SC(543).
Deuteromycotina - Hyphomycetes
Cercospora sequoiae. Leaf blight and spot. **1**:FL(35). **3**:FL(35).
Chaetopsina illinii. On wood. **1**:IL(719).
Geotrichum sp. Pink wood stain. **3**:Gulf States(82).
Phymatotrichopsis omnivora. Root rot. **3**:Unk.(82).
Sporidesmium taxodii. On decaying leaves. **1**:IL(720).
Deuteromycotina - Coelomycetes
Diplodia sp. Leaf blight. **1**:FL(35).
Pestalotiopsis funerea. Twig blight. **3**:TX(82).
Phomopsis occulta. On twigs. **3**:NC(82).
Phyllosticta sp. Leaf spot. **1**:FL(35).
Deuteromycotina - Other
Rhizoctonia sp. Leaf blight, root and basal stem rot. **1**:FL(35). **3**:FL(35).

Theaceae

Camellia L.
1. *C. japonica* L. [common camellia]
2. *C. sasanqua* Thunb. [sasanqua camellia]
3. *C. sinensis* (L.) Kuntze (=*C. sinensis* (L.) Kuntze var. *sinensis*, *Thea bohea* L., *Thea sinensis* L., *Thea viridis* L., *C. thea* Link, *C. theifera* Griff.) [tea-plant]
4. *Camellia* sp.

Myxomycetes
Physarum gyrosum. Slime mold. **1**:MS(2720).
Oomycetes
Phytophthora cinnamomi. Dieback, root rot. **1**:AL(3624), CA(3624,4010), FL(35,3624), LA(3624), NC(1303). **2**:FL(35,3624), NC(1303).
Pythium splendens. Root rot. **1**:FL(35). **2**:FL(35).
Pythium sp. Damping-off. **2**:NC(1303).
Ascomycotina
Botryosphaeria dothidea. Limb canker. **1**:FL(35).
B. ribis. Limb canker. **1**:MS(82,2720).
Chaetomium sp. On flowers. **4**:SE states(3787).
Ciborinia camelliae. Flower and petal blight. **1**:CA(82), FL(35), GA(82,1215), LA(82), NC(82,1303), OR(82,3202), SC(224), TX(2831). **2**:FL(35). **4**:GA(3787), LA(42,1262,2823,2828), NC(3787), OR(3989).
Diaporthe eres. On flowers. **4**:SE states(3787).
Diaporthe sp. Cutting dieback. **1**:FL(35).
Didymosphaeria sp. **1**:GA(1438).
Elsinoe leucospila. White scab on leaves. **1**:FL(35,2525). **2**:FL(35,2524,2525). **4**:FL(2525).
Glomerella cingulata. Anthracnose, canker, dieback, on flowers and leaves. **1**:Range of host(82), FL(35), GA(1438), MS(2720), NC(1303), OR(3202). **2**:FL(35), NC(1303). **3**:FL(35,82), SC(82), TX(82). **4**:S states(225), SE states(3787).
Guignardia camelliae. On leaves. **3**:CA(82).

Laestadia sp. On flowers. **4**:SE states(3787).
Leptosphaeria sp. Twig dieback. **2**:FL(35).
Meliola camelliae. Black mildew. **1**:FL(35). **2**:FL(35).
Metasphaeria sp. Stem canker, twig dieback. **1**:FL(35). **2**:FL(35).
Mycosphaerella sp. Leaf spot. **1**:FL(35).
Physalospora sp. Stem canker. **1**:FL(35).
Sclerotinia sclerotiorum. On flowers. **4**:GA(3787), NC(3787).

Basidiomycotina - Other

Armillaria mellea. **1**:CA(2898).
A. tabescens. Dieback, trunk girdle, root rot. **1**:FL(35), LA(2815,2829).
Exobasidium camelliae. Flower, leaf and stem gall. **1**:FL(35,82), LA(82), MS(82,2720), NC(1303), TX(82). **2**:FL(35,82), GA(1439), LA(82), MS(82), NC(1303), TX(82,2440). **3**:FL(35).
E. camelliae var. *gracilis*. On young leaves. **2**:NC(3945).
E. monosporum. Leaf curl. **1**:AL(82).
E. vaccinii. Leaf gall. **2**:FL(35).
Heterobasidion annosum. **2**:NC(1303).
Septobasidium castaneum. **1**:SC(82).
S. pseudopedicellatum. Felt fungus. **1**:SC(82).

Deuteromycotina - Hyphomycetes

Alternaria alternata. On flowers. **4**:SE states(3787).
A. camelliae. Leaf spot. **1**:MS(2720), OK(2872). **2**:GA(82).
Alternaria sp. Flower blight. **1**:FL(35). **2**:FL(35).
[*Arthrinium phaeospermum*]. On flowers. **4**:SE states(3787).
Botrytis cinerea. Bud and flower blight, flower rot. **1**:Range of host(82), AL(3533), FL(35), NC(1303). **2**:FL(35), LA(82). **4**:SE states(3787), AL(3534).
Cercoseptoria ocellata. Leaf spot. **2**:LA(2833).
Cercospora sp. Leaf spot. **1**:FL(35), GA(82), NC(1303).
Cladosporium herbarum. On flowers. **4**:SE states(3787).
Curvularia sp. On flowers. **4**:SE states(3787).
Cylindrocladiella peruviana. Seed damp-off. **3**:FL(35).
Cylindrocladium sp. Root rot. **1**:FL(35).
Epicoccum sp. On flowers. **4**:SE states(3787).
Fusarium oxysporum. Root rot. **1**:FL(35).
F. poae. On flowers. **4**:SE states(3787).
Fusarium sp. Stem canker, root rot. **2**:FL(35). **4**:OR(197).
Helminthosporium sp. On flowers. **4**:SE states(3787).
Nigrospora oryzae. On flowers. **4**:SE states(3787).
Penicillium implicatum. Flower rot. **1**:AL(3533).
P. rugulosum. Flower rot. **1**:AL(3533).
Penicillium sp. Flower blight. **4**:SE states(3787), AL(3534).
Pseudocercospora theae. Leaf spot. **1**:FL(35). **2**:FL(35). **3**:FL(35).
Trichoderma sp. On flowers. **4**:SE states(3787).

Deuteromycotina - Coelomycetes

Colletotrichum gloeosporioides. Anthracnose, twig dieback, on flowers and leaves. **1**:CA(81), FL(35), MS(82,2720). **2**:FL(35). **3**:FL(35). **4**:SE states(3787).
Colletotrichum sp. Bud and flower blight. **1**:NC(1303).
Coniothyrium sp. Stem canker. **2**:FL(35).
Cylindrosporium camelliae. On flowers. **4**:SE states(3787).
Diplodia sp. Twig dieback. **1**:FL(35). **2**:FL(35).
Discosia artocreas. On flowers. **4**:SE states(3787).
Heteropatella sp. On flowers. **4**:SE states(3787).
Macrophoma sp. Canker. **1**:FL(35), MS(82). **4**:SE states(3787).
Pestalotia sp. Flower and petal blight, flower rot, leaf spot. **1**:AL(3533), FL(35), OR(3202). **2**:FL(35). **4**:AL(3531,3534).
Pestalotiopsis karstenii. On leaves. **1**:GA(82), MS(82,2720).
P. maculans. Twig blight, leaf spot, on flowers and leaves. **1**:Range of host(82), FL(35), GA(82), MS(2720), NC(1303), OK(2872), WA(3202). **2**:NC(1303). **3**:SC(82). **4**:SE states(3787).
Phoma camelliae. On flowers. **4**:SE states(3787).
P. tenuis. On flowers. **4**:SE states(3787).
Phoma sp. Purple blotch, dieback. **1**:NC(1303). **3**:FL(35).
Phomopsis sp. Dieback on flowers and stems. **1**:Unk.(82), FL(35). **2**:FL(35). **4**:SE states(3787), OR(3202).
Phyllosticta camelliae. Leaf spot. **1**:SE states(82), AL(285), FL(35), MS(2720), NC(1303), WA(3202). **2**:FL(35).
P. erratica. Leaf spot. **3**:AL(82), FL(35).
Phyllosticta sp. On flowers. **4**:SE states(3787).
Phyllostictina sp. On flowers. **4**:SE states(3787).
Pilidium concavum. On flowers. **4**:SE states(3787).
Sphaceloma sp. White scab. **1**:FL(82), GA(82), LA(82), MS(2720), OR(3202). **3**:LA(82).
Sphaeropsis sp. On decayed roots. **1**:NC(1303).
Sporonema camelliae. Leaf spot. **1**:AL(82), VA(82), WI(1341).
Sporonema sp. On flowers. **4**:SE states(3787).

Deuteromycotina - Other

Rhizoctonia solani. Root rot. **1**:FL(35). **2**:FL(35).
Rhizoctonia sp. Damping-off. **2**:NC(1303).

Cleyera Thunb.

1. *C. japonica* Thunb.

Oomycetes

Pythium splendens. Root rot. **1**:FL(35).

Ascomycotina

Botryosphaeria sp. Leaf spot. **1**:FL(35).
Diatrype sp. Leaf spot. **1**:FL(35).

Deuteromycotina - Hyphomycetes

Cercospora sp. Leaf spot. **1**:FL(35).
Fusarium oxysporum. Root rot. **1**:FL(35).

Deuteromycotina - Coelomycetes

Colletotrichum sp. Anthracnose. **1**:FL(35).
Entomosporium mespili. Leaf spot. **1**:FL(35).
Macrophoma sp. Leaf spot. **1**:FL(35).
Phyllosticta sp. Leaf spot. **1**:FL(35).

Deuteromycotina - Other

Rhizoctonia sp. Leaf necrosis. **1**:FL(35).

Gordonia J. Ellis

1. *G. lasianthus* (L.) J. Ellis [black laurel, loblolly-bay]

Ascomycotina

Antennularia robinsonii. Leaf spot. **1**:FL(35).
Irenopsis cryptocarpa. Black mildew. **1**:FL(35,82,1452), LA(82).
Mycosphaerella gordoniae. Leaf spot. **1**:FL(35), GA(82,1438).
Nectria leucorrhodina. **1**:FL(3085).

Deuteromycotina - Hyphomycetes

Cercospora sp. Leaf spot. **1**:FL(35).
Cylindrocladium sp. Leaf spot. **1**:FL(35).
Phymatotrichopsis omnivora. Root rot. **1**:TX(82).

Deuteromycotina - Coelomycetes

Cytospora gordoniae. On flower stalks. **1**:SC(82).
Phyllosticta gordoniae. Leaf spot. **1**:FL(35,82), TX(82).

Ternstroemia Mutis ex L. fil.

1. *T. gymnanthera* (Wight & Arnott) Sprague

Ascomycotina

Elsinoe leucospila. Spot anthracnose. **1**:FL(35,2524).

Deuteromycotina - Hyphomycetes

Cercospora sp. Leaf spot. **1**:FL(35).

Deuteromycotina - Coelomycetes

Phyllosticta sp. Leaf spot. **1**:NC(1303).

Thea L.

Ascomycotina

Theophrastaceae

Jacquinia L.

1. *Jacquinia* sp.

Ascomycotina

Asterina paupercula. Black spot. **1**:FL(35,82).

Deuteromycotina - Coelomycetes

Phyllosticta sp. Leaf spot. **1**:FL(35).

Thymelaeaceae

Daphne L.

1. *D. cneorum* L. [garland-flower, rose daphne]
2. *D. mezereum* L. [February daphne, mezereon]
3. *D. odora* Thunb. [winter daphne]
4. *Daphne* sp.

Oomycetes

Phytophthora cactorum. **2**:CA(82). **4**:CA(82), OR(1429).
P. nicotianae var. *parasitica*. Root and stem rot. **3**:CA(3655).
Phytophthora sp. Collar, crown and stem rot. **2**:NY(82). **4**:NC(1303), NY(82).
Pythium sp. Root rot. **4**:NC(1303).

Ascomycotina
Nectria cinnabarina. On twigs. **2**:NJ(82). **4**:NJ(82).
Basidiomycotina - Other
Athelia rolfsii. **3**:OR(3202).
Deuteromycotina - Hyphomycetes
Botrytis sp. Twig blight. **2**:NE states(82), NW states(82). **4**:NE states(82), NW states(82).
Fusarium sp. Dieback. **4**:NJ(82).
Verticillium albo-atrum. Wilt. **1**:IL(1558).
Deuteromycotina - Coelomycetes
Marssonina daphnes. Leaf spot. **2**:WA(82). **4**:WA(82,3202).
Deuteromycotina - Other
Rhizoctonia solani. Stem rot. **2**:NY(82). **4**:NY(82).
Sclerotium rolfsii. Wilt. **2**:FL(82). **4**:FL(82).

Dirca L.
1. *D. palustris* L. [leatherwood, moosewood, ropebark, wicopy]

Basidiomycotina - Uredinales
Aecidium hydnoideum. **1**:NC states(82), NE states(82), Unk.(82), AL(82), MO(82).
Deuteromycotina - Coelomycetes
Gloeodes pomigena. On twigs. **1**:IN(82).
Phyllosticta dircae. **1**:WI(1348).

Wikstroemia Endl.
1. *W. elongata* A. Gray
2. *W. oahuensis* (A. Gray) Rock
3. *W. phillyreifolia* A. Gray
4. *W. uva-ursi* A. Gray
5. *W. viridiflora* Meisn. (=*W. foetida* A. Gray)
6. *Wikstroemia* sp.

Ascomycotina
Amazonia wikstroemiae. Black mildew. **1**:HI(1272,1452,2903). **2**:HI(2903). **3**:HI(1272,1452,2903). **5**:HI(1272,1452). **6**:HI(1268,1272,1452,2903).
Botryosphaeria ribis. **3**:HI(2903).
Calloriopsis gelatinosa. **6**:HI(2903).
Diaporthe sp. On limbs. **3**:HI(2903).
Didymosphaeria wikstroemiae. On decaying stems. **6**:HI(2784).
Physalospora abdita. **3**:HI(2903).
Seuratia millardetii. On leaves. **6**:HI(2337).
Basidiomycotina - Uredinales
Uredo wikstroemiae. **4**:HI(2903).
Deuteromycotina - Hyphomycetes
Cladosporium avellaneum. On leaves. **6**:HI(2903).
C. cladosporioides. On leaves. **6**:HI(2903).
Epicoccum nigrum. **6**:HI(2903).

Tiliaceae

Tilia L.
1. *T. americana* L. (=*T. glabra* Venten.) [American basswood, American linden, whitewood]
2. *T. cordata* Mill. [European linden, littleleaf linden]
3. *T. heterophylla* Venten. [white basswood]
4. *T. heterophylla* Venten. var. *michauxii* (Nutt.) Sarg.
5. *T. neglecta* Spach
6. *T. pubescens* Aiton
7. *Tilia* sp.

Ascomycotina
Apiognomonia tiliae. Leaf spot. **1**:NE states(82), IA(82), MN(82), TX(82).
Apiosporella sp. **4**:GA(1438).
Bombardia fasciculata. **7**:NY(2456).
Botryosphaeria dothidea. **4**:GA(1438).
B. obtusa. On limbs. **1**:NY(82), VA(82).
B. ribis. On limbs. **1**:MD(82), NJ(82), NY(82).
Botryosphaeria sp. Trunk canker. **5**:MD(82).
Calycellina populina. On leaves. **7**:NY(2181).
Cercophora appalachianensis. **1**:VA(1554). **7**:VA(1554).
Ciboria tiliae. On fallen fruits. **1**:IA(82).
Daldinia concentrica. Wood rot. **1**:MN(82), NY(82).
D. grandis. **1**:MN(82).
D. occidentalis. **1**:MN(82).
Diaporthe eres. On limbs. **1**:MI(82).
Diatrypella favacea. **4**:GA(1438).
D. quercina. **4**:GA(1438).
Dothivalsaria megalospora. **1**:VT(185).
Elsinoe tiliae. Anthracnose. **5**:VA(726).
Eutypella juglandicola. **4**:GA(1438).
E. leprosa. **1**:IA(3629), VA(2925).
E. prunastri. **1**:IA(3629).
E. stellulata. On branches. **1**:OK(82,2873).
Hercospora tiliae. **4**:GA(1438).
Hypocrea sulphurea. **1**:ND(377).
Hypoxylon cinereo-lilacinum. On wood. **7**:IA(1225).
H. fuscum. On wood. **7**:IA(1225).
H. howeianum. On wood. **7**:IA(1225).
H. mediterraneum. On wood. **7**:IA(1225).
H. rubiginosum. On wood. **7**:IA(1225).
H. serpens. On decaying wood. **7**:IA(1225).
Lachnella tiliae. On limbs and trunks. **1**:E & C states(456), ND(377). **7**:MI(2945), MO(2945), NY(2945).
Lasiosphaeria chrysentera. **1**:VA(1554). **7**:NY(1554).
Lepteutypa fuckelii. **1**:NC(3256).
Lulworthia rotunda. On wood submerged in sea water. **7**:NC(1791).
Massarina corticola. On twigs. **7**:MA(319).
Massariovalsa sudans. **4**:GA(1438).
Melanconis juglandis var. *tiliae*. On limbs. **1**:MI(82).
Microsphaera penicillata. Powdery mildew. **1**:MN(82).
Mycosphaerella microsora. **1**:IA(82).
M. punctiformis. On leaves. **4**:GA(1438). **7**:Unk.(3943).
Nectria cinnabarina. Twig dieback. **1**:Range of host(82), AK(526), FL(35), OK(2873). **4**:GA(1438).
N. coccinea. On limbs and trunks. **1**:NY(82), PA(82), VA(82).
N. galligena. Trunk canker. **1**:MN(2240), NY(82), PA(82), VA(82).
Nitschkia broomeiana. **4**:GA(1438).
N. calyculus. **7**:FL(2782).
Phorcys tiliae. **3**:GA(1438).
Phragmodiaporthe tiliacea. On limbs. **1**:IA(82), ME(82), MI(82), NY(82), PA(82).
Phyllactinia guttata. Powdery mildew. **1**:MN(82).
Pseudomassaria chondrospora. On petioles and twigs. **4**:GA(175,1438).
Pyrenopeziza minuta. **4**:GA(1438).
Schizoxylon insigne. **4**:GA(1438).
Scutellinia erinaceus. On rotten wood. **7**:NY(848).
Sphaerangium sp. **4**:GA(1438).
Splanchnonema ampullaceum. On limbs. **4**:GA(1438). **7**:DC(188), GA(188), MA(188), ME(188), MI(188), NY(188).
Tympanis alnea. **1**:NY(2670).
Uncinula clintonii. Powdery mildew. **1**:Range of host(82), FL(35), MS(2720), ND(377).
Ustulina deusta. Root rot. **1**:NY(82).
Valsa ambiens. Canker, on dead twigs. **1**:NY(82), OK(2873). **4**:GA(1438).
V. ceratosperma. **4**:GA(1438).
Vibrissea truncorum. On fruit bracts and dead leaves. **7**:NY(638).
Winterella wehmeyeriana. **1**:DC(2948), NY(2948).
Basidiomycotina - Other
Albatrellus peckianus. On buried wood. **7**:Unk.(1204).
Aleurodiscus oakesii. On bark. **1**:TN(2083).
Basidiodendron cinereum. On wood. **1**:IA(128).
Bjerkandera adusta. Wood rot. **1**:MN(82), MO(82), NY(82), VT(82), WI(82).
B. fumosa. Wood rot. **1**:NY(82).
Cerrena unicolor. **1**:IA(82), ND(82).
Climacodon septentrionalis. White spongy rot. **1**:AL(82), MI(82), TX(82).
Crepidotus cinereipallens. On bark. **1**:MI(1544).
C. cinnabarinus. On wood. **1**:MI(2192).
C. obscurus. On wood. **1**:MI(1544).
Cystostereum murrayi. **1**:PA(82).
Daedaleopsis confragosa. White spongy rot. **1**:VT(82).
Datronia mollis. Wood rot. **1**:VT(82).
Dendrothele acerina. On bark. **1**:IA(82), MO(82), PA(82).
D. alliacea. On bark. **1**:IA(2084).
D. griseo-cana. On bark. **1**:IA(82), MO(82), PA(82).
D. microspora. On bark. **1**:WI(2084).

Dentocorticium sulphurellum. White rot. **1**:MN(1197), NY(2043).
Flammulina velutipes. Sapwood rot. **1**:Unk.(82).
Fomitopsis pinicola. Brown cubical rot. **1**:NY(82).
Ganoderma applanatum. White butt. **1**:VT(82).
Gloeocystidiellum clavuligerum. White rot. **1**:MN(2553).
G. porosum. **1**:MN(128,1197).
Gymnopilus praecox. On bark and wood. **1**:IN(2738).
Hyphoderma mutatum. **1**:NY(2321). **7**:NY(2321).
Hyphodontia albicans. **1**:MN(1197).
Ischnoderma resinosum. Wood rot. **1**:MN(82), NY(82), VT(82).
Junghuhnia nitida. **1**:MN(128).
Lenzites betulina. Wood rot. **1**:NY(82), VT(82).
Lopharia cinerascens. **1**:IA(82), MN(82). **7**:VT(543).
Oxyporus corticola. **1**:MN(128).
O. populinus. On wood. **1**:NE states(82), FL(35,82).
Peniophora cinerea. **1**:MN(1197).
Perenniporia medulla-panis. Wood rot. **1**:MI(82).
Phanerochaete filamentosa. White rot. **1**:NY(433). **7**:NY(128).
P. rimosa. White rot. **1**:MI(433).
Phellinus gilvus. Wood rot. **1**:Unk.(82).
P. punctatus. **1**:ND(378).
Phlebia rufa. **1**:WI(128).
Pholiota adiposa. Brown-mottled heart rot. **1**:E states(82), MA(82). **7**:WA(3202).
Pleurotus ostreatus. Sapwood rot. **1**:Range of host(82), MN(63,3774).
Plicaturopsis crispa. **1**:MN(1197).
Polyporus mori. **1**:MN(1197).
P. squamosus. Wood rot. **1**:Unk.(82).
P. varius. Wood rot. **1**:MO(82), NY(82), OH(82).
Rhodotus palmatus. On wood. **1**:MI(2422).
Rigidoporus ulmarius. On wood. **1**:FL(35,82).
Schizophyllum commune. Wood rot. **1**:MN(82).
Sebacina incrustans. On logs. **1**:NY(128).
Septobasidium fumigatum. Parasitic on scale insects. **1**:FL(35,82).
Simocybe tiliophila. On dead limbs and trunks. **1**:NY(1544).
Sistotrema raduloides. **1**:MN(1197).
Spongipellis spumeus. Wood rot. **1**:NY(82).
Stereum hirsutum. **1**:CT(543).
S. ostrea. **1**:VT(82). **7**:NC(1307), SC(1307).
Tomentella ferruginea. **7**:MI(2039).
T. rubiginosa. **7**:PA(2039).
T. subvinosa. **7**:NY(2039).
T. umbrinospora. **7**:NY(2039).
Trametes hirsuta. White rot. **1**:MA(82), NC(1303), NY(82), VT(82). **4**:GA(1439). **7**:NC(1307), SC(1307).
T. versicolor. Wood rot. **1**:IN(82), VT(82).
Trichaptum biforme. Wood rot. **1**:NY(82), VT(82). **7**:NC(1307), SC(1307).
Tyromyces galactinus. Wood rot. **1**:MN(82).
Deuteromycotina - Hyphomycetes
Cacumisporium capitulatum. **7**:NY(1264).
Cercospora microsora. Leaf blight and spot. **1**:Range of host(82), FL(35), NC(1303), SD(2243). **2**:NJ(82), PA(82), VA(82).
C. tragopogonis. Leaf spot. **1**:OK(2872).
Coniosporium olivaceum. **7**:DE(3543).
Conoplea olivacea. **7**:NY(1672), WI(1672).
Corynespora olivacea. **1**:NY(3547).
Exosporium tiliae. On limbs. **1**:E states(82), DE(3543), MI(82), MS(2720).
Fusarium sp. **1**:WI(1341).
Haplotrichum ramosissimum. On limbs. **7**:MA(2131), NY(2131).
Helicomyces roseus. On bark. **7**:NY(1270).
Mycoleptodiscus minimus. On leaves. **6**:FL(634).
Phymatotrichopsis omnivora. Root rot. **1**:TX(82).
Strumella sp. Trunk canker. **1**:NJ(82).
[*Torula ligniperda*]. Gray sapwood stain. **1**:Unk.(82).
Verticillium albo-atrum. Wilt. **1**:IL(82,992,1558). **2**:IL(1558).
Deuteromycotina - Coelomycetes
Aplosporella tiliacea. On branches. **1**:NY(82), OK(2873). **2**:NY(82).
Asteroma tiliae. On leaves. **1**:IN(82), WI(82,1356).
Colletotrichum dematium. On samaras. **7**:MD(651).
Colletotrichum sp. **1**:NY(2456).
Dendrophoma tiliae. On limbs. **1**:NY(82).
Discosia artocreas. **1**:NY(2456).
Discula umbrinella. Leaf spot. **2**:NY(82).
Lasiodiplodia theobromae. Canker. **7**:DC(3862).
Macrophoma tiliacea. On limbs. **1**:OH(82), OK(82), WI(1337).
Myxosporium fumosum. On dead twigs. **1**:IL(82,971).
M. tiliae. On dead twigs. **1**:NY(82,832).
Phlyctema tiliae. Leaf spot. **1**:TN(82).
P. verrucarioides. **1**:NY(3081).
Phomopsis communis. **1**:AK(526).
Phyllosticta bacterioides. **1**:WI(1354).
P. praetervisa. Leaf spot. **1**:WI(82).
P. tiliae. Leaf spot. **1**:NE states(82), FL(35), NY(2456), WI(1339).
Rabenhorstia tiliae. On limbs. **1**:IA(82), ME(82), NC(82). **7**:NY(2783).
Sphaeropsis sp. Leaf spot. **1**:OK(82,2873).

Triumfetta L.

1. *T. semitriloba* (L.) Jacq. [Sacramento bur]

Basidiomycotina - Uredinales
Pucciniosira pallidula. Autoecious. **1**:FL(2285).

Trapaceae

Trapa L.

1. *T. natans* L. [caltrop, jesuit-nut, ling, saligot, trapa-nut, water-chestnut]

Deuteromycotina - Other
Sclerotium hydrophilum. **1**:NY(2888).

Tropaeolaceae

Tropaeolum L.

1. *T. majus* L. [garden nasturtium, Indian cress, tall nasturtium]
2. *Tropaeolum* sp.

Oomycetes
Pythium sp. Root rot. **1**:HI(2903).
Ascomycotina
Pleospora tropaeoli. Leaf spot. **1**:OK(2872).
Pleospora sp. Leaf spot. **1**:MS(82,2720), NJ(82), OH(82).
Basidiomycotina - Uredinales
[*Puccinia aristidae*]. **1**:UT(82).
Deuteromycotina - Hyphomycetes
Acroconidiella tropaeoli. Leaf spot. **1**:CA(81,82,147), HI(2899), NY(82). **2**:HI(2903).
Botrytis cinerea. Gray mold. **1**:AK(82,526).
Cercospora tropaeoli. Leaf spot. **1**:AL(82), TX(82).
Cercospora sp. Leaf spot. **1**:HI(2903).
Deuteromycotina - Coelomycetes
Phyllosticta sp. **1**:WI(1341).
Deuteromycotina - Other
Rhizoctonia solani. Root rot. **1**:FL(35).

Turneraceae

Turnera L.

1. *T. ulmifolia* L. [yellow-alder]

Deuteromycotina - Hyphomycetes
Pseudocercospora turnerae. Leaf spot. **1**:FL(35).

Typhaceae

Typha L.

1. *T. angustifolia* L. [narrowleaf cattail, small bulrush, soft-flag]
2. *T. latifolia* L. [bullrush, common cattail, Cossack asparagus, nail-rod]
3. *Typha* sp.

Oomycetes
Pythium helicoides. Leaf rot. **2**:OH(82).
Ascomycotina
Didymosphaeria futilis. On culms. **2**:NY(82,195).
Epichloe typhina. Leaf spot, on culms. **2**:NE(82), NY(82), OR(82), TX(82), WI(82).
Guignardia sp. **2**:GA(82,1438).
Leptosphaeria hydrophila. On culms and leaves. **1**:NY(82).

Lewia infectoria. **2**:CA(2439).
Lophodermium typhinum. On leaves. **2**:GA(82,1438), NY(82).
Massarina gloeospora. **3**:FL(2782).
Mycosphaerella typhae. On leaves. **2**:GA(82,1438), ID(3202), MD(82), ND(82), NY(82), PA(82).
Ophiobolus sp. Culm rot. **2**:AR(82).
Phaeosphaeria luctuosa. On culms and leaves. **2**:MO(82).
P. typharum. On culms and leaves. **2**:ID(3202), MO(82), MT(82,3202), ND(377), NY(82), WA(3202).
Pleospora herbarum. On leaves. **2**:CA(82).
Basidiomycotina - Other
Cellypha goldbachii. **2**:MI(2945).
Hyphoderma typhicola. **2**:NY(3315).
Psathyrella typhae. On leaf fragments in ponds and streams. **1**:Unk.(1766). **2**:Unk.(1766). **3**:ME(2931).
Typhula latissima. On dead culms. **2**:NY(82,264,2955).
Deuteromycotina - Hyphomycetes
Cladosporium herbarum. **1**:Range of host(82). **2**:Range of host(82), CA(2439), OK(2872).
C. typhae. **1**:Range of host(82). **2**:Range of host(82).
C. typharum. **1**:Range of host(82). **2**:Range of host(82), DE(3543), ID(3202), MT(3202), ND(378).
Dictyosporium elegans. On culms. **2**:CA(637).
Epicoccum nigrum. **2**:DE(3543).
Heterosporium maculatum. On leaves. **2**:CO(82), ID(3202), MO(82), MT(3202), ND(82), WA(3202).
Scolicotrichum typhae. Leaf spot. **2**:CO(82).
Deuteromycotina - Coelomycetes
Colletotrichum typhae. On living leaves. **2**:WI(1349).
Hymenopsis hydrophila. On leaves. **2**:NY(82).
H. typhae. On culms and leaves. **1**:NY(82). **2**:CA(81), MA(82,1394), NY(82).
Phoma orthosticha. On leaves. **2**:ME(82).
P. typharum. **2**:ID(3202), WA(3202).
P. typhicola. **2**:CA(2439).
Phyllosticta typhina. **2**:ID(3202), OR(3202), WA(3202), WI(64). **3**:NY(64).
Phyllosticta sp. Leaf spot. **2**:FL(35).
Scolecosporiella typhae. On leaves. **1**:WI(82,1336). **2**:MS(2720).
Stagonospora typhoidearum. Leaf spot. **2**:FL(35), TX(82), WI(82).
Deuteromycotina - Other
Sclerotium hydrophilum. On culms. **2**:AR(82).

Ulmaceae

Celtis L.

1. *C. aculeata* Sw. (=*Mertensia laevigata* H. B. K.)
2. *C. georgiana* Small (=*C. occidentalis* L. var. *georgiana* (Small) E. Murray)
3. *C. laevigata* Willd. [Mississippi hackberry, southern hackberry, sugarberry]
4. *C. occidentalis* L. [common hackberry, nettle-tree, sugarberry]
5. *C. pallida* Torr. [desert hackberry, granjeno]
6. *C. reticulata* Torr. (=*C. douglasii* Planch.) [netleaf hackberry]
7. *Celtis* sp.

Oomycetes
Pseudoperonospora celtidis. Downy mildew. **3**:GA(82), MD(82). **4**:MD(82), WI(1355).
Ascomycotina
Acrospermum compressum. **2**:GA(1438). **3**:GA(1438).
Botryosphaeria obtusa. On dead limbs. **3**:Unk.(82). **4**:Unk.(82).
Cucurbitaria celtidis. On dead limbs. **4**:KS(82,3209).
Diatrype whitmanensis. **7**:WA(3019).
Diatrypella favacea. **4**:IA(3629).
D. subfulva. **2**:GA(1438).
Didymella celtidis. On limbs. **3**:SC(82).
Didymosphaeria celtidis. On dead limbs. **4**:KS(82), OK(660).
Eutypella leprosa. **4**:IA(3629).
E. stellulata. On limbs and bark. **7**:MS(2720).
Hypoxylon jecorinum. On bark. **7**:LA(128).
H. rubiginosum. **2**:GA(1438). **3**:GA(82).
Lecanidion atratum. **6**:ID(3202), WA(3202).
Leptosphaeria erigerontis. **1**:WA(3814).
Lojkania nuda. **7**:KS(189).
Lopadostoma caespitosum. On dead limbs. **4**:KS(82).
Lophiostoma quadrinucleatum. **6**:ID(3202).
L. quadrinucleatum var. *triseptatum*. **2**:GA(1438). **4**:OK(660).
Mycosphaerella punctiformis. On fallen leaves. **4**:KS(82).
Niesslia pulchriseta. **2**:GA(1438).
Nitschkia callista. **7**:OK(660).
Nummularia signata. On trunks and larger limbs. **7**:SE states(1810), LA(1810).
N. tinctor. **7**:FL(2782).
Physalospora abdita. On dead limbs. **3**:Unk.(82).
Pleochaeta polychaeta. Powdery mildew. **3**:S states(82), MS(2720), TX(82). **4**:S states(82), MS(2720).
Rhytidhysteron rufulum. **3**:AL(82).
Sphaerotheca phytoptophila. Powdery mildew, on galls. **3**:FL(35,82). **4**:C states(82), NC(1303), OK(2873). **7**:OK(660).
Splanchnonema foedans. On limbs. **2**:GA(1438). **7**:KS(188).
Thyronectria pyrrhochlora. **2**:GA(1438).
Uncinula macrospora. Powdery mildew. **2**:FL(82), GA(82). **3**:FL(35).
U. parvula. Powdery mildew. **2**:GA(1438). **3**:Range of host(82), FL(35). **4**:MS(2720). **6**:WA(3202). **7**:OK(2872).
Uncinula sp. Powdery mildew. **3**:FL(35).
Valsa ambiens. **2**:GA(1438). **4**:IA(1226), KS(3415).
V. ceratosperma. **2**:GA(1438).
Valsaria celtidis. Twig dieback, wood rot. **2**:GA(1438). **3**:OK(82). **4**:OK(618,2872).
Basidiomycotina - Uredinales
Puccinia recondita. **1**:ID(82), OR(82).
Basidiomycotina - Other
Armillaria mellea. Root rot. **3**:TX(82). **4**:Unk.(82).
A. tabescens. Root rot. **3**:FL(35).
Arrhytidia involuta. On decorticated wood. **7**:MS(128).
Auricularia auricula. **7**:AR(2675), LA(2675).
Bjerkandera adusta. On wood. **3**:Range of host(82). **7**:AR(2675), LA(2675), MS(2675).
Cerrena unicolor. Wood rot. **7**:Unk.(82), LA(2675).
Chondrostereum purpureum. Silver leaf. **3**:OK(82). **7**:OK(2872).
Corticium sp. **6**:ID(3202).
Daedaleopsis confragosa. Wood rot. **3**:Unk.(82). **7**:Unk.(82).
Dendrophora albobadia. **6**:AZ(1202).
Eichleriella leucophaea. **5**:AZ(1202).
Fomes fasciatus. Butt rot, heart rot. **3**:S states(82), FL(35). **7**:LA(2675).
Fuscocerrena portoricensis. On branches. **4**:OH(638).
Ganoderma colossum. On wood. **3**:FL(1204).
G. lucidum. Butt rot. **3**:S states(82), FL(35). **6**:AZ(1202). **7**:LA(2675).
Helicobasidium brebissonii. Seedling blight. **3**:TX(82).
Hericium erinaceus. Usually on living trees. **3**:Unk.(82).
Inonotus cuticularis. On wood. **3**:Range of host(82). **7**:AR(2675), LA(2675).
Irpex lacteus. Sapwood rot. **4**:IN(654), MN(82).
Laetiporus sulphureus. On wood. **3**:Range of host(82).
Laxitextum bicolor. White rot. **3**:Range of host(82), FL(35). **7**:AR(2675), LA(543,2675).
Lentinus tigrinus. **3**:Unk.(82).
Meruliopsis corium. **6**:AZ(1202).
Oxyporus latemarginatus. Root rot. **3**:SC states(82), TX(82). **7**:LA(2675).
Pellicularia koleroga. Thread blight. **3**:FL(82).
Peniophora nuda. **6**:AZ(1202).
P. tamaricicola. White rot. **6**:AZ(1180,1202).
Perenniporia robiniophila. On wood. **3**:Range of host(82), TX(128). **4**:MO(128), OH(128). **7**:Unk.(2681), MS(128,2675), OH(128).
Phanerochaete flavido-alba. White rot. **7**:LA(433).
P. tuberculata. **6**:AZ(1202).
Phellinus melleoporus. On wood. **7**:Gulf states(1205).
P. texanus. **5**:AZ(1202).
Phellinus sp. **4**:IN(654).
Phlebia celtidis. **6**:ID(3202), WA(3202). **7**:ID(643).
Pleurotus ostreatus. On fire-scarred trees. **3**:LA(82), MS(82,2720). **4**:LA(82), MS(82). **7**:LA(82), MS(82).
Polyporus arcularius. **6**:AZ(1202).
P. tenuiculus. On wood. **7**:SE states(2681).
Rigidoporus lineatus. On wood. **3**:Range of host(82), AL(2311). **7**:AR(2675), LA(2675).

R. ulmarius. Butt rot, heart rot. **3:**S states(82), FL(35).
Steccherinum laeticolor. **6:**AZ(1202).
Stereum ostrea. On wood. **3:**MS(2720).
Trametes elegans. Wood rot. **7:**Unk.(82), AR(2675), LA(2675), MS(2675).
T. hirsuta. On wood. **3:**Range of host(82). **4:**OK(660). **6:**WA(3202). **7:**LA(2675).
T. versicolor. On wood. **3:**Range of host(82).
Trichaptum biforme. On wood. **3:**Range of host(82). **4:**OK(660).

Deuteromycotina - Hyphomycetes

Berkleasmium concinnum. **6:**WA(3202).
Cercospora spegazzinii. Leaf spot. **3:**FL(35). **4:**IA(82).
Cristulariella moricola. **2:**AL(2056), GA(2056). **3:**AL(2056), GA(2056).
Dendrostilbella sp. **7:**LA(2675).
Fusarium lateritium. On galls. **4:**MS(2720).
Phymatotrichopsis omnivora. Root rot. **3:**TX(82). **4:**OK(82,2872), TX(82).
Ramularia celtidis. Leaf spot. **3:**AL(82), TX(82). **4:**C states(82), OK(2872), WI(1355).
Sirosporium antenniforme. On leaves. **3:**AL(82), TX(82). **4:**AR(82), KS(82), MO(82), WV(82).
Tubercularia vulgaris. **7:**WV(3178).

Deuteromycotina - Coelomycetes

Aplosporella celtidis. On limbs. **4:**KS(82), WV(82).
Ascochyta sp. On leaves. **4:**WI(1353).
Botryodiplodia celtidis. **4:**MO(82), NE(82). **7:**LA(2783).
Camarosporium celtidis. On limbs. **4:**KS(82).
Coniothyrium sp. **4:**WI(1342).
Cylindrosporium celtidis. Leaf spot. **3:**AL(82), MS(2720), TX(82).
C. defoliatum. Leaf blight and spot. **3:**TX(82). **4:**OK(2872), TX(82). **6:**OK(2872).
Cytospora leucosperma. On limbs. **4:**KS(82), WV(82).
Dinemasporium decipiens. On dead twigs. **4:**KS(82), PA(82).
Diplodia celtidigena. On dead limbs. **4:**KS(82), MO(82), OK(660).
Discula umbrinella. On leaves. **4:**NE(82), WI(1333).
Dothiorella celtidis. On dead twigs. **4:**NY(82).
Hendersonia celtidis. On dead twigs. **4:**IL(82), NJ(82), PA(82,965).
H. celtifolia. On leaves. **4:**Unk.(82), WI(1339).
Libertella sp. Twig dieback. **4:**OK(618).
Melanconium celtidis. On dead limbs. **4:**KS(82).
Myxosporium ulmi. On limbs. **4:**TX(82).
Phyllosticta celtidis. Leaf spot. **3:**FL(35), TX(82). **4:**Range of host(82), OK(2872), SD(2243), WI(1355). **7:**SD(2243).
Phyllosticta sp. On leaves. **4:**WI(1345).
Placosphaeria celtidis. On twigs. **4:**NY(82).
Polynema radiatum. On dead twigs. **4:**KS(82), PA(82,965).
Seiridium unicorne. **4:**OK(660).
Septogloeum celtidis. Leaf spot. **4:**NY(82,832).
Septoria poseyi. On living leaves. **1:**OR(82,3421).
Septoria sp. Leaf spot. **3:**TX(82).
Sphaceloma sp. **7:**FL(2525).
Stagonospora celtidis. Leaf spot. **4:**C states(82), NE states(82), KS(82).
S. macromaculans. On living leaves. **4:**WI(1339).

Deuteromycotina - Other

Rhizoctonia solani. Thread blight. **3:**FL(35).

Planera J. F. Gmelin

1. *P. aquatica* (Walter) J. F. Gmelin [planertree, water-elm]
2. *Planera* sp.

Basidiomycotina - Other

Asterostroma cervicolor. **2:**LA(2675).
Ganoderma lucidum. **2:**LA(2675).
Hyphoderma setigerum. **2:**LA(2675).
Trametes elegans. Wood rot. **1:**SC(82).

Trema Lour.

1. *T. micrantha* (L.) Blume

Deuteromycotina - Hyphomycetes

Cercospora trematis. Leaf spot. **1:**FL(35).

Ulmus L.

1. *U. alata* Michx. [Wahoo elm, winged elm]
2. *U. americana* L. [American elm, water elm, white elm]
3. *U. americana* L. cv. 'ascendens'
4. *U. americana* L. cv. 'moline' [Moline elm]
5. *U. americana* L. cv. 'pendula'
6. *U. campestris* L.
7. *U. crassifolia* Nutt. [cedar elm]
8. *U. davidiana* Planch. var. *japonica* (Sarg. ex Rehd.) Nakai (=*U. japonica* (Sarg. ex Rehd.) Sarg.) [Japanese elm]
9. *U. glabra* Huds. [Scotch elm, Wych elm]
10. *U. minor* Mill. (=*U. carpinifolia* Gled.)
11. *U. parvifolia* Jacq. [Chinese elm, leatherleaf elm]
12. *U. procera* Salisb. [English elm]
13. *U. procera* Salisb. cv. 'argenteo-variegata'
14. *U. pumila* L. [Chinese elm, dwarf elm, Siberian elm]
15. *U. rubra* Muhl. (=*U. fulva* Michx., *U. pubescens* Walter) [red elm, slippery elm]
16. *U. serotina* Sarg. [red elm, September elm]
17. *U. thomasii* Sarg. (=*U. racemosa* D. Thomas) [cork elm, rock elm]
18. *Ulmus* sp.

Chytridiomycetes

Carpenterella molinea. **2:**WI(82).
Synchytrium ulmi. Leaf gall. **18:**LA(627).

Oomycetes

Phytophthora inflata. Pit canker of trunks. 'Phytophthora canker.' **2:**CT(82), MA(82), NY(82), PA(82). **15:**CT(82), MA(82), NY(82), PA(82).
Phytophthora sp. Root rot. **2:**FL(35), MD(1867). **11:**FL(35).
Pythium sp. Damping-off, root rot. **2:**FL(35), MD(1867), OR(3202). **11:**FL(35). **14:**NGP(82), FL(35), OK(2873), OR(3202).

Ascomycotina

Acrospermum compressum. **1:**GA(1438), NC(3944). **2:**GA(1438).
Apioporthe apiospora. Twig canker. **2:**IA(82).
Biscogniauxia marginata. On wood. **2:**OK(618).
B. nummularia. On wood. **18:**IA(1225).
B. repanda. On dead limbs. **2:**IA(1225), IN(82), KS(82), MO(82), NE(82), WI(82). **15:**DE(82).
Botryosphaeria dothidea. 'Botryosphaeria canker.' **2:**GA(1438). **15:**GA(1438).
B. obtusa. On limbs. **2:**CT(82), GA(82), IA(82), KS(82). **15:**GA(82). **17:**AL(82).
B. ribis. Stem canker of seedlings. **2:**GA(82,2195). **14:**AR(82).
Botryosphaeria sp. On stem galls. **11:**FL(35).
Camarops microspora. On branches. **2:**AK(82), ID(82), NY(82).
C. polysperma. **18:**FL(2561).
Daldinia concentrica. Wood rot. **2:**Range of host(82). **18:**IA(1225).
Diaporthe eres. Twig blight. **2:**MI(82), OK(82,2873).
Diatrype albopruinosa. **2:**IA(3629). **18:**IA(3629).
D. stigma. **18:**IA(3629).
Diatrypella eutypaeformis. **18:**IA(3629).
D. eutypoides. On dead limbs. **2:**MA(1394).
D. favacea. **2:**GA(1438). **14:**GA(1438).
D. nitschkei. **2:**IA(3629). **15:**IA(3629). **18:**IA(3629), OK(660).
D. subfulva. **2:**GA(1438).
Didymosphaeria sp. On twigs. **2:**GA(1438). **14:**TX(82).
Dothidotthia ramulicola. On twigs. **2:**NY(195).
Endothia gyrosa. On limbs and exposed roots. **2:**GA(82).
Eutypa lata. **2:**IA(3629).
Eutypella dissepta. **15:**IA(3629). **18:**IA(3629).
E. juglandicola. On dead limbs. **2:**NY(195). **14:**GA(1438). **15:**GA(82,1438), IN(82).
E. leprosa. On dead limbs. **2:**CO(82), DC(82), DE(82), GA(82,1438), IA(3629), IL(82), KY(82), MD(82), MO(82), ND(377), NY(82), OH(82), RI(82), TN(82), VT(82). **14:**GA(1438). **15:**IA(3629). **18:**IA(3629).
E. scoparia. On dead limbs. **2:**CO(82), DC(82), DE(82), GA(82), IA(3629), IL(82), KY(82), MD(82), MO(82,2925), NY(82,2925), OH(82), RI(82), TN(82), VT(82). **11:**GA(1438). **14:**AL(82), GA(82), IA(3629), LA(82). **15:**GA(82), IA(3629), IN(82), MO(2925). **18:**IA(3629).
E. stellulata. On limbs. **2:**CO(82), DC(82), DE(82), GA(82,1438), IL(82), KY(82), MD(82), MO(82), NY(82), OH(82), RI(82), TN(82), VT(82). **14:**AL(82), GA(82), LA(82).
Eutypella sp. On dead limbs. **2:**CO(82), DC(82), DE(82), GA(82), IL(82), KY(82), MD(82), MO(82), NY(82), OH(82), RI(82), TN(82), VT(82). **15:**GA(82), IN(82).
Gloniopsis praelonga. On dead limbs. **2:**MA(1394).
Hapalocystis ulmi. On limbs and twigs. **2:**MI(82). **15:**IA(82).

Hypoxylon albocinctum. On wood. **18:**IA(1225).
H. cinereo-lilacinum. On wood. **2:**IA(1225).
H. mediterraneum. On wood. **2:**IN(82), KS(82), MO(82), NE(82), WI(82). **17:**IL(82). **18:**IA(1225).
H. rubiginosum. On wood. **2:**GA(1438). **15:**GA(1438). **18:**IA(1225).
H. serpens var. *effusum*. On wood. **18:**IA(1225).
H. truncatum. On wood. **2:**IA(1225).
Hysteropatella prostii. **1:**GA(1438). **2:**GA(1438). **16:**GA(1438). **18:**OK(660).
Kluyveromyces sp. Isolated from slime flux. **18:**LA(658).
Lachnella tiliae. On bark. **18:**E & C states(456).
Lophiostoma asperum. **2:**OK(2872).
Massaria ulmi. **15:**GA(1438).
Massariovalsa sudans. On limbs. **2:**MI(82), PA(82).
Microsphaera neglecta. **2:**IL(366). **15:**MO(366).
M. penicillata. Powdery mildew. **2:**IA(82), IL(82), MS(82,2720), OH(82). **9:**GA(1438). **18:**MS(2720).
Mycosphaerella punctiformis. **2:**GA(1438). **14:**GA(1438).
M. ulmi. Leaf spot. **1:**MS(2720). **2:**E states(82), GA(1438), KS(82), MS(2720), NC(3944), WI(82). **6:**E states(82), KS(82), WI(82). **12:**CT(82), NY(82). **15:**E states(82), KS(82), MS(2720), NC(3944), OK(2872), WI(82). **17:**IL(82), MS(82,2720), NC(82), TX(82).
Mycosphaerella sp. Leaf spot. **1:**GA(1438). **16:**GA(82,1438).
Mycothyridium marylandicum. **18:**MD(2777).
Nectria aurantiaca. **18:**WA(3202).
N. cinnabarina. Twig blight, coral spot. 'Nectria canker.' **2:**Range of host(82), FL(35). **5:**OK(2872). **9:**WA(3202). **12:**MA(82), NJ(82), NY(82). **14:**Range of host(82), OK(2872), OR(3202), WA(3202). **15:**MO(82). **18:**OK(2872).
N. coccinea. 'Nectria canker.' **2:**MA(82), NJ(82), NY(82). **18:**WA(3202).
N. flammea. **1:**GA(1438).
N. galligena. 'Nectria canker.' **2:**E states(82).
Nectriella pironii. Stem canker. **2:**FL(35).
Nummularia signata. On trunks and larger limbs. **18:**SE states(1810).
Ophiostoma ulmi. 'Dutch elm disease.' **1:**AL(405), AR(3922), LA(21), MS(20,3947), NC(1303,1320,1869), SC(3934), VA(3493). **2:**NC states(82), NE states(82), AR(3922,3923,3966), AZ(3966), CO(82,865,1614,2060,3966), CT(1616), DE(1614,1615,1616), GA(3073), IA(1616,3966), ID(2722,2723,3202,3966), IL(479,480,514,1614,1615,1616,3492), KS(1616,1968,2685,3966), KY(1614,1616,1776,1779,1780), LA(21), MA(1614,1781), MD(1614,1616,1729), ME(1614,1615,1616), MI(1479,1614,1615,1616,3966), MN(1114,3156), MO(1614), MS(1616,3947), MT(582,3966), NC(1303,1320,3492), ND(3155,3966), NE(3966), NH(1614,1615,1616), NJ(1616,3492), NM(3966), NY(1614,1616), OH(1616), OK(2733,3610,3966), OR(2725), PA(1614,1616), RI(1616), SD(2243,2919,3966), TN(1616,2942,2943), TX(806,3966), UT(142,3966), VA(1614,1615,1616,3492,3493), VT(1614,1615,1616), WI(1614,1616), WV(1614,1615,1616), WY(386,3966). **7:**LA(21). **12:**CA(1807), CT(82), NY(82). **14:**CO(2060), MN(141), NY(82). **15:**IN(82), LA(21), NC(1303), NY(82), OH(82), WI(1360). **18:**AR(1342,3920), CA(2851), ID(2724), KS(3912,3913), KY(1777,1778), NE(2752).
Phyllactinia angulata. Powdery mildew. **1:**GA(1441).
P. guttata. 'Powdery mildew.' **1:**SC states(82), SE states(82), GA(1438), IL(82), NC(1303,3944), SC(634). **2:**SC states(82), SE states(82), FL(35), IA(82), MS(2720). **11:**NC(1303). **12:**NC(82). **17:**IA(82).
Physalospora abdita. On limbs. **2:**AL(82).
Platystomum compressum. **18:**OK(660).
Pleospora sp. **2:**GA(1438).
Proliferodiscus pulveraceus. **14:**GA(1438). **16:**GA(1438).
Rebentischia massalongii. On limbs. **2:**MA(186).
Rhytisma ulmi. Anthracnose. **2:**MN(82), NJ(82), PA(82). **11:**FL(35).
Saccharomyces sp. Isolated from a slime flux. **18:**LA(658).
Splanchnonema foedans. On limbs. **18:**MA(188).
Stegophora ulmea. Anthracnose. 'Black spot.' **1:**OK(2873). **2:**Range of host(82), FL(35), GA(1438,2405), IL(1052), IN(987), MS(2720,3780), NC(1303,3944), ND(377), OK(618,2872), SC(1533), SD(2243), TN(1533), VA(3493). **7:**LA(82), MS(82,2720), TX(82). **10:**WI(1358). **11:**FL(35), WI(1358). **12:**NJ(82). **14:**Range of host(82), FL(35), MS(2720), OK(2873), SD(2243), WI(1358). **15:**Range of host(82), WI(1342). **16:**MS(82,2720). **17:**IL(82), OK(82), TX(82), VA(82). **18:**AL(285), MI(665), MS(2720), NC(1303), OK(2872).
Taphrina ulmi. 'Leaf blister.' **1:**AL(82,2454), AR(82,2454), KY(1967). **2:**E states(82), KS(1967,3361), MS(82,2720), OK(82,2873), WI(82). **7:**LA(1967), TX(2457). **12:**MA(82), WI(82). **14:**FL(35), OK(2872). **15:**KS(82,1967,2454,2458), OK(2872), VA(82,2454). **17:**KS(2457).
Thyridium vestitum. **2:**IA(1226).
Thyronectria chlorinella. On inner bark. **2:**AL(82), NC(82,3173).
T. chrysogramma. **2:**KS(82,3173). **18:**NY(3173).
Uncinula macrospora. 'Powdery mildew.' **1:**GA(1438), IL(2568), MS(2720), NC(1303,1533,3944), OK(2872). **2:**Range of host(82), FL(35), MS(2720), ND(377), OK(2873). **7:**TX(82). **15:**NC states(82), NE states(82), MO(82), OK(2872). **17:**Range of host(82). **18:**NC(1303).
U. parvula. Powdery mildew. **1:**GA(2405).
Uncinula sp. Mildew. **18:**AL(285).
Ustulina deusta. White sapwood rot. **2:**MD(82).
Valsa ambiens. On dead twigs. **1:**GA(1438). **2:**Range of host(82), IA(1226), ND(377). **15:**IA(1226).
V. ceratosperma. **2:**IA(1226).
V. modesta. **2:**IA(1226).
V. sordida. Canker. **2:**MN(82), OK(618). **14:**OK(618).
Valsaria exasperans. **2:**IA(1226).
Xylaria hypoxylon. Root rot. **2:**IL(82), MS(82), VA(82).
X. mali. Root rot. **2:**IL(82), MS(82), VA(82).
X. polymorpha. Root rot. **2:**IL(82), MS(82), VA(82). **18:**MS(2720).
Xylaria sp. Root rot. **2:**IL(82), MS(82), VA(82).

Basidiomycotina - Other

Abortiporus biennis. **18:**OR(3202).
Aleurodiscus oakesii. On bark. **2:**IA(82), MO(82), NY(82). **18:**NC(2083).
Armillaria mellea. 'Shoestring root rot.' **2:**CA(2891,2898), MO(82). **11:**CA(2900). **18:**OK(2872).
A. tabescens. Root rot. **2:**FL(35), GA(2684). **11:**FL(35), MS(1055). **14:**FL(35), OK(618).
Auricularia auricula. On wood. **18:**GA(914).
Bjerkandera adusta. White rot. **2:**IN(82), NY(82), VA(82), WI(82). **14:**OK(618).
B. fumosa. White rot. **2:**IN(82), MA(82), NY(82), OH(82), WI(82).
Ceratobasidium anceps. On seedlings. **2:**WI(1359).
Cerrena unicolor. Butt rot. **2:**NY(82), VT(82).
Chondrostereum purpureum. **2:**Unk.(82), CT(543), ME(543). **18:**NY(543).
Clavicorona pyxidata. **18:**OH(884).
Clitocybula familia. At base of trees. **2:**KS(3193).
Corticium roseum. **18:**AZ(128).
Crepidotus cinnabarinus. On wood. **2:**IA(2192).
C. subluteus. On wood. **18:**MI(1544).
C. subnidulans. On wood. **2:**OK(618).
Cylindrobasidium laeve. **2:**VT(128).
Cystostereum australe. White rot. **18:**FL(2554).
C. murrayi. **2:**NY(543).
Dacrymyces minor. On wood. **18:**IA(128).
Daedaleopsis confragosa. White rot. **2:**Range of host(82). **17:**TN(82).
Datronia scutellata. **2:**VA(82).
Dendrophora versiformis. **2:**NY(543).
Dendrothele candida. On bark. **1:**LA(2084), NC(2084).
D. griseo-cana. On bark. **2:**IA(82), MO(82), NY(82), OH(653). **18:**NY(2084).
D. microspora. On bark. **2:**CT(2084), NC(2084).
Dichostereum pallescens. **2:**VT(128).
Efibulobasidium rolleyi. On wood. **18:**IA(3852).
Exidiopsis glaira. **18:**IA(128).
E. podlachica. On wood. **18:**GA(2642).
Flammulina velutipes. Brown heart rot. 'Decay.' **2:**Range of host(82). **14:**WA(3202).
Fomes fasciatus. On wood. **2:**TX(82). **18:**LA(2675), MS(2720).
Fomitella supina. **18:**LA(2675).
Ganoderma applanatum. White-mottled butt rot. 'Decay.' **2:**ME(82), MI(82), NY(82), OH(82).
G. lucidum. White spongy heart rot. **2:**NY(82), OH(82), OR(3202).
Gloeocystidiellum clavuligerum. White rot. **2:**MN(2553).
Gloeophyllum sepiarium. On wood. **2:**OK(618).

Gloeoporus dichrous. **18:**AR(2675), LA(2675).
Grifola frondosa. Wood rot. **2:**CT(82), OH(82).
Helicobasidium brebissonii. 'Violet root rot.' **2:**TX(82).
Hexagonia hydnoides. **18:**LA(2675).
Hyphoderma setigerum. **18:**LA(2675).
Hypsizygus ulmarius. Brown ring rot. **2:**Range of host(82), SD(2243). **18:**TN(2421), VA(2421), WA(3202), WV(2421).
Inonotus cuticularis. **14:**AZ(1181). **18:**AR(2675), LA(2675).
I. dryadeus. Wood rot. **2:**NY(82).
I. munzii. **14:**AZ(1202).
Irpex lacteus. On wood. **2:**NY(82).
Ischnoderma resinosum. Wood rot. **2:**ME(82), VA(82).
Junghuhnia nitida. **18:**GA(2168), NY(2168).
Laetiporus sulphureus. Wood rot. **2:**Unk.(82).
Laxitextum bicolor. **2:**CT(543). **18:**AR(2675), LA(2675).
Lentinellus vulpinus. On wounds of living trees. **18:**N states(2423).
Lenzites betulina. On wood. **2:**IN(82), MA(82), MD(82).
Lopharia cinerascens. White rot. **2:**Unk.(82), CT(543), NY(543), OH(543), RI(543). **8:**FL(35). **18:**NY(543), OH(543), OR(3202,4007).
Marasmius sp. On outer bark of living trees. **2:**Unk.(82).
Mycena ulmicola. On wood. **18:**NC(3326).
Mycorrhaphium adustum. On dead limbs. **2:**NC(1303).
Oxyporus latemarginatus. On wood. **18:**LA(2166).
O. populinus. White spongy rot. **2:**NE states(82). **15:**CT(82), MA(82).
Pellicularia koleroga. **18:**LA(3640).
Perenniporia fraxinea. **2:**Gulf states(82), AR(82).
P. fraxinophila. **2:**Gulf states(82), AR(82).
P. medulla-panis. Wood rot. **2:**MI(82).
Phaeosolenia brenckleana. On wood. **18:**KS(646), ND(646).
Phanerochaete affinis. White rot. **2:**WI(433).
P. burtii. White rot. **1:**FL(433).
P. chrysosporium. **2:**NY(433,435).
P. corymbata. White rot. **15:**LA(433).
P. filamentosa. White rot. **2:**NY(433), TN(433).
P. rimosa. White rot. **2:**WI(433).
Phellinus everhartii. **15:**WV(82).
P. gilvus. White rot. **2:**NY(82), WI(82). **14:**OK(82,2873). **18:**MS(2720).
P. igniarius. White spongy heart rot. **2:**NE states(82), NY(2456), WI(82). **17:**TN(82).
Phlebia rufa. **18:**NJ(128).
P. tremellosus. **18:**NY(128).
Pleurotus ostreatus. White sapwood rot. **1:**GA(1439). **2:**Range of host(82), MN(63,3774), OH(63), SD(2243).
Polyporus admirabilis. Wood rot. **2:**NY(82).
P. arcularius. On wood. **2:**OK(618).
P. squamosus. White spongy heart rot. 'Decay.' **2:**NC states(82), NE states(82).
Psathyrella incerta. On roots around stump. **2:**NY(3491).
Rhodotus palmatus. On wood. **2:**MI(2422).
Rigidoporus ulmarius. On wood. **2:**Gulf states(82), AR(82), FL(35). **18:**LA(2675), MS(2675,2720).
Schizophyllum commune. On wood. **2:**Range of host(82), MS(2720). **14:**CA(82).
Sebacina sublilacina. On wood. **18:**IA(128).
Septobasidium pseudopedicellatum. Brown felt, parasitic on scale insects. **1:**NC(3944). **2:**FL(35), NC(82,3944). **17:**NC(82).
Spongipellis delectans. Wood rot. **2:**NY(82).
S. spumeus. White sapwood rot. **2:**NY(82), OH(82).
Steccherinum laeticolor. On wood. **2:**AL(1175). **18:**MI(1175).
Stereum hirsutum. **2:**CT(543). **18:**OH(543).
S. ostrea. White rot. **2:**MA(543), OK(82,2873), WA(543).
Thanatephorus cucumeris. **2:**OR(3202). **14:**OR(3202).
Tomentella botryoides. **18:**NY(2039).
T. lateritia. **18:**IL(2039).
T. pilosa. **18:**NY(2039).
T. punicea. **18:**IL(2039).
T. umbrinospora. **18:**NY(2039).
Trametes conchifer. On wood. **1:**GA(491,1439), NC(3944). **2:**Range of host(82), GA(491,1439), MA(656), MT(3202,3832), NC(1303,3944), NY(2456), OK(2872). **18:**NY(2456).
T. elegans. On wood. **18:**OK(2872).
T. hirsuta. Wood rot. **2:**MO(82), NY(82).
T. maxima. **18:**FL(2681), LA(2675,2681).
T. versicolor. 'Decay.' **2:**NY(82), OK(618). **14:**OK(618).
Tulasnella bifrons. On wood. **18:**NC(2640).
T. pruinosa. **18:**IA(128).
Tyromyces chioneus. On wood. **2:**NY(82).
T. galactinus. Wood rot. **2:**ME(82), NY(82).
Wolfiporia cocos. On wood. **2:**WI(793).
Xylobolus subpileatus. **2:**Unk.(82).

Deuteromycotina - Hyphomycetes

Alternaria sp. Leaf spot. **11:**FL(35). **14:**FL(35).
Bispora occulta. Inside old fence post. **18:**KS(3149).
Botrytis sp. **18:**WA(3202).
Cephalosporium sp. Dieback, Dutch elm disease, wilt. **1:**MS(20). **7:**AR(1056), MS(1056). **12:**VA(3493). **18:**KY(1777).
Cephalothecium sp. **18:**OK(2872).
Cercospora sphaeriaeformis. Leaf spot. **2:**FL(35), LA(82), TX(82). **14:**FL(35).
Chalara thielavioides. Seedling root rot. 'Chalara root rot.' **14:**W states(82), NJ(82), NY(82), OK(2872), WY(82).
Cladosporium herbarum. Sooty mold associated with scale insect infestation. **2:**CA(2362). **11:**CA(2362).
Conoplea olivacea. **18:**WI(1672).
Cristulariella moricola. Leaf spot. **11:**FL(35).
Cylindrocladium scoparium. Leaf spot. **11:**FL(35).
Dendrodochium sepultum. On dead limbs. **15:**KS(971).
Dictyodesmium ulmicola. Leaf spot. **15:**KS(82), MS(2720), NE(82), SD(2243). **18:**MS(2720).
Fumago sp. On leaves. **2:**IL(1052).
Fusarium equiseti. On bark. **2:**CT(82), IL(82), MA(82).
F. oxysporum. On bark. **2:**CT(82), IL(82), MA(82).
Fusarium sp. Root rot. **2:**OR(3202). **11:**FL(35). **14:**OR(3202).
Helicoma limpidum. On wood. **18:**OH(1271).
H. minutissimum. On bark. **18:**MA(1271).
Oidium candicans. On wood. **18:**MA(2131).
Paradidymobotryum oblongum. On rotten wood. **2:**NY(3772).
Phaeostalagmus altissimus. On rotten wood. **2:**NY(3771).
Phymatotrichopsis omnivora. 'Phymatotrichum root rot.' **2:**TX(82). **7:**TX(82). **14:**AZ(82), OK(82,2872), TX(82). **17:**TX(82). **18:**OK(2872).
Rhinocladium dingleyae. On rotten wood. **2:**NY(3771).
Spadicoides catenulata. On rotten wood. **2:**NY(3771).
Stigmina compacta. 'Thyrostroma canker.' **14:**IL(82).
Thielaviopsis basicola. Root rot. **2:**MD(1867).
Tubercularia grayana. On limbs. **2:**NY(3178).
T. hymenuloides. On twigs. **14:**OK(82).
T. nigricans. On dead twigs. **14:**MD(82), ND(82).
T. ulmea. 'Tubercularia canker.' **14:**IL(82), MN(1986), MT(1986), ND(1986), SD(1986).
T. vulgaris. **18:**NY(2456).
Tubercularia sp. Canker. **2:**OK(618).
Verticillium albo-atrum. 'Verticillium wilt.' **2:**IL(992,1558), KS(1558,3022), VA(3493). **3:**IL(992). **4:**IL(992). **11:**IL(992,1558). **12:**CT(82), IL(992,1558), MA(82), NJ(82), NY(82). **15:**IL(992,1558), PA(82), WI(82).
V. rhizophagum. On roots, probably secondary, on trees injured by phloem necrosis. **2:**OH(82).
Verticillium sp. Wilt. **1:**MS(20). **2:**NE states(82), OR(82,3202), WI(82).

Deuteromycotina - Coelomycetes

Actinopelte sp. Leaf spot. **11:**FL(35).
Aplosporella ulmea var. *rhizophila*. On dead, exposed roots. **18:**NE(3209).
Asteroma inconspicuum. Twig blight, leaf spot. 'Anthracnose.' **1:**OK(2872). **2:**E states(82), MN(82), OK(82,618,2872). **12:**MA(82). **14:**OK(2872). **18:**OK(2872).
A. ulmeum. Anthracnose, leaf spot. **2:**FL(35), NC(3944), OK(2872). **11:**FL(35). **14:**FL(35). **18:**NC(1303).
Colletotrichum sp. Anthracnose. **2:**FL(35). **11:**FL(35). **14:**FL(35).
Coniothyrium celtidis-australis. Leaf spot, on living leaves. **2:**OH(82), OK(2872). **6:**TX(3597), WI(1361). **7:**OK(2872). **11:**NC(1303). **12:**TX(82). **14:**WV(82).
C. fuckelii. **11:**WI(1342).
C. wisconsinensis. On living leaves. **2:**WI(1361).
Coniothyrium sp. Stem and twig canker, dieback. **2:**IL(82), MA(82), MI(82), WI(82,1360). **11:**FL(35).
Coryneum sp. Leaf spot. **18:**OK(2872).

Cryptostictis insidens. **2**:NY(1393).
Cylindrosporium tenuisporum. Leaf spot. **1**:TX(652). **7**:TX(82,652). **17**:TX(82).
C. ulmicola. **2**:Unk.(82).
Cytospora carbonacea. Dieback. **2**:MA(82), TX(82).
C. leucosperma. On twigs. 'Cytospora canker.' **2**:Range of host(82), MA(1391).
Cytospora sp. Canker, dieback. **2**:Range of host(82). **11**:FL(35), TX(3780), WY(3780). **14**:W states(82), OR(82,3202). **18**:TX(3780), WY(3780).
Cytosporina ludibunda. 'Cytosporina canker.' **2**:IL(82), KS(82).
Dinemasporium decipiens. **18**:OK(660).
Diplodia ulmi. On cankers. **2**:IL(82).
Diplodia sp. Twig dieback, on stem galls. **2**:Unk.(82). **7**:TX(82). **11**:FL(35).
Dothiorella ulmi. Dieback, wilt. 'Dothiorella canker and wilt.' **2**:NE states(82), Range of host(82), MT(82,3202), OK(82,2872), VA(3493). **15**:NE states(82), OK(82,2872).
Gloeodes pomigena. On twigs. **15**:IN(82).
Gloeosporium ulmicola. Leaf spot, on fruits and leaves. 'Anthracnose.' **2**:CT(82), IL(82,1052), IN(82), WI(82). **11**:KS(3780), TN(3780). **13**:IL(50). **14**:CT(82), GA(82), KS(82), OK(2872), TN(82), TX(82). **15**:NY(82). **17**:IL(50).
Libertella sp. Canker. **11**:CO(3780). **14**:CO(82).
Macrophoma sp. On stem galls. **11**:FL(35).
Melasmia samaricola. On green samaras. **10**:WI(1354).
M. ulmicola. On leaves. **2**:NJ(82). **18**:WI(1361).
Monochaetia monochaeta. Zonate leaf spot. **1**:GA(1509). **17**:GA(82).
Myxosporella ulmicola. On dead twigs. **2**:WV(128,331).
Phloeospora ulmi. Leaf spot. **1**:NC(3944). **2**:NC(3944). **15**:NC(3944). **18**:OK(2872).
Phoma cincta. On leaves. **2**:SC(82).
Phoma sp. Twig canker, stem gall. **2**:IL(82). **11**:FL(35).
Phomopsis oblonga. Twig blight, dieback. 'Phomopsis canker.' **2**:NJ(3780). **12**:MA(82).
Phomopsis sp. Twig blight and canker, dieback. **2**:E states(82), IL(82), MN(82). **11**:FL(35), VA(3493). **14**:WI(82).
Phyllosticta confertissima. Leaf spot. **2**:PA(82), WI(1350). **15**:KS(82).
P. erratica. Leaf spot. **2**:AL(82), OK(2872), TX(82). **15**:AL(82).
P. melaleuca. Leaf spot. **1**:MS(2720). **2**:E states(82), NC(1303), OK(82,2872), WI(82). **12**:VT(82). **15**:IN(82), MI(82), WV(82). **17**:MS(82,2720), WI(82).
P. ulmi-rubrae. On living leaves. **15**:WI(1358).
Phyllosticta sp. Blight leaf spot. **2**:WI(1344). **11**:FL(35). **14**:MS(2720), NJ(82), SD(2243). **15**:OK(2872), WI(1354). **18**:KY(1777).
Sacidium ulmi-gallae. On leaf galls. **2**:KS(82), TX(82), WI(82). **14**:TX(82). **15**:WI(82).
Seimatosporium pestalozzioides. Leaf spot. **2**:NY(82).
Septogloeum parasiticum. Twig blight. **2**:MI(82).
S. profusum. Leaf spot. **2**:IL(2568), MS(82,2720). **17**:AL(82).
Sphaeropsis ulmicola. Twig canker. **2**:OK(2873). **11**:VA(3493). **12**:PA(82). **14**:CT(82), MN(1986), MT(1986), ND(1986), OK(618), SD(1986,2668). **18**:OK(2872).
Sphaeropsis sp. Twig canker, stem gall. **1**:NC(1303). **2**:E states(82), MS(82), OK(82), WI(82). **11**:FL(35).
Tubakia dryina. On leaves. **1**:IL(1246). **17**:IL(82).

Deuteromycotina - Other

Rhizoctonia solani. Root rot. 'Rhizoctonia damping-off.' **2**:Range of host(82), FL(35), OK(2873). **14**:FL(35), OK(2873).

Zelkova Spach

1. *Z. carpinifolia* (Pall.) K. Koch [elmleaf zelkova]
2. *Z. serrata* (Thunb.) Makino [Japanese zelkova, saw-leaf zelkova]
3. *Zelkova* sp.

Basidiomycotina - Other

Armillaria mellea. Root rot. **2**:CA(2897).

Deuteromycotina - Hyphomycetes

Cercospora zelkowae. Leaf spot. **3**:FL(35).

Deuteromycotina - Coelomycetes

Melasmia ulmicola. **1**:WI(1355).

Urticaceae

Boehmeria Jacq.

1. *B. cylindrica* (L.) Sw. [bog hemp, false nettle]
2. *B. nivea* (L.) Gaud. [China-grass, Chinese silk-plant, ramie]

Chytridiomycetes

Synchytrium cellulare. Leaf gall. **1**:WI(82).

Ascomycotina

Acrospermum compressum. **1**:GA(1438).
Calloria neglecta. **1**:GA(1438).
Hymenoscyphus scutula. **1**:GA(1438).
Leptosphaeria doliolum. **1**:GA(1438).
Metasphaeria sp. **1**:GA(1438).

Basidiomycotina - Uredinales

Aecidium boehmeriae. **1**:IN(82), MD(82), NY(82).

Deuteromycotina - Hyphomycetes

Alternaria sp. Leaf spot. **1**:FL(35).
Cercospora boehmeriae. Leaf spot. **1**:AL(82), DE(3543), NC(82,3944), NE(82), NY(82), PA(82), TN(82), WI(82), WV(82).
Fusarium sp. Leaf spot. **1**:FL(35).

Deuteromycotina - Coelomycetes

Phoma exigua. **1**:WI(1348).
Phyllosticta boehmeriicola. Leaf spot. **1**:WI(82).
Phyllosticta sp. Leaf spot. **2**:FL(35).
Septoria tenuissima. Leaf spot. **1**:KS(82), MO(82).

Laportea Gaud. in Freyc.

1. *L. aestuans* (L.) Chew
2. *L. canadensis* (L.) Wedd. (=*Urtica canadensis* L.)

Ascomycotina

Cistella grevillei. On stems. **2**:NY(849).

Deuteromycotina - Hyphomycetes

Corynespora cassiicola. Leaf spot. **1**:FL(35).

Deuteromycotina - Coelomycetes

Colletotrichum urticae. **2**:WI(1354).
Epistigme laporteae. Leaf spot. **2**:OK(2872).
Phyllosticta boehmeriicola. **2**:WI(1338).
Septoria urticae. Leaf spot. **2**:OK(2873).

Parietaria L.

1. *P. floridana* Nutt.
2. *P. pensylvanica* Muhl. ex Willd. [Pennsylvania pellitory]

Oomycetes

Plasmopara illinoensis. Downy mildew. **2**:WI(82).

Ascomycotina

Erysiphe cichoracearum. Powdery mildew. **1**:FL(35). **2**:C states(82), MT(3202), OR(3202).

Deuteromycotina - Hyphomycetes

Ramularia parietariae. Leaf spot. **2**:MO(82).

Deuteromycotina - Coelomycetes

Septoria parietariae. Leaf spot. **2**:IA(82), SD(2243), WI(82).

Pellionia Gaud. in Freyc.

1. *P. daveauana* (Godefroy-Leb.) N. E. Brown [trailing watermelon-begonia]
2. *P. pulchra* N. E. Brown [rainbow-vine, satin pellionia]

Oomycetes

Pythium splendens. Root rot. **2**:FL(35).

Deuteromycotina - Hyphomycetes

Fusarium sp. Leaf necrosis. **2**:FL(35).
Oidium sp. Powdery mildew. **2**:FL(35).

Deuteromycotina - Coelomycetes

Phyllosticta sp. Leaf spot. **1**:FL(35). **2**:FL(35).

Deuteromycotina - Other

Rhizoctonia solani. Leaf rot. **2**:FL(35).
Sclerotium rolfsii. Southern blight. **2**:FL(35).

Pilea Lindl.

1. *P. cadierei* Gagnep. & Guillaumin [aluminum-plant, watermelon pilea]
2. *P. depressa* (Sw.) Blume
3. *P. involucrata* (Sims) Urb. [friendship-plant, panamiga]
4. *P. microphylla* (L.) Liebm. [artillery-plant, artillery-weed]
5. *P. nummulariifolia* (Sw.) Wedd. [creeping-Charlie]
6. *P. pumila* (L.) A. Gray [clearweed, coolwort, richweed]

7. *P. spruceana* Wedd.
8. *Pilea* sp.

Oomycetes

Phytophthora nicotianae var. *parasitica*. Bud, leaf and stem blight. 'Phytophthora blight.' **4:**FL(35). **8:**FL(2308).
Pythium aphanidermatum. Stem rot. **4:**FL(35).
P. dissotocum. Rootlet rot. **6:**MD(82).
P. splendens. Cutting and root rot. **1:**FL(35). **2:**FL(35). **3:**FL(35). **4:**FL(35). **5:**FL(35). **7:**FL(35).
Pythium sp. Root rot. **4:**HI(2903). **5:**HI(2903).

Ascomycotina

Calloria neglecta. **6:**GA(1438).
Erysiphe cichoracearum. Powdery mildew. **1:**FL(35). **6:**IA(82), IL(82), IN(82).

Deuteromycotina - Hyphomycetes

Alternaria sp. Leaf necrosis. **1:**FL(35). **3:**FL(35). **7:**FL(35).
Botrytis cinerea. Gray mold. **1:**FL(35). **3:**FL(35). **4:**FL(35). **5:**FL(35).
Cercospora pileae. Leaf spot. **1:**FL(35). **4:**FL(35).
Corynespora cassiicola. Leaf spot. **1:**FL(35).
Curvularia sp. Leaf spot. **2:**HI(2903).
Fusarium sp. Root and stem rot. **1:**FL(35). **3:**FL(35). **5:**FL(35).
Helminthosporium sp. Leaf spot. **1:**FL(35).
Myrothecium sp. Leaf necrosis. **1:**FL(35). **3:**FL(35).

Deuteromycotina - Coelomycetes

Colletotrichum gloeosporioides. Anthracnose. **4:**HI(2903).
Colletotrichum sp. Anthracnose. **3:**FL(35). **5:**FL(35).
Phyllosticta sp. Leaf spot. **1:**FL(35).
Septoria pileae. Leaf spot. **4:**IA(82), IL(82), IN(82), MI(82), MO(82), NY(82), WI(82). **6:**IA(82), IL(82), IN(82), MI(82), MO(82), NY(82), WI(82,1351). **8:**IA(82), IL(82), IN(82), MI(82), MO(82), NY(82), WI(82).

Deuteromycotina - Other

Rhizoctonia solani. Cutting, root and stem rot. **1:**FL(35). **2:**FL(35). **3:**FL(35). **4:**FL(35). **5:**FL(35). **7:**FL(35).
Rhizoctonia sp. Root rot. **2:**HI(2903). **5:**HI(2903).
Sclerotium rolfsii. 'Southern blight.' **1:**FL(35). **8:**FL(35).

Pipturus Wedd.

1. *P. albidus* (Hook. & Arnott) A. Gray
2. *Pipturus* sp.

Oomycetes

Phytophthora cinnamomi. Root rot. **1:**HI(2903).

Ascomycotina

Asteridiella pipturi. Black mildew. **1:**HI(1272,1452,2903).
A. triloba. Black mildew. **1:**HI(82).
Botryosphaeria ribis. On limbs. **1:**HI(82,2903).
Diaporthe sp. On limbs. **1:**HI(2903).
Lachnella villosa. On dead limbs. **1:**HI(2903).
Rhynchosphaeria shearii. On dead leaves. **1:**HI(2781).
Seuratia millardetii. On leaves. **1:**HI(2337,2784).

Basidiomycotina - Other

Armillaria mellea. **1:**HI(2011,2903).

Deuteromycotina - Hyphomycetes

Cercospora pipturi. Leaf spot. **1:**HI(82,2903).
Chaetotrichum sandwicense. On living leaves. **2:**HI(2784).

Soleirolia Gaud. in Freyc.

1. *S. repens* Kuntze
2. *S. soleirolii* (Req.) Dandy [angel's-tears, baby's-tears, Corsican carpet-plant, Corsican-curse, Irish-moss, Japanese-moss, mind-your-own-business-plant, peace-in-the-home, pollyana-vine]

Oomycetes

Pythium splendens. Foliar blight, basal stem rot, root rot. **2:**FL(35).
Pythium sp. Root rot. **1:**HI(2903).

Deuteromycotina - Other

Rhizoctonia solani. Basal stem rot. **2:**FL(35).
Rhizoctonia sp. Root rot. **1:**HI(2903).

Urtica L.

1. *U. cardiophylla* Rydb.
2. *U. chamaedryoides* Pursh
3. *U. dioica* L. (=*U. procera* Muhl. ex Willd.) [European nettle, stinging nettle]
4. *U. dioica* L. subsp. *gracilis* (Aiton) Selander (=*U. gracilis* Aiton)
5. *U. gracilenta* Greene
6. *U. holosericea* Nutt. (=*U. gracilis* Aiton var. *holosericea* Jeps.)
7. *U. lyallii* S. Wats.
8. *U. urens* L. [burning nettle, dog nettle]
9. *Urtica* sp.

Chytridiomycetes

Synchytrium chamaedryoidis. Galls on flowers, leaves and stems. **2:**LA(626), TX(1837).

Oomycetes

Peronospora debaryi. **4:**WI(1338).
Pseudoperonospora urticae. Downy mildew. **4:**WI(82).

Ascomycotina

Brunnipila clandestina. **9:**WA(3202).
Comoclathris permunda. **4:**ND(377).
Crocicreas cyathoideum. **4:**ND(377), SD(505).
Dasyscyphus spiraeicolus. **9:**WA(3202).
Didymella eupyrena. **7:**AK(526), WA(3202).
Erysiphe cichoracearum. Powdery mildew. **4:**IL(82). **9:**CA(82).
Leptosphaeria acuta. On stems. **3:**NY(3249). **4:**CA(82). **7:**AK(526), CA(3249), WA(3249). **9:**AK(82), CA(82), WA(82,3202,3249).
L. conoidea. **4:**ND(377).
Mollisia urticae. At base of living stems. **7:**CA(527).
Mycosphaerella superflua. On stems. **4:**CA(82), MT(82,3202). **6:**CA(2439).
Nectria tubercular iformis. On dead stems. **4:**ND(377). **9:**ND(3084).
Sclerotinia sclerotiorum. Stem rot. **9:**CA(82).
Sclerotinia sp. Stem rot. **9:**FL(35).

Basidiomycotina - Uredinales

Aecidium libertum. **2:**OK(2872). **9:**OK(82).
Puccinia caricina. Heteroecious: aecial host. **1:**MT(3202). **3:**Range of host(82), CO(1195). **6:**CA(81,661). **8:**OK(2873). **9:**Range of host(82), AK(82), FL(35).

Basidiomycotina - Other

Ceratobasidium anceps. On leaves. **3:**WI(128).

Deuteromycotina - Hyphomycetes

Cercospora sp. Leaf spot. **3:**FL(35).
Dendryphiella vinosum. On stems. **4:**ND(82,378), NY(82), TX(82).
Macrosporium toruloides. **7:**WA(3202).
Phymatotrichopsis omnivora. Root rot. **3:**TX(82). **4:**TX(82). **9:**TX(82).
Ramularia urticae. Leaf spot. **3:**SD(2243), WI(1333), WY(82). **4:**E states(82), SC states(82), CA(82), MT(3202), ND(378), WA(82). **6:**WA(3202). **7:**AK(526), ID(3202), OR(3202), WA(3202). **9:**UT(82), WA(82).

Deuteromycotina - Coelomycetes

Colletotrichum urticae. On living leaves. **3:**WI(1354). **4:**WI(1346).
Cylindrosporium urticae. Leaf spot. **4:**WY(1195). **7:**WA(3202). **9:**WA(82).
Phoma exigua. **3:**WI(1353).
P. nebulosa. **3:**TX(82). **4:**TX(82). **6:**CA(2439). **9:**TX(82).
Phyllosticta boehmeriicola. **3:**WI(1357).
Septoria urticae. Leaf spot. **4:**IL(82), MN(82), MT(82,3202), SD(82), TX(82), WI(82,1330), WY(82). **5:**AZ(1195,3378). **7:**WA(3202). **9:**AK(82,526), LA(82), TX(82), WA(82).
S. urticaria. On living leaves. **2:**TX(3597). **9:**TX(82).

Deuteromycotina - Other

Rhizoctonia sp. Root and stem rot. **3:**FL(35).

Valerianaceae

Centranthus Neck. ex Lam. & A. DC.

1. *C. ruber* (L.) DC. [fox's-brush, Jupiter's-beard, red valerian]

Deuteromycotina - Hyphomycetes

Ramularia centranthi. Leaf spot. **1:**CA(82).

Plectritis (Lindl.) DC.

1. *P. macrocera* Torr. & A. Gray

Basidiomycotina - Uredinales

Aecidium valerianellae. **1:**CA(81).

Valeriana L.

1. *V. capitata* Pall. ex Link
2. *V. edulis* Nutt. ex Torr. & A. Gray (=*V. ceratophylla* (Hook.) Piper, *V. ciliata* Torr. & A. Gray, *V. furfurescens* A. Nels.) [edible valerian,

tobacco-root]
3. *V. occidentalis* A. Heller
4. *V. officinalis* L. [common valerian, garden heliotrope]
5. *V. sitchensis* Bong.
6. *V. sitchensis* Bong. var. *scouleri* (Rydb.) G. Mey.
7. *Valeriana* sp.
Chytridiomycetes
Synchytrium perforatum. **5**:MT(3202), WA(3202).
Oomycetes
Peronospora valerianae. **5**:MT(3202).
Ascomycotina
Apiosporella alpina. **5**:WA(3202,3815).
Erysiphe cichoracearum. Powdery mildew. **2**:CO(82), UT(82), WY(1195). **6**:CA(1155). **7**:CO(82).
Hyponectria sceptri. **5**:WA(3202,3815).
Leptosphaeria drabae. **5**:WA(3202,3814).
L. erigerontis. **5**:WA(3202,3814).
Lewia scrophulariae. **5**:WA(3202,3814).
Lophiostoma herbarum. **5**:WA(3202,3815).
Mycosphaerella delphiniicola. **5**:WA(3202,3815).
Pleospora helvetica. **5**:WA(3202,3814).
P. penicillus. **5**:WA(3202,3814).
Wettsteinina duplex. **5**:WA(3252).
Basidiomycotina - Uredinales
Aecidium sp. **2**:WY(1195).
Puccinia commutata. Autoecious. **1**:AK(526). **3**:OR(3202). **5**:OR(3202), WA(3202). **7**:AK(82), NY(82), OR(82).
P. dioicae. Heteroecious: aecial host. **3**:WY(1195). **7**:UT(82).
Deuteromycotina - Hyphomycetes
Cercoseptoria valerianae. **2**:WY(1195).
Cercospora sp. **2**:WI(1356).
Ramularia centranthi. Leaf spot. **4**:CA(82).
R. eamesii. **7**:NY(82).
R. valerianae. **4**:MI(82). **5**:WA(3202). **7**:MT(3202).
Deuteromycotina - Coelomycetes
Ascochyta valerianae. On leaves. **4**:WI(1341,1342).
Heteropatella conglutinata. **5**:WA(3202).
H. umbilicata. **5**:WA(3202,3816).
Leptothyrella graminis. **5**:WA(3816).
Phoma exigua. **4**:WI(1340).
P. pedicularis. **5**:WA(3202).
Phyllosticta sp. **2**:WI(1356).
Septoria valerianae. Leaf spot. **2**:WI(1332). **4**:WI(1342). **7**:WI(82).
Deuteromycotina - Other
Rhizoctonia solani. Root rot. **4**:NY(82).
Sclerotium rolfsii var. *delphinii*. Stem rot. **4**:CT(82), NJ(82).

Valerianella Mill.
1. *V. radiata* (L.) Dufr. [beaked cornsalad]
2. *Valerianella* sp.
Chytridiomycetes
Synchytrium aureum. Leaf gall. **1**:MS(82,2720).
Oomycetes
Peronospora sp. Downy mildew. **2**:OK(2872).
Deuteromycotina - Hyphomycetes
Phymatotrichopsis omnivora. Root rot. **2**:TX(82).
Deuteromycotina - Coelomycetes
Septoria valerianellae. Leaf spot. **1**:MS(2720), OK(2872). **2**:TX(82).

Verbenaceae

Aloysia Juss.
1. *A. gratissima* (Gillies & Hook.) Tronc. (=*Lippia lycioides* (Cham.) Steud., *A. lycioides* Cham.) [whitebrush]
Basidiomycotina - Uredinales
Prospodium lippiae. Autoecious. **1**:TX(742).

Avicennia L.
1. *A. germinans* (L.) L. (=*A. nitida* Jacq.) [black mangrove, blackwood]
Oomycetes
Pythium sp. Root rot. **1**:FL(35).
Ascomycotina
Hydronectria tethys. On dead roots. **1**:FL(1942).
Leptosphaeria avicenniae. In pneumatophores. **1**:FL(1943).
Mycosphaerella pneumatophorae. In pneumatophores. **1**:FL(1943,1945).
Deuteromycotina - Hyphomycetes
Fusarium sp. Root rot. **1**:FL(35).
Deuteromycotina - Coelomycetes
Phyllosticta hibiscina. Leaf spot. **1**:FL(35).
Rhabdospora avicenniae. In pneumatophores. **1**:FL(1943).

Callicarpa L.
1. *C. americana* L. [American beautyberry, French mulberry]
2. *C. dichotoma* (Lour.) K. Koch [purple beautyberry]
3. *Callicarpa* sp.
Chytridiomycetes
Synchytrium callicarpae. Leaf gall. **1**:LA(627).
Ascomycotina
Botryosphaeria callicarpae. On stems. **1**:GA(82), SC(82).
B. dothidea. **1**:GA(1438).
B. obtusa. On stems. **1**:SC(82). **2**:SC(82).
Didymella sp. **2**:OK(2872).
Meliola amphitricha. On leaves. **1**:FL(634).
M. cookeana. Black mildew. **1**:FL(35,82,2782), LA(82), TX(82). **2**:FL(82), LA(82), TX(82). **3**:FL(1452).
Nectria cinnabarina. On stems. **1**:AL(82). **2**:AL(82).
Valsa ceratosperma. **1**:GA(1438).
Basidiomycotina - Other
Mycena sp. **3**:TX(652).
Deuteromycotina - Hyphomycetes
Atractilina parasitica. Leaf spot. **1**:FL(35,82,832). **2**:FL(82).
Cercospora callicarpae. Leaf spot. **1**:SC states(82), SE states(82), FL(35), MS(2720), OK(2872). **2**:SC states(82), SE states(82). **3**:FL(634).
C. missouriensis. Leaf spot. **1**:FL(35).
Corynespora sp. Leaf spot. **1**:FL(35).
Deuteromycotina - Coelomycetes
Coniothyrium callicarpae. On stems. **1**:SC(82). **2**:SC(82).
Phyllosticta sp. Leaf spot. **1**:FL(35).

Clerodendrum L.
1. *C. bungei* Steud.
2. *C. thomsoniae* Balf. [bagflower, bleeding glory-bower, glory-tree, tropical bleeding-heart]
3. *Clerodendrum* sp.
Ascomycotina
Diaporthe sp. Stem canker. **1**:FL(35).
Nectriella pironii. Stem canker. **1**:FL(35). **3**:FL(35).
Deuteromycotina - Hyphomycetes
Cercospora apii f. sp. *clerodendri*. Leaf spot. **2**:FL(35). **3**:FL(35).
Corynespora sp. Leaf spot. **2**:FL(35).
Kutilakesa pironii. Stem cankers. **1**:FL(34).
Deuteromycotina - Coelomycetes
Dothiorella sp. Leaf spot. **3**:FL(35).
Phyllosticta sp. Leaf spot. **2**:FL(35).
Septoria phlyctanoides. Leaf spot. **3**:SC(82).

Duranta L.
1. *D. repens* L. [Brazilian skyflower, garden dewdrop, golden dewdrop, pigeonberry, skyflower]
Ascomycotina
Phyllachora fusicarpa. Anthracnose. **1**:FL(35,82).
Deuteromycotina - Hyphomycetes
Cercospora durantae. Leaf spot. **1**:FL(35).
Deuteromycotina - Other
Sclerotium rolfsii. Seedling blight. **1**:FL(35,82).

Junellia Mold.
1. *J. ligustrina* (Lag.) Mold. (=*Lippia ligustrina* (Lag.) Britton)
Deuteromycotina - Coelomycetes
Cylindrosporium lippiae. Leaf spot. **1**:TX(82).

Lantana L.
1. *L. camara* L. [lantana, yellow-sage]
2. *L. horrida* H. B. K.
3. *L. involucrata* L.
4. *L. montevidensis* (Spreng.) Briq. [pole-cat geranium, trailing lantana, weeping lantana]

5. *Lantana* sp.
Oomycetes
Pythium sp. Root rot. **5**:FL(35).
Ascomycotina
Botryosphaeria obtusa. **5**:AL(82).
Diaporthe sp. On limbs. **1**:HI(2903).
Diatrypella lantanae. On stems. **1**:FL(35). **5**:FL(82).
Godronia lantanae. On dead limbs. **1**:HI(2903,3167).
Lecanidion atratum. **5**:HI(2903).
Meliola cookeana. Black mildew. **1**:FL(35). **3**:FL(35). **4**:FL(35). **5**:FL(82).
Physalospora abdita. On stems. **1**:HI(2903). **5**:HI(82).
Schizoxylon insigne. On stems. **5**:HI(2903).
Basidiomycotina - Uredinales
Puccinia lantanae. Autoecious. **1**:FL(35). **3**:FL(35). **4**:FL(35). **5**:FL(35,82).
Deuteromycotina - Hyphomycetes
Alternaria sp. Leaf spot. **5**:FL(35), TX(82).
Fusarium sp. Wilt. **5**:NJ(82).
Mycovellosiella lantanae. Leaf spot. **1**:FL(35). **4**:FL(35). **5**:FL(35).
Phymatotrichopsis omnivora. Root rot. **2**:TX(82).
Ramularia sp. Leaf spot. **5**:FL(35).
Deuteromycotina - Coelomycetes
Asteromella lantanae. On dead twigs. **1**:HI(2784).
Colletotrichum sp. Leaf spot. **5**:FL(35).
Phyllosticta sp. Leaf spot. **5**:FL(35).
Deuteromycotina - Other
Rhizoctonia solani. Leaf necrosis, root rot. **1**:FL(35). **5**:FL(35).
Rhizoctonia sp. Root rot. **1**:HI(2903).
Sclerotium rolfsii. Basal stem rot. **3**:FL(35).

Lippia L.
1. *L. micromera* Schauer [Spanish thyme]
2. *L. stoechadifolia* H. B. K.
3. *Lippia* sp.
Ascomycotina
Meliola lippiae. Black mildew. **2**:FL(35). **3**:FL(82).
Basidiomycotina - Uredinales
Puccinia lantanae. Autoecious. **2**:FL(35).
Deuteromycotina - Hyphomycetes
Cercospora lippiae. Leaf spot. **3**:TX(82).
Phymatotrichopsis omnivora. Root rot. **3**:TX(82).
Deuteromycotina - Coelomycetes
Sphaceloma lippiae. **3**:FL(2525).
Deuteromycotina - Other
Sclerotium rolfsii. Southern blight. **1**:HI(2903).

Phryma L.
1. *P. leptostachya* L.
Basidiomycotina - Uredinales
Puccinia dioicae. Heteroecious: aecial host. **1**:NC states(82), NE states(82).
Deuteromycotina - Hyphomycetes
Cercosporella exilis. Leaf spot. **1**:WI(82).
Sclerotiomyces colchicus. **1**:WI(1356).
Deuteromycotina - Coelomycetes
Septoria leptostachya. Leaf spot. **1**:E & C states(82), MN(82), MS(2720), OK(2872), SD(2243).

Phyla Lour.
1. *P. lanceolata* (Michx.) Greene (=*Lippia lanceolata* Michx.) [northern frogfruit]
2. *P. nodiflora* (L.) Greene (=*Lippia canescens* H. B. K., *Lippia nodiflora* (L.) Michx.) [capeweed, frogfruit, mat-grass, turkey-tangle]
Ascomycotina
Meliola lippiae. Black mildew. **2**:AZ(358), FL(35).
Deuteromycotina - Hyphomycetes
Cercospora lippiae. Leaf spot. **1**:Range of host(82), IL(650), MS(2720), OK(2872). **2**:FL(35), LA(962).
Oidium sp. Powdery mildew. **2**:FL(35).
Deuteromycotina - Coelomycetes
Sphaceloma lippiae. Spot anthracnose, scab. **1**:IN(82,1745). **2**:FL(35,2525).

Deuteromycotina - Other
Sclerotium rolfsii. Southern blight. **2**:CA(82), HI(2299,2903).

Stachytarpheta Vahl
1. *S. jamaicensis* (L.) Vahl
2. *Stachytarpheta* sp.
Basidiomycotina - Uredinales
Puccinia urbaniana. Autoecious. **2**:FL(35).
Deuteromycotina - Hyphomycetes
Cercospora stachytarphetae. Leaf spot. **2**:FL(35,82).
Deuteromycotina - Coelomycetes
Phoma herbarum. **1**:HI(2903).
Deuteromycotina - Other
Rhizoctonia sp. Root rot. **1**:HI(2903).

Tectona L. fil.
1. *T. grandis* L. fil. [teak]
Deuteromycotina - Hyphomycetes
Cercospora tectoniae. Leaf spot. **1**:HI(82,2903).

Verbena L.
1. *V. bipinnatifida* Nutt. [Dakota vervain]
2. *V. bracteata* Lag. & Rodriguez (=*V. bracteosa* Michx.) [prostrate vervain]
3. *V. canadensis* (L.) Britton [clump verbena, creeping vervain, rose verbena, rose vervain]
4. *V. carnea* Medik.
5. *V. caroliniana* Hort. ex Schau
6. *V. ciliata* Benth.
7. *V. hastata* L. [blue verbena, blue vervain, simpler's-joy]
8. *V.* × *hybrida* Voss. Probably a hybrid of *V. peruviana* and other species. [garden verbena]
9. *V. peruviana* (L.) Britton
10. *V. prostrata* R. Br. in Aiton
11. *V. pumila* Rydb.
12. *V. scabra* Vahl
13. *V. simplex* Lehm. (=*V. angustifolia* Michx.)
14. *V. stricta* Venten. [hoary vervain, woolly verbena]
15. *V. urticifolia* L. [white verbena, white vervain]
16. *V. xutha* Lehm.
17. *Verbena* sp.
Oomycetes
Phytophthora sp. Crown and root rot. **8**:FL(35).
Plasmopara halstedii. Downy mildew. **17**:NM(82).
Pythium aphanidermatum. Root rot. **9**:HI(2903).
Pythium sp. Damping-off. **8**:FL(35).
Ascomycotina
Erysiphe cichoracearum. Powdery mildew. **2**:WA(3202). **7**:Range of host(82), MT(3202). **14**:Range of host(82), MS(2720), SD(2243). **15**:Range of host(82), NC(1303), OK(2872), SD(2243). **8**:Range of host(82), FL(35), NC(1303). **17**:Unk.(82), GA(1438).
Pleospora herbarum. On stems. **10**:CA(2439).
Stictis stellata var. *philippensis*. **9**:HI(2903).
Basidiomycotina - Uredinales
[*Puccinia aristidae*]. **6**:AZ(1203). **17**:AZ(82).
P. subnitens. **6**:AZ(3981).
P. vilfae. Heteroecious: aecial host. **2**:NE(82,753). **3**:OK(82,753,2872). **7**:C states(82), KS(753), SD(82). **13**:KS(753). **14**:C states(82), IL(753), KS(753), NE(753), OK(2872). **15**:C states(82), KS(753), NE(753), OK(2872). **17**:KS(82), MO(82).
Basidiomycotina - Other
Ceratobasidium anceps. **15**:WI(1354).
Deuteromycotina - Hyphomycetes
Alternaria sp. Leaf spot. **8**:FL(35).
Botrytis cinerea. Flower and stem blight. **8**:IL(312), MA(82).
Cercospora papillosa. Leaf spot. **17**:AL(581).
C. septatissima. Leaf spot. **4**:MS(2720). **5**:MS(581).
C. truncatella. **17**:AL(82).
C. verbenicola. Leaf spot. **1**:TX(82). **14**:IL(82), KS(82), WI(1334). **16**:LA(962). **17**:AL(82), LA(82), TX(82).
Cercospora sp. Leaf spot. **8**:FL(35).
Oidium sp. Powdery mildew. **17**:HI(2903).
Phymatotrichopsis omnivora. Root rot. **1**:TX(82). **8**:TX(82). **17**:TX(82).

Thielaviopsis basicola. Root rot. **8:**PA(82).
Deuteromycotina - Coelomycetes
Ascochyta verbenae. Leaf spot. **14:**WI(82,1332). **15:**WI(1358).
Ascochyta sp. On leaves. **15:**WI(1347).
Macrophomina phaseolina. Charcoal rot. **8:**OK(82,2872).
Phoma exigua. **14:**WI(1337).
Phyllosticta texensis. Leaf spot. **14:**TX(82).
P. verbenicola. On living leaves. **1:**TX(64,3597). **7:**NJ(82).
Septoria verbenae. Leaf spot. **1:**TX(82). **2:**ID(82,3202), KS(82), OK(2872), SD(82), WI(82). **3:**OK(2872). **7:**E states(82), MS(82), OK(2873), SD(82,2243), TX(82). **10:**CA(2439). **11:**OK(2872). **12:**AZ(1195). **13:**OK(2873). **14:**E states(82), MS(82), SD(82), TX(82), WI(1329). **15:**E states(82), MS(82), OK(2872), SD(82), TX(82). **17:**E states(82), MS(82), SD(82), TX(82).
Stagonospora verbenae. On living leaves. **14:**WI(1343).
Stagonospora sp. Leaf spot. **7:**WI(1353). **14:**WI(1345).
Deuteromycotina - Other
Rhizoctonia solani. Root rot. **8:**FL(35), NY(82).
Rhizoctonia sp. Root rot. **17:**HI(2903).

Vitex L.

1. *V. agnus-castus* L. [chaste-tree, hemp-tree, Indian spice, monk's-pepper-tree, sage-tree, wild pepper]
2. *V. negundo* L. var. *incisa* (Lam.) C. B. Clarke
3. *Vitex* sp.

Oomycetes
Pythium sp. Root rot. **1:**FL(35).
Basidiomycotina - Other
Armillaria mellea. Root rot. **1:**CA(2900).
Phellinus badius. **1:**AZ(1202).
P. texanus. **1:**AZ(1202).
Deuteromycotina - Hyphomycetes
Alternaria sp. Leaf spot. **1:**FL(35).
Cercospora viticis. Leaf spot. **1:**FL(35), LA(82,962), OK(82,2872), TX(82). **2:**OK(2873). **3:**FL(35).
Corynespora sp. Leaf spot. **3:**FL(35).
Cristulariella moricola. Zonate leaf spot. **3:**FL(35).
Fusarium sp. Leaf necrosis, root rot. **1:**FL(35).
Helminthosporium viticis. Leaf spot. **1:**FL(35).
Phymatotrichopsis omnivora. Root rot. **1:**TX(82).
Stemphylium sp. Leaf spot. **1:**FL(35).
Deuteromycotina - Coelomycetes
Phoma sp. **1:**OK(2872).
Phyllosticta sp. Leaf spot. **1:**FL(35).

Violaceae

Hybanthus Jacq.

1. *H. concolor* (T. F. Forster) Spreng. (=*Ionidium concolor* (T. F. Forster) Benth. & J. D. Hook. ex S. Wats.)

Deuteromycotina - Hyphomycetes
Cercospora tandilensis. Leaf spot. **1:**MO(581,962).

Serresia Montrouz.

1. *S. japonica* (Thunb.) Thunb.

Oomycetes
Pythium splendens. Root rot. **1:**FL(35).

Viola L.

1. *V. adunca* Sm. [hook-spur violet, western dog violet]
2. *V. affinis* Leconte
3. *V. arvensis* J. A. Murray [European field pansy, field violet]
4. *V. blanda* Willd. [sweet white violet]
5. *V. canadensis* L. [Canada violet, tall white violet]
6. *V. canadensis* L. var. *rugulosa* (Greene) C. Hitchc. (=*V. rugulosa* Greene, *V. rydbergii* Greene)
7. *V. canina* L. (=*V. longipes* Nutt. in Torr. & A. Gray)
8. *V. conspersa* Rchb. [American dog violet]
9. *V. cornuta* L. [bedding pansy, horned violet, tufted pansy, viola]
10. *V. cucullata* Aiton [marsh blue violet]
11. *V. glabella* Nutt. [stream violet]
12. *V. hastata* Michx. [halberb-leaved violet]
13. *V. howellii* A. Gray
14. *V. incognita* Brainerd [large-leaved white violet]
15. *V. jooi* Janka
16. *V. lanceolata* L. [lance-leaved violet]
17. *V. langloisii* Greene
18. *V. langsdorfii* Fisch. ex Ging. [Alaska violet]
19. *V. lobata* Benth. [yellow wood violet]
20. *V. lunellii* Greene
21. *V. macloskeyi* J. Lloyd [western sweet white violet]
22. *V. missouriensis* Greene [Missouri violet]
23. *V. nephrophylla* Greene [northern bog violet]
24. *V. nuttallii* Pursh (=*V. linguaefolia* Nutt. ex Torr. & A. Gray, *V. nuttallii* Pursh var. *linguaefolia* (Nutt. ex Torr. & A. Gray) Piper, *V. vallicola* A. Nels.) [yellow prairie violet]
25. *V. occidentalis* Howell
26. *V. ocellata* Torr. & A. Gray [two-eyed violet]
27. *V. odorata* L. [English violet, florist's violet, sweet violet]
28. *V. orbiculata* Geyer ex Holz. [western round-leaved violet]
29. *V. pallens* (Banks) Brainerd
30. *V. palmata* L. [early blue violet, wild-okra]
31. *V. palustris* L. (=*V. epipsila* Ledeb.) [alpine marsh violet]
32. *V. papilionacea* Pursh
33. *V. patrinii* Ging.
34. *V. pedata* L. [birdsfoot violet]
35. *V. pedatifida* G. Don [larkspur violet, purple prairie violet]
36. *V. primulifolia* L. [primrose-leaved violet]
37. *V. pubescens* Aiton (=*V. eriocarpa* Schwein., *V. pensylvanica* Michx.) [downy yellow violet, smooth yellow violet]
38. *V. purpurea* Kellogg
39. *V. rafinesquii* Greene (=*V. kitaibeliana* Roem. & Schult. var. *rafinesquii* (Greene) Fernald) [field pansy]
40. *V. renifolia* A. Gray [northern white violet]
41. *V. renifolia* A. Gray var. *brainerdii* (Greene) Fernald
42. *V. rostrata* Pursh [long-spurred violet]
43. *V. rotundifolia* Michx.
44. *V. rupestris* F. W. Schmidt (=*V. arenaria* DC.)
45. *V. sagittata* Aiton [arrow-leaved violet]
46. *V. scabriuscula* Shafer
47. *V. selkirkii* Pursh ex Goldie [great-spurred violet]
48. *V. sororia* Willd. (=*V. priceana* Pollard) [woolly blue violet]
49. *V. striata* Aiton [cream violet, pale violet, stripped violet]
50. *V. tricolor* L. (=*V. tricolor* L. var. *hortensis* DC.) [European wild pansy, field pansy, hearts-ease, Johnny-jump-up, miniature pansy, pansy, wild violet]
51. *Viola* sp.
52. *Viola* × sp. (unnamed hybrids)

Chytridiomycetes
Synchytrium aureum. Leaf gall. **29:**WI(1345). **37:**WI(82). **45:**WI(1331). **51:**NY(82), WA(3202), WI(82), WY(2924).
S. violae. Leaf gall. **51:**LA(627).
Synchytrium sp. On leaves and stems. **35:**WI(1338).
Oomycetes
Aphanomyces sp. Root rot. **50:**MD(82), NC(1303).
Bremiella megasperma. Downy mildew. **3:**IL(606), MO(606). **27:**FL(82), NJ(82). **39:**AL(606), KS(606). **50:**AL(606), FL(35). **51:**IA(82), IL(82), MN(82).
Peronospora violae. Downy mildew. **39:**AL(82), MS(82). **50:**IL(82), MS(82,2720), NE(82).
Phytophthora nicotianae var. *parasitica*. Leaf blight, crown and root rot. **9:**FL(35). **27:**FL(35). **50:**FL(35).
Pythium debaryanum. Damping-off. **50:**CT(82), MO(82), NJ(82).
P. mamillatum. Damping-off. **50:**CA(82).
P. ultimum. Damping-off. **50:**CA(82).
Pythium sp. Damping-off, root rot. **9:**FL(35). **27:**FL(35). **50:**Unk.(82), FL(35), NC(1303). **51:**HI(2903).
Ascomycotina
Ciborinia violae. **32:**PA(208). **37:**PA(208). **43:**NY(208), PA(208).
Elsinoe sp. Scab. **51:**MD(1736).
Nectria galligena. Canker. **51:**OR(82).
Pleosphaerulina violae. On leaves. **27:**AL(82).
Scorias spongiosa. Sooty mold. **51:**ME(82).
Sphaerotheca fuliginea. Powdery mildew. **9:**IA(82). **10:**CO(82). **50:**IA(82), KS(82), WA(82). **51:**CA(82), CO(82), ND(82), OR(82), WI(82), WY(82).
S. macularis. Powdery mildew. **1:**OR(3202). **5:**ND(377), WY(1195). **6:**WY(1195). **9:**FL(35). **10:**CO(82). **11:**WA(3202).

24:WA(3202). **27:**FL(35). **38:**CA(638,642). **50:**FL(35), ID(3202), OR(3202), WA(3202). **51:**CA(82), CO(82), IL(312), ND(82), OR(82), WI(82), WY(82).
S. violae. On leaves and stems. **5:**WY(371).

Basidiomycotina - Uredinales

Phragmidium sp. **1:**ID(2774,3202). **51:**CA(2774).
Puccinia canadensis. Autoecious. **28:**WA(3105).
P. ellisiana. Heteroecious: aecial host. **10:**WI(1341). **20:**ND(378). **32:**NC states(82), NE states(82), ND(378). **34:**NC states(82), NE states(82), WI(1353). **35:**WI(1338). **37:**WI(1342). **50:**KS(82), NE(82). **51:**E states(82), NC states(82), NM(82), WY(82).
P. fergussonii. Autoecious. **4:**AK(526). **18:**AK(526,3106). **31:**AK(60,526), MT(3202). **51:**AK(82), CO(82), MT(82), UT(82).
P. ornatula. Autoecious. **11:**AK(3107).
P. violae. Autoecious. **1:**Range of host(82), AK(60,526), CA(81,642), ID(2774,3202), MT(3202), OR(3202), WA(3202), WI(1356). **4:**Range of host(82), CA(81), MA(1394), WI(1330). **5:**ID(3202), MT(3202), ND(378), NM(1195), OR(3202), SD(2243), UT(1195), WA(3202). **6:**WY(1195). **8:**WI(1342). **9:**FL(35). **10:**Range of host(82), SD(2243). **11:**AK(60,526), CA(81), ID(3202), MT(3202), OR(3202), WA(3202). **12:**NC(1303). **13:**WA(3202). **14:**WI(1342). **16:**FL(35), GA(1439). **18:**AK(60,526). **19:**CA(642). **21:**WA(3202). **22:**OK(2872). **23:**OR(3202), WA(3202). **24:**WA(3202). **25:**CA(81). **26:**CA(81). **27:**CT(82), FL(35). **30:**FL(35), MS(2720). **31:**AK(526), CA(81), ID(3202), WA(3202). **32:**Range of host(82), NC(1303), ND(378), OK(2872). **34:**GA(1439). **36:**GA(1439), MS(2720). **37:**Range of host(82), MS(2720). **41:**AK(526). **44:**ND(378). **46:**ND(378). **48:**FL(35), ND(378). **50:**CT(82), FL(35,82), KS(82), NC(1303), ND(82), NJ(82), SC(82). **51:**Range of host(82), AK(526), CA(82,661), CO(1195), IN(647), MI(665), MS(2720), NC(1307,3944), OK(2872), SC(634), SD(2243), VT(1224), WY(2924).
Uromyces andropogonis. Heteroecious: aecial host. **32:**NC(82,1303), TN(82), WV(82). **34:**GA(1439). **36:**GA(1439), MS(2720). **49:**NC(1303). **50:**CT(82). **51:**SE states(82), GA(1439).

Basidiomycotina - Ustilaginales

Urocystis kmetiana. **39:**AR(82,1077), GA(1077), KS(1077), MO(82,1077), TN(82,1077,1539).
U. violae. Leaf smut. **1:**UT(1077). **7:**UT(1077). **11:**AK(526,1077). **24:**CA(1077), UT(1077). **51:**AK(82), CA(82), MN(82), TX(82), UT(82).

Basidiomycotina - Other

Sebacina incrustans. On leaves and stems. **51:**MD(128), NC(1307), NY(128).

Deuteromycotina - Hyphomycetes

Alternaria tenuissima. Leaf spot. **10:**IN(82). **27:**CT(82), MD(82), TX(82). **50:**NJ(82), NY(82), PA(82), WA(3202). **51:**E states(82), MS(2720), TX(82), WI(82,1330).
Alternaria sp. Leaf spot. **9:**FL(35). **50:**FL(35). **51:**MA(82), MD(82), MI(82), MN(82), NJ(82), NY(82), TX(82).
Botrytis cinerea. Gray mold blight. **9:**FL(35). **27:**AK(82), MD(82), OH(82). **50:**AK(82,526), FL(35), HI(2903), NC(1303), WA(3202).
Cercospora granuliformis. Leaf spot. **10:**E states(82), DE(3543), OK(82), SD(82). **22:**OK(2873). **27:**MD(82). **32:**E states(82), OK(82,2872), SD(82). **34:**WI(1347). **37:**E states(82), OK(82), SD(82). **51:**IL(82), NC(1303), SD(2243), WI(1356).
C. violae. Leaf spot. **1:**WA(3202). **4:**E states(82), MS(2720), ND(82), TX(82). **5:**WI(1352). **9:**FL(35), IN(82). **10:**E states(82), DE(3543), ND(82), TX(82). **13:**WA(3202). **16:**FL(35). **22:**OK(2872). **27:**AL(82), CT(82), FL(35), MI(82), PA(82), TX(82). **30:**FL(35), MS(2720). **32:**E states(82), IL(82), ND(82,378), OK(2872), TX(82). **34:**E states(82), ND(82), TX(82), WI(1331). **35:**WI(1338). **37:**WI(1341). **42:**WI(1356). **45:**WI(1334). **47:**WI(1343). **48:**FL(35). **50:**CT(82), FL(35), MI(82), NC(1303), NY(82), TX(82), WI(82,1330).
Fusarium oxysporum. Root rot, wilt. **9:**FL(35). **27:**FL(35,82), MS(82,2720), OH(82). **50:**FL(35), MI(82), NC(1303), NE(82), NJ(82), NY(82), OH(82), TX(82).
Fusarium sp. Damping-off, wilt. **50:**HI(2903), NC(1303).
Heterosporium sp. Leaf spot. **27:**AK(82).
Mycocentrospora acerina. In seed. **50:**GA(1209,1214), WA(3202). **51:**AK(82,526), CA(82), NC(3944).
Mycovellosiella murina. **32:**IN(82), KS(82). **37:**WI(1361).
Myrothecium sp. Leaf necrosis. **50:**FL(35).
Phymatotrichopsis omnivora. Root rot. **27:**TX(82). **51:**TX(82).
Ramularia agrestis. Leaf spot. **39:**OR(82). **50:**OR(3202). **51:**NE(82), OR(82).
R. ionophila. On leaves. **51:**AK(82), CO(82), WI(82).
R. lactea. Leaf spot, white spot. **1:**WA(3202). **11:**ID(3202), WA(3202). **27:**AK(82). **38:**OR(3202), WA(3202). **50:**NC(1303), WA(82,3202). **51:**AK(526), CO(82), MS(82,2720), MT(82,3202).
Sclerotiomyces colchicus. **51:**WI(1356).
Stemphylium sp. Leaf spot. **9:**FL(35). **50:**FL(35).
Thielaviopsis basicola. Root rot. **27:**CT(82), MA(82), MD(82), NY(82). **50:**CT(82). **51:**E & C states(82), KS(82), MS(2720), OH(82).

Deuteromycotina - Coelomycetes

Ascochyta violae. Leaf spot. **10:**IN(82). **35:**WI(1338). **37:**IA(82), WI(82). **48:**WI(1360). **51:**PA(82).
A. violicola. Leaf spot. **27:**AK(82). **51:**AK(526).
Colletotrichum dematium. Anthracnose. **1:**WI(1348). **2:**WI(1356). **10:**WI(1342). **22:**OK(2872). **34:**WI(1340). **35:**WI(1338). **37:**WI(1342). **40:**WI(1359). **48:**WI(1330).
C. gloeosporioides. Anthracnose, leaf spot. **9:**FL(35). **27:**CT(82), FL(35), MA(82). **32:**OK(2872). **39:**PA(82). **50:**E states(82), DE(2159), FL(35), IN(82), MI(82), NC(1303), WA(82,3202). **51:**MS(2720).
Cryptostictis violae. Leaf spot. **51:**IL(82).
Cylindrosporium violae. Leaf spot. **51:**MT(82,3202), WA(3202).
Marssonina violae. Leaf spot. **10:**IA(82), IL(82), NY(82). **27:**NJ(82). **32:**NY(82). **51:**E states(82), IA(82), MN(82).
Phyllosticta nigrescens. Leaf spot. **38:**CA(638,642). **51:**CA(82), WA(3202).
P. rafinesquii. Leaf spot. **39:**AL(82), IL(82).
P. violae. Leaf spot. **10:**NY(82), WI(1341). **14:**WI(1354). **16:**FL(35). **27:**CT(82), MA(82), NY(82). **30:**FL(35). **35:**WI(1338). **37:**IA(82). **47:**WI(1356). **48:**FL(35). **50:**MI(82). **51:**E states(82), CA(82), HI(82,2903), KS(82), MN(82), WI(1334).
Septoria hyalina. **31:**AK(526). **51:**ID(3202).
S. violae-palustris. Leaf spot. **4:**NC states(82), NE states(82). **10:**NC states(82), NE states(82). **11:**ID(3202). **14:**WI(1341). **16:**FL(35). **29:**MA(664). **30:**FL(35). **32:**NC states(82), NE states(82). **37:**NC states(82), NE states(82), WI(1329). **40:**WI(1359). **45:**WI(1334). **48:**FL(35). **51:**E & C states(82), AK(82), KS(82), LA(82), MS(2720), NC(1307), NM(1195), WI(1341).
Sphaceloma violae. Spot anthracnose, scab on leaves and scapes. **5:**NY(1745). **9:**FL(35). **10:**E states(82), CT(1745), DC(1745), MD(1745), MS(82,2720), VA(1745). **15:**NY(1745). **16:**FL(35). **17:**LA(1745). **27:**FL(35,82,1745), GA(82,1745), MA(82), MD(82,1736,1745), MS(82,965,2720), NJ(82), NY(82,1744,1745), PA(82,1745), TX(82), VA(82,1745). **30:**FL(35), NY(1745). **32:**E states(82), CT(1745), MS(82,2720), NY(1745), VA(1745). **33:**KS(2818,3749). **34:**E states(82), MS(82,2720). **48:**FL(35), MD(1745), NY(1745). **49:**CT(1745), NY(1745). **50:**FL(35), KS(82,2818,3749), MD(82,1745), NJ(82), NY(1745). **52:** E states(82), CT(1745), FL(2525), HI(2903), KS(82,2818,3749), MD(1733,1745), MS(82,2720), NC(1745), NH(1745), NY(1744), VA(1745), WA(3202).

Deuteromycotina - Other

Rhizoctonia crocorum. Root rot. **51:**TX(82).
R. solani. Damping-off, root and stem rot, wilt. **9:**FL(35). **16:**FL(35). **27:**FL(35). **30:**FL(35). **48:**FL(35). **50:**DE(82), FL(35), IL(82), MN(82), NC(1303), NY(82). **51:**FL(82), MN(82), MS(82,2720), NY(82).
Rhizoctonia sp. Root rot. **50:**HI(2903).
Sclerotium rolfsii. Southern blight, root and stem rot. **9:**FL(35). **10:**MS(2720). **16:**FL(35). **27:**AL(82,285), AR(82), MD(82), TX(82). **30:**FL(35). **32:**IL(312). **48:**FL(35). **50:**FL(35,82), VA(82). **51:**SE states(82), TX(82).
S. rolfsii var. *delphinii*. Southern blight. **51:**CA(82), CT(82), VT(82).

Viscaceae

Arceuthobium M. Bieb.

1. *A. abietinum* Engelm. ex Munz (=*A. campylopodum* Engelm. in A. Gray f. *abietinum* (Engelm. ex Munz) L. S. Gill)
2. *A. americanum* Nutt. ex Engelm.
3. *A. campylopodum* Engelm. in A. Gray (=*A. campylopodum* Engelm. in A. Gray f. *campylopodum*)
4. *A. douglasii* Engelm.
5. *A. laricis* (Piper) St. John (=*A. campylopodum* Engelm. in A. Gray f. *laricis* (Piper) L. S. Gill)
6. *A. pusillum* Peck
7. *A. tsugense* (Rosend.) G. N. Jones (=*A. campylopodum* Engelm. in A. Gray f. *tsugensis* (Rosend.) L. S. Gill)
8. *Arceuthobium* sp.

Ascomycotina

Caliciopsis arceuthobii. On floral parts of pistillate shoots. **2**:ID(3202), WA(3202,3889). **4**:ID(3202,3889), OR(1933), WA(3202,3889). **6**:NY(1933). **8**:MT(3202).
Glomerella cingulata. **2**:WA(3202). **3**:WA(3202). **4**:ID(3202), WA(3202).

Deuteromycotina - Hyphomycetes

Alternaria alternata. **2**:Co(2247).
Aureobasidium pullulans. **2**:CO(2247).
Botrytis sp. **2**:CO(2247).
Cephalosporium sp. **2**:CO(2247).
Cladosporium herbarum. **2**:CO(2247).
Cylindrocarpon gillii. Anthracnose, on all parts of aerial shoots. **1**:WA(3889). **2**:CO(2247), MT(1221,3202). **3**:OR(3202,3889), WA(3202,3889). **5**:WA(3889). **7**:WA(3889).
Epicoccum nigrum. **2**:CO(2247).
Gonatobotrys sp. **2**:CO(2247).
Penicillium sp. **2**:CO(2247).
Torula herbarum. **2**:CO(2247).
Trichoderma viride. **3**:ID(3202).

Deuteromycotina - Coelomycetes

Colletotrichum gloeosporioides. Shoot blight, dieback. **1**:CA(2719). **2**:WA(3889). **3**:WA(3889). **4**:ID(3889), WA(3889). **5**:WA(3889).
Pestalotiopsis maculiformans. **3**:WA(3202).
Phoma sp. **2**:CO(2247).
Phomopsis sp. **3**:WA(3202).

Korthalsella Tiegh.

1. *K. complanata* (Tiegh.) Engl.
2. *K. latissima* (Tiegh.) Danser
3. *Korthalsella* sp.

Ascomycotina

Cucurbitaria obducens. Stem lesions. **1**:HI(1922,2903).
Echidnodes visci. On living stems. **3**:HI(2784).
Meliola visci. **3**:HI(2903).
Pleospora sp. On stems. **1**:HI(1922). **2**:HI(2903).

Phoradendron Nutt.

1. *P. juniperinum* Engelm.
2. *P. libocedri* (Engelm.) Howell
3. *P. serotinum* (Raf.) M. Johnston (=*P. flavescens* (Pursh) Nutt.) [American mistletoe]
4. *P. tomentosum* (DC.) A. Gray subsp. *macrophyllum* (Engelm.) Wiens (=*P. flavescens* (Pursh) Nutt. var. *macrophyllum* Engelm.)
5. *P. villosum* Nutt.
6. *Phoradendron* sp.

Ascomycotina

Asterina phoradendricola. Black mildew. **3**:FL(35).
Nectria cinnabarina. Twig blight. **6**:TX(82).
Valsaria phoradendri. On stems. **6**:Unk.(82).

Basidiomycotina - Uredinales

Uredo phoradendri. **1**:AZ(1497,3122,3981), NM(1497,3122). **4**:CA(81,3122). **5**:CA(3122), OR(1497,3122,3202). **6**:CA(82), OR(82).

Basidiomycotina - Other

Septobasidium pseudopedicellatum. Parasitic on scale insects. **3**:FL(35).

Deuteromycotina - Hyphomycetes

Cercospora struthanthi. Leaf spot. **3**:FL(35,450).
Exosporium phoradendri. Leaf spot. **3**:TX(3597). **6**:TX(82).

Deuteromycotina - Coelomycetes

Colletotrichum gloeosporioides. **3**:CA(81). **4**:CA(81). **5**:CA(81).
Phyllosticta phoradendri. Leaf blight. **4**:CA(82,325).
Sphaeropsis visci. Leaf blight. **2**:CA(642). **3**:FL(35,2782). **6**:FL(82), SC(82), TX(82).

Viscum L.

1. *V. articulatum* N. L. Burm.

Ascomycotina

Meliola visci. Black mildew. **1**:HI(1272,1452,2903).

Vitaceae

Ampelopsis Michx.

1. *A. aconitifolia* Bunge [monk's-hood-vine]
2. *A. arborea* (L.) Koehne [peppervine]
3. *A. cordata* Michx. [heartleaf ampelopsis]
4. *Ampelopsis* sp.

Oomycetes

Plasmopara viticola. Downy mildew. **4**:NY(82), WI(82).

Ascomycotina

Botryosphaeria obtusa. **4**:AL(82), MD(82).
Guignardia bidwellii. Leaf spot. **2**:FL(35), MS(82,2720), NJ(82). **4**:Range of host(82).
Linospora psederae. Leaf spot. **4**:WV(82).
Nectria cinnabarina. Dieback. **4**:Unk.(82).
Rhytidhysteron rufulum. **4**:GA(82).
Uncinula necator. Powdery mildew. **2**:FL(35). **4**:Range of host(82).

Basidiomycotina - Other

Pellicularia koleroga. Thread blight. **2**:LA(82,3640).

Deuteromycotina - Hyphomycetes

Cercospora arboriae. Leaf spot. **2**:FL(35), TX(82,3597).
C. truncata. Leaf spot. **3**:LA(82).
Pseudocercospora vitis. Leaf spot. **2**:LA(82). **3**:AL(82), LA(82).
Tubercularia nigricans. Dieback. **1**:NH(82), TX(82).

Deuteromycotina - Coelomycetes

Phyllosticta ampelicida. **2**:Unk.(82).
Septoria sp. **2**:LA(82).

Deuteromycotina - Other

Rhizoctonia crocorum. Root rot. **4**:TX(82).

Cissus L.

1. *C. incisa* (Nutt. ex Torr. & A. Gray) Des Moul. [marine ivy, marine-vine, possum-grape]
2. *C. rhombifolia* Vahl [grape ivy, Venezuela treebine]
3. *C. sicyoides* L. [princess-vine]
4. *Cissus* sp.

Oomycetes

Phytophthora nicotianae var. *parasitica*. Leaf and stem blight. **2**:FL(35).
Plasmopara viticola. Downy mildew. **2**:FL(35). **4**:FL(35).

Basidiomycotina - Uredinales

Aecidium mexicanum. **1**:OK(2872). **4**:OK(82).
Endophyllum circumscriptum. Autoecious. **4**:FL(35).
Uromyces commelinae. Heteroecious: aecial host. **1**:TX(741).

Basidiomycotina - Ustilaginales

Mycosyrinx cissi. Smut. **3**:FL(1077). **4**:FL(35,82).

Deuteromycotina - Hyphomycetes

Alternaria sp. Leaf spot. **4**:FL(35).
Botrytis sp. Gray mold. **4**:FL(35).
Cercospora arboriae. **4**:TX(82).
Cercospora sp. Leaf spot. **2**:FL(35). **4**:FL(35).
Cylindrocladiella peruviana. Foliage blight. **2**:FL(35).
Cylindrocladium ilicicola. Stem gall. **2**:FL(35).
C. pteridis. Stem gall. **2**:FL(35).
Fusarium sp. Wilt. **4**:FL(35).
Phymatotrichopsis omnivora. Root rot. **4**:TX(82).
Pseudocercospora vitis. Leaf spot. **4**:LA(82).

Deuteromycotina - Coelomycetes

Colletotrichum sp. Anthracnose. **4**:FL(35).
Phyllosticta cissicola. Leaf spot. **4**:TX(82).

Deuteromycotina - Other

Rhizoctonia solani. Root rot. **2**:'Rhizoctonia aerial blight.' FL(35). **4**:FL(35).

Parthenocissus Planch.

1. *P. quinquefolia* (L.) Planch. (=*Ampelopsis quinquefolia* (L.) Michx., *Psedera quinquefolia* (L.) Greene) [American ivy, five-leaved ivy, Virginia creeper, woodbine]
2. *P. tricuspidata* (Siebold & Zucc.) Planch. (=*Psedera tricuspidata* (Siebold & Zucc.) Rehd.) [Boston ivy, Japanese ivy]
3. *P. vitacea* (Knerr) Hitchc. (=*Psedera vitacea* (Knerr) Greene)
4. *Parthenocissus* sp.

Chytridiomycetes

Physoderma sp. On leaves. **1:**LA(1844).
Synchytrium parthenocissi. Leaf gall. **1:**LA(627). **4:**LA(627).

Oomycetes

Plasmopara viticola. Downy mildew. **1:**AL(82), IA(82), ME(82), MN(82), NY(82), TX(82), WI(82). **2:**NJ(82), TX(82).

Ascomycotina

Botryosphaeria obtusa. **1:**AL(82), NJ(82).
B. quercuum. On bark. **1:**WV(82).
Diaporthe ampelopsidis. On stems. **1:**GA(1438), MA(82,3807), NJ(82,3807), OH(82).
Discohainesia oenotherae. On stems. **1:**VA(82).
Elsinoe parthenocissi. Scab, on leaves and stems. **1:**FL(35,82,2524,2525), MO(82), NH(82,1745), PA(82).
Elsinoe sp. Leaf scab. **1:**NH(3780).
Guignardia bidwellii. Black rot, leaf spot. **1:**Range of host(82), FL(35), GA(2405), IN(647), MS(2720), NC(3944). **2:**Range of host(82), MS(2720), OK(2872).
G. bidwellii f. *parthenocissi*. **1:**GA(1438). **2:**GA(1438).
Lophiostoma vigheffulensis. **1:**OK(660).
Plowrightia neomexicana. On stems. **1:**NM(82).
Tubeufia pezizula. **4:**FL(2782).
Uncinula necator. Powdery mildew on leaves. **1:**Range of host(82), FL(35), ID(3202), MA(1390), MS(2720), ND(377), SD(2243), WA(3202). **2:**TX(82), WA(3202).
Valsa ampelopsidis. **1:**NJ(82).
V. vitigera. **1:**IA(1226).

Basidiomycotina - Other

Meruliopsis corium. **4:**FL(128).
Pellicularia koleroga. Thread blight. **1:**FL(82).
Thanatephorus cucumeris. Stem rot. **2:**CT(82).

Deuteromycotina - Hyphomycetes

Alternaria sp. Leaf spot. **1:**FL(35).
Cercospora ampelopsidis. Leaf spot. **1:**Range of host(82), DE(3543), FL(35), IL(82), MS(2720), SD(2243), TX(82). **2:**MS(82), TX(82).
Cladosporium herbarum. Dieback. **2:**IN(82), NJ(82), OH(82), PA(82).
Cladosporium sp. Dieback. **1:**NJ(82).
Cristulariella moricola. **1:**AL(2056), GA(2056).
Fusarium sp. Leaf spot. **1:**FL(35).
Helminthosporium sp. Leaf spot. **1:**FL(35).
Phymatotrichopsis omnivora. Root rot. **1:**TX(82). **2:**TX(82).

Deuteromycotina - Coelomycetes

Aplosporella hedericola. Leaf spot. **2:**NJ(82).
Aplosporella sp. **2:**NJ(82).
Colletotrichum dematium. On petioles. **3:**WI(1348,1350).
Colletotrichum sp. Anthracnose. **1:**FL(35).
Coniothyrium fuckelii. Stem canker. **1:**SD(2243), WV(82).
Macrophoma ampelopsidis. On stems. **1:**IL(82).
Phloeospora ampelopsidis. Leaf spot. **1:**IA(82), IL(82), NE(82), WI(82,966). **2:**IA(82). **3:**WI(1335).
Phomopsis viticola. On stems. **1:**NJ(82).
Phyllosticta ampelicida. Leaf spot. **2:**IN(663), NY(64), WI(1334). **3:**WI(64).
Stagonospora heterospora. On stems. **1:**NY(82).

Deuteromycotina - Other

Rhizoctonia crocorum. Root rot. **1:**TX(82).
R. solani. Thread blight. **1:**FL(35).

Tetrastigma (Miq.) Planch. in A. DC. & C. DC.

1. *T. voinieranum* (Baltet) Pierre ex Gagnep. [chestnut-vine, lizard-plant]

Deuteromycotina - Coelomycetes

Phyllosticta sp. Leaf spot. **1:**FL(35).

Vitis L.

1. *V. aestivalis* Michx. [bunch grape, pigeon grape, summer grape]
2. *V. arizonica* Engelm. [canyon grape]
3. *V. baileyana* Munson [possum grape]
4. *V. berlandieri* Planch. [Spanish grape, winter grape]
5. *V. californica* Benth. [California wild grape, Western wild grape]
6. *V. cinerea* Engelm. ex Millard. [grayback grape]
7. *V. indivisa* Willd.
8. *V. labrusca* L. (=*V. labruscana* L. H. Bailey) [American grape, fox grape, skunk grape, stunt grape]
9. *V. munsoniana* J. H. Simson ex Planch. [bird grape]
10. *V. palmata* Vahl [cat grape, catbird grape, red grape]
11. *V. riparia* Michx. [frost grape, riverbank grape]
12. *V. rotundifolia* Michx. [bullace grape, muscadine grape, scupernong, southern fox grape]
13. *V. rupestris* Scheele [bush grape, mountain grape, rock grape, sand grape, sugar grape]
14. *V. vinifera* L. [European grape, wine grape]
15. *V. vulpina* L. (=*V. cordifolia* Michx.) [chicken grape, frost grape, winter grape]
16. *Vitis* sp.

Oomycetes

Phytophthora sp. Root rot. **8:**CA(1110). **14:**CA(1110).
Plasmopara viticola. 'Downy mildew.' **1:**Range of host(82), SC(634). **5:**CA(82,1110). **8:**Range of host(82), FL(35), MS(2720), NC(1303). **12:**AL(82), MI(82), MS(2720). **13:**IL(82), IN(82), LA(82), OK(2872). **14:**CA(82), FL(35), MI(82), NC(1303,3944), NY(82), SD(2243). **15:**Range of host(82). **16:**FL(35), HI(2903), MS(2720), NC(1303), OK(2872), SD(2243).
Pythium sp. Root rot. **12:**FL(35). **14:**FL(35).

Zygomycotina

Mucor sp. **16:**WA(3202).
Rhizopus arrhizus. 'Fruit rot. Raisin mold.' **14:**CA(1110).
R. stolonifer. **16:**ID(3202), WA(3202).

Ascomycotina

Botryosphaeria dothidea. Canker, dieback. 'Botryosphaeria rot. Botryosphaeria necrosis.' **8:**FL(35), GA(1438). **12:**FL(35), GA(1438), NC(1303). **14:**FL(35), GA(1438), NC(1303). **16:**FL(35).
B. obtusa. On canes and limbs. **8:**E states(82), MI(82). **12:**AL(82), GA(82). **15:**VA(82).
B. rhodina. **12:**SC(82).
B. ribis. On canes. **12:**FL(82), GA(82).
Coniochaeta pulveracea. On wood. **16:**IA(1225).
Cryptosporella viticola. Dead arm, canker, limb necrosis. **8:**NE states(82), Range of host(82), AR(282), CA(282), DE(282), FL(35), GA(282), IA(82), IL(282), IN(282), KS(282), MI(82,282), MS(282), NC(282), NE(282), NJ(282), NY(282), OH(282), PA(282), SC(282), VA(282), WV(282). **12:**MS(82,2720), NC(82,1303), SC(82). **14:**Unk.(82), CA(1548), NC(1303). **16:**FL(35), NC(1303).
Cryptovalsa ampelina. **16:**FL(2782).
Daldinia concentrica. On stumps. **8:**MD(82), VA(82).
D. vernicosa. On stumps. **8:**MD(82), VA(82).
Diaporthe rudis. On stems. **8:**VA(82).
Diaporthe sp. **12:**GA(1438).
Didymosphaeria sarmentii. On twigs. **14:**CA(82,1110).
Discohainesia oenotherae. On stems. **8:**VA(82). **15:**VA(82).
Elsinoe ampelina. On fruits, leaves and stems. 'Anthracnose. Bird's-eye rot.' **1:**FL(82). **8:**Range of host(82), FL(35). **12:**FL(35). **14:**FL(35), NC(1303), TX(82,1745). **16:**FL(35,1745), MD(1736,1737), MI(1745), MS(1745,2720), NC(1303), OK(2872), SC(1745), WI(1343).
Endothia gyrosa. On canes. **8:**NC(82).
Erysiphe graminis f. sp. *avenae*. Powdery mildew. **16:**AR(3029).
Eutypa armeniacae. Drying arm disease, canker. 'Eutypa dieback.' **8:**MI(3677). **14:**CA(2459), NY(3716).
E. lata. **14:**CA(1110).
Eutypella aequilinearis. On bark. **1:**PA(82), SC(82). **8:**MD(82), MI(82), PA(82). **12:**AL(82), GA(82), NC(82). **13:**OK(82,2873). **14:**NC(82), PA(82). **15:**GA(2925), NJ(82), PA(82). **16:**IL(971), OK(2872), TX(2925).
E. juglandicola. **12:**GA(1438).
E. leprosa. **8:**IA(3629).
Glomerella cingulata. 'Ripe rot.' **8:**Range of host(82), FL(35).

12:FL(35), NC(1303). 14:FL(35), NC(1303), TX(82). 16:FL(35).
Glonium lineare. 12:GA(1438).
Glonium sp. 12:GA(1438).
Guignardia bidwellii. Leaf spot. 'Black rot.' 1:Range of host(82), MS(2720). 4:OK(2872). 8:Range of host(82), FL(35), GA(1438). 9:OK(2872). 12:FL(35), MS(2720), NC(1303,3944), SC(1307). 13:TX(82), WV(82). 14:AL(82), FL(35), MA(82), NC(1303), NE(82). 15:Range of host(82). 16:AL(285), FL(35), MS(2720), NC(1303), OK(2872), WA(3202).
G. bidwellii f. *muscadinii*. Muscadine black rot, leaf spot. 12:SE states(82), FL(35), GA(1438), IA(82), MD(82).
Hainesia lythri. 11:WI(1357).
Herpotrichia sp. 12:GA(1438).
Hypocrea gelatinosa. Rotten wood. 8:GA(1438).
Hypoxylon hypophlaeum. 8:GA(1438).
H. mediterraneum. On stems. 8:AL(82).
H. rubiginosum. 12:GA(1438).
H. tinctor. 12:GA(1438).
Hysterographium flexuosum. On stems. 12:GA(1438), MS(2720). 15:NC(82).
H. mori. On stems. 8:MD(82), NJ(82), VA(82). 12:NC(3944).
Massarina microcarpa. 3:GA(1438).
Metschnikowia pulcherrima. 8:CA(204).
Mollisia cinerea. 12:GA(1438).
M. melaleuca. 12:GA(1438).
M. pullata. 16:NY(850).
Monilinia fructicola. Brown rot. 16:OK(2872).
Mycosphaerella angulata. 'Angular leaf spot.' 12:AL(82), GA(82,1438), MS(2720), NC(1303), VA(82). 14:NC(1303). 16:NC(1303).
M. personata. 'Isariopsis blight.' 5:CA(82,1110). 8:Range of host(82), FL(35), GA(1438), MS(2720). 12:FL(35), GA(82), MS(2720), NC(82). 13:KS(82). 14:FL(82), GA(82), IA(82), LA(82), NC(82). 15:GA(82), IL(82), LA(82), MI(82), MO(82), MS(2720). 16:FL(35), MS(2720).
Nectria coccinea. On canes following crown gall. 8:OR(82). 16:OR(3202).
N. viticola. On limbs. 14:AL(82).
Ostreichnion curtisii. 12:GA(1438), MS(2720).
Phyllachora picea. On stems. 1:NY(82).
Physalospora abdita. On canes, limbs and stems. 12:SC(82). 15:SC(82).
Physalospora sp. Dieback. 12:FL(35).
Pleospora herbarum. 16:WA(3202).
P. penicillus. 3:GA(1438).
Pseudopezicula tetraspora. On leaves. 16:NY(1959).
Rhytidhysteron rufulum. Dieback. 12:FL(35,82).
Rhytisma vitis. Tar spot. 15:WI(1340).
Roesleria pallida. Root rot. 8:E states(82), IA(82), MO(82).
Rosellinia aquila. 12:GA(35).
R. necatrix. 'Dematophora root rot.' 8:AL(82), IN(82), MI(82), NY(82), OH(82).
Rosellinia sp. 8:WA(3202).
Schizothyrium pomi. Fly speck. 8:PA(82), WV(82).
Schizoxylon insigne. 12:GA(1438).
Sclerotinia sclerotiorum. Shoot blight. 14:CA(82,1110).
Stictis viticola. 12:AL(3225).
Thyridaria viticola. On dead stems. 14:KS(82).
Thyridium vitis. On stems. 15:KS(82).
Tubeufia pezizula. 12:GA(1438).
Uncinula necator. 'Powdery mildew.' 1:Range of host(82). 5:CA(82,1110). 8:Range of host(82), CA(1110), FL(35), GA(1438), HI(2903). 12:FL(35), IA(82), NC(1303). 13:CA(1110), IL(82). 14:Unk.(3976), CA(82,1110), HI(2903), IA(82), ID(82), MD(82), NC(1303), OR(82), SD(2243), WA(82,3202). 15:Range of host(82), MS(2720), WI(1336). 16:FL(35), HI(2903), ID(3202), NC(1303), OK(2872), OR(3202), SD(2243), WA(3202).
Valsa ceratosperma. 12:GA(1438).
V. vitigera. 16:IA(1226).
Xylaria polymorpha. 16:VA(3018).

Basidiomycotina - Uredinales

Physopella ampelopsidis. Autoecious. 1:FL(82). 8:FL(35). 12:AL(82), FL(35,82), GA(1439), NC(587,1303). 14:FL(35,82), NC(1303), SC(82). 16:FL(35).

Basidiomycotina - Other

Aleurobotrys botryosus. On dead stems. 16:AL(128,2083), MA(128,2083), PA(2083).
Armillaria mellea. Root rot. 'Shoestring root rot.' 8:AR(82), FL(35), MO(82), TX(82). 12:FL(35,82), MS(82,2720), NC(82,1303). 14:CA(82,1110), FL(35), NC(1303), WA(82). 16:CA(2898), FL(35), WA(3202).
A. tabescens. Root rot. 8:SC states(82), SE states(82). 12:FL(35). 14:NC(1303). 16:OK(2872).
Chondrostereum purpureum. On wood. 8:FL(35,82). 16:FL(35).
Corticium appalachiensis. 16:TN(431,2043).
C. cremeo-albidum. On wood. 16:FL(2051).
Corticium sp. Root rot. 3:TX(82).
Crepidotus amarus. On dead vines. 16:FL(1544).
Dendrophora albobadia. 2:AZ(1194). 12:GA(1439).
D. erumpens. 12:AL(543).
Grandinia pruni. 2:AZ(1194,1202).
Hapalopilus rutilans. White rot. 2:AZ(1202). 16:Unk.(2681).
Irpex lacteus. Wood rot. 8:VA(82). 15:ND(378).
Pachykytospora papyracea. Wood rot. 8:NC(82), VA(82).
Perenniporia medulla-panis. 16:GA(1439).
Phanerochaete crassa. 8:VA(82).
P. flavido-alba. 16:KY(128).
P. viticola. White rot. 16:MD(128,433), MS(128,433).
Phellinus gilvus. Wood rot. 8:TN(82).
P. igniarius. Black measles. 'Wood rot.' 16:CA(573).
P. viticola. Wood rot. 8:MD(82), VA(82).
Polyporus tenuiculus. On wood. 16:SE states(2681).
Pulcherricium caeruleum. On wood. 12:GA(491,1439).
Ramaricium polyporoideum. On wood. 16:GA(430), TN(430).
Schizophyllum commune. Wood rot. 8:VA(82).
Schizopora paradoxa. Wood rot. 8:MI(82), VA(82).
Sebacina incrustans. At base of living tree. 16:LA(2675).
Stereum hirsutum. White rot. 'Wood rot.' 16:CA(573), DC(543).
Stigmatolemma poriaeforme. On wood. 8:GA(491,1439). 12:GA(491,1439).
Trametes hirsuta. Wood rot. 8:VA(82).
T. versicolor. White rot. 8:AL(82), GA(1439), MI(82), VA(82), WI(82). 15:NE(82).
Typhula viticola. 1:NY(264).
Vararia pectinata. 16:MD(128).

Deuteromycotina - Hyphomycetes

Alternaria alternata. 'Fruit rot. Raisin mold.' 14:CA(1110). 16:WA(3202).
A. gomphrenae. Leaf spot. 12:FL(35).
Alternaria sp. Leaf and stem blotch, storage decay, fruit rot, leaf spot. 1:FL(82). 8:FL(35). 14:CA(82), FL(35). 16:CA(1487), FL(35), OK(2872).
Aspergillus niger. Bunch mold. 'Fruit rot. Raisin mold.' 14:CA(82,1110). 16:WA(3202).
Asperisporium minutulum. On leaves. 5:OR(82,3202), WA(82,3202). 16:OR(3202).
Berkleasmium corticola. On dead stems. 16:NC(1303,1307).
Botrytis cinerea. 'Botrytis bunch rot. Botrytis cane blight.' 8:Range of host(82). 14:CA(82,1397), HI(2903), OR(82). 16:WA(3202).
Botrytis sp. 16:ID(3202), WA(3202).
Briosia ampelophaga. 'Leaf blotch.' 8:TX(82). 11:WI(1349).
Cephalosporium sp. Black measles. 16:CA(573).
Cercospora brachypus. Leaf spot. 12:FL(35), NC(1303). 14:FL(35).
C. truncata. On living leaves. 7:LA(962).
C. vulpinae. Leaf spot. 15:KS(82,581), WI(1334).
Cercospora sp. Leaf spot. 11:WI(1350). 12:NC(1303). 14:NC(1303). 16:HI(2899,2903), NC(1303).
Cladosporium herbarum. 'Fruit rot. Raisin mold.' 14:CA(1110).
Cladosporium sp. Green-mold rot. 14:CA(82).
Corynespora sp. Leaf spot. 12:FL(35).
Cristulariella moricola. 'Target spot.' 8:MD(2848), WV(158). 10:IL(2572). 14:NC(1303).
Cylindrocarpon sp. Root rot. 12:FL(35). 16:FL(35).
Cylindrocladium sp. Root rot. 12:FL(35).
Epicoccum nigrum. 16:WA(3202).
Fumago sp. Sooty mold, following insects. 16:NC(1303).
Fusarium equiseti. 16:WA(3202).
F. poae. 16:WA(3202).
Fusarium sp. Root rot. 12:FL(35). 16:FL(35).

Helminthosporium siliquosum. On twigs. **15**:SC(82).
Helminthosporium sp. Leaf spot. **1**:FL(82). **8**:FL(35). **12**:FL(35). **14**:FL(35).
Myrothecium sp. Leaf spot. **12**:FL(35).
Oospora sp. **16**:WA(3202).
Penicillium expansum. **16**:WA(3202).
P. rolfsii. **16**:WA(3202).
P. variabile. **16**:WA(3202).
Penicillium sp. Blue mold rot. **8**:Range of host(82), FL(35). **14**:Range of host(82), CA(1110), FL(35). **16**:ID(3202).
Periconia byssoides. On dead herbaceous stems. **12**:NC(1303).
Phaeoramularia heterospora. Leaf spot. **5**:CA(81,82), OR(81).
Phymatotrichopsis omnivora. 'Phymatotrichum root rot.' **8**:AZ(82), NV(82), TX(82). **13**:TX(82). **14**:NV(82), TX(82).
Pseudocercospora vitis. Leaf spot. **1**:WI(1354). **8**:FL(35). **12**:FL(35). **14**:NC(1303). **16**:FL(35), NC(1303), TX(3597).
Ramularia sp. **16**:WA(3202).
Sporidesmium rauii. **16**:PA(2468).
Stemphylium sp. Storage decay. **16**:CA(1487).
Trichothecium roseum. **16**:WA(3202).
Tubercularia vulgaris. **16**:PA(3178).
Verticillium dahliae. 'Verticillium wilt.' **14**:CA(1110,3140). **16**:CA(2881).

Deuteromycotina - Coelomycetes

Aplosporella fabaeformis. On stems. **8**:KS(82), MD(82), NY(82). **15**:KS(82), MD(82), NJ(82), NY(82).
Ascochyta ampelina. On leaves. **15**:WI(82).
Colletotrichum crassipes. **16**:DE(3543).
C. gloeosporioides. Fruit rot, ripe rot. **8**:NY(82). **12**:NC(822,823,1303).
Coniella diplodiella. Dieback. 'White rot.' **8**:E states(82), FL(35), TX(82). **12**:FL(35). **16**:FL(35).
Coniothyrium ampelinum. On leaves. **1**:SC(634).
C. berlandieri. Leaf spot. **3**:TX(82).
Cryptostictis inaequalis. On leaves. **12**:IL(82).
Cytospora vitis. On twigs. **8**:KS(82), VA(82).
Cytosporina sp. Dieback. **8**:WA(1251).
Dichomera viticola. On dead stems. **5**:CA(82,1110).
Diplodia ampelina. On bark. **16**:SC(634).
D. viticola. Dieback, on canes, stems and twigs. **1**:PA(82). **8**:MD(82), MI(82), NJ(82), VA(82). **12**:FL(35). **14**:NJ(82).
Discosia artocreas. **16**:OK(660).
D. artocreas var. *viticola*. On leaves. **1**:SC(634).
Gloeodes pomigena. On stems. **15**:IN(82).
Greeneria uvicola. Stem fleck, scab and leaf spot. 'Bitter rot.' **8**:Range of host(82), FL(35). **12**:FL(35,82), GA(82,2199), MS(2720), NC(1303). **14**:FL(35), GA(82), NC(1303). **16**:FL(35), NC(1303), OK(2872), SC(634).
Hendersonia sarmentorum. On twigs. **8**:NJ(82), TX(82).
Lasiodiplodia theobromae. Dieback. 'Cane-tip blight. Summer bunch rot.' **12**:FL(35). **14**:CA(1110), FL(35).
Macrophoma farlowiana. On leaves. **1**:WI(1360). **8**:NY(82). **15**:NY(82).
M. longispora. On stems. **1**:SC(82).
M. reniformis. On stems. **8**:VA(82).
M. viticola. On leaves. **1**:SC(82).
Macrophoma sp. Dieback. **12**:FL(35).
Macrophomina phaseolina. Charcoal and root rot. **8**:TX(82). **14**:HI(2903).
Micropera ampelina. On limbs. **14**:NY(82).
Myxosporium viticola. On stems. **15**:NY(82).
Pestalotia menezesiana. Fruit rot, leaf and stem spot. **8**:CT(82), FL(35,82). **14**:FL(35).
P. pezizoides. Fruit rot, on stems. **8**:MD(82), NJ(82), VA(82). **14**:SC(82). **15**:KS(82), MD(82), SC(82).
[*Pestalotiopsis funerea*]. Leaf spot. **16**:MS(2720).
P. uvicola. Fruit rot. **8**:Range of host(82).
Phoma ampelina. On stems. **1**:PA(82).
P. vitis. On stems. **5**:CA(82,1110).
Phoma sp. Dieback, on fruit. **12**:FL(35). **14**:CA(1110).
Phomopsis viticola. Dead arm. 'Phomopsis cane blight. Phomopsis leaf spot.' **1**:NY(82). **8**:FL(35). **12**:FL(35). **14**:CA(1110), FL(35). **16**:FL(35).
Phomopsis sp. On stems. **8**:VA(82).
Phyllosticta ampelicida. Black rot of fruit, leaf spot. **6**:OK(2872). **8**:Unk.(82), FL(35), NJ(2). **11**:IL(2568). **12**:FL(35). **13**:OK(2872). **14**:FL(35), SD(2243). **15**:IA(64), ND(378), SC(2). **16**:FL(35), LA(658,2675), NC(1303).
P. spermoides. Leaf spot. **5**:CA(82,1110). **12**:NC(1303), VA(82). **15**:IL(82), KS(82), NY(82), WI(82).
Pilidium concavum. **11**:WI(1361).
Pyrenochaeta vitis. Leaf spot. **1**:FL(82), MS(82,2720), SC(82). **3**:TX(82). **8**:FL(35,82), MA(82), NC(82), NY(82), PA(82), TX(82), VA(82). **15**:NC(82), PA(82), TX(82). **16**:FL(35).
Rhabdospora mueggenburgii. On stems. **1**:LA(82). **15**:LA(82).
Seiridium unicorne. Leaf spot. **16**:MS(2720).
Septoria ampelina. Leaf spot. 'Melanose complex.' **1**:NY(82), VA(82). **8**:NY(82), TX(82), VA(82). **11**:WI(1354). **12**:NC(82), TX(82). **15**:IA(82), SC(82,634), TX(82).
S. kellermaniana. **15**:KS(82).
S. viticola. On leaves. **16**:SC(634).
Sphaceloma ampelinum. Spot anthracnose, bird's-eye rot. **8**:FL(35). **12**:FL(35). **14**:FL(35), TX(3556). **16**:FL(35,3556), MI(3556), MS(3556), SC(3556).

Deuteromycotina - Other

Rhizoctonia solani. Root rot. **12**:FL(35). **14**:FL(35).
Rhizoctonia sp. Damping-off, root rot. **12**:MS(2720). **16**:HI(2903).
Sclerotium rolfsii. On stems, buds, growing tips. **14**:CA(1110).

Xyridaceae

Xyris L.

1. *X. ambigua* Beyrich ex Kunth
2. *X. elata* Chapm.
3. *X. fimbriata* Elliott

Deuteromycotina - Hyphomycetes

Cercospora xyridis. Leaf spot. **2**:MS(581,2720).
Cladosporium xyridis. **3**:MS(2720).
Curvularia sp. Leaf spot. **1**:FL(35).

Zamiaceae

Dioon Lindl.

1. *Dioon* sp.

Basidiomycotina - Other

Ganoderma lucidum. Butt rot. **1**:FL(35).

Deuteromycotina - Hyphomycetes

Cercospora sp. Leaf spot. **1**:FL(35).
Penicillium sp. Stem rot. **1**:FL(35).

Deuteromycotina - Coelomycetes

Phoma sp. Dieback. **1**:FL(35).

Encephalartos Lehm.

1. *Encephalartos* sp.

Oomycetes

Pythium sp. Root rot. **1**:FL(35).

Deuteromycotina - Hyphomycetes

Fusarium sp. Root rot. **1**:FL(35).
Mycoleptodiscus sp. Leaf necrosis. **1**:FL(35).

Deuteromycotina - Coelomycetes

Coniothyrium sp. Leaf spot. **1**:FL(35).
Phomopsis sp. Leaf spot. **1**:FL(35).

Deuteromycotina - Other

Rhizoctonia sp. Root rot. **1**:FL(35).

Macrozamia Miq.

1. *Macrozamia* sp.

Oomycetes

Pythium sp. Root rot. **1**:FL(35).

Deuteromycotina - Hyphomycetes

Fusarium sp. Root rot. **1**:FL(35).
Mycoleptodiscus sp. Leaf necrosis. **1**:FL(35).

Zamia L.

1. *Z. debilis* Aiton
2. *Z. fisheri* Miq.
3. *Z. loddigesii* Miq.
4. *Z. muricata* Willd.

5. *Z. pumila* L. (=*Z. floridana* A. DC., *Z. furfuracea* L. fil., *Z. umbrosa* Small) [comptie, coontie, Florida arrowroot, sago-cycas, seminole-bread]
6. *Zamia* sp.

Oomycetes

Pythium sp. Root rot. **6**:FL(35).

Deuteromycotina - Hyphomycetes

Cercospora sp. Leaf spot. **2**:FL(35). **5**:FL(35).
Fusarium lateritium. Leaf necrosis. **3**:FL(35).
F. moniliforme. Leaf necrosis. **4**:FL(35).
Fusarium sp. Root rot. **5**:FL(35).
Mycoleptodiscus indicus. Leaf spot. **1**:FL(35). **6**:FL(35).
Penicillium sp. Isolated from ovules. **5**:FL(134).

Deuteromycotina - Coelomycetes

Diplodia sp. Leaf necrosis. **5**:FL(35).
Pestalotia cycadis. On leaves. **5**:FL(82), IN(82).
Phyllosticta sp. Leaf spot. **6**:FL(35).

Deuteromycotina - Other

Rhizoctonia solani. Root rot. **6**:FL(35).

Zingiberaceae

Alpinia Roxb.

1. *A. purpurata* (Vieill.) K. Schum. [red ginger]
2. *A. zerumbet* (Pers.) B. L. Burtt & R. M. Smith [pink porcelain-lily, shell ginger, shellflower]

Oomycetes

Pythium sp. Root rot. **1**:HI(2903).

Deuteromycotina - Hyphomycetes

Fusarium sp. Root rot. **2**:FL(35).

Deuteromycotina - Other

Rhizoctonia sp. Root rot. **1**:HI(2903).

Costus L.

1. *C. malortieanus* H. Wendl. [spiral ginger, spiral-flag, stepladder-plant]

Deuteromycotina - Hyphomycetes

Helminthosporium sp. Leaf spot. **1**:FL(35).

Curcuma L.

1. *C. roscoeana* Wallich

Deuteromycotina - Hyphomycetes

Alternaria sp. Leaf spot. **1**:FL(35).
Cercospora sp. Leaf spot. **1**:FL(35).

Elettaria Maton

1. *E. cardomomum* (L.) Maton [cardamom, Ceylon cardamom, Malabar cardamom]

Oomycetes

Pythium sp. Root rot. **1**:FL(35).

Deuteromycotina - Other

Rhizoctonia solani. Root rot. **1**:FL(35).

Hedychium J. König in Retz.

1. *H. coronarium* J. König ex Retz. [buterfly ginger, butterfly-lily, cinnamon-jasmine, garland-flower, ginger-lily, white ginger]
2. *H. gardneranum* Roscoe [Kahili ginger]
3. *Hedychium* sp.

Oomycetes

Pythium sp. Root rot. **2**:HI(2903). **3**:FL(35).

Ascomycotina

Mycosphaerella hedychii. On living leaves. **1**:HI(2903,3480).

Deuteromycotina - Coelomycetes

Colletotrichum sp. Anthracnose. **3**:FL(35).
Phomopsis sp. Leaf necrosis. **3**:FL(35).

Deuteromycotina - Other

Rhizoctonia sp. Root rot. **1**:HI(2903). **2**:HI(2903). **3**:FL(35).

Tapeinocheilos Miq.

1. *T. pungens* (Teijsm. & Binnend.) Miq.

Deuteromycotina - Hyphomycetes

Helminthosporium sp. Leaf spot. **1**:FL(35).

Zingiber Boehmer in Ludwig

1. *Z. officinale* Roscoe (=*Amomum zingiber* L.) [Canton ginger, common ginger, true ginger]
2. *Z. zerumbet* (L.) Sm. [zerumbet ginger]

Oomycetes

Pythium aphanidermatum. Root rot. **1**:HI(82,2903).
P. graminicola. **1**:HI(2903).
Pythium sp. **1**:HI(2903).

Deuteromycotina - Hyphomycetes

Cercospora zingiberi. Leaf spot. **1**:FL(35).
Fusarium oxysporum f. sp. *zingiberi*. **1**:HI(2903).
F. solani. **1**:HI(2903).
Fusarium sp. Rhizome rot. **1**:HI(82,2903).
Myrothecium sp. Rhizome rot. **1**:HI(2899,2903).

Deuteromycotina - Coelomycetes

Colletotrichum gloeosporioides. Anthracnose. **1**:FL(35).
Coniothyrium zingiberi. Leaf spot. **1**:HI(82,2903).
Phyllosticta zingiberis. Leaf spot. **2**:HI(2903).
Phyllosticta sp. **1**:HI(2903).

Deuteromycotina - Other

Rhizoctonia solani. Leaf blight, rhizome and root rot. **1**:FL(35), HI(2899).
Rhizoctonia sp. Root and rhizome rot. **1**:HI(2903).
Sclerotium rolfsii. Southern blight. **1**:FL(35).

Zosteraceae

Zostera L.

1. *Z. marina* L. [eelgrass]

Ascomycotina

Corollospora maritima. **1**:MA(1170).

Deuteromycotina - Hyphomycetes

Alternaria sp. **1**:MA(1170).
Asteromyces cruciatus. On leaves. **1**:CA(357), MA(1170).
Dendryphiella salina. **1**:MA(1170).
Sigmoidea marina. On leaves. **1**:VA(357).
Varicosporina ramulosa. Isolated from leaves. **1**:NC(357).

Zygophyllaceae

Guaiacum L.

1. *G. sanctum* L. [holy-wood lignum, lignum-vitae]

Oomycetes

Pythium splendens. Root rot. **1**:FL(35).

Basidiomycotina - Other

Ganoderma sp. Butt rot. **1**:FL(35).

Deuteromycotina - Hyphomycetes

Fusarium sp. Root rot. **1**:FL(35).

Larrea Cav.

1. *L. mexicana* Moric. (=*L. tridentata* (Sessé & Moc. ex DC.) J. M. Coulter) [creosote-bush]

Basidiomycotina - Other

Omphalia sp. On leaves. **1**:CA(82).
Phellinus texanus. **1**:AZ(1202).

Tribulus L.

1. *T. terrestris* L. [burnut, puncture-vine]

Deuteromycotina - Coelomycetes

Phyllosticta sp. Leaf spot. **1**:OK(2872).

Deuteromycotina - Other

Rhizoctonia solani. Root rot. **1**:TX(82).

Fungus List

Abortiporus Murrill. Basidiomycetes, Aphyllophorales

Abortiporus biennis (Bull.:Fr.) Singer
Heteroporus biennis (Bull.:Fr.) Lázaro
Polyporus biennis (Bull.:Fr.) Fr.
Polyporus abortivus Peck
Polyporus balloui C.G. Lloyd
Abortiporus distortus (Schwein.:Fr.) Murrill
Polyporus distortus (Schwein.:Fr.) Fr.
Polyporus oleae Panizzi, sensu Bonar
Distr.: Widespread.
Notes: On wood, usually of hardwoods. Gilbertson & Ryvarden (North American Polypores 1:85. 1986) stated that the fungus reported as *Polyporus oleae* on olive trees in CA is *A. biennis*.
Hosts: Acer, Carya, Magnolia, Malus, Melia, Nyssa, Olea, Quercus, Ulmus

Abortiporus fractipes (Berk. & M.A. Curtis) R.L. Gilbertson & Ryvarden
Polyporus fractipes Berk. & M.A. Curtis
Heteroporus fractipes (Berk. & M.A. Curtis) Fidalgo
Polyporus humilis Peck
Distr.: Central and eastern United States.
Notes: On hardwoods.
Hosts: Acer, Betula, Fagus

Absidia Tiegh. Zygomycetes, Mucorales

Absidia californica J.J. Ellis & Hesseltine
Distr.: CA.
Notes: Isolated primarily from dung but also from various organic substrates (Ellis & Hesseltine. Mycologia 57:230. 1965).
Host: Carya

Absidia corymbifera (Cohn) Sacc. & Trott.
Absidia ramosa (Lindt) Lendn.
Distr.: Widespread.
Notes: Isolated primarily from humans and other warm-blooded animals. A study of this species is presented by Nottebrock et al. (Sabouraudia 12:64. 1974). Lunn (C.M.I. Descr. 521. 1977).
Hosts: Carya, Zea

Absidia repens Tiegh.
Distr.: Widespread.
Notes: Isolated from soil and various kinds of organic matter (Hesseltine & Ellis. Mycologia 58:773. 1966).
Host: Zea

Absidia spinosa Lendn.
Distr.: Widespread.
Notes: Isolated primarily from soil and various organic substrates (Hesseltine & Ellis. Mycologia 56:580. 1964).
Host: Comandra

Acantharia Theiss. & Syd. Loculoascomycetes, Dothideales

Acantharia echinata (Ellis & Everh.) Theiss. & Syd.
Dimerosporium echinata Ellis & Everh.
Venturia echinata (Ellis & Everh.) Theiss.
Distr.: CA, OR.
Notes: On living leaves. The anamorph forms on the surface of the stroma (Sivanesan. Trans. Brit. Mycol. Soc. 82:507. 1984).
Host: Quercus

Acanthonitschkea Speg. Pyrenomycetes, Sordariales

Acanthonitschkea tristis (Pers.:Fr.) Nannf.
Calyculosphaeria tristis (Pers.:Fr.) Fitzp.
Distr.: FL; Europe.
Notes: Reported from North America and the Caribbean; however, Nannfeldt (Svensk Bot. Tidskr. 69:321. 1975) stated that this species is restricted to Europe.
Host: Magnolia

Acanthophiobolus Berl. Loculoascomycetes, Dothideales

Notes: Walker (Mycotaxon 11:2. 1980) reviewed this genus.

Acanthophiobolus helicosporus (Berk. & Broome) J. Walker
Ophiobolus helicosporus (Berk. & Broome) Sacc.
Leptospora helminthospora Rehm
Distr.: Temperate regions.
Notes: On various plant parts, mostly of monocotyledons, in damp places and on cloth.
Host: Sorghum

Acanthophysium (Pilát) G. Cunn. Basidiomycetes, Aphyllophorales

Aleurodiscus Rabenh. ex J. Schröt.

Acanthophysium bertii (C.G. Lloyd) Boidin in Boidin et al.
Aleurodiscus bertii C.G. Lloyd
Aleurodiscus cremeus Burt
Distr.: NM; Mexico.
Notes: On wood.
Hosts: Cercocarpus, Juniperus, Quercus

Acanthophysium canadense (Skolko) Parmasto
Aleurodiscus canadensis Skolko
Distr.: North America.
Notes: On wood.
Host: Pinus

Acanthophysium cerussatum (Bres.) Boidin in Boidin et al.
Aleurodiscus cerussatus (Bres.) Höhn. & Litsch.
Distr.: Temperate northern hemisphere.
Notes: On wood.
Hosts: Abies, Cercocarpus, Fraxinus, Juniperus, Physocarpus, Prunus, Quercus, Serenoa, Thuja

Acanthophysium diffissum (Sacc.) Parmasto
Aleurodiscus diffissus (Sacc.) Burt
Distr.: Western North America; northeastern Asia.
Notes: On dead twigs and branches, mostly of Ericaceae.
Hosts: Arbutus, Arctostaphylos, Ceanothus, Juglans, Juniperus, Lycium, Quercus

Acanthophysium fennicum (Laurila) Parmasto
Aleurodiscus fennicus Laurila
Distr.: Northeastern North America; northern Europe.
Notes: On conifers, mainly *Picea* spp.
Host: Picea

Acanthophysium fruticetorum (W.B. Cooke) Boidin in Boidin et al.
Aleurodiscus fruticetorum W.B. Cooke

Distr.: CA.
Notes: On twigs and stems.
Hosts: Arctostaphylos, Ceanothus
Acanthophysium lapponicum (Litsch.) Boidin in Boidin et al.
Aleurodiscus lapponicus Litsch.
Distr.: MO; Canada(Manitoba); Sweden.
Notes: On wood of hardwoods.
Hosts: Maclura, Populus
Acanthophysium lividocaeruleum (P. Karst.) Boidin in Boidin et al.
Corticium lividocaeruleum P. Karst.
Aleurodiscus lividocaeruleus (P. Karst.) P.A. Lemke
Gloeocystidiellum lividocaeruleum (P. Karst.) Donk
Distr.: Temperate northern hemisphere.
Notes: On conifers.
Hosts: Abies, [Acer], Cupressus, Juniperus, Larix, Picea, Pinus, [Populus], Pseudotsuga, Thuja, Tsuga
Acanthophysium macrocystidiatum (P.A. Lemke) Boidin in Boidin et al.
Aleurodiscus macrocystidiatus P.A. Lemke
Distr.: Pacific Northwest.
Notes: On recently dead branches.
Host: Arctostaphylos
Acanthophysium mirabile (Berk. & M.A. Curtis) Parmasto
Aleurodiscus mirabilis (Berk. & M.A. Curtis) Höhn.
Aleurodiscus apiculatus Burt
Aleurodiscus spinulosus Henn.
Distr.: Widespread.
Notes: On hardwood limbs.
Hosts: [Larix], Metrosideros, Rhododendron
Acanthophysium spiniger (D.P. Rogers & P.A. Lemke) Parmasto
Aleurodiscus spiniger D.P. Rogers & P.A. Lemke in P.A. Lemke
Distr.: Northwestern North America; China.
Notes: On conifers.
Host: Pseudotsuga
Acanthophysium succineum (Bres.) Boidin in Boidin et al.
Aleurodiscus succineus Bres.
Distr.: Pacific Northwest.
Notes: On wood.
Host: Arbutus
Acanthophysium weirii (Burt) Parmasto
Aleurodiscus weirii Burt
Distr.: Northern North America.
Notes: On conifers.
Hosts: Abies, Larix, Picea, Pseudotsuga, Thuja, Tsuga

Acanthostigma De Not. Loculoascomycetes, Dothideales

Notes: Arx & Müller (Stud. Mycol. 9:109. 1975) considered *Acanthostigma* to be congeneric with *Nematostoma* but did not formally transfer the species.
Acanthostigma berenice (Berk. & M.A. Curtis) Sacc.
Venturia saccardioides Ellis & G. Martin
Distr.: Southeastern United States.
Host: Magnolia
Acanthostigma decastylum (Cooke) Sacc.
Lasiosphaeria subvelutina Ellis & Everh.
Distr.: Eastern United States.
Hosts: Pinus, Quercus, Rubus
Acanthostigma sequoiae (Plowr.) Sacc.
Laestadia consociata (Ellis & Harkn.) Cooke
Physalospora consociata (Ellis & Harkn.) Sacc.
Wallrothiella consociata (Ellis & Harkn.) Ellis & Everh.
Distr.: CA.
Notes: This may be a synonym of *Seynesiella sequoiae* (Cooke & Harkness) E. Müller.
Host: Sequoiadendron

Acarosporina M.A. Sherwood. Discomycetes, Ostropales

Notes: Sherwood (Mycotaxon 6:215. 1977).
Acarosporina hormospora (Speg.) M.A. Sherwood
Schizoxylon hormosporum Speg.
Schizoxylon abutilonis Cash
Distr.: HI; Argentina.
Notes: On woody plants, associated with cankers.
Host: Abutilon
Acarosporina microspora (R.W. Davidson & Lorentz) M.A. Sherwood
Schizoxylon microspora R.W. Davidson & Lorentz
Distr.: Northeastern North America.
Notes: Bark canker of maples. An unnamed pycnidial stage is also associated with the cankers.
Host: Acer
Acarosporina monilifera (Ellis & Everh.) M.A. Sherwood
Schizoxylon moniliferum Ellis & Everh.
Distr.: Northeastern United States.
Notes: On bark cankers.
Hosts: Fagus, Quercus

Acerbia (Sacc.) Sacc. & Syd. Loculoascomycetes, Dothideales

Notes: This genus is considered a synonym of *Ophiobolus*, but not all names have been redisposed.
Acerbia bacillata (Cooke) Berl.
Ophioceras bacillata (Cooke) Sacc.
Distr.: CA; Europe.
Host: Penstemon

Achrotelium Syd. in Syd. & Petr. Basidiomycetes, Uredinales

Achrotelium lucumae Cummins
Distr.: FL; Cuba.
Notes: On Sapotaceae. Cummins (Mycologia 48:601. 1956); Cummins (Bull. Torrey Bot. Club 67:70. 1940).
Host: Pouteria

Acremoniella Sacc. Hyphomycetes

Notes: This name is considered illegitimate according to Holubová-Jechová (Folia Geobot. Phytotax. 9:315. 1974) and is replaced by *Harzia*, but not all names in *Acremoniella* have been redisposed.
Acremoniella alascensis R. Sprague
Distr.: AK.
Host: Agrostis

Acremonium Link:Fr. Hyphomycetes

Notes: Gams (*Cephalosporium*-artige Schimmelpilze [Hyphomycetes]. 1971).
Acremonium bacillisporum (Onions & Barron)W. Gams
Paecilomyces bacillisporum Onions & Barron
Distr.: Widespread.
Notes: Isolated from soil, stored grain and various kinds of debris.
Host: Acer
Acremonium chisosum J.F. White & Morgan-Jones
Distr.: TX.
Notes: White & Morgan-Jones (Mycotaxon 28: 179. 1987).
Host: Stipa
Acremonium coenophialum Morgan-Jones & W. Gams
Distr.: Southeastern United States; possibly New Zealand.
Notes: Morgan-Jones & W. Gams (Mycotaxon 15:311. 1982).
Hosts: Festuca, Poa
Acremonium crotocinigenum (Schol-Schwarz) W. Gams
Cephalosporium crotocinigenum Schol-Schwarz
Distr.: Temperate and subtropical regions.
Notes: Leaf spot. Also isolated from fungi and dead wood.
Hosts: Philodendron, Syngonium
Acremonium curvulum W. Gams
Distr.: Widespread.
Notes: Isolated from soil, leaves and various kinds of debris.
Hosts: Cheirodendron, Metrosideros
Acremonium diospyri (Crandall)W. Gams

Cephalosporium diospyri Crandall
Distr.: Southern United States.
Notes: Persimmon wilt.
Host: Diospyros
Acremonium huerfanum J.F. White, J.R. Cole & Morgan-Jones
Distr.: CO.
Notes: Endophytic in grasses (White et al. Mycologia 79:150. 1987; White & Morgan-Jones. Mycotaxon 30:92. 1987).
Host: Festuca
Acremonium implicatum (Gilman & E. Abbott) W. Gams
Monilia implicatua Gilman & E. Abbott
Acremonium terricola (J.H. Miller, J.E. Giddens & A.A. Foster) W. Gams
Fusidium terricola J.H. Miller, J.E. Giddens & A.A. Foster
Distr.: Temperate regions(rare); subtropical and tropical regions.
Notes: Isolated from soil, fungi, sugar cane, sunflower seeds and various kinds of debris.
Host: Helianthus
Acremonium sclerotigenum (Valenta)W. Gams
Cephalosporium sclerotigenum Valenta
Distr.: Widespread.
Notes: Isolated from soil, fungi, sunflower seeds and various kinds of debris.
Host: Helianthus
Acremonium starrii J.F. White & Morgan-Jones
Distr.: TX, western North America.
Notes: Endophytic in grasses (White & Morgan-Jones. Mycotaxon 30:88. 1987).
Hosts: Bromus, Festuca
Acremonium strictum W. Gams
Cephalosporium acremonium Auct. non Corda
Distr.: Cosmopolitan.
Notes: Common in soil and on plant surfaces.
Hosts: Acacia, Alnus, Carya, Chrysanthemum, Ficus, Glycine, Gossypium, Metrosideros, Sorghum, Triticum, Zea
Acremonium typhinum Morgan-Jones & W. Gams
Epichloe typhina (Pers.:Fr.) Tul. in Tul. & C. Tul. (teleomorph)
Distr.: Temperate regions.
Notes: Endophytic in grasses (White & Morgan-Jones. Mycotaxon 29:489. 1987). White & Morgan-Jones (Mycotaxon 30:94. 1987) provided a key to species in the *albo-lanosa* section of *Acremonium*.
Hosts: Agrostis, Elymus, Sphenopholis
Acremonium zonatum (Sawada) W. Gams
Cephalosporium zonatum Sawada
Cephalosporium fici Tims & Olive
Distr.: Subtropical and tropical regions.
Notes: Zonate leaf spot of figs (Hawksworth. C.M.I. Descr. 502. 1976).
Hosts: Eichhornia, Ficus

Acroconidiella Lindquist & Alippi. Hyphomycetes

Acroconidiella eschscholtziae (Harkn.) M.B. Ellis
Heterosporium eschscholtziae Harkn.
Distr.: CA(type); Europe; Mauritius.
Notes: On leaves and capsules (Ellis. More dematiaceous Hyphomycetes. p. 407. 1972).
Host: Eschscholzia
Acroconidiella tropaeoli (Bondartsev) Lindquist & Alippi
Heterosporium tropaeoli Bondartsev
Distr.: North and South America; Africa; Australia; East Indies.
Notes: Leaf spot of nasturtium (Ellis. C.M.I. Descr. 161. 1968).
Host: Tropaeolum

Acrodictys M.B. Ellis. Hyphomycetes

Acrodictys atroapicula Wang & Sutton
Distr.: Northeastern United States.
Notes: On wood (Wang & Sutton. Mycologia 74:492. 1982).
Hosts: Acer, Betula, Quercus
Acrodictys erecta (Ellis & Everh.) M.B. Ellis
Mystrosporium erectum Ellis & Everh.
Macrosporium erectum (Ellis & Everh.) Pound & Clements
Piricauda serendipita R.T. Moore
Distr.: Central and southern United States; Venezuela.
Notes: On *Arundo* and maize (Ellis. Dematiaceous Hyphomycetes. p. 130. 1971).
Host: Zea
Acrodictys fimicola M.B. Ellis & Gunnell in Ellis
Distr.: HI; Malaya; Sierra Leone.
Notes: On elephant dung, bamboo, and *Argyroxiphium* (Ellis. Dematiaceous Hyphomycetes. p. 130. 1971).
Host: Argyroxiphium

Acrodontium De Hoog. Hyphomycetes

Notes: De Hoog (Stud. Mycol. 1:23. 1972).
Acrodontium intermissum De Hoog & V.G. Rao
Distr.: Eastern United States; Norway(type).
Notes: Isolated from wood and soil (De Hoog & Rao. Persoonia 8:207. 1975).
Host: Pinus

Acrospermoides J.H. Miller & G.E. Thompson. Loculoascomycetes, Dothideales

Acrospermoides subulata J.H. Miller & G.E. Thompson
Distr.: GA(type)
Host: Morus

Acrospermum Tode:Fr. Loculoascomycetes, Dothideales

Acrospermum album Peck
Distr.: NY.
Host: Aralia
Acrospermum compressum Tode:Fr.
Dactylaria sp. (anamorph)
Distr.: Temperate North America; Europe.
Hosts: Agrostis, Alnus, Betula, Boehmeria, Calamagrostis, Celtis, Cinna, Morus, Prunella, Ulmus
Acrospermum foliicola Berk.
Acrospermum compressum Tode:Fr. var. *foliicola* (Berk.) Riddle
Distr.: Southeastern United States.
Hosts: Acer, Aster, Baptisia, Broussonetia, Cercis, Cirsium, Coreopsis, Cornus, Crataegus, Eupatorium, Fraxinus, Hamamelis, Morus, Ostrya, Polymnia, Quercus, Rhododendron, Solidago, Vaccinium
Acrospermum graminum Lib.
Virgariella sp. (anamorph)
Distr.: Temperate regions.
Notes: Webster (Trans. Brit. Mycol. Soc. 39:361-366. 1956) described the conidial state and compared it with that of *A. compressum*.
Hosts: Carex, Dactylis, Erianthus, Lolium, Panicum, Paspalum
Acrospermum urceolatum M.E. Olson
Distr.: MN, WA.
Host: Selaginella

Acrothecium (Corda) G. Preuss. Hyphomycetes

Acrothecium melanopus (Schwein.) Sacc.
Dactylium melanopus Schwein.
Distr.: Eastern United States.
Host: Allium

Actidium Fr. Loculoascomycetes, Dothideales

Notes: Zogg (Beitr. Kryptogamenfl. Schweiz 11(3):122. 1962).
Actidium nitidum (Cooke & Ellis) Zogg
Glonium nitidum Ellis ex Cooke & Ellis
Distr.: NJ(type), WA.
Notes: On dead wood and bark of conifers.
Host: Juniperus

Actinomucor Schostakowitsch. Zygomycetes, Mucorales

Actinomucor elegans (Eidam) C.R. Benjamin & Hesseltine
Rhizopus elegans Eidam
Mucor glomerula Lendn.
Distr.: Widespread.
Notes: Isolated from soil and various kinds of organic substrates (Benjamin & Hesseltine. Mycologia 49:241. 1957). This fungus is used to ferment several foods.
Hosts: Zea, Zebrina

Actinonema Pers.:Fr. Coelomycetes

Notes: Vassiljevsky & Karakulin (Fungi Imperfecti Parasitici. Pars II. Melanconiales. p. 377. 1950.) reduced *Actinonema* to synonymy with *Marssonina*. Lectotypification by Hughes (Canad. J. Bot. 36:733. 1958) placed *Actinonema* in the Mycelia Sterilia, but not all names have been redisposed.
Actinonema psoraleae Ellis & Everh.
Distr.: KS(type).
Host: Pediomelum

Actinothecium Ces. Coelomycetes

Actinothecium juglandis Dearn. & House
Distr.: NY(type).
Host: Juglans

Aecidium Pers.:Pers. Basidiomycetes, Uredinales

Aecidium ageratinae Savile
Distr.: Eastern North America, south to Mississippi.
Notes: Savile (Canad. J. Bot. 50:2590. 1972).
Host: Ageratina
Aecidium arcularium Arth.
Distr.: AR, CO, NM.
Host: Brickellia
Aecidium avocense Cummins & H.C. Greene in H.C. Greene
Puccinia avocensis Cummins & H.C. Greene in H.C. Greene (teleomorph)
Distr.: WI.
Notes: The connection with *P. avocensis* is tentative.
Host: Callirhoe
Aecidium batesii Arth.
Distr.: NE.
Host: Rudbeckia
Aecidium bethelii Arth.
Distr.: CA.
Host: Ludwigia
Aecidium boehmeriae Arth.
Distr.: IN, MD, NY.
Host: Boehmeria
Aecidium borrichiae Syd. & P. Syd.
Distr.: AL.
Host: Borrichia
Aecidium bouvardiae Dietel & Holw.
Puccinia diplachnis Arth. (teleomorph)
Distr.: Southwestern United States.
Notes: Heteroecious.
Host: Bouvardia
Aecidium campanulastri G.W. Wils.
Distr.: Central United States.
Host: Campanula
Aecidium chamaecristae Arth.
Distr.: Central United States.
Host: Cassia
Aecidium columbiense Ellis & Everh.
Distr.: Western North America, NY.
Host: Hieracium
Aecidium cyrillae Arth.
Distr.: Southeastern United States.
Host: Cyrilla
Aecidium eupatorii-rotundifolii Savile
Distr.: Eastern United States.
Notes: Savile (Canad. J. Bot. 50:2591. 1972).
Host: Eupatorium
Aecidium graebnerianum Henn.
Puccinia praegracilis Arth. (teleomorph)
Distr.: Northern North America.
Notes: Aecial host - Orchidaceae; telial host - Poaceae.
Hosts: Habenaria, Orchis
Aecidium hesleri Arth.
Distr.: TN.
Host: Lithospermum
Aecidium humuli Hotson
Distr.: WA(type).
Host: Humulus
Aecidium hydnoideum Berk. & M.A. Curtis
Distr.: AL, MO; southern Canada.
Host: Dirca
Aecidium insulsum Arth.
Distr.: UT.
Host: Collinsia
Aecidium ivae H. Jacks.
Distr.: DE to LA(coastal).
Host: Iva
Aecidium libertum Arth.
Distr.: OK.
Host: Urtica
Aecidium lycii Arth.
Distr.: AZ.
Host: Lycium
Aecidium mesadeniae Arth.
Distr.: Central United States.
Host: Cacalia
Aecidium mexicanum Dietel & Holw. in Holw.
Distr.: OK; Mexico.
Host: Cissus
Aecidium minutulum H. Jacks. in Barth.
Distr.: TX(type).
Host: Rumex
Aecidium modestum Arth.
Distr.: TX.
Host: Manfreda
Aecidium physalidis Burrill
Distr.: Central and western North America.
Notes: Compare with the aecial state of *Puccinia kansensis* Ellis & Barth. (Parmelee & de Carteret. Fungi Canadenses 186. 1980).
Host: Physalis
Aecidium plenum Arth.
Distr.: TX.
Host: Argemone
Aecidium ranunculacearum DC.
Distr.: Circumboreal.
Notes: This species has been mistaken for aecia of *Puccinia eatoniae*.
Hosts: Anemone, Ranunculus
Aecidium renatum Arth.
Distr.: NM.
Host: Polygala
Aecidium residuum Arth.
Distr.: OK, TX.
Host: Sisyrinchium
Aecidium rubromaculans E. West
Distr.: FL.
Host: Viburnum
Aecidium subsimulans Arth. & Mains
Puccinia biporula J.W. Baxter (teleomorph)
Distr.: Southwestern United States; Mexico.
Host: Salvia
Aecidium tithymali Arth.
Distr.: Central and eastern United States.
Host: Euphorbia
Aecidium tracyanum Syd. & P. Syd.
Distr.: FL; Mexico.
Host: Dyschoriste
Aecidium triostei Arth.
Distr.: MO.
Host: Triosteum

Aecidium valerianellae Biv.-Bern.
Puccinia gladioli Castagne (teleomorph)
Distr.: North America; Europe.
Notes: The telia are not known to occur in the United States (Cummins. Suppl. to Arthur's Manual of the rusts. p. 2A. 1962).
Host: Plectritis

Agaricus L., nom. cons. Basidiomycetes, Agaricales

Agaricus subrufescens Peck
Pholiota fulvosquamosa Peck
Distr.: Northeastern United States.
Notes: Redhead (Sydowia 37:250. 1984).
Host: Acer

Aglaospora De Not. Loculoascomycetes, Pyrenulales

Aglaospora profusa (Fr.:Fr.) De Not.
Aglaospora anomia (Fr.:Fr.) Lambotte
Distr.: Eastern North America; Europe.
Notes: Barr (Mycotaxon 9:20. 1979); Shoemaker & Kokko (Fungi Canadenses 101. 1977).
Hosts: Carpinus, Gleditsia, Robinia

Agrocybe Fayod. Basidiomycetes, Agaricales

Agrocybe aegerita (V. Briganti) Singer
Pholiota aegerita (V. Briganti) Quél.
Distr.: ID; Europe.
Host: Thuja
Agrocybe molesta (Lasch) Singer
Agrocybe dura Auct. non (Bolton:Fr.) Singer
Pholiota dura Auct. non (Bolton:Fr.) P. Kumm.
Distr.: Worldwide in grasslands.
Hosts: Avena, Hordeum, Triticum
Agrocybe praecox (Pers.:Fr.) Fayod
Pholiota praecox (Pers.:Fr.) P. Kumm.
Distr.: Temperate regions.
Hosts: Avena, Hordeum, Triticum

Alatospora Ingold. Hyphomycetes

Alatospora acuminata Ingold
Distr.: Widespread.
Notes: On leaf litter and in foam in aquatic habitats (Nilsson. Symb. Bot. Upsal. 18:78. 1964).
Hosts: Acer, Quercus

Albatrellus S.F. Gray. Basidiomycetes, Aphyllophorales

Albatrellus peckianus (Cooke) Niemelä
Polyporus peckianus Cooke
Distr.: Central and eastern North America.
Notes: Gilbertson & Ryvarden (North American Polypores 1:103. 1986) reported that *A. peckianus* may be terrestrial or inhabit buried wood.
Hosts: Fagus, Tilia

Albotricha Raitviir. Discomycetes, Helotiales

Albotricha washingtonensis (Dennis) Raitviir
Dasyscyphus washingtonensis Dennis
Lachnella pteridicola Seaver
Hyalopeziza pteridis Kanouse
Distr.: Temperate regions.
Host: Pteridium

Albugo (Pers.) Roussel. Oomycetes, Peronosporales

Albugo amaranthi (Schwein.) Kuntze
Distr.: FL.
Notes: The place of publication of this name could not be found.
Host: Acnida
Albugo bliti (Biv.-Bern.) Kuntze
Cystopus bliti (Biv.-Bern.) Lév.
Albugo froelichiae G.W. Wils.
Distr.: Cosmopolitan.
Hosts: Acnida, Amaranthus, Chenopodium, Froelichia, Gomphrena
Albugo candida (Pers.) Kunze
Cystopus candidus (Pers.) Lév.
Albugo cruciferarum (DC.) S.F. Gray
Distr.: Cosmopolitan.
Notes: White rust and staghead of Brassicaceae (Mukerji. C.M.I. Descr. 460. 1975).
Hosts: Arabidopsis, Arabis, Armoracia, Barbarea, Brassica, Camelina, Capparis, Capsella, Cardamine, Cheiranthus, Cochlearia, Conringia, Dentaria, Descurainia, Draba, Erysimum, Hesperis, Iberis, Lepidium, Nasturtium, Raphanus, Rorippa, Sibara, Sinapis, Sisymbrium, Stanleya, Thelypodium, Thlaspi, Thysanocarpus
Albugo caryophyllacearum (Wallr.) Cif. & Biga in Biga
Albugo lepigoni (de Bary) Kuntze
Distr.: North and South America; Europe.
Hosts: Spergula, Spergularia
Albugo ipomoeae-panduratae (Schwein.) Swingle
Cystopus ipomoeae-panduratae (Schwein.) F. Stevens & Swingle
Cystopus ipomoeae-panduranae (Schwein.) F. Stevens & Swingle, orth. var.
Distr.: Cosmopolitan.
Notes: White rust of Convolvulaceae (Mukerji & Critchett. C.M.I. Descr. 459. 1975)
Hosts: Calystegia, Convolvulus, Ipomoea, Jacquemontia, Merremia
Albugo occidentalis G.W. Wils.
Distr.: North America; Europe.
Hosts: Chenopodium, Spinacia
Albugo platensis (Speg.) Swingle
Distr.: Cosmopolitan.
Hosts: Allionia, Boerhavia, Mirabilis
Albugo portulacae (DC.) Kuntze
Cystopus portulacae (DC.) Lév.
Albugo quadrata (Wallr.) S.D. Baker
Distr.: Cosmopolitan.
Hosts: Portulaca, Portulacaria
Albugo tragopogonis (Pers.) S.F. Gray
Cystopus tragopogonis (Pers.) J. Schröt.
Distr.: Cosmopolitan.
Notes: White rust of *Tragopogon* and other Asteraceae (Mukerji. C.M.I. Descr. 458. 1975).
Hosts: Ambrosia, Antennaria, Artemisia, Centaurea, Cirsium, Helianthus, Iva, Matricaria, Parthenium, Scorzonera, Senecio, Tragopogon, Xanthium
Albugo trianthemae G.W. Wils.
Distr.: Southwestern United States.
Hosts: Sesuvium, Trianthema

Aleurobotrys Boidin, Lanquetin, Gilles, Candoussau & Hugueney. Basidiomycetes, Aphyllophorales

Notes: Boidin et al. (Bull. Soc. Mycol. France 101: 355. 1985).
Aleurobotrys botryosus (Burt) Boidin, Lanquetin, Gilles, Candoussau & Hugueney
Aleurodiscus botryosus Burt
Acanthophysium botryosum (Burt) G. Cunn.
Distr.: Widespread.
Notes: On limbs of conifers and hardwoods.
Hosts: Juniperus, [Lonicera], Myrica, Ostrya, [Rubus], [Symphoricarpos], Thuja, Vitis

Aleurocystidiellum P.A. Lemke. Basidiomycetes, Aphyllophorales

Aleurocystidiellum disciforme (DC.:Fr.) Boidin, P. Terra & Lanquetin
Aleurodiscus disciformis (DC.:Fr.) Pat.
Distr.: Western North America; Europe; Soviet Union.
Notes: On bark.
Host: Quercus
Aleurocystidiellum subcruentatum (Berk. & M.A. Curtis) P.A. Lemke
Aleurodiscus subcruentatus (Berk. & M.A. Curtis) Burt
Distr.: Temperate northern hemisphere.
Notes: On conifers.
Hosts: Abies, Picea, Pseudotsuga

Aleurodiscus Rabenh. ex J. Schröt., nom. cons. Basidiomycetes, Aphyllophorales

Notes: Lemke (Canad. J. Bot. 42:213. 1964).
Aleurodiscus abietis H. Jacks. & P.A. Lemke in P.A. Lemke
Distr.: Northern North America.
Notes: On limbs of conifers.
Hosts: Abies, Tsuga
Aleurodiscus amorphus (Pers.:Fr.) J. Schröt. in Cohn
Distr.: Temperate northern hemisphere.
Notes: On suppressed, injured, or recently dead conifers, especially *Abies* spp.
Hosts: Abies, Larix, Picea, Pinus, Pseudotsuga, Thuja, Tsuga
Aleurodiscus aurantius (Pers.:Fr.) J. Schröt. in Cohn
Distr.: Northwestern North America; South Americ; Europe; Japan; New Zealand.
Notes: On dead twigs and stems.
Hosts: Arctostaphylos, Lithocarpus, Rhododendron, Ribes, Rubus, Umbellularia
Aleurodiscus farlowii Burt
Distr.: North America.
Notes: On twigs and limbs of conifers.
Hosts: Abies, Tsuga
Aleurodiscus grantii C.G. Lloyd
Distr.: Western North America.
Notes: Usually on small, dead limbs of living trees or on dead trees.
Hosts: Abies, Picea, Pinus, Pseudotsuga, Thuja, Tsuga
Aleurodiscus mesaverdensis Lindsey
Distr.: CO.
Notes: On bark.
Host: Pinus
Aleurodiscus oakesii (Berk. & M.A. Curtis) Höhn. & Litsch.
Distr.: Temperate northern hemisphere.
Notes: On bark, often of living trees.
Hosts: Betula, Carpinus, Carya, Fraxinus, Ostrya, Quercus, Salix, Tilia, Ulmus
Aleurodiscus penicillatus Burt
Distr.: Northern North America.
Notes: On limbs of conifers.
Hosts: Picea, Pinus, Pseudotsuga, Tsuga
Aleurodiscus piceinus Lyon & P.A. Lemke in P.A. Lemke
Distr.: Northeastern North America.
Host: Picea
Aleurodiscus tsugae Yasuda in C.G. Lloyd
Distr.: Northern North America; Japan.
Notes: On limbs of conifers.
Host: Thuja
Aleurodiscus utahensis Lindsey & R.L. Gilbertson
Distr.: UT(type).
Notes: On wood of conifers.
Host: Pinus

Allantonectria Earle. Pyrenomycetes, Hypocreales

Allantonectria yuccae Earle
Distr.: Southwestern United States; Mexico.
Host: Yucca

Allantoporthe Petr. Pyrenomycetes, Diaporthales

Notes: Barr (Mycol. Mem. 7:152. 1978).
Allantoporthe decedens (Fr.:Fr.) Barr
Diaporthe decedens (Fr.:Fr.) Fuckel
Phomopsis sp. (anamorph)
Distr.: Temperate North America; Europe.
Hosts: Corylus, Umbellularia
Allantoporthe tessella (Pers.:Fr.) Petr.
Diaporthe tessella (Pers.:Fr.) Rehm
Distr.: Temperate North America; Europe.
Notes: On limbs.
Host: Salix

Allantospora Wakk. Hyphomycetes

Notes: This genus is considered a synonym of *Cylindocarpon* (Booth. Mycol. Pap. 104. 1966), but not all names have been redisposed.
Allantospora radicicola Wakk.
Distr.: North America(questionable); Java(type).
Notes: Booth (Mycol. Pap. 104. 1966) indicated this name is probably based on a *Cylindrocarpon* species.
Host: Saccharum

Allescheriella Henn. Hyphomycetes

Allescheriella crocea (Mont.) S.J. Hughes in R.E.D. Baker & W.T. Dale
Gymnosporium fulvum Berk. & M.A. Curtis
Rhinotrichum fulvum Berk. & M.A. Curtis in Berk.
Botryobasidium croceum Lentz (teleomorph)
Distr.: Warm temperate and tropical regions.
Notes: On woody debris (Hughes. Mycol. Pap. 41:1. 1951). The teleomorph was discovered by Lentz (Mycopath. Mycol. Appl. 32:1. 1967).
Host: Casuarina

Allomyces E.J. Butler. Chytridiomycetes, Blastocladiales

Allomyces arbuscula E.J. Butler
Distr.: Widespread.
Notes: Isolated from various kinds of organic debris especially insect cadavers.
Host: Colocasia

Alpakesa Subramanian & Ramakrishan. Coelomycetes

Notes: Morgan-Jones et al. (Canad. J. Bot. 50:879. 1972); Sutton (The Coelomycetes. p. 56. 1980).
Alpakesa nolinae (Pollack) Morgan-Jones, Nag Raj, & Kendrick
Bartalinia nolinae Pollack
Distr.: AZ.
Notes: On leaves.
Host: Nolina
Alpakesa uniseptata Morgan-Jones, Nag Raj, & Kendrick
Distr.: TX.
Notes: On leaves.
Hosts: Nolina, Yucca
Alpakesa yuccifolia (J.G. Hall) Subramanian & Ramakrishan
Neottiospora yuccifolia J.G. Hall
Distr.: Temperate North America.
Notes: On leaves.
Host: Yucca

Alternaria Nees, nom. cons. Hyphomycetes

Notes: Ellis (Dematiaceous Hyphomycetes. p. 464. 1971; More Dematiaceous Hyphomycetes. p. 411. 1976).
Alternaria abutilonis (Speg.) C.A. Schwarze
Macrosporium abutilonis Speg.
Distr.: WI; Italy(type).
Host: Abutilon

Alternaria alternantherae G.E. Holcomb & A.A. Antonopoulos
Distr.: Southeastern United States.
Host: Alternanthera
Alternaria alternata (Fr.:Fr.) Keissl.
Alternaria fasciculata (Cooke & Ellis) L. Jones & Grout
Macrosporium fasciculatum Cooke & Ellis
Macrosporium maydis Cooke & Ellis
Alternaria tenuis Nees
Distr.: Cosmopolitan.
Hosts: Acacia, Acer, Agropyron, Allium, Alnus, Antirrhinum, Arachis, Arceuthobium, Asclepias, Asparagus, Avena, Beta, Brassica, Bromus, Calathea, Callistephus, Calocedrus, Camellia, Cannabis, Carica, Carya, Chrysalidocarpus, Chrysanthemum, Clarkia, Cucumis, Cucurbita, Dahlia, Daucus, Dendrobium, Dichondra, Dimorphotheca, Fagopyrum, Ficus, Gladiolus, Glycine, Gossypium, Hedera, Helianthus, Hibiscus, Hordeum, Ilex, Juncus, Juniperus, Lathyrus, Lens, Ligustrum, Linum, Lycopersicon, Magnolia, Malus, Medicago, Metrosideros, Nicotiana, Oenothera, Pandanus, Pastinaca, Pelargonium, Petroselinum, Petunia, Phaseolus, Pieris, Pinus, Pisum, Plectranthus, Populus, Prunus, Pseudotsuga, Ptelea, Pyrus, Rhododendron, Rhus, Robinia, Secale, Spartina, Syagrus, Syringa, Tephrosia, Tragopogon, Trifolium, Trigonella, Triticum, Vaccinium, Vicia, Vigna, Vitis, Yucca, Zea
Alternaria amaranthi (Peck) Van Hook
Distr.: Eastern United States.
Hosts: Amaranthus, Chenopodium, Froelichia
Alternaria angustiovoidea E. Simmons
Distr.: ND; Canada(Manitoba, type).
Notes: Simmons (Mycotaxon 25:198. 1986) reviewed several species of *Alternaria* on *Euphorbia*.
Host: Euphorbia
Alternaria araliae H.C. Greene
Distr.: WI(type).
Host: Aralia
Alternaria atrans W. Gibson
Distr.: AZ.
Host: Glycine
Alternaria brassicae (Berk.) Sacc.
Macrosporium brassicae Berk.
Alternaria herculea (Ellis & G. Martin) J.A. Elliott
Macrosporium herculeum Ellis & G. Martin
Distr.: Cosmopolitan.
Notes: Ellis (C.M.I. Descr. 162. 1968).
Hosts: Abelmoschus, Armoracia, Brassica, Lepidium, Lunaria, Raphanus, Ricinus, Syringa, Triticum, Zea
Alternaria brassicicola (Schwein.) Wiltshire
Helminthosporium brassicicola Schwein.
Alternaria brassicae (Berk.) Sacc. var. *microspora* C. Brun
Alternaria circinans (Berk. & M.A. Curtis) P.C. Bolle
Alternaria oleracea Milbrath
Distr.: Cosmopolitan.
Notes: Leafspot of cabbage and cauliflower (Ellis. C.M.I. Descr. 163. 1968).
Hosts: Armoracia, Brassica, Crambe, Lunaria, Raphanus, [Rosa]
Alternaria camelliae (Cooke & Massee) Joly
Macrosporium camelliae Cooke & Massee
Distr.: Southeastern United States; England(type).
Host: Camellia
Alternaria cassiae A.M.M. Jurair & A. Khan
Distr.: Southeastern United States; India; Pakistan.
Notes: Simmons (Mycotaxon 14:24. 1982).
Hosts: Senna, Vicia
Alternaria catalpae (Ellis & G. Martin) Joly
Macrosporium catalpae Ellis & G. Martin
Distr.: Eastern United States.
Host: Catalpa
Alternaria cheiranthi (Lib.) P.C. Bolle
Helminthosporium cheiranthi Lib.
Macrosporium cheiranthi (Lib.) Fr.
Distr.: OK; Europe.
Notes: Common on wallflower, reported from throughout Europe (Ellis. Dematiaceous Hyphomycetes. p. 472. 1971). The report from United States is questionable.
Host: [Spinacia]
Alternaria cinerariae S. Hori & Enjoji
Alternaria senecionis Neergaard
Distr.: CA, FL; Europe; Japan.
Notes: Leaf spot (Ellis. More dematiaceous Hyphomycetes. p. 418. 1976).
Host: Senecio
Alternaria citri Ellis & N. Pierce in N. Pierce
Distr.: Worldwide.
Notes: Ellis & Holliday (C.M.I. Descr. 242. 1970).
Hosts: Citrofortunella, Citrus, Fortunella, Murraya, Phoenix, Prunus, Psidium
Alternaria compacta (Cooke) McClellen
Distr.: Eastern North America; Europe.
Host: Ricinus
Alternaria crassa (Sacc.) Rands
Cercospora daturae Peck
Distr.: North America; Africa; Asia; Europe.
Notes: Leaf spots and lesions (Ellis. Dematiaceous Hyphomycetes. p. 485. 1971; Ellis & Holliday. C.M.I. Descr. 243. 1970).
Host: Datura
Alternaria cucumerina (Ellis & Everh.) J.A. Elliott
Macrosporium cucumerinum Ellis & Everh.
Distr.: Cosmopolitan.
Notes: Ellis & Holliday (C.M.I. Descr. 244. 1970).
Hosts: Citrullus, Cucumis, Cucurbita, Cyamopsis
Alternaria dauci (Kühn) Groves & Skolko
Alternaria carotae (Ellis & Langl.) J.A. Stevenson & Wellman
Distr.: Cosmopolitan.
Hosts: Daucus, Petroselinum
Alternaria dianthicola Neergaard
Distr.: Widespread.
Notes: Flower bud rot of carnation.
Host: Dianthus
Alternaria euphorbiae (Barth.) Aragaki & Uchida
Alternaria euphorbiae (Barth.) E. Simmons & S. Francis, nom. illeg.
Distr.: KS(type).
Notes: The cause of poinsettia leaf spot (Pl. Dis. 70:73. 1986) from Hawaii is actually *A. euphorbiicola*. Simmons (Mycotaxon 25:195. 1986) also made this new combination and discussed *Alternaria* on poinsettia.
Host: Euphorbia
Alternaria euphorbiicola E. Simmons & Engelhard
Macrosporium euphorbiae Reichert non Barth., hom. illeg.
Distr.: FL,HI; Egypt(type).
Notes: According to Simmons (pers. comm.), this is the cause of poinsettia leaf spot (Mycotaxon 25:196. 1986).
Host: Euphorbia
Alternaria flagelloideum (Atk.) Luttrell
Helminthosporium flagelloideum Atk.
Bipolaris flagelloidea (Atk.) Shoemaker
Distr.: AL(type), WI; Dominican Republic.
Notes: Luttrell (Mycologia 47:268. 1955).
Hosts: Carex, Panicum, Scirpus
Alternaria forsythiae Harter
Distr.: DC.
Notes: According to Simmons (pers. comm.), the type is lost.
Host: Forsythia
Alternaria gerberae Raabe et al., nom. nud.
Distr.: HI.
Notes: This name appeared in the list of Hawaiian fungi but lacks a description and is, therefore, not validly published.
Host: Gerbera
Alternaria gomphrenae Togashi
Distr.: Widespread in tropical regions.
Notes: Leaf spot (Ellis. More dematiaceous Hyphomycetes. p. 423. 1976).
Hosts: Gomphrena, Vitis
Alternaria gossypii (Jacz.) Nisikado, K. Kimura & Miyawaki
Macrosporium gossypii Jacz.
Distr.: FL; Europe(type); Japan.
Host: Gossypium
Alternaria graminum (Cooke) Joly
Macrosporium graminum Cooke
Distr.: HI, SC(type).
Host: Saccharum
Alternaria helianthi (Hansf.) Tubaki & Nishihara

Helminthosporium helianthi Hansf.
Helminthosporium helianthi Pavgi non Hansf., hom. illeg.
Distr.: Widespread.
Notes: Leaf spot, blossom blight, and head blight of sunflower (Anahosur. C.M.I. Descr. 582. 1978). Simmons (Mycotaxon 25:210. 1986) presented a recent discussion of this species.
Host: Helianthus

Alternaria helianthinficiens E. Simmons, I. Walcz & Roberts
Distr.: ND; Canada(Manitoba); Hungary.
Notes: Simmons (Mycotaxon 25:204. 1986) discussed several *Alternaria* species on *Helianthus*.
Host: Helianthus

Alternaria inconspicua H.C. Greene
Distr.: WI.
Notes: According to Simmons (pers. comm.), this is not an *Alternaria*.
Host: Fraxinus

Alternaria iridicola (Ellis & Everh.) J.A. Elliott
Distr.: Central United States.
Host: Iris

Alternaria leucanthemi E.S. Nelen
Alternaria chrysanthemi E. Simmons & Crosier in E. Simmons
Distr.: FL; Austria; India; The Netherlands; Soviet Union.
Notes: Simmons (Mycotaxon 25:211. 1986); Ellis (C.M.I. Descr. 164. 1968 as *Alternaria chrysanthemi*).
Host: Chrysanthemum

Alternaria longipes (Ellis & Everh.) E. Mason
Distr.: Cosmopolitan.
Notes: Brown spot (Ellis & Holliday. C.M.I. Descr. 245. 1970). The report on *Daucus* is probably erroneous.
Hosts: [Daucus], Nicotiana

Alternaria longissima Deighton & MacGarvie
Distr.: Subtropical and tropical regions.
Notes: On pollen grains of corn, husks and grain of rice and *Sorghum*, and on leaves of various plants, often mixed with other fungi (Mycol. Pap. 113:1. 1968). The report on *Juncus* is probably erroneous.
Host: [Juncus]

Alternaria macrospora A. Zimmerm.
Distr.: Subtropical and tropical regions.
Notes: Leaf spot of cotton (Ellis & Holliday. C.M.I. Descr. 246. 1970).
Hosts: Anoda, Gossypium

Alternaria mali Roberts
Distr.: Eastern United States.
Notes: Joly (Le genre *Alternaria*. p. 158. 1964) considered this taxon to be problematical.
Hosts: Cydonia, Malus, Prunus, Pyrus

Alternaria maritima G.K. Sutherland
Distr.: NC; England(type).
Notes: Kohlmeyer & Kohlmeyer (Marine mycology. p. 552. 1979).
Host: Spartina

Alternaria multirostrata E. Simmons & C.R. Jackson
Distr.: GA.
Host: Richardia

Alternaria negundinicola (Ellis & Barth.) Joly
Macrosporium negundinicola Ellis & Barth.
Distr.: KS(type).
Host: Acer

Alternaria nelumbii Enlows & F.V. Rand
Distr.: Central and eastern North America.
Notes: Enlows & Rand (Phytopathology 11:135. 1921).
Host: Nelumbo

Alternaria panax Whetzel in Whetzel & Rosenb.
Distr.: HI, eastern United States.
Hosts: Aralia, Dizygotheca, Eleutherococcus, Panax, Polyscias, Schefflera, Tupidanthus

Alternaria passiflorae J.H. Simmonds
Distr.: HI; Canada; subtropical and tropical regions.
Notes: Brown fruit spot (Ellis & Holliday. C.M.I. Descr. 247. 1970).
Host: Passiflora

Alternaria peglionii Curzi
Distr.: ND; Europe.
Host: Triticum

Alternaria pomicola A.S. Horne
Distr.: WA; England(type).
Host: Malus

Alternaria porri (Ellis) Cif.
Macrosporium porri Ellis
Alternaria alli Nolla
Distr.: Cosmopolitan.
Notes: Purple blotch (Ellis & Holliday. C.M.I. Descr. 248. 1970).
Hosts: Allium, [Gerbera]

Alternaria radicina Meier, Drechs., & E.D. Eddy
Pseudostemphylium radicina (Meier, Drechs., & E.D. Eddy) Subramanian
Stemphylium radicinum (Meier, Drechs., & E.D. Eddy) Neergaard
Thyrospora radicina (Meier, Drechs., & E.D. Eddy) Neergaard
Distr.: Temperate northern hemisphere.
Notes: Black rot and damping-off of carrots and other Apiaceae (Ellis & Holliday. C.M.I. Descr. 346. 1972).
Hosts: Cucurbita, Daucus, Petroselinum

Alternaria ramulosa (Sacc.) Joly
Stemphylium ramulosum Sacc.
Distr.: FL; Europe(type).
Notes: Ellis (More dematiaceous Hyphomycetes. p. 413. 1976).
Host: Apium

Alternaria raphani Groves & Skolko
Distr.: Widespread in the northern hemisphere.
Notes: On seeds and pods of *Rhaphanus* and other Brassicaceae.
Hosts: Brassica, [Carya], Matthiola, Raphanus

Alternaria ricini (Yoshii) Hansf.
Distr.: Southeastern United States; Africa; southern Asia.
Notes: Ellis & Holliday (C.M.I. Descr. 249. 1970).
Host: Ricinus

Alternaria saponariae (Peck) Neergaard
Macrosporium saponariae Peck
Alternaria dianthi F. Stevens & J.G. Hall
Macrosporium nobile Vize ex Cooke
Distr.: North America; Europe; Jamaica; Brazil.
Hosts: Dianthus, Lychnis, Saponaria, Silene, Vaccaria

Alternaria scorzonerae (Aderhold) Loerakker
Sporidesmium scorzonerae Aderhold
Distr.: Eastern United States; Europe.
Notes: Leaf spot disease of scorzonera or black salsify (Loerakker. Neth. J. Pl. Pathol. 90:35. 1984).
Host: Tragopogon

Alternaria sesami (S. Kawamura) Mohanty & Behera
Distr.: Southeastern United States; South America; Africa; Asia.
Notes: Ellis & Holliday (C.M.I. Descr. 250. 1970).
Host: Sesamum

Alternaria solani Sorauer
Macrosporium solani Ellis & G. Martin
Distr.: Temperate and tropical regions.
Notes: Ellis & Gibson (C.M.I. Descr. 475. 1975).
Hosts: Amaranthus, Capsicum, Datura, Galinsoga, Hyoscyamus, Ipomoea, Lycopersicon, Petunia, Physalis, Solanum

Alternaria sonchi J.J. Davis in J.A. Elliott
Distr.: United States; Africa; Europe.
Hosts: Cichorium, Emilia, Lactuca, Sonchus

Alternaria stemphylioides D.E. Bliss
Alternaria chartarum G. Preuss f. *stemphylioides* (D.E. Bliss) Joly
Distr.: CA.
Notes: Simmons (Mycologia 59:73. 1967) transferred *A. chartarum* to *Ulocladium* but did not mention the taxonomic status of this form. It is possible that this form represents a *Ulocladium* species.
Host: Phoenix

Alternaria tagetica Shome & Mustafee
Distr.: SC; England; India.
Notes: Simmons (Mycotaxon 14:20. 1982) provided a recent discussion of this species.
Host: Tagetes

Alternaria tenuissima (Kunze:Fr.) Wiltshire
Alternaria brassicae (Berk.) Sacc. f. *phaseoli* C. Brun
Alternaria citri Ellis & N. Pierce var. *cerasi* B.A. Rudolph
Alternaria tabacina (Ellis & Everh.) S. Hori
Macrosporium tabacinium Ellis & Everh.
Alternaria tomato (Cooke) G.F. Weber
Macrosporium tomato Cooke
Alternaria violae B.T. Galloway & Dorsett
Distr.: Cosmopolitan.
Hosts: Amaranthus, Cajanus, Cichorium, Fragaria, Glycine, Lycopersicon, Nicotiana, Passiflora, Phaseolus, Pittosporum, Prunus,

Santolina, Tragopogon, Vaccinium, Viola
Alternaria tenuissima (Kunze:Fr.) Wiltshire f. sp. ***euphorbiae*** J.M. Krupinsky & R.J. Lorenz
Distr.: Central United States.
Notes: Useful as a biological control agent of leafy spurge (Krupinsky & Lorenz. Weed Science 31:86. 1983).
Host: Euphorbia
Alternaria zinniae M.B. Ellis
Distr.: Cosmopolitan.
Hosts: Eupatorium, Helianthus, Zinnia

Alveolaria Lagerh. Basidiomycetes, Uredinales

Alveolaria cordiae Lagerh.
Distr.: CA; Central and South America.
Host: Cordia

Amarenographium O. Eriksson. Coelomycetes

Amarenographium metableticum (Trail) O. Eriksson
Camarosporium metableticum Trail
Camarographium metableticum (Trail) Grove
Camarosporium graminicola Ellis & Everh.
Distr.: MA; northern Europe.
Host: Ammophila

Amazonia Theiss., nom. cons. Pyrenomycetes, Meliolales

Amazonia alyxiae Hansf.
Distr.: HI.
Host: Alyxia
Amazonia asterinoides (G. Wint.) Theiss.
Distr.: HI; West Indies; South Africa.
Hosts: Gouldia, Lobelia
Amazonia clermontiae Hansf.
Distr.: HI(type).
Host: Clermontia
Amazonia polypoda Syd.
Amazonia psychotriae (Henn.) Theiss. var. *straussiae* Hansf.
Distr.: HI.
Hosts: Psychotria, Straussia
Amazonia psychotriae (Henn.) Theiss.
Distr.: Widespread in tropical regions.
Hosts: Clermontia, Coprosma, Euphorbia, Labordia, Lobelia, Psychotria, Scaevola
Amazonia psychotriae (Henn.) Theiss. var. ***labordiae*** Hansf.
Distr.: HI.
Host: Labordia
Amazonia scaevolae Hansf.
Distr.: HI.
Host: Scaevola
Amazonia stevensii Hansf.
Actinodothis perrottetiae F. Stevens
Distr.: HI.
Host: Perrottetia
Amazonia suttoniae (F. Stevens) Hansf.
Actinodothis suttoniae F. Stevens
Distr.: HI.
Host: Myrsine
Amazonia tetrorchidii Hansf.
Distr.: HI; Africa.
Host: Euphorbia
Amazonia wikstroemiae Hansf.
Distr.: HI.
Host: Wikstroemia

Ambrosiella Arx & Hennebert. Hyphomycetes

Notes: These fungi are associated with wood-inhabiting beetles (Batra. Mycologia 59:976. 1967).
Ambrosiella brunnea Batra
Monilia brunnea (Verrall) Batra non Gilman & E. Abbott, hom. illeg.
Distr.: Eastern and southern United States.
Notes: This new name was based on a later homonym.
Host: Acer
Ambrosiella ferruginea Batra
Monilia ferruginea Mathiesen-Kaarik non Pers., hom. illeg.
Distr.: Western United States; Canada; Europe.
Notes: This new name was based on a later homonym.
Hosts: Picea, Pinus
Ambrosiella xylebori L. Brader
Distr.: AR, GA; Africa; India.
Host: Acer

Amerosporium Speg. Coelomycetes

Notes: Sutton (The Coelomycetes. p. 618. 1980) limited the generic concept to only two species, but not all names have been redisposed.
Amerosporium atrum (Fuckel) Höhn.
Chaetomella atra Fuckel
Amerosporium caricinum (Lib.) Sacc.
Distr.: Eastern United States; Europe.
Notes: Stolk (Trans. Brit. Mycol. Soc. 46:409. 1963).
Hosts: Carex, Muhlenbergia, Panicum, Sporobolus
Amerosporium ilicinum Ellis & Everh.
Distr.: LA(type).
Host: Ilex
Amerosporium macrochaeta Ellis & Everh.
Distr.: LA(type).
Host: Rhynchospora
Amerosporium sabalinum Ellis & Everh.
Distr.: LA(type).
Host: Sabal
Amerosporium subclausum Ellis & Kellerm.
Distr.: KS(type).
Host: Gymnocladus

Amphinema P. Karst. Basidiomycetes, Aphyllophorales

Amphinema arachispora Burdsall & Nakasone
Distr.: GA.
Notes: Burdsall & Nakasone (Mycologia 73:454. 1981).
Host: Cornus
Amphinema byssoides (Pers.:Fr.) J. Eriksson
Coniophora byssoidea (Pers.:Fr.) P. Karst.
Peniophora byssoidea (Pers.:Fr.) Höhn. & Litsch.
Amphinema tomentellum (Bres.) M. Christiansen
Distr.: Widespread.
Notes: On wood and plant debris.
Hosts: Abies, Acer, Betula, Picea, Pinus, Populus, Quercus, Selaginella, Tsuga

Amphiporthe Petr. Pyrenomycetes, Diaporthales

Notes: Barr (Mycol. Mem. 7:140. 1978).
Amphiporthe aculeans (Schwein.) Barr
Cryptodiaporthe aculeans (Schwein.) Wehmeyer
Valsa aculeans (Schwein.) Berk.
Valsa rufescens (Schwein.) Peck
Stilbum rhois Berk. & M.A. Curtis (anamorph)
Distr.: Eastern North America; Japan.
Notes: On limbs.
Host: Rhus
Amphiporthe castanea (Tul.) Barr
Cryptodiaporthe castanea (Tul.) Wehmeyer
Fusicoccum castaneum Sacc. (anamorph)
Distr.: Eastern North America; Europe; Japan.
Notes: On limbs. According to Arx (Biblioth. Mycol. 24:90. 1970) the anamorph is a *Discella* sp.
Host: Castanea
Amphiporthe raveneliana (Thuem. & Rehm) Barr
Diaporthe raveneliana Thuem. & Rehm
Cryptodiaporthe raveneliana (Thuem. & Rehm) T. Kobayashi
Diaporthe leiphaemia (Fr.:Fr.) Sacc. var. *raveneliana* (Thuem. &

Rehm) Wehmeyer
Cryptodiaporthe densissima (Ellis) Wehmeyer
Distr.: Eastern North America; Japan.
Notes: On limbs.
Host: Quercus

Amphisphaeria Ces. & De Not., nom. cons. Pyrenomycetes, Amphisphaeriales

Amphisphaeria depressa Petr.
Distr.: HI(type).
Host: Senna
Amphisphaeria granulosa Ellis & Everh.
Distr.: NY(type).
Host: Quercus
Amphisphaeria juniperi Tracy & Earle
Distr.: Western United States.
Host: Tsuga
Amphisphaeria succinea (Roberge in Desmaz.) J. Krug
Charonectria succinea (Roberge in Desmaz.) Sacc.
Nectriella succinea (Roberge in Desmaz.) Weese
Amphisphaeria cupula (Ellis) Schrantz
Cainia cupula (Ellis) E. Müller
Didymosphaeria cupula (Ellis) Sacc.
Distr.: Eastern United States; Europe.
Notes: On acorn cupules and leaves.
Host: Quercus

Amphobotrys Hennebert. Hyphomycetes

Notes: Hennbert (Persoonia 7:192. 1973).
Amphobotrys ricini (Buchwald) Hennebert
Botrytis ricini Buchwald
Botrytis bifurcata J.H. Miller, J.E. Giddens, & A.A. Foster
Botryotinia ricini (Godfrey) Whetzel (teleomorph)
Distr.: Southeastern United States; West Indies.
Notes: Gray mold of castor bean.
Host: Caperonia

Amyloathelia Hjortstam & Ryvarden. Basidiomycetes, Aphyllophorales

Amyloathelia amylacea (Bourd. & Galzin) Hjortstam & Ryvarden
Aleurodiscus amylaceus (Bourd. & Galzin) D.P. Rogers
Distr.: North America; Europe.
Notes: On wood.
Hosts: Picea, Thuja

Amylocorticium Pouzar. Basidiomycetes, Aphyllophorales

Amylocorticium canadense (Burt) J. Eriksson & Weresub in Weresub
Corticium canadense Burt
Distr.: Temperate northern hemisphere.
Notes: On wood (Weresub. Fungi Canadenses 45. 1974).
Host: Tsuga
Amylocorticium subincarnatum (Peck) Pouzar
Peniophora subincarnata (Peck) Litsch.
Distr.: Temperate northern hemisphere.
Notes: On wood (Weresub. Fungi Canadenses 91. 1977).
Hosts: Acer, Picea, Pinus
Amylocorticium subsulphureum (P. Karst.) Pouzar
Corticium subsulphureum P. Karst.
Peniophora subsulphurea (P. Karst.) Höhn. & Litsch.
Distr.: Temperate northern hemisphere.
Notes: On wood.
Hosts: Larix, Picea, Pinus

Amylocystis Bondartsev & Singer ex Singer. Basidiomycetes, Aphyllophorales

Notes: Gilbertson & Ryvarden (North American Polypores 1:111. 1986).
Amylocystis lapponica (Romell) Singer, comb. inval.
Polyporus lapponicus Romell
Polyporus ursinus C.G. Lloyd non (Link) Fr., hom. illeg.
Distr.: Northeastern and western North America; Europe.
Notes: On conifers, especially *Picea* spp.
Hosts: Abies, Larix, Picea, Pinus, Pseudotsuga, Taxus, Thuja, Tsuga

Amylosporus Ryvarden. Basidiomycetes, Aphyllophorales

Amylosporus campbellii (Berk.) Ryvarden
Polyporus campbellii Berk.
Tyromyces graminicola Murrill
Distr.: AZ, CA, TX; subtropical regions.
Notes: On living hardwoods, dead conifer wood, and grass roots.
Hosts: Citrus, Cynodon, Melia

Amylostereum Boidin. Basidiomycetes, Aphyllophorales

Amylostereum chailletii (Pers.:Fr.) Boidin
Stereum chailletii (Pers.:Fr.) Fr.
Peniophora atkinsonii Ellis & Everh.
Hymenochaete simulans Ellis & Everh., nom. nud.
Distr.: Widespread.
Notes: On wood of conifers; reports on hardwoods are questionable.
Hosts: Abies, [Acer], Chamaecyparis, Juniperus, Larix, Picea, Pinus, Pseudotsuga, Sequoia, Thuja, Tsuga
Amylostereum laevigatum (Fr.:Fr.) Boidin
Corticium laevigatum (Fr.:Fr.) Fr.
Peniophora laevigata (Fr.:Fr.) Massee
Distr.: Northeastern United States and adjacent Canada; Europe.
Notes: On bark. Rare in North America.
Host: Thuja

Angelina Fr. Discomycetes, Helotiales

Angelina rufescens (Schwein.:Fr.) Duby
Distr.: Eastern United States.
Hosts: Liriodendron, Quercus

Anguillospora Ingold. Hyphomycetes

Notes: Includes both aquatic and plant parasitic fungi.
Anguillospora coryli Redhead & G.P. White
Distr.: WI.
Notes: Redhead & White (Canad. J. Bot. 63:1431. 1985).
Host: Corylus
Anguillospora longissima (Sacc. & P. Syd.) Ingold
Fusarium longissimum Sacc. & P. Syd.
Distr.: Widespread.
Notes: On leaf litter and in foam in aquatic habitats (Nilsson. Symb. Bot. Upsal. 18:80. 1964).
Hosts: Acer, Quercus
Anguillospora vermiformis (J.J. Davis) Redhead & G.P. White
Cylindrosporium vermiforme J.J. Davis
Cercoseptoria vermiformis (J.J. Davis) J.J. Davis
Distr.: WI.
Notes: Redhead & White (Canad. J. Bot. 63:1431. 1985).
Hosts: Alnus, Corylus

Aniptodera C.A. Shearer & M. Miller. Pyrenomycetes, Halosphaeriales

Aniptodera chesapeakensis C.A. Shearer & M. Miller
Distr.: East coast of the United States.
Notes: Shearer & Miller (Mycologia 69: 894. 1977).
Hosts: Juncus, Spartina

Anisogramma Theiss. & Syd. Pyrenomycetes, Diaporthales

Anisogramma anomala (Peck) E. Müller in E. Müller & Arx
Apioporthe anomala (Peck) Höhn.

Cryptosporella anomala (Peck) Sacc.
Distr.: Eastern and northwestern North America.
Notes: Canker disease. Gottwald & Cameron (Mycologia 71:1107. 1979) described the life history of populations infecting *Corylus* orchards in the Pacific Northwest.
Host: Corylus
Anisogramma virgultorum (Fr.:Fr.) Theiss. & Syd.
Distr.: Northeastern North America; Europe.
Notes: On branches.
Host: Betula

Annellophora S.J. Hughes. Hyphomycetes

Annellophora phoenicis M.B. Ellis
Distr.: TX; Malaya; New Guinea; Sierra Leone.
Notes: This fungus was recently reported from the United States by Vann & Taber (Pl. Dis. 69:903. 1985).
Host: Phoenix

Anomoporia Pouzar. Basidiomycetes, Aphyllophorales

Anomoporia albolutescens (Romell) Pouzar
Poria albolutescens (Romell) Egeland
Poria grandis Overh.
Distr.: Temperate northern hemisphere.
Notes: On conifers.
Hosts: Abies, Larix, Picea, [Populus], Pseudotsuga, Tsuga
Anomoporia bombycina (Fr.:Fr.) Pouzar
Poria bombycina (Fr.:Fr.) Cooke
Poria coniferarum D. Baxter
Poria fulvella Bres.
Distr.: Circumboreal.
Notes: On conifers.
Hosts: Abies, Larix, Picea, Pinus, [Populus], Pseudotsuga, Tsuga
Anomoporia myceliosa (Peck) Pouzar
Poria myceliosa Peck
Poria perextensa Murrill
Distr.: Circumboreal on conifers.
Hosts: [Alnus], [Betula], Picea, Tsuga

Anomothallus F. Stevens. Loculoascomycetes, Dothideales

Notes: The identity of this genus is uncertain (Petrak. Sydowia 5:353. 1951).
Anomothallus erraticus F. Stevens
Distr.: HI(type).
Host: Rubus

Antennaria Link non Gaertn., hom. illeg. Hyphomycetes

Notes: Because no sporulating structures were described by the original author, Hughes (New Zealand J. Bot. 12:299. 1970) regarded *Antennaria* as a *nomen ambiguum*.
Antennaria rectangularis Sacc. in Sacc., Peck, & Trel.
Distr.: AK(type).
Notes: Saccardo's original description appears to be based solely on vegetative structures.
Hosts: Cassiope, Phyllodoce

Antennellina Mendoza. Loculoascomycetes, Dothideales

Notes: This genus is considered a synonym of *Scorias* (Arx & Müller. Stud. Mycol. 9:142. 1975); however, disposition of the type species is still uncertain.
Antennellina hawaiiensis Mendoza
Distr.: HI(type).
Host: Mangifera

Antennularia Rchb. Loculoascomycetes, Dothideales

Notes: This genus is considered a synonym of *Protoventuria* by Barr (Canad. J. Bot. 49:1959. 1971), although Arx & Müller (Stud. Mycol. 9:101. 1975) disagree.
Antennularia robinsonii Berk. & Mont.
Distr.: FL.
Hosts: Gordonia, Lyonia

Antennulariella Woronichin. Loculoascomycetes, Dothideales

Antennulariella alpina (W.B. Cooke) Barr
Dimerium alpinum W.B. Cooke
Distr.: Western United States.
Notes: Forming black crusts usually on lower surface of leaves (Cooke. Mycologia 41:609. 1949). Barr (Mycotaxon 29:501. 1987).
Host: Penstemon

Anthostoma Nitschke. Pyrenomycetes, Xylariales

Notes: Many species have been transferred to other genera.
Anthostoma amygdalinum (Cooke) Sacc.
Distr.: AL, SC.
Notes: Known only from two records.
Host: Liquidambar
Anthostoma juglandinum Rehm var. ***caryae*** Rehm
Distr.: IL.
Host: Carya
Anthostoma melanotes (Berk. & Broome) Sacc. var. ***symphoricarpi*** Brenckle
Distr.: ND.
Host: Symphoricarpos
Anthostoma quercina (Petr.) Arx & E. Müller
Leptomassaria quercina Petr.
Distr.: MD(type).
Host: Quercus
Anthostoma saprophilum Ellis
Distr.: NJ(type), eastern United States.
Host: Acer
Anthostoma simplex (G. Otth) Sacc.
Leptomassaria simplex (G. Otth) Petr.
Distr.: GA; Europe.
Host: Tillaea

Anthostomella Sacc. Pyrenomycetes, Xylariales

Notes: Francis (Mycol. Pap. 139. 1975) provided a comprehensive account of this genus. Anamorphs of some species belong to the form-genera *Nodulisporium* and *Virgariella*.
Anthostomella baptisiae (Cooke) Sacc.
Distr.: SC.
Host: Baptisia
Anthostomella brachystoma Ellis & Everh.
Xylosphaeria brachystoma (Ellis & Everh.) Cooke
Distr.: WA.
Host: Tsuga
Anthostomella cacti (Schwein.) Sacc.
Distr.: CA, NJ.
Host: Opuntia
Anthostomella chionostoma (Durieu & Mont.) Sacc.
Distr.: NJ; Algeria; Europe; India.
Notes: On herbaceous debris.
Host: Andropogon
Anthostomella consanguinea (Ces.) Sacc.
Sordaria consanguinea Ces.
Distr.: FL; Italy; Portugal.
Notes: On leaves and petioles.
Host: Serenoa
Anthostomella contaminans (Durieu & Mont.) Sacc.
Distr.: Warmer regions.
Notes: On various monocotyledonous hosts.

Host: Phoenix
Anthostomella cornicola Ellis & Everh.
Entosordaria cornicola (Ellis & Everh.) Höhn.
Distr.: WA(type).
Host: Cornus
Anthostomella leucobasis (Ellis & G. Martin) Sacc.
Distr.: FL.
Host: Sabal
Anthostomella ludoviciana Ellis & Langl. in Ellis & Everh.
Distr.: LA.
Host: Smilax
Anthostomella melanosticta Ellis & Everh.
Distr.: Southeastern United States.
Notes: On dead leaf stalks and blades.
Host: Sabal
Anthostomella minor Ellis & Everh.
Distr.: FL.
Host: Sabal
Anthostomella nigro-annulata (Berk. & M.A. Curtis) Sacc.
Distr.: Southern United States; Cuba.
Host: Yucca
Anthostomella nigrotecta (Berk. & Ravenel) Sacc.
Sphaeria nigrotecta Berk. & Ravenel
Distr.: SC(type).
Host: Ilex
Anthostomella picaceum (Cooke & Ellis) Sacc.
Anthostoma picaceum (Cooke & Ellis) Cooke
Distr.: CO, ND.
Host: Fraxinus
Anthostomella ratibidae Solheim
Distr.: NM.
Host: Ratibida
Anthostomella rhizomorphae (Kunze) Berl. & Voglino
Distr.: Puerto Rico, Europe.
Host: Rhizophora
Anthostomella rostraspora (W. Gerard) Sacc.
Distr.: NY; Europe.
Host: Inula
Anthostomella sabalensioides (Ellis & G. Martin) Sacc.
Distr.: Southeastern United States.
Notes: On dead leaf stalks and blades of woody monocotyledons.
Host: Sabal
Anthostomella sabiniana S. Francis
Distr.: CA(type); Scotland.
Host: Pinus
Anthostomella sepelibilis (Berk. & M.A. Curtis) Sacc.
Distr.: Eastern and southeastern United States; Argentina; Soviet Union.
Notes: On dead stems.
Host: Smilax
Anthostomella tomicoides Sacc.
Anthostoma mortuosum (Ellis) Sacc.
Distr.: NJ; Europe.
Notes: On herbaceous debris.
Host: Eupatorium
Anthostomella unguiculata (Mont.) Sacc.
Anthostomella magnoliae Ellis & Everh.
Distr.: LA; Chile; Belgium.
Notes: On fallen leaves.
Host: Magnolia

Anthracoidea Bref. Basidiomycetes, Ustilaginales

Notes: Nannfeldt (Bot. Not. 130:351. 1977); Nannfeldt (Symb. Bot. Upsal. 22:1. 1979).
Anthracoidea americana (Nannf. & Lindeberg) Kukk.
Cintractia americana Nannf. & Lindeberg
Distr.: Northern and western North America.
Host: Carex
Anthracoidea aspera (Liro) Kukk.
Cintractia aspera Liro
Distr.: Temperate northern hemisphere.
Host: Carex
Anthracoidea atratae (Savile) Kukk.
Cintractia atratae Savile
Distr.: Northern and western North America.
Host: Carex
Anthracoidea buxbaumii Kukk.
Distr.: Northern and western North America; Europe.
Host: Carex
Anthracoidea calderi (Savile) Kukk.
Cintractia calderi Savile
Distr.: Northern North America.
Host: Carex
Anthracoidea caricis (Pers.:Pers.) Bref.
Cintractia caricetorum Speg.
Cintractia caricis (Pers.:Pers.) Magnus
Ustilago caricis (Pers.:Pers.) Unger
Cintractia californica Zundel in W.B. Cooke
Distr.: North America; Europe.
Notes: Reports on genera other than *Carex* are erroneous.
Hosts: Carex, [Juncus], [Kobresia], [Scirpus]
Anthracoidea caricis (Pers.:Pers.) Bref. var. ***intermedia*** Savile
Cintractia caricis (Pers.:Pers.) Magnus var. *intermedia* Savile
Distr.: Northern and western North America.
Host: Carex
Anthracoidea caryophylleae Kukk.
Distr.: Northern North America; Europe.
Host: Carex
Anthracoidea douglasii Nannf.
Ustilago caricis (Pers.:Pers.) Unger var. *douglasii* Shear in Ellis & Everh., nom. nud.
Distr.: Western United States.
Notes: According to Nannfeldt (Bot. Not. 130:366. 1977), this species has been misidentified as *Cintractia limosa* var. *minor*.
Host: Carex
Anthracoidea eleocharidis Kukk.
Distr.: WY; Canada.
Notes: Kukkonen (Trans. Brit. Mycol. Soc. 47:273. 1964).
Host: Carex
Anthracoidea elynae (Syd.) Kukk.
Cintractia carpophila (Schumach.) Liro var. *elynae* (Syd.) Savile
Distr.: Temperate northern hemisphere.
Hosts: [Carex], Kobresia
Anthracoidea externa (D. Griffiths) Kukk.
Cintractia externa (D. Griffiths) G.P. Clinton
Distr.: North America.
Host: Carex
Anthracoidea fischeri (P. Karst.) Kukk.
Cintractia fischeri (P. Karst.) Liro
Distr.: Temperate northern hemisphere.
Host: Carex
Anthracoidea heterospora (Lindeberg in Nannf. & Lindeberg) Kukk.
Cintractia caricis (Pers.) Magnus var. *acutarum* Savile
Cintractia carpophila (Schumach.) Liro
Uredo carpophila Schumach., nom. illeg.
Distr.: Temperate northern hemisphere; South America; New Zealand.
Notes: See notes under *A. karii* and *Cintractia carpophila* var. *kenaica*.
Host: Carex
Anthracoidea karii (Liro) Nannf.
Cintractia karii Liro
Distr.: Temperate northern hemisphere.
Notes: Formerly considered a synonym of *Cintractia (Anthracoidea) carpophila* (= *A. heterospora*) (Fischer. Manual of the North American smut fungi. p. 54. 1953). Thus some records under the latter may actually be *A. karii*.
Host: Carex
Anthracoidea lasiocarpae Lindeberg ex Kukk.
Distr.: AK; Finland.
Host: Carex
Anthracoidea limosa (Syd.) Kukk.
Cintractia limosa Syd.
Distr.: Circumpolar.
Host: Carex
Anthracoidea limosa (Syd.) Kukk. var. ***gigantissima*** (Lehtola) Savile

Cintractia gigantissima Lehtola
Cintractia limosa Syd. var. *gigantissima* (Lehtola) Savile
Distr.: Western North America.
Host: Carex
Anthracoidea limosa (Syd.) Kukk. var. ***minor*** Savile
Cintractia limosa Syd. var. *minor* Savile
Distr.: Western North America.
Notes: According to Nannfeldt (Bot. Not. 130:366. 1977) these records refer to *Anthracoidea douglassii.*
Host: Carex
Anthracoidea lindebergiae (Kukk.)Kukk.
Cintractia lindebergiae Kukk.
Distr.: Temperate northern hemisphere.
Host: Kobresia
Anthracoidea paniceae Kukk.
Distr.: Northern North America; Europe.
Host: Carex
Anthracoidea scirpi (Kühn) Kukk.
Ustilago scirpi Kühn
Cintractia scirpi (Kühn) Schellenb.
Distr.: North America; Scotland.
Notes: On Cyperaceae.
Host: Scirpus
Anthracoidea scirpoideae Kukk.
Distr.: Northern and western North America.
Host: Carex
Anthracoidea subinclusa (Körn.) Bref.
Ustilago subinclusa Körn. in Rabenh.
Cintractia subinclusa (Körn.) Magnus
Distr.: North and South America; Europe.
Host: Carex
Anthracoidea verrucosa (Savile) Nannf.
Cintractia carpophila (Schumach.) Liro var. *verrucosa* Savile
Distr.: Western North America; Greenland.
Host: Carex

Anthracophyllum Ces. Basidiomycetes, Agaricales

Anthracophyllum lateritium (Berk. & M.A. Curtis) Singer
Distr.: Subtropical and tropical western hemisphere.
Host: Juniperus

Antrodia P. Karst. Basidiomycetes, Aphyllophorales

Antrodia albida (Fr.:Fr.) Donk
Coriolellus sepium (Berk.) Murrill
Trametes sepium Berk.
Antrodia serpens (Fr.:Fr.) P. Karst.
Distr.: Widespread.
Notes: On wood, often on fence posts and structural timbers.
Hosts: Acer, Arbutus, Carya, Castanea, Catalpa, Ceanothus, Cupressus, Fagus, Fraxinus, Juniperus, Picea, Pinus, Platanus, Quercus, Rhus, Salix, Sassafras, Thuja, Tsuga, Viburnum
Antrodia albobrunnea (Romell) Ryvarden
Polyporus albobrunneus Romell
Poria albobrunnea (Romell) D. Baxter
Poria dichroa Bres.
Distr.: Temperate northern hemisphere.
Notes: On conifers, especially *Pinus* spp.
Hosts: Abies, Chamaecyparis, Larix, Picea, Pinus, Pseudotsuga, Tsuga
Antrodia alpina (Litsch.) R.L. Gilbertson & Ryvarden
Poria alpina Litsch.
Distr.: Western North America; European Alps.
Notes: On conifers.
Hosts: Larix, Picea
Antrodia carbonica (Overh.) Ryvarden & R.L. Gilbertson
Poria carbonica Overh.
Distr.: Western North America; Central America; North Africa; Himalayas.
Notes: On conifers, especially *Pseudotsuga* in North America.
Hosts: Abies, Larix, Picea, Pinus, Pseudotsuga, Thuja
Antrodia crassa (P. Karst.) Ryvarden
Physisporus crassus P. Karst.
Poria crassa (P. Karst.) Sacc.
Distr.: Western North America; Europe.
Notes: On conifers.
Hosts: [Alnus], [Betula], Pinus, Pseudotsuga, [Quercus], Tsuga
Antrodia ferox (W.H. Long & D. Baxter) R.L. Gilbertson & Ryvarden
Poria ferox W.H. Long & D. Baxter
Distr.: Southcentral and southwestern United States.
Hosts: Juniperus, [Pinus], [Pseudotsuga]
Antrodia gossypina (Speg.) Ryvarden
Poria gossypina Speg.
Leptoporus resupinatus (Bourd. & Galzin) Pilát
Distr.: Northwestern United States; temperate regions.
Notes: On conifers.
Host: Pseudotsuga
Antrodia heteromorpha (Fr.:Fr.) Donk
Lenzites heteromorpha (Fr.:Fr.) Fr.
Coriolellus heteromorphus (Fr.:Fr.) Bondartsev & Singer
Trametes heteromorpha (Fr.:Fr.) Bres.
Trametes alaskana D. Baxter
Coriolellus alaskanus (D. Baxter) K.J. Martin & R.L. Gilbertson
Trametes lacerata C.G. Lloyd
Distr.: Circumboreal.
Notes: On wood, usually of conifers.
Hosts: Abies, Alnus, Betula, Condalia, Fraxinus, Juniperus, Larix, Picea, Pinus, Platanus, Populus, Prosopis, Prunus, Pseudotsuga, Quercus, Salix, Thuja, Tsuga
Antrodia juniperina (Murrill) Niemelä & Ryvarden
Daedalea juniperina (Murrill) Murrill
Poria juniperina (Murrill) Sacc. & Trott.
Distr.: United States; East Africa; Europe.
Notes: On living trees.
Hosts: Juniperus, Taxodium
Antrodia malicola (Berk. & M.A. Curtis) Donk
Trametes malicola Berk. & M.A. Curtis
Coriolellus malicola (Berk. & M.A. Curtis) Murrill
Distr.: Eastern North America(common); northern hemisphere.
Notes: On hardwoods.
Hosts: Acer, Ailanthus, Betula, Carya, Fraxinus, Lindera, Malus, Metrosideros, Platanus, Populus, Quercus, Salix
Antrodia odora (Peck ex Sacc.) R.L. Gilbertson & Ryvarden
Poria odora Peck ex Sacc.
Polyporus odorus Peck non Fr., hom. illeg.
Distr.: Northeastern United States(not common) and southwestern United States.
Notes: On conifers.
Hosts: Picea, Pinus
Antrodia oleracea (R.W. Davidson & Lombard) Ryvarden
Poria oleracea R.W. Davidson & Lombard
Distr.: AZ, central and eastern United States.
Notes: On hardwoods, especially *Quercus* spp.
Hosts: Fraxinus, Quercus
Antrodia radiculosa (Peck) R.L. Gilbertson & Ryvarden
Polyporus radiculosus Peck
Poria radiculosa (Peck) Sacc.
Poria flavida Murrill
Poria luteofimbriata D. Baxter
Poria subradiculosa Murrill
Distr.: North and Central America.
Notes: On wood, usually of conifers.
Hosts: Fraxinus, Pinus, Platanus, Pseudotsuga, Quercus, Tsuga
Antrodia serialis (Fr.:Fr.) Donk
Polyporus serialis Fr.:Fr.
Coriolellus serialis (Fr.:Fr.) Murrill
Trametes serialis (Fr.:Fr.) Fr.
Poria callosa Auct. non (Fr.:Fr.) Cooke
Distr.: Widespread.
Notes: On wood, usually of conifers.
Hosts: Abies, Juniperus, Larix, Picea, Pinus, Populus, Prunus, Pseudotsuga, Thuja, Tsuga
Antrodia sinuosa (Fr.:Fr.) P. Karst.
Poria sinuosa (Fr.:Fr.) Cooke
Poria sylvestris Romell ex D. Baxter
Poria vaporaria (Pers.:Fr.) Cooke
Polyporus vaporarius Pers.:Fr.

Distr.: Temperate northern hemisphere.
Notes: On wood, usually of conifers; sometimes on burnt wood.
Hosts: Abies, Alnus, Citrus, Juniperus, Kalmia, Larix, Magnolia, Picea, Pinus, Prunus, Pseudotsuga, Rhus, Salix, Thuja, Tsuga
Antrodia sitchensis (D. Baxter) R.L. Gilbertson & Ryvarden
Poria sitchensis D. Baxter
Distr.: NY, western North America.
Notes: On conifers.
Hosts: Picea, Pinus, Pseudotsuga, Tsuga
Antrodia sordida Ryvarden & R.L. Gilbertson
Poria oleagina Overh., nom. inval.
Distr.: Eastern North America.
Notes: On conifers.
Hosts: Picea, Pinus
Antrodia vaillantii (DC.:Fr.) Ryvarden
Polyporus vaillantii (DC.:Fr.) Fr.
Poria vaillantii (DC.:Fr.) Cooke
Poria bergii Speg.
Distr.: Widespread.
Notes: On wood of conifers, usually on structural timbers, in mines, etc.
Hosts: Abies, Larix, Picea, Pinus, Pseudotsuga, Thuja, Tsuga
Antrodia variiformis (Peck) Donk
Polyporus variiformis Peck
Coriolellus variiformis (Peck) A. Sarkar
Trametes variiformis (Peck) Peck
Distr.: Temperate northern hemisphere.
Notes: On wood, usually of conifers.
Hosts: Betula, Larix, Picea, Pinus, Populus, Quercus, Tsuga
Antrodia xantha (Fr.:Fr.) Ryvarden
Amyloporia xantha (Fr.:Fr.) Bondartsev & Singer
Poria xantha (Fr.:Fr.) Cooke
Distr.: North America, cooler parts; Jamaica; Europe; New Zealand.
Notes: On wood, usually of conifers.
Hosts: Abies, Castanea, Chamaecyparis, Larix, Picea, Pinus, Populus, Prunus, Pseudotsuga, Thuja, Tsuga

Antrodiella Ryvarden & I. Johansen. Basidiomycetes, Aphyllophorales

Antrodiella americana Ryvarden & R.L. Gilbertson
Poria aestivale Overh., nom. inval.
Distr.: Northern hardwood forests of North America.
Hosts: Fagus, [Picea]
Antrodiella incrustans (Berk. & M.A. Curtis ex Cooke) Ryvarden
Poria incrustans Berk. & M.A. Curtis ex Cooke
Poria distorta Murrill
Poria humilis Murrill
Distr.: Eastern North America.
Notes: On wood, usually of hardwoods.
Hosts: Lyonothamnus, Quercus
Antrodiella overholtsii Ryvarden & R.L. Gilbertson
Polyporus canadensis Overh., nom. inval.
Tyromyces canadensis (Overh.) J. Lowe, comb. inval.
Distr.: Northern United States and adjacent Canada.
Notes: On wood.
Host: Picea
Antrodiella romellii (Donk) Niemelä
Poria romellii Donk
Flaviporus romellii (Donk) Ginns
Poria byssina Romell
Distr.: Western and southwestern United States.
Notes: On wood, usually of conifers.
Hosts: Larix, Picea, Pinus, Populus, Pseudotsuga
Antrodiella semisupina (Berk. & M.A. Curtis) Ryvarden
Polyporus semisupinus Berk. & M.A. Curtis
Flaviporus semisupinus (Berk. & M.A. Curtis) Ginns
Tyromyces semisupinus (Berk. & M.A. Curtis) Murrill
Polyporus pachycheiles Ellis & Everh.
Distr.: Widespread.
Notes: On wood.
Hosts: Acer, Alnus, Betula, Pinus, Quercus
Antrodiella versicutis (Berk. & M.A. Curtis) R.L. Gilbertson & Ryvarden
Polyporus versicutis Berk. & M.A. Curtis
Coriolus ochrotinctellus Murrill
Polyporus ochrotinctellus (Murrill) Overh.
Distr.: Southeastern United States to Brazil.
Notes: On hardwoods.
Host: Quercus

Anungitea Sutton & C.S. Hodges. Hyphomycetes

Anungitea globosa Sutton & C.S. Hodges
Distr.: HI; New Zealand.
Notes: On leaves.
Host: Eucalyptus

Aphanomyces de Bary. Oomycetes, Saprolegniales

Notes: Scott (Virginia Agr. Exp. Sta. Tech. Bull. 151:1. 1961) provided a comprehensive account of this genus.
Aphanomyces camptostylus Drechs.
Distr.: Northern United States.
Hosts: Avena, Festuca, Setaria, Stipa
Aphanomyces cladogamus Drechs.
Distr.: Eastern North America.
Hosts: Linum, Lycopersicon, Spinacia
Aphanomyces cochlioides Drechs.
Distr.: Northern North America.
Hosts: Beta, Chenopodium
Aphanomyces euteiches Drechs.
Distr.: Northern North America; Europe; Japan; Tasmania.
Notes: The isolates causing diseases on beans and peas have been segregated into *formae speciales*. Stamps (C.M.I. Descr. 600. 1978).
Hosts: Apium, Echinodorus, Hordeum, Lathyrus, Medicago, Melilotus, Pisum, Vicia
Aphanomyces euteiches Drechs. f. sp. ***phaseoli*** W.F. Pfender & D.J. Hagedorn
Distr.: WI.
Notes: Bean root and hypocotyl rot (Pfender & Hagedorn. Phytopathology 72:306. 1982).
Hosts: Phaseolus, Pisum
Aphanomyces euteiches Drechs. f. sp. ***pisi*** W.F. Pfender & D.J. Hagedorn
Distr.: WI.
Notes: Pfender & Hagedorn (Phytopathology 72:306. 1982).
Host: Pisum
Aphanomyces raphani Kendrick
Distr.: Eastern North America; Europe.
Host: Raphanus

Aphanostigme Syd. Loculoascomycetes, Dothideales

Notes: This genus is considered a synonym of *Nematostoma* Syd. (Rossman. Mycol. Pap. 157:48. 1987), but not all names have been redisposed.
Aphanostigme lanugine V. Miller & Bonar
Distr.: CA.
Host: Garrya

Apiocarpella Melnik. Coelomycetes

Apiocarpella thermopsidis (Solheim) Melnik
Ascochyta thermopsidis Solheim
Distr.: WY(type).
Host: Thermopsis

Apiognomonia Höhn. Pyrenomycetes, Diaporthales

Notes: Barr (Mycol. Mem. 7:23. 1978); Monod (Beih. Sydowia 9:47. 1983).
Apiognomonia errabunda (Roberge) Höhn.
Gnomonia errabunda (Roberge) Auersw.

Gnomonia quercina Kleb.
Discula umbrinella (Berk. & Broome) Sutton (anamorph)
Distr.: Temperate North America; Europe.
Notes: Anthracnose of hardwood trees.
Hosts: Fagus, Platanus, Quercus
Apiognomonia tiliae (Rehm) Höhn.
Gnomonia tiliae Kleb.
Discula sp. (anamorph)
Distr.: Eastern North America; Europe.
Notes: On leaves.
Host: Tilia
Apiognomonia veneta (Sacc. & Speg.) Höhn.
Gnomonia veneta (Sacc. & Speg.) Kleb.
Gnomonia platani Kleb.
Discula platani (Peck) Sacc. (anamorph)
Distr.: Temperate North America; Europe.
Notes: Anthracnose.
Hosts: Platanus, Quercus

Apioplagiostoma Barr. Pyrenomycetes, Diaporthales

Notes: Barr (Mycol. Mem. 7:102. 1978).
Apioplagiostoma populi (Cash & A.M. Waterman) Barr
Plagiostoma populi Cash & A.M. Waterman
Distr.: Northeastern United States.
Notes: On leaves.
Host: Populus

Apioporthe Höhn. Pyrenomycetes, Diaporthales

Notes: This genus is considered a synonym of *Anisogramma*, but not all names have been redisposed.
Apioporthe apiospora (Ellis & Everh.) Wehmeyer
Diaporthe apiospora Ellis & Everh.
Distr.: Eastern North America.
Notes: Barr (Mycol. Mem. 7:96. 1978) suggested that this species may be better placed in the genus *Cryptodiaporthe*.
Host: Ulmus
Apioporthe utahensis (Sacc.) Wehmeyer
Chorostate utahensis Sacc.
Diaporthe utahensis (Sacc.) Petr.
Distr.: UT.
Notes: Possibly a species of *Amphiporthe* (Barr. Mycol. Mem. 7:142. 1978).
Host: Quercus

Apiospora Sacc. Pyrenomycetes, Amphisphaeriales

Apiospora montagnei Sacc.
Arthrinium arundinis Dyko & Sutton (anamorph)
Distr.: Cosmopolitan.
Notes: Saprobic in soil, on grasses and other organic debris.
Hosts: Andropogon, Aristida, Arundinaria, Arundo, Chasmanthium, Cinna, Cortaderia, Cynodon, Digitaria, Elymus, Erianthus, Gymnopogon, Helianthus, Panicum, Paspalum, Saccharum, Schizachyrium, Smilax, Sorghastrum, Sorghum, Sporobolus, Tridens, Tripsacum, Uniola
Apiospora parallela (P. Karst.) Sacc.
Sphaerella parallela P. Karst.
Distr.: AK; Europe.
Notes: On grasses (Arx. Ber. Schweiz. Bot. Ges. 62:344. 1952).
Host: Arctagrostis
Apiospora spinulosa Petr.
Distr.: HI(type).
Host: Panicum

Apiosporella Höhn. ex Theiss. Pyrenomycetes, Amphisphaeriales

Apiosporella alpina Wehmeyer
Distr.: WA.
Notes: Probably a *Didymella* sp. (Barr. Mycologia 56:860. 1964).
Hosts: Arnica, Lupinus, Pedicularis, Valeriana

Apiosporina Höhn. Loculoascomycetes, Dothideales

Notes: Barr (Canad. J. Bot. 46:855. 1968).
Apiosporina collinsii (Schwein.) Höhn.
Dimerosporium collinsii (Schwein.) Thuem.
Fusicladium sp. (anamorph)
Distr.: North America.
Notes: Witches' broom (Corlett. Fungi Canadenses 76. 1975).
Host: Amelanchier
Apiosporina morbosa (Schwein.:Fr.) Arx
Dibotryon morbosum (Schwein.:Fr.) Theiss. & Syd.
Plowrightia morbosa (Schwein.:Fr.) Sacc.
Fusicladium sp. (anamorph)
Distr.: North America.
Notes: Black knot (Corlett. Fungi Canadenses 84. 1976; Sivanesan. The bitunicate Ascomycetes. p. 599. 1984; Sutton & Waterston. C.M.I. Descr. 224. 1970).
Hosts: [Plumeria], Prunus

Apiosporium Kunze:Fr. Agonomycetes

Notes: A synonym of *Sclerotium* based on Höhnel's examination of the type species, *A. salicis* Kunze.
Apiosporium abietis Cooke
Meliola abietis (Cooke) Sacc.
Distr.: Pacific Northwest.
Notes: Hansford (Beih. Sydowia 2:765. 1961).
Host: Abies
Apiosporium pinophilum Fuckel
Distr.: Pacific Northwest.
Host: Juniperus

Aplopsora Mains. Basidiomycetes, Uredinales

Aplopsora nyssae Mains
Distr.: Southeastern United States.
Notes: Only uredinial and telial stages known.
Host: Nyssa

Aplosporella Speg. Coelomycetes

Podosporium Bonord. non Schwein., hom. illeg.
Podosporium Sacc. & Schulzer non Schwein., hom. illeg.
Pleosphaeropsis Died.
Epicyta Petr.
Notes: *Haplosporella* is an orthographic variant. There is no modern treatment of this genus. Petrak & Sydow (Gatt. Repert. Spec. Nov. Regni Veg. p. 16-113. 1927) provided the most comprehensive account of the genus.
Aplosporella aesculi Cooke & Massee
Distr.: Eastern North America; Great Britain.
Host: Aesculus
Aplosporella ailanthi Ellis & Everh.
Sphaeropsis ailanthi Ellis & Barth.
Haplosporella dothideoides Ellis & Barth.
Sphaeropsis melanconioides Peck
Distr.: Temperate North America.
Host: Ailanthus
Aplosporella alnicola (Peck) Petr. & Syd.
Sphaeropsis alnicola Peck
Distr.: NY.
Hosts: Alnus, Betula
Aplosporella amygdalina Dearn.
Distr.: KS(type).
Host: Prunus
Aplosporella araliae Ellis & Everh.
Distr.: WV(type).
Host: Aralia
Aplosporella asiminae (Ellis & Everh.) Petr.

Sphaeropsis asiminae Ellis & Everh.
Distr.: NC(type).
Host: Asimina
Aplosporella baccharidicola (Dearn.) Petr. & Syd.
Sphaeropsis baccharidicola Dearn.
Distr.: NY.
Host: Baccharis
Aplosporella baptisiae Petr. & Syd.
Distr.: SC(type).
Host: Baptisia
Aplosporella calycanthi Fairm.
Distr.: NY(type).
Host: Calycanthus
Aplosporella celtidis Ellis & Everh.
Sphaeropsis celtidis Ellis & Everh.
Distr.: KS(type).
Host: Celtis
Aplosporella cercidis Ellis & Barth.
Distr.: KS.
Host: Cercis
Aplosporella chamaecristae Petr.
Distr.: FL(type).
Host: Cassia
Aplosporella clematidis (Dearn. & House) Petr. & Syd.
Macroplodia clematidis Dearn. & House
Distr.: NY.
Host: Clematis
Aplosporella clintonii (Peck) Petr. & Syd.
Sphaeropsis clintonii Peck
Haplosporella clintonii (Peck) Petr. & Syd., orth. var.
Haplosporella acerina (Ellis & Barth.) Ellis & Everh.
Sphaeropsis acerina Ellis & Barth.
Sphaeropsis albescens Ellis & Everh.
Sphaeropsis grandinea Ellis & Everh.
Sphaeropsis lineata Ellis & Dearn.
Distr.: Eastern North America.
Host: Acer
Aplosporella cydoniae (Cooke & Ellis) Petr.
Sphaeropsis cydoniae Cooke & Ellis
Distr.: VA(type).
Host: Chaenomeles
Aplosporella diospyri (Dearn. & Barth.) Petr. & Syd.
Sphaeropsis diospyri Dearn. & Barth.
Distr.: Southeastern United States.
Host: Diospyros
Aplosporella elaeagnina (Fairm.) Petr. & Syd.
Sphaeropsis elaeagnina Fairm.
Distr.: ND.
Host: Elaeagnus
Aplosporella euonymi Ellis & Everh.
Distr.: NY(type).
Host: Euonymus
Aplosporella fabaeformis (Pass. & Thuem.) Petr. & Syd.
Diplodia fabaeformis Pass. & Thuem.
Sphaeropsis fabaeformis Sacc.
Sphaeropsis fabaeformis Sacc. var. *lignicola* Maia
Sphaeropsis viticola Pass.
Sphaeropsis vitigena Ellis & Everh.
Distr.: Central and eastern North America.
Host: Vitis
Aplosporella gleditschiicola (Cooke) Petr. & Syd.
Sphaeropsis gleditschiicola Cooke
Aplosporella gleditschiae Grove
Sphaeropsis triacanthi Ellis & Barth.
Distr.: KS(type).
Host: Gleditsia
Aplosporella hedericola Speg.
Diplodia hedericola Speg.
Sphaeropsis hedericola (Speg.) Sacc.
Distr.: Eastern United States; Argentina.
Hosts: Hedera, Parthenocissus
Aplosporella hibisci (Berk.) Petr. & Syd.
Sphaeria hibisci Berk.
Sphaeropsis hibisci (Berk.) Cooke
Sphaeropsis lyndonvillae Sacc.
Distr.: Eastern United States.
Host: Hibiscus
Aplosporella jasmini Ellis & Everh.
Distr.: AL(type).
Host: Jasminum
Aplosporella juniperi (Peck) Petr.
Sphaeropsis juniperi Peck
Macroplodia sabina House
Distr.: Southeastern United States.
Hosts: Chamaecyparis, Juniperus
Aplosporella lathamii Dearn.
Distr.: NY(type).
Host: Myrica
Aplosporella linderae (Peck) Petr.
Sphaeropsis linderae Peck
Distr.: Eastern United States.
Host: Lindera
Aplosporella linearis (Peck) Petr.
Sphaeropsis linearis Peck
Distr.: Central and eastern United States; Europe.
Hosts: Carya, Quercus
Aplosporella liquidambaris (Dearn. & House) Petr. & Syd.
Sphaeropsis liquidambaris Dearn. & House
Distr.: Eastern North America.
Host: Liquidambar
Aplosporella mori (Berl.) Petr. & Syd.
Sphaeropsis mori Berl.
Sphaeropsis sepulta Ellis & Everh.
Distr.: Central and eastern United States; Europe.
Host: Morus
Aplosporella negundinis Ellis & Barth.
Haplosporella negundinis Ellis & Barth., orth. var.
Distr.: KS(type).
Notes: Possibly a synonym of *Aplosporella clintonii* (Peck) Petr. & Syd.
Host: Acer
Aplosporella neilliae (Ellis & Everh.)
Sphaeropsis neilliae Ellis & Everh.
Distr.: NY(type).
Host: Sorbaria
Aplosporella nerii Sacc.
Distr.: Eastern North America; Europe.
Host: Nerium
Aplosporella novae-hollandiae (Speg.) Speg.
Diplodia novae-hollandiae Speg.
Sphaeropsis novae-hollandiae (Speg.) Sacc.
Distr.: FL(type); Argentina.
Host: Eucalyptus
Aplosporella pini Peck
Distr.: NY.
Host: Pinus
Aplosporella prosopidis (Dearn. & Barth.) Petr. & Syd.
Sphaeropsis prosopidis Dearn. & Barth.
Distr.: TX.
Host: Prosopis
Aplosporella punctum (Cooke & Ellis) Petr.
Sphaeropsis punctum Cooke & Ellis
Distr.: DC(type).
Host: Lonicera
Aplosporella rumicicola (Sacc.) Petr. & Syd.
Sphaeropsis rumicicola Sacc.
Distr.: WA; Italy(type).
Host: Rumex
Aplosporella salicicola (Pass.) Petr. & Syd.
Sphaeropsis salicicola Pass.
Sphaeropsis salicis Ellis & Barth.
Distr.: Central and eastern United States; Europe.
Host: Salix
Aplosporella sambuci (Peck) Petr.
Sphaeropsis sambuci Peck
Distr.: North America; Europe.
Host: Sambucus
Aplosporella smilacis (Ellis & Everh.) Petr. & Syd.
Sphaeropsis smilacis Ellis & Everh.
Distr.: NY(type).
Host: Smilax

Aplosporella squieriae (G.W. Clinton) Petr.
Sphaeropsis squieriae G.W. Clinton in Peck
Distr.: MA(type).
Host: Aristolochia
Aplosporella tecomae (Dearn.) Petr. & Syd.
Sphaeropsis tecomae Dearn.
Distr.: Eastern United States.
Host: Campsis
Aplosporella thalictri (Ellis & Fairm.) Petr. & Syd.
Sphaeropsis thalictri Ellis & Fairm.
Distr.: NY.
Host: Thalictrum
Aplosporella tiliacea (Peck) Petr.
Sphaeropsis tiliacea Peck
Aplosporella guttifera Petr.
Distr.: Central and eastern United States; Europe.
Host: Tilia
Aplosporella ulmea (Ellis & Barth.) Petr. & Syd. var. **rhizophila** (Shear) Petr. & Syd.
Aplosporella rhizophila Shear
Distr.: Central United States.
Host: Ulmus
Aplosporella wistariae Ellis & Barth.
Sphaeropsis wistariana Fairm.
Distr.: Central and eastern North America.
Host: Wisteria

Apodothina Petr. Loculoascomycetes, Dothideales

Apodothina pringlei (Peck) Petr.
Dothidea pringlei Peck
Sphaerodothis pringlei (Peck) Theiss. & Syd.
Distr.: Southwestern United States.
Notes: On leaves.
Host: Yucca

Aporpium Bondartsev & Singer ex Singer. Basidiomycetes, Tremellales

Aporpium caryae (Schwein.) Teix. & D.P. Rogers
Polyporus caryae Schwein.
Poria caryae (Schwein.) Cooke
Poria argillacea (Cooke) Cooke
Polyporus argillaceus Cooke
Aporpium canescens (P. Karst.) Bondartsev & Singer, nom. inval.
Poria canescens P. Karst.
Poria cordylina G. Cunn.
Aporpium pilati (Bourd.) Bondartsev & Singer ex Bondartsev
Poria pilati Bourd.
Distr.: Widespread.
Notes: On wood. Teixeira & Rogers (Mycologia 47:408-415. 1955) listed a number of specimens but do not indicate hosts.
Hosts: Alnus, Fagus, Juglans, Liquidambar, Liriodendron, Populus, Quercus

Aposphaeria Sacc., nom. cons. Coelomycetes

Notes: According to Sutton (The Coelomycetes. p. 407. 1980.), there are about 200 taxa with no revisionary treatments available.
Aposphaeria caricicola R. Sprague
Distr.: WA(type).
Host: Carex
Aposphaeria fuscomaculans Sacc.
Plenodomus fuscomaculans (Sacc.) Coons
Distr.: CA, MI; Europe(type).
Host: Malus
Aposphaeria major Syd.
Distr.: UT.
Notes: On stems.
Host: Rubus
Aposphaeria mollis (Lév.) Sacc.
Distr.: CA(type).
Host: Juncus

Apostrasseria Nag Raj. Coelomycetes

Apostrasseria lunata (Shear) Nag Raj
Ceuthospora lunata Shear
Neottiospora lycopodina Höhn.
Strasseria lycopodina (Höhn.) Höhn.
Phacidium lunatum DiCosmo, Nag Raj, & Kendrick (teleomorph)
Distr.: Northeastern North America.
Notes: On leaves and fruits of Ericaceae, pine cones, and organic debris.
Hosts: Gaultheria, Gaylussacia, Taxus, Vaccinium
Apostrasseria robusta Nag Raj
Phacidium dearnessii DiCosmo, Nag Raj, & Kendrick (teleomorph)
Distr.: Pacific Northwest.
Notes: On leaves.
Host: Taxus

Appendiculella Höhn. Pyrenomycetes, Meliolales

Appendiculella araliae (Spreng.) Hansf.
Amphitrichum araliae Spreng.
Irene araliae (Spreng.) F. Stevens
Meliola araliae (Spreng.) Mont.
Distr.: MS, PR; Guyana.
Notes: On Araliaceae (Hansford. Beih. Sydowia 2:480. 1961).
Host: [Magnolia]
Appendiculella calostroma (Desmaz.) Höhn.
Sphaeria calostroma Desmaz.
Irene calostroma (Desmaz.) Höhn.
Meliola autumnalis Syd.
Irene puiggarii (Speg.) Doidge
Meliola rubicola Henn.
Irenina sanguinea (Ellis & Everh.) F. Stevens
Distr.: Cosmopolitan.
Notes: On living leaves. Additional synonyms listed by Hansford (Beih. Sydowia 2:236. 1961).
Hosts: Myrica, Osteomeles, Rubus
Appendiculella kalalauensis Hansf.
Distr.: HI.
Host: Perrottetia
Appendiculella pinicola (Dearn.) Pirozynski & Shoemaker
Meliola pinicola Dearn.
Asteridiella pinicola (Dearn.) Hansf.
Irenina pinicola (Dearn.) F. Stevens
Distr.: Northern and temperate North America.
Notes: On living needles of Pinaceae.
Host: Pinus
Appendiculella splendens (F. Stevens) Hansf.
Irene splendens F. Stevens
Distr.: HI.
Host: Alphitonia

Arachnopeziza Fuckel. Discomycetes, Helotiales

Arachnopeziza aurata Fuckel
Arachnopeziza arctostaphyli Cash
Belonidium phlegmaceum (Ellis) Sacc.
Belonium phlegmaceum (Ellis) Seaver
Distr.: Widespread, most common in North America; Europe.
Notes: On various kinds of decaying plant material, particularly on rotted wood of *Quercus* (Korf. Lloydia 14:153. 1951).
Host: Arctostaphylos
Arachnopeziza aurelia (Pers.:Fr.) Fuckel
Distr.: Widespread.
Notes: On rotten wood, leaves and especially on fruits and involucres. Korf (Lloydia 14:160. 1951).
Host: Quercus

Arcyria A. Wigg. Myxomycetes

Arcyria nutans (Bull.) Grev.
Distr.: Cosmopolitan.
Host: Quercus

Arcyria occidentalis (T. Macbr.) G. List.
Lachnobolus occidentalis T. Macbr.
Distr.: Temperate North America, rare elsewhere.
Host: Pinus

Aristastoma Tehon emend. Sutton. Coelomycetes

Aristastoma oeconomicum (Ellis & Tracy) Tehon
Amerosporium oeconomicum Ellis & Tracy
Aristastoma concentricum Tehon
Distr.: Central and eastern United States.
Notes: On living leaves of Fabaceae.
Hosts: Phaseolus, Pisum, Vicia, Vigna

Armillaria (Fr.:Fr.) Staude. Basidiomycetes, Agaricales

Armillariella (P. Karst.) P. Karst.
Armillaria mellea (Vahl:Fr.) P. Kumm.
Armillariella mellea (Vahl:Fr.) P. Karst.
Rhizomorpha subcorticalis Pers. (anamorph)
Distr.: Cosmopolitan.
Notes: A wide number of possible taxa and/or ecotypes have been included under this name. Many of these entities utilize different ecological niches, including wood and litter decomposition, primary and secondary root pathogens, mycorrhiza formation, and mycoparasitism of *Entoloma abortivum* (Watling et al. Trans. Brit. Mycol. Soc. 78:271. 1983; Roll-Hansen. Eur. J. Forest Pathol. 15:22. 1985; Pegler & Gibson. C.M.I. Descr. 321. 1972).
Hosts: Abies, Abutilon, Acacia, Acer, Actinidia, Aesculus, Agave, Ailanthus, Albizia, Alnus, Alsophila, Annona, Arbutus, Arundo, Asparagus, Baptisia, Begonia, Bergenia, Betula, Broussonetia, Buxus, Callistemon, Calocedrus, Calothamnus, Camellia, Carpinus, Carya, Castanea, Castanopsis, Casuarina, Catalpa, Ceanothus, Cedrus, Celtis, Cephalanthus, Ceratonia, Chamaecyparis, Chimonanthus, Cinnamomum, Cistus, Citrus, Coprosma, Cornus, Corylus, Cotinus, Cotoneaster, Crassula, Crataegus, Cupressus, Dahlia, Dianthus, Diospyros, Eriobotrya, Erythrina, Eucalyptus, Euonymus, Euryops, Exochorda, Fabiana, Fagus, Feijoa, Ficus, Fontanesia, Fragaria, Fraxinus, Fuchsia, Gardenia, Gouldia, Grevillea, Hebe, Helianthus, Hibiscadelphus, Hippeastrum, Hovenia, Humulus, Hydrangea, Ilex, Impatiens, Jacaranda, Juglans, Juniperus, Kalmia, Lagerstroemia, Larix, Lavandula, Leptospermum, Ligustrum, Liquidambar, Liriodendron, Lonicera, Lupinus, Magnolia, Malus, Maytenus, Medicago, Melaleuca, Metrosideros, Morus, Myoporum, Myrsine, Narcissus, Nerium, Nyssa, Olea, Opuntia, Osmanthus, Ostrya, Paeonia, Pelargonium, Persea, Peumus, Phlomis, Phoenix, Phormium, Photinia, Picea, Pieris, Pinus, Pipturus, Pistacia, Pittosporum, Platanus, Populus, Prunus, Pseudotsuga, Psychotria, Pterocarya, Punica, Pyracantha, Pyrus, Quercus, Rhaphiolepis, Rheum, Rhododendron, Rhus, Ribes, Robinia, Rosa, Rubus, Salix, Sapium, Sassafras, Schinus, Sequoia, Sequoiadendron, Sesbania, Solanum, Sophora, Sorbus, Straussia, Styphelia, Syringa, Taxus, Tecomaria, Thuja, Tsuga, Ulmus, Vaccinium, Viburnum, Vitex, Vitis, Watsonia, Xanthosoma, Xylosma, Zantedeschia, Zanthoxylum, Zelkova
Armillaria putrida (Scop.) Murrill
Distr.: CA; Europe(type).
Notes: A doubtful name (Termorshuizen & Arnolds. Mycotaxon 30:112. 1987).
Hosts: Quercus, Sambucus
Armillaria tabescens (Scop.) Dennis, Orton & Hora
Armillariella tabescens (Scop.) Singer
Clitocybe tabescens (Scop.) Bres.
Clitocybe monadelphus (Morg.) Sacc.
Clitocybe parasitica E.M. Wilcox
Distr.: Eastern North America; southern Europe.
Notes: On roots and wood.
Hosts: Abelia, Acacia, Acalypha, Acer, Acokanthera, Albizia, Aleurites, Annona, Antigonon, Ardisia, Arecaceae, Bauhinia, Bucida, Buddleja, Butia, Caesalpinia, Calliandra, Callistemon, Camellia, Carya, Cassia, Castanea, Casuarina, Cedrus, Celtis, Ceratonia, Cercis, Chamaecyparis, Cinnamomum, Citrus, Coccoloba, Cochlospermum, Cordia, Cornus, Cotoneaster, Crataegus, Crotalaria, Cuphea, Cupressus, Cycas, Dalbergia, Delonix, Diospyros, Dombeya, Elaeagnus, Eriobotrya, Erythrina, Eucalyptus, Eugenia, Euphorbia, Feijoa, Ficus, Fragaria, Gardenia, Gaylussacia, Grevillea, Hamelia, Hibiscus, Ilex, Ixora, Jacaranda, Jasminum, Jatropha, Juglans, Juniperus, Koelreuteria, Lagerstroemia, Ligustrum, Liquidambar, Litchi, Lonicera, Magnolia, Malus, Malvaviscus, Manilkara, Melaleuca, Melia, Melicoccus, Michelia, Mimosa, Monstera, Morus, Musa, Myrica, Myrtus, Nerium, Osmanthus, Ostrya, Parkinsonia, Persea, Phoenix, Photinia, Pinus, Pithecellobium, Pittosporum, Plumbago, Podocarpus, Poncirus, Prunus, Psidium, Pterygota, Punica, Pyracantha, Pyrus, Quercus, Rhaphiolepis, Rhododendron, Rhus, Ricinus, Rosa, Rosmarinus, Rumohra, Salix, Sanchezia, Sapium, Schinus, Schlumbergera, Senecio, Sida, Spiraea, Syzygium, Tabernaemontana, Tecoma, Tecomaria, Tetrazygia, Thuja, Thunbergia, Tibouchina, Tithonia, Trachelospermum, Ulmus, Viburnum, Vitis, Washingtonia, Wisteria

Arnium Nitschke. Pyrenomycetes, Sordariales

Notes: Eleven species (Krug & Cain. Canad. J. Bot. 50:367. 1972).
Arnium apiculatum (D. Griffiths) N. Lundqvist
Delitschia apiculata D. Griffiths
Distr.: IL, SD; England.
Notes: Shearer et al. (Mycotaxon 11:169. 1980).
Hosts: Equisetum, Salsola

Arrhytidia Berk. & M.A. Curtis. Basidiomycetes, Dacrymycetales

Arrhytidia involuta (Schwein.) Coker
Ceracea corticioides (Ellis & Everh.) Pat.
Dacrymyces corticioides Ellis & Everh.
Distr.: North America.
Notes: On wood.
Hosts: Betula, Celtis, Pinus, Taxodium, Tsuga

Arthothelium C. Massal. Loculoascomycetes, Arthoniales

Notes: Most of these species are lichenized.
Arthothelium lichenalis (Peck) Barr in Barr et al.
Pleospora lichenalis (Peck) Sacc.
Distr.: VT(type).
Notes: Barr et al. (New York State Mus. Bull. 459:27. 1986).
Host: Betula

Arthrinium Kunze:Fr. in Kunze & J.C. Schmidt. Hyphomycetes

Arthrinium arundinis Dyko & Sutton
Papularia arundinis (Corda) Fr.
Apiospora montagnei Sacc. (teleomorph)
Distr.: Cosmopolitan.
Notes: On dead and decaying grass leaves.
Hosts: Alnus, Arundinaria, Cortaderia, Cucumis, Cynodon, Phragmites
Arthrinium caricicola Kunze:Fr. in Kunze & J.C. Schmidt
Distr.: Eastern United States; Europe.
Host: Carex
Arthrinium cuspidatum (Cooke & Harkn.) Höhn.
Camptoum cuspidatum Cooke & Harkn.
Arthrinium bicorne Rostr.
Tureenia juncoidea J.G. Hall
Distr.: Widespread in Europe and North America.
Hosts: Carex, Juncus, Scirpus
Arthrinium phaeospermum (Corda) M.B. Ellis
Coniosporium inquinans Durieu & Mont.
Epicoccum simplex Berk. & M.A. Curtis in Berk.
Papularia sphaerosperma (Pers.) Höhn.
Papularia sphaerosperma (Pers.) Höhn. var. *inquinans* (Durieu & Mont.) Grove
Distr.: Cosmopolitan.

Notes: On culms of Poaceae and other monocots (Ellis. Dematiaceous Hyphomycetes. p. 571. 1971; Ellis. Mycol. Pap. 103:8. 1965).
Hosts: Agropyron, Andropogon, Arundinaria, Arundo, Bambusa, [Camellia], [Carya], Cortaderia, Cynodon, Phragmites, Phyllostachys, Poaceae, Saccharum, Spartina, Zea
Arthrinium sporophleum Kunze
Distr.: Widespread.
Hosts: Carex, Metrosideros

Arthrobotrys Corda. Hyphomycetes
Notes: Haard (Mycologia 60:1140. 1968).
Arthrobotrys oligospora Fresen.
Distr.: Central and eastern United States; Europe.
Notes: Isolated from soil, rotten wood and various kinds of debris.
Host: Antirrhinum

Arthrobotryum Ces. Hyphomycetes
Arthrobotryum atrocephalum (Ellis) Sacc.
Stilbum atrocephalum Ellis
Distr.: NJ.
Host: Kalmia

Arthrographis Sigler & J.W. Carmichael. Hyphomycetes
Notes: Sigler & Carmichael (Mycotaxon 4:359. 1976).
Arthrographis cuboidea (Sacc. & Ellis) Sigler
Oospora cuboidea Sacc. & Ellis
Distr.: Eastern United States.
Notes: Pink stain of wood.
Hosts: Pinus, Quercus

Arthuriomyces Cummins & Y. Hiratsuka. Basidiomycetes, Uredinales
Arthuriomyces peckianus (E. Howe) Cummins & Y. Hiratsuka
Puccinia peckiana E. Howe in Peck
Gymnoconia peckiana (E. Howe) Trott.
Gymnoconia interstitialis Lagerh.
Distr.: Central and eastern North America; Eurasia.
Notes: Autoecious. Long-cycled orange rust.
Host: Rubus

Articularia Höhn. Hyphomycetes
Articularia quercina (Peck) Höhn.
Distr.: IL, MS.
Notes: On fallen leaves.
Hosts: [Carya], Quercus
Articularia quercina (Peck) Höhn. var. ***minor*** V. Charles
Distr.: MO, TX.
Notes: On leaves.
Hosts: Carya, Quercus

Articulophora Wang & Sutton. Hyphomycetes
Notes: Monotypic (Mycologia 74:489. 1982).
Articulophora variispora Wang & Sutton
Distr.: NY.
Notes: On dead wood.
Host: Castanea

Arwidssonia B. Eriksson. Pyrenomycetes, Phyllachorales
Notes: Two species (Eriksson. Svensk Bot. Tidskr. 68:199. 1974).
Arwidssonia empetri (Rehm) B. Eriksson
Heterosphaeria empetri Rehm
Sphaeropezia empetri (Rehm) Rehm
Distr.: Northern temperate regions.
Host: Empetrum

Arxiomyces P. Cannon & D. Hawksworth. Pyrenomycetes, Sordariales
Arxiomyces vitis (Fuckel) P. Cannon & D. Hawksworth
Phaeostoma vitis (Fuckel) Arx & E. Müller, nom. illeg.
Ceratostoma graphioides Sacc.
Chaetoceratostoma graphioides (Sacc.) C. Booth & Dennis in Dennis
Distr.: ND(questionable); temperate Europe.
Notes: On bark of woody shrubs and trees. Cannon & Hawksworth (Trans. Brit. Mycol. Soc. 81:644. 1983) provided a replacement for *Phaeostoma* Arx & E. Müller, a later homonym.
Host: Crataegus

Ascocalyx Naumov. Discomycetes, Helotiales
Ascocalyx abietina (Lagerberg) Schläpfer
Lagerbergia abietina (Lagerberg) J. Reid in Dennis
Gremmeniella abietina (Lagerberg) Morelet
Scleroderris abietina (Lagerberg) Gremmen
Scleroderris lagerbergii Gremmen
Brunchorstia pinea (P. Karst.) Höhn. (anamorph)
Distr.: Northern North America; Europe.
Notes: Die-back and canker (Punithalingam. C.M.I. Descr. 369. 1973).
Host: Pinus

Ascochyta Lib. Coelomycetes
Notes: Originally spelled "*Ascoxyta*". Melnik (Key to fungi of the genus *Ascochyta* Lib. 1977).
Ascochyta abelmoschi Harter
Distr.: Eastern North America.
Hosts: Abelmoschus, Hibiscus
Ascochyta achlyicola Ellis & Everh.
Distr.: Western United States.
Host: Achlys
Ascochyta actaeae (Bres.) J.J. Davis
Marssonia actaeae Bres.
Actinonema actaeae Allesch.
Marssonina actaeae (Bres.) Magnus
Distr.: Eastern North America; Europe.
Hosts: Actaea, Cimicifuga
Ascochyta agropyri-repentis (R. Sprague) Punithalingam
Apiocarpella agropyri R. Sprague in H.C. Greene
Distr.: WI.
Host: Agropyron
Ascochyta agropyrina (Fairm.) Trott.
Ascochytula agropyrina Fairm.
Distr.: Temperate United States.
Notes: Restricted to *Agropyron* (Punithalingam. Mycol. Pap. 142:39. 1979).
Hosts: Agropyron, [Elymus]
Ascochyta ampelina Sacc.
Pseudodiplodia ampelina (Sacc.) Petr.
Distr.: WI; Italy(type).
Host: Vitis
Ascochyta aquilegiae (Roum. & Pat.) Sacc.
Phyllosticta aquilegiae Roum. & Pat.
Ascochyta aquilegiae (Rabenh.) Höhn.
Diplodina delphinii Laskaris
Distr.: North America; Europe; New Zealand.
Notes: Buchanan (Mycol. Pap. 156:17. 1987).
Hosts: Aquilegia, Clematis, Delphinium
Ascochyta armoraciae Fuckel
Distr.: Temperate regions.
Notes: Brandenburger (Parasitiche Pilze an Gefarbpflanzen in Europa. p. 182. 1985) suggested that this might be a synonym of *Septoria armoraciae* Sacc.
Hosts: Armoracia, Brassica
Ascochyta asclepiadicola Ellis & Everh.
Distr.: WI
Notes: A doubtful name; probably refers to *A. asclepiadis*.
Host: Asclepias
Ascochyta asclepiadis Ellis & Everh.
Distr.: DE(type), WI.
Host: Asclepias

Ascochyta asparagina (Petr.) P.K. Buchanan
Ascochytula asparagina Petr.
Distr.: Southeastern United States; Europe.
Notes: Buchanan (Mycol. Pap. 156:22. 1987).
Host: Asparagus
Ascochyta aspidistrae Massee
Distr.: FL, MN, NJ; Europe.
Notes: A doubtful species (Grove. British stem and leaf fungi. Vol. I. p. 322. 1935).
Host: Aspidistra
Ascochyta avenae (Petr.) R. Sprague & A.G. Johnson
Ascochytella avenae Petr.
Pseudodiplodia avenae (Petr.) Petr.
Distr.: Western North America; Europe; New Zealand.
Notes: Leaf blotch of oats and other grasses (Punithalingam. C.M.I. Descr. 731. 1982).
Hosts: Avena, Helictotrichon
Ascochyta babylonica H.C. Greene
Distr.: WI; Europe.
Notes: Melnik (Key to fungi of the genus *Ascochyta* Lib. p. 158. 1977) indicated that this name may be synonymous with *Ascochyta translucens*.
Host: Salix
Ascochyta biguttulata E.Y. Daniels
Distr.: Temperate United States.
Host: Polygonum
Ascochyta bohemica Kab. & Bubák
Stagonospora bohemica (Kab. & Bubák) Tobisch
Distr.: North and South America; Asia; Europe.
Notes: Synonymy according to Melnik (Key to fungi of the genus *Ascochyta* Lib. p. 92. 1977). Boerema & Bollen (Persoonia 8:138. 1975) discussed the morphological characteristics of this species.
Host: Campanula
Ascochyta boutelouae Fairm.
Distr.: ND(type), NM.
Notes: A weak pathogen (Punithalingam. Mycol. Pap. 142:59. 1979).
Host: Bouteloua
Ascochyta brachypodii (Syd.) R. Sprague & A.G. Johnson
Ascochyta graminicola Sacc. var. *brachypodii* Trail
Distr.: North America(questionable); Europe.
Notes: Sprague and Johnson's concept of *A. graminicola* var. *brachypodii* was inconsistent with the type specimen; therefore specimens identified as that species based on R. Sprague and Johnson's descriptions could be a variety of fungi (Punithalingam. Mycol. Pap. 142:60. 1979).
Hosts: Agrohordeum, Andropogon, Avena, Bromus, Chloris, Festuca, Lolium, Melica, Phalaris, Pleuropogon, Poa, Schedonnardus, Schizachyrium, Sorghastrum, Stipa
Ascochyta bresadolae Sacc. & P. Syd. in Sacc.
Ascochyta fagopyri Bres.
Distr.: Temperate North America, Asia and Europe.
Host: Fagopyrum
Ascochyta cannabis Lasch
Phyllosticta cannabis (L. Kirchner) Speg.
Septoria cannabis (Lasch) Sacc.
Didymella arcuata Röder (teleomorph)
Distr.: Central and southeastern United States; Europe.
Notes: Sivanesan (The bitunicate Ascomycetes. p. 416. 1984); Punithalingam (C.M.I. Descr. 668. 1980 as *Septoria cannabis*).
Host: Cannabis
Ascochyta caricae Rabenh.
Distr.: OR; Asia; Europe.
Host: Ficus
Ascochyta caricicola Melnik
Ascochyta caricis W.B. Cooke & C.G. Shaw non Lambotte & Fautrey, hom. illeg.
Distr.: AK, CA; Europe; Asia.
Notes: This is the name used by Melnik (Key to fungi of the genus *Ascochyta* Lib. p. 107. 1977) but it appears to be a later name for *Ascochyta caricis-arenariae* Melnik.
Host: Carex
Ascochyta carthaginensis Sacc.
Distr.: Southeastern United States.
Host: Manihot
Ascochyta caryae H.C. Greene
Distr.: WI.
Host: Carya
Ascochyta cassandrae Peck
Distr.: Temperate North America.
Notes: According to Melnik (Key to fungi of the genus *Ascochyta* Lib. p. 180. 1977), this is a *Discella*.
Host: Chamaedaphne
Ascochyta catalpae Tassi
Distr.: IL, PA; Italy(type).
Host: Catalpa
Ascochyta cephalanthi Ellis & Everh. in Sacc.
Distr.: Southern United States.
Host: Cephalanthus
Ascochyta cheiranthi Bres.
Ascochyta brassicae-rapae Bondartzeva-Monteverde
Ascochyta hesperidis Died.
Distr.: AK; Asia; Europe.
Host: Cheiranthus
Ascochyta clematidina Thuem. var. **thalictri** J.J. Davis
Distr.: WI.
Notes: According to Melnik (Key to fungi of the genus *Ascochyta* Lib. p. 181. 1977), this is a *Phyllosticta*.
Host: Thalictrum
Ascochyta compositarum J.J. Davis
Ascochyta compositarum J.J. Davis var. *parva* J.J. Davis
Distr.: Temperate North America.
Hosts: Ambrosia, Aster, Cacalia, Echinacea, Erechtites, Eupatorium, Helenium, Helianthus, Heliopsis, Parthenium, Prenanthes, Ratibida, Rudbeckia, Senecio, Silphium, Solidago, Tithonia
Ascochyta confusa Ellis & Everh.
Distr.: Eastern North America.
Host: Smilax
Ascochyta conicola Dearn. & House, nom. nud.
Distr.: NY.
Notes: This name was published without a latin diagnosis.
Host: Tsuga
Ascochyta cornicola Sacc.
Distr.: Temperate North America.
Host: Cornus
Ascochyta cycadina Scalia
Distr.: MO, TX; Italy(type).
Notes: According to Melnik (Key to fungi of the genus *Ascochyta* Lib. p. 182. 1977), this species is a *Hendersonia*.
Host: Cycas
Ascochyta cypripedii H.C. Greene
Distr.: WI.
Host: Cypripedium
Ascochyta deformis (P. Karst.) P.K. Buchanan
Diplodia deformis P. Karst.
Diplodina deformis (P. Karst.) Sacc.
Pseudodiplodia deformis (P. Karst.) H. Rupprecht
Distr.: WA; Europe.
Host: Sambucus
Ascochyta desmazieresii Cavara
Diplodina lolii A. Zimmerm.
Distr.: North America; Europe; Japan.
Notes: Leaf spot. Punithalingam (Mycol. Pap. 142 :69. 1979); Punithalingam (C.M.I. Descr. 661. 1980).
Host: Lolium
Ascochyta dianthi (Albertini & Schwein.) Lib.
Distr.: North America; Europe.
Hosts: Dianthus, Stellaria
Ascochyta doronici Allesch.
Ascochyta gerberae Maffei
Ascochyta kuhniae H.C. Greene
Ascochyta lappae Kab. & Bubák
Gloeosporium lappae Dearn. & House
Ascochyta rudbeckiae (Ellis & Everh.) H.C. Greene non Bondartzeva-Monteverde, hom. illeg.
Ascochyta taraxaci (Hollós) Grove
Phyllosticta taraxaci Hollós
Distr.: Cosmopolitan.
Notes: On Asteraceae. The epithet *doronici* does not appear to be the earliest available name of the synonyms listed by Melnik (Key to fungi of the genus *Ascochyta* Lib. p. 99. 1977).

Hosts: Arctium, Gerbera, Helianthus, Kuhnia, Rudbeckia, Taraxacum
Ascochyta equiseti (Desmaz.) Grove
Phoma equiseti Desmaz.
Distr.: North America; temperate Europe.
Host: Equisetum
Ascochyta fremontiae Harkn.
Distr.: CA.
Host: Fremontodendron
Ascochyta garrettiana Syd. & P. Syd.
Distr.: Western United States.
Hosts: Orthocarpus, Penstemon
Ascochyta gossypii Woronichin
Distr.: Worldwide in cotton-growing regions.
Notes: Wet weather blight of cotton. Holliday and Punithalingam (C.M.I. Descr. 271. 1970). Boerema et al. (Persoonia 7:133. 1973) provided a discussion of the differences between this and some plurivorous *Phoma* spp.
Hosts: Gossypium, Hibiscus, [Lupinus]
Ascochyta graminea (Sacc.) R. Sprague & A.G. Johnson
Diplodina graminea Sacc.
Distr.: Temperate regions.
Notes: Punithalingam (Mycol. Pap. 142:84. 1979) indicated that most of Sprague's specimens from the U.S. have slightly smaller conidia than the type and designated *Ascochyta graminea* var. *minuscula* Punithalingam for those fungi. Variety *graminea* occurs on *Cynodon dactylon*.
Hosts: [Bouteloua], [Buchloe], Cynodon, [Glyceria], [Hilaria], [Hordeum], [Muhlenbergia]
Ascochyta graminicola Sacc.
Distr.: Widespread.
Notes: This name has been applied to several graminicolous *Ascochyta* spp. Examination of the type revealed that the name is based on the mycoparasite *Sphaerellopsis filum* (Biv.-Bern.:Fr.) Sutton (Punithalingam. Mycol. Pap. 142:178. 1979).
Hosts: Agropyron, Agrostis, Andropogon, Arrhenatherum, Avena, Bromus, Calamagrostis, Dactylis, Deschampsia, Elymus, Festuca, Hierochloe, Holcus, Hordeum, Hystrix, Muhlenbergia, Oryzopsis, Panicum, Poa, Schizachne, Schizachyrium, Secale, Setaria, Sphenopholis, Stenotaphrum, Triticum
Ascochyta graminicola Sacc. var. **holci** Sacc.
Distr.: OR; WA.
Notes: Punithalingam (Mycol. Pap. 142:188. 1979) indicated that the status of this variety is uncertain.
Hosts: Holcus, Sitanion
Ascochyta greenei Melnik
Ascochyta solidaginis H.C. Greene
Distr.: WI.
Host: Solidago
Ascochyta grossulariae Oudem. non Lib., hom. illeg.
Diplodina grossulariae Sacc. & Briard
Hendersonia grossulariae Oudem.
Diplodina oudemansii Allesch.
Distr.: NY, OR, VA; Europe.
Notes: Buchanan (Mycol. Pap. 156:40. 1987).
Host: Ribes
Ascochyta hansenii Ellis & Everh.
Distr.: Southern and western North America.
Host: Arbutus
Ascochyta hordei K. Hara
Distr.: Temperate northern hemisphere.
Notes: Ring spot disease of barley and other grasses. Punithalingam (Mycol. Pap. 142:88. 1979) recognized 3 varieties. Most records in older literature cannot be ascribed to a specific variety.
Hosts: Bromus, Festuca, Holcus, Hordeum, Phleum, Secale, Stipa
Ascochyta hordei K. Hara var. **americana** Punithalingam
Distr.: Temperate North America.
Notes: Also on *Triticum aestivum* and *Secale cereale* (Punithalingam. Mycol. Pap. 142:96. 1979).
Host: Hordeum
Ascochyta hyalospora (Cooke & Ellis) Boerema, Mathur, & Neergaard
Diplodia hyalospora Cooke & Ellis
Phloeospora chenopodii Ellis & Kellerm.
Diplodina ellisii Sacc.
Distr.: Central and eastern United States; Bolivia
Notes: On leaves, stems, and seeds (Boerema et al. Neth. J. Pl. Pathol. 83:153. 1977).
Host: Chenopodium
Ascochyta hydrophylli-virginiani H.C. Greene
Distr.: WI.
Host: Hydrophyllum
Ascochyta idaei Oudem.
Ascochyta osmaroniae (Dearn.) Arx
Gloeosporium osmaroniae Dearn.
Distr.: WA; Europe.
Host: Oemleria
Ascochyta infuscans Ellis & Everh.
Ascochyta anemones Kab. & Bubák
Distr.: Temperate North America; Europe.
Host: Ranunculus
Ascochyta iridis Oudem.
Distr.: Northeastern North America; Europe.
Host: Iris
Ascochyta ischaemi Sacc.
Ascochyta zeae G.L. Stout
Distr.: CA, IL; Europe; Mauritius.
Notes: On Poaceae.
Host: Zea
Ascochyta juglandis Boltshauser
Distr.: North America; Asia; Europe.
Host: Juglans
Ascochyta lathyri Trail
Distr.: North America; Europe.
Host: Lathyrus
Ascochyta ledi Rostr.
Distr.: WA, WI; Greenland(type).
Host: Ledum
Ascochyta lentis Vassiljevsky
Distr.: Western North America; widespread in lentil-growing regions.
Notes: Gossen et al. (Canad. J. Pl. Pathol. 8:154. 1986) considered this a *forma specialis* of *Ascochyta fabae* Speg. Boerema (Versl. Meded. Plziektenk. Dienst Wageningen 162:25. 1984) provided a discussion of the differences between the two species.
Host: Lens
Ascochyta leonuri Ellis & Dearn.
Ascochyta nepetae J.J. Davis
Distr.: North America; Asia; Europe.
Hosts: Glechoma, Leonurus, Nepeta
Ascochyta leptospora (Trail) K. Hara
Ascochyta elymi Tehon & E.Y. Daniels non Brunaud, hom. illeg.
Ascochyta graminicola Sacc. var. *diedickeana* Baudys & Picb.
Distr.: Widespread but actual distribution difficult to determine due to misidentifications.
Notes: On many grasses. Punithalingam (Mycol. Pap. 142:103. 1979).
Hosts: Bromus, Elymus
Ascochyta lethalis Ellis & Barth. in Barth.
Ascochyta caulicola Laubert
Ascochyta meliloti (Trel.) J.J. Davis
Gloeosporium meliloti Trel.
Didymella lethalis (R. Stone) Sivanesan (teleomorph)
Distr.: Central and eastern North America.
Notes: Sivanesan (The bitunicate Ascomycetes. p. 424. 1984). *Ascochyta viciae* Lib. would be an earlier name for this species if the two are synonymous as suggested by Melnik (Key to fungi of the genus *Ascochyta* Lib. p. 129. 1977).
Host: Melilotus
Ascochyta lomatii W.B. Cooke
Distr.: WA.
Host: Lomatium
Ascochyta lonicerae-canadensis H.C. Greene
Distr.: WI.
Host: Lonicera
Ascochyta lophanthi J.J. Davis
Distr.: MA, WI.
Hosts: Agastache, Lycopus
Ascochyta lophanthi J.J. Davis var. **lycopina** J.J. Davis
Distr.: MA, WI.
Host: Lycopus
Ascochyta lophanthi J.J. Davis var. **osmophila** J.J. Davis
Distr.: WI.

Notes: Melnik (Key to fungi of the genus *Ascochyta* Lib. p. 70. 1977) considered this a synonym of *Ascochyta lagochili* Byz. which occurs in Asia and North America.
Host: Agastache
Ascochyta luzulicola R. Sprague
Distr.: AK.
Host: Luzula
Ascochyta macrospora (Speg.) Melnik
Apiosporella macrospora Speg.
Apiocarpella macrospora (Speg.) Syd. & P. Syd.
Apiocarpella bromi R. Sprague
Distr.: Western North America; Argentina.
Notes: On leaves.
Hosts: Bromus, Poa
Ascochyta madisonensis H.C. Greene
Distr.: WI.
Notes: Melnik (Key to fungi of the genus *Ascochyta* Lib. p. 90. 1977) considered this a synonym of the European *Ascochyta boraginis* Brezhn.
Host: Mertensia
Ascochyta majalis C. Massal.
Mycosphaerella convallariae W.E. McKeen & Zimmer (teleomorph)
Distr.: Eastern North America; Europe.
Notes: Lily-of-the-valley leaf blotch (McKeen & Zimmer. Canad. J. Bot. 42:665. 1964).
Host: Convallaria
Ascochyta mali Ellis & Everh.
Distr.: Midwestern United States.
Host: Malus
Ascochyta marginata J.J. Davis
Distr.: North America; Asia; Europe.
Host: Aralia
Ascochyta maydis G.L. Stout
Distr.: IL; Europe.
Host: Zea
Ascochyta menyanthicola Melnik
Ascochyta menyanthis Oudem.
Distr.: NY; Europe.
Host: Menyanthes
Ascochyta minima (P. Karst. & Har.) Arx
Gloeosporium minimum P. Karst. & Har.
Distr.: OK; Brazil(type).
Notes: Arx (Biblioth. Mycol. 24:141. 1970).
Hosts: Anthurium, Dieffenbachia
Ascochyta minor (R. Sprague) Punithalingam
Apiocarpella minor R. Sprague in H.C. Greene
Distr.: WI.
Host: Calamagrostis
Ascochyta missouriensis R. Sprague & A.G. Johnson
Distr.: ND; Great Britain.
Host: Panicum
Ascochyta necans (Ellis & Everh.) J.J. Davis
Gloeosporium necans Ellis & Everh.
Ascochyta pteridis Bres.
Distr.: North America; Europe.
Notes: Arx (Biblioth. Mycol. 24:145. 1970) confirmed the synonymy of *Ascochyta pteridis* with *A. necans*.
Hosts: Pteridium, Rumohra
Ascochyta negundinis Bres.
Distr.: IL; Italy(type).
Notes: A poorly known species.
Host: Acer
Ascochyta oryzae Cattaneo
Distr.: VA; Italy.
Notes: A doubtful species (Punithalingam. Mycol. Pap. 142:191. 1979).
Host: Oryza
Ascochyta osmundae H.C. Greene
Distr.: WI.
Host: Osmunda
Ascochyta oxybaphi Trel.
Ascochyta abroniae R. Sprague
Ascochyta boerhaviae Tharp
Distr.: Temperate United States.
Notes: Synonymy according to Melnik (Key to fungi of the genus *Ascochyta* Lib. p. 138. 1977). On Nyctaginaceae.
Hosts: Abronia, Boerhavia, Mirabilis
Ascochyta paliuri Sacc.
Ascochyta rhamni W.B. Cooke & C.G. Shaw
Distr.: ID; Europe.
Host: Rhamnus
Ascochyta paspali (Syd.) Punithalingam
Ascochytula paspali Syd.
Scolecosporiella spraguei Sutton & Alcorn
Distr.: Southeastern United States; South America; Oceania.
Notes: On leaves (Buchanan. Mycol. Pap. 156:51. 1987; Punithalingam. C.M.I. Descr. 821. 1985).
Host: Paspalum
Ascochyta paucisporula R. Sprague
Distr.: ID.
Host: Luzula
Ascochyta paulowniae Sacc. & Brunaud in Roum.
Distr.: MD.
Notes: This is not an *Ascochyta* according to Melnik (Key to fungi of the genus *Ascochyta* Lib. p. 190. 1977).
Host: Paulownia
Ascochyta pellucida Bubák
Distr.: WI; Europe.
Host: Calla
Ascochyta petuniae Speg.
Ascochyta solani-nigri Died.
Distr.: North America; Europe.
Hosts: Petunia, Solanum
Ascochyta philadelphi Sacc. & Speg.
Ascochyta ribicola H.C. Greene
Distr.: North America; Europe.
Notes: Synonymy according to Melnik (Key to fungi of the genus *Ascochyta* Lib. p. 160. 1977).
Hosts: Philadelphus, Ribes
Ascochyta phleina R. Sprague
Distr.: Central and western North America.
Host: Phleum
Ascochyta phlogis Voglino
Ascochyta phlogis Voglino var. *phlogina* Fairm.
Distr.: North America; Asia; Europe.
Host: Phlox
Ascochyta pinodes L.K. Jones
Mycosphaerella pinodes (Berk. & Bloxam) Vestergr. (teleomorph)
Distr.: Temperate and subtropical regions.
Notes: Stem, leaf, pod, and foot rot of Fabaceae (Punithalingam & Holliday. C.M.I. Descr. 340. 1972). Warmelo (Bothalia 9:183. 1966) discussed the taxonomy of the teleomorph.
Host: Pisum
Ascochyta pisi Lib.
Ascochyta pisicola (Berk.) Sacc.
Distr.: Cosmopolitan.
Notes: Leaf and pod spot of peas and other legumes (Punithalingam & Holliday. C.M.I. Descr. 334. 1972).
Hosts: Crotalaria, Lathyrus, Lupinus, Melilotus, Pisum, Trifolium, [Vicia], Vigna
Ascochyta plumbaginicola Henn.
Distr.: IA; Europe.
Host: Limonium
Ascochyta rhei (Ellis & Everh.) Ellis & Everh.
Phyllosticta rhei Ellis & Everh.
Distr.: HI, temperate North America.
Host: Rheum
Ascochyta rhododendri Lind
Distr.: WA.
Notes: Melnik (Key to fungi of the genus *Ascochyta* Lib. p. 192. 1977) indicated that this is not an *Ascochyta*.
Host: Rhododendron
Ascochyta rhodotypi H.W. Anderson
Distr.: IL.
Host: Rhodotypos
Ascochyta rhynchosiae (Thuem.) Sacc.
Distr.: SC.
Notes: Melnik (Key to fungi of the genus *Ascochyta* Lib. p. 192. 1977) indicated that this is a *Phyllosticta*.
Host: Rhynchosia
Ascochyta salicis Bonar

Distr.: CA.
Host: Salix
Ascochyta silenes Ellis & Everh.
Distr.: North America; Asia; Europe.
Host: Silene
Ascochyta skagwayensis (R. Sprague) Punithalingam
Ascochyta hordei K. Hara f. *skagwayensis* R. Sprague
Distr.: AK.
Notes: Punithalingham (Mycol. Pap. 142:142. 1979).
Host: Bromus
Ascochyta socialis Sacc.
Distr.: ND; Europe.
Host: Carex
Ascochyta sodalis Naumov
Ascochyta plantaginella Tehon
Ascochyta plantaginicola Melnik
Ascochyta plantaginis Sacc. & Speg. non Ces., hom. illeg.
Distr.: North America; Asia; Europe.
Notes: Synonymy according to Melnik (Key to fungi of the genus *Ascochyta* Lib. p. 145. 1977) who used the superfluous name *Ascochyta plantaginicola* for this species.
Host: Plantago
Ascochyta sorghi Sacc.
Distr.: Warm-temperate to tropical regions.
Notes: Rough leaf spot of *Sorghum* spp. and other grasses (Punithalingam. C.M.I. Descr. 632. 1979).
Hosts: Agrohordeum, Agropyron, Agrostis, Alopecurus, Anthoxanthum, Arrhenatherum, Avena, Beckmannia, Bromus, Buchloe, Calamagrostis, Dactylis, Deschampsia, Elymus, Festuca, Hierochloe, Hordeum, Hystrix, Melica, Muhlenbergia, Panicum, Phalaris, Phleum, Poa, Schizachne, Sclerochloa, Secale, Setaria, Sitanion, Sorghum, Sphenopholis, Stipa, Trisetum, Triticum
Ascochyta sorghina Sacc.
Distr.: Europe; Asia.
Notes: A doubtful species; most reports are probably *A. sorghi* Sacc. (Punithalingam. Mycol. Pap. 142:193. 1979).
Host: Sorghum
Ascochyta stipae Died.
Distr.: Europe.
Notes: Punithalingam (Mycol. Pap. 142:194. 1979) considered this an unrecognizable species as no type or authentic material has been located.
Hosts: Sporobolus, Stipa
Ascochyta subalpina R. Sprague & A.G. Johnson
Distr.: North America; Australia; Europe.
Notes: Punithalingam (Mycol. Pap. 142:154. 1979).
Hosts: Cynosurus, Deschampsia, Distichlis, Oryzopsis
Ascochyta syringae Bres.
Ascochyta fraxini Kab. & Bubák
Distr.: North America; Asia; Europe.
Host: Syringa
Ascochyta tecomae Sacc.
Distr.: NY.
Host: Campsis
Ascochyta tenerrima Sacc. & Roum.
Ascochyta vulgaris Kab. & Bubák
Distr.: North America; Europe.
Hosts: Lonicera, Symphoricarpos
Ascochyta teretiuscula Sacc. & Roum.
Distr.: ID; Europe.
Host: Carex
Ascochyta treleasei Berl. & Voglino
Distr.: WI.
Hosts: Silphium, Vernonia
Ascochyta tritici S. Hori & Enjoji
Distr.: Central and eastern United States.
Notes: The status of *A. tritici* is uncertain (Punithalingam. Mycol. Pap. 142:195. 1979) but it may be a synonym of *A. hordei*.
Hosts: Triticum, Zea
Ascochyta utahensis R. Sprague
Distr.: Western North America.
Hosts: Agropyron, Elymus
Ascochyta valerianae A.L. Sm. & Ramsb.
Distr.: WI; Asia; Europe.
Host: Valeriana
Ascochyta veratri Cavara
Gloeosporium veratrinum Allesch.
Distr.: WI; Asia; Europe.
Host: Zigadenus
Ascochyta veratrina Ellis & Everh.
Distr.: WA(type).
Notes: Melnik (Key to fungi of the genus *Ascochyta* Lib. p. 198. 1977) considered this a synonym of *Diplodia cylindrospora* Bubák.
Host: Veratrum
Ascochyta verbenae Siemaszko
Ascochyta cuneomaculata H.C. Greene
Distr.: WI; Europe; Asia.
Host: Verbena
Ascochyta versicolor Bubák
Distr.: ID; Europe.
Notes: On Aristolochiaceae.
Host: Asarum
Ascochyta viciae Lib.
Distr.: North America; central Asia; Europe.
Notes: See comments under *A. lethalis*.
Host: Vicia
Ascochyta violae Sacc. & Speg.
Distr.: North America; Europe.
Host: Viola
Ascochyta violicola McAlpine
Distr.: AK; Australia.
Host: Viola
Ascochyta vodakii Bubák
Ascochyta hepaticae Died.
Gloeosporium hepaticae Peck
Didymella vodakii E. Müller (teleomorph)
Distr.: WI.
Host: Anemone
Ascochyta volubilis Sacc. & Malbr.
Ascochyta polygoni (Dearn. & House) Arx
Gloeosporium polygoni Dearn. & House
Ascochyta polygonicola Kab. & Bubák
Distr.: Northeastern United States; Europe.
Host: Polygonum
Ascochyta wisconsina J.J. Davis
Distr.: North America; Europe.
Host: Sambucus
Ascochyta zeicola Ellis & Everh.
Ascochyta diedickei Staritz
Distr.: NJ; Cuba; Venezuela; Pakistan.
Notes: Punithalingam (Mycol. Pap. 142:164. 1979) listed this on many graminaceous hosts.
Hosts: [Carex], [Luzula], Zea

Ascochytella Tassi. Coelomycetes

Ascochytella pentstemonii Jenk.
Distr.: CA(type).
Host: Penstemon

Ascochytulina Petr. Coelomycetes

Ascochytulina deflectens (P. Karst.) Petr.
Diplodia deflectens P. Karst.
Distr.: AR; Europe.
Notes: Sutton (The Coelomycetes. p. 429. 1980).
Host: Lonicera

Ascoconidium Seaver. Coelomycetes

Ascoconidium castanae Seaver
Distr.: PA(type).
Notes: On dead limbs.
Host: Castanea

Ascocoryne Groves & D.E. Wilson. Discomycetes, Helotiales

Ascocoryne sarcoides (Jacq.) Groves & D.E. Wilson

Coryne sarcoides (Jacq.) Tul. & C. Tul.
Coryne dubia (Pers.) S.F. Gray (synanamorph)
Phialophora sp. (synanamorph)
Distr.: Temperate northern hemisphere.
Notes: On wood (Roll-Hansen & Roll-Hansen. Norw. J. Bot. 46:193. 1979).
Hosts: Aleurites, Pinus, Quercus

Ascodichaena Butin. Discomycetes, Rhytismatales

Ascodichaena rugosa Butin
Dichaena rugosum (L.:Fr.) Fr.
Dichaena faginea (Pers.:Fr.) Sacc., nom. illeg.
Polymorphum quercinum (Pers.) Chev. (anamorph)
Distr.: Northern North America; Europe.
Notes: A bark parasite. Hawksworth (Taxon 32:212. 1983).
Host: Fagus

Ascospora Fr. Loculoascomycetes, Dothideales

Notes: This genus is considered a synonym of *Omphalospora*, but not all names have been redisposed.
Ascospora graminis Lindau
Distr.: OR; Greenland(type).
Host: Agropyron
Ascospora ruborum Zeller
Coryneum sp. (anamorph)
Distr.: OR(type).
Host: Rubus

Ascotricha Berk. Pyrenomycetes, Xylariales

Notes: Eight species, often on organic debris (Hawksworth. Mycol. Pap. 126:1. 1971).
Ascotricha xylina L. Ames
Dicyma sp. (anamorph)
Distr.: GA; India; Pakistan; The Philippines.
Notes: Recently isolated from sunflower seeds (Roberts. Mycologia 76:963. 1984).
Host: Helianthus

Ashbya Guilliermond. Hemiascomycetes, Endomycetales

Ashbya gossypii (S. Ashby & W. Nowell) Guilliermond
Nematospora gossypii S. Ashby & W. Nowell
Distr.: Worldwide in cotton-growing regions.
Notes: Stigmatomycosis and internal boll rot of cotton. The fungus is transmitted by insects (Mukerji. C.M.I. Descr. 185. 1968).
Hosts: Citrus, Gossypium

Aspergillus P. Mich. ex Link:Fr. Hyphomycetes

Notes: Raper & Fennell (The genus *Aspergillus*. 1965); Samson & Pitt (Advances in *Penicillium* and *Aspergillus* systematics. 1985).
Aspergillus alliaceus Thom & Church
Distr.: Cosmopolitan.
Hosts: Allium, Cereus, Citrus, Echinocactus, Opuntia, Zea
Aspergillus caespitosus Raper & Thom
Distr.: Southwestern United States.
Host: Zea
Aspergillus candidus Link
Distr.: Cosmopolitan.
Notes: In soil and on a variety of plant materials.
Hosts: Arachis, Carya, Helianthus, Nicotiana, Triticum, Zea
Aspergillus carbonarius (Bainier) Thom
Distr.: Widespread.
Host: Zea
Aspergillus chevalieri (L. Mangin) Thom & Church var. ***intermedius*** Thom & Raper, nom. illeg.
Distr.: Widespread.
Host: Zea
Aspergillus clavatus Desmaz.
Distr.: Cosmopolitan.
Hosts: Malus, Pseudotsuga, Zea
Aspergillus duricaulis Raper & Fennell
Distr.: Widespread.
Host: Carya
Aspergillus echinulatus (Delacr.) Thom & Church, nom. illeg.
Eurotium echinulatum Delacr. (teleomorph)
Distr.: Widespread.
Host: Zea
Aspergillus elegans Gasp.
Distr.: Widespread.
Hosts: Cucumis, Malus, Zea
Aspergillus equitis R.A. Samson & W. Gams
Aspergillus chevalieri (L. Mangin) Thom & Church, nom. illeg.
Eurotium chevalieri L. Mangin (teleomorph)
Distr.: Cosmopolitan.
Hosts: Carya, Zea
Aspergillus ficuum (Reichardt) Henn.
Distr.: Widespread.
Hosts: Carya, Prunus
Aspergillus fischerianus R.A. Samson & W. Gams
Aspergillus fischeri Wehmer, nom. illeg.
Neosartorya fischeri (Wehmer) Malloch & Cain (teleomorph)
Distr.: Widespread.
Host: Pelargonium
Aspergillus flavipes (Bainier & Sartory) Thom & Church
Distr.: Widespread.
Hosts: Helianthus, Zea
Aspergillus flavus Link:Fr.
Distr.: Cosmopolitan.
Notes: Usually a saprophyte or spoilage organism but may become pathogenic to seedlings or animals (Onions. C.M.I. Descr. 91. 1966).
Hosts: Arachis, Avena, Bertholletia, Cannabis, Carya, Citrus, Cotinus, Eichhornia, Glycine, Gossypium, Helianthus, Hordeum, Lycopersicon, Narcissus, Nicotiana, Phoenix, Pistacia, Prunus, Pyrus, Saccharum, Trifolium, Triticum, Zea
Aspergillus foetidus (Nakazawa) Thom & Raper
Distr.: Widespread.
Host: Citrus
Aspergillus fumigatus Fresen.
Distr.: Cosmopolitan.
Notes: Raper & Fennell (The genus *Aspergillus*. p. 242. 1965) provided a list of probable synonyms. Dennis (British Ascomycetes. p. 385. 1978) cited *A. fumigatus* as the anamorph of *Satorya fumigata* Vuill., but Subramanian (Curr. Sci. 41:755. 1972). considered *A. fischeri* to be its anamorph. Onions (C.M.I. Descr. 92. 1966).
Hosts: Arachis, Carya, Cucumis, Glycine, Gossypium, Helianthus, Malus, Saccharum, Zea
Aspergillus glaucus Link:Fr.
Aspergillus herbariorum (A. Wigg.:Fr.) E. Fisch.
Aspergillus minor (L. Mangin) Thom & Raper
Aspergillus umbrosus Bainier & Sartory
Eurotium herbariorum Link:Fr. (teleomorph)
Distr.: Cosmopolitan.
Notes: Isolated from soil and various kinds of organic substrates (Domsch et al. Compendium of soil fungi. p. 293. 1980).
Hosts: Arachis, Avena, Glycine, Gossypium, Helianthus, Nicotiana, Prunus, Saccharum, Triticum, Zea
Aspergillus hollandicus R.A. Samson & W. Gams
Aspergillus amstelodami (L. Mangin) Thom & Church, nom. illeg.
Aspergillus montevidensis Talice & Mackinnon, nom. illeg.
Eurotium amstelodami L. Mangin (teleomorph)
Distr.: Widespread, most common in subtropical and tropical regions.
Notes: Isolated from soil and various kinds of organic debris.
Hosts: Helianthus, Hordeum, Nicotiana, Platanus, Zea
Aspergillus luteo-niger (M.L. Lutz) Thom & Church
Distr.: TX; Africa(type).
Notes: Possibly a synonym of *A. niger* var. *usamii* (Al-Musallam. Revision of the black *Aspergillus* species. p. 79. 1980).
Host: Lycopersicon
Aspergillus mangini Thom & Raper
Distr.: Widespread.
Host: Zea

Aspergillus nidulellus R.A. Samson & W. Gams
Aspergillus nidulans (Eidam) G. Wint.
Emericella nidulans (Eidam) Vuill. (teleomorph)
Distr.: Cosmopolitan.
Notes: Raper & Fennell (The genus *Aspergillus*. p. 495. 1965); Onions (C.M.I. Descr. 93. 1966 as *Aspergillus nidulans*).
Hosts: Gossypium, Lycopersicon, Saccharum, Zea
Aspergillus niger Tiegh.
Distr.: Cosmopolitan.
Notes: Extensive synonymy in Raper & Fennell (The genus *Aspergillus*. p. 309. 1965); Onions (C.M.I. Descr. 94. 1966).
Hosts: Allium, Alnus, Arachis, Argyroxiphium, Cannabis, Carya, Cheirodendron, Citrus, Cucumis, Dracaena, Eichhornia, Ficus, Glycine, Gossypium, Helianthus, Hordeum, Ipomoea, Juncus, Malus, Medicago, Narcissus, Nicotiana, Oenothera, Phoenix, Platanus, Prunus, Pseudotsuga, Punica, Pyrus, Saccharum, Solanum, Sorghum, Spartina, Triticum, Tulipa, Vitis, Zea
Aspergillus niger Tiegh. var. ***awamorii*** (Nakazawa) Al-Musallam
Aspergillus awamorii Nakazawa
Aspergillus luchuensis Inui
Distr.: Widespread.
Hosts: Gossypium, Helianthus
Aspergillus niger Tiegh. var. ***phoenicis*** (Corda) Al-Musallam
Aspergillus phoenicis (Corda) Thom
Aspergillus pyri English, nom. nud.
Distr.: Widespread.
Host: Pyrus
Aspergillus ochraceus K. Wilh.
Distr.: Cosmopolitan.
Hosts: Carya, Glycine, Gossypium, Helianthus, Lycopersicon, Platanus, Prunus, Triticum, Zea
Aspergillus oryzae (Ahlburg) Cohn
Eurotium oryzae Ahlburg
Aspergillus flavus Link var. *oryzae* (Ahlburg) K. Saito, nom. inval.
Distr.: Cosmopolitan.
Notes: This fungus is used extensively in fermenting soybeans and and other legumes (Hesseltine & Wang. Mycol. Mem. 11:332. 1986). Problems in identification of this "domesticated" species are enumerated by Christensen (Mycologia 73:1056. 1981) and Kozlowski & Stepien. (J. Gen. Microbiol. 128:471. 1982).
Hosts: Carya, Nicotiana
Aspergillus panamensis Raper & Thom
Distr.: Widespread.
Host: Metrosideros
Aspergillus parasiticus Speare
Distr.: Cosmopolitan.
Notes: Generally an active aflatoxin producer. Raper & Fennell (The genus *Aspergillus*. p. 369. 1965) listed four probable synonyms.
Hosts: Cannabis, Carya, Helianthus, Saccharum, Zea
Aspergillus penicillioides Speg.
Distr.: Cosmopolitan.
Hosts: Gossypium, Helianthus, Saccharum
Aspergillus reptans R.A. Samson & W. Gams
Aspergillus repens (de Bary) E. Fisch. in Engl. & Prantl
Eurotium repens de Bary (teleomorph)
Distr.: Cosmopolitan.
Hosts: Carya, Festuca, Gossypium, Nicotiana, Platanus, Saccharum, Zea
Aspergillus restrictus G. Sm.
Distr.: Cosmopolitan.
Hosts: Avena, Carya, Nicotiana, Triticum, Zea
Aspergillus rubrobrunneus R.A. Samson & W. Gams
Aspergillus ruber (König, Spieckermann, & W. Bremer) Thom & Church, nom. inval.
Eurotium rubrum König, Spieckermann, & W. Bremer (teleomorph)
Distr.: Cosmopolitan.
Hosts: Carya, Nicotiana, Platanus, Zea
Aspergillus sclerotiorum G.A. Huber
Distr.: Worldwide, most common in subtropical and tropical regions.
Host: Malus
Aspergillus stellifer R.A. Samson & W. Gams
Aspergillus variecolor (Berk. & Broome) Thom & Raper, nom. inval.
Emericella variecolor Berk. & Broome (teleomorph)
Distr.: Widespread.
Host: Zea
Aspergillus sulphureus (Fresen.) Wehmer
Distr.: Cosmopolitan.
Notes: In soil and on a variety of organic substrates.
Host: Zea
Aspergillus sydowii (Bainier & Sartory) Thom & Church
Distr.: Cosmopolitan.
Notes: Raper & Fennell (The genus *Aspergillus*. p. 450. 1965) listed several possible synonyms.
Hosts: Cucumis, Saccharum, Zea
Aspergillus tamarii Kita
Distr.: Cosmopolitan.
Hosts: Arachis, Carya, Cotinus, Helianthus, Lycopersicon, Nicotiana, Zea
Aspergillus terreus Thom in Thom & Church
Distr.: Cosmopolitan.
Notes: Onions (C.M.I. Descr. 95. 1966).
Hosts: Carya, Cucumis, Helianthus, Lycopersicon, Saccharum, Spartina, Trifolium, Zea
Aspergillus terricola Ém. Marchal
Distr.: Widespread.
Notes: Isolated from sunflower seeds.
Host: Helianthus
Aspergillus terricola Ém. Marchal var. ***indicus*** (Mehrotra & Agnihotri) Raper & Fennell
Distr.: Widespread.
Host: Argyroxiphium
Aspergillus unguis (Emile-Weil & L. Gaudin) Thom & Raper
Distr.: Widespread.
Hosts: Cucumis, Zea
Aspergillus ustus (Bainier) Thom & Church
Distr.: Cosmopolitan, most common in subtropical and tropical regions.
Hosts: Carya, Gossypium, Metrosideros, Zea
Aspergillus versicolor (Vuill.) Tiraboschi
Aspergillus lateralis (Harkn.) C.A. Peek & Solheim
Theclospora lateralis Harkn.
Distr.: Cosmopolitan.
Hosts: Carya, Cucumis, Gossypium, Helianthus, Nerium, Nicotiana, Zea
Aspergillus wentii Wehmer
Distr.: Widespread.
Notes: In soil and on a variety of organic substrates.
Hosts: Carya, Helianthus, Malus, Prunus, Zea

Asperisporium Maubl. Hyphomycetes

Asperisporium asclepiadis (Ellis & Everh.) Deighton in M.B. Ellis
Scolicotrichum asclepiadis Ellis & Everh.
Passalora asclepiadis (Ellis & Everh.) Höhn.
Distr.: CA(type), MN.
Notes: Leaf spot.
Host: Asclepias
Asperisporium caricae (Speg.) Maubl.
Distr.: Subtropical and tropical western hemisphere.
Notes: Ellis & Holliday (C.M.I. Descr. 347. 1972).
Host: Carica
Asperisporium minutulum (Sacc.) Deighton
Fusicladium minutulum Sacc.
Distr.: Pacific Northwest.
Notes: Leaf spot of grapes.
Host: Vitis
Asperisporium punctulatum (Tracy & Earle) Maubl.
Scolicotrichum punctulatum Tracy & Earle
Distr.: MS.
Hosts: Ipomoea, Iris

Asterella (Sacc.) Sacc. Loculoascomycetes, Dothideales

Notes: This genus is considered a synonym of *Asterina* (Arx & Müller. Stud. Mycol. 9:42. 1975), but not all names have been redisposed.
Asterella bignoniae (Ellis & Everh.) Sacc.
Asterina bignoniae Ellis & Everh.
Distr.: LA.

Notes: Theissen (Abh. K.K. Zool.-Bot. Ges. Wien 7:21. 1913).
Host: Bignonia
Asterella myricae L.E. Miles
Distr.: MS(type).
Host: Myrica

Asteridiella McAlpine. Pyrenomycetes, Meliolales

Asteridiella acervata (Ellis & Everh.) Hansf.
Meliola acervata Ellis & Everh.
Irene acervata (Ellis & Everh.) Hansf.
Distr.: HI.
Host: Physalis
Asteridiella americana Hansf.
Distr.: FL, GA.
Notes: On living leaves (Hansford. Beih. Sydowia 2:529. 1961).
Host: Osmanthus
Asteridiella cheirodendronis (F. Stevens) Hansf.
Meliola cheirodendronis F. Stevens
Irenina cheirodendronis (F. Stevens) F. Stevens
Distr.: HI.
Hosts: Cheirodendron, Panax, Vaccinium
Asteridiella coprosmae Hansf.
Distr.: HI(type).
Notes: On living leaves (Hansford. Beih. Sydowia 2:579. 1961).
Host: Coprosma
Asteridiella cyrtandrae (F. Stevens)Hansf.
Irenina cyrtandrae F. Stevens
Distr.: HI.
Host: Cyrtandra
Asteridiella exilis (Syd.) Hansf.
Meliola exilis Syd.
Irene exilis (Syd.) F. Stevens
Irenina exilis (Syd.) F. Stevens
Distr.: HI.
Host: Vaccinium
Asteridiella inermis (Kalchbr. & Cooke) Hansf.
Meliola inermis Kalchbr. & Cooke
Irene inermis (Kalchbr. & Cooke) Theiss. & Syd.
Meliola quinqueseptata Rehm
Meliola quinquespora Thuem.
Distr.: HI; Brazil; South Africa.
Notes: Hansford (Beih. Sydowia 2:523. 1961).
Host: Physalis
Asteridiella manca (Ellis & G. Martin) Hansf.
Meliola manca Ellis & G. Martin
Irenina manca (Ellis & G. Martin) F. Stevens
Meliola manca Ellis & G. Martin var. *tenuis* G. Wint.
Distr.: Southeastern United States.
Hosts: Myrica, Quercus
Asteridiella negundinis J.A. Stevenson
Distr.: AL, FL, GA.
Host: Acer
Asteridiella ohiana (F. Stevens) Hansf.
Amazonia ohianus F. Stevens
Distr.: HI.
Host: Metrosideros
Asteridiella perrottetiae (F. Stevens) Hansf.
Amazonia perrottetiae F. Stevens
Irene perrottetiae (F. Stevens) Hansf.
Distr.: HI.
Host: Perrottetia
Asteridiella perseae (F. Stevens) Hansf.
Irene perseae (F. Stevens) Toro
Distr.: FL; West Indies; South America.
Host: Persea
Asteridiella pipturi (Hansf.) Hansf.
Irenina pipturi Hansf.
Distr.: HI; The Philippines.
Host: Pipturus
Asteridiella pitya (Sacc.) Hansf.
Irenina pitya (Sacc.) F. Stevens
Asteridiella taxi (Sawada) F. Stevens
Irenina taxi Sawada
Distr.: Temperate North America; eastern Asia.
Notes: On living needles.
Host: Taxus
Asteridiella sheariana Hansf. ex Hansf.
Distr.: HI(type).
Host: Cryptocarya
Asteridiella triloba (G. Wint.) Hansf.
Meliola triloba G. Wint.
Irene triloba (G. Wint.) Theiss. & Syd.
Irenina triloba (G. Wint.) F. Stevens
Distr.: HI; Africa.
Host: Pipturus
Asteridiella waimeana Hansf.
Distr.: HI(type).
Notes: On living leaves (Hansford. Beih. Sydowia 2:342. 1961).
Host: Perrottetia

Asteridium (Sacc.) Speg. ex Sacc. Pyrenomycetes, Meliolales

Notes: This genus is considered a synonym of *Meliola*, but not all names have been redisposed.
Asteridium illicii Tracy & Earle
Distr.: MS.
Host: Illicium
Asteridium lepidigenoides (Ellis & Everh.) Sacc.
Asterina lepidigenoides Ellis & Everh.
Distr.: FL.
Host: Capparis
Asteridium lepidigenum (Ellis & G. Martin) Sacc.
Asterina lepidigena Ellis & G. Martin
Distr.: FL.
Host: Lyonia
Asteridium xerophylli (Ellis) Sacc.
Asterina xerophylli Ellis
Distr.: NJ.
Host: Xerophyllum

Asterina Lév. Loculoascomycetes, Dothideales

Asterina aspidii (Henn.) Theiss.
Asterella aspidii Henn.
Opeasterina aspidii (Henn.) Speg.
Distr.: HI; Japan.
Notes: Stevens & Ryan (Illinois Biol. Monogr. 17:48. 1939).
Host: Diospyros
Asterina asterophora Ellis & G. Martin, nom. nud.
Distr.: FL(type), GA.
Host: Osmanthus
Asterina caricarum Rehm
Asterostomella caricae Henn. (anamorph)
Distr.: FL; West Indies; South America.
Host: Carica
Asterina clavuligera Cooke
Distr.: Eastern United States.
Hosts: Andromeda, Vaccinium
Asterina clermontiae F. Stevens & Ryan
Distr.: HI.
Host: Clermontia
Asterina congesta Cooke
Distr.: HI; India.
Notes: Stevens & Ryan (Illinois Biol. Monogr. 17:66. 1939).
Host: Santalum
Asterina conglobata Berk. & M.A. Curtis
Distr.: ME.
Host: Arctostaphylos
Asterina cupressina (Rehm) Cooke
Venturia cupressina Rehm
Asterinella cupressina (Rehm) Theiss.
Distr.: Eastern North America.
Notes: Barr (Canad. J. Bot. 46:799. 1968) stated this was an *Arnaudiella* sp.

Hosts: Chamaecyparis, Juniperus
Asterina delitescens Ellis & G. Martin
Asterina tenella Cooke
Distr.: HI, southeastern United States.
Hosts: Persea, Vaccinium
Asterina diplodioides Berk. & M.A. Curtis
Distr.: AL.
Host: Leucothoe
Asterina gouldiae F. Stevens & Ryan
Distr.: HI.
Host: Gouldia
Asterina graminicola Ellis & Everh.
Asterella graminicola (Ellis & Everh.) Sacc.
Distr.: MI.
Notes: Theissen (Abh. K.K. Zool.-Bot. Ges. Wien 7:111. 1913).
Host: Oryzopsis
Asterina ildefonsiae (Rehm) Theiss.
Seynesia balansae Speg. var. *ildefonsiae* Rehm
Asterella gardoquiae Syd.
Distr.: HI; South America.
Notes: On living leaves of Cucurbitaceae, Euphorbiaceae, Lamiaceae and Scrophulariaceae (Stevens & Ryan. Illinois Biol. Monogr. 17:50. 1939).
Hosts: Cladium, Claoxylon
Asterina lepiniana Mont.
Asterina pelliculosa Berk.
Distr.: Southern United States; Tahiti.
Host: Ilex
Asterina lobeliae F. Stevens & Ryan
Parasterina lobeliae (F. Stevens & Ryan) F. Stevens
Distr.: HI.
Host: Lobelia
Asterina orbicularis Berk. & M.A. Curtis
Englerulaster orbicularis (Berk. & M.A. Curtis) Höhn.
Asterina asperulispora Gaillard
Englerulaster asperulispora (Gaillard) Theiss.
Asterella opulenta Henn.
Asterina opulenta (Henn.) Sacc.
Distr.: North and South America.
Notes: Müller & Arx (Beitr. Kryptogamenfl. Schweiz 11(2):107. 1962).
Host: Ilex
Asterina paupercula Ellis & Everh.
Asterella paupercula (Ellis & Everh.) Sacc.
Distr.: FL(type).
Host: Jacquinia
Asterina phoradendricola J.A. Stevenson & Pollack
Distr.: FL.
Notes: Common on living leaves of American mistletoe.
Host: Phoradendron
Asterina phyllostegiae F. Stevens & Ryan
Distr.: HI.
Host: Phyllostegia
Asterina punctiformis Lév. var. ***fimbriata*** (Kalchbr. & Cooke) Theiss.
Asterina fimbriata Kalchbr. & Cooke
Distr.: HI; South Africa.
Host: Lobelia
Asterina radians Ellis
Distr.: FL.
Notes: A doubtful species (Theissen. Abh. K.K. Zool.-Bot. Ges. Wien 7:115. 1913).
Host: Capparis
Asterina rickii Theiss.
Distr.: HI; Brazil.
Host: Metrosideros
Asterina sabalicola Earle
Distr.: FL.
Notes: Theissen (Abh. K.K. Zool.-Bot. Ges. Wien 7:115. 1913).
Host: Sabal
Asterina subcyanea Ellis & G. Martin
Dictyothyrium subcyaneum (Ellis & G. Martin) Theiss.
Distr.: FL(type).
Host: Quercus
Asterina suttoniae F. Stevens & Ryan
Distr.: HI.
Hosts: Myrsine, Rapanea
Asterina tantalina Petr.
Distr.: HI.
Host: Ilex

Asterinella Theiss. Loculoascomycetes, Dothideales

Asterinella humiriae (Henn.) Theiss.
Seynesia humiriae Henn.
Distr.: HI; Brazil.
Host: Ilex
Asterinella intensa (Cooke & Massee) Theiss.
Asterina intensa Cooke & Massee
Distr.: HI; New Zealand(type).
Host: Osmanthus
Asterinella mabae F. Stevens & Ryan
Distr.: HI(type).
Host: Diospyros
Asterinella puiggarii (Speg.) Theiss.
Asterina puiggarii Speg.
Distr.: Subtropical and tropical regions.
Hosts: Eugenia, Syzygium

Asterodon Pat. Basidiomycetes, Aphyllophorales

Asterodon ferruginosum Pat.
Distr.: Temperate northern hemisphere.
Notes: On wood.
Hosts: Betula, Picea, Pinus, Thuja, Tsuga

Asteroma DC.:Fr. Coelomycetes

Notes: Sutton (The Coelomycetes. p. 494. 1980) indicated that about 150 species have been described in this genus; few of them are congeneric with the type species of the genus, *A. padi* DC.
Asteroma alneum (Pers.:Fr.) Sutton
Cylindrosporella alnea (Pers.:Fr.) Höhn.
Gloeosporium alneum Westend.
Ophiodothis alneum (Pers.:Fr.) Ellis & Everh.
Gloeosporium alnicola Dearn. & House
Gloeosporium cylindrospermum (Bonord.) Sacc.
Gnomoniella tubiformis (Tode:Fr.) Sacc. (teleomorph)
Distr.: United States; Europe; Great Britain.
Notes: Leaf lesions.
Host: Alnus
Asteroma capreae Desmaz.
Distr.: CT; Europe.
Notes: The North American report is based on a single collection determined by G. P. Clinton.
Host: Salix
Asteroma caryae (Peck) Sutton
Phyllosticta caryae Peck
Cylindrosporella caryae (Ellis & Dearn.) Petr.
Gloeosporium caryae Ellis & Everh. non Ellis & Dearn., hom. illeg.
Gloeosporium caryae Ellis & Dearn.
Gnomonia caryae F.A. Wolf (teleomorph)
Distr.: Eastern North America.
Notes: Barr (Mycol. Mem. 7:45. 1978).
Host: Carya
Asteroma dianthii Cooke & Harkn.
Distr.: CA(type).
Host: Dianthus
Asteroma fraserae Ellis & Everh.
Distr.: CO, ID.
Host: Frasera
Asteroma garrettianum Syd. in Syd. & P. Syd.
Distr.: CO, UT.
Host: Primula
Asteroma graminis Westend.
Distr.: KS, MO; western Europe(type).
Notes: A poorly known species. The Kansas and Missouri reports are uncertain.

Host: Andropogon
Asteroma inconspicuum (Cavara) Sutton
Gloeosporium inconspicuum Cavara
Distr.: Eastern North America; Europe.
Notes: Lesions caused by this species are similar to those of *Gloeosporium ulmicola*.
Host: Ulmus
Asteroma lactucae J.J. Davis
Distr.: MD, SD, TN, TX, WI.
Notes: A poorly known species.
Host: Lactuca
Asteroma liriodendri Cooke
Distr.: DC, MD, SC.
Notes: A poorly known species.
Host: Liriodendron
Asteroma microspermum (Peck) Sutton
Septoria microsperma Peck
Gloeosporium betulae-luteae Sacc. & Dearn.
Distr.: Eastern North America; Great Britain.
Host: Betula
Asteroma parkinsoniae Ellis & Everh.
Distr.: AZ(type).
Host: Parkinsonia
Asteroma ribicola Ellis & Everh.
Distr.: MT, NY, WI.
Host: Ribes
Asteroma saxifragae Ellis & Everh.
Distr.: AK(type).
Host: Saxifraga
Asteroma solidaginis Cooke
Distr.: IA; Great Britain.
Host: Solidago
Asteroma tenerrimum Grognot
Distr.: Temperate North America.
Notes: A poorly known species.
Hosts: Erythronium, Fritillaria
Asteroma tennerrimum Grognot var. **erythronii** Sacc.
Distr.: ID, MT, WA.
Host: Erythronium
Asteroma tiliae Rudolphi
Distr.: Eastern North America; Europe.
Host: Tilia
Asteroma ulmeum (L.E. Miles) Sutton
Gloeosporium ulmeum L.E. Miles
Cylindrosporella ulmea (L.E. Miles) Arx
Stegophora ulmea (Schwein.:Fr.) Syd. & P. Syd. (teleomorph)
Distr.: North America.
Host: Ulmus
Asteroma venulosum (Wallr.) Fuckel
Distr.: TX, WI; Europe.
Host: Iris

Asteromassaria Höhn. Loculoascomycetes, Dothideales

Notes: Barr (Mycotaxon 15:349. 1982).
Asteromassaria distincta (Schwein.) Barr
Massaria distincta (Schwein.) Cooke
Distr.: Eastern North America.
Host: Sambucus
Asteromassaria macrospora (Desmaz.) Höhn.
Massaria macrospora (Desmaz.) Sacc.
Scolicosporium macrosporium (Berk.) Sutton (anamorph)
Distr.: Eastern North America; Europe.
Notes: On branches.
Host: Fagus
Asteromassaria minor (Peck) Barr
Caryospora minor Peck
Distr.: Northeastern North America.
Notes: On pericarps.
Host: Carya
Asteromassaria olivaceohirta (Schwein.) Barr
Massaria olivaceohirta (Schwein.) Cooke
Massaria epileuca Berk. & M.A. Curtis
Trematosphaeria epileuca (Berk. & M.A. Curtis) Shoemaker & P.M. LeClair
Massaria olivacea Cooke
Distr.: Temperate northern hemisphere.
Notes: On branches.
Host: Morus
Asteromassaria pulchra (Harkn.) Shoemaker & P.M. LeClair
Massaria pulchra Harkn.
Distr.: Western North America.
Host: Umbellularia
Asteromassaria seriata (Cooke) Shoemaker & P.M. LeClair
Massariella seriata (Cooke) Sacc.
Distr.: Eastern North America.
Notes: On branches.
Hosts: Carya, Cydonia

Asteromella Pass. & Thuem. Coelomycetes

Asteromella andrewsii Petr. in J.J. Davis
Phyllosticta gentianicola (DC.) Ellis & Everh.
Mycosphaerella andrewsii (Sacc.) J.J. Davis (teleomorph)
Distr.: Eastern North America; Europe.
Host: Gentiana
Asteromella asteris Peck
Distr.: KS, NE.
Notes: J.J. Davis (Trans. Wisconsin Acad. Sci. 21:281. 1924) suggested this may be a form of *A. astericola*.
Host: Aster
Asteromella brassicae (Chev.) Boerema & Van Kesteren
Phyllosticta brassicicola McAlpine
Phyllosticta napi Sacc.
Mycosphaerella brassicicola (Duby) Lindau in Engl. & Prantl (teleomorph)
Distr.: Temperate Europe and North America, especially in cool moist climates.
Notes: Ring spot (Punithalingam. C.M.I. Descr. 468. 1975; Boerema & Van Kesteren. Persoonia 3:17. 1964; Corlett. Mycotaxon 31:63. 1988).
Host: Brassica
Asteromella castanicola (Ellis & Everh.) Petr.
Phyllosticta castanicola Ellis & Everh.
Distr.: North America, most common in the West.
Host: Castanopsis
Asteromella lantanae Petr.
Distr.: HI(type).
Host: Lantana
Asteromella scaevolae Petr.
Distr.: HI(type).
Host: Scaevola

Asteromyces Moreau & Mme. Moreau ex Hennebert. Hyphomycetes

Asteromyces cruciatus Moreau & Mme. Moreau ex Hennebert
Distr.: Temperate marine habitats.
Host: Zostera

Asterosporium Kunze. Coelomycetes

Asterosporium asterospermum (Pers.:Fr.) S.J. Hughes
Asterosporium hoffmannii Kunze
Botryosphaeria hoffmannii (Kunze) Höhn., nom. illeg.
Distr.: Eastern North America; Eurasia.
Notes: On bark and twigs.
Host: Fagus
Asterosporium betulinum Peck
Distr.: Temperate North America.
Notes: On limbs.
Host: Betula

Asterostomella Speg. Coelomycetes

Asterostomella saccardoi (Thuem.) Petr.
Phyllosticta saccardoi Thuem.
Distr.: Temperate North America.
Notes: Pirozynski (Fungi Canadenses 21. 1974 as *Botryosphaeria rhodorae*).
Host: Rhododendron

Asterostroma Massee. Basidiomycetes, Aphyllophorales

Asterostroma cervicolor (Berk. & M.A. Curtis) Massee
Distr.: North America; West Indies; Japan.
Hosts: Abies, Pinus, Planera, Quercus
Asterostroma laxum Bres. in Bourd. & Galzin
Distr.: Temperate northern hemisphere.
Hosts: Abies, Quercus
Asterostroma medium Bres.
Distr.: Temperate northern hemisphere.
Notes: On wood.
Host: Pinus
Asterostroma muscicola (Berk. & M.A. Curtis) Massee
Distr.: Northern hemisphere.
Notes: On wood.
Hosts: Fouquieria, Juniperus, Pseudotsuga
Asterostroma ochroleucum Bres. in Torrend
Distr.: Temperate northern hemisphere.
Notes: On wood.
Host: Abies

Atelocauda Arth. & Cummins. Basidiomycetes, Uredinales

Atelocauda digitata (G. Wint.) Cummins & Y. Hiratsuka
Uromyces digitatus G. Wint.
Distr.: HI; Australia; Java; New Zealand.
Notes: Autoecious.
Host: Acacia
Atelocauda koae (Arth.) Cummins & Y. Hiratsuka
Uromyces koae Arth.
Distr.: HI.
Notes: Autoecious.
Host: Acacia

Athelia Pers. Basidiomycetes, Aphyllophorales

Notes: Jülich (Willdenowia Beih. 7:42. 1972).
Athelia arachnoidea (Berk.) Jülich
Corticium arachnoideum Berk.
Athelia bispora (J. Schröt.) Donk
Corticium bisporum (J. Schröt.) Höhn. & Litsch.
Corticium consimile Bres.
Distr.: Northern hemisphere.
Notes: On wood.
Hosts: Cibotium, Dasylirion, Elephantopus, Heteromeles, Larix, Pinus, Prosopis, Rosa, Salix
Athelia bombacina Pers.
Distr.: Temperate northern hemisphere.
Notes: On wood and plant debris.
Hosts: Malus, Populus, Quercus
Athelia coprophila (Wakef.) Jülich
Corticium coprophilum Wakef.
Distr.: Temperate northern hemisphere.
Notes: On wood, litter, and dung.
Hosts: Acer, Populus, Prosopis, Quercus
Athelia decipiens (Höhn. & Litsch.) J. Eriksson
Corticium decipiens Höhn. & Litsch.
Distr.: Temperate northern hemisphere.
Notes: On wood, usually of conifers.
Hosts: Abies, Alnus, Carnegiea, Fouquieria, Picea, Pinus
Athelia epiphylla Pers.:Fr.
Corticium epiphyllum (Pers.:Fr.) Massee
Distr.: Northern hemisphere.
Notes: On wood, bark, dead leaves, and debris of conifers and hardwoods.
Hosts: Alnus, Cupressus, Eucalyptus, Pinus, Populus, Quercus
Athelia fibulata M. Christiansen
Distr.: Northern hemisphere.
Notes: On wood.
Host: Populus
Athelia laxa (Burt) Jülich
Peniophora laxa Burt
Distr.: North America.
Notes: On bark and wood of conifers.
Host: Tsuga
Athelia neuhoffii (Bres.) Donk
Distr.: Northern hemisphere.
Notes: On wood.
Hosts: Picea, Populus
Athelia rolfsii (Curzi) Tu & Kimbrough
Corticium rolfsii Curzi
Pellicularia rolfsii E. West
Sclerotium rolfsii Sacc. (anamorph)
Distr.: Cosmopolitan.
Notes: On roots of a wide variety of plants (Mordue. C.M.I. Descr. 410. 1974).
Hosts: Arrhenatherum, Avena, Beta, Citrullus, Cynodon, Daphne, Daucus, Ficus, Hordeum, Iris, Lilium, Narcissus, Nicotiana, Oryza, Pennisetum, Saccharum, Scilla, Sorghum, Triticum, Tulipa, Vigna, Zea, Zinnia
Athelia salicum Pers.
Athelia incrustata M. Christiansen
Distr.: North America, most common in the west, northern hemisphere.
Notes: On wood and bark; occasionally on basidiomata of *Hymenochaete tabacina*.
Hosts: Acer, Betula, Quercus
Athelia scutellaris (Berk. & M.A. Curtis) R.L. Gilbertson
Corticium scutellare Berk. & M.A. Curtis
Distr.: North America.
Notes: On wood.
Hosts: Arundinaria, Liriodendron, Morus, Pinus, Rhododendron
Athelia teutoburgensis (Brinkmann) Jülich
Corticium teutoburgense Brinkmann
Hyphoderma teutoburgense (Brinkmann) J. Eriksson
Corticium flavescens Bres.
Athelia macrospora (Bourd. & Galzin) M. Christiansen
Distr.: Northern hemisphere.
Notes: On wood.
Hosts: Pinus, Quercus

Atichia Flot. Hyphomycetes

Seuratia Pat.
Notes: Anamorph of *Seuratia* Pat. Meeker (Canad. J. Bot. 53:2483. 1975).
Atichia glomerulosa (Achar. ex H. Mann) Stein in Cohn
Collema glomerulosum Achar.
Synalissa glomerulosa Nyl.
Atichia mosigii Flot.
Seuratia millardetii (Racib.) Meeker (teleomorph)
Distr.: Tropical regions.
Hosts: Abies, Pinus, Pseudotsuga
Atichia solaridiscoidea Meeker
Distr.: HI.
Notes: On Myrtaceae.
Hosts: Metrosideros, Psidium

Atkinsonella Diehl. Pyrenomycetes, Clavicipitales

Notes: Diehl (Agric. Monogr. U.S.D.A. 4:48. 1950) discussed this species.
Atkinsonella hypoxylon (Peck) Diehl
Epichloe hypoxylon Peck
Balansia hypoxylon (Peck) Atk.
Hypocrella hypoxylon (Peck) Sacc.

Ephelis borealis Ellis & Everh. (anamorph)
Distr.: Eastern North America.
Notes: Leuchtmann & Clay (Mycologia 80:192. 1988).
Hosts: Danthonia, Stipa

Atopospora Petr. Loculoascomycetes, Dothideales

Atopospora betulina (Fr.:Fr.) Petr.
Dothidea betulina Fr.:Fr.
Euryachora betulina (Fr.:Fr.) J. Schröt. in Cohn
Dothidella betulina (Fr.:Fr.) Sacc.
Phyllachora betulina (Fr.:Fr.) Fuckel
Rehmiodothis betulina (Fr.:Fr.) Arx
Distr.: North America; northern Europe.
Notes: On leaves (Corlett & Carter. Fungi Canadenses 88. 1976).
Hosts: Alnus, Betula

Atractilina Dearn. & Barth. Hyphomycetes

Atractilina parasitica (G. Wint.) Deighton & Pirozynski
Atractilina callicarpae Dearn. & Barth.
Isariopsis penicillata Ellis & Everh.
Distr.: Pantropical.
Notes: Reported growing on leaves but is parasitizing species of black mildews, i.e. *Meliola*, *Irenoposis*, *Amazonia*, etc.
Host: Callicarpa

Atractogloea Oberwinkler & Bandoni. Basidiomycetes, Auriculariales

Notes: Oberwinkler & Bandoni (Mycologia 74:634. 1982).
Atractogloea stillata Oberwinkler & Bandoni
Distr.: CA.
Notes: On spaths.
Host: Phoenix

Atropellis Zeller & Goodd. Discomycetes, Helotiales

Atropellis apiculata Lohman, Cash & R.W. Davidson
Distr.: NC, VA.
Host: Pinus
Atropellis pinicola Zeller & Goodd.
Godronia zelleri Seaver
Distr.: Pacific coastal North America.
Hosts: Pinus, Ribes
Atropellis piniphila (Weir) Lohman & Cash
Cenangium piniphilum Weir
Atropellis arizonica Lohman & Cash
Distr.: Western North America, occasionally reported in the Southeast.
Host: Pinus
Atropellis tingens Lohman & Cash
Distr.: Eastern North America.
Host: Pinus

Auerswaldiella Theiss. & Syd. Loculoascomycetes, Dothideales

Auerswaldiella puccinioides (Speg.) Theiss. & Syd.
Auerswaldia puccinioides Speg.
Distr.: FL, GA; South America.
Host: Prunus

Aulacostroma Syd. & P. Syd. Loculoascomycetes, Dothideales

Aulacostroma osmanthi F. Stevens & Ryan
Distr.: HI(type).
Host: Osmanthus

Aulographella Höhn. Loculoascomycetes, Dothideales

Notes: This genus is considered a synonym of *Morenoina* (Arx & Müller. Stud. Mycol. 9:32. 1975), but not all names have been redisposed.
Aulographella baumeae F. Stevens
Distr.: HI(type).
Host: Cladium

Aulographina Arx & E. Müller. Loculoascomycetes, Dothideales

Notes: Two species. Arx & Müller (Sydowia 14:330. 1960).
Aulographina eucalypti (Cooke & Massee) Arx & E. Müller
Aulographum eucalypti Cooke & Massee
Lembosiopsis eucalyptina Petr. & Syd.
Thyrinula eucalypti Petr. & Syd. (anamorph)
Distr.: HI; southern Africa; Australia; Great Britain.
Notes: Wall & Keane (Trans. Brit. Mycol. Soc. 82:257. 1984).
Host: Eucalyptus

Aulographum Lib. Loculoascomycetes, Dothideales

Notes: The generic status of many species described in this genus is uncertain.
Aulographum arundinariae Cooke
Distr.: GA(type).
Host: Arundinaria
Aulographum hederae Lib.
Aulographum vagum Desmaz.
Distr.: NJ; Europe.
Notes: On leaves of various shrubs (Müller & Arx. Beitr. Kryptogamenfl. Schweiz 11(2):127. 1962).
Host: Ilex

Aureobasidium Viala & Boyer. Hyphomycetes

Notes: Hermanides-Nijhof (Stud. Mycol. 15:141. 1977).
Aureobasidium apocryptum (Ellis & Everh.) Hermanides-Nijhof
Gloeosporium apocryptum Ellis & Everh.
Kabatiella apocrypta (Ellis & Everh.) Arx
Microstroma apocryptum (Ellis & Everh.) W.B. Cooke
Distr.: Temperate northern hemisphere.
Host: Acer
Aureobasidium caulivorum (Kirchn.) W.B. Cooke
Gloeosporium caulivorum Kirchn.
Kabatiella caulivora (Kirchn.) Karakulin
Distr.: Worldwide in red clover-growing regions.
Notes: Clover scorch or clover anthracnose.
Hosts: Kummerowia, Melilotus, Trifolium
Aureobasidium lini (Lafferty) Hermanides-Nijhof
Polyspora lini Lafferty
Aureobasidium pullulans (de Bary) G. Arnaud var. *lini* (Lafferty) W.B. Cooke
Kabatiella lini (Lafferty) Karakulin in Vassiljevsky & Karakulin
Distr.: Worldwide in flax-growing regions.
Notes: Parasitic.
Host: Linum
Aureobasidium microstictum (Bubák) W.B. Cooke
Kabatiella microsticta Bubák
Distr.: OR; Europe.
Notes: Leaf spot.
Hosts: Convallaria, Iris, Lilium
Aureobasidium nigricans (Atk. & Edgerton) W.B. Cooke
Protocoronospora nigricans Atk. & Edgerton
Kabatiella nigricans (Atk. & Edgerton) Karakulin
Distr.: North America.
Notes: Leaf spot.
Host: Vicia
Aureobasidium prunicola (Ellis & Everh.) Hermanides-Nijhof
Gloeosporium prunicola Ellis & Everh.
Kabatiella prunicola (Ellis & Everh.) Arx
Microstroma prunicola (Ellis & Everh.) W.B. Cooke

Microstroma tonellianum Ferraris
Distr.: Northern United States.
Notes: Leaf spot.
Host: Prunus
Aureobasidium pullulans (de Bary) G. Arnaud
Dematium pullulans de Bary
Pullularia pullulans (de Bary) Berkhout
Aureobasidium vitis Viala & Boyer
Exobasidium vitis (Viala & Boyer) Prill. & Delacr.
Distr.: Cosmopolitan.
Notes: Saprophytic on a wide variety of substrates.
Hosts: Abies, Acacia, Acer, Alnus, Arceuthobium, Carya, Cheirodendron, Cucumis, Fagus, Gossypium, Hibiscus, Linum, Lycopersicon, Malus, Medicago, Metrosideros, Phaseolus, Pinus, Pisum, Platanus, Prunus, Pseudotsuga, Pyrus, Sequoia, Shorea, Thuja, Trifolium, Triticum, Vicia, Zea
Aureobasidium sanguinariae (Ellis & Everh.) Hermanides-Nijhof
Gloeosporium sanguinariae Ellis & Everh.
Kabatiella sanguinariae (Ellis & Everh.) Arx
Microstroma sanguinariae (Ellis & Everh.) W.B. Cooke
Distr.: Eastern United States.
Notes: Leaf spot.
Host: Sanguinaria
Aureobasidium umbellulariae (J.M. Harvey) Hermanides-Nijhof
Kabatiella phoradendri (L. Darling) J.M. Harvey f. *umbellulariae* J.M. Harvey
Microstroma umbellulariae (J.M. Harvey) W.B. Cooke
Distr.: CA.
Notes: Leaf spot.
Host: Umbellularia
Aureobasidium zeae (Narita & Hiratsuka) J.M. Dingley
Kabatiella zeae Narita & Hiratsuka
Distr.: Temperate regions.
Notes: Isolated from *Zea* (Hermanides-Nijhof. Stud. Mycol. 15:165. 1977).
Host: Zea

Auricularia Bull. Basidiomycetes, Auriculariales

Auricularia auricula (L.:Fr.) Underw.
Auricularia auricula-judae (Bull.:Fr.) Wettst.
Auricularia auricularis (S.F. Gray) G.W. Martin
Distr.: Widespread.
Notes: On wood, usually of conifers.
Hosts: Abies, Acer, Betula, Carya, Celtis, Gleditsia, Juglans, Liquidambar, Picea, Pseudotsuga, Psidium, Quercus, Thuja, Ulmus
Auricularia cornea Ehrenb.:Fr.
Exidia cornea (Ehrenb.:Fr.) Fr.
Auricularia polytricha (Mont.) Sacc.
Auricularia tenuis (Lév.) Farl.
Distr.: United States(warm areas) to the neotropics; Australasia.
Notes: Formerly widely known as *A. polytricha* (Wong & Wells. Mycologia 79:847. 1987). On wood; used for food, particularly in the Orient.
Hosts: Acacia, Aleurites, Dracaena, Hibiscus, Liquidambar, Metrosideros
Auricularia fuscosuccinea (Mont.) Farl.
Distr.: MD, TN; neotropical region.
Notes: On wood.
Host: Acer

Auriporia Ryvarden. Basidiomycetes, Aphyllophorales

Auriporia aurea (Peck) Ryvarden
Poria aurea Peck
Distr.: Infrequent in northeastern and western United States.
Notes: On wood.
Hosts: Acer, Pinus, Tsuga

Auriscalpium S.F. Gray. Basidiomycetes, Aphyllophorales

Notes: About 5 species.
Auriscalpium vulgare S.F. Gray
Auriscalpium auriscalpium (L.:Fr.) Banker
Hydnum auriscalpium L.:Fr.
Distr.: Northern hemisphere.
Notes: On cones of various conifers. Although the oldest name for this species is *Hydnum auriscalpium* L., the combination in the genus *Auriscalpium* cannot be used because it contradicts Art. 23.4 of the ICBN.
Hosts: Pinus, Pseudotsuga

Bacillispora S. Nilsson. Hyphomycetes

Bacillispora aquatica S. Nilsson
Distr.: IL; northern Europe.
Notes: On submerged, decaying leaves (Nilsson. Bot. Not. 115:77. 1962; Nilsson. Symb. Bot. Upsal. 18:83. 1964).
Hosts: Acer, Quercus

Bactridium Kunze. Hyphomycetes

Notes: Most species are saprophytic (Hughes. New Zealand J. Bot. 4:522. 1966).
Bactridium flavum Kunze
Distr.: HI.
Notes: On dead wood and bark.
Host: Aleurites

Bactrodesmium Cooke. Hyphomycetes

Bactrodesmium cedricola M.B. Ellis
Distr.: Northwestern North America; Pakistan.
Notes: Hughes & White (Fungi Canadenses 255. 1983).
Host: Juniperus
Bactrodesmium pithoideum (Dearn. & House) Sutton
Coryneum pithoideum Dearn. & House
Distr.: NY.
Notes: Hughes & White (Fungi Canadenses 252. 1983).
Host: Celastrus

Badhamia Berk. Myxomycetes

Badhamia panicea (Fr.) Rostr.
Distr.: Cosmopolitan.
Host: Solenostemon

Baeodromus Arth. Basidiomycetes, Uredinales

Baeodromus californicus Arth.
Distr.: CA.
Notes: Autoecious.
Host: Senecio

Baeospora Singer. Basidiomycetes, Agaricales

Baeospora myosura (Fr.:Fr.) Singer
Collybia myosura (Fr.:Fr.) Quél.
Distr.: Temperate northern hemisphere.
Notes: On cones of conifers.
Host: Pinus

Bagnisiopsis Theiss. & Syd. Pyrenomycetes, Phyllachorales

Notes: Arx & Müller (Beitr. Kryptogamenfl. Schweiz 11(1):254. 1954) considered *Bagnisiopsis* to be a synonym of *Coccostroma*.
Bagnisiopsis eucalypti Dearn. & Barth.
Distr.: CA.
Host: Eucalyptus

Bahusakala Subramanian. Hyphomycetes
Bahusakala cookei M.B. Ellis
Distr.: CA.
Notes: Ellis (More dematiaceous Hyphomycetes. p. 18. 1976).
Host: Salix

Balansia Speg. Pyrenomycetes, Clavicipitales
Notes: Diehl (Agric. Monogr. U.S.D.A. 4:1. 1950).
Balansia aristidae (Atk.) Diehl
Dothichloe aristidae Atk.
Distr.: Southeastern United States; West Indies; northern South America.
Notes: Often confused with *Dothidiella aristidae* (Schwein.) Ell.& Everh.
Host: Aristida
Balansia claviceps Speg.
Ephelis mexicana Fr. in Berk. (anamorph)
Distr.: Widespread in tropical regions.
Host: Cenchrus
Balansia cyperacearum (Berk. & M.A. Curtis) Diehl
Distr.: Southeastern United States; northern South America.
Host: Cyperus
Balansia cyperi Edgerton
Distr.: Southeastern United States; South America.
Notes: Leuchtmann & Clay (Mycologia 80:192. 1988).
Host: Cyperus
Balansia epichloe (Weese) Diehl
Dothichloe epichloe Weese
Dothichloe limitata Diehl
Distr.: Eastern North America; West Indies; South America.
Hosts: Agrostis, Andropogon, Calamagrostis, Chloris, Ctenium, Eragrostis, Gymnopogon, Oryzopsis, Panicum, Paspalum, Schizachyrium, Sporobolus, Tridens
Balansia hemicrypta Diehl
Distr.: TX to Central and South America.
Host: Aristida
Balansia henningsiana (A. Möller) Diehl
Distr.: KS, NY; Mexico; West Indies; South America.
Notes: Often reported under the name *Dothichloe atramentosa*.
Hosts: Andropogon, Chloris, Gymnopogon, Panicum, Schizachyrium
Balansia obtecta Diehl
Distr.: FL; Mexico; West Indies.
Host: Cenchrus
Balansia oryzae-sativae Hashioka
Balansia oryzae (Syd.) Narasimhan & Thirumalachar, nom. inval.
Ephelis oryzae Syd. (anamorph)
Distr.: LA; southwestern China; East Indies; India.
Notes: Udbatta disease of rice (Booth. C.M.I. Descr. 640. 1979). Ou (Rice diseases. p. 315. 1985).
Host: Oryza
Balansia strangulans (Mont.) Diehl
Epichloe strangulans (Mont.) Sacc.
Dothichloe nigricans (Speg.) Seaver in J.A. Stevenson
Distr.: Central and eastern United States; West Indies; Central and South America.
Hosts: Axonopus, Bouteloua, Dichanthelium, Panicum, Phleum
Balansia strangulans (Mont.) Diehl f. ***discoidea*** (Henn.) Diehl
Balansia discoidea Henn.
Dothichloe granulosa Chárdon
Distr.: Central United States; Central and South America.
Hosts: Andropogon, Bouteloua

Balansiopsis Höhn. Pyrenomycetes, Clavicipitales
Balansiopsis pilulaeformis (Berk. & M.A. Curtis) Diehl
Balansia pilulaeformis (Berk. & M.A. Curtis) Diehl in J.H. Miller
Balansia clavula (Berk. & M.A. Curtis) C.G. Lloyd
Distr.: Southeastern United States; South America.
Notes: Black crust.
Hosts: Paspalum, Uniola

Basidiobotrys Höhn. Hyphomycetes
Basidiobotrys pallida (Berk. & M.A. Curtis) S.J. Hughes
Botrytis pallida Berk. & M.A. Curtis
Streptothrix glauca Ellis & Everh.
Distr.: NJ, OH, SC.
Hosts: Trifolium, Zea

Basidiodendron Rick. Basidiomycetes, Tremellales
Notes: Luck-Allen (Canad. J. Bot. 41:1025. 1963); Wells & Raitviir (Mycologia 67:909. 1975).
Basidiodendron caesiocinereum (Höhn. & Litsch.) Luck-Allen
Bourdotia caesiocinerea (Höhn. & Litsch.) Bourd. & Galzin ex Pilát
Sebacina caesiocinerea (Höhn. & Litsch.) D.P. Rogers
Distr.: Temperate northern hemisphere.
Notes: On wood.
Hosts: Fraxinus, Pinus, Platanus, Quercus
Basidiodendron cinereum (Bres.) Luck-Allen
Sebacina cinerea Bres.
Bourdotia cinerea (Bres.) Bourd. & Galzin
Distr.: Widespread.
Notes: On wood.
Hosts: Fagus, Pinus, Platanus, Salix, Tilia
Basidiodendron eyrei (Wakef.) Luck-Allen
Sebacina eyrei Wakef.
Bourdotia eyrei (Wakef.) Bourd. & Galzin
Bourdotia deminuta (Bourd.) Bourd. & Galzin
Sebacina deminuta Bourd.
Basidiodendron deminutum (Bourd.) Luck-Allen
Corticium involucrum Burt
Distr.: Widespread.
Notes: On wood.
Hosts: Juglans, Picea, Pinus, Populus, Prunus, Quercus, Thuja
Basidiodendron grandinioides (Bourd. & Galzin) Luck-Allen
Bourdotia grandinioides Bourd. & Galzin
Sebacina grandinioides (Bourd. & Galzin) D.P. Rogers
Distr.: Temperate northern hemisphere.
Notes: On hardwoods.
Hosts: Magnolia, Populus

Basidiophora Roze & Cornu. Oomycetes, Peronosporales
Basidiophora entospora Roze & Cornu
Plasmopara entospora (Roze & Cornu) J. Schröt.
Peronospora simplex Peck
Distr.: Eastern North America; Ecuador; Europe; Iraq; New Zealand.
Notes: Downy mildew of asters (Francis. C.M.I. Descr. 681. 1983).
Hosts: Aster, Callistephus, Chrysopsis, Conyza, Erigeron, Solidago
Basidiophora kellermanii (Swingle ex Sacc.) G.W. Wils.
Distr.: Central United States.
Host: Iva

Basidioradulum Nobles. Basidiomycetes, Aphyllophorales
Notes: Nobles (Mycologia 59:192. 1967).
Basidioradulum radula (Fr.:Fr.) Nobles
Hyphoderma radula (Fr.:Fr.) Donk
Corticium colliculosum Berk. & M.A. Curtis
Corticium hydnans (Schwein.) Burt
Distr.: Northern hemisphere.
Notes: On wood.
Hosts: Alnus, Castanea, Eucalyptus, Picea, Prunus, Pseudotsuga, Taxus, Tsuga

Beelia F. Stevens & Ryan. Loculoascomycetes, Dothideales
Notes: Monotypic. Petrak (Sydowia 7:351. 1953).
Beelia suttoniae F. Stevens & Ryan
Distr.: HI(type).
Host: Myrsine

Belaina Batista & Peres. Coelomycetes

Belaina asclepiadis Batista & Peres.
Crucellisporium selaginellae M.L. Farr in M.L. Farr & H.T. Horner
Host: Selaginella

Belonidium Mont. & Durieu. Discomycetes, Helotiales

Notes: A heterogeneous assemblage of discomycetes, many of which have been transferred to other genera, e.g., *Dasyscyphus*, *Hymenoscyphus*, *Lachnellula*, etc.
Belonidium elegantulum (P. Karst.) Raitviir
Lachnella elegantulum (P. Karst.) P. Karst.
Dasyscyphus elegantulus (P. Karst.) Rehm
Helotium elegantulum (P. Karst.) P. Karst.
Trichopeziza elegantula (P. Karst.) Sacc.
Distr.: WA; Europe.
Notes: Dennis (Mycol. Pap. 62:186. 1956).
Host: Delphinium
Belonidium eupatorii (Schwein.) Raitviir
Dasyscyphus eupatorii (Schwein.) Massee
Lachnella eupatorii (Schwein.) Seaver
Trichopeziza eupatorii (Schwein.) Sacc.
Distr.: PA(type).
Host: Eupatorium
Belonidium sulphureum (Pers.:Fr.) Raitviir
Dasyscyphus sulphureus (Pers.:Fr.) Massee
Lachnella sulphurea (Pers.:Fr.) Quél.
Lachnum sulphureum (Pers.:Fr.) Rehm
Distr.: Temperate North America; Europe.
Hosts: Rubus, Sambucus

Belonium Sacc. Discomycetes, Helotiales

Belonium aggregatum Cash
Distr.: CO(type).
Notes: Cash (Mycologia 28:303. 1936).
Host: Populus
Belonium arabicola Ellis & Everh.
Distr.: WA(type).
Host: Arabis
Belonium consanguineum Ellis & Everh.
Distr.: AL(type).
Notes: According to Seaver (North Amer. cup fungi. [Inoperculates]. p. 176. 1951), this is close to *Belonium bicolor*.
Host: Ilex
Belonium inconspicuum Cash
Distr.: CO(type).
Notes: Cash (Mycologia 28:302. 1936).
Host: Abies

Belonopsis (Sacc.) Rehm. Discomycetes, Helotiales

Belonopsis junciseda (P. Karst.) Le Gal & G. Mangenot
Belonidium juncisedum (P. Karst.) Rehm
Mollisia junciseda (P. Karst.) P. Karst.
Distr.: North America(questionable); Europe.
Hosts: Juncus, Scirpus

Beltrania Penz. Hyphomycetes

Notes: Ellis (Dematiaceous Hyphomycetes. p. 237. 1971).
Beltrania querna Harkn.
Distr.: CA; Italy.
Notes: On dead leaves.
Hosts: Eucalyptus, Juncus, Quercus, Sassafras
Beltrania rhombica Penz.
Beltrania indica Subramanian
Beltrania multispora H.J. Swart
Distr.: Pantropical.
Notes: Saprobic on plant debris and in soil.
Hosts: Eucalyptus, Liquidambar, Podocarpus, Quercus, Thuja

Beltraniella Subramanian. Hyphomycetes

Beltraniella portoricensis (F. Stevens) Pirozynski & S.D. Patil
Ellisiella portoricensis F. Stevens
Ellisiopsis portoricensis (F. Stevens) Pirozynski
Pseudomassaria carolinensis Barr & C.S. Hodges (teleomorph)
Distr.: Worldwide in warm or tropical regions.
Notes: Common in leaf litter (Pirozynski & Patil. Canad. J. Bot. 48:575. 1975).
Host: Eucalyptus

Beniowskia Racib. Hyphomycetes

Beniowskia sphaeroidea (Kalchbr. & Cooke) E. Mason
Sporotrichum peribebuyense Speg.
Clathrotrichum subcarneum Pat.
Botrytis uredinicola Peck
Distr.: Warm-temperate to tropical regions.
Notes: Parasitic on grasses (Hanlin. Mycotaxon 28:229. 1987).
Hosts: Panicum, Setaria

Berkleasmium J. Zobel. Hyphomycetes

Notes: Most species occur on dead wood (Moore. Mycologia 51:734. 1959).
Berkleasmium concinnum (Berk.) R.T. Moore
Sporidesmium concinnum Berk.
Berkleasmium cordeanum J. Zobel
Distr.: Temperate regions.
Notes: On dead wood.
Hosts: Celtis, Juglans, Liriodendron, Pinus, Quercus, Tsuga
Berkleasmium conglobatum (Cooke & Ellis) R.T. Moore
Sporidesmium conglobatum Cooke & Ellis
Distr.: NJ(type).
Notes: On dead wood.
Host: Quercus
Berkleasmium corticola (P. Karst.) R.T. Moore
Sporidesmium moriforme Peck subsp. *corticola* P. Karst.
Distr.: NC; Finland(type).
Notes: On dead wood.
Hosts: Abies, Vitis
Berkleasmium moriforme (Peck) R.T. Moore
Sporidesmium moriforme Peck
Distr.: NC, NY; Italy.
Notes: On dead wood.
Hosts: Acer, Malus
Berkleasmium opacum R.T. Moore
Sporidesmium opacum Sacc. non Corda, hom. illeg.
Distr.: NY.
Notes: On dead wood.
Host: Juglans

Bertia De Not. Pyrenomycetes, Sordariales

Bertia moriformis (Tode:Fr.) De Not.
Distr.: Temperate northern hemisphere.
Hosts: Abies, Acer, Alnus, Betula, Calocedrus, Fagus, Picea, Populus, Quercus, Rhododendron, Salix, Thuja, Tsuga, Vaccinium
Bertia moriformis (Tode:Fr.) De Not. var. ***latispora*** Corlett & J. Krug
Distr.: North America; Switzerland (Corlett & Krug. Canad. J. Bot. 62:2564. 1984).
Hosts: Abies, Pinus, Tsuga

Bertiella (Sacc.) Sacc. & P. Syd. Loculoascomycetes, Dothideales

Bertiella rhodospila (Berk. & M.A. Curtis) Barr in Barr et al.
Herpotrichia rhodospila (Berk. & M.A. Curtis) Sacc.
Melogramma effusum Peck
Metasphaeria rubida Ellis & Everh.
Distr.: Eastern North America.
Notes: On wood (Barr et al. New York State Mus. Bull. 459:19. 1986).
Hosts: Liriodendron, Populus, Quercus

Bifusella Höhn. Discomycetes, Rhytismatales

Bifusella linearis (Peck) Höhn.
Lophodermium lineare (Peck) Ellis & Everh.
Hypodermium effusum Schwein. (anamorph)
Distr.: Northern North America.
Notes: *Bifusella* blight, tar spot needle cast, needle blight of 5-needle pines. Minter & Millar (C.M.I. Descr. 782. 1984).
Host: Pinus
Bifusella pini (Dearn.) Darker
Hypoderma pini (Dearn.) Darker
Distr.: CA, NV.
Host: Pinus
Bifusella saccata (Darker) Darker
Hypoderma saccatum Darker
Distr.: CO, NM.
Host: Pinus
Bifusella vaccinii Tehon
Distr.: NH, NJ.
Host: Vaccinium

Bipolaris Shoemaker. Hyphomycetes

Notes: Sivanesan (Mycol. Pap. 158. 1987).
Bipolaris arizonica R. Sprague
Distr.: AZ(type).
Host: Muhlenbergia
Bipolaris australiensis (M.B. Ellis) Tsuda & Ueyama
Drechslera australiensis (Bugnicourt) Subramanian & P.C. Jain ex M.B. Ellis
Helminthosporium australiensis Bugnicourt, nom. inval.
Cochliobolus australiensis (Tsuda & Ueyama) Alcorn (teleomorph)
Distr.: HI; Asia; Africa; Australia.
Host: Acacia
Bipolaris buchloes (Lefebvre & A.G. Johnson) Shoemaker
Helminthosporium buchloes Lefebvre & A.G. Johnson
Drechslera buchloes (Lefebvre & A.G. Johnson) Subramanian & P.C. Jain
Helminthosporium inconspicuum Cooke & Ellis var. *buchloes* Ellis & Everh., nom. nud.
Distr.: Central and western North America.
Hosts: Bouteloua, Buchloe
Bipolaris cactivora (Petr.) Alcorn
Helminthosporium cactivorum Petr.
Drechslera cactivora (Petr.) M.B. Ellis
Distr.: CA, FL; Europe.
Hosts: Astrophytum, Cephalocereus, Cereus, Echinopsis, Espostoa, Ferocactus, Lemaireocereus, Lithops, Lophocereus, Mammillaria, Rhipsalidopsis, Schlumbergera, Selenicereus, Stapelia, Trichocereus
Bipolaris cookei (Sacc.) Shoemaker
Helminthosporium cookei Sacc.
Distr.: SC, FL.
Host: Sorghum
Bipolaris cynodontis (Marignoni) Shoemaker
Helminthosporium cynodontis Marignoni
Drechslera cynodontis (Marignoni) Subramanian & P.C. Jain
Cochliobolus cynodontis R.R. Nelson (teleomorph)
Distr.: Cosmopolitan.
Hosts: Cynodon, Eleusine, Muhlenbergia
Bipolaris hadrotrichoides (Ellis & Everh.) Luttrell
Septosporium hadrotrichoides Ellis & Everh. in Sumstine, nom. nud.
Distr.: Central and eastern United States.
Notes: This listing follows Sivanesan (Mycol. Pap. 158:82. 1987). Shoemaker (Canad. J. Bot. 37:886. 1959) excluded this species from the *Bipolaris-Drechslera* complex. Although the specific epithet was originally published as "*hadotrichoides*", the name should have been "*hadrotrichoides*".
Hosts: Eleusine, Eragrostis
Bipolaris hawaiiensis (M.B. Ellis) Uchida & Aragaki
Drechslera hawaiiensis M.B. Ellis
Helminthosporium hawaiiense Bugnicourt, nom. inval.
Distr.: Subtropical and tropical regions.
Notes: On a variety of plant materials. Sivanesan & Holliday (C.M.I. Descr. 728. 1982).
Hosts: Chloris, Juncus, Oryza, Saccharum
Bipolaris incurvata (C. Bernard) Alcorn
Helminthosporium incurvatum C. Bernard
Drechslera incurvata (C. Bernard) M.B. Ellis
Distr.: FL; Jamaica; Asia; Australia; Oceania; the Seychelles.
Notes: Leaf spot of young coconut (Ellis & Holliday. C.M.I. Descr. 342. 1972).
Host: Cocos
Bipolaris iridis (Oudem.) C.H. Dickinson
Clasterosporium iridis Oudem.
Drechslera iridis (Oudem.) M.B. Ellis
Mystrosporium adustum Massee
Distr.: North America; Europe.
Notes: Ink disease (Ellis & Waller. C.M.I. Descr. 434. 1974).
Host: Iris
Bipolaris kusanoi (Nisikado) Shoemaker
Helminthosporium kusanoi Nisikado
Drechslera kusanoi (Nisikado) Subramanian & P.C. Jain
Cochliobolus kusanoi (Nisikado) Drechs. ex Dastur (teleomorph)
Distr.: KS; Japan.
Notes: Sivanesan (The bitunicate Ascomycetes. p. 376. 1984).
Host: Eragrostis
Bipolaris leersiae (Atk.) Shoemaker
Helminthosporium leersiae Atk.
Drechslera leersiae (Atk.) Subramanian & P.C. Jain
Distr.: Central and eastern United States.
Hosts: Bromus, Leersia
Bipolaris maydis (Nisikado & Miyake) Shoemaker
Helminthosporium maydis Nisikado & Miyake
Drechslera maydis (Nisikado & Miyake) Subramanian & P.C. Jain
Cochliobolus heterostrophus (Drechs.) Drechs. (teleomorph)
Distr.: Warm temperate, subtropical, and tropical regions.
Notes: Southern leaf blight of maize (Ellis & Holliday. C.M.I. Descr. 301. 1971).
Hosts: Antirrhinum, Dianthus, Pelargonium, Sorghum, Zea
Bipolaris micropus (Drechs.) Shoemaker
Helminthosporium micropus Drechs.
Distr.: FL.
Host: Paspalum
Bipolaris nodulosa (Berk. & M.A. Curtis) Shoemaker
Helminthosporium nodulosum Berk. & M.A. Curtis in Sacc.
Drechslera nodulosa (Berk. & M.A. Curtis) Subramanian & P.C. Jain
Bipolaris leucostyla (Drechs.) Shoemaker
Helminthosporium leucostylum Drechs.
Cochliobolus nodulosus Luttrell (teleomorph)
Distr.: Southeastern United States; subtropical and tropical regions.
Notes: Leaf blight, seedling blight, and foot rot of finger millet and other grasses (Ellis & Holliday. C.M.I. Descr. 341. 1972 as *Cochliobolus nodulosus*).
Hosts: Eleusine, Eragrostis
Bipolaris oryzae (Breda de Haan) Shoemaker
Helminthosporium oryzae Breda de Haan
Drechslera oryzae (Breda de Haan) Subramanian & P.C. Jain
Cochliobolus miyabeanus (Ito & Kuribayashi) Drechs. ex Dastur (teleomorph)
Distr.: Cosmopolitan.
Notes: Brown spot and seedling blight of rice (Ellis & Holliday. C.M.I. Descr. 302. 1971).
Hosts: Oryza, Zizania
Bipolaris palousensis R. Sprague
Distr.: Northwestern North America.
Notes: Sprague (Res. Stud. Wash. State Univ. 29:77. 1961).
Host: Juncus
Bipolaris ravenelii (M.A. Curtis) Shoemaker
Helminthosporium ravenelii M.A. Curtis
Drechslera ravenelii (M.A. Curtis) Subramanian & P.C. Jain
Cochliobolus ravenelii Alcorn (teleomorph)
Distr.: Southern United States; warm temperate to tropical regions.
Notes: False smut or head mold of grasses, especially *Sporobolus* spp. (Sivanesan. C.M.I. Descr. 884. 1987).
Hosts: Axonopus, Bromus, Eragrostis, Sporobolus
Bipolaris sacchari (E.J. Butler) Shoemaker
Helminthosporium sacchari E.J. Butler in E.J. Butler & Hafiz Khan
Drechslera sacchari (E.J. Butler) Subramanian & P.C. Jain
Distr.: Subtropical and tropical regions.

Notes: Eye spot and seedling blight of sugarcane (Ellis & Holliday. C.M.I. Descr. 305. 1971).
Hosts: Cymbopogon, Panicum, Pennisetum, Saccharum, Zea
Bipolaris setariae **(Sawada) Shoemaker**
Helminthosporium setariae Sawada
Drechslera setariae (Sawada) Subramanian & P.C. Jain
Cochliobolus setariae (Ito & Kuribayashi in Ito) Drechs. ex Dastur (teleomorph)
Distr.: North America; South America; Africa; Asia.
Notes: Spot blotch and secondary root rot of *Setaria* (Ellis & Gibson. C.M.I. Descr. 473. 1975).
Hosts: Antirrhinum, Brachiaria, Calathea, Caryota, Chamaedorea, Chrysalidocarpus, Chrysanthemum, Dendrobium, Dianthus, Maranta, Panicum, Pelargonium, Phoenix, Rosa, Setaria
Bipolaris sorghicola **(Lefebvre & Sherwin) Alcorn**
Helminthosporium sorghicola Lefebvre & Sherwin
Drechslera sorghicola (Lefebvre & Sherwin) M.J. Richardson & E.M. Fraser
Distr.: Worldwide in sorghum-growing regions.
Notes: Target spot of *Sorghum* (Ellis & Holliday. C.M.I. Descr. 491. 1976).
Hosts: Sorghum, Zea
Bipolaris sorokiniana **(Sacc.) Shoemaker**
Helminthosporium sorokinianum Sacc. in Sorokin
Drechslera sorokiniana (Sacc.) Subramanian & P.C. Jain
Helminthosporium californicum Mackie & G. Paxton
Helminthosporium sativum Pammel, C.M. King & Bakke
Cochliobolus sativus (Ito & Kuribayashi) Drechs. ex Dastur (teleomorph)
Distr.: Temperate regions.
Notes: Foot rot or spot blotch of grasses (Sivanesan & Holliday. C.M.I. Descr. 701. 1981).
Hosts: Agrohordeum, Agropyron, Agrostis, Ammophila, Andropogon, Arthraxon, Avena, Bouteloua, Bromus, Buchloe, Calamagrostis, Calamovilfa, Cenchrus, Chloris, Cynodon, Dactylis, Dendrobium, Dichanthelium, Digitaria, Echinochloa, Elymus, Eragrostis, Eremopyrum, Festuca, Hordeum, Hystrix, Koeleria, Linum, Lolium, Medicago, Muhlenbergia, Oryzopsis, Panicum, Phalaris, Phleum, Poa, Secale, Setaria, Sorghum, Stipa, Trifolium, Triticum, Vulpia, Zea, Zizania
Bipolaris spicifera **(Bainier) Subramanian**
Brachycladium spiciferum Bainier
Curvularia spicifera (Bainier) Boedijn
Drechslera spicifera (Bainier) Arx
Helminthosporium spiciferum (Bainier) J. Nicot
Bipolaris tetramera (McKinney) Shoemaker
Helminthosporium tetramera McKinney
Cochliobolus spicifer R.R. Nelson (teleomorph)
Distr.: Cosmopolitan.
Notes: Sivanesan & Holliday (C.M.I. Descr. 702. 1981).
Hosts: Agropyron, Bromus, Cynodon, Festuca, Platanus, Poa, Prunus, Trifolium, Triplasis, Triticum, Vicia
Bipolaris stenospila **(Drechs.) Shoemaker**
Helminthosporium stenospilum Drechs.
Drechslera stenospila (Drechs.) Subramanian & P.C. Jain
Cochliobolus stenospilus Matsumoto & W. Yamamoto, nom. inval. (teleomorph)
Distr.: Worldwide in sugarcane-growing regions.
Notes: Brown stripe of sugarcane (Ellis & Holliday. C.M.I. Descr. 306. 1971).
Hosts: Cynodon, Pennisetum, Saccharum
Bipolaris urochloae **(Putterill) Shoemaker**
Helminthosporium urochloae Putterill
Drechslera urochloae (Putterill) Subramanian & P.C. Jain
Distr.: HI; Australia; South Africa; Zimbabwe.
Hosts: Dendrobium, Melinis
Bipolaris victoriae **(F. Meehan & Murphy) Shoemaker**
Helminthosporium victoriae F. Meehan & Murphy
Drechslera victoriae (F. Meehan & Murphy) Subramanian & P.C. Jain
Cochliobolus victoriae R.R. Nelson (teleomorph)
Distr.: Cosmopolitan.
Notes: Victoria blight of oats (Sivanesan & Holliday. C.M.I. Descr. 703. 1981).
Hosts: Agropyron, Avena, Chloris, Dactylis, Glycine, Hordeum, Oryza, Paspalum, Phalaris, Phleum, Setaria, Sorghum, Zea
Bipolaris yamadai **(Shoemaker) Nisikado**
Helminthosporium yamadai Nisikado
Distr.: WI; Japan(type).
Host: Panicum
Bipolaris zeicola **(G.L. Stout) Shoemaker**
Helminthosporium zeicola G.L. Stout
Drechslera zeicola (G.L. Stout) Subramanian & P.C. Jain
Drechslera carbonum (Ullstrup) Sivanesan
Helminthosporium carbonum Ullstrup
Cochliobolus carbonum R.R. Nelson (teleomorph)
Distr.: Widespread in maize-growing regions of the world.
Notes: Leaf spot of maize (Ellis & Holliday. C.M.I. Descr. 349. 1972).
Hosts: Sorghum, Tripsacum, Zea

Biscogniauxia Kuntze. Pyrenomycetes, Xylariales

Notes: *Biscogniauxia* was published to replace *Nummularia*, a later homonym. Whalley & Edwards (Trans. Brit. Mycol. Soc. 85:385. 1985) provided a comprehensive account of this genus.
Biscogniauxia marginata **(Fr.) Pouzar**
Nummularia discincola (Schwein.) Cooke
Nummularia discreta (Schwein.) Tul. & C. Tul.
Distr.: Eastern North America.
Notes: Apple blister canker and on a variety of hardwoods. The cultural characters and anamorph were described by Callan & Rogers (Canad. J. Bot. 64:842. 1986).
Hosts: Amelanchier, Carya, Gleditsia, Magnolia, Malus, Pyrus, Sorbus, Ulmus
Biscogniauxia nummularia **(Bull.:Fr.) Kuntze**
Hypoxylon nummularium Bull.:Fr.
Nummularia bulliardii Tul. & C. Tul.
Distr.: Cosmopolitan.
Notes: On hardwoods. According to Miller (A monograph of the world species of *Hypoxylon*. p. 121. 1961), the type variety only occurs in Europe.
Hosts: Acer, Alnus, Arbutus, Holodiscus, Quercus, Ulmus
Biscogniauxia nummularia **(Bull.:Fr.) Kuntze var. *pseudopachyloma* (Speg.) J.H. Miller**
Hypoxylon nummularium Bull.:Fr. var. *pseudopachyloma* (Speg.) J.H. Miller
Distr.: Southern United States; China; southern hemisphere.
Notes: On hardwoods.
Host: Crataegus
Biscogniauxia nummularia **(Bull.:Fr.) Kuntze var. *rumpens* (Cooke) J.H. Miller**
Hypoxylon nummularium Bull.:Fr. var. *rumpens* (Cooke) J.H. Miller
Hypoxylon rumpens (Cooke) J.H. Miller
Periconiella sp. (anamorph)
Distr.: Southeastern United States.
Notes: On hardwoods.
Hosts: Berchemia, Crataegus, Fraxinus
Biscogniauxia repanda **(Fr.:Fr.) Kuntze**
Nummularia repanda (Fr.:Fr.) Nitschke
Nummularia pezizoides Ellis & Everh.
Distr.: Central and eastern North America; Europe.
Notes: On hardwoods.
Hosts: Sorbus, Ulmus

Bispora Corda. Hyphomycetes

Bispora betulina **(Corda) S.J. Hughes**
Torula aequalis Berk. & M.A. Curtis in Berk.
Torula binale Cooke & Ellis
Torula dissita Berk. & M.A. Curtis in Berk.
Distr.: Temperate northern hemisphere.
Notes: On dead wood. Ellis (Dematiaceous Hyphomycetes. p. 91. 1971).
Hosts: Fagus, Salix
Bispora effusa **Peck**
Distr.: NY; Canada(Manitoba).
Host: Acer
Bispora occulta **(Ellis & Barth.) J.L. Crane & Schoknecht**
Torula occulta Ellis & Barth.

Distr.: KS(type).
Notes: Crane & Schoknecht (Mycologia 69:533. 1977).
Host: Ulmus

Bisporella Sacc. Discomycetes, Helotiales

Bisporella citrina (Batsch:Fr.) Korf & S. Carpenter
Calycella citrina (Batsch:Fr.) Boud.
Helotium citrinum (Hedw.) Fr.
Distr.: Temperate northern hemisphere.
Hosts: Acer, Alnus, Betula, Crataegus, Pinus, Populus, Prunus, Pseudotsuga, Quercus, Salix, Sorbus
Bisporella drosodes (Rehm) S. Carpenter
Allophylaria drosodes (Rehm) Nannf.
Distr.: ID; Europe.
Notes: Carpenter (Mem. New York Bot. Gard. 33:209. 1981).
Host: Senecio
Bisporella sulfurina (Quél.) S. Carpenter
Helotium sulfurinum Quél.
Distr.: HI; Europe.
Host: Aleurites

Bjerkandera P. Karst. Basidiomycetes, Aphyllophorales

Bjerkandera adusta (Willd.:Fr.) P. Karst.
Polyporus adustus (Willd.:Fr.) Fr.
Polyporus crispus (Pers.:Fr.) Fr.
Polyporus halesiae Berk. & M.A. Curtis
Distr.: Widespread.
Notes: On wood, usually of hardwoods.
Hosts: Abies, Acer, Alnus, Betula, Carpinus, Carya, Castanea, Catalpa, Celtis, Cercis, Citrus, Cornus, Fagus, Fraxinus, Gleditsia, Halesia, Ilex, Juglans, Juniperus, Larix, Liquidambar, Liriodendron, Malus, Nyssa, Ostrya, Picea, Pinus, Platanus, Populus, Prosopis, Prunus, Pseudotsuga, Quercus, Rhus, Salix, Sambucus, Thuja, Tilia, Tsuga, Ulmus
Bjerkandera fumosa (Pers.:Fr.) P. Karst.
Polyporus fumosus Pers.:Fr.
Polyporus fragrans Peck
Distr.: Eastern North America and the Pacific Coast; Australasia; Europe.
Notes: On wood of hardwoods.
Hosts: Acer, Betula, Liriodendron, Populus, Quercus, Salix, Tilia, Ulmus

Blakeslea Thaxt. Zygomycetes, Mucorales

Blakeslea trispora Thaxt.
Distr.: Warm temperate to tropical regions.
Notes: Saprobic in soil and on plant debris.
Hosts: Brassica, Dionaea, Gossypium, Juncus, Sorghum, Zea

Blastocladia Reinsch. Chytridiomycetes, Blastocladiales

Blastocladia ramosa Thaxt.
Distr.: Temperate regions.
Notes: On twigs and fruits in aquatic environments.
Hosts: Betula, Populus

Blumeriella Arx. Discomycetes, Helotiales

Blumeriella jaapii (Rehm) Arx
Pseudopeziza jaapii Rehm
Coccomyces hiemalis Higgins
Higginsia hiemalis (Higgins) Nannf.
Coccomyces lutescens Higgins
Higginsia lutescens (Higgins) Nannf.
Coccomyces prunophorae Higgins
Phloeosporella padi (Lib.) Arx (synanamorph)
Microgloeum pruni Petr. in Syd. & Petr. (synanamorph)
Distr.: North America; temperate Europe.
Notes: Leaf spot and shot-hole (Arx. Phytopathol. Z. 42:161. 1961).
Host: Prunus
Blumeriella kerriae (V.B. Stewart) Korf
Coccomyces kerriae V.B. Stewart
Higginsia kerriae (V.B. Stewart) Nannf.
Cylindrosporium kerriae V.B. Stewart, nom. nud. (anamorph)
Distr.: Eastern United States.
Notes: Leaf and twig blight.
Host: Kerria

Bolbitius Fr. Basidiomycetes, Agaricales

Bolbitius reticulatus (Pers.:Fr.) Ricken
Pluteolus reticulatus (Pers.:Fr.) Gill.
Distr.: Temperate North America; Europe.
Host: Quercus

Bombardia Fr. Pyrenomycetes, Sordariales

Bombardia fasciculata Fr.
Distr.: Temperate regions.
Notes: On dead wood.
Host: Tilia
Bombardia solaris (Cooke & Ellis) Höhn.
Sphaeria solaris Cooke & Ellis
Trichosphaeria solaris (Cooke & Ellis) Ellis & Everh.
Distr.: Western United States.
Hosts: Abies, Picea

Bondarzewia Singer. Basidiomycetes, Aphyllophorales

Bondarzewia berkeleyi (Fr.) Bondartsev & Singer
Polyporus berkeleyi Fr.
Distr.: North America, most common in the East; Africa; Australasia; Europe.
Notes: On buried or exposed wood around trees and stumps of hardwoods.
Hosts: Abies, Castanea, Juniperus, Larix, Picea, Pinus, Prunus, Pseudotsuga, Quercus
Bondarzewia montana (Quél.) Singer
Polyporus montanus (Quél.) Ferry
Distr.: Western North America; Europe.
Notes: On buried or exposed wood around trees and stumps of conifers.
Hosts: Abies, Pinus, Salix, Tsuga

Boreostereum Parmasto. Basidiomycetes, Aphyllophorales

Boreostereum radiatum (Peck) Parmasto
Stereum radiatum Peck
Distr.: Circumboreal.
Notes: On wood of conifers.
Hosts: [Malus], Picea, Pinus, Tsuga

Bostrichonema Ces. Hyphomycetes

Bostrichonema polygonia (Unger) J. Schröt.
Bostrichonema alpestre Ces.
Diplocarpon polygoni E. Müller (teleomorph)
Distr.: Europe; distribution elsewhere is questionable.
Host: Polygonum

Bothrodiscus Shear. Coelomycetes

Bothrodiscus berenice (Berk. & M.A. Curtis) Groves
Fusisporium berenice Berk. & M.A. Curtis
Bothrodiscus pinicola Shear
Ascocalyx abietis Naumov (teleomorph)
Distr.: Northeastern North America.
Notes: On branches.
Host: Abies

Botryobasidium Donk. Basidiomycetes, Aphyllophorales

Botryobasidium ansosum (H. Jacks. & D.P. Rogers) Parmasto
Pellicularia ansosa H. Jacks. & D.P. Rogers in D.P. Rogers
Distr.: Southwestern United States.
Hosts: Abies, Picea, Pinus
Botryobasidium botryosum (Bres.) J. Eriksson
Distr.: Temperate northern hemisphere.
Notes: On wood.
Hosts: Abies, Picea, Quercus
Botryobasidium candicans J. Eriksson
Distr.: AZ; Europe.
Notes: Eriksson & Ryvarden (The Corticiaceae of North Europe 2:157. 1973).
Host: Pinus
Botryobasidium flavescens (Bonord.) D.P. Rogers
Hypochnus flavescens Bonord.
Pellicularia flavescens (Bonord.) D.P. Rogers
Distr.: CA; Europe.
Notes: This species probably does not belong to the genus *Botryobasidium*. Jülich (Kleine Kryptogamenflora IIb/1 pp. 453-454. 1984) puts *H. flavescens* in questionable synonymy *U. fusisporum* and *U. ochraceum* (Rogers. Farlowia 1:105. 1943). Since the basionym of *B. flavescens* is the oldest name, it has priority.
Hosts: Abies, Populus
Botryobasidium laeve (J. Eriksson) Parmasto
Distr.: North temperate regions.
Notes: Eriksson & Ryvarden (The Corticiaceae of North Europe 2:165. 1973).
Host: Pinus
Botryobasidium medium J. Eriksson
Haplotrichum medium (Holubová-Jechová) Holubová-Jechová (anamorph)
Distr.: Southwestern United States; Canada; Europe.
Notes: Eriksson & Ryvarden (The Corticiaceae of North Europe 2:167. 1973).
Host: Pinus
Botryobasidium pruinatum (Bres.) J. Eriksson
Distr.: Temperate northern hemisphere.
Notes: On wood and litter.
Hosts: Alnus, Picea, Pinus, Populus
Botryobasidium subcoronatum (Höhn. & Litsch.) Donk
Corticium subcoronatum Höhn. & Litsch.
Pellicularia subcoronata (Höhn. & Litsch.) D.P. Rogers
Distr.: Widespread in temperate regions.
Notes: Eriksson & Ryvarden (The Corticiaceae of North Europe 2:173. 1973).
Hosts: Abies, Alnus, Arbutus, Picea, Pinus, Pseudotsuga, Tsuga
Botryobasidium vagum (Berk. & M.A. Curtis) D.P. Rogers
Pellicularia vaga (Berk. & M.A. Curtis) D.P. Rogers ex Linder
Haplotrichum curtisii (Berk.) Holubová-Jechová (anamorph)
Distr.: Widespread.
Notes: On wood.
Hosts: Abies, Picea, Pinus, Populus, Sambucus, Tsuga

Botryodiplodia (Sacc.) Sacc. Coelomycetes

Notes: Zambettakis (Bull. Soc. Mycol. France 70:219. 1955).
Botryodiplodia acerina Ellis & Everh.
Distr.: WV; Canada(Ontario).
Host: Acer
Botryodiplodia ailanthi (Cooke) Sacc.
Diplodia ailanthi Cooke
Distr.: Southeastern United States.
Notes: Zambettakis (Bull. Soc. Mycol. France 70:219. 1955) treated this taxon as a synonym of *B. congesta*.
Host: Ailanthus
Botryodiplodia celtidis (Ellis & Everh.) Petr. & Syd.
Macrophoma celtidicola Dearn. & House
Macrophoma celtidis Ellis & Everh., nom. nud.
Distr.: North America.
Host: Celtis
Botryodiplodia circinans (Berk. & Broome) Petr.
Diplodia circinans Berk. & Broome
Distr.: Southern United States; Sri Lanka.
Host: Yucca
Botryodiplodia compressa (Cooke) Sacc. f. ***toxicodendri*** Dearn.
Distr.: NY(type).
Host: Rhus
Botryodiplodia congesta (Lév.) Sacc.
Diplodia congesta Lév.
Distr.: OR; Europe.
Host: Juglans
Botryodiplodia diplocarpa Ellis & Everh., nom. nud.
Distr.: FL, LA(type); Trinidad(dubious).
Host: Poncirus
Botryodiplodia gallae (Schwein.) Petr. & Syd.
Dothiorella gallae Starb.
Dothiorella gallae Ellis & Everh.
Sphaeropsis gallae (Schwein.) W.A. Archer non Berk. & M.A. Curtis, hom. illeg.
Sphaeropsis gallae Berk. & M.A. Curtis
Dothiorella quercina (Cooke & Ellis) Sacc.
Sphaeropsis quercina Cooke & Ellis, hom. illeg.
Distr.: Temperate North America.
Host: Quercus
Botryodiplodia gallicola (Sacc.) Petr.
Macrophoma gallicola Sacc.
Distr.: ND.
Host: Solidago
Botryodiplodia juniperina (Peck) Petr. & Syd.
Macrophoma juniperina Peck
Distr.: NY.
Host: Juniperus
Botryodiplodia lecanidion (Speg.) Petr. & Syd.
Pseudopatella lecanidion Speg.
Metadiplodia lecanidion (Speg.) Zambettakis
Nothopatella lecanidium Sacc.
Distr.: FL(type); Brazil.
Host: Citrus
Botryodiplodia nerii Syd.
Distr.: FL(type).
Host: Nerium
Botryodiplodia oncidii (Henn.) Petr. & Syd.
Macrophoma oncidii Henn.
Distr.: Reported from unknown localities in the United States; Europe.
Hosts: Cattleya, Dendrobium, Laelia, Oncidium, Phalaenopsis
Botryodiplodia pallida Ellis & Everh.
Distr.: AL(type).
Host: Glycine
Botryodiplodia palmarum (Cooke) Petr. & Syd.
Sphaeropsis palmarum Cooke
Macrophoma palmarum (Cooke) Berl. & Voglino
Phoma palmarum Sacc.
Striodiplodia palmarum (Cooke) Zambettakis
Distr.: FL; Guyana(type).
Host: Serenoa
Botryodiplodia ravenelii Sacc.
Distr.: OH, SC.
Host: Quercus

Botryohypochnus Donk. Basidiomycetes, Aphyllophorales

Botryohypochnus chordulatus (D.P. Rogers) Burdsall & Nakasone
Pellicularia chordulata D.P. Rogers
Distr.: Southeastern United States.
Notes: Burdsall & Nakasone (Mycologia 73:457. 1981).
Host: Quercus
Botryohypochnus isabellinus (Fr.) J. Eriksson
Hypochnus isabellinus Fr.
Pellicularia isabellina (Fr.) D.P. Rogers
Distr.: Northern hemisphere.
Notes: On wood.
Hosts: Abies, Pinus, Populus, Quercus
Botryohypochnus verrucisporus Burdsall & R.L. Gilbertson

Distr.: AZ.
Notes: On wood. Burdsall & Gilbertson (Mycotaxon 15:333. 1982).
Host: Quercus

Botryosphaeria Ces. & De Not. Loculoascomycetes, Dothideales

Notes: Some authorities consider *Guignardia* a synonym.
Botryosphaeria abietina (Prill. & Delacr.) Maubl.
Physalospora abietina Prill. & Delacr.
Cryptosporina abietina (Prill. & Delacr.) Theiss.
Distr.: North America; northern Europe.
Host: Abies
Botryosphaeria abrupta Berk. & M.A. Curtis
Distr.: NC(type).
Host: Cyrilla
Botryosphaeria apocyni (Ellis & Everh.) Barr
Laestadia apocyni Ellis & Everh.
Distr.: Eastern North America.
Host: Apocynum
Botryosphaeria arctostaphyli (Plowr.) Sacc.
Distr.: Western United States.
Host: Arctostaphylos
Botryosphaeria arundinariae Earle
Distr.: AL, LA.
Host: Arundinaria
Botryosphaeria callicarpae (Cooke) Cooke
Melogramma callicarpae Cooke
Distr.: GA, SC.
Host: Callicarpa
Botryosphaeria calycanthi (Schwein.) Sacc.
Distr.: NC, VA.
Notes: Arx & Müller (Beitr. Kryptogamenfl. Schweiz 11(1):33. 1954) treated this name as a synonym of *Botryosphaeria quercuum*.
Host: Calycanthus
Botryosphaeria castaneae (Schwein.:Fr.) Sacc.
Distr.: NC(type).
Notes: Arx & Müller (Beitr. Kryptogamenfl. Schweiz 11(1):33. 1954) synonymized this species with *B. quercuum*.
Host: Castanea
Botryosphaeria corticis (Demaree & M.S. Wilcox) Arx & E. Müller
Physalospora corticis Demaree & M.S. Wilcox
Distr.: Southeastern United States.
Notes: Blueberry cane canker (Demaree & Wilcox. Phytopathology 32:1068. 1942).
Host: Vaccinium
Botryosphaeria dasylirii (Peck) Theiss. & Syd.
Dothidea dasylirii Peck
Phyllachora dasylirii (Peck) Sacc.
Distr.: Southwestern United States.
Host: Dasylirion
Botryosphaeria disrupta (Berk. & M.A. Curtis) Arx & E. Müller
Phomatospora disrupta (Berk. & M.A. Curtis) Cooke
Physalospora disrupta (Berk. & M.A. Curtis) Sacc.
Desmotascus portoricensis F. Stevens
Distr.: PR; Central and South America; central Africa.
Notes: Arx & Müller (Beitr. Kryptogamenfl. Schweiz 11(1):37. 1954) redescribed this species.
Hosts: Acacia, Carya, Cocos, Ficus, Liquidambar, Magnolia, Mangifera, Manihot, Persea, Poncirus, Rhododendron, Rosa, Smilax, Yucca, Zea
Botryosphaeria dothidea (Moug.:Fr.) Ces. & De Not.
Botryosphaeria berengeriana De Not.
Botryosphaeria dothidea (Moug.:Fr.) Ces. & De Not. f. *chromogena*
Physalospora suberumpens Ellis & Everh.
Fusicoccum aesculi Corda (anamorph)
Distr.: Temperate regions.
Notes: Arx & Müller (Beitr. Kryptogamenfl. Schweiz 11(1):37. 1954); Barr (Contr. Univ. Michigan Herb. 9:557. 1972). Morgan-Jones & White (Mycotaxon 30:117. 1987) and Pennycook & Samuels (Mycotaxon 24:445. 1985) discussed the anamorph.
Hosts: Acer, Aesculus, Ailanthus, Albizia, Aleurites, Amorpha, Aralia, Aronia, Arundinaria, Baccharis, Baptisia, Berberis, Berchemia, Betula, Broussonetia, Callicarpa, Calycanthus, Camellia, Campsis, Carya, Castanea, Ceanothus, Cephalanthus, Cercis, Chamaecyparis, Cinnamomum, Cirsium, Citrus, Cocculus, Cornus, Cotinus, Cotoneaster, Crataegus, Croton, Cyrilla, Diospyros, Elaeagnus, Eucalyptus, Euonymus, Ficus, Foeniculum, Fraxinus, Gelsemium, Gleditsia, Gossypium, Hibiscus, Ilex, Itea, Juglans, Juniperus, Lespedeza, Ligustrum, Lindera, Liquidambar, Liriodendron, Lonicera, Ludwigia, Lyonia, Lysimachia, Magnolia, Malus, Melia, Metasequoia, Myrica, Nerium, Oenothera, Ostrya, Oxydendrum, Paulownia, Persea, Photinia, Physocarpus, Pieris, Pinus, Pistacia, Platanus, Poncirus, Populus, Prunus, Pseudotsuga, Psidium, Pueraria, Punica, Pyracantha, Pyrus, Quercus, Rhamnus, Rhizophora, Rhododendron, Rhus, Ribes, Ricinus, Robinia, Rosa, Rubus, Rumex, Salix, Sambucus, Sequoia, Sequoiadendron, Solidago, Spiraea, Thuja, Tilia, Ulmus, Vaccinium, Viburnum, Vitis, Wisteria, Xylosma
Botryosphaeria festucae (Lib.) Arx & E. Müller
Physalospora festucae (Lib.) Sacc.
Physalospora conica Ellis & Everh.
Physalospora oxystoma Sacc. & Ellis
Rostrosphaeria phlei Tehon & E.Y. Daniels
Physalospora zeicola Ellis & Everh.
Diplodia frumenti Ellis & Everh. (anamorph)
Distr.: Eastern and southern North America; Europe.
Hosts: Agropyron, Arundinaria, Panicum, Phleum, Zea
Botryosphaeria ficus (Cooke) Sacc.
Melogramma ficus Cooke
Distr.: NC(type).
Host: Ficus
Botryosphaeria graphidea (Berk. & Ravenel) Sacc.
Melogramma graphideum Berk. & Ravenel
Distr.: NC.
Host: Myrica
Botryosphaeria hibisci (Schwein.) Sacc.
Melogramma hibisci (Schwein.) Berk.
Distr.: LA, SC.
Host: Hibiscus
Botryosphaeria marconii V. Charles & Jenk.
Distr.: MD, VA.
Host: Cannabis
Botryosphaeria melanops (Tul. & C. Tul.) G. Wint.
Botryosphaeria advena Sacc. non Ces. & De Not., hom. illeg.
Dothiorella advena Sacc. (anamorph)
Distr.: Temperate northern hemisphere.
Host: Quercus
Botryosphaeria melathroa Cooke
Distr.: MD, PA.
Host: Crataegus
Botryosphaeria minor Ellis & Everh.
Melanops minor (Ellis & Everh.) Weese
Distr.: LA(type).
Host: Sesbania
Botryosphaeria muriculata Ellis & Everh.
Melanops muriculata (Ellis & Everh.) Weese
Distr.: AL(type).
Host: Smilax
Botryosphaeria obtusa (Schwein.) Shoemaker
Physalospora obtusa (Schwein.) Cooke
Physalospora corni Ellis & Everh. non Sacc., hom. illeg.
Physalospora everhartii Sacc.
Sphaeropsis sp. (anamorph)
Distr.: Temperate regions.
Notes: Punithalingam & Waller (C.M.I. Descr. 394. 1973).
Hosts: Acer, Aesculus, Ailanthus, Alcea, Alnus, Amorpha, Ampelopsis, Aristolochia, Aronia, Artemisia, Baccharis, Baptisia, Betula, Bignonia, Broussonetia, Callicarpa, Calycanthus, Calystegia, Campsis, Canna, Carissa, Carpinus, Carya, Castanea, Catalpa, Ceanothus, Celastrus, Celtis, Cephalanthus, Cercis, Chaenomeles, Chamaecyparis, Citrus, Clematis, Clethra, Cocculus, Cornus, Corylus, Cotinus, Cotoneaster, Crataegus, Cydonia, Cytisus, Dahlia, Diospyros, Elaeagnus, Eriobotrya, Euonymus, Ficus, Fraxinus, Gelsemium, Genista, Gleditsia, Gossypium, Hamamelis, Hedera, Hibiscus, Ilex, Jasminum, Juglans, Juniperus, Koelreuteria, Lagerstroemia, Lantana, Larix, Lespedeza, Leucaena, Ligustrum, Lindera, Liquidambar, Liriodendron, Lonicera, Ludwigia, Lyonia, Maclura, Magnolia, Malus, Melia, Morus, Myrica, Nerium,

Oenothera, Opuntia, Osteomeles, Ostrya, Oxydendrum, Parthenocissus, Paulownia, Persea, Physocarpus, Pinus, Pistacia, Platanus, Polygonum, Poncirus, Populus, Pouteria, Prosopis, Prunus, Pueraria, Pyracantha, Pyrus, Quercus, Rhamnus, Rhododendron, Rhus, Ribes, Ricinus, Robinia, Rosa, Rubus, Rumex, Salix, Sambucus, Sassafras, Smilax, Solidago, Sorbus, Spiraea, Syringa, Tabebuia, Thuja, Tilia, Ulmus, Vaccinium, Vinca, Vitis, Wisteria

Botryosphaeria pyriospora (Ellis) Sacc.
Distr.: NJ, WV.
Notes: Arx & Müller (Beitr. Kryptogamenfl. Schweiz 11(1):34. 1954) synonymized this species with *B. quercuum*.
Host: Chionanthus

Botryosphaeria quercuum (Schwein.) Sacc.
Botryosphaeria ambigua (Schwein.) Sacc.
Melogramma ambiguum (Schwein.) Berk.
Physalospora cupressi (Berk. & M.A. Curtis) Sacc.
Physalospora cydoniae G. Arnaud
Wallrothiella eunotiaespora (Cooke & Harkn.) Berl. & Voglino
Physalospora glandicola (Schwein.) N. Stevens
Amerodothis ilicis (Cooke) Theiss. & Syd.
Physalospora thuyoidea (Cooke & Ellis) Sacc.
Distr.: Worldwide on branches of woody dicotyledons
Notes: Shoemaker (Canad. J. Bot. 42:1299. 1962) and Sivanesan (The bitunicate Ascomycetes. p. 136. 1964) included descriptions of the unnamed anamorph.
Hosts: Aleurites, Arachis, Beta, Cajanus, Carissa, Carya, Castanea, Chamaecyparis, Citrullus, Citrus, Cocos, Crataegus, Cydonia, Eriobotrya, Ficus, Fortunella, Gleditsia, Gossypium, Grevillea, Ilex, Ipomoea, Juglans, Juniperus, Lagerstroemia, Liquidambar, Malus, Mangifera, Manihot, Melia, Myrciaria, Myrica, Opuntia, Parthenocissus, Persea, Phoenix, Pittosporum, Poncirus, Pyrus, Quercus, Rhus, Ricinus, Rosa, Rubus, Salix, Viburnum, Zea, Ziziphus

Botryosphaeria rhizogena (Berk.) Sacc.
Melogramma rhizogenum Berk.
Distr.: NC, OH.
Host: Gleditsia

Botryosphaeria rhodina (Cooke) Arx
Physalospora rhodina Cooke
Distr.: Widespread in the northern hemisphere.
Notes: Punithalingam (Biblioth. Mycol. 71:13. 1980) suggested this species might be the teleomorph of *Lasiodiplodia theobromae* (Pat.) Griff. & Maubl. Arx & Müller (Beitr. Kryptogamenfl. Schweiz 11(1):34. 1954) considered this species a synonym of *B. quercuum* (Schwein.) Sacc.
Hosts: Albizia, Aleurites, Arachis, Baccharis, Beta, Carissa, Carya, Cassia, Catalpa, Cinnamomum, Citrullus, Citrus, Cocos, Cornus, Cucumis, Datura, Diospyros, Dracaena, Eucalyptus, Ficus, Fortunella, Gelsemium, Gleditsia, Gossypium, Ilex, Liquidambar, Malus, Mangifera, Manihot, Melia, Oenothera, Opuntia, Persea, Phoenix, Pittosporum, Platanus, Prunus, Pyrus, Quercus, Rhododendron, Rhus, Ricinus, Rosa, Rubus, Salix, Sambucus, Sorghum, Vitis, Zea

Botryosphaeria rhodorae (Cooke) Barr
Guignardia rhodorae (Cooke) B.H. Davis
Laestadia rhodorae (Cooke) Berl. & Voglino
Phyllosticta maxima Ellis & Everh. (synanamorph macroconidial)
Asteromella saccardoi (Thuem.) Petr. (synanamorph microconidial)
Distr.: Cosmopolitan.
Notes: Based on Barr (Mycologia 62:381. 1970), this species is recognized as separate from *Discochora philoprina* (Aa. Stud. Mycol. 5:44. 1973). Pirozynski (Fungi Canadenses 21. 1974).
Host: Rhododendron

Botryosphaeria ribis Gross. & Duggar
Botryosphaeria ribis Gross. & Duggar f. *achromogena* Gross. & Duggar
Botryosphaeria ribis Gross. & Duggar f. *chromogena* Gross. & Duggar
Fusicoccum sp. (anamorph)
Distr.: Temperate northern hemisphere.
Notes: On a wide variety of woody plants; also causes fruit rots. Morgan-Jones & White (Mycotaxon 30:117. 1987) discussed the anamorph. Sivanesan (The bitunicate Ascomycetes. p. 137. 1984); Punithalingam & Holliday (C.M.I. Descr. 395. 1973).
Hosts: Acalypha, Acer, Aesculus, Ailanthus, Albizia, Aleurites, Alnus, Aralia, Berberis, Caesalpinia, Camellia, Carya, Castanea, Catalpa, Ceanothus, Celastrus, Cephalanthus, Ceratonia, Cercis, Chamaecyparis, Cinnamomum, Citrus, Cladrastis, Croton, Cydonia, Diospyros, Eriobotrya, Euphorbia, Fagus, Ficus, Firmiana, Forsythia, Fraxinus, Gaultheria, Gleditsia, Gossypium, Grevillea, Hibiscus, Hypericum, Ilex, Juglans, Juniperus, Kalanchoe, Kalmia, Laburnum, Lagerstroemia, Leucaena, Lindera, Liquidambar, Liriodendron, Malus, Mangifera, Melia, Morus, Myrica, Nyssa, Ostrya, Pandanus, Persea, Picea, Pinus, Pipturus, Platanus, Populus, Prunus, Psidium, Pueraria, Pyracantha, Pyrus, Quercus, Rhamnus, Rhododendron, Rhus, Ribes, Ricinus, Robinia, Rosa, Salix, Sambucus, Schinus, Smilax, Taxus, Tilia, Ulmus, Vaccinium, Vitis, Wikstroemia

Botryosphaeria stevensii Shoemaker
Physalospora malorum Shear, N. Stevens, & M.S. Wilcox
Physalospora mutila N. Stevens, nom. inval.
Diplodia mutila (Fr.:Fr.) Mont. (anamorph)
Distr.: Eastern North America; Europe.
Notes: Shoemaker (Canad. J. Bot. 42:1299. 1964).
Hosts: Acer, Malus, Prosopis, Rhus

Botryosphaeria tamaricis (Cooke) Theiss. & Syd.
Dothidea tamaricis Cooke
Bagnisiella tamaricis (Cooke) Sacc.
Distr.: NC(type).
Host: Tamarix

Botryosphaeria trames (Berk. & M.A. Curtis) Sacc.
Lasiosphaeria trames (Berk. & M.A. Curtis) Cooke
Trichosphaeria trames (Berk. & M.A. Curtis) Ellis & Everh.
Distr.: SC(type), WA.
Host: Acer

Botryosphaeria vaccinii (Shear) Barr
Guignardia vaccinii Shear
Mycosphaerella andromedae L.E. Miles
Mycosphaerella oxycocci Dearn. & House
Phyllosticta elongata G.J. Weidemann in G.J. Weidemann, D.M. Boone & Burdsall (anamorph)
Distr.: North America.
Notes: Flower and fruit blast, leaf spot, fruit rot of cranberries and other Ericaceae. The recent article by Weidemann et al. (Mycologia 74:59. 1982) proves that the anamorph of this species is *Phyllosticta elongata*, rather than *P. vaccinii* as suggested by Van der Aa (Stud. Mycol. 5:85. 1973).
Hosts: Gaultheria, Gaylussacia, Kalmia, Lyonia, Vaccinium

Botryosphaeria vanillae (Stoneman) Petch & Ragunathan
Gnomoniopsis vanillae Stoneman
Distr.: Subtropical and tropical America; Sri Lanka.
Hosts: Oncidium, Orchidaceae, Vanilla

Botryosphaeria wistariae (Rehm) Sacc.
Melogramma wistariae (Rehm) Cooke
Distr.: Eastern North America.
Host: Wisteria

Botryosphaeria zeae (G.L. Stout) Arx & E. Müller
Physalospora zeae G.L. Stout
Distr.: United States; eastern and southern Africa; France.
Notes: Gray ear rot of maize. Sivanesan & Holliday (C.M.I. Descr. 774. 1983) suggested that the anamorph belongs in *Dothiorella*.
Hosts: Triticum, Zea

Botryosporium Corda. Hyphomycetes

Botryosporium madrasense Raghukumar
Distr.: FL; Great Britain; India.
Host: Nuphar

Botryosporium pulchrum Corda
Distr.: North America; Europe.
Hosts: Lycopersicon, Nicotiana, Pelargonium, Selaginella

Botryotinia Whetzel. Discomycetes, Helotiales

Botryotinia calthae Hennebert & M.E. Elliott in Hennebert & Groves
Botrytis calthrae Hennebert (anamorph)
Distr.: Northeastern North America; Belgium; Denmark.
Host: Caltha

Botryotinia convoluta (Drayton) Whetzel

Sclerotinia convoluta Drayton
Distr.: Northern North America; Europe.
Host: Iris
Botryotinia fuckeliana (de Bary) Whetzel
Sclerotinia fuckeliana (de Bary) Fuckel
Botrytis cinerea Pers.:Fr. (anamorph)
Distr.: Cosmopolitan.
Notes: Ellis & Waller (C.M.I. Descr. 431. 1974 as *Sclerotinia fuckeliana*).
Hosts: Amaryllis, Coffea, Daucus, Lilium, Phaseolus, Picea
Botryotinia narcissicola (C.T. Gregory) Buchwald
Sclerotinia narcissicola C.T. Gregory
Distr.: Widespread in Europe and North America.
Host: Narcissus
Botryotinia polyblastis (C.T. Gregory) Buchwald
Sclerotinia polyblastis C.T. Gregory
Distr.: Widespread in Europe and North America.
Hosts: Iris, Narcissus
Botryotinia ranunculi Hennebert & Groves
Botrytis ranunculi Hennebert (anamorph)
Distr.: Northeastern North America.
Host: Ranunculus
Botryotinia ricini (Godfrey) Whetzel
Sclerotinia ricini Godfrey
Distr.: Southeastern United States; West Indies.
Notes: Gray mold of castor bean.
Hosts: Caladium, Phaseolus, Ricinus
Botryotinia squamosa Vien.-Bourg.
Sclerotinia squamosa (Vien.-Bourg.) Dennis
Botrytis squamosa J.C. Walker (anamorph)
Distr.: Widespread in Europe and North America.
Host: Allium

Botrytis P. Mich. ex Pers. Hyphomycetes

Botrytis aclada Fresen.
Botrytis allii Munn
Distr.: Cosmopolitan.
Notes: Neck rot of onions and other *Allium* spp. (Ellis & Waller. C.M.I. Descr. 433. 1974 as *B. allii*; Hennebert. Persoonia 7:183. 1973).
Host: Allium
Botrytis anthophila Bondartsev
Sclerotinia spermophila M. Noble (teleomorph)
Distr.: Northeastern North America; Europe.
Notes: Noble (Trans. Brit. Mycol. Soc. 30:84. 1948) provided verification of the anamorph-teleomorph connection.
Host: Trifolium
Botrytis byssoidea J.C. Walker
Botryotinia allii (Sawada) W. Yamamoto (teleomorph)
Distr.: Temperate North America; Japan.
Notes: Yamamoto et al. (Rep. Hygo Univ. Agric. 2:17. 1956).
Host: Allium
Botrytis cinerea Pers.:Fr.
Haplaria grisea Link
Botrytis vulgaris Link:Fr.
Botryotinia fuckeliana (de Bary) Whetzel (teleomorph)
Distr.: Cosmopolitan.
Notes: An aggregate of morphologically indistinct conidial states, many of which are plurivorous, while others have restricted host ranges.
Hosts: Abies, Actinidia, Aechmea, Aeschynanthus, Agave, Allium, Alnus, Alsophila, Amaryllis, Amelanchier, Amorphophallus, Anemone, Angelica, Anthurium, Antirrhinum, Apium, Aquilegia, Arabis, Arachis, Aristolochia, Artemisia, Artocarpus, Asparagus, Aster, Athyrium, Avena, Begonia, Bellis, Berberis, Beta, Billbergia, Brassica, Brassolaeliocattleya, Bromus, Calathea, Calceolaria, Calendula, Callistephus, Calluna, Camassia, Camellia, Campanula, Cannabis, Capsicum, Carthamus, Castilleja, Cattleya, Cereus, Chamaedorea, Cheiranthus, Chrysanthemum, Cichorium, Cicuta, Citrus, Clarkia, Conioselinum, Coreopsis, Cornus, Crataegus, Crossandra, Crotalaria, Cryptanthus, Cucumis, Cucurbita, Cuphea, Cyclamen, Cydonia, Cymbalaria, Cynara, Cynoglossum, Cytisus, Dahlia, Daucus, Davallia, Delphinium, Dendrobium, Dianthus, Dimorphotheca, Diospyros, Dracaena, Epidendrum, Epilobium, Epipremnum, Episcia, Erigeron, Erythronium, Eschscholzia, Eucalyptus, Eucharis, Euonymus, Eupatorium, Euphorbia, Exacum, Fagus, Feijoa, Felicia, Ficus, Foeniculum, Fragaria, Fuchsia, Gardenia, Gentiana, Geranium, Gerbera, Gilia, Gladiolus, Glycine, Gynura, Gypsophila, Helianthus, Heliotropium, Helleborus, Heuchera, Hibiscus, Hippeastrum, Hosta, Hydrangea, Hylocereus, Iberis, Impatiens, Ipomoea, Iris, Iva, Kalanchoe, Lactuca, Laelia, Laeliocattleya, Lathyrus, Lens, Leucospermum, Lilium, Limonium, Linaria, Linum, Lithospermum, Lobelia, Lonicera, Lupinus, Lychnis, Lycopersicon, Macadamia, Maclura, Magnolia, Malus, Mandevilla, Mangifera, Matthiola, Medicago, Melissa, Mimulus, Myosotis, Narcissus, Nepeta, Nephrolepis, Nicotiana, Oenothera, Omphalodes, Oncidium, Oplopanax, Orchidaceae, Orobanche, Oxalis, Paeonia, Panax, Papaver, Parthenium, Pastinaca, Pelargonium, Peperomia, Petroselinum, Petunia, Phalaenopsis, Phanerophlebia, Phaseolus, Philadelphus, Philodendron, Phlox, Picea, Pilea, Pinus, Pistacia, Pisum, Plectranthus, Poa, Podophyllum, Polianthes, Potamogeton, Primula, Prunus, Pseudotsuga, Pteris, Punica, Pyracantha, Pyrola, Pyrus, Ranunculus, Rheum, Rhinanthus, Rhododendron, Ribes, Ricinus, Rosa, Rubus, Ruellia, Rumex, Saintpaulia, Sambucus, Sarcococca, Saxifraga, Scutellaria, Sedum, Senecio, Sequoia, Sequoiadendron, Sinningia, Solanum, Solenostemon, Staphylea, Stokesia, Syringa, Tagetes, Taraxacum, Thuja, Thymus, Tradescantia, Trifolium, Trigonella, Triodanis, Tropaeolum, Tsuga, Tulipa, Vaccinium, Vanda, Veratrum, Verbena, Viburnum, Vicia, Vigna, Vinca, Viola, Vitis, Xanthium, Zantedeschia, Zea, Zinnia
Botrytis convoluta Whetzel & Drayton
Botryotinia convoluta (Drayton) Whetzel (teleomorph)
Distr.: Northern North America; Europe.
Host: Iris
Botrytis douglasii Tub.
Botrytis cinerea Pers.:Fr. f. *douglasii* Tub.
Distr.: Western United States; Europe.
Notes: Jarvis (Canad. Dept. Agric. Monogr. 15:16 1977) mentioned that this was a doubtful species. Klebahn (Z. Bot. 23:251. 1930) regarded this species as morphologically identical with but physiologically distinct from *B. cinerea*.
Hosts: Larix, Sequoia, Sequoiadendron
Botrytis elliptica (Berk.) Cooke
Distr.: Widespread in Europe and North America.
Notes: Usually on liliaceous plants.
Hosts: Clematis, Colchicum, Erythronium, Gladiolus, Lilium, Polianthes, Stephanotis
Botrytis fusca (Cooke & Harkn.) Sacc.
Polyactis fusca Cooke & Harkn.
Distr.: CA(type).
Host: Eucalyptus
Botrytis galanthina (Berk. & Broome) Sacc.
Polyactis galanthina Berk. & Broome
Distr.: Europe. The sclerotial stage has not yet been imported into North America.
Notes: *Sclerotinia galanthi* Ludwig was based on the anamorph of this fungus. The teleomorph is imperfectly known.
Host: Galanthus
Botrytis gladioli Kleb.
Distr.: WA; Germany(type).
Notes: According to Moore (Diseases of bulbs. p. 103. 1979) this is probably a synonym of *Botrytis gladiolorum*.
Host: Gladiolus
Botrytis gladiolorum Timmermans
Botryotinia draytonii (Buddin & Wakef.) Seaver (teleomorph)
Distr.: North America; Europe.
Host: Gladiolus
Botrytis hyacinthi Westend. & Van Beyma
Distr.: WA; The Netherlands.
Notes: Fire disease of hyacinth (Moore. Diseases of bulbs. p. 2. 1979).
Host: Hyacinthus
Botrytis mali Ruehle
Distr.: WA(type).
Notes: A doubtful species.
Host: Malus
Botrytis narcissicola Kleb.
Botryotinia narcissicola (C.T. Gregory) Buchwald (teleomorph)
Distr.: Widespread in Europe and North America.
Host: Narcissus
Botrytis paeoniae Oudem.

Distr.: Eastern North America; South America; Europe.
Hosts: Allium, Convallaria, Paeonia
Botrytis polyblastis Dowson
Botryotinia polyblastis (C.T. Gregory) Buchwald (teleomorph)
Distr.: Widespread in Europe and North America.
Host: Narcissus
Botrytis squamosa J.C. Walker
Botryotinia squamosa Vien.-Bourg. (teleomorph)
Distr.: Widespread in Europe and North America.
Notes: Leaf rot and neck rot of onions.
Host: Allium
Botrytis tulipae (Lib.) Lind
Sclerotium tulipae Lib.
Distr.: Cosmopolitan.
Hosts: Allium, Lilium, Tulipa
Botrytis viciae H.C. Greene
Distr.: WI(type).
Host: Linaria
Botrytis vulgaris Link:Fr. var. **interrupta** Fr.
Distr.: ND; Germany(type).
Notes: The type variety is a synonym of *B. cinerea*.
Host: Populus

Bourdotia (Bres.) Trott. Basidiomycetes, Tremellales
Notes: Monotypic.
Bourdotia galzinii (Bres.) Trott.
Distr.: Temperate northern hemisphere.
Notes: On wood.
Hosts: Fraxinus, Populus

Brachysporiella Batista. Hyphomycetes
Brachysporiella turbinata (Cooke & Harkn.) M.B. Ellis
Mystrosporium turbinatum Cooke & Harkn.
Distr.: CA.
Notes: Ellis (Mycol. Pap. 72:18. 1959).
Host: Sequoia

Brachysporium Sacc. Hyphomycetes
Brachysporium nigrum (Link:Fr.) S.J. Hughes
Cladotrichum nigrum (Link:Fr.) Sacc.
Acrothecium apicale (Berk. & M.A. Curtis) Höhn.
Brachysporium apicale (Berk. & M.A. Curtis) Sacc.
Helminthosporium apicale Berk. & M.A. Curtis
Distr.: Temperate North America; Europe.
Notes: On hardwood debris (Hughes & Kokko. Fungi Canadenses 69. 1975).
Hosts: Acer, Fagus
Brachysporium oosporum (Corda) Sacc.
Helminthosporium oosporum Corda
Distr.: CA; Europe(type).
Host: Sambucus

Brasiliomyces Viégas. Pyrenomycetes, Erysiphales
Notes: Hanlin & Tortolero (Mycologia 76:439. 1984); Zheng (Mycotaxon 19:281. 1984).
Brasiliomyces setosus C.S. Hodges
Distr.: HI.
Notes: On living leaves (Hodges. Mycologia 77:978. 1985).
Host: Sapindus
Brasiliomyces trina (Harkn.) Zheng
Erysiphe trina Harkn.
Californiomyces trina (Harkn.) U. Braun
Distr.: AZ, CA.
Notes: Powdery mildew of oaks.
Hosts: Lithocarpus, Quercus

Bremia Regel. Oomycetes, Peronosporales
Bremia lactucae Regel
Peronospora gangliformis Berk. ex de Bary
Distr.: Cosmopolitan.
Notes: Downy mildew of *Lactuca* spp. and other Asteraceae (Morgan. C.M.I. Descr. 682. 1981).
Hosts: Agoseris, Centaurea, Cichorium, Erechtites, Gnaphalium, Helichrysum, Hieracium, Krigia, Lactuca, Prenanthes, Sonchus

Bremiella G.W. Wils. Oomycetes, Peronosporales
Bremiella megasperma (Berl.) G.W. Wils.
Distr.: Temperate North America.
Notes: Downy mildew.
Host: Viola

Bricookea Barr. Loculoascomycetes, Dothideales
Notes: Barr (Mycotaxon 15:345. 1982).
Bricookea sepalorum (Vleugel) Barr
Metasphaeria sepalorum Vleugel
Leptosphaeria sepalorum (Vleugel) Lind
Distr.: Pacific coastal range of North America; Europe.
Hosts: Carex, Juncus

Briosia Cavara. Hyphomycetes
Briosia ampelophaga Cavara
Distr.: Central and eastern North America; Brazil; Europe.
Notes: Leaf spot.
Host: Vitis

Broomella Sacc. Pyrenomycetes, Sphaeriales
Broomella montaniensis (Ellis & Everh.) E. Müller & Ahmad
Lophiostoma montaniensis Ellis & Everh.
Keissleria montaniensis (Ellis & Everh.) Höhn.
Pestalotia pestalozzioides Dearn. & Fairm. in Fairm. (anamorph)
Distr.: Western United States.
Notes: Shoemaker & Müller (Canad. J. Bot. 41:1237. 1963).
Host: Clematis

Brunnipila Baral & Krieglsteiner. Discomycetes, Helotiales
Brunnipila calyculiformis (Schumach.:Fr.) Rehm
Dasyscyphus calyculiformis (Schumach.:Fr.) Rehm
Lachnum calyculiforme (Schumach.:Fr.) P. Karst.
Distr.: AK, CO, WA; Europe.
Host: Salix
Brunnipila clandestina (Bull.:Fr.) Baral & Krieglsteiner
Dasyscyphus clandestinus (Bull.:Fr.) Fuckel
Lachnum clandestinum (Bull.:Fr.) P. Karst.
Distr.: Temperate northern hemisphere.
Notes: Commonly on stems of *Rubus* spp. and herbaceous plants.
Hosts: Acer, Aruncus, Rubus, Salix, Urtica

Buergenerula Syd. Pyrenomycetes, Sphaeriales
Buergenerula spartinae Kohlmeyer & R.V. Gessner
Distr.: Coastal regions of the Atlantic Ocean.
Host: Spartina

Bulbillomyces Jülich. Basidiomycetes, Aphyllophorales
Bulbillomyces farinosa (Bres.) Jülich
Peniophora farinosa (Bres.) Höhn. & Litsch.
Aegerita candida Pers.:Fr. (anamorph)
Distr.: Temperate northern hemisphere.

Notes: On wood.
Host: Pinus

Bulgaria Fr.:Fr. Discomycetes, Helotiales

Bulgaria inquinans (Pers.:Fr.) Fr.
Phaeobulgaria inquinans (Pers.:Fr.) Nannf.
Distr.: Temperate North America; Europe.
Notes: Erumpent through bark of hardwoods, especially oaks.
Hosts: Acer, Lithocarpus, Quercus

Bullera Derx. Hyphomycetes

Notes: About 6 species. This genus of yeasts appears to have affinities with the Basidiomycotina.
Bullera alba (Hanna) Derx
Sporobolomyces albus Hanna in Bisby et al.
Distr.: KS.
Hosts: Althaea, Euonymus, Impatiens, Ribes

Burrillia Setch. Basidiomycetes, Ustilaginales

Burrillia decipiens (G. Wint.) G.P. Clinton
Distr.: Eastern North America.
Host: Nymphoides
Burrillia echinodori G.P. Clinton
Distr.: CA, FL, TX.
Host: Echinodorus
Burrillia limnanthemi Cif.
Distr.: Central United States; Canada; Dominican Republic(type).
Host: Sagittaria
Burrillia pustulata Setch.
Distr.: Central United States; Canada(Ontario).
Host: Sagittaria

Butlerelfia Weresub & Illman. Basidiomycetes, Aphyllophorales

Notes: Monotypic.
Butlerelfia eustacei Weresub & Illman
Corticium centrifugum (Lév.) Bres., sensu auct.
Distr.: North America; Europe; Asia.
Notes: Fisheye rot of stored apples (Weresub & Illman. Canad. J. Bot. 58:137. 1980).
Host: Malus

Byssochlamys Westling. Pyrenomycetes, Eurotiales

Byssochlamys nivea Westling
Paecilomyces niveus A.C. Stolk & R.A. Samson (anamorph)
Distr.: Temperate regions.
Notes: Isolated from soil and various organic substrates including applejuice, barley grain, silage and wood (Stolk & Samson. Persoonia 6:351. 1971).
Hosts: Acer, Zea

Byssocorticium Bondartsev & Singer ex Singer. Basidiomycetes, Aphyllophorales

Notes: Jülich (Willdenowia Beih. 7:133. 1972).
Byssocorticium atrovirens (Fr.:Fr.) Bondartsev & Singer ex Singer
Corticium atrovirens (Fr.:Fr.) Fr.
Distr.: Northern hemisphere.
Notes: On wood and bark.
Hosts: Castanea, Fagus, Pinus, Quercus
Byssocorticium californicum Jülich
Distr.: CA.
Notes: On wood.
Host: Pinus
Byssocorticium neomexicanum R.L. Gilbertson & Budington
Distr.: NM.
Notes: On wood.
Host: Pinus

Byssoporia M. Larsen & Zak. Basidiomycetes, Aphyllophorales

Notes: Monotypic.
Byssoporia terrestris (DC.:Fr.) M. Larsen & Zak
Byssocorticium terrestre (DC.:Fr.) Bondartsev & Singer
Poria terrestris (DC.:Fr.) Sacc.
Poria parksii Murrill
Distr.: North America; possibly Asia and Europe.
Notes: This species fruits on wood but is mycorrhizal according to Larsen & Zak (Canad. J. Bot. 56:1122. 1978) who described five varieties.
Hosts: Larix, Picea, Pinus, Populus, Pseudotsuga, Quercus

Byssosphaeria Cooke. Loculoascomycetes, Dothideales

Notes: Barr (Mycotaxon 20:28. 1984).
Byssosphaeria alnea (Peck) Barr
Cucurbitaria alnea Peck
Gibberidea alnea (Peck) Wehmeyer
Massarina alnea (Peck) L. Holm
Distr.: Northeastern North America.
Notes: On limbs.
Host: Alnus
Byssosphaeria diffusa (Schwein.) Cooke
Herpotrichia diffusa (Schwein.) Ellis & Everh.
Neopeckia diffusa (Schwein.) Starb.
Herpotrichia parietalis (Berk. & M.A. Curtis) Ellis & Everh.
Pyrenochaeta sp. (anamorph)
Distr.: Cosmopolitan.
Notes: On wood and limbs of various dicotyledons.
Hosts: Nyssa, Robinia
Byssosphaeria rhodomphala (Berk.) Cooke
Herpotrichia diffusa (Schwein.) Ellis & Everh. var. *rhodomphala* (Berk.) Ellis & Everh.
Macbridella chaetostroma (Ellis & T. Macbr.) Seaver
Nectria chaetostroma Ellis & T. Macbr.
Herpotrichia lanuginosa (Berk. & M.A. Curtis) Ellis & Everh.
Herpotrichia rhodospiloides Peck
Herpotrichia rhodosticta (Berk. & Broome) Sacc.
Neopeckia rhodosticta (Berk. & Broome) Sacc.
Amphisphaeria subiculosa Ellis & Everh.
Pyrenochaeta sp. (anamorph)
Distr.: Cosmopolitan.
Notes: On wood and bark of various trees. Samuels & Müller (Sydowia 31:157. 1978) described the anamorph.
Hosts: Populus, Robinia
Byssosphaeria salebrosa (Cooke & Peck) Barr
Amphisphaeria salebrosa (Cooke & Peck) Barr
Trematosphaeria salebrosa (Cooke & Peck) Sivanesan
Distr.: NY.
Host: Vaccinium
Byssosphaeria schiedermayeriana (Fuckel) Barr
Herpotrichia schiedermayeriana Fuckel
Neopeckia rhodostoma Syd.
Pyrenochaeta sp. (anamorph)
Distr.: Worldwide in warm regions.
Notes: On various substrates, rotting logs and branches, endocarps of coconuts, culms and petioles, on wood or cord in greenhouses
Host: Hibiscus

Byssothecium Fuckel. Loculoascomycetes, Dothideales

Byssothecium circinans Fuckel
Trematosphaeria circinans (Fuckel) G. Wint.
Distr.: SD; Europe.
Notes: A saprobe or weak parasite on woody substrates (Boise. Mycologia 75:666. 1983).
Host: Medicago

Cacumisporium G. Preuss. Hyphomycetes

Cacumisporium capitulatum (Corda) S.J. Hughes
Helminthosporium capitulatum Corda
Distr.: Widespread in Europe and North America.
Notes: On dead wood and bark of dicotyledonous trees.
Hosts: Acer, Tilia

Caeoma Link. Basidiomycetes, Uredinales

Caeoma torreyae Bonar
Distr.: CA(type).
Host: Torreya

Calathella D. Reid. Basidiomycetes, Agaricales

Calathella ellisii Agerer
Lachnella campanula (Ellis) Sacc.
Flagelloscypha coloradensis W.B. Cooke
Lachnella oregonensis W.B. Cooke
Distr.: Western United States.
Host: Salix
Calathella eruciformis (Batsch:Fr.) D. Reid
Cyphella eruciformis (Batsch:Fr.) Fr.
Flagelloscypha eruciformis (Batsch:Fr.) Singer
Lachnella eruciformis (Batsch:Fr.) W.B. Cooke
Cyphella albissima Pat. & Doassans in Pat.
Cyphella albocarnea Quél.
Cyphella eumorpha P. Karst.
Stromatocyphella lataensis W.B. Cooke
Distr.: North America; Europe; Turkey.
Notes: On fallen branches.
Hosts: Populus, Salix

Caliciopsis Peck. Pyrenomycetes, Coryneliales

Caliciopsis arceuthobii (Peck) Barr
Wallrothiella arceuthobii (Peck) Sacc.
Distr.: Temperate North America.
Host: Arceuthobium
Caliciopsis calicioides (Ellis & Everh.) Fitzp.
Hypsotheca calicioides Ellis & Everh.
Caliciopsis ellisii Sacc.
Distr.: Western North America.
Host: Populus
Caliciopsis nigra (Schrad.:Fr.) Fitzp.
Lagenula nigra (Schrad.:Fr.) G. Arnaud
Ceratostoma juniperinum Ellis & Everh.
Phaeostoma juniperinum (Ellis & Everh.) Arx & E. Müller, nom. illeg.
Ceratostoma stromaticum Delacr.
Distr.: Eastern North America; Africa; Europe.
Notes: Cannon & Hawksworth (Bot. J. Linn. Soc. 84:138. 1982).
Host: Juniperus
Caliciopsis pinea Peck
Caliciopsis stenocyboides (Nyl.) Rehm
Distr.: Eastern United States; Germany.
Notes: Funk (Canad. J. Bot. 41:506. 1963).
Hosts: Pinus, Tsuga
Caliciopsis pseudotsugae Fitzp.
Distr.: Western United States.
Host: Pseudotsuga
Caliciopsis thujina (Ellis & Everh.) Fitzp.
Hypsotheca thujina Ellis & Everh.
Distr.: NJ(type).
Host: Chamaecyparis

Callistosporium Singer. Basidiomycetes, Agaricales

Callistosporium luteo-olivaceum (Berk. & M.A. Curtis) Singer
Collybia colorea (Peck) Sacc.
Callistosporium graminicolor J.W. Lennox
Callistosporium luteofuscum Singer
Callistosporium psilocybe Murrill & Singer
Distr.: Temperate and tropical forested regions.
Notes: Redhead (Sydowia 35:223. 1982) provided a taxonomic revision and complete synonymy.
Host: Pinus

Calloria Fr. Discomycetes, Helotiales

Notes: Hein (Willdenowia Beih. 9:52. 1976).
Calloria caricinella (Peck) Seaver
Helotium caricinellum Peck
Distr.: NY(type).
Host: Carex
Calloria cremea (Cash) Seaver
Helotium cremeum Cash
Distr.: CA, HI.
Hosts: Cibotium, Dicranopteris, Gleichenia
Calloria lithospermi (Ellis & Everh.) Seaver
Niptera lithospermi Ellis & Everh.
Distr.: MT(type).
Host: Lithospermum
Calloria neglecta (Lib.) B. Hein
Calloria fusarioides (Berk.) Fr.
Cylindrocolla urticae (Fr.) Bonord. (anamorph)
Distr.: GA; Europe.
Hosts: Boehmeria, Pilea

Calloriopsis Syd. & P. Syd. Discomycetes, Helotiales

Calloriopsis gelatinosa (Sacc.) Syd.
Coryne gelatinosa (Sacc.) Rehm
Distr.: Widespread in tropical regions.
Notes: On the mycelium of *Meliola* and related fungi on living leaves (Pfister. Mycotaxon 4:341. 1976).
Hosts: Persea, Scaevola, Wikstroemia

Calocera (Fr.) Fr. Basidiomycetes, Dacrymycetales

Calocera cornea (Batsch:Fr.) Fr.
Clavaria cornea Batsch:Fr.
Calopposis damae-cornis C.G. Lloyd
Calopposis nodulosa C.G. Lloyd
Calocera palmata (Schumach.) Fr.
Calocera rufa C.G. Lloyd
Calocera vermicularis C.G. Lloyd
Distr.: Cosmopolitan.
Notes: Usually on hardwoods.
Hosts: Acer, Alnus, Gleditsia, Juniperus, Pinus, Populus, Pseudotsuga, Quercus, Tsuga

Calonectria De Not. Pyrenomycetes, Hypocreales

Notes: About 10 species, all with a *Cylindrocladium* anamorph. Rossman (Mycotaxon 8:485, 1979; Mycol. Pap. 150:50. 1983).
Calonectria crotalariae (C.A. Loos) D.K. Bell & Sobers
Calonectria theae C.A. Loos var. *crotalariae* C.A. Loos
Distr.: Warm-temperate and tropical regions.
Notes: *Cylindrocladium* black rot of peanut. Seedling diseases. Peerally (C.M.I. Descr. 430. 1974).
Hosts: Acacia, Arachis, Carica, Crotalaria, Eucalyptus, Glycine, Leea, Medicago, Nerium, Vaccinium
Calonectria kyotensis Terashita
Calonectria floridana Sobers
Calonectria uniseptata Gerlach
Cylindrocladium scoparium Morg. (anamorph)
Distr.: Widespread.
Notes: Saprophytic and pathogenic on a variety of hosts and substrates. Peerally (C.M.I. Descr. 421. 1974) reviewed the diseases caused by this fungus. Sobers (Phytopathology 62:485. 1972) compared the morphology and pathogenicity of this species and its synonyms.

Booth & Gibson (C.M.I. Descr. 362. 1973 as *Cylindrocladium scoparium*).
Hosts: Araucaria, Liquidambar
Calonectria pyrochroa (Desmaz.) Sacc.
Calonectria hederae G. Arnaud ex C. Booth
Cylindrocladium ilicicola (Hawley) Boedijn & Reitsma (anamorph)
Distr.: Warm-temperate and tropical regions.
Notes: Pathogenic and saprophytic on a variety of hosts and substrates. *Cercosporella* disease of tea bushes (Peerally. C.M.I. Descr. 426. 1974 as *Calonectria hederae*; Rossman. Mycol. Pap. 150:50. 1983).
Host: Swietenia
Calonectria theae C.A. Loos
Cylindrocladium theae (Petch) Subramanian (anamorph)
Distr.: HI, southeastern United States; Sri Lanka.
Notes: Peerally (C.M.I. Descr. 424. 1974).
Hosts: Acacia, Metrosideros, Rhododendron

Calosphaeria Tul. & C. Tul. Pyrenomycetes, Calosphaeriales

Notes: Barr (Mycologia 77:549. 1985) reviewed the genus.
Calosphaeria abnormis Petr.
Distr.: MD(type); Austria.
Host: Sambucus
Calosphaeria arctica Otani
Distr.: AK(type).
Host: Salix
Calosphaeria barbirostris (Fr.:Fr.) Ellis & Everh.
Valsa lasiostoma Ellis & Everh.
Eutypa scabriseta (Schwein.) Berk. & M.A. Curtis
Distr.: Eastern North America; Europe.
Notes: On hardwood limbs.
Host: Quercus
Calosphaeria cornicola Ellis & Everh.
Togninia cornicola (Ellis & Everh.) Berl.
Distr.: KS, NY.
Host: Cornus
Calosphaeria fagicola (Ellis & Everh.) Ellis & Everh.
Valsa fagicola Ellis & Everh.
Distr.: PA(type).
Host: Fagus
Calosphaeria herbicola Ellis & Everh.
Jattaea herbicola (Ellis & Everh.) Berl.
Distr.: NJ, TX.
Host: Lactuca
Calosphaeria microsperma Ellis & Everh.
Circinostoma microsperma (Ellis & Everh.) House
Distr.: Northeastern North America.
Host: Carpinus
Calosphaeria pulchella (Pers.:Fr.) J. Schröt. in Cohn
Circinostoma pulchellum (Pers.:Fr.) S.F. Gray
Valsa pulchella (Pers.:Fr.) Fr.
Calosphaeria princeps Tul. & C. Tul.
Distr.: Temperate North America; Europe.
Notes: On limbs.
Hosts: Aronia, Prunus
Calosphaeria pulchelloidea (Cooke & Ellis) Ellis & Everh.
Valsa pulchelloidea Cooke & Ellis
Distr.: OR, eastern North America.
Host: Quercus
Calosphaeria subcuticularis (Cooke & Ellis) Ellis & Everh.
Valsa subcuticularis Cooke & Ellis
Distr.: NJ(type).
Host: Ilex

Calothyriella Höhn. Loculoascomycetes, Dothideales

Notes: Synonym of *Microthyrium* (Müller & Arx. Beitr. Kryptogamenfl. Schweiz 11(2):509. 1962).
Calothyriella osmanthi F. Stevens & Ryan
Distr.: HI.
Host: Osmanthus

Calothyriopeltis F. Stevens & Ryan. Loculoascomycetes, Dothideales

Calothyris Clements & Shear
Notes: A confused name (Petrak. Sydowia 5:183. 1951).
Calothyriopeltis clermontiae F. Stevens & Ryan
Distr.: HI.
Host: Clermontia
Calothyriopeltis metrosideri F. Stevens & Ryan
Distr.: HI.
Hosts: Lobelia, Metrosideros
Calothyriopeltis scaevolae F. Stevens & Ryan
Distr.: HI.
Host: Scaevola

Calothyrium Theiss. Loculoascomycetes, Dothideales

Notes: Synonym of *Asterinella* (Müller & Arx. Beitr. Kryptogamenfl. Schweiz 11(2):514. 1962).
Calothyrium osmanthi F. Stevens & Ryan
Distr.: HI.
Host: Osmanthus
Calothyrium pustulatum (Ellis & G. Martin) Theiss.
Asterina pustulata Ellis & G. Martin
Asterella pustulata (Ellis & G. Martin) Sacc.
Microthyrium pustulatum (Ellis & G. Martin) Theiss.
Distr.: FL.
Host: Quercus
Calothyrium ryanae F. Stevens
Calothyrium psychotriae Ryan non Doidge, hom. illeg.
Microthyrium psychotriae (Ryan) Toro
Distr.: FL, PR.
Host: Psychotria
Calothyrium stomatophorum (Ellis & G. Martin) Theiss.
Asterina stomatophora Ellis & G. Martin
Asterella stomatophora (Ellis & G. Martin) Sacc.
Distr.: FL.
Hosts: [Gelsemium], Quercus
Calothyrium suttoniae F. Stevens & Ryan
Distr.: HI.
Host: Myrsine

Calycella (Sacc.) Sacc. Discomycetes, Helotiales

Calycella subochracea (Cooke & Peck) Dennis
Lachnella subochracea (Cooke & Peck) Seaver
Trichopeziza subochracea (Cooke & Peck) Sacc.
Distr.: NY, WA; Canada(Newfoundland).
Notes: Dennis (Kew Bull. 17:366. 1963) provided a recent description.
Host: Rubus

Calycellina Höhn. Discomycetes, Helotiales

Calycellina carolinensis Nag Raj & Kendrick
Chaetochalara aspera Pirozynski & C.S. Hodges (anamorph)
Distr.: NC; China; New Zealand.
Notes: Korf & Zhuang (Mycotaxon 22:498. 1985).
Host: Myrica
Calycellina castanea (Sacc. & Ellis) Dennis
Helotium castaneum Sacc. & Ellis
Distr.: Eastern United States.
Host: Quercus
Calycellina leucella (P. Karst.) Dennis ex E. Müller
Peziza leucella P. Karst.
Helotium leucellum (P. Karst.) P. Karst.
Distr.: AK; Europe.
Host: Alnus
Calycellina osmundae (Cooke & Ellis) Dennis
Peziza osmundae Cooke & Ellis
Trichopeziza osmundae (Cooke & Ellis) Sacc.
Distr.: NJ.
Host: Osmunda

Calycellina populina (Fuckel) Höhn.
Helotium populinum Fuckel
Pezizella populina (Fuckel) Rehm in Rabenh.
Distr.: Temperate North America; Europe.
Hosts: Betula, Robinia, Tilia
Calycellina pulviscula (Cooke) Dennis
Orbilia pulviscula (Cooke) Seaver
Distr.: Northeastern United States.
Host: Phytolacca
Calycellina punctata (Fr.) R. Lowen & Dumont
Distr.: Northern hemisphere; Colombia.
Notes: On fallen leaves of hardwoods (Lowen & Dumont. Mycologia 76:1006. 1984).
Host: Quercus
Calycellina sadleriae (F. Stevens & P.A. Young) Dennis
Dasyscyphus sadleriae F. Stevens & P.A. Young
Distr.: HI.
Host: Sadleria

Calycina Nees. Discomycetes

Notes: This genus is considered a synonym of *Hymenoscyphus* (Korf & Carpenter. Mycotaxon 1:51. 1974), but not all names have been redisposed.
Calycina herbarum (Pers.:Fr.) S.F. Gray
Helotium herbarum (Pers.:Fr.) Fr.
Hymenoscyphus herbarum (Pers.:Fr.) Dennis
Distr.: Temperate regions.
Notes: On dead stems of dicotyledonous plants.
Hosts: Ambrosia, Cirsium
Calycina maculosa (Ellis & G. Martin) Kuntze
Helotium maculosum Ellis & G. Martin
Distr.: FL(type).
Notes: On leaves (Dennis. Persoonia 3:52. 1964).
Host: Persea

Calyptella Quél. Basidiomycetes, Agaricales

Calyptella capula (Holmsk.:Fr.) Quél.
Cyphella capula (Holmsk.:Fr.) Fr.
Distr.: North America; Europe.
Notes: Redhead & Traquair (Fungi Canadenses 202. 1981).
Hosts: Foeniculum, Oxydendrum

Calyptronectria Speg. Pyrenomycetes, Hypocreales

Calyptronectria ohiensis (Ellis & Everh.) Barr
Teichospora ohiensis Ellis & Everh.
Strickeria ohiensis (Ellis & Everh.) Kuntze
Distr.: GA, OH(type).
Notes: Barr (Mycotaxon 18:155. 1983).
Host: Salix

Camarops P. Karst. Pyrenomycetes, Sphaeriales

Notes: Nannfeldt (Svensk Bot. Tidskr. 66:335-376. 1972) reviewed the genus.
Camarops microspora (P. Karst.) Shear
Anthostoma microspora P. Karst.
Diatrype microspora Ellis
Valsa exudans Peck
Distr.: North America; northern Europe.
Notes: On wood.
Hosts: Alnus, Ulmus
Camarops ohiensis (Ellis & Everh.) Nannf.
Hypoxylon ohiense Ellis & Everh.
Distr.: Eastern United States.
Notes: On wood.
Hosts: Picea, Quercus
Camarops petersii (Berk. & M.A. Curtis) Nannf.
Hypoxylon petersii Berk. & M.A. Curtis
Peridoxylon petersii (Berk. & M.A. Curtis) Shear
Distr.: Eastern United States; Cuba; Japan; Soviet Union.
Notes: On wood of Fagaceae.
Hosts: Castanea, Quercus
Camarops polysperma (Mont.) J.H. Miller
Hypoxylon polysperma Mont.
Hypoxylon cylindrophorum Ellis & Everh.
Distr.: Temperate regions.
Notes: On hardwoods, especially common on *Alnus* spp.
Hosts: Acer, Citrus, Liriodendron, Magnolia, Quercus, Rhododendron, Ulmus
Camarops pugillus (Schwein.) Shear
Distr.: Eastern United States.
Notes: On hardwoods.
Hosts: Acer, Fagus, Liriodendron, Rhododendron
Camarops tubulina (Albertini & Schwein.:Fr.) Shear
Bolinia tubulina (Albertini & Schwein.:Fr.) Sacc.
Distr.: LA; Europe.
Notes: Old reports from North America probably refer to *C. ohiensis* or *C. lutea* (Nannfeldt. Svensk Bot. Tidskr. 66:370. 1972).
Host: Gleditsia

Camaropycnis Cash. Coelomycetes

Camaropycnis libocedri Cash
Distr.: CA.
Hosts: Calocedrus, Pinus

Camarosporellum Tassi. Coelomycetes

Camarosporellum cercocarpi Sutton & Pollack
Distr.: Western United States.
Notes: Sutton & Pollack (Mycopath. Mycol. Appl. 52:337. 1974).
Host: Cercocarpus

Camarosporium Schulzer. Coelomycetes

Camarosporulum Tassi
Pseudodichomera Höhn.
Notes: Generic synonymy is discussed in Sutton (Mycol. Pap. 141:29. 1977). Several species are treated in Kohlmeyer & Kohlmeyer (Marine mycology. p. 517. 1979).
Camarosporium abnormale (Peck & Cooke) Sacc.
Hendersonia abnormalis Peck & Cooke
Distr.: NY.
Host: Celastrus
Camarosporium berkeleyanum (Lév.) Sacc.
Hendersonia berkeleyanum Lév.
Distr.: NY; Europe.
Host: Ailanthus
Camarosporium ceanothi Dearn.
Distr.: MD(type).
Host: Ceanothus
Camarosporium celtidis Ellis & Everh.
Distr.: KS(type).
Host: Celtis
Camarosporium cerei (Peck) Sacc.
Hendersonia cerei Peck
Distr.: AZ.
Host: Carnegiea
Camarosporium coluteae (Peck & Cooke) Sacc.
Hendersonia coluteae Peck & Cooke
Distr.: NY.
Host: Colutea
Camarosporium compositarum (Cooke & Harkn.) Sacc.
Dichomera compositarum Cooke & Harkn.
Distr.: Western North America.
Hosts: Achillea, Artemisia
Camarosporium elaeagnellum Fairm.
Distr.: Temperate North America.
Host: Elaeagnus
Camarosporium ellipticum Cooke & Harkn.
Distr.: Western North America.
Host: Mesembryanthemum
Camarosporium eriocryptum Fairm.

Distr.: CA(type).
Host: Salvia
Camarosporium eriogoni Ellis & Everh.
Distr.: CO(type), FL.
Host: Callistephus
Camarosporium kriegeri Bres.
Distr.: Western North America.
Host: Achillea
Camarosporium laburni Sacc. & Roum.
Cucurbitaria laburni (Pers.:Fr.) De Not. (teleomorph)
Distr.: Eastern United States; Europe.
Hosts: Cytisus, Laburnum
Camarosporium linderae Ellis & Everh. in Millsp. & L. Nutt.
Distr.: WV.
Host: Lindera
Camarosporium lyndonvillae Fairm.
Distr.: NY.
Host: Hibiscus
Camarosporium palliatum Kohlmeyer & E. Kohlmeyer
Distr.: NC; Argentina.
Host: Salicornia
Camarosporium parasiticum H.C. Greene
Distr.: WI.
Host: Grindelia
Camarosporium pini (Westend.) Sacc.
Hendersonia pini Westend.
Distr.: PA; Europe.
Host: Pinus
Camarosporium pteridis H.C. Greene
Distr.: WI.
Host: Pteridium
Camarosporium rosellinioides Ellis & Everh.
Distr.: CO(type).
Host: Chrysothamnus
Camarosporium roumeguerii Sacc.
Camarosporium obiones Jaap
Distr.: North and South America; Europe.
Notes: On stems of halophytic plants (Kohlmeyer & Kohlmeyer. Marine mycology. p. 522. 1979).
Hosts: Bassia, Salicornia, Salsola
Camarosporium umbonatum Brenckle
Distr.: SD(type).
Host: Symphoricarpos
Camarosporium wistarianum Fairm.
Distr.: NY.
Host: Wisteria

Campanella Henn. Basidiomycetes, Agaricales

Campanella floridana Singer
Distr.: FL.
Host: Magnolia
Campanella junghuhnii (Mont.) Singer
Campanella cucullata (Jungh.) C.G. Lloyd
Distr.: Subtropical and tropical regions.
Hosts: Dodonaea, Hibiscus

Camposporium Harkn. Hyphomycetes

Camposporium antennatum Harkn.
Distr.: Subtropical and tropical regions.
Host: Eucalyptus

Camptomeris Syd. Hyphomycetes

Camptomeris floridana E.A. Bessey
Distr.: FL; West Indies.
Host: Pithecellobium

Candelabrochaete Boidin. Basidiomycetes, Aphyllophorales

Candelabrochaete langloisii (Pat.) Boidin
Botryobasidium langloisii (Pat.) R.L. Gilbertson & Budington
Pellicularia langloisii (Pat.) D.P. Rogers
Phanerochaete insolita Burdsall & Nakasone
Peniophora magnahypha Burt
Distr.: Widespread.
Notes: On wood.
Hosts: Liquidambar, Pinus
Candelabrochaete septocystidia (Burt) Burdsall
Peniophora septocystidia Burt
Phanerochaete septocystidia (Burt) J. Eriksson & Ryvarden
Distr.: Temperate northern hemisphere.
Notes: On wood.
Hosts: Acer, Liriodendron, Pinus, Populus

Candida Berkhout, nom. cons. Hyphomycetes

Notes: Lodder (The yeasts. p. 893. 1970).
Candida albicans (C.P. Robin) Berkhout
Oidium albicans C.P. Robin
Distr.: Cosmopolitan.
Notes: Isolated from a wide variety of substrates including humans (Davis. C.M.I. Descr. 871. 1986).
Hosts: Prunus, Zea
Candida deserticola Phaff, Starmer, Tredick & M. Miranda
Distr.: AZ; Mexico; Caribbean Islands.
Notes: Isolated from necrotic stems of Cactaceae.
Host: Carnegiea
Candida guilliermondii (Castellani) Langeron & Guerra
Endomyces guilliermondii Castellani
Distr.: Cosmopolitan.
Notes: Isolated from various substrates (Davis. C.M.I. Descr. 872. 1986).
Host: Zea
Candida intermedia (Cif. & Ashford) Langeron & Guerra
Blastodendrion intermedium Cif. & Ashford
Distr.: Widespread, but uncommon.
Notes: Isolated from varied and unusual substrates.
Host: Zea
Candida krusei (A. Castellani) Berkhout
Saccharomyces krusei A. Castellani
Distr.: Widespread.
Notes: Isolated from various substrates (Davis. C.M.I. Descr. 874. 1986).
Hosts: Citrus, Zea
Candida parapsilosis (Ashford) Langeron & Talice
Monilia parapsilosis Ashford
Distr.: Widespread.
Notes: Isolated from a variety of substrates (Davis. C.M.I. Descr. 875. 1986).
Host: Zea
Candida pseudotropicalis (A. Castellani) Basgal
Endomyces pseudotropicalis A. Castellani
Distr.: Cosmopolitan.
Notes: On a variety of substrates.
Host: Zea

Capitotricha (Raitviir) Baral & Krieglsteiner. Discomycetes, Helotiales

Capitotricha bicolor (Bull.:Fr.) Baral & Krieglsteiner
Dasyscyphus bicolor (Bull.:Fr.) Fuckel
Erinella bicolor (Bull.:Fr.) Quél.
Lachnum bicolor (Bull.:Fr.) P. Karst.
Distr.: Temperate northern hemisphere.
Notes: Baral & Krieglsteiner (Beih. Z. Mykol. 6:60. 1985).
Hosts: Acer, Alnus, Arctostaphylos, Castanopsis, Ceanothus, Populus, Ribes, Rubus, Salix, Shepherdia

Capnocybe S.J. Hughes. Hyphomycetes

Capnocybe spongiosa (Hoerl) S.J. Hughes
Arthrobotryum spongiosum Hoerl
Distr.: CA; Corsica; Portugal.
Notes: Hughes (New Zealand J. Bot. 4:349. 1966).
Hosts: Calocedrus, Chamaecyparis, Pseudotsuga

Capnodium Mont. Loculoascomycetes, Dothideales

Capnodium caespitosum Ellis & Everh.
Distr.: CA.
Host: Eriobotrya
Capnodium citri Mont.
Meliola citri Sacc.
Pleosphaeria citri (Mont.) G. Arnaud
Distr.: Gulf Coast States; Australia; France.
Notes: Additional synonyms were included by Fraser (Proc. Linn. Soc. New South Wales 58:384. 1933) and Hughes (Mycologia 68:738. 1976).
Hosts: Citrus, Gardenia, Ilex
Capnodium citricola McAlpine
Limacinula citricola (McAlpine) Speg.
Distr.: Gulf Coast States; Australia.
Host: Citrus
Capnodium elongatum Berk. & Desmaz.
Polychaetella elongata Speg. (anamorph)
Distr.: Southern United States, cosmopolitan.
Notes: A doubtful species, according to Arnaud (Ann. Ecole Natl. Agric. Montpellier n.s. 10:260. 1911)
Hosts: Bignonia, Ilex, Liriodendron, Nerium, Osmanthus, Pittosporum
Capnodium pelliculosum Berk. & Ravenel
Limacinia pelliculosa (Berk. & Ravenel) Sacc.
Distr.: Southern United States.
Notes: Reynolds (Mycotaxon 14:199. 1982) discussed this species. Most specimens are misidentifications of *Capnodium grandisporum*.
Hosts: Magnolia, Prunus
Capnodium pini Berk. & M.A. Curtis
Distr.: Southern United States.
Notes: A doubtful species (Arnaud. Ann. Ecole Natl. Agric. Montpellier n.s. 10:261. 1911).
Host: Pinus
Capnodium rhamni Cooke & Harkn.
Distr.: CA.
Notes: A doubtful species (Arnaud. Ann. Ecole Natl. Agric. Montpellier n.s. 10:260. 1911).
Host: Rhamnus
Capnodium salicinum Mont.
Polychaeton salicinum (Mont.) Kuntze
Fumagospora capnodioides G. Arnaud (anamorph)
Distr.: Temperate northern hemisphere.
Notes: The *Fumagospora* pycnidia are often intimately associated with the ascomata. Fraser (Proc. Linn. Soc. New South Wales 58:384. 1933) provided additional synonyms.
Hosts: Acer, Baccharis, Populus, Salix
Capnodium tuba Cooke & Harkn., nom. inval.
Distr.: CA.
Notes: A coelomycete.
Host: Umbellularia

Capronia Sacc. Loculoascomycetes, Dothideales

Notes: This genus is reviewed by Müller et al. (Trans. Brit. Mycol. Soc. 88:63. 1987).
Capronia mansonii (Schol-Schwarz) E. Müller et al.
Dictyotrichiella mansonii Schol-Schwarz
Exophiala mansonii (A. Castellani) De Hoog (anamorph)
Distr.: Cosmopolitan.
Notes: This fungus in the anamorphic state is isolated from various substrates including human beings.
Host: Metrosideros

Carpenterella Tehon & H.A. Harris. Chytridiomycetes, Chytridiales

Carpenterella molinea Tehon & H.A. Harris
Distr.: WI(type).
Host: Ulmus

Caryospora De Not. Loculoascomycetes, Dothideales

Notes: Barr (Mycotaxon 9:17. 1979); Hawksworth (Trans. Brit. Mycol. Soc. 79:69. 1982).
Caryospora langloisii Ellis & Everh.
Distr.: LA(type).
Host: Arundinaria
Caryospora minima Jeffers
Distr.: MD.
Notes: On peach endocarps.
Host: Prunus
Caryospora putaminum (Schwein.:Fr.) De Not.
Distr.: North America; Europe.
Notes: On peach endocarps.
Hosts: Carya, Prunus

Catacauma Theiss. & Syd. Pyrenomycetes, Phyllachorales

Notes: This genus is considered a synonym of *Phyllachora* Nitschke ex Fuckel (Petrak. Ann. Mycol. 25:295. 1927), but not all names have been redisposed.
Catacauma sabal Chárdon
Distr.: Southeastern United States.
Notes: On Palmaceae.
Hosts: Acoelorrhaphe, Sabal

Catenophora Luttrell. Coelomycetes

Catenophora pruni Luttrell
Distr.: NC(type).
Notes: Luttrell (Mycologia 32:535. 1940).
Host: Prunus
Catenophora yuccae Nag Raj
Distr.: Southwestern United States.
Notes: Nag Raj (Canad. J. Bot. 55:757. 1977).
Host: Yucca

Catenularia Grove. Hyphomycetes

Catenularia fuliginea K. Saito
Distr.: AZ, CA; Japan(type).
Notes: Hughes (New Zealand J. Bot. 3:148. 1965) did not include this species in *Catenularia*.
Host: Phoenix

Catosphaeropsis Tehon. Coelomycetes

Catosphaeropsis caulivora Tehon
Distr.: IL.
Host: Kummerowia

Caulorhiza J.W. Lennox. Basidiomycetes, Agaricales

Caulorhiza umbonata (Peck) J.W. Lennox
Collybia umbonata Peck
Hydropus umbonata (Peck) Singer
Distr.: West coast of the United States.
Hosts: Pseudotsuga, Sequoia

Cellypha Donk. Basidiomycetes, Aphyllophorales

Cellypha goldbachii (Weinm.) Donk
Cyphella goldbachii Weinm.

Cyphella caricina Peck
Cyphella dumetorum E. Bommer & M. Rousseau
Cyphella lactea Bres.
Cyphella malbranchei Pat.
Cyphella ochroleuca Berk. & Broome
Cyphella rubi Fuckel
Distr.: Temperate northern hemisphere.
Notes: On woody twigs, dead herbaceous stems, and remains of grasses and sedges.
Hosts: Carex, Typha
Cellypha musaecola (Berk. & M.A. Curtis) W.B. Cooke
Cyphella fumosa Cooke
Distr.: SC; Caribbean Islands.
Notes: On debris of monocotyledons (Cooke. Beih. Sydowia 4:56. 1961).
Host: Gladiolus
Cellypha subgelatinosa (Berk. & M.A. Curtis) W.B. Cooke
Cyphella subgelatinosa Berk. & M.A. Curtis
Distr.: SC(type).
Host: Alnus

Cenangella Sacc. Discomycetes, Helotiales

Notes: Most members of this genus belong to *Encoeliopsis*, but not all species have been redisposed.
Cenangella helotioides (Mont. & Sacc.) Rehm
Distr.: OR; Europe.
Host: Juniperus

Cenangium Fr.:Fr. Discomycetes, Helotiales

Cenangium acicola (Fuckel) Rehm
Encoelia acicola (Fuckel) Höhn.
Cenangium acuum Cooke & Peck
Mollisia pinastri (Cooke & Peck) Sacc.
Phibalis pinastri (Cooke & Peck) Korf & L.M. Kohn
Distr.: Temperate North America; Europe.
Notes: On needles. Torkelsen & Eckblad (Norw. J. Bot. 24:134. 1977).
Host: Pinus
Cenangium alboatrum Ellis & Everh.
Distr.: WV(type).
Notes: Seaver (North Amer. cup fungi. [Inoperculates]. p. 304. 1951) suggested that this is not a *Cenangium*.
Host: Castanea
Cenangium alpinum Ellis & Everh.
Distr.: CO(type).
Host: Pinus
Cenangium atropurpureum Cash & R.W. Davidson
Distr.: Eastern North America.
Notes: On needles.
Host: Pinus
Cenangium aureum Ellis & Everh.
Distr.: CA, CO.
Host: Ceanothus
Cenangium cassandrae Peck
Distr.: NY(type).
Host: Chamaedaphne
Cenangium castaneae Schwein.
Distr.: PA(type).
Host: Castanea
Cenangium crataegi Schwein.
Distr.: PA(type).
Notes: Seaver (North Amer. cup fungi. [Inoperculates]. p. 305. 1951) could not locate the type collection.
Host: Crataegus
Cenangium episphaeria Schwein.
Distr.: GA.
Notes: A doubtful species according to Seaver (North Amer. cup fungi. [Inoperculates]. p. 305. 1951).
Host: Acer
Cenangium ferruginosum Fr.:Fr.
Cenangium abietis (Pers.) Duby
Dermea pini W. Phillips & Harkn.
Distr.: Circumboreal on Pinaceae.
Notes: Saprobic, a wound parasite or causing die-back of shoots and twigs.
Hosts: Abies, Picea, Pinus
Cenangium griseum Dearn. & House
Distr.: PA; Canada(Ontario, type).
Notes: A poorly known species.
Host: Acer
Cenangium leptospermum Berk. & M.A. Curtis
Distr.: OR.
Notes: Seaver (North Amer. cup fungi. [Inoperculates]. p. 305. 1951) considered the description inadequate.
Host: Pseudotsuga
Cenangium magnoliae Berk. & M.A. Curtis
Dermatella magnoliae (Berk. & M.A. Curtis) Seaver
Distr.: Eastern United States.
Notes: Seaver (North Amer. cup fungi. [Inoperculates]. p. 314. 1951) recognized this species in *Dermatella*, a genus now considered a synonym of *Dermea*.
Host: Persea
Cenangium pezizoides Peck
Distr.: NY(type).
Notes: Seaver (North Amer. cup fungi. [Inoperculates]. p. 306. 1951) stated that the spores of this species suggested *Dermea*.
Host: Chamaedaphne
Cenangium populneum (Pers.) Rehm var. ***prunicola*** Rehm
Distr.: ND.
Notes: The record from ND is *Encoelia fascicularis* according to Seaver (North Amer. cup fungi. [Inoperculates]. p. 301. 1951).
Host: Prunus
Cenangium purpureum (Hedw.:Fr.) Sacc.
Distr.: ID.
Host: Pinus
Cenangium tetrasporum (Ellis) Sacc.
Dermatea tetraspora Ellis
Phaeangium tetrasporum (Ellis) Sacc. & Syd.
Distr.: NJ(type).
Host: Quercus
Cenangium tryblidioides Ellis
Distr.: WA(type).
Notes: Seaver (North Amer. cup fungi. [Inoperculates]. p. 306. 1951) stated that the type specimen is unsatisfactory.
Host: Salix
Cenangium tuberculiforme Ellis & Everh.
Distr.: Eastern United States.
Hosts: Ilex, Malus
Cenangium yuccae Clements
Distr.: Western United States.
Host: Yucca

Cephaliophora Thaxt. Hyphomycetes

Cephaliophora tropica Thaxt.
Distr.: Subtropical and tropical regions.
Notes: Isolated from various substrates (Ellis. Dematiaceous Hyphomycetes. p. 178. 1971; White & Darbyshire. Fungi Canadenses 290. 1984).
Host: Helianthus

Cephalosporium Corda. Hyphomycetes

Notes: This genus is considered a synonym of *Acremonium*, but not all names have been redisposed.
Cephalosporium bertholletianum J.A. Spencer
Distr.: Reported from an unknown locality, probably imported.
Notes: According to Gams (*Cephalosporium*-artige Schimmelpilze [Hyphomycetes]. p. 206. 1971) this species belongs to *Fusarium* sect. *Martiella*.
Host: Bertholletia
Cephalosporium carpogenum Ruehle
Distr.: PA, northwestern North America.
Notes: Apple rot fungi (Ruehle. Phytopathology 21:1141. 1931).
Hosts: Malus, Pyrus
Cephalosporium cinnamomeum M.B. Linn, nom. nud.
Distr.: FL, NY(type, from a greenhouse).
Notes: According to Gams (*Cephalosporium*-artige Schimmelpilze [Hyphomycetes]. p. 207. 1971), this may be a *Cylindrocarpon*.

Host: Syngonium
Cephalosporium dieffenbachiae M.B. Linn, nom. nud.
Distr.: NY(greenhouse, type).
Notes: According to Gams (*Cephalosporium*-artige Schimmelpilze [Hyphomycetes]. p. 208. 1971), this is a *Cylindrocarpon*.
Host: Dieffenbachia

Ceraceomyces Jülich. Basidiomycetes, Aphyllophorales

Notes: Eriksson & Ryvarden (The Corticiaceae of North Europe 2:202 1973); Jülich (Willdenowia Beih. 7:146. 1972).
Ceraceomyces borealis (Romell) J. Eriksson & Ryvarden
Merulius borealis Romell
Athelia borealis (Romell) Parmasto
Distr.: Northern hemisphere.
Notes: On wood.
Hosts: Betula, Picea, Pinus, Populus
Ceraceomyces serpens (Fr.:Fr.) Ginns
Merulius serpens Fr.:Fr.
Byssomerulius serpens (Fr.:Fr.) Parmasto
Ceraceomerulius serpens (Fr.:Fr.) J. Eriksson
Byssomerulius ceracellus (Berk. & M.A. Curtis) Parmasto
Merulius ceracellus Berk. & M.A. Curtis
Byssomerulius farlowii (Burt) Parmasto
Merulius farlowii Burt
Athelia porinoides (Fr.:Fr.) Parmasto
Distr.: Temperate northern hemisphere.
Notes: On wood (Ginns. Canad. J. Bot. 54:147. 1976).
Hosts: Abies, Acer, Larix, Magnolia, Picea, Pinus, Platanus, Populus, Quercus, Thuja
Ceraceomyces sublaevis (Bres.) Jülich
Corticium sublaeve Bres.
Athelia sublaevis (Bres.) Parmasto
Peniophora sublaevis (Bres.) Höhn. & Litsch.
Distr.: Northern hemisphere.
Notes: On wood. This species also causes spoilage of flax straw (*Linum usitatissimum*) (Brown & Sharma, Trans. Brit. Mycol. Soc. 86:173-175. 1986).
Host: Abies
Ceraceomyces sulphurinus (P. Karst.) J. Eriksson & Ryvarden
Peniophora sulphurina (P. Karst.) Höhn. & Litsch.
Phanerochaete sulphurina (P. Karst.) Budington & R.L. Gilbertson
Distr.: Temperate northern hemisphere; India.
Notes: On conifers.
Host: Pinus
Ceraceomyces tessulatus (Cooke) Jülich
Corticium tessulatum Cooke
Athelia tessulata (Cooke) Donk
Distr.: Northern hemisphere.
Notes: On wood.
Hosts: Alnus, Picea, Pinus, Pseudotsuga

Ceratobasidium D.P. Rogers. Basidiomycetes, Tulasnellales

Notes: Talbot (Persoonia 3:371. 1965).
Ceratobasidium anceps (Bres. & Syd.) H. Jacks.
Tulasnella anceps Bres. & Syd. in Syd.
Corticium anceps (Bres. & Syd.) M. Gregor
Sclerotium deciduum J.J. Davis (anamorph)
Distr.: Central and northern North America, cooler parts; Europe.
Notes: Web-blight on leaves and petioles of ferns, monocots, and dicots.
Hosts: Adiantum, Aralia, Aster, Clintonia, Cornus, Cryptotaenia, Diervilla, Eupatorium, Fragaria, Galium, Hydrophyllum, Lysimachia, Matteuccia, Monarda, Onoclea, Pteridium, Ranunculus, Rubus, Rudbeckia, Salix, Senecio, Solidago, Thelypteris, Trientalis, Triosteum, Ulmus, Urtica, Verbena, Veronicastrum
Ceratobasidium cornigerum (Bourd.) D.P. Rogers
Corticium cornigerum Bourd.
Distr.: Northern hemisphere.
Notes: On wood.
Hosts: Quercus, Rhus
Ceratobasidium ramicola Tu, D.A. Roberts & Kimbrough
Rhizoctonia ramicola G.F. Weber & D.A. Roberts (anamorph)
Distr.: FL.
Notes: Threadblight disease.
Host: Pittosporum

Ceratocystiopsis Upadhyay & Kendrick. Pyrenomycetes, Ophiostomatales

Notes: Upadhyay (A monograph of *Ceratocystis* and *Ceratocystiopsis*. 1981).
Ceratocystiopsis alba (DeVay, R.W. Davidson, & W.J. Moller) Upadhyay
Ceratocystis alba DeVay, R.W. Davidson, & W.J. Moller
Distr.: Central and western North America.
Hosts: Juglans, Picea, Populus, Prunus
Ceratocystiopsis minuta (Siemaszko) Upadhyay & Kendrick
Ceratocystis minuta (Siemaszko) J. Hunt
Ceratocystis dolominuta H.D. Griffin
Hyalorhinocladiella sp. (anamorph)
Distr.: Temperate North America; Europe.
Hosts: Pinus, Pseudotsuga
Ceratocystiopsis minuta-bicolor (R.W. Davidson) Upadhyay & Kendrick
Ceratocystis minuta-bicolor R.W. Davidson
Ceratocystis pallida H.D. Griffin
Hyalorhinocladiella minuta-bicolor Upadhyay & Kendrick (anamorph)
Distr.: Western United States.
Host: Pinus
Ceratocystiopsis ranaculosus J.R. Bridges & T.J. Perry
Distr.: Southern United States.
Notes: Isolated from beetle-infested pine trees (Bridges & Perry. Mycologia 79:630. 1987).
Host: Pinus
Ceratocystiopsis retusi (R.W. Davidson & T.E. Hinds) Upadhyay
Ceratocystis retusi R.W. Davidson & T.E. Hinds
Distr.: CO.
Host: Populus

Ceratocystis Ellis & Halst. Pyrenomycetes, Ophiostomatales

Notes: Griffin (Canad. J. Bot. 46:689. 1968); Upadhyay (A monograph of *Ceratocystis* and *Ceratocystiopsis*. 1981).
Ceratocystis adiposa (E.J. Butler) C. Moreau
Ceratostomella adiposa (E.J. Butler) Sartoris
Endoconidiophora adiposa (E.J. Butler) R.W. Davidson
Ophiostoma adiposa (E.J. Butler) Nannf.
Ceratocystis major (Van Beyma) C. Moreau
Chalara sp. (anamorph)
Distr.: Warm-temperate to tropical regions.
Host: Saccharum
Ceratocystis angusticollis E. Wright & H.D. Griffin
Sporothrix sp. (anamorph)
Distr.: Northern North America.
Host: Pseudotsuga
Ceratocystis brunnea R.W. Davidson
Distr.: CO.
Notes: A doubtful species (Upadhyay. A monograph of *Ceratocystis* and *Ceratocystiopsis*. p. 137. 1981).
Host: Abies
Ceratocystis cana (Münch) Moreau
Ceratostomella cana Münch
Ophiostoma canum (Münch) Syd. & P. Syd.
Pachnodium canum Upadhyay & Kendrick (anamorph)
Distr.: MN; Europe.
Host: Populus
Ceratocystis coerulescens (Münch) Bakshi
Endoconidiophora coerulescens Münch
Ophiostoma coerulescens (Münch) Syd. & P. Syd.
Endoconidiophora coerulescens Münch f. *douglasii* R.W. Davidson
Ceratocystis virescens (R.W. Davidson) C. Moreau
Endoconidiophora virescens R.W. Davidson

Chalara ungeri Sacc. (anamorph)
Distr.: Temperate North America; Europe.
Hosts: Acer, Fagus, Liquidambar, Liriodendron, Magnolia, Nyssa, Picea, Pinus, Pseudotsuga, Quercus
Ceratocystis denticulata R.W. Davidson
Sporothrix sp. (anamorph)
Distr.: CO.
Notes: R.W. Davidson (Mycologia 71:1088. 1979).
Host: Pinus
Ceratocystis fagacearum (T.W. Bretz) J. Hunt
Endoconidiophora fagacearum T.W. Bretz
Chalara quercina B.W. Henry (anamorph)
Distr.: Eastern North America.
Notes: Oak wilt.
Host: Quercus
Ceratocystis fimbriata Ellis & Halst.
Ceratostomella fimbriata (Ellis & Halst.) J.A. Elliott
Endoconidiophora fimbriata (Ellis & Halst.) R.W. Davidson
Ophiostoma fimbriata (Ellis & Halst.) Nannf.
Ceratocystis variospora (R.W. Davidson) C. Moreau
Endoconidiophora variospora R.W. Davidson
Ophiostoma variospora (R.W. Davidson) Arx
Chalara sp. (anamorph)
Distr.: Cosmopolitan.
Notes: Causes a variety of diseases (Morgan-Jones. C.M.I. Descr. 141. 1967).
Hosts: Ipomoea, Populus, Prunus, Quercus, Syngonium
Ceratocystis fimbriata Ellis & Halst. f. **platani** J.M. Walter
Endoconidiophora fimbriata (Ellis & Halst.) R.W. Davidson f. *platani* J.M. Walter in J.M. Walter et al.
Distr.: Eastern United States.
Notes: Sycamore canker.
Host: Platanus
Ceratocystis fraxinopennsylvanica T.E. Hinds in T.E. Hinds & R.W. Davidson
Calosphaeria fraxinopennsylvanica (T.E. Hinds) Upadhyay, nom. inval.
Distr.: ND.
Host: Fraxinus
Ceratocystis hyalothecium R.W. Davidson
Distr.: WY.
Host: Pinus
Ceratocystis leucocarpa R.W. Davidson
Sporothrix sp. (anamorph)
Distr.: Western North America.
Hosts: Abies, Pinus
Ceratocystis microspora (Arx) R.W. Davidson
Ophiostoma microspora Arx
Ceratostomella microspora R.W. Davidson non Ellis & Everh., hom. illeg.
Ceratocystis perparvispora J. Hunt
Sporothrix sp. (anamorph)
Distr.: Central and eastern United States.
Hosts: Castanea, Fagus, Quercus
Ceratocystis moniliformis (Hedgc.) C. Moreau
Ceratostomella moniliformis Hedgc.
Endoconidiophora moniliformis (Hedgc.) R.W. Davidson
Ophiostoma moniliforme (Hedgc.) Syd. & P. Syd.
Chalara sp. (anamorph)
Distr.: North America; Africa; West Indies.
Notes: On cacao and a variety of hardwoods (Morgan-Jones. C.M.I. Descr. 142. 1967).
Hosts: Liquidambar, Nyssa, Pinus, Populus
Ceratocystis olivaceapinii R.W. Davidson
Graphium sp. (anamorph)
Distr.: Western United States.
Notes: According to the guidelines proposed by De Hoog & Scheffer (Mycologia 76:292. 1984), this taxon should be transferred to *Ophiostoma*.
Host: Pinus
Ceratocystis paradoxa (Dade) C. Moreau
Ceratostomella paradoxa Dade
Endoconidiophora paradoxa (De Seyn.) R.W. Davidson
Ophiostoma paradoxum (Dade) Nannf.
Chalara paradoxa (De Seyn.) Sacc. (anamorph)
Distr.: Subtropical and tropical regions.
Notes: Morgan-Jones (C.M.I. Descr. 143. 1967).
Hosts: Ananas, Arecaceae, Cocos, Ficus, Hemerocallis, Musa, Phoenix, Sabal, Saccharum
Ceratocystis piceaperda (Rumbold) Moreau
Ceratostomella piceaperda Rumbold
Grosmannia piceaperda (Rumbold) Goidanich
Ophiostoma piceaperdum (Rumbold) Arx
Ceratocystis europhioides E. Wright & Cain
Leptographium sp. (anamorph)
Distr.: Northern North America.
Hosts: Picea, Pseudotsuga
Ceratocystis radicicola (D.E. Bliss) C. Moreau
Ceratostomella radicicola D.E. Bliss
Ophiostoma radicicola (D.E. Bliss) Arx
Chalara sp. (anamorph)
Distr.: CA; Mexico.
Notes: Wilt and root rot of date palm.
Host: Phoenix
Ceratocystis sagmatospora E. Wright & Cain
Phialographium sagmatosporum Upadhyay & Kendrick (anamorph)
Distr.: North America.
Notes: In wood of conifers. According to the criteria proposed by De Hoog & Scheffer (Mycologia 76:292. 1984), this taxon should be transferred to *Ophiostoma*.
Host: Pinus
Ceratocystis serpens (Goidanich) Moreau
Leptographium sp. (anamorph)
Distr.: MN.
Notes: Upadhyay (A monograph of *Ceratocystis* and *Ceratocystiopsis*. p. 137. 1981) considered this a doubtful species.
Host: [Populus]
Ceratocystis seticollis R.W. Davidson
Hyalorhinocladiella sp. (anamorph)
Distr.: NY.
Host: Tsuga
Ceratocystis tenella R.W. Davidson
Ceratocystis capitata H.D. Griffin
Ceratocystis coronata Olchowecki & J. Reid
Sporothrix sp. (anamorph)
Distr.: Temperate North America.
Notes: According to criteria presented by De Hoog & Scheffer (Mycologia 76:292. 1984), this taxon should be transferred to *Ophiostoma*.
Host: Picea

Ceratophacidium J. Reid & Pirozynski. Discomycetes, Rhytismatales

Ceratophacidium aristosporum (Bonar) J. Reid & Pirozynski
Pezizella aristospora Bonar
Distr.: CA.
Notes: Reid & Pirozynski (Canad. J. Bot. 44:645. 1966).
Host: Sequoiadendron

Ceratophorum Sacc. Hyphomycetes

Ceratophorum epiphyllum (Schwein.) Sacc.
Coryneum epiphyllum Schwein.
Sporidesmium epiphyllum (Schwein.) Berk. & M.A. Curtis
Distr.: Central and eastern North America.
Notes: On leaves (Sutton. Mycol. Pap. 138:107. 1975).
Hosts: Carya, Castanea
Ceratophorum uncinatum (G.W. Clinton) Sacc.
Clasterosporium uncinatum G.W. Clinton in Peck
Distr.: Central and eastern North America.
Notes: On leaves of hardwoods.
Hosts: Carya, Quercus

Ceratosporella Höhn. Hyphomycetes

Ceratosporella bicornis (Morg.) Höhn.
Triposporium bicorne Morg.
Distr.: OH(type).

Host: Zea

Ceratostoma Fr. Pyrenomycetes, Sordariales

Notes: This name has been rejected, but not all names have been redisposed.
Ceratostoma marylandicum Petr.
Distr.: FL, MD.
Host: Pinus
Ceratostoma subrufum Ellis & Everh.
Distr.: NJ(type).
Host: Rubus

Ceratostomella Sacc. Pyrenomycetes, Ophiostomatales

Notes: This genus is a synonym of *Endoxyla*. Most species have been transferred to *Ceratocystis* or *Ophiostoma*.
Ceratostomella hystricina (Cooke) Sacc.
Distr.: SC.
Host: Ficus
Ceratostomella microspora Ellis & Everh.
Distr.: Northeastern United States.
Notes: Not a *Ceratostomella* (Hunt. Lloydia 19:55. 1956).
Hosts: Castanea, Fagus, Quercus
Ceratostomella nyssicola (Berk. & M.A. Curtis) Sacc.
Distr.: PA.
Host: Nyssa

Cercophora Fuckel. Pyrenomycetes, Sordariales

Cercophora appalachianensis O. Hilber & R. Hilber
Distr.: IL, VA.
Notes: On bark and decorticated wood (Hilber et al. Mycotaxon 30:269. 1987).
Host: Tilia

Cercoseptoria Petr. Hyphomycetes

Notes: Leaf spot diseases. Deighton (Mycol. Pap. 140:1. 1976; Mycol. Pap. 151:1. 1983); Evans (Mycol. Pap. 153:1. 1984).
Cercoseptoria andropogonis H.C. Greene
Distr.: WI(type).
Host: Schizachyrium
Cercoseptoria blephiliae H.C. Greene
Distr.: WI(type).
Hosts: Blephilia, Pycnanthemum
Cercoseptoria cerastii H.C. Greene
Distr.: WI(type).
Host: Cerastium
Cercoseptoria crataegi (Ellis & Everh.) J.J. Davis
Cylindrosporium crataegi Ellis & Everh.
Phloeospora crataegi (Ellis & Everh.) Höhn.
Distr.: Temperate North America.
Host: Crataegus
Cercoseptoria crepidis Solheim
Distr.: WY.
Host: Crepis
Cercoseptoria iridis (Ellis & Halst.) H.C. Greene
Cylindrosporium iridis Ellis & Halst.
Distr.: Central and eastern United States.
Host: Iris
Cercoseptoria ocellata Deighton
Cercospora theae Breda de Haan
Distr.: LA; Indonesia; Nigeria; Sri Lanka; Tanzania.
Host: Camellia
Cercoseptoria prunicola (Ellis & Everh.) J. Yen
Cercospora prunicola Ellis & Everh.
Cercospora pruni-yedoensis Sawada
Distr.: Southeastern United States; central and eastern Asia.
Host: Prunus
Cercoseptoria scirpi H.C. Greene
Distr.: WI(type).
Host: Scirpus
Cercoseptoria septoriopsidis (Dearn. & Overh.) O. Constantinescu
Septoria septoriopsidis Dearn. & Overh.
Distr.: PA.
Host: Betula
Cercoseptoria sesami (Hansf.) Deighton
Cylindrosporium sesami Hansf.
Cercospora sesamicola Mohanty
Distr.: Widespread in tropical regions.
Host: Sesamum
Cercoseptoria valerianae Solheim
Distr.: WY.
Host: Valeriana

Cercospora Fresen. Hyphomycetes

Notes: Most of the about 2,000 described species of *Cercospora* cause leaf spots (Chupp. A monograph of the fungus genus *Cercospora*. 1953; Pollack. Mycol. Mem. 12. 1987).
Cercospora abeliae Plakidas non Katsuki, hom. illeg.
Distr.: LA(type).
Host: Abelia
Cercospora abutilonis Tehon & E.Y. Daniels
Distr.: IL(type).
Notes: According to Chupp (A monograph of the fungus genus *Cercospora*. p. 368. 1953), the type specimen is a *Colletotrichum*.
Host: Abutilon
Cercospora acalyphae Peck
Cercospora acalypharum Tharp
Distr.: Central and eastern United States; Brazil; Japan.
Host: Acalypha
Cercospora acetosellae Ellis
Distr.: Central and eastern North America.
Host: Rumex
Cercospora aciculina Chupp
Distr.: HI.
Host: Ageratum
Cercospora acnidae Ellis & Everh.
Distr.: DE, WI.
Host: Amaranthus
Cercospora adusta Heald & F.A. Wolf
Distr.: Southeastern United States.
Host: Ligustrum
Cercospora aesculina Ellis & Everh.
Distr.: FL, WV.
Notes: According to Chupp (A monograph of the fungus genus *Cercospora*. p. 509. 1953), this species is a *Pseudocercospora*.
Host: Aesculus
Cercospora ageratoides Ellis & Everh.
Distr.: Central and eastern United States.
Host: Eupatorium
Cercospora agnostoica Speg.
Distr.: SD; Brazil(type).
Host: Borago
Cercospora agrostidis Atk.
Distr.: Central and eastern United States.
Hosts: Agrostis, Sphenopholis
Cercospora alismatis Ellis & Holw.
Distr.: Central and eastern United States.
Hosts: Alisma, Sagittaria
Cercospora alternantherae Ellis & Langl.
Distr.: LA; West Indies.
Host: Alternanthera
Cercospora althaeina Sacc.
Cercospora kellermanii Bubák
Cercospora ramularia Siemaszko
Distr.: Temperate regions.
Notes: Leaf spot of Malvaceae.
Hosts: Abutilon, Alcea, Callirhoe, Gossypium, Hibiscus, Malva, Modiola, Napaea
Cercospora amaryllidis Ellis & Everh.
Distr.: Southeastern United States; West Indies; Europe.
Hosts: Amaryllis, Hippeastrum, Hymenocallis, Manfreda, Zephyranthes

Cercospora ammanniae Tharp
Distr.: TX(type).
Host: Ammannia
Cercospora ampelopsidis Peck
Cercospora psedericola Tehon
Cercospora pustula Cooke
Distr.: Central and eastern North America; Europe; Japan.
Hosts: Campsis, Parthenocissus
Cercospora angelicae (Sacc. & Scalia) Chupp
Cercospora apii Fresen. var. *angelicae* Sacc. & Scalia
Distr.: AK.
Host: Angelica
Cercospora angraeci Feuillebois & Roum.
Distr.: FL; France(type).
Hosts: Cattleya, Laelia, Odontoglossum, Oncidium, Orchidaceae
Cercospora angulata G. Wint.
Distr.: Central and eastern United States; China.
Hosts: Philadelphus, Ribes
Cercospora annulata Cooke
Distr.: FL; India(type).
Host: Ficus
Cercospora anomala Ellis & Halst.
Distr.: IA.
Notes: Chupp (A monograph of the fungus genus *Cercospora*. p. 121. 1953) indicated that this fungus is a *Cladosporium* spp.
Host: Verbesina
Cercospora antirrhini Muller & Chupp
Distr.: HI, central and southern United States; Guatemala.
Notes: *Cercospora* blight of snapdragon.
Host: Antirrhinum
Cercospora apii Fresen.
Distr.: Worldwide in celery-growing regions.
Notes: Ellis (Dematiaceous Hyphomycetes. p. 278. 1971) suggested that several other *Cercospora* spp. were synonyms.
Hosts: Apium, Lactuca, Nicotiana
Cercospora apii Fresen. f. sp. **clerodendri** Sobers & A.P. Martinez
Distr.: FL(type).
Host: Clerodendrum
Cercospora apocyni Ellis & Kellerm.
Cercosporella apocyni (Ellis & Kellerm.) Trel.
Distr.: Central and eastern United States; West Indies.
Host: Apocynum
Cercospora aquilegiae Kellerm. & Swingle
Distr.: Central United States; Europe.
Host: Aquilegia
Cercospora arachidicola S. Hori
Mycosphaerella arachidis Deighton (teleomorph)
Distr.: Throughout range of host.
Notes: Early leaf spot of peanuts (Mulder & Holliday. C.M.I. Descr. 411. 1974).
Host: Arachis
Cercospora arboriae Tharp
Distr.: TX.
Hosts: Ampelopsis, Cissus
Cercospora arcti-ambrosiae Halst.
Cercospora arctii F. Stevens
Distr.: HI, central United States; Japan.
Hosts: Ambrosia, Arctium
Cercospora argythamniae Dearn. & House
Distr.: OK(type).
Host: Argythamnia
Cercospora ariae Fuckel
Distr.: FL; Germany(type).
Host: Rhaphiolepis
Cercospora armoraciae Sacc.
Distr.: MN, WV; Italy(type).
Host: Armoracia
Cercospora asclepiadis Ellis
Cercospora maculans Pat.
Distr.: United States; Tunisia.
Host: Asclepias
Cercospora asclepiodorae Ellis & Kellerm.
Distr.: Central and southeastern United States.
Notes: Reports of this fungus on *Asclepias tuberosa* and *A. jamesii* were probably misdeterminations of *Cercospora asclepiodis* Ellis.
Host: Asclepias
Cercospora asparagi Sacc.
Distr.: Temperate North America; West Indies; temperate Europe; Africa; eastern Asia.
Host: Asparagus
Cercospora asplenii Jaap
Mycosphaerella asplenii Thirumalachar & Govindu non (Auersw.) Lindau, hom. illeg. (teleomorph)
Distr.: FL; India; Yugoslavia(type).
Host: Asplenium
Cercospora asprellae Ellis & B.T. Galloway
Distr.: IL.
Notes: A doubtful taxon (Chupp. A monograph of the fungus genus *Cercospora*. p. 256. 1954).
Host: Hystrix
Cercospora asterata Atk.
Distr.: Central and southeastern United States.
Host: Aster
Cercospora astragali Woronichin
Cercospora astragali Sawada
Distr.: Central United States; Japan; Soviet Union.
Host: Astragalus
Cercospora atrogrisea Ellis & Everh.
Distr.: NJ.
Host: Raphanus
Cercospora atropae Kvashnina
Distr.: VA; Soviet Union.
Host: Atropa
Cercospora averrhoae Petch
Distr.: FL; Sri Lanka(type).
Host: Averrhoa
Cercospora avicennae Chupp
Distr.: Central and eastern United States; China.
Host: Abutilon
Cercospora avicularis G. Wint.
Distr.: North America; Venezuela; Europe.
Notes: Pirozynski (Fungi Canadenses 62. 1975).
Host: Polygonum
Cercospora baccharidis Ellis & Everh.
Distr.: CA.
Notes: This species may be a *Cercosporella* or a *Ramularia* (Chupp. A monograph of the fungus genus *Cercospora*. p. 122. 1953).
Host: Baccharis
Cercospora bacilligera (Berk. & Broome) Fresen.
Fusarium bacilligerum (Berk. & Broome) Sacc.
Cercospora aeruginosa Cooke
Distr.: Eastern North America; Asia; Temperate Europe.
Host: Rhamnus
Cercospora barbarea (Sacc.) Chupp in H.C. Greene
Distr.: Northcentral and northeastern United States; Italy.
Hosts: Arabis, Barbarea
Cercospora begoniae S. Hori
Distr.: FL; Japan(type); Poland.
Host: Begonia
Cercospora bellynckii (Westend.) Niessl
Cladosporium bellynckii Westend.
Distr.: Europe, southern Soviet Union, China.
Notes: Reports of this fungus in North America are probably misdeterminations (Chupp. A monograph of the fungus genus *Cercospora*. p. 70. 1953).
Host: Vincetoxicum
Cercospora beticola Sacc.
Cercospora anthelmintica Atk.
Cercospora chenopodiicola Bres.
Cercospora spinaciae Oudem.
Distr.: Cosmopolitan.
Notes: Kirk (C.M.I. Descr. 721. 1982).
Hosts: Bassia, Beta, Chenopodium, Cycloloma, Malva, Polygonum, Proboscidea, Spinacia
Cercospora bicolor G. Wint.
Distr.: FL; Bermuda; Portugal(type).
Host: Coccoloba
Cercospora bidentis Tharp
Distr.: Cosmopolitan.
Hosts: Bidens, Coreopsis

Cercospora biformis Peck
Distr.: Southeastern United States; West Indies.
Host: Passiflora
Cercospora bischofiae W. Yamamoto
Distr.: FL; Taiwan(type).
Host: Bischofia
Cercospora bixae Allesch. & F. Noack
Distr.: FL; West Indies; South America; Taiwan.
Host: Bixa
Cercospora blephiliae Chupp & H.C. Greene
Distr.: WI(type).
Hosts: Blephilia, Lycopus
Cercospora boehmeriae Peck
Cercospora boehmeriae Fukui
Distr.: Eastern United States; Bermuda; Japan; The Philippines; Soviet Union.
Host: Boehmeria
Cercospora boutelouae Chupp & H.C. Greene
Distr.: Central and western United States.
Hosts: Berberis, Bouteloua
Cercospora brachiata Ellis & Everh.
Distr.: Southeastern United States; West Indies; eastern Africa; eastern Asia; Soviet Union.
Host: Amaranthus
Cercospora brachypoda Speg.
Distr.: LA, PR; Argentina(type).
Host: Abutilon
Cercospora brachypus Ellis & Everh.
Mycosphaerella angulata W.A. Jenkins (teleomorph)
Distr.: Southeastern United States; China.
Notes: Angular leaf spot of Muscadine grapes (Jenkins. Phytopathology 32:71. 1942).
Host: Vitis
Cercospora brassicicola Henn.
Cercospora bloxami Berk. & Broome
Distr.: Central and eastern United States; West Indies; Africa; southern Asia; East Indies.
Notes: Kirk (C.M.I. Descr. 722. 1982).
Hosts: Brassica, Sinapis
Cercospora briareus Ellis & Everh.
Distr.: DE, MD, WI.
Host: Asclepias
Cercospora broussonetiae Chupp & Linder
Distr.: FL; China.
Host: Broussonetia
Cercospora brunkii Ellis & B.T. Galloway
Distr.: Eastern North America, especially common in the southeastern United States; The Philippines.
Hosts: Geranium, Pelargonium
Cercospora bunchosiae Chupp & Muller
Distr.: FL, HI; Venezuela(type).
Host: Malpighia
Cercospora caespitosa Ellis & Everh.
Distr.: FL, MS.
Host: Chloris
Cercospora calatheae Viégas & Chupp
Distr.: FL; Brazil(type).
Host: Calathea
Cercospora calendulae Sacc.
Distr.: Central and eastern United States; Africa; Europe; India.
Host: Calendula
Cercospora callae Peck & G.W. Clinton
Cercospora pachyspora Ellis & Everh.
Distr.: Central and eastern North America.
Hosts: Alisma, Calla, Peltandra, Zantedeschia
Cercospora callicarpae Cooke
Distr.: Southern United States; eastern Asia.
Host: Callicarpa
Cercospora calotropidis Ellis & Everh.
Distr.: Widespread in tropical regions.
Notes: Ellis (More dematiaceous Hyphomycetes. p. 245. 1976).
Host: Vaccinium
Cercospora calystegiae Speg.
Cercospora tuberculella J.J. Davis
Distr.: WI; Argentina.
Hosts: Calystegia, Convolvulus
Cercospora camptosori J.J. Davis
Distr.: WI.
Hosts: Camptosorus, Dryopteris
Cercospora camptothecae F.L. Tai
Distr.: FL; China(type).
Host: Camptotheca
Cercospora canescens Ellis & G. Martin
Cercospora vignicaulis Tehon
Distr.: Eastern United States; subtropcial and tropical regions.
Notes: Mulder & Holliday (C.M.I. Descr. 462. 1975).
Hosts: Amaranthus, Glycine, Lablab, Lycopersicon, Phaseolus, Ricinus, Vicia, Vigna
Cercospora cannabis K. Hara & Fukui
Distr.: MO, WI; India; Japan(type).
Host: Cannabis
Cercospora capitati Tharp
Distr.: Southeastern United States.
Host: Croton
Cercospora capsici Heald & F.A. Wolf
Distr.: Southern United States; subtropical and tropical regions.
Notes: Frog-eye leaf spot, fruit rot (Kirk. C.M.I. Descr. 723. 1982).
Host: Capsicum
Cercospora caracallae (Speg.) Vassiljevsky & Karakulin
Cercosporina caracallae Speg.
Cercospora caracallae (Speg.) Chupp
Cercospora caracallae (Speg.) H.C. Greene
Distr.: WI; South America.
Host: Vigna
Cercospora caricis Oudem.
Cercospora caricina Ellis & Dearn.
Cercospora caricis Dearn. & House non Oudem.
Cercospora microstigma Sacc.
Distr.: Central and eastern North America; western Europe.
Notes: Chupp (A monograph of the fungus genus *Cercospora*. p. 191. 1953) suggested that records on *Cyperus* spp. may be misdeterminations, but Blaney et al. (Mycologia 80:418. 1988) considered this species for biological control of *Cyperus*.
Hosts: Carex, Cyperus
Cercospora carolinensis Tharp
Distr.: TX.
Host: Solanum
Cercospora carotae (Pass.) Solheim
Distr.: Temperate regions.
Host: Daucus
Cercospora carveriana Sacc. & D. Sacc.
Cercospora richardsoniae Ellis & Everh.
Distr.: Southeastern United States.
Host: Richardia
Cercospora catalpae G. Wint.
Distr.: Central and eastern United States; China.
Host: Catalpa
Cercospora caulophylli Peck
Distr.: Central and eastern United States.
Hosts: Anemonella, Caulophyllum
Cercospora ceanothi Kellerm. & Swingle
Distr.: Central United States.
Host: Ceanothus
Cercospora celosiae Syd.
Distr.: Southeastern United States; South America; eastern Africa; eastern Asia.
Host: Celosia
Cercospora centaureae Died.
Distr.: FL; Bermuda; Venezuela; Germany.
Host: Centaurea
Cercospora cephalanthi Ellis & Kellerm.
Ramularia cephalanthi (Ellis & Kellerm.) Heald
Cercospora perniciosa Heald & F.A. Wolf
Distr.: Central and eastern United States.
Host: Cephalanthus
Cercospora cercidicola Ellis
Cercospora cercidicola Ellis var. *coremioides* Tehon
Mycosphaerella cercidicola (Ellis & Kellerm.) F.A. Wolf (teleomorph)
Distr.: Temperate North America.

Host: Cercis
Cercospora chamaecrista Ellis & Kellerm.
Distr.: KS.
Host: Cassia
Cercospora chenopodii Fresen.
Distr.: Temperate northern hemisphere; India; Kenya.
Host: Chenopodium
Cercospora chionanthi Ellis & Everh. in Millsp. & L. Nutt.
Distr.: Eastern United States.
Host: Chionanthus
Cercospora chionea Ellis & Everh.
Cercosporella chionea (Ellis & Everh.) Sacc.
Cercospora cercidis W.W. Ray
Distr.: Central and eastern United States; China
Host: Cercis
Cercospora chrysanthemi Heald & F.A. Wolf
Distr.: Central and eastern United States; South America; eastern Asia; probably wherever *Chrysanthemum* is cultivated.
Host: Chrysanthemum
Cercospora chrysobalani Ellis & Everh.
Distr.: Southeastern United States; West Indies; western Africa.
Host: Chrysobalanus
Cercospora cichorii J.J. Davis
Cercospora cichorii-intybi Woronichin
Distr.: Northeastern North America; Bermuda; Argentina; Soviet Union.
Host: Cichorium
Cercospora cinchonae Ellis & Everh.
Distr.: LA; central Africa.
Host: Cinchona
Cercospora circumscissa Sacc.
Cercospora cerasella Sacc.
Mycosphaerella cerasella Aderhold (teleomorph)
Distr.: Temperate and subtropical regions
Notes: Shot hole or leaf spot (Wilson. C.M.I. Descr. 911. 1987).
Host: Prunus
Cercospora cirsii Ellis & Everh.
Distr.: WA.
Host: Cirsium
Cercospora citrullina Cooke
Cercospora cucurbitae Ellis & Everh.
Distr.: Eastern United States; subtropical and tropical regions.
Notes: Leaf spot of Cucurbitaceae (Little. C.M.I. Descr. 917. 1987).
Hosts: Citrullus, Cucumis, Cucurbita, Lagenaria, Luffa, Melothria, Momordica, Sechium
Cercospora clavata (W. Gerard) Cooke
Helminthosporium clavatum W. Gerard
Cercospora clavata (W. Gerard) Peck
Virgasporium clavatum (W. Gerard) Cooke
Cercospora incarnata Ellis & Everh.
Distr.: Eastern North America to UT.
Host: Asclepias
Cercospora clavicarpa Ellis & Everh.
Distr.: CA.
Host: Stephanomeria
Cercospora clemensiae Graff
Distr.: HI.
Notes: The place of publication of this name could not be found.
Host: Eragrostis
Cercospora cleomes Ellis & Halst.
Distr.: Northeastern United States; Venezuela; India.
Host: Cleome
Cercospora cocculicola W.W. Ray
Distr.: OK.
Host: Cocculus
Cercospora cochlospermi R.E.D. Baker & W.T. Dale
Distr.: FL; Trinidad(type); Venezuela.
Host: Cochlospermum
Cercospora coffeicola Berk. & Cooke
Cercospora coffeae A. Zimmerm.
Distr.: Subtropical and tropical regions.
Notes: Brown eye spot of coffee (Mulder & Holliday. C.M.I. Descr. 415. 1974).
Host: Coffea
Cercospora coleosanthi Ellis & Everh.
Distr.: Western United States.
Host: Brickellia
Cercospora coleroides Sacc.
Distr.: FL; Mexico(type); Guatemala.
Host: Casimiroa
Cercospora comandrae Ellis & Dearn.
Distr.: Northeastern North America.
Host: Comandra
Cercospora comari Peck
Distr.: Eastern North America; Europe.
Host: Potentilla
Cercospora commonsii Sacc.
Cercospora stylosanthis Ellis & Everh. non Speg., hom. illeg.
Distr.: DE.
Host: Stylosanthes
Cercospora condensata Ellis & Kellerm.
Distr.: Central and eastern United States.
Host: Gleditsia
Cercospora confluens Lieneman
Cercospora crataegi Heald & F.A. Wolf non Sacc. & C. Massal., hom. illeg.
Distr.: TX(type).
Notes: Chupp (A monograph of the fungus genus *Cercospora*. p. 474. 1953) suggested this might be a *Pseudocercospora*.
Host: Crataegus
Cercospora conocarpi Chupp & Muller
Distr.: FL; Venezuela(type).
Host: Conocarpus
Cercospora consociata G. Wint.
Distr.: Central and southeastern United States; China.
Hosts: Dyschoriste, Ruellia
Cercospora cordylines Henn.
Cercospora cordylines Speg.
Distr.: FL, HI; Brazil.
Host: Cordyline
Cercospora coreopsidis W.W. Ray
Distr.: OK.
Host: Coreopsis
Cercospora corni J.J. Davis
Distr.: Central United States.
Notes: Possibly a *Helminthosporium* (Chupp. A monograph of the fungus genus *Cercospora*. p. 175. 1953).
Host: Cornus
Cercospora cornicola Tracy & Earle
Distr.: Southeastern United States; Japan.
Host: Cornus
Cercospora corylina W.W. Ray
Distr.: OK.
Host: Corylus
Cercospora cotizensis Muller & Chupp
Cercospora crotalariae Syd. non Sacc., hom. illeg.
Distr.: FL; Venezuela.
Host: Crotalaria
Cercospora crassoides J.J. Davis
Distr.: Central and eastern United States.
Notes: Probably an *Alternaria* (Chupp. A monograph of the fungus genus *Cercospora*. p.33. 1953).
Host: Froelichia
Cercospora crataegi Sacc. & C. Massal.
Cercospora apiifoliae Tharp
Distr.: TX; southern Europe; Japan.
Host: Crataegus
Cercospora crinospora Atk.
Distr.: AL.
Host: Rhynchospora
Cercospora crotonicola Ellis & Barth.
Distr.: TX.
Host: Croton
Cercospora crotonifolia Cooke
Distr.: AL, SC.
Host: Croton
Cercospora crotonophila Speg.
Distr.: WI; South America.
Host: Croton
Cercospora cruciferarum Ellis & Everh.

Distr.: Southeastern United States; South America.
Notes: Previous reports on *Sisymbrium officinale* were actually based on a *Septoria* sp. (Chupp. A monograph of the fungus genus *Cercospora*. p. 182. 1953).
Hosts: Raphanus, [Sisymbrium]
Cercospora cryptotaeniae Dearn.
Distr.: NY.
Host: Cryptotaenia
Cercospora cydoniae Ellis & Everh.
Cercospora cydoniae Rangel
Distr.: Southern United States; South America.
Host: Chaenomeles
Cercospora cynoglossi Van Hook
Distr.: Northcentral United States.
Hosts: Cynoglossum, Hackelia
Cercospora cypericola Chupp & H.C. Greene
Distr.: WI.
Host: Cyperus
Cercospora cypripedii Ellis & Dearn.
Distr.: Northeastern North America.
Hosts: Cypripedium, Orchidaceae
Cercospora daemonicola R. Sprague
Distr.: OR.
Notes: Probably not a *Cercospora* (Chupp. A monograph of the fungus genus *Cercospora*. p. 63. 1953).
Host: Oplopanax
Cercospora daleae Ellis & Kellerm.
Distr.: KS.
Notes: Probably not a *Cercospora* (Chupp. A monograph of the fungus genus *Cercospora*. p. 300. 1953).
Host: Dalea
Cercospora daturicola (Speg.) W.W. Ray
Cercosporina daturicola Speg.
Distr.: OK; Argentina; China; Soviet Union.
Host: Datura
Cercospora davisii Ellis & Everh.
Mycosphaerella davisii F.R. Jones (teleomorph)
Distr.: Central North America; Pakistan.
Host: Melilotus
Cercospora decumariae Tracy & Earle
Distr.: MS.
Host: Decumaria
Cercospora delphinii Thuem.
Distr.: Temperate regions.
Host: Delphinium
Cercospora demetrioniana G. Wint.
Distr.: HI, southeastern United States; West Indies; France; Uganda.
Host: Crotalaria
Cercospora dendrobii H.C. Burnett
Distr.: FL(type).
Host: Dendrobium
Cercospora depazeoides (Desmaz.) Sacc.
Cercospora sambuci F. Stevens & King in F. Stevens
Cercospora sambucina Ellis & Kellerm.
Distr.: Temperate northern hemisphere.
Host: Sambucus
Cercospora desmodii Ellis & Kellerm.
Distr.: Central and eastern United States; West Indies; France.
Host: Desmodium
Cercospora desmodiicola Atk.
Cercospora melaleuca Ellis & Everh.
Distr.: Central and eastern United States; West Indies.
Host: Desmodium
Cercospora destructiva (Ravenel) Ellis & Everh.
Distr.: Southeastern United States; eastern Asia.
Host: Euonymus
Cercospora deutziae Ellis & Everh.
Distr.: Central and eastern United States; Japan.
Hosts: Abelia, Deutzia
Cercospora diantherae Ellis & Kellerm.
Cercospora jacobinae Mendoza
Distr.: Central and eastern United States; Venezuela; The Philippines.
Hosts: Dianthera, Justicia
Cercospora didymospora Ellis & Barth.
Distr.: KS.
Notes: Chupp (A monograph of the fungus genus *Cercospora*. p. 420. 1953) indicated that this species may be a *Coryneum*.
Host: Oenothera
Cercospora diervillae Ellis & Everh.
Distr.: ME(type), TX.
Notes: Possibly a *Pseudocercospora* (Chupp. A monograph of the fungus genus *Cercospora*. p. 101. 1953).
Host: Diervilla
Cercospora diffusa Ellis & Everh.
Distr.: Central and eastern North America; Central America; northern South America.
Hosts: Lycopersicon, Physalis
Cercospora diodiae Cooke
Distr.: Central and eastern United States; Brazil.
Host: Diodia
Cercospora diodiae-virginianae Atk.
Distr.: Southeastern United States.
Host: Diodia
Cercospora dulcamarae (Peck) Ellis & Everh.
Ramularia dulcamarae Peck
Cercospora dulcarmaraecola Hollós
Distr.: Temperate North America; Europe.
Host: Solanum
Cercospora duplicata Ellis & Everh.
Cercospora capreolata Ellis & Everh.
Cercospora langloisii Sacc.
Cercospora pallida Ellis & Everh.
Distr.: Southeastern United States.
Hosts: Bignonia, Campsis
Cercospora durantae Chupp & Muller
Distr.: FL; Venezuela(type).
Host: Duranta
Cercospora echii G. Wint.
Distr.: Eastern North America; southern Europe.
Host: Echium
Cercospora echinochloae J.J. Davis
Distr.: WI.
Host: Echinochloa
Cercospora echinocystis Ellis & G. Martin
Distr.: HI, central and eastern United States.
Hosts: Benincasa, Echinocystis, Sicyos
Cercospora effusa (Berk. & M.A. Curtis) Ellis
Didymaria effusa (Berk. & M.A. Curtis) Solheim
Cercospora hydropiperis (Thuem.) Speg.
Helminthosporium hydropiperis Thuem.
Cercospora polygonorum Cooke
Distr.: Central and eastern United States; West Indies; South America; China.
Notes: Deighton (Trans. Brit. Mycol. Soc. 86:637. 1986).
Hosts: Lobelia, Polygonum
Cercospora elaeagni Heald & F.A. Wolf
Cercospora carrii Barth., nom. nud.
Distr.: Southeastern United States.
Host: Elaeagnus
Cercospora elaeochroma Sacc.
Distr.: Central and eastern United States.
Host: Asclepias
Cercospora elasticae A. Zimmerm.
Distr.: FL; Venezuela; Java(type).
Host: Ficus
Cercospora eleocharidis J.J. Davis
Distr.: ID, LA, WI(type).
Host: Eleocharis
Cercospora elephantopi Ellis & Everh.
Distr.: Southern United States; neotropical regions.
Host: Elephantopus
Cercospora elongata Peck
Cercosporina elongata (Peck) Speg.
Distr.: Eastern North America; Argentina; Europe.
Notes: On leaves (Sutton & Pons. Mycotaxon 12:208. 1980).
Host: Dipsacus
Cercospora epicoccoides Cooke & Massee
Distr.: FL; Australia(type).
Host: Eucalyptus
Cercospora epidendronis Bolick in Alfieri et al., nom. nud.

Distr.: FL.
Notes: The place of publication of this name could not be found.
Host: Epidendrum
Cercospora epigaeae Ellis & Dearn.
Cercospora epigaeina J.J. Davis
Distr.: Central and eastern North America.
Host: Epigaea
Cercospora epipactidis C. Massal.
Distr.: FL; Germany; Italy(type); Soviet Union.
Host: Orchidaceae
Cercospora erechtitis Atk.
Distr.: Southern United States; West Indies; Central America; northern South America.
Hosts: Erechtites, Hieracium
Cercospora eriobotryae (Enjoji) Sawada
Cercosporina eriobotryae Enjoji
Distr.: FL; Japan; Taiwan(type).
Host: Eriobotrya
Cercospora eriogoni Ellis & Everh.
Distr.: CA.
Host: Eriogonum
Cercospora erysimi J.J. Davis
Distr.: Northern United States.
Host: Erysimum
Cercospora erythrinae Ellis & Everh.
Distr.: Southeastern United States; South America; Africa.
Host: Erythrina
Cercospora erythrinicola Tharp
Distr.: Southeastern United States; Brazil.
Host: Erythrina
Cercospora erythrogena Atk.
Distr.: Southeastern United States; Colombia.
Hosts: Miconia, Rhexia
Cercospora eucalypti Cooke & Massee
Distr.: Subtropical and tropical regions.
Host: Eucalyptus
Cercospora eugeniae (Rangel) Chupp non Sawada, hom. illeg.
Cercosporina eugeniae Rangel
Distr.: FL; Bermuda; Brazil; Taiwan.
Notes: This name is a later homonym of *C. eugeniae* Sawada (Chupp. A monograph of the fungus genus *Cercospora*. p. 406. 1953).
Host: Eugenia
Cercospora euonymi Ellis
Distr.: FL, KY(type).
Notes: According to Chupp (A monograph of the fungus genus *Cercospora*. p. 109. 1953), this may be a *Ramularia*.
Host: Euonymus
Cercospora eupatorii Peck
Distr.: Eastern United States.
Host: Eupatorium
Cercospora euphorbiae Kellerm. & Swingle
Distr.: Central United States; West Indies; South America; eastern Asia; India.
Host: Euphorbia
Cercospora euphorbiaecola Atk.
Distr.: AL(type), NE.
Host: Euphorbia
Cercospora euphorbiaecola Atk. var. ***tragiae*** Tharp
Distr.: Southeastern United States.
Host: Tragia
Cercospora eustomae Peck
Cercospora nepheloides Ellis & Holw.
Distr.: Central and western United States.
Host: Eustoma
Cercospora fagopyri Chupp & E. Müller non Nakata & Takimoto, hom. illeg.
Distr.: IL; Venezuela(type); China; Japan; Uganda.
Notes: This name is illegitimate for several reasons: no latin description, invalid publication, and later homonym of a species described by Nakata & Takimoto.
Host: Fagopyrum
Cercospora festucae Hardison
Distr.: Central and western United States.
Hosts: Bromus, Festuca
Cercospora fici Heald & F.A. Wolf
Distr.: Subtropical and tropical regions.
Host: Ficus
Cercospora ficina Tharp
Distr.: TX(type).
Host: Ficus
Cercospora filiformis (J.J. Davis) Chupp
Cercosporella filiformis J.J. Davis
Cercospora pulsatillae Clements, nom. nud
Distr.: CO, WI.
Hosts: Anemone, Thalictrum
Cercospora fingens J.J. Davis
Distr.: Central United States.
Host: Thalictrum
Cercospora flagellaris Ellis & G. Martin
Distr.: Central and eastern United States, PR; Japan.
Notes: Reports of this species on *Rivina* represent another *Cercospora* species (Chupp. A monograph of the fungus genus *Cercospora*. p.437. 1953).
Hosts: Phytolacca, [Rivina]
Cercospora flagellifera Atk.
Distr.: Southeastern United States; West Indies.
Hosts: Galactia, Gaura
Cercospora floricola Heald & F.A. Wolf
Distr.: TX.
Host: Yucca
Cercospora fraxinites Ellis & Everh.
Distr.: Southeastern United States; Japan.
Host: Fraxinus
Cercospora fuchsiae Chupp & Muller
Distr.: FL; Guatemala; Brazil; Venezuela(type).
Host: Fuchsia
Cercospora fukushiana (Matsuura) W. Yamamoto
Distr.: Central and southeastern United States; West Indies; Central America; Africa; eastern and southeastern Asia; India.
Host: Impatiens
Cercospora fuligniosa Ellis & Kellerm.
Cercospora atra Ellis & Everh.
Distr.: Central and eastern United States; South America.
Host: Diospyros
Cercospora fulvella Heald & F.A. Wolf
Distr.: TX.
Host: Verbesina
Cercospora furfurella Speg.
Cercospora boerhaviae Canonaco
Distr.: TX; West Indies; Central and South America; Central Africa; India; Pakistan.
Host: Boerhavia
Cercospora fusco-virens Sacc.
Distr.: Southern United States; Trinidad.
Host: Passiflora
Cercospora galactiae Ellis & Everh.
Distr.: FL.
Host: Galactia
Cercospora garryae Harkn.
Distr.: CA.
Host: Garrya
Cercospora gaultheriae Ellis & Everh.
Cercospora arctostaphyli J.J. Davis
Distr.: Central and eastern North America.
Hosts: Arctostaphylos, Gaultheria
Cercospora gaurae Kellerm. & Swingle
Distr.: Central and eastern United States.
Host: Gaura
Cercospora gentianae Peck
Distr.: Northcentral and northeastern United States.
Host: Gentiana
Cercospora gentianicola Ellis & Everh.
Distr.: DE(type), WI.
Host: Gentiana
Cercospora geranii Kellerm. & Swingle
Distr.: United States to Mexico.
Host: Geranium
Cercospora gerardiae Ellis & Dearn.
Distr.: Eastern North America.
Host: Aureolaria

Cercospora gerberae Chupp & Viégas
Distr.: Subtropical and tropical regions.
Host: Gerbera
Cercospora gigantea F.E. Fisher
Distr.: FL(type).
Notes: Associated with tar spot of citrus (Fisher. Phytopathology 51:300. 1961).
Host: Citrus
Cercospora glandulosa Ellis & Kellerm.
Distr.: Central and southern United States; South America; China.
Host: Ailanthus
Cercospora glottidiicola Tracy & Earle
Distr.: Southeastern United States.
Host: Sesbania
Cercospora gnaphalii Harkn.
Distr.: CA(type), TX.
Host: Gnaphalium
Cercospora gomphrenae W.W. Ray non Sawada, hom. illeg.
Distr.: Southeastern United States.
Host: Gomphrena
Cercospora gonolobi W.W. Ray
Distr.: Central United States.
Host: Cynanchum
Cercospora gossypina Cooke
Mycosphaerella gossypina (Atk.) Earle (teleomorph)
Distr.: Worldwide in cotton-growing regions.
Notes: Leaf spot of cotton (Little. C.M.I. Descr. 914. 1987).
Host: Gossypium
Cercospora granuliformis Ellis & Holw.
Cercospora sororiae Tehon
Distr.: Central and eastern North America; Brazil
Host: Viola
Cercospora graphioides Ellis
Distr.: Central and eastern North America.
Host: Prunus
Cercospora grindeliae Ellis & Everh.
Distr.: CA(type), TX, WI.
Notes: Chupp (A monograph of the fungus genus *Cercospora*. p. 139. 1953) indicated that this species might be a *Cercosporella*.
Host: Grindelia
Cercospora grisea Cooke & Ellis
Cercospora polygalae Henn.
Distr.: Central and eastern United States; Brazil
Host: Polygala
Cercospora guttulata Ellis & Kellerm.
Distr.: MO, WV(type).
Host: Aristolochia
Cercospora gymnocladi Ellis & Kellerm.
Cercospora superflua Ellis & Holw.
Distr.: Central and eastern United States.
Notes: The host of *C. superflua* Ellis & Kellerm. was misdetermined as *Fraxinus*. Possibly a *Pseudocercospora* or *Helminthosporium* (Chupp. A monograph of the fungus genus *Cercospora*. p. 419. 1953).
Hosts: [Fraxinus], Gymnocladus
Cercospora haleniae Chupp & Bisby
Distr.: MI, WI; Canada(Manitoba, type).
Host: Halenia
Cercospora halesiae Dearn., nom. inval.
Distr.: TN.
Host: Halesia
Cercospora halstedii Ellis & Everh. in Seym. & Earle
Distr.: Central and eastern United States.
Host: Carya
Cercospora hamamelidis (Peck) Ellis & Everh.
Ramularia hamamelidis Peck
Distr.: Eastern North America; Japan.
Host: Hamamelis
Cercospora hansenii Ellis & Everh.
Distr.: CA(type), DE, MO.
Host: Asclepias
Cercospora hederae Katsuki & Togashi
Distr.: FL; Japan(type).
Host: Hedera
Cercospora helenii Tharp
Distr.: AL, TX(type).
Host: Helenium
Cercospora helianthi Ellis & Everh.
Distr.: Central and eastern United States; Soviet Union.
Host: Helianthus
Cercospora heliotropii Ellis & Everh.
Distr.: Western United States.
Host: Heliotropium
Cercospora hemerocallidis Tehon
Distr.: IL(type); Bermuda.
Host: Hemerocallis
Cercospora heterospora Ellis & Everh.
Distr.: WI.
Host: Euphorbia
Cercospora heucherae Ellis & G. Martin
Distr.: Central and eastern United States.
Host: Heuchera
Cercospora hieracii Ellis & Everh.
Distr.: Southeastern United States.
Host: Hieracium
Cercospora houstoniae Ellis & Everh.
Distr.: DE.
Host: Hedyotis
Cercospora huricola Chupp
Distr.: FL; Brazil(type); Venezuela.
Host: Hura
Cercospora hyalina Muller & Chupp
Distr.: FL, OK; Brazil(type); Colombia; Venezuela; India.
Host: Rosa
Cercospora hyalospora Muller & Chupp
Distr.: IL; Brazil(type).
Host: Sida
Cercospora hydrangeae Ellis & Everh. in Atk.
Cercospora hydrangeana Tharp
Cercosporina hydrangeicola Speg.
Cercospora arborescentis Tehon & E.Y. Daniels
Distr.: Southcentral and southeastern United States; Argentina; central Africa; eastern and southeastern Asia.
Host: Hydrangea
Cercospora hydrocotyles Ellis & Everh.
Distr.: Southeastern United States; West Indies; Central America; Japan.
Host: Hydrocotyle
Cercospora hyperici Tehon & E.Y. Daniels
Distr.: IL(type), WI.
Notes: Possibly a *Didymaria* (Chupp. A monograph of the fungus genus *Cercospora*. p. 260. 1953).
Host: Hypericum
Cercospora ichthyomethiae Dearn. & Barth.
Distr.: FL.
Host: Piscidia
Cercospora ilicis Ellis
Distr.: Eastern United States.
Host: Ilex
Cercospora ilicis-opacae Chupp
Distr.: Southeastern United States.
Host: Ilex
Cercospora insulana Sacc.
Distr.: Worldwide, scattered reports.
Host: Limonium
Cercospora ipomoeae G. Wint.
Cercospora alabamensis Atk.
Cercospora stuckertiana Syd. & P. Syd.
Cercospora viridula Ellis & Everh.
Distr.: Central and eastern United States; tropical and subtropical regions.
Hosts: Ipomoea, Jacquemontia
Cercospora isanthi Ellis & Kellerm.
Distr.: KS(type), MS, WI.
Host: Isanthus
Cercospora ithacensis Chupp
Distr.: Central and western United States.
Notes: The name *C. geranii* Kellerm. & Swingle has been misapplied to this species in the past (Chupp. A monograph of the fungus genus *Cercospora*. p. 241. 1953).
Host: Geranium

Cercospora jacquiniana Thuem.
Cercospora senecionis Ellis & Everh.
Distr.: Eastern North America; Argentina; Europe.
Host: Senecio
Cercospora jahnii Syd.
Distr.: FL; West Indies; northern South America.
Host: Tabebuia
Cercospora jatrophae Atk.
Distr.: AL(type); Martinique.
Host: Cnidoscolus
Cercospora jatropharum Speg.
Distr.: FL; Argentina(type).
Notes: Sutton & Pons (Mycotaxon 12:210. 1980) suggested that this species belongs in *Pseudocercospora*.
Host: Jatropha
Cercospora jatrophicola (Speg.) Chupp
Distr.: TX; Argentina(type).
Host: Cnidoscolus
Cercospora juglandis Kellerm. & Swingle
Distr.: Central and eastern United States; Brazil
Host: Juglans
Cercospora jujubae Chowdhury
Distr.: Subtropical and tropical regions (Ellis. More Dematiaceous Hyphomycetes. p. 280. 1976).
Host: Ziziphus
Cercospora juncina Sacc.
Cercospora junci J.J. Davis
Distr.: Temperate North America.
Notes: Sprague (Res. Stud. Wash. State Univ. 29:82. 1961).
Host: Juncus
Cercospora justiciae F.L. Tai
Distr.: FL; China(type).
Host: Justicia
Cercospora justiciicola F.L. Tai
Distr.: FL; China(type).
Hosts: Justicia, Pachystachys
Cercospora kaki Ellis & Everh.
Distr.: Southern United States; Brazil; eastern Asia; India.
Host: Diospyros
Cercospora kalmiae Ellis & Everh.
Distr.: Eastern United States.
Hosts: Andromeda, Kalmia
Cercospora kansensis Syd. & P. Syd.
Distr.: KS(type), TX.
Host: Cirsium
Cercospora kikuchii (Matsumoto & Tomoyasu) M.W. Gardner
Cercosporina kikuchii Matsumoto & Tomoyasu
Distr.: Worldwide in soybean-growing regions.
Notes: Brown leaf spot, purple stain of soybeans (Mulder & Holliday. C.M.I. Descr. 466. 1975).
Hosts: Cyamopsis, Glycine
Cercospora kolkwitziae W.W. Ray
Distr.: AL(type), OK.
Host: Kolkwitzia
Cercospora lanuginosa Heald & F.A. Wolf
Distr.: MO, TX(type).
Notes: A doubtful species (Chupp. A monograph of the fungus genus *Cercospora*. p. 513. 1953).
Host: Sideroxylon
Cercospora latens Ellis & Everh.
Cercospora lespedezae Ellis & Dearn.
Distr.: Central and eastern North America; central and eastern Asia.
Hosts: Kummerowia, Lespedeza, Pediomelum, Psoralidium
Cercospora lateritia Ellis & Halst.
Distr.: Central United States.
Host: Sambucus
Cercospora lathyri Dearn. & House
Distr.: Northcentral and northeastern North America; Romania.
Host: Lathyrus
Cercospora lathyrina Ellis & Everh.
Distr.: NJ to southeastern United States.
Hosts: Lathyrus, Pisum
Cercospora lecheae Chupp & H.C. Greene
Distr.: WI.
Host: Lechea
Cercospora leonotidis Cooke
Distr.: LA; Africa(type).
Host: Leonotis
Cercospora leonuri F. Stevens & Solheim
Distr.: WI; Costa Rica(type).
Host: Leonurus
Cercospora leptandrae J.J. Davis
Distr.: WI.
Host: Veronicastrum
Cercospora leptosperma Peck
Cercosporella leptospermum (Peck) J.J. Davis
Cercoseptoria leptosperma (Peck) Petr.
Septoriopsis leptospermum (Peck) J.J. Davis
Cylindrosporium leptospermum (Peck) J.J. Davis
Distr.: Northcentral and northeastern United States.
Host: Aralia
Cercospora ligusticicola Bonar
Distr.: CA(type).
Hosts: Clematis, Ligusticum
Cercospora ligustricola F.L. Tai
Distr.: FL; China(type).
Notes: According to Chupp (A monograph of the fungus genus *Cercospora*. p. 417. 1953), this is not a *Cercospora*.
Host: Ligustrum
Cercospora lilacis (Desmaz.) Sacc.
Distr.: Temperate northern hemisphere to northern South America.
Notes: Leaf spot of lilac.
Hosts: Ligustrum, Syringa
Cercospora limnobii K. Conway
Distr.: FL.
Notes: Conway (Trans. Brit. Mycol. Soc. 71:521. 1978).
Host: Limnobium
Cercospora lini Ellis & Everh.
Distr.: DE(type), TX.
Host: Linum
Cercospora lippiae Ellis & Everh.
Distr.: Central and southern United States; West Indies; Africa; southern Asia; India; Pakistan.
Notes: Leaf spot.
Hosts: Lippia, Phyla
Cercospora liquidambaris Cooke & Ellis
Cercospora liquidambaris Sawada
Distr.: Southeastern United States; Mexico; Taiwan.
Host: Liquidambar
Cercospora liriopes F.L. Tai
Distr.: FL; China(type).
Host: Liriope
Cercospora lithospermi Chupp & H.C. Greene
Distr.: WI(type).
Host: Lithospermum
Cercospora lobeliae Kellerm. & Swingle
Distr.: Central and eastern United States; Brazil.
Host: Lobelia
Cercospora longipes E.J. Butler
Distr.: Worldwide in sugar cane-growing regions.
Notes: Brown spot of sugarcane (Mulder & Holliday. C.M.I. Descr. 418. 1974). Distinct from *Mycovellosiella koepkei* (Krüger) Deighton (Deighton. Mycol. Pap. 144:22. 1979).
Host: Saccharum
Cercospora longispora Peck
Cercoseptoria longispora (Peck) J.J. Davis
Cylindrosporium longisporium Ellis & Dearn.
Cercospora filispora Peck in Ellis & Everh.
Distr.: Central and eastern North America.
Host: Lupinus
Cercospora longissima Cugini ex Traverso non Cooke & Ellis, hom. illeg.
Cercospora lactucae C. Welles non J.A. Stevenson, hom. illeg.
Cercospora lactucae J.A. Stevenson
Distr.: Cosmopolitan.
Notes: Leaf spot of lettuce.
Host: Lactuca
Cercospora lonicerae Chupp
Distr.: Southeastern United States; Bermuda.
Host: Lonicera

Cercospora loti Hollós
Distr.: GA; Europe.
Host: Lotus
Cercospora ludwigiae Atk.
Distr.: AL(type), WI.
Host: Ludwigia
Cercospora lumbricoides Turconi & Maffei
Distr.: Mexico(type); distribution elsewhere is questionable.
Host: Fraxinus
Cercospora lupinicola Lieneman
Cercospora texensis Tharp non Ellis & B.T. Galloway, hom. illeg.
Distr.: TX.
Host: Lupinus
Cercospora lychnidis H.C. Greene
Distr.: WI(type).
Host: Lychnis
Cercospora lycii Ellis & Halst.
Distr.: Central and eastern United States; Poland
Host: Lycium
Cercospora lyoniae Katsuki & T. Kobayashi
Distr.: FL; Japan(type).
Host: Lyonia
Cercospora lythracearum Heald & F.A. Wolf
Cercospora lagerstroemiae Syd. & P. Syd.
Distr.: TX; West Indies; eastern Africa; eastern Asia; India.
Hosts: Lagerstroemia, Punica
Cercospora macclatchieana Sacc. & Syd.
Cercospora fuliginosa Ellis & Everh.
Distr.: CA(type), WI.
Hosts: Ceanothus, Diospyros
Cercospora maclurae Ellis & Everh.
Distr.: AL.
Notes: Possibly a *Pseudocercospora* (Chupp. A monograph of the fungus genus *Cercospora*. p. 398. 1953).
Host: Maclura
Cercospora macroguttata Atk.
Distr.: Southeastern United States.
Host: Chrysopsis
Cercospora macromaculans Heald & F.A. Wolf
Distr.: TX(type), WI.
Host: Syringa
Cercospora madisonensis Chupp & H.C. Greene in H.C. Greene
Distr.: WI.
Host: Aureolaria
Cercospora maesae Hansf.
Distr.: UT(type).
Host: Myrsine
Cercospora malachrae Heald & F.A. Wolf
Distr.: Southern United States; West Indies; Venezuela; southeastern Asia.
Hosts: Malachra, Urena
Cercospora malayensis F. Stevens & Solheim
Distr.: Worldwide.
Notes: Brown leaf spot (Little. C.M.I. Descr. 916. 1987).
Hosts: Abelmoschus, Hibiscus
Cercospora malkoffii Bubák
Distr.: Bulgaria(type); distribution elsewhere is unclear.
Host: Pimpinella
Cercospora mallotii Ellis & Everh.
Distr.: MS(type); eastern Asia.
Host: Mallotus
Cercospora malvarum Sacc.
Distr.: Temperate North America; Europe.
Host: Malva
Cercospora malvicola Ellis & G. Martin
Cercospora polymorpha Bubák
Distr.: Northeastern United States; Yugoslavia.
Host: Malva
Cercospora mamaonis Viégas & Chupp
Distr.: HI; Brazil(type).
Host: Carica
Cercospora maritima Tracy & Earle
Distr.: Southeastern United States; West Indies; northern South America.
Host: Croton
Cercospora marrubii Tharp
Cercospora marrubii Savul. & Sandu
Distr.: TX, OK; Romania.
Host: Marrubium
Cercospora medicaginis Ellis & Everh.
Distr.: Worldwide.
Hosts: Medicago, Trifolium
Cercospora megalopotamica Speg.
Distr.: HI; Argentina(type); China; Uganda.
Host: Bidens
Cercospora meliae Ellis & Everh.
Cercospora leucosticta Ellis & Everh.
Distr.: Southeastern United States; eastern Asia; India.
Host: Melia
Cercospora melongenae C. Welles
Distr.: Cosmopolitan.
Notes: Williams (C.M.I. Descr. 936. 1987).
Host: Solanum
Cercospora menthicola Tehon & E.Y. Daniels
Distr.: IL(type), TX.
Host: Mentha
Cercospora microsora Sacc.
Cercospora exitiosa Syd. & P. Syd.
Cercospora tiliae Peck
Mycosphaerella sp. (teleomorph)
Distr.: Temperate northern hemisphere.
Notes: According to Chupp (A monograph of the fungus genus *Cercospora*. p. 565. 1953), there is some disagreement over the identity of the teleomorph.
Host: Tilia
Cercospora mikaniacola F. Stevens
Distr.: FL, PR(type).
Host: Mikania
Cercospora mikaniae Ellis & Everh.
Cercospora lemnischea Cif.
Distr.: Eastern United States; West Indies.
Host: Mikania
Cercospora mimuli Ellis & Everh.
Distr.: MO(type).
Host: Mimulus
Cercospora minuta (J.J. Davis) Chupp
Cercoseptoria minuta J.J. Davis
Distr.: WI(type).
Host: Campanula
Cercospora mirabilis Tharp
Distr.: TX(type); France.
Host: Mirabilis
Cercospora mississippiensis Tracy & Earle
Distr.: Central and eastern United States.
Host: Smilax
Cercospora missouriensis G. Wint.
Cercospora pulvinulata Sacc. & G. Wint.
Distr.: MS; eastern Asia; Australia; central Europe.
Hosts: Callicarpa, Morus
Cercospora mitellae G.H. Hicks in Seym., nom. nud.
Distr.: MI.
Notes: The name has never been published validly (Chupp. A monograph of the fungus genus *Cercospora*. p. 518. 1953).
Host: Mitella
Cercospora modiolae Tharp
Distr.: AL, TX(type).
Host: Modiola
Cercospora molleriana G. Wint.
Distr.: FL; Portugal(type).
Host: Arbutus
Cercospora molluginis Halst.
Cercospora mollugini̇cola Lieneman
Cercospora molluginis J.J. Davis, hom. illeg.
Distr.: Central and eastern United States.
Host: Mollugo
Cercospora moricola Cooke
Distr.: Central and eastern North America; Brazil; Africa; Europe.
Host: Morus
Cercospora morindicola Jenk. & Chupp
Distr.: FL; West Indies.

Host: Morinda
Cercospora morongiae Tracy & Earle
Distr.: MS.
Host: Schrankia
Cercospora muhlenbergiae Atk.
Distr.: Central and eastern United States.
Notes: Sprague & Rogerson (Mycologia 50:639. 1958).
Host: Muhlenbergia
Cercospora namae Dearn. & House
Cercosporina namae (Dearn. & House) Sacc.
Distr.: OK(type).
Host: Hydrolea
Cercospora nandinae Nagatomo
Distr.: Southeastern United States; Japan.
Host: Nandina
Cercospora nasturtii Pass.
Distr.: Temperate northern hemisphere.
Hosts: Berteroa, Descurainia, Nasturtium, Rorippa, Sisymbrium, Stanleya
Cercospora negundinis Ellis & Everh.
Distr.: NE(type), central and eastern United States.
Host: Acer
Cercospora nepetae Tehon
Distr.: IL(type), TX.
Hosts: Glechoma, Nepeta
Cercospora nesaeae Ellis & Everh.
Cercospora decodontis Tehon & E.Y. Daniels
Distr.: Central and eastern United States.
Host: Decodon
Cercospora nicotianae Ellis & Everh.
Cercospora raciborskii Sacc. & Syd.
Distr.: Worldwide in tobacco-growing regions.
Notes: Frog-eye leaf spot, barn spot of tobacco (Mulder & Holliday. C.M.I. Descr. 416. 1974).
Hosts: Lycopersicon, Nicotiana
Cercospora noveboracensis Ellis & Everh.
Distr.: Central and eastern United States.
Host: Vernonia
Cercospora nudiflorae Chupp
Distr.: FL; Bermuda(type).
Host: Commelina
Cercospora nyssae Tharp
Distr.: TX(type).
Host: Nyssa
Cercospora nyssae-sylvaticae H.C. Greene
Distr.: WI(type).
Host: Nyssa
Cercospora obesa Ellis & Everh.
Cercospora ditissima Ellis & Everh.
Distr.: KS.
Host: Cirsium
Cercospora obscura Heald & F.A. Wolf
Distr.: TX(type); Venezuela.
Host: Cynara
Cercospora oculata Ellis & Kellerm.
Distr.: Central United States.
Host: Vernonia
Cercospora odontoglossi Prill. & Delacr.
Distr.: Eastern United States; Europe; New Zealand.
Hosts: Brassolaeliocattleya, Cattleya, Dendrobium, Epidendrum, Laelia, Laeliocattleya, Odontoglossum, Oncidium, Orchidaceae
Cercospora oenotherae Ellis & Everh.
Distr.: Central and eastern United States; Japan.
Hosts: Calylophus, Oenothera
Cercospora oenotherae-sinuatae Atk.
Distr.: Southeastern United States.
Host: Oenothera
Cercospora omphacodes Ellis & Holw.
Cercospora phlogina Peck
Distr.: Central and eastern North America; Japan; South Africa.
Hosts: Phlox, Polemonium
Cercospora opuli (Fuckel) Höhn.
Cercospora penicillata (Ces.) Fresen. var. *opuli* Fuckel
Distr.: Temperate northern hemisphere.
Host: Viburnum
Cercospora oryzae Miyake
Sphaerulina oryzina K. Hara (teleomorph)
Distr.: Southeastern United States; subtropical and tropical rice-growing regions.
Notes: Narrow brown leaf spot of rice (Mulder & Holliday. C.M.I. Descr. 420. 1974).
Host: Oryza
Cercospora osmorrhizae Ellis & Everh.
Distr.: Central and eastern North America.
Host: Osmorhiza
Cercospora oxalidiphila Chupp & Muller
Distr.: WI; West Indies; South America; eastern Africa.
Host: Oxalis
Cercospora oxalidis Muller & Chupp
Distr.: FL; Brazil(type).
Host: Oxalis
Cercospora oxybaphi Ellis & Halst.
Distr.: Central and eastern United States.
Hosts: Allionia, Mirabilis
Cercospora oxydendri Tracy & Earle
Cercospora oxydendri Ellis & Everh.
Distr.: Southeastern United States.
Host: Oxydendrum
Cercospora pachypus Ellis & Kellerm.
Distr.: Central and southeastern United States.
Hosts: Helianthus, Tithonia
Cercospora paeoniae Tehon & E.Y. Daniels
Distr.: Central United States; China.
Host: Paeonia
Cercospora palustrium Chupp
Distr.: FL; Bermuda(type).
Host: Baccharis
Cercospora pancratii Ellis & Everh.
Cercospora hymenocallidis Pat.
Distr.: Southeastern United States; West Indies; Central America; India; Japan.
Hosts: Crinum, Hymenocallis
Cercospora papavericola Chupp
Cercospora papaveri Muller & Chupp non Nakata, hom. illeg.
Distr.: Southeastern United States; Brazil.
Host: Papaver
Cercospora papayae Hansf.
Distr.: Widespread in tropical regions.
Notes: Ellis (More dematiaceous Hyphomycetes. p. 247. 1976).
Host: Carica
Cercospora papillosa Atk.
Distr.: AL, PR.
Host: Verbena
Cercospora partheniphila Chupp & H.C. Greene
Distr.: WI(type); San Domingo.
Host: Parthenium
Cercospora passaloroides G. Wint.
Cylindrosporium passaloroides (G. Wint.) Gilman & W.A. Archer
Distr.: Central and eastern United States.
Host: Amorpha
Cercospora passiflorae Muller & Chupp
Distr.: FL, PR; Dominican Republic; Brazil(type).
Host: Passiflora
Cercospora pastinacae (Sacc.) Peck
Cercospora apii Fresen. var. *pastinacae* Sacc.
Distr.: Central and eastern United States; Soviet Union.
Host: Pastinaca
Cercospora peckiana Chupp
Cercospora acetosellae Peck var. *maculosa* Peck
Distr.: NY.
Host: Rumex
Cercospora penicillus Ellis & Everh.
Cercospora dispersa Ellis & Everh.
Cercospora myricae Tracy & Earle
Distr.: Eastern United States.
Host: Myrica
Cercospora penniseti Chupp
Distr.: GA.
Host: Pennisetum
Cercospora pentstemonis Ellis & Kellerm.

Distr.: Temperate North America.
Host: Penstemon
Cercospora penzigii Sacc.
Cercospora aurantia Heald & F.A. Wolf
Cercospora fumosa Penz. non Speg., hom. illeg.
Distr.: Citrus-growing regions of southwestern United States; Brazil; Italy.
Hosts: Citrus, Poncirus
Cercospora peregrina Chupp
Distr.: FL, TX(type).
Host: Tabernaemontana
Cercospora peristeriae H.C. Burnett
Distr.: FL(type).
Host: Orchidaceae
Cercospora petersii (Berk. & M.A. Curtis) Atk.
Helminthosporium petersii Berk. & M.A. Curtis
Distr.: Central and eastern United States to Mexico.
Host: Smilax
Cercospora petuniae (K. Saito) Chupp
Cercosporina petuniae K. Saito
Distr.: Central and southern United States; West Indies; South America; Africa; eastern Asia.
Host: Petunia
Cercospora phormii Bolick in Alfieri et al., nom. nud.
Distr.: FL.
Notes: The place of publication of this name could not be found.
Host: Phormium
Cercospora photiniae-serrulatae Anzalone & Plakidas
Distr.: Southeastern United States.
Host: Photinia
Cercospora phyllitidis H. Hume in Ellis & Everh.
Distr.: Fl(type); England.
Hosts: Angiopteris, Davallia, Nephrolepis, Osmunda, Polypodium, Pteris, Rumohra, Thelypteris
Cercospora physalidis Ellis
Cercospora physalicola Ellis & Barth.
Distr.: Central and eastern United States; Central and South America; Africa; eastern Asia; India.
Notes: On leaves. Misidentified as occurring on tomato.
Host: Physalis
Cercospora physocarpi W.W. Ray
Distr.: IL, OK.
Hosts: Physocarpus, Spiraea
Cercospora piaropi Tharp
Distr.: Southeastern United States.
Host: Eichhornia
Cercospora pileae F.L. Tai
Distr.: FL; China(type).
Host: Pilea
Cercospora pinnulaecola Atk.
Distr.: AL(type).
Host: Cassia
Cercospora pipturi F. Stevens & Glick
Distr.: HI.
Host: Pipturus
Cercospora pisa-sativae J.A. Stevenson
Distr.: Southeastern United States; West Indies; China; India; East Indies.
Host: Pisum
Cercospora pittospori Plakidas
Distr.: Southeastern United States; Japan.
Host: Pittosporum
Cercospora plantaginis Sacc.
Cercospora plantaginella Tehon
Distr.: HI; northern hemisphere to Brazil.
Host: Plantago
Cercospora platanicola Ellis & Everh.
Mycosphaerella platanifolia (Cooke) F.A. Wolf (teleomorph)
Distr.: Central and eastern North America.
Notes: Sivanesan (The bitunicate Ascomycetes. p. 192. 1984).
Host: Platanus
Cercospora platycerii Chupp
Distr.: FL, NJ(type).
Host: Platycerium
Cercospora plumeriae Chupp
Distr.: FL; India(type).
Host: Plumeria
Cercospora podophylli Tehon & E.Y. Daniels
Distr.: IL.
Host: Podophyllum
Cercospora polygonacea Ellis & Everh.
Cercospora avicularis G. Wint. var. *sagittati* Atk.
Distr.: Central and eastern United States; Japan.
Host: Polygonum
Cercospora polygonacea Ellis & Everh. var. **avicularis** Sumstine, nom. nud.
Distr.: ND.
Notes: The place of publication of this name could not be found; the name may be based on *C. avicularis*.
Host: Polygonum
Cercospora polytaeniae Ellis & Kellerm.
Distr.: KS(type), WI.
Host: Polytaenia
Cercospora polytricha Cooke
Cercospora macrochaeta Ellis & Everh.
Distr.: CA, southeastern United States.
Host: Quercus
Cercospora populicola Tharp
Distr.: TX(type); Brazil.
Host: Populus
Cercospora potentillae Chupp & Greene
Distr.: WI.
Host: Potentilla
Cercospora praecincta (J.J. Davis) Chupp
Cercospora althaeina Sacc. var. *praecincta* J.J. Davis
Distr.: Central United States.
Host: Callirhoe
Cercospora prenanthis Ellis & Kellerm.
Cercospora brunnea Peck
Distr.: Central and southeastern United States.
Host: Prenanthes
Cercospora prosopidis Heald & F.A. Wolf
Distr.: TX.
Host: Prosopis
Cercospora psidii Rangel
Distr.: FL; Brazil(type).
Host: Psidium
Cercospora psychotriae Chupp & Viégas non Sawada, hom. illeg.
Distr.: FL; Bermuda; Brazil(type).
Host: Psychotria
Cercospora pteleae G. Wint.
Cercospora afflata G. Wint.
Distr.: Central United States.
Hosts: Harpullia, Ptelea
Cercospora pteridis Siemaszko
Distr.: WI; Colombia; Soviet Union; Spain.
Host: Pteridium
Cercospora pulcherrimae Tharp
Cercospora pulcherrimae Tharp var. *minima* Tharp
Distr.: Subtropical and tropical regions.
Host: Euphorbia
Cercospora pulvinula Cooke & Ellis
Distr.: NJ to southeastern United States.
Host: Ilex
Cercospora punctiformis Sacc. & Roum.
Cercospora vincetoxici Ellis & Everh.
Distr.: Southeastern United States; northern Africa; eastern Asia; southern Europe.
Hosts: Matelea, Morrenia
Cercospora pyracanthae Katsuki
Distr.: FL; Japan(type).
Host: Pyracantha
Cercospora pyri Farl.
Cercosporella pyri (Farl.) Karakulin in Vassiljevsky & Karakulin
Distr.: Northeastern North America.
Hosts: Aronia, Malus, Pyrus
Cercospora pyrostegiae Viégas
Distr.: FL; Brazil(type).
Host: Pyrostegia

Cercospora quarta Chupp & H.C. Greene
Distr.: WI(type).
Host: Aster
Cercospora querci Chupp
Distr.: LA.
Host: Quercus
Cercospora rafinesquiae Harkn.
Distr.: CA(type).
Host: Rafinesquia
Cercospora ranunculi Ellis & Holw.
Distr.: Central United States.
Host: Ranunculus
Cercospora rapaneae Syd.
Distr.: FL; Ecuador(type).
Host: Rapanea
Cercospora ratibidae Ellis & Barth.
Distr.: Central United States.
Host: Ratibida
Cercospora rautensis C. Massal.
Distr.: PA; Europe.
Host: Coronilla
Cercospora regalis Tharp
Distr.: TX.
Host: Passiflora
Cercospora repens Ellis & Everh.
Distr.: Southeastern United States.
Hosts: Nerium, Trachelospermum
Cercospora resedae Fuckel
Distr.: North America; South Africa; temperate Europe.
Host: Reseda
Cercospora rhamni Fuckel
Mycosphaerella vogelii (Syd. & P. Syd.) Tomilin (teleomorph)
Distr.: Temperate North America; Europe.
Host: Rhamnus
Cercospora rhapontici Tehon & E.Y. Daniels
Distr.: IL.
Host: Rheum
Cercospora rhizophorae D.B. Creager
Distr.: FL.
Host: Rhizophora
Cercospora ribicola Ellis & Everh.
Distr.: Pacific Northwest.
Host: Ribes
Cercospora ribis Earle
Distr.: Central and southeastern United States.
Host: Ribes
Cercospora richardiaecola Atk.
Distr.: Southeastern United States; West Indies; Central America; South Africa.
Hosts: Anthurium, Epipremnum, Zantedeschia
Cercospora ricinella Sacc. & Berl.
Cercospora albido-maculans G. Wint.
Cercospora ricini Speg.
Distr.: Warm-temperate, subtropical and tropical regions.
Host: Ricinus
Cercospora rodmanii K. Conway
Distr.: FL.
Notes: Conway (Canad. J. Bot. 54:1079. 1976).
Host: Eichhornia
Cercospora rosicola Pass.
Cercospora rosae Van Hook
Cercospora rosae-indiananensis Van Hook
Cercospora rosicola Pass. var. *undosa* J.J. Davis
Cercospora rosigena Tharp
Mycosphaerella rosicola B.H. Davis ex Deighton (teleomorph)
Distr.: Worldwide in rose-growing regions.
Host: Rosa
Cercospora rubella Cooke
Distr.: SC(type), TX.
Host: Eriogonum
Cercospora rubigo Cooke & Harkn.
Distr.: Central and western United States.
Notes: Previous reports of this species on *Clematis* were based on misdeterminations (Chupp. A monograph of the fungus genus *Cercospora*. p. 488. 1953).
Host: Spiraea
Cercospora rubrotincta Ellis & Everh.
Cercospora consobrina Ellis & Everh.
Distr.: Eastern United States; Africa; Europe; India; Middle East.
Notes: Williams (C.M.I. Descr. 937. 1987).
Host: Prunus
Cercospora rumicis Ellis & Langl.
Distr.: IA, LA(type), TX.
Host: Rumex
Cercospora sabbatiae Ellis & Everh.
Cercospora erythraeae Hollós
Distr.: Southeastern United States; Hungary.
Host: Sabatia
Cercospora saccharini Liberta & Boewe
Distr.: IL.
Host: Acer
Cercospora sagittariae Ellis & Kellerm.
Distr.: North America, scattered reports elsewhere.
Host: Sagittaria
Cercospora salicis Chupp & H.C. Greene
Distr.: WI.
Host: Salix
Cercospora salviicola Tharp
Distr.: Central and eastern United States; Central America.
Host: Salvia
Cercospora sanguinariae Peck
Distr.: Central and eastern United States.
Host: Sanguinaria
Cercospora saniculae J.J. Davis
Distr.: WI.
Host: Sanicula
Cercospora saururi Ellis & Everh.
Cercospora saururicola Sawada
Ramularia saururi (Ellis & Everh.) Tharp
Distr.: NY to southeastern United States; Taiwan.
Host: Saururus
Cercospora saximontanensis Deighton
Distr.: Western United States.
Notes: This species has previously been reported as *Cercospora ribicola* Ellis & Everh. and *Cercoseptoria ribis* (J.J. Davis) Dearn. & House (Deighton. Mycol. Pap. 151:7. 1983).
Host: Ribes
Cercospora schinii Syd. & P. Syd.
Distr.: FL; Argentina(type).
Host: Schinus
Cercospora scutellariae Ellis & Everh.
Distr.: Central United States.
Host: Scutellaria
Cercospora secalis Chupp
Distr.: VA(type), IL.
Hosts: Avena, Hordeum, Secale, Triticum
Cercospora sechii J.A. Stevenson
Distr.: FL; West Indies.
Host: Sechium
Cercospora sedi Ellis & Everh.
Distr.: AL(type).
Host: Sedum
Cercospora sedoides Ellis & Everh.
Distr.: Central United States.
Host: Penthorum
Cercospora selaginellarum Cour & Joly
Host: Selaginella
Cercospora selini-gmelini (Sacc. & Scalia) Chupp
Cercospora apii Fresen. var. *selini-gmelini* Sacc. & Scalia
Distr.: AK; Japan.
Host: Conioselinum
Cercospora seminalis Ellis & Everh.
Distr.: Central and western United States.
Hosts: Buchloe, Cynodon
Cercospora senecionicola J.J. Davis
Distr.: WI.
Host: Senecio
Cercospora septatissima Tracy & Earle
Distr.: MS(type).
Hosts: Verbascum, Verbena

Cercospora sequoiae Ellis & Everh.
Cercospora thujina Plakidas
Distr.: Eastern United States(nurseries).
Notes: Hodges (Mycologia 54:62. 1962); Mulder & Gibson (C.M.I. Descr. 366. 1973).
Hosts: Chamaecyparis, Cupressus, Juniperus, Sequoia, Sequoiadendron, Taxodium, Thuja
Cercospora sequoiae Ellis & Everh. var. ***juniperi*** Ellis & Everh.
Distr.: Central and eastern North America.
Notes: Needle blight of eastern red cedar (Hodges. Mycologia 54:64. 1962).
Hosts: Cupressus, Juniperus
Cercospora seriata Atk.
Distr.: AL(type).
Host: Sporobolus
Cercospora serpentaria Ellis & Everh.
Distr.: Eastern United States; West Indies; Brazil.
Host: Aristolochia
Cercospora sesami A. Zimmerm.
Distr.: Subtropical and tropical regions.
Notes: Leaf and pod spot of sesame (Waller & Sutton. C.M.I. Descr. 627. 1979).
Host: Sesamum
Cercospora setariae Atk.
Cercospora setariicola Tehon & E.Y. Daniels
Cercospora paspali W.W. Ray
Distr.: Central and eastern United States; West Indies; Central America; eastern Africa; central and eastern Asia.
Hosts: Paspalum, Setaria, Sporobolus
Cercospora seymouriana G. Wint.
Cercospora berkeleyi Cooke
Helminthosporium olivaceum Berk. & Ravenel
Distr.: Southeastern United States.
Hosts: Gleditsia, Litsea
Cercospora sidaecola Ellis & Everh.
Cercospora densissima Speg.
Distr.: Southeastern United States; West Indies; South America.
Host: Sida
Cercospora silphii Ellis & Everh.
Cercospora silphii Ellis & Everh. var. *laciniati* Tehon & E.Y. Daniels
Distr.: Central and eastern United States.
Host: Silphium
Cercospora simulans Ellis & Kellerm.
Cercosporidium simulans (Ellis & Kellerm.) X. Liu & Guo
Distr.: Southeastern United States; China.
Host: Amphicarpaea
Cercospora smilacis Thuem.
Cercospora smilacina Sacc.
Distr.: Worldwide.
Host: Smilax
Cercospora sojina K. Hara
Cercosporidium sojinum (K. Hara) X. Liu & Guo
Distr.: Central and eastern United States; subtropical regions.
Notes: Frog-eye leaf spot of soybeans.
Host: Glycine
Cercospora solani Thuem.
Distr.: Europe, Africa.
Notes: North American reports have been based on misdeterminations (Chupp. A monograph of the fungus genus *Cercospora*. p. 550. 1953).
Host: Solanum
Cercospora solanicola Atk.
Distr.: Southern United States; Venezuela; Central Africa; eastern Asia.
Host: Solanum
Cercospora solidaginis Chupp & H.C. Greene
Distr.: WI(type); Japan.
Host: Solidago
Cercospora sorghi Ellis & Everh.
Cercospora sorghi Ellis & Everh. var. *maydis* Ellis & Everh.
Distr.: Cosmopolitan.
Notes: Gray leaf spot of sorghum, sudan grass, and broom corn (Mulder & Holliday. C.M.I. Descr. 419. 1974).
Hosts: Echinochloa, Sorghum, Zea
Cercospora sparsa Cooke
Distr.: PA to southeastern United States.
Hosts: Kalmia, Vaccinium
Cercospora speculariae Ellis & Langl., nom. nud.
Distr.: LA.
Host: Triodanis
Cercospora spegazzinii Sacc.
Cercospora dubia Speg.
Distr.: IA; Argentina(type).
Host: Celtis
Cercospora sphaeriaeformis Cooke
Cercospora ulmi Syd.
Distr.: Southeastern United States; eastern Asia.
Host: Ulmus
Cercospora spiraeae Thuem.
Distr.: Austria(type); distribution elsewhere is questionable.
Host: Physocarpus
Cercospora spiraeicola Muller & Chupp
Distr.: FL; Guatemala(type).
Host: Spiraea
Cercospora squalidula Peck
Distr.: Temperate North America; Ethiopia; Soviet Union.
Host: Clematis
Cercospora stachydis Ellis & Everh.
Distr.: Central and eastern United States.
Host: Stachys
Cercospora stachytarphetae Ellis & Everh.
Distr.: FL; West Indies; northern South America.
Host: Stachytarpheta
Cercospora staphyleae W.W. Ray & J.H. McLaughlin
Distr.: OK(type).
Host: Staphylea
Cercospora stevensonii Chupp
Distr.: FL, TX(type).
Host: Codiaeum
Cercospora stomatica Ellis & J.J. Davis
Cercospora parvimaculans J.J. Davis
Distr.: Central and western United States; Brazil
Host: Solidago
Cercospora streptopi Dearn. & Barth.
Distr.: Temperate North America.
Hosts: Disporum, Streptopus
Cercospora stromatis Clements, nom. nud.
Distr.: CO.
Notes: This name was used on an exsiccati specimen but was never described.
Host: Crepis
Cercospora struthanthi Chupp & Muller
Distr.: FL; Venezuela(type).
Host: Phoradendron
Cercospora stylismae Tracy & Earle
Distr.: AL, MS.
Host: Bonamia
Cercospora styracis Chupp
Distr.: MS(type).
Host: Styrax
Cercospora symplocarpi Peck
Distr.: Northeastern North America.
Host: Symplocarpus
Cercospora tabacina Ellis & Everh.
Cercospora rudbeckiae Peck
Distr.: Central and eastern United States.
Hosts: Echinacea, Prenanthes, Rudbeckia
Cercospora tageticola Ellis & Everh.
Distr.: Southeastern United States; Sierra Leone.
Host: Tagetes
Cercospora tandilensis Speg.
Cercospora columbiensis Ellis & Everh.
Distr.: MO; South America.
Host: Hybanthus
Cercospora tectoniae F. Stevens
Distr.: HI; West Indies; India.
Host: Tectona
Cercospora tenuis Peck
Distr.: Northcentral and northeastern North America.
Host: Galium
Cercospora tephrosiae Atk.

Distr.: AL(type), WI.
Host: Tephrosia
Cercospora ternateae Petch
Distr.: Southern United States; subtropical and tropical regions.
Notes: On leaves and pods.
Hosts: Canavalia, Clitoria
Cercospora tessellata Atk.
Distr.: AL(type).
Host: Dactyloctenium
Cercospora texensis Ellis & B.T. Galloway
Distr.: TX(type).
Host: Fraxinus
Cercospora thaliae Ellis & Langl.
Distr.: LA.
Host: Thalia
Cercospora thaspii Ellis & Everh. in Atk.
Distr.: Eastern United States; China.
Hosts: Angelica, Thaspium
Cercospora thaspiicola J.J. Davis
Cercospora cordatae Chupp & H.C. Greene
Cylindrosporium ziziae Ellis & Everh.
Distr.: Northcentral United States.
Hosts: Thaspium, Zizia
Cercospora thlaspiae Chupp & H.C. Greene
Distr.: WI(type).
Host: Thlaspi
Cercospora torta Tracy & Earle
Distr.: MS(type), TX.
Host: Mitreola
Cercospora tortipes J.J. Davis
Distr.: Northeastern United States.
Host: Veronica
Cercospora toxicodendri Ellis
Cercospora bartholomei Ellis & Kellerm.
Cercospora infuscans Ellis & Everh.
Distr.: Eastern North America; Japan.
Host: Rhus
Cercospora tragopogonis Ellis & Everh.
Distr.: Central and western United States.
Hosts: Tilia, Tragopogon
Cercospora traversiana Sacc.
Distr.: IA; Africa; western Asia; Europe.
Host: Trigonella
Cercospora trematis (F. Stevens & Solheim) Chupp
Ragnhildiana trematis F. Stevens & Solheim
Distr.: FL; South Africa, scattered throughout the tropical region.
Host: Trema
Cercospora triostei Chupp & H.C. Greene
Distr.: WI.
Host: Triosteum
Cercospora tropaeoli Atk.
Distr.: Southeastern United States; Brazil; eastern Africa.
Host: Tropaeolum
Cercospora truncata Ellis & Everh.
Distr.: LA, PR.
Hosts: Ampelopsis, Vitis
Cercospora truncatella Atk.
Distr.: Southeastern United States.
Hosts: Passiflora, Verbena
Cercospora tuberculans Ellis & Everh.
Distr.: Southeastern United States.
Host: Liquidambar
Cercospora tuberosa Ellis & Kellerm.
Cercospora glaucescens G. Wint.
Distr.: Central and eastern North America.
Host: Apios
Cercospora umbrata Ellis & Holw.
Cercospora umbrata Ellis & Holw. var. *maculata* Dearn.
Distr.: Central and eastern North America.
Host: Bidens
Cercospora uramensis Chupp & Muller
Distr.: WI; Venezuela(type).
Host: Cleome
Cercospora uromycestri Pollack
Distr.: FL(type).
Notes: On urediniospores of *Uromyces cestri* (Pollack. Mycologia 63:691. 1971).
Host: Cestrum
Cercospora varia Peck
Distr.: Northcentral and northeastern North America.
Notes: Reports from the southeastern United States are suspect (Chupp. A monograph of the fungus genus *Cercospora*. p. 105. 1953).
Hosts: Lonicera, Viburnum
Cercospora variicolor G. Wint.
Distr.: Central and eastern North America.
Host: Paeonia
Cercospora velutina Ellis & Kellerm.
Distr.: Central United States.
Host: Baptisia
Cercospora venturioides Peck
Cercospora illinoensis Barth.
Distr.: Central and eastern North America, PR.
Host: Asclepias
Cercospora verbascicola Ellis & Everh.
Distr.: Central and eastern United States.
Host: Verbascum
Cercospora verbenicola Ellis & Everh.
Cercospora verbenae-strictae Peck
Distr.: Central and eastern United States.
Host: Verbena
Cercospora vernoniae Ellis & Kellerm.
Distr.: Central and eastern United States.
Host: Vernonia
Cercospora vexans C. Massal.
Distr.: CA, northeastern North America; Europe.
Host: Fragaria
Cercospora viburnicola W.W. Ray
Distr.: OK.
Host: Viburnum
Cercospora violae Sacc.
Cercospora violae-tricoloris Briosi & Cavara
Cercospora difformis Tehon
Cercospora kiusana Sawada
Distr.: Temperate northern hemisphere; South Africa.
Host: Viola
Cercospora violamaculans Fukui
Distr.: FL; Japan(type).
Host: Rhaphiolepis
Cercospora viticis Ellis & Everh.
Cercospora viticis Sawada
Distr.: Southeastern United States; eastern Asia.
Host: Vitex
Cercospora vulpinae Ellis & Kellerm.
Distr.: Central United States; Soviet Union.
Host: Vitis
Cercospora websteri Van der Zwet, nom. nud.
Mycosphaerella websteri Wiehe (teleomorph)
Distr.: LA; Malawi(Nyasaland, type).
Notes: Wiehe (Mycol. Pap. 55:1. 1953). This name was used for the anamorph of *Mycosphaerella websteri* without valid publication or description.
Host: Aleurites
Cercospora wisconsinensis Chupp & H.C. Greene
Distr.: WI(type).
Host: Prenanthes
Cercospora wisteriae Muller & Chupp
Distr.: FL; Brazil.
Host: Wisteria
Cercospora wulffiae Muller & Chupp
Distr.: FL; Dominican Republic; Brazil(type).
Host: Wedelia
Cercospora xanthicola Heald & F.A. Wolf
Distr.: TX(type), WI.
Host: Xanthium
Cercospora xanthoxyli Cooke
Cercospora fagaricola Sawada
Distr.: Southeastern United States; Taiwan.
Host: Zanthoxylum
Cercospora xyridis L.E. Miles
Distr.: MS.

Host: Xyris
Cercospora yuccae Cooke
Distr.: Southeastern United States.
Notes: Deighton (Trans. Brit. Mycol. Soc. 61:111. 1973).
Host: Yucca
Cercospora zeae-maydis Tehon & E.Y. Daniels
Distr.: Southeastern United States; West Indies; South America.
Notes: Gray leaf spot of corn.
Host: Zea
Cercospora zebrina Pass.
Distr.: Northern hemisphere to Africa; South America.
Notes: Williams (C.M.I. Descr. 938. 1987).
Hosts: Agropyron, Medicago, Melilotus, Trifolium, Zebrina
Cercospora zelkowae S. Hori
Distr.: FL; Japan(type).
Host: Zelkova
Cercospora zingiberi Togashi & Katsuki
Distr.: FL; Japan(type).
Host: Zingiber
Cercospora zinniae Ellis & G. Martin
Distr.: Cosmopolitan.
Hosts: Cosmos, Zinnia
Cercospora ziziae Ellis & Everh.
Distr.: Northcentral and northeastern United States.
Hosts: Thaspium, Zizia
Cercospora zonata G. Wint.
Cercospora fabae Fautrey
Cercospora viciae Ellis & Holw.
Distr.: Temperate and subtropical regions
Notes: Williams (C.M.I. Descr. 939. 1987).
Hosts: Lathyrus, Vicia

Cercosporella Sacc. Hyphomycetes

Cercosporella anethi Sacc.
Distr.: ND.
Host: Anethum
Cercosporella arachnoidea F.A. Wolf
Mycosphaerella arachnoidea F.A. Wolf (teleomorph)
Distr.: Southeastern United States.
Notes: False mildew of red mulberry.
Host: Morus
Cercosporella bouteIouae R. Sprague & C.T. Rogerson
Distr.: KS.
Notes: Sprague & Rogerson (Mycologia 50:638. 1958).
Host: Bouteloua
Cercosporella californica Bonar
Distr.: CA(type).
Host: Rhus
Cercosporella exilis J.J. Davis
Distr.: WI(type).
Host: Phryma
Cercosporella fraserae (Ellis & Everh.) Sacc.
Cercospora fraserae Ellis & Everh.
Distr.: Rocky Mountains.
Host: Frasera
Cercosporella geranii W.B. Cooke & C.G. Shaw
Distr.: ID, WA.
Host: Geranium
Cercosporella helianthellae Ellis & Everh.
Distr.: CO(type), WA.
Host: Helianthella
Cercosporella holci R. Sprague
Cercospora holci (R. Sprague) R. Sprague
Distr.: Pacific Northwest.
Host: Holcus
Cercosporella idahoensis Sacc.
Distr.: ID, WA.
Hosts: Disporum, Smilacina
Cercosporella inconspicua (G. Wint.) Höhn.
Cylindrosporium inconspicua G. Wint.
Cercosporella lilii Dearn.
Distr.: Temperate North America; Europe.
Notes: Savile (Mycologia 53:49. 1961).
Host: Lilium
Cercosporella mirabilis Peck
Distr.: CO(type), NY, WI.
Host: Crataegus
Cercosporella mori Peck
Distr.: Central and eastern United States.
Hosts: Broussonetia, Morus
Cercosporella palustris R. Sprague
Distr.: WA.
Host: Carex
Cercosporella pastinacae P. Karst.
Cercospora pastinacina Solheim
Ramularia pastinacae Bubák non (P. Karst.) Lindr. & Vestergr., hom. illeg.
Ramularia pastinacae (P. Karst.) Lindr. & Vestergr.
Distr.: Temperate North America; Europe.
Notes: Guba (Pl. Dis. Reporter 39:747. 1955) provided a discussion of the synonymy.
Host: Pastinaca
Cercosporella poagena R. Sprague
Distr.: OR.
Host: Poa
Cercosporella pontederiae (Ellis & Dearn.) J.J. Davis
Cercospora pontederiae Ellis & Dearn.
Distr.: WI, TX; Canada(Ontario, type).
Notes: Savile (Canad. J. Bot. 35:201. 1957).
Host: Pontederia
Cercosporella prolificans (Ellis & Holw.) Sacc.
Cercospora prolificans Ellis & Holw.
Distr.: CA(type), NM, OR.
Host: Sambucus
Cercosporella pycnanthemi Atk.
Distr.: AL(type).
Host: Pycnanthemum
Cercosporella pyrina Ellis & Everh.
Cercospora pirina Ellis & Everh., nom. nud.
Distr.: IL, MI, WI(type).
Hosts: Aronia, Malus
Cercosporella rubi (G. Wint.) Plakidas
Fusisporium rubi G. Wint.
Distr.: Eastern North America.
Host: Rubus
Cercosporella saxifragae Rostr.
Distr.: WI; Norway(type).
Host: Saxifraga
Cercosporella scirpina J.J. Davis
Distr.: WI.
Host: Scirpus
Cercosporella trichophila J.J. Davis
Distr.: WI.
Host: Fraxinus
Cercosporella trollii (Jacz.) Bubák
Didymaria trollii Jacz.
Ramularia trollii (Jacz.) Lindr.
Distr.: CO; Finland; Soviet Union.
Host: Trollius
Cercosporella veratri Peck
Distr.: NY(type); Colombia.
Notes: Savile (Mycologia 53:50. 1961).
Host: Veratrum
Cercosporella virgaureae (Thuem.) Allesch.
Ramularia virgaureae Thuem.
Cercospora virgaureae (Thuem.) Allesch. ex Oudem.
Ovularia virgaureae (Thuem.) Sacc.
Cercospora cana Sacc.
Cercosporella cana (Sacc.) Sacc.
Cercosporella cana (Sacc.) Sacc. var. *gracilis* J.J. Davis
Cercosporella dearnessii Bubák & Sacc.
Cercospora reticulata (Peck) Ellis & Everh.
Cercospora viminei Tehon
Distr.: Temperate northern hemisphere.
Notes: Leaf spot of Asteraceae (Deighton. Mycol. Pap. 133:4. 1973; Pirozysnki. Fungi Canadenses 61. 1975).
Hosts: Aster, Conyza, Erigeron, Eriophyllum, Grindelia, Solidago

Cercosporidium Earle. Hyphomycetes

Notes: Deighton (Mycol. Pap. 112:27. 1967); Ellis (Dematiaceous Hyphomycetes. p. 279. 1971); Ellis (More dematiaceous Hyphomycetes. p. 294. 1976).

Cercosporidium angelicae (Ellis & Everh.) Deighton
Fusicladium angelicae Ellis & Everh.
Distr.: Temperate North America; Europe.
Notes: Leaf spot.
Hosts: Angelica, [Heracleum], [Osmorhiza], [Taenidia]

Cercosporidium bougainvilleae (Muntañola) Sobers & C.P. Seymour
Cercospora bougainvilleae Muntañola
Distr.: FL; West Indies; South America.
Notes: Leaf spot (Ellis. More dematiaceous Hyphomycetes. p. 297. 1976).
Host: Bougainvillea

Cercosporidium cassiae (Henn.) Deighton
Cercospora cassiae Henn.
Berteromyces aeneus Cif.
Distr.: Central United States; West Indies; eastern Africa.
Notes: Leaf spot.
Host: Senna

Cercosporidium cassiocarpum (Sacc.) L.G. Brown & Morgan-Jones
Cercospora occidentalis Cooke var. *cassiocarpa* Sacc.
Cercospora cassiocarpa (Sacc.) Chupp
Cercospora europa (Gonz. Frag.) O. Constantinescu
Cercospora psoraleae W.W. Ray
Cercospora psoraleae-bituminosae Savul. & Sandu
Distr.: OK; South America; southern Europe; The Philippines.
Notes: Leaf and pod spot (Brown & Morgan-Jones. Mycotaxon 6:274. 1977).
Host: Pediomelum

Cercosporidium chaetomium (Cooke) Deighton
Cladosporium chaetomium Cooke
Cercosporidium euphorbiae (Tracy & Earle) Earle, nom. illeg.
Pyricularia euphorbiae (Tracy & Earle) Atk.
Scolicotrichum euphorbiae Tracy & Earle
Distr.: Central and southern United States; Cuba.
Notes: On stems, leaves, and flowers.
Host: Euphorbia

Cercosporidium compactum (Berk. & M.A. Curtis) Deighton
Cladosporium compactum Berk. & M.A. Curtis
Cercospora scolecotrichoides Atk.
Distr.: Southeastern United States.
Notes: On leaves.
Host: Arundinaria

Cercosporidium depressum (Berk. & Broome) Deighton
Cladosporium depressum Berk. & Broome
Fusicladium depressum (Berk. & Broome) Roum.
Cercospora depressa (Berk. & Broome) Vassiljevsky in Vassiljevsky & Karakulin
Megacladosporium depresssum (Berk. & Broome) Vien.-Bourg.
Scolicotrichum depressum (Berk. & Broome) J. Schröt. in Cohn
Passalora depressa (Berk. & Broome) Sacc.
Passalora polythrincioides Fuckel
Distr.: Temperate northern hemisphere.
Notes: On leaves.
Hosts: Angelica, Osmorhiza, Oxypolis, Perideridia, Sanicula

Cercosporidium desmanthi (Ellis & Kellerm.) Dearn.
Cercospora condensata Ellis & Kellerm. var. *desmanthi* Ellis & Kellerm.
Cercospora desmanthi (Ellis & Kellerm.) Ellis & Everh.
Distr.: Central, eastern and southwestern United States; Jamaica.
Notes: Leaf spot.
Host: Desmanthus

Cercosporidium galii (Ellis & Holw.) Deighton
Cercospora galii Ellis & Holw.
Cercospora concinna Syd.
Passalora ruthenica (Petr.) Petr.
Fusicladium ruthenicum Petr.
Scolicotrichum ruthenicum (Petr.) Karakulin & Vassiljevsky
Distr.: Eastern North America; Europe.
Notes: Leaf spot.
Host: Galium

Cercosporidium graminis (Fuckel) Deighton
Scolicotrichum graminis Fuckel
Cercospora graminicola Tracy & Earle
Napicladium gramineum Peck
Cercospora graminis (Fuckel) Horsfall
Passalora graminis (Fuckel) Höhn.
Distr.: Temperate regions.
Notes: Leaf streak of grasses.
Hosts: Agrohordeum, Agropyron, Agrositanion, Agrostis, Alopecurus, Ammophila, Arctagrostis, Arrhenatherum, Arundinaria, Avena, Beckmannia, Bromus, Calamagrostis, Cinna, Cynosurus, Dactylis, Danthonia, Deschampsia, Digitaria, Elymus, Elysitanion, Eragrostis, Festuca, Glyceria, Hordeum, Hystrix, Koeleria, Leersia, Leucopoa, Lolium, Melica, Milium, Muhlenbergia, Oryzopsis, Phleum, Phragmites, Poa, Puccinellia, Secale, Sitanion, Spartina, Stipa, Trisetum

Cercosporidium guanicense (F. Stevens) Deighton
Cladosporium guanicense F. Stevens
Distr.: Neotropical region.
Host: Argemone

Cercosporidium henningsii (Allesch.) Deighton
Cercospora henningsii Allesch.
Cercospora cassavae Ellis & Everh.
Cercospora manihotis Henn. in De Wild.
Septogloeum manihotis A. Zimmerm.
Mycosphaerella henningsii Sivanesan (teleomorph)
Distr.: Subtropical and tropical regions.
Notes: Brown leaf spot (Sivanesan. Trans. Brit. Mycol. Soc. 84:551. 1985; Wilson C.M.I. Descr. 912. 1987).
Host: Manihot

Cercosporidium magnoliae (Ellis & Harkn.) Sivanesan
Cercospora magnoliae Ellis & Harkn.
Isariopsis magnoliae Plakidas
Phaeoisariopsis magnoliae (Ellis & Harkn.) S.C. Jong & E.F. Morris
Mycosphaerella milleri C.S. Hodges & Haasis (teleomorph)
Distr.: Southeastern United States.
Notes: Angular leaf spot (Hodges & Haasis. Mycologia 56:53. 1964; Sivanesan. The bitunicate Ascomycetes. p. 220. 1984).
Host: Magnolia

Cercosporidium manitobanum (J.J. Davis) Sutton
Cercospora manitobana J.J. Davis
Distr.: ID, MT; Canada(Manitoba, Saskatchewan).
Notes: Sutton (Mycol. Pap. 132:25. 1973).
Host: Elaeagnus

Cercosporidium punctiforme (G. Wint.) Deighton
Fusicladium punctiforme G. Wint. in Rabenh.
Fusicladium depressum (Berk. & Broome) Sacc. var. *platysporum* (Ellis & Holw.) J.J. Davis
Cercospora platyspora Ellis & Holw.
Didymaria platyspora (Ellis & Holw.) Ellis & Everh.
Distr.: Central and eastern North America.
Notes: On leaves.
Hosts: Oxypolis, Perideridia, Pimpinella, Taenidia

Cercosporidium punctum (Lacr.) Deighton
Cercospora anethi Sacc.
Cercospora apii Fresen. var. *petroselini* Sacc.
Fusicladium depresseum (Berk. & Broome) Sacc. f. *petroselini* Sacc.
Cercospora foeniculi Magnus
Cercospora petroselini Sacc.
Mycosphaerella anethi (Pers.:Fr.) Petr. (teleomorph)
Phoma anethi (Pers.:Fr.) Sacc. (synanamorph)
Distr.: Northern hemisphere to Kenya.
Notes: On leaves and stems of Apiaceae (Deighton. Mycol. Pap. 112:47. 1967). The *Phoma anethi* synanamorph is probably an *Asteromella*.
Hosts: Anethum, Petroselinum

Cercosporidium sii (Ellis & Everh.) Deighton
Cercospora sii Ellis & Everh.
Fusicladium depressum (Berk. & Broome) Sacc. var. *sii* (Ellis & Everh.) J.J. Davis
Distr.: Central United States.
Notes: Deighton (Mycol. Pap. 112:56. 1967).
Host: Sium

Cerebella Ces. in Rabenh. Hyphomycetes

Notes: Langdon (Mycol. Pap. 61:8. 1955).
Cerebella andropogonis Ces.
Epicoccum andropogonis (Ces.) Schol-Schwarz
Cerebella anthaenantiae Tracy & Earle
Stigmella graminicola Linder
Cerebella panici Tracy & Earle
Cerebella paspali Cooke & Massee
Cerebella sorghi Tracy & Earle
Cerebella spartinae Ellis & Everh.
Distr.: Cosmopolitan.
Notes: Head mold of grasses, often associated with ergot.
Hosts: Andropogon, Anthaenantia, Axonopus, Bothriochloa, Brachiaria, Distichlis, Erianthus, Heteropogon, Panicum, Paspalum, Rhynchelytrum, Saccharum, Schizachyrium, Sorghastrum

Cerinomyces G.W. Martin. Basidiomycetes, Dacrymycetales

Cerinomyces ceraceus Ginns
Distr.: MS.
Notes: On limbs. Ginns (Canad. J. Bot. 60:519. 1982).
Host: Magnolia
Cerinomyces pallidus G.W. Martin
Distr.: North and South America.
Notes: On wood.
Hosts: Quercus, Tsuga

Ceriospora Niessl. Pyrenomycetes, Amphisphaeriales

Ceriospora dubyi Niessl
Distr.: NM; Europe(type).
Host: Humulus
Ceriospora sandwicensis Petr.
Distr.: HI(type).
Host: Gunnera

Ceriosporopsis Linder in Barghoorn & Linder. Pyrenomycetes, Halosphaeriales

Ceriosporopsis halima Linder in Barghoorn & Linder
Distr.: Worldwide in marine habitats and inland salt lakes.
Hosts: Pinus, Spartina

Ceriporia Donk. Basidiomycetes, Aphyllophorales

Notes: Gilbertson & Ryvarden (North American Polypores 1:177. 1986).
Ceriporia alachuana (Murrill) Hallenberg
Poria alachuana Murrill
Distr.: Warmer parts of North America.
Notes: On hardwoods.
Hosts: Carya, Quercus
Ceriporia excelsa (S. Lundell) Parmasto
Poria excelsa S. Lundell in S. Lundell & Nannf.
Distr.: Widespread.
Notes: On wood.
Host: Quercus
Ceriporia purpurea (Fr.:Fr.) Donk
Poria purpurea (Fr.:Fr.) Cooke
Ceriporia bresadolae (Bourd. & Galzin) Bondartsev & Singer
Poria bresadolae Bourd. & Galzin
Poria purpurea (Fr.:Fr.) Cooke var. *roseo-lilacina* Bres.
Distr.: Temperate northern hemisphere.
Notes: On wood.
Hosts: Acer, Alnus, Arbutus, Fouquieria, Juglans, Juniperus, Picea, Pinus, Platanus, Populus, Pseudotsuga, Quercus
Ceriporia reticulata (Pers.: Fr.) Dománski
Poria reticulata (Pers.:Fr.) Cooke
Distr.: Widespread.
Notes: On wood.
Hosts: Alnus, Betula, Juglans, Picea, Populus, Quercus, Salix
Ceriporia spissa (Schwein.) Rajchenberg
Physisporinus spissus (Schwein.) Murrill
Poria spissa (Schwein.) Cooke
Poria crocipora (Berk. & M.A. Curtis) Sacc.
Polyporus crociporus Berk. & M.A. Curtis in M.A. Curtis
Distr.: Widespread.
Notes: Usually on hardwoods.
Hosts: Acer, Alnus, Betula, Cercis, Malus, Pinus, Platanus, Populus, Pseudotsuga, Quercus
Ceriporia tarda (Berk.) Ginns
Poria tarda (Berk.) Cooke
Poria semitincta (Peck) Cooke
Distr.: North America.
Notes: On wood.
Hosts: Abies, Arctostaphylos, Chilopsis, Cornus, Fagus, Fouquieria, Fraxinus, Juniperus, Nyssa, Pinus, Populus, Prosopis, Quercus, Yucca
Ceriporia viridans (Berk. & Broome) Donk
Polyporus viridans Berk. & Broome
Poria viridans (Berk. & Broome) Cooke
Poria rhodella (Fr.:Fr.) Cooke, nom. ambig.
Distr.: Widespread.
Notes: On wood.
Hosts: Alnus, Ceanothus, Fraxinus, Heteromeles, Juglans, Lithocarpus, Pinus, Platanus, Populus, Sambucus
Ceriporia xylostromatoides (Berk.) Ryvarden & I. Johansen
Poria xylostromatoides (Berk.) Cooke
Poria cremeicolor Murrill
Poria flaccida Overh.
Poria subcollapsa Murrill
Poria subcorticola Murrill
Poria taxodium D. Baxter
Distr.: Southern United States; pantropical and subtropical.
Hosts: Carya, Citrus, Fouquieria, Fraxinus, Juniperus, Liquidambar, Myrtus, Nandina, Nerium, Phoenix, Platanus, Populus, Quercus, Rosa, Syringa, Taxodium, Washingtonia

Ceriporiopsis Dománski. Basidiomycetes, Aphyllophorales

Notes: Gilbertson & Ryvarden (North American Polypores 1:187. 1986).
Ceriporiopsis aneirina (Sommerf.:Fr.) Dománski
Poria aneirina (Sommerf.:Fr.) Cooke
Distr.: Temperate northern hemisphere.
Notes: Usually on aspen in North America.
Hosts: Betula, Populus
Ceriporiopsis carnegiea (D. Baxter) R.L. Gilbertson & Ryvarden
Poria carnegiea D. Baxter
Distr.: North America(Sonoran Desert).
Notes: On Cactaceae.
Host: Carnegiea
Ceriporiopsis gilvescens (Bres.) Dománski
Poria gilvescens Bres.
Distr.: Northeastern and central North America; Europe.
Notes: On hardwoods.
Hosts: Alnus, Pinus
Ceriporiopsis mucida (Pers.:Fr.) R.L. Gilbertson & Ryvarden
Poria mucida Pers.:Fr.
Distr.: Temperate northern hemisphere.
Notes: On wood.
Hosts: Alnus, Quercus
Ceriporiopsis pannocincta (Romell) R.L. Gilbertson & Ryvarden
Gloeoporus pannocinctus (Romell) J. Eriksson
Poria pannocincta (Romell) J. Lowe
Distr.: Temperate northern hemisphere.
Notes: On hardwoods, especially aspen.
Hosts: Betula, Malus, Pinus, Populus
Ceriporiopsis rivulosa (Berk. & M.A. Curtis) R.L. Gilbertson & Ryvarden
Poria rivulosa (Berk. & M.A. Curtis) Cooke
Poria albipellucida D. Baxter
Distr.: North America, most common in the West.
Notes: On conifers.
Hosts: Abies, Larix, [Platanus], Pseudotsuga, Sequoia, Thuja, Tsuga

Ceriporiopsis subrufa (Ellis & Dearn.) Ginns
Poria subrufa Ellis & Dearn.
Distr.: Eastern and northcentral United States and adjacent Canada.
Notes: On hardwoods.
Host: Pinus
Ceriporiopsis subvermispora (Pilát) R.L. Gilbertson & Ryvarden
Poria subvermispora Pilát
Spongiporus subvermisporus (Pilát) Ryvarden & R.L. Gilbertson
Poria notata Overh., nom. inval.
Distr.: Temperate northern hemisphaere.
Notes: On wood.
Hosts: Abies, Picea, Pinus, Pseudotsuga

Cerocorticium Henn. & Warb. Basidiomycetes, Aphyllophorales

Cerocorticium confluens (Fr.:Fr.) Jülich & Stalpers
Corticium confluens (Fr.:Fr.) Fr.
Radulomyces confluens (Fr.:Fr.) M. Christiansen
Coniophora avellanea Burt
Corticium rubellum Burt
Distr.: Widespread.
Notes: On wood of hardwoods.
Hosts: Alnus, Fouquieria, Juglans, Populus, Quercus, Robinia
Cerocorticium cremoricolor (Berk. & M.A. Curtis) Ginns
Corticium cremoricolor Berk. & M.A. Curtis
Distr.: North America.
Notes: On wood of hardwoods.
Hosts: Abies, Corylus
Cerocorticium fuscum (C.G. Lloyd) Jülich & Stalpers
Merulius fuscus C.G. Lloyd
Radulomyces fuscus (C.G. Lloyd) Ginns
Serpula fusca (C.G. Lloyd) W.B. Cooke
Distr.: PA; Tasmania.
Notes: On wood.
Host: Quercus
Cerocorticium hiemale (Laurila) Jülich & Stalpers
Corticium hiemale Laurila
Aleurodiscus hiemalis (Laurila) J. Eriksson ex R. Sprague, comb. inval.
Radulomyces hiemalis (Laurila) Parmasto
Distr.: Temperate northern hemisphere.
Notes: On wood of conifers.
Hosts: Picea, Thuja
Cerocorticium notabile (H. Jacks.) Jülich & Stalpers
Corticium notabile H. Jacks.
Distr.: Northern hemisphere.
Notes: On wood.
Host: Castanea
Cerocorticium sulfureo-isabellinum (Litsch.) Jülich & Stalpers
Corticium sulfureo-isabellinum Litsch. in Pilát
Phlebia sulfureo-isabellina (Litsch.) Parmasto
Distr.: North America.
Notes: On conifers.
Host: Abies

Cerotelium Arth. Basidiomycetes, Uredinales

Cerotelium dicentrae Mains & H.W. Anderson in Mains
Distr.: Central and eastern United States.
Notes: Aecial host - *Dicentra*; telial host - *LaPortea*.
Host: Dicentra
Cerotelium fici (E.J. Butler) Arth.
Physopella fici Arth.
Distr.: Subtropical and tropical regions.
Notes: Common fig rust (Laundon & Rainbow. C.M.I. Descr. 281. 1971). Aecia unknown.
Hosts: Ficus, Maclura, Morus
Cerotelium tanakae Ito in Ito & Homma
Distr.: IN; Japan.
Notes: Aecia unknown.
Host: Amphicarpaea

Cerrena S.F. Gray. Basidiomycetes, Aphyllophorales

Cerrena unicolor (Bull.:Fr.) Murrill
Daedalea unicolor (Bull.:Fr.) Fr.
Distr.: Temperate northern hemisphere and high mountains of Africa.
Notes: On hardwoods.
Hosts: Abies, Acer, Ailanthus, Alnus, Amelanchier, Betula, Carya, Castanea, Celtis, Fagus, Liquidambar, Liriodendron, Malus, Nyssa, Ostrya, Pinus, Populus, Prunus, Quercus, Rhamnus, Robinia, Salix, Tilia, Tsuga, Ulmus

Ceuthospora Grev., nom. cons. Coelomycetes

Notes: Sutton (The Coelomycetes. p. 471. 1980).
Ceuthospora brevispora Cooke & Harkn.
Distr.: CA(type).
Host: Heteromeles
Ceuthospora foliicola (Lib.) Jaap
Cytospora foliicola Lib.
Ceuthospora feurichii Bubák
Phacidium vincae Fuckel (teleomorph)
Distr.: Widespread.
Notes: Dicosmo et al. (Mycotaxon 21:162. 1984).
Hosts: Garrya, Kalmia, Rhamnus
Ceuthospora latitans (Fr.:Fr.) Höhn.
Dothidea latitans Fr.:Fr.
Dothiopsis latitans (Fr.:Fr.) P. Karst.
Dothiorella latitans (Fr.:Fr.) Sacc.
Phyllachora latitans (Fr.:Fr.) Sacc.
Phacidium vaccinii Fr. (teleomorph)
Distr.: AK; Finland; Sweden.
Notes: DiCosmo et al. (Mycotaxon 21:157. 1984).
Host: Vaccinium
Ceuthospora lycopodii Lind
Distr.: Eastern North America; Europe.
Host: Lycopodium
Ceuthospora minima Cooke & Harkn.
Distr.: CA.
Host: Vaccinium

Chaetapiospora Petr. Pyrenomycetes, Amphisphaeriales

Notes: This genus is considered a synonym of *Pseudomassaria* (Barr. Mycologia 56:843. 1964), but not all names have been redisposed.
Chaetapiospora rhododendri (Tengwall) Arx
Venturia rhododendri Tengwall
Distr.: MD, VA, WA; The Netherlands(type).
Notes: Bigelow & Barr (Rhodora 71:201. 1969) compared this species with *Pseudomassaria thistletonia*.
Host: Rhododendron

Chaetasbolisia Speg. Coelomycetes

Chaetasbolisia falcata V. Miller & Bonar
Distr.: CA, NC; Caribbean Islands; Brazil; China.
Notes: Sooty mold on numerous hosts.
Hosts: Lithocarpus, Polystichum, Rhamnus, Sequoia, Umbellularia, Woodwardia

Chaetoblastophorum Morgan-Jones. Hyphomycetes

Chaetoblastophorum ingramii Morgan-Jones
Distr.: AL.
Host: Quercus

Chaetocalathus Singer. Basidiomycetes, Agaricales

Chaetocalathus bicolor (Pat. & Demange) Singer
Lachnella bicolor (Pat. & Demange) Locquin
Distr.: CA; Asia.
Host: Castanopsis

Chaetochalara Sutton & Pirozynski. Hyphomycetes

Chaetochalara aspera Pirozynski & C.S. Hodges
Calycellina carolinensis Nag Raj & Kendrick (teleomorph)
Distr.: Southeastern United States; China; New Zealand.
Notes: On leaves. Korf & Zhuang (Mycotaxon 22:498. 1985).
Hosts: Myrica, Persea
Chaetochalara setosa (Harkn.) Nag Raj & Kendrick
Chalara setosa Harkn.
Distr.: CA(type).
Host: Lithocarpus

Chaetoconis Clements. Coelomycetes

Chaetoconis polygoni (Ellis & Everh.) Clements
Kellermania polygoni Ellis & Everh.
Amphorula polygoni (Ellis & Everh.) Petr.
Kellermania rumicis Fautrey & Lambotte
Ceriospora polygonacearum (Petr.) Pirozynski & Morgan-Jones (teleomorph)
Distr.: Temperate North America; Europe.
Notes: On dead stems.
Host: Polygonum

Chaetoderma Parmasto. Basidiomycetes, Aphyllophorales

Chaetoderma luna (Romell ex D.P. Rogers & H. Jacks.) Parmasto
Peniophora luna Romell ex D.P. Rogers & H. Jacks.
Peniophora odorata (P. Karst.) Burt, sensu Burt
Distr.: Temperate northern hemisphere.
Notes: On wood of conifers.
Hosts: Abies, Larix, Picea, Pinus, Pseudotsuga, Thuja

Chaetomella Fuckel. Coelomycetes

Notes: Stolk (Trans. Brit. Mycol. Soc. 46:409. 1963).
Chaetomella oblonga Fuckel
Chaetomella oblonga Fuckel var. *major* Sacc.
Amerosporium patellarioides A.L. Sm. & Ramsb.
Distr.: CA; Europe; Great Britain.
Host: Chamaecyparis

Chaetomium Kunze:Fr. Pyrenomycetes, Sordariales

Notes: Ames (U.S. Army Res. Developm., Ser. 2:1. 1961); Arx Persoonia 12:169. 1984).
Chaetomium aterrimum Ellis & Everh.
Distr.: KS.
Host: Triticum
Chaetomium aureum Chivers
Chaetomium minimum Van Beyma
Distr.: NJ, MI; Canada; Mexico.
Notes: Also on paper and dung.
Hosts: Capsicum, Cucumis
Chaetomium bostrychodes Zopf
Distr.: Widespread in the northern hemisphere.
Hosts: Capsicum, Lycopersicon, Zea
Chaetomium brasiliense Batista & Pontual
Distr.: Widespread.
Host: Zea
Chaetomium convolutum Chivers
Distr.: ID; Canada; Germany.
Notes: Also on decaying cotton and mouse dung.
Host: Tragopogon
Chaetomium cupreum L. Ames
Distr.: Widespread.
Host: Glycine
Chaetomium dolichotrichum L. Ames
Distr.: Midwestern United States; Canada.
Notes: Also on paper.
Host: Zea
Chaetomium elatum J.C. Schmidt & Kunze:Fr.
Distr.: Cosmopolitan.
Hosts: Gossypium, Pseudotsuga
Chaetomium erectum Skolko & Groves
Distr.: CT(type);Britain..
Host: Pinus
Chaetomium funicola Cooke
Distr.: Cosmopolitan.
Hosts: Alnus, Arachis, Gossypium, Malus, Pseudotsuga, Spartina, Zea
Chaetomium globosum Kunze:Fr.
Chaetomium cochliodes Palliser
Chaetomium olivaceum Cooke & Ellis
Distr.: Cosmopolitan.
Notes: *Chaetomium olivaceum* Cooke & Ellis is sometimes considered to be a distinct species.
Hosts: Alnus, Arachis, Beta, Brassica, Cucumis, Cucurbita, Glycine, Gossypium, Helianthus, Lens, Lycopersicon, Malus, Metrosideros, Pastinaca, Phaseolus, Pinus, Platanus, Pyrus, Zea
Chaetomium incomptum L. Ames
Distr.: CA, GA.
Notes: Also on vegetable detritus. Arx. et al. (Beih. Nova Hedwigia 84:62. 1986).
Host: Helianthus
Chaetomium indicum Corda
Chaetomium melioloides Cooke & Peck
Distr.: Cosmopolitan.
Hosts: Helianthus, Zea
Chaetomium megalocarpum Bainier
Chaetomium atrosporum Skolko & Groves
Distr.: Widespread in Europe and North America.
Host: Daucus
Chaetomium murorum Corda
Distr.: Widespread in the northern hemisphere.
Hosts: Abies, Beta, Cucurbita, Pastinaca, Petroselinum, Populus, Zea
Chaetomium spirale Zopf
Distr.: Widespread.
Hosts: Carya, Helianthus
Chaetomium succineum L. Ames
Distr.: CA.
Notes: Isolated from rotting organic material.
Hosts: Apium, Beta, Petroselinum
Chaetomium torulosum Bainier
Distr.: New England; Canada(Ontario); Japan.
Notes: Also on paper and dung. A doubtful species (Arx. et al. Beih. Nova Hedwigia 84:67. 1986).
Host: Zea
Chaetomium trilaterale Chivers
Distr.: Cosmopolitan.
Hosts: Arachis, Glycine, Malus

Chaetophoma Cooke. Coelomycetes

Chaetophoma pellicula Sacc. & Syd.
Distr.: CA; Chile(type).
Host: Acer
Chaetophoma quercifolia Cooke
Distr.: CA, SC(type).
Host: Quercus
Chaetophoma setigera Peck
Chaetophomella setigera (Peck) Speg.
Distr.: CA.
Hosts: Heteromeles, Quercus

Chaetoplea (Sacc.) Clements. Loculoascomycetes, Dothideales

Chaetoplea oblongispora (Ellis & Everh.) Barr
Teichospora oblongispora Ellis & Everh.
Distr.: CO, NM.
Notes: Barr (Mycotaxon 29:504. 1987).
Host: Populus

Chaetoporellus Bondartsev & Singer ex Singer. Basidiomycetes, Aphyllophorales

Chaetoporellus latitans (Bourd. & Galzin) Bondartsev & Singer ex Singer
Poria latitans Bourd. & Galzin
Distr.: Northern hemisphere.
Notes: On wood of conifers.
Host: Pinus

Chaetopsina Rambelli. Hyphomycetes

Notes: Kirk & Sutton (Trans. Brit. Mycol. Soc. 85:709. 1985); Wingfield (Mycotaxon 30:149. 1987).
Chaetopsina auburnensis Morgan-Jones
Distr.: AL.
Notes: Morgan-Jones (Mycotaxon 8:411. 1979).
Host: Quercus
Chaetopsina fulva Rambelli
Nectria chaetopsinae Samuels (teleomorph)
Distr.: HI; Canada; Italy(type); New Zealand.
Notes: The connection with the *Nectria* teleomorph is tentative (Samuels. Mycotaxon 22:18. 1985).
Host: Eucalyptus
Chaetopsina illinii (J.L. Crane) M.J. Wingfield
Phialocephala illinii J.L. Crane
Distr.: IL.
Host: Taxodium
Chaetopsina ramifera T. Matsushima
Distr.: FL; Brazil; Papua-New Guinea(type).
Host: Quercus
Chaetopsina splendida Sutton & C.S. Hodges
Distr.: HI; Brazil(type).
Host: Eucalyptus

Chaetopsis Grev. Hyphomycetes

Chaetopsis grisea (Ehrenb.:Pers.) Sacc.
Distr.: MI; Czechoslovakia(type); Great Britain.
Notes: On wood and bark of various deciduous trees (Ellis. Dematiaceous Hyphomycetes. p. 564. 1971).
Host: Quercus

Chaetoseptoria Tehon. Coelomycetes

Chaetoseptoria vignae Tehon
Distr.: IL.
Notes: Sutton (Mycol. Pap. 97:11. 1964).
Host: Vigna

Chaetospermum Sacc. Coelomycetes

Chaetospermum carneum Tassi
Chaetospermum camelliae Agnihothrudu
Chaetospermum elasticae Koord.
Chaetospermum elasticae Koord. var. *artocarpi* Nag Raj
Ciliospora gelatinosa A. Zimmerm.
Chaetospermella populinum Naumov
Distr.: Subtropical and tropical regions.
Notes: On leaves, twigs, and stems (Sutton. The Coelomycetes. p. 152. 1980).
Host: Carya
Chaetospermum chaetosporum (Pat.) A.L. Sm. & Ramsb.
Chaetospermum tubercularioides Sacc.
Distr.: North America; Africa; Europe; India.
Host: Paspalum

Chaetosphaeria Tul. & C. Tul. Pyrenomycetes, Trichosphaeriales

Notes: Booth (Mycol. Pap. 68:1. 1957) described several species and their anamorphs, but many names have not been considered in modern studies. Hughes (New Zealand J. Bot. 6:323. 1968).
Chaetosphaeria atrobarbum (Cooke & Ellis) Sacc.
Acanthostigma atrobarbum (Cooke & Ellis) Ellis & Everh.
Distr.: NJ.
Host: Quercus
Chaetosphaeria dingleyae S.J. Hughes, Kendrick & Shoemaker
Codinaea sp. (anamorph)
Distr.: HI; New Zealand(type).
Notes: Hughes & Kendrick (New Zealand J. Bot. 6:343. 1968).
Host: Eucalyptus
Chaetosphaeria ornata Harkn.
Distr.: CA(type).
Host: Eucalyptus
Chaetosphaeria pannicola (Berk. & M.A. Curtis) Sacc.
Distr.: GA, NJ(type).
Hosts: Ostrya, Platanus, Salix
Chaetosphaeria parvicapsa (Cooke) Sacc.
Distr.: GA, SC(type).
Host: Juniperus
Chaetosphaeria pulviscula (Curr.) C. Booth
Zignoella pulviscula (Curr.) Sacc.
Menispora caesia G. Preuss (anamorph)
Distr.: Temperate regions.
Notes: On dead wood. Booth (Mycol. Pap. 68:10. 1957).
Hosts: Betula, Quercus

Chaetosticta Petr. & Syd. Coelomycetes

Chaetosticta perforata (Ellis & Everh.) Petr. & Syd.
Chaetomella perforata Ellis & Everh.
Distr.: Central United States.
Notes: Crane (Canad. J. Bot. 49:31. 1971).
Hosts: Artemisia, Cirsium

Chaetostroma Corda. Hyphomycetes

Chaetostroma aterrima (Cooke) Sacc.
Psilonia aterrima Cooke
Distr.: SC(type).
Host: Cortaderia

Chaetothyrina Theiss. Loculoascomycetes, Dothideales

Chaetopeltopsis Theiss.
Plochmopeltidella Mendoza in F. Stevens & Manter
Setopeltis Batista & Vital
Chaetothyrina asterinoides (Ellis & G. Martin) Barr
Venturia asterinoides Ellis & G. Martin
Distr.: FL.
Host: Quercus
Chaetothyrina musarum (Speg.) Theiss.
Chaetothyrium musarum Speg.
Plochmopeltidella smilacina Mendoza
Distr.: HI; South America.
Host: Musa

Chaetothyrium Speg. Loculoascomycetes, Dothideales

Merismella Syd.
Chaetothyrium hawaiiense Mendoza in F. Stevens
Distr.: HI.
Hosts: Citrus, Morinda
Chaetothyrium mangiferae Mendoza in F. Stevens
Distr.: HI.
Host: Mangifera
Chaetothyrium straussiae Mendoza in F. Stevens
Distr.: HI; Guyana.
Notes: Batista & Ciferri (Beih. Sydowia 3:13. 1962).
Host: Psychotria

Chaetotrichum Syd. non Rabenh., hom. illeg. Hyphomycetes

Notes: A synonym of *Anellophora* S.J. Hughes, but not all names have been redisposed.

Chaetotrichum macrosporum Bonar
Distr.: CA.
Notes: Hughes (Trans. Brit. Mycol. Soc. 34:549. 1951).
Hosts: Gaultheria, Physocarpus
Chaetotrichum sandwicense Petr.
Distr.: HI(type).
Host: Pipturus

Chalara (Corda) Rabenh. Hyphomycetes
Notes: Nag Raj & Kendrick (A monograph of *Chalara* and allied genera. 1975).
Chalara acuaria Cooke & Ellis
Distr.: Eastern United States; New Zealand.
Notes: On needles.
Hosts: Abies, Pinus
Chalara cibotti (Plunkett) Nag Raj & Kendrick
Excioconidium cibotti Plunkett in F. Stevens
Distr.: HI.
Notes: On dead leaves and rachi.
Host: Cibotium
Chalara longipes (G. Preuss) Cooke
Cylindrosporium longipes G. Preuss
Distr.: FL; Germany(type).
Host: Serenoa
Chalara paradoxa (De Seyn.) Sacc.
Thielaviopsis paradoxa (De Seyn.) Höhn.
Thielaviopsis ethacetica Went
Ceratocystis paradoxa (Dade) C. Moreau (teleomorph)
Distr.: Subtropical and tropical regions.
Hosts: Ananas, Arecaceae, Ficus, Musa, Plumeria, Saccharum
Chalara quercina B.W. Henry
Ceratocystis fagacearum (T.W. Bretz) J. Hunt (teleomorph)
Distr.: Eastern North America.
Notes: Oak wilt.
Hosts: Castanea, Quercus
Chalara selaginellae M.L. Farr in M.L. Farr & H.T. Horner
Distr.: FL, NC.
Notes: Farr & Horner (Nova Hedwigia 15:269. 1968).
Host: Selaginella
Chalara thielavioides (Peyronel) Nag Raj & Kendrick
Chalaropsis thielavioides Peyronel
Distr.: Temperate and subtropical regions
Hosts: Cryptomeria, Daucus, Euphorbia, Ilex, Juniperus, Rosa, Thuja, Ulmus
Chalara ungeri Sacc.
Ceratocystis coerulescens (Münch) Bakshi (teleomorph)
Distr.: Temperate North America; Europe.
Notes: On wood; often causes blue stain.
Host: Pseudotsuga

Chapeckia Barr. Pyrenomycetes, Diaporthales
Notes: Barr (Mycol. Mem. 7:164. 1978).
Chapeckia nigrospora (Peck) Barr
Melanconis nigrospora (Peck) Wehmeyer
Distr.: Northeastern North America.
Host: Betula
Chapeckia ribesia (Sacc. & Scalia) Barr
Pseudovalsa ribesia Sacc. & Scalia in Sacc., Peck, & Trel.
Distr.: Western North America.
Notes: On limbs.
Host: Ribes

Charomyces Seifert. Hyphomycetes
Notes: Seifert (Canad. J. Bot. 65:230. 1987).
Charomyces amphimelas Seifert
Distr.: HI.
Notes: On fallen debris.
Host: Poaceae

Cheilaria Lib. Coelomycetes
Cheilaria agrostis Lib.
Labrella agrostis (Lib.) Sacc.
Placosphaeria graminis Sacc. & Roum.
Septogloeum oxysporum E. Bommer, M. Rousseau, & Sacc.
Distr.: Temperate regions.
Notes: Tar spot or blotch of Poaceae (Gibson & Sutton. C.M.I. Descr. 488. 1976).
Hosts: Agropyron, Agrostis, Alopecurus, Arrhenatherum, Bromus, Calamagrostis, Elymus, Festuca, Glyceria, Hordeum, Muhlenbergia, Poa, Puccinellia, Trisetum

Cheimonophyllum Singer. Basidiomycetes, Agaricales
Cheimonophyllum candidissimum (Berk. & M.A. Curtis) Singer
Pleurotus candidissimum (Berk. & M.A. Curtis) Sacc.
Geopetalum haedinum (Berk. & M.A. Curtis) Murrill
Pleurotus haedinus (Berk. & M.A. Curtis) Sacc.
Geopetalum oregonense Murrill
Pleurotus oregonense (Murrill) Murrill
Geopetalum subelatinum Murrill
Pleurotus subelatinum (Murrill) Murrill
Distr.: Cosmopolitan.
Notes: On wood and plant debris.
Host: Acer

Cheiromyces Berk. & M.A. Curtis. Hyphomycetes
Cheiromyces stellatus Berk. & M.A. Curtis
Distr.: Eastern United States.
Notes: Sutton (Proc. Indian Acad. Sci. 94:229. 1985).
Host: Scirpus

Chionosphaera Cox. Basidiomycetes, Filobasidiales
Notes: Cox (Mycologia 68:481. 1976).
Chionosphaera apobasidialis Cox
Distr.: Central United States.
Notes: On hardwood limbs.
Hosts: Carpinus, Quercus

Chlorencoelia J.R. Dixon. Discomycetes, Helotiales
Notes: Dixon (Mycotaxon 1:224. 1974).
Chlorencoelia versiformis (Pers.:Fr.) J.R. Dixon
Chlorociboria versiformis (Pers.:Fr.) Ramamurthi, Korf, & Batra
Midotis versiformis (Pers.:Fr.) Seaver
Distr.: Eastern North America to Mexico; Europe.
Hosts: Castanea, Quercus

Chlorociboria Seaver ex Ramamurthi, Korf, & Batra, nom. cons. Discomycetes, Helotiales
Notes: Dixon (Mycotaxon 1:193. 1974).
Chlorociboria aeruginosa (Pers.:Fr.) Seaver ex Ramamurthi, Korf, & Batra
Chlorosplenium aeruginosum (Pers.:Fr.) Auct.
Distr.: Cosmopolitan.
Hosts: Abies, Acer, Alnus, Arctostaphylos, Betula, Castanea, Fraxinus, Liriodendron, Lithocarpus, Populus, Pseudotsuga, Quercus, Salix

Chloroscypha Seaver. Discomycetes, Helotiales
Kriegeria Rabenh.
Notes: Weak pathogens of Cupressaceae. Petrini (Sydowia 35:206. 1982).
Chloroscypha alutipes (W. Phillips) Dennis

Kriegeria alutipes (W. Phillips) Seaver, nom. inval.
Distr.: CA, OR.
Hosts: Calocedrus, Chamaecyparis
Chloroscypha cedrina (Cooke) Seaver
Distr.: NC, NY.
Host: Juniperus
Chloroscypha chloromela (W. Phillips & Harkn.) Seaver
Kriegeria chloromela (W. Phillips & Harkn.) Seaver, nom. inval.
Distr.: CA.
Host: Sequoia
Chloroscypha enterochroma (Peck) Petrini
Kriegeria enterochroma (Peck) Seaver, nom. inval.
Ombrophila enterochroma (Peck) Sacc.
Chloroscypha jacksonii Seaver
Distr.: North America.
Host: Calocedrus
Chloroscypha sabinae (Fuckel) Dennis
Dermatea juniperina Ellis
Kriegeria juniperina (Ellis) Seaver, nom. inval.
Distr.: Widespread in Europe and North America.
Host: Juniperus
Chloroscypha seaveri Rehm ex Seaver
Distr.: North America; Europe.
Notes: Petrini (Sydowia 35:216. 1982) recognized two forms.
Host: Thuja

Chlorosplenium Fr. Discomycetes, Helotiales

Notes: Dixon (Mycotaxon 1:65. 1974).
Chlorosplenium chlora (Schwein.:Fr.) M.A. Curtis in C. Sprague
Lachnella chlorascens (Schwein.) Morg.
Distr.: Cosmopolitan.
Notes: On decorticated and decayed wood.
Hosts: Castanea, Quercus

Choanephora Curr. Zygomycetes, Mucorales

Notes: Kirk (Mycol. Pap. 152:29. 1984).
Choanephora cucurbitarum (Berk. & Ravenel) Thaxt.
Choanephora americana A. Möller
Distr.: Temperate to tropical regions.
Notes: Fruit rot and blossom blight of cucurbits and many other plants.
Hosts: Abelmoschus, Amaranthus, Capsicum, Carica, Citrullus, Cucumis, Cucurbita, Dahlia, Desmodium, Ficus, Gossypium, Hemerocallis, Hibiscus, Ipomoea, Malvaviscus, Nasturtium, Pisum, Sicana, Solanum, Tabernaemontana, Tagetes, Vicia, Vigna, Zinnia
Choanephora infundibulifera (Curr.) Sacc.
Choanephora conjuncta Couch
Distr.: Warm-temperate to tropical regions.
Notes: Saprobic in soil and on plant debris.
Hosts: Abelmoschus, Gossypium, Hibiscus, Ipomoea, Jasminum, Petunia, Pisum

Chondrostereum Pouzar. Basidiomycetes, Aphyllophorales

Chondrostereum purpureum (Pers.:Fr.) Pouzar
Stereum purpureum Pers.:Fr.
Stereum rugosiusculum Berk. & M.A. Curtis
Distr.: Widespread.
Notes: Saprobic on recently dead or on living hardwoods. Silver-leaf disease of fruit trees especially *Prunus* spp.
Hosts: Abies, Acer, Aesculus, Alnus, Amelanchier, Arbutus, Betula, Carpinus, Carya, Castanea, Celtis, Cornus, Cydonia, Fagus, Liriodendron, Malus, Nyssa, Ostrya, Platanus, Populus, Prunus, Pyrus, Quercus, Salix, Syringa, Tsuga, Ulmus, Vitis

Chromelosporium Corda. Hyphomycetes

Chromelosporium fulvum (Link:Fr.) McGinty, Korf & Hennebert in Hennebert & Korf
Botrytis fulva Link:Fr.
Chromelosporium ollare (Pers.) Hennebert
Dematium ollare Pers.
Distr.: Temperate regions.
Notes: Peat mould. On debris associated with heat-sterilized soil, sand, and peat moss in heated greenhouses (Hennebert & Korf. Mycologia 67:214. 1975).
Host: Crambe

Chrysocyclus Syd. Basidiomycetes, Uredinales

Chrysocyclus cestri (Dietel & Henn.) Syd.
Puccinia cestri Dietel & Henn.
Distr.: Central and South America, possibly from FL, but no voucher specimen was deposited to verify this report.
Host: Cestrum

Chrysomyxa Unger. Basidiomycetes, Uredinales

Notes: Ziller (The tree rusts of western Canada. p. 45. 1974).
Chrysomyxa arctostaphyli Dietel
Peridermium coloradense (Dietel) Arth. & F. Kern (anamorph)
Distr.: Northern North America.
Notes: Spruce broom rust. Heteroecious; aecial host - *Picea* spp.; telial host - *Arctostaphylos*.
Hosts: Arctostaphylos, Picea
Chrysomyxa chiogenis Dietel
Distr.: Northeastern North America; Canada(British Columbia).
Notes: Heteroecious; aecial host - *Picea* spp.; telial host - *Gaultheria hispidula*.
Host: Gaultheria
Chrysomyxa empetri J. Schröt. ex Cummins
Distr.: Northern hemisphere; Falkland Islands.
Notes: Heteroecious; aecial host - *Picea* spp.; telial host - *Empetrum nigrum*.
Hosts: Empetrum, Picea
Chrysomyxa ilicina (Arth.) Arth.
Distr.: Southern Appalachian Mountains.
Notes: Heteroecious; aecial host - unknown; telial host - *Ilex opaca*.
Host: Ilex
Chrysomyxa ledi de Bary
Chrysomyxa ledi de Bary var. *glandulosi* Savile
Distr.: Widespread in the northern hemisphere.
Notes: Heteroecious; aecial host - *Picea* spp.; telial host - *Ledum* spp.
Hosts: Ledum, Picea, [Rhododendron]
Chrysomyxa ledi de Bary var. ***cassandrae*** (Peck & G.W. Clinton) Savile
Chrysomyxa cassandrae (Peck & G.W. Clinton) Tranzschel
Melampsoropsis cassandrae (Tranzschel) Arth.
Distr.: Temperate northern hemisphere.
Notes: Heteroecious; aecial host - *Picea* spp.; telial host - *Chamaedaphne calyculata*.
Hosts: [Andromeda], Chamaedaphne, Picea
Chrysomyxa ledi de Bary var. ***rhododendri*** (de Bary) Savile
Chrysomyxa rhododendri de Bary
Distr.: Temperate regions.
Notes: Heteroecious; aecial host - *Picea* spp. telial host - *Rhododendron* spp. (Only uredinia known in North America).
Host: Rhododendron
Chrysomyxa ledicola Lagerh.
Distr.: Northern North America; Japan; Soviet Union.
Notes: Heteroecious; aecial host - *Picea* spp.; telial host - *Ledum* spp.
Hosts: Ledum, Picea, [Rhododendron]
Chrysomyxa monesis Ziller
Distr.: Pacific Northwest.
Notes: Heteroecious; aecial host - *Picea sitchensis*; telial host - *Moneses uniflora*.
Hosts: Moneses, Picea
Chrysomyxa piperiana Sacc. & Trott. ex Cummins
Melampsoropsis piperiana Arth.
Distr.: Pacific Northwest.
Notes: Heteroecious; aecial host - *Picea sitchensis*; telial host - *Rhododendron macrophyllum* and its hybrids.
Hosts: Picea, Rhododendron

Chrysomyxa pirolata G. Wint. in Rabenh.
Chrysomyxa pyrolae (DC.) Rostr., nom. nud.
Distr.: Temperate northern hemisphere.
Notes: Heteroecious; aecial host - *Picea* spp.; telial host - *Pyrola* spp. and *Moneses uniflora*.
Hosts: Moneses, Picea, Pyrola
Chrysomyxa roanensis (Arth.) Arth.
Distr.: TN.
Notes: Heteroecious; aecial host - unknown, but possibly *Picea rubens* (Savile. Canad. J. Res. Sect. C, Bot. Sci. 28:323. 1950); telial host - *Rhododendron* spp.
Hosts: Picea, Rhododendron
Chrysomyxa weirii H. Jacks.
Distr.: Temperate North America; southcentral Asia.
Notes: Autoecious. Sutton (C.M.I. Descr. 221. 1970).
Host: Picea

Chrysonilia Arx. Hyphomycetes

Chrysonilia sitophilia (Mont.) Arx
Monilia sitophila (Mont.) Sacc.
Neurospora sitophila Shear & B.O. Dodge (teleomorph)
Distr.: Cosmopolitan.
Hosts: Cucumis, Saccharum, Zea

Ciboria Fuckel. Discomycetes, Helotiales

Ciboria acerina Whetzel & Buchwald in Groves & M.E. Elliott
Distr.: Northeastern North America.
Notes: On catkins.
Hosts: Acer, Myrica, Salix
Ciboria alni (O. Rostr.) Buchwald
Sclerotinia alni O. Rostr.
Helotium seminicola (Kienholz & Cash) Seaver
Phialea seminicola Kienholz & Cash
Distr.: Northern North America; Europe.
Host: Alnus
Ciboria americana E.J. Durand
Rutstroemia americana (E.J. Durand) W.L. White
Distr.: Eastern North America.
Notes: On involucres.
Host: Castanea
Ciboria aschersoniana (Henn. & Plöttner) Whetzel
Sclerotinia aschersoniana Henn. & Plöttner
Distr.: CO, NY; Europe.
Host: Carex
Ciboria betulae (Woronin) W.L. White
Distr.: Circumboreal.
Notes: On sclerotified fruits. Schumacher (Norw. J. Bot. 25:147.1978).
Host: Betula
Ciboria carunculoides (Siegler & Jenk.) Whetzel
Sclerotinia carunculoides Siegler & Jenk.
Distr.: Southeastern United States.
Notes: On fruits of mulberry (Whetzel & Wolf. Mycologia 37:476.1945).
Host: Morus
Ciboria caucus (Rebent.:Fr.) Fuckel
Ciboria amentacea (Balb.:Fr.) Fuckel
Distr.: Temperate regions.
Notes: On decaying catkins of various trees. Schumacher (Norw. J. Bot. 25:149. 1978).
Hosts: Alnus, Populus, Salix
Ciboria nyssaegena (Ellis) Sacc.
Peziza nyssaegena Ellis
Sclerotinia nyssaegena (Ellis) Rehm
Distr.: NJ(type).
Host: Nyssa
Ciboria rufo-fusca (O. Weberb.) Sacc.
Distr.: Western United States; Canada.
Notes: On cone scales.
Host: Abies
Ciboria tiliae (Reade) Buchwald
Sclerotinia tiliae Reade
Distr.: IA(type).
Host: Tilia

Ciborinia Whetzel. Discomycetes, Helotiales

Notes: Batra (Amer. J. Bot. 47:819. 1960); Batra & Korf (Amer. J. Bot. 46:441. 1959).
Ciborinia camelliae L.M. Kohn
Sclerotinia camelliae K. Hara
Distr.: CA; Japan.
Notes: On fallen flowers (Kohn. Trans. Mycol. Soc. Japan 25:149. 1984).
Host: Camellia
Ciborinia candolleana (Lév.) Whetzel
Phialea candolleana (Lév.) Quél.
Sclerotinia candolleana (Lév.) Fuckel
Distr.: Eastern North America; Europe.
Notes: On leaves.
Host: Quercus
Ciborinia davidsoniana Whetzel ex Groves & Bowerman
Distr.: CO.
Host: Populus
Ciborinia erythronii (Whetzel) Whetzel
Sclerotinia erythronii Whetzel
Distr.: Central and eastern North America.
Host: Erythronium
Ciborinia foliicola (Cash & R.W. Davidson) Whetzel
Sclerotinia foliicola Cash & R.W. Davidson
Distr.: CO(type); Great Britain.
Host: Salix
Ciborinia gracilipes (Cooke) Seaver
Sclerotinia gracilipes (Cooke) Sacc.
Distr.: NJ(type).
Notes: Parasitic on petals.
Host: Magnolia
Ciborinia gracilis (Clements) Whetzel
Sclerotinia gracilis Clements
Distr.: IL, NE.
Host: Erythronium
Ciborinia pseudobifrons Whetzel ex Groves & Bowerman
Distr.: Northeastern North America.
Notes: On leaves and twigs.
Host: Populus
Ciborinia seaveri Groves & Bowerman
Ciborinia confundens Whetzel
Sclerotinia confundens Whetzel
Distr.: Western North America.
Notes: Ink spot disease.
Host: Populus
Ciborinia trillii Batra & Korf
Distr.: Eastern North America.
Host: Trillium
Ciborinia violae Batra & Korf
Distr.: NY, PA.
Host: Viola
Ciborinia whetzelii (Seaver) Seaver
Sclerotinia whetzelii Seaver
Ciborinia bifrons Whetzel
Sclerotinia bifrons Whetzel non Seaver & Shope, hom. illeg.
Sclerotium bifrons Ellis & Everh. ex Sacc. & Syd. (anamorph sclerotial)
Distr.: Eastern North America.
Notes: Ink spot disease.
Host: Populus
Ciborinia wisconsinensis Batra
Distr.: WI.
Notes: On leaves.
Host: Salix

Cilioplea Munk. Loculoascomycetes, Dothideales

Notes: Crivelli (Ueber die heterogene Ascomycetengattung *Pleospora* Rabh.; Vorschlag für eine Aufteilung. p. 170. 1983).
Cilioplea coronata (Niessl) Munk
Pleospora coronata Niessl
Teichospora nepetae Ellis & Everh.

Distr.: Cosmopolitan.
Notes: On herbaceous stems of various dicots.
Host: Salsola
Cilioplea kansensis (Ellis & Everh.) P.G. Crivelli
Pleospora kansensis Ellis & Everh.
Distr.: Temperate North America; Europe.
Host: Melilotus

Cintractia Cornu. Basidiomycetes, Ustilaginales

Notes: Many species formerly treated in *Cintractia* have been placed in *Anthracoidea* or *Ustilago*.
Cintractia affinis Peck
Distr.: NY.
Host: Rhynchospora
Cintractia axicola (Berk.) Cornu
Distr.: Widespread.
Notes: On Cyperaceae (except *Carex* spp.).
Hosts: Fimbristylis, Scirpus
Cintractia carpophila (Schumach.) Liro var. ***kenaica*** Savile
Distr.: AK(type).
Notes: The type variety is now a synonym of *Anthracoidea heterospora*.
Host: Carex
Cintractia cyperi G.P. Clinton
Distr.: Eastern North America.
Host: Cyperus
Cintractia farlowii G.P. Clinton in Zundel
Distr.: Eastern North America.
Host: Rhynchospora
Cintractia fuirenae Hasija & F.A. Wolf
Distr.: NE(type).
Notes: Hasija & Wolf (Mycologia 61:680. 1969).
Host: Fuirena
Cintractia junci (Schwein.) Trel.
Distr.: Western hemisphere.
Host: Juncus
Cintractia leucoderma (Berk.) Henn.
Distr.: Widespread.
Notes: On Cyperaceae (except *Carex* spp.).
Hosts: Cyperus, Rhynchospora
Cintractia limitata G.P. Clinton
Distr.: Warmer parts of the western hemisphere and Africa.
Notes: On Cyperaceae (except *Carex* spp.).
Host: Cyperus
Cintractia pachyderma Syd.
Distr.: FL.
Host: Rhynchospora
Cintractia peribebuyensis (Speg.) Sawada
Cintractia minor (G.P. Clinton) H. Jacks.
Distr.: Widespread.
Notes: On Cyperaceae (except *Carex* spp.).
Host: Cyperus
Cintractia psilocaryae (Tracy & Earle) G.P. Clinton
Distr.: Central and eastern United States.
Notes: On Cyperaceae (except *Carex* spp.).
Host: Rhynchospora
Cintractia striata G.P. Clinton & Zundel
Cintractia leucoderma (Berk.) Henn. var. *striata* (G.P. Clinton & Zundel) Ling
Distr.: Eastern United States.
Host: Rhynchospora
Cintractia taubertiana (Henn.) G.P. Clinton
Distr.: Widespread.
Hosts: Fimbristylis, Rhynchospora
Cintractia utriculicola (Henn.) G.P. Clinton
Distr.: Pantropical and subtropical.
Notes: On Cyperaceae (except *Carex* spp.).
Host: Rhynchospora

Circinella Tiegh. & A. LeMonnier. Zygomycetes, Mucorales

Circinella muscae (Sorokin) Berl. & De Toni
Distr.: Widespread.
Notes: Isolated from soil, dung, nuts and other organic substrates (Hesseltine & Fennell. Mycologia 47:201. 1955).
Host: Zea

Circinotrichum Nees. Hyphomycetes

Notes: Ellis (Dematiaceous Hyphomycetes. p. 356. 1971); Ellis. (More dematiaceous Hyphomycetes. p. 361. 1976); Pirozynski. (Mycol. Pap. 84:1. 1962).
Circinotrichum fertile Pirozynski & C.S. Hodges
Distr.: SC(type).
Notes: Pirozynski & Hodges (Canad. J. Bot. 51:160. 1973).
Host: Persea
Circinotrichum maculiforme Nees
Distr.: FL; Europe.
Notes: On fallen leaves and dead twigs.
Host: Juncus
Circinotrichum olivaceum (Speg.) Pirozynski
Helicotrichum olivaceum Speg.
Distr.: Subtropical and tropical regions.
Host: Eucalyptus
Circinotrichum papakurae S.J. Hughes & Pirozynski
Distr.: HI; New Zealand(type).
Host: Eucalyptus

Cirrenalia S.P. Meyers & R.T. Moore. Hyphomycetes

Notes: Ellis (More dematiaceous Hyphomycetes. p. 38. 1976); Kohlmeyer & Kohlmeyer (Marine mycology. p. 476. 1979).
Cirrenalia macrocephala (Kohlmeyer) S.P. Meyers & R.T. Moore
Helicoma macrocephala Kohlmeyer
Distr.: Worldwide in marine habitats, less common in tropical regions than in temperate regions.
Notes: Saprobic on driftwood and intertidal wood and bark.
Hosts: Juncus, Pinus, Spartina
Cirrenalia pseudomacrocephala Kohlmeyer
Distr.: Atlantic Ocean.
Notes: Saprobic on driftwood under mangroves.
Host: Juncus

Cistella Quél., nom. cons. Discomycetes, Helotiales

Cistella coarctata (Ellis & Everh.) Dennis
Trichopeziza coarctata Ellis & Everh.
Distr.: WA.
Notes: On decorticated wood (Dennis. Kew Bull. 17:338. 1963).
Hosts: Arctostaphylos, Vaccinium
Cistella dentata (Pers.:Fr.) Quél.
Clavidisculum dentatum (Pers.:Fr.) Raitviir
Cyathicula dentata (Pers.:Fr.) Sacc.
Hyaloscypha dentata (Pers.:Fr.) Boud.
Distr.: Possibly North America; Europe.
Notes: Seaver (North Amer. cup fungi. [Inoperculates]. p. 113. 1951) considered the occurrence of this species in North America as doubtful.
Host: Acer
Cistella grevillei (Berk.) Raitviir
Dasyscyphus grevillei (Berk.) Massee
Trichopeziza urticina (Peck) Sacc.
Distr.: NY; Europe.
Notes: On herbaceous stems (Dennis. Kew Bull. 17:373. 1963).
Host: Laportea
Cistella parksii Cash
Distr.: CA(type).
Hosts: Picea, Pinus, Tsuga
Cistella xylita (P. Karst.) Nannf.
Distr.: WA; Europe.
Host: Alnus

Cladobotryum Nees. Hyphomycetes

Notes: De Hoog (Persoonia 10:33. 1978).
Cladobotryum verticillatum (Link) S.J. Hughes
Acremonium verticillatum Link
Distr.: Temperate regions.
Notes: Frequently associated with other fungi.
Host: Pseudotsuga

Cladochytrium Nowakowski. Chytridiomycetes, Chytridiales

Cladochytrium replicatum Karling
Cladochytrium nowakowskii Sparrow
Distr.: North and South America; Europe.
Hosts: Eriocaulon, Tradescantia

Cladosporium Link:Fr. Hyphomycetes

Notes: De Vries (Contribution to the knowledge of the genus *Cladosporium* Link ex Fr. 1952).
Cladosporium aecidiicola Thuem.
Distr.: AZ; Europe(type).
Notes: On aecia of rusts on various plants (Ellis. More dematiaceous Hyphomycetes. p. 330. 1976).
Host: Helenium
Cladosporium allii (Ellis & G. Martin) P.M. Kirk & J.G. Crompton
Heterosporium allii Ellis & G. Martin
Cladosporium allii-porri (Sacc. & Briard) Boerema
Heterosporium allii Ellis & G. Martin var. *allii-porri* (Sacc. & Briard) Sacc.
Heterosporium ornithogali Klotzsch var. *allii-porri* Sacc. & Briard
Distr.: Temperate regions.
Notes: Leaf blotch (Kirk & Crompton. Pl. Pathol. 33:317. 1984; C.M.I. Descr. 841. 1986).
Hosts: Allium, Sisyrinchium, Triteleia
Cladosporium americanum H.C. Greene
Distr.: WI.
Host: Prunus
Cladosporium aromaticum Ellis & Everh.
Distr.: United States.
Host: Rhus
Cladosporium artemisiae H.C. Greene
Distr.: WI.
Host: Artemisia
Cladosporium arthrinioides Thuem.
Distr.: TX; Italy(type).
Host: Bougainvillea
Cladosporium astericola J.J. Davis
Distr.: IA, WA, WI(type).
Hosts: Aster, Euthamia, Machaeranthera, Solidago
Cladosporium atriellum Cooke
Distr.: SC(type).
Host: Yucca
Cladosporium avellaneum G.A. De Vries
Distr.: HI; The Netherlands(type).
Notes: De Vries (Contribution to the knowledge of the genus *Cladosporium* Link ex Fr. p. 56. 1952) described four cultural forms.
Host: Wikstroemia
Cladosporium baptisiae H.C. Greene
Distr.: WI.
Host: Baptisia
Cladosporium brachyelytri H.C. Greene
Distr.: WI.
Host: Brachyelytrum
Cladosporium brassicae (Ellis & Barth.) M.B. Ellis
Cladotrichum brassicae Ellis & Barth.
Distr.: KS.
Host: Brassica
Cladosporium brevipes Peck
Distr.: NY(type).
Host: Quercus
Cladosporium brunneolum Sacc.
Cladosporium brunnea Cooke & Harkn. non Corda, hom. illeg.
Distr.: CA(type).
Host: Hedera
Cladosporium caducum J.J. Davis
Distr.: WI(type).
Host: Betula
Cladosporium carpophilum Thuem.
Fusicladium carpophilum (Thuem.) Oudem.
Venturia carpophila E.E. Fisher (teleomorph)
Distr.: Cosmopolitan.
Notes: Scab and freckle (Sivanesan. C.M.I. Descr. 402. 1974; Sivanesan. The bitunicate Ascomycetes. p. 609. 1984).
Hosts: Acacia, Cheirodendron, Metrosideros, Prunus
Cladosporium caryigenum (Ellis & Langl.) Gottwald
Fusicladium caryigenum G. Wint. & Langl.
Cladosporium effusum (G. Wint.) Demaree non Berk. & M.A. Curtis, hom. illeg.
Fusicladium effusum G. Wint.
Distr.: Central and eastern North America.
Notes: Scab and leaf spot (Gottwald. Mycologia 74:382. 1982).
Hosts: Carpinus, Carya
Cladosporium caryigenum (Ellis & Langl.) Gottwald var. ***carpinum*** Ellis & Everh.
Fusicladium effusum G. Wint. var. *carpinum* Ellis & Everh.
Distr.: Eastern North America.
Notes: Leaf spot (Gottwald. Mycologia 74:389. 1982).
Hosts: Carpinus, [Carya]
Cladosporium citri Massee
Distr.: AL, FL(type), LA.
Host: Citrus
Cladosporium cladosporioides (Fresen.) G.A. De Vries
Penicillium cladosporioides Fresen.
Hormodendrum cladosporioides (Fresen.) Sacc.
Distr.: Cosmopolitan.
Notes: A secondary pathogen of a wide variety of plants; a common saprobe on dead plant material.
Hosts: Acacia, Acer, Alnus, Carya, Cheirodendron, Cucumis, Helianthus, Juncus, Liriodendron, Malus, Medicago, Metrosideros, Pinus, Pisum, Platanus, Prunus, Pseudotsuga, Tragopogon, Trifolium, Vicia, Wikstroemia, Zea
Cladosporium cladosporioides (Fresen.) G.A. De Vries f. sp. ***pisicola*** (W.C. Snyder) G.A. De Vries
Cladosporium pisicola W.C. Snyder
Distr.: Widespread throughout pea-growing regions.
Notes: Necrotic spotting of leaves, stems, and fruits of peas.
Host: Pisum
Cladosporium coreopsidis H.C. Greene
Distr.: WI.
Host: Coreopsis
Cladosporium cucumerinum Ellis & Arth.
Cladosporium scabies Cooke
Distr.: North America; West Indies; eastern and southern Asia; Europe.
Notes: Scab or gummosis of curcubits (Ellis & Holliday. C.M.I. Descr. 348. 1972).
Hosts: Citrullus, Cucumis, Cucurbita, Helianthus
Cladosporium dracaenatum Thuem.
Distr.: SC(type).
Hosts: Cordyline, Dracaena
Cladosporium echinulatum (Berk.) G.A. De Vries
Heterosporium echinulatum (Berk.) Cooke
Mycosphaerella dianthi (C.C. Burt) Jørst. (teleomorph)
Distr.: Worldwide on leaves and flowers.
Hosts: Cerastium, Comandra, Dianthus
Cladosporium elatum (C. Harz) Nannf.
Cadophora elatum (C. Harz) Nannf. in Melin & Nannf.
Distr.: United States; Europe.
Hosts: Malus, Medicago, Pisum, Trifolium, Vicia
Cladosporium erianthi Thuem.
Distr.: SC(type).
Host: Saccharum
Cladosporium extorre Sacc.
Distr.: WA(type).
Host: Malus
Cladosporium fumagineum Sacc.
Distr.: WA(type).
Host: Quercus

Cladosporium gloeosporioides Atk.
Distr.: AL(type), MS, NJ, NY, WI.
Host: Hypericum
Cladosporium graminum Corda f. ***caricicola*** Sacc.
Distr.: ND.
Notes: The type form is a synonym of *C. herbarum*.
Host: Carex
Cladosporium grumosum Link
Distr.: NY; Europe(type).
Host: Pteridium
Cladosporium heliotropii Eriks.
Distr.: AK; Scandinavia(type).
Host: Heliotropium
Cladosporium herbarum (Pers.:Fr.) Link
Cladosporium caricicola Corda
Cladosporium epiphyllum (Pers.:Fr.) Fr.
Cladosporium fasciculatum Corda
Helminthosporium flexuosum Corda
Cladosporium fuscatum Link
Cladosporium graminum (Pers.:Fr.) Link
Mycosphaerella tassiana (De Not.) Johans. (teleomorph)
Distr.: Cosmopolitan.
Notes: Ellis (Dematiaceous Hyphomycetes. p. 313. 1971); Prasil & De Hoog (Trans. Brit. Mycol. Soc. 90:49. 1988).
Hosts: Abutilon, Acer, Agropyron, Agrostis, Alnus, Amaryllis, Anthoxanthum, Arachis, Arceuthobium, Arrhenatherum, Asclepias, Asparagus, Avena, Bambusa, Bromus, Buddleja, Buxus, Calamagrostis, Calamovilfa, Camellia, Capsicum, Carex, Citrus, Clematis, Cleome, Dianthus, Echinochloa, Elymus, Euonymus, Festuca, Ficus, Foeniculum, Fragaria, Gladiolus, Glycine, Gossypium, Helianthus, Hordeum, Iris, Lathyrus, Lens, Leucothoe, Lolium, Lonicera, Lupinus, Lycopersicon, Magnolia, Malus, Narcissus, Nicotiana, Oryza, Paeonia, Parthenocissus, Phalaris, Phaseolus, Philadelphus, Phleum, Phyllostachys, Pinus, Pisum, Poa, Poaceae, Prunus, Pseudotsuga, Puccinellia, Quercus, Rhododendron, Ribes, Robinia, Rosa, Rubus, Saccharum, Salix, Secale, Setaria, Sinningia, Sorghastrum, Sorghum, Stipa, Syringa, Tragopogon, Trigonella, Trisetum, Triticum, Tsuga, Typha, Ulmus, Vaccinium, Vinca, Vitis, Zea
Cladosporium herbarum (Pers.:Fr.) Link var. ***citricola*** Farl.
Distr.: FL.
Hosts: Citrus, Poncirus
Cladosporium heugelinianum Thuem.
Distr.: VA; Middle East(type).
Host: Buddleja
Cladosporium humile J.J. Davis
Venturia acerina Plakidas ex Barr (teleomorph)
Distr.: Northeastern North America.
Notes: Leaf spot.
Host: Acer
Cladosporium iridis (Fautrey & Roum.) G.A. De Vries
Heterosporium iridis (Fautrey & Roum.) J. Jacques
Brachysporium gracile (Wallr.) Sacc.
Helminthosporium gracile Wallr.
Heterosporium gracile (Wallr.) Sacc.
Mycosphaerella macrospora (Kleb.) Jørst. (teleomorph)
Distr.: Cosmopolitan.
Notes: Leaf spot and blotch of *Iris* and other Liliaceae (Ellis & Waller. C.M.I. Descr. 435. 1975).
Hosts: Allium, Belamcanda, Chlorogalum, Freesia, Iris, [Pandanus]
Cladosporium letiferum Peck
Distr.: NY(type).
Host: Populus
Cladosporium lysimachiae H.C. Greene non Guba, hom. illeg.
Distr.: WI(type).
Host: Lysimachia
Cladosporium lysimachiae Guba
Distr.: MA(type).
Host: Lysimachia
Cladosporium macrocarpum G. Preuss
Distr.: Cosmopolitan, more prevalent in temperate regions.
Notes: Saprobic on plant debris or in soil. Ellis (Dematiaceous Hyphomycetes. p. 315. 1971).
Hosts: Alnus, Pseudotsuga, Spinacia
Cladosporium malorum Ruehle
Distr.: WA(type).
Host: Malus
Cladosporium molle Cooke
Distr.: SC(type).
Host: Asclepias
Cladosporium monardae H.C. Greene
Distr.: WI(type).
Host: Monarda
Cladosporium myriosporum Ellis & Dearn.
Distr.: Northern North America.
Host: Pisum
Cladosporium nigrellum Ellis & Everh.
Distr.: WV.
Host: Robinia
Cladosporium oxycocci Shear
Distr.: MA, NJ, WA; Canada(Nova Scotia).
Host: Vaccinium
Cladosporium oxysporum Berk. & M.A. Curtis
Distr.: Subtropical and tropical regions.
Hosts: Alnus, Bambusa, Citrus, Helianthus, Pseudotsuga
Cladosporium paeoniae Pass.
Distr.: North America; Europe.
Notes: Leaf spot (De Vries. Contribution to the knowledge of the genus *Cladosporium* Link ex Fr. p. 94. 1952).
Host: Paeonia
Cladosporium pericarpium Cooke
Distr.: SC(type).
Host: Juglans
Cladosporium phlei (C.T. Gregory) G.A. De Vries
Heterosporium phlei C.T. Gregory
Distr.: Western United States; Europe; New Zealand.
Notes: Eye spot of timothy.
Hosts: Agropyron, Arctagrostis, Bromus, Calamagrostis, Elymus, Festuca, Phleum, Poa
Cladosporium piricularioides Dearn. & House, nom. inval.
Distr.: NY.
Host: Panicum
Cladosporium pisi Cugini & Macch.
Distr.: OR; Europe(type).
Notes: The status of this taxon is uncertain.
Host: Pisum
Cladosporium punctulatum Sacc. & Ellis
Distr.: CA, NJ(type).
Host: Hedera
Cladosporium rambosum Desmaz.
Distr.: IL.
Notes: The place of publication of this name could not be found.
Host: Populus
Cladosporium salicis-sitchensis Dearn. & Barth.
Distr.: WA(type).
Host: Salix
Cladosporium sitchense (Dearn.) Barth., nom. nud.
Distr.: WA.
Host: Cyperus
Cladosporium sphaerospermum Penz.
Distr.: Cosmopolitan.
Notes: Isolated from air, soil, foodstuffs, occasionally from humans and animals; also a secondary invader on many different plants. Ellis (Dematiaceous Hyphomycetes. p. 315. 1971).
Hosts: Alnus, Carya, Helianthus, Juncus, Pseudotsuga
Cladosporium spongiosum Berk. & M.A. Curtis
Distr.: Subtropical and tropical regions.
Notes: On inflorescences of grasses.
Host: Cenchrus
Cladosporium stipae H.C. Greene
Distr.: WI.
Host: Stipa
Cladosporium subnodosum Cooke
Distr.: DE, SC(type).
Host: Spinacia
Cladosporium tenuissimum Cooke
Distr.: Cosmopolitan, especially common in tropical regions.
Notes: On many plant species; also isolated from air, bread, and soil (Ellis. More dematiaceous Hyphomycetes. p. 326. 1976).
Host: Zea

Cladosporium trichophilum H.C. Greene
Distr.: WI.
Host: Lonicera
Cladosporium triostei Peck
Distr.: NY(type), WV.
Notes: Leaf spot. De Vries (Contribution to the knowledge of the genus *Cladosporium* Link ex Fr. p. 99. 1952) considered this species a *Napicladium*.
Host: Triosteum
Cladosporium typhae Schwein.
Distr.: PA(type).
Host: Typha
Cladosporium typharum Desmaz.
Distr.: United States; Europe.
Host: Typha
Cladosporium variabile (Cooke) G.A. De Vries
Heterosporium variabile Cooke
Helminthosporium variabile Cooke, nom. nud.
Distr.: Cosmopolitan.
Notes: Leaf spot and leaf mold of spinach. Ellis (Dematiaceous Hyphomycetes. p. 315. 1971).
Hosts: [Brassica], [Sorghum], Spinacia
Cladosporium velutinum Ellis & Tracy
Distr.: MS(type).
Host: Phalaris
Cladosporium vignae M.W. Gardner
Distr.: Eastern United States.
Hosts: Pisum, Vicia, Vigna
Cladosporium xyridis Tracy & Earle
Distr.: MS(type).
Host: Xyris
Cladosporium zeae Peck
Distr.: NY(type).
Host: Zea
Cladosporium ziziphi P. Karst. & Roum.
Distr.: FL; Finland(type).
Host: Ziziphus

Cladotrichum Corda. Hyphomycetes

Notes: This genus is considered a synonym of *Oedemium*, but not all names have been redisposed.
Cladotrichum leersiae Atk.
Distr.: AL(type), WI.
Host: Leersia

Clasterosporium Schwein. Hyphomycetes

Notes: Ellis (Mycol. Pap. 70:1. 1958).
Clasterosporium caricinum Schwein.
Distr.: Temperate North America; Europe.
Notes: Black mold.
Hosts: Carex, Scirpus
Clasterosporium putrefaciens (Fuckel) Sacc.
Sporidesmium putrefaciens Fuckel
Distr.: CO, MI, WA; Germany(type).
Notes: Wehmeyer (A world monograph of the genus *Pleospora* and its segregates. p. 372. 1961) stated that this teleomorph is not distinguishable from *Pleospora scropulariae*.
Host: Beta

Clathrospora Rabenh., nom. rej. Loculoascomycetes, Dothideales

Clathrospora bakeri Wehmeyer
Distr.: CO, NY.
Host: Juncus
Clathrospora elynae Rabenh.
Pleospora elynae (Rabenh.) Ces. & De Not.
Distr.: North America; Europe.
Hosts: Carex, Juncus
Clathrospora heterospora (De Not.) Wehmeyer
Pleospora heterospora De Not.
Distr.: Western North America; eastern Arctic Canada; Europe.
Host: Carex
Clathrospora heterospora (De Not.) Wehmeyer var. ***simmonsii*** (Wehmeyer) Wehmeyer
Clathrospora simmonsii Wehmeyer
Distr.: WA.
Host: Carex
Clathrospora juncicola (Ellis & Everh.) Wehmeyer
Pleospora juncicola Ellis & Everh.
Distr.: Northern temperate regions.
Host: Juncus
Clathrospora multiseptata E. Müller
Clathrospora cookei Wehmeyer
Distr.: Northwestern North America; Switzerland.
Hosts: Muhlenbergia, Stipa
Clathrospora pentamera (P. Karst.) Berl.
Pleospora pentamera P. Karst.
Platyspora pentamera (P. Karst.) Wehmeyer
Clathrospora aurea (Ellis) Berl.
Pleospora aurea Ellis
Pleospora platyspora Sacc.
Pleospora quadriseptata Cooke & Harkn.
Distr.: Temperate northern hemisphere.
Notes: On herbaceous stems and debris.
Hosts: Andropogon, Arenaria, Astragalus, Carex, Cerastium, Cinna, Clematis, Delphinium, Diapensia, Dupontia, Erysimum, Lupinus, Papaver, Poa, Puccinellia, Secale, Triticum, Wyethia
Clathrospora platysporioides M.L. Farr & H.T. Horner
Host: Selaginella

Claudopus Gill. Basidiomycetes, Agaricales

Notes: A segregate of *Entoloma*. Nooderloos (Persoonia 11:121. 1981) reduced it to a subgenus of *Entoloma*.
Claudopus affinis Kauffm.
Distr.: OR.
Notes: Largent (Brittonia 23:244. 1971) examined the holotype and indicated that this taxon is probably a *Pleurotus* sp.
Host: Quercus
Claudopus cortinicinctus Kauffm.
Distr.: OR.
Notes: Largent (Brittonia 23:238. 1971) indicated that this name was based on a *Crepidotus* sp.
Host: Acer
Claudopus subargillaceus Kauffm.
Distr.: OR.
Notes: Largent (Brittonia 23:238. 1971) indicated this is probably a *Pleurotus* sp.
Host: Quercus

Claussenomyces Kirschst. Discomycetes, Helotiales

Notes: Korf & Abawi (Canad. J. Bot. 49:1879. 1971).
Claussenomyces atrovirens (Pers.:Fr.) Korf & Abawi
Corynella atrovirens (Pers.:Fr.) Boud.
Distr.: FL; Europe.
Host: Sambucus
Claussenomyces olivaceus (Fuckel) M.A. Sherwood
Retinocyclus olivaceus Fuckel
Tromera olivaceus (Fuckel) Sacc.
Tympanis olivacea (Fuckel) Rehm
Distr.: CO; Europe.
Notes: On conifer resin. An unnamed pycnidial anamorph forms in culture. Hawksworth & Sherwood (Canad. J. Bot. 59:367. 1981) provided a description and a nomenclatural discussion.
Host: Picea
Claussenomyces prasinula (P. Karst.) Korf & Abawi
Peziza prasinula P. Karst.
Corynella prasinula (P. Karst.) Boud.
Helotium albovirens Cooke
Distr.: North America; Europe.
Notes: On decaying wood (Dennis. Persoonia 3:34. 1964).
Host: Acer
Claussenomyces pseudotsugae (Groves) Ouellette & Pirozynski

Tympanis pseudotsugae Groves
Distr.: Western North America.
Notes: Ouellette & Pirozynski (Canad. J. Bot. 52:1909. 1974).
Host: Pseudotsuga

Clavariadelphus Donk. Basidiomycetes, Aphyllophorales

Clavariadelphus ligula (Fr.) Donk
Distr.: Northern temperate regions.
Notes: On coniferous debris (Wells & Kempton. Michigan Bot. 7:53. 1968).
Host: Picea

Clavariopsis De Wild. Hyphomycetes

Clavariopsis aquatica De Wild.
Distr.: Temperate regions.
Notes: On fallen leaves and various kinds of organic debris in aquatic habitats (Nilsson. Symb. Bot. Upsal. 18:86. 1964).
Hosts: Acer, Quercus
Clavariopsis bulbosa Anastasiou
Corollospora pulchella Kohlmeyer, J.C. Schmidt & N.B. Nair (teleomorph)
Distr.: Widespread in marine habitats.
Notes: Saprobic on submerged wood, mangrove roots, and rotting algae (Kohlmeyer & Kohlmeyer. Marine mycology. p. 276. 1979).
Host: Tamarix

Claviceps Tul. Pyrenomycetes, Clavicipitales

Notes: Brady (Lloydia 25:1. 1962); Langdon (Univ. Queensland Dept. Bot. Pap. 3:651. 1954).
Claviceps cinerea D. Griffiths
Distr.: Southwestern United States.
Hosts: Hilaria, Melica
Claviceps grohii Groves
Distr.: North America.
Notes: Groves (Mycologia 35:608. 1943).
Host: Carex
Claviceps junci J.F. Adams
Distr.: WI; Ireland(type).
Host: Juncus
Claviceps nigricans Tul.
Sphacelia nigricans Sacc. (anamorph)
Distr.: Temperate regions.
Host: Eleocharis
Claviceps paspali F. Stevens & J.G. Hall
Claviceps rolfsii F. Stevens & J.G. Hall
Distr.: Cosmopolitan.
Notes: Ergot.
Host: Paspalum
Claviceps purpurea (Fr.:Fr.) Tul.
Claviceps microcephala (Wallr.) Tul.
Sphacelia segetum Lév. (anamorph)
Distr.: Worldwide on grasses.
Notes: Ergot. Occurs on about 200 species of grasses in about 50 genera.
Hosts: Agrohordeum, Agropyron, Agrostis, Alopecurus, Ammophila, Andropogon, Arctagrostis, Arctophila, Arrhenatherum, Avena, Bothriochloa, Bouteloua, Brachiaria, Bromus, Calamagrostis, Cinna, Dactylis, Danthonia, Deschampsia, Distichlis, Elymus, Festuca, Glyceria, Hierochloe, Holcus, Hordeum, Hystrix, Koeleria, Leucopoa, Leymus, Lolium, Paspalum, Phalaris, Phleum, Phragmites, Poa, Puccinellia, Saccharum, Schizachyrium, Secale, Setaria, Sitanion, Sorghastrum, Spartina, Stipa, Tripsacum, Trisetum, Triticum, Vulpia, Zizania
Claviceps ranunculoides A. Möller
Distr.: Southwestern United States; Brazil.
Host: Setaria
Claviceps tripsaci F. Stevens & J.G. Hall
Distr.: Southeastern United States.
Host: Tripsacum
Claviceps yanagawaensis Togashi
Distr.: Eastern United States(introduced); Japan.
Host: Zoysia
Claviceps zizaniae (Fyles) Pantidou
Spermoedia zizaniae Fyles
Distr.: Northern North America.
Notes: Pantidou (Canad. J. Bot. 37:1233. 1959) reviewed this species.
Host: Zizania

Clavicorona Doty. Basidiomycetes, Aphyllophorales

Clavicorona cristata (Kauffm.) Doty
Distr.: Northwestern North America; Canada(Quebec).
Notes: On conifer logs.
Host: Pseudotsuga
Clavicorona divaricata Leathers & A.H. Sm.
Distr.: North America.
Notes: On wood.
Hosts: Abies, Populus
Clavicorona piperata (Kauffm.) Leathers & A.H. Sm.
Clavaria piperata Kauffm.
Clavicorona avellanea Leathers & A.H. Sm.
Distr.: Northwestern North America; Australia.
Notes: On conifer logs (Dodd. Mycologia 64:753. 1972).
Hosts: Pinus, Tsuga
Clavicorona pyxidata (Pers.:Fr.) Doty
Clavaria pyxidata Pers.:Fr.
Clavaria coronata Schwein.
Distr.: Northern hemisphere.
Notes: On wood.
Hosts: Acer, Alnus, Betula, Carya, Cornus, Fraxinus, Liquidambar, Populus, Quercus, Rhododendron, Salix, Tsuga, Ulmus

Clavispora Rodrigues de Miranda. Hemiascomycetes, Endomycetales

Clavispora opuntiae Phaff, M. Miranda, Starmer, Tredick & J. Barker
Distr.: AZ, TX; Jamaica; Venezuela; Australia.
Notes: Isolated from necrotic tissue.
Host: Opuntia

Cleistophoma Petr. & Syd. Coelomycetes

Cleistophoma dryina (Berk. & M.A. Curtis) Petr. & Syd.
Macrophoma dryina (Berk. & M.A. Curtis) Berl. & Voglino
Distr.: IL, PA.
Host: Quercus

Climacocystis Kotlaba & Pouzar. Basidiomycetes, Aphyllophorales

Climacocystis borealis (Fr.:Fr.) Kotlaba & Pouzar
Polyporus borealis Fr.:Fr.
Abortiporus borealis (Fr.:Fr.) Singer
Tyromyces borealis (Fr.:Fr.) Imazeki
Spongipellis borealis (Fr.:Fr.) Pat.
Polyporus pacificus Kauffm.
Distr.: Temperate northern hemisphere.
Notes: On stumps, logs, and roots of conifers, rarely on hardwoods.
Hosts: Abies, Acer, Larix, Picea, Pinus, Pseudotsuga, Tsuga

Climacodon P. Karst. Basidiomycetes, Aphyllophorales

Climacodon pulcherrimus (Berk. & M.A. Curtis) Nikolaeva
Hydnum pulcherrimum Berk. & M.A. Curtis
Donkia pulcherrima (Berk. & M.A. Curtis) Pilát
Steccherinum pulcherrimum (Berk. & M.A. Curtis) Banker
Distr.: Central and eastern United States and adjacent Canada., most common in the South; Africa; western Asia; Europe.
Notes: On wood.
Hosts: Abies, Acer, Betula, Fagus, Liquidambar, Pinus, Platanus,

Populus, Quercus
Climacodon septentrionalis (Fr.:Fr.) P. Karst.
Hydnum septentrionale Fr.:Fr.
Steccherinum septentrionale (Fr.:Fr.) Banker
Distr.: Temperate northern hemisphere, cooler parts.
Notes: On living hardwoods.
Hosts: Acer, Carya, Liquidambar, Malus, Populus, Quercus, Tilia

Clintamra Cordas & Durán. Basidiomycetes, Ustilaginales

Clintamra nolinae (G.P. Clinton) Cordas & Durán
Tolyposporella nolinae G.P. Clinton
Melanotaenium nolinae (G.P. Clinton) Thirumalachar, M.D. Whitehead & M. O'Brien
Distr.: Southwestern United States.
Host: Nolina

Clithris (Fr.) P. Karst. Discomycetes, Rhytismatales

Notes: This genus is considered a synonym of *Colpoma*, but not all names have been redisposed.
Clithris andromedae (Schwein.) Ellis & Everh.
Sporomega andromedae (Schwein.) Duby
Distr.: Eastern United States.
Host: Leucothoe
Clithris graphis Rehm
Distr.: WA.
Hosts: Abies, Pinus, Tsuga
Clithris leucothoicola Tehon
Distr.: PA.
Host: Leucothoe
Clithris sequoiae Bonar
Distr.: CA(type).
Host: Sequoia

Clitocybe P. Kumm. Basidiomycetes, Agaricales

Clitocybe eccentrica Peck
Distr.: Eastern North America.
Notes: On woody debris.
Host: Quercus
Clitocybe isabellina (A.H. Sm. & D.E. Stuntz) H.E. Bigelow
Omphalina isabellina A.H. Sm. & D.E. Stuntz
Distr.: Pacific Northwest.
Host: Pseudotsuga
Clitocybe martiorum J. Favre
Distr.: Temperate North America; Europe.
Notes: On conifer needles.
Host: Pinus
Clitocybe setiseda (Schwein.) Sacc.
Distr.: NC.
Notes: The status of this taxon is uncertain.
Host: Quercus

Clitocybula (Singer) Métrod. Basidiomycetes, Agaricales

Clitocybula abundans (Peck) Singer
Collybia abundans (Peck) Sacc.
Distr.: Temperate North America.
Notes: On wood.
Host: Tsuga
Clitocybula familia (Peck) Singer
Collybia familia (Peck) Sacc.
Distr.: Eastern North America.
Notes: On wood.
Host: Ulmus

Clonostachys Corda. Hyphomycetes

Clonostachys araucaria Corda
Distr.: OR; Europe(type).
Notes: Hawksworth & Punithalingam (Trans. Brit. Mycol. Soc. 62:91. 1975).
Host: Pyrus
Clonostachys araucaria Corda var. ***rosea*** G. Preuss
Distr.: Reported from an unknown locality in the United States; Germany; Italy.
Notes: Tuber rot of potatoes (Hooker. Compendium of potato diseases. p. 66. 1981).
Host: Solanum

Clypeolella Höhn. Loculoascomycetes, Dothideales

Clypeolella clermontiae F. Stevens & Ryan
Distr.: HI.
Host: Clermontia
Clypeolella leemingii (Ellis & Everh.) Theiss.
Asterina leemingii Ellis & Everh.
Distr.: MS, NC, VA.
Host: Galax

Clypeoporthe Höhn. Pyrenomycetes, Diaporthales

Notes: Barr (Mycol. Mem. 7:78. 1978).
Clypeoporthe andropogonis (Ellis & Everh.) Barr
Gnomonia andropogonis Ellis & Everh.
Distr.: Central and eastern United States.
Notes: On culms.
Hosts: Andropogon, Schizachyrium
Clypeoporthe iliau (Lyon) Barr
Gnomonia iliau Lyon
Distr.: Southeastern United States; West Indies; South America; Australia; Mauritius; The Philippines; South Pacific Islands.
Notes: Iliau disease of sugarcane (Sivanesan & Holliday. C.M.I. Descr. 705. 1981).
Hosts: Axonopus, Saccharum

Clypeoporthella Petr. Pyrenomycetes, Diaporthales

Clypeoporthella brencklei Petr.
Diaporthopsis brencklei (Petr.) Arx & E. Müller
Phomopsis brencklei Petr. (anamorph)
Distr.: ND.
Host: Aster

Clypeopycnis Petr. Coelomycetes

Clypeopycnis aeruginascens Petr.
Distr.: AK, SD.
Host: Ribes

Clypeoseptoria F. Stevens & P.A. Young. Coelomycetes

Clypeoseptoria rockii F. Stevens & P.A. Young
Distr.: HI(type).
Host: Platydesma

Clypeosphaeria Fuckel. Pyrenomycetes, Amphisphaeriales

Clypeosphaeria mamillana (Fr.) Lambotte
Distr.: MT; Europe.
Notes: On hardwoods.
Host: Cornus
Clypeosphaeria stevensii Syd. in F. Stevens
Distr.: HI(type).
Host: Freycinetia

Clypeostagonospora Punithalingam. Coelomycetes

Clypeostagonospora muhlenbergiae Punithalingam
Distr.: MI.
Host: Muhlenbergia

Coccodiella K. Hara. Loculoascomycetes, Dothideales

Notes: Species in this genus have been confused with those in *Coccoidella* (Katumoto. J. Jap. Bot. 43:277. 1968).
Coccodiella nuda (F. Stevens) Hino & Katumoto
Yoshinagella nuda F. Stevens
Bagnisiopsis nuda (F. Stevens) Petr.
Coccostroma nudum (F. Stevens) Arx & E. Müller
Distr.: HI(type).
Notes: On leaves. Arx & Müller (Beitr. Kryptogamenfl. Schweiz 11(1):266. 1954) discussed this species.
Host: Cibotium
Coccodiella polymorpha (F. Stevens) Hino & Katumoto
Bagnisiopsis polymorphum (F. Stevens) Petr.
Coccostroma polymorphum (F. Stevens) Arx & E. Müller
Distr.: HI(type).
Host: Cibotium

Coccodinium C. Massal. Loculoascomycetes, Dothideales

Notes: Eriksson (Opera Bot. 60:43. 1981) reviewed this genus.
Coccodinium corticola (Ellis & Everh.) Barr
Pleosphaeria corticola Ellis & Everh.
Distr.: ID, WV.
Notes: Barr (Mycotaxon 29:502. 1987).
Host: Larix
Coccodinium laricis W.B. Cooke & C.G. Shaw
Distr.: ID.
Host: Larix

Coccodothis Theiss. & Syd. Loculoascomycetes, Dothideales

Notes: Müller & Arx (Beitr. Kryptogamenfl. Schweiz 11(2):83. 1962).
Coccodothis sphaeroidea (Cooke) Theiss. & Syd.
Dothidea sphaeroidea Cooke
Dothidella sphaeroidea (Cooke) Ellis & Everh.
Keithia juniperi J.K. Miller
Distr.: Southeastern United States.
Host: Juniperus

Coccoidella Höhn. Loculoascomycetes, Dothideales

Coccoidella scutula (Berk. & M.A. Curtis) Höhn.
Distr.: Southeastern United States.
Notes: On leaves.
Host: Persea

Coccomyces De Not. Discomycetes, Rhytismatales

Notes: Sherwood (Occas. Pap. Farlow Herb. 15:1. 1980) provided a comprehensive account of this genus.
Coccomyces arctostaphyli (Rehm) B. Eriksson
Distr.: Western North America; central and northern Europe.
Host: Arctostaphylos
Coccomyces bipartitus (Kauffm.) M.A. Sherwood
Schizoxylon bipartitus Kauffm.
Distr.: Northeastern United States.
Notes: On conifer wood.
Host: Picea
Coccomyces castanopsidis (Seaver) M.A. Sherwood
Godronia castanopsidis Seaver
Colpoma californicum Cash
Distr.: Western North America.
Notes: On dead bark and wood.
Hosts: Castanopsis, Quercus
Coccomyces coronatus (Schumach.: Fr.) De Not.
Distr.: North America; temperate Europe.
Notes: On dead leaves of hardwoods.
Hosts: Berberis, Betula, Fagus, Gaultheria, Ledum, Quercus, Rhododendron, Vaccinium
Coccomyces dentatus (J.C. Schmidt & Kunze:Fr.) Sacc.
Distr.: Warm-temperate regions.
Notes: On dead leaves of angiosperms, especially Fagaceae and Ericaceae.
Hosts: Arbutus, Berberis, Gaultheria, Lithocarpus, Quercus, Rhododendron
Coccomyces farlowii M.A. Sherwood
Distr.: MA.
Notes: On fallen leaves.
Host: Quercus
Coccomyces gaultheriae (Dearn.) DiCosmo, Nag Raj & Kendrick
Phacidium gaultheriae Dearn.
Distr.: Western United States.
Host: Gaultheria
Coccomyces heterophyllae Funk
Distr.: Pacific Northwest.
Notes: On twigs.
Host: Tsuga
Coccomyces irretitus M.A. Sherwood
Distr.: ME.
Notes: On bark.
Host: Picea
Coccomyces leptideus (Fr.:Fr.) B. Eriksson
Coccomyces quadratus (J.C. Schmidt & Kunze) P. Karst.
Distr.: Western North America; Europe.
Notes: On leaves and twigs of Ericaceae.
Hosts: Arbutus, Gaultheria, Rhododendron
Coccomyces mertensianae M.A. Sherwood
Distr.: OR.
Notes: On decorticated twigs.
Host: Tsuga
Coccomyces parvulus M.A. Sherwood
Distr.: OR, WA.
Notes: Often misdetermined in the past as various *Clithris* species.
Hosts: Pinus, Pseudotsuga
Coccomyces petersii (Berk. & M.A. Curtis) M.A. Sherwood
Hysterium petersii Berk. & M.A. Curtis
Clithris juniperi (P. Karst.) Rehm, pro parte
Coccomyces juniperi P. Karst., pro parte
Colpoma juniperinum Cooke & Peck
Distr.: Eastern North America.
Notes: Specimens of this species have been misdetermined as *Coccomyces juniperi* (= *Clithris juniperi*), a species which only occurs in Europe.
Host: Juniperus
Coccomyces pseudotsugae Funk
Distr.: Pacific Northwest.
Notes: Leader die-back. The name *Clithris juniperi* has been misapplied to this species (Sherwood. Occas. Pap. Farlow Herb. 15:81. 1980).
Host: Abies
Coccomyces radiatus M.A. Sherwood
Distr.: Appalachian Mountains; possibly South America.
Notes: On leaves.
Host: Rhododendron
Coccomyces strobi J. Reid & Cain
Coccophacidium cembrae Rehm, pro parte
Distr.: North America.
Notes: Reid & Cain (Canad. J. Bot. 39:1127. 1961) described this species for various misdetermined specimens including one of *Coccophacidium cembrae* from Massachusetts.
Host: Pinus
Coccomyces triangularis (Schwein.:Fr.) Sacc.
Colpoma triangularis (Schwein.:Fr.) Petr.
Distr.: Eastern North America.
Notes: On dead twigs.
Hosts: Oxydendrum, Quercus
Coccomyces tumidus (Fr.:Fr.) De Not.

Coccomyces comitialis (Batsch) Dearn. & House
Distr.: North America; temperate Europe.
Notes: On fallen leaves.
Hosts: Amelanchier, Fagus, Quercus

Coccomycetella Höhn. Discomycetes, Rhytismatales

Notes: Monotypic (Sherwood-Pike. Mycotaxon 28:144. 1987).
Coccomycetella richardsonii (Leight.) M.A. Sherwood
Odontotrema belonosporum Nyl.
Distr.: North America; Scandinavia.
Notes: On bleached, decorticated conifer wood.
Host: Pinus

Cochliobolus Drechs. Loculoascomycetes, Dothideales

Notes: Sivanesan (Mycol. Pap. 158. 1987) included species with known anamorphs.
Cochliobolus bouteLouae R. Sprague
Distr.: KS, NM.
Host: Bouteloua
Cochliobolus carbonum R.R. Nelson
Bipolaris zeicola (G.L. Stout) Shoemaker (anamorph)
Distr.: Worldwide in maize-growing regions.
Notes: Leaf spot of maize (Ellis & Holliday. C.M.I. Descr. 349. 1972).
Host: Zea
Cochliobolus heterostrophus (Drechs.) Drechs.
Ophiobolus heterostrophus Drechs.
Distr.: Warm temperate, subtropical, and tropical regions.
Notes: Southern leaf blight of maize (Ellis & Holliday. C.M.I. Descr. 301. 1971).
Host: Zea
Cochliobolus lunatus R.R. Nelson & Haasis
Pseudocochliobolus lunatus (R.R. Nelson & Haasis) Tsuda, Ueyama & Nishihara
Curvularia lunata (Wakk.) Boedijn (anamorph)
Distr.: Cosmopolitan, most common in tropical regions.
Notes: Ellis & Gibson (C.M.I. Descr. 474. 1975); Sivanesan (Mycol. Pap. 158:117. 1987).
Host: Saccharum
Cochliobolus miyabeanus (Ito & Kuribayashi in Ito) Drechs. ex Dastur
Ophiobolus miyabeanus Ito & Kuribayashi in Ito
Bipolaris oryzae (Breda de Haan) Shoemaker (anamorph)
Distr.: Widespread.
Notes: Brown spot and seedling blight of rice.
Host: Oryza
Cochliobolus nodulosus Luttrell
Bipolaris nodulosa (Berk. & M.A. Curtis) Shoemaker (anamorph)
Distr.: Southeastern United States; subtropical and tropical regions.
Notes: Leaf blight, seedling blight, and foot rot of finger millet (Ellis & Holliday. C.M.I. Descr. 341. 1972).
Host: Eleusine
Cochliobolus sativus (Ito & Kuribayashi) Drechs. ex Dastur
Ophiobolus sativus Ito & Kuribayashi
Distr.: Temperate regions.
Notes: Foot rot or spot blotch of grasses (Sivanesan & Holliday. C.M.I. Descr. 701. 1981).
Hosts: Aira, Alopecurus, Bromus, Cynosurus, Danthonia, Deschampsia, Elymus, Festuca, Glyceria, Holcus, Hordeum, Lolium, Phleum, Triticum, Vulpia, Zea
Cochliobolus spicifer R.R. Nelson
Bipolaris spicifera (Bainier) Subramanian (anamorph)
Distr.: Cosmopolitan.
Notes: Sivanesan & Holliday (C.M.I. Descr. 702. 1981).
Host: Cynodon
Cochliobolus stenospilus Matsumoto & W. Yamamoto, nom. inval.
Bipolaris stenospila (Drechs.) Shoemaker (anamorph)
Distr.: Worldwide in sugarcane-growing regions.
Notes: Brown stripe of sugarcane (M.B. Ellis & Holliday. C.M.I. Descr. 306. 1971). According to Sivanesan (Mycol. Pap. 158:99. 1987), this teleomorph name was published without a latin diagnosis and is therefore not valid.
Host: Saccharum

Codinaea Maire. Hyphomycetes

Notes: Hewings & Crane (Mycotaxon 13:419. 1981); Hughes & Kendrick (New Zealand J. Bot. 6:354. 1968); Morgan-Jones & Ingram (Mycotaxon 4:504. 1976).
Codinaea cylindrospora Morgan-Jones & E.G. Ingram
Distr.: AL(type).
Host: Quercus
Codinaea parva S.J. Hughes & Kendrick
Distr.: HI; New Zealand(type).
Host: Eucalyptus
Codinaea simplex S.J. Hughes & Kendrick
Distr.: North America; Asia; Europe; New Zealand.
Notes: On bark, leaves, and acorns.
Hosts: Eucalyptus, Quercus
Codinaea unisetula Morgan-Jones & E.G. Ingram
Distr.: AL(type).
Host: Quercus

Codinaeopsis Morgan-Jones. Hyphomycetes

Codinaeopsis gonytrichodes (C.A. Shearer & J.L. Crane) Morgan-Jones
Codinaea gonytrichodes C.A. Shearer & J.L. Crane
Distr.: Eastern United States; Japan.
Notes: Morgan-Jones (Mycotaxon 4:166. 1976).
Hosts: Carya, Rubus

Coleophoma Höhn. Coelomycetes

Coleophoma cylindrospora (Desmaz.) Höhn.
Macrophoma cylindrospora (Desmaz.) Berl. & Voglino
Distr.: CA; Europe.
Host: Phlox
Coleophoma empetri (Rostr.) Petr.
Septoria empetri Rostr.
Sporonema oxycocci Shear
Distr.: Widespread.
Hosts: Arctostaphylos, Empetrum, Galax, Gaultheria, Gaylussacia, Kalmia, Malus, Paxistima, Rhododendron, Vaccinium
Coleophoma oleae (DC.:Fr.) Petr. & Syd.
Macrophoma oleae (DC.:Fr.) Berl. & Voglino
Distr.: United States; Europe.
Host: Olea
Coleophoma sandwicensis Petr.
Distr.: HI(type).
Host: Clermontia
Coleophoma taxi (Petr.) Nag Raj
Xenodomus taxi Petr.
Distr.: ID(type).
Notes: Nag Raj (Canad. J. Bot. 56:696. 1978).
Host: Taxus

Coleosporium Lév. Basidiomycetes, Uredinales

Notes: Kaneko (Rep. Tottori Mycol. Inst. 19. 1981).
Coleosporium adenocaulonis H. Jacks.
Uredo adenocaulonis (H. Jacks.) Cummins (anamorph)
Distr.: Northwestern United States.
Host: Adenocaulon
Coleosporium apocynaceum Cooke
Distr.: Southeastern United States.
Notes: Heteroecious; aecial host - *Pinus* spp.; telial host - *Amsonia* spp.
Hosts: Amsonia, Pinus
Coleosporium arizonicum Cummins
Coleosporium aridum H. Jacks. in Arth., nom. illeg.
Uredo arida (H. Jacks. in Arth.) Cummins (anamorph)
Distr.: Southwestern United States.
Notes: Autoecious.

Host: Brickellia
Coleosporium asterum (Dietel) Syd. & P. Syd.
Coleosporium heterothecae Hedgc. & N. Hunt
Coleosporium solidaginis Thuem., nom. illeg.
Distr.: Widespread in the northern hemisphere.
Notes: Heteroecious; aecial host - *Pinus* spp.; telial host - Asteraceae.
Hosts: Aster, Callistephus, Chrysopsis, Erigeron, Euthamia, Gaillardia, Grindelia, Gutierrezia, Haplopappus, Heterotheca, Machaeranthera, Pinus, Sericocarpus, Solidago
Coleosporium bletiae Dietel
Distr.: FL; eastern Asia; India.
Notes: Heteroecious; aecial host - *Pinus* spp.; telial host - Orchidaceae.
Host: Orchidaceae
Coleosporium crowellii Cummins
Distr.: AZ, CO, NM.
Notes: Autoecious.
Host: Pinus
Coleosporium delicatulum Arth.
Distr.: Central and eastern United States.
Notes: Heteroecious; aecial host - *Pinus* spp. telial host - *Euthamia* spp. (sometimes as *Solidago*).
Hosts: Euthamia, Pinus, Solidago
Coleosporium helianthi (Schwein.) Arth.
Coleosporium terebinthinaceae Arth.
Distr.: Central and eastern United States.
Notes: Heteroecious; aecial host - *Pinus* spp.; telial host - Asteraceae.
Hosts: Helianthus, Heliopsis, [Iva], Madia, Parthenium, Pinus, Silphium
Coleosporium inconspicuum Arth.
Distr.: Southeastern United States.
Notes: Heteroecious; aecial host - *Pinus* spp.; telial host - *Coreopsis* spp.
Hosts: Coreopsis, Pinus
Coleosporium ipomoeae (Schwein.) Burrill
Distr.: NJ to southern United States; West Indies.
Notes: Heteroecious; aecial host - *Pinus* spp.; telial host - Convolvulaceae. Orange rust of sweet potato (Laundon & Rainbow. C.M.I. Descr. 282. 1971).
Hosts: Argyreia, Calystegia, Convolvulus, Ipomoea, Jacquemontia, Pinus
Coleosporium jonesii (Peck) Arth.
Coleosporium ribicola Arth.
Distr.: Central and western North America.
Notes: Heteroecious; aecial host - *Pinus* spp.; telial host - *Ribes* spp.
Hosts: Pinus, Ribes
Coleosporium laciniariae Arth.
Distr.: AR, FL, NJ.
Notes: Heteroecious; aecial host - *Pinus* spp.; telial host - *Liatris* spp.
Hosts: Liatris, Pinus
Coleosporium minutum Hedgc. & N. Hunt
Distr.: FL.
Notes: Heteroecious; aecial host - *Pinus* spp.; telial host - *Forestiera ligustrina*.
Hosts: Forestiera, Pinus
Coleosporium pacificum Cummins
Coleosporium madiae Arth. nom. illegit.
Coleosporium madiae Cooke, nom. illeg.
Stichopsora madiae Syd.
Peridermium californicum Arth. & F. Kern (anamorph)
Distr.: Western North America.
Notes: Heteroecious; aecial host - *Pinus radiata*; telial host - Asteraceae.
Hosts: Arnica, Haplopappus, Helianthus, Madia, Pinus, Tagetes
Coleosporium paederiae Dietel ex Hiratsuka
Distr.: HI; eastern Asia.
Notes: Heteroecious; aecial host - *Pinus* spp.; telial host - Rubiaceae.
Host: Paederia
Coleosporium pinicola Arth.
Coleosporium pini B.T. Galloway non Lagerh., hom. illeg.
Distr.: Eastern United States; Soviet Union.
Notes: Autoecious.
Host: Pinus
Coleosporium plumierae Pat.
Coleosporium domingensis (Berk.) Arth., nom. illeg.
Distr.: FL; West Indies; Central America.
Notes: Autoecious.
Host: Plumeria
Coleosporium steviae Arth.
Coleosporium eupatorii Cummins non Hiratsuka, hom. illeg.
Distr.: Southwestern United States; Central America.
Notes: Aecial host unknown; telial hosts are *Brickellia*, *Eupatorium* and *Stevia* spp.
Host: Brickellia
Coleosporium tussilaginis (Pers.) Lév. in C. d'Orb.
Coleosporium campanulae Lév. ex J. Kickx fil.
Coleosporium occidentale Arth.
Coleosporium senecionis J. Kickx fil.
Coleosporium sonchi (F. Strauss) Tul.
Coleosporium sonchi-arvensis Lév. in Berk., nom. illeg.
Distr.: Widespread in temperate regions; West Indies; New Zealand.
Notes: Heteroecious; aecial host - *Pinus* spp. telial host - *Campanula* spp. Parmelee & de Carteret (Fungi Canadenses 218. 1982 as *C. campanulae*).
Hosts: Campanula, [Lysimachia], Pinus, [Senecio], [Sonchus], Triodanis
Coleosporium vernoniae Berk. & M.A. Curtis in Berk.
Coleosporium carneum H. Jacks.
Coleosporium elephantopodis Thuem.
Distr.: Eastern United States; West Indies; Central and South America.
Notes: Heteroecious; aecial host - *Pinus* spp.; telial host - *Elephantopus* and *Vernonia* spp.
Hosts: Elephantopus, Pinus, Vernonia
Coleosporium viburni Arth.
Distr.: Central and eastern North America; Central and South America; Japan.
Notes: Heteroecious; aecial host - *Pinus* spp.; telial host - *Viburnum* spp.
Host: Viburnum
Coleosporium viguierae Dietel & Holw. in Holw.
Distr.: Southwestern United States; Mexico.
Notes: Autoecious.
Hosts: Verbesina, Viguiera

Coleroa (Fr.) Rabenh. Loculoascomycetes, Dothideales

Aphysa Theiss. & Syd.
Coleroa chaetomium (Kunze:Fr.) Rabenh.
Venturia kunzei Sacc.
Coccomyces rubi (Fr.) P. Karst.
Phacidium rubi Fr.
Distr.: Northwestern North America; Europe; Japan; The Philippines.
Notes: On living leaves (Corlett. Fungi Canadenses 20. 1974).
Host: Rubus
Coleroa circinans (Fr.:Fr.) G. Wint. in Rabenh.
Venturia circinans (Fr.:Fr.) Sacc.
Spilosticta glomerata (Cooke) Munk
Venturia glomerata Cooke
Distr.: Temperate North America; Europe.
Notes: On living leaves. Müller & Arx (Beitr. Kryptogamenfl. Schweiz 11(2):415. 1962).
Host: Geranium
Coleroa sporoboli (H.C. Greene) Barr
Venturia sporoboli H.C. Greene
Distr.: WI.
Notes: On leaves.
Hosts: Oryzopsis, Schizachyrium, Sporobolus

Collecephalus J.A. Spencer. Hyphomycetes

Collecephalus hemerocalli J.A. Spencer
Distr.: Southeastern United States.
Notes: Leaf-streak.
Host: Hemerocallis
Collecephalus hemerocalli J.A. Spencer var. ***macrosporum*** J.N. Lunsford & J.A. Spencer
Distr.: MD.
Host: Hemerocallis

Colletogloeum Petr. Coelomycetes

Notes: Arx (Phytopathol. Z. 29:413. 1957).

Colletogloeum fouquieriae Pollack & Sutton

Distr.: AZ.

Notes: Pollack & Sutton (Mycologia 64:702. 1972).

Host: Fouquieria

Colletogloeum rhois Sutton

Coryneum rhois Halst. non Sacc., hom. illeg.

Distr.: NJ.

Host: Rhus

Colletotrichum Corda in Sturm. Coelomycetes

Notes: Arx (Phytopathol. Z. 19:413. 1957); Sutton (The Coelomycetes. p. 523. 1980).

Colletotrichum acutatum J.H. Simmonds

Distr.: North America; Australia; New Zealand.

Notes: Fruit rot of a variety of plants. Many older reports of *C. gloeosporioides* may refer to this species. Dyko & Mordue (C.M.I. Descr. 630. 1979).

Host: Fragaria

Colletotrichum aquatilis R. Sprague

Distr.: Central and western North America.

Hosts: Catabrosa, Puccinellia

Colletotrichum arisaematis H.C. Greene

Distr.: WI.

Host: Arisaema

Colletotrichum bromeliacearum Birge in Heald & F.A. Wolf, nom. nud.

Distr.: FL, TX(type).

Hosts: Billbergia, Tillandsia

Colletotrichum capsici (Syd.) E.J. Butler & Bisby

Vermicularia capsici Syd.

Distr.: Subtropical and tropical regions.

Notes: Mordue (C.M.I. Descr. 317. 1971); Maiello (Mycologia 80:410. 1988).

Hosts: Canavalia, Capsicum, Gossypium, Rosa

Colletotrichum caudatum (Sacc.) Peck

Ellisiella caudata Sacc.

Ellisiellina caudata (Sacc.) Camara

Distr.: Cosmopolitan.

Notes: Arx (Phytopathol. Z. 29:455. 1957) synonymized this species with *C. graminicola* (Ces.) Wilson, but Sutton retained them as separate species.

Hosts: Agropyron, Andropogon, Aristida, Dichanthium, Eragrostis, Koeleria, Panicum, Puccinellia, Rottboellia, Schizachyrium, Setaria, Sorghastrum, Sporobolus

Colletotrichum cereale T.F. Manns in A. Selby & T.F. Manns

Distr.: Central and eastern United States.

Hosts: Dactylis, Zea

Colletotrichum circinans (Berk.) Voglino

Vermicularia circinans Berk.

Colletotrichum dematium (Pers.) Grove f. *circinans* (Berk.) Arx

Distr.: Cosmopolitan.

Notes: Smudge.

Hosts: Allium, Nolina, Nothoscordum

Colletotrichum coccodes (Wallr.) S.J. Hughes

Colletotrichum atramentarium (Berk. & Broome) Taubenhaus

Distr.: Cosmopolitan.

Notes: Anthracnose of fruits, black dot of tomato and potato, brown rot of stems of Solanaceae (Mordue. C.M.I. Descr. 131. 1967; Kokko & Shoemaker. Fungi Canadenses 51. 1974).

Hosts: Abutilon, Dahlia, Lycopersicon, Solanum

Colletotrichum coffeanum F. Noack

Distr.: Subtropical and tropical regions.

Notes: Probably a synonym of *C. gloeosporioides* but the name has been applied to the causal agent of coffee berry disease or to any *Colletotrichum* isolate from coffee (Firman & Waller. Phytopathol. Pap. 20:1. 1977).

Host: Coffea

Colletotrichum crassipes (Speg.) Arx

Gloeosporium crassipes Speg.

Gloeosporium agaves Syd.

Colletotrichum lobeliae F. Stevens

Fusarium schweinitzii Ellis & Harkn.

Colletotrichum sordidum J.J. Davis

Distr.: Cosmopolitan.

Hosts: Agave, Menispermum, Vitis

Colletotrichum dematium (Pers.) Grove

Vermicularia dematium (Pers.) Fr.

Colletotrichum carpophilum Kellerm. & Swingle

Vermicularia compacta Cooke & Ellis

Colletotrichum erumpens Sacc.

Colletotrichum fructus (F. Stevens & J.G. Hall) Sacc.

Vermicularia herbarum Westend.

Colletotrichum liliacearum Ferraris

Colletotrichum liliacearum (Westend.) M.M. Duke, hom. illeg.

Vermicularia liliacearum Westend.

Colletotrichum lineola Corda in Sturm

Schizotrichella lunata E.F. Morris

Colletotrichum omnivorum Halst.

Colletotrichum peckii (Sacc.) J.J. Davis

Colletotrichum silphii J.J. Davis

Colletotrichum smilacis Tehon

Colletotrichum solitarium Ellis & Barth.

Colletotrichum uncinatum (Berk. & M.A. Curtis) M.M. Duke

Colletotrichum violae-rotundifoliae (House) J.J. Davis

Colletotrichum violarum J.J. Davis

Distr.: Temperate and subtropical regions.

Notes: Arx (Phytopathol. Z. 29:457. 1957) listed 84 synonyms.

Hosts: Aeschynanthus, Amaryllis, Andropogon, Angelica, Asparagus, Aspidistra, Aster, Astragalus, Caulophyllum, Colchicum, Convallaria, Coptis, Dahlia, Desmodium, Dianthus, Erythronium, Fragaria, Gloriosa, Glycine, Hemerocallis, Heracleum, Hosta, Ipomoea, Iris, Kniphofia, Lilium, Lycopersicon, Maianthemum, Malus, Medicago, Monotropa, Osmorhiza, Panax, Parthenocissus, Pinus, Polygonatum, Polyscias, Rheum, Rumex, Silphium, Smilacina, Smilax, Solidago, Streptopus, Teramnus, Tilia, Trillium, Uvularia, Veratrum, Viola, Yucca, Zephyranthes, Zornia

Colletotrichum dematium (Pers.) Grove f. ***ipomoeae*** Arx nom. nud.

Distr.: NC.

Notes: The place of publication of this name could not be found.

Host: Ipomoea

Colletotrichum dematium (Pers.) Grove f. ***spinaciae*** (Ellis & Halst.) Arx

Colletotrichum spinaciae Ellis & Halst.

Distr.: Temperate regions.

Notes: Leaf spot and anthracnose of spinach and other Chenopodiaceae.

Host: Spinacia

Colletotrichum destructivum O'Gara

Colletotrichum sativum Horn

Glomerella glycines F. Lehm. & F.A. Wolf (teleomorph)

Distr.: North America.

Notes: Manandhar et al. (Phytopathology 76:282. 1986).

Hosts: Agropyron, Bromus, Coronilla, Cuscuta, Festuca, Glycine, Medicago, Melilotus, Nicotiana, Phalaris, Trifolium

Colletotrichum dianallae F. Stevens & P.A. Young

Distr.: HI(type).

Host: Dianella

Colletotrichum falcatum Went

Glomerella tucumanensis (Speg.) Arx & E. Müller (teleomorph)

Distr.: Subtropical and tropical regions.

Notes: Red rot of sugarcane (Mordue. C.M.I. Descr. 133. 1967).

Hosts: Carica, Dichorisandra, Eremochloa, Erianthus, Ophiopogon, Pisum, Saccharum, Sorghum, Syngonium, Zebrina, Zoysia

Colletotrichum fusarioides (Ellis & Kellerm.) O'Gara in Dearn.

Gloeosporium fusarioides Ellis & Kellerm.

Colletotrichum salmonicolor O'Gara

Distr.: Central and eastern North America.

Notes: Leaf spot.

Host: Asclepias

Colletotrichum fuscum Laubert

Distr.: Temperate regions.

Hosts: Arctium, Digitalis

Colletotrichum georgius-fischeri R. Sprague

Distr.: Western North America; Great Britain.

Hosts: Muhlenbergia, Poa

Colletotrichum gloeosporioides (Penz.) Penz. & Sacc. in Penz.

Gloeosporium affine Sacc.
Colletotrichum agaves Cavara
Gloeosporium ailanthi Dearn. & Barth. in Dearn.
Gloeosporium aletridis Henn.
Gloeosporium aleuriticum Sacc.
Gloeosporium alleschertanum Wollenweb.
Gloeosporium amaranthi Ellis & Everh.
Gloeosporium amygdalinum Britzelm.
Colletotrichum antirrhini F.C. Stewart
Gloeosporium apocryptum Ellis & Everh. var. *ramicola* Ellis & Everh.
Colletotrichum artocarpi Delacr.
Colletotrichum azaleae Ellis & Everh.
Gloeosporium barringtoniae F. Stevens & E. Young
Colletotrichum bletiae Halst.
Gloeosporium cactorum Stoneman
Colletotrichum cajani Rangel
Gloeosporium callae Oudem.
Colletotrichum camelliae Massee
Gloeosporium camphorae Sacc.
Gloeosporium canavaliae Syd.
Colletotrichum caricae Syd.
Gloeosporium catalpae Ellis & Everh.
Macrophoma cattleyicola Henn.
Colletotrichum cerei Earle
Gloeosporium cerei Pass.
Colletotrichum cinctum Stoneman
Gloeosporium cinctum Berk. & M.A. Curtis
Gloeosporium cingulatum Atk.
Colletotrichum cinnamomi Tharp
Volutella citrulli Stoneman
Colletotrichum commelinae Ellis & Everh.
Vermicularia concentrica Lév.
Gloeosporium corni H.C. Greene non Woronichin, hom. illeg.
Colletotrichum crotalariae Petch
Colletotrichum curvisetum F. Stevens
Colletotrichum cyclamenae Halst.
Gloeosporium dendrobii Maubl.
Colletotrichum derridis Van Hoof
Colletotrichum dioscoreae Tehon
Colletotrichum dracaenae Allesch.
Colletotrichum elasticae Tassi
Gloeosporium elasticae Cooke & Massee
Gloeosporium epidendri Henn.
Colletotrichum erythrinae Ellis & Everh.
Gloeosporium eucalypti McAlpine
Gloeosporium foliicola Nishida
Colletotrichum fragariae A.N. Brooks
Gloeosporium fructigenum Berk.
Gloeosporium glottidii Ellis & G. Martin
Colletotrichum griseum Heald & F.A. Wolf
Gloeosporium hawaiense Thuem.
Colletotrichum heucherae Ellis & B.T. Galloway
Colletotrichum hibisci Pollacci
Gloeosporium hysterioides Ellis & Everh.
Gloeosporium intermedium Sacc.
Colletotrichum ipomoeae Camara
Colletotrichum janiphae Grove
Colletotrichum jussiaeae Earle
Colletotrichum kentiae Camara & Luz
Gloeosporium laeliae Henn.
Gloeosporium leguminum (Cooke) Sacc.
Gloeosporium limetticola R.E. Clausen
Gloeosporium liriodendri Ellis & Everh.
Gloeosporium mangiferae Henn.
Gloeosporium melongenae Ellis & Halst.
Colletotrichum nigrum Ellis & Halst.
Gloeosporium ochraceum F. Patterson
Gloeosporium oleae F. Patterson
Gloeosporium oleandri Sacc.
Gloeosporium olivarum J.V. Almeida
Gloeosporium oncidii Oudem.
Gloeosporium opuntiae Ellis & Everh.
Colletotrichum orchidearum Allesch.
Colletotrichum papayae Henn.
Colletotrichum passiflorae F. Stevens & E. Young
Macrophoma peckiana (Thuem.) Berl. & Voglino
Gloeosporium peleae Henn.
Colletotrichum peregrinum Pass.
Colletotrichum philodendri Henn.
Colletotrichum phomoides F. Chester
Kabatiella phoradendri (L. Darling) Harvey
Protocoronospora phoradendri L. Darling
Colletotrichum phyllocacti Ellis & Everh.
Gloeosporium piperatum Ellis & Everh. in Halst.
Colletotrichum pisi Pat.
Colletotrichum pollaccii Magnaghi
Gloeosporium polymorphum Trinchieri
Colletotrichum primulae Halst.
Gloeosporium revolutum Ellis & Everh.
Gloeosporium rubicola Ellis & Everh.
Gloeosporium rufo-maculans (Berk.) Thuem.
Gloeosporium sansevieriae Verwoerd & Plessis
Gloeosporium schixanthi C.N. Jensen & V.B. Stewart
Vermicularia sparsipila Ellis & Kellerm.
Gloeosporium syringae Allesch.
Gloeosporium tabernaemontanae Speg.
Colletotrichum tertium (Cooke) Grove
Gloeosporium thumenii Sacc.
Colletotrichum trillii Tehon
Colletotrichum vanillae Scalia
Gloeosporium vanillae Cooke
Colletotrichum vermicularioides Halst.
Gloeosporium vexans Atk.
Colletotrichum violae-tricoloris R.E. Sm.
Colletotrichum xanthii Halst.
Gloeosporium yuccaegenum Ellis & Everh.
Glomerella cingulata (Stoneman) Spauld. & H. Schrenk (teleomorph)

Distr.: Cosmopolitan.

Notes: Arx (Phytopathol. Z. 29:413. 1957) listed about 600 synonyms.

Hosts: Abelmoschus, Abutilon, Acer, Aechmea, Aeschynomene, Agapanthus, Agave, Ailanthus, Aletris, Aleurites, Allamanda, Alysicarpus, Amaranthus, Amorphophallus, Ananas, Annona, Anthurium, Antirrhinum, Arceuthobium, Arecaceae, Aristolochia, Artocarpus, Asparagus, Aucuba, Averrhoa, Barringtonia, Billbergia, Brassica, Brassolaeliocattleya, Cactus, Cajanus, Calendula, Callirhoe, Callistephus, Camellia, Canavalia, Canna, Capsicum, Carica, Carissa, Casimiroa, Catalpa, Catha, Cattleya, Centrosema, Cereus, Chrysalidocarpus, Chrysobalanus, Cinnamomum, Citrofortunella, Citrullus, Citrus, Clidemia, Clitoria, Codiaeum, Commelina, Convolvulus, Cornus, Crotalaria, Cryptanthus, Cryptbergia, Cymbidium, Dalbergia, Dendrobium, Desmodium, Dichondra, Dieffenbachia, Dionaea, Dioscorea, Diospyros, Dovyalis, Dracaena, Echinopsis, Epidendrum, Epiphyllum, Eriobotrya, Erythrina, Eucalyptus, Eugenia, Euonymus, Euphorbia, Ficus, Flacourtia, Fortunella, Fragaria, Galactia, Gardenia, Glycine, Guardiola, Guzmania, Gynura, Heuchera, Hibiscus, Hoya, Hylocereus, Iberis, Indigofera, Jasminum, Jatropha, Juglans, Kalanchoe, Laelia, Laeliocattleya, Lathyrus, Limonium, Linaria, Liriodendron, Lobelia, Lonchocarpus, Ludwigia, Lycopersicon, Macroptilium, Magnolia, Malpighia, Malus, Mammillaria, Mangifera, Manihot, Matthiola, Medicago, Melilotus, Mucuna, Musa, Neoregelia, Nerium, Noronhia, Odontoglossum, Olea, Oncidium, Opuntia, Orchidaceae, Osmanthus, Pandanus, Passiflora, Pelea, Persea, Phalaenopsis, Philodendron, Phoenix, Phoradendron, Phytolacca, Pilea, Pisum, Pithecellobium, Plumeria, Podocarpus, Polyscias, Poncirus, Populus, Potentilla, Pouteria, Primula, Prosopis, Prunus, Psidium, Punica, Pyracantha, Quercus, Rhaphiolepis, Rhododendron, Robinia, Roystonea, Rubus, Sansevieria, Sarracenia, Sassafras, Schlumbergera, Senna, Sesbania, Smilax, Solanum, Spathiphyllum, Stapelia, Strelitzia, Strongylodon, Stylosanthes, Swietenia, Syngonium, Syringa, Syzygium, Tabernaemontana, Tagetes, Tecomaria, Teramnus, Thunbergia, Trigonella, Trillium, Umbellularia, Vaccinium, Vanda, Vanilla, Viola, Vitis, Vriesea, Xanthium, Yucca, Zantedeschia, Zingiber

Colletotrichum gloeosporioides (Penz.) Penz. & Sacc. in Penz. f. sp. ***aeschynomene***

Distr.: Central and eastern United States.

Notes: Useful as a biological control agent on northern jointvetch (Daniel et al. Weed Science 21:303. 1973).
Host: Aeschynomene
Colletotrichum gossypii Southworth
Glomerella gossypii Edgerton (teleomorph)
Distr.: Worldwide in cotton-growing regions.
Notes: Some have considered this a synonym of *C. gloeosporioides*.
Host: Gossypium
Colletotrichum graminicola (Ces.) G.W. Wils.
Vermicularia graminicola Westend.
Ellisiella mutica G. Wint.
Colletotrichum sublineola Henn. in Kab. & Bubák
Glomerella graminicola Politis (teleomorph)
Distr.: Cosmopolitan.
Notes: Red stalk rot, red leaf spot, leaf anthracnose, and seedling blight of *Sorghum*, other cereals, and legumes (Mordue. C.M.I. Descr. 132. 1967). However, Sutton (The Coelomycetes. p. 534. 1980) applied this name only to a fungus pathogenic on *Zea*.
Hosts: [Agropyron], Agrostis, Alopecurus, Ammophila, Andropogon, Anthaenantia, Anthoxanthum, Aristida, Arrhenatherum, Arundinaria, Avena, Beckmannia, Bouteloua, Brachypodium, Bromus, Buchloe, Calamagrostis, Calamovilfa, Carex, Catabrosa, Cenchrus, Chasmanthium, Chloris, Cynodon, Dactylis, Danthonia, Diarrhena, Dichanthelium, Digitaria, Echinochloa, Eleusine, Elymus, Eragrostis, Eremochloa, Festuca, Glyceria, Helictotrichon, Hierochloe, Holcus, Hordeum, Koeleria, Lolium, Medicago, Melica, Milium, Muhlenbergia, Oryzopsis, Panicum, Paspalum, Phalaris, Phleum, Poa, Polypogon, Puccinellia, Saccharum, Schizachne, Schizachyrium, Secale, Silphium, Sisyrinchium, Sorghastrum, Sorghum, Sphenopholis, Sporobolus, Stipa, Tridens, Trifolium, Tripsacum, Trisetum, Triticum, Vulpia, Zea, Zizania
Colletotrichum helianthi J.J. Davis
Distr.: WI(type).
Host: Helianthus
Colletotrichum helianthi J.J. Davis var. ***macromaculans*** H.C. Greene
Distr.: WI(type).
Host: Helianthus
Colletotrichum hepaticae Peck
Distr.: NY(type), WI.
Host: Hepatica
Colletotrichum higginsianum Sacc. in Higgins
Distr.: Cosmopolitan.
Host: Brassica
Colletotrichum hysteriiformis (Peck) H.C. Greene
Vermicularia hysteriiformis Peck
Distr.: Northeastern United States.
Host: Caulophyllum
Colletotrichum lilii Plakidas ex Boerema & Hamers
Notes: See Boerema & Hamers (Neth. J. Pl. Pathol. 94:12. 1988).
Hosts: Hemerocallis, Lilium
Colletotrichum lindemuthianum (Sacc. & Magnus) Lams.-Scrib.
Glomerella lindemuthiana Shear (teleomorph)
Distr.: Cosmopolitan.
Notes: Causes anthracnose of leaves, pods, and stems of *Phaseolus* spp. and other legumes (Mordue. C.M.I. Descr. 316. 1971). The connection with *Glomerella lindemuthiana* is tentative. The author citation follows the recommendation of Stevenson (Pl. Dis. Reporter 40:831. 1956).
Hosts: Phaseolus, Vigna
Colletotrichum lineola Corda var. ***halepense*** Heald & F.A. Wolf
Distr.: TX(type).
Notes: Leaf spot of Johnson grass (Heald & Wolf. U.S.D.A. Bur. Pl. Industr. Bull. 226:51. 1912).
Host: Sorghum
Colletotrichum lini (Westerd.) Tochinai
Gloeosporium lini Westerd.
Colletotrichum lini T.F. Manns & Bolley
Distr.: Temperate northern hemisphere.
Host: Linum
Colletotrichum linicola Pethybr. & Lafferty
Distr.: Temperate North America; Europe.
Notes: Anthracnose disease of flax (Pethyb. & Lafferty. Sci. Proc. Roy. Dublin Soc. 15:359. 1918).
Host: Linum
Colletotrichum lucidae H.C. Greene
Distr.: WI.
Hosts: Populus, Salix
Colletotrichum madisonensis H.C. Greene
Distr.: WI.
Notes: On leaves.
Host: Carex
Colletotrichum malvarum (A. Braun & Casp.) Southworth
Steirochaete malvarum A. Braun & Casp.
Colletotrichum athaeae Southworth
Colletotrichum magnusianum Bres.
Distr.: Temperate northern hemisphere.
Notes: Leaf spot of Malvaceae.
Hosts: Abutilon, Alcea, Althaea, Lavatera, Malva, Sida
Colletotrichum musae (Berk. & M.A. Curtis) Arx
Myxosporium musae Berk. & M.A. Curtis
Gloeosporium musarum Cooke & Massee
Distr.: Subtropical and tropical regions.
Notes: Anthracnose fruit and stem rot, leaf spot of banana (Sutton & Waterson. C.M.I. Descr. 222. 1970).
Host: Musa
Colletotrichum orbiculare (Berk. & Mont.) Arx
Gloeosporium orbiculare Berk. & Mont.
Colletotrichum lagenarium (Pass.) Ellis & Halst.
Gloeosporium lagenarium (Pass.) Sacc.
Glomerella lagenarium F. Stevens (teleomorph)
Distr.: Temperate, subtropical and tropical regions.
Notes: Anthracnose and fruit rot of Cucurbitaceae.
Hosts: Benincasa, Citrullus, Cucumis, Cucurbita, Echinocystis, Lagenaria, Luffa, Momordica, Sechium, Sicana, Sicyos, Trichosanthes
Colletotrichum paludosum (Ellis & B.T. Galloway) Arx
Gloeosporium paludosum Ellis & B.T. Galloway
Distr.: Eastern United States.
Host: Peltandra
Colletotrichum phyllachoroides (Ellis & Everh.) Arx
Gloeosporium phyllachoroides Ellis & Everh.
Distr.: CA.
Host: Artemisia
Colletotrichum prunicola H.C. Greene
Distr.: WI(type).
Host: Prunus
Colletotrichum psoraleae (Peck) Arx
Gloeosporium psoraleae Peck
Distr.: FL, NE(type).
Notes: Leaf necrosis.
Hosts: Pediomelum, Psoralea
Colletotrichum pyrolae (Trel.) Parmelee
Ovularia pyrolae Trel.
Ramularia pyrolae (Trel.) Ellis & Everh.
Distr.: DE, NY, WI.
Notes: On living leaves (Parmelee. Canad. J. Bot. 36:872. 1958).
Host: Pyrola
Colletotrichum rhexiae Ellis & Everh.
Distr.: DE(type).
Hosts: Miconia, Rhexia
Colletotrichum rudbeckiae Peck
Distr.: NY(type).
Host: Rudbeckia
Colletotrichum schizanthi C.N. Jensen & V.B. Stewart
Distr.: NY(type).
Notes: Anthracnose (Jensen & Stewart. Phytopathology 1:125. 1911).
Host: Schizanthus
Colletotrichum trichellum (Fr.:Fr.) M.M. Duke
Amerosporium trichellum (Fr.:Fr.) Lind
Distr.: FL; Australia; Europe; India; Turkey.
Host: Hedera
Colletotrichum trifolii Bain & Essary
Distr.: United States.
Notes: Southern anthracnose of clover and other legumes.
Hosts: Coronilla, Kummerowia, Lespedeza, Medicago, Melilotus, Trifolium, Vicia
Colletotrichum truncatum (Schwein.) Andrus & W.D. Moore
Colletotrichum dematium (Pers.) Grove f. *truncatum* (Schwein.) Arx
Colletotrichum caulicola Heald & F.A. Wolf

Colletotrichum glycines S. Hori
Vermicularia polytricha Cooke
Colletotrichum viciae Dearn. & Overh.
Colletotrichum villosum J.L. Weimer
Distr.: Cosmopolitan.
Notes: On various plants, especially common on Fabaceae.
Hosts: Abutilon, Aeschynomene, Alysicarpus, Amaranthus, Ambrosia, Apocynum, Asclepias, Canavalia, Centrosema, Chenopodium, Clitoria, Cyperus, Datura, Desmanthus, Desmodium, Galactia, Glycine, Hibiscus, Indigofera, Ipomoea, Lotus, Macroptilium, Macrotyloma, Medicago, Phaseolus, Polygonum, Senna, Solanum, Strelitzia, Stylosanthes, Vicia, Vigna, Xanthium
Colletotrichum typhae H.C. Greene
Distr.: WI(type).
Host: Typha
Colletotrichum urticae H.C. Greene
Distr.: WI.
Hosts: Laportea, Urtica

Collybia (Fr.) Staude, nom. cons. Basidiomycetes, Agaricales

Collybia acervata (Fr.:Fr.) P. Kumm.
Distr.: Circumboreal on wood and debris of conifers.
Hosts: Abies, Picea
Collybia bakerensis A.H. Sm.
Distr.: In CA, typically on *Abies*. In CO, ID and WA, typically on *Tsuga*.
Hosts: Abies, Tsuga
Collybia dichrous (Berk. & M.A. Curtis) Gilliam
Marasmius dichrous Berk. & M.A. Curtis
Marasmius cespitosus Peck
Marasmius fagineus Morg.
Distr.: Eastern North America.
Notes: On wood of various hardwoods, especially common on *Fagus*.
Host: Quercus
Collybia dryophila (Bull.:Fr.) P. Kumm.
Distr.: Temperate northern hemisphere.
Notes: On litter and debris, occasionally on wood.
Host: Rubus

Colpoma Wallr. Discomycetes, Rhytismatales

Colpoma azaleae (Schwein.) Cooke
Distr.: Eastern United States.
Hosts: Lyonia, Rhododendron
Colpoma crispum (Pers.:Fr.) Sacc.
Clithris crispa (Pers.:Fr.) Rehm
Distr.: Temperate regions.
Notes: On conifer twigs.
Hosts: Abies, Larix, Pinus, [Populus], Tsuga
Colpoma leucothoicola Fr.
Distr.: NC.
Host: Leucothoe
Colpoma morbida (Peck) Sacc.
Triblidium morbidum Peck
Clithris morbida (Peck) Ellis & Everh.
Distr.: NY, OR.
Host: Pseudotsuga
Colpoma quercinum (Pers.) Wallr.
Hysterium quercinum Pers.
Cenangium quercinum (Pers.) Fr.:Fr.
Clithris quercina (Pers.) P. Karst.
Conostroma didymum (Fautrey & Roum.) Moesz (anamorph)
Distr.: Temperate regions.
Notes: Die-back of twigs and young branches (Cannon & Minter. Mycol. Pap. 155:35. 1986).
Hosts: Castanea, Quercus

Coltriciella Murrill. Basidiomycetes, Aphyllophorales

Coltriciella dependens (Berk. & M.A. Curtis) Murrill
Polyporus dependens Berk. & M.A. Curtis
Polyporus deceptivus C.G. Lloyd
Distr.: Widespread.
Notes: On wood.
Host: Pinus

Columnocystis Pouzar. Basidiomycetes, Aphyllophorales

Columnocystis abietina (Pers.:Fr.) Pouzar
Stereum abietinum (Pers.:Fr.) Fr.
Hymenochaete rugispora Ellis & Everh.
Stereum rugisporum (Ellis & Everh.) Burt
Distr.: Temperate northern hemisphere; in North America, common in the West, occasional in the East.
Notes: On wood of conifers.
Hosts: Abies, Larix, Picea, Pinus, Pseudotsuga, Thuja, Tsuga
Columnocystis ambigua (Peck) Pouzar
Stereum ambiguum Peck
Distr.: Temperate northern hemisphere.
Notes: On wood of conifers.
Hosts: Abies, Picea, Pinus, Tsuga
Columnocystis pimeriensis R.L. Gilbertson
Distr.: AZ(type).
Notes: On wood.
Host: Pinus

Comatricha G. Preuss. Myxomycetes

Comatricha nigra (Pers.) J. Schröt.
Distr.: Cosmopolitan.
Host: Quercus
Comatricha typhoides (Bull.) Rostr.
Distr.: Cosmopolitan.
Host: Acer

Comoclathris Clements. Loculoascomycetes, Dothideales

Comoclathris permunda (Cooke) E. Müller
Clathrospora permunda (Cooke) Berl.
Platyspora permunda (Cooke) Wehmeyer
Pleospora permunda (Cooke) Sacc.
Pleospora baccata Ellis
Pleospora compressa Harkn.
Pyrenophora depressa Peck
Clathrospora diplospora (Ellis & Everh.) Wehmeyer
Pleospora diplospora Ellis & Everh.
Pleospora silenes Earle
Pleospora utahensis Ellis & Everh.
Pleospora xerophylli Petr.
Distr.: Temperate North America; South America; Eurasia.
Notes: On a variety of herbaceous hosts.
Hosts: Amorpha, Anemone, Arabis, Asclepias, Aster, Balsamorhiza, Brickellia, Carum, Castilleja, Clematis, Cycladenia, Cymopterus, Delphinium, Guardiola, Hedysarum, Lewisia, Ligusticum, Linaria, Linum, Lomatium, Lupinus, Monardella, Muhlenbergia, Penstemon, Petroselinum, Polygonum, Sambucus, Senecio, Sphenosciadium, Stephanomeria, Synthyris, Urtica, Xerophyllum

Completoria Lohde. Zygomycetes, Entomophthorales

Completoria complens Lohde
Distr.: North America; Europe.
Notes: On fern prothalli.
Hosts: Phanerophlebia, Pteris

Confertobasidium Jülich. Basidiomycetes, Aphyllophorales

Confertobasidium montanum (Jülich) Jülich & Stalpers
Confertobasidium olivaceo-album (Bourd. & Galzin) Jülich var. *montanum* Jülich
Distr.: Northern hemisphere(rare).

Notes: On wood.
Host: Abies
Confertobasidium olivaceo-album (Bourd. & Galzin) Jülich
Athelia fuscostrata (Burt) Donk
Corticium fuscostratum Burt
Distr.: Northern hemisphere.
Notes: On wood.
Hosts: Abies, Acer, Picea, Pinus, Populus, Pseudotsuga

Confertopeltis Tehon. Coelomycetes

Confertopeltis asparagi Tehon
Distr.: IL.
Notes: A doubtful species (Sutton. Mycol. Pap. 141:46. 1977).
Host: Asparagus

Confistulina Stalpers in Stalpers & Vlug. Hyphomycetes

Notes: Monotypic (Stalpers & Vlug. Canad. J. Bot. 61:1660. 1983).
Confistulina hepatica (Sacc.) Stalpers in Stalpers & Vlug
Ceriomyces hepaticus Sacc.
Ptychogaster hepaticus (Sacc.) C.G. Lloyd
Fistulina hepatica (Schaeff.:Fr.) Sibth. (teleomorph)
Distr.: Widespread in temperate regions.
Notes: On wood.
Hosts: Castanea, Pinus, Quercus

Conidioxyphium Batista & Cif. Coelomycetes

Conidioxyphium gardeniorum Batista & Cif.
Distr.: LA.
Host: Gardenia

Coniella Höhn. Coelomycetes

Coniella castaneicola (Ellis & Everh.) Sutton
Gloeosporium castaneicola Ellis & Everh.
Pilidiella castaneicola (Ellis & Everh.) Arx
Coniella eucalypticola Nag Raj
Pilidiella quercicola (Oudem.) Petr.
Schizoparme straminea Shear (teleomorph)
Distr.: Warm-temperate to tropical regions.
Notes: On a variety of hosts. The anamorph-teleomorph connection is based on the synonymy of Sutton (The Coelomycetes. p. 420. 1980) and the study by Maas et al. (Mycologia 71:92. 1979).
Hosts: Castanea, Eucalyptus, Fragaria
Coniella diplodiella (Speg.) Petr. & Syd.
Phoma diplodiella Speg.
Coniothyrium diplodiella (Speg.) Sacc.
Distr.: Cosmopolitan.
Notes: White rot of vine. Sutton & Waterston (C.M.I. Descr. 82. 1966).
Host: Vitis
Coniella granati (Sacc.) Petr. & Syd.
Distr.: Widespread.
Host: Punica

Coniochaeta (Sacc.) Cooke. Pyrenomycetes, Sordariales

Notes: Mahoney & LaFavre (Mycologia 73:931. 1981) included a discussion of many species of *Coniochaeta*.
Coniochaeta ligniaria (Grev.) Cooke
Rosellinia ligniaria (Grev.) Nitschke ex Fuckel
Coniochaeta discospora (Auersw. ex Niessl) Cain
Phialophora sp. (anamorph)
Distr.: Temperate northern hemisphere.
Hosts: Abies, Acer, Alnus, Ceanothus, Cercocarpus, Crataegus, Larix, Oplopanax, Ostrya, Physocarpus, Pinus, Populus, Sambucus, Tsuga
Coniochaeta malacotricha (Auersw. in Niessl) Cooke
Phialophora sp. (anamorph)
Distr.: Cosmopolitan.
Notes: On wood. Rogers & Grand (Canad. J. Bot. 49:2239. 1971) discussed the anamorph of this species.
Host: Pinus
Coniochaeta pulveracea (Ehrenb.:Fr.) Munk
Rosellinia pulveracea (Ehrenb.:Fr.) Nitschke ex Fuckel
Anthostomella myricae Grove
Distr.: Temperate northern hemisphere.
Notes: On wood of many hardwoods.
Hosts: Acer, Alnus, Amelanchier, Magnolia, Physocarpus, Platanus, Populus, Quercus, Sambucus, Vitis
Coniochaeta sordaria (Fr.) Petr.
Distr.: Temperate regions.
Host: Sambucus

Coniophora DC. Basidiomycetes, Aphyllophorales

Coniophora arida (Fr.:Fr.) P. Karst.
Distr.: Widespread.
Notes: On wood.
Hosts: Abies, Betula, Chamaecyparis, Chilopsis, Juglans, Juniperus, Picea, Pinus, Platanus, Populus, Prunus, Pseudotsuga, Rhododendron, Taxodium, Thuja, Tsuga
Coniophora arida (Fr.:Fr.) P. Karst. var. **_suffocata_** (Peck) Ginns
Coniophora suffocata (Peck) Massee
Distr.: Northern hemisphere.
Notes: On wood.
Hosts: Abies, Fraxinus, Pinus, Platanus, Populus, Salix, Thuja, Tsuga
Coniophora eremophila Lindsey & R.L. Gilbertson
Distr.: North America(Sonoran Desert); Chile.
Notes: On wood.
Hosts: Carnegiea, Juniperus, Prosopis
Coniophora fusispora (Cooke & Ellis) Sacc.
Corticium fusisporum Cooke & Ellis
Distr.: Northern hemisphere.
Notes: On wood, occasionally causes butt rot of living conifers.
Hosts: Juniperus, Pinus
Coniophora olivacea (Pers.:Fr.) P. Karst.
Coniophorella olivacea (Pers.:Fr.) P. Karst.
Coniophora atrocinerea P. Karst.
Coniophora ellisii (Berk. & M.A. Curtis) Sacc.
Coniophora leucothrix (Berk. & M.A. Curtis) Sacc.
Distr.: Widespread.
Notes: On wood, usually of conifers, especially *Picea* species.
Hosts: Abies, Chamaecyparis, Picea, Pinus, Platanus, Populus, Pseudotsuga, Quercus, Rhamnus
Coniophora puteana (Schumach.:Fr.) P. Karst.
Coniophora cerebella (Pers.) Pers.
Coniophora laxa (Fr.:Fr.) Quél.
Distr.: Widespread.
Notes: On wood, usually of conifers, sometimes associated with decay of wood in service or with brown butt rot of living trees.
Hosts: Abies, Alnus, Betula, Carya, Castanea, Crataegus, Cupressus, Fraxinus, Larix, Liriodendron, Malus, Picea, Pinus, Populus, Prunus, Pseudotsuga, Quercus, Thuja, Tsuga
Coniophora submembranacea (Berk. & Broome) Sacc.
Coniophora inflata Burt
Distr.: AZ; Mexico; Singapore; Sri Lanka.
Notes: On wood. Ginns (Opera Bot. 61:52. 1982) reported collections Mexico, Singapore, and Sri Lanka.
Hosts: Olneya, Prosopis

Coniosporium Link: Fr. Hyphomycetes

Coniosporium diversum (Cooke) J.L. Crane & Schoknecht in Schoknecht & J.L. Crane
Torula diversa Cooke
Sirodesmium diversum (Cooke) S.J. Hughes
Trichocladium olivaceum Petch
Distr.: Southeastern United States; Sri Lanka.
Notes: Schoknecht & Crane (Mycologia 69:535. 1977).
Host: Agave
Coniosporium gramineum Ellis & Everh.
Distr.: LA(type).
Host: Arundinaria
Coniosporium harknessioides (Ellis & Holw.) Sacc.

Gymnosporium harknessioides Ellis & Holw.
Distr.: IA(type).
Host: Teucrium
Coniosporium limoniiforme Syd.
Distr.: UT(type).
Host: Rosa
Coniosporium nucifoedum Fairm.
Distr.: NY(type).
Host: Carya
Coniosporium olivaceum Link
Septonema bicolor Peck
Sirodesmium granulosum De Not.
Sporidesmium peziza Cooke & Ellis
Septonema spilomeum Berk.
Distr.: United States; Europe.
Notes: On dead wood (Ellis. The Dematiaceous Hyphomycetes. p. 40. 1971).
Hosts: Acer, Cornus, Crataegus, Morus, Populus, Quercus, Tilia
Coniosporium rhizophilum (G. Preuss) Sacc.
Gymnosporium rhizophilum G. Preuss
Distr.: OK; Europe.
Notes: On grasses.
Host: Cynodon
Coniosporium subgranulosum Ellis & Everh.
Distr.: MT(type).
Host: Populus

Coniothecium Corda. Hyphomycetes

Notes: Hughes (Canad. J. Bot. 36:754. 1958) was unable to locate a type specimen for the the type species, *C. atrum*, and therefore considered it a doubtful genus. Many taxa have been transferred to various genera of the Hyphomycetes, but not all names have been redisposed.
Coniothecium agaves Petr.
Distr.: HI(type).
Host: Agave
Coniothecium applanatum Sacc.
Distr.: WA; Europe(type).
Host: Acer
Coniothecium mollerianum Thuem.
Distr.: ND; Portugal(type).
Host: Aster
Coniothecium richardiae (W.B. Mercer) Jauch
Phoma richardiae W.B. Mercer
Distr.: United States; Argentina(type).
Host: Zantedeschia

Coniothyrium Corda, nom. cons. Coelomycetes

Notes: Most species described in *Coniothyrium* need to be transferred to other genera such as *Microsphaeropsis*. Sutton (Mycol. Pap. 123:32. 1971) provided a discussion of the genus. Biga et al. (Sydowia 12:258. 1958).
Coniothyrium ampelinum Cooke
Distr.: SC.
Host: Vitis
Coniothyrium arctostaphyli (Vize) Tassi
Sphaeropsis arctostaphyli (Vize) Sacc.
Distr.: Western United States.
Host: Arctostaphylos
Coniothyrium asterinum (Cooke & Harkn.) Tassi
Distr.: Western United States.
Host: Arbutus
Coniothyrium baptisiae (Thuem.) Tassi
Sphaeropsis baptisiae Thuem.
Distr.: SC.
Host: Baptisia
Coniothyrium bartholomaei Dearn. & Barth.
Distr.: CA.
Host: Yucca
Coniothyrium berlandieri Viala & Sauv.
Distr.: TX.
Host: Vitis
Coniothyrium callicarpae Cooke
Distr.: SC(type).
Host: Callicarpa
Coniothyrium caryogenum F.V. Rand
Distr.: Southeastern United States.
Host: Carya
Coniothyrium celtidis-australis (Sacc.) Biga
Coniothyrium olivaceum Bonord. f. *celtidis-australis* Sacc.
Coniothyrium radicicola Tehon
Coniothyrium ulmi Tharp non (P. Karst.) Tassi, hom. illeg.
Distr.: Widespread in temperate regions.
Host: Ulmus
Coniothyrium cephalanthi Ellis & Everh.
Distr.: Southern United States.
Host: Cephalanthus
Coniothyrium chionanthi Fairm.
Distr.: NY(type).
Host: Chionanthus
Coniothyrium convolutum A.S. Horne
Distr.: WA.
Host: Malus
Coniothyrium cydoniae Brunaud
Distr.: WA.
Notes: Biga et al. (Sydowia 12:284. 1958) considered this a synonym of *Coniothyrium cystotricha* (Schulzer) Sacc. & Traverso and indicated that it has also been considered synonymous with *Microsphaeropsis olivacea* (as *Coniothyrium*).
Host: Malus
Coniothyrium decipiens Cooke & Harkn.
Coniothyrium punctum Cooke & Harkn.
Distr.: CA.
Host: Acacia
Coniothyrium dracaenae F. Stevens & Weedon in F. Stevens
Distr.: HI.
Host: Dracaena
Coniothyrium fagi Tehon
Distr.: IL.
Host: Fagus
Coniothyrium fuckelii Sacc.
Diapleella coniothyrium (Fuckel) Barr (teleomorph)
Distr.: Cosmopolitan.
Notes: Stem blight and canker of Rosaceae, especially *Rubus* spp. (Punithalingam. C.M.I. Descr. 663. 1980).
Hosts: Acer, Carya, Euonymus, Fragaria, Malus, Parthenocissus, Prunus, Rosa, Rubus, Smilax, Ulmus
Coniothyrium fuligineum (P. Karst.) Sacc.
Distr.: Circumpolar.
Host: Salix
Coniothyrium fuscidulum Sacc.
Distr.: AK; Europe.
Host: Sambucus
Coniothyrium hellebori Cooke & Massee
Distr.: North America; Europe.
Notes: Ellis & Ellis (Microfungi on land plants. p. 367. 1985) provided a description and illustration.
Host: Helleborus
Coniothyrium hibisci (Berk.) Tassi
Distr.: SC.
Notes: Probably not a *Coniothyrium* (Biga et al. Sydowia 12:294. 1958).
Host: Hibiscus
Coniothyrium ilicinum Ellis & F.W. Anderson
Distr.: Eastern United States.
Host: Ilex
Coniothyrium junci Ellis & Everh.
Distr.: OR.
Host: Juncus
Coniothyrium marrubii Fairm.
Distr.: CA.
Host: Marrubium
Coniothyrium microscopicum Sacc.
Coniothyrium truncisedum Vestergr.
Distr.: IL.
Host: Quercus
Coniothyrium negundinis Tehon & E.Y. Daniels

Distr.: IL(type), distribution elsewhere is questionable.
Notes: Wollenweber & Hochapfel (Z. Parasitol. 9:614. 1937) considered this to be a synonym of *C. fuckelii* Sacc.
Host: Acer
Coniothyrium orni Henn.
Distr.: CA.
Notes: This species does not belong in *Coniothyrium* (Biga et al. Sydowia 12:307. 1958).
Host: Fraxinus
Coniothyrium ovispermum (Fuckel) Sacc.
Coniothyrium oospermum (Fuckel) Sacc.
Distr.: WA.
Notes: Biga et al. (Sydowia 12:266. 1958) indicated that the epithet *oospermum* is incorrect and that *ovispermum* is the correct epithet for this fungus.
Host: Cryptomeria
Coniothyrium pallidospermum H.C. Greene
Distr.: WI.
Host: Carya
Coniothyrium pandanicola Petr.
Distr.: HI(type).
Host: Pandanus
Coniothyrium penstemonis Earle in H.C. Greene
Distr.: Western United States.
Notes: Biga et al. (Sydowia 12:304. 1958 as "*pentastemonis*") considered this a doubtful species.
Host: Penstemon
Coniothyrium phyllogenum Sacc.
Distr.: Western United States; Europe.
Host: Rhododendron
Coniothyrium poavora R. Sprague in H.C. Greene
Distr.: WI.
Host: Poa
Coniothyrium psammae Oudem.
Distr.: OR; Europe.
Notes: Ellis & Ellis (Microfungi on land plants. p. 481. 1985) provided a description and illustration.
Host: Calamagrostis
Coniothyrium pyrinum (Sacc.) J. Sheld.
Distr.: Temperate northern hemisphere.
Hosts: Crataegus, Cydonia, Malus, Pyrus
Coniothyrium rhamnigenum (Sacc.) Bubák
Phyllosticta rhamnigena Sacc.
Distr.: Northwestern North America; Europe.
Host: Rhamnus
Coniothyrium rhododendri Henn.
Distr.: WA; Germany.
Notes: Biga et al. (Sydowia 12:303. 1958) suggested that this is a doubtful species.
Host: Rhododendron
Coniothyrium rhois-radicantis (Sacc.) Biga, Cif. & Bestagno
Coniothyrium rhois Tharp
Distr.: Southern United States; Europe.
Host: Rhus
Coniothyrium rosarum Cooke & Harkn.
Distr.: United States.
Notes: Rose graft canker. This species is similar to *Coniothyrium fuckelii* but appears to be distinct (Horst. Westcott's plant disease handbook. p. 158. 1979).
Host: Rosa
Coniothyrium rupestris M.L. Farr
Host: Selaginella
Coniothyrium sambucinum (Cooke) Tassi
Coniothyrium sambuci Earle
Distr.: Western United States.
Host: Sambucus
Coniothyrium scirpi Trail
Coniothyrium zeae G.L. Stout
Paraphaeosphaeria michotii (Westend.) O. Eriksson (teleomorph)
Distr.: Temperate northern hemisphere.
Notes: Mostly on grasses and rushes (Shoemaker & Babcock. Canad. J. Bot. 63: 1286. 1985.)
Host: Zea
Coniothyrium scrophulariae (Fuckel) Sacc.
Distr.: CA; Europe(type).
Notes: Biga et al. (Sydowia 12:304. 1958) considered this a doubtful species.
Host: Scrophularia
Coniothyrium sepium Fairm.
Distr.: NM.
Host: Calystegia
Coniothyrium spokanense Sacc.
Distr.: WA.
Host: Salix
Coniothyrium vaccinicola (Schwein.) Starb.
Distr.: Northeastern United States.
Host: Vaccinium
Coniothyrium wernsdorffiae Laubert
Distr.: North America; Europe; Soviet Union.
Notes: Brand canker of rose. Ellis & Ellis (Microfungi on land plants. p. 230. 1985) provided a description.
Host: Rosa
Coniothyrium wisconsinensis H.C. Greene
Distr.: WI.
Host: Ulmus
Coniothyrium zingiberi F. Stevens & Atienza
Distr.: HI; The Philippines.
Host: Zingiber

Conocybe Fayod, nom. cons. Basidiomycetes, Agaricales

Conocybe blattaria (Fr.:Fr.) Kühner
Pholiotina blattaria (Fr.:Fr.) Fayod
Distr.: Northern hemisphere.
Host: Acer

Conohypha Jülich. Basidiomycetes, Aphyllophorales

Conohypha albocremea (Höhn. & Litsch.) Jülich
Corticium albocremeum Höhn. & Litsch.
Hyphoderma albocremeum (Höhn. & Litsch.) J. Eriksson & A. Strid in J. Eriksson & Ryvarden
Distr.: North America (rare); Europe.
Notes: On wood.
Host: Juniperus
Conohypha terricola (Burt) Jülich
Peniophora terricola Burt
Hyphoderma terricola (Burt) K.J. Martin & R.L. Gilbertson
Distr.: United States.
Notes: On wood. Jülich (Persoonia 8:442. 1976).
Host: Populus

Conoplea Pers.:Fr. Hyphomycetes

Notes: Hughes (Canad. J. Bot. 38:659. 1960).
Conoplea abietina (Peck) S.J. Hughes
Distr.: NY(type), northwestern United States.
Hosts: Abies, Picea
Conoplea elegantula (Cooke) Ellis
Botrytis elegantula Cooke
Distr.: SC(type).
Notes: On conifer needles (Ellis. Dematiaceous Hyphomycetes. p. 236. 1971).
Host: Pinus
Conoplea fusca Pers.
Streptothrix fusca Corda
Trichosporum crispulum Sacc. & Malbr. in Sacc.
Streptothrix mounceae Sumstine
Botrytis torta Ellis & Everh.
Acremonium vaccinii (Fuckel) Sacc.
Distr.: Temperate North America; Europe.
Notes: On wood.
Hosts: Cymophyllus, Eucalyptus, Eupatorium, Fagus, Juniperus, Ostrya, Pseudotsuga, Quercus
Conoplea geniculata (Corda) S.J. Hughes
Botrytis geniculata Corda
Distr.: Northeastern North America; Europe.

Notes: On decorticated branches of hardwoods and conifers (Hughes. Canad. J. Bot. 38:691. 1960).
Hosts: Abies, Carya
Conoplea globosa (Schwein.:Fr.) S.J. Hughes
Streptothrix globosa (Schwein.:Fr.) S.J. Hughes
Strumella coryneoidea Sacc. & G. Wint.
Cladosporium polysporum Link
Urnula craterium (Schwein.:Fr.) Fr. (teleomorph)
Distr.: Eastern North America.
Notes: Canker of oaks and occasionally other hardwoods. Wolf (Mycologia 50:8 37. 1959) provided the anamorph-teleomorph connection.
Hosts: Acer, Betula, Carpinus, Carya, Castanea, Juglans, Nyssa, Ostrya, Quercus, Rhus
Conoplea juniperi S.J. Hughes
Streptothrix atra Berk. & M.A. Curtis
Distr.: Eastern North America.
Notes: Apparently restricted to *Juniperus*.
Hosts: [Abies], Juniperus, [Vaccinium]
Conoplea juniperi S.J. Hughes var. ***robusta*** S.J. Hughes
Distr.: Pacific Northwest.
Host: Abies
Conoplea olivacea Fr.:Fr.
Conoplea sphaerica (Pers.) Pers.
Distr.: Eastern North America; Europe.
Notes: On dead branches of various hardwoods. Often misdetermined as *Streptothrix fusca* Pers. in old literature. Ellis (Dematiaceous Hyphomycetes. p. 236. 1971).
Hosts: Asimina, Carya, Crataegus, Fagus, Fraxinus, Hamamelis, Juglans, Nemopanthus, Ostrya, Prunus, Quercus, Ribes, Taxus, Tilia, Ulmus

Cookella Sacc. Loculoascomycetes, Dothideales

Cookella microscopica Sacc.
Ascomycetella quercina Peck
Cookella quercina (Peck) Sacc.
Distr.: Temperate northern hemisphere.
Host: Quercus

Coprinus Pers. Basidiomycetes, Agaricales

Coprinus alnivorus Van De Bogart
Distr.: WA.
Host: Alnus
Coprinus atramentarius (Bull.:Fr.) Fr.
Distr.: Temperate northern hemisphere.
Notes: White rot of hardwoods. Older literature indicated that this fungus caused brown rots.
Hosts: Acer, Citrus, Populus
Coprinus disseminatus (Pers.:Fr.) S.F. Gray
Psathyrella disseminata (Pers.:Fr.) Quél.
Pseudocoprinus disseminatus (Pers.:Fr.) Kühner
Distr.: Temperate northern hemisphere.
Hosts: Betula, Cornus, Liquidambar, Quercus
Coprinus lagopus (Fr.:Fr.) Fr. var. ***sphaerosporus*** Kühner & Josserand
Distr.: Temperate North America; Europe.
Host: Populus
Coprinus martinii J. Favre ex P.D. Orton
Distr.: Alpine tundra and boreal North America; Europe.
Notes: On culms of reeds and sedges.
Hosts: Carex, Eriophorum
Coprinus micaceus (Bull.:Fr.) Fr.
Distr.: Temperate and tropical regions.
Notes: White rot of living or dead hardwoods.
Hosts: Acer, Juglans, Liriodendron, Populus, Quercus
Coprinus papagoensis Lindsey & R.L. Gilbertson
Distr.: AZ.
Hosts: Carnegiea, Fouquieria
Coprinus psychromorbidus Redhead & J.A. Traquair
Distr.: Northern North America.
Notes: Redhead & Traquair (Mycotaxon 8:373. 1981); Traquair (Canad. J. Bot. 65:928. 1987).
Hosts: Calamagrostis, Deschampsia, Hordeum, Phleum, Poa, Trisetum
Coprinus radians (Desmaz.) Fr.
Distr.: Temperate North America; Europe.
Host: Populus

Cordana G. Preuss. Hyphomycetes

Cordana musae (A. Zimmerm.) Höhn.
Scolicotrichum musae A. Zimmerm.
Distr.: Worldwide in banana-growing regions.
Notes: Leaf blotch or spot of banana (Ellis & Holliday. C.M.I. Descr. 350. 1972).
Host: Musa

Coremium Link. Hyphomycetes

Notes: This genus is considered a synonym of *Penicillium*, but not all names have been redisposed.
Coremium triostei H.C. Greene
Distr.: WI.
Host: Triosteum

Coriolopsis Murrill. Basidiomycetes, Aphyllophorales

Coriolopsis byrsina (Mont.) Ryvarden
Polyporus byrsinus Mont.
Distr.: Gulf Coast States; tropical America; Africa.
Notes: On wood, usually of hardwoods.
Host: Populus
Coriolopsis gallica (Bull.:Fr.) Ryvarden
Funalia gallica (Bull.:Fr.) Bondartsev & Singer
Trametes hispida Bagl.
Distr.: Temperate northern hemisphere.
Notes: On wood.
Hosts: Acer, Aesculus, Chilopsis, Citrus, Fraxinus, Malus, Morus, Pinus, Platanus, Populus, Prosopis, Prunus, Pseudotsuga, Salix, Schinus, Thevetia
Coriolopsis rigida (Berk. & Mont.) Murrill
Trametes rigida Berk. & Mont.
Distr.: Southeastern United States; neotropical regions.
Notes: On wood of hardwoods, rarely conifers.
Hosts: Acer, Carya, Fraxinus, Juniperus, Lespedeza, Liquidambar, Nyssa, Quercus, Salix

Corniculariella P. Karst. Coelomycetes

Notes: DiCosmo (Canad. J. Bot. 56:1667. 1978).
Corniculariella harpographoidea Dearn. ex DiCosmo
Cornularia harpographoides Dearn., nom. inval.
Durandiella lenticellicola Groves (teleomorph)
Distr.: Northeastern North America.
Notes: On branches.
Host: Crataegus
Cornicuiariella hystricina (Ellis) DiCosmo
Sphaeronaema hystricinum Ellis
Sphaerographium hystricinum (Ellis) Sacc.
Dermea viburni Groves (teleomorph)
Distr.: Northeastern United States.
Hosts: Rhododendron, Viburnum
Cornicuiariella populi DiCosmo
Cornularia populi Dearn., nom. inval.
Distr.: MI.
Host: Populus
Cornicuiariella pseudotsugae (W.L. White) DiCosmo
Chondropodium pseudotsugae W.L. White
Distr.: OR.
Host: Pseudotsuga
Cornicuiariella spina (Berk. & Ravenel) DiCosmo
Sphaeronaema spina Berk. & Ravenel
Cornularia spina (Berk. & Ravenel) Sacc. & Syd.
Sphaerographium fraxinii (Peck) Sacc.
Sphaeronaema fraxinii Peck

Distr.: Eastern North America.
Host: Fraxinus

Cornularia Sacc. Coelomycetes

Notes: This genus is considered a synonym of *Cornulariella* P. Karst., but not all names have been redisposed.
Cornularia sphaeroidea (Ellis) Sacc.
Sphaeronaema sphaeroidea Ellis
Distr.: Western United States.
Notes: The generic placement of this species is uncertain, possibly in *Heteropatella* Fuckel (DiCosmo. Canad. J. Bot. 56:1680. 1978).
Host: Arabis

Corollospora Werderm. Pyrenomycetes, Halosphaeriales

Corollospora maritima Werderm.
Distr.: Widespread in marine habitats.
Hosts: Spartina, Zostera
Corollospora pulchella Kohlmeyer, J.C. Schmidt & N.B. Nair
Clavariopsis bulbosa Anastasiou (anamorph)
Distr.: Widespread in marine habitats.
Notes: Saprobic on submerged wood, mangrove roots, and rotting algae (Kohlmeyer & Kohlmeyer. Marine mycology. p. 276. 1979).
Host: Tamarix

Coronellaria P. Karst. Discomycetes, Helotiales

Coronellaria castanopsidis Kanouse
Distr.: OR(type).
Host: Castanopsis

Coronicium J. Eriksson & Ryvarden. Basidiomycetes, Aphyllophorales

Coronicium alboglaucum (Bourd. & Galzin) Jülich
Corticium albo-glaucum Bourd. & Galzin
Xenasma albo-glaucum (Bourd. & Galzin) Liberta
Distr.: Temperate northern hemisphere.
Notes: On wood.
Host: Acer

Coronophora Fuckel. Pyrenomycetes, Sordariales

Coronophora annexa (Nitschke) Fuckel
Cryptosphaerella annexa (Nitschke) Höhn.
Distr.: Temperate regions.
Host: Betula

Corticium Pers. Basidiomycetes, Aphyllophorales

Laeticorticium Donk
Notes: This genus was recently defined in a narrow sense with *C. roseum* as its type. Boidin & Lanquetin (Bull. Soc. Mycol. France 99:269. 1983); Jülich (Int. J. Mycol. Lichenol. 1:27. 1982). Some species listed here belong in other genera.
Corticium appalachiensis (Burdsall & M. Larsen) M. Larsen in Nakasone
Laeticorticium appalachiensis Burdsall & M. Larsen in Burdsall
Distr.: TN.
Hosts: Nyssa, Vitis
Corticium argentatum Burt
Distr.: KS.
Notes: Apparently not a member of *Corticium sensu stricto*; according to Liberta (Nova Hedwigia 18:216. 1969), the type specimen is sterile.
Host: Salix
Corticium canfieldii (M. Larsen & R.L. Gilbertson) Boidin & Lanquetin
Laeticorticium canfieldii M. Larsen & R.L. Gilbertson
Distr.: AZ.
Notes: On hardwoods.
Host: Juglans
Corticium cremeo-albidum (M. Larsen & Nakasone) M. Larsen in Nakasone
Laeticorticium cremeo-albidum M. Larsen & Nakasone
Distr.: FL(type).
Host: Vitis
Corticium efibulatum (M. Larsen & Nakasone) M. Larsen in Nakasone
Laeticorticium efibulatum M. Larsen & Nakasone
Distr.: MS(type).
Host: Vaccinium
Corticium floridensis (M. Larsen & Nakasone) M. Larsen in Nakasone
Laeticorticium floridensis M. Larsen & Nakasone
Distr.: FL(type).
Host: Myrica
Corticium lombardiae (M. Larsen & R.L. Gilbertson) Boidin & Lanquetin
Laeticorticium lombardiae M. Larsen & R.L. Gilbertson
Distr.: WI.
Notes: On wood. Larsen & Gilbertson (Mycologia 70:206. 1978).
Host: Populus
Corticium maculare Lair
Distr.: NC.
Notes: The generic placement of this species is uncertain.
Host: Quercus
Corticium microsclerotium G.F. Weber
Distr.: Southern United States.
Hosts: Echinochloa, Eleusine, Ficus, Phaseolus, Pueraria, Sambucus
Corticium minnsiae (H. Jacks.) Boidin & Lanquetin
Aleurodiscus minnsiae H. Jacks.
Laeticorticium minnsiae (H. Jacks.) Donk
Distr.: North America.
Notes: On conifers, especially *Tsuga canadensis*.
Hosts: Pinus, Pseudotsuga, Tsuga
Corticium pini (H. Jacks.) Boidin & Lanquetin
Aleurodiscus pini H. Jacks.
Laeticorticium pini (H. Jacks.) Donk
Distr.: Eastern North America.
Host: Pinus
Corticium roseum Pers.:Fr.
Aleurodiscus roseus (Pers.:Fr.) Höhn. & Litsch.
Laeticorticium roseum (Pers.:Fr.) Donk
Distr.: Northern hemisphere.
Notes: On wood. Some reports of this species on *Populus* spp. may be *C. lombardiae*.
Hosts: Abies, Alnus, Arctostaphylos, Juglans, Liriodendron, Nyssa, Populus, Quercus, Rhus, Tsuga, Ulmus
Corticium rubicundum Burt
Distr.: WA.
Notes: Apparently not a member of *Corticium sensu stricto*; according to Liberta (Nova Hedwigia 18:225. 1969), the type specimen is sterile.
Host: Pinus
Corticium spretum Burt
Distr.: WA.
Notes: Apparently not a member of *Corticium sensu stricto*; according to Liberta (Nova Hedwigia 18:227. 1969), the type specimen is sterile.
Host: Fraxinus
Corticium stevensii Burt
Distr.: Southern United States, WA.
Notes: According to Talbot (Persoonia 3:401. 1965), this species belongs in or near *Ceratobasidium*.
Hosts: Aleurites, Aucuba, Cephalanthus, Citrus, Cornus, Dianthus, Eugenia, Feijoa, Ilex, Liquidambar, Malus, Punica, Pyrus

Corynascus Arx. Pyrenomycetes, Sordariales

Corynascus sepedonium (C.W. Emmons) Arx
Thielavia sepedonium C.W. Emmons
Distr.: Cosmopolitan.
Notes: In soil.
Host: Zea

Coryne Nees. Hyphomycetes

Coryne sarcoides (Jacq.) Tul. & C. Tul. var. ***viridescens*** Rehm
Distr.: MI; Germany(type).
Notes: This variety is described as an ascomycete and should be placed in *Ascocoryne*.
Host: Betula

Corynespora H.T. Güssow. Hyphomycetes

Notes: Ellis (Dematiaceous Hyphomycetes. p. 372. 1971).
Corynespora arctespora (Cooke & Ellis) Carris
Helminthosporium arctesporum Cooke & Ellis
Distr.: DE, NJ.
Notes: Carris (Mycotaxon 30:127. 1987).
Host: Vaccinium
Corynespora aterrima (Berk. & M.A. Curtis ex Cooke) M.B. Ellis
Mystrosporium aterrimum Berk. & M.A. Curtis ex Cooke
Distr.: Southeastern United States; South Africa.
Host: Smilax
Corynespora cassiicola (Berk. & M.A. Curtis) C.T. Wei
Helminthosporium cassiicola Berk. & M.A. Curtis
Cercospora melonis Cooke
Helminthosporium vignae Olive in Olive, Bain, & Lefebvre
Cercospora vignicola E. Kawamura
Helminthosporium vignicola (E. Kawamura) Olive
Distr.: Cosmopolitan.
Notes: Target spot of a wide variety of plants (Ellis & Holliday. C.M.I. Descr. 303. 1971).
Hosts: Adiantum, Aeschynanthus, Ajuga, Allamanda, Aphelandra, Begonia, Broussonetia, Capsicum, Carica, Ceiba, Citrullus, Columnea, Cucumis, Cucurbita, Dracaena, Euphorbia, Fatshedera, Ficus, Glycine, Hevea, Hydrangea, Ipomoea, Jasminum, Justicia, Laportea, Ligustrum, Lonicera, Lycopersicon, Moluccella, Momordica, Nematanthus, Pachystachys, Passiflora, Peperomia, Pilea, Pisum, Polyscias, Psidium, Pyrus, Rhododendron, Russelia, Saintpaulia, Sechium, Sedum, Sesamum, Tabebuia, Tolmiea, Trachelospermum, Vicia, Vigna, Wisteria, Zantedeschia
Corynespora foveolata (Pat.) S.J. Hughes
Helminthosporium foveolatum Pat.
Distr.: Subtropical and tropical regions; Great Britain.
Notes: Most common on bamboos, also on palms.
Host: Bambusa
Corynespora obclavata Dyko & Sutton
Distr.: FL.
Notes: On dead petioles (Dyko & Sutton. Mycotaxon 8:121. 1979).
Host: Serenoa
Corynespora olivacea (Wallr.) M.B. Ellis
Clasterosporium olivaceum (Wallr.) Sacc.
Coryneum clavisporum Peck
Exosporium clavisporum (Peck) Sacc.
Coryneum cupulatum Ellis & Everh.
Helminthosporium oligosporum (Corda) S.J. Hughes
Distr.: Eastern North America; Europe.
Notes: Hughes (Fungi Canadenses 245. 1983 as *Helminthosporium oligosporum*).
Hosts: Fraxinus, Liriodendron, Tilia, Tsuga
Corynespora pruni (Berk. & M.A. Curtis) M.B. Ellis
Helminthosporium pruni Berk. & M.A. Curtis in Berk.
Helminthosporium persistens Cooke in Cooke & Ellis
Distr.: Eastern North America; Great Britain.
Notes: On wood and bark of hardwoods (Ellis. Mycol. Pap. 76:30. 1960).
Hosts: Gaultheria, Prunus, Quercus

Coryneum Nees. Coelomycetes

Notes: Sutton (Mycol. Pap. 138. 1975).
Coryneum betulinum Schulzer
Distr.: Eastern North America.
Notes: On bark.
Host: Betula
Coryneum castaneicola Berk. & M.A. Curtis in Berk.
Stegonsporium castaneicola (Berk. & M.A. Curtis) Sacc.
Distr.: OR, eastern United States.
Host: Castanea
Coryneum japonicum (Sacc.) Sutton
Coryneum kunzei Corda var. *japonicum* Sacc.
Distr.: Temperate North America.
Notes: On bark.
Host: Quercus
Coryneum longestipitatum Berl. & Bres.
Distr.: Italy(type); distribution elsewhere is questionable.
Notes: Sutton (Mycol. Pap. 138:194. 1975) was unable to establish the identity of this fungus. Possibly the same as *Seimatosporium cassiopes* (Rostr.) Sutton.
Hosts: Amelanchier, Malus, Pyrus
Coryneum longistipitatum Berl. & Bres.
Distr.: Temperate North America; Italy(type).
Hosts: Amelanchier, Malus
Coryneum megaspermum Syd.
Distr.: Temperate North America; Pakistan.
Notes: On bark.
Host: Quercus
Coryneum modonium (Sacc.) Griffon & Maubl.
Coryneum castaneae (Sacc. & Roum.) Petr.
Pseudovalsa modonia (Tul.) Höhn. (teleomorph)
Distr.: Temperate northern hemisphere.
Notes: On limbs.
Host: Castanea
Coryneum salicinum (Corda) Sacc.
Distr.: AK, ND; Europe.
Notes: A doubtful species. Reports from North Dakota are *Seimatosporium lichenicola* (Corda) Shoemaker & E. Müller.
Host: Salix
Coryneum stromatoideum (Dearn.) Sutton
Helminthosporium stromatoideum Dearn.
Sporidesmium stromatoideum (Dearn.) Morgan-Jones
Stigmina stromatoidea (Dearn.) S.J. Hughes
Distr.: NY(type).
Notes: On bark (Morgan-Jones. Canad. J. Bot. 49:1015. 1971).
Host: Tsuga
Coryneum umbonatum Nees
Coryneum disciforme J.C. Schmidt in Kunze & J.C. Schmidt
Coryneum kunzei Corda
Coryneum pustulatum Peck
Pseudovalsa longipes (Tul.) Sacc. (teleomorph)
Distr.: Temperate North America; Europe.
Notes: According to Sutton (Mycol. Pap. 138:50. 1975) this species is restricted to *Quercus*.
Hosts: [Betula], [Castanea], [Forsythia], [Psidium], Quercus

Costantinella Matr. Hyphomycetes

Notes: Ellis (Dematiaceous Hyphomycetes. p. 364. 1971).
Costantinella micheneri (Berk. & M.A. Curtis) S.J. Hughes
Botrytis micheneri Berk. & M.A. Curtis
Distr.: North America; Europe.
Notes: On woody debris, rarely on bare soil.
Hosts: Pinus, Pseudotsuga
Costantinella terrestris (Link:Fr.) S.J. Hughes
Stachylidium terrestre Link:Fr.
Costantinella cristata Matr.
Costantinella tilletii (Desmaz.) E. Mason & S.J. Hughes
Distr.: North America; Europe.
Notes: On decaying wood and other plant debris.
Hosts: Abies, Salix

Crandallia Ellis & Sacc. Coelomycetes

Crandallia juncicola Ellis & Sacc.
Duplicaria acuminata Ellis & Everh. (teleomorph)
Distr.: Western North America.
Notes: Often misdetermined as *Leptothyrium juncinum* Cooke & Harkn. (Powell. Mycologia 65:1356. 1973).
Host: Juncus

Craterium Trentep. Myxomycetes

Craterium leucocephalum (Pers.) Ditmar
Distr.: Cosmopolitan.
Host: Salix

Cremasteria S.P. Meyers & R.T. Moore. Hyphomycetes

Notes: Kohlmeyer & Kohlmeyer (Marine mycology. p. 485. 1979).
Cremasteria cymatilis S.P. Meyers & R.T. Moore
Distr.: Temperate marine habitats.
Notes: Saprobic on intertidal wood and test panels.
Host: Juncus

Creothyriella Batista & C.A. Costa. Coelomycetes

Creothyriella sideroxyfoliae Batista & C.A. Costa
Distr.: ID; India(type).
Host: Xerophyllum

Crepidotus (Fr.) Staude. Basidiomycetes, Agaricales

Notes: Hesler & Smith (North American species of *Crepidotus*. 1965).
Crepidotus alabamensis Murrill
Distr.: HI, southeastern United States.
Hosts: Aleurites, Diospyros, Hibiscus, Psidium, Syzygium
Crepidotus albescens (Murrill) Redhead
Geopetalum albescens Murrill
Crepidotus phaseoliformis Hesler & A.H. Sm.
Distr.: NY; Canada(Ontario).
Notes: On wood.
Host: Betula
Crepidotus alnicola Hesler & A.H. Sm.
Distr.: MI.
Host: Alnus
Crepidotus amarus Murrill
Distr.: FL.
Host: Vitis
Crepidotus amygdalosporus Kühner
Distr.: HI, temperate North America; Europe.
Host: Syzygium
Crepidotus appalachianensis Hesler & A.H. Sm.
Distr.: HI, eastern United States.
Host: Pluchea
Crepidotus applanatus (Pers.) P. Kumm. var. ***globigera*** (Berk.) Sacc.
Distr.: HI, temperate North America; Australia.
Host: Hibiscus
Crepidotus bakerae Ueki & C.W. Sm.
Distr.: HI.
Host: Mangifera
Crepidotus cinereipallens Hesler & A.H. Sm.
Distr.: MI(type).
Host: Tilia
Crepidotus cinnabarinus Peck
Crepidotus decurrens J. States
Distr.: Central and eastern North America.
Hosts: Fagus, Populus, Tilia, Ulmus
Crepidotus cinnamomeus Hesler & A.H. Sm.
Distr.: ID(type).
Host: Betula
Crepidotus citrinus Petch
Distr.: HI; Japan; Sri Lanka.
Notes: Pegler (Kew Bull., Add. Ser. 12:446. 1986).
Host: Hibiscus
Crepidotus coloradensis Hesler & A.H. Sm.
Distr.: CO, MI, NM.
Host: Populus
Crepidotus contortus Hesler & A.H. Sm.
Distr.: WA(type).
Hosts: Acer, Populus
Crepidotus flexuosus Hesler & A.H. Sm.
Distr.: MI(type).
Host: Populus
Crepidotus fusisporus Hesler & A.H. Sm. var. ***abietinus*** Hesler & A.H. Sm.
Distr.: MI(type).
Host: Abies
Crepidotus hamulatus Hesler & A.H. Sm.
Distr.: MI(type).
Host: Acer
Crepidotus kauffmanii Hesler & A.H. Sm.
Crepidotus calolepsis (Fr.) P. Karst., sensu Kauffm.
Distr.: Temperate North America.
Host: Salix
Crepidotus latifolius Peck
Distr.: Temperate North America.
Host: Betula
Crepidotus maximus Hesler & A.H. Sm.
Distr.: TN(type).
Host: Carya
Crepidotus mollis (Schaeff.:Fr.) Staude
Crepidotus calolepioides Murrill
Distr.: Northern hemisphere; South America; Africa.
Notes: Usually on hardwoods but occasionally reported on conifers.
Hosts: Epipremnum, Hibiscus, Populus, Syzygium
Crepidotus mollis (Schaeff.:Fr.) Staude var. ***cystidiosus*** Hesler & A.H. Sm.
Distr.: WA(type).
Host: Alnus
Crepidotus montanensis Hesler & A.H. Sm.
Distr.: MT.
Host: Betula
Crepidotus nephrodes (Berk. & M.A. Curtis) Sacc.
Crepidotus fulvifibrillosus Murrill
Distr.: HI, temperate North America.
Hosts: Crataegus, Juglans, Populus, Psidium
Crepidotus nyssicola (Murrill) Singer
Pleuropus nyssicola Murrill
Distr.: FL, MI.
Host: Nyssa
Crepidotus obscurus Hesler & A.H. Sm.
Distr.: MI(type).
Host: Tilia
Crepidotus occidentalis Hesler & A.H. Sm.
Distr.: Western North America.
Hosts: Alnus, Populus
Crepidotus ochraceus Hesler & A.H. Sm.
Distr.: OR.
Host: Alnus
Crepidotus paxilloides Singer
Paxillus reniformis Berk. & Ravenel
Distr.: NC.
Host: Quercus
Crepidotus payettensis Hesler & A.H. Sm.
Distr.: ID(type).
Host: Pinus
Crepidotus praelatifolius Murrill
Distr.: FL(type).
Host: Magnolia
Crepidotus pseudoflammeus Hesler & A.H. Sm.
Distr.: WA(type).
Host: Alnus
Crepidotus ramosus Hesler & A.H. Sm.
Distr.: CO(type).
Host: Picea
Crepidotus smithii Hesler
Distr.: TN.
Host: Quercus
Crepidotus sphaerosporus (Pat.) J.E. Lange
Distr.: AZ; Europe.
Notes: The occurrence of this taxon in North America is questionable.
Host: Picea
Crepidotus subapplanatus Hesler & A.H. Sm.
Distr.: Eastern United States.
Host: Castanea
Crepidotus subfibrillosus Hesler & A.H. Sm.

Distr.: MI(type).
Host: Fraxinus
Crepidotus subluteus Hesler & A.H. Sm.
Distr.: MI(type).
Host: Ulmus
Crepidotus submollis Murrill
Distr.: Central and western North America.
Host: Alnus
Crepidotus subnidulans (Overh.) Hesler & A.H. Sm.
Claudopus subnidulans Overh.
Distr.: Central United States.
Host: Ulmus
Crepidotus subverrucisporus Hesler & A.H. Sm. var. **megalosporus** Hesler & A.H. Sm.
Distr.: CO.
Hosts: Alnus, Populus
Crepidotus tennesseensis Hesler
Distr.: TN.
Host: Quercus
Crepidotus variabilis (Pers.:Fr.) P. Kumm.
Distr.: North America, scattered reports; Europe.
Host: Amelanchier
Crepidotus versutus (Peck) Sacc.
Distr.: Eastern and northwestern North America(rare).
Notes: On bark and wood of hardwoods.
Hosts: Abies, Acer, Populus

Cribropeltis Tehon. Coelomycetes

Cribropeltis citrullina Tehon
Distr.: IL.
Notes: Fruit speck of watermelon.
Host: Citrullus

Crinipellis Pat. Basidiomycetes, Agaricales

Crinipellis foliicola Singer
Distr.: FL.
Host: Smilax
Crinipellis piceae Singer
Distr.: Western North America; Asia.
Host: Picea
Crinipellis zonata (Peck) Pat.
Collybia zonata Peck
Distr.: Eastern North America.
Hosts: Betula, Carya, Quercus

Crinula Fr.:Fr. Hyphomycetes

Holwaya Sacc.
Crinula caliciiformis Fr.:Fr.
Ceratopodium giganteum (Peck) Kuntze
Graphium giganteum (Peck) Sacc.
Stilbum giganteum Peck
Holwaya mucida (Schulzer) Korf & Abawi (teleomorph)
Distr.: North America.
Notes: On wood. Korf & Abawi (Canad. J. Bot. 49:1880. 1971) presented a complete account of the synonyms of this species.
Host: Acer

Cristinia Parmasto. Basidiomycetes, Aphyllophorales

Cristinia gallica (Pilát) Jülich
Hydnum mucidum Pers., sensu Bourd. & Galzin, non Fr., hom. illeg.
Distr.: FL; Europe.
Hosts: Malus, Persea
Cristinia helvetica (Pers.) Parmasto
Hydnum helveticum Pers.
Distr.: Temperate northern hemisphere.
Notes: On wood.
Host: Populus
Cristinia sonorae Nakasone & R.L. Gilbertson
Distr.: AZ(type).
Notes: On wood. Nakasone & Gilbertson (Mycologia 70:271. 1978).
Host: Fouquieria

Cristulariella Höhn. emend. Redhead. Hyphomycetes

Notes: Redhead (Canad. J. Bot. 53:700. 1975).
Cristulariella depraedans (Cooke) Höhn.
Distr.: Temperate North America; Europe.
Hosts: Acer, Bucida
Cristulariella moricola (Hino) Redhead
Cristulariella pyramidalis A.M. Waterman & R.P. Marshall
Sclerotium cinnamomi Sawada (synanamorph)
Grovesinia pyramidalis M. Cline, J.L. Crane, & S. Cline (teleomorph)
Distr.: Temperate northern hemisphere.
Notes: Bull's eye or zonate leaf spot on at least 73 species in 36 families of vascular plants (Trolinger et al. Pl. Dis. Reporter 62:710. 1978; Cline et al. Mycologia 75:988. 1983).
Hosts: Acer, Ailanthus, Aleurites, Amelanchier, Arachis, Asimina, Bidens, Campsis, Carya, Catalpa, Celtis, Chenopodium, Commelina, Cornus, Cuphea, Desmodium, Eupatorium, Fraxinus, Hibiscus, Impatiens, Ipomoea, Juglans, Lindera, Lobelia, Lycopersicon, Magnolia, Malus, Menispermum, Oxydendrum, Parthenocissus, Perilla, Persea, Phaseolus, Phytolacca, Platanus, Polygonum, Prunus, Rhus, Rumex, Sassafras, Sida, Smilax, Solanum, Solidago, Ulmus, Vaccinium, Viburnum, Vitex, Vitis

Crocicreas Fr. Discomycetes, Helotiales

Notes: Carpenter (Mem. New York Bot. Gard. 33:1. 1981).
Crocicreas amenti (Batsch) S. Carpenter
Ciboria amenti (Batsch) Whetzel
Pezizella amenti (Batsch) Dennis
Helotium amenti (Batsch) Fuckel
Distr.: North America; northern Europe.
Notes: On catkins.
Host: Salix
Crocicreas coronatum (Bull.) S. Carpenter
Cyathicula coronata (Bull.) De Not.
Helotium coronatum (Bull.) P. Karst.
Hymenoscypha coronata (Bull.) W. Phillips
Distr.: Temperate Northern Hemisphere.
Notes: On a wide variety of herbaceous angiosperms.
Hosts: Achillea, Eupatorium, Senecio, Solidago
Crocicreas culmicola (Desmaz.) S. Carpenter
Belonium culmicola (Desmaz.) Seaver
Cyathicula culmicola (Desmaz.) S. Carpenter & Dumont
Cyathicula incarnata (Quél.) Sacc.
Belonium subgibbosum (Ellis) Sacc.
Belonioscypha vexata (De Not.) Rehm
Distr.: Temperate North America; Europe.
Notes: On plant debris, especially culms of grasses.
Hosts: Andropogon, Panicum, Phalaris
Crocicreas cyathoideum (Bull.:Fr.) S. Carpenter
Hymenoscyphus cyathoideus (Bull.:Fr.) S.F. Gray
Cyathicula cyathoidea (Bull.:Fr.) Thuem.
Helotium cyathoideum (Bull.:Fr.) P. Karst.
Phialea cyathoidea (Bull.:Fr.) Gill.
Lachnella canadense (Ellis & Dearn.) Seaver
Pseudohelotium canadense Ellis & Dearn.
Phialea pallida Kanouse
Cyathicula petasata (P. Karst.) Dennis
Phialea umbellulariae Cash
Phialea urticae (Pers.) Sacc.
Distr.: Cosmopolitan.
Notes: Saprobic on herbaceous plant debris.
Hosts: Arctium, Boschniakia, Corydalis, Delphinium, Epilobium, Equisetum, Erechtites, Eupatorium, Gleditsia, Heracleum, Impatiens, Iva, Marah, Rubus, Sambucus, Solanum, Streptopus, Urtica, Veratrum, Zea
Crocicreas cyathoideum (Bull.:Fr.) S. Carpenter var. **cacaliae** (Pers.) S. Carpenter
Cyathicula cacaliae (Pers.) Dennis

Helotium cacaliae (Pers.) Sacc.
Cyathicula alpina Ellis & Everh.
Distr.: Temperate Northern Hemisphere.
Notes: On dead herbaceous stems.
Hosts: Actaea, Agastache, Aquilegia, Arnica, Artemisia, Aster, Boschniakia, Castilleja, Clematis, Delphinium, Elymus, Epilobium, Geum, Ivesia, Lathyrus, Lomatium, Osmorhiza, Pedicularis, Phytolacca, Rudbeckia, Sambucus, Scrophularia, Senecio, Silene
Crocicreas cyathoideum (Bull.) S. Carpenter var. ***olympianum*** (Kanouse) S. Carpenter
Phialea olympiana Kanouse
Distr.: WA.
Host: Acer
Crocicreas eschscholtziae (W. Phillips & Harkn.) S. Carpenter
Peziza eschscholtziae W. Phillips & Harkn.
Helotium eschscholtziae (W. Phillips & Harkn.) Seaver
Hymenoscyphus eschscholtziae (W. Phillips & Harkn.) Kuntze
Phialea eschscholtziae (W. Phillips & Harkn.) Sacc.
Distr.: CA.
Host: Eschscholzia
Crocicreas fuscum (W. Phillips & Harkn.) S. Carpenter
Belonidium fuscum W. Phillips & Harkn.
Distr.: CA.
Notes: Carpenter erroneously listed this basionym as *Belonium*.
Host: Sanicula
Crocicreas gramineum (Fr.:Fr.) Fr.
Allophylaria pusiola (P. Karst.) Nannf.
Cyathicula stipae (Fuckel) E. Müller
Phialea stipae (Fuckel) Rehm
Distr.: Circumboreal.
Notes: On culms and leaves of Poaceae.
Hosts: Phleum, Poa, Stipa
Crocicreas helios (Penz. & Sacc.) S. Carpenter
Davincia helios Penz. & Sacc.
Distr.: Widespread in tropical regions.
Host: Eupatorium
Crocicreas hysterioides (Rehm) S. Carpenter
Helotium hysterioides Rehm
Cyathicula hysterioides (Rehm) E. Müller
Distr.: Alpine and arctic North America; Europe.
Host: Xerophyllum
Crocicreas melanosporum (Rehm) S. Carpenter
Belonium melanosporum (Rehm) Sacc.
Scelobelonium melanosporum (Rehm) Höhn.
Distr.: Alpine and arctic North America; Europe.
Notes: On dead herbaceous plants.
Hosts: Agrostis, Anemone, Bromus, Elymus, Juncus, Penstemon, Poa, Sitanion
Crocicreas nigrescens (Cooke) S. Carpenter
Helotium nigrescens (Cooke) Rehm
Phialea nigrescens (Cooke) Dennis
Peziza heterocarpa Ellis
Distr.: Northeastern United States.
Notes: On stems of herbaceous plants.
Hosts: Conyza, Daucus, Erigeron, Verbascum
Crocicreas nigrofuscum (Rehm) S. Carpenter
Phialea nigrofusca Rehm
Helotium nigrofusum (Rehm) Boud.
Distr.: Alpine North America; Europe.
Hosts: Erigeron, Heracleum, Veratrum
Crocicreas nigrofuscum (Rehm) S. Carpenter var. ***allantosporum*** S. Carpenter
Distr.: Rocky Mountains.
Notes: On dead herbaceous stems.
Hosts: Agastache, Delphinium, Mertensia, Osmorhiza, Rudbeckia, Sidalcea
Crocicreas panici (Höhn.) S. Carpenter
Phialea panici Höhn.
Belonioscypha miniata Kanouse
Distr.: NY, OR; Java.
Host: Carex
Crocicreas phaeoconium (Fairm.) S. Carpenter
Phialea phaeoconia Fairm.
Helotium phaeoconium (Fairm.) Seaver
Distr.: NY.
Notes: On leaves.
Host: Antennaria
Crocicreas starbaeckii (Rehm) S. Carpenter
Phialea starbaeckii Rehm
Distr.: Cosmopolitan.
Notes: On various herbaceous plants.
Host: Ranunculus

Cronartium Fr. Basidiomycetes, Uredinales

Notes: Petersen (Rep. Tottori Mycol. Inst. 10:203. 1973).
Cronartium appalachianum Hepting
Peridermium appalachianum Hepting & Cummins
Distr.: Southern Appalachian Mountains.
Notes: Heteroecious; aecial host - *Pinus virginiana*; telial host - *Bucklyea distichophylla*.
Hosts: Buckleya, Pinus
Cronartium arizonicum Cummins
Distr.: AZ.
Notes: Heteroecious; aecial host - *Pinus ponderosa*; telial host - *Castilleja patriotica*.
Hosts: Castilleja, Pinus
Cronartium coleosporioides Arth.
Cronartium filamentosum Hedgc., nom. nud.
Peridermium stalactiforme Arth. & F. Kern (anamorph)
Distr.: Western North America.
Notes: Stalactiform blister rust. Heteroecious; aecial host - *Pinus* spp.; telial host - Scrophulariaceae. Mordue & Gibson (C.M.I. Descr. 577. 1978).
Hosts: Castilleja, Orthocarpus, Pedicularis, Pinus, Rhinanthus
Cronartium comandrae Peck
Cronartium asclepiadeum (Willd.) Fr. var. *thesii* Berk.
Cronartium pyriforme Hedgc. & W.H. Long
Distr.: Southern United States to northern Canada.
Notes: Comandra blister rust. Heteroecious; aecial host - *Pinus* spp.; telial host - *Comandra* and *Geocaulon* spp. Mordue & Gibson (C.M.I. Descr. 578. 1978).
Hosts: Comandra, Geocaulon, Pinus
Cronartium comptoniae Arth.
Peridermium comptoniae Orton & J.F. Adams (anamorph)
Distr.: Temperate North America.
Notes: Sweet fern blister rust. Heteroecious; aecial host - *Pinus* spp.; telial host - *Comptonia* and *Myrica*. Mordue & Gibson (C.M.I. Descr. 579. 1978).
Hosts: Comptonia, Myrica, Pinus
Cronartium conigenum Hedgc. & N. Hunt
Distr.: Southwestern United States; Mexico; El Salvador.
Notes: Arizona pine cone rust. Heteroecious; aecial host - *Pinus* spp.; telial host - *Quercus* spp. The distinction between this species and *C. quercuum* (Berk.) Miyabe ex Shirai is questionable (Peterson. Rep. Tottori Mycol. Inst. 10:213. 1973).
Hosts: Pinus, Quercus
Cronartium occidentale Hedgc., Bethel, & N. Hunt
Peridermium occidentale Hedgc., Bethel, & N. Hunt (anamorph)
Distr.: Western United States.
Notes: Piñon pine blister rust. Heteroecious; aecial host - species of *Pinus* subg. *strobus*; telial host - *Ribes* spp.
Hosts: Pinus, Ribes
Cronartium quercuum (Berk.) Miyabe ex Shirai
Cronartium asclepiadeum (Willd.) Fr. var. *quercuum* Berk.
Cronartium cerebrum (Peck) Hedgc. & W.H. Long
Peridermium cerebrum Peck (anamorph)
Distr.: Central and eastern North America; Asia; India; Soviet Union.
Notes: Eastern gall rust. Heteroecious; aecial host - *Pinus* spp.; telial host - Fagaceae. Burdsall & Snow (Mycologia 69:503. 1979) recognized 5 *formae speciales*.
Hosts: Castanea, Lithocarpus, Pinus, Quercus
Cronartium quercuum (Berk.) Miyabe ex Shirai f. sp. ***banksianae*** Burdsall & G. Snow
Distr.: Northcentral United States.
Notes: Primarily pathogenic on jack pine.
Host: Pinus
Cronartium quercuum (Berk.) Miyabe ex Shirai f. sp. ***echinatae*** Burdsall & G. Snow
Distr.: Southeastern United States.

Notes: Primarily pathogenic on shortleaf pine.
Host: Pinus
Cronartium quercuum (Berk.) Miyabe ex Shirai f. sp. **fusiforme** (Hedgc. & N. Hunt) Burdsall & G. Snow
Cronartium fusiforme Hedgc. & N. Hunt
Cronartium fusiforme Cummins
Peridermium fusiforme Arth. & F. Kern (anamorph)
Distr.: Southeastern United States.
Notes: Fusiform rust of southern pines. Heteroecious; aecial host - *Pinus* spp.; telial host - Fagaceae.
Hosts: Pinus, Quercus
Cronartium quercuum (Berk.) Miyabe ex Shirai f. sp. **virginianae** Burdsall & G. Snow
Distr.: Southeastern United States.
Notes: Primarily pathogenic on Virginia pine.
Host: Pinus
Cronartium ribicola J.C. Fisch.
Distr.: Temperate northern hemisphere.
Notes: Heteroecious; aecial host - *Pinus* subg. *strobus*; telial host - Grossulariaceae. White pine blister rust, currant rust (Laundon & Rainbow. C.M.I. Descr. 283. 1971).
Hosts: Pinus, Ribes
Cronartium strobilinum Hedgc. & Hahn
Peridermium strobilinum (Arth.) R.S. Peterson (anamorph)
Distr.: Southeastern United States.
Notes: Southern cone rust. Heteroecious; aecial host - *Pinus* spp.; telial host - *Quercus* spp.
Hosts: Pinus, Quercus

Crossopsora Syd. & P. Syd. Basidiomycetes, Uredinales

Crossopsora bixae Buriticá
Uredo bixae Arth. (anamorph)
Distr.: FL, PR; South America.
Notes: Autoecious.
Host: Bixa

Crucibulum Tul. Basidiomycetes, Nidulariales

Crucibulum laeve (Bull.) Kambly in Kambly & R.E. Lee
Crucibulum vulgare Tul.
Distr.: Cosmopolitan.
Notes: On wood and plant debris.
Hosts: Fagus, Juniperus, Pseudotsuga
Crucibulum parvulum Brodie
Distr.: Western North America.
Host: Tetradymia

Crumenulopsis Groves. Discomycetes, Helotiales

Crumenulopsis lacrimiformia Funk
Distr.: FL.
Notes: Funk (Mycotaxon 27:286. 1986).
Host: Pinus
Crumenulopsis sororia (P. Karst.) Groves
Crumenula sororia P. Karst.
Godronia sororia (P. Karst.) P. Karst.
Digitosporium piniphilum Gremmen (anamorph)
Distr.: NM. Although reported, Groves (Canad. J. Bot. 47:50. 1969) stated that there are no genuine records of this species in North America.
Host: Pinus

Crustoderma Parmasto. Basidiomycetes, Aphyllophorales

Crustoderma carolinense Nakasone
Distr.: NC.
Host: Castanea
Crustoderma corneum (Bourd. & Galzin) Nakasone
Phlebia cornea (Bourd. & Galzin) Parmasto
Distr.: MT; Canada; northern Europe.
Hosts: Pinus, Quercus
Crustoderma dryinum (Berk. & M.A. Curtis) Parmasto
Corticium dryinum Berk. & M.A. Curtis in Berk.
Peniophora dryina (Berk. & M.A. Curtis) D.P. Rogers & H. Jacks.
Peniophora tabacina Burt
Peniophora weiri Bres.
Distr.: Northern hemisphere.
Hosts: Picea, Pinus, Populus, Pseudotsuga, Quercus
Crustoderma flavescens Nakasone & R.L. Gilbertson
Distr.: North and Central America.
Hosts: Castanea, Quercus
Crustoderma marianum Nakasone
Distr.: MD.
Hosts: Pinus, Robinia
Crustoderma opuntiae Nakasone & R.L. Gilbertson
Distr.: AZ.
Notes: On wood.
Host: Opuntia
Crustoderma resinosum (H. Jacks. & Dearden) R.L. Gilbertson
Peniophora resinosa H. Jacks. & Dearden
Hyphoderma resinosum (H. Jacks. & Dearden) K.J. Martin & R.L. Gilbertson
Distr.: Western North America.
Notes: On wood. Gilbertson (Mycotaxon 12:377. 1981).
Host: Pinus

Cryphonectria (Sacc.) Sacc. Pyrenomycetes, Diaporthales

Notes: Barr (Mycol. Mem. 7:142. 1978).
Cryphonectria coccolobii (Vizioli) Micales & Stipes
Endothia coccolobii Vizioli
Distr.: Widespread in tropical regions.
Notes: Micales & Stipes (Mycotaxon 26:99. 1986).
Host: Coccoloba
Cryphonectria cubensis (Bruner) C.S. Hodges
Diaporthe cubensis Bruner
Cryptosporella eugeniae F.J. Nutman & F.M. Roberts
Endothia eugeniae (F.J. Nutman & F.M. Roberts) J. Reid & C. Booth
Distr.: Subtropical and tropical regions.
Notes: Canker of *Eucalyptus* spp., die-back of *Syzygium aromaticum*. Hodges et al. (Mycologia 78:343. 1986); Micales et al. (Mycologia 79:707. 1987); Booth & Gibson (C.M.I. Descr. 363. 1973 as *Endothia eugeniae*).
Host: Eucalyptus
Cryphonectria havanensis (Bruner) Barr
Endothia havanensis Bruner
Distr.: Widespread in tropical regions.
Notes: Although Hodges (Mycologia 72:542. 1980) suggested that this species is a synonym of *C. gyrosa*, Walker et al. (Mycotaxon 23:367. 1985) presented arguments to retain these as distinct species.
Host: Eucalyptus
Cryphonectria parasitica (Murrill) Barr
Diaporthe parasitica Murrill
Endothia parasitica (Murrill) P.J. Anderson & H.W. Anderson
Distr.: Temperate regions.
Notes: Mostly on *Castanea* and *Quercus* but reported from a few other woody hosts (Sivanesan & Holliday. C.M.I. Descr. 704. 1981).
Hosts: Carya, Castanea, Castanopsis, Quercus
Cryphonectria radicalis (Schwein.:Fr.) Barr
Endothia radicalis (Schwein.:Fr.) Ces. & De Not.
Distr.: Temperate northern hemisphere.
Notes: On bark of hardwoods.
Hosts: Castanea, Quercus

Cryptendoxyla Malloch & Cain. Pyrenomycetes, Eurotiales

Notes: Malloch & Cain (Canad. J. Bot. 48:1816. 1970).
Cryptendoxyla hypophloia Malloch & Cain
Distr.: Northeastern North America.
Host: Acer

Cryptocline Petr. Coelomycetes
Notes: Morgan-Jones (Canad. J. Bot. 51:309. 1973).
Cryptocline abietina **Petr.**
Distr.: OR(type).
Host: Abies
Cryptocline betularum **(Ellis & G. Martin) Arx**
Gloeosporium betularum Ellis & G. Martin
Gloeosporidium betularum (Ellis & G. Martin) Petr.
Distr.: Temperate North America.
Host: Betula
Cryptocline ceanothi **(H.C. Greene) Arx**
Gloeosporium ceanothi H.C. Greene
Distr.: WI.
Host: Ceanothus
Cryptocline cinerescens **(Bubák) Arx**
Gloeosporium cinerescens Bubák
Gloeosporium bicolor J.J. Davis
Distr.: Temperate North America; Europe.
Notes: On leaves.
Host: Quercus
Cryptocline effusa **Petr.**
Distr.: OR(type).
Host: Abies
Cryptocline hypophyllum **(D.E. Ellis & Gill.) Morgan-Jones & Nag Raj in Morgan-Jones**
Rhabdogloeum hypophyllum D.E. Ellis & Gill.
Distr.: Southwestern United States.
Host: Pseudotsuga
Cryptocline taxicola **(Allesch.) Petr.**
Gloeosporium taxicola Allesch.
Distr.: Northeastern North America; Europe.
Host: Taxus

Cryptococcus Kütz. Hyphomycetes
Cryptococcus albidus **(K. Saito) C.E. Skinner**
Distr.: Cosmopolitan.
Hosts: Trifolium, Triticum
Cryptococcus laurentii **(Kuff.) C.E. Skinner**
Torula laurentii Kuff.
Distr.: Widespread.
Notes: Isolated from various kinds of substrates (Lodder. The yeasts. p. 1120. 1970).
Hosts: Triticum, Zea

Cryptocoryneum Fuckel. Coelomycetes
Cryptocoryneum condensatum **(Wallr.) E. Mason & S.J. Hughes**
Hormiscium uniforme (Peck) Sacc.
Torula uniformis Peck
Distr.: Temperate northern hemisphere.
Notes: On dead wood and woody debris (Ellis. Dematiaceous Hyphomycetes. p. 109. 1971).
Host: Acer
Cryptocoryneum simmonsii **Sacc.**
Distr.: Western North America.
Host: Pinus

Cryptodiaporthe Petr. Pyrenomycetes, Diaporthales
Notes: Barr (Mycol. Mem. 7:135. 1978); Butin (Phytopathol. Z. 32:399. 1958).
Cryptodiaporthe acerinum **J. Reid & Cain**
Distr.: Eastern North America.
Notes: On branches. Sometimes misdetermined as *Valsa myinda* Cooke.
Host: Acer
Cryptodiaporthe aesculi **(Fuckel) Petr.**
Diplodina aesculi (Sacc.) Sutton (anamorph)
Distr.: Temperate North America; Europe.
Notes: On branches and twigs.
Host: Aesculus
Cryptodiaporthe aubertii **(Westend.) Wehmeyer var. *comptoniae*** (Schwein.) Wehmeyer
Uniseta flagellifera (Ellis & Everh.) Ciccarone (anamorph)
Distr.: Eastern North America.
Host: Comptonia
Cryptodiaporthe calosphaerioides **(Ellis & Everh.) Wehmeyer**
Diaporthe calosphaerioides Ellis & Everh.
Distr.: AK; Canada(Ontario).
Host: Sambucus
Cryptodiaporthe corni **(Wehmeyer) Petr.**
Apioporthe corni Wehmeyer
Distr.: Northeastern North America.
Hosts: Cornus, Sassafras
Cryptodiaporthe galericulata **(Tul.) Wehmeyer**
Distr.: Eastern North America; Europe.
Notes: On limbs.
Host: Fagus
Cryptodiaporthe hicoriae **(Atk.) Wehmeyer**
Melanconis hicoriae Atk.
Distr.: AL.
Host: Carya
Cryptodiaporthe lebiseyi **(Desmaz.) Wehmeyer**
Phomopsis lebiseyi (Sacc.) Died. (anamorph)
Distr.: CA; Europe.
Notes: Although reported occasionally, no reports from North America have been verified.
Host: Acer
Cryptodiaporthe liquidambaris **Petr.**
Distr.: MD(type).
Host: Liquidambar
Cryptodiaporthe macounii **(Dearn.) Wehmeyer**
Distr.: Temperate North America.
Notes: On limbs.
Host: Spiraea
Cryptodiaporthe petiolophila **(Peck) Barr**
Gnomonia petiolophila (Peck) Berl. & Voglino
Gnomoniella petiolophila (Peck) J.H. Miller
Cryptodiaporthe densissima (Ellis) Wehmeyer var. *spicata* (Ellis & Everh.) Wehmeyer
Diaporthe spicata Ellis & Everh.
Distr.: Eastern North America.
Notes: On twigs.
Host: Acer
Cryptodiaporthe phomaspora **(Cooke & Ellis) Barr**
Valsa phomaspora Cooke & Ellis
Apioporthe phomaspora (Cooke & Ellis) Wehmeyer
Distr.: Eastern North America.
Host: Myrica
Cryptodiaporthe pyrrhocystis **(Berk. & Broome) Wehmeyer**
Diaporthe pyrrhocystis Berk. & Broome
Distr.: ID; Europe.
Notes: Barr (Mycol. Mem. 7:153. 1978) stated that this species belongs in *Allantoporthe*.
Host: Alnus
Cryptodiaporthe salicella **(Fr.:Fr.) Petr.**
Cryptodiaporthe salicina (Curr.) Wehmeyer
Discella microsperma (G. Johnst.) Sutton (anamorph)
Distr.: Circumboreal.
Notes: On limbs.
Hosts: Populus, Salix
Cryptodiaporthe vepris **(Delacr.) Petr.**
Apioporthe vepris (Delacr.) Wehmeyer
Distr.: Eastern North America; Europe.
Host: Rubus

Cryptodiscus Corda. Discomycetes, Ostropales
Distr.: Sherwood (Mycotaxon 5:88. 1977).
Cryptodiscus foveolaris **(Rehm) Rehm**
Distr.: North America; Europe.
Hosts: Castanea, Populus, Quercus
Cryptodiscus niveopurpureus **(Ellis & Everh.) Sacc.**
Stictis niveopurpureus Ellis & Everh.
Distr.: NJ(type).

Host: Quercus
Cryptodiscus pallidus (Pers.:Fr.) Corda
Stictis pallida Pers.:Fr.
Distr.: NY; Europe; Canary Islands.
Notes: On decorticated wood.
Hosts: Acer, Castanea
Cryptodiscus sambuci Cash
Distr.: FL(type).
Host: Sambucus

Cryptomycella Höhn. Coelomycetes

Cryptomycella pteridis (Kalchbr.:Fr.) Höhn.
Fusidium pteridis Kalchbr.:Fr.
Cylindrosporium pteridis Gilman & W.A. Archer
Gloeosporium pteridis Harkn.
Gloeosporium pteridis (Kalchbr.:Fr.) Bubák & Kab. non Harkn., hom. illeg.
Gloeosporium obtegens Syd.
Cryptomycina pteridis (Rebent.:Fr.) Höhn. (teleomorph)
Distr.: Temperate northern hemisphere.
Notes: Leaf-roll disease of ferns (DiCosmo et al. Mycotaxon 11:35. 1984).
Host: Pteridium

Cryptomyces Grev. Discomycetes, Rhytismatales

Cryptomyces maximus (Fr.) Rehm
Rhytisma maximum Fr.
Distr.: Southwestern United States; Europe.
Notes: On living branches and twigs.
Host: Salix

Cryptomycina Höhn. Discomycetes, Rhytismatales

Cryptomycina filicina (Fr.:Fr.) L. Holm & K. Holm
Leptostroma filicinum Fr.:Fr.
Cryptomycina osmundae Syd.
Hysterium osmundae Schwein.
Distr.: Northeastern United States; Europe.
Notes: On culms of ferns (Holm & Holm. Bot. Not. 131:97. 1978).
Host: Osmunda
Cryptomycina pteridis (Rebent.:Fr.) Höhn.
Cryptomyces pteridis (Rebent.:Fr.) Rehm
Phyllachora pteridis (Rebent.:Fr.) Fuckel
Cryptomycella pteridis (Kalchbr.:Fr.) Höhn. (anamorph)
Distr.: Temperate regions.
Notes: Leaf-roll disease of ferns (DiCosmo et al. Mycotaxon 11:32. 1984).
Hosts: Dryopteris, Pteridium

Cryptophiale Pirozynski. Hyphomycetes

Notes: Farr (Mycotaxon 11:181. 1980).
Cryptophiale kakombensis Pirozynski
Distr.: HI; Tanzania.
Host: Eucalyptus
Cryptophiale udagawae Pirozynski & Ichin.
Distr.: HI; Japan.
Host: Eucalyptus

Cryptoporus (Peck) Shear. Basidiomycetes, Aphyllophorales

Cryptoporus volvatus (Peck) Shear
Polyporus volvatus Peck
Distr.: Northeastern and western North America; East Asia.
Notes: On recently killed conifers, especially *Pinus* sp.
Hosts: Abies, Calocedrus, Larix, Picea, Pinus, Pseudotsuga, Tsuga

Cryptosphaerella Sacc. Pyrenomycetes, Sordariales

Notes: This genus is considered a synonym of *Coronophora*, but not all names have been redisposed.
Cryptosphaerella shearii Petr.
Distr.: HI.
Host: Phytolacca

Cryptosphaeria Grev. Pyrenomycetes, Diatrypales

Notes: Glawe (Mycotaxon 20:415. 1984) included species known from the Pacific Northwest. Rappaz (Mycol. Helv. 2:380. 1987).
Cryptosphaeria eunomia (Fr.:Fr.) Fuckel
Valsa eunomia (Fr.:Fr.) Nitschke
Cryptosphaeria millepunctata Grev.
Cryptosphaeria sepulta (Nitschke) Sacc.
Cytosporina millepunctata Sacc. (anamorph)
Distr.: Temperate North America; Europe.
Hosts: Fraxinus, Robinia
Cryptosphaeria eunomia (Fr.:Fr.) Fuckel var. ***fraxini*** (Richon) Rappaz
Cryptosphaerina fraxini (Ellis & Everh.) Traverso
Endoxyla fraxini (Ellis & Everh.) Ellis & Everh.
Thyridaria fraxini Ellis & Everh.
Distr.: North America; Europe.
Host: Fraxinus
Cryptosphaeria lignyota (Fr.:Fr.) Auersw.
Diatrype lignyota (Fr.:Fr.) Fr.
Cryptosphaeria populina Sacc.
Eutypa populina (Sacc.) L.C. Tiffany & Gilman
Cytosporina sp. (anamorph)
Distr.: North temperate regions.
Notes: Jacobs et al. (Mycologia 80:307. 1988).
Hosts: Populus, Salix
Cryptosphaeria pullmanensis Glawe
Distr.: WA(type).
Host: Populus
Cryptosphaeria subcutanea (Wahlenberg:Fr.) Rappaz
Eutypa subcutanea (Wahlenberg:Fr.) Sacc.
Valsa subcutanea (Wahlenberg:Fr.) Nitschke
Anthostoma ontariensis Ellis & Everh.
Eutypa ontariensis (Ellis & Everh.) L.C. Tiffany & Gilman
Eutypa salicicola (Allesch.) Sacc.
Cryptosphaeria vicinula (Nyl.) P. Karst.
Eutypa vicinula (Nyl.) L.C. Tiffany & Gilman
Distr.: North temperate regions.
Host: Salix
Cryptosphaeria vexata (Cooke & Ellis) Sacc.
Distr.: NJ.
Notes: According to Rappaz (Mycol. Helv. 2:569. 1987), this species belongs in *Valsa*.
Hosts: Lyonia, Vaccinium

Cryptospora Tul. & C. Tul. Coelomycetes

Cryptospora leiphaemoides Dearn. & House
Distr.: NY.
Notes: This type specimen is too poor to allow accurate determination of its identity (Reid & Booth, Canad. J. Bot. 65:1340).
Host: Quercus
Cryptospora pulviniceps (Peck) Sacc.
Distr.: Eastern United States.
Host: Sambucus

Cryptosporella Sacc. Pyrenomycetes, Diaporthales

Cryptosporella acerina Wehmeyer in W.B. Cooke
Distr.: CA(type).
Notes: According to Barr (Mycol. Mem. 7:182. 1978), this fungus belongs in *Pseudocryptosporella*.
Host: Acer
Cryptosporella araucariae Voglino

Melanops araucariae (Voglino) Petr.
Distr.: WA; Italy(type).
Host: Araucaria
Cryptosporella beltsvillensis Petr.
Harknessia liquidambaris (Berk. & M.A. Curtis) Nag Raj & DiCosmo (anamorph)
Distr.: Eastern United States.
Notes: On leaves and twigs (Nag Raj & DiCosmo. Biblioth. Mycol. 80:42. 1981) .
Host: Liquidambar
Cryptosporella compta (Tul. & C. Tul.) Sacc.
Valsa compta Tul. & C. Tul.
Cryptospora compta (Tul. & C. Tul.) Ellis & Everh.
Distr.: Eastern United States; Europe(type).
Host: Fagus
Cryptosporella farinosa (Ellis) Sacc.
Valsa farinosa Ellis
Cryptospora farinosa (Ellis) Ellis & Everh.
Harknessia caudata Ellis & Everh. (anamorph)
Distr.: Eastern United States.
Notes: On leaves and twigs.
Host: Quercus
Cryptosporella hypodermia (Fr.:Fr.) Sacc.
Valsa hypodermia (Fr.:Fr.) Fr.
Distr.: Eastern North America; Europe.
Host: Acer
Cryptosporella paucispora (Peck) Berl. & Voglino
Valsa paucispora Peck
Cryptospora paucispora (Peck) Ellis & Everh.
Distr.: Eastern North America.
Notes: Reid & Cain (Canad. J. Bot. 38:945. 1960) redescribed this species.
Hosts: Alnus, Quercus
Cryptosporella thujina Nag Raj & DiCosmo
Harknessia thujina Ellis & Everh. (anamorph)
Distr.: Eastern North America.
Notes: On needles and bark.
Hosts: Chamaecyparis, Juniperus
Cryptosporella umbrina (Jenk.) Jenk. & Wehmeyer
Diaporthe umbrina Jenk.
Distr.: Worldwide in rose-growing regions.
Notes: Brown canker.
Host: Rosa
Cryptosporella viticola Shear
Distr.: Central and eastern United States.
Notes: Although described as the cause of dead-arm of grape, Punithalingam (C.M.I. Descr. 635. 1979) doubted the connection with *Phomopsis viticola* and suggested use of the latter name for this disease.
Host: Vitis

Cryptosporiopsis Bubák & Kab. Coelomycetes

Notes: Sutton (The Coelomycetes. p. 509. 1980).
Cryptosporiopsis cornina (Peck) Petr. & Syd.
Sphaeropsis cornina Peck
Macrophoma cornina (Peck) Sacc.
Macrophoma paniculata (Ellis & Dearn.) Sacc. & Syd.
Pezicula corni Petr. (teleomorph)
Distr.: Northern North America.
Notes: On twigs.
Host: Cornus
Cryptosporiopsis corticola (Edgerton) Nannf.
Myxosporium corticola Edgerton
Distr.: Temperate North America; Europe.
Notes: Bark canker.
Hosts: Malus, Pyrus
Cryptosporiopsis curvispora (Peck) Gremmen in Boerema & Gremmen
Macrophoma curvispora Peck
Cryptosporiopsis malicorticis (Cordley) Nannf.
Gloeosporium malicorticis Cordley
Cryptosporiopsis perennans (Zeller & Childs) Wollenweb.)
Gloeosporium perennans Zeller & Childs
Pezicula malicorticis (H. Jacks.) Nannf. (teleomorph)
Distr.: Temperate regions.
Notes: Anthracnose, bulls' eye rot, and perennial canker.
Hosts: Amelanchier, Malus
Cryptosporiopsis fasciculata (Tode ex Tul.) Petr.
Tubercularia fasciculata Tode ex Tul.
Pezicula carpinae (Pers.:Pers.) Tul. ex Fuckel
Distr.: Eastern North America; Europe; Great Britain.
Host: Carpinus
Cryptosporiopsis pruinosa (Peck) Wollenweb.
Sphaeronaema pruinosum Peck
Lagynodella pruinosa (Peck) Petr.
Glutinium macrosporum Zeller
Pezicula pruinosa Farl. (teleomorph)
Distr.: Temperate North America; Europe.
Notes: On twigs of Rosaceae.
Hosts: Amelanchier, Malus, Pyrus

Cryptosporium Kunze:Fr. Coelomycetes

Cryptosporium candidum Dearn.
Cryptosporium falcatum Dearn. non Cooke & Harkn., hom. illeg.
Distr.: Northwestern North America; Canada(Ontario, type).
Hosts: Abies, Arctostaphylos
Cryptosporium eucalypti Cooke & Harkn.
Distr.: CA(type), FL.
Host: Eucalyptus
Cryptosporium lunasporum Linder
Distr.: ID(type).
Host: Pinus
Cryptosporium lupini Cooke
Distr.: CA.
Host: Lupinus
Cryptosporium macrospermum Peck
Distr.: NY(Adirondack Mountains;type).
Host: Abies
Cryptosporium minimum Laubert
Distr.: Temperate North America; Germany(type).
Host: Rosa
Cryptosporium punctiforme Cooke & Harkn.
Distr.: CA(type).
Host: Arbutus
Cryptosporium robiniae Dearn. & House
Distr.: NY.
Host: Robinia

Cryptostictis Fuckel. Coelomycetes

Notes: *Cryptostictis* is considered a synonym of *Seimatosporium*, but not all names have been redisposed.
Cryptostictis inaequalis Tehon & G.L. Stout
Distr.: IL.
Host: Vitis
Cryptostictis insidens (Zabr.) Guba
Pestalotia insidens Zabr.
Monochaetia insidens (Zabr.) Sacc.
Distr.: NY.
Host: Ulmus
Cryptostictis libertiana (Roum. & Sacc.) Guba
Monochaetia monochaeta (Desmaz.) Allesch. var. *libertiana* Roum. & Sacc.
Pestalozzia monochaeta Desmaz. var. *libertiana* Sacc.
Distr.: AK; Europe.
Host: Rubus
Cryptostictis oenotherae (Ellis & Barth.) Guba
Pestalotia oenotherae Ellis & Barth.
Distr.: Central and southern United States.
Hosts: Calylophus, Oenothera
Cryptostictis paeoniae Tehon & E.Y. Daniels
Distr.: IL.
Host: Paeonia
Cryptostictis violae Tehon & E.Y. Daniels
Distr.: IL.
Host: Viola

Cryptostroma P. Gregory & S. Waller. Hyphomycetes

Cryptostroma corticale (Ellis & Everh.) P. Gregory & S. Waller
Coniosporium corticale Ellis & Everh.
Distr.: Temperate North America(introduced); Germany; Great Britain; Norway.
Notes: Sutton & Gibson (C.M.I. Descr. 539. 1977).
Host: Acer

Cryptovalsa Ces. & De Not. ex Fuckel. Pyrenomycetes, Diaporthales

Cryptovalsa ampelina (Nitschke) Fuckel
Distr.: FL; Europe.
Host: Vitis
Cryptovalsa deusta (Ellis & G. Martin) Petr.
Diatrypella deusta Ellis & G. Martin
Distr.: FL (Petrak. Sydowia 7:103. 1953).
Host: Serenoa

Cucurbidothis Petr. Loculoascomycetes, Dothideales

Cucurbidothis pithyophila (J.C. Schmidt & Kunze:Fr.) Petr.
Cucurbitaria pithyophila (J.C. Schmidt & Kunze:Fr.) De Not.
Melogramma spraguei Berk. & M.A. Curtis
Coniothyrium pithyophilum (Höhn.) Petr. & Syd. (anamorph)
Distr.: North America; Europe.
Host: Pinus

Cucurbitaria S.F. Gray. Loculoascomycetes, Dothideales

Notes: Mirza (Nova Hedwigia 16:161. 1968); Sivanesan (The bitunicate Ascomycetes. p. 388. 1984).
Cucurbitaria ailanthi Rabenh.
Cucurbitaria negundinis G. Wint. in Rabenh.
Distr.: North America(rare); Germany.
Host: Acer
Cucurbitaria amorphae (Wallr.) Fuckel
Camarosporium amorphae Sacc. (synanamorph)
Diplodia amorphae (Wallr.) Sacc. (synanamorph)
Distr.: KS, NM; Europe.
Host: Amorpha
Cucurbitaria arizonica Ellis & Everh.
Distr.: Central and western United States.
Host: Acacia
Cucurbitaria berberidis (Pers.: Fr.) S.F. Gray
Pyrenochaeta berberidis (Sacc.) Brunaud (anamorph)
Distr.: Temperate northern hemisphere.
Host: Berberis
Cucurbitaria caraganae P. Karst.
Camarosporium caraganae P. Karst. (anamorph)
Distr.: Central United States; Eurasia.
Notes: Sivanesan (The bitunicate Ascomycetes. p. 385. 1984) mentioned both *Camarosporium caraganae* Karst. and *Pseudodichomera elaeagni* (Karst.) Höhn. as possible anamorphs.
Host: Caragana
Cucurbitaria cassandrae Ellis, nom. nud.
Distr.: Northeastern United States.
Host: Chamaedaphne
Cucurbitaria ceanothi Dearn. & House
Distr.: NY(type).
Host: Ceanothus
Cucurbitaria celtidis Shear
Distr.: KS(type).
Host: Celtis
Cucurbitaria comptoniae Cooke & Ellis
Distr.: NJ; Canada.
Host: Comptonia
Cucurbitaria congesta Cooke & Ellis
Distr.: NJ(type).
Host: Magnolia
Cucurbitaria conglobata (Fr.) Ces. & De Not.
Distr.: AK; Europe.
Hosts: Alnus, Corylus
Cucurbitaria conglobata (Fr.) Ces. & De Not. var. ***crataegi*** Rehm
Distr.: ND.
Notes: The place of publication of this name could not be found.
Host: Crataegus
Cucurbitaria crataegi (Schwein.) Ellis & Everh.
Distr.: Temperate North America; Europe.
Host: Crataegus
Cucurbitaria elaeagni Mirza
Cucurbitaria caraganae P. Karst. var. *elaeagni* Rehm, nom. nud.
Distr.: ND.
Notes: Mirza (Nova Hedwigia 16:199. 1968) described this species for the invalid name *Cucurbitaria caraganae* var. *elaeagni*.
Host: Elaeagnus
Cucurbitaria elongata (Fr.:Fr.) Grev.
Camarosporium sp. (anamorph)
Distr.: Temperate North America; Europe.
Hosts: Colutea, Gleditsia, Robinia
Cucurbitaria fraxini Ellis & Everh.
Distr.: North America; Europe.
Hosts: Alnus, Fraxinus
Cucurbitaria interstitialis (Cooke & Peck) Barr in Barr et al.
Teichospora interstitialis (Cooke & Peck) Sacc.
Distr.: NY.
Host: Prunus
Cucurbitaria laburni (Pers.:Fr.) De Not.
Camarosporium laburni Sacc. & Roum. (anamorph)
Distr.: NY, OH; Europe.
Host: Laburnum
Cucurbitaria longitudinalis Peck
Distr.: Northeastern United States.
Host: Lyonia
Cucurbitaria obducens (Schumach.:Fr.) Fuckel
Strickeria obducens (Schumach.:Fr.) G. Wint.
Teichospora obducens (Schumach.:Fr.) Fuckel
Distr.: Temperate regions.
Notes: On wood (Dennis. British Ascomycetes. p. 463. 1978).
Hosts: Artemisia, Chamaebatiaria, Crataegus, Korthalsella, Physocarpus, Ribes
Cucurbitaria occultata Oudem.
Distr.: ND; Europe.
Host: Shepherdia
Cucurbitaria recuperata Theiss.
Botryosphaeria gleditsiae (Schwein.) Sacc.
Distr.: SC.
Host: Gleditsia
Cucurbitaria ribis (Fr.) Niessl
Diplodia ribis Sacc. (anamorph)
Distr.: United States; Europe; Pakistan.
Notes: The anamorph connection is tentative.
Host: Ribes
Cucurbitaria rosae G. Wint. & Sacc.
Distr.: NM, NY; Europe.
Host: Rosa
Cucurbitaria shepherdiae (Ellis & Everh.) Ellis & Everh.
Curreya shepherdiae Ellis & Everh.
Distr.: MT, ND, WA.
Notes: Barr (Mycologia 73:602. 1981) suggested that this species belongs in *Strickeria*.
Host: Shepherdia
Cucurbitaria sorbi P. Karst.
Karstenula sorbicola Petr.
Distr.: Temperate North America; Europe.
Notes: On limbs.
Host: Sorbus
Cucurbitaria staphula Dearn. ex R.H. Arnold & Russell
Dichomera sp. (anamorph)
Distr.: North America.
Notes: Arnold & Russell (Mycologia 52:499. 1960) reported the association of this species with galls formed by *Macrophoma tumefaciens*. Arnold (Fungi Canadenses 17. 1974).
Host: Populus
Cucurbitaria umbilicata Ellis
Distr.: ND, UT.

Host: Chrysothamnus

Cumminsiella Arth. Basidiomycetes, Uredinales

Notes: McCain & Hennen (Syst. Bot. 7:48-59. 1982).
Cumminsiella mirabilissima (Peck) Nannf. in S. Lundell & Nannf.
Puccinia mirabilissima Peck
Cumminsiella sanguinea (Peck) Arth.
Uropyxis sanguinea Arth.
Distr.: Western North America, introduced into MI and eastern Canada; northern Europe.
Notes: Autoecious. Sivanesan (C.M.I. Descr. 261. 1970); Parmelee & de Carteret (Fungi Canadenses 288. 1984).
Host: Berberis
Cumminsiella texana (Holw. & W.H. Long) Arth.
Distr.: TX; Mexico.
Notes: Autoecious.
Hosts: Berberis, Mahonia
Cumminsiella wootoniana (Arth.) Arth.
Uropyxis wootoniana Arth.
Distr.: Southwestern United States; Mexico.
Notes: Autoecious.
Hosts: Berberis, Mahonia

Cunninghamella Matr. Zygomycetes, Mucorales

Cunninghamella echinulata (Thaxt.) Thaxt.
Distr.: Widespread.
Notes: Isolated from soil, dung and various organic substrates (Samson. Proc. Koninkl. Nederl. Akad. Wetens., Ser. C, 72:322. 1969).
Hosts: Cucumis, Gossypium
Cunninghamella elegans Lendn.
Distr.: Widespread.
Notes: Isolated from various organic substrates.
Hosts: Daucus, Gossypium, Tetragonia

Curreya Sacc. Loculoascomycetes, Dothideales

Curreya peckiana Sacc.
Distr.: NY(type).
Notes: Barr (Mycologia 73:606. 1981) examined the type specimen and suggested that this is a species of *Dothiora*.
Host: Nemopanthus

Curvularia Boedijn. Hyphomycetes

Notes: The teleomorph state of these species belong to *Cochliobolus*. Sivanesan (Mycol. Pap. 158. 1987) provided descriptions of species on Poaceae.
Curvularia affinis Boedijn
Distr.: Cosmopolitan.
Host: Eichhornia
Curvularia brachyspora Boedijn
Distr.: Subtropical and tropical regions.
Host: Oryza
Curvularia crepinii (Westend.) Boedijn
Distr.: Temperate North America; Europe.
Notes: On living leaves.
Host: Ophioglossum
Curvularia cymbopogonis (C.W. Dodge) Groves & Skolko
Helminthosporium cymbopogonis C.W. Dodge
Distr.: Subtropical and tropical regions.
Notes: Seed and seedling blight, leaf spot (Sivanesan. Mycol. Pap. 158:110. 1987; Sivanesan & Holliday. C.M.I. Descr. 726. 1982 as *Cochliobolus cymbopogonis*).
Host: Rottboellia
Curvularia eragrostidis (Henn.) J.A. Meyer
Curvularia maculans (Bancroft) Boedijn
Cochliobolus eragrostidis (Tsuda & Ueyama) Sivanesan (teleomorph)
Distr.: Subtropical and tropical regions.
Hosts: Alysicarpus, Gossypium, Opuntia, Oryza, Pennisetum, Tradescantia, Zea
Curvularia fallax Boedijn
Distr.: Subtropical and tropical regions.
Notes: On a variety of hosts and debris.
Host: Populus
Curvularia geniculata (Tracy & Earle) Boedijn
Helminthosporium geniculatum Tracy & Earle
Distr.: Cosmopolitan, most common in tropical regions.
Notes: Sivanesan & Holliday (C.M.I. Descr. 727. 1982).
Hosts: Agropyron, Bouteloua, Brachypodium, Brassica, Bromus, Digitaria, Elymus, Eragrostis, Eremochloa, Hordeum, Koeleria, Linum, Muhlenbergia, Oryzopsis, Panicum, Pennisetum, Poa, Pyrus, Sambucus, Setaria, Sorghum, Sporobolus, Stipa, Striga, Tridens, Trifolium, Triticum, Zea
Curvularia gladioli Van Weerdt, unpublished name
Distr.: FL.
Notes: Leaf spot (Weber & Van Weerdt. Phytopathology 46:30. 1956).
Host: Gladiolus
Curvularia gudauskasii Morgan-Jones & G.W. Karr
Distr.: AL; Tanzania.
Notes: On leaves (Morgan-Jones & Karr. Mycotaxon 3:559. 1976).
Host: Zea
Curvularia inaequalis (Shear) Boedijn
Helminthosporium inaequale Shear
Acrothecium arenarium Moreau & Mme. Moreau
Distr.: Temperate regions.
Notes: Commonly isolated from grain and roots of wheat; occurs on a wide variety of plants.
Hosts: Arachis, Brassica, Dactylis, Festuca, Linum, Pyrus, Secale, Spinacia, Triticum, Vaccinium, Zea
Curvularia intermedia Boedijn
Cochliobolus intermedius R.R. Nelson (teleomorph)
Distr.: North America; Africa; Asia.
Hosts: Cupressus, Cynodon, Juniperus, Platanus, Zea
Curvularia lunata (Wakk.) Boedijn
Cochliobolus lunatus R.R. Nelson & Haasis (teleomorph)
Distr.: Cosmopolitan, most common in tropical regions.
Notes: Ellis & Gibson (C.M.I. Descr. 474. 1975).
Hosts: Agrostis, Allium, Alnus, Brachiaria, Capsicum, Caryota, Cocos, Cynodon, Digitaria, Eragrostis, Eremochloa, Erioneuron, Gladiolus, Glycine, Gossypium, Lolium, Ophiopogon, Oryza, Pennisetum, Phalaenopsis, Pinus, Polygala, Ravenala, Sabal, Saccharum, Stenotaphrum, Zea, Zoysia
Curvularia lunata (Wakk.) Boedijn var. ***aeria*** (Batista, Lima & Vasconcelos) M.B. Ellis
Distr.: Subtropical and tropical regions.
Notes: Plurivorous.
Host: Glycine
Curvularia pallescens Boedijn
Curvularia leonensis M.B. Ellis
Cochliobolus pallescens (Tsuda & Ueyama) Sivanesan (teleomorph)
Distr.: IA; tropical regions.
Notes: On grasses and otehr substrates.
Host: Zea
Curvularia penniseti (M. Mitra) Boedijn
Acrothecium penniseti M. Mitra
Distr.: FL; tropical regions.
Notes: On grasses.
Host: Eichhornia
Curvularia protuberata R.R. Nelson & C.S. Hodges
Distr.: North and South America; Africa; Europe.
Hosts: Juncus, Panicum, Zea
Curvularia richardiae Alcorn
Distr.: NC; Australia; Kenya; Mauritius.
Notes: On living leaves (Ellis. More dematiaceous Hyphomycetes. p. 404. 1976).
Host: Richardia
Curvularia robusta Kilpatr. & Luttrell
Distr.: TX.
Notes: Pathogenic on leaves.
Host: Dichanthium
Curvularia senegalensis (Speg.) Subramanian
Brachysporium senegalense Speg.
Curvularia falcata (Tehon) Boedijn

Acrothecium falcatum Tehon
Distr.: Subtropical and tropical regions.
Hosts: Carya, Saccharum
Curvularia sociata (Ellis & Everh.) Morgan-Jones & Kendrick
Exosporium sociatum Ellis & Everh.
Distr.: LA.
Host: Acer
Curvularia trifolii (Kauffm.) Boedijn
Brachysporium trifolii Kauffm. in Bonar
Blennoria trifolii Bonar, nom. nud.
Distr.: Cosmopolitan.
Notes: Leaf blight and seed mold of *Trifolium*, but occurs on a wide variety of plants.
Hosts: Agropyron, Cucumis, Cucurbita, Digitaria, Echinochloa, Gladiolus, Trifolium
Curvularia trifolii (Kauffm.) Boedijn f. sp. ***gladioli*** Parmelee & Luttrell in Parmelee
Distr.: Cosmopolitan.
Notes: Various rots (Laundon. C.M.I. Descr. 307. 1971).
Host: Gladiolus
Curvularia tuberculata P.C. Jain
Cochliobolus tuberculatus Sivanesan (teleomorph)
Distr.: FL; France; tropical regions.
Notes: On grasses and various other substrates.
Host: Juncus

Cyanoderma Höhn. Loculoascomycetes, Dothideales

Cyanoderma viridulum (Berk. & M.A. Curtis) Höhn.
Acrospermum viridulum Berk. & M.A. Curtis
Distr.: North America; Europe.
Notes: Eriksson (Ark. Botanik 6:388. 1967).
Host: Quercus

Cyanodermella O. Eriksson. Loculoascomycetes, Dothideales

Notes: Eriksson (Opera Bot. 60:155. 1981) accepted 2 species.
Cyanodermella candidum (Setch.) O. Eriksson
Acrospermum candidum Setch.
Cyanoderma candidum (Setch.) O. Eriksson
Distr.: Southwestern United States; Central America; Scandinavia.
Notes: Eriksson (Ark. Botanik 6:389. 1967).
Host: Alsophila

Cyanodiscus E. Müller & M.L. Farr. Loculoascomycetes, Dothideales

Notes: Müller & Farr (Mycologia 63:1080. 1971).
Cyanodiscus occidentalis E. Müller & M.L. Farr
Distr.: LA(type).
Host: Chamaecyparis

Cyathus Haller:Pers. Basidiomycetes, Nidulariales

Notes: Brodie (Bot. Not. 130:453. 1977).
Cyathus olla Batsch:Pers.
Cyathus vernicosus DC.
Distr.: Temperate regions.
Hosts: Juniperus, Pseudotsuga
Cyathus striatus (Huds.:Pers.) Pers.
Distr.: Cosmopolitan.
Notes: On wood, plant debris, and plant-derived fabrics.
Hosts: Abies, Acer, Carya, Quercus

Cyclaneusma DiCosmo, Peredo & Minter. Discomycetes, Rhytismatales

Notes: Two species on pines (DiCosmo et al. Eur. J. Forest Pathol. 13:206. 1983).
Cyclaneusma minus (Butin) DiCosmo, Peredo & Minter
Naemacyclus minor Butin
Distr.: Widespread on pine needles.
Notes: Millar & Minter (C.M.I. Descr. 659. 1980 as *Naemacylus minor*).
Host: Pinus
Cyclaneusma niveum (Pers.:Fr.) DiCosmo, Peredo, & Minter
Naemacyclus niveus (Pers.:Fr.) Sacc.
Distr.: Widespread on pine needles.
Notes: Butin (Eur. J. Forest Pathol. 3:146. 1973); Minter & Millar (C.M.I. Descr. 660. 1980 as *Naemcyclus niveus*).
Hosts: Pinus, Primula

Cyclodomus Höhn. Coelomycetes

Notes: Monotypic (Sutton. The Coelomycetes. p. 483. 1980).
Cyclodomus umbellulariae Höhn.
Distr.: CA(type).
Host: Umbellularia

Cyclomyces Fr. Basidiomycetes, Aphyllophorales

Cyclomyces iodinus (Mont.) Pat.
Polyporus iodinus Mont.
Distr.: FL, TX, neotropical region.
Notes: On wood of hardwoods.
Host: Quercus

Cylindrobasidium Jülich. Basidiomycetes, Aphyllophorales

Cylindrobasidium corrugum (Burt) Ginns
Coniophora corrugis Burt
Corticium corruge (Burt) Burt, nom. inval.
Distr.: Western North America.
Notes: On conifers.
Hosts: Abies, Acer, Juniperus, Picea, Pinus, Populus, Pseudotsuga, Ribes, Sambucus, Tsuga
Cylindrobasidium laeve (Pers.:Fr.) Chamuris
Corticium laeve Pers.:Fr., sensu Jülich
Hyphoderma arizonicum Lindsey & R.L. Gilbertson
Corticium evolvens Fr.:Fr.
Cylindrobasidium evolvens (Fr.:Fr.) Jülich
Distr.: Widespread.
Notes: On wood, usually of recently dead hardwoods. Jülich (Persoonia 8:59. 1974) had previously selected a neotype from Persoon's herbarium which preserved the historial application of the name. Eriksoon et al. (The Corticiaceae of North Europe 56:1007. 1978) changed the established concept of the species.
Hosts: Acer, Betula, Fagus, Liquidambar, Magnolia, Morus, Pinus, Populus, Pseudotsuga, Quercus, Salix, Ulmus
Cylindrobasidium torrendii (Bres.) Hjortstam
Peniophora torrendii Bres. in Torrend
Peniophora albula Atk. & Burt
Corticium albulum (Atk. & Burt) D.P. Rogers & H. Jacks.
Cylindrobasidium albulum (Atk. & Burt) J. Eriksson & Ryvarden
Distr.: North America; Europe.
Notes: On wood.
Hosts: Abies, Acer, Fraxinus, Liriodendron, Pinus, Populus, Quercus, Salix

Cylindrocarpon Wollenweb., nom. cons. Hyphomycetes

Notes: Booth (Mycol. Pap. 104. 1966).
Cylindrocarpon album (Sacc.) Wollenweb.
Fusarium album Sacc.
Cylindrocarpon album (Sacc.) Wollenweb. var. *majus* Wollenweb.
Nectria punicea (J.C. Schmidt:Fr.) Rabenh. (teleomorph)
Distr.: North temperate regions; West Indies; Malaysia.
Notes: A weak pathogen of various trees and other plants.
Hosts: Aesculus, Fagus
Cylindrocarpon angustum Wollenweb. in Zeller
Distr.: OR(type).

Notes: Booth (Mycol. Pap. 104:50. 1966) was not able to characterize these species which is known only from the initial report.
Host: Malus
Cylindrocarpon cylindroides Wollenweb.
Nectria macrospora (Wollenweb.) Ouellette (teleomorph)
Distr.: Northwestern North America; Europe.
Notes: Canker of balsam fir and other Pinaceae (Booth. C.M.I. Descr. 623. 1979).
Host: Abies
Cylindrocarpon destructans (Zinssmeister) Scholten
Ramularia destructans Zinssmeister
Cylindrocarpon radicicola Wollenweb.
Distr.: Cosmopolitan.
Notes: In soil, associated with the roots of a wide variety of plants (Booth. C.M.I. Descr. 148. 1967).
Hosts: Aster, Beta, Campanula, Conyza, Erigeron, Fragaria, Hieracium, Lilium, Malus, Narcissus, Panax, Pyrus, Rhododendron, Rosa, Rumex, Saccharum, Saintpaulia, Solanum
Cylindrocarpon didymum (Harting) Wollenweb.
Fusisporium didymum Harting
Distr.: Widespread.
Notes: Root rots and seedling blights. Also isolated from various kinds of plant debris and soil (Domsch et al. Compendium of soil fungi. p. 244. 1980; Brayford. C.M.I. Descr. 923. 1987).
Host: Rhizophora
Cylindrocarpon gillii (D.E. Ellis) J.A. Muir
Septogloeum gillii D.E. Ellis
Distr.: Western North America.
Notes: Parasitic on dwarf mistletoe (Muir. Canad. J. Bot. 51:1997. 1973).
Hosts: Arceuthobium, Pinus
Cylindrocarpon liriodendri J.D. MacDonald & E.E. Butler
Distr.: CA(type).
Notes: Isolated from diseased roots (MacDonald & Butler. Pl. Dis. 65:154. 1981).
Host: Liriodendron
Cylindrocarpon magnusianum Wollenweb.
Ramularia magnusiana (Wollenweb.) Lindau
Septocylindrium magnusianum Sacc.
Cylindrocarpon ehrenbergii Wollenweb.
Nectria ramulariae (Wollenweb.) E. Müller (teleomorph)
Distr.: Widespread.
Notes: On woody substrates and isolated from soil (Domsch et al. Compendium of soil fungi. p. 245. 1980).
Hosts: Aesculus, Euphrasia, Malus, Medicago, Poa, Pyrus, Solanum, Trientalis
Cylindrocarpon obtusisporum (Cooke & Harkn.) Wollenweb.
Fusarium obtusisporum Cooke & Harkn.
Distr.: Temperate and warm-temperate regions.
Notes: Root rot (Brayford. C.M.I. Descr. 928. 1987).
Hosts: Acacia, Medicago, Pseudotsuga, Pyrus

Cylindrocladiella Boesewinkel. Hyphomycetes

Notes: Boesewinkel (Canad. J. Bot. 60:2288. 1982) established this genus for short-spored *Cylindrocladium*.
Cylindrocladiella camelliae (Venkataramani & Venkata Ram) Boesewinkel
Cylindrocladium camelliae Venkataramani & Venkata Ram
Distr.: Subtropical and tropical regions.
Notes: On fruits and roots of various plants (Rossman. Mycol. Pap. 150:45. 1983; Peerally. C.M.I. Descr. 428. 1974 as *Cylindrocladium camelliae*).
Host: Garcinia
Cylindrocladiella parva (P.J. Anderson) Boesewinkel
Cylindrocladium parvum P.J. Anderson
Distr.: Cosmopolitan.
Hosts: Allium, Cupressus, Juglans, Rhapis, Rhizophora
Cylindrocladiella peruviana (Batista, Bezerra & Herrera) Boesewinkel
Cylindrocladium peruviana Batista, Bezerra & Herrera
Distr.: Subtropical and tropical regions.
Hosts: Aglaonema, Camellia, Cissus, Cupressus, Cyclamen, Phoenix, Rhododendron

Cylindrocladium Morg. Hyphomycetes

Cylindrocladium avesiculatum D.L. Gill, Alfieri & Sobers
Distr.: Southeastern United States.
Notes: Parasitic on *Ilex* and other plants (Gill et al. Phytopathology 61:58. 1971).
Hosts: Ilex, Leucothoe, Rhododendron
Cylindrocladium citri (H. Fawc. & L.J. Klotz) Boedijn & Reitsma
Candelospora citri H. Fawc. & L.J. Klotz
Distr.: CA, FL.
Notes: On decaying fruits (Boedijn & Reitsma. Reinwardtia 1:57. 1950).
Host: Citrus
Cylindrocladium clavatum C.S. Hodges & L.C. May
Distr.: FL, HI; Brazil; Mauritius.
Notes: Peerally (C.M.I. Descr. 422. 1974).
Hosts: Araucaria, Callistemon, Medicago, Parkinsonia, Rhododendron
Cylindrocladium crotalariae (C.A. Loos) D.K. Bell & Sobers
Cercosporella theae Petch
Calonectria crotolariae (C.A. Loos) D.K. Bell & Sobers (teleomorph)
Distr.: Subtropical and tropical regions.
Notes: Black rot of peanuts, causes various seedling diseases (Peerally. C.M.I. Descr. 429. 1974).
Hosts: Actinidia, Arachis, Glycine, Liquidambar, Medicago, Trifolium, Vaccinium
Cylindrocladium ellipticum Alfieri, C.P. Seymour & Sobers
Distr.: FL.
Notes: Isolated from leaves (Alfieri et al. Phytopathology 60:1212. 1970).
Host: Berberis
Cylindrocladium heptaseptatum Sobers, Alfieri & J.F. Knauss
Distr.: FL; Honduras.
Notes: Leaf spot of leatherleaf fern. A microconidial state was recently discovered (El-Gholl et al. Canad. J. Bot. 65:1733. 1987).
Host: Rumohra
Cylindrocladium ilicicola (Hawley) Boedijn & Reitsma
Candelospora ilicicola Hawley
Distr.: Worldwide, most common in tropical regions.
Notes: On decaying organic debris, occasionally causing diseases (Peerally. C.M.I. Descr. 425. 1974; Rossman. Mycol. Pap. 150:50. 1983).
Hosts: Araucaria, Cissus, Vaccinium
Cylindrocladium pteridis F.A. Wolf
Cylindrocladium macrosporum Sherb.
Distr.: Subtropical and tropical regions.
Notes: On ferns, palms and various dicotyledonous hosts (Sobers. Phytopathology 58:1265. 1968).
Hosts: Chamaedorea, Cissus, Cocos, Crassula, Dryopteris, Eucalyptus, Howeia, Melaleuca, Nephrolepis, Rumohra, Strelitzia, Syagrus, Synsepalum, Washingtonia
Cylindrocladium scoparium Morg.
Diplocladium cylindrosporum Ellis & Everh.
Cylindrocladium floridanum Sobers & C.P. Seymour
Calonectria kyotensis Terashita (teleomorph)
Distr.: Cosmopolitan.
Notes: Associated with root decline of peaches (Sobers & Seymour. Phytopathology 57:389. 1967). Rossman (Mycol. Pap. 150:47. 1983).
Hosts: Abies, Acacia, Arecaceae, Asimina, Berberis, Beta, Bucida, Callistemon, Cercis, Chrysanthemum, Citrullus, Conocarpus, Dianthus, Eucalyptus, Euphorbia, Feijoa, Ficus, Fragaria, Galphimia, Gardenia, Howeia, Ilex, Juglans, Juniperus, Lagerstroemia, Larix, Leucothoe, Liquidambar, Liriodendron, Magnolia, Malus, Melaleuca, Pelargonium, Persea, Photinia, Picea, Pimenta, Pinus, Prunus, Quercus, Rhododendron, Rosa, Rumohra, Syringa, Trifolium, Tsuga, Ulmus
Cylindrocladium spathiphylli Schoulties, El-Gholl & Alfieri
Distr.: FL(type).
Notes: Isolated from petioles and roots (Schoulties et al. Mycotaxon 16:265. 1982).
Host: Spathiphyllum
Cylindrocladium theae (Petch) Subramanian
Calonectria theae C.A. Loos (teleomorph)
Distr.: HI, southeastern United States; Sri Lanka.

Notes: Peerally (C.M.I. Descr. 423. 1974).
Hosts: Eucalyptus, Leucothoe, Rhododendron

Cylindrocolla Bonord. Hyphomycetes

Cylindrocolla bigeloviae Ellis & Everh.
Distr.: CO(type).
Host: Chrysothamnus

Cylindrogloeum Petr. Coelomycetes

Cylindrogloeum trillii (Ellis & Everh.) Arx
Gloeosporium trillii Ellis & Everh.
Distr.: Northwestern United States.
Notes: Leaf spot (Sutton. The Coelomycetes. p. 596. 1980).
Host: Trillium

Cylindrosporella Höhn. Coelomycetes

Notes: This genus is considered a synonym of *Asteroma* DC.:Fr., but not all names have been redisposed (Sutton. The Coelomycetes. p. 494. 1980).
Cylindrosporella betulae-papyriferae (Dearn. & Overh.) Arx
Gloeosporium betulae-papyriferae Dearn. & Overh.
Distr.: Northeastern North America.
Host: Betula
Cylindrosporella caricina H.C. Greene
Distr.: WI(type).
Host: Carex
Cylindrosporella conspicua H.C. Greene
Distr.: WI(type).
Host: Salix
Cylindrosporella leptothyrioides (Kab. & Bubák) Arx
Gloeosporium leptothyrioides Kab. & Bubák
Distr.: WI.
Host: Betula

Cylindrosporium Grev. Coelomycetes

Notes: According to Sutton (The Coelomycetes. p. 503. 1980), many of these species should be transferred to other genera.
Cylindrosporium acerinum Tracy & Earle
Distr.: UT(type).
Notes: Possibly a synonym of *Septoria aceris* (Lib.) Berk. & Broome.
Host: Acer
Cylindrosporium acicola Bres.
Distr.: CA; Germany(type).
Host: Pinus
Cylindrosporium alni Dearn. & Barth.
Distr.: Western North America.
Host: Alnus
Cylindrosporium andropogoni R. Sprague & C.T. Rogerson
Distr.: KS.
Host: Andropogon
Cylindrosporium apocyni Ellis & Everh.
Distr.: Temperate North America.
Host: Apocynum
Cylindrosporium aroniae Sacc.
Distr.: Temperate North America.
Host: Amelanchier
Cylindrosporium artemisiae Dearn. & Barth.
Distr.: Temperate North America.
Host: Artemisia
Cylindrosporium balsamorhizae Solheim
Distr.: WY.
Host: Balsamorhiza
Cylindrosporium bambusae Miyake & K. Hara
Distr.: GA; Japan.
Notes: The report from Georgia is apparently based on a single specimen determined by V.K. Charles.
Hosts: Bambusa, Poaceae
Cylindrosporium betulae J.J. Davis
Distr.: Temperate North America.
Notes: The generic placement of this species is uncertain (Constantinescu. Trans. Brit. Mycol. Soc. 83:394. 1984).
Host: Betula
Cylindrosporium brevispina Dearn.
Distr.: Western North America.
Host: Crataegus
Cylindrosporium burkii W.B. Cooke & C.G. Shaw
Distr.: ID.
Host: Lupinus
Cylindrosporium calamagrostidis Ellis & Everh.
Distr.: UT, WI.
Hosts: Calamagrostis, Muhlenbergia
Cylindrosporium camelliae Gonz. Frag.
Distr.: Southeastern United States; Portugal(type).
Host: Camellia
Cylindrosporium campicola C.T. Rogerson & R. Sprague
Distr.: KS.
Notes: Sprague & Rogerson (Mycologia 50:637. 1958).
Hosts: Chloris, Cynodon
Cylindrosporium castaneae (Lév.) Krenner
Distr.: Eastern North America; Europe.
Notes: The place of publication of this name could not be found.
Host: Castanea
Cylindrosporium celtidis Earle
Distr.: Southeastern United States.
Host: Celtis
Cylindrosporium cercosporoides Ellis & Everh.
Distr.: Eastern United States.
Host: Liriodendron
Cylindrosporium chrysanthemi Ellis & Dearn.
Distr.: Temperate North America.
Host: Chrysanthemum
Cylindrosporium cicutae Ellis & Everh.
Distr.: Temperate North America.
Host: Cicuta
Cylindrosporium circinans G. Wint.
Distr.: Temperate North America.
Host: Sanguinaria
Cylindrosporium clematidis Ellis & Everh.
Distr.: Eastern United States.
Host: Clematis
Cylindrosporium consociatum Dearn.
Distr.: ID, WA.
Host: Acer
Cylindrosporium convolvulicola Vassiljevsky in Vassiljevsky & Karakulin
Septogloeum convolvuli Ellis & Everh.
Distr.: CA(type).
Hosts: Calystegia, Convolvulus
Cylindrosporium corni Solheim
Distr.: WY.
Host: Cornus
Cylindrosporium crescentum Barth.
Distr.: CO, TX, WI.
Host: Pastinaca
Cylindrosporium defoliatum Heald & F.A. Wolf
Distr.: Southeastern United States.
Host: Celtis
Cylindrosporium eminens J.J. Davis
Distr.: Temperate North America.
Host: Helianthemum
Cylindrosporium eryngii Ellis & Kellerm.
Distr.: Central United States.
Host: Eryngium
Cylindrosporium fairmanianum Sacc.
Distr.: NY.
Host: Spiraea
Cylindrosporium filipendulae Thuem.
Ascochyta salicifoliae Trel.
Cylindrosporium salicifoliae (Trel.) J.J. Davis
Phloeospora salicifoliae (Trel.) Petr.
Septoria salicifoliae (Trel.) Ellis & Everh.
Distr.: Temperate North America; Europe.
Hosts: Aruncus, Holodiscus, Spiraea
Cylindrosporium fraserae Wehmeyer
Distr.: CO, WY.
Host: Frasera

Cylindrosporium fraxini (Ellis & Kellerm.) Ellis & Everh.
Cercosporella fraxini Ellis & Kellerm.
Phloeochora fraxini (Ellis & Kellerm.) Höhn.
Phloeospora fraxini (Ellis & Kellerm.) Höhn.
Septoria fraxini Desmaz.
Mycosphaerella fraxini (Niessl) Lindau (teleomorph)
Distr.: Central and eastern North America.
Notes: The relationship of this species to *Piggotia fraxini* Berk. & M.A. Curtis is in need of clarification.
Host: Fraxinus
Cylindrosporium gei Farl.
Distr.: Northcentral and northeastern United States.
Host: Geum
Cylindrosporium geranii Ellis & Everh.
Distr.: LA.
Host: Geranium
Cylindrosporium glyceriae Ellis & Everh.
Distr.: NY, WI.
Hosts: Agrostis, Glyceria
Cylindrosporium glycyrrhizae Harkn.
Distr.: CA.
Host: Glycyrrhiza
Cylindrosporium gnaphalicola Atk.
Distr.: Southern United States.
Host: Gnaphalium
Cylindrosporium griseum Heald & F.A. Wolf
Distr.: OK, TX.
Host: Sapindus
Cylindrosporium guttatum G. Wint.
Distr.: Temperate North America; Brazil(type).
Host: Hypoxis
Cylindrosporium humuli Ellis & Everh.
Distr.: Central and eastern North America.
Host: Humulus
Cylindrosporium interstitialis H.C. Greene
Distr.: WI(type).
Host: Spartina
Cylindrosporium juglandis F.A. Wolf
Mycosphaerella juglandis K.J. Kessler (teleomorph)
Distr.: Central and eastern United States.
Notes: Leaf spot. Kessler (Mycologia 76:364. 1984) suggested that this anamorph state is a *Septoria*.
Host: Juglans
Cylindrosporium kelloggii Ellis & Everh.
Distr.: CA, OR; Brazil.
Notes: Black-oak leaf spot.
Host: Quercus
Cylindrosporium kerriae V.B. Stewart, nom. nud.
Blumeriella kerriae (V.B. Stewart) Korf (teleomorph)
Distr.: NC.
Notes: This name was never validated by Stewart (Phytopathology 7:399. 1917).
Host: Kerria
Cylindrosporium lactucicola Ellis & Everh.
Distr.: Temperate North America.
Host: Lactuca
Cylindrosporium lippiae Heald & F.A. Wolf
Distr.: TX.
Host: Junellia
Cylindrosporium lupini Ellis & Everh.
Distr.: Western North America.
Host: Lupinus
Cylindrosporium microspilum Sacc. & G. Wint.
Distr.: Central North America.
Host: Quercus
Cylindrosporium montenegrinum Bubák
Distr.: WY.
Host: Trollius
Cylindrosporium negundinis Ellis & Everh.
Distr.: KS(type).
Notes: Possibly a synonym of *Septoria aceris* (Lib.) Berk. & Broome, but satisfactory studies are unavailable.
Host: Acer
Cylindrosporium nuttallii (Harkn.) Dearn.
Septogloeum nuttallii Harkn.
Cylindrosporium conservans Peck
Distr.: Western North America.
Hosts: Oemleria, Prunus, Salix
Cylindrosporium oculatum Ellis & Everh.
Distr.: United States.
Host: Populus
Cylindrosporium officinale Ellis & Everh.
Distr.: IN(type).
Host: Saponaria
Cylindrosporium parkinsoniae Heald, nom. nud.
Distr.: TX.
Host: Parkinsonia
Cylindrosporium phaceliae Ellis & Everh.
Distr.: Temperate North America.
Host: Phacelia
Cylindrosporium populinum (Peck) Vassiljevsky
Marssonia rhabdospora Ellis & Everh.
Marssonina rhabdospora (Ellis & Everh.) Magnus
Distr.: North America.
Host: Populus
Cylindrosporium pulchrum Speg.
Distr.: Temperate North America; Argentina(type).
Host: Rumex
Cylindrosporium quercinum J.C. Carter
Distr.: IL.
Notes: Oak dieback (Carter. Illinois Nat. Hist. Survey Bull. 21:221. 1941).
Host: Quercus
Cylindrosporium ranunculi (Bonord.) Sacc.
Fusidium ranunculi Bonord.
Distr.: WY; Italy(type).
Host: Ranunculus
Cylindrosporium rhamni Ellis & Everh.
Distr.: Western United States.
Host: Rhamnus
Cylindrosporium ribis J.J. Davis
Cercoseptoria ribis (J.J. Davis) Dearn. & House
Cercospora septoriopsis Chupp
Distr.: Central and western United States.
Notes: Deighton (Mycol. Pap. 151:6. 1983).
Host: Ribes
Cylindrosporium rubi Ellis & Morg.
Sphaerulina rubi Demaree & M.S. Wilcox (teleomorph)
Distr.: Central and eastern North America; Europe.
Notes: On leaves.
Host: Rubus
Cylindrosporium saccharinum Ellis & Everh.
Distr.: WI(type).
Notes: Possibly a synonym of *Septoria aceris* (Lib.) Berk. & Broome.
Host: Acer
Cylindrosporium salicinum (Peck) Dearn.
Septoria salicina Peck
Septoria albaniensis Thuem.
Distr.: United States.
Host: Salix
Cylindrosporium saximontanense Solheim
Distr.: CO.
Host: Populus
Cylindrosporium scirpivora R. Sprague
Distr.: Western North America.
Host: Scirpus
Cylindrosporium scrophulariae Sacc. & Ellis
Distr.: Temperate North America.
Host: Scrophularia
Cylindrosporium smilacinae Ellis & Everh.
Distr.: Pacific Northwest.
Notes: Savile (Mycologia 53:48. 1961).
Host: Smilacina
Cylindrosporium smilacis Ellis & Everh.
Distr.: Southern United States
Host: Smilax
Cylindrosporium solitarium Heald & F.A. Wolf
Distr.: TX.
Host: Robinia
Cylindrosporium sparganii (Pass.) Ellis & Everh.

Septoria sparganii Pass.
Distr.: WI; Europe.
Host: Sparganium
Cylindrosporium stachydis Ellis
Distr.: IL(type).
Host: Stachys
Cylindrosporium steironematis Atk.
Distr.: AL.
Host: Lysimachia
Cylindrosporium tenuisporum Heald & F.A. Wolf
Distr.: TX.
Host: Ulmus
Cylindrosporium thalictri (Ellis & Everh.) J.J. Davis
Cylindrosporium ranunculi Bonord. f. *thalictri* Ellis & Everh.
Distr.: Central and western United States.
Host: Thalictrum
Cylindrosporium triflori H.C. Greene
Distr.: WI, WY.
Hosts: Geum, Potentilla
Cylindrosporium triostei Kellerm. & Swingle
Distr.: Temperate North America.
Host: Triosteum
Cylindrosporium ulmicola Ellis & Everh.
Distr.: Temperate North America; Europe.
Hosts: Morus, Ulmus
Cylindrosporium umbelliferarum Wehmeyer
Ascochyta heraclei Lib.
Cylindrosporium heraclei (Lib.) Höhn. non Ellis & Everh., hom. illeg.
Cylindrosporium heraclei Ellis & Everh.
Septoria heraclei (Lib.) Desmaz.
Distr.: Western United States.
Hosts: Heracleum, Ligusticum
Cylindrosporium urticae Dearn.
Distr.: WA(type).
Host: Urtica
Cylindrosporium vagnerae H.C. Greene
Distr.: WI.
Host: Smilacina
Cylindrosporium veratrinum Sacc. & G. Wint.
Cercosporella terminalis Peck
Distr.: Temperate North America; Europe.
Notes: Savile (Mycologia 53:50. 1961).
Host: Veratrum
Cylindrosporium violae Sacc.
Distr.: Temperate North America; Italy(type).
Host: Viola

Cylindroxyphium Batista & Cif. Coelomycetes

Cylindroxyphium virginianum Batista & Cif.
Distr.: FL(type).
Notes: On leaves (Sutton. The Coelomycetes. p. 365. 1980).
Host: Quercus

Cymadothea F.A. Wolf. Loculoascomycetes, Dothideales

Cymadothea trifolii (C. Killian) F.A. Wolf
Plowrightia trifolii C. Killian
Dothidella trifolii Bayliss-Elliott & Stansfield
Phyllachora trifolii (Pers.) Fuckel
Mycosphaerella killianii Petr.
Placosphaeria trifolii (Pers.: Fr.) Traverso (anamorph microconidial)
Polythrincium trifolii Kunze in J.C. Schmidt & Kunze (anamorph)
Distr.: Temperate regions.
Notes: Sooty blotch disease of clover (Boerema & Verhoeven. Neth. J. Pl. Pathol. 82:162. 1979; Sivanesan. The bitunicate Ascomycetes. 1984; Booth & O'Rourke. C.M.I. Descr. 393. 1973).
Host: Trifolium

Cyphellathelia Jülich. Basidiomycetes, Aphyllophorales

Cyphellathelia pezizoidea (Ellis & Everh.) Jülich
Corticium pezizoideum Ellis & Everh.
Distr.: AL, NJ.
Host: Arundinaria

Cyphellopsis Donk. Basidiomycetes, Agaricales

Cyphellopsis anomala (Pers.:Fr.) Donk
Cyphella anomala (Pers.:Fr.) Pat.
Merismodes anomalus (Pers.:Fr.) Singer
Henningsomyces anomalus (Pers.:Fr.) Kuntze
Solenia anomala (Pers.:Fr.) Fuckel
Distr.: Widespread.
Notes: On wood.
Hosts: Abies, Acer, Alnus, Betula, Corylus, Fraxinus, Juglans, Lindera, Liriodendron, Populus, Quercus, Robinia, Salix, Umbellularia
Cyphellopsis confusa (Bres.) D. Reid
Solenia confusa Bres.
Distr.: Temperate northern hemisphere.
Notes: On branches and twigs of hardwoods.
Host: Alnus
Cyphellopsis mellea (Burt) D. Reid
Cyphella mellea Burt
Merismodes melleus (Burt) Singer
Distr.: LA(type).
Host: Salix

Cyptotrama Singer. Basidiomycetes, Agaricales

Cyptotrama asprata (Berk.) Redhead & Ginns
Armillaria asprata (Berk.) Petch
Xerulina asprata (Berk.) Pegler
Collybia chrysopepla (Berk. & M.A. Curtis) A. Pearson
Xerulina chrysopepla (Berk. & M.A. Curtis) Singer
Tricholomopsis streetsii R.L. Gilbertson
Distr.: Pantropical, extending northward into eastern Canada.
Notes: On woody debris.
Hosts: Abies, Alnus, Picea, Pinus, Psidium, Quercus

Cystobasidium (Lagerh.) Neuhoff. Basidiomycetes, Auriculariales

Cystobasidium proliferans Olive
Distr.: Eastern North America.
Notes: On insect excreta and inner bark.
Host: Pinus

Cystostereum Pouzar. Basidiomycetes, Aphyllophorales

Crustomyces Jülich
Cystostereum australe Nakasone
Distr.: Subtropical and tropical America.
Notes: On wood.
Hosts: Carya, Ulmus
Cystostereum murrayi (Berk. & M.A. Curtis) Pouzar
Stereum murrayi (Berk. & M.A. Curtis) Burt
Corticium effusum Overh.
Distr.: Widespread.
Notes: On hardwoods in North America, especially *Betula* and *Acer* spp.; on conifers in Europe. Heart rot and stem cankers of living trees.
Hosts: Acer, Betula, Carpinus, Fagus, Ficus, Fraxinus, Malus, Ostrya, Picea, Populus, Quercus, Tilia, Ulmus
Cystostereum pini-canadense (Schwein.) Parmasto
Radulum pini-canadensis Schwein.
Peniophora piceina Overh.
Cystostereum piceinum (Overh.) Lindsey & R.L. Gilbertson
Distr.: Central and eastern United States and adjacent Canada.

Notes: On wood.
Hosts: Abies, Alnus, Betula, Larix, Liriodendron, Picea, Pinus, Populus, Quercus, Thuja, Tsuga
Cystostereum pini-canadense (Schwein.) Parmasto subsp. ***subabruptum*** (Bourd. & Galzin) Chamuris
Odontia subabrupta Bourd. & Galzin
Distr.: Western North America; Europe.
Notes: On wood.
Hosts: Acer, Alnus, Chamaecyparis, Lithocarpus

Cystotheca Berk. & M.A. Curtis. Pyrenomycetes, Erysiphales

Cystotheca lanestris (Harkn.) Miyabe in Ideta
Sphaerotheca lanestris Harkn.
Distr.: Central and southern United States; Mexico; Japan.
Notes: Powdery mildew of oak.
Hosts: Eriogonum, Lithocarpus, Quercus

Cytidia Quél. Basidiomycetes, Aphyllophorales

Cytidia flocculenta (Fr.:Fr.) Höhn. & Litsch.
Cyphella ampla Lév.
Stereum pubescens Burt
Distr.: Widespread.
Notes: On hardwood limbs.
Hosts: Alnus, Salix
Cytidia lanata W.B. Cooke
Distr.: ID.
Notes: On bark.
Host: Betula
Cytidia patelliformis (Burt) A.L. Welden
Stereum patelliforme Burt
Distr.: North America.
Notes: On wood.
Hosts: Acer, Corylus, Crataegus, Fraxinus, Purshia, Rhus, Rosa
Cytidia salicina (Fr.:Fr.) Burt
Distr.: Widespread.
Notes: On hardwood limbs.
Hosts: Populus, Salix
Cytidia stereoides W.B. Cooke
Distr.: CA.
Notes: On hardwood limbs.
Hosts: Ceanothus, Cercocarpus, Purshia, Quercus

Cytoplea Bizz. & Sacc. in Bizz. Coelomycetes

Cytoplea insitiva (Sacc.) Petr.
Coniothyrium insitivum Sacc.
Distr.: Temperate North America; Europe.
Hosts: Ailanthus, Albizia, Berberis

Cytospora Ehrenb.:Fr. Coelomycetes

Notes: *Cytispora* is an orthographic variant. Spielman (Canad. J. Bot. 63:1355. 1985).
Cytospora abietis Sacc.
Valsa abietis (Fr.:Fr.) Fr. (teleomorph)
Distr.: Circumboreal on conifers.
Host: Abies
Cytospora albiceps Ellis & Kellerm.
Distr.: KS.
Host: Juglans
Cytospora annulata Ellis & Everh.
Cytospora negundinis Ellis & Everh.
Valsa ambiens (Pers.:Fr.) Fr. subsp. *leucostomoides* (Peck) Spielman (teleomorph)
Distr.: Northcentral and northeastern North America; northern Europe.
Host: Acer
Cytospora carbonacea Fr.
Distr.: MA, TX.
Host: Ulmus
Cytospora ceanothi Schwein.
Distr.: PA.
Notes: Spielman (Canad. J. Bot. 63:1368. 1985) provisionally assigns this species to *Pseudosclerophoma* Petr.
Host: Ceanothus
Cytospora celastri Clements
Distr.: Temperate North America.
Notes: This species belongs in *Phomopsis* (Spielman. Canad. J. Bot. 63:1368. 1985).
Host: Celastrus
Cytospora cenisia Zentmyer f. ***littoralis*** Zentmyer
Distr.: CA.
Notes: Canker of Italian cypress.
Host: Cupressus
Cytospora cercidis Ellis & Everh., nom. nud.
Distr.: WV.
Host: Cercis
Cytospora chrysosperma (Pers.:Fr.) Fr.
Cytospora pulcherrima Dearn. & Hansbrough
Valsa sordida Nitschke (teleomorph)
Distr.: Circumboreal.
Notes: Mostly on *Populus* and *Salix*, but also on other hardwoods (Sutton. The Coelomycetes. p. 580. 1980).
Hosts: Acer, Betula, Malus, Picea, Populus, Prunus, Quercus, Salix, Sambucus, Sorbus
Cytospora cincta Sacc.
Leucostoma cincta (Fr.:Fr.) Höhn. (teleomorph)
Distr.: Temperate North America; Europe.
Notes: Perennial canker of peach (Dhanvantari. Canad. J. Pl. Pathol. 4:221. 1982).
Host: Prunus
Cytospora coccinea (Rebent.) Fr.
Naemaspora coccinea Rebent. (teleomorph)
Distr.: Temperate North America; Europe.
Host: Robinia
Cytospora corni Westend.
Valsa fallax Nitschke (teleomorph)
Distr.: Temperate North America; Belgium(type).
Host: Cornus
Cytospora decipiens Sacc.
Distr.: MN; Europe; Great Britain.
Host: Betula
Cytospora elaeagni Allesch.
Distr.: Temperate North America; Europe(type).
Host: Elaeagnus
Cytospora eriogoni W.B. Cooke in W.B. Cooke & C.G. Shaw
Distr.: CA.
Notes: According to Spielman (Canad. J. Bot. 63:1370. 1985), this may belong in *Chaetophoma*.
Host: Eriogonum
Cytospora euonymi Cooke
Distr.: Temperate North America; Great Britain.
Host: Euonymus
Cytospora fugax (Bull.:Fr.) Fr.
Valsa salicina (Pers.:Fr.) Fr. (teleomorph)
Distr.: United States; Europe.
Hosts: Fraxinus, Salix
Cytospora globifera Fr.
Distr.: ID; Europe(type).
Host: Physocarpus
Cytospora gordoniae Berk.
Distr.: SC(type).
Host: Gordonia
Cytospora grossulariae Laubert
Distr.: MD; Germany(type).
Host: Ribes
Cytospora halesiae Ellis & Everh.
Distr.: WV.
Notes: This species belongs in *Phomopsis* (Spielman. Canad. J. Bot. 63:1371. 1985).
Host: Halesia
Cytospora kunzei Sacc.
Leucocytospora kunzei (Sacc.) Z. Urban
Leucostoma kunzei (Fr.:Fr.) Munk (teleomorph)

Distr.: Temperate northern hemisphere.
Notes: Canker and branch die-back of conifers.
Hosts: Abies, Picea, Pinus, Pseudotsuga, Tsuga
Cytospora laxa **Berk. & M.A. Curtis**
Distr.: Eastern North America.
Host: Calycanthus
Cytospora leucosperma **(Pers.:Fr.) Fr.**
Cytospora ambiens Sacc.
Cytospora celastrina Ellis & Everh.
Cytospora celtidis Ellis & Everh.
Cytospora exasperans Ellis & Everh.
Cytospora juglandicola Ellis & Everh.
Cytospora nyssae Fairm.
Cytospora sambucicola Tehon & G.L. Stout
Cytospora sambucina Tehon & E.Y. Daniels
Cytospora sambucina Ellis & Barth.
Cytospora sassafras Ellis & Everh.
Valsa ambiens (Pers.:Fr.) Fr. (teleomorph)
Distr.: Temperate northern hemisphere.
Notes: On a variety of woody angiosperms.
Hosts: Celtis, Juglans, Malus, Populus, Prunus, Robinia, Sambucus, Sassafras, Ulmus
Cytospora leucosticta **Ellis & Barth.**
Distr.: KS(type).
Notes: This is an anamorph of *Leucostoma* or *Valsella* (Spielman. Canad. J. Bot. 63:1372. 1985).
Host: Malus
Cytospora leucostoma **Sacc.**
Leucostoma persoonii Höhn. (teleomorph)
Distr.: Temperate northern hemisphere.
Notes: On limbs of Rosaceae and other hardwoods.
Hosts: Liriodendron, Malus, Picea, Prunus, Sorbus
Cytospora microspora **(Corda) Rabenh.**
Naemaspora microspora Corda (teleomorph)
Distr.: Temperate North America; Europe.
Hosts: Crataegus, Malus, Sorbus
Cytospora minuta **Thuem.**
Distr.: Eastern United States; Europe.
Host: Fraxinus
Cytospora pallida **Ellis & Everh.**
Distr.: NJ.
Host: Quercus
Cytospora palmarum **Cooke**
Distr.: FL; Great Britain.
Hosts: Arecaceae, Cocos
Cytospora persicae **Schwein.**
Distr.: Temperate North America.
Notes: This species is an anamorph of *Leucostoma* or *Valsella* (Spielman. Canad. J. Bot. 63:1374. 1985).
Host: Prunus
Cytospora personata **Fr.**
Valsa auerswaldii Nitschke (teleomorph)
Distr.: Temperate North America; Europe.
Host: Malus
Cytospora pinastri **Fr.:Fr.**
Cytospora dubyi Sacc.
Valsa friesii (Duby) Fuckel (teleomorph)
Distr.: Circumboreal on conifers.
Hosts: Abies, Pinus, Sequoia
Cytospora platani **Fuckel**
Distr.: Temperate North America; Europe.
Host: Platanus
Cytospora pruinosa **(Fr.:Fr.) Sacc.**
Cytophoma pruinosa (Fr.:Fr.) Höhn.
Cytospora annularis Ellis & Everh.
Cytospora chionanthi Ellis & Everh.
Valsa cypri (Tul.) Tul. & C. Tul. (teleomorph)
Distr.: North America; Europe.
Notes: On hardwoods.
Host: Fraxinus
Cytospora pustulata **Berk. & M.A. Curtis**
Distr.: Temperate North America.
Host: Fagus
Cytospora rhizophorae **Kohlmeyer & E. Kohlmeyer**
Distr.: FL, HI; the Bahamas; Central America; Africa.
Notes: A halotolerant and facultative marine species (Kohlmeyer & Kohlmeyer. Mycologia 63: 847. 1971).
Host: Rhizophora
Cytospora rhuina **Fr.**
Distr.: Temperate North America.
Host: Rhus
Cytospora rubescens **Fr.**
Eutypella sorbi (J.C. Schmidt:Fr.) Sacc. (teleomorph)
Distr.: Temperate North America; Europe.
Hosts: Malus, Sorbus
Cytospora rubi **Schwein.**
Distr.: Temperate North America.
Notes: This is a species of *Phomopsis* (Spielman. Canad. J. Bot. 63:1375. 1985).
Host: Rubus
Cytospora sacchari **E.J. Butler**
Distr.: Worldwide in sugarcane-growing regions.
Notes: Sheath rot of sugarcane (Sivanesan. C.M.I. Descr. 777. 1983).
Host: Saccharum
Cytospora sacculus **(Schwein.) Gvritischvili**
Cytospora ailanthi Berk. & M.A. Curtis
Cytospora amorphae Ellis & Barth.
Cytospora caryigena Ellis & Everh.
Cytospora ceratophora Sacc.
Cytospora gleditsiae Ellis & Barth.
Cytospora grandis Peck
Cytospora maclurae Ellis & Barth.
Cytospora smilacis Cooke
Cytospora sphaerocephala (Schwein.) Sacc.
Cytospora tumulosa Ellis & Everh.
Valsa ceratosperma (Tode:Fr.) Maire (teleomorph)
Distr.: Cosmopolitan.
Hosts: Ailanthus, Amorpha, Carya, Fraxinus, Gleditsia, Hydrangea, Magnolia, Rhus, Smilax
Cytospora sassafrasicola **Tehon & E.Y. Daniels**
Distr.: IL.
Notes: This is a species of *Phomopsis* (Spielman. Canad. J. Bot. 63:1376. 1985).
Host: Sassafras
Cytospora sophorae **Bres.**
Distr.: Temperate North America; Europe.
Host: Sophora
Cytospora thujae **Sacc. & Ellis**
Distr.: Temperate North America.
Host: Thuja
Cytospora translucens **Sacc.**
Valsa translucentis Nitschke (teleomorph)
Distr.: Temperate North America; Europe.
Host: Populus
Cytospora umbrina **(Bonord.) Sacc.**
Cytosporopsis umbrina (Bonord.) Höhn.
Cytospora melanodiscus Höhn.
Cytospora truncata Sacc.
Valsa melanodiscus G. Otth (teleomorph)
Distr.: Temperate North America; Europe.
Host: Alnus
Cytospora vitis **Mont.**
Eutypella aequilinearis (Schwein.:Fr.) Starb. (teleomorph)
Distr.: Temperate North America; Europe.
Host: Vitis

Cytosporella Sacc. Coelomycetes

Cytosporella carnea **Ellis & Everh.**
Distr.: VA, WV.
Host: Castanea

Cytosporina (Sacc.) Sacc. Coelomycetes

Notes: This genus is considered a synonym of *Dumortiera* according to Sutton (Mycol. Pap. 141:59. 1977), but not all names have been redisposed.
Cytosporina ludibunda **Sacc.**
Distr.: United States; Europe.
Hosts: Prunus, Ulmus

Dacrymyces Nees:Fr. Basidiomycetes, Dacrymycetales

Notes: McNabb (New Zealand J. Bot. 11:461. 1973); Reid (Trans. Brit. Mycol. Soc. 62:449. 1974).

Dacrymyces capitatus Schwein.
Dacrymyces azaleae Schwein.
Dacrymyces concavus C.G. Lloyd in J.A. Stevenson & Cash
Dacryomitra cystidiata Brasfield
Dacryomitra depallens C.G. Lloyd
Dacrymyces digressus C.G. Lloyd
Dacrymyces ellisii Coker
Dacrymyces involutus Schwein.
Dacrymyces stipitatus (Bourd. & Galzin) Neuhoff
Dacrymyces viticola Schwein.
Distr.: Widespread.
Notes: On wood.
Hosts: Acer, Alnus, Gleditsia, Platanus, Quercus

Dacrymyces chrysospermus Berk. & M.A. Curtis
Dacrymyces flabellum Ellis & Everh.
Tremella palmata Schwein. non Schumach., hom. illeg.
Dacrymyces palmatus (Schwein.) Bres. in Höhn., nom. inval.
Distr.: Widespread.
Notes: On wood.
Hosts: Abies, Alnus, Larix, Picea, Pinus, Pseudotsuga

Dacrymyces cokeri McNabb
Dacrymyces pallidus Coker non C.G. Lloyd, hom. illeg.
Distr.: NC.
Notes: On wood.
Host: Pinus

Dacrymyces dictyosporus G.W. Martin
Distr.: North and Central America.
Notes: On wood and bark.
Host: Pinus

Dacrymyces minor Peck
Distr.: North America.
Notes: On hardwoods.
Hosts: Baccharis, Fraxinus, Malus, Ostrya, Parkinsonia, Pinus, Prosopis, Quercus, Ulmus

Dacrymyces minutus (Olive) McNabb
Guepiniopsis minuta Olive
Distr.: Southeastern United States.
Notes: On wood.
Host: Tsuga

Dacrymyces pedunculatus (Berk. & M.A. Curtis) Coker
Exidia pedunculata Berk. & M.A. Curtis in Berk.
Distr.: SC.
Notes: On wood.
Host: Pinus

Dacrymyces stillatus Nees:Fr.
Dacrymyces abietinus (Pers.) J. Schröt.
Dacrymyces deliquescens (Bull.) Duby, nom. dub.
Dacrymyces flavus C.G. Lloyd in J.A. Stevenson & Cash
Dacrymyces intermedius Olive
Dacrymyces puro-alba C.G. Lloyd
Arrhytidia pustulata Brasfield
Distr.: Temperate northern hemisphere.
Notes: On wood, usually of conifers. The names *stillatus*, *deliquescens* and *abietinus* have been used by various American authors. *Dacrymyces stillatus*, which may form arthroconidial pustules, has been confused with *D. variisporus* in the past. Reid (Trans. Brit. Mycol. Soc. 62:433-493. 1974) and Donk (Proc. Koninkl. Nederl. Akad. Wetens., Ser. C, 67:1-18. 1964) provided discussions.
Hosts: Abies, Alnus, Arctostaphylos, Calocedrus, Cercocarpus, Eucalyptus, Larix, Picea, Pinus, Populus, Prosopis, Purshia, Quercus, Tsuga

Dacrymyces tortus (Willd.:Fr.) Fr.
Dacrymyces punctiformis Neuhoff
Distr.: Temperate northern hemisphere.
Notes: On wood, mostly of conifers.
Hosts: Picea, Pinus

Dacrymyces variisporus McNabb
Distr.: Temperate northern hemisphere.
Notes: On wood, usually of conifers. This species has been called *D. stillatus* (Martin. Revision of the North Central Tremellales. p. 30. 1952) and *D. abietinus* (Kennedy. Mycologia 50:902. 1958).
Host: Pinus

Dacryobolus Fr. emend. Oberwinkler. Basidiomycetes, Aphyllophorales

Dacryobolus karstenii (Bres.) Oberwinkler ex Parmasto
Stereum karstenii Bres.
Peniophora crassa Burt in Peck
Distr.: Temperate northern hemisphere.
Notes: On wood.
Hosts: Abies, Chamaecyparis, Corylus, Picea, Pinus, Pseudotsuga, Quercus, Tsuga

Dacryobolus sudans (Albertini & Schwein.:Fr.) Fr.
Odontia sudans (Albertini & Schwein.:Fr.) Bres.
Distr.: Temperate northern hemisphere.
Notes: On wood of conifers.
Hosts: Juniperus, Larix, Pinus, [Platanus], [Populus], [Quercus]

Dacryopinax G.W. Martin. Basidiomycetes, Dacrymycetales

Dacryopinax elegans (Berk. & M.A. Curtis) G.W. Martin
Guepinia elegans Berk. & M.A. Curtis
Guepinia biformis Peck
Distr.: On wood.
Host: Acer

Dacryopinax spathularia (Schwein.:Fr.) G.W. Martin
Guepinia spathularia (Schwein.:Fr.) Fr.
Distr.: HI, North and South America; Asia.
Notes: On wood, reported on railroad ties.
Hosts: Abies, Acer, Castanea, Citrus, Fagus, Lagerstroemia, Liquidambar, Liriodendron, Malus, Picea, Pinus, Platanus, Prunus, Quercus, Taxodium, Tsuga

Dactylaria Sacc. Hyphomycetes

Notes: Bhatt & Kendrick (Canad. J. Bot. 46:1253. 1968).

Dactylaria constricta (E. Abbott) D.M. Dixon & Salkin
Scolecobasidium constrictum E. Abbott
Heterosporium terrestre Atk.
Distr.: North America; Asia; Australia; South America.
Notes: Isolated from soil and humans (Dixon & Salkin. J. Clin. Microbiol. 24:12. 1986; Dixon & Salkin. Mycotaxon 29:377. 1987).
Host: Cucumis

Dactylaria graminum (Schwein.) Sacc.
Dactylium graminum Schwein.
Distr.: Northeastern United States.
Notes: A doubtful species (De Hoog & Van Oorschot. Stud. Mycol. 26:111. 1985).
Host: Leersia

Dactylaria higginsii (Luttrell) M.B. Ellis
Pyricularia higginsii Luttrell
Distr.: Southern United States; West Indies; South America; Sudan.
Notes: On leaves (Ellis. More dematiaceous Hyphomycetes. p. 173. 1976).
Host: Cyperus

Dactylaria humicola G.C. Bhatt & Kendrick
Distr.: FL; Canada(Ontario,type).
Host: Philodendron

Dactylospora Koerb. Discomycetes, Lecanorales

Kymadiscus Kohlmeyer & E. Kohlmeyer
Notes: Hafellner (Beih. Nova Hedwigia 62:90. 1979).

Dactylospora bloxamii (Berk.) Hafellner
Patellaria bloxamii Berk.
Buellia bloxamii (Berk.) D. Hawksworth
Karschia bloxamii (Berk.) Sacc.
Distr.: North America; Central and South America; Europe.
Notes: On rotten wood.
Host: Betula

Dactylospora haliotrepha (Kohlmeyer & E. Kohlmeyer) Hafellner

Buellia haliotrepha Kohlmeyer & E. Kohlmeyer
Kymadiscus haliotrephus (Kohlmeyer & E. Kohlmeyer) Kohlmeyer & E. Kohlmeyer
Distr.: Atlantic and Pacific Oceans.
Notes: On intertidal wood and bark (Kohlmeyer & Kohlmeyer. Marine mycology. p. 395. 1979).
Hosts: Hibiscus, Rhizophora
Dactylospora imperfecta (Ellis) Hafellner
Patellaria imperfecta Ellis
Buellia imperfecta (Ellis) E. Müller
Karschia imperfecta (Ellis) Sacc.
Distr.: NJ.
Notes: On rotting corn stalks.
Host: Zea
Dactylospora stygia (Berk. & M.A. Curtis) Hafellner
Patellaria stygia Berk. & M.A. Curtis
Karschia stygia (Berk. & M.A. Curtis) Massee
Karschia elaeospora Fairm.
Karschia nigerrima (Sacc.) Sacc.
Patellaria nigerrima Sacc.
Karschia taveliana Rehm
Distr.: Widespread.
Notes: On rotten wood.
Hosts: Aleurites, Betula, Carya, Fraxinus, Liriodendron, Myrica, Populus

Daedalea Fr.:Fr. Basidiomycetes, Aphyllophorales

Daedalea quercina (L.:Fr.) Fr.
Distr.: Widespread.
Notes: On hardwoods, rarely on living trees.
Hosts: Acer, Betula, Carya, Castanea, Fagus, Juglans, Quercus

Daedaleopsis J. Schröt. Basidiomycetes, Aphyllophorales

Daedaleopsis confragosa (Bolton:Fr.) J. Schröt.
Daedalea confragosa (Bolton:Fr.) Fr.
Daedalea corrugata Klotzsch
Lenzites corrugata (Klotzsch) Weiss & M. O'Brien
Distr.: Widespread.
Notes: On wood.
Hosts: Abies, Acer, Alnus, Amelanchier, Betula, Carpinus, Carya, Castanea, Celtis, Chionanthus, Cornus, Crataegus, Fagus, Fraxinus, Juglans, Ledum, Liquidambar, Liriodendron, Malus, Nyssa, Ostrya, Pinus, Platanus, Populus, Prunus, Pseudotsuga, Quercus, Rhus, Salix, Sassafras, Tilia, Ulmus

Daldinia Ces. & De Not., nom. cons. Pyrenomycetes, Xylariales

Notes: Petrini & Müller (Mycol. Helv. 1:567. 1986).
Daldinia concentrica (Bolton:Fr.) Ces. & De Not.
Hypoxylon concentricum (Bolton:Fr.) Grev.
Nodulisporium sp. (anamorph)
Distr.: Temperate regions.
Notes: On hardwoods.
Hosts: Acer, Ailanthus, Alnus, Betula, Carpinus, Citrus, Diospyros, Fagus, Fraxinus, Hamamelis, Liquidambar, Maclura, Magnolia, Malus, Persea, Poaceae, Quercus, Salix, Tilia, Ulmus, Umbellularia, Vitis
Daldinia eschscholzii (Ehrenb.) Rehm
Nodulisporium sp. (anamorph)
Distr.: Cosmopolitan.
Host: Platanus
Daldinia grandis Child
Distr.: Central and western North America; South America; Australia; New Zealand.
Notes: Originally spelled "*grande*". On various hardwoods.
Hosts: Acer, Salix, Tilia
Daldinia loculata (Lév.) Sacc.
Nodulisporium sp. (anamorph)
Distr.: North America; Asia; southwestern Europe.
Notes: On hardwoods.
Hosts: Carpinus, Ostrya
Daldinia occidentalis Child
Nodulisporium sp. (anamorph)
Distr.: Temperate North America; Europe.
Notes: On hardwoods.
Hosts: Alnus, Betula, Malus, Populus, Salix, Tilia
Daldinia vernicosa (Schwein.) Ces. & De Not.
Nodulisporium sp. (anamorph)
Distr.: North and South America; Europe.
Hosts: Acer, Alnus, Betula, Cornus, Diospyros, Fagus, Fraxinus, Ilex, Liriodendron, Magnolia, Malus, Nyssa, Populus, Quercus, Robinia, Sassafras, Vitis

Darkera H.S. Whitney, J. Reid, & Pirozynski. Discomycetes, Rhytismatales

Darkera abietis H.S. Whitney, J. Reid, & Pirozynski
Tiarosporella abietis H.S. Whitney, J. Reid, & Pirozynski (anamorph)
Distr.: Northern North America.
Notes: Whitney et al. (Canad. J. Bot. 53:3052. 1975).
Host: Abies

Dasyscyphella Tranzschel. Discomycetes, Helotiales

Dasyscyphella crystallina (Fuckel) Raitviir
Dasyscyphus crystallinus (Fuckel) Sacc.
Distr.: North America; Europe.
Host: Quercus
Dasyscyphella dryinus (P. Karst.) Raitviir
Dasyscyphus dryinus (P. Karst.) Sacc.
Helotium dryinum (P. Karst.) P. Karst.
Lachnella dryinum (P. Karst.) P. Karst.
Distr.: WA; Finland.
Notes: Dennis (Mycol. Pap. 62:186. 1956) redescribed this species.
Host: Alnus

Dasyscyphus S.F. Gray. Discomycetes, Helotiales

Dasyscyphus acerinus (Cooke & Ellis) Cash
Trichopeziza acerina (Cooke & Ellis) Sacc.
Distr.: NJ(type), eastern United States.
Notes: Seaver (North Amer. cup fungi. [Inoperculates]. p. 273. 1951) considered this a doubtful species.
Host: Acer
Dasyscyphus albolabra (Ellis & Everh.) Dennis
Lachnella albolabra Ellis & Everh.
Distr.: CO.
Notes: Dennis (Kew Bull. 17:322. 1963).
Host: Ribes
Dasyscyphus albopileatus (Cooke) Sacc.
Distr.: GA.
Notes: Seaver (North Amer. cup fungi. [Inoperculates]. p. 273. 1951) stated that this species suggests *Lachnella virginea*, now *Lachnum virgineum*.
Host: Magnolia
Dasyscyphus aplopappi Cash
Distr.: CA(type).
Host: Haplopappus
Dasyscyphus atropurpureus (E.J. Durand) Dennis
Lachnum atropurpureum E.J. Durand
Lachnella atropurpurea (E.J. Durand) Seaver
Distr.: CA.
Notes: On bark (Dennis. Kew Bull. 17:328. 1963).
Host: Eucalyptus
Dasyscyphus bicolor (Bull.:Fr.) Fuckel var. ***indicus*** E. Müller & Dennis
Distr.: CA; India.
Notes: The type variety has been transferred to *Capitotricha*.
Host: Rubus
Dasyscyphus cerinus (Pers.:Fr.) Fuckel
Belonidium cerinum (Pers.:Fr.) Raitviir

Distr.: Temperate regions.
Hosts: Ceanothus, Heteromeles
Dasyscyphus citrinoalbus (Penz. & Sacc.) Cash
Trichopeziza citrinoalba Penz. & Sacc.
Distr.: HI; Java(type).
Hosts: Metrosideros, Rhus
Dasyscyphus corticalis (Pers.:Fr.) Massee
Lachnum corticale (Pers.:Fr.) Nannf.
Lachnella lonicerae (Albertini & Schwein.:Fr.) Fuckel
Lachnella rhizophila Ellis & Everh.
Distr.: CA, WA; Europe.
Hosts: Baccharis, Physocarpus, Rhus, Rubus
Dasyscyphus distinctus (Peck) Cash
Distr.: GA.
Hosts: Andropogon, Erianthus
Dasyscyphus echinophilus Cash
Distr.: CA(type).
Host: Castanopsis
Dasyscyphus ellisianus (Rehm) Sacc.
Lachnella ellisiana (Rehm) Seaver
Distr.: Central and eastern United States.
Notes: On bark of various conifers.
Hosts: Larix, Picea, Pinus, Pseudotsuga
Dasyscyphus hystriculus (Ellis & Everh.) Sacc.
Lachnella hystricula Ellis & Everh.
Distr.: LA(type).
Host: Magnolia
Dasyscyphus ivae Rehm
Lachnella ivae (Rehm) Seaver
Distr.: ND(type).
Host: Iva
Dasyscyphus labrosus W. Phillips & Harkn.
Distr.: CA(type).
Host: Arctostaphylos
Dasyscyphus oblongosporus Hahn & Ayers
Lachnella oblongospora (Hahn & Ayers) Seaver
Distr.: Central and northeastern United States.
Notes: On various conifers.
Hosts: Larix, Picea, Pinus, Pseudotsuga
Dasyscyphus oncospermatis (Berk. & Broome) Sacc.
Dasyscyphus javanicus Penz. & Sacc.
Distr.: HI; Australia; Java; The Philippines; Sri Lanka.
Notes: On large ferns of the Cyatheaceae (Haines. Mycotaxon 11:212. 1980).
Host: Cibotium
Dasyscyphus patulus (Pers.:Fr.) Sacc.
Lachnella patula (Pers.:Fr.) W. Phillips
Distr.: ID, PA; Europe.
Notes: Dennis (Mycol. Pap. 32:37. 1949) provided a description and suggested that this species is a foliicolous form of *D. bicolor*.
Hosts: Holodiscus, Philadelphus, Physocarpus
Dasyscyphus phlogis Cash
Distr.: CA.
Host: Phlox
Dasyscyphus populicolus (Seaver) McKnight
Lachnella populicola Seaver
Distr.: Western United States.
Notes: Haines & McKnight (Mycotaxon 5:425. 1977).
Host: Populus
Dasyscyphus pteridis (Albertini & Schwein.:Fr.) Massee
Urceolella pteridis (Albertini & Schwein.:Fr.) Boud.
Lachnella frondicola (Ellis & Everh.) Sacc.
Pirottaea frondicola (Ellis & Everh.) Sacc.
Distr.: North America; Europe.
Notes: Dennis (Kew Bull. 17:345. 1963) included a description.
Hosts: Onoclea, Osmunda
Dasyscyphus puberulus Sacc.
Peziza puberula Berk. & M.A. Curtis non Lasch, hom. illeg.
Distr.: SC.
Notes: On fallen leaves (Dennis. Kew Bull. 17:360. 1963). The status of this name is uncertain.
Host: Fraxinus
Dasyscyphus salmoneus Cash
Distr.: CA.
Host: Haplopappus
Dasyscyphus scabrovillosus (W. Phillips) Sacc.
Distr.: Western United States.
Hosts: Alnus, Rubus
Dasyscyphus serinellus (Quél.) Sacc.
Erinella serinella Quél.
Distr.: CA, OR; France(type).
Hosts: Abies, Pinus, Tsuga
Dasyscyphus setigerus (W. Phillips) Dennis
Lachnum setigerus (W. Phillips) Rehm
Trichopeziza setigera (W. Phillips) Sacc.
Distr.: CA.
Notes: On dead stems (Dennis. Kew Bull. 17:364. 1963).
Host: Aralia
Dasyscyphus solenia (Peck) Dennis
Lachnella solenia (Peck) Seaver
Solenopezia solenia (Peck) Sacc.
Distr.: NY.
Notes: On dead stems (Dennis. Kew Bull. 17:364. 1963).
Host: Eupatorium
Dasyscyphus spiraeicolus (P. Karst.) Sacc.
Lachnum spiraeicola (P. Karst.) Rehm
Distr.: MI, WA; Europe(type).
Host: Urtica
Dasyscyphus subauratus (Ellis) Dennis
Lachnella albopileata (Cooke) W. Phillips var. *subaurata* Ellis in W. Phillips
Distr.: NJ.
Notes: On leaves (Dennis. Kew Bull. 17:323. 1963).
Host: Clethra
Dasyscyphus ulei (G. Wint.) Sacc.
Distr.: HI, PR; Brazil; Peru.
Notes: Haines (Mycotaxon 11:191. 1980).
Hosts: Dicranopteris, Gleichenia, Hicriopteris
Dasyscyphus venturioides (Ellis & Everh.) Dennis
Lachnella venturioides Ellis & Everh.
Lachnum venturioides (Sacc. & Romell) Nannf.
Pirottaea venturioides Sacc. & Romell
Trichopeziza venturioides (Ellis & Everh.) Sacc.
Distr.: NJ.
Notes: On fallen leaves (Dennis. Kew Bull. 17:373. 1963).
Host: Gaylussacia

Datronia Donk. Basidiomycetes, Aphyllophorales

Datronia mollis (Sommerf.:Fr.) Donk
Trametes mollis (Sommerf.:Fr.) Fr.
Trametes serpens Fr.
Distr.: Temperate northern hemisphere.
Notes: On hardwoods.
Hosts: Acer, Alnus, Betula, Carpinus, Carya, Fagus, Fraxinus, Ostrya, Populus, Quercus, Salix, Sambucus, Thuja, Tilia
Datronia scutellata (Schwein.) R.L. Gilbertson & Ryvarden
Fomes scutellatus (Schwein.) Cooke
Fomitopsis scutellata (Schwein.) Bondartsev & Singer
Distr.: Widespread.
Notes: On wood.
Hosts: Acer, Alnus, Amelanchier, Castanea, Cornus, Hamamelis, Platanus, Quercus, Ulmus
Datronia stereoides (Fr.:Fr.) Ryvarden
Polyporus stereoides Fr.:Fr.
Antrodia stereoides (Fr.:Fr.) Bondartsev & Singer
Datronia epilobii (P. Karst.) Donk
Trametes epilobii P. Karst.
Coriolus planellus Murrill
Polyporus planellus (Murrill) Overh.
Polyporus planus Peck non Wallr., hom. illeg.
Distr.: Northern hemisphere.
Notes: On hardwoods, especially *Populus* and *Betula* spp., rarely on conifers.
Hosts: Acer, Alnus, Betula, Corylus, Populus, Ribes, Salix, Sambucus

Davisiella Petr. Coelomycetes

Davisiella elymina **(J.J. Davis) Petr.**
Cytodiplospora elymina J.J. Davis
Distr.: WI.
Notes: Associated with ascomata of *Phyllachora* spp.
Hosts: Andropogon, Bouteloua, Elymus, Hystrix, Muhlenbergia, Oryzopsis

Davisomycella Darker. Discomycetes, Rhytismatales

Notes: Darker (Contr. Arnold Arbor. 1:40. 1932).
Davisomycella ampla **(J.J. Davis) Darker**
Hypodermella ampla (J.J. Davis) Darker
Lophodermium ampla (J.J. Davis) J.J. Davis
Distr.: Northern and western North America.
Notes: Needle cast of 2-3 needle pines (Minter & Gibson. C.M.I. Descr. 561. 1978).
Host: Pinus
Davisomycella lacrimiformis **(Darker) Darker**
Hypodermella lacrimiformis Darker
Distr.: CA, OR.
Notes: Pathogenic on pine needles.
Host: Pinus
Davisomycella limitata **(Darker) Darker**
Hypodermella limitata Darker
Distr.: CA.
Notes: On needles.
Host: Pinus
Davisomycella medusa **(Darker) Darker**
Hypodermella medusa Darker
Distr.: CA, CO.
Notes: On needles.
Host: Pinus
Davisomycella montana **(Darker) Darker**
Hypodermella montana Darker
Distr.: CA, ID, OR.
Notes: On needles.
Host: Pinus
Davisomycella ponderosae **(Staley) Dubin**
Lophodermium ponderosae Staley
Distr.: CO.
Notes: Needle cast of ponderosa pine (Staley. Mycologia 56:757. 1964; Dubin & Staley. Mycologia 61:396. 1969).
Host: Pinus

Deightoniella S.J. Hughes. Hyphomycetes

Deightoniella argemonensis **Pollack & Matthews**
Distr.: FL.
Notes: On living leaves (Pollack & Matthews. Mycologia 68:1095. 1976).
Host: Argemone
Deightoniella arundinacea **(Corda) S.J. Hughes**
Helminthosporium arundinaceum Corda
Napicladium arundinaceum (Corda) Sacc.
Distr.: North America; Asia; Australia; Europe.
Host: Phragmites
Deightoniella torulosa **(Syd.) M.B. Ellis**
Brachysporium torulosum Syd.
Helminthosporium torulosum (Syd.) S. Ashby
Distr.: Subtropical and tropical regions.
Notes: Black-tip and leaf spot (Subramanian. C.M.I. Descr. 165. 1968).
Hosts: Musa, Ravenala

Delortia Pat. & Gaillard. Hyphomycetes

Delortia palmicola **Pat.**
Distr.: Widespread in tropical regions.
Notes: On sodden, decaying palm litter (Goos. Mycologia 72:596. 1980).
Host: Arecaceae

Delphinella (Sacc.) Kuntze. Loculoascomycetes, Dothideales

Delphinella abietis **(Rostr.) E. Müller in E. Müller & Arx**
Mycosphaerella abietis (Rostr.) Lindau
Rehmiellopsis abietis (Rostr.) O. Rostr.
Rehmiellopsis bohemica Bubák & Kab.
Dothiorella sp. (anamorph)
Distr.: Circumboreal.
Notes: Parasitic on shoots and needles.
Host: Abies
Delphinella balsameae **(A.M. Waterman) E. Müller in E. Müller & Arx**
Rehmiellopsis balsameae A.M. Waterman
Distr.: Northeastern North America.
Notes: On living needles.
Host: Abies
Delphinella peckii **(Lindau) Barr**
Mycosphaerella peckii Lindau in Engl. & Prantl
Delphinella tsugae (House) Barr
Mycosphaerella tsugae (House) House
Distr.: Eastern North America.
Notes: On cone scales.
Host: Tsuga

Dendrina Fr. Hyphomycetes

Dendrina diospyri **Berk. & M.A. Curtis**
Distr.: DE.
Notes: The place of publication of this name could not be found.
Host: Diospyros

Dendrocorticium M. Larsen & R.L. Gilbertson. Basidiomycetes, Aphyllophorales

Notes: Larsen & Gilbertson (Norw. J. Bot. 24:110. 1977).
Dendrocorticium piceinum **P.A. Lemke in M. Larsen & R.L. Gilbertson**
Distr.: SC; Canada(Ontario).
Notes: On wood of conifers.
Host: Taxodium
Dendrocorticium roseocarneum **(Schwein.) M. Larsen & R.L. Gilbertson**
Laeticorticium roseocarneum (Schwein.) Boidin
Laxitextum roseocarneum (Schwein.) Lentz
Stereum roseocarneum (Schwein.) Fr.
Corticium lilacino-fuscum Berk. & M.A. Curtis
Stereum lilacino-fuscum (Berk. & M.A. Curtis) C.G. Lloyd
Distr.: Eastern North America; Brazil; Japan.
Notes: On limbs, usually of hardwoods.
Hosts: Acer, Alnus, Crataegus, Liriodendron, Malus, Nyssa, Prunus, Quercus

Dendrodochium Bonord. Hyphomycetes

Notes: The generic concept is heterogeneous and in need of clarification.
Dendrodochium affine **Sacc.**
Distr.: DE; Europe(type).
Host: Euphorbia
Dendrodochium berolinense **Wollenweb.**
Thyronectria berolinensis (Sacc.) Seaver (teleomorph)
Distr.: Widespread in Europe and North America.
Notes: On dead branches of *Ribes*, also causes limited stem canker of cultivated currants (Seeler. J. Arnold Arboretum 21:445. 1940).
Host: Ribes
Dendrodochium densipes **Sacc. & Ellis**
Distr.: DE.
Host: Salix
Dendrodochium nigrescens **Ellis & Everh.**
Distr.: MT(type).
Host: Acer
Dendrodochium pezizoides **Ellis & Everh. in Sumstine, nom. nud.**
Distr.: DE.

Host: Triticum
Dendrodochium rubellum Sacc.
Distr.: DE; Europe(type).
Host: Platanus
Dendrodochium sepultum Ellis & Everh.
Distr.: KS.
Host: Ulmus

Dendrophoma Sacc. Coelomycetes

Notes: This genus is considered a synonym of *Dinemasporium*, but not all names have been redisposed (Sutton. Trans. Brit. Mycol. Soc. 48:611. 1965).
Dendrophoma albomaculans (Schwein.) Starb.
Sphaeria albomaculans Schwein.
Distr.: Eastern United States.
Host: Syringa
Dendrophoma azaleae Dearn. & House
Distr.: NY.
Host: Rhododendron
Dendrophoma caespitosa Sacc.
Distr.: WA; Soviet Union(type).
Host: Salix
Dendrophoma cephalanthi Peck
Tympanis cephalanthi Dearn. & House (teleomorph)
Distr.: NY.
Host: Cephalanthus
Dendrophoma clypeata D. Sacc.
Distr.: FL; Europe.
Host: Cycas
Dendrophoma gouldiae F. Stevens & Plunkett
Distr.: HI.
Host: Gouldia
Dendrophoma hispalensis Gonz. Frag.
Distr.: CA; Spain.
Host: Eucalyptus
Dendrophoma longipes (Berk. & M.A. Curtis) Grove
Phoma longipes Berk. & M.A. Curtis
Phomopsis longipes (Berk. & M.A. Curtis) Camara
Distr.: Eastern United States.
Host: Morus
Dendrophoma marconii Cavara
Distr.: MD, VA; Italy.
Host: Cannabis
Dendrophoma nigrescens Fairm.
Distr.: NY.
Hosts: Ilex, Rhus
Dendrophoma phyllogena Sacc.
Distr.: NY.
Host: Chamaedaphne
Dendrophoma pruinosa (Fr.) Sacc.
Sphaeria pruinosa Fr.
Distr.: Central and eastern United States; Europe.
Host: Fraxinus
Dendrophoma rhododendri (Roum.) Sacc.
Cytospora rhododendri Roum.
Distr.: WA; Europe.
Hosts: Arbutus, Rhododendron
Dendrophoma syringae Dearn.
Distr.: NY.
Host: Syringa
Dendrophoma tiliae Peck
Distr.: NY.
Host: Tilia
Dendrophoma zeae Tehon
Distr.: IL.
Host: Zea

Dendrophora (Parmasto) Chamuris. Basidiomycetes, Aphyllophorales

Peniophora Cooke
Notes: Chamuris (Mycol. Mem. 14. 1988).
Dendrophora albobadia (Schwein.:Fr.) Chamuris
Peniophora albobadia (Schwein.:Fr.) Boidin
Stereum albobadium (Schwein.:Fr.) Fr.
Peniophora albomarginata (Schwein.) Massee
Distr.: Central and eastern North America; South America.
Notes: On limbs, usually of hardwoods.
Hosts: Acer, Albizia, Alnus, Baccharis, Brassica, Carpinus, Carya, Castanea, Catalpa, Celtis, Cercis, Condalia, Diospyros, Forsythia, Fortunella, Fouquieria, Fraxinus, Haplopappus, Hibiscus, Juglans, Juniperus, Ligustrum, Liriodendron, Lycium, Maclura, Magnolia, Malus, Ostrya, Philadelphus, Platanus, Prosopis, Prunus, Ptelea, Quercus, Rhus, Sambucus, Simmondsia, Syringa, Umbellularia, Viburnum, Vitis
Dendrophora erumpens (Burt) Chamuris
Stereum erumpens Burt
Peniophora erumpens (Burt) Boidin
Distr.: North America.
Notes: On trunks and branches of hardwoods.
Hosts: Acer, Alnus, Betula, Carpinus, Castanea, Catalpa, Malus, Prunus, Salix, Sassafras, Vitis
Dendrophora versiformis (Berk. & M.A. Curtis) Chamuris
Stereum versiforme Berk. & M.A. Curtis
Peniophora versiformis (Berk. & M.A. Curtis) Boidin
Distr.: Widespread.
Notes: On branches, most often on hardwoods.
Hosts: Acer, Alnus, Amelanchier, Betula, Carpinus, Carya, Castanea, Cornus, Crataegus, Diospyros, Fraxinus, Hamamelis, Liriodendron, Malus, Myrica, Populus, Prunus, Quercus, Rhus, Salix, Ulmus

Dendropleella Munk. Loculoascomycetes, Dothideales

Notes: Munk (Dansk Bot. Ark. 17:392. 1957).
Dendropleella hirta (Fr.) Munk
Massaria hirta (Fr.) Fuckel
Distr.: CO; Europe.
Host: Sambucus

Dendrosporium Plakidas & Edgerton ex J.L. Crane. Hyphomycetes

Dendrosporium lobatum Plakidas & Edgerton ex J.L. Crane
Distr.: Central United States; India
Notes: Crane (Trans. Brit. Mycol. Soc. 58:423. 1972).
Hosts: Juglans, Pyrus

Dendrothele Höhn. & Litsch. Basidiomycetes, Aphyllophorales

Aleurocorticium P.A. Lemke
Notes: Lempke (Canad. J. Bot. 42:723. 1964).
Dendrothele acerina (Pers.:Fr.) P.A. Lemke
Aleurocorticium acerinum (Pers.:Fr.) P.A. Lemke
Aleurodiscus acerinus (Pers.:Fr.) Höhn. & Litsch.
Stereum acerinum (Pers.:Fr.) Fr.
Distr.: Widespread.
Notes: On bark of hardwoods, usually of living trees.
Hosts: Acer, Castanea, Fraxinus, Mortonia, Morus, Quercus, Tilia
Dendrothele alliacea (Quél.) P.A. Lemke
Aleurocorticium alliaceum (Quél.) P.A. Lemke
Aleurodiscus alliaceus (Quél.) Boidin
Distr.: Widespread.
Notes: On bark of hardwoods, usually of living trees.
Hosts: Acer, Carya, Quercus, Tilia
Dendrothele candida (Schwein.:Fr.) P.A. Lemke
Acanthophysium candidum (Schwein.:Fr.) G. Cunn.
Aleurocorticium candidum (Schwein.:Fr.) P.A. Lemke
Aleurodiscus candidus (Schwein.:Fr.) Burt
Stereum candidum (Schwein.:Fr.) Fr.
Aleurodiscus candidus (Schwein.:Fr.) Burt var. *sphaerosporus* Coker
Distr.: Widespread.
Notes: On bark of living and dead hardwoods.
Hosts: Acer, Carya, Castanea, Fraxinus, Lithocarpus, Quercus, Ulmus
Dendrothele dryina (Pers.) P.A. Lemke
Aleurocorticium dryinum (Pers.) P.A. Lemke

Aleurodiscus dryinus (Pers.) Bourd.
Distr.: Eastern North America; Europe.
Notes: On bark of living trees, especially *Acer* spp.
Hosts: Acer, Arbutus
Dendrothele griseo-cana (Bres.) Bourd. & Galzin
Aleurocorticium griseo-canum (Bres.) P.A. Lemke
Aleurodiscus griseo-canus (Bres.) Höhn. & Litsch.
Distr.: Eastern North America; Mexico; Europe.
Notes: On bark and wood.
Hosts: Acer, Carya, Fraxinus, Juniperus, Ligustrum, Liriodendron, Lonicera, Ostrya, Platanus, Quercus, Salix, Thuja, Tilia, Ulmus
Dendrothele incrustans (P.A. Lemke) P.A. Lemke
Aleurocorticium incrustans P.A. Lemke
Distr.: Pacific Northwest.
Notes: On twigs.
Hosts: Ceanothus, Holodiscus, Juniperus, Picea, Pseudotsuga, Ribes, Spiraea, Tsuga
Dendrothele itihummensis R.L. Gilbertson & Blackwell
Distr.: LA, MS.
Notes: On bark. Gilbertson & Blackwell (Mycotaxon 24:331. 1985).
Host: Juniperus
Dendrothele macrodens (Coker) P.A. Lemke
Aleurodiscus macrodens Coker
Aleurocorticium macrodens (Coker) P.A. Lemke
Distr.: North America.
Notes: On bark of living *Fraxinus* spp., rarely on other hardwoods.
Hosts: Fraxinus, Nyssa, Quercus, Salix
Dendrothele maculata (H. Jacks. & P.A. Lemke) P.A. Lemke
Aleurocorticium maculatum H. Jacks. & P.A. Lemke in P.A. Lemke
Distr.: Northeastern North America.
Notes: On hardwoods.
Host: Prunus
Dendrothele microspora (H. Jacks. & P.A. Lemke) P.A. Lemke
Aleurocorticium microsporum H. Jacks. & P.A. Lemke in P.A. Lemke
Distr.: North America.
Notes: On bark of hardwoods, usually of living trees.
Hosts: Acer, Carya, Fraxinus, Liquidambar, Nyssa, Platanus, Populus, Quercus, Salix, [Taxodium], Tilia, Ulmus
Dendrothele nivosa (Berk. & M.A. Curtis ex Höhn. & Litsch.) P.A. Lemke
Acanthophysium nivosum (Berk. & M.A. Curtis ex Höhn. & Litsch.) G. Cunn.
Aleurocorticium nivosum (Berk. & M.A. Curtis ex Höhn. & Litsch.) P.A. Lemke
Aleurodiscus nivosus (Berk. & M.A. Curtis ex Höhn. & Litsch.) Höhn. & Litsch.
Distr.: Widespread.
Notes: On bark.
Hosts: Chamaecyparis, Juniperus, Thuja
Dendrothele pachysterigmata (H. Jacks. & P.A. Lemke) P.A. Lemke
Aleurocorticium pachysterigmatum H. Jacks. & P.A. Lemke in P.A. Lemke
Distr.: Eastern North America; Mexico.
Notes: On bark.
Hosts: Carya, Juniperus, Thuja
Dendrothele seriata (Berk. & M.A. Curtis) P.A. Lemke
Aleurocorticium seriatum (Berk. & M.A. Curtis) P.A. Lemke
Aleurodiscus seriatus (Berk. & M.A. Curtis) Burt
Distr.: Subtropical and tropical America.
Notes: On bark of hardwoods.
Host: Fraxinus
Dendrothele strumosa (Fr.) P.A. Lemke
Aleurocorticium strumosum (Fr.) P.A. Lemke
Aleurodiscus strumosus (Fr.) Burt
Distr.: Southeastern United States; subtropical and tropical America.
Notes: On bark of living hardwoods.
Hosts: Carya, Fraxinus, Liquidambar, Ostrya, Quercus
Dendrothele subfusispora Burdsall & Nakasone
Distr.: FL(type).
Notes: On wood.
Host: Ostrya

Dendryphiella Bubák & Ranoj. Hyphomycetes

Dendryphiella arenaria J. Nicot
Scolecobasidium arenaria (J. Nicot) M.B. Ellis
Distr.: Worldwide in marine habitats of the Atlantic Ocean.
Notes: On decaying marine or estuarine plants (Kohlmeyer & Kohlmeyer. Marine mycology. p. 488. 1979).
Host: Spartina
Dendryphiella infuscans (Thuem.) M.B. Ellis
Cladosporium infuscans Thuem.
Distr.: Temperate North America; Europe.
Notes: On dead herbaceous stems.
Hosts: Desmodium, Eichhornia
Dendryphiella salina (G.K. Sutherland) Pugh & J. Nicot
Cercospora salina G.K. Sutherland
Scolecobasidium salinum (G.K. Sutherland) M.B. Ellis
Distr.: Worldwide in marine habitats of all oceans except the Pacific Ocean.
Notes: Saprobic on decaying marine or estuarine plants (Kohlmeyer & Kohlmeyer. Marine mycology. p. 488. 1979).
Hosts: Juncus, Spartina, Zostera
Dendryphiella vinosum (Berk. & M.A. Curtis) O. Reisinger
Helminthosporium vinosum Berk. & M.A. Curtis
Dendryphion vinosum (Berk. & M.A. Curtis) S.J. Hughes
Helminthosporium fuscum Fuckel
Dendryphiella interseminata (Berk. & Ravenel) Bubák & Ranoj.
Dendryphion interseminatum (Berk. & Ravenel) S.J. Hughes
Helminthosporium interseminatum Berk. & Ravenel
Dendryphion nodulosum Sacc.
Helminthosporium urticae Peck
Distr.: Cosmopolitan.
Notes: On dead herbaceous material.
Hosts: Cicuta, Collinsonia, Phoenix, Phytolacca, Urtica

Dendryphion Wallr. Hyphomycetes

Dendryphion nanum (Nees) S.J. Hughes
Helminthosporium nanum Nees
Distr.: North America; temperate Europe.
Notes: Saprophytic on herbaceous stems, roots, tree stumps (Hughes & Cooke. Fungi Canadenses 147. 1979).
Host: Lesquerella
Dendryphion nubilosum Ellis & Everh.
Distr.: Rocky Mountains.
Host: Astragalus

Dendryphiopsis S.J. Hughes. Hyphomycetes

Dendryphiopsis atra (Corda) S.J. Hughes
Dendryphion atrum Corda
Kirschsteiniothelia aethiops (Berk. & M.A. Curtis) D. Hawksworth (teleomorph)
Distr.: Temperate regions.
Notes: On dead wood (Hughes. New Zealand J. Bot. 16:360. 1978).
Host: Salix

Dennisiodiscus Svrček. Discomycetes, Helotiales

Notes: This genus was proposed by Svrček (Ceska Mykol. 30:9. 1976) for species of *Trichodiscus*.
Dennisiodiscus crossotus (Ellis) Svrček
Trichodiscus crossotus (Ellis) Dennis
Trichopeziza crossota (Ellis) Sacc.
Distr.: Western United States.
Notes: On weathered wood (Dennis. Kew Bull. 17:341. 1963).
Host: Arctostaphylos

Denticularia Deighton. Hyphomycetes

Notes: De Hoog (Persoonia 10:51. 1978).
Denticularia tertia (Chupp & Greene) Deighton
Cercospora tertia Chupp & Greene
Distr.: Central North America.
Notes: Sometimes misdetermined as *Cercosporella virgaureae* (Thüm.)

Allesch.
Host: Aster

Dentipellis Donk. Basidiomycetes, Aphyllophorales

Dentipellis fragilis (Pers.:Fr.) Donk
Distr.: Temperate northern hemisphere.
Notes: On wood. According to Jülich & Stalpers (The resupinate non-poroid Aphyllophorales of the temperate northern hemisphere. p. 95. 1980), *Hydnum macrodon* Pers.:Fr. (*Mycoacia macrodon* (Pers.:Fr.) L. Miller & Boyle) is a possible synonym. If this is the case, then *H. macrodon* would be the correct basionym because it is an older name.
Host: Populus

Dentocorticium M. Larsen & R.L. Gilbertson. Basidiomycetes, Aphyllophorales

Notes: Larsen & Gilbertson (Norw. J. Bot. 24:117. 1977).
Dentocorticium sulphurellum (Peck) M. Larsen & R.L. Gilbertson
Hydnum sulphurellum Peck
Grandinia sulphurellum (Peck) Burt
Laeticorticium sulphurellum (Peck) R.L. Gilbertson
Distr.: Central and eastern North America.
Notes: On wood.
Hosts: Acer, Liriodendron, Tilia

Depazea Fr. Coelomycetes

Notes: This genus is considered a synonym of *Asteroma* according to Sutton (Mycol. Pap. 141:63. 1977), but not all names have been redisposed.
Depazea carpinea (Schwein.) Fr.
Xyloma carpineum Schwein.
Distr.: NC.
Host: Carpinus

Dermatella P. Karst. Discomycetes, Helotiales

Notes: This genus is considered a synonym of *Dermea*, but not all names have been redisposed.
Dermatella cucurbitaria Cooke
Distr.: NJ(type).
Host: Nyssa
Dermatella populina Petr.
Distr.: ID(type).
Host: Populus

Dermateopsis Nannf. Discomycetes, Helotiales

Dermateopsis tabacina (Cooke) Nannf.
Dermatea tabacina Cooke
Godronia tabacina (Cooke) Seaver
Distr.: NJ.
Host: Quercus

Dermea Fr. Discomycetes, Helotiales

Notes: Groves (Mycologia 38:351. 1946).
Dermea acerina (Peck) Rehm
Tympanis acerina Peck
Sphaeronema acerinum Peck (anamorph)
Distr.: Eastern North America.
Notes: Groves (Mycologia 38:409. 1946).
Host: Acer
Dermea ariae (Pers.:Fr.) Tul. ex P. Karst.
Distr.: Northeastern North America; Europe.
Notes: On limbs.
Host: Sorbus
Dermea balsamea (Peck) Seaver in B.O. Dodge
Cenangium balsameum Peck
Distr.: Boreal and temperate regions of North America.
Notes: Weakly parasitic on branches.
Hosts: Abies, Tsuga
Dermea bicolor (Ellis) Groves
Tympanis bicolor Ellis
Dermatea rosae Rehm
Pezicula rosae Sacc.
Distr.: North America; Europe.
Notes: On limbs.
Hosts: Amelanchier, Rosa
Dermea cerasi (Pers.:Fr.) Fr.
Cenangium cerasi (Pers.:Fr.) Fr.
Dermatea cerasi (Pers.:Fr.) Fr.
Foveostroma drupacearum (Lév.) DiCosmo (anamorph)
Distr.: Temperate northern hemisphere.
Notes: On limbs and bark.
Host: Prunus
Dermea chionanthi Ellis & Everh.
Distr.: DE, MD.
Host: Chionanthus
Dermea crypta Cooke
Scleroderris crypta (Cooke) Sacc.
Distr.: GA, NJ(type).
Notes: Groves (Mycologia 38:415. 1946) stated that this species belongs in *Pezicula*.
Hosts: Clethra, Rhododendron
Dermea hamamelidis (Peck) Groves
Patellaria hamamelidis Peck
Distr.: Eastern North America.
Host: Hamamelis
Dermea libocedri Groves
Distr.: CA.
Host: Calocedrus
Dermea molliuscula (Schwein.) Cash
Gelatinosporium fulvum Peck (anamorph)
Distr.: Northeastern North America.
Notes: On limbs.
Host: Betula
Dermea mori Peck
Distr.: KS(type).
Host: Morus
Dermea olivacea Ellis non G. Otth, hom. illeg.
Distr.: NJ.
Notes: Probably a *Pezicula* (Groves. Mycologia 38:382. 1946).
Host: Ilex
Dermea padi (Albertini & Schwein.) Fr.
Micropera padina Sacc. (anamorph)
Distr.: Northeastern United States; Europe.
Host: Prunus
Dermea peckiana (Rehm) Groves
Cenangium peckianum Rehm
Micropera stellata (Ellis) Jacz. (anamorph)
Distr.: Northeastern United States.
Notes: Groves (Mycologia 38:380. 1946).
Hosts: Ilex, Nemopanthus
Dermea piceae (Pers.) Rehm
Distr.: ID; Europe.
Notes: According to Groves (Mycologia 38:421. 1946) this species does not belong in *Dermea* but a correct generic disposition was not determined.
Host: Abies
Dermea prunastri (Pers.:Fr.) Fr.
Cenangium prunastri (Pers.:Fr.) Fr.
Micropera spuria (Fr.) Höhn. (anamorph)
Distr.: Temperate regions.
Host: Prunus
Dermea pseudotsugae Funk
Foveostroma boycei (Dearn.) Funk (anamorph)
Distr.: Pacific Northwest.
Notes: Bark and phloem necrosis, cankers, or die-back on young conifers.
Host: Pseudotsuga
Dermea spiraeae Schwein.
Distr.: NC, PA.

Notes: According to Groves (Mycologia 38:424. 1946) the type specimen of this species is too scanty to determine its identity and this name should be disregarded.
Host: Physocarpus
Dermea xanthoxyli Peck
Distr.: NY(type).
Notes: According to Groves (Mycologia 38:425. 1946) the type specimen of this species has no *Dermea*-like fungus and the name should be disregarded.
Host: Zanthoxylum

Desmazierella Lib. Discomycetes, Pezizales
Desmazierella acicola Lib.
Verticillium trifidum G. Preuss (anamorph)
Distr.: Northwestern North America; China.
Notes: On fallen needles of conifers (Dennis. British Ascomycetes. p. 36. 1978).
Host: Tsuga

Desmella Syd. & P. Syd. Basidiomycetes, Uredinales
Desmella aneimiae Syd.
Desmella superficialis Syd. & P. Syd.
Distr.: FL; West Indies.
Host: Nephrolepis

Diachea Fr. Myxomycetes
Diachea leucopodia (Bull.) Rostr.
Distr.: Cosmopolitan.
Hosts: Fragaria, Phaseolus, Smilax

Diapleella Munk. Pyrenomycetes, Amphisphaeriales
Diapleella coniothyrium (Fuckel) Barr
Leptosphaeria coniothyrium (Fuckel) Sacc.
Melanomma coniothyrium (Fuckel) L. Holm
Clypeosphaeria hendersoniae (Ellis) Sacc.
Sphaeria melantera Peck
Coniothyrium fuckelii Sacc. (anamorph)
Distr.: Cosmopolitan.
Notes: Stem blight and canker of Rosaceae, especially *Rubus* spp. (Punithalingam. C.M.I. Descr. 663. 1980). Barr et al. (New York State Mus. Bull. 459:30. 1986) established this synonymy.
Hosts: Baptisia, Ceanothus, Epilobium, Fragaria, Gaultheria, Ilex, Malus, Prunus, Rhododendron, Ribes, Rosa, Rubus, Silphium, Vaccinium

Diaporthe Nitschke. Pyrenomycetes, Diaporthales
Notes: Kobayashi (Bull. Gov. Forest Exp. Sta. 226:1. 1970); Wehmeyer (Univ. Michigan Stud., Sci. Ser. 9:1. 1933).
Diaporthe acerina (Peck) Sacc.
Valsa acerina Peck
Valsa albocincta Cooke & Peck
Phomopsis sp. (anamorph)
Distr.: Eastern North America.
Host: Acer
Diaporthe actinidiae N.F. Sommer & L. Beraha
Phomopsis sp. (anamorph)
Distr.: United States(introduced); New Zealand.
Notes: Stem-end decay of fruits in storage.
Host: Actinidia
Diaporthe aliena Ellis & Everh.
Distr.: Eastern North America.
Notes: On limbs.
Host: Crataegus
Diaporthe alleghaniensis R.H. Arnold
Phomopsis sp. (anamorph)
Distr.: Northeastern North America; Japan.
Notes: Canker and die-back (Arnold. Canad. J. Bot. 45:783; Canad. J. Bot. 45:1525. 1970; Fungi Canadenses 70. 1975).
Host: Betula
Diaporthe amorphae Ellis & Everh.
Distr.: Central and southeastern United States; Japan.
Host: Amorpha
Diaporthe ampelopsidis (Ellis) Ellis & Everh.
Distr.: Eastern United States.
Host: Parthenocissus
Diaporthe apocrypta (Cooke & Ellis) Sacc.
Distr.: NJ.
Notes: A doubtful species (Wehmeyer. Univ. Michigan Stud., Sci. Ser. 9:236. 1933).
Host: Carya
Diaporthe araliae Ellis & Everh.
Distr.: WA, eastern United States.
Hosts: Aralia, Oplopanax
Diaporthe arctii (Lasch) Nitschke
Phomopsis sp. (anamorph)
Distr.: Temperate northern hemisphere.
Notes: Usually on stems of herbaceous plants.
Hosts: Abelmoschus, Ambrosia, Arctium, Asclepias, Aureolaria, Baptisia, Bidens, Callistephus, Cassia, Chelone, Chenopodium, Chrysanthemum, Cirsium, Conyza, Daucus, Decumaria, Delphinium, Desmodium, Elephantopus, Erigeron, Eupatorium, Euphorbia, Firmiana, Helianthus, Heracleum, Ipomoea, Lactuca, Lespedeza, Ligusticum, Magnolia, Manfreda, Monarda, Oenothera, Pastinaca, Phaseolus, Phleum, Phytolacca, Platanus, Polygonum, Polymnia, Pycnanthemum, Ricinus, Rudbeckia, Rumex, Senna, Sesbania, Silphium, Smallanthus, Solanum, Solidago, Sonchus, Vernonia, Xanthium, Yucca
Diaporthe arctii (Lasch) Nitschke var. ***achilleae*** (Auersw.) Wehmeyer
Phomopsis achilleae (Sacc.) Höhn. (anamorph)
Distr.: IN; Europe(type).
Host: Verbesina
Diaporthe austro-americana Speg.
Distr.: FL; Argentina.
Host: Ludwigia
Diaporthe bakeri Wehmeyer
Distr.: Eastern United States.
Host: Carpinus
Diaporthe beckhausii Nitschke
Diaporthe elaeagni Rehm
Diaporthe halesiae Ellis & Everh.
Diaporthe menispermi Dearn. & House
Phomopsis beckhausii (Cooke) Traverso (anamorph)
Distr.: Temperate North America; Europe.
Notes: On woody stems.
Hosts: Elaeagnus, Halesia, Menispermum
Diaporthe biglobosa (Cooke & Ellis) Sacc.
Distr.: NJ.
Host: Sassafras
Diaporthe binoculata (Ellis) Sacc.
Distr.: Eastern United States.
Notes: On limbs.
Host: Magnolia
Diaporthe ceanothi Dearn. & House
Diaporthe minuta Dearn. & House non Nitschke, hom. illeg.
Distr.: NY.
Notes: A doubtful species (Wehmeyer. Univ. Michigan Stud., Sci. Ser. 9:261. 1933).
Host: Ceanothus
Diaporthe celastrina Ellis & Barth.
Distr.: Central and eastern United States.
Host: Celastrus
Diaporthe citri F.A. Wolf
Phomopsis citri H. Fawc. non (Sacc.) Traverso & Spessa, hom. illeg. (anamorph)
Distr.: Worldwide in citrus-growing regions.
Notes: Melanose and stem-end rot (Punithalingam & Holliday. C.M.I. Descr. 396. 1973).
Hosts: Citrofortunella, Citrus, Fortunella, Poncirus
Diaporthe conorum (Desmaz.) Niessl
Phomopsis conorum (Sacc.) Died. (anamorph)

Distr.: Circumboreal.
Notes: Saprobic on conifers. Wehmeyer (Univ. Michigan Stud., Sci. Ser. 9:70. 1933) considered this species to be a synonym of *D. eres* Nitschke.
Host: Abies
Diaporthe crotalariae G.F. Weber
Phomopsis crotalariae G.F. Weber (anamorph)
Distr.: FL.
Host: Crotalaria
Diaporthe dakotensis Wehmeyer
Distr.: ND.
Host: Polygonum
Diaporthe decorticans (Lib.) Sacc. & Roum.
Diaporthe padi G. Otth
Phomopsis padina (Sacc. & Roum.) Died. (anamorph)
Distr.: Temperate northern hemisphere.
Notes: On limbs.
Host: Prunus
Diaporthe detrusa (Fr.:Fr.) Fuckel
Phomopsis detrusa (Sacc.) Traverso (anamorph)
Distr.: Temperate northern hemisphere.
Notes: On limbs.
Host: Berberis
Diaporthe dubia Nitschke
Diaporthe congenera Ellis & Everh.
Diaporthe ontariensis Ellis & Everh.
Diaporthe robusta Peck
Diaporthe subcongrua Ellis & Everh.
Phomopsis sp. (anamorph)
Distr.: Eastern North America.
Host: Acer
Diaporthe eres Nitschke
Diaporthe ambigua Nitschke
Diaporthe fallaciosa Nitschke
Diaporthe hydrangeae Ellis & Everh.
Diaporthe pithya Sacc.
Diaporthe pungens Nitschke
Diaporthe rhododendri Feltgen
Phomopsis oblonga (Desmaz.) Traverso (anamorph)
Distr.: Temperate northern hemisphere.
Notes: Usually on woody branches. *Diaporthe perniciosa* is sometimes regarded as a distinct species on apple.
Hosts: Acer, Aesculus, Camellia, Castanea, Cornus, Cryptomeria, Cupressus, Geranium, Hibiscus, Hydrangea, Ilex, Ligustrum, Liriodendron, Lonicera, Magnolia, Ostrya, Physocarpus, Prunus, Pseudotsuga, Pyrus, Rhododendron, Ribes, Rosa, Tilia, Ulmus
Diaporthe eucalypti Harkn.
Distr.: CA.
Host: Eucalyptus
Diaporthe euonymi Dearn.
Distr.: DE; Canada(Ontario).
Host: Euonymus
Diaporthe fagi Wehmeyer
Phomopsis sp. (anamorph)
Distr.: Northeastern North America.
Host: Fagus
Diaporthe fagi Wehmeyer var. ***longispora*** Wehmeyer
Distr.: MI.
Host: Fagus
Diaporthe helianthi M. Muntanola-Cvetkovic, M. Mihaljcevic, & M. Petrov
Phomopsis helianthi M. Muntanola-Cvetkovic, M. Mihaljcevic, & M. Petrov (anamorph)
Distr.: Temperate North America; Europe.
Notes: Stem canker and stalk rot.
Host: Helianthus
Diaporthe hicoriae Wehmeyer
Distr.: MI.
Host: Carya
Diaporthe ilicis (Ellis & Everh.) Wehmeyer
Distr.: Northeastern North America.
Host: Ilex
Diaporthe impulsa (Cooke & Peck) Sacc.
Diaporthe congesta Ellis & Everh.
Phomopsis sp. (anamorph)
Distr.: Temperate northern hemisphere.
Notes: On limbs.
Host: Sorbus
Diaporthe inaequalis (Curr.) Nitschke
Phomopsis inaequalis Traverso (anamorph)
Distr.: Temperate North America; Europe.
Notes: On shrubby Fabaceae.
Host: Cytisus
Diaporthe kalmiae F.A. Wolf & A.R. Cavaliere
Phomopsis kalmiae Enlows (anamorph)
Distr.: Eastern United States.
Host: Kalmia
Diaporthe kokiae Petr.
Distr.: HI.
Host: Kokia
Diaporthe ligustrina Ellis & Everh.
Distr.: NJ.
Notes: Wehmeyer (Univ. Michigan Stud., Sci. Ser. 9:85. 1933) considered this species a synonym of *D. eres* Nitschke.
Hosts: Ligustrum, Lyonia
Diaporthe leariformis Petr.
Apioporthe leariformis (Petr.) Wehmeyer
Distr.: ND.
Host: Solidago
Diaporthe linearis (Fr.:Fr.) Nitschke
Diaporthe aorista Ellis & Everh.
Diaporthe exercitalis (Peck) Sacc.
Phomopsis linearis (Sacc.) Traverso (anamorph)
Distr.: Temperate North America; Europe.
Notes: On Asteraceae.
Hosts: Aster, Euthamia, Solidago
Diaporthe lokoyae Funk
Phomopsis lokoyae Hahn (anamorph)
Distr.: Western North America.
Notes: Dieback of douglas fir (Funk. Canad. J. Bot. 46:601. 1968).
Host: Pseudotsuga
Diaporthe lupini Harkn.
Distr.: CA.
Host: Lupinus
Diaporthe magnoliae Ellis & Everh.
Distr.: Eastern United States.
Notes: Considered a synonym of *D. eres* Nitschke (Wehmeyer. Univ. Michigan Stud., Sci. Ser. 9:85. 1933).
Host: Magnolia
Diaporthe megalospora Ellis & Everh.
Phomopsis sp. (anamorph)
Distr.: Northeastern North America.
Host: Sambucus
Diaporthe oncostoma Fuckel
Phomopsis oncostoma (Thuem.) Traverso (anamorph)
Distr.: Temperate northern hemisphere.
Notes: On limbs of hardwoods.
Host: Robinia
Diaporthe oxyspora (Peck) Sacc.
Phomopsis oxyspora Petr. (anamorph)
Distr.: Eastern North America.
Host: Ilex
Diaporthe padi G. Otth var. ***patria*** (Speg.) Wehmeyer
Distr.: Southeastern United States.
Notes: On limbs.
Host: Aesculus
Diaporthe pardalota (Mont.) Fuckel
Phomopsis sp. (anamorph)
Distr.: Temperate North America; Europe.
Notes: On stems of herbaceous and soft-woody plants.
Hosts: Cornus, Sassafras
Diaporthe paulula (Cooke & Ellis) Sacc.
Valsa paulula Cooke & Ellis
Cryptodiaporthe paulula (Cooke & Ellis) Wehmeyer
Phomopsis sp. (anamorph)
Distr.: Eastern United States.
Host: Nyssa
Diaporthe peckii Sacc.
Diaporthe sparsa Peck, Niessl
Phomopsis sp. (anamorph)

Distr.: Eastern North America.
Host: Rhus
Diaporthe pennsylvanica (Berk. & M.A. Curtis) Wehmeyer
Distr.: Central and eastern North America.
Notes: On limbs.
Host: Prunus
Diaporthe perniciosa Ém. Marchal
Phomopsis mali Roberts, nom. illeg. (anamorph)
Distr.: Temperate regions.
Notes: This is considered a synonym of *Diaporthe eres* by some workers.
Host: Malus
Diaporthe phaseolorum (Cooke & Ellis) Sacc.
Diaporthe batatatis Harter & E.C. Field
Diaporthe phaseolorum (Cooke & Ellis) Sacc. var. *batatatis* (Harter & E.C. Field) Wehmeyer
Diaporthe phaseolorum (Cooke & Ellis) Sacc. var. *caulivora* K.L. Athow & R.M. Caldwell
Diaporthe phaseolorum (Cooke & Ellis) Sacc. var. *sojae* (S.G. Lehman) Wehmeyer
Diaporthe sojae S.G. Lehman
Phomopsis phaseoli (Desmaz.) Sacc. (anamorph)
Distr.: Cosmopolitan.
Notes: Stem canker, blight, and pod rot of soybeans and other legumes (Punithalingam & Holliday. C.M.I. Descr. 336. 1972; Kulik. Mycologia 76:274. 1984); also on a wide variety of other hosts.
Hosts: Abelmoschus, Allium, Arachis, Capsicum, Desmanthus, Glycine, Ipomoea, Lespedeza, Lupinus, Lycopersicon, Phalaris, Phaseolus, Strophostyles, Vigna, Zea
Diaporthe pruni Ellis & Everh.
Phomopsis pruni (Ellis & Dearn.) Wehmeyer (anamorph)
Distr.: Central and eastern North America.
Notes: On limbs.
Host: Prunus
Diaporthe prunicola (Peck) Wehmeyer
Valsa prunicola Peck
Distr.: Central and eastern North America.
Notes: On limbs.
Host: Prunus
Diaporthe rudis (Fr.:Fr.) Nitschke
Diaporthe faginea (Curr.) Sacc.
Diaporthe medusaea Nitschke
Phomopsis rudis (Fr.) Höhn. (anamorph)
Distr.: Temperate North America; Europe.
Notes: On a variety of woody hosts.
Hosts: Acacia, Ailanthus, Citrus, Cytisus, Eucalyptus, Fortunella, Genista, Laburnum, Pelargonium, Syringa, Vitis
Diaporthe sarothamni Auersw. ex Nitschke
Phomopsis sarothamni (Sacc.) Höhn. (anamorph)
Distr.: Europe; distribution elsewhere is questionable.
Host: Cytisus
Diaporthe sarothamni Auersw. ex Nitschke var. ***baccharidis*** (Cooke) Wehmeyer
Distr.: Eastern United States.
Host: Baccharis
Diaporthe sheariana Petr.
Distr.: HI.
Host: Acacia
Diaporthe sociabilis Nitschke var. ***sambuci*** (Ellis & Everh.) Wehmeyer
Diaporthe sambuci Ellis & Everh.
Diaporthe callicarpae Peck
Distr.: Temperate North America.
Notes: On limbs.
Host: Sambucus
Diaporthe sociata (Cooke & Ellis) Sacc.
Valsa sociata Cooke & Ellis
Distr.: Eastern United States.
Host: Lindera
Diaporthe spiculosa (Albertini & Schwein.:Fr.) Nitschke
Phomopsis sp. (anamorph)
Distr.: Temperate northern hemisphere.
Notes: On woody stems.
Hosts: Juglans, Rhus, Sambucus
Diaporthe staphylina Ellis & Everh.
Distr.: Northeastern North America.
Host: Staphylea
Diaporthe stewartii A.L. Harrison
Phomopsis stewartii Peck (anamorph)
Distr.: United States.
Notes: Stem blight (Harrison. Mycologia 27:521. 1935).
Host: Cosmos
Diaporthe strumella (Fr.:Fr.) Fuckel
Dothiorella ribis (Fuckel) Sacc. (synanamorph)
Phomopsis sp. (synanamorph)
Distr.: Temperate North America; Europe.
Notes: On woody limbs, especially of *Ribes* spp.
Host: Ribes
Diaporthe syngenesia (Fr.:Fr.) Nitschke ex Fuckel
Diaporthe berlesiana Sacc. & Roum.
Phomopsis syngenesia (Brunaud) Höhn. (anamorph)
Distr.: Temperate North America; Europe.
Notes: On stems of Rhamnaceae.
Host: Rhamnus
Diaporthe tetraptera Ellis & Everh.
Distr.: Southeastern United States.
Host: Halesia
Diaporthe tuberculosa (Ellis) Sacc.
Phomopsis sp. (anamorph)
Distr.: Northeastern North America.
Notes: On limbs.
Host: Amelanchier
Diaporthe tuberculosa (Ellis) Sacc. var. ***corymbosa*** (Cooke & Ellis) Wehmeyer
Diaporthe corymbosa (Cooke & Ellis) Sacc.
Distr.: Eastern United States.
Host: Clethra
Diaporthe tuberculosa (Ellis) Sacc. var. ***pruni*** Dearn. & House
Distr.: Northeastern United States.
Host: Prunus
Diaporthe vaccinii Shear in Shear, N. Stevens, & H. Bain
Phomopsis vaccinii Shear in Shear, N. Stevens, & H. Bain (anamorph)
Distr.: Temperate North America.
Notes: On Ericaceae. Fruit rot and twig blight. Jacobs et al. (Mycologia 80:307. 1988).
Hosts: Arctostaphylos, Gaultheria, Rhododendron, Vaccinium
Diaporthe viburni Dearn. & Bisby
Distr.: Temperate North America.
Notes: On limbs.
Host: Viburnum
Diaporthe viburni Dearn. & Bisby var. ***spiraeicola*** Wehmeyer
Phomopsis sp. (anamorph)
Distr.: MA.
Host: Spiraea

Diaporthella Petr. Pyrenomycetes, Diaporthales

Diaporthella aristata (Fr.:Fr.) Petr.
Diaporthella platasca (Peck) Wehmeyer
Diatrype platasca Peck
Distr.: Temperate North America; Europe.
Notes: On limbs.
Host: Betula
Diaporthella sphendamnina (Berk. & M.A. Curtis) Wehmeyer
Distr.: Northeastern United States.
Host: Acer

Diaporthopsis Fabre. Pyrenomycetes, Diaporthales

Diaporthopsis apiculosa (Ellis) Wehmeyer
Diaporthe apiculosa Ellis
Distr.: NJ.
Host: Conyza
Diaporthopsis sclerophila Bonar
Distr.: CA.
Host: Baccharis
Diaporthopsis trinucleata (Niessl) Höhn.

Diaporthe trinucleata Niessl
Distr.: GA; Canada(Ontario); Europe.
Host: Gnaphalium

Diatrype Fr. Pyrenomycetes, Diatrypales

Notes: Rappaz (Mycol. Helv. 2:394. 1987).
Diatrype albopruinosa (Schwein.) Cooke
Diatrype americana Berl.
Phaeotrype brencklei Sacc.
Diatrype verrucoides Peck
Distr.: Temperate North America.
Notes: On dead twigs and limbs of hardwoods (Pirozynski. Fungi Canadenses 72. 1975).
Hosts: Acer, Amelanchier, Betula, Carpinus, Carya, Cornus, Corylus, Fraxinus, Juglans, Ostrya, Populus, Prunus, Quercus, Rhus, Ulmus
Diatrype americana Ellis & Everh. f. ***ostryae*** Rehm
Distr.: MD, ND(type).
Notes: Although the type variety is a synonym of *D. albopruinosa*, this form has never been synonymized.
Host: Ostrya
Diatrype americana Ellis & Everh. f. ***quercus*** Rehm
Distr.: ND(type).
Notes: Although *D. americana* is a synonym of *D. albopruinosa*, this form has never been synonymized.
Host: Quercus
Diatrype arundinariae (Berl.) Rappaz
Eutypella arundinariae Berl.
Distr.: LA.
Host: Arundinaria
Diatrype asterostoma Berk. & M.A. Curtis
Distr.: Eastern United States.
Notes: On wood.
Hosts: Acer, Liquidambar, Magnolia, Nyssa, Prunus
Diatrype azedarachtae Cooke
Eutypella azedarachtae (Cooke) Berl.
Distr.: SC.
Host: Melia
Diatrype bullata (Hoffm.:Fr.) Fr.
Diatrype macounii Ellis & Everh.
Distr.: Temperate northern hemisphere.
Hosts: Acer, Alnus, Populus, Quercus, Salix
Diatrype concolor (Schwein.) Cooke
Distr.: NJ.
Host: Vaccinium
Diatrype dakotensis Rehm
Distr.: ND.
Host: Salix
Diatrype diffidens Kauffm.
Distr.: OR(type).
Host: Pseudotsuga
Diatrype disciformis (Hoffm.:Fr.) Fr.
Libertella disciformis Höhn. (anamorph)
Distr.: Temperate northern hemisphere.
Notes: On limbs of hardwoods.
Hosts: Alnus, Betula, Fagus, Magnolia, Salix
Diatrype durieui Mont.
Distr.: Eastern United States; Algeria(type).
Host: Magnolia
Diatrype flavovirens (Pers.:Fr.) Fr.
Eutypa flavovirens (Pers.:Fr.) Tul. & C. Tul.
Valsa flavovirens (Pers.:Fr.) Nitschke
Diatrype hullensis Ellis & Everh.
Eutypella virescens Wehmeyer
Cytosporina flavovirens (Sacc.) Grove (anamorph)
Distr.: Cosmopolitan.
Notes: On various hardwoods.
Hosts: Acer, Alnus, Amelanchier, Ceanothus, Populus, Quercus, Sambucus
Diatrype friabilis (Pers.:Fr.) Cooke
Diatrypella friabilis (Pers.:Fr.) Ellis & Everh.
Distr.: Eastern United States; Europe.
Notes: Rappaz (Mycol. Helv. 2:559. 1987) examined two possible type specimens and was unable to characterize this species.
Host: Ilex
Diatrype linearis Ellis & Everh. f. ***salicis*** Rehm
Distr.: ND(type).
Notes: Although *D. linearis* is a synonym of *D. prominens*, this form has never been synonymized.
Host: Salix
Diatrype oregonensis (Wehmeyer) Rappaz
Eutypella oregonensis Wehmeyer in Kauffm.
Distr.: OR.
Host: Acer
Diatrype princeps Penz. & Sacc.
Distr.: HI; Java(type).
Notes: According to Rappaz (Mycol. Helv. 2:568. 1987), this species belongs in *Valsa* sp.
Host: Metrosideros
Diatrype prominens Cooke & Harkn. non E. Howe, hom. illeg.
Diatrype baccharidis Earle
Diatrype ceanothi Cooke & Harkn.
Eutypella ceanothi (Cooke & Harkn.) Berl.
Diatrype eucalypti Cooke & Harkn.
Diatrype linearis Ellis & Everh.
Diatrype microstega Ellis & Everh.
Distr.: CA.
Notes: On various hardwoods. Although Rappaz (Mycol. Helv. 2:430. 1987) used the name *D. prominens* Cooke & Harkn. 1881, this epithet is a later homonym of *D. prominens* Howe 1874.
Hosts: Baccharis, Lupinus, Quercus, Umbellularia
Diatrype sambucivora (Schwein.) Cooke
Distr.: PA(type).
Notes: According to Rappaz (Mycol. Helv. 2:568. 1987), the type specimen is a mix of *Eutypella* and *Diatrypella* spp.
Host: Sambucus
Diatrype standleyi Fairm.
Eutypella brunaudiana Sacc. var. *ribis-aurei* Fairm.
Distr.: NM.
Host: Cercocarpus
Diatrype stigma (Hoffm.:Fr.) Fr.
Diatrype stigmoides Kauffm.
Libertella betulina Desmaz. (anamorph)
Distr.: Temperate northern hemisphere.
Notes: Several groups have been recognized within the broad species concept of *D. stigma*, based on stromal morphology, ascospore size, and conidial size (Rappaz. Mycotaxon 30:209. 1987).
Hosts: Acer, Alnus, Betula, Carpinus, Carya, Ceanothus, Cercis, Crataegus, Fagus, Lithocarpus, Magnolia, Ostrya, Quercus, Salix, Ulmus
Diatrype stigmaoides Kauffm.
Distr.: OR, WA.
Notes: On fallen branches. Glawe (Mycotaxon 20:423. 1984) suggested that this species is close to or synonymous with *D. stigma*; Rappaz (Mycol. Helv. 2:402. 1987) recognized this species.
Host: Quercus
Diatrype tremellophora Ellis ex Ellis & Everh.
Diatrype disciformis (Hoffm.) Fr. var. *magnoliae* Thuem.
Distr.: Eastern United States.
Host: Magnolia
Diatrype tumida Ellis & Everh. f. ***populi*** Rehm
Distr.: ND(type).
Host: Populus
Diatrype tumida Ellis & Everh. f. ***pruni*** Rehm
Distr.: ND(type).
Host: Prunus
Diatrype undulata (Pers.:Fr.) Fr.
Distr.: NJ; Europe.
Notes: On bark (Rappaz. Mycotaxon 30:209. 1987).
Host: Betula
Diatrype virescens (Schwein.) M.A. Curtis
Distr.: Eastern North America.
Notes: On twigs and limbs of Fagaceae (Pirozynski. Fungi Canadenses 73. 1975).
Hosts: Fagus, Populus, Quercus
Diatrype whitmanensis J.D. Rogers & Glawe
Distr.: WA(type).
Notes: Rogers & Glawe (Mycotaxon 18:73. 1983).
Host: Celtis

Diatrypella (Ces. & De Not.) Sacc. Pyrenomycetes, Diatrypales

Diatrypella annulans (Schwein.) Ellis & Everh.
Distr.: Eastern United States.
Host: Acer
Diatrypella betulina (Peck) Sacc.
Diatrype betulina Peck
Distr.: Eastern North America.
Host: Betula
Diatrypella cephalanthi (Schwein.) Sacc.
Distr.: Central and eastern United States.
Notes: Tiffany & Gilman (Iowa State J. Sci. 40:147. 1965).
Host: Cephalanthus
Diatrypella decipiens Ellis & Everh.
Distr.: OR.
Notes: Glawe (Mycotaxon 20:432. 1984) examined the type and one other specimen but could not characterize this species.
Hosts: Corylus, Umbellularia
Diatrypella decorata Nitschke
Diatrypella discoidea (Cooke & Peck) Sacc.
Distr.: Temperate North America; Europe.
Hosts: Alnus, Betula, Carpinus, Quercus, Salix
Diatrypella eutypaeformis (Sacc.) L.C. Tiffany & Gilman
Cryptovalsa eutypaeformis Sacc.
Cryptovalsa pustulata (Ellis & Everh.) Ellis & Everh.
Cryptovalsa sparsa Ellis & Everh.
Distr.: Temperate regions (Tiffany & Gilman. Iowa State J. Sci. 40:146. 1965).
Hosts: Amelanchier, Carpinus, Juglans, Quercus, Ulmus
Diatrypella eutypoides Niessl
Distr.: MA; Czechoslovakia(type).
Host: Ulmus
Diatrypella favacea (Fr.:Fr.) De Not.
Diatrype favacea (Fr.:Fr.) Fr.
Diatrypella tocciaeana De Not.
Diatrypella verruciformis (Ehrenb.:Fr.) Nitschke
Libertella favacea Traverso (anamorph)
Distr.: Eastern North America; Europe.
Notes: Generally on dead wood of hardwoods.
Hosts: Acer, Alnus, Baccharis, Berchemia, Betula, Broussonetia, Castanea, Celtis, Cephalanthus, Cercis, Corylus, Crataegus, Cyrilla, Eupatorium, Fagus, Fraxinus, Hibiscus, Ilex, Juniperus, Lonicera, Maclura, Magnolia, Morus, Nyssa, Populus, Prunus, Quercus, Smilax, Tilia, Ulmus
Diatrypella fraxini Ellis & Everh.
Distr.: GA, KS.
Host: Fraxinus
Diatrypella frostii Peck
Cytosporina sp. (anamorph)
Distr.: North America.
Notes: Common on branches of sugar maple and other species of *Acer*. Glawe (Mycologia 75:913. 1983) presented details of the anamorph.
Hosts: Acer, Corylus
Diatrypella herbacea Ellis & Everh.
Distr.: FL, LA.
Host: Solidago
Diatrypella lantanae Earle
Distr.: FL(type).
Host: Lantana
Diatrypella minutispora Dearn.
Distr.: NY(type).
Host: Corylus
Diatrypella missouriensis Ellis & Everh.
Distr.: MO(type).
Host: Corylus
Diatrypella nigro-angulata (Grev.) Nitschke
Sphaeria nigro-angulata Grev.
Distr.: MS, TX; Europe.
Host: Ilex
Diatrypella nitschkei (Fuckel) L.C. Tiffany & Gilman
Cryptovalsa nitschkei Fuckel
Distr.: Temperate regions.
Notes: Tiffany & Gilman (Iowa State J. Sci. 40:141. 1965).
Hosts: Carpinus, Crataegus, Magnolia, Morus, Populus, Prunus, Pyrus, Quercus, Sabal, Ulmus
Diatrypella opaca Cooke
Distr.: Southeastern United States.
Host: Ilex
Diatrypella placenta Rehm
Diatrypella discoidea Cooke var. *alni* Ravenel ex Ellis & Everh. (as var. *alni* Cooke)
Distr.: Temperate regions.
Notes: Common on recently fallen or still-attached dead branches (Glawe. Mycotaxon 20:433. 1984; Glawe. Mycotaxon 25:19. 1986).
Host: Alnus
Diatrypella populi Ellis & Holw.
Distr.: IA.
Host: Populus
Diatrypella prominens (E. Howe) Ellis & Everh.
Diatrype prominens E. Howe
Distr.: United States.
Hosts: Arbutus, Platanus
Diatrypella pulvinata Nitschke
Distr.: Northwestern North America; Europe.
Notes: On dead branches (Glawe. Mycotaxon 20:435. 1984).
Host: Quercus
Diatrypella quercina (Pers.:Fr.) Cooke
Libertella quercina (Sacc.) Grove (anamorph)
Distr.: Eastern North America; Europe.
Notes: On hardwood limbs.
Hosts: Acer, Berchemia, Crataegus, Fraxinus, Ilex, Magnolia, Morus, Populus, Quercus, Salix, Tilia
Diatrypella rhois (Schwein.:Fr.) Ellis & Everh.
Diatrype rhois (Schwein.:Fr.) Cooke
Distr.: CA, PA; Europe.
Host: Rhus
Diatrypella rimosa Shear
Distr.: OR.
Notes: Glawe (Mycotaxon 20:436. 1984) suggested that this species is possibly conspecific with *D. discoidea* var. *alni*, now *D. placenta*.
Hosts: Alnus, Tsuga
Diatrypella sassafras Ellis & Everh.
Cryptovalsa sassafras (Ellis & Everh.) Berl.
Distr.: NJ(type).
Host: Sassafras
Diatrypella subfulva (Berk. & M.A. Curtis) Sacc.
Distr.: Southeastern United States.
Notes: On hardwood branches.
Hosts: Acer, Celtis, Crataegus, Nyssa, Quercus, Ulmus
Diatrypella underwoodii Peck
Distr.: TX.
Host: Prosopis

Dicarpella Syd. & P. Syd. Pyrenomycetes, Diaporthales

Notes: Barr (Mycol. Mem. 7:156. 1978).
Dicarpella bina (Harkn.) Syd. & P. Syd.
Physalospora bina Harkn.
Disperma bina (Harkn.) Theiss.
Distr.: CA.
Notes: Leaf spot.
Host: Quercus
Dicarpella georgiana (J.H. Miller & G.E. Thompson) Barr
Gnomoniella georgiana J.H. Miller & G.E. Thompson
Distr.: Southeastern United States.
Notes: On leaves.
Hosts: Liquidambar, Nyssa
Dicarpella quercifolia (Ellis & Everh.) Barr
Physalospora quercifolia Ellis & Everh.
Pseudomassaria quercifolia (Ellis & Everh.) Arx
Mastigosporella hyalina (Ellis & Everh.) Höhn. (anamorph)
Distr.: Eastern United States.
Host: Quercus

Dicellomyces Olive. Basidiomycetes, Dacrymycetales

Dicellomyces gloeosporus L. Olive
Distr.: Southeastern United States.
Notes: Parasitic on leaves.
Host: Arundinaria

Dichaena Fr. Coelomycetes

Notes: This genus is considered a synonym of *Polymorphum* (Hawksworth. Taxon 32:212. 1983), but not all names have been redisposed.
Dichaena strumosa Fr.
Distr.: Eastern United States.
Notes: This name is probably a synonym of *Polymorphum quercinum*.
Host: Quercus

Dichomera Cooke. Coelomycetes

Dichomera juglandis Ellis & Everh.
Distr.: OH, MD.
Hosts: Acer, Juglans
Dichomera viticola Cooke & Harkn.
Distr.: CA.
Host: Vitis

Dichomitus D. Reid. Basidiomycetes, Aphyllophorales

Dichomitus campestris (Quél.) Dománski & Orlicz
Coriolellus campestris (Quél.) Bondartsev
Trametes campestris (Quél.) Quél.
Distr.: Southwestern United States; Europe.
Hosts: Alnus, Quercus
Dichomitus squalens (P. Karst.) D. Reid
Polyporus anceps Peck
Polyporus ellisianus (Murrill) Sacc. & Trott.
Poria stenospora Overh., nom. inval.
Distr.: Circumboreal.
Notes: On wood of conifers.
Hosts: Abies, Larix, Picea, Pinus, Pseudotsuga, Thuja, Tsuga

Dichostereum Pilát. Basidiomycetes, Aphyllophorales

Notes: Boidin & Lanquetin (Mycotaxon 6:277. 1977).
Dichostereum effuscatum (Cooke & Ellis) Boidin & Lanquetin
Corticium effuscatum Cooke & Ellis
Vararia effuscata (Cooke & Ellis) D.P. Rogers & H. Jacks.
Distr.: North America; Europe.
Notes: On wood, especially of hardwoods in contact with wet soil.
Hosts: Acer, Alnus, Fraxinus, Juglans, Populus, Quercus, Tsuga
Dichostereum pallescens (Schwein.) Boidin & Lanquetin
Vararia pallescens (Schwein.) D.P. Rogers & H. Jacks.
Corticium thelephoroides Ellis & Everh.
Distr.: United States to Central America.
Notes: On wood.
Hosts: Alnus, Betula, Castanea, Pinus, Pseudotsuga, Quercus, Rubus, Ulmus

Dichotomophthora Mehrlich & Fitzp. ex M.B. Ellis. Hyphomycetes

Notes: DeHoog & Van Oorschot (Proc. Koninkl. Nederl. Akad. Wetens., Ser. C, 86:55. 1983).
Dichotomophthora lutea (Routien) De Hoog & Van Oorschot
Dactylaria lutea Routien
Dichotomophthora indica P.N. Rao, nom. inval.
Distr.: Eastern North America; West Indies; Europe; India.
Notes: Stem and leaf blight of purslane.
Host: Portulaca
Dichotomophthora portulacae Mehrlich & Fitzp. ex M.B. Ellis
Distr.: CA, HI; West Indies; South America; northern Africa.
Notes: Leaf spot of purslane.
Host: Portulaca

Dichotomophthoropsis M.B. Ellis. Hyphomycetes

Dichotomophthoropsis nymphaearum (F.V. Rand) M.B. Ellis
Helicoceras nymphaearum (F.V. Rand) Linder
Distr.: Eastern North America; India.
Notes: Ellis (More dematiaceous Hyphomycetes. p. 386. 1976).
Hosts: Brasenia, Nuphar, Nymphaea

Dicranidion Harkn. Hyphomycetes

Dicranidion fragile Harkn.
Dicranidion inaequalis Tubaki & Yokoyama
Orbilia sp. (teleomorph)
Distr.: Worldwide, although not common.
Notes: On rotten wood (Butterfield. Canad. J. Bot. 51:795. 1973).
Host: Nerium

Dictydium Schrad. Myxomycetes

Dictydium cancellatum (Batsch) T. Macbr.
Dictydium cancellatum (Batsch) T. Macbr. f. *anomalum* (E. Jahn) G. List.
Dictydium cancellatum (Batsch) T. Macbr. var. *purpureum* T. Macbr.
Distr.: Cosmopolitan.
Host: Pinus

Dictyocatenulata Finley & E.F. Morris in E.F. Morris & Finley. Hyphomycetes

Dictyocatenulata alba Finley & E.F. Morris in E.F. Morris & Finley
Stilbum glomerulisporum Ellis & Everh., nom. nud.
Distr.: Northern hemisphere.
Notes: On bark, often of living trees.
Hosts: Betula, Fagus, Picea

Dictyochaeta Speg. Hyphomycetes

Dictyochaeta fertilis (S.J. Hughes & Kendrick) Holubová-Jechová
Codinaea fertilis S.J. Hughes & Kendrick
Distr.: North America; Europe; New Zealand.
Notes: On a variety of dead plant materials.
Hosts: Coronilla, Glycine, Kummerowia, Lespedeza, Medicago, Melilotus, Phaseolus, Pisum, Trifolium, Vicia, Zea

Dictyochora Theiss. & Syd. Loculoascomycetes, Dothideales

Dictyochora gambellii Fairm.
Distr.: NY(type).
Host: Zea

Dictyodesmium S.J. Hughes. Hyphomycetes

Dictyodesmium ulmicola (Ellis & Kellerm.) S.J. Hughes
Ceratophorum ulmicola Ellis & Kellerm.
Distr.: Central United States.
Host: Ulmus

Dictyoporthe Petr. Pyrenomycetes, Diaporthales

Notes: Barr (Mycol. Mem. 7:187. 1978).
Dictyoporthe acerophila Barr
Hendersonula sp. (anamorph)
Distr.: NH(type).
Host: Acer

Dictyosporium Corda. Hyphomycetes

Dictyosporium elegans Corda
Dictyosporium opacum Cooke & Harkn.
Distr.: North Ameria; West Africa; Europe.
Notes: On dead wood, herbaceous plants, and stubble of grasses.
Host: Typha
Dictyosporium pelagicum (Linder) G.C. Hughes ex T.W. Johnson & Sparrow
Speira pelagica Linder
Distr.: Worldwide in marine habitats.
Notes: On submerged leaves, driftwood and drifting bark, intertidal bark and wood, and test panels (Kohlmeyer & Kohlmeyer. Marine mycology. p. 490. 1979).
Host: Spartina

Dicyma Boul. Hyphomycetes

Hansfordia S.J. Hughes
Basifimbria Subramanian & Lodha
Gonytrichella Y. Emoto & Tubaki
Dicyma pulvinata (Berk. & M.A. Curtis) Arx
Polyactis pulvinata Berk. & M.A. Curtis
Hansfordia pulvinata (Berk. & M.A. Curtis) S.J. Hughes
Hansfordia grewiae (Hansf.) S.J. Hughes
Verticicladium grewiae Hansf.
Distr.: Southern United States, common in warmer areas.
Notes: Possibly a mycoparasite (Arx. Proc. Koninkl. Nederl. Akad. Wetens., Ser. C, 85:21. 1982).
Hosts: Acacia, Magnolia, Yucca

Didymaria Corda. Hyphomycetes

Notes: This genus is considered a synonym of *Ramularia*, but not all names have been redisposed.
Didymaria anethiana W.B. Cooke
Distr.: OR(type).
Host: Anethum
Didymaria clematidis Cooke & Harkn.
Distr.: CA(type).
Host: Clematis
Didymaria conferta Syd. & P. Syd.
Distr.: Western United States.
Notes: On living leaves.
Host: Wyethia
Didymaria fulva Ellis & Everh.
Distr.: IN(type).
Host: Dioscorea
Didymaria hydrophylli (Ellis & Everh.) Solheim
Ramularia hydrophylli Ellis & Everh.
Distr.: Western United States.
Notes: Solheim (Univ. Wyoming Publ. 24:46. 1960).
Host: Hydrophyllum
Didymaria puncta J.J. Davis
Distr.: WI.
Host: Sisyrinchium
Didymaria spissa Harkn.
Distr.: CA(type).
Host: Euthamia
Didymaria wyethiae (Ellis & Everh.) Solheim
Marssonia wyethiae Ellis & Everh.
Marssonina wyethiae (Ellis & Everh.) Magnus
Distr.: Western North America.
Hosts: Helianthella, Wyethia

Didymascella Maire & Sacc. Discomycetes, Helotiales

Notes: On conifers.
Didymascella chamaecyparidis (J.F. Adams) Maire
Distr.: NJ(type).
Host: Chamaecyparis
Didymascella tetramicrospora Pantidou & Darker
Keithia tetramicrospora Shope, nom. inval.
Distr.: CO.
Notes: On fallen foliage of juniper (Pantidou & Darker. Mycologia 55:417. 1963).
Host: Juniperus
Didymascella thujina (E.J. Durand) Maire
Keithia thujina E.J. Durand
Distr.: WI.
Host: Thuja

Didymella Sacc. Loculoascomycetes, Dothideales

Didymella andropogonis Ellis & Everh.
Distr.: LA, MS.
Host: Vetiveria
Didymella applanata (Niessl) Sacc.
Didymosphaeria applanata Niessl
Mycosphaerella rubina (Peck) House
Phoma sp. (anamorph)
Distr.: Temperate regions.
Notes: Sour blight or gray bark (Punithalingam. C.M.I. Descr. 735. 1982; Corlett. Fungi Canadenses 49. 1974).
Host: Rubus
Didymella bryoniae (Auersw.) Rehm
Mycosphaerella citrullina (C.O. Sm.) Gross.
Didymella melonis Pass.
Mycosphaerella melonis (Pass.) Chiu & J.C. Walker
Phoma cucurbitacearum (Fr.:Fr.) Sacc. (anamorph)
Distr.: Cosmopolitan.
Notes: Gummy stem blight, blackrot of fruit, and various other symptoms on Cucurbitaceae (Corlett et al., Fungi Canadenses 303. 1986; Punithalingam & Holliday. C.M.I. Descr. 332. 1972).
Hosts: Citrullus, Cucumis, Cucurbita
Didymella castanella (Cooke & Ellis) Sacc.
Distr.: NJ(type).
Host: Castanea
Didymella castillejae Wehmeyer
Distr.: ID, WY(type).
Notes: Corlett (Canad. J. Bot. 59:2024. 1981).
Hosts: Castilleja, Thermopsis
Didymella catariae (Cooke & Ellis) Ellis
Cercidospora catariae (Cooke & Ellis) Kuntze
Distr.: NJ(type).
Host: Nepeta
Didymella celtidis Berk. & Cooke
Distr.: SC(type).
Host: Celtis
Didymella cornuta Ellis & Everh.
Distr.: NJ; Europe.
Host: Asclepias
Didymella culmigena Sacc.
Distr.: MS; Italy(type).
Host: Muhlenbergia
Didymella delphinii Earle
Distr.: Western North America.
Notes: On stems (Corlett. Fungi Canadenses 131. 1979).
Hosts: Arnica, Delphinium, Frasera
Didymella dioscoreae (Berk. & M.A. Curtis) Sacc.
Distr.: FL(type).
Host: Dioscorea
Didymella douglasii Ellis & Everh.
Mycosphaerella andersonii (Sacc.) Lindau
Sphaerella andersonii Sacc. non Ellis & Everh., hom. illeg.
Sphaerella conigena Ellis & Everh. non Peck, hom. illeg.
Distr.: MT(type).
Host: Pseudotsuga
Didymella effusa (Niessl) Sacc.
Didymosphaeria effusa Niessl
Didymella sambucina Feltgen
Distr.: ND, UT; Europe.
Notes: Sivanesan (The bitunicate Ascomycetes. p. 420. 1984).
Host: Artemisia
Didymella eumorpha (Berk. & M.A. Curtis) Sacc.
Distr.: Southern United States.
Notes: On monocotyledonous plants.

Hosts: Arundinaria, Zizaniopsis
Didymella eupyrena Sacc.
Didymella superflua (Auersw.) Sacc. var. *macedonica* Petr.
Phoma sp. (anamorph)
Distr.: AK, WA; Europe.
Notes: On stems; type described from *Aconitum* (Corlett. Canad. J. Bot. 59:2026. 1981).
Host: Urtica
Didymella eurotiae Fairm.
Distr.: NM(type).
Host: Ceratoides
Didymella exigua (Niessl) Sacc.
Didymosphaeria exigua Niessl
Distr.: Temperate and subtropical regions
Notes: On various dicotyledonous substrates. Sivanesan (The bitunicate Ascomycetes. p. 420. 1984).
Hosts: Carya, Dipsacus, Eichhornia, Heracleum
Didymella festucae (H. Wegelin) L. Holm
Didymosphaeria festucae H. Wegelin
Distr.: Temperate regions.
Notes: Stem eyespot disease (Corlett. Canad. J. Bot. 59:2028. 1981; Smith & Shoemaker. Canad. J. Bot. 52:2061. 1974).
Host: Festuca
Didymella fuchsiae Cooke & Ellis
Distr.: CA(type).
Host: Fuchsia
Didymella juncina (Berk. & Ravenel) Sacc.
Distr.: SC(type).
Host: Juncus
Didymella leguminis-cytisis (Desmaz.) Höhn.
Mycosphaerella leguminis-cytisi (Desmaz.) Lindau
Sphaerella leguminis-cytisi (Desmaz.) Ces. & De Not.
Distr.: CA; Europe.
Host: Cercis
Didymella lethalis (R. Stone) Sivanesan
Mycosphaerella lethalis R. Stone
Distr.: Central and eastern North America.
Notes: Sivanesan (The bitunicate Ascomycetes. p. 424. 1984).
Host: Melilotus
Didymella ligulicola (K. Baker, Dimock, & L.H. Davis) Arx in E. Müller & Arx
Mycosphaerella ligulicola K. Baker, Dimock, & L.H. Davis
Distr.: North America; Australia; Europe; Japan.
Notes: Ray blight (Walker & Baker. Trans. Brit. Mycol. Soc. 80:31. 1983); Punithalingam. C.M.I. Descr. 662. 1980 as *Didymella chrysanthemi*). Boerema & Bollen (Persoonia 8:114. 1975) provided a discussion of the anamorph synonymy.
Host: Chrysanthemum
Didymella lupini (Cooke & Harkn.) Berl. & Voglino
Distr.: CA(type).
Host: Lupinus
Didymella megarrhizae Cooke & Harkn.
Distr.: CA.
Host: Marah
Didymella nigrificans P. Karst.
Distr.: NM; Finland(type).
Host: Rosa
Didymella nucis-hicoriae Fairm.
Distr.: NY(type).
Host: Carya
Didymella obscura Rehm
Distr.: MD; Germany(type).
Host: Melilotus
Didymella phacidiomorpha (Ces.) Sacc.
Distr.: FL; Italy(type)
Hosts: Arecaceae, Coccothrinax
Didymella prominens Ellis & Everh.
Distr.: GA, LA.
Hosts: Ambrosia, Hypericum
Didymella proximella (P. Karst.) Sacc.
Didymella refracta (Cooke) Sacc.
Stagonospora sp. (anamorph)
Distr.: OR; Europe.
Notes: On culms and leaves (Corlett. Canad. J. Bot. 59:2033. 1981; Corlett & Smith. Canad. J. Bot. 56:2818. 1978).
Host: Carex
Didymella ramonae Fairm.
Distr.: CA(type).
Host: Salvia
Didymella ricini Ellis & Everh.
Cercidospora ricini (Ellis & Everh.) Kuntze
Distr.: LA(type).
Host: Ricinus
Didymella segna (Cooke & Ellis) Sacc.
Distr.: NJ(type).
Host: Nyssa
Didymella stromatica C.T. Rogerson & Barr
Distr.: UT(type).
Notes: On dead stems (Rogerson & Barr. Mycotaxon 17:251. 1983).
Host: Thermopsis
Didymella trifolii (Fuckel) Sacc.
Phoma sp. (anamorph)
Distr.: MS, western North America; Europe.
Notes: On dead stems (Corlett. Canad. J. Bot. 59:2034. 1981).
Host: Trifolium
Didymella wehmeyeri Dearn. & House, nom. inval.
Distr.: NY.
Host: Ceanothus

Didymellina W. Höhnk. Loculoascomycetes, Dothideales
Notes: This genus is considered a synonym of *Mycosphaerella*, but not all names have been redisposed.
Didymellina poecilospora McWhorter
Cladosporium sp. (anamorph)
Distr.: OR(type).
Host: Iris

Didymium Schrad. Myxomycetes
Didymium iridis (Ditmar) Fr.
Distr.: Cosmopolitan.
Host: Zea

Didymobotryopsis Henn. Hyphomycetes
Didymobotryopsis cerastii Solheim
Distr.: WY.
Host: Cerastium

Didymochaeta Sacc. & Ellis. Coelomycetes
Didymochaeta fraserae (Ellis & Everh.) Sutton
Ascochyta fraserae Ellis & Everh.
Diplodina fraserae (Ellis & Everh.) Tracy & Earle
Didymochaeta americana Sacc. & Ellis
Distr.: CO.
Notes: Sutton (The Coelomycetes. p. 410. 1980).
Host: Frasera

Didymolepta Munk. Loculoascomycetes, Dothideales
Didymolepta mimuli (Wehmeyer) Barr
Apiosporella mimuli Wehmeyer
Distr.: Western United States.
Notes: Barr (Mycotaxon 29:503. 1987) transferred this to *Didymolepta*. *Macrophoma mimuli* (Wehmeyer. Mycologia 38:323. 1946) may be the anamorph.
Hosts: Allium, Mimulus, Senecio, Veratrum

Didymopeltis Batista & Lima. Loculoascomycetes, Dothideales
Notes: This genus is considered a synonym of *Schizothyrium*, but not all names have been redisposed.
Didymopeltis pini Batista
Distr.: CA; Austria(type).

Notes: The identity of this species is uncertain.
Host: Pinus

Didymosamarospora T.W. Johnson & H.S. Gold. Loculoascomycetes, Dothideales

Didymosamarospora euryhalina T.W. Johnson & H.S. Gold
Distr.: NC.
Notes: A doubtful species (Kohlmeyer & Kohlmeyer. Marine mycology. p. 547. 1979).
Host: Juncus

Didymosphaeria Fuckel. Loculoascomycetes, Dothideales

Notes: Scheinpflug (Ber. Schweiz. Bot. Ges. 68:325. 1958).
Didymosphaeria andropogonis Ellis & Langl.
Microthelia andropogonis (Ellis & Langl.) Kuntze
Distr.: LA; Bermuda.
Host: Vetiveria
Didymosphaeria arenaria V. Mouton var. ***macrospora*** Sacc. & Scalia
Distr.: AK(type), WA.
Hosts: Deschampsia, Elymus
Didymosphaeria arundinariae Ellis & Everh.
Microthelia arundinariae (Ellis & Everh.) Kuntze
Distr.: GA, LA.
Host: Arundinaria
Didymosphaeria catalina Ellis & Everh.
Distr.: CA(type).
Host: Artemisia
Didymosphaeria catalpae J.B. Parker
Phyllosticta catalpae Ellis & G. Martin (anamorph)
Distr.: OH(type).
Host: Catalpa
Didymosphaeria ceanothi (Cooke & Harkn.) Berl. & Voglino
Distr.: CA(type).
Host: Ceanothus
Didymosphaeria celtidis Ellis & Everh.
Distr.: KS, OK.
Host: Celtis
Didymosphaeria chionanthi J.H. Miller & G.E. Thompson
Distr.: GA(type).
Host: Chionanthus
Didymosphaeria circinans Harkn.
Distr.: CA(type).
Host: Eucalyptus
Didymosphaeria clementsii Sacc. & D. Sacc.
Phorcys minuta Clements
Distr.: CO, WY.
Host: Yucca
Didymosphaeria cryptosphaerioides Rehm
Distr.: TX(type).
Host: Prosopis
Didymosphaeria decolorans Rehm var. ***americana*** Sacc.
Distr.: ND(type).
Notes: The type variety is considered a synonym of *D. futilis*.
Host: Cornus
Didymosphaeria empetri (DC.) Sacc.
Distr.: AK; Europe.
Host: Empetrum
Didymosphaeria enalia Kohlmeyer
Distr.: Widespread in marine habitats.
Host: Rhizophora
Didymosphaeria futilis (Berk. & Broome) Rehm
Didymosphaeria albescens Niessl
Didymosphaeria brunneola Niessl
Didymosphaeria clematidis Fautrey
Didymosphaeria crastophila (Niessl) G. Wint.
Didymosphaeria decolorans Rehm
Didymosphaeria epidermidis (Fr.) Fuckel
Didymosphaeria typhae Peck
Dendrophoma sp. (synanamorph)
Fusicladiella sp. (synanamorph)
Distr.: Cosmopolitan.
Notes: Sivanesan (The bitunicate Ascomycetes. p. 342. 1984).
Hosts: Apocynum, Asparagus, Aster, Clematis, Conyza, Eucalyptus, Pinus, Pluchea, Salix, Sambucus, Schizachyrium, Symphoricarpos, Typha, Viburnum
Didymosphaeria graminicola Ellis & Everh.
Microthelia graminicola (Ellis & Everh.) Kuntze
Distr.: GA, KS.
Notes: This is probably a synonym of *D. futilis*.
Host: Zea
Didymosphaeria grumata (Cooke) Rehm
Distr.: NJ(type).
Host: Lyonia
Didymosphaeria housei Dearn.
Distr.: NY(type).
Notes: This is probably a synonym of *D. futilis*.
Host: Ceanothus
Didymosphaeria johansenii Dearn.
Distr.: IA(type).
Host: Mertensia
Didymosphaeria magnoliae J.H. Miller & G.E. Thompson
Distr.: GA(type).
Host: Magnolia
Didymosphaeria massarioides Sacc. & Brunaud
Distr.: WA; France(type).
Host: Sitanion
Didymosphaeria nana Rostr.
Distr.: AK; Greenland(type).
Host: Alnus
Didymosphaeria nana Rostr. var. ***brachyspora*** Sacc.
Distr.: AK(type).
Host: Alnus
Didymosphaeria oregonensis Goodd.
Distr.: AK, western United States.
Notes: Canker on alder (Goodding. Phytopathology 21:913. 1931).
Host: Alnus
Didymosphaeria polysticta (Berk. & M.A. Curtis) Sacc.
Distr.: AL(type), MD.
Host: Smilax
Didymosphaeria saccata Clements, nom. nud.
Distr.: CO.
Host: Iris
Didymosphaeria sarmentii (Cooke & Harkn.) Berl. & Voglino
Distr.: CA(type).
Host: Vitis
Didymosphaeria thalictri Ellis & Dearn.
Distr.: GA; Canada(Ontario); Europe.
Host: Thalictrum
Didymosphaeria wikstroemiae Petr.
Distr.: HI(type).
Host: Wikstroemia
Didymosphaeria yuccogena (Cooke) Sacc.
Distr.: CA(type).
Host: Yucca

Didymosporina Höhn. Coelomycetes

Didymosporina aceris (Lib.) Höhn.
Marssonina truncatula (Sacc.) Magnus
Gloeosporium truncatulum Sacc.
Distr.: Temperate northern hemisphere.
Host: Acer

Didymosporium Sacc. Coelomycetes

Notes: According to Sutton (Mycol. Pap. 141:67. 1977) this is a doubtful name, but not all names have been redisposed.
Didymosporium arbuticola Zeller
Distr.: OR, WA.
Host: Arbutus
Didymosporium concinnum Petr.
Distr.: VA(type).
Host: Quercus

Diedickea Syd. & P. Syd. Coelomycetes
Diedickea piceae Bonar
Distr.: CA.
Notes: Yellowing and discoloration of leaves.
Host: Picea

Dilophospora Desmaz. Coelomycetes
Dilophospora alopecuri (Fr.:Fr.) Fr.
Dilophospora graminis Desmaz.
Dilophospora holci Fuckel
Distr.: Temperate regions.
Notes: Twist of cereals and grasses (Gibson & Sutton. C.M.I. Descr. 490. 1976).
Hosts: Agropyron, Calamagrostis, Dactylis, Deschampsia, Elymus, Festuca, Holcus, Hordeum, Leersia, Leucopoa, Melica, Phalaris, Poa, Secale, Sitanion, Triticum

Dimeriella Speg. Loculoascomycetes, Dothideales
Notes: Farr (Mycologia 71:243. 1979).
Dimeriella sacchari (Breda de Haan) Hansf. ex E. Abbott in S.J. Hughes et al.
Coleroa sacchari Breda de Haan
Eriosphaeria sacchari (Breda de Haan) Went
Venturia sacchari (Breda de Haan) Sacc.
Distr.: Widespread in tropical regions.
Notes: Red leaf spot of sugarcane, a disease of minor importance (Hughes, Abbott, and Wismer. Sugar Cane Dis. 2:42. 1964; Sivanesan. C.M.I. Descr. 775. 1983).
Host: Saccharum

Dimerium (Sacc. & Syd.) Sacc. & Syd. Loculoascomycetes, Dothideales
Dimerium balsamicola (Peck) Shoemaker
Meliola balsamicola Peck
Dimeriella balsamicola (Peck) Petr. in Terrier
Dimerosporium balsamicola (Peck) Ellis & Everh.
Distr.: Northeastern North America.
Hosts: Abies, Picea
Dimerium langloisii (Ellis & G. Martin) Sacc. & Syd.
Dimerosporium langloisii Ellis & G. Martin
Dimerosporium galactis Ellis & Everh.
Distr.: Southern United States.
Notes: Mycoparasite of *Asterina* spp. in the tropics and subtropics (Müller & Arx. Beitr. Kryptogamenfl. Schweiz 11(2):477. 1962).
Hosts: Galax, Justicia
Dimerium nigrosporum L.E. Miles
Distr.: MS(type).
Notes: On *Meliola* on leaves.
Host: Ilex
Dimerium piceum Berk. & M.A. Curtis
Phaeostigme picea (Berk. & M.A. Curtis) Syd.
Coniothyrium glabroides F. Stevens (anamorph)
Distr.: GA.
Host: Magnolia

Dimerosporiopsis Henn. Loculoascomycetes, Dothideales
Notes: This genus is a synonym of *Antennularia* (Müller & Arx. Beitr. Kryptogamenfl. Schweiz 11(2):429. 1962), but not all names have been redisposed.
Dimerosporiopsis macrospora Batista & Bezerra
Distr.: ID.
Host: Pinus

Dimerosporium Fuckel. Loculoascomycetes, Dothideales
Notes: This genus is a synonym of *Asterina* Lév., but not all names have been redisposed.
Dimerosporium ilicinum Cooke
Distr.: NC, SC.
Host: Ilex
Dimerosporium nimbosum Ellis & G. Martin
Distr.: FL, SC.
Notes: Theissen (Beih. Bot. Centralbl. 29:63. 1912).
Host: Smilax
Dimerosporium populi Ellis & Everh., nom. nud.
Distr.: MT, NJ.
Host: Populus

Dimorphospora Tubaki. Hyphomycetes
Dimorphospora foliicola Tubaki
Distr.: North temperate regions; Tanzania.
Notes: On submerged leaves and other organic matter in streams and ponds (Nilsson. Symb. Bot. Upsal. 18:92. 1964).
Hosts: Acer, Quercus

Dinemasporium Lév. Coelomycetes
Dinemasporium decipiens (De Not.) Sacc.
Excipula decipiens De Not.
Distr.: United States; Europe.
Hosts: Celtis, Ulmus
Dinemasporium graminum (Lib.) Lév. var. ***strigosulum*** P. Karst.
Distr.: ND; Europe.
Host: Agropyron
Dinemasporium strigosum (Pers.:Fr.) Sacc.
Dinemasporium graminum (Lib.) Lév.
Dinemasporium longisetum Syd.
Phomatospora dinemasporium J. Webster (teleomorph)
Distr.: Cosmopolitan.
Notes: Common on many graminicolous hosts, recorded on at least 46 different host genera (Sutton. The Coelomycetes. p. 458. 1980).
Hosts: Poa, Schizachyrium, Secale

Diplocarpon F.A. Wolf. Hyphomycetes
Diplocarpon earlianum (Ellis & Everh.) F.A. Wolf
Peziza earliana Ellis & Everh.
Fabraea earliana (Ellis & Everh.) Seaver
Marssonina fragariae (Lib.) Kleb. (anamorph)
Distr.: Widespread.
Notes: Strawberry leaf scorch (Sivanesan & Gibson. C.M.I. Descr. 486. 1976).
Host: Fragaria
Diplocarpon mespili (Sorauer) Sutton
Fabraea maculata Atk.
Diplocarpon maculatum (Atk.) Jørst.
Entomosporium mespili (DC.) Sacc. (anamorph)
Distr.: Cosmopolitan.
Notes: Leaf and rose spot of Rosaceae (Sivanesan & Gibson. C.M.I. Descr. 481. 1976 as *D. maculatum*).
Hosts: Amelanchier, Chaenomeles, Cotoneaster, Crataegus, Cydonia, Eriobotrya, Heteromeles, Malus, Mespilus, Photinia, Prunus, Pyrus, Sorbus
Diplocarpon rosae F.A. Wolf
Marssonina rosae (Lib.) Died. (anamorph)
Distr.: Cosmopolitan.
Notes: Sivanesan & Gibson (C.M.I. Descr. 485. 1976).
Host: Rosa

Diplodia Fr. in Mont. Coelomycetes
Diplodia abrotani Fuckel
Distr.: ND; temperate Europe.
Host: Artemisia
Diplodia acericola Tehon & G.L. Stout
Distr.: IL.
Host: Acer
Diplodia aesculi Lév.
Distr.: Eastern North America; Europe.
Host: Aesculus
Diplodia akebiae Fairm.

Distr.: NY.
Host: Akebia
Diplodia amelanchieris Sacc.
Distr.: France(type); distribution elsewhere is questionable.
Host: Amelanchier
Diplodia amorphae (Wallr.) Sacc.
Camarosporium amorphae Sacc. (synanamorph)
Cucurbitaria amorphae (Wallr.) Fuckel (teleomorph)
Distr.: MS; Europe.
Notes: Sivanesan (The bitunicate Ascomycetes. p. 393. 1984).
Host: Amorpha
Diplodia ampelina Cooke
Distr.: SC.
Host: Vitis
Diplodia asclepiadea Cooke & Ellis
Distr.: NJ(type).
Host: Asclepias
Diplodia asparagi C. Brun
Distr.: United States; Europe.
Host: Asparagus
Diplodia atrata (Desmaz.) Sacc.
Distr.: North America(questionable); Europe.
Host: Acer
Diplodia atrocoerulea Ellis & Everh.
Distr.: Southern United States.
Host: Datura
Diplodia australiae Speg.
Distr.: CA; Argentina.
Host: Eucalyptus
Diplodia bambusae Ellis & Langl.
Distr.: LA.
Hosts: Bambusa, Poaceae
Diplodia berberidina Sacc.
Distr.: NY.
Host: Berberis
Diplodia cacaoicola Henn.
Distr.: FL; Africa.
Host: Cajanus
Diplodia camphorae Tassi
Distr.: MS; Italy.
Host: Cinnamomum
Diplodia caryigena Ellis & Everh.
Distr.: Temperate North America.
Host: Carya
Diplodia celtidigena Ellis & Barth.
Distr.: Central United States.
Host: Celtis
Diplodia cercidis Ellis & Everh.
Distr.: Eastern United States.
Host: Cercis
Diplodia coluteae Schnabl
Distr.: PA; Europe.
Notes: Mirza (Nova Hedwigia 16:202. 1968).
Host: Colutea
Diplodia convolvuli Dearn. & House
Distr.: NY.
Host: Calystegia
Diplodia coryli Fuckel
Otthia coryli Fuckel (teleomorph)
Distr.: MI; Europe.
Host: Corylus
Diplodia cowdellii Berk. & Broome
Diplodiella cowdellii (Berk. & Broome) Sacc.
Distr.: ME; Great Britain.
Host: Gossypium
Diplodia crassulae Cooke & Harkn.
Distr.: CA.
Host: Crassula
Diplodia crataegi Westend. in J. Kickx fil.
Distr.: Eastern United States; Europe.
Host: Pyracantha
Diplodia cyparissa Cooke & Harkn.
Distr.: CA.
Host: Cupressus
Diplodia epicocos Cooke
Distr.: FL; Tahiti.
Host: Cocos
Diplodia eriobotryae Sacc.
Distr.: FL; Italy.
Host: Eriobotrya
Diplodia eucalypti Cooke & Harkn.
Distr.: CA(type).
Host: Eucalyptus
Diplodia euonymi Westend. in J. Kickx fil.
Distr.: AL, NY; Europe.
Host: Euonymus
Diplodia fairmanii Ellis & Everh.
Distr.: NY.
Host: Menispermum
Diplodia frangulae Fuckel
Cucurbitaria rhamni Fuckel (teleomorph)
Distr.: CA; France
Host: Rhamnus
Diplodia frumenti Ellis & Everh.
Botryosphaeria festucae (Lib.) Arx & E. Müller (teleomorph)
Distr.: Eastern and southern North America; Europe.
Hosts: Arachis, Zea
Diplodia fuchsiae Cooke & Harkn.
Distr.: CA.
Host: Fuchsia
Diplodia hamamelidis Fairm.
Distr.: NY.
Host: Hamamelis
Diplodia hederae Fuckel
Distr.: North and South America; Europe.
Host: Hedera
Diplodia herbarum (Corda) Lév.
Sporocadus herbarum Corda
Distr.: United States; Algeria; Europe.
Host: Marrubium
Diplodia heteromelina Fairm. in Millsp. & L. Nutt.
Distr.: CA.
Host: Heteromeles
Diplodia hibiscina Cooke & Ellis
Distr.: HI, NJ.
Host: Hibiscus
Diplodia holodisci Barth. nom. nud.
Distr.: WA.
Notes: The place of publication of this name could not be found.
Host: Holodiscus
Diplodia hortensis Sacc.
Distr.: Eastern United States.
Host: Clematis
Diplodia juglandis (Fr.) Fr.
Sphaeria juglandis Fr.
Distr.: United States; Europe.
Host: Juglans
Diplodia julibrissin Speg.
Distr.: WA; southern Europe.
Host: Albizia
Diplodia koelreuteriae Sacc.
Distr.: NJ; France.
Host: Koelreuteria
Diplodia laelio-cattleyae Sibilia
Distr.: FL; Europe.
Host: Cattleya
Diplodia langloisii Sacc. & Syd.
Botryodiplodia meliae (Ellis & Everh.) Ellis & Everh.
Diplodia meliae Ellis & Everh. non Tassi, hom. illeg.
Distr.: LA(type).
Host: Melia
Diplodia leptodactyli Earle
Distr.: TX(type).
Host: Gilia
Diplodia linderae Ellis & Everh.
Microdiplodia linderae (Ellis & Everh.) Tassi
Distr.: Northeastern North America.
Host: Lindera
Diplodia liriodendri Peck
Distr.: Temperate North America.

Host: Liriodendron
Diplodia longispora Cooke & Ellis
Distr.: Central and eastern North America.
Notes: This species has been placed in synonymy with several species various genera; its correct placement remains uncertain.
Hosts: Castanea, Quercus
Diplodia lonicerae Fuckel
Distr.: North America; Europe.
Host: Lonicera
Diplodia lupini Cooke & Harkn.
Distr.: CA.
Host: Lupinus
Diplodia maculata Cooke & Harkn.
Distr.: Western United States.
Host: Arbutus
Diplodia malorum Fuckel
Diplodia maura Cooke & Ellis
Diplodia pseudodiplodia Fuckel
Distr.: Cosmopolitan.
Hosts: Malus, Pyrus
Diplodia mamillana Fr.
Distr.: WA; Europe.
Host: Cornus
Diplodia maydis (Berk.) Sacc.
Sphaeria maydis Berk.
Distr.: North America; Europe.
Notes: Sutton & Waterston (C.M.I. Descr. 84. 1966).
Host: Zea
Diplodia megalospora Berk. & M.A. Curtis
Distr.: Eastern and southern United States.
Host: Pinus
Diplodia mutila (Fr.:Fr.) Mont.
Sphaeropsis malorum Peck
Sphaeropsis malorum Berk.
Diplodia pittospori Cooke & Harkn.
Botryosphaeria stevensii Shoemaker (teleomorph)
Distr.: Eastern North America; Europe.
Notes: Sivanesan (The bitunicate Ascomycetes. 1984; Shoemaker. Canad. J. Bot. 42:1299. 1964; Sutton. The Coelomycetes. p. 76. 1980; Vajna. Eur. J. Forest Path. 16:233. 1986).
Hosts: Ficus, Gossypium, Malus, Pinus, Pittosporum, Prosopis, Pyrus
Diplodia nerii Speg.
Distr.: CA; Italy.
Host: Nerium
Diplodia officinalis Ellis & Everh.
Distr.: MI, NY.
Host: Sassafras
Diplodia opuntiae Sacc.
Distr.: HI, eastern and central United States; Italy.
Host: Opuntia
Diplodia oryzae Miyake
Distr.: LA; Japan.
Host: Zizania
Diplodia paraphysaria Sacc.
Distr.: MD, MO; Brazil.
Host: Cattleya
Diplodia persicina (Berk. & M.A. Curtis) Grove
Sphaeropsis persicina Berk. & M.A. Curtis
Distr.: LA, PA; Great Britain.
Notes: According to Petrak & Sydow (Repert. Spec. Nov. Regni Veg. 42:51. 1927) this species may be synonymous with *Haplosporella demersa* var. *missouriensis* (Bubák) Petr. & Syd.
Host: Prunus
Diplodia phoenicum (Sacc.) H. Fawc. & L.J. Klotz
Botryodiplodia phoenicum Sacc.
Distr.: CA, FL.
Host: Phoenix
Diplodia pinastri Grove
Distr.: United States; Europe.
Host: Pinus
Diplodia polygonicola Peck non Hollós, hom. illeg.
Distr.: KS(type).
Host: Polygonum
Diplodia populina Fuckel
Distr.: PA; Europe(type).
Host: Populus
Diplodia propullans (Schwein.) Peck
Sphaeria propullans Schwein.
Diplodia celastri Cooke
Diplodia celastrina Ellis & Everh.
Sphaeropsis celastrina Peck
Distr.: United States.
Host: Celastrus
Diplodia pruni Fuckel in Sacc.
Distr.: FL; Europe.
Hosts: Prunus, Vicia
Diplodia puerariae Barth.
Distr.: MD, TX.
Host: Pueraria
Diplodia punctipetiolae Cooke
Distr.: SC.
Notes: Grove (British stem and leaf fungi. Vol. II. p. 47. 1937.) suggested that this species may be synonymous with *Diplodia magnoliae* Westendorp.
Host: Magnolia
Diplodia quercina Westend.
Distr.: CA; Europe.
Host: Quercus
Diplodia radicicola Tassi
Distr.: Reported from an unknown locality in the United States.
Host: Aristolochia
Diplodia resurgens Cooke & Harkn.
Microdiplodia resurgens (Cooke & Harkn.) Tassi
Distr.: CA(type).
Host: Rhus
Diplodia rosarum Fr.
Diplodia rosae Westend. non Berk. & M.A. Curtis, hom. illeg.
Otthia rosae Fuckel (teleomorph)
Distr.: CA; Europe.
Host: Rosa
Diplodia rubi Fr.
Macrophoma conica Pass.
Didymosphaeria diplospora Rehm (teleomorph)
Distr.: VA, WA; Europe.
Host: Rubus
Diplodia salicina Lév.
Cucurbitaria salicina Fuckel (teleomorph)
Distr.: Eastern and southern United States; Europe.
Host: Salix
Diplodia sambuci Tehon & E.Y. Daniels
Distr.: IL.
Host: Sambucus
Diplodia sambucina Sacc.
Distr.: Central and eastern United States; Europe.
Host: Sambucus
Diplodia sarmentorum (Fr.:Fr.) Fr.
Diplodia alni-rubrae Peck
Otthia spiraeae (Fuckel) Fuckel (teleomorph)
Distr.: Northwestern North America; Europe.
Notes: Stevens (Mycologia 28:334. 1936) discussed the distribution of this fungus in North America and provided an extensive list of synonyms. Booth (Trans. Brit. Mycol. Soc. 41:335. 1958).
Hosts: Acer, Alnus, Corylus, Malus, Menispermum, Pyrus, Rosa
Diplodia sarothamni Cooke & Harkn.
Distr.: CA; Europe.
Host: Spartium
Diplodia sassafras Tracy & Earle
Distr.: MS.
Host: Sassafras
Diplodia sedicola Cooke & Harkn.
Distr.: CA.
Host: Sedum
Diplodia shearii Petr.
Otthia shearii Petr. (teleomorph)
Distr.: HI(type).
Host: Leucaena
Diplodia smilacina Berk.
Distr.: Central and eastern United States.
Host: Smilax
Diplodia sophorae Speg. & Sacc.

Distr.: OH; Europe.
Host: Sophora
Diplodia subcuticularis Dearn. & House
Distr.: NY.
Hosts: Acacia, Sassafras
Diplodia suberina Durieu & Mont.
Distr.: Eastern North America; Europe.
Host: Quercus
Diplodia subtecta Fr.
Distr.: Europe; distribution elsewhere is questionable.
Notes: Zambettakis (Bull. Soc. Mycol. France 70:219. 1954) considered this a synonym of *Diplodia acerina* Lév.
Host: Acer
Diplodia sycina Mont. in Castagne
Distr.: OR; Europe.
Host: Ficus
Diplodia sycina Mont. in Castagne var. ***syconophila*** Sacc.
Distr.: NC,OR, TX; Europe.
Host: Ficus
Diplodia tamariscina Sacc.
Distr.: NY; Europe.
Host: Tamarix
Diplodia taxi (Sowerby) De Not.
Distr.: OR, WI; Europe.
Notes: On branches and leaves.
Host: Taxus
Diplodia tenuis Cooke & Harkn.
Distr.: CA(type).
Host: Eucalyptus
Diplodia thalictri Ellis & Dearn.
Distr.: Northeastern North America.
Host: Thalictrum
Diplodia thalictricola Syd.
Distr.: Germany(type); distribution elsewhere is uncertain.
Host: Clematis
Diplodia thujina Peck & G.W. Clinton
Distr.: NY(type).
Host: Thuja
Diplodia thyoidea Cooke & Ellis
Distr.: NJ(type).
Host: Chamaecyparis
Diplodia tumefaciens (Shear) Zalasky
Macrophoma tumefaciens Shear in Hubert
Distr.: North America; northern Europe.
Host: Populus
Diplodia ulmi Dearn.
Distr.: IL; Canada(Ontario).
Host: Ulmus
Diplodia umbellulariae Ellis & Everh.
Distr.: CA(type).
Host: Umbellularia
Diplodia vaccinii Berl. & Roum.
Distr.: France(type), distribution elsewhere is questionable.
Host: Vaccinium
Diplodia veratri Earle
Ascochytella veratri (Earle) Petr.
Distr.: NV(type).
Host: Veratrum
Diplodia vincae Sacc. in Rabenh.
Distr.: OH, MD.
Host: Vinca
Diplodia virginiana Cooke & Ravenel
Distr.: SC(type).
Host: Juniperus
Diplodia viticola Desmaz.
Distr.: Eastern North America(rare); Europe.
Host: Vitis
Diplodia vulgaris Lév.
Distr.: FL, NJ; Great Britain.
Host: Prunus
Diplodia weigelae Sacc.
Distr.: NY.
Host: Weigela

Diplodiella Petr. Coelomycetes

Notes: According to Sutton (Mycol. Pap. 141:69. 1977) this name is illegitimate, but not all names have been redisposed.
Diplodiella tecomae (Cooke) Sacc.
Diplodia tecomae Cooke non Pass., hom. illeg.
Distr.: SC.
Host: Campsis

Diplodina Westend. Coelomycetes

Diplodina acerina (Pass.) Sutton
Gloeosporium acerinum Pass. in Thuem. non Westend., hom. illeg.
Discella acerina (Pass.) Arx
Gloeosporium acerinum Westend.
Marssonina acerina (Westend.) Magnus
Septogloeum acerinum (Pass.) Sacc.
Hendersonia collapsa Cooke & Ellis
Stagonospora collapsa (Cooke & Ellis) Sacc.
Marssonina tulasnei (Sacc.) Died.
Cryptodiaporthe hyareix (Tode:Fr.) Petr. (teleomorph)
Distr.: Temperate North America; Europe.
Notes: Sutton (The Coelomycetes. p. 604. 1980).
Hosts: Acer, Hibiscus
Diplodina aesculi (Sacc.) Sutton
Septomyxa aesculi Sacc.
Gloeosporium carpigenum Cooke
Cryptodiaporthe aesculi (Fuckel) Petr. (teleomorph)
Distr.: Temperate North America; Europe.
Notes: On branches and twigs. Sutton (The Coelomycetes. p. 606. 1980).
Host: Aesculus
Diplodina andromedae (Rostr.) Jørst.
Diplodina rostrupii Vestergr.
Encoeliopsis rhododendri (Ces.) Nannf. var. *harrimanellae* L. Holm & K. Holm (teleomorph)
Distr.: Circumboreal.
Notes: On capsules of *Cassiope* spp. and other Ericaceae (Holm & Holm. Norw. J. Bot. 27:180. 1979).
Hosts: Cassiope, Phyllodoce
Diplodina coloradensis Ellis & Everh.
Diplodinula coloradensis (Ellis & Everh.) Tassi
Distr.: CO(type).
Host: Gutierrezia
Diplodina dahliae Hollós
Distr.: WA; Europe.
Host: Dahlia
Diplodina eurhododendri G. Voss
Encoeliopsis rhododendri (Ces. ex De Not.) Nannf. (teleomorph)
Distr.: Temperate northern hemisphere.
Notes: On leaves, twigs, and capsules.
Host: Rhododendron
Diplodina fructigena P. Karst.
Distr.: Western United States; Europe.
Host: Ledum
Diplodina fusispora Peck
Distr.: IL(type), reports from elsewhere need confirmation.
Host: Acer
Diplodina lyngei Lind
Distr.: AK; Norway.
Host: Potentilla
Diplodina pedicularidis (Fuckel) Lind
Distr.: Alpine and boreal North America; Europe.
Notes: Leaf spot.
Host: Pedicularis
Diplodina persicae Horn & Hawthorne, nom. illeg.
Distr.: Southern United States.
Host: Prunus
Diplodina psoraleae Ellis & Barth.
Diplodinula psoraleae (Ellis & Barth.) Tassi
Distr.: KS(type).
Host: Psoralidium
Diplodina ramulorum Ellis & Everh.
Distr.: NJ.
Hosts: Lycium, Smilax
Diplodina silenes W.B. Cooke

Distr.: CA.
Host: Silene
Diplodina smilacis Ellis & Everh.
Diplodinula smilacis (Ellis & Everh.) Tassi
Mycosphaerella smilacina (Ellis & Everh.) Tomilin (teleomorph)
Distr.: NJ.
Host: Smilax
Diplodina stevensii Sacc., nom. nud.
Distr.: ND.
Host: Mentzelia
Diplodina tridentatae W.B. Cooke & C.G. Shaw
Distr.: UT.
Host: Artemisia

Diplomitoporus Dománski. Basidiomycetes, Aphyllophorales

Diplomitoporus crustulinus (Bres.) Dománski
Poria crustulina Bres.
Poria chromatica Overh. non (Berk. & Cooke) Cooke, hom. illeg.
Poria conwayana Pilát
Distr.: Temperate northern hemisphere.
Notes: On wood, usually of conifers.
Hosts: Abies, Picea, Populus, Tsuga
Diplomitoporus lenis (P. Karst.) R.L. Gilbertson & Ryvarden
Poria lenis (P. Karst.) Sacc.
Poria earlei Murrill
Poria montana Murrill
Poria tenuiporia Murrill
Distr.: Temperate northern hemisphere.
Notes: On wood, usually of conifers.
Hosts: Abies, Alnus, Chamaecyparis, Juniperus, Larix, Liquidambar, Picea, Pinus, Pseudotsuga, Thuja, Tsuga
Diplomitoporus linbladii (Berk.) R.L. Gilbertson & Ryvarden
Polyporus cinerascens Bres., nom. inval.
Poria cinerascens (Bres.) Sacc. & Syd., nom. inval.
Poria subfuscoflavida (Rostk.) Cooke
Distr.: Widespread.
Notes: On wood, usually of conifers.
Hosts: Abies, Acer, Alnus, Betula, Larix, Picea, Pinus, Populus, Pseudotsuga, Thuja, Tsuga
Diplomitoporus overholtsii (Pilát) R.L. Gilbertson & Ryvarden
Poria overholtsii Pilát
Distr.: AZ, eastern North America.
Notes: On wood.
Host: Platanus
Diplomitoporus rimosus (Murrill) R.L. Gilbertson & Ryvarden
Poria rimosa Murrill
Distr.: Western North America.
Notes: On *Juniperus* spp., rarely on *Cupressus* spp.
Hosts: Cupressus, Juniperus

Diplonaevia Sacc. Discomycetes, Helotiales

Notes: This genus is considered a synonym of *Merostictis* (Hein. Willdenowia Beih. 9:10. 1976), but not all names have been redisposed.
Diplonaevia melaleuca Ellis & Everh.
Distr.: MT(type).
Notes: Seaver (North Amer. cup fungi. [Inoperculates]. p. 276. 1951) suggested that this is a synonym *Solenopeziza fimbriata* Ellis & Barth.
Host: Populus

Diplopeltis Pass. Coelomycetes

Notes: This genus was renamed *Pycnoseynesia* Kuntze (Sutton. Mycol. Pap. 141:226. 1977), but not all names have been redisposed.
Diplopeltis sassafrasicola Tehon & G.L. Stout
Distr.: IL.
Host: Sassafras

Diplosporonema Höhn. Coelomycetes

Diplosporonema delastrei (Delacr.) Petr.
Gloeosporium delastrei Delacr. in Mont.
Marssonina delastrei (Delacr.) Magnus
Distr.: North America; temperate Europe.
Notes: On leaves and stems of Caryophyllaceae.
Hosts: Agrostemma, Silene

Diplotheca Starb. Loculoascomycetes, Dothideales

Diplotheca tunae (Spreng.:Fr.) Starb.
Perisporiopsis wrightii (Berk. & M.A. Curtis) F. Stevens
Perisporium wrightii Berk. & M.A. Curtis
Stevensea wrightii (Berk. & M.A. Curtis) Trott.
Distr.: Subtropical and tropical North America.
Notes: On Cactaceae.
Hosts: Cereus, Opuntia

Diporotheca C.C. Gordon & C.G. Shaw. Pyrenomycetes, Meliolales

Notes: Monotypic.
Diporotheca rhizophila C.C. Gordon & C.G. Shaw
Distr.: WA(type).
Notes: On roots (Gordon & Shaw. Mycologia 52:327. 1960).
Host: Solanum

Discella Berk. & Broome. Coelomycetes

Discella arida Peck
Distr.: NY.
Host: Chamaedaphne
Discella leguminum Cooke
Distr.: TX.
Host: Prosopis
Discella microsperma (G. Johnst.) Sutton
Discella carbonacea (Fr.:Fr.) Berk. & Broome, sensu Wehmeyer
Discella salicis (Westend.) Boerema
Cryptodiaporthe salicella (Fr.:Fr.) Petr. (teleomorph)
Distr.: Circumboreal.
Notes: On limbs.
Host: Salix
Discella ochroleuca (Berk. & M.A. Curtis) Arx
Septoria ochroleuca Berk. & M.A. Curtis
Gloeosporium ochroleuca (Berk. & M.A. Curtis) Ellis & Everh.
Marssonina ochroleuca (Berk. & M.A. Curtis) Lentz
Distr.: Eastern North America.
Notes: Leaf spot (Lentz. Mycologia 42:259. 1950).
Host: Castanea
Discella pilosula Ellis & Everh.
Distr.: CO, NY(type).
Host: Acer
Discella populina Sacc.
Distr.: IA; Europe.
Host: Populus
Discella quercina (G. Wint.) Arx
Marssonia quercina G. Wint.
Marssonina quercina (G. Wint.) Lentz
Marssonia quercina G. Wint. var *major* Overh.
Gloeosporium septorioides Sacc.
Gloeosporium septorioides Sacc. var. *major* Ellis & Everh.
Distr.: Central and eastern North America.
Notes: Leaf spot (Lentz. Mycologia 42:259. 1950).
Hosts: Festuca, Quercus
Discella tenuispora Cooke & Harkn.
Distr.: CA.
Host: Juncus

Discina Fr. Discomycetes, Pezizales

Notes: McKnight (Mycologia 61:614. 1969) reviewed the species known from North America. They occur primarily on soil and have not been reported on a vascular plant host.

Discina ancilis (Pers.) Sacc.
Discina perlata (Fr.) Fr.
Distr.: Temperate regions.
Notes: On soil or decaying wood in conifer and hardwood forests.
Hosts: Abies, Pinus

Discocainia J. Reid & Funk. Discomycetes, Rhytismatales

Discocainia arctica (Ehrenb.:Fr.) L. Holm
Cenangium arcticum Ehrenb.:Fr.
Distr.: Arctic region.
Notes: Torkelson & Eckblad (Norw. J. Bot. 24:137. 1977).
Host: Cassiope
Discocainia treleasei (Sacc. in Sacc., Peck, & Trel.) J. Reid & Funk
Atropellis treleasei (Sacc.) Zeller & Goodd.
Godronia treleasei (Sacc.) Seaver
Scleroderris treleasei Sacc.
Distr.: Pacific Northwest.
Notes: Reid & Funk (Mycologia 58:432. 1966).
Hosts: Picea, Tsuga

Discochora Höhn. Loculoascomycetes, Dothideales

Notes: *Discochora* was proven by Bissett (Mycotaxon 25:519. 1986) to be the correct name for *Guignardia*.
Discochora philoprina (Berk. & M.A. Curtis) Bissett
Botryosphaeria philoprina (Berk. & M.A. Curtis) Arx & E. Müller
Physalospora philoprina (Berk. & M.A. Curtis) Sacc.
Botryosphaeria foliorum (Sacc.) Arx & E. Müller
Discochora ilicis (Schleich.) Höhn.
Physalospora ilicis (Schleich.) Sacc.
Phyllosticta concentrica Sacc. (anamorph)
Leptodothiorella sp. (anamorph spermatia)
Distr.: Eastern United States; Europe; India.
Notes: Aa (Stud. Mycol. 5:44. 1973) included the extensive synonymy of this species. As *Physalospora ilicis*, this fungus is reported to cause a disease of holly in the southeastern United States.
Hosts: Ilex, Yucca
Discochora smilacinina (Peck) Bissett
Anthostomella smilacinina (Peck) Sacc.
Botryosphaeria smilacinina (Peck) Barr
Guignardia smilacinae (Dearn. & House) Dearn. & House
Distr.: Eastern North America.
Notes: Bissett (Canad. J. Bot. 64:1721. 1986).
Host: Smilacina
Discochora yuccae Bissett
Leptodothiorella notabilis Petr. & Cif. (synanamorph)
Phyllosticta yuccae Bissett (anamorph)
Distr.: FL.
Notes: Bissett (Canad. J. Bot. 64:1720. 1986).
Host: Yucca

Discocurtisia Nannf. Discomycetes, Helotiales

Discocurtisia arundinariae (Berk. & M.A. Curtis) Nannf.
Belonium arundinariae (Berk. & M.A. Curtis) Massee
Scleroderris arundinariae Groves
Pyrenopeziza arundinariae (Berk. & M.A. Curtis) Sacc.
Belonium eustegiaeforme (Berk. & M.A. Curtis) Sacc.
Distr.: Southeastern United States.
Notes: Nannfeldt (Mycologia 75:307. 1983).
Hosts: Arundinaria, Spartina

Discogloeum Petr. Coelomycetes

Discogloeum concentricum (Heald & F.A. Wolf) Morgan-Jones
Exosporium concentricum Heald & F.A. Wolf
Distr.: Southern United States.
Notes: Morgan-Jones (Canad. J. Bot. 49:1461. 1971).
Hosts: Euonymus, Ligustrum
Discogloeum veronicae (Lib.) Petr.
Leptothyrium veronicae Lib.
Gloeosporium veronicae Dearn. & House
Distr.: Temperate North America; Europe.
Host: Veronica

Discohainesia Nannf. Discomycetes, Helotiales

Discohainesia oenotherae (Cooke & Ellis) Nannf.
Peziza oenotherae Cooke & Ellis
Pezizella oenotherae (Cooke & Ellis) Sacc.
Pezizella lythri Shear & B.O. Dodge
Hainesia lythri (Desmaz.) Höhn. (anamorph)
Distr.: Cosmopolitan.
Notes: Sutton & Gibson (C.M.I. Descr. 535. 1977).
Hosts: Acer, Betula, Carya, Castanea, Cercis, Cornus, Cotinus, Duchesnea, Epilobium, Eucalyptus, Eugenia, Fragaria, Galax, Gaultheria, Gaura, Gaylussacia, Ligustrum, Liriodendron, Ludwigia, Lysimachia, Lythrum, Magnolia, Malus, Nyssa, Oenothera, Paeonia, Parthenocissus, Pelargonium, Pinus, Podophyllum, Populus, Potentilla, Prunus, Quercus, Rhus, Ribes, Rosa, Rubus, Salix, Shortia, Smilax, Vaccinium, Vitis

Discosia Lib. Coelomycetes

Discosia artocreas (Tode:Fr.) Fr.
Cryptostictis glandicola Starb.
Distr.: Widespread in temperate regions.
Notes: The taxonomic position of this species is uncertain (Sutton. The Coelomycetes. p. 148. 1980).
Hosts: Betula, Camellia, Carya, Castanea, Cercis, Cornus, Crataegus, Cydonia, Eucalyptus, Fagus, Hamamelis, Hepatica, Ilex, Liquidambar, Paeonia, Platanus, Polygonatum, Quercus, Rhododendron, Rhus, Rosa, Rubus, Tilia, Vaccinium, Vitis
Discosia artocreas (Tode:Fr.) Fr. var. ***viticola*** M.A. Curtis
Distr.: SC(type).
Host: Vitis
Discosia baarnensis Vanev
Distr.: CA(type).
Notes: On dead leaves.
Host: Photinia
Discosia blumencronii Bubák
Distr.: OR, WA.
Host: Gaultheria
Discosia deflectens Sacc.
Distr.: SC(type).
Host: Ilex
Discosia maculicola W. Gerard
Distr.: Eastern and southern United States.
Hosts: Epigaea, Gaultheria, Impatiens, Smilax
Discosia pini Heald
Distr.: NE(type).
Host: Pinus
Discosia poikilimera Fairm.
Distr.: CA.
Hosts: Heteromeles, Laurocerasus
Discosia potentillae Tehon
Distr.: IL(type).
Host: Potentilla
Discosia rugulosa Berk. & M.A. Curtis
Distr.: AL(type), FL.
Host: Carya
Discosia smilacina De Not.
Distr.: MS; Italy(type).
Host: Smilax

Discosphaerina Höhn. Loculoascomycetes, Dothideales

Notes: Barr (Contr. Univ. Michigan Herb. 9:579. 1972).
Discosphaerina boltoniae (Dearn. & Barth.) Barr
Guignardia boltoniae Dearn. & Barth. in Dearn.
Distr.: Central United States.
Host: Boltonia

Discosphaerina fagi (H.J. Hudson) Barr
Guignardia fagi H.J. Hudson
Distr.: Northeastern North America; Europe.
Notes: On overwintered leaves of hardwoods (Sivanesan. The bitunicate Ascomycetes. p. 148. 1984).
Host: Acer

Discosporium Höhn. Coelomycetes

Discosporium populeum (Sacc.) Sutton
Phoma populea Sacc.
Chondroplea populea (Sacc.) Kleb.
Dothichiza populea Sacc. & Briard in Sacc.
Melanconium hyalinum Ellis
Distr.: Temperate North America; Europe.
Notes: The type species is a pathogen that has been known as *Chondroplea populea* (Sutton. The Coelomycetes. p. 573. 1980). Booth et al. (C.M.I. Descr. 364. 1973 as *Cryptodiaporthe populea*).
Host: Populus
Discosporium tremuloides (Ellis & Everh.) Sutton
Gloeosporium tremuloides Ellis & Everh.
Cryptocline tremuloides (Ellis & Everh.) Sutton
Leptomelanconium tremuloides (Ellis & Everh.) Arx
Distr.: WI(type).
Host: Populus

Discostroma Clements. Pyrenomycetes, Amphisphaeriales

Notes: Brockmann (Sydowia 28:275. 1975).
Discostroma corticola (Fuckel) I. Brockmann
Clethridium corticola (Fuckel) Shoemaker & E. Müller
Griphosphaeria corticola (Fuckel) Höhn.
Pleosphaerulina corticola (Fuckel) Rehm
Metasphaeria leiostega (Ellis) Sacc.
Metasphaeria staphyleae Dearn. & House
Seimatosporium lichenicola (Corda) Shoemaker & E. Müller (anamorph)
Distr.: Temperate North America; Europe.
Notes: On limbs of woody plants.
Hosts: Carpinus, Fraxinus, Rhododendron, Rosa, Sassafras, Staphylea, Vaccinium
Discostroma massarina (Sacc.) I. Brockmann
Metasphaeria massarina Sacc.
Clethridium massarina (Sacc.) Berl.
Homostegia kelseyi Ellis & Everh.
Phragmodothella kelseyi (Ellis & Everh.) Theiss. & Syd.
Curreya rehmii Schnabl
Curreyella rehmii (Schnabl) Lindau
Discostroma rehmii (Schnabl) Clements
Seimatosporium ribis-alpini (Fautrey) Shoemaker & E. Müller (anamorph)
Distr.: Temperate North America; Europe.
Notes: On stems.
Host: Ribes
Discostroma muricata Ellis & Everh.
Strickeria muricata (Ellis & Everh.) Kuntze
Teichospora muricata Ellis & Everh.
Distr.: Western United States.
Notes: Barr (Mycotaxon 18:151. 1983) provided a description.
Hosts: Acer, Opuntia

Discula Sacc. Coelomycetes

Discula betulina (Westend.) Arx
Gloeosporium betulinum Westend. in J. Kickx fil.
Gloeosporium betulicola Sacc. & Dearn.
Gloeosporium betulinum (Westend.) Petr.
Distr.: Temperate North America; Europe.
Host: Betula
Discula brenckleana (Sacc. & Syd.) Petr.
Macrophoma brenckleana Sacc. & Syd.
Macrophoma salicis Defago & Barth.
Distr.: Western United States.
Host: Salix
Discula fraxinea (Peck) Redlin & Stack
Gloeosporium fraxineum Peck
Gloeosporium aridum Ellis & Holw.
Gnomoniella fraxini Redlin & Stack (teleomorph)
Distr.: Central and eastern United States.
Notes: Anthracnose of ash (Redlin & Stack. Mycotaxon 32:175. 1988).
Host: Fraxinus
Discula graminum (Rostr.) Arx
Gloeosporium graminum Rostr.
Distr.: Temperate North America; Europe.
Notes: Anthracnose.
Host: Poa
Discula microsperma (Berk. & Broome) Sacc.
Discella microsperma Berk. & Broome
Distr.: CA, MS; Europe.
Host: Salix
Discula platani (Peck) Sacc.
Macrophoma petiolata (Cooke) Berl. & Voglino
Sphaeropsis petiolata Cooke
Apiognomonia veneta (Sacc. & Speg.) Höhn. (teleomorph)
Distr.: North America; Europe.
Notes: Anthracnose of hardwoods.
Host: Platanus
Discula runcinata Ellis & Everh.
Distr.: MT.
Host: Stephanomeria
Discula sassafras (Cooke) Arx
Phyllosticta sassafras Cooke
Gloeosporium sassafras (Cooke) Ellis & Everh.
Gloeosporium affine (Ellis & Kellerm.) Ellis & Everh. non Sacc., hom. illeg.
Phyllosticta affine Ellis & Kellerm.
Distr.: Eastern North America.
Notes: Leaf spot.
Hosts: Acer, Brassolaeliocattleya, Cattleya, Cymbidium, Dendrobium, Epidendrum, Laelia, Laeliocattleya, Odontoglossum, Oncidium, Orchidaceae, Phalaenopsis, Sassafras, Vanda
Discula umbrinella (Berk. & Broome) Sutton
Gloeosporium umbrinellum Berk. & Broome
Gloeosporidium umbrinellum (Berk. & Broome) Petr.
Gloeosporium canadense Ellis & Everh.
Gloeosporium celtidis Ellis & Everh.
Gloeosporium divergens Peck
Discula fagi Oudem.
Gloeosporium fagi (Desmaz. & Roberge) Westend.
Gloeosporium fagicola Pass. in Brunaud
Gloeosporium irregulare Peck
Gloeosporium nervisequum (Fuckel) Sacc.
Myxosporium platanicola Ellis & Everh.
Discula quercina (Westend.) Arx non (Cooke) Sacc., hom. illeg.
Gloeosporium quercinum Westend.
Gloeosporium quernum Harkn.
Gloeosporium tiliae Oudem.
Gloeosporium valsoideum Sacc.
Myxosporium valsoideum (Sacc.) Allesch.
Apiognomonia errabunda (Roberge) Höhn. (teleomorph)
Distr.: Temperate North America; Europe.
Notes: Anthracnose of hardwood trees.
Hosts: Celtis, Fagus, Liquidambar, Platanus, Quercus, Tilia
Discula xanthoxyli Ellis & Everh.
Distr.: LA.
Host: Zanthoxylum

Ditiola Fr.:Fr. Basidiomycetes, Dacrymycetales

Femsjonia Fr.
Notes: Two species (Reid. Trans. Brit. Mycol. Soc. 62:473. 1974).
Ditiola pezizaeformis (Lév.) D. Reid
Exidia pezizaeformis Lév.
Femsjonia pezizaeformis (Lév.) P. Karst.
Femsjonia radiculata (Sowerby:Fr.) G.W. Martin, sensu G.W. Martin
Distr.: Widespread.

Notes: On wood, usually of hardwoods.
Host: Liriodendron
Ditiola radicata (Albertini & Schwein.:Fr.) Fr.
Helotium radicatum Albertini & Schwein.:Fr.
Dacrymyces radicatus (Albertini & Schwein.:Fr.) Donk
Dacryomitra nuda (Berk. & Broome) Pat.
Dacryomitra stipitata (Peck) Burt
Distr.: Southeastern and southwestern United States; Europe.
Notes: Reid (Trans. Brit. Mycol. Soc. 62:479. 1974) doubted that this species occurred in North America.
Hosts: Pinus, Quercus

Ditopella J. Schröt. in Cohn. Pyrenomycetes, Diaporthales

Ditopella asclepiadea Bonar
Distr.: CA.
Notes: Barr (Mycol. Mem. 7:161. 1978) suggested that this species belongs in *Delphinella* or *Plurisperma*.
Host: Asclepias

Ditopellopsis J. Reid & C. Booth. Pyrenomycetes, Diaporthales

Notes: Reid & Booth (Canad. J. Bot. 45:1479. 1967); Barr (Mycol. Mem. 7:90. 1978).
Ditopellopsis alni (G.E. Thompson & J.H. Miller) Barr
Mamiania alni G.E. Thompson & J.H. Miller
Distr.: GA.
Host: Alnus
Ditopellopsis clethrae J. Reid & C. Booth
Distr.: MD.
Host: Clethra
Ditopellopsis racemula (Cooke & Peck) Barr
Diaporthe racemula (Cooke & Peck) Sacc.
Distr.: Temperate North America.
Host: Epilobium

Doassansia Cornu. Basidiomycetes, Ustilaginales

Doassansia alismatis (Nees) Cornu
Distr.: North America; possibly Africa; Asia; Europe.
Notes: On Alismataceae.
Hosts: Alisma, Echinodorus, Sagittaria
Doassansia callitriches H. Jacks. & Linder in Linder
Distr.: CA(type).
Host: Callitriche
Doassansia deformans Cornu
Distr.: Central and eastern North America.
Host: Sagittaria
Doassansia downingiae Liro
Distr.: ID(type).
Host: Downingia
Doassansia epilobii Farl.
Distr.: Northern hemisphere.
Host: Epilobium
Doassansia limosellae (J. Kunze) J. Schröt.
Protomyces limosellae J. Kunze
Burrillia limosellae (J. Kunze) Liro
Distr.: MT, OR.
Notes: According to Vánky (Symb. Bot. Upsal. 24:43) this fungus is only reported from Europe but is easily overlooked.
Host: Limosella
Doassansia opaca Setch.
Distr.: Eastern North America.
Host: Sagittaria
Doassansia ranunculina J.J. Davis
Distr.: Central United States; Canada.
Host: Ranunculus
Doassansia sagittariae (Fuckel) C. Fisch
Distr.: Widespread.
Notes: On Alismataceae.
Host: Sagittaria
Doassansia zizaniae J.J. Davis
Distr.: WI(type).
Notes: This fungus is probably the sclerotium of an ascomycete.
Host: Zizania

Doassansiopsis (Setch.) Dietel. Basidiomycetes, Ustilaginales

Doassansiopsis furva (J.J. Davis) K. Vánky
Doassansia furva J.J. Davis
Distr.: WI; Canada(Manitoba).
Host: Sagittaria
Doassansiopsis hydrophila (A. Dietr.) Lavrov
Doassansia hydrophila (A. Dietr.) Lindeberg
Doassansia domingensis Cif.
Doassansia martianoffiana (Thuem.) J. Schröt.
Doassansiopsis martianoffiana (Thuem.) Dietel
Distr.: North America; Europe.
Notes: Vánky (Symb. Bot. Upsal. 24:46. 1985) provided a recent description and synonymy.
Host: Potamogeton
Doassansiopsis intermedia (Setch.) K. Vánky
Doassansia intermedia Setch.
Doassansia affinis Ellis & Dearn.
Distr.: North America.
Notes: Vánky (Sydowia 34:171. 1981) reviewed this species.
Host: Sagittaria
Doassansiopsis occulta (H. Hoffm.) Dietel
Doassansia occulta (H. Hoffm.) Cornu
Doassansia occulta (H. Hoffm.) Cornu var. *farlowii* (Cornu) Setch.
Doassansiella aquatilis (Peck) Zambettakis (synanamorph)
Ramularia aquatilis Peck (synanamorph)
Distr.: North America; Europe.
Notes: Vánky (Symb. Bot. Upsal. 24:47. 1985).
Host: Potamogeton

Donkioporia Kotlaba & Pouzar. Basidiomycetes, Aphyllophorales

Donkioporia expansa (Desmaz.) Kotlaba & Pouzar
Distr.: OH; Canada(Ontario); central Europe.
Notes: On structural timbers, especially of *Quercus* spp.
Host: Quercus

Doratomyces Corda. Hyphomycetes

Notes: Morton & Smith (Mycol. Pap. 86. 1963).
Doratomyces eichhornius K. Conway & Kimbrough
Distr.: FL.
Notes: Saprobic on leaves (Conway & Kimbrough. Mycotaxon 2:127. 1975).
Host: Eichhornia
Doratomyces stemonitis (Pers.:Fr.) F.J. Morton & G. Sm.
Cephalotrichum stemonitis (Pers.:Fr.) Link
Stysanus stemonitis (Pers.:Fr.) Corda
Stysanus tubericola Ellis & Dearn.
Distr.: Cosmopolitan.
Notes: Saprobic on many organic substrates. Associated with rot of potato tubers.
Hosts: Pinus, Solanum, Zea

Dothichiza Lib. ex Roum. Coelomycetes

Dothichiza caroliniana Demaree & M.S. Wilcox
Distr.: NC.
Notes: Leaf disease of blueberry (Taylor & Clayton. Phytopathology 49:65. 1959).
Host: Vaccinium
Dothichiza cassandrae Ellis & Everh.
Distr.: MI.
Host: Chamaedaphne
Dothichiza garryae Bonar
Distr.: CA.
Host: Garrya

Dothichiza serotina Atk.
Distr.: AL, WV.
Host: Prunus
Dothichiza sorbi Lib.
Micropera sorbi Thuem.
Dothiorella pyrenophora Sacc.
Dothiora pyrenophora (Fr.:Fr.) Fr. (teleomorph)
Distr.: Temperate North America; Europe.
Notes: On branches. Sivanesan (The bitunicate Ascomycetes. p. 158. 1984).
Host: Sorbus

Dothidasteromella Höhn. Loculoascomycetes, Dothideales

Dothidasteromella magnoliae (Tracy & Earle) M.L. Farr
Dimerosporium magnoliae Tracy & Earle
Distr.: MS.
Notes: Farr (Sydowia 38:67. 1985)
Host: Magnolia

Dothidea Fr.:Fr. Loculoascomycetes, Dothideales

Notes: Barr (Contr. Univ. Michigan Herb. 9:608. 1972).
Dothidea acerva Barr
Distr.: Northeastern North America.
Host: Juniperus
Dothidea atramentaria Berk.
Distr.: SC.
Host: Poa
Dothidea berberidis (Wahlenberg) De Not.
Sphaeria berberidis Wahlenberg
Dothidella berberidis (Wahlenberg) Theiss. & Syd.
Plowrightia berberidis (Wahlenberg) Sacc.
Distr.: MA, northwestern United States; Asia; Europe.
Host: Berberis
Dothidea bigeloviae Ellis & Everh.
Dothidea montaniensis Ellis & Everh.
Distr.: MT.
Host: Chrysothamnus
Dothidea conspicua D. Griffiths
Phragmodothis conspicua (D. Griffiths) Theiss. & Syd.
Thyridaria conspicua (D. Griffiths) Petr.
Dothidea yuccae Earle
Dothidea yuccae Ellis & Everh. non Earle, hom. illeg.
Phaeodothis yuccae (Ellis & Everh.) Sacc.
Piptarthron yuccae (Ellis & Everh.) Sutton (anamorph)
Distr.: Western United States.
Host: Yucca
Dothidea flabella Berk. & M.A. Curtis
Distr.: SC.
Host: Pteris
Dothidea linderae W. Gerard
Distr.: Northeastern United States.
Host: Lindera
Dothidea puccinioides Fr.:Fr.
Systremma artemisiae (Schwein.) Theiss. & Syd.
Systremma baccharidis (Cooke) Theiss. & Syd.
Dothidea collecta (Schwein.) Ellis & Everh.
Dothidea coluteae Berk. & M.A. Curtis
Dothidea crystallophora Berk. & M.A. Curtis
Dothidea tetraspora Berk. & Broome
Distr.: United States; Asia; Europe.
Notes: On dead branches and attached leaves of most deciduous woody plants. Luttrell (Amer. J. Bot. 38:460. 1951) described the developmental morphology of *D. collecta*. Loeffler (Phytopathol. Z. 30:357. 1957) considered Luttrell's fungus to be *Dothidea puccinioides* Fr.:Fr.
Hosts: Acer, Alnus, Artemisia, Asimina, Baccharis, Bignonia, Colutea, Cytisus, Foeniculum, Fraxinus, Genista, Gleditsia, Hibiscus, Ipomoea, Lespedeza, Lonicera, Maclura, Melia, Morus, Prunus, Rhododendron, Rhus, Sambucus, Sida, Solidago
Dothidea rugodisca Cooke & Harkn.
Homostegia rugodisca (Cooke & Harkn.) Sacc.
Distr.: CA(type).
Host: Arbutus
Dothidea sambuci Fr.:Fr.
Systremma lonicerae (Cooke) Theiss. & Syd.
Systremma natans (Tode) Theiss. & Syd.
Distr.: North America; Asia; Europe.
Notes: On dead branches of numerous deciduous woody plants. Loeffler (Phytopathol. Z. 30:352. 1957) gave extensive synonymy.
Hosts: Garrya, Lonicera, Sambucus, Spiraea
Dothidea solidaginis Fr.:Fr.
Xyloma solidaginis Fr.
Distr.: Eastern and midwestern United States; Canada.
Host: Solidago
Dothidea tuberculiformis Ellis
Plowrightia tuberculiformis (Ellis) Sacc.
Syncarpella tuberculiformis (Ellis) Theiss. & Syd.
Distr.: Western United States.
Host: Artemisia

Dothidella Speg. Loculoascomycetes, Dothideales

Dothidella adusta (Fuckel) Lind
Distr.: AK; Soviet Union.
Host: Epilobium
Dothidella aristidae (Schwein.) Ellis & Everh.
Distr.: North America.
Notes: Char spot.
Hosts: Aristida, Distichlis
Dothidella calystegiae (Cooke & Harkn.) Theiss. & Syd.
Dothidea calystegiae Cooke & Harkn.
Plowrightia calystegiae (Cooke & Harkn.) Berl. & Voglino
Distr.: CA.
Host: Calystegia
Dothidella caricina Dearn. & House
Distr.: NY.
Host: Carex
Dothidella kaduae (F. Stevens) Arx
Polystomella kaduae F. Stevens
Distr.: HI(type).
Host: Kadua
Dothidella longissima (Pers.:Fr.) Ellis & Everh.
Sphaeria longissima Pers.:Fr.
Distr.: Midwest United States; Germany.
Host: Chenopodium
Dothidella minima Sacc. & P. Syd.
Distr.: LA.
Host: Arundinaria
Dothidella parryi (Cooke) Theiss. & Syd.
Endothia parryi Cooke
Distr.: Southwestern United States.
Host: Agave
Dothidella rubiacearum (G. Arnaud) Arx
Polystomella rubiacearum G. Arnaud
Pluriporus gouldiae F. Stevens & Ryan
Distr.: HI; South America.
Notes: Müller & Arx (Beitr. Kryptogamenfl. Schweiz 11(2):536. 1962).
Host: Gouldia
Dothidella sphaerelloides Dearn.
Distr.: AK; arctic Canada.
Host: Saxifraga

Dothideovalsa Speg. Pyrenomycetes, Diatrypales

Dothideovalsa diantherae (I. Lewis) Theiss. & Syd.
Bagnisiella diantherae I. Lewis
Eutypa diantherae (I. Lewis) Petr.
Distr.: TX.
Notes: Black knot disease (Lewis. Mycologia 4:66. 1912; Rappaz. Mycol. Helv. 2:551. 1987).
Host: Justicia

Dothidotthia Höhn. Loculoascomycetes, Dothideales

Dothidotthia andersonii (Shoemaker) Barr
Gibbera andersonii Shoemaker
Gibbera distegiae (Tracy & Earle) Barr
Otthia distegiae Tracy & Earle
Dibotryon symphoricarpi (Rehm) Höhn.
Gibbera symphoricarpi (Ellis & Everh.) Ellis & Everh.
Otthia symphoricarpi (Ellis & Everh.) Ellis & Everh.
Plowrightia symphoricarpi Ellis & Everh.
Pseudotthia symphoricarpi (Ellis & Everh.) Rehm
Valsaria symphoricarpi (Ellis & Everh.) Theiss. & Syd.
Distr.: Central and western North America.
Notes: Barr (Mycotaxon 29:503. 1987); Barr (Canad. J. Bot. 46:838. 1968); Shoemaker (Canad. J. Bot. 41:1421. 1963).
Hosts: Lonicera, Symphoricarpos
Dothidotthia ramulicola (Peck) Barr
Leptosphaeria ramulicola (Peck) Sacc.
Neodeightonia ramulicola (Peck) Barr in Barr et al.
Distr.: NY.
Notes: Barr (Mycotaxon 29:503. 1987); Barr et al. (New York State Mus. Bull. 459:38. 1986).
Host: Ulmus

Dothiora Fr., nom. cons. Loculoascomycetes, Dothideales

Notes: Barr (Contr. Univ. Michigan Herb. 9:572. 1972); Froidevaux (Nova Hedwigia 23:679. 1972).
Dothiora pyrenophora (Fr.:Fr.) Fr.
Dothidea pyrenophora Fr.:Fr.
Distr.: Temperate North America; Europe.
Notes: On branches (Sivanesan. The bitunicate Ascomycetes. p. 158 1984).
Host: Pyrus
Dothiora ribesia (Fr.:Fr.) Barr
Dothidea ribesia Fr.:Fr.
Phragmodothella ribesia (Fr.:Fr.) Petr.
Plowrightia ribesia (Fr.:Fr.) Sacc.
Aureobasidium sp. (anamorph)
Distr.: Temperate North America; Europe.
Notes: On branches (Sivanesan & Holliday. C.M.I. Descr. 707. 1981).
Host: Ribes
Dothiora sambucina (Peck) Barr
Sphaerulina sambucina Peck
Distr.: Eastern North America.
Notes: On branches (Barr. Contr. Univ. Michigan Herb. p. 574. 1972).
Host: Sambucus
Dothiora schizospora Luttrell
Distr.: MO(type).
Notes: On stems (Luttrell. Mycologia 52:64. 1960).
Host: Symphoricarpos
Dothiora slippii Barr
Distr.: ID(type).
Host: Pinus
Dothiora staphylina (Peck) Barr
Hysterium staphylina (Peck) Dearn. & House
Metasphaeria staphylina (Peck) Sacc.
Distr.: Eastern North America.
Host: Staphylea
Dothiora taxicola (Peck) Barr
Leptosphaeria taxicola (Peck) Sacc.
Metasphaeria taxicola (Peck) Peck
Sphaerulina taxicola (Peck) Berl.
Sphaerulina taxi (Cooke) Massee
Dothichiza sp. (anamorph)
Distr.: Temperate North America; Europe.
Notes: On leaves. The anamorph was originally described as *Cytospora taxifolia* Cooke & Massee.
Host: Taxus
Dothiora wolfii Barr
Sphaerulina polyspora F.A. Wolf
Distr.: Eastern United States.
Notes: Twig blight.
Host: Oxydendrum

Dothiorella Sacc. Coelomycetes

Dothiorella aberrans Peck
Distr.: OH.
Host: Asimina
Dothiorella agavicola Petr.
Distr.: HI(type).
Host: Agave
Dothiorella asiminae Ellis & Everh.
Distr.: WV.
Host: Asimina
Dothiorella barringtoniae (Cooke & Massee) Petr. & Syd.
Phoma barringtoniae Cooke & Massee
Distr.: HI; Sri Lanka(type).
Host: Barringtonia
Dothiorella berengeriana Sacc.
Botryosphaeria dothidea (Moug.:Fr.) Ces. & De Not. (teleomorph)
Distr.: NJ, NY, OH; Europe.
Notes: On various trees.
Hosts: Liquidambar, Syringa
Dothiorella botrya Sacc.
Distr.: SC.
Host: Manihot
Dothiorella celastri Peck
Distr.: IL.
Host: Celastrus
Dothiorella celtidis Peck
Distr.: NY.
Host: Celtis
Dothiorella concaviuscula Ellis & Barth.
Distr.: KS.
Host: Fraxinus
Dothiorella diatrypea (Cooke & Ellis) Grove
Sphaeropsis diatrypea Cooke & Ellis
Phoma diatrypea (Cooke & Ellis) Grove
Distr.: NJ(type).
Host: Chionanthus
Dothiorella eucalypti (Berk. & Broome) Sacc.
Sphaeropsis eucalypti Berk. & Broome
Distr.: FL; Australia.
Host: Eucalyptus
Dothiorella everhartii Sacc. & Syd.
Dothiorella fraxini Ellis & Everh. non (Sacc.) Roum., hom. illeg.
Distr.: Reported from an unknown locality in the United States; Canada(Ontario).
Host: Fraxinus
Dothiorella fraxinicola Ellis & Everh.
Distr.: IA, NE.
Host: Fraxinus
Dothiorella glandulosa (Cooke) Sacc.
Sphaeropsis glandulosa Cooke
Distr.: SC.
Hosts: Ailanthus, Robinia
Dothiorella gleditschiae Grove
Distr.: MS, SC.
Host: Gleditsia
Dothiorella gregaria Sacc.
Distr.: CA; France; Italy.
Notes: On branches of various trees.
Hosts: Juglans, Persea, Salix, Syagrus
Dothiorella hawaiiensis Petr.
Distr.: HI(type).
Host: Asclepias
Dothiorella hicoriae Dearn. & House
Distr.: NY.
Host: Carya
Dothiorella inversa (Fr.:Fr.) Höhn.
Sphaeria inversa Fr.:Fr.
Distr.: AK; Europe.
Host: Alnus
Dothiorella macrospora (Sacc. & Briard) Petr. & Syd.
Fusicoccum macrosporum Sacc. & Briard
Sphaeropsis fagicola G. Otth

Distr.: MI, PA; France.
Host: Sambucus
Dothiorella mali P. Karst. var. ***fructus*** Dearn., nom. inval.
Distr.: TN.
Host: Malus
Dothiorella minor Ellis & Everh.
Distr.: MD, WV.
Host: Liriodendron
Dothiorella mori Berl.
Distr.: NJ, TX; Europe.
Host: Morus
Dothiorella nelumbii Ellis & F.W. Anderson
Distr.: Central Atlantic United States.
Host: Nelumbo
Dothiorella nucis Fairm.
Distr.: NY.
Host: Carya
Dothiorella opuntiae Siemaszko ex Petr.
Distr.: HI; Soviet Union.
Host: Cereus
Dothiorella phomiformis (Sacc.) Petr. & Syd.
Phyllosticta phomiformis Sacc.
Distr.: Central and eastern North America.
Host: Quercus
Dothiorella phomopsis Fairm.
Distr.: NM.
Host: Clematis
Dothiorella pinastri Linder
Distr.: ID.
Host: Pinus
Dothiorella pyrenophora (P. Karst.) Sacc. var. ***salicis*** P. Karst.
Distr.: AK; Finland.
Host: Salix
Dothiorella radicans Ellis & Everh.
Distr.: CO, IL, NJ.
Host: Rhus
Dothiorella rhoina Ellis & Everh.
Distr.: CO, IL, northeastern United States.
Host: Rhus
Dothiorella ribis (Fuckel) Sacc.
Podosporium ribis Fuckel
Distr.: OK, NY; Germany.
Hosts: Ribes, Salix
Dothiorella scopulina Dearn. & Barth.
Distr.: WY.
Host: Sorbus
Dothiorella stramonii Petr.
Distr.: HI(type).
Host: Datura
Dothiorella toxica Ellis & Everh.
Distr.: CO, IL, NJ.
Host: Rhus
Dothiorella ulmi Verrall & May
Distr.: United States.
Host: Ulmus

Dothistroma Hulbary. Coelomycetes

Notes: Sutton (The Coelomycetes. p. 172. 1980).
Dothistroma septospora (Doroguine) Morelet
Cytosporina septospora Doroguine
Septoria septospora (Doroguine) Arx
Septoriella septospora (Doroguine) Sacc. in Trott.
Dothistroma pini Hulbary
Mycosphaerella pini Rostr. in Munk (teleomorph)
Distr.: Cosmopolitan.
Notes: Red band needle blight (Evans. Mycol. Pap. 153:36. 1984).
Host: Pinus
Dothistroma septospora (Doroguine) Morelet var. ***lineare*** (Thyr & C.G. Shaw) Sutton
Dothistroma pini Hulbary var. *lineare* Thyr & C.G. Shaw
Distr.: ID(type).
Notes: Thyr & Shaw (Mycologia 56:103. 1964).
Host: Pinus

Dothivalsaria Petr. Loculoascomycetes, Pyrenulales

Notes: Barr (Mycotaxon 9:26. 1978) reviewed this monotypic genus.
Dothivalsaria megalospora (Auersw.) Petr.
Valsaria megalospora Auersw.
Massariovalsa megalospora (Auersw.) E. Müller in E. Müller & Arx
Myrmaecium megalospora (Auersw.) Niessl
Diatrype angulare Peck
Distr.: Temperate regions.
Notes: On woody branches of dicotyledonous trees.
Hosts: Alnus, Sapindus, Tilia, Ungnadia

Drechslera Ito, nom. cons. Hyphomycetes

Notes: Sivanesan (Mycol. Pap. 158:1. 1967).
Drechslera avenacea (M.A. Curtis ex Cooke) Shoemaker
Helminthosporium avenaceum M.A. Curtis ex Cooke
Distr.: North America; temperate Europe.
Hosts: Avena, Poa, Trisetum
Drechslera avenae (Eidam) Scharif
Helminthosporium avenae Eidam
Distr.: Cosmopolitan.
Notes: Leaf spot and seedling blight of oats and other grasses (Ellis & Waller. C.M.I. Descr. 389. 1973 as *Pyrenophora avenae*). Shoemaker (Canad. J. Bot. 40:815. 1962) and Sivanesan (Mycol. Pap. 158:162. 1987) regarded *D. avenae* as a synonym of *D. avenacea* and *P. avenae* as a synonym of *P. chaetomioides*. Some authors (Alcorn. Mycotaxon 17:1. 1983) retain them as separate species.
Hosts: Arrhenatherum, Avena, Hordeum, Koeleria
Drechslera biseptata (Sacc. & Roum.) M.J. Richardson & E.M. Fraser
Helminthosporium biseptatum Sacc. & Roum.
Distr.: North America; Australia; Europe.
Hosts: Dactylis, Metrosideros
Drechslera brizae (Nisikado) Subramanian & P.C. Jain
Helminthosporium brizae Nisikado
Bipolaris brizae (Nisikado) Shoemaker
Distr.: CA; Japan.
Host: Gastridium
Drechslera bromi (Died.) Shoemaker
Helminthosporium bromi (Died.) Died.
Pyrenophora bromi (Died.) Drechs. (teleomorph)
Distr.: North America; Australia; temperate Europe.
Notes: Leaf spot of *Bromus* and other grasses.
Host: Bromus
Drechslera campanulata (Lév.) Sutton
Angiopoma campanulata Lév.
Bipolaris cyclops (Drechs.) R. Sprague
Helminthosporium cyclops Drechs.
Drechslera verticillata (O'Gara) Shoemaker
Podosporiella verticillata O'Gara
Pyrenophora semeniperda (Brittlebank & D.B. Adam) Shoemaker (teleomorph)
Distr.: Temperate North America; South Africa; Australia; New Zealand.
Notes: Leaf spot of wheat, oats, and other grasses.
Hosts: Danthonia, Poa, Secale, Trisetum, Triticum
Drechslera catenaria (Drechs.) Ito
Helminthosporium catenarium Drechs.
Distr.: Northern North America.
Hosts: Agrostis, Beckmannia, Cinna, Phalaris, Sorghum
Drechslera dactylidis Shoemaker
Pyrenophora dactylidis Ammon (teleomorph)
Distr.: PA.
Notes: Shoemaker (Canad. J. Bot. 40:820. 1964).
Host: Dactylis
Drechslera dematioidea (Bubák & Wróbl.) Subramanian & P.C. Jain
Helminthosporium dematioideum Bubák & Wróbl.
Distr.: Southern North America; Africa; temperate Europe.
Hosts: Agrostis, Anthoxanthum
Drechslera dictyoides (Drechs.) Shoemaker
Helminthosporium dictyoides Drechs.
Distr.: North America; Australia; temperate Europe.

Notes: Net blotch of fescues and other grasses.
Hosts: Festuca, Lolium, Poa, Vulpia, Zoysia
Drechslera erythrospila (Drechs.) Shoemaker
Helminthosporium erythrospilum Drechs.
Distr.: Eastern North America; Australia, Europe.
Host: Agrostis
Drechslera fugax (Wallr.) Shoemaker in S.J. Hughes
Helminthosporium fugax Wallr.
Drechslera stenacra (Drechs.) Shoemaker
Helminthosporium stenacrum Drechs.
Distr.: Temperate North America; Europe.
Notes: Leaf mold.
Hosts: Agrostis, Apera, Eleusine
Drechslera gigantea (Heald & F.A. Wolf) Ito
Helminthosporium giganteum Heald & F.A. Wolf
Distr.: North and South America.
Notes: Zonate eye spot of grasses; also causes eye spot of banana.
Hosts: Agropyron, Agrostis, Bromus, Caulophyllum, Cynodon, Echinochloa, Eleusine, Elymus, Eragrostis, Leersia, Muhlenbergia, Musa, Panicum, Pennisetum, Phalaris, Phleum, Poa, Sporobolus, Zizania
Drechslera graminea (Rabenh.) Shoemaker
Helminthosporium gramineum Rabenh.
Distr.: Cosmopolitan.
Notes: Leaf stripe of barley and other grasses (Ellis & Waller. C.M.I. Descr. 388. 1973).
Hosts: Chloris, Hordeum, Pennisetum
Drechslera heveae (Petch) M.B. Ellis
Helminthosporium heveae Petch
Distr.: Subtropical and tropical regions.
Notes: Bird's-eye spot of rubber (Ellis & Holliday. C.M.I. Descr. 343. 1972).
Host: Hevea
Drechslera nobleae D. Matthews & E. McKenzie
Distr.: AL; Europe; New Zealand.
Notes: Sivanesan (C.M.I. Descr. 890. 1986).
Host: Lolium
Drechslera phlei (J.H. Graham) Shoemaker
Helminthosporium dictyoides Drechs. var. *phlei* J.H. Graham
Distr.: North America; temperate Europe.
Host: Phleum
Drechslera poae (Baudys) Shoemaker
Helminthosporium poae Baudys
Drechslera vagans (Drechs.) Shoemaker
Helminthosporium vagans Drechs.
Distr.: Temperate North America; Australia; Europe.
Notes: Leaf spot and foot rot of grasses.
Hosts: Koeleria, Lolium, Poa
Drechslera portulacae (Rader) De Hoog & Van Oorschot
Helminthosporium portulacae Rader
Distr.: NY.
Host: Portulaca
Drechslera siccans (Drechs.) Shoemaker
Helminthosporium siccans Drechs.
Distr.: North America; Australia; temperate Europe.
Notes: Leaf spot, brown blight, and foot rot of rye grasses (Ellis & Holliday. C.M.I. Descr. 492. 1976).
Hosts: Lolium, Poa
Drechslera teres (Sacc.) Shoemaker
Helminthosporium teres Sacc.
Helminthosporium hordei Eidam
Pyrenophora teres Drechs. (teleomorph)
Distr.: Cosmopolitan.
Notes: Net blotch of barley and other grasses. Ellis & Waller (C.M.I. Descr. 390. 1973).
Hosts: Hordeum, Lolium
Drechslera triseptata (Drechs.) Subramanian & P.C. Jain
Helminthosporium triseptatum Drechs.
Distr.: Temperate North America.
Hosts: Agrostis, Dactylis, Holcus, Phleum, Poa
Drechslera tritici-repentis (Died.) Shoemaker
Helminthosporium tritici-repentis Died.
Drechslera tritici-vulgaris (Nisikado) Ito
Helminthosporium tritici-vulgaris Nisikado
Pyrenophora tritici-repentis (Died.) Drechs. (teleomorph)
Distr.: Cosmopolitan.
Notes: Yellow leaf spot of cereals and grasses (Ellis & Waller. C.M.I. Descr. 494. 1976).
Hosts: Agropyron, Elymus, Hordeum, Secale, Stenotaphrum, Triticum
Drechslera tuberosa (Atk.) Shoemaker
Helminthosporium tuberosum Atk.
Drechslera japonica (Ito & Kuribayashi) Shoemaker
Helminthosporium japonicum Ito & Kuribayashi
Helminthosporium secalis M.D. Whitehead & Dickson
Pyrenophora japonica Ito & Kuribayashi (teleomorph)
Distr.: Temperate northern hemisphere.
Notes: Parasitic on cereals and grasses.
Host: Secale

Drepanopeziza (Kleb.) Höhn. Discomycetes, Helotiales

Notes: Gremmen (Nova Hedwigia 9:170. 1965); Rimpau (Phytopathol. Z. 43:257. 1961).
Drepanopeziza populi-albae (Kleb.) Nannf.
Pseudopeziza populi-albae Kleb.
Distr.: Temperate northern hemisphere.
Notes: Pirozynski (Fungi Canadenses 14. 1974 as *Marssonina castagnei*).
Host: Populus
Drepanopeziza populorum (Desmaz.) Höhn.
Marssonina populi (Lib.) Magnus (anamorph)
Distr.: Temperate regions.
Notes: Pirozynski (Fungi Canadenses 15. 1974 as *Marssonina populi*).
Host: Populus
Drepanopeziza ribis (Kleb.) Höhn.
Pseudopeziza ribis Kleb.
Distr.: Temperate regions.
Notes: Leaf spot (Booth & Waller. C.M.I. Descr. 638. 1979).
Host: Ribes
Drepanopeziza salicis (Tul. & C. Tul.) Höhn.
Pseudopeziza salicis (Tul. & C. Tul.) Potebnia
Monostichella salicis (Westend.) Arx (anamorph)
Distr.: Temperate North America; Europe.
Host: Salix

Drudeola Kuntze. Coelomycetes

Drudeola sarraceniae (Peck & G.W. Clinton) Kuntze
Peckia sarraceniae Peck & G.W. Clinton
Distr.: MI, NY.
Notes: The fungus name *Peckia* is a later homonym which was replaced by *Drudeola* with *D. sarraceinaea* as the lectotype (Sutton. Mycol. Pap. 151:78. 1977).
Host: Sarracenia

Ductifera C.G. Lloyd. Basidiomycetes, Tremellales

Ductifera sucina (A. Möller) K. Wells
Exidia cystidiata Olive
Sebacina lactescens Burt
Distr.: North and Central America.
Notes: On wood.
Hosts: Cornus, Quercus

Duplicaria Fuckel. Discomycetes, Rhytismatales

Duplicaria acuminata Ellis & Everh.
Bifusella acuminata (Ellis & Everh.) W.B. Cooke & C.G. Shaw
Crandallia juncicola Ellis & Sacc. (anamorph)
Distr.: Western North America.
Notes: Powell (Mycologia 65:1356. 1973).
Hosts: Carex, Juncus

Durandiella Seaver. Discomycetes, Helotiales

Notes: Groves (Canad. J. Bot. 32:116. 1954).
Durandiella andromedae (Schwein.:Fr.) Groves
Cenangium andromedae Schwein.:Fr.
Distr.: NC.
Hosts: Andromeda, Oxydendrum
Durandiella fraxini (Schwein.:Fr.) Seaver
Godronia fraxini (Schwein.:Fr.) Höhn.
Distr.: Eastern North America.
Host: Fraxinus
Durandiella lenticellicola Groves
Corniculariella harpographoidea Dearn. ex DiCosmo (anamorph)
Distr.: Northeastern North America.
Host: Crataegus
Durandiella nemopanthis (Peck) Groves
Tympanis nemopanthis Peck
Godronia nemopanthis (Peck) Sacc.
Distr.: Western North America.
Host: Nemopanthus
Durandiella pseudotsugae Funk
Corniculariella pseudotsugae (W.L. White) DiCosmo (anamorph)
Distr.: Western North America.
Notes: Funk (Canad. J. Bot. 40:332. 1962).
Host: Pseudotsuga
Durandiella rosae (Schwein.) Groves
Cenangium rosae Schwein.
Distr.: PA(type).
Host: Rosa
Durandiella rugosa (Ellis & Everh.) Groves
Godronia rugosa Ellis & Everh.
Distr.: Southeastern United States.
Host: Oxydendrum
Durandiella seriata (Fr.) Groves
Cenangium seriatum Fr.
Scleroderris seriata (Fr.) Rehm
Gelatinosporium betulinum Peck (anamorph)
Distr.: Temperate regions.
Host: Betula
Durandiella viburnicola (Seaver) Groves
Godronia viburnicola Seaver
Distr.: Eastern North America.
Notes: On limbs.
Host: Viburnum

Durella Tul. & C. Tul. Discomycetes, Helotiales

Notes: Dennis (Mycol. Pap. 62. 1956); Sherwood (Mycotaxon 38:137. 1987).
Durella atrocyanea (Fr.) Höhn.
Distr.: Western North America; Europe.
Notes: On decorticated wood and woody herbaceous stems.
Hosts: Amelanchier, Arbutus

Earliella Murrill. Basidiomycetes, Aphyllophorales

Earliella scabrosa (Pers.) R.L. Gilbertson & Ryvarden
Trametes scabrosa (Pers.) G. Cunn.
Trametes corrugata (Pers.) Bres.
Polyporus persoonii Mont.
Trametes persoonii (Mont.) Pat.
Distr.: FL, LA, pantropical.
Notes: On hardwoods.
Hosts: Aleurites, Ficus, Mangifera, Pandanus, Psidium

Echidnodella Theiss. & Syd. Loculoascomycetes, Dothideales

Echidnodella angustiformis (Tracy & Earle) F. Stevens
Lembosia angustiformis Tracy & Earle
Aulographum angustiforme (Tracy & Earle) Theiss. & Syd.
Morenoella angustiformis (Tracy & Earle) Theiss.
Morenoella ilicis (Tracy & Earle) Theiss.
Distr.: MS.
Host: Ilex
Echidnodella camphorae F. Stevens & Ryan
Lembosia camphorae Earle
Morenoella camphorae (Earle) Theiss.
Distr.: FL.
Host: Cinnamomum
Echidnodella mabae F. Stevens & Ryan
Lembosia mabae (F. Stevens & Ryan) G. Arnaud
Distr.: HI.
Host: Diospyros
Echidnodella prinoides F. Stevens & Ryan
Lembosia prinoides Tracy & Earle
Morenoella prinoides (Tracy & Earle) Theiss.
Distr.: MS.
Host: Ilex
Echidnodella raillardiae F. Stevens & Ryan
Distr.: HI.
Hosts: Dubautia, Railliardia
Echidnodella rugispora (Tracy & Earle) F. Stevens & Ryan
Lembosia rugispora Tracy & Earle
Morenoella rugispora (Tracy & Earle) Theiss.
Distr.: MS, NC.
Host: Persea

Echidnodes Theiss. & Syd. Loculoascomycetes, Dothideales

Echidnodes pisoniae F. Stevens & Ryan
Maurodothella pisoniae (F. Stevens & Ryan) G. Arnaud
Distr.: HI.
Hosts: Pisonia, Rockia
Echidnodes sandwicensis Petr.
Distr.: HI(type).
Host: Pandanus
Echidnodes visci Petr.
Distr.: HI(type).
Notes: Petrak (Sydowia 7:386. 1953) reported the host as *Viscum* sp.
Host: Korthalsella

Echinodontium Ellis & Everh. Basidiomycetes, Aphyllophorales

Echinodontium ballouii (Banker) H. Gross
Steccherinum ballouii Banker
Distr.: NJ.
Notes: On wood.
Host: Chamaecyparis
Echinodontium tinctorium (Ellis & Everh.) Ellis & Everh.
Fomes tinctorius Ellis & Everh.
Distr.: Western North America.
Notes: Heart rot of living trees, especially *Picea* and *Abies* spp.
Hosts: Abies, Larix, Picea, Pseudotsuga, Thuja, Tsuga

Ectostroma Fr. Hyphomycetes

Ectostroma afflatum (Schwein.) Fr.
Xyloma afflata Schwein.
Distr.: PA, VA.
Host: Cimicifuga
Ectostroma liriodendri (Schwein.) Fr.
Distr.: Eastern United States.
Notes: Sterile mycelium in leaf spots.
Host: Liriodendron

Efibulobasidium K. Wells. Basidiomycetes, Tremellales

Efibulobasidium rolleyi (Olive) K. Wells
Exidia rolleyi Olive
Distr.: IA; Tahiti.
Notes: On corticated wood and leaves.
Host: Ulmus

Eichleriella Bres. Basidiomycetes, Tremellales

Eichleriella deglubens (Berk. & Broome) D. Reid
Radulum deglubens Berk. & Broome
Distr.: Temperate northern hemisphere.
Notes: On wood. Previously confused with *E. spinulosa*. Reid (Trans. Brit. Mycol. Soc. 55:436-437. 1970).
Host: Populus
Eichleriella leucophaea Bres.
Exidiopsis leucophaea (Bres.) K. Wells
Eichleriella schrenkii Burt
Distr.: Temperate northern hemisphere.
Notes: On wood.
Hosts: Acacia, Celtis, Prosopis
Eichleriella leveilliana (Berk. & M.A. Curtis) Burt
Corticium leveillianum Berk. & M.A. Curtis
Distr.: Eastern United States to South America.
Notes: On wood.
Host: Crataegus
Eichleriella macrospora (Ellis & Everh.) G.W. Martin
Exidiopsis macrospora (Ellis & Everh.) K. Wells
Sebacina macrospora (Ellis & Everh.) Burt
Sebacina monticola Burt
Distr.: North America.
Notes: On wood.
Host: Picea
Eichleriella pulvinata Coker
Distr.: NC(type).
Notes: On twigs. According to Wells (Mycologia 53:363. 1961), this species may be close to *Exidia* spp.
Host: Carya

Ellisiodothis Theiss. Loculoascomycetes, Dothideales

Ellisiodothis inquinans (Ellis & Everh.) Theiss.
Asterina inquinans Ellis & Everh.
Distr.: Southeastern United States.
Hosts: Campsis, Desmodium, Lespedeza, Phaseolus, Sabal

Elsinoe Racib. Loculoascomycetes, Dothideales

Elsinoe alyxiae A.M. Watson & Jenk.
Distr.: HI(type).
Host: Alyxia
Elsinoe ampelina Shear
Sphaceloma ampelinum de Bary (anamorph)
Distr.: Worldwide in grape-growing regions.
Notes: Anthracnose, bird's-eye rot of grapes (Sivanesan & Critchett. C.M.I. Descr. 439. 1974).
Host: Vitis
Elsinoe batatas Viégas & Jenk.
Sphaceloma batatas Sawada (anamorph)
Distr.: HI; Brazil; Taiwan.
Host: Ipomoea
Elsinoe cinnamomi Pollack & Jenk.
Sphaceloma sp. (anamorph)
Distr.: MS.
Notes: Limber et al. (Mycologia 38:470. 1946).
Host: Cinnamomum
Elsinoe corni Jenk. & Bitancourt
Distr.: Eastern United States.
Host: Cornus
Elsinoe diospyri Bitancourt & Jenk.
Distr.: FL; Brazil.
Host: Diospyros
Elsinoe euonymi-japonici Jenk. & Bitancourt
Sphaceloma euonymi-japonici Kurosawa & Katsuki (anamorph)
Distr.: FL; Japan.
Host: Euonymus
Elsinoe fawcettii Bitancourt & Jenk.
Sphaceloma fawcettii Jenk. (anamorph)
Distr.: Worldwide in citrus-growing regions.
Notes: Common citrus scab (Sivanesan & Critchett. C.M.I. Descr. 438. 1974).
Hosts: Citrofortunella, Citrus, Clausena, Fortunella, Poncirus
Elsinoe ilicis Plakidas
Distr.: FL, LA.
Notes: Spot anthracnose on leaves, stems, and berries of holly.
Host: Ilex
Elsinoe jasmini Bitancourt & Jenk.
Distr.: FL; Brazil.
Host: Jasminum
Elsinoe ledi (Peck) Zeller
Aulographum ledi Peck
Distr.: Temperate North America.
Hosts: Gaultheria, Ledum, Leucothoe
Elsinoe lepagei Bitancourt & Jenk.
Distr.: FL; Brazil.
Hosts: Manilkara, Pouteria
Elsinoe leucospila Bitancourt & Jenk.
Sphaceloma theae Kurosawa ex Bitancourt & Jenk. (anamorph)
Distr.: FL; Antigua; Central America; Peru; Japan.
Notes: White scab of tea.
Hosts: Camellia, Ternstroemia
Elsinoe magnoliae J.H. Miller & Jenk.
Sphaceloma magnoliae Jenk. & J.H. Miller (anamorph)
Distr.: Southern United States.
Host: Magnolia
Elsinoe mangiferae Bitancourt & Jenk.
Sphaceloma mangiferae Bitancourt & Jenk. (anamorph)
Distr.: FL; Caribbean.
Host: Mangifera
Elsinoe mattirolianum G. Arnaud & Bitancourt
Sphaceloma mattirolianum (Sacc. & D. Sacc.) Jenk. (anamorph)
Distr.: CA; Italy.
Host: Arbutus
Elsinoe mulleri Bitancourt & Jenk.
Distr.: HI; Venezuela.
Host: Polyscias
Elsinoe panici L.C. Tiffany & Mathre
Distr.: IA.
Notes: Gabel & Tiffany (Mycologia 79:737. 1987).
Host: Panicum
Elsinoe parthenocissi Jenk. & Bitancourt
Distr.: FL, NH.
Host: Parthenocissus
Elsinoe phaseoli Jenk. in Bruner & Jenk.
Sphaceloma sp. (anamorph)
Distr.: Southeastern United States; West Indies; Central America; northern South America; East Africa.
Notes: Lima bean scab (Sivanesan & Holliday. C.M.I. Descr. 314. 1971).
Host: Phaseolus
Elsinoe piri (Woronichin) Jenk.
Sphaceloma pirinum (Peglion) Jenk. (anamorph)
Distr.: Temperate northern hemisphere.
Notes: Anthracnose of apple and pear.
Hosts: Cydonia, Malus, Pyrus
Elsinoe quercicola Bitancourt & Jenk.
Distr.: FL.
Host: Quercus
Elsinoe quercus-falcatae J.H. Miller
Sphaceloma sp. (anamorph)
Distr.: Southeastern United States.
Host: Quercus
Elsinoe randii Jenk. & Bitancourt
Sphaceloma randii Jenk. & Bitancourt (anamorph)
Distr.: FL, GA; Brazil.
Host: Carya
Elsinoe rosarum Jenk. & Bitancourt
Sphaceloma rosarum (Pass.) Jenk. (anamorph)
Distr.: Temperate regions.
Notes: Rose anthracnose.
Host: Rosa
Elsinoe sacchari Lo
Sphaceloma sacchari Lo (anamorph)
Distr.: Subtropical and tropical regions.

Notes: White rash of sugarcane.
Host: Saccharum
Elsinoe sesbaniae **Limber & Jenk.**
Sphaceloma sp. (anamorph)
Distr.: SC.
Notes: Limber et al. (Mycologia 38:465. 1946).
Host: Sesbania
Elsinoe solidaginis **Jenk. & Ukkelberg**
Distr.: FL, GA, WI.
Hosts: Aster, Solidago
Elsinoe tiliae **Creelman**
Distr.: United States; Canada(Nova Scotia); Argentina.
Host: Tilia
Elsinoe vaccinii **D.E. Gardner & C.S. Hodges**
Distr.: HI.
Notes: Leaf anthracnose (Gardner & Hodges. Mycologia 78:506. 1986).
Host: Vaccinium
Elsinoe veneta **(Burkholder) Jenk.**
Plectodiscella veneta Burkholder
Distr.: Temperate regions.
Notes: Cane spot or anthracnose of raspberry (Sivanesan & Critchett. C.M.I. Descr. 484. 1974).
Host: Rubus
Elsinoe wisconsinensis **H.C. Greene**
Sphaceloma wisconsinensis H.C. Greene (anamorph)
Distr.: WI.
Host: Desmodium

Elytroderma Darker. Discomycetes, Rhytismatales

Elytroderma deformans **(Weir) Darker**
Hypoderma deformans Weir
Distr.: North America.
Notes: Witches' broom of pine (Minter. C.M.I. Descr. 655. 1980).
Host: Pinus

Embellisia E. Simmons. Hyphomycetes

Embellisia allii **(Campanile) E. Simmons**
Helminthosporium allii Campanile
Distr.: Cosmopolitan.
Notes: On bulbs (Simmons. Mycologia 63:380. 1971).
Host: Allium
Embellisia chlamydospora **(J.A. Hoes, G.W. Bruehl & C.G. Shaw) E. Simmons**
Pseudostemphylium chlamydosporum J.A. Hoes, G.W. Bruehl & C.G. Shaw
Distr.: Widespread.
Notes: Isolated from various plants, air and soil (Ellis. More dematiaceous Hyphomycetes. p. 435. 1976).
Host: Triticum

Emericella Berk. & Broome in Berk. Pyrenomycetes, Eurotiales

Notes: *Emericella* is the teleomorph of some *Aspergillus* species (Christensen & Raper. Trans. Brit. Mycol. Soc. 71:177. 1978).
Emericella nidulans **(Eidam) Vuill.**
Aspergillus nidulellus R.A. Samson & W. Gams (anamorph)
Distr.: Cosmopolitan.
Hosts: Alnus, Helianthus
Emericella rugulosa **(Thom & Raper) C.R. Benjamin**
Aspergillus rugulosa Thom & Raper (anamorph)
Distr.: Widespread.
Notes: Isolated from a variety of substrates (Domsch et al. Compendium of soil fungi. p. 270. 1980).
Host: Carya
Emericella variecolor **Berk. & Broome**
Theclospora bifida Harkn.
Aspergillus stellifer R.A. Samson & W. Gams (anamorph)
Distr.: Widespread.
Notes: Peek & Solheim (Mycologia 50:851. 1958).
Host: Eucalyptus

Enchnoa Fr. Pyrenomycetes, Calosphaeriales

Enchnoa canavaliae **F. Stevens & E. Young, nom. nud.**
Distr.: HI.
Notes: The place of publication of this name could not be found.
Host: Canavalia
Enchnoa subcorticalis **(Peck) Barr in H.E. Bigelow & Barr**
Trichosphaeria subcorticalis (Peck) Sacc.
Distr.: Eastern North America.
Notes: On limbs of hardwoods. Bigelow & Barr (Rhodora 71:798. 1969).
Host: Carpinus

Encoelia (Fr.:Fr.) P. Karst., nom. cons. Discomycetes, Helotiales

Encoelia fascicularis **(Albertini & Schwein.:Fr.) P. Karst.**
Phibalis fascicularis (Albertini & Schwein.:Fr.) Wallr.
Cenangium populneum (Pers.) Rehm
Distr.: Temperate North America; Europe.
Hosts: Populus, Salix
Encoelia furfuracea **(Roth:Pers.) P. Karst.**
Cenangium furfuraceum (Roth:Pers.) Sacc.
Distr.: North temperate regions.
Notes: Torkelson & Eckblad (Norw. J. Bot. 24:139. 1977).
Hosts: Alnus, Corylus
Encoelia pruinosa **(Ellis & Everh.) Torkelson & Eckblad**
Dermea pruinosa Ellis & Everh.
Cenangium pruinosa (Ellis & Everh.) Seaver
Phibalis pruinosa (Ellis & Everh.) L.M. Kohn & Korf
Cenangium singulare (Rehm) R.W. Davidson & Cash
Distr.: Temperate North America; Europe.
Notes: Sooty bark canker.
Host: Populus

Encoeliopsis Nannf. Discomycetes, Helotiales

Encoeliopsis laricina **(Ettlinger) Groves**
Crumenula laricina Ettlinger
Scleroderris laricina (Ettlinger) Gremmen
Brunchorstia laricina Ettlinger (anamorph)
Distr.: Western North America; European Alps.
Notes: Shoot-blight disease of larch (Funk. Canad. J. Bot. 47:1509. 1969).
Host: Larix
Encoeliopsis oricostata **(Cash) Groves**
Cenangella oricostata Cash
Distr.: Southwestern United States.
Host: Ribes
Encoeliopsis rhododendri **(Ces. ex De Not.) Nannf.**
Cenangella rhododendri (Ces. ex De Not.) Rehm
Diplodina eurhododendri G. Voss (anamorph)
Distr.: Temperate northern hemisphere.
Notes: On leaves, twigs, and capsules.
Host: Rhododendron

Endobotrya Berk. & M.A. Curtis in Berk. Coelomycetes

Endobotrya elegans **Berk. & M.A. Curtis in Berk.**
Distr.: Eastern United States.
Notes: Sutton (The Coelomycetes. p. 202. 1980).
Host: Fagus

Endocalyx Berk. & Broome. Hyphomycetes

Endocalyx melanoxanthus **(Berk. & Broome) Petch**
Distr.: Subtropical and tropical regions.
Notes: Especially common on palms (Ellis. Dematiaceous Hyphomycetes. p. 580. 1971).
Hosts: Cocos, Sabal, Serenoa, Smilax

Endoconidium Prill. & Delacr. Hyphomycetes

Notes: Nag Raj & Kendrick (A monograph of *Chalara* and allied genera. p. 45. 1975).

Endoconidium temulentum Prill. & Delacr.
Gloeotinia granigena (Quél.) T. Schumacher (teleomorph)
Distr.: Temperate North America; Europe.
Notes: Blind seed disease of grasses.
Hosts: Agrostis, Alopecurus, Festuca, Glyceria, Holcus, Phleum

Endocronartium Y. Hiratsuka. Basidiomycetes, Uredinales

Endocronartium harknessii (J.P. Moore) Y. Hiratsuka
Cronartium harknessii E. Meinecke, nom. nud.
Peridermium cerebroides E. Meinecke
Peridermium harknessii J.P. Moore (anamorph)
Distr.: Temperate North America.
Notes: Autoecious. Western gall rust, Woodgate rust. According to Epstein & Buurlage (Mycologia 80:235. 1988) this fungus is an asexual rust.
Host: Pinus

Endomycopsis Dekker. Hemiascomycetes, Endomycetales

Endomycopsis burtonii (Boidin, Pignal, Lehodey, Vey & Abadie) Kreger-Van Rij
Pichia burtonii Boidin, Pignal, Lehodey, Vey & Abadie
Distr.: Widespread in warmer regions.
Host: Triticum

Endomycopsis mali (I. Lewis) Dekker
Endomyces mali I. Lewis
Distr.: ME(type).
Notes: This name could not be characterized (Lodder. The yeasts. p. 205. 1970).
Host: Malus

Endophragmiella Sutton. Coelomycetes

Notes: Hughes (New Zealand J. Bot. 17:139. 1979).

Endophragmiella biseptata (Peck) S.J. Hughes
Monotospora biseptata Peck
Endophragmia nannfeldtii M.B. Ellis
Distr.: Temperate North America; Europe; New Zealand.
Notes: On decaying wood and bark of hardwoods (Hughes. Fungi Canadenses 125. 1978).
Host: Acer

Endophragmiella boewei (J.L. Crane) S.J. Hughes
Endophragmia boewei J.L. Crane
Distr.: FL(type); Asia.
Notes: Crane (Mycologia 64:658. 1972).
Host: Podocarpus

Endophragmiella boothii (M.B. Ellis) S.J. Hughes
Endophragmia boothii M.B. Ellis
Distr.: FL; England(type); New Zealand.
Host: Serenoa

Endophragmiella dimorphospora (Awao & Udagawa) S.J. Hughes
Endophragmia dimorphospora Awao & Udagawa
Selenosporella sp. (synanamorph)
Distr.: FL; Asia.
Notes: Isolated from soil, leaves and other debris (Sutton. Mycologia 70:787. 1978).
Host: Serenoa

Endophragmiella taxi (M.B. Ellis) S.J. Hughes
Endophragmia taxi M.B. Ellis
Distr.: WV(type).
Host: Taxus

Endophyllum Lév. Basidiomycetes, Uredinales

Endophyllum alaskanum Savile
Distr.: AK.
Notes: Savile (Canad. J. Bot. 40:1393. 1962).
Host: Epilobium

Endophyllum circumscriptum Whetzel & Olive in Olive & Whetzel
Distr.: FL; Mexico to Argentina including the West Indies.
Notes: Buriticá & Hennen (Flora Neotropica 24:13. 1980).
Host: Cissus

Endophyllum sempervivi de Bary
Distr.: Eastern United States; Europe.
Host: Sempervivum

Endophyllum tuberculatum (Ellis & Kellerm.) Arth. & Fromme
Aecidium tuberculatum Ellis & Kellerm. (anamorph)
Distr.: Central and western North America.
Notes: Autoecious; on Malvaceae.
Hosts: Alcea, Callirhoe, Sidalcea

Endoraecium C.S. Hodges & D.E. Gardner. Basidiomycetes, Uredinales

Notes: Hodges & Gardner (Mycologia 76:339. 1984).

Endoraecium acaciae C.S. Hodges & D.E. Gardner
Distr.: HI.
Notes: Autoecious.
Host: Acacia

Endoraecium hawaiiense C.S. Hodges & D.E. Gardner
Distr.: HI.
Notes: Autoecious.
Host: Acacia

Endothia Fr. Pyrenomycetes, Diaporthales

Notes: Some species have been placed in *Cryphonectria* (Barr. Mycol. Mem. 7:142. 1978).

Endothia gyrosa (Schwein.:Fr.) Fr.
Endothiella gyrosa Sacc. (anamorph)
Distr.: Eastern North America; Australia.
Notes: On limbs and trunks of hardwoods, especially *Quercus* spp. Snow et al. (C.M.I. Descr. 449. 1975).
Hosts: Acer, Castanea, Corylus, Fagus, Ilex, Liquidambar, Quercus, Ulmus, Vitis

Endothia metrosideri (M.K. Roane & Fosb.) Barr
Diaporthopsis metrosideri M.K. Roane & Fosb.
Distr.: HI.
Host: Metrosideros

Endothia singularis Shear & N. Stevens
Endothiella singularis (Syd. & P. Syd.) Shear & N. Stevens (anamorph)
Distr.: Western North America; eastern Asia; Europe.
Notes: On limbs of Fagaceae.
Host: Quercus

Endothia viridistroma Wehmeyer
Distr.: GA.
Host: Cercis

Endothiella Sacc. Coelomycetes

Notes: These are anamorphs of *Cryphonectria* and *Endothia*.

Endothiella gyrosa Sacc.
Endothia gyrosa (Schwein.:Fr.) Fr. (teleomorph)
Distr.: Southeastern North America.
Notes: Blight of pin oak.
Host: Quercus

Endotryblidium Petr. Discomycetes, Helotiales

Endotryblidium insculptum (Cooke) Petr.
Triblidium insculptum Cooke
Dothidea corylina Cooke & Harkn.
Distr.: CA, NJ.
Host: Corylus

Endoxyla Fuckel. Pyrenomycetes, Sordariales

Endoxyla cirrhosa (Pers.:Fr.) Arx & E. Müller
Ceratostomella cirrhosa (Pers.:Fr.) Sacc.
Distr.: Eastern North America; Europe.
Notes: On hardwoods.
Host: Nyssa
Endoxyla inusta (Cooke) Ellis & Everh.
Distr.: GA(type).
Host: Juniperus
Endoxyla operculata (Albertini & Schwein.:Fr.) Sacc.
Distr.: Western United States; Europe.
Notes: On dead wood of conifers. Arx & Müller (Beitr. Kryptogamenfl. Schweiz 11(1):354. 1954) provided an extensive synonymy.
Host: Pinus
Endoxyla parallela (Fr.) Fuckel
Distr.: Western United States; Europe.
Notes: On dead wood of conifers.
Host: Pinus
Endoxyla rostrata (Tode) Munk, nom. illeg.
Ceratostomella rostrata (Fr.) Sacc.
Distr.: Eastern North America; Europe.
Host: Nyssa

Endoxylina Romell. Pyrenomycetes, Sphaeriales

Endoxylina allantospora (Ellis & Everh.) Shoemaker & Egger
Valsaria allantospora Ellis & Everh.
Distr.: Central and western North America.
Notes: On dead branches (Shoemaker & Egger. Fungi Canadenses 277. 1982).
Host: Acer
Endoxylina moroides (Cooke & Peck) Shoemaker & Egger
Diatrype moroides Cooke & Peck
Valsaria moroides (Cooke & Peck) Sacc.
Distr.: IA, NY.
Hosts: Alnus, Carpinus

Enthallopycnidium F. Stevens. Coelomycetes

Enthallopycnidium gouldiae F. Stevens
Distr.: HI.
Host: Gouldia

Entodesmium Riess. Loculoascomycetes, Dothideales

Entodesmium niessleanum (Rabenh. ex Niessl) L. Holm
Leptosphaeria niessleana Rabenh. ex Niessl
Distr.: Western United States; Europe.
Hosts: Hieracium, Ligusticum, Pedicularis

Entomocorticium H.S. Whitney, Bandoni & Oberwinkler. Basidiomycetes, Aphyllophorales

Entomocorticium dendroctoni H.S. Whitney
Distr.: MT; Canada(British Columbia).
Notes: In bark beetle galleries, associated with *Dendroctonus ponderosae*.
Host: Pinus

Entomosporium Lév. Coelomycetes

Entomosporium mespili (DC.) Sacc.
Entomosporium cydoniae (Cooke & Ellis) Sacc.
Entomosporium maculatum Lév.
Entomosporium maculatum Lév. var. *cydoniae* (Sacc.) Grove
Entomosporium maculatum Lév. var. *domesticum* (Sacc.) Grove
Entomosporium mespili (DC.) Sacc. var. *cydoniae* Briosi & Cavara
Entomosporium thuemenii (Cooke) Sacc.
Diplocarpon mespili (Sorauer) Sutton (teleomorph)
Distr.: Cosmopolitan.
Notes: Leaf and fruit spot of Rosaceae (Sutton. The Coelomycetes. p. 150. 1980; Sivanesan & Gibson. C.M.I. Descr. 481. 1976 as *Diplocarpon maculatum*).
Hosts: Amelanchier, Cercis, Cleyera, Crataegus, Cydonia, Eriobotrya, Feijoa, Heteromeles, Malus, Photinia, Pyrus, Rhaphiolepis, Sorbus

Entophlyctis A. Fischer. Chytridiomycetes, Chytridiales

Entophlyctis texana Karling
Distr.: TX(type).
Notes: Barr (Canad. J. Bot. 49:2217. 1971) suggested that this may be a pigmented form of *E. confervae-glomeratae* (Cienkowski) Sparrow.
Host: Eriocaulon

Entyloma de Bary. Basidiomycetes, Ustilaginales

Notes: Vánky (Symb. Bot. Upsal. 24:52. 1985).
Entyloma alsines Halst.
Distr.: NJ(type), TX.
Host: Stellaria
Entyloma arnicale Ellis & Everh.
Entylomella arnicale (Ellis & Everh.) Cif.
Entyloma circaeae Dearn.
Distr.: Western United States.
Notes: Savile (Fungi Canadenses 112. 1978).
Hosts: Arnica, Circaea
Entyloma australe Speg.
Entyloma physalidis (Kalchbr. & Cooke) G. Wint.
Distr.: North America; South America; Africa; eastern Asia.
Notes: On Solanaceae.
Hosts: Lycopersicon, Physalis, Solanum
Entyloma calendulae (Oudem.) de Bary
Ramularia arnicalis Ellis & Everh.
Entyloma holwayi Syd.
Distr.: Temperate regions.
Notes: On Asteraceae (Mordue. C.M.I. Descr. 801. 1984).
Hosts: Antennaria, Arnica, Calendula, Cosmos, Gaillardia, Hieracium, Senecio
Entyloma clintonianum Zundel & A.A. Dunlap
Distr.: Western United States.
Host: Mimulus
Entyloma compositarum Farl.
Cercosporella colubrina Ellis & Everh.
Entyloma madiae Cif.
Entyloma wisconsiniense Cif.
Distr.: North America; Great Britain.
Notes: On Asteraceae.
Hosts: Achillea, Agoseris, Ambrosia, [Amsinckia], Arnica, Aster, Bidens, Boltonia, Calendula, Chrysopsis, Echinacea, Erigeron, Eupatorium, Gaillardia, Gnaphalium, Helenium, Helianthella, Heterotheca, Hieracium, Lactuca, Lagophylla, Madia, Ratibida, Rudbeckia, Senecio, Silphium, Viguiera
Entyloma dactylidis (Pass.) Cif.
Thecaphora dactylidis Pass. in A. Fisch. v. Waldh.
Entyloma catabrosae Johans.
Entyloma crastophilum Sacc. & Roum. in Roum. & Sacc.
Entyloma holci Liro
Entyloma irregulare I. Johansen
Entyloma speciosum J. Schröt. & Henn.
Distr.: Widespread.
Notes: On Poaceae.
Hosts: Agrostis, Alopecurus, Catabrosa, Deschampsia, Glyceria, Holcus, Muhlenbergia, Oryza, Panicum, Phleum, Poa, Zizania
Entyloma dahliae Syd. & P. Syd.
Distr.: Cosmopolitan.
Notes: Mordue (C.M.I. Descr. 802. 1984 as *Entyloma calendulae* f. *dahliae*).
Host: Dahlia
Entyloma davisii Cif.
Distr.: Central United States; Canada.
Notes: Zundel (The Ustilaginales of the world. p. 244. 1953) suggested that this species is a synonym of *E. compositarum*, but other authors

disagree.
Host: Rudbeckia
Entyloma ellisii Halst.
Distr.: NJ(type), WA.
Host: Spinacia
Entyloma eryngii (Corda) de Bary
Entyloma saniculae Peck
Distr.: Northern hemisphere.
Notes: On Apiaceae.
Hosts: Eryngium, Sanicula
Entyloma eschscholtziae Harkn.
Distr.: CA.
Hosts: Dendromecon, Eschscholzia
Entyloma ficariae A. Fisch. v. Waldh.
Entyloma ranunculi (Bonord.) J. Schröt.
Distr.: North America; Europe.
Notes: On Ranunculaceae.
Hosts: Anemone, Ranunculus, Thalictrum
Entyloma fimbriata G. Fisch.
Distr.: CA(type).
Host: Hydrocotyle
Entyloma floerkeae Holw. in Ellis & Everh.
Distr.: Eastern United States.
Host: Floerkea
Entyloma fuscum J. Schröt. in Cohn
Distr.: North America; Africa; Asia; Europe.
Notes: Mordue (C.M.I. Descr. 803. 1984).
Host: Papaver
Entyloma gratiolae (J.J. Davis) Cif.
Entyloma linariae J. Schröt. var. *gratiolae* J.J. Davis
Distr.: WI(type).
Host: Gratiola
Entyloma hydrocotyles Speg.
Distr.: CA; Argentina(type).
Host: Hydrocotyle
Entyloma lapponicum Liro
Entyloma montis-rainieri G. Fisch. & W.B. Cooke
Distr.: North America; Europe.
Notes: On Polemoniaceae.
Host: Polemonium
Entyloma linariae J. Schröt. in Cohn
Entyloma collinsiae Harkn.
Entyloma linariae J. Schröt. var. *veronicae* G. Wint. in Rabenh.
Entyloma veronicae (G. Wint.) Lagerh. in Pat. & Lagerh.
Distr.: North America; North Africa; Asia; Europe.
Notes: On Scrophulariaceae.
Hosts: Collinsia, Linaria, Tonella, Veronica
Entyloma lineatum (Cooke) J.J. Davis
Distr.: Eastern North America.
Hosts: Poa, Zizania
Entyloma lithophragmae G. Fisch. & C.G. Shaw
Distr.: WA(type).
Host: Lithophragma
Entyloma lobeliae Farl.
Distr.: Eastern North America.
Host: Lobelia
Entyloma meliloti McAlpine
Entyloma trigonellae J.A. Stevenson
Distr.: Southern and western United States; Australia(type).
Hosts: Melilotus, Trigonella
Entyloma menispermi Farl. & Trel.
Distr.: Central and eastern North America.
Host: Menispermum
Entyloma microsporum (Unger) J. Schröt. in Rabenh.
Distr.: Widespread.
Host: Ranunculus
Entyloma oryzae Syd. & P. Syd.
Distr.: Worldwide in rice-growing regions.
Notes: Leaf smut of rice (Mulder & Holliday. C.M.I. Descr. 296. 1971).
Host: Oryza
Entyloma parvum J.J. Davis
Distr.: MA, WI(type).
Host: Eleocharis
Entyloma polysporum (Peck) Farl.
Entyloma aster-sericeanum Zundel
Entyloma guaraniticum Speg.
Distr.: Western hemisphere; Japan.
Notes: On Asteraceae.
Hosts: Agoseris, Ambrosia, Aster, Bidens, Echinacea, Gaillardia, Helenium, Helianthus, Ratibida, Rudbeckia
Entyloma serotinum J. Schröt. in Cohn
Distr.: North America; Europe.
Notes: On Boraginaceae.
Hosts: Amsinckia, Hackelia, Mertensia, Myosotis
Entyloma thalictri J. Schröt. in Cohn
Distr.: Temperate northern hemisphere.
Host: Thalictrum
Entyloma winteri Linh.
Entyloma wyomingense Zundel
Distr.: Northern hemisphere.
Host: Delphinium

Entylomella Höhn. Hyphomycetes

Entylomella armoraciae (Fuckel) Cif.
Ramularia armoraciae Fuckel
Entyloma armoraciae Cif.
Distr.: Eastern North America; temperate Europe.
Notes: On Brassicaceae.
Hosts: Armoracia, Brassica, Capsella, Thlaspi

Ephelis Fr. Hyphomycetes

Ephelis borealis Ellis & Everh.
Atkinsonella hypoxylon (Peck) Diehl (teleomorph)
Distr.: Temperate North America.
Notes: Leuchtmann & Clay (Mycologia 80:192. 1988).
Host: Danthonia
Ephelis mexicana Fr. in Berk.
Balansia claviceps Speg. (teleomorph)
Distr.: Southern United States; tropical regions.
Notes: Rykard et al. (Mycologia 76:1095).
Host: Cynodon

Epichloe (Fr.) Tul. & C. Tul. Pyrenomycetes, Clavicipitales

Epichloe typhina (Pers.:Fr.) Tul. in Tul. & C. Tul.
Typhodium typhinum (Pers.:Fr.) Seaver
Acremonium typhinum Morgan-Jones & W. Gams (anamorph)
Distr.: Temperate northern hemisphere.
Notes: Choke disease of grasses, cat-tail disease (Booth. C.M.I. Descr. 639. 1979). Traditionally, the anamorph was referred to *Sphacelia typhina* Sacc., however compare Morgan-Jones & Gams (Mycotaxon 15:311. 1982) and Rykard et al. (Phytopathology 75:950. 1985).
Hosts: Agropyron, Agrostis, Alopecurus, Andropogon, Bouteloua, Bromus, Calamagrostis, Calamovilfa, Cinna, Dactylis, Danthonia, Elymus, Festuca, Glyceria, Hystrix, Juncus, Koeleria, Leersia, Melica, Muhlenbergia, Phleum, Poa, Schizachyrium, Sphenopholis, Typha

Epicoccum Link:Fr. Hyphomycetes

Epicoccum nigrum Link
Epicoccum asterinum Pat.
Epicoccum duriaeanum Mont.
Epicoccum granulatum Penz.
Epicoccum neglectum Desmaz.
Epicoccum negundinis Ellis & Everh., nom. nud.
Epicoccum purpurascens Ehrenb.
Epicoccum vulgare Corda
Distr.: Cosmopolitan.
Notes: Cannon (Microbiol. Sci. 3:168. 1986) explained the use of *E. nigrum*, rather than *E. purpurascens*. Booth (C.M.I. Descr. 680. 1980) presented a recent description of this species as *E. purpurascens*. Schol-Schwarz (Trans. Brit. Mycol. Soc. 42:149. 1959) reduced most described *Epicoccum* names to synonymy with *E. nigrum*.
Hosts: Acacia, Acer, Agropyron, Alnus, Amaryllis, Amphicarpaea, Arceuthobium, Artemisia, Avena, Carya, Catalpa, Cocos, Echinochloa, Eichhornia, Epilobium, Erythrina, Erythronium,

Euonymus, Fagus, Ginkgo, Glycine, Gossypium, Helianthus, Iris, Juncus, Lens, Lotus, Lupinus, Magnolia, Malus, Medicago, Metrosideros, Microsteris, Oryza, Phaseolus, Pinus, Pisum, Platanus, Prunus, Pseudotsuga, Roystonea, Rubus, Scirpus, Setaria, Spartina, Stipa, Trifolium, Tripsacum, Triticum, Typha, Vicia, Vitis, Wikstroemia, Yucca, Zea

Episphaerella Petr. Loculoascomycetes, Dothideales

Episphaerella dodonaeae (F. Stevens) M.L. Farr, Schoknecht, & J.L. Crane
Dimerina dodonaeae F. Stevens
Distr.: FL; Costa Rica; Ecuador.
Notes: Farr et al. (Canad. J. Bot. 63:1983. 1985).
Host: Dodonaea

Epistigme Syd. Coelomycetes

Epistigme laporteae (Peck) Arx
Gloeosporium laporteae Peck
Distr.: OK(type).
Host: Laportea

Epithele Pat. Basidiomycetes, Aphyllophorales

Notes: Boquiren (Mycologia 63:937. 1971) provided a broad generic concept.
Epithele hydnoides Burt
Distr.: HI.
Host: Cibotium
Epithele sulphurea Burt
Distr.: FL, HI.
Host: Sabal

Epochnium Link. Hyphomycetes

Notes: This genus is considered a synonym of *Monilia*, but not all names have been redisposed.
Epochnium isthmophorum Sacc.
Distr.: Western United States.
Hosts: Artemisia, Chrysothamnus

Eriopezia (Sacc.) Rehm. Discomycetes, Helotiales

Eriopezia microspora (Kanouse) Dennis
Lachnella tricolor (Sowerby ex Pers.:Fr.) W. Phillips var. *microspora* Kanouse
Lachnella microspora (Kanouse) Seaver
Peziza alboviridis Cooke non Saut., hom. illeg.
Trichopeziza alboviridis Sacc.
Distr.: Central and eastern United States.
Notes: According to Korf (Mycotaxon 7:482. 1978) this species belongs in *Lachnellula*, but he did not propose a new combination.
Hosts: Myrica, Quercus

Eriosphaeria Sacc. Pyrenomycetes, Trichosphaeriales

Eriosphaeria vermicularia (Nees:Fr.) Sacc.
Venturia lanea Dearn.
Distr.: Temperate North America; Europe.
Hosts: Chamaecyparis, Picea, Pinus

Eriosporella Höhn. Coelomycetes

Eriosporella calami (Niessl) Höhn.
Cryptosporium calami Niessl in Rabenh.
Distr.: HI.
Host: Eucalyptus

Erysiphe R. Hedw. ex DC.:Fr. Pyrenomycetes, Erysiphales

Erysiphe aggregata (Peck) Farl.
Erysiphella aggregata Peck
Erysiphe vernalis P. Karst. var. *aggregata* (Peck) Junell
Distr.: Temperate North America.
Notes: On twigs and female catkins.
Host: Alnus
Erysiphe americana U. Braun
Distr.: North America.
Notes: On Malvaceae. Braun (Mycotaxon 18:124. 1983).
Host: Napaea
Erysiphe brunneopunctata U. Braun
Distr.: CA.
Notes: A segregate of *E.cichoracearum* DC. (Braun. Mycotaxon 20:494. 1984).
Host: Mimulus
Erysiphe californica U. Braun
Distr.: Western United States.
Notes: A segregate of *E. cichoracearum* (Braun. Mycotaxon 20:492. 1984).
Host: Eriogonum
Erysiphe cichoracearum DC.
Erysiphe asterum Schwein.
Erysiphe lamprocarpa (Wallr.:Fr.) Duby
Oidium asteris-punicei Peck (anamorph)
Distr.: Cosmopolitan.
Notes: Powdery mildew of chicory and other Asteraceae (Kapoor. C.M.I. Descr. 152. 1967).
Hosts: Abelmoschus, Acaena, Acalypha, Achillea, Ageratum, Ajuga, Alcea, Althaea, Ambrosia, Amsinckia, Angelonia, Antennaria, Antirrhinum, Arctium, Arnica, Artemisia, Asclepias, Aster, Baccharis, Balsamorhiza, Begonia, Bidens, Boltonia, Cacalia, Calendula, Callistephus, Campanula, Campsis, Capsicum, Carica, Carthamus, Centaurea, Chelone, Chrysanthemum, Chrysopsis, Chrysothamnus, Cichorium, Cirsium, Citrullus, Clarkia, Conyza, Cordia, Coreopsis, Cosmos, Crepis, Cryptantha, Cucumis, Cucurbita, Cynara, Cynoglossum, Dahlia, Delphinium, Doronicum, Dracopis, Dyssodia, Echinocystis, Erechtites, Erigeron, Eriogonum, Eucalyptus, Eupatorium, Euthamia, Fraxinus, Gaillardia, Galeopsis, Galium, Gerbera, Gloxinia, Grindelia, Gutierrezia, Hackelia, Haplopappus, Hebe, Hedera, Helenium, Helianthus, Heliopsis, Hesperochiron, Heuchera, Hibiscus, Hieracium, Humulus, Hydrophyllum, Hypericum, Inula, Iva, Lactuca, Lagenaria, Lamium, Lappula, Layia, Liatris, Linaria, Lithospermum, Luffa, Machaeranthera, Madia, Mangifera, Matricaria, Meconopsis, Melothria, Mentha, Mertensia, Microsteris, Mimulus, Momordica, Monarda, Monardella, Myosotis, Navarretia, Nemophila, Nicotiana, Parietaria, Parthenium, Pectocarya, Penstemon, Phacelia, Phlox, Pilea, Piqueria, Plagiobothrys, Plantago, Polemonium, Populus, Prenanthes, Prunella, Ratibida, Rhus, Rudbeckia, Salvia, Satureja, Scrophularia, Scutellaria, Sechium, Senecio, Sicyos, Silphium, Solanum, Solidago, Sonchus, Stachys, Stokesia, Tanacetum, Taraxacum, Teucrium, Tragopogon, Trifolium, Urtica, Valeriana, Vauquelinia, Verbascum, Verbena, Verbesina, Vernonia, Veronicastrum, Viguiera, Xanthium, Zauschneria, Zinnia
Erysiphe cichoracearum DC. var. ***latispora*** U. Braun
Erysiphe ambrosiae Schwein.
Oidium latisporum U. Braun (anamorph)
Distr.: Temperate northern hemisphere.
Notes: On Asteraceae. Distinguished on the basis of conidial shape (Braun. Mycotaxon 18:117. 1983).
Host: Helianthus
Erysiphe communis (Wallr.) Link
Distr.: Cosmospolitan.
Notes: According to Junell (Trans. Brit. Mycol. Soc. 48:541. 1965), this name has been used in many different senses and should thus be regarded as a *nomen ambiguum*. Fungi prevsiouly placed in *E. communis* are now placed in several species (Junell. Svensk Bot. Tidskr. 61:209. 1967).
Hosts: Achillea, Anemone, Aquilegia, Artemisia, Clematis, Dahlia, Glycyrrhiza, Lotus, Lupinus, Mertensia, Pelargonium, Solidago
Erysiphe cumminsiana U. Braun
Distr.: Central and southwestern United States.

Notes: On Asteraceae.
Hosts: Heliopsis, Senecio
Erysiphe cynoglossi (Wallr.) U. Braun
Alphitomorpha cynoglossi Wallr.
Erysiphe asperifoliorum (Wallr.) Rabenh.
Erysiphe horridula (Wallr.) Lév.
Distr.: Temperate regions.
Notes: Powdery mildew of Boraginaceae.
Hosts: Amsinckia, Cryptantha, Lappula
Erysiphe euphorbiae Peck
Distr.: Eastern United States.
Notes: In the past, this name was considered synonymous with *Microsphaera euphorbiae*. According to Braun (Mycotaxon 20:498. 1984), it is a distinct species.
Host: Euphorbia
Erysiphe galeopsidis DC.
Erysiphe cichoracearum DC. f. *galeopsidis* (DC.) E.S. Salmon
Erysiphe chelones Schwein.
Erysiphe gerardiae Schwein.
Oidium leucas-javanicae Sawada (anamorph)
Distr.: Temperate regions.
Notes: Powdery mildew of Lamiaceae. Braun (Nova Hedwigia 34:698. 1981) suggested that *E. gerardiae* is a synonym of *E. cichoracearum* or *E. galeopsidis*.
Hosts: Aureolaria, Chelone, [Eupatorium], [Heliopsis], Lamium, Mentha, Physostegia, Scutellaria, Stachys, Teucrium
Erysiphe graminis DC.
Blumeria graminis (DC.) E.O. Speer
Oidium monilioides (Nees) Link (anamorph)
Distr.: Cosmopolitan.
Notes: Powdery mildew of grasses. Several physiologically specialized races exist including at least four named *formae speciales* (Kapoor. C.M.I. Descr. 153. 1967; Parmelee. Fungi Canadenses 71. 1975; Speer. Sydowia 27:1. 1975). Hiura in Spencer (The powdery mildews. p. 101. 1978) discussed the genetic basis of *formae speciales*.
Hosts: Agropyron, Agrostis, Andropogon, Avena, Beckmannia, Bromus, Buchloe, Calamagrostis, Catabrosa, Cinna, Cynodon, Dactylis, Deschampsia, Digitaria, Elymus, Elysitanion, Festuca, Glyceria, Hordeum, Hystrix, Koeleria, Lolium, Melica, Milium, Phalaris, Phleum, Poa, Polypogon, Puccinellia, Secale, Sitanion, Sphenopholis, Sporobolus, Stipa, Trisetum, Triticum
Erysiphe graminis DC. f. sp. ***avenae*** Ém. Marchal
Distr.: Widespread.
Hosts: Antirrhinum, Arrhenatherum, Avena, Dactylis, Fragaria, Hordeum, Lagerstroemia, Lespedeza, Malus, Phaseolus, Rosa, Syringa, Trifolium, Triticum, Vigna, Vitis, Zinnia
Erysiphe graminis DC. f. sp. ***hordei*** Ém. Marchal
Distr.: Widespread.
Host: Hordeum
Erysiphe graminis DC. f. sp. ***secalis*** Ém. Marchal
Distr.: Southern United States.
Host: Secale
Erysiphe graminis DC. f. sp. ***tritici*** Ém. Marchal
Distr.: Widespread.
Host: Triticum
Erysiphe heraclei DC.
Erysiphe umbelliferaum de Bary
Distr.: Widespread.
Notes: Powdery mildew of Apiaceae (Kapoor. C.M.I. Descr. 154. 1967).
Host: Daucus
Erysiphe immersa U. Braun
Distr.: CA.
Notes: A segregate of *E. cicoracearum* (Braun. Mycotaxon 20:494. 1984).
Host: Eremocarpus
Erysiphe lagerstroemiae E. West
Distr.: Southern United States.
Notes: A doubtful species. According to Braun (Beih. Nova Hedwigia 89:622. 1987), all specimens of this species were actually the anamorph of *Uncinula astraliana*.
Host: Lagerstroemia
Erysiphe polygoni DC.
Distr.: Worldwide on Polygonaceae.
Notes: In the past, this name was used for powdery mildews on a wide variety of hosts. Recently, the name has been restricted to the pathogen of Polygonaceae (Junell. Symb. Bot. Upsal. 19:41. 1976; Sivanesan. C.M.I. Descr. 509. 1976). Parmelee (Canad. J. Bot. 55:1940. 1977) presented a table explaining the redisposition of *E. polygoni sensu* lato.
Hosts: Acacia, Aconitum, Althaea, Alysicarpus, Amelanchier, Amphicarpaea, Anemone, Anemonella, Apios, Aquilegia, Arabis, Arenaria, Armoracia, Astilbe, Astragalus, Baptisia, Begonia, Beta, Brassica, Calendula, Callistephus, Caltha, Cassia, Castilleja, Ceanothus, Chelone, Chrysothamnus, Clematis, Colutea, Crotalaria, Cucumis, Cucurbita, Cuphea, Dahlia, Daucus, Delphinium, Desmanthus, Desmodium, Duboisia, Epilobium, Erechtites, Erica, Eriogonum, Erysimum, Eschscholzia, Fagopyrum, Galium, Gardenia, Gaura, Genista, Geranium, Geum, Glycine, Glycyrrhiza, Gutierrezia, Helianthus, Homalocladium, Hydrangea, Hydrophyllum, Iberis, Iliamna, Kummerowia, Lathyrus, Lespedeza, Liriodendron, Lobularia, Lonicera, Lotus, Lupinus, Lycium, Lycopersicon, Medicago, Melilotus, Mentha, Nothochelone, Oenothera, Oxytropis, Pachycarpus, Paeonia, Papaver, Parnassia, Passiflora, Phaseolus, Pisum, Polygonum, Populus, Potentilla, Primula, Psoralidium, Quercus, Ranunculus, Raphanus, Rhododendron, Robinia, Rumex, Scabiosa, Sedum, Sesbania, Silene, Sinapis, Stellaria, Strophostyles, Thalictrum, Thermopsis, Trifolium, Vicia, Vigna, Whipplea, Zizia
Erysiphe pseudosepulta U. Braun
Distr.: Western United States.
Notes: On Asteraceae. A segregate of *E.cichoracearum* DC. (Braun. Mycotaxon 20:495. 1984).
Host: Chrysothamnus
Erysiphe sparsa U. Braun
Distr.: WI(type).
Notes: A segregate of *E.cichoracearum* DC. (Braun. Mycotaxon 20:495. 1984).
Host: Acalypha

Erythricium J. Eriksson & Hjortstam. Basidiomycetes, Aphyllophorales

Erythricium chaparralum Burdsall & R.L. Gilbertson
Distr.: AZ.
Notes: On herbaceous plants (Burdsall & Gilbertson. Mycotaxon 15:335. 1982).
Host: Yucca
Erythricium laetum (P. Karst.) J. Eriksson & Hjortstam
Distr.: North America; Europe.
Notes: On wood.
Hosts: Alnus, Betula, Populus
Erythricium salmonicolor (Berk. & Broome) Burdsall
Corticium salmonicolor Berk. & Broome
Phanerochaete salmonicolor (Berk. & Broome) Jülich
Necator decretus Massee (anamorph)
Distr.: Widespread.
Notes: Mordue & Gibson (C.M.I. Descr. 511. 1976 as *Corticium salmonicolor*).
Hosts: Aleurites, Buxus, Cercis, Euonymus, Ficus, Gardenia, Ilex, Jasminum, Ligustrum, Malus, Pittosporum, Pyrus, Severinia

Euantennaria Speg. Loculoascomycetes, Dothideales

Euantennaria alaskensis (Sacc. & Scalia) Speg.
Limacinia alaskensis Sacc. & Scalia
Aithalomyces alaskensis (Sacc. & Scalia) Barr
Morfea alaskensis (Sacc. & Scalia) Batista & Cif.
Distr.: Northwestern North America.
Notes: Sooty mold on conifers and alder (Barr. Canad. J. Bot. 33:501. 1955). Possibly synonymous with *E. rhododendri* (Woronichin) S.J. Hughes.
Host: Abies
Euantennaria rhododendri (Woronichin) S.J. Hughes
Aithalomyces rhododendri Woronichin
Distr.: Circumboreal.
Notes: Hughes (New Zealand J. Bot. 12:299. 1974).
Host: Arctostaphylos

Eudimeriolum Speg. Loculoascomycetes, Dothideales

Eudimeriolum plantaginis (Tehon & E.Y. Daniels) Barr
Stigmatea plantaginis Tehon & E.Y. Daniels
Distr.: IL.
Notes: Barr (Mycotaxon 29:504. 1987).
Host: Plantago
Eudimeriolum septentrionale M.L. Farr & Luttrell
Distr.: GA(type).
Notes: Farr & Luttrell (Mycologia 71:1269. 1979).
Host: Betula

Eupenicillium C.A. Ludw. Hyphomycetes

Eupenicillium cinnamopurpureum D.B. Scott & A.C. Stolk
Eupenicillium idahoense Paden
Penicillium dierckxii Biourge (anamorph)
Distr.: Widespread.
Notes: Isolated from various kinds of organic substrates (Stolk & Samson. Stud. Mycol. 23:61. 1983).
Host: Helianthus

Eupropolella Höhn. Discomycetes, Helotiales

Notes: Eriksson (Symb. Bot. Upsal. 19:22. 1970).
Eupropolella arundinariae (Cash) Dennis
Sphaeropezia arundinariae Cash
Distr.: GA; Great Britain.
Host: Arundinaria
Eupropolella oxycocci (Dearn. ex Cash) B. Eriksson
Naevia oxycocci Dearn. ex Cash in Shear, N. Stevens, & H. Bain
Distr.: WA, northeastern North America; northern Europe.
Host: Vaccinium
Eupropolella vaccinii (Rehm) Höhn.
Phacidium arctostaphyli P. Karst.
Sphaeropezia arctostaphyli (P. Karst.) Rehm
Phacidium vaccinii Fr. var. *oxycoccini* Dearn. & House
Distr.: Temperate regions.
Hosts: Arctostaphylos, Vaccinium

Eurotium Link:Fr. Hyphomycetes

Notes: *Eurotium* includes the teleomorphs of the *Aspergillus glaucus* group. Blaser (Sydowia 28:1. 1976).
Eurotium amstelodami L. Mangin
Aspergillus hollandicus R.A. Samson & W. Gams (anamorph)
Distr.: Widespread, most common in subtropical and tropical regions.
Notes: Isolated from soil and various kinds of organic debris (Domsch, et al. Compendium of soil fungi. p. 289. 1980).
Host: Helianthus
Eurotium chevalieri L. Mangin
Aspergillus equitis R.A. Samson & W. Gams (anamorph)
Distr.: Subtropical and tropical regions.
Notes: Isolated from air, soil, and various kinds of organic debris (Domsch et al. Compendium of soil fungi. p. 291. 1980).
Hosts: Carya, Helianthus
Eurotium echinulatum Delacr.
Aspergillus echinulatus (Delacr.) Thom & Church, nom. illeg. (anamorph)
Distr.: Widespread.
Notes: On harvested grain and other substrates (Niesh & McGregor. Fungi Canadenses 315. 1987).
Host: Helianthus
Eurotium herbariorum Link:Fr.
Eurotium umbrosum (Bainier & Sartory) Malloch & Cain
Aspergillus glaucus Link:Fr. (anamorph)
Distr.: Worldwide, most common in subtropical and tropical regions.
Notes: Isolated from soil and various kinds of organic substrates (Domsch et al. Compendium of soil fungi. p. 293. 1980).
Hosts: Coptis, Geum, Helianthus, Rubus
Eurotium repens de Bary
Aspergillus reptans R.A. Samson & W. Gams (anamorph)
Distr.: Widespread.
Notes: Isolated from various kinds of organic substrates.
Hosts: Carya, Helianthus
Eurotium rubrum König, Spieckermann & W. Bremer
Aspergillus rubrobrunneus R.A. Samson & W. Gams (anamorph)
Distr.: Widespread.
Notes: Samson & Pitt (Advances in *Penicillium* and *Aspergillus* systematics. p. 389. 1985).
Hosts: Carya, Helianthus, Zea
Eurotium tonophilum Ohtsuki
Aspergillus tonophilus Ohtsuki (anamorph)
Distr.: GA; Japan(type).
Notes: Isolated from optical glass and *Helianthus*.
Host: Helianthus

Euryachora Fuckel. Loculoascomycetes, Dothideales

Notes: Obrist (Phytopathol. Z. 35:382. 1959).
Euryachora aristidae (Schwein.) Theiss. & Syd.
Distr.: United States.
Notes: On grasses.
Hosts: Agrostis, Aristida, Distichlis, Elymus
Euryachora neowashingtonianae Dearn.
Distr.: FL(type).
Host: Washingtonia

Eustilbum Rabenh. Hyphomycetes

Eustilbum aureum (Pers.:Fr.) Seifert & S. Carpenter
Stilbum resinarium Peck
Bisporella resinicola (Baranyay & Funk) S. Carpenter & Seifert in Seifert & S. Carpenter (teleomorph)
Distr.: Temperate northern hemisphere.
Notes: On resins of various conifers. Seifert & Carpenter (Canad. J. Bot. 65:1262. 1987).
Host: Abies

Eutryblidiella (Rehm) Höhn. Loculoascomycetes, Patellariales

Eutryblidiella macrospora (Bonar & Cash) Hafellner
Tryblidiella macrospora Bonar & Cash
Distr.: CA.
Notes: Hafellner (Beih. Nova Hedwigia 62:170. 1979).
Host: Calocedrus
Eutryblidiella viburni (Schwein.:Fr.) Groves
Cenangium viburni (Fr.:Fr.) Groves
Cenangella ravenelii (Berk.) Sacc.
Distr.: Southeastern United States.
Notes: On bark.
Host: Ilex

Eutypa Tul. & C. Tul. Pyrenomycetes, Diatrypales

Notes: Rappaz (Mycol. Helv. 2:334. 1987).
Eutypa armeniacae Hansf. & M.V. Carter
Distr.: Cosmopolitan.
Notes: Determinations of this fungus as a pathogen of grapes in the Pacific Northwest have often been based on isolations of the anamorph state. Rappaz (Mycol. Helv. 2:356. 1987) considered this species to be conspecific with *E. lata* (Pers.:Fr.) Tul . & C. Tul. See also Carter & Talbot (C.M.I. Descr. 436. 1974).
Hosts: Ceanothus, Prunus, Vitis
Eutypa consobrina (Mont.) Rappaz
Diatrype consobrina Mont.
Distr.: Southern United States; Algeria.
Host: Arundinaria
Eutypa laevata (Nitschke) Sacc.
Valsa laevata Nitschke
Eutypa nitida (Nitschke) Sacc.
Distr.: ND; Europe.
Host: Salix
Eutypa lata (Pers.:Fr.) Tul. & C. Tul.

Eutypa fraxini (Nitschke) Sacc.
Eutypa milliaria (Fr.:Fr.) Sacc.
Distr.: Temperate northern hemisphere.
Hosts: Actinidia, Cornus, Crossosoma, Fraxinus, Liriodendron, Lonicera, Populus, Prunus, Quercus, Ulmus, Vitis
Eutypa lata **(Pers.) Tul. & C. Tul. var. *aceri* Rappaz**
Eutypa velutina (Wallr.) Sacc.
Distr.: Eastern United States; Europe.
Notes: On wood.
Hosts: Acer, Morus
Eutypa limaeformis **(Schwein.) Cooke**
Eutypa confusa (Schwein.) Cooke
Distr.: Central and eastern United States.
Notes: On wood and bark of hardwoods.
Hosts: Acer, Quercus
Eutypa maura **(Fr.:Fr.) Fuckel**
Diatrype maura (Fr.:Fr.) Ces. & De Not.
Valsa maura (Fr.:Fr.) Nitschke
Eutypa acharii Tul. & C. Tul.
Cytosporina acharii (Sacc.) Grove (anamorph)
Distr.: Europe; distribution elsewhere is questionable.
Hosts: Acer, Cornus
Eutypa mela **(Schwein.) Cooke**
Eutypa quercina (Sacc.) Berl.
Eutypa velutina (Wallr.) Sacc. var. *quercina* Sacc.
Distr.: Central and eastern United States.
Host: Quercus
Eutypa polycocca **(Fr.:Fr.) P. Karst.**
Valsa polycocca (Fr.:Fr.) Nitschke
Eutypa aspera (Nitschke) Fuckel
Valsa aspera Nitschke
Distr.: ND; Europe.
Host: Quercus
Eutypa rivulosa **(Schwein.) Ellis & Everh.**
Distr.: PA.
Host: Lindera
Eutypa scabrosa **(Bull.:Fr.) Fuckel**
Distr.: ND; Europe(type).
Notes: Rappaz (Mycol. Helv. 2:563. 1987) was unable to characterize this species.
Hosts: Fraxinus, Salix
Eutypa spinosa **(Pers.:Fr.) Tul. & C. Tul.**
Distr.: Temperate North America; Europe.
Hosts: Crataegus, Quercus

Eutypella (Nitschke) Sacc. Pyrenomycetes, Diatrypales

Notes: Rappaz (Mycol. Helv. 2:452. 1987).
Eutypella aequilinearis **(Schwein.:Fr.) Starb.**
Sphaeria aequilinearis Schwein.
Diatrype capnostroma Berk. & Ravenel ex Ellis & Everh.
Diatrype nigerrima Ellis & Everh.
Eutypella vitis (Schwein.:Fr.) Ellis & Everh.
Valsa vitis (Schwein.:Fr.) M.A. Curtis
Cytospora vitis Mont. (anamorph)
Distr.: Eastern United States.
Notes: On hardwoods.
Hosts: Berberis, Morus, Vitis
Eutypella angulosa **(Nitschke) Sacc. f. *fraxini* Rehm**
Distr.: ND(type).
Notes: Although *E. angulosa* is a synonym of *E. sorbi*, this form has not been synonymized.
Host: Fraxinus
Eutypella angulosa **(Nitschke) Sacc. f. *negundinis* Rehm**
Distr.: ND(type).
Notes: Although *E. angulosa* is a synonym of *E. sorbi*, this form has not been synonymized.
Host: Acer
Eutypella canodisca **(Ellis & Holw.) Sacc.**
Valsa canodisca Ellis & Holw. in Ellis & Everh.
Diatrype albopruinosa (Schwein.) Cooke var. *salicina* Rehm
Distr.: Central North America.
Host: Salix
Eutypella ceranata **Fairm.**
Distr.: CA(type).
Host: Heteromeles
Eutypella cerviculata **(Fr.:Fr.) Sacc.**
Eutypa cerviculata (Fr.:Fr.) L.C. Tiffany & Gilman
Valsa cerviculata (Fr.:Fr.) Fr.
Eutypella alnifraga (Wahlenberg) Sacc.
Eutypella halseyana (Schwein.) Berl. & Voglino
Valsa halseyana (Schwein.) Cooke
Diatrype megastoma Ellis & Everh.
Distr.: Temperate northern hemisphere.
Notes: Glawe & Rogers (Mycotaxon 14:334. 1982) described an anamorph from culture.
Hosts: Alnus, Amelanchier, Betula, Carpinus, Corylus, Ostrya, Prunus, Quercus
Eutypella citricola **Speg.**
Distr.: FL; Argentina(type).
Hosts: Citrus, Poncirus
Eutypella conseptata **(Schwein.) Ellis & Everh.**
Valsa conseptata (Schwein.) W.C. Stevenson
Distr.: PA.
Notes: On bark.
Host: Gleditsia
Eutypella dissepta **(Fr.:Fr.) Rappaz**
Eutypa dissepta (Fr.:Fr.) Berl.
Quaternaria dissepta (Fr.:Fr.) Tul. & C. Tul.
Valsa dissepta (Fr.:Fr.) Fr.
Cryptosphaeria juglandina Ellis & Holw.
Libertella dissepta Traverso (anamorph)
Distr.: North America; Europe.
Notes: On hardwoods.
Hosts: Acer, Juglans, Ulmus
Eutypella domicalis **Fairm.**
Distr.: CA(type).
Host: Adenostoma
Eutypella elevans **(Schwein.) Berl.**
Eutypa elevans Schwein.
Distr.: PA.
Host: Rhus
Eutypella glandulosa **(Cooke) Ellis & Everh.**
Valsa glandulosa Cooke
Distr.: Eastern United States.
Hosts: Ailanthus, Alnus, Sambucus
Eutypella goniostoma **(Schwein.) Sacc.**
Eutypella goniostoma (Schwein.) Ellis & Everh.
Eutypella berchemiae (Cooke) Sacc.
Valsa berchemiae Cooke
Eutypella herbicola Ellis & Everh.
Eutypella indistincta (Schwein.) Ellis & Everh.
Eutypella linderae (Peck) Berl.
Valsa linderae Peck
Valsa monticulosa Berk. & M.A. Curtis ex Cooke
Diatrype plagia Berk. & M.A. Curtis
Eutypella tumidula (Cooke & Peck) Sacc.
Valsa tumidula Cooke & Peck
Distr.: Central and eastern United States.
Notes: On hardwoods.
Hosts: Aster, Atriplex, Baccharis, Berchemia, Crataegus, Lindera, Liriodendron, Magnolia, Sassafras
Eutypella juglandicola **(Schwein.:Fr.) Ellis & Everh.**
Valsa juglandicola (Schwein.:Fr.) Cooke
Eutypa allostoma (Schwein.) Cooke
Valsa allostoma (Schwein.) Cooke
Diatrype callicarpae Berk. & Ravenel ex Ellis & Everh.
Eutypella constellata (Berk. & M.A. Curtis ex Cooke) Berl. & Voglino
Valsa constellata Berk. & M.A. Curtis ex Cooke
Eutypella fraxinicola (Cooke & Peck) Sacc.
Valsa fraxinicola Cooke & Peck
Eutypella innumerabilis (Peck) Sacc.
Valsa innumerabilis Peck
Eutypella juglandicola (Schwein.) Ellis & Everh. var. *juglandina* (Cooke & Ellis) Ellis & Everh.
Eutypella juglandina (Cooke & Ellis) Sacc.
Valsa juglandina Cooke & Ellis

Valsa ventriosa Cooke & Ellis
Distr.: Eastern United States.
Notes: On hardwoods.
Hosts: Acer, Aesculus, Ailanthus, Amelanchier, Carya, Catalpa, Cercis Diospyros, Elaeagnus, Fagus, Fraxinus, Gleditsia, Hydrangea, Ilex, Juglans, Lespedeza, Lonicera, Magnolia, Melia, Morus, Nyssa, Rhododendron, Rhus, Rosa, Salix, Smilax, Staphylea, Tilia, Ulmus, Viburnum, Vitis
Eutypella leprosa (Pers.:Fr.) Berl.
Valsa leprosa (Pers.:Fr.) Nitschke
Diatrype celastrina Ellis & Everh.
Valsa celastrina (Ellis & Everh.) Kuntze
Diatrype daldiniana De Not.
Eutypa daldiniana (De Not.) L.C. Tiffany & Gilman
Diatrype hochelagae Ellis & Everh.
Eutypa ludibunda (Sacc.) Thuem.
Valsa ludibunda Sacc.
Diatrype radiata Ellis
Diatrype tumida Ellis & Everh.
Eutypella tumida (Ellis & Everh.) Wehmeyer
Distr.: Eastern United States; Europe.
Notes: On wood of hardwoods.
Hosts: Acer, Ailanthus, Amelanchier, Amorpha, Betula, Caragana, Carya, Celastrus, Celtis, Cercis, Cornus, Crataegus, Euonymus, Fraxinus, Juglans, Magnolia, Malus, Morus, Nyssa, Physocarpus, Picea, Pinus, Platanus, Populus, Prunus, Quercus, Ribes, Robinia, Salix, Sambucus, Smilax, Tilia, Ulmus, Vitis, Zanthoxylum
Eutypella minuta Berl. & F. Sacc.
Distr.: OR; Portugal(type).
Notes: Glawe (Mycotaxon 20:421. 1984) suggested that the report of this species from Oregon is erroneous and may be *Diatrype microstega*.
Host: [Umbellularia]
Eutypella parasitica R.W. Davidson & Lorentz
Libertella sp. (anamorph)
Distr.: North America(east of the Rocky Mountains), most common in the Great Lakes region.
Notes: Canker disease (Johnson & Kuntz. Canad. J. Bot. 56:1518. 1978).
Host: Acer
Eutypella platani (Schwein.) Sacc.
Valsa platani (Schwein.) Cooke
Eutypella aleurina (Berk. & M.A. Curtis) Berl. & Voglino
Valsa aleurina Berk. & M.A. Curtis
Distr.: Eastern United States.
Host: Platanus
Eutypella populi Ellis & Everh.
Distr.: CA; Canada(type).
Host: Populus
Eutypella prunastri (Pers.:Fr.) Sacc.
Eutypa prunastri (Pers.:Fr.) L.C. Tiffany & Gilman
Distr.: Temperate regions.
Hosts: Malus, Prunus, Tilia
Eutypella quadrifida (Schwein.) Ellis & Everh.
Diatrype infuscans Ellis & Everh.
Distr.: NC, NJ, TX.
Host: Smilax
Eutypella quaternata (Pers.:Fr.) Rappaz
Eutypa quaternata (Pers.:Fr.) L.C. Tiffany & Gilman
Quaternaria quaternata (Pers.:Fr.) J. Schröt.
Valsa quaternata (Pers.:Fr.) Fr.
Quaternaria persoonii Tul. & C. Tul., nom. illeg.
Libertella faginea Desmaz. (anamorph)
Distr.: North America; Europe.
Notes: On hardwoods.
Hosts: Liriodendron, Salix
Eutypella sabalina (Cooke) Ellis & Everh.
Valsa sabalina Cooke
Distr.: Southern United States.
Host: Sabal
Eutypella sarcobati Ellis & Everh.
Distr.: CO(type).
Host: Sarcobatus
Eutypella scoparia (Schwein.:Fr.) Ellis & Everh.
Eutypa scoparia (Schwein.:Fr.) L.C. Tiffany & Gilman
Valsa scoparia (Schwein.:Fr.) M.A. Curtis
Eutypella aesculina Ellis & Everh.
Valsa atomaespora Cooke
Eutypella capillata (Ellis & Everh.) Sacc.
Valsa capillata Ellis & Everh.
Eutypella clavulata (Cooke) Sacc.
Valsa clavulata Cooke
Eutypella corynostoma (Berk. & Ravenel) Sacc.
Valsa corynostoma Berk. & Ravenel
Eutypella deusta (Ellis & Everh.) Ellis & Everh.
Valsa deusta Ellis & Everh.
Diatrype dimorpha Schwein. ex Berk. & M.A. Curtis
Eutypa echinata Ellis & Everh.
Eutypella exigua Ellis & Everh. ex Berl.
Eutypella fici Ellis & Everh.
Eutypa heteracantha (Sacc.) Sacc.
Peroneutypa heteracantha (Sacc.) Berl.
Valsa heteracantha Sacc.
Valsa hylodes Ellis & Everh.
Eutypella longirostris Peck
Eutypella microcarpa Sacc.
Eutypa sepulta (Berk. & M.A. Curtis) Sacc.
Eutypella sepulta (Berk. & M.A. Curtis) Dearn.
Distr.: Cosmopolitan.
Notes: On woody tissue of various plants.
Hosts: Acer, Aesculus, Ailanthus, Albizia, Betula, Broussonetia, Carpinus, Carya, Cornus, Elaeagnus, Ficus, Fraxinus, Gleditsia, Lyonia, Magnolia, Malus, Melia, Morus, Pueraria, Robinia, Sambucus, Smilax, Ulmus
Eutypella sorbi (J.C. Schmidt:Fr.) Sacc.
Valsa sorbi (Albertini & Schwein.:Fr.) Fr.
Eutypella angulosa (Nitschke) Sacc.
Valsa angulosa Nitschke
Cytospora rubescens Fr. (anamorph)
Distr.: Eastern United States; Europe.
Notes: Apparently restricted to *Sorbus* spp. but reported on other hardwoods.
Hosts: [Betula], Sorbus
Eutypella staphyleae Dearn. & House
Distr.: NY(type).
Host: Staphylea
Eutypella stellulata (Fr.:Fr.) Sacc.
Valsa stellulata (Fr.:Fr.) Fr.
Eutypella radula (Pers.:Fr.) Ellis & Everh.
Cytosporina stellulata Sacc. (anamorph)
Distr.: North America; Europe.
Hosts: Albizia, Alnus, Carya, Celtis, Cornus, Fagus, Liquidambar, Liriodendron, Melia, Populus, Sambucus, Tilia, Ulmus
Eutypella tetraploa (Berk. & M.A. Curtis ex Berk. & Broome) Sacc.
Valsa tetraploa Berk. & M.A. Curtis ex Berk. & Broome
Eutypella amorphae Ellis & Everh.
Eutypella coryli Ellis & Everh.
Eutypella microspora (Cooke & Plowr.) Sacc.
Valsa microspora Cooke & Plowr.
Distr.: North America; Europe.
Notes: On hardwoods.
Hosts: Corylus, Maclura

Eversia J.L. Crane & Schoknecht in Schoknecht & J.L. Crane. Hyphomycetes

Eversia subopaca (Cooke & Ellis) J.L. Crane & Schoknecht in Schoknecht & J.L. Crane
Helminthosporium subopacum Cooke & Ellis
Cheiromycella annulata T. Matsushima
Cryptocoryneum bigeminum (Cooke & Ellis) S.J. Hughes
Dictyosporium minor (Sacc.) Damon
Distr.: Eastern United States; Japan, probably widespread.
Notes: Schoknecht & Crane (Mycologia 69:539. 1977).
Hosts: Magnolia, Pinus, Quercus

Exarmidium P. Karst. Pyrenomycetes, Phyllachorales

Notes: Barr & Boise (Mycotaxon 23:233. 1985).

Exarmidium diaphanum (Cooke & Ellis) Barr & Boise
Zignoella diaphana (Cooke & Ellis) Sacc.
Distr.: North America.
Hosts: Abies, Phyllodoce

Exarmidium fusariisporum (Ellis & Everh.) Theiss. & Syd.
Rhopographus fusariisporus Ellis & Everh.
Distr.: KS(type).
Notes: Not included in *Exarmidium*; this species may belong to *Chaetosphaeria*.
Host: Panicum

Exarmidium hysteriiforme (P. Karst.) P. Karst.
Dothidea hysteriiforme P. Karst.
Rhopographus hysteriiforme (P. Karst.) Sacc.
Distr.: Circumboreal.
Notes: On decorticated wood.
Hosts: Abies, Thuja

Exarmidium morthieri (Fuckel) Barr & Boise
Trematosphaeria morthieri Fuckel
Trematostoma morthieri (Fuckel) Shear
Zignoella albocincta (Cooke & Ellis) Sacc.
Clypeothecium weirii Petr.
Exarmidium weirii (Petr.) Petr.
Distr.: Northern North America; Europe.
Notes: On decorticated coniferous wood.
Hosts: Abies, Thuja

Exidia Fr.:Fr. Basidiomycetes, Tremellales

Exidia candida C.G. Lloyd
Distr.: WA(type).
Host: Corylus

Exidia cokeri Olive
Distr.: Southeastern United States.
Hosts: Carya, Quercus

Exidia glandulosa Fr.:Fr.
Tremella myricae Berk. & Cooke
Exidia spiculata Schwein.
Exidia spiculosa (Pers.) Sommerf.
Distr.: Temperate northern hemisphere.
Notes: On wood.
Hosts: Alnus, Betula, Carya, Fagus, Holodiscus, Myrica, Persea, Pinus, Populus, Pseudotsuga, Quercus, Salix

Exidia nucleata (Schwein.:Fr.) Burt
Naematelia nucleata (Schwein.:Fr.) Fr.
Naematelia atrata Peck
Distr.: North America.
Notes: On wood. This species has been cited by some American workers as *Myxarium nucleatum* Wallr. 1833, a European species. Most authors (e.g. Donk, Persoonia 4:232-235. 1966) regard *E. nucleata* as a member of *Myxarium* based on basidial characters; however, the combination in *Myxarium* is preoccupied by the Wallroth species.
Hosts: Aralia, Magnolia, Populus, Quercus, Sambucus

Exidia recisa Fr.:Fr.
Exidia gelatinosa (Bull.) Duby
Distr.: Temperate northern hemisphere.
Notes: On wood (Ginns. Fungi Canadenses 86. 1976).
Hosts: Acer, Adenostoma, Amelanchier, Betula, Liquidambar, Quercus

Exidia repanda Fr.:Fr.
Distr.: Temperate northern hemisphere.
Notes: On wood.
Hosts: Alnus, Salix

Exidia saccharina (Albertini & Schwein.:Fr.) Fr.
Tremella spiculosa Pers. var. *saccharina* Albertini & Schwein.
Distr.: Temperate northern hemisphere.
Notes: On wood.
Hosts: Pinus, Quercus

Exidia tremelloides Olive
Distr.: LA.
Notes: On bark and wood.
Hosts: Magnolia, Quercus, Rhamnus

Exidia umbrinella Bres.
Distr.: OR; Europe.
Notes: On wood.
Hosts: Amelanchier, Salix

Exidia zelleri C.G. Lloyd
Distr.: OR(type).
Host: Sambucus

Exidiopsis (Johan-Olsen in Bref.) A. Möller. Basidiomycetes, Tremellales

Exidiopsis calcea (Pers.) K. Wells
Sebacina calcea (Pers.) Bres.
Distr.: Temperate northern hemisphere.
Notes: On living limbs or dead wood.
Hosts: Abies, Acer, Cercocarpus, Juniperus, Lycium, Paulownia, Philadelphus, Picea, Pinus, Populus, Prosopis, Pseudotsuga, Robinia, Sambucus, Tsuga

Exidiopsis fugacissima (Bourd. & Galzin) Sacc. & Trott. in Sacc.
Sebacina fugacissima Bourd. & Galzin
Distr.: Temperate northern hemisphere.
Notes: On wood.
Hosts: Pinus, Salix

Exidiopsis fuliginea Rick
Sebacina adusta Burt
Sebacina variseptata Olive
Distr.: ID, WA, warm parts of the United States to South America.
Notes: On wood.
Hosts: Betula, Populus, Tsuga

Exidiopsis glaira (C.G. Lloyd) K. Wells
Exidiopsis opalea (Bourd. & Galzin) D. Reid
Sebacina opalea Bourd. & Galzin
Distr.: Temperate northern hemisphere.
Notes: On wood and basidiomata of various polypores.
Hosts: Fraxinus, Pinus, Ulmus

Exidiopsis grisea (Pers.) Bourd. & Maire
Exidiopsis plumbescens (Burt) K. Wells
Sebacina plumbescens Burt
Sebacina umbrina D.P. Rogers
Distr.: Widespread.
Notes: On wood.
Hosts: Alnus, Arctostaphylos, Betula, Fraxinus, Liriodendron, Lithocarpus, Populus, Quercus, Umbellularia

Exidiopsis molybdea (J.M. McGuire) Ervin
Sebacina molybdea J.M. McGuire
Distr.: IA, eastern United States.
Notes: On wood.
Hosts: Acer, Diospyros, Populus, Prunus, Quercus, Salix

Exidiopsis podlachica (Bres.) Ervin
Sebacina podlachica Bres.
Distr.: Widespread.
Notes: According to Wells (Mycologia 53:363. 1961) this species is related to *Stypella* or *Myxarium* spp.
Hosts: Elaeagnus, Pinus, Prunus, Quercus, Ulmus

Exidiopsis sordida (Olive) K. Wells
Sebacina sordida Olive
Distr.: NC.
Notes: Wells (Mycologia 53:357-358. 1961).
Host: Pinus

Exobasidium Woronin. Basidiomycetes, Exobasidiales

Notes: Nannfeldt (Symb. Bot. Upsal. 23:1. 1981); Savile (Canad. J. Bot. 6:218. 1984).

Exobasidium angustisporum Linder
Distr.: Arctic North America.
Host: Arctostaphylos

Exobasidium burtii Zeller
Distr.: NJ; northwestern United States and adjacent Canada.
Notes: On living leaves.
Host: Rhododendron

Exobasidium camelliae Shirai

Distr.: United States; Australasia; Europe; Great Britain.
Notes: Parasitic on flower buds and shoots.
Host: Camellia
Exobasidium camelliae Shirai var. ***gracilis*** Shirai
Distr.: SC; Japan.
Host: Camellia
Exobasidium canadense Savile
Distr.: Northeastern United States and adjacent Canada.
Notes: On living leaves.
Host: Rhododendron
Exobasidium cordilleranum Savile var. ***minor*** Savile
Distr.: Western North America.
Notes: On living leaves.
Host: Vaccinium
Exobasidium decolorans Harkn.
Distr.: CA.
Notes: On living leaves.
Host: Rhododendron
Exobasidium dimorphosporum Savile
Distr.: Western United States and adjacent Canada.
Notes: On living leaves.
Host: Vaccinium
Exobasidium monosporum Sawada
Distr.: LA(on introduced host); Taiwan.
Host: Camellia
Exobasidium myrtilli Siegm.
Exobasidium vaccinii (Fuckel) Woronin var. *myrtilli* Thuem.
Distr.: Widespread in arctic-alpine regions.
Host: Cassiope
Exobasidium perenne N. Nickerson
Distr.: WI; Canada(Nova Scotia, Newfoundland), probably throughout northcentral and northeastern United States and adjacent Canada.
Notes: Red shoot disease of cranberries (Nickerson. Canad. J. Pl. Pathol. 6:218. 1984).
Host: Vaccinium
Exobasidium symploci Ellis & G. Martin
Distr.: Southern United States.
Notes: Parasitic on *Symplocos tinctoria*. According to Savile (Canad. J. Bot. 37:645. 1959), known only from conidia, but probably an *Exobasidium*.
Host: Symplocos
Exobasidium uvae-ursi (Maire) Juel
Distr.: Western North America; Europe.
Notes: On living leaves.
Host: Arctostaphylos
Exobasidium vaccinii (Fuckel) Woronin
Exobasidium andromedae Peck
Exobasidium azaleae Peck
Exobasidium cassandrae Peck
Exobasidium discoideum Ellis
Exobasidium parvifolii Hotson
Exobasidium peckii Halst.
Distr.: Widespread.
Notes: Parasitic on Ericaceae (Booth. C.M.I. Descr. 778. 1983).
Hosts: Andromeda, Arbutus, Arctostaphylos, Camellia, Cassiope, Chamaedaphne, Gaultheria, Gaylussacia, Kalmia, Ledum, Leiophyllum, Leucothoe, Lyonia, Menziesia, Pieris, Rhododendron, Vaccinium
Exobasidium vaccinii (Fuckel) Woronin var. ***arctostaphyli*** (Harkn.) Savile
Exobasidium arctostaphyli Harkn.
Distr.: CA; Canada(British Columbia).
Host: Arctostaphylos
Exobasidium vaccinii-uliginosi Boud. in Boud. & E. Fisch.
Distr.: Western North America; Europe.
Notes: Parasitic on Ericaceae.
Hosts: Arctostaphylos, Phyllodoce, Rhododendron, Vaccinium

Exophiala J.W. Carmichael. Hyphomycetes

Notes: De Hoog (Stud. Mycol. 15:110. 1977).
Exophiala jeanselmei (Langeron) McGinnis & Padhye
Torula jeanselmei Langeron
Distr.: Widespread.
Notes: Isolated from humans as well as various kinds of other organic substrates (Domsch et al. Compendium of soil fungi. p. 298. 1980; Brady. C.M.I. Descr. 743. 1983).
Host: Quercus
Exophiala jeanselmei (Langeron) McGinnis & Padhye var. ***heteromorpha*** (Nannf.) De Hoog
Trichosporum heteromorphum Nannf. in Melin & Nannf.
Phialophora heteromorpha (Nannf.) Wang
Distr.: Widespread.
Notes: Isolated from wood, humans, and various organic substrates. This synonymy is not accepted by Domsch et al. (Compendium of soil fungi. p. 297. 1980) in which "*Phialophora heteromorpha*" *sensu* Wang (Canad. J. Bot. 42:1011. 1964) is keyed out separately from *Exophiala jeanselmei* var. *heteromorpha*.
Host: Pinus
Exophiala mansonii (A. Castellani) De Hoog
Microsporum mansonii A. Castellani
Rhinocladiella mansonii (A. Castellani) Schol-Schwarz
Capronia mansonii (Schol-Schwarz) E. Müller et al. (teleomorph)
Distr.: Widespread.
Notes: Isolated from various substrates including humans.
Hosts: Metrosideros, Pinus

Exophoma Weedon. Coelomycetes

Exophoma astericola Tehon & G.L. Stout
Distr.: IL.
Host: Aster
Exophoma magnoliae Weedon
Distr.: Southern United States.
Host: Magnolia

Exosporium Link. Hyphomycetes

Exosporium phoradendri Tharp
Distr.: TX.
Host: Phoradendron
Exosporium platani G. Otth
Distr.: MS; Switzerland(type).
Host: Platanus
Exosporium tiliae Link
Helminthosporium tiliae (Link) Fr.
Distr.: Eastern North America; Europe.
Notes: On limbs and twigs (Hughes. Canad. J. Bot. 38:664. 1960).
Host: Tilia

Exserohilum K.J. Leonard & E.G. Suggs. Hyphomycetes

Notes: Sivanesan (Mycol. Pap. 158:1. 1987).
Exserohilum holmii (Luttrell) K.J. Leonard & E.G. Suggs
Helminthosporium holmii Luttrell
Drechslera holmii (Luttrell) Subramanian & P.C. Jain
Setosphaeria holmii (Luttrell) K.J. Leonard & E.G. Suggs (teleomorph)
Distr.: United States; Africa; India.
Host: Dactyloctenium
Exserohilum monoceras (Drechs.) K.J. Leonard & E.G. Suggs
Helminthosporium monoceras Drechs.
Bipolaris monoceras (Drechs.) Shoemaker
Setosphaeria monoceras Alcorn (teleomorph)
Distr.: Temperate North America; Asia; Australia; Israel.
Notes: Leaf blotch of grasses (Sivanesan. C.M.I. Descr. 886. 1986).
Hosts: Echinochloa, Panicum, Setaria
Exserohilum pedicellatum (A.W. Henry) K.J. Leonard & E.G. Suggs
Helminthosporium pedicellatum A.W. Henry
Drechslera pedicellata (A.W. Henry) Subramanian & P.C. Jain
Setosphaeria pedicellata (R.R. Nelson) K.J. Leonard & E.G. Suggs (teleomorph)
Distr.: Temperate North America; Egypt; India; Pakistan.
Hosts: Triticum, Zea
Exserohilum prolatum K.J. Leonard & E.G. Suggs
Drechslera prolata (K.J. Leonard & E.G. Suggs) Sivanesan
Setosphaeria prolata K.J. Leonard & E.G. Suggs (teleomorph)

Distr.: Southeastern United States; Central America; eastern Africa.
Notes: Leaf spot of grasses (Sivanesan. C.M.I. Descr. 888. 1986).
Hosts: Eichhornia, Zea

Exserohilum rostratum (Drechs.) K.J. Leonard & E.G. Suggs
Helminthosporium rostratum Drechs.
Bipolaris rostrata (Drechs.) Shoemaker
Drechslera rostrata (Drechs.) M.J. Richardson & E.M. Fraser
Bipolaris halodes (Drechs.) Shoemaker
Drechslera halodes (Drechs.) Subramanian & P.C. Jain
Exserohilum halodes (Drechs.) K.J. Leonard & E.G. Suggs
Setosphaeria rostrata K.J. Leonard (teleomorph)
Distr.: Cosmopolitan.
Notes: Leaf spot and foot rot of wheat and other grasses (Anahosur & Sivanesan. C.M.I. Descr. 587. 1978).
Hosts: Aechmea, Agropyron, [Aloe], [Bromelia], [Caryota], [Chamaedorea], Chrysalidocarpus, Cynodon, Dendrobium, Distichlis, Eragrostis, Geonoma, [Juncus], Muhlenbergia, Oryza, Panicum, Paspalum, Pennisetum, Rhapis, [Roystonea], Saccharum, Schedonnardus, Senna, Setaria, Sorghum, Spartina, Sporobolus, Stenotaphrum, Uniola, [Washingtonia], Zea

Exserohilum turcicum (Pass.) K.J. Leonard & E.G. Suggs
Helminthosporium turcicum Pass.
Bipolaris turcica (Pass.) Shoemaker
Drechslera turcica (Pass.) Subramanian & P.C. Jain
Helminthosporium inconspicuum Cooke & Ellis
Setosphaeria turcica (Luttrell) K.J. Leonard & E.G. Suggs (teleomorph)
Distr.: Cosmopolitan.
Notes: Northern leaf blight of maize and sorghum (Ellis & Holliday. C.M.I. Descr. 304. 1971).
Hosts: Bouteloua, Buchloe, Cinna, Panicum, Pennisetum, Setaria, Sorghum, Zea

Extrawettsteinina Barr. Loculoascomycetes, Dothideales

Notes: Barr (Contr. Univ. Michigan Herb. 9:538. 1972).

Extrawettsteinina andromedae (Auersw.) Barr
Leptosphaeria andromedae (Auersw.) Sacc.
Wettsteinina andromedae (Auersw.) Barr
Distr.: Northwestern North America.
Notes: A recent description is provided by Shoemaker & Babcock (Canad. J. Bot. 65:374. 1987). Barr (Mycotaxon 29:502. 1987) transferred this species to *Extrawettsteinina*.
Host: Cassiope

Extrawettsteinina minuta Barr
Distr.: VT(type).
Host: Juniperus

Fabraea Sacc. Discomycetes, Helotiales

Fabraea cincta Sacc. & Scalia in Sacc., Peck, & Trel.
Distr.: AK; Canada (British Columbia).
Notes: On leaves (Pirozynski. Fungi Canadenses 81. 1976).
Host: Rubus

Fabraea thuemenii H.C. Greene
Distr.: WI(type).
Notes: Leaf blight of English hawthorn. This genus is considered a synonym of *Leptotrochila*, but *F. thuemnii* has not been redisposed.
Host: Crataegus

Fabrella Kirschst. Discomycetes, Helotiales

Notes: Korf (Mycologia 54:28. 1962).

Fabrella tsugae (Farl.) Kirschst.
Didymascella tsugae (Farl.) Maire
Distr.: Temperate North America.
Host: Tsuga

Fairmaniella Petr. & Syd. Coelomycetes

Fairmaniella leprosa (Fairm.) Petr. & Syd.
Coniothyrium leprosum Fairm. in Millsp. & L. Nutt.
Melanconium eucalypticola Hansf.
Distr.: CA, HI; Australia; Chile; New Zealand; Zambia.
Notes: On leaves (Sutton. Mycol. Pap. 123:39. 1971).
Host: Eucalyptus

Farrowia D. Hawksworth. Pyrenomycetes, Sordariales

Notes: Hawksworth (Persoonia 8:167. 1975).

Farrowia seminuda (L. Ames) D. Hawksworth
Chaetomium seminudum L. Ames
Distr.: Widespread in the northern hemisphere.
Host: Lycopersicon

Farysia Racib. Basidiomycetes, Ustilaginales

Farysia thuemenii (A. Fisch. v. Waldh.) Nannf. in Lindeberg
Farysia olivacea (DC.) Syd. non (Jaap) Höhn., hom. illeg.
Ustilago olivacea (DC.) Tul. & C. Tul.
Distr.: Widespread.
Host: Carex

Favolaschia (Pat.) Pat. & Lagerh. Basidiomycetes, Agaricales

Favolaschia sabalensis (V. Charles) Singer
Laschia sabalensis V. Charles
Distr.: FL; Caribbean Islands, Central America.
Notes: On dead leaves, petioles and inflorescences (Singer. Beih. Nova Hedwigia 50:36. 1974).
Host: Sabal

Fenestella Tul. & C. Tul. Loculoascomycetes, Dothideales

Fenestella erratica (Peck) Barr
Cucurbitaria erratica Peck
Distr.: OH.
Host: Aesculus

Fenestella phaeospora Sacc.
Distr.: GA, ID, WA; Italy(type).
Hosts: Castanea, Crataegus, Quercus

Fenestella subcaespitosa (G. Otth) Barr
Cucurbitaria subcaespitosa G. Otth
Distr.: North America; Europe.
Notes: On branches (Barr. Rep. Kevo Subarctic Res. Sta. 11:14. 1974).
Host: Sorbus

Fibricium J. Eriksson. Basidiomycetes, Aphyllophorales

Fibricium rude Jülich
Corticium rude P. Karst. non Pat., hom. illeg.
Peniophora rudis Bourd. & Galzin
Peniophora alba Burt
Corticium greschikii Bres.
Fibricium greschikii (Bres.) J. Eriksson
Peniophora greschikii (Bres.) Bourd. & Galzin
Peniophora subcremea Höhn. & Litsch.
Distr.: Temperate northern hemisphere.
Notes: On conifers (Hallenberg. Mycotaxon 30:206. 1987).
Hosts: Abies, [Alnus], [Juglans], Picea, Pinus, [Populus], Pseudotsuga, Thuja

Fibulobasidium Bandoni. Basidiomycetes, Tremellales

Notes: Monotypic.

Fibulobasidium inconspicuum Bandoni

Distr.: LA.
Notes: On corticate limbs.
Host: Quercus

Fibulomyces Jülich. Basidiomycetes, Aphyllophorales

Fibulomyces mutabilis (Bres.) Jülich
Corticium mutabile Bres.
Athelia mutabilis (Bres.) Donk
Corticium pelliculare (P. Karst.) P. Karst.
Athelia pellicularis (P. Karst.) Donk
Distr.: Temperate northern hemisphere.
Notes: On wood (Jülich. Willdenowia Beih. 7:182. 1972).
Hosts: Abies, Carya, Pseudotsuga, Quercus
Fibulomyces septentrionalis (J. Eriksson) Jülich
Athelia septentrionalis J. Eriksson
Distr.: Northern hemisphere.
Notes: On wood.
Host: Picea

Filosporella Nawawi. Hyphomycetes

Filosporella annelidica (C.A. Shearer & J.L. Crane) C.A. Shearer & J.L. Crane
Rogersia annelidica C.A. Shearer & J.L. Crane
Distr.: IL, NC.
Notes: On organic matter in streams (Dyko & Sutton. Mycotaxon 7:325. 1978).
Hosts: Acer, Platanus, Quercus

Fistulina Bull.:Fr. Basidiomycetes, Aphyllophorales

Fistulina hepatica (Schaeff.:Fr.) Sibth.
Confistulina hepatica (Sacc.) Stalpers in Stalpers & Vlug (anamorph)
Distr.: Northern hemisphere.
Hosts: Castanea, Quercus

Flagelloscypha Donk in Singer. Basidiomycetes, Agaricales

Flagelloscypha libertiana (Cooke) Agerer
Cyphella libertiana Cooke
Lachnella libertiana (Cooke) W.B. Cooke
Lachnella rosae W.B. Cooke
Distr.: ID.
Host: Rosa
Flagelloscypha minutissima (Burt) Donk
Cyphella minutissima Burt
Distr.: NH(type).
Host: Populus

Flagellospora Ingold. Hyphomycetes

Flagellospora curvula Ingold
Distr.: Cosmopolitan.
Notes: On decaying leaves and in foam in aquatic habitats (Nilsson. Symb. Bot. Upsal. 18:93. 1964).
Hosts: Acer, Quercus

Flammulina P. Karst. Basidiomycetes, Agaricales

Flammulina velutipes (Curtis:Fr.) Singer
Collybia velutipes (Curtis:Fr.) P. Kumm.
Distr.: Cosmopolitan.
Notes: On wood. An excellent edible fungus.
Hosts: Abies, Acer, Aesculus, Catalpa, Liriodendron, Lupinus, Malus, Picea, Populus, Quercus, Salix, Tilia, Ulmus

Flosculomyces Sutton. Hyphomycetes

Flosculomyces floridaensis Sutton
Distr.: FL; Japan.
Notes: On dead leaves (Sutton. Mycologia 70:789. 1978).
Host: Podocarpus

Fomes (Fr.) Fr. Basidiomycetes, Aphyllophorales

Notes: The modern, narrow concept of *Fomes* is accepted here, including only *F. fomentarius* and *F. fasciatus*. Other species require disposition in other genera.
Fomes fasciatus (Sw.:Fr.) Cooke
Fomes marmoratus (Berk. & M.A. Curtis) Cooke
Fomes sclerodermeus (Lév.) Cooke
Distr.: AZ, southeastern United States; tropical western hemisphere.
Notes: On living or dead hardwoods.
Hosts: Acer, Carya, Celtis, Diospyros, Gleditsia, Liquidambar, Magnolia, Prunus, Quercus, Salix, Ulmus
Fomes fasciculatus Burt
Distr.: HI(type).
Notes: The generic placement of this species is uncertain.
Host: Acacia
Fomes fomentarius (L.:Fr.) J. Kickx fil.
Distr.: Temperate northern hemisphere.
Notes: On dead or living hardwoods.
Hosts: Acer, Alnus, Betula, Fagus, Malus, Populus, Prunus, Pseudotsuga, Quercus, Salix, Umbellularia
Fomes fullageri (Berk.) Cooke
Polyporus fullageri Berk.
Distr.: HI(type).
Notes: The generic placement of this species is uncertain.
Host: Acacia
Fomes hawaiiensis C.G. Lloyd
Distr.: CA, HI(type).
Notes: The generic placement of this species is uncertain.
Hosts: Acacia, Aleurites, Osmanthus

Fomitella Murrill. Basidiomycetes, Aphyllophorales

Fomitella supina (Sw.:Fr.) Murrill
Polyporus supinus (Sw.:Fr.) Fr.
Polyporus hemileucus Berk. & M.A. Curtis
Distr.: Southeastern United States.
Notes: On hardwoods; rarely causes heart rot of living *Quercus* spp.
Hosts: Carpinus, Carya, Diospyros, Fraxinus, Gleditsia, Liquidambar, Nyssa, Persea, Prunus, Quercus, Salix, Ulmus

Fomitopsis P. Karst. Basidiomycetes, Aphyllophorales

Fomitopsis cajanderi (P. Karst.) Kotlaba & Pouzar
Fomes cajanderi Weir
Trametes carnea C.G. Lloyd
Trametes subrosea Weir
Fomes subroseus (Weir) Overh.
Distr.: Temperate northern hemisphere.
Notes: On wood, usually of conifers.
Hosts: Abies, Acer, Alnus, Arbutus, Betula, Chamaecyparis, Juniperus, Larix, Picea, Pinus, Populus, Prunus, Pseudotsuga, Taxodium, Thuja, Tsuga
Fomitopsis durescens (Overh. ex J. Lowe) R.L. Gilbertson & Ryvarden
Polyporus durescens Overh. ex J. Lowe
Distr.: Eastern North America.
Notes: On hardwoods.
Host: Acer
Fomitopsis feei (Fr.) Kreisel
Polyporus feei Fr.
Fomes feei (Fr.) J. Lowe
Distr.: FL; neotropical regions.
Notes: On hardwoods.
Host: Juniperus

Fomitopsis meliae (Underw.) R.L. Gilbertson & Ryvarden
Polyporus meliae Underw.
Fomes meliae (Underw.) Murrill
Polyporus submurinus (Murrill) C.G. Lloyd
Trametes subvinosa Murrill
Distr.: North America, most common in central and eastern states.
Notes: On wood of hardwoods, rarely conifers. Causes heart rot of living peach trees and decays wood in service.
Hosts: Acer, Carya, Fraxinus, Gleditsia, Larix, Melia, Populus, Prunus, Quercus, Sequoia
Fomitopsis officinalis (Villars:Fr.) Bondartsev & Singer
Fomes officinalis (Villars:Fr.) Faull
Laricifomes officinalis (Villars:Fr.) Kotlaba & Pouzar
Fomes laricis Jacq.
Distr.: Central and western North America; Europe.
Notes: Heart rot of living conifers, especially *Larix* and *Pseudotsuga* spp.
Hosts: Abies, Larix, Picea, Pinus, Pseudotsuga, Tsuga
Fomitopsis palustris (Berk. & M.A. Curtis) R.L. Gilbertson & Ryvarden
Polyporus palustris Berk. & M.A. Curtis
Tyromyces palustris (Berk. & M.A. Curtis) Murrill
Distr.: Southeastern United States.
Notes: On wood.
Hosts: Pinus, Prunus, Pseudotsuga
Fomitopsis pinicola (Sw.:Fr.) P. Karst.
Fomes pinicola (Sw.:Fr.) Cooke
Distr.: Temperate northern hemisphere.
Notes: On wood, usually conifers; heart rot of living conifers and black cherry; important in decay of lumber and wood in service.
Hosts: Abies, Acer, Alnus, Betula, Calocedrus, Carya, Castanea, Chamaecyparis, Fagus, Fraxinus, Larix, Magnolia, Malus, Picea, Pinus, Platanus, Populus, Prunus, Pseudotsuga, Pyrus, Thuja, Tilia, Tsuga
Fomitopsis rosea (Albertini & Schwein.:Fr.) P. Karst.
Fomes roseus (Albertini & Schwein.:Fr.) Cooke
Distr.: Northern hemisphere.
Notes: On wood, usually of conifers.
Hosts: Abies, Fagus, Juniperus, Picea, Pinus, Populus, Prunus, Pseudotsuga, Taxus, Thuja, Tsuga
Fomitopsis spraguei (Berk. & M.A. Curtis) R.L. Gilbertson & Ryvarden
Polyporus spraguei Berk. & M.A. Curtis
Trametes spraguei (Berk. & M.A. Curtis) Ryvarden
Tyromyces spraguei (Berk. & M.A. Curtis) Murrill
Distr.: Central and eastern North America, rare in the northwest; Europe.
Notes: Heart rot, also on dead wood.
Hosts: Acer, Betula, Castanea, Cladrastis, Diospyros, Fagus, Fraxinus, Liquidambar, Paulownia, Prunus, Quercus, Tsuga

Foveostroma DiCosmo. Coelomycetes

Micropera Lév.
Notes: DiCosmo (Canad. J. Bot. 56:1665. 1978) designated *Foveostroma* as a *nomen novum* for *Micropera*.
Foveostroma abietinum (Peck) DiCosmo
Gelatinosporium abietinum Peck
Micropera abietina (Peck) Höhn.
Dermea balsamea (Peck) Seaver in B.O. Dodge (teleomorph)
Distr.: Northeastern North America.
Hosts: Abies, Pseudotsuga, Tsuga
Foveostroma boycei (Dearn.) Funk
Cryptosporium boycei Dearn.
Brunchorstia boycei (Dearn.) Hahn
Micropera boycei (Dearn.) Groves
Brunchorstia gibbosa Wollenweb.
Dermea pseudotsugae Funk (teleomorph)
Distr.: Pacific Northwest.
Notes: Groves (Mycologia 60:718.1968) provided a history of the name.
Host: Pseudotsuga
Foveostroma drupacearum (Lév.) DiCosmo
Micropera drupacearum Lév.
Distr.: Temperate northern hemisphere.
Notes: On limbs and twigs.
Host: Prunus

Frommeella Cummins & Y. Hiratsuka. Basidiomycetes, Uredinales

Frommeella duchesneae (Arth.) Yohem, Cummins & R.L. Gilbertson
Kuehneola duchesneae Arth.
Frommea obtusa (F. Strauss) Arth. var. *duchesneae* (Arth.) Arth.
Distr.: Eastern and southern United States.
Host: Duchesnea

Fuligo Haller. Myxomycetes

Fuligo septica (L.) A. Wigg.
Fuligo violacea Pers.
Distr.: Cosmopolitan.
Notes: On lawns and numerous substrates in wet weather.
Hosts: Agrostis, Festuca, Fragaria, Ipomoea

Fulvia Cif. Hyphomycetes

Fulvia fulva (Cooke) Cif.
Cladosporium fulvum Cooke
Distr.: Cosmopolitan.
Notes: Leaf mold of tomato (Holliday & Mulder. C.M.I. Descr. 487. 1976).
Host: Lycopersicon

Fumago Pers., nom. conf. Hyphomycetes

Notes: Friend (Trans. Brit. Mycol. Soc. 48:371. 1965); Hughes. (Mycologia 68:773. 1976).
Fumago salicina (Pers.:Fr.) Tul. & C. Tul.
Sporotrichum salicinum Pers.:Fr.
Dematium salicinum Pers.
Cladosporium fumago Link
Distr.: SC; Europe.
Hosts: Osmanthus, Salix
Fumago vagans Pers., nom. conf.
Distr.: Cosmopolitan.
Notes: This species is composed of *Aureobasidium pullulans* and *Cladosporium herbarum*. No connection to *Capnodium* has been proven (Friend. Trans. Brit. Mycol. Soc. 48:371. 1965).
Hosts: Ficus, Humulus, Ilex, Malus, Nicotiana, Prunus, Quercus, Rhododendron, Ribes

Furcaspora Bonar. Coelomycetes

Furcaspora pinicola Bonar
Distr.: CA.
Notes: Bonar (Mycologia 57:379. 1965); Sutton (The Coelomycetes. p. 218. 1980).
Hosts: Abies, Pinus

Fusamen (Sacc.) P. Karst. Coelomycetes

Fusamen fructigenum (P. Karst.)
Gloeosporium weirianum Sacc.
Platycarpium weirianum (Sacc.) Petr.
Distr.: Circumboreal.
Notes: Parasitic on catkins and leaves (Sutton. The Coelomycetes. p. 214. 1980).
Host: Salix

Fusariella Sacc. Hyphomycetes

Fusariella kansensis (Ellis & Barth.) M.B. Ellis
Clasterosporium kansense Ellis & Barth.
Distr.: KS(type).
Host: Populus
Fusariella obstipa (Pollack) S.J. Hughes
Dendryphion obstipum Pollack

Distr.: Widespread.
Notes: Isolated from various plants and soil (Ellis. Dematiaceous Hyphomycetes. p. 529. 1971).
Host: Juncus

Fusarium Link:Fr. Hyphomycetes

Fusarium acuminatum **Ellis & Everh.**
Fusarium scirpi Lambotte & Fautrey var. *acuminatum* (Ellis & Everh.) Wollenweb.
Gibberella acuminata Wollenweb. (teleomorph)
Distr.: Cosmopolitan.
Notes: Foot rot and root rot of many plants, stem rot of maize; also isolated from insects.
Hosts: Aegilops, Aesculus, Agrohordeum, Agropyron, Agrostis, Allium, Andropogon, Aristida, Bouteloua, Brachypodium, Bromus, Buchloe, Calamagrostis, Calamovilfa, Carex, Cenchrus, Chloris, Cucumis, Dactylis, Danthonia, Deschampsia, Dianthus, Dichanthelium, Digitaria, Distichlis, Echinochloa, Elymus, Eragrostis, Erioneuron, Festuca, Glycine, Gossypium, Hilaria, Hordeum, Ipomoea, Koeleria, Leymus, Linum, Lolium, Malus, Medicago, Melica, Muhlenbergia, Oryzopsis, Panax, Panicum, Phalaris, Phleum, Phragmites, Picea, Pisum, Poa, Pyrus, Ricinus, Schedonnardus, Schizachyrium, Secale, Setaria, Sitanion, Solanum, Sorghum, Spartina, Sporobolus, Stipa, Trisetum, Triticum, Vulpia, Zea

Fusarium annuum **Leonian**
Distr.: United States; West Indies; Central and South America; Yugoslavia; China.
Notes: Vascular wilt and fruit rot. Possibly a synonym of *Fusarium oxysporum* Schlechtend.
Host: Capsicum

Fusarium aquaeductuum **(Radlk. & Rabenh.) Lagerh.**
Selenosporium aquaeductuum Radlk. & Rabenh.
Fusarium oxydendri Ellis & Everh.
Rhabdospora oxydendri (Ellis & Everh.) Sherb.
Distr.: Cosmopolitan.
Notes: Commonly isolated from flowing water and water pipes. Also associated with sap flow from trees, dead twigs, and wood-inhabiting sphaeriaceous fungi.
Host: Oxydendrum

Fusarium arthrosporioides **Sherb.**
Distr.: Cool temperate regions.
Notes: Reported as a pathogen of clover and of conifer seedlings. Possibly confused with *Fusarium avenaceum* (Fr.:Fr.) Sacc. Nelson et al. (*Fusarium* species: an illustrated manual for identification. p. 80. 1983) treated this provisionally as a synonym of *F. avenaceum*; other authors considered it to be a distinct species.
Hosts: Alnus, Hordeum, Pinus, Solanum, Triticum

Fusarium avenaceum **(Fr.:Fr.) Sacc.**
Fusisporium avenaceum Fr.:Fr.
Fusarium anguioides Sherb.
Fusarium granulosum Ellis & Everh.
Fusarium roseum Link f. *cerealis* (Cooke) W.C. Snyder & E.M. Hans.
Fusarium viticola Thuem.
Gibberella avenacea R.J. Cooke (teleomorph)
Distr.: Cosmopolitan, most common in temperate regions.
Notes: Damping-off, seedling blights, and rots of a wide variety of plants (Booth & Waterson. C.M.I. Descr. 25. 1964). The synonym *F. roseum* f. *cerealis* is a *nomen confusum*.
Hosts: Agrostis, Avena, Beta, Bromus, Carex, Castilleja, Citrus, Cucurbita, Dactylis, Daucus, Dianthus, Echinochloa, Elymus, Festuca, Hordeum, Leucopoa, Malus, Matthiola, Medicago, Muhlenbergia, Phleum, Picea, Pinus, Poa, Prunus, Pseudotsuga, Pyrus, Ribes, Robinia, Secale, Smilax, Solanum, Triticum, Zea

Fusarium buxicola **Sacc.**
Fusarium buxi Sacc. non Spreng., hom. illeg.
Nectria desmazierii Beccari & De Not. (teleomorph)
Distr.: Temperate North America; Europe.
Notes: Associated with premature leaf fall. Not to be confused with *Fusarium lateritium* Nees var. *buxi* C. Booth.
Host: Buxus

Fusarium carpineum **J.J. Davis**
Distr.: WI.
Hosts: Carpinus, Carya

Fusarium chlamydosporum **Wollenweb. & Reinking**
Fusarium sporotrichioides Sherb. var. *chlamydosporum* (Wollenweb. & Reinking) Joffe
Dactylium fusarioides Gonz. Frag. & Cif.
Fusarium fusarioides (Gonz. Frag. & Cif.) C. Booth
Distr.: Cosmopolitan.
Notes: In soil and on a wide variety of plant materials.
Hosts: Carya, Gossypium, Helianthus

Fusarium coccophilum **(Desmaz.) Wollenweb. & Reinking**
Nectria flammea (Tul. & C. Tul.) J.M. Dingley (teleomorph)
Distr.: Subtropical and tropical regions.
Notes: Parasitic on scale insects.
Host: Citrus

Fusarium concolor **Reinking**
Distr.: United States; Europe(type).
Notes: Predominantly a soil organism.
Host: Helianthus

Fusarium crookwellense **L.W. Burgess, P.E. Nelson & T.A. Toussoun**
Distr.: Widespread.
Notes: Isolated from various kinds of plant debris (Burgess et al. Trans. Brit. Mycol. Soc. 79:497. 1982).
Host: Potamogeton

Fusarium culmorum **(Wm.G. Sm.) Sacc.**
Fusarium culmorum (Wm.G. Sm.) Sacc. var. *leteius* Sherb.
Distr.: Cosmopolitan.
Notes: Seedling blight, foot rot, ear blight, stalk rot, and other diseases of cereals, grasses, and a wide variety of monocots and dicots (Booth & Waterson. C.M.I. Descr. 26. 1964).
Hosts: Agropyron, Agrostis, Avena, Beta, Bromus, Callistephus, Cucumis, Dactylis, Dianthus, Echinochloa, Elymus, Festuca, Hordeum, Ipomoea, Lolium, Poa, Secale, Solanum, Sorghum, Triticum, Vicia, Zea

Fusarium decemcellulare **C. Brick**
Nectria rigidiuscula Berk. & Broome (teleomorph)
Distr.: Subtropical and tropical regions.
Notes: An important pathogen of *Theobroma cacao*. Also a wound pathogen and associated with die-back and cankers of many woody hosts.
Hosts: Araucaria, Carya, Theobroma

Fusarium episphaeria **(Tode) W.C. Snyder & H.N. Hans.**
Distr.: Widespread.
Notes: According to Booth (The genus *Fusarium*. p. 40, 70. 1971) this name has been applied to *Microdochium dimerum*, *Fusarium aquaeductuum*, and *F. merismoides*.
Hosts: Cucumis, Metrosideros, Zea

Fusarium equiseti **(Corda) Sacc.**
Fusarium equiseti (Corda) Sacc. var. *bullatum* Wollenweb.) Wollenweb.
Fusarium roseum Link f. *gibbosum* W.C. Snyder & H.N. Hans.
Fusarium scirpi Lambotte & Fautrey
Fusarium scirpi Lambotte & Fautrey var. *compactum* Wollenweb.
Gibberella intricans Wollenweb. (teleomorph)
Distr.: Cosmopolitan, most common in tropical regions.
Notes: Rots of a wide variety of plants, including stalk rot of maize (Booth. C.M.I. Descr. 571. 1978).
Hosts: Agropyron, Andropogon, Avena, Beta, Bouteloua, Bromus, Buchloe, Citrullus, Citrus, Cucumis, Dianthus, Echinochloa, Elymus, Euonymus, Glycine, Gossypium, Helianthus, Hordeum, Koeleria, Linum, Malus, Muhlenbergia, Oryzopsis, Panax, Panicum, Phalaris, Phleum, Pisum, Poa, Pyrus, Secale, Setaria, Solanum, Sorghum, Stipa, Taraxacum, Triticum, Ulmus, Vitis, Zea

Fusarium flocciferum **Corda**
Distr.: Temperate North America; Europe.
Notes: Frequently isolated from a wide variety of plant materials and soil. Associated with black rot of potatoes in the northwestern United States. *Gibberella heterochroma* Wollenw. may be the teleomorph.
Hosts: Beta, Solanum

Fusarium graminearum **Schwabe**
Gibberella zeae (Schwein.) Petch (teleomorph)
Distr.: Cosmopolitan.
Notes: Diseases of cereals and grasses (Booth. C.M.I. Descr. 384. 1973).
Hosts: Agropyron, Agrostis, Avena, Bromus, Cenchrus, Cortaderia,

Cucumis, Echinochloa, Hordeum, Lolium, Medicago, Poa, Secale, Setaria, Sorghum, Trifolium, Triticum, Zea

Fusarium heterosporum Nees:Fr.

Fusarium graminum Corda, pro parte
Fusarium negundi Sherb.
Fusarium reticulatum Mont.
Fusarium reticulatum Mont. var. *negundi* (Sherb.) Wollenweb.
Gibberella gordonia C. Booth (teleomorph)

Distr.: North America; Africa; Australia; Europe.
Notes: Booth (C.M.I. Descr. 572. 1978). The synonym *Fusarium negundi* has erroneously been reported as *Fusarium negundinis* and as *Fusarium reticulatum* var. *negundinis*.
Hosts: Acer, Axonopus, Beta, Digitaria, Elymus, Festuca, Glyceria, Malus, Paspalum, Phleum, Poa, Secale

Fusarium javanicum Koord.

Fusarium javanicum Koord. var. *radicicola* (Wollenweb.) Wollenweb.
Fusarium radicicola Wollenweb.
Fusarium solani (Mart.) Sacc. f. sp. *radicicola* (Wollenweb.) W.C. Snyder & H.N. Hans.
Nectria ipomoeae Halst. (teleomorph)

Distr.: Widespread, but most common in warm regions and greenhouses.
Notes: Recognition of this species follows Gerlach & Nirenberg (Mitt. Biol. Bundesanst. Land-Forstw. Berlin-Dahlem 209:359. 1982)
Hosts: Beta, Cucumis, Hibiscus, Ipomoea, Persea, Solanum

Fusarium juruanum Henn.

Nectria diploa Berk. & M.A. Curtis (teleomorph)

Distr.: Subtropical and tropical regions.
Notes: Parasitic on scale insects.
Host: Citrus

Fusarium lateritium Nees:Fr.

Fusarium celtidis Ellis & Tracy
Fusarium cinnabarinum (Berk. & M.A. Curtis) Sacc.
Fusarium episphaericum (Cooke & Ellis) Sacc.
Fusarium illosporioides Sacc.
Fusarium lateritium Nees:Fr. var. *fructigenum* (Fr.) Wollenweb.
Gibberella lateritium (Wallr.) Sacc. (teleomorph)

Distr.: Cosmopolitan.
Notes: Wilt, die-back, and cankering of a wide variety of woody plants.
Hosts: Acacia, Acer, Ailanthus, Albizia, Amorphophallus, Anoda, Arachis, Araucaria, Buxus, Carya, Celtis, Citrus, Euonymus, Ficus, Gardenia, Helianthus, Hibiscus, Juglans, Laburnum, Leucophyllum, Malus, Melia, Metrosideros, Morus, Musa, Ostrya, Pinus, Prunus, Pyrus, Ribes, Sida, Torreya, Yucca, Zamia

Fusarium lateritium Nees:Fr. f. sp. ***mori*** (Desmaz.) Matuo & K. Sato

Fusarium lateritium Nees:Fr. var. *mori* Desmaz.

Distr.: Northern hemisphere; Tasmania.
Notes: Twig blight and die-back of mulberry, probably throughout the range of the host.
Host: Morus

Fusarium lateritium Nees:Fr. f. sp. ***pini*** Hepting in W.C. Snyder, Toole, Hepting

Distr.: Southeastern United States; Haiti.
Notes: Although initially thought to be the cause of pitch canker of pines, that organism has now been identified as *Fusarium subglutinans*. The reports of *F. lateritium* f. sp. *pini* are probably misdeterminations of *F. subglutinans*.
Host: Pinus

Fusarium melanochlorum (Casp.) Sacc.

Fusisporium melanochlorum Casp.

Distr.: Temperate regions.
Notes: On woody plants.
Host: Quercus

Fusarium merismoides Corda

Fusarium rimosum (Peck) Sacc.

Distr.: Cosmopolitan.
Notes: Usually considered a soil saprobe but commonly isolated from a wide variety of plant materials.
Hosts: Acacia, Bambusa, Beta, Carya, Pisum, Platanus, Pyrus, Zea

Fusarium moniliforme J. Sheld.

Sporotrichum atropurpureum Peck
Gibberella fujikori (Sawada) Ito in Ito & K. Kimura (teleomorph)

Distr.: Cosmopolitan, rare in cooler regions.
Notes: Causal agent of Bakanae disease of rice; pathogenic on grasses and many other plants.
Hosts: Agave, Agropyron, Albizia, Allium, Arachis, Ardisia, Asparagus, Aspidistra, Astragalus, Avena, Bidens, Buxus, Calamovilfa, Carya, Chamaedorea, Chrysanthemum, Cocos, Cordyline, Cucumis, Daucus, Dichanthelium, Dieffenbachia, Dracaena, Elymus, Euphorbia, Festuca, Ficus, Glycine, Gossypium, Helianthus, Hordeum, Liquidambar, Lupinus, Lycopersicon, Mangifera, Medicago, Narcissus, Nolina, Oryza, Oryzopsis, Panicum, Pennisetum, Philodendron, Phoenix, Pinus, Pisum, Platanus, Pseudotsuga, Rhapis, Rhododendron, Saccharum, Sansevieria, Schlumbergera, Setaria, Solanum, Sorghum, Syagrus, Trifolium, Triticum, Zamia, Zea

Fusarium oxysporum Schlechtend.:Fr.

Fusarium angustum Sherb.
Fusarium aurantiacum (Link) Sacc.
Fusarium bulbigenum Cooke & Massee
Fusarium dianthi Prill. & Delacr.
Fusarium hyperoxysporum Wollenweb.
Fusarium orthoceras Appel & Wollenweb.
Fusarium orthoceras Appel & Wollenweb. var. *longius* (Sherb.) Wollenweb.
Fusarium oxysporum Schlechtend.:Fr. var. *aurantiacum* (Link) Wollenweb.
Fusarium oxysporum Schlechtend.:Fr. var. *orthoceras* (Appel & Wollenweb.) Bilai
Fusarium vasinfectum Atk. var. *lutulatum* (Sherb.) Wollenweb.

Distr.: Cosmopolitan.
Notes: The most frequently recorded *Fusarium* in soil. Many *formae speciales* have been recognized, many of which cause vascular wilts. We follow the list provided by Armstrong & Armstrong in Nelson et al. (*Fusarium*: diseases, biology, and taxonomy. p. 391. 1981). Booth (C.M.I. Descr. 211. 1970).
Hosts: Abelmoschus, Abies, Acacia, Aglaonema, Agropyron, Ajuga, Albizia, Alcea, Aloe, Alternanthera, Ambrosia, Apium, Araucaria, Ardisia, Asparagus, Avena, Beta, Bouteloua, Brassica, Bromus, Caladium, Callistemon, Camellia, Capsicum, Carnegiea, Carthamus, Carya, Cereus, Chamaedorea, Chrysanthemum, Citrus, Cleyera, Cocos, Codiaeum, Cornus, Crocus, Crotalaria, Cucumis, Cycas, Dahlia, Dieffenbachia, Dizygotheca, Dracaena, Echinochloa, Echinopsis, Elymus, Eucalyptus, Euphorbia, Fragaria, Freesia, Gardenia, Gerbera, Glycine, Gossypium, Gynura, Gypsophila, Hedera, Helianthus, Hibiscus, Hilaria, Hordeum, Hyacinthus, Ipomoea, Iris, Juglans, Juniperus, Lactuca, Lagerstroemia, Lens, Leucophyllum, Ligustrum, Lithops, Lophocereus, Lycopersicon, Malus, Mammillaria, Mandevilla, Maranta, Medicago, Melica, Muhlenbergia, Musa, Myrica, Nolina, Opuntia, Oryzopsis, Panicum, Phanerophlebia, Phaseolus, Phleum, Picea, Pinus, Pistacia, Pisum, Pittosporum, Plantago, Plectranthus, Podocarpus, Polyscias, Prunus, Pseudotsuga, Pyracantha, Pyrus, Quercus, Raphanus, Rhizophora, Rhododendron, Rosa, Roystonea, Rubus, Saccharum, Schlumbergera, Senna, Setaria, Solanum, Sorghum, Sporobolus, Stapelia, Syngonium, Tagetes, Trifolium, Triticum, Tulipa, Ulmus, Vigna, Viola, Zea, Zigadenus, Zinnia

Fusarium oxysporum Schlechtend.:Fr. f. sp. ***apii*** (R.R. Nelson & Sherb.) W.C. Snyder & H.N. Hans.

Distr.: Widespread.
Hosts: Apium, Petroselinum

Fusarium oxysporum Schlechtend.:Fr. f. sp. ***asparagi*** S.I. Cohen

Distr.: Widespread.
Host: Asparagus

Fusarium oxysporum Schlechtend.:Fr. f. sp. ***batatas*** (Wollenweb.) W.C. Snyder & H.N. Hans.

Fusarium batatatis Wollenweb.

Distr.: Cosmopolitan, more prevalent in temperate regions.
Notes: Wilt of sweet potatoes. Booth (C.M.I. Descr. 212. 1970).
Hosts: Cassia, Ipomoea, Richardia, Solanum

Fusarium oxysporum Schlechtend.:Fr. f. sp. ***callistephi*** (Beach) W.C. Snyder & H.N. Hans.

Fusarium conglutinans Wollenweb. var. *callistephi* Beach

Distr.: United States; New Zealand; Switzerland.
Hosts: Callistephus, Centaurea, Chrysanthemum

Fusarium oxysporum Schlechtend.:Fr. f. sp. ***cannabis*** Noviello & W.C. Snyder

Host: Cannabis

Fusarium oxysporum Schlechtend.:Fr. f. sp. ***carthami*** Klisiewicz &

Houston
Distr.: CA.
Host: Carthamus
Fusarium oxysporum Schlechtend.:Fr. f. sp. ***cattleyae*** V. Foster
Distr.: CA.
Hosts: Brassolaeliocattleya, Cattleya, Cymbidium, Dendrobium, Epidendrum, Laelia, Laeliocattleya, Oncidium, Orchidaceae, Phalaenopsis, Vanda
Fusarium oxysporum Schlechtend.:Fr. f. sp. ***cepae*** (H.N. Hans.) W.C. Snyder & H.N. Hans.
Distr.: ID.
Host: Allium
Fusarium oxysporum Schlechtend.:Fr. f. sp. ***chrysanthemi*** G.M. Armstrong, J.K. Armstrong & R.H. Littrell
Distr.: Eastern United States.
Host: Chrysanthemum
Fusarium oxysporum Schlechtend.:Fr. f. sp. ***ciceris*** (Padwick) Matuo & K. Sato
Fusarium lateritium Nees:Fr. f. sp. *ciceri* (Padwick) D.C. Erwin
Distr.: CA; India.
Host: Cicer
Fusarium oxysporum Schlechtend.:Fr. f. sp. ***citri*** Timmer et al.
Distr.: FL(greenhouses); possibly Brazil.
Host: Citrus
Fusarium oxysporum Schlechtend.:Fr. f. sp. ***conglutinans*** (Wollenweb.) W.C. Snyder & H.N. Hans.
Fusarium conglutinans Wollenweb.
Distr.: Cosmopolitan.
Notes: Yellow of Brassicaceae (Subramanian. C.M.I. Descr. 213. 1970).
Hosts: [Beta], Brassica, Callistephus, [Chrysanthemum], Crambe, [Solanum]
Fusarium oxysporum Schlechtend.:Fr. f. sp. ***cubense*** (E.F. Sm.) W.C. Snyder & H.N. Hans.
Fusarium cubense E.F. Sm.
Fusarium oxysporum Schlechtend.:Fr. var. *cubense* (E.F. Sm.) Wollenweb.
Distr.: Cosmopolitan.
Notes: Vascular wilt of bananas and abaca. Also occurs on grasses (Subramanian. C.M.I. Descr. 214. 1970).
Host: Musa
Fusarium oxysporum Schlechtend.:Fr. f. sp. ***cucumerinum*** J.H. Owen
Distr.: Cosmopolitan.
Notes: Wilt and damping-off of cucumbers (Holliday. C.M.I. Descr. 215. 1970).
Host: Cucumis
Fusarium oxysporum Schlechtend.:Fr. f. sp. ***cyclaminis*** Gerlach
Distr.: CA; Europe.
Host: Cyclamen
Fusarium oxysporum Schlechtend.:Fr. f. sp. ***delphinii*** Laskaris
Distr.: FL, WA.
Host: Delphinium
Fusarium oxysporum Schlechtend.:Fr. f. sp. ***dianthi*** (Prill. & Delacr.) W.C. Snyder & H.N. Hans.
Fusarium oxysporum Schlechtend.:Fr. f. sp. *barbati* W.C. Snyder
Distr.: Widespread.
Host: Dianthus
Fusarium oxysporum Schlechtend.:Fr. f. sp. ***gladioli*** (L. Massey) W.C. Snyder & H.N. Hans.
Fusarium orthoceras Appel & Wollenweb. var. *gladioli* L. McCulloch
Distr.: Widespread.
Notes: Corm rot, root rot, vascular wilt.
Hosts: Freesia, Gladiolus, Ixia, Tigridia, Tritonia
Fusarium oxysporum Schlechtend.:Fr. f. sp. ***koae*** D.E. Gardner
Distr.: HI.
Host: Acacia
Fusarium oxysporum Schlechtend.:Fr. f. sp. ***lilii*** Imle
Distr.: Widespread.
Notes: Vascular wilt.
Host: Lilium
Fusarium oxysporum Schlechtend.:Fr. f. sp. ***lini*** (Bolley) W.C. Snyder & H.N. Hans.
Fusarium lini Bolley
Distr.: Widespread.
Host: Linum
Fusarium oxysporum Schlechtend.:Fr. f. sp. ***lycopersici*** (Sacc.) W.C. Snyder & H.N. Hans.
Fusarium oxysporum Schlechtend.:Fr. ssp. *lycopersici* Sacc.
Fusarium lycopersici D. Bruschi
Distr.: Widespread.
Notes: Vascular wilt of tomato (Subramanian. C.M.I. Descr. 217. 1970).
Host: Lycopersicon
Fusarium oxysporum Schlechtend.:Fr. f. sp. ***mathioli*** K. Baker
Distr.: AZ, CA.
Host: Matthiola
Fusarium oxysporum Schlechtend.:Fr. f. sp. ***medicaginis*** (J.L. Weimer) W.C. Snyder & H.N. Hans.
Distr.: Widespread.
Hosts: Medicago, Pisum, Vicia
Fusarium oxysporum Schlechtend.:Fr. f. sp. ***melonis*** W.C. Snyder & H.N. Hans.
Distr.: North America; Asia; Australia; Europe.
Notes: Wilt of cantaloupe and melons (Holliday. C.M.I. Descr. 218. 1970).
Host: Cucumis
Fusarium oxysporum Schlechtend.:Fr. f. sp. ***narcissi*** W.C. Snyder & H.N. Hans.
Distr.: Widespread.
Host: Narcissus
Fusarium oxysporum Schlechtend.:Fr. f. sp. ***nicotianae*** (J. Johnson) W.C. Snyder & H.N. Hans.
Distr.: Widespread.
Notes: Armstrong & Armstrong (Phytopathology 58:1242. 1968) considered this to be a doubtful form.
Host: Nicotiana
Fusarium oxysporum Schlechtend.:Fr. f. sp. ***niveum*** (E.F. Sm.) W.C. Snyder & H.N. Hans.
Fusarium niveum E.F. Sm.
Fusarium bulbigenum Cooke & Massee var. *niveum* (E.F. Sm.) Wollenweb.
Distr.: Cosmopolitan.
Notes: Wilt and damping-off of watermelon (Holliday. C.M.I. Descr. 219. 1970).
Hosts: Citrullus, Cucumis, Cucurbita
Fusarium oxysporum Schlechtend.:Fr. f. sp. ***perniciosum*** (Hepting) Toole
Fusarium perniciosum Hepting
Distr.: Eastern and southern United States.
Host: Albizia
Fusarium oxysporum Schlechtend.:Fr. f. sp. ***phaseoli*** J.B. Kendrick & W.C. Snyder
Distr.: Widespread.
Notes: *Fusarium* yellows of beans.
Host: Phaseolus
Fusarium oxysporum Schlechtend.:Fr. f. sp. ***pisi*** (J.C. Hall) W.C. Snyder & Hanna
Fusarium orthoceras Appel & Wollenweb. var. *pisi* Linford
Distr.: Widespread.
Host: Pisum
Fusarium oxysporum Schlechtend.:Fr. f. sp. ***radici-lupini*** J.L. Weimer
Distr.: Widespread.
Notes: Root rot of *Lupinus*. spp.
Host: Lupinus
Fusarium oxysporum Schlechtend.:Fr. f. sp. ***radicis-lycopersici*** W.R. Jarvis & Shoemaker
Distr.: Widespread.
Host: Lycopersicon
Fusarium oxysporum Schlechtend.:Fr. f. sp. ***raphani*** J.B. Kendrick & W.C. Snyder
Distr.: CA, WI.
Host: Raphanus
Fusarium oxysporum Schlechtend.:Fr. f. sp. ***rhois*** W.C. Snyder, Toole, & Hepting
Distr.: CT, VA.
Host: Rhus
Fusarium oxysporum Schlechtend.:Fr. f. sp. ***spinaciae*** (Sherb.) W.C. Snyder & H.N. Hans.
Fusarium conglutinans Wollenweb. var. *betae* D. Stewart

Fusarium oxysporum Schlechtend.:Fr. f. sp. *betae* (D. Stewart) W.C. Snyder & H.N. Hans.
Distr.: Widespread.
Notes: *Fusarium* yellows.
Hosts: Beta, Spinacia

Fusarium oxysporum Schlechtend.:Fr. f. sp. ***tracheiphilum*** (E.F. Sm.) W.C. Snyder & H.N. Hans.
Fusarium tracheiphilum E.F. Sm.
Fusarium bulbigenum Cooke & Massee var. *tracheiphilum* (E.F. Sm.) Wollenweb.
Distr.: Cosmopolitan.
Notes: Pathogenic on legumes (Holliday. C.M.I. Descr. 220. 1970).
Hosts: [Chrysanthemum], [Chrysobalanus], Crotalaria, Glycine, Vicia, Vigna

Fusarium oxysporum Schlechtend.:Fr. f. sp. ***vanillae*** (Tucker) W.L. Gordon
Distr.: FL; Puerto Rico; Asia.
Host: Orchidaceae

Fusarium oxysporum Schlechtend.:Fr. f. sp. ***vasinfectum*** (Atk.) W.C. Snyder & H.N. Hans.
Fusarium vasinfectum Atk.
Distr.: Cosmopolitan.
Notes: Vascular wilt of cotton and other plants (Booth & Waterston. C.M.I. Descr. 28. 1964).
Hosts: Abelmoschus, Arachis, Crotalaria, Gossypium, Lathyrus, Musa, Phaseolus, Pinus, Vigna

Fusarium oxysporum Schlechtend.:Fr. f. sp. ***zingiberi*** E.E. Trujillo
Distr.: HI.
Host: Zingiber

Fusarium pallidoroseum (Cooke) Sacc.
Fusisporium pallidoroseum Cooke
Pseudofusarium fusarioideum T. Matsushima
Fusarium semitectum auct. non Berk. & Ravenel
Pseudofusarium semitectum (Berk. & Ravenel) T. Matsushima
Pseudohansfordia venezuelensis (Deighton & Pirozynski) De Hoog
Sympodiophora venezuelensis Deighton & Pirozynski
Distr.: Subtropical and tropical regions, common.
Notes: Storage rots and damping-off (Booth & Sutton. Trans. Brit. Mycol. Soc. 83:702. 1984).
Hosts: Acer, Beta, Carex, Carya, Castilleja, Chenopodium, Cucumis, Glycine, Gossypium, Helianthus, Oryza, Phoenix, Zea

Fusarium poae (Peck) Wollenweb.
Sporotrichum poae Peck
Fusarium tricinctum (Corda) Sacc. f. sp. *poae* (Peck) W.C. Snyder & H.N. Hans.
Distr.: Temperate regions.
Notes: Common on grasses and herbaceous plants. White cob rot and silver top of corn, white head or silver top of cereals and grasses, and bud rot of chrysanthemums and carnations (Booth. C.M.I. Descr. 308. 1971; Neish. Fungi Canadenses 234. 1982).
Hosts: Avena, Bouteloua, Bromus, Camellia, Chrysanthemum, Dactylis, Dianthus, Elymus, Festuca, Helianthus, Hordeum, Muhlenbergia, Panicum, Phleum, Poa, Setaria, Sitanion, Solanum, Triticum, Tulipa, Vitis, Vulpia, Zea

Fusarium proliferatum (T. Matsushima) Nirenberg
Cephalosporium proliferatum T. Matsushima
Distr.: Widespread.
Notes: Foot rots, fruit rots and leaf spots on a variety of hosts. Redescribed in Gerlach & Nirenberg (Mitt. Biol. Bundesanst. Land-Forstw. Berlin-Dahlem 209:309. 1982) as a segregate of *F. moniliforme*.
Host: Zea

Fusarium redolens Wollenweb.
Fusarium oxysporum Schlechtend.:Fr. var. *redolens* W.L. Gordon
Distr.: Temperate regions.
Notes: Vascular wilts and damping-off (Booth & Waterson. C.M.I. Descr. 27. 1964).
Hosts: Pyrus, Solanum

Fusarium retusum Wellman
Distr.: IN(type).
Notes: Vascular wilt of tomato. Booth (The genus *Fusarium*. 1971) suggested this may be synonym of *Fusarium oxysporum* Schlecht.:Fr.
Host: Lycopersicon

Fusarium roseum Link:Fr.
Distr.: Widespread.
Notes: A confused name that has been applied to *F. lateritium*, *F. sambucinum*, *F. graminearum*, and *F. concolor*.
Hosts: Asparagus, Beta, Carya, Chamaedorea, Cheirodendron, Chrysanthemum, Citrullus, Colocasia, Coronilla, Cucumis, Cucurbita, Dianthus, Elaeagnus, Eremochloa, Fragaria, Gardenia, Glycine, Gossypium, Humulus, Ipomoea, Lens, Lotus, Lycopersicon, Malus, Medicago, Mentha, Metrosideros, Phaseolus, Phytolacca, Picea, Pinus, Pisum, Platanus, Poa, Prunus, Pseudotsuga, Rhizophora, Solanum, Striga, Trifolium, Triticum, Tulipa, Vicia, Zea

Fusarium sacchari (E.J. Butler) W. Gams
Cephalosporium sacchari E.J. Butler in E.J. Butler & Hafiz Khan
Distr.: Subtropical and tropical regions.
Host: Zea

Fusarium sambucinum Fuckel
Fusarium sarcochroum (Desmaz.) Sacc.
Gibberella pulicaris (Fr.:Fr.) Sacc. (teleomorph)
Distr.: Temperate northern hemisphere; northern Africa; Australia; New Zealand.
Notes: Die-back, canker, root rot, and storage rot of a wide variety of plants (Booth C.M.I. Descr. 385. 1973).
Hosts: Acer, Beta, Medicago, Melia, Pinus, Pyrus, Robinia, Solanum, Taraxacum

Fusarium semitectum Berk. & Ravenel var. ***majus*** Wollenweb.
Fusarium diversisporum Sherb.
Distr.: Temperate North America; Europe.
Hosts: Beta, Ipomoea, Juglans

Fusarium solani (Mart.) Sacc.
Fusarium lathyri Taubenhaus
Fusarium malli Taubenhaus
Fusarium martii Appel & Wollenweb. var. *phaseoli* Burkholder
Fusarium solani (Mart.) Sacc. var. *aduncisporum* (J.L. Weimer & Harter) Wollenweb. & Reinking
Fusarium solani (Mart.) Sacc. var. *martii* (Appel & Wollenweb.) Wollenweb. & Reinking
Fusarium solani (Mart.) Sacc. var. *striatum* (Sherb.) Wollenweb.
Nectria haematococca Berk. & Broome (teleomorph)
Distr.: Cosmopolitan in soil and plant debris.
Notes: On a wide variety of plants. Many physiological races adapted to specific hosts have been recognized (Booth & Waterson. C.M.I. Descr. 29. 1964)
Hosts: Abies, Acer, Actinidia, Allium, Ambrosia, Araucaria, Asparagus, Beta, Billbergia, Broussonetia, Buxus, Caladium, Carica, Carnegiea, Cercis, Chamaedorea, Chrysanthemum, Citrofortunella, Citrullus, Citrus, Coccoloba, Cocos, Codiaeum, Colocasia, Cornus, Cucumis, Cupressus, Dieffenbachia, Dracaena, Euphorbia, Eustoma, Fragaria, Freesia, Gentiana, Gerbera, Glycine, Gossypium, Hedera, Helianthus, Ilex, Ipomoea, Juglans, Lagerstroemia, Lathyrus, Linum, Liquidambar, Liriodendron, Liriope, Lycopersicon, Malus, Medicago, Murraya, Nerium, Nolina, Onobrychis, Parthenium, Persea, Phaseolus, Phoenix, Pinus, Podocarpus, Poncirus, Populus, Prunus, Pseudotsuga, Punica, Pyrus, Ravenala, Rhizophora, Rhododendron, Sabal, Saintpaulia, Schefflera, Sedum, Senna, Setaria, Solanum, Spinacia, Striga, Syngonium, Thuja, Tillandsia, Trevesia, Trifolium, Xanthosoma, Zea, Zingiber

Fusarium solani (Mart.) Sacc. f. sp. ***cucurbitae*** W.C. Snyder & H.N. Hans.
Distr.: Cosmopolitan.
Hosts: Cucumis, Cucurbita

Fusarium solani (Mart.) Sacc. f. sp. ***eumartii*** (C. Carpenter) W.C. Snyder & H.N. Hans.
Fusarium eumartii C. Carpenter
Distr.: Temperate regions.
Host: Solanum

Fusarium solani (Mart.) Sacc. f. sp. ***lupini*** J.L. Weimer
Distr.: FL.
Host: Lupinus

Fusarium solani (Mart.) Sacc. f. sp. ***phaseoli*** (Burkholder) W.C. Snyder & H.N. Hans.
Distr.: Cosmopolitan.
Notes: Bean root rot, bean yellows.
Hosts: Arachis, Phaseolus, Vicia, Vigna

Fusarium solani (Mart.) Sacc. f. sp. ***pisi*** (F.R. Jones) W.C. Snyder & H.N. Hans.
Fusarium martii Appel & Wollenweb. var. *pisi* F.R. Jones
Distr.: Cosmopolitan.

Notes: Root rot and vascular wilt of peas.
Hosts: Cicer, Pisum
Fusarium solani (Mart.) Sacc. var. ***coeruleum*** (Lib. ex Sacc.) C. Booth
Fusarium coeruleum Lib. ex Sacc.
Distr.: Cosmopolitan.
Notes: Dry rot of stored potatoes.
Hosts: Pisum, Solanum
Fusarium sporotrichioides Sherb.
Fusarium sporotrichiella Bilai var. *sporotrichioides* (Sherb.) Bilai
Distr.: Temperate and tropical regions.
Notes: Usually considered a soil fungus but frequently isolated from seeds of a wide variety of plants (Neish. Fungi Canadenses 235. 1982).
Hosts: Agropyron, Bouteloua, Bromus, Elymus, Helianthus, Hordeum, Oryzopsis, Panicum, Pinus, Solanum, Triticum
Fusarium subglutinans (Wollenweb. & Reinking) P.E. Nelson, T.A. Toussoun, & Marasas
Fusarium moniliforme J. Sheld. var. *subglutinans* Wollenweb. & Reinking
Distr.: Cosmopolitan.
Notes: Pitch canker of southern pines, stalk rot of maize, and several other diseases (Kuhlman et al. Mycologia 70:1131. 1978; Booth & Waterston. C.M.I. Descr. 23. 1964).
Hosts: Ananas, Asparagus, Cucumis, Helianthus, Hordeum, Musa, Pinus, Zea
Fusarium trichothecioides Wollenweb. in Jamieson & Wollenweb.
Distr.: North America; Australia; temperate Europe.
Notes: White powdery dry rot of stored potatoes and seed potatoes.
Hosts: Beta, Solanum, Vicia
Fusarium tricinctum (Corda) Sacc.
Selenosporium tricinctum Corda
Fusarium citriforme Jamalainen
Fusarium sporotrichiella Bilai var. *tricintum* (Corda) Bilai
Distr.: Cosmopolitan.
Notes: A common soil-inhabiting fungus.
Hosts: Astragalus, Glycine, Pisum, Sorghum, Trifolium, Zea
Fusarium udum E.J. Butler f. sp. ***crotalariae*** (G.S. Kulkarni) Subramanian
Fusarium lateritium Nees:Fr. f. sp. *crotalariae* (Padwick) W.L. Gordon
Fusarium udum E.J. Butler var. *crotalariae* (G.S. Kulkarni) Padwick
Distr.: North America; Brazil, India.
Notes: Vascular wilt.
Hosts: Crotalaria, Glycine
Fusarium ventricosum Appel & Wollenweb.
Fusarium argillaceum (Fr.:Fr.) Sacc., sensu Wollenweb.
Fusarium cuneiforme Sherb.
Nectria ventricosa C. Booth (teleomorph)
Distr.: Temperate northern hemisphere.
Notes: Associated with rot of discarded potatoes or sugar beets.
Hosts: Beta, Helianthus, Solanum

Fuscocerrena Ryvarden. Basidiomycetes, Aphyllophorales

Fuscocerrena portoricensis (Fr.:Fr.) Ryvarden
Polyporus portoricensis Fr.:Fr.
Irpex coriaceus Berk. & Ravenel
Daedalea farinacea (Fr.) Overh.
Irpex farinaceus Fr.
Irpex griseofuscus Mont.
Cerrenella subcoriacea Murrill
Distr.: Eastern North America, rare in the West.
Notes: On hardwoods.
Hosts: Acer, Alnus, Celtis, Quercus

Fusicladium Bonord. Hyphomycetes

Fusicladium alopecuri Ellis & Everh.
Distr.: MO.
Hosts: Alopecurus, Malus
Fusicladium aplectri Ellis & Everh.
Distr.: DE(type).
Host: Aplectrum
Fusicladium ascyrinum Ellis & Everh.
Distr.: LA(type).
Host: Hypericum
Fusicladium brevipes Ellis & Everh.
Distr.: Western United States.
Host: Astragalus
Fusicladium cerasi (Rabenh.) Sacc.
Venturia cerasi Aderhold (teleomorph)
Distr.: North America; Brazil; Europe; Middle East; New Zealand.
Notes: Cherry scab (Sivanesan & Holliday. C.M.I. Descr. 706. 1981).
Host: Prunus
Fusicladium destruens Peck
Distr.: NY.
Host: Avena
Fusicladium fasciculatum Cooke & Ellis
Cercosporidium fasciculatum (Cooke & Ellis) Höhn.
Passalora fasciculata (Cooke & Ellis) Earle
Scolicotrichum fasciculatum (Cooke & Ellis) Shear
Distr.: Central and eastern United States.
Notes: On leaves and stems (Deighton. Mycol. Pap. 112:19. 1967).
Hosts: [Alopecurus], [Ammophila], Euphorbia
Fusicladium fasciculatum Cooke & Ellis var. ***didymum*** Deighton
Distr.: Central and southern United States.
Notes: On stems and leaves (Deighton. Mycol. Pap. 112:21. 1967).
Host: Euphorbia
Fusicladium gnaphaliatum Bonar
Distr.: CA.
Notes: Parasitic on leaves (Bonar. Mycologia 57:392. 1965).
Host: Gnaphalium
Fusicladium lathyrinum (Ellis & B.T. Galloway) S.J. Hughes & Pirozynski
Dicoccum lathyrinum Ellis & B.T. Galloway
Distr.: MT(type).
Notes: Hughes & Pirozynski (Canad. J. Bot. 50:2528. 1972).
Host: Lathyrus
Fusicladium levieri Magnus in Sommier & Levier
Distr.: CT, FL, MS; Europe(type).
Notes: On leaves.
Host: Diospyros
Fusicladium pisicola Linford
Distr.: UT.
Notes: Black-leaf of peas (Linford. Phytopathology 16:549. 1926).
Host: Pisum
Fusicladium poricola Bonar
Distr.: CA.
Notes: On pore surface of *Poria ferrea* on logs of *Alnus* (Bonar. Mycologia 57:393. 1965).
Host: Alnus
Fusicladium pyrorum (Lib.) Fuckel
Venturia pyrina Aderhold (teleomorph)
Distr.: Temperate regions.
Notes: Scab or black spot of pear (Sivanesan. The bitunicate Ascomycetes. p. 620. 1984; Sivanesan & Waller. C.M.I. Descr. 404. 1974; Corlett. Fungi Canadenses 36. 1974 as *Venturia pirina*).
Hosts: Chaenomeles, Pyrus
Fusicladium staticis Ellis & Everh.
Distr.: DE, NJ, TX.
Host: Limonium
Fusicladium tremulae A.B. Frank
Distr.: AK.
Notes: No account of this species could be located.
Host: Populus

Fusicoccum Corda in Sturm. Coelomycetes

Fusicoccum amygdali Delacr.
Distr.: Eastern United States; Europe.
Host: Prunus
Fusicoccum canavaliae Lyon
Distr.: HI(type).
Host: Canavalia
Fusicoccum dakotense Sacc. & Syd.
Distr.: ND.
Host: Prunus

Fusicoccum ellisianum (Ellis & Everh.) Sacc. & Syd.
Fusicoccum quercinum Ellis & Everh.
Distr.: Eastern United States.
Host: Quercus
Fusicoccum ilicinum Ellis & Everh.
Micropera ilicina (Ellis & Everh.) Höhn.
Distr.: WV(type).
Host: Ilex
Fusicoccum persicae Ellis & Everh.
Distr.: Eastern and southern United States.
Host: Prunus
Fusicoccum putrefaciens Shear
Distr.: Central and eastern United States.
Notes: End rot of cranberries.
Host: Vaccinium
Fusicoccum pyrorum Chupp & Clapp
Distr.: OR(type).
Notes: Canker of apple.
Host: Malus

Fusicolla Bonord. Hyphomycetes

Fusicolla foliicola P. Karst.
Distr.: CA; Finland(type).
Host: Lithocarpus

Fusidium Link. Hyphomycetes

Notes: Although Booth (Mycol. Pap. 104. 1966) considered *Fusidium* a synonym of *Cylindrocarpon*, the change in the starting point for fungi legitimizes this genus with *F. griseum* as the lectotype (Hughes. Canad. J. Bot. 36:768. 1958).
Fusidium griseum Link
Distr.: CA, IA; Europe(type).
Notes: The identity of this species is uncertain.
Hosts: Lithocarpus, Zea

Fusisporium Link. Hyphomycetes

Notes: This genus is considered a synonym of *Fusarium*, but not all species have been redisposed.
Fusisporium cerealis Cooke
Distr.: FL(type).
Host: Zea
Fusisporium leguminum Cooke
Distr.: SC(type).
Host: Acacia
Fusisporium pallido-roseum Cooke
Distr.: SC(type).
Host: Chenopodium

Gaeumannomyces Arx & D. Olivier. Pyrenomycetes, Diaporthales

Notes: Walker (Mycotaxon 11:1. 1980).
Gaeumannomyces cylindrosporus D. Hornby, D. Slope, R. Gutteridge & Sivanesan
Phialophora graminicola (Deacon) J. Walker (anamorph)
Distr.: RI; England.
Notes: Hornby et al. (Trans. Brit. Mycol. Soc. 69:21. 1977); Jackson & Landschoot (Phytopathology 76:655. 1986).
Hosts: Lolium, Poa
Gaeumannomyces graminis (Sacc.) Arx & D. Olivier
Ophiobolus graminis (Sacc.) Sacc. in Roum. & Sacc.
Distr.: Cosmopolitan.
Notes: Crown and sheath rot of Poaceae (Walker. C.M.I. Descr. 381. 1973).
Hosts: Agropyron, Agrostis, Avena, Bromus, [Carex], Cinna, Deschampsia, Elymus, Festuca, [Glycine], Holcus, Hordeum, [Luzula], Oryza, Panicum, Phleum, Poa, Secale, Setaria, Trisetum, Triticum, Vulpia, Zea, Zizania
Gaeumannomyces graminis (Sacc.) Arx & D. Olivier var. ***avenae*** (E.M. Turner) Dennis
Ophiobolus graminis (Sacc.) Sacc. var. *avenae* E.M. Turner
Distr.: Temperate regions.
Notes: Take-all of oats and other cereals (Walker. C.M.I. Descr. 382. 1973).
Hosts: Agrostis, Festuca
Gaeumannomyces graminis (Sacc.) Arx & D. Olivier var. ***tritici*** J. Walker
Distr.: Temperate regions.
Notes: Take-all of cereals (Walker. C.M.I. Descr. 383. 1973; Shoemaker. Fungi Canadenses 37. 1974).
Hosts: Bromus, Triticum

Galerina Earle. Basidiomycetes, Agaricales

Notes: Smith & Singer (A monograph of the genus *Galerina* Earle. 1964).
Galerina autumnalis (Peck) A.H. Sm. & Singer
Pholiota autumnalis Peck
Distr.: Temperate northern hemisphere.
Notes: Usually on wood of hardwoods, occasionally on conifers.
Hosts: Abies, Pinus
Galerina castaneipes A.H. Sm. & Singer
Distr.: OR.
Host: Quercus
Galerina filiformis A.H. Sm. & Singer
Distr.: CA.
Host: Sequoia
Galerina laticeps A.H. Sm.
Distr.: Temperate North America.
Host: Pseudotsuga
Galerina marginata (Batsch:Fr.) Kühner
Pholiota marginata (Batsch:Fr.) Quél.
Distr.: Temperate northern hemisphere.
Notes: On wood.
Host: Quercus
Galerina megalocystis A.H. Sm. & Singer
Distr.: Central and western North America.
Host: Abies
Galerina nigripes A.H. Sm. & Singer
Distr.: OR.
Host: Quercus
Galerina olympiana A.H. Sm.
Distr.: WA.
Host: Populus
Galerina pallidispora A.H. Sm.
Distr.: WA.
Host: Alnus
Galerina perangusta A.H. Sm. & Singer
Distr.: CA.
Host: Sequoia
Galerina reflexa (Murrill) A.H. Sm. & Singer
Galerula reflexa Murrill
Distr.: HI; Mexico.
Host: Hibiscus
Galerina subbadia A.H. Sm. & Singer
Distr.: Temperate North America.
Host: Abies
Galerina thujina A.H. Sm. & Singer
Distr.: MI.
Host: Thuja
Galerina tsugae A.H. Sm. & Singer
Distr.: MI.
Host: Tsuga

Galiella Nannf. & Korf. Discomycetes, Pezizales

Galiella rufa (Schwein.) Korf
Bulgaria rufa Schwein.
Distr.: Central and eastern North America.
Notes: On dead wood.
Host: Quercus

Galzinia Bourd. Basidiomycetes, Aphyllophorales

Galzinia cymosa D.P. Rogers
Distr.: MA.
Notes: On wood.
Host: Pinus
Galzinia occidentalis D.P. Rogers
Distr.: Western United States.
Notes: On wood.
Hosts: Pinus, Pseudotsuga
Galzinia pedicellata Bourd.
Distr.: Temperate northern hemisphere.
Notes: On wood.
Host: Pinus

Gampsonema Nag Raj. Coelomycetes

Gampsonema exile (Tassi) Nag Raj
Pestalotia exilis Tassi
Distr.: HI; Brazil; India.
Notes: On dead bark (Nag Raj. Canad. J. Bot. 53:1621. 1975; Sutton. The Coelomycetes. p. 230. 1980).
Host: Eucalyptus

Ganoderma P. Karst. Basidiomycetes, Aphyllophorales

Ganoderma annularis (Fr.) R.L. Gilbertson
Fomes annularis (Fr.) C.G. Lloyd
Distr.: CA, FL.
Notes: On wood.
Hosts: Casuarina, Quercus, Umbellularia
Ganoderma applanatum (Pers.) Pat.
Fomes applanatus (Pers.) Gill.
Polyporus applanatus (Pers.) Wallr.
Fomes leucophaeus Mont.
Elfvingia megaloma (Lév.) Murrill
Fomes megaloma (Lév.) Cooke
Ganoderma megaloma (Lév.) Bres.
Distr.: Northern hemisphere as far south as India.
Notes: Usually on dead but sometimes on living trees, usually hardwoods. Root and butt rot of *Populus* spp. Steyaert (C.M.I. Descr. 443. 1975).
Hosts: Abies, Acacia, Acer, Aesculus, Alnus, Baccharis, Betula, Carya, Castanea, Casuarina, Cercis, Chrysalidocarpus, Citrus, Crataegus, Eucalyptus, Fagus, Fraxinus, Gleditsia, Juglans, Ligustrum, Liquidambar, Liriodendron, Magnolia, Malus, Morus, Nyssa, Ostrya, Picea, Pinus, Platanus, Populus, Prunus, Pseudotsuga, Quercus, Robinia, Salix, Schinus, Spiraea, Syringa, Taxodium, Thuja, Tilia, Tsuga, Ulmus, Umbellularia
Ganoderma australe (Fr.:Fr.) Pat.
Fomes australis (Fr.:Fr.) Cooke
Distr.: Southern United States; pantropical.
Notes: On hardwoods. According to Steyaert (Trans. Brit. Mycol. Soc. 65:451. 1975), *G. australe* is a synonym of *G. tornatum* (Pers.) Bres.
Hosts: Acacia, Diospyros, Eucalyptus, Fraxinus, Ilex, Quercus
Ganoderma brownii (Murrill) R.L. Gilbertson
Elfvingia brownii Murrill
Distr.: CA.
Notes: On wood. According to Steyaert (Trans. Brit. Mycol. Soc. 65:451-467. 1975), *G. brownii* is a synonym of *Ganoderma tornatum* (Pers.) Bres. but Gilbertson & Ryvarden (North American Polypores 1:293. 1987) recognize this species as distinct.
Hosts: Citrus, Hakea, Quercus, Schinus, Umbellularia
Ganoderma colossum (Fr.) C. Baker
Distr.: Subtropical and tropical regions.
Notes: On wood.
Host: Celtis
Ganoderma lobatum (Schwein.) Atk.
Polyporus lobatus Schwein.
Fomes lobatus (Schwein.) Cooke
Distr.: Eastern and southwestern North America.
Notes: On wood.
Hosts: Acer, Prunus, Quercus
Ganoderma lucidum (Curtis:Fr.) P. Karst.
Polyporus lucidus (Curtis:Fr.) Fr.
Ganoderma curtisii (Berk.) Murrill
Polyporus curtisii Berk.
Ganoderma polychromum (Copeland) Murrill
Ganoderma sessile Murrill
Distr.: North America; Asia; Europe.
Notes: The relationship with many tropical species described within the "*Ganoderma lucidum*" complex is unclear. Steyaert (C.M.I. Descr. 445. 1975).
Hosts: Acer, Albizia, Betula, Carya, Cassia, Castanopsis, Casuarina, Celtis, Cercis, Citrofortunella, Citrus, Dioon, Diospyros, Eriobotrya, Fraxinus, Gleditsia, Gymnocladus, Limonia, Liquidambar, Macadamia, Malus, Olea, Parkinsonia, Picea, Pinus, Planera, Platanus, Poncirus, Populus, Prosopis, Prunus, Pseudotsuga, Quercus, Rhus, Robinia, Salix, Schinus, Tsuga, Ulmus
Ganoderma oregonense Murrill
Polyporus oregonensis (Murrill) Kauffm.
Ganoderma sequoiae Murrill
Distr.: Western North America.
Notes: Possibly conspecific with *G. tsugae*. Butt and root rot of living and dead trees, especially *Abies* and *Tsuga* spp.
Hosts: Abies, Castanopsis, Picea, Pinus, Pseudotsuga, Sequoia, Tsuga
Ganoderma tsugae Murrill
Polyporus tsugae (Murrill) Overh.
Distr.: Eastern North America.
Notes: On wood, usually of conifers.
Hosts: Abies, Betula, Picea, Pinus, Pseudotsuga, Tsuga
Ganoderma zonatum Murrill
Polyporus lucidus (Curtis:Fr.) Fr. var. *zonatus* (Murrill) Overh.
Ganoderma sulcatum Murrill
Distr.: Southern United States.
Notes: Steyaert (C.M.I. Descr. 448. 1975).
Hosts: Acacia, Aphanes, Arecaceae, Butia, Chrysalidocarpus, Cocos, Grevillea, Leucaena, Lonchocarpus, Persea, Phoenix, Prosopis, Prunus, Quercus, Roystonea, Sabal, Syagrus, Washingtonia

Gelasinospora Dowding. Pyrenomycetes, Sordariales

Gelasinospora autosteira Alexopoulos & Sung
Distr.: MS(type).
Host: Tillandsia
Gelasinospora tetrasperma Dowding
Distr.: Cosmopolitan.
Notes: Isolated from dung, also occasionally from seeds and other organic matter.
Host: Alnus

Gelatinodiscus Kanouse & A.H. Sm. Discomycetes, Helotiales

Gelatinodiscus flavidus Kanouse & A.H. Sm.
Distr.: Pacific Northwest.
Host: Chamaecyparis

Gelatinosporium Peck. Coelomycetes

Gelatinosporium betulinum Peck
Gelatinosporis betulina (Peck) Clements & Shear
Durandiella seriata (Fr.) Groves (teleomorph)
Distr.: Northeastern North America.
Notes: On branches (DiCosmo. Canad. J. Bot. 56:1685. 1978).
Host: Betula
Gelatinosporium fulvum Peck
Dermea molliuscula (Schwein.) Cash (teleomorph)
Distr.: Northeastern North America.
Notes: On limbs.
Host: Betula
Gelatinosporium pinicola (Linder) Funk
Cryptosporium pinicola Linder
Micropera pinicola (Linder) Funk
Distr.: Pacific Northwest.
Notes: Stem and branch canker. Morgan-Jones (Canad. J. Bot. 49:1935.

1971) illustrated the type material.
Hosts: Abies, Pinus

Geniculodendron G.A. Salt. Hyphomycetes

Geniculodendron pyriforme G.A. Salt
Caloscypha fulgens (Pers.) Boud. (teleomorph)
Distr.: Western North America.
Notes: A psychrophilic fungus pathogenic on seeds of various conifers (Salt. Trans. Brit. Mycol. Soc. 63:339. 1974; Paden et al. Canad. J. Bot. 56:2375. 1978;Ginns. Fungi Canadenses 66. 1975).
Host: Picea

Geniculosporium Chesters & Greenhalgh. Hyphomycetes

Geniculosporium serpens Chesters & Greenhalgh
Hypoxylon serpens (Pers.:Fr.) J. Kickx fil. (teleomorph)
Distr.: Widespread.
Notes: On dead wood and bark (Ellis. Dematiaceous Hyphomycetes. p. 229. 1971; Petrini & Rogers. Mycotaxon 26:401. 1986).
Host: Carya

Geotrichum Link. Hyphomycetes

Geotrichum candidum Link
Oospora lactis (Fresen.) Sacc.
Oospora lactis (Fresen.) Sacc. var. *parasitica* Pritchard & Porte
Galactomyces geotrichum (E.E. Butler & L.J. Petersen) Redhead & Malloch (teleomorph)
Distr.: Cosmopolitan.
Notes: On a variety of organic substrates (De Hoog et al. Stud. Mycol. 29:83. 1986).
Hosts: Citrus, Cucumis, Daucus, Lycopersicon, Prunus
Geotrichum citri-aurantii (Ferraris) E.E. Butler
Oidium citri-aurantii Ferraris
Geotrichum candidum Link var. *citri-aurantii* (Ferraris) Cif.
Oospora citri-aurantii (Ferraris) Sacc.
Galactomyces citri-aurantii E.E. Butler (teleomorph)
Distr.: Worldwide.
Notes: Sour rot of *Citrus*. Butler et al. (Mycotaxon 33:197. 1988).
Hosts: Citrus, Poncirus

Gerronema Singer. Basidiomycetes, Agaricales

Gerronema chrysophyllum (Fr.:Fr.) Singer
Omphalina chrysophylla (Fr.:Fr.) Murrill
Clitocybe luteophylla H.E. Bigelow & Hesler
Distr.: Temperate northern hemisphere.
Notes: On conifer wood.
Host: Picea
Gerronema subclavatum (Peck) Singer
Omphalina subclavata Peck
Clitocybe subclavata (Peck) H.E. Bigelow & Hesler
Omphalina subclavata (Peck) Murrill
Distr.: Eastern North America.
Notes: On hardwood debris (Bigelow. Mycologia 62:1. 1970).
Host: Quercus

Gerwasia Racib. Basidiomycetes, Uredinales

Gerwasia epiphylla (Arth.) Cummins
Mainsia epiphylla Arth.
Distr.: TX.
Host: Rubus

Gibbago E. Simmons. Hyphomycetes

Gibbago trianthemae E. Simmons
Distr.: TX; Cuba; Venezuela.
Notes: Leaf spot of horse-purslane (Simmons. Mycotaxon 17:107. 1986).
Host: Trianthema

Gibbera Fr. Loculoascomycetes, Dothideales

Notes: Barr (Canad. J. Bot. 46:835. 1968).
Gibbera andromedae (Rehm) E. Müller
Coleroa andromedae (Rehm) G. Wint. in Rabenh.
Distr.: Northeastern North America.
Host: Andromeda
Gibbera cassandrae (Peck) Barr
Venturia cassandrae Peck
Distr.: Northeastern North America.
Host: Andromeda
Gibbera cercocarpi (Ellis & Everh.) Barr
Dothidea cercocarpi Ellis & Everh.
Distr.: CA.
Host: Cercocarpus
Gibbera kalmiae (Peck) Barr
Venturia kalmiae Peck
Coleroa kalmiae (Peck) Rehm
Distr.: Temperate North America.
Notes: On leaves, twigs, and fruits.
Host: Kalmia
Gibbera myrtilli (Cooke) Petr.
Venturia myrtilli Cooke
Distr.: Temperate North America.
Notes: On leaves. The name *Venturia kalmiae* Peck has been misapplied to this species.
Host: Vaccinium
Gibbera selaginellae M.L. Farr & H.T. Horner
Distr.: OR, WA, on *Selaginella oregana*.
Host: Selaginella
Gibbera vaccinii (Sowerby:Fr.) Fr.
Brachysporium vaccinii (Fr.) Sacc. (anamorph)
Distr.: Circumboreal.
Host: Vaccinium

Gibberella Sacc. Pyrenomycetes, Hypocreales

Notes: Booth (The genus *Fusarium*. 1971); Gerlach and Nirenberg (Mitt. Biol. Bundesanst. Land-Forstw. Berlin-Dahlem 209. 1982).
Gibberella acervalis (Moug.:Fr.) Sacc.
Distr.: Temperate regions.
Host: Salix
Gibberella avenacea R.J. Cooke
Fusarium avenaceum (Fr.:Fr.) Sacc. (anamorph)
Distr.: Cosmopolitan, most common in temperate regions.
Notes: See comments under *Fusarium avenaceum*.
Host: Triticum
Gibberella baccata (Wallr.) Sacc.
Gibberella moricola (De Not.) Sacc.
Fusarium lateritium Nees:Fr. (anamorph)
Distr.: Cosmopolitan.
Notes: Wilt, die-back, and cankering of woody plants. Booth (C.M.I. Descr. 310. 1971).
Hosts: Carpinus, Citrus, Cotoneaster, Ficus, Juglans, Malus, Morus, Prunus, Pyrus, Robinia
Gibberella cyanogena (Desmaz.) Sacc.
Sphaeria cyanogena Desmaz.
Gibberella saubinetii (Mont.) Sacc.
Fusarium sulphureum Schlechtend. (anamorph)
Distr.: Widespread as a facultative pathogen.
Notes: Isolated from soil, debris, and a wide range of herbaceous plants. Storage rot of potatoes (Booth. The genus *Fusarium*. p. 183. 1971; Booth. C.M.I. Descr. 574. 1978 as *Fusarium sulphureum*).
Hosts: Asparagus, Cannabis, Carya, Hordeum, Mesembryanthemum, Triticum, Zea
Gibberella fujikuroi (Sawada) Ito in Ito & K. Kimura
Lisea fujikuroi Sawada
Gibberella moniliforme (J. Sheld.) Wineland
Fusarium moniliforme J. Sheld. (anamorph)
Distr.: Warm-temperate and tropical regions.
Notes: Parasitic on Poaceae and a wide range of other hosts. Bakanae disease of rice, associated with seedling blight, root rot and pink boll of cotton, associated with seed rot, seedling blight, root rot and stalk rot of maize and sorghum, pokkah boeng disease of sugarcane, and

various other diseases (Booth. The genus *Fusarium*. p. 123. 1971). Gerlach & Nirenberg (Mitt. Biol. Bundesanst. Land-Forstw. Berlin-Dahlem 209:301. 1982) do not agree with this synonymy. Booth & Waterston (C.M.I. Descr. 22. 1964).
Hosts: Saccharum, Sorghum, Triticum, Zea
Gibberella lagerheimii Rehm
Distr.: HI; Ecuador(type).
Host: Freycinetia
Gibberella pulicaris (Fr.:Fr.) Sacc.
Fusarium sambucinum Fuckel (anamorph)
Distr.: Temperate northern hemisphere; northern Africa; Australia; New Zealand.
Notes: Die-back, canker, root rot, or storage rot of a wide variety of plants (Booth. C.M.I. Descr. 385. 1973).
Hosts: Sambucus, Solanum, Zea
Gibberella quinqueseptata Sherb.
Distr.: IN(type).
Host: Cannabis
Gibberella subglutinans (E. Edwards) P.E. Nelson, T.A. Toussoun, & Marasas
Gibberella fujikuroi (Sawada) Ito var. *subglutinans* E. Edwards
Distr.: Cosmopolitan.
Hosts: Saccharum, Zea
Gibberella zeae (Schwein.) Petch
Gibberella roseum (Link) W.C. Snyder & E.M. Hans.
Gibberella roseum (Link) W.C. Snyder & E.M. Hans. f. sp. *cerealis* (Cooke) W.C. Snyder & H.N. Hans.
Fusarium graminearum Schwabe (anamorph)
Distr.: Cosmopolitan.
Notes: Diseases of cereals and grasses (Booth. C.M.I. Descr. 384. 1973).
Hosts: Avena, Bromus, Cenchrus, Hordeum, Lycopersicon, Poa, Schizachyrium, Secale, Setaria, Sorghum, Triticum, Zea

Gilbertella Hesseltine. Zygomycetes, Mucorales

Gilbertella persicaria (E.D. Eddy) Hesseltine
Choanephora persicaria E.D. Eddy
Distr.: North and South America; India.
Notes: Fruit rot of peaches and mulberries (Sarbhoy. C.M.I. Descr. 104. 1966).
Host: Prunus

Gilmaniella Barron. Hyphomycetes

Gilmaniella humicola Barron
Distr.: Cosmopolitan.
Notes: Isolated from soil and dung (Ellis. Dematiaceous Hyphomycetes. p. 61. 1971).
Host: Lens

Glabrocyphella W.B. Cooke. Basidiomycetes, Aphyllophorales

Glabrocyphella ailanthi W.B. Cooke
Distr.: NJ.
Notes: On bark.
Host: Ailanthus
Glabrocyphella bananae (Cooke) W.B. Cooke
Cyphella bananae Cooke
Distr.: FL.
Notes: On leaves.
Host: Musa
Glabrocyphella ellisiana W.B. Cooke
Distr.: NJ.
Hosts: Oenothera, Populus, Zea

Gliocephalotrichum J.J. Ellis & Hesseltine. Hyphomycetes

Gliocephalotrichum bulbilium J.J. Ellis & Hesseltine
Distr.: HI; Africa; India.
Notes: Isolated from wood, soil, and various plants (Ellis. Dematiaceous Hyphomycetes. p. 552. 1971).
Host: Psidium
Gliocephalotrichum simplex (J.A. Meyer) B. Wiley & E. Simmons
Cylindrocladium simplex J.A. Meyer
Distr.: Subtropical and tropical regions.
Notes: Isolated from soil and plant debris (Wiley & Simmons. Mycologia 63:579. 1971).
Host: Cornus

Gliocladium Corda. Hyphomycetes

Notes: Domsch et al. (Compendium of soil fungi. p. 368. 1980).
Gliocladium catenulatum Gilman & E. Abbott
Gliocladium roseum Bainier var. *viride* Rall
Gliocladium varians Pidoplichko
Distr.: Cosmopolitan.
Notes: In soil.
Host: Abies
Gliocladium roseum Bainier
Penicillium roseum Link
Gliocladium aureum Rader
Isaria clonostachoides Pritchard & Porte
Gliocladium verticilloides Pidoplichko
Nectria ochroleuca (Schwein.) Berk. (teleomorph)
Distr.: Cosmopolitan.
Notes: In soil and on plant debris, also mycoparasitic.
Hosts: Acacia, Acer, Beta, Buxus, Citrus, Cucumis, Daucus, Fragaria, Glycine, Liquidambar, Lycopersicon, Metrosideros, Phaseolus, Phoenix, Pinus, Poncirus, Psilotum, Quercus, Rhizophora, Trifolium
Gliocladium vermoeseni (Biourge) Thom
Penicillium vermoeseni Biourge
Distr.: Widespread.
Hosts: Chamaedorea, Chrysalidocarpus, Howeia, Phoenix, Syagrus, Washingtonia
Gliocladium virens J.H. Miller, J.E. Giddens, & A.A. Foster
Hypocrea sp. (teleomorph)
Distr.: Widespread.
Notes: In soil and on a variety of plant materials.
Hosts: Pinus, Pseudotsuga
Gliocladium viride Matr.
Gliocladium deliquescens Sopp
Distr.: Northern hemisphere.
Notes: In soil or on plant debris.
Hosts: Malus, Metrosideros

Gliomastix Guég. Hyphomycetes

Gliomastix murorum (Corda) S.J. Hughes var. ***felina*** (Ém. Marchal) S.J. Hughes
Distr.: Widespread.
Notes: Common on seeds, wood and dead herbaceous stems and isolated from air, cardboard, paint, rope and soil (Ellis. Dematiaceous Hyphomycetes. p. 521. 1971).
Host: Metrosideros

Globifomes Murrill. Basidiomycetes, Aphyllophorales

Globifomes graveolens (Schwein.:Fr.) Murrill
Polyporus graveolens (Schwein.:Fr.) Fr.
Polyporus conglobatus Berk.
Distr.: Central and eastern North America.
Notes: On trunks or logs, sometimes on wounds of living trees.
Hosts: Acer, Fagus, Liriodendron, Quercus

Globulina Speg. Pyrenomycetes, Clavicipitales

Globulina antennariae Hasselbring
Distr.: NY(type).
Host: Antennaria

Gloeocephalus J.A. Spencer, nom. illeg. Hyphomycetes

Notes: This genus and its one species were published without a latin description or type designation.
Gloeocephalus hemerocalli J.A. Spencer, nom. illeg.
Distr.: MS(type).
Notes: Leaf-streak disease (Spencer. Pl. Dis. Reporter 52:751. 1968).
Host: Hemerocallis

Gloeocercospora Bain & Edgerton ex Deighton. Hyphomycetes

Gloeocercospora inconspicua Demaree & M.S. Wilcox ex Deighton
Distr.: Southeastern United States.
Notes: Demaree & Wilcox (Phytopathology 37:496. 1947) provided the original account of this leaf spot dieases. Deighton (Trans. Brit. Mycol. Soc. 57:360. 1971) commented that this fungus is probably not a *Gloeocercospora*.
Host: Vaccinium
Gloeocercospora sorghi Bain & Edgerton ex Deighton
Distr.: Southern United States; Central and South America; tropical Africa; Australia; India; Japan.
Notes: Zonate leaf spot or copper spot of grasses. Bain & Edgarton (Phytopathology 33:220. 1943) and Luttrell (Pl. Dis. Reporter 34:45.1950) provided accounts of the disease on sorghum. Mulder & Holliday (C.M.I. Descr. 300. 1971).
Hosts: Agrostis, Pennisetum, Saccharum, Sorghum, Zea

Gloeocoryneum J. Weindlymayr. Coelomycetes

Gloeocoryneum cinereum (Dearn.) J. Weindlymayr
Coryneum cinereum Dearn.
Leptomelanconium cinereum (Dearn.) Morgan-Jones
Distr.: Western North America; Central America.
Notes: On needles (Evans. Mycol. Pap. 153:85. 1984).
Host: Pinus
Gloeocoryneum hawaiiense Sutton & C.S. Hodges
Distr.: HI(type).
Notes: Sutton & Hodges (Mycologia 75:280. 1983).
Host: Acacia

Gloeocystidiellum Donk. Basidiomycetes, Aphyllophorales

Notes: Burdsall et al. (Syst. Bot. 6:422. 1981).
Gloeocystidiellum clavuligerum (Höhn. & Litsch.) Nakasone
Corticium vesiculosum Burt
Distr.: Widespread.
Notes: On wood. Nakasone (Mycotaxon 14:316. 1982).
Hosts: Acer, Ostrya, Populus, Tilia, Ulmus
Gloeocystidiellum convolvens (P. Karst.) Donk
Distr.: North America; Europe.
Notes: On wood.
Host: Populus
Gloeocystidiellum fimbriatum Burdsall, Nakasone & G. Freeman
Distr.: Southeastern United States.
Notes: On hardwoods.
Hosts: Carpinus, Ostrya, Quercus
Gloeocystidiellum heimii Boidin
Distr.: MN.
Host: Populus
Gloeocystidiellum karstenii (Bourd. & Galzin) Donk
Gloeocystidium karstenii Bourd. & Galzin
Distr.: Northern hemisphere.
Notes: On wood.
Hosts: Populus, Pseudotsuga
Gloeocystidiellum marianum Burdsall, Nakasone & G. Freeman
Distr.: MD.
Notes: On wood.
Host: Quercus
Gloeocystidiellum ochraceum (Fr.:Fr.) Donk
Distr.: Northern hemisphere.
Notes: On wood.
Hosts: Picea, Pinus, Populus
Gloeocystidiellum porosum (Berk. & M.A. Curtis) Donk
Corticium porosum Berk. & M.A. Curtis
Corticium stramineum Bres. in Brinkmann
Distr.: Widespread.
Notes: On wood. Nakasone (Mycotaxon 14:316. 1982).
Hosts: Acer, Alnus, Liquidambar, Picea, Pinus, Populus, Prunus, Quercus, Salix, Tilia
Gloeocystidiellum sinuosum G. Freeman in Burdsall, Nakasone & G. Freeman
Distr.: Southeastern United States.
Notes: On wood.
Hosts: Pinus, Quercus, Salix
Gloeocystidiellum tropicale Burdsall, Nakasone & G. Freeman
Distr.: FL.
Notes: On wood.
Host: Sideroxylon
Gloeocystidiellum turpe G. Freeman in Burdsall, Nakasone & G. Freeman
Distr.: FL.
Notes: On wood.
Hosts: Liquidambar, Ostrya
Gloeocystidiellum wakullum Burdsall, Nakasone & G. Freeman
Distr.: Southeastern United States.
Notes: On wood.
Host: Taxodium

Gloeodes Colby. Coelomycetes

Gloeodes pomigena (Schwein.) Colby
Dothidea pomigena Schwein.
Phyllachora pomigena (Schwein.) Sacc.
Distr.: Temperate northern hemisphere.
Notes: On twigs and fruits of at least 22 plant species, mostly trees and shrubs.
Hosts: Acer, Asimina, Brassolaeliocattleya, Cattleya, Celastrus, Cercis, Citrus, Cornus, Crataegus, Dendrobium, Dirca, Epidendrum, Fraxinus, Gaultheria, Gymnocladus, Laeliocattleya, Lindera, Liriodendron, Magnolia, Malus, Orchidaceae, Platanus, Poncirus, Prunus, Pyrus, Quercus, Rhus, Rubus, Salix, Sassafras, Smilax, Sorbus, Staphylea, Ulmus, Vanda, Vitis

Gloeodontia Boidin. Basidiomycetes, Aphyllophorales

Notes: About 2 species.
Gloeodontia columbiensis Burdsall & Lombard
Distr.: Western North America.
Notes: On wood.
Hosts: Acer, Alnus, Populus
Gloeodontia discolor (Berk. & M.A. Curtis) Boidin
Irpex discolor Berk. & M.A. Curtis
Distr.: Southern United States; probably Africa.
Notes: On wood.
Hosts: Bursera, Liquidambar, Quercus, Salix

Gloeomucro R. Petersen. Basidiomycetes, Aphyllophorales

Notes: Petersen (Mycologia 72:303. 1980).
Gloeomucro nodosus (Linder) R. Petersen
Myxomycidium nodosum Linder
Distr.: TN(type), WA.
Host: Pseudotsuga

Gloeophyllum P. Karst. Basidiomycetes, Aphyllophorales

Gloeophyllum carbonarium (Berk. & M.A. Curtis) Ryvarden
Hexagonia carbonaria Berk. & M.A. Curtis
Poria carbonaria (Berk. & M.A. Curtis) D. Baxter
Coriolellus carbonarius (Berk. & M.A. Curtis) Bondartsev & Singer
Griseoporia carbonaria (Berk. & M.A. Curtis) Ginns
Trametes carbonaria (Berk. & M.A. Curtis) Overh.

Trametes sequoiae Copeland
Distr.: Northern hemisphere.
Notes: On burnt or charred wood of conifers.
Hosts: Abies, Chamaecyparis, Larix, Picea, Pinus, Pseudotsuga, Sequoia, Thuja, Tsuga
Gloeophyllum mexicanum (Mont.) Ryvarden
Lenzites mexicana Mont.
Daedalea berkeleyi Sacc.
Daedalea rhabarbarina Berk. & Cooke
Distr.: United States; subtropical and tropical regions.
Notes: On wood of conifers, often on wood in service.
Host: Pinus
Gloeophyllum odoratum (Wulfen:Fr.) Imazeki
Polyporus odoratus (Wulfen:Fr.) Fr.
Osmoporus odoratus (Wulfen:Fr.) Singer
Trametes odorata (Wulfen:Fr.) Fr.
Trametes americana Overh., nom. inval.
Distr.: Northern hemisphere.
Notes: On conifer wood; often on structural timber. Gilbertson & Ryvarden (North American Polypores 1:318. 1986) state that records of this species are in question due to confusion with *G. protractum*. See also Ryvarden & Gilbertson (Mycotaxon 19:137-144. 1984).
Hosts: Abies, Alnus, Juniperus, Larix, Picea, Pinus, Pseudotsuga, Sequoia, Thuja, Tsuga
Gloeophyllum protractum (Fr.) Imazeki
Trametes protracta Fr.
Distr.: Northern hemisphere.
Notes: On wood of conifers, especially *Pinus* spp.
Host: Pinus
Gloeophyllum sepiarium (Wulfen:Fr.) P. Karst.
Lenzites sepiaria (Wulfen:Fr.) Fr.
Gloeophyllum hirsutum (Schaeff.) Murrill
Distr.: Temperate northern hemisphere.
Notes: On wood, usually of conifers, including wood in service such as structural timbers, railroad ties, and utility poles. Also spelled "*saepiaria*".
Hosts: Abies, Acer, Alnus, Arbutus, Betula, Calocedrus, Chamaecyparis, Cupressus, Juniperus, Larix, Liquidambar, Liriodendron, Malus, Picea, Pinus, Populus, Prunus, Pseudotsuga, Quercus, Salix, Sequoiadendron, Taxodium, Thuja, Tsuga, Ulmus
Gloeophyllum striatum (Sw.:Fr.) Murrill
Lenzites striata (Sw.:Fr.) Fr.
Distr.: AZ, FL; pantropical.
Notes: On wood.
Hosts: Cupressus, Taxodium
Gloeophyllum trabeum (Pers.:Fr.) Murrill
Lenzites trabea (Pers.:Fr.) Fr.
Trametes trabea (Pers.:Fr.) Bres.
Lenzites vialis Peck
Distr.: Widespread.
Notes: Usually on hardwoods, often on conifers; decays structural timbers, especially in buildings.
Hosts: Acer, Alnus, Betula, Castanea, Chilopsis, Fagus, Juniperus, Liquidambar, Malus, Picea, Pinus, Populus, Prunus, Pseudotsuga, Quercus, Salix, Tsuga

Gloeoporus Mont. Basidiomycetes, Aphyllophorales

Caloporus P. Karst.
Gloeoporus dichrous (Fr.:Fr.) Bres.
Polyporus dichrous Fr.:Fr.
Distr.: Widespread.
Notes: On wood, usually of hardwoods.
Hosts: Abies, Acer, Alnus, Arbutus, Betula, Carya, Cephalanthus, Chamaecyparis, Cornus, Fagus, Fraxinus, Juniperus, Liquidambar, Liriodendron, Picea, Pinus, Platanus, Populus, Prunus, Pseudotsuga, Quercus, Rhus, Salix, Thuja, Tsuga, Ulmus

Gloeosporidiella Petr. Coelomycetes

Notes: Sutton (The Coelomycetes. p. 538. 1980).
Gloeosporidiella bartholomaei (Dearn.) Arx
Gloeosporium bartholomaei Dearn.
Distr.: Pacific Northwest.
Notes: Leaf spot.
Host: Ribes
Gloeosporidiella cercidis H.C. Greene
Distr.: WI.
Host: Cercis
Gloeosporidiella desmodii (H.C. Greene) Arx
Gloeosporium desmodii H.C. Greene
Distr.: WI.
Host: Desmodium
Gloeosporidiella ramosa (Ellis & Everh.) Arx
Gloeosporium ramosum Ellis & Everh.
Distr.: Central and eastern United States.
Notes: Anthracnose.
Host: Polygala
Gloeosporidiella ribis (Lib.) Petr.
Leptothyrium ribis Lib.
Gloeosporium ribis (Lib.) Mont. & Desmaz.
Gloeosporium curvatum Oudem.
Drepanopeziza ribis (Kleb.) Höhn. (teleomorph)
Distr.: Temperate regions.
Notes: Leaf spot.
Host: Ribes
Gloeosporidiella variabilis (Laubert) Nannf.
Gloeosporium variabile Laubert
Distr.: WI; Europe.
Host: Ribes

Gloeosporidina Petr. Coelomycetes

Gloeosporidina cercocarpi (Ellis & Everh.) Sutton & Pollack
Gloeosporium cercocarpi Ellis & Everh.
Melanophora cercocarpi (Ellis & Everh.) Arx
Distr.: CA.
Notes: Sutton & Pollack (Mycologia 65:1133. 1973).
Host: Cercocarpus

Gloeosporium Desmaz. & Mont. Coelomycetes

Notes: This genus is considered a synonym of *Marssonina*, but not all names have been redisposed (Arx. Biblioth. Mycol. 24:3. 1970).
Gloeosporium aceris Cooke
Distr.: NC.
Notes: This species may belong in *Cladosporium* or *Fusicladium* (Arx. Biblioth. Mycol. 24:90. 1970).
Host: Acer
Gloeosporium amaranthicola Dearn., nom. nud.
Distr.: NC.
Notes: This name was apparently never published.
Host: Amaranthus
Gloeosporium aquifolii Penz. & Sacc.
Distr.: Italy(type); distribution elsewhere is questionable.
Notes: A doubtful species (Arx. Biblioth. Mycol. 24:97. 1970).
Host: Ilex
Gloeosporium arctostaphyli W.B. Cooke & C.G. Shaw
Distr.: Western United States.
Notes: A doubtful species. Arx & Bezerra (Proc. Koninkl. Nederl. Akad. Wetens., Ser. C, 66:177. 1963) found the type specimen to be an ascomycete and described it as *Mycosphaerella shawii* Arx & Bezerra.
Host: Arctostaphylos
Gloeosporium argemonis Ellis & Everh.
Septoria argemones Tharp
Distr.: Central United States.
Notes: A doubtful species.
Host: Argemone
Gloeosporium brunneo-maculatum H.C. Greene
Distr.: WI.
Host: Trillium
Gloeosporium capsularum Cooke & Harkn.
Distr.: CA(type).
Notes: A poorly known species, possibly the same as *G. carpigenum* Cooke (Arx. Biblioth. Mycol. 24:106. 1970).
Host: Eucalyptus
Gloeosporium carthami (Fukui) S. Hori & Hemmi

Marssonia carthami Fukui
Distr.: Southeastern United States; eastern Asia.
Host: Carthamus
Gloeosporium castanopsidis Dearn. & House
Distr.: CA(type).
Host: Castanopsis
Gloeosporium cattleyae Henn.
Distr.: VA; Brazil(type).
Notes: A poorly known species, possibly a *Botryodiplodia* (Arx. Biblioth. Mycol. 24:108. 1970).
Host: Cattleya
Gloeosporium confluens Ellis & Dearn.
Distr.: Central and eastern North America.
Notes: The generic placement of this species is uncertain (Arx. Biblioth. Mycol. 24:112. 1970.)
Host: Sagittaria
Gloeosporium convallariae Allesch.
Distr.: Germany(type).
Notes: A poorly known species, possibly the same as *Aureobasidium microstictum* (Bubák) W.B. Cooke (Arx. Biblioth. Mycol. 24:113. 1970).
Host: Convallaria
Gloeosporium decolorans Ellis & Everh.
Distr.: MA.
Notes: A doubtful species according to Arx (Biblioth. Mycol. 24:117. 1970).
Host: Acer
Gloeosporium eragrostidis H.C. Greene
Distr.: WI.
Host: Eragrostis
Gloeosporium ferrugineum Dearn., nom. inval.
Distr.: NC.
Notes: Published without a latin diagnosis (Dearness. Mycologia 33:363. 1941).
Host: Rhododendron
Gloeosporium ficariae (Berk.) Cooke
Cylindrosporium ficariae Berk.
Distr.: WA; Europe(type).
Host: Ranunculus
Gloeosporium hemerocallidis Ellis & Everh., nom. nud.
Distr.: TX.
Host: Hymenocallis
Gloeosporium hydrophylli Dearn. & House
Ascochyta hydrophylli R. Sprague & F.D. Bailey
Distr.: NY, OR.
Host: Hydrophyllum
Gloeosporium ilicis Dearn., nom. inval.
Distr.: TN(type), southeastern United States, scattered reports.
Notes: Published without a latin diagnosis (Dearness. Mycologia 33:363. 1941).
Host: Ilex
Gloeosporium impatientis H.W. Anderson non Petch, hom. illeg.
Distr.: IL, IN.
Host: Impatiens
Gloeosporium intermedium Sacc. var. ***poinsettiae*** Sacc.
Distr.: CA; Europe.
Host: Euphorbia
Gloeosporium lagenarium (Pass.) Sacc. & Roum. var. ***musarum*** Ellis & Everh.
Distr.: NE.
Host: Musa
Gloeosporium lupinicola Dearn., nom. nud.
Distr.: ID.
Host: Lupinus
Gloeosporium minus Shear
Distr.: Eastern United States.
Host: Vaccinium
Gloeosporium niveum J.J. Davis
Distr.: WI(type).
Notes: A doubtful species, based on the conidial state of a *Rhytisma* (Arx. Biblioth. Mycol. 24:146. 1970).
Host: Ilex
Gloeosporium papulatum Dearn., nom. inval.
Distr.: TN(type).
Notes: Published without a latin diagnosis. A doubtful species (Arx. Biblioth. Mycol. 24:152. 1970).
Host: Rubus
Gloeosporium pentstemonicola H.C. Greene
Distr.: WI.
Host: Penstemon
Gloeosporium quercuum Miura
Distr.: CA; China(type).
Host: Quercus
Gloeosporium rosae Halst., nom. nud.
Distr.: NJ(type).
Notes: Although Halsted mentioned this fungus, the name has not been validly described nor has the fungus been characterized in recent studies.
Host: Rosa
Gloeosporium rumicis Ellis & Everh.
Distr.: WV(type).
Notes: A doubtful species (Arx. Biblioth. Mycol. 24:167. 1970).
Host: Rumex
Gloeosporium saururi Ellis & Dearn.
Distr.: North America.
Host: Saururus
Gloeosporium senecionis Ellis & Everh.
Distr.: CA(type).
Notes: A doubtful species (Arx. Biblioth. Mycol. 24:169. 1970).
Host: Senecio
Gloeosporium thalictri J.J. Davis
Distr.: WI(type).
Host: Thalictrum
Gloeosporium toxicodendri Ellis & G. Martin
Cylindrosporium toxicodendri (Ellis & G. Martin) Ellis & Everh.
Marssonina toxicodendri (Ellis & G. Martin) Magnus
Marssonia toxicodendri (Ellis & G. Martin) Sacc.
Septoria toxicodendri M.A. Curtis in G. Martin
Cylindrosporium irregulare (Peck) Dearn., nom. nud.
Phloeospora irregularis (Peck) Bubák
Septoria irregularis Peck
Distr.: Central and western United States.
Hosts: Pisum, Rhus
Gloeosporium ulmicola L.E. Miles
Distr.: Central and eastern North America.
Notes: Probably a synonym of *Discula umbrinella* (Berk. & Broome) Sutton (Arx. Biblioth. Mycol. 24:180. 1970).
Host: Ulmus
Gloeosporium veratri (Allesch.) Höhn.
Fusoma veratri Allesch.
Marssonia veratri Ellis & Everh.
Marssonina veratri (Ellis & Everh.) Magnus
Septogloeum veratri (Allesch.) Wollenweb.
Distr.: Western United States; Europe.
Notes: According to Arx (Biblioth. Mycol. 24:181. 1970) this species is not placed in the correct genus.
Host: Veratrum

Gloeotinia M. Wilson, M. Noble, & E. Gray. Discomycetes, Helotiales

Gloeotinia granigena (Quél.) T. Schumacher
Phialea granigena Quél.
Gloeotinia temulenta (Prill. & Delacr.) M. Wilson, M. Noble, & E. Gray
Phialea temulenta Prill. & Delacr.
Endoconidium temulentum Prill. & Delacr. (anamorph)
Distr.: Temperate North America; Europe.
Notes: Blind seed disease of grasses (Schumacher. Mycotaxon 8:125. 1979).
Hosts: Agrostis, Aira, Alopecurus, Bromus, Cynosurus, Dactyloctenium, Danthonia, Deschampsia, Festuca, Glyceria, Holcus, Hordeum, Lolium, Phleum, Vulpia

Gloiodon P. Karst. Basidiomycetes, Aphyllophorales

Notes: Monotypic.
Gloiodon strigosus (Sw.:Fr.) P. Karst.

Hydnum strigosum Sw.:Fr.
Distr.: Temperate northern hemisphere.
Notes: On wood.
Host: Abies

Glomerella Spauld. & H. Schrenk. Pyrenomycetes, Phyllachorales

Glomerella cingulata (Stoneman) Spauld. & H. Schrenk
Physalospora cattleyae Maubl. & Lasnier
Glomerella cinctum Berk. & M.A. Curtis
Glomerella cingulata (Stoneman) Spauld. & H. Schrenk var. *vaccinii* Shear
Physalospora dracaenae J. Sheld.
Glomerella mayor Tunstall
Physalospora miyabeana Fukushi
Physalospora orchidearum Henn.
Glomerella phomoides Swank
Glomerella piperata (Stoneman) Spauld. & H. Schrenk
Glomerella rubicola (Stoneman) Spauld. & H. Schrenk
Venturia rubicola Ellis & Everh.
Glomerella rufomaculans Spauld. & H. Schrenk
Physalospora uvispora (Cooke) Sacc.
Glomerella vignicaulis Tehon
Colletotrichum gloeosporioides (Penz.) Penz. & Sacc. in Penz. (anamorph)
Distr.: Cosmopolitan.
Notes: About 750 names exist for either the teleomorph or anamorph. Arx & Müller (Beitr. Kryptogamenfl. Schweiz 11(1):188. 1954) provided a list of synonyms. Mordue (C.M.I. Descr. 315. 1971).
Hosts: Abutilon, Aesculus, Agave, Aglaonema, Aleurites, Annona, Arceuthobium, Arecaceae, Asclepias, Aucuba, Begonia, Betula, Buxus, Calathea, Camellia, Capsicum, Carica, Carya, Caryota, Cassia, Castanea, Cattleya, Celastrus, Chrysobalanus, Cinnamomum, Citrofortunella, Citrus, Clausena, Clematis, Coccoloba, Coffea, Cordyline, Crataegus, Cryptanthus, Cucurbita, Cyclamen, Cydonia, Dendrobium, Dieffenbachia, Dracaena, Epidendrum, Epipremnum, Eriobotrya, Euonymus, Fatshedera, Feijoa, Ficus, Garcinia, Ginkgo, Gleditsia, Glycine, Gossypium, Hedera, Hevea, Homalomena, Humulus, Illicium, Ipomoea, Iris, Kummerowia, Lathyrus, Lespedeza, Ligustrum, Lindera, Lupinus, Lycopersicon, Magnolia, Malus, Mangifera, Maranta, Melia, Monstera, Musa, Nandina, Odontoglossum, Olea, Oncidium, Orchidaceae, Pandanus, Peperomia, Persea, Phalaenopsis, Philodendron, Phormium, Podophyllum, Poncirus, Prunus, Psidium, Pyrus, Quercus, Rhododendron, Ribes, Rollinia, Rosa, Rubus, Salix, Sansevieria, Sapindus, Sassafras, Sechium, Smilax, Solanum, Sorbus, Stephanotis, Syagrus, Syngonium, Syzygium, Thespesia, Vaccinium, Vanda, Vanilla, Vigna, Vitis, Yucca, Zantedeschia

Glomerella glycines F. Lehm. & F.A. Wolf
Colletotrichum destructivum O'Gara (anamorph)
Distr.: Temperate and subtropical regions
Notes: Anthracnose of soybeans, usually affecting aging plants at harvest or soybean stubble (Manandhar et al. Phytopathology 76:282. 1986).
Host: Glycine

Glomerella gossypii Edgerton
Colletotrichum gossypii Southworth (anamorph)
Distr.: Worldwide in cotton-growing regions.
Notes: Some have considered this a synonym of *G. cingulata*.
Host: Gossypium

Glomerella graminicola Politis
Colletotrichum graminicola (Ces.) G.W. Wils. (anamorph)
Distr.: Cosmopolitan in the anamorph state.
Notes: This teleomorph was described by Politis (Mycologia 67:56. 1975) who considered it distinct from *Glomerella tucumanensis*. See comments on the disease under the anamorph.
Host: Zea

Glomerella lagenaria F. Stevens
Glomerella cingulata (Stoneman) Spauld. & Schenk var. *orbiculare* S.F. Jenkins & Winstead
Colletotrichum orbiculare (Berk. & Mont.) Arx (anamorph)
Distr.: Widespread.
Notes: The teleomorph is rarely encountered. See comments under the anamorph.
Hosts: Citrullus, Cucumis, Cucurbita

Glomerella lindemuthiana Shear
Colletotrichum lindemuthianum (Sacc. & Magnus) Lams.-Scrib. (anamorph)
Distr.: Cosmopolitan.
Notes: See comments under the anamorph.
Host: Phaseolus

Glomerella magna S.F. Jenkins & Winstead
Colletotrichum sp. (anamorph)
Distr.: NC.
Notes: Anthracnose of cucurbits (Jenkins & Winstead. Phytopathology 54:452. 1964). The relationship of this species with other *Glomerella* species has not been determined.
Hosts: Citrullus, Cucumis, Cucurbita, Trichosanthes

Glomerella nephrolepis Faris, nom. inval.
Colletotrichum sp. (anamorph)
Distr.: NY(type).
Notes: This name was published without a latin description. It is probably a synonym of *G. cingulata* but this has not been proven (Faris. Mycologia 15:92. 1923).
Host: Nephrolepis

Glomerella pezizoidea Petr.
Distr.: HI(type).
Host: Cheirodendron

Glomerella tucumanensis (Speg.) Arx & E. Müller
Physalospora tucumanensis Speg.
Distr.: Subtropical and tropical regions.
Notes: Red rot of sugarcane (Mordue. C.M.I. Descr. 133. 1967).
Hosts: Erianthus, Leptochloa, Saccharum, Sorghum, Zea

Glomopsis D.M. Henderson. Hyphomycetes

Glomerularia Peck non P. Karst., hom. illeg.
Notes: Donk (Persoonia 4:213. 1966).

Glomopsis corni (Peck) D.M. Henderson
Glomerularia corni Peck
Distr.: Northern North America.
Notes: Sutton (Mycol. Pap. 132:65. 1973).
Host: Cornus

Glomosporium Kochman. Basidiomycetes, Ustilaginales

Glomosporium leptideum (Syd. & P. Syd.) Kochman
Tolyposporium leptideum Syd. & P. Syd.
Distr.: North America(rare); Australia; Europe.
Host: Chenopodium

Gloniella Sacc. Loculoascomycetes, Dothideales

Gloniella abietina Syd. in Syd. & Petr.
Distr.: ID(type).
Notes: On decorticated wood.
Host: Abies

Gloniella lapponica (P. Karst.) Sacc.
Hysterium lapponicum P. Karst.
Distr.: Western United States; Lapland(type).
Notes: A doubtful species.
Host: Arctostaphylos

Gloniella vaccinicola Dearn. & House
Distr.: NY(type).
Host: Vaccinium

Gloniopsis De Not. Loculoascomycetes, Dothideales

Gloniopsis curvata (Fr.:Fr.) Sacc.
Gloniopsis ellisii Cash
Gloniopsis gerardiana Sacc.
Gloniopsis smilacis (Schwein.) Ellis & Everh.
Hysterium smilacis Schwein.
Distr.: Cosmopolitan.

Notes: On wood, bark, and limbs of woody species.
Hosts: Campsis, Pieris, Quercus, Smilax
Gloniopsis lineolata **(Cooke) Sacc.**
Hysterium lineolatum Cooke
Distr.: North and Central America.
Notes: On dead wood. Lohman (Pap. Michigan Acad. Sci. 17:229. 1933) suggested that this is a species of *Tryblidaria*.
Hosts: Persea, Serenoa
Gloniopsis praelonga **(Schwein.) Zogg**
Hysterium praelongum Schwein.
Hysterographium praelongum (Schwein.) Ellis & Everh.
Gloniopsis brevisaccata Lohman
Gloniopsis cookeana (W. Gerard) Sacc.
Gloniopsis insignis (Cooke & Harkn.)Berl. & Voglino
Gloniopsis lonicerae (W. Phillips & Harkn.) Berl. & Voglino
Gloniopsis verbasci (Schwein.) Rehm
Distr.: Cosmopolitan.
Notes: On wood and bark of many woody species.
Hosts: Arctostaphylos, Berchemia, Buddleja, Heteromeles, Lespedeza, Lyonia, Malus, Sabal, Ulmus, Vaccinium

Glonium Mühlenberg. Loculoascomycetes, Dothideales

Notes: Zogg (Beitr. Kryptogamenfl. Schweiz 11(3):57. 1962).
Glonium abbreviatum **(Schwein.) Lohman**
Hysterium abbreviatum Schwein.
Glonium parvulum (W. Gerard) Sacc.
Distr.: Cosmopolitan.
Notes: On dead wood and bark of various trees.
Hosts: Cercocarpus, Heteromeles, Quercus
Glonium lineare **(Fr.) De Not.**
Hysterium lineare Fr.
Glonium clavisporum Seaver
Glonium simulans W. Gerard
Distr.: Widespread in temperate regions.
Notes: On dead wood and bark of various trees.
Hosts: Liquidambar, Quercus, Vitis
Glonium stellatum **Mühlenberg**
Distr.: North America.
Notes: On dead wood.
Hosts: Acer, Betula, Prunus, Quercus, Tsuga
Glonium strobilarum **(P. Karst.) P. Karst.**
Hysterium strobilarum P. Karst.
Distr.: Western United States; Finland(type).
Notes: According to Lohman (Mycologia 31:354. 1939) this fungus is not a member of the Hysteriales, but he does not suggest a correct genus.
Host: Picea
Glonium tryblidioides **Ellis & Everh.**
Distr.: Western United States.
Notes: A doubtful species.
Host: Pseudotsuga
Glonium vestigiale **Fairm.**
Distr.: CA(type).
Host: Nicotiana

Glyphium Nitschke ex F. Lehm. Loculoascomycetes, Dothideales

Notes: The two species of *Glyphium* occurring in western Canada and the United States were described by Goree (Canad. J. Bot. 52:1265. 1974).
Glyphium corrugatum **(Ellis) H. Goree**
Acrospermum corrugatum Ellis
Lophionema apoclastosporum Solheim
Glyphium leptothecium (Earle in Greene) Sutton
Lophium leptothecium Earle in Greene
Peyronelia sp. (anamorph)
Distr.: Western North America.
Notes: Possibly the most common hysteriaceous fungus in western North America .
Hosts: Acer, Alnus, Amelanchier, Artemisia, Ceanothus, Cornus, Corylus, Holodiscus, Malus, Populus, Rhus, Salix, Sorbus
Glyphium elatum **(Grev.:Fr.) Zogg**
Lophium elatum Grev.:Fr.
Glyphium dolabriforme (Wallr.) F. Lehm.
Lophium dolabriforme Wallr.
Acrospermum fultum Harkn.
Distr.: Western North America; possibly Chile; Europe.
Hosts: Acer, Alnus, Arctostaphylos, Eucalyptus, Juncus, Populus, Salix
Glyphium schizosporum **(Maire) Zogg**
Distr.: Europe.
Notes: This fungus occurs only in Europe; reports from the U.S. are erroneous.
Hosts: Lonicera, Populus, Salix
Glyphium tillandsiae **(Cash) Zogg**
Lophium tillandsiae Cash
Distr.: FL(type).
Host: Tillandsia

Gnomonia Ces. & De Not. Pyrenomycetes, Diaporthales

Notes: Barr (Mycol. Mem. 7:33. 1978); Monod (Beih. Sydowia 9:101. 1983).
Gnomonia alni **Plowr.**
Hypospila californica Dearn. & Barth.
Leptothyrium alni Zeller (anamorph)
Distr.: Western North America.
Notes: Parasitic on leaves.
Host: Alnus
Gnomonia caryae **F.A. Wolf**
Gnomonia setacea (Pers.:Fr.) Ces. & De Not. var. *macrospora* Ellis & Everh.
Distr.: Eastern North America.
Notes: On dead leaves. Barr (Mycol. Mem. 7:45. 1978) indicated the anamorph is *Asteroma caryae* (Peck) Sutton (= *Gloeosporium caryae* Ellis & Dearn).
Host: Carya
Gnomonia clavulata **Ellis**
Didymella clavulata (Ellis) Sacc.
Distr.: NJ.
Host: Carya
Gnomonia comari **P. Karst.**
Gnomonia fragariae Kleb. f. *fructicola* G. Arnaud
Gnomonia fructicola (G. Arnaud) J. Fall
Zythia fragariae Laibach (anamorph)
Distr.: Temperate regions.
Notes: Leaf blotch and stem-end rot.
Host: Fragaria
Gnomonia dispora **Demaree & J.R. Cole ex Barr**
Distr.: Southeastern United States.
Notes: Pecan leaf blotch.
Host: Carya
Gnomonia fasciculata **Fuckel**
Gnomoniella fasciculata (Fuckel) Sacc.
Distr.: Temperate North America; Europe.
Notes: On dead leaves.
Host: Quercus
Gnomonia fragariae **Kleb.**
Distr.: Europe.
Notes: Reports of this species in North America probably refer to *Gnomonia comari* P. Karst. (Bolay. Ber. Schweiz. Bot. Ges. 81:398. 1971).
Host: Fragaria
Gnomonia gnomon **(Tode:Fr.) J. Schröt. in Cohn**
Gnomoniella gnomon (Tode:Fr.) Magnus
Gnomonia vulgaris Ces. & De Not.
Gnomoniella vulgaris (Ces. & De Not.) Sacc.
Distr.: Temperate North America; Europe.
Notes: On dead leaves (Barr. Mycol. Mem. 7:33. 1978).
Hosts: Alnus, Corylus, Ostrya
Gnomonia leptostyla **(Fr.:Fr.) Ces. & De Not.**
Marssonina juglandis (Lib.) Magnus (anamorph)
Distr.: Temperate regions.
Notes: On leaves.
Host: Juglans

Gnomonia linnaeae Auersw.
Distr.: North America; northern Europe.
Host: Linnaea
Gnomonia myricae Cooke & Ellis
Apiognomonia myricae (Cooke & Ellis) Dennis
Distr.: Temperate North America.
Notes: On dead leaves.
Host: Myrica
Gnomonia nerviseda J.R. Cole
Gnomonia caryae F.A. Wolf var. *pecanae* J.R. Cole
Gnomonia setacea (Pers.:Fr.) Ces. & De Not. var. *caryae* Dearn. & House
Leptothyrium nerviseda J.R. Cole (anamorph)
Distr.: Central and eastern North America.
Notes: On dead leaves.
Host: Carya
Gnomonia petiolorum (Schwein.:Fr.) Cooke
Gnomoniella amoena (Nees:Fr.) Ces. & De Not. var. *petiolorum* (Schwein.:Fr.) Sacc.
Distr.: Eastern North America.
Notes: On overwintered leaves.
Host: Liquidambar
Gnomonia rauii (Ellis & Everh.) Petr.
Didymella rauii (Ellis & Everh.) Ellis
Distr.: Eastern United States.
Notes: On dead limbs.
Host: Rosa
Gnomonia ribicola Barr
Distr.: Temperate North America.
Notes: On leaves and twigs.
Host: Ribes
Gnomonia rostellata (Fr.:Fr.) Bref.
Distr.: Temperate North America; Europe.
Notes: On leaves.
Host: Rubus
Gnomonia rubi (Rehm) G. Wint.
Distr.: Temperate North America; Europe.
Notes: On leaves and stems.
Host: Rubus
Gnomonia salina E.B.G. Jones
Distr.: CT; Great Britain (Kohlmeyer. Marine mycology. p. 244. 1979).
Host: Spartina
Gnomonia setacea (Pers.:Fr.) Ces. & De Not.
Distr.: Cosmopolitan.
Notes: On overwintered leaves of deciduous trees and shrubs.
Hosts: Alnus, Betula, Carya, Castanea, Corylus, Quercus, Rubus, Vaccinium
Gnomonia triostei (Dearn. & House) Dearn. in Wehmeyer
Diaporthe triostei Dearn. & House
Distr.: NY.
Host: Triosteum
Gnomonia waldsteiniae Dearn. & House, nom. inval.
Distr.: Northeastern North America.
Notes: Published without a latin diagnosis.
Host: Waldsteinia

Gnomoniella Sacc. Pyrenomycetes, Diaporthales

Notes: Barr (Mycol. Mem. 7:19. 1978).
Gnomoniella destruens Barr & C.S. Hodges
Distr.: HI.
Host: Eucalyptus
Gnomoniella tubiformis (Tode:Fr.) Sacc.
Gnomonia tubiformis (Tode:Fr.) Ces. & De Not.
Asteroma alneum (Pers.:Fr.) Sutton (anamorph)
Distr.: North America; Europe.
Notes: On dead leaves.
Host: Alnus

Godronia Moug. & Lév. Discomycetes, Helotiales

Notes: Groves (Canad. J. Bot. 43:1195. 1965).
Godronia cassandrae Peck
Distr.: Central and eastern North America; northern Europe.
Notes: On stems and leaves of Ericaceae (Smerlis. Canad. J. Bot. 46:597. 1968; Pl. Dis. Reporter 53:807. 1969).
Hosts: Chamaedaphne, Rhododendron, Vaccinium
Godronia cassandrae Peck f. ***ribicola*** Groves
Distr.: CO, NH.
Host: Ribes
Godronia cassandrae Peck f. ***vaccinii*** Groves
Distr.: Temperate North America.
Notes: On stems, leaves, and fruits.
Host: Vaccinium
Godronia cephalanthi (Schwein.:Fr.) Dearn. & House
Scleroderris cephalanthi (Schwein.:Fr.) Farl.
Distr.: Eastern North America.
Notes: The generic placement of this species is uncertain (Groves. Canad. J. Bot. 43:1258. 1965).
Host: Cephalanthus
Godronia confertus (Hone) Groves
Godronia urceolus (J.C. Schmidt:Fr.) P. Karst. var. *confertus* Hone
Distr.: North America.
Host: Prunus
Godronia davidsonii Cash
Distr.: Western North America.
Host: Ribes
Godronia fuliginosa (Fr.:Fr.) Seaver
Scleroderris fuliginosa (Fr.:Fr.) P. Karst.
Topospora proboscidea (Fr.) Fr. (anamorph)
Distr.: Temperate North America; Europe.
Host: Salix
Godronia grossulariae Groves
Distr.: OR.
Host: Ribes
Godronia kalmiae (Rehm) Seaver
Gorgoniceps kalmiae Rehm
Distr.: Reported from an unknown locality in the United States(type).
Host: Vaccinium
Godronia lantanae (Cash) Seaver
Scleroderris lantanae Cash
Distr.: HI.
Notes: The generic placement of this species is uncertain (Groves. Canad. J. Bot. 43:1260. 1965).
Host: Lantana
Godronia linnaeae (P. Karst.) P. Karst.
Distr.: Circumboreal.
Host: Linnaea
Godronia lobata (Cash) Seaver
Scleroderris lobata Cash
Distr.: CA.
Host: Ribes
Godronia rhois Groves
Distr.: Northeastern North America.
Host: Rhus
Godronia ribis (Fr.) Seaver
Cenangium ribis Fr.
Peziza ribesia Pers.
Fuckelia ribesia (Link) Sutton (anamorph)
Distr.: North America(questionable); Europe.
Notes: Groves (Canad. J. Bot. 43:1236. 1965).
Hosts: Arctostaphylos, Juniperus, Ribes
Godronia spiraeae (Rehm) Seaver
Belonidium spiraeae Dearn. & House
Cenangium spiraeae (Schwein.) Berk.
Belonidium macounii Dearn. & House
Distr.: Temperate North America; Europe.
Notes: On limbs.
Host: Spiraea
Godronia tumoricola (Cash) Seaver
Scleroderris tumoricola Cash
Distr.: CO(type).
Notes: Groves (Canad. J. Bot. 43:1264. 1965) stated that this species is

not a *Godronia* but could not be placed satisfactorily at present.
Host: Ribes
Godronia turbinata (Schwein.) Farl. in Thaxt.
Distr.: Northeastern North America.
Host: Diervilla
Godronia uberiformis Groves
Topospora uberiformis (Fr.) Fr. (anamorph)
Distr.: Temperate North America; Europe.
Host: Ribes
Godronia urceolata (Ellis) Höhn.
Cenangella urceolata (Ellis) Sacc.
Sphaeronaema clethrincola Ellis (anamorph)
Distr.: Eastern North America.
Hosts: Clethra, Vaccinium
Godronia urceolus (J.C. Schmidt:Fr.) P. Karst.
Distr.: Temperate North America; Europe.
Hosts: Acer, Clethra, Ribes

Godroniopsis Diehl & Cash. Discomycetes, Helotiales

Godroniopsis nemopanthis Groves
Sphaeronaema peckii Sacc. & Syd. (anamorph)
Distr.: MI; Canada.
Notes: Groves (Mycologia 29:71. 1937).
Hosts: Ilex, Nemopanthus
Godroniopsis quernea (Schwein.) Diehl & Cash
Dichaena quernea (Schwein.) Diehl & Cash
Dichaenopsella quernea Petr. (anamorph)
Distr.: Eastern United States.
Host: Quercus

Gonatobotrys Corda. Hyphomycetes

Gonatobotrys simplex Corda
Distr.: Cosmopolitan.
Notes: A mycoparasite.
Host: Zea
Gonatobotrys zeae M.C. Futrell & D. Bain, nom. nud.
Distr.: MS.
Notes: This name was published without a description or type specimen (Futrell & Bain. Phytopathology 58:728. 1968).
Host: Zea

Gonatobotryum Sacc. Hyphomycetes

Gonatobotryum apiculatum (Peck) S.J. Hughes
Haplographium apiculatum Peck
Gonatobotrys maculicola G. Wint.
Gonatobotryum maculicola (G. Wint.) Sacc.
Distr.: Eastern North America; India.
Notes: On leaves of *Hamamelis* and Anacardiaceae and from soil (Kendrick et al. Canad. J. Bot. 46:591. 1968).
Host: Hamamelis

Gonatophragmium Deighton. Hyphomycetes

Gonatophragmium mori (Sawada) Deighton
Ormathodium fici Tims & Olive
Acrospermum viticola Ikata in Ikata & Hitomi (teleomorph)
Distr.: Widespread.
Notes: Deighton (Mycol. Pap. 117:13. 1969).
Host: Ficus

Gorgoniceps P. Karst. Discomycetes, Helotiales

Gorgoniceps aridula (P. Karst.) P. Karst.
Apostemidium aridulum (P. Karst.) Boud.
Arachnopeziza rhaphidospora (Ellis) Rehm
Erinella rhaphidospora (Ellis) Sacc.
Distr.: Widespread.
Host: Pinus
Gorgoniceps iowensis Rehm
Distr.: Central and eastern United States.
Host: Acer

Grammothele Berk. & M.A. Curtis. Basidiomycetes, Aphyllophorales

Notes: About 5 species, mostly tropical.
Grammothele fuligo (Berk. & Broome) Ryvarden
Poria ravenalae (Berk. & Broome) Sacc.
Distr.: AZ, CA; pantropical.
Host: Phoenix
Grammothele lineata Berk. & Broome
Distr.: FL, pantropical.
Host: Quercus

Grandinia Fr. Basidiomycetes, Aphyllophorales

Hyphodontia J. Eriksson
Notes: Formerly a large and heterogeneous genus, now restricted. Jülich (Int. J. Mycol. Lichenol. 1:27. 1982) demonstrated that *Grandinia* is an earlier name for *Hyphodontia*.
Grandinia abieticola (Bourd. & Galzin) Jülich
Odontia barba-jovis (Fr.:Fr.) Fr. subsp. *abieticola* Bourd. & Galzin
Hyphodontia abieticola (Bourd. & Galzin) J. Eriksson
Distr.: Temperate northern hemisphere.
Notes: On wood.
Hosts: Abies, Pinus, Pseudotsuga
Grandinia alutaria (Burt) Jülich
Peniophora alutaria Burt
Hyphodontia alutaria (Burt) J. Eriksson
Distr.: Temperate northern hemisphere.
Notes: On wood.
Hosts: Picea, Pinus, Populus, Thuja
Grandinia arguta (Fr.:Fr.) Jülich
Hyphodontia arguta (Fr.:Fr.) J. Eriksson
Odontia arguta (Fr.:Fr.) Quél.
Distr.: Northern hemisphere.
Notes: On wood.
Hosts: Abies, Alnus, Arbutus, Juglans, Juniperus, Larix, Picea, Pinus, Populus, Quercus, Salix, Thuja
Grandinia barba-jovis (Fr.:Fr.) Jülich
Hyphodontia barba-jovis (Fr.:Fr.) J. Eriksson
Odontia barba-jovis (Fr.:Fr.) Fr.
Distr.: Temperate northern hemisphere.
Notes: On wood.
Hosts: Abies, Carya, Castanea, Picea, Pinus
Grandinia breviseta (P. Karst.) Jülich
Hyphodontia breviseta (P. Karst.) J. Eriksson
Distr.: Temperate northern hemisphere.
Notes: On wood.
Hosts: Abies, Arbutus, Picea, Pinus, Populus, Pseudotsuga
Grandinia crustosa (Pers.:Fr.) Fr.
Hyphodontia crustosa (Pers.:Fr.) J. Eriksson
Odontia crustosa (Pers.:Fr.) Quél.
Grandinia burtii Peck
Distr.: Temperate northern hemisphere.
Notes: On wood.
Hosts: Alnus, Arctostaphylos, Betula, Juniperus, Picea, Pinus, Populus, Quercus
Grandinia floccosa (Bourd. & Galzin) Jülich
Hyphodontia floccosa (Bourd. & Galzin) J. Eriksson
Distr.: Temperate northern hemisphere.
Notes: On wood.
Hosts: Cowania, Larix, Pinus
Grandinia granulosa (Pers.:Fr.) Fr.
Grandinia aspera Fr.
Hyphodontia aspera (Fr.) J. Eriksson
Distr.: Temperate northern hemisphere.
Notes: On wood. Most reports of *G. granulosa* refer to *Vararia borealis*. Jülich (Int. J. Mycol. Lichenol. 1:27. 1982); Pouzar (Ceska Mykol. 36:72. 1982).
Hosts: Acer, Picea, Pinus, Populus, Robinia
Grandinia hastata (Litsch.) Jülich

Peniophora hastata Litsch.
Hyphodontia hastata (Litsch.) J. Eriksson
Distr.: Temperate northern hemisphere.
Notes: On wood, usually of conifers.
Hosts: Abies, Pinus, Thuja
Grandinia juniperi (Bourd. & Galzin) Jülich
Hyphodontia juniperi (Bourd. & Galzin) J. Eriksson & Hjortstam in J. Eriksson et al.
Distr.: Northern hemisphere.
Notes: On Cupressaceae.
Host: Juniperus
Grandinia nespori (Bres.) Cejp
Odontia nespori Bres.
Hyphodontia nespori (Bres.) J. Eriksson & Hjortstam in J. Eriksson et al.
Odontia papillosa Fr.:Fr.
Distr.: Northern hemisphere.
Notes: On wood. Eriksson & Ryvarden (The Corticiaceae of North Europe 4:656. 1976) provided a discussion on the use of *G. nespori* over *Odontia papillosa*.
Hosts: Acer, Juniperus
Grandinia pallidula (Bres.) Jülich
Hyphodontia pallidula (Bres.) J. Eriksson
Peniophora pallidula (Bres.) Bres. in Bourd. & Galzin
Distr.: Temperate northern hemisphere.
Notes: On wood.
Hosts: Abies, Alnus, Picea, Pinus, Populus, Pseudotsuga, Thuja, Tsuga
Grandinia pruni (Lasch) Jülich
Odontia pruni Lasch in Rabenh.
Hyphoderma pruni (Lasch) Jülich
Hyphodontia pruni (Lasch) J. Eriksson
Hyphodontia bugellensis (Ces.) J. Eriksson
Odontia bugellensis Ces. in Rabenh.
Distr.: Northern hemisphere.
Notes: On wood.
Hosts: Acer, Baccharis, Chilopsis, Cowania, Fraxinus, Lycium, Olneya, Platanus, Populus, Prosopis, Prunus, Quercus, Rubus, Vitis
Grandinia quercina (Fr.:Fr.) Jülich
Hyphodontia quercina (Fr.:Fr.) J. Eriksson
Distr.: Temperate northern hemisphere.
Notes: On hardwoods.
Hosts: Alnus, Prosopis
Grandinia spathulata (Schrad.:Fr.) Jülich
Hyphodontia spathulata (Schrad.:Fr.) Parmasto
Odontia spathulata (Schrad.:Fr.) Litsch.
Hydnum pithyophilum Berk. & M.A. Curtis
Distr.: Northern hemisphere.
Notes: On wood.
Hosts: Alnus, Fraxinus, Juglans, Juniperus, Malus, Pinus, Platanus, Populus, Prunus, Quercus, Tsuga
Grandinia stenospora (P. Karst.) Jülich
Hyphodontia alutacea (Fr.:Fr.) J. Eriksson
Odontia alutacea (Fr.:Fr.) Bres.
Hydnum alutaceum Fr.:Fr.
Distr.: Temperate northern hemisphere.
Notes: On wood. Although *H. alutacea* is an older name than *G. stenospora* and is sanctioned, the combination in *Grandinia* is preoccupied by *G. alutacea* Berk. & Ravenel in Berk. 1873.
Hosts: Picea, Pinus, Populus, Robinia
Grandinia subalutacea (P. Karst.) Jülich
Corticium subalutaceum P. Karst.
Hyphodontia subalutacea (P. Karst.) J. Eriksson
Peniophora subalutacea (P. Karst.) Höhn. & Litsch.
Distr.: Widespread.
Notes: On wood.
Hosts: Acer, Picea, Pinus, Quercus

Graphiola Poit. Basidiomycetes, Ustilaginales

Notes: Formerly placed in the Hyphomycetes, recently shown to belong in the Ustilaginales. Oberwinkler et al. (Pl. Syst. Evol. 140:251. 1982); Cole (Mycologia 75:93. 1983).
Graphiola congesta Berk. & Ravenel
Distr.: Southern United States.
Host: Sabal
Graphiola phoenicis (Moug.) Poit.
Distr.: Subtropical and tropical regions; elsewhere in greenhouses.
Notes: On Palmaceae.
Hosts: Acoelorrhaphe, Arecaceae, Arenga, Butia, Chrysalidocarpus, Cocos, Phoenix, Roystonea, Sabal, Syagrus, Washingtonia
Graphiola thaxteri E. Fisch.
Distr.: FL.
Host: Sabal

Graphium Corda. Hyphomycetes

Graphium album (Corda) Sacc.
Ceratopodium album Corda
Distr.: AR; Europe(type).
Host: Fagus
Graphium dubautiae F. Stevens & Weedon
Distr.: HI(type).
Host: Dubautia
Graphium gracile Peck
Distr.: Central and eastern United States.
Host: Rubus
Graphium hamamelidis Van Hook
Stilbum hamamelidis (Van Hook) Overh.
Distr.: Eastern United States.
Notes: According to Seifert (Stud. Mycol. 27:198. 1985) the only available material contains *Gonatobotryum apiculatum* which does not agree with the diagnosis.
Host: Hamamelis
Graphium penicillioides Corda
Distr.: LA; Europe.
Notes: Also reported on wood of *Populus* (Ellis. Dematiaceous Hyphomycetes. p. 333. 1971).
Host: Zea
Graphium rigidum (Pers.) Sacc.
Stilbum rigidum Pers.
Distr.: United States.
Notes: According to Seifert (Stud. Mycol. 27:206. 1985) this name has been used for a variety of black synnematous fungi and the type specimen is inadequate to characterize the name.
Hosts: Liquidambar, Pinus, Quercus
Graphium rubrum Rumbold
Distr.: Central and southeastern United States.
Notes: Lumber stain (Rumbold. Phytopathology 24:300. 1934).
Hosts: Liquidambar, Liriodendron, Pinus, Populus, Quercus
Graphium sessile Dearn. & House, nom. inval.
Distr.: NY.
Notes: Published without a latin diagnosis.
Host: Sanguisorba
Graphium sorbi Peck
Distr.: NY, WI.
Host: Sorbus

Graphostroma Pirozynski. Pyrenomycetes, Calosphaeriales

Graphostroma platystoma (Schwein.) Pirozynski
Diatrype platystoma (Schwein.) Berk.
Diatrype afflata (Schwein.) Cooke
Diatrype discostoma Cooke
Eutypa micropuncta Cooke
Diatrype tenuissima Cooke
Nodulisporium sp. (anamorph)
Distr.: Central and eastern North America.
Notes: On limbs and bark of hardwoods (Pirozynski. Canad. J. Bot. 52:2131. 1974).
Hosts: Acer, Carpinus, Ostrya, Persea, Quercus, Rhododendron

Graphyllium Clements. Loculoascomycetes, Dothideales

Notes: Barr (Mycotaxon 29:504. 1987).
Graphyllium chloes Clements

Pleospora chloes (Clements) Petr.
Distr.: NE(type), ND.
Hosts: Aristida, Bouteloua
Graphyllium dakotense Rehm
Clathrospora dakotensis (Rehm) Höhn.
Pleospora dakotensis (Rehm) Petr.
Distr.: ND(type).
Hosts: Artemisia, Phragmites, Schizachyrium
Graphyllium hysterioides (Ellis & Everh.) Barr
Pleospora hysterioides Ellis & Everh.
Hysterographium hysterioides (Ellis & Everh.) Wehmeyer
Graphyllium graminis (Ellis & Everh.) Rehm
Distr.: GA, KS, ND.
Notes: Zogg (Beitr. Kryptogamenfl. Schweiz 11(3):154. 1962).
Hosts: Andropogon, Aristida, Panicum, Phragmites, Schizachyrium, Sporobolus
Graphyllium ipomoeae (Clements) Barr
Comoclathris ipomoeae Clements
Pleospora ipomoeae (Clements) Wehmeyer
Pyrenophora salsolae D. Griffiths
Distr.: CO, ND; Europe.
Host: Salsola
Graphyllium planispora (Ellis) Barr
Pleospora planispora Ellis
Comoclathris planispora (Ellis) J. Harr
Platyspora planispora (Ellis) Wehmeyer
Distr.: Western United States; Europe.
Hosts: Elymus, Penstemon

Greeneria Lams.-Scrib. & Viala. Coelomycetes

Greeneria uvicola (Berk. & M.A. Curtis) Punithalingam
Phoma uvicola Berk. & M.A. Curtis
Phyllosticta frankiana Sacc. & Syd.
Greeneria fuligineum Lams.-Scrib. & Viala
Melanconium fuligineum Lams.-Scrib. & Viala
Distr.: Temperate, subtropical and tropical regions.
Notes: Bitter rot of grapes, necrotic fleck of grape vines; also on a variety of fruit crops (Sutton & Gibson. C.M.I. Descr. 538. 1970).
Hosts: Malus, Vitis

Grifola S.F. Gray. Basidiomycetes, Aphyllophorales

Grifola frondosa (Dickson:Fr.) S.F. Gray
Polyporus frondosus (Dickson:Fr.) Fr.
Distr.: Temperate northern hemisphere.
Notes: On roots at base of living hardwoods, especially *Quercus* and *Castanea* spp., rarely on conifers. Butt rot of heartwood; sometimes at base of dead trees and stumps. A good edible fungus.
Hosts: Acer, Castanea, Pinus, Prunus, Pseudotsuga, Quercus, Ulmus

Griphosphaeria Höhn. Pyrenomycetes, Amphisphaeriales

Notes: This genus is considered a synonym of *Discostroma*, but not all names have been redisposed.
Griphosphaeria stevensonii (Petr.) E. Müller & Arx
Griphosphaerella stevensonii Petr.
Distr.: FL(type).
Host: Panicum

Griphosphaerioma Höhn. Pyrenomycetes, Amphisphaeriales

Griphosphaerioma kansensis (Ellis & Everh.) Shoemaker
Cryptospora kansensis Ellis & Everh.
Calospora kansensis (Ellis & Everh.) Berl.
Clethridium symphoricarpi (Rehm ex Höhn.) Petr.
Curreyella symphoricarpi (Rehm ex Höhn.) Petr.
Labridella cornu-cervae Brenckle (anamorph)
Distr.: Northcentral North America.
Notes: On hardwood twigs (Shoemaker. Canad. J. Bot. 41:1419. 1968).
Host: Symphoricarpos

Grovesiella Morelet. Discomycetes, Helotiales

Grovesiella abieticola (Zeller & Goodd.) Morelet & Gremmen
Scleroderris abieticola Zeller & Goodd.
Godronia abieticola (Zeller & Goodd.) Seaver
Distr.: Northern North America; Europe.
Notes: Gremmen & Morelet (Eur. J. Forest Pathol. 1:80. 1971).
Hosts: Abies, Pinus

Grovesinia M. Cline, J.L. Crane, & S. Cline. Discomycetes, Helotiales

Notes: Cline et al. (Mycologia 75:988. 1983).
Grovesinia pyramidalis M. Cline, J.L. Crane, & S. Cline
Sclerotium cinnamomi Sawada (synanamorph)
Cristulariella moricola (Hino) Redhead (synanamorph)
Distr.: Known only to form in culture, but the anamorphs occur throughout the northern hemisphere.
Host: Juglans

Guepiniopsis Pat. Basidiomycetes, Dacrymycetales

Heterotextus C.G. Lloyd
Guepiniopsis alpina (Tracy & Earle) Brasfield
Heterotextus alpinus (Tracy & Earle) G.W. Martin
Distr.: Widespread.
Notes: On wood of conifers.
Hosts: Abies, Picea, Pinus, Tsuga
Guepiniopsis buccina (Pers.:Fr.) L. Kennedy
Guepiniopsis torta (Pers.:Fr.) Pat.
Distr.: Western United States.
Notes: On wood.
Hosts: Abies, Acer, Artemisia, Pinus, Pseudotsuga
Guepiniopsis chrysocomus (Bull.:Fr.) Brasfield
Peziza chrysocoma Bull.:Fr.
Distr.: GA, WA.
Notes: This name may be misapplied to the species reported from North America; see Donk (Proc. Koninkl. Nederl. Akad. Wetens., Ser. C, 67:1. 1964).
Hosts: Pinus, Tsuga

Guignardia Viala & Ravaz. Loculoascomycetes, Dothideales

Notes: According to Bissett (Mycotaxon 25:519. 1986) *Discochora* has precedence over *Guignardia*. *Guignardia* is also sometimes considered a synonym of *Botryosphaeria*.
Guignardia aesculi (Peck) V.B. Stewart
Laestadia aesculi Peck
Botryosphaeria aesculi (Peck) Barr
Leptodothiorella aesculicola (Sacc.) Sivanesan (synanamorph macroconidial)
Phyllosticta sphaeropsoidea Ellis & Everh. (synanamorph microconidial)
Distr.: Eastern North America; Europe.
Notes: Buckeye blotch (Sivanesan. The bitunicate Ascomycetes. p. 165. 1984).
Host: Aesculus
Guignardia ailanthi (Grove) Sacc.
Laestadia ailanthi Grove
Distr.: SC(type).
Host: Ailanthus
Guignardia alyxiae F. Stevens
Distr.: HI.
Host: Alyxia
Guignardia bambusae Miyake & K. Hara
Distr.: MD; Asia(type).
Notes: The validity of this name is in question.
Hosts: Bambusa, Poaceae
Guignardia bidwellii (Ellis) Viala & Ravaz

Botryosphaeria bidwellii (Ellis) Petr.
Physalospora bidwellii (Ellis) Sacc.
Phyllosticta ampelicida (Engelm.) van der Aa (anamorph pycnidial)
Leptodothiorella sp. (anamorph spermatial)
Distr.: Temperate North America; Europe.
Notes: Sivanesan & Holliday (C.M.I. Descr. 710. 1981).
Hosts: Ampelopsis, Parthenocissus, Vitis
Guignardia bidwellii (Ellis) Viala & Ravaz f. ***muscadinii*** Luttrell
Distr.: Central and eastern United States.
Notes: Luttrell (Phytopathology 36:905. 1946) differentiated this form based primarily on the host.
Host: Vitis
Guignardia bidwellii (Ellis) Viala & Ravaz f. ***parthenocissi*** Luttrell
Distr.: GA.
Notes: Luttrell (Phytopathology 38:716. 1948) differentiated this form primarily on the host.
Host: Parthenocissus
Guignardia bumeliae J.H. Miller & G.E. Thompson
Distr.: GA(type).
Host: Sideroxylon
Guignardia camelliae (Cooke) E.J. Butler
Distr.: CA; Sri Lanka.
Host: Camellia
Guignardia caryophyllea (Cooke & Harkn.) Wehmeyer
Sphaerella caryophyllea Cooke & Harkn.
Distr.: Western United States.
Hosts: Artemisia, Penstemon
Guignardia citricarpa Kiely
Phyllosticta citricarpa (McAlpine) van der Aa (synanamorph)
Leptodothiorella sp. (synanamorph)
Distr.: Worldwide in citrus-growing regions.
Notes: Citrus black spot (Sutton & Waterston. C.M.I. Descr. 85. 1966).
Host: Mangifera
Guignardia galactina (Dearn. & House) Sacc.
Laestadia galactina Dearn. & House
Distr.: NC.
Host: Galax
Guignardia haydenii (Berk. & M.A. Curtis) Arx & E. Müller
Dothidea haydenii Berk. & M.A. Curtis
Ophiodothis haydenii (Berk. & M.A. Curtis) Sacc.
Ascospora pseudohimantia Rehm, nom. nud.
Discosphaerina pseudohimantia (Rehm) Petr., nom. inval.
Placosphaeria haydenii (Berk. & M.A. Curtis) Petr. (anamorph)
Distr.: North America.
Notes: On stems of Asteraceae (Arx & Müller. Beitr. Kryptogamenfl. Schweiz 11(1):53. 1954).
Host: Aster
Guignardia jussiaeae F. Stevens
Distr.: HI(type).
Host: Ludwigia
Guignardia musae Racib.
Guignardia musae F. Stevens
Phyllosticta musarum (Cooke) van der Aa (anamorph)
Distr.: HI; widespread in southeastern Asia; East Indies; South Pacific; eastern Africa.
Notes: Punithalingam & Holliday (C.M.I. Descr. 467. 1975).
Host: Musa
Guignardia polygonati (Schwein.) Lindau
Distr.: PA(type).
Host: Polygonatum
Guignardia reticulata (DC.:Fr.) van der Aa
Dothidea reticulata DC.:Fr.
Distr.: North America; Europe.
Notes: Leaf spot of Liliaceae (van der Aa. Stud. Mycol. 5:23. 1973). Although reported on various hosts from North America, these records appear to be misdeterminations (Bissett. Canad. J. Bot. 57:2085. 1979). The anamorph name was conserved against *P. cruenta*.
Host: [Lomatium]
Guignardia scabiosae (Lambotte & Fautrey) Arx & E. Müller
Laestadia scabiosae Lambotte & Fautrey
Distr.: North America; Europe.
Host: Aster
Guignardia spinicola (Ellis & Everh.) Lindau
Distr.: PA(type).
Host: Rosa

Gymnoconia Lagerh. Basidiomycetes, Uredinales

Gymnoconia nitens (Schwein.) F. Kern & H.W. Thurston
Kunkelia nitens (Schwein.) Arth.
Distr.: Temperate regions.
Notes: Autoecious. Short-cycled orange rust (Laundon & Rainbow. C.M.I. Descr. 201. 1969).
Host: Rubus

Gymnopilus P. Karst. Basidiomycetes, Agaricales

Notes: Hesler (Mycol. Mem. 3:1. 1969).
Gymnopilus amarissimus Murrill
Distr.: FL.
Host: Pinus
Gymnopilus armillatus Murrill
Distr.: FL.
Host: Liquidambar
Gymnopilus bellulus (Peck) Murrill
Distr.: Temperate North America.
Notes: On conifer logs and stumps.
Host: Abies
Gymnopilus croceoluteus Hesler
Distr.: Temperate North America.
Notes: On conifers.
Host: Pinus
Gymnopilus decurrens Hesler
Distr.: CA.
Host: Pinus
Gymnopilus fuscosquamulosus Hesler
Distr.: NC.
Hosts: Aesculus, Rhododendron
Gymnopilus liquiritiae (Pers.:Fr.) P. Karst.
Distr.: Circumboreal.
Notes: Usually on conifer wood, occasionally on hardwoods.
Host: Pinus
Gymnopilus luteoviridis Thiers
Distr.: TX.
Host: Quercus
Gymnopilus melleus Hesler
Distr.: AL.
Host: Pinus
Gymnopilus naucorioides Hesler
Distr.: TN.
Host: Tsuga
Gymnopilus oregonensis Murrill
Distr.: OR.
Host: Abies
Gymnopilus pacificus Hesler
Distr.: HI.
Host: Metrosideros
Gymnopilus penetrans (Fr.:Fr.) Murrill
Flammula penetrans (Fr.:Fr.) Quél.
Distr.: North America; Jamaica; Europe.
Notes: On conifers and hardwoods.
Hosts: Abies, Pinus
Gymnopilus praecox (Peck) Murrill
Flammula praecox Peck
Distr.: IN(type).
Notes: Probably a *Pholiota* sp. (Hesler. Mycol. Mem. 3:87. 1969).
Host: Tilia
Gymnopilus praefloccosus Murrill
Distr.: FL.
Host: Liquidambar
Gymnopilus pulchrifolius (Peck) Murrill
Distr.: Eastern United States.
Host: Tsuga
Gymnopilus rufosquamulosus Hesler
Distr.: TX.
Host: Quercus

Gymnopilus sapineus (Fr.:Fr.) Maire
Flammula sapinea (Fr.:Fr.) P. Kumm.
Distr.: Temperate North America; Europe.
Notes: On conifers and hardwoods.
Hosts: Abies, Pinus, Quercus
Gymnopilus spectabilis (Fr.:Fr.) A.H. Sm.
Pholiota spectabilis (Fr.:Fr.) Gill.
Distr.: Widespread in the northern hemisphere.
Notes: On wood of conifers and hardwoods.
Hosts: Populus, Pseudotsuga, Salix
Gymnopilus subdryophilus Murrill
Distr.: FL.
Host: Quercus
Gymnopilus subtropicus Hesler
Distr.: HI, southeastern United States.
Host: Quercus
Gymnopilus underwoodii (Peck) Murrill
Distr.: Southeastern United States.
Notes: On wood.
Host: Pinus

Gymnosporangium R. Hedw. ex DC. in Lam. & DC. Basidiomycetes, Uredinales

Notes: Kern (A revised taxonomic account of the genus *Gymnosporangium*. 1973); Ziller (The tree rusts of western Canada. p. 101. 1974).
Gymnosporangium asiaticum Miyabe ex G. Yamada
Gymnosporangium haraeanum Syd. & P. Syd.
Distr.: United States(introduced to the east and west coasts); eastern Asia.
Notes: Japanese pear rust. Heteroecious; aecial host - Rosaceae; telial host - *Juniperus* spp.
Hosts: Juniperus, Pyrus
Gymnosporangium bermudianum Earle in Seym. & Earle
Distr.: Southeastern United States; Bahamas; Bermuda.
Notes: Autoecious.
Host: Juniperus
Gymnosporangium bethelii F. Kern
Gymnosporangium connersii Parmelee
Gymnosporangium tubulatum F. Kern
Distr.: Northern and western North America.
Notes: Heteroecious; aecial host - *Crataegus* spp.; telial host - *Juniperus* spp. Parmelee (Fungi Canadenses 28. 1974 as *G. connersii*; Fungi Canadenses 134. 1978; Canad. J. Bot. 43:245. 1965).
Hosts: Crataegus, Juniperus
Gymnosporangium biseptatum Ellis
Distr.: East coast of the United States.
Notes: Heteroecious; aecial host - *Amelanchier* spp.; telial host - *Chamaecyparis thyoides*.
Hosts: Amelanchier, Calocedrus, Chamaecyparis
Gymnosporangium clavariiforme (Wulfen in Jacq.:Pers.) DC.
Distr.: Temperate regions.
Notes: Hawthorn rust, claviform juniper rust. Heteroecious; aecial host - Rosaceae, especially *Crataegus* spp.; telial host - *Juniperus* spp. Laundon (C.M.I. Descr. 542. 1977); Parmelee (Fungi Canadenses 115. 1978).
Hosts: Amelanchier, Aronia, Crataegus, Cydonia, Juniperus, Pyrus
Gymnosporangium clavipes (Cooke & Peck) Cooke & Peck in Peck
Gymnosporangium germinale F. Kern
Distr.: North America.
Notes: Quince rust. Heteroecious; aecial host - Rosaceae; telial host - *Juniperus* spp. Laundon (C.M.I. Descr. 543. 1977); Parmelee (Fungi Canadenses 116. 1978).
Hosts: Amelanchier, Aronia, Chaenomeles, Crataegus, Cydonia, Juniperus, Malus, Mespilus, Photinia, Pyrus
Gymnosporangium confusum Plowr.
Distr.: CA(introduced); northern Africa; Asia; Europe.
Notes: Medlar rust. Heteroecious; aecial host - Rosaceae; telial host - *Juniperus* spp. Laundon (C.M.I. Descr. 544. 1977).
Hosts: Crataegus, Juniperus
Gymnosporangium cornutum Arth. ex F. Kern
Roestelia cornuta Fr.
Gymnosporangium aurantiacum Chev., sensu Arth.
Gymnosporangium juniperinum (L.) Fr.
Distr.: Temperate northern hemisphere.
Notes: Heteroecious; aecial host - *Sorbus* and *Malus* spp.; telial host - *Juniperus* spp. Parmelee (Fungi Canadenses 117. 1978).
Hosts: Juniperus, Sorbus
Gymnosporangium cunninghamianum Barclay
Gymnosporangium cupressi W.H. Long & Goodd. in W.H. Long
Gymnosporangium cupressi W.H. Long & Goodd. in W.H. Long var. *cascadense* R.S. Peterson
Distr.: Southwestern North America; central and eastern Asia.
Notes: Heteroecious; aecial host - *Amelanchier* spp. in North America, *Pyrus* spp. in Asia; telial host - *Cupressus* spp.
Hosts: Amelanchier, Cupressus
Gymnosporangium davisii F. Kern
Distr.: Northeastern North America.
Notes: Heteroecious; aecial host - *Aronia* spp.; telial host - *Juniperus* spp.
Hosts: Aronia, Juniperus
Gymnosporangium effusum F. Kern
Distr.: East coast of the United States.
Notes: Aecial host unknown; telial host - *Juniperus* spp.
Host: Juniperus
Gymnosporangium ellisii (Berk.) Ellis
Distr.: East coast of the United States.
Notes: Heteroecious; aecial host - Myricaceae; telial host - *Chamaecyparis thyoides*.
Hosts: Chamaecyparis, Comptonia, Myrica
Gymnosporangium exiguum F. Kern
Distr.: Southwestern United States.
Notes: Heteroecious; aecial host - *Crataegus* and *Heteromeles* spp.; telial host - *Juniperus* spp.
Hosts: Crataegus, Heteromeles, Juniperus
Gymnosporangium exterum Arth. & F. Kern in Arth.
Distr.: Southcentral and southeastern United States.
Notes: Heteroecious; aecial host - *Gillenia* spp.; telial host - *Juniperus* spp.
Hosts: Gillenia, Juniperus
Gymnosporangium floriforme Thaxt. in F. Kern
Distr.: Southeastern United States.
Notes: Heteroecious; aecial host - *Crataegus* spp.; telial host - *Juniperus* spp.
Hosts: Crataegus, Juniperus
Gymnosporangium fraternum F. Kern
Gymnosporangium transformans (Ellis) F. Kern
Roestelia transformans Ellis in Peck
Distr.: East coast of the United States.
Notes: Heteroecious; aecial host - *Aronia* spp.; telial host - *Chamaecyparis thyoides*.
Hosts: Aronia, Chamaecyparis
Gymnosporangium fuscum R. Hedw. in DC.
Gymnosporangium sabinae (Dickson) G. Wint.
Distr.: CA; Canada (British Columbia, introduced); northern Africa; Asia; Europe.
Notes: European pear rust, trellis rust. Heteroecious; aecial host - *Pyrus* spp.; telial host - *Juniperus* spp. Laundon (Mycotaxon 3:140. 1975) provided nomenclatural comments. Laundon (C.M.I. Descr. 545. 1977); Parmelee (Fungi Canadenses 43. 1974).
Host: Pyrus
Gymnosporangium globosum (Farl.) Farl.
Distr.: AK, eastern North America.
Notes: American hawthorn rust. Heteroecious; aecial host - Rosacae; telial host - *Juniperus* spp. Laundon (C.M.I. Descr. 546. 1977); Parmelee (Fungi Canadenses 135. 1979).
Hosts: Amelanchier, Crataegus, Juniperus, Malus, Pyrus, Sorbus
Gymnosporangium gracile Pat.
Distr.: TX; northern Mexico; northern Africa; southern Europe.
Notes: Heteroecious; aecial host - Rosaceae; telial host - *Juniperus* spp.
Host: Juniperus
Gymnosporangium harknessianum F. Kern ex Arth.
Distr.: Western United States.
Notes: Heteroecious; aecial host - *Amelanchier* spp. telial host - *Juniperus* spp.
Hosts: Amelanchier, Juniperus
Gymnosporangium hyalinum F. Kern ex Cummins

Roestelia hyalina Cooke (anamorph)
Distr.: Southeastern United States.
Notes: Heteroecious; aecial host - *Crataegus* spp.: telial host - *Chamaecyparis thyoides*.
Hosts: Chamaecyparis, Crataegus, [Pyrus]
Gymnosporangium inconspicuum F. Kern
Distr.: Western North America.
Notes: Heteroecious; aecial host - *Amelanchier* and *Peraphyllum* spp.; telial host - *Juniperus* spp. Parmelee (Fungi Canadenses 136. 1979).
Hosts: Amelanchier, Juniperus, Peraphyllum
Gymnosporangium japonicum P. Syd.
Distr.: United States(coastal); eastern Asia.
Notes: Heteroecious; aecial host - *Photinia* spp.; telial host - *Juniperus* spp.
Hosts: Heteromeles, Juniperus
Gymnosporangium juniperi-virginianae Schwein.
Gymnosporangium macropus Link
Distr.: CA, central and eastern North America.
Notes: Cedar-apple rust. Heteroecious; aecial host - *Malus* spp. telial host - *Juniperus* spp. Laundon (C.M.I. Descr. 547. 1977); Parmelee (Fungi Canadenses 137. 1979).
Hosts: [Cedrus], [Crataegus], Juniperus, Malus
Gymnosporangium kernianum Bethel
Distr.: Western United States.
Notes: Heteroecious; aecial host - Rosaceae; telial host - *Juniperus* spp.
Hosts: Amelanchier, Juniperus
Gymnosporangium libocedri (Henn.) F. Kern
Gymnosporangium blasdaleanum F. Kern
Distr.: Western United States.
Notes: Heteroecious; aecial host - Rosaceae; telial host - *Calocedrus decurrens*. Laundon (C.M.I. Descr. 548. 1977).
Hosts: Amelanchier, Calocedrus, Chaenomeles, Crataegus, Cydonia, [Juglans], Malus, Pyrus, Sorbus
Gymnosporangium multiporum F. Kern
Distr.: Southwestern United States.
Notes: Aecia unknown, but see Kern & Keener (Science 131:298. 1960). Telial host - *Juniperus* spp.
Host: Juniperus
Gymnosporangium nelsonii Arth.
Gymnosporangium corniculans F. Kern in Arth.
Distr.: Temperate North America.
Notes: Heteroecious; aecial host - Rosaceae; telial host - *Juniperus* spp. Savile (Fungi Canadenses 138. 1979).
Hosts: Amelanchier, Crataegus, Cydonia, Juniperus, Malus, Peraphyllum, Pyrus, Sorbus
Gymnosporangium nidus-avis Thaxt.
Gymnosporangium juvenescens F. Kern
Distr.: Temperate North America.
Notes: Heteroecious; aecial host - Rosaceae; telial host - *Juniperus* spp. Parmelee (Fungi Canadenses 139. 1979).
Hosts: Amelanchier, Cydonia, Juniperus, Malus
Gymnosporangium nootkatense Arth.
Distr.: Northwestern North America.
Notes: Heteroecious; aecial host - Rosaceae; telial host - *Chamaecyparis nootkatensis*.
Hosts: Amelanchier, Chamaecyparis, Malus, Sorbus
Gymnosporangium speciosum Peck
Distr.: Southwestern United States.
Notes: Heteroecious; aecial host - *Fendlera* and *Philadelphus* spp.; telial host - *Juniperus* spp.
Hosts: Fendlera, Juniperus, Philadelphus
Gymnosporangium trachysorum F. Kern ex Arth.
Distr.: Central and eastern United States.
Notes: Heteroecious; aecial host - *Crataegus* spp.; telial host - *Juniperus* spp.
Hosts: Crataegus, Juniperus
Gymnosporangium tremelloides R. Hartig
Distr.: North America; western Africa; eastern Asia; Europe.
Notes: Heteroecious; aecial host - Rosaceae; telial host - *Juniperus* spp. Laundon (C.M.I. Descr. 549. 1977); Parmelee (Fungi Canadenses 119. 1978).
Hosts: Juniperus, [Philadelphus], Sorbus
Gymnosporangium vauqueliniae W.H. Long & Goodd.
Distr.: AZ.
Notes: Heteroecious; aecial host - *Vauquelinia californica*; telial host - *Juniperus monosperma*.
Hosts: Juniperus, Vauquelinia

Gyromitra Fr., nom. cons. Discomycetes, Pezizales

Notes: All species occur on soil, but some have been reported in association with vascular plants.
Gyromitra infula (Schaeff.) Quél.
Distr.: Temperate regions.
Notes: On soil in conifer woods (Ginns. Fungi Canadenses 52. 1974).
Host: Pinus

Gyrostroma Naumov. Coelomycetes

Gyrostroma austroamericana Seeler
Kaskaskia gleditsiae G.L. Born & J.L. Crane
Thyronectria austroamericana (Speg.) Seeler (teleomorph)
Distr.: North America; Brazil, Paraguay.
Notes: Canker of honey locust (Bedker & Wingfield. Trans. Brit. Mycol. Soc. 81:179. 1983).
Host: Gleditsia
Gyrostroma missouriense Seeler
Nectria missouriensis Ellis & Everh. (anamorph)
Distr.: Eastern United States.
Hosts: Acer, Carya

Gyrothrix (Corda) Corda. Hyphomycetes

Gyrothrix circinata (Berk. & M.A. Curtis) S.J. Hughes
Campsotrichum circinatum Berk. & M.A. Curtis
Distr.: Widespread in tropical regions.
Notes: On fallen leaves of various plants (Ellis. Dematiaceous Hyphomycetes. p. 360. 1971).
Host: Eucalyptus
Gyrothrix microsperma (Höhn.) Pirozynski
Circinotrichum microspermum Höhn.
Distr.: FL; Java(type).
Notes: On dry leaves (Ellis. Dematiaceous Hyphomycetes. p. 360. 1971).
Host: Serenoa
Gyrothrix pediculata J.L. Cunningham
Distr.: MD.
Notes: On fallen leaves (Cunningham. Mycologia 66:122. 1974).
Host: Robinia

Hadrotrichum Fuckel. Hyphomycetes

Hadrotrichum blasdalei Sacc. in Ellis & Everh.
Distr.: CA(type).
Host: Vicia
Hadrotrichum chenopodinum Peck
Distr.: NY(type).
Host: Chenopodium
Hadrotrichum globiferum (Ellis & Everh.) J.J. Davis
Ovularia globifera Ellis & Everh.
Distr.: Temperate North America.
Notes: Leaf blight.
Host: Lupinus
Hadrotrichum heteromelis Ellis & Everh.
Distr.: CA(type).
Host: Heteromeles
Hadrotrichum lineare Peck
Distr.: AK, NY.
Host: Calamagrostis
Hadrotrichum triseti R. Sprague
Distr.: ID.
Notes: Sprague (Mycologia 54:44. 1962).
Host: Trisetum

Haematomyxa Sacc. Loculoascomycetes, Patellariales

Haematomyxa sequoiae (Plowr.) Bonar
Distr.: CA.
Host: Sequoiadendron
Haematomyxa tetraspora Cash
Distr.: CA(type).
Host: Physocarpus
Haematomyxa vinosa (Cooke & Ellis) Sacc.
Haematomyxa rufa Rehm
Distr.: CA, NJ.
Host: Crataegus

Haglundia Nannf. Discomycetes, Helotiales

Haglundia perelegans Nannf.
Distr.: CA; Europe.
Host: Arctostaphylos

Hainesia Ellis & Sacc. Discomycetes, Helotiales

Hainesia borealis Ellis & Everh.
Distr.: Western North America.
Host: Galium
Hainesia lythri (Desmaz.) Höhn.
Sphaeronaema corneum Cooke & Ellis
Hainesia rhoina (Sacc.) Sacc. & Ellis
Tubercularia rhois Halst.
Discohainesia oenotherae (Cooke & Ellis) Nannf. (teleomorph)
Distr.: Cosmopolitan.
Notes: Sutton & Gibson (C.M.I. Descr. 535. 1977).
Hosts: Acer, Cornus, Fragaria, Geum, Hamamelis, Ludwigia, Lysimachia, Oenothera, Penthorum, Potentilla, Rhus, Rosa, Rubus, Vitis

Halbaniella Theiss. Loculoascomycetes, Dothideales

Notes: According to Arx & Müller (Stud. Mycol. 9:126. 1975), this genus is a synonym of *Actinopeltis*, but not all names have been redisposed.
Halbaniella linnaeae Dearn. in Bisby, Buller, & Dearn.
Distr.: NY; Canada(Manitoba).
Host: Linnaea

Haligena Kohlmeyer. Pyrenomycetes, Halosphaeriales

Haligena spartinae E.B.G. Jones
Distr.: Coastal regions of the Atlantic Ocean.
Host: Spartina
Haligena viscidula Kohlmeyer & E. Kohlmeyer
Distr.: Widespread in marine habitats.
Notes: Saprobic on woody substrates (Kohlmeyer & Kohlmeyer. Marine mycology. p. 284. 1979).
Host: Pinus

Halocyphina Kohlmeyer & E. Kohlmeyer. Basidiomycetes, Aphyllophorales

Notes: Monotypic.
Halocyphina villosa Kohlmeyer & E. Kohlmeyer
Distr.: FL; Liberia.
Notes: Marine. Ginns & Malloch (Mycologia 69:53. 1977).
Host: Rhizophora

Halosphaeria Linder in Barghoorn & Linder. Pyrenomycetes, Sphaeriales

Notes: Kohlmeyer & Kohlmeyer (Marine mycology. p. 286. 1979).
Halosphaeria cucullata (Kohlmeyer) Kohlmeyer
Remispora cucullata Kohlmeyer
Distr.: Worldwide in marine and inland saline habitats.
Notes: On woody substrates.
Hosts: Pinus, Pluchea
Halosphaeria hamata (W. Höhnk) Kohlmeyer
Remispora hamata W. Höhnk
Distr.: Widespread in marine habitats.
Hosts: Juncus, Spartina
Halosphaeria mediosetigera A.B. Cribb & J.W. Cribb
Halosphaeriopsis mediosetigera (A.B. Cribb & J.W. Cribb) T.W. Johnson
Halosphaeria mediosetigera A.B. Cribb & J.W. Cribb var. *grandispora* Kohlmeyer
Trichocladium achrasporum (S.P. Meyers & R.T. Moore) M. Dixon (anamorph)
Distr.: Worldwide in oceans and inland salt lakes.
Notes: On driftwood, bark, test panels and other organic debris.
Host: Pinus
Halosphaeria quadricornuta A.B. Cribb & J.W. Cribb
Distr.: Widespread in coastal regions of the tropics.
Host: Hibiscus
Halosphaeria salina (S.P. Meyers) Kohlmeyer
Arenariomyces salina S.P. Meyers
Remispora salina (S.P. Meyers) Kohlmeyer
Distr.: In oceans throughout the world.
Notes: On intertidal and drifting wood, test panels and woody debris.
Host: Cocos

Hansenula Syd. & P. Syd. Hemiascomycetes, Endomycetales

Hansenula anomala (E. Hans.) Syd. & P. Syd.
Distr.: Widespread.
Notes: Isolated from soil and various kinds of organic debris.
Hosts: Triticum, Zea
Hansenula canadensis Wickerham
Distr.: Eastern North America.
Notes: Isolated from frass associated with pine trees.
Host: Pinus
Hansenula capsulata Wickerham
Distr.: Temperate regions.
Notes: Isolated from frass associated with coniferous trees.
Hosts: Picea, Pinus
Hansenula polymorpha J. Morais & Maia
Hansenula angusta Teunisson, H.H. Hall & Wickerham
Distr.: Widespread.
Notes: Isolated from various kinds of organic debris.
Hosts: Liquidambar, Nyssa, Populus, Quercus

Hapalocystis Auersw. ex Fuckel. Pyrenomycetes, Diaporthales

Notes: Barr (Mycol. Mem. 7:191. 1978).
Hapalocystis berkeleyi Auersw. ex Fuckel
Pseudovalsa hapalocystis (Berk. & Broome) Sacc.
Distr.: Temperate North America; Europe.
Notes: On limbs.
Host: Platanus
Hapalocystis corni (Wehmeyer) Barr
Melanconis corni Wehmeyer
Melanconiella corni (Wehmeyer) Petr.
Distr.: GA.
Host: Cornus
Hapalocystis corticale (Schwein.) Barr
Prosthecium corticale (Schwein.) Wehmeyer
Distr.: Eastern United States.
Host: Platanus
Hapalocystis ulmi (Wehmeyer) Barr
Pseudovalsa ulmi Wehmeyer
Prosthecium ulmi (Wehmeyer) Wehmeyer
Distr.: Central and eastern North America.
Notes: On limbs.
Host: Ulmus

Hapalopilus P. Karst. Basidiomycetes, Aphyllophorales

Notes: Gilbertson & Ryvarden (North American Polypores 1:334. 1986).

Hapalopilus croceus (Pers.:Fr.) Donk
Polyporus croceus (Pers.:Fr.) Fr.
Tyromyces croceus (Pers.:Fr.) J. Lowe
Polyporus castanophilus Atk.
Polyporus pilotae Schwein.
Distr.: Northern hemisphere.
Notes: On hardwoods.
Hosts: Castanea, Quercus

Hapalopilus mutans (Peck) R.L. Gilbertson & Ryvarden
Polyporus mutans Peck
Poria mutans (Peck) Peck
Distr.: Eastern North America.
Notes: On hardwoods.
Hosts: Castanea, Prunus

Hapalopilus rutilans (Pers.:Fr.) Murrill
Polyporus rutilans Pers.:Fr.
Hapalopilus nidulans (Fr.:Fr.) P. Karst.
Polyporus nidulans Fr.:Fr.
Polyporus pallido-cervinus Schwein.
Distr.: Temperate northern hemisphere.
Notes: On wood.
Hosts: Abies, Acer, Alnus, Betula, Carya, Castanea, Fagus, Juglans, Morus, Picea, Pinus, Platanus, Populus, Prunus, Pseudotsuga, Pyrus, Quercus, Rosa, Salix, Sorbus, Tsuga, Vitis

Hapalosphaeria Syd. in Died. & Syd. Coelomycetes

Hapalosphaeria deformans (Syd.) Syd.
Paepalopsis deformans Syd.
Distr.: Western United States; Germany.
Notes: On anthers (Sutton. The Coelomycetes. p. 377. 1980).
Host: Rubus

Haplobasidion Eriks. Hyphomycetes

Notes: Ellis (Dematiaceous Hyphomycetes. p. 353. 1971).

Haplobasidion lelebae Sawada ex M.B. Ellis
Distr.: FL; Taiwan(type).
Host: Juncus

Haplobasidion thalictri Eriks.
Haplobasidion pavoninum Höhn.
Distr.: Northern North America; Europe; India.
Notes: On leaves.
Hosts: Aquilegia, Thalictrum

Haplographium Berk. & Broome. Hyphomycetes

Haplographium atrobrunneum (Cooke) Sacc.
Penicillium atrobrunneum Cooke
Distr.: FL.
Host: Musa

Haplotrichum Link:Fr. Hyphomycetes

Acladium (Link) Pers.

Haplotrichum conspersum (Link:Fr.) Holubová-Jechová
Oidium conspersum (Link:Fr.) Linder
Botryobasidium conspersum J. Eriksson (teleomorph)
Distr.: MA; Europe.
Hosts: Betula, Fagus, Malus, Quercus

Haplotrichum curtisii (Berk.) Holubová-Jechová
Acladium curtisii (Berk.) M.B. Ellis
Oidium curtisii (Berk.) Linder
Botryobasidium vagum (Berk. & M.A. Curtis) D.P. Rogers (teleomorph)
Distr.: Widespread.
Hosts: Ostrya, Pinus, Quercus

Haplotrichum magnisporum (Linder) Holubová-Jechová
Oidium magnisporum Linder
Distr.: CA(type).
Host: Abies

Haplotrichum ramosissimum (Berk. & M.A. Curtis) Holubová-Jechová
Acladium ramosissimum (Berk. & M.A. Curtis) M.B. Ellis
Oidium ramosissimum (Berk. & M.A. Curtis) Linder
Distr.: Eastern United States.
Hosts: Pinus, Quercus, Robinia, Sassafras, Tilia, Tsuga

Harknessia Cooke in Cooke & Harkn. Coelomycetes

Notes: Nag Raj & DiCosmo (Biblioth. Mycol. 80:1. 1981); Sutton (The Coelomycetes. p. 177. 1980).

Harknessia aggregata Syd. in Syd. & P. Syd.
Distr.: UT(type).
Notes: On old leaves. Sutton & DiCosmo (Biblioth. Mycol. 80:55. 1981) were unable to comment on its generic status.
Host: Ceanothus

Harknessia americana (Mont.) Sutton
Distr.: Southern United States; Chile.
Notes: Although Sutton (Mycol. Pap. 123:4. 1971) synonymized *H. liquidambaris* with this species, Nag Raj & DiCosmo (Biblioth. Mycol. 80:17. 1983) considered these species distinct.
Host: Liquidambar

Harknessia arctostaphyli Cooke & Harkn.
Distr.: Western United States.
Notes: On leaves.
Hosts: Arbutus, Arctostaphylos

Harknessia caudata Ellis & Everh.
Mastigonetron caudatum (Ellis & Everh.) Höhn.
Cryptosporella farinosa (Ellis) Sacc. (teleomorph)
Distr.: Eastern United States to TX.
Notes: On leaves and stems.
Host: Quercus

Harknessia eucalypti Cooke in Cooke & Harkn.
Distr.: CA; Australia; Great Britain; Italy; New Zealand.
Notes: On leaves.
Host: Eucalyptus

Harknessia globosa Sutton
Distr.: CA; New Zealand.
Notes: On leaves.
Host: Eucalyptus

Harknessia gunnerae F. Stevens & E. Young
Distr.: HI.
Notes: On leaves.
Host: Gunnera

Harknessia hawaiiensis F. Stevens & E. Young
Distr.: FL, HI; Zambia.
Notes: On leaves.
Host: Eucalyptus

Harknessia insueta Sutton
Distr.: United States; Brazil; Mauritius.
Host: Eucalyptus

Harknessia liquidambaris (Berk. & M.A. Curtis) Nag Raj & DiCosmo
Seiridium liquidambaris Berk. & M.A. Curtis
Monochaetia liquidambaris (Berk. & M.A. Curtis) Died.
Mastigonetron affine (Ellis & Everh.) Petr.
Harknessia affinis Ellis & Everh.
Cryptosporella beltsvillensis Petr. (teleomorph)
Distr.: Eastern United States.
Notes: On leaves and twigs.
Host: Liquidambar

Harknessia rhoina Ellis & Everh.
Distr.: CA.
Notes: On leaves.
Host: Rhus

Harknessia shearii Petr.
Distr.: HI.
Notes: On dead leaves.
Host: Metrosideros

Harknessia thujina Ellis & Everh.
Harknessia foeda Sacc. & Dearn.

Cryptosporella thujina Nag Raj & DiCosmo (teleomorph)
Distr.: Eastern North America.
Notes: On needles and bark.
Hosts: Chamaecyparis, Juniperus, Thuja
Harknessia uromycoides (Speg.) Speg.
Harknessia longipes Harkn.
Sphaeropsis stictoides Earle
Cryptosporella sp. (teleomorph)
Distr.: CA; northern Africa; Argentina; Australia; southern Europe.
Notes: On leaves.
Host: Eucalyptus
Harknessia ventricosa Sutton & C.S. Hodges in Sutton
Distr.: HI.
Notes: On dead leaves.
Host: Eucalyptus

Harknessiella Sacc. Loculoascomycetes, Dothideales
Notes: The identity of this genus is in doubt.
Harknessiella purpurea (W. Phillips & Harkn.) Sacc.
Phillipsiella purpurea W. Phillips & Harkn.
Distr.: CA(type).
Host: Garrya

Harpographium Sacc. Hyphomycetes
Harpographium fasciculatum Sacc.
Distr.: North America; Europe; India.
Notes: On wood.
Hosts: Carya, Platanus

Harzia Costantin. Hyphomycetes
Harzia acremonioides (C. Harz) Costantin
Acremoniella atra (Corda) Sacc.
Distr.: Cosmopolitan.
Notes: On a variety of plant materials.
Hosts: Vicia, Zea

Helicascus Kohlmeyer. Loculoascomycetes, Dothideales
Notes: Kohlmeyer & Kohlmeyer (Marine mycology. p. 405. 1979).
Helicascus kanaloanus Kohlmeyer
Distr.: HI(type).
Host: Rhizophora

Helicobasidium Pat. Basidiomycetes, Auriculariales
Helicobasidium brebissonii (Desmaz.) Donk
Helicobasidium purpureum Pat.
Rhizoctonia crocorum (Pers.:Fr.) DC. (anamorph)
Distr.: Widespread.
Notes: Violet root rot (Boerema. Neth. J. Pl. Pathol. 93. Suppl. 1:9. 1987).
Hosts: Beta, Carya, Celtis, Daucus, Fraxinus, Ipomoea, Medicago, Melia, Morus, Rubus, Salix, Sapindus, Sideroxylon, Smilax, Solanum, Ulmus
Helicobasidium candidum G.W. Martin
Distr.: AZ; Canada(Quebec).
Notes: On wood.
Host: Quercus
Helicobasidium corticioides Bandoni
Distr.: Western United States.
Notes: On wood of conifers.
Hosts: Abies, Picea, Pinus
Helicobasidium holospirum Bourd.
Distr.: NY; Europe.
Notes: The generic placement of this species is uncertain (Donk. Persoonia 4:157. 1966).
Host: Dryopteris

Helicoceras Linder. Hyphomycetes
Notes: A confused name (Goos. Mycologia 79:15. 1987).
Helicoceras oryzae Linder & Tullis
Distr.: Southeastern United States; eastern Asia.
Host: Oryza

Helicogloea Pat. in Pat. & Lagerh. Basidiomycetes, Auriculariales
Helicogloea contorta G. Baker
Distr.: IA.
Notes: On wood.
Host: Quercus
Helicogloea lagerheimii Pat. in Pat. & Lagerh.
Distr.: North America; Europe.
Notes: On wood.
Hosts: Fraxinus, Quercus
Helicogloea longispora G. Baker
Distr.: OR.
Notes: On wood.
Host: Pseudotsuga
Helicogloea terminalis Olive
Distr.: NC.
Notes: On wood.
Host: Betula

Helicoma Corda. Hyphomycetes
Notes: Goos (Mycologia 78:744. 1986).
Helicoma ambiens Morg.
Helicosporium ambiens (Morg.) Sacc.
Distr.: HI, North America; Africa.
Host: Acer
Helicoma inflatum Linder
Distr.: MO(type).
Host: Quercus
Helicoma limpidum Morg.
Distr.: OH.
Host: Ulmus
Helicoma machaerinae Goos
Distr.: HI.
Host: Machaerina
Helicoma microscopicum (Ellis) Linder
Helicosporium microscopicum Ellis
Distr.: NJ.
Host: Alnus
Helicoma minutissimum Linder
Distr.: MA.
Host: Ulmus
Helicoma monilipes Ellis & L.N. Johnson in Ellis & Everh.
Distr.: Northeastern United States; Great Britain; Japan.
Host: Quercus
Helicoma morganii Linder
Distr.: United States.
Hosts: Juglans, Populus
Helicoma muelleri Corda
Helicosporium muelleri (Corda) Sacc.
Helicoma curtisii Berk.
Helicosporium tiliae Peck
Tubeufia pezizula (Berk. & M.A. Curtis) Barr (teleomorph)
Distr.: HI, North America; Africa.
Notes: On wood and plant debris.
Hosts: Arecaceae, Fraxinus
Helicoma olivaceum (P. Karst.) Linder
Helicopsis punctata Peck
Distr.: Northern hemisphere.
Host: Prunus
Helicoma palmigenum (Penz. & Sacc.) Linder
Helicoma westonii Linder
Distr.: Widespread in tropical regions.
Hosts: Arecaceae, Roystonea, Syagrus
Helicoma perelegans Thaxt. ex Linder
Distr.: MA, Japan.
Host: Platanus

Helicoma phaeosporium Fresen.
Helicoma velutinum Ellis
Helicosporium velutinum (Ellis) Sacc.
Distr.: Northeastern North America; Europe.
Host: Magnolia
Helicoma polysporum Morg.
Distr.: OH.
Host: Acer
Helicoma repens Morg.
Distr.: OH.
Host: Acer
Helicoma taenia R.T. Moore
Distr.: IA.
Host: Populus
Helicoma tenuifilum Linder
Distr.: MO, VA.
Hosts: Carya, Juglans

Helicomyces Link. Hyphomycetes

Helicomyces ambiguus (Morg.) Linder
Helicoma ambiguum Morg.
Distr.: OH; Japan.
Host: Platanus
Helicomyces bellus Morg.
Distr.: Eastern North America.
Notes: On dead wood.
Hosts: Fraxinus, Juglans
Helicomyces colligatus R.T. Moore
Distr.: Central United States; India.
Host: Salix
Helicomyces roseus Link
Helicomyces fuscopes Linder
Tubeufia cylindrothecia (Seaver) Höhn. (teleomorph)
Distr.: Cosmopolitan.
Notes: On dead wood and plant debris.
Hosts: Acer, Arecaceae, Bambusa, Carya, Platanus, Populus, Spathodea, Tilia
Helicomyces scandens Morg.
Distr.: Central and eastern North America.
Hosts: Castanea, Populus

Helicosporium Nees. Hyphomycetes

Helicosporium aureum (Corda) Linder
Helicosporium pilosum Ellis & Everh.
Tubeufia scopula (Cooke & Peck) Barr (teleomorph)
Distr.: Temperate regions.
Notes: On decorticated coniferous wood.
Host: Pinus
Helicosporium lumbricoides Sacc.
Distr.: Widespread.
Notes: On dead wood and bark (Goos. Mycologia 72:605. 1980).
Host: Populus
Helicosporium panacheum R.T. Moore
Distr.: Widespread.
Notes: On rotten woody plant parts (Goos. Mycologia 72:605. 1980).
Hosts: Casuarina, Liquidambar, Syagrus, Zea
Helicosporium pannosum (Berk. & M.A. Curtis) R.T. Moore
Drepanospora pannosa Berk. & M.A. Curtis
Helicosporium elinorae Linder
Helicosporium nematosporum Linder
Helicosporium serpentinum Linder
Tubeufia helicoma (W. Phillips & Plowr.) Pirozynski (anamorph)
Distr.: Widespread.
Notes: On dead, decaying woody branches and logs, on sawdust, and on large monocot culms (Barr. Mycotaxon 12:161. 1980; Moore. Mycologia 49:582. 1957).
Host: Eucalyptus
Helicosporium phragmitis Höhn.
Tubeufia helicomyces Höhn. (teleomorph)
Distr.: Temperate regions.
Notes: Common on decaying grasses, leaves and other substrates, always in damp situations (Ellis. Dematiaceous Hyphomycetes. p. 193. 1971).
Host: Carex
Helicosporium viride (Corda) Sacc.
Helicoma proliferans Linder
Distr.: Widespread.
Notes: On well-rotted wood (Pirozynski. Mycol. Pap. 129:35. 1972).
Hosts: Acer, Castanea

Heliocybe Redhead & Ginns. Basidiomycetes, Agaricales

Heliocybe sulcata (Berk.) Redhead & Ginns
Lentinus sulcatus Berk.
Pleurotus sulcatus (Berk.) D. Reid
Panus fulvidus Bres.
Lentinus pholiotoides Ellis & F.W. Anderson
Distr.: Temperate and subtropical regions
Notes: Brown rot, often on decorticated wood in xeric habitats.
Hosts: Arbutus, Artemisia, Carnegiea, Chilopsis, Fouquieria, Fraxinus, Juniperus, Platanus, Populus, Prosopis

Heliscus Sacc. & Therry. Hyphomycetes

Heliscus lugdunensis Sacc. & Therry
Heliscus aquaticus Ingold
Nectria lugdunensis J. Webster (teleomorph)
Distr.: Temperate and cold regions.
Notes: On organic debris in aquatic habitats (Nilsson. Symb. Bot. Upsal. 18:97. 1964).
Host: Potamogeton

Helminthosporium Link:Fr., nom. cons. Hyphomycetes

Notes: Originally spelled "*Helmisporium*". Most species have been transferred to *Biploria, Curvularia, Drechslera* and *Exserohilum*.
Helminthosporium ahmadii M.B. Ellis
Distr.: NY; Pakistan(type).
Host: Zea
Helminthosporium asterinum Cooke
Distr.: FL(type).
Host: Liquidambar
Helminthosporium cibotii F. Stevens & Weedon
Distr.: HI(type).
Host: Cibotium
Helminthosporium fumosum Ellis & G. Martin
Brachysporium fumosum (Ellis & G. Martin) Sacc.
Distr.: FL(type).
Host: Persea
Helminthosporium gleicheniae F. Stevens & Glick
Distr.: HI(type).
Hosts: Dicranopteris, Gleichenia
Helminthosporium gossypii Tucker
Distr.: SC, PR(type).
Host: Gossypium
Helminthosporium lunariae Pollacci
Distr.: MA; Italy(type).
Host: Lunaria
Helminthosporium macrocarpum Grev.
Distr.: DE, WA; Europe.
Notes: On dicotyledonous trees.
Hosts: Alnus, Cornus, Liriodendron, Magnolia
Helminthosporium macrocarpum Grev. var. ***caudatum*** Berk. & M.A. Curtis
Distr.: SC(type).
Host: Quercus
Helminthosporium papulosum A. Berg
Distr.: Eastern United States.
Notes: Black pox of apples.
Hosts: Malus, Pyrus
Helminthosporium papulosum A. Berg f. ***pyri*** Acedo & Agrios
Distr.: MA.
Notes: Bark disease of pear trees.
Host: Pyrus
Helminthosporium parvulum Cooke

Distr.: SC(type).
Host: Cortaderia
Helminthosporium sarraceniae MacMill.
Distr.: MN(type).
Host: Sarracenia
Helminthosporium sesami Miyake
Distr.: TX; China(type).
Host: Sesamum
Helminthosporium siliquosum Berk. & M.A. Curtis
Distr.: SC(type).
Host: Vitis
Helminthosporium solani Durieu & Mont.
Spondylocladium atrovirens (C. Harz) C. Harz ex Sacc.
Distr.: Cosmopolitan.
Notes: Silver scurf of potatoes (Ellis. C.M.I. Descr. 166. 1968; Hughes. Fungi Canadenses 236. 1982).
Host: Solanum
Helminthosporium sorghi Schwein.
Distr.: SC(type).
Notes: The type specimen is in poor condition, thus the modern characterization of this name is not possible (Sivanesan. Mycol. Pap. 158:238. 1987).
Host: Sorghum
Helminthosporium spiculiferum Ellis & Everh.
Distr.: LA(type).
Host: Sabal
Helminthosporium tomato Ellis & Barth.
Brachysporium tomato (Ellis & Barth.) Hiroe & T. Watanabe
Distr.: KS(type), TX; Japan.
Host: Lycopersicon
Helminthosporium variegatum Ellis & Everh., nom. nud.
Distr.: DE.
Notes: This names was never validly published.
Host: Quercus
Helminthosporium velutinum Link:Fr.
Septosporium fuliginosum Ellis
Distr.: Cosmopolitan.
Notes: Usually on dead and decaying plant material (Ellis. Dematiaceous Hyphomycetes. p. 390. 1971; Hughes. Fungi Canadenses 163. 1980).
Hosts: Acer, Buxus, Castanea, Cornus, Staphylea
Helminthosporium viticis Syd.
Distr.: FL.
Host: Vitex

Helotiella Sacc. Discomycetes, Helotiales

Helotiella microspora Burt in Millsp. & L. Nutt.
Distr.: CA(type).
Host: Heteromeles

Helotium Pers. non Tode, hom. illeg. Discomycetes, Helotiales

Notes: A large heterogeneous assemblage of inoperculate discomycetes. The generic name has been proposed for rejection (Hawksworth. Taxon 33:730. 1984) because the name has also been applied to agarics (Redhead. Canad. J. Bot. 60:1998. 1982). Most well known species have been transferred to other genera, e.g., *Hymenoscyphus*, but not all names have been redisposed.
Helotium consanguineum Ellis & Everh. ex Seaver
Distr.: NJ(type).
Notes: On dead stems (Dennis. Persoonia 3:39. 1964).
Host: Tephrosia
Helotium fastidiosum (Kuntze) Peck
Calycina fastidiosa Peck
Distr.: North America (White. Mycologia 34:165. 1942).
Host: Alnus
Helotium furfuraceum W. Phillips & Harkn.
Distr.: CA(type).
Host: Quercus
Helotium polystichi Cash
Distr.: CA(type).
Host: Polystichum
Helotium rhizicola Seaver
Distr.: NY.
Notes: On roots (Dennis. Persoonia 3:61. 1964).
Host: Collinsonia

Helvella L. Discomycetes, Pezizales

Notes: Dissing (Dansk Bot. Ark. 25:1. 1966).
Helvella lacunosa Afzel.
Helvella sulcata Afzel.
Distr.: Temperate regions.
Notes: In soil.
Host: Quercus

Hemibeltrania Pirozynski. Hyphomycetes

Hemibeltrania navicularis Sutton
Distr.: Subtropical and tropical regions.
Notes: On fallen, leathery leaves (Sutton. Mycologia 70:791. 1978).
Hosts: Eucalyptus, Podocarpus

Hemimycena Singer. Basidiomycetes, Agaricales

Notes: A synonym of *Helotium* Tode (Redhead. Canad. J. Bot. 60:1998. 1982; Gulden et al. Taxon 33:331. 1984).
Hemimycena albida (Murrill) Singer
Gymnopus albidus Murrill
Helotium floridanum Redhead
Distr.: Southeastern United States.
Host: Juniperus
Hemimycena lactea (Pers.:Fr.) Singer
Mycena delicatella (Peck) A.H. Sm.
Helotium delicatellum (Peck) Redhead
Distr.: Temperate North America; Europe.
Notes: On coniferous debris.
Host: Rosa

Hemiphacidium Korf. Discomycetes, Helotiales

Distr.: Korf (Mycologia 54:26. 1962).
Hemiphacidium convexum (Dearn.) Korf
Phacidium convexum Dearn.
Distr.: NC.
Host: Pinus
Hemiphacidium planum (J.J. Davis) Korf
Phacidium planum J.J. Davis
Distr.: Central and western North America (Reid & Cain. Mycologia 54:495. 1962).
Host: Pinus

Hemitrichia Rostr. Myxomycetes

Hemitrichia clavata (Pers.) Rostr.
Distr.: Temperate regions.
Host: Quercus
Hemitrichia serpula (Scop.) Rostr.
Distr.: Cosmopolitan.
Host: Quercus

Hendersonia Berk., nom. rej. Coelomycetes

Notes: This genus is a synonym of *Stagonospora*, but not all names have been redisposed.
Hendersonia agropyri-repentis Oudem.
Distr.: WA; The Netherlands(type).
Host: Elymus
Hendersonia arundinacea (Desmaz.) Sacc.
Sphaeria arundinacea Desmaz.
Distr.: United States; Europe; Great Britain.
Host: Phragmites
Hendersonia asparagi Pass.
Distr.: FL; Italy.

Host: Asparagus
Hendersonia bliti Clements, nom. nud.
Distr.: CO(type).
Notes: This species was issued as an exsiccati number without an adequate diagnosis.
Host: Chenopodium
Hendersonia calamovilfae Petr.
Distr.: Central and western United States.
Hosts: Calamagrostis, Calamovilfa
Hendersonia calospora Fautrey
Distr.: ND; Europe.
Host: Poa
Hendersonia celtidis Ellis & Everh.
Sporocadus celtidis (Ellis & Everh.) Kuntze
Distr.: IL, NJ, PA.
Host: Celtis
Hendersonia celtifolia Cooke
Distr.: WI.
Host: Celtis
Hendersonia concentrica Ellis & Everh.
Distr.: NC(type).
Hosts: Gaultheria, Rhododendron
Hendersonia corynoidea Cooke & Harkn.
Distr.: CA.
Host: Eucalyptus
Hendersonia crataegicola Atk.
Distr.: AL, TX.
Host: Crataegus
Hendersonia culmicola Sacc. var. ***minor*** Sacc.
Distr.: United States; Europe; Great Britain.
Hosts: Bromus, Calamagrostis, Deschampsia, Festuca, Poa
Hendersonia culmiseda Sacc.
Hendersonia culmicola Cooke non Sacc., hom. illeg.
Distr.: OR; Great Britain.
Host: Cortaderia
Hendersonia cydoniae Cooke & Ellis
Distr.: Central and eastern United States.
Hosts: Cydonia, Malus, Pyrus
Hendersonia davisii Ellis & Everh.
Distr.: WI.
Host: Carya
Hendersonia distans Brunaud
Distr.: CA; France.
Host: Carex
Hendersonia donacis Sacc.
Distr.: AL; Italy.
Host: Saccharum
Hendersonia effusa Berk. & M.A. Curtis
Distr.: Southern United States.
Host: Aristida
Hendersonia erianthi Atk.
Distr.: AL.
Host: Saccharum
Hendersonia eriogoni Fairm.
Distr.: NM.
Host: Eriogonum
Hendersonia fiedleri Westend.
Hendersonia corni Fuckel
Leptosphaeria fiedleri (Niessl) Sacc. (teleomorph)
Distr.: NY; Europe; Great Britain.
Host: Cornus
Hendersonia findens Cooke
Hendersonula findens (Cooke) Tassi
Distr.: GA(type).
Host: Ficus
Hendersonia foliorum Fuckel var. ***hamamelidina*** Fairm.
Distr.: NY.
Host: Hamamelis
Hendersonia foliorum Fuckel var. ***viburni*** Sacc.
Distr.: Eastern United States.
Host: Viburnum
Hendersonia fuscispora M.L. Farr
Host: Selaginella
Hendersonia grantii Dearn.
Distr.: WA.
Host: Phragmites
Hendersonia heterophragmia Ellis & Everh.
Distr.: Western United States.
Host: Sarcobatus
Hendersonia hortilecta Fairm.
Distr.: NY(type).
Host: Clematis
Hendersonia hydrangeae Fairm.
Distr.: NY.
Host: Hydrangea
Hendersonia hypocarpa Fairm.
Distr.: NY(type).
Host: Rosa
Hendersonia jonesiana Petr.
Distr.: NV(type).
Host: Prunus
Hendersonia kalmicola Ellis & Barth.
Distr.: NJ(type).
Host: Kalmia
Hendersonia leucelenes Fairm.
Distr.: NM(type).
Host: Aster
Hendersonia linderae Sacc.
Distr.: Northeastern North America.
Host: Lindera
Hendersonia lineolans (Schwein.) Starb.
Sphaeria lineolans Schwein.
Distr.: PA.
Host: Salix
Hendersonia lupini Cooke & Harkn.
Distr.: CA.
Host: Lupinus
Hendersonia magnoliae Sacc.
Distr.: VA; Europe.
Host: Magnolia
Hendersonia mali Thuem.
Pleospora mali Hesler non G.A. Newton, hom. illeg. (teleomorph)
Distr.: United States; Europe.
Host: Malus
Hendersonia mouiensis Petr.
Distr.: HI(type).
Host: Trisetum
Hendersonia opuntiae Ellis & Everh.
Leptosphaeria opuntiae B.O. Dodge (teleomorph)
Distr.: Subtropical and tropical western hemisphere.
Hosts: Cereus, Echinocactus, Opuntia
Hendersonia oudemansii Sacc. & Syd.
Hendersonia rhododendri Oudem. non Thuem., hom. illeg.
Distr.: WA; The Netherlands(type).
Host: Rhododendron
Hendersonia pallida Berk. & M.A. Curtis
Stagonopsis pallida (Berk. & M.A. Curtis) Sacc.
Distr.: AL.
Host: Cornus
Hendersonia pallidispora M.L. Farr
Host: Selaginella
Hendersonia panicicola Petr.
Distr.: ND.
Host: Panicum
Hendersonia petalostemonis Fairm.
Distr.: NM.
Host: Dalea
Hendersonia petiolorum (Ellis & Everh.) Kuntze
Stagonospora petiolorum Ellis & Everh.
Distr.: WV.
Host: Aralia
Hendersonia pinicola Wehmeyer
Distr.: Western United States.
Notes: Stahl et al. (Mycotaxon 31:323. 1988) studied this fungus in detail.
Hosts: Pinus, [Pyrus]
Hendersonia rockiana Petr.
Distr.: HI(type).
Host: Trisetum
Hendersonia sabaleos Ces.

Distr.: MS; Italy.
Host: Sabal
Hendersonia sarmentorum Westend.
Distr.: North America; Europe; Great Britain.
Hosts: Gossypium, Rhododendron, Vitis
Hendersonia scirpicola Cooke & Harkn.
Distr.: CA.
Hosts: Juncus, Scirpus
Hendersonia septem-septata Vestergr.
Distr.: KS; Europe.
Host: Caragana
Hendersonia spartinae Greene
Distr.: WI.
Host: Spartina
Hendersonia spinaciae (Ellis & Everh.) Kuntze
Stagonospora spinaciae Ellis & Everh.
Distr.: SD.
Host: Spinacia
Hendersonia stanleyellae Fairm.
Distr.: NM.
Host: Thelypodium
Hendersonia staphyleae Ellis & Everh.
Distr.: Eastern United States.
Host: Staphylea
Hendersonia stipae-pennatae Fautrey
Distr.: NV; France.
Host: Bromus
Hendersonia tenella J. Schröt.
Distr.: North America; Great Britain.
Host: Arenaria
Hendersonia thyoides Cooke & Ellis
Distr.: NJ, OR.
Hosts: Chamaecyparis, Thuja
Hendersonia tini Ellis & Langl.
Distr.: LA(type).
Host: Viburnum
Hendersonia varians Cooke & Harkn.
Distr.: CA.
Host: Sphacele
Hendersonia wistariae Cooke
Distr.: SC.
Host: Wisteria
Hendersonia xerophylli Ellis
Distr.: NJ.
Host: Xerophyllum

Hendersonula Speg. Coelomycetes

Hendersonula pinicola Dearn.
Distr.: Southern United States.
Host: Pinus
Hendersonula toruloidea Nattras
Torula dimidiata Penz.
Exosporina fawcettii E.E. Wilson
Scytalidium sp. (synanamorph)
Distr.: Cosmopolitan.
Notes: Punithalingam & Waterston (C.M.I. Descr. 274. 1970).
Hosts: Arbutus, Castanea, Citrus, Ficus, Juglans, Musa, Populus, Prunus, Rhus, Sequoiadendron

Hendersonulina Tassi. Coelomycetes

Notes: Sutton (Mycol. Pap. 141:98. 1977).
Hendersonulina fraxini (Ellis & Barth.) Tassi
Hendersonia fraxini Ellis & Barth.
Distr.: KS.
Host: Fraxinus
Hendersonulina pustulata (Ellis & Everh.) Tassi
Hendersonia pustulata Ellis & Everh.
Distr.: Eastern United States.
Host: Carya

Henningsomyces Kuntze. Basidiomycetes, Aphyllophorales

Henningsomyces candidus (Pers.:Fr.) Kuntze
Solenia candida Pers.:Fr.
Cyphella candida (Pers.:Fr.) Pat.
Solenia polyporoidea (Peck) Burt
Distr.: Widespread.
Notes: On wood and bark.
Hosts: Abies, Betula, Calocedrus, Fraxinus, Juniperus, Pinus, Populus, Prosopis, Pseudotsuga, Quercus, Taxodium, Tsuga, Washingtonia

Heptameria Rehm & Thuem. Loculoascomycetes, Dothideales

Notes: Lucas & Sutton (Trans. Brit. Mycol. Soc. 57:283. 1971).
Heptameria arvensis (Speg.) Cooke
Leptosphaeria arvensis Speg.
Distr.: CA; Italy.
Host: Equisetum
Heptameria obesa (Durieu & Mont.) Sacc.
Sphaeria obesa Durieu & Mont.
Leptosphaeria mesoedema (Berk. & M.A. Curtis) M.B. Ellis & Everh.
Distr.: Southern United States; Europe.
Hosts: Aster, Conyza, Gnaphalium, Solidago, Verbesina
Heptameria tenera (Ellis) Cooke
Leptosphaeria tenera Ellis
Distr.: Western United States.
Hosts: Arabis, Carex, Cinna, Clematis

Hercospora Fr. Pyrenomycetes, Diaporthales

Notes: Barr (Mycol. Mem. 7:194. 1978).
Hercospora taleola (Fr.:Fr.) E. Müller in E. Müller & Arx
Diaporthe taleola (Fr.:Fr.) Sacc.
Hercospora concinna Petr.
Distr.: Eastern North America; Europe.
Notes: On limbs.
Host: Quercus
Hercospora tiliae (Pers.:Fr.) Fr. ex Tul. & C. Tul.
Rabenhorstia tiliae Fr. (anamorph)
Distr.: Eastern North America; Europe.
Notes: On wood.
Host: Tilia

Hericium Pers. Basidiomycetes, Aphyllophorales

Notes: Ginns (Canad. J. Bot. 63:1551. 1985).
Hericium abietis (Weir ex Hubert) K.A. Harrison
Hydnum abietis Weir ex Hubert
Hericium weirii A.H. Sm.
Distr.: Pacific Northwest.
Notes: On wood of conifers.
Hosts: Abies, Picea, Pseudotsuga, Tsuga
Hericium americanum Ginns
Hericium caput-ursi (Fr.) Banker
Distr.: Eastern North America.
Notes: On wood, especially of *Carya* and *Fagus* spp. Formerly known in North America as *H. coralloides* (Ginns. Mycotaxon 20:39. 1984). The European species *H. alpestre* is similar morphologically to *H. americanum* but reproductively isolated.
Hosts: Abies, Acer, Carya, Fagus, Picea, Platanus, Populus, Pseudotsuga, Quercus, Tsuga
Hericium coralloides (Scop.:Fr.) S.F. Gray
Hericium laciniatum (Leers) Banker
Hericium ramosum (Bull.) Letell.
Distr.: Eastern North America, rare in the West; Europe.
Notes: On wood, especially of *Carya* and *Fagus* spp. Commonly known as *H. ramosum* or *H. laciniatum* in North America (Ginns. Mycotaxon 20:39. 1984).
Hosts: Acer, Alnus, Betula, Fagus, Populus, Quercus
Hericium erinaceus (Bull.:Fr.) Pers.
Hydnum erinaceus Bull.:Fr.
Distr.: Temperate northern hemisphere.

Notes: On wood, especially of *Carya* and *Fagus* spp. Commonly known as *H. ramosum* or *H. laciniatum* in North America (Ginns. Mycotaxon 20:39. 1984).
Hosts: Acer, Carya, Celtis, Diospyros, Fagus, Fraxinus, Liquidambar, Liriodendron, Nyssa, Platanus, Pseudotsuga, Quercus, Salix
Hericium erinaceus (Bull.:Fr.) Pers. subsp. ***erinaceo-abietis*** Burdsall, O.K. Miller & Nishijima
Distr.: VA(type).
Notes: On living trees.
Host: Quercus

Herpobasidium Lind, emend. Oberwinkler & K. Wells. Basidiomycetes, Auriculariales

Herpobasidium abnorme Oberwinkler & K. Wells
Distr.: ID.
Host: Polystichum
Herpobasidium filicinum (Rostr.) Lind
Distr.: North America; Europe.
Hosts: Gymnocarpium, Plantago

Herpotrichia Fuckel. Loculoascomycetes, Dothideales

Notes: Barr (Mycotaxon 20:1. 1984).
Herpotrichia graminea Dearn. & House, nom. inval.
Distr.: NY.
Notes: A doubtful species.
Host: Calamagrostis
Herpotrichia herpotrichoides (Fuckel) P. Cannon
Distr.: North America; Europe; India.
Hosts: Carya, Epilobium, Ribes
Herpotrichia juniperi (Duby) Petr.
Herpotrichia nigra R. Hartig
Pyrenochaeta sp. (anamorph)
Distr.: Northern North America; Europe.
Notes: Snow mold of conifers (Sivanesan & Gibson. C.M.I. Descr. 328. 1972).
Hosts: Abies, Calocedrus, Juniperus, Phyllodoce, Picea, Pinus, Pseudotsuga, Taxus, Thuja, Tsuga
Herpotrichia macrotricha (Berk. & Broome) Sacc.
Didymella agrostidis Dearn. & House
Distr.: Temperate northern hemisphere.
Notes: On various plant parts.
Hosts: Acer, Agropyron, Agrostis, Arundinaria, Carex, Fraxinus, Solidago, Thuja
Herpotrichia parasitica (R. Hartig) Rostr.
Trichosphaeria parasitica R. Hartig
Acanthostigma parasitica (R. Hartig) Sacc.
Distr.: North America(questionable); Europe.
Notes: On needles of Pinaceae (Freyer & Aa. Eur. J. Forest Pathol. 5:177. 1975). The fungus may be transferred to *Nematostoma* in the future (M. E. Barr, pers. comm.).
Host: Abies
Herpotrichia purpurea Ellis & Everh.
Distr.: WA.
Notes: Probably belongs to the Helotiales.
Host: Calamagrostis
Herpotrichia symphoricarpi (Tracy & Earle) Barr in L. Holm
Gibberidea symphoricarpi Tracy & Earle
Distr.: CO.
Host: Symphoricarpos

Heterobasidion Bref. Basidiomycetes, Aphyllophorales

Heterobasidion annosum (Fr.:Fr.) Bref.
Fomes annosus (Fr.:Fr.) Cooke
Fomitopsis annosa (Fr.:Fr.) P. Karst.
Spiniger meineckellum (A. Olson) Stalpers (anamorph)
Distr.: Widespread.
Notes: Usually on stumps and trunk bases of conifers, occasionally hardwoods; an important forest pathogen (Pegler & Waterston. C.M.I. Descr. 192. 1968).
Hosts: Abies, Acer, Alnus, Arbutus, Arctostaphylos, Artemisia, Calocedrus, Camellia, Carya, Castanea, Cercocarpus, Chamaecyparis, Diospyros, Juniperus, Kalmia, Larix, Lonicera, Malus, Paxistima, Picea, Pinus, Populus, Prunus, Pseudotsuga, Purshia, Quercus, Rhododendron, Sequoia, Sequoiadendron, Thuja, Tsuga

Heterochaete Pat. Basidiomycetes, Tremellales

Heterochaete shearii (Burt) Burt
Sebacina shearii Burt
Distr.: LA.
Notes: On wood.
Host: Juniperus
Heterochaete spinulosa (Berk. & M.A. Curtis) D. Reid
Radulum spinulosum Berk. & M.A. Curtis
Eichleriella spinulosa (Berk. & M.A. Curtis) Burt
Distr.: Subtropical and tropical western hemisphere.
Notes: On wood.
Host: Cephalanthus

Heteroconium Petr. Hyphomycetes

Heteroconium chaetospira (Grove) M.B. Ellis
Septocylindrium chaetospira Grove
Septonema chaetospira (Grove) S.J. Hughes
Distr.: North temperate regions.
Notes: On rotten wood, often overgrowing other fungi; also isolated from millipede droppings (Ellis. More dematiaceous Hyphomycetes. p. 65. 1976).
Host: Pseudotsuga

Heteropatella Fuckel. Coelomycetes

Notes: Sutton (The Coelomycetes. p. 230. 1980).
Heteropatella cercosperma Lind
Septoria cercosperma Rostr.
Distr.: TX; Europe.
Hosts: Gnaphalium, Hieracium
Heteropatella conglutinata (Ellis & Everh.) Nannf.
Excipula conglutinata Ellis & Everh.
Distr.: Western United States.
Hosts: Gilia, Lupinus, Pedicularis, Valeriana
Heteropatella umbilicata (Pers.) Jaap
Discosia acuta Dearn.
Distr.: Temperate regions.
Notes: On herbaceous and woody debris (Sutton. The Coelomycetes. p. 230. 1980).
Hosts: Achillea, Agastache, Agoseris, Arabis, Artemisia, Aster, Astragalus, Castilleja, Heracleum, Juncus, Ligusticum, Lomatium, Machaeranthera, Monardella, Pedicularis, Penstemon, Ranunculus, Senecio, Silene, Thelypodium, Valeriana, Veratrum
Heteropatella valtellinensis (Traverso) Wollenweb.
Distr.: WA; Australia; Kenya.
Notes: Sutton (The Coelomycetes. p. 233. 1980).
Host: Dianthus

Heterosphaeria Grev. Discomycetes, Helotiales

Notes: Leuchtmann (Mycotaxon 28:261. 1987).
Heterosphaeria patella (Tode:Fr.) Grev.
Pyrenopeziza fuscoatra Hazsl.
Heterosphaeria linariae (Rabenh.) Rehm
Heteropatella lacera Fuckel (anamorph)
Distr.: Temperate regions.
Hosts: Anethum, Heracleum, Linaria, Veratrum

Heterosporium Klotzsch. Hyphomycetes

Heterosporium abroniae Harkn.
Distr.: CA, TX.
Host: Abronia
Heterosporium asperatum Massee

Distr.: United States.
Notes: On herbaceous plants.
Hosts: Smilacina, Smilax
Heterosporium avenae Oudem.
Distr.: Temperate North America; Europe.
Notes: Possibly a *Cladosporium* spp.
Hosts: Agropyron, Avena, Bouteloua, Calamagrostis, Carex, Danthonia, Elymus, Hordeum, Koeleria, Oryza, Phalaris, Poa, Triticum
Heterosporium betae Dowson
Distr.: WA(type).
Host: Beta
Heterosporium californicum Ellis & Everh., nom. nud.
Heterosporium eucalypti Ellis & Everh. var. *maculicola* Ellis & Everh., nom. nud.
Distr.: CA(type).
Notes: These names were based on a collection issued as exsiccati specimens without descriptions.
Host: Eriodictyon
Heterosporium caulicola Ellis & Everh.
Distr.: CO(type).
Host: Rumex
Heterosporium equiseti H.C. Greene
Distr.: WI.
Host: Equisetum
Heterosporium eucalypti Ellis & Everh.
Distr.: CA(type).
Host: Eucalyptus
Heterosporium fraxini Ferd. & Winge
Distr.: WA; Denmark(type).
Host: Fraxinus
Heterosporium hordei Bubák
Distr.: AK; Hungary(type).
Host: Hordeum
Heterosporium hybridum Ellis & Everh.
Distr.: IA, MT.
Host: Cleome
Heterosporium interseminatum Berk. & Ravenel
Distr.: WV.
Host: Sambucus
Heterosporium laricis Cooke & Massee
Distr.: WA; England(type).
Host: Larix
Heterosporium maculatum Klotzsch in Cooke
Distr.: Western United States; Europe(type).
Notes: On decaying stems and leaves of monocotyledonous plants, also reported on apples.
Hosts: Malus, Typha
Heterosporium oxybaphi F. Patterson
Distr.: NE(type).
Host: Allionia
Heterosporium petuniae R. Sprague
Distr.: WA(type).
Notes: On overwintering leaves (Sprague. Mycologia 57:658. 1965).
Host: Petunia
Heterosporium robiniae Kab. & Bubák
Distr.: NY; Europe(type).
Notes: Jacques (Contr. Inst. Bot. Univ. Montreal 39:38. 1941).
Host: Robinia
Heterosporium sambuci Earle
Distr.: AL(type), OK.
Host: Sambucus
Heterosporium selaginellarum M.L. Farr
Host: Selaginella
Heterosporium sphaeriiforme Ellis & Everh.
Distr.: CO(type).
Host: Eriogonum
Heterosporium spiraeae Petr.
Distr.: AK; Austria(type).
Host: Spiraea
Heterosporium syringae Oudem.
Distr.: MS, NC, NJ; The Netherlands(type).
Host: Syringa
Heterosporium trillii Ellis & Everh.
Distr.: Western United States.
Host: Trillium
Heterosporium tuberculans Ellis & Everh.
Distr.: ID, ND, TX.
Host: Iva

Hexagonella F. Stevens & Guba. Loculoascomycetes, Dothideales

Hexagonella peleae F. Stevens & Guba
Distr.: HI(type).
Host: Pellaea

Hexagonia Fr. Basidiomycetes, Aphyllophorales

Notes: As pointed out by Gilbertson & Ryvarden (North American Polypores 1:345. 1986), this name is a later homonym of *Hexagonia* Pollini 1816. Because of the historical use of *Hexagonia sensu* Fr., this name should be proposed for conservation.
Hexagonia hydnoides (Sw.:Fr.) M. Fidalgo
Polyporus hydnoides (Sw.:Fr.) Fr.
Pogonomyces hydnoides (Sw.:Fr.) Murrill
Trametes hydnoides (Sw.:Fr.) Fr.
Distr.: Gulf coastal states, pantropical.
Notes: On hardwoods.
Hosts: Carya, Citrus, Delonix, Fraxinus, Gleditsia, Liquidambar, Mangifera, Quercus, Sideroxylon, Ulmus

Heyderia (Fr.) Link. Discomycetes, Helotiales

Notes: A genus related to *Mitrula* (Knudsen. Bot. Tiddskr. 69:248. 1978; Maas Geesteranus. Persoonia 3:87. 1964).
Heyderia abietis (Fr.) Link
Mitrula abietis Fr.
Mitrula cucullata (Batsch) Fr.
Distr.: North temperate regions.
Notes: In wet places.
Hosts: Picea, Pseudotsuga, Tsuga

Himantia Pers. Agonomycetes

Himantia stellifera J.R. Johnston
Distr.: FL, HI; Caribbean Islands.
Notes: According to Stevenson (Fungi of Puerto Rico and American Virgin Islands. 1975), the teleomorph is assumed to be *Odontia sacchari* Burt.
Host: Saccharum

Hirudinaria Ces. Hyphomycetes

Hirudinaria macrospora Ces.
Clasterosporium curvatum (Berk. & M.A. Curtis) Sacc.
Sporidesmium curvatum Berk. & M.A. Curtis
Distr.: Eastern North America; Europe.
Notes: On leaves (Ellis. Dematiaceous Hyphomycetes. p. 53. 1971; Hughes. Mycol. Pap. 39:11. 1951).
Host: Crataegus

Hobsonia Berk. in Massee. Hyphomycetes

Hobsonia mirabilis (Peck) Linder
Helicomyces mirabilis Peck
Distr.: Worldwide, most common in tropical regions.
Notes: On rotting woody and herbaceous debris (Martin. Brittonia 11:98. 1959).
Hosts: Arecaceae, Brassica, Machaerina, Pandanus

Hohenbuehelia Schulzer. Basidiomycetes, Agaricales

Hohenbuehelia angustata (Berk.) Singer
Panus angustatus Berk.
Distr.: Central and eastern North America.

Host: Carya
Hohenbuehelia approximans (Peck) Singer
Nematoctonus robustus F.R. Jones (anamorph)
Distr.: Eastern North America.
Host: Juniperus
Hohenbuehelia atrocaerulea (Fr.) Singer
Panus albido-tomentosus Cooke
Distr.: GA; Europe.
Notes: On wood.
Host: Quercus
Hohenbuehelia cyphelliformis (Berk. & M.A. Curtis) O.K. Miller
Pleurotus cyphelliformis Berk. & M.A. Curtis
Pleurotus campanulatus Peck
Distr.: Eastern North America.
Host: Solanum
Hohenbuehelia elegans (Coker) Singer
Pleurotus elegans Coker
Distr.: NC.
Host: Juniperus
Hohenbuehelia inversa (Kauffm. & A.H. Sm.) O.K. Miller
Pleurotus inversus Kauffm. & A.H. Sm.
Distr.: MI.
Host: Fagus
Hohenbuehelia mastrucata (Fr.:Fr.) Singer
Pleurotus mastrucatus (Fr.:Fr.) Sacc.
Distr.: Eastern North America; Japan.
Host: Quercus
Hohenbuehelia nigra (Schwein.:Fr.) Singer
Pleurotus niger (Schwein.:Fr.) Sacc.
Nematoctonus robustus F.R. Jones (anamorph)
Distr.: Temperate North America.
Hosts: Acer, Catalpa
Hohenbuehelia petalodes (Bull.:Fr.) Schulzer
Pleurotus petalodes (Bull.:Fr.) Quél.
Pleurotus spathulatus (Pers.:Fr.) Peck
Nematoctonus geogenius R.G. Thorn & Barron (anamorph)
Distr.: North America; Venezuela; Europe; Japan; New Zealand.
Hosts: Abies, Quercus
Hohenbuehelia unguicularis (Fr.:Fr.) O.K. Miller
Pleurotus unguicularis Fr.:Fr.
Resupinatus unguicularis (Fr.:Fr.) Singer
Distr.: Widespread in Europe and North America.
Host: Pinus

Holmiella Petrini, Samuels, & E. Müller. Loculoascomycetes, Patellariales

Holmiella sabina (De Not.) Petrini, Samuels, & E. Müller
Caldesia sabina (De Not.) Rehm
Eutryblidiella sabina (De Not.) Höhn.
Cenangella deformata (Peck) Sacc.
Dermatella deformata (Peck) Seaver
Cenangium deformatum Peck
Diplodia kansensis Ellis & Everh.
Tryblidiopsis occidentalis Earle in Greene
Cornicularіella sp. (anamorph)
Distr.: Temperate North America; Europe; Pakistan.
Notes: On wood and twigs (Petrini et al. Rep. Swiss Bot. Soc. 89:80. 1979; Pirozynski. Fungi Canadenses 11. 1974 as *Eutryblidiella sabina*).
Host: Juniperus

Holwaya Sacc. Discomycetes, Helotiales

Crinula Fr.:Fr.
Notes: Korf & Abawi (Canad. J. Bot. 49:1880. 1971).
Holwaya mucida (Schulzer) Korf & Abawi
Ditiola mucida Schulzer
Chlorosplenium canadense Ellis & Everh.
Holwaya leptosperma (Peck) E.J. Durand
Patellaria leptosperma Peck
Bulgaria ophiobolus Ellis
Holwaya tiliacea Ellis & Everh.
Crinula caliciiformis Fr.:Fr. (anamorph)
Distr.: North America.
Notes: On wood.
Hosts: Acer, Fagus

Hormiactella Sacc. Hyphomycetes

Hormiactella sacchari J.R. Johnston in J.R. Johnston & J.A. Stevenson non Went, hom. illeg.
Distr.: FL, HI.
Host: Saccharum

Hormiscium Kunze. Hyphomycetes

Notes: This genus is considered a synonym of *Torula*, but not all names have been redisposed.
Hormiscium gelatinosum Hedgc.
Distr.: Reported from an unknown locality in the United States(type).
Notes: De Hoog & Hermanides-Nijhof (Stud. Mycol. 15:202. 1977) suggested that this may be a synonym of *Aureobasidium pullulans*.
Host: Pinus

Hormoconis Arx & G.A. De Vries. Hyphomycetes

Hormoconis resinae (Lindau) Arx & G.A. De Vries
Hormodendrum resinae Lindau
Cladosporium resinae (Lindau) G.A. De Vries
Amorphotheca resinae D.G. Parbery (teleomorph)
Distr.: Cosmopolitan.
Notes: Kerosene fungus; also occurs on a variety of organic substrates (Ellis. Dematiaceous Hyphomycetes. p. 309. 1971).
Host: Pinus

Hormodendrum Bonord. Hyphomycetes

Notes: This genus is considered a synonym of *Cladosporium*, but not all names have been redisposed.
Hormodendrum hordei Bruhne in Zopf
Distr.: MT; Germany(type).
Host: Hordeum
Hormodendrum microsporum Lagerberg & Melin
Distr.: MT; Sweden(type).
Host: Thuja
Hormodendrum pyri English, nom. nud.
Distr.: WA(type).
Host: Pyrus
Hormodendrum viride (Fresen.) Sacc.
Penicillium viride Sacc.
Distr.: Western United States; Europe(type).
Notes: According to Thom (The *Penicillia*. p. 589. 1930), this is a "member of the *Cladosporium-Hormodendrum* series".
Host: Tragopogon

Hormotheca Bonord. Loculoascomycetes, Dothideales

Notes: Corlett & Barr (Mycotaxon 25:256. 1986).
Hormotheca plantaginis (Ellis) Corlett & Barr
Asterina plantaginis Ellis
Aphysa plantaginis (Ellis) Theiss.
Coleroa plantaginis (Ellis) Barr
Mycosphaerella plantaginis (Ellis) Theiss. non (Sollm.) Vestergr., hom. illeg.
Mycosphaerella theissenii Tomilin
Distr.: Northeastern North America.
Notes: Corlett & Kokko (Fungi Canadenses 120. 1978 as *Coleroa plantaginis*).
Host: Plantago
Hormotheca robertiani (Fr.:Fr.) Höhn.
Dothidea robertiani Fr.:Fr.
Coleroa robertiani (Fr.:Fr.) E. Müller in E. Müller & Arx
Stigmatea robertiani (Fr.:Fr.) Fr.
Hormotheca geranii (Wallr.) Bonord.
Sphaeria geranii Wallr.
Cryptosphaeria nitida Grev.

Distr.: Northeastern North America; Asia; Europe.
Notes: On leaves (Corlett & Kokko. Fungi Canadenses 93. 1977 as *Coleroa robertiani*).
Host: Geranium
Hormotheca rubicola (Ellis & Everh.) Corlett & Barr
Asterina rubicola Ellis & Everh.
Coleroa rubicola (Ellis & Everh.) E. Müller & Arx
Stigmatea rubicola (Ellis & Everh.) Theiss.
Coleroa chaetomium (Kunze:Fr.) Rabenh. var. *americana* Petr. in Syd. & Petr.
Microthelia manitobiensis (Ellis & Everh.) Kunze
Distr.: Temperate North America.
Notes: On leaves (Corlett. Fungi Canadenses 92. 1977 as *Coleroa rubicola*).
Host: Rubus

Humaria Fuckel. Discomycetes, Pezizales

Humaria hemisphaerica (A. Wigg.) Fuckel
Patella albida (Schaeff.) Seaver
Distr.: Temperate regions.
Notes: On soil and wet wood.
Host: Quercus
Humaria lusatiae (Cooke) Kanouse
Peziza lusatiae Cooke
Distr.: WA; Germany(type).
Host: Populus

Humicola Traaen. Hyphomycetes

Humicola alopallonella S.P. Meyers & R.T. Moore
Distr.: Worldwide in marine habitats.
Notes: Saprobic on intertidal wood and organic debris (Kohlmeyer & Kohlmeyer. Marine mycology. p. 494. 1979).
Host: Pluchea
Humicola brevis (Gilman & E. Abbott) Gilman
Acremoniella brevis Gilman & E. Abbott
Monotospora brevis (Gilman & E. Abbott) E. Mason
Distr.: HI, LA.
Notes: Isolated from soil and on *Acacia*.
Host: Acacia
Humicola fuscoatra Traaen
Distr.: Cosmopolitan.
Notes: Saprobic in soil and on plant debris.
Hosts: Alnus, Cucumis, Gossypium, Metrosideros, Pseudotsuga

Hyalinia Boud. Discomycetes, Helotiales

Hyalinia crenato-marginata Höhn.
Distr.: WA; Europe.
Hosts: Acer, Alnus
Hyalinia crystallina (Quél.) Boud.
Helotium crystallinum Quél.
Cyathicula crystallina (Quél.) Sacc.
Distr.: NY; Europe.
Host: Quercus

Hyalopeziza Fuckel. Discomycetes, Helotiales

Notes: Raschle (Sydowia 29:192. 1977).
Hyalopeziza ciliata Fuckel
Dasyscyphus ciliatus (Fuckel) Höhn.
Lachnella ciliata (Fuckel) Boud.
Lachnum ciliata (Fuckel) Rehm
Trichopeziza ciliata (Fuckel) Sacc.
Distr.: Temperate regions.
Hosts: Delphinium, Pseudotsuga, Rhododendron
Hyalopeziza tautilla (W. Phillips & Harkn.) Dennis
Dasyscyphus tautillus (W. Phillips & Harkn.) Sacc.
Lachnella tautilla (W. Phillips & Harkn.) Seaver
Distr.: CA.
Notes: Among hairs on lower surface of living leaves (Dennis. Kew Bull. 17:369. 1963).
Host: Garrya

Hyalopsora Magnus. Basidiomycetes, Uredinales

Hyalopsora aspidiotus (Magnus) Magnus
Melampsorella aspidiotus Magnus
Distr.: North America, cooler parts; western Asia; Europe; Japan.
Notes: Heteroecious; aecial host - *Abies* spp.; telial host - Polypodiaceae.
Hosts: Abies, Gymnocarpium
Hyalopsora cheilanthis Arth.
Distr.: Central and western United States; Argentina; Chile; Ecuador.
Notes: Heteroecious; aecial host - unknown; telial host - Polypodiaceae.
Hosts: Cheilanthes, Cryptogramma, Notholaena, Pellaea, Pityrogramma
Hyalopsora polypodii (Dietel) Magnus
Distr.: Temperate northern hemisphere.
Notes: Heteroecious; aecial host - unknown; telial host - Polypodiaceae.
Hosts: Cystopteris, Woodsia

Hyaloscypha Boud. Discomycetes, Helotiales

Hyaloscypha alniseda Velen.
Distr.: OR; Europe.
Host: Alnus
Hyaloscypha hyalina (Pers.:Fr.) Boud.
Lachnella hyalina (Pers.:Fr.) W. Phillips
Distr.: Temperate regions.
Notes: On dead wood and bark.
Hosts: Alnus, Rubus, Salix
Hyaloscypha stevensonii (Berk. & Broome) Nannf.
Hyaloscypha atomaria (Starb.) Nannf.
Distr.: Temperate regions.
Notes: On dead conifer wood.
Hosts: Abies, Pinus, Pseudotsuga, Tsuga

Hyalothyridium Tassi. Coelomycetes

Notes: A synonym of *Camarosporium* (Sutton. Mycol. Pap. 141:101. 1977).
Hyalothyridium calamagrostidis H.C. Greene
Distr.: WI.
Host: Calamagrostis

Hyalotia Guba. Coelomycetes

Notes: This genus is considered a synonym of *Bartalinia* (Nag Raj. Canad. J. Bot. 53:1615. 1975), but not all names have been redisposed.
Hyalotia pistacina J.L. Maas
Distr.: MD.
Notes: Maas (Mycologia 63:663. 1971).
Host: Pistacia

Hyalotrochophora Finley & E.F. Morris in E.F. Morris & Finley. Hyphomycetes

Notes: According to Carmichael et al. (Genera of Hyphomycetes. p. 112. 1980), this genus may be a synonym of *Delortia* Pat. & Gaill.
Hyalotrochophora lignitalis (Thaxt.) Finley & E.F. Morris in E.F. Morris & Finley
Everhartia lignitalis Thaxt.
Distr.: IA, MA.
Notes: On wood.
Hosts: Catalpa, Pinus

Hydnochaete Bres. Basidiomycetes, Aphyllophorales

Hydnochaete olivacea (Schwein.:Fr.) Banker
Irpex cinnamomeus Fr.:Fr.

Hydnoporia fuscescens (Schwein.) Murrill
Irpex quisquilaris Pat.
Distr.: Central and eastern North America; Jamaica; Ecuador.
Notes: On wood.
Hosts: Alnus, Betula, Carpinus, Populus, Prunus, Pseudotsuga, Quercus, Salix
Hydnochaete tabacina (Berk. & M.A. Curtis) Ryvarden
Irpex tabacinus Berk. & M.A. Curtis
Irpex ravenelii (Berk.) Overh., comb. inval.
Distr.: Southeastern United States.
Notes: On wood.
Host: Quercus

Hydronectria Kirschst. Pyrenomycetes, Hypocreales

Notes: Kohlmeyer & Kohlmeyer (Marine mycology. p. 327. 1979).
Hydronectria tethys Kohlmeyer & E. Kohlmeyer
Distr.: Worldwide in marine habitats.
Hosts: Avicennia, Hibiscus, Rhizophora

Hydropus (Kühner) Singer. Basidiomycetes, Agaricales

Hydropus marginellus (Pers.:Fr.) Singer
Mycena marginella (Pers.:Fr.) Quél.
Distr.: North America; Europe.
Notes: On conifer wood.
Hosts: Abies, Picea
Hydropus sabalis Singer
Distr.: FL.
Host: Sabal
Hydropus serifluus (Murrill) Singer
Galactopus serifluus Murrill
Mycena seriflua (Murrill) Murrill
Distr.: FL.
Host: Magnolia
Hydropus taxodii (Murrill) Singer
Prunulus taxodii Murrill
Mycena taxodii (Murrill) Murrill
Distr.: FL(type).
Host: Taxodium

Hymenochaete Lév., nom. cons. Basidiomycetes, Aphyllophorales

Hymenochaete agglutinans Ellis
Distr.: Central and eastern North America; Cuba.
Notes: On living limbs, seedlings or saplings, usually where two or more limbs come into contact; stem canker. According to Reeves & Welden (Mycologia 59:1034. 1967) this species is a sterile stage of *H. corrugata*; they considered it a synonym of the latter. *Hymenochaete agglutinans* is maintained as a separate species here because of its notoriety as a pathogen.
Hosts: Acer, Alnus, Betula, Corylus, Fagus, Fraxinus, Lindera, Liquidambar, Malus, Morus, Nyssa, Pinus, Prunus, Rhododendron, Salix, Sambucus, Sassafras, Syringa, Tsuga
Hymenochaete arida P. Karst.
Distr.: Temperate northern hemisphere.
Notes: On hardwoods.
Hosts: Ceanothus, Prosopis, Quercus
Hymenochaete badio-ferruginea (Mont.) Lév.
Distr.: Eastern North America.
Notes: On wood. Probably a form of *H. tabacina*.
Hosts: Betula, Quercus, Vaccinium
Hymenochaete borealis Burt
Distr.: NJ, VY; Canada(Ontario).
Notes: On hardwoods, rare.
Host: Carya
Hymenochaete cinnamomea (Pers.:Fr.) Bres.
Distr.: Widespread.
Notes: On wood.
Hosts: Acer, Ceanothus, Corylus, Fagus, Pinus, Quercus, Salix
Hymenochaete corrugata (Fr.:Fr.) Lév.
Distr.: Widespread.
Notes: On wood, usually of hardwoods. See notes under *H. agglutinans*.
Hosts: Abies, Acer, Alnus, Carpinus, Castanea, Fagus, Magnolia, Quercus, Thuja
Hymenochaete corticolor Berk. & Ravenel
Distr.: Eastern United States to warm regions of the western hemisphere.
Notes: On wood.
Host: Magnolia
Hymenochaete curtisii (Berk.) Morg.
Distr.: Central and eastern United States.
Notes: On hardwoods.
Hosts: Carpinus, Castanea, Quercus
Hymenochaete episphaeria (Schwein.:Fr.) Massee
Distr.: Central and eastern United States.
Notes: On hardwoods.
Hosts: Acer, Castanea
Hymenochaete fusca (P. Karst.) Sacc.
Hymenochaete fuliginosa (Pers.) Bres. non Lév., hom. illeg.
Distr.: Eastern North America; Europe.
Notes: On hardwoods.
Hosts: Acer, Tsuga
Hymenochaete leonina Berk. & M.A. Curtis
Distr.: Southern United States; Mexico; Cuba.
Notes: On hardwoods.
Hosts: Fraxinus, Quercus
Hymenochaete pinnatifida Burt
Distr.: Southeastern United States; warmer parts of the western hemisphere.
Notes: On wood.
Hosts: Alnus, Elaeagnus, Pinus, Prunus
Hymenochaete rubiginosa (Dickson:Fr.) Lév.
Distr.: Temperate northern hemisphere.
Hosts: Castanea, Juglans, Myrica, Prosopis, Quercus, Rhamnus, Sambucus
Hymenochaete sallei Berk. & M.A. Curtis
Distr.: Southeastern United States to South America.
Notes: On wood.
Host: Quercus
Hymenochaete spreta Peck
Distr.: North America.
Notes: On wood.
Hosts: Acer, Alnus, Ceanothus, Pseudotsuga, Quercus, Thuja
Hymenochaete tabacina (Sowerby:Fr.) Lév.
Distr.: Temperate northern hemisphere.
Notes: On wood.
Hosts: Abies, Acer, Alnus, Amelanchier, Arctostaphylos, Aristolochia, Betula, Castanea, Castanopsis, Ceanothus, Fagus, Holodiscus, Larix, Lithocarpus, Lyonothamnus, Paxistima, Physocarpus, Picea, Pinus, Populus, Prunus, Pseudotsuga, Quercus, Rosa, Salix, Sequoia, Sorbus, Symphoricarpos, Taxus, Thuja, Vaccinium
Hymenochaete tenuis Peck
Distr.: North America.
Notes: On wood.
Hosts: Picea, Pinus, Pseudotsuga, Thuja

Hymenopsis Sacc. Coelomycetes

Notes: Sutton (The Coelomycetes. p. 331. 1980).
Hymenopsis hydrophila Sacc.
Distr.: NY(type).
Host: Typha
Hymenopsis trochiloides (Sacc.) Sacc.
Myrothecium trochiloides Sacc.
Distr.: DE; France; India
Notes: On leaves and stems.
Host: Scirpus
Hymenopsis typhae (Peck) Sutton
Melanconium typhae Peck
Cryptomela typhae (Peck) Died.
Distr.: Eastern North America; Europe; Great Britain.
Host: Typha

Hymenoscyphus S.F. Gray. Discomycetes, Helotiales

Hymenoscyphus alniellus (Nyl.) Kuntze
Phialea alniella (Nyl.) Sacc.
Distr.: Temperate regions.
Host: Alnus

Hymenoscyphus calyculus (Sowerby) W. Phillips
Helotium virgultorum (Vahl) Fr.
Distr.: Temperate regions.
Notes: On dead wood and twigs.
Hosts: Alnus, Ribes, Rosa, Salix, Sambucus, Symphoricarpos

Hymenoscyphus caudatus (P. Karst.) Dennis
Helotium caudatum (P. Karst.) Velen.
Helotium naviculasporum Ellis
Idriella sp. (anamorph)
Distr.: Temperate regions (White. Farlowia 1:151. 1943).
Hosts: Acer, Alnus, Berberis

Hymenoscyphus conscriptus (P. Karst.) Korf
Phialea ampla Ellis & Everh.
Helotium amplum (Ellis & Everh.) Seaver
Distr.: Western United States; Europe.
Notes: Carpenter (Mem. New York Bot. Gard. 33:189. 1981) suggested this synonymy.
Hosts: Populus, Salix

Hymenoscyphus dearnessii (Ellis & Everh.) Kuntze
Phialea dearnessii Ellis & Everh.
Helotium dearnessii (Ellis & Everh.) W.L. White
Distr.: Eastern North America.
Host: Lysimachia

Hymenoscyphus epiphyllus (Pers.:Fr.) Rehm ex Kauffm.
Helotium epiphyllum (Pers.:Fr.) Fr.
Distr.: Temperate regions.
Notes: White (Farlowia 1:139. 1943).
Hosts: Acer, Alnus, Quercus, Silphium

Hymenoscyphus fructigenus (Bull.) W. Phillips
Helotium fructigenum (Bull.) Fuckel
Helotium virgultorum (Vahl:Fr.) P. Karst. var. *fructigenum* (Bull.) Rehm
Distr.: Widespread in Europe and North America.
Notes: On fallen fruits of various trees.
Hosts: Carya, Quercus

Hymenoscyphus imberbis (Bull.:Fr.) Dennis
Helotium imberbe (Bull.:Fr.) Fr.
Distr.: MI, WA, WI; Europe.
Host: Alnus

Hymenoscyphus lutescens (Hedw.:Fr.) W. Phillips
Helotium lutescens (Hedw.:Fr.) Fr.
Distr.: NY, WA; Europe.
Host: Pinus

Hymenoscyphus phiala (Vahl:Fr.) E. Arnolds ex Baral
Helotium phiala (Vahl:Fr.) Fr.
Distr.: WA; Europe.
Notes: Carpenter (Mem. New York Bot. Gard. 33:243. 1981) stated that the taxonomic status of this species was unknown.
Host: Sambucus

Hymenoscyphus phyllophilus (Desmaz.) Kuntze
Helotium phyllophilum (Desmaz.) Fr.
Phialea phyllophila (Desmaz.) Gill.
Distr.: GA; France(type).
Notes: White (Farlowia 1:157. 1943).
Host: Fagus

Hymenoscyphus salicellus (Fr.) Dennis
Helotium salicellum (Fr.) Fr.
Distr.: Temperate regions.
Notes: Dennis (Persoonia 3:73. 1964).
Hosts: Populus, Salix

Hymenoscyphus scutula (Pers.:Fr.) W. Phillips
Helotium scutula (Pers.:Fr.) P. Karst.
Phialea scutula (Pers.:Fr.) Gill.
Belonioscypha ciliatospora (Fuckel) Rehm
Distr.: Temperate regions.
Notes: On dicotyledonous debris.
Hosts: Aster, Boehmeria, Dennstaedtia, Dryas, Epilobium, Impatiens, Lysimachia, Rubus

Hymenoscyphus serotinus (Pers.:Fr.) W. Phillips
Helotium serotinum (Pers.:Fr.) Fr.
Distr.: ID, OH; Europe.
Notes: On plant stems in wet places.
Host: Cornus

Hymenoscyphus sphaerophoroides W. Phillips & Harkn.
Distr.: CA(type).
Host: Sequoia

Hymenula Fr. Hyphomycetes

Notes: A superfluous substitute name for *Hymenella* Fr.:Fr.

Hymenula affinis (Fautrey & Lambotte) Wollenweb.
Fusarium affine Fautrey & Lambotte
Distr.: HI.
Notes: This is an obscure species in a doubtful genus, now typified and recognized as *Hymenella* (Tulloch. Mycol. Pap. 130:6. 1972). The report of *Fusarium affine* as *F. udum* goes back to Larsen (Rep. Exp. Sta. Hawaiian Sugar Planter's Assoc. 10:35. 1910). Booth (The genus *Fusarium*. p. 187. 1971) stated that *Fusarium affine* is in part also a synonym of *Fusarium solani*.
Host: Ananas

Hymenula cerealis Ellis & Everh.
Cephalosporium gramineum Nisikado & Ikata in Nisikado et al.
Distr.: North America; southern Africa; temperate Europe; Japan.
Notes: Commonly known as "Cephalosporium" stripe of cereals and grasses. As pointed out by Hawksworth & Waller (C.M.I. Descr. 501. 1976), the generic position of this species is in need of revision.
Hosts: Agropyron, Bromus, Dactylis, Elymus, Secale, Triticum

Hyphochytrium Zopf. Hyphochytridiomycetes, Hyphochytriales

Hyphochytrium catenoides Karling
Distr.: NY.
Host: Zea

Hyphoderma Wallr. emend. Donk. Basidiomycetes, Aphyllophorales

Hyphoderma amoenum (Burt) Donk
Peniophora amoena Burt
Peniophora subalba (Burt) D.P. Rogers & H. Jacks.
Distr.: North America.
Notes: On hardwoods.
Hosts: Agave, Alnus, Garrya, Holodiscus, Picea, Prosopis, Prunus, Quercus, Sageretia

Hyphoderma anasaziense Lindsey
Distr.: CO(type).
Notes: On wood.
Host: Quercus

Hyphoderma argillaceum (Bres.) Donk
Peniophora argillacea Bres.
Peniophora fusca Burt
Distr.: Temperate northern hemisphere.
Notes: On wood.
Hosts: Abies, Juniperus, Picea, Pinus, Populus, Tsuga

Hyphoderma assimile (H. Jacks. & Dearden) Donk
Peniophora assimilis H. Jacks. & Dearden
Distr.: Temperate northern hemisphere.
Notes: On wood.
Host: Purshia

Hyphoderma baculorubrense R.L. Gilbertson & Blackwell
Distr.: Gulf Coast States.
Notes: On bark. Gilbertson & Blackwell (Mycotaxon 20:85. 1984).
Hosts: Juniperus, Quercus

Hyphoderma budingtonii Lindsey & R.L. Gilbertson
Distr.: AZ.
Notes: On wood. Lindsey & Gilbertson (Mycotaxon 5:311. 1977).
Host: Populus

Hyphoderma clavigerum (Bres.) Donk
Peniophora clavigera (Bres.) Bourd. & Galzin
Distr.: Temperate northern hemisphere.
Notes: On wood.

Hosts: Betula, Carnegiea, Chilopsis, Quercus
Hyphoderma cristulatum (Fr.:Fr.) Donk
Odontia cristulata (Fr.:Fr.) Fr.
Distr.: Temperate northern hemisphere.
Notes: On wood. Jülich & Stalpers (The resupinate non-poroid Aphyllophorales of the temperate northern hemisphere. p. 110. 1980) considered this species to be a synonym of *H. setigerum*. If this is accepted, then *H. cristulatum* must be used because its basionym is older (1821) than that of *H. setigerum* (1828).
Hosts: Alnus, Pinus, Quercus
Hyphoderma definitum (H. Jacks.) Donk
Corticium definitum H. Jacks.
Distr.: Northern hemisphere.
Notes: On wood.
Host: Pinus
Hyphoderma deserticola R.L. Gilbertson & Lindsey
Distr.: AZ.
Notes: Gilbertson & Lindsey (Great Basin Naturalist 35:288. 1975).
Host: Juniperus
Hyphoderma echinocystis J. Eriksson & A. Strid in J. Eriksson & Ryvarden
Distr.: Northern hemisphere.
Host: Sabal
Hyphoderma fouquieriae Nakasone & R.L. Gilbertson
Distr.: AZ.
Notes: On wood. Nakasone & Gilbertson (Mycologia 70:266. 1978).
Host: Fouquieria
Hyphoderma guttuliferum (P. Karst.) Donk
Peniophora guttulifera (P. Karst.) Sacc.
Distr.: North America; Europe.
Notes: On wood.
Hosts: Alnus, Platanus, Populus, Quercus
Hyphoderma heterocystidia (Burt) Donk
Peniophora heterocystidia Burt
Distr.: Northern hemisphere.
Notes: On hardwoods.
Hosts: Acer, Carpinus, Fagus, Juglans, Robinia
Hyphoderma karstenii Jülich
Metulodontia nivea (P. Karst.) Parmasto
Peniophora nivea (P. Karst.) Bourd. & Galzin
Distr.: Temperate northern hemisphere.
Notes: On wood. Although the basionym of *Metulodnotia nivea* (*Kneiffia nivea* P. Karst. 1896) is older than *H. karstenii* Jülich 1974, the combination in *Hyphoderma* is preoccupied by *H. niveum* Fuckel 1869.
Host: Populus
Hyphoderma leoninum Burdsall & Nakasone
Distr.: Eastern North America.
Notes: On hardwoods. Burdsall & Nakasone (Mycotaxon 17:253. 1983).
Host: Acer
Hyphoderma litschaueri (Burt) J. Eriksson & A. Strid in J. Eriksson & Ryvarden
Corticium litschaueri Burt
Basidioradulum alienum Parmasto
Hyphoderma bresadolae Jülich
Distr.: North America; Europe.
Notes: On wood.
Hosts: Alnus, Malus
Hyphoderma medioburiense (Burt) Donk
Peniophora medioburiensis Burt
Distr.: Northern hemisphere.
Notes: On wood.
Hosts: Picea, Pinus, Populus, Quercus
Hyphoderma mutatum (Peck) Donk
Corticium mutatum Peck
Peniophora mutata (Peck) Höhn. & Litsch.
Peniophora allescheri (Bres.) Sacc. & Syd.
Distr.: Probably widespread.
Notes: On hardwoods.
Hosts: Castanea, Corylus, Heteromeles, Juglans, Liquidambar, Liriodendron, Populus, Quercus, Tilia, Viburnum
Hyphoderma obtusiforme J. Eriksson & A. Strid in J. Eriksson & Ryvarden
Distr.: North America (rare); Europe.
Notes: On wood.
Host: Juniperus
Hyphoderma pallidum (Bres.) Donk
Corticium pallidum Bres.
Corticium ochrofarctum Burt
Distr.: Northern hemisphere.
Notes: On wood.
Hosts: Juniperus, Pinus, Populus
Hyphoderma pilosum (Burt) R.L. Gilbertson & Budington
Peniophora pilosa Burt
Distr.: North America.
Notes: On conifers.
Hosts: Abies, [Alnus], [Arbutus], Pinus, [Populus], [Quercus]
Hyphoderma populneum (Peck) Donk
Stereum populneum Peck
Peniophora populnea (Peck) Burt
Distr.: North America.
Notes: On hardwoods.
Host: Populus
Hyphoderma praetermissum (P. Karst.) J. Eriksson
Peniophora praetermissa P. Karst.
Sebacina africana Burt
Peniophora albugo Burt
Peniophora taxodii Burt
Hyphoderma tenue (Pat.) Donk
Distr.: Temperate northern hemisphere.
Notes: On wood.
Hosts: Abies, Alnus, Juniperus, Picea, Pinus, Platanus, Populus, Pseudotsuga, Quercus, Taxodium
Hyphoderma puberum (Fr.:Fr.) Wallr.
Peniophora pubera (Fr.:Fr.) Sacc.
Peniophora tenella Burt
Peniophora tenuissima Peck
Distr.: Widespread.
Hosts: Acer, Alnus, Pinus, Populus, Quercus, Thuja
Hyphoderma rimosum Burdsall & Nakasone
Distr.: Eastern North America.
Notes: On hardwoods. Burdsall & Nakasone (Mycotaxon 17:253. 1983).
Hosts: Acer, Cornus, Ilex, Lagerstroemia, Liquidambar
Hyphoderma roseocremeum (Bres.) Donk
Corticium roseocremeum Bres.
Distr.: Northern hemisphere.
Notes: On wood.
Hosts: Fraxinus, Platanus, Prunus, Quercus
Hyphoderma sambuci (Pers.) Jülich
Lyomyces sambuci (Pers.) P. Karst.
Hyphodontia sambuci (Pers.) J. Eriksson
Peniophora sambuci (Pers.) Burt
Corticium cretaceum (Fr.) Cooke ex Sacc.
Corticium hariotii Bres.
Hyphodontia hariotii (Bres.) Parmasto
Peniophora irregularis Burt
Peniophora thujae Burt
Distr.: Widespread.
Notes: On wood.
Hosts: Alnus, Fouquieria, Juglans, Juniperus, Maclura, Morus, Picea, Pinus, Populus, Prosopis, Quercus, Robinia, Sambucus, Thuja
Hyphoderma setigerum (Fr.:Fr.) Donk
Kneiffia setigera (Fr.:Fr.) Fr.
Odontia setigera (Fr.:Fr.) L.W. Miller
Peniophora setigera (Fr.:Fr.) Höhn. & Litsch.
Odontia acerina Peck
Peniophora aspera (Pers.) Sacc.
Corticium berkeleyi Cooke
Peniophora subtestacea Litsch.
Hyphoderma subtestaceum (Litsch.) Donk
Peniophora trachytricha Ellis & Everh.
Odontia vesiculosa Burt in Povah
Distr.: Widespread.
Notes: On wood.
Hosts: Abies, Acer, Alnus, Betula, Chilopsis, Corylus, Larix, Liquidambar, Maclura, Magnolia, Persea, Pinus, Planera, Platanus, Populus, Prunus, Quercus, Rhus, Ribes, Ulmus
Hyphoderma tsugae (Burt) J. Eriksson & A. Strid in J. Eriksson et

al.
Corticium tsugae Burt
Distr.: Temperate northern hemisphere.
Notes: On wood of conifers.
Host: Tsuga
Hyphoderma typhicola (Burt) Donk
Peniophora typhicola Burt
Distr.: North America.
Host: Typha

Hyphodermella J. Eriksson & Ryvarden. Basidiomycetes, Aphyllophorales

Hyphodermella corrugata (Fr.) J. Eriksson & Ryvarden
Grandinia corrugata Fr.
Odontia corrugata (Fr.) Bourd. & Galzin
Distr.: Temperate northern hemisphere.
Notes: On wood.
Hosts: Artemisia, Baccharis, Populus, Prunus, Ribes

Hyphodermopsis Jülich. Basidiomycetes, Aphyllophorales

Notes: Monotypic.
Hyphodermopsis polonensis (Bres.) Jülich
Hyphoderma polonense (Bres.) Donk
Peniophora polonensis (Bres.) Höhn. & Litsch.
Peniophora canadensis Burt
Distr.: Temperate northern hemisphere.
Notes: On wood.
Hosts: Abies, Fraxinus

Hyphodontia J. Eriksson. Basidiomycetes, Aphyllophorales

Notes: As pointed out by Jülich (Int. J. Mycol. Lichenol. 1:27. 1982), the earliest name for this genus is *Grandinia*. A few residual species are listed here under *Hyphodontia*.
Hyphodontia albicans (Pers.) Parmasto
Distr.: MN; Europe.
Notes: The generic placement of this species is uncertain.
Hosts: Populus, Tilia
Hyphodontia lanata Burdsall & Nakasone
Distr.: Southern United States; New Zealand.
Notes: Burdsall & Nakasone (Mycologia 73:454. 1981).
Hosts: Ligustrum, Liriodendron, Sabal
Hyphodontia magnacystidiata Lindsey & R.L. Gilbertson
Distr.: NY.
Notes: On wood. Lindsey & Gilbertson (Mycotaxon 5:311. 1977).
Host: Populus
Hyphodontia rimosissima (Peck) R.L. Gilbertson
Odontia rimosissima Peck
Distr.: Northeastern United States.
Notes: On wood.
Host: Alnus
Hyphodontia stipata (Fr.:Fr.) R.L. Gilbertson
Hydnum stipatum Fr.:Fr.
Odontia stipata (Fr.:Fr.) Quél.
Hydnum combinans Peck
Hydnum rimulosum Peck
Distr.: Temperate northern hemisphere.
Notes: On wood.
Hosts: Abies, Acer, Alnus, Fraxinus, Pinus, Populus

Hypholoma (Fr.) Quél., nom. cons. Basidiomycetes, Agaricales

Hypholoma capnoides (Fr.:Fr.) P. Kumm.
Naematoloma capnoides (Fr.:Fr.) P. Karst.
Distr.: Northern hemisphere.
Hosts: Abies, Tsuga
Hypholoma fasciculare (Huds.:Fr.) P. Kumm.
Naematoloma fasciculare (Huds.:Fr.) P. Karst.
Distr.: Temperate northern hemisphere.
Hosts: Acer, Arbutus, Fagus, Juniperus, Liriodendron, Malus, Pinus, Pseudotsuga, Quercus, Tsuga
Hypholoma sublateritium (Fr.) Quél.
Naematoloma sublateritium (Fr.) P. Karst.
Hypholoma perplexum (Peck) Sacc.
Distr.: Temperate northern hemisphere.
Notes: On dead hardwoods.
Hosts: Acer, Betula, Castanea, Malus, Pinus, Quercus, Ribes

Hyphosoma Syd. Hyphomycetes

Notes: Hughes (New Zealand J. Bot. 8:153. 1970) considered this a confused generic name because the type species was based on a mixed collection. Taxa still remaining under *Hyphosoma* are in need of revision.
Hyphosoma lumbricoideum (Dearn.) S.J. Hughes
Helminthosporium lumbricoideum Dearn.
Helminthosporium abietis W.B. Cooke & C.G. Shaw
Distr.: WA(type).
Hosts: Abies, Vaccinium
Hyphosoma piceae (Funck) S.J. Hughes
Monilia piceae Funck
Antennatula pinophila (Nees) F. Strauss (synanamorph)
Distr.: OR, WA; central Europe.
Notes: On living twigs (Hughes. New Zealand J. Bot. 8:180. 1970).
Host: Juniperus
Hyphosoma resinacea (Cooke) S.J. Hughes
Helmisporium resinaceum Cooke
Helminthosporium pseudotsugae W.B. Cooke
Strigopodia resinae (Sacc. & Bres.) S.J. Hughes (teleomorph)
Distr.: Western United States; Europe.
Notes: Hughes (Canad. J. Bot. 46:1100. 1968).
Hosts: Larix, Pseudotsuga

Hypocenia Berk. & M.A. Curtis. Coelomycetes

Notes: According to Sutton (The Coelomycetes. p. 416. 1980), this genus is a synonym of *Topospora*, but not all names have been redisposed.
Hypocenia caespitosa Petr.
Distr.: MD.
Host: Viburnum

Hypochniciellum Hjortstam & Ryvarden. Basidiomycetes, Aphyllophorales

Hypochniciellum molle (Fr.:Fr.) Hjortstam
Leucogyrophana mollis (Fr.:Fr.) Parmasto
Peniophora mollis (Fr.:Fr.) Bourd. & Galzin
Distr.: Temperate northern hemisphere.
Notes: On wood. Ginns (Canad. J. Bot. 56:1953. 1978) did not include this species in *Leucogyrophana*.
Host: Pinus
Hypochniciellum subillaqueatum (Litsch.) Hjortstam
Corticium subillaqueatum Litsch.
Leucogyrophana subillaqueatum (Litsch.) Jülich
Trechispora subillaqueata (Litsch.) R.L. Gilbertson & Budington
Distr.: Southwestern United States.
Hosts: Pinus, Quercus

Hypochnicium J. Eriksson. Basidiomycetes, Aphyllophorales

Hypochnicium analogum (Bourd. & Galzin) J. Eriksson
Corticium analogum (Bourd. & Galzin) Burt
Distr.: Temperate northern hemisphere.
Notes: On wood.
Hosts: Carya, Fraxinus, Populus, Quercus, Salix
Hypochnicium bombycinum (Sommerf.:Fr.) J. Eriksson
Corticium bombycinum (Sommerf.:Fr.) P. Karst.
Distr.: Northern hemisphere.
Notes: On wood.
Hosts: Alnus, Betula, Juniperus, Picea, Pinus, Populus, Prosopis,

Prunus, Quercus, Salix
Hypochnicium eichleri (Bres.) J. Eriksson & Ryvarden
Peniophora eichleri Bres. in Sacc. & Syd.
Distr.: Temperate northern hemisphere.
Notes: On wood.
Host: Juniperus
Hypochnicium geogenium (Bres.) J. Eriksson
Corticium geogenium Bres.
Distr.: Temperate northern hemisphere.
Notes: On wood.
Hosts: Alnus, Juglans, Pinus, Pseudotsuga, Thuja
Hypochnicium lundellii (Bourd.) J. Eriksson
Distr.: Temperate northern hemisphere.
Notes: On wood.
Hosts: Fraxinus, Juniperus, Pinus, Populus
Hypochnicium prosopidis Burdsall in R.L. Gilbertson et al.
Distr.: Southwestern United States.
Notes: On wood.
Host: Prosopis
Hypochnicium punctulatum (Cooke) J. Eriksson
Corticium punctulatum Cooke
Distr.: Widespread.
Notes: On wood.
Hosts: Acer, Alnus, Ceanothus, Juniperus, Pinus, Populus, Pseudotsuga, Sageretia, Simmondsia, Thuja
Hypochnicium sphaerosporum (Höhn. & Litsch.) J. Eriksson
Peniophora sphaerospora Höhn. & Litsch.
Distr.: Widespread.
Notes: On wood. Some authors (Rogers & Jackson. Farlowia 1:321. 1943) considered this species to be synonymous with *H. punctulatum*.
Hosts: Acer, Alnus, Pinus
Hypochnicium stratosum Burdsall & Nakasone
Distr.: FL(type).
Notes: On wood.
Host: Fagus
Hypochnicium vellereum (Ellis & Cragin) Parmasto
Corticium vellereum Ellis & Cragin
Distr.: Northern hemisphere.
Notes: On wood.
Hosts: Acer, Populus, Thuja

Hypochnopsis P. Karst. Basidiomycetes, Aphyllophorales

Notes: Monotypic.
Hypochnopsis mustialaensis (P. Karst.) P. Karst.
Coniophora mustialaensis (P. Karst.) Massee
Distr.: Temperate northern hemisphere.
Notes: On wood.
Hosts: Abies, Alnus, Pinus

Hypocrea Fr. Pyrenomycetes, Hypocreales

Notes: About 100 species.
Hypocrea ceramica Ellis & Everh.
Chromocrea ceramica (Ellis & Everh.) Seaver
Trichoderma koningii Oudem. (anamorph)
Distr.: Cosmopolitan.
Notes: On decorticated wood.
Host: Juniperus
Hypocrea chionea Ellis & Everh.
Distr.: North America.
Host: Pinus
Hypocrea citrina (Pers.:Fr.) Fr.
Acremonium sp. (anamorph)
Distr.: North America; Europe.
Notes: On polypores. Canham (Mycologia 61:315. 1969).
Host: Acer
Hypocrea gelatinosa (Tode) Fr.
Chromocrea gelatinosa (Tode) Seaver
Creopus gelatinosus (Tode) Link
Gliocladium sp. (anamorph)
Distr.: Cosmopolitan.
Notes: On dead wood and decaying matter.
Hosts: Betula, Salix, Vitis
Hypocrea patella Cooke & Peck
Verticillium sp. (anamorph)
Distr.: Cosmopolitan.
Notes: On rotten wood and other fungi.
Hosts: Alnus, Liriodendron
Hypocrea rufa (Pers.:Fr.) Fr.
Trichoderma sp. Pers.:Fr. (anamorph)
Distr.: Cosmopolitan.
Notes: On rotting wood.
Hosts: Quercus, Rhododendron
Hypocrea schweinitzii (Fr.) Sacc.
Hypocrea lenta Auth. non (Tode) Berk. & Broome
Hypocrea melaleuca Ellis & Everh.
Hypocrea olivacea Cooke & Ellis
Trichoderma sp. (anamorph)
Distr.: North and South America.
Notes: On rotten wood. The type specimen of *Hypocrea lenta* proved this species to be xylariaceous, but the name *H. lenta* has been frequently misapplied and these records probably refer to *H. schweinitzii*.
Hosts: Acer, Betula, Carya, Liquidambar, Pinus, Quercus
Hypocrea sulphurea (Schwein.) Sacc.
Acremonium sp. (anamorph)
Distr.: United States.
Notes: On rotting wood.
Hosts: Liquidambar, Quercus, Tilia

Hypocrella Sacc. Pyrenomycetes, Clavicipitales

Hypocrella tuberiformis (Berk. & Ravenel) Atk.
Hypocrea tuberiformis Berk. & Ravenel
Echinodothis tuberiformis (Berk. & Ravenel) Atk.
Distr.: Southern United States; Caribbean region; South America.
Notes: Patouillard (Bull. Soc. Mycol. France 6:107. 1890) suggested that this species belonged in *Dussiella*.
Host: Arundinaria

Hypocreopsis P. Karst. Pyrenomycetes, Hypocreales

Hypocreopsis consimilis (Ellis) Seaver
Hypocrea consimilis Ellis
Distr.: CA, NJ(type).
Host: Eucalyptus

Hypoderma De Not. Discomycetes, Rhytismatales

Hypoderma cassandrae Ellis & Everh.
Hypodermopsis cassandrae (Ellis & Everh.) Kuntze
Distr.: MI(type).
Host: Chamaedaphne
Hypoderma commune (Fr.:Fr.) Duby
Leptothyrium vulgare (Fr.:Fr.) Sacc. (anamorph)
Distr.: United States; Europe.
Hosts: Amsonia, Aster, Bidens, Bignonia, Chelone, Conyza, Desmodium, Eupatorium, Fraxinus, Helenium, Helianthus, Lespedeza, Magnolia, Rudbeckia, Silphium, Solidago, Tofieldia, Vernonia, Xerophyllum
Hypoderma eucalypti Cooke & Harkn.
Distr.: CA(type).
Host: Eucalyptus
Hypoderma ilicinum De Not.
Distr.: Southern United States; Europe.
Notes: On dead leaves.
Host: Quercus
Hypoderma nitidum De Not.
Distr.: MS; Europe(type).
Host: Solidago
Hypoderma pacificensis Tehon
Distr.: OR(type).
Host: Acer

Hypoderma rubi (Pers.) DC. ex Chev.
Hysterium rubi Pers.
Hypoderma virgultorum DC., nom. illeg.
Leptostroma virgultorum Sacc.
Distr.: Temperate regions.
Notes: Minter (C.M.I. Descr. 781. 1984).
Host: Rubus
Hypoderma rufilabrum (Berk. & M.A. Curtis) Duby
Hysterium rufilabrum Berk. & M.A. Curtis
Distr.: Eastern United States.
Host: Acer
Hypoderma scirpinum (Fr.) DC.:Fr.
Hysterium scirpinum Fr.
Distr.: United States; Europe.
Notes: Sutton (The Coelomycetes. p. 213. 1980) included a brief description of the anamorph.
Hosts: Arundinaria, Scirpus
Hypoderma smilacis (Schwein.) Rehm
Hypodermopsis smilacis (Schwein.) Rehm
Distr.: Eastern United States.
Notes: Cash (Mycologia 31:289. 1939) clarified the identity of this name which is based on a mixed type collection.
Host: Smilax
Hypoderma tunicatum Ellis & Everh.
Hypodermopsis tunicatum (Ellis & Everh.) Kuntze
Distr.: WA.
Host: Arctostaphylos
Hypoderma variegatum (Berk. & M.A. Curtis) Duby
Distr.: NC, SC.
Hosts: Leucothoe, Lyonia, Oxydendrum

Hypodermella Tub. Discomycetes, Rhytismatales

Notes: Many species were transferred to *Lirula* Darker (Darker. Canad. J. Bot. 45:1399. 1967).
Hypodermella laricis Tub.
Hypodermella laricis Tub. var. *octospora* Dearn.
Leptothyrella laricis Dearn. (anamorph)
Distr.: Western United States; Europe.
Host: Larix

Hypomyces (Fr.) Tul. Pyrenomycetes

Notes: Although most species occur on other fungi, they have been reported as occurring on vascular plant substrates.
Hypomyces aurantius (Pers.:Fr.) Tul.
Cladobotryum varium Nees (anamorph)
Distr.: Temperate regions.
Notes: On old fruiting bodies of *Polyporus* and other fungi.
Host: Abies
Hypomyces papyraceus (Ellis & Holw.) Seaver
Hypocrea papyraceus Ellis & Holw.
Distr.: IA(type), OH, WY.
Host: Picea
Hypomyces rosellus (Albertini & Schwein.) Tul.
Cladobotryum dendroides (Bull.:Fr.) W. Gams & Hoozemans (anamorph)
Distr.: Temperate regions.
Notes: On old fruiting bodies of *Stereum*.
Hosts: Fraxinus, Sambucus

Hyponectria Sacc. Pyrenomycetes, Amphisphaeriales

Notes: Barr (Mycologia 69:959. 1977).
Hyponectria auripuncta (Harkn.) Barr
Laestadia auripunctum Harkn.
Distr.: Southwestern North America.
Host: Quercus
Hyponectria betulina Barr
Distr.: Eastern North America.
Host: Betula
Hyponectria buxi (DC. ex Desmaz.) Sacc.
Laestadia buxifolia (Cooke) Sacc.
Distr.: North America; Europe.
Host: Buxus
Hyponectria dakotensis Seaver
Distr.: ND(type).
Notes: This species does not belong to *Hyponectria*, rather it is hypocreaceous, probably belonging to *Nectriella*.
Host: Ambrosia
Hyponectria depressa (Peck) Barr
Guignardia depressa (Peck) Dearn. & House
Mycosphaerella depressa (Peck) House
Distr.: Eastern United States.
Hosts: Lactuca, Rheum
Hyponectria gossypii (Schwein.) Sacc.
Distr.: SC(type).
Notes: This is probably not a *Hyponectria* but has not been placed in a suitable genus.
Host: Gossypium
Hyponectria gregaria Barr
Distr.: CA.
Host: Haplopappus
Hyponectria lonicerae (Dearn. & Barth.) Barr
Guignardia lonicerae Dearn. & Barth.
Distr.: Western North America.
Notes: On leaves.
Host: Lonicera
Hyponectria magnoliae (Schwein.) Barr
Guignardia magnoliae (Schwein.) J.H. Miller
Mycosphaerella magnoliae (Ellis) Petr.
Sphaerella magnoliae Ellis
Laestadia magnoliae (Schwein.) Sacc.
Distr.: Eastern North America.
Notes: On dead leaves.
Host: Magnolia
Hyponectria mohavensis Bonar
Distr.: Southwestern United States.
Notes: This species does not belong in *Hyponectria*, rather it is hypocreaceous, probably belonging to *Nectriella*.
Host: Yucca
Hyponectria populi (G.E. Thompson) Barr
Guignardia populi G.E. Thompson
Distr.: Eastern North America.
Host: Populus
Hyponectria rubescens (Ellis & Everh.) Barr
Laestadia rubescens Ellis & Everh.
Anisostomula rubescens (Ellis & Everh.) Barr
Distr.: Eastern North America.
Host: Castanea
Hyponectria sceptri (P. Karst.) Samuels et al.
Nectriella sceptri (P. Karst.) A.Y. Rossman
Calloria oleosa (Ellis) Sacc.
Nectriella pedicularis (Tracy & Earle) Seaver
Distr.: Temperate North America; Europe.
Notes: On a variety of herbaceous plants. Samuels et al. (Canad. J. Bot. 61:1896. 1984).
Hosts: Aquilegia, Cirsium, Helianthella, Heracleum, Ligusticum, Lomatium, Pedicularis, Petroselinum, Senecio, Valeriana

Hypoxylon Bull.:Fr., nom. cons. Pyrenomycetes, Xylariales

Notes: A monograph was provided by Miller (A monograph of the world species of *Hypoxylon*. 1961). Jong & Rogers (Washington Agr. Exp. Sta. Tech. Bull. 71. 1972) described and illustrated the anamorphs of 18 *Hypoxylon* spp.
Hypoxylon albocinctum Ellis & Everh.
Hypoxylon hydnicola (Schwein.) Sacc.
Distr.: Eastern United States.
Notes: On wood.
Hosts: Acer, Betula, Crataegus, Fraxinus, Ostrya, Ulmus
Hypoxylon argillaceum (Pers.) Berk.
Distr.: Central and eastern North America; Europe.
Notes: On twigs of hardwoods.
Hosts: Chionanthus, Fraxinus

Hypoxylon atropunctatum (Schwein.:Fr.) Cooke
Albocrustum atropunctatum (Schwein.:Fr.) C.G. Lloyd
Numulariola atropunctatum (Schwein.:Fr.) House
Distr.: Central and eastern United States.
Notes: On bark of hardwoods, especially common on dying and recently dead *Quercus* spp.
Hosts: Acer, Fagus, Malus, Ostrya, Quercus
Hypoxylon atropurpureum (Fr.:Fr.) Fr.
Hypoxylon reticulatum P. Karst.
Distr.: Temperate regions.
Notes: On wood (Petrini & Rogers. Mycotaxon 26:414. 1986).
Host: Rhamnus
Hypoxylon bartholomaei Peck
Distr.: Pacific Northwest.
Host: Alnus
Hypoxylon caries (Schwein.) Sacc.
Hypoxylon balansae Speg.
Geniculosporium sp. (anamorph)
Distr.: HI, continental United States; Brazil; The Philippines.
Notes: On hardwoods (Petrini & Rogers. Mycotaxon 26:401. 1986).
Hosts: Acer, Liriodendron
Hypoxylon cinereo-lilacinum J.H. Miller
Distr.: Central and eastern North America; Hungary.
Notes: On wood.
Hosts: Acer, Crataegus, Quercus, Salix, Tilia, Ulmus
Hypoxylon cohaerens (Pers.:Fr.) Fr.
Virgariella sp. (anamorph)
Distr.: Temperate North America; Europe.
Notes: On hardwoods, especially *Fagus* spp.
Hosts: Corylus, Fagus, Quercus
Hypoxylon croceum J.H. Miller
Distr.: Southeastern United States.
Notes: On hardwoods.
Hosts: Liriodendron, Quercus
Hypoxylon crocopeplum Berk. & M.A. Curtis
Hypoxylon pumilo Sacc. & Fairm.
Distr.: Eastern United States; West Indies; South America; India; East Indies.
Hosts: Acer, Carya, Lindera, Quercus
Hypoxylon culmorum Cooke
Distr.: Southeastern United States; West Indies; Central and South America; The Philippines.
Notes: On *Arundinaria* and other grasses.
Host: Arundinaria
Hypoxylon fragiforme (Scop.:Fr.) J. Kickx fil.
Hypoxylon coccineum Bull.
Hypoxylon majusculum Cooke
Nodulisporium sp. (anamorph)
Distr.: North and South America; Europe.
Notes: On a variety of woody plants, especially *Fagus*.
Hosts: Alnus, Fagus, Picea, Quercus
Hypoxylon fuscum (Pers.:Fr.) Fr.
Hypoxylon pruinatoides Kauffm.
Virgariella sp. (anamorph)
Distr.: Northern hemisphere.
Notes: On dead wood, especially of Betulaceae.
Hosts: Acer, Alnus, Betula, Carpinus, Carya, Castanea, Corylus, Fraxinus, Ilex, Nyssa, Ostrya, Prunus, Quercus, Salix, Tilia
Hypoxylon grenadense J.H. Miller var. ***macrospora*** J.H. Miller
Rhinocladiella sp. (anamorph)
Distr.: WA; West Indies.
Notes: On wood.
Host: Acer
Hypoxylon howeianum Peck
Nodulisporium sp. (anamorph)
Distr.: Cosmopolitan.
Notes: On hardwoods.
Hosts: Acer, Betula, Castanea, Fagus, Liriodendron, Malus, Ostrya, Oxydendrum, Populus, Quercus, Tilia
Hypoxylon hypophlaeum (Berk. & Ravenel) J.H. Miller
Hypoxylon cinereum Ellis & Everh.
Distr.: Southern United States, subtropical and tropical regions.
Notes: On hardwoods.
Hosts: Acer, Alnus, Baccharis, Carpinus, Cercis, Fraxinus, Liriodendron, Magnolia, Ostrya, Persea, Quercus, Vitis
Hypoxylon illitum (Schwein.) M.A. Curtis
Distr.: Central and eastern United States; China; Italy.
Notes: On hardwoods.
Hosts: Carya, Castanea, Fagus, Fraxinus, Liquidambar, Platanus, Quercus
Hypoxylon investiens (Schwein.) M.A. Curtis
Geniculosporium sp. (anamorph)
Distr.: Western hemisphere; Africa; Asia.
Hosts: Acer, Fagus, Fraxinus, Quercus
Hypoxylon investiens (Schwein.) M.A. Curtis var. ***epiphaeum*** (Berk. & M.A. Curtis) J.H. Miller
Hypoxylon epiphaeum Berk. & M.A. Curtis
Distr.: Southeastern United States.
Notes: Apparently restricted to *Magnolia virginiana*.
Host: Magnolia
Hypoxylon jecorinum Berk. & Ravenel
Distr.: Southeastern United States; West Indies; Central and South America; Uganda.
Notes: On a variety of woods.
Hosts: Acer, Celtis, Quercus
Hypoxylon laurus J.H. Miller
Distr.: OH, VA.
Host: Gleditsia
Hypoxylon maculum (Schwein.) J.H. Miller
Distr.: Central and southeastern United States.
Host: Quercus
Hypoxylon mammatum (Wahlenberg) J.H. Miller
Hypoxylon blakei Berk. & M.A. Curtis
Entoleuca callimorpha Syd.
Hypoxylon morsei Berk. & M.A. Curtis
Hypoxylon pruinatum (Klotzsch) Cooke
Geniculosporium sp. (anamorph)
Distr.: Temperate North America.
Notes: On various hardwoods, but most frequently on *Populus* and *Salix* spp. Miller (A monograph of the world species of *Hypoxylon*. p. 64. 1961) reviewed the synonyms of this species. Hawksworth (C.M.I. Descr. 356. 1972).
Hosts: Acer, Alnus, Betula, Carpinus, Malus, Populus, Quercus, Salix
Hypoxylon mediterraneum (De Not.) J.H. Miller
Diatrype clypeus (Schwein.) Berk.
Nummularia clypeus (Schwein.) Cooke
Periconiella sp. (anamorph)
Distr.: Eastern and southeastern United States(common); temperate northern hemisphere.
Notes: On hardwoods (Hawksworth. C.M.I. Descr. 359. 1972).
Hosts: Acer, Alnus, Betula, Ceanothus, Cornus, Diospyros, Fagus, Malus, Ostrya, Paulownia, Populus, Quercus, Salix, Tilia, Ulmus, Vitis
Hypoxylon mediterraneum (De Not.) J.H. Miller var. ***microspora*** J.H. Miller
Periconiella sp. (anamorph)
Distr.: Western North America; Africa; Asia; New Zealand.
Host: Alnus
Hypoxylon microplacum (Berk. & M.A. Curtis) J.H. Miller
Nummularia microplaca (Berk. & M.A. Curtis) Cooke
Hansfordia sp. (anamorph)
Distr.: North America; West Indies; Central America; eastern Asia; Maritius.
Notes: Often on wood of Lauraceae and Myrtaceae.
Hosts: Magnolia, Persea, Quercus, Sassafras
Hypoxylon morganii Ellis & Everh.
Distr.: OH.
Host: Carya
Hypoxylon multiforme (Fr.:Fr.) Fr.
Nodulisporium sp. (anamorph)
Distr.: Temperate northern hemisphere, high elevations in tropical regions.
Hosts: Acer, Alnus, Betula, Quercus
Hypoxylon notatum Berk. & M.A. Curtis
Hypoxylon pallidum Ellis & Everh.
Distr.: Southeastern United States.
Notes: On wood.
Host: Quercus
Hypoxylon nummularium Bull.:Fr. var. ***exutans*** (Cooke) J.H. Miller
Nummularia exutans Cooke

Distr.: Warm temperate and tropical regions.
Notes: Although the type variety has been placed in *Biscogniauxia*, this variety has not been transferred.
Host: Sabal
Hypoxylon nummularium Bull.:Fr. var. **macrosporum** Gilman & L.C. Tiffany
Distr.: IA.
Host: Quercus
Hypoxylon papillatum Ellis & Everh.
Nodulisporium sp. (anamorph)
Distr.: Central and eastern North America to Nicaragua.
Notes: On a variety of woods. Rogers (Mycotaxon 23:429. 1985) described the anamorph.
Hosts: Betula, Carya, Quercus
Hypoxylon punctulatum (Berk. & Ravenel) Cooke
Distr.: Central and eastern North America.
Notes: On hardwoods. The anamorph has been assigned to *Basidiobotrys* (Barnett. Mycologia 49:588. 1957) or, more recently, *Xylocladium* (Jong & Rogers. Washington Agr. Exp. Sta. Tech. Bull. 71:40. 1972).
Hosts: Castanea, Halesia, Ilex, Quercus
Hypoxylon quadratum (Schwein.) Ellis & Everh.
Distr.: OH, PA.
Notes: On hardwoods.
Host: Juglans
Hypoxylon regale Morg.
Distr.: Eastern United States.
Notes: On hardwoods.
Hosts: Acer, Fraxinus, Quercus
Hypoxylon rubiginosum (Pers.:Fr.) Fr.
Hypoxylon bicolor Ellis & Everh.
Hypoxylon botrys Nitschke
Hypoxylon fuscopurpureum (Schwein.) Berk.
Nummularia lateritia Ellis & Everh.
Hypoxylon nuttallii Ellis & Everh.
Hypoxylon perforatum Schwein.:Fr.
Virgariella sp. (anamorph)
Distr.: Cosmopolitan.
Notes: Probably the most common *Hypoxylon* sp. (Hawksworth. C.M.I. Descr. 357. 1972).
Hosts: Acer, Aesculus, Ailanthus, Alnus, Amelanchier, Arundinaria, Betula, Carpinus, Carya, Castanea, Catalpa, Celtis, Cercis, Cercocarpus, Cinnamomum, Cliftonia, Cornus, Corylus, Crataegus, Dichrostachys, Fagus, Fraxinus, Garrya, Gleditsia, Heteromeles, Ilex, Juglans, Ligustrum, Lindera, Liriodendron, Lonicera, Magnolia, Malus, Myrica, Nyssa, Ostrya, Paulownia, Platanus, Populus, Prunus, Pyrus, Quercus, Rhamnus, Rhododendron, Rhus, Salix, Sambucus, Swietenia, Taxodium, Tilia, Ulmus, Umbellularia, Viburnum, Vitis
Hypoxylon rubiginosum (Pers.:Fr.) Fr. var. **ferrugineum** (G. Otth) J.H. Miller
Hypoxylon ferrugineum G. Otth
Distr.: Temperate North America; Europe.
Host: Acer
Hypoxylon rubrostromaticum J.H. Miller
Distr.: GA; Trinidad; Brazil; New Zealand.
Notes: On wood.
Hosts: Acer, Fraxinus, Quercus
Hypoxylon sassafras (Schwein.:Fr.) M.A. Curtis
Hypoxylon callostroma (Schwein.) Berk.
Rosellinia linderae Peck
Rosellinia prinicola (Berk. & M.A. Curtis) Sacc.
Distr.: North and South America; China; Europe.
Notes: On wood of Lauraceae.
Hosts: Lindera, [Magnolia], Persea, Sassafras, Umbellularia
Hypoxylon sclerophaeum Berk. & M.A. Curtis
Distr.: Temperate North America to the southern hemisphere.
Hosts: Liquidambar, Sassafras
Hypoxylon serpens (Pers.:Fr.) J. Kickx fil.
Hypoxylon subluteum Ellis & Everh.
Geniculosporium serpens Chesters & Greenhalgh (anamorph)
Distr.: Temperate and subtropical regions
Notes: On hardwoods (Hawksworth. C.M.I. Descr. 358. 1972; Petrini & Rogers. Mycotaxon 26:401-436. 1986). *H. serpens* has been resolved into several species, e.g. *H. aeneum*, *H. atropurpureum*, *H. bipapillatum*, and *H. atropurpureum*.
Hosts: Acer, Berberis, Carya, Ceanothus, Cornus, Fagus, Liriodendron, Malus, Populus, Quercus, Rhododendron, Tilia
Hypoxylon serpens (Pers.:Fr.) J. Kickx fil. var. **effusum** (Nitschke) J.H. Miller
Hypoxylon effusum Nitschke
Distr.: HI, temperate North America; Europe.
Hosts: Magnolia, Mangifera, Ulmus
Hypoxylon serpens (Pers.:Fr.) J. Kickx fil. var. **macrospora** J.H. Miller
Geniculosporium sp. (anamorph)
Distr.: North America, most common in the West.
Notes: On hardwoods.
Hosts: Acer, Cornus, Fagus, Populus, Prunus, Quercus
Hypoxylon smilacicola E. Howe
Distr.: Eastern United States.
Notes: On dead stems.
Host: Smilax
Hypoxylon sphaeriostomum (Schwein.) Sacc.
Hypoxylon regale Morg. var. *macrosporum* J.H. Miller
Distr.: Eastern United States.
Notes: On hardwoods.
Hosts: Fagus, Quercus
Hypoxylon stygium (Lév.) Sacc.
Hypoxylon platystomum Ellis & Everh.
Nummularia rufa Ellis & Everh.
Distr.: Warm-temperate to tropical regions.
Notes: On a variety of woods.
Hosts: Fagus, Fraxinus, Myrica, Persea, Quercus
Hypoxylon thouarsianum (Lév.) C.G. Lloyd
Daldinia malleola (Berk. & Ravenel) Kauffm.
Hypoxylon malleolus Berk. & Ravenel
Hypoxylon occidentale Ellis & Everh.
Distr.: Southern United States; West Indies; Central and South America; South Pacific Islands.
Notes: On a variety of woody plants.
Hosts: Quercus, Sequoia
Hypoxylon tinctor (Berk.) Cooke
Xylocladium sp. (anamorph)
Distr.: Central and eastern United States; Central and South America.
Notes: On wood of hardwoods.
Hosts: Acer, Carpinus, Fraxinus, Liriodendron, Magnolia, Nyssa, Platanus, Quercus, Vitis
Hypoxylon truncatum (Schwein.:Fr.) J.H. Miller
Hypoxylon annulatum (Schwein.) Mont. in Gay
Hypoxylon marginatum (Schwein.) Berk.
Hypoxylon vernicosum Ellis & Everh.
Nodulisporium sp. (anamorph)
Distr.: Cosmopolitan.
Notes: On hardwoods, especially common on white oak.
Hosts: Acacia, Acer, Betula, Cercis, Fagus, Heteromeles, Magnolia, Myrica, Ostrya, Quercus, Ulmus
Hypoxylon vogesiacum (Pers.) Sacc.
Hypoxylon oregonense Kauffm.
Nodulisporium sp. (anamorph)
Distr.: Western North America; Europe.
Notes: On hardwoods.
Hosts: Acer, Alnus, Amelanchier, Arbutus, Quercus
Hypoxylon vogesiacum (Pers.) Sacc. var. **macrospora** J.H. Miller
Nodulisporium sp. (anamorph)
Distr.: Western North America; Europe (Whalley & Petrini. Trans. Brit. Mycol. Soc. 82:364. 1984).
Host: Salix
Hypoxylon vogesiacum (Pers.) Sacc. var. **microspora** J.H. Miller
Distr.: Temperate North America.
Notes: On hardwoods.
Host: Quercus
Hypoxylon weldenii J.D. Rogers
Distr.: LA(type).
Notes: Rogers (Canad. J. Bot. 55:372. 1977).
Host: Nyssa

Hypsizygus Singer. Basidiomycetes, Agaricales

Hypsizygus tessellatus (Bull.:Fr.) Singer
Hypsizygus elongatipes (Peck) H.E. Bigelow
Pleurotus elongatipes Peck
Agaricus marmoreus Peck
Hypsizygus marmoreus (Peck) H.E. Bigelow
Distr.: Temperate northern hemisphere.
Notes: Many old reports of *Pleurotus ulmarius* may refer to this taxon (Redhead. Mycologia 78:522. 1986).
Hosts: Acer, Alnus, Liriodendron, Populus, Quercus, Tsuga
Hypsizygus ulmarius (Bull.:Fr.) Redhead
Lyophyllum ulmarium (Bull.:Fr.) Kühner
Pleurotus ulmarius (Bull.:Fr.) P. Kumm.
Distr.: Temperate northern hemisphere.
Notes: Redhead (Trans. Mycol. Soc. Japan 25:1. 1984; Mycologia 78:522. 1986).
Hosts: Acer, Liriodendron, Malus, Platanus, Quercus, Salix, Ulmus

Hysterium Pers.:Fr. Loculoascomycetes, Dothideales

Notes: Zogg (Beitr. Kryptogamenfl. Schweiz 11(3):21. 1962).
Hysterium angustatum Albertini & Schwein.:Fr.
Hysterium acuminatum Fr.:Fr.
Hysterium acuminatum Fr.:Fr. var. *alpinum* Rehm
Hysterium eucalypti W. Phillips & Harkn.
Distr.: Cosmopolitan.
Notes: On bark and wood.
Hosts: Abies, Betula, Lonicera, Pinus, Populus, Quercus, Tsuga, Vaccinium
Hysterium insidens Schwein.
Hysterium depressum Berk. & M.A. Curtis
Hysterium fusiger Berk. & M.A. Curtis
Coniosporium sp. (anamorph)
Distr.: Cosmopolitan.
Notes: On dead wood of trees. For the anamorph, see Ellis (Dematiaceous Hyphomycetes. p. 41. 1971).
Hosts: Quercus, Robinia
Hysterium macrosporum W. Gerard in Peck
Distr.: FL, GA, NY.
Notes: On dead wood.
Hosts: Leucothoe, Pinus, [Smilax], Vaccinium
Hysterium magnosporum W. Gerard
Distr.: CA, GA, NC.
Hosts: Athyrium, Castanea, Quercus, Salix
Hysterium pulicare Pers.:Fr.
Hysterium acervulatum Schwein.
Hysterium betuliginum Schwein.
Hysterium teres Schwein.
Distr.: Cosmopolitan.
Notes: On bark of living trees and dead wood.
Hosts: Betula, Cornus, Eucalyptus, Malus, Quercus
Hysterium thujae (Roberge) House
Hysterium pinastri Schrad.:Fr. var. *thujae* Roberge
Distr.: NY; Europe (type).
Notes: This fungus may belong in *Lophodermium*.
Host: Thuja

Hysterographium Corda emend. De Not. Loculoascomycetes, Dothideales

Notes: Zogg (Beitr. Kryptogamenfl. Schweiz 11(3):32. 1962).
Hysterographium arctostaphyli Kauffm.
Distr.: North America.
Notes: On dead wood.
Host: Arctostaphylos
Hysterographium bakeri Tracy & Earle
Distr.: CA(type).
Host: Heteromeles
Hysterographium flexuosum (Schwein.:Fr.) Sacc.
Hysterographium nova-caesariense (Ellis) Ellis & Everh.
Hysterium prominens W. Phillips & Plowr.
Hysterographium prominens (W. Phillips & Plowr.) Berl. & Voglino
Hysterographium vulvatum (Schwein.) Rehm
Distr.: North America; Europe.
Hosts: Acer, Adenostoma, Arbutus, Arctostaphylos, Castanea, Heteromeles, Lyonothamnus, Pinus, Prunus, Quercus, Rhus, Salix, Sassafras, Vitis
Hysterographium fraxini (Pers.:Fr.) De Not.
Hysterographium acerinum Peck
Hysterographium elongatum (Wahlenberg) Corda
Distr.: Widespread on a variety of woody hosts.
Hosts: Acer, Alnus, Cercocarpus, Fraxinus, Salix
Hysterographium mori (Schwein.) Rehm
Hysterographium lesquereuxii (Duby) Sacc.
Hysterographium rousselii De Not.
Hysterographium viticola (Cooke & Peck) Rehm
Distr.: Worldwide on a variety of woody hosts.
Hosts: Catalpa, Cercocarpus, Chionanthus, Gleditsia, Leucothoe, Lyonothamnus, Morus, Pinus, Quercus, Salix, Viburnum, Vitis
Hysterographium nucicola (Schwein.) Ellis & Everh.
Hysterium nucicola Schwein.
Distr.: NJ(type).
Host: Carya
Hysterographium opuntiae J. Brown
Distr.: AZ.
Notes: A doubtful species.
Host: Opuntia
Hysterographium subrugosum (Cooke & Ellis) Sacc.
Hysterographium kansense Ellis & Everh.
Distr.: North America; Europe.
Notes: On wood and bark.
Hosts: Castanea, Cercocarpus, Quercus

Hysteromyxa Sacc. & Ellis. Ascomycetes

Notes: According to Petrak (Sydowia 5:196. 1951), this genus cannot be adequately characterized.
Hysteromyxa effugiens Sacc. & Ellis
Distr.: NJ(type).
Host: Chamaecyparis

Hysteropatella Rehm. Loculoascomycetes, Dothideales

Notes: Zogg (Phytopathol. Z. 14:310. 1962) excluded this genus from the Hysteriaceae.
Hysteropatella prostii (Duby) Rehm
Hysterium prostii Duby
Distr.: Central and southern United States.
Hosts: Abelmoschus, Cornus, Hibiscus, Salix, Ulmus

Hysteropezizella Höhn. Discomycetes, Helotiales

Hysteropezizella diminuens (P. Karst.) Nannf.
Hysteropezizella subvelata (Rehm) Höhn.
Distr.: Arctic and north temperate regions.
Host: Carex
Hysteropezizella ignobilis (P. Karst.) Lind
Naevia ignobilis P. Karst.
Distr.: Arctic and subarctic regions.
Host: Carex
Hysteropezizella multipuncta (Peck) Nannf.
Pyrenopeziza multipuncta (Peck) Sacc.
Distr.: NY(type).
Host: Carex

Hysterostoma Theiss. Loculoascomycetes, Dothideales

Notes: This genus is a synonym of *Dothidasteromella*, but not all names have been redisposed.
Hysterostoma weberi Westend.
Distr.: FL.
Notes: The correct placement of this species is uncertain.
Host: Lyonia

Hysterostomella Speg. Loculoascomycetes, Dothideales

Hysterostomella floridana Tracy & Earle
Distr.: FL(type).
Host: Ardisia
Hysterostomella sparsa (Peck & G.W. Clinton) Barr in Barr et al.
Rhytisma sparsa Peck & G.W. Clinton
Cocconia sparsa (Peck & G.W. Clinton) Sacc.
Distr.: FL.
Host: Sabal

Idriella P.E. Nelson & K. Wilh. Hyphomycetes

Notes: Morgan-Jones (Mycotaxon 8:402. 1979); Nelson & Wilhelm (Mycologia 48:547. 1956).
Idriella angustispora Morgan-Jones
Distr.: AL(type).
Host: Quercus
Idriella bambusae Morgan-Jones
Distr.: AL(type).
Host: Bambusa
Idriella lunata P.E. Nelson & K. Wilh.
Distr.: CA, MD, probably widespread.
Notes: Root rot of strawberry.
Host: Fragaria

Illosporium Mart. Hyphomycetes

Notes: Hawksworth (Bull. Brit. Mus. Nat. Hist., Bot. 6:181. 1979) noted that the name has been employed for a large number of fungi which are not congeneric with the type species, *I. carneum* Fr.:Fr.
Illosporium coccinellum Cooke
Distr.: SC(type).
Host: Pinus
Illosporium commelinae F. Stevens
Distr.: PR(type), TX.
Host: Commelina
Illosporium conicola Ellis & Everh. in Shear
Distr.: DC, MD.
Host: Pinus
Illosporium malifoliorum J. Sheld.
Distr.: Eastern United States.
Notes: Leaf spot of apple.
Host: Malus
Illosporium pallidum Cooke
Distr.: SC(type).
Host: Zea

Incrucipulum Baral & Krieglsteiner. Discomycetes, Helotiales

Notes: Baral & Krieglsteiner (Beih. Z. Mykol. 6:71. 1985).
Incrucipulum capitatum (Peck) Baral & Krieglsteiner
Lachnella capitata (Peck) Seaver
Distr.: Eastern United States; Europe.
Host: Quercus
Incrucipulum ciliare (Schrad.:Fr.) Baral & Krieglsteiner
Lachnella ciliaris (Schrad.:Fr.) W. Phillips
Lachnum ciliare (Schrad.:Fr.) Rehm
Distr.: Eastern North America; Europe.
Notes: On leaves.
Host: Quercus
Incrucipulum sulphurellum (Peck) Baral & Krieglsteiner
Dasyscyphus sulphurellus (Peck) Sacc.
Dasyscyphus crucifera (W. Phillips) Sacc.
Distr.: NC; Europe.
Hosts: Abies, Betula, Gaylussacia

Incrupila Raitviir. Discomycetes, Helotiales

Incrupila aspidii (Lib.) Raitviir
Peziza aspidii Lib.
Dasyscyphus aspidii (Lib.) Cash
Distr.: AK; Europe.
Notes: On fern fronds.
Host: Dryopteris

Inonotus P. Karst. Basidiomycetes, Aphyllophorales

Notes: Gilbertson & Ryvarden (North American Polypores 1:356. 1986).
Inonotus andersonii (Ellis & Everh.) Cerny
Mucronoporus andersonii Ellis & Everh.
Poria andersonii (Ellis & Everh.) J.J. Neuman
Distr.: Temperate northern hemisphere.
Notes: On hardwoods, especially *Quercus* and *Carya* spp.
Hosts: Carya, Quercus, Salix
Inonotus arizonicus R.L. Gilbertson
Distr.: Southwestern United States.
Notes: On wood.
Host: Platanus
Inonotus circinatus (Fr.) R.L. Gilbertson
Trametes circinatus Fr.
Onnia circinata (Fr.) P. Karst.
Polyporus tomentosus Fr.:Fr. var. *circinatus* (Fr.) Sartory & Maire
Polyporus circinatus (Fr.) Fr.
Polyporus dualis Peck
Distr.: Temperate northern hemisphere.
Notes: On living trunks or roots of conifers, especially *Pinus* spp.; also on dead wood.
Hosts: Abies, Larix, Picea, Pinus, Pseudotsuga, Tsuga
Inonotus crocitinctus (Berk. & M.A. Curtis) Ryvarden
Distr.: TX; Cuba.
Host: Juniperus
Inonotus cuticularis (Bull.:Fr.) Fr.
Polyporus cuticularis (Bull.:Fr.) Fr.
Distr.: Temperate northern hemisphere.
Notes: On hardwoods, developing from wounds on living trees, or on dead wood.
Hosts: Acer, Carya, Celtis, Fagus, Ficus, Morus, Ostrya, Populus, Quercus, Salix, Sambucus, Schinus, Ulmus
Inonotus dryadeus (Pers.:Fr.) Murrill
Polyporus dryadeus (Pers.:Fr.) Fr.
Distr.: Widespread.
Notes: Usually at base of living *Quercus* species, sometimes on dead wood in eastern North America; on *Abies* spp. in the West.
Hosts: Abies, Acer, Picea, Pinus, Quercus, Tsuga, Ulmus
Inonotus dryophilus (Berk.) Murrill
Polyporus dryophilus Berk.
Distr.: Temperate northern hemisphere.
Notes: Usually on living *Quercus* spp; occasionally on other hardwoods.
Hosts: [Abies], Acer, Populus, Prunus, Quercus, Salix, Schinus
Inonotus farlowii (C.G. Lloyd) R.L. Gilbertson
Polyporus farlowii C.G. Lloyd
Distr.: Southwestern United States.
Notes: Heart rot of living trees.
Hosts: Morus, Populus, Salix, Sambucus, Schinus
Inonotus glomeratus (Peck) Murrill
Polyporus glomeratus Peck
Distr.: Eastern North America(common), western North America(rare); Australasia; Europe.
Notes: Heart rot of living hardwoods, associated with sterile conks, commonly on *Acer*, *Fagus*, and *Populus* spp.
Hosts: Acer, Betula, Castanea, Fagus, Populus, Quercus
Inonotus hispidus (Bull.:Fr.) P. Karst.
Polyporus hispidus (Bull.:Fr.) Fr.
Distr.: Eastern and western North America; Europe.
Notes: Usually on living trees (Pegler & Waterston. C.M.I. Descr. 193. 1968).
Hosts: Abies, Acer, Betula, Carya, Fagus, Fraxinus, Juglans, Morus, Quercus, Rhamnus
Inonotus jamaicensis Murrill
Distr.: AZ; Jamaica.
Notes: On wood.

Host: Quercus
Inonotus juniperinus Murrill
Distr.: TX.
Notes: On roots.
Host: Juniperus
Inonotus ludovicianus (Pat.) Murrill
Polyporus ludovicianus (Pat.) Sacc. & Trott.
Distr.: Southeastern United States.
Notes: On hardwoods.
Hosts: Liquidambar, Quercus
Inonotus munzii (C.G. Lloyd) R.L. Gilbertson
Polyporus munzii C.G. Lloyd
Inonotus schinii J. Brown
Distr.: Southwestern United States; Mexico.
Notes: On hardwoods.
Hosts: Morus, Platanus, Populus, Quercus, Salix, Schinus, Ulmus
Inonotus obliquus (Pers.:Fr.) Pilát
Polyporus obliquus (Pers.:Fr.) Fr.
Poria obliqua (Pers.:Fr.) P. Karst.
Distr.: Temperate northern hemisphere.
Notes: On hardwoods.
Hosts: Alnus, Betula, Carya, Fagus, Ostrya
Inonotus patouillardii (Rick) Imazeki
Distr.: AZ; tropical regions.
Host: Quercus
Inonotus porrectus Murrill
Polyporus porrectus (Murrill) Sacc. & Trott.
Distr.: LA; Bahamas.
Hosts: Gleditsia, Quercus
Inonotus quercustris Blackwell & R.L. Gilbertson
Distr.: LA.
Notes: On living trees. Blackwell & Gilbertson (Mycotaxon 23:285. 1985).
Host: Quercus
Inonotus radiatus (Sowerby:Fr.) P. Karst.
Polyporus radiatus (Sowerby:Fr.) Fr.
Polyporus illinoisensis D. Baxter
Polyporus radiatus (Sowerby:Fr.) Fr. var. *cephalanthi* Overh.
Distr.: Eastern North America, rare in the West; Asia; Europe.
Notes: On hardwoods.
Hosts: Alnus, Betula, Cephalanthus, Corylus, Fagus, Salix
Inonotus rheades (Pers.) Bondartsev & Singer
Polyporus rheades Pers.
Polyporus dryophilus Berk. var. *vulpinus* (Fr.) Overh.
Polyporus vulpinus Fr.
Distr.: Temperate northern hemisphere.
Hosts: Fagus, Populus, Quercus
Inonotus rickii (Pat.) D. Reid
Distr.: Southern United States.
Hosts: Casuarina, Cercidium, Myrica, Parkinsonia, Quercus
Inonotus subiculosus (Peck) J. Eriksson & A. Strid
Poria subiculosa (Peck) Cooke
Distr.: Temperate northern hemisphere.
Notes: On conifers.
Host: Thuja
Inonotus texanus Murrill
Polyporus texanus (Murrill) Sacc. & Trott.
Distr.: Southwestern United States.
Hosts: Acacia, Prosopis
Inonotus tomentosus (Fr.:Fr.) S. Teng
Polyporus tomentosus Fr.:Fr.
Coltricia tomentosa (Fr.:Fr.) Murrill
Onnia tomentosa (Fr.:Fr.) P. Karst.
Distr.: Temperate northern hemisphere.
Notes: On dead wood of conifers or on living trunks or roots.
Hosts: Abies, Larix, Picea, Pinus, Pseudotsuga, Thuja, Tsuga

Insolibasidium Oberwinkler & Bandoni. Basidiomycetes, Auriculariales

Notes: Oberwinkler & Bandoni (Trans. Brit. Mycol. Soc. 83:639. 1984).
Insolibasidium deformans (C.J. Gould) Oberwinkler & Bandoni
Herpobasidium deformans C.J. Gould
Distr.: North America.
Notes: Parasitic on leaves.
Host: Lonicera

Intextomyces J. Eriksson & Ryvarden. Basidiomycetes, Aphyllophorales

Intextomyces contiguus (P. Karst.) J. Eriksson & Ryvarden
Corticium contiguum P. Karst.
Corticium subcinereum Burt
Distr.: Temperate northern hemisphere.
Notes: On wood.
Host: Fraxinus

Ionomidotis E.J. Durand. Discomycetes, Helotiales

Ionomidotis fulvotingens (Berk. & M.A. Curtis) Cash
Cenangium fulvotingens Berk. & M.A. Curtis
Distr.: GA, PA.
Host: Quercus
Ionomidotis plicata (W. Phillips & Harkn.) E.J. Durand
Midotis plicata W. Phillips & Harkn.
Distr.: CA(type).
Host: Umbellularia

Irenopsis F. Stevens. Pyrenomycetes, Meliolales

Notes: Hansford (Beih. Sydowia 2. 1961).
Irenopsis cryptocarpa (Ellis & G. Martin) Hansf.
Meliola cryptocarpa Ellis & G. Martin
Irenina cryptocarpa (Ellis & G. Martin) Hansf.
Distr.: FL.
Host: Gordonia
Irenopsis quercifolia Hansf.
Distr.: FL(type).
Host: Quercus
Irenopsis tehoniana (Trott.) Hansf.
Meliola tehoniana Trott.
Irenopsis compacta (Earle) F. Stevens
Meliola compacta Earle
Irenopsis conferta (Tehon) F. Stevens
Meliola conferta Tehon
Meliola earliana Cif.
Distr.: FL, PR; Dominican Republic; Brazil.
Host: Crossopetalum

Irpex Fr.:Fr. Basidiomycetes, Aphyllophorales

Irpex lacteus (Fr.:Fr.) Fr.
Irpex rimosus Peck
Irpex tulipiferae (Schwein.) Schwein.
Polyporus tulipiferae (Schwein.) Overh.
Distr.: Widespread.
Notes: On wood.
Hosts: Abies, Acer, Albizia, Alnus, Aralia, Baccharis, Betula, Carpinus, Carya, Castanea, Celtis, Cornus, Fagus, Fraxinus, Ginkgo, Gleditsia, Liriodendron, Magnolia, Malus, Picea, Pinus, Populus, Prunus, Pseudotsuga, Quercus, Rhus, Sabal, Salix, Symphoricarpos, Tsuga, Ulmus, Vitis

Isariopsis Fresen. Hyphomycetes

Isariopsis albo-rosella (Desmaz.) Sacc.
Distr.: Temperate regions.
Notes: On dead herbarceous plant material.
Hosts: Cerastium, Stellaria
Isariopsis bulbigera (Fuckel) Savile
Scolicotrichum bulbigerium Fuckel
Ovularia bulbigera (Fuckel) Sacc.
Mycosphaerella pseudomaculiformis (Desmaz.) J. Schröt. (teleomorph)
Distr.: Temperate North America; Europe.

Notes: Leaf blotch (Savile. Canad. J. Bot. 35:201. 1957).
Host: Sanguisorba
Isariopsis carnea Ellis & G. Martin, nom. nud.
Distr.: FL.
Host: Osmanthus
Isariopsis linderae (Ellis & Everh.) Sacc.
Graphium linderae Ellis & Everh.
Exosporium linderae (Ellis & Everh.) Höhn.
Distr.: Eastern United States.
Hosts: Diospyros, Lindera
Isariopsis subulata Ellis & Everh.
Distr.: LA.
Host: Zea

Ischnoderma P. Karst. Basidiomycetes, Aphyllophorales

Ischnoderma resinosum (Schrad.:Fr.) P. Karst.
Polyporus resinosus (Schrad.:Fr.) Fr.
Trametes benzoina (Wahlenberg:Fr.) Fr.
Polyporus benzoinus (Wahlenberg:Fr.) Fr.
Ischnoderma fuliginosum (Scop.) Murrill
Polyporus fuliginosus Scop.
Polyporus fuscus (Pers.) C.G. Lloyd
Distr.: Northern hemisphere.
Notes: On wood.
Hosts: Abies, Acer, Betula, Castanea, Fagus, Fraxinus, Juniperus, Larix, Liriodendron, Malus, Nyssa, Picea, Pinus, Populus, Pseudotsuga, Quercus, Thuja, Tilia, Tsuga, Ulmus

Isothea Fr. Pyrenomycetes, Phyllachorales

Isothea nyssae Berk. & M.A. Curtis
Distr.: SC(type).
Host: Nyssa

Isthmiella Darker. Discomycetes, Rhytismatales

Notes: Funk (Foliar fungi of western trees. p. 54. 1985).
Isthmiella abietis (Dearn.) Darker
Bifusella abietis Dearn.
Distr.: Canada to NM, Rocky Mountains.
Notes: *Istmiella quadrispora* Ziller is a similar species occurring on subalpine fir in western Canada.
Host: Abies
Isthmiella crepidiformis (Darker) Darker
Bifusella crepidiformis Darker
Distr.: MT; Canada(Ontario).
Notes: On needles.
Host: Picea
Isthmiella faullii (Darker) Darker
Bifusella faullii Darker
Distr.: Northeastern North America.
Notes: Cannon & Minter (C.M.I. Descr. 793. 1984).
Host: Abies

Isthmospora F. Stevens. Hyphomycetes

Notes: The ascomycetous state is *Trichothyrium* (Hughes. Mycol. Pap. 50:77. 1953).
Isthmospora glabra F. Stevens
Trichothyrium reptans (Berk. & M.A. Curtis) S.J. Hughes (teleomorph)
Distr.: Widespread in tropical regions.
Notes: Associated with *Meliola* on living leaves.
Host: Desmodium

Itersonilia Derx. Hyphomycetes

Notes: Sowell & Korf (Mycologia 52:934. 1960).
Itersonilia perplexans Derx
Itersonilia pyriformis Nyland
Distr.: Temperate North America; Australia.
Hosts: Acer, Chrysanthemum

Jahnoporus I. Nuss. Basidiomycetes, Aphyllophorales

Jahnoporus hirtus (Quél. ex Cooke) Donk
Fomes hirtus Quél. ex Cooke
Polyporus hirtus Quél., nom. inval.
Polyporus hispidellus Peck
Distr.: North America(rare).
Notes: Around trees or stumps, on buried wood or roots of uptorn trees, usually associated with conifers. Gilbertson & Ryvarden (North American Polypores 1:408. 1986).
Hosts: Abies, Picea, Pseudotsuga

Janetia M.B. Ellis. Hyphomycetes

Notes: Hughes (New Zealand J. Bot. 21:177. 1983).
Janetia bonarii (M.B. Ellis) S.J. Hughes
Sporidesmium bonarii M.B. Ellis
Distr.: CA.
Host: Torreya
Janetia garryae (Bonar) S.J. Hughes
Sporidesmium garryae Bonar
Distr.: CA.
Host: Garrya

Jattaea Berl. Pyrenomycetes, Calosphaeriales

Notes: Barr (Mycologia 77:549. 1985).
Jattaea ceanothina (Peck) Barr
Physalospora ceanothina (Peck) Sacc.
Distr.: NY.
Host: Ceanothus
Jattaea curvicolla (Peck) Barr
Eutypa curvicolla (Peck) Barr
Gnomoniella curvicolla (Peck) Sacc.
Distr.: NY.
Host: Polygonella
Jattaea microtheca (Cooke & Ellis) Berl.
Calosphaeria microtheca (Cooke & Ellis) Sacc.
Distr.: Eastern North America; Europe.
Notes: On limbs of hardwoods.
Hosts: Ailanthus, Andromeda, Malus, Populus, Quercus, Vaccinium

Johansonia Sacc. Loculoascomycetes, Dothideales

Johansonia cupulata (Ellis & G. Martin) Barr
Venturia cupulata Ellis & G. Martin
Distr.: FL.
Notes: Barr (Mycotaxon 29:501. 1987).
Host: Quercus

Johncouchia S.J. Hughes & Cavalcanti. Hyphomycetes

Notes: Hughes & Cavalcanti (Canad. J. Bot. 61:2224. 1983).
Johncouchia mangiferae (Batista) S.J. Hughes & Cavalcanti
Septobasidium pilosum Boedijn & B.A. Steinmann (teleomorph)
Distr.: Southern United States to Brazil; Java; Taiwan.
Notes: Parasitic on scale insects on limbs and leaves.
Hosts: Magnolia, Mangifera

Julella Fabre. Loculoascomycetes, Dothideales

Notes: Barr (Sydowia 38:11. 1985).
Julella vitrispora (Cooke & Harkn.) Barr
Pleospora vitrispora Cooke & Harkn.
Peltosphaeria vitrispora (Cooke & Harkn.) Berl.
Teichosporella lonicera Fairm. in Millsp. & L. Nutt.

Distr.: CA; Cuba.
Notes: On dead twigs.
Host: Lonicera

Junghuhnia Corda emend. Ryvarden. Basidiomycetes, Aphyllophorales

Notes: Gilbertson & Ryvarden (North American Polypores 1:410. 1986).
Junghuhnia collabens (Fr.) Ryvarden
Polyporus collabens Fr.
Poria rixosa (P. Karst.) P. Karst.
Distr.: Temperate northern hemisphere.
Notes: On wood, usually of conifers.
Hosts: Abies, Acer, Corylus, Larix, Picea, Pinus, Pseudotsuga, Thuja, Tsuga
Junghuhnia fimbriatella (Peck) Ryvarden
Polyporus fimbriatellus Peck
Chaetoporus fimbriatella (Peck) Parmasto
Poria fimbriatella (Peck) Sacc.
Distr.: Northeastern and northwestern North America; northern hemisphere.
Notes: On hardwoods, especially *Populus* spp.
Hosts: Acer, Alnus, Populus
Junghuhnia luteoalba (P. Karst.) Ryvarden
Poria luteoalba (P. Karst.) Sacc.
Distr.: Temperate northern hemisphere.
Notes: On conifers, especially *Pinus* spp.
Hosts: Abies, Picea, Pinus, Pseudotsuga, Tsuga
Junghuhnia nitida (Pers.:Fr.) Ryvarden
Poria nitida (Pers.:Fr.) Cooke
Poria attenuata (Peck) Peck
Polyporus attenuatus Peck
Poria eupora (P. Karst.) Cooke
Distr.: Widespread.
Notes: On wood.
Hosts: Acer, Alnus, Castanea, Fagus, Fraxinus, Hamamelis, Larix, Liquidambar, Liriodendron, Lithocarpus, Ostrya, Picea, Pinus, Populus, Prunus, Quercus, Salix, Tilia, Ulmus, Viburnum
Junghuhnia separabilima (Pouzar) Ryvarden
Chaetoporus separabilima Pouzar
Distr.: North America(rare); Europe.
Notes: On wood.
Host: Acer
Junghuhnia subfimbriata (Romell) Ginns
Polyporus radula (Pers.:Fr.) Fr.
Poria radula Pers.:Fr.
Distr.: Widespread in temperate regions.
Hosts: Picea, Pinus, Populus
Junghuhnia zonata (Bres.) Ryvarden
Poria zonata Bres.
Distr.: Western North America.
Notes: On conifers.
Hosts: Abies, Larix, Pseudotsuga, Tsuga

Kabatia Bubák. Coelomycetes

Notes: Sutton (The Coelomycetes. p. 166. 1980).
Kabatia fragariae Solheim
Distr.: WY.
Hosts: Fragaria, Geum
Kabatia lonicerae (Harkn.) Höhn.
Marssonia lonicerae Harkn.
Marssonina lonicerae (Harkn.) Magnus
Distr.: CA.
Notes: Leaf lesions.
Hosts: Lonicera, Physocarpus
Kabatia lonicerae (Harkn.) Höhn. var. ***involucratae*** Conners
Distr.: MT(type).
Notes: Leaf lesions.
Host: Lonicera
Kabatia mirabilis Bubák
Guignardia mirabilis E. Müller (teleomorph)
Distr.: United States; Europe.
Notes: Leaf lesions.
Host: Lonicera
Kabatia mirabilis Bubák var. ***oblongifoliae*** Conners
Distr.: NY, WI; Canada (Ontario).
Notes: Leaf lesions.
Host: Lonicera
Kabatia periclymeni (Desmaz.) Morelet
Labrella periclymeni Desmaz.
Colletotrichella periclymeni (Desmaz.) Höhn.
Leptothyrium periclymeni (Desmaz.) Sacc.
Leptothyrium pictum Berk. & Broome
Phyllosticta vossii Thuem.
Depazea xylostei Pass.
Distr.: MT; Europe.
Host: Lonicera

Kabatiella Bubák. Hyphomycetes

Notes: De Hoog & Hermanides-Nijhof (Stud. Mycol. 15:188. 1977).
Kabatiella berberidis (Cooke) C.G. Shaw & Arx in Arx
Gloeosporium berberidis Cooke
Distr.: WA.
Notes: Probably an *Aureobasidium*.
Host: Berberis
Kabatiella boreale (Ellis & Everh.) Arx
Gloeosporium boreale Ellis & Everh.
Microstroma boreale (Ellis & Everh.) W.B. Cooke
Distr.: VT(type).
Notes: Probably an *Aureobasidium*.
Host: Salix
Kabatiella iliamnae (W.B. Cooke & C.G. Shaw) Arx
Gloeosporium iliamnae W.B. Cooke & C.G. Shaw
Microstroma iliamnae (W.B. Cooke & C.G. Shaw) W.B. Cooke
Distr.: OR(type).
Notes: Probably an *Aureobasidium*.
Host: Iliamna

Kabatina R. Schneider & Arx. Coelomycetes

Kabatina juniperi R. Schneider & Arx
Distr.: Central and eastern United States; Europe.
Notes: Hermanides-Nijhof (Stud. Mycol. 15:168. 1977).
Hosts: Calocedrus, Chamaecyparis, Juniperus, Thuja

Kaleidosporium Van Warmelo & Sutton. Coelomycetes

Kaleidosporium fenestratum (Ellis & Everh.) Van Warmelo & Sutton
Stilbospora fenestrata Ellis & Everh.
Stegonsporium fenestratum (Ellis & Everh.) Sacc.
Distr.: NJ.
Notes: Van Warmelo & Sutton (Mycol. Pap. 145:24. 1981).
Host: Clethra

Kalmusia Niessl. Loculoascomycetes, Dothideales

Kalmusia clivensis (Berk. & Broome) Barr
Diapleella clivensis (Berk. & Broome) Munk
Leptosphaeria clivensis (Berk. & Broome) Sacc.
Leptosphaeria steironematis Ellis & Everh.
Distr.: Northeastern North America; Europe.
Notes: Barr (Mycotaxon 29:504. 1987).
Hosts: Clematis, Dipsacus, Lysimachia
Kalmusia pinicola Bonar
Distr.: Western United States.
Notes: On fallen needles of pines (Bonar. Mycologia 57:386. 1965).
Host: Pinus

Karschia Koerb. Loculoascomycetes, Dothideales

Karschia melaspileoides Rehm
Distr.: WA; Europe(type).
Notes: Hafellner (Beih. Nova Hedwigia 62:216. 1979) stated that the type specimen no longer exists.
Host: Populus

Karstenula Speg. Loculoascomycetes, Dothideales

Karstenula alnicola Petr.
Distr.: MD(type).
Host: Alnus
Karstenula shepherdiae (Peck) Barr
Pleospora shepherdiae Peck
Karstenula carpinicola (Ellis & Everh.) Berl.
Pleospora carpinicola Ellis & Everh.
Coniothyrium sp. (anamorph)
Distr.: Temperate North America; Great Britain.
Host: Shepherdia

Kavinia Pilát. Basidiomycetes, Aphyllophorales

Kavinia alboviridis (Morg.) R.L. Gilbertson & Budington
Mycoacia alboviride (Morg.) L.W. Miller & Boyle
Oxydontia alboviride (Morg.) L.W. Miller
Distr.: Temperate northern hemisphere.
Notes: On wood.
Hosts: Abies, Pinus, Populus, Pseudotsuga
Kavinia himantia (Schwein.) J. Eriksson
Clavaria himantia (Schwein.) Bourd. & Galzin
Mycoacia himantia (Schwein.) L.W. Miller & Boyle
Oxydontia himantia (Schwein.) L.W. Miller
Odontia himantia (Schwein.) Bres.
Hydnum subfuscum Peck
Distr.: Temperate northern hemisphere.
Notes: On wood.
Hosts: Abies, Arbutus, Ostrya, Pinus, Quercus

Keissleriella Höhn. Loculoascomycetes, Dothideales

Keissleriella blepharospora Kohlmeyer & E. Kohlmeyer
Distr.: Atlantic and Pacific Oceans.
Notes: On bark of seedlings and roots (Kohlmeyer & Kohlmeyer. Marine mycology. p. 407. 1979).
Host: Rhizophora
Keissleriella culmifida (P. Karst.) Bose
Leptosphaeria culmifida P. Karst.
Trichometasphaeria culmifida (P. Karst.) L. Holm
Leptosphaeria poae Niessl
Metasphaeria poae (Niessl) Sacc.
Distr.: OK; Europe.
Notes: On dead grass culms (Bose. Phytopathol. Z. 41:188. 1961).
Host: Chasmanthium

Kellermania Ellis & Everh. Coelomycetes

Notes: Sutton (The Coelomycetes. p. 54. 1980); Morgan-Jones et al. (Canad. J. Bot. 50:1641. 1972).
Kellermania alpina Ellis & Everh.
Heteropatella alpina (Ellis & Everh.) W.B. Cooke in R. Sprague & W.B. Cooke
Distr.: Western United States.
Hosts: Agastache, Aquilegia, Aster, Juncus, Ligusticum, Penstemon
Kellermania anomala (Cooke) Höhn.
Discella anomala Cooke
Discula anomala (Cooke) Sacc.
Kellermania major Dearn. & Barth.
Distr.: Southern and western United States.
Hosts: Dracaena, Yucca
Kellermania yuccigena Ellis & Everh.
Distr.: Central and western United States.
Notes: Also spelled "*yuccaegena*." (Morgan-Jones et al. Canad. J. Bot. 50: 1646. 1972).
Host: Yucca

Kernella Thirumalachar. Basidiomycetes, Uredinales

Kernella lauricola (Thirumalachar) Thirumalachar
Kernia lauricola Thirumalachar
Distr.: HI; India.
Host: Pelea

Khuskia H.J. Hudson. Pyrenomycetes, Sordariales

Khuskia oryzae H.J. Hudson
Monotospora oryzae Berk. & Broome
Basisporium gallarum Molliard
Nigrospora oryzae (Berk. & Broome) Petch (anamorph)
Distr.: Warm regions.
Notes: Cob and stalk rot of maize, discoloration of rice, and on various kinds of organic debris (Sivanesan & Holliday. C.M.I. Descr. 311. 1971).
Hosts: Acacia, Metrosideros, Pseudotsuga

Kirschsteiniella Petr. Coelomycetes

Notes: Hawksworth (Bot. J. Linn. Soc. 82:35. 1981).
Kirschsteiniella applanata (Fr.:Fr.) Petr.
Amphisphaeria applanata (Fr.:Fr.) Ces. & De Not.
Microthelia applanata (Fr.:Fr.) E. Müller
Trematosphaeria applanata (Fr.:Fr.) Fuckel
Distr.: Temperate regions.
Notes: On wood of various trees. According to Hawksworth (Bot. J. Linn. Soc. 82:54. 1981), this name belongs in the coelomycete genus *Cyclothyrium*. The fungi often reported under this name are actually *Kirschsteiniothelia aethiops*.
Hosts: [Acer], [Alnus], [Pinus], [Thuja], [Tsuga]

Kirschsteiniothelia D. Hawksworth. Loculoascomycetes, Dothideales

Notes: Hawksworth (Bot. J. Linn. Soc. 91:181. 1985).
Kirschsteiniothelia acerina (A.Y. Rossman & H. Wilcox) D. Hawksworth
Microthelia acerina A.Y. Rossman & H. Wilcox
Distr.: NY.
Notes: On mycorrhizal rootlets (Rossman & Wilcox. Mycologia 77:165. 1985).
Host: Acer
Kirschsteiniothelia aethiops (Berk. & M.A. Curtis) D. Hawksworth
Amphisphaeria aethiops (Berk. & M.A. Curtis) Sacc.
Amphisphaeria incrustans Ellis & Everh.
Microthelia incrustans (Ellis & Everh.) Corlett & S.J. Hughes
Dendryphiopsis atra (Corda) S.J. Hughes (anamorph)
Distr.: Temperate regions.
Notes: On wood. According to Hawksworth (Bot. J. Linn. Soc. 91:191. 1985), this species has been confused with *Kirschsteiniella applanata*; this latter name was actually based on a *Cyclothyrium* sp.
Host: Populus
Kirschsteiniothelia recessa (Cooke & Peck) D. Hawksworth
Melanopsamma recessa (Cooke & Peck) Sacc.
Distr.: NY.
Host: Acer
Kirschsteiniothelia thujina (Peck) D. Hawksworth
Amphisphaeria thujina (Peck) Sacc.
Kirschsteiniella thujina (Peck) Pomerleau & D.E. Etheridge
Distr.: Northeastern North America.
Notes: Blue stain in conifer wood.
Hosts: Abies, Thuja

Kmetia Bres. & Sacc. Hyphomycetes

Kmetia corticola W.B. Cooke
Distr.: CA.
Notes: On dead bark (Cooke. Mycologia 44:808. 1952).
Host: Abies

Konenia K. Hara. Pyrenomycetes, Sphaeriales

Konenia bambusae K. Hara
Distr.: NC; Japan(type).
Notes: The disposition of this monotypic genus is uncertain.
Host: Arundinaria

Korfia J. Reid & Cain. Discomycetes, Helotiales

Notes: Monotypic. Reid & Cain (Mycologia 55:781. 1963).
Korfia tsugae (Cash & R.W. Davidson) J. Reid & Cain
Phacidium tsugae Cash & R.W. Davidson
Distr.: NC(type).
Host: Tsuga

Kretzschmaria Fr. Pyrenomycetes, Xylariales

Kretzschmaria clavus (Fr.) Sacc.
Hypoxylon clavus (Fr.) Mont.
Kretzschmaria pusilla Ellis & Everh.
Distr.: Subtropical and tropical regions.
Notes: On woody substrates (Dennis. Kew Bull. 1957:309. 1957).
Host: Macadamia

Kriegeria Brès. Basidiomycetes, Auriculariales

Xenogloea Syd. & P. Syd.
Notes: Donk (Taxon 7:206. 1958) provided a discussion of the use of *Kriegeria* over *Xenogloea*.
Kriegeria eriophori Bres.
Platygloea eriophori (Bres.) Höhn.
Xenogloea eriophori (Bres.) Syd. & P. Syd.
Distr.: Temperate northern hemisphere.
Notes: Parasitic on grasses.
Hosts: Brachyelytrum, Scirpus

Kuehneola Magnus. Basidiomycetes, Uredinales

Kuehneola malvicola Arth.
Uredo malvicola Speg.
Distr.: Southern United States to South America.
Hosts: Alcea, Hibiscus, Malvaviscus
Kuehneola uredinis (Link) Arth.
Chrysomyxa albida Kühn
Distr.: Temperate regions.
Notes: Autoecious. Parmelee & de Carteret (Fungi Canadenses 307. 1986).
Host: Rubus

Kuehneromyces Singer & A.H. Sm. Basidiomycetes, Agaricales

Kuehneromyces lignicola (Peck) Redhead
Kuehneromyces vernalis (Peck) Singer & A.H. Sm.
Pholiota vernalis (Peck) A.H. Sm. & Hesler
Distr.: Boreal and temperate regions of North America.
Notes: More frequent on conifer wood but also on hardwoods.
Host: Abies
Kuehneromyces mutabilis (Schaeff.:Fr.) Singer & A.H. Sm.
Pholiota mutabilis (Schaeff.:Fr.) P. Kumm.
Distr.: Boreal and temperate North America.
Notes: On logs, stumps, or buried wood of conifers and hardwoods.
Hosts: Abies, Alnus, Populus, Rubus, Tsuga

Kutilakesa Subramanian. Hyphomycetes

Kutilakesa pironii Alfieri
Nectriella pironii Alfieri & Samuels (teleomorph)
Distr.: FL.
Notes: Parasitic on woody ornamentals. Carmichael et al. (Genera of Hyphomycetes. p. 132. 1980) indicated this may be a species of *Myrothecium*.
Hosts: Aphelandra, Clerodendrum, Codiaeum

Labrella Fr. Coelomycetes

Notes: According to Sutton (Mycol. Pap. p. 106. 1977), this is a doubtful name, but not all species have been redisposed.
Labrella aspidistrae Tehon & E.Y. Daniels
Distr.: IL, LA.
Host: Aspidistra
Labrella celastri Dearn. & House
Distr.: Northeastern United States.
Host: Celastrus

Labridella Brenckle. Coelomycetes

Labridella cornu-cervae Brenckle
Pestalotia cornu-cervae (Brenckle) Guba f. *longiseta* Dearn.
Griphosphaerioma kansensis (Ellis & Everh.) Shoemaker (teleomorph)
Distr.: Northcentral North America.
Notes: On twigs (Shoemaker. Canad. J. Bot. 41.1419. 1963).
Host: Symphoricarpos

Lachnaster Höhn. Discomycetes, Helotiales

Notes: This genus is now considered a synonym of *Psilachnum*, but not all names have been redisposed.
Lachnaster miniatus Kanouse
Distr.: WA(type).
Host: Pteridium

Lachnella Fr. emend. Donk. Discomycetes, Helotiales

Lachnella alboviolascens (Albertini & Schwein.:Fr.) Fr.
Cyphella alboviolascens (Albertini & Schwein.:Fr.) P.L. Crouan & H. Crouan
Cyphellopsis alboviolascens (Albertini & Schwein.:Fr.) Donk
Distr.: Widespread.
Notes: On woody and herbaceous stems.
Hosts: Arctostaphylos, Ceanothus, Fouquieria, Garrya, Juniperus, Lycium, Platanus
Lachnella cenangioides (Ellis) Sacc.
Distr.: Western United States.
Notes: On herbaceous stems (Dennis. Kew Bull. 17:337. 1963).
Host: Mertensia
Lachnella tiliae (Peck) Donk in Singer
Cyphella tiliae (Peck) Cooke
Trichopeziza tiliae (Peck) Sacc.
Distr.: North America.
Notes: On branches and trunks.
Hosts: Tilia, Ulmus
Lachnella villosa (Pers.:Fr.) Gill.
Peziza villosa Pers.:Fr.
Cyphella villosa (Pers.:Fr.) H. Crouan & P.L. Crouan
Distr.: Cosmopolitan
Hosts: Ammophila, Artemisia, Encelia, Helianthus, Pinus, Pipturus, Solidago

Lachnellula P. Karst. Discomycetes, Helotiales

Notes: Dharne (Phytopathol. Z. 53:101. 1964).
Lachnellula abietis (P. Karst.) Dennis

Dasyscyphus abietis (P. Karst.) Sacc.
Distr.: NY, VT; Europe.
Host: Picea
Lachnellula agassizii (Berk. & M.A. Curtis) Dennis
Dasyscyphus agassizii (Berk. & M.A. Curtis) Sacc.
Lachnella agassizii (Berk. & M.A. Curtis) Seaver
Naemaspora sp. (anamorph)
Distr.: Northern North America.
Notes: On bark and wood of Pinaceae.
Hosts: Abies, Picea, Pinus, Pseudotsuga, Tsuga
Lachnellula arida (W. Phillips) Dennis
Dasyscyphus aridus (W. Phillips) Sacc.
Lachnella arida (W. Phillips) Seaver
Trichoscyphella arida (W. Phillips) E. Müller & Ahmad
Lachnum engelmannii Earle
Distr.: Circumboreal.
Notes: On dead branches of Pinaceae.
Hosts: Abies, Larix, Picea, Pinus, Pseudotsuga, Tsuga
Lachnellula calyciformis (Batsch) Dharne
Dasyscyphus calyciformis (Batsch) Rehm
Distr.: Circumboreal; New Zealand.
Notes: On dead branches of Pinaceae.
Hosts: Abies, Larix, Pinus
Lachnellula cervina (Ellis & Everh.) Dennis
Erinella cervina Ellis & Everh.
Erinellina cervina (Ellis & Everh.) Seaver
Distr.: CO.
Notes: On decaying branches (Dennis. Kew Bull. 17:338. 1963).
Host: Betula
Lachnellula ciliata Dennis
Dasyscyphus ciliatus Hahn non (Fuckel) Höhn., hom. illeg.
Distr.: Western North America.
Notes: Dharne (Phytopathol. Z. 53:101. 1964).
Hosts: Pseudotsuga, Rhododendron
Lachnellula fuscosanguinea (Rehm) Dennis
Dasyscyphus fuscosanguineus Rehm
Dasyscyphus pini Hahn & Ayers
Distr.: Northwestern North America; Europe.
Host: Pinus
Lachnellula laricis (Cooke) Dharne
Distr.: VT; Europe.
Host: Larix
Lachnellula occidentalis (Hahn & Ayers) Dharne
Dasyscyphus occidentalis Hahn & Ayers
Lachnella hahniana Seaver
Lachnellula hahniana (Seaver) Dennis
Trichoscyphella hahniana (Seaver) Manners
Distr.: Circumboreal.
Notes: Pathogenic or saprobic on branches of Pinaceae.
Host: Larix
Lachnellula pini (Brunch.) Dennis
Lachnella pini Brunch.
Distr.: Northwestern North America; Europe.
Notes: Canker of 5-needle pines.
Host: Pinus
Lachnellula pseudotsugae (Hahn) Dennis
Dasyscyphus pseudotsugae Hahn
Lachnella pseudotsugae (Hahn) Seaver
Naemospora sp. (anamorph)
Distr.: Pacific Northwest.
Notes: Target canker.
Host: Pseudotsuga
Lachnellula resinaria (Cooke & W. Phillips) Rehm
Dasyscyphus resinarius (Cooke & W. Phillips) Rehm
Trichoscypha resinaria (Cooke & W. Phillips) Boud.
Trichoscyphella resinaria (Cooke & W. Phillips) Dennis
Naemaspora sp. (anamorph)
Distr.: North America, scattered reports; Europe.
Notes: On dead branches of Pinaceae.
Host: Abies
Lachnellula subtilissima (Cooke) Dennis
Dasyscyphus subtilissimus (Cooke) Sacc.
Lachnella subtilissima (Cooke) W. Phillips
Dasyscyphus calycinus (Schumach.:Fr.) Fuckel
Naemaspora strobi Allesch. (anamorph)
Distr.: Circumboreal on Pinaceae.
Hosts: Abies, Larix, Picea, Pinus, Pseudotsuga, Tsuga
Lachnellula suecica (de Bary ex Fuckel) Nannf.
Lachnellula chrysophthalma (Pers.) P. Karst., sensu Nyl.
Distr.: North America; northern Europe.
Notes: On bark and twigs of conifers.
Hosts: Abies, Picea, Pinus, Tsuga
Lachnellula theiodea (Cooke & Ellis) Sacc.
Distr.: NJ.
Notes: Dennis (Kew Bull. 17:370. 1963) suggested that this species might be a *Pithyella*.
Host: Rhus
Lachnellula willkommii (R. Hartig) Dennis
Dasyscyphus willkommii (R. Hartig) Rehm
Trichoscyphella willkommii (R. Hartig) Nannf.
Distr.: Northeastern North America; Europe; Japan.
Notes: Larch canker (Buczaki. C.M.I. Descr. 450. 1975 as *Trichoscyphella willkommii*).
Hosts: Larix, Pseudolarix

Lachnum Retz. Discomycetes, Helotiales

Lachnum abnorme (Mont.) J. Haines & Dumont
Peziza abnorme Mont.
Erinella longispora P. Karst.
Dasyscyphus longisporus (P. Karst.) Dennis
Distr.: Widespread in tropical regions.
Notes: On wood and bark of trees (Haines & Dumont. Mycotaxon 19:10. 1984).
Host: Mangifera
Lachnum alneum Velen.
Distr.: WA; Czechoslovakia(type).
Notes: A questionable taxon (J. Haines. pers. comm.).
Host: Alnus
Lachnum arundinariae Cash
Lachnella arundinariae (Cash) Seaver
Dasyscyphus milleri Dennis
Distr.: GA.
Host: Arundinaria
Lachnum carneolum (Sacc.) Rehm
Dasyscyphus carneolus (Sacc.) Sacc.
Distr.: AK, MI; Europe.
Host: Calamagrostis
Lachnum corni Cash
Lachnella corni (Cash) Seaver
Distr.: GA.
Notes: According to Dennis (Kew Bull. 17:340. 1963), this taxon is doubtfully distinct from *Dasyscyphus fuscescens* (Pers.:Fr.) S.F. Gray.
Host: Cornus
Lachnum crystalligerum Sacc.
Distr.: UT.
Host: Rubus
Lachnum flavidulum (Rehm) J. Haines in Korf & Zhuang
Dasyscyphus gleicheniae (Cash) Dennis
Lachnella gleicheniae (Cash) Seaver
Lachnum gleicheniae Cash
Distr.: Pantropical (Haines. Mycotaxon 11:203. 1980).
Notes: On ferns.
Hosts: Cibotium, Dicranopteris, Gleichenia
Lachnum flavofuligineum (Albertini & Schwein.:Fr.) Rehm
Dasyscyphus flavofuligineus (Albertini & Schwein.:Fr.) Fuckel
Distr.: Eastern North America; Europe.
Notes: The report from Washington on *Sambucus* based on a collection by Kanouse is questionable (J. Haines. pers. comm.).
Host: Sambucus
Lachnum fuscescens (Pers.:Fr.) P. Karst.
Dasysyphus fuscescens (Pers.:Fr.) Rehm
Dasyscyphus fuscidulus (Cooke) Sacc.
Trichopeziza marginata (Cooke) Sacc.
Distr.: NJ; Europe.
Notes: On old leaves (Dennis. Kew Bull. 17:346. 1963).
Hosts: Andromeda, Quercus
Lachnum fuscofloccosum Rehm
Distr.: WA.
Host: Lonicera

Lachnum gaultheriae (Ellis & Everh.) Zeller
Dasyscyphus gaultheriae Ellis & Everh.
Lachnella gaultheriae (Ellis & Everh.) Seaver
Distr.: Pacific Northwest.
Host: Gaultheria
Lachnum halesiae Cash
Dasyscyphus halesiae (Cash) Dennis
Distr.: GA.
Host: Halesia
Lachnum hyalinellum Rehm
Distr.: WA; Europe.
Notes: The report from Washington was based on a misdetermination of *Lachnum virgineum* (Batsch) P. Karst. (J. Haines. pers. comm.).
Host: Alnus
Lachnum morthieri (Cooke) Rehm
Lachnella morthieri Cooke
Distr.: GA.
Notes: A doubtful taxon (J. Haines. pers. comm.)
Host: Hypericum
Lachnum nardi Rehm
Distr.: WA; Europe.
Notes: The report from Washington by Kanouse, as well as the taxon, are doubtful (J. Haines. pers. comm.).
Host: Carex
Lachnum niveum (Hedw.:Fr.) P. Karst.
Dasyscyphus niveus (Hedw.:Fr.) Sacc.
Dasyscyphella nivea (Hedw.:Fr.) Raitviir
Lachnum niveum (Hedw.:Fr.) P. Karst.
Distr.: Temperate regions.
Notes: On dead wood of dicotyledonous trees.
Hosts: Abies, Arctostaphylos, Leucothoe, Oxydendrum, Quercus, Rubus
Lachnum pollinarium (Cooke) Cash
Dasyscyphus pollinarius (Cooke) Dennis
Lachnella pollinaria (Cooke) Seaver
Distr.: Eastern United States.
Host: Quercus
Lachnum pygmaeum (Fr.) Bres.
Dasyscyphus pygmaeus (Fr.) Sacc.
Helotium affinissimum Peck
Helotium rhizogenum Ellis & Everh.
Distr.: Temperate regions.
Notes: On roots and rhizomes of grasses and other herbaceous plants and on limbs of trees (White. Amer. Midl. Naturalist 28:512. 1942).
Host: Andropogon
Lachnum sclerotii (A.L. Sm.) J. Haines & Dumont
Belonidium sclerotii A.L. Sm.
Dasyscyphella subcorticalis (Pat. in Duss) Cash
Erinella subcorticalis Pat. in Duss
Distr.: Widespread in tropical regions.
Notes: On various hardwood substrates (Haines & Dumont. Mycotaxon 19:17. 1984).
Host: Zanthoxylum
Lachnum virginellum (Sacc.) Zeller
Dasyscyphus virginellus Seaver
Lachnella virginella (Sacc.) Seaver
Dasyscyphus albocitrinus (Cooke) Sacc.
Distr.: Eastern North America.
Host: Vaccinium
Lachnum virgineum (Batsch:Fr.) S.F. Gray
Dasyscyphus virgineus (Batsch:Fr.) S.F. Gray
Distr.: Cosmopolitan.
Notes: On woody and herbaceous debris.
Hosts: Alnus, Castanopsis, Populus, Rhododendron, Salix, Taxodium, Vaccinium

Laestadia Auersw. non Kunth ex Less., hom. illeg. Pyrenomycetes, Diaporthales

Notes: This genus is considered a synonym of *Guignardia*, but not all names have been redisposed.
Laestadia asarifolia (Cooke) Sacc.
Distr.: SC(type).
Host: Asarum
Laestadia berberidis Delacr.
Distr.: MD; France(type).
Host: Berberis
Laestadia brunnea (Berk. & M.A. Curtis) Sacc.
Depazea brunnea Berk. & M.A. Curtis
Sphaerella brunnea (Berk. & M.A. Curtis) Cooke
Distr.: NC(questionable), SC(type).
Notes: An imperfectly known species.
Host: Acer
Laestadia caelata Harkn.
Distr.: CA(type).
Host: Lithocarpus
Laestadia cucurbitacearum (Schwein.:Fr.) Sacc.
Distr.: PA(type).
Host: Lagenaria
Laestadia fraxinicola (M.A. Curtis & Peck) Sacc.
Depazea fraxinicola M.A. Curtis & Peck
Distr.: NY(type).
Host: Fraxinus
Laestadia glaucescens (Cooke) Sacc.
Sphaerella glaucescens Cooke
Distr.: SC.
Host: Acer
Laestadia illiciicola Tracy & Earle
Distr.: MS(type).
Host: Illicium
Laestadia saxifragae Sacc. & Scalia
Distr.: AK.
Hosts: Leptarrhena, Saxifraga

Laeticorticium Donk. Basidiomycetes, Aphyllophorales

Notes: This genus is considered a synonym of *Corticium*, but not all names have been redisposed.
Laeticorticium simplicibasidium Lindsey & R.L. Gilbertson
Distr.: AZ(type).
Notes: According to Larsen (Mycologia 76:353. 1984) this species is a member of the Auriculariales.
Host: Populus

Laetinaevia Nannf. Discomycetes, Helotiales

Notes: Hein (Willdenowia Beih. 9:28. 1976).
Laetinaevia caulophylli (Ellis & Everh.) Korf
Orbilia caulophylli Ellis & Everh.
Calloria caulophylli (Ellis & Everh.) Rehm
Distr.: NY.
Notes: Hein (Willdenowia Beih. 9:114. 1976) considered this species to be a *Merostictis*.
Host: Caulophyllum
Laetinaevia veratri Cash
Distr.: CA, OR.
Host: Veratrum

Laetiporus Murrill. Basidiomycetes, Aphyllophorales

Laetiporus persicinus (Berk. & M.A. Curtis) R.L. Gilbertson
Polyporus persicinus Berk. & M.A. Curtis
Meripilus persicinus (Berk. & M.A. Curtis) Ryvarden
Distr.: Southeastern United States.
Notes: On living roots and base.
Host: Quercus
Laetiporus sulphureus (Bull.:Fr.) Murrill
Polyporus sulphureus (Bull.:Fr.) Fr.
Sporotrichum versisporum (C.G. Lloyd) Stalpers (anamorph)
Distr.: Widespread.
Notes: Most commonly on wood of *Quercus* spp., also on other hardwoods and conifers. Heart rot of living trees.
Hosts: Abies, Acacia, Acer, Betula, Castanea, Celtis, Eucalyptus, Fraxinus, Gleditsia, Juglans, Larix, Liriodendron, Malus, Persea, Picea, Pinus, Populus, Prunus, Pseudotsuga, Pyrus, Quercus, Robinia,

Salix, Schinus, Sequoia, Tamarix, Thuja, Tsuga, Ulmus

Laetisaria Burdsall. Basidiomycetes, Aphyllophorales

Notes: Stalpers & Loerakker (Canad. J. Bot. 60:529. 1982).
Laetisaria agavei Burdsall & R.L. Gilbertson
Distr.: AZ.
Notes: On dead leaves.
Host: Agave
Laetisaria fuciformis (McAlpine) Burdsall
Hypochnus fuciformis McAlpine
Corticium fuciforme (Berk.) Wakef., comb. inval.
Phanerochaete fuciformis (Berk.) Jülich, comb. inval.
Isaria fuciformis Berk. (anamorph)
Distr.: Northeastern United States; Australia; Europe.
Notes: On grasses. Although *Corticium fuciforme* is listed here as a synonym, Wakefield's specimen serves as the type of a new species, *Athelia fuciformis* (Wakefield) Burdsall (Burdsall. Trans. Brit. Mycol. Soc. 72:419. 1979).
Hosts: Agrostis, Briza, Cynodon, Festuca, Poa, Vulpia

Lagenidium Schenk. Oomycetes, Lagenidiales

Notes: Most species occur as parasites of algae, fungi, rotifers and their eggs, and other aquatic organisms.
Lagenidium muenscheri V.M. Cutter
Distr.: NY.
Notes: Parasitic in and causing hypertrophy on root hairs (Karling. Predominantly holocarpic & eucarpic simple biflagellate Phycomycetes. p. 127. 1981).
Host: Potamogeton

Lageniformia Plunkett. Pyrenomycetes, Diatrypales

Notes: This monotypic genus may be a synonym of *Eutypa*.
Lageniformia bambusae Plunkett
Distr.: HI(type).
Host: Bambusa

Lahmia Koerb. Pyrenomycetes, Lahmiales

Parkerella Funk
Notes: *Parkerella* was recently shown to be a synonym of *Lahmia* (Eriksson. Mycotaxon 27:347. 1986; Funk. Canad. J. Bot. 54:868. 1976).
Lahmia kunzei Koerb.
Parkerella populi Funk
Distr.: WI; Canada(British Columbia).
Notes: Associated with declining hybrid poplar.
Host: Populus

Lambertella Höhn. Discomycetes, Helotiales

Lambertella cephalanthi Whetzel & Dumont in Dumont
Distr.: Northeastern United States.
Host: Cephalanthus
Lambertella corni-maris Höhn.
Distr.: North and South America; Australia; Europe.
Notes: On stromatized fruits.
Host: Carya
Lambertella garryae (Cash) Dumont
Phaeociboria garryae Cash
Distr.: CA.
Host: Garrya
Lambertella jasmini Seaver, Whetzel, & Dumont in Dumont
Distr.: OR; Bermuda.
Host: Prunus
Lambertella pruni Whetzel, Zeller, & Dumont in Dumont
Distr.: Western United States.
Notes: On fruits and seedlings.
Host: Prunus
Lambertella viburni Whetzel & Dumont in Dumont
Distr.: NH.
Host: Viburnum

Lamprospora De Not. Discomycetes, Pezizales

Notes: These species occur on soil and have been reported in association with vascular plants.
Lamprospora crec'hqueraultii (H. Crouan) Boud.
Distr.: Temperate regions.
Notes: On wet clay soil.
Host: Betula

Lanzia Sacc. Discomycetes, Helotiales

Lanzia echinophila (Bull.:Fr.) Korf
Ciboria echinophila (Bull.:Fr.) Sacc.
Rutstroemia echinophila (Bull.:Fr.) Höhn.
Sclerotinia echinophila (Bull.:Fr.) Rehm
Distr.: GA; Europe(type).
Notes: On involucres.
Host: Castanea
Lanzia longipes (Cooke & Peck) Dumont & Korf in Korf & S. Gruff
Phialea longipes (Cooke & Peck) Sacc.
Phialea anomala Peck
Distr.: NY.
Host: Fraxinus
Lanzia luteovirescens (Roberge) Dumont & Korf in Korf & S. Gruff
Ciboria luteovirescens (Roberge) Sacc.
Rutstroemia luteovirescens (Roberge) W.L. White
Ciboria sulfurella (Ellis & Everh.) Rehm
Helotium sulfurellum Ellis & Everh.
Distr.: Eastern and midwestern North America; Europe.
Notes: On leaves and petioles.
Hosts: Acer, Fraxinus
Lanzia rufescens (Kanouse) Dumont
Ciboria rufescens Kanouse
Ciboriella rufesens (Kanouse) Seaver
Distr.: Pacific Northwest.
Hosts: Acer, Alnus

Lasiobelonium Ellis & Everh. Discomycetes, Helotiales

Lasiobelonium boreale (Ellis & Holw.) Raitviir
Dasyscyphus borealis (Ellis & Holw.) Sacc.
Lasiobelonium subflavidum Ellis & Everh.
Distr.: Temperate North America.
Notes: Korf (Mycotaxon 7:399. 1978) reviewed the genus *Lasiobelonium* of which *L. subflavidum* is the type. Raitviir (Scripta Mycol. 9:99. 1980).
Host: Salix

Lasiobotrys Kuntze. Loculoascomycetes, Dothideales

Notes: Müller & Arx (Beitr. Kryptogamenfl. Schweiz 11(2):460. 1962) presented a synopsis of the genus.
Lasiobotrys lonicerae Kunze
Lasiobotrys affinis Harkn.
Distr.: Temperate regions.
Notes: On leaves and small twigs.
Hosts: Lonicera, [Symphoricarpos]

Lasiodiplodia Ellis & Everh. Coelomycetes

Lasiodiplodia theobromae (Pat.) Griffon & Maubl.
Botryodiplodia theobromae Pat.
Diplodia theobromae (Pat.) W. Nowell
Botryodiplodia gossypii Ellis & Barth.
Diplodia gossypina Cooke
Diplodia natalensis Pole-Evans
Lasiodiplodia triflorae Higgins

Diplodia tubericola (Ellis & Everh.) Taubenhaus
Lasiodiplodia tubericola Ellis & Everh. in Clendenin
Distr.: Worldwide in warm regions.
Notes: On at least 280 genera of vascular plants (Punithalingam. C.M.I. Descr. 519. 1976 as *Botryodiplodia theobromae*).
Hosts: Abelmoschus, Actinidia, Ailanthus, Albizia, Aleurites, Allium, Annona, Arachis, Araucaria, Arecaceae, Bucida, Buddleja, Carica, Carissa, Carya, Cassia, Cattleya, Chamaecyparis, Chenopodium, Cinnamomum, Citrofortunella, Citrus, Cocos, Cucumis, Cucurbita, Diospyros, Dracaena, Epidendrum, Eranthemum, Eugenia, Euphorbia, Feijoa, Ficus, Fortunella, Geranium, Glycine, Gossypium, Hibiscus, Indigofera, Ipomoea, Kalanchoe, Ligustrum, Liquidambar, Lupinus, Lycopersicon, Maclura, Medicago, Musa, Orchidaceae, Pandanus, Parthenium, Persea, Pinus, Platanus, Polyscias, Poncirus, Prenanthes, Prunus, Ricinus, Rosa, Solanum, Sorghum, Theobroma, Tilia, Torreya, Viburnum, Vigna, Vitis, Xanthium, Zanthoxylum

Lasiosphaeria Ces. & De Not. Pyrenomycetes, Sordariales

Lasiosphaeria canescens (Pers.:Fr.) P. Karst.
Leptospora canescens (Pers.:Fr.) G. Wint.
Lasiosphaeria hystrix Ellis & Everh.
Lasiosphaeria stictochaetophora Fairm.
Distr.: Temperate regions.
Notes: On wood. Hilber & Hilber (Hoppea, Denkschr. Regensb. Bot. Ges. 37:16. 1978) provided a modern description of this species.
Host: Populus
Lasiosphaeria caudata (Fuckel) Sacc.
Leptospora caudata Fuckel
Distr.: Temperate regions.
Notes: On wood and herbaceous stems.
Host: Populus
Lasiosphaeria chrysentera Carroll & Munk
Distr.: ID; Costa Rica(type).
Notes: Carroll & Munk (Mycologia 56:83. 1964); Hilber et al. (Mycotaxon 30:275. 1987).
Hosts: Alnus, Tilia
Lasiosphaeria foliicola O. Hilber & R. Hilber
Distr.: VA.
Notes: On dead leaves (Hilber et al. Mycotaxon 30:281. 1987).
Host: Quercus
Lasiosphaeria hispida (Tode:Fr.) Fuckel
Phialophora sp. (anamorph)
Distr.: Temperate regions.
Notes: On wood.
Hosts: Liriodendron, Quercus, Robinia
Lasiosphaeria ovina (Pers.:Fr.) Ces. & De Not.
Phialophora sp. (anamorph)
Distr.: Temperate regions.
Notes: On wood.
Host: Liriodendron
Lasiosphaeria spermoides (Hoffm.:Fr.) Ces. & De Not.
Distr.: Temperate northern hemisphere; South America.
Hosts: Acer, Rubus
Lasiosphaeria strigosa (Albertini & Schwein.:Fr.) Sacc.
Distr.: Temperate North America; Europe.
Hosts: Acer, Alnus, Liriodendron, Quercus
Lasiosphaeria stuppea Ellis & Everh.
Distr.: WA(type).
Host: Tsuga

Lasiostemma Theiss., Syd. & P. Syd. Loculoascomycetes, Dothideales

Lasiostemma erysipheoides (Ellis & Everh.) M.L. Farr
Dimerosporium erysipheoides Ellis & Everh.
Dimeriella erysipheoides (Ellis & Everh.) M.L. Farr
Epipolaeum erysipheoides (Ellis & Everh.) M.L. Farr
Lojkania cynodonifolii Batista
Distr.: Southern United States; West Indies; India.
Notes: On grasses (Farr. Mycologia 71:267. 1979).
Hosts: Cynodon, Panicum, Uniola
Lasiostemma fimbriatum (Dearn. & House) Barr
Venturia fimbriata Dearn. & House
Wentiomyces fimbriatus (Dearn. & House) Barr
Distr.: Temperate North America.
Notes: A description is provided by Barr (Contr. Inst. Bot. Univ. Montreal 73:35. 1959 as *Venturia*). Barr (Mycotaxon 29:504. 1987) transferred this species to *Lasiostemma*.
Host: Antennaria
Lasiostemma melioloides (Berk. & M.A. Curtis) Theiss., Syd., & P. Syd.
Asterina melioloides Berk. & M.A. Curtis
Dimeriella melioloides (Berk. & M.A. Curtis) Theiss.
Wentiomyces melioloides (Berk. & M.A. Curtis) E. Müller
Dimerosporium baccharidis (Berk. & Ravenel) Sacc.
Meliola baccharidis Berk. & Ravenel
Dimeriella erigeronicola F. Stevens
Distr.: Southeastern United States; West Indies; South America.
Notes: Farr (Mycologia 71:248. 1979).
Host: Baccharis

Lasmeniella Petr. & Syd. Coelomycetes

Notes: Sutton (The Coelomycetes. p. 326. 1980).
Lasmeniella alpina (Ellis & Everh.) Petr. & Syd.
Aplosporella alpina Ellis & Everh.
Distr.: Western North America.
Host: Sambucus
Lasmeniella cocoes (Henn.) Petr. & Syd.
Coniothyrium cocoes Henn.
Distr.: Widespread in tropical regions.
Notes: On leaves.
Host: Cocos

Laurilia Pouzar. Basidiomycetes, Aphyllophorales

Laurilia sulcata (Burt) Pouzar
Stereum sulcatum Burt
Echinodontium sulcatum (Burt) H. Gross
Spiniger sp. (anamorph)
Distr.: Temperate northern hemisphere, especially north of 40 degrees latitude.
Notes: On coniferous wood.
Hosts: Abies, Larix, Picea, Pinus, Pseudotsuga, Thuja, Tsuga
Laurilia taxodii (Lentz & McKay) Parmasto
Stereum taxodii Lentz & McKay in R.W. Davidson, Lentz & McKay
Distr.: Southeastern United States; Japan; Taiwan.
Notes: Wood decay known as "pecky cypress."
Host: Taxodium

Laxitextum Lentz. Basidiomycetes, Aphyllophorales

Notes: Chamuris (Mycol. Mem. 14. 1988).
Laxitextum bicolor (Pers.:Fr.) Lentz
Stereum bicolor (Pers.:Fr.) Fr.
Stereum fuscum (Schrad.) P. Karst.
Distr.: Cosmopolitan.
Notes: On wood.
Hosts: Acer, Alnus, Betula, Castanea, Celtis, Fagus, Fraxinus, Liquidambar, Liriodendron, Magnolia, Malus, Nyssa, Pinus, Platanus, Populus, Prunus, Quercus, Salix, Ulmus
Laxitextum incrustatum Hjortstam & Ryvarden
Distr.: LA; Africa.
Notes: On wood (Hjorstam & Ryvarden. Mycotaxon 13:35. 1981).
Host: Quercus

Lazulinospora Burdsall & M. Larsen. Basidiomycetes, Aphyllophorales

Notes: Burdsall & Larsen (Mycologia 66:96. 1974).
Lazulinospora cinnamomea Burdsall & Nakasone
Distr.: FL.
Notes: Burdsall & Nakasone (Mycologia 73:454. 1981).

Hosts: Magnolia, Serenoa
Lazulinospora wakefieldii **Burdsall & M. Larsen**
Distr.: FL.
Notes: On hardwoods.
Host: Quercus

Leandria Rangel. Hyphomycetes

Leandria momordicae **Rangel**
Stemphylium cucurbitacearum G.A. Osner
Distr.: Central and eastern United States; Brazil.
Notes: *Stemphylium* leaf spot of cucumbers.
Hosts: Cucumis, Cucurbita, Echinocystis, Momordica

Lecanidiella M.A. Sherwood. Discomycetes

Notes: Sherwood-Pike (Sydowia 38:272. 1985) described this genus whose aspect in the field is "that of a bark-inhabiting *Coccomyces*".
Lecanidiella contortae **M.A. Sherwood**
Distr.: Western North America.
Notes: On bark.
Host: Pinus

Lecanidion Endl. Loculoascomycetes, Patellariales

Notes: Butler (Mycologia 32:791. 1940).
Lecanidion atratum **(Hedw.) Rabenh.**
Patellaria atrata (Hedw.) Fr.
Distr.: Temperate regions.
Notes: On dry wood.
Hosts: Celtis, Erythrina, Hibiscus, Juglans, Lantana, Liriodendron, Morus, Prunus, Sambucus, Symphoricarpos, Zea
Lecanidion simile **(W. Gerard) E.T. Butler**
Patellaria similis W. Gerard
Patellaria atrofusca Berk. & M.A. Curtis
Patellaria fuscoatra Rehm
Lecanidion fuscoatrum Ellis
Distr.: Central and eastern United States.
Hosts: Cercis, Quercus

Lecanosticta Syd. in Syd. & Petr. Coelomycetes

Lecanosticta acicola **(Thuem.) Syd. in Syd. & Petr.**
Cryptosporium acicola Thuem.
Septoria acicola (Thuem.) Sacc.
Lecanosticta pini Syd. in Syd. & Petr.
Mycosphaerella dearnessii Barr (teleomorph)
Distr.: North and South America; Europe.
Notes: Brown needle spot (Evans. Mycol. Pap. 153:7. 1984).
Host: Pinus

Lembosia Lév. Loculoascomycetes, Dothideales

Morenoella Speg.
Lembosia acicola **(Harkn.) Sacc.**
Aulographum acicola Harkn.
Distr.: CA.
Host: Pinus
Lembosia brevis **Tracy & Earle**
Lembosiopsis brevis (Tracy & Earle) Theiss.
Distr.: FL(type).
Host: Ilex
Lembosia cactorum **Tracy & Earle**
Lembosiopsis cactorum (Tracy & Earle) Theiss.
Distr.: FL.
Host: Opuntia
Lembosia cliftoniae **Tracy & Earle**
Morenoella cliftoniae (Tracy & Earle) Theiss.
Distr.: MS.
Host: Cliftonia
Lembosia eucalypti **F. Stevens & M. Dixon**
Distr.: HI.
Host: Eucalyptus
Lembosia illiciicola **Tracy & Earle**
Distr.: Southern United States.
Host: Illicium
Lembosia oleae **Tracy & Earle**
Lembosiopsis oleae (Tracy & Earle) Theiss.
Distr.: MS.
Host: Osmanthus
Lembosia rolfsii **A.S. Horne**
Distr.: FL.
Host: Vanilla

Lembosina Theiss. Loculoascomycetes, Dothideales

Lembosina cocculi **(F. Stevens & Ryan) Arx**
Echidnodella cocculi F. Stevens & Ryan
Maurodothella cocculi (F. Stevens & Ryan) G. Arnaud
Distr.: HI.
Host: Cocculus
Lembosina liturae **(Cooke) Arx**
Aulographum liturae Cooke
Echidnodes liturae (Cooke) Theiss. & Syd.
Lembosia liturae (Cooke) Sacc.
Distr.: Southeastern United States.
Notes: Müller & Arx (Beitr. Kryptogamenfl. Schweiz 11(2):121. 1962).
Host: Quercus
Lembosina quercina **(Ellis & G. Martin) Arx**
Aulographum quercinum Ellis & G. Martin
Lembosia quercina (Ellis & G. Martin) Tracy & Earle
Echidnodes quercina (Ellis & G. Martin) Hansf.
Morenoella quercina (Ellis & G. Martin) Theiss.
Distr.: FL.
Notes: Müller & Arx (Beitr. Kryptogamenfl. Schweiz 11(2):122. 1962).
Host: Quercus

Lembosiopsis Theiss. Loculoascomycetes, Dothideales

Notes: Synonym of *Aulographum*, according to Müller & Arx (Beitr. Kryptogamenfl. Schweiz 11(2):127. 1962).
Lembosiopsis andromedae **(Tracy & Earle) Theiss.**
Lembosia andromedae Tracy & Earle
Distr.: Southeastern United States.
Host: Lyonia

Lemonniera De Wild. Hyphomycetes

Notes: Descals et al. (Trans. Brit. Mycol. Soc. 69:89. 1977).
Lemonniera aquatica **De Wild.**
Distr.: Temperate regions.
Notes: On plant debris in freshwater habitats.
Hosts: Acer, Quercus
Lemonniera filiforme **R. Petersen**
Distr.: NY; Germany; Soviet Union.
Notes: In foam and on dicotyledonous leaves in streams.
Host: Acer
Lemonniera pseudofloscula **Dyko in E. Descals, J. Webster & Dyko**
Distr.: NC.
Notes: On rotting leaves in streams.
Hosts: Acer, Quercus
Lemonniera terrestris **Tubaki**
Lemonniera brachycladia Ingold
Distr.: Temperate regions.
Notes: On plant debris in streams.
Hosts: Acer, Quercus

Lentaria Corner. Basidiomycetes, Aphyllophorales

Lentaria byssiseda **Corner**
Distr.: Widespread in north temperate regions.

Notes: On wood.
Host: Acer

Lentinellus P. Karst. Basidiomycetes, Agaricales

Lentinellus cochleatus (Fr.:Fr.) P. Karst.
Lentinellus umbilicatus (Peck) Singer
Lentinus umbilicatus Peck
Distr.: North America; temperate Europe.
Notes: On wood.
Hosts: Alnus, Betula, Castanea, Fagus, Pinus, Populus, Pseudotsuga, Quercus
Lentinellus flabelliformis (Bolton:Fr.) Ito
Distr.: North America; Africa; Asia; Europe.
Notes: On wood.
Hosts: Alnus, Arctostaphylos, Betula, Populus
Lentinellus micheneri (Berk. & M.A. Curtis) Pegler
Lentinellus omphalodes (Fr.) P. Karst.
Lentinus omphalodes Fr.
Distr.: Temperate northern hemisphere.
Notes: On wood.
Hosts: Abies, Acer, Alnus, Fagus, Picea, Rosa
Lentinellus montanus O.K. Miller
Distr.: Western North America.
Notes: On conifer logs and branches, often fruiting near receding snowbanks.
Hosts: Abies, Picea, Pinus, Tsuga
Lentinellus ursinus (Fr.:Fr.) Kühner
Lentinus ursinus (Fr.:Fr.) Fr.
Distr.: Temperate northern hemisphere.
Notes: On wood.
Hosts: Abies, Acer, Fagus, Fraxinus, Juglans, Liquidambar, Nyssa, Pinus, Populus, Pseudotsuga, Quercus, Salix, Tsuga
Lentinellus vulpinus (Fr.:Fr.) Kühner & Maire
Lentinus vulpinus (Fr.:Fr.) Fr.
Distr.: North America; temperate Europe.
Notes: On wood.
Hosts: Betula, Fagus, Populus, Ulmus

Lentinus Fr. Basidiomycetes, Agaricales

Notes: Redhead & Ginns (Trans. Mycol. Soc. Japan 26:349. 1985).
Lentinus crinitus (L.:Fr.) Fr.
Distr.: Subtropical and tropical western hemisphere.
Notes: On wood.
Host: Nyssa
Lentinus strigosus (Schwein.:Fr.) Fr.
Panus fragilis O.K. Miller
Lentinus lecomtei Fr.
Panus rudis Fr.
Distr.: Cosmopolitan.
Notes: On a wide variety of hardwoods.
Hosts: Acer, Betula, Castanea, Fouquieria, Liquidambar, Liriodendron, Pinus, Populus, Prunus, Quercus
Lentinus tigrinus (Bull.:Fr.) Fr.
Panus tigrinus (Bull.:Fr.) Singer
Lentinus ravenelii Berk. & M.A. Curtis
Lentodium squamulosum Morg.
Distr.: Cosmopolitan.
Notes: On hardwoods.
Hosts: Acer, Celtis, Diospyros, Fraxinus, Liquidambar, Nyssa, Platanus, Quercus, Salix
Lentinus torulosus (Pers.:Fr.) C.G. Lloyd
Panus torulosus (Pers.:Fr.) Fr.
Distr.: Temperate northern hemisphere.
Notes: On wood.
Host: Quercus
Lentinus velutinus Fr.
Panus fulvus (Berk.) Pegler & R.W. Rayner
Distr.: Subtropical and tropical regions.
Host: Quercus

Lenzites Fr. Basidiomycetes, Aphyllophorales

Notes: Gilbertson & Ryvarden (North American Polypores 1:424. 1986).
Lenzites betulina (L.:Fr.) Fr.
Distr.: Widespread.
Notes: On wood.
Hosts: Acer, Alnus, Betula, Castanea, Cornus, Fagus, Liquidambar, Liriodendron, Malus, Nyssa, Populus, Prunus, Quercus, Salix, Tilia, Tsuga, Ulmus, Umbellularia

Leocarpus Link. Myxomycetes

Leocarpus fragilis (Dicks.) Rostr.
Distr.: Cosmopolitan.
Hosts: Juniperus, Quercus

Lepteutypa Petr. Pyrenomycetes, Amphisphaeriales

Notes: Nag Raj & Kendrick (Sydowia 38:190. 1985).
Lepteutypa fuckelii (Nitschke ex Fuckel) Petr.
Massaria fuckelii Nitschke ex Fuckel
Distr.: NC; Europe.
Notes: On wood (Shoemaker & Müller. Canad. J. Bot. 43:1459. 1965).
Host: Tilia
Lepteutypa fusispora Petr.
Distr.: HI.
Notes: According to Nag Raj & Kendrick (Sydowia 38:193. 1985), this species does not belong in *Lepteutypa*.
Host: Wisteria

Leptodermella Höhn. Coelomycetes

Leptodermella opuntiae B.O. Dodge
Distr.: WA; Bermuda.
Notes: This species probably does not belong in *Leptodermella* (Sutton. The Coelomycetes. p. 160. 1980).
Host: Opuntia

Leptodontium De Hoog. Hyphomycetes

Notes: De Hoog (Stud. Mycol. 15:44. 1977).
Leptodontium elatius (G. Mangenot) De Hoog
Rhinocladiella elatior G. Mangenot
Distr.: Temperate regions; Nigeria.
Notes: Isolated from wood.
Host: Pinus

Leptographium Lagerberg & Melin. Hyphomycetes

Notes: Wingfield (Trans. Brit. Mycol. Soc. 85:81. 1985).
Leptographium abietinum (Peck) M.J. Wingfield
Sporocybe abietina Peck
Verticicladiella abietina (Peck) S.J. Hughes
Distr.: Temperate North America.
Host: Pinus
Leptographium antibioticum (Kendrick) M.J. Wingfield
Verticicladiella antibiotica Kendrick
Distr.: Western North America.
Host: Pinus
Leptographium engelmanii R.W. Davidson
Distr.: CO.
Notes: On sapwood and bark.
Hosts: Picea, Pinus
Leptographium lundbergii Lagerberg & Melin
Distr.: Temperate regions.
Notes: On conifer wood often associated with bark beetles (Ellis. Dematiaceous Hyphomycetes. p. 333. 1971).
Host: Pinus
Leptographium microsporum R.W. Davidson
Distr.: Southern United States.
Notes: Isolated from wood.

Hosts: Fagus, Liquidambar
Leptographium penicillatum Grosmann
Verticicladiella penicillata (Grosmann) Kendrick
Distr.: Temperate North America; Europe.
Hosts: Abies, Larix, Pinus, Pseudotsuga
Leptographium procerum (Kendrick) M.J. Wingfield
Verticicladiella procera Kendrick
Distr.: Central and eastern North America.
Notes: Associated with root decline.
Host: Pinus
Leptographium pyrinum R.W. Davidson
Distr.: NM.
Host: Pinus
Leptographium terebrantis S.J. Barras & T.J. Perry
Distr.: Central and western United States, LA, MA.
Notes: Associated with blue stain and mortality of *Pinus* and black turpentine beetles (Highley & Tattar. Pl. Dis. 69:528. 1985).
Host: Pinus
Leptographium wageneri (Kendrick) M.J. Wingfield
Verticicladiella wageneri Kendrick
Distr.: Western North America.
Notes: In roots of conifers. These records have not been designated as variety *wageneri*; for the most part no variety was reported.
Hosts: Abies, Pinus, Pseudotsuga, Tsuga
Leptographium wageneri (Kendrick) M.J. Wingfield var. ***ponderosum*** (T.C. Harrington & F.W. Cobb) T.C. Harrington & F.W. Cobb
Verticicladiella wageneri Kendrick var. *ponderosa* T.C. Harrington & F.W. Cobb
Distr.: Western North America.
Notes: Black stain root disease of conifers (Harrington & Cobb. Mycologia 78:562. 1986). Harrington & Cobb (Mycotaxon 30:505. 1987) transferred this variety to *Leptographium* and compared it with the other described varieties. Based on host associations *Ophiostoma wageneri* is presumed to be the teleomorph of this variety.
Hosts: Pinus, Tsuga
Leptographium wageneri (Kendrick) M.J. Wingfield var. ***pseudotsugae*** T.C. Harrington & F.W. Cobb
Distr.: Western North America.
Notes: Harrington & Cobb (Mycotaxon 30:501. 1987).
Host: Pseudotsuga

Leptopeltis Höhn. Loculoascomycetes, Dothideales

Leptopeltis litigiosa (Sacc.) L. Holm & K. Holm
Microthyrium litigiosum Sacc.
Distr.: North America; Europe.
Notes: On dead petioles (Holm & Holm. Bot. Not. 130:221. 1977).
Host: Pteridium

Leptophyma Sacc. Loculoascomycetes, Dothideales

Notes: Müller & Arx (Beitr. Kryptogamenfl. Schweiz 11(2):205. 1962).
Leptophyma aurantiaca (Ellis & Mart.) Sacc.
Ascomycetella aurantiaca Ellis & Mart.
Distr.: FL.
Notes: On leaves.
Host: Quercus

Leptoporus Quél. Basidiomycetes, Aphyllophorales

Notes: Gilbertson & Ryvarden (North American Polypores 1:427. 1986).
Leptoporus mollis (Pers.:Fr.) Pilát
Polyporus mollis (Pers.:Fr.) Fr.
Tyromyces mollis (Pers.:Fr.) Kotlaba & Pouzar
Distr.: Northern hemisphere.
Notes: On wood of conifers.
Hosts: Abies, Larix, Picea, Pinus

Leptosphaeria Ces. & De Not., nom. cons. Loculoascomycetes, Dothideales

Notes: Müller (Sydowia 4:185. 1950); Shoemaker (Canad. J. Bot. 62:2688. 1984).
Leptosphaeria acuta (Fuckel) P. Karst.
Pleospora acuta Fuckel
Phoma piskorzii (Petr.) Boerema & Loerakker (anamorph)
Distr.: Temperate North America; Europe.
Notes: Boerema & Loerakker (Persoonia 11:311. 1981).
Host: Urtica
Leptosphaeria agnita (Desmaz.) Ces. & De Not.
Phoma sp. (anamorph)
Distr.: Europe.
Notes: The occurrence of this species in North America is uncertain.
Hosts: Castilleja, Hibiscus, Lactuca, Lupinus
Leptosphaeria albopunctata (Westend.) Sacc.
Phoma sp. (anamorph)
Distr.: Widespread in coastal regions.
Notes: On culms. Although Kohlmeyer & Kohlmeyer (Marine mycology. p. 410. 1979) listed *L. sticta* and *L. spartinae* as synonyms, Shoemaker (pers. comm.) disagrees.
Host: Spartina
Leptosphaeria anthelmintica (Cooke) Sacc.
Sphaeria anthelmintica Cooke
Metasphaeria anthelmintica (Cooke) Dearn. in House
Distr.: ND, SC.
Host: Chenopodium
Leptosphaeria artemisiae (Fuckel) Auersw. in Rabenh.
Distr.: Temperate North America; Europe.
Notes: On Asteraceae. Lucas & Webster (Trans. Brit. Mycol. Soc. 50:89. 1967) reported a conidial state for which no name exists.
Hosts: Artemisia, Rudbeckia
Leptosphaeria arthrophyma (Fairm.) L. Holm
Gibberidea arthrophyma Fairm.
Distr.: NM.
Notes: Holm (Svensk Bot. Tidskr. 62:228. 1968).
Host: Chrysothamnus
Leptosphaeria arunci Zeller
Distr.: OR(type).
Host: Aruncus
Leptosphaeria asclepiadis Shoemaker
Distr.: Northeastern North America.
Host: Asclepias
Leptosphaeria asparagina P. Karst.
Distr.: FL, WA; Europe.
Hosts: Agrostis, Asparagus, Calamagrostis, Elymus, Festuca, Heracleum, Phleum
Leptosphaeria astericola Ellis & Everh.
Distr.: KS, ND.
Host: Aster
Leptosphaeria aulica (Cooke & Ellis) Ellis
Sphaeria aulica Cooke & Ellis
Metasphaeria aulica (Cooke & Ellis) Sacc.
Distr.: GA, NJ.
Notes: On herbaceous stems.
Host: Impatiens
Leptosphaeria australiensis (A.B. Cribb & J.W. Cribb) G.C. Hughes
Metasphaeria australiensis A.B. Cribb & J.W. Cribb
Distr.: Subtropical and tropical marine habitats.
Notes: On wood.
Hosts: Hibiscus, Juncus, Rhizophora
Leptosphaeria avicenniae Kohlmeyer & E. Kohlmeyer
Distr.: FL.
Notes: Along the Atlantic Ocean in tropical and subtropical regions (Kohlmeyer & Kohlmeyer. Marine mycology. p. 413. 1979).
Host: Avicennia
Leptosphaeria berberidis Richon
Distr.: Central and eastern United States; France.
Hosts: Berberis, Mahonia
Leptosphaeria bicuspidata (Cooke & Harkn.) Berl. & Voglino
Sphaeria bicuspidata Cooke & Harkn.
Distr.: CA(type).
Host: Baccharis

Leptosphaeria borealis Ellis & Everh.
Distr.: Central United States.
Host: Salix
Leptosphaeria brightonensis Shoemaker
Distr.: UT.
Host: Agastache
Leptosphaeria bupleuri P. Syd.
Distr.: WA; France.
Host: Pedicularis
Leptosphaeria byssincola Earle ex Shoemaker
Distr.: CO.
Host: Lupinus
Leptosphaeria cacuminispora Fairm.
Distr.: Northern North America.
Host: Carya
Leptosphaeria californica (Cooke & Harkn.) Berl. & Voglino
Sphaeria californica Cooke & Harkn.
Distr.: CA.
Host: Araucaria
Leptosphaeria cassiaecola Ellis & Everh.
Distr.: NY, TX.
Host: Cassia
Leptosphaeria castillejae Shoemaker
Distr.: MT.
Host: Castilleja
Leptosphaeria castrensis Shoemaker
Distr.: CA.
Host: Lupinus
Leptosphaeria ceanothi (Cooke & Harkn.) Berl. & Voglino
Sphaeria ceanothi Cooke & Harkn.
Distr.: CA(type).
Host: Ceanothus
Leptosphaeria cercocarpi Syd. & P. Syd.
Distr.: UT.
Host: Cercocarpus
Leptosphaeria clavigera (Cooke & Ellis) Sacc.
Distr.: Southeastern United States.
Hosts: Cimicifuga, Lespedeza, Phytolacca, Yucca
Leptosphaeria collinsoniae Dearn. & House
Distr.: Northeastern North America.
Hosts: Aralia, Arctium, Collinsonia
Leptosphaeria complanata (Tode) Ces. & De Not.
Sphaeria complanata Tode
Metasphaeria complanata (Tode) Sacc.
Distr.: GA; England.
Notes: On Asteraceae and *Polygonum*.
Hosts: Ambrosia, Aster, Polygonum, Rudbeckia
Leptosphaeria compressa (Rehm) L. Holm
Ophiobolus compressa Rehm
Distr.: IL; Europe.
Host: Erigeron
Leptosphaeria concinna Ellis & Everh.
Distr.: Western United States.
Host: Castilleja
Leptosphaeria conoidea (De Not.) Sacc.
Leptosphaeria doliolum (Pers.:Fr.) Ces. & De Not. var. *conoidea* (De Not.) Sacc.
Distr.: Temperate North America; Europe.
Hosts: Ambrosia, Aralia, Aster, Baccharis, Celastrus, Chelone, Impatiens, Lactuca, Monarda, Physocarpus, Thalictrum, Urtica, Xanthium
Leptosphaeria consessa (Cooke & Ellis) Sacc.
Sphaeria consessa Cooke & Ellis
Distr.: Temperate North America.
Notes: On stems of various dicotyledonous plants.
Hosts: Ambrosia, Helianthus, Scrophularia
Leptosphaeria corallorhizae Peck
Phaeosphaeria corallorhizae (Peck) Barr in Barr et al.
Leptosphaeria solani Ellis & Everh. non Romell, hom. illeg.
Distr.: Northcentral and northeastern North America.
Hosts: Bidens, Corallorrhiza, Orchidaceae
Leptosphaeria cylindrospora Auersw. & Niessl ex Sacc.
Distr.: Circumboreal.
Host: Epilobium
Leptosphaeria darkeri Shoemaker
Distr.: UT.
Host: Agastache
Leptosphaeria dearnessii Shoemaker
Distr.: Northeastern North America.
Host: Asclepias
Leptosphaeria distributa (Cooke & Ellis) Sacc.
Metasphaeria distributa (Cooke & Ellis) Berl.
Distr.: Eastern North America.
Hosts: Desmodium, Lespedeza
Leptosphaeria doliolum (Pers.:Fr.) Ces. & De Not.
Leptosphaeria subconica (Cooke & Peck) Sacc.
Phoma hoehnelii Van Kesteren (anamorph)
Distr.: Temperate North America; Europe.
Notes: Holm (Symb. Bot. Upsal. 14:19. 1957) recognized two varieties.
Hosts: Abelmoschus, Ambrosia, Aralia, Arnica, Asclepias, Aster, Aureolaria, Boehmeria, Chelone, Chrysopsis, Cirsium, Datura, Draba, Eupatorium, Foeniculum, Heracleum, Lactuca, Passiflora, Pedicularis, Petasites, Phytolacca, Ricinus, Saponaria, Senecio, Solidago, Verbesina, Vernonia
Leptosphaeria drabae (Nyl.) P. Karst.
Sphaeria drabae Nyl.
Distr.: North temperate regions.
Host: Valeriana
Leptosphaeria dracaenae Camara
Distr.: HI; Guatemala; Portugal.
Notes: The place of publication of this name could not be found.
Host: Dracaena
Leptosphaeria dryophila (Cooke & Harkn.) Sacc.
Sphaerella dryophila Cooke & Harkn.
Distr.: CA, NC.
Host: Quercus
Leptosphaeria ellisiana Berl.
Leptosphaeria onagrae Rehm
Distr.: Northeastern North America.
Notes: On Onagraceae.
Host: Oenothera
Leptosphaeria elongata Wehmeyer
Distr.: WA(type).
Hosts: Calamagrostis, Elymus
Leptosphaeria elymi Atk.
Distr.: Central and southern United States.
Host: Elymus
Leptosphaeria epicarecta (Cooke) Sacc.
Sphaeria epicarecta Cooke
Distr.: GA; Great Britain.
Host: Carex
Leptosphaeria equiseti P. Karst.
Distr.: NY; Europe.
Host: Equisetum
Leptosphaeria erigerontis (Berl.) Berl.
Leptosphaeria agnita (Desmaz.) Ces. & De Not. var. *erigerontis* Berl.
Distr.: Temperate North America; Europe.
Hosts: Agoseris, Aquilegia, Aster, Carum, Celtis, Conyza, Delphinium, Erigeron, Heracleum, Ligusticum, Lupinus, Mertensia, Pedicularis, Valeriana
Leptosphaeria eumorpha (Berk. & M.A. Curtis) Earle
Sphaeria eumorpha Berk. & M.A. Curtis
Distr.: Central and eastern United States.
Host: Arundinaria
Leptosphaeria fiedleri (Niessl) Sacc.
Cryptosphaeria fiedleri Niessl
Metasphaeria fiedleri (Niessl) Sacc.
Hendersonia fiedleria Westend. (anamorph)
Distr.: NY; Italy.
Host: Cornus
Leptosphaeria filamentosa Ellis & Everh.
Distr.: NC, NJ, TX.
Host: Yucca
Leptosphaeria foeniculacea Fabre subsp. **lupina** Sacc. & Scalia
Distr.: AK(type).
Host: Lupinus
Leptosphaeria folliculata Ellis & Everh. var. **oxyspora** J.J. Davis
Distr.: WI(type).
Host: Carex
Leptosphaeria fraxini Ellis & Everh.

Distr.: MO(type).
Host: Fraxinus
Leptosphaeria fulgida (Cooke & Peck) Barr
Ophiobolus fulgidus (Cooke & Peck) Sacc.
Distr.: Temperate North America.
Notes: On Asteraceae.
Hosts: Ambrosia, [Phytolacca], Sonchus, Vernonia
Leptosphaeria fusispora Niessl
Distr.: NY, VA; Europe.
Host: Polygonum
Leptosphaeria galiorum (Roberge ex Desmaz.) Ces. & De Not.
Sphaeria galiorum Roberge ex Desmaz.
Metasphaeria galiorum (Roberge ex Desmaz.) Sacc.
Leptosphaeria aparines (Fuckel) Sacc.
Pleospora aparines Fuckel
Distr.: CA; Europe.
Hosts: Galium, Lupinus
Leptosphaeria galiorum (Roberge ex Desmaz.) Ces. & De Not. var. **gnaphaliana** Fairm.
Distr.: CA(type).
Host: Gnaphalium
Leptosphaeria gaultheriae Dearn.
Distr.: Pacific Northwest.
Host: Gaultheria
Leptosphaeria georgius-fischeri R. Sprague
Distr.: OR(type).
Host: Muhlenbergia
Leptosphaeria halima T.W. Johnson
Distr.: East coast of North America.
Notes: On culms of *Spartina* and on dead wood.
Host: Spartina
Leptosphaeria hamamelidis Fairm.
Distr.: NY(type).
Host: Hamamelis
Leptosphaeria helianthi Ellis & Everh., nom. nud.
Distr.: ND, NM.
Notes: This species was published without a description.
Host: Helianthus
Leptosphaeria heliopsidis (Schwein.:Fr.) L. Holm
Gibberidea heliopsidis (Schwein.:Fr.) Shear
Montagnella heliopsidis (Schwein.:Fr.) Sacc.
Rosenscheldia heliopsidis (Schwein.:Fr.) Theiss. & Syd.
Distr.: Temperate North America.
Notes: On stems of Asteraceae. Holm (Svensk Bot. Tidskr. 62:230. 1968) discussed the relationships of this species.
Hosts: Aster, Helianthus, Heliopsis, [Hypericum], Lactuca, Solidago
Leptosphaeria helminthospora (Ces.) Ces. & De Not.
Leptosphaeria clavispora J.H. Miller & M.G. Burton
Distr.: Eastern North America; Europe.
Notes: On dead stems.
Host: Eupatorium
Leptosphaeria helvetica Sacc. & Speg.
Host: Selaginella
Leptosphaeria hydrophila Sacc.
Distr.: NY; Italy.
Host: Typha
Leptosphaeria jacksonensis Shoemaker
Distr.: WY(type).
Host: Pedicularis
Leptosphaeria korrae J.C. Walker & A.M. Sm.
Distr.: NY; Australia(type).
Notes: Spring dead spot. A root and stolon parasite of various grasses. Smiley & Fowler (Pl. Dis. 68:440. 1984).
Hosts: Cynodon, Poa
Leptosphaeria leersiana Sacc.
Leptosphaeria leersiae Pass.
Distr.: AK; Italy.
Host: Agrostis
Leptosphaeria lethalis Ellis & Everh.
Distr.: Western United States.
Host: Penstemon
Leptosphaeria licatensis Sacc.
Distr.: IL; Europe.
Host: Aster
Leptosphaeria lindquistii Frezzi
Phoma macdonaldii Boerema (anamorph)
Distr.: Central United States; Argentina.
Notes: Black stem disease of sunflower.
Host: Helianthus
Leptosphaeria longipedicellata J.H. Miller & M.G. Burton
Distr.: GA.
Hosts: Conyza, Daucus, Polymnia, Smallanthus, Solidago
Leptosphaeria lupini (Earle) Boise
Trematosphaeria lupini Earle
Distr.: CA, CO.
Host: Lupinus
Leptosphaeria lythri Peck
Distr.: KS(type).
Host: Lythrum
Leptosphaeria maclurae Ellis & Everh.
Metasphaeria maclurae (Ellis & Everh.) Sacc.
Distr.: MO(type).
Host: Maclura
Leptosphaeria macrospora (Fuckel) Thuem.
Scolecosporiella bernardiana (Sacc.) Sivanesan (anamorph)
Distr.: Temperate North America; Europe.
Notes: On herbaceous stems.
Hosts: Chenopodium, Eupatorium, Lupinus, Rumex, Solidago, Zea
Leptosphaeria maculans (Desmaz.) Ces. & De Not.
Phoma lingam (Tode:Fr.) Desmaz. (anamorph)
Distr.: Temperate regions.
Notes: Black leg, canker, and dry rot of Brassicaceae. Reports on *Lupinus* spp. are doubtful (Shoemaker. Canad. J. Bot. 62:2707. 1984). Punithalingam & Holliday (C.M.I. Descr. 331. 1972).
Hosts: Brassica, Lepidium, [Lupinus]
Leptosphaeria marina Ellis & Everh.
Distr.: East coast of North America.
Notes: On culms.
Hosts: Juncus, Spartina
Leptosphaeria maydis G.L. Stout
Distr.: IL(type).
Host: Zea
Leptosphaeria mellispora M.L. Farr & H.T. Horner
Host: Selaginella
Leptosphaeria millefolii (Fuckel) Niessl
Pleospora millefolii Fuckel
Distr.: North America; temperate Europe.
Hosts: Achillea, Artemisia, [Lupinus]
Leptosphaeria monticola Ellis & Everh.
Distr.: CO(type).
Host: Trifolium
Leptosphaeria muhlenbergiae Rehm
Distr.: ND, OK.
Host: Muhlenbergia
Leptosphaeria muirensis R. Sprague
Distr.: AK(type).
Host: Poa
Leptosphaeria multiseptata G. Wint.
Distr.: Western United States; Europe.
Host: Achillea
Leptosphaeria muricata Ellis & Everh.
Distr.: Southern United States.
Host: Andropogon
Leptosphaeria myricae Dearn. & House
Distr.: NY(type).
Host: Myrica
Leptosphaeria neomaritima R.V. Gessner & Kohlmeyer
Leptosphaeria maritima (Cooke & Plowr.) Sacc.
Distr.: East coast of North America; Great Britain.
Notes: On culms.
Hosts: Juncus, Spartina
Leptosphaeria nigricans P. Karst. var. **inculta** Sacc.
Distr.: ND.
Notes: The place of publication of this name could not be found.
Host: Grindelia
Leptosphaeria norvegica Rostr.
Distr.: Western United States; northern Europe.
Host: Scirpus
Leptosphaeria noveboracensis Shoemaker
Distr.: NY.

Host: Solidago
Leptosphaeria obiones (P.L. Crouan & H. Crouan) Sacc.
Pleospora obiones P.L. Crouan & H. Crouan
Leptosphaeria discors Sacc. & Ellis
Passeriniella discors (Sacc. & Ellis) Apinis & Chesters
Didymosphaeria spartinae Grove
Distr.: Cosmopolitan.
Notes: On dead culms of Poaceae.
Host: Spartina
Leptosphaeria occidentalis Ellis & Everh.
Distr.: KS(type).
Host: Echinochloa
Leptosphaeria odora (Cooke & Harkn.) Berl. & Voglino
Distr.: CA.
Host: Umbellularia
Leptosphaeria ogilviensis (Berk. & Broome) Ces. & De Not.
Distr.: North America; temperate Europe.
Notes: On herbaceous stems.
Hosts: Aconitum, Aster, Eupatorium, Haplopappus, Lupinus, Senecio
Leptosphaeria olericola (Berk. & M.A. Curtis) Sacc.
Distr.: Northeastern United States.
Host: Brassica
Leptosphaeria ophiopogonis Sacc. f. **graminum** Sacc.
Distr.: AK; Europe.
Host: Festuca
Leptosphaeria opuntiae B.O. Dodge
Hendersonia opuntiae Ellis & Everh. (anamorph)
Distr.: Subtropical and tropical western hemisphere.
Host: Opuntia
Leptosphaeria oraemaris Linder in Barghoorn & Linder
Distr.: Widespread in marine habitats.
Notes: On culms of grasses and driftwood.
Host: Spartina
Leptosphaeria oreophila Sacc.
Distr.: WY.
Hosts: Aquilegia, Castilleja, Delphinium, Hedysarum, Pedicularis, Saxifraga
Leptosphaeria orthogramma (Berk. & M.A. Curtis) Sacc.
Sphaeria orthogramma Berk. & M.A. Curtis
Distr.: Central and eastern United States.
Hosts: Erianthus, Saccharum
Leptosphaeria pacifica Rehm
Distr.: CA.
Host: Yucca
Leptosphaeria pelagica E.B.G. Jones
Distr.: Worldwide in marine habitats.
Notes: On culms of grasses and driftwood.
Host: Spartina
Leptosphaeria phaseolorum Ellis & Everh.
Distr.: NJ(type).
Host: Phaseolus
Leptosphaeria phlogis Oudem.
Distr.: Reported from an unknown locality in the United States; The Netherlands(type).
Host: Phlox
Leptosphaeria physalidis Ellis & Everh.
Distr.: KY(type).
Host: Physalis
Leptosphaeria planiuscula (Riess in Rabenh.) Ces. & De Not.
Leptosphaeria asteris J.H. Miller & M.G. Burton
Leptosphaeria virginica (Cooke & Ellis) Sacc.
Distr.: Temperate North America; Europe.
Notes: On stems of Asteraceae and Brassicaceae (Shoemaker. Canad. J. Bot. 62:2709. 1984).
Hosts: Aster, Conyza, Eupatorium, Lepidium, Solidago
Leptosphaeria platanicola (E. Howe) Sacc.
Sphaeria platanicola E. Howe
Distr.: Eastern United States.
Host: Platanus
Leptosphaeria pratensis Sacc. & Briard
Phoma meliloti Allesch. (synanamorph)
Stagonospora meliloti (Lasch) Petr. (synanamorph)
Distr.: North America, north Europe.
Notes: Leaf spot and root rot of alfalfa.
Hosts: Medicago, Melilotus, Trifolium
Leptosphaeria preandina (Speg.) L. Holm
Gibberidea preandina Speg.
Gibberidea artemisiae Earle
Distr.: NV.
Notes: Holm (Svensk Bot. Tidskr. 62:228. 1968).
Host: Artemisia
Leptosphaeria proteispora Speg.
Distr.: HI; Argentina(type).
Host: Paspalum
Leptosphaeria rostrata M.L. Farr & H.T. Horner
Host: Selaginella
Leptosphaeria rubicunda Rehm ex G. Wint.
Distr.: North America; Hungary.
Host: Leonurus
Leptosphaeria rubrotincta Ellis & Everh.
Distr.: PA(type).
Host: Staphylea
Leptosphaeria sacchari Breda de Haan
Phyllosticta sp. (anamorph)
Distr.: Subtropical and tropical regions.
Notes: Ring spot of sugarcane (Morgan-Jones. C.M.I. Descr. 145. 1967; Sivanesan & Waller. Phytopathol. Pap. 29:30. 1986).
Hosts: [Pennisetum], Saccharum
Leptosphaeria salsolae Hollós
Distr.: Western United States; Hungary.
Hosts: Castilleja, Hieracium, Lupinus, Phacelia
Leptosphaeria sambuci Fautrey
Distr.: Western United States; Europe.
Hosts: Holodiscus, Sambucus
Leptosphaeria sambucina Ellis & Everh.
Distr.: CO(type).
Host: Sambucus
Leptosphaeria senecionis (Fuckel) G. Wint.
Pleospora senecionis Fuckel
Distr.: Eastern North America; Europe.
Host: Solidago
Leptosphaeria shastensis Shoemaker
Distr.: CA.
Host: Lupinus
Leptosphaeria simmonsii Sacc.
Distr.: WA(type).
Host: Heracleum
Leptosphaeria spectabilis Niessl
Distr.: TX; Europe.
Host: Elymus
Leptosphaeria sporoboli Ellis & B.T. Galloway
Distr.: MT.
Host: Muhlenbergia
Leptosphaeria stellariae Rostr.
Distr.: Alpine, arctic and subarctic North America; Europe.
Hosts: Arenaria, Stellaria
Leptosphaeria subcutanea (Cooke & Ellis) Ellis
Sphaeria subcutanea Cooke & Ellis
Conisphaeria subcutanea (Cooke & Ellis) Cooke
Metasphaeria subcutanea (Cooke & Ellis) Sacc.
Distr.: GA, NJ.
Notes: On decorticated limbs.
Host: Styrax
Leptosphaeria substerilis Peck
Phyllosticta sp. (anamorph)
Distr.: NY.
Notes: On living leaves.
Host: Mentha
Leptosphaeria suffulta (Nees:Fr.) Niessl
Distr.: ID(questionable); Europe.
Host: Sorbus
Leptosphaeria tamaricis (Grev.) Sacc.
Sphaeria tamaricis Grev.
Distr.: OK; Great Britain.
Host: Tamarix
Leptosphaeria tetonensis (Ellis & Everh.) Rehm
Melanomma occidentale (Ellis) Sacc. var. *tetonense* Ellis & Everh.
Distr.: Temperate North America.
Hosts: Artemisia, [Lupinus], [Pedicularis]
Leptosphaeria thalictri G. Wint.

Distr.: ND; Europe.
Host: Thalictrum
Leptosphaeria thomasiana Sacc. & Roum.
Distr.: OR; France.
Host: Rubus
Leptosphaeria tini Ellis & Everh.
Distr.: LA; Portugal.
Host: Viburnum
Leptosphaeria trimeroides Rehm
Distr.: ND; Germany.
Host: Liatris
Leptosphaeria tumefaciens (Ellis & Harkn.) Petr.
Montagnella tumefaciens (Ellis & Harkn.) Berl. & Voglino
Syncarpella tumefaciens (Ellis & Harkn.) Theiss. & Syd.
Distr.: Western North America.
Notes: Holm (Svensk Bot. Tidskr. 62:222. 1968).
Hosts: Artemisia, Haplopappus
Leptosphaeria variiseptata G.L. Stout
Distr.: IL(type).
Host: Zea
Leptosphaeria veratri Earle
Distr.: CO(type).
Host: Veratrum
Leptosphaeria wehmeyeri Shoemaker
Distr.: Western North America.
Hosts: Lupinus, Thermopsis
Leptosphaeria xerophylli Ellis
Metasphaeria xerophylli (Ellis) Sacc.
Sphaerulina xerophylli (Ellis) Cooke
Leptosphaeria hysterioides Ellis & Everh.
Distr.: Temperate North America.
Host: Xerophyllum

Leptosphaerulina McAlpine. Loculoascomycetes, Dothideales

Leptosphaerulina americana (Ellis & Everh.) J.H. Graham & Luttrell
Pleospora americana Ellis & Everh.
Pleosphaerulina hyalospora (Ellis & Everh.) Berl.
Pleospora hyalospora Ellis & Everh. non Speg., hom. illeg.
Distr.: Eastern United States.
Hosts: Lysimachia, Pisum, Trifolium, Vicia
Leptosphaerulina argentinensis (Speg.) J.H. Graham & Luttrell
Pleosphaerulina argentinensis Speg.
Distr.: North America; Australia.
Notes: Although Booth & Pirozynski (C.M.I. Descr. 146. 1967) considered this species a synonym of *L. trifolii*, Irwin & Davis (Austral. J. Bot. 33:233. 1985) demonstrated that *L. argentinensis* should be recognized as a separate species.
Hosts: Melilotus, Phleum, Pueraria, Sorghum, Trifolium
Leptosphaerulina crassiasca (Sechet) C.R. Jackson & D.K. Bell
Pleospora crassiasca Sechet
Leptosphaerulina arachidicola J. Yen, Chen & K. Huang
Distr.: North America; Australia; Taiwan.
Notes: Jackson & Bell (Oleagineux 23:387. 1968) listed the synonymy for this species. Although Booth & Pirozynski (C.M.I. Descr. 146. 1967) considered the synonym *L. arachidicola* to be *L. trifolii*, Irwin & Davis (Austral. J. Bot. 33:233. 1985) demonstrated that *L. arachidicola*, thus *L. crassiasca*, is a distinct species.
Host: Arachis
Leptosphaerulina sidalceae Zeller
Distr.: OR(type).
Host: Sidalcea
Leptosphaerulina trifolii (Rostr.) Petr.
Sphaerulina trifolii Rostr.
Pseudoplea trifolii (Rostr.) Petr.
Leptosphaerulina australis McAlpine
Leptosphaerulina briosiana (Pollacci) J.H. Graham & Luttrell
Pleosphaerulina briosiana Pollacci
Pseudoplea briosiana (Pollacci) Höhn.
Pseudoplea medicaginis L.E. Miles
Pleosphaerulina oryzae Miyake
Leptosphaerulina vignae Tehon & G.L. Stout
Pleosphaerulina zeicola G.L. Stout
Distr.: Cosmopolitan.
Notes: Irwin & Davis (Austral. J. Bot. 33:233. 1985) confirmed the synonyms of this species.
Hosts: Agropyron, Arachis, Glycine, Lespedeza, Medicago, Oryza, Panicum, Phaseolus, Poa, Trifolium, Vigna, Zea

Leptospora Rabenh. Loculoascomycetes, Dothideales

Notes: Walker (Mycotaxon 11:34. 1980).
Leptospora implexa (Ellis & Everh.) J. Walker
Lophiostoma implexum Ellis & Everh.
Ophiobolus implexus (Ellis & Everh.) Ellis & Everh.
Distr.: LA; Sri Lanka.
Notes: On dead herbaceous material (Walker. Mycotaxon 11:35. 1980).
Host: Sorghum
Leptospora rubella (Pers.:Fr.) Rabenh.
Sphaeria rubella Pers.:Fr.
Ophiobolus rubellus (Pers.:Fr.) Sacc.
Ophiobolus porphyrogonus (Tode) Sacc.
Ophiobolus tenellus (Auersw.) Sacc.
Ophiobolus trichosporus Ellis & Everh.
Distr.: Temperate regions.
Hosts: Abelmoschus, Amaranthus, Arctium, Cryptotaenia, Erechtites, Lactuca, Lappula, Leonurus, Lycopersicon, Medicago, Sinapis, Veratrum, Verbesina

Leptosporomyces Jülich. Basidiomycetes, Aphyllophorales

Notes: Eriksson & Ryvarden (The Corticiaceae of North Europe 4:800. 1976).
Leptosporomyces galzinii (Bourd.) Jülich
Corticium galzinii Bourd.
Athelia galzinii (Bourd.) Donk
Corticium canum Burt
Athelia grisea M. Christiansen
Distr.: Northern hemisphere.
Notes: On wood and plant debris.
Hosts: Abies, Alnus, Juniperus, Picea, Pinus, Pseudotsuga, Quercus, Tsuga
Leptosporomyces juniperinus R.L. Gilbertson & Lindsey
Distr.: Southwestern United States.
Notes: On wood (Gilbertson & Lindsey. Great Basin Naturalist 38:45. 1978).
Host: Juniperus
Leptosporomyces mundus (H. Jacks. & Dearden) Jülich
Peniophora munda H. Jacks. & Dearden
Athelia munda (H. Jacks. & Dearden) M. Christiansen
Distr.: North America.
Notes: On wood.
Hosts: Abies, Picea, Tsuga
Leptosporomyces ovoideus Jülich
Distr.: North America; Europe.
Notes: On plant debris.
Host: Acer
Leptosporomyces raunkiaerii (M. Christiansen) Jülich
Athelia raunkiaerii M. Christiansen
Distr.: NJ, IA; Canada(Ontario); northern hemisphere.
Notes: On hardwoods.
Hosts: Abies, Populus

Leptostroma Fr.:Fr. Coelomycetes

Notes: Species on pine needles are reviewed by Minter (Canad. J. Bot. 58:906. 1980).
Leptostroma caricinum Fr.:Fr.
Distr.: WA; Europe.
Host: Carex
Leptostroma decipiens Petr.
Lophodermium baculiferum Mayr (teleomorph)
Distr.: Western North America.
Host: Pinus

Leptostroma donacis **Schwein.**
Distr.: NC.
Host: Arundo
Leptostroma durissimum **Cooke**
Lophodermium australe Dearn. (teleomorph)
Distr.: Eastern United States.
Host: Pinus
Leptostroma hedgcockii **Dearn.**
Ploioderma hedgcockii (Dearn.) Darker (teleomorph)
Distr.: Southern United States.
Notes: Needle blight of pines (Cannon & Minter. C.M.I. Descr. 799. 1984).
Host: Pinus
Leptostroma herbarum **(Fr.:Fr.) Link**
Sclerotium herbarum Fr.:Fr.
Distr.: AK; Europe.
Notes: On stems and twigs of many hosts.
Hosts: Heuchera, Polemonium
Leptostroma hyperici **Schwein.**
Distr.: PA.
Host: Hypericum
Leptostroma juncacearum **Sacc.**
Distr.: Pacific Northwest; Europe.
Notes: On dead culms of *Juncus conglomeratus* and on leaves of *Luzula sylvatica*. May be synonymous with *Leptostroma juncinum* Fr.
Host: Juncus
Leptostroma lupini **Wehmeyer**
Distr.: WY.
Hosts: Lupinus, Pedicularis
Leptostroma pinorum **Sacc. var. *graupianum* Minter**
Distr.: Temperate regions.
Notes: Minter (Mycol. Pap. 147:29. 1981) described this variety occurring on *Pinus* species other than *P. pinaster* and *P. cembra*.
Host: Pinus
Leptostroma pteridis **Ehrenb.**
Distr.: ME; Europe.
Notes: On stems.
Host: Pteridium
Leptostroma querci **Tehon**
Distr.: IL.
Host: Quercus
Leptostroma rostrupii **Minter**
Lophodermium seditiosum Minter, Staley & Millar (teleomorph)
Distr.: WA.
Host: Pinus
Leptostroma sequoiae **Cooke & Harkn.**
Distr.: CA.
Hosts: Chamaecyparis, Sequoia
Leptostroma smilacis **Cooke**
Distr.: NY, SC.
Host: Smilax
Leptostroma xerophylli **Petr.**
Distr.: ID.
Host: Xerophyllum

Leptostromella (Sacc.) Sacc. Coelomycetes

Leptostromella andropogonis **Dearn. & House**
Distr.: AL; Panama.
Hosts: Andropogon, Schizachyrium
Leptostromella angelicae **Dearn. & House**
Distr.: NY.
Host: Angelica
Leptostromella bignoniae **Dearn., nom. inval.**
Distr.: TN.
Host: Bignonia
Leptostromella chenopodii **Dearn. & House**
Distr.: NY.
Host: Chenopodium
Leptostromella conigena **Dearn.**
Distr.: Temperate North America.
Host: Picea
Leptostromella cynodontis **Sacc.**
Distr.: AL; southern Europe.
Host: Cynodon
Leptostromella elastica **Ellis & Everh.**
Distr.: Northeastern United States.
Host: Ficus
Leptostromella eriogoni **Earle**
Distr.: NV.
Host: Eriogonum
Leptostromella filicina **(Berk. & M.A. Curtis) Sacc.**
Cryptosporium filicinum Berk. & M.A. Curtis
Distr.: Central and eastern United States.
Hosts: Dryopteris, Osmunda, Pteridium
Leptostromella hysterioides **(Fr.) Sacc. var. *graminicola* De Not.**
Distr.: CA; Italy
Host: Sitanion
Leptostromella panici **Dearn.**
Distr.: SC.
Host: Panicum
Leptostromella scirpina **Peck**
Distr.: NE, NY.
Host: Scirpus

Leptothyrella Sacc. Coelomycetes

Asterinula Ellis & Everh.
Leptothyrella aceris **Dearn. & House**
Distr.: NY(type).
Host: Acer
Leptothyrella caricis **Dearn. & Barth.**
Distr.: KS.
Host: Carex
Leptothyrella graminis **Wehmeyer**
Distr.: WA.
Hosts: Aster, Festuca, Heracleum, Lupinus, Muhlenbergia, Pedicularis, Phleum, Valeriana
Leptothyrella langloisii **(Ellis & Everh.) Sacc.**
Asterinula langloisii Ellis & Everh.
Distr.: LA.
Host: Magnolia

Leptothyrium Kunze in Kunze & J.C. Schmidt. Coelomycetes

Notes: The generic limits have been redefined (Sutton. The Coelomycetes. p. 212. 1980).
Leptothyrium alneum **(Lév.) Sacc.**
Melasmia alnea Lév.
Distr.: AK; Italy.
Notes: On wilting leaves.
Host: Alnus
Leptothyrium arcticum **(Fuckel) Lind**
Labrella arctica Fuckel
Distr.: AR; Europe.
Host: Potentilla
Leptothyrium astericola **(Ellis & Everh.) H.C. Greene**
Phyllosticta astericola Ellis & Everh.
Distr.: TX, WI.
Host: Aster
Leptothyrium avenae **Tehon**
Distr.: IL.
Host: Avena
Leptothyrium brunnichiae **Tehon**
Distr.: IL.
Host: Brunnichia
Leptothyrium californicum **Bubák**
Distr.: CA.
Host: Quercus
Leptothyrium carpophilum **Pass.**
Distr.: AL; Italy.
Host: Pyrus
Leptothyrium castaneae **(Spreng.) Sacc.**
Leptostroma castaneae Spreng.
Xyloma geographicum Pers.
Coccomyces dentatus (Kunze & J.C. Schmidt) Sacc. f. *castaneae* Sacc. (teleomorph)
Distr.: FL, WV; Europe.

Notes: On fallen leaves.
Host: Castanea
Leptothyrium castaneae (Spreng.) Sacc. var. ***quercus*** C. Massal.
Distr.: VA; Italy.
Host: Quercus
Leptothyrium chenopodii Dearn. & Barth.
Distr.: KS.
Host: Chenopodium
Leptothyrium clypeosphaerioides Sacc.
Distr.: AK; France.
Host: Rubus
Leptothyrium conspicuum Dearn. & House
Distr.: AK.
Host: Vaccinium
Leptothyrium cylindricum Atk.
Distr.: AL.
Host: Arundinaria
Leptothyrium dearnessii Bubák
Distr.: TX.
Host: Erigeron
Leptothyrium doellingeriae Dearn. & House, nom. nud.
Distr.: NY.
Notes: Published without a latin diagnosis.
Host: Aster
Leptothyrium foraminulatum Sacc. & Ellis
Distr.: Southeastern United States.
Host: Ilex
Leptothyrium gentianaecola (Fr.) Bäumler
Depazea gentianaecola Fr.
Distr.: ND; Europe.
Host: Gentiana
Leptothyrium gleicheniae F. Stevens & P.A. Young
Distr.: HI.
Hosts: Dicranopteris, Gleichenia, Hicriopteris
Leptothyrium kellermanii Bubák
Distr.: OH.
Host: Sassafras
Leptothyrium lactucae Dearn. & Barth.
Distr.: KS.
Host: Lactuca
Leptothyrium liriodendri Cooke
Distr.: SC, WV.
Host: Liriodendron
Leptothyrium litigiosum (Desmaz.) Sacc.
Leptostroma litigiosum Desmaz.
Leptopeltis litigiosa (Sacc.) L. Holm & K. Holm (teleomorph)
Distr.: North America; Europe.
Notes: On dead petioles (Holm & Holm. Bot. Not. 130:221. 1977).
Hosts: Osmunda, Pteridium
Leptothyrium lunariae Kunze
Microthyrium lunariae Fuckel (teleomorph)
Distr.: MA; Austria; Germany.
Host: Lunaria
Leptothyrium lychnidis Berk. & Cooke
Distr.: AL.
Host: Lychnis
Leptothyrium maximum Tehon & E.Y. Daniels
Distr.: IL(type).
Host: Acer
Leptothyrium nervisedum J.R. Cole
Distr.: Southern United States.
Notes: Vein spot of pecan leaves.
Host: Carya
Leptothyrium parvulum Dearn., nom. inval.
Distr.: NC.
Host: Rhododendron
Leptothyrium petiolorum (Cooke & Ellis) Sacc.
Leptostroma petiolorum Cooke & Ellis
Distr.: NJ, OH.
Host: Ailanthus
Leptothyrium petrakii Solheim
Distr.: WY.
Host: Elymus
Leptothyrium pothi Weedon in F. Stevens
Distr.: HI.
Host: Pothos
Leptothyrium protuberans Sacc.
Distr.: North America; Europe.
Notes: On herbaceous stems.
Host: Epilobium
Leptothyrium pseudotsugae Dearn.
Distr.: CO, ID.
Host: Pseudotsuga
Leptothyrium pulchrum Dearn.
Distr.: Arctic region.
Host: Salix
Leptothyrium punctiforme Berk. & M.A. Curtis
Distr.: WI, eastern United States.
Host: Erigeron
Leptothyrium rosarum Cooke
Distr.: FL, SC.
Host: Rosa
Leptothyrium salicicola H.C. Greene
Distr.: WI.
Host: Salix
Leptothyrium sidae F. Stevens & P.A. Young
Distr.: HI.
Host: Sida
Leptothyrium similisporum (Ellis & J.J. Davis) J.J. Davis
Phyllosticta similispora Ellis & J.J. Davis
Leptothyrium tumidulum Sacc.
Distr.: WI; Canada(Ontario).
Hosts: Aster, Solidago
Leptothyrium smilacis Dearn.
Distr.: NY.
Host: Smilax
Leptothyrium stenosporum Dearn.
Distr.: GA.
Host: Pinus
Leptothyrium symploci (Cooke) Tassi
Sacidium symploci Cooke
Distr.: SC.
Host: Symplocos
Leptothyrium vulgare (Fr.:Fr.) Sacc.
Leptostroma vulgare Fr.:Fr.
Distr.: AK; Europe.
Notes: On dead stems of many herbaceous plants.
Host: Rubus
Leptothyrium vulgare (Fr.:Fr.) Sacc. f. ***parryae*** Sacc. in Sacc., Peck, & Trel.
Distr.: AK.
Host: Parrya
Leptothyrium zeae G.L. Stout
Distr.: IL.
Host: Zea

Leptotrochila P. Karst. Discomycetes, Helotiales

Notes: Schuepp (Phytopathol. Z. 36:236. 1959).
Leptotrochila cerastiorum (Wallr.) H. Schüepp
Pseudopeziza cerastiorum (Wallr.) Fuckel
Distr.: OR; Europe.
Notes: Leaf spot.
Host: Cerastium
Leptotrochila medicaginis (Fuckel) H. Schüepp
Pyrenopeziza medicaginis Fuckel
Pseudopeziza jonesii Nannf.
Sporonema phacidioides Desmaz. (anamorph)
Distr.: Cosmopolitan.
Notes: Yellow leaf blotch of alfalfa.
Host: Medicago
Leptotrochila ranunculi (Fr.) H. Schüepp
Fabraea ranunculi (Fr.) P. Karst.
Pseudopeziza ranunculi (Fr.) Fuckel
Distr.: Central and eastern North America; Europe.
Notes: Shoemaker & Egger (Fungi Canadenses 228. 1982).
Host: Ranunculus
Leptotrochila repanda (Fr.) P. Karst.

Pseudopeziza repanda (Fr.) P. Karst.
Fabraea dehnii (Rabenh.) Nannf.
Mollisia dehnii (Rabenh.) P. Karst.
Distr.: Western United States; Europe.
Hosts: Fragaria, Galium, Potentilla
Leptotrochila sanguisorbae (Jaap) H. Schüepp
Fabraea sanguisorbae Jaap
Distr.: AK; Europe.
Host: Sanguisorba

Leptoxyphium Speg. Hyphomycetes

Astragoxyphium Batista, Nascim., & Cif.
Leptoxyphium axillatum (Cooke) S.J. Hughes
Capnodium axillatum Cooke
Polychaeton axillatum (Cooke) Kuntze
Distr.: LA, SC.
Host: Catalpa

Leucocytospora Höhn. Coelomycetes

Leucocytospora massariana (Sacc.) Z. Urban
Cytospora massariana Sacc.
Leucostoma massariana (De Not.) Höhn. (teleomorph)
Distr.: ID; Europe.
Host: Sorbus

Leucogyrophana Pouzar. Basidiomycetes, Aphyllophorales

Notes: Ginns (Canad. J. Bot. 56:1953. 1978).
Leucogyrophana arizonica Ginns
Distr.: AZ.
Notes: On wood of conifers.
Hosts: Pinus, Pseudotsuga
Leucogyrophana mollusca (Fr.:Fr.) Pouzar
Merulius molluscus Fr.:Fr.
Serpula mollusca (Fr.:Fr.) P. Karst.
Leucogyrophana pseudomollusca (Parmasto) Parmasto
Merulius pseudomolluscus Parmasto
Distr.: Northern hemisphere.
Notes: On wood.
Hosts: Abies, Betula, Castanea, Larix, Picea, Pinus, Populus, Pseudotsuga, Rhododendron, Thuja
Leucogyrophana montana (Burt) Dománski
Merulius montanus Burt
Distr.: ID, NY; Germany.
Notes: On conifers.
Host: Pinus
Leucogyrophana olivascens (Berk. & M.A. Curtis) Ginns & Weresub
Corticium olivascens Berk. & M.A. Curtis
Coniophora olivascens (Berk. & M.A. Curtis) Massee
Coniophora subochracea Peck
Distr.: Northern hemisphere.
Notes: On wood and woody debris.
Hosts: Picea, Pinus, Platanus, Populus
Leucogyrophana pinastri (Fr.:Fr.) Ginns & Weresub
Gyrophana pinastri (Fr.:Fr.) Bourd. & Galzin
Serpula pinastri (Fr.:Fr.) W.B. Cooke
Merulius pinastri (Fr.:Fr.) Burt
Merulius atrovirens Burt
Merulius byssoideus Burt
Merulius irpicinus Peck
Distr.: Widespread.
Notes: On wood, usually of conifers.
Hosts: Abies, Alnus, Liriodendron, Picea, Pinus, Populus
Leucogyrophana sororia (Burt) Ginns
Merulius sororius Burt
Distr.: United States; Great Britain; New Zealand; Sweden.
Notes: On conifers.
Host: Pinus

Leucostoma (Nitschke) Höhn. Pyrenomycetes, Diaporthales

Leucostoma cincta (Fr.:Fr.) Höhn.
Valsa cincta (Fr.:Fr.) Fr.
Cytospora cincta Sacc. (anamorph)
Distr.: Eastern North America; Europe.
Notes: Perennial canker of peach (Dhanvantari. Canad. J. Pl. Pathol. 4:221. 1982).
Hosts: Crataegus, Prunus, Robinia
Leucostoma kunzei (Fr.:Fr.) Munk
Valsa kunzei (Fr.:Fr.) Fr.
Cytospora kunzei Sacc. (anamorph)
Distr.: Temperate northern hemisphere.
Notes: Canker and branch die-back of conifers (Funk. Parasitic microfungi of western trees. p. 90. 1981).
Hosts: Abies, Picea, Pinus, Pseudotsuga, Thuja
Leucostoma massariana (De Not.) Höhn.
Valsa massariana De Not.
Distr.: ID; Asia; Europe.
Notes: Kobayshi (Bull. Gov. Forest Exp. Sta. 226:129. 1970).
Host: Sorbus
Leucostoma nivea (Hoffm.:Fr.) Höhn.
Valsa nivea (Hoffm.:Fr.) Fr.
Cytospora nivea Sacc. (anamorph)
Distr.: Temperate northern hemisphere.
Notes: Canker.
Hosts: Populus, Salix
Leucostoma persoonii Höhn.
Valsa leucostoma (Pers.:Fr.) Fr.
Cytospora leucostoma Sacc. (anamorph)
Distr.: Temperate northern hemisphere.
Notes: On limbs of Rosaceae and other hardwoods.
Hosts: Cydonia, Malus, Populus, Prunus, Pyrus, Sorbus
Leucostoma translucens (De Not.) Höhn.
Valsa translucens De Not.
Valsa abrupta Cooke
Valsa orbicula Berk. & M.A. Curtis
Cytospora sp. (anamorph)
Distr.: Temperate regions.
Hosts: Populus, Salix

Leveillula G. Arnaud. Pyrenomycetes, Erysiphales

Leveillula taurica (Lév.) G. Arnaud
Erysiphe taurica Lév.
Distr.: Cosmopolitan.
Notes: Powdery mildew of cotton and many other plants (Mukerji. C.M.I. Descr. 182. 1968).
Hosts: Capsicum, Cyamopsis, Heliopsis, Hibiscus, Lycopersicon, Prosopis

Lewia Barr & E. Simmons in E. Simmons. Loculoascomycetes, Dothideales

Notes: Simmons (Mycotaxon 25:287. 1986).
Lewia infectoria (Fuckel) Barr & E. Simmons in E. Simmons
Pleospora infectoria Fuckel
Alternaria infectoria E. Simmons
Distr.: Widespread in temperate regions.
Hosts: Bromus, Danthonia, Lupinus, Poa, Typha
Lewia scrophulariae (Desmaz.) Barr & E. Simmons in E. Simmons
Pleospora scrophulariae (Desmaz.) Höhn.
Pyrenophora alternarina M.D. Whitehead & Dickson
Pleospora media Niessl var. *obtusa* Wehmeyer
Pleospora vulgaris Niessl
Alternaria conjuncta E. Simmons (anamorph)
Distr.: Cosmopolitan.
Notes: On various grasses and dicots.
Hosts: Aquilegia, Avena, Brassica, Bupleurum, Castilleja, Draba, Erysimum, Festuca, Iva, Lloydia, Lomatium, Pedicularis, Penstemon, Salix, Senecio, Silene, Valeriana

Libertella Desmaz. Coelomycetes

Libertella acerina Westend.
Distr.: Europe; distribution elsewhere is uncertain.
Host: Acer
Libertella faginea Desmaz.
Eutypella quaternata (Pers.:Fr.) Rappaz (teleomorph)
Distr.: United States; Europe; Great Britain.
Notes: Sutton (The Coelomycetes. p. 221. 1980).
Host: Fagus
Libertella gleditschiae G. Wint.
Distr.: MO, OK.
Host: Gleditsia
Libertella kokkiae Petr.
Distr.: HI(type).
Host: Kokia
Libertella lonicerae Cooke & Harkn.
Distr.: CA.
Host: Lonicera
Libertella olivacea F. Patterson
Distr.: NE.
Host: Rhus
Libertella platani Ellis & Everh., nom. nud.
Distr.: MA, OH, WV.
Host: Platanus

Licea Schrad. Myxomycetes

Licea operculata (Wingate) G.W. Martin
Orcadella operculata Wingate
Distr.: Widespread.
Hosts: Larix, Pinus

Licrostroma P.A. Lemke. Basidiomycetes, Aphyllophorales

Licrostroma subgiganteum (Berk.) P.A. Lemke
Corticium subgiganteum Berk.
Aleurodiscus subgiganteus (Berk.) Höhn.
Michenera artocreas Berk. & M.A. Curtis (anamorph)
Distr.: Eastern North America; Cuba; Japan.
Hosts: Acer, Fagus, Magnolia

Ligniera Maire & Tison. Plasmodiophoromycetes, Plasmodiophorales

Notes: Karling (The Plasmodiophorales. p. 76. 1968).
Ligniera hypogeae (Borzi) Karling
Rhizomyxa hypogeae Borzi
Distr.: CA; Europe.
Notes: Parasitic in cortical cells of young roots and root hairs.
Host: Stellaria
Ligniera junci (E.J. Schwartz) Maire & Tison
Sorosphaera junci E.J. Schwartz
Distr.: Eastern United States; England.
Notes: Parasitic in root hairs and roots of various plants.
Hosts: Alisma, Amaranthus, Eupatorium, Panicum, Physalis, Polygonum, Stellaria, Zea
Ligniera verrucosa Maire & Tison
Distr.: Northern hemisphere.
Notes: Parasitic in roots and root hairs.
Host: Veronica

Lignincola W. Höhnk. Pyrenomycetes, Halosphaeriales

Notes: Kohlmeyer & Kohlmeyer (Marine mycology. p. 307. 1979).
Lignincola laevis W. Höhnk
Distr.: Widespread in coastal regions.
Notes: On intertidal wood.
Hosts: Hibiscus, Pinus, Pluchea, Rhizophora, Spartina

Limacinia Neger. Loculoascomycetes, Dothideales

Notes: According to Hughes (Mycologia 68:782. 1976), this genus is a *nomen dubium*.
Limacinia lithocarpi Bonar & V. Miller
Scolecobonaria lithocarpi (Bonar & V. Miller) Batista
Distr.: Western United States.
Hosts: Lithocarpus, Pseudotsuga

Limaciniella Mendoza in F. Stevens. Loculoascomycetes, Dothideales

Notes: Arx & Müller (Stud. Mycol. 9:117. 1975) listed *Limaciniella* as a synonym of *Actinocymbe* Höhn. Hughes (Mycologia 68:783. 1976).
Limaciniella psidii Mendoza in F. Stevens
Distr.: HI.
Host: Psidium

Limaciniopsis Mendoza in F. Stevens. Loculoascomycetes, Dothideales

Notes: A synonym of *Capnodium* or *Phragmocapnias* (Hughes. Mycologia 68:783. 1976).
Limaciniopsis rollandiae Mendoza in F. Stevens
Distr.: HI.
Host: Rollandia

Limacinula (Sacc.) Höhn. emend. Reynolds. Loculoascomycetes, Dothideales

Shanoriella Batista & Cif.
Notes: Reynolds (Mycologia 63:1173. 1971).
Limacinula anomala (Cooke & Harkn.) Reynolds
Asterina anomala Cooke & Harkn.
Chaetasterina anomala (Cooke & Harkn.) Bubák
Phaeosaccardinula anomala (Cooke & Harkn.) V. Miller & Bonar
Limacinia heteromeles (Cooke & Harkn.) Batista & Cif.
Phaeosaccardinula heteromeles (Cooke & Harkn.) V. Miller & Bonar
Limacinia quitensis Syd.
Distr.: CA; Asia.
Hosts: Heteromeles, Lithocarpus, Umbellularia
Limacinula theae Syd. & E.J. Butler
Ainsworthia morindae (Mendoza) Batista
Phaeosaccardinula morindae Mendoza in F. Stevens
Distr.: HI; tropical India.
Host: Morinda

Lindtneria Pilát. Basidiomycetes, Aphyllophorales

Notes: Gilbertson & Ryvarden (North American Polypores 1:429. 1986; North American Polypores 2:810. 1987).
Lindtneria baboquivariensis (R.L. Gilbertson) R.L. Gilbertson & Ryvarden
Poria baboquivariensis R.L. Gilbertson in R.L. Gilbertson et al.
Distr.: AZ.
Notes: On wood.
Host: Prosopis
Lindtneria flava Parmasto
Distr.: NY; Europe.
Notes: On wood.
Host: Fagus
Lindtneria leucobryophila (Henn.) Jülich
Corticium leucobryophilum (Henn.) Bourd. & Galzin
Trechispora leucobryophila (Henn.) Liberta
Corticium petrophilum Bourd. & Galzin
Trechispora variecolor (Bourd. & Galzin) Liberta
Distr.: Temperate northern hemisphere.
Notes: On bark and plant debris.
Host: Pinus
Lindtneria thujatsugina M. Larsen
Distr.: ID.
Notes: On duff.

Hosts: Thuja, Tsuga

Linocarpon Syd. & P. Syd. Pyrenomycetes, Diaporthales

Notes: Barr (Mycol. Mem. 7:72. 1978).
Linocarpon palmetto **(Ellis & Everh.) Barr**
Linospora palmetto Ellis & Everh.
Distr.: LA.
Host: Sabal
Linocarpon pandani **(Syd. & P. Syd.) Syd. & P. Syd.**
Linospora pandani Syd. & P. Syd.
Distr.: HI, PR; The Philippines.
Notes: On leaves.
Host: Pandanus
Linocarpon versisporum **(Ellis & G. Martin) Petr.**
Ophiobolus versisporus Ellis & G. Martin
Distr.: FL.
Hosts: Roystonea, Sabal, Serenoa

Linodochium Höhn. Hyphomycetes

Notes: Dyko & Sutton (Canad. J. Bot. 57:382. 1979).
Linodochium hyalinum **(Lib.) Höhn.**
Dacrymyces hyalinus Lib.
Distr.: Temperate regions.
Notes: On submerged, decaying leaves and needles.
Host: Quercus

Linospora Fuckel. Pyrenomycetes, Diaporthales

Notes: Barr (Mycol. Mem. 7:128. 1978).
Linospora capreae **(DC.:Fr.) Fuckel**
Linospora saligna (Ehrh.:Fr.) Traverso
Distr.: North America; Europe.
Host: Salix
Linospora gleditschiae **J.H. Miller & F.A. Wolf**
Melasmia hypophylla (Berk. & Ravenel) Sacc. (anamorph)
Distr.: Central and southeastern United States.
Host: Gleditsia
Linospora psederae **Dearn.**
Distr.: WV(type).
Host: Ampelopsis
Linospora tetraspora **G.E. Thompson**
Distr.: Northern North America.
Notes: On leaves.
Host: Populus

Lirula Darker. Discomycetes, Rhytismatales

Notes: Funk (Foliar fungi of western trees. p. 62. 1985).
Lirula abietis-concoloris **(Mayor ex Dearn.) Darker**
Hypodermella abietis-concoloris Mayor ex Dearn.
Distr.: NC(questionable), Rocky Mountains to the Pacific Coast.
Notes: Needle blight. A pycnidial anamorph forms on needles.
Host: Abies
Lirula macrospora **(R. Hartig) Darker**
Lophodermium macrosporum (R. Hartig) Rehm
Lophodermium filiforme Darker
Hypodermina hartigii Hilitzer (anamorph)
Distr.: United States; Europe.
Notes: Needle disease of spruce. Cannon & Minter (C.M.I. Descr. 794. 1984) state that this fungus was introduced into Vermont. Other reports may be erroneous.
Host: Picea
Lirula mirabilis **(Darker) Darker**
Hypodermella mirabilis Darker
Distr.: Northeastern North America.
Host: Abies
Lirula nervata **(Darker) Darker**
Hypodermella nervata Darker
Distr.: Northeastern North America.
Host: Abies
Lirula nervisequia **(DC.:Fr.) Darker var. *conspicua* Darker**
Hypodermina nervisequa (Link) Höhn. (anamorph)
Distr.: CA; Italy(type).
Notes: Needle cast (Darker. Canad. J. Bot. 45:1420. 1967).
Host: Abies
Lirula punctata **(Darker) Darker**
Hypodermella punctata Darker
Distr.: Northern North America.
Notes: On needles.
Host: Abies

Lobatopedis P.M. Kirk. Hyphomycetes

Lobatopedis elegans **P.M. Kirk**
Distr.: HI; Tanzania.
Notes: Kirk (Mycologia 74:875. 1982).
Host: Eucalyptus

Loculohypoxylon Barr. Loculoascomycetes, Dothideales

Notes: Barr (Mycotaxon 3:325. 1976).
Loculohypoxylon grandineum **(Berk. & Ravenel) Barr**
Hypoxylon grandineum (Berk. & Ravenel) J.H. Miller
Distr.: Eastern United States.
Host: Quercus

Lojkania Rehm. Loculoascomycetes, Dothideales

Notes: Barr (Mycotaxon 20:14. 1984).
Lojkania decorticata **(Cooke & Harkn.) Barr**
Amphisphaeria decorticata (Cooke & Harkn.) Berl. & Voglino
Distr.: CA.
Host: Yucca
Lojkania nuda **(Ellis & Everh.) Barr**
Amphisphaeria nuda Ellis & Everh.
Distr.: Central and eastern North America; South Africa.
Notes: On bark or wood.
Host: Celtis
Lojkania separans **(Ellis & Everh.) Barr**
Amphisphaeria separans Ellis & Everh.
Herpotrichia separans (Ellis & Everh.) Sivanesan
Distr.: North America; India; Pakistan.
Notes: On woody limbs.
Host: Populus
Lojkania utahensis **(Petr.) Barr**
Melanopsammina utahensis Petr.
Distr.: Western United States.
Notes: On woody limbs.
Host: Atriplex

Lopadostoma (Nitschke) Traverso. Pyrenomycetes, Xylariales

Lopadostoma caespitosum **(Ellis & Everh.) P. Martin**
Rosellinia caespitosa Ellis & Everh.
Distr.: KS.
Host: Celtis
Lopadostoma gastrinum **(Fr.:Fr.) Traverso**
Anthostoma gastrinum (Fr.:Fr.) Sacc.
Distr.: Temperate North America; Europe.
Hosts: Alnus, Betula, Quercus
Lopadostoma sphinctrinum **(Fr.:Fr.) P. Martin**
Anthostoma dryophilum (Curr.) Sacc.
Distr.: Eastern North America; Europe.
Hosts: Castanea, Quercus
Lopadostoma turgidum **(Pers.:Fr.) Traverso**
Anthostoma turgidum (Pers.:Fr.) Nitschke
Distr.: Temperate regions.
Notes: On branches.
Host: Fagus

Lopharia Kalchbr. & MacOwan. Basidiomycetes, Aphyllophorales

Notes: Welden (Mycologia 67:530. 1975).
Lopharia cinerascens (Schwein.) G. Cunn.
Stereum cinerascens (Schwein.) Massee
Distr.: Widespread.
Notes: On wood, especially of *Ulmus* spp. and *Acer negundo* in eastern and central North America.
Hosts: Acer, Carya, Elaeagnus, Euonymus, Morus, Platanus, Populus, Prunus, Quercus, Rhamnus, Robinia, Salix, Sambucus, Tilia, Ulmus, Umbellularia
Lopharia papyrina (Mont.) Boidin
Stereum papyrinum Mont.
Distr.: Neotropical and adjacent subtropical region.
Notes: On wood. With the transfer of *L. crassa* to *Phanerochaete*, the placement of *L. papyrina* in *Lopharia* is in question.
Host: Quercus

Lophidium P. Karst. Loculoascomycetes, Dothideales

Notes: This genus is considered a synonym of *Lophium*, but not all names have been redisposed.
Lophidium trifidum Ellis & Everh.
Distr.: WA(type).
Notes: Chesters & Bell (Mycol. Pap. 120:54. 1970) examined the type specimen and excluded this species from the Lophiostomataceae.
Host: Salix

Lophiosphaera Trevis. Loculoascomycetes, Dothideales

Notes: Arx & Müller (Stud. Mycol. 9:64. 1975) considered this genus a synonym of *Lophiostoma*, but not all names have been redisposed.
Lophiosphaera fluviatilis Ellis & Everh.
Distr.: MT(type).
Host: Salix
Lophiosphaera gloniospora Ellis & Everh.
Distr.: MT(type).
Host: Salix
Lophiosphaera lophospora (Sacc. & Speg.) Corlett
Didymosphaeria lophospora Sacc. & Speg.
Didymella lophospora (Sacc. & Speg.) Sacc.
Didymella lophospora (Sacc. & Speg.) Sacc. var. *acetosellae* Ellis & Sacc.
Distr.: Temperate regions.
Notes: On various dicotyledons (Corlett. Canad. J. Bot. 59:2029. 1981). Pirozynski & Morgan-Jones (Trans. Brit. Mycol. Soc. 51:196. 1968) considered this species to be a synonym of *Lophiosphaera fuckelii*.
Hosts: Quercus, Rumex
Lophiosphaera querceti (Sacc. & Speg.) Sacc.
Distr.: CA; Italy(type).
Host: Sambucus
Lophiosphaera velata (Ellis & Everh.) Lohman
Glonium velatum Ellis & Everh.
Distr.: Southern United States.
Notes: On decorticated wood. Lohman (Amer. J. Bot. 21:314. 1934).
Host: Quercus
Lophiosphaera zeicola Ellis & Everh.
Distr.: KS(type).
Host: Zea

Lophiostoma Ces. & De Not., nom. cons. Loculoascomycetes, Dothideales

Notes: Chesters & Bell (Mycol. Pap. 120:1. 1970).
Lophiostoma alpigenum Fuckel
Lophiotrema alpigenum (Fuckel) Sacc.
Distr.: North America; Europe.
Host: Populus
Lophiostoma angustilabrum (Berk. & Broome) Cooke var. ***crenatum*** (Pers.:Fr.) Chesters & A. Bell
Lophionema crenatum (Pers.:Fr.) Ellis & Everh.
Lophiotrema crenatum (Pers.:Fr.) Sacc.
Lophiostoma crenatum (Pers.:Fr.) Fuckel
Distr.: Eastern North America; Europe.
Host: Acer
Lophiostoma arundinis (Pers.:Fr.) Ces. & De Not.
Distr.: Temperate North America; Europe.
Notes: On Poaceae.
Hosts: Agropyron, Phragmites, Scirpus, Zea
Lophiostoma asperum Ellis & Everh.
Lophiosphaera asperum (Ellis & Everh.) Sacc.
Distr.: KS, OK.
Notes: On dead wood.
Host: Ulmus
Lophiostoma caulium (Fr.:Fr.) Ces. & De Not.
Lophiostoma appendiculatum Niessl
Lophiostoma insidiosum (Desmaz.) Ces. & De Not.
Distr.: Eastern North America; Europe.
Notes: On a variety of herbaceous and woody hosts.
Hosts: Amaranthus, Aster, Chrysothamnus, Guardiola, Haplopappus, Helianthus, Hypericum, Lespedeza, Monarda, Platanus, Populus, Prunella, Rumex, Solidago, Stachys, Verbascum, Yucca
Lophiostoma caulium (Fr.) Ces. & De Not. var. ***congregatum*** (Harkn.) Chesters & A. Bell
Lophiostoma congregatum Harkn.
Lophiostoma brenckleanum Sacc.
Distr.: Western United States.
Hosts: Chrysothamnus, Haplopappus, Sambucus
Lophiostoma herbarum (Wehmeyer) Shoemaker & Babcock
Eriosphaeria herbarum Wehmeyer
Distr.: Western North America.
Notes: On dead stems of herbaceous plants (Shoemaker & Babcock. Canad. J. Bot. 67:389. 1987).
Hosts: Achillea, Lupinus, Valeriana
Lophiostoma hysterioides (Schwein.) Sacc.
Distr.: North America.
Notes: On woody parts of various dicotyledons.
Hosts: Amelanchier, Sambucus
Lophiostoma insidiosum (Desmaz.) Ces. & De Not. f. ***salicis*** Rehm
Distr.: ND(type).
Notes: The type form is now a synonym of *Lophiostoma caulium*.
Host: Salix
Lophiostoma macrostomoides (De Not.) Ces. & De Not.
Distr.: North America; Europe.
Notes: Mostly on wood of *Populus* and *Salix*.
Hosts: Cercis, Populus, Salix, Symphoricarpos
Lophiostoma prominens Peck
Distr.: NY.
Host: Cephalanthus
Lophiostoma quadrinucleatum P. Karst.
Distr.: North America; Europe.
Hosts: Amelanchier, Celtis, Morus, Paxistima, Populus, Rhamnus, Salix
Lophiostoma quadrinucleatum P. Karst. var. ***triseptatum*** (Peck) Chesters & A. Bell
Lophiostoma triseptatum Peck
Lophiostoma pruni Ellis & Everh.
Distr.: Cosmopolitan.
Notes: On dead wood.
Hosts: Celtis, Prunus, Rhamnus
Lophiostoma ribis (Tracy & Earle) L. Holm
Gibberidea ribis Tracy & Earle
Distr.: Western North America.
Notes: Holm (Svensk Bot. Tidskr. 62:237. 1968).
Host: Ribes
Lophiostoma sieversiae Peck
Leptosphaeria sieversiae (Clements) Petr.
Massaria sieversiae Clements
Wettsteinina sieversiae (Peck) Barr
Wettsteinina andromedae (Auersw.) Barr var. *cassiopes* (Dearn. & House ex Barr) Barr
Phaeospora cassiopes Dearn. & House ex Barr
Distr.: Western United States.
Notes: Shoemaker & Babcock (Canad. J. Bot. 65:389. 1987).
Hosts: Cassiope, Geum
Lophiostoma spiraeae Peck

Lophiotrema spiraeae (Peck) Sacc.
Distr.: NY.
Notes: Barr et al. (New York State Mus. Bull. 459:43. 1986) retained this species separate from *L. alpigenum*.
Host: Physocarpus
Lophiostoma starbackii P. Karst.
Distr.: ND; Finland(type).
Host: Prunus
Lophiostoma submagnum Kauffm.
Distr.: OR(type).
Host: Philadelphus
Lophiostoma thujae Ellis & Everh.
Navicella thujae (Ellis & Everh.) Sacc.
Distr.: MI, NY; Canada(Ontario).
Host: Thuja
Lophiostoma triseptatum Peck in Ellis & Everh. f. ***fraxini*** Rehm
Distr.: ND(type).
Notes: The type form is a synonym of *Lophiostoma quadrinucleatum* var. *triseptatum*.
Host: Fraxinus
Lophiostoma vagabundum (Sacc.) Chesters & A. Bell
Lophiotrema vagabundum Sacc.
Distr.: North America; Europe.
Hosts: Cornus, Salix, Solidago
Lophiostoma vigheffulensis (Pass.) Chesters & A. Bell
Lophiosphaera vigheffulensis Pass.
Lophiostoma stenostomum Ellis & Everh.
Distr.: North America; Europe.
Host: Parthenocissus
Lophiostoma viridarium Cooke
Lophiostoma desmazierii Sacc. & Speg.
Distr.: North America; Europe.
Host: Quercus

Lophiotrema Sacc. Loculoascomycetes, Dothideales

Lophiotrema halesiae Fairm.
Distr.: NY(type).
Host: Halesia

Lophodermella Höhn. Discomycetes, Rhytismatales

Notes: Darker (Canad. J. Bot. 45:1425. 1967).
Lophodermella arcuata (Darker) Darker
Hypodermella arcuata Darker
Distr.: OR.
Notes: On needles.
Host: Pinus
Lophodermella cerina (Darker) Darker
Hypodermella cerina Darker
Distr.: CA.
Notes: On needles.
Host: Pinus
Lophodermella concolor (Dearn.) Darker
Hypodermella montivaga (Petr.) Dearn. f. *concolor* Dearn.
Hypodermella concolor (Dearn.) Darker
Distr.: Central and western North America.
Notes: On needles.
Host: Pinus
Lophodermella montivaga Petr.
Hypodermella montivaga (Petr.) Darker
Distr.: Western United States.
Notes: On needles.
Host: Pinus
Lophodermella morbida Staley & Bynum
Distr.: CO, OR.
Notes: On needles (Staley & Bynum. Mycologia 72:722. 1972).
Host: Pinus

Lophodermium Chev. Discomycetes, Rhytismatales

Notes: Species on pines are reviewed by Minter (Mycol. Pap. 147:1. 1981). Darker (Canad. J. Bot. 45:1399. 1967).
Lophodermium acacicola Tehon
Distr.: HI(type).
Notes: Tehon (Illinois Biol. Monogr. 13:40. 1935).
Host: Acacia
Lophodermium alpinum (Rehm) Terrier
Lophodermium arundinaceum (Schrad.:Fr.) Chev. var. *alpinum* Rehm
Ophiobolus festucae Tracy & Earle
Distr.: Temperate North America; Europe.
Host: Festuca
Lophodermium apiculatum (M. Wormsk.:Fr.) Sacc.
Actinothyrium graminis Kunze (anamorph)
Distr.: CO; Europe.
Notes: On grasses.
Host: Calamagrostis
Lophodermium arundinaceum (Schrad.:Fr.) Chev.
Distr.: Temperate regions.
Notes: On grasses.
Hosts: Agropyron, Agrostis, Ammophila, Arundinaria, Bouteloua, Calamagrostis, Cladium, Danthonia, Elymus, Festuca, Hordeum, Muhlenbergia, Oryzopsis, Phleum, Poa, Puccinellia, Sitanion, Trisetum
Lophodermium arundinaceum (Schrad.) Chev. var. ***gramineum*** Duby
Distr.: Eastern United States.
Notes: This variety has not been treated in recent studies.
Host: Phleum
Lophodermium aucupariae (Schleich.) Darker
Lophodermium tumidum Rehm ex Nannf.
Leptostroma sorbicola Hilitzer (anamorph)
Distr.: Temperate regions.
Notes: On Roasaceae.
Host: Amelanchier
Lophodermium australe Dearn.
Leptostroma durissimum Cooke (anamorph)
Distr.: Warm-temperate, subtropical and tropical regions.
Notes: On needles (Cannon & Minter. Mycol. Pap. 155:54. 1986; Minter & Millar. C.M.I. Descr. 563. 1978).
Host: Pinus
Lophodermium baculiferum Mayr
Leptostroma decipiens Petr. (anamorph)
Distr.: Western North America.
Notes: On 2- and 3-needle pines (Cannon & Minter. C.M.I. Descr. 795. 1984).
Host: Pinus
Lophodermium caricinum (Roberge ex Desmaz.) Duby
Lophodermina caricina (Roberge ex Desmaz.) Tehon
Distr.: Temperate northern hemisphere.
Host: Carex
Lophodermium consociatum Darker
Dermascia consociata (Darker) Tehon
Distr.: Pacific Northwest.
Host: Abies
Lophodermium culmigenum (Fr.:Fr.) De Not.
Lophodermium arundinaceum (Schrad.:Fr.) Chev. var. *culmigenum* (Fr.) P. Karst.
Distr.: Eastern North America; Europe.
Notes: On various grasses.
Hosts: Poa, Secale, Triticum
Lophodermium cyrillicola Tracy & Earle
Distr.: Southern United States.
Host: Cyrilla
Lophodermium danthoniae Tehon
Distr.: WI(type).
Host: Danthonia
Lophodermium decorum Darker
Distr.: MT, OR.
Host: Abies
Lophodermium dracaenae W. Phillips & Harkn.
Dermascia dracaenae (W. Phillips & Harkn.) Tehon
Distr.: CA(type).

Host: Dracaena
Lophodermium durilabrum Darker
Distr.: Central and western North America.
Notes: Cannon & Minter (C.M.I. Descr. 796. 1984).
Host: Pinus
Lophodermium exaridum Cooke & Peck in Cooke
Hysterium exaridum Cooke & Peck in Peck
Distr.: Temperate North America.
Hosts: Gaultheria, Kalmia, Pyrola
Lophodermium foliicola (Fr.:Fr.) P. Cannon & Minter
Lophodermellina hysterioides (Pers.) Höhn.
Lophodermium hysterioides (Pers.) Sacc.
Leptostroma crataegi Nannf. (anamorph)
Distr.: North America; temperate Europe.
Hosts: Amelanchier, Myrica
Lophodermium gramineum (Fr.) Chev.
Hysterium gramineum Pers., nom. nud.
Distr.: Temperate regions. Cannon & Minter (Mycol. Pap. 155:61. 1986).
Host: Poa
Lophodermium heteromelis (W. Phillips & Harkn.) Ellis & Everh.
Hypoderma heteromelis W. Phillips & Harkn.
Distr.: CA(type).
Host: Heteromeles
Lophodermium intermissum Starb.
Distr.: HI; Europe.
Notes: Eriksson (Symb. Bot. Upsal. 19:39. 1970).
Host: Acacia
Lophodermium juniperinum (Fr.) De Not.
Lophodermina juniperina (Fr.) Tehon
Lophodermium juniperi (Grev.) Darker
Distr.: Temperate regions.
Notes: On needles (Cannon & Minter. C.M.I. Descr. 797. 1984)
Hosts: Calocedrus, Juniperus
Lophodermium juniperinum (Fr.) De Not. f. **cupressi-thyoidis** Sacc.
Lophodermina cupressi-thyoides (Sacc.) Tehon
Distr.: NJ(type).
Notes: Tehon (Illinois Biol. Monogr. 13:94. 1935) considered this a separate species.
Host: Chamaecyparis
Lophodermium lacerum Darker
Dermascia lacera (Darker) Tehon
Distr.: Northern North America.
Notes: On needles.
Host: Abies
Lophodermium maculare (Fr.:Fr.) De Not.
Hysterium maculare Fr.:Fr.
Distr.: Circumboreal.
Notes: On leaves (Shoemaker & Egger. Fungi Canadenses 196. 1981).
Hosts: Arctostaphylos, Chamaedaphne, Ledum, Magnolia, Pyrola, Quercus, Salix, Vaccinium
Lophodermium melaleucum (Fr.:Fr.) De Not.
Lophodermium melaleucum (Fr.:Fr.) De Not. var. *epiphyllum* Zeller
Distr.: Temperate northern hemisphere.
Notes: On leaves (Shoemaker & Egger. Fungi Canadenses 195. 1981).
Hosts: Rhododendron, Vaccinium
Lophodermium molitoris Minter
Distr.: Widespread in Canada to southeastern United States.
Notes: On 2- and 3-needle pines.
Host: Pinus
Lophodermium nitens Darker
Lophodermina nitens (Darker) Tehon
Distr.: Temperate North America.
Notes: On secondary pine needles. Ascocarp development was studied by Morgan-Jones & Hulton (Canad. J. Bot. 55:2605. 1977). Minter & Millar (C.M.I. Descr. 566. 1978).
Host: Pinus
Lophodermium orbiculare (Ehrenb.:Fr.) Sacc.
Lophodermium gracile (Ehrenb.:Fr.) Sacc.
Distr.: Temperate North America.
Notes: Tehon (Illinois Biol. Monogr. 13:103. 1935) proposed this synonymy.
Hosts: Cassiope, Chamaedaphne, Ledum, Leucothoe
Lophodermium oxycocci (Fr.) P. Karst.
Hysterium oxycocci Fr.
Lophodermium hypophyllum (Dearn. & House) Shear
Lophodermium oxycocci (Fr.) P. Karst. var. *hypophyllum* Dearn. & House
Distr.: North America; Europe.
Notes: On leaves (Eriksson. Symb. Bot. Upsal. 19:42. 1970).
Host: Vaccinium
Lophodermium petiolicola Fuckel
Distr.: OK; Europe.
Notes: On petioles. Tehon (Illinois Biol. Monogr. 32:336) considered this species a synonym of *Lophodermium punctiforme*.
Host: Rhus
Lophodermium phlei Tehon & G.L. Stout
Distr.: NJ(type).
Host: Phleum
Lophodermium phloxii Bonar & W.B. Cooke
Lophodermium melaleuca (Fr.) De Not. var. *epiphyllum* Zeller
Distr.: CA(type).
Host: Phlox
Lophodermium piceae (Fuckel) Höhn.
Lophodermium abietis Rostr.
Leptostroma abietis Darker (anamorph)
Distr.: Circumboreal.
Notes: Found to occur as an endophyte (Barklund. Trans. Brit. Mycol. Soc. 89:307. 1987).
Hosts: Abies, Picea
Lophodermium pinastri (Schrad.:Fr.) Chev.
Hysterium pinastri Schrad.:Fr.
Lophodermium laricis Dearn.
Lophodermium pinicola Tehon
Leptostroma sp. (anamorph)
Distr.: Worldwide.
Notes: This name has been used for any *Lophodermium*-like fungus on pines. Minter & Millar (C.M.I. Descr. 567. 1978).
Hosts: Larix, Pinus
Lophodermium punctiforme (Fr.:Fr.) Fuckel
Distr.: Eastern North America; Europe.
Hosts: Acer, Fraxinus, Liriodendron
Lophodermium pyrolae Parmelee
Distr.: Northern North America.
Notes: Leaf spot (Parmelee. Canad. J. Bot. 36:865. 1958).
Host: Pyrola
Lophodermium ravenelii Minter
Distr.: SC, TX.
Host: Pinus
Lophodermium rubiicola Earle
Dermascia rubiicola Earle
Distr.: AL, CA, GA.
Notes: Tehon (Illinois Biol. Monogr. 13:303. 1935).
Host: Rubus
Lophodermium sacchari Lyon
Distr.: HI(type).
Host: Saccharum
Lophodermium schweinitzii M. Wilson & N.F. Robertson
Coccomyces rhododendri (Schwein.) Sacc.
Hysterium rhododendri Schwein.
Lophodermium rhododendri (Schwein.) Ellis & Everh., nom. inval.
Distr.: Temperate North America.
Hosts: Gaultheria, Rhododendron, Vaccinium
Lophodermium seditiosum Minter, Staley, & Millar
Leptostroma rostrupii Minter (anamorph)
Distr.: Circumboreal.
Notes: Much of the literature on *L pinastri* was actually based on this fungus. Minter & Millar (C.M.I. Descr. 568. 1978).
Host: Pinus
Lophodermium sphaerioides (Albertini & Schwein.:Fr.) Duby
Distr.: Circumboreal.
Notes: On leaves (Shoemaker & Egger. Fungi Canadenses 197. 1981).
Host: Ledum
Lophodermium staleyi Minter
Distr.: OR(type).
Notes: On secondary needles.
Host: Pinus
Lophodermium thujae J.J. Davis
Lophodermina thujae (J.J. Davis) Tehon

Distr.: Central and eastern North America.
Host: Thuja
Lophodermium typhinum (Fr.) Lambotte
Distr.: Temperate regions.
Host: Typha
Lophodermium uncinatum Darker
Lophodermina uncinata (Darker) Tehon
Distr.: Pacific Northwest.
Notes: On needles.
Host: Abies
Lophodermium versicolor (Wahlenberg:Fr.) Rehm
Distr.: CO, PA; Europe.
Notes: On dead leaves.
Host: Salix

Lophomerum Ouellette & Magasi. Discomycetes, Rhytismatales

Notes: Ouellette & Magasi (Mycologia 58:275. 1966); Darker (Canad. J. Bot. 45:1399. 1967).
Lophomerum autumnale (Darker) Magasi in Ouellette & Magasi
Lophodermium autumnale Darker
Distr.: MI to the Pacific Coast.
Notes: A secondary pathogen.
Host: Abies
Lophomerum septatum Ouellette in Ouellette & Magasi
Lophodermina septata Tehon
Lophodermium septatum (Tehon) Terrier
Distr.: OR(type).
Host: Picea

Lophophacidium Lagerberg. Discomycetes, Helotiales

Notes: DiCosmo et al. (Mycotaxon 21:83. 1984).
Lophophacidium hyperboreum Lagerberg
Distr.: Temperate regions.
Host: Picea

Loramyces W. Weston. Discomycetes, Helotiales

Loramyces juncicola W. Weston
Distr.: East coast of the United States; Europe.
Notes: On dead aquatic monocots (Digby & Goos. Mycologia 79:821. 1987).
Host: Juncus

Lulworthia G.K. Sutherland. Pyrenomycetes, Halosphaeriales

Notes: Kohlmeyer & Kohlmeyer (Marine mycology. p. 313. 1979).
Lulworthia attenuata T.W. Johnson
Distr.: NC(coastal).
Host: Liriodendron
Lulworthia conica T.W. Johnson
Distr.: NC(coastal).
Host: Picea
Lulworthia floridana S.P. Meyers
Distr.: Widespread in marine habitats.
Host: Spartina
Lulworthia grandispora S.P. Meyers
Lulworthia grandispora S.P. Meyers var. *apiculata* T.W. Johnson
Distr.: Worldwide in marine habitats.
Host: Acer
Lulworthia medusa (Ellis & Everh.) A.B. Cribb & J.W. Cribb
Ophiobolus medusa Ellis & Everh.
Linocarpon medusa (Ellis & Everh.) Petr.
Distr.: Widespread in marine habitats.
Hosts: Spartina, Vetiveria
Lulworthia rotunda T.W. Johnson
Distr.: NC(coastal).
Host: Tilia
Lulworthia submersa T.W. Johnson
Distr.: Coastal regions of the Atlantic Ocean.
Hosts: Acer, Picea

Lycoperdon Tourn.:Pers. Basidiomycetes, Lycoperdales

Lycoperdon pyriforme Schaeff.:Pers.
Distr.: Temperate northern hemisphere.
Notes: On wood.
Hosts: Quercus, Tsuga

Lylea Morgan-Jones. Hyphomycetes

Lylea catenulata Morgan-Jones
Distr.: AL.
Notes: Morgan-Jones (Mycotaxon 3:129. 1975).
Host: Pinus

Lyonella Syd. Pyrenomycetes, Sphaeriales

Lyonella neurophila Syd.
Distr.: HI(type).
Host: Psychotria

Macroderma Höhn. Discomycetes, Rhytismatales

Notes: DiCosmo et al. (Mycotaxon 21:48. 1984).
Macroderma curtisii (Berk. & Ravenel) Höhn.
Rhytisma curtisii Berk. & Ravenel
Phacidium curtisii (Berk. & Ravenel) Luttrell
Distr.: Southeastern United States.
Host: Ilex

Macrodiaporthe Petr. Pyrenomycetes, Diaporthales

Notes: Barr (Mycol. Mem. 7:167. 1978).
Macrodiaporthe everhartii (Ellis) Barr
Melanconis everhartii Ellis
Melanconis dasycarpa Ellis & Kellerm.
Cryptodiaporthe magnispora (Ellis & Everh.) Wehmeyer
Valsa magnispora Ellis & Everh.
Distr.: Eastern North America.
Notes: On limbs.
Host: Acer
Macrodiaporthe occulta (Nitschke) Petr.
Melanconis occulta (Nitschke) Sacc.
Distr.: Temperate North America; Europe.
Notes: On limbs.
Host: Populus

Macrodiplodiopsis Petr. Coelomycetes

Macrodiplodiopsis desmazieresii (Mont.) Petr.
Hendersonia platani Peck
Stegonsporium platani G. Preuss
Splanchnonema platani (Ces.) Barr (teleomorph)
Distr.: North America; Europe.
Notes: On branches.
Host: Platanus

Macrophoma (Sacc.) Berl. & Voglino. Coelomycetes

Notes: *Macrophoma* and related genera *Botryodiplodia* and *Dothiorella* are poorly defined and in need of revision. Sutton (The Coelomycetes. p. 118. 1980) considered *Macrophoma* to be a synonym of *Sphaeropsis*. Many of the 600+ species have yet to be transferred to other genera.
Macrophoma ampelopsidis (Cooke & Ellis) Berl. & Voglino
Sphaeropsis ampelopsidis Cooke & Ellis
Distr.: IL.
Host: Parthenocissus

Macrophoma astericola (Atk.) Tassi
Phoma astericola Atk.
Distr.: AL.
Host: Aster
Macrophoma candollei (Berk. & Broome) Berl. & Voglino
Ludwigiella candollei (Berk. & Broome) Petr.
Distr.: United States; Europe.
Host: Buxus
Macrophoma cercis Tehon & G.L. Stout
Distr.: Central and eastern United States.
Host: Cercis
Macrophoma chrysothamni W.B. Cooke in W.B. Cooke & C.G. Shaw
Distr.: CA.
Host: Chrysothamnus
Macrophoma cupressi (Cooke & Harkn.) Berl. & Voglino
Sphaeropsis cupressi Cooke & Harkn.
Distr.: CA, southern United States.
Hosts: Chamaecyparis, Cupressus
Macrophoma diospyri Earle
Distr.: Southern United States.
Notes: Probably a synonym of *Phyllosticta arxii* Aa (Stud. Mycol. 5:34. 1973).
Host: Diospyros
Macrophoma elongata (Berk. & M.A. Curtis) Berl. & Voglino
Dothiorella lanceolata (Cooke & Ellis) Petr.
Macrophoma lanceolata (Cooke & Ellis) Berl. & Voglino
Distr.: Eastern United States.
Hosts: Asparagus, Gladiolus
Macrophoma farlowiana (Viala & Sauv.) Tassi
Phoma farlowiana Viala & Sauv.
Distr.: NY, WI.
Host: Vitis
Macrophoma fici J.V. Almeida & Camara
Distr.: Southern United States; Europe.
Notes: May be a *Botryodiplodia* according to Petrak & Sydow (Beih. Repert. Spec. Nov. Regni Veg. 42:122. 1926).
Host: Ficus
Macrophoma gladioli (Cooke) Berl. & Voglino
Distr.: SC.
Host: Gladiolus
Macrophoma halstedii (Ellis & Everh.) Tassi
Distr.: Eastern North America.
Host: Syringa
Macrophoma hyalina (Berk. & Cooke) Berl. & Voglino
Distr.: PA(type).
Host: Fraxinus
Macrophoma ilicella (Sacc. & Penz.) Berl. & Voglino
Dothiorella ilicella (Sacc. & Penz.) Petr. & Syd.
Distr.: FL; Europe.
Host: Ilex
Macrophoma longispora (Thuem.) Berl. & Voglino
Distr.: SC; Italy.
Host: Vitis
Macrophoma lupini (Cooke & Harkn.) Berl. & Voglino
Phoma lupini Cooke & Harkn.
Distr.: Western United States.
Host: Lupinus
Macrophoma maculiformis (Berk. & M.A. Curtis) Berl. & Voglino
Distr.: CA.
Host: Arbutus
Macrophoma magnifructa (Peck) Sacc.
Phoma magnifructa Peck
Dothiorella magnifructa (Peck) Petr. & Syd.
Distr.: NY(type).
Hosts: Calocedrus, Pinus
Macrophoma mamillaris (Berk. & M.A. Curtis) Tassi
Distr.: SC.
Host: Gleditsia
Macrophoma mimuli Wehmeyer
Distr.: WY.
Notes: *Apiosporella mimuli* may be the teleomorph (Wehmeyer. Mycologia 38:323. 1946).
Host: Mimulus
Macrophoma molleriana (Thuem.) Berl. & Voglino
Distr.: CA, FL; Europe.
Host: Eucalyptus
Macrophoma negundinis Ellis & Everh.
Distr.: CO, a few unconfirmed reports from the Midwest.
Host: Acer
Macrophoma nervicola Ellis & Everh.
Distr.: OK, WI(type).
Host: Quercus
Macrophoma numerosa Peck
Distr.: Eastern United States.
Host: Robinia
Macrophoma oblongata Tehon
Distr.: IL.
Host: Poa
Macrophoma oenotherae-biennis Dearn.
Dothiorella oenotherae (Tassi) Petr. & Syd.
Macrophoma oenotherae Tassi
Distr.: North America; Europe.
Host: Oenothera
Macrophoma opunticola (Speg.) P. Syd.
Phoma opunticola Speg.
Distr.: HI; Argentina(type).
Host: Opuntia
Macrophoma pachysandrae W.G. Hutchinson
Distr.: VA.
Host: Pachysandra
Macrophoma pandani (Lév.) Berl. & Voglino
Distr.: CA, FL; Europe.
Host: Pandanus
Macrophoma persicina (Berk. & M.A. Curtis) Berl. & Voglino
Distr.: Central and eastern United States.
Host: Prunus
Macrophoma philodendri Peck
Distr.: FL, MI.
Host: Monstera
Macrophoma phlei Tehon & G.L. Stout
Distr.: Temperate United States.
Hosts: Agrostis, Danthonia, Elymus, Phleum
Macrophoma pulchrispora (Peck & G.W. Clinton) Sacc.
Distr.: NY.
Host: Polygonum
Macrophoma ramulicola (Desmaz.) Höhn.
Distr.: CA, LA; Europe.
Host: Euonymus
Macrophoma reniformis (Viala & Ravaz) Cavara
Dothiorella reniformis (Viala & Ravaz) Petr. & Syd.
Distr.: VA; Europe.
Host: Vitis
Macrophoma rubi Tehon
Distr.: IL.
Host: Rubus
Macrophoma sacchari (Cooke) Berl. & Voglino
Distr.: GA.
Host: Saccharum
Macrophoma secalina Tehon
Distr.: IL.
Hosts: Avena, Hordeum, Secale, Triticum
Macrophoma seminalis (Berk. & M.A. Curtis) Berl. & Voglino
Distr.: SC.
Host: Citrullus
Macrophoma smilacina (Peck) Berl. & Voglino
Sphaeropsis smilacina Peck
Phoma smilacina (Peck) Sacc.
Dothiorella smilacina (Peck) Petr. & Syd.
Macrophoma smilacinae Tehon & G.L. Stout
Phyllosticta smilacis Ellis & G. Martin in Ellis & Everh.
Phyllosticta smilacis Ellis & Everh.
Phyllosticta smilacina (Peck) Dearn.
Phyllosticta oakesiae Dearn. & House
Phyllosticta yuccaegena Ellis & Everh.
Distr.: Temperate North America.
Notes: Bissett (Canad. J. Bot. 57:2071-2081. 1979; Canad. J. Bot. 64:1721. 1986).
Hosts: Smilax, Uvularia, Yucca
Macrophoma smilacina (Peck) Berl. & Voglino var. **smilacis** (Ellis

& G. Martin) Bissett
Ascochyta smilacis Ellis & G. Martin
Stagonospora smilacis (Ellis & G. Martin) Sacc.
Diplodia uvulariae J.J. Davis
Distr.: Central and eastern North America.
Notes: Bissett (Canad. J. Bot. 57:2071. 1979).
Hosts: Asparagus, Polygonatum, Smilax, Uvularia
Macrophoma sorghicola Speg.
Distr.: MT; Argentina(type).
Host: Agropyron
Macrophoma spartiicola Berl. & Voglino
Distr.: MI; Europe.
Host: Agrostis
Macrophoma sphaeropsispora (Ellis & Everh.) Tassi
Distr.: CA.
Host: Solidago
Macrophoma sporoboli R. Sprague
Distr.: Western United States.
Hosts: Bouteloua, Sporobolus
Macrophoma straminella (Bres.) Died.
Phyllosticta straminella Died.
Distr.: Temperate North America; Europe.
Host: Rheum
Macrophoma subconica Ellis & Everh.
Distr.: Southern United States, IL.
Hosts: Canna, Melia, Solanum, Vigna
Macrophoma taxi (Berk.) Berl. & Voglino
Distr.: WA; Europe.
Host: Taxus
Macrophoma thermopsidis (Ellis & B.T. Galloway) Sacc. & Syd.
Distr.: MT, NM.
Host: Thermopsis
Macrophoma tiliacea Peck
Distr.: Central United States.
Host: Tilia
Macrophoma triacanthi (Sacc.) Berl. & Voglino
Phoma triacanthi Sacc.
Distr.: SC.
Host: Gleditsia
Macrophoma trichosanthis Syd.
Distr.: AL; The Philippines(type).
Host: Trichosanthes
Macrophoma vincae (Curr.) Berl. & Voglino
Distr.: Central and eastern United States; Europe.
Host: Vinca
Macrophoma viticola (Cooke) Berl. & Voglino
Distr.: SC.
Host: Vitis
Macrophoma xanthoxyli Ellis & Everh.
Distr.: LA.
Host: Zanthoxylum
Macrophoma zeae Tehon & E.Y. Daniels
Botryosphaeria zeae (G.L. Stout) Arx & E. Müller (teleomorph)
Distr.: Central United States; eastern and southern Africa; France.
Notes: Gray ear rot of maize. Sivanesan & Holliday (C.M.I. Descr. 774. 1983) suggested that this anamorph belongs in *Dothiorella*.
Host: Zea

Macrophomina Petr. Coelomycetes

Macrophomina phaseolina (Tassi) Goidanich
Macrophoma phaseolina Tassi
Macrophoma phaseoli Maubl.
Botryodiplodia phaseoli (Maubl.) Thirumalachar
Macrophomina phaseoli (Maubl.) S. Ashby
Rhizoctonia bataticola (Taubenhaus) E.J. Butler
Sclerotium bataticola Taubenhaus
Distr.: Cosmopolitan, most common in subtropical and tropical regions.
Notes: Reported on at least 284 host species (Holliday & Punithalingam. C.M.I. Descr. 275. 1970; Punithalingam. Nova Hedwigia 36:266. 1982).
Hosts: Abelmoschus, Abies, Abutilon, Acer, Allium, Amaranthus, Ambrosia, Antirrhinum, Apocynum, Arachis, Arrhenatherum, Asclepias, Asparagus, Avena, Begonia, Beta, Bidens, Bouteloua, Brassica, Cajanus, Campanula, Cannabis, Capsicum, Cassia, Catalpa, Celosia, Chamaecyparis, Chenopodium, Chrysanthemum, Cicer, Cirsium, Citrullus, Citrus, Conyza, Cornus, Crotalaria, Cucumis, Cucurbita, Cupressus, Cyamopsis, Dahlia, Datura, Dichondra, Elymus, Erigeron, Eryngium, Eupatorium, Euphorbia, Fagopyrum, Fragaria, Glycine, Gossypium, Hedera, Helianthus, Hibiscus, Ipomoea, Juniperus, Koelreuteria, Kummerowia, Lactuca, Lespedeza, Ligustrum, Lilium, Lotus, Lupinus, Lycopersicon, Malva, Medicago, Melilotus, Muhlenbergia, Nicandra, Nicotiana, Nyssa, Oenothera, Opuntia, Parthenium, Phaseolus, Phlox, Picea, Pinus, Pisum, Polygonum, Prunus, Pseudotsuga, Pueraria, Pyracantha, Quercus, Rhododendron, Ricinus, Robinia, Roystonea, Rudbeckia, Salvia, Santolina, Schefflera, Senna, Sequoiadendron, Sesamum, Sesbania, Setaria, Sida, Solanum, Solidago, Sorghum, Strophostyles, Tagetes, Thuja, Trifolium, Tristania, Verbena, Vicia, Vigna, Vitis, Zea, Zinnia

Macrosporium Fr. Hyphomycetes

Notes: According to Hughes (Canad. J. Bot. 36:782. 1958) this genus is considered a synonym of *Alternaria*, but not all names have been redisposed.
Macrosporium avicennae Ellis & Everh., nom. nud.
Distr.: DE(type).
Host: Abutilon
Macrosporium baccatum Ellis & Kellerm.
Distr.: KS(type).
Host: Aesculus
Macrosporium clematidis Peck
Distr.: KS(type).
Host: Clematis
Macrosporium cocos Pass.
Distr.: FL; Italy(type); distribution elsewhere is uncertain.
Host: Cocos
Macrosporium cydoniae Ellis & Everh. in Sumstine, nom. nud.
Distr.: NJ(type).
Host: Cydonia
Macrosporium heteronemum (Desmaz.) Sacc.
Septonema heteronemum Desmaz.
Distr.: ND; Europe(type).
Host: Robinia
Macrosporium hybridum Ellis & Everh.
Distr.: KS(type).
Host: Sambucus
Macrosporium iridis Cooke & Ellis
Distr.: DE, ID, NJ.
Host: Iris
Macrosporium leguminum Cooke
Distr.: SC(type).
Host: Senna
Macrosporium maculatum Cooke & Ellis in Sumstine, nom. nud.
Distr.: DE(type).
Host: Zea
Macrosporium nerii Cooke
Distr.: AL, GA.
Host: Nerium
Macrosporium nigricantium Atk.
Distr.: AL, FL.
Host: Gossypium
Macrosporium pinguedinis Berk.
Distr.: MI, OH.
Host: Calamagrostis
Macrosporium rosarium Penz.
Distr.: WA; Italy(type).
Host: Eremocarpus
Macrosporium subglobosum Cooke & Ravenel
Distr.: SC(type).
Host: Quercus
Macrosporium toruloides Ellis & Everh.
Distr.: WA(type).
Host: Urtica

Macroventuria van der Aa. Loculoascomycetes, Dothideales

Notes: van der Aa (Persoonia 6:359. 1971).
Macroventuria wentii van der Aa
Distr.: NV(type).
Notes: Isolated from air, insect debris and plant litter.
Host: Ambrosia

Magnaporthe R. Krause & R.E. Webster. Pyrenomycetes, Phyllachorales

Magnaporthe salvinii (Cattaneo) R. Krause & R.K. Webster
Leptosphaeria salvinii Cattaneo
Distr.: Worldwide in rice-growing regions.
Notes: Stem rot of rice (Krause & Webster. Mycologia 64:103. 1972; Ellis & Holliday. C.M.I. Descr. 344. 1972).
Hosts: Echinochloa, Oryza, Zizaniopsis

Maireina W.B. Cooke. Basidiomycetes, Aphyllophorales

Notes: Cooke (Beih. Sydowia 4:90. 1961).
Maireina cinerea (Burt) W.B. Cooke
Solenia cinerea Burt in Millsp. & L. Nutt.
Distr.: CA.
Host: Quercus
Maireina marginata (McAlpine) W.B. Cooke
Cyphella marginata McAlpine
Distr.: OR; Australia(type).
Notes: On dead twigs of fruit trees, usually on living plants.
Hosts: Cytisus, Malus, Prunus
Maireina texensis (Berk. & M.A. Curtis ex Cooke) W.B. Cooke
Cyphella texensis Berk. & M.A. Curtis ex Cooke
Distr.: TX.
Notes: On wood.
Host: Quercus

Malbranchea Sacc. Hyphomycetes

Malbranchea sulfurea (Miehe) Sigler & J.W. Carmichael
Thermoideum sulfureum Miehe
Malbranchea pulchella Sacc. & Penz. var. *sulfurea* (Miehe) Cooney & R. Emerson
Distr.: Temperate regions.
Notes: Isolated from various kinds of rotting, organic debris (Sigler & Carmichael. Mycotaxon 4:441. 1976).
Host: Helianthus

Mamiania Ces. & De Not. Pyrenomycetes, Diaporthales

Notes: Barr (Mycol. Mem. 7:88. 1978).
Mamiania fimbriata (Pers.:Fr.) Ces. & De Not.
Gnomoniella fimbriata (Pers.:Fr.) Sacc.
Distr.: Temperate northern hemisphere.
Notes: On leaves.
Hosts: Carpinus, Ostrya

Mamianiella Höhn. Pyrenomycetes, Diaporthales

Notes: Barr (Mycol. Mem. 7:86. 1978).
Mamianiella coryli (Batsch:Fr.) Höhn.
Gnomoniella coryli (Batsch:Fr.) Höhn.
Distr.: Temperate North America; Europe.
Notes: On leaves.
Host: Corylus
Mamianiella coryli (Batsch:Fr.) Höhn. var. ***spiralis*** (Peck) Barr
Gnomoniella coryli (Batsch:Fr.) Höhn. var. *spiralis* Peck
Distr.: Central and eastern North America.
Notes: On leaves.
Host: Corylus

Marasmiellus Murrill. Basidiomycetes, Agaricales

Notes: Singer (Beih. Nova Hedwigia 44:1. 1973).
Marasmiellus candidus (Bolton:Fr.) Singer
Marasmius candidus Bolton:Fr.
Marasmiellus albuscorticis (Secr.) Singer, nom. inval.
Marasmius magnisporus Murrill
Distr.: Temperate northern hemisphere.
Hosts: Juglans, Umbellularia
Marasmiellus filopes (Peck) Redhead
Marasmius filopes Peck
Distr.: Eastern North America.
Notes: Redhead (Fungi Canadenses 179. 1980).
Hosts: Abies, Tsuga
Marasmiellus juniperinus Murrill
Distr.: LA; West Indies to Patagonia.
Notes: On bark of living branches (Gilbertson & Ryvarden. Mycotaxon 24:339. 1985).
Host: Juniperus
Marasmiellus mesosporus Singer in Singer et al.
Distr.: NC; Pakistan.
Host: Ammophila
Marasmiellus nigripes (Schwein.) Singer
Marasmius nigripes (Schwein.) Fr.
Marasmius caesius Murrill
Marasmius subcinereus Berk. & Broome
Distr.: Cosmopolitan.
Notes: On plant debris of all kinds.
Host: Quercus
Marasmiellus opacus (Berk. & M.A. Curtis) Singer
Distr.: Eastern North America.
Notes: Common saprobe in litter of eastern deciduous forest.
Host: Acer
Marasmiellus pigmentatus (D.E. Bliss) Singer
Omphalia pigmentata D.E. Bliss
Distr.: CA.
Notes: According to Singer (Beih. Nova Hedwigia 44:319. 1973) this does not belong in *Marasmiellus*.
Hosts: Phoenix, Washingtonia
Marasmiellus ramealis (Bull.:Fr.) Singer
Marasmius ramealis (Bull.:Fr.) Fr.
Distr.: Temperate northern hemisphere.
Host: Jasminum
Marasmiellus ramealis (Bull.:Fr.) Singer var. ***californicus*** Desjardin
Distr.: California(type).
Notes: Desjardin (Mycologia 79:132. 1987).
Host: Quercus
Marasmiellus semiustus (Berk. & M.A. Curtis) Singer
Marasmius semiustus Berk. & M.A. Curtis
Collybia semiustus (Berk. & M.A. Curtis) Dennis
Distr.: Subtropical and tropical western hemisphere.
Host: Musa
Marasmiellus stenophyllus (Mont.) Singer
Marasmius stenophyllus Mont.
Marasmius subsynodicus Murrill
Distr.: Subtropical and tropical western hemisphere.
Notes: Reports of this taxon as a sugarcane pathogen refer to other *Marasmiellus* spp. (Singer. Beih. Nova Hedwigia 44:37. 1973).
Host: Saccharum

Marasmius Fr., nom. cons. Basidiomycetes, Agaricales

Notes: Gilliam (Mycotaxon 4:1. 1976).
Marasmius albogriseus (Peck) Singer
Collybia albogrisea Peck
Distr.: CA, OR.
Notes: Halling (Brittonia 35:318. 1985).
Host: Pseudotsuga
Marasmius androsaceus (L.:Fr.) Fr.
Distr.: Boreal and temperate North America.
Notes: On needles and debris of conifers.
Host: Pinus

Marasmius atlanticus Singer
Distr.: FL(type).
Host: Quercus
Marasmius capillaris Morg.
Distr.: Central and eastern North America.
Notes: Usually on fallen leaves of hardwoods.
Host: Quercus
Marasmius caricis (P. Karst.) P. Karst.
Gloiocephala caricis (P. Karst.) Bas
Marasmius caricicola Kauffm.
Distr.: Temperate North America; Europe.
Host: Carex
Marasmius cohortalis Berk. var. ***alachuanus*** (Murrill) Singer
Marasmius alachuanus Murrill
Distr.: FL(type).
Host: Quercus
Marasmius copelandii Peck
Distr.: California.
Notes: Desjardin (Mycologia 79:127. 1987).
Hosts: Berberis, Castanopsis, Lithocarpus, Quercus
Marasmius copelandii Peck var. ***olidus*** (Gilliam) Desjardin
Marasmius olidus Gilliam
Distr.: MI.
Notes: Desjardin (Mycologia 79:129. 1987).
Host: Quercus
Marasmius corrugatus (Pat.) Sacc. var. ***aurantiacus*** (Murrill) Singer
Marasmius aurantiacus (Murrill) Singer
Distr.: FL; South America.
Host: Quercus
Marasmius epidryas Kühner
Distr.: Circumboreal.
Notes: Redhead et al. (Fungi Canadenses 213. 1982).
Host: Dryas
Marasmius epifagus Gilliam
Distr.: Eastern North America.
Notes: On fallen leaves.
Host: Fagus
Marasmius epiphyllus (Pers.:Fr.) Fr.
Distr.: Temperate northern hemisphere.
Notes: On hardwood litter.
Host: Potentilla
Marasmius graminum (Lib.) Berk. & Broome
Marasmius tritici P.A. Young
Distr.: Cosmopolitan.
Notes: Saprobic or facultatively parasitic on stems, leaves, and roots of grasses (Young. Phytopathology 15:115. 1925).
Hosts: Agrostis, Avena, Hordeum, Secale, Triticum
Marasmius heliomyces Murrill
Distr.: FL.
Host: Quercus
Marasmius insipidus Gilliam
Distr.: OH.
Host: Quercus
Marasmius magnoliae Singer
Distr.: FL.
Notes: Singer (Mycologia 37:435. 1945).
Host: Magnolia
Marasmius minutus Peck
Marasmius pyrinus Ellis
Distr.: Eastern North America; Europe.
Host: Malus
Marasmius pallidocephalus Gilliam
Distr.: Central and eastern North America.
Hosts: Picea, Tsuga
Marasmius plicatulus Peck
Distr.: Pacific coastal North America.
Hosts: Cupressus, Picea, Pinus, Pseudotsuga, Quercus
Marasmius quercophilus Pouzar
Distr.: California; Europe(type).
Notes: Desjardin (Mycologia 79:124. 1987).
Hosts: Lithocarpus, Quercus
Marasmius robinianus Gilliam
Distr.: MI(type).
Host: Robinia
Marasmius rotula (Scop.:Fr.) Fr.
Distr.: Eastern North America; Europe.
Notes: Usually on partially decayed hardwood logs and stumps.
Hosts: Acer, Betula, Fagus, Pinus, Quercus
Marasmius sacchari Wakk.
Distr.: Southern United States, HI.
Notes: A doubtful species. Reports of this taxon are probably based on *Marasmiellus dealbatus* (Berk. & M.A. Curtis) Singer, *M. inoderma* (Berk.) Singer, or *M. semiustus* (Berk. & M.A. Curtis) Singer (Singer. Beih. Nova Hedwigia 44:36. 1973).
Host: Saccharum
Marasmius salalis Desjardin & Redhead
Distr.: OR, WA; Canada(coastal British Columbia).
Notes: Desjardin & Redhead (Mycotaxon 29:3307. 1987).
Hosts: Berberis, Gaultheria
Marasmius spadiceus Gilliam
Distr.: MI.
Host: Quercus
Marasmius thiersii Desjardin
Distr.: California(type).
Notes: Desjardin (Mycologia 79:123. 1987).
Hosts: Lithocarpus, Pinus

Maravalia Arth. Basidiomycetes, Uredinales
Scopella Mains
Notes: Ono (Mycologia 76:892. 1984).
Maravalia sapotae (Mains) Y. Ono
Scopella sapotae Mains in Cummins
Distr.: FL; West Indies; Central and South America.
Notes: Autoecious.
Host: Manilkara

Mariannaea G. Arnaud ex R.A. Samson. Hyphomycetes
Mariannaea elegans (Corda) R.A. Samson
Penicillium elegans Corda
Paecilomyces elegans (Corda) E. Mason & S.J. Hughes
Spicaria elegans (Corda) C. Harz var. *sorghina* Sacc.
Distr.: Northern temperate regions.
Notes: On rotting coniferous bark and isolated from forest soils and other organic debris (Samson. Stud. Mycol. 6:75. 1974; Bissett. Fungi Canadenses 160. 1979).
Host: Zea

Marssonia J.C. Fisch. Coelomycetes
Notes: *Marssonina* has replaced the later homonym *Marssonia* (Sutton. Mycol. Pap. 141:120. 1977), but not all names have been redisposed.
Marssonia adunca Sacc.
Distr.: Western United States.
Host: Geum
Marssonia callae Dearn. & House
Distr.: NY(type).
Host: Calla
Marssonia coronariae Sacc. & Dearn.
Distr.: Temperate North America.
Host: Malus
Marssonia galii Solheim
Distr.: AZ(type).
Host: Galium
Marssonia potentillae (Desmaz.) J.C. Fisch. var. ***helleri*** Peck
Distr.: CA.
Host: Potentilla

Marssonina Magnus. Coelomycetes
Marssonina apicalis (Ellis & Everh.) Magnus
Marssonia apicalis Ellis & Everh.
Distr.: Central and western United States.
Host: Salix
Marssonina bracteosa Dearn. & Barth.

Distr.: Western North America.
Host: Ribes
Marssonina californica (Ellis & Everh.) Magnus
Marssonia californica Ellis & Everh.
Distr.: CA.
Host: Juglans
Marssonina carpogena (Cooke & Harkn.) Arx
Gloeosporium carpogenum Cooke & Harkn.
Distr.: CA(type).
Host: Populus
Marssonina castagnei (Desmaz. & Mont.) Magnus
Gloeosporium castagnei Desmaz. & Mont.
Distr.: Temperate northern hemisphere.
Notes: Pirozynski (Fungi Canadenses 14. 1974).
Hosts: Acer, Populus
Marssonina chrysothamni Solheim
Distr.: CO, WY.
Host: Chrysothamnus
Marssonina coronaria (Ellis & J.J. Davis) J.J. Davis
Ascochyta coronaria Ellis & J.J. Davis
Marssonia coronaria Sacc. & Dearn.
Distr.: Temperate northern hemisphere.
Notes: Leaf spot of apple (Parmelee. Canad. Pl. Dis. Surv. 51:91. 1971).
Host: Malus
Marssonina daphnes (Desmaz. & Roberge) Magnus
Marssonia daphnes (Desmaz. & Roberge) Sacc.
Gloeosporium mezerei Cooke
Distr.: WA; Europe.
Host: Daphne
Marssonina fragariae (Lib.) Kleb.
Leptothyrium fragariae Lib.
Ascochyta fragariae Sacc.
Ascochyta colorata Peck
Marssonia potentillae (Desmaz.) J.C. Fisch. in Rabenh.
Marssonina potentillae (Desmaz.) Magnus
Phyllosticta potentillae Desmaz.
Diplocarpon earlianum (Ellis & Everh.) F.A. Wolf (teleomorph)
Distr.: Widespread.
Notes: Strawberry leaf scorch (Sivanesan & Gibson. C.M.I. Descr. 486. 1976).
Hosts: Fragaria, Geum, Potentilla
Marssonina fraserae (Ellis & Everh.) Magnus
Marssonia fraserae Ellis & Everh.
Distr.: Western United States.
Host: Frasera
Marssonina gloeodes (H.C. Greene) H.C. Greene
Marssonia gloeodes H.C. Greene
Distr.: WI.
Host: Fraxinus
Marssonina grossulariae (Oudem.) Magnus
Marssonia grossulariae Oudem.
Distr.: OH, WI; Europe.
Host: Ribes
Marssonina juglandis (Lib.) Magnus
Marssonia juglandis (Lib.) Sacc.
Marssoniella juglandis (Lib.) Höhn.
Gnomonia leptostyla (Fr.:Fr.) Ces. & De Not. (teleomorph)
Distr.: Central and eastern North America; Europe.
Notes: On leaves.
Hosts: Carya, Juglans
Marssonina kriegeriana (Bres.) Magnus
Marssonia kriegeriana Bres.
Distr.: United States; Europe.
Host: Salix
Marssonina martinii (Sacc. & Ellis) Magnus
Marssonia martinii Sacc. & Ellis
Gloeosporium martinii (Sacc. & Ellis) Ellis & Everh.
Marssonia quercus Peck
Marssonina quercus (Peck) Magnus
Distr.: Central and eastern North America.
Notes: Leaf spot of oaks (Lentz. Mycologia 42:259. 1950).
Host: Quercus
Marssonina melonis Dolan
Distr.: NY.
Hosts: Citrullus, Cucumis
Marssonina neilliae (Harkn.) Magnus
Marssonia neilliae Harkn.
Distr.: CA.
Hosts: Neillia, Physocarpus
Marssonina nigricans (Ellis & Everh.) Magnus
Marssonia nigricans Ellis & Everh.
Distr.: CA.
Host: Salix
Marssonina piriformis (Riess) Magnus
Didymosporium piriforme Riess
Marssonia piriformis (Riess) Sacc.
Distr.: OR; Europe.
Host: Populus
Marssonina populi (Lib.) Magnus
Gloeosporium populi (Lib.) Mont. & Desmaz.
Drepanopeziza populorum (Desmaz.) Höhn. (teleomorph)
Distr.: Temperate northern hemisphere.
Notes: Leaf spot (Pirozynski. Fungi Canadenses 15. 1974).
Hosts: Populus, Salix
Marssonina potentillae (Desmaz.) Magnus var. ***tormentillae*** (Trail) Magnus
Marssonia potentillae (Desmaz.) E. Fisch. var. *tormentillae* Trail
Distr.: WI; Europe.
Hosts: Fragaria, Rubus
Marssonina punctiformis (Ellis & Everh.) Arx
Gloeosporium punctiforme Ellis & Everh.
Gloeosporium everhartii Ellis
Distr.: Eastern United States.
Notes: Leaf spot.
Hosts: Fraxinus, Vicia
Marssonina rhamni (Ellis & Everh.) Magnus
Marssonia rhamni Ellis & Everh.
Distr.: WA.
Host: Rhamnus
Marssonina ribicola (Ellis & Everh.) Magnus
Marssonia ribicola Ellis & Everh.
Distr.: CO, IL.
Host: Ribes
Marssonina rosae (Lib.) Died.
Asteroma rosae Lib.
Actinonema rosae (Lib.) Fr.
Diplocarpon rosae F.A. Wolf (teleomorph)
Distr.: Cosmopolitan.
Notes: Sivanesan & Gibson (C.M.I. Descr. 485. 1975 as *D. rosae*).
Host: Rosa
Marssonina rubiginosa (Ellis & Everh.) Magnus
Marssonia rubiginosa Ellis & Everh.
Distr.: ID, WI.
Host: Salix
Marssonina salicicola (Bres.) Magnus
Marssonia salicicola Bres.
Distr.: Eastern United States; Europe.
Host: Salix
Marssonina salicina Tehon
Distr.: IL.
Host: Salix
Marssonina salicis (Trail) Magnus
Marssonia salicis Trail
Distr.: MA; Europe.
Host: Salix
Marssonina saxifragae Arx
Distr.: WA.
Host: Saxifraga
Marssonina sennenis (Gonz. Frag.) Vassiljevsky & Karakulin
Marssonia sennensis Gonz. Frag.
Distr.: AK; Europe.
Host: Sanguisorba
Marssonina stenospora (Ellis & Kellerm.) Magnus
Gloeosporium stenospora Ellis & Kellerm.
Distr.: United States.
Host: Populus
Marssonina tremulae (Lib.) Kleb.
Leptothyrium tremulae Lib.
Gloeosporium brunnea Ellis & Everh.
Marssonia brunnea (Ellis & Everh.) Sacc.

Marssonina brunnea (Ellis & Everh.) Magnus
Gloeosporium brunneum Ellis & Everh.
Drepanopeziza tremulae Rimpau (teleomorph)
Distr.: NJ; Europe.
Notes: Pirozynski (Fungi Canadenses 13. 1974 as *Marssonina brunnea*).
Host: Populus
Marssonina violae (Pass.) Magnus
Distr.: Central and eastern United States; Europe.
Host: Viola

Massaria De Not. Loculoascomycetes, Pyrenulales

Notes: Barr (Mycotaxon 9:17. 1979); Shoemaker & LeClair (Canad. J. Bot. 53:1568. 1975).
Massaria cleistotheca Harkn.
Distr.: CA(type).
Host: Dendromecon
Massaria conspurcata (Wallr.) Sacc.
Distr.: United States; Europe.
Notes: On dead wood of Rosaceae.
Hosts: Amelanchier, Prunus
Massaria inquinans (Tode:Fr.) De Not.
Massaria gigaspora (Desmaz.) Berk.
Massaria vomitoria Berk. & M.A. Curtis
Distr.: Temperate northern hemisphere.
Notes: A mild pathogen of *Acer* spp., especially common on *A. saccharum* (Shoemaker & Kokko. Fungi Canadenses 104. 1977).
Hosts: Acer, Amelanchier, Castanea, Fagus
Massaria lantanae (G. Otth) Shoemaker & P.M. LeClair
Massaria plumigera Ellis & Everh.
Massaria plumigera Ellis & Everh. var. *tetraspora* Dearn. & House in House
Massarina plumigera (Ellis & Everh.) Sacc. & Trott. var. *tetraspora* Dearn. & House in House
Distr.: Eastern North America; Europe.
Notes: On branches of hardwoods (Shoemaker & Kokko. Fungi Canadenses 105. 1979).
Hosts: Acer, Cephalanthus, Morus, Viburnum
Massaria pyri G. Otth in Tul. & C. Tul.
Aglaospora occulta (Ellis) Farl.
Massaria pruni Wehmeyer
Distr.: Northeastern North America.
Notes: On limbs of Rosaceae.
Host: Malus
Massaria shearii Petr.
Distr.: GA(type).
Host: Magnolia
Massaria ulmi Fuckel
Distr.: GA; Germany(type).
Notes: According to Shoemaker & LeClair (Canad. J. Bot. 53:1593. 1975), this is an ambiguous name.
Host: Ulmus
Massaria xanthoxyli (Peck) Petr.
Massariella xanthoxyli Peck
Distr.: Central and eastern North America.
Notes: Shoemaker & Kokko (Fungi Canadenses 106. 1977 as *M. zanthoxyli*).
Hosts: Morus, Zanthoxylum

Massarina Sacc. Loculoascomycetes, Dothideales

Massarina arundinacea (Sowerby:Fr.) Leuchtmann
Leptosphaeria arundinacea (Sowerby:Fr.) Sacc.
Distr.: Central and eastern United States; Europe.
Notes: Leuchtmann(Sydowia 37:179. 1984).
Hosts: Arundinaria, Spartina, Zea
Massarina berchemiae Petr.
Distr.: LA(type).
Host: Berchemia
Massarina coffeicola (Speg.) Bose
Oraniella coffeicola Speg.
Distr.: HI, tropical regions.
Notes: On dead twigs (Bose. Phytopathol. Z. 41:172. 1961).
Host: Coffea
Massarina corticola (Fuckel) L. Holm
Trematosphaeria corticola Fuckel
Distr.: Temperate regions.
Notes: On numerous woody hosts.
Hosts: Sambucus, Tilia
Massarina floridana Petr.
Distr.: FL(type).
Host: Sabal
Massarina gloeospora (Berk. & Curr.) Barr in Barr et al.
Leptosphaeria gloeospora (Berk. & Curr.) Sacc.
Trichometasphaeria gloeospora (Berk. & Curr.) L. Holm
Leptosphaeria asparagi Peck
Leptosphaeria comatella (Cooke & Ellis) Sacc.
Leptosphaeria defodiens (Ellis) Ellis
Metasphaeria defodiens (Ellis) Sacc.
Sphaeria defodiens Ellis
Leptosphaeria stictostoma (Berk. & M.A. Curtis) Sacc.
Leptosphaeria trichostoma Pass. in Thuem.
Metasphaeria trichostoma (Pass.) Sacc.
Distr.: Eastern North America; Europe.
Notes: On a variety of herbaceous plants. Holm (Symb. Bot. Upsal. 14:144. 1957).
Hosts: Asparagus, Aster, Aureolaria, Bidens, Chrysopsis, Coreopsis, Daucus, Desmodium, Erigeron, Eupatorium, Foeniculum, Galium, Haplopappus, Helianthus, Juncus, Lespedeza, Monarda, Panicum, Polygonum, Ricinus, Rudbeckia, Saccharum, Saponaria, Solidago, Sorghastrum, Typha, Verbesina, Xanthium
Massarina lacustris (Fuckel) Leuchtmann
Leptosphaeria lacustris (Fuckel) G. Wint. in Rabenh.
Metasphaeria lacustris (Fuckel) Sacc.
Stagonospora sp. (anamorph)
Distr.: Eastern United States; Europe.
Notes: On herbaceous debris (Leuchtmann. Sydowia 37:181. 1984).
Host: Spartina
Massarina microcarpa (Fuckel) Sacc.
Massaria microcarpa Fuckel
Leptosphaeria vagabunda Sacc.
Distr.: United States; Europe.
Hosts: Clematis, Conium, Hydrangea, Ribes, Viburnum, Vitis
Massarina myricae (Peck) Berl.
Metasphaeria myricae Peck
Distr.: NY.
Notes: Barr et al. (New York State Mus. Bull. 459:31. 1986).
Host: Myrica
Massarina staphyleae Petr.
Massarina floridana Petr. 1952 non Petr. 1951, hom. illeg.
Distr.: FL(type).
Host: Staphylea

Massarinula Géneau de Lamarlière. Loculoascomycetes, Dothideales

Notes: This genus is considered a synonym of *Massarina*, but not all names have been redisposed.
Massarinula lignorum W.B. Cooke
Distr.: CA(type).
Host: Acer

Massariosphaeria (E. Müller) P.G. Crivelli. Loculoascomycetes, Dothideales

Notes: Crivelli (Ueber die heterogene Ascomycetengattung *Pleospora* Rabh.; Vorschlag für eine Aufteilung. p. 141. 1983).
Massariosphaeria rubelloides (Plowr. ex Cooke) P.G. Crivelli
Pleospora rubelloides (Plowr. ex Cooke) J. Webster
Pleospora alismatis Ellis & Everh.
Pleospora rubicunda Niessl var. *americana* Wehmeyer
Phoma sp. (anamorph)
Distr.: Temperate North America; Europe.
Hosts: Alisma, Pedicularis, Potentilla
Massariosphaeria rubicunda (Niessl) P.G. Crivelli
Pleospora rubicunda Niessl

Distr.: Western United States; Europe.
Notes: On various monocots and dicots.
Hosts: Epilobium, Pedicularis, Potentilla
Massariosphaeria scirpina (G. Wint.) Leuchtmann
Leptosphaeria scirpina G. Wint.
Distr.: Eastern United States; Europe.
Host: Spartina

Massariothea Syd. Coelomycetes
Notes: Sutton (The Coelomycetes. p. 638. 1980).
Massariothea paspali (Ellis & Everh.) Sutton
Coryneum paspali Ellis & Everh.
Apocoryneum paspali (Ellis & Everh.) Sutton
Distr.: LA.
Host: Axonopus

Massariovalsa Sacc. Pyrenomycetes, Diaporthales
Notes: Barr (Mycol. Mem. 7:189. 1978).
Massariovalsa californica (Wehmeyer) Barr
Melanconis californica Wehmeyer
Distr.: CA.
Host: Arbutus
Massariovalsa sudans (Berk. & M.A. Curtis) Sacc.
Melanconis sudans (Berk. & M.A. Curtis) Wehmeyer
Valsa apatosa Cooke & Ellis
Valsaria apatosa (Cooke & Ellis) Sacc.
Melanconis nyssagena Ellis & Everh.
Melanconiopsis inquinans Ellis & Everh. (anamorph)
Distr.: Eastern North America.
Notes: On branches of hardwoods.
Hosts: Acer, Aesculus, Carya, Cornus, Nyssa, Tilia, Ulmus

Mastigosporella Höhn. Coelomycetes
Mastigosporella hyalina (Ellis & Everh.) Höhn.
Harknessia hyalina Ellis & Everh.
Dicarpella quercifolia (Ellis & Everh.) Barr (teleomorph)
Distr.: Eastern United States.
Host: Quercus
Mastigosporella nyssae Nag Raj & DiCosmo
Dicarpella georgiana (J.H. Miller & G.E. Thompson) Barr (teleomorph)
Distr.: Southeastern United States.
Notes: On leaves.
Host: Nyssa

Mastigosporium Riess in Fresen. Hyphomycetes
Notes: Hughes (Mycol. Pap. 36:1. 1951).
Mastigosporium album Riess
Distr.: NY; Europe.
Notes: Leaf spot.
Host: Alopecurus
Mastigosporium cylindricum R. Sprague
Distr.: OR(type).
Host: Bromus
Mastigosporium rubricosum (Dearn. & Barth.) Nannf.
Fusoma rubricosum Dearn. & Barth.
Mastigosporium rubricosum (Dearn. & Barth.) R. Sprague
Amastigosporium graminicola Bondartzeva-Monteverde
Distr.: Western North America; Europe.
Notes: Leaf fleck of grasses.
Hosts: Agrostis, Calamagrostis, Dactylis, Deschampsia, Phleum, Trisetum

Maurodothina G. Arnaud ex Pirozynski & Shoemaker. Loculoascomycetes, Dothideales
Notes: Arx & Müller (Stud. Mycol. 9:44. 1975) considered this genus a synonym of *Eupelte*, but not all names have been redisposed. Pirozynski & Shoemaker (Canad. J. Bot. 48:1327. 1970).
Maurodothina dothideoides (Ellis & Everh.) Pirozynski & Shoemaker
Asteridium dothideoides Ellis & Everh.
Morenoella dothideoides (Ellis & Everh.) Höhn.
Distr.: FL.
Host: Lyonia
Maurodothina farrae Pirozynski & Shoemaker
Eupelte farrae (Pirozynski & Shoemaker) Arx & E. Müller
Pirozynskia farriae Subramanian (anamorph)
Distr.: Northern and temperate North America.
Notes: On living needles (Ellis. More dematiaceous Hyphomycetes. p. 365. 1976).
Host: Abies

Mazzantia Mont. Pyrenomycetes, Diaporthales
Notes: Monod (Beih. Sydowia 9:198. 1983).
Mazzantia borealis (Sacc.) M. Monod
Physalospora borealis Sacc.
Heteropera borealis (Sacc.) Theiss.
Distr.: AK; Europe.
Host: Anemone
Mazzantia galii (Fr.:Fr.) Mont.
Physalospora galii (Fr.:Fr.) Rostr.
Distr.: North America; Europe.
Notes: On overwintered stems.
Host: Galium

Medeolaria Thaxt. Discomycetes, Medeolariales
Medeolaria farlowii Thaxt.
Distr.: Northeastern United States.
Host: Medeola

Megachytrium Sparrow. Chytridiomycetes, Chytridiales
Megachytrium westonii Sparrow
Distr.: NY.
Host: Elodea

Megacollybia Kotlaba & Pouzar. Basidiomycetes, Agaricales
Megacollybia platyphylla (Fr.:Fr.) Kotlaba & Pouzar
Collybia platyphylla (Fr.:Fr.) P. Kumm.
Tricholomopsis platyphylla (Fr.:Fr.) Singer
Distr.: Temperate northern hemisphere.
Notes: Usually on hardwood logs, stumps, or debris.
Hosts: Abies, Alnus, Pinus, Populus, Pseudotsuga, Quercus

Megalocystidium Jülich. Basidiomycetes, Aphyllophorales
Megalocystidium lactescens (Berk.) Jülich
Corticium lactescens (Berk.) Berk.
Gloeocystidiellum lactescens (Berk.) Boidin
Distr.: Northern hemisphere.
Notes: On wood.
Hosts: Agave, Alnus, Juglans, Populus, Quercus
Megalocystidium leucoxanthum (Bres.) Jülich
Corticium leucoxanthum Bres.
Gloeocystidiellum leucoxanthum (Bres.) Boidin
Distr.: Northern hemisphere.
Notes: On wood, usually of hardwoods.
Hosts: Alnus, Picea, Populus, Quercus
Megalocystidium luridum (Bres.) Jülich
Corticium luridum Bres.
Gloeocystidiellum luridum (Bres.) Boidin

Distr.: North America; Europe.
Notes: On wood.
Hosts: Abies, Alnus, Picea

Melampsora Castagne. Basidiomycetes, Uredinales

Notes: Ziller (The tree rusts of western Canada. p. 138. 1974).
Melampsora abieti-capraearum Tub.
Melampsora americana Arth.
Melampsora humboldtiana Speg.
Distr.: Western hemisphere; western Europe.
Notes: See remarks under *M. epitea*. Heteroecious; aecial host - *Abies* spp.; telial host - *Salix* spp.
Hosts: Abies, Salix
Melampsora abietis-canadensis C.A. Ludw. ex Arth.
Caeoma abietis-canadensis Farl.
Distr.: Northern North America.
Notes: Aecial host - Pinaceae; telial host - Salicaceae.
Hosts: Populus, Tsuga
Melampsora arctica Rostr.
Distr.: Circumboreal.
Notes: See remarks under *M. epitea*. Heteroecious; aecial host - *Saxifraga* spp.; telial host - *Salix* spp.
Hosts: Salix, Saxifraga
Melampsora cumminsii Bagyanarayana & Ramachar
Distr.: NH(type).
Notes: Heteroecious; aecial host - unknown; telial host - *Populus* sp.
Host: Populus
Melampsora epitea Thuem.
Distr.: Widespread in temperate regions.
Notes: According to Hylander et al. (Opera Bot. 1:1. 1953), the following species are morphologically indistinguishable from *M. epitea*, and should be considered synonymous: *M. arctica*, *M. abieti-capraearum* , *M. ribesii-purpureae*. Here we follow Ziller (The tree rusts of western Canada. p. 138. 1974) in treating these species separately. Reports of *M. epitea* from the United States refer to one or another of these species, distinguished on aecial host.
Hosts: Abies, Larix, Ribes, Salix, Tsuga
Melampsora epitea Thuem. f. sp. ***tsugae*** Ziller
Caeoma dubium C.A. Ludw.
Distr.: Pacific Northwest.
Notes: Heteroecious; aecial host - *Tsuga* spp.; telial host - *Salix* spp.
Host: Tsuga
Melampsora euphorbiae Castagne
Distr.: Widespread.
Notes: On Euphorbiaceae.
Host: Euphorbia
Melampsora euphorbiae-gerardianae W. Muell. Stettin
Distr.: Central and estern United States; Asia; Europe.
Notes: Aecia unknown; uredinia and telia on Euphorbiaceae.
Host: Euphorbia
Melampsora farlowii (Arth.) J.J. Davis
Necium farlowii Arth.
Distr.: Northeastern North America.
Notes: Aecia and uredinia lacking; telia on Pinaceae.
Host: Tsuga
Melampsora hypericorum G. Wint. in Rabenh.
Mesopsora hypericorum Dietel
Uredo hypericorum DC. (anamorph)
Distr.: Western United States.
Host: Hypericum
Melampsora lini (Ehrenb.) Desmaz.
Melampsora lini (Ehrenb.) Desmaz. var. *liniperda* Körn.
Distr.: Cosmopolitan.
Notes: Autoecious. Laundon & Waterston (C.M.I. Descr. 51. 1965).
Host: Linum
Melampsora medusae Thuem.
Melampsora albertensis Arth.
Caeoma faulliana L.M. Hunter
Distr.: North America; Argentina; Australia; Japan.
Notes: Heteroecious; aecial host - Pinaceae; telial host - *Populus* spp. Walker (C.M.I. Descr. 480. 1975); Hubbes et al. (Pl. Dis. 67:217. 1983).
Hosts: Abies, Larix, Populus, Pseudotsuga
Melampsora monticola Mains
Distr.: Central and western North America.
Notes: Pycnia and aecia unknown; telial host - Euphorbiaceae.
Host: Euphorbia
Melampsora occidentalis H. Jacks.
Distr.: Western North America.
Notes: Heteroecious; aecial host - Pinaceae; telial host - *Populus* spp.
Hosts: Populus, Pseudotsuga
Melampsora paradoxa Dietel & Holw. in Dietel
Melampsora bigelowii Thuem.
Distr.: Northern North America.
Notes: Heteroecious; aecial host - *Larix* spp.; telial host - *Salix* spp.
Hosts: Larix, Salix
Melampsora populnea (Pers.) P. Karst.
Melampsora aecidioides (DC.) J. Schröt.
Distr.: Western United States, rare in eastern United States; Europe; India.
Notes: Aecia unknown; uredinia on Salicaceae.
Host: Populus
Melampsora ribesii-purpureae Kleb. in Pringsh.
Melampsora confluens H. Jacks.
Distr.: Temperate northern hemisphere.
Notes: See remarks under *M. epitea*. Heteroecious; aecial host - *Ribes* spp.; telial host - *Salix* spp.
Hosts: Ribes, Salix

Melampsorella J. Schröt. Basidiomycetes, Uredinales

Melampsorella caryophyllacearum J. Schröt.
Melampsorella cerastii J. Schröt. in Cohn
Melampsorella elatina Arth.
Distr.: Temperate northern hemisphere; Falkland Island(introduced).
Notes: Fir broom rust. Heteroecious; aecial host - *Abies* spp.; telial host - Carophyllaceae.
Hosts: Abies, Cerastium, [Picea], Stellaria

Melampsoridium Kleb. Basidiomycetes, Uredinales

Melampsoridium betulinum Kleb.
Melampsora betulina G. Wint.
Melampsoridium alni Dietel
Distr.: Temperate northern hemisphere; New Zealand.
Notes: Birch rust. Heteroecious; aecial host - *Larix* spp.; telial host - *Betula* and *Alnus* spp. Kaneko & Hiratsuka (Trans. Mycol. Soc. Japan 22:463. 1981) reidentified a specimen on *Ostrya* purported to be *M. carpini* as *M. betulinum*.
Hosts: Alnus, Betula, Larix, [Ostrya]
Melampsoridium carpini (Fuckel) Dietel in Engl. & Prantl
Melampsora carpini Fuckel
Distr.: Temperate regions.
Host: [Ostrya]

Melanconiopsis Ellis & Everh. Coelomycetes

Melanconiopsis inquinans Ellis & Everh.
Massariovalsa sudans (Berk. & M.A. Curtis) Sacc. (teleomorph)
Distr.: Temperate North America.
Notes: Sutton (The Coelomycetes. p. 330. 1980).
Host: Carya

Melanconis Tul. & C. Tul. Pyrenomycetes, Diaporthales

Notes: Wehmeyer (Univ. Michigan Stud., Sci. Ser. 14. 1941).
Melanconis alni Tul. & C. Tul.
Melanconium sphaeroideum Link:Fr. (anamorph)
Distr.: Temperate regions.
Notes: On limbs.
Host: Alnus
Melanconis apocrypta Ellis
Melanconium sp. (anamorph)

Distr.: Central and eastern North America.
Notes: On limbs.
Host: Populus
Melanconis chrysostroma (Fr.) Tul. & C. Tul.
Valsa chrysostroma (Fr.) Tul. & C. Tul.
Diaporthe hyperopta Nitschke in G. Otth
Melanconis hyperopta (Nitschke) Wehmeyer
Distr.: North America; Europe.
Notes: On limbs of Betulaceae.
Hosts: Carpinus, [Quercus]
Melanconis chrysostroma (Fr.) Tul. & C. Tul. var. ***ellisii*** (Rehm) Wehmeyer
Melanconis carpinigera (Berk. & M.A. Curtis) Petr.
Distr.: Eastern North America.
Host: Carpinus
Melanconis decorahensis Ellis
Melanconiella decorahensis (Ellis) Sacc.
Melanconium subviridis Dearn. & House (anamorph)
Distr.: Temperate North America.
Notes: On limbs.
Hosts: Betula, Corylus
Melanconis flavovirens (G. Otth) Wehmeyer
Melanconium sp. (anamorph)
Distr.: Temperate North America; Europe.
Notes: On limbs.
Host: Corylus
Melanconis juglandis (Ellis & Everh.) Graves
Diaporthe juglandis Ellis & Everh.
Distr.: Eastern North America.
Host: Juglans
Melanconis juglandis (Ellis & Everh.) Graves var. ***caryae*** Wehmeyer
Melanconium sp. (anamorph)
Distr.: Southeastern United States.
Host: Carya
Melanconis juglandis (Ellis & Everh.) Graves var. ***tiliae*** Wehmeyer
Melanconium sp. (anamorph)
Distr.: Northeastern North America; Sweden.
Notes: On limbs.
Host: Tilia
Melanconis marginalis (Peck) Wehmeyer
Diaporthe marginalis Peck
Melanconis alni Tul. & C. Tul. var. *marginalis* (Peck) Wehmeyer
Melanconium marginale Wehmeyer (anamorph)
Distr.: Temperate North America; Japan.
Notes: On limbs.
Host: Alnus
Melanconis ostryae (Dearn.) Wehmeyer
Melanconium sp. (anamorph)
Distr.: Northeastern North America.
Host: Ostrya
Melanconis pallida (Rehm) Wehmeyer
Melanconiella pallida Rehm
Distr.: Eastern North America.
Notes: On limbs.
Host: Carya
Melanconis platystroma Wehmeyer
Melanconium sp. (anamorph)
Distr.: MI.
Host: Carpinus
Melanconis stilbostoma (Fr.:Fr.) Tul.
Melanconium bicolor Nees:Fr. (anamorph)
Distr.: Temperate North America; Europe.
Notes: On limbs (Kobayashi. Bull. Gov. Forest Exp. Sta. 226:13. 1970). Barr (Mycol. Mem. 7:183. 1978).
Host: Betula
Melanconis xanthostroma (Mont.) J. Schröt.
Valsa xanthostroma (Mont.) Tul.
Cryptosporella aurea Fuckel
Distr.: GA; Europe.
Notes: The synonymy of *Cryptosporella aurea* follows Wehmeyer (Revision of *Melanconis, Pseudovalsa, Prosthecium,* and *Titania*. p. 74. 1941).
Host: Liquidambar

Melanconium Link:Fr. Coelomycetes

Melanconium americanum Peck & G.W. Clinton
Distr.: NY.
Host: Agave
Melanconium angustum Ellis & Everh.
Distr.: Central and eastern United States.
Host: Carya
Melanconium apiocarpum Link
Distr.: Western North America; Europe.
Host: Alnus
Melanconium atrum Link
Distr.: Temperate northern hemisphere.
Notes: Sutton (The Coelomycetes. p. 330. 1980).
Hosts: Acer, Populus
Melanconium betulinum Schumach. & Kunze
Distr.: United States; Europe.
Host: Betula
Melanconium bicolor Nees:Fr.
Melanconium zonatum Ellis & Everh.
Melanconis stilbostoma (Fr.:Fr.) Tul. (teleomorph)
Distr.: North America; Asia; Europe.
Notes: Associated with dieback (Kobayashi. Bull. Gov. Forest Exp. Sta. 226:13. 1970).
Hosts: Betula, Carya
Melanconium candidum (Peck) Zeller
Melanconium bicolor Nees var. *candidum* Peck
Distr.: Pacific Northwest.
Notes: Zeller (Mycologia 27:465. 1935).
Host: Alnus
Melanconium celtidis Ellis & Everh.
Distr.: KS.
Host: Celtis
Melanconium cerasinum Peck
Distr.: Central and western United States.
Host: Prunus
Melanconium cinctum Berk. & M.A. Curtis
Distr.: Southern United States.
Hosts: Castanea, Castanopsis
Melanconium globosum Cooke & Harkn.
Distr.: CA.
Host: Eucalyptus
Melanconium gracile Ellis & Everh.
Distr.: Central and eastern United States.
Host: Carya
Melanconium juncinum (Cooke & Harkn.) P.E. Powell
Leptothyrium juncinum Cooke & Harkn.
Distr.: Western North America.
Host: Juncus
Melanconium magnoliae Ellis & Everh.
Distr.: NJ.
Host: Magnolia
Melanconium magnum Berk.
Distr.: Central and eastern United States.
Host: Carya
Melanconium oblongum Berk.
Melanconis juglandis (Ellis & Everh.) Graves (teleomorph)
Distr.: Eastern North America.
Notes: On limbs.
Host: Juglans
Melanconium pallidum Peck
Distr.: Central and eastern United States.
Host: Carya
Melanconium palmarum Cooke & Ravenel
Melanconium sabal Cooke, nom. nud.
Distr.: Southern United States.
Hosts: Roystonea, Sabal
Melanconium pandani Lév.
Distr.: FL, HI, MD; Europe.
Host: Pandanus
Melanconium salicinum Ellis & Everh.
Distr.: LA.
Host: Salix
Melanconium sphaeroideum Link:Fr.
Melanconis alni Tul. & C. Tul. (teleomorph)
Distr.: Temperate regions.

Notes: Associated with branch dieback (Funk. Parasitic microfungi of western trees. p. 93. 1981).
Host: Alnus
Melanconium viscosum Schwein.
Distr.: PA(type).
Host: Robinia

Melanocarpus Arx. Pyrenomycetes, Sordariales

Melanocarpus albomyces (Cooney & R. Emerson) Arx
Myriococcum albomyces Cooney & R. Emerson
Thielavia albomyces (Cooney & R. Emerson) Malloch & Cain
Distr.: NV, OK.
Notes: On rotting debris.
Host: Arachis

Melanodothis R.H. Arnold. Loculoascomycetes, Dothideales

Melanodothis caricis R.H. Arnold
Hyalodothis caricis Pat. & Har., nom. dub.
Distr.: Northwestern North America; China; Japan; New Guinea.
Notes: On inflorescences. Arnold (Canad. J. Bot. 49:2188. 1971); Arnold (Fungi Canadenses 30. 1974).
Host: Carex

Melanographium Sacc. Hyphomycetes

Melanographium cookei M.B. Ellis
Monotospora fasciculata Cooke & Massee
Distr.: Widespread in tropical regions.
Notes: On dead wood and bark of various trees (Ellis. Dematiaceous Hyphomycetes. p. 196. 1971).
Host: Pritchardia

Melanomma Nitschke ex Fuckel. Loculoascomycetes, Dothideales

Melanomma alpestre Ellis & Everh.
Distr.: WA.
Host: Arctostaphylos
Melanomma caryophagum (Schwein.) Fairm.
Distr.: Eastern United States.
Host: Carya
Melanomma clypeatum (Sacc. & Pavgi) Berk.
Distr.: HI.
Host: Freycinetia
Melanomma conjunctum (Petr.) L. Holm
Cucurbidothis conjuncta Petr.
Gibberidea conjuncta (Petr.) Arx
Distr.: ID.
Notes: Holm (Svensk Bot. Tidskr. 62:229. 1968).
Host: Thuja
Melanomma cupulata Ellis & Everh.
Distr.: WA(type).
Host: Salix
Melanomma dothideasporum Cooke
Distr.: CA(type).
Host: Mimulus
Melanomma dryadis Johans.
Distr.: Circumboreal.
Notes: On flowers (Holm. Bot. Not. 132:87. 1979).
Host: Dryas
Melanomma minutum Berl.
Distr.: ND; Europe(type).
Host: Salix
Melanomma occidentale (Ellis) Sacc.
Cucurbitaria occidentale Ellis
Distr.: ND, UT.
Host: Chrysothamnus
Melanomma pulvis-pyrius (Pers.:Fr.) Fuckel
Aposphaeria agminalis (Sacc.) Sacc. (anamorph)
Distr.: Temperate North America; Europe.
Hosts: Alnus, Betula, Populus, Salix, Sambucus
Melanomma pulvis-pyrius (Pers.:Fr.) Fuckel f. ***fraxini*** Rehm
Distr.: ND(type).
Host: Fraxinus
Melanomma rhododendri Rehm
Gibberidea kalmiae (Peck) Barr
Leptosphaeria kalmiae Peck
Gibberidea turfosa Syd.
Distr.: Temperate northern hemisphere.
Notes: On Ericaceae.
Hosts: Kalmia, Ledum, Rhododendron
Melanomma sambuci Earle
Distr.: AK, NV.
Host: Sambucus
Melanomma trochus Penz. & Sacc.
Distr.: HI; Java.
Host: Coix

Melanoporia Murrill. Basidiomycetes, Aphyllophorales

Melanoporia nigra (Berk.) Murrill
Polyporus niger Berk.
Nigroporus niger (Berk.) Ryvarden
Poria nigra (Berk.) Cooke
Distr.: Central and eastern United States.
Notes: On wood of hardwoods.
Host: Quercus

Melanopsamma Niessl. Pyrenomycetes, Trichosphaeriales

Melanopsamma improvisa (P. Karst.) Sacc.
Distr.: ND; Europe.
Host: Malus
Melanopsamma pomiformis (Pers.:Fr.) Sacc.
Zignoella exigua (Cooke & Peck) Sacc.
Stachybotrys socia (Sacc.) Sacc. (anamorph)
Distr.: Temperate regions.
Notes: On dead wood.
Hosts: Juncus, Malus
Melanopsamma salicaria (P. Karst.) Sacc.
Distr.: ND; Europe.
Host: Salix
Melanopsamma salicaria (P. Karst.) Sacc. subsp. ***fallax*** Sacc. & Briard
Distr.: ND; France(type).
Host: Salix

Melanopsichium Beck. Basidiomycetes, Ustilaginales

Melanopsichium austro-americanum (Speg.) Beck
Distr.: South America.
Notes: According to Fischer (Manual of the North American smut fungi. p. 107. 1953), reports of this fungus occurring in North America are erroneous and are actually specimens of *M. pennsylvanicum*.
Host: Polygonum
Melanopsichium pennsylvanicum Hirschhorn
Melanopsichium missouriense M.D. Whitehead & Thirumalachar
Melanopsichium pennsylvanicum Hirschhorn var. *besseyanum* Zundel
Melanopsichium pennsylvanicum Hirschhorn var. *caulicola* Zundel
Distr.: Widespread.
Notes: Reports on *Glycine max* were shown to be erroneous (Schiller & Sinclair. Phytopathology 69:605. 1979).
Host: Polygonum

Melanospora Corda. Pyrenomycetes, Sordariales

Notes: Cannon & Hawksworth (Bot. J. Linn. Soc. 84:115. 1982); Doguet (Le Botaniste 39:1. 1955).
Melanospora interna Tehon & G.L. Stout

Distr.: IL(type).
Host: Lycopersicon
Melanospora multiformis (G.H. Martin & V. Charles) Arx & E. Müller
Erostrotheca multiformis G.H. Martin & V. Charles
Distr.: Eastern United States(greenhouses).
Host: Lathyrus
Melanospora zamiae Corda
Melanospora pampeana Speg.
Melanospora townei D. Griffiths
Distr.: Warm-temperate and tropical regions.
Notes: Common on rotting vegetation, possibly parasitic on other fungi.
Hosts: Eichhornia, Salsola, Zea

Melanotaenium de Bary. Basidiomycetes, Ustilaginales

Melanotaenium euphorbiae (L. Lenz) M.D. Whitehead & Thirumalachar
Tilletia euphorbiae L. Lenz
Distr.: Southern United States.
Host: Euphorbia
Melanotaenium oxalidis S. Dietz & G. Fisch.
Distr.: WA(type).
Notes: Dietz & Fischer (Mycologia 62:405. 1970).
Host: Oxalis

Melanotus Pat. Basidiomycetes, Agaricales

Melanotus flavolivens (Berk. & M.A. Curtis) Singer
Crepidotus flavolivens (Berk. & M.A. Curtis) Sacc.
Distr.: Pacific basin.
Notes: Horak (Persoonia 9:323. 1977).
Host: Epipremnum
Melanotus textilis Redhead & P. Kroeger
Distr.: West coast of North America.
Notes: Redhead & Kroeger (Mycologia 76:868. 1984).
Host: Pinus

Melasmia Lév. Coelomycetes

Notes: Sutton (The Coelomycetes. p. 563. 1980).
Melasmia acerina Lév.
Rhytisma acerina (Pers.:Fr.) Fr. (teleomorph)
Distr.: Temperate northern hemisphere.
Host: Acer
Melasmia galii Ellis & Everh., nom. nud.
Distr.: Temperate North America.
Host: Galium
Melasmia hypophylla (Berk. & Ravenel) Sacc.
Leptostroma hypophyllum Berk. & Ravenel
Melasmia gleditschiae Ellis & Everh.
Linospora gleditschiae J.H. Miller & F.A. Wolf (teleomorph)
Distr.: Central and eastern United States.
Host: Gleditsia
Melasmia imitans Peck
Distr.: WY.
Host: Pteridium
Melasmia menziesiae Dearn. & Barth.
Distr.: Western North America.
Hosts: Menziesia, Rhododendron
Melasmia rhododendri Henn. & Shirai
Distr.: AK; Japan.
Host: Rhododendron
Melasmia samaricola H.C. Greene
Distr.: WI.
Host: Ulmus
Melasmia setariae Atk.
Distr.: AL.
Host: Setaria
Melasmia ulmicola Berk. & M.A. Curtis
Distr.: NJ, WI.
Hosts: Ulmus, Zelkova

Meliola Fr. Pyrenomycetes, Meliolales

Notes: Black mildews. Hansford (Beih. Sydowia 2:1. 1961).
Meliola amphitricha Fr.:Fr.
Distr.: Cosmopolitan.
Notes: An ambiguous name (Hansford. Beih. Sydowia 2:765. 1961).
Hosts: Acer, Callicarpa, Cladium, Magnolia, Osmanthus, Persea, Sabal.
Meliola argentina Speg.
Distr.: Subtropical and tropical regions.
Notes: The host, *Baumea meyenii*, has been erroneously reported as *Brumea meyeri* (Hansford. Beih. Sydowia 2:733. 1961).
Hosts: Baumea, Cladium, Cyperus, Gahnia, Machaerina, Rhynchospora, Scirpus
Meliola argentina Speg. var. ***hawaiiensis*** Hansf.
Distr.: HI.
Host: Gahnia
Meliola atkinsonii Hansf.
Distr.: AL(type).
Host: Vaccinium
Meliola bidentata Cooke
Distr.: HI, PR, southern United States; South America.
Hosts: Acacia, Bignonia, Capparis, Cynanchum, Psychotria, Sabal, Zea
Meliola bignoniacearum F. Stevens
Distr.: FL; Central and South America.
Host: Bignonia
Meliola bumeliae Hansf.
Distr.: FL.
Hosts: Bumelia, Sideroxylon
Meliola camelliae (Cattaneo) Sacc.
Fumago camelliae Cattaneo
Distr.: FL; Italy(type).
Notes: Hansford (Beih. Sydowia 2:785. 1961) excluded this species from *Meliola* but did not transfer it elsewhere.
Host: Camellia
Meliola castelae J.A. Stevenson
Distr.: TX.
Host: Castela
Meliola circinans Earle
Distr.: Southeastern United States; West Indies.
Hosts: Cyperus, Rhynchospora
Meliola circinans Earle var. ***rhynchosporae*** Hansf.
Distr.: FL; West Indies.
Host: Rhynchospora
Meliola condaliae J.A. Stevenson
Distr.: TX(type).
Host: Coccocypselum
Meliola cookeana Speg.
Meliola rizalensis Syd.
Distr.: FL; Asia; Sierra Leone.
Hosts: Callicarpa, Lantana
Meliola cyperi Pat.
Meliola argentina Speg. var. *africana* Hansf.
Distr.: FL, HI; Africa.
Hosts: Cladium, Gahnia, Rhynchospora
Meliola dracaenae F. Stevens
Distr.: HI.
Host: Dracaena
Meliola floridensis Hansf.
Distr.: FL(type).
Host: Persea
Meliola furcata Lév.
Distr.: FL; Central and South America; South Africa; The Philippines.
Hosts: Bignonia, Sabal
Meliola garryae Hansf.
Distr.: TX(type).
Host: Garrya
Meliola gouldiae Hansf.
Distr.: HI.
Host: Gouldia
Meliola gregoriana F. Stevens
Distr.: HI.
Host: Dianella
Meliola hawaiiensis F. Stevens
Distr.: HI.
Host: Eugenia
Meliola heteroseta Höhn.

Meliola leptidea Syd.
Meliola leptidea Syd. var. *major* Hansf.
Meliola leptoclada Syd.
Distr.: HI; Africa; The Philippines.
Notes: Hansford (Beih. Sydowia 2:482. 1961).
Host: Vaccinium
Meliola intricata Syd.
Distr.: FL; West Indies; The Philippines.
Hosts: Cladium, Cyperus
Meliola juddiana F. Stevens
Distr.: HI.
Hosts: Freycinetia, Pelea, Pellaea
Meliola kaduae F. Stevens
Distr.: HI.
Hosts: Gouldia, Kadua, Psychotria, Straussia
Meliola kauaiensis F. Stevens
Distr.: HI.
Host: Kadua
Meliola kisubiensis Hansf. var. ***peleicola*** Hansf.
Distr.: HI.
Host: Pelea
Meliola koae F. Stevens
Meliola acaciae-confusae Sawada
Distr.: HI.
Host: Acacia
Meliola krugiodendri Cif.
Distr.: FL, PR; Dominican Republic.
Host: Krugiodendron
Meliola lippiae Maubl.
Distr.: AZ, FL; Sierra Leone(type).
Hosts: Lippia, Phyla
Meliola lobeliae F. Stevens
Distr.: HI.
Hosts: Clermontia, Lobelia
Meliola lyoni F. Stevens
Meliola cookeana Speg. var. *major* Gaillard
Distr.: HI; Jamaica; Brazil.
Hosts: Dodonaea, Pritchardia
Meliola magnoliae F. Stevens
Distr.: Southeastern United States.
Host: Magnolia
Meliola malacotricha Speg.
Meliola hewittiae Rehm
Meliola ipomoeae Rehm
Meliola ipomoeae Earle
Meliola lepistemonis Hansf.
Meliola merremiae Rehm
Distr.: FL; pantropical.
Host: Bonamia
Meliola martiniana Gaillard
Irenopsis martiniana (Gaillard) F. Stevens
Meliola perseae F. Stevens forma *setulifera* Speg.
Meliola setulifera (Speg.) F. Stevens
Distr.: Southern United States; Dominican Republic.
Host: Persea
Meliola mitchellae Cooke
Distr.: FL, SC(questionable).
Host: Mitchella
Meliola moerenhoutiana Mont. in Sagra
Meliola alyxiae F. Stevens
Distr.: HI; Chile; Tahiti.
Notes: Stevens (Ann. Mycol. 26:173. 1928).
Host: Alyxia
Meliola morbosa F. Stevens
Distr.: HI.
Hosts: Cladium, Claoxylon
Meliola nashii Hansf.
Distr.: Southeastern United States.
Host: Magnolia
Meliola nidulans (Schwein.) Cooke
Distr.: Southeastern United States.
Notes: Reports of this species on *Vaccinium* may be *M. niessleana* Winter (Hansford. Beih. Sydowia 2:478. 1961).
Hosts: Cornus, Gaultheria, [Vaccinium]
Meliola niessleana G. Wint.
Meliola ellisii Roum.
Meliola nidulans (Schwein.) Cooke var. *germanica* Rehm
Distr.: Western United States, FL; Dominican Republic; Europe; Formosa.
Hosts: Gaultheria, Vaccinium
Meliola osmanthi Syd.
Distr.: HI; Japan; Taiwan.
Host: Osmanthus
Meliola osmanthi Syd. var. ***hawaiiensis*** Hansf.
Distr.: HI.
Host: Osmanthus
Meliola osmanthicola Hansf.
Distr.: HI.
Host: Osmanthus
Meliola osmanthina Hansf.
Distr.: HI.
Host: Osmanthus
Meliola palmicola G. Wint.
Meliola contigua P. Karst. & Roum.
Distr.: Southern United States; Cuba; Brazil.
Notes: On Palmaceae.
Hosts: Arecaceae, Pritchardia, Sabal, Serenoa, Thrinax
Meliola panici Earle
Meliola chaetochloae F. Stevens
Meliola stenotaphri F. Stevens
Meliola substenospora Höhn.
Distr.: Pantropical.
Host: Andropogon
Meliola peleae F. Stevens
Distr.: HI.
Hosts: Cryptocarya, Pelea, Pellaea
Meliola praetervisa Gaillard
Distr.: HI; Australia; Java.
Host: Cryptocarya
Meliola psychotriae Earle
Distr.: HI; Central and South America; Africa; Asia.
Host: Guettarda
Meliola psychotriae Earle var. ***chiococcae*** Hansf.
Distr.: PR; Barbados.
Host: Chiococca
Meliola recurvipoda Hansf.
Distr.: HI.
Host: Pelea
Meliola sabalidis Sacc., nom. nud.
Distr.: FL.
Host: Sabal
Meliola sandwicensis Ellis & Everh.
Distr.: HI.
Hosts: Gouldia, Kadua
Meliola sandwicensis Ellis & Everh. var. ***gouldiae*** Hansf.
Distr.: HI.
Hosts: Gouldia, Kadua
Meliola sandwicensis Ellis & Everh. var. ***major*** Hansf. ex Hansf.
Distr.: HI.
Host: Gouldia
Meliola scaevolae Syd.
Distr.: HI; Asia.
Host: Scaevola
Meliola scaevolicola (F. Stevens) Hansf.
Irene scaevolicola F. Stevens
Irenopsis scaevolicola (F. Stevens) F. Stevens
Distr.: HI.
Host: Scaevola
Meliola serjaniae F. Stevens
Distr.: TX; Central America; Brazil.
Host: Serjania
Meliola sideroxyli F. Stevens
Distr.: HI.
Notes: This species has sometimes been reported using the orthographic variant, *M. sideroxylonis*.
Host: Sideroxylon
Meliola sideroxylicola Hansf.
Distr.: FL.
Host: Sideroxylon
Meliola straussiae Hansf.

Distr.: HI.
Hosts: Psychotria, Straussia
Meliola subdentata Pat.
Distr.: HI; The Philippines.
Host: Dracaena
Meliola sydowiana F. Stevens & Larson
Distr.: HI; The Philippines.
Host: Metrosideros
Meliola tenuis Berk. & M.A. Curtis
Distr.: Southeastern United States.
Host: Arundinaria
Meliola vaccinii F. Stevens
Distr.: HI.
Host: Vaccinium
Meliola venezuelana Orejuela var. ***floridensis*** Hansf.
Distr.: FL(type).
Host: Pithecellobium
Meliola visci F. Stevens
Distr.: HI.
Hosts: Korthalsella, Viscum

Meliolina Syd. Loculoascomycetes, Dothideales

Notes: The taxonomic position of this genus is uncertain (Arx & Müller. Stud. Mycol. 9:109. 1975).
Meliolina haplochaeta Syd.
Distr.: HI.
Host: Metrosideros
Meliolina mollis (Berk. & Broome) Höhn.
Meliola mollis Berk. & Broome
Distr.: HI.
Host: Metrosideros
Meliolina novae-zealandiae Hansf.
Distr.: HI; New Zealand.
Host: Metrosideros
Meliolina sydowiana F. Stevens
Distr.: HI.
Host: Metrosideros

Melittosporiella Höhn. Discomycetes, Ostropales

Notes: Sherwood-Pike (Sydowia 38:268. 1985).
Melittosporiella macrospora M.A. Sherwood
Distr.: OR.
Notes: On dead wood.
Hosts: Juniperus, Salix

Melittosporium Corda. Loculoascomycetes, Dothideales

Melittosporium hysterinum (Fr.:Fr.) Gill.
Stictis hysterina Fr.:Fr.
Xylogramma hysterinum (Fr.:Fr.) Rehm
Distr.: NY, OR, WA; Europe.
Host: Abies

Meloderma Darker. Discomycetes, Rhytismatales

Meloderma desmazieresii (Duby) Darker
Hypoderma desmazieresii Duby
Hypoderma brachysporum (Rostr.) Tub.
Distr.: Temperate northern hemisphere.
Notes: On needles (Cannon & Minter. Mycol. Pap. 155:87. 1986; Minter & Gibson. C.M.I. Descr. 569. 1978 as *Meloderma desmazierii*).
Host: Pinus

Melogramma Fr. Pyrenomycetes, Diaporthales

Notes: This genus was placed in the Diaporthales by LaFlamme(Sydowia 28:237. 1975) whereas Barr (Mycol. Mem. 7:207. 1978) suggested that is is allied with *Phyllachora*.
Melogramma campylosporum Fr.
Melogramma fuscosporum (Schwein.) Sacc.
Diatrype lateritia Ellis
Hypoxylon myriangioides Berk. & Ravenel
Thyridaria myriangioides (Berk. & Ravenel) Sacc.
Broomella ravenelii Sacc.
Hypocrea ravenelii Berk.
Melogramma vagans De Not.
Cytosporina bulliardii Höhn. (anamorph)
Distr.: Temperate North America; Europe.
Notes: On bark of *Carpinus*, occasionally on other trees (LaFlamme. Sydowia 28:240. 1975).
Host: Carpinus
Melogramma meliae M.A. Curtis in Cooke
Distr.: Southeastern United States.
Host: Melia
Melogramma patens Morg.
Distr.: IN, OH.
Host: Carpinus

Melomastia Nitschke ex Sacc. Pyrenomycetes, Trichosphaeriales

Melomastia shastensis Earle
Distr.: CA.
Host: Abies

Melzericium Hauerslev. Basidiomycetes, Aphyllophorales

Notes: About 2 species, on wood.
Melzericium udicola (Bourd.) Hauerslev
Corticium udicola Bourd.
Distr.: Northwestern United States; Europe.
Notes: On wood.
Hosts: Crataegus, Rosa, Sambucus

Membranosorus Ostenf. & H.E. Petersen. Plasmodiophoromycetes, Plasmodiophorales

Notes: Karling (The Plasmodiophorales. p. 67. 1968).
Membranosorus heterantherae Ostenf. & H.E. Petersen
Distr.: Eastern North America.
Host: Heteranthera

Memnoniella Höhn. Hyphomycetes

Memnoniella echinata (Rivolta) B.T. Galloway
Stachybotrys echinata (Rivolta) G. Sm.
Distr.: Cosmopolitan.
Notes: A major cause of fabric deterioration, especially in the tropics (Jong & Davis. Mycotaxon 3:464. 1976).
Hosts: Alnus, Bambusa, Gossypium, Juncus, Pseudotsuga, Tragopogon
Memnoniella subsimplex (Cooke) Deighton
Stachybotrys subsimplex Cooke
Distr.: Subtropical and tropical regions.
Notes: Ellis (Dematiaceous Hyphomycetes. p. 603. 1971).
Host: Eichhornia

Menispora Pers. Hyphomycetes

Menispora uncinata S.J. Hughes & Kendrick
Distr.: FL; New Zealand.
Notes: On rotting wood and leaves (Sutton. Mycologia 70:791. 1978).
Host: Serenoa

Menisporopsis S.J. Hughes. Hyphomycetes

Menisporopsis profusa Pirozynski & C.S. Hodges
Distr.: SC.
Notes: Ellis (More dematiaceous Hyphomycetes. p. 451. 1976).
Host: Persea

Meria Vuill. Hyphomycetes

Meria laricis Vuill.
Distr.: North America; Europe.
Notes: Needle cast of larch (Funk. Foliar fungi of western trees. p. 74. 1985).
Host: Larix

Meria parkeri M.A. Sherwood, J.K. Stone & Carroll
Rhabdocline parkeri M.A. Sherwood, J.K. Stone & Carroll (teleomorph)
Distr.: OR.
Notes: On galled needles of Douglas fir (Sherwood-Pike et al. Canad. J. Bot. 64:1852. 1986).
Host: Pseudotsuga

Meripilus P. Karst. Basidiomycetes, Aphyllophorales

Notes: Larsen and Lombard (Mycologia 80:619. 1988) provide a key to four species.

Meripilus sumstinei (Murrill) M. Larsen & Lombard
Grifola sumstinei Murrill
Distr.: North America.
Notes: At base of stumps or trees. In the past this species was erroneously referred to the European *Meripilus giganteus* (Pers.:Fr.) P. Karst. See Larsen and Lombard (Mycologia 80:612. 1988) for discussion of this and related species.
Hosts: Pinus, Pseudotsuga, Quercus

Merismodes Earle. Basidiomycetes, Aphyllophorales

Merismodes fasciculatus (Schwein.) Donk in Singer
Cyphella fasciculata (Schwein.) Berk. & M.A. Curtis in Berk.
Cyphella fulva Berk. & Ravenel
Distr.: North America; Europe; Japan.
Notes: On limbs.
Hosts: Alnus, Salix

Merismodes ochraceus (Hoffm.) D. Reid
Solenia ochracea Hoffm.
Distr.: Northern hemisphere.
Hosts: Alnus, Betula, Carya, Lonicera, Salix

Meruliopsis Bondartsev in Parmasto. Basidiomycetes, Aphyllophorales

Byssomerulius Parmasto
Ceraceomerulius (Parmasto) J. Eriksson & Ryvarden
Notes: Ginns (Canad. J. Bot. 54:100. 1976).

Meruliopsis albostramineus (Torrend) Jülich & Stalpers
Merulius albostramineus Torrend
Ceraceomerulius albostramineus (Torrend) Ginns
Merulius armeniacus Bres.
Merulius atropurpureus W.B. Cooke
Distr.: Widespread.
Notes: On wood.
Hosts: Abies, Acer, Pinus, Populus, Tsuga

Meruliopsis ambiguus (Berk.) Ginns
Merulius ambiguus Berk.
Byssomerulius ambiguus (Berk.) R.L. Gilbertson & Budington
Merulius succineus C.G. Lloyd
Distr.: Temperate northern hemisphere.
Notes: On wood.
Hosts: Larix, Pinus

Meruliopsis bellus (Berk. & M.A. Curtis) Ginns
Merulius bellus Berk. & M.A. Curtis
Distr.: North America.
Notes: On wood. This name may be a synonym of *M. hirtellus*; specimens of *Ceraceomyces serpens* have been misdetermined as *M. bellus* (Ginns. Canad. J. Bot. 54:122. 1976).
Hosts: Abies, Cercocarpus, Pinus, Tsuga

Meruliopsis corium (Fr.:Fr.) Ginns
Byssomerulius corium (Fr.:Fr.) Parmasto
Merulius corium (Fr.:Fr.) Fr.
Byssomerulius confluens (Schwein.) Lindsey & R.L. Gilbertson
Merulius confluens Schwein.
Distr.: Widespread.
Notes: On wood.
Hosts: Abies, Acer, Alnus, Baccharis, Betula, Carya, Celtis, Citrus, Cytisus, Encelia, Ilex, Larix, Ligustrum, Liquidambar, Nerium, Nyssa, Ostrya, Parthenocissus, Picea, Pinus, Populus, Prosopis, Prunus, Quercus, Rhamnus, Salix, Simmondsia, Viburnum

Meruliopsis hirtellus (Burt) Ginns
Merulius hirtellus Burt
Byssomerulius hirtellus (Burt) Parmasto
Byssomerulius armeniacus Parmasto
Distr.: Northern hemisphere.
Notes: On wood.
Hosts: Abies, Acer, Picea, Pinus, Populus, Pseudotsuga

Meruliopsis taxicola (Pers.) Bondartsev in Parmasto
Caloporus taxicola (Pers.) Ryvarden
Gloeoporus taxicola (Pers.) R.L. Gilbertson & Ryvarden
Merulius taxicola (Pers.) Duby
Poria taxicola Pers.) Bres.
Merulius ravenelii Berk.
Poria rufa (Schrad.) Cooke
Distr.: Widespread.
Notes: On conifers.
Hosts: Abies, Chamaecyparis, Picea, Pinus

Meruliporia Murrill. Basidiomycetes, Aphyllophorales

Meruliporia incrassata (Berk. & M.A. Curtis) Murrill
Merulius incrassatus Berk. & M.A. Curtis
Poria incrassata (Berk. & M.A. Curtis) Burt
Serpula incrassata (Berk. & M.A. Curtis) Donk
Distr.: Widespread.
Notes: On wood, usually of conifers; decays wood in service.
Hosts: Magnolia, Pinus, Pseudotsuga, Quercus, Robinia, Sequoia, Sequoiadendron, Taxodium, Tsuga

Metacoleroa Petr. Loculoascomycetes, Dothideales

Metacoleroa dickiei (Berk. & Broome) Petr.
Gibbera dickiei (Berk. & Broome) Arx
Venturia dickiei (Berk. & Broome) Ces. & De Not.
Distr.: Northern North America.
Notes: Leaf spot (Corlett & Kokko. Fungi Canadenses 114. 1978).
Host: Linnaea

Metasphaeria Sacc. Loculoascomycetes, Dothideales

Notes: This genus is considered a synonym of *Leptosphaeria* (Arx & Müller. Stud. Mycol. 9:75. 1975), but not all names have been redisposed.

Metasphaeria ambrosiaecola Atk.
Distr.: AL, GA, ND.
Hosts: Ambrosia, Iva

Metasphaeria ambrosiaecola Atk. var. ***praetans*** Rehm, nom. nud.
Distr.: ND.
Notes: The place of publication of this name could not be found.
Host: Ambrosia

Metasphaeria anisometra (Cooke & Harkn.) Sacc.
Sphaeria anisometra Cooke & Harkn.
Distr.: CA, ND.
Notes: On branches of a wide variety of plants.
Hosts: Baccharis, Eucalyptus, Galium, Holodiscus, Malacothamnus, Mimulus, Nicotiana, Rhus, Solanum, Suaeda

Metasphaeria annae Oudem.
Distr.: AK; Soviet Union.
Host: Ranunculus

Metasphaeria apiosporoides Petr.
Distr.: FL.
Host: Sabal

Metasphaeria asparagi Tehon & G.L. Stout
Distr.: IL.

Host: Asparagus
Metasphaeria bifoveolata Speg.
Distr.: FL; Costa Rica.
Host: Psychotria
Metasphaeria cavernosa (Ellis & Everh.) Sacc.
Sphaeria cavernosa Ellis & Everh.
Distr.: GA.
Host: Taxodium
Metasphaeria hawaiiensis F. Stevens & P.A. Young
Distr.: HI.
Host: Metrosideros
Metasphaeria ilicis Ellis & Everh.
Distr.: CA.
Host: Ilex
Metasphaeria infuscans Ellis & Everh.
Distr.: Southern United States.
Host: Andropogon
Metasphaeria palmetta (Cooke) Sacc.
Sphaeria palmetta Cooke
Distr.: FL, GA.
Host: Sabal
Metasphaeria panicorum (Cooke) Sacc.
Sphaerella panicum Cooke
Distr.: GA, NH.
Notes: The specific epithet is an orthographic correction of the basionym.
Hosts: Panicum, Sorghum
Metasphaeria punctulata Ellis & Everh.
Distr.: Southern United States.
Host: Panicum
Metasphaeria puyae Speg.
Distr.: CA; Chile.
Host: Puya
Metasphaeria rimularum (Cooke) Sacc.
Sphaeria rimularum Cooke
Distr.: FL, GA.
Host: Arundinaria
Metasphaeria sassafrasicola Tehon & G.L. Stout
Distr.: IL.
Host: Sassafras
Metasphaeria stenotheca (Ellis & Everh.) Ellis & Everh. in Langl.
Sphaeria stenotheca Ellis & Everh.
Distr.: LA.
Host: Panicum
Metasphaeria subseriata Ellis & Everh.
Distr.: KS.
Host: Panicum
Metasphaeria washingtoniae Earle
Distr.: CA.
Host: Washingtonia

Metatrichia Ing. Myxomycetes

Metatrichia vesparium (Batsch) Nann.-Brem.
Hemitrichia vesparium (Batsch) T. Macbr.
Distr.: North temperate regions, less common in tropical regions and the southern hemisphere.
Host: Quercus

Metschnikowia T. Kamienski. Hemiascomycetes, Endomycetales

Metschnikowia pulcherrima P.I. Pitt & M.W. Miller
Distr.: North America; Europe.
Notes: Associated with flowers, fruits, and nectar-feeding insects.
Hosts: Medicago, Vitis

Microascus Zukal. Pyrenomycetes, Microascales

Notes: Arx (Persoonia 8:191. 1975); Morton & Smith (Mycol. Pap. 86:25. 1963). The anamorphs belongs in *Scopulariopsis* and *Wardomyces*.
Microascus cinereus (Emile-Weil & L. Gaudin) Curzi
Scopulariopsis cinereus Emile-Weil & L. Gaudin (anamorph)
Distr.: Temperate regions.
Notes: Often isolated from soil, decaying matter, flowers and seeds.
Hosts: Helianthus, Zea
Microascus cirrosus Curzi
Distr.: Temperate regions.
Notes: Isolated from soil and organic debris, most recently from seeds.
Hosts: Helianthus, Zea
Microascus desmosporus (Lechmere) Curzi
Distr.: IA; Europe.
Notes: Morton & Smith (Mycol. Pap. 85:29. 1963).
Host: Zea
Microascus longirostris Zukal
Microascus variabilis Massee & E.S. Salmon
Scopulariopsis sp. (anamorph)
Distr.: Temperate regions.
Notes: Isolated from soil and stored grains (Corlett & Neish. Fungi Canadenses 180. 1980).
Host: Zea
Microascus senegalensis Arx
Distr.: GA; Senegal.
Notes: Originally isolated from mangrove soil, now also known from sunflower achenes.
Host: Helianthus
Microascus trigonosporus C.W. Emmons & B.O. Dodge
Scopulariopsis sp. (anamorph)
Distr.: Cosmopolitan.
Notes: Isolated from various plant and animal debris.
Hosts: Carya, Glycine, Helianthus, Triticum

Microdiplodia Allesch. Coelomycetes

Notes: Zambettakis (Bull. Soc. Mycol. France 70:243. 1954).
Microdiplodia bonnayae Petr.
Distr.: CA.
Host: Boerhavia
Microdiplodia ceanothi Dearn. & House
Distr.: NY.
Host: Ceanothus
Microdiplodia conigena Allesch.
Distr.: CA.
Host: Pinus
Microdiplodia cupressina (Cooke) Tassi
Diplodia cupressina Cooke
Distr.: CA, GA(type), WA.
Hosts: Juniperus, Sequoiadendron
Microdiplodia diervillae Fairm.
Distr.: NY.
Host: Diervilla
Microdiplodia hosackiae (Cooke & Harkn.) Bonar
Phoma hosackiae Cooke & Harkn.
Distr.: CA.
Host: Lotus
Microdiplodia ilicigena Fairm.
Distr.: NY.
Host: Ilex
Microdiplodia ivicola (Ellis & Everh.) Sacc. & D. Sacc.
Diplodia ivicola Ellis & Everh.
Distr.: SD.
Host: Iva
Microdiplodia laurina Grove
Diplodia laurina Cooke & Harkn. non Sacc., hom. illeg.
Diplodia harknessii Sacc.
Distr.: CA; Europe.
Host: Umbellularia
Microdiplodia leucelenes Fairm.
Distr.: NM.
Host: Aster
Microdiplodia microsporella Allesch.
Diplodia microsporella Sacc.
Diplodia microspora Sacc.
Distr.: United States; Great Britain.
Hosts: Eucalyptus, Philadelphus
Microdiplodia mimuli Fairm.
Distr.: CA.

Host: Mimulus

Microdiplodia minuta (Ellis & Tracy) Zambettakis
Diplodia minuta Ellis & Tracy
Distr.: MS.
Host: Campsis

Microdiplodia paupercula (Berk. & Broome) Zambettakis
Diplodia paupercula Berk. & Broome
Distr.: Eastern United States.
Host: Sambucus

Microdiplodia perpusila (Desmaz.) Allesch.
Microdiplodia subtecta Allesch.
Distr.: Europe; distribution elsewhere is uncertain.
Host: Acer

Microdiplodia ramonae Fairm.
Distr.: CA.
Host: Salvia

Microdiplodia spiraeicola (Ellis & Everh.) Tassi
Diplodia spiraeicola Ellis & Everh. in Fairm.
Distr.: NY; Europe.
Host: Spiraea

Microdiplodia viciae Buchwald
Distr.: Central and western United States; Faeröe Islands.
Host: Vicia

Microdochium Syd. Hyphomycetes

Notes: Species previously placed in *Fusarium* section *Arachnites* are now included here based on conidiogenesis. Arx (Sydowia 34:30. 1981).

Microdochium bolleyi (R. Sprague) De Hoog & Hermanides-Nijhof
Gloeosporium bolleyi R. Sprague
Aureobasidium bolleyi (R. Sprague) Arx
Distr.: Central and western North America.
Notes: Root necrosis and decay of grasses. Recorded on at least 120 species of grasses. DeHoog & Hermanides-Nijhof (Stud. Mycol. 15:194. 1977).
Hosts: Agrohordeum, Agropyron, Agrostis, Andropogon, Aristida, Avena, Bouteloua, Bromus, Buchloe, Calamagrostis, Calamovilfa, Cenchrus, Cynodon, Dactylis, Dichanthelium, Distichlis, Echinochloa, Elymus, Eragrostis, Festuca, Hilaria, Hordeum, Koeleria, Leymus, Melica, Muhlenbergia, Oryzopsis, Panicum, Phalaris, Phleum, Poa, Schedonnardus, Schizachyrium, Secale, Setaria, Sitanion, Sorghastrum, Sorghum, Sporobolus, Stipa, Triticum, Vulpia, Zea

Microdochium caespitosum Sutton, Pirozynski & Deighton
Distr.: HI; Tanzania(type).
Notes: On dead leaves (Sutton et al. Canad. J. Bot. 50:1904. 1972).
Host: Eucalyptus

Microdochium cylindricum Sutton & C.S. Hodges
Distr.: HI; Brazil(type).
Notes: On dead leaves (Sutton & Hodges. Nova Hedwigia 27:216. 1976).
Host: Eucalyptus

Microdochium dimerum (Penz. in Sacc.) Arx
Fusarium aquaeductuum (Radlk. & Rabenh.) Lagerh. var. *dimerum* (Penz.) Raillo
Fusarium dimerum Penz. in Sacc.
Distr.: Cosmopolitan.
Notes: A common soil fungus. Isolated from a wide variety of plant materials. Associated with storage rots of vegetables.
Hosts: Acacia, Beta, Solanum

Microdochium falcatum Sutton & C.S. Hodges
Distr.: HI; Brazil(type).
Notes: On dead leaves (Sutton & Hodges. Nova Hedwigia 27:217. 1976).
Host: Eucalyptus

Microdochium griseum Sutton, Pirozynski & Deighton
Distr.: HI; Tanzania(type).
Notes: On dead leaves (Sutton et al. Canad. J. Bot. 50:1904. 1972).
Host: Eucalyptus

Microdochium lunatum (Ellis & Everh.) Arx
Gloeosporium lunatum Ellis & Everh.
Fusarium lunatum (Ellis & Everh.) Arx
Monographella opuntiae (Ellis & Everh.) Arx (teleomorph)
Distr.: Central and eastern United States; West Indies; Canary Islands.
Notes: Brown spotting.
Host: Opuntia

Microdochium nivale (Fr.) Samuels & I.C. Hallett
Fusarium nivale Ces. ex Berl. & Voglino
Gerlachia nivalis (Ces. ex Berl. & Voglino) W. Gams & E. Müller
Fusarium nivale (Fr.) Sorauer
Monographella nivalis (Schaffnit) E. Müller (teleomorph)
Distr.: Cosmopolitan.
Notes: Cereal and turf diseases (Booth. C.M.I. Descr. 309. 1971).
Hosts: Agrohordeum, Agropyron, Agrostis, Avena, Bromus, Calamagrostis, Carex, Carya, Cinna, Cynodon, Dactylis, Danthonia, Elymus, Festuca, Helianthus, Holcus, Hordeum, Hystrix, Koeleria, Lolium, Metrosideros, Oryzopsis, Phalaris, Phleum, Poa, Secale, Setaria, Sitanion, Stipa, Stiporyzopsis, Triticum, Vulpia, Zea

Microdochium oryzae (Hashioka & Yokogi) Samuels & I.C. Hallett
Rhynchosporium oryzae Hashioka & Yokogi
Distr.: Worldwide in rice-growing regions.
Notes: Leaf-scald of rice (Sivanesan. C.M.I. Descr. 729. 1982 as *Monographella albescens*).
Host: Oryza

Microdochium panattonianum (Berl.) Sutton, Galea & Price in Galea, Price, & Sutton
Marssonia panattoniana Berl.
Marssonina panattoniana (Berl.) Magnus
Marssonia perforans Ellis & Everh. in A. Selby
Distr.: North America; Europe; Australia.
Notes: Galea et al. (Trans. Brit. Mycol. Soc. 86:620. 1986).
Hosts: Cichorium, Lactuca

Microdochium tabacinum (Van Beyma) Arx
Cephalosporium tabacinum Van Beyma
Fusarium tabacinum (Van Beyma) W. Gams
Monographella cucumerina (Lindfors) Arx (teleomorph)
Distr.: Cosmopolitan.
Notes: Common in soil, often associated with decaying plant material.
Host: Pseudotsuga

Microdothella Syd. & P. Syd. Loculoascomycetes, Dothideales

Microdothella ramularis (Ellis) Theiss. & Syd.
Asterina ramularis Ellis
Distr.: Eastern United States.
Notes: On twigs.
Host: Lindera

Microgloeum Petr. Coelomycetes

Microgloeum pruni Petr. in Syd. & Petr.
Blumeriella jaapii (Rehm) Arx (teleomorph)
Phloeosporella padi (Lib.) Arx (synanamorph)
Distr.: North America; temperate Europe.
Notes: Leaf lesions (shot-hole) (Nag Raj. Canad. J. Bot. 56:696. 1978; Sutton. The Coelomycetes. p. 224. 1980).
Host: Prunus

Micromphale Nees. Basidiomycetes, Agaricales

Micromphale arbuticola Desjardin
Distr.: CA.
Host: Arbutus

Micromphale foetidum (Sowerby:Fr.) Singer
Marasmius foetidus (Sowerby:Fr.) Fr.
Distr.: Temperate northern hemisphere.
Notes: On dead wood of hardwoods.
Host: Acer

Micromphale perforans (Hoffm.:Fr.) Singer
Marasmius perforans (Hoffm.:Fr.) Fr.
Distr.: Temperate northern hemisphere.
Notes: On conifer needles and debris.
Hosts: Pinus, Tsuga

Micromphale sequoiae Desjardin
Distr.: CA.

Host: Sequoia

Micropeltis Mont. Loculoascomycetes, Dothideales

Micropeltis alabamensis Earle
Distr.: AL(type).
Notes: Black leaf spot.
Host: Magnolia

Micropera Lév. Coelomycetes

Notes: This genus is a later homonym for which *Foveostroma* was established (DiCosmo. Canad. J. Bot. 56:1665. 1978), but not all the names have been redisposed.
Micropera ampelina Sacc. & Fairm.
Distr.: NY.
Host: Vitis
Micropera fraxini Ellis & Everh.
Distr.: OH; Canada (Ontario).
Host: Fraxinus
Micropera spuria (Fr.) Höhn.
Ceratostoma spurium Fr.
Sphaeronema spurium (Fr.) Sacc.
Dermea prunastri (Pers.:Fr.) Fr. (teleomorph)
Distr.: North America; Europe.
Notes: Groves (Mycologia 38:406. 1946).
Host: Prunus
Micropera tenella Sacc. & Ellis
Distr.: NJ.
Host: Thuja
Micropera vaccinii Ellis & Everh.
Distr.: IN.
Host: Vaccinium

Microporellus Murrill. Basidiomycetes, Aphyllophorales

Microporellus obovatus (Jungh.) Ryvarden
Polyporus obovatus Jungh.
Polyporus mutabilis Berk. & M.A. Curtis
Distr.: Subtropical and tropical regions.
Hosts: Diospyros, Myrica, Nyssa, Pinus, Quercus

Microsphaera Lév. Pyrenomycetes, Erysiphales

Notes: Powdery mildews (Braun. Beih. Nova Hedwigia 89:1. 1987).
Microsphaera calocladophora Atk.
Microsphaera alni (Wallr.) G. Wint. var. *calocladophora* (Atk.) E.S. Salmon
Distr.: Southern United States.
Host: Quercus
Microsphaera cinnamomi (Olive) U. Braun
Microsphaera alni (Wallr.) G. Wint. var. *cinnamomi* Olive
Distr.: LA(type).
Notes: Braun (Mycotaxon 15:132. 1982).
Host: Cinnamomum
Microsphaera diffusa Cooke & Peck
Distr.: Temperate regions.
Notes: Powdery mildew of Fabaceae. Previously referred to as *Erysiphe polygoni* (Paxon & Rogers. Mycologia 66:894. 1974).
Hosts: Apios, Asclepias, Bauhinia, Crotalaria, Desmodium, Glycine, Glycyrrhiza, Kummerowia, Lathyrus, Lespedeza, Lupinus, Lycium, Phaseolus, Psoralea, Robinia, Symphoricarpos, Vicia
Microsphaera diffusa Cooke & Peck f. ***elongata*** U. Braun
Distr.: North America.
Notes: Braun (Mycotaxon 20:492. 1984).
Host: Psoralea
Microsphaera elevata Burrill
Distr.: Central and eastern United States.
Notes: On Bignoniaceae.
Host: Catalpa
Microsphaera euonymi (DC.) Sacc.
Distr.: Temperate regions.
Host: Euonymus
Microsphaera euphorbiae Berk. & M.A. Curtis
Distr.: Eastern North America; Central and South America.
Notes: In the past, this species was thought to be based on *Erysiphe euphorbiae* Peck, but the two names represent distinct species (Braun. Mycotaxon 20:498. 1984).
Hosts: Astragalus, Euphorbia, Hibiscus, Macrotyloma, Phaseolus
Microsphaera extensa Cooke & Peck
Microsphaera alni Schwein. var. *extensa* (Cooke & Peck) E.S. Salmon
Microsphaera quercina (Schwein.) Burrill var. *extensa* (Cooke & Peck) Atk.
Microsphaera penicillata (Wallr.:Fr.) Lév. var. *extensa* (Cooke & Peck) W.B. Cooke
Distr.: Temperate North America.
Hosts: Platanus, Quercus
Microsphaera friesii Lév.
Microsphaera polonica Siemaszko
Distr.: North America; Europe.
Notes: Powdery mildew of hydrangea (Spencer. The powdery mildews. p. 434. 1978).
Host: Hydrangea
Microsphaera grossulariae (Wallr.) Lév.
Distr.: Temperate regions.
Notes: Powdery mildew of gooseberry (Sivanesan. C.M.I. Descr. 252. 1970).
Hosts: Ribes, Sambucus

Pseudomerulius Jülich. Basidiomycetes, Aphyllophorales

Pseudomerulius curtisii (Berk.) Redhead & Ginns
Paxillus curtisii Berk. in Berk. & M.A. Curtis
Paxillus corrugatus Atk.
Distr.: Eastern North America.
Notes: On dead wood.
Hosts: Pinus, Tristania, Tsuga

Microsphaera Lév. Pyrenomycetes, Erysiphales

Notes: Powdery mildews (Braun. Beih. Nova Hedwigia 89:1. 1987).
Microsphaera ludens (E.S. Salmon) S. Blumer
Microsphaera alni (Wallr.) G. Wint. var. *ludens* E.S. Salmon
Distr.: North America.
Notes: On Fabaceae.
Hosts: Desmodium, Vicia
Microsphaera magnifica U. Braun
Distr.: WA, eastern United States; Japan.
Host: Magnolia
Microsphaera neglecta U. Braun
Distr.: Central United States.
Notes: On Ulmaceae.
Host: Ulmus
Microsphaera neomexicana U. Braun
Distr.: AZ.
Notes: On Oleaceae.
Host: Forestiera
Microsphaera penicillata (Wallr.: Fr.) Lév.
Erysiphe penicillata Wallr.:Fr.) Fr.
Microsphaera alni (Wallr.:Fr.) G. Wint.
Distr.: Temperate regions.
Notes: Powdery mildew of alder and lilac (Mukerji. C.M.I. Descr. 183. 1968).
Hosts: Acer, Alnus, Apios, Baptisia, Betula, Campsis, Carpinus, Carya, Castanea, Castanopsis, Catalpa, Ceanothus, Celastrus, Cephalanthus, Cornus, Corylus, Diervilla, Euonymus, Fagus, Forestiera, Fraxinus, Gaultheria, Gaylussacia, Gleditsia, Hydrangea, Ilex, Juglans, Kalmia, Lathyrus, Ledum, Leucothoe, Ligustrum, Lithocarpus, Lobularia, Lonicera, Magnolia, Menispermum, Nemopanthus, Ostrya, Platanus, Quercus, Rhamnus, Rhododendron, Sambucus, Senna, Sophora, Spiraea, Symphoricarpos, Syringa, Tilia, Trifolium, Ulmus, Vaccinium, Viburnum, Vicia, Vigna
Microsphaera pusilla U. Braun

Distr.: Central and eastern United States.
Notes: On Celastraceae.
Host: Euonymus
Microsphaera quercina (Schwein.) Burrill
Erysiphe quercina Schwein.
Distr.: Cosmopolitan.
Notes: According to Braun (Beih. Nova Hedwigia 89. 1987), this name is ambiguous and has been used for the following species: *Microsphaera abbreviata, M. alphitoides, M. calocladophora,* and *M. extensa.*
Host: Quercus
Microsphaera ravenelii Berk.
Distr.: Central and eastern United States.
Notes: On Fabaceae (Braun. Nova Hedwigia 39:239. 1984).
Hosts: Gleditsia, Lathyrus, Vicia
Microsphaera russellii G.W. Clinton in Peck
Oidium oxalidis McAlpine (anamorph)
Distr.: Eastern North America. Introduced into various parts of the world
Host: Oxalis
Microsphaera sydowiana U. Braun
Distr.: WA.
Host: Ceanothus
Microsphaera symphoricarpi E. Howe
Distr.: North America.
Notes: Braun (Nova Hedwigia 39:239. 1984).
Host: Symphoricarpos
Microsphaera syringae (Schwein.) Magnus
Microsphaera fraxini U. Braun
Distr.: Temperate North America.
Host: Fraxinus
Microsphaera tortilis (Wallr.:Fr.) E.O. Speer
Erysiphe tortilis Wallr.:Fr.
Erysiphe corni Duby
Distr.: Temperate northern hemisphere.
Host: Cornus
Microsphaera trifolii (Grev.) U. Braun
Erysiphe trifolii Grev.
Microsphaera caulicola Petr.
Erysiphe martii S. Blumer
Distr.: Cosmopolitan.
Notes: Powdery mildew of Fabaceae (Kapoor. C.M.I. Descr. 156. 1967).
Host: Trifolium
Microsphaera vaccinii (Schwein.) Cooke & Peck
Erysiphe vaccinii Schwein.
Microsphaera alni (Wallr.) E.S. Salmon var. *vaccinii* (Schwein.) E.S. Salmon
Microsphaera penicillata (Wallr.:Fr.) Lév. var. *vaccinii* (Schwein.) W.B. Cooke
Distr.: Temperate North America.
Notes: Powdery mildew of Ericaceae (Parmelee. Canad. J. Bot. 55:1958. 1977).
Hosts: Andromeda, [Catalpa], Epigaea, Gaylussacia, Kalmia, Ledum, Lyonia, Menziesia, Rhododendron, Vaccinium, [Vicia]

Microsphaeropsis Höhn. Coelomycetes

Notes: Many species described in *Coniothyrium* have been placed in this genus (Sutton. Mycol. Pap. 123:32. 1971).
Microsphaeropsis callista (Syd.) Sutton
Coniothyrium callistum Syd.
Distr.: CA; Australia.
Notes: Sutton (The Coelomycetes. p. 423. 1980).
Host: Eucalyptus
Microsphaeropsis clidemiae Sutton
Distr.: CA(type).
Host: Clidemia
Microsphaeropsis concentrica (Desmaz.) Morgan-Jones
Coniothyrium concentricum (Desmaz.) Sacc.
Coniothyrium hysterioideum P. Karst. & Har.
Distr.: North America; Europe.
Notes: Morgan-Jones & White (Mycotaxon 30:177. 1987) provided a description, discussion and illustrations.
Hosts: Agave, Aloe, Dasylirion, Furcraea, Yucca
Microsphaeropsis olivacea (Bonord.) Höhn.
Coniothyrium olivaceum Bonord. in Fuckel
Distr.: Cosmopolitan.
Notes: Additional hosts are listed in Sutton (The Coelomycetes. p. 425. 1980).
Hosts: Atriplex, Magnolia, Prunus, Rubus, Sambucus

Microstroma Niessl. Basidiomycetes, Exobasidiales

Notes: Formerly considered a genus of Hyphomycetes, the "conidiophores" are actually basidia(Arx et al., Mycologia 74:285. 1982; Blanz, Z. Mykol. 44:91. 1978; Oberwinkler, Z. Mykol. 44:13. 1978).
Microstroma album (Desmaz.) Sacc.
Fusisporum album Desmaz.
Distr.: Temperate regions.
Notes: On living leaves (Arx. Biblioth. Mycol. 24:79. 1970).
Host: Quercus
Microstroma juglandis (Berenger) Sacc.
Exobasidium juglandis (Berenger) Pat.
Microstroma brachysporum (Peck) Vestergr.
Distr.: Temperate northern hemisphere.
Notes: Parasitic on leaves.
Hosts: Carya, Juglans
Microstroma juglandis (Berenger) Sacc. var. ***robustum*** Higgins
Distr.: Central and southeastern United States.
Notes: Catkin blight.
Host: Carya

Microthecium Corda. Pyrenomycetes, Sordariales

Notes: Cannon & Hawksworth (Bot. J. Linn. Soc. 84:118. 1982) considered *Microthecium* to be a synonym of *Melanospora.*
Microthecium levitum Udagawa & Cain
Distr.: RI; Japan.
Notes: Udagawa & Cain (Canad. J. Bot. 47:1915. 1969) described this species, a non-ostiolate relative of *Melanospora.* Cannon & Hawksworth (Bot. J. Linn. Soc. 84:144. 1982) suggested this species belongs in *Sphaerodes.*
Host: Spartina

Microthyriella Höhn. Loculoascomycetes, Dothideales

Notes: According to Müller & Arx (Beitr. Kryptogamenfl. Schweiz 11(2):195. 1962), this genus is considered a synonym of *Schizothyrium*, but not all names have been redisposed.
Microthyriella celastri (Ellis & Kellerm.) Theiss.
Asterina celastri Ellis & Kellerm.
Asteridium celastri (Ellis & Kellerm.) Sacc.
Distr.: KS, ME, MI.
Host: Celastrus
Microthyriella cuticulosa (Cooke) Höhn.
Asterina cuticulosa Cooke
Distr.: Southeastern United States.
Host: Ilex
Microthyriella hibisci F. Stevens
Distr.: HI.
Host: Hibiscus

Microthyrium Desmaz. Loculoascomycetes, Dothideales

Notes: Ellis (Trans. Brit. Mycol. Soc. 67:381. 1976).
Microthyrium harrimanii Sacc.
Distr.: AK.
Notes: Stevens & Ryan (Illinois Biol. Monogr. 17:18. 1939).
Host: Tsuga
Microthyrium michelianum Tognini
Distr.: FL; Italy.
Host: Castanea
Microthyrium microscopicum Desmaz.
Microthyrium moravicum Petr.

Microthyrium pinophyllum (Höhn.) Petr.
Distr.: Cosmopolitan.
Notes: On dead leaves of Fagaceae and numerous other hosts.
Hosts: Arabis, Carex, Fraxinus, Magnolia, Pinus, Quercus
Microthyrium mitchellae **Cooke**
Distr.: FL.
Host: Mitchella
Microthyrium rubicola **H.C. Greene**
Distr.: WI.
Host: Rubus
Microthyrium thujae **Dearn.**
Distr.: CA, OR.
Notes: On Cupressaceae.
Hosts: Chamaecyparis, Thuja

Microxiphium (Harvey ex Berk. & Desmaz.) Thuem. Hyphomycetes

Notes: *Microxyphium* is an alternate spelling (Hughes. Mycologia 68:791. 1976).
Microxiphium fagi **(Pers.) S.J. Hughes**
Fumago fagi Pers.
Capnodium footii Harvey ex Berk. & Desmaz., nom. illeg.
Microxiphium footii (Harvey ex Berk. & Desmaz.) Thuem.
Dennisiella babingtonii (Berk.) Batista & Cif. (teleomorph)
Distr.: MS; Canada; Europe.
Host: Illicium

Milesina Magnus. Basidiomycetes, Uredinales

Milesia F.B. White
Notes: *Milesia* is based on a *Uredo* (anamorphic) stage of *Milesina dieteliana* Magnus and is therefore an illegitimate name, but not all names have been redisposed.
Milesina darkeri **(Faull) Hiratsuka**
Milesia darkeri Faull
Distr.: Pacific Northwest.
Notes: Heteroecious; aecial host - unknown; telial host - *Cryptogramma crispa* var. *acrostichoides*.
Host: Cryptogramma
Milesina dilatata **(Faull) Hiratsuka**
Milesia dilatata Faull
Distr.: Pacific Northwest.
Notes: Heteroecious; aecial host - unknown; telial host - *Dryopteris austriaca*.
Host: Dryopteris
Milesina fructuosa **(Faull) Hiratsuka**
Milesia fructuosa Faull
Distr.: Northeastern North America.
Notes: Heteroecious; aecial host - *Abies* spp.; telial host - *Dryopteris* spp.
Hosts: Abies, Dryopteris
Milesina laeviuscula **(Dietel) Hiratsuka**
Milesia laeviuscula (Dietel) Faull
Distr.: Western North America.
Notes: Heteroecious; aecial host - *Abies grandis*; telial host - *Polypodium*.
Hosts: Phanerophlebia, Polypodium
Milesina marginalis **(Faull & A.M. Watson) Faull & A.M. Watson ex Hiratsuka**
Milesia marginalis Faull & A.M. Watson in Faull
Distr.: Northeastern United States and adjacent Canada.
Notes: Heteroecious; aecial host - *Abies balsamea*; telial host - *Dryopteris marginalis*.
Hosts: Abies, Dryopteris
Milesina pycnograndis **(Arth.) Hiratsuka**
Milesia pycnograndis Arth.
Milesia polypodophila (H.P. Bell) Faull
Milesina polypodophila Hiratsuka
Uredinopsis polypodophila H.P. Bell
Distr.: Eastern North America.
Notes: Heteroecious; aecial host - *Abies* spp. telial host - *Polypodium virginianum*.
Hosts: Abies, Polypodium
Milesina vogesiaca **(Faull) Hiratsuka**
Milesia vogesiaca Faull
Distr.: Western North America; northern Africa; Europe; Japan.
Notes: Heteroecious; aecial host - *Abies* spp. (not known in North America); telial host - Polypodiaceae.
Host: Polystichum
Milesina winelandii **Hiratsuka**
Milesia polystichi Wineland ex Faull
Distr.: Western North America.
Notes: Heteroecious; aecial host - unknown; telial host - *Polystichum munitum*.
Host: Polystichum

Mitrula Fr. Discomycetes, Helotiales

Mitrula paludosa **Fr.**
Distr.: Reported from GA.
Notes: According to Redhead (Canad. J. Bot. 55:320. 1977), this species only occurs in Europe, thus these reports are probably erroneous.
Hosts: Pinus, Quercus

Miuraea K. Hara. Hyphomycetes

Miuraea persica **(Sacc.) K. Hara**
Cercospora persica Sacc.
Cercosporella persica (Sacc.) Sacc.
Fusarium persicum (Sacc.) Atk.
Mycosphaerella pruni-persicae Deighton (teleomorph)
Distr.: Warm temperate regions of the northern hemisphere.
Notes: Frosty mildew of peaches.
Host: Prunus

Mixia C.L. Kramer. Hemiascomycetes, Taphrinales

Mixia osmundae **(Nishida) C.L. Kramer**
Taphrina osmundae Nishida
Gloeosporium osmundae Ellis & Everh.
Taphrina higginsii Mix
Distr.: GA, MI; Japan.
Notes: Mix (Mycologia 39:71. 1947); Kramer (Mycologia 50:916. 1958).
Host: Osmunda

Moellerodiscus Henn. Discomycetes, Helotiales

Moellerodiscus lentus **(Berk. & Broome) Dumont**
Helotium lentum Berk. & Broome
Ciboriopsis simulata (Ellis) Dennis
Peziza simulata Ellis
Distr.: NJ(type); tropical region.
Notes: On decaying parts of dicotyledonous plants (Dumont. Mycologia 68:245. 1976).
Host: Acer
Moellerodiscus tenuistipes **(J. Schröt.) Dumont**
Ciboria tenuistipes J. Schröt.
Ciboriopsis tenuistipes (J. Schröt.) J.A. Palmer
Ciboriopsis bramleyi Dennis
Distr.: WA(questionable); Europe(type).
Notes: Dumont (Mycologia 68:260. 1976) did not locate any specimens from the United States.
Host: Alnus

Moesziomyces K. Vánky. Basidiomycetes, Ustilaginales

Notes: Vánky (Nord. J. Bot. 6:67. 1986).
Moesziomyces bullatus **(J. Schröt.) K. Vánky**
Sorosporium bullatum J. Schröt.
Tolyposporium bullatum (J. Schröt.) J. Schröt. in Cohn
Tolyposporium globuligerum (Berk. & Broome) Rick
Moesziomyces globuligerus (Berk. & Broome) K. Vánky

Tolyposporium penicillariae Bref.
Distr.: Central and eastern United States and adjacent Canada.
Notes: On Poaceae.
Hosts: Echinochloa, Leersia, Panicum, Pennisetum
Moesziomyces eriocauli (G.P. Clinton) K. Vánky
Tolyposporium eriocauli G.P. Clinton
Distr.: Eastern United States.
Host: Eriocaulon

Mollisia (Fr.) P. Karst. Discomycetes, Helotiales

Mollisia amenticola (Sacc.) Rehm
Distr.: WA; Europe.
Host: Alnus
Mollisia atrata (Pers.) P. Karst.
Pyrenopeziza atrata (Pers.) Fuckel
Distr.: Temperate regions.
Notes: Dennis (British Ascomycetes. p. 209. 1978) suggested that this is a composite species of which most specimens are referable to *Pyrenopeziza*.
Hosts: Ambrosia, Aster, Bidens, Chelone, Conyza, Eupatorium, Hypericum, Sarracenia, Sonchus
Mollisia atrata (Pers.:Fr.) P. Karst. var. ***liliaceara*** W. Phillips & Harkn.
Distr.: CA(type).
Host: Lilium
Mollisia atriella (Cooke) Sacc.
Distr.: Eastern and southern United States; Europe.
Hosts: Andropogon, Muhlenbergia
Mollisia benesuada (Tul.) W. Phillips
Pyrenopeziza benesuada (Tul.) Gremmen
Distr.: Temperate regions.
Notes: LeGal & Mangenot (Rev. Mycol. 25:178. 1960).
Host: Alnus
Mollisia caespiticia (P. Karst.) P. Karst.
Pyrenopeziza caespiticia (P. Karst.) Gremmen
Distr.: WA; Europe.
Hosts: Alnus, Malus
Mollisia cinerea (Batsch) P. Karst.
Distr.: Temperate regions.
Notes: On decaying wood.
Hosts: Abies, Acer, Aster, Ceanothus, Paulownia, Physocarpus, Pinus, Populus, Pritchardia, Quercus, Vitis
Mollisia clavigera (Ellis & Everh.) Sacc.
Peziza clavigera Ellis & Everh.
Distr.: MT(type).
Host: Calamovilfa
Mollisia complicatula Rehm
Distr.: MI, WA; Austria(type).
Host: Lonicera
Mollisia conigena (Pers.) Boud.
Distr.: OR; Europe.
Notes: On decaying cones of conifers.
Host: Abies
Mollisia culmina (Sacc.) Rehm
Distr.: GA; Europe.
Hosts: Andropogon, Carex, Chasmanthium, Leersia, Schizachyrium, Uniola
Mollisia dakotensis (Rehm) Dennis
Pezizella dakotensis Rehm
Helotium dakotense (Rehm) Seaver
Distr.: ND(type).
Notes: On dead fallen twigs (Dennis. Persoonia 3:42. 1964).
Host: Symphoricarpos
Mollisia emergens W. Phillips & Harkn. non P. Karst., hom. illeg.
Distr.: CA(type).
Host: Eucalyptus
Mollisia exigua (Cooke) Seaver
Peziza exigua Cooke
Pezizella exigua (Cooke) Sacc.
Distr.: NJ(type).
Notes: On old stems. Dennis (Persoonia 3:46. 1964) suggested that this is an *Orbilia*.
Host: Conyza
Mollisia fumigata (Ellis & Everh.) Sacc.
Distr.: NJ(type).
Host: Tsuga
Mollisia gaultheriae Ellis & Everh.
Distr.: WA(type).
Host: Gaultheria
Mollisia hysteroides Cash
Distr.: CA(type).
Hosts: Haplopappus, Phlox
Mollisia iridis (Rehm) Sacc.
Micropeziza iridis Rehm
Distr.: NY; Europe.
Notes: Forming large, reddish, discolored areas on leaves.
Host: Iris
Mollisia ligni (Desmaz.) P. Karst.
Pyrenopeziza tamaricis (Roum.) Sacc.
Distr.: AK, NY; Europe.
Hosts: Sambucus, Tamarix
Mollisia lithocarpi Cash
Distr.: CA(type).
Host: Lithocarpus
Mollisia lycopodii Lebreton & Malbr.
Distr.: MA; France(type).
Notes: Holm & Holm (Karstenia 21:60. 1981) reported that they were unable to locate the type specimen.
Host: Lycopodium
Mollisia melaleuca (Fr.) Sacc.
Niptera melaleuca (Fr.) Fuckel
Patella melaleuca (Fr.) Quél.
Distr.: Temperate regions.
Hosts: Corylus, Prunus, Quercus, Salix, Vaccinium, Vitis
Mollisia nigritella W. Phillips & Harkn.
Distr.: CA(type).
Host: Galium
Mollisia petiolorum Cash
Distr.: HI.
Hosts: Aleurites, Freycinetia, Hibiscus, Terminalia
Mollisia piceae (Pers.) Quél.
Distr.: CA(type).
Host: Abies
Mollisia propinqua (Sacc. & Ellis) Dennis
Helotium propinquum Sacc. & Ellis
Calycina propinqua (Sacc. & Ellis) Kuntze
Distr.: PA(type).
Notes: On dead twigs (Dennis. Persoonia 3:60. 1964).
Host: Cornus
Mollisia pteridina (Nyl.) P. Karst.
Cyathicula aquilina (Rehm) Sacc.
Distr.: Widespread.
Notes: The synonym of *Cyathicula aquilina* is tentative (Carpenter. Mem. New York Bot. Gard. 33:190. 1981).
Host: Gaultheria
Mollisia pullata (W. Gerard in Cooke) Dennis
Helotium pullatum W. Gerard in Cooke
Distr.: NY(type).
Notes: Dennis (Persoonia 3:61. 1964).
Host: Vitis
Mollisia sabalidis (Ellis & Mart.) Seaver
Dermatea sabalidis Ellis & Mart.
Cenangium sabalidis (Ellis & Mart.) Sacc.
Distr.: FL(type).
Notes: The generic disposition of this species remains unsettled (Groves. Mycologia 38:424. 1946; Seaver. North Amer. cup fungi. [Inoperculates]. p. 214. 1951).
Host: Sabal
Mollisia scirpina (Peck) Sacc.
Distr.: NY(type).
Host: Scirpus
Mollisia shastensis Cash
Distr.: CA(type).
Hosts: Arctostaphylos, Eriogonum
Mollisia subcornea W. Phillips & Harkn.
Distr.: CA(type).
Host: Eucalyptus

Mollisia sublividula (Nyl.) P. Karst.
Distr.: AK.
Host: Salix
Mollisia urticae Cash
Distr.: CA(type).
Host: Urtica

Mollisina Höhn. Discomycetes, Helotiales

Mollisina acerina (V. Mouton) Höhn.
Pezizella acerina V. Mouton
Urceolella acerina (V. Mouton) Boud.
Distr.: CO; Europe.
Notes: On leaves,.
Host: Populus
Mollisina ammophilae Cash
Distr.: CA(type).
Host: Ammophila

Monascostroma Höhn. Loculoascomycetes, Dothideales

Notes: Barr (Contr. Univ. Michigan Herb. 9:539. 1972).
Monascostroma euryasca (Ellis & B.T. Galloway) Barr
Didymosphaeria euryasca Ellis & B.T. Galloway
Distr.: Western North America.
Host: Pinus

Monascus Tiegh. Discomycetes, Pezizales

Notes: This genus is tentatively placed in the Pezizales. Hawksworth & Pitt (Austral. J. Bot. 31:55. 1983).
Monascus floridanus E. Barnard & P. Cannon
Distr.: FL.
Notes: Isolated from roots.
Host: Pinus
Monascus purpureus Went
Distr.: Cosmopolitan.
Hosts: Oryza, Sorghum, Zea
Monascus ruber Tiegh.
Basipetospora rubra J.R. Cole & Kendrick (anamorph)
Distr.: Widespread.
Notes: Isolated from silage, deteriorating grain and soil.
Host: Helianthus

Monilia Bonord., nom. cons. Hyphomycetes

Notes: Arx (Sydowia 34:13. 1981).
Monilia angustior (Sacc.) Reade
Monilia peckiana Sacc. & Voglino var. *angustior* Sacc.
Distr.: North America; Finland(teleomorph, type).
Notes: Parasitic on twigs, leaves and fruits.
Host: Prunus
Monilia crataegi Died.
Monilinia johnsonii (Ellis & Everh.) Honey (teleomorph)
Distr.: Northeastern North America; Europe.
Notes: On leaves.
Host: Crataegus
Monilia implicata Gilman & E. Abbott
Distr.: GA, LA.
Notes: Isolated from soil, reported to cause sour rot of peaches (Wells. Phytopathology 67:404. 1977).
Host: Prunus
Monilia laxa (Ehrenb.) Sacc. & Voglino
Monilia cinerea Bonord.
Monilinia laxa (Aderhold & Ruhland) Honey (teleomorph)
Distr.: North temperate regions.
Notes: Brown fruit rot of Rosaceae and other families (Mordue. C.M.I. Descr. 619. 1979).
Host: Prunus
Monilia peckiana Sacc. & Voglino
Distr.: NY.
Hosts: Gaylussacia, Vaccinium
Monilia seaveri Reade
Monilinia seaveri (Rehm) Honey (teleomorph)
Distr.: Central and eastern North America.
Notes: On leaves, twigs, and fruits.
Host: Prunus

Monilinia Honey. Discomycetes, Helotiales

Notes: Byrde & Willetts (The brown-rot fungi of fruit: their biology and control. 1977); Harada (Bull. Fac. Agric. Hirosaki Univ. 27:30. 1977).
Monilinia amelanchieris Honey
Monilia amelanchieris Reade (anamorph)
Distr.: NY.
Notes: On leaves and fruits.
Host: Amelanchier
Monilinia azaleae Honey
Distr.: Eastern United States.
Host: Rhododendron
Monilinia corni (Reade) Hone, nom. inval.
Sclerotinia corni Reade
Distr.: Northeastern North America.
Notes: An invalid name based on the anamorph only (Kohn. Mycotaxon 9:404. 1979).
Host: Cornus
Monilinia demissa (Dana) Honey
Sclerotinia demissa Dana
Distr.: WA.
Notes: The validity of this name is uncertain.
Host: Prunus
Monilinia fructicola (G. Wint.) Honey
Sclerotinia fructicola (G. Wint.) Rehm
Monilia sp. (anamorph)
Distr.: Temperate regions.
Notes: Brown rot of stone fruits (Mordue. C.M.I. Descr. 616. 1979 as *Sclerotinia fructicola*; Elliott. Fungi Canadenses 38. 1974).
Hosts: Amelanchier, Chaenomeles, Cydonia, Gaylussacia, Malus, Prunus, Pyrus, Rubus, Vitis
Monilinia fructigena Honey in Whetzel
Sclerotinia fructigena Aderhold ex Sacc.
Monilia fructigena Pers.:Fr. (anamorph)
Distr.: Western hemisphere, rare; northern Africa; Asia; Europe.
Notes: Brown rot of pome and stone fruits (Mordue. C.M.I. Descr. 617. 1979 as *Sclerotinia fructigena*).
Hosts: Prunus, Pyrus
Monilinia gregaria (Dana) Honey
Sclerotinia gregaria Dana
Distr.: WA(type).
Notes: On leaves and fruits. The validity of this name is uncertain.
Host: Amelanchier
Monilinia johnsonii (Ellis & Everh.) Honey
Monilia crataegi Died. (anamorph)
Distr.: Northeastern North America; Europe.
Notes: On leaves.
Host: Crataegus
Monilinia laxa (Aderhold & Ruhland) Honey
Sclerotinia laxa Aderhold & Ruhland
Distr.: Temperate northern hemisphere.
Notes: Brown rot, flower and twig blight of *Prunus* spp. and other Rosaceae (Mordue. C.M.I. Descr. 619. 1979 as *Sclerotinia laxa*).
Hosts: Chaenomeles, Cydonia, Malus, Prunus, Pyrus
Monilinia ledi (Nawaschin) Whetzel
Sclerotinia ledi Nawaschin
Distr.: NY; Europe.
Notes: Batra (Mycologia 75:175. 1983).
Host: Vaccinium
Monilinia megalospora (Woronin) Whetzel
Sclerotinia megalospora Woronin
Distr.: NH; Europe.
Notes: Batra (Mycologia 75:146. 1983).
Host: Vaccinium
Monilinia oxycocci (Woronin) Honey
Sclerotinia oxycocci Woronin
Distr.: Worldwide in cranberry-growing regions.
Host: Vaccinium
Monilinia padi (Woronin) Honey

Monilia angustior (Sacc.) Reade (anamorph)
Distr.: Eastern and northwestern North America; Finland(type).
Host: Prunus
Monilinia polycodii (Reade) Honey
Monilia polycodii Reade (anamorph)
Distr.: NY.
Host: Vaccinium
Monilinia seaveri (Rehm) Honey
Sclerotinia seaveri Rehm
Distr.: Central and eastern North America.
Notes: On leaves, twigs, and fruits.
Host: Prunus
Monilinia urnula (Weinm.) Whetzel
Sclerotinia urnula (Weinm.) Rehm
Distr.: WA; Europe.
Host: Vaccinium
Monilinia vaccinii-corymbosi (Reade) Honey
Sclerotinia vaccinii-corymbosi Reade
Distr.: Eastern United States.
Notes: Blight of twigs, leaves, and fruits. Batra (Mycologia 75:131. 1983).
Hosts: [Prunus], Vaccinium

Monilochaetes Halst. ex Harter. Hyphomycetes

Monilochaetes infuscans Halst. ex Harter
Distr.: Widespread.
Notes: Sweet potato scurf.
Host: Ipomoea

Monochaetia (Sacc.) Allesch. Coelomycetes

Notes: Guba (Monograph of *Monochaetia* and *Pestalotia*. p. 21. 1966); Sutton (The Coelomycetes. p. 267. 1980).
Monochaetia concentrica (Berk. & Broome) Sacc.
Pestalotia concentrica Berk. & Broome
Distr.: Eastern North America.
Notes: On dead leaves of hardwoods.
Hosts: Castanea, Malus, Quercus
Monochaetia coryneoidea (Harkn.) Sacc.
Pestalotia coryneoidea Harkn.
Distr.: CA.
Host: Umbellularia
Monochaetia cryptomeriae M. Wilson
Distr.: WA; Scotland.
Host: Rhododendron
Monochaetia ellisiana Sacc. & D. Sacc.
Pestalotia monochaetoidea Sacc. & Ellis
Distr.: NJ(type), WV.
Notes: Guba (Monograph of *Monochaetia* and *Pestalotia*. p. 28. 1961) suggested that this is probably a synonym of *Cryptostictia lonicerae*, now *Seimatosporium lonicerae*, but this has not been confirmed.
Host: Physocarpus
Monochaetia hysteriiformis (Berk. & M.A. Curtis) Guba
Pestalotia hysteriiformis Berk. & M.A. Curtis
Distr.: Eastern North America.
Hosts: Lithocarpus, Quercus
Monochaetia jefferisii (Ellis) Sacc.
Pestalotia jefferisii Ellis
Distr.: PA.
Host: Viburnum
Monochaetia kansensis (Ellis & Barth.) Sacc.
Pestalotia kansensis Ellis & Barth.
Monochaetia flagellata (Earle) Sacc.
Monochaetia pachyspora Bubák
Distr.: Southeastern United States; southern Europe.
Notes: Usually on leaves of *Castanea* and *Quercus* but occasionally on other hardwoods. Often misdetermined as *Monochaetia desmazierii* Sacc. in older literature.
Hosts: Acer, Castanea
Monochaetia monochaeta (Desmaz.) Allesch. in Rabenh.
Pestalotia monochaeta Desmaz.
Pestalotia affinis Ellis & Everh., hom. illeg.
Monochaetia desmazierii Sacc.
Pestalotia nervalis Ellis & Everh.
Pestalotia pallida Ellis & Everh.
Monochaetia taphrinicola (Ellis & Everh.) Sacc.
Pestalotia taphrinicola Ellis & Everh.
Distr.: United States; Europe.
Hosts: Acer, Aesculus, Carya, Castanea, Eucalyptus, Hamamelis, Quercus, Rhododendron, Rubus, Ulmus
Monochaetia paeoniae (Maubl.) Sacc. & D. Sacc.
Pestalotia paeoniae Maubl.
Distr.: Eastern United States; Europe.
Host: Paeonia
Monochaetia turgida (Atk.) Sacc.
Pestalotia turgida Atk.
Monochaetia crataegi (Ellis & Everh.) Sacc.
Distr.: Southern United States; Canada(Ontario).
Host: Crataegus

Monochytrium Griggs. Chytridiomycetes, Chytridiales

Notes: This genus is considered a synonym of *Olpidium* but not all species have been redisposed.
Monochytrium stevensianum Griggs
Distr.: NC(type).
Host: Ambrosia

Monodictys S.J. Hughes. Hyphomycetes

Monodictys antiqua (Corda) S.J. Hughes
Sporidesmium antiquum Corda
Distr.: DE; Europe.
Notes: On dead wood (Ellis. More dematiaceous Hyphomycetes. p. 43. 1976).
Host: Platanus
Monodictys austrina Tubaki in Tubaki & I. Asano
Distr.: FL; Anarctica(type).
Notes: Isolated from soil, algae, and *Juncus*.
Host: Juncus
Monodictys castaneae (Wallr.) S.J. Hughes
Acrospeira macrosporoidea (Berk.) Wiltshire
Distr.: Temperate North America; Europe.
Notes: On wood and organic debris. Ellis (Dematiaceous Hyphomycetes. p. 68. 1971).
Host: Pseudotsuga
Monodictys glauca (Cooke & Harkn.) S.J. Hughes
Epochnium glaucum Cooke & Harkn.
Stemphylium glaucum (Cooke & Harkn.) Sacc.
Stemphylium subsphaericum Fairm.
Distr.: HI, North America.
Notes: Ellis (Dematiaceous Hyphomycetes. p. 70. 1971).
Host: Metrosideros
Monodictys pelagica (T.W. Johnson) E.B.G. Jones
Piricauda pelagica E.B.G. Jones
Piricauda arcticoceanorum R.T. Moore
Distr.: Worldwide in marine and inland saline habitats.
Host: Spartina

Monographella Petr. Pyrenomycetes, Amphisphaeriales

Notes: The genus is based on teleomorphs of *Fusarium* section *Arachnites* Wollenw. (Arx. Trans. Brit. Mycol. Soc. 82:373. 1984).
Monographella nivalis (Schaffnit) E. Müller
Calonectria nivalis Schaffnit
Griphosphaeria nivalis (Schaffnit) E. Müller & Arx
Micronectriella nivalis (Schaffnit) C. Booth
Calonectria graminicola Wollenweb.
Microdochium nivale (Fr.) Samuels & I.C. Hallett (anamorph)
Distr.: Cosmopolitan.
Notes: Causes diseases of cereal and turf (Booth. C.M.I. Descr. 309. 1971).
Hosts: Hordeum, Secale, Triticum
Monographella opuntiae (Ellis & Everh.) Arx

Sphaerella opuntiae Ellis & Everh.
Mycosphaerella opuntiae (Ellis & Everh.) Dearn.
Microdochium lunatum (Ellis & Everh.) Arx (anamorph)
Distr.: Central and eastern United States; West Indies; Canary Islands.
Notes: Brown spot of fleshy joints of Cactaceae.
Hosts: Cereus, Echinocactus, Mammillaria, Opuntia

Monosporascus Pollack & Uecker. Pyrenomycetes, Sordariales

Notes: Pollack & Uecker (Mycologia 66:346. 1974); Hawksworth & Ciccarone (Mycopath. Mycol. Appl. 66:147. 1978).
Monosporascus cannonballus Pollack & Uecker
Distr.: AZ; India.
Notes: Associated with roots of canteloupe.
Host: Cucumis

Monostichella Höhn. Coelomycetes

Notes: Arx (Biblioth. Mycol. 24:14. 1970) accepted 10 species, but Sutton (The Coelomycetes. p. 516. 1980) indicated that Arx's concept of the genus was heterogeneous.
Monostichella alni (Ellis & Everh.)
Gloeosporium alni Ellis & Everh.
Distr.: Eastern United States.
Host: Alnus
Monostichella chamaedaphnes (Dearn.) Arx
Gloeosporium chamaedaphnes Dearn.
Distr.: Northeastern North America.
Host: Chamaedaphne
Monostichella falcata (Dearn. & House) Arx
Gloeosporium falcatum Dearn. & House
Distr.: NY(type).
Host: Lindera
Monostichella hysteroidea (Dearn. & Barth.) Arx
Gloeosporium hysteroideum Dearn. & Barth.
Distr.: OH(type).
Host: Acer
Monostichella robergei (Desmaz.) Höhn.
Gloeosporium robergei Desmaz.
Distr.: Temperate North America; Europe.
Hosts: Carpinus, Ostrya
Monostichella salicis (Westend.) Arx
Gloeosporium salicis Westend.
Gloeosporidiella salicis (Westend.) Nannf.
Drepanopeziza salicis (Tul. & C. Tul.) Höhn. (teleomorph)
Distr.: Temperate North America; Europe.
Host: Salix

Montagnula Berl. Loculoascomycetes, Dothideales

Notes: Crivelli (Ueber die heterogene Ascomycetengattung *Pleospora* Rabh.; Vorschlag für eine Aufteilung. p. 151. 1983).
Montagnula obtusa (Fuckel) P.G. Crivelli
Teichospora obtusa Fuckel
Pleospora obtusa (Fuckel) Höhn.
Distr.: Temperate regions.
Notes: On woody plants.
Hosts: Carya, Vaccinium
Montagnula spinosella (Rehm) P.G. Crivelli
Pleospora spinosella Rehm
Pleospora scrophulariae (Desmaz.) Höhn. var. *spinosella* (Rehm) Wehmeyer
Distr.: Cosmopolitan.
Notes: On grasses.
Host: Chasmanthium
Montagnula thuemeniana (Sacc.) P.G. Crivelli
Pleospora thuemeniana Sacc.
Distr.: GA, SC.
Host: Yucca

Morenoina Theiss. Loculoascomycetes, Dothideales

Morenoina lucens (Harkn.) Theiss.
Aulographum lucens Harkn.
Lembosia lucens (Harkn.) Sacc.
Distr.: CA.
Host: Garrya
Morenoina selaginellae M.L. Farr
Host: Selaginella

Morrisographium Morelet. Hyphomycetes

Notes: Illman & White (Canad. J. Bot. 63:419. 1984).
Morrisographium hispidulum (Ellis) Illman & G.P. White
Cornularia hispidula (Ellis) Sacc.
Pseudographium hispidulum (Ellis) Jacz.
Sphaeronaema hispidulum Ellis non Corda, hom. illeg.
Distr.: Eastern United States.
Host: Nyssa
Morrisographium persicae (Schwein.:Fr.) Illman & G.P. White
Periconia persicae Schwein.
Cornularia persicae (Schwein.:Fr.) Sacc.
Sphaeronaema persicae (Schwein.:Fr.) Ellis
Harpographium magnum Sacc.
Isariopsis pilosa Earle
Phragmographium pilosum (Earle) E.F. Morris
Distr.: Central and eastern North America.
Notes: On limbs and twigs.
Host: Prunus

Mortierella Coem. Zygomycetes, Mucorales

Notes: Domsch et al. (Compendium of soil fungi. p. 433. 1980).
Mortierella isabellina Oudem. & Koning
Distr.: Widespread.
Notes: Isolated from soil, rotten wood and various substrates.
Hosts: Acer, Metrosideros
Mortierella ramanniana (A. Möller) Linnemann var. ***angulispora*** (Naumov) Linnemann
Distr.: Widespread.
Notes: Isolated from soil and various organic substrates.
Hosts: Alnus, Pseudotsuga

Mucilago P. Mich. ex Battarra. Myxomycetes

Mucilago crustacea A. Wigg.
Mucilago spongiosa (Leyss.) Morg.
Distr.: Cosmopolitan.
Hosts: Cynodon, Fragaria, Hypoxis, Trifolium

Mucor P. Mich.:Fr. Zygomycetes, Mucorales

Notes: Domsch et al. (Compendium of soil fungi. p. 461. 1980).
Mucor circinelloides Tiegh.
Mucor ambiguus Vuill.
Distr.: Cosmopolitan.
Notes: In soil and on a variety of organic substrates.
Hosts: Lycopersicon, Zea
Mucor fragilis Bainier
Distr.: Widespread.
Notes: Isolated from soil and various organic substrates. Schipper (Stud. Mycol. 4:36. 1973; Stud. Mycol. 12:37. 1976) was unable to characterize this name but suspected that it belonged to the *M. hiemalis*-group or the *M. circinelloides*-group *sensu* Zycha.
Hosts: Helianthus, Zea
Mucor hiemalis Wehmer
Distr.: Cosmopolitan.
Notes: Common soil saprobe.
Hosts: Carya, Helianthus, Lycopersicon, Psidium, Zea
Mucor hiemalis Wehmer f. ***silvaticus*** (Hagem) M.A.A. Schipper
Mucor silvaticus Hagem
Distr.: Temperate regions.
Host: Cucumis
Mucor mucedo P. Mich. ex Saint-Amans

Distr.: Cosmopolitan.
Hosts: Malus, Nicotiana, Zea
Mucor paronychia Suth.-Camp. & Plunkett
Distr.: CA.
Notes: This name was published in an obscure journal and has not been considered by those who have studied the genus *Mucor*.
Host: Citrus
Mucor piriformis E. Fisch.
Distr.: Temperate regions.
Hosts: Malus, Prunus, Pyrus
Mucor plumbeus Bonord.
Mucor spinosus Tiegh.
Distr.: Cosmopolitan.
Hosts: Carya, Zea
Mucor racemosus Fresen.
Distr.: Cosmopolitan.
Notes: Storage rot of fruit and vegetables (Lunn. C.M.I. Descr. 529. 1977).
Hosts: Carya, Citrus, Ipomoea, Malus, Nicotiana, Zea
Mucor recurvus E.E. Butler
Distr.: Widespread.
Host: Fragaria

Mucronella Fr. Basidiomycetes, Aphyllophorales

Mucronella aggregata Fr.
Distr.: Temperate northern hemisphere.
Notes: On wood.
Hosts: Abies, Betula, Picea, Pinus, Populus, Pseudotsuga

Muirella R. Sprague. Hyphomycetes

Muirella alascensis R. Sprague
Gloeocercospora alascensis R. Sprague, nom. illeg.
Distr.: AK.
Hosts: Elymus, Trisetum

Mundkurella Thirumalachar. Basidiomycetes, Ustilaginales

Notes: Savile (Mycologia 67:275. 1975).
Mundkurella mossii Savile
Distr.: SD; Canada.
Host: Aralia

Murangium Seaver. Loculoascomycetes, Patellariales

Murangium sequoiae (Plowr. ex W. Phillips) Seaver
Cenangium sequoiae Plowr. ex W. Phillips
Scleroderris sequoiae (Plowr. ex W. Phillips) Sacc.
Distr.: CA, SC.
Hosts: Juniperus, Sequoiadendron

Mycena (Pers.) Roussel. Basidiomycetes, Agaricales

Notes: Smith (North American species of *Mycena*. 1947).
Mycena albicolor A.H. Sm.
Distr.: OR.
Host: Thuja
Mycena albidula (Peck) A.H. Sm.
Helotium californiense (Murrill) Redhead
Omphalopsis californiensis Murrill
Distr.: Temperate North America.
Notes: On hardwood debris.
Host: Rosa
Mycena albissima A.H. Sm.
Distr.: CA.
Host: Sequoia
Mycena alcalina (Fr.:Fr.) P. Kumm.
Distr.: Temperate North America; Europe.
Notes: On wood and debris of conifers.
Host: Salix
Mycena alnicola A.H. Sm.
Distr.: WA.
Host: Alnus
Mycena austinii (Peck) Kühner
Distr.: NY.
Host: Picea
Mycena borealis A.H. Sm.
Distr.: Eastern North America.
Notes: On wood of conifers.
Hosts: Abies, Picea
Mycena capillaris (Schumach.:Fr.) P. Kumm.
Distr.: Temperate North America; Europe.
Notes: On hardwood debris.
Host: Rosa
Mycena carolinensis A.H. Sm. & Hesler
Distr.: Southern Appalachian Mountains.
Hosts: Leucothoe, Rhododendron
Mycena cinerella (P. Karst.) P. Karst.
Mycena brownii A.H. Sm.
Distr.: Temperate North America; Europe.
Host: Rubus
Mycena citrinomarginata Gill.
Distr.: Temperate North America; Europe.
Notes: On conifer needles.
Host: Pinus
Mycena clavata (Peck) Redhead
Omphalia clavata Peck
Mycena thujina A.H. Sm.
Distr.: Temperate North America; Europe.
Notes: On conifer wood (Redhead. Mycologia 78:522. 1986).
Hosts: Sequoia, Thuja
Mycena clavicularis (Fr.:Fr.) Gill.
Distr.: Temperate northern hemisphere.
Notes: On needles of conifers.
Host: Pinus
Mycena collybiiformis Murrill
Distr.: NY.
Host: Quercus
Mycena corticalis A.H. Sm.
Distr.: OR.
Host: Thuja
Mycena cucullata (Ellis) Redhead
Marasmius cucullatus Ellis
Distr.: Northeastern United States.
Notes: Redhead (Sydowia 37:252. 1984).
Host: Vaccinium
Mycena culmigena Maas G.
Mycena juncicola (Fr.) Gill., sensu A.H. Sm.
Distr.: Pacific Northwest.
Notes: On sedges (Maas Geesteranus. Proc. Koninkl. Nederl. Akad. Wetens., Ser. C, 89:164. 1986).
Host: Carex
Mycena cylindrospora A.H. Sm.
Distr.: TN.
Host: Rhododendron
Mycena elegantula Peck var. ***inflata*** D.H. Mitchel & A.H. Sm.
Distr.: Western North America.
Hosts: Abies, Picea
Mycena excisa (Lasch) Gill.
Distr.: Temperate North America; Europe.
Host: Juglans
Mycena fallax A.H. Sm.
Distr.: OR.
Host: Rubus
Mycena filiformis A.H. Sm.
Distr.: MI.
Host: Fagus
Mycena fragillima A.H. Sm.
Distr.: Pacific Northwest.
Host: Polystichum
Mycena fuliginella A.H. Sm.
Distr.: WA.
Host: Pseudotsuga
Mycena fusco-ocula A.H. Sm.

Distr.: Pacific Northwest.
Hosts: Sequoia, Tsuga
Mycena galericulata (Scop.:Fr.) S.F. Gray
Distr.: Temperate North America; Europe.
Host: Quercus
Mycena gaultheri A.H. Sm.
Distr.: WA(type).
Host: Gaultheria
Mycena griseoviridis A.H. Sm. var. ***cascadensis*** A.H. Sm.
Distr.: WA.
Host: Abies
Mycena haematopoda (Pers.:Fr.) P. Kumm.
Mycena haematopus (Pers.:Fr.) Quél.
Distr.: Temperate northern hemisphere.
Notes: On hardwoods.
Host: Fagus
Mycena haematopoda (Pers.:Fr.) P. Kumm. var. ***cuspidata*** D.H. Mitchel & A.H. Sm.
Distr.: CO.
Hosts: Alnus, Populus
Mycena incarnatifolia A.H. Sm.
Distr.: MI.
Host: Prunus
Mycena intertexta (Berk. & M.A. Curtis) Sacc.
Mycena avellanea Murrill
Distr.: Eastern United States.
Host: Pinus
Mycena leaiana (Berk.) Sacc.
Collybia leaiana Fairm.
Prunulus leaianus Murrill
Distr.: Central and eastern North America.
Notes: On hardwoods.
Host: Fagus
Mycena lilacifolia (Peck) A.H. Sm.
Distr.: Temperate North America.
Notes: On conifer wood.
Host: Pseudotsuga
Mycena lohwagii Singer
Distr.: Northwestern North America; temperate Eurasia.
Host: Athyrium
Mycena longiseta Höhn.
Distr.: Temperate North America; Europe.
Notes: On conifer needles.
Host: Picea
Mycena luteopallens (Peck) Sacc.
Marasmius nucicola W. McDougall
Distr.: Eastern United States.
Hosts: Carya, Juglans
Mycena madronicola A.H. Sm.
Distr.: OR.
Host: Arbutus
Mycena meliigena (Berk. & M.A. Curtis) Sacc.
Mycena corticola Auct. non (Pers.:Fr.) S.F. Gray
Distr.: Temperate North America; Europe.
Notes: On bark of living trees.
Hosts: Diospyros, Melia, Quercus, Sciadopitys
Mycena metata (Fr.) P. Kumm.
Distr.: Temperate North America; Europe.
Notes: On conifer debris.
Host: Tsuga
Mycena mucor (Fr.) Quél.
Distr.: Eastern North America; Europe.
Host: Quercus
Mycena overholtsii A.H. Sm. & Solheim
Distr.: Western North America.
Notes: On conifer wood.
Hosts: Abies, Tsuga
Mycena paucilamellata A.H. Sm.
Distr.: CA.
Host: Sequoia
Mycena plicosa (Fr.) Gill.
Distr.: Eastern United States; Europe.
Hosts: Picea, Pinus, Tsuga
Mycena plumbea (Fr.) Sacc.
Distr.: Temperate North America; Europe.
Host: Picea
Mycena pterigena (Fr.:Fr.) P. Kumm.
Distr.: Temperate Northern Hemisphere.
Hosts: Athyrium, Onoclea
Mycena radicatella (Peck) Sacc.
Distr.: Northeastern North America to NC.
Host: Acer
Mycena rugulosiceps (Kauffm.) A.H. Sm.
Distr.: Pacific Northwest.
Hosts: Alnus, Vaccinium
Mycena sabali Murrill
Distr.: LA.
Host: Sabal
Mycena sanguinolenta (Albertini & Schwein.:Fr.) Quél.
Distr.: Temperate northern hemisphere.
Notes: On conifer needles.
Host: Picea
Mycena setulosa A.H. Sm.
Distr.: TN.
Host: Liriodendron
Mycena subcana A.H. Sm.
Distr.: Pacific Northwest.
Hosts: Chamaecyparis, Pseudotsuga
Mycena subepipterygia Murrill
Distr.: FL.
Host: Quercus
Mycena subincarnata (Peck) Sacc.
Distr.: Northeastern North America.
Host: Pinus
Mycena subinclinata Murrill
Distr.: FL.
Host: Quercus
Mycena subplicosa P. Karst.
Distr.: Temperate North America; Europe.
Host: Picea
Mycena subvitrea A.H. Sm.
Distr.: OR.
Host: Pseudotsuga
Mycena tenax A.H. Sm.
Distr.: Temperate North America.
Notes: On conifer needles.
Hosts: Abies, Picea
Mycena tenuiceps A.H. Sm.
Distr.: TN.
Host: Fagus
Mycena texensis A.H. Sm.
Distr.: Southeastern United States.
Host: Quercus
Mycena ulmicola A.H. Sm.
Distr.: Eastern United States.
Host: Ulmus

Mycoacia Donk. Basidiomycetes, Aphyllophorales

Notes: Eriksson & Ryvarden (The Corticiaceae of North Europe 4:873. 1976).
Mycoacia aurea (Fr.:Fr.) J. Eriksson & Ryvarden
Mycoacia stenodon (Pers.) Donk
Distr.: Temperate northern hemisphere.
Notes: On wood.
Host: Populus
Mycoacia austro-occidentale E.R. Canfield
Distr.: Southwestern United States.
Notes: On wood.
Hosts: Fouquieria, Prosopis
Mycoacia fuscoatra (Fr.:Fr.) Donk
Hydnum fuscoatrum Fr.:Fr.
Odontia fuscoatra (Fr.:Fr.) Bres.
Steccherinum fuscoatrum (Fr.:Fr.) R.L. Gilbertson
Hydnum carbonarium Peck
Distr.: Temperate northern hemisphere.
Notes: On wood.
Hosts: Acer, Castanea, Populus, Quercus

Mycoacia meridionalis Burdsall & Nakasone
Distr.: Southern United States.
Notes: Burdsall & Nakasone (Mycologia 73:454. 1981).
Hosts: Carpinus, Carya, Persea, Pinus, Quercus
Mycoacia uda (Fr.:Fr.) Donk
Odontia uda (Fr.:Fr.) Bres.
Distr.: Temperate northern hemisphere.
Notes: On wood.
Hosts: Alnus, Populus, Pseudotsuga

Mycocalia J.T. Palmer. Basidiomycetes, Nidulariales

Notes: Brodie (The bird's nest fungi. p. 139. 1975).
Mycocalia reticulata (Petch) J.T. Palmer
Nidularia reticulata Petch
Distr.: HI, LA; France(greenhouse); tropical regions.
Host: Cecropia

Mycocentrospora Deighton. Hyphomycetes

Centrospora Neergaard non Trevis., hom. illeg.
Notes: This name replaced *Centrospora*, a later hymonym.
Mycocentrospora acerina (R. Hartig) Deighton
Cercospora acerina R. Hartig
Centrospora acerina (R. Hartig) A.G. Newhall
Ansatospora macrospora (Osterwalder) A.G. Newhall
Cercospora macrospora Osterwalder
Cercospora praegrandis R. Sprague
Distr.: Temperate North America; Europe; Australia; New Zealand.
Notes: A low temperature plant pathogen that is persistant in the soil. On a wide range of host plants and on submerged leaves in streams and soils (Sutton & Gibson. C.M.I. Descr. 537. 1977).
Hosts: Apium, Daucus, Osmorhiza, Potamogeton, Viola
Mycocentrospora asiminae (Ellis & Kellerm.) Deighton
Cercospora asiminae Ellis & Kellerm.
Centrospora asiminae (Ellis & Kellerm.) Deighton
Distr.: Southeastern United States.
Host: Asimina
Mycocentrospora cladosporioides (Sacc.) P. Costa ex Deighton
Cercospora cladosporioides Sacc.
Distr.: CA; northern Africa; southern Europe.
Hosts: Olea, Prunus
Mycocentrospora verrucosa Pollack & C.W. Ellett
Distr.: OH.
Notes: Foliar shot-hole (Pollack & Ellett. Mycologia 66:170. 1974).
Host: Euonymus

Mycoenterolobium Goos. Hyphomycetes

Mycoenterolobium platysporum Goos
Distr.: HI.
Notes: Goos (Mycologia 62:171. 1970).
Host: Araucaria

Mycogloea L. Olive. Basidiomycetes, Auriculariales

Mycogloea carnosa Olive
Distr.: Southern United States.
Notes: On wood.
Host: Acer

Mycoleptodiscus Ostazeski. Hyphomycetes

Notes: Ostazeski (Mycologia 59:970. 1967).
Mycoleptodiscus brasiliensis Sutton & C.S. Hodges
Distr.: HI; Brazil.
Notes: On dead leaves (Sutton & Hodges. Nova Hedwigia 27:694. 1976).
Host: Eucalyptus
Mycoleptodiscus indicus (Sahni) Sutton
Distr.: Subtropical and tropical regions.
Notes: Usually on monocots (Sutton. Trans. Brit. Mycol. Soc. 60:528. 1973).
Hosts: Hippeastrum, Ilex, Sabal, Serenoa, Zamia
Mycoleptodiscus minimus (Berk. & M.A. Curtis) Vanev
Discosia minima Berk. & M.A. Curtis
Distr.: Southern United States.
Notes: Leaf spot of holly (Vanev. Proc. Koninkl. Nederl. Akad. Wetens., Ser. C, 86:433. 1983).
Hosts: Arbutus, Ilex, Tilia
Mycoleptodiscus sphaericus Ostazeski
Distr.: MD, VA.
Notes: Associated with root and crown rot.
Host: Lotus
Mycoleptodiscus terrestris J.W. (Gerdemann) Ostazeski
Leptodiscus terrestris J.W. Gerdemann
Distr.: Southeastern United States.
Hosts: Arachis, Capsicum, Eichhornia, Glycine, Lotus, Medicago, Melilotus, Phaseolus, Pisum, Trifolium

Mycomicrothelia Keissl. Loculoascomycetes, Dothideales

Notes: Hawksworth (Bull. Brit. Mus. Nat. Hist., Bot. 14:64. 1985).
Mycomicrothelia dothideaspora (Cooke & Harkn.) D. Hawksworth
Amphisphaeria dothideaspora Cooke & Harkn.
Distr.: CA(type).
Host: Mimulus
Mycomicrothelia macularis (Hampe ex C. Massal.) Keissl.
Distr.: CA, but according to Hawksworth (Bull. Brit. Mus. Nat. Hist., Bot. 14:92. 1985) this species is known reliably only from the central European mountains.
Host: Holodiscus

Mycopappus Redhead & G.P. White. Hyphomycetes

Notes: Redhead & White (Canad. J. Bot. 63:1429. 1985).
Mycopappus aceris (Dearn. & Barth.) Redhead & G.P. White
Cercosporella aceris Dearn. & Barth.
Distr.: Pacific Northwest.
Notes: This fungus has been misdetermined and reported under the name *Illosporium maculicola* Sacc. Funk and Dorworth (Canad. J. Bot. 66:295. 1988) provided a description and suggested that *Mycosphaerella mycopappi* and *Stigmina zilleri* are morphs of this fungus.
Host: Acer
Mycopappus alni (Dearn. & Barth.) Redhead & G.P. White
Cercosporella alni Dearn. & Barth.
Distr.: Pacific Northwest.
Host: Alnus

Mycorrhaphium Maas G. Basidiomycetes, Aphyllophorales

Mycorrhaphium adustum (Schwein.:Fr.) Maas G.
Steccherinum adustum (Schwein.:Fr.) Banker
Distr.: Eastern North America; Asia.
Notes: On wood.
Hosts: Carpinus, Carya, Castanea, Fagus, Ostrya, Pinus, Quercus, Ulmus

Mycosphaerella Johans. Loculoascomycetes, Dothideales

Notes: Tomilin (Classification key of the fungi of the genus *Mycosphaerella*. 1979); Barr (Contr. Univ. Michigan Herb. 9:596. 1972); Sivanesan (The bitunicate Ascomycetes. p. 182. 1984).
Mycosphaerella acaciae (Cooke & Harkn.) Tomilin
Sphaerella acaciae Cooke & Harkn.
Distr.: CA.
Host: Acacia
Mycosphaerella acervata (Ellis & Everh.) Barr
Diatrype acervata Ellis & Everh.
Diaporthe acervata (Ellis & Everh.) Ellis & Everh.

Distr.: North America.
Notes: On leaves.
Host: Yucca
Mycosphaerella acicola (Cooke & Harkn.) Lindau
Sphaerella acicola Cooke & Harkn.
Distr.: CA(type).
Host: Pinus
Mycosphaerella aesculi (Cocc. & Morini) Tomilin
Mycosphaerella maculiformis (Pers.:Fr.) J. Schröt. var. *hippocastani* Jaap
Distr.: CA; Europe.
Host: Aesculus
Mycosphaerella aggregata Earle non (Schwein.) J.A. Stevenson, hom. illeg.
Distr.: TX; Central America.
Host: Hymenocallis
Mycosphaerella ailanthi House
Sphaerella ailanthi Ellis & Barth. non Cooke, hom. illeg.
Distr.: NY(type).
Host: Ailanthus
Mycosphaerella aleuritis Ou
Pseudocercospora aleuritis (Miyake) Deighton (anamorph)
Distr.: Throughout range of host.
Notes: On leaves.
Host: Aleurites
Mycosphaerella allicina (Fr.:Fr.) Vestergr.
Sphaerella allicina (Fr.:Fr.) Auersw.
Distr.: Temperate North America; Europe.
Notes: On leaves.
Host: Allium
Mycosphaerella alnicola (Peck) Jaap
Distr.: Temperate North America; Europe.
Host: Alnus
Mycosphaerella alpina Arx
Distr.: WA; Europe.
Host: Luzula
Mycosphaerella altera (Pass.) House
Distr.: NY; Europe(type).
Host: Equisetum
Mycosphaerella andrewsii (Sacc.) J.J. Davis
Asteromella andrewsii Petr. in J.J. Davis (anamorph)
Distr.: Central United States; Italy(type).
Host: Gentiana
Mycosphaerella angulata W.A. Jenkins
Cercospora brachypus Ellis & Everh. (anamorph)
Distr.: Southeastern United States; China.
Notes: Angular leaf spot of Muscadine grapes (Jenkins. Phytopathology 32:71. 1942).
Host: Vitis
Mycosphaerella annulata (Cooke) L.E. Miles
Sphaerella annulata Cooke
Distr.: Southeastern United States.
Host: Magnolia
Mycosphaerella applanata (Ellis & Everh.) Lind
Sphaerella applanata Ellis & Everh.
Distr.: MT, TX.
Host: Clematis
Mycosphaerella aquatica (Cooke) J.H. Miller
Distr.: Southeastern United States.
Host: Quercus
Mycosphaerella aquilegiae-jonesii Tomilin
Sphaerella aquilegiae Ellis & B.T. Galloway
Distr.: MT, WA.
Hosts: Aquilegia, Lupinus
Mycosphaerella arachidis Deighton
Mycosphaerella arachidicola Jenk. non Chochrjakov, hom. illeg.
Cercospora arachidicola S. Hori (anamorph)
Distr.: Worldwide in peanut-growing regions.
Notes: Early leaf spot of peanuts (Mulder & Holliday. C.M.I. Descr. 411. 1974).
Host: Arachis
Mycosphaerella arachnoidea F.A. Wolf
Cercosporella arachnoidea F.A. Wolf (anamorph)
Distr.: Southeastern United States.
Notes: False mildew of red mulberry.
Host: Morus
Mycosphaerella arbuticola (Peck) Barr
Distr.: Western North America.
Hosts: Arbutus, Umbellularia
Mycosphaerella arbutifoliae House, nom. nud.
Distr.: GA, NY(type).
Host: Aronia
Mycosphaerella areola J. Ehrlich & F.A. Wolf
Ramularia gossypii (Speg.) Cif. (anamorph)
Distr.: Worldwide in cotton-growing regions.
Notes: Areolate mildew of cotton.
Host: Gossypium
Mycosphaerella aronici Volkart
Distr.: WA; Europe.
Host: Aster
Mycosphaerella artocarpi F. Stevens & P.A. Young in F. Stevens
Distr.: HI.
Host: Artocarpus
Mycosphaerella arundinariae (Atk.) Earle in C. Mohr
Distr.: AL.
Host: Arundinaria
Mycosphaerella asiminae (Ellis & Everh.) Tomilin
Distr.: OH(type).
Host: Asimina
Mycosphaerella asterinoides (Ellis & Everh.) Fairm. non (Pat.) Lindau, hom. illeg.
Sphaerella asterinoides Ellis & Everh.
Distr.: NY.
Host: Dipsacus
Mycosphaerella audibertiae Rehm
Sphaerella audibertiae (Rehm) Sacc.
Distr.: CA(type).
Host: Salvia
Mycosphaerella bacillifera (P. Karst.) Lind
Sphaerella bacillifera P. Karst.
Distr.: WI; Europe.
Host: Scheuchzeria
Mycosphaerella balsamorhizae Earle
Sphaerella balsamorhizae (Earle) Sacc. & Trott.
Distr.: NV(type).
Host: Balsamorhiza
Mycosphaerella baptisiicola (Cooke) Earle in C. Mohr
Sphaerella baptisiicola (Cooke) Berl. & Voglino in Sacc.
Distr.: Eastern United States.
Host: Baptisia
Mycosphaerella berberidis (Auersw.) Lindau
Septoria berberidis Niessl (anamorph)
Distr.: Temperate North America; Europe; Middle East.
Notes: On leaves.
Host: Berberis
Mycosphaerella berkeleyi Jenk.
Phaeoisariopsis personata (Berk. & M.A. Curtis) Arx (anamorph)
Distr.: Worldwide in peanut-growing regions.
Notes: Late leaf spot of peanuts (Mulder & Holliday. C.M.I. Descr. 412. 1974).
Host: Arachis
Mycosphaerella bolleana Higgins
Sphaerella bolleana (Higgins) Trott. in Sacc.
Pseudocercospora bolleana (Thuem.) Sivanesan (anamorph)
Distr.: HI, southern United States; Middle East; southern Europe.
Notes: Leaf spot.
Host: Ficus
Mycosphaerella borealis (Bubák & Vleugel) Barr
Distr.: Northeastern North America; Europe.
Notes: On leaves.
Host: Alnus
Mycosphaerella botrychii (Rostr.) Savile
Distr.: Boreal North America; Europe.
Host: Botrychium
Mycosphaerella brassicicola (Duby) Lindau in Engl. & Prantl
Asteromella brassicae (Chev.) Boerema & Van Kesteren (anamorph)
Distr.: Temperate North America; Europe, most common in cool, moist climates.
Notes: Ring spot (Punithalingam & Holliday. C.M.I. Descr. 468. 1975; Boerema & Kesteren. Persoonia 3:17. 1984; Corlett. Mycotaxon

31:63. 1988).
Host: Brassica
Mycosphaerella caespitosa (Ellis & Everh.) Diehl in W.B. Cooke
Distr.: Southern and western United States.
Host: Quercus
Mycosphaerella campanulae (Ellis & Kellerm.) Petr.
Distr.: Eastern North America; Europe.
Host: Campanula
Mycosphaerella caricae Syd. & P. Syd.
Phoma caricae-papayae (Tarr) Punithalingam (anamorph)
Distr.: FL, HI; southeast Asia; India.
Notes: Fruit rot, trunk rot, and leaf spot of papaya.
Host: Carica
Mycosphaerella caricis (Dearn. & House) Petr. & Syd.
Laestadia caricis Dearn. & House
Guignardia caricis (Dearn. & House) Dearn. & House
Distr.: NY.
Host: Carex
Mycosphaerella caroliniana (F.A. Wolf) J.H. Miller
Distr.: Southeastern United States.
Host: Oxydendrum
Mycosphaerella caryigena Demaree & J.R. Cole
Pseudocercosporella caryigena (Ellis & Everh.) Sivanesan (anamorph)
Distr.: Central and eastern North America.
Notes: On Juglandaceae.
Host: Carya
Mycosphaerella caryophylli (Pass.) Cruchet
Sphaerella caryophylli Pass.
Distr.: AK; Europe.
Host: Dianthus
Mycosphaerella catesbeyi (Cooke) J.H. Miller
Distr.: Southeastern United States(rare).
Host: Quercus
Mycosphaerella cerasella Aderhold
Sphaerella cerasella (Aderhold) Sacc. & Syd.
Cercospora circumscissa Sacc. (anamorph)
Distr.: Central and eastern United States; central Europe.
Notes: Leaf spot (Jenkins. Phytopathology 20:329. 1920).
Host: Prunus
Mycosphaerella cercidicola (Ellis & Kellerm.) F.A. Wolf
Sphaerella cercidicola Ellis & Kellerm.
Distr.: Temperate North America.
Notes: On leaves.
Host: Cercis
Mycosphaerella chenopodii Dearn. & Barth. in Dearn.
Distr.: KS.
Host: Chenopodium
Mycosphaerella chimaphilae (Ellis & Everh.) Höhn.
Mycosphaerella chimaphilina (Peck in Sacc.) House
Distr.: Northeastern North America; northern Europe.
Notes: Parmelee (Canad. J. Bot. 36:870. 1958).
Hosts: Chimaphila, Pyrola
Mycosphaerella chlorogali Fairm. in Millsp. & L. Nutt.
Distr.: CA.
Host: Chlorogalum
Mycosphaerella cinnafolia R. Sprague
Distr.: OR, possibly AK.
Host: Cinna
Mycosphaerella citri Whiteside
Stenella citri-grisea (F.E. Fisher) Sivanesan (anamorph)
Distr.: FL.
Notes: Greasy spot of citus (Sivanesan & Holliday. C.M.I. Descr. 510.1976; Fisher. Phytopathology 51:297. 1961; Whiteside. Phytopathology 62:260. 1972).
Hosts: Citrofortunella, Citrus, Fortunella, Murraya, Poncirus
Mycosphaerella clallamensis R. Sprague
Distr.: WA.
Host: Bromus
Mycosphaerella clintoniana (House) House
Sphaerella clintoniana House
Sphaerella rhododendri Cooke
Distr.: Temperate North America; Europe.
Host: Rhododendron
Mycosphaerella clymenia (Sacc.) Johans. ex Oudem.
Distr.: Central and eastern United States; Europe.
Notes: On living leaves.
Host: Lonicera
Mycosphaerella coerulea (Ellis & Everh.) Tracy & Earle in H.C. Greene
Distr.: CO.
Host: Aquilegia
Mycosphaerella colorata (Peck) Earle in C. Mohr
Distr.: Eastern North America.
Notes: Leaf spot.
Host: Kalmia
Mycosphaerella columbi Rehm
Distr.: OH.
Host: Plantago
Mycosphaerella confusa F.A. Wolf
Mycosphaerella dubia F.A. Wolf non L.E. Miles, hom. illeg.
Pseudocercospora rubi (Sacc.) Deighton (anamorph)
Distr.: North America; Brazil.
Host: Rubus
Mycosphaerella convexula (Schwein.) F.V. Rand
Distr.: Eastern United States.
Notes: Pecan leaf blotch (Rand. Phytopathology 1:135. 1911).
Host: Carya
Mycosphaerella coptis (Schwein.) House
Laestadia coptis (Schwein.) Ellis & Everh.
Sphaerella coptis (Schwein.) Farl.
Septoria coptis Berk. & M.A. Curtis (anamorph)
Distr.: Northeastern North America.
Host: Coptis
Mycosphaerella corni (Schwein.) J.H. Miller
Distr.: Southeastern United States.
Host: Cornus
Mycosphaerella cornicola Tehon & E.Y. Daniels
Distr.: IL.
Host: Cornus
Mycosphaerella cruciferarum (Fr.:Fr.) Lindau in Engl. & Prantl
Distr.: Temperate North America; Europe.
Notes: On overwintered stems and leaves of Brassicaceae (Corlett. Mycotaxon 31:64. 1988).
Hosts: Arabis, Cheiranthus
Mycosphaerella cruenta D.H. Latham
Pseudocercospora cruenta (Sacc.) Deighton (anamorph)
Distr.: North America; Solomon Islands; Uganda.
Hosts: Glycine, Mucuna, Phaseolus, Vigna
Mycosphaerella crus-galli (Ellis & Kellerm.) Lindau in Engl. & Prantl
Sphaerella crus-galli Ellis & Kellerm.
Distr.: KS(type), NY.
Hosts: Echinochloa, Tridens
Mycosphaerella cyaneae F. Stevens & P.A. Young in F. Stevens
Distr.: HI.
Host: Cyanea
Mycosphaerella cypripedii (Peck) House
Distr.: NY.
Host: Cypripedium
Mycosphaerella dahliae (Cooke & Ellis) Coons
Distr.: Eastern United States.
Host: Dahlia
Mycosphaerella davisii F.R. Jones
Cercospora davisii Ellis & Everh. (anamorph)
Distr.: Central North America; Pakistan.
Notes: Leaf blight.
Host: Melilotus
Mycosphaerella dearnessii Barr
Dothidea acicola (Dearn.) Morelet
Oligostroma acicola Dearn.
Scirrhia acicola (Dearn.) Siggers
Systremma acicola (Dearn.) F.A. Wolf & Barbour
Lecanosticta acicola (Thuem.) Syd. in Syd. & Petr. (anamorph)
Distr.: North and South America; Europe.
Notes: Brown needle spot (Evans. Mycol. Pap. 153:7. 1984; Punithalingam & Gibson. C.M.I. Descr. 367. 1973 as *Scirrhia acicola*).
Host: Pinus
Mycosphaerella decidua (Ellis & Kellerm.) Tomilin
Distr.: KS.
Hosts: Scrophularia, Vernonia

Mycosphaerella degeneri Petr.
Distr.: HI.
Host: Euphorbia
Mycosphaerella delphiniicola Earle in Greene
Distr.: Western United States.
Hosts: Delphinium, Ligusticum, Valeriana
Mycosphaerella dendroides (Cooke) Demaree & J.R. Cole
Distr.: Southeastern United States.
Notes: Leaf blotch (Demaree & Cole. J. Agric. Res. 40:777. 1930).
Host: Carya
Mycosphaerella deschampsiae R. Sprague
Distr.: AK.
Host: Deschampsia
Mycosphaerella dianellae F. Stevens & Weedon in F. Stevens
Distr.: HI.
Host: Dianella
Mycosphaerella dolichospora (Sacc. & Fautrey) Wehmeyer
Distr.: Temperate North America; Europe.
Hosts: Carum, Clematis
Mycosphaerella effigurata (Schwein.) House
Sphaerella effigurata (Schwein.) Cooke
Sphaerella fraxinea Peck non Niessl, hom. illeg.
Sphaerella quadrangulata Ellis & Everh.
Mycosphaerella sapindi (Ellis & Everh.) Lindau
Sphaerella sapindi Ellis & Everh.
Piggotia fraxini Berk. & M.A. Curtis (anamorph)
Distr.: Central and eastern North America.
Notes: On leaves.
Hosts: Fraxinus, [Sapindus]
Mycosphaerella eriophila (Niessl) Lindau
Distr.: AK; Europe(type).
Host: Artemisia
Mycosphaerella erysiphina (Berk. & Broome) Kirchn.
Distr.: CA; Great Britain(type).
Host: Humulus
Mycosphaerella eugeniae Rehm
Distr.: HI; Brazil(type).
Host: Eugenia
Mycosphaerella exutans (Cooke) L.E. Miles
Distr.: TX.
Host: Persea
Mycosphaerella fagi (Auersw.) Lindau in Engl. & Prantl
Distr.: ME; Europe.
Host: Fagus
Mycosphaerella fendleri Tracy & Earle in Greene
Distr.: CO.
Host: Thalictrum
Mycosphaerella fijiensis Morelet
Paracercospora fijiensis (Morelet) Deighton (anamorph)
Distr.: HI; southeast Asia; Pacific Basin.
Notes: Black leaf streak (Mulder & Punithalingam. C.M.I. Descr. 413. 1974).
Host: Musa
Mycosphaerella filicum (Desmaz.) Starb.
Distr.: FL; Europe.
Notes: On ferns.
Host: Blechnum
Mycosphaerella fragariae (Tul.) Lindau
Sphaerella fragariae (Tul.) Sacc.
Ramularia brunnea Peck (anamorph)
Distr.: Worldwide on *Fragaria* spp.
Notes: Leaf blight of strawberries (Sivanesan & Holliday. C.M.I. Descr. 708. 1981).
Host: Fragaria
Mycosphaerella fraxini (Niessl) Lindau
Sphaerella fraxini Niessl
Distr.: Temperate regions.
Host: Fraxinus
Mycosphaerella fraxinicola (Schwein.) House
Phyllosticta viridis Ellis & Kellerm. (anamorph)
Distr.: Central and eastern North America.
Notes: Leaf spot (Wolf. Mycologia 31:258. 1939).
Host: Fraxinus
Mycosphaerella freycinetiae F. Stevens
Distr.: HI.
Host: Freycinetia
Mycosphaerella fructicum Starb.
Distr.: AK; Scandinavia(type).
Notes: A poorly known species.
Host: Rubus
Mycosphaerella gardeniae (Cooke) Henn.
Sphaerella gardeniae Cooke
Distr.: GA, SC(type).
Host: Gardenia
Mycosphaerella gaultheriae (Cooke & Ellis) House
Distr.: Temperate North America.
Notes: On leaves.
Host: Gaultheria
Mycosphaerella gentianae (Niessl) Lindau
Distr.: CO; Europe.
Host: Gentiana
Mycosphaerella glauca (Cooke) L.E. Miles non Woronichin, hom. illeg.
Sphaerella glauca Cooke
Distr.: Eastern United States.
Host: Magnolia
Mycosphaerella glycosmae Tracy & Earle in Greene
Distr.: CO.
Host: Osmorhiza
Mycosphaerella gordoniae (Cooke) J.H. Miller
Sphaerella gordoniae Cooke
Distr.: GA.
Host: Gordonia
Mycosphaerella gossypina (Atk.) Earle
Sphaerella gossypina Atk.
Distr.: Cotton-growing regions of United States and Africa.
Notes: Little (C.M.I. Descr. 914. 1987).
Host: Gossypium
Mycosphaerella greenei Tomilin
Mycosphaerella calamagrostidis H.C. Greene non Petr., hom. illeg.
Distr.: OR, WI(type).
Host: Calamagrostis
Mycosphaerella harthensis (Auersw.) P. Larsen
Sphaerella harthensis Auersw.
Mycosphaerella perparva (Sacc.) House
Sphaerella perparva Sacc.
Distr.: North America; Europe.
Hosts: Alnus, Betula
Mycosphaerella hawaiiensis F. Stevens & P.A. Young in F. Stevens
Distr.: HI.
Host: Gunnera
Mycosphaerella hedychii F. Stevens & P.A. Young in F. Stevens
Distr.: HI.
Host: Hedychium
Mycosphaerella hermione (Sacc.) Lindau in Ranoj.
Sphaerella hermione Sacc.
Mycosphaerella bumeliae (Cooke) J.H. Miller
Sphaerella bumeliae Cooke
Mycosphaerella punctiformis (Fr.) J. Schröt. var. *clematidis* Jaap
Distr.: GA, WY; Europe.
Notes: On decaying stems of herbaceous plants.
Hosts: Bumelia, Epilobium, Sideroxylon
Mycosphaerella heucherae (Ellis & Everh.) Petr.
Dothidella heucherae Ellis & Everh.
Distr.: Western North America.
Host: Heuchera
Mycosphaerella holci Tehon
Distr.: Cosmopolitan.
Notes: Anahosur & Sivanesan (C.M.I. Descr. 584. 1978). Although *Phoma sorghina* (Sacc.) Boerema, Dorenbosch & Van Kesteren is often mentioned as the anamorph, White & Morgan-Jones (Mycotaxon 18:12. 1983) stated that this connection has not been confirmed.
Hosts: Avena, Hordeum, Sorghum
Mycosphaerella horii K. Hara
Distr.: Japan, mistakenly reported from the United States.
Notes: This species was thought to be the cause of greasy spot in the United States but apparently does not occur outside of Japan (Wellings. Trans. Brit. Mycol. Soc. 76:495. 1981.)

Hosts: Citrofortunella, Citrus
Mycosphaerella hosackiae (Cooke & Harkn.) Bonar
Sphaerella hosackiae Cooke & Harkn.
Ditopella hosackiae (Cooke & Harkn.) Sacc.
Microdiplodia hosackiae (Cooke & Harkn.) Bonar (anamorph)
Distr.: CA.
Host: Lotus
Mycosphaerella hypericina (Ellis) Tomilin
Sphaerella hypericina Ellis
Distr.: NJ.
Host: Hypericum
Mycosphaerella hypodermellae Wehmeyer
Distr.: WY.
Host: Pinus
Mycosphaerella hypsicola (Ellis & Everh.) Lindau in Engl. & Prantl
Sphaerella hypsicola Ellis & Everh.
Distr.: CO.
Host: Trollius
Mycosphaerella ignobilis (Auersw.) Lindau
Sphaerella ignobilis Auersw.
Distr.: AK; Europe; Soviet Union.
Notes: On grasses.
Host: Agrostis
Mycosphaerella ilicella (Cooke) House
Sphaerella ilicella Cooke
Distr.: NY.
Host: Ilex
Mycosphaerella ilicis (Ellis) Lindau in Engl. & Prantl
Distr.: Eastern North America.
Host: Ilex
Mycosphaerella immersa Dearn.
Distr.: Arctic region.
Host: Cassiope
Mycosphaerella impatiens (Peck & G.W. Clinton) House
Septoria sp. (anamorph)
Distr.: Central and eastern North America.
Notes: Leaf spot.
Host: Impatiens
Mycosphaerella infuscans (Ellis & Everh.) Barr
Sphaerella infuscans Ellis & Everh.
Distr.: WV.
Host: Liriodendron
Mycosphaerella isariphora (Desmaz.) Johans.
Sphaerella isariphora Desmaz.
Distr.: North America; Asia; Europe.
Host: Pyrola
Mycosphaerella janus (Berk. & M.A. Curtis) Petr.
Dothidella janus (Berk. & M.A. Curtis) Höhn.
Leptosphaeria janus (Berk. & M.A. Curtis) Sacc.
Dothidella castanicola Bonar
Dothidella castanopsidis Dearn.
Mycosphaerella castanopsidis (Dearn.) Petr.
Asteromella castanicola (Ellis & Everh.) Petr. (anamorph)
Distr.: Southern and western North America.
Notes: On leaves.
Hosts: Castanopsis, Quercus
Mycosphaerella juglandis K.J. Kessler
Cylindrosporium juglandis F.A. Wolf (anamorph)
Distr.: Central and eastern United States.
Notes: Leaf spot.
Host: Juglans
Mycosphaerella juniperina (Ellis) Tomilin
Sphaerella juniperina Ellis
Dothidella juniperina (Ellis) Höhn.
Dothidella juniperi (Desmaz.) Höhn.
Mycosphaerella juniperi (Fautrey & Roum.) Petr.
Laestadia juniperina (Ellis) Sacc.
Mycosphaerella sylvatica (Sacc.) Petr. & Syd.
Distr.: Temperate North America; Europe.
Notes: On needles (Holm & Holm. Symb. Bot. Upsal. 21:34. 1977).
Host: Juniperus
Mycosphaerella kaduae F. Stevens & P.A. Young in F. Stevens
Distr.: HI.
Host: Kadua
Mycosphaerella koae Petr.
Distr.: HI.
Host: Acacia
Mycosphaerella krigiae (Ellis & Everh.) H.C. Greene
Sphaerella krigiae Ellis & Everh.
Distr.: IL, WI.
Host: Krigia
Mycosphaerella lactucae (Ellis & Kellerm.) Tomilin
Sphaerella lactucae Ellis & Kellerm.
Distr.: Temperate North America; Europe.
Host: Lactuca
Mycosphaerella lageniformis Rehm
Sphaerella lageniformis (Rehm) Trott. in Sacc.
Distr.: CA.
Host: Citrus
Mycosphaerella laricina (R. Hartig) Mig. in Thomé
Cercoseptoria sp. (anamorph)
Distr.: Worldwide.
Host: Larix
Mycosphaerella leptospora (Sacc. & Scalia) Tomilin
Sphaerella leptospora Sacc. & Scalia
Distr.: AK.
Host: Carex
Mycosphaerella ligea (Sacc.) Zeller
Distr.: North America; Asia; Europe.
Host: Rubus
Mycosphaerella lineolata (Roberge ex Desmaz.) J. Schröt. in Cohn
Distr.: Temperate North America; Europe.
Host: Carex
Mycosphaerella linhartiana (Niessl) Starb.
Distr.: NJ; Europe.
Host: Melilotus
Mycosphaerella linicola Naumov
Mycosphaerella linorum (Wollenweb.) Garcia-Rada
Septoria linicola (Speg.) Garassini (anamorph)
Distr.: Worldwide in flax-growing regions.
Notes: Pasmo disease of flax (Sivanesan & Holliday. C.M.I. Descr. 709. 1981).
Host: Linum
Mycosphaerella liriodendri (Cooke) Woronichin
Distr.: Eastern United States.
Host: Liriodendron
Mycosphaerella longissima (Fuckel) Lindau
Distr.: AK; Europe.
Host: Bromus
Mycosphaerella louisianae Plakidas
Distr.: Southeastern United States.
Notes: Purple leaf spot of strawberry.
Host: Fragaria
Mycosphaerella lycopodii (Peck) House
Distr.: Northern North America.
Host: Lycopodium
Mycosphaerella lythracearum F.A. Wolf
Pseudocercospora punicae (Henn.) Deighton (anamorph)
Distr.: Southeastern United States; Bermuda.
Notes: Pomegranate blotch. Chupp (A monograph of the fungus genus *Cercospora*. p. 460. 1952) suggested this anamorph.
Host: Punica
Mycosphaerella maclurae (Ellis & Everh.) Tomilin
Sphaerella maclurae Ellis & Everh.
Distr.: MO.
Host: Maclura
Mycosphaerella macrospora (Kleb.) Jørst.
Didymellina macrospora Kleb.
Didymellina iridis (Desmaz.) Höhn.
Mycosphaerella iridis (Desmaz.) J. Schröt. in Cohn
Distr.: Cosmopolitan.
Notes: Leaf spot or blotch of several liliaceous genera.
Hosts: [Ipomoea], Iris, Triteleia
Mycosphaerella metrosideri F. Stevens & P.A. Young in F. Stevens
Distr.: HI.
Hosts: Acacia, Metrosideros
Mycosphaerella microsora Syd.
Distr.: IA; Europe.

Host: Tilia
Mycosphaerella millepunctata (Desmaz.) Barr
Laestadia millepunctata (Desmaz.) Sacc.
Guignardia millepunctata (Desmaz.) Lind
Phyllachora millepunctata (Desmaz.) Sacc.
Distr.: Eastern North America; Europe.
Notes: On leaves.
Host: Rhododendron
Mycosphaerella minor (P. Karst.) Johans.
Distr.: Alpine and arctic regions.
Hosts: Arctagrostis, Arctostaphylos, Campanula, Cerastium, Chaenactis, Epilobium, Gentianella, Pyrola, Salix, Saxifraga, Spiraea, Tofieldia
Mycosphaerella molleriana (Thuem.) Lindau in Engl. & Prantl
Distr.: Warm-temperate to tropical regions.
Host: Eucalyptus
Mycosphaerella mori (Fuckel) F.A. Wolf
Phloeospora maculans (Berenger) Allesch. (anamorph)
Distr.: Worldwide.
Host: Morus
Mycosphaerella muhlenbergiae (Ellis) Wehmeyer
Sphaerella muhlenbergiae Ellis
Distr.: NJ(type), widespread.
Hosts: Festuca, Juncus, Muhlenbergia, Poa
Mycosphaerella musae (Speg.) Syd. & P. Syd.
Distr.: HI; Australia; Pacific Basin.
Notes: Leaf speckling of bananas.
Host: Musa
Mycosphaerella musicola J.L. Mulder in J.L. Mulder & Stover
Pseudocercospora musae (A. Zimmerm.) Deighton (anamorph)
Distr.: Throughout banana-growing regions of the world, including HI.
Notes: Sigatoka disease of banana (Mulder & Holliday. C.M.I. Descr. 414. 1974).
Host: Musa
Mycosphaerella myricae L.E. Miles
Distr.: Southeastern United States.
Host: Myrica
Mycosphaerella nemoseridis Fairm.
Distr.: CA(type).
Host: Rafinesquia
Mycosphaerella nigerristigma Higgins
Sphaerella nigerristigma (Higgins) Trott. in Sacc.
Distr.: Central and eastern United States.
Host: Prunus
Mycosphaerella nigredo (Schwein.) Lindau
Sphaerella nigredo (Schwein.) Cooke
Distr.: PA.
Host: Rhus
Mycosphaerella nigrita (Cooke) J.H. Miller
Distr.: Eastern United States.
Host: Quercus
Mycosphaerella nigromaculans Shear
Ramularia nigromaculans Shear (anamorph)
Distr.: Eastern North America.
Host: Vaccinium
Mycosphaerella nyssicola (Cooke) F.A. Wolf
Sphaerella nyssicola Cooke
Distr.: Eastern United States.
Notes: Leaf spot.
Host: Nyssa
Mycosphaerella oenotherae (Ellis & Everh.) Lindau in Engl. & Prantl
Sphaerella oenotherae Ellis & Everh.
Mycosphaerella granulata (Ellis & Everh.) Lindau in Engl. & Prantl
Distr.: Central and eastern North America.
Hosts: Baptisia, Oenothera
Mycosphaerella oleina (Cooke) J.H. Miller
Distr.: GA.
Host: Osmanthus
Mycosphaerella olindensis Petr.
Distr.: HI.
Host: Pteris
Mycosphaerella ootheca (Sacc.) Magnus
Distr.: Arctic North America.
Host: Dryas
Mycosphaerella operculata (Sacc.) Barr
Sphaerella operculata Sacc.
Distr.: Western North America.
Host: Quercus
Mycosphaerella ornithogali (J. Jacques) G.A. De Vries
Didymellina ornithogali J. Jacques
Distr.: Northern United States; Europe.
Notes: Known primarily in the anamorphic state.
Host: Ornithogalum
Mycosphaerella oryzae (Garov. & Cattaneo) Jacz.
Distr.: TX; Europe; Japan.
Host: Oryza
Mycosphaerella pachyasca (Rostr.) Vestergr.
Distr.: Temperate North America; Europe.
Hosts: Astragalus, Diapensia, Draba, Plantago, Thymus
Mycosphaerella pachystimae Dearn.
Distr.: Western North America.
Host: Paxistima
Mycosphaerella pardalota (Cooke & Ellis) L.E. Miles
Distr.: NJ.
Host: Myrica
Mycosphaerella pentstemonis Tracy & Earle in Greene
Distr.: CO.
Host: Penstemon
Mycosphaerella perseae L.E. Miles
Distr.: Fl; West Indies; Mexico.
Hosts: Persea, Vaccinium
Mycosphaerella personata Higgins
Pseudocercospora vitis (Lév.) Speg. (anamorph)
Distr.: Cosmopolitan.
Host: Vitis
Mycosphaerella phlogina (Ellis & Everh.) Earle in Greene
Distr.: MT.
Host: Phlox
Mycosphaerella physostegiae Jenk.
Distr.: Southwestern United States.
Host: Dracocephalum
Mycosphaerella pini Rostr. in Munk
Scirrhia pini Funk & A.K. Parker
Dothistroma septospora (Doroguine) Morelet (anamorph)
Distr.: Cosmopolitan.
Notes: Red band needle blight (Evans. Mycol. Pap. 153:36. 1984; Punithalingam & Gibson. C.M.I. Descr. 368. 1973 as *Scirrhia pini*).
Host: Pinus
Mycosphaerella pinodes (Berk. & Bloxam) Vestergr.
Didymella pinodes (Berk. & Bloxam) Petr.
Ascochyta pinodes L.K. Jones (anamorph)
Distr.: Temperate and subtropical regions.
Notes: Leaf, stem, pod, and foot rot of Fabaceae (Punithalingam & Holliday. C.M.I. Descr. 340. 1972). Warmelo (Bothalia 9:183. 1966) provided a discussion of this fungus.
Hosts: Lathyrus, Lupinus, Piptanthus, Pisum, Vicia, Vigna
Mycosphaerella pinsapo (Thuem.) Lindau
Distr.: NY; Europe.
Host: Thuja
Mycosphaerella pittospori (Cooke) Weiss
Distr.: CA.
Host: Pittosporum
Mycosphaerella plantaginicola (Schwein.) Dearn. & House
Distr.: Northeastern North America.
Host: Plantago
Mycosphaerella platanifolia (Cooke) F.A. Wolf
Sphaerella platanifolia Cooke
Distr.: Central and eastern North America.
Host: Platanus
Mycosphaerella pneumatophorae Kohlmeyer
Distr.: Coastal regions of the tropical Atlantic Ocean.
Host: Avicennia
Mycosphaerella polifoliae (Ellis & Everh.) Bubák
Distr.: Northeastern North America.
Host: Andromeda
Mycosphaerella polypodii (Rabenh.) Johans. ex Oudem.
Distr.: MD; Eurasia.
Host: Lycopodium
Mycosphaerella pomi (Pass.) Lindau

Sphaerella pomi Pass.
Cylindrosporium pomi C. Brooks (anamorph)
Distr.: Temperate northern hemisphere.
Notes: Brook's fruit spot.
Hosts: Chaenomeles, Cydonia, Malus
Mycosphaerella pontederiae (Peck) House
Distr.: Central and eastern North America.
Hosts: Nuphar, Nymphaea, Pontederia, Sarracenia
Mycosphaerella populi (Auersw.) J. Schröt. in Cohn
Septoria populi Desmaz. (anamorph)
Distr.: Temperate North America; Europe.
Host: Populus
Mycosphaerella populicola G.E. Thompson
Septoria populicola Peck (anamorph)
Distr.: Temperate North America.
Notes: Thompson (Phytopathology 31:241. 1941).
Host: Populus
Mycosphaerella populifolia (Cooke) House
Sphaerella populifolia Cooke
Distr.: Eastern United States.
Host: Populus
Mycosphaerella populnea (Sacc.) House
Sphaerella populnea Sacc.
Distr.: NY.
Host: Populus
Mycosphaerella populorum G.E. Thompson
Septoria musiva Peck (anamorph)
Distr.: Eastern North America.
Notes: On leaves.
Host: Populus
Mycosphaerella prenanthis (Ellis & Everh.) Barr
Laestadia prenanthis Ellis & Everh.
Distr.: AL.
Notes: On leaves.
Host: Prenanthes
Mycosphaerella primulae (Auersw. & Heufl.) J. Schröt. in Cohn
Stigmatea primulae Auersw. & Heufl.
Distr.: Temperate regions.
Host: Androsace
Mycosphaerella pruni-persicae Deighton
Mycosphaerella persica Higgins & F.A. Wolf non Syd., hom. illeg. & P. Syd.
Miuraea persica (Sacc.) K. Hara (anamorph)
Distr.: Warm temperate regions of the northern hemisphere.
Notes: Frosty mildew of peaches.
Host: Prunus
Mycosphaerella pseudacaciae (Auersw.) Lindau
Sphaerella pseudacaciae Auersw.
Distr.: CA; Europe; Soviet Union.
Host: Robinia
Mycosphaerella psilospora Gilman & Wadley
Septoria querceti Thuem. (anamorph)
Distr.: IA.
Notes: Leaf spot (Gilman & Wadley. Mycologia 44:216. 1952).
Host: Quercus
Mycosphaerella pteridicola Dearn. & House
Distr.: Northeastern North America.
Host: Pteridium
Mycosphaerella pteridis (Desmaz.) J. Schröt. in Cohn
Sphaerella pteridis Desmaz.
Mycosphaerella indistincta (Peck) Lindau in Engl. & Prantl
Distr.: Temperate North America; Europe.
Hosts: Adiantum, Blechum, Pteridium
Mycosphaerella puerariicola J.L. Weimer & Luttrell
Pseudocercospora puerariicola (W. Yamamoto) Deighton (anamorph)
Distr.: GA; Japan.
Notes: Angular leaf spot of kudzu. Originally spelled "*pueraricola*".
Host: Pueraria
Mycosphaerella punctata Dearn. & House ex Barr
Distr.: NY.
Host: Thalictrum
Mycosphaerella punctiformis (Pers.:Fr.) Starb.
Sphaerella punctiformis (Pers.:Fr.) Rabenh.
Mycosphaerella maculiformis (Pers.:Fr.) J. Schröt.
Sphaerella maculiformis (Pers.:Fr.) Auersw.
Ramularia sp. (anamorph)
Distr.: Cosmopolitan.
Notes: On dead leaves of many plant species.
Hosts: Acer, Aesculus, Amelanchier, Asimina, Baptisia, Betula, Carpinus, Carya, Cassiope, Castanea, Celtis, Chionanthus, Corylus, Crataegus, Diospyros, Fagus, Firmiana, Fraxinus, Halesia, Hamamelis, Ledum, Liquidambar, Liriodendron, Lupinus, Magnolia, Ostrya, Platanus, Populus, Prunus, Quercus, Styrax, Tilia, Ulmus, Vaccinium, Viburnum
Mycosphaerella pusilla (Auersw.) Johans.
Distr.: AK; northern Europe.
Host: Arctophila
Mycosphaerella pyrenaica (Speg.) Arx
Distr.: AK; Europe.
Notes: On Brassicaceae.
Host: Draba
Mycosphaerella pyri (Auersw.) Boerema
Mycosphaerella sentina (Fr.:Fr.) J. Schröt. in Cohn
Septoria pyricola (Desmaz.) Desmaz. (anamorph)
Distr.: Temperate northern hemisphere.
Notes: Leaf-fleck. Boerema (Neth. J. Pl. Pathol. 76:166. 1970) provided a nomenclatural discussion.
Hosts: Malus, Pyrus
Mycosphaerella pyrina (Ellis & Everh.) J.H. Miller
Distr.: Eastern United States.
Hosts: Chaenomeles, Pyrus
Mycosphaerella rabiei Kovachevski
Phoma rabiei (Pass.) Khune & J.N. Kapoor (anamorph)
Distr.: Worldwide in chick-pea growing regions.
Notes: Boerema (pers. comm.) agreed with the transfer of the anamorph of this species from *Ascochyta* to *Phoma* (Khune & Kapoor. Indian Phytopathol. 33:120. 1980). Kaiser & Hannan (Pl. Dis. 71:192. 1987).
Host: Cicer
Mycosphaerella ranunculi (P. Karst.) Lind
Sphaerella ranunculi P. Karst.
Mycosphaerella innumerella (P. Karst.) Starb.
Cercosporella sp. (anamorph)
Distr.: Northern hemisphere, cooler parts.
Hosts: Anemone, Potentilla, Ranunculus
Mycosphaerella recutita (Fr.: Fr.) Johans.
Distr.: North America; northern Europe.
Notes: On Poaceae and other monocots.
Hosts: Arctagrostis, Bromus, Carex, Chasmanthium, Dupontia
Mycosphaerella ribis Feltgen
Mycosphaerella aurea R. Stone
Mycosphaerella grossulariae (Fr.) Lindau in Engl. & Prantl
Phyllosticta grossulariae Sacc. (synanamorph)
Septoria ribis (Lib.) Desmaz. (synanamorph)
Distr.: Temperate northern hemisphere.
Notes: Leaf spot. Boerema (Neth. J. Pl. Pathol. 78:30. 1972) and Jørstad (Skr. Norske Vidensk.-Akad. Oslo, Mat.-Naturvidensk. Kl. 24:64. 1965) provided nomenclatural discussions.
Host: Ribes
Mycosphaerella rosicola B.H. Davis ex Deighton
Phaeosphaerella rosicola (B.H. Davis ex Deighton) Tomilin
Cercospora rosicola Pass. (anamorph)
Distr.: Cosmopolitan.
Notes: Leaf spot.
Host: Rosa
Mycosphaerella rosigena (Ellis & Everh.) S.M. McMurran in F.C. Stewart
Distr.: HI, eastern United States; Central America.
Host: Rosa
Mycosphaerella rubi Roark
Septoria rubi Westend. (anamorph)
Distr.: Temperate northern hemisphere.
Host: Rubus
Mycosphaerella rudbeckiae Clements
Distr.: CO.
Host: Rudbeckia
Mycosphaerella salicina (Ellis & Everh.) Tomilin
Distr.: KS.
Host: Salix

Mycosphaerella salicorniae (Auersw.) Lindau
Distr.: Worldwide in marine, saline, or alkaline habitats.
Host: Salicornia
Mycosphaerella sarraceniae (Schwein.: Fr.) Henn.
Distr.: Northern North America.
Hosts: Darlingtonia, Sarracenia
Mycosphaerella sassafras (Ellis & Everh.) Bubák & Kab. in Seym.
Sphaerella sassafras Ellis & Everh.
Distr.: Central and eastern United States.
Host: Sassafras
Mycosphaerella scaevolae F. Stevens & P.A. Young in F. Stevens
Distr.: HI.
Host: Scaevola
Mycosphaerella scirrhoides Barr
Scirrhia lineata Dearn. & House, nom. inval.
Distr.: NY.
Host: Carex
Mycosphaerella septorioides (Desmaz.) Lindau
Distr.: Europe; distribution elsewhere is questionable.
Host: Acer
Mycosphaerella sequoiae Bonar
Distr.: CA.
Host: Sequoia
Mycosphaerella serrulata (Ellis & Everh.) Diehl
Sphaerella serrulata Ellis & Everh.
Distr.: Southeastern United States.
Host: Serenoa
Mycosphaerella shawii Arx & Bezerra
Distr.: ID(type).
Host: Arctostaphylos
Mycosphaerella sicyicola (Ellis & Everh.) Tomilin
Sphaerella sicyicola Ellis & Everh.
Distr.: Central United States.
Notes: Possibly a synonym of *Didymella bryoniae* (Auersw.) Rehm.
Hosts: Echinocystis, Sicyos
Mycosphaerella sidaecola (Ellis & Everh.) Tomilin
Sphaerella sidaecola Ellis & Everh.
Distr.: CA.
Host: Sidalcea
Mycosphaerella smilacicola (Schwein.) Overh.
Sphaerella smilacicola (Schwein.) Cooke
Distr.: Temperate northern hemisphere.
Host: Smilax
Mycosphaerella smilacina (Ellis & Everh.) Tomilin
Sphaerella smilacina Ellis & Everh.
Distr.: NJ.
Host: Smilax
Mycosphaerella solani (Ellis & Everh.) Wollenweb.
Didymella solani (Ellis & Everh.)W. Gams & Gerlach
Distr.: Eastern United States.
Host: Solanum
Mycosphaerella spartinae (Ellis & Everh.) Tomilin
Sphaerella spartinae Ellis & Everh.
Distr.: NE(type).
Host: Spartina
Mycosphaerella sphaerellula (Peck) Barr
Didymella sphaerellula (Peck) Sacc.
Distr.: Eastern North America.
Host: Acer
Mycosphaerella spleniata (Cooke & Peck) House
Distr.: Temperate North America.
Notes: Leaf spot.
Host: Quercus
Mycosphaerella staphylina (Ellis & Everh.) J.H. Miller
Sphaerella staphylina Ellis & Everh.
Distr.: GA, KS.
Host: Staphylea
Mycosphaerella stellarinearum (Rabenh.) Johans.
Distr.: North America; temperate Europe.
Hosts: Arenaria, Stellaria
Mycosphaerella stigmina-platani F.A. Wolf
Mycosphaerella polymorpha D.J. Sm. & C.O. Sm.
Stigmina platani (Fuckel) Sacc. (anamorph)
Distr.: Central and eastern North America; Europe.
Notes: Leaf blight.
Host: Platanus
Mycosphaerella striatiformans N.A. Cobb
Distr.: HI; Fiji; Indonesia.
Notes: Leaf splitting disease of sugarcane.
Host: Saccharum
Mycosphaerella subcongregata (Ellis & Everh.) Tomilin
Sphaerella subcongregata Ellis & Everh.
Distr.: WA.
Host: Erigeron
Mycosphaerella superflua (Auersw.) Petr.
Didymella superflua (Auersw.) Sacc.
Didymosphaeria superflua (Auersw.) Niessl
Ramularia urticae Ces. (anamorph)
Distr.: North America; Europe; India.
Notes: On leaves.
Host: Urtica
Mycosphaerella tabifica (Prill. & Delacr.) Lind
Distr.: Northwestern North America; Europe.
Host: Beta
Mycosphaerella taraxaci (P. Karst.) Lind
Distr.: Northern North America; Europe.
Notes: On various Asteraceae.
Hosts: Erigeron, Gnaphalium, Machaeranthera
Mycosphaerella tassiana (De Not.) Johans.
Sphaerella tassiana De Not.
Mycosphaerella bakeri Rehm
Mycosphaerella confinis (P. Karst.) Dearn. & House
Physalospora leptosperma Rostr.
Phaeosphaerella schoenoprasii (Rabenh.) Petr.
Sphaerella schoenoprasii (Rabenh.) Auersw.
Didymella sphaerelloides Sacc.
Mycosphaerella sphaerelloides (Sacc.) Petr.
Mycosphaerella washingtoniae Rehm
Cladosporium herbarum (Pers.:Fr.) Link (anamorph)
Distr.: Cosmopolitan.
Notes: On a variety of plants. Synonymy is according to Arx (Sydowia 3:40. 1949).
Hosts: Aconitum, Agastache, Agropyron, Allium, Alopecurus, Anaphalis, Androsace, Antennaria, Aquilegia, Arabis, Arenaria, Artemisia, Astragalus, Avena, Calamagrostis, Calamovilfa, Carex, Castanea, Cerastium, Chaenactis, Chimaphila, Chrysanthemum, Cirsium, Draba, Eichhornia, Elymus, Epilobium, Eriogonum, Festuca, Geum, Gnaphalium, Heuchera, Hierochloe, Holodiscus, Juncus, Lomatium, Lupinus, Monarda, Muhlenbergia, Oxytropis, Papaver, Parnassia, Pedicularis, Phleum, Phlox, Poa, Polemonium, Polygonum, Puccinellia, Sagina, Saxifraga, Senecio, Setaria, Sibbaldia, Silene, Sitanion, Stellaria, Thlaspi, Tofieldia, Veratrum, Washingtonia, Xerophyllum, Yucca, Zigadenus
Mycosphaerella tassiana (De Not.) Johans. var. **arctica** (Rostr.) Barr
Laestadia arctica Rostr.
Distr.: Alpine and arctic regions.
Hosts: Arabis, Diapensia, Silene, Veratrum
Mycosphaerella taxi (Cooke) Lind
Distr.: Western United States; Europe.
Notes: On dead leaves.
Host: Taxus
Mycosphaerella taxodii (Cooke) Lindau in Engl. & Prantl
Distr.: SC.
Host: Taxodium
Mycosphaerella tecomae F.A. Wolf
Pseudocercospora sordida (Sacc.) Deighton (anamorph)
Distr.: Southeastern United States.
Notes: Leaf spot.
Host: Campsis
Mycosphaerella tetraspora Seaver
Distr.: PR, TX.
Host: Commelina
Mycosphaerella thalictri (Ellis & Everh.) Lindau in Engl. & Prantl
Sphaerella thalictricola Sacc. & Syd.
Sphaerella septorioides Peck non (Desmaz.) Niessl, hom. illeg.
Distr.: Temperate North America.
Notes: Leaf spot.
Host: Thalictrum
Mycosphaerella tingens (Niessl) Lindau

Distr.: NM; Europe.
Host: Arenaria
Mycosphaerella tulasnei (Jancz.) Lindau
Distr.: Temperate North America; Europe.
Notes: Possibly a synonym of *M. tassiana* (De Not.) Johan.
Hosts: Agropyron, Agrostis, Calamagrostis, Cinna, Elymus, Festuca, Holcus, Hordeum, Juncus, Lathyrus, Malus, Phleum, Poa, Pyrus, Trisetum, Triticum
Mycosphaerella tulipifera (Schwein.) Higgins
Sphaerella fuscans Ellis & Everh.
Phaeoisariopsis liriodendri (Ellis & Harkn.) Morgan-Jones & L.G. Brown (anamorph)
Distr.: Southeastern United States.
Host: Liriodendron
Mycosphaerella tussilaginis (Rehm) Lindau
Distr.: Eastern North America; Europe.
Host: Tussilago
Mycosphaerella typhae (Lasch) Lindau in Engl. & Prantl
Distr.: Central and eastern North America; Europe.
Host: Typha
Mycosphaerella ulmi Kleb.
Phyllosticta bellunensis C. Mart. (synanamorph spermatial)
Phloeospora ulmi (Fr.:Fr.) Wallr. (synanamorph)
Distr.: Temperate northern hemisphere.
Host: Ulmus
Mycosphaerella vaccinii (Cooke) J. Schröt. in Cohn
Carlia vaccinii (Cooke) Höhn.
Distr.: Eastern North America; Europe.
Host: Vaccinium
Mycosphaerella vagnerae Earle
Distr.: NV.
Host: Smilacina
Mycosphaerella verbascicola (Ellis & Everh.) Fairm.
Distr.: Eastern United States.
Host: Verbascum
Mycosphaerella weigelae Fairm.
Distr.: Eastern United States.
Host: Weigela
Mycosphaerella wichuriana (J. Schröt.) Johans.
Distr.: Temperate North America; Europe.
Notes: On Poaceae and other monocots.
Host: Carex
Mycosphaerella xanthicola (Cooke & Harkn.) Lindau
Distr.: CA(type).
Host: Xanthium
Mycosphaerella xerophylli Petr. in Syd. & Petr.
Distr.: ID.
Host: Xerophyllum
Mycosphaerella yuccae (Ellis & Everh.) Tomilin
Sphaerella yuccae Ellis & Everh.
Distr.: Central and eastern United States.
Host: Yucca
Mycosphaerella zeae (Sacc.) Woronow
Mycosphaerella zeicola G.L. Stout
Distr.: Temperate northern hemisphere.
Notes: Mukunya & Boothroyd (Phytopathology 63:530. 1973) compared three *Mycosphaerella* spp. on corn.
Hosts: Secale, Zea
Mycosphaerella zeae-maydis Mukunya & Boothroyd
Phyllosticta maydis D.C. Arny & R.R. Nelson (anamorph)
Distr.: Central and eastern United States; southern Canada.
Notes: Yellow leaf blight of corn.
Host: Zea
Mycosphaerella zizaniae (Schwein.) Lindau in Engl. & Prantl
Mycosphaerella spraguei Tomilin
Distr.: Eastern United States.
Host: Zizania

Mycosyrinx Beck. Basidiomycetes, Ustilaginales

Mycosyrinx cissi (DC.) Beck
Ustilago cissi Tul.
Distr.: Widespread in tropical regions.
Host: Cissus
Mycosyrinx osmundae Beck
Distr.: Central and eastern United States.
Notes: This species is thought to be a hyphomycete (Zambettakis & Foko. Rev. Mycol. 35:304. 1970).
Host: Osmunda

Mycothyridium Petr. Loculoascomycetes, Dothideales

Mycothyridium ceanothi (Dearn. & House) Petr.
Thyridium ceanothi Dearn. & House
Distr.: NY(type).
Host: Ceanothus
Mycothyridium deviatum (Petr.) Petr.
Xylosphaeria deviata Petr.
Distr.: FL(type).
Host: Sabal
Mycothyridium floridanum (Petr.) Petr.
Xylosphaeria floridana Petr.
Distr.: FL(type).
Host: Quercus
Mycothyridium lividum (Pers.:Fr.) Petr.
Thyridium lividum (Pers.:Fr.) Sacc.
Distr.: North America(questionable); Europe.
Host: Acer
Mycothyridium marylandicum (Petr.) Petr.
Xylosphaeria marylandica Petr.
Distr.: MD(type).
Host: Ulmus

Mycotribulus Nag Raj & Kendrick. Coelomycetes

Mycotribulus mirabilis Nag Raj & Kendrick
Distr.: HI; Brazil; India.
Notes: On rotten leaves (Nag Raj & Kendrick. Canad. J. Bot. 48:2219. 1970; Sutton. The Coelomycetes. p. 66. 1980).
Host: Eucalyptus

Mycovellosiella Rangel. Hyphomycetes

Notes: Deighton (Mycol. Pap. 137:1. 1974; Mycol. Pap. 144:1. 1979).
Mycovellosiella ambrosiae (Olive) Deighton
Ormathodium ambrosiae Olive
Cercospora ferruginea Fuckel var. *ambrosiae* (Seym. & Earle) J.J. Davis
Cercospora racemosa Ellis & G. Martin var. *ambrosiae* Seym. & Earle
Cercospora trifidae Ellis & G. Martin
Distr.: Central and eastern United States.
Notes: Leaf spot.
Host: Ambrosia
Mycovellosiella concors (Casp.) Deighton
Cercospora concors (Casp.) Sacc.
Cercospora heterosperma Bres.
Distr.: Eastern united States; East Africa; Asia; Europe.
Notes: Leaf spot (Kirk. C.M.I. Descr. 724. 1982).
Host: Solanum
Mycovellosiella ferruginea (Fuckel) Deighton
Cercospora ferruginea Fuckel in Fresen.
Cercospora absinthii (Peck) Sacc.
Helminthosporium absinthii Peck
Cercosporidium artemisiae Sawada
Cercospora olivacea G. Otth
Distr.: Central and eastern North America; Europe; India; Taiwan.
Notes: On leaves.
Hosts: Ambrosia, Artemisia, Lobelia
Mycovellosiella koepkei (Krüger) Deighton
Cercospora koepkei Krüger
Pseudocercospora miscanthi Katsuki
Distr.: Southeastern United States; Africa; southeastern Asia; Australia; Sri Lanka; South Pacific Islands.
Notes: Yellow spot of sugarcane (Mulder & Holliday. C.M.I. Descr. 417. 1974.).

Host: Saccharum
Mycovellosiella lantanae (Chupp) Deighton
Cercospora lantanae Chupp
Distr.: FL, neotropical regions.
Host: Lantana
Mycovellosiella lobeliae-cardinalis (Schwein.) Deighton
Cercospora lobeliicola Solheim
Cercospora ochracea Sacc. & Malbr.
Distr.: Central and eastern United States; France.
Notes: On leaves (Deighton. Trans. Brit. Mycol. Soc. 86:637. 1986).
Host: Lobelia
Mycovellosiella maclurae (Ellis & Langl.) Deighton
Ovularia maclurae Ellis & Langl.
Distr.: Southeastern United States.
Notes: Cottony leaf spot of Osage orange.
Host: Maclura
Mycovellosiella minax (J.J. Davis) Deighton
Ramularia minax J.J. Davis
Distr.: Central North America.
Notes: Leaf spot.
Host: Solidago
Mycovellosiella murina (Ellis & Kellerm.) Deighton
Cercospora murina Ellis & Kellerm.
Cercospora ii Trail
Cercospora lilacina Bres.
Distr.: Eastern United States; Europe.
Host: Viola
Mycovellosiella perfoliati (Ellis & Everh.) Muntañola
Cercospora perfoliati Ellis & Everh.
Cercospora agerati F. Stevens
Ragnhildiana agerati (F. Stevens) F. Stevens & Solheim
Cercosporella coorgica Muthappa
Cladosporium versicolor Bondartsev
Distr.: HI, central United States; tropical and subtropical regions.
Hosts: Ageratum, Eupatorium
Mycovellosiella pongamiae (A.K. Kar & M. Mandal) Deighton
Cercospora pongamiae A.K. Kar & M. Mandal
Distr.: FL; India.
Host: Pongamia
Mycovellosiella sidarum (Petr. & Cif.) Deighton
Ramularia sidarum Petr. & Cif.
Distr.: Gulf coastal states; West Indies.
Host: Sida
Mycovellosiella teucrii (Schwein.) Deighton
Cercospora teucrii (Schwein.) Arth. & Bisby non Ellis & Kellerm., hom. illeg.
Cercospora racemosa Ellis & G. Martin
Distr.: Central and eastern United States.
Notes: Leaf spot.
Hosts: Ambrosia, Stachys, Teucrium
Mycovellosiella vaginae (Krüger) Deighton
Cercospora vaginae Krüger
Distr.: Subtropical and tropical regions.
Notes: On leaf sheaths (Kirk. C.M.I. Descr. 725. 1982).
Host: Saccharum

Myiocopron Speg. Loculoascomycetes, Dothideales

Haplopeltis Theiss.
Peltella Syd. & P. Syd.
Peltopsis Batista
Myiocopron freycinetiae (F. Stevens & Ryan) G. Arnaud
Peltella freycinetiae F. Stevens & Ryan
Distr.: HI.
Notes: Arx & Müller (Beitr. Kryptogamenfl. Schweiz 11(1):90. 1954) reviewed this species.
Host: Freycinetia
Myiocopron smilacis (De Not.) Sacc.
Microthyrium smilacis De Not.
Peltella smilacis (De Not.) F. Stevens in F. Stevens & Ryan
Distr.: PA, SC; Europe.
Hosts: Abrus, Smilax

Mylittopsis Pat. Basidiomycetes, Auriculariales

Mylittopsis marmorata (Berk. & M.A. Curtis) D.P. Rogers & G.W. Martin
Tremella marmorata Berk. & M.A. Curtis
Distr.: Southeastern United States; Malaya.
Notes: On wood.
Host: Quercus

Myriangium Mont. & Berk. in Berk. Loculoascomycetes, Dothideales

Notes: Arx (Persoonia 2:421. 1963); Miller (Mycologia 32:587. 1940).
Myriangium catalinae Fairm.
Distr.: CA(type).
Host: Salvia
Myriangium citri Henn.
Myriangium floridanum Rehm
Distr.: Subtropical and tropical regions.
Notes: On scale insects on *Citrus*, also reported on *Gleditschia*.
Host: Citrus
Myriangium duriaei Mont. & Berk. in Berk.
Myriangium curtisii Mont. & Berk.
Distr.: Subtropical and tropical regions.
Hosts: Acer, Carya, Citrus, Crataegus, Diospyros, Heteromeles, Liquidambar, Nyssa, Quercus
Myriangium sabaleos Weedon
Myrianginella sabaleos (Weedon) Jenk. & Limber
Distr.: FL, GA.
Notes: Limber & Jenkins (Mycologia 41:550. 1949) conducted a detailed a study and placed this species in *Myrianginella*, but that genus is now considered a synonym of *Uleomyces*.
Host: Sabal
Myriangium superficialis (Peck & G.W. Clinton) Barr in Barr et al.
Melogramma superficialis Peck & G.W. Clinton
Fenestella superficialis (Peck & G.W. Clinton) Sacc.
Cenangium asterinosporum Ellis & Everh.
Myriangium asterinosporum (Ellis & Everh.) J.H. Miller
Distr.: GA, NY.
Notes: On twigs of various woody plants.
Hosts: Crataegus, Cyrilla, Ilex, Malus, Sorbus, Vaccinium
Myriangium tuberculans L.E. Miles
Distr.: IL, southern United States.
Notes: On scale insects.
Hosts: Carya, Crataegus

Myriellina Höhn. Coelomycetes

Notes: Sutton (The Coelomycetes. p. 610. 1980).
Myriellina cydoniae (Desmaz.) Höhn.
Gloeosporium cydoniae Mont. in Castagne
Septogloeum cydoniae (Mont.) Peglion
Phyllosticta cydoniae (Desmaz.) Sacc.
Distr.: FL, MS; Europe.
Host: Cotoneaster

Myrioconium Syd. Coelomycetes

Notes: Anamorphic states of *Myriosclerotinia*.
Myrioconium comitatum J.J. Davis
Distr.: WI.
Notes: On leaves. According to Buchwald (Contr. Dept. Pl. Pathol. Roy. Veterin. Agric. Coll. Copenhagen 32:154. 1949) this species should not be referred to *Myrioconium*, but it has not been redisposed.
Host: Populus
Myrioconium comitatum J.J. Davis var. ***salicarium*** J.J. Davis
Distr.: WI.
Notes: On leaves.
Host: Salix

Myriogenospora Atk. Pyrenomycetes, Clavicipitales

Notes: Luttrell & Bacon (Canad. J. Bot. 55:2090. 1977).
Myriogenospora aciculispora Vizioli
Distr.: LA; Argentina, Brazil.
Notes: Leaf binding or tangle top of sugarcane (Sivanesan & Waller. Phytopathol. Pap. 29:35. 1986).
Host: Saccharum
Myriogenospora atramentosa (Berk. & M.A. Curtis) Diehl
Hypocrea atramentosa Berk. & M.A. Curtis
Dothichloe atramentosa (Berk. & M.A. Curtis) Atk.
Myriogenospora paspali Atk.
Ephelis sp. (anamorph)
Distr.: Southeastern United States.
Notes: A systemic parasite of pasture grasses (Rykard et al. Mycologia 74:648. 1982).
Hosts: Andropogon, Chloris, Cymbopogon, Eremochloa, Panicum, Paspalum, Schizachyrium, Sorghastrum, Sporobolus, Tridens
Myriogenospora bresadoleana Henn.
Distr.: FL; Brazil(type).
Host: Panicum

Myriophacidium M.A. Sherwood. Discomycetes, Rhytismatales

Notes: Sherwood (Mycologia 66:690. 1974).
Myriophacidium aphyophyllicum M.A. Sherwood
Distr.: OR.
Hosts: Castanopsis, Quercus, Rhododendron

Myriosclerotinia Buchwald. Discomycetes, Helotiales

Myriosclerotinia borealis (Bubák & Vleugel) L.M. Kohn
Sclerotinia borealis Bubák & Vleugel
Distr.: Northern hemisphere, cooler parts.
Notes: Snow scald or snow mold of cereals and grasses.
Hosts: Secale, Triticum
Myriosclerotinia caricis-ampullaceae (Nyberg) Buchwald
Sclerotinia caricis-ampullaceae Nyberg
Distr.: North America; northern Europe.
Host: Carex
Myriosclerotinia curreyana (Berk.) Buchwald
Ciboria juncigena Ellis & Everh.
Sclerotinia juncigena (Ellis & Everh.) Whetzel
Distr.: United States(rare); Europe.
Notes: Kohn (Mycotaxon 9:414. 1979).
Host: Juncus
Myriosclerotinia longisclerotialis (Whetzel) Buchwald
Sclerotinia longisclerotialis Whetzel
Distr.: Northeastern North America.
Host: Carex
Myriosclerotinia sulcata (Whetzel) Buchwald, nom. illeg.
Sclerotinia sulcata Whetzel
Distr.: Temperate regions.
Notes: Kohn (Mycotaxon 9:428. 1979) discussed the nomenclatural problems for this species; there is no validly published name.
Host: Carex

Myrmaecium Nitschke ex Fuckel. Pyrenomycetes, Calosphaeriales

Notes: This genus is considered to be a synonym of *Valsaria*, but not all names have been redisposed.
Myrmaecium harperianum Rehm
Distr.: WA.
Host: Cornus

Myrothecium Tode:Fr. Hyphomycetes

Notes: Tulloch (Mycol. Pap. 130:1. 1972).
Myrothecium cinctum (Corda) Sacc.
Fusarium cintum Corda
Distr.: Cosmopolitan.
Notes: On dead plant material, particularly of grasses, and isolated from soil (Domsch et al. Compendium of soil fungi. p. 482. 1980).
Hosts: Eichhornia, Zea
Myrothecium gramineum Lib.
Distr.: Widespread.
Notes: On grasses and occasionally on other plants, also isolated from soil (Ellis. Dematiaceous Hyphomycetes. p. 553. 1971).
Host: Zea
Myrothecium leucotrichum (Peck) Tulloch
Excipula leucotricha Peck
Myrothecium jollymannii N.C. Preston
Distr.: Widespread.
Notes: On various kinds of plant debris (Tulloch. Mycol. Pap. 130:14. 1972; Ellis. Dematiaceous Hyphomycetes. p. 553. 1971 as *M. jollymannii*).
Host: Juncus
Myrothecium roridum Tode:Fr.
Gliocladium nigrum Moreau & Moreau
Distr.: Cosmopolitan, most common in warmer regions.
Notes: Fitton & Holliday (C.M.I. Descr. 253. 1970).
Hosts: Aglaonema, Ajuga, Alcea, Antirrhinum, Asimina, Begonia, Campsis, Cucumis, Dieffenbachia, Eremurus, Ficus, Gardenia, Glycine, Gossypium, Juncus, Leucophyllum, Lycopersicon, Matthiola, Peperomia, Rudbeckia, Sinningia, Vigna
Myrothecium verrucaria (Albertini & Schwein.) Ditmar:Fr.
Gliocladium fibriatum Gilman & E. Abbott
Metarhizium glutinosum S. Pope
Starkeyomyces koorchalomoides Agnihothrudu
Distr.: Cosmopolitan, especially common in soil.
Hosts: Cucumis, Gossypium, Lotus, Zea

Mystrosporium Corda. Hyphomycetes

Mystrosporium consors Thuem.
Distr.: SC(type).
Host: Cortaderia

Mytilinidion Sacc. Loculoascomycetes, Dothideales

Notes: Zogg (Beitr. Kryptogamenfl. Schweiz 11(3):1. 1962) as *Mytilidion*.
Mytilinidion acicola G. Wint.
Distr.: Most common in Europe.
Notes: Bark of living twigs and branches.
Host: Juniperus
Mytilinidion decipiens (P. Karst.) Sacc.
Lophium decipiens P. Karst.
Distr.: Central and eastern United States; European Alps.
Notes: Living branches, bark and dying needles.
Host: Juniperus
Mytilinidion gemmigenum Fuckel
Mytilidion fusisporum (Cooke) Sacc.
Lophium fusisporum Cooke
Herpotrichia quinqueseptata Weir
Distr.: Most common in Europe.
Notes: On woody debris of conifers.
Hosts: Abies, Larix, Picea, Pinus
Mytilinidion mytilinellum (Fr.) Zogg
Lophium mytilinellum Fr.
Distr.: North temperate regions.
Notes: Dead wood and bark of conifers.
Hosts: Abies, Pinus
Mytilinidion oblongisporum S. Teng
Distr.: OR; China(type).
Host: Thuja
Mytilinidion rhenanum Fuckel
Mytilinidion karstenii Sacc.
Distr.: GA; Europe.
Notes: Dead wood, bark and branches.
Host: Pinus
Mytilinidion thujarum (Cooke & Peck) Lohman
Hysterium thujarum Cooke & Peck
Distr.: Northeastern United States.

Host: Thuja
Mytilinidion tortile (Schwein.) Ellis & Everh.
Hysterium tortile Schwein.
Mytilidion juniperi Ellis & Everh. .
Distr.: Western United States, GA.
Notes: On woody debris of conifers.
Hosts: Juniperus, Pinus, Pseudotsuga

Myxocyclus Riess in Fresen. Coelomycetes

Myxocyclus polycistis (Berk. & Broome) Sacc.
Hendersonia polycistis Berk. & Broome
Coryneum irregulare Berk. & M.A. Curtis in Berk.
Splanchnonema argus (Berk. & Broome) Kuntze (teleomorph)
Distr.: Eastern North America; Europe.
Notes: On branches.
Host: Betula

Myxofusicoccum Died. Coelomycetes

Notes: This genus is considered a synonym of *Pseudodiscula* (Sutton. Mycol. Pap. 141:132. 1977), but not all names have been redisposed.
Myxofusicoccum peckianum (Dearn. & House) Petr.
Macrophoma ceanothi Dearn. & House
Distr.: NY.
Host: Ceanothus
Myxofusicoccum salicis Died.
Distr.: Western United States; Europe.
Host: Salix

Myxosporella Sacc. Coelomycetes

Myxosporella ulmicola Bonar
Distr.: WV.
Host: Ulmus

Myxosporium Link, nom. conf. Coelomycetes

Notes: A heterogeneous assemblage of fungi, many of which have been placed in a variety of genera; others are in need of revision.
Myxosporium acerinum Peck
Distr.: Eastern North America.
Host: Acer
Myxosporium alboluteum Dearn. & Barth.
Distr.: KS.
Host: Populus
Myxosporium carpini (Grove) Grove
Myxosporium carneum Lib. var. *carpini* Grove
Distr.: Eastern United States; Great Britain.
Host: Carpinus
Myxosporium castaneum Peck
Distr.: NY.
Host: Castanea
Myxosporium castaneum Peck var. ***quercus*** Peck
Distr.: NY.
Host: Quercus
Myxosporium coloratum (Peck) Sacc.
Melanconium coloratum Peck
Distr.: NY(type).
Host: Liriodendron
Myxosporium cytosporeum Sacc.
Distr.: Northwestern United States.
Host: Salix
Myxosporium depressum Sacc.
Distr.: NY.
Host: Syringa
Myxosporium diedickei Syd.
Distr.: TX, WA; Europe.
Host: Morus
Myxosporium everhartii Sacc. & Syd.
Myxosporium corni Ellis & Everh. non Allesch., hom. illeg.
Distr.: MI, TN.
Host: Cornus
Myxosporium fumosum Ellis & Everh.
Distr.: IL.
Host: Tilia
Myxosporium liriodendri Dearn. & House
Distr.: NY.
Host: Liriodendron
Myxosporium longisporum Edgerton
Distr.: NY.
Host: Liriodendron
Myxosporium luteum Ellis & Everh.
Distr.: WV.
Host: Carya
Myxosporium negundicola Dearn. & Barth.
Distr.: KS(type).
Host: Acer
Myxosporium nitidum Berk. & M.A. Curtis
Distr.: Reported from an unknown locality, probably SC.
Host: Cornus
Myxosporium pruni (Peck) Dearn. & House
Phoma pruni Peck
Sphaeropsis pruni (Peck) Kuntze
Distr.: NY.
Host: Prunus
Myxosporium rhois (Berk. & M.A. Curtis) Sacc.
Naemaspora rhois Berk. & M.A. Curtis
Distr.: Eastern North America.
Host: Rhus
Myxosporium rosae Fuckel
Distr.: Reported from an unknown locality in the United States; Europe.
Host: Rosa
Myxosporium salicinum Sacc. & Roum.
Distr.: MA; Europe.
Host: Salix
Myxosporium sambuci Dearn. nom. nud.
Distr.: WA.
Host: Sambucus
Myxosporium seriatum Ellis & Everh.
Distr.: NY, WV(type).
Host: Acer
Myxosporium stellatum Dearn.
Distr.: NY.
Host: Quercus
Myxosporium tiliae Dearn.
Distr.: NY.
Host: Tilia
Myxosporium tulipiferae Died.
Distr.: IA; Germany(type).
Host: Liriodendron
Myxosporium ulmi (Oudem.) Sacc.
Discella ulmi Oudem.
Distr.: TX; Europe.
Host: Celtis
Myxosporium viticola Dearn. & House, nom. inval.
Distr.: NY.
Host: Vitis

Myxotrichum Kunze. Plectomycetes, Gymnoascales

Myxotrichum setosum (Eidam) G.F. Orr & Plunkett
Gymnoascus setosus Eidam
Distr.: Temperate regions.
Notes: Usually associated with bee hives (Orr. Mycol. Pap. 96:20. 1964).
Host: Phoenix

Naemacyclus Fuckel. Discomycetes, Rhytismatales

Notes: Most species have been transferred to *Cyclaneusma* (DiCosmo et al. Eur. J. Forest Pathol. 13:206. 1983).
Naemacyclus culmigenus Ellis & Langl.
Distr.: Southern United States.
Hosts: Andropogon, Panicum, Paspalum, Saccharum, Schizachyrium, Sorghastrum

Naemacyclus fimbriatus (Schwein.) DiCosmo, Peredo & Minter
Stictis fimbriata Schwein.
Lasiostictis fimbriata (Schwein.) Bäumler
Distr.: Eastern United States; Europe.
Notes: A saprobe on scales of fallen cones, rarely on needles (Sherwood. Mycotaxon 1:41. 1974).
Host: Pinus
Naemacyclus stictiformis Petr.
Distr.: FL(type).
Host: Pinus

Naemosphaera (Sacc.) Sacc. Coelomycetes
Notes: This genus is considered a synonym of *Rabenhorstia* (Sutton. Mycol. Pap. 141:134. 1977), but not all names have been redisposed.
Naemosphaera shastensis R. Sprague & W.B. Cooke
Distr.: CA.
Host: Streptanthus

Naevia Fr. Discomycetes, Helotiales
Notes: This genus is a synonym of *Naevala* (Hein. Willdenowia Beih. 9:8. 1976).
Naevia stenospora Sacc.
Distr.: WA.
Notes: Hein (Willdenowia Beih. 9:34. 1976.) suggested that this species belongs in *Merostictis*, but the transfer was not made.
Host: Heracleum

Nais Kohlmeyer. Pyrenomycetes, Halosphaeriales
Notes: Monotypic.
Nais inornata Kohlmeyer
Distr.: Worldwide along ocean shores and saline lakes.
Notes: On intertidal and drifting wood and bark (Kohlmeyer & Kohlmeyer. Marine mycology. p. 320. 1979; Shearer & Crane. Mycotaxon 7:443. 1978).
Host: Spartina

Nakataea K. Hara. Hyphomycetes
Nakataea sigmoidea (Cavara) K. Hara
Helminthosporium sigmoideum Cavara
Curvularia sigmoidea (Cavara) K. Hara
Vakrabeeja sigmoidea (Cavara) Subramanian
Sclerotium oryzae Cattaneo (synanamorph)
Magnaporthe salvinii (Cattaneo) R. Krause & R.K. Webster (teleomorph)
Distr.: Worldwide in rice-growing regions.
Notes: Stem rot of rice (Ellis & Holliday. C.M.I. Descr. 344. 1972).
Hosts: Echinochloa, Zizania, Zizaniopsis
Nakataea sigmoidea (Cavara) K. Hara var. ***irregulare*** Cralley & Tullis
Helminthosporium sigmoideum Cavara var. *irregulare* Cralley & Tullis
Nakatea irregulare (Cralley & Tullis) K. Hara
Vakrabeeja sigmoideum (Cavara) Subramanian var. *irregulare* (Cralley & Tullis) Shoemaker
Distr.: Southeastern United States; eastern Asia.
Notes: Stem rot of rice. According to Ou (Rice diseases. p. 253. 1985) this variety has never been validly transferred into the genus *Nakatea*.
Host: Oryza

Nannfeldtiomyces K. Vánky. Basidiomycetes, Ustilaginales
Nannfeldtiomyces anomalus (Crowell) K. Vánky
Burrillia anomala Crowell
Distr.: North temperate regions.
Notes: Vánky (Sydowia 34:173. 1981).
Host: Sparganium

Napicladium Thuem. Hyphomycetes
Notes: This genus is considered a synonym of *Spilocaea*, but not all names have been redisposed.
Napicladium astragali Ellis & Everh.
Distr.: MT(type), OK.
Notes: On leaves.
Host: Astragalus
Napicladium prosopodium Tharp
Distr.: TX(type).
Host: Prosopis

Naucoria (Fr.) P. Kumm. Basidiomycetes, Agaricales
Notes: A large, heterogeneous assemblage of brown-spored agarics. Most species have now been transferred to other genera.
Naucoria cerealis Boewe
Distr.: IL.
Notes: On culms of cereals and grasses.
Hosts: Hordeum, Secale, Triticum

Nectria (Fr.) Fr. Pyrenomycetes, Hypocreales
Notes: Booth (Mycol. Pap. 73. 1959); Perrin (Bull. Soc. Mycol. France 92:335. 1976); Samuels (Mem. New York Bot. Gard. 26. 1976); Seifert (Stud. Mycol. 27. 1985).
Nectria apocyni Peck
Dendrodochium sp. (anamorph)
Distr.: NY; Venezuela.
Notes: On stems and bark.
Host: Apocynum
Nectria aquifolii (Fr.) Berk.
Distr.: Temperate regions.
Notes: On woody substrates.
Host: Ilex
Nectria atrofusca (Schwein.) Ellis & Everh.
Otthia seriata (Peck) Sacc.
Otthia staphylina (Ellis & Everh.) Ellis & Everh.
Plowrightia staphylina Ellis & Everh.
Fusarium staphyleae Samuels & C.T. Rogerson (anamorph)
Distr.: IA, eastern North America.
Notes: Twig blight of bladder nut. Samuels & Rogerson (Brittonia 36:81. 1984) provided a description and the teleomorph-anamorph connection.
Hosts: Euonymus, Staphylea
Nectria aurantiaca (Tul.) Jacz.
Sphaerostilbe aurantiaca Tul.
Distr.: WA; Europe.
Host: Ulmus
Nectria aureo-fulva Cooke & Ellis
Dendrodochium sp. (anamorph)
Distr.: Widespread.
Notes: On woody substrates, rarely on herbaceous tissue.
Hosts: Magnolia, Rhus
Nectria brassicae Ellis & Sacc.
Fusarium sp. (anamorph)
Distr.: NJ.
Host: Solanum
Nectria bulbicola Henn.
Tubercularia bulbicola (Henn.) Sacc. (anamorph)
Distr.: Reported from an unknown locality in the United States; Germany.
Host: Phalaenopsis
Nectria canadensis Ellis & Everh.
Scoleconectria canadensis (Ellis & Everh.) Seaver
Tubercularia grayana (Sacc. & Ellis) Seifert (anamorph)
Distr.: Northeastern North America.
Notes: On wood.
Host: Carya
Nectria cancri Rutgers
Hypomyces cancri (Rutgers) Wollenweb.
Distr.: MD; Java(type).
Host: Cannabis

Nectria cinnabarina (Tode:Fr.) Fr.
Nectria offuscata Berk. & M.A. Curtis
Creonectria purpurea (L.) Seaver
Nectria ribis (Tode:Fr.) Rabenh.
Tubercularia vulgaris Tode:Fr. (anamorph)
Distr.: Temperate regions.
Notes: Coral spot fungus (Booth. C.M.I. Descr. 531. 1977).
Hosts: Acer, Aesculus, Ailanthus, Albizia, Aleurites, Alnus, Althaea, Amelanchier, Ampelopsis, Aralia, Asimina, Betula, Bougainvillea, Broussonetia, Buxus, Caesalpinia, Callicarpa, Calycanthus, Carya, Celastrus, Chaenomeles, Chrysanthemum, Citrus, Cornus, Cotoneaster, Cydonia, Cytisus, Daphne, Elaeagnus, Epilobium, Eriobotrya, Erythrina, Euonymus, Fagus, Ficus, Firmiana, Fraxinus, Gleditsia, Gossypium, Hibiscus, Holodiscus, Hydrangea, Indigofera, Juglans, Kerria, Koelreuteria, Ligustrum, Maclura, Magnolia, Malus, Melia, Mimosa, Morus, Philadelphus, Phoradendron, Populus, Prunus, Pyrus, Quercus, Rhododendron, Rhodotypos, Rhus, Ribes, Robinia, Rosa, Roystonea, Rubus, Salix, Sambucus, Schinus, Solanum, Sophora, Sorbus, Spiraea, Symphoricarpos, Tilia, Ulmus, Umbellularia, Vaccinium, Viburnum, Wisteria
Nectria coccinea (Pers.:Fr.) Fr.
Cylindrocarpon candidum (Link) Wollenweb. (anamorph)
Distr.: Cosmopolitan.
Notes: Booth (C.M.I. Descr. 532. 1977).
Hosts: Acer, Ailanthus, Albizia, Alnus, Betula, Carya, Ceanothus, Cornus, Cytisus, Fagus, Fraxinus, Ilex, Juglans, Malus, Melia, Populus, Pyrus, Quercus, Robinia, Salix, Sambucus, Tilia, Ulmus, Umbellularia, Vitis
Nectria coccinea (Pers.:Fr.) Fr. var. **faginata** Lohman, A.M. Watson, & Ayers
Cylindrocarpon faginatum C. Booth (anamorph)
Distr.: Northeastern North America.
Notes: Booth (C.M.I. Descr. 533. 1977).
Hosts: Carya, Fagus
Nectria coryli Fuckel
Creonectria coryli (Fuckel) Seaver
Distr.: Temperate North America; Europe.
Notes: On stems of hardwoods.
Hosts: Betula, Corylus, Populus, Salix, Symphoricarpos
Nectria decora (Wallr.) Fuckel
Calonectria dearnessii Ellis & Everh.
Fusarium sp. (anamorph)
Distr.: United States; Europe.
Notes: On carbonaceous stromata of *Massaria inquinans* on dead branches of dicotylendonous trees (Pirozynski. Fungi Canadenses 74. 1975 as *Calonectria decora*).
Host: Robinia
Nectria depauperata Cooke
Distr.: OR; England.
Host: Howeia
Nectria diploa Berk. & M.A. Curtis
Calonectria diploa (Berk. & M.A. Curtis) Wollenweb.
Fusarium juruanum Henn. (anamorph)
Distr.: Subtropical and tropical regions.
Notes: Parasitic on scale insects.
Host: Citrus
Nectria discophora (Mont.) Mont.
Sphaeria discophora Mont.
Nectria mammoidea W. Phillips & Plowr.
Cylindrocarpon ianothele Wollenweb. (anamorph)
Distr.: Cosmopolitan.
Notes: On woody substrates both as a pathogen and a saprobe.
Hosts: Betula, Quercus
Nectria ditissima Tul. & C. Tul.
Cylindrocarpon willkommii (Lindau) Wollenweb. (anamorph)
Distr.: Temperate North America; Europe.
Notes: On woody plant tissue.
Hosts: Acacia, Alnus, Juglans, Malus, Ribes
Nectria episphaeria (Tode:Fr.) Fr.
Fusarium aquaeductuum Lagerh. var. *medium* Wollenweb. (anamorph)
Distr.: Temperate northern hemisphere.
Notes: On wood-inhabiting stromatic Ascomycetes but often reported on the vascular plant host.
Hosts: Alnus, Corylus, Gleditsia, Lithocarpus, Lupinus, Pinus, Quercus, Sambucus
Nectria erubescens (Desmaz.) W. Phillips & Plowr.
Calonectria erubescens (Desmaz.) Sacc.
Cephalosporiopsis sp. (anamorph)
Distr.: Subtropical and tropical regions.
Hosts: Myrica, Osmanthus
Nectria eucalypti (Cooke & Harkn.) Berl. & Voglino
Distr.: CA.
Hosts: Eucalyptus, Lupinus
Nectria filicina (Cooke & Harkn.) Berl. & Voglino
Distr.: CA.
Host: Cyathea
Nectria flammea (Tul. & C. Tul.) J.M. Dingley
Sphaerostilbe flammea Tul. & C. Tul.
Nectria coccophila (Tul.) Wollenweb. & Reinking
Nectria laeticolor Berk. & M.A. Curtis
Nectria subcoccinea Sacc. & Ellis
Fusarium coccophilum (Desmaz.) Wollenweb. & Reinking (anamorph)
Distr.: Subtropical and tropical regions.
Notes: Parasitic on scale insects (Booth. C.M.I. Descr. 715. 1981).
Hosts: Acer, Citrus, Fraxinus, Oemleria, Quercus, Ulmus
Nectria fuckeliana C. Booth
Creonectria curcubitula (Sacc.) Seaver
Cylindrocarpon cylindroides Wollenweb. var. *tenue* Wollenweb. (anamorph)
Distr.: Circumboreal.
Notes: Wound parasite and dieback of Pinaceae (Booth. C.M.I. Descr. 624. 1979).
Hosts: Abies, Pinus
Nectria galii Plowr. & Harkn.
Distr.: CA(type).
Notes: This species may belong in *Nectriella*.
Host: Galium
Nectria galligena Bres. in Strass.
Cylindrocarpon heteronemum (Berk. & Broome) Wollenweb. (anamorph)
Distr.: Temperate northern hemisphere.
Notes: Booth (C.M.I. Descr. 147. 1967).
Hosts: [Acer], Alnus, Amelanchier, Betula, Carpinus, Carya, Cornus, Crataegus, Cydonia, Fagus, Fraxinus, Ilex, Juglans, Malus, Nyssa, Populus, Pyrus, Quercus, Rhus, Salix, Tilia, Ulmus, Viola
Nectria gracilipes (Tul. & C. Tul.) Wollenweb.
Sphaerostilbe gracilipes Tul. & C. Tul.
Distr.: Warm-temperate to tropical regions.
Notes: On wood and bark.
Hosts: Hibiscus, Melia
Nectria grammicospora Ferd. & Winge
Clonostachys sp. (anamorph)
Distr.: Subtropical and tropical regions.
Notes: On woody substrates.
Host: Cajanus
Nectria haematococca Berk. & Broome
Hypomyces haematococcus (Berk. & Broome) Wollenweb.
Fusarium solani (Mart.) Sacc. (anamorph)
Distr.: Worldwide on a variety of hosts.
Notes: Both a pathogen and a saprobe.
Hosts: Albizia, Cornus, Exacum, Ipomoea
Nectria hippocastani Allesch.
Distr.: Germany(type).
Notes: A doubtful species.
Host: Aesculus
Nectria inventa Pethybr.
Verticillium tenerum (Pers.:Fr.) Link (anamorph)
Distr.: Cosmopolitan.
Notes: Commonly encountered as the anamorph. Domsch et al. (Compendium of soil fungi. p. 501. 1980) presented a useful account.
Hosts: Solanum, Tulipa
Nectria ipomoeae Halst.
Hypomyces ipomoeae (Halst.) Wollenweb.
Fusarium javanicum Koord. (anamorph)
Distr.: Widespread, but most common in warm regions and greenhouses.
Notes: Recognition of this species follows Gerlach & Nirenberg (Mitt. Biol. Bundesanst. Land-Forstw. Berlin-Dahlem 209:362. 1982).

Hosts: Ipomoea, Solanum, Staphylea
Nectria lateritia (P. Karst.) A.Y. Rossman
Stilbonectria lateritia P. Karst.
Distr.: Tropical regions.
Notes: On dead bark and wood (Rossman. Mycol. Pap. 150:22. 1983).
Host: Acacia
Nectria leucorrhodina (Mont.) Samuels
Peziza leucorrhodina Mont.
Pseudomeliola collapsa Earle
Pseudonectria pipericola F. Stevens
Distr.: Subtropical and tropical regions.
Notes: Hyperparasites on meliolaceous leaf parasites and on leaves of dicotyledonous plants (Samuels. Mem. New York Bot. Gard. 26:90. 1976).
Host: Gordonia
Nectria magnoliae Lohman & Hepting in Lohman & A.M. Watson
Distr.: Eastern United States.
Hosts: Liriodendron, Magnolia
Nectria mauritiicola (Henn.) Seifert & Samuels in Seifert
Corallomyces mauritiicola Henn.
Corallomyces elegans Berk. & M.A. Curtis
Sphaerostilbe repens Berk. & M.A. Curtis
Rhizostilbella hibisci (Pat.) Seifert (anamorph)
Distr.: Widespread in tropical regions.
Notes: Mildly parasitic or saprophytic on roots or bark of trees; also isolated from soil (Booth & Holliday. C.M.I. Descr. 391. 1973 as *Sphaerostilbe repens*).
Host: Citrus
Nectria missouriensis Ellis & Everh.
Pleonectria missouriensis (Ellis & Everh.) Sacc.
Thyronectria missouriensis (Ellis & Everh.) Seaver
Gyrostroma missouriense Seeler (anamorph)
Distr.: Eastern United States.
Host: Carya
Nectria ochroleuca (Schwein.) Berk.
Nectria bainii Massee
Nectria gliocladioides Smalley & E.M. Hans.
Nectria pallida Ellis & Everh.
Nectria saccharina Berk. & M.A. Curtis
Nectria vulgaris Speg.
Gliocladium roseum Bainier (anamorph)
Distr.: Cosmopolitan.
Notes: Among the most common *Nectria* species encountered in tropical, subtropical, and temperate regions.
Hosts: Acer, Andromeda, Arecaceae, Carya, Citrus, Clethra, Cupressus, Lilium, Lindera, Liriodendron, Magnolia, Platanus, Theobroma, Veratrum
Nectria oropensoides Rehm in Bref. & Tavel
Nectria lactea Ellis & Morg.
Acremonium sp. (anamorph)
Distr.: Widespread.
Notes: On bark, *Stereum*, and other fungi.
Hosts: Acer, Melia
Nectria pallidula Cooke
Distr.: Eastern United States; England.
Notes: On wood.
Hosts: Amelanchier, Lindera
Nectria peziza (Tode:Fr.) Fr.
Neuronectria peziza (Tode:Fr.) Munk
Nectria umbellulariae Plowr. & Harkn.
Acremonium sp. (anamorph)
Distr.: Cosmopolitan.
Notes: On decorticated wood, less frequently on bark, dung, decaying cloth, herbaceous tissues and basidiocarps (Samuels. New Zealand J. Bot. 14:231-260. 1976).
Hosts: Acer, Betula, Carya, Fagus, Morus, Platanus, Populus, Quercus, Salix, Umbellularia
Nectria pinea J.M. Dingley
Cylindrocarpon pineum C. Booth (anamorph)
Distr.: North America; Great Britain; New Zealand.
Notes: On conifers (Booth. Mycol. Pap. 104:26. 1966).
Host: Abies
Nectria pithoides Ellis & Everh.
Distr.: WA.
Hosts: Alnus, Pinus
Nectria pseudotrichia Berk. & M.A. Curtis
Megalonectria pseudotrichia (Berk. & M.A. Curtis) Speg.
Sphaerostilbe pseudotrichia (Berk. & M.A. Curtis) Berk. & Broome
Thyronectria pseudotrichia (Berk. & M.A. Curtis) Seeler
Sphaerostilbe lateritia Berk. & M.A. Curtis
Calonectria polythalama (Berk.) Sacc.
Nectria polythalama Berk.
Scoleconectria polythalama (Berk.) Seaver
Tubercularia lateritia (Berk.) Seifert (anamorph)
Distr.: Subtropical and tropical regions.
Notes: On bark and wood.
Hosts: Carya, Chionanthus, Citrus, Fraxinus, Grevillea, Jasminum, Medicago
Nectria punicea (J.C. Schmidt:Fr.) Rabenh.
Nectria punicea (J.C. Schmidt:Fr.) Rabenh. var. *ilicis* C. Booth
Cylindrocarpon album (Sacc.) Wollenweb. (anamorph)
Distr.: Cosmopolitan.
Notes: On woody plant tissue.
Hosts: Juglans, Quercus
Nectria purtonii (Grev.) Berk.
Nectria applanata Fuckel
Fusarium aquaeductuum (Radlk. & Rabenh.) Lagerh. (anamorph)
Distr.: Cosmopolitan.
Notes: Usually found on dead wood, often associated with sphaeriaceous fungi.
Host: Acer
Nectria radicicola Gerlach & L. Nilsson
Cylindrocarpon destructans (Zinssmeister) Scholten (anamorph)
Distr.: Cosmopolitan.
Notes: Associated with roots of a wide variety of plants (Booth. C.M.I. Descr. 148. 1967).
Hosts: Lilium, Narcissus, Rhododendron
Nectria ramulariae (Wollenweb.) E. Müller
Neonectria ramulariae Wollenweb.
Cylindrocarpon magnusianum Wollenweb. (anamorph)
Distr.: United States; Germany(type).
Host: Salix
Nectria rubicarpa Cooke
Distr.: Southeastern United States.
Notes: This species is close to *N. coccinea*.
Hosts: Gelsemium, Sida
Nectria rubricarpa Cooke
Distr.: Southeastern United States.
Notes: This species is close to *Nectria coccinea*.
Host: Ilex
Nectria sanguinea Bolton:Fr.
Distr.: Temperate regions.
Notes: No type specimen exists. Most specimens identified as this species have been reidentified as other *Nectria* species.
Hosts: Ceanothus, Liquidambar, Lupinus, Malus, Salix
Nectria secalina Ellis & Everh.
Distr.: AL(type).
Host: Secale
Nectria squamulosa Ellis
Byssonectria chrysocoma Cooke & Harkn.
Nectria chrysocoma (Cooke & Harkn.) Berl. & Voglino
Distr.: CA, NJ, NY.
Notes: On decorticated wood.
Host: Eucalyptus
Nectria subquaternata Berk. & Broome f. **farinosa** Syd.
Distr.: HI(type).
Host: Aleurites
Nectria sulphurata Ellis & Everh.
Pseudonectria sulphurata (Ellis & Everh.) Seaver
Distr.: MT(type).
Host: Populus
Nectria swieteniae-mahagoni Chen
Distr.: FL; Taiwan(type).
Host: Swietenia
Nectria sylvana V. Mouton
Distr.: Temperate regions.
Notes: On dead herbaceous stems.
Host: Equisetum
Nectria thujana (Rehm) Sacc.
Nectriella thujana Rehm

Distr.: NJ(type).
Host: Chamaecyparis
Nectria truncata Ellis
Distr.: NJ(type).
Host: Chamaecyparis
Nectria tuberculariformis (Rehm ex Sacc.) G. Wint.
Hypocrea tuberculariformis Rehm ex Sacc.
Creonectria tuberculariformis (Rehm ex Sacc.) Seaver
Distr.: Boreal regions.
Notes: This and three other low-temperature tolerant, alpine-boreal fungal antagonists were discussed by Samuels et al. (Canad. J. Bot. 62:1898. 1984).
Host: Urtica
Nectria verrucosa (Schwein.) Sacc.
Creonectria verrucosa (Schwein.) Seaver
Distr.: Eastern North America.
Notes: On branches of hardwoods.
Hosts: Acer, Morus, Populus, Sassafras
Nectria viticola Berk. & M.A. Curtis
Distr.: AL(type).
Notes: Booth (Mycol. Pap. 73:73. 1959) considered this species a synonym of *N. episphaeria*; however, an examination of the type specimen at NY suggested that this species is distinct.
Host: Vitis
Nectria vulpina (Cooke) Ellis
Peziza vulpina Cooke
Distr.: North America; Europe.
Notes: On decorticated wood and *Stereum*.
Hosts: Acer, Malus

Nectriella Nitschke ex Fuckel. Pyrenomycetes, Hypocreales

Nectriella cacti Ellis & Everh.
Hyponectria cacti (Ellis & Everh.) Seaver
Distr.: AL(type).
Host: Opuntia
Nectriella pironii Alfieri & Samuels
Kutilakesa pironii Alfieri (anamorph)
Distr.: FL.
Notes: Parasitic on ornamental shrubs (Alfieri & Samuels. Mycologia 71:1178. 1979).
Hosts: Albizia, Aphelandra, Barleria, Bauhinia, Callistemon, Chrysanthemum, Clerodendrum, Codiaeum, Cuphea, Cyphomandra, Dieffenbachia, Dombeya, Dracaena, Eupatorium, Euphorbia, Fatsia, Ficus, Forsythia, Gelsemium, Hedera, Hydrangea, Jasminum, Lansium, Leea, Leucophyllum, Ligustrum, Ludwigia, Lysiloma, Mahonia, Melia, Osmanthus, Parkinsonia, Philodendron, Pittosporum, Plumbago, Polyscias, Psychotria, Salix, Salvia, Sambucus, Solenostemon, Trevesia, Ulmus
Nectriella sambuci (Höhn.) Weese
Charonectria sambuci Höhn.
Distr.: CA; Germany(type).
Host: Sambucus

Nematogonum Desmaz. Hyphomycetes

Notes: Walker & Minter (Trans. Brit. Mycol. Soc. 77:299. 1981).
Nematogonum ferrugineum (Pers.) S.J. Hughes
Nematogonum aurantiacum Desmaz.
Distr.: Northern temperate regions.
Notes: A contact mycoparasite on various fungi; reported on dead wood, bark and roots of various trees and on corms.
Hosts: Alnus, Malus

Nematospora Peglion. Hemiascomycetes, Endomycetales

Nematospora coryli Peglion
Distr.: Warm-temperate to subtropical regions.
Notes: Mukerji (C.M.I. Descr. 184. 1968).
Hosts: Capsicum, Carya, Citrus, Corylus, Glycine, Gossypium, Lycopersicon, Maclura, Punica, Vigna
Nematospora lycopersici A. Schneid.
Distr.: Warm temperate and tropical regions.
Notes: Isolated from various crop plants.
Hosts: Glycine, Lycopersicon, Phaseolus
Nematospora phaseoli Wingard
Distr.: Central and southeastern United States.
Notes: Possibly a synonym of *N. coryli* Peglion (Batra. U.S.D.A. Tech. Bull. 1469:24. 1973).
Hosts: Phaseolus, Vigna

Nematostoma Syd. & P. Syd. Loculoascomycetes, Dothideales

Nematostoma occidentalis (Ellis & Everh.) Barr
Venturia occidentalis Ellis & Everh.
Acanthostigma occidentale (Ellis & Everh.) Sacc.
Distr.: Eastern United States.
Hosts: Amorpha, Antennaria, Artemisia, Cirsium, Dalea

Neobulgaria Petr. Discomycetes, Helotiales

Neobulgaria pura (Fr.) Petr.
Coryne microspora Ellis & Everh.
Ombrophila microspora (Ellis & Everh.) Sacc. & Speg.
Distr.: Temperate regions.
Notes: On decaying wood. The synonymy was determined by Roll-Hansen (Norw. J. Bot. 26:155. 1979). Ginns (Fungi Canadenses 85. 1976).
Host: Populus

Neocosmospora E.F. Sm. Pyrenomycetes, Hypocreales

Neocosmospora vasinfecta E.F. Sm.
Neocosmospora ornamentata Barbosa, nom. inval.
Acremonium sp. (anamorph)
Distr.: Warm temperate to tropical regions.
Notes: Cannon & Hawksworth (Trans. Brit. Mycol. Soc. 82:673. 1984) also recognized variety *africana*. Records from the United States appear to be variety *vasinfecta*.
Hosts: Arachis, Capsicum, Crotalaria, Glycine, Gossypium, Lablab, Medicago, Saccharum, Solanum, Trifolium

Neofuckelia Zeller & Goodd. Coelomycetes

Neofuckelia pinicola Zeller & Goodd.
Distr.: Northwestern North America.
Notes: Associated with *Atropellis pinicola* on cankered branches of living trees (Zeller. Mycologia 27:464. 1935).
Host: Pinus

Neohendersonia Petr. Coelomycetes

Neohendersonia kickxii (Westend.) Sutton & Pollack
Hendersonia pyriformis G. Otth
Neohendersonia pyriformis (G. Otth) Petr.
Distr.: Eastern United States; Europe.
Host: Fagus

Neokeissleria Petr. Pyrenomycetes, Diaporthales

Neokeissleria ribis (Henn. & Plöttner) Petr.
Ceriospora ribis Henn. & Plöttner
Melanconis ribis (Henn. & Plöttner) Sacc. & Trott.
Distr.: AK; Europe.
Host: Ribes

Neolentinus Redhead & Ginns. Basidiomycetes, Agaricales

Notes: Redhead & Ginns (Trans. Mycol. Soc. Japan 26:357. 1985).
Neolentinus adhaerens (Albertini & Schwein.:Fr.) Redhead & Ginns
Lentinus adhaerens (Albertini & Schwein.:Fr.) Fr.

Distr.: Northeastern North America; Europe.
Notes: On wood of conifers.
Host: Tsuga

Neolentinus kauffmanii (A.H. Sm.) Redhead & Ginns
Lentinus kauffmanii A.H. Sm. in Bier & Nobles
Panus kauffmanii (A.H. Sm.) Corner
Distr.: Pacific Coast of North America; Japan.
Notes: Brown rot of conifers.
Hosts: Picea, Tsuga

Neolentinus lepideus (Fr.:Fr.) Redhead & Ginns
Lentinus lepideus (Fr.:Fr.) Fr.
Distr.: Temperate regions.
Notes: Causes brown rot in wood-in-service. Especially common on coniferous wood, but occasionally reported from hardwoods (Pegler. Kew Bull., Add. Ser. 10:182. 1983).
Hosts: Abies, Calocedrus, Larix, Liquidambar, Picea, Pinus, Pseudotsuga, Quercus, Thuja, Tsuga

Neolentinus ponderosus (O.K. Miller) Redhead & Ginns
Lentinus ponderosus O.K. Miller
Distr.: Western North America.
Notes: On stumps and logs of conifers; associated with a brown cubical rot.
Host: Pinus

Neopeckia Sacc. Loculoascomycetes, Dothideales

Notes: Barr (Mycotaxon 20:10. 1984).

Neopeckia coulteri (Peck) Sacc. in Peck
Herpotrichia coulteri (Peck) Bose
Lasiosphaeria coulteri (Peck) Ellis & Everh.
Amphisphaeria acicola (Cooke) Sacc.
Pyrenochaeta sp. (anamorph)
Distr.: Western North America; Europe.
Notes: Snow mold, usually on *Pinus* spp., rarely on *Picea* spp. Sivanesan & Gibson (C.M.I. Descr. 327. 1972 as *Herpotrichia coulteri*).
Hosts: Picea, Pinus

Neophoma Petr. & Syd. Coelomycetes

Neophoma graminella (Sacc.) Petr. & Syd.
Phoma graminella Sacc.
Macrophoma graminella (Sacc.) Berl. & Voglino
Distr.: MS; Europe.
Host: Bromus

Neottiospora Desmaz. Coelomycetes

Neottiospora umbelliferarum Greene
Distr.: WI(type).
Notes: Pirozynski & Shoemaker (Canad. J. Bot. 49:532. 1971) suggested that this species belongs in the *Robillarda* complex.
Hosts: Cicuta, Oxypolis

Neottiosporina Subramanian. Coelomycetes

Notes: Sutton & Alcorn (Austral. J. Bot. 22:517. 1974).

Neottiosporina paspali (Atk.) Sutton & Alcorn
Stagonospora paspali Atk.
Distr.: AL.
Notes: Forming lesions on leaves. Sutton & Alcorn. (Austral. J. Bot. 22:519. 1974) discussed the problems of misidentified specimens.
Hosts: Andropogon, Paspalum, Tridens

Neovossia Körn. Basidiomycetes, Ustilaginales

Notes: Some species previously placed in *Neovossia* are now considered in *Tilletia*.

Neovossia iowensis H. Hume & E.R. Hodgson in E.R. Hodgson
Distr.: Northern hemisphere.
Host: Phragmites

Neurospora Shear & B.O. Dodge. Pyrenomycetes, Sordariales

Neurospora sitophila Shear & B.O. Dodge
Chrysonilia sitophila (Mont.) Arx (anamorph)
Distr.: Cosmopolitan.
Notes: Generally occurring on burned plant parts.
Hosts: Alnus, Pyrus, Saccharum

Nia R.T. Moore & S.P. Meyers. Basidiomycetes, Melanogastrales

Nia vibrissa R.T. Moore & S.P. Meyers
Distr.: Temperate regions.
Notes: A marine species, on wood and grasses. Kohlmeyer & Kohlmeyer (Marine mycology. p. 460. 1979).
Host: Spartina

Nidula V.S. White. Basidiomycetes, Nidulariales

Nidula candida (Peck) V.S. White
Nidularia candida Peck
Distr.: West coast of North America.
Notes: Brodie (The bird's nest fungi. p. 144. 1975).
Hosts: Pteridium, Rubus

Nidularia Fr., nom. cons. Basidiomycetes, Nidulariales

Nidularia griseolazulina Lindsey & R.L. Gilbertson
Distr.: AZ.
Notes: On ribs of fallen saguaro. Lindsey & Gilbertson (Mycotaxon 2:99. 1975).
Host: Carnegiea

Niesslia Auersw. Pyrenomycetes, Trichosphaeriales

Niesslia coloradensis (Cash & R.W. Davidson) W. Gams
Acanthonitschkea coloradensis Cash & R.W. Davidson
Distr.: North America; northern Europe.
Notes: Gams (*Cephalosporium*-artige Schimmelpilze [Hyphomycetes]. p. 163. 1971). Nannfeldt (Svensk Bot. Tidskr. 69:62. 1975) suggested that this fungus belongs in the Diapothales. Barr (Canad. J. Bot. 31:820. 1953).
Hosts: Abies, Pinus

Niesslia exilis (Albertini & Schwein.) G. Wint.
Niesslia pusilla (Fr.:Fr.) J. Schröt.
Venturia vaccinii Ellis & Everh.
Monocillium sp. (anamorph)
Distr.: Temperate regions.
Hosts: Abies, Ledum, Penstemon, Pinus, Vaccinium

Niesslia pulchriseta (Peck) Barr in Barr et al.
Lohwagiella pulchriseta (Peck) Petr.
Trichosphaeria pulchriseta (Peck) Ellis & Everh.
Trichosphaeria flavida Ellis & Everh.
Distr.: Eastern North America.
Notes: On wood.
Hosts: Acer, Celtis, Quercus

Nigroporus Murrill. Basidiomycetes, Aphyllophorales

Nigroporus vinosus (Berk.) Murrill
Polyporus vinosus Berk.
Distr.: Southeastern United States; pantropical.
Notes: On wood.
Hosts: Liquidambar, Pinus

Nigrospora A. Zimmerm. Hyphomycetes

Nigrospora oryzae (Berk. & Broome) Petch
Khuskia oryzae H.J. Hudson (teleomorph)
Distr.: Subtropical and tropical regions.
Notes: On many different kinds of plants; also isolated from air and soil (Ellis. Dematiaceous Hyphomycetes. p. 320. 1971).
Hosts: Acacia, Camellia, Carya, Eichhornia, Helianthus, Lycopersicon, Metrosideros, Oryza, Panicum, Phyllostachys, Saccharum, Sorghum, Trifolium, Zea
Nigrospora sphaerica (Sacc.) E. Mason
Trichosporum sphaerica Sacc.
Distr.: Cosmopolitan.
Notes: Saprobic or weakly parasitic on a wide variety of plants. Ear and stalk rot of grasses.
Hosts: Acacia, Agropyron, Avena, Beta, Carya, Cynodon, Eichhornia, Gossypium, Helianthus, Juncus, Malus, Quercus, Secale, Setaria, Sorghum, Stenotaphrum, Tridens, Triticum, Zea

Nimbomollisia Nannf. Discomycetes, Helotiales

Notes: Nannfeldt (Mycologia 75:292. 1983).
Nimbomollisia aberrans (Peck) Nannf.
Belonium aberrans (Peck) Sacc.
Belonium andropogonis (Berk. & M.A. Curtis) Sacc.
Hysteropezizella andropogonis (Berk. & M.A. Curtis) Nannf.
Distr.: Eastern United States.
Notes: On Poaceae.
Hosts: Andropogon, Aristida, Erianthus, Muhlenbergia, Saccharum, Schizachyrium, Sporobolus
Nimbomollisia stictoidea (Cooke & Ellis) Nannf.
Mollisia stictoidea (Cooke & Ellis) Sacc.
Distr.: NJ.
Host: Juncus

Niptera Fr. Discomycetes, Helotiales

Niptera guestphalica (Rehm) Boud.
Belonopsis guestphalicum (Rehm) Nannf.
Distr.: Temperate regions.
Notes: In semi-aquatic habitats. Aebi (Nova Hedwigia 23:103. 1972); Dennis (Kew Bull. 26:442. 1972).
Hosts: Carex, Juncus
Niptera melanophaea Rehm
Distr.: NY; Germany(type).
Host: Carex
Niptera stictella Sacc. & Speg.
Mollisia stictella Sacc. & Speg., nom. inval.
Distr.: ID; Europe.
Host: Rosa

Nipterella Starb. ex Dennis. Discomycetes, Helotiales

Notes: Dennis (Persoonia 2:188. 1962).
Nipterella parksii (Cash) Dennis
Belonidium parksii Cash
Belonium parksii (Cash) Seaver
Distr.: CA(type).
Notes: On decorticated stems of various woody shrubs.
Host: Ribes

Nitschkia G. Otth ex P. Karst. Pyrenomycetes, Sordariales

Scortechinia Sacc.
Scortechiniella Arx & E. Müller
Notes: Nannfeldt (Svensk Bot. Tidskr. 69:289. 1975).
Nitschkia acanthostroma (Mont.) Nannf.
Scortechinia acanthostroma (Mont.) Sacc. & Berl.
Scortechinia culicitella (Berk. & Ravenel) Speg.
Distr.: Southeastern North America; subtropical and tropical regions.
Hosts: Liquidambar, Quercus
Nitschkia broomeiana (Berk.) Nannf.
Fracchiaea americana Berl.
Fracchiaea heterogenea Sacc.
Fracchiaea moricarpa (Cooke) Sacc.
Nitschkia paurida (Berk. & M.A. Curtis) Cooke
Fracchiaea subcongregata (Berk. & M.A. Curtis) Ellis & Everh.
Fracchiaea subconnata (Berk. & M.A. Curtis) Cooke
Distr.: Cosmopolitan.
Notes: On woody plants.
Hosts: Cercis, Ficus, Liquidambar, Morus, Myrica, Pinus, Prunus, Robinia, Salix, Tilia
Nitschkia callista (Berk. & M.A. Curtis) Nannf.
Cucurbitaria callista Berk. & M.A. Curtis
Cryptosphaerella callista (Berk. & M.A. Curtis) Fitzp.
Fracchiaea callista (Berk. & M.A. Curtis) Sacc.
Distr.: Eastern United States.
Hosts: Carpinus, Celtis, Cornus, Sorbus
Nitschkia calyculus (Mont.) Kuntze
Calyculosphaeria calyculus (Mont.) Fitzp.
Byssosphaeria calyculus (Mont.) Cooke
Winterina calyculus (Mont.) Höhn.
Distr.: FL; South America; Africa.
Host: Tilia
Nitschkia confertula (Schwein.) Nannf.
Tympanopsis euomphala (Berk. & M.A. Curtis) Starb.
Distr.: Eastern United States; Europe; scattered reports from other regions.
Notes: Usually associated with *Hypoxylon rubiginosum*.
Hosts: Carpinus, Cornus, Fraxinus, Quercus
Nitschkia crustacea (P. Karst.) Sacc. & Trott.
Coelosphaeria crustacea P. Karst.
Distr.: ND; Finland(type).
Notes: Fitzpatrick (Mycologia 15:35. 1923) stated that the specimen from North Dakota was "certainly not a *Nitschkia*".
Host: Prunus
Nitschkia grevillii (Rehm) Nannf.
Melanopsamma grevillii Rehm
Winteria tuberculifera Ellis & Everh.
Distr.: Cosmopolitan.
Host: Alnus
Nitschkia polygoni E.Y. Daniels
Distr.: IL(type).
Host: Polygonum

Nivatogastrium Singer & A.H. Sm. Basidiomycetes, Agaricales

Nivatogastrium nubigenum (Harkn.) Singer & A.H. Sm.
Secotium nubigenum Harkn.
Distr.: Western North America.
Notes: On wood of conifers (Singer & Smith. Brittonia 11:224. 1959). Allied with the genus *Pholiota*.
Hosts: Abies, Tsuga

Nodulisporium G. Preuss. Hyphomycetes

Notes: Many species are anamorphs of *Hypoxylon*.
Nodulisporium gregarium (Berk. & M.A. Curtis) J.A. Meyer
Stachylidium gregarium Berk. & M.A. Curtis
Distr.: Widespread in tropical regions.
Notes: On dead wood (Ellis. Dematiaceous Hyphomycetes. p. 232. 1971).
Hosts: Cheirodendron, Metrosideros
Nodulisporium hinnuleum G. Sm.
Distr.: KS.
Notes: Isolated from seeds of millet (Smith. Trans. Brit. Mycol. Soc. 45:390. 1962).
Hosts: Carya, Platanus

Nodulosphaeria Rabenh. Loculoascomycetes, Dothideales

Notes: Shoemaker (Canad. J. Bot. 62:2730. 1984).
Nodulosphaeria dolioloides (Auersw. in Rabenh.) Rabenh.
Leptosphaeria dolioloides Auersw. in Rabenh.

Distr.: North temperate and arctic regions.
Notes: On stems of Asteraceae.
Host: Petasites
Nodulosphaeria erythrospora (Riess) L. Holm
Ophiobolus erythrosporus (Riess) G. Wint.
Distr.: North America; Europe.
Notes: On herbaceous stems.
Hosts: Sambucus, Scrophularia
Nodulosphaeria modesta (Desmaz.) Munk ex L. Holm
Sphaeria modesta Desmaz.
Leptosphaeria modesta (Desmaz.) Auersw.
Distr.: Temperate regions.
Hosts: Actaea, Castilleja, Cirsium, Helenium, Lupinus, Pedicularis, Scrophularia
Nodulosphaeria octoseptata (Wehmeyer) L. Holm
Leptosphaeria octoseptata Wehmeyer
Distr.: Western United States.
Hosts: Senecio, Veratrum
Nodulosphaeria olivacea (Ellis) L. Holm
Leptosphaeria olivacea Ellis
Ophiobolus olivaceus (Ellis) Sacc.
Distr.: Western North America.
Notes: On dead stems of herbaceous plants.
Hosts: Agastache, Aquilegia, Carum, Erigeron, Ligusticum, Lomatium, Mertensia, Senecio
Nodulosphaeria winteri (Niessl) P.G. Crivelli
Leptosphaeria winteri Niessl
Wettsteinina winteri (Niessl) E. Müller
Wettsteinina engadinensis E. Müller
Leptosphaeria morthieriana Sacc.
Distr.: CA; Switzerland.
Notes: Crivelli (Ueber die heterogene Ascomycetengattung *Pleospora* Rabh.; Vorschlag für eine Aufteilung. p. 168. 1983).
Host: Plantago

Nothophacidium J. Reid & Cain. Discomycetes, Helotiales

Notes: Smerlis (Canad. J. Bot. 44:563. 1966).
Nothophacidium abietinellum (Dearn.) J. Reid & Cain
Phacidium abietinellum Dearn.
Distr.: Northeastern North America.
Notes: Snow blight.
Host: Abies

Nummularia Tul. & C. Tul. non J. Hill, hom. illeg. Pyrenomycetes, Xylariales

Notes: This generic name is replaced by *Biscogniauxia* Kuntze, but not all names have been transferred. Whalley & Edwards (Trans. Brit. Mycol. Soc. 85:385. 1985); Jong & Benjamin (Mycologia 63:862. 1971).
Nummularia broomeiana (Berk. & M.A. Curtis) J.H. Miller
Hypoxylon broomeiana Berk. & M.A. Curtis
Jongiella broomeiana (Berk. & M.A. Curtis) Morelet
Nummularia emergens C.G. Lloyd
Xylocladium sp. (anamorph)
Distr.: Southern United States.
Notes: On hardwoods (Rogers. Amer. J. Bot. 62:761. 1975).
Hosts: Fraxinus, Magnolia, Mangifera, Quercus
Nummularia guaranitica Speg.
Distr.: HI; South America(type).
Notes: Pouzar (Ceska Mykol. 33:217. 1979) discussed the generic placement of this species.
Host: Acacia
Nummularia hypophloea (Berk. & Ravenel) Cooke
Diatrype hypophloea Berk. & Ravenel
Anthostoma hypophloea (Berk. & Ravenel) Sacc.
Distr.: FL, SC(type).
Host: Citrus
Nummularia mauritanica Berk. & Cooke
Distr.: HI; Mauritius(type).
Host: Metrosideros
Nummularia repanda (Fr.:Fr.) Nitschke f. ***querceti*** Rehm
Distr.: ND(type).
Host: Quercus
Nummularia signata S.C. Jong & C.R. Benjamin
Jongiella signata (S.C. Jong & C.R. Benjamin) Morelet
Distr.: Southeastern United States.
Notes: On hardwoods.
Hosts: Celtis, Gleditsia, Quercus, Ulmus
Nummularia subconcava (Schwein.) Ellis & Everh.
Sphaeria subconcava Schwein.
Cytoplea subconcava (Schwein.) Starb.
Distr.: ID, PA.
Hosts: Oplopanax, Wisteria
Nummularia tinctor (Berk.) Ellis & Everh.
Distr.: FL.
Hosts: Carya, Celtis

Nyssopsora Arth. Basidiomycetes, Uredinales

Nyssopsora clavellosa (Berk.) Arth.
Triphragmium clavellosum Berk.
Distr.: Temperate North America; eastern Asia.
Notes: Autoecious; on Araliaceae. Parmelee & de Carteret (Fungi Canadenses 221. 1982).
Host: Aralia
Nyssopsora echinata (Lév.) Arth.
Triphragmium echinatum Lév.
Distr.: Temperate North America; Europe.
Notes: Autoecious; on Apiaceae.
Hosts: Cicuta, Coelopleurum, Conioselinum, Ligusticum, Lomatium, Oenanthe

Octospora Hedw. Discomycetes, Pezizales

Octospora wrightii (Berk. & M.A. Curtis) J. Moravec
Humaria wrightii (Berk. & M.A. Curtis) Boud.
Lamprospora wrightii (Berk. & M.A. Curtis) Seaver
Distr.: NE, southern United States; Europe.
Notes: On bark of trees among mosses. According to Döbbeler (Nova Hedwigia 31:844. 1979) this fungus causes rhizoid galls on the moss *Amblystegium*.
Host: Juniperus

Odontia Pers. Basidiomycetes, Aphyllophorales

Notes: The genus is considered a synonym of *Tomentella* nom. cons., but not all names have been redisposed. Jülich (Int. J. Mycol. Lichenol. 1:27. 1982).
Odontia crustula L.W. Miller
Distr.: Central and western United States.
Notes: On wood.
Host: Mortonia
Odontia lateritia Berk. & M.A. Curtis
Distr.: Western United States.
Notes: Ginns (Mycologia 62:243. 1970).
Host: Castanea
Odontia saccharicola Burt
Distr.: HI(type).
Host: Saccharum

Odonticium Parmasto. Basidiomycetes, Aphyllophorales

Odonticium romellii (S. Lundell) Parmasto
Odontia romellii S. Lundell in J. Eriksson
Distr.: Temperate northern hemisphere.
Notes: On wood.
Hosts: Abies, Larix, Pinus, Pseudotsuga, Thuja

Odontotrema Nyl. Discomycetes, Ostropales

Notes: Sherwood-Pike (Mycotaxon 28:137. 1987).

Odontotrema majusculum Rehm

Odontotrema minus Nyl. f. *salicella* Kauffm.

Distr.: Western North America; Europe.

Notes: Usually on decorticated wood of conifers.

Hosts: Abies, Populus

Odontotrema minus Nyl.

Distr.: European Alps; Scandinavia.

Notes: Sherwood-Pike (Mycotaxon 28:158. 1987) stated that reports of this fungus from North America are misdeterminations of *Xylopezia hemisphaerica*.

Hosts: Abies, Pinus, Tsuga

Odontotrema oregonense M.A. Sherwood

Distr.: OR.

Notes: On dry decorticated coniferous wood and desert shrubs at high elevations.

Hosts: Artemisia, Pinus, Purshia

Oidiodendron Robak. Hyphomycetes

Oidiodendron griseum Robak in Melin & Nannf.

Distr.: Temperate regions; Trinidad.

Notes: Isolated from litter, soil and wood pulp (Ellis. Dematiaceous Hyphomycetes. p. 35. 1971).

Host: Pinus

Oidiodendron truncatum Barron

Distr.: Temperate regions.

Notes: Isolated from soil, the rhizosphere, and wheat (Domsch et al. Compendium of soil fungi. p. 524. 1980).

Host: Abies

Oidiopsis Scalia. Hyphomycetes

Oidiopsis macrospora (Uppal, M.K. Patel & M.N. Kamat) Mundk. & Thirumalachar

Oidiopsis taurica E.S. Salmon var. *macrospora* Uppal, M.K. Patel & M.N. Kamat

Distr.: FL; India(type).

Notes: On living leaves.

Host: Hibiscus

Oidiopsis sicula Scalia

Oidiopsis taurica E.S. Salmon, nom. illeg.

Leveillula taurica (Lév.) G. Arnaud (teleomorph)

Distr.: Cosmopolitan.

Notes: Powdery mildew of cotton and many other plants (Mukerji. C.M.I. Descr. 182. 1968 as *L. taurica*).

Hosts: Allium, Gossypium, Lycopersicon

Oidium Link, nom. cons. Hyphomycetes

Notes: This name is used for conidial states of many species in the Erysiphaceae (Braun. Beih. Nova Hedwigia 89. 1987).

Oidium ambrosiae Thuem.

Distr.: AZ, SC(type).

Host: Ambrosia

Oidium bixae Viégas

Distr.: FL; Brazil(type).

Host: Bixa

Oidium candicans (Sacc.) Linder

Monilia candicans Sacc.

Distr.: North America; Europe.

Hosts: Quercus, Ulmus

Oidium caricae F. Noack

Oidium caricicola J. Yen & Wang

Sphaerotheca caricae-papayae S. Tanda & U. Braun (teleomorph)

Distr.: Subtropical and tropical regions.

Notes: Powdery mildew of papaya (Tanda & Braun. Trans. Mycol. Soc. Japan 26:316. 1985).

Host: Carica

Oidium erysiphoides Fr.

Distr.: NM; Europe.

Notes: According to Braun (Beih. Nova Hedwigia 89:625. 1987) this is a doubtful name.

Host: Anoda

Oidium erysiphoides Fr. f. ***crotalariae*** Cif. & Gonz. Frag.

Distr.: Subtropical and tropical Americas.

Host: Crotalaria

Oidium euonymi-japonici (Arcang.) Sacc. in E.S. Salmon

Microsphaera euonymi-japonici Vien.-Bourg. (teleomorph)

Distr.: Temperate regions.

Host: Euonymus

Oidium laevisporum (Cooke) Linder

Zygodesmus laevisporus Cooke

Rhinotrichum laevisporum (Cooke) Sumstine

Distr.: FL, NC.

Host: Magnolia

Oidium obductum Ellis & Langl.

Acrosporium obductum (Ellis & Langl.) Sumstine

Distr.: Eastern and southern United States.

Notes: Described on *Quercus*.

Host: Platanus

Oidium pirinum Ellis & Everh.

Distr.: FL, WI(type).

Hosts: Crataegus, Malus

Oidium tingitaninum C.N. Carter

Acrosporium tingitaninum (C.N. Newton) Subramanian, nom. illeg.

Distr.: CA(type), worldwide on citrus.

Notes: *Acrosporium* Nees:Fr. is a rejected name, thus the combination is illegitimate.

Hosts: Citrus, Poncirus

Oligoporus Bref. Basidiomycetes, Aphyllophorales

Notes: Gilbertson & Ryvarden (North American Polypores 2:486. 1987).

Oligoporus subpendulus (Atk.) R.L. Gilbertson & Ryvarden

Tyromyces subpendulus Atk.

Distr.: NY.

Notes: On wood.

Host: Tsuga

Oligostroma Syd. & P. Syd. Loculoascomycetes, Dothideales

Notes: This genus is considered a synonym of *Mycosphaerella*, but not all names have been redisposed.

Oligostroma suttoniae F. Stevens

Distr.: HI(type).

Host: Myrsine

Oliveonia Donk. Basidiomycetes, Tulasnellales

Notes: Talbot (Persoonia 3:381. 1965).

Oliveonia subviolacea (Peck) M. Larsen

Oliveonia atrata (Bres.) Talbot

Ceratobasidium atratum (Bres.) D.P. Rogers in G.W. Martin

Corticium atratum Bres.

Distr.: Western hemisphere; possibly Europe.

Notes: On wood.

Hosts: Pinus, Tsuga

Olpidium (A. Braun) J. Schröt. Chytridiomycetes, Chytridiales

Olpidium brassicae (Woronin) P.A. Dang.

Distr.: Cosmopolitan.

Notes: Obligate root parasite on a wide variety of plants (Barr. Fungi Canadenses 176. 1980).

Hosts: Agrostis, Brassica, Fragaria, Lactuca, Lycopersicon, Nicotiana, Spinacia, Triticum

Olpidium trifolii J. Schröt.

Distr.: Widespread.

Notes: Parasitic on epidermal cells.

Host: Trifolium

Olpitrichum Atk. Hyphomycetes

Olpitrichum macrosporum (Farl.) Sumstine
Rhinotrichum macrosporum Farl.
Oidium macrosporum (Farl.) Linder
Plectrothrix globosa Shear
Distr.: Eastern North America; Australia.
Notes: Often associated with cotton-boll decay.
Hosts: Buxus, Eucalyptus, Ficus, Gossypium, Liriodendron, Vaccinium
Olpitrichum tenellum (Berk. & M.A. Curtis) Holubová-Jechová
Rhinotrichum tenellum Berk. & M.A. Curtis
Acladium tenellum (Berk. & M.A. Curtis) Subramanian
Oidium tenellum (Berk. & M.A. Curtis) Linder
Gonatobotrys cucumerinum (Berk. & M.A. Curtis) Sumstine
Distr.: Temperate North America; Europe.
Notes: On various plants; reported to be a mycoparasite (Kuykendall et al., Mycologia 75:656. 1983).
Hosts: Allium, Gossypium, Pennisetum, Zea

Ombrophila Fr. Discomycetes, Helotiales

Ombrophila lysichitonis Kanouse
Distr.: WA(type).
Notes: Seaver (North Amer. cup fungi. [Inoperculates]. p. 233. 1951) stated that the description strongly suggests *Ombrophila clavus*.
Host: Lysichiton

Omphalia (Pers.) S.F. Gray. Basidiomycetes, Agaricales

Notes: This genus is no longer recognized, but not all names have been redisposed.
Omphalia tralucida D.E. Bliss
Distr.: CA.
Notes: The type of this specimen is apparently lost.
Hosts: Phoenix, Washingtonia

Omphalina Quél. Basidiomycetes, Agaricales

Omphalina epichysium (Pers.:Fr.) Quél.
Clitocybe epichysium (Pers.:Fr.) H.E. Bigelow
Distr.: Temperate North America; Europe.
Notes: On wood.
Host: Tsuga

Omphalospora Theiss. & Syd. Loculoascomycetes, Dothideales

Omphalospora stellariae (Lib.) Theiss. & Syd.
Euryachora stellariae (Lib.) Fuckel
Phyllachora stellariae (Lib.) J. Schröt.
Distr.: North America; northern Europe.
Notes: On leaves and stems.
Hosts: Paeonia, Stellaria

Omphalotus Fayod. Basidiomycetes, Agaricales

Omphalotus olearius (DC.:Fr.) Singer
Clitocybe illudens (Schwein.) Sacc.
Omphalotus illudens (Schwein.) Bresinsky & Besl
Monodelphus subilludens Murrill
Distr.: Worldwide in temperate and tropical regions.
Notes: Lignicolous; especially common on hardwood logs and stumps in eastern North America. Notable for its luminescent lamellae and ability to cause gastrointestinal poisoning when eaten.
Hosts: Abies, Castanea, Malus, Olea, Quercus, Robinia
Omphalotus olivascens H.E. Bigelow, O.K. Miller, & Thiers
Distr.: CA.
Notes: On wood of living or dead hardwoods.
Hosts: Arctostaphylos, Cercis, Lithocarpus, Quercus

Oncopodium Sacc. Hyphomycetes

Oncopodium simmonsii Sutton
Distr.: FL.
Notes: On dead branches (Sutton. Mycologia 70:793. 1978).
Host: Serenoa

Oncospora Kalchbr. & Cooke. Coelomycetes

Oncospora abietina Oudem. & Fautrey
Distr.: CA; France.
Host: Sequoia

Oospora Wallr. Hyphomycetes

Notes: This name is illegitimate (Hughes. Canad. J. Bot. 36:791. 1958), but not all names have been redisposed.
Oospora mali M.N. Kidd & A. Beaumont
Distr.: WA; England(type).
Host: Malus
Oospora nicotianae Penz. & Sacc.
Toruloidea nicotianae (Penz. & Sacc.) Sumstine
Distr.: KY, OH; Italy(type).
Host: Nicotiana
Oospora otophila C. Harz
Distr.: WA; Europe(type).
Host: Malus
Oospora tulipiferae Ellis & G. Martin
Alysidium tulipiferae (Ellis & G. Martin) Pound & Clements
Toruloidea tulipiferae (Ellis & G. Martin) Sumstine
Distr.: NC, PA.
Host: Liriodendron

Ophiobolus Riess. Loculoascomycetes, Dothideales

Notes: Shoemaker (Canad. J. Bot. 54:2365. 1976); Walker (Mycotaxon 11:50. 1980).
Ophiobolus acuminatus (Sowerby:Fr.) Duby
Sphaeria acuminatus Sowerby:Fr.
Ophiobolus disseminans Riess
Coniothyrium sp. (anamorph)
Distr.: Temperate regions.
Notes: On stems of Asteraceae.
Hosts: Ambrosia, Cirsium, [Heracleum], [Panicum], [Rumex], [Scirpus], [Setaria], Sorghum, Xanthium, [Zea]
Ophiobolus anguillidus (Cooke) Sacc.
Sphaeria anguillidus Cooke
Distr.: Temperate North America.
Notes: On Asteraceae.
Hosts: Ambrosia, Arctium, Aster, [Cannabis], Cirsium, Coreopsis, Erigeron, Helianthus, Monarda, Rudbeckia, Xanthium
Ophiobolus anthrisci (L. Holm) L. Holm
Ophiobolus nigromaculatus Rehm f. *anthrisci* L. Holm
Distr.: UT.
Host: Iva
Ophiobolus brachyascus (G. Wint.) Sacc.
Rhaphidospora brachyascus G. Wint.
Distr.: NJ; Europe.
Notes: Described from *Lavendula*.
Host: Panicum
Ophiobolus byssicola Harkn.
Distr.: CA, CO.
Host: Sambucus
Ophiobolus cariceti (Berk. & Broome) Sacc.
Sphaeria cariceti Berk. & Broome
Distr.: Southern United States.
Notes: Walker (Trans. Brit. Mycol. Soc. 58:446. 1972) stated that this name is based on several disparate elements and therefore cannot be used.
Hosts: Elymus, Leersia
Ophiobolus castillejae Tracy & Earle
Distr.: Western United States.
Host: Castilleja
Ophiobolus cirsii (P. Karst.) Sacc. f. ***furcatus*** Rehm

Distr.: NJ(type).
Notes: The type form was redescribed by Shoemaker (Canad. J. Bot. 54:2373. 1976) but this form was not mentioned.
Host: Cirsium
Ophiobolus cirsii-altissima J.H. Miller & M.G. Burton
Distr.: GA(type).
Notes: Possibly a synonym of *Leptosphaeria compressa* (Rehm) L. Holm according to Holm (Symb. Bot. Upsal. 14:29. 1957).
Host: Cirsium
Ophiobolus claviger Harkn.
Distr.: CA, ND.
Hosts: Artemisia, Iva, Salvia
Ophiobolus collapsus Ellis & Sacc.
Distr.: Temperate North America.
Hosts: Chenopodium, Lappula, Trifolium
Ophiobolus consimilis Ellis & Everh.
Distr.: Southern United States.
Host: Abelmoschus
Ophiobolus crassus Shoemaker
Distr.: NC.
Hosts: Aster, Verbesina
Ophiobolus drechsleri Shoemaker
Distr.: Central and eastern North America.
Notes: On stems of Asteraceae.
Hosts: Ambrosia, Helianthus, Heliopsis, Verbesina
Ophiobolus gnaphalii (Sacc. & Briard) Fairm. var. ***lanaria*** Fairm.
Distr.: Central and eastern United States.
Notes: On stems of Asteraceae.
Host: Antennaria
Ophiobolus heveae Henn.
Distr.: FL; Brazil.
Host: Hevea
Ophiobolus instabilis Ellis & Everh.
Distr.: WI(type).
Host: Artemisia
Ophiobolus junci J.H. Miller & M.G. Burton
Distr.: GA(type).
Notes: Walker (Mycotaxon 11:54. 1980) stated that this species may belong in *Ophiosphaerella*.
Host: Juncus
Ophiobolus lonicerae Fabre
Distr.: CA; France.
Host: Lonicera
Ophiobolus niesslii Bäumler
Distr.: North America; Europe.
Hosts: Aralia, Lythrum
Ophiobolus nigro-clypeata J.H. Miller & M.G. Burton
Distr.: GA.
Hosts: Achillea, Ambrosia, Amsonia, Cassia, Cimicifuga, Coreopsis, Daucus, Desmodium, Firmiana, Helianthus, Ligusticum, Lonicera, Oxypolis, Sanicula, Smilax, Solidago
Ophiobolus ophioboloides (Sacc.) L. Holm
Leptosphaeria ophioboloides Sacc.
Leptosphaeria chrysanthemi Clements
Distr.: North America; Europe.
Notes: On stems of Asteraceae.
Host: Cirsium
Ophiobolus prunellae (Ellis & Everh.) Shoemaker
Linospora prunellae Ellis & Everh.
Ceuthocarpon brunellae (Ellis & Everh.) Berl.
Linospora brunellae Ellis & Everh., orth. var.
Leptosphaeria hesperia Barr
Leptosporopsis rostrupii (Ferd. & Winge) Höhn.
Ophiobolus rostrupii Ferd. & Winge
Distr.: Western North America.
Notes: Shoemaker (Canad. J. Bot. 54:2388. 1976) discussed this species and the problems with the orthographic variant. Walker (Mycotaxon 11:37. 1980) considered this a synonym of *Leptosporopsis rostrupii* (Ferd. & Winge) Höhn.
Host: Prunella
Ophiobolus stictisporus (Cooke & Ellis) Sacc.
Sphaeria stictispora Cooke & Ellis
Distr.: GA, NJ.
Notes: Walker (Trans. Brit. Mycol. Soc. 58:451. 1972) suggested that this species may belong in the Ostropales. The epithet has occasionally been misspelled as *strictispora*.
Hosts: Arundinaria, Danthonia
Ophiobolus tanaceti (Fuckel) Sacc.
Leptosphaeria tanaceti (Fuckel) L. Holm
Distr.: North America; Europe.
Notes: On stems of Asteraceae.
Host: Artemisia
Ophiobolus vermisporus (Ellis) Shoemaker
Lophiostoma vermisporum Ellis
Lophionema vermisporum (Ellis) Sacc.
Distr.: Central and eastern United States.
Host: Oenothera

Ophiocapnocoma Batsch & Cif. Loculoascomycetes, Dothideales

Ophiocapnocoma phloiophilia (E.E. Fisher) S.J. Hughes
Limacinia phloiophilia E.E. Fisher
Limacinia multiseptata Barr
Ophiocapnocoma multiseptata (Barr) Batista & Cif.
Hormiokrypsis libocedri Batsch & Nascim. (synanamorph)
Capnophialophora sp. (synanamorph)
Distr.: CA; Australia; New Zealand.
Notes: On bark of trunks (Hughes. New Zealand J. Bot. 10:233. 1972).
Host: Arctostaphylos

Ophiodothella (Henn.) Höhn. Pyrenomycetes, Phyllachorales

Ophiodothella ferruginea (Ellis & G. Martin) Barr
Linospora ferruginea Ellis & G. Martin
Ceuthocarpon ferrugineum (Ellis & G. Martin) Berl.
Distr.: FL.
Host: Lyonia
Ophiodothella fici E.A. Bessey
Distr.: FL.
Host: Ficus
Ophiodothella floridana Chárdon
Distr.: FL.
Host: Ficus
Ophiodothella leucospila (Berk. & M.A. Curtis) J.H. Miller & G.E. Thompson
Linospora leucospila (Berk. & M.A. Curtis) Sacc.
Distr.: Southeastern United States.
Notes: Miller & Thompson (Mycologia 32:10. 1940).
Host: Platanus
Ophiodothella vaccinii Boyd
Distr.: Central, eastern, and southern United States.
Notes: Parasitic on leaves of ericaceous plants (Boyd. Mycologia 26:465. 1934).
Hosts: Gaylussacia, Vaccinium, Zenobia

Ophiosphaerella Speg. Loculoascomycetes, Dothideales

Ophiosphaerella herpotricha (Fr.:Fr.) J.C. Walker
Ophiobolus herpotrichus (Fr.:Fr.) Sacc. & Roum.
Phaeosphaeria herpotricha (Fr.:Fr.) L. Holm
Ophiobolus medusae Ellis & Everh. f. *bromi* Brenckle
Ophiobolus oryzae Miyabe
Scolecosporiella sp. (anamorph)
Distr.: Cosmopolitan.
Notes: On Poaceae. Walker (Mycotaxon 11:76. 1980) reviewed the nomenclature and synonyms of this species.
Hosts: Agropyron, Oryza, Vetiveria, Zea

Ophiostoma Syd. & P. Syd. Pyrenomycetes, Ophiostomatales

Notes: The generic limits between *Ceratocystis* and *Ophiostoma* are based on anamorphic states and biochemical characteristics as proposed by De Hoog & Scheffer (Mycologia 76:292. 1984). The *Sporothrix* anamorph of several species are described by De Hoog (Stud. Mycol. 7:12. 1974). Harrington (Mycotaxon 23:39. 1987).

Ophiostoma abiocarpum (R.W. Davidson) T.C. Harrington
Ceratocystis abiocarpa R.W. Davidson
Distr.: Southwestern United States.
Host: Picea
Ophiostoma adjuncti (R.W. Davidson) T.C. Harrington
Ceratocystis adjuncti R.W. Davidson
Distr.: NM.
Notes: Upadhyay (Monograph of *Ceratocystis* and *Ceratocystopsis*. p. 79. 1981) treated this taxon as a synonym of *Ceratocystis ips* (Rumb.) C. Moreau.
Host: Pinus
Ophiostoma aureum (Robinson-Jeffrey & R.W. Davidson) T.C. Harrington
Europhium aureum Robinson-Jeffrey & R.W. Davidson
Ceratocystis aurea (Robinson-Jeffrey & R.W. Davidson) Upadhyay
Leptographium aureum M.J. Wingfield (anamorph)
Distr.: Western North America.
Host: Pinus
Ophiostoma bicolor R.W. Davidson & D. Wells in R.W. Davidson
Ceratocystis bicolor (R.W. Davidson & D. Wells) R.W. Davidson
Hyalorhinocladiella sp. (anamorph)
Distr.: Western United States.
Hosts: Picea, Pinus
Ophiostoma brevicolla (R.W. Davidson) De Hoog & R.J. Scheffer
Ceratocystis brevicollis R.W. Davidson
Distr.: Western United States.
Host: Populus
Ophiostoma clavigerum (Robinson-Jeffrey & R.W. Davidson) T.C. Harrington
Europhium clavigerum Robinson-Jeffrey & R.W. Davidson
Ceratocystis clavigera (Robinson-Jeffrey & R.W. Davidson) Upadhyay
Graphiocladiella clavigerum Upadhyay (anamorph)
Distr.: Western North America.
Host: Pinus
Ophiostoma crassivaginatum (H.D. Griffin) T.C. Harrington
Ceratocystis crassivaginata H.D. Griffin
Ceratocystiopsis crassivaginata (H.D. Griffin) Upadhyay
Leptographium sp. (anamorph)
Distr.: Northern North America.
Host: Populus
Ophiostoma distortum (R.W. Davidson) De Hoog & R.J. Scheffer
Ceratocystis distorta R.W. Davidson
Distr.: Western North America.
Hosts: Abies, Picea
Ophiostoma grandifoliae (R.W. Davidson) T.C. Harrington
Ceratocystis grandifoliae R.W. Davidson
Distr.: IA.
Host: Fagus
Ophiostoma huntii (Robinson-Jeffrey) De Hoog & R.J. Scheffer
Ceratocystis huntii Robinson-Jeffrey in Robinson-Jeffrey & Grinchenko
Distr.: Northwestern North America.
Host: Pinus
Ophiostoma ips (Rumbold) Nannf.
Ceratostomella ips Rumbold
Ceratocystis ips (Rumbold) C. Moreau
Ceratocystis montia (Rumbold) J. Hunt
Graphilbum sp. (synanamorph)
Hyalorhinocladiella sp. (synanamorph)
Distr.: North America; South Africa; Europe.
Notes: In bark of beetle-infested conifer wood.
Hosts: Larix, Picea, Pinus
Ophiostoma leptographioides (R.W. Davidson) Arx
Ceratostomella leptographioides R.W. Davidson
Ceratocystis leptographioides (R.W. Davidson) J. Hunt
Leptographium sp. (anamorph)
Distr.: Eastern United States.
Host: Quercus
Ophiostoma longirostellatum (Bakshi) Arx & E. Müller
Ceratocystis longirostellata Bakshi
Ceratocystis californica DeVay, R.W. Davidson, & W.J. Moller
Sporothrix sp. (anamorph)
Distr.: Temperate North America; Europe.
Hosts: Pinus, Prunus
Ophiostoma megalobrunneum (R.W. Davidson & Toole) De Hoog & R.J. Scheffer
Ceratocystis megalobrunnea R.W. Davidson & Toole in R.W. Davidson, T.E. Hinds & Toole
Distr.: MS.
Host: Quercus
Ophiostoma minus (Hedgc.) Syd. & P. Syd.
Ceratostomella minor Hedgc.
Ceratocystis minor (Hedgc.) J. Hunt
Ceratostomella pini Münch
Ceratostomella pseudotsugae Rumbold
Hyalorhinocladiella sp. (anamorph)
Distr.: Temperate northern hemisphere.
Notes: Blue stain of conifer lumber.
Hosts: Abies, Larix, Picea, Pinus, Pseudotsuga
Ophiostoma multiannulatum (Hedgc. & R.W. Davidson) Arx
Ceratostomella multiannulata Hedgc. & R.W. Davidson in R.W. Davidson
Ceratocystis multiannulata (Hedgc. & R.W. Davidson) J. Hunt
Sporothrix sp. (anamorph)
Distr.: Southeastern United States.
Notes: Common on lumber.
Hosts: Liriodendron, Pinus
Ophiostoma narcissi Limber
Ceratocystis narcissi (Limber) J. Hunt
Sporothrix sp. (anamorph)
Distr.: Temperate North America; Europe; New Zealand.
Notes: On bulbs.
Host: Narcissus
Ophiostoma nigrocarpum (R.W. Davidson) De Hoog
Ceratocystis nigrocarpa R.W. Davidson
Distr.: Temperate North America.
Hosts: Abies, Pinus, Pseudotsuga, Tsuga
Ophiostoma nigrum (R.W. Davidson) De Hoog & R.J. Scheffer
Ceratocystis nigra R.W. Davidson
Distr.: Western United States.
Host: Abies
Ophiostoma obscurum (R.W. Davidson) Arx
Ceratostomella obscura R.W. Davidson
Ceratocystis obscura (R.W. Davidson) J. Hunt
Leptographium sp. (anamorph)
Distr.: Southern United States.
Notes: A doubtful species (Upadhyay. A monograph of *Ceratocystis* and *Ceratocystiopsis*. p. 137. 1981).
Host: Pinus
Ophiostoma olivaceum Mathiesen
Ceratocystis olivacea (Mathiesen) J. Hunt
Ceratocystis vesca R.W. Davidson
Phialographium sp. (anamorph)
Distr.: Western North America; Sweden.
Hosts: Abies, Picea, Pinus
Ophiostoma penicillatum (Grosmann) Siemaszko
Ceratostomella penicillata Grosmann
Ceratocystis penicillata (Grosmann) C. Moreau
Ceratocystis pseudoeurophioides Olchowecki & J. Reid
Ceratocystis truncicolor (R.W. Davidson) H.D. Griffin
Ophiostoma truncicolor R.W. Davidson
Leptographium penicillatum Grosmann (anamorph)
Distr.: Western United States.
Hosts: Abies, Picea
Ophiostoma piceae (Münch) Syd. & P. Syd.
Ceratostomella piceae Münch
Ceratocystis piceae (Münch) Bakshi
Pesotum piceae J.L. Crane & Schoknecht (synanamorph)
Sporothrix sp. (synanamorph)
Distr.: Temperate North America; Europe.
Hosts: Abies, Acer, Fagus, Picea, Populus, Pseudotsuga, Quercus, Tsuga
Ophiostoma piliferum (Fr.:Fr.) Syd. & P. Syd.
Ceratocystis pilifera (Fr.:Fr.) C. Moreau
Ceratostomella pilifera (Fr.:Fr.) G. Wint.
Ceratocystis ambrosia Bakshi
Ceratocystis capillifera (Hedgc.) C. Moreau
Ceratostomella coerulea Münch
Ceratostomella echinella Ellis & Everh.

Ceratocystis schrenkiana (Hedgc.) C. Moreau
Sporothrix sp. (anamorph)
Distr.: Temperate North America; Europe; Australia.
Notes: Wood stain of conifer and hardwood lumber.
Hosts: Acer, Aesculus, Carya, Eucalyptus, Fagus, Liquidambar, Picea, Pinus, Platanus, Populus
Ophiostoma pluriannulatum (Hedgc.) Syd. & P. Syd.
Ceratostomella pluriannulata Hedgc.
Ceratocystis pluriannulata (Hedgc.) C. Moreau
Hyalodendron sp. (anamorph)
Distr.: Eastern North America; Europe.
Notes: Stain of hardwood and conifer lumber.
Hosts: Aesculus, Carya, Liquidambar, Liriodendron, Pinus, Populus, Quercus
Ophiostoma populinum (T.E. Hinds & R.W. Davidson) De Hoog & R.J. Scheffer
Ceratocystis populina T.E. Hinds & R.W. Davidson
Distr.: North America.
Notes: Stem and branch canker.
Hosts: Pinus, Populus
Ophiostoma robustum (Robinson-Jeffrey & R.W. Davidson) T.C. Harrington
Europhium robustum Robinson-Jeffrey & R.W. Davidson
Ceratocystis robusta (Robinson-Jeffrey & R.W. Davidson) Upadhyay
Leptographium robustum M.J. Wingfield (anamorph)
Distr.: Western North America.
Host: Pinus
Ophiostoma rostrocoronatum (R.W. Davidson & Eslyn) De Hoog & R.J. Scheffer
Ceratocystis rostrocoronata R.W. Davidson & Eslyn in Eslyn & R.W. Davidson
Distr.: ME.
Hosts: Acer, Betula, Fagus
Ophiostoma rostrocylindricum (R.W. Davidson) Arx
Ceratostomella rostrocylindrica R.W. Davidson
Ceratocystis rostrocylindrica (R.W. Davidson) J. Hunt
Leptographium sp. (anamorph)
Distr.: Northeastern North America.
Host: Quercus
Ophiostoma sparsum (R.W. Davidson) De Hoog & R.J. Scheffer
Ceratocystis sparsa R.W. Davidson
Distr.: AK.
Host: Picea
Ophiostoma stenoceras (Robak) Melin & Nannf.
Ceratostomella stenoceras Robak
Ceratocystis stenoceras (Robak) C. Moreau
Ceratocystis eucastaneae R.W. Davidson
Ceratocystis gossypina R.W. Davidson
Ceratocystis gossypina R.W. Davidson var. *robusta* R.W. Davidson
Ophiostoma gossypinum (R.W. Davidson) J.J. Taylor
Sporothrix pallidum (Tubaki) T. Matsushima (anamorph)
Distr.: Cosmopolitan.
Notes: A common saprobe of wood.
Hosts: Abies, Betula, Castanea, Pinus, Quercus
Ophiostoma subannulatum Livingston & R.W. Davidson
Sporothrix subannulata Livingston & R.W. Davidson (anamorph)
Distr.: ID(type).
Notes: Livingston & Davidson (Mycologia 79:145. 1987).
Host: Abies
Ophiostoma tremulo-aureum (R.W. Davidson & T.E. Hinds) De Hoog & R.J. Scheffer
Ceratocystis tremulo-aurea R.W. Davidson & T.E. Hinds
Distr.: Western North America.
Host: Populus
Ophiostoma ulmi (Buisman) Nannf.
Ceratostomella ulmi Buisman
Ceratocystis ulmi (Buisman) C. Moreau
Sporothrix sp. (anamorph)
Pesotum ulmi (M.B. Schwartz) J.L. Crane & Schoknecht (anamorph)
Distr.: Central and eastern North America; Europe.
Notes: Dutch elm disease (Booth & Gibson. C.M.I. Descr. 361. 1973).
Host: Ulmus
Ophiostoma wageneri (D.J. Goheen & F.W. Cobb) T.C. Harrington
Ceratocystis wageneri D.J. Goheen & F.W. Cobb
Distr.: Western North America.
Notes: In roots of conifers. The anamorph is presumed to be *Leptographium wageneri* var. *ponderosum* based on circumstantial evidence (Harrington & Cobb. Mycotaxon 30:507. 1987).
Hosts: Abies, Pinus, Pseudotsuga, Tsuga

Oramasia Urries. Hyphomycetes

Notes: Sutton (Mycologia 70:796. 1978).
Oramasia hirsuta Urries
Distr.: Widespread.
Notes: On decaying bark and fallen leathery leaves.
Host: Eucalyptus
Oramasia hirsuta Urries var. ***ramosa*** Sutton
Distr.: FL.
Notes: On dead leaves.
Host: Quercus

Orbilia Fr. Discomycetes, Helotiales

Orbilia abutilonis Cash
Distr.: HI(type).
Host: Abutilon
Orbilia chrysocoma (Bull.) Sacc.
Calloria chrysocoma (Bull.) Fr.
Distr.: Cosmopolitan.
Notes: On old wood and pasteboard.
Host: Sambucus
Orbilia coccinella (Sommerf.) Fr.
Calloria coccinella (Sommerf.) W. Phillips
Mollisia coccinella (Sommerf.) Gill.
Distr.: Temperate regions.
Notes: On wood, bark and occasionally on fungi.
Hosts: Oxydendrum, Purshia
Orbilia cruenta (Schwein.) Seaver
Calloria occulta Rehm
Orbilia occulta (Rehm) Sacc.
Distr.: Temperate regions.
Host: Salix
Orbilia epipora (Nyl.) P. Karst.
Distr.: Temperate regions.
Notes: On wood, bark and occasionally fungi (polypores).
Hosts: Artocarpus, Mangifera
Orbilia eucalypti (W. Phillips & Harkn.) Sacc.
Calloria eucalypti W. Phillips & Harkn.
Distr.: CA, possibly WA.
Hosts: Eucalyptus, Salix
Orbilia leucostigma (Fr.) Fr.
Calloria leucostigma (Fr.) W. Phillips
Mollisia leucostigma (Fr.) Gill.
Distr.: Temperate regions.
Notes: On rotting wood.
Host: Psidium
Orbilia luteorubella (Nyl.) P. Karst.
Calloria luteorubella (Nyl.) W. Phillips
Distr.: Cosmopolitan.
Notes: On decorticated wood of dicotyledonous trees. Benny et al. (Canad. J. Bot. 56:2006. 1978) described the ultrastructural morphology of this species.
Host: Quercus
Orbilia myriospora (W. Phillips & Harkn.) Seaver
Calloria myriospora W. Phillips & Harkn.
Distr.: CA.
Host: Psoralea
Orbilia paradoxa Velen.
Distr.: AK; Austria(type).
Host: Alnus
Orbilia rubrococcinea (Rehm) Sacc.
Distr.: GA, KY(type).
Host: Albizia
Orbilia vinosa (Albertini & Schwein.) P. Karst.
Calloria vinosa (Albertini & Schwein.) Fr.
Mollisia vinosa (Albertini & Schwein.) Gill.
Distr.: Temperate regions.
Host: Morus

Orbilia xanthostigma (Fr.) Fr.
Calloria xanthostigma (Fr.) W. Phillips
Mollisia xanthostigma (Fr.) Gill.
Distr.: Temperate regions.
Host: Tsuga

Orphanomyces Savile. Basidiomycetes, Ustilaginales

Orphanomyces arcticus (Rostr.) Savile
Cintractia arctica (Rostr.) Lagerh.
Distr.: North America; Asia; northern Europe.
Notes: Savile (Canad. J. Bot. 52:341. 1974).
Host: Carex

Ossicaulis Redhead & Ginns. Basidiomycetes, Agaricales

Ossicaulis lignatilis (Pers.:Fr.) Redhead & Ginns
Pleurotus lignatilis (Pers.:Fr.) P. Kumm.
Pleurotus abscondens Peck
Distr.: Northern hemisphere; Japan.
Notes: On trunks of living or recently fallen trees (Redhead & Ginns. Trans. Mycol. Soc. Japan 26:362. 1985).
Host: Abies

Osteina Donk. Basidiomycetes, Aphyllophorales

Osteina obducta (Berk.) Donk
Oligoporus obductus (Berk.) R.L. Gilbertson & Ryvarden
Polyporus osseus Kalchbr.
Distr.: North America; Europe.
Notes: On wood.
Hosts: Abies, Alnus, Betula, Larix, Picea, Pinus, Pseudotsuga, Tsuga

Ostreichnion Duby. Loculoascomycetes, Dothideales

Notes: Barr (Mycotaxon 3:81. 1975).
Ostreichnion curtisii (Duby) Barr
Hysterium curtisii Duby
Glonium curtisii (Duby) Lohman
Glonium macrosporium Tracy & Earle
Distr.: Southern United States.
Notes: On dead wood of various woody substrates.
Hosts: Berchemia, Carpinus, Castanea, Cyrilla, Persea, Quercus, Smilax, Vaccinium, Vitis
Ostreichnion sassafras (Schwein.) Barr
Lophium sassafras Schwein.
Mytilidion sassafras (Schwein.) Zogg
Ostreichnion americanum Duby
Ostreion americanum (Duby) Sacc.
Distr.: Eastern United States; China.
Host: Quercus

Ostreola Darker. Loculoascomycetes, Dothideales

Ostreola formosa (Cooke) Barr
Hysterium formosum Cooke
Hysterographium formosum (Cooke) Sacc.
Distr.: CA.
Notes: Barr (Mycotaxon 29:505. 1987).
Hosts: Abies, Pinus

Ostropa Fr. Discomycetes, Ostropales

Ostropa cinerea (Pers.) Fr.
Distr.: WA.
Notes: Sherwood (Mycotaxon 5:70. 1977) suggested that this species occurs only in Europe, possibly also Algeria and the Canary Islands.
Host: Rubus

Ostropella (Sacc.) Höhn. Loculoascomycetes, Dothideales

Notes: Müller & Arx (Beitr. Kryptogamenfl. Schweiz 11(2):339. 1962).
Ostropella leve (Syd.) E. Müller in E. Müller & Arx
Xenolophium leve Syd.
Distr.: HI(type).
Host: Metrosideros
Ostropella verrucosa (Syd.) E. Müller in E. Müller & Arx
Xenolophium verrucosum Syd.
Distr.: HI(type).
Host: Metrosideros

Otthia Nitschke ex Fuckel. Loculoascomycetes, Dothideales

Otthia amica Sacc., E. Bommer & M. Rousseau
Hendersonia diplodioides Ellis & Everh. (anamorph)
Distr.: Western United States; Europe.
Hosts: Malus, Sambucus
Otthia diminuta P. Karst.
Distr.: AK; Finland(type).
Host: Salix
Otthia fruticola (Ellis & Everh.) Ellis & Everh.
Parodiella fruticola Ellis & Everh.
Distr.: Western United States.
Notes: On stems of *Clematis*, confused with *Scirrhia insculpta* as discussed by Barr (Contr. Univ. Michigan Herb. 9:566. 1972).
Host: Clematis
Otthia rubicola Petr.
Distr.: HI(type).
Host: Rubus
Otthia shearii Petr.
Diplodia shearii Petr. (anamorph)
Distr.: HI(type).
Host: Leucaena
Otthia spiraeae (Fuckel) Fuckel
Cucurbitaria spiraeae Fuckel
Distr.: Northwestern North America; Europe.
Notes: Booth (Trans. Brit. Mycol. Soc. 41:335. 1958).
Host: Crataegus

Oudemansiella Speg. Basidiomycetes, Agaricales

Oudemansiella canarii (Jungh.) Höhn.
Distr.: FL, pantropical.
Notes: On living and recently dead woody species.
Hosts: Bursera, Carya, Coccoloba, Ficus, Liquidambar, Nectandra, Quercus

Ovularia Sacc. Hyphomycetes

Notes: According to Hughes (Trans. Brit. Mycol. Soc. 32:34. 1939; Canad. J. Bot. 36:792. 1958), the distinctions between *Didymaria*, *Ovularia*, and *Ramularia* are unclear. Arx (Proc. Koninkl. Nederl. Akad. Wetens., Ser. C, 86:52.1983) provided a recent circumscription.
Ovularia aristolochiae Dearn., nom. nud.
Distr.: WV.
Host: Aristolochia
Ovularia asperifolii Sacc. var. **lappulae** J.J. Davis
Distr.: WI(type).
Notes: On living leaves.
Host: Hackelia
Ovularia asteris Solheim
Distr.: WY.
Hosts: Aster, Chrysopsis
Ovularia avicularis Peck
Distr.: NE(type).
Host: Polygonum
Ovularia bullata Ellis & Everh.
Distr.: CA.
Host: Stachys
Ovularia carletonii Ellis & Kellerm.

Distr.: Central United States.
Host: Lactuca
Ovularia cercidis Camara
Distr.: MO; Portugal(type).
Host: Cercis
Ovularia compacta Ellis & Everh.
Distr.: Western United States.
Host: Agoseris
Ovularia decipiens Sacc.
Distr.: Temperate North America; Europe.
Host: Ranunculus
Ovularia delphinii Solheim
Distr.: WY.
Host: Delphinium
Ovularia edwiniae Clements, nom. nud.
Distr.: CO.
Notes: Published in an exsiccati set without a description.
Host: Jamesia
Ovularia hughesiana Sacc.
Distr.: MT.
Host: Arnica
Ovularia isarioides Sacc.
Oidium irregulare Peck, nom. nud.
Distr.: Central and eastern North America.
Host: Staphylea
Ovularia lolii Volkart
Distr.: OR; Argentina; Europe.
Notes: Sprague (Diseases of cereals and grasses in North America. p. 409. 1950).
Host: Lolium
Ovularia lotophaga Ellis & Everh.
Distr.: Central and western United States.
Hosts: Cyphia, Lotus
Ovularia lupini W.B. Cooke & C.G. Shaw
Distr.: CO(type).
Host: Lupinus
Ovularia lupinicola Pollack in J.A. Stevenson
Distr.: Western United States.
Notes: Stevenson (Mycologia 38:531. 1946).
Host: Lupinus
Ovularia rhamnigena Ellis & Everh.
Distr.: OR.
Host: Rhamnus
Ovularia sphaeroidea Sacc.
Ovularia schwarziana Magnus
Distr.: Temperate North America; Europe.
Host: Vicia
Ovularia stachydis-ciliatae Peck
Distr.: WA.
Host: Stachys

Ovulariopsis Pat. & Har. Hyphomycetes

Ovulariopsis farinosa Syd.
Distr.: FL; Costa Rica(type).
Notes: On living leaves.
Host: Zanthoxylum

Ovulinia Weiss. Discomycetes, Helotiales

Ovulinia azaleae Weiss
Sclerotinia azaleae (Weiss) Dennis
Ovulitis azaleae Buchwald (anamorph)
Distr.: Temperate regions.
Notes: Petal blight.
Hosts: Kalmia, Rhododendron

Oxydothis Penz. & Sacc. Pyrenomycetes, Amphisphaeriales

Oxydothis pandanicola (Syd. & P. Syd.) Petr.
Didymella pandanicola Syd. & P. Syd.
Schizochora pandani F. Stevens
Distr.: HI; The Philippines.
Hosts: Pandanus, Pritchardia
Oxydothis sabalensis (Cooke) Petr.
Dilophia sabalensis (Cooke) Sacc.
Metasphaeria sabalensis (Cooke) Cooke
Distr.: Southeastern United States.
Host: Sabal

Oxyporus (Bourd. & Galzin) Donk. Basidiomycetes, Aphyllophorales

Notes: Gilbertson & Ryvarden (North American Polypores 2:491. 1987).
Oxyporus corticola (Fr.:Fr.) Ryvarden
Poria corticola (Fr.:Fr.) Cooke
Poria separans Murrill
Poria vesiculosus (Berk. & M.A. Curtis in Berk.) Cooke
Poria vicina Bres.
Distr.: Western United States.
Notes: The name *Poria vesiculosus* is based on an abnormal growth form of this species.
Hosts: Abies, Alnus, Fouquieria, Pinus, Platanus, Populus, Prosopis, Prunus, Pseudotsuga, Salix, Tilia, Tsuga
Oxyporus cuneatus (Murrill) Aoshima
Coriolellus cuneatus Murrill
Polyporus cuneatus (Murrill) Zeller
Distr.: Western North America.
Notes: On wood and bark of conifers.
Hosts: Thuja, Tsuga
Oxyporus latemarginatus (Durieu & Mont.) Donk
Poria latemarginata (Durieu & Mont.) Cooke
Poria ambigua Bres.
Distr.: Widespread.
Notes: On wood.
Hosts: Acer, Albizia, Aralia, Baccharis, Carya, Celtis, Citrus, Cladrastis, Fagus, Fraxinus, Gleditsia, Juglans, Liquidambar, Magnolia, Malus, Melia, Nyssa, Ostrya, Phoenix, Populus, Prosopis, Prunus, Quercus, Robinia, Saccharum, Salix, Sambucus, Syringa, Ulmus, Umbellularia
Oxyporus nobilissimus W.B. Cooke
Fomes nobilissimus (W.B. Cooke) J. Lowe
Distr.: Northwestern United States.
Notes: On wood of conifers.
Hosts: Abies, Tsuga
Oxyporus pellucida (Jungh.) Ryvarden
Poria proxima Bres.
Oxyporus proximus (Bres.) Ryvarden
Distr.: Central and eastern United States; tropical Africa; Asia.
Notes: On hardwoods.
Host: Carya
Oxyporus populinus (Schumach.:Fr.) Donk
Fomes populinus (Schumach.:Fr.) Cooke
Rigidoporus populinus (Schumach.:Fr.) Pouzar
Fomitopsis connata (Weinm.) P. Karst.
Flaviporus connatus (Weinm.) G. Cunn.
Fomes connatus (Weinm.) Gill.
Oxyporus connatus (Weinm.) Kühner
Distr.: Widespread.
Notes: Usually on living hardwoods.
Hosts: Acer, Betula, Carya, Fagus, Ginkgo, Gleditsia, Nyssa, Ostrya, Platanus, Populus, Prunus, Quercus, Salix, Tilia, Ulmus
Oxyporus similis (Bres.) Ryvarden
Poria similis Bres.
Distr.: Northern United States; Canada.
Notes: On hardwoods, especially *Populus* spp.
Hosts: Populus, Thuja, Tsuga

Ozonium Link. Hyphomycetes

Ozonium auriconum Link
Distr.: Worldwide on wood.
Notes: A sterile mycelial form of several *Coprinus* spp.
Hosts: Gossypium, Salix
Ozonium texanum D.C. Neal & R.E. Webster
Distr.: TX(type).

Notes: Isolated from decayed cotton roots. The differences between this fungus and *Phymatotrichum omnivorum* are elaborated by Streets & Bloss (Amer. Phytopth. Soc. Monographs 9:14. 1973).
Host: Gossypium

Pachnocybe Berk. Hyphomycetes

Pachnocybe ferruginea (Sowerby:Fr.) Berk.
Distr.: OR; Europe.
Notes: On floor boards and other timber (Ellis. Dematiaceous Hyphomycetes. p. 184. 1971).
Host: Pseudotsuga

Pachyella Boud. Discomycetes, Pezizales

Pachyella clypeata (Schwein.) Le Gal
Peziza clypeata Schwein.
Distr.: Temperate North America; South America(rare).
Notes: Pfister (Canad. J. Bot. 51:2009. 1973).
Host: Oxydendrum

Pachykytospora Kotlaba & Pouzar. Basidiomycetes, Aphyllophorales

Notes: Gilbertson & Ryvarden (North American Polypores 2:501. 1987).
Pachykytospora alabamae (Berk. & Cooke) Ryvarden
Polyporus alabamae Berk. & Cooke
Poria alabamae (Berk. & Cooke) Cooke
Distr.: Southern United States.
Notes: On wood.
Hosts: Juniperus, Myrica
Pachykytospora papyracea (Schwein.) Ryvarden
Poria papyracea (Schwein.) Cooke
Distr.: Widespread.
Notes: On wood.
Hosts: Chamaecyparis, Thuja, Vitis
Pachykytospora tuberculosa (Fr.:Fr.) Kotlaba & Pouzar
Polyporus tuberculosus Fr.:Fr.
Trametes colliculosa (Pers.) S. Lundell & Nannf.
Trametes morganii C.G. Lloyd
Distr.: Temperate northern hemisphere.
Notes: On wood.
Hosts: Acer, Quercus

Paecilomyces Bainier. Hyphomycetes

Notes: Samson (Mycol. Pap. 6:1. 1974).
Paecilomyces carneus (Duche & R. Heim) A.H.S. Brown & G. Sm.
Distr.: Cosmopolitan.
Notes: In soil (Bissett. Fungi Canadenses 152. 1979).
Host: Acer
Paecilomyces farinosus (Holmsk.) A.H.S. Brown & G. Sm.
Cordyceps memorabilis Ces. (teleomorph)
Distr.: Cosmopolitan.
Notes: A ubiquitous insect parasite; also isolated from soil and wood (Onions. C.M.I. Descr. 613. 1979; Bissett. Fungi Canadenses 153. 1979).
Hosts: Acer, Pinus
Paecilomyces lilacinus (Thom) R.A. Samson
Penicillium lilacinum Thom
Distr.: Widespread.
Notes: Isolated from insects and a variety of plant materials (Bissett. Fungi Canadenses 156. 1979).
Host: Carya
Paecilomyces marquandii (Massee) S.J. Hughes
Distr.: Cosmopolitan.
Notes: In soil (Bissett. Fungi Canadenses 157. 1979).
Hosts: Acer, Pinus
Paecilomyces variotii Bainier
Distr.: Southern and tropical regions.
Notes: This is a thermotolerant fungus isolated from soil (Bissett. Fungi Canadenses 151. 1979).
Hosts: Carya, Helianthus, Pinus

Panellus P. Karst. Basidiomycetes, Agaricales

Panellus longinquus (Berk.) Singer subsp. ***pacificus*** S.D. Libonati-Barnes & Redhead
Distr.: Pacific Northwest.
Notes: Libonati-Barnes & Redhead (Mycotaxon 20:205. 1984).
Host: Alnus
Panellus mitis (Pers.:Fr.) Singer
Distr.: Temperate North America; Europe.
Notes: Usually on conifer logs and sticks.
Host: Pseudotsuga
Panellus pusillus (Pers.) Burdsall & O.K. Miller
Favolus rhipidium (Berk.) Sacc.
Polyporus rhipidium Berk.
Distr.: Central and eastern United States; Central and South America; eastern Asia; Australasia.
Notes: On hardwoods.
Hosts: Acer, Carpinus, Liquidambar, Magnolia, Nyssa, Ostrya, Pinus, Quercus
Panellus serotinus (Pers.:Fr.) Kühner
Pleurotus serotinus (Pers.:Fr.) Gill.
Distr.: Temperate northern hemisphere.
Notes: On hardwood and conifer logs and branches.
Hosts: Abies, Alnus, Betula, Fagus, Malus, Populus, Quercus, Salix, Sorbus, Tsuga
Panellus stipticus (Bull.:Fr.) P. Karst.
Panus stipticus (Bull.:Fr.) Fr.
Distr.: Temperate North America; Europe.
Notes: On logs, stumps, branches, and wood in service of hardwoods.
Hosts: Acer, Betula, Castanea, Malus, Picea, Prunus, Quercus
Panellus violaceofulvus (Batsch:Fr.) Singer
Pleurotus violaceofulvus (Batsch:Fr.) Pilát
Distr.: North America(rare); Europe.
Notes: This species is restricted to conifers. The report on *Rhododendron* is probably *P. ringens* (Miller. Michigan Bot. 9:28. 1970).
Hosts: Abies, Rhododendron

Papularia Fr. Hyphomycetes

Papularia vinosa (Berk. & M.A. Curtis) E. Mason
Gymnosporium vinosum Berk. & M.A. Curtis
Melanconium lineolatum Sacc.
Melanconium saccharinum Penz. & Sacc.
Apiospora camptospora Penz. & Sacc. (teleomorph)
Distr.: Widespread in tropical regions.
Notes: On grasses. This anamorph belongs in *Pteroconium* (Ellis. Dematiaceous Hyphomycetes. p. 579. 1971), but the name has not been transferred to that genus.
Hosts: Andropogon, Arundinaria, Arundo, Bambusa, Phyllostachys, Poaceae, Saccharum

Papulaspora G. Preuss. Hyphomycetes

Notes: Weresub & LeClair (Canad. J. Bot. 49:2203. 1971).
Papulaspora appendicularis Hotson
Distr.: NY(type).
Host: Gladiolus
Papulaspora aspera Bernhardt & Duniway
Distr.: CA.
Host: Potamogeton
Papulaspora brachiata (Ellis & Barth.) J.L. Crane & Schoknecht
Torula brachiata Ellis & Barth.
Distr.: CO.
Notes: On dead wood (Crane & Schoknecht. Mycologia 67:666. 1975).
Host: Symphoricarpos
Papulaspora coprophila (Zukal) Hotson
Helicosporangium coprophilum Zukal
Distr.: United States; Europe.
Notes: On dung and various kinds of organic debris.
Hosts: Gladiolus, Solanum
Papulaspora halima Anastasiou
Distr.: Worldwide in marine habitats.
Notes: Saprobic on submerged wood and leaves (Kohlmeyer &

Kohlmeyer. Marine mycology. p. 465. 1979).
Host: Juncus
Papulaspora rubida Hotson
Distr.: PA; Argentina(type).
Notes: On dung and *Gladiolus*.
Host: Gladiolus
Papulaspora selaginellae M.L. Farr
Host: Selaginella
Papulaspora sepedonioides G. Preuss
Papulaspora dodgei Conners ex Linder
Distr.: Temperate North America.
Notes: On corms and tubers (Weresub. Fungi Canadenses 27. 1974).
Host: Gladiolus

Paradidymobotryum Wang & Sutton. Hyphomycetes

Paradidymobotryum oblongum Wang & Sutton
Distr.: NY.
Notes: On rotten wood (Wang & Sutton. Mycologia 76:572. 1984).
Host: Ulmus

Paraphaeosphaeria O. Eriksson. Loculoascomycetes, Dothideales

Notes: Shoemaker & Babcock (Canad. J. Bot. 63:1284. 1985).
Paraphaeosphaeria michotii (Westend.) O. Eriksson
Leptosphaeria michotii Westend.
Leptosphaeria monilispora (Fuckel) Sacc.
Pleospora monilispora Fuckel
Leptosphaeria zeae G.L. Stout
Coniothyrium scirpi Trail (anamorph)
Distr.: Temperate northern hemisphere.
Notes: On leaves of monocots, mainly grasses.
Hosts: Andropogon, Juncus, Zea
Paraphaeosphaeria obtusispora (Speg.) Hedjaroude
Leptosphaeria obtusispora Speg.
Leptosphaeria aloes Dias & Camara
Leptosphaeria draconis Urries
Leptosphaeria transluscens G. Wint.
Coniothyrium henriquesii Thuem. (anamorph)
Distr.: Southern United States; Argentina; Canary Islands; Portugal.
Notes: On leaves of Agavaceae.
Host: Yucca

Paratomenticola M.B. Ellis. Hyphomycetes

Paratomenticola lanceolatus (Cooke) M.B. Ellis
Helminthosporium lanceolatum Cooke
Sporidesmium lanceolatum (Cooke) S.J. Hughes
Distr.: SC.
Host: Berchemia

Parksia Cash. Discomycetes, Helotiales

Notes: Monotypic.
Parksia libocedri Cash
Distr.: CA(type).
Host: Calocedrus

Parmastomyces Kotlaba & Pouzar. Basidiomycetes, Aphyllophorales

Parmastomyces transmutans (Overh.) Ryvarden & R.L. Gilbertson
Polyporus transmutans Overh.
Tyromyces transmutans (Overh.) J. Lowe
Parmastomyces kravtzevianus (Bondartsev & Parmasto) Kotlaba & Pouzar
Tyromyces kravtzevianus (Overh.) Dománski, nom. inval.
Polyporus subcartilagineus Overh., nom. inval.
Tyromyces subcartilagineus Overh., nom. inval.
Distr.: Northern North America; Europe.
Notes: On dead conifers and hardwoods (Gilbertson & Ryvarden. North American Polypores 2:506. 1987).
Hosts: Abies, Pinus, Populus, Prunus, Pseudotsuga

Parodiella Speg. Loculoascomycetes, Dothideales

Parodiella hedysari (Schwein.) S.J. Hughes
Didymosporium hedysari Schwein.
Coryneum hedysari (Schwein.) Schwein.
Parodiella grammodes (Kuntze) Cooke
Parodiella griffithsii Theiss. & Syd.
Parodiella paraguayensis Speg.
Dothidea perisporioides Berk. & M.A. Curtis
Parodiella perisporioides (Berk. & M.A. Curtis) Speg.
Dothidea seminata Berk. & M.A. Curtis
Stigmatea seminata (Berk. & M.A. Curtis) Sacc.
Ascochytopsis leguminosarum (Petr.) Sutton (anamorph)
Distr.: North America; South America; Africa; Asia, subtropical and tropical regions.
Notes: Parasitic on Fabaceae (Sutton. Mycol. Pap. 138. p. 117. 1975).
Hosts: Amphicarpaea, [Convolvulus], Crotalaria, Desmodium, Gleditsia, Hedysarum, Indigofera, Lablab, Lespedeza, Pediomelum, Psoralidium, Rhynchosia, Strophostyles, Tephrosia

Paschelkiella M.A. Sherwood. Discomycetes, Ostropales

Paschelkiella pini (Romell) M.A. Sherwood
Distr.: Western North America; Europe.
Notes: On decorticated coniferous wood. Sherwood (Mycotaxon 28:166. 1987).
Host: Libocedrus

Passalora Fr. Hyphomycetes

Passalora alni (Chupp & H.C. Greene) Deighton
Cercospora alni Chupp & H.C. Greene in H.C. Greene
Passalora bacilligera (Mont. & Fr.) Mont. & Fr. f. *alnobetulae* Jaap
Passalora bacilligera (Mont. & Fr.) Mont. & Fr. var. *alnobetulae* (Jaap) Lindau in Rabenh.
Distr.: Circumboreal.
Notes: Leaf spot. Ellis (Dematiaceous Hyphomycetes. p. 267. 1971).
Host: Alnus
Passalora melioloides Tracy & Earle
Distr.: FL, SC.
Host: Quercus

Patellaria Fr.:Fr. non Hoffm., hom. illeg. Loculoascomycetes, Patellariales

Notes: The legitimate name is now *Lecanidion* Endl. (Hawksworth. Lichenologist 14:131. 1982).
Patellaria lecideola (Fr.:Fr.) Fr.
Durella lecideola (Fr.:Fr.) Rehm
Distr.: WA; Northern Europe.
Notes: This poorly known taxon was reported by Kanouse (Mycologia 39:676. 1947).
Host: Acer
Patellaria nigrocyanea W. Phillips & Harkn.
Distr.: CA(type).
Host: Salvia
Patellaria signata Ellis & Everh.
Blitridium signatum (Ellis & Everh.) Sacc.
Triblidium signatum (Ellis & Everh.) Kuntze
Distr.: WA(type).
Hosts: Pseudotsuga, Tsuga

Patellariopsis Dennis. Discomycetes, Helotiales

Patellariopsis clavispora (Berk. & Broome) Dennis
Lecanidion clavisporum (Berk. & Broome) Sacc.

Distr.: GA; Europe.
Notes: On decorticated wood.
Hosts: Crataegus, Liriodendron, Oxydendrum, Quercus

Patellea (Fr.) Sacc. Discomycetes, Helotiales

Notes: This genus is considered a synonym of *Durella* (Dennis. Mycol. Pap. 61:12. 1956), but not all names have been redisposed.
Patellea californica Rehm
Distr.: Western United States.
Notes: This species undoubtedly belongs in *Durella*.
Hosts: Arbutus, Arctostaphylos, Pseudotsuga, Salix

Patinella Sacc. Discomycetes, Helotiales

Notes: This is an obscure genus; most names have not been treated by modern taxonomists.
Patinella abietina (Cooke) Sacc.
Patellaria abietina Cooke
Distr.: CA.
Host: Abies
Patinella aloysii-sabaudiae Sacc.
Distr.: CA(type).
Host: Veratrum
Patinella atroviridis Rehm
Distr.: Western United States; Switzerland(type).
Host: Pinus
Patinella flavobrunnea Petr.
Distr.: WA(type).
Host: Alnus

Pauahia F. Stevens. Pyrenomycetes, Meliolales

Notes: Monotypic.
Pauahia sideroxyli F. Stevens
Distr.: HI(type).
Host: Sideroxylon

Paullicorticium J. Eriksson. Basidiomycetes, Aphyllophorales

Paullicorticium niveocremeum (Höhn. & Litsch.) Oberwinkler
Distr.: Temperate northern hemisphere.
Notes: On wood.
Host: Quercus
Paullicorticium pearsonii (Bourd.) J. Eriksson
Corticium subinvisibile D.P. Rogers
Distr.: Northern hemisphere.
Notes: On conifer wood.
Hosts: Fraxinus, Pinus, Tsuga

Paxillus Fr. Basidiomycetes, Agaricales

Notes: Redhead & Ginns (Trans. Mycol. Soc. Japan 26:349. 1985).
Paxillus atrotomentosus (Batsch:Fr.) Fr.
Distr.: Temperate northern hemisphere.
Notes: Trophic role is unclear; reported to cause brown rots or to form mycorrhizae with conifers.
Hosts: Pinus, Tsuga

Pellicularia Cooke. Basidiomycetes, Aphyllophorales

Notes: The type species, *P. koleroga*, is a *nomen confusum* (Talbot. Persoonia 3:371. 1965).
Pellicularia koleroga Cooke
Corticium koleroga (Cooke) Höhn.
Distr.: Southern United States.
Notes: This is a confused name; the type specimen is a mixture of the vegetative hyphae of a resupinate basidiomycete and the conidia of a deuteromycete. Reports probably refer to various species of *Botryobasidium*, *Ceratobasidium*, and *Thanatephorus*.
Hosts: Acer, Aleurites, Ampelopsis, Buxus, Carya, Celtis, Cephalanthus, Chaenomeles, Cinnamomum, Citrus, Cocos, Cornus, Cydonia, Dianthus, Diospyros, Erythrina, Euonymus, Ficus, Hibiscus, Ilex, Ipomoea, Jacquemontia, Kummerowia, Lagenaria, Lagerstroemia, Laurus, Ligustrum, Lonicera, Magnolia, Malus, Melia, Michelia, Nyssa, Parthenocissus, Pistacia, Pittosporum, Prunus, Psidium, Punica, Pyrus, Rhododendron, Ribes, Rosa, Rubus, Sambucus, Solidago, Syringa, Ulmus, Vigna

Pellionella Sacc. Coelomycetes

Pellionella macrospora E.R. Spencer
Distr.: NY(type).
Notes: On imported Brazil nuts.
Host: Bertholletia
Pellionella tetonensis Wehmeyer
Distr.: WY.
Host: Lupinus

Peltosphaeria Berl. Loculoascomycetes, Dothideales

Notes: Riedl (Sydowia 15:301. 1961).
Peltosphaeria sandwicensis Petr.
Distr.: HI(type).
Host: Myoporum

Penicillium Link:Fr. Hyphomycetes

Notes: *Penicillium* spp. are commonly isolated from soil, air, and various kinds of organic matter, occasionally causing storage rots and food spoilage. Several works helpful in identification are available (Pitt. The genus *Penicillium* and its teleomorphic states *Eupenicillium* and *Talaromyces*. 1979; Pitt. A laboratory guide to the common *Penicillium* species. 1985; Samson & Pitt. Advances in *Penicillium* and *Aspergillus* systematics. 1985; Samson et al. Stud. Mycol. 11:1. 1976; Stolk & Samson. Stud. Mycol. 23:1. 1983).
Penicillium aculeatum Raper & Fennell
Distr.: Tropical regions.
Host: Carya
Penicillium atramentosum Thom
Distr.: North America; Australia; Europe.
Host: Cucumis
Penicillium aurantiogriseum Dierckx
Penicillium cyclopium Westling
Penicillium martensii Biourge
Penicillium solitum Westling
Distr.: Cosmopolitan.
Notes: In soil and on a variety of plant materials.
Hosts: Asparagus, Beta, Carya, Hordeum, Lilium, Malus, Nicotiana, Pyrus, Tulipa, Zea
Penicillium brevicompactum Dierckx
Penicillium stoloniferum Thom
Distr.: Cosmopolitan.
Notes: On a variety of organic materials.
Hosts: Acer, Carya, Citrus, Platanus, Zea
Penicillium camembertii Thom
Penicillium caseicola Bainier
Distr.: Cosmopolitan.
Notes: Associated with manufacture of soft cheeses.
Hosts: Carya, Metrosideros
Penicillium canescens Sopp
Distr.: Cosmopolitan.
Hosts: Acacia, Cheirodendron, Metrosideros, Zea
Penicillium capsulatum Raper & Fennell
Distr.: Widespread in tropical regions.
Host: Metrosideros
Penicillium chermesinum Biourge
Distr.: North America; Africa; India.
Host: Cucumis
Penicillium chrysogenum Thom
Penicillium notatum Westling
Distr.: Cosmopolitan.
Notes: On a variety of organic materials.

Hosts: Carya, Hordeum, Pinus, Pyrus, Zea
Penicillium citreonigrum Dierckx
Penicillium citreoviride Biourge
Distr.: Widespread.
Hosts: Carya, Metrosideros
Penicillium citrinum Thom
Penicillium steckii Zaleski
Distr.: Cosmopolitan.
Notes: In soil and on a variety of organic materials.
Hosts: Acer, Arachis, Carya, Cucumis, Metrosideros, Rhizophora, Zea
Penicillium claviforme Bainier
Distr.: Worldwide, but rare.
Notes: Primarily in soil.
Hosts: Beta, Zea
Penicillium corylophilum Dierckx
Distr.: Widespread.
Host: Carya
Penicillium crustosum Thom
Penicillium pseudocasei Abe ex G. Sm.
Penicillium terrestre C.N. Jensen, sensu Raper & Thom
Distr.: Cosmopolitan.
Notes: Associated with spoilage of stored plant products, especially pome fruits.
Hosts: Carya, Pyrus, Zea
Penicillium cyaneum (Bainier & Sartory) Biourge
Distr.: Worldwide, but rare.
Hosts: Carya, Metrosideros, Zea
Penicillium dangeardii J.I. Pitt
Penicillium vermiculatum P.A. Dang.
Talaromyces flavus (Klöcker) A.C. Stolk & R.A. Samson (teleomorph)
Distr.: Cosmopolitan.
Host: Zea
Penicillium decumbens Thom
Distr.: Cosmopolitan.
Host: Pinus
Penicillium dierckxii Biourge
Penicillium cinnamopurpureum Abe, nom. inval.
Penicillium phoeniceum Van Beyma
Eupenicillium cinnamopurpureum D.B. Scott & A.C. Stolk (teleomorph)
Distr.: Cosmopolitan.
Notes: In soil and on a variety of plant materials.
Host: Zea
Penicillium digitatum (Pers.:Fr.) Sacc.
Distr.: Cosmopolitan.
Notes: Rot of *Citrus* fruits, on other plant materials and in soil (Onions. C.M.I. Descr. 96. 1966).
Hosts: Beta, Citrofortunella, Citrus, Iris, Poncirus, Zea
Penicillium diversum Raper & Fennell
Distr.: Widespread.
Notes: On a variety of organic substrates.
Host: Metrosideros
Penicillium duclauxii Delacr.
Distr.: Worldwide, but rare.
Host: Cucumis
Penicillium echinulatum Fassatiová
Penicillium cyclopium Westling var. *echinulatum* Raper & Thom, nom. nud.
Distr.: Probably widely distributed.
Host: Carya
Penicillium expansum Link
Coremium glaucum Link
Penicillium glaucum Link
Distr.: Cosmopolitan.
Notes: Commonly associated with fruit; on a variety of plant materials (Onions. C.M.I. Descr. 97. 1966).
Hosts: Agropyron, Avena, Carya, Cucumis, Cydonia, Daucus, Diospyros, Eucalyptus, Feijoa, Hordeum, Malus, Panicum, Persea, Prunus, Punica, Pyrus, Saccharum, Sorghum, Triticum, Vitis, Zea
Penicillium fellutanum Biourge
Penicillium charlesii G. Sm.
Distr.: Cosmopolitan.
Notes: In soil and on plant material.
Hosts: Carya, Zea
Penicillium funiculosum Thom
Distr.: Widespread.
Notes: Common in soil.
Hosts: Ananas, Beta, Carya, Cucumis, Gladiolus, Metrosideros, Prunus, Quercus, Zea
Penicillium glabrum (Wehmer) Westling
Penicillium frequentans Westling
Distr.: Cosmopolitan.
Notes: In soil and on a variety of plant materials.
Hosts: Acacia, Acer, Carya, Cheirodendron, Cucumis, Metrosideros, Zea
Penicillium gladioli L. McCulloch & Thom
Eupenicillium crustaceum F. Ludw. (teleomorph)
Distr.: Widespread.
Notes: Corm rot (Onions. C.M.I. Descr. 98. 1966).
Hosts: Cucumis, Gladiolus, Iris, Scilla, Tigridia
Penicillium granulatum Bainier
Distr.: Widespread.
Hosts: Acer, Zea
Penicillium griseofulvum Dierckx
Penicillium urticae Bainier
Distr.: Cosmopolitan.
Notes: On a variety of plant materials.
Hosts: Carya, Cucumis, Zea
Penicillium herquei Bainier & Sartory
Distr.: Widespread.
Hosts: Acer, Cucumis, Metrosideros, Zea
Penicillium hirsutum Dierckx
Penicillium corymbiferum Westling
Distr.: Widespread.
Hosts: Allium, Armoracia, Hyacinthus, Iris, Lilium, Tulipa
Penicillium implicatum Biourge
Distr.: Widespread.
Hosts: Acer, Camellia, Carya, Cucumis, Zea
Penicillium islandicum Sopp
Distr.: Worldwide, most common in tropical regions.
Host: Carya
Penicillium italicum Wehmer
Distr.: Widespread.
Notes: Fruit rot of *Citrus* spp.; common in soil and on other plant materials (Onions. C.M.I. Descr. 99. 1966).
Hosts: Citrus, Poncirus, Prunus
Penicillium janczewskii Zaleski
Penicillium nigricans Bainier in Thom
Distr.: Cosmopolitan.
Notes: Primarily a soil-inhabiting organism.
Hosts: Carya, Cucumis
Penicillium janthinellum Biourge
Distr.: Cosmopolitan.
Hosts: Acacia, Arachis, Cucumis, Metrosideros, Zea
Penicillium jensenii Zaleski
Penicillium nalgiovense Laxa
Distr.: Worldwide, but rare.
Notes: Usually in soil.
Host: Carya
Penicillium lividum Westling
Distr.: Temperate North America; Europe.
Notes: Associated with forest soils.
Host: Acer
Penicillium miczynskii Zaleski
Distr.: Worldwide, but rare.
Host: Metrosideros
Penicillium oxalicum Currie & Thom
Distr.: Cosmopolitan.
Notes: In soil and on decaying plant material.
Hosts: Cucumis, Sorghum, Zea
Penicillium pinophilum Hedgc.
Distr.: Cosmopolitan.
Notes: In soil and on a variety of organic materials.
Host: Ananas
Penicillium primulinum J.I. Pitt
Penicillium diversum Raper & Fennell var. *aureum* Raper & Fennell
Distr.: MD.
Host: Quercus
Penicillium puberulum Bainier

Penicillium commune Thom
Penicillium lannosum Westling
Distr.: Worldwide, but rare.
Hosts: Beta, Carya, Malus, Pyrus, Zea
Penicillium purpurogenum O. Stoll
Penicillium rubrum O. Stoll
Distr.: Cosmopolitan.
Notes: Primarily in soil.
Hosts: Carya, Pinus, Zea
Penicillium raistrickii G. Sm.
Distr.: Widespread.
Notes: Important in biodeterioration of organic materials.
Host: Acacia
Penicillium resticulosum Birkinshaw, Raistrick & G. Sm.
Penicillium japonicum G. Sm.
Distr.: Great Britain; Japan; distribution elsewhere is questionable.
Host: Carya
Penicillium restrictum Gilman & E. Abbott
Distr.: Cosmopolitan.
Notes: Primarily a soil-inhabiting organism.
Host: Cucumis
Penicillium rolfsii Thom
Distr.: FL, WA.
Host: Vitis
Penicillium roquefortii Thom
Distr.: Widespread.
Notes: On a variety of organic substrates.
Hosts: Carya, Pyrus, Zea
Penicillium rugulosum Thom
Penicillium chrysitis Biourge
Distr.: Widespread.
Hosts: Camellia, Cucumis, Pyrus, Zea
Penicillium sclerotiorum Van Beyma
Penicillium multicolor Grigorieva-Manoilova & Poradielova *sensu* Raper & Thom
Distr.: Worldwide, but rare.
Hosts: Acer, Zea
Penicillium simplicissimum (Oudem.) Thom
Penicillium piscarium Westling
Distr.: Cosmopolitan.
Notes: In soil and on a variety of organic substrates.
Hosts: Cheirodendron, Cucumis, Metrosideros
Penicillium spinulosum Thom
Distr.: Cosmopolitan.
Notes: In soil and on a variety of organic substrates.
Host: Metrosideros
Penicillium sublateritium Biourge
Distr.: Widespread.
Host: Cucumis
Penicillium thomii Maire
Distr.: Cosmopolitan.
Notes: In soil and on a variety of organic substrates.
Hosts: Acer, Carya, Cucumis, Zea
Penicillium variabile Sopp
Distr.: Cosmopolitan.
Hosts: Carya, Cucumis, Vitis, Zea
Penicillium verrucosum Dierckx
Distr.: Widespread.
Notes: Common in foodstuffs, also in soil.
Hosts: Malus, Zea
Penicillium viridicatum Westling
Penicillium olivinoviride Biourge
Penicillium palitans Westling
Distr.: Cosmopolitan.
Notes: Associated with grains and grain products.
Hosts: Malus, Zea
Penicillium waksmanii Zaleski
Distr.: Widespread.
Host: Zea

Peniophora Cooke. Basidiomycetes, Aphyllophorales

Peniophora aurantiaca (Bres.) Höhn. & Litsch.
Peniophora shearii Burt
Aleurodiscus zelleri Burt
Distr.: Temperate northern hemisphere.
Notes: On wood, usually of hardwoods.
Hosts: Abies, Alnus, Betula, Menziesia, Salix
Peniophora cinerea (Pers.:Fr.) Cooke
Peniophora obscura (Pers.) Bres.
Distr.: Widespread.
Notes: On wood.
Hosts: Abies, Acer, Alnus, Betula, Carpinus, Castanea, Cercis, Cercocarpus, Corylus, Fagus, Fraxinus, Ligustrum, Liriodendron, Malus, Melia, Picea, Pinus, Populus, Purshia, Quercus, Rhus, Salix, Symphoricarpos, Tilia
Peniophora decorticans Burt
Distr.: Western North America.
Notes: On hardwoods.
Hosts: Acer, Quercus, Rhus
Peniophora erikssonii Boidin
Distr.: Temperate northern hemisphere.
Notes: On hardwoods, especially attached limbs of *Alnus* spp.
Host: Alnus
Peniophora exima H. Jacks. & Dearden
Distr.: Western North America.
Notes: On conifers.
Host: Abies
Peniophora fuscomarginata Burt
Phanerochaete fuscomarginata (Burt) R.L. Gilbertson in R.L. Gilbertson, E.R. Canfield & Cummins
Distr.: AZ.
Hosts: Fraxinus, Juglans, Platanus, Quercus, Rubus
Peniophora hydnoidea (Pers.:Fr.) Donk
Peniophora laeta (Fr.:Fr.) Donk
Distr.: CA; Europe.
Notes: Formerly known as *P. laeta*. *P. hydnoidea* is an older, sanctioned name (Donk. Fungus 27:17. 1957).
Host: Amelanchier
Peniophora incarnata (Pers.:Fr.) P. Karst.
Corticium incarnatum (Pers.:Fr.) Fr.
Peniophora aemulans P. Karst.
Distr.: Widespread.
Notes: On wood.
Hosts: Acer, Alnus, Arctostaphylos, Baccharis, Crataegus, Fagus, Magnolia, Platanus, Populus, Quercus, Salix, Sambucus, Symphoricarpos
Peniophora isabellina Burt
Distr.: VA; Jamaica(type).
Notes: The generic placement of this species is uncertain (Liberta. Mycologia 60:827. 1968).
Host: Rubus
Peniophora junipericola J. Eriksson
Distr.: Gulf Coast; Europe.
Notes: On wood.
Host: Juniperus
Peniophora malenconii Boidin & Lanquetin subsp. ***americana*** Chamuris
Distr.: AZ, CA, UT.
Notes: On hardwoods. Formerly known as *Stereum heterosporum* Burt (Chamuris. Mycotaxon 28:543. 1987).
Hosts: Acacia, Acer, Adenostoma, Aesculus, Baccharis, Chrysothamnus, Citrus, Crossosoma, Encelia, Eucalyptus, Laurocerasus, Platanus, Quercus, Rosa, Umbellularia
Peniophora nuda (Fr.:Fr.) Bres.
Peniophora ochracea Massee
Peniophora overholtsii Burt
Distr.: Widespread.
Notes: On wood.
Hosts: Betula, Celtis, Cercocarpus, Corylus, Fraxinus, Juglans, Platanus, Prosopis, Prunus, Quercus, Ribes
Peniophora perexigua H. Jacks.
Distr.: AZ; Canada (Ontario).
Notes: On wood and bark. The generic placement of this species is uncertain.
Host: Pseudotsuga
Peniophora piceae (Pers.) J. Eriksson
Peniophora separans Burt

Distr.: Temperate northern hemisphere; infrequent in North America.
Notes: On conifers.
Host: Abies
Peniophora pini (Schleich.:Fr.) Boidin subsp. **duplex** (Burt) Weresub & S. Gibson
Peniophora duplex Burt
Distr.: Eastern North America.
Notes: On conifers, especially *Pinus* spp. (Weresub & Gibson. Canad. J. Bot. 38:861. 1960).
Host: Pinus
Peniophora pithya (Pers.) J. Eriksson
Distr.: Circumboreal.
Notes: On wood.
Hosts: Abies, Picea
Peniophora polygonia (Pers.:Fr.) Bourd. & Galzin
Corticium polygonium Pers.:Fr.
Cryptochaete polygonia (Pers.:Fr.) P. Karst.
Distr.: Western North America; Europe.
Notes: On wood, usually reported on *Populus* spp.
Hosts: Acer, Populus
Peniophora pseudopini Weresub & S. Gibson
Distr.: Temperate North America.
Notes: On conifers, especially *Pinus* spp.
Host: Pinus
Peniophora pseudoversicolor Boidin
Distr.: Temperate northern hemisphere.
Notes: On wood.
Host: Quercus
Peniophora pusilla H. Jacks.
Distr.: NH; Canada(Ontario, Quebec).
Notes: On wood and bark. The generic placement of this species is uncertain.
Host: Tsuga
Peniophora quercina (Pers.:Fr.) Cooke
Peniophora corticalis (Bull.) Bres.
Peniophora pezizoides Massee
Distr.: Temperate northern hemisphere.
Notes: On hardwoods.
Host: Mortonia
Peniophora rufa (Fr.:Fr.) Boidin
Cryptochaete rufa (Fr.:Fr.) P. Karst.
Sterellum rufum (Fr.:Fr.) J. Eriksson
Stereum rufum (Fr.:Fr.) Fr.
Distr.: Temperate northern hemisphere.
Notes: On wood, usually on *Populus tremuloides* in North America.
Hosts: Populus, Salix
Peniophora sphaerocystidiata Burdsall & Nakasone
Distr.: MS(type).
Notes: On wood. Burdsall & Nakasone (Mycotaxon 17:253. 1983).
Host: Vaccinium
Peniophora tamaricicola Boidin & Malencon in Boidin
Distr.: Temperate northern hemisphere.
Notes: On wood.
Hosts: Baccharis, Carnegiea, Celtis, Condalia, Fouquieria, Fraxinus, Juglans, Parkinsonia, Prosopis
Peniophora violaceolivida (Sommerf.) Massee
Peniophora syringae P. Karst.
Distr.: Temperate northern hemisphere.
Notes: On hardwoods.
Hosts: Abies, Populus, Quercus, Thuja

Penzigia Sacc. Pyrenomycetes, Xylariales

Penzigia frustulosa (Berk. & M.A. Curtis) J.H. Miller in Chárdon & Toro
Hypoxylon exiguum Cooke
Distr.: Subtropical and tropical regions.
Notes: On wood (Jong & Rogers. Mycologia 62:851. 1970).
Hosts: Cornus, Quercus, Salix

Perenniporia Murrill. Basidiomycetes, Aphyllophorales

Notes: Gilbertson & Ryvarden (North American Polypores 2:508. 1987).
Perenniporia amylodextrinoidea R.L. Gilbertson & Ryvarden
Distr.: AZ.
Notes: On wood.
Host: Alnus
Perenniporia compacta (Overh.) Ryvarden & R.L. Gilbertson
Polyporus compactus Overh.
Poria compacta (Overh.) Ginns
Distr.: Northeastern and northcentral North America.
Notes: On dead standing trunks of *Quercus alba*; occasionally other hardwoods.
Hosts: Acer, Quercus, Robinia
Perenniporia ellipsospora Ryvarden & R.L. Gilbertson
Poria conferta Overh., nom. inval.
Distr.: Central and eastern United States.
Notes: On wood.
Host: Alnus
Perenniporia ellisiana (F.W. Anderson) R.L. Gilbertson & Ryvarden
Fomes ellisianus F.W. Anderson
Fomes fraxinophilus (Peck) Sacc. f. *ellisianus* (F.W. Anderson) D. Baxter
Distr.: Western United States and adjacent Canada.
Notes: On wood.
Host: Shepherdia
Perenniporia fergusii R.L. Gilbertson & Ryvarden
Distr.: IL.
Notes: On wood.
Host: Carya
Perenniporia fissiliformis (Pilát) R.L. Gilbertson & Ryvarden
Poria fissiliformis Pilát
Distr.: Central and eastern North America.
Notes: On hardwoods.
Host: Acer
Perenniporia fraxinea (Bull.:Fr.) Ryvarden
Fomes fraxineus (Bull.:Fr.) Cooke
Distr.: Northern hemisphere.
Notes: On wood.
Hosts: Acer, Fraxinus, Liquidambar, Nyssa, Quercus, Ulmus
Perenniporia fraxinophila (Peck) Ryvarden
Polyporus fraxinophilus Peck
Fomes fraxinophilus (Peck) Sacc.
Poria fraxinophila (Peck) J. Wright
Distr.: North America.
Notes: Usually on living or dead hardwoods.
Hosts: Acer, Fagus, Fraxinus, Garrya, Juniperus, Prunus, Quercus, Salix, Shepherdia, Ulmus
Perenniporia medulla-panis (Jacq.:Fr.) Donk
Poria medulla-panis (Jacq.:Fr.) Bres.
Fomitopsis unita (Pers.:Fr.) Bondartsev
Poria unita (Pers.:Fr.) P. Karst.
Fomes unitus (Pers.:Fr.) J. Lowe
Polyporus unitus Pers.:Fr.
Distr.: Temperate northern hemisphere.
Notes: On wood.
Hosts: Abies, Acacia, Acer, Arbutus, Betula, Carya, Castanea, Cercocarpus, Chilopsis, Cornus, Cupressus, Fagus, Juglans, Larix, Opuntia, Pinus, Populus, Prosopis, Prunus, Pseudotsuga, Quercus, Rhus, Robinia, Salix, Thuja, Tilia, Tsuga, Ulmus, Vitis
Perenniporia narymica (Pilát) Pouzar
Perenniporia amylohypha Ryvarden & R.L. Gilbertson
Poria elongata Overh., nom. inval.
Distr.: Eastern North America.
Notes: On hardwoods.
Hosts: Acer, Populus
Perenniporia ohiensis (Berk.) Ryvarden
Trametes ohiensis Berk.
Fomes ohiensis (Berk.) Murrill
Fomitopsis ohiensis (Berk.) Bondartsev & Singer
Truncospora ohiensis (Berk.) Pilát
Distr.: Central and eastern North America.
Notes: On hardwoods.
Hosts: Acacia, Acer, Arbutus, Castanea, Fagus, Fraxinus, Magnolia, Ostrya, Parkinsonia, Quercus, Robinia
Perenniporia phloiophila R.L. Gilbertson & Blackwell

Distr.: Gulf Coast States.
Notes: On bark. Gilbertson & Blackwell (Mycotaxon 20:85. 1984).
Host: Quercus
Perenniporia robiniophila (Murrill) Ryvarden
Trametes robiniophila Murrill
Polyporus robiniophilus (Murrill) C.G. Lloyd
Distr.: Widespread.
Notes: On hardwoods.
Hosts: Acer, Broussonetia, Celtis, Liriodendron, Morus, Paulownia, Quercus, Robinia
Perenniporia subacida (Peck) Donk
Poria subacida (Peck) Sacc.
Poria colorea Overh.
Poria homaena (Berk.) Cooke
Distr.: Temperate northern hemisphere.
Notes: On wood.
Hosts: Abies, Acer, Alnus, Betula, Castanea, Cupressus, Juniperus, Larix, Liriodendron, Picea, Pinus, Populus, Pseudotsuga, Thuja, Tsuga
Perenniporia tenuis (Schwein.) Ryvarden
Poria tenuis (Schwein.) Cooke
Polyporus dryinus Berk. & Cooke ex Cooke
Poria pulchella (Schwein.) Cooke
Poria tenuis (Schwein.) Cooke var. *pulchella* (Schwein.) J. Lowe
Poria tomentocincta (Berk. & Ravenel) Cooke
Distr.: Widespread.
Notes: On wood.
Hosts: Acer, Alnus, Betula, Castanea, Cornus, Cupressus, Fagus, Fraxinus, Gleditsia, Gymnocladus, Hamamelis, Juglans, Liriodendron, Malus, Myrica, Picea, Pinus, Platanus, Populus, Prunus, Pseudotsuga, Quercus, Tsuga

Perichaena Fr. Myxomycetes
Perichaena vermicularis (Schwein.) Rostr.
Distr.: Cosmopolitan.
Host: Zea

Periconia Tode:Fr. Hyphomycetes
Periconia byssoides Pers.
Sporocybe byssoides (Pers.) Fr.
Periconia pycnospora Fresen.
Distr.: Cosmopolitan.
Notes: Saprobic on plant debris (Ellis. Dematiaceous Hyphomycetes. p. 345. 1971).
Hosts: Acacia, Carya, Cheirodendron, Commelina, Cytisus, Eichhornia, Hibiscus, Impatiens, Liriodendron, Magnolia, Pisum, Silphium, Vitis
Periconia circinata (L. Mangin) Sacc.
Distr.: Central North America; Europe.
Notes: Milo root rot (Ellis. C.M.I. Descr. 167. 1968).
Hosts: Agropyron, Agrostis, Andropogon, Arthraxon, Avena, Bromus, Calamovilfa, Dactylis, Elymus, Eragrostis, Festuca, Glycine, Hordeum, Koeleria, Leymus, Medicago, Panicum, Phleum, Poa, Schizachyrium, Setaria, Sorghastrum, Sorghum, Triticum, Vulpia, Zea
Periconia cookei E. Mason & M.B. Ellis
Distr.: Eastern North America; Europe.
Notes: On dead herbaceous stems (Ellis. Dematiaceous Hyphomycetes. p. 345. 1971).
Hosts: Carya, Juncus, Sorghum
Periconia digitata (Cooke) Sacc.
Periconia paludosa E. Mason & M.B. Ellis
Distr.: Subtropical and tropical regions.
Notes: On dead leaves and culms of monocotyledonous plants (Ellis. Dematiaceous Hyphomycetes. p. 349. 1971).
Hosts: Juncus, Maranta
Periconia echinochloae (Batista) M.B. Ellis
Periconiella echinochloae Batista
Distr.: Widespread in tropical regions.
Notes: On grasses, also isolated from air and soil (Ellis. Dematiaceous Hyphomycetes. p. 347. 1971).
Host: Juncus
Periconia hispidula (Pers.) E. Mason & M.B. Ellis
Conoplea hispidula Pers.
Distr.: Temperate regions.
Notes: On dead leaves of grasses, sedges, and similar plants (Mason. Mycol. Pap. 56:112. 1953).
Host: Thuja
Periconia igniaria E. Mason & M.B. Ellis
Distr.: Widespread.
Notes: On plants which have been scorched or prematurely killed by burning (Mason & Ellis. Mycol. Pap. 56:104. 1953).
Host: Juncus
Periconia lateralis Ellis & Everh.
Harpocephalum dematioides Atk.
Distr.: Cosmopolitan.
Notes: On a variety of plants and in soil.
Hosts: Cynodon, Tridens, Zanthoxylum
Periconia macrospinosa Lefebvre & A.G. Johnson in Lefebvre et al.
Distr.: Temperate regions.
Notes: Ellis (C.M.I. Descr. 168. 1968).
Hosts: Sorghum, Triticum
Periconia maculans (Cooke) J.L. Crane & Schoknecht
Torula maculans Cooke
Distr.: SC, WY.
Notes: On dead leaves (Crane & Schoknecht. Mycologia 67:667. 1975).
Host: Yucca
Periconia minutissima Corda
Distr.: Cosmopolitan.
Notes: Saprobic on plant debris.
Hosts: Arundinaria, Cynodon, Juncus, Morus
Periconia nigriceps (Peck) Sacc.
Sporocybe nigriceps Peck
Distr.: AK, NY(type).
Host: Carex
Periconia prolifica Anastasiou
Halosphaeria cucullata (Kohlmeyer) Kohlmeyer (teleomorph)
Distr.: Widespread in marine habitats.
Notes: Saprobic on intertidal and drifting wood.
Host: Pluchea
Periconia sacchari J.R. Johnston in J.R. Johnston & J.A. Stevenson
Distr.: Southeastern United States; West Indies; South America; Africa.
Notes: Saprobic on leaves and stalks.
Host: Saccharum

Periconiella Sacc. Hyphomycetes
Notes: Ellis (Dematiaceous Hyphomycetes. p. 295. 1971).
Periconiella angusiana M.B. Ellis
Distr.: HI; Africa.
Hosts: Acacia, Metrosideros
Periconiella verrucosa E. Stewart & M. Corden
Distr.: OR.
Notes: Isolated from utility poles and pilings of douglas fir (Stewart & Corden. Mycotaxon 8:302. 1979).
Host: Pseudotsuga

Peridermium (Link) J.C. Schmidt & Kunze, nom. cons. Basidiomycetes, Uredinales
Notes: Ziller (The tree rusts of western Canada. p. 45. 1974).
Peridermium californicum Arth. & F. Kern
Coleosporium pacificum Cummins (teleomorph)
Distr.: Canada(British Columbia) to CA.
Notes: Heteroecious; aecial host - *Pinus radiata*; telial host - Asteraceae.
Host: [Ephedra]
Peridermium cerebrum Peck
Cronartium quercuum (Berk.) Miyabe & Shirai (teleomorph)
Distr.: Central and eastern North America; Asia; India; Soviet Union.
Notes: Anamorph of eastern gall rust. Heteroecious; aecial host - *Pinus* spp.; telial host - Fagaceae.
Host: Pinus
Peridermium coloradense (Dietel) Arth. & F. Kern

Chrysomyxa arctostaphyli Dietel (teleomorph)
Distr.: Northern North America.
Notes: Anamorph of spruce broom rust. Heteroecious; aecial host - *Picea*; telial host - *Arctostaphylos*.
Host: Picea
Peridermium comptoniae Orton & J.F. Adams
Cronartium comptoniae Arth. (teleomorph)
Distr.: Temperate North America.
Notes: Anamorph of sweet fern blister rust. Heteroecious; aecial host - *Pinus* spp.; telial host - *Comptonia* and *Myrica*.
Host: Pinus
Peridermium ephedrae Cooke
Distr.: Southwestern United States; Mexico.
Notes: Kern & Keener (Science 131:298. 1960).
Host: Ephedra
Peridermium filamentosum Peck
Distr.: Western United States.
Notes: Limb rust. One race is heteroecious, aecial host - *Pinus* spp.; telial host - *Castilleja* spp.; the other races are autoecious (Peterson. Phytopathology 58:309. 1968).
Host: Pinus
Peridermium fusiforme Arth. & F. Kern
Cronartium quercuum (Berk.) Miyabe & G. Snow f. sp. *fusiforme* (Hedgc. & N. Hunt) Burdsall & G. Snow (teleomorph)
Distr.: Southeastern United States.
Notes: Anamorph of fusiform rust of southern pines. Heteroecious; aecial host - *Pinus* spp.; telial host - Fagaceae.
Host: Pinus
Peridermium guatemalense Arth. & F. Kern
Peridermium floridanum Hedgc. & Hahn
Distr.: FL; Mexico; Central America.
Host: Pinus
Peridermium harknessii J.P. Moore
Endocronartium harknessii (J.P. Moore) Y. Hiratsuka (teleomorph)
Distr.: Western North America.
Notes: Western gall rust. According to Epstein & Buurlage (Mycologia 80:235. 1988) this fungus is an asexual rust.
Host: Pinus
Peridermium holwayi Syd.
Pucciniastrum goeppertianum (Kühn) Kleb. (teleomorph)
Peridermium ornamentale Arth. (synanamorph)
Distr.: North America; Europe; Japan.
Notes: Witches' broom rust on blueberries. Heteroecious; aecial host- *Abies* spp.; telial host- *Vaccinium* spp. Hiratsuka et al. (Canad. J. Bot. 45:1913. 1967) provided a discussion of the *Peridermium* states.
Host: Abies
Peridermium occidentale Hedgc., Bethel & N. Hunt
Cronartium occidentale Hedgc., Bethel & N. Hunt (teleomorph)
Distr.: Western United States.
Notes: Anamorph of pinon pine blister rust. Heteroecious; aecial host - species of *Pinus* subg. *strobus*; telial host - *Ribes* and *Grossularia* spp.
Host: Pinus
Peridermium ornamentale Arth.
Pucciniastrum goeppertianum (Kühn) Kleb. (teleomorph)
Peridermium holwayi Syd. (synanamorph)
Distr.: North America; Europe; Japan.
Notes: Witches' broom of blueberries. Heteroecious; aecial host - *Abies* spp.; telia host - *Vaccinium* spp. Hiratsuka et al. (Canad. J. Bot. 45:1913. 1967) provided a discussion of the *Peridermium* states.
Host: Abies
Peridermium rugosum H. Jacks.
Distr.: Western North America.
Notes: Probably heteroecious, aecial host - *Abies* spp.; telial connection unknown.
Host: Abies
Peridermium stalactiforme Arth. & F. Kern
Cronartium coleosporioides Arth. (teleomorph)
Distr.: Western North America.
Notes: Stalactiform blister rust. Heteroecious; aecial host - *Pinus* spp.; telial host - Scrophulariaceae.
Host: Pinus
Peridermium strobilinum (Arth.) R.S. Peterson
Cronartium strobilinum Hedgc. & Hahn (teleomorph)
Distr.: Southeastern United States.
Notes: Southern cone rust. Heteroecious; aecial host - *Pinus* spp.; telial host - *Quercus* spp.
Host: Pinus
Peridermium weirii Arth.
Distr.: ID.
Notes: Aecial host - *Pinus*.
Host: Pinus

Perisporium Fr. Loculoascomycetes, Dothideales

Notes: This genus is considered a synonym of *Preussia*, but not all names have been redisposed.
Perisporium zeae Berk. & M.A. Curtis
Distr.: SC(type).
Hosts: Panicum, Zea

Peroneutypa Berl. Pyrenomycetes, Diatrypales

Peroneutypa corniculata (Ehrh. ex Pers.) Berl.
Distr.: ND; Europe.
Host: Prunus

Peronosclerospora (Ito) Shirai & K. Hara. Oomycetes, Peronosporales

Peronosclerospora sorghi (W. Weston & Uppal) C.G. Shaw
Sclerospora sorghi W. Weston & Uppal
Distr.: North, Central and South America; Africa; Asia; Australia.
Notes: Downy mildew of sorghum (Francis & Williams. C.M.I. Descr. 761. 1983).
Hosts: Sorghum, Zea

Peronospora Corda. Oomycetes, Peronosporales

Peronospora alsinearum Casp.
Distr.: North America; Europe.
Notes: This species is not well-defined and has been considered synonymous *pro parte* with *P. media* Gäumann and *P. vernalis* Gäumann.
Hosts: Cerastium, Spergula, Stellaria
Peronospora alta Fuckel
Distr.: Temperate regions.
Notes: Downy mildew of plantain (Francis. C.M.I. Descr. 683. 1981).
Host: Plantago
Peronospora americana Gäum.
Distr.: Central United States.
Host: Polygonum
Peronospora antirrhini J. Schröt.
Distr.: Cosmopolitan.
Notes: Downy mildew of snapdragon (Francis. C.M.I. Descr. 685. 1981).
Host: Antirrhinum
Peronospora arborescens (Berk.) Casp.
Distr.: Temperate and warm-temperate regions.
Notes: Downy mildew of poppy (Francis. C.M.I. Descr. 686. 1981).
Host: Argemone
Peronospora arthurii Farl.
Distr.: Worldwide.
Notes: On Onagraceae.
Hosts: Camissonia, Clarkia, Epilobium, Gaura, Oenothera
Peronospora calotheca de Bary
Distr.: Worldwide.
Hosts: Galium, Hedyotis, Hydrangea
Peronospora candida Fuckel
Peronospora anagallidis J. Schröt.
Distr.: KS; Europe.
Host: Androsace
Peronospora claytoniae Farl.
Distr.: North America.
Hosts: Calandrinia, Claytonia, Montia
Peronospora cochleariae Gäum.

Distr.: AK; Denmark(type).
Host: Cochlearia
Peronospora conglomerata Fuckel
Distr.: Worldwide.
Host: Geranium
Peronospora corydalis de Bary
Distr.: Worldwide.
Hosts: Corydalis, Dicentra
Peronospora cynoglossi (Burrill) Swingle
Distr.: Central and eastern United States.
Host: Cynoglossum
Peronospora cyperi Ideta
Phytophthora cyperi (Ideta) Ito
Distr.: Southern United States; Asia.
Notes: The identity of this species is obscure. The name *Peronospora cyperi* Miyabe & Ideta is a later homonym which has been placed in *Phytophthora*. Waterhouse (Mycol. Pap. 122:2. 1970) explained the confusion.
Host: Cyperus
Peronospora davisii C.G. Shaw
Distr.: WI.
Host: Dracocephalum
Peronospora debaryi P. Karst.
Distr.: WI; Europe.
Notes: The occurrence of this species in North America is uncertain (Constantinescu. Mycotaxon 24:303. 1985).
Host: Urtica
Peronospora destructor (Berk.) Casp. in Berk.
Peronospora schleideni Unger
Peronospora schleideniana Wm.G. Sm.
Distr.: Cosmopolitan.
Notes: Downy mildew of onion (Palti as "Mukerji". C.M.I. Descr. 456. 1975).
Host: Allium
Peronospora dianthicola Barthelet
Distr.: CA; Colombia; Asia Minor; Europe.
Notes: Downy mildew of carnation (Francis. C.M.I. Descr. 764. 1983).
Host: Dianthus
Peronospora dicentrae Gäum.
Distr.: Central and eastern United States.
Host: Dicentra
Peronospora dipsaci Tul. ex de Bary
Distr.: Temperate North America; Europe.
Host: Dipsacus
Peronospora echinospermi (Swingle) Swingle
Peronospora cynoglossi (Burrill) Swingle var. *echinospermi* Swingle
Distr.: Central and western United States; Korea; Soviet Union..
Hosts: Amsinckia, Hackelia, Lappula
Peronospora effusa (Grev.) Tul.
Distr.: Worldwide on Chenopodiaceae.
Hosts: Chenopodium, Monolepis, Spinacia, Tetragonia
Peronospora eriogoni Solheim & R.L. Gilbertson
Distr.: AZ.
Host: Eriogonum
Peronospora erodii Fuckel
Pseudoperonospora erodii (Fuckel) G.W. Wils.
Distr.: Widespread.
Notes: Waterhouse & Brothers (Mycol. Pap. 148:21. 1981).
Host: Erodium
Peronospora euphorbiae Fuckel
Peronospora chamaesyces G.W. Wils.
Distr.: Central United States; Europe.
Host: Euphorbia
Peronospora farinosa (Fr.:Fr.) Fr.
Peronospora amaranthi Gäum.
Peronospora chenopodii-glauci Gäum.
Peronospora farinosa (Fr.:Fr.) Fr. f. sp. *betae* Byford
Peronospora schachtii Fuckel
Distr.: Central and western United States; Europe; India.
Notes: Although this name has been considered a synonym of various other species, *P. farinosa* is recognized and redescribed by Solheim & Gilbertson (Mycopath. Mycol. Appl. 49:158. 1973). The synonymy was proposed by Yerkes & Shaw (Phytopathology 49:503. 1959). Francis & Byford (C.M.I. Descr. 765. 1983) discussed the species as *P. farinosa* f.sp. *betae*.
Hosts: Amaranthus, Atriplex, Beta, Chenopodium, Spinacia
Peronospora ficariae (Nees) Tul.
Distr.: Worldwide on Ranunculaceae.
Hosts: Delphinium, Ranunculus, Trautvetteria
Peronospora floerkeae Kellerm.
Distr.: Central and western United States.
Host: Floerkea
Peronospora fragariae Roze & Cornu
Distr.: IA; Europe.
Host: Fragaria
Peronospora grisea (Unger) Unger
Distr.: Temperate regions.
Notes: Downy mildew of *Hebe* (Francis & Berrie. C.M.I. Descr. 766. 1983).
Hosts: [Collinsia], Veronica
Peronospora hedeomae Kellerm. & Swingle
Distr.: Central United States.
Hosts: Dracocephalum, Hedeoma
Peronospora hiemalis Gäum.
Distr.: Temperate regions.
Host: Ranunculus
Peronospora hydrophylli M.B. Waite
Distr.: Central and western United States.
Hosts: Ellisia, Hydrophyllum, Phacelia
Peronospora hyoscyami de Bary
Distr.: CA; Europe.
Host: Nicotiana
Peronospora jacksonii C.G. Shaw
Distr.: Worldwide.
Host: Mimulus
Peronospora lamii A. Braun
Peronospora swinglei Ellis & Everh.
Distr.: Worldwide, except Australia.
Notes: Downy mildew of mints (Francis. C.M.I. Descr. 688. 1981).
Hosts: Lamium, Salvia
Peronospora lathyri-palustris Gäum.
Distr.: Worldwide.
Host: Lathyrus
Peronospora leptosperma de Bary
Distr.: Central and eastern North America; Europe.
Hosts: Artemisia, Chrysanthemum
Peronospora linariae Fuckel
Distr.: Central and eastern North America; Europe.
Host: Linaria
Peronospora lini J. Schröt.
Distr.: IA; Europe(type).
Host: Linum
Peronospora lophanthi Farl.
Distr.: Central and eastern United States.
Host: Agastache
Peronospora lophanthi Farl. var. ***moldavicae*** Dearn. & Barth.
Distr.: WY(type).
Host: Dracocephalum
Peronospora lotorum Syd.
Distr.: Northwestern North America; Europe.
Host: Lotus
Peronospora manshurica (Naumov) Syd. in Gäum.
Peronospora sojae F. Lehm. & F.A. Wolf
Distr.: Cosmopolitan.
Notes: Downy mildew of soybean (Francis. C.M.I. Descr. 689. 1981).
Host: Glycine
Peronospora mayorii Gäum.
Distr.: Worldwide.
Host: Vicia
Peronospora myosotidis de Bary
Distr.: Widespread on Boraginaceae.
Hosts: Amsinckia, Cryptantha, Hackelia, Mertensia, Myosotis, Plagiobothrys
Peronospora narbonensis Gäum.
Distr.: Central and western United States.
Host: Vicia
Peronospora nemophilae C.G. Shaw
Distr.: Pacific Northwest.
Host: Nemophila
Peronospora obovata Bonord. in Rabenh.

Distr.: Worldwide.
Host: Spergula
Peronospora oxybaphi Ellis & Everh.
Distr.: Central United States.
Hosts: Abronia, Mirabilis
Peronospora parasitica (Pers.:Fr.) Fr.
Peronospora lepidii (McAlpine) G.W. Wils.
Peronospora lepidii-sativi Gäum.
Peronospora lepidii-virginici Gäum.
Distr.: Cosmopolitan.
Notes: Downy mildew of Brassicaceae. Many of the synonyms were proposed by Yerkes & Shaw (Phytopathology 49:499. 1959) who conducted inoculation experiments.
Hosts: Arabidopsis, Arabis, Armoracia, Astragalus, Barbarea, Brassica, Capsella, Cardamine, Cleome, Coronopus, Dentaria, Descurainia, Draba, Erysimum, Hesperis, Iberis, Lepidium, Lesquerella, Lobularia, Lomatium, Matthiola, Nasturtium, Raphanus, Rorippa, Sinapis, Sisymbrium, Streptanthus, Thelypodium, Thysanocarpus
Peronospora pennsylvanica Gäum.
Distr.: Eastern United States.
Host: Ranunculus
Peronospora phlogina Dietel & Holw.
Peronospora giliae Ellis & Everh.
Distr.: Central and western United States.
Hosts: Collomia, Gilia, Microsteris, Navarretia, Phlox
Peronospora pisi Syd.
Distr.: Worldwide in pea-growing regions.
Host: Pisum
Peronospora plantaginis Underw.
Distr.: Central United States; India.
Notes: Rao (Nova Hedwigia 16:277. 1968).
Host: Plantago
Peronospora polygoni Halst.
Distr.: Temperate regions.
Host: Polygonum
Peronospora potentillae de Bary
Distr.: Widespread on herbaceous members of the Rosaceae.
Hosts: Agrimonia, Duchesnea, Geum, Potentilla, Rubus
Peronospora ranunculi Gäum.
Distr.: Temperate regions.
Host: Ranunculus
Peronospora rubi Rabenh.
Distr.: Temperate regions.
Host: Rubus
Peronospora rumicis Corda
Distr.: Temperate regions.
Host: Rheum
Peronospora saxifragae Bubák
Distr.: Western United States; Europe.
Host: Lithophragma
Peronospora seymourii Burrill in Underw.
Distr.: Central and southern United States.
Host: Hedyotis
Peronospora silenes G.W. Wils.
Distr.: Central and western United States.
Host: Silene
Peronospora sordida Berk. & Broome
Distr.: United States; Europe.
Hosts: Orthocarpus, Scrophularia, Verbascum
Peronospora sparsa Berk.
Distr.: Cosmopolitan.
Notes: Downy mildew of rose (Francis. C.M.I. Descr. 690. 1981).
Hosts: Feijoa, Rosa, Rubus
Peronospora stigmaticola Raunk.
Distr.: WI; Europe.
Host: Mentha
Peronospora tabacina D.B. Adam
Distr.: Temperate regions.
Notes: Downy mildew of tobacco and other solanaceous plants (Spencer. The downy mildews. p. 577. 1981).
Hosts: Capsicum, Lycopersicon, Nicotiana, Solanum
Peronospora trifoliorum de Bary
Distr.: Cosmopolitan.
Notes: Downy mildew of alfalfa (Francis. C.M.I. Descr. 768. 1983).
Hosts: Astragalus, Hedysarum, Lathyrus, Lupinus, Medicago, Melilotus, Trifolium, Vigna
Peronospora valerianae Trail
Distr.: MT; Europe.
Host: Valeriana
Peronospora viciae (Berk.) Casp.
Distr.: Cosmopolitan.
Notes: Downy mildew of *Pisum* spp. and other Fabaceae. Mukerji (C.M.I. Descr. 455. 1975).
Hosts: Medicago, Pisum, Trifolium, Vicia
Peronospora viciae (Berk.) Casp. var. ***americana*** J.J. Davis
Distr.: NC, WI(type).
Host: Vicia
Peronospora vicicola L. Campbell
Distr.: Northwestern United States.
Host: Vicia
Peronospora violae de Bary
Distr.: Worldwide.
Notes: Often confused with *Bremiella megasperma* (Berl.) G.W. Wilson (Constantinescu. Trans. Brit. Mycol. Soc. 72:510. 1979).
Host: Viola
Peronospora whippleae Ellis & Everh.
Distr.: CA(type).
Host: Whipplea

Perrotia Boud. Discomycetes, Helotiales

Perrotia flammea (Albertini & Schwein.:Fr.) Boud.
Dasyscyphus flammeus (Albertini & Schwein.:Fr.) Fr.
Helotium flammea (Albertini & Schwein.:Fr.) P. Karst.
Lachnella flammea (Albertini & Schwein.:Fr.)Fr.
Lachnella rhoina Earle in H.C. Greene
Distr.: Temperate regions.
Notes: Dennis (British Ascomycetes. p. 172. 1978).
Hosts: Alnus, Arbutus, Cercocarpus, Haplopappus, Quercus, Rhus, Salix
Perrotia populina (Seaver) Dennis
Lachnella populina Seaver
Distr.: Western North America.
Notes: On limbs.
Host: Populus
Perrotia succina (W. Phillips) Dennis
Dasyscyphus succina (W. Phillips) Sacc.
Lachnella succina (W. Phillips) Seaver
Distr.: CA, UT.
Notes: Dennis (Kew Bull. 17:367. 1963).
Hosts: Ceanothus, Prunus, Quercus

Pestalopezia Seaver. Discomycetes, Helotiales

Notes: Barr (Mycologia 67:192. 1975) suggested that this genus is related to *Velutarina* and that a *Pestalotia* anamorph is unlikely.
Pestalopezia brunneopruinosa (Zeller) Seaver
Dermatea bruneopruinosa Zeller
Distr.: Western United States.
Hosts: Arbutus, Arctostaphylos, Gaultheria, Vaccinium
Pestalopezia rhododendri Seaver
Distr.: Southeastern United States.
Host: Rhododendron

Pestalosphaeria Barr. Pyrenomycetes, Amphisphaeriales

Notes: Nag Raj (Mycotaxon 22:57. 1985).
Pestalosphaeria concentrica Barr
Pestalotiopsis guepinii (Desmaz.) Steyaert var. *macrotricha* (Kleb.) Sutton (anamorph)
Distr.: Eastern United States.
Notes: On leaves and twigs (Barr. Mycologia 67.187. 1975).
Host: Rhododendron
Pestalosphaeria varia Nag Raj
Pestalotiopsis besseyii (Guba) Nag Raj (anamorph)
Distr.: HI.
Host: Acacia

Pestalotia De Not. Coelomycetes

Notes: *Pestalozzia* is a later spelling. Sutton (Canad. J. Bot. 48:2083. 1969) argued that the generic concept should be monotypic and that most of the species should be redisposed in *Pestalotiopsis*.

Pestalotia adusta Ellis & Everh.
Pestalotia kalmicola Ellis & Everh.
Pestalotia vaccinicola Guba
Distr.: Cosmopolitan.
Notes: On leaves of a variety of woody dicotyledons.
Hosts: Acer, Carya, Kalmia, Prunus, Vaccinium

Pestalotia avenae R. Sprague
Distr.: WA(type).
Notes: Sprague (Mycologia 54:47. 1962).
Host: Avena

Pestalotia baarnensis (Steyaert) Guba
Distr.: WA.
Host: Rhododendron

Pestalotia batatae Ellis & Everh.
Distr.: AL; South Africa.
Notes: On tubers.
Host: Ipomoea

Pestalotia brevipes Cooke
Distr.: FL.
Host: Aralia

Pestalotia breviseta Sacc.
Pestalotia quercina Guba
Distr.: United States; Italy; Japan.
Hosts: Acacia, Quercus

Pestalotia cibotii R.P. White
Distr.: NJ; Scotland(on seeds).
Host: Cibotium

Pestalotia clavata Cooke & Ellis
Distr.: United States; Italy.
Host: Smilax

Pestalotia cryptomeriae Cooke
Distr.: SC.
Host: Cryptomeria

Pestalotia cycadis Allesch.
Distr.: Eastern United States; Germany.
Hosts: Cycas, Zamia

Pestalotia foedans Sacc. & Ellis
Pestalotia peregrina Ellis & G. Martin
Distr.: Widespread.
Host: Pinus

Pestalotia gracilis Kleb.
Pestalotia sassafras Ellis & Everh.
Distr.: Eastern United States.
Host: Gossypium

Pestalotia hartigii Tub.
Distr.: Western United States; Europe.
Hosts: Malus, Pyrus

Pestalotia langloisii Guba
Distr.: Southern United States; Japan.
Host: Gardenia

Pestalotia laurocerasi Westend.
Monochaetia rosenwaldii Khazanoff
Pestalotia truncata Lév. var. *rubi* P. Karst.
Distr.: Widespread.
Hosts: Prunus, Rubus

Pestalotia leprogena Speg.
Distr.: HI, PR; South America.
Host: Musa

Pestalotia longiaristata Maubl.
Distr.: Widespread.
Host: Eriobotrya

Pestalotia longisetula Guba
Distr.: Eastern United States; Israel.
Notes: Fruit rot of strawberry.
Hosts: Fragaria, Rosa

Pestalotia mangifolia Guba
Distr.: Widespread.
Host: Carya

Pestalotia matildae Richatt
Distr.: Reported from an unknown locality in the United States(type); Chile.
Host: Carya

Pestalotia maura Ellis & Everh.
Distr.: Several reports from the southeastern United States.
Notes: The identity of this species is uncertain (Nag Raj. Mycotaxon 26: 216. 1986).
Hosts: Persea, Quercus

Pestalotia menezesiana Bres. & Torr.
Distr.: Widespread.
Host: Vitis

Pestalotia microspora Speg.
Pestalotia nucicola Ellis & Everh.
Distr.: Widespread.
Hosts: Alnus, Carya

Pestalotia moorei Harkn.
Distr.: CA; Europe.
Host: Lotus

Pestalotia multiseta (Speg.) Guba
Pestalotia funerea Desmaz. var. *multiseta* Speg.
Distr.: CA; Italy.
Host: Gaultheria

Pestalotia neglecta Thuem.
Distr.: North, Central and South America; Europe.
Host: Carya

Pestalotia oleandri Guba
Pestalotia stellata Berk. & M.A. Curtis var. *nerii* Cooke
Distr.: Southern United States; Haiti.
Host: Rosa

Pestalotia pestalozzioides (Dearn. & Fairm.) Shoemaker & E. Müller
Arthrobotryum pestalozzioides Dearn. & Fairm.
Pestalotia berberidis Guba
Broomella montaniensis (Ellis & Everh.) E. Müller & Ahmad (teleomorph)
Distr.: CO, MA, NM.
Notes: On bark and stems (Shoemaker & Müller. Canad. J. Bot. 41:1237. 1963).
Hosts: Berberis, Clematis

Pestalotia pezizoides De Not.
Pestalotia compacta Berk. & M.A. Curtis
Hendersonia longipes Berk. & M.A. Curtis
Pestalotia tremelloides Ellis & Everh.
Distr.: Temperate North America; Europe.
Notes: On vines (Sutton. Canad. J. Bot. 47:2084. 1969).
Hosts: Rosa, Vitis

Pestalotia planimi Vize
Distr.: CA; Europe.
Hosts: Euonymus, Spartina

Pestalotia rosae Westend.
Distr.: United States; Europe.
Host: Rosa

Pestalotia scirpina Ellis & G. Martin
Distr.: North and South America; Malta.
Host: Scirpus

Pestalotia scirrofaciens N.A. Brown
Distr.: Southern United States.
Host: Manilkara

Pestalotia sorbi Pat.
Pestalotia swieteniae Gonz. Frag. & Cif.
Distr.: Widespread.
Host: Swietenia

Pestalotia sphaerelloides Ellis & Langl. in Ellis & Everh., nom. nud.
Distr.: Southern United States.
Host: Carya

Pestalotia stellata Berk. & M.A. Curtis
Pestalotia myricae Ellis & G. Martin
Distr.: United States.
Hosts: Ilex, Myrica

Pestalotia stevensonii Peck
Distr.: United States; Europe.
Hosts: Acacia, Metrosideros, Picea

Pestalotia subcuticularis Guba
Pestalotia aucubae R.P. White, hom. illeg.
Distr.: NJ.
Host: Aucuba

Pestalotia suffocata Ellis & Everh.
Distr.: TX; Brazil.
Host: Rosa
Pestalotia sydowiana Bres.
Distr.: North and South America; Europe; Japan.
Hosts: Gaultheria, Rhododendron
Pestalotia thujicola J.L. Maas
Pestalotia thujae Sawada non Hollós, hom. illeg.
Distr.: Northeastern United States.
Notes: On cones, stems, and needles of conifers.
Hosts: Picea, Thuja
Pestalotia unicolor Berk. & M.A. Curtis
Pestalotia decolorata Speg.
Distr.: Widespread.
Notes: Sutton (Mycol. Pap. 141:143. 1977) provided a discussion of placement of this species in a new genus .
Host: Myrtus
Pestalotia vaccinii (Shear) Guba
Pestalotia guepini Desmaz. var. *vaccinii* Shear
Distr.: United States; Europe.
Hosts: Gaylussacia, Kalmia, Lyonia, Pseudotsuga, Rhododendron, Vaccinium
Pestalotia versicolor Speg.
Pestalotia cliftoniae Tracy & Earle
Pestalotia coccolobae Ellis & Everh.
Distr.: Widespread.
Hosts: Cliftonia, Coccoloba, Nerium, Sciadopitys
Pestalotia zahlbruckneriana Henn.
Distr.: Cosmopolitan.
Notes: On leaves.
Host: Acer

Pestalotiopsis Steyaert. Coelomycetes

Notes: Steyaert (Darwiniana 12:157. 1961).
Pestalotiopsis annulata (Berk. & M.A. Curtis) Steyaert
Pestalotia annulata Berk. & M.A. Curtis
Distr.: FL, eastern United States; Great Britain(Kew).
Host: Ilex
Pestalotiopsis aquatica (Ellis & Everh.) Steyaert
Pestalotia aquatica Ellis & Everh.
Distr.: Central and eastern United States.
Notes: Saprophytic on living leaves.
Hosts: Peltandra, Sarracenia
Pestalotiopsis besseyii (Guba) Nag Raj
Pestalotia besseyii Guba
Distr.: HI.
Notes: Nag Raj (Mycotaxon 22:54. 1985).
Host: Acacia
Pestalotiopsis clavispora (Atk.) Steyaert
Pestalotia clavispora Atk.
Distr.: AL, IL, OR; South Africa.
Hosts: Pseudotsuga, Quercus
Pestalotiopsis dichaeta (Speg.) Steyaert
Pestalotia dichaeta Speg.
Pestalotia micheneri Guba
Distr.: Cosmopolitan.
Notes: Leaf blight and die-back of *Araucaria* spp. and other hosts (Mordue. C.M.I. Descr. 675. 1980).
Host: Araucaria
Pestalotiopsis disseminata (Thuem.) Steyaert
Pestalotia disseminata Thuem.
Distr.: North and South America; Europe.
Hosts: Rhizophora, Terminalia
Pestalotiopsis funerea (Desmaz.) Steyaert
Pestalotia funerea Desmaz.
Pestalotia polychaetia Cooke & Harkn.
Distr.: Cosmopolitan.
Notes: On stems and leaves of conifers (Mordue. C.M.I. Descr. 514. 1976).
Hosts: Araucaria, Chamaecyparis, [Chrysobalanus], Cryptomeria, Cupressus, [Cytisus], [Eucalyptus], Juniperus, Picea, Pinus, Pithecellobium, Sequoiadendron, [Spartium], Taxodium, Taxus, Thuja, [Vitis]
Pestalotiopsis glandicola (Castagne) Steyaert
Robillarda glandicola Castagne
Pestalotia caffra Syd.
Distr.: Widespread.
Notes: On a variety of hosts (Steyaert. Bothalia 6:381. 1954).
Host: Mimusops
Pestalotiopsis guepini (Desmaz.) Steyaert var. ***macrotricha*** (Kleb.) Sutton
Pestalotia macrotricha Kleb.
Pestalotiopsis macrotricha (Kleb.) Steyaert in F.L. Raymond et al., comb. inval.
Distr.: Eastern North America.
Notes: On leaves and twigs (Barr. Mycologia 67:187. 1975).
Hosts: Cucumis, Rhododendron
Pestalotiopsis karstenii (Sacc. & Syd.) Steyaert
Pestalotia karstenii Sacc. & Syd.
Monochaetia karstenii (Sacc. & Syd.) Sutton
Monochaetia camelliae L.E. Miles
Pestalotia inquinans P. Karst. non Cooke & Harkn., hom. illeg.
Distr.: Southeastern United States; Europe.
Notes: On leaves (Nag Raj. Mycotaxon 32:122. 1988; Sutton. The Coelomycetes. p. 272. 1980).
Host: Camellia
Pestalotiopsis leucothoes (R.P. White) Steyaert
Pestalotia leucothoes R.P. White
Distr.: NJ.
Host: Leucothoe
Pestalotiopsis maculans (Corda) Nag Raj
Sporocadus maculans Corda
Hendersonia maculans (Corda) Lév.
Pestalotia maculans (Corda) S.J. Hughes
Pestalotia funerea Desmaz. var. *camelliae* Brunaud
Pestalotia guepini Desmaz.
Pestalotiopsis guepini (Desmaz.) Steyaert
Distr.: United States; Europe.
Notes: Nag Raj (Mycotaxon 22:47. 1985) discussed the synonymy accepted here. Mordue (C.M.I. Descr. 320. 1971 as *Pestalotiopsis guepini*).
Hosts: Camellia, Cercis, Citrus, Cocculus, Diospyros, Magnolia, Poncirus, Pyrus, Rhododendron, Vaccinium
Pestalotiopsis maculiformans (Guba & Zeller) Steyaert
Pestalotia maculiformans Guba & Zeller
Distr.: Western United States; Africa.
Hosts: Arceuthobium, Ledum, Vaccinium
Pestalotiopsis mangiferae (Henn.) Steyaert
Pestalotia mangiferae Henn.
Distr.: Subtropical and tropical regions.
Notes: Gray leaf spot of mango (Mordue. C.M.I. Descr. 676. 1980).
Host: Mangifera
Pestalotiopsis montellica (Sacc. & Voglino) T. Kobayashi
Pestalotia montellica Sacc. & Voglino
Distr.: Temperate North America; Great Britain(Kew); Italy.
Notes: California collections of this species have been mistakenly called *P. castagnei* Desm.
Hosts: Lithocarpus, Quercus
Pestalotiopsis palmarum (Cooke) Steyaert
Pestalotia palmarum Cooke
Distr.: Subtropical and tropical regions.
Notes: Mordue & Holliday (C.M.I. Descr. 319. 1971).
Hosts: Arecaceae, Bismarckia, Butia, Caryota, Chamaedorea, Chrysalidocarpus, Cocos, Pandanus, Phoenix, Rhapis, Roystonea, Sabal, Syagrus, Veitchia, Washingtonia
Pestalotiopsis smilacis (Schwein.) Sutton
Coryneum smilacis Schwein.
Pestalotia smilacis (Schwein.) S.J. Hughes
Distr.: Eastern North America.
Notes: On stems (Sutton. Mycol. Pap. 138:171. 1975).
Host: Smilax
Pestalotiopsis sydowiana (Bres.) Sutton
Pestalotia rhododendri Guba non Westend., hom. illeg.
Distr.: Eastern United States; Europe.
Notes: On leaves of Ericaceae and other plants (Sutton. Mycol. Pap. 80:14. 1961).
Hosts: Carya, Rhododendron
Pestalotiopsis uvicola (Speg.) Bissett
Pestalotia uvicola Speg.

Pestalotia gaurae Guba
Distr.: United States; Brazil; Europe.
Notes: Nag Raj (Mycotaxon 26:218. 1986; Mycotaxon 32:126. 1988).
Hosts: Carya, Gaura, Vitis

Pestalozziella Sacc. & Ellis in Sacc. Coelomycetes

Pestalozziella andersonii Ellis & Everh. in Ellis & B.T. Galloway
Distr.: Western United States.
Notes: Nag Raj & Kendrick (Canad. J. Bot. 50:608. 1972).
Host: Apocynum
Pestalozziella gossypina Atk.
Distr.: AL(type), southern United States.
Notes: Not a *Pestalozziella*, possibly a *Chaetospermum* (Nag Raj & Kendrick. Canad. J. Bot. 50:615. 1972).
Host: Gossypium
Pestalozziella subsessilis Sacc. & Ellis
Dilophospora geranii J. Schröt.
Neottiospora geranii (J. Schröt.) H.C. Greene
Pestalozziella geranii-pusilli C. Massal.
Distr.: Eastern North America; Europe.
Notes: Leaf spot (Nag Raj & Kendrick. Canad. J. Bot. 50:611. 1972).
Host: Geranium
Pestalozziella yuccae P. Karst. & Har.
Distr.: TX(type).
Notes: A doubtful species (Nag Raj & Kendrick. Canad. J. Bot. 50:615. 1972).
Host: Yucca

Petrakia Syd. Hyphomycetes

Petrakia echinata (Peglion) Syd.
Epicoccum echinatum Peglion
Distr.: WI; Europe(type).
Notes: Leaf spot (Ellis. Dematiaceous Hyphomycetes. p. 152. 1971).
Host: Acer

Petriella Curzi. Pyrenomycetes, Microascales

Notes: This genus is considered to be a synonym or closely related to *Microascus*.
Petriella guttulata Barron & Cain
Graphium sp. (synanamorph)
Sporotrichum sp. (synanamorph)
Distr.: North America; Europe.
Notes: Isolated from dung and various other substrates (Barron et al., Canad. J. Bot. 39:841. 1961).
Host: Carya

Peyronelia Cif. & Gonz. Frag. Hyphomycetes

Peyronelia sirodesmioides Cif. & Gonz. Frag.
Distr.: Western United States; Dominican Republic.
Hosts: Holodiscus, Philadelphus

Pezicula Tul. & C. Tul. Discomycetes, Helotiales

Notes: Parasitic and saprobic on woody substrates (Dennis. Kew Bull. 29:157. 1974).
Pezicula acericola (Peck) Sacc.
Dermatea acericola (Peck) Cooke
Dermatea simillima Ellis & Everh.
Cryptosporiopsis sp. (anamorph)
Distr.: Northeastern North America; Great Britain; Japan.
Host: Acer
Pezicula alni Rehm
Distr.: Eastern North America; Europe.
Notes: On twigs.
Host: Alnus
Pezicula australis Rehm
Distr.: MA(dubious); Europe (type).
Host: Gleditsia
Pezicula carnea (Cooke & Ellis) Rehm
Cryptosporiopsis sp. (anamorph)
Distr.: Northeastern North America; Great Britain.
Host: Acer
Pezicula carpinea (Pers.) Tul. ex Fuckel
Cryptosporium fasiculata (Tode:Fr.) Petr. (anamorph)
Distr.: Eastern North America; Europe.
Notes: Twig canker of hardwoods.
Host: Carpinus
Pezicula cinnamomea (DC.) Sacc.
Cryptosporiopsis quercina Petr. (anamorph)
Distr.: Temperate North America; Europe.
Notes: On twigs of hardwoods, especially *Quercus*.
Hosts: Castanea, Castanopsis, Quercus
Pezicula corni Petr.
Cryptosporiopsis cornina (Peck) Petr. & Syd. (anamorph)
Distr.: Northern North America.
Notes: On twigs.
Host: Cornus
Pezicula cornicola Seaver
Myxosporium sp. (anamorph)
Distr.: NY.
Host: Cornus
Pezicula corylina Groves
Cryptosporiopsis coryli (Peck) Sutton (anamorph)
Distr.: Temperate regions.
Host: Corylus
Pezicula frangulae (Fr.) Fuckel
Tympanis frangulae Fr.
Dermatea frangulae (Fr.) Tul.
Cryptosporiopsis versiformis (Albertini & Schwein.) Wollenweb. (anamorph)
Distr.: Western United States; Europe.
Host: Rhamnus
Pezicula livida (Berk. & Broome) Rehm
Dermea livida (Berk. & Broome) W. Phillips
Durella livida (Berk. & Broome) Sacc.
Cryptosporiopsis abietina Petr. (anamorph)
Distr.: Temperate northern hemisphere.
Notes: On conifer wood, needles, and cones.
Hosts: Abies, Pinus, Sequoia
Pezicula malicorticis (H. Jacks.) Nannf.
Neofabraea malicorticis H. Jacks.
Neofabraea perennans Kienholz
Distr.: Temperate regions.
Notes: Anthracnose, bull's eye rot, and perennial canker of apples and pears.
Hosts: Cydonia, Malus, Pyrus
Pezicula myrtillina P. Karst.
Distr.: CA; Europe.
Host: Vaccinium
Pezicula ocellata (Pers.:Fr.) Seaver
Ocellaria ocellata (Pers.:Fr.) Petr.
Ocellaria aurea Tul. & C. Tul., nom. illeg.
Pezicula eximia Rehm
Cryptosporiopsis scutellata (G. Otth) Petr. (anamorph)
Distr.: Temperate North America; Europe.
Notes: On twigs and branches of Salicaceae.
Hosts: [Cornus], Populus, Salix
Pezicula pruinosa Farl.
Cryptosporiopsis pruinosum (Peck) Wollenweb. (anamorph)
Distr.: Temperate North America; Europe.
Notes: On twigs.
Hosts: Amelanchier, Malus, Pyrus
Pezicula purpurascens (Ellis & Everh.) Seaver
Dermatea purpurascens Ellis & Everh.
Distr.: Eastern North America.
Notes: On dead limbs of chestnut.
Host: Castanea
Pezicula rhododendricola Rehm
Distr.: PA; Europe(type).
Host: Rhododendron
Pezicula rubi (Lib.) Rabenh.
Myxosporium phaeosporum (Sacc.) Allesch. (anamorph)

Distr.: Temperate regions.
Host: Rubus
Pezicula sepium (Desmaz.) Dennis
Dermatea crataegicola E.J. Durand
Pezicula crataegicola (E.J. Durand) Groves
Distr.: Temperate regions.
Notes: Dennis (Kew Bull. 29:161. 1974).
Host: Crataegus
Pezicula spiculata Seaver
Sphaeronaema sp. (anamorph)
Distr.: NY(type).
Host: Acer
Pezicula subcarnea Groves
Cryptosporiopsis sp. (anamorph)
Distr.: Northeastern North America.
Notes: Groves (Mycologia 33:317. 1941).
Host: Acer

Peziza L. Discomycetes, Pezizales

Notes: Most species occur on soil and are occasionally reported in association with a vascular plant.
Peziza repanda Pers.
Plicaria repanda (Pers.) Rehm
Peziza pallida Cooke & Peck
Oedocephalum sp. (anamorph)
Distr.: Temperate regions.
Hosts: Acer, Betula, Malus

Pezizella Fuckel. Discomycetes, Helotiales

Pezizella aurantiaca Cash
Distr.: CO(type).
Host: Picea
Pezizella carneorosea (Cooke & Harkn.) Sacc.
Peziza carneorosea Cooke & Harkn.
Distr.: CA(type).
Notes: On twigs (Dennis. Persoonia 3:38. 1964).
Host: Eucalyptus
Pezizella helotioides Starb.
Distr.: CA.
Host: Allium
Pezizella minuta Dearn.
Distr.: Southeastern United States.
Host: Pinus
Pezizella pteridina (Nyl.) Rehm
Distr.: CA, PA.
Hosts: Dennstaedtia, Polystichum
Pezizella regalis (Cooke & Ellis) Sacc.
Peziza regalis Cooke & Ellis
Distr.: NJ(type).
Notes: On bark. Dennis (Persoonia 3:61. 1964) suggested that this is an *Orbilia*.
Host: Malus
Pezizella vulgaris (Fr.) Sacc.
Allophylaria vulgaris (Fr.) Nannf.
Hyalinia vulgaris (Fr.) Boud.
Mollisia vulgaris (Fr.) Gill.
Calycina albuminum (Cooke & Peck) Kuntze
Helotium albuminum (Cooke & Peck) Sacc.
Dasyscyphus translucida (Berk. & M.A. Curtis) Sacc.
Chalara cylindrica P. Karst. (anamorph)
Distr.: PA, WA; Europe.
Notes: Carpenter (Mem. New York Bot. Gard. 33:270. 1981) was unable to determine the correct generic disposition of this species. Nannfelt (Nova Acta Regiae Soc. Sci. Upsal., Ser. 4, 8:291. 1932) listed *Chalara minima* as the anamorph but according to Nag Raj & Kendrick (A monograph of *Chalara* and allied genera. p. 179. 1975) this is a synonym of *C. cylindrica*.
Hosts: Acer, Castanea, Salix
Pezizella vulgaris (Fr.) Sacc. var. ***sanguinella*** (Berk. & M.A. Curtis) Sacc.
Peziza vulgaris Fr. var. *sanguinella* Berk. & M.A. Curtis
Distr.: SC(type).
Notes: Dennis (Persoonia 3:70. 1964).
Host: Liquidambar

Pezizellaster Höhn. Discomycetes, Helotiales

Pezizellaster herbarum Cash
Distr.: CA(type).
Host: Scrophularia

Pezoloma Clements. Discomycetes, Helotiales

Pezoloma laricina (Ellis & Everh.) Korf
Pseudohelotium laricinum Ellis & Everh.
Sphagnicola laricina (Ellis & Everh.) Dennis
Distr.: MI(type).
Notes: On fallen needles (Dennis. Persoonia 3:50. 1964).
Host: Larix

Phacidina Höhn. Loculoascomycetes, Dothideales

Phacidina gracilis (Niessl) Höhn.
Phacidium gracile Niessl
Distr.: Northeastern North America; Europe.
Notes: DiCosmo et al. (Mycotaxon 21:61. 1984).
Host: Lycopodium

Phacidiopycnis Potebnia. Coelomycetes

Phacidiopycnis boycei (Hahn) Funk
Phomopsis boycei Hahn
Distr.: Northwestern North America.
Notes: The anamorph of *Phacidium balsamicola* (=*Potebniamyces balsamicola* var. *balsamicola*) was transferred to the genus *Apostrasseria* by DiCosmo et al. (Mycotaxon 21:105. 1984) as *Apostrasseria balsamicola* (=*Phaciopycnis balsamicola*) but *Phacidiopynis boycei*, the anamorph of *Potebniamyces balsamicola* var. *boycei*, was never transferred to that genus.
Host: Abies
Phacidiopycnis pseudotsugae (M. Wilson) G. Hahn
Phomopsis pseudotsugae M. Wilson
Distr.: Northern hemisphere; New Zealand.
Notes: According to DiCosmo (Mycotaxon 21:114. 1984) this anamorph belongs in *Apostrasseria*.
Hosts: Abies, Larix, Pinus, Pseudotsuga, Tsuga
Phacidiopycnis tuberivora (H.T. Güssow & W.R. Foster) Sutton
Phomopsis tuberivora H.T. Güssow & W.R. Foster
Distr.: Temperate regions.
Notes: Stem-end hard rot, dry rot, or corky rot of potatoes (Punithalingam. C.M.I. Descr. 823. 1985).
Host: Solanum

Phacidium Fr., nom. cons. Discomycetes, Helotiales

Notes: DiCosmo et al. (Mycotaxon 21:1. 1984).
Phacidium abietis (Dearn.) J. Reid & Cain
Phacidium infestans P. Karst. var. *abietis* Dearn.
Distr.: Western United States; Canada.
Notes: Populations in western North America were referred to as *P. infestans* in some of the older literature.
Hosts: Abies, Pseudotsuga
Phacidium arbuti Cooke & Harkn.
Ceuthospora sp. (anamorph)
Distr.: CA(type).
Host: Arbutus
Phacidium balsamicola (Smerlis) DiCosmo, Nag Raj, & Kendrick
Potebniamyces balsamicola Smerlis
Distr.: ID; Canada.
Notes: On branches and twigs.
Host: Abies
Phacidium coniferarum (Hahn) DiCosmo, Nag Raj, & Kendrick
Phacidiella coniferarum Hahn

Potebniamyces coniferarum (Hahn) Smerlis
Phacidium pulverulentum J.C. Schmidt in Kunze & J.C. Schmidt:Fr.
Phacidiopycnis pseudotsugae (M. Wilson) Hahn (anamorph)
Distr.: Coniferous forests of the northern hemisphere; New Zealand.
Notes: According to DiCosmo et al. (Mycotaxon 21:1. 1984), the anamorph belongs to the genus *Apostrasseria*. Punithalingam & Gibson (C.M.I. Descr. 517. 1976 as *Potebniamyces coniferarum*).
Host: Pinus
Phacidium elegantissimum Berk. & M.A. Curtis
Distr.: AL.
Notes: DiCosmo et al. (Mycotaxon 21:178. 1984) reported that this name could not be characterized.
Host: Ilex
Phacidium heteromelis W. Phillips & Harkn.
Ceuthospora sp. (anamorph)
Distr.: CA(type).
Host: Heteromeles
Phacidium infestans P. Karst.
Gremmenia gigaspora (Gremmen) Korf
Phragmonaevia gigaspora Gremmen
Distr.: Northern United States; Europe.
Notes: The records from North America are questionable. Reid & Pirozynski (Mycologia 60:526. 1968); Minter & Millar (C.M.I. Descr. 652. 1980).
Hosts: Abies, Picea, Pinus, Pseudotsuga, Thuja
Phacidium lacerum Fr.:Fr.
Ceuthospora pinastri (Fr.) Höhn. (anamorph)
Distr.: Temperate regions.
Notes: On fallen needles of conifers.
Host: Arctostaphylos
Phacidium negundinis Tehon & E.Y. Daniels
Distr.: IL(type).
Notes: DiCosmo et al. (Mycotaxon 21:1. 1984) reexamined the type and found it to be a coelomycete similar to *Cryptocline abietina*.
Host: Acer
Phacidium nigrum Cooke
Distr.: GA(type).
Notes: DiCosmo et al. (Mycotaxon 21:193. 1984) suggested that this species probably belongs in *Rhytisma*.
Hosts: Leucothoe, Lyonia
Phacidium sherwoodiae DiCosmo, Nag Raj & Kendrick
Ceuthospora sp. (anamorph)
Distr.: OR.
Hosts: Gaultheria, Thuja
Phacidium sphaeroideum Cooke & Ellis
Distr.: NJ(type).
Host: Ilex
Phacidium taxi (Fr.) Fr.
Distr.: WI; Europe.
Notes: DiCosmo et al. (Mycotaxon 21:203. 1984) were unable to determine the characteristics of this taxon.
Host: Taxus
Phacidium taxicola Dearn. & House in Dearn.
Apostrasseria fusiformis Nag Raj (anamorph)
Distr.: Northeastern North America.
Notes: Snow blight on leaves of *Taxus canadensis*. Smerlis (Pl. Dis. Reporter 52:403. 1968) demonstrated its pathogenicity.
Host: Taxus
Phacidium vaccinii Fr.
Ceuthospora latitans (Fr.:Fr.) Höhn. (anamorph)
Distr.: Western United States; Europe.
Hosts: Arctostaphylos, Lyonia, Rhododendron, Vaccinium

Phaeangellina Dennis. Discomycetes, Helotiales

Phaeangellina empetri (W. Phillips) Dennis
Cenangium empetri W. Phillips
Distr.: Circumboreal in north temperate in arctic regions.
Host: Empetrum

Phaeangium (Sacc.) Sacc. Discomycetes, Helotiales

Notes: This genus is considered a synonym of *Pezicula*, but not all names have been redisposed.
Phaeangium magnisporum Cash & J.A. Stevenson
Distr.: GA(type).
Host: Betula
Phaeangium sphaeroides Ellis & Everh.
Distr.: CA(type).
Hosts: Arctostaphylos, Rhamnus, Rhus

Phaeocryptopus Naumov. Loculoascomycetes, Dothideales

Phaeocryptopus gaeumannii (T. Rohde) Petr.
Adelopus gaeumannii T. Rohde
Distr.: North America; Europe; Japan; New Zealand; Australia.
Notes: On needles.
Host: Pseudotsuga
Phaeocryptopus nudus (Peck) Petr.
Adelopus nudus (Peck) Theiss. in Höhn.
Adelopus balsamicola Theiss., nom. inval.
Distr.: Northern hemisphere.
Notes: On needles.
Hosts: Abies, Tsuga
Phaeocryptopus pinastri (Ellis & Sacc.) Petr.
Asterina pinastri Ellis & Sacc.
Distr.: NJ(type).
Host: Pinus

Phaeocytostroma Petr. Coelomycetes

Phaeocytostroma ambiguum (Mont.) Petr. in Petr. & Syd.
Sphaeropsis ambigua Mont.
Phaeocytosporella zeae G.L. Stout
Distr.: Central United States; Australia; Europe.
Host: Zea
Phaeocytostroma sacchari (Ellis & Everh.) Sutton
Trullula sacchari Ellis & Everh. in Cockerell
Melanconium sacchari Massee in Speg.
Pleocyta sacchari (Massee) Petr. & Syd.
Distr.: Subtropical and tropical regions.
Notes: Rind disease of sugarcane (Sutton & Waterson. C.M.I. Descr. 87. 1966).
Host: Saccharum

Phaeodiscula Cuboni. Coelomycetes

Phaeodiscula texensis R. Sprague
Distr.: TX(type).
Host: Bouteloua

Phaeodochium M.L. Farr. Hyphomycetes

Distr.: Farr & Horner (Nova Hedwigia 15:268. 1968)
Phaeodochium myrothecioides M.L. Farr
Host: Selaginella

Phaeoisariopsis Ferraris. Hyphomycetes

Notes: Ellis (More dematiaceous Hyphomycetes. p. 230. 1976).
Phaeoisariopsis bambusicola Morgan-Jones
Distr.: AL.
Notes: Morgan-Jones (Mycotaxon 7:130. 1978).
Host: Bambusa
Phaeoisariopsis gnaphaliaceae (Cooke) Morgan-Jones
Cercospora gnaphaliaceae Cooke
Distr.: Central and eastern North America; West Indies; South America; Japan.
Hosts: Filago, Gnaphalium
Phaeoisariopsis griseola (Sacc.) Ferraris
Isariopsis griseola Sacc.
Cercospora griseola (Sacc.) Ragunathan & Ramakrishan
Cercospora columnaris Ellis & Everh.

Pseudocercospora columnaris (Ellis & Everh.) J. Yen in J. Yen & G. Lim
Isariopsis laxa (Ellis) Sacc.
Graphium laxum Ellis
Distr.: Cosmopolitan.
Notes: Angular leaf spot of beans (Anon. C.M.I. Descr. 847. 1986).
Hosts: Lathyrus, Phaseolus
Phaeoisariopsis liriodendri (Ellis & Harkn.) Morgan-Jones & L.G. Brown
Cercospora liriodendri Ellis & Harkn.
Distr.: Southeastern United States.
Notes: Morgan-Jones & Brown (Mycotaxon 4:494. 1976).
Host: Liriodendron
Phaeoisariopsis melanochaeta (Ellis & Everh.) Deighton
Cercospora melanochaeta Ellis & Everh.
Distr.: Central United States; southern Africa.
Notes: Anon. (C.M.I. Descr. 848. 1986).
Host: Celastrus
Phaeoisariopsis menispermi (Ellis & Holw.) Deighton
Cercospora menispermi Ellis & Holw.
Cercosporiopsis menispermi (Ellis & Holw.) Miura
Distr.: Central and eastern North America; China; India.
Hosts: Calycocarpum, Cocculus, Menispermum
Phaeoisariopsis nigricans (Cooke) L.G. Brown & Morgan-Jones
Cercospora nigricans Cooke
Pseudocercospora nigricans (Cooke) Deighton
Cercospora atromaculans Ellis & Everh.
Cercospora torae Tharp
Distr.: Eastern United States; West Indies; East Africa; Japan.
Notes: Leaf spot (Brown & Morgan-Jones. Mycotaxon 6:267. 1977).
Hosts: Aralia, Cassia, Senna
Phaeoisariopsis personata (Berk. & M.A. Curtis) Arx
Cladosporium personatum Berk. & M.A. Curtis
Cercospora personata (Berk. & M.A. Curtis) Ellis & Everh.
Cercosporidium personatum (Berk. & M.A. Curtis) Deighton
Cercospora arachidis Henn.
Mycosphaerella berkeleyi Jenk. (teleomorph)
Distr.: Worldwide in peanut-growing regions.
Notes: Late leaf spot of peanuts (Mulder & Holliday. C.M.I. Descr. 412. 1974).
Hosts: Arachis, Senna
Phaeoisariopsis pubescens (Cooke & Ellis) M.B. Ellis
Graphium pubescens Cooke & Ellis
Didymobotryum pubescens (Cooke & Ellis) Sacc.
Distr.: NJ.
Host: Smilax
Phaeoisariopsis robiniae (Shear) Deighton in M.B. Ellis
Fusicladium robiniae Shear
Camptomeris robiniae (Shear) Cif.
Passalora robiniae (Shear) S.J. Hughes
Distr.: Eastern United States.
Notes: Leaf spot of black locust (C.M.I. Descr. 849. 1986).
Host: Robinia
Phaeoisariopsis simulata (Ellis & Everh.) L.G. Brown & Morgan-Jones
Cercospora simulata Ellis & Everh.
Distr.: Eastern United States; West Indies; South America; East Africa.
Notes: Leaf spot (C.M.I. Descr. 850. 1986).
Hosts: Cassia, Senna

Phaeolus Pat. Basidiomycetes, Aphyllophorales
Notes: Gilbertson & Ryvarden (North American Polypores 2:539. 1987).
Phaeolus schweinitzii (Fr.:Fr.) Pat.
Polyporus schweinitzii Fr.:Fr.
Distr.: Temperate northern hemisphere.
Notes: Butt rot of conifers.
Hosts: Abies, Acacia, Betula, Eucalyptus, Larix, Liquidambar, Picea, Pinus, Prunus, Pseudotsuga, Quercus, Taxus, Thuja, Tsuga

Phaeomarasmius Scherff. Basidiomycetes, Agaricales
Notes: Many species are placed in *Flammulaster* Earle.
Phaeomarasmius distans (Peck) Singer
Distr.: NY(type).
Host: Crataegus
Phaeomarasmius erinaceus (Fr.) Kühner in Singer
Crinipellis alnicola Murrill
Naucoria badia Murrill
Distr.: Temperate North America; Europe.
Notes: On wood (Redhead. Sydowia 37:261. 1984; Redhead. Fungi Canadenses 175. 1980).
Hosts: Alnus, Quercus
Phaeomarasmius rufolateritius (Bres.) Singer
Distr.: MO(type).
Host: Crataegus

Phaeophacidium Henn. & Lindau. Discomycetes, Rhytismatales
Phaeophacidium abietinum Sacc.
Distr.: WA(type).
Host: Abies

Phaeoramularia Muntañola. Hyphomycetes
Notes: Deighton (Mycol. Pap. 144:1. 1979); Ellis (More dematiaceous Hyphomycetes. p. 316. 1976).
Phaeoramularia antigoni (M.L. Farr) Deighton
Cercospora antigoni M.L. Farr
Distr.: TX; Cuba.
Host: Antigonon
Phaeoramularia antipus (Ellis & Holw.) Deighton
Cercospora antipus Ellis & Holw.
Distr.: Central North America.
Notes: Leaf spot.
Host: Lonicera
Phaeoramularia capsicicola (Vassiljevsky) Deighton
Cercospora capsicicola Vassiljevsky
Cladosporium capsici (Ém. Marchal & Steyaert) Kovachevski
Cercospora unamunoi Castellani
Phaeoramularia unamunoi (Castellani) Muntañola, nom. inval.
Distr.: Subtropical and tropical regions.
Notes: Leaf spot (Deighton. Trans. Brit. Mycol. Soc. 67:140. 1976).
Host: Capsicum
Phaeoramularia catenospora (Atk.) Deighton
Cercospora catenospora Atk.
Distr.: Central and eastern North America.
Host: Sambucus
Phaeoramularia cucurbiticola (Henn.) Deighton in M.B. Ellis
Cercospora cucurbiticola Henn.
Distr.: FL; South America.
Host: Cucurbita
Phaeoramularia dioscoreae (Ellis & G. Martin) Deighton
Cercospora dioscoreae Ellis & G. Martin
Distr.: Central and eastern United States; India.
Host: Dioscorea
Phaeoramularia fusimaculans (Atk.) X. Liu & Guo
Cercospora fusimaculans Atk.
Cercospora panici J.J. Davis
Cercosporina panici (J.J. Davis) Sacc.
Cercospora panici-miliacei Sawada
Distr.: United States; West Indies; Central America; northern South America; Africa; southeast Asia; Australia; India.
Notes: On leaves of many grasses.
Hosts: Dichanthelium, Digitaria, Echinochloa, Panicum, Stenotaphrum
Phaeoramularia gayophyti (Ellis & Everh.) Deighton
Cercospora gayophyti Ellis & Everh.
Ramularia gayophyti (Ellis & Everh.) Solheim
Distr.: Western United States.
Notes: Solheim (Univ. Wyoming Publ. 24:48. 1960) redescribed this fungus based on collections from Wyoming.
Host: Gayophytum

Phaeoramularia heterospora (Ellis & B.T. Galloway) Deighton
Septosporium heterosporum Ellis & B.T. Galloway
Stigmina vitis Dearn. & Barth.
Distr.: CA; Israel.
Notes: On leaves (Ellis. More dematiaceous Hyphomycetes. p. 325. 1976).
Host: Vitis
Phaeoramularia maculicola (Roum. & Sacc.) Sutton
Torula maculicola Roum. & Sacc.
Cladosporium subsessile Ellis & Barth.
Distr.: Temperate North America; Europe.
Notes: Leaf spot (Ellis. Dematiaceous Hyphomycetes. p. 322. 1976).
Host: Populus
Phaeoramularia manihotis (F. Stevens & Solheim) M.B. Ellis
Ragnhildiana manihotis F. Stevens & Solheim
Cercospora caribaea Cif. in Muller & Chupp
Distr.: Subtropical and tropical regions.
Host: Manihot
Phaeoramularia occidentalis (Cooke) Deighton
Cercospora occidentalis Cooke
Cercospora occidentalis Ellis & Kellerm. in Heald & F.A. Wolf non Cooke, hom. illeg.
Cercosporina occidentalis (Cooke) Sacc.
Ramularia cassiaecola Heald & F.A. Wolf
Cercospora somalensis Curzi
Distr.: Southern United States; West Indies; Central and South America.
Hosts: Cassia, Senna
Phaeoramularia periclymeni (G. Wint.) Deighton
Cercospora periclymeni G. Wint.
Distr.: TX; Germany; Malta.
Notes: On leaves.
Host: Lonicera
Phaeoramularia symphoricarpi (Ellis & Everh.) Deighton
Cercospora symphoricarpi Ellis & Everh.
Distr.: Central and eastern North America.
Notes: Leaf spot (Ellis. More dematiaceous Hyphomycetes. p. 317. 1976).
Host: Symphoricarpos

Phaeosaccardinula Henn. Loculoascomycetes, Dothideales

Phaeosaccardinula dematia V. Miller & Bonar
Distr.: Western United States.
Hosts: Abies, Baccharis, Umbellularia
Phaeosaccardinula martinii (Ellis & Sacc.) Höhn.
Saccardia martini Ellis & Sacc.
Distr.: Southern United States.
Notes: On living leaves of trees.
Host: Quercus

Phaeoseptoria Speg. Coelomycetes

Notes: Punithalingam (Nova Hedwigia 32:585. 1980).
Phaeoseptoria airae (Grove) R. Sprague
Septoria alopecuri Syd. var. *airae* Grove
Phaeoseptoria festucae R. Sprague
Distr.: Temperate regions.
Notes: On a wide variety of grasses.
Hosts: Agrostis, Bromus, Calamagrostis, Carex, Catabrosa, Danthonia, Deschampsia, Festuca, Hordeum, Muhlenbergia, Phalaris, Phleum, Poa, Puccinellia, Secale
Phaeoseptoria calamagrostidis R. Sprague
Distr.: Midwest and western United States.
Hosts: Agrostis, Calamagrostis
Phaeoseptoria caricicola (Sacc.) R. Sprague
Septoria caricicola Sacc.
Distr.: Western United States; Europe.
Host: Carex
Phaeoseptoria caricis Tehon & E.Y. Daniels
Distr.: IL.
Notes: According to Petrak (Ann. Mycol. 39:292. 1941) this species belongs in *Hendersonia*.
Host: Carex
Phaeoseptoria elymi R. Sprague
Distr.: NE.
Host: Elymus
Phaeoseptoria eucalypti Hansf.
Distr.: HI; Australia.
Notes: Although *Phaeoseptoria* is now restricted to graminicolous species, this name has not been transferred elsewhere (Walker. Proc. Linn. Soc. New South Wales 87:171. 1962).
Host: Eucalyptus
Phaeoseptoria festucae R. Sprague var. ***andropogonis*** R. Sprague in Greene
Distr.: Midwest and western North America.
Hosts: Andropogon, Festuca
Phaeoseptoria festucae R. Sprague var. ***muhlenbergiae*** R. Sprague ex Punithalingam
Distr.: Midwest and western United States.
Hosts: Agropyron, Elymus, Muhlenbergia
Phaeoseptoria phalaridis (Trail) R. Sprague
Septoria bromi Sacc. var. *phalaridis* Trail
Distr.: OR.
Host: Phalaris
Phaeoseptoria poae R. Sprague
Distr.: Western United States.
Hosts: Phleum, Poa
Phaeoseptoria scirpi Wehmeyer
Hendersonia scirpi (Wehmeyer) Petr.
Distr.: SD, WY.
Host: Scirpus

Phaeosolenia Speg. Basidiomycetes, Agaricales

Phaeosolenia brenckleana (Sacc.) W.B. Cooke
Distr.: Central United States.
Host: Ulmus
Phaeosolenia ravenelii W.B. Cooke
Cyphella ravenelii Berk. & M.A. Curtis non Sacc., hom. illeg.
Distr.: SC(type).
Host: Carya

Phaeosphaerella P. Karst. Loculoascomycetes, Dothideales

Notes: This genus is considered a synonym of *Venturia*, but not all names have been redisposed.
Phaeosphaerella dianellae F. Stevens
Distr.: HI.
Host: Dianella
Phaeosphaerella mangiferae F. Stevens & Weedon in F. Stevens
Distr.: HI.
Host: Mangifera
Phaeosphaerella paspali Tehon
Distr.: OK; Puerto Rico.
Host: Paspalum
Phaeosphaerella scirpicola Earle
Distr.: CA.
Notes: A confused name (Barr. Canad. J. Bot. 46:862. 1968).
Host: Scirpus

Phaeosphaeria Miyake. Loculoascomycetes, Dothideales

Notes: Eriksson (Ark. Botanik 6:381. 1967); Leuchtmann (Sydowia 37:75. 1984).
Phaeosphaeria acori H.C. Greene
Distr.: WI(type).
Host: Acorus
Phaeosphaeria agminalis (Sacc. & Morthier) Tomilin
Leptosphaeria agminalis Sacc. & Morthier
Distr.: OR; Europe.
Host: Clarkia
Phaeosphaeria avenaria (G.F. Weber) O. Eriksson
Leptosphaeria avenaria G.F. Weber

Distr.: Cosmopolitan.
Notes: Speckle blotch of oats (Sivanesan. C.M.I. Descr. 312. 1971 as *Leptosphaeria avenaria*).
Hosts: Avena, Hordeum
Phaeosphaeria avenaria (G.F. Weber) O. Eriksson f. sp. ***triticea*** T. Johnson
Leptosphaeria avenaria G.F. Weber f. sp. *triticea* T. Johnson
Stagonospora avenae (A.B. Frank) Bissett f. sp. *triticea* T. Johnson (anamorph)
Distr.: Cosmopolitan.
Notes: A physiolgical form which occurs on a wide range of festucoid grasses (Bissett. Fungi Canadenses 239. 1982).
Host: Triticum
Phaeosphaeria caricinella (P. Karst.) O. Eriksson
Leptosphaeria caricinella P. Karst.
Distr.: North temperate regions.
Notes: On Juncaceae and Poaceae.
Hosts: Agrostis, Carex, Danthonia, Eriophorum, Festuca, Juncus, Phleum, Poa, Sitanion, Spartina, Trisetum
Phaeosphaeria clavispora (Cooke & Peck) Barr
Dothidea clavispora (Cooke & Peck) Peck
Rhopographus clavisporus (Cooke & Peck) Sacc.
Leptosphaeria clavicarpa Ellis & Everh.
Distr.: Eastern North America.
Notes: Barr et al. (New York State Mus. Bull. 459:14. 1986).
Hosts: Andropogon, Phragmites, Rhynchospora, Setaria
Phaeosphaeria culmorum (Auersw.) Leuchtmann
Leptosphaeria culmorum Auersw.
Phaeosphaeria microscopica P. Karst. var. *culmorum* (Auersw.) O. Eriksson
Distr.: North temperate regions.
Notes: On grasses.
Hosts: Agropyron, Andropogon, Dactylis, Elymus, Erianthus, Tridens
Phaeosphaeria eustoma (Fuckel) L. Holm
Pleospora eustoma Fuckel
Leptosphaeria eustoma (Fuckel) Sacc.
Stagonospora sp. (anamorph)
Distr.: Central and southern United States; Europe.
Notes: On leaves of grasses and other monocots.
Hosts: [Aquilegia], Dieffenbachia, [Draba], [Thalictrum], Trisetum, Zea
Phaeosphaeria fuckelii (Niessl) L. Holm
Leptosphaeria fuckelii Niessl
Leptosphaeria lycopodiicola Peck
Distr.: Central and eastern United States; Europe.
Host: Lycopodium
Phaeosphaeria herpotrichoides (De Not.) L. Holm
Leptosphaeria herpotrichoides De Not.
Trematosphaeria herpotrichoides (De Not.) G. Wint.
Leptosphaeria culmifraga Auct.
Leptosphaeria sparsa (Fuckel) Sacc.
Stagonospora sp. (anamorph)
Distr.: Cool temperate northern hemisphere.
Notes: Secondary root rot, node canker of cereals and grasses.
Hosts: Agropyron, Agrostis, Calamagrostis, Chasmanthium, Cinna, Elymus, Festuca, Glyceria, Muhlenbergia, Phleum, Poa, Secale, Setaria, Spartina, Triticum
Phaeosphaeria heterospora (De Not.) Boise
Leptosphaeria heterospora (De Not.) Niessl
Trematosphaeria heterospora (De Not.) G. Wint.
Phoma sp. (anamorph)
Distr.: Temperate regions.
Notes: On rhizomes (Boise. Mycologia 77:236. 1985).
Hosts: Arnica, Iris, Lupinus, Poa, Stipa, Trisetum
Phaeosphaeria insignis (P. Karst.) L. Holm
Leptosphaeria insignis P. Karst.
Distr.: CA; Europe.
Notes: On grasses.
Hosts: Arctagrostis, [Carex]
Phaeosphaeria juncicola (Rehm) L. Holm
Leptosphaeria juncicola Rehm
Distr.: Temperate regions.
Host: Scirpus
Phaeosphaeria juncina (Auersw.) L. Holm
Leptosphaeria juncina Auersw.
Distr.: FL; Europe.
Host: Juncus
Phaeosphaeria luctuosa (Niessl) Otani & Mikawa
Leptosphaeria luctuosa Niessl
Distr.: Cosmopolitan.
Host: Typha
Phaeosphaeria lycopodina (Mont.) Hedjaroude
Leptosphaeria lycopodina (Mont.) Sacc.
Leptosphaeria crepini (Westend.) De Not.
Distr.: Temperate regions.
Host: Lycopodium
Phaeosphaeria marcyensis (Peck) L. Holm & K. Holm
Leptosphaeria marcyensis (Peck) Sacc.
Distr.: NY.
Host: Lycopodium
Phaeosphaeria microscopica (P. Karst.) O. Eriksson
Leptosphaeria microscopica P. Karst.
Distr.: Temperate regions.
Notes: On a wide variety of grasses.
Hosts: Agrostis, Secale, Trisetum
Phaeosphaeria nigrans (Roberge ex Desmaz.) L. Holm
Leptosphaeria nigrans (Roberge ex Desmaz.) Ces. & De Not.
Leptosphaeria culmicola Auersw.
Metasphaeria leersiae (Pass.) Sacc.
Sphaerella leersiae Pass.
Stagonospora sp. (anamorph)
Distr.: Cosmopolitan.
Notes: On grasses and other herbaceous substrates.
Hosts: Aloe, Arundinaria, Carex, Chasmanthium, Leersia, Panicum, Phragmites
Phaeosphaeria nodorum (E. Müller) Hedjaroude
Leptosphaeria nodorum E. Müller
Distr.: Cosmopolitan.
Notes: Glume blotch of wheat and other grasses (Sutton & Waterson. C.M.I. Descr. 86. 1966; Bissett. Fungi Canadenses 240. 1982).
Hosts: Bromus, Triticum
Phaeosphaeria silenes-acaulis (De Not.) L. Holm
Leptosphaeria silenes-acaulis De Not.
Distr.: Arctic-alpine North American and Europe.
Hosts: Silene, Stellaria
Phaeosphaeria tritici (Garov.) Hedjaroude
Pleospora tritici Garov.
Leptosphaeria tritici (Garov.) Pass.
Hendersonia sp. (anamorph)
Distr.: United States; Europe.
Notes: Leuchtmann (Sydowia 37:105. 1984).
Host: Triticum
Phaeosphaeria typharum (Desmaz.) L. Holm
Leptosphaeria typharum (Desmaz.) P. Karst.
Pleospora typharum (Desmaz.) Fuckel
Leptosphaeria kunzeana Berl.
Scolecosporiella typhae (Oudem.) Petr. (anamorph)
Distr.: North America; temperate Europe; India.
Hosts: Agrostis, Carex, Festuca, Muhlenbergia, Phleum, Poa, Spartina, Typha
Phaeosphaeria vagans (Niessl) O. Eriksson
Pleospora vagans Niessl
Distr.: Temperate regions.
Notes: On herbaceous stems of various plants.
Hosts: Geranium, Oxytropis, Poa, Spartina
Phaeosphaeria viridella (Peck) Sacc.
Leptosphaeria viridella (Peck) Sacc.
Stagonospora sp. (anamorph)
Distr.: Eastern North America; Europe.
Notes: Frequently misdetermined as *L. pratensis* Sacc. & Briard (Shoemaker. Canad. J. Bot. 62:2715. 1984).
Host: Melilotus

Phaeostalagmus W. Gams. Hyphomycetes

Notes: Wang & Sutton (Mycologia 74:489. 1982).
Phaeostalagmus altissimus Wang & Sutton
Distr.: NY.
Notes: Lignicolous.
Hosts: Acer, Ulmus

Phaeostalagmus arbusculus Wang & Sutton
Distr.: NY.
Notes: Lignicolous.
Hosts: Acer, Fagus, Populus

Phaeotrichoconis Subramanian. Hyphomycetes

Phaeotrichoconis crotalariae (M.A. Salam & P.N. Rao) Subramanian
Trichoconis crotalariae M.A. Salam & P.N. Rao
Distr.: Widespread in tropical regions.
Notes: On various plants, isolated from air and soil (Ellis. Dematiaceous Hyphomycetes. p. 463. 1971).
Hosts: Caryota, Chamaedorea, Chrysalidocarpus, Gronophyllum, Rhapis, Strelitzia

Phakopsora Dietel. Basidiomycetes, Uredinales

Phakopsora apoda (Har. & Pat.) Mains
Puccinia apoda Har. & Pat.
Distr.: HI; Africa; Australia.
Host: Pennisetum
Phakopsora cherimoliae Cummins ex Cummins
Distr.: FL, TX; Mexico; Cuba; Central America.
Host: Annona
Phakopsora crotonis (Burrill) Arth.
Melampsora crotonis Burrill
Bubakia crotonis (Burrill) Arth.
Pucciniastrum crotonis De Toni in Sacc.
Distr.: Southern United States; Mexico; South America.
Notes: On Euphorbiaceae.
Hosts: Codiaeum, Croton, Crotonopsis
Phakopsora gossypii (Lagerh.) Hiratsuka
Doassansia gossypii Lagerh.
Kuehneola gossypii Arth.
Phakopsora gossypii (Arth.) W.T. Dale non (Lagerh.) Hiratsuka, hom. illeg.
Aecidium desmium Berk. & Broome
Cerotelium desmium Arth.
Phakopsora desmium Cummins
Distr.: Cosmopolitan.
Notes: Cotton rust. Autoecious on Malvaceae. Punithalingham (C.M.I. Descr. 172. 1968).
Host: Gossypium
Phakopsora jatrophicola Cummins ex Cummins
Uredo jatrophicola Arth. (anamorph)
Distr.: FL, TX; Mexico; West Indies.
Host: Jatropha
Phakopsora mexicana (Arth.) Arth.
Bubakia mexicana Arth.
Distr.: TX; Mexico.
Host: Croton
Phakopsora tecta H. Jacks. & Holw. in H. Jacks.
Distr.: HI, PR; South America.
Host: Commelina
Phakopsora zizyphi-vulgaris Dietel
Uredo zizyphi-vulgaris Henn. (anamorph)
Distr.: FL; Asia.
Host: Ziziphus

Phanerochaete P. Karst. Basidiomycetes, Aphyllophorales

Notes: Burdsall (Mycol. Mem. 10:1. 1985).
Phanerochaete affinis (Burt) Parmasto
Peniophora affinis Burt
Peniophora gilvidula Bres.
Distr.: Widespread.
Notes: On wood. Burdsall (Mycol. Mem. 10:90. 1985).
Hosts: Acer, Alnus, Betula, Carpinus, Castanea, Cornus, Fagus, Pinus, Populus, Prunus, Quercus, Salix, Thuja, Ulmus
Phanerochaete allantospora Burdsall & R.L. Gilbertson
Grandiniella allantospora (Burdsall & R.L. Gilbertson) Burdsall
Distr.: AZ.
Notes: On hardwoods.
Hosts: Fouquieria, Fraxinus, Juglans, Pinus, Platanus, Prosopis
Phanerochaete arizonica Burdsall & R.L. Gilbertson
Grandiniella arizonica (Burdsall & R.L. Gilbertson) Burdsall
Distr.: Southwestern United States.
Notes: On hardwoods.
Hosts: Fouquieria, Lemaireocereus, Mimosa, Mortonia, Opuntia, Parkinsonia, Platanus, Populus, Prosopis, Prunus, Quercus
Phanerochaete avellanea (Bres.) J. Eriksson & Hjortstam in J. Eriksson et al.
Corticium avellaneum Bres. in Bourd. & Galzin
Distr.: Cosmopolitan.
Notes: On hardwoods, sometimes on herbaceous stems.
Hosts: Alnus, Platanus, Quercus
Phanerochaete burtii (Romell) Parmasto
Peniophora burtii Romell in Burt
Distr.: United States, especially in the South; Australia.
Notes: On wood, usually of hardwoods.
Hosts: Alnus, Fagus, Juglans, Liquidambar, Pinus, Platanus, Populus, Prunus, Pseudotsuga, Quercus, Sabal, Sequoia, Ulmus
Phanerochaete cacaina (Bourd. & Galzin) Burdsall & R.L. Gilbertson
Peniophora cacaina Bourd. & Galzin
Distr.: AZ; France.
Notes: On conifers, especially *Pinus* spp.
Host: Pinus
Phanerochaete carnosa (Burt) Parmasto
Peniophora carnosa Burt
Distr.: Circumboreal.
Notes: On wood, usually of conifers.
Hosts: Abies, Acer, Alnus, Betula, Juniperus, Larix, Picea, Pinus, Platanus, Populus, Pseudotsuga, Quercus, Robinia, Thuja
Phanerochaete chrysorhiza (Torr.) Budington & R.L. Gilbertson
Hydnum chrysorhizon Torr. in Eaton
Mycoacia chrysorhiza (Torr.) Aoshima & H. Furukawa
Hydnum fragillissima Berk. & M.A. Curtis
Mycoacia fragillissima (Berk. & M.A. Curtis) L.W. Miller & Boyle
Odontia fragillissima (Berk. & M.A. Curtis) C.A. Brown
Distr.: Eastern North America.
Notes: On wood of hardwoods, rarely conifers.
Hosts: Acer, Carnegiea, Carpinus, Citrus, Cornus, Fouquieria, Fraxinus, Gossypium, Haplopappus, Juglans, Liquidambar, Lycium, Nectandra, Olneya, Ostrya, Pinus, Platanus, Populus, Prosopis, Quercus
Phanerochaete chrysosporium Burdsall in Burdsall & Eslyn
Phanerochaete macrocystidiata Hallenberg
Sporotrichum pruinosum Gilman & E. Abbott (anamorph)
Distr.: Temperate northern hemisphere.
Hosts: Heteromeles, Liriodendron, Platanus, Ulmus
Phanerochaete corymbata (G. Cunn.) Burdsall
Corticium corymbatum G. Cunn.
Distr.: LA; New Zealand.
Notes: On hardwoods.
Host: Ulmus
Phanerochaete crassa (Lév.) Burdsall
Laxitextum crassum (Lév.) Lentz
Lopharia crassa (Lév.) Boidin
Stereum umbrinum Berk. & M.A. Curtis non Fr., hom. illeg.
Distr.: United States, most common in the South; Africa; Asia; Australia.
Notes: On hardwoods.
Hosts: Acacia, [Agave], Canotia, Carpinus, Castanea, [Dasylirion], Fagus, Fouquieria, Ilex, Liquidambar, Liriodendron, Platanus, Prosopis, Prunus, Quercus, Vitis
Phanerochaete ericina (Bourd.) J. Eriksson & Ryvarden in J. Eriksson et al.
Peniophora ericina Bourd.
Distr.: Southeastern United States; southern Europe.
Notes: On hardwoods.
Hosts: Acer, Carya, Fagus, Liriodendron, Quercus, Robinia
Phanerochaete filamentosa (Berk. & M.A. Curtis) Burdsall
Corticium filamentosum Berk. & M.A. Curtis
Peniophora filamentosa (Berk. & M.A. Curtis) Burt in Coker
Phanerochaete borneensis Jülich

Peniophora unicolor Peck
Distr.: Widespread.
Notes: On wood.
Hosts: Abies, Acer, Betula, Castanea, Fagus, Platanus, Populus, Prunus, Quercus, Robinia, Tilia, Ulmus
Phanerochaete flavido-alba (Cooke) S.S. Rattan
Peniophora flavido-alba Cooke
Peniophora texana Burt
Peniophora vernicosa Ellis & Everh. in Burt
Distr.: Southeastern United States; Caribbean; South America; Asia.
Notes: On wood.
Hosts: Carpinus, Celtis, Citrus, Fagus, Gleditsia, Juniperus, Liquidambar, Liriodendron, Magnolia, Malus, Pinus, Populus, Quercus, Salix, Vitis
Phanerochaete gigantea (Fr.:Fr.) S.S. Rattan et al. in S.S. Rattan
Peniophora gigantea (Fr.:Fr.) Massee
Phlebia gigantea (Fr.:Fr.) Donk
Phlebiopsis gigantea (Fr.:Fr.) Jülich
Peniophora globulifera Ellis & Everh.
Distr.: Northern Hemisphere.
Notes: On wood, usually of conifers.
Hosts: Abies, Larix, Picea, Pinus, Populus, Thuja, Tsuga
Phanerochaete hiulca (Burt) A.L. Welden
Peniophora hiulca Burt
Distr.: Southeastern United States; Bermuda.
Notes: On wood.
Host: Quercus
Phanerochaete magnoliae (Berk. & M.A. Curtis) Burdsall
Radulum magnoliae Berk. & M.A. Curtis
Phanerochaete cumulodentata (Nikolaeva) Parmasto, nom. illeg.
Radulum cumulodentatum Nikolaeva, nom. nud.
Phanerochaete raduloides J. Eriksson & Ryvarden in J. Eriksson et al.
Distr.: Eastern United States; Iran.
Notes: On hardwoods.
Hosts: Betula, Populus, Quercus, Salix
Phanerochaete omnivora (Shear) Burdsall & Nakasone
Hydnum omnivorum Shear
Distr.: Southwestern United States.
Notes: On hardwoods.
Hosts: Acacia, Baccharis, Carnegiea, Chilopsis, Fouquieria, Maclura, Platanus, Prosopis
Phanerochaete ravenelii (Cooke) Burdsall
Peniophora ravenelii Cooke
Peniophora molleriana Bres. in Sacc.
Metulodontia roumeguerii (Bres.) Parmasto
Peniophora roumeguerii (Bres.) Höhn. & Litsch.
Phlebia roumeguerii (Bres.) Donk
Phlebiopsis roumeguerii (Bres.) Jülich & Stalpers
Peniophora stratosa Burt non Petch, hom. illeg.
Distr.: Southern United States; southern Europe.
Notes: On hardwoods.
Hosts: Acer, Alnus, Carya, Crataegus, Diospyros, Ligustrum, Lithocarpus, Quercus
Phanerochaete rimosa (Cooke) Burdsall
Peniophora rimosa Cooke
Scopuloides rimosa (Cooke) Jülich
Odontia hydnoides (Cooke & Massee) Höhn.
Peniophora hydnoides Cooke & Massee
Phlebia hydnoides (Cooke & Massee) M. Christiansen
Scopuloides hydnoides (Cooke & Massee) Hjortstam & Ryvarden
Distr.: Eastern United States; Europe.
Notes: On hardwoods.
Hosts: Acer, Fagus, Fraxinus, Nyssa, Populus, Tilia, Ulmus
Phanerochaete salmoneolutea Burdsall & R.L. Gilbertson
Distr.: Southern United States.
Notes: On hardwoods.
Hosts: Carya, Quercus
Phanerochaete sanguinea (Fr.:Fr.) Pouzar
Peniophora sanguinea (Fr.:Fr.) Bres.
Corticium calotrichum P. Karst.
Peniophora limonia Burt
Peniophora miniata (Berk. in Hook.) Burt
Distr.: Temperate northern hemisphere.
Notes: On wood.
Hosts: Abies, Acer, Alnus, Betula, Castanea, Castanopsis, Pinus, Populus, Prunus, Pseudotsuga, Quercus, Robinia, Sequoia, Thuja, Tsuga
Phanerochaete sordida (P. Karst.) J. Eriksson & Ryvarden
Corticium sordidum P. Karst.
Peniophora sordida (P. Karst.) Burt
Peniophora arachnoidea Burt
Peniophora cremea (Bres.) Sacc. & Syd.
Phanerochaete cremea (Bres.) J. Eriksson
Distr.: Widespread.
Notes: On wood.
Hosts: Acer, Alnus, Betula, Carya, Crataegus, Delonix, Heteromeles, Juglans, Larix, Liriodendron, Nyssa, Picea, Pinus, Platanus, Populus, Quercus, Viburnum
Phanerochaete subceracea (Burt) Burdsall
Corticium subceraceum Burt
Distr.: Eastern United States.
Notes: On wood.
Host: Quercus
Phanerochaete tuberculata (P. Karst.) Parmasto
Corticium tuberculatum P. Karst.
Grandiniella tuberculata (P. Karst.) Burdsall
Distr.: Widespread.
Notes: On wood.
Hosts: Agave, Betula, Carnegiea, Celtis, Chilopsis, Condalia, Fouquieria, Juglans, Lycium, Platanus, Populus, Prosopis, Quercus, Vauquelinia, Yucca
Phanerochaete velutina (DC.:Fr.) P. Karst.
Peniophora velutina (DC.:Fr.) Cooke
Peniophora phosphorescens Burt
Distr.: Temperate northern hemisphere.
Notes: On wood.
Hosts: Acer, Alnus, Amelanchier, Arbutus, Arctostaphylos, Carya, Castanea, Corylus, Crossosoma, Fagus, Juniperus, Liquidambar, Liriodendron, Magnolia, Picea, Pinus, Populus, Pseudotsuga, Quercus, Thuja
Phanerochaete viticola (Schwein.) Parmasto
Peniophora viticola (Schwein.) Höhn. & Litsch.
Corticium subaurantiacum Peck
Distr.: Temperate northern hemisphere.
Notes: On wood.
Hosts: Abies, Acer, Betula, Fagus, Picea, Thuja, Vitis
Phanerochaete xerophila Burdsall
Distr.: Southwestern United States.
Notes: On fallen limbs of desert shrubs and cacti.
Hosts: Baccharis, Carnegiea, Opuntia, Platanus, Prosopis, Quercus

Phellinus Quél. Basidiomycetes, Aphyllophorales

Notes: Gilbertson (Mycotaxon 9:51. 1979); Gilbertson & Ryvarden (North American Polypores 2:541. 1987).
Phellinus arctostaphyli (W.H. Long) Niemelä
Fomes arctostaphyli W.H. Long
Distr.: Western North America.
Notes: Heart rot of living *Arctostaphylos*; also on *Cercocarpus*. Sometimes considered a synonym of *F. igniarius*, but see Niemelä (Ann. Bot. Fenn. 11:202. 1974).
Hosts: Arctostaphylos, Cercocarpus
Phellinus badius (Berk.) G. Cunn.
Fomes badius Berk. ex Cooke
Distr.: Subtropical and tropical regions.
Notes: On wood, especially living *Acacia* and *Prosopis*.
Hosts: Acacia, Chilopsis, Prosopis, Vitex
Phellinus bicuspidatus Lombard & M. Larsen
Distr.: LA.
Notes: On wood.
Host: Quercus
Phellinus chrysoloma (Fr.) Donk
Polyporus chrysoloma Fr.
Distr.: Western North America; Europe.
Notes: On conifers. Sometimes considered a synonym of *P. pini*; many of the older reports of *P. pini* may refer to *P. chrysoloma*.
Host: Picea
Phellinus conchatus (Pers.:Fr.) Quél.

Fomes conchatus (Pers.:Fr.) Gill.
Polyporus conchatus (Pers.:Fr.) Fr.
Fomes densus C.G. Lloyd
Distr.: Widespread.
Notes: Usually on dead hardwoods, rarely living trees.
Hosts: Acer, Alnus, Baccharis, Betula, Carya, Crataegus, Fagus, Fraxinus, Gleditsia, Juglans, Liquidambar, Ostrya, Physocarpus, Populus, Salix
Phellinus contiguus (Pers.:Fr.) Pat.
Poria contigua (Pers.:Fr.) P. Karst.
Distr.: Widespread.
Notes: On wood.
Hosts: Quercus, Thuja
Phellinus crocatus (Fr.) Ryvarden
Polyporus crocatus Fr.
Distr.: Southern United States; Mexico.
Notes: On wood.
Hosts: Magnolia, Populus, Salix
Phellinus dependens (Murrill) Ryvarden
Pyropolyporus dependens Murrill
Fomes dependens (Murrill) Sacc. & Trott.
Distr.: Southern United States to Brazil.
Notes: On wood.
Host: Lysiloma
Phellinus everhartii (Ellis & B.T. Galloway) A. Ames
Mucronoporus everhartii Ellis & B.T. Galloway
Fomes everhartii (Ellis & B.T. Galloway) H. Schrenk & Spauld.
Pyropolyporus everhartii (Ellis & B.T. Galloway) Murrill
Fomes praerimosus (Murrill) Sacc. & D. Sacc.
Distr.: North America.
Notes: Usually on living trunks of *Quercus* spp., occasionally on dead wood.
Hosts: Betula, Castanea, Fagus, Juglans, Ostrya, Populus, Prosopis, Quercus, Ulmus
Phellinus extensus (Lév.) Pat.
Polyporus extensus Lév.
Fomes extensus (Lév.) Cooke
Distr.: Subtropical and tropical regions.
Notes: On hardwoods.
Hosts: Exothea, Lysiloma, Taxodium
Phellinus ferreus (Pers.) Bourd. & Galzin
Polyporus ferreus Pers.
Poria ferrea (Pers.) Overh.
Poria fulvida Ellis & Everh., nom. nud.
Distr.: Cosmopolitan.
Notes: On wood.
Hosts: Abies, Acer, Alnus, Amelanchier, Arctostaphylos, Betula, Castanea, Castanopsis, Cornus, Corylus, Fraxinus, Gaultheria, Lithocarpus, Myrica, Populus, Prunus, Pseudotsuga, Quercus, Rubus, Salix, Thuja, Tsuga, Umbellularia, Vaccinium
Phellinus ferrugineo-velutinus (Henn.) Ryvarden
Poria flavomarginata (Murrill) Sacc. & Trott.
Distr.: Southern United States to Brazil.
Notes: On wood.
Host: Liriodendron
Phellinus ferrugineofuscus (P. Karst.) Bourd.
Poria ferrugineofusca P. Karst.
Poria marginella (Peck) Sacc.
Polyporus marginellus Peck
Distr.: Temperate northern hemisphere.
Notes: On wood.
Hosts: Abies, Larix, Picea, Pinus, Pseudotsuga, Thuja, Tsuga
Phellinus ferruginosus (Schrad.:Fr.) Pat.
Polyporus ferruginosus (Schrad.:Fr.) Fr.
Poria ferruginosa (Schrad.:Fr.) P. Karst.
Fomes salicinus (Pers.:Fr.) J. Kickx fil.
Poria umbrina (Fr.) Cooke, nom. nud.
Distr.: Temperate northern hemisphere.
Notes: On wood.
Hosts: Abies, Acer, Alnus, Arbutus, Arctostaphylos, Betula, Castanea, Corylus, Fagus, Fraxinus, Liquidambar, Maclura, Malus, Ostrya, Picea, Pinus, Populus, Prosopis, Prunus, Pseudotsuga, Quercus, Robinia, Salix, Sassafras, Thuja, Tsuga, Umbellularia, Vaccinium
Phellinus fragrans M. Larsen & Lombard
Distr.: MI, NY.
Host: Acer
Phellinus gilvus (Schwein.:Fr.) Pat.
Hapalopilus gilvus (Schwein.:Fr.) Murrill
Polyporus gilvus (Schwein.:Fr.) Fr.
Polyporus licnoides Mont.
Fomes taxodii Murrill
Distr.: Widespread.
Notes: On wood, usually of hardwoods.
Hosts: Acer, Albizia, Aleurites, Alnus, Betula, Carpinus, Carya, Castanea, Cephalanthus, Cercis, Citrus, Crataegus, Diospyros, Eucalyptus, Fagus, Fraxinus, Gleditsia, Hamamelis, Juglans, Juniperus, Liquidambar, Malus, Mangifera, Morus, Persea, Pinus, Pithecellobium, Platanus, Populus, Prosopis, Prunus, Pyrus, Quercus, Rhus, Robinia, Salix, Sassafras, Syringa, Taxodium, Tilia, Tsuga, Ulmus, Vitis
Phellinus grenadensis (Murrill) Ryvarden
Pyropolyporus grenadensis Murrill
Distr.: LA; tropical region.
Notes: On wood.
Host: Quercus
Phellinus hartigii (Allesch. & Schnabl) Pat.
Polyporus hartigii Allesch. & Schnabl
Fomes hartigii (Allesch. & Schnabl) Bres.
Distr.: Western North America; Europe.
Notes: On conifers.
Host: Taxus
Phellinus igniarius (L.:Fr.) Quél.
Fomes igniarius (L.:Fr.) J. Kickx fil.
Distr.: Temperate northern hemisphere.
Notes: On living or dead hardwoods. A variable species probably composed of biologically distinct varieties or species (Pegler & Waterston. C.M.I. Descr. 194. 1968).
Hosts: Acer, Alnus, Amelanchier, Arctostaphylos, Betula, Carpinus, Carya, Castanopsis, Cornus, Fagus, Fraxinus, Juglans, Liriodendron, Malus, Ostrya, Populus, Prunus, Pseudotsuga, Pyrus, Quercus, Rhamnus, Robinia, Salix, Sambucus, Tsuga, Ulmus, Umbellularia, Vitis, Zanthoxylum
Phellinus inermis (Ellis & Everh.) G. Cunn.
Poria inermis Ellis & Everh.
Distr.: HI, North America, most common in the East; Australasia.
Notes: On hardwoods.
Hosts: Ilex, Metrosideros, Nemopanthus
Phellinus johnsonianus (Murrill) Ryvarden
Fomes johnsonianus (Murrill) J. Lowe
Distr.: North America.
Notes: On hardwoods.
Host: Fraxinus
Phellinus kawakamii M. Larsen, Lombard & C.S. Hodges
Distr.: HI.
Notes: Larsen et al. (Mycologia 77:345. 1985).
Hosts: Acacia, Casuarina
Phellinus laevigatus (Fr.) Bourd. & Galzin
Polyporus laevigatus Fr.
Fomes igniarius (L.:Fr.) J. Kickx fil. var. *laevigatus* (Fr.) Overh.
Fuscoporia laevigata (Fr.) G. Cunn.
Poria laevigata (Fr.) P. Karst.
Fomitiporella betulina Murrill
Fuscoporella ludoviciana Murrill
Fomitiporia pereffusa Murrill
Poria pereffusa (Murrill) Sacc. & Trott.
Distr.: Widespread.
Notes: On hardwoods. Formerly considered to be a variety of *P. igniarius* (Overholts, The Polyporaceae of the United States, Alaska and Canada. p. 63. 1953).
Hosts: Alnus, Betula, Fagus, Gleditsia, Liquidambar, Picea, Populus, Prunus, Quercus, Rhododendron
Phellinus linteus (Berk. & M.A. Curtis) S. Teng
Polyporus linteus Berk. & M.A. Curtis
Distr.: Pantropical and subtropical.
Notes: On wood.
Host: Quercus
Phellinus lundellii Niemelä
Distr.: Temperate northern hemisphere.
Notes: On wood.
Hosts: Alnus, Betula

Phellinus melleoporus (Murrill) Ryvarden
Fomitiporella melleopora Murrill
Distr.: Southern United States; South America.
Notes: On wood.
Hosts: Acer, Celtis, Liquidambar, Quercus

Phellinus merrillii (Murrill) Ryvarden
Pyropolyporus merrillii Murrill
Distr.: Southeastern United States.
Notes: On wood.
Host: Tsuga

Phellinus nigricans (Fr.:Fr.) P. Karst.
Fomes igniarius (L.:Fr.) J. Kickx fil. var. *nigricans* Fr.:Fr.
Fomes nigricans (Fr.:Fr.) Fr.
Distr.: Western United States, Europe.
Hosts: Alnus, Betula, Salix

Phellinus nigrolimitatus (Romell) Bourd. & Galzin
Polyporus nigrolimitatus Romell
Fomes nigrolimitatus (Romell) Egeland
Fomes putearius Weir
Distr.: Western North America; Europe.
Notes: On wood of conifers, sometimes causes root and butt rot of living trees.
Hosts: Abies, Larix, Picea, Pinus, Pseudotsuga, Taxus, Thuja, Tsuga

Phellinus occidentalis (Overh. ex Lombard, R.W. Davidson & R.L. Gilbertson) R.L. Gilbertson in Lombard, R.W. Davidson & R.L. Gilbertson
Fomes occidentalis Overh. ex Lombard, R.W. Davidson & R.L. Gilbertson
Distr.: Western United States.
Notes: On dead and living trees.
Host: Crataegus

Phellinus palmicolus (Berk. & M.A. Curtis) Ryvarden
Polyporus palmicolus Berk. & M.A. Curtis
Distr.: FL; Mexico; Cuba.
Host: Arecaceae

Phellinus pini (Thore:Fr.) A. Ames
Fomes pini (Thore:Fr.) P. Karst.
Porodaedalea pini (Thore:Fr.) Murrill
Trametes pini (Thore:Fr.) Fr.
Distr.: Temperate northern hemisphere.
Notes: On wood of conifers, often causes heart rot of living trees. Larsen et al. (Canad. J. For. Res. 9:31. 1979) recognized the variety *cancriformans* M. Larsen, Lombard & Aho on *Abies concolor* in Oregon and California.
Hosts: Abies, Acer, Alnus, Betula, Calocedrus, Cedrus, Chamaecyparis, Crataegus, Cupressus, Juniperus, Larix, Picea, Pinus, Pseudotsuga, Quercus, Taxus, Thuja, Tsuga

Phellinus pomaceus (Pers.) Maire
Polyporus pomaceus Pers.
Fomes pomaceus (Pers.) C.G. Lloyd
Fomes fulvus (Scop.) Gill.
Distr.: Northern hemisphere.
Notes: On limbs and trunks of hardwoods, especially *Prunus* spp.
Hosts: Crataegus, Malus, Prunus

Phellinus prunicola (Murrill) R.L. Gilbertson
Fomitiporia prunicola Murrill
Poria prunicola (Murrill) Sacc. & Trott.
Distr.: Western North America.
Notes: On wood.
Host: Prunus

Phellinus punctatus (Fr.) Pilát
Polyporus punctatus Fr.
Poria punctata (Fr.) P. Karst.
Fomitiporia earleae Murrill
Poria laminata (Murrill) Sacc. & Trott.
Fomitiporia langloisii Murrill
Fomitiporia tsugina Murrill
Poria tsugina (Murrill) Sacc. & Trott.
Distr.: Widespread.
Notes: On wood.
Hosts: Acer, Ailanthus, Alnus, Asimina, Berberis, Betula, Buxus, Caragana, Carpinus, Castanea, Cercis, Fagus, Fraxinus, Juglans, Kalmia, Liquidambar, Liriodendron, Maclura, Melia, Nyssa, Ostrya, Oxydendrum, Picea, Prunus, Quercus, Rhamnus, Rhus, Salix, Syringa, Tilia, Tsuga

Phellinus repandus (Overh.) R.L. Gilbertson
Fomes repandus Overh.
Distr.: Western North America.
Notes: On conifers.
Hosts: Abies, Larix, Picea, Pinus, Pseudotsuga, Tsuga

Phellinus robiniae (Murrill) A. Ames
Pyropolyporus robiniae Murrill
Fomes robiniae (Murrill) Sacc. & D. Sacc.
Fulvifomes robiniae (Murrill) Murrill
Fomes rimosus (Berk.) Cooke
Polyporus rimosus Berk.
Distr.: Widespread.
Notes: On living trees or dead wood of *Robinia* spp., occasionally of other Fabaceae.
Hosts: Gleditsia, Prosopis, Robinia

Phellinus robustus (P. Karst.) Bourd. & Galzin
Fomes robustus P. Karst.
Poria robusta (P. Karst.) Weiss & M. O'Brien
Fomes abramsianus (Murrill) Sacc. & Trott. in Trott.
Fomes calkinsii (Murrill) Sacc. & D. Sacc.
Pyropolyporus calkinsii Murrill
Distr.: Widespread.
Notes: On living or dead hardwoods (Pegler & Waterston. C.M.I. Descr. 197. 1968).
Hosts: [Abies], Acer, Betula, Cercocarpus, Dodonaea, Eucalyptus, Fraxinus, Nyssa, [Opuntia], [Pinus], [Pseudotsuga], Quercus, Robinia, Salix, [Taxodium], [Taxus], [Tsuga]

Phellinus sonorae R.L. Gilbertson
Distr.: AZ.
Host: Dodonaea

Phellinus spiculosus (W.A. Campbell & R.W. Davidson) Niemelä
Poria spiculosa W.A. Campbell & R.W. Davidson
Distr.: Eastern North America.
Notes: On hardwoods.
Hosts: Carya, Gleditsia, Quercus

Phellinus texanus (Murrill) A. Ames
Pyropolyporus texanus Murrill
Fomes texanus (Murrill) Hedgc. & W.H. Long
Distr.: Southwestern United States.
Notes: Heart rot of living desert hardwoods and *Juniperus* spp.
Hosts: Canotia, Carnegiea, Celtis, Cercocarpus, Cocculus, Cowania, Fouquieria, Juniperus, Larrea, Morus, Olneya, Opuntia, Parkinsonia, Prunus, Vauquelinia, Vitex

Phellinus torulosus (Pers.) Bourd. & Galzin
Polyporus torulosus Pers.
Fomes torulosus (Pers.) C.G. Lloyd
Distr.: Southern United States; Europe.
Notes: Overholts (The Polyporaceae of the United States, Alaska and Canada. p. 72. 1953) listed a number of hosts and states. However, Gilbertson (Mycotaxon 9:81. 1979) stated that *P. torulosus* is known in North America only from southern Arizona on *Pinus* spp.
Hosts: [Fraxinus], Pinus, [Pseudotsuga]

Phellinus tremulae (Bondartsev) Bondartsev & Borisov in Bondartsev
Fomes igniarius (L.:Fr.) J. Kickx fil. f. *tremulae* Bondartsev
Distr.: Temperate northern hemisphere.
Notes: On wood.
Host: Populus

Phellinus vaninii Ljub.
Distr.: MN; Soviet Union.
Notes: On wood.
Host: Populus

Phellinus viticola (Schwein.:Fr.) Donk
Polyporus viticola Schwein.:Fr.
Fomes viticola (Schwein.:Fr.) J. Lowe
Poria viticola (Schwein.:Fr.) Cooke
Trametes gilvoides C.G. Lloyd
Poria isabellina (Fr.) Overh. non (Pat.) Sacc., hom. illeg.
Trametes isabellina Fr.
Phellinus isabellinus (Fr.) Bourd. & Galzin
Trametes setosus Weir
Fomes tenuis P. Karst.
Trametes tenuis (P. Karst.) Overh.
Distr.: North America; Europe; Soviet Union.
Hosts: Abies, Acer, Alnus, Arbutus, Asimina, Betula, Calocedrus,

Chamaecyparis, Fagus, Larix, Picea, Pinus, Populus, Prunus, Pseudotsuga, Quercus, Rhododendron, Salix, Sequoia, Thuja, Tsuga, Vitis

Phellinus weirianus (Bres.) R.L. Gilbertson
Fomes weirianus Bres.
Distr.: Southwestern United States.
Notes: White heart rot of living trees.
Host: Juglans

Phellinus weirii (Murrill) R.L. Gilbertson
Fomitiporia weirii Murrill
Inonotus weirii (Murrill) Kotlaba & Pouzar
Poria weirii (Murrill) Murrill
Distr.: Western North America; Japan.
Notes: On conifers (Pegler & Gibson. C.M.I. Descr. 323. 1972 as *Inonotus weirii*).
Hosts: Abies, Chamaecyparis, Larix, Picea, Pinus, Pseudotsuga, Thuja, Tsuga

Phialea (Fr.) Gill. Discomycetes, Helotiales

Notes: Carpenter (Mem. New York Bot. Gard. 33:8. 1981) rejected this genus, but not all names have been redisposed.

Phialea acicularum (Rolland) Rehm
Helotium acicularum Rolland
Distr.: WI; France(type).
Notes: Carpenter (Mem. New York Bot. Gard. 33. 1981) was unable to ascertain the identity of this fungus.
Host: Larix

Phialea campanula (Nees:Fr.) Gill.
Belonioscypha campanula (Nees:Fr.) Rehm
Distr.: Western United States; Europe(type).
Hosts: Eriogonum, Hordeum

Phialea rhodoleuca (Fr.) Sacc.
Hymenoscyphus rhodoleuca (Fr.) W. Phillips
Distr.: CO; Europe(type).
Notes: According to Carpenter (Mem. New York Bot. Gard. 33:249. 1981), the identity of this fungus is uncertain.
Host: Equisetum

Phialocephala Kendrick. Hyphomycetes

Notes: Jong & Davis (Mycologia 64:1351. 1972).

Phialocephala dimorphospora Kendrick
Distr.: Eastern North America.
Notes: On rotten deciduous wood, isolated from pine litter (Ellis. More dematiaceous Hyphomycetes. p. 445. 1976).
Host: Pinus

Phialocephala repens (Cooke & Ellis) Kendrick
Penicillium repens Cooke & Ellis
Distr.: Northeastern North America.
Host: Magnolia

Phialocephala truncata Sutton
Distr.: HI; England(type).
Notes: On woody plant parts (Sutton. Trans. Brit. Mycol. Soc. 64:414. 1975).
Host: Eucalyptus

Phialophora Medlar. Hyphomycetes

Notes: Schol-Schwarz (Persoonia 6:59. 1970).

Phialophora bubakii (Laxa) Schol-Schwarz
Margarinomyces bubakii Laxa
Cadophora obscura Nannf.
Phialophora obscura (Nannf.) Conant
Distr.: North temperate regions.
Notes: Isolated from soft wood and margarine (Ellis. More dematiaceous Hyphomycetes. p. 454. 1976).
Host: Quercus

Phialophora cinerescens (Wollenweb.) Van Beyma
Verticillium cinerescens Wollenweb.
Distr.: Temperate regions.
Notes: Carnation wilt (Hawksworth & Gibson. C.M.I. Descr. 503. 1976).
Hosts: Dianthus, Fragaria

Phialophora decumbens (Van Beyma) Schol-Schwarz
Margarinomyces decumbens Van Beyma
Distr.: OR; Europe.
Notes: This species is a member of the *P. hoffmannii*-group.
Hosts: Calocedrus, Juniperus, Thuja

Phialophora fastigiata (Lagerberg & Melin) Conant
Cadophora fastigiata Lagerberg & Melin
Distr.: Temperate regions.
Host: Pinus

Phialophora graminicola (Deacon) J. Walker
Phialophora radicicola Cain var. *graminicola* Deacon
Distr.: North America; Australia; Europe.
Notes: Previously considered the causal agent of summer patch disease of turf (Smiley. Pl. Dis. 71:779. 1987), this fungus appears to be a weak saprophyte with some biocontrol potential (Jackson & Landschoot. Phytopathology 76:654. 1986).
Hosts: Poa, Triticum

Phialophora gregata (Allington & D.W. Chamberlain)W. Gams
Cephalosporium gregatum Allington & D.W. Chamberlain
Distr.: Central United States.
Notes: Brown stem rot of soybeans.
Hosts: Glycine, Phaseolus

Phialophora hoffmannii (Van Beyma) Schol-Schwarz
Margarinomyces hoffmannii Van Beyma
Distr.: Widespread.
Notes: Isolated from dead wood and other organic substrates.
Hosts: Alnus, Juniperus, Pseudotsuga, Thuja

Phialophora malorum (M.N. Kidd & A. Beaumont) McColloch
Sporotrichum malorum M.N. Kidd & A. Beaumont
Sporotrichum carpogenum Rühle
Distr.: Temperate regions; New Guinea.
Notes: Isolated from wood, soil, various chemicals, air, and soil (Domsch et al. Compendium of soil fungi. p. 627. 1980).
Hosts: Malus, Pyrus

Phialophora melinii (Nannf.) Conant
Cadophora melinii Nannf.
Distr.: North temperate regions.
Notes: Isolated from soft- and hardwood and wood pulp (Ellis. More dematiaceous Hyphomycetes. p. 455. 1976).
Host: Acer

Phialophora olivacea W. Gams
Distr.: Eastern United States; Germany(type).
Notes: Isolated from stalactites and various organic substrates (Gams. Stud. Mycol. 13:65. 1976).
Host: Pinus

Phialophora parasitica Ajello, L.K. Georg, & Wang
Distr.: North America; Central America; northern Africa; Middle East.
Notes: Vascular wilt and die-back woody plant, phaeohyphomycosis in humans (Ajello et al. Mycologia 66:490. 1974; Hawksworth & Gibson. C.M.I. Descr. 504. 1976).
Host: Quercus

Phialophora repens (R.W. Davidson) Conant
Cadophora repens R.W. Davidson
Distr.: Southeastern United States.
Host: Pinus

Phialophora richardsiae (Nannf.) Conant
Cadophora richardsiae Nannf. in Melin & Nannf.
Cadophora brunnescens R.W. Davidson
Phialophora brunnescens (R.W. Davidson) Conant
Distr.: Cosmopolitan.
Notes: In soil, on wood debris, and a variety of organic substrates.
Host: Pinus

Phialophora verrucosa Medlar
Cadophora americana Nannf.
Phialophora americana (Nannf.) S.J. Hughes
Distr.: North America.
Notes: Isolated from wood, wood pulp, air and humans (Ellis. More dematiaceous Hyphomycetes. p. 454. 1976).
Host: Alnus

Phillipsiella Cooke. Loculoascomycetes, Patellariales

Microphyma Speg.
Phillipsiella atra Cooke
Asterina erysiphoides Ellis & G. Martin, nom. nud.
Asterina patelloides Ellis & G. Martin
Microthyriella patelloides (Ellis & G. Martin) Theiss.
Allosoma quercifoliae Batista & Nascim.
Distr.: FL.
Notes: Müller & Arx (Beitr. Kryptogamenfl. Schweiz 11(2):216. 1962).
Host: Quercus
Phillipsiella ilicis (Ellis) Höhn.
Asterina ilicis Ellis
Henningsiella ilicis (Ellis) Höhn.
Microphyma ilicis (Ellis) Speg.
Distr.: NJ.
Host: Ilex
Phillipsiella nigella W. Phillips & Harkn.
Microphyma nigellum (W. Phillips & Harkn.) Sacc.
Distr.: CA.
Host: Quercus

Phlebia Fr.:Fr. Basidiomycetes, Aphyllophorales

Merulius Fr.:Fr.
Phlebia albida Fr.
Phlebia centrifuga P. Karst.
Distr.: Northern hemisphere.
Notes: On wood.
Hosts: Abies, Picea, Pinus, Populus, Quercus, Salix, Tsuga
Phlebia brevispora Nakasone in Nakasone & Eslyn
Distr.: Eastern United States.
Notes: On wood.
Host: Pinus
Phlebia celtidis W.B. Cooke
Distr.: Western United States.
Notes: On wood.
Host: Celtis
Phlebia chrysocrea (Berk. & M.A. Curtis) Burdsall in Lombard, Burdsall & R.L. Gilbertson
Corticium chrysocreas Berk. & M.A. Curtis in Berk.
Distr.: Temperate northern hemisphere; Sri Lanka.
Notes: On hardwoods (Berry. Pl. Dis. Reporter 60:308. 1976; Lombard et al. Mycologia 67:495. 1975).
Hosts: Betula, Castanea, Gleditsia, Liquidambar, Liriodendron, Prunus, Quercus, Robinia, Sassafras
Phlebia coccineo-fulva Schwein.
Peniophora coccineo-fulva (Schwein.) Burt
Peniophora martiana (Berk. & M.A. Curtis) Burt
Phlebia martiana (Berk. & M.A. Curtis) Parmasto
Corticium martianum Berk. & M.A. Curtis
Peniophora rhodochroa Bres.
Distr.: Temperate northern hemisphere.
Notes: On wood.
Hosts: Alnus, Betula
Phlebia concentrica (Cooke & Ellis) Kropp & Nakasone
Radulum concentricum Cooke & Ellis
Pirex concentricum (Cooke & Ellis) Hjortstam & Ryvarden in Hallenberg, Hjortstam & Ryvarden
Irpex owensii C.G. Lloyd
Radulum owensii (C.G. Lloyd) C.G. Lloyd
Distr.: Pacific Northwest.
Notes: On wood (Kropp & Nakasone. Mycotaxon 24:423. 1985; Hallenberg et al. Mycotaxon 24:287. 1985 as *Pirex concentricus*).
Hosts: Acer, Alnus, Corylus, Lithocarpus, Myrtus, Prunus, Quercus, Salix
Phlebia incarnata (Schwein.) Nakasone & Burdsall in Nakasone, Burdsall & Noll
Merulius incarnatus Schwein.
Byssomerulius incarnatus (Schwein.) R.L. Gilbertson
Merulius rubellus Peck
Distr.: Temperate northern hemisphere.
Notes: On wood, usually of hardwoods (Nakasone et al. Mycotaxon 14:3. 1982).
Hosts: Abies, Pinus, Quercus
Phlebia lilascens (Bourd.) J. Eriksson & Hjortstam in J. Eriksson et al.
Corticium lilascens Bourd.
Distr.: Temperate northern hemisphere.
Notes: On wood.
Host: Quercus
Phlebia livida (Pers.:Fr.) Bres.
Corticium lividum Pers.:Fr.
Distr.: Western United States; Europe.
Hosts: Picea, Pinus, Thuja, Tsuga
Phlebia ludoviciana (Burt) Nakasone & Burdsall in Nakasone, Burdsall & Noll
Peniophora ludoviciana Burt
Hyphoderma ludovicianum (Burt) K.J. Martin & R.L. Gilbertson
Distr.: North America.
Notes: On wood. Nakasone et al. (Mycotaxon 14:3. 1982).
Hosts: Acer, Alnus, Baccharis, Betula, Fraxinus, Liquidambar, Lycium, Picea, Platanus, Populus, Quercus
Phlebia merismoides (Fr.:Fr.) Fr.
Merulius merismoides Fr.:Fr.
Phlebia cinnabarina Schwein.
Phlebia cystidiata H. Jacks. in W.B. Cooke
Distr.: Widespread.
Notes: On wood. Formerly known as *P. radiata*.
Hosts: Abies, Acer, Alnus, Betula, Corylus, Malus, Pinus, Platanus, Populus, Pseudotsuga, Quercus, Salix, Tsuga
Phlebia murrillii W.B. Cooke
Distr.: FL.
Notes: On wood.
Host: Pinus
Phlebia ochraceofulva (Bourd. & Galzin) Donk
Distr.: AZ, NY; Europe.
Hosts: Acer, Prosopis
Phlebia romellii (Litsch.) Parmasto
Peniophora romellii Litsch. in Bourd.
Distr.: Temperate northern hemisphere.
Notes: On wood.
Host: Thuja
Phlebia rufa (Fr.:Fr.) M. Christiansen
Merulius rufus Fr.:Fr.
Phlebia acerina Peck
Phlebia castanea C.G. Lloyd
Phlebia erecta Rea
Merulius interruptus Bres.
Phlebia merulioidea C.G. Lloyd
Merulius pallens Berk.
Merulius pilosus Burt
Merulius pruni Peck
Distr.: Widespread.
Notes: On wood.
Hosts: Acer, Castanea, Fraxinus, Juglans, Liriodendron, Malus, Nyssa, Pinus, Populus, Prunus, Quercus, Tilia, Tsuga, Ulmus
Phlebia subalbida W.B. Cooke
Distr.: CA.
Notes: On wood.
Host: Abies
Phlebia subochracea (Bres.) J. Eriksson & Ryvarden
Grandinia subochracea Bres.
Mycoacia subochracea (Bres.) Parmasto
Corticium granulatum Burt
Distr.: Temperate northern hemisphere.
Notes: On hardwoods. Nakasone et al. (Mycotaxon 14:7. 1982).
Hosts: Platanus, Populus, Quercus, Salix
Phlebia subserialis (Bourd. & Galzin) Donk
Corticium subseriale Bourd. & Galzin
Peniophora subserialis (Bourd. & Galzin) Slysh
Peniophora phlebioides H. Jacks. & Dearden
Phlebia phlebioides (H. Jacks. & Dearden) R.L. Gilbertson & Budington
Distr.: Temperate northern hemisphere.
Notes: On wood, mainly of conifers. Nakasone et. al. (Mycotaxon 14:8. 1982).
Hosts: Abies, Acer, Juniperus, Picea, Pinus, Populus, Pseudotsuga

Phlebia tremellosus (Schrad.:Fr.) Nakasone & Burdsall
Merulius tremellosus Schrad.:Fr.
Distr.: Widespread.
Notes: On wood.
Hosts: Abies, Acer, Alnus, Betula, Carya, Castanea, Malus, Pinus, Populus, Quercus, Sorbus, Tsuga, Ulmus
Phlebia umbrata (Bourd. & Galzin) Parmasto
Phlebia lilacea M. Christiansen
Distr.: Northern hemisphere.
Host: Quercus

Phlebiella P. Karst. Basidiomycetes, Aphyllophorales

Notes: Hjortstam & Larsson. (Mycotaxon 29:315. 1987).
Phlebiella filicina (Bourd.) K. Larsson & Hjortstam in Hjortstam & K. Larsson
Corticium filicinum Bourd.
Xenasma filicinum (Bourd.) M. Christiansen
Aphanobasidium filicinum (Bourd.) Jülich
Xenasmatella filicina (Bourd.) Oberwinkler
Distr.: Northern hemisphere.
Notes: On plant materials.
Hosts: Abies, Larix, Osmunda, Picea, Pseudotsuga
Phlebiella ralla (H. Jacks.) K. Larsson & Hjortstam in Hjortstam & K. Larsson
Corticium rallum H. Jacks.
Aphanobasidium rallum (H. Jacks.) Jülich
Xenasma rallum (H. Jacks.) Liberta
Distr.: North America.
Notes: On wood and bark.
Hosts: Acer, Picea
Phlebiella subnitens (Bourd. & Galzin) K. Larsson & Hjortstam in Hjortstam & K. Larsson
Corticium subnitens Bourd. & Galzin
Aphanobasidium subnitens (Bourd. & Galzin) Jülich
Xenasma subnitens (Bourd. & Galzin) Liberta
Distr.: IA, OR; France.
Notes: On wood.
Hosts: Pseudotsuga, Quercus
Phlebiella tulasnelloidea (Höhn. & Litsch.) Oberwinkler
Corticium tulasnelloideum Höhn. & Litsch.
Xenasma tulasnelloideum (Höhn. & Litsch.) Donk
Corticium incanum Burt
Distr.: Northern hemisphere.
Notes: On wood and bark.
Hosts: Acacia, Acer, Lindera, Pinus, Quercus
Phlebiella vaga (Fr.:Fr.) P. Karst.
Phlebia vaga Fr.:Fr.
Grandinia vaga (Fr.:Fr.) P. Karst.
Hypochnus vagus (Fr.:Fr.) Kauffm.
Trechispora vaga (Fr.:Fr.) Liberta
Odontia vaga (Fr.:Fr.) P. Karst.
Cristella donkii Parmasto
Trechispora donkii (Parmasto) Liberta
Hypochnus fumosus Fr.
Athelia sericea Pers.
Cristella sulphurea (Pers.:Fr.) Donk
Corticium sulphureum (Pers.:Fr.) Fr.
Hyphoderma sulphureum (Pers.:Fr.) Wallr.
Odontia tenuis Peck
Distr.: Temperate northern hemisphere.
Notes: On wood and litter. *Corticium sulphureum* 1796 is an accepted synonym of *Phlebia vaga* 1821 (Liberta, Canad. J. Bot. 51:1871. 1973). Since both names are sanctioned by Fries, the earlier name should take priority over the later one. *Phlebiella vaga* is used here because there is no extant combination for *C. sulphureum* in *Phlebiella*.
Hosts: Abies, Acer, Alnus, Betula, Juglans, Picea, Pinus, Populus, Pseudotsuga, Thuja, Tsuga

Phleogena Link. Basidiomycetes, Auriculariales

Phleogena faginea (Fr.:Fr.) Link
Distr.: Temperate northern hemisphere.
Notes: On wood and bark.
Hosts: Carpinus, Quercus

Phloeospora Wallr. Coelomycetes

Phloeospora adusta Heald & F.A. Wolf
Distr.: TX.
Host: Clematis
Phloeospora ampelopsidis (Ellis & Everh.) Bubák
Gloeosporium ampelopsidis Ellis & Everh.
Septogloeum ampelopsidis (Ellis & Everh.) Sacc.
Phloeospora ampelopsidis (Ellis) Bubák
Rhabdospora ampelopsidis (Ellis) Kunze
Septoria ampelopsidis Ellis
Distr.: Central United States.
Hosts: Ampelopsis, Parthenocissus
Phloeospora balsamorhizae Solheim & K.S. Wilson
Distr.: WY.
Host: Balsamorhiza
Phloeospora caraganae Jacz.
Distr.: WY; Soviet Union(type).
Host: Caragana
Phloeospora caricis Ellis & Everh. in G. Martin
Septoria caricis (Ellis & Everh.) Kuntze
Distr.: Eastern United States.
Host: Carex
Phloeospora cercocarpi (Bonar) Sutton & Pollack
Septogloeum cercocarpi Bonar
Distr.: CA.
Notes: Sutton (The Coelomycetes. p. 306. 1980).
Host: Cercocarpus
Phloeospora graminearum R. Sprague & Hardison
Distr.: MI.
Hosts: Agropyron, Elymus
Phloeospora hansenii Bubák
Distr.: CA.
Host: Quercus
Phloeospora idahoensis R. Sprague
Didymella festucae (H. Wegelin) L. Holm (teleomorph)
Distr.: ID.
Notes: Stem eyespot disease.
Host: Festuca
Phloeospora lunelliana (Sacc.) Solheim
Septoria lunelliana Sacc.
Distr.: WY.
Host: Carex
Phloeospora maculans (Berenger) Allesch.
Cercosporella maculans (Berenger) F.A. Wolf
Cylindrosporium mori (Lév.) Berl.
Phloeospora mori (Lév.) Sacc.
Septoria mori Lév.
Mycosphaerella mori (Fuckel) F.A. Wolf (teleomorph)
Distr.: Worldwide.
Notes: Sutton (The Coelomycetes. p. 309. 1980).
Host: Morus
Phloeospora megarrhizae Ellis & Everh.
Distr.: Western United States.
Host: Echinocystis
Phloeospora moricola (Pass.) Sacc.
Septoria moricola Pass.
Distr.: Eastern United States.
Host: Morus
Phloeospora muhlenbergiae R. Sprague & Solheim in Solheim
Distr.: AZ; possibly Mexico.
Host: Muhlenbergia
Phloeospora multimaculans Heald & F.A. Wolf
Distr.: IN, OK, TX.
Hosts: Juglans, Platanus
Phloeospora osmorrhizae (Peck) House
Septoria osmorrhizae Peck

Distr.: United States.
Hosts: Osmorhiza, Pseudocymopterus
Phloeospora oxytropidis **Ellis & B.T. Galloway**
Septoria oxytropidis (Ellis & B.T. Galloway) Kuntze
Distr.: Western United States.
Host: Oxytropis
Phloeospora panici **H.C. Greene**
Distr.: WI(type).
Host: Panicum
Phloeospora plucheae **J.A. Stevenson**
Distr.: FL, TX.
Host: Pluchea
Phloeospora pseudoplatani **Bubák**
Distr.: CO; Europe(type).
Host: Acer
Phloeospora pteleae **Tharp**
Distr.: TX.
Host: Ptelea
Phloeospora robiniae **(Lib.) Höhn.**
Ascochyta robiniae Lib.
Distr.: Central and eastern United States; Europe.
Host: Robinia
Phloeospora skagwayensis **R. Sprague**
Distr.: AK.
Host: Carex
Phloeospora spiraeicola **(Ellis & Everh.) Bubák**
Cylindrosporium spiraeicola Ellis & Everh.
Distr.: Central and western United States.
Host: Spiraea
Phloeospora ulmi **(Fr.:Fr.) Wallr.**
Septoria ulmi Fr.: Fr.
Phyllosticta bellunensis C. Mart. (synanamorph)
Mycosphaerella ulmi Kleb. (teleomorph)
Distr.: Temperate North America; Europe.
Notes: On leaves.
Host: Ulmus
Phloeospora vagnerae **Petr.**
Distr.: MT.
Host: Smilacina

Phloeosporella Höhn. Coelomycetes
Notes: Sutton (The Coelomycetes. p. 240. 1980).
Phloeosporella ariaefoliae **(Ellis & Everh.) Sutton**
Cylindrosporium ariaefoliae Ellis & Everh.
Phloeospora ariaefoliae (Ellis & Everh.) Höhn.
Distr.: Western United States.
Notes: On leaves.
Hosts: Holodiscus, Spiraea
Phloeosporella ceanothi **(Ellis & Everh.) Höhn.**
Cylindrosporium ceanothi Ellis & Everh.
Distr.: Central and western United States.
Notes: Leaf spot.
Host: Ceanothus
Phloeosporella hedysari **(Solheim) Sutton**
Septogloeum hedysari Solheim
Distr.: WY.
Host: Hedysarum
Phloeosporella padi **(Lib.) Arx**
Ascochyta padi Lib.
Cylindrosporium padi (Lib.) P. Karst. ex Sacc.
Cylindrosporium hiemalis Higgins
Cylindrosporium lutescens Higgins
Cylindrosporium prunophorae Higgins
Blumeriella jaapii (Rehm) Arx (teleomorph)
Microgloeum pruni Petr. in Syd. & Petr. (synanamorph)
Distr.: North America; temperate Europe.
Notes: Leaf spot and shot-hole.
Host: Prunus

Phloeosporina Höhn. Coelomycetes
Phloeosporina fraxini **(Harkn.) Arx**
Septogloeum fraxini Harkn.
Gloeosporium fraxini (Harkn.) Ellis & Everh.
Phloeosporina minor (Ellis & Kellerm.) Höhn.
Cylindrosporium minus Ellis & Kellerm.
Cylindrosporium viride Ellis & Everh.
Distr.: Temperate North America.
Notes: On leaves. Arx (Biblioth. Mycol. 24:124. 1970).
Host: Fraxinus

Phlyctema Desmaz. Coelomycetes
Phlyctema albocincta **Ellis & Everh.**
Distr.: NJ.
Host: Rhus
Phlyctema andersonii **Ellis**
Distr.: MT.
Hosts: Arabis, Draba
Phlyctema arcuata **Berk.**
Distr.: CA, TX.
Hosts: Arctium, Solanum
Phlyctema campanulae **Clements, nom. nud.**
Distr.: CO.
Host: Campanula
Phlyctema complanata **(Berk. & M.A. Curtis) Sacc.**
Septoria complanata Berk. & M.A. Curtis
Distr.: Eastern United States.
Host: Polygonum
Phlyctema ficuum **Henn.**
Distr.: FL; Brazil.
Host: Ficus
Phlyctema gossypii **Sacc.**
Distr.: Southern United States.
Host: Gossypium
Phlyctema ipomoeae **Ellis & Everh.**
Distr.: WV.
Host: Ipomoea
Phlyctema linicola **Speg.**
Distr.: United States; Argentina(type).
Host: Linum
Phlyctema simulans **(Berk. & M.A. Curtis) Sacc.**
Septoria simulans Berk. & M.A. Curtis
Distr.: Northeastern United States.
Host: Prenanthes
Phlyctema smilacis **Cooke**
Distr.: Southern United States.
Host: Smilax
Phlyctema tiliae **Dearn., nom. nud.**
Distr.: TN.
Notes: Published without a latin diagnosis.
Host: Tilia
Phlyctema vagabunda **Desmaz.**
Gloeosporium allantoideum Peck
Gloeosporium allantosporum Fautrey
Gloeosporium diervillae Grove
Gloeosporium frigidum Sacc.
Gloeosporium tineum Sacc.
Pezicula alba Guthrie (teleomorph)
Distr.: Temperate regions.
Notes: Sutton (The Coelomycetes. p. 580. 1980).
Hosts: Euonymus, Malus, Rubus, Sambucus
Phlyctema verrucarioides **Sacc.**
Distr.: NY.
Host: Tilia

Phlyctochytrium J. Schröt. Chytridiomycetes, Chytridiales
Notes: Barr (Canad. J. Bot. 47:991. 1969).
Phlyctochytrium lackeyi **Sparrow**
Distr.: FL(type).
Host: Pinus
Phlyctochytrium macrosporum **J. Schröt.**
Distr.: FL(type).
Host: Pinus

Pholiota (Fr.) P. Kumm., nom. cons. Basidiomycetes, Agaricales

Notes: Smith & Hesler (The North American species of *Pholiota*. 1968).

Pholiota abieticola A.H. Sm. & Hesler
Distr.: NC.
Notes: On dead wood in high-elevation spruce-fir forest.
Host: Abies

Pholiota adiposa (Fr.:Fr.) P. Kumm.
Pholiota aurivella (Fr.) P. Kumm.
Distr.: Temperate northern hemisphere.
Notes: On wood.
Hosts: Abies, Acer, Alnus, Castanea, Fagus, Malus, Picea, Pinus, Populus, Pseudotsuga, Quercus, Salix, Sorbus, Tilia, Tsuga

Pholiota albocrenulata (Peck) Sacc.
Distr.: Northeastern and western North America.
Notes: Mostly on *Acer*, but occasionally on other hardwoods and conifers.
Host: Acer

Pholiota alnicola (Fr.:Fr.) Singer
Flammula alnicola (Fr.:Fr.) P. Kumm.
Distr.: Temperate North America; Europe.
Notes: On wood.
Hosts: Abies, Alnus, Tsuga

Pholiota alniphila (Zeller) Redhead
Naucoria alniphila Zeller
Pholiota occidentalis A.H. Sm. & Hesler
Distr.: Pacific Northwest.
Notes: Redhead (Sydowia 37:264. 1984).
Host: Alnus

Pholiota aurivelloides Overh.
Distr.: Western North America.
Notes: On wood.
Hosts: Alnus, Salix

Pholiota bakerensis A.H. Sm. & Hesler
Distr.: Western North America.
Notes: On conifer wood.
Hosts: Picea, Pinus, Pseudotsuga

Pholiota bridgii A.H. Sm. & Hesler
Distr.: CA(type).
Host: Abies

Pholiota decorata (Murrill) A.H. Sm. & Hesler
Distr.: Western North America.
Notes: On wood.
Hosts: Abies, Picea, Pinus, Populus, Pseudotsuga

Pholiota flammans (Fr.:Fr.) P. Kumm.
Pholiota kauffmaniana A.H. Sm.
Distr.: Circumboreal.
Hosts: Abies, Tsuga

Pholiota granulosa (Peck) A.H. Sm. & Hesler
Flammula granulosa Peck
Distr.: Eastern North America.
Notes: On wood of hardwoods and conifers.
Host: Fagus

Pholiota graveolens (Peck) A.H. Sm. & Hesler
Flammula graveolens Peck
Distr.: Temperate North America.
Notes: On conifer wood.
Host: Abies

Pholiota limonella (Peck) Sacc.
Pholiota abietis A.H. Sm. & Hesler
Pholiota connata A.H. Sm. & Hesler
Pholiota squarroso-adiposa J.E. Lange
Pholiota subvelutipes A.H. Sm.
Distr.: Temperate North America; Europe.
Notes: On wood. Often misdetermined as *P. adiposa* (Fr.:Fr.) Kummer. Farr et al. (Canad. J. Bot. 55:1172. 1977).
Hosts: Abies, Acer, Populus, Pseudotsuga, Salix

Pholiota malicola (Kauffm.) A.H. Sm. var. ***macropoda*** A.H. Sm. & Hesler
Distr.: NC, western North America.
Notes: On wood.
Host: Abies

Pholiota olympiana (A.H. Sm.) A.H. Sm. & Hesler
Hypholoma olympianum A.H. Sm.
Naematoloma olympianum (A.H. Sm.) A.H. Sm.
Distr.: WA(type).
Host: Pseudotsuga

Pholiota oregonensis Murrill
Distr.: OR.
Host: Salix

Pholiota populnea (Pers.:Fr.) Kuyper & Tjall.-Beukers
Pholiota destruens (Brond.) Gill.
Distr.: Temperate North America; Europe.
Notes: On wood (Kuyper & Tjallingii-Beukers. Persoonia 13:81. 1986).
Host: Populus

Pholiota spumosa (Fr.:Fr.) Singer
Distr.: Temperate North America; Europe.
Notes: On wood and woody debris.
Hosts: Pinus, Pseudotsuga

Pholiota squarrosa (Fr.:Fr.) P. Kumm.
Distr.: Temperate North America; Europe.
Notes: On wood.
Hosts: Betula, Castanea, Malus, Populus, Salix

Pholiota squarrosoides (Peck) Sacc.
Distr.: Temperate North America.
Notes: On trunks and stumps of various hardwoods.
Hosts: Abies, Acer, Betula, Fagus, Tsuga

Pholiota stratosa A.H. Sm. & Hesler
Distr.: MI(type).
Host: Acer

Phoma Sacc., nom. cons. Coelomycetes

Notes: Boerema (Trans. Brit. Mycol. Soc. 67:289. 1976); Boerema & Dorenbosch (Stud. Mycol. 3:1. 1973); Boerema et al. (Trans. Brit. Mycol. Soc. 77:61. 1981); Johnston (New Zealand J. Bot. 19:173. 1981); Sutton (The Coelomycetes. p. 378. 1980).

Phoma acuum Cooke & Ellis
Distr.: NJ, NY.
Host: Pinus

Phoma agapanthi (Thuem.) Sacc.
Sphaeropsis agapanthi Thuem.
Distr.: HI; Portugal.
Host: Agapanthus

Phoma akebiae Dearn.
Distr.: NY.
Host: Akebia

Phoma allicola Sacc. & Roum.
Distr.: Europe; distribution elsewhere is questionable.
Host: Allium

Phoma amaranthi Halst. in Seym. & Earle non Brunaud, hom. illeg.
Distr.: NJ.
Host: Amaranthus

Phoma amaranthicola Brunaud
Distr.: OR; Europe.
Host: Amaranthus

Phoma americana Morgan-Jones & J.F. White
Distr.: Southeastern United States.
Notes: Morgan-Jones & White (Mycotaxon 16:403. 1983).
Hosts: Triticum, Zea

Phoma ampelina Berk. & M.A. Curtis
Phomopsis ampelina (Berk. & M.A. Curtis) Grove
Distr.: PA.
Host: Vitis

Phoma anceps Sacc.
Distr.: MT; Europe.
Host: Medicago

Phoma anemone Kauffm. in Dearn.
Distr.: Eastern United States.
Host: Anemone

Phoma anethi (Pers.:Fr.) Sacc.
Mycosphaerella anethi (Pers.:Fr.) Petr. (teleomorph)
Cercosporidium punctum (Lacr.) Deighton (synanamorph)
Distr.: North America; northern Africa; temperate Europe.
Notes: Boerema (pers. comm.) suggested that this species is an *Asteromella*.
Host: Anethum

Phoma antennariae Clements, nom. nud.

Distr.: CO.
Host: Antennaria
Phoma apiicola Kleb.
Distr.: Temperate North America; Europe.
Host: Apium
Phoma apocrypta Ellis & Everh.
Distr.: Southeastern United States.
Host: Phytolacca
Phoma arachidicola Marasas, G.D. Pauer, & Boerema
Ascochyta adzamethica Schoschiaschvili
Didymosphaeria arachidicola (Chochrjakov) Alcorn, Punithalingam, G. McCarthy (teleomorph)
Distr.: Worldwide in peanut-growing regions.
Notes: Net blotch or web blotch of peanuts (Alcorn et al. Trans. Brit. Mycol. Soc. 66:351. 1976; Punithalingam. C.M.I. Descr. 736. 1982).
Host: Arachis
Phoma asclepiadea Ellis & Everh. in Millsp. & L. Nutt.
Distr.: Eastern and southwestern United States.
Host: Asclepias
Phoma asparagina Tehon & G.L. Stout
Distr.: IL.
Host: Asparagus
Phoma astragali Cooke & Harkn.
Distr.: Central and western North America.
Host: Astragalus
Phoma atrocincta Sacc.
Distr.: FL; Southern Europe.
Host: Ficus
Phoma bacilliformis Wehmeyer
Distr.: WY.
Host: Senecio
Phoma bacteriophila Peck
Distr.: NY.
Host: Pinus
Phoma bakeriana Sacc.
Distr.: MS; East Indies; The Philippines.
Host: Vigna
Phoma baptisiae Henn.
Distr.: SC.
Host: Baptisia
Phoma betae A.B. Frank
Gloeosporium betae Dearn. & Barth.
Phyllosticta betae Oudem.
Phoma spinaciae Bubák & W. Krieg. in Bubák
Phyllosticta spinaciae A. Zimmerm.
Pleospora betae (Berl.) Nevodovsky (teleomorph)
Distr.: Cosmopolitan.
Notes: Black leg, root rot, storage rot, and leaf spot of beets (Booth. C.M.I. Descr. 149. 1967). Boerema et al. (Neth. J. Pl. Pathol. 93:12. 1987).
Host: Beta
Phoma boltoniae Dearn. & Barth.
Discosphaerina boltoniae (Dearn. & Barth.) Barr (teleomorph)
Selenophoma boltoniae (Dearn. & Barth.) Sivanesan (synanamorph)
Distr.: Central United States.
Host: Boltonia
Phoma bresadolae Sacc.
Distr.: Southeastern United States; southern Europe.
Host: Cycas
Phoma bumeliae Tassi
Distr.: Southeastern United States; southern Europe.
Host: Sideroxylon
Phoma camelliae Pass. non Cooke, hom. illeg.
Distr.: Southern United States; southern Europe.
Host: Camellia
Phoma caricae-papayae (Tarr) Punithalingam
Ascochyta caricae-papayae Tarr
Phoma caricae (Pat.) Punithalingam
Mycosphaerella caricae Syd. & P. Syd. (teleomorph)
Distr.: FL, HI; southeast Asia; India.
Notes: Fruit rot, trunk rot, and leaf spot of papaya (Sivanesan. The bitunicate Ascomycetes. p. 260. 1984).
Host: Carica
Phoma caricis (Fr.) Sacc.
Macrophoma caricis (Fr.) Berl. & Voglino
Distr.: CA; Europe.
Host: Carex
Phoma castanea Peck
Distr.: Eastern United States.
Host: Castanea
Phoma cava Schulzer
Phoma aposphaerioides Briard & Har.
Phoma collabens Schulzer & Sacc.
Aposphaeria collabescens Schulzer & Sacc.
Distr.: United States; Europe.
Hosts: Acer, Quercus
Phoma cephalanthi Henn.
Distr.: NY.
Notes: The status of this taxon is unknown.
Host: Cephalanthus
Phoma cercidicola Fairm. non Henn., hom. illeg.
Distr.: NY.
Host: Cercis
Phoma chrysanthemi Voglino
Ascochyta chrysanthemi F. Stevens
Didymella ligulicola (K. Baker, Dimock, & L.H. Davis) Arx in E. Müller & Arx (teleomorph)
Distr.: North America; Australia; Europe; Japan.
Notes: Ray blight (Walker & Baker. Trans. Brit. Mycol. Soc. 80:31. 1983; Boerema & Bollen. Persoonia 8:114. 1975).
Host: Chrysanthemum
Phoma cincta Berk. & M.A. Curtis
Distr.: SC.
Host: Ulmus
Phoma clematidina (Thuem.) Boerema
Ascochyta clematidina Thuem. emend. Gloyer
Distr.: North America; Europe.
Notes: Boerema (Versl. Meded. Plziektenk. Dienst Wageningen 153:17. 1979).
Host: Clematis
Phoma clitoricarpa (Cooke) Sacc.
Sphaeropsis clitoricarpa Cooke
Distr.: TX; India(type).
Host: Sesbania
Phoma coloradoensis Earle in Greene
Distr.: CO.
Host: Pedicularis
Phoma complanata (Tode:Fr.) Desmaz.
Distr.: Temperate North America; Europe.
Notes: On herbaceous stems.
Hosts: Angelica, Heracleum
Phoma corallorhizae Ellis & Everh. in Coons, nom. nud.
Distr.: Northeastern United States.
Hosts: Corallorrhiza, Orchidaceae
Phoma coryphae Cooke & Massee
Distr.: FL; Sri Lanka.
Host: Pandanus
Phoma cucurbitacearum (Fr.:Fr.) Sacc.
Ascochyta citrullina (F. Chester) C.O. Sm.
Diplodina citrullina (F. Chester) Gross.
Phyllosticta citrullina F. Chester
Ascochyta cucumis Fautrey & Roum.
Ascochyta melonis Potebnia
Didymella bryoniae (Auersw.) Rehm (teleomorph)
Distr.: Worldwide on cucurbits.
Notes: Gummy stem blight, black rot of fruit, and various other symptoms on Cucurbitaceae (Punithalingam & Holliday. C.M.I. Descr. 332. 1972; Corlett et al. Fungi Canadenses 303. 1986 as *Didymella bryoniae*). Often referred to as *Ascochyta cucumis*.
Hosts: Citrullus, Cucumis, Cucurbita
Phoma cydoniae Sacc. & Schulzer
Distr.: Midwest United States; Europe.
Host: Cydonia
Phoma cynoglossi Dearn.
Distr.: NY, TX.
Host: Cynoglossum
Phoma dahliae Berk. in Sm.
Distr.: Eastern United States; Europe.
Notes: A doubtful species according to Grove (British stem and leaf fungi. Vol. I. p. 79. 1935).

Host: Dahlia
Phoma delphiniicola Tracy & Earle in Greene
Distr.: CO, TX.
Host: Delphinium
Phoma dendritica Thuem.
Distr.: SC.
Host: Quercus
Phoma destructiva Plowr.
Diplodina destructiva (Plowr.) Petr.
Phyllosticta lycopersici Peck
Distr.: Cosmopolitan.
Notes: Black spot of tomato and other Solanaceae. Morgan-Jones & Burch (Mycotaxon 32:253. 1988).
Hosts: Capsicum, Lycopersicon, Solanum
Phoma dioscoreae Fairm.
Distr.: NY.
Host: Dioscorea
Phoma discosiiformis Cooke & Harkn.
Distr.: CA.
Host: Quercus
Phoma douglasii Oudem.
Distr.: CO; Europe.
Host: Pseudotsuga
Phoma dura Sacc.
Distr.: CA; Great Britain(Kew); Italy.
Host: Abies
Phoma eriogoni W.B. Cooke in W.B. Cooke & C.G. Shaw
Distr.: CA.
Host: Eriogonum
Phoma erysiphoides Ellis & Everh.
Distr.: CO, TX.
Hosts: Achillea, Gnaphalium
Phoma erythrinicola Tracy & Earle
Distr.: MS.
Notes: The status of this name is uncertain.
Host: Erythrina
Phoma estrelti Fairm.
Distr.: NM.
Host: Haplopappus
Phoma euphorbiicola (Schwein.) Starb.
Distr.: PA, TX.
Host: Euphorbia
Phoma eupyrena Sacc.
Distr.: Cosmopolitan.
Notes: On a variety of organic substrates. Morgan-Jones & Burch (Mycotaxon 31:427. 1988).
Hosts: Abies, Pseudotsuga, Rhizophora, Solanum
Phoma exigua Desmaz.
Ascochyta althaeina Sacc. & Bizz. in Sacc.
Phyllosticta althaeina Sacc.
Ascochyta asteris (Bres.) Gloyer
Phyllosticta asteris Bres.
Ascochyta capsici Bondartzeva-Monteverde
Phyllosticta decidua Ellis & Kellerm.
Phyllosticta hortorum Speg.
Ascochyta hydrangeae M. Arnaud & G. Arnaud
Gloeosporium melleum Dearn. & Overh.
Phyllosticta mulgedii J.J. Davis
Ascochyta nicotianae Pass.
Ascochyta parasitica Fautrey
Ascochyta phaseolorum Sacc.
Ascochyta pirina Peglion
Phyllosticta sambuci Desmaz.
Phoma solanicola Prill. & Delacr.
Ascochyta sonchi (Sacc.) Grove
Phyllosticta sonchi Sacc.
Phoma tuberosa Melhus, Rosenb., & E.S. Schultz
Distr.: Cosmopolitan.
Notes: A plurivorous soil-borne, weakly parasitic fungus (Boerema & Höweler. Persoonia 5:15. 1967; Boerema. Trans. Brit. Mycol. Soc. 67:292. 1976). A number of subspecific taxa have been described in *Phoma*. Morgan-Jones & Burch (Mycotaxon 32:477 1988) provided a treatment of *Phoma exigua* var. *exigua*. Sutton & Waterston (C.M.I. Descr. 81. 1966 as *Ascochyta phaseolorum*).
Hosts: Abelmoschus, Abutilon, Acer, Agastache, Agrimonia, Alcea, Althaea, Aralia, Armoracia, Aronia, Bidens, Boehmeria, Callistephus, Capsicum, Ceratoides, Chenopodium, Circaea, Citrullus, Clinopodium, Cucumis, Cynoglossum, Datura, Eupatorium, Fraxinus, Galeopsis, Geum, Glechoma, Gossypium, Hackelia, Hieracium, Humulus, Hydrangea, Lactuca, Leonurus, Lycopersicon, Lycopus, Lysimachia, Malva, Melissa, Mentha, Monarda, Myosotis, Nepeta, Nerium, Nicotiana, Phaseolus, Pisum, Plantago, Pyrus, Ratibida, Ricinus, Rudbeckia, Sambucus, Scutellaria, Solanum, Solidago, Sonchus, Stachys, Teucrium, Urtica, Valeriana, Verbena, Vernonia, Veronicastrum, Vigna, Vinca
Phoma exigua Desmaz. var. ***inoxydabilis*** Boerema & Vegh in Vegh et al.
Phyllosticta vincae-majoris Allesch.
Distr.: Temperate North America; Europe.
Notes: Stem blight and leaf spot (Boerema. Trans. Brit. Mycol. Soc. 67:292. 1976).
Host: Vinca
Phoma exocarpina Peck
Distr.: Eastern United States.
Host: Carya
Phoma florida Dearn. & House
Distr.: NY.
Host: Cornus
Phoma fuliginea M.N. Kidd & A. Beaumont
Distr.: WA; Great Britain.
Host: Malus
Phoma fusispora Wehmeyer
Distr.: WY.
Host: Linum
Phoma genista Brunaud
Distr.: CA; Europe.
Host: Genista
Phoma gentianae Kühn
Distr.: TX(questionable); Europe.
Host: Campanula
Phoma glandulosum Cooke
Distr.: SC.
Host: Ailanthus
Phoma glomerata (Corda) Wollenweb. & Hochapfel
Phoma conidiogena Schegg
Aposphaeria fibricola (Berk.) Sacc.
Phoma fibricola Berk.
Distr.: Cosmopolitan.
Notes: Blights, rots, and other diseases (Morgan-Jones. C.M.I. Descr. 134. 1967; White & Morgan-Jones. Mycotaxon 28:437. 1987).
Hosts: Buxus, Elaeagnus, Juniperus, Metrosideros, Pseudotsuga
Phoma glumarum Ellis & Tracy
Phyllosticta glumarum (Ellis & Tracy) Miyake
Distr.: Subtropical and tropical regions.
Notes: According to Boerema, Dorenbosch & Van Kesteren (Persoonia 7:131. 1973), this name has been misapplied. The type specimen revealed that this is a species of *Coniothyrium*. These records probably refer to *Phoma sorghina*.
Hosts: Chasmanthium, Oryza, Uniola
Phoma gossypii Sacc.
Distr.: Worldwide in cotton-growing regions.
Host: Gossypium
Phoma graminis Westend.
Distr.: AK; Europe.
Host: Poa
Phoma gramma (Schwein.) Starb.
Sphaeria gramma Schwein.
Distr.: Southeastern United States.
Host: Stylosanthes
Phoma gyneriicola Cooke & Massee
Distr.: CA, FL; Great Britain(Kew).
Host: Cortaderia
Phoma halesiae Fairm.
Distr.: NY.
Host: Halesia
Phoma harknessii Sacc.
Phoma pini Cooke & Harkn.
Distr.: Eastern and western United States.
Hosts: Pinus, Triticum
Phoma henningsii Sacc.

Distr.: HI; Southern Europe.
Host: Albizia
Phoma heraclei Earle in Greene
Distr.: CO, TX.
Host: Heracleum
Phoma herbarum Westend.
Phoma oleracea Sacc.
Phoma oleracea Sacc. var. *dipsaci* Sacc.
Distr.: Cosmopolitan.
Notes: An ubiquitous saprophyte. Boerema (Trans. Brit. Mycol. Soc. 67:303. 1976). Morgan-Jones (Mycotaxon 33:81. 1988).
Hosts: Agrimonia, Arenaria, Cardamine, Carya, Castilleja, Cerastium, Dipsacus, Heracleum, Pseudotsuga, Rumex, Stachytarpheta
Phoma herbicola Wehmeyer
Distr.: WY.
Hosts: Carex, Pedicularis, Synthyris
Phoma hieracii Rostr.
Distr.: TX; Europe.
Host: Hieracium
Phoma houseana Sacc.
Distr.: Northeastern United States.
Host: Vaccinium
Phoma ilicina Ellis & F.W. Anderson
Sphaeropsis ilicina (Ellis & F.W. Anderson) Kuntze
Distr.: Central and eastern United States.
Host: Ilex
Phoma infossa Ellis & Everh.
Distr.: Northeastern United States.
Host: Fraxinus
Phoma iowana Sacc.
Distr.: Central and eastern United States.
Host: Aster
Phoma iridina Sacc.
Distr.: MT.
Host: Iris
Phoma jejuna Wehmeyer
Distr.: WY.
Hosts: Draba, Penstemon
Phoma jolyana Pirozynski & Morgan-Jones
Phoma musae (Joly) Boerema, Dorenbosch, & Van Kesteren non (Cooke) Sacc., hom. illeg.
Distr.: Subtropical and tropical regions.
Notes: Morgan-Jones & Burch (Mycotaxon 30: 239. 1987).
Hosts: Musa, Prosopis
Phoma junci G. Preuss
Distr.: Eastern United States; Europe.
Host: Juncus
Phoma karstenii (P. Karst.) Berl. & Voglino
Phoma microsperma P. Karst.
Distr.: CA; Europe.
Host: Lomatium
Phoma krigiae Tassi
Distr.: Reported from an unknown locality in the United States.
Notes: The status of this name is uncertain.
Host: Krigia
Phoma lagenariae (Thuem.) Sacc.
Sphaeropsis lagenariae Thuem.
Distr.: NY; Europe.
Host: Cucurbita
Phoma lagerstroemiae Speg. var. ***foliicola*** Ellis & Everh.
Distr.: LA.
Host: Lagerstroemia
Phoma lanuginis Fairm.
Distr.: NY, TX.
Host: Marrubium
Phoma leprosa Peck
Distr.: NY.
Host: Crataegus
Phoma leptospora Sacc. non Sacc. & Briard, hom. illeg.
Distr.: ND.
Host: Grindelia
Phoma ligustrina Sacc.
Distr.: NY.
Host: Ligustrum
Phoma linariae Dearn. & House
Distr.: NY, TX.
Host: Linaria
Phoma linearis (Thuem.) Sacc. non Sacc., hom. illeg.
Phoma lineata Sacc.
Distr.: CA.
Host: Cortaderia
Phoma linearispora Wehmeyer
Distr.: WY.
Host: Leptodactylon
Phoma lineolata Desmaz.
Distr.: MT; Europe.
Host: Picea
Phoma lingam (Tode:Fr.) Desmaz.
Plenodomus lingam (Tode) Höhn.
Leptosphaeria maculans (Desmaz.) Ces. & De Not. (teleomorph)
Distr.: Temperate regions.
Notes: Black leg, canker, and dry rot of Brassicaceae. Boerema (Trans. Brit. Mycol. Soc. 67:311. 1976); Sivanesan (The bitunicate Ascomycetes. p. 463. 1984).
Hosts: Brassica, Cardamine, Iberis, Lobularia, Raphanus
Phoma lomatiae Speg.
Distr.: OR; Chile.
Host: Lomatium
Phoma longissima (Pers.:Fr.) Westend.
Distr.: North America, scattered reports; Europe.
Host: Amaranthus
Phoma lophanthi Bubák
Distr.: OH.
Host: Agastache
Phoma lupini Ellis & Everh.
Asteromella lupini (Ellis & Everh.) Petr.
Distr.: Western North America.
Host: Lupinus
Phoma lupinicola Earle in Greene
Distr.: CO.
Host: Lupinus
Phoma lycopersici Cooke
Ascochyta lycopersici (Plowr.) Brunaud
Diplodina lycopersici Hollós
Phoma lycopersici (Plowr.) Jacz.
Didymella lycopersici Kleb. (teleomorph)
Distr.: Worldwide in tomato-growing regions.
Notes: Stem and fruit rot of tomato (Holliday & Punithalingam. C.M.I. Descr. 272. 1970). Morgan-Jones & Burch (Mycotaxon 32:133. 1988) provided a description, illustrations, and a discussion of this species.
Hosts: Lycopersicon, Solanum
Phoma macdonaldii Boerema
Leptosphaeria lindquistii Frezzi (teleomorph)
Distr.: HI, central North America.
Notes: Black stem of sunflower. Boerema et al. (Trans. Brit. Mycol. Soc. 77:64. 1981).
Host: Helianthus
Phoma macrostoma Mont.
Phoma cicatricum Pass.
Phoma limitata (Peck) Boerema in Boerema & Dorenbosch
Phyllosticta limitata Peck
Phoma mali Sacc. & Schulzer
Phyllosticta mali Prill. & Delacr.
Phoma mororum Sacc.
Phoma phyllostictoides Desmaz.
Aposphaeria pomi (Schulzer & Sacc.) Sacc.
Phoma pomi Schulzer & Sacc.
Phyllosticta taxi Hollós
Distr.: Cosmopolitan.
Notes: On a variety of hosts, especially Rosaceae (White & Morgan-Jones. Mycotaxon 20:197. 1984).
Hosts: Carya, Cydonia, Malus, Pyrus, Taxus
Phoma macularis Desmaz.
Distr.: HI; Europe.
Host: Hibiscus
Phoma maculata (Cooke & Harkn.) Sacc.
Sphaeropsis maculata Cooke & Harkn.
Distr.: CA.
Host: Arecaceae
Phoma majanthemi Peck

Distr.: Northern North America.
Host: Maianthemum
Phoma mamillariae Ellis & Everh.
Sphaeropsis mamillariae (Ellis & Everh.) Kuntze
Distr.: MT.
Hosts: Coryphantha, Mammillaria, Opuntia
Phoma mamillaris (Berk. & M.A. Curtis) Sacc.
Sphaeropsis mammillaris Berk. & M.A. Curtis
Distr.: SC.
Host: Gleditsia
Phoma mariae G.W. Clinton
Distr.: Temperate North America; Europe.
Host: Lonicera
Phoma media Ellis & Everh.
Distr.: Central and eastern United States.
Host: Asparagus
Phoma medicaginis Malbr. & Roum. in Roum.
Phoma herbarum Westend. f. *medicaginum* Westend. in Rabenh.
Ascochyta imperfecta Peck
Distr.: Cosmopolitan.
Notes: Black stem of forage legumes and on a wide variety of plants (Morgan-Jones & Burch. Mycotaxon 29:477. 1987).
Hosts: Medicago, Melilotus, Trifolium
Phoma megarrhizae Fairm.
Distr.: CA.
Host: Marah
Phoma melaleuca Berk. & M.A. Curtis
Phomopsis melaleuca (Berk. & M.A. Curtis) Grove
Distr.: Eastern United States.
Hosts: Aralia, Magnolia
Phoma microspora Berk. & M.A. Curtis, nom. nud. non Sacc. non Pat., hom. illeg.
Distr.: ND, SC.
Host: Asparagus
Phoma microsporella P. Karst. & Har.
Distr.: IN, OH.
Host: Asimina
Phoma minuta Wehmeyer
Distr.: WY.
Hosts: Heracleum, Lupinus, Pedicularis, Polemonium
Phoma nebulosa (Pers.:Fr.) Berk.
Distr.: Temperate North America; Europe.
Notes: An ubiquitous saprophyte commonly found on *Urtica* stems. Boerema (Trans. Brit. Mycol. Soc. 67:306. 1976).
Hosts: Arabis, Pastinaca, Thalictrum, Urtica
Phoma necatrix Thuem.
Distr.: Southeastern United States; Europe.
Host: Oryza
Phoma negundicola Thuem. var. **ramicola** Ellis & Everh.
Distr.: WV.
Host: Acer
Phoma negundinicola Thuem.
Distr.: ID, NJ.
Host: Acer
Phoma nyssicarpa Cooke
Distr.: Eastern United States.
Host: Nyssa
Phoma okra Cooke
Distr.: Southeastern United States.
Host: Abelmoschus
Phoma oncidii-sphacelati Tassi
Distr.: FL; Italy.
Host: Cattleya
Phoma orthosticha Ellis & Everh.
Distr.: ME.
Host: Typha
Phoma pallens Berk. & M.A. Curtis
Distr.: NY, SC(type).
Notes: Grove (Kew Bull. 4:184. 1919) suggested that this species belongs in *Phomopsis* but did not transfer the name.
Host: Celastrus
Phoma pedicularis Wehmeyer non Fuckel, hom. illeg.
Distr.: WY.
Hosts: Aster, Lupinus, Pedicularis, Valeriana
Phoma pedunculi Ellis & Everh.
Distr.: WV.
Host: Magnolia
Phoma pennsylvanica Ellis & Everh.
Distr.: WV.
Host: Acer
Phoma persicae Schulzer & Sacc.
Distr.: Eastern United States; Europe.
Host: Prunus
Phoma pezizoides (Ellis & Everh.) Boerema & Van Kesteren
Aposphaeria pezizoides Ellis & Everh.
Aposphaeria salicum Sacc. in Syd.
Distr.: Eastern United States; central and southern Europe.
Notes: On hardwoods. Boerema & Van Kesteren (Persoonia 11:322. 1981).
Hosts: Acer, Fraxinus, Platanus, Salix
Phoma phacidiella Sacc.
Macrophoma phacidiella (Sacc.) Berl. & Voglino
Sphaeropsis phacidioides Cooke & Ellis
Distr.: United States.
Host: Ilex
Phoma philadelphi Cooke
Distr.: OR; England(type).
Host: Philadelphus
Phoma piceina Peck
Distr.: Eastern North America.
Host: Picea
Phoma pinodella (L.K. Jones) Morgan-Jones & K.B. Burch
Ascochyta pinodella L.K. Jones
Phoma medicaginis Malbr. & Roum. var. *pinodella* (L.K. Jones) Boerema
Phoma trifolii E.M. Johnson & Valleau
Distr.: Cosmopolitan.
Notes: Black stem and foot rot of Fabaceae (White & Morgan-Jones. Mycotaxon 28:241. 1987; Morgan-Jones & Burch. Mycotaxon 29:477. 1987; Punithalingam & Gibson. C.M.I. Descr. 518. 1976 as *Phoma medicaginis* var. *pinodella*).
Hosts: Lens, Medicago, Pisum, Trifolium
Phoma platysperma Peck
Distr.: IL, WV.
Host: Salix
Phoma polygalae Cooke & Harkn.
Distr.: CA.
Host: Polygala
Phoma polygramma (Fr.:Fr.) Sacc. var. **plantaginis** Sacc.
Distr.: Eastern United States; Europe.
Host: Plantago
Phoma pomorum Thuem.
Phoma bismarckii M.N. Kidd & A. Beaumont
Phyllosticta pruni-avium Allesch.
Depazea prunicola Opiz, nom. nud.
Phoma prunicola (Opiz) Wollenweb. & Hochapfel non Schwein., hom. illeg.
Phyllosticta prunicola Sacc.
Phyllosticta pyrina Sacc.
Distr.: Cosmopolitan.
Notes: Leaf spot of Rosaceae, but isolated from a wide variety of plants (White & Morgan-Jones. Mycotaxon 25:461. 1986).
Hosts: Malus, Prunus, Pyrus
Phoma potentillica Allesch.
Distr.: TX; Europe.
Host: Potentilla
Phoma pulchella (Berk. & M.A. Curtis) Sacc.
Sphaeropsis pulchella Berk. & M.A. Curtis
Distr.: Eastern United States.
Host: Rhus
Phoma pulchellicola Wehmeyer
Distr.: WY.
Host: Lactuca
Phoma punctiformis Desmaz.
Phyllosticta punctiformis (Desmaz.) Allesch.
Sclerophoma punctiformis (Desmaz.) Höhn.
Distr.: North America, scattered reports; Europe.
Hosts: Lesquerella, Lychnis
Phoma quercina (Peck) Sacc.
Sphaeropsis quercina Peck

Distr.: Central and eastern North America.
Host: Quercus
Phoma rabiei (Pass.) Khune & J.N. Kapoor
Ascochyta rabiei (Pass.) Labrousse
Mycosphaerella rabiei Kovachevski (teleomorph)
Distr.: Worldwide in chick-pea growing regions.
Notes: This fungus was transferred to *Phoma* (Khune & Kapoor. Indian Phytopathol. 33:120. 1980). Punithalingam & Holliday (C.M.I. Descr. 337. 1972 as *Ascochyta rabiei*); Derie et al. (Pl. Dis. 69:268. 1985); Tu (Pl. Dis. 68:826. 1984).
Host: Cicer
Phoma regina Fairm.
Distr.: NY.
Host: Anemone
Phoma rhamnicola Cooke & Harkn.
Distr.: CA.
Host: Rhamnus
Phoma rhodocarpa Sacc.
Distr.: NY.
Host: Rosa
Phoma rubicola Sacc.
Distr.: OK; Europe.
Host: Rubus
Phoma sclerotioides G. Preuss ex Sacc.
Plenodomus meliloti Dearn. & G.B. Sanford
Distr.: North America; northern Europe.
Notes: On roots of herbaceous plants; pathogenic on forage legumes (Boerema & Van Kesteren. Persoonia 11:325. 1981).
Hosts: Festuca, Medicago, Poa, Trifolium
Phoma selenophomoides Wehmeyer
Distr.: WY.
Host: Erigeron
Phoma sepincola (J. Kickx fil.) Sacc.
Sphaeropsis sepincola J. Kickx fil.
Distr.: NE; Europe.
Host: Rosa
Phoma sidalceae Fairm.
Distr.: NM.
Host: Sidalcea
Phoma smilacis G. Boyer & Jacz.
Distr.: NY; Europe.
Host: Smilax
Phoma socia F.A. Wolf
Distr.: FL.
Notes: Associated with cankers caused by *Xanthomonas citri*.
Hosts: Citrus, Fortunella
Phoma solidaginis Cooke var. **longispora** Sacc.
Distr.: NY.
Notes: The status of this name is uncertain.
Host: Solidago
Phoma sorghina (Sacc.) Boerema, Dorenbosch, & Van Kesteren
Phyllosticta sorghina Sacc.
Phoma glumicola Speg.
Phyllosticta glumicola (Speg.) K. Hara
Phoma insidiosa Tassi
Phyllosticta sacchari Speg.
Distr.: Cosmopolitan.
Notes: Glume blight of rice and other grasses. White & Morgan-Jones (Mycotaxon 18:3. 1983) stated that the teleomorph connection with *Mycosphaerella holci* Tehon has not been confirmed. Punithalingam & Holliday (C.M.I. Descr. 333. 1972 as *Phoma insidiosa*); Punithalingam (C.M.I. Descr. 825. 1985).
Hosts: Brachiaria, Chloris, Oryza, Panicum, Phragmites, Rhynchelytrum, Saccharum, Setaria, Sorghastrum, Sorghum
Phoma spermoides Dearn.
Distr.: NY, PA.
Host: Thalictrum
Phoma strasseri Moesz
Distr.: OR; Europe.
Host: Mentha
Phoma subvelata Sacc.
Distr.: TX; Europe.
Hosts: Cucurbita, Lagenaria
Phoma tenuis Pass.
Distr.: Southeastern United States; southern Europe.
Host: Camellia
Phoma terrestris E.M. Hans.
Pyrenochaeta terrestris (E.M. Hans.) Gorenz, J.C. Walker, & R.H. Larson
Distr.: Worldwide, most common in subtropical and tropical regions.
Notes: Pink root of *Allium*; frequently isolated from soil (Punithalingam & Holliday. C.M.I. Descr. 397. 1973). Schneider & Schwarz (Mitt. Biol. Bundesanst. Land-Forstw. Berlin-Dahlem 189:1. 1979).
Hosts: Agropyron, Agrostis, Allium, Andropogon, Arctostaphylos, Artemisia, Arthraxon, Avena, Bouteloua, Bromus, Calamovilfa, Cenchrus, Chrysothamnus, Cucumis, Cucurbita, Cynodon, Dactylis, Distichlis, Echinochloa, Elymus, Eragrostis, Festuca, Fragaria, Holcus, Hordeum, Ipomoea, Juniperus, Koeleria, Lepidium, Linum, Lycopersicon, Medicago, Melilotus, Muhlenbergia, Oryzopsis, Panicum, Phaseolus, Phleum, Pinus, Pisum, Poa, Populus, Purshia, Ribes, Rumex, Saccharum, Salix, Schedonnardus, Setaria, Sitanion, Solanum, Sorghum, Spinacia, Stipa, Triticum, Verbascum, Vigna, Vulpia, Zea
Phoma thapsi Ellis & Everh.
Distr.: NJ, TX.
Host: Verbascum
Phoma thermopsidicola Henn.
Distr.: CA; Europe.
Host: Thermopsis
Phoma tulipiferae Schwein.
Distr.: Eastern United States.
Host: Liriodendron
Phoma typharum Sacc.
Distr.: Western United States; Europe.
Host: Typha
Phoma typhicola Oudem.
Distr.: CA; Europe.
Host: Typha
Phoma urens Ellis & Everh.
Distr.: MA.
Host: Populus
Phoma vaccinii Dearn. & House
Distr.: Northeastern North America.
Host: Vaccinium
Phoma verbascicarpa Fairm.
Distr.: NY.
Host: Verbascum
Phoma verbascicola (Schwein.) Cooke
Distr.: Eastern United States.
Host: Verbascum
Phoma viburni (Roum. ex Sacc.) Boerema & M.J. Griffin
Ascochyta viburni Roum. ex Sacc.
Phyllosticta viburnicola Roum.
Phyllosticta lantanoides Peck
Rhabdospora lantanoides (Peck) Jacz.
Phyllosticta punctata Ellis & Dearn.
Distr.: Temperate North America; Europe.
Notes: Leaf spot and shoot blight (Boerema & Griffin. Trans. Brit. Mycol. Soc. 63:109. 1974).
Host: Viburnum
Phoma vitis Bonord.
Distr.: Temperate North America; Europe.
Host: Vitis
Phoma vulgaris Sacc.
Distr.: WV; Europe.
Host: Medicago
Phoma weigeliae Speg.
Distr.: WA; Europe.
Host: Weigela
Phoma wyomingensis Wehmeyer
Distr.: WY.
Host: Penstemon
Phoma xylostei Cooke & Harkn.
Distr.: CA; Great Britain.
Host: Lonicera
Phoma zeicola Ellis & Everh.
Distr.: Central and eastern United States.
Host: Zea

Phomatospora Sacc. Pyrenomycetes, Amphisphaeriales

Phomatospora argyrostigma (Berk.) Sacc.
Distr.: Central and southern United States.
Notes: Although this name has been included in *Physalospora*, the transfer was never validly published.
Host: Yucca

Phomatospora berkeleyi Sacc.
Physalospora minutella (Peck) Sacc.
Distr.: CA; Europe.
Notes: Kirk (Mycotaxon 19:318. 1984).
Hosts: Lupinus, Polygala

Phomatospora cylindrotheca Petr.
Distr.: HI(type).
Host: Pandanus

Phomatospora datiscae Harkn.
Distr.: CA(type).
Host: Datisca

Phomatospora dinemasporium J. Webster
Dinemasporium strigosum (Pers.:Fr.) Sacc. (anamorph)
Distr.: Cosmopolitan.
Notes: On grasses (Webster. Trans. Brit. Mycol. Soc. 38:360. 1955).
Host: [Carex]

Phomatospora gelatinospora Barr
Distr.: NH.
Host: Rhododendron

Phomatospora leptasca (Peck & G.W. Clinton) J. Reid & C. Booth
Valsa leptasca Peck & G.W. Clinton
Cryptospora leptasca (Peck & G.W. Clinton) Ellis & Everh.
Cryptosporella leptasca (Peck & G.W. Clinton) Sacc.
Flageoletia leptasca (Peck & G.W. Clinton) Höhn.
Distr.: Northeastern United States.
Notes: Reid & Booth (Canad. J. Bot. 44:448. 1966).
Host: Rhus

Phomatospora luzulae Cocc. & Morini
Distr.: WA; Italy(type).
Host: Elymus

Phomatospora pandani (Ellis & Everh.) Theiss.
Physalospora pandani Ellis & Everh.
Distr.: Subtropical and tropical regions.
Host: Pandanus

Phomatospora rosae Rehm
Distr.: ND(type).
Host: Rosa

Phomatospora therophila (Desmaz.) Sacc.
Phyllachora therophila (Desmaz.) Arx & E. Müller
Distr.: WY; Canada(Quebec); Europe.
Host: Carex

Phomatospora wistariae Ellis & Everh.
Distr.: AL(type).
Host: Wisteria

Phomopsis (Sacc.) Bubák. Coelomycetes

Notes: Uecker (Mycol. Mem. 13. 1988).

Phomopsis abietina (R. Hartig) Höhn.
Phoma abietina R. Hartig
Fusicoccum abietinum Prill. & Delacr.
Distr.: Reported from the western United States; Europe.
Notes: This name was used by early researchers for the fungus now known as *Phacidiopycnis boycei* Funk. Apparently *Phomopsis abietina* is known only on *Abies* spp. in Europe.
Host: Abies

Phomopsis acerina Pirone & J.C. Carter
Distr.: NJ.
Host: Acer

Phomopsis achilleae (Sacc.) Höhn.
Diaporthe arctii (Lasch) Nitschke var. *achilleae* (Auersw.) Wehmeyer (teleomorph)
Distr.: HI; Europe(type).
Hosts: Dahlia, Hemerocallis

Phomopsis ailanthi (Sacc.) Died.
Phoma ailanthi Sacc.
Distr.: TX; southern Europe.
Host: Ailanthus

Phomopsis alnea (Sacc.) Höhn.
Phoma alnea Sacc.
Distr.: KY; France(type).
Host: Alnus

Phomopsis ambigua (Sacc.) Traverso
Phoma ambigua Sacc.
Distr.: Temperate North America; Europe.
Hosts: Malus, Pyrus

Phomopsis archeri Sutton
Phomopsis pittospori S.A. Archer non (Cooke & Harkn.) Grove, hom. illeg.
Distr.: CA; Great Britain.
Notes: Sutton (The Coelomycetes. p. 571. 1980).
Host: Pittosporum

Phomopsis arecae Syd. & P. Syd.
Distr.: Temperate North America; The Philippines(type).
Hosts: Livistona, Trachycarpus

Phomopsis arnoldiae Sutton
Fusicoccum elaeagni J.C. Carter & Sacamano
Phomopsis elaeagni (J.C. Carter & Sacamano) R.H. Arnold & J.C. Carter
Distr.: Temperate North America; Great Britain.
Notes: Canker or dieback of Russian olive (Sutton. The Coelomycetes. p. 571. 1980).
Hosts: Elaeagnus, Juglans

Phomopsis asparagi (Sacc.) Bubák
Phoma asparagi Sacc.
Distr.: Eastern United States; China; Europe.
Host: Asparagus

Phomopsis asteriscus (Berk. & Broome) Grove
Phoma asteriscus Berk. & Broome
Distr.: CA(type); Great Britain.
Hosts: Conium, Heracleum

Phomopsis aucubae (Westend.) Traverso
Phoma aucubae Westend.
Distr.: NC; Europe(type).
Host: Aucuba

Phomopsis berberina (Sacc. & Roum.) Traverso & Spessa
Phoma berberina Sacc. & Roum.
Distr.: Temperate North America; Europe.
Host: Berberis

Phomopsis bertholletianum J.A. Spencer
Distr.: Reported from an unknown locality in the Untied States, probably imported; Brazil.
Notes: Decay of Brazil nuts.
Host: Bertholletia

Phomopsis buddleiae Grove
Distr.: NC; Great Britain(type).
Host: Buddleja

Phomopsis cacti (Berk.) Sacc.
Phoma cacti Berk.
Phyllosticta cacti (Berk.) W.A. Archer
Distr.: NM; Europe.
Host: Opuntia

Phomopsis callistephi Tehon & E.Y. Daniels
Distr.: HI, temperate North America.
Notes: Stem canker.
Host: Callistephus

Phomopsis cannabina Curzi
Distr.: IL; Italy(type).
Host: Cannabis

Phomopsis caraganae Bondartsev
Diaporthe caraganae Jacz. (teleomorph)
Distr.: OH; Soviet Union(type).
Host: Caragana

Phomopsis caricae-papayae Petr. & Cif.
Distr.: HI; Dominican Republic(type).
Notes: Punithalingam (C.M.I. Descr. 827. 1985).
Host: Carica

Phomopsis carposchiza Fairm.
Distr.: NY.
Notes: Common nut and pit fungus.
Host: Aesculus

Phomopsis cassiae (Sacc.) Ratiani non Camara, hom. illeg.

Phoma cassiae Sacc.
Distr.: Southeastern United States; Soviet Union.
Hosts: Cassia, Senna
Phomopsis cinerascens (Sacc.) Bubák in Bubák & Kab.
Phoma cinerascens Sacc.
Distr.: MD; Europe(type).
Notes: Boerema & Dorenbosch (Stud. Mycol. 3:8. 1973) suggested that this may be a synonym of *Phomopsis macrostoma* Mont.
Host: Ficus
Phomopsis citri H. Fawc. non (Sacc.) Traverso & Spessa, hom. illeg.
Diaporthe citri F.A. Wolf (teleomorph)
Distr.: Worldwide in citrus-growing regions.
Notes: Melanose and stem-end rot (Punithalingam & Holliday. C.M.I. Descr. 396. 1973).
Host: Citrus
Phomopsis coluteae (Sacc. & Roum.) Died.
Phoma coluteae Sacc. & Roum.
Distr.: NJ; France(type).
Host: Colutea
Phomopsis communis (Roberge) Höhn.
Phoma communis Roberge in Desmaz.
Distr.: AK, TX; Europe.
Hosts: Rhamnus, Tilia
Phomopsis conorum (Sacc.) Died.
Phoma conorum Sacc.
Distr.: Temperate North America; Europe.
Notes: On cone scales.
Hosts: Picea, Pinus
Phomopsis corni (Sacc.) Traverso
Phoma corni Sacc.
Distr.: Reported from an unknown locality in the United States; Europe.
Host: Cornus
Phomopsis cuscutae H.C. Greene
Distr.: WI(type).
Host: Cuscuta
Phomopsis depressa (Lév.) Traverso
Distr.: Temperate North America; Europe.
Host: Syringa
Phomopsis diachenii Sacc.
Cyphellopycnis pastinacae Tehon & G.L. Stout
Distr.: Temperate North America; Great Britain.
Notes: On stems, dried seed and fruit (Sutton. The Coelomycetes. p. 571. 1980).
Host: Pastinaca
Phomopsis diospyri (Sacc.) Traverso & Spessa
Phoma diospyri Sacc.
Distr.: FL; southern Europe.
Host: Diospyros
Phomopsis dulcamarina (Sacc.) Traverso in Trott.
Phoma dulcamarina Sacc.
Phoma dulcamarae Thuem. non Nitschke, hom. illeg.
Distr.: OR; Europe.
Host: Solanum
Phomopsis ericaceana Fairm.
Distr.: CA(type).
Host: Rhododendron
Phomopsis excelsior Tapy in H.C. Greene, nom. nud.
Distr.: WI.
Host: Quercus
Phomopsis fagopyri Traverso & Spessa
Distr.: DE; Portugal(type)
Host: Fagopyrum
Phomopsis filicina H.C. Greene
Distr.: WI(type).
Host: Athyrium
Phomopsis foveolaris (Fr.:Fr.) Traverso
Sphaeria foveolaris Fr.:Fr.
Distr.: Temperate North America; Europe.
Host: Euonymus
Phomopsis ganjae McPartland
Distr.: IL; India.
Notes: McPartland (Mycotaxon 18:527. 1983).
Host: Cannabis
Phomopsis gardeniae Buddin & Wakef.
Distr.: United States; Europe(type).
Notes: Stem canker of gardenia.
Host: Gardenia
Phomopsis glandicola (Lév.) Gonz. Frag.
Phoma glandicola Lév.
Distr.: Eastern North America; Europe.
Host: Quercus
Phomopsis hibisci Dearn.
Distr.: NY.
Host: Hibiscus
Phomopsis hieracii H.C. Greene
Distr.: WI.
Host: Hieracium
Phomopsis ilicis (Desmaz.) Höhn.
Phoma ilicis Desmaz.
Distr.: Eastern North America; Europe.
Notes: An ambiguous name (Van Der Aa. Stud. Mycol. 6:47. 1973) which has been confused with *Phyllosticta concentrica* Sacc.
Host: Ilex
Phomopsis impatientis Dearn. & House
Distr.: NY.
Host: Impatiens
Phomopsis imperialis (Sacc. & Roum.) Grove
Phoma imperialis Sacc. & Roum. in Sacc.
Distr.: NY; France; Soviet Union.
Host: Paulownia
Phomopsis ipomeae-batatas Punithalingam
Depazea batatas Thuem.
Phyllosticta batatas (Thuem.) Cooke
Distr.: HI, southcentral and southeastern United States; worldwide in warm sweet potato-growing regions.
Notes: Leaf blight of sweet potato. Punithalingam (C.M.I. Descr. 739. 1982).
Hosts: Calystegia, Ipomoea
Phomopsis japonica (Sacc.) Traverso
Phoma japonica Sacc.
Distr.: Temperate North America; Italy(type).
Host: Kerria
Phomopsis juglandina (Sacc.) Höhn.
Phoma juglandina Sacc.
Distr.: CA; Europe.
Host: Juglans
Phomopsis juniperovora Hahn
Distr.: Temperate North America.
Notes: Needle blight of Cupressaceae.
Hosts: Cedrus, Cephalotaxus, Chamaecyparis, Cupressus, Juniperus, Sequoiadendron, Taxus, Thuja
Phomopsis kalmiae Enlows
Diaporthe kalmiae F.A. Wolf & A.R. Cavaliere (teleomorph)
Distr.: Eastern United States.
Host: Kalmia
Phomopsis landeghemiae (Sacc.) Höhn.
Phoma landeghemiae Sacc.
Distr.: Eastern United States; Europe.
Host: Philadelphus
Phomopsis lavaterae (Westend.) Died.
Phoma lavaterae Westend.
Phomopsis malvacearum (Westend.) Died.
Distr.: MD; Europe.
Host: Hibiscus
Phomopsis lebiseyi (Sacc.) Died.
Cryptodiaporthe lebiseyi (Desmaz.) Wehmeyer (teleomorph)
Distr.: IA; Europe.
Notes: Reports from North America are in need of verification.
Host: Acer
Phomopsis leptostromiformis (Kühn) Bubák in Kab. & Bubák
Cryptosporium leptostromiforme Kühn
Distr.: Southern United States; Europe.
Notes: Stem rot of yellow lupine (Punithalingam & Gibson. C.M.I. Descr. 476. 1975 as *Diaporthe woodii*).
Host: Lupinus
Phomopsis liatridis H.C. Greene
Distr.: WI(type).
Host: Liatris

Phomopsis ligustri-vulgaris Petr.
Distr.: PA; Soviet Union(type).
Host: Ligustrum
Phomopsis lirella (Desmaz.) Grove
Phoma lirella Desmaz.
Distr.: Temperate North America; Europe.
Host: Vinca
Phomopsis lokoyae Hahn
Diaporthe lokoyae Funk (teleomorph)
Distr.: Western North America.
Notes: Dieback of douglas fir (Funk. Canad. J. Bot. 46:601. 1968).
Host: Pseudotsuga
Phomopsis longicolla T.W. Hobbs
Distr.: Central and eastern United States.
Notes: Seed decay of soybeans. Hobbs et al. (Mycologia 77:535. 1985).
Host: Glycine
Phomopsis macrospora T. Kobayashi & Chiba
Diaporthe santonensis Sacc. (teleomorph)
Distr.: MS; Japan.
Notes: Cottonwood canker.
Host: Populus
Phomopsis magnoliicola (Syd. & P. Syd.) Died.
Phoma magnoliicola Syd. & P. Syd.
Distr.: FL; Europe.
Host: Magnolia
Phomopsis majuscula Sacc.
Distr.: NY.
Host: Campsis
Phomopsis mali Roberts non (Schulzer & Sacc.) Died., hom. illeg.
Diaporthe perniciosa Ém. Marchal (teleomorph)
Distr.: Temperate North America.
Notes: Apple and pear canker. Rosenberger & Burr (Pl. Dis. 66:1073. 1982).
Hosts: Malus, Prunus, Pyrus
Phomopsis menispermi (Peck) Grove
Phoma menispermi Peck
Distr.: KS; Great Britain(Kew).
Host: Menispermum
Phomopsis missouriensis Bubák
Diaporthe asclepiadis Ellis & Everh. (teleomorph)
Distr.: MO.
Host: Asclepias
Phomopsis montanensis Hahn
Distr.: ID, MT.
Notes: Known only from two specimens (Hahn. Trans. Brit. Mycol. Soc. 15:32. 1930).
Host: Abies
Phomopsis moricola (Sacc.) Sacc.
Phoma moricola Sacc.
Distr.: WI; Italy(type).
Notes: The status of this taxon is uncertain (Wehmeyer. Univ. Michigan Stud., Sci. Ser. 9:308. 1933).
Host: Morus
Phomopsis nicotianae Fairm. in Millsp. & L. Nutt.
Distr.: CA.
Host: Nicotiana
Phomopsis oblita Sacc.
Distr.: CA; Europe.
Host: Artemisia
Phomopsis oblonga (Desmaz.) Traverso
Phoma oblongum Desmaz.
Distr.: Eastern United States; Europe.
Notes: Elm canker. Webber & Gibbs (Trans. Brit. Mycol. Soc. 82:348. 1984).
Host: Ulmus
Phomopsis obscurans (Ellis & Everh.) Sutton
Phoma obscurans Ellis & Everh.
Dendrophoma obscurans (Ellis & Everh.) H.W. Anderson
Phyllosticta obscurans (Ellis & Everh.) Tassi
Sphaeronaemella fragariae F. Stevens & A. Peterson
Distr.: Cosmopolitan.
Notes: Lesions. The synonymy of *Sphaeronaemella fragariae* was presented in Maas (Compendium of strawberry diseases p. 76. 1984). Sutton & Waterston (C.M.I. Descr. 227. 1970). Sutton (The Coelomycetes. p. 571. 1980).
Host: Fragaria
Phomopsis occulta (Sacc.) Traverso
Phoma occulta Sacc. non Desmaz., hom. illeg.
Diaporthe occulta (Fuckel) Nitschke (teleomorph)
Distr.: Circumboreal.
Notes: Saprobic on conifers.
Hosts: Abies, Cupressus, Juniperus, Larix, Picea, Pseudotsuga, Sequoia, Sequoiadendron, Taxodium, Taxus, Thuja, Thujopsis, Tsuga
Phomopsis oenotherae (Dearn.) Petr.
Myxosporium oenotherae Dearn.
Distr.: NY.
Host: Oenothera
Phomopsis oncostoma (Thuem.) Traverso
Phoma oncostoma Thuem.
Distr.: NJ; Europe.
Host: Robinia
Phomopsis ophites (Sacc.) Traverso
Phoma ophites Sacc.
Distr.: PA; Europe.
Host: Hibiscus
Phomopsis oxalina (Ellis & Everh.) Syd.
Phyllachora oxalina Ellis & Everh.
Distr.: Eastern United States.
Host: [Oxalis]
Phomopsis oxyspora Petr.
Diaporthe oxyspora (Peck) Sacc. (teleomorph)
Distr.: MD.
Host: Ilex
Phomopsis padina (Sacc. & Roum.) Died.
Phoma padina Sacc. & Roum.
Distr.: United States; Europe.
Notes: Cherry canker.
Host: Prunus
Phomopsis pallida (Fuckel) Sacc. & D. Sacc.
Libertella pallida Fuckel
Distr.: WA; Austria.
Host: Salix
Phomopsis palmicola (G. Wint.) Sacc.
Phoma palmicola G. Wint.
Distr.: Subtropical and tropical regions.
Hosts: Livistona, Washingtonia
Phomopsis phaseoli (Desmaz.) Sacc.
Phomopsis batatae Harter & E.C. Field
Phomopsis glycines Petr.
Phomopsis sojae S.G. Lehman
Phoma subcircinata Ellis & Everh.
Diaporthe phaseolorum (Cooke & Ellis) Sacc. (teleomorph)
Distr.: Cosmopolitan.
Notes: Stem canker and blight, pod rot of soybeans and other legumes (Punithalingam & Holliday. C.M.I. Descr. 336. 1972; Kulik. Mycologia 76:274. 1984).
Hosts: Abutilon, Glycine, Phaseolus, Strophostyles
Phomopsis phoenicola Traverso & Spessa
Distr.: CA; Portugal(Bot. Garden).
Host: Phoenix
Phomopsis phytolaccae (Berk. & M.A. Curtis) Grove
Distr.: Eastern United States; Pakistan.
Host: Phytolacca
Phomopsis pimpinellae (Ellis) H.C. Greene
Septoria pimpinellae Ellis
Rhabdospora pimpinellae (Ellis) Kuntze
Distr.: Northcentral United States.
Hosts: Polytaenia, Taenidia
Phomopsis pittospori (Cooke & Harkn.) Grove
Phoma pittospori Cooke & Harkn.
Distr.: CA.
Host: Pittosporum
Phomopsis pritchardiae (Cooke & Harkn.) Sacc.
Phoma pritchardiae Cooke & Harkn.
Macrophoma pritchardiae (Cooke & Harkn.) Sacc.
Distr.: CA, WA.
Notes: Sutton (The Coelomycetes. p. 569. 1980).
Hosts: Pritchardia, Washingtonia
Phomopsis prunorum (Cooke) Grove
Phoma prunorum Cooke

Distr.: WA; Great Britain.
Host: Prunus
Phomopsis psidii Nag Raj & Ponnappa
Distr.: HI; India.
Host: Psidium
Phomopsis quercina Höhn.
Distr.: OK, VA: Italy.
Notes: According to Höhnel the basionym is *Cytispora (Libertella) quercina* Sacc., but according to Saccardo the correct basionym is *Cytispora (Libertella) quercina* Westendorp. Both are later homonyms of *Cytispora quercina* Schwein.
Host: Quercus
Phomopsis ricini Grove
Diaporthe ricini Speg. (teleomorph)
Distr.: SC.
Host: Ricinus
Phomopsis rosae (Schulzer & Sacc.) Traverso & Spessa
Phoma rosae Schulzer & Sacc.
Distr.: CA; Yugoslavia.
Host: Rosa
Phomopsis rubiseda Fairm.
Distr.: NY.
Host: Rubus
Phomopsis rudis (Fr.) Höhn.
Diaporthe rudis (Fr.:Fr.) Nitschke (teleomorph)
Distr.: OH; Europe; Great Britain.
Notes: Wehmeyer (Univ. Michigan Stud., Sci. Ser. 9:104. 1933).
Hosts: Caragana, Genista, Laburnum
Phomopsis salicina (Westend.) Died.
Phoma salicina Westend.
Distr.: Temperate North America; Europe.
Host: Salix
Phomopsis sambucina (Sacc.) Traverso
Phoma sambucina Sacc.
Distr.: Reported from an unknown locality in the United States; Europe; Great Britain.
Host: Sambucus
Phomopsis sarothamni (Sacc.) Höhn.
Phoma sarothamni Sacc.
Distr.: CA; Europe.
Host: Spartium
Phomopsis scabra (Sacc.) Traverso
Phoma scabra Sacc.
Distr.: United States; Europe.
Host: Platanus
Phomopsis seposita (Sacc.) Traverso
Phoma seposita Sacc.
Distr.: CA, MA; Europe.
Host: Wisteria
Phomopsis spiraeae (Desmaz.) Grove
Phoma spiraeae Desmaz.
Distr.: MD; Europe.
Host: Spiraea
Phomopsis stictica (Berk. & Broome) Traverso
Phoma stictum Berk. & Broome
Distr.: MD; Europe.
Host: Buxus
Phomopsis strobi Syd. in Syd. & Petr.
Distr.: ME.
Host: Pinus
Phomopsis subordinaria (Desmaz.) Traverso
Phoma subordinaria Desmaz.
Distr.: United States; Europe.
Host: Plantago
Phomopsis thalictri H.C. Greene
Distr.: WI(type).
Host: Thalictrum
Phomopsis thalictrina (Sacc. & Malbr.) Died.
Phoma thalictrina Sacc. & Malbr.
Distr.: NM; Europe.
Host: Thalictrum
Phomopsis trollii Fairm.
Distr.: NY.
Host: Trollius
Phomopsis vaccinii Shear in Shear, N. Stevens, & H. Bain
Diaporthe vaccinii Shear in Shear, N. Stevens, & H. Bain (teleomorph)
Distr.: Temperate North America.
Notes: Fruit rot and twig blight. Milholland & Daykin (Pl. Dis. 67:325. 1983).
Host: Vaccinium
Phomopsis venenosa (Sacc.) Traverso & Spessa
Phoma venensoa Sacc.
Distr.: Eastern United States; Europe.
Host: Datura
Phomopsis vexans (Sacc. & Syd.) Harter
Distr.: North America; Africa; Asia.
Notes: Blight and fruit rot of eggplant (Punithalingam & Holliday. C.M.I. Descr. 338. 1972). The name *Diaporthe vexans* has been applied to the teleomorph but was invalidly published.
Host: Solanum
Phomopsis viridarii (Sacc.) Traverso & Spessa
Phoma viridarii Sacc.
Distr.: Reported from an unknown locality in the United States; Europe.
Host: Magnolia
Phomopsis viticola (Sacc.) Sacc.
Fusicoccum viticola Reddick
Phomopsis viticola (Sacc.) Sacc. var. *ampelopsidis* Grove
Distr.: Temperate regions.
Notes: Dead arm or necrosis of grape (Punithalingam. C.M.I. Descr. 635. 1979).
Hosts: Parthenocissus, Vitis

Phorcys Niessl. Pyrenomycetes, Amphisphaeriales

Notes: This genus is considered a synonym of *Amphisphaeria*, but not all names have been redisposed.
Phorcys seriata (Cooke) J.H. Miller, unpublished name
Distr.: GA(type).
Notes: This name is apparently unpublished.
Host: Cydonia
Phorcys tiliae (M.A. Curtis) J. Schröt.
Distr.: GA.
Host: Tilia

Phragmidium Link. Basidiomycetes, Uredinales

Notes: Hiratsuka et al. (Rep. Tottori Mycol. Inst. 18:53. 1980); Peterson (Mycologia 54:389. 1962).
Phragmidium alaskanum (Arth.) P. Syd. & Syd.
Distr.: AK.
Notes: Autoecious.
Host: Rubus
Phragmidium americanum (Peck) Dietel
Distr.: Eastern North America; Japan.
Notes: Autoecious.
Host: Rosa
Phragmidium andersonii Shear
Distr.: Northern North America; Asia; Europe.
Notes: Autoecious.
Hosts: [Cowania], Potentilla
Phragmidium biloculare Dietel & Holw.
Distr.: Western North America.
Notes: Autoecious.
Host: Potentilla
Phragmidium boreale Tranzschel
Phragmidium ivesiae Syd. & Syd. subsp. *wyomingense* R.S. Peterson
Distr.: Western United States.
Notes: Autoecious. Savile (Canad. J. Bot. 54:1693. 1976).
Host: Potentilla
Phragmidium fusiforme J. Schröt.
Phragmidium rosae-acicularis Liro
Distr.: Temperate northern hemisphere.
Notes: Autoecious. Savile (Fungi Canadenses 54. 1974) segregated populations in northwestern North America as var. *novi-boreale* Savile.

Host: Rosa
Phragmidium fusiforme J. Schröt. var. **novi-boreale** Savile
Distr.: Arctic region.
Notes: Autoecious. Savile (Fungi Canadenses 54. 1974).
Host: Rosa
Phragmidium horkeliae Garrett
Distr.: UT.
Notes: Autoecious.
Host: Ivesia
Phragmidium ivesiae Syd. & P. Syd.
Uredo acaenae Ellis & Everh. (anamorph)
Distr.: Central and western North America.
Notes: Autoecious. Savile (Canad. J. Bot. 54:1690. 1976).
Hosts: Acaena, Horkelia, Ivesia, Potentilla
Phragmidium jonesii Dietel
Distr.: Western United States.
Notes: Autoecious.
Hosts: Ivesia, Sanguisorba
Phragmidium montivagum Arth.
Distr.: Temperate North America and Asia.
Notes: Autoecious.
Host: Rosa
Phragmidium mucronatum (Pers.:Pers.) Schlechtend.
Phragmidium disciflorum (Tode) J. James
Phragmidium subcorticium G. Wint.
Distr.: Cosmopolitan.
Notes: Autoecious. Some records may be misidentifications of *Phragmidium rosae-pimpinellifoliae*. Laundon & Rainbow (C.M.I. Descr. 204. 1969); Bedlan (Pflanzenschutzberichte 46:33. 1984).
Host: Rosa
Phragmidium occidentale Arth. in H.C. Greene
Phragmidium rubi-odorati Dietel
Distr.: North America.
Notes: Autoecious. Savile (Fungi Canadenses 80. 1975).
Host: Rubus
Phragmidium peckianum Arth.
Distr.: Western United States.
Notes: On Rosaceae.
Host: Rubus
Phragmidium potentillae (Pers.:Pers.) P. Karst.
Frommea obtusa (F. Strauss) Arth.
Distr.: Circumboreal.
Notes: Autoecious. Savile (Fungi Canadenses 41. 1974); Bedlan (Pflanzenschutzberichte 46:33. 1984).
Hosts: [Duchesnea], [Fragaria], Potentilla
Phragmidium rosae-arkansanae Dietel
Distr.: Central North America.
Notes: Autoecious.
Host: Rosa
Phragmidium rosae-californicae Dietel
Distr.: Northwestern North America.
Notes: Autoecious.
Host: Rosa
Phragmidium rosae-pimpinellifoliae Dietel
Distr.: North America; temperate Europe.
Notes: Autoecious. Laundon & Rainbow (C.M.I. Descr. 205. 1969).
Host: Rosa
Phragmidium rosicola (Ellis & Everh.) Arth.
Distr.: Central North America.
Notes: Autoecious.
Host: Rosa
Phragmidium rubi-idaei (DC.) P. Karst.
Phragmidium gracile Arth.
Phragmidium imitans Arth.
Distr.: Temperate regions.
Notes: Autoecious. Laundon & Rainbow (C.M.I. Descr. 207. 1969).
Host: Rubus
Phragmidium speciosum (Fr.) Cooke
Aregma speciosa Fr.
Earlea speciosa Arth.
Distr.: North America.
Notes: Autoecious.
Host: Rosa

Phragmocapnias Theiss. & Syd. Loculoascomycetes, Dothideales

Notes: This genus is considered to be a synonym of *Capnodium*, but not all names have been redisposed.
Phragmocapnias smilacina Mendoza
Distr.: HI.
Hosts: Cyrtandra, Kadua, Pelea, Smilax

Phragmodiaporthe Wehmeyer. Pyrenomycetes, Diaporthales

Notes: Barr (Mycol. Mem. 7:154. 1978).
Phragmodiaporthe caryae (Peck) Wehmeyer
Cryptospora caryae Peck
Valsa apatela Ellis & Holw.
Pseudovalsa fairmani Ellis & Everh.
Distr.: Eastern United States.
Host: Carya
Phragmodiaporthe tiliacea (Ellis) Barr
Diatrype tiliacea Ellis
Diaporthe tiliacea (Ellis) Höhn.
Melanconis tiliacea (Ellis) Ellis & Everh.
Diaporthe atropuncta Peck
Distr.: Eastern North America.
Hosts: Alnus, Tilia

Phragmodothidea Dearn. & Barth. Loculoascomycetes, Dothideales

Notes: This genus is considered to be a synonym of *Metameris*, but not all names have been redisposed.
Phragmodothidea eucalypti Dearn. & Barth.
Distr.: CA(type).
Host: Eucalyptus

Phragmodothis Theiss. & Syd. Loculoascomycetes, Dothideales

Phragmodothis berberidis (J.V. Almeida & Camara) Theiss. & Syd.
Montagnella berberidis J.V. Almeida & Camara
Distr.: CA; Portugal(type).
Notes: On dead stems of Berberidaceae.
Host: Vancouveria
Phragmodothis circumscissa (Tracy & Earle) Theiss. & Syd.
Plowrightia circumscissa Tracy & Earle
Distr.: CA.
Notes: According to Barr (Contr. Univ. Michigan Herb. 9:610. 1972), this may be a synonym of *Dothidea conspicua*.
Host: Yucca

Phragmopeltis Henn. Coelomycetes

Phragmopeltis phragmitis Dearn.
Distr.: WA.
Host: Phragmites

Phragmopyxis Dietel in Engl. & Prantl. Basidiomycetes, Uredinales

Phragmopyxis acuminata (W.H. Long) P. Syd. & Syd.
Distr.: AZ; probably northern Mexico.
Notes: Autoecious.
Host: Coursetia
Phragmopyxis deglubens (Berk. & M.A. Curtis) Dietel in Engl. & Prantl
Distr.: AZ to Ecuador.
Notes: Autoecious.
Host: Cracca
Phragmopyxis noelii J.W. Baxter
Distr.: AZ; Mexico.
Notes: Autoecious.
Host: Coursetia

Phragmotrichum Kunze in Kunze & J.C. Schmidt. Coelomycetes

Notes: Sutton (The Coelomycetes. p. 35. 1980); Sutton & Sandhu (Trans. Brit. Mycol. Soc. 52:67. 1969).
Phragmotrichum chailletii Kunze in Kunze & Schumach.
Distr.: NC, NY; Europe.
Host: Picea
Phragmotrichum pini (W.B. Cooke) Sutton & Sandhu
Trimmatostroma pini W.B. Cooke
Distr.: WA.
Host: Pinus

Phycomyces Kunze. Zygomycetes, Mucorales

Phycomyces nitens Kunze
Distr.: Widespread.
Notes: Isolated from soil, dung and various organic substrates.
Host: Zea

Phyllachora Nitschke ex Fuckel, nom. cons. Pyrenomycetes, Phyllachorales

Notes: Orton (Mycologia 36:18. 1944); Parbery (Austral. J. Bot. 15:271. 1967; Austral. J. Bot. 19:207. 1971).
Phyllachora acaciae Henn.
Phyllachora texana Tharp
Distr.: Southern United States; Central America; South America.
Host: Acacia
Phyllachora ambrosiae (Berk. & M.A. Curtis) Sacc.
Dothidea ambrosiae Berk. & M.A. Curtis
Physalospora arthuriana Sacc.
Physalospora lepachydis Ellis & Everh.
Distr.: PR; temperate North America; Colombia.
Hosts: Ambrosia, Baccharis, Helianthus, Iva, Ratibida
Phyllachora americana D.G. Parbery
Phyllachora andropogonis (Schwein.) Ellis & Everh.
Phyllachora nervisequia (Schwein.) Orton
Sphaeria nervisequia Schwein.
Distr.: Temperate North America; West Indies; possibly South America.
Host: Andropogon
Phyllachora ammophilae Orton
Phyllachora calamagrostis Sawada
Distr.: Eastern United States; Uruguay; Taiwan.
Host: Ammophila
Phyllachora asterigena Ellis & Everh.
Distr.: KS, NE.
Host: Aster
Phyllachora beaumontii (Berk. & M.A. Curtis) Cooke
Dothidea beaumontii Berk. & M.A. Curtis
Distr.: AL(type).
Host: Prunus
Phyllachora bonariensis Speg.
Phyllachora congruens Rehm
Distr.: Widespread.
Host: Panicum
Phyllachora canafistulae F. Stevens & N. Dalbey
Distr.: Southern United States; West Indies; Central America.
Host: Cassia
Phyllachora canaliculata (Schwein.) Sacc.
Dothidea canaliculata (Schwein.) Berk.
Distr.: Temperate North America.
Host: Cyperus
Phyllachora caricis (Fr.) Sacc.
Dothidea caricis Fr.
Distr.: Boreal regions.
Host: Carex
Phyllachora catesbyana Chárdon
Distr.: FL.
Host: Ocotea
Phyllachora cibotii C.S. Hodges
Distr.: HI.
Host: Cibotium
Phyllachora coloradensis Orton
Distr.: CO(type).
Host: Muhlenbergia
Phyllachora cynodontis (Sacc.) Niessl
Phyllachora graminis (Pers.:Fr.) Fuckel var. *cynodontis* Sacc.
Physalospora cynodontis Delacr.
Phyllachora boutelouae Rehm
Phyllachora serialis Ellis & Everh.
Distr.: Widespread.
Hosts: Bouteloua, Buchloe, Chloris, Distichlis, Elymus, Festuca, Spartina
Phyllachora cyperi Rehm
Distr.: Southern United States; Mexico; West Indies.
Host: Cyperus
Phyllachora dalibardae (Peck) Sacc.
Dothidea dalibardae Peck
Distr.: Northeastern United States.
Host: Dalibarda
Phyllachora dimeriae Theiss. & Syd.
Phyllachora erianthi Orton
Distr.: Southeastern United States; Cyprus; Japan.
Hosts: Erianthus, Saccharum
Phyllachora domingensis Chárdon
Distr.: FL; Dominican Republic.
Host: Picramnia
Phyllachora epicampis Orton
Distr.: Southwestern United States; Central America.
Host: Muhlenbergia
Phyllachora eragrostidis Chárdon
Distr.: PR, southern United States; Mexico.
Hosts: Eragrostis, Tridens
Phyllachora eucalypti (Cooke & Massee) Theiss. & Syd.
Physalospora latitans Sacc.
Distr.: Cosmopolitan.
Host: Eucalyptus
Phyllachora ficuum Niessl in Rabenh.
Trabutia ficuum (Niessl) Theiss. & Syd.
Distr.: FL; India; The Philippines.
Notes: Arx & Müller (Beitr. Kryptogamenfl. Schweiz 11(1):219. 1954).
Host: Ficus
Phyllachora filicina Sacc. & Scalia
Trabutiella filicina (Sacc. & Scalia) Theiss. & Syd.
Distr.: Northwestern North America.
Notes: Tar spot (Savile. Mycologia 51:296. 1959).
Host: Polystichum
Phyllachora flabella (Schwein.) Thuem.
Sphaeria flabella Schwein.
Catacauma flabellum (Schwein.) Theiss. & Syd.
Distr.: WA, eastern United States.
Host: Pteridium
Phyllachora freycinetiae F. Stevens
Distr.: HI.
Host: Freycinetia
Phyllachora fusicarpa Seaver in Britton & Millsp.
Distr.: FL; Mexico; West Indies; Venezuela.
Notes: Tar spot and defoliation.
Host: Duranta
Phyllachora graminis (Pers.:Fr.) Nitschke
Linochora graminis (Grove) D.G. Parbery (anamorph)
Distr.: Temperate regions.
Notes: Black leaf spot or tar spot of grasses.
Hosts: Agrohordeum, Agropyron, Agrostis, Ammophila, Andropogon, Avena, Brachyelytrum, Bromus, Calamagrostis, Cinna, Danthonia, Digitaria, Elymus, Eragrostis, Festuca, Heteropogon, Hordeum, Hystrix, Lasiacis, Panicum, Pappophorum, Paspalum, Phleum, Saccharum, Schizachne, Spartina, Sporobolus, Triticum, Uniola
Phyllachora halstediana Petr.
Halstedia portoricensis F. Stevens
Distr.: FL, PR.
Notes: Petrak (Ann. Mycol. 32:318. 1934).
Host: Sideroxylon
Phyllachora heraclei (Fr.) Fuckel
Distr.: Temperate northern hemisphere.
Host: Heracleum
Phyllachora junci (Albertini & Schwein.:Fr.) Fuckel
Dothidella junci (Albertini & Schwein.:Fr.) Sacc.

Endodothella junci (Albertini & Schwein.:Fr.) Theiss. & Syd.
Distr.: Temperate northern hemisphere.
Notes: On leaves and stems.
Host: Juncus
Phyllachora kalmiae (Peck) Petr.
Dothidella kalmiae (Peck) Sacc.
Distr.: Temperate North America.
Notes: On twigs.
Host: Kalmia
Phyllachora lespedezae (Schwein.) Cooke
Sphaeria lespedezae Schwein.
Distr.: Central and eastern United States; Canada(Ontario).
Hosts: Kummerowia, Lespedeza
Phyllachora luteo-maculata (Schwein.) Orton
Phyllachora brevifolia Chárdon
Distr.: North and South America.
Notes: On living leaves.
Hosts: Andropogon, Bothriochloa, Schizachyrium, Sorghastrum
Phyllachora melanoplaca (Desmaz.) Sacc.
Distr.: Temperate northern hemisphere.
Host: Veratrum
Phyllachora minutissima (Welw. & Curr.) G.W. Sm.
Isothea minutissima Welw. & Curr.
Phyllachora cornispora Atk.
Distr.: Subtropical and tropical regions.
Hosts: Panicum, Paspalum
Phyllachora monanthochloes Petr.
Distr.: CA.
Notes: According to Parbery (Austral. J. Bot. 15:271. 1967), this is a doubtful species.
Host: Monanthochloe
Phyllachora orbiculata (Schwein.) Sacc.
Sphaeria orbiculata Schwein.
Dothidea orbiculata (Schwein.) Fr.
Distr.: SC, TN.
Notes: Rare.
Host: Diospyros
Phyllachora oryzopsidis (Rehm) Theiss. & Syd.
Distr.: Widespread.
Host: Oryzopsis
Phyllachora palmicola (F. Stevens) J.A. Stevenson
Catacauma palmicola F. Stevens
Distr.: FL, PR.
Host: Arecaceae
Phyllachora paspalicola Henn.
Phyllachora chardonii Orton
Phyllachora insularis Chárdon
Phyllachora wilsonii Orton
Distr.: Widespread.
Hosts: Digitaria, Muhlenbergia, Panicum, Paspalum
Phyllachora perseae C.S. Hodges
Distr.: Southeastern United States.
Host: Persea
Phyllachora phalaridis Orton
Distr.: Eastern United States.
Host: Phalaris
Phyllachora picea (Berk. & M.A. Curtis) Sacc.
Dothidea picea Berk. & M.A. Curtis
Distr.: NY.
Host: Vitis
Phyllachora polemonii Harkn.
Distr.: CO.
Host: Polemonium
Phyllachora polygalae (Schwein.) Sacc.
Dothidea polygalae Schwein.
Distr.: FL(type).
Host: Polygala
Phyllachora punctum (Schwein.) Orton & F. Stevens
Sphaeria punctum Schwein.
Phyllachora guianensis F. Stevens
Phyllachora panici Sacc.
Distr.: Widespread.
Hosts: Dichanthelium, Digitaria, Oplismenus, Panicum, Paspalum
Phyllachora quadraspora Tehon
Distr.: Subtropical and tropical regions.
Hosts: Andropogon, Rottboellia
Phyllachora rosae (Schwein.) Sacc.
Distr.: MS.
Host: Rosa
Phyllachora scleriicola L.E. Miles
Distr.: MS(type).
Host: Scleria
Phyllachora shiraiana Syd.
Phyllachora arundinariae Orton
Distr.: Widespread.
Host: Arundinaria
Phyllachora simillima (Berk. & Ravenel) Sacc.
Dothidea simillima Berk. & Ravenel
Distr.: SC(type).
Host: Desmodium
Phyllachora simplex Starb.
Distr.: FL; West Indies; Paraguay.
Notes: Rare.
Host: Coccoloba
Phyllachora solidaginum (Schwein.) Sacc.
Sphaeria solidaginum Schwein.
Distr.: United States.
Hosts: Euthamia, Solidago
Phyllachora spartinae Orton
Distr.: Eastern United States.
Host: Spartina
Phyllachora sphaerosperma G. Wint.
Sphaerodothis sphaerosperma (G. Wint.) F. Stevens & W.D. Moore
Phyllachora cenchricola Speg.
Phyllachora sphaerospora Pat.
Distr.: Subtropical and tropical regions.
Notes: The type locality, erroneously cited as San Francisco, California, is actually San Francisco, Brazil.
Host: Cenchrus
Phyllachora subcuticularis (Schwein.) Sacc.
Dothidea subcuticularis Schwein.
Distr.: NY, PA.
Host: Cornus
Phyllachora swieteniae Petr. & Cif.
Distr.: FL; Dominican Republic.
Notes: Mahogany tar spot.
Host: Swietenia
Phyllachora sylvatica Sacc. & Speg.
Homostegia diplocarpa Ellis & Everh.
Phyllachora diplocarpa (Ellis & Everh.) Ellis & Everh.
Phyllachora nuttalliana Fairm.
Phyllachora pammelii Orton
Phyllachora phari-latifolia Chárdon
Dothidella tracyi (Ellis & Everh.) Sacc.
Endodothella tracyi (Ellis & Everh.) Theiss. & Syd.
Phyllachora tracyi Ellis & Everh.
Distr.: Temperate regions; Dominican Republic; Peru.
Notes: Black leaf spot.
Hosts: Distichlis, Festuca, Spartina, Vulpia
Phyllachora texensis Orton
Distr.: TX(type).
Host: Muhlenbergia
Phyllachora urbaniana Allesch. & Henn.
Catacauma urbanianum (Allesch. & Henn.) Theiss. & Syd.
Distr.: FL; Brazil.
Notes: On Myrtaceae.
Host: Calyptranthes
Phyllachora viequesensis Orton & Toro
Distr.: TX; Dominican Republic.
Notes: On living leaves.
Host: Cyperus
Phyllachora vulgata Theiss. & Syd.
Distr.: North America; Mexico; Uruguay.
Hosts: Blepharoneuron, Muhlenbergia, Sporobolus

Phyllactinia Lév. Pyrenomycetes, Erysiphales

Phyllactinia angulata (E.S. Salmon) S. Blumer

Phyllactinia corylea (Pers.) P. Karst. var. *angulata* E.S. Salmon
Distr.: Temperate regions.
Notes: On various trees (Braun. Mycotaxon 22:93. 1985).
Host: Ulmus
Phyllactinia eleagni Linder
Distr.: WY(type).
Host: Elaeagnus
Phyllactinia guttata (Wallr.:Fr.) Lév.
Erysiphe betulae DC.
Phyllactinia corylea (Pers.) P. Karst.
Erysiphe liriodendri Schwein.
Phyllactinia suffulta (Rebent.) Sacc.
Phyllactinia suffulta (Rebent.) Sacc. var. *macrospora* Atk.
Oidium sp. (anamorph)
Distr.: Temperate regions.
Notes: Powdery mildew of hardwoods, especially Betulaceae. At least six *formae speciales* described.
Hosts: [Acer], Aesculus, Alnus, Amelanchier, Aralia, Asclepias, Berberis, Betula, Calycanthus, Carpinus, Carya, Castanea, Catalpa, Celastrus, Cephalanthus, Chelone, Chionanthus, Cirsium, Cladrastis, Cornus, Corylus, Crataegus, Cydonia, Desmodium, Dipsacus, Elaeagnus, Elymus, Epilobium, Erigeron, Fagus, Fouquieria, Fragaria, Fraxinus, Hamamelis, Heuchera, Holodiscus, Ilex, Juglans, Lactuca, Lagerstroemia, Liriodendron, Magnolia, Malus, Melia, Morus, Oemleria, Ostrya, Philadelphus, Philibertia, Physocarpus, Plantago, Platanus, Polygonum, Populus, Prunus, Quercus, Rhus, Ribes, Robinia, Rosa, Rubus, Rudbeckia, Rumex, Salix, Sambucus, Sarcostemma, Sassafras, Shepherdia, Smilax, Solidago, Sorbus, Taraxacum, Tilia, Ulmus, Vaccinium, Zanthoxylum

Phylleutypa Petr. Loculoascomycetes, Dothideales

Phylleutypa wittrockii (Eriks.) Petr.
Dothidea wittrockii Eriks.
Phyllachora wittrockii (Eriks.) Sacc.
Distr.: Temperate regions.
Host: Linnaea

Phylloporia Murrill. Basidiomycetes, Aphyllophorales

Phylloporia frutica (Berk. & M.A. Curtis) Ryvarden
Polyporus fruticum Berk. & M.A. Curtis
Inonotus amplectens Murrill
Polyporus amplectens (Murrill) Sacc. & Trott.
Distr.: Southeastern United States; neotropical regions; Africa.
Notes: On wood.
Hosts: Asimina, Citrus
Phylloporia ribis (Schumach.:Fr.) Ryvarden
Fomes ribis (Schumach.:Fr.) Gill.
Phellinus ribis (Schumach.:Fr.) Quél.
Polyporus ribis (Schumach.:Fr.) Fr.
Fomes langloisii (Murrill) Sacc. & D. Sacc.
Distr.: Temperate northern hemisphere.
Notes: At bases of stems of living shrubs.
Hosts: Crataegus, Lonicera, Ribes, Sassafras, Symphoricarpos

Phyllosticta Pers., nom. cons. Coelomycetes

Phyllostictina Syd.
Phyllosticta acanthospermi Ellis & Everh.
Distr.: FL.
Host: Acanthospermum
Phyllosticta acetosellae A.L. Sm. & Ramsb.
Distr.: WI; Europe.
Host: Rumex
Phyllosticta adoxae Clements in Seaver
Distr.: CO.
Host: Adoxa
Phyllosticta agrifolia Ellis & Everh.
Distr.: CA.
Host: Quercus
Phyllosticta ailanthi Sacc.
Distr.: VA; Europe.
Host: Ailanthus
Phyllosticta alcides Sacc.
Distr.: Temperate North America; Europe.
Host: Populus
Phyllosticta allegheniensis Tehon & G.L. Stout
Distr.: Eastern United States.
Host: Rubus
Phyllosticta alliariifoliae Allesch.
Distr.: North America, scattered reports; Europe.
Host: Campanula
Phyllosticta allii Tehon & E.Y. Daniels
Distr.: United States.
Notes: Leaf blight.
Host: Allium
Phyllosticta alnea Oudem.
Distr.: AK; Europe.
Host: Alnus
Phyllosticta alpinicola Solheim
Distr.: WY.
Host: Trifolium
Phyllosticta amaranthi Ellis & Kellerm.
Distr.: Central and eastern North America; Europe.
Hosts: Alternanthera, Amaranthus
Phyllosticta ambrosiae J.J. Davis
Distr.: CA, WI.
Hosts: Ambrosia, Balsamorhiza, Wyethia
Phyllosticta ambrosioides Thuem.
Distr.: Temperate North America; Europe.
Host: Chenopodium
Phyllosticta amicta Ellis & Everh.
Distr.: Western United States.
Host: Arctostaphylos
Phyllosticta ampelicida (Engelm.) van der Aa
Phyllosticta ampelopsidis Ellis & G. Martin
Phyllosticta labruscae Thuem.
Phyllosticta viticola Thuem.
Guignardia bidwellii (Ellis) Viala & Ravaz (teleomorph)
Leptodothiorella sp. (synanamorph)
Distr.: Temperate North America; Europe.
Notes: Black rot of Vitaceae.
Hosts: Ampelopsis, Parthenocissus, Vitis
Phyllosticta anacardiacearum van der Aa
Phyllostictina mangiferae Batista & Vital
Guignardia mangiferae A.J. Roy (teleomorph)
Distr.: Subtropical and tropical regions.
Host: Mangifera
Phyllosticta andromedae Westend.
Distr.: NJ; Europe.
Host: Pieris
Phyllosticta andropogonivora R. Sprague & C.T. Rogerson
Distr.: KS.
Notes: Sprague & Rogerson (Mycologia 50:639. 1958).
Host: Andropogon
Phyllosticta anemonicola Sacc. & Syd. in Sacc.
Phyllosticta anemones Ellis & Everh. non Fuckel, hom. illeg.
Distr.: Central and eastern United States.
Host: Anemone
Phyllosticta angelicae Sacc.
Distr.: Temperate North America; Europe.
Hosts: Angelica, Cymopterus, Ligusticum
Phyllosticta anserinae Tehon
Distr.: IL.
Host: Potentilla
Phyllosticta antennariae Ellis & Everh.
Distr.: Central and eastern United States.
Host: Antennaria
Phyllosticta anthoxella R. Sprague
Distr.: OR.
Host: Anthoxanthum
Phyllosticta antirrhini P. Syd.
Distr.: Eastern North America; Europe.
Hosts: Antirrhinum, Penstemon
Phyllosticta apicalis J.J. Davis

Distr.: Temperate North America.
Host: Salix
Phyllosticta apii Halst.
Distr.: DE, NJ.
Host: Apium
Phyllosticta aplectri Ellis & Everh.
Distr.: DE(type).
Notes: Van der Aa (Stud. Mycol. 5:32. 1973).
Host: Aplectrum
Phyllosticta apocyni Trel.
Distr.: Temperate North America.
Host: Apocynum
Phyllosticta aquilegicola Brunaud
Distr.: WA; Europe.
Host: Aquilegia
Phyllosticta arachidis-hypogaea V.G. Rao
Distr.: FL; India(type).
Host: Arachis
Phyllosticta araucariae Woronichin
Distr.: FL; Soviet Union.
Host: Araucaria
Phyllosticta arbutifoliae Ellis & G. Martin
Distr.: NJ, OK.
Host: Aronia
Phyllosticta argillacea Bres.
Ascochyta argillacea (Bres.) Bondartzeva-Monteverde
Distr.: WI; Europe.
Host: Rubus
Phyllosticta argyraea Speg.
Distr.: North and South America; Asia; Europe.
Host: Elaeagnus
Phyllosticta aricola Bubák
Distr.: HI; southern Europe.
Hosts: Epipremnum, Pothos
Phyllosticta arida Earle
Distr.: AL(type).
Host: Acer
Phyllosticta aristolochiae Tassi
Distr.: NJ; Italy(type).
Host: Aristolochia
Phyllosticta armeriae Allesch.
Distr.: WI; Europe.
Host: Limonium
Phyllosticta arnicae Fuckel
Distr.: Western North America; Europe.
Host: Arnica
Phyllosticta artocarpi Speg.
Distr.: Subtropical and tropical regions.
Host: Artocarpus
Phyllosticta arxii van der Aa
Dothiorella diospyri (Ellis & Everh.) Arx
Gloeosporium diospyri Ellis & Everh.
Phyllostictina diospyri (Ellis & Everh.) Arx
Distr.: Central and eastern United States.
Notes: Anthracnose and leaf spot of paw paw (Van Der Aa. Stud. Mycol. 5:34. 1978).
Host: Diospyros
Phyllosticta asiminae Ellis & Kellerm.
Distr.: Eastern United States.
Host: Asimina
Phyllosticta astragali Peck
Distr.: Temperate North America.
Host: Astragalus
Phyllosticta atomata Tehon
Distr.: IL.
Host: Plantago
Phyllosticta aucubae Sacc. & Speg. in Sacc.
Distr.: Southern North America; southern Europe.
Host: Aucuba
Phyllosticta aucubicola Sacc.
Distr.: CT; Europe.
Host: Aucuba
Phyllosticta aurantiicola (Berk. & Cooke) Sacc.
Distr.: Southern United States; southern Europe.
Host: Citrus
Phyllosticta avenophila Tehon & E.Y. Daniels
Distr.: MS.
Host: Avena
Phyllosticta azedarachis Thuem.
Distr.: AL; Europe.
Host: Melia
Phyllosticta baccharidis Dearn. & House in House
Distr.: FL, NY.
Host: Baccharis
Phyllosticta bacterioides Vuill.
Distr.: WI; Europe.
Host: Tilia
Phyllosticta baldensis C. Massal.
Distr.: North America, scattered reports; Europe.
Host: Paeonia
Phyllosticta barssii R. Sprague
Distr.: OR.
Host: Saussurea
Phyllosticta berberidis Rabenh. non Westend., hom. illeg.
Distr.: Temperate North America; Europe.
Host: Berberis
Phyllosticta berolinensis Henn.
Phyllosticta rhododendri-flavi Bubák & Kab.
Chaetapiospora rhododendri (Tengwall) Arx (teleomorph)
Distr.: North America; northern Europe.
Host: Rhododendron
Phyllosticta betulina Sacc.
Distr.: OK; Europe.
Host: Betula
Phyllosticta biformis Heald & F.A. Wolf
Distr.: Southeastern United States.
Host: Diospyros
Phyllosticta boehmeriicola J.J. Davis
Distr.: WI.
Hosts: Boehmeria, Laportea, Urtica
Phyllosticta bridgesii Speg.
Distr.: IN; South America.
Host: Lobelia
Phyllosticta bromivora R. Sprague
Distr.: NM.
Host: Bromus
Phyllosticta brunellae Ellis & Everh.
Distr.: TX, WA; Europe.
Host: Prunella
Phyllosticta brunnea Dearn. & Barth.
Distr.: Temperate North America.
Host: Populus
Phyllosticta bumeliifolia Heald & F.A. Wolf
Distr.: Southeastern United States.
Host: Sideroxylon
Phyllosticta cacaliae H.C. Greene
Distr.: WI.
Hosts: Cacalia, Cirsium, Prenanthes, Senecio, Silphium, Solidago
Phyllosticta cajani Rangel
Distr.: FL; West Indies; South America.
Host: Cajanus
Phyllosticta camelliae Westend. in Sacc.
Phyllosticta camelliaecola Brunaud
Distr.: North America; southern Europe.
Host: Camellia
Phyllosticta canescens Ellis & Everh.
Distr.: ID.
Host: Ribes
Phyllosticta capitalensis Henn.
Phyllostictina pyriformis Cash & A.M. Watson
Distr.: Cosmopolitan.
Notes: Van Der Aa (Stud. Mycol. 5:38. 1973).
Hosts: Brassolaeliocattleya, Cattleya, Cymbidium, Dendrobium, Epidendrum, Guzmania, Laelia, Laeliocattleya, Nidularium, Odontoglossum, Oncidium, Orchidaceae, Phalaenopsis, Vanda
Phyllosticta caricae-papayae Allesch.
Distr.: Subtropical and tropical regions.
Host: Carica
Phyllosticta caricicola Sacc. & Scalia
Distr.: Northwestern North America.

Host: Carex
Phyllosticta caricis (Fuckel) Sacc.
Distr.: Temperate North America; Europe.
Host: Carex
Phyllosticta carpini Schulzer & Sacc.
Distr.: NM; Europe.
Host: Alnus
Phyllosticta carpogena (Shear) van der Aa
Phyllostictina carpogena Shear
Distr.: Eastern North America.
Notes: Fruit rot. van der Aa (Stud. Mycol. 5:89. 1973).
Host: Rubus
Phyllosticta casimiroae F. Stevens & Weedon
Distr.: HI.
Host: Casimiroa
Phyllosticta cassiicola Kellerm. & Swingle
Distr.: KS.
Host: Cassia
Phyllosticta castaneae Ellis & Everh.
Distr.: Central and eastern United States.
Host: Castanea
Phyllosticta catalpae Ellis & G. Martin
Didymosphaeria catalpae J.B. Parker (teleomorph)
Distr.: Eastern United States.
Host: Catalpa
Phyllosticta ceanothi L.E. Miles
Distr.: MS.
Host: Ceanothus
Phyllosticta celastri Ellis & Everh. in Millsp. & L. Nutt.
Distr.: Eastern United States.
Host: Celastrus
Phyllosticta celtidis Ellis & Kellerm.
Distr.: Central and eastern North America.
Host: Celtis
Phyllosticta cephalanthi Tharp
Distr.: FL, TX.
Host: Cephalanthus
Phyllosticta cercidicola Ellis & Everh. in Millsp. & L. Nutt.
Distr.: Eastern United States.
Host: Cercis
Phyllosticta chamaenerii Allesch.
Distr.: OR; Europe.
Host: Epilobium
Phyllosticta chenopodii Westend.
Septoria westendorpii G. Wint.
Distr.: Temperate North America; Europe.
Notes: The status of this taxon is uncertain (Seaver. North Amer. Fl. 6:83. 1922).
Hosts: Chenopodium, Spinacia
Phyllosticta chenopodii-albi Siemaszko
Distr.: WI; Europe.
Hosts: Atriplex, Chenopodium
Phyllosticta chionanthi Thuem.
Distr.: Eastern United States; Europe.
Host: Chionanthus
Phyllosticta chrysanthemi Ellis & Dearn.
Distr.: Eastern North America.
Host: Chrysanthemum
Phyllosticta cinerea (Desmaz.) Allesch. non Pass., hom. illeg.
Distr.: AZ; Europe.
Host: Rhamnus
Phyllosticta circuligerens Tehon & E.Y. Daniels
Distr.: IL, OK.
Host: Rumex
Phyllosticta circumscissa Cocke
Distr.: Worldwide.
Host: Prunus
Phyllosticta cirsii Desmaz.
Distr.: Temperate North America; Europe.
Host: Cirsium
Phyllosticta cissicola Speg.
Distr.: Subtropical and tropical western hemisphere.
Host: Cissus
Phyllosticta citricarpa (McAlpine) van der Aa
Phoma citricarpa McAlpine
Phyllostictina citricarpa (McAlpine) Petr.
Guignardia citricarpa Kiely (teleomorph)
Leptodothiorella sp. (synanamorph)
Distr.: Worldwide in citrus-growing regions.
Notes: Black spot (Sutton & Waterson. C.M.I. Descr. 85. 1966).
Host: Citrus
Phyllosticta citricola S. Hori ex K. Hara
Distr.: MS; Asia.
Host: Fortunella
Phyllosticta clematidis Ellis & Dearn.
Distr.: Temperate North America.
Host: Clematis
Phyllosticta clethrae P. Syd.
Distr.: North America, scattered reports; Europe.
Host: Clethra
Phyllosticta clethricola Ellis & G. Martin
Distr.: Central and eastern United States.
Host: Clethra
Phyllosticta clypeata Ellis & Everh.
Distr.: OR.
Host: Malus
Phyllosticta coccolobae Ellis & Everh.
Distr.: FL; West Indies.
Host: Coccoloba
Phyllosticta colocasiophila Weedon in F. Stevens
Distr.: HI.
Host: Colocasia
Phyllosticta commelinicola E. Young
Distr.: OK, TX; West Indies.
Host: Commelina
Phyllosticta commonsii Ellis & Everh.
Distr.: Eastern United States.
Host: Paeonia
Phyllosticta concava Seaver
Distr.: Central and eastern United States; Bermuda.
Host: Opuntia
Phyllosticta concentrica Sacc.
Phyllosticta decipiens Ellis & Everh.
Phoma hysterella Sacc.
Phyllostictina hysterella (Sacc.) Petr.
Phoma ilicicola (Cooke & Ellis) Sacc.
Phyllosticta ilicicola (Cooke & Ellis) Ellis & Everh.
Sphaeropsis ilicicola Cooke & Ellis
Discochora philoprina (Berk. & M.A. Curtis) Bissett (teleomorph)
Leptodothiorella sp. (synanamorph)
Distr.: Temperate northern hemisphere.
Notes: Van Der Aa (Stud. Mycol. 5:44. 1973).
Hosts: Fatshedera, Hedera, Ilex, Taxus
Phyllosticta concomitans Ellis & Everh.
Distr.: LA.
Host: Ilex
Phyllosticta confertissima Ellis & Everh.
Distr.: Central and eastern United States.
Host: Ulmus
Phyllosticta congesta Heald & F.A. Wolf
Distr.: Eastern United States.
Host: Prunus
Phyllosticta convallariae Pers.:Fr.
Phyllosticta cruenta (Kunze:Fr.) J. Kickx fil.
Phyllostictina cruenta (Kunze:Fr.) Petr. & Syd.
Sphaeropsis cruenta (Kunze:Fr.) Gilman & W.A. Archer
Guignardia reticulata (DC.:Fr.) van der Aa (teleomorph)
Distr.: One questionable report from the United States; Europe.
Notes: Leaf spot of Liliaceae. van der Aa (Stud. Mycol. 5:23. 1973) discussed this fungus as *P. cruenta*. Although widely reported from North America in the past, no occurrences could be verified (Bissett. Canad. J. Bot. 57:2085. 1979). Previous reports appear to be based on misdeterminations of *Macrophoma smilacina* (Peck) Berl. & Vogl. or of *Phyllosticta pallidior* Peck.
Hosts: [Iris], Maianthemum, Polygonatum, Smilacina, Smilax, Uvularia
Phyllosticta convexula Bubák
Distr.: Central and eastern United States.
Host: Carya
Phyllosticta cookei Sacc.

Distr.: North America; Europe.
Host: Magnolia
Phyllosticta cordylinophila P.A. Young in F. Stevens
Distr.: HI.
Host: Cordyline
Phyllosticta coreopsidis H.C. Greene
Distr.: WI.
Host: Coreopsis
Phyllosticta cornicola (DC.) Rabenh.
Distr.: Temperate North America; Europe.
Host: Cornus
Phyllosticta cornuti Ellis & Kellerm.
Distr.: Central and eastern United States.
Host: Asclepias
Phyllosticta corydalis (Ellis & J.J. Davis) H.C. Greene
Distr.: TX, WI.
Host: Corydalis
Phyllosticta coryli Westend.
Distr.: Temperate North America; Europe.
Host: Corylus
Phyllosticta cotoneastri Allesch. in P. Syd.
Distr.: North America, scattered reports; Europe.
Host: Cotoneaster
Phyllosticta crataegi (Cooke) Sacc.
Distr.: Temperate North America.
Host: Crataegus
Phyllosticta crataegicola Sacc.
Distr.: North America, scattered reports; Europe.
Host: Crataegus
Phyllosticta crustosa Bonar & W.B. Cooke
Distr.: CA.
Notes: According to Petrak (Sydowia 10:302. 1956), a synonym of *Sporonema punctiforme* (Fuckel) Höhn.
Host: Kelloggia
Phyllosticta crypta Bissett
Distr.: Eastern United States.
Notes: Leaf spot (Bissett. Canad. J. Bot. 57:2085. 1979).
Host: Smilax
Phyllosticta cryptomeriae S. Kawamura
Guignardia sawadae T. Kobayashi (teleomorph)
Distr.: Eastern United States; Japan.
Notes: Kobayashi & Sasaki (Trans. Mycol. Soc. Japan 16:230. 1975).
Host: Cryptomeria
Phyllosticta cucurbitacearum Sacc.
Distr.: North America; West Indies; Europe.
Hosts: Citrullus, Cucumis, Cucurbita
Phyllosticta cumminsii Bissett
Distr.: Central and eastern United States.
Notes: Leaf spot (Bissett. Canad. J. Bot. 57:2088. 1979).
Host: Smilax
Phyllosticta cunninghamii Allesch. in P. Syd.
Distr.: WA; Europe.
Host: Rhododendron
Phyllosticta curtisii (Sacc.) Ellis & Everh. in H. Hume
Phoma curtisii Sacc.
Phoma maculans (Berk. & M.A. Curtis) Sacc.
Sphaeropsis maculans Berk. & M.A. Curtis
Distr.: Southeastern United States.
Host: Sideroxylon
Phyllosticta cyanococci Dearn. & House in House
Distr.: NC, NY.
Host: Vaccinium
Phyllosticta cycadina Pass.
Distr.: FL(as *Phomopsis*); Italy(type).
Host: Cycas
Phyllosticta cyclaminicola Trel.
Distr.: Central United States.
Host: Cyclamen
Phyllosticta cyclaminis Brunaud
Distr.: VA; Europe.
Host: Cyclamen
Phyllosticta cyrillae Ellis & G. Martin
Distr.: FL.
Host: Cyrilla
Phyllosticta cystopteridis H.C. Greene
Distr.: WI.
Host: Cystopteris
Phyllosticta dahliaecola Brunaud
Distr.: FL; Brazil; Europe; India.
Host: Dahlia
Phyllosticta davisii (Ellis & Everh.) van der Aa
Gloeosporium davisii Ellis & Everh.
Phyllostictina davisii (Ellis & Everh.) Arx
Distr.: Central United States.
Hosts: Astragalus, Lathyrus, Vicia
Phyllosticta dearnessii Sacc.
Distr.: Temperate North America.
Host: Rubus
Phyllosticta delphinii Clements in Seaver
Distr.: Western North America.
Host: Delphinium
Phyllosticta desmodii Ellis & Everh.
Distr.: Central and eastern North America.
Host: Desmodium
Phyllosticta destruens Desmaz.
Distr.: United States; Europe(type).
Host: Prunus
Phyllosticta deutziae Ellis & Everh.
Distr.: Eastern United States.
Host: Deutzia
Phyllosticta dictamni Fairm.
Distr.: NY.
Host: Dictamnus
Phyllosticta diervillae J.J. Davis
Distr.: WI.
Host: Diervilla
Phyllosticta digitalis Bellynck
Distr.: Northern North America; Europe.
Host: Digitalis
Phyllosticta digitariae R. Sprague
Distr.: UT.
Host: Digitaria
Phyllosticta dioscoreae Cooke
Phyllostictina dioscoreae (Cooke) Petr.
Septoria dioscoreae Cooke
Guignardia dioscoreae Pande (teleomorph)
Distr.: North America; Asia; Europe.
Notes: Leaf spot (Pande. Sydowia 22:366. 1968).
Host: Dioscorea
Phyllosticta dircae Ellis & Dearn.
Distr.: Northeastern North America.
Host: Dirca
Phyllosticta discincola Ellis & Everh.
Distr.: DC, MD.
Host: Forsythia
Phyllosticta divergens Sacc.
Distr.: FL, PR; Europe.
Host: Albizia
Phyllosticta dodecathei Trel.
Distr.: TX, WI.
Host: Dodecatheon
Phyllosticta dracaenae Henn.
Distr.: Subtropical and tropical regions.
Hosts: Cordyline, Dracaena
Phyllosticta dracocephali Dearn. & Bisby
Distr.: Central North America.
Host: Dracocephalum
Phyllosticta draconis Berk.
Distr.: PA; Europe.
Host: Dracaena
Phyllosticta dryopteridis H.C. Greene
Distr.: WI(type).
Host: Thelypteris
Phyllosticta dulcamarae Sacc.
Distr.: Central and eastern North America; Europe.
Host: Solanum
Phyllosticta ellisiana Lambotte & Fautrey
Distr.: VT; Europe.
Host: Anemone
Phyllosticta ellisii Sacc. & Syd. in Sacc.

Phyllosticta ribis Ellis & Everh. in Millsp. & L. Nutt. non Speg., hom. illeg.
Distr.: WV.
Host: Ribes
Phyllosticta elongata G.J. Weidemann
Botryosphaeria vaccinii (Shear) Barr (teleomorph)
Distr.: North America.
Notes: Fruit and leaf spot, speckle and rot of cranberry (Weidemann et al. Mycologia 74:62. 1982).
Host: Vaccinium
Phyllosticta eminens H.C. Greene
Distr.: WI.
Host: Salix
Phyllosticta entylomicola H.C. Greene
Distr.: WI.
Host: Ratibida
Phyllosticta epigaeae Peck
Distr.: Northeastern United States.
Host: Epigaea
Phyllosticta erechtitis F. Stevens & P.A. Young
Distr.: HI.
Host: Erechtites
Phyllosticta eriobotryae Thuem.
Distr.: Southeastern United States; Europe.
Host: Eriobotrya
Phyllosticta eriogoni W.B. Cooke in W.B. Cooke & C.G. Shaw
Distr.: WA.
Host: Eriogonum
Phyllosticta erratica Ellis & Everh.
Distr.: Southeastern United States.
Hosts: Camellia, Citrus, Rosa, Ulmus
Phyllosticta erysimi Westend.
Distr.: WI; Europe.
Host: Erysimum
Phyllosticta erysiphoides Sacc.
Distr.: France(type); distribution elsewhere is questionable.
Host: Chilopsis
Phyllosticta erythronii W.B. Cooke & C.G. Shaw
Distr.: WI, WY.
Host: Erythronium
Phyllosticta eugeniae E. Young
Distr.: FL, PR.
Hosts: Eugenia, Syzygium
Phyllosticta euonymi Sacc.
Distr.: Widespread in North America; Europe.
Host: Euonymus
Phyllosticta everhartii Sacc. & Syd.
Phyllosticta araliae Ellis & Everh.
Distr.: Southeastern United States.
Host: Aralia
Phyllosticta excavata Sacc.
Distr.: AK, ID.
Host: Heuchera
Phyllosticta eximia Bubák
Distr.: Western United States; Europe.
Host: Crepis
Phyllosticta extensa Sacc. & Syd. in Sacc.
Distr.: CA, FL.
Host: Eucalyptus
Phyllosticta faginea Peck
Distr.: Central and eastern North America.
Host: Fagus
Phyllosticta fatiscens Peck
Distr.: Temperate North America.
Hosts: Nuphar, Nymphaea
Phyllosticta favillensis H.C. Greene
Distr.: WI.
Host: Silphium
Phyllosticta ferax Ellis & Everh.
Distr.: Central and western North America.
Host: Lupinus
Phyllosticta fimbriata Ellis & Everh.
Distr.: West coast of the United States.
Host: Arbutus
Phyllosticta forsythiae Sacc.
Distr.: CT; Europe.
Host: Forsythia
Phyllosticta fragaricola Desmaz. & Roberge
Distr.: Eastern North America; Europe.
Hosts: Fragaria, Potentilla
Phyllosticta fraserae Ellis & Everh.
Distr.: CO.
Host: Frasera
Phyllosticta fraxinicola (Curr.) Ellis & Everh.
Distr.: Temperate North America; Europe.
Host: Fraxinus
Phyllosticta fritillariae Bonar & W.B. Cooke
Distr.: Pacific Northwest.
Hosts: Fritillaria, Triteleia
Phyllosticta fusispora Ellis & Everh.
Distr.: AL, TX.
Host: Castanea
Phyllosticta galactis (Cooke) Ellis & Everh. in Millsp. & L. Nutt.
Phoma galactis Cooke
Distr.: Southeastern United States.
Host: Galax
Phyllosticta gallicola Ellis & Everh.
Distr.: CO, ID.
Host: Solidago
Phyllosticta gallorum Thuem.
Distr.: AK; Soviet Union(type).
Host: Caragana
Phyllosticta garrettii Syd. & P. Syd.
Distr.: Western North America.
Host: Senecio
Phyllosticta garryae Cooke & Harkn.
Distr.: CA, TX; Great Britain.
Host: Garrya
Phyllosticta gaultheriae Ellis & Everh.
Phyllosticta gaultheriae Ellis & Everh. f. *shallon* Dearn.
Guignardia gaultheriae van der Aa (teleomorph)
Leptodothiorella sp. (synanamorph)
Distr.: Temperate North America.
Host: Gaultheria
Phyllosticta gelsemii Ellis & Everh.
Distr.: NJ, TX.
Host: Gelsemium
Phyllosticta geranii Ellis & Everh.
Distr.: LA, TX.
Host: Geranium
Phyllosticta ghaesembillae Koord.
Phyllosticta codiaei Sacc.
Phyllosticta codiaei Died.
Phyllostictina euphorbiae Petr.
Distr.: Subtropical and tropical regions.
Notes: Leaf spot of Euphorbiaceae (Van Der Aa. Stud. Mycol. 6:58. 1973).
Hosts: Codiaeum, Euphorbia
Phyllosticta ginkgo Brunaud
Distr.: Eastern North America; Europe.
Host: Ginkgo
Phyllosticta gladioli Ellis & Everh.
Distr.: Southeastern United States.
Host: Gladiolus
Phyllosticta glauca Cooke ex Ellis & Everh.
Distr.: Southeastern United States.
Host: Magnolia
Phyllosticta globifera Ellis & Everh. in Millsp. & L. Nutt.
Distr.: Central and eastern United States.
Host: Cornus
Phyllosticta globigera Sacc.
Distr.: Pacific Northwest.
Host: Sorbus
Phyllosticta gordoniae Ellis & G. Martin
Distr.: FL, TX.
Host: Gordonia
Phyllosticta gossypina Ellis & G. Martin
Distr.: Southeastern United States.
Host: Gossypium
Phyllosticta grisea Peck

Distr.: NY.
Host: Crataegus
Phyllosticta grossulariae Sacc.
Mycosphaerella ribis Feltgen (teleomorph)
Septoria ribis (Lib.) Desmaz. (synanamorph)
Distr.: Temperate North America; Europe.
Host: Ribes
Phyllosticta guttulatae Halst. in Seym. & Earle
Distr.: Northeastern North America.
Host: Oxalis
Phyllosticta gymnocladi Tehon & E.Y. Daniels
Distr.: IL.
Host: Gymnocladus
Phyllosticta halstedii Ellis & Everh.
Distr.: NJ(type).
Host: Syringa
Phyllosticta hamamelidis Cooke in Ravenel & Cooke
Gloeosporium hamamelidis Cooke
Phyllosticta hamamelidis Peck
Phyllostictina hamamelidis (Cooke) Petr. & Syd.
Distr.: Eastern North America.
Host: Hamamelis
Phyllosticta hawaiiensis Caum
Distr.: HI.
Hosts: Mimosa, Saccharum
Phyllosticta hawaiiensis Petr. non Caum, hom. illeg.
Distr.: HI(type).
Notes: Petrak (Sydowia 7:381. 1953).
Hosts: Mimosa, Saccharum
Phyllosticta haynaldii Roum. & Sacc. in Sacc.
Distr.: WV; Europe.
Host: Ilex
Phyllosticta healdii R. Sprague
Distr.: NE.
Host: Dichanthelium
Phyllosticta hederae Sacc. & Roum.
Distr.: Eastern North America; northern Africa; Europe.
Host: Hedera
Phyllosticta hedericola Durieu & Mont. in Mont.
Distr.: North America; northern Africa; Europe.
Host: Hedera
Phyllosticta helenae R. Sprague
Distr.: ID.
Host: Glyceria
Phyllosticta helianthi Ellis & Everh. ex H.C. Greene
Distr.: WI.
Host: Helianthus
Phyllosticta heliconiae F. Stevens & P.A. Young
Distr.: HI.
Host: Heliconia
Phyllosticta heliopsidis H.C. Greene
Distr.: WI.
Host: Heliopsis
Phyllosticta helleboricola C. Massal. var. ***coptidis*** Sacc. & Scalia
Distr.: AK.
Notes: The status of this taxon is uncertain (Seaver. North Amer. Fl. 6:84. 1922).
Host: Coptis
Phyllosticta heraclei Ellis & Dearn.
Distr.: Northern North America.
Host: Heracleum
Phyllosticta hesleri Dearn., nom. nud.
Distr.: TN.
Notes: Published without a latin description.
Host: Acer
Phyllosticta hesperidearum (Cattaneo) Penz.
Distr.: North America; Europe.
Hosts: Carpinus, Citrofortunella, Citrus, Poncirus
Phyllosticta heteromeles Cooke & Harkn.
Distr.: CA.
Host: Heteromeles
Phyllosticta heucherae Ellis & Everh. non Brunaud, hom. illeg.
Asteromella heucherae (Ellis & Everh.) Petr.
Stictochorella heucherae (Ellis & Everh.) Petr.
Mycosphaerella heucherae (Ellis & Everh.) Petr. (teleomorph)
Distr.: Western North America.
Host: Heuchera
Phyllosticta hibiscina Ellis & Everh.
Distr.: Central and eastern North America.
Hosts: Abelmoschus, Avicennia, Hibiscus
Phyllosticta hieracii Allesch.
Distr.: WI; Europe.
Host: Hieracium
Phyllosticta hispida Ellis & Dearn.
Distr.: Northeastern North America.
Host: Smilax
Phyllosticta hosackiae W.B. Cooke in W.B. Cooke & C.G. Shaw
Distr.: WA.
Host: Lotus
Phyllosticta humuli Sacc. & Speg. in Sacc.
Phyllosticta humuli Sacc. & Speg. var. *major* Ellis & Everh.
Distr.: Central and eastern North America; Europe.
Host: Humulus
Phyllosticta hydrangeae Ellis & Everh.
Distr.: Eastern North America.
Host: Hydrangea
Phyllosticta hymenocallidis Seaver
Distr.: Southeastern United States; Bermuda; West Indies.
Hosts: Hymenocallis, Manfreda
Phyllosticta illinoensis Tehon & E.Y. Daniels
Distr.: IL, MA.
Host: Sassafras
Phyllosticta indianensis Van Hook
Distr.: IN.
Host: Rosa
Phyllosticta innumera Cooke & Harkn.
Distr.: CA.
Host: Fraxinus
Phyllosticta innumerabilis Peck
Distr.: Central North America.
Host: Amelanchier
Phyllosticta intermixta Seaver
Phyllosticta populina Sacc. non Pers., hom. illeg.
Distr.: Temperate North America; Europe.
Host: Populus
Phyllosticta ipomoeae Ellis & Kellerm.
Distr.: Eastern United States; West Indies.
Host: Ipomoea
Phyllosticta iridis Ellis & G. Martin in Ellis & Everh.
Distr.: Temperate North America.
Host: Iris
Phyllosticta iteae Dearn., nom. inval.
Distr.: TN.
Host: Itea
Phyllosticta ivaecola Ellis & Everh.
Distr.: Central United States.
Host: Iva
Phyllosticta ixorae Rangel
Distr.: FL; Brazil(type).
Host: Ixora
Phyllosticta jambosicola van der Aa
Phyllostictina jambosae Petr. non Batista & Vittal, hom. illeg.
Distr.: HI(type).
Host: Syzygium
Phyllosticta japonica Thuem. non Fautrey, hom. illeg.
Distr.: Central and eastern North America; Europe.
Hosts: Berberis, Mahonia, Oryza
Phyllosticta juglandina Sacc.
Distr.: OR; Europe.
Host: Juglans
Phyllosticta juglandis (DC.) Sacc.
Distr.: Temperate North America; Europe.
Host: Juglans
Phyllosticta juliflora Ellis & Barth.
Distr.: OK, TX.
Host: Prosopis
Phyllosticta kalmicola (Schwein.) Ellis & Everh.
Distr.: Eastern United States.
Host: Kalmia
Phyllosticta lactucae Atk.

Distr.: Southeastern United States.
Host: Lactuca
Phyllosticta laeliae Keissl.
Distr.: Subtropical and tropical regions.
Host: Laelia
Phyllosticta lagerstroemiae Ellis & Everh.
Distr.: Southeastern United States.
Host: Lagerstroemia
Phyllosticta lappae Sacc.
Distr.: Eastern North America; Europe.
Host: Arctium
Phyllosticta latifolia Ellis & Everh.
Distr.: Eastern United States.
Host: Kalmia
Phyllosticta laurocerasi Sacc. & Speg. in Sacc.
Distr.: CA, FL; Europe.
Hosts: Laurocerasus, Prunus
Phyllosticta lentisci (Pass.) Allesch.
Phoma lentisci Pass.
Distr.: TX; Europe.
Host: Pistacia
Phyllosticta liatridis J.J. Davis
Distr.: OK, WI.
Host: Liatris
Phyllosticta linderae Ellis & Everh.
Distr.: Eastern United States.
Host: Lindera
Phyllosticta lindericola Ellis & Everh.
Distr.: WV.
Host: Lindera
Phyllosticta liriodendrica Sacc.
Phyllosticta liriodendri Cooke non Thuem., hom. illeg.
Phyllosticta macrospora Ellis & Everh.
Distr.: Eastern United States.
Host: Liriodendron
Phyllosticta livida Ellis & Everh.
Distr.: Central and eastern North America.
Host: Quercus
Phyllosticta longispora McAlpine
Distr.: Southeastern United States; West Indies; Australia.
Host: Citrus
Phyllosticta ludoviciana Ellis & G. Martin
Distr.: Southeastern United States.
Host: Quercus
Phyllosticta ludwigiae Peck
Distr.: IA, NY, WI.
Host: Ludwigia
Phyllosticta lupini Bonar
Distr.: CA.
Host: Lupinus
Phyllosticta lychnidis (Kunze & J.C. Schmidt) Ellis & Everh.
Distr.: IA, TX.
Notes: North American reports are doubtful (Seaver. North Amer. Fl. 6:84. 1922).
Hosts: Lychnis, Silene
Phyllosticta lycii Ellis & Kellerm.
Distr.: NY, OH.
Host: Lycium
Phyllosticta lysimachiae Allesch.
Distr.: NY; Europe.
Host: Lysimachia
Phyllosticta maclurae Ellis & Everh.
Distr.: Central and eastern United States.
Host: Maclura
Phyllosticta macroguttata Earle
Distr.: AL.
Host: Desmodium
Phyllosticta maculans Ellis & Everh.
Distr.: Central and eastern North America.
Host: Populus
Phyllosticta maculicola Halst.
Distr.: North America, scattered reports; Europe.
Hosts: Cordyline, Dracaena, Yucca
Phyllosticta maculiformis Sacc.
Distr.: North America, scattered reports; Europe.
Host: Alnus
Phyllosticta magnoliae Sacc.
Distr.: Eastern United States; Europe.
Host: Magnolia
Phyllosticta mahoniicola Pass. in Brunaud
Phoma mahoniana Sacc.
Phyllosticta mahoniana (Sacc.) Allesch.
Distr.: Temperate North America; Europe.
Hosts: Berberis, Mahonia
Phyllosticta maianthemi W.B. Cooke
Distr.: WA.
Host: Maianthemum
Phyllosticta malkoffii Bubák
Distr.: TX, PR; Europe.
Host: Gossypium
Phyllosticta malvavisci J.A. Stevenson
Distr.: FL, TX.
Host: Malvaviscus
Phyllosticta marantaceae Henn.
Distr.: HI; South America.
Host: Maranta
Phyllosticta maxima Ellis & Everh.
Phyllostictina maxima (Ellis & Everh.) Petr.
Botryosphaeria rhodorae (Cooke) Barr (teleomorph)
Asterostomella saccardoi (Thuem.) Petr. (synanamorph microconidial)
Distr.: Temperate North America.
Notes: Pirozynski (Fungi Canadenses 21. 1974 as *Botryosphaeria rhodorae*).
Hosts: Pieris, Rhododendron
Phyllosticta maydis D.C. Arny & R.R. Nelson
Mycosphaerella zeae-maydis Mukunya & Boothroyd (teleomorph)
Distr.: Central and eastern United States; southern Canada.
Notes: Yellow leaf blight of corn.
Host: Zea
Phyllosticta medeolae Dearn. & House in House
Distr.: NY.
Host: Medeola
Phyllosticta meibomiae Seaver
Distr.: IL, southeastern United States.
Host: Desmodium
Phyllosticta melaleuca Ellis & Everh.
Distr.: Central and eastern North America.
Host: Ulmus
Phyllosticta melanoplaca Thuem.
Distr.: Temperate northern hemisphere.
Host: Veratrum
Phyllosticta meliae Ellis & Everh.
Distr.: LA, TX.
Host: Melia
Phyllosticta menispermicola Tehon & E.Y. Daniels
Distr.: IL, WI.
Host: Menispermum
Phyllosticta mentzeliae Ellis & Kellerm.
Distr.: Central United States.
Host: Mentzelia
Phyllosticta micropuncta Cooke
Phyllosticta perseae Ellis & G. Martin
Distr.: Southeastern United States.
Host: Persea
Phyllosticta minima (Berk. & M.A. Curtis) Underw. & Earle
Phyllosticta acericola Cooke & Ellis
Phyllosticta saccharina Ellis & G. Martin
Distr.: Eastern North America.
Host: Acer
Phyllosticta minor Ellis & Everh.
Distr.: Eastern North America; Europe.
Host: Vinca
Phyllosticta minutaspora R. Sprague
Distr.: ID.
Host: Muhlenbergia
Phyllosticta minutissima Ellis & Everh.
Distr.: Western North America.
Notes: Associated with leaf spots caused by *Septoria* sp.
Host: Acer

Phyllosticta mitellae Peck
Distr.: NY, WI.
Host: Mitella
Phyllosticta miyakei Syd.
Distr.: Southeastern United States; Japan.
Notes: Possibly a synonym of *Phoma sorghina* (Sacc.) Boerema, Doenbosch, Van Kesteren (Ou. Rice diseases. p. 322. 1985).
Host: Oryza
Phyllosticta molluginis Ellis & Halst.
Distr.: NJ.
Host: Mollugo
Phyllosticta monardae Ellis & Barth.
Distr.: Central United States.
Hosts: Lycopus, Mentha, Monarda, Pycnanthemum
Phyllosticta monardellae W.B. Cooke
Asteromella monardellae (W.B. Cooke) Petr.
Distr.: CA.
Host: Monardella
Phyllosticta moricola Ellis & Everh.
Distr.: KS, SC.
Host: Morus
Phyllosticta mortonii Fairm.
Distr.: Southern United States; Mexico; West Indies.
Host: Mangifera
Phyllosticta mucunae Ellis & Everh.
Distr.: HI, southeastern United States.
Host: Mucuna
Phyllosticta musae F. Stevens & P.A. Young in F. Stevens
Distr.: HI.
Host: Musa
Phyllosticta musarum (Cooke) van der Aa
Sphaeropsis musarum Cooke
Phyllostictina musarum (Cooke) Petr.
Macrophoma musae (Cooke) Berl. & Voglino
Guignardia musae Racib. (teleomorph)
Distr.: HI; southeastern Asia; East Indies; South Pacific; eastern Africa.
Host: Musa
Phyllosticta musicola F. Stevens & P.A. Young in F. Stevens
Distr.: HI.
Host: Musa
Phyllosticta myricae Cooke
Distr.: Temperate United States.
Host: Myrica
Phyllosticta nandinae Tassi
Distr.: OK; Europe.
Host: Nandina
Phyllosticta nebulosa Sacc.
Phyllosticta pallida Halst. in Seym. & Earle
Distr.: Eastern North America; Europe.
Hosts: Arenaria, Lychnis, Silene
Phyllosticta negundinis Sacc. & Speg.
Distr.: Temperate North America; Europe.
Host: Acer
Phyllosticta neomexicana Kab. & Bubák
Distr.: FL; Europe.
Host: Robinia
Phyllosticta nerii Westend.
Distr.: HI, North America; Europe.
Host: Nerium
Phyllosticta nesaeae Peck
Distr.: Eastern United States.
Host: Decodon
Phyllosticta neuroterigallicola Tehon
Distr.: IL.
Host: Quercus
Phyllosticta nicotiana Ellis & Everh.
Distr.: Eastern United States.
Host: Nicotiana
Phyllosticta nigrescens Bonar & W.B. Cooke
Distr.: CA.
Host: Viola
Phyllosticta nigromaculans Sacc.
Distr.: Brazil; distribution elsewhere is questionable.
Host: Oncidium
Phyllosticta numerospora H.C. Greene
Distr.: WI.
Host: Potentilla
Phyllosticta nymphaeacea Ellis & Everh.
Distr.: Southeastern United States.
Hosts: Nuphar, Nymphaea
Phyllosticta nymphaeicola Tehon & E.Y. Daniels
Distr.: IL.
Host: Nuphar
Phyllosticta nyssae Cooke
Mycosphaerella nyssicola (Cooke) F.A. Wolf (teleomorph)
Distr.: Eastern United States.
Host: Nyssa
Phyllosticta occulta Bubák
Distr.: WA; Europe.
Host: Rhododendron
Phyllosticta oleae Ellis & G. Martin
Distr.: Southeastern United States.
Hosts: Olea, Osmanthus
Phyllosticta olympica R. Sprague
Distr.: WA.
Host: Carex
Phyllosticta opaca Ellis & Everh. in Millsp. & L. Nutt.
Distr.: Temperate North America.
Host: Ilex
Phyllosticta opulasteris Petr.
Distr.: ID.
Host: Physocarpus
Phyllosticta orbicula Ellis & Everh.
Distr.: Northeastern North America; Europe.
Host: Armoracia
Phyllosticta orbicularis Ellis & Everh.
Distr.: Northeastern United States.
Hosts: Cucurbita, Sicyos
Phyllosticta orobella Sacc.
Distr.: Temperate North America; Europe.
Host: Lathyrus
Phyllosticta orontii Ellis & G. Martin
Distr.: Eastern North America.
Host: Orontium
Phyllosticta osmanthi Tassi
Distr.: LA; Europe.
Host: Osmanthus
Phyllosticta osteospora Sacc.
Distr.: AK; Europe.
Host: Populus
Phyllosticta ovalifolii Brunaud
Distr.: North America, scattered reports; Europe.
Host: Ligustrum
Phyllosticta owensii R. Sprague
Distr.: Pacific Northwest.
Hosts: Dactylis, Panicum
Phyllosticta oxalidis Sacc.
Distr.: TX, WI; Europe.
Host: Oxalis
Phyllosticta oxydendri Ellis & Everh. in Millsp. & L. Nutt.
Distr.: Southeastern United States.
Host: Oxydendrum
Phyllosticta pachysandrae Dearn. & House in House
Distr.: Eastern United States.
Host: Pachysandra
Phyllosticta pachystimae Wehmeyer
Distr.: WY.
Host: Paxistima
Phyllosticta pallens Ellis & Everh.
Distr.: Southeastern United States.
Host: Euonymus
Phyllosticta pallidior Peck
Phyllosticta cruenta (Kunze: Fr.) J. Kickx fil. var. *pallidior* (Peck) J.J. Davis
Phyllostictina pallidior (Peck) Petr. & Syd.
Distr.: Central and eastern North America.
Notes: Leaf spot (Bissett. Canad. J. Bot. 57:2089. 1979).
Hosts: Polygonatum, Smilacina
Phyllosticta palmetto Ellis & Everh.

Distr.: Southeastern United States.
Hosts: Cocos, Sabal
Phyllosticta palustris Ellis & Dearn.
Distr.: Central and eastern North America.
Host: Stachys
Phyllosticta pandanicola E. Young
Distr.: Southeastern United States; West Indies.
Host: Pandanus
Phyllosticta panici E. Young
Distr.: Southern United States, PR.
Host: Panicum
Phyllosticta parkinsoniae Ellis & Everh.
Distr.: FL, TX.
Host: Parkinsonia
Phyllosticta paulowniae Sacc.
Distr.: Central and eastern North America; Europe.
Host: Paulownia
Phyllosticta paupercula Peck
Distr.: KS, MT.
Host: Amelanchier
Phyllosticta pedicularidis Solheim
Distr.: WY.
Host: Pedicularis
Phyllosticta pennsylvanica H.C. Greene
Distr.: WI.
Host: Prunus
Phyllosticta pentstemonicola H.C. Greene
Distr.: WI.
Host: Cacalia
Phyllosticta persicae Sacc.
Distr.: Central and eastern North America; Europe.
Host: Prunus
Phyllosticta petroselini Rothers
Distr.: NY; Soviet Union(type).
Host: Petroselinum
Phyllosticta phaseolina Sacc.
Distr.: Temperate North America; Europe.
Hosts: Amphicarpaea, Apios, Glycine, Phaseolus, Strophostyles, Vicia, Vigna
Phyllosticta phoradendri Bonar
Distr.: CA.
Host: Phoradendron
Phyllosticta physaleos Sacc.
Distr.: CT; Europe.
Host: Physalis
Phyllosticta physocarpi H.C. Greene
Distr.: WI.
Host: Physocarpus
Phyllosticta physopellae Dearn. & Barth.
Distr.: FL.
Host: Ficus
Phyllosticta phytolaccae Cooke
Distr.: Eastern United States.
Host: Phytolacca
Phyllosticta pitcheriana Fairm.
Distr.: NY.
Host: Heliopsis
Phyllosticta pithecolobii E. Young
Distr.: HI; West Indies.
Host: Albizia
Phyllosticta plantaginella Sacc.
Distr.: NY.
Host: Plantago
Phyllosticta plantaginicola Tehon & E.Y. Daniels
Distr.: IL.
Host: Plantago
Phyllosticta plantaginis Sacc.
Distr.: AK; Europe.
Host: Plantago
Phyllosticta platani Sacc. & Speg. in Sacc.
Distr.: Eastern North America; Europe.
Host: Platanus
Phyllosticta platanoidis Sacc.
Distr.: North America, scattered reports; Europe.
Host: Acer
Phyllosticta podophylli (M.A. Curtis) G. Wint.
Phyllostictina podophylli (M.A. Curtis) Petr.
Distr.: Central and eastern North America.
Host: Podophyllum
Phyllosticta podophyllina Tehon & G.L. Stout
Distr.: IL.
Host: Podophyllum
Phyllosticta polygonorum Sacc.
Distr.: Eastern North America; Europe.
Hosts: Fagopyrum, Polygonum
Phyllosticta porteri Tehon & E.Y. Daniels
Distr.: IL.
Host: Syringa
Phyllosticta pothicola Weedon in F. Stevens
Distr.: HI.
Hosts: Epipremnum, Pothos
Phyllosticta praetervisa Bubák
Distr.: WI; Europe.
Host: Tilia
Phyllosticta primulicola Desmaz.
Distr.: WI; Europe.
Host: Primula
Phyllosticta pruni-virginianae H.C. Greene
Distr.: WI.
Host: Prunus
Phyllosticta pseudocapsici Roum.
Distr.: LA; Europe.
Host: Solanum
Phyllosticta psidiicola (Petr.) van der Aa
Phyllostictina psidiicola Petr.
Distr.: HI(type).
Host: Psidium
Phyllosticta psoraleae (Cooke) Tassi
Phoma psoraleae Cooke
Distr.: OR, SC(type).
Host: Psoralea
Phyllosticta pteleicola Tehon & E.Y. Daniels
Distr.: IL.
Host: Ptelea
Phyllosticta pteridis Halst.
Distr.: Northeastern United States.
Hosts: Pteridium, Pteris
Phyllosticta putrefaciens Shear
Distr.: WA, northeastern United States.
Notes: The status of this taxon is uncertain (Weidemann et al. Mycologia 74:65. 1982).
Host: Vaccinium
Phyllosticta pyrolae Ellis & Everh.
Phyllostictina pyrolae (Ellis & Everh.) Petr.
Leptodothiorella sp. (synanamorph)
Distr.: Temperate North America; Europe.
Host: Pyrola
Phyllosticta pyrorum Cooke
Distr.: Southeastern United States.
Host: Pyrus
Phyllosticta quercea J.J. Davis
Septogloeum querceum J.J. Davis (synanamorph)
Distr.: WI.
Host: Quercus
Phyllosticta quercus Sacc. & Speg. in Sacc.
Distr.: WI; Europe.
Host: Quercus
Phyllosticta quercus-ilicis Sacc.
Distr.: MS; Europe.
Host: Quercus
Phyllosticta quercus-prini Ellis & Everh.
Distr.: FL, WV.
Host: Quercus
Phyllosticta rafinesquii H.W. Anderson
Distr.: AL, IL.
Host: Viola
Phyllosticta ragnhildae Solheim
Distr.: WY.
Host: Antennaria
Phyllosticta raui (Peck) Dearn. & House

Sphaeropsis raui Peck
Macrophoma raui (Peck) Berl. & Voglino
Distr.: NY, western United States.
Host: Artemisia
Phyllosticta rhamni Westend.
Distr.: WI; Europe.
Host: Rhamnus
Phyllosticta rhexiae Dearn. & House in Dearn.
Distr.: FL.
Hosts: Miconia, Rhexia
Phyllosticta rhododendri Westend.
Distr.: WA; Europe.
Host: Rhododendron
Phyllosticta rhododendricola C. Brun
Distr.: Western United States; Europe.
Host: Rhododendron
Phyllosticta rhodorae (Cooke) Tassi
Phoma rhodorae Cooke
Distr.: WA; England(type).
Host: Rhododendron
Phyllosticta rhoicola Ellis & Everh.
Distr.: Temperate North America.
Host: Rhus
Phyllosticta rhoiseda Fairm.
Distr.: CA.
Host: Rhus
Phyllosticta rhynchosiae L.E. Miles
Distr.: MS.
Host: Rhynchosia
Phyllosticta ribesicida Speg.
Distr.: North America, scattered reports; Argentina(type).
Host: Ribes
Phyllosticta richardsoniae Ellis & Everh.
Distr.: AL.
Host: Richardia
Phyllosticta robertii Boyer & Jacz.
Distr.: Subtropical and tropical regions.
Host: Ficus
Phyllosticta robiniae Sacc.
Distr.: LA; Europe.
Host: Robinia
Phyllosticta rogleri R. Sprague
Distr.: IA.
Host: Digitaria
Phyllosticta rosae Desmaz.
Distr.: Worldwide in rose-growing regions.
Host: Rosa
Phyllosticta rosae-setigerae Van Hook
Distr.: IN.
Host: Rosa
Phyllosticta rosicola C. Massal.
Distr.: North America, scattered reports; Europe.
Host: Rosa
Phyllosticta rubicola Rabenh.
Distr.: North America, scattered reports; Europe.
Host: Rubus
Phyllosticta ruborum Sacc.
Distr.: North America; Australia; Europe.
Host: Rubus
Phyllosticta rudbeckiae Ellis & Everh.
Distr.: Central and eastern United States.
Host: Rudbeckia
Phyllosticta salicicola Thuem.
Distr.: Temperate North America; Europe.
Host: Salix
Phyllosticta salicis Kellerm. & Swingle
Distr.: KS.
Host: Salix
Phyllosticta sambucicola Kalchbr.
Distr.: IN; Europe.
Host: Sambucus
Phyllosticta sanguinariae G. Wint.
Distr.: Central and eastern United States.
Host: Sanguinaria
Phyllosticta sapindii Henn.
Distr.: FL; Brazil(type).
Host: Sapindus
Phyllosticta saxifragarum Allesch.
Distr.: WY; Europe.
Host: Saxifraga
Phyllosticta scaevolae Ellis & Everh.
Distr.: HI.
Host: Scaevola
Phyllosticta scariolicola Tehon
Distr.: IL.
Host: Lactuca
Phyllosticta serotina Cooke
Distr.: Eastern North America.
Host: Prunus
Phyllosticta siccata H.C. Greene
Distr.: WI.
Host: Solidago
Phyllosticta sicyna Sacc.
Distr.: NY, TX.
Host: Sicyos
Phyllosticta silenes Peck
Distr.: Central and eastern United States.
Hosts: Dianthus, Silene
Phyllosticta siliquastri Sacc. & Speg. in Sacc.
Distr.: Southeastern United States; Europe.
Host: Cercis
Phyllosticta sinuosa Ellis & G. Martin
Distr.: Southeastern United States.
Host: Osmanthus
Phyllosticta smilacinae Solheim
Distr.: CO.
Host: Smilacina
Phyllosticta smilacinae-trifoliae H.C. Greene
Distr.: WI.
Host: Smilacina
Phyllosticta sojicola C. Massal.
Phyllosticta glycinea Tehon & E.Y. Daniels
Pleosphaerulina sojicola Miura (teleomorph)
Distr.: Worldwide in soybean-growing regions.
Hosts: Clitoria, Glycine
Phyllosticta solani Ellis & G. Martin
Distr.: Southeastern United States.
Host: Solanum
Phyllosticta solidaginicola Tehon & E.Y. Daniels
Distr.: IL.
Host: Solidago
Phyllosticta solidaginis Bres.
Distr.: Central United States; Europe.
Hosts: Petradoria, Solidago
Phyllosticta solidaginis-serotinae Petr.
Distr.: WI.
Host: Solidago
Phyllosticta solitaria Ellis & Everh.
Distr.: Central and eastern North America.
Notes: Fruit blotch, leaf spot, and twig canker of pome fruits.
Hosts: Crataegus, Malus, Pyrus
Phyllosticta sophorae Ellis & Everh.
Distr.: OK, TX.
Host: Sophora
Phyllosticta sorbariae H.C. Greene
Distr.: WI.
Host: Sorbaria
Phyllosticta sorbi Westend.
Distr.: Central and eastern North America; Europe.
Host: Sorbus
Phyllosticta spermoides Peck
Distr.: North America.
Host: Vitis
Phyllosticta sphaeropsoidea Ellis & Everh.
Phyllosticta paviae Desmaz.
Guignardia aesculi (Peck) V.B. Stewart (teleomorph)
Leptodothiorella aesculicola (Sacc.) Sivanesan (synanamorph)
Distr.: Eastern North America; Europe.
Hosts: Aesculus, Hamamelis, Thuja
Phyllosticta spinosa Ellis & Kellerm.

Distr.: FL, KS, TX.
Host: Sida
Phyllosticta staphyleae Dearn. in Ellis & Everh.
Distr.: Northeastern North America.
Host: Staphylea
Phyllosticta starbaeckii Sacc. & Syd. in Sacc.
Distr.: CA, OK, PA.
Host: Cornus
Phyllosticta staticis Petr.
Distr.: NY; Albania(type).
Host: Limonium
Phyllosticta steironematis Dearn. & House
Phyllostictina steironematis (Dearn. & House) Petr.
Distr.: Northeastern North America.
Host: Lysimachia
Phyllosticta stillingiae Ellis & Everh.
Distr.: Southeastern United States.
Host: Sapium
Phyllosticta subeffusa (Ellis & Everh.) Tehon & G.L. Stout
Phyllosticta smilacis Ellis & G. Martin var. *subeffusa* Ellis & Everh. in Millsp. & L. Nutt.
Distr.: Central and eastern United States.
Notes: Leaf spot (Bissett. Canad. J. Bot. 57:2091. 1979).
Hosts: Smilacina, Smilax
Phyllosticta subtilis Peck
Distr.: Eastern United States.
Host: Carya
Phyllosticta succinosa H.C. Greene
Distr.: WI.
Host: Ribes
Phyllosticta suecica Bubák & Vleugel
Distr.: CT; Sweden(type).
Host: Lathyrus
Phyllosticta symphoricarpi Westend.
Distr.: North America, scattered reports; Europe.
Host: Symphoricarpos
Phyllosticta syriaca Sacc.
Distr.: NY; Europe.
Host: Hibiscus
Phyllosticta syringae Westend.
Phyllosticta syringella (Fuckel) Rabenh.
Distr.: Temperate North America; Europe.
Host: Syringa
Phyllosticta tabaci Pass.
Distr.: FL; Europe.
Host: Nicotiana
Phyllosticta tecomae Sacc.
Distr.: MS; Europe.
Host: Campsis
Phyllosticta tenerrima Ellis & Everh.
Distr.: Eastern North America.
Host: Saponaria
Phyllosticta terminalis Ellis & G. Martin
Phyllosticta leucothoes Ellis & G. Martin
Distr.: Eastern United States.
Hosts: Forsythia, Ilex, Leucothoe, Magnolia, Osmanthus
Phyllosticta tetonensis Solheim
Distr.: WY.
Host: Clematis
Phyllosticta teucrii Sacc. & Speg. in Sacc.
Distr.: IA; Europe.
Host: Teucrium
Phyllosticta texensis Seaver
Distr.: TX.
Host: Verbena
Phyllosticta thermopsidis Thuem.
Distr.: ID; Soviet Union(type).
Host: Thermopsis
Phyllosticta tiliae Sacc. & Speg. in Sacc.
Distr.: Temperate North America; Europe.
Host: Tilia
Phyllosticta tinea Sacc.
Distr.: North America; Australia; Europe.
Host: Viburnum
Phyllosticta tineola Sacc.
Distr.: MN, WI; Europe.
Host: Viburnum
Phyllosticta toxica Ellis & G. Martin
Distr.: Central and eastern North America.
Notes: Bissett & Darbyshire (Fungi Canadenses 281. 1984).
Host: Rhus
Phyllosticta toxicodendri Thuem.
Distr.: SC.
Host: Rhus
Phyllosticta trillii Ellis & Everh.
Distr.: Temperate North America.
Host: Trillium
Phyllosticta trollii Trail
Distr.: Rocky Mountains; Europe.
Host: Trollius
Phyllosticta tuberosa Ellis & G. Martin
Distr.: Central and eastern United States.
Host: Asclepias
Phyllosticta tumoricola Peck
Distr.: NY.
Host: Quercus
Phyllosticta typhina Sacc. & Malbr.
Phyllosticta renouana Sacc. & Roum.
Distr.: Temperate North America; Europe.
Host: Typha
Phyllosticta ulmi-rubrae H.C. Greene
Distr.: WI.
Host: Ulmus
Phyllosticta umbrino-fumosa H.C. Greene
Distr.: WI.
Host: Eupatorium
Phyllosticta uvulariae B.T. Galloway ex F. Patterson in J.J. Davis
Phyllosticta cruenta (Kunze:Fr.) J. Kickx fil. var. *discincta* (J.J. Davis) J.J. Davis
Phyllosticta discincta J.J. Davis
Distr.: Central and eastern United States.
Notes: Leaf spot (Bissett. Canad. J. Bot. 57:2094. 1979).
Host: Uvularia
Phyllosticta vaccinii Earle
Phyllostictina vaccinii Demaree & M.S. Wilcox non (Earle) Shear, hom. illeg.
Phyllostictina vaccinii (Earle) Shear
Phyllosticta sparsa Bonar
Distr.: Temperate North America.
Notes: Fruit rot and leaf spot of cranberry. According to Weidemann et al. (Mycologia 74:59. 1982) no teleomorph for this species is known.
Host: Vaccinium
Phyllosticta vagans Peck
Distr.: Temperate North America.
Hosts: Convallaria, Smilacina
Phyllosticta variabilis Peck
Phyllosticta bicolor Peck
Distr.: Temperate North America.
Host: Rubus
Phyllosticta verbascicola Ellis & Kellerm.
Distr.: Central and eastern United States.
Host: Verbascum
Phyllosticta verbenicola G. Martin
Distr.: NJ, TX.
Host: Verbena
Phyllosticta verbesinae Heald & F.A. Wolf
Distr.: TX.
Host: Verbesina
Phyllosticta vesicatoria Thuem.
Distr.: SC.
Host: Quercus
Phyllosticta villosa Ellis & Everh., nom. nud.
Distr.: Eastern United States.
Host: Rubus
Phyllosticta violae Desmaz.
Distr.: Eastern North America; Europe.
Host: Viola
Phyllosticta virens Ellis & Langl.
Distr.: LA, MS.

Host: Quercus
Phyllosticta virginiana **(Ellis & Halst.) Ellis**
Phoma virginiana Ellis & Halst.
Distr.: Temperate North America.
Hosts: Amelanchier, Prunus
Phyllosticta viridis **Ellis & Kellerm.**
Mycosphaerella fraxinicola (Schwein.) House (teleomorph)
Distr.: Central and eastern North America.
Host: Fraxinus
Phyllosticta vulgaris **Desmaz.**
Phyllosticta lonicerae Westend.
Distr.: North America, scattered reports; Europe.
Hosts: Lonicera, Prunus
Phyllosticta wisconsinensis **H.C. Greene**
Distr.: WI.
Host: Helianthus
Phyllosticta wislizenii **Ellis & Everh.**
Distr.: CA.
Host: Quercus
Phyllosticta wistariae **Sacc.**
Distr.: Eastern United States; Europe.
Host: Wisteria
Phyllosticta wulfeniae **H.C. Greene**
Distr.: WI.
Host: Besseya
Phyllosticta wyomingensis **Solheim**
Distr.: WY.
Host: Epilobium
Phyllosticta xanthorrhizae **Ellis & L. Nutt. in Millsp. & L. Nutt.**
Distr.: NC, WV.
Host: Xanthorhiza
Phyllosticta zanthoxyli **H.C. Greene**
Distr.: WI.
Host: Zanthoxylum
Phyllosticta zeae **G.L. Stout**
Distr.: Eastern United States.
Host: Zea
Phyllosticta zingiberis **F. Stevens & Ryan**
Distr.: HI.
Host: Zingiber
Phyllosticta zonata **Ellis & Everh.**
Distr.: IA.
Host: Malus

Phyllostictina Syd. Coelomycetes

Notes: A synonym of *Phyllosticta* Pers.
Phyllostictina divergens **Petr. & Syd.**
Distr.: FL; The Philippines(type).
Notes: Probably a *Phyllosticta* sp. (Van Der Aa. Stud. Mycol. 5:190. 1973).
Host: Bambusa

Phyllotopsis E.J. Gilbert & Donk ex Singer. Basidiomycetes, Agaricales

Phyllotopsis nidulans **(Pers.:Fr.) Singer**
Claudopus nidulans (Pers.:Fr.) P. Karst.
Distr.: Temperate regions.
Notes: White rots of hardwoods and conifers.
Hosts: Abies, Acer, Liriodendron, Pinus, Populus, Quercus

Phymatotrichopsis Hennebert. Hyphomycetes

Phymatotrichopsis omnivora **(Duggar) Hennebert**
Phymatotrichum omnivorum Duggar
Ozonium omnivorum Shear
Distr.: Southwestern United States; Mexico; reports from elsewhere need confirmation.
Notes: A serious root pathogen on many (>2000) plant species. The identification of the basidial teleomorph (Baniecki & Bloss. Mycologia 61:1054. 1969) as *Sistotrema brinkmanii* (Bres.) J. Eriks. has been considered doubtful(Hennebert. Persoonia 7:183. 1973). *Phymatotrichum fungicola* Zeller might be a synonym.
Hosts: Abelia, Abelmoschus, Abutilon, Acacia, Acalypha, Acer, Achillea, Acnida, Aconitum, Aesculus, Agrimonia, Agrostemma, Ailanthus, Alcea, Aleurites, Allionia, Alnus, Amaranthus, Ambrosia, Amelanchier, Amorpha, Anethum, Angelica, Annona, Anthemis, Antirrhinum, Apios, Apocynum, Aquilegia, Arabis, Arachis, Aralia, Araucaria, Arbutus, Arctium, Arctostaphylos, Arctotis, Arenaria, Argemone, Argyreia, Armoracia, Aronia, Artemisia, Asclepias, Asimina, Astragalus, Atriplex, Baccharis, Baptisia, Bassia, Bellis, Berberis, Beta, Betula, Bidens, Boerhavia, Brassica, Brickellia, Broussonetia, Buddleja, Caesalpinia, Cajanus, Calendula, Callirhoe, Callistephus, Calocedrus, Calystegia, Camassia, Campsis, Cannabis, Capsicum, Caragana, Carica, Carissa, Carpinus, Carya, Cassia, Castanea, Catalpa, Cedrus, Celastrus, Celtis, Centaurea, Cerastium, Ceratonia, Cercis, Cereus, Chaenomeles, Chamaecyparis, Chelidonium, Chenopodium, Chilopsis, Chionanthus, Chrysanthemum, Cichorium, Cicuta, Cinnamomum, Cissus, Citrullus, Citrus, Claytonia, Clematis, Clethra, Clitoria, Cnidoscolus, Cocculus, Codiaeum, Convolvulus, Conyza, Cordia, Coreopsis, Cornus, Corylus, Cosmos, Cotinus, Cotoneaster, Crataegus, Crotalaria, Croton, Cucumis, Cucurbita, Cupressus, Cyamopsis, Cydonia, Cynara, Cynodon, Cynoglossum, Dahlia, Dalea, Datura, Daucus, Delphinium, Descurainia, Desmodium, Dianthus, Dichondra, Diervilla, Dioscorea, Diospyros, Dipsacus, Echinacea, Echinocactus, Elaeagnus, Eleutherococcus, Ephedra, Epilobium, Erechtites, Erigeron, Eriobotrya, Eriogonum, Eryngium, Erythrina, Eucalyptus, Euonymus, Eupatorium, Euphorbia, Evolvulus, Eysenhardtia, Feijoa, Ficus, Firmiana, Foeniculum, Forestiera, Forsythia, Fouquieria, Fraxinus, Froelichia, Gaillardia, Galium, Gardenia, Garrya, Gaura, Gelsemium, Geranium, Geum, Gilia, Ginkgo, Gleditsia, Glycine, Gnaphalium, Gordonia, Gossypium, Grevillea, Grindelia, Gutierrezia, Gymnocladus, Haplopappus, Hedera, Hedyotis, Helenium, Helianthemum, Helianthus, Heliopsis, Hemerocallis, Heracleum, Heteromeles, Hibiscus, Hieracium, Hydrangea, Hydrastis, Hymenopappus, Hypochaeris, Ilex, Indigofera, Ipomoea, Iva, Jacaranda, Jatropha, Juglans, Juniperus, Justicia, Kalmia, Kerria, Koelreuteria, Krameria, Krigia, Kuhnia, Lablab, Lactuca, Lagenaria, Lagerstroemia, Lantana, Lappula, Larix, Lathyrus, Lavatera, Lepidium, Lespedeza, Lesquerella, Leucophyllum, Liatris, Ligustrum, Limonium, Linaria, Lindera, Linum, Lippia, Liquidambar, Liriodendron, Lithospermum, Lobelia, Lonicera, Lotus, Lupinus, Lycium, Lycopersicon, Lysimachia, Maackia, Macfadyena, Maclura, Magnolia, Malus, Malva, Malvastrum, Malvaviscus, Malvella, Mammillaria, Mangifera, Manihot, Manilkara, Matthiola, Medicago, Melia, Melilotus, Mentzelia, Mimosa, Mirabilis, Mollugo, Morus, Mucuna, Myrica, Nandina, Nerium, Nicandra, Nicotiana, Nyssa, Oenothera, Olea, Oncoba, Onosmodium, Osmanthus, Ostrya, Oxydendrum, Paeonia, Parkinsonia, Parthenium, Parthenocissus, Passiflora, Pastinaca, Paulownia, Pavonia, Penstemon, Persea, Petroselinum, Phacelia, Phaseolus, Philadelphus, Phlox, Phoenix, Photinia, Physalis, Physocarpus, Phytolacca, Picea, Pimpinella, Pinus, Pistacia, Pisum, Pittosporum, Plantago, Platanus, Platycodon, Plumeria, Polygonum, Poncirus, Populus, Portulaca, Potentilla, Primula, Proboscidea, Prosopis, Prunus, Pseudotsuga, Psidium, Psoralidium, Ptelea, Pueraria, Punica, Pyracantha, Pyrus, Quercus, Ranunculus, Raphanus, Ratibida, Rhamnus, Rheum, Rhododendron, Rhus, Ribes, Ricinus, Rivina, Robinia, Rosa, Rubus, Rudbeckia, Rumex, Salix, Salsola, Salvia, Sambucus, Sapindus, Sapium, Saponaria, Sassafras, Scabiosa, Schinus, Scutellaria, Senecio, Senna, Sequoia, Sesbania, Sesuvium, Shepherdia, Sida, Sideroxylon, Silene, Silphium, Solanum, Solidago, Sonchus, Sophora, Sorbus, Spergula, Sphaeralcea, Spinacia, Spiraea, Stellaria, Stillingia, Strophostyles, Symplocos, Syringa, Tamarix, Taraxacum, Taxodium, Tecoma, Tephrosia, Thuja, Tilia, Tragia, Tragopogon, Trifolium, Triodanis, Tsuga, Ulmus, Ungnadia, Urtica, Vaccinium, Valerianella, Verbascum, Verbena, Verbesina, Vernonia, Veronica, Veronicastrum, Vicia, Vigna, Viguiera, Vincetoxicum, Viola, Vitex, Vitis, Washingtonia, Weigela, Wisteria, Xanthium, Zea, Zinnia

Physalacria Peck. Basidiomycetes, Agaricales

Physalacria luttrellii **G. Baker**
Distr.: GA(type).

Notes: On dead stems.
Host: Lespedeza

Physalospora Niessl. Pyrenomycetes, Amphisphaeriales

Notes: Barr (Mycologia 62:386. 1970).
Physalospora abdita (Berk. & M.A. Curtis) N. Stevens in Voorhees
Anthostomella abdita (Berk. & M.A. Curtis) Sacc.
Physalospora fusca N. Stevens
Distr.: Southeastern United States; Cuba.
Hosts: Acacia, Acalypha, Acer, Alnus, Baccharis, Beta, Brunnichia, Campsis, Carya, Cassia, Celtis, Citrullus, Citrus, Cocos, Diospyros, Ficus, Gossypium, Hibiscus, Juniperus, Lantana, Liquidambar, Magnolia, Mangifera, Manihot, Melia, Persea, Platanus, Prunus, Pyrus, Rhododendron, Rhus, Ricinus, Rosa, Salix, Sambucus, Sassafras, Ulmus, Vitis, Wikstroemia, Zea
Physalospora dearnessii Barr
Distr.: Eastern United States.
Host: Rhododendron
Physalospora empetri Rostr.
Botryosphaeria empetri (Rostr.) Arx & E. Müller
Physalospora crepiniana Sacc. & Él. Marchal in Rehm
Distr.: Boreal region of North America; Europe.
Notes: On dead leaves.
Host: Empetrum
Physalospora erratica (Cooke & Ellis) Sacc.
Distr.: CA, NJ(type).
Host: Nicotiana
Physalospora eucalyptina Fairm.
Distr.: CA(type).
Host: Eucalyptus
Physalospora geranii Cooke & Harkn.
Distr.: CA; Portugal.
Notes: This species belongs in *Botryosphaeria* but has not yet been transferred.
Host: Geranium
Physalospora gregaria Sacc.
Distr.: Western United States, WV.
Notes: This species has been confused with specimens of *Botryosphaeria dothidea* and *B. quercuum*.
Hosts: Salix, Taxus
Physalospora heteromelina Fairm.
Distr.: CA(type).
Host: Heteromeles
Physalospora kalmiae F.A. Wolf & A.R. Cavaliere
Distr.: Eastern United States.
Host: Kalmia
Physalospora minima Ellis & Everh.
Distr.: AL(type).
Host: Rubus
Physalospora orontii Ellis & Everh.
Distr.: NJ(type).
Host: Orontium
Physalospora potentillae Rostr.
Distr.: TX; Greenland(type).
Host: Potentilla
Physalospora rhododendri (De Not.) Rehm
Distr.: Europe.
Notes: Although reported from the United States, Barr (Mycologia 62:390. 1970) revealed that the type variety occurs only in Europe and that the fungus reported from North America is *P. rhododendri* var. *albimontana*.
Host: Rhododendron
Physalospora rhododendri (De Not.) Rehm var. ***albimontana*** Barr
Distr.: Northeastern United States.
Host: Rhododendron
Physalospora sesbaniae Ellis & Everh.
Phomatospora sesbaniae (Ellis & Everh.) Sacc.
Distr.: LA(type).
Host: Sesbania
Physalospora vaccinii (Shear) Arx & E. Müller
Acanthorhynchus vaccinii Shear
Distr.: Eastern North America; Europe.
Hosts: Chamaedaphne, Vaccinium

Physarella Peck. Myxomycetes

Physarella oblonga (Berk. & M.A. Curtis) Morg.
Distr.: Temperate North America; Europe(rare); tropical regions.
Host: Pandanus

Physarum Pers. Myxomycetes

Physarum cinereum (Batsch) Pers.
Physarum plumbeum Fr.
Distr.: Cosmopolitan.
Hosts: Agrostis, Axonopus, Cynodon, Eremochloa, Euonymus, Festuca, Fragaria, Ipomoea, Lactuca, Medicago, Plantago, Poa, Rhododendron, Stenotaphrum, Taraxacum, Trifolium
Physarum citrinum Schumach.
Distr.: Temperate areas, rare in North America.
Host: Pinus
Physarum gyrosum Rostr.
Distr.: Eastern United States; western Europe(mostly in greenhouses); tropical regions.
Host: Camellia
Physarum leucophaeum Fr.
Physarum nutans Pers. var. *leucophaeum* List.
Distr.: Temperate and tropical regions.
Hosts: Fragaria, Quercus
Physarum nutans Pers.
Distr.: Cosmopolitan.
Host: Quercus
Physarum pulcherripes Peck
Distr.: North America; rare throughout the world.
Host: Pinus
Physarum pusillum (Berk. & M.A. Curtis) G. List.
Distr.: Cosmopolitan.
Host: Zea
Physarum vernum Sommerf. in Fr.
Distr.: Widespread but questionably known from North America.
Host: Medicago

Physisporinus P. Karst. Basidiomycetes, Aphyllophorales

Notes: Gilbertson & Ryvarden (North American Polypores 2:628. 1987).
Physisporinus sanguinolentus (Albertini & Schwein.:Fr.) Pilát
Poria sanguinolenta (Albertini & Schwein.:Fr.) Cooke
Rigidoporus sanguinolentus (Albertini & Schwein.:Fr.) Donk
Poria decolorans (Schwein.) Cooke
Distr.: Northern hemisphere.
Notes: On wood.
Hosts: Abies, Picea, Pinus, Populus, Pseudotsuga, Quercus, Tsuga
Physisporinus vitreus (Pers.:Fr.) P. Karst.
Poria vitrea Pers.:Fr.
Rigidoporus vitreus (Pers.:Fr.) Donk
Poria alborosea Murrill
Poria holoseparans Murrill
Poria undata (Pers.:Fr.) Quél.
Rigidoporus undatus (Pers.:Fr.) Donk
Distr.: Temperate North America; Europe.
Hosts: Abies, Carpinus, Fraxinus, Pinus, Quercus

Physoderma Wallr. Chytridiomycetes, Blastocladiales

Notes: Karling (Lloydia 13:29. 1950).
Physoderma alfalfae (Pat. & Lagerh.) Karling
Cladochytrium alfalfae Pat. & Lagerh.
Urophlyctis alfalfae (Pat. & Lagerh.) Magnus
Distr.: Cosmopolitan.
Notes: Crownwart of alfalfa (Walker. C.M.I. Descr. 751. 1983).
Host: Medicago
Physoderma alpina R. Sprague

Distr.: UT.
Host: Poa
Physoderma australasica (McAlpine) J. Walker
Physoderma claytoniana H.C. Greene var. *sparrowii* Savile & Parmelee
Distr.: North America; Australia.
Notes: Walker (Austral. J. Bot. Suppl. Ser. 10:112. 1983).
Host: Claytonia
Physoderma beckmanniae R. Sprague
Distr.: Western North America.
Host: Beckmannia
Physoderma butomi J. Schröt.
Distr.: MI; northern Africa; Europe.
Host: Butomus
Physoderma claytoniana H.C. Greene
Distr.: Northcentral North America.
Host: Claytonia
Physoderma comari (Berk. & F.B. White in Berk. & Broome) Lagerh.
Protomyces comari Berk. & F.B. White in Berk. & Broome
Distr.: MI; Europe.
Host: Potentilla
Physoderma dulichii R.M. Johns
Distr.: MI.
Host: Dulichium
Physoderma gerhardtii J. Schröt.
Entyloma spectabilis P. Karst.
Entyloma spragueanum Zundel
Distr.: United States; Europe.
Hosts: Glyceria, Poa
Physoderma graminis (Büsgen) De Wild.
Cladochytrium graminis Büsgen
Distr.: MT, WI; Germany(type).
Notes: A doubtful taxon (Sparrow. Mycologia 57:624. 1965).
Hosts: Agropyron, Agrostis, Elymus
Physoderma heleocharidis (Fuckel) J. Schröt.
Distr.: HI, North America; Europe.
Hosts: Elaeocarpus, Eleocharis
Physoderma hydrocotylidis Viégas & Teix.
Distr.: North and South America.
Host: Hydrocotyle
Physoderma johnsii Sparrow
Distr.: MI(type).
Host: Caltha
Physoderma lathyri (B. Palm) Karling
Urophlyctis lathyri B. Palm
Distr.: OR; Sweden.
Host: Lathyrus
Physoderma leproides (Trab.) Karling
Urophlyctis leproides (Trab.) Magnus
Distr.: Temperate regions.
Notes: Leaf gall or crown wart of beets (Walker. C.M.I. Descr. 752. 1983).
Host: Beta
Physoderma lycopi Sparrow
Distr.: MI.
Host: Lycopus
Physoderma maculare Wallr.
Cladochytrium maculare (Wallr.) Graff
Distr.: Temperate North America; Europe.
Host: Alisma
Physoderma majus J. Schröt.
Cladochytrium majus (J. Schröt.) E. Fisch.
Urophlyctis major J. Schröt.
Distr.: Northcentral United States; Chile; Europe.
Host: Rumex
Physoderma marsiliae M.S. Brewster
Distr.: KS.
Host: Marsilea
Physoderma maydis (Miyabe) Miyabe
Physoderma zeae-maydis F.J. Shaw in Syd., P. Syd., & E.J. Butler
Distr.: Cosmopolitan.
Notes: Brown spot of maize (Walker. C.M.I. Descr. 753. 1983).
Host: Zea
Physoderma menthae J. Schröt.
Distr.: MI; Europe.
Notes: Parasitic in the stems and leaves.
Host: Mentha
Physoderma menyanthis de Bary
Distr.: Temperate North America; Europe.
Host: Menyanthes
Physoderma palustris Sparrow
Distr.: WI(type).
Host: Sium
Physoderma pluriannulatum (Berk. & M.A. Curtis) Karling
Synchytrium pluriannulatum Berk. & M.A. Curtis
Urophlyctis pluriannulata (Berk. & M.A. Curtis) Farl.
Distr.: Temperate North America.
Hosts: Angelica, Ligusticum, Osmorhiza, Sanicula, Zizia
Physoderma potteri A.W. Bartlett
Distr.: OR; Great Britain.
Host: Lotus
Physoderma pulposum Wallr.
Urophlyctis pulposa (Wallr.) J. Schröt.
Distr.: Temperate regions.
Notes: Leaf and crown gall of Chenopodiaceae (Walker. C.M.I. Descr. 754. 1983).
Hosts: Atriplex, Chenopodium
Physoderma trifolii (Pass.) Karling
Urophlyctis trifolii (Pass.) Magnus
Distr.: CA; Europe; India.
Host: Trifolium
Physoderma vagans J. Schröt.
Distr.: Northcentral United States; Europe.
Host: Sium

Physopella Arth. Basidiomycetes, Uredinales

Angiopsora Mains
Notes: Cummins & Ramachar (Mycologia 50:741. 1958).
Physopella ampelopsidis (Dietel & P. Syd.) Cummins & Ramachar
Physopella vitis Arth.
Distr.: North America; South America; eastern Asia.
Notes: Autoecious. Punithalingam (C.M.I. Descr. 173. 1968).
Host: Vitis
Physopella cameliae (Mayor) Cummins & Ramachar
Angiopsora cameliae (Arth.) Mains
Distr.: Southern United States to Brazil.
Notes: Aecia unknown.
Hosts: Panicum, Setaria
Physopella compressa (Mains) Cummins & Ramachar
Angiopsora compressa Mains
Distr.: Southern United States to Brazil; Bolivia.
Notes: Aecia unknown.
Hosts: Axonopus, Paspalum
Physopella pallescens (Arth.) Cummins & Ramachar
Angiopsora pallescens (Arth.) Mains
Uredo pallida Dietel & Holw.
Distr.: FL; Mexico to Colombia.
Notes: Aecia unknown.
Hosts: Tripsacum, Zea
Physopella zeae (Mains) Cummins & Ramachar
Angiopsora zeae Mains
Distr.: FL; West Indies; Central and South America.
Notes: Aecia unknown. Seymour & Miller (Florida State Hort. Soc. 84:124. 1974) reported that this fungus was eradicated from Florida. Laundon & Waterston (C.M.I. Descr. 5. 1964).
Host: Zea

Phytophthora de Bary. Oomycetes, Peronosporales

Notes: Erwin et al. (*Phytophthora*: its biology, taxonomy, ecology, and pathology. 1983); Newhook et al. (Mycol. Pap. 143:1. 1978); Waterhouse (Mycol. Pap. 122:1. 1970).
Phytophthora cactorum (Lebert & Cohn) J. Schröt.
Distr.: Temperate regions.

Notes: On at least 154 genera of vascular plants in 54 families. Waterhouse & Waterston (C.M.I. Descr. 111. 1966).
Hosts: Abies, Acer, Antirrhinum, Arbutus, Avena, Betula, Brassolaeliocattleya, Caragana, Cattleya, Centaurea, Ceratonia, Citrullus, Citrus, Clarkia, Colutea, Cornus, Cucumis, Cucurbita, Daphne, Dendrobium, Echinochloa, Elaeagnus, Epidendrum, Eriobotrya, Fagus, Fortunella, Fragaria, Gypsophila, Heteromeles, Hordeum, Humulus, Juglans, Juniperus, Laeliocattleya, Lathyrus, Lilium, Lolium, Lychnis, Lycopersicon, Malus, Melilotus, Nyssa, Oncidium, Orchidaceae, Paeonia, Panax, Panicum, Persea, Picea, Pinus, Platycodon, Poa, Prunus, Pseudotsuga, Pyracantha, Pyrus, Quercus, Rheum, Rhododendron, Syringa, Trachymene, Tulipa, Vanda, Vicia, Vigna, Zea

Phytophthora cambivora (Petri) Buisman
Distr.: Cosmopolitan.
Notes: Root rots and cankers of various hardwoods, often in conjunction with other *Phytophthora* spp. Waterhouse & Waterston (C.M.I. Descr. 112. 1966).
Hosts: Acer, Castanea, Malus, Prunus, Pyrus

Phytophthora capsici Leonian
Distr.: Western hemisphere; Asia; Europe.
Notes: Fruit and stem rot of peppers, tomatoes, and cucurbits (Stamps. C.M.I. Descr. 836. 1985).
Hosts: Capsicum, Carica, Citrullus, Cucumis, Cucurbita, Lycopersicon, Macadamia

Phytophthora cinnamomi Rands
Distr.: Cosmopolitan.
Notes: A serious root pathogen, recorded on over 900 species and varieties of plants (Waterhouse & Waterston. C.M.I. Descr. 113. 1966; Zentmeyer. Phytopathology Soc. Monogr. 10. 1980).
Hosts: Abies, Acacia, Aleurites, Allium, Ananas, Antidesma, Araucaria, Aucuba, Banksia, Betula, Broussaisia, Buxus, Cajanus, Calluna, Calocedrus, Camellia, Castanea, Cedrus, Cercis, Chamaecyparis, Cheirodendron, Cibotium, Citrus, Comptonia, Coprosma, Cornus, Cotoneaster, Cryptomeria, Cupressus, Dicranopteris, Dracaena, Erica, Eucalyptus, Ficus, Flindersia, Forsythia, Freycinetia, Gaultheria, Gleichenia, Gouldia, Hudsonia, Ilex, Juglans, Juniperus, Kalmia, Larix, Leucospermum, Lithocarpus, Lycopersicon, Lycopodium, Macadamia, Melastoma, Metrosideros, Mucuna, Myrsine, Myrtus, Pandanus, Perrottetia, Persea, Picea, Pieris, Pinus, Pipturus, Platanus, Polystichum, Protea, Prunus, Pseudotsuga, Psidium, Psychotria, Pteris, Pyracantha, Pyrus, Quercus, Rhododendron, Richardia, Ricinus, Robinia, Rubus, Sciadopitys, Sophora, Styphelia, Taxus, Thuja, Toona, Tsuga, Vaccinium, Viburnum

Phytophthora citricola Sawada
Phytophthora cactorum (Lebert & Cohn) J. Schröt. var. *applanata* F. Chester
Distr.: Cosmopolitan.
Notes: Rot and die-back of various crop plants (Waterhouse & Waterston. C.M.I. Descr. 114. 1966).
Hosts: Abies, Aucuba, Humulus, Juglans, Persea, Pieris, Prunus, Psidium, Quercus, Rhododendron

Phytophthora citrophthora (R.E. Sm. & E.H. Sm.) Leonian
Distr.: Cosmopolitan.
Notes: Waterhouse & Waterston (C.M.I. Descr. 33. 1964).
Hosts: Actinidia, Ananas, Citrullus, Citrus, Cucurbita, Gleditsia, Juglans, Juniperus, Persea, Pieris, Poncirus, Prunus, Pyrus

Phytophthora colocasiae Racib.
Distr.: Warmer regions.
Hosts: Alocasia, Catharanthus, Colocasia

Phytophthora cryptogea Pethybr. & Lafferty
Distr.: Temperate regions.
Notes: This fungus attacks a variety of plants (Ho & Jong. Mycotaxon 27:289. 1986; Stamps. C.M.I. Descr. 592. 1978).
Hosts: Abies, Actinidia, Antirrhinum, Callistephus, Carpobrotus, Carthamus, Chrysanthemum, Cichorium, Gerbera, Helianthus, Juniperus, Lycopersicon, Malus, Matthiola, Medicago, Nasturtium, Osteospermum, Prunus, Pseudotsuga, Sinningia, Solanum, Tagetes, Zinnia

Phytophthora drechsleri Tucker
Distr.: Cosmopolitan.
Notes: A recent description is provided by Stamps (C.M.I. Descr. 840. 1985). Based on a detailed study of morphological and physiological characteristics, Ho & Jong (Mycotaxon 27:289. 1986) suggested that *P. drechsleri* is a synonym of *P. cryptogea*.
Hosts: Abies, Allium, Ananas, Beta, Cajanus, Carthamus, Citrus, Cucumis, Euphorbia, Leucaena, Lycopersicon, Malus, Parthenium, Pinus, Prunus, Pseudotsuga, Solanum, Spinacia, Zea

Phytophthora epistomium Fell & Master
Distr.: FL.
Notes: Fell & Master (Canad. J. Bot. 53:2914. 1975).
Host: Rhizophora

Phytophthora erythroseptica Pethybr.
Distr.: Cosmopolitan.
Notes: Rots of various crops (Stamps. C.M.I. Descr. 593. 1978).
Hosts: Rubus, Saccharum, Solanum, Trifolium, Zantedeschia

Phytophthora fragariae C.J. Hickman
Distr.: Western United States; England; Scotland.
Notes: Red stele root rot of strawberry. Converse (Phytopathology 57:173. 1967) distinguished physiologic races. Ho & Jong (Mycotaxon 31:305. 1988) redefined the species.
Host: Fragaria

Phytophthora heveae A. Thompson
Distr.: Warm regions.
Notes: Rot, leaf blight and stem canker of various crops (Stamps. C.M.I. Descr. 594. 1978).
Host: Rhododendron

Phytophthora hibernalis Carne
Distr.: Widespread in citrus growing regions.
Notes: Waterhouse & Waterston (C.M.I. Descr. 31. 1964).
Host: Citrus

Phytophthora ilicis Buddenhagen & R.A. Young
Distr.: Northwestern United States.
Notes: Leaf and twig disease of English holly (Phytopathology 47:95. 1957).
Host: Ilex

Phytophthora infestans (Mont.) de Bary
Distr.: Cosmopolitan.
Notes: Late blight of tomato and potato (Stamps. C.M.I. Descr. 838. 1985).
Hosts: Lycopersicon, Solanum

Phytophthora infestans (Mont.) de Bary f. sp. ***thalictri*** (G.W. Wils. & J.J. Davis) G.M. Waterhouse
Phytophthora thalictri G.W. Wils. & J.J. Davis
Distr.: Central and eastern United States.
Host: Thalictrum

Phytophthora inflata Caroselli & Tucker
Distr.: Northeastern United States.
Notes: Pit canker of elm. Caroselli & Tucker (Phytopathology 39:481. 1949).
Host: Ulmus

Phytophthora katsurae W.H. Ko & H.S. Chang
Phytophthora castaneae Katsura & Uchida
Distr.: HI; Japan; Taiwan.
Notes: Disease of chestnut, also isolated from coconut and soil (Ko & Chang. Mycologia 71:840. 1979; Stamps. C.M.I. Descr. 837. 1985).
Host: Cocos

Phytophthora lateralis Tucker & Milbrath
Distr.: Northwestern North America.
Notes: Root rot (Tucker & Milbraith. Mycologia 34:94. 1942).
Host: Chamaecyparis

Phytophthora megasperma Drechs.
Distr.: Cosmopolitan.
Notes: Waterhouse & Waterston (C.M.I. Descr. 115. 1966).
Hosts: Actinidia, Alcea, Brassica, Citrus, Cynara, Daucus, Glycine, Juglans, Malus, Matthiola, Medicago, Onobrychis, Prunus, Pseudotsuga, Pyrus, Rosa, Saccharum, Spinacia, Trifolium

Phytophthora megasperma Drechs. f. sp. ***glycinea*** T. Kuan & D.C. Erwin
Phytophthora megasperma Drechs. var. *sojae* A.A. Hildebrand
Phytophthora sojae M.J. Kaufmann & J.W. Gerdemann
Distr.: Central and eastern United States; northeastern Canada.
Notes: Root and stem rot of soybean (Jones & Johnson. Phytopathology 59:504. 1969). Kuan & Erwin (Phytopathology 70:333. 1980) discussed the forms of this organism describing a new name for a previously published taxa. Because *formae speciales* do not fall under the ICBN, this name can be used. If recognized at the varietal level, the name must be based on the earliest epithet, namely that described by Kaufmann & Gerdemann.

Hosts: Glycine, Lupinus
Phytophthora megasperma Drechs. f. sp. ***medicaginis*** T. Kuan & D.C. Erwin
Distr.: Temperate regions.
Notes: Kuan & Erwin (Phytopathology 70:333. 1980) established this *forma specialis* for isolates that attack alfalfa.
Host: Medicago
Phytophthora nicotianae Breda de Haan
Phytophthora parasitica Dastur var. *nicotianae* (Breda de Haan) Tucker
Phytophthora terrestris Sherb.
Distr.: Warm regions.
Notes: Black shank of tobacco, also reported from other plants (Waterhouse & Waterston. C.M.I. Descr. 34. 1964).
Hosts: Castanea, Fremontodendron, Lycopersicon, Nicotiana, Philodendron, Pinus, Ricinus, Saintpaulia, Sinningia
Phytophthora nicotianae Breda de Haan var. ***parasitica*** (Dastur) G.M. Waterhouse
Phytophthora parasitica Dastur
Distr.: Cosmopolitan.
Notes: Damping-off, crown rot, leaf blight, fruit rot, etc., of a wide variety of plants (Waterhouse & Waterston. C.M.I. Descr. 35. 1964).
Hosts: Abelmoschus, Allium, Aloe, Ananas, Antirrhinum, Aphelandra, Atropa, Bidens, Bougainvillea, Buxus, Cajanus, Callistemon, Capsicum, Carica, Carissa, Carthamus, Catharanthus, Celosia, Chrysanthemum, Cissus, Citrofortunella, Citrullus, Citrus, Cordyline, Coriandrum, Cornus, Crotalaria, Cucurbita, Daphne, Dianthus, Dracaena, Euphorbia, Fatshedera, Fatsia, Fortunella, Fragaria, Fuchsia, Gossypium, Gypsophila, Hedera, Hibiscus, Ilex, Jasminum, Juniperus, Lilium, Lupinus, Lycopersicon, Macadamia, Manihot, Melia, Mucuna, Murraya, Nicotiana, Opuntia, Pastinaca, Peperomia, Persea, Petroselinum, Petunia, Phaseolus, Philodendron, Pilea, Pinus, Pistacia, Poncirus, Prunus, Pyrus, Rheum, Rhododendron, Richardia, Ricinus, Robinia, Saintpaulia, Schefflera, Schlumbergera, Sempervivum, Sinningia, Solanum, Spathiphyllum, Trifolium, Vinca, Viola, Washingtonia, Zea
Phytophthora palmivora (E.J. Butler) E.J. Butler
Kawakamia carica K. Hara
Phytophthora carica (K. Hara) S. Hori
Distr.: Warm-temperate, subtropical and tropical regions.
Notes: Stamps (C.M.I. Descr. 831. 1985).
Hosts: Allium, Ananas, Arecaceae, Cajanus, Carica, Catharanthus, Cattleya, Chamaedorea, Cocos, Cornus, Crotalaria, Dendrobium, Dieffenbachia, Epidendrum, Ficus, Hedera, Lycopersicon, Macadamia, Morrenia, Mucuna, Nolina, Peperomia, Persea, Phoenix, Richardia, Ricinus, Roystonea, Syagrus, Vanda, Washingtonia
Phytophthora phaseoli Thaxt.
Distr.: North America; West Indies; central Africa; Asia; Europe.
Host: Phaseolus
Phytophthora pseudotsugae P.B. Hamm & E.M. Hans.
Distr.: Northwestern United States.
Notes: Root rot of Douglas fir (Canad. J. Bot. 61:2626. 1983).
Host: Pseudotsuga
Phytophthora richardiae Buisman
Phytophthora cryptogea Pethybr. & Lafferty var. *richardiae* (Buisman) S. Ashby
Distr.: Widespread.
Notes: Root rots of various plants.
Host: Zantedeschia
Phytophthora spinosa Fell & Master
Distr.: Warm oceanic regions.
Notes: Isolated from leaves of mangroves and other debris (Fell & Master. Canad. J. Bot. 53:2917. 1975).
Host: Rhizophora
Phytophthora syringae (Kleb.) Kleb.
Distr.: Temperate regions.
Notes: Twig blight and wilt of lilac and other hardwoods (Waterhouse & Waterston. C.M.I. Descr. 32. 1964).
Hosts: Citrus, Prunus, Pyrus, Syringa
Phytophthora vesicula Anastasiou & Churchland
Distr.: Worldwide in marine environments.
Notes: Fell & Master (Canad. J. Bot. 53:2919. 1975).
Host: Rhizophora

Pichia E. Hans. Hemiascomycetes, Endomycetales

Pichia amethionina Starmer, Phaff, M. Miranda & M.W. Miller var. ***pachycereana*** Starmer, Phaff, M. Miranda & M.W. Miller
Distr.: AZ; Mexico.
Notes: Associated with stem decay (Starmer et al. Int. J. Systematic Bacteriol. 28:433. 1978).
Host: Carnegiea
Pichia cactophila Starmer, Phaff, M. Miranda & M.W. Miller
Distr.: AZ, HI; Mexico.
Notes: Isolated from decaying plant tissue.
Hosts: Carnegiea, Opuntia
Pichia deserticola Phaff, Starmer, Tredick & M. Miranda
Distr.: AZ, TX.
Notes: Isolated from necrotic stems (Phaff et al. Int. J. Systematic Bacteriol. 35:211. 1985).
Host: Opuntia
Pichia farinosa (Lindner) E. Hans.
Distr.: Widespread.
Notes: Isolated from various organic substrates.
Host: Triticum
Pichia fermentans Lodder
Distr.: Widespread.
Host: Triticum
Pichia heedii Phaff, Starmer, M. Miranda & M.W. Miller
Distr.: AZ; Mexico.
Notes: Isolated from necrotic tissues.
Host: Carnegiea

Piggotia Berk. & Broome. Coelomycetes

Notes: Sutton (The Coelomycetes. p. 336. 1980).
Piggotia coryli (Desmaz.) Sutton
Gloeosporium coryli (Desmaz.) Sacc.
Labrella coryli (Desmaz.) Sacc.
Monostichella coryli (Desmaz.) Höhn.
Distr.: Temperate North America; Europe.
Notes: On leaves.
Host: Corylus
Piggotia depressa Dearn.
Distr.: Temperate North America.
Host: Angelica
Piggotia fraxini Berk. & M.A. Curtis
Asterostromella fraxini (Berk. & M.A. Curtis) Petr.
Marssonia fraxini Ellis & J.J. Davis
Marssonina fraxini (Ellis & J.J. Davis) J.J. Davis
Septoria besseyi Peck
Gloeosporium decipiens Ellis & Everh.
Mycosphaerella effigurata (Schwein.) House (teleomorph)
Distr.: Central and eastern North America.
Notes: The relationship of this fungus to *Cylindrosporium fraxini* (Ellis & Kellerm.) Ellis & Everh. and *Phloeosporina fraxini* (Harkn.) Arx is in need of clarification (Wolf & Davidson. Mycologia 33:526. 1941; Arx. Biblioth. Mycol. 24:124. 1970).
Host: Fraxinus
Piggotia negundinis Ellis & Dearn.
Distr.: Central North America; Canada(Ontario, type).
Notes: A poorly known species.
Host: Acer
Piggotia vaccinii J.J. Davis
Distr.: WI.
Host: Vaccinium

Pileolaria Castagne. Basidiomycetes, Uredinales

Pileolaria brevipes Berk. & Ravenel
Uromyces brevipes (Berk. & Ravenel) Pazschke
Pileolaria toxicodendri (Berk. & Ravenel) Arth.
Uromyces toxicodendri Berk. & Ravenel
Distr.: Eastern and northwestern North America; Japan.
Notes: Autoecious. Parmelee & Elliott (Fungi Canadenses 50. 1974).
Host: Rhus
Pileolaria cotini-coggygriae F.L. Tai & Cheo

Distr.: AR, GA; China.
Notes: Smoke-tree rust (Thompson. Mycologia 52:321. 1960).
Host: Cotinus
Pileolaria effusa **Peck**
Distr.: AZ, CO, OK.
Notes: Microcyclic, pycnia and telia on *Rhus* (Leon-Gallegos & Cummins. Uredinales [Royas] de Mexico, vol. 2:325. 1981).
Host: Rhus
Pileolaria patzcuarensis **(Holw.) Arth.**
Distr.: CO, NM; Canada (Ontario); Mexico.
Notes: On Anacardiaceae.
Host: Rhus

Pilidium Kunze in Kunze & J.C. Schmidt. Coelomycetes
Notes: Sutton (The Coelomycetes. p. 563. 1980).
Pilidium acerinum **Kunze in Kunze & J.C. Schmidt**
Coccomyces acerinus (Kunze) Quél.
Leptothyrium acerinum (Kunze) Corda
Distr.: Cosmopolitan.
Notes: Leaf spot on various hardwoods.
Hosts: Acer, Eucalyptus
Pilidium concavum **(Desmaz.) Höhn.**
Ceuthospora concava Desmaz.
Sclerotiopsis concava (Desmaz.) Shear & B.O. Dodge
Sporonema pulvinatum Shear
Distr.: Temperate North America; Europe.
Hosts: Acer, Camellia, Corylus, Quercus, Vitis

Piloderma Jülich. Basidiomycetes, Aphyllophorales
Notes: Jülich (Willdenowia Beih. 7:211. 1972).
Piloderma bicolor **(Peck) Jülich**
Corticium bicolor Peck
Athelia bicolor (Peck) Parmasto non Pers., hom. illeg.
Distr.: Temperate northern hemisphere.
Notes: On wood, plant debris in litter (especially around conifers); apparently mycorrhizal.
Hosts: Abies, Picea, Pinus, Populus, Tsuga
Piloderma byssinum **(P. Karst.) Jülich**
Athelia byssina (P. Karst.) Parmasto
Corticium byssinum (P. Karst.) Massee
Distr.: Temperate northern hemisphere.
Notes: On wood and litter.
Hosts: Picea, Pinus, Populus, Symphoricarpos
Piloderma reticulatum **Jülich**
Athelia reticulata (Litsch.) Parmasto
Corticium reticulatum Litsch., nom. illeg.
Distr.: North America; Europe.
Notes: On wood.
Host: Picea
Piloderma sphaerosporum **Jülich**
Distr.: WA; Europe.
Notes: On wood.
Host: Acer

Piptarthron Mont. ex Höhn. Coelomycetes
Piptarthron macrosporum **(Durieu & Mont.) Höhn.**
Septoria macrospora Durieu & Mont.
Stagonospora macrospora (Durieu & Mont.) Sacc.
Hendersonia montagnei Cooke
Kellermania multiseptata Morgan-Jones, Nag Raj, & Kendrick
Hendersonia piptarthra Sacc.
Distr.: Western United States; southern Europe.
Notes: Morgan-Jones et al. (Canad. J. Bot. 50:1644 as *Kellermania multise ptata*) provided a description of North American specimens.
Hosts: Agave, Yucca
Piptarthron yuccae **(Ellis & Everh.) Sutton**
Phyllachora yuccae Ellis & Everh.
Kellermania attenuata Morgan-Jones, Nag Raj, & Kendrick
Septoplaca limbata Petr.
Piptarthron limbatum (Petr.) Sutton
Dothidea conspicua D. Griffiths (teleomorph)
Distr.: Southwestern North America.
Notes: Sutton (The Coelomycetes. p. 54. 1980 as *Piptarthron limbatum*; Trans. Brit. Mycol. Soc. 81:407. 1983).
Hosts: Agave, Yucca

Piptoporus P. Karst. Basidiomycetes, Aphyllophorales
Piptoporus betulinus **(Bull.:Fr.) P. Karst.**
Polyporus betulinus (Bull.:Fr.) Fr.
Distr.: Northern hemisphere.
Notes: On wood.
Host: Betula
Piptoporus soloniensis **(Dubois:Fr.) Pilát**
Distr.: Southeastern United States; Europe.
Notes: On wood.
Host: Quercus

Pirostoma Fr. Coelomycetes
Pirostoma circinans **(Nees) Fr.**
Coniosporium circinans Nees
Distr.: WI; Europe.
Hosts: Schizachyrium, Sorghastrum
Pirostoma dianellae **F. Stevens & P.A. Young**
Distr.: HI.
Host: Dianella

Pithoascus Arx. Pyrenomycetes, Microascales
Notes: Arx (Persoonia 7:372. 1973).
Pithoascus intermedius **(C.W. Emmons & B.O. Dodge) Arx**
Microascus intermedius C.W. Emmons & B.O. Dodge
Distr.: Temperate regions.
Notes: Isolated from roots, more recently from seeds (Corlett & MacLatchy. Fungi Canadenses 305. 1986).
Hosts: Fragaria, Helianthus, Zea
Pithoascus nidicola **(Massee & E.S. Salmon) Arx**
Microascus nidicola Massee & E.S. Salmon
Distr.: Temperate regions.
Notes: Isolated from bees' nest and seeds.
Host: Helianthus
Pithoascus schumacheri **(E. Hans.) Arx**
Microascus schumacheri (E. Hans.) Curzi
Rosellinia schumacheri (E. Hans.) Sacc.
Distr.: Temperate regions.
Notes: Isolated from stored organic matter as well as dung.
Host: Zea

Pithomyces Berk. & Broome. Hyphomycetes
Notes: Rao & de Hoog (Stud. Mycol. 28:23. 1986).
Pithomyces atro-olivaceus **(Cooke & Harkn.) M.B. Ellis**
Helminthosporium atro-olivaceum Cooke & Harkn.
Distr.: CA; India.
Notes: On bark (Ellis. Dematiaceous Hyphomycetes. p. 44. 1971).
Hosts: Acacia, Juncus
Pithomyces chartarum **(Berk. & M.A. Curtis) M.B. Ellis**
Sporidesmium chartarum Berk. & M.A. Curtis
Piricauda chartarum (Berk. & M.A. Curtis) R.T. Moore
Leptosphaerulina chartarum Roux (teleomorph)
Distr.: Cosmopolitan.
Notes: Usually saprophytic; causes glume blotch of rice and sorghum, also produces mycotoxins (Sutton & Gibson. C.M.I. Descr. 540. 1977). Roux (Trans. Brit. Mycol. Soc. 86:319. 1986).
Hosts: Argyroxiphium, Cynodon, Dactylis, Eichhornia, Glycine, Gossypium, Juncus, Metrosideros
Pithomyces maydicus **(Sacc.) M.B. Ellis**
Clasterosporium maydicum Sacc.
Distr.: Widespread in tropical regions.
Notes: On leaves and other parts of various plants (Ellis.

Dematiaceous Hyphomycetes. p. 44. 1971).
Hosts: Juncus, Zea
Pithomyces quadrata (Atk.) M.B. Ellis
Sporidesmium quadratum Atk.
Piricauda quadrata (Atk.) R.T. Moore
Scheleobrachea quadrata (Atk.) S.J. Hughes
Stemphylium crataegi (Ellis & Everh.) Höhn.
Stigmella crataegi Ellis & Everh.
Distr.: Eastern North America.
Notes: Ellis (Mycol. Pap. 76:18. 1960).
Host: Crataegus

Pithya Fuckel. Discomycetes, Pezizales

Notes: Denison (Mycologia 64:609. 1972).
Pithya cupressina (Pers.) Fuckel
Pithya cupressi (Batsch) Rehm
Helotium thujinum Peck
Molliardomyces cupressina Paden (anamorph)
Distr.: North America; Bermuda; Europe.
Notes: On newly-fallen foliage of Cupressaceae.
Hosts: [Abies], Calocedrus, Chamaecyparis, Cupressus, Juniperus, Pseudotsuga, Thuja
Pithya vulgaris Fuckel
Distr.: CA; Europe.
Notes: On needles. Although Eckblad (Nytt Mag. Bot. 15:105. 1968) considered this a synonym of *P. cupressina*, Denison (Mycologia 64:616. 1972) recognized the distinct characteristics of *P. vulgaris*.
Host: Pseudotsuga

Placosphaeria (De Not.) Sacc. Coelomycetes

Placosphaeria arctostaphyli Ellis & Everh.
Distr.: Western United States.
Host: Arctostaphylos
Placosphaeria baccharidis Dearn. & House
Distr.: NY.
Host: Baccharis
Placosphaeria celtidis Dearn. & House
Distr.: NY.
Host: Celtis
Placosphaeria cornicola Dearn.
Distr.: Western North America.
Host: Cornus
Placosphaeria decipiens Fairm.
Distr.: NM.
Host: Aster
Placosphaeria galii Sacc.
Distr.: ND; Europe.
Host: Galium
Placosphaeria haydenii (Berk. & M.A. Curtis) Petr.
Guignardia haydeni (Berk. & M.A. Curtis) Arx & E. Müller (teleomorph)
Distr.: North America.
Notes: On stems of Asteraceae (Arx & Müller. Beitr. Kryptogamenfl. Schweiz 11(1):53. 1954).
Hosts: Aster, Solidago
Placosphaeria junci Bubák
Distr.: Western United States; Europe.
Host: Juncus
Placosphaeria lupini Syd.
Stictochorella lupini (Syd.) Syd.
Distr.: Western United States.
Host: Lupinus
Placosphaeria medicaginis Tehon
Distr.: IL.
Host: Medicago
Placosphaeria punctiformis (Fuckel) Massee & Sacc.
Phyllachora punctiformis Fuckel
Distr.: United States; Europe.
Host: Galium

Placostroma Theiss. & Syd. Pyrenomycetes, Phyllachorales

Placostroma bambusae (Turconi) R. Sprague
Scirrhodothis bambusae (Turconi) Trott.
Distr.: Southeastern United States; southern Europe.
Hosts: Bambusa, Poaceae
Placostroma sporoboli (Atk.) R. Sprague
Scirrhia sporoboli Atk.
Distr.: Southeastern United States.
Hosts: Muhlenbergia, Sporobolus

Plagiostoma Fuckel. Pyrenomycetes, Diaporthales

Notes: Barr (Mycol. Mem. 7:104. 1978).
Plagiostoma acerophilum (Dearn. & House) Barr
Gnomoniopsis acerophila Dearn. & House
Gnomonia acerophila (Dearn. & House) Barr in H.E. Bigelow & Barr
Distr.: Northeastern North America.
Notes: On overwintered petioles.
Host: Acer
Plagiostoma alneum (Fr.:Fr.) Arx
Gnomonia perversa Rehm
Distr.: North America; Europe.
Host: Alnus
Plagiostoma campylostylum (Auersw.) Barr
Gnomonia campylostyla Auersw.
Distr.: Eastern North America; Europe.
Notes: On dead leaves.
Host: Betula
Plagiostoma campylostylum (Auersw.) Barr var. **mirabilis** (Peck) Barr
Gnomonia mirabilis (Peck) M. Monod
Gnomoniella mirabilis (Peck) Sacc.
Distr.: NY.
Host: Betula
Plagiostoma conradii (Ellis) Barr
Diaporthe conradii Ellis
Gnomonia conradii (Ellis) Wehmeyer
Distr.: Eastern North America.
Notes: On dead limbs.
Host: Hudsonia
Plagiostoma devexum (Desmaz.) Fuckel
Gnomoniella eccentrica (Cooke & Peck) Sacc.
Distr.: Eastern North America; Europe.
Notes: On dead stems.
Host: Polygonum
Plagiostoma magnoliae (Ellis) Barr
Gnomonia magnoliae Ellis
Distr.: Eastern United States.
Host: Magnolia
Plagiostoma solidaginis J.C. Cooke & Barr
Distr.: CT.
Notes: Cooke & Barr (Mycotaxon 18:87. 1983).
Host: Solidago

Plasmodiophora Woronin. Plasmodiophoromycetes, Plasmodiophorales

Plasmodiophora brassicae Woronin
Asterocystis radicis De Wild.
Distr.: Worldwide on crucifers, most common in temperate regions.
Notes: Clubroot (Buczacki. C.M.I. Descr. 621. 1979).
Hosts: Alyssum, Armoracia, Brassica, Camelina, Capsella, Erysimum, Iberis, Lepidium, Lobularia, Lunaria, Matthiola, Nasturtium, Nicotiana, Raphanus, Rorippa, Sinapis, Sisymbrium, Thlaspi
Plasmodiophora diplantherae (Ferd. & Winge) M.T. Cook
Distr.: FL; West Indies(type).
Notes: Large galls on stems.
Host: Halodule

Plasmopara J. Schröt. Oomycetes, Peronosporales

Plasmopara acalyphae (G.W. Wils.) G.W. Wils.
Rhysotheca acalyphae G.W. Wils.
Distr.: WI(type).
Host: Acalypha
Plasmopara australis (Speg.) Swingle
Peronospora australis Speg.
Distr.: Worldwide on Cucurbitaceae.
Hosts: Echinocystis, Sicyos
Plasmopara brassicae Woronin
Distr.: NJ.
Notes: No account of this name was located.
Host: Hesperis
Plasmopara cephalophora J.J. Davis
Distr.: WI(type).
Host: Physostegia
Plasmopara cercidis C.G. Shaw
Distr.: TN.
Host: Cercis
Plasmopara crustosa (Fr.:Fr.) Jørst.
Plasmopara nivea Auct.
Distr.: Worldwide on Apiaceae.
Notes: Boerema & Verhoeven (Neth. J. Pl. Pathol. 82:206. 1976).
Hosts: Angelica, Cicuta, Ligusticum, Lomatium
Plasmopara epilobii (G. Otth) Sacc. & Syd. in Sacc.
Distr.: Temperate northern hemisphere.
Host: Epilobium
Plasmopara geranii (Peck) Berl. & De Toni
Peronospora geranii Peck
Distr.: Worldwide.
Host: Geranium
Plasmopara gonolobi (Lagerh.) Swingle
Peronospora gonolobi Lagerh.
Distr.: Central and eastern United States; Ecuador(type).
Hosts: Cynanchum, Matelea, Vincetoxicum
Plasmopara halstedii (Farl.) Berl. & De Toni in Sacc.
Peronospora halstedii Farl.
Distr.: North and South America; Africa; Eurasia.
Notes: Downy mildew of Asteraceae.
Hosts: Acanthospermum, Ambrosia, Arctotis, Bidens, Cacalia, Carthamus, Centaurea, Coreopsis, Dimorphotheca, Erechtites, Erigeron, Eupatorium, Gnaphalium, Helianthus, Hymenopappus, Iva, Lactuca, Madia, Ratibida, Rudbeckia, Senecio, Silphium, Solidago, Verbena, Verbesina, Vernonia, Xanthium
Plasmopara illinoensis (Farl.) J.J. Davis
Peronospora illinoensis Farl.
Distr.: IL, WI.
Host: Parietaria
Plasmopara latifolii Savile
Distr.: Boreal and lower arctic regions.
Notes: Distinguished from *P. epilobii* by larger sporangia (Savile. Canad. J. Bot. 40:1387. 1962).
Host: Epilobium
Plasmopara myosotidis C.G. Shaw
Distr.: WI.
Host: Myosotis
Plasmopara obducens (J. Schröt.) J. Schröt. in Cohn
Peronospora obducens J. Schröt.
Distr.: Worldwide.
Host: Impatiens
Plasmopara pygmaea (Unger) J. Schröt.
Peronospora pygmaea Unger
Distr.: Temperate regions.
Hosts: Aconitum, Anemone, Hepatica, Isopyrum
Plasmopara pygmaea (Unger) J. Schröt. var. ***fusca*** (Peck) J.J. Davis
Protomyces fusca Peck
Distr.: NY, WI.
Host: Hepatica
Plasmopara ribicola J. Schröt. ex J.J. Davis
Distr.: United States; Europe.
Host: Ribes
Plasmopara viburni Peck
Distr.: Central and eastern United States.
Host: Viburnum
Plasmopara viticola (Berk. & M.A. Curtis) Berl. & De Toni in Sacc.
Peronospora viticola (Berk. & M.A. Curtis) de Bary
Distr.: Worldwide in grape-growing regions.
Notes: Downy mildew of Vitaceae.
Hosts: Ampelopsis, Cissus, Parthenocissus, Vitis

Platycarpium P. Karst. Coelomycetes

Notes: Although this name has been considered a synonym of *Fusamen*, Sutton (The Coelomycetes. p. 214. 1980) suggested that it be retained for sympodial species.
Platycarpium caulicola Wehmeyer
Distr.: WA.
Host: Anaphalis

Platychora Petr. Loculoascomycetes, Dothideales

Platychora alni (Peck) Petr.
Dothidella alni Peck
Distr.: Northern North America.
Notes: On leaves.
Host: Alnus

Platygloea J. Schröt. Basidiomycetes, Auriculariales

Notes: The correct generic name for these species is *Achroomyces* (Donk. Persoonia 4:145. 1966), but not all names have been redisposed.
Platygloea acanthophysa Burdsall
Distr.: WI(type).
Notes: On wood (Burdsall. Mycotaxon 27:499. 1986).
Host: Tsuga
Platygloea caroliniana Coker
Distr.: NC.
Notes: On dead limbs.
Host: Lagerstroemia
Platygloea lagerstroemiae Coker
Distr.: NC.
Notes: On dead limbs.
Host: Lagerstroemia
Platygloea laplata Lindsey
Distr.: CO.
Host: Quercus
Platygloea sphaerospora G.W. Martin
Distr.: NJ.
Notes: On wood.
Host: Quercus
Platygloea unispora L. Olive
Distr.: Southeastern United States.
Notes: On dead limbs.
Hosts: Chamaecyparis, Juniperus

Platystomum Trevis. Loculoascomycetes, Dothideales

Platystomum compressum (Pers.:Fr.) Trevis.
Lophidium compressum (Pers.:Fr.) Sacc.
Distr.: Temperate northern hemisphere.
Notes: On decorticated wood, especially *Salix*.
Hosts: Acer, Ulmus
Platystomum compressum (Pers.:Fr.) Trevis. var. ***pseudomacrostomum*** (Sacc.) Chesters & A. Bell
Lophiostoma pseudomacrostomum Sacc.
Lophiostoma erosum Ellis & Everh.
Lophiostoma rudi Ellis & Everh.
Lophiostoma scelestum Cooke & Ellis
Distr.: North America.
Hosts: Acer, Clematis, Rhamnus

Plectania Fuckel. Discomycetes, Pezizales

Plectania melastoma (Sowerby) Fuckel
Bulgaria melastoma (Sowerby) Seaver
Distr.: Widespread.
Notes: On debris in woods.
Hosts: Juniperus, Larix
Plectania nannfeldtii Korf
Paxina nigrella Seaver
Distr.: Western United States.
Notes: On rotten wood in coniferous woods.
Hosts: Juniperus, Pinus, Tsuga

Plectosphaera Theiss. Pyrenomycetes, Phyllachorales

Plectosphaera selaginellae M.L. Farr
Distr.: NC(type).
Notes: Farr & Horner (Nova Hedwigia 15:244. 1968)
Host: Selaginella

Pleiochaeta (Sacc.) S.J. Hughes. Hyphomycetes

Pleiochaeta setosa (Kirchn.) S.J. Hughes
Ceratophorum setosum Kirchn.
Distr.: Cosmopolitan.
Notes: Leaf, stem, and pod lesions of legumes (Ellis & Holliday. C.M.I. Descr. 495. 1976; Pirozynski. Fungi Canadenses 12. 1974).
Hosts: Cytisus, Lupinus

Pleiopatella Rehm. Discomycetes, Helotiales

Pleiopatella harperi Rehm
Distr.: WA(type).
Host: Thuja

Plenodomus G. Preuss. Coelomycetes

Notes: This genus is now considered to be a section of *Phoma*.
Plenodomus destruens Harter
Distr.: Southern United States.
Notes: Foot rot of sweet potato. Despite several major treatments (Boerema & Van Kesteren. Persoonia 11:317. 1981; Boerema et al. Trans. Brit. Mycol. Soc. 77:61. 1981), *Plenodomus destruens* has not been transferred to *Phoma*.
Host: Ipomoea

Pleochaeta Sacc. & Speg. Pyrenomycetes, Erysiphales

Pleochaeta polychaeta (Berk. & M.A. Curtis) Kimbrough & Korf
Erysiphe polychaeta Berk. & M.A. Curtis in Berk.
Uncinula polychaeta (Berk. & M.A. Curtis) Massee
Uncinulopsis polychaeta (Berk. & M.A. Curtis) Homma
Uncinula petersii Seym.
Ovulariopsis sp. (anamorph)
Distr.: Southeastern United States; Central and South America; India; Pakistan.
Notes: Powdery mildew of hackberry. Kimbrough & Korf (Mycologia 55:619. 1963) provided the nomenclatural history.
Host: Celtis
Pleochaeta prosopidis (Speg.) U. Braun
Uncinula prosopidis Speg.
Distr.: TX; Argentina(type).
Host: Prosopis

Pleomassaria Speg. Loculoascomycetes, Dothideales

Notes: Barr (Mycotaxon 15:367. 1982).
Pleomassaria acericola Petr.
Distr.: Eastern North America.
Notes: On branches.
Host: Acer
Pleomassaria carpini (Fuckel) Sacc.
Massaria carpini Fuckel
Distr.: Eastern North America; Europe.
Notes: On branches.
Host: Carpinus
Pleomassaria maxima Ellis & Everh.
Massaria magnoliae Ellis & Everh.
Pleomassaria magnoliae Shear
Shearia fusca (Berk. & M.A. Curtis) Barr (anamorph)
Distr.: Eastern North America.
Notes: On branches.
Host: Magnolia
Pleomassaria siparia (Berk. & Broome) Sacc.
Sphaeria siparia Berk. & Broome
Massaria siparia (Berk. & Broome) C. Tul. & Tul.
Prosthemium betulinum Kunze (anamorph)
Distr.: Eastern North America; Europe.
Notes: On branches.
Host: Betula

Pleosphaeria Speg. Pyrenomycetes, Sphaeriales

Pleosphaeria echinata (Ellis & Everh.) J.H. Miller
Cucurbitaria echinata Ellis & Everh.
Distr.: Eastern United States.
Notes: On old stromata of *Hypoxylon rubiginosum* (Miller. Mycologia 33:78. 1941).
Hosts: Acer, Crataegus, Fraxinus
Pleosphaeria modesta Harkn.
Distr.: CA(type).
Host: Eucalyptus
Pleosphaeria salicina (Mont.) G. Arnaud
Distr.: CA.
Host: Baccharis

Pleosphaerulina Pass. Loculoascomycetes, Dothideales

Notes: This genus is considered a synonym of *Pringsheimia*, but not all species have been redisposed..
Pleosphaerulina californica Berl.
Distr.: CA(type).
Host: Dendromecon
Pleosphaerulina corticola (Fuckel) Rehm f. ***crataegi*** Rehm
Distr.: ND(type).
Notes: The type is now placed in *Discostroma*.
Host: Crataegus
Pleosphaerulina corticola (Fuckel) Rehm f. ***rosae*** Rehm
Distr.: ND(type).
Notes: The type is now placed in *Discostroma*.
Host: Rosa
Pleosphaerulina sojicola Miura
Phyllosticta sojicola C. Massal. (anamorph)
Distr.: Worldwide in soybean-growing regions.
Host: Glycine
Pleosphaerulina violae Nagornyi
Distr.: AL; Soviet Union.
Host: Viola

Pleospora Rabenh. ex Ces. & De Not., nom. cons. Loculoascomycetes, Dothideales

Notes: A comprehensive account is provided by Wehmeyer (A world monograph of the genus *Pleospora* and its segregates. 1961). See also Eriksson (Ark. Botanik 6:339. 1967), Crivelli (Ueber die heterogene Ascomycetengattung *Pleospora* Rabh.; Vorschlag für eine Aufteilung. 1983) and Müller (Sydowia 5:248. 1951).
Pleospora abscondita Sacc. & Roum.
Distr.: Western United States; Europe.
Host: Silene
Pleospora allii (Rabenh.) Ces. & De Not.

Sphaeria allii Rabenh.
Stemphylium vesicarium (Wallr.) E. Simmons (anamorph)
Distr.: Cosmopolitan.
Notes: Sivanesan (The bitunicate Ascomycetes. p. 537. 1984) listed this connection tenuously.
Hosts: Asparagus, Erigeron
Pleospora amelanchieris Wehmeyer
Distr.: CO, WY.
Host: Amelanchier
Pleospora amplispora Ellis & Everh.
Distr.: Northwestern United States.
Host: Lupinus
Pleospora anthyllidis Auersw. & Niessl
Distr.: Western United States; Europe.
Notes: On various herbaceous plants.
Host: Arnica
Pleospora anthyllidis Auersw. & Niessl var. ***abbreviata*** (Fuckel) Wehmeyer
Pleospora abbreviata Fuckel
Pyrenophora ampla Syd.
Pyrenophora polyphragmoides Sacc. & Scalia
Distr.: Western United States.
Hosts: Anemone, Polemonium
Pleospora argyrospora Harkn.
Distr.: CA(type).
Host: Dendromecon
Pleospora bataanensis Petr.
Distr.: CA; subtropical and tropical regions.
Host: Agave
Pleospora calvescens (Fr. ex Desmaz.) Tul.
Chaetoplea calvescens (Fr. ex Desmaz.) Clements in Clements & Shear
Pyrenophora calvescens (Fr. ex Desmaz.) Sacc.
Pyrenophora echinella (Cooke) Berl.
Leptosphaeria eutypoides Peck
Ascochyta caulina (P. Karst.) van der Aa & Van Kesteren (anamorph)
Distr.: Eastern North America; Europe.
Notes: Aa & Van Kesteren (Persoonia 10:271. 1979) presented a detailed account of the *Ascochyta* anamorph.
Hosts: Chenopodium, Suaeda
Pleospora cassiae Ellis & Everh.
Distr.: TX(type).
Host: Cassia
Pleospora cerastii Oudem.
Pyrenophora cerastii (Oudem.) Lind
Distr.: AK; Greenland(type).
Notes: Wehmeyer (A world monograph of the genus *Pleospora* and its segregates. p. 343. 1961) stated that this name should not be used for lack of a type specimen and adequate definition.
Hosts: Oxyria, Pedicularis, Salix, Silene
Pleospora chlamydospora Sacc.
Pleospora balsamorhizae Tracy & Earle
Distr.: Cosmopolitan.
Notes: On herbaceous stems of various dicots.
Hosts: Alopecurus, Artemisia, Balsamorhiza, Bupleurum, Lupinus, Myosotis
Pleospora chlorogalli Fairm. in Millsp. & L. Nutt.
Distr.: CA(type).
Host: Chlorogalum
Pleospora coloradensis Ellis & Everh.
Distr.: Western North America.
Hosts: Arabis, Balsamorhiza, Potentilla
Pleospora comata Auersw. & Niessl
Pyrenophora comata (Auersw. & Niessl) Sacc.
Distr.: Western United States; Europe.
Notes: On herbaceous stems of various dicots.
Hosts: Anemone, Arenaria, Astragalus, Cardamine, Dalea, Draba, Erigeron, Frasera, Lupinus, Oxyria, Phacelia, Phlox, Polemonium, Ranunculus, Silene
Pleospora dichromatricha (Speg.) Wehmeyer
Pyrenophora dichromatricha Speg.
Pleospora clementsii Wehmeyer
Distr.: CO; Argentina.
Notes: This is a doubtful species according to Wehmeyer (A world monograph of the genus *Pleospora* and its segregates. p. 77. 1961).
Hosts: Mentha, Mentzelia
Pleospora discors (Durieu & Mont.) Ces. & De Not.
Sphaeria discors Durieu & Mont.
Distr.: Cosmopolitan.
Notes: On Cyperaceae.
Host: Carex
Pleospora ellisii Wehmeyer
Distr.: NC, NM.
Host: Yucca
Pleospora eocoronis (Clements) Wehmeyer
Pyrenophora ciliata (Ellis) Sacc. var. *eocoronis* Clements
Distr.: AK, CO, UT.
Host: Trisetum
Pleospora farlowiana Rehm
Host: Selaginella
Pleospora gaudefroyi Pat.
Distr.: Atlantic Ocean shore.
Notes: On driftwood and marine woody plants (Kohlmeyer & Kohlmeyer. Marine mycology. p. 438. 1979).
Host: Salicornia
Pleospora gigaspora P. Karst.
Stemphylium sp. (anamorph)
Distr.: Alpine and arctic regions.
Notes: On herbaceous stems of various plants.
Hosts: Carex, Lupinus
Pleospora heleocharidis P. Karst. var. ***arctica*** (P. Karst.) Wehmeyer
Pleospora arctica P. Karst.
Pleospora karstenii Berl. & Voglino
Distr.: Arctic and temperate regions.
Notes: On grasses.
Hosts: Dupontia, Hierochloe, Lycopodium, Poa
Pleospora helvetica Niessl
Pleospora chrysospora Niessl ex G. Wint.
Pyrenophora chrysospora (Niessl ex G. Wint.) Sacc.
Distr.: Western United States; temperate Europe.
Hosts: Achillea, Agastache, Arenaria, Arnica, Artemisia, Aster, Bupleurum, Castilleja, Claytonia, Cystopteris, Gilia, Hieracium, Iliamna, Ligusticum, Lloydia, Lupinus, Oxytropis, Papaver, Parrya, Pedicularis, Sagina, Saussurea, Saxifraga, Silene, Valeriana, Xerophyllum
Pleospora herbarum (Pers.:Fr.) Rabenh.
Pleospora frangulae Fuckel
Pleospora fructicola Ruehle
Pleospora herbarum (Pers.:Fr.) Rabenh. var. *citrorum* Sacc.
Pleospora labiatarum Cooke & Harkn.
Pleospora leguminum (Wallr.) Rabenh.
Pleospora lycopersici Él. Marchal & Ém. Marchal
Pleospora meliloti Rabenh.
Pleospora salsolae Fuckel
Pleospora typhae Pass.
Stemphylium herbarum E. Simmons (anamorph)
Distr.: Cosmopolitan.
Notes: Saprobic or rarely weakly parasitic on a wide variety of plants. Booth & Pirozynski (C.M.I. Descr. 150. 1967), Simmons (Mycologia 61:6. 1969) and Corlett et al. (Fungi Canadenses 232. 1982) provided traditional interpretations. Simmons (Sydowia 38:284. 1985) redefined the teleomorph-anamorph connection.
Hosts: Acacia, Agrostis, Allium, Ambrosia, Arnica, Artemisia, Asparagus, Astragalus, Balsamorhiza, Callistephus, Chrysanthemum, Cirsium, Citrus, Crepis, Cymopterus, Dactylis, Epilobium, Eupatorium, Foeniculum, Haplopappus, Iliamna, Lactuca, Linum, Lonicera, Lotus, Lupinus, Lychnis, Lycopersicon, Malus, Marrubium, Medicago, Melilotus, Oxytropis, Papaver, Pedicularis, Penstemon, Phaseolus, Phleum, Phoenix, Pisum, Potentilla, Primula, Pyrus, Rafinesquia, Ranunculus, Rhamnus, Rhododendron, Rosa, Rubus, Salicornia, Salix, Salvia, Saxifraga, Silene, Sitanion, Spartina, Stephanomeria, Syringa, Trifolium, Triticum, Typha, Vaccinium, Verbena, Vitis
Pleospora herbarum (Pers.:Fr.) Rabenh. var. ***occidentalis*** Wehmeyer
Pyrenophora eriogoni Earle
Pleospora lepidiicola Earle
Distr.: Cosmopolitan.

Hosts: Agastache, Artemisia, Eriogonum, Iliamna, Ivesia, Lepidium, Lupinus, Penstemon, Senecio
Pleospora lactucicola Ellis & Everh.
Pleospora diaporthoides Ellis & Everh.
Distr.: Temperate North America; Chile; temperate Europe.
Notes: On herbaceous stems.
Hosts: Ambrosia, Helianthus, Hieracium, Lactuca, Pastinaca
Pleospora laricina Rehm
Distr.: Temperate regions.
Notes: On various woody substrates.
Host: Cornus
Pleospora lecanora (Fabre) Rehm
Leptosphaeria lecanora Fabre
Pleospora pellita (Fr.) Rabenh. var. *bardanae* (Niessl) Wehmeyer
Distr.: Cosmopolitan.
Notes: On herbaceous stems of dicots.
Host: Salsola
Pleospora longispora Speg.
Pleospora gigantasca Rostr.
Pleospora laxa Ellis & Holw.
Distr.: Northern North America; Iceland; Tierra del Fuego.
Hosts: Carex, Elymus, Hesperochloa, Potentilla
Pleospora lutea Wehmeyer
Distr.: AK.
Hosts: Agropyron, Alopecurus
Pleospora magnifica Peck
Distr.: Western United States.
Notes: Barr et al. (New York State Mus. Bull. 459:29. 1986).
Hosts: Lupinus, Phlox
Pleospora magnusiana Berl.
Distr.: Arctic region.
Notes: On grasses.
Host: Poa
Pleospora mali Hesler
Hendersonia mali Thuem. (anamorph)
Distr.: NY.
Host: Malus
Pleospora montana Wehmeyer
Distr.: Western North America.
Hosts: Arenaria, Lupinus
Pleospora moravica (Petr.) Wehmeyer
Pyrenophora moravica Petr.
Distr.: AK; Algeria; Austria.
Host: Salix
Pleospora nigricantia Atk.
Distr.: AL(type).
Host: Gossypium
Pleospora nivalis Niessl
Host: Selaginella
Pleospora njegusensis Bubák
Distr.: Western United States; Europe.
Notes: On various dicots. North American species are probably *Pleospora magnifica* Peck (Barr et al. New York State Mus. Bull. 459:29. 1986).
Hosts: Castilleja, Lupinus
Pleospora paucitricha Fuckel
Pyrenophora paucitricha (Fuckel) Berl. & Voglino
Distr.: AK; Greenland(type).
Notes: Wehmeyer (A world monograph of the genus *Pleospora* and its segregates. p. 369. 1961) stated that this species is similar to *P. richtophensis*.
Host: Papaver
Pleospora pelagica T.W. Johnson
Distr.: East coast of North America.
Host: Spartina
Pleospora pellita (Fr.:Fr.) Rabenh. ex Ces. & De Not. var. ***oligostachyae*** (Ellis & Everh.) Wehmeyer
Pleospora oligostachyae Ellis & Everh.
Distr.: Central United States; Europe; India.
Host: Bouteloua
Pleospora penicillus (J.C. Schmidt) Fuckel
Sphaeria penicillus J.C. Schmidt
Pleospora ambigua (Berl. & Bres.) Wehmeyer
Pleospora hispida Niessl
Pyrenophora hispida (Niessl) Sacc.
Pleospora media Niessl
Pleospora phaeocomoides (Berk. & Broome) G. Wint.
Sphaeria phaeocomoides Berk. & Broome
Camarosporium sp. (anamorph)
Distr.: Temperate regions.
Notes: On various herbaceous stems.
Hosts: Androsace, Anemone, Antennaria, Artemisia, Aster, Astragalus, Chrysanthemum, Daucus, Draba, Erigeron, Eupatorium, Foeniculum, Linnaea, Lupinus, Machaeranthera, Papaver, Petasites, Rudbeckia, Saxifraga, Trisetum, Valeriana, Verbesina, Vitis
Pleospora phaeospora (Duby) Ces. & De Not.
Pyrenophora phaeospora (Duby) Sacc.
Distr.: WY; Europe.
Host: Arenaria
Pleospora pleosphaerioides Wehmeyer
Distr.: WY(type).
Host: Leptodactylon
Pleospora pluriseptata Wehmeyer
Distr.: Western United States.
Notes: On various dicots.
Hosts: Arabis, Lupinus
Pleospora potentillae (Rostr.) E. Müller
Pleospora oligasca Bubák
Distr.: Western United States; Europe.
Host: Geum
Pleospora punctata Wehmeyer
Distr.: WY; Switzerland.
Host: Scirpus
Pleospora rainierensis Wehmeyer
Pleospora asymmetrica Wehmeyer
Distr.: Western United States.
Notes: On various herbaceous stems. These species were synonymized by Wehmeyer (A world monograph of the genus *Pleospora* and its segregates. p. 104. 1961).
Hosts: Agrostis, Festuca, Lupinus, Muhlenbergia, Poa
Pleospora rehmiana Staritz
Distr.: ID, VA; Germany.
Notes: This species may be *Pleospora scrophulariae* (Desmaz.) Höhn. (Wehmeyer. A world monograph of the genus *Pleospora* and its segregates. p. 372. 1961).
Hosts: Medicago, Melilotus
Pleospora richtophensis Ellis & Everh.
Distr.: Western United States.
Hosts: Achillea, Agastache, Aster, Castilleja, Erigeron, Monardella, Osmorhiza, Ranunculus, Senecio
Pleospora richtophensis Ellis & Everh. var. ***pallida*** Wehmeyer
Distr.: Western United States.
Notes: On various dicots.
Hosts: Cirsium, Ranunculus
Pleospora rudis Berl.
Distr.: Cosmopolitan.
Host: Astragalus
Pleospora scaevolae F. Stevens & E. Young
Distr.: HI.
Host: Scaevola
Pleospora scrophulariae (Desmaz.) Höhn. var. ***compositarum*** (Earle) Wehmeyer
Pleospora compositarum Earle
Distr.: Temperate regions.
Hosts: Agastache, Castilleja, Helenium, Kuhnia, Penstemon, Rudbeckia
Pleospora stenospora J. Schröt.
Distr.: WY; Europe.
Notes: According to Wehmeyer (A world monograph of the genus *Pleospora* and its segregates. p. 377. 1961), this species in the United States may be the same as *Pleospora coloradensis*.
Hosts: Hedysarum, Hymenoxys, Linum, Senecio
Pleospora straminis Sacc. & Speg.
Distr.: Eastern North America; Europe.
Host: Zea
Pleospora togwotiensis Wehmeyer
Distr.: WY; Switzerland.
Host: Lupinus
Pleospora tomentosa Wehmeyer
Distr.: WY.

Hosts: Castilleja, Rudbeckia
Pleospora tragacanthae Rabenh.
Pyrenophora fenestrata Peck
Distr.: Cosmopolitan.
Notes: On herbaceous stems of dicots.
Hosts: Arenaria, Arnica, Astragalus, Bupleurum, Castilleja, Cerastium, Eupatorium, Lupinus, Oxytropis, Phlox, Silene, Umbellularia
Pleospora tropaeoli Halst.
Distr.: OK(type).
Notes: This species may be a *Leptosphaerulina* (Wehmeyer. A world monograph of the genus *Pleospora* and its segregates. p. 333. 1961).
Host: Tropaeolum
Pleospora vitalbae (De Not.) Berl.
Leptosphaeria vitalbae (De Not.) G. Wint.
Teichospora vitalbae (De Not.) Sacc.
Pleospora clematidis Fuckel
Leptosphaeria pleosporoides Auersw.
Stagonospora sp. (anamorph)
Distr.: Temperate regions.
Notes: Sivanesan (The bitunicate Ascomycetes. p. 533. 1984).
Host: Clematis
Pleospora vulgaris Niessl f. ***astericola*** Rehm
Distr.: ND(type).
Notes: The type variety is now considered a synonym of *P. scrophulariae*.
Hosts: Aster, Prunus

Pleuroceras Riess. Pyrenomycetes, Diaporthales

Notes: Barr (Mycol. Mem. 7:117. 1978); Monod (Beih. Sydowia 9:161. 1983).
Pleuroceras pleurostylum (Auersw.) Barr
Gnomonia pleurostyla Auersw.
Distr.: Temperate regions.
Host: Salix
Pleuroceras populi G.E. Thompson
Septomyxa populina (Peck) Arx (anamorph)
Distr.: Central and eastern North America.
Host: Populus
Pleuroceras sassafras (Ellis & Everh.) Barr
Gnomonia sassafras Ellis & Everh.
Distr.: Eastern United States.
Host: Sassafras
Pleuroceras tenellum (Ellis & Everh.) Barr
Gnomonia tenella Ellis & Everh.
Gnomoniella tenella (Ellis & Everh.) Sacc.
Distr.: Eastern North America.
Notes: Often occuring on the same leaves as *Plagiostoma acerophila* (Dearn. & House) Barr.
Host: Acer

Pleurocybella Singer. Basidiomycetes, Agaricales

Notes: Sometimes considered to be a synonym of *Nothopanus* Singer.
Pleurocybella porrigens (Pers.:Fr.) Singer
Nothopanus porrigens (Pers.:Fr.) Singer
Pleurotus porrigens (Pers.:Fr.) P. Kumm.
Distr.: Temperate northern hemisphere.
Notes: Especially common on conifer wood.
Hosts: Abies, Pseudotsuga

Pleuroflammula Singer in Singer & A.H. Sm. Basidiomycetes, Agaricales

Pleuroflammula flammea (Murrill) Singer in Singer & A.H. Sm.
Crepidotus flammeus Murrill
Distr.: Eastern North America to Colombia.
Notes: Horak (Persoonia 9:441. 1978).
Host: Hibiscus

Pleurophomopsis Petr. Coelomycetes

Pleurophomopsis eucalypti Petr.
Distr.: HI(type).
Host: Eucalyptus

Pleurophragmium Costantin. Hyphomycetes

Notes: Ellis (Mycol. Pap. 114:42. 1968).
Pleurophragmium flumeanum (Sacc.) S.J. Hughes
Helminthosporium flumeanum Sacc.
Distr.: FL; The Philippines(type).
Host: Eremochloa

Pleurostoma Tul. Pyrenomycetes, Diatrypales

Pleurostoma ootheca (Berk. & M.A. Curtis) Barr
Coronophora ootheca (Berk. & M.A. Curtis) Sacc.
Neoarcangelia ootheca (Berk. & M.A. Curtis) Berl.
Distr.: Eastern United States.
Notes: On decorticated wood (Barr. Mycologia 77:564. 1985).
Host: Quercus

Pleurotellus Fayod. Basidiomycetes, Agaricales

Pleurotellus hypnophilus (Berk.) Sacc.
Claudopus commixtus Bres. in Sacc.
Crepidotus herbarum (Peck) Sacc.
Distr.: Temperate northern hemisphere.
Hosts: Acer, Cornus, Pinus

Pleurothyrium Bubák. Coelomycetes

Notes: This genus is monotypic.
Pleurothyrium longissimum (Lib.) Bubák
Leptostroma longissimum Lib.
Distr.: CA; Austria.
Host: Athyrium

Pleurotus (Fr.) P. Kumm., nom. cons. Basidiomycetes, Agaricales

Notes: Hilber (Biblioth. Mycol. 87. 1982).
Pleurotus albolanatus (Peck) Kauffm.
Distr.: MI.
Notes: According to Smith (Beih. Sydowia 1:55. 1957) this species is close to *P. porrigens*.
Host: Pseudotsuga
Pleurotus cystidiosus O.K. Miller
Antromycopsis broussonetiae Pat. & Trab. (anamorph)
Distr.: Temperate North America; Europe.
Hosts: Acer, Liquidambar, Populus, Quercus, Salix
Pleurotus dryinus (Pers.:Fr.) P. Kumm.
Pleurotus corticatus (Pers.:Fr.) Quél.
Distr.: Temperate northern hemisphere.
Hosts: Carya, Liquidambar, Malus, Quercus
Pleurotus levis (Berk. & M.A. Curtis) Singer
Panus levis Berk. & M.A. Curtis
Lentinus levis (Berk. & M.A. Curtis) Murrill
Pleurotus strigosus (Berk. & M.A. Curtis) Singer
Distr.: Eastern North America; Honduras.
Notes: On logs and stumps of various hardwoods.
Hosts: Acer, Betula, Juglans, Liquidambar, Salix
Pleurotus minutus Peck
Crepidopus minutus (Peck) Murrill
Distr.: NC, NY(type).
Hosts: Betula, Carya, Quercus
Pleurotus ostreatus (Jacq.:Fr.) P. Kumm.
Pleurotus limpidus (Fr.) Gill.
Pleurotus salignus (Pers.:Fr.) P. Kumm.
Pleurotus sapidus Kalchbr.
Pleurotus similis Peck
Distr.: Northern hemisphere.

Notes: Also commercially cultivated as an edible mushroom.
Hosts: Abies, Acacia, Acer, Alnus, Betula, Carpinus, Carya, Castanea, Celtis, Diospyros, Eucalyptus, Fagus, Fraxinus, Ilex, Juglans, Larix, Laurocerasus, Liquidambar, Liriodendron, Lupinus, Magnolia, Malus, Morus, Nyssa, Ostrya, Pandanus, Picea, Pistacia, Populus, Pseudotsuga, Quercus, Salix, Tilia, Ulmus, Wisteria
Pleurotus septicus Fr.:Fr.
Distr.: CA; Europe.
Notes: The status of this taxon is uncertain (Hilber. Biblioth. Mycol. 87:266. 1982).
Host: Heteromeles

Plicatura Peck. Basidiomycetes, Aphyllophorales

Plicatura nivea (Fr.:Fr.) P. Karst.
Merulius niveus Fr.:Fr.
Distr.: North America; Europe.
Notes: On wood, usually of hardwoods.
Hosts: Abies, Alnus, Betula, Populus

Plicaturopsis D. Reid. Basidiomycetes, Aphyllophorales

Plicaturopsis crispa (Pers.:Fr.) D. Reid
Cantharellus crispus Pers.:Fr.
Merulius crispus (Pers.:Fr.) Pers.
Plicatura crispa (Pers.:Fr.) Rea
Trogia crispa (Pers.:Fr.) Fr.
Plicatura faginea (Schrad.) P. Karst.
Merulius fagineus Schrad.
Distr.: Temperate North America; central Asia; Europe.
Notes: Commonly on twigs and branches of *Betula*, *Alnus*, and *Fagus*, but also on many other hardwoods and conifers.
Hosts: Acer, Alnus, Betula, Campsis, Fagus, Platanus, Populus, Prunus, Quercus, Salix, Tilia

Plochmopeltis Theiss. Loculoascomycetes, Dothideales

Plochmopeltis intricata (Ellis & G. Martin) Theiss.
Asterina intricata Ellis & G. Martin
Microthyriella intricata (Ellis & G. Martin) Petr.
Distr.: Southeastern United States.
Host: Quercus

Ploioderma Darker. Discomycetes, Rhytismatales

Notes: Darker (Canad. J. Bot. 45:1399. 1967).
Ploioderma hedgcockii (Dearn.) Darker
Hypoderma hedgcockii Dearn.
Distr.: Southern United States.
Notes: Needle blight of pines (Cannon & Minter. C.M.I. Descr. 799. 1984).
Host: Pinus
Ploioderma lethale (Dearn.) Darker
Hypoderma lethale Dearn.
Distr.: Eastern United States.
Notes: Needle blight of pines (Darker. Contr. Arnold Arbor. 1:32. 1932; Minter & Gibson. C.M.I. Descr. 570. 1978).
Host: Pinus
Ploioderma pedatum (Darker) Darker
Hypoderma pedatum Darker
Distr.: CA.
Notes: On needles.
Host: Pinus

Plowrightia Sacc. Loculoascomycetes, Dothideales

Plowrightia neomexicana Earle
Distr.: NM(type).
Host: Parthenocissus

Pluteus Fr. Basidiomycetes, Agaricales

Notes: Homola (Mycologia 64:1211. 1972).
Pluteus atricapillus (Batsch) Fayod
Pluteus cervinus (Schaeff.) P. Kumm.
Distr.: Temperate northern hemisphere.
Notes: On wood. Vellinga & Schreurs (Persoonia 12:342. 1985).
Hosts: Abies, Betula, Gleditsia, Picea, Pinus, Populus, Pseudotsuga, Quercus
Pluteus aurantiorugosus (Trog) Sacc.
Pluteus caloceps Atk.
Pluteus coccineus (Massee) Lange
Distr.: Eastern North America; Europe.
Notes: On dead hardwood stumps and logs.
Host: Acer
Pluteus fuliginosus Murrill
Distr.: NY.
Host: Pinus
Pluteus latifolius Murrill
Distr.: WA.
Host: Alnus
Pluteus lilacinus (Mont.) Singer
Distr.: Eastern United States.
Host: Quercus
Pluteus longistriatus Peck
Distr.: Temperate North America.
Notes: On hardwoods.
Host: Platanus
Pluteus petasatus (Fr.) Gill.
Pluteus curtisii Auct. non (Berk.) Sacc.
Distr.: Eastern North America; Europe.
Notes: On wood.
Host: Quercus

Pocillum De Not. Discomycetes, Helotiales

Pocillum americanum Cooke
Distr.: FL, GA(type).
Notes: On recently fallen oak leaves.
Host: Quercus

Pocosphaeria (Sacc.) Berl. Loculoascomycetes, Dothideales

Notes: This genus is considered a synonym of *Leptosphaeria*, but not all species have been redisposed.
Pocosphaeria dendromeconis Earle
Distr.: CA(type).
Host: Dendromecon

Poculum Velen. Discomycetes, Helotiales

Notes: This name was rejected in favor of the conserved name, *Rutstroemia* Karst., but not all names have been redisposed.
Poculum fraternum (Peck) Dumont
Helotium fraternum Peck
Hymenoscyphus fraternus (Peck) Dennis
Distr.: Eastern United States.
Notes: Dumont (Mycotaxon 12:273. 1980).
Host: Acer
Poculum petiolorum (Roberge) Dumont & Korf in Korf & S. Gruff
Calycina petiolorum (Roberge) Seaver
Rutstroemia petiolorum (Roberge) W.L. White
Distr.: Eastern North America; temperate Europe.
Notes: On petioles and veins of Fagaceae.
Host: Quercus
Poculum sydowiana (Rehm) Dumont
Ciboria sydowiana Rehm
Ombrophila sydowiana Rehm, nom. inval.
Rutstroemia sydowiana (Rehm) W.L. White
Distr.: OR; Europe.
Notes: Dumont (Mycologia 68:872. 1976).

Hosts: Acer, Quercus

Pododimeria E. Müller. Loculoascomycetes, Dothideales

Notes: Luttrell & Barr (Amer. J. Bot. 65:251. 1978).
Pododimeria gelatinosa Luttrell & Barr
Distr.: VA(type).
Notes: On living needles.
Host: Juniperus
Pododimeria juniperi (Batista & Peres) Luttrell & Barr
Lojkania juniperi Batista & Peres
Distr.: Eastern United States.
Notes: On living needles.
Host: Juniperus

Podonectria Petch. Loculoascomycetes, Dothideales

Notes: On scale insects (Rossman. Mycotaxon 7:164. 1978), but often reported on their plant hosts.
Podonectria coccicola (Ellis & Everh.) Petch
Tetracrium coccicola Höhn. (anamorph)
Distr.: Subtropical and tropical regions.
Host: Citrus

Podosphaera Kunze. Pyrenomycetes, Erysiphales

Notes: Braun (Beih. Nova Hedwigia 89. 1987; Mycotaxon 19:370. 1984).
Podosphaera biuncinata Cooke & Peck
Distr.: Eastern North America.
Notes: Powdery mildew of witch hazel.
Host: Hamamelis
Podosphaera clandestina (Wallr.:Fr.) Lév.
Podosphaera oxyacanthae (DC.) de Bary
Oidium sp. (anamorph)
Distr.: Temperate northern hemisphere.
Notes: Powdery mildew of hawthorn (Khairi & Preece. C.M.I. Descr. 478. 1975). On Rosaceae.
Hosts: Amelanchier, Crataegus, Cydonia, Diospyros, Holodiscus, Malus, Prunus, Pyracantha, Pyrus, Sanguisorba, Spiraea, Symphoricarpos, Vaccinium
Podosphaera leucotricha (Ellis & Everh.) E.S. Salmon
Sphaerotheca leucotricha Ellis & Everh.
Distr.: Cosmopolitan.
Notes: Powdery mildew of apple (Kapoor. C.M.I. Descr. 158. 1967.)
Hosts: Cydonia, Malus, Photinia, Prunus, Pyrus
Podosphaera tridactyla (Wallr.) de Bary
Podosphaera oxyacanthae (DC.) de Bary var. *tridactyla* (Wallr.) E.S. Salmon
Oidium passerinii G. Bertol. (anamorph)
Distr.: Temperate regions.
Notes: Powdery mildew of plum (Mukerji. C.M.I. Descr. 187. 1968).
Hosts: Prunus, Sorbus, Spiraea

Podospora Ces., nom. cons. Pyrenomycetes, Sordariales

Notes: Most species occur on dung (Lundqvist. Symb. Bot. Upsal. 20:1. 1972).
Podospora curvicolla (G. Wint.) Niessl
Sordaria curvicolla G. Wint.
Distr.: Cosmopolitan.
Notes: Usually on dung.
Host: Salsola
Podospora minor Ellis & Everh.
Pleurage minor (Ellis & Everh.) D. Griffiths
Sordaria minor (Ellis & Everh.) Sacc. & Syd.
Distr.: KS(type).
Notes: Mirza & Cain (Canad. J. Bot. 47:2032. 1969).
Host: Zea

Podosporiella Ellis & Everh. Hyphomycetes

Podosporiella glomerata (Harkn.) Bonar
Cercospora glomerata Harkn.
Podosporium humile (Ellis & Everh.) Pound & Clements
Podosporiella humilis Ellis & Everh.
Distr.: CA.
Notes: Leaf spot (Ellis. Dematiaceous Hyphomycetes. p. 292. 1971).
Host: Garrya

Podosporium Schwein. Hyphomycetes

Podosporium rigidum Schwein.
Distr.: DE, MS.
Notes: On dead stems and branches (Ellis. Dematiaceous Hyphomycetes. p. 380. 1971).
Host: Rhus

Poliomopsis A.W. Ramaley. Basidiomycetes, Uredinales

Notes: Ramaley (Mycotaxon 28:361. 1987).
Poliomopsis thermopsidis A.W. Ramaley
Distr.: CO.
Notes: Only the telial stage is known.
Host: Thermopsis

Pollaccia Baldacci & Cif. Hyphomycetes

Notes: Ellis (More dematiaceous Hyphomycetes. p. 109. 1976).
Pollaccia peucedani (Ellis & Holw.) Deighton
Fusicladium peucedani Ellis & Holw.
Asperisporium peucedani (Ellis & Holw.) Maubl.
Distr.: Western United States.
Hosts: Glehnia, Lomatium
Pollaccia radiosa (Lib.) Baldacci & Cif.
Fusicladium radiosum (Lib.) Lind
Clasterosporium populi Ellis & Everh.
Stigmina populi (Ellis & Everh.) Peck
Dicoccum populinum Ellis & Everh.
Napicladium tremulae (A.B. Frank) Sacc.
Venturia macularis (Fr.:Fr.) E. Müller & Arx (teleomorph)
Distr.: Temperate North America; Europe.
Notes: Shoot tip or scab (Sivanesan. C.M.I. Descr. 403. 1974). Morelet (Crypt. Mycol. 6:101. 1985).
Host: Populus
Pollaccia saliciperda (Allesch. & Tub.) Arx
Fusicladium saliciperdum (Allesch. & Tub.) Tub.
Venturia saliciperda J. Nüesch (teleomorph)
Distr.: Temperate North America; Europe.
Notes: On twigs and leaves (Sivanesan. C.M.I. Descr. 482. 1976; Corlett. Fungi Canadenses 247. 1983 as *Venturia saliciperda*).
Host: Salix

Polychaeton (Pers.) Lév. Coelomycetes

Morfea Roze
Notes: Hughes (Mycologia 68:802. 1976); Sutton (Mycol. Pap. 141:160. 1977).
Polychaeton quercinum (Pers.) Kuntze
Fumago quercina Pers.
Capnodium quercinum (Pers.) Berk. & Desmaz., nom. inval.
Distr.: SC.
Host: Quercus

Polymorphum Chev. Coelomycetes

Polymorphum quercinum (Pers.) Chev.
Dichaena quercina (Pers.) Sacc.
Ascodichaena rugosa Butin (teleomorph)
Distr.: Temperate regions.
Notes: Beech bark fungus (Hawksworth. Taxon 32:212. 1983).
Host: Quercus

Polymyxa Ledingham. Plasmodiophoromycetes, Plasmodiophorales

Polymyxa betae Keskin
Distr.: Temperate regions.
Notes: Parasitic on roots of Chenopodiaceae (Barr. Fungi Canadenses 200. 1981). Responsible for transmitting beet necrotic yellow vein virus, associated with rhizomania disease of sugar beet.
Hosts: Amaranthus, Beta

Polymyxa graminis Ledingham
Distr.: Temperate North America; New Zealand, possibly more widely distributed.
Notes: Parasitic on roots of grasses (Barr. Fungi Canadenses 199. 1981).
Hosts: Agrostis, Avena, Cynodon, Trifolium, Triticum

Polynema Lév. Coelomycetes

Polynema radiatum (Ellis & Everh.) Nag Raj
Dinemasporium radiatum Ellis & Everh.
Distr.: PA.
Notes: Sutton (The Coelomycetes. p. 468. 1980).
Host: Celtis

Polynema tricristatum (Pat.) Sacc.
Distr.: FL; western Africa(type).
Host: Chloris

Polyporus P. Mich. ex Adans.:Fr., sensu Donk. Basidiomycetes, Aphyllophorales

Notes: Gilbertson & Ryvarden (North American Polypores 2:639. 1987).

Polyporus admirabilis Peck
Polyporus underwoodii Murrill
Distr.: Northeastern North America.
Notes: On living hardwoods or on stumps and roots.
Hosts: Acer, Betula, Fraxinus, Juglans, Malus, Salix, Ulmus

Polyporus arcularius (Batsch:Fr.) Fr.
Distr.: Temperate northern hemisphere.
Notes: On hardwoods.
Hosts: Acacia, Acer, Alnus, Celtis, Cowania, Fouquieria, Gleditsia, Juniperus, Liquidambar, Metrosideros, Morus, Platanus, Populus, Prosopis, Prunus, Quercus, Salix, Ulmus

Polyporus badius (Pers.) Schwein.
Polyporus fissus Berk.
Polyporus picipes Fr.
Distr.: Widespread.
Notes: On wood.
Hosts: Abies, Acer, Alnus, Betula, Castanea, Castanopsis, Picea, Pinus, Populus, Pseudotsuga, Quercus, Salix, Sambucus, Thuja, Tsuga

Polyporus brumalis (Pers.:Fr.) Fr.
Distr.: Temperate northern hemisphere.
Notes: On wood.
Hosts: Acer, Alnus, Betula, Populus, Salix

Polyporus coronadensis R.L. Gilbertson & K.J. Martin
Distr.: AZ.
Notes: On wood.
Host: Quercus

Polyporus melanopus Fr.:Fr.
Distr.: Western North America; Europe.
Notes: On wood, sometimes buried.
Hosts: Abies, Acer, Alnus, Ostrya, Picea, Pinus, Populus, Thuja, Tsuga

Polyporus mori (Pollini:Fr.) Fr.
Favolus alveolaris (DC.:Fr.) Quél.
Polyporus alveolaris (DC.:Fr.) Bondartsev & Singer, non (Bosc:Fr.) Fr., hom. illeg.
Distr.: Temperate northern hemisphere.
Notes: On hardwood limbs. Previously known in North America as *Favolus alveolaris*. Although the basionym of *F. alveolaris* (*Merulius alveolaris* DC. 1815) is older than that of *P. mori* (*Hexagonia mori* Pollini 1816) , *P. mori* is the correct name because the combination *Polyporus alveolaris* is occupied by another species (Donk. Check list of European Polypores. p. 138. 1974). In a genus other than *Polyporus*, the DeCandolle name would have priority.
Hosts: Acer, Betula, Carya, Castanea, Fagus, Fraxinus, Populus, Quercus, Salix, Tilia

Polyporus squamosus (Huds.:Fr.) Fr.
Agaricus squamosus P. Mich. ex Huds.
Polyporus westii Murrill
Distr.: Widespread.
Notes: On living or dead hardwoods.
Hosts: Acer, Aesculus, Alnus, Fagus, Fraxinus, Magnolia, Populus, Pyrus, Salix, Tilia, Ulmus

Polyporus tenuiculus (P. Beauv.:Fr.) Fr.
Favolus brasiliensis (Fr.:Fr.) Fr.
Distr.: Subtropical and tropical regions.
Notes: On wood of hardwoods.
Hosts: Carya, Celtis, Quercus, Vitis

Polyporus umbellatus (Pers.:Fr.) Fr.
Dendropolyporus umbellatus (Pers.:Fr.) Jülich
Grifola umbellata (Pers.:Fr.) Pilát
Polypilus umbellatus (Pers.:Fr.) P. Karst.
Distr.: Northern hemisphere.
Notes: At base of trees and stumps or from underground wood.
Hosts: Castanea, Picea

Polyporus varius Fr.:Fr.
Polyporus elegans (Bull.) Fr.
Distr.: Widespread.
Notes: On wood.
Hosts: Abies, Acer, Alnus, Amelanchier, Betula, Ceanothus, Corylus, Fagus, Liriodendron, Picea, Pinus, Populus, Prunus, Pseudotsuga, Rubus, Salix, Sambucus, Symphoricarpos, Thuja, Tilia, Tsuga

Polyporus virgatus Berk. & M.A. Curtis
Polyporus amygdalinus Berk. & Ravenel
Distr.: Southeastern United States; tropical regions.
Notes: On wood of hardwoods.
Host: Quercus

Polyschema Upadhyay. Hyphomycetes

Polyschema bicellularis C.A. Shearer
Distr.: IL(type).
Notes: On twigs (Shearer. Mycotaxon 14:91. 1982).
Host: Platanus

Polyschema olivacea (Ellis & Everh.) M.B. Ellis
Clasterosporium olivacea Ellis & Everh.
Clasterosporium elaeodes Pound & Clements
Torula elaeodes (Pound & Clements) S.J. Hughes
Clasterosporium zeae Sacc. & Syd.
Distr.: Eastern United States.
Notes: Ellis (Dematiaceous Hyphomycetes. p. 372. 1971).
Host: Zea

Polyscytalum Riess. Hyphomycetes

Polyscytalum cylindroides Sacc. & Ellis
Distr.: LA(type).
Host: Quercus

Polyscytalum pustulans (M.N. Owen & Wakef.) M.B. Ellis
Oospora pustulans M.N. Owen & Wakef.
Distr.: Temperate regions.
Notes: Parasitic on tubers of potato causing skin spot disease, also on roots, stolons and stems (Ellis. More dematiaceous Hyphomycetes. p. 159. 1976).
Host: Solanum

Polyscytalum sericeum Sacc.
Distr.: DE, FL; Europe.
Hosts: Magnolia, Quercus

Polyscytalum truncatum Sutton & C.S. Hodges
Distr.: HI; Brazil(type).
Notes: On fallen leaves (Sutton & Hodges. Nova Hedwigia 28: 487. 1977).
Host: Eucalyptus

Polystigma DC. Pyrenomycetes, Phyllachorales

Polystigma astragali (Lasch) Höhn.
Physalospora astragali (Lasch) Sacc.

Physalospora aurantia Ellis & Everh.
Physalospora megastoma (Peck) Sacc.
Distr.: Cosmopolitan.
Notes: On leaves and twigs of Fabaceae.
Host: Astragalus
Polystigma bumeliae Schwein.
Distr.: AR(type).
Host: Sideroxylon

Polythrincium Kunze. Hyphomycetes

Polythrincium trifolii Kunze in J.C. Schmidt & Kunze
Cymadothea trifolii (C. Killian) F.A. Wolf (teleomorph)
Distr.: Widespread.
Notes: Black blotch or sooty blotch of clover (Sivanesan. The bitunicate Ascomycetes. p. 143. 1984).
Host: Trifolium

Porodisculus Murrill. Basidiomycetes, Aphyllophorales

Porodisculus pendulus (Schwein.) Murrill
Polyporus pocula (Schwein.) Berk. & M.A. Curtis
Distr.: Eastern United States to Central America.
Notes: On wood, usually of hardwoods.
Hosts: Castanea, Fraxinus, Juglans, Prunus, Quercus, Rhus

Porotheleum Fr.:Fr. Basidiomycetes, Aphyllophorales

Stromatoscypha D. Reid
Porotheleum fimbriatum (Pers.:Fr.) Fr.
Poria fimbriata Pers.:Fr.
Stromatoscypha fimbriata (Pers.:Fr.) Donk
Porotheleum pezizoides Schwein.
Cyphella porioides (Albertini & Schwein.) Quél.
Distr.: Widespread.
Notes: On wood.
Hosts: Abies, Alnus, Betula, Juglans, Lithocarpus, Pinus, Populus, Quercus
Porotheleum perenne W.B. Cooke
Distr.: NY.
Notes: On wood.
Host: Acer

Postia Fr. Basidiomycetes, Aphyllophorales

Notes: Most species listed here under *Postia* are treated under *Oligoporus* by Gilbertson & Ryvarden (North American Polypores 2:456. 1987). These authors rejected *Postia* P. Karst. 1881 as a later homonym of *Postia* Boiss & Blanch. 1875; however *Postia* Fr. 1874 takes priority.
Postia amara (Hedgc.) M. Larsen & Lombard
Polyporus amarus Hedgc.
Fomes amarus (Hedgc.) Murrill
Oligoporus amarus (Hedgc.) R.L. Gilbertson & Ryvarden
Tyromyces amarus (Hedgc.) J. Lowe
Distr.: Western North America, not common.
Notes: On living or dead conifers, especially *Calocedrus decurrens*.
Hosts: Abies, Calocedrus
Postia angulopora (M. Larsen & Lombard) M. Larsen & Lombard
Fibroporia angulopora M. Larsen & Lombard
Distr.: OR.
Notes: On wood.
Host: Pseudotsuga
Postia balsamea (Peck) Jülich
Polyporus balsameus Peck
Oligoporus balsameus (Peck) R.L. Gilbertson & Ryvarden
Tyromyces balsameus (Peck) Murrill
Polyporus basilaris Overh. in H.E. Bailey
Tyromyces basilaris (Overh.) K.J. Martin
Polyporus carbonarius (Murrill) Murrill
Tyromyces carbonarius Murrill
Polyporus cutifractus (Murrill) Murrill
Tyromyces cutifractus Murrill
Distr.: Temperate northern hemisphere.
Notes: On conifers, rarely on hardwoods.
Hosts: Abies, Acer, Cupressus, Larix, Picea, Pinus, Populus, Prunus, Pseudotsuga, Thuja, Tsuga
Postia caesia (Schrad.:Fr.) P. Karst.
Polyporus caesius (Schrad.:Fr.) Fr.
Oligoporus caesius (Schrad.:Fr.) R.L. Gilbertson & Ryvarden
Tyromyces caesius (Schrad.:Fr.) Murrill
Distr.: Temperate northern hemisphere.
Notes: On hardwoods.
Hosts: Abies, Acer, Alnus, Betula, Calocedrus, Castanea, Juniperus, Larix, Liriodendron, Malus, Picea, Pinus, Populus, Pseudotsuga, Quercus, Rhododendron, Salix, Sambucus, Thuja, Tsuga
Postia ceriflua (Berk. & M.A. Curtis) Jülich
Polyporus cerifluus Berk. & M.A. Curtis in Berk.
Tyromyces cerifluus (Berk. & M.A. Curtis) Murrill
Oligoporus minusculoides (Pilát) R.L. Gilbertson & Ryvarden
Polyporus minusculoides (Pilát) J. Lowe
Tyromyces minusculoides (Pilát) Bondartsev
Distr.: SC.
Notes: On wood.
Hosts: Pinus, Populus, Pseudotsuga, Quercus, Tsuga
Postia floriformis (Quél.) Jülich
Polyporus floriformis Quél. in Bres.
Oligoporus floriformis (Quél.) R.L. Gilbertson & Ryvarden
Tyromyces floriformis (Quél.) Bondartsev & Singer
Distr.: Widespread.
Notes: On wood, usually of conifers.
Hosts: Acer, Larix, Pinus
Postia fragilis (Fr.:Fr.) Jülich
Polyporus fragilis Fr.:Fr.
Oligoporus fragilis (Fr.:Fr.) R.L. Gilbertson & Ryvarden
Tyromyces fragilis (Fr.:Fr.) Donk
Spongipellis fragilis (Fr.:Fr.) Murrill
Distr.: Temperate northern hemisphere.
Notes: On wood, mainly of conifers.
Hosts: Abies, Calocedrus, Chamaecyparis, Larix, Picea, Pinus, Populus, Pseudotsuga, Taxus, Thuja, Tsuga
Postia guttulata (Peck) Jülich
Polyporus guttulatus Peck in Sacc.
Oligoporus guttulatus (Peck) R.L. Gilbertson & Ryvarden
Tyromyces guttulatus (Peck) Murrill
Distr.: Widespread.
Notes: On wood, usually of conifers.
Hosts: Abies, Acer, Larix, Picea, Pinus, Populus, Pseudotsuga, Quercus, Thuja, Tsuga
Postia hibernica (Berk. & Broome) Jülich
Oligoporus hibernicus (Berk. & Broome) R.L. Gilbertson & Ryvarden
Poria simanii (Pilát) R.L. Gilbertson & J. Lowe
Postia simanii (Pilát) Jülich
Distr.: AZ; Europe.
Notes: On wood of conifers.
Hosts: Abies, Picea
Postia johnstonii (Murrill) Jülich
Poria johnstonii Murrill
Oligoporus johnstonii (Murrill) R.L. Gilbertson & Ryvarden
Tyromyces johnstonii (Murrill) Ryvarden
Distr.: CA.
Notes: On conifers.
Host: Tsuga
Postia leucomallela (Murrill) Jülich
Tyromyces leucomallelus Murrill
Oligoporus leucomallelus (Murrill) R.L. Gilbertson & Ryvarden
Distr.: Western North America.
Notes: On conifers.
Hosts: Larix, Picea, Pinus
Postia leucospongia (Cooke & Harkn.) Jülich
Polyporus leucospongia Cooke & Harkn. in Cooke
Spongiporus leucospongia (Cooke & Harkn.) Murrill
Oligoporus leucospongia (Cooke & Harkn.) R.L. Gilbertson & Ryvarden
Tyromyces leucospongia (Cooke & Harkn.) Bondartsev & Singer
Distr.: Western North America.
Notes: On conifers.

Hosts: Abies, Arctostaphylos, Larix, Picea, Pinus, Populus, Pseudotsuga, Tsuga

Postia mappa M. Larsen & Lombard
Oligoporus mappus (Overh. & J. Lowe) R.L. Gilbertson & Ryvarden
Poria mappa Overh. & J. Lowe
Distr.: ID, NY; Canada(British Columbia).
Notes: On wood.
Host: Picea

Postia perdelicata (Murrill) M. Larsen & Lombard
Tyromyces perdelicatus Murrill
Oligoporus perdelicatus (Murrill) R.L. Gilbertson & Ryvarden
Polyporus perdelicatus (Murrill) Murrill
Distr.: Western North America.
Notes: On conifer wood.
Hosts: Abies, Acer, Larix, Picea, Pinus, Populus, Tsuga

Postia placenta (Fr.) M. Larsen & Lombard
Polyporus placentus Fr.
Oligoporus placentus (Fr.) R.L. Gilbertson & Ryvarden
Poria placenta (Fr.) Cooke
Poria carnicolor D. Baxter
Poria microspora Overh. in Nobles
Poria monticola Murrill
Bjerkandera roseomaculata P. Karst.
Distr.: Northern North America; Europe.
Notes: Hirt & Lowe (Phytopathology 35:217. 1945; J. Forest. [Washington] 43:717. 1945); Davidson et al. (Mycologia 39:313. 1947 as *P. microspora*).
Hosts: Abies, Larix, Picea, Pinus, Populus, Pseudotsuga, Quercus, Tsuga, Zizia

Postia rancida (Bres.) M. Larsen & Lombard
Poria rancida Bres.
Oligoporus rancidus (Bres.) R.L. Gilbertson & Ryvarden
Poria cognata Overh.
Distr.: Eastern United States.
Notes: On wood.
Hosts: Abies, Fraxinus, Juglans

Postia salmonicolor (Berk. & M.A. Curtis) M. Larsen & Lombard
Polyporus salmonicolor Berk. & M.A. Curtis
Hapalopilus salmonicolor (Berk. & M.A. Curtis) Pouzar
Poria salmonicolor (Berk. & M.A. Curtis) Cooke
Poria rubens Overh. & J. Lowe
Distr.: Temperate northern hemisphere.
Notes: On conifers.
Host: Pinus

Postia sequoiae (Bonar) M. Larsen & Lombard
Poria sequoiae Bonar
Oligoporus sequoiae (Bonar) R.L. Gilbertson & Ryvarden
Distr.: CA.
Notes: On wood.
Host: Sequoia

Postia sericeomollis (Romell) Jülich
Polyporus sericeomollis Romell
Poria sericeomollis (Romell) Egeland
Oligoporus sericeomollis (Romell) Pouzar
Tyromyces sericeomollis (Romell) Bondartsev & Singer
Poria asiatica (Pilát) Overh.
Distr.: North America; Europe; Asia.
Hosts: Abies, Juniperus, Larix, Picea, Pinus, Prunus, Pseudotsuga, Thuja, Tsuga

Postia stiptica (Pers.:Fr.) Jülich
Oligoporus stipticus (Pers.:Fr.) R.L. Gilbertson & Ryvarden
Polyporus stipticus (Pers.:Fr.) Fr.
Polyporus immitis Peck
Tyromyces immitis (Peck) Bondartsev
Distr.: Temperate northern hemisphere.
Notes: On wood, usually of hardwoods. Lowe (Mycotaxon 2:55. 1975).
Hosts: Abies, Acer, Alnus, Cupressus, Fraxinus, Malus, Picea, Pinus, Platanus, Pseudotsuga, Quercus, Thuja, Tsuga

Postia tephroleuca (Fr.:Fr.) Jülich
Polyporus tephroleucus Fr.:Fr.
Oligoporus tephroleucus (Fr.:Fr.) R.L. Gilbertson & Ryvarden
Tyromyces tephroleucus (Fr.:Fr.) Donk
Distr.: Temperate northern hemisphere.
Notes: On wood of hardwoods, occasionally conifers.
Hosts: Acer, Alnus, Malus, Picea, Pinus, Platanus, Prunus, Pseudotsuga, Pyrus

Postia undosa (Peck) Jülich
Polyporus undosus Peck
Oligoporus undosus (Peck) R.L. Gilbertson & Ryvarden
Tyromyces undosus (Peck) Murrill
Distr.: North America; Europe.
Hosts: Abies, Acer, Betula, Fagus, Picea, Pinus, Populus, Pseudotsuga, Quercus, Tsuga

Pragmopora C. Massal. Discomycetes, Helotiales

Notes: Groves (Canad. J. Bot. 45:169. 1967).

Pragmopora pini Groves
Distr.: Western North America.
Host: Pinus

Pragmopora pithya (Fr.) Groves
Phacidium pithya (Fr.) Fr.
Pragmopycnis pithya Sutton & Funk (anamorph)
Distr.: Temperate regions.
Host: Pinus

Pragmopycnis Sutton & Funk. Coelomycetes

Notes: Sutton & Funk (Canad. J. Bot. 53:522. 1975).

Pragmopycnis pithya Sutton & Funk
Pragmopora pithya (Fr.) Groves (teleomorph)
Distr.: Temperate regions.
Notes: On conifers.
Hosts: Pinus, Pseudotsuga

Preussia Fuckel. Loculoascomycetes, Dothideales

Preussia vulgare (Corda) Cain
Perisporium vulgare Corda
Distr.: HI; Europe(type).
Notes: Cain (Canad. J. Bot. 39:1642. 1961).
Hosts: Acacia, Cheirodendron, Metrosideros

Pringsheimia Schulzer. Loculoascomycetes, Dothideales

Pringsheimia sepincola (Fr.) Höhn.
Saccothecium sepincola (Fr.) Fr.
Pleosphaerulina intermixta (Berk. & Broome) Berl.
Sphaerulina intermixta (Berk. & Broome) Sacc.
Dothichiza sp. (synanamorph)
Hormonema sp. (synanamorph)
Distr.: Temperate regions; Pakistan.
Notes: On dead twigs of Rosaceae. Sivanesan (The bitunicate Ascomycetes. p. 266. 1984).
Host: Rubus

Proliferodiscus J. Haines & Dumont. Discomycetes, Helotiales

Notes: Haines & Dumont (Mycologia 75:535. 1983).

Proliferodiscus earoleucus (Berk. & Broome) J. Haines & Dumont
Trichopeziza earoleuca (Berk. & Broome) Sacc.
Distr.: AK; tropical regions.
Host: Lupinus

Proliferodiscus pulveraceus (Albertini & Schwein.:Fr.) Baral
Dasyscyphus pulveraceus (Albertini & Schwein.:Fr.) Höhn.
Lachnellula pulveracea (Albertini & Schwein.:Fr.) Dennis
Distr.: Eastern North America; Europe; Macronesia.
Notes: Baral & Krieglsteiner (Beih. Z. Mykol. 6:84. 1985).
Hosts: Melia, Prunus, Quercus, Ulmus

Propolidium Sacc. Discomycetes, Ostropales

Notes: Sherwood (Mycotaxon 5:74) restricted this genus to two species, but not all names have been redisposed.

Propolidium californicum Cash

Distr.: CA(type).
Host: Salvia
Propolidium glaucum (Ellis) Sacc.
Propolis glauca Ellis
Distr.: Eastern North America.
Notes: On corticate oak twigs.
Host: Quercus
Propolidium shearii Cash
Distr.: CA(type).
Host: Quercus

Propolis (Fr.) Corda. Discomycetes, Rhytismatales

Notes: Sherwood (Mycotaxon 5:322. 1977).
Propolis decidua (Ellis & Everh.) Sacc.
Stictis decidua Ellis & Everh.
Distr.: NJ.
Host: Pinus
Propolis emarginata (Cooke & Massee) M.A. Sherwood
Coccomyces martinae Hansf.
Distr.: Widespread.
Notes: Common on fallen leaves.
Host: Eucalyptus
Propolis faginea (Schrad.) P. Karst. var. ***atra*** Cash
Distr.: CO(type).
Host: Populus
Propolis leonis (Tul. & C. Tul.) Rehm
Stictis leonis Tul. & C. Tul.
Distr.: Temperate regions.
Notes: Sherwood (Mycotaxon 5:203. 1977) suggested that this species belongs in *Propolomyces*.
Hosts: Larix, Pinus, Pseudotsuga, Tsuga
Propolis phacidioides (Fr.:Fr.) Corda
Stictis phacidioides Fr.:Fr.
Naemacyclus phacidioides (Fr.:Fr.) B. Eriksson
Phacidium albidum W. Phillips & Harkn.
Coccomyces albidus (W. Phillips & Harkn.) Sacc.
Stictis myricae Cash
Stictis vaccinii Ellis & Everh.
Distr.: North America; temperate Europe.
Notes: On leaves of Ericaceae.
Hosts: Myrica, Vaccinium
Propolis quercifolia (Cooke & Ellis) M.A. Sherwood
Stictis quericfolia Cooke & Ellis
Stictis quercina Peck
Distr.: Eastern North America.
Notes: On dead leaves of red oaks.
Host: Quercus
Propolis rhòdoleuca (Sommerf.) Fr.
Stictis rhodoleuca Sommerf.
Pseudopatellina conigena Höhn.
Stictis fulva Peck
Distr.: Temperate regions.
Notes: On cones. Sherwood (Mycotaxon 5:239. 1977) suggested that this species belongs in *Propolomyces*.
Host: Pinus
Propolis strobilina (Cooke) Sacc.
Stictis strobilina Cooke
Distr.: GA, SC(type).
Host: Pinus

Propolomyces M.A. Sherwood. Discomycetes, Rhytismatales

Propolomyces versicolor (Fr.:Fr.) Dennis
Stictis versicolor Fr.:Fr.
Propolis versicolor (Fr.:Fr.) Fr.
Propolis faginea (Schrad.) P. Karst.
Distr.: North America; Europe.
Notes: Sherwood (Mycotaxon 5:320. 1977).
Hosts: Acer, Alnus, Amelanchier, Arctostaphylos, Betula, Catalpa, Holodiscus, Laurocerasus, Liriodendron, Lonicera, Pinus, Prunus, Pseudotsuga, Quercus, Rhododendron, Rosa, Salix

Prospodium Arth. Basidiomycetes, Uredinales

Prospodium appendiculatum (G. Wint.) Arth.
Puccinia appendiculatum G. Wint.
Distr.: FL, TX; Mexico; Cuba; South America.
Notes: On Bignoniaceae.
Host: Tecoma
Prospodium lippiae (Speg.) Arth.
Puccinia lippiae Speg.
Distr.: AZ, TX; Mexico; Central and South America.
Notes: On Verbenaceae.
Host: Aloysia
Prospodium plagiopus (Mont.) Arth.
Puccinia plagiopus Mont.
Distr.: FL, PR; Cuba.
Notes: On Bignoniaceae.
Host: Tabebuia

Prosthecium Fresen. Pyrenomycetes, Diaporthales

Notes: Barr (Mycol. Mem. 7:184. 1978).
Prosthecium acrocystis (Peck) Barr
Valsa acrocystis Peck
Melanconis acrocystis (Peck) Ellis & Everh.
Distr.: Northeastern North America.
Notes: On limbs.
Host: Betula
Prosthecium appendiculatum (G. Otth) Barr
Diaporthe appendiculata G. Otth
Melanconis appendiculata (G. Otth) Wehmeyer
Distr.: Central and eastern North America; Europe.
Host: Acer
Prosthecium ellipsosporum Fresen.
Melanconis macrosperma Tul.
Distr.: MI, NE, NJ; Europe.
Notes: On Betulaceae (Wehmeyer. Univ. Michigan Stud., Sci. Ser. 14:96. 1941).
Host: Carya
Prosthecium stylosporum (Ellis & Everh.) Wehmeyer
Pseudovalsa stylospora Ellis & Everh.
Distr.: Northeastern North America.
Host: Acer

Protodontia Höhn. Basidiomycetes, Tremellales

Protodontia piceicola (Bourd.) G.W. Martin
Distr.: Temperate northern hemisphere.
Notes: On wood.
Host: Tsuga

Protomyces Unger. Hemiascomycetes, Taphrinales

Notes: Reddy & Kramer (Mycotaxon 3:1. 1975).
Protomyces andinus Pat.
Distr.: Central and South America.
Notes: Reports in temperate North America are based on misdeterminations of *P. grandisporus* Reddy & Kramer or *P. gravidus* J.J. Davis.
Hosts: Ambrosia, Bidens
Protomyces conglomeratus Peck
Distr.: NY(type).
Notes: Probably not a *Protomyces* (Reddy & Kramer. Mycotaxon 3:29. 1975).
Host: Salicornia
Protomyces gravidus J.J. Davis
Distr.: North America; temperate Europe.
Notes: Galls.
Hosts: Ambrosia, Bidens
Protomyces macrosporus Unger
Distr.: Widespread in the northern hemisphere.
Notes: Galls on Apiaceae.

Hosts: Cicuta, Torilis
Protomyces pachydermus Thuem.
Distr.: North America; temperate Europe.
Notes: Swellings and galls on Asteraceae.
Host: Taraxacum

Protopeltis Syd. Loculoascomycetes, Dothideales

Notes: This genus is considered a synonym of *Myriangiella* Zimm., but not all names have been redisposed.
Protopeltis lithocarpi V. Miller & Bonar
Distr.: CA(type).
Host: Lithocarpus

Protostroma Batista. Coelomycetes

Protostroma hyphaenea Batista
Distr.: FL.
Host: Hyphaene

Protoventuria Berl. & Sacc. Loculoascomycetes, Dothideales

Notes: Barr (Canad. J. Bot. 49:1959. 1971) considered this the correct genus for species of *Antennularia*.
Protoventuria alpina (Sacc.) Barr
Venturia alpina Sacc.
Antennularia alpina (Sacc.) E. Müller
Venturia arctostaphyli Cooke & Harkn.
Distr.: Circumboreal.
Notes: On leaves of Ericaceae.
Hosts: Andromeda, Arctostaphylos, Chamaedaphne, Gaultheria, Leucothoe
Protoventuria curviseta (Peck) Barr
Venturia curviseta Peck
Antennularia curviseta (Peck) Barr
Distr.: Northeastern North America.
Host: Nemopanthus
Protoventuria latispora (Barr) Barr
Antennularia latispora Barr
Distr.: Temperate North America.
Host: Arctostaphylos
Protoventuria maculosa (Ellis) Barr
Meliola maculosa Ellis
Antennularia maculosa (Ellis) Barr
Dimerosporium ellisii Sacc.
Venturia pezizoidea Sacc. & Ellis, pro parte
Distr.: Northeastern North America.
Notes: On leaves of Ericaceae.
Hosts: Gaylussacia, Leucothoe
Protoventuria quercina (Pers.) E. Müller
Antennularia quercina (Pers.) E. Müller
Phaeodimeriella quercina (Pers.) E. Müller in S.J. Hughes
Venturia orbicula (Schwein.) Cooke & Peck
Distr.: Central and eastern North America.
Notes: On overwintered leaves.
Host: Quercus
Protoventuria rhododendri (Höhn.) Barr
Antennularia rhododendri Höhn.
Distr.: Northeastern North America; Europe.
Host: Kalmia

Psathyrella (Fr.) Quél. Basidiomycetes, Agaricales

Psathyrella alboalutacea A.H. Sm.
Distr.: WA.
Host: Alnus
Psathyrella atomatoides (Peck) A.H. Sm.
Psilocybe atomatoides (Peck) Sacc.
Distr.: Eastern United States.
Hosts: Saccharum, Tsuga
Psathyrella candidissima A.H. Sm.
Distr.: MI, WA.
Host: Alnus
Psathyrella incerta (Peck) A.H. Sm.
Hypholoma incertum (Peck) Sacc.
Distr.: Temperate North America.
Notes: On stumps, logs, and debris of hardwoods.
Hosts: Acer, Carya, Malus, Ulmus
Psathyrella maculata (C.S. Parker) A.H. Sm.
Hypholoma maculata C.S. Parker
Distr.: Temperate North America; Europe.
Host: Alnus
Psathyrella naucoria A.H. Sm.
Distr.: ID, WA.
Host: Alnus
Psathyrella piluliformis (Bull.:Fr.) P.D. Orton
Hypholoma hydrophila (Bull.) Quél.
Psathyrella hydrophila (Bull.) Maire
Distr.: Temperate North America; Europe.
Host: Pseudotsuga
Psathyrella rubicola A.H. Sm.
Distr.: MI, WA.
Host: Rubus
Psathyrella sublateritia A.H. Sm. var. ***intermedia*** D.H. Mitchel & A.H. Sm.
Distr.: CO.
Host: Populus
Psathyrella typhae (Kalchbr.) A. Pearson & Dennis
Naucoria scirpicola Peck
Distr.: Temperate North America; Europe.
Notes: Redhead (Fungi Canadenses 133. 1979). Johnson et al. (Mycologia 67:879. 1975) misidentified this fungus as *Galerina hypnorum*, a species which occurs only on mosses.
Host: Typha

Pseudobeltrania Henn. Hyphomycetes

Pseudobeltrania chumrungensis Sutton
Distr.: HI; Nepal(type).
Notes: On fallen leathery leaves (Sutton. Trans. Brit. Mycol. Soc. 55:506. 1970).
Host: Eucalyptus

Pseudocenangium P. Karst. Coelomycetes

Notes: Dyko & Sutton (Canad. J. Bot. 57:370. 1979).
Pseudocenangium succineum (Sprée) Dyko & Sutton
Dacrymyces succineus Sprée
Phialea fumosella (Cooke & Ellis) Sacc.
Distr.: Temperate North America; Europe.
Host: Pinus

Pseudocercospora Speg. Hyphomycetes

Notes: Deighton (Mycol. Pap. 140. 1976; Mycol. Pap. 144. 1979).
Pseudocercospora abelmoschi (Ellis & Everh.) Deighton
Cercospora abelmoschi Ellis & Everh.
Cercospora hibisci Tracy & Earle
Distr.: Southeastern United States; subtropical and tropical regions.
Notes: Kirk (C.M.I. Descr. 678. 1980).
Hosts: Abelmoschus, Hibiscus
Pseudocercospora aleuritis (Miyake) Deighton
Cercospora aleuritis Miyake
Cercosporina aleuritis (Miyake) Sacc.
Mycosphaerella aleuritis Ou (teleomorph)
Distr.: Throughout range of host.
Notes: On leaves (Sivanesan. The bitunicate Ascomycetes. p. 204. 1984).
Host: Aleurites
Pseudocercospora atromarginalis (Atk.) Deighton
Cercospora atromarginalis Atk.
Cercospora nigri Tharp
Distr.: Subtropical and tropical regions.
Notes: Leafspot disease (Williams. C.M.I. Descr. 940. 1987).
Host: Solanum
Pseudocercospora bauhiniae (Syd. & P. Syd.) Deighton

Cercospora bauhiniae Syd. & P. Syd.
Distr.: Subtropical and tropical regions.
Host: Bauhinia
Pseudocercospora bolleana (Thuem.) Sivanesan
Cercospora bolleana (Thuem.) Speg.
Cercosporidium bolleana (Thuem.) X. Liu & Guo
Septosporium bolleana Thuem.) Speg.
Mycosphaerella bolleana Higgins (teleomorph)
Distr.: HI, southern United States; Middle East; southern Europe.
Notes: Leaf spot (Sivanesan. The bitunicate Ascomycetes. p. 205. 1984).
Host: Ficus
Pseudocercospora borreriae (Ellis & Everh.) Deighton
Cercospora borreriae Ellis & Everh.
Distr.: Southeastern United States; West Indies; Central America; eastern Africa.
Hosts: Borreria, Spermacoce
Pseudocercospora cannabina (Wakef.) Deighton
Cercospora cannabina Wakef.
Distr.: MS, WI; Uganda(type).
Host: Cannabis
Pseudocercospora caperoniae (L. Olive) Deighton
Helicomina caperoniae L. Olive
Distr.: Southeastern United States; West Indies; South America.
Host: Caperonia
Pseudocercospora ceratoniae (Pat. & Trab.) Deighton
Cercospora ceratoniae Pat. & Trab.
Distr.: FL; northern Africa; southern Europe; Taiwan.
Host: Ceratonia
Pseudocercospora clitoriae (Atk.) Deighton
Cercospora clitoriae Atk.
Distr.: AL(type).
Hosts: Centrosema, Clitoria
Pseudocercospora conspicua (Earle) Deighton
Cercospora conspicua Earle
Distr.: Southeastern United States; West Indies.
Host: Cleome
Pseudocercospora crotonis (Ellis & Everh.) Deighton
Cercospora crotonis Ellis & Everh.
Distr.: CA; southeastern United States.
Host: Croton
Pseudocercospora cruenta (Sacc.) Deighton
Cercospora cruenta Sacc.
Cercospora phaseoli Dearn. & Barth.
Cercospora phaseolorum Cooke
Cercospora vignae Ellis & Everh.
Mycosphaerella cruenta D.H. Latham (teleomorph)
Distr.: North America; British Soloman Islands; Uganda.
Notes: On decaying leaves (Sivanesan. The bitunicate Ascomycetes. p. 208. 1984; Mulder & Holliday. C.M.I. Descr. 463. 1975).
Hosts: Clitoria, Glycine, Lycopersicon, Phaseolus, Strophostyles, Vicia, Vigna
Pseudocercospora dolichi (Ellis & Everh.) J. Yen
Cercospora dolichi Ellis & Everh.
Distr.: Southeastern United States; West Indies; eastern Asia; India.
Notes: Yen (Bull. Soc. Mycol. France 97:152. 1981).
Host: Vigna
Pseudocercospora fuligena (Roldan) Deighton
Cercospora fuligena Roldan
Distr.: FL; India; Ivory Coast; Japan; The Philippines.
Notes: Cercospora leaf mold of tomato. Early reports of this disease (Solheim & Stevens. Mycologia 23:389. 1931) were misdetermined as *C. canescans* Ellis & G. Martin. Mulder & Holliday (C.M.I. Descr. 465. 1975 as *Cercospora fuligena*).
Host: Lycopersicon
Pseudocercospora glycines (Cooke) Deighton
Cercospora glycines Cooke
Distr.: FL; Australia(type).
Host: Sesbania
Pseudocercospora handelii (Bubák) Deighton
Cercospora handelii Bubák
Cercoseptoria handelii (Bubák) Deighton
Cercospora rhododendri Ém. Marchal & Verplancke
Distr.: Temperate northern hemisphere.
Notes: Leaf spot (Pirozynski. Fungi Canadenses 22. 1974 as *Cercospora handelii*).
Hosts: Actinidia, Rhododendron
Pseudocercospora ixorae (Solheim) Deighton
Cercospora ixorae Solheim
Distr.: FL; India; Taiwan.
Host: Ixora
Pseudocercospora jamaicensis (Chupp) Deighton
Cercospora jamaicensis Chupp
Distr.: FL; West Indies; Central America.
Host: Datura
Pseudocercospora jasminicola Muller & Chupp ex Deighton
Cercospora jasminicola Muller & Chupp, nom. inval.
Distr.: Subtropical and tropical regions.
Host: Jasminum
Pseudocercospora jussiaeae (Atk.) Deighton
Cercospora jussiaeae Atk.
Distr.: Southern United States; northern South America; South Africa.
Host: Ludwigia
Pseudocercospora laburni (W.W. Ray) Deighton
Cercospora laburni W.W. Ray
Distr.: OK; Japan.
Host: Laburnum
Pseudocercospora leucothoes (B.H. Davis) Deighton
Cercospora leucothoes B.H. Davis
Distr.: Eastern United States.
Host: Leucothoe
Pseudocercospora ligustri Deighton
Cercospora ligustri Auct. non Roum.
Distr.: Southern United States.
Notes: In the older literature, this fungus was misdetermined as *Cercospora ligustri* Roum.
Host: Ligustrum
Pseudocercospora lupini (Cooke) Deighton
Cercospora lupini Cooke
Chaetotrichum lupini (Cooke) Petr.
Septosporium lupini Thuem.
Distr.: OR, southeastern United States.
Host: Lupinus
Pseudocercospora lycopodis (Ellis & Everh.) Deighton
Cercospora lycopodis Ellis & Everh.
Cercospora hypticola Chupp & Muller
Distr.: Southeastern United States; northern South America.
Host: Lycopus
Pseudocercospora lysimachiae (Ellis & Halst.) Deighton
Cercospora lysimachiae Ellis & Halst.
Distr.: Northeastern North America.
Host: Lysimachia
Pseudocercospora mali (Ellis & Everh.) Deighton
Cercospora mali Ellis & Everh.
Cercospora minima Tracy & Earle
Distr.: Southeastern United States; South America; India.
Hosts: Aronia, Malus, Pyrus
Pseudocercospora monoicae (Ellis & Holw.) Deighton
Cercospora monoicae Ellis & Holw.
Distr.: Central and eastern North America.
Host: Amphicarpaea
Pseudocercospora muntingiae (Petr. & Cif.) Deighton
Cercospora muntingiae Petr. & Cif.
Distr.: FL; West Indies.
Host: Muntingia
Pseudocercospora musae (A. Zimmerm.) Deighton
Cercospora musae A. Zimmerm.
Cercospora musae Massee
Mycosphaerella musicola J.L. Mulder in J.L. Mulder & Stover (teleomorph)
Distr.: Worldwide in banana-growing regions.
Notes: Sigatoka disease of banana (Mulder & Holliday. C.M.I. Descr. 414. 1974).
Host: Musa
Pseudocercospora myrticola (Speg.) Deighton
Cercospora myrticola Speg.
Distr.: North and South America; Eurasia.
Host: Myrtus
Pseudocercospora neriella (Sacc.) Deighton
Cercospora neriella Sacc.

Cercoseptoria neriella (Sacc.) J.M. Dingley
Cercospora nerii-indici W. Yamamoto
Distr.: Widespread in tropical regions.
Host: Nerium
Pseudocercospora nymphaeacea (Cooke & Ellis) Deighton
Cercospora nymphaeacea Cooke & Ellis
Cercoseptoria nymphaeacea (Cooke & Ellis) Deighton
Cercospora nymphaeacea Ellis & Everh. in Atk. non Cooke & Ellis, hom. illeg.
Cercospora exotica Ellis & Everh.
Cercospora nelumbonis Tharp
Cercospora nymphaeae Ellis & Everh. in Underw. & Earle
Distr.: United States; West Indies; Brazil; Japan; India.
Hosts: Nelumbo, Nuphar, Nymphaea, Nymphoides
Pseudocercospora puderi B.H. Davis ex Deighton
Cercospora puderi B.H. Davis, nom. inval.
Distr.: Southeastern United States; subtropical and tropical regions.
Host: Rosa
Pseudocercospora puerariicola (W. Yamamoto) Deighton
Cercospora puerariicola W. Yamamoto
Distr.: GA; Japan.
Notes: Angular leaf spot of kudzu (Sivanesan. The bitunicate Ascomycetes. p. 211. 1984). First described in the United States by Weimer & Luttrell (Phytopathology 38:348.1948).
Host: Pueraria
Pseudocercospora punicae (Henn.) Deighton
Cercospora punicae Henn.
Distr.: Subtropical and tropical regions.
Notes: Leaf spot of pomegranate.
Host: Punica
Pseudocercospora purpurea (Cooke) Deighton
Cercospora purpurea Cooke
Distr.: Southeastern United States; West Indies; Central and South America.
Host: Persea
Pseudocercospora rhoina (Cooke & Ellis) Deighton
Cercospora rhoina Cooke & Ellis
Cercospora copallina Cooke
Cercospora rhoina Cooke & Ellis var. *nigromaculans* Peck
Distr.: Central and eastern North America.
Hosts: Cotinus, Rhus
Pseudocercospora rubi (Sacc.) Deighton
Fusarium rubi G. Wint.
Cercospora rubi G. Wint.
Ramularia rubi (G. Wint.) Wollenweb.
Cercospora bliti Tharp
Cercospora septorioides Ellis & Everh.
Mycosphaerella confusa F.A. Wolf (teleomorph)
Distr.: North America; Brazil.
Notes: Leaf spot (Sivanesan. The bitunicate Ascomycetes. p. 207. 1984).
Hosts: Rosa, Rubus
Pseudocercospora salicina (Ellis & Everh.) Deighton
Cercospora salicina Ellis & Everh.
Cercospora reducta Syd. & P. Syd.
Cercospora sessilis Ellis & Everh. non Sorokin, hom. illeg.
Mycosphaerella togashiana Ito & T. Kobayashi (teleomorph)
Distr.: Temperate North America; Asia.
Notes:
Hosts: Populus, Salix
Pseudocercospora sordida (Sacc.) Deighton
Cercospora sordida Sacc.
Distr.: Southeastern United States.
Notes: Wolf (Mycologia 35:503. 1943).
Host: Campsis
Pseudocercospora spilosticta (Syd.) Deighton
Cercospora spilosticta Syd.
Distr.: FL; South America; Taiwan.
Host: Pithecellobium
Pseudocercospora stillingiae (Ellis & Everh.) J. Yen, A.K. Kar, & B.K. Das
Cercospora stillingiae Ellis & Everh.
Distr.: Southeastern United States.
Notes: Yen et al. (Mycotaxon 16:66. 1982).
Hosts: Sapium, Stillingia
Pseudocercospora stizolobii (Syd. & P. Syd.) Deighton
Cercospora stizolobii Syd. & P. Syd.
Distr.: Subtropical and tropical regions.
Host: Mucuna
Pseudocercospora subsessilis (Syd. & P. Syd.) Deighton
Cercospora subsessilis Syd. & P. Syd.
Cercoseptoria domingensis Cif.
Distr.: Southeastern United States; subtropical and tropical regions.
Hosts: Melia, Swietenia
Pseudocercospora tabernaemontanae (Syd. & P. Syd.) Deighton
Cercospora tabernaemontane Syd. & P. Syd.
Distr.: FL; southeastern Asia; India.
Host: Tabernaemontana
Pseudocercospora theae (Cavara) Deighton
Septoria theae Cavara
Cercoseptoria theae (Cavara) Curzi
Distr.: Warm-temperate to subtropical regions.
Host: Camellia
Pseudocercospora turnerae (Ellis & Everh.) Deighton
Cercospora turnerae Ellis & Everh.
Distr.: Neotropical region.
Host: Turnera
Pseudocercospora vitis (Lév.) Speg.
Cercospora vitis (Lév.) Sacc.
Cercospora viticola (Ces.) Sacc.
Cladosporium vitis (Lév.) Sacc.
Isariopsis clavispora (Berk. & Cooke) Sacc.
Mycosphaerella personata Higgins (teleomorph)
Distr.: Cosmopolitan.
Hosts: Ampelopsis, Cercis, Cissus, Vitis, Ziziphus
Pseudocercospora weigeliae (Ellis & Everh.) Deighton
Cercospora weigeliae Ellis & Everh.
Distr.: Eastern United States; Japan.
Host: Weigela

Pseudocercosporella Deighton. Hyphomycetes

Pseudocercosporella capsellae (Ellis & Everh.) Deighton
Cylindrosporium capsellae Ellis & Everh.
Cercosporella albomaculans (Ellis & Everh.) Sacc.
Cercosporella brassicae (Fautrey & Roum.) Höhn.
Distr.: Widespread.
Notes: Deighton (Mycol. Pap. 133:42. 1973).
Hosts: Brassica, Capsella, Lepidium, Sinapis, Sisymbrium
Pseudocercosporella caryigena (Ellis & Everh.) Sivanesan
Cylindrosporium caryigenum Ellis & Everh.
Cercoseptoria caryigena (Ellis & Everh.) J.J. Davis
Cercosporella caryigena (Ellis & Everh.) Höhn.
Mycosphaerella caryigena Demaree & J.R. Cole (teleomorph)
Distr.: Central and eastern North America.
Notes: Sivanesan (The bitunicate Ascomycetes. p. 201. 1984).
Hosts: Carpinus, Carya
Pseudocercosporella herpotrichoides (Fron) Deighton
Cercosporella herpotrichoides Fron
Distr.: Temperate regions.
Notes: Eyespot of cereals and grasses (Booth & Waller. C.M.I. Descr. 386. 1973).
Hosts: Aegilops, Agropyron, Avena, Balsamorhiza, Bromus, Delphinium, Festuca, Hordeum, Koeleria, Lithospermum, Lomatium, Poa, Secale, Sitanion, Triticum
Pseudocercosporella nivea (Ellis & Barth.) Deighton
Cercosporella nivea Ellis & Barth.
Distr.: Central United States.
Notes: Deighton (Mycol. Pap. 133:53. 1973).
Host: Solidago

Pseudociboria Kanouse. Discomycetes, Helotiales

Pseudociboria umbrina Kanouse
Helotium umbrinum (Kanouse) Seaver
Distr.: WA(type); Brazil; Malaysia.
Notes: Dumont (Mycologia 66:706. 1974) stated that many records of

this species are misidentifications of *Lambertella chromoflava*.
Host: Alnus

Pseudoclitocybe (Singer) Singer. Basidiomycetes, Agaricales

Pseudoclitocybe cyathiformis (Fr.:Fr.) Singer
Clitocybe cyathiformis (Fr.:Fr.) P. Kumm.
Distr.: Temperate North America; Europe.
Notes: On wood and woody debris.
Hosts: Fagus, Nyssa, Quercus

Pseudodictya Tehon & G.L. Stout. Coelomycetes

Notes: According to Sutton (Mycol. Pap. 141:164. 1977) this genus is of doubtful application.
Pseudodictya sassafrasicola Tehon & G.L. Stout
Distr.: IL(type).
Host: Sassafras

Pseudodoassansia (Setch.) K. Vánky. Basidiomycetes, Ustilaginales

Pseudodoassansia obscura (Setch.) K. Vánky
Doassansia obscura Setch.
Distr.: North America.
Host: Sagittaria

Pseudofavolus Pat. Basidiomycetes, Aphyllophorales

Pseudofavolus cucullatus (Mont.) Pat.
Favolus cucullatus Mont.
Distr.: Southeastern United States; tropical regions.
Notes: On wood of hardwoods.
Hosts: Persea, Taxodium

Pseudofistulina Fidalgo & M. Fidalgo. Basidiomycetes, Aphyllophorales

Pseudofistulina radicata (Schwein.) Burdsall
Fistulina radicata Schwein.
Fistulina brasiliensis Fidalgo & M. Fidalgo
Pseudofistulina brasiliensis (Fidalgo & M. Fidalgo) Fidalgo & M. Fidalgo
Fistulina firma Peck
Fistulina pallida Berk. & Ravenel
Fistulina spathulata Berk. & M.A. Curtis
Distr.: Eastern United States to Brazil.
Notes: On wood.
Hosts: Castanea, Quercus

Pseudographis Nyl., nom. cons. Discomycetes, Ostropales

Pseudographis elatina (Achar.) Nyl.
Distr.: Temperate regions.
Notes: On bark of conifers (Dennis. British Ascomycetes. p. 162. 1978).
Host: Pseudotsuga
Pseudographis intermedia Rehm
Distr.: WA(type).
Host: Thuja
Pseudographis phragmitis Dearn. & House
Distr.: NY(type).
Host: Phragmites
Pseudographis tetraspora Cummins
Distr.: MT(type).
Host: Thuja

Pseudohelotium Fuckel. Discomycetes, Helotiales

Notes: According to Dennis (Mycol. Pap. 62:63. 1976), this genus only occurs in Europe.
Pseudohelotium sacchariferum (Berk.) Sacc.
Distr.: AL(type), GA.
Notes: Seaver (North Amer. cup fungi. [Inoperculates]. p. 217. 1951) suggested that this is a synonym of *Orbilia cruenta*.
Host: Liquidambar

Pseudohydnum P. Karst. Basidiomycetes, Tremellales

Pseudohydnum gelatinosum (Scop.:Fr.) P. Karst.
Tremellodon gelatinosum (Scop.:Fr.) Fr.
Distr.: Temperate northern hemisphere.
Notes: On wood, usually of conifers.
Hosts: Abies, Picea, Pinus, Pseudotsuga, Quercus, Tsuga

Pseudolachnea Ranoj. Coelomycetes

Pseudolachnea hispidula (Schrad.:Fr.) Sutton
Dinemasporium hispidulum (Schrad.:Fr.) M.A. Curtis
Dinemasporium herbarum (Cooke) Grove
Dinemasporium patellum Cooke & Ellis
Distr.: Temperate North America; Europe.
Notes: On a wide variety of woody substrates.
Hosts: Acer, Leonurus

Pseudomassaria Jacz. Pyrenomycetes, Amphisphaeriales

Notes: Barr (Mycologia 56:841. 1964).
Pseudomassaria agrifolia (Ellis & Everh.) Arx
Physalospora agrifolia Ellis & Everh.
Distr.: CA.
Host: Quercus
Pseudomassaria carolinensis Barr & C.S. Hodges
Beltraniella portoricensis (F. Stevens) Pirozynski & S.D. Patil (anamorph)
Distr.: SC.
Notes: Barr & Hodges (Mycologia 63:562. 1971).
Host: Persea
Pseudomassaria chondrospora (Ces.) Jacz.
Physalospora citrispora (Berk. & M.A. Curtis) Sacc.
Distr.: Eastern North America; Europe.
Notes: On twigs and petioles.
Host: Tilia
Pseudomassaria corni (Sowerby) Arx
Apiosporella cornina Fairm.
Beltraniella sp. (anamorph)
Distr.: Temperate North America; Europe.
Notes: On twigs (Corlett. Fungi Canadenses 121. 1978).
Host: Cornus
Pseudomassaria erumpens Barr
Distr.: Northeastern North America.
Host: Kalmia
Pseudomassaria foliicola Barr
Distr.: Northeastern North America.
Host: Cornus
Pseudomassaria inversa Barr
Distr.: NH.
Notes: Bigelow & Barr (Rhodora 71:200. 1969).
Host: Rhododendron
Pseudomassaria leucothoes (Cooke) Petr. & Syd.
Laestadia leucothoes (Cooke) Sacc.
Guignardia leucothoes (Cooke) Lindau
Mycosphaerella leucothoes L.E. Miles
Distr.: Southeastern United States.
Host: Leucothoe
Pseudomassaria leucothoes (Cooke) Petr. & Syd. var. ***gaultheriae*** Barr
Distr.: MA.
Host: Gaultheria

Pseudomassaria lycopodina (P. Karst.) Arx
Mycosphaerella lycopodina (P. Karst.) Starb.
Pseudapiospora lycopodina (P. Karst.) L. Holm
Distr.: Eastern North America; Europe.
Notes: On dead leaves.
Host: Lycopodium
Pseudomassaria oxydendri (F.A. Wolf) Barr
Venturia oxydendri F.A. Wolf
Distr.: Southeastern United States.
Host: Oxydendrum
Pseudomassaria polystigma (Ellis & Everh.) Arx
Laestadia polystigma (Ellis & Everh.) Berl. & Voglino in Sacc.
Sphaerognomonia polystigma (Ellis & Everh.) J.H. Miller & G.E. Thompson
Laestadia celata Harkn.
Distr.: Eastern North America.
Notes: On overwintered leaves.
Host: Quercus
Pseudomassaria sepincolaeformis (De Not.) Arx
Didymella sepincolaeformis (De Not.) Sacc.
Distr.: MD, MN; Italy(type).
Host: Rosa
Pseudomassaria thistletonia (Cooke) Arx
Physalospora thistletonia Cooke
Distr.: Eastern United States; Europe.
Host: Rhododendron

Pseudombrophila Boud. Discomycetes, Pezizales

Pseudombrophila deerata (P. Karst.) Seaver
Ascophanus deerata (P. Karst.) Boud.
Humaria deerata (P. Karst.) Sacc.
Distr.: ND; Europe.
Host: Brassica

Pseudomerulius Jülich. Basidiomycetes, Aphyllophorales

Pseudomerulius aureus (Fr.:Fr.) Jülich
Merulius aureus Fr.:Fr.
Plicatura aurea (Fr.:Fr.) Parmasto
Distr.: Temperate northern hemisphere.
Notes: On conifers, especially *Pinus* spp.
Hosts: Larix, Picea, Pinus, Pseudotsuga, Tsuga
Pseudomerulius curtisii (Berk.) Redhead & Ginns
Paxillus curtisii Berk. in Berk. & M.A. Curtis
Paxillus corrugatus Atk.
Distr.: Eastern North America.
Notes: On dead wood.
Hosts: Pinus, Tristania, Tsuga

Pseudonectria Seaver. Pyrenomycetes, Hypocreales

Pseudonectria aquifoliae (Cooke & Ellis) Dennis
Peziza aquifoliae Cooke & Ellis
Hymenoscyphus aquifoliae (Cooke & Ellis) Kuntze
Pezizella aquifoliae (Cooke & Ellis) Sacc.
Distr.: NJ(type).
Notes: On leaves (Dennis. Persoonia 3:35. 1964).
Host: Ilex
Pseudonectria pachysandricola B.O. Dodge
Volutella pachysandricola B.O. Dodge (anamorph)
Distr.: Eastern United States.
Notes: Leaf blight (Dodge. Mycologia 36:532. 1944).
Host: Pachysandra
Pseudonectria rousseliana (Mont.) Seaver
Volutella buxi (DC.) Berk. (anamorph)
Distr.: Temperate regions.
Notes: Boxwood canker (Samuels. Mycologia 69:256. 1977).
Host: Buxus

Pseudoperonospora Rostovzev. Oomycetes, Peronosporales

Notes: Waterhouse & Brothers (Mycol. Pap. 148. 1981).
Pseudoperonospora celtidis (M.B. Waite) G.W. Wils.
Distr.: Eastern United States; Japan.
Notes: Downy mildew.
Host: Celtis
Pseudoperonospora cubensis (Berk. & M.A. Curtis) Rostovzev
Peronospora cubensis Berk. & M.A. Curtis
Peronoplasmopara cubensis (Berk. & M.A. Curtis) G.W. Clinton
Distr.: Cosmopolitan.
Notes: Downy mildew of Cucurbitaceae (Palti. C.M.I. Descr. 457. 1975).
Hosts: Benincasa, Brassica, Bryonopsis, Citrullus, Cucumis, Cucurbita, Echinocystis, Lagenaria, Luffa, Melothria, Momordica, Sicyos, Trichosanthes
Pseudoperonospora humuli (Miyabe & Takah.) G.W. Wils.
Distr.: Temperate regions of the world including northwestern United States.
Notes: Downy mildew of hops. (Francis. C.M.I. Descr. 769. 1983).
Host: Humulus
Pseudoperonospora urticae (Lib. ex Berk.) E.S. Salmon & W.N. Ware
Distr.: Temperate North America; Europe.
Notes: Constantinescu (Mycotaxon 24:303. 1985).
Host: Urtica

Pseudopezicula Korf in Korf, R.C. Pearson, Zhuang & B. Dubos. Discomycetes, Helotiales

Notes: Korf et al. (Mycotaxon 26:457. 1986) described this genus for two species causing diseases on grapes.
Pseudopezicula tetraspora Korf, R.C. Pearson & Zhuang
Phialophora sp. (anamorph)
Distr.: NY.
Notes: This is the American counterpart of the organism causing grape disease called "Rotbrenner" in Europe.
Host: Vitis

Pseudopeziza Fuckel. Discomycetes, Helotiales

Pseudopeziza calthae (W. Phillips) Massee
Fabraea rousseauana Sacc. & E. Bommer
Distr.: Temperate regions.
Host: Caltha
Pseudopeziza caricina (Lib.) Sacc.
Pyrenopeziza caricina (Lib.) Rehm
Distr.: CO, ND; Europe.
Notes: Nannfeldt (Nova Acta Regiae Soc. Sci. Upsal., Ser. 4, 8:178. 1932) suggested that this species belongs in *Hysteropezizella*.
Host: Carex
Pseudopeziza cerastiorum (Wallr.) J. Schröt. var. **arenariae** Sacc.
Distr.: AK(type).
Notes: The type variety is now placed in *Leptotrochila*.
Host: Arenaria
Pseudopeziza medicaginis (Lib.) Sacc.
Distr.: Worldwide in *Medicago*-growing regions.
Notes: Common leaf spot (Booth & Waller. C.M.I. Descr. 637. 1979).
Hosts: Medicago, Vicia
Pseudopeziza pyri Peck
Distr.: NY(type).
Host: Sorbus
Pseudopeziza singularis (Peck) J.J. Davis
Mollisia singularis (Peck) Sacc.
Distr.: Northeastern and northcentral North America.
Notes: On leaves. Shoemaker & Egger (Fungi Canadenses 229. 1982).
Host: Ranunculus
Pseudopeziza trifolii (Biv.-Bern.:Fr.) Fuckel
Phacidium trifolii (Biv.-Bern.:Fr.) Bonord.
Pseudopeziza meliloti Syd.
Distr.: Temperate regions.

Notes: Leaf spot of clover (Booth & Waller. C.M.I. Descr. 636. 1979).
Hosts: Melilotus, Trifolium
Pseudopeziza trifolii (Biv.-Bern.:Fr.) Fuckel f. sp. ***medicaginis-sativae*** Schmiedeknecht
Distr.: Temperate regions.
Notes: Meyer & Luttrell (Mycologia 78:529. 1986) presented a detailed account of the ascoma morphology.
Host: Medicago

Pseudophacidium P. Karst. Discomycetes, Rhytismatales

Notes: Egger (Sydowia 20:288. 1968).
Pseudophacidium ledi (Albertini & Schwein.:Fr.) P. Karst.
Pseudophacidium callunae (P. Karst.) P. Karst.
Myxofusicoccum ericeti (Sacc.) Petr. (anamorph)
Distr.: Temperate regions.
Notes: On ericaceous hosts (Eriksson. Symb. Bot. Upsal. 19:55. 1970).
Host: Vaccinium

Pseudoplectania Fuckel. Discomycetes, Pezizales

Pseudoplectania melaena (Fr.) Sacc.
Melascypha melaena (Fr.) Boud.
Pseudoplectania vogesiaca (Pers.) Seaver
Distr.: Northern North America; Europe.
Notes: On rotting wood in coniferous forests.
Host: Thuja
Pseudoplectania nigrella (Pers.:Fr.) Fuckel
Distr.: Temperate regions; Jamaica.
Notes: On rotting wood in coniferous forests.
Host: Thuja

Pseudorhytisma Juel. Discomycetes, Rhytismatales

Pseudorhytisma bistortae (Lib.) Juel
Pseudopeziza bistortae (Lib.) Fuckel
Distr.: AK; Europe.
Host: Polygonum

Pseudorobillarda Morelet. Coelomycetes

Pseudorobillarda agrostidis (R. Sprague) Nag Raj, Morgan-Jones, & Kendrick
Robillarda agrostidis R. Sprague
Robillarda lathyrii H. Rupprecht
Distr.: Western North America; Europe.
Notes: Nag Raj et al. (Canad. J. Bot. 50:863. 1972).
Hosts: Agrostis, Muhlenbergia
Pseudorobillarda sojae Uecker & Kulik
Distr.: Southeastern United States.
Notes: Uecker & Kulik (Mycologia 78:449. 1986).
Host: Glycine

Pseudoseptoria Speg. Coelomycetes

Notes: Sutton (The Coelomycetes. p. 97. 1980).
Pseudoseptoria bromigena (Sacc.) Sutton
Septoria bromigena Sacc.
Selenophoma bromigena (Sacc.) R. Sprague & A.G. Johnson
Distr.: Temperate northern hemisphere.
Notes: Bissett (Fungi Canadenses 237. 1982).
Host: Bromus
Pseudoseptoria donacis (Pass.) Sutton
Septoria donacis Pass.
Selenophoma donacis (Pass.) R. Sprague & A.G. Johnson
Lunospora oxyspora (Penz. & Sacc.) Frandsen
Septoria oxyspora Penz. & Sacc.
Distr.: Temperate regions.
Notes: Halo spot of cereals and grasses (Punithalingam & Waller. C.M.I. Descr. 400. 1973).
Hosts: Agropyron, Agrostis, Arundo, Avena, Bambusa, Bouteloua, Dactylis, Elymus, Elysitanion, Festuca, Hordeum, Leymus, Oryzopsis, Panicum, Phalaris, Phleum, Phragmites, Poa, Poaceae, Schizachyrium, Secale, Sitanion, Sporobolus, Stipa, Triticum
Pseudoseptoria everhartii (Sacc. & Syd.) Sutton
Septoria everhartii Sacc. & Syd.
Selenophoma everhartii (Sacc. & Syd.) R. Sprague & A.G. Johnson
Distr.: Temperate North America; northern Africa; Soviet Union.
Notes: Eye spot of grasses.
Hosts: Agropyron, Agrostis, Alopecurus, Aristida, Bouteloua, Calamagrostis, Danthonia, Deschampsia, Dichanthelium, Festuca, Hystrix, Leucopoa, Muhlenbergia, Poa, Puccinellia, Sphenopholis, Trisetum, Vulpia
Pseudoseptoria obtusa (R. Sprague & A.G. Johnson) Sutton
Selenophoma obtusa R. Sprague & A.G. Johnson
Distr.: Central and western United States.
Notes: Eye spot of grasses.
Hosts: Agropyron, Elymus, Elysitanion, Poa, Sitanion, Stipa
Pseudoseptoria stomaticola (Bäumler) Sutton
Phyllosticta stomaticola Bäumler
Selenophoma donacis (Pass.) R. Sprague & A.G. Johnson var. *stomaticola* (Bäumler) R. Sprague & A.G. Johnson
Septoria culmifida Lind
Septoria lunata Grove
Distr.: Northern and temperate North America; Europe.
Notes: Leaf spot of grasses (Bissett. Fungi Canadenses 238. 1982).
Hosts: Agropyron, Aristida, Arrhenatherum, Calamovilfa, Dactylis, Danthonia, Deschampsia, Elymus, Elysitanion, Festuca, Hordeum, Koeleria, Leucopoa, Melica, Oryzopsis, Phleum, Poa, Puccinellia, Secale, Sitanion, Sporobolus, Stipa, Taeniatherum, Trisetum, Vulpia

Pseudosphaeria Höhn. Loculoascomycetes, Dothideales

Notes: This genus is considered a synonym of *Wettsteinina*, but not all names have been redisposed.
Pseudosphaeria wallowaensis R. Sprague
Distr.: OR(type).
Host: Carex

Pseudospiropes M.B. Ellis. Hyphomycetes

Pseudospiropes lotorus Morgan-Jones
Distr.: AL; India.
Notes: Isolated from roots (Morgan-Jones. Mycotaxon 5:381. 1977) and on rotten woody debris (Rao & De Hoog. Stud. Mycol. 28:45. 1986).
Host: Lotus
Pseudospiropes nodosus (Wallr.) M.B. Ellis
Helminthosporium nodosum Wallr.
Pleurophragmium nodosum (Wallr.) S.J. Hughes
Distr.: North temperate regions.
Notes: On dead wood and bark, occasionally on dead, herbaceous stems (Ellis. Dematiaceous Hyphomycetes. p. 258. 1971; Hughes & Cooke. Fungi Canadenses 144. 1979).
Hosts: Alnus, Sambucus

Pseudostegia Bubák. Coelomycetes

Pseudostegia nubilosa (Ellis & Everh.) Bubák
Cryptosporium nubilosum Ellis & Everh.
Distr.: Temperate North America.
Host: Carex

Pseudotomentella Svrček. Basidiomycetes, Aphyllophorales

Notes: Larsen (Mycologia 66:167. 1974).
Pseudotomentella atrofusca M. Larsen
Distr.: AZ.
Notes: On wood.
Host: Pinus
Pseudotomentella flavovirens (Höhn. & Litsch.) Svrček
Tomentella flavovirens Höhn. & Litsch.
Distr.: Temperate northern hemisphere.

Notes: On hardwoods.
Hosts: Acer, Betula
Pseudotomentella fumosa M. Larsen
Distr.: OR.
Notes: On wood.
Host: Pseudotsuga
Pseudotomentella griseopergamacea M. Larsen
Distr.: NY; Canada(Ontario).
Notes: On wood.
Hosts: Acer, Pinus, Tsuga
Pseudotomentella griseoveneta M. Larsen
Distr.: OR.
Notes: On conifers.
Hosts: Pseudotsuga, Tsuga
Pseudotomentella kaniksuensis M. Larsen
Distr.: ID.
Notes: On wood.
Host: Pseudotsuga
Pseudotomentella molybdea M. Larsen
Distr.: MT.
Notes: On wood.
Host: Pseudotsuga
Pseudotomentella mucidula (P. Karst.) Svrček
Hypochnus mucidulus P. Karst.
Distr.: Temperate northern hemisphere.
Notes: On wood.
Hosts: Picea, Pinus
Pseudotomentella nigra (Höhn. & Litsch.) Svrček
Tomentella nigra Höhn. & Litsch.
Distr.: Temperate northern hemisphere.
Notes: On wood of conifers.
Hosts: Picea, Pinus
Pseudotomentella tristis (P. Karst.) M. Larsen
Hypochnus tristis (P. Karst.) P. Karst.
Tomentella tristis (P. Karst.) Höhn. & Litsch.
Pseudotomentella umbrina (Fr.) M. Larsen
Distr.: Temperate northern hemisphere.
Notes: On wood.
Hosts: Abies, Picea, Pinus, Populus, Tsuga

Pseudotrichia Kirschst. Loculoascomycetes, Dothideales

Notes: Barr (Mycotaxon 20:19. 1984).
Pseudotrichia mutabilis (Pers.:Fr.) Wehmeyer
Herpotrichia mutabilis (Pers.:Fr.) G. Wint.
Scoleconectria atkinsonii (Rehm) Seaver
Pseudotrichia aurata (Rehm) Wehmeyer
Pseudotrichia viridicoma (Cooke & Peck) Wehmeyer
Distr.: North America; Europe.
Host: Acer

Pseudovalsa Ces. & De Not. Pyrenomycetes, Diaporthales

Notes: Barr (Mycol. Mem. 7:169. 1978).
Pseudovalsa lanciformis (Fr.:Fr.) Ces. & De Not.
Diatrype elliptica (Peck) Peck
Coryneum brachyanum Link (anamorph)
Distr.: Eastern North America; Europe.
Notes: On limbs.
Hosts: Betula, Quercus
Pseudovalsa longipes (Tul.) Sacc.
Pseudovalsa sigmoidea (Cooke & Ellis) Sacc.
Coryneum umbonatum Nees (anamorph)
Distr.: Temperate North America; Europe.
Notes: On limbs.
Host: Quercus
Pseudovalsa modonia (Tul.) Höhn.
Melanconis modonia Tul.
Distr.: Temperate northern hemisphere.
Notes: On limbs.
Host: Castanea

Pseudovalsella Höhn. Pyrenomycetes, Diaporthales

Notes: Wehmeyer. (Univ. Michigan Stud., Sci. Ser. 14:52. 1941).
Pseudovalsella thelebola (Fr.:Fr.) Höhn.
Melanconis thelebola (Fr.:Fr.) Sacc.
Cryptodiaporthe konseiensis T. Kobayashi
Hendersonula thelebola (Sacc.) T. Kobayashi (anamorph)
Distr.: Temperate northern hemisphere.
Notes: On limbs (Kobayashi. Bull. Gov. Forest Exp. Sta. 226:31. 1970; Barr. Mycol. Mem. 7:184. 1978).
Host: Alnus

Psilachnum Höhn. Discomycetes, Helotiales

Psilachnum cassandrae (Kanouse) Shoemaker, Egger, & Kokko
Phialea cassandrae Kanouse
Helotium cassandrae (Kanouse) Seaver
Distr.: Northeastern North America.
Notes: Shoemaker et al. (Fungi Canadenses 190. 1980).
Host: Chamaedaphne
Psilachnum chrysostigmum (Fr.) Raitviir
Pezizella chrysostigma (Fr.) Sacc.
Helotium aspidiicola (Berk. & Broome) Rehm
Mollisia aspidiicola (Berk. & Broome) Quél.
Pezizella aspidiicola (Berk. & Broome) Rehm
Distr.: HI, MA, MI; Europe.
Notes: On fallen stems of ferns.
Hosts: Adiantum, Cibotium, Osmunda, Sadleria
Psilachnum crinellum (Ellis & Everh.) Dennis
Dasyscyphus crinella (Ellis & Everh.) Sacc.
Distr.: PA.
Notes: On dead leaves (Dennis. Kew Bull. 17:340. 1963).
Host: Carex
Psilachnum lanceolato-paraphysatum (Rehm) Dennis
Pezizella lanceolata-paraphysa Rehm
Distr.: NY(type).
Notes: On dead stems (Dennis. Persoonia 3:50. 1964).
Host: Filipendula

Psilocybe (Fr.) P. Kumm. Basidiomycetes, Agaricales

Psilocybe agrariella Atk. var. ***vaccinii*** V. Charles
Distr.: MA, NJ.
Notes: This species belongs in *Psathyrella* (S.A. Redhead. pers. comm.).
Host: Vaccinium

Pteridiospora Penz. & Sacc. Coelomycetes

Pteridiospora spinosispora Filer
Distr.: MS.
Notes: Isolated from tree roots (Filer. Mycologia 61:167. 1969).
Hosts: Fraxinus, Liquidambar

Ptychogaster Corda. Hyphomycetes

Notes: Anamorphs of various Aphyllophorales (Sigler & Carmichael. Mycotaxon 4:394. 1976).
Ptychogaster cubensis Pat.
Ceriomyces cubensis (Pat.) Sacc. & Syd.
Distr.: Southern United States; Cuba.
Hosts: Myrica, Parkinsonia, Quercus
Ptychogaster rubescens Boud.
Punctularia atropurpurascens (Berk. & Broome) Petch (teleomorph)
Distr.: Worldwide.
Notes: Usually on wood or in basidiomata of teleomorph. Variously reported as the anamorph of *Punctularia atropurpurascens* and *Tyromyces* spp. *sensu lato*. Arx (Persoonia 7:127. 1973).
Hosts: Pinus, Thuja

Puccinia Pers.:Pers. Basidiomycetes, Uredinales

Notes: Cummins (Rust fungi of cereals, grasses and bamboos. 1971); Cummins (Rust fungi on legumes and composites in North America. 1978).

Puccinia aberrans Peck
Distr.: Western North America.
Notes: Autoecious.
Host: Smelowskia

Puccinia abnormis Henn.
Puccinia flaccida Auct. non Berk. & Broome
Distr.: Western hemisphere.
Notes: Aecia unknown; telial host - *Echinochloa* spp.
Host: Echinochloa

Puccinia abramsii Parmelee
Distr.: CA.
Notes: Autoecious. Parmelee (Canad. J. Bot. 45:2291. 1967).
Host: Geraea

Puccinia abrupta Dietel & Holw. in Dietel
Distr.: Southwestern United States; Central and South America.
Notes: Autoecious. Reports on *Viguiera* spp. in the United States refer to variety *abrupta*. Variety *partheniicola* occurs on *Parthenium* in Central and South America.
Hosts: [Verbesina], Viguiera

Puccinia acetosae Körn.
Distr.: Temperate northern hemisphere.
Notes: Autoecious.
Host: Rumex

Puccinia acrophila Peck
Distr.: Western United States.
Notes: Autoecious.
Hosts: Besseya, Synthyris

Puccinia acroptili P. Syd. & Syd.
Distr.: Western North America.
Notes: Autoecious.
Host: Centaurea

Puccinia adoxae R. Hedw. ex DC.
Distr.: Temperate northern hemisphere.
Notes: Autoecious.
Host: Adoxa

Puccinia agnita Arth.
Distr.: CO.
Notes: Autoecious.
Host: Claytonia

Puccinia agoseridis Parmelee & Savile
Distr.: Western North America.
Notes: Autoecious.
Host: Agoseris

Puccinia albulensis Magnus
Distr.: Western North America; Europe.
Notes: Autoecious.
Hosts: Hebe, Veronica

Puccinia aletridis Berk. & M.A. Curtis in Berk.
Distr.: Eastern United States.
Notes: Autoecious.
Host: Aletris

Puccinia allii F. Rudolphi
Puccinia mixta Fuckel
Puccinia porri G. Wint.
Distr.: Cosmopolitan.
Notes: Autoecious. Laundon & Waterston (C.M.I. Descr. 52. 1965).
Host: Allium

Puccinia altissimorum Savile
Distr.: Central and eastern North America.
Notes: Autoecious.
Host: Cirsium

Puccinia altissimorum Savile var. ***horriduli*** Savile
Distr.: LA.
Notes: Autoecious.
Host: Cirsium

Puccinia amphigena Dietel
Distr.: Central and western North America.
Notes: Heteroecious; aecial host - *Smilax* spp.; telial host - *Calamovilfa longifolia*.
Hosts: [Ammophila], Calamovilfa, [Smilacina], Smilax

Puccinia amphispilusa Dietel & Holw.
Distr.: CA.
Notes: Autoecious.
Host: Polygonum

Puccinia andina Dietel & Neger
Distr.: Central United States; Mexico; South America.
Notes: Autoecious.
Host: Ranunculus

Puccinia andropogonis Schwein.
Puccinia andropogonis Schwein. var. *gerardiae* Arth.
Puccinia andropogonis Schwein. var. *melampyri* Arth.
Puccinia andropogonis Schwein. var. *micropuncta* Arth.
Puccinia andropogonis Schwein. var. *onobrychidis* Arth.
Puccinia andropogonis Schwein. var. *oxalidis* Arth.
Puccinia andropogonis Schwein. var. *pentstemonis* Arth.
Puccinia andropogonis Schwein. var. *polygalina* Arth.
Puccinia andropogonis Schwein. var. *pustulata* (Arth.) Arth.
Puccinia andropogonis Schwein. var. *xanthoxyli* (Peck) Arth.
Puccinia pustulata Arth.
Aecidium aesculi Ellis & Kellerm. (anamorph)
Distr.: North America.
Notes: Heteroecious; aecial host - on several families of angiosperms; telial host - *Andropogon* spp. The connection with *Aecidium aesculi* was proven by Baxter (Pl. Dis. Reporter 46:706. 1962).
Hosts: Aesculus, Amphicarpaea, Andropogon, Apios, Aureolaria, Baptisia, Castilleja, Chelone, Comandra, Dalea, Desmodium, Lupinus, Mimulus, Orthocarpus, Oxalis, Penstemon, Polygala, Psoralea, Schizachyrium, Zanthoxylum

Puccinia andropogonis-hirti Beltrán
Distr.: FL; Mediterranean region.
Notes: Telial host - *Hyparrhenia* spp.
Host: Hyparrhenia

Puccinia anemones-virginianae Schwein.
Distr.: North America.
Notes: Autoecious. Parmelee (Fungi Canadenses 262. 1984).
Host: Anemone

Puccinia angelicae (Schumach.) Fuckel
Distr.: Central, eastern and northwestern North America; Asia; Europe.
Notes: Autoecious; on Apiaceae.
Hosts: Angelica, Taenidia

Puccinia angulata Dietel & Neger
Distr.: AZ; South America.
Notes: Autoecious.
Host: Sisyrinchium

Puccinia angustata Peck
Distr.: Widespread.
Notes: Heteroecious; aecial host - Lamiaceae; telial host - Cyperaceae.
Hosts: Cyperus, Eriophorum, Lycopus, Mentha, Monarda, Pycnanthemum, Scirpus, Senecio

Puccinia angustatoides R. Stone
Puccinia angustata Peck var. *angustatoides* (R. Stone) Arth.
Distr.: Southcentral United States; subtropical and tropical America.
Notes: Aecia unknown; telial host- *Rhynchospora* spp.
Host: Rhynchospora

Puccinia anisacanthi Dietel & Holw.
Puccinia bouvardiae D. Griffiths
Distr.: AZ; Mexico.
Notes: Autoecious; on Acanthaceae. Cummins & Baxter (Madroño 16:210. 1962) stated that the type specimen of *P. bouvardiae* is actually on *Anisacanthus thurberi*, not *Bouvardia triphylla* as previously reported.
Hosts: Anisacanthus, [Bouvardia], Carlowrightia

Puccinia antirrhini Dietel & Holw.
Distr.: Cosmopolitan.
Notes: Autoecious. Sivanesan (C.M.I. Descr. 262. 1970).
Hosts: Antirrhinum, [Linaria]

Puccinia arachidis Speg.
Distr.: Southern United States; Central and South America.
Notes: Autoecious. Laundon & Waterston (C.M.I. Descr. 53. 1965).
Host: Arachis

Puccinia arachidis Speg. var. ***offuscata*** (Arth.) Cummins
Puccinia offuscata Arth.
Distr.: Southern United States; West Indies; South America.

Notes: Autoecious.
Host: Zornia
Puccinia araliae Ellis & Everh.
Distr.: Northeastern North America.
Notes: Autoecious. Parmelee & de Carteret (Fungi Canadenses 222. 1982).
Host: Panax
Puccinia arechavaletae Speg.
Distr.: FL, TX; Mexico; West Indies; Central and South America.
Notes: On Sapindaceae.
Host: Serjania
Puccinia arenariae (Schumach.) G. Wint.
Puccinia dianthi DC.
Puccinia spergulae DC.
Distr.: North and South America; Asia; Europe.
Notes: Autoecious; on Caryophyllaceae.
Hosts: Arenaria, Cerastium, Dianthus, Lychnis, Spergula, Stellaria
Puccinia areolata Dietel & Holw. in Dietel
Puccinia nephrophyllidii Mains
Distr.: Western North America.
Notes: Autoecious. Parmelee & de Carteret (Fungi Canadenses 269. 1984).
Hosts: Caltha, [Nephrophyllidium]
Puccinia argentata (C.F. Schultz) G. Wint.
Distr.: Central North America; Europe; India; The Philippines.
Notes: Heteroecious; aecial host - *Adoxa* spp.; telial host - *Impatiens* spp.
Hosts: Adoxa, Impatiens
Puccinia aristidae Tracy
Aecidium caspicum Jacz. (anamorph)
Distr.: Northern Africa; Middle East.
Notes: Heteroecious; aecial host - *Heliotropium europaeum*; telial host - *Aristida* spp. The name *P. aristidae* has been misapplied to collections of *P. subnitens* Dietel in North America.
Hosts: [Abronia], [Allionia], [Amaranthus], [Anemopsis], Aristida, [Atriplex], [Bassia], [Beta], [Boerhavia], [Brassica], [Camelina], [Camissonia], [Capsella], [Chenopodium], [Cleome], [Corydalis], [Cryptantha], [Datura], [Descurainia], [Distichlis], [Draba], [Eriogonum], [Erysimum], [Gilia], Heliotropium, [Hilaria], [Lappula], [Lepidium], [Lesquerella], [Linanthus], [Mentzelia], [Mirabilis], [Monolepis], [Nasturtium], [Nicotiana], [Oenothera], [Pectocarya], [Petunia], [Phacelia], [Physalis], [Plantago], [Polygonum], [Primula], [Raphanus], [Rorippa], [Rumex], [Salsola], [Sarcobatus], [Scleropogon], [Sesuvium], [Silene], [Sinapis], [Sisymbrium], [Spergularia], [Spinacia], [Stanleya], [Stellaria], [Thelypodium], [Thlaspi], [Triglochin], [Tropaeolum], [Vaccaria], [Verbena]
Puccinia aristidae Tracy var. ***chaetariae*** Cummins & S.M. Husain
Aecidium pupaliae Prasad, Sharma, & R.D. Singh (anamorph)
Distr.: North and South America; Africa; India.
Notes: Heteroecious; aecial host - Amaranthaceae and Nyctaginaceae; telial host - Poaceae. Confused with *P. subnitens* Dietel in the older literature.
Hosts: Aristida, Boerhavia, Hilaria
Puccinia arnicalis Peck
Distr.: Western North America.
Notes: Autoecious. Savile (Fungi Canadenses 111. 1978).
Host: Arnica
Puccinia artemisiae-norvegicae Tranzschel & Woronichin
Distr.: AK, Soviet Union.
Notes: Autoecious.
Host: Artemisia
Puccinia arundinariae Schwein.
Aecidium smilacis Schwein. (anamorph)
Distr.: Southern United States.
Notes: Heteroecious; aecial host - *Smilax* spp.; telial host - *Arundinaria tecta*.
Hosts: Arundinaria, Smilax
Puccinia asarina Kunze in Kunze & J.C. Schmidt
Distr.: Western North America; Japan; Soviet Union.
Notes: On Aristolochiaceae.
Host: Asarum
Puccinia asparagi DC. in Lam. & DC.
Distr.: Worldwide in asparagus-growing regions.
Notes: Waterston (C.M.I. Descr. 54. 1965).
Hosts: Allium, Asparagus
Puccinia asperior Ellis & Everh.
Distr.: Western United States.
Notes: On Apiaceae.
Host: Lomatium
Puccinia atrofusca (W.R. Dudley & C.H. Thomps.) Holw.
Puccinia universalis Arth.
Distr.: North America; Asia.
Notes: Aecial host - Asteraceae; telial host - Cyperaceae.
Hosts: Artemisia, Aster, Carex, Hieracium
Puccinia atropuncta Peck & G.W. Clinton in Peck
Distr.: Central and eastern North America.
Notes: Heteroecious; aecial host - Asteraceae; telial host - Liliaceae.
Hosts: Prenanthes, Stenanthium, Veratrum, Xerophyllum, Zigadenus
Puccinia avocensis Cummins & H.C. Greene in H.C. Greene
Aecidium avocense Cummins & H.C. Greene in H.C. Greene (anamorph)
Distr.: WI.
Notes: Aecial host possibly *Callirhoe triangulata*; telial host - *Stipa spartea*.
Host: Stipa
Puccinia baccharidis Dietel & Holw.
Eriosporangium punctato-striatum Arth.
Distr.: Southwestern United States; Central and South America.
Notes: Autoecious.
Host: Baccharis
Puccinia ballotaeflorae W.H. Long
Distr.: TX.
Notes: Aecial host is unknown; telial host is *Salvia ballotaeflora*.
Host: Salvia
Puccinia balsamorhizae Peck
Puccinia wyethiae Peck in Harkn.
Distr.: Western North America.
Notes: Autoecious. Parmelee (Fungi Canadenses 89. 1976).
Hosts: Balsamorhiza, [Borrichia], Wyethia
Puccinia batesiana Arth.
Distr.: Central and eastern United States.
Notes: Autoecious.
Host: Heliopsis
Puccinia biocellata Vestergr.
Puccinia plucheae (Syd.) Arth.
Distr.: FL; West Indies; Central and South America.
Notes: Autoecious.
Host: Pluchea
Puccinia biporula J.W. Baxter
Aecidium subsimulans Arth. & Mains (anamorph)
Distr.: Southwestern United States; Mexico.
Notes: Autoecious.
Host: Salvia
Puccinia bistortae (F. Strauss) DC.
Distr.: Northern regions of the northern hemisphere.
Notes: Heteroecious; aecial host - Apiaceae; telial host - Polygonaceae.
Hosts: Angelica, Conioselinum, Ligusticum, Polygonum
Puccinia blasdalei Dietel & Holw.
Distr.: Western North America; Asia; Europe.
Notes: Autoecious.
Host: Allium
Puccinia blyttiana Lagerh.
Puccinia ranunculi A. Blytt
Distr.: North America; temperate Europe.
Notes: Autoecious. Parmelee & de Carteret (Fungi Canadenses 263. 1984).
Host: Ranunculus
Puccinia bolleyana Schwein.
Puccinia sambuci Arth.
Distr.: Central and eastern North America.
Notes: Heteroecious; aecial host - *Sambucus* spp.; telial host - *Carex* spp.
Hosts: Carex, Sambucus
Puccinia boutelouae (H.S. Jennings) Holw.
Distr.: Southwestern United States; Central America; Brazil.
Notes: Aecia unknown; telial host - Poacae.
Host: Bouteloua
Puccinia brachypodii G. Otth
Distr.: Europe; Japan; reported from WY before the recognition of the

two varieties.
Notes: Heteroecious; aecial host - *Berberis* spp; telial host - *Brachypodium* spp.
Hosts: [Deschampsia], [Phleum], [Poa]
Puccinia brachypodii G. Otth var. **arrhenatheri** (Kleb.) Cummins & H.C. Greene
Puccinia deschampsiae Arth.
Puccinia koeleriae Arth.
Distr.: Temperate regions.
Notes: Heteroecious; aecial host - *Berberis* spp.; telial host - Poaceae.
Hosts: Berberis, Deschampsia, Festuca, Hordeum, Koeleria, Poa, Trisetum
Puccinia brachypodii G. Otth var. **poae-nemoralis** (G. Otth) Cummins & H.C. Greene
Puccinia poae-nemoralis G. Otth
Puccinia poae-sudeticae Jørst.
Puccinia poae-sudeticae Jørst. var. *airae* (Lagerh.) Arth.
Distr.: Temperate regions, high elevations in tropical regions.
Notes: Heteroecious; aecial host - *Berberis* spp.; telial host - Poaceae.
Hosts: Agrostis, Alopecurus, Anthoxanthum, Calamagrostis, Catabrosa, Deschampsia, Festuca, Glyceria, Melica, Phleum, Poa, Puccinellia, Trisetum
Puccinia brachypodii-phoenicoidis Guyot & Malencon var. **chisosana** (Cummins) Cummins & H.C. Greene
Puccinia pygmaea Eriks. var. *chisosana* Cummins
Distr.: Southwestern United States; Mexico; Brazil.
Notes: Aecial host unknown.
Host: Bromus
Puccinia brachypodii-phoenicoidis Guyot & Malencon var. **davisii** Cummins & H.C. Greene
Distr.: Great Lakes region of the United States.
Notes: Aecial host unknown; telial host *Oryzopsis asperifolia*.
Host: Oryzopsis
Puccinia brandegei Peck
Distr.: CO, WA.
Notes: On Fumariaceae.
Host: Corydalis
Puccinia buchnerae Cummins
Uredo cumula Arth. (anamorph)
Distr.: FL; New Guinea.
Notes: This species was reported from Florida as *Uredo cumula*.
Host: Buchnera
Puccinia bupleuri F. Rudolphi
Distr.: AK; Canada(Yukon).
Host: Bupleurum
Puccinia burnettii D. Griffiths
Distr.: Western North America; Iran; Soviet Union.
Notes: Heteroecious; aecial host - *Eurotia* spp.; telial host - Poaceae.
Hosts: Ceratoides, Oryzopsis, Stipa
Puccinia cacabata Arth. & Holw. in Arth.
Puccinia stakmanii Presley in Presley & C.J. King
Distr.: Southwestern United States; West Indies; Central and South America.
Notes: Heteroecious; aecial host - *Gossypium* spp.; telial host - Poaceae. Mulder & Holliday (C.M.I. Descr. 294. 1971).
Hosts: Bouteloua, Chloris, Gossypium
Puccinia calanticariae Parmelee
Distr.: AZ; Mexico.
Notes: Autoecious.
Host: Viguiera
Puccinia calcitrapae DC.
Distr.: CA, WY; Europe.
Notes: Autoecious; on Asteraceae. The type variety does not occur in North America (Cummins. Rust fungi on legumes and composites in North America. p. 141. 1978).
Hosts: [Arctium], [Cirsium]
Puccinia calcitrapae DC. var. **bardanae** (Wallr.) Cummins
Puccinia bardanae (Wallr.) Corda
Distr.: North America; Europe.
Notes: Autoecious.
Host: Arctium
Puccinia calcitrapae DC. var. **centaureae** (DC.) Cummins
Puccinia carthami Corda
Puccinia cirsii Lasch in Rabenh. non L. Kirchner, hom. illeg.
Puccinia irrequisita H. Jacks.
Puccinia laschii Lagerh.
Distr.: North America; Asia; Australia; Europe; New Zealand.
Notes: Autoecious; on Asteraceae.
Hosts: Carthamus, Centaurea, Cirsium
Puccinia caleae Arth.
Distr.: TX; Mexico.
Notes: Autoecious.
Host: Viguiera
Puccinia californica Dietel & Holw. in Dietel
Distr.: Southwestern United States.
Notes: Autoecious.
Host: Cirsium
Puccinia californica Dietel & Holw. in Dietel var. **deserticola** Savile
Distr.: AZ, UT.
Notes: On Asteraceae (Savile. Canad. J. Bot. 48:1576. 1970).
Host: Cirsium
Puccinia callaquensis Neger
Distr.: HI; Chile.
Host: Geranium
Puccinia calochorti Peck
Distr.: Western United States; Mexico.
Notes: On Liliaceae.
Host: Calochortus
Puccinia calthae Link in L.
Distr.: Temperate northern hemisphere.
Notes: Autoecious. Parmelee & de Carteret (Fungi Canadenses 270. 1984).
Host: Caltha
Puccinia calthicola J. Schröt. in Cohn
Distr.: Northern North America.
Notes: Autoecious.
Host: Caltha
Puccinia campanulae Carmich. ex Berk. in Sm.
Distr.: Temperate northern hemisphere.
Notes: Autoecious. Parmelee & de Carteret (Fungi Canadenses 219. 1982).
Host: Campanula
Puccinia canadensis Arth.
Distr.: Pacific Northwest.
Notes: Autoecious. Savile (Fungi Canadenses 56. 1974).
Host: Viola
Puccinia canaliculata (Schwein.) Lagerh.
Distr.: HI, North America; West Indies.
Notes: Heteroecious; aecial host - Asteraceae; telial host - *Cyperus*.
Hosts: Ambrosia, Bahia, Cosmos, Cyperus, Heliopsis, Senecio, Xanthium
Puccinia carduorum Jacky
Distr.: CA, OR; Eurasia.
Notes: On Asteraceae.
Host: Carduus
Puccinia caricina DC.
Puccinia caricis (Schumach.) J. Schröt. non Rebent., hom. illeg.
Puccinia caricis (Schumach.) J. Schröt. var. *caricis-strictae* (Dietel) Arth.
Uredo caricis Schumach. non Pers., hom. illeg.
Puccinia caricis (Schumach.) J. Schröt. var. *grossulariata* Arth.
Puccinia caricis (Schumach.) J. Schröt. var. *quadriporula* Arth.
Puccinia caricis (Schumach.) J. Schröt. var. *uniporula* (Orton) Arth.
Puccinia caricis (Schumach.) J. Schröt. var. *urticata* (F. Kern) Arth.
Puccinia grossulariae Lagerh.
Puccinia ludibunda Ellis & Everh.
Puccinia urticae Lagerh.
Puccinia urticata F. Kern
Distr.: Northern hemisphere.
Notes: Heteroecious; aecial host - Grossulariaceae and Urticaceae; telial host - *Carex* spp.
Hosts: Carex, [Juncus], Ribes, Urtica
Puccinia caricina DC. var. **uliginosa** (Juel) Jørst.
Puccinia uliginosa Juel
Distr.: AK; Europe.
Notes: Autoecious.
Host: Parnassia
Puccinia caricis-shepherdiae J.J. Davis
Aecidium allenii G.W. Clinton

Puccinia allenii (G.W. Clinton) Barth., nom. inval.
Distr.: Temperate North America.
Notes: Heteroecious; aecial host - Elaeagnaceae; telial host - *Carex* spp.
Hosts: Carex, Elaeagnus, Shepherdia
Puccinia carnegiana Arth.
Distr.: AZ.
Notes: Autoecious; on Liliaceae.
Hosts: Brodiaea, Dichelostemma
Puccinia cascadensis (Savile) Cummins
Puccinia albulensis Magnus subsp. *cascadensis* Savile
Distr.: Western North America.
Notes: Autoecious. Arthur (Manual of the rusts in the United States and Canada. p. 335. 1934) mistakenly used the name *P. rhaetica* for this fungus (Savile, Canad. J. Bot. 46:633. 1968). Some reports of *P. rhaetica* from western North America actually represent mixed collections of *P. cascadensis* and *P. rhaetica* subsp. *americana*.
Host: Veronica
Puccinia castillejae Arth. in Blasdale
Distr.: CA, UT.
Notes: Autoecious; on Scrophulariaceae.
Host: Castilleja
Puccinia caulicola Tracy & B.T. Galloway
Distr.: Central and southern United States; Mexico.
Notes: Autoecious; on Lamiaceae.
Host: Salvia
Puccinia cenchri Dietel & Holw.
Distr.: Southern United States; West Indies; Central and South America; Pacific Islands.
Notes: Aecia unknown; telial host - *Cenchrus*.
Host: Cenchrus
Puccinia cesatii J. Schröt.
Distr.: Southwestern United States; Mexico; China; Egypt; southern Europe; India; Iran.
Notes: Autoecious; on Poaceae.
Host: Bothriochloa
Puccinia chaetochloae Arth.
Distr.: Southern United States; West Indies; Central and South America.
Notes: Autoecious; on Poaceae
Hosts: Paspalum, Setaria
Puccinia chamaesarachae P. Syd. & Syd.
Distr.: CA, NV.
Notes: Autoecious.
Host: Leucophysalis
Puccinia chelonis Dietel & Holw. in Dietel
Distr.: Pacific Northwest.
Notes: Autoecious.
Hosts: Chelone, Nothochelone
Puccinia chisosensis Cummins
Distr.: TX; Mexico.
Notes: Autoecious.
Host: Piptochaetium
Puccinia chloracae Parmelee
Distr.: Southwestern United States and adjacent Mexico.
Notes: Autoecious.
Host: Viguiera
Puccinia chloridis Speg.
Puccinia bartholomaei Dietel
Distr.: Southern United States to Argentina.
Notes: Heteroecious; aecial host - Asclepiadaceae; telial host - Poaceae.
Hosts: Asclepias, Bouteloua, Bouvardia, Chloris, Leptochloa, Philibertia, Sarcostemma
Puccinia chondrillina Bubák & Syd.
Distr.: Eastern United States; Europe.
Notes: Autoecious. This species was introduced from Europe as a biological control agent of skeleton weed (Parmelee & Savile. Canad. J. Bot. 59:1085. 1981).
Host: Chondrilla
Puccinia cicutae Lasch in Klotzsch
Distr.: North America; Asia; Europe; Japan.
Notes: Autoecious.
Hosts: Cicuta, Perideridia
Puccinia circaeae Pers.:Pers.
Distr.: Widespread in the northern hemisphere.
Notes: Autoecious.
Host: Circaea
Puccinia cladii Ellis & Tracy in Ellis & Everh.
Distr.: Southern United States.
Notes: Aecia unknown; telia on Cyperaceae.
Hosts: Cladium, Rhynchospora
Puccinia claytoniicola Cummins
Distr.: WY.
Notes: Probably autoecious.
Host: Claytonia
Puccinia clintonii Peck
Distr.: Temperate North America; British Isles.
Notes: Autoecious.
Host: Pedicularis
Puccinia clintonii Peck var. ***albae*** Savile
Distr.: WY.
Notes: Autoecious.
Host: Pedicularis
Puccinia clintonii Peck var. ***bracteosae*** Savile
Distr.: Western North America.
Notes: Autoecious.
Host: Pedicularis
Puccinia clintonii Peck var. ***groenlandicae*** Savile
Distr.: Western United States.
Notes: Autoecious.
Host: Pedicularis
Puccinia clintonii Peck var. ***racemosae*** Savile
Distr.: OR.
Notes: Autoecious.
Host: Pedicularis
Puccinia cnici H. Mart.
Puccinia cirsii-lanceolati J. Schröt. ex Cohn
Distr.: North America; temperate Europe.
Notes: Autoecious.
Host: Cirsium
Puccinia cnici-oleracei Pers.
Puccinia asteris Duby
Puccinia columbiensis Ellis & Everh.
Puccinia conferta Dietel & Holw. in Dietel
Puccinia millefolii Willd.
Puccinia rudbeckiae Barth. in Arth.
Distr.: North and South America; Asia; Europe.
Notes: Autoecious; on Asteraceae.
Hosts: Achillea, Agoseris, Artemisia, Aster, Erigeron, Hieracium, Krigia, Rudbeckia
Puccinia cockerelliana Bethel ex Arth.
Distr.: Central and western North America.
Notes: Heteroecious; aecial host - *Thalictrum fendleri*; telial host - *Festuca* spp.
Hosts: Festuca, Thalictrum
Puccinia coelopleuri Arth.
Distr.: AK, northwestern North America.
Notes: Autoecious; on Apiaceae.
Hosts: Angelica, Conioselinum
Puccinia cognata P. Syd. & Syd.
Distr.: Southern United States; Central America.
Notes: Autoecious.
Host: Verbesina
Puccinia collinsiae Henn.
Aecidium collinsiae Ellis & Everh. (anamorph)
Distr.: Western United States.
Hosts: Collinsia, Tonella
Puccinia comandrae Peck
Distr.: Western United States east along Canadian border.
Notes: Autoecious.
Host: Comandra
Puccinia commelinae Holw.
Distr.: TX.
Host: Commelina
Puccinia commutata Syd.
Distr.: OR, NY; Canada(British Columbia); Europe.
Notes: Autoecious; on Valerianaceae.
Host: Valeriana
Puccinia concinna Arth.

Distr.: AZ; possibly northwestern Mexico.
Notes: Autoecious; on Asteraceae.
Host: Eupatorium
Puccinia confraga Arth. & Cummins
Distr.: AZ.
Notes: Autoecious; on Scrophulariaceae.
Hosts: Keckiella, Penstemon
Puccinia congdonii Savile
Distr.: CA.
Notes: Aecia unknown; telia on *Scirpus congdonii.*
Host: Scirpus
Puccinia conglomerata (F. Strauss) Röhl.
Distr.: North America; temperate Europe.
Notes: Autoecious. Savile (Fungi Canadenses 110. 1977).
Host: Petasites
Puccinia conoclinii Seym. ex Burrill
Distr.: Central United States to Costa Rica; South America.
Notes: Autoecious; on Asteraceae.
Hosts: Ageratum, Ambrosia, Eupatorium
Puccinia consimilis Ellis & Everh.
Distr.: MT, UT.
Notes: Autoecious; on Brassicaceae.
Hosts: Erysimum, Sisymbrium
Puccinia convolvuli Castagne
Distr.: Cosmopolitan.
Notes: Autoecious.
Hosts: Calystegia, Convolvulus, Ipomoea
Puccinia cooperiae W.H. Long
Distr.: FL, NC, TX.
Notes: Autoecious.
Host: Zephyranthes
Puccinia cordiae Arth.
Distr.: FL; West Indies; Central and South America.
Notes: Autoecious.
Host: Cordia
Puccinia coronata Corda
Puccinia coronata Corda var. *bromi* W.P. Fraser & Ledingham
Puccinia coronata Corda var. *calamagrostis* W.P. Fraser & Ledingham
Puccinia coronata Corda var. *elaeagni* W.P. Fraser & Ledingham
Dicaeoma rhamni Kuntze
Puccinia rhamni (Pers.) Wettst.
Distr.: Cosmopolitan.
Notes: Heteroecious; aecial host - Rhamnaceae and Eleagnaceae; telial host - Poaceae.
Hosts: Agropogon, Agropyron, Agrostis, Alopecurus, Ammophila, Arrhenatherum, Arundo, Avena, Beckmannia, Berchemia, Bromus, Calamagrostis, Cinna, Dactylis, Deschampsia, Elaeagnus, Elymus, Festuca, Glyceria, Helictotrichon, Hierochloe, Holcus, Hordeum, Hystrix, Koeleria, Lagurus, Lamarckia, Limnodea, Lolium, Panicum, Paspalum, Phalaris, Phleum, Poa, Polypogon, Puccinellia, Rhamnus, Schizachne, Scolochloa, Shepherdia, Trisetum
Puccinia coronata Corda var. ***avenae*** W.P. Fraser & Ledingham
Distr.: Worldwide in oat-growing regions.
Notes: Heteroecious; aecial host - Rhamnaceae; telial host - *Avena* spp. Savile (Fungi Canadenses 311. 1987).
Host: Avena
Puccinia coronopsora Cummins
Distr.: FL; China(type).
Notes: Autoecious.
Host: Lindera
Puccinia crandallii Pammel & H. Hume in H. Hume
Puccinia abundans H. Jacks.
Aecidium abundans Peck (anamorph)
Distr.: Western United States.
Notes: Heteroecious; aecial host - *Symphoricarpos* spp.; telial host - Poaceae.
Hosts: Festuca, Koeleria, Leucopoa, Poa, Symphoricarpos
Puccinia crassipes Berk. & M.A. Curtis
Distr.: Southern United States; West Indies; Central and South America.
Notes: Autoecious.
Host: Ipomoea
Puccinia crepidis-montanae Magnus
Distr.: Western North America; Europe.
Notes: Autoecious.
Hosts: Agoseris, Crepis
Puccinia cruciferarum Rudolphi
Distr.: Circumpolar and at high elevations in the United States and Europe.
Notes: Autoecious.
Host: Cardamine
Puccinia cruciferarum Rudolphi subsp. ***borealis*** Savile
Distr.: Subarctic regions.
Notes: Savile (Mycologia 56:241. 1964).
Host: Cardamine
Puccinia cruciferarum Rudolphi subsp. ***wyomingensis*** Savile
Distr.: UT, WY.
Notes: Savile (Mycologia 56:243. 1964).
Host: Cardamine
Puccinia cryptandri Ellis & Barth.
Puccinia simulans (Peck) Barth.
Distr.: Central and western North America to northern Mexico.
Notes: Aecia unknown; telial host- *Sporobolus* spp.
Hosts: Sporobolus, [Triodia]
Puccinia cryptandri Ellis & Barth. var. ***luxurians*** (Arth.) Cummins & H.C. Greene
Puccinia luxuriosa Syd.
Distr.: Central and western North America to Mexico.
Notes: Heteroecious; aecial host - *Sarcobatus*; telial host - *Sporobolus* spp.
Hosts: [Muhlenbergia], Sarcobatus, Sporobolus
Puccinia cryptanthes Dietel & Holw.
Distr.: CA, WA.
Notes: Autoecious.
Host: Cryptantha
Puccinia cryptotaeniae Peck
Distr.: Northern United States and adjacent Canada.
Notes: Autoecious.
Host: Cryptotaenia
Puccinia cyani Pass. in Rabenh.
Distr.: Warm-temperate to tropical regions.
Notes: Autoecious.
Host: Centaurea
Puccinia cynodontis Lacr. ex Desmaz.
Aecidium plantaginis Ces. (anamorph)
Distr.: Cosmopolitan.
Notes: Heteroecious; aecial host - several families of angiosperms; telial host - *Cynodon* spp. Hennen & Cummins (Mycologia 48:126. 1956; Mulder & Holliday. C.M.I. Descr. 292. 1971).
Host: Cynodon
Puccinia cyperi Arth.
Distr.: Central and eastern United States; West Indies; Central and South America.
Notes: Heteroecious; aecial host - Asteraceae; telial host - *Cyperus* spp.
Hosts: Chrysopsis, Conyza, Cyperus, Erigeron, Heterotheca
Puccinia cyperi-tagetiformis F. Kern
Distr.: TX; West Indies; Asia.
Notes: Aecia unknown; telial host- *Cyperus* spp.
Host: Cyperus
Puccinia cypripedii Arth. & Holw.
Distr.: Northern United States; Canada (Saskatchewan).
Notes: Autoecious; on Orchidaceae.
Hosts: Calopogon, Cypripedium
Puccinia dayi G.W. Clinton
Distr.: Central and eastern North America.
Notes: Autoecious. Parmelee (Fungi Canadenses 204. 1981).
Host: Lysimachia
Puccinia delphinii Dietel & Holw. in Dietel
Distr.: CA.
Notes: Autoecious.
Host: Delphinium
Puccinia dentariae (Albertini & Schwein.) Fuckel
Distr.: OR; Europe.
Notes: Autoecious.
Host: Dentaria
Puccinia dichelostemmae Dietel & Holw. in Dietel
Distr.: CA, WA.
Notes: Autoecious.
Hosts: Brodiaea, Dichelostemma, Triteleia

Puccinia dichondrae Mont. in Gay
Distr.: CA, coastal areas along eastern and southern United States; West Indies; Central and South America; Australia; New Zealand.
Notes: Autoecious.
Host: Dichondra
Puccinia dietelii Sacc. & Syd.
Distr.: Southern United States to Argentina.
Notes: Heteroecious; aecial host - *Acalypha*; telial host - Poaceae.
Hosts: Acalypha, Chloris
Puccinia difformis Kunze in Kunze & J.C. Schmidt
Puccinia ambigua Lagerh.
Distr.: North and South America; Europe.
Notes: Autoecious.
Host: Galium
Puccinia dioicae Magnus
Puccinia asterum F. Kern
Puccinia caricis-asteris Arth.
Puccinia extensicola Plowr.
Puccinia extensicola Plowr. var. *asteris* (Thuem.) Arth.
Puccinia extensicola Plowr. var. *erigerontis* Arth.
Puccinia extensicola Plowr. var. *euthamii* Arth.
Puccinia extensicola Plowr. var. *hieraciata* (Schwein.) Arth.
Puccinia extensicola Plowr. var. *oenotherae* (Mont.) Arth.
Puccinia extensicola Plowr. var. *phrymae* (Halst.) Arth.
Puccinia extensicola Plowr. var. *rudbeckiae* Arth.
Puccinia extensicola Plowr. var. *valerianae* Arth.
Puccinia peckii Kellerm.
Puccinia vulpinoideae Dietel
Aecidium boltoniae Arth. (anamorph)
Distr.: Northern hemisphere.
Notes: Heteroecious; aecial host - Asteraceae, Onagraceae, Phrymaceae, and Valerianceae; telial host - Cyperaceae.
Hosts: Agoseris, Apargidium, Artemisia, Aster, Boltonia, Calylophus, Carex, Conyza, Crepis, Epilobium, Erigeron, Euthamia, Gaura, Hieracium, Krigia, Lactuca, Microseris, Oenothera, Phryma, Prenanthes, Rudbeckia, Senecio, Solidago, Valeriana
Puccinia diplachnis Arth.
Aecidium bouvardiae Dietel & Holw. (anamorph)
Distr.: AZ, TX; Mexico.
Notes: Heteroecious; aecial host - *Bouvardia*; telial host - Poaceae.
Hosts: Bouteloua, Bouvardia, Leptochloa
Puccinia distichlidis Ellis & Everh.
Distr.: Temperate North America.
Notes: Heteroecious; aecial host - Primulaceae; telial host - *Spartina* spp. Parmelee & de Carteret (Fungi Canadenses 205. 1981).
Hosts: Glaux, Lysimachia, Spartina
Puccinia distorta Holw.
Distr.: AZ, CA; Mexico; South America.
Notes: Autoecious.
Host: Hyptis
Puccinia dochmia Berk. & M.A. Curtis
Distr.: TX; Mexico.
Notes: Autoecious.
Host: Muhlenbergia
Puccinia dolosa Arth. & Fromme var. ***circumdata*** (Mains) Ramachar & Cummins
Puccinia circumdata Mains
Distr.: Southern United States; West Indies; Central and South America.
Notes: Aecia unknown; telial host - *Panicum* spp.
Host: Panicum
Puccinia douglasii Ellis & Everh.
Distr.: Western North America.
Notes: Autoecious. Parmelee (Mycologia 78:454. 1986) recognized two varieties.
Host: Phlox
Puccinia douglasii Ellis & Everh. var. ***hoodii*** Parmelee
Distr.: Western North America.
Notes: Autoecious.
Host: Phlox
Puccinia dovrensis A. Blytt var. ***leptoticha*** (J. Hennen) J. Hennen
Puccinia russa Arth. & Cummins var. *leptoticha* J. Hennen
Distr.: CO, WY.
Notes: Autoecious. Cummins (Rust fungi on legumes and composites in North America. p. 175. 1978).
Host: Erigeron
Puccinia dovrensis A. Blytt var. ***russa*** (Arth. & Cummins) J. Hennen in J. Hennen & J.W. Baxter
Puccinia russa Arth. & Cummins
Distr.: CA.
Notes: Cummins (Rust fungi on legumes and composites in North America. p. 175. 1978).
Host: Erigeron
Puccinia drabae F. Rudolphi
Distr.: North America; northern Europe.
Notes: Autoecious.
Host: Draba
Puccinia dulichii Syd.
Puccinia extensicola Syd. var. *solidaginis* (Schwein.) Arth.
Distr.: Central and eastern United States.
Notes: Heteroecious; aecial host - Asteraceae; telial host - Cyperaceae. Savile (Canad. J. Bot. 50:2582. 1972).
Hosts: Aster, Carex, Dulichium, Grindelia, Helenium, Sericocarpus, Solidago
Puccinia durangensis Cummins
Distr.: AZ, CO.
Notes: Aecia unknown; telial host- Poaceae.
Hosts: Piptochaetium, Stipa
Puccinia dyssodiae Cummins
Distr.: AZ; Mexico.
Notes: Autoecious.
Host: Dyssodia
Puccinia eatoniae Arth.
Puccinia eatoniae Arth. var. *myosotidis* Mains
Puccinia eatoniae Arth. var. *ranunculi* Mains
Distr.: North America; West Indies.
Notes: Heteroecious; aecial host - *Ranunculus* spp. or *Myosotis virginica*; telial host - *Sphenopholis* spp.
Hosts: Myosotis, Ranunculus, Sphenopholis
Puccinia echeveriae Linder
Distr.: CA.
Notes: Autoecious.
Host: Echeveria
Puccinia electrae Dietel & Holw.
Distr.: TX; Mexico; Central America.
Notes: Autoecious.
Host: Zexmenia
Puccinia electrae Dietel & Holw. var. ***depressiporosa*** Parmelee
Distr.: TX.
Notes: Autoecious.
Host: Zexmenia
Puccinia electrae Dietel & Holw. var. ***robusta*** Parmelee
Distr.: TX.
Notes: Autoecious.
Host: Zexmenia
Puccinia eleocharidis Arth.
Distr.: United States; Mexico; West Indies.
Notes: Heteroecious; aecial host - Asteraceae; telial host - *Eleocharis* spp.
Hosts: Eleocharis, Eupatorium
Puccinia ellisiana Thuem.
Puccinia sagittata W.H. Long
Distr.: Central and eastern North America to Mexico.
Notes: Heteroecious; aecial host - *Viola* spp.; telial host - *Andropogon* spp.
Hosts: Andropogon, Schizachyrium, Viola
Puccinia ellisii De Toni in Sacc.
Distr.: Western North America.
Notes: Autoecious; on Apiaceae.
Host: Angelica
Puccinia emaculata Schwein.
Puccinia pammelii Arth.
Puccinia panici Dietel
Distr.: Central and eastern North America to Mexico; possibly Chile.
Notes: Heteroecious; aecial host - *Euphorbia* spp.; telial host - Poaceae.
Hosts: Euphorbia, Panicum, Paspalum
Puccinia emiliae Henn.
Distr.: FL; West Indies; Central America.
Notes: Autoecious.

Host: Emilia
Puccinia enceliae Dietel & Holw.
Puccinia aemulans P. Syd. & Syd.
Puccinia tithoniae Dietel & Holw. in Holw.
Distr.: Western United States; Central and South America.
Notes: Autoecious; on Asteraceae.
Hosts: Encelia, Tithonia, Viguiera
Puccinia epilobii DC.
Distr.: Northern North America; Europe; Japan.
Notes: Autoecious.
Host: Epilobium
Puccinia erigeniae (Orton) Arth.
Distr.: OH; Canada(Ontario).
Notes: Autoecious.
Host: Erigenia
Puccinia eriophori Thuem.
Puccinia angustata Peck var. *eriophori* (Thuem.) Arth.
Distr.: Boreal regions.
Notes: Heteroecious; aecial host - *Senecio*; telial host - *Eriophorum*. Savile (Canad. J. Bot. 50:2584. 1972).
Hosts: Eriophorum, Senecio
Puccinia eriophori-alpini Allesch.
Distr.: Boreal regions.
Notes: Heteroceious; aecial host - *Solidago*; telial host - *Scirpus*. Savile (Canad. J. Bot. 50:2585. 1972).
Host: Scirpus
Puccinia esclavensis Dietel & Holw. in Holw.
Aecidium mirabilis Dietel & Holw. (anamorph)
Distr.: Southwestern United States; West Indies; Central and South America.
Notes: Heteroecious; aecial host - *Mirabilis* spp.; telial host - Poaceae.
Hosts: Digitaria, Mirabilis, Panicum, Setaria
Puccinia esclavensis Dietel & Holw. in Holw. var. ***panicophila*** (Speg.) Ramachar & Cummins
Puccinia panicophila Speg.
Puccinia atra Dietel & Holw. in Holw.
Distr.: Southwestern United States; West Indies; Central America; Argentina.
Notes: Aecial host unknown.
Hosts: Digitaria, Panicum, Paspalum, Setaria
Puccinia eumacrospora Cummins in Cummins & D. Baxter
Puccinia macrospora Arth. non Spreng., hom. illeg.
Distr.: KS, eastern United States.
Notes: Heteroecious; aecial host - *Smilax* spp.; telial host - *Carex* spp.
Hosts: Carex, Smilax
Puccinia eurotiae D. Griffiths
Distr.: NV, UT.
Notes: Autoecious.
Host: Ceratoides
Puccinia evadens Harkn.
Eriosporangium evadens (Harkn.) Arth.
Distr.: Southern United States; Central and South America.
Notes: Autoecious. Mordue (C.M.I. Descr. 620. 1979).
Host: Baccharis
Puccinia exasperans Holw.
Distr.: Southern United States; Mexico.
Notes: Aecia unknown; telial host - *Bouteloua* spp.
Host: Bouteloua
Puccinia excursionis Savile
Distr.: CO.
Notes: Autoecious.
Host: Erigeron
Puccinia farinacea W.H. Long
Distr.: NE to TX; Mexico; Central America.
Notes: Autoecious.
Host: Salvia
Puccinia farinacea W.H. Long var. ***azurea*** J.W. Baxter
Distr.: Central and southern United States.
Notes: Autoecious.
Host: Salvia
Puccinia fergussonii Berk. & Broome
Distr.: Western North America; Eurasia.
Notes: Autoecious. Savile (Fungi Canadenses 64. 1975).
Host: Viola
Puccinia festucae Plowr.
Distr.: AK; Eurasia.
Notes: Heteroecious; aecial host - *Lonicera* spp.; telial host - *Festuca* spp.
Hosts: Festuca, Lonicera
Puccinia fimbristylidis Arth.
Distr.: TX; Mexico; West Indies; South America.
Notes: Aecia unknown; telial host - *Fimbristylis* spp.
Host: Fimbristylis
Puccinia fragilis Tracy & B.T. Galloway
Distr.: Western North America.
Notes: Autoecious; on Polemoniaceae. Parmelee (Mycologia 78:460. 1986).
Hosts: Linanthus, Phlox
Puccinia fragilis Tracy & B.T. Galloway var. ***divaricatae*** Parmelee
Distr.: Central and eastern North America.
Notes: Autoecious. Parmelee (Mycologia 78:461. 1986).
Host: Phlox
Puccinia franseriae Syd. & P. Syd.
Puccinia caborcensis Parmelee
Distr.: Western United States; Mexico; western South America.
Notes: Autoecious; on Asteraceae.
Hosts: Ambrosia, Hymenoclea
Puccinia fraseri Arth.
Distr.: Northcentral and northeastern North America.
Notes: Autoecious.
Host: Hieracium
Puccinia fuirenae Cooke
Distr.: Southern United States.
Notes: Aecia unknown; telia on Cyperaceae.
Host: Fuirena
Puccinia gallula J. Hennen & Cummins
Distr.: TX; Mexico.
Notes: Autoecious.
Host: Porophyllum
Puccinia gemella Dietel & Holw. in P. Syd. & Syd.
Distr.: Western North America.
Notes: Autoecious. Parmelee & de Carteret (Fungi Canadenses 264. 1984).
Host: Caltha
Puccinia gentianae (F. Strauss) Röhl.
Puccinia probabilis Arth. & Cummins
Distr.: Widespread in the northern hemisphere.
Notes: Autoecious. Parmelee & de Carteret (Fungi Canadenses 295. 1985).
Hosts: Gentiana, [Veronica]
Puccinia geranii-silvatici P. Karst.
Distr.: HI; Finland(type).
Host: Geranium
Puccinia gibberulosa J. Schröt.
Distr.: AZ.
Host: Ranunculus
Puccinia gigantea P. Karst.
Distr.: Western North America; northern Europe.
Notes: Autoecious.
Host: Epilobium
Puccinia gigantispora Bubák
Distr.: Northern North America.
Notes: Autoecious.
Host: Anemone
Puccinia giliae Harkn.
Distr.: Western United States.
Notes: Autoecious; on Polemoniaceae. Parmelee (Mycologia 78:461. 1986).
Hosts: Collomia, Eriastrum, Gilia, Linanthus, Navarretia
Puccinia giliicola Henn.
Distr.: CA.
Notes: Autoecious. Parmelee (Mycologia 78:454. 1986).
Host: Leptodactylon
Puccinia glabella Holw.
Distr.: Western United States.
Notes: Autoecious.
Host: Boisduvalia
Puccinia globosipes Peck
Distr.: Western United States.
Notes: Autoecious.

Host: Lycium
Puccinia glomerata Grev.
Puccinia expansa Link in Willd.
Distr.: Temperate northern hemisphere.
Notes: Autoecious.
Host: Senecio
Puccinia gnaphaliicola Henn.
Puccinia gnaphalii Henn.
Distr.: Central and southern United States; South America; Australia; New Zealand.
Notes: Autoecious; on Asteraceae.
Host: Gnaphalium
Puccinia graminella Dietel & Holw. in Dietel
Distr.: CA; southern and western South America.
Notes: Autoecious; on Poaceae
Host: Stipa
Puccinia graminis Pers.:Pers.
Puccinia graminis Pers.:Pers. f. sp. *avenae* Eriks. & E. Henn.
Puccinia graminis Pers.:Pers. f. sp. *poae* Eriks. & Henn.
Puccinia graminis Pers.:Pers. f. sp. *secalis* Eriks. & E. Henn.
Puccinia graminis Pers.:Pers. f. sp. *tritici* Eriks. & E. Henn.
Puccinia graminis Pers.:Pers. var. *agropyri-repentis* Guyot
Puccinia graminis Pers.:Pers. var. *stakmanii* Guyot, Massenot, & Saccas ex Z. Urban
Puccinia graminis Pers.:Pers. var. *tritici* Guyot, Massenot, & Saccas ex Z. Urban
Puccinia poculiformis Wettst.
Distr.: Widespread.
Notes: Heteroecious; aecial host - *Berberis* spp.; telial host - Poaceae. Cummins (Rust fungi of cereals, grasses and bamboos. p. 208. 1971) and Urban (Ceska Mykol. 21:12. 1967) recognized two subspecies based on urediniospore size.
Hosts: [Aegilops], Agrohordeum, Agropyron, Agrostis, Alopecurus, Ammophila, Anthoxanthum, Arrhenatherum, Avena, Beckmannia, Berberis, Bouteloua, Briza, Bromus, Buchloe, Calamagrostis, Calamovilfa, Catabrosa, Cinna, Cynodon, Dactylis, Dactyloctenium, Danthonia, Deschampsia, Distichlis, [Echinochloa], Elymus, Festuca, Gastridium, Glyceria, Helictotrichon, Hierochloe, Holcus, Hordeum, Hystrix, Koeleria, Lamarckia, Limnodea, Lolium, Mahonia, Melica, Milium, Molinia, Muhlenbergia, Oryzopsis, Panicum, Phalaris, Phleum, Poa, Polypogon, Puccinellia, Secale, Sitanion, Sphenopholis, Sporobolus, Stipa, Tridens, Trisetum, Triticum, Vulpia
Puccinia graminis Pers.:Pers. subsp. ***graminicola*** Z. Urban
Puccinia dactylidis Gäum.
Puccinia graminis Pers.:Pers. f. sp. *phlei-pratensis* (Eriks. & E. Henn.) Stakman & Piemeisel
Distr.: Widespread.
Notes: Heteroecious; aecial host - *Berberis* spp.; telial host - Poaceae.
Hosts: Bromus, Cinna, Dactylis, Elymus, Festuca, Koeleria, Lolium, Phleum
Puccinia granulispora Ellis & B.T. Galloway ex Ellis & Everh.
Distr.: Western North America.
Notes: Autoecious.
Host: Allium
Puccinia grindeliae Peck
Distr.: Central and western North America to Mexico.
Notes: Autoecious; on Asteraceae.
Hosts: Aster, Baileya, Chrysopsis, Chrysothamnus, Dyssodia, Erigeron, Grindelia, Gutierrezia, Haplopappus, Hymenopappus, Hymenoxys, Lygodesmia, Machaeranthera, Petradoria, Psilostrophe, Senecio, Solidago, Tetradymia
Puccinia grindeliae Peck subsp. ***riddellii*** (D. Griffiths) Cummins
Gymnoconia riddelliae D. Griffiths
Distr.: AZ.
Notes: Autoecious.
Hosts: Baileya, Psilostrophe
Puccinia grumosa Syd. & Holw.
Distr.: Subarctic North America; Soviet Union.
Notes: Autoecious; on Liliaceae. Ono (Trans. Mycol. Soc. Japan 26:493. 1985).
Hosts: Stenanthium, Zigadenus
Puccinia guillemineae Dietel & Holw.
Distr.: Southwestern United States; Mexico.
Notes: Autoecious; on Amaranthaceae.
Host: Guilleminea
Puccinia gulosa H. Jacks. in Barth.
Distr.: CA.
Notes: Autoecious. Parmelee (Mycologia 78:456. 1986).
Host: Polemonium
Puccinia gymnandrae Tranzschel
Distr.: AK.
Host: Lagotis
Puccinia haleniae Arth. & Holw. in Arth.
Distr.: Central and northern North America; eastern Asia.
Notes: Autoecious. Parmelee & de Carteret (Fungi Canadenses 296. 1985).
Hosts: Gentiana, Halenia
Puccinia helianthellae Arth.
Puccinia gaillardiae (Dietel & Holw.) Barth.
Uredo gaillardiae Dietel & Holw. in Dietel (anamorph)
Distr.: Western United States.
Notes: Autoecious.
Hosts: Gaillardia, Helianthella
Puccinia helianthi Schwein.
Puccinia helianthorum Schwein.
Puccinia heliopsidis Schwein.
Puccinia xanthifoliae Ellis & Everh.
Distr.: Cosmopolitan.
Notes: Autoecious; on Asteraceae (Laundon & Waterston. C.M.I. Descr. 55. 1965; Parmelee. Fungi Canadenses 95. 1977).
Hosts: Helianthus, Heliopsis, Iva, Krigia
Puccinia helicalis Savile
Distr.: Northwestern North America.
Notes: Autoecious.
Host: Pedicularis
Puccinia heterospora Berk. & M.A. Curtis
Distr.: HI, southern United States; West Indies; South America; Africa; Taiwan.
Notes: Autoecious; on Malvaceae.
Hosts: Abutilon, Alcea, Herissantia, Hibiscus, Malvastrum, Malvaviscus, Sida
Puccinia heucherae (Schwein.) Dietel
Puccinia congregata Ellis & Harkn.
Distr.: Temperate northern hemisphere.
Notes: Autoecious; on Saxifragaceae.
Hosts: Elmera, Heuchera, Lithophragma, Mitella, Saxifraga, Tellima, Tiarella, Tolmiea
Puccinia heucherae (Schwein.) Dietel var. ***austroberingiana*** Savile
Distr.: Northwestern North America; northeastern Asia.
Notes: Autoecious; on Saxifragaceae.
Hosts: Mitella, Saxifraga, Tellima
Puccinia heucherae (Schwein.) Dietel var. ***lithophragmae*** (Holw.) Savile
Puccinia lithophragmae Holw.
Distr.: Western North America.
Host: Lithophragma
Puccinia heucherae (Schwein.) Dietel var. ***saxifragae*** (Schlechtend.) Savile
Puccinia saxifragae Schlechtend.
Puccinia curtipes E. Howe
Puccinia striatospora Peck
Distr.: Circumboreal.
Notes: Autoecious; on Saxifragaceae (Savile. Canad. J. Bot. 32:408. 1954).
Hosts: Heuchera, Saxifraga
Puccinia hieracii (Röhl.) H. Mart.
Puccinia eriophylli H. Jacks.
Puccinia flosculosorum (Albertini & Schwein.) Röhl.
Puccinia sejuncta Syd.
Puccinia taraxaci Plowr.
Distr.: Cosmopolitan.
Notes: Autoecious; on Asteraceae.
Hosts: Agoseris, Cichorium, Cnicus, Crepis, Eriophyllum, Hieracium, Hypochaeris, Inula, Krigia, Lactuca, Leontodon, Microseris, Pyrrhopappus, Taraxacum
Puccinia hieracii (Röhl.) H. Mart. var. ***harknessii*** (Vize) Cummins
Puccinia harknessii Vize
Distr.: Western United States.
Notes: Autoecious; on Asteraceae. Parmelee & Savile (Canad. J. Bot.

59:1078. 1981) treated this taxon as a variety of *P. harknessii*. We follow Cummins (Rust fungi on legumes and composites in North America. p. 121. 1978).
Hosts: Lygodesmia, Malacothrix, Rafinesquia
Puccinia hieracii (Röhl.) H. Mart. var. ***stephanomeriae*** (P. Syd. & Syd.) Cummins
Puccinia stephanomeriae P. Syd. & Syd.
Puccinia harknessii Vize var. *stephanomeriae* (P. Syd. & Syd.) Parmelee & Savile
Puccinia harknessii Vize var. *major* Arth.
Distr.: CA.
Notes: Autoecious. Parmelee & Savile (Canad. J. Bot. 59:1078. 1981) treated this taxon as a variety of *P. harknessii*. We follow Cummins (Rust fungi on legumes and composites in North America. p. 121. 1978).
Host: Stephanomeria
Puccinia hordei G. Otth
Puccinia hordei Fuckel non G. Otth, hom. illeg.
Puccinia hordei-murini Buchwald
Puccinia anomala Rostr.
Puccinia holcina Eriks.
Puccinia rubigo-vera G. Wint. f. sp. *holcina* (Eriks.) Mains
Distr.: Cosmopolitan.
Notes: Heteroecious; aecial host - *Allium*, *Ornithogalum* and *Sedum*; telial host - Poaceae. Savile (Fungi Canadenses 312. 1987) included a key to cereal leaf rusts.
Hosts: Holcus, Hordeum
Puccinia horiana Henn.
Distr.: United States; Argentina; Africa; Asia; Europe.
Notes: Autoecious. Punithalingam (C.M.I. Descr. 176. 1968).
Host: Chrysanthemum
Puccinia huberi Henn.
Distr.: HI, PR; Central and South America.
Notes: Aecial host unknown; telial host - *Panicum* spp.
Host: [Paspalum]
Puccinia hydrocotyles Cooke
Distr.: HI, North and South America; Europe; New Zealand.
Notes: Autoecious.
Host: Hydrocotyle
Puccinia hydrophylli Peck & G.W. Clinton
Distr.: North America.
Notes: Autoecious.
Hosts: Hackelia, Hydrophyllum, Myosotis
Puccinia hydrophylli Peck & G.W. Clinton subsp. ***mertensiae*** (Peck) J. Hennen
Puccinia mertensiae Peck
Distr.: Western United States.
Notes: Autoecious; on Boraginaceae.
Hosts: Hackelia, Lappula, Mertensia, Myosotis
Puccinia hypochaeridis Oudem.
Distr.: CA, WA.
Notes: Autoecious.
Host: Hypochaeris
Puccinia hyptidis Tracy & Earle
Distr.: Southeastern United States; West Indies.
Notes: Autoecious.
Host: Hyptis
Puccinia hyptidis-mutabilis Mayor
Distr.: FL; Central and South America; Africa.
Notes: Autoecious.
Host: Hyptis
Puccinia hyssopi Schwein.
Distr.: United States.
Notes: Autoecious; on Lamiaceae.
Host: Agastache
Puccinia inanipes Dietel & Holw.
Distr.: AZ; Mexico.
Notes: Autoecious.
Host: Eupatorium
Puccinia inclusa P. Syd. & Syd.
Distr.: Western North America to central Mexico.
Notes: Autoecious.
Host: Cirsium
Puccinia intermixta Peck
Distr.: Central and western North America.
Notes: Autoecious. Parmelee & de Carteret (Fungi Canadenses 96. 1977).
Host: Iva
Puccinia interveniens Bethel in Blasdale
Distr.: Western United States; Mexico; western South America.
Notes: Heteroecious; aecial host - Malvaceae; telial host - *Nassella* and *Stipa* spp.
Hosts: Callirhoe, Iliamna, Malacothamnus, Sidalcea, Stipa
Puccinia invaginata Arth. & J.R. Johnston
Distr.: FL; Central and South America.
Notes: Autoecious; on Rhamnaceae.
Host: Gouania
Puccinia invelata H. Jacks. in Arth.
Puccinia parthenices H. Jacks.
Distr.: AZ; Mexico.
Notes: Autoecious; on Asteraceae.
Host: Parthenice
Puccinia investita Schwein.
Distr.: Central and eastern North America; Central and South America.
Notes: Autoecious.
Host: Gnaphalium
Puccinia iridis Wallr. in Rabenh.
Distr.: Temperate regions.
Notes: Heteroecious; aecial host - *Urtica* and *Valeriana* spp. (known only in Eurasia); telial host - Iridaceae. Parmelee (Fungi Canadenses 320. 1987).
Hosts: [Belamcanda], [Dietes], [Ipomoea], Iris
Puccinia iridis Wallr. in Rabenh. f. sp. ***australis*** Mains
Distr.: CA, FL.
Hosts: Belamcanda, Iris
Puccinia iridis Wallr. in Rabenh. f. sp. ***floridanum*** Sobers
Distr.: FL.
Host: Iris
Puccinia iridis Wallr. in Rabenh. f. sp. ***septentrionalis*** Mains
Distr.: CA, FL.
Host: Iris
Puccinia jonesii Peck
Puccinia cymopteri Dietel & Holw. in Dietel
Puccinia jonesii Peck var. *cymopteri* (Dietel & Holw.) Arth.
Puccinia lindrothii Syd. ex Lindr.
Distr.: Western North America.
Notes: Autoecious; on Apiaceae.
Hosts: Cymopterus, Ligusticum, Lomatium, Musineon, Tauschia
Puccinia jussiaeae Speg.
Distr.: Eastern and southern United States; West Indies; South America.
Notes: Autoecious; on Onagraceae.
Host: Ludwigia
Puccinia justiciae A. Puttemans
Puccinia diantherae Syd.
Distr.: TX; West Indies; Central America.
Notes: Autoecious.
Host: Justicia
Puccinia kansensis Ellis & Barth.
Distr.: Central United States to Mexico.
Notes: Heteroecious; aecial host - *Physalis* spp.; telial host - *Buchloe* spp. Compare aecial stage with *Aecidium physalidis* Burrill.
Host: Buchloe
Puccinia karelica Tranzschel
Distr.: Circumboreal.
Notes: Heteroecious; aecial host - *Trientalis* spp.; telial host - *Carex* spp. Savile (Canad. J. Bot. 43:231. 1965) recognized two subspecies. Parmelee & de Carteret (Fungi Canadenses 206. 1981).
Hosts: Carex, Trientalis
Puccinia karelica Tranzschel subsp. ***laurentina*** Savile
Distr.: Northeastern North America.
Notes: Heteroecious; aecial host - *Trientalis borealis*; telial host - *Carex* spp.
Host: Trientalis
Puccinia kuhniae Schwein.
Distr.: Central and western United States.
Notes: Autoecious; on Asteraceae.
Hosts: Brickellia, Kuhnia
Puccinia kuhniae Schwein. var. ***brickelliae*** (Peck) Cummins

Puccinia brickelliae Peck
Distr.: AZ; Mexico.
Notes: Autoecious; on Asteraceae.
Host: Brickellia
Puccinia kuhniae Schwein. var. ***robusta*** Cummins, M.P. Britton, & J.W. Baxter
Distr.: Western United States; Central America.
Notes: Autoecious.
Hosts: Brickellia, Kuhnia
Puccinia lantanae Farl.
Distr.: FL, TX; Japan; Taiwan; subtropical and tropical western hemisphere.
Notes: Autoecious; on Verbenaceae and Acanthaceae.
Hosts: Lantana, Lippia, Ruellia
Puccinia lapathicola Hyl., Jørst., & Nannf.
Puccinia suaveolens Rostr.
Distr.: CA, FL.
Notes: This name is used for the rust on *Rumex*, formerly known as the later homonym, *P. punctiformis* Dietel & Holw.
Host: Rumex
Puccinia lapsanae Fuckel
Puccinia variabilis Grev. var. *lapsanae* (Fuckel) Cummins
Distr.: Temperate northern hemisphere.
Notes: Autoecious.
Host: Lapsana
Puccinia lateripes Berk. & Ravenel in Berk.
Puccinia agrimoniae (Arth.) Arth.
Puccinia ruelliae Auct. non Syd. & P. Syd.
Distr.: Southcentral and southeastern United States; West Indies; Central America; northern South America.
Notes: Autoecious; on Acanthaceae. Laundon (Mycol. Pap. 89:46. 1963) recognized four varieties.
Hosts: Blechum, Justicia, Ruellia
Puccinia lateritia Berk. & M.A. Curtis
Distr.: Subtropical and tropical western hemisphere; India; Taiwan.
Notes: Autoecious; on Rubiaceae.
Hosts: Borreria, Diodia, Ernodea, Hedyotis
Puccinia laurentiana Trel.
Distr.: AK(type).
Host: Saxifraga
Puccinia leonotidicola Henn.
Puccinia leonotidis Arth., nom. inval.
Distr.: FL; West Indies; Central and South America; Africa.
Notes: Autoecious.
Host: Leonotis
Puccinia leptochloae Arth. & Fromme
Distr.: Southern United States; West Indies; Central and South America.
Notes: Heteroecious; aecial host - Portulacaceae; telial host - *Leptochloa* spp.
Hosts: Leptochloa, Talinum
Puccinia levata J. Hennen & C.S. Hodges
Puccinia velata Arth. in Cummins & J.A. Stevenson, nom. nud.
Uredo velata Ellis & Everh. (anamorph)
Distr.: FL, HI; West Indies; Central and South America.
Notes: Autoecious.
Host: Euphorbia
Puccinia leveillei Mont. in Gay
Distr.: North and South America; Africa; Asia; Europe.
Notes: Autoecious. Sivanesan (C.M.I. Descr. 264. 1970).
Host: Geranium
Puccinia levis (Sacc. & Bizz.) Magnus
Puccinia paspali Tracy & Earle
Distr.: Southern United States to Argentina; possibly Australia.
Notes: Aecia unknown; telial host - Poaceae.
Hosts: Paspalum, Rhynchelytrum
Puccinia liberta F. Kern
Distr.: West Indies; Central America.
Notes: Aecia unknown; telial host - *Eleocharis* spp. Reports from California need confirmation (Savile. Canad. J. Bot. 50:2588. 1972).
Host: Eleocharis
Puccinia ligustici Ellis & Everh.
Distr.: Western North America.
Notes: Autoecious; on Apiaceae.
Hosts: Angelica, Conioselinum, Ligusticum, Lomatium, Oxypolis
Puccinia liliacearum Duby
Distr.: Northeastern North America(rare); South Africa; Eurasia.
Notes: Autoecious.
Host: Ornithogalum
Puccinia limnodeae Cummins
Distr.: TX.
Notes: Autoecious.
Host: Limnodea
Puccinia limosae Magnus
Puccinia caricina DC. var. *limosae* (Magnus) Jørst.
Distr.: Northcentral and northeastern United States.
Notes: Heteroecious; aecial host - *Lysimachia* spp.; telial host - *Carex* spp. Parmelee (Fungi Canadenses 207. 1981).
Hosts: Carex, Lysimachia
Puccinia linkii Klotzsch
Distr.: Northern North America.
Notes: Autoecious.
Host: Viburnum
Puccinia lithospermi Ellis & Kellerm.
Distr.: Southwestern United States; West Indies; Central and South America; southwestern Africa; Japan.
Notes: Autoecious.
Host: Evolvulus
Puccinia lobata Berk. & M.A. Curtis in Berk.
Distr.: CA, TX; Mexico.
Notes: Autoecious; on Malvaceae.
Hosts: Alcea, Malvella
Puccinia lobeliae W. Gerard ex Peck
Distr.: Central and eastern North America.
Notes: Autoecious. Parmelee & de Carteret (Fungi Canadenses 220. 1982).
Host: Lobelia
Puccinia longiana Syd. & P. Syd.
Distr.: TX; West Indies; Central America.
Notes: Autoecious.
Host: Ruellia
Puccinia longipes Lagerh.
Puccinia bullata Schwein.
Puccinia vernoniae Schwein.
Distr.: Central and eastern North America.
Notes: Autoecious. Parmelee (Fungi Canadenses 97. 1977).
Hosts: [Conioselinum], Vernonia
Puccinia longipes Lagerh. var. ***brevipes*** (Dietel) Z. Urban
Distr.: Central and southern North America; Mexico..
Notes: Autoecious. Urban (Acta. Univ. Carol. - Biol. 1971:70. 1973).
Host: Vernonia
Puccinia ludovicianae Fahrendorff
Distr.: Western North America.
Notes: Autoecious.
Host: Artemisia
Puccinia magnusiana Körn.
Puccinia simillima Arth.
Distr.: Cosmopolitan.
Notes: Heteroecious; aecial host - Ranunculaceae; telial host - Poaceae.
Hosts: Anemone, Phragmites
Puccinia malvacearum Bertero ex Mont. in Gay
Distr.: Cosmopolitan.
Notes: Autoecious; on Malvaceae. Sivanesan (C.M.I. Descr. 265. 1970); Parmelee & de Carteret (Fungi Canadenses 171. 1980).
Hosts: Abutilon, Alcea, Althaea, Hibiscus, Lavatera, Malva, Malvastrum, Malvella, Sida, Sphaeralcea
Puccinia mariae-wilsoniae G.W. Clinton in Peck
Puccinia claytoniatum Peck
Distr.: Central and eastern North America.
Notes: Autoecious.
Host: Claytonia
Puccinia mariae-wilsoniae G.W. Clinton in Peck var. ***montiae*** Savile in Savile & Parmelee
Distr.: Western North America.
Notes: Autoecious. Savile & Parmelee (Mycologia 48:573. 1956).
Host: Montia
Puccinia marianae P. Syd. & Syd.
Distr.: FL.
Notes: Autoecious.

Hosts: Chrysopsis, Haplopappus
Puccinia marylandica Lindr.
Distr.: Eastern North America; southwestern Canada(British Columbia).
Notes: Autoecious; on Apiaceae.
Host: Sanicula
Puccinia massalis Arth.
Distr.: Southwestern United States.
Notes: Autoecious.
Host: Helianthus
Puccinia mcclatchieana Dietel & Holw. in Dietel
Distr.: Northern North America.
Notes: Aecia unknown; telial host - *Scirpus* spp.
Host: Scirpus
Puccinia melampodii Dietel & Holw. in Holw.
Puccinia flaveriae H. Jacks.
Puccinia ripulae Mains
Distr.: Central United States to South America.
Notes: Autoecious; on Asteraceae.
Hosts: Baccharis, Calendula, Dimorphotheca, Emilia, Parthenium, Tridax
Puccinia melanconioides Ellis & Harkn.
Distr.: CA, OR.
Notes: Autoecious.
Host: Dodecatheon
Puccinia melanocephala Syd. & P. Syd.
Distr.: Southern United States; West Indies; South America; Asia.
Notes: Sugarcane rust (Purdy. Pl. Dis. 67:1292. 1983).
Hosts: [Bambusa], [Phyllostachys], [Poaceae], Saccharum
Puccinia mellifera Dietel & Holw. in Dietel
Distr.: CA, NV.
Notes: Autoecious.
Host: Salvia
Puccinia menthae Pers.:Pers.
Distr.: Cosmopolitan.
Notes: Autoecious; on Lamiaceae (Laundon & Waterston. C.M.I. Descr. 7. 1964).
Hosts: Blephilia, Clinopodium, Cunila, Hedeoma, Mentha, Monarda, Monardella, Pycnanthemum, Satureja, Teucrium
Puccinia mesnieriana Thuem.
Puccinia digitata Ellis & Harkn.
Distr.: CA; western Asia; southern Europe.
Notes: Autoecious.
Host: Rhamnus
Puccinia mesomajalis Berk. & M.A. Curtis in Peck
Micropuccinia mesomajalis (Berk. & M.A. Curtis) Arth. & H. Jacks.
Distr.: Northern and western North America.
Notes: Autoecious.
Host: Clintonia
Puccinia mexicensis H.C. Greene & Cummins
Distr.: Southwestern United States; Mexico.
Notes: Aecia unknown; telial host - *Stipa*.
Host: Stipa
Puccinia micrantha D. Griffiths
Distr.: Central and western United States.
Notes: Heteroecious; aecial host - *Ribes* spp.; telial host - *Oryzopsis micrantha*.
Hosts: Oryzopsis, Ribes
Puccinia microsora Körn.
Distr.: LA, OR, WI, eastern United States(mountainous regions); northern Europe.
Notes: Aecia unknown; telial host - *Carex* spp.
Host: Carex
Puccinia microspora Dietel
Distr.: Southwestern United States; eastern Asia; Borneo.
Notes: Aecia unknown; telial host - Poaceae.
Host: Imperata
Puccinia minussensis Thuem.
Puccinia hemisphaerica Ellis & Everh.
Distr.: Temperate northern hemisphere.
Notes: Autoecious.
Host: Lactuca
Puccinia minuta Dietel
Distr.: AL, FL.
Notes: Aecia unknown.
Host: Carex
Puccinia minutissima Arth.
Distr.: Central and eastern United States.
Notes: Heteroecious; aecial host - *Decodon verticillatus*; telial host - *Carex* spp.
Hosts: Carex, Decodon
Puccinia mirifica Dietel & Holw.
Puccinia triannulata (Berk. & M.A. Curtis) H. Jacks.
Distr.: Southern United States; Mexico.
Notes: Autoecious.
Host: Borrichia
Puccinia molokaiensis Cummins
Distr.: HI.
Notes: Aecia unknown; telial host - *Carex* spp.
Host: Carex
Puccinia monardae J.W. Baxter
Distr.: WV.
Notes: Autoecious.
Host: Monarda
Puccinia monoica Arth.
Distr.: Central and western North America.
Notes: Heteroecious; aecial host - Brassicaceae; telial host - Poaceae.
Hosts: Arabis, Descurainia, Draba, Koeleria, Lepidium, Oryzopsis, Phoenicaulis, Poa, Sisymbrium, Smelowskia, Stipa, Thlaspi, Trisetum
Puccinia montanensis Ellis
Aecidium fendleri Tracy & Earle (anamorph)
Distr.: Central and western North America.
Notes: Heteroecious; aecial host - *Berberis* spp.; telial host - Poaceae.
Hosts: Agrohordeum, Agropyron, Arrhenatherum, Berberis, Elymus, Eremopyrum, Hordeum, Hystrix, [Lithospermum], Lolium, Melica, [Onosmodium], Sitanion, [Thalictrum]
Puccinia moreniana W.R. Dudley & C.H. Thomps.
Distr.: CA.
Notes: Autoecious.
Host: Dichelostemma
Puccinia musenii Ellis & Everh.
Distr.: Western United States.
Notes: Autoecious; on Apiaceae.
Hosts: Cymopterus, Lomatium
Puccinia mutabilis Ellis & B.T. Galloway
Distr.: Western North America.
Notes: Autoecious.
Host: Allium
Puccinia nakanishikii Dietel
Distr.: HI; Asia.
Notes: Aecia unknown.
Host: Cymbopogon
Puccinia neocoronata H.C. Greene & Cummins
Distr.: AZ; Mexico.
Notes: Aecia unknown.
Hosts: Piptochaetium, Stipa
Puccinia nodosa Ellis & Harkn.
Distr.: Central to southern CA.
Notes: Autoecious.
Hosts: Brodiaea, Dichelostemma, [Dodecatheon]
Puccinia nuda Ellis & Everh.
Puccinia hemizoniae Ellis & Tracy
Distr.: Pacific Coast of United States; South America.
Notes: Autoecious; on Asteraceae.
Hosts: Calycadenia, Hemizonia, Lagophylla, Madia, Malacothrix
Puccinia oahuensis Ellis & Everh.
Distr.: Warm regions.
Notes: Aecia unknown; telial host - *Digitaria* spp. Mordue & Critchett (C.M.I. Descr. 516. 1976).
Host: Digitaria
Puccinia obliqua Berk. & M.A. Curtis in Berk.
Distr.: Southern United States; West Indies; Central and South America.
Notes: Autoecious; on Asclepiadaceae.
Hosts: Asclepias, Calotropis, Cynanchum, Matelea, Morrenia, Philibertia, Sarcostemma, Vincetoxicum
Puccinia obscura J. Schröt. in Pass.
Distr.: Temperate northern hemisphere.
Notes: Heteroecious; aecial host - *Bellis perennis* in Europe; telial host - *Luzula* spp.

Host: Luzula
Puccinia obtecta Peck
Distr.: Temperate North America.
Notes: Heteroecious; aecial host - *Bidens* spp.; telial host - *Scirpus* spp.
Hosts: Bidens, Scirpus
Puccinia oenotherae Vize
Puccinia oenotherae Vize var. *heteranthae* (Ellis & Everh.) Arth.
Puccinia zauschneriae Syd.
Distr.: Western North America.
Notes: Autoecious; on Onagraceae.
Hosts: Boisduvalia, Camissonia, Clarkia, Epilobium, Oenothera, Zauschneria
Puccinia orbicula Peck & G.W. Clinton
Distr.: Temperate North America.
Notes: Autoecious.
Host: Prenanthes
Puccinia ornata Arth. & Holw. in Arth.
Distr.: Northern North America; Europe; Japan.
Notes: Autoecious.
Host: Rumex
Puccinia ornatula Holw.
Distr.: AK; Canada(British Columbia).
Notes: Autoecious. Savile (Fungi Canadenses 77. 1975).
Host: Viola
Puccinia ortonii H. Jacks.
Distr.: Western North America.
Notes: Autoecious. Parmelee (Fungi Canadenses 208. 1981).
Host: Dodecatheon
Puccinia osoyoosensis Savile
Distr.: Western North America to Mexico.
Notes: Heteroecious; aecial host - *Xanthium strumarium* var. *canadense*; telial host - *Scirpus americanus*. Savile (Canad. J. Bot. 50:2587. 1972).
Host: Scirpus
Puccinia oudemansii Tranzschel
Distr.: UT; Norway; Turkey.
Notes: Autoecious.
Hosts: Parrya, Thlaspi
Puccinia oxalidis Dietel & Ellis in Dietel
Distr.: Southern United States; West Indies; Central and South America.
Notes: Heteroecious; aecial host - *Berberis* spp.; telial host - *Oxalis* spp.
Hosts: Berberis, Oxalis
Puccinia oxyriae Fuckel
Distr.: Alpine and arctic regions.
Notes: Autoecious.
Host: Oxyria
Puccinia pacifica (Blasdale) Arth.
Distr.: CA.
Notes: Autoecious.
Host: Plantago
Puccinia pagana Arth.
Distr.: CO.
Notes: Autoecious.
Host: Lloydia
Puccinia pallidissima Speg.
Distr.: Southeastern United States; South America.
Notes: Autoecious.
Host: Stachys
Puccinia palmeri Dietel & Holw. in Dietel
Distr.: Western North America.
Notes: Autoecious.
Host: Penstemon
Puccinia paradoxica Ricker
Distr.: MI.
Notes: Aecia unknown.
Host: Melica
Puccinia parca Arth.
Distr.: AK, eastern United States.
Notes: Autoecious.
Host: Polygonum
Puccinia parkerae Dietel & Holw. in Dietel
Distr.: Western North America; eastern Asia.
Notes: Autoecious.
Host: Ribes
Puccinia parnassiae Arth.
Distr.: Western North America.
Notes: Autoecious. Savile (Fungi Canadenses 19. 1974).
Host: Parnassia
Puccinia paroselae Cummins
Distr.: CA.
Notes: Autoecious.
Host: Dalea
Puccinia parthenii Arth.
Distr.: TX; Mexico; South America.
Notes: Autoecious.
Host: Parthenium
Puccinia pattersoniana Arth.
Distr.: Western North America.
Notes: Heteroecious; uredinioid aecia on *Brodiaea*; telial host - Poaceae.
Hosts: Agropyron, Brodiaea, Elymus, Elysitanion, Sitanion, Triteleia
Puccinia pazschkei Dietel
Distr.: Northern North America.
Notes: Autoecious; on Saxifragaceae.
Host: Saxifraga
Puccinia pazschkei Dietel var. ***heterisiae*** (H. Jacks.) Savile
Puccinia heterisiae H. Jacks. in Arth.
Puccinia aspera Dietel & Holw. in Arth.
Distr.: Western North America.
Host: Saxifraga
Puccinia pazschkei Dietel var. ***tricuspidatae*** Savile
Puccinia turrita Arth.
Distr.: CO, UT.
Notes: Autoecious.
Host: Saxifraga
Puccinia pelargonii-zonalis Doidge
Distr.: Cosmopolitan.
Notes: Autoecious. Sivanesan (C.M.I. Descr. 266. 1970).
Host: Pelargonium
Puccinia pentstemonis Peck
Distr.: Western North America.
Notes: Autoecious.
Host: Penstemon
Puccinia phaceliae Syd. & Holw.
Distr.: WA.
Notes: Autoecious.
Host: Phacelia
Puccinia phragmitis (Schumach.) Körn.
Puccinia rubella Arth.
Distr.: Cosmopolitan.
Notes: Heteroecious; aecial host - Polygonaceae; telial host - *Phragmites* spp.
Hosts: Phragmites, Rheum, Rumex
Puccinia phyllostachydis Kusano
Distr.: HI, southeastern United States; China; Japan.
Notes: Aecia unknown; telial host - *Bambusa* and *Phyllostachys* spp.
Host: Phyllostachys
Puccinia physalidis Peck
Distr.: Temperate North America.
Notes: Autoecious. Parmelee & de Carteret (Fungi Canadenses 187. 1980).
Host: Physalis
Puccinia physostegiae Peck & G.W. Clinton in Peck
Distr.: IN, MT, NY.
Notes: Autoecious.
Hosts: Dracocephalum, Physostegia
Puccinia pimpinellae (F. Strauss) Mart.
Puccinia osmorrhizae G.W. Clinton & Peck in Peck
Distr.: Temperate northern hemisphere.
Notes: Autoecious; on Apiaceae.
Hosts: Osmorhiza, Pimpinella
Puccinia pinaropappi P. Syd. & Syd.
Distr.: TX.
Notes: Autoecious.
Host: Pinaropappus
Puccinia piperi Ricker
Distr.: CA, OR.
Notes: Aecia unknown.
Hosts: Festuca, Vulpia

Puccinia pistorica Arth.
Distr.: FL.
Notes: Autoecious.
Host: Baccharis
Puccinia plumasensis J. Hennen
Distr.: CA, CO.
Notes: Autoecious.
Host: Hydrophyllum
Puccinia plumbaria Peck
Distr.: Western North America.
Notes: Autoecious; on Polemoniaceae. Parmelee (Mycologia 78:456. 1986).
Hosts: Collomia, [Gaura], Gilia, Microsteris, Phlox
Puccinia poarum P. Nielsen
Puccinia conspicua Mains
Puccinia liatridis Bethel in Arth.
Aecidium tussilaginis Pers. (anamorph)
Distr.: Northern hemisphere to South America.
Notes: Heteroecious; aecial host - Asteraceae; telial host - Poaceae.
Hosts: Agrostis, Brickellia, Festuca, Helenium, Koeleria, Liatris, Petasites, Poa
Puccinia podophylli Schwein.
Distr.: Central and eastern North America.
Notes: Autoecious. Parmelee & de Carteret (Fungi Canadenses 289. 1984).
Host: Podophyllum
Puccinia polemonii Dietel & Holw.
Micropuccinia polemonii Arth. & H. Jacks.
Distr.: Western United States; Asia; northern Europe.
Notes: Autoecious. Parmelee (Mycologia 78:455. 1986).
Host: Polemonium
Puccinia polygoni-alpini Cruchet & Mayor in Cruchet
Distr.: Western North America.
Host: Polygonum
Puccinia polygoni-amphibii Pers.:Pers.
Puccinia polygoni-amphibii Pers.:Pers. var. *convolvuli* Arth.
Puccinia polygoni-amphibii Pers.:Pers. var. *persicariae* Arth.
Puccinia polygoni-amphibii Pers.:Pers. var. *tovariae* Arth.
Distr.: Cosmopolitan.
Notes: Heteroecious; aecial host - *Geranium* spp.; telial host - *Polygonum* spp.
Hosts: Geranium, Polygonum
Puccinia polysora Underw.
Distr.: United States; Central and South America; Central Africa; southeastern Asia.
Notes: Aecia unknown; telial host - Poaceae. Laundon & Waterston (C.M.I. Descr. 4. 1964).
Hosts: Saccharum, Tripsacum, Zea
Puccinia poromera Holw.
Distr.: UT.
Notes: Autoecious.
Host: Angelica
Puccinia porphyrogenita M.A. Curtis in Thuem.
Distr.: Northern North America; northeastern Asia.
Notes: Autoecious.
Host: Cornus
Puccinia posadensis Sacc. & Trott.
Puccinia kaernbachii Arth.
Puccinia venustula Arth.
Distr.: Southern United States; Central and South America.
Notes: Aecia unknown; telial host - Poaceae.
Hosts: Andropogon, Imperata, Schizachyrium
Puccinia praegracilis Arth.
Aecidium graebnerianum Henn. (anamorph)
Distr.: Northern North America.
Notes: Aecial host - Orchidaceae; telial host - Poaceae. Several varieties are recognized by Savile (Canad. J. Bot. 35:199. 1957).
Hosts: Habenaria, Orchis
Puccinia procera Dietel & Holw.
Distr.: CA(coastal areas); Black Sea region.
Notes: Heteroecious; aecial host - *Phacelia* spp.; telial host - *Elymus* spp.
Host: Elymus
Puccinia proserpinacae Farl.
Distr.: Central and eastern United States.
Notes: Autoecious.
Host: Proserpinaca
Puccinia pseudocymopteri Holw.
Distr.: Western United States.
Notes: Autoecious.
Hosts: Cymopterus, Pseudocymopterus
Puccinia psidii G. Wint.
Distr.: Subtropical and tropical western hemisphere.
Notes: Autoecious; on Myrtaceae (Laundon & Waterston. C.M.I. Descr. 56. 1965).
Hosts: Melaleuca, Pimenta
Puccinia pulsatillae Kalchbr.
Puccinia debaryana Thuem.
Distr.: Western United States; Canada; eastern Asia; Europe.
Notes: Autoecious; on Ranunculaceae. Parmelee & Savile (Fungi Canadenses 265. 1984).
Hosts: Anemone, Clematis, Trautvetteria
Puccinia pulverulenta Grev.
Puccinia epilobii-tetragoni G. Wint. in Rabenh.
Puccinia gayophyti J.S. Billings
Puccinia vagans Arth.
Puccinia vagans Arth. var. *epilobii-tetragoni* Arth. non DC., hom. illeg.
Puccinia vagans Arth. var. *gayophyti* (J.S. Billings) Arth.
Distr.: Temperate northern hemisphere.
Notes: Autoecious; on Onagraceae.
Hosts: Boisduvalia, Camissonia, Clarkia, Epilobium, Gayophytum, Oenothera
Puccinia punctata Link
Distr.: North America; Central and South America; Africa; Asia; Australia; Europe; New Zealand.
Notes: Autoecious; on Rubiaceae.
Host: Galium
Puccinia punctata Link var. ***troglodytes*** (Lindr.) Arth.
Distr.: United States.
Notes: Autoecious; on Rubiaceae.
Host: Galium
Puccinia punctiformis (F. Strauss) Röhl.
Puccinia obtegens Tul. ex Fuckel
Distr.: Widespread.
Notes: Autoecious.
Host: Cirsium
Puccinia puritanica Cummins
Distr.: MS, WI.
Host: Carex
Puccinia purpurea Cooke
Distr.: Warm regions.
Notes: Uredinia and telia on species of *Sorghum*. *Oxalis corniculata* was reported as an aecial host (LeRoux & Dickson. Phytopathology 47:101. 1957). Laundon & Waterston (C.M.I. Descr. 8. 1964).
Host: Sorghum
Puccinia pygmaea Eriks.
Distr.: Temperate regions.
Notes: Heteroecious; aecial host - *Berberis* spp.; telial host - Poaceae.
Hosts: Agrostis, Ammophila, Arctagrostis, Berberis, Calamagrostis, Festuca, Hordeum, Koeleria, Melica, Milium, Oryzopsis
Puccinia pygmaea Eriks. var. ***ammophilina*** (Mains ex Cummins) Cummins & H.C. Greene
Puccinia ammophilina Mains ex Cummins
Distr.: Pacific coast of United States; Europe.
Notes: Aecial host unknown.
Host: Ammophila
Puccinia pyrolae Cooke
Distr.: Northeastern North America.
Notes: Autoecious.
Host: Polygala
Puccinia raunkaerii Ferd. & Winge
Distr.: FL, TX; West Indies; South America.
Notes: Autoecious.
Host: Rivina
Puccinia recedens P. Syd. & Syd.
Distr.: Northern United States; Canada.
Notes: Autoecious.
Host: Senecio
Puccinia recondita Roberge ex Desmaz.

Puccinia aconiti-elymi Mayor
Puccinia aconiti-rubrae Lüdi
Puccinia agropyri Ellis & Everh.
Puccinia agropyrina Eriks.
Puccinia alternans Arth.
Puccinia apocrypta Ellis & Tracy
Puccinia bromi-maximi Guyot
Puccinia bromicola Guyot
Puccinia cinerea Arth.
Dicaeoma clematidis Arth.
Puccinia clematidis Lagerh.
Puccinia hordei Fuckel
Puccinia impatientis Arth.
Puccinia rubigo-vera G. Wint. in Rabenh.
Puccinia rubigo-vera G. Wint. f. *aconiti-elymi* (Mayor) Mains
Puccinia rubigo-vera G. Wint. f. *aconiti-rubrae* (Lüdi) Mains
Puccinia rubigo-vera G. Wint. f. *alternans* (Arth.) Mains
Puccinia rubigo-vera G. Wint. f. *arthuri* Mains
Puccinia rubigo-vera G. Wint. f. *bromicola* Mains
Puccinia rubigo-vera G. Wint. f. *phaceliae* Mains
Puccinia rubigo-vera G. Wint. f. *triseti* (Eriks.) Mains
Puccinia rubigo-vera G. Wint. var. *agropyri* (Ellis & Everh.) Arth.
Puccinia rubigo-vera G. Wint. var. *agropyrina* (Eriks.) Arth.
Puccinia rubigo-vera G. Wint. var. *apocrypta* (Ellis & Tracy) Arth.
Puccinia rubigo-vera G. Wint. var. *impatientis* (Arth.) Mains
Puccinia rubigo-vera G. Wint. var. *secalis* (Eriks.) Carleton
Puccinia rubigo-vera G. Wint. var. *tritici* (Eriks.) Carleton
Puccinia triseti Eriks.
Aecidium clematidis DC. (anamorph)
Distr.: Temperate regions.
Notes: Heteroecious; aecial host - Ranunculaceae, Boraginaceae, Hydrophyllaceae, and Balsaminaceae; telial host - Poaceae. Cummins (Rust fungi of cereals, grasses and bamboos. p. 320. 1971) recognized this as a variable species. Recent studies have supported this concept (Mercé. Crypt. Mycol. 7:319. 1986). Savile (Fungi Canadenses 310. 1986).
Hosts: Aconitum, Actaea, Aegilops, Agropyron, Agrostis, Alopecurus, Anchusa, Anemone, Anemonella, Aquilegia, Avena, Bromus, Celtis, Cimicifuga, Cinna, Clematis, Dactylis, Delphinium, Ellisia, Elymus, Festuca, Glyceria, Holcus, Hordeum, Hydrophyllum, Hystrix, Impatiens, Isopyrum, Lithospermum, Lolium, Melica, Mertensia, Nemophila, Onosmodium, Phacelia, Phalaris, Poa, Puccinellia, Ranunculus, Scolochloa, Secale, Sitanion, Thalictrum, Trisetum, Triticum

Puccinia redfieldiae Tracy
Aecidium anograe Arth. (anamorph)
Distr.: Central and western United States.
Notes: Heteroecious; aecial host - *Oenothera nuttallii*; telial host - *Redfieldia*.
Hosts: Oenothera, Redfieldia

Puccinia retecta Syd. & P. Syd.
Distr.: CO, WY.
Notes: Autoecious. Reports from Alaska are erroneous (Hennen. Mycologia 59:270. 1967).
Host: Anemone

Puccinia rhaetica E. Fisch. subsp. ***americana*** Savile
Distr.: Western North America.
Notes: Autoecious. See notes under *P. cascadensis*.
Host: Veronica

Puccinia rhytismoides Johans.
Distr.: Widespread in arctic regions.
Host: Thalictrum

Puccinia ribis DC.
Distr.: Temperate North America; Europe.
Host: Ribes

Puccinia romanzoffiae H. Jacks.
Distr.: OR.
Notes: Autoecious.
Host: Romanzoffia

Puccinia rubefaciens Johans.
Distr.: Northern North America; Asia; Europe.
Notes: Autoecious.
Host: Galium

Puccinia ruelliae-parviflorae G. Laundon
Distr.: TN.
Notes: Autoecious.
Host: Ruellia

Puccinia rufescens Dietel & Holw.
Distr.: Southwestern United States.
Notes: Autoecious.
Host: Pedicularis

Puccinia rydbergii Garrett
Distr.: UT; Canada(British Columbia).
Notes: Autoecious.
Host: Sedum

Puccinia salviicola Dietel & Holw.
Distr.: FL, TX; Mexico; South America.
Notes: Autoecious.
Host: Salvia

Puccinia scandica Johans.
Distr.: UT, northwestern North America; Sweden.
Notes: Autoecious.
Host: Epilobium

Puccinia schedonnardi Kellerm. & Swingle
Puccinia epicampis Arth.
Puccinia muhlenbergiae Arth. & Holw.
Puccinia tosta Arth.
Aecidium hibisciatum Schwein. (anamorph)
Distr.: Western hemisphere; Japan; The Philippines.
Notes: Heteroecious; aecial host - Malvaceae; telial host - Poaceae. Mulder & Holliday (C.M.I. Descr. 293. 1971); Parmelee (Fungi Canadenses 172. 1980).
Hosts: Alcea, Callirhoe, Gossypium, Hibiscus, Lycurus, Malacothamnus, Malvella, Melica, Monanthochloe, Muhlenbergia, Napaea, Schedonnardus, Sidalcea, Sphaeralcea, Sporobolus, Triplasis

Puccinia scirpi DC.
Distr.: FL; West Indies; Europe.
Notes: Heteroecious; aecial host - *Nymphoides* spp.; telial host - *Scripus* spp.
Host: Nymphoides

Puccinia scleriicola Arth.
Distr.: GA; West Indies.
Notes: Aecia unknown; telial host - *Scleria* spp.
Host: Scleria

Puccinia scleropogonis Cummins
Distr.: Southwestern United States; Mexico.
Notes: Heteroecious; aecial host - *Chaemaesaracha*; telial host - *Scleropogon*.
Host: Scleropogon

Puccinia senecionis Lib.
Puccinia subcircinata Ellis & Everh.
Distr.: Central and western North America; Europe.
Notes: Autoecious.
Host: Senecio

Puccinia septentrionalis Juel
Aecidium sommerfelti Johans. (anamorph)
Distr.: Circumboreal.
Notes: Heteroecious; aecial host - *Thalictrum* spp.; telial host - *Polygonum* spp.
Hosts: Polygonum, Thalictrum

Puccinia sessilis W.G. Schneid. in J. Schröt.
Puccinia majanthemi Dietel
Aecidium majanthae Schumach. (anamorph)
Distr.: Temperate northern hemisphere.
Notes: Heteroecious; aecial host - Araceae, Iridaceae, Liliaceae and Orchidaceae; telial host - Poaceae.
Hosts: Iris, Maianthemum, Phalaris, Polygonatum, Smilacina, Streptopus, Uvularia

Puccinia sessilis W.G. Schneid. in J. Schröt. var. ***minor*** Cummins
Distr.: Southcentral United States.
Notes: Aecial host unknown.
Host: Phalaris

Puccinia setariae Dietel & Holw.
Distr.: Southern United States; Central and South America.
Notes: Aecia unknown; telial host - *Setaria geniculata*.
Host: Setaria

Puccinia seymouriana Arth.
Puccinia cephalanthi Barth.
Distr.: Central and eastern North America.
Notes: Heteroecious; aecial host - Apocynaceae, Asclepiadaceae, and

Caprifoliaceae; telial host - *Spartina* spp.
Hosts: Amsonia, Apocynum, Asclepias, Cephalanthus, Spartina
Puccinia sherardiana Körn.
Puccinia malvastri Peck
Distr.: Central and western North America; South America; eastern Asia.
Notes: Autoecious; on Malvaceae. Parmelee (Fungi Canadenses 173. 1980).
Hosts: Alcea, Iliamna, Malacothamnus, Sidalcea, Sphaeralcea
Puccinia sieversiae Arth.
Puccinia tatrensis Z. Urban
Distr.: Rocky Mountains; Czechoslovakia(Carpathian Mountains).
Notes: Autoecious.
Host: Geum
Puccinia silphii Schwein.
Distr.: Central and eastern North America.
Notes: Autoecious. Parmelee (Fungi Canadenses 98. 1977).
Host: Silphium
Puccinia similis Ellis & Everh.
Distr.: Central and western United States.
Notes: Autoecious.
Host: Artemisia
Puccinia smilacis Arth.
Distr.: Central and eastern United States; West Indies; Central America.
Notes: Heteroecious; aecial host - *Apocynum* spp.; telial host - *Smilax* spp.
Hosts: Apocynum, Smilax
Puccinia solheimii Cummins
Distr.: WY.
Host: Dodecatheon
Puccinia sonorica Cummins & S.M. Husain
Distr.: Southwestern United States; Mexico.
Notes: Aecia unknown; telial host - *Aristida* spp.
Host: Aristida
Puccinia sorghi Schwein.
Aecidium oxalidis Thuem. (anamorph)
Distr.: World-wide in maize-growing regions.
Notes: Heteroecious; aecial host - *Oxalis* spp.; telial host - *Euchlaena* and *Zea* (Parmelee & Savile. Fungi Canadenses 302. 1986; Laundon & Waterston. C.M.I. Descr. 3. 1964).
Hosts: Oxalis, Zea
Puccinia sparganioides Ellis & Barth.
Puccinia fraxinata Arth.
Puccinia peridermiospora Arth.
Distr.: Central and eastern North America; Brazil.
Notes: Heteroecious; aecial host - Oleaceae; telial host - *Spartina* spp. McCain & Hennen (Proc. Indiana Acad. Sci. 91:504).
Hosts: Forestiera, Fraxinus, Spartina
Puccinia spegazzinii De Toni in Sacc.
Distr.: Southern United States; Central and South America.
Notes: Autoecious.
Host: Mikania
Puccinia sphaeralceoides Cummins
Puccinia sphaeralceae Gäum. non Ellis & Everh., hom. illeg.
Distr.: CA, ID.
Notes: Autoecious. Cummins (Bull. Torrey Bot. Club 79:224. 1952).
Host: Sphaeralcea
Puccinia splendens Vize
Distr.: Southwestern United States; Mexico.
Notes: Autoecious.
Hosts: Ambrosia, Hymenoclea
Puccinia sporoboli Arth.
Aecidium lilii G.W. Clinton (anamorph)
Distr.: Northcentral North America.
Notes: Heteroecious; aecial host - Liliaceae; telial host - *Sporobolus* spp.
Hosts: Lilium, Sporobolus
Puccinia sporoboli Arth. var. **robusta** Cummins & H.C. Greene
Aecidium yuccae Arth. (anamorph)
Distr.: Central and western North America.
Notes: Heteroecious; aecial host - Liliaceae; telial host - Poaceae.
Hosts: Calamovilfa, Leucocrinum, Smilacina, Sporobolus, Yucca
Puccinia stenotaphricola J. Walker
Puccinia stenotaphri Cummins non (P. Syd. & Syd.) Cummins, hom. illeg.
Distr.: FL, HI, PR; Old World tropical regions.
Notes: Aecia unknown. Walker (Mycotaxon 31:118. 1988).
Hosts: Pennisetum, Stenotaphrum
Puccinia stipae Arth.
Aecidium grindeliae D. Griffiths non Syd., hom. illeg. (anamorph)
Distr.: Central North America to Mexico; Bolivia.
Notes: Heteroecious; aecial host - Asteraceae; telial host - *Stipa* spp.
Hosts: Agoseris, Aster, Chrysopsis, Chrysothamnus, Crepis, Erigeron, Grindelia, Gutierrezia, Haplopappus, [Koeleria], Lygodesmia, [Oryzopsis], Senecio, Solidago, Stipa
Puccinia striiformis Westend.
Puccinia glumarum Eriks. & E. Henn.
Uredo glumarum J.C. Schmidt (anamorph)
Distr.: Western North America; South America; North Africa; Europe; Japan.
Notes: Probably heteroecious; telial host - Poaceae. (Savile. Fungi Canadenses 250. 1983; Mulder & Booth. C.M.I. Descr. 291. 1971).
Hosts: Aegilops, Agropyron, Agrostis, Bromus, Chloris, Cinna, Elymus, Elysitanion, Hierochloe, Hordeum, Hystrix, Koeleria, Leymus, Lolium, Phalaris, Poa, Puccinellia, Secale, Sitanion, Trisetum, Triticum, Vulpia
Puccinia stromatica Berk. & M.A. Curtis
Distr.: AL.
Notes: Autoecious.
Host: Clematis
Puccinia subangulata Holw.
Distr.: WA.
Notes: Autoecious.
Hosts: Dichelostemma, Triteleia
Puccinia subdecora Syd., P. Syd., & Holw.
Distr.: Western United States.
Notes: Autoecious.
Host: Brickellia
Puccinia subnitens Dietel
Peridermium gracile Harkn.
Puccinia sarcobati Arth.
Distr.: Western hemisphere.
Notes: Heteroecious; aecial host - several angiosperm families; telial host - Poaceae. Parmelee & de Carteret (Fungi Canadenses 209. 1981).
Hosts: Atriplex, Camissonia, Chenopodium, Cleome, Cryptantha, Datura, Distichlis, Draba, Eriogonum, Gilia, Lappula, Lepidium, Lesquerella, Linanthus, Mirabilis, Nasturtium, Nicotiana, Pectocarya, Petunia, Phacelia, Plantago, Raphanus, Rumex, Salsola, Sarcobatus, Silene, Spinacia, Verbena
Puccinia substerilis Ellis & Everh.
Puccinia scaber (Ellis & Everh.) Barth.
Distr.: Central and western North America.
Notes: Autoecious.
Hosts: Oryzopsis, Stipa
Puccinia substerilis Ellis & Everh. var. **oryzopsidis** H.C. Greene & Cummins
Distr.: Western United States.
Notes: Autoecious; on Poaceae
Hosts: Oryzopsis, Stipa
Puccinia substerilis Ellis & Everh. var. **scribneri** H.C. Greene & Cummins
Distr.: CO.
Notes: Autoecious.
Host: Stipa
Puccinia substriata Ellis & Barth.
Puccinia penniseti A. Zimmerm.
Distr.: HI, southern United States to South America; possibly Uganda.
Notes: Heteroecious; aecial host - *Solanum* spp.; telial host - Poaceae.
Hosts: Digitaria, Panicum, Paspalum, Pennisetum, Physalis, Setaria, Solanum
Puccinia substriata Ellis & Barth. var. **imposita** (Arth.) Ramachar & Cummins
Puccinia imposita Arth.
Distr.: Southern United States; West Indies; Central and South America.
Notes: Heteroecious; aecial host - *Solanum* spp.; telial host - *Digitaria* spp.
Host: Digitaria

Puccinia substriata Ellis & Barth. var. **indica** Ramachar & Cummins
Distr.: Southeastern United States; India.
Notes: Heteroecious; aecial host - *Solanum* spp.; telial host - Poaceae.
Host: Pennisetum
Puccinia suksdorfii Ellis & Everh.
Distr.: Central and western United States.
Notes: Autoecious.
Hosts: Agoseris, Microseris
Puccinia swertiae G. Wint. in Rabenh.
Distr.: Western North America; Asia; Europe.
Notes: Autoecious.
Host: Swertia
Puccinia symphoricarpi Harkn.
Distr.: Western North America.
Notes: Autoecious.
Host: Symphoricarpos
Puccinia tageticola Dietel & Holw. in Holw.
Distr.: TX, PR; Central and South America.
Notes: Aecial host - unknown; telial host - *Tagetes* spp.
Host: Tagetes
Puccinia tanaceti DC.
Puccinia absinthii DC.
Puccinia chrysanthemi Roze
Distr.: Cosmopolitan.
Notes: Autoecious. Punithalingam (C.M.I. Descr. 175. 1968 as *Puccinia chrysanthemi*).
Hosts: [Artemisia], Chrysanthemum
Puccinia tanaceti DC. var. **dracunculina** (Fahrendorff) Cummins
Puccinia dracunculina Fahrendorff
Distr.: Western North America; Europe.
Notes: Autoecious.
Host: Artemisia
Puccinia tardissima Garrett
Distr.: Western United States.
Notes: Autoecious.
Host: Arenaria
Puccinia tenuis Burrill
Distr.: Central and eastern United States.
Notes: Autoecious.
Host: Eupatorium
Puccinia tetramerii Seym.
Distr.: AZ; Central America.
Notes: Autoecious; on Acanthaceae.
Hosts: Dicliptera, Justicia, Tetramerium
Puccinia thaliae Dietel
Puccinia cannae (G. Wint.) Henn.
Distr.: Warm-temperate to tropical regions.
Notes: Autoecious; on Cannaceae and Marantaceae. Sivanesan (C.M.I. Descr. 267. 1970).
Hosts: Canna, Maranta, Thalia
Puccinia thelypodii Cummins
Puccinia lupinicola Gäum.
Distr.: CA.
Notes: Probably autoecious. The host of *P. lupinicola* is probably a *Thelypodium*, not *Lupinus benthami* as previously reported. (Cummins & Stevenson. Pl. Dis. Reporter Suppl. 240:167. 1956).
Hosts: [Lupinus], Thelypodium
Puccinia thlaspeos C. Schub.
Puccinia holboellii (Hornem.) Rostr.
Distr.: Western North America.
Notes: Autoecious; on Brassicaceae. Savile (Fungi Canadenses 47. 1974 as *P. holboellii*).
Hosts: Arabis, Descurainia, Draba, Erysimum, Lepidium, Sisymbrium, Smelowskia, Thlaspi
Puccinia tillandsiae Cummins & Pollack
Distr.: FL; West Indies; Central America.
Notes: Autoecious.
Host: Tillandsia
Puccinia tolimensis Mayor
Distr.: NY; Central and South America.
Notes: Autoecious.
Host: Eupatorium
Puccinia tomipara Trel.
Distr.: Western Great Lakes region to Saskatchewan.
Notes: Heteroecious; aecial host - *Thalictrum* spp.; telial host - *Bromus* spp.
Host: Bromus
Puccinia trautvetteriae P. Syd., Syd., & Holw.
Distr.: Pacific Northwest.
Notes: Autoecious. Parmelee & Savile (Fungi Canadenses 266. 1984).
Host: Trautvetteria
Puccinia treleasiana Pazschke in Rabenh.
Distr.: Western North America(montane regions).
Notes: Autoecious. Parmelee & de Carteret (Fungi Canadenses 267. 1984).
Host: Caltha
Puccinia tripsaci Dietel & Holw.
Puccinia ceanothi Arth.
Distr.: Central North America; Mexico.
Notes: Heteroecious; aecial host - *Ceanothus* spp.; telial host - Poaceae.
Hosts: Andropogon, Ceanothus, Tripsacum
Puccinia triticina Eriks.
Puccinia dispersa Eriks. & E. Henn. f. sp. *tritici* Eriks. & E. Henn.
Puccinia recondita Roberge ex Desmaz. f. sp. *tritici* (Eriks. & E. Henn.) D.M. Henderson
Distr.: Temperate North America; Europe.
Notes: Heteroecious; aecial host - *Thalictrum*; telial host - *Aegilops*, *Secale*, and *Triticum* (Savile. Fungi Canadenses 309. 1986).
Host: Triticum
Puccinia troximontis Peck
Distr.: Western North America.
Notes: Autoecious.
Hosts: Agoseris, Microseris
Puccinia tumamocensis Arth.
Distr.: AZ.
Notes: Autoecious.
Host: Dichelostemma
Puccinia tumidipes Peck
Distr.: Central and eastern North America.
Notes: Autoecious. Parmelee (Fungi Canadenses 188. 1980).
Host: Lycium
Puccinia turgidipes H. Jacks.
Distr.: Southwestern United States; Mexico.
Notes: Autoecious.
Host: Viguiera
Puccinia tuyutensis Speg.
Puccinia cressae (DC.) Lagerh.
Distr.: Southwestern United States; Mexico; Central and South America.
Host: Cressa
Puccinia umbilici Guépin in Duby
Distr.: Rocky Mountains; Europe.
Notes: Autoecious; on Crassulaceae.
Host: Sedum
Puccinia unica Holw. in Arth. & Fromme
Distr.: Southwestern United States; Mexico.
Notes: Aecia unknown; telial host - *Aristida* spp.
Host: Aristida
Puccinia urbaniana Henn.
Distr.: FL; West Indies; Central and South America.
Notes: Autoecious.
Host: Stachytarpheta
Puccinia ustalis Berk.
Distr.: Alpine and subalpine Northern Hemisphere.
Notes: Autoecious. Parmelee & Savile (Fungi Canadenses 268. 1984).
Host: Ranunculus
Puccinia variabilis Grev.
Distr.: FL, SD(questionable); Canada(Nova Scotia, Quebec); Europe.
Notes: Autoecious.
Host: Taraxacum
Puccinia variabilis Grev. var. **insperata** (H. Jacks.) Cummins
Puccinia insperata H. Jacks.
Distr.: AK, northwestern North America.
Notes: Autoecious. This species was reported from Alaska as *P. prenanthis*.
Host: Prenanthes
Puccinia veratri Duby
Distr.: Temperate northern hemisphere.

Notes: Heteroecious; aecial host - *Epilobium* spp.; telial host - *Veratrum* spp.
Hosts: Epilobium, Veratrum
Puccinia verbesinae Schwein.
Distr.: Southeastern United States.
Notes: Autoecious.
Host: Verbesina
Puccinia veronicarum DC.
Distr.: Temperate northern hemisphere.
Notes: Autoecious.
Hosts: Veronica, Veronicastrum
Puccinia versicolor Dietel & Holw.
Aecidium plectroniae Cooke (anamorph)
Distr.: Subtropical and tropical regions.
Notes: Heteroecious; aecial host - Verbenaceae; telial host - Poaceae.
Host: Heteropogon
Puccinia vertisepta Tracy & B.T. Galloway
Distr.: AZ, NM; Mexico.
Notes: Autoecious.
Host: Salvia
Puccinia vesiculosa Schlechtend.
Distr.: AK; Europe.
Host: Anemone
Puccinia vexans Farl.
Aecidium cannonii D. Griffiths (anamorph)
Distr.: Central and western North America.
Notes: Heteroecious; aecial host - Asclepiadaceae and Fouquieriaceae; telial host - *Bouteloua* spp.
Hosts: Asclepias, Bouteloua, Fouquieria
Puccinia vilfae Arth. & Holw.
Puccinia verbenicola Arth.
Distr.: Central and eastern North America; South Africa.
Notes: Heteroecious; aecial host - *Verbena* spp.; telial host - *Sporobolus* spp.
Hosts: Sporobolus, Verbena
Puccinia vincae Berk. in Sm.
Distr.: MA to MI(greenhouses); Europe.
Notes: Autoecious.
Hosts: Catharanthus, Vinca
Puccinia violae (Schumach.) DC.
Puccinia effusa Dietel & Holw.
Aecidium violae Schumach. (anamorph)
Distr.: Cosmopolitan.
Notes: Autoecious. Savile (Fungi Canadenses 75. 1975) segregated North America populations into subsp. *americana* Savile and indicated that subsp. *violae* may have been introduced into North America via contaminated seed. Ono (Trans. Mycol. Soc. Japan 27:175. 1986).
Host: Viola
Puccinia virgae-aurea (DC.) Lib.
Distr.: Temperate northern hemisphere.
Notes: Autoecious.
Host: Solidago
Puccinia virgata Ellis & Everh.
Puccinia clavispora Ellis & Barth.
Distr.: Northern United States to Brazil.
Notes: Autoecious.
Hosts: Saccharum, Sorghastrum
Puccinia vitata J. Hennen & C.S. Hodges
Distr.: HI.
Notes: Autoecious.
Host: Euphorbia
Puccinia volkartiana E. Fisch.
Distr.: AK; Canada(Yukon); Europe.
Notes: Autoecious. Parmelee (Fungi Canadenses 210. 1981).
Host: Androsace
Puccinia waldsteiniae M.A. Curtis ex Peck
Distr.: VT to ID; Japan.
Notes: Autoecious.
Host: Waldsteinia
Puccinia windsoriae Schwein.
Aecidium pteleae Berk. & M.A. Curtis (anamorph)
Distr.: Central and eastern United States.
Notes: Heteroecious; aecial host - *Ptelea trifoliata*; telial host - *Tridens flavus*.
Hosts: [Ptelea], [Stenotaphrum], [Stipa], Tridens, [Triodia]
Puccinia wulfeniae Dietel & Holw. in Dietel
Distr.: Western United States.
Notes: Autoecious.
Hosts: Besseya, Synthyris, Veronica
Puccinia xanthii Schwein.
Distr.: Cosmopolitan.
Notes: Autoecious. Parmelee (Fungi Canadenses 99. 1977).
Hosts: Ambrosia, Xanthium
Puccinia ximenesiae W.H. Long
Distr.: Southwestern United States; Mexico.
Notes: Autoecious.
Host: Verbesina
Puccinia yosemitana Blasdale
Distr.: CA, CO.
Notes: Autoecious. Parmelee (Mycologia 78:464. 1986).
Hosts: Gilia, Leptodactylon
Puccinia zexmeniae Dietel & Holw. in Holw.
Distr.: AZ; Mexico.
Notes: Autoecious.
Host: Lasianthaea
Puccinia ziziae Ellis & Everh.
Distr.: WA; Canada (Saskatchewan).
Notes: Autoecious.
Host: Zizia
Puccinia zoysiae Dietel
Distr.: HI, central and southern United States; eastern Asia.
Notes: Heteroecious; aecial host - *Paederia* spp.; telial host - *Zoysia* spp.
Host: Zoysia

Pucciniastrum G. Otth. Basidiomycetes, Uredinales

Notes: Hiratsuka (Revision of taxonomy of the Pucciniastreae. 1958).
Pucciniastrum agrimoniae (Dietel) Tranzschel
Distr.: Cosmopolitan.
Notes: Agrimony rust. Heteroecious; aecial host - unknown; telial host - *Agrimonia* spp.
Host: Agrimonia
Pucciniastrum americanum (Farl.) Arth.
Pucciniastrum arcticum Tranzschel var. *americanum* Farl.
Distr.: Temperate North America.
Notes: Heteroecious; aecial host - *Picea glauca*; telial host - *Rubus* spp. Needle rust of white spruce, late leaf rust of raspberry (Laundon & Rainbow. C.M.I. Descr. 210. 1969).
Hosts: Picea, Rubus
Pucciniastrum arcticum Tranzschel
Distr.: Circumboreal.
Notes: Heteroecious; aecial host - *Picea* spp.; telial host - *Rubus* spp.
Hosts: Picea, Rubus
Pucciniastrum epilobii G. Otth
Pucciniastrum fuchsiae Hiratsuka
Uredo fuchsiae Arth. & Holw. non Henn., hom. illeg. (anamorph)
Distr.: North America; Asia; Australia; Europe; New Zealand.
Notes: Heteroecious; aecial host - *Abies* spp.; telial host - *Epilobium* spp. and other Onagraceae.
Hosts: Abies, Epilobium, Fuchsia
Pucciniastrum goeppertianum (Kühn) Kleb.
Calyptospora goeppertiana Kühn
Distr.: North America; Europe; Japan.
Notes: Heteroecious; aecial host - *Abies* spp. telial host - *Vaccinium* spp. Witches' broom rust on blueberries. Some mycologists regard *Calyptospora* as a valid segregated genus. For comparison with *P. epilobii* and a discussion of the *Peridermium* states, see Hiratsuka et al. (Canad. J. Bot. 45:1913. 1967).
Hosts: Abies, Thibaudia, Vaccinium
Pucciniastrum guttatum (J. Schröt.) Hyl., Jørst., & Nannf.
Pucciniastrum galii E. Fisch.
Distr.: Temperate northern hemisphere.
Notes: Heteroecious; aecial host - unknown; telial host - Rubiaceae. Hylander et al. (Opera Bot. 1:81. 1953).
Host: Galium
Pucciniastrum hydrangeae (Magnus) Arth.

Thekopsora hydrangeae Magnus
Distr.: Eastern North America.
Notes: Heteroecious; aecial host - *Tsuga* spp.; telial host - *Hydrangea* spp.
Hosts: Hydrangea, Tsuga
Pucciniastrum potentillae Kom. ex Jacz., Kom., & Tranzschel
Distr.: Temperate northern hemisphere; New Guinea.
Notes: Heteroecious; aecial host - unknown; telial host - *Potentilla* spp.
Host: Potentilla
Pucciniastrum pustulatum Dietel in Engl. & Prantl
Distr.: Northern North America.
Notes: Heteroecious; aecial host - *Abies* spp.; telial host - on Onagraceae. Savile (Canad. J. Bot. 40:1392. 1962).
Hosts: Abies, Clarkia, Epilobium, Fuchsia, Oenothera
Pucciniastrum pyrolae Dietel ex Arth.
Distr.: Circumboreal.
Notes: Heteroecious; aecial host - unknown; telial host - Pyrolaceae.
Hosts: Chimaphila, Pyrola
Pucciniastrum sparsum (G. Wint.) E. Fisch.
Thekopsora sparsa (G. Wint.) Magnus
Distr.: Temperate northern hemisphere.
Notes: Heteroecious; aecial host - *Picea* spp.; telial host - Ericaceae (Hiratsuka. Canad. J. Bot. 48:433. 1970).
Hosts: Arbutus, Arctostaphylos
Pucciniastrum vaccinii (G. Wint.) Jørst.
Thekopsora vaccinii (G. Wint.) Hiratsuka
Uraecium holwayi (Arth.) Arth.
Pucciniastrum myrtilli Arth.
Distr.: Widespread in temperate regions, occasionally encountered in tropical areas.
Notes: Heteroecious; aecial host - *Tsuga* spp.; telial host - Ericaceae.
Hosts: Gaylussacia, Lyonia, Menziesia, Rhododendron, Thibaudia, Tsuga, Vaccinium

Pucciniosira Lagerh. Basidiomycetes, Uredinales

Notes: Buriticá & Hennen (Flora Neotropica 24. 1980).
Pucciniosira pallidula (Speg.) Lagerh.
Distr.: FL; Central and South America.
Notes: Autoecious.
Host: Triumfetta

Pulcherricium Parmasto. Basidiomycetes, Aphyllophorales

Pulcherricium caeruleum (Schrad.:Fr.) Parmasto
Corticium caeruleum (Schrad.:Fr.) Fr.
Distr.: Northern hemisphere.
Notes: On hardwoods.
Hosts: Acer, Castanea, Cornus, Juglans, Liriodendron, Malus, Quercus, Vitis

Pulveria Malloch & C.T. Rogerson. Pyrenomycetes, Xylariales

Notes: Monotypic (Malloch & Rogerson. Canad. J. Bot. 55:1505. 1977).
Pulveria porrecta Malloch & C.T. Rogerson
Distr.: NY; Canada(Ontario).
Notes: On wood of dead standing hardwood trees.
Host: Fraxinus

Punctularia Pat. in Pat. & Lagerh. Basidiomycetes, Aphyllophorales

Phaeophlebia W.B. Cooke
Punctularia atropurpurascens (Berk. & Broome) Petch
Corticium conigenum Shear & R.W. Davidson
Ptychogaster rubescens Boud. (anamorph)
Distr.: Northern hemisphere.
Notes: On wood.
Host: Quercus
Punctularia strigoso-zonata (Schwein.) Talbot
Merulius strigoso-zonatus Schwein.
Phaeophlebia strigoso-zonata (Schwein.) W.B. Cooke
Phlebia strigoso-zonata (Schwein.) C.G. Lloyd
Phlebia reflexa Berk.
Distr.: Widespread.
Notes: On hardwoods, especially *Populus* spp. in North America.
Host: Populus

Pycnopeziza W.L. White & Whetzel. Discomycetes, Helotiales

Pycnopeziza sympodialis W.L. White & Whetzel
Acarosporium sympodiale Bubák & Vleugel (anamorph)
Distr.: NY(type).
Hosts: Acer, Populus

Pycnoporellus Murrill. Basidiomycetes, Aphyllophorales

Pycnoporellus alboluteus (Ellis & Everh.) Kotlaba & Pouzar
Aurantioporellus alboluteus (Ellis & Everh.) Murrill
Phaeolus alboluteus (Ellis & Everh.) Pilát
Polyporus alboluteus (Ellis & Everh.) Ellis & Everh.
Distr.: North America, most common in cooler parts and in the West; Europe.
Notes: On wood, usually of conifers.
Hosts: Abies, Larix, Picea, Pinus, Populus, Pseudotsuga, Tsuga
Pycnoporellus fulgens (Fr.) Donk
Hydnum fulgens Fr.
Polyporus aurantiacus Peck
Phaeolus fibrillosus (P. Karst.) A. Ames
Polyporus fibrillosus P. Karst.
Distr.: Temperate northern hemisphere.
Notes: Usually on conifers, occasionally hardwoods.
Hosts: Abies, Acer, Alnus, Larix, Picea, Pinus, Prunus, Pseudotsuga, Thuja, Tsuga

Pycnoporus P. Karst. Basidiomycetes, Aphyllophorales

Notes: Gilbertson & Ryvarden (North American Polypores 2:687. 1987).
Pycnoporus cinnabarinus (Jacq.:Fr.) P. Karst.
Polyporus cinnabarinus Jacq.:Fr.
Coriolus cinnabarinus (Jacq.:Fr.) G. Cunn.
Phellinus cinnabarinus (Jacq.:Fr.) Quél.
Polystictus cinnabarinus (Jacq.:Fr.) Sacc.
Trametes cinnabarina (Jacq.:Fr.) Fr.
Distr.: Temperate northern hemisphere.
Notes: On wood.
Hosts: Abies, Acer, Alnus, Betula, Castanea, Fagus, Fouquieria, Fraxinus, Juglans, Liquidambar, Liriodendron, Magnolia, Nyssa, Opuntia, Picea, Pinus, Populus, Prunus, Pseudotsuga, Quercus, Salix, Thuja, Tsuga
Pycnoporus sanguineus (L.:Fr.) Murrill
Polyporus sanguineus L.:Fr.
Polystictus sanguineus (L.:Fr.) Fr.
Trametes cinnabarina (Jacq.:Fr.) Fr. var. *sanguinea* (L.:Fr.) Pilát
Coriolus sanguineus (L.:Fr.) G. Cunn.
Trametes sanguinea (L.:Fr.) C.G. Lloyd
Distr.: Central and eastern North America, especially common in the South; South America; Africa; India; New Guinea.
Notes: On hardwoods.
Hosts: Acer, Betula, Campsis, Carpinus, Castanea, Juglans, Liquidambar, Mangifera, Myrica, Nyssa, Pinus, Platanus, Prunus, Quercus, Salix

Pycnostysanus Lindau. Hyphomycetes

Pycnostysanus azaleae (Peck) E. Mason
Periconia azaleae Peck
Briosia azaleae (Peck) Dearn.
Sporocybe azaleae (Peck) Sacc.
Distr.: Temperate regions.

Notes: Twig blight and bud blast.
Host: Rhododendron

Pyrenobotrys Theiss. & Syd. Loculoascomycetes, Dothideales

Pyrenobotrys compacta (Peck) B. Eriksson
Venturia compacta Peck
Gibbera compacta (Peck) Shear
Distr.: Northeastern North America; Scandinavia.
Notes: On leaves and twigs (Eriksson. Svensk Bot. Tidskr. 68:192. 1974).
Host: Vaccinium
Pyrenobotrys conferta (Fr.:Fr.) Theiss. & Syd.
Sphaeria conferta Fr.:Fr.
Gibbera conferta (Fr.:Fr.)Petr.
Stigmatea conferta (Fr.:Fr.) Fr.
Venturia atramentaria Cooke
Dothidella vaccinii Rostr.
Distr.: Alpine and arctic regions.
Notes: On living and overwintered leaves. Müller & Arx. (Beitr. Kryptogamenfl. Schweiz 11(2):412. 1962); Barr (Canad. J. Bot. 46:827. 1968).
Host: Vaccinium
Pyrenobotrys pulchella (Cooke & Peck) Barr
Venturia pulchella Cooke & Peck in Peck
Gibbera pulchella (Cooke & Peck) Petr.
Stigmatea pulchella (Cooke & Peck) Barr
Distr.: Northcentral and northeastern North America.
Host: Chamaedaphne

Pyrenochaeta De Not. Coelomycetes

Notes: Schneider & Schwartz (Mitt. Biol. Bundesanst. Land-Forstw. Berlin-Dahlem 189. 1979).
Pyrenochaeta berberidis (Sacc.) Brunaud
Phoma berberidis Sacc.
Distr.: Temperate northern hemisphere.
Notes: Sivanesan (The bitunicate Ascomycetes. p. 393. 1984).
Host: Berberis
Pyrenochaeta elymi R. Sprague
Distr.: UT(type).
Notes: According to Schneider (In: Subramanian. Ed. Taxonomy of Fungi. Part 2. p. 520. 1984), this is an *Ameropsorium*.
Host: Elymus
Pyrenochaeta fraxinina Fairm.
Distr.: NY.
Host: Fraxinus
Pyrenochaeta graminis Ellis & Everh.
Distr.: KS(type), TX.
Notes: A doubtful species (Schneider. In: Subramanian. Ed. Taxonomy of Fungi. Part 2. p. 521. 1984).
Hosts: Chloris, Distichlis
Pyrenochaeta lycopersici R. Schneider & Gerlach
Distr.: United States; Europe; New Zealand.
Notes: Punithalingam & Holliday (C.M.I. Descr. 398. 1973); Sutton (The Coelomycetes. p. 395. 1980).
Hosts: Carthamus, Lycopersicon, Solanum, Spinacia
Pyrenochaeta mali M.A. Sm.
Distr.: United States.
Notes: Apple rot.
Host: Malus
Pyrenochaeta minuta J.C. Carter
Distr.: IL.
Notes: *Pyrenochaeta* oak dieback (Carter. Illinois Nat. Hist. Survey Bull. 21:218. 1941). A doubtful species (Schneider. In: Subramanian. Ed. Taxonomy of Fungi. Part 2. p. 522. 1984).
Host: Quercus
Pyrenochaeta nucinata Fairm.
Distr.: NY.
Host: Carya
Pyrenochaeta oryzae Shirai ex Miyake
Distr.: Worldwide in rice-growing regions.
Notes: Sheath blotch of rice (Punithalingam. C.M.I. Descr. 666. 1980).
Host: Oryza
Pyrenochaeta phlogis Massee
Distr.: NY; Great Britain(type).
Notes: Probably a *Phoma* (Schneider. In: Subramanian. Ed. Taxonomy of Fungi. Part 2. p. 518. 1984).
Host: Phlox
Pyrenochaeta setariae H.C. Greene
Distr.: WI.
Host: Setaria
Pyrenochaeta vitis Viala & Sauv.
Distr.: Central and eastern United States; France(type).
Notes: Possibly a *Dothiorella* (Schneider. In: Subramanian. Ed. Taxonomy of Fungi. Part 2. p. 521. 1984).
Host: Vitis

Pyrenopeziza Fuckel. Discomycetes, Helotiales

Notes: Hütter (Phytopathol. Z. 33:1. 1958).
Pyrenopeziza absinthii (Lasch) Rehm
Distr.: ND; Europe.
Notes: Gremmen (Fungus 25:11. 1955) stated that this species does not belong in *Pyrenopeziza*.
Host: Iva
Pyrenopeziza artemisiae (Lasch) Rehm
Distr.: Temperate North America; Europe.
Host: Artemisia
Pyrenopeziza artemisiae (Lasch) Rehm var. ***solidaginis*** Rehm
Distr.: NY; Canada(Ontario,type).
Host: Solidago
Pyrenopeziza californica Sacc.
Pezizella californica (Sacc.) Rehm
Distr.: CA(type).
Host: Linum
Pyrenopeziza carduorum Rehm
Mollisia carduorum (Rehm) Gremmen
Distr.: ID; Europe.
Notes: On decaying stems.
Host: Cirsium
Pyrenopeziza coloradensis Ellis & Everh.
Distr.: CO(type).
Host: Potentilla
Pyrenopeziza compressula Rehm
Distr.: CA, CO; Europe.
Notes: On dead stems of legumes.
Host: Lotus
Pyrenopeziza ellisii (Rehm) Massee
Niptera ellisii Rehm
Distr.: MT, NJ.
Notes: This name may be a *nomen nudem* fide Seaver (North Amer. cup fungi. [Inoperculates]. p. 224. 1951).
Host: Calamovilfa
Pyrenopeziza foliicola (P. Karst.) Sacc.
Distr.: GA; Europe.
Notes: Seaver (North Amer. cup fungi. [Inoperculates]. p. 229. 1951) suggested that the report from Georgia is erroneous.
Host: Alnus
Pyrenopeziza heteromeles (W. Phillips & Harkn.) Sacc.
Distr.: CA(type).
Host: Heteromeles
Pyrenopeziza karstenii Sacc.
Hysteropezizella leporinae Nannf.
Distr.: North temperate regions.
Hosts: Agropyron, Agrostis, Carex, Festuca, Poa
Pyrenopeziza lacerata (Cooke & Peck) Sacc.
Distr.: NY, WV.
Host: Rubus
Pyrenopeziza latebrosa (W. Phillips & Harkn.) Dennis
Lachnella latebrosa W. Phillips & Harkn.
Trichopeziza latebrosa (W. Phillips & Harkn.) Sacc.
Distr.: CA.
Notes: On dead leaves (Dennis. Kew Bull. 17:351. 1963).
Host: Arctostaphylos
Pyrenopeziza laurocerasi (Desmaz.) Nannf.

Trochila laurocerasi (Desmaz.) Fr.
Distr.: CA; Europe.
Notes: Greenhalgh & Morgan-Jones (Trans. Brit. Mycol. Soc. 47:315. 1964) presented details of the morphology of this species under the name *Trochila laurocerasi*.
Host: Prunus
Pyrenopeziza leucodermis Cash
Distr.: GA.
Host: Acer
Pyrenopeziza minuta Cash
Distr.: GA(type).
Host: Tilia
Pyrenopeziza multipunctoidea Dearn. & House
Distr.: NY(type).
Host: Carex
Pyrenopeziza polygoni (Lasch) Gremmen
Mollisia polygoni (Lasch) Rehm
Mollisia atrocinerea (Cooke) W. Phillips
Distr.: Eastern United States; Europe.
Hosts: Chrysopsis, Polygonum
Pyrenopeziza potentillae (Rostr.) Nannf.
Trochila potentillae Rostr.
Distr.: North temperate and arctic regions.
Host: Potentilla
Pyrenopeziza prinicola (Ellis & Everh.) Cash
Mollisia prinicola (Ellis & Everh.) Sacc.
Distr.: Southeastern United States.
Host: Quercus
Pyrenopeziza rigidae (Nannf.) Defago
Hysteropezizella rigidae Nannf.
Distr.: North temperate regions.
Host: Carex
Pyrenopeziza rubi (Fr.) Rehm
Mollisia rubi (Fr.) P. Karst.
Distr.: Temperate regions.
Notes: On dead canes.
Host: Rubus
Pyrenopeziza smilacicola Dearn. & House
Distr.: NY(type).
Host: Smilax
Pyrenopeziza stictoidea Sacc.
Distr.: AK; Europe.
Host: Hierochloe
Pyrenopeziza symphoricarpi (Ellis & Everh.) Dennis
Lachnella symphoricarpi Ellis & Everh.
Distr.: CO.
Notes: On dead stems (Dennis. Kew Bull. 17:367. 1963).
Host: Symphoricarpos
Pyrenopeziza thalictri (Peck) Sacc.
Distr.: Temperate regions.
Host: Thalictrum
Pyrenopeziza yogoensis (Ellis & B.T. Galloway) Sacc.
Distr.: MT(type).
Host: Carex

Pyrenophora Fr. Loculoascomycetes, Dothideales

Notes: Sivanesan (Mycol. Pap. 158:161. 1987).
Pyrenophora avenae Ito & Kuribayashi
Drechslera avenae (Eidam) Scharif (anamorph)
Distr.: Cosmopolitan.
Notes: Leaf spot and seedling blight of oats and other grasses (Ellis & Waller. C.M.I. Descr. 389. 1973).
Hosts: Acacia, Avena
Pyrenophora bromi (Died.) Drechs.
Pleospora bromi Died.
Distr.: Temperate North America; Australia; Europe.
Notes: Leaf spot of *Bromus* and other grasses.
Host: Bromus
Pyrenophora japonica Ito & Kuribayashi in Ito
Pyrenophora secalis M.D. Whitehead & Dickson
Drechslera tuberosa (Atk.) Shoemaker (anamorph)
Distr.: Temperate northern hemisphere.
Notes: Parasitic on cereals and grasses (Sivanesan. The bitunicate Ascomycetes. p. 547. 1984; Whitehead & Dickson. Mycologia 44:752. 1952).
Host: Secale
Pyrenophora leucelenes Fairm.
Distr.: NM(type).
Notes: Wehmeyer (A world monograph of the genus *Pleospora* and its segregates. p. 360. 1961) suggested that this is a synonym of *Pleospora tracanthae*.
Host: Aster
Pyrenophora phaeocomes Fuckel non Fr., hom. illeg.
Pleospora phaeocomes (Rebent.) G. Wint. non Niessl, hom. illeg.
Distr.: North America; Europe.
Host: Dactylis
Pyrenophora schroeteri Barr
Pleospora macrospora J. Schröt. non Fuckel, hom. illeg.
Distr.: North America; Europe.
Notes: On culms of Poaceae (Barr. Contr. Univ. Michigan Herb. 9:551. 1972).
Hosts: Arctophila, [Helianthella]
Pyrenophora scirpi (Rabenh.) Wehmeyer
Pleospora aquatica D. Griffiths
Distr.: SD; Europe.
Host: Eleocharis
Pyrenophora teres Drechs.
Pleospora teres Died., nom. nud.
Drechslera teres (Sacc.) Shoemaker (anamorph)
Distr.: Cosmopolitan.
Notes: Net blotch of barley and other grasses (Ellis & Waller. C.M.I. Descr. 390. 1973).
Hosts: Avena, Hordeum
Pyrenophora trichostoma (Fr.) Fuckel
Pleospora trichostoma (Fr.) Ces. & De Not.
Pleospora culmorum (Cooke) Sacc.
Pyrenophora relicina (Fr.) Fuckel
Distr.: United States; Europe.
Notes: This name has erroneously been considered a synonym of *Pyrenophora tritici-repentis*.
Hosts: Bromus, Elymus, Hierochloe, Juncus, Leymus, Poa, Puccinellia, Secale, Triticum
Pyrenophora tritici-repentis (Died.) Drechs.
Pleospora tritici-repentis Died.
Pyrenophora tritici-vulgaris Dickson
Drechslera tritici-repentis (Died.) Shoemaker (anamorph)
Distr.: Cosmopolitan.
Notes: Yellow leaf spot of cereals and grasses (Ellis & Waller. C.M.I. Descr. 494. 1976).
Hosts: Bromus, Hordeum, Secale, Triticum

Pyricularia Sacc. Hyphomycetes

Notes: Ellis (Dematiaceous Hyphomycetes. p. 218. 1971).
Pyricularia grisea (Cooke) Sacc.
Trichothecium griseum Cooke
Magnaporthe grisea (T.T. Hebert) Yaegashi & Udagawa (teleomorph)
Distr.: North and South America; Africa; Asia.
Notes: Blast or gray spot of grasses (Yaegashi & Udagawa. Canad. J. Bot. 56:180. 1978).
Hosts: Agrostis, Andropogon, Cladium, Commelina, Cynodon, Digitaria, Echinochloa, Eragrostis, Eremochloa, Glycine, Hystrix, Leersia, Muhlenbergia, Oryza, Panicum, Paspalum, Poa, Scolochloa, Setaria, Sorghum, Stenotaphrum
Pyricularia oryzae Cavara
Distr.: Southern North America; South America; Africa; Asia.
Notes: Rice blast disease (Subramanian. C.M.I. Descr. 169. 1968). *Pyricularia grisea* (Cooke) Sacc. is an older name for *P. oryzae*; the latter should be conserved (Kirk. Mycotaxon 18:289. 1983) but this has not been done.
Hosts: Eremochloa, Oryza
Pyricularia parasitica Ellis & Everh.
Distr.: WI.
Hosts: Elymus, Hystrix, Muhlenbergia

Pyrofomes Kotlaba & Pouzar. Basidiomycetes, Aphyllophorales

Notes: Gilbertson & Ryvarden (North American Polypores 2:691. 1987).
Pyrofomes demidoffii (Lév.) Kotlaba & Pouzar
Polyporus demidoffii Lév. in Demidoff
Fomes demidoffii (Lév.) Cooke
Fomes earlei (Murrill) Sacc. & D. Sacc.
Fomes juniperinus (H. Schrenk) Sacc. & Syd.
Fulvifomes juniperinus (H. Schrenk) Murrill
Polyporus juniperinus H. Schrenk
Distr.: Widespread.
Notes: On wood of *Juniperus* and *Cupressus* spp.
Host: Juniperus

Pyronema Carus. Discomycetes, Pezizales

Notes: Moore & Korf (Bull. Torrey Bot. Club 90:33. 1963).
Pyronema omphalodes (Bull.:Fr.) Fuckel
Pyronema confluens (Pers.:Pers.) Tul. & C. Tul.
Distr.: Cosmopolitan.
Notes: On burned areas, charcoal beds, and steamed soil in greenhouses.
Hosts: Sequoia, Zea

Pythium Pringsh., nom. cons. Oomycetes, Peronosporales

Notes: Plaats-Niterink (Stud. Mycol. 21:1. 1981).
Pythium acanthicum Drechs.
Distr.: Cosmopolitan.
Notes: On roots, fruits, and seedlings of a wide variety of plants.
Hosts: Citrullus, Fragaria, Solanum, Zea
Pythium acanthophoron Sideris
Distr.: HI; The Phillipines.
Host: Ananas
Pythium adhaerens Sparrow
Distr.: United States; Malaysia.
Notes: Damping-off, root rots, or fruits rots.
Host: Zea
Pythium anandrum Drechs.
Distr.: Cosmopolitan.
Notes: On roots, seedlings, and fruits of a wide variety of plants.
Hosts: Cucumis, Rheum, Spinacia
Pythium angustatum Sparrow
Distr.: Widespread.
Notes: Generally not pathogenic but found to cause a root-rot of corn under certain environmental conditions (Plaats-Niterink. Stud. Mycol. 21:34. 1981).
Host: Zea
Pythium aphanidermatum (Edson) Fitzp.
Rheosporangium aphanidermatum Edson
Nematosporangium aphanidermatum (Edson) Fitzp.
Pythium butleri L. Subramanian
Distr.: Cosmopolitan, especially common in warm regions.
Notes: Waterhouse & Waterston (C.M.I. Descr. 36. 1964; C.M.I. Descr. 37. 1964 as *Pythium butleri*).
Hosts: [Agrostis], Amaranthus, Ananas, Aphelandra, Avena, Beta, Capsicum, Carica, Carpobrotus, Chrysanthemum, Chrysobalanus, Citrullus, Citrus, Codiaeum, Colocasia, Cordyline, Cucumis, Cucurbita, Cynodon, Delphinium, Dracaena, Echeveria, Euphorbia, Geranium, Glycine, Gossypium, Gypsophila, Hordeum, Ipomoea, Lavandula, Leea, Linum, Lolium, Lycopersicon, Medicago, Momordica, Musa, Nicotiana, Opuntia, Oryza, Parthenium, Passiflora, Pelargonium, Peperomia, Petroselinum, Phaseolus, Phlox, Pilea, Pinus, Pisum, Raphanus, Rosa, Saccharum, Schlumbergera, Secale, Solanum, Spinacia, Triticum, Verbena, Zea, Zebrina, Zingiber, Zoysia
Pythium aristosporum Vanterpool
Distr.: Temperate North America.
Notes: Usually on grasses.
Hosts: Agropyron, Agrostis, Arrhenatherum, Avena, Bromus, Dactylis, Echinochloa, Hordeum, Phaseolus, Triticum
Pythium arrhenomanes Drechs.
Nematosporangium arrhenomaes (Drechs.) Sideris
Distr.: Cosmopolitan.
Notes: Common on grasses; also on a variety of other plants (Waterhouse & Waterston. C.M.I. Descr. 39. 1964).
Hosts: Aegilops, Agropyron, Agrostis, Ammophila, Ananas, Andropogon, Arrhenatherum, Avena, Billbergia, Bouteloua, Bromus, Cajanus, Calamovilfa, Commelina, Echinochloa, Elymus, Festuca, Hordeum, Ipomoea, Lycopersicon, Muhlenbergia, Musa, Oryzopsis, Panicum, Saccharum, Secale, Setaria, Solanum, Sorghastrum, Sorghum, Stipa, Triticum, Vicia, Vigna, Vulpia, Zea
Pythium carolinianum Matthews
Distr.: CA, HI; Costa Rica; Japan.
Notes: Nursery root rot of *Macadamia integrifolia* in Costa Rica (Plaats-Niterink. Stud. Mycol. 21:188. 1981).
Hosts: Colocasia, Myriophyllum, Potamogeton
Pythium catenulatum Matthews
Distr.: Cosmopolitan.
Notes: On plant debris and in soil. Seedling blight and root rots.
Hosts: Lactuca, Phaseolus
Pythium chamaehyphon Sideris
Distr.: HI(type).
Host: Carica
Pythium coloratum Vaartaja
Distr.: Cosmopolitan.
Host: Allium
Pythium debaryanum Auct. non R. Hesse
Distr.: Widespread.
Notes: According to Plaats-Niterink (Stud. Mycol. 21:189. 1981), Hesse's original description was based on a mixture of fungi. *P. debaryanum* has been applied to several common *Pythium* spp., including *P. intermedium*, *P. ultimum*, and probably *P. sylvaticum*.
Hosts: Aegilops, Agrohordeum, Agropyron, Agrostis, Amaranthus, Amaryllis, Ananas, Andropogon, Antirrhinum, Apium, Arrhenatherum, Atropa, Avena, Bassia, Beckmannia, Begonia, Berberis, Beta, Bouteloua, Bromus, Cajanus, Calamovilfa, Canavalia, Capsicum, Carica, Centaurea, Chrysanthemum, Cichorium, Citrullus, Citrus, Clarkia, Colocasia, Cucumis, Cucurbita, Cynodon, Dactylis, Dahlia, Distichlis, Echinochloa, Elymus, Epilobium, Eremochloa, Euphorbia, Festuca, Glycine, Gossypium, Gypsophila, Helianthus, Hippeastrum, Hordeum, Ipomoea, Koeleria, Lactuca, Lathyrus, Lavatera, Lepidium, Lilium, Linum, Lobelia, Lolium, Lupinus, Lycopersicon, Medicago, Melilotus, Nasturtium, Nicotiana, Opuntia, Oryzopsis, Panax, Panicum, Pelargonium, Petroselinum, Phalaris, Phaseolus, Phleum, Phragmites, Picea, Pinus, Pisum, Poa, Polianthes, Pyrus, Ranunculus, Raphanus, Rhododendron, Saccharum, Salsola, Salvia, Schedonnardus, Schizachyrium, Secale, Setaria, Sitanion, Solanum, Solenostemon, Sorghum, Spinacia, Sporobolus, Stipa, Triticum, Vicia, Vigna, Viola, Vulpia, Zea
Pythium deliense Meurs
Distr.: Warmer regions.
Notes: Pathogenic on a variety of crop plants.
Host: Beta
Pythium dictyosporum Racib.
Nematosporangium dictyosporum (Racib.) J. Schröt.
Distr.: WA; Europe.
Host: Pyrus
Pythium dimorphum J.W. Hendrix & W.A. Campbell
Distr.: LA(type).
Notes: Hendrix & Campbell (Mycologia 63:978. 1971).
Hosts: Pinus, Rhododendron
Pythium dissotocum Drechs.
Pythium araiosporum Sideris
Pythium oryzae Ito & Tokunaga
Pythium perigynosum Sparrow
Distr.: Cosmopolitan.
Notes: Frequently isolated from soil and from a wide variety of plant roots.
Hosts: Beta, Fragaria, Lactuca, Medicago, Phaseolus, Pilea, Pisum, Saccharum, Spinacia
Pythium graminicola Subramanian
Distr.: Cosmopolitan.
Notes: Pathogenic on many grasses; also isolated from a wide variety of dicotyledonous plants (Waterhouse & Waterston. C.M.I. Descr. 38. 1964).
Hosts: Aegilops, Agrohordeum, Agropyron, Agrostis, Ammophila,

Ananas, Andropogon, Aristida, Arrhenatherum, Avena, Bouteloua, Bromus, Calamovilfa, Cenchrus, Chasmanthium, Colocasia, Dactylis, Echinochloa, Elymus, Eragrostis, Eremopyrum, Festuca, Hordeum, Leucopoa, Leymus, Lolium, Lupinus, Melica, Muhlenbergia, Oryza, Oryzopsis, Panicum, Phleum, Phragmites, Pisum, Poa, Redfieldia, Saccharum, Sarracenia, Schedonnardus, Scolochloa, Secale, Setaria, Sorghastrum, Sorghum, Sphenopholis, Sporobolus, Stipa, Triticum, Vicia, Vulpia, Zea, Zingiber

Pythium grandisporangium Fell & Master
Distr.: FL.
Notes: On mangroves (Fell & Master. Canad. J. Bot. 53:2920. 1975).
Host: Rhizophora

Pythium helicandrum Drechs.
Distr.: MD(type), scattered reports from soil isolates in the United States; The Netherlands.
Host: Rumex

Pythium helicoides Drechs.
Phytophthora fagopyri Takimoto
Distr.: North America; Australia; Malaysia.
Notes: Implicated in several root diseases and incidences of damping-off.
Hosts: Buxus, Citrullus, Phaseolus, Saccharum, Spinacia, Typha

Pythium heterothallicum W.A. Campbell & J.W. Hendrix
Distr.: Cosmopolitan.
Notes: Generally thought to be saprobic but pathogenicity has not been studied.
Host: Triticum

Pythium hydnosporum (Mont.) J. Schröt.
Pythium artotrogus de Bary
Pythium artotrogus de Bary var. *macracanthum* Sideris
Distr.: Cosmopolitan.
Notes: Associated with damping-off, root and stem rot.
Hosts: Ananas, Cajanus, Canavalia, Citrullus, Ipomoea, Matthiola, Panicum, Saccharum, Vigna

Pythium hypogynum Middleton
Distr.: Central United States.
Hosts: Avena, Fragaria, Hordeum, Triticum

Pythium indigoferae E.J. Butler
Distr.: HI; Brazil; India(type).
Host: Ananas

Pythium inflatum Matthews
Distr.: Cosmopolitan.
Host: Saccharum

Pythium intermedium de Bary
Distr.: Cosmopolitan.
Notes: Ubiquitous in soil and plant species debris and isolated from many living plant species (Waterhouse & Waterston. C.M.I. Descr. 40. 1964).
Hosts: Ananas, Begonia, Pteris, Ricinus

Pythium irregulare Buisman
Pythium dactyliferum Drechs. in Rands
Pythium fabae G. Cheney
Pythium polymorphon Sideris
Distr.: Cosmopolitan.
Notes: A common soil and plant-inhabiting species.
Hosts: Agropyron, Allium, Ananas, Andropogon, Antirrhinum, Apium, Arrhenatherum, Atropa, Avena, Begonia, Bouteloua, Bromus, Buxus, Cajanus, Caladium, Canavalia, Capsicum, Carya, Chrysanthemum, Citrullus, Citrus, Cornus, Daucus, Dizygotheca, Elymus, Festuca, Fragaria, Helianthus, Hordeum, Ipomoea, Juniperus, Lactuca, Ligustrum, Limonium, Medicago, Nicotiana, Ocimum, Pinus, Pisum, Pittosporum, Podocarpus, Primula, Ranunculus, Rheum, Rhododendron, Saccharum, Schlumbergera, Solanum, Spinacia, Sporobolus, Stipa, Triticum, Vicia, Vigna, Zea

Pythium iwayamai Ito
Distr.: Northwestern North America; Japan.
Notes: On cereals and conifer seedlings.
Host: Triticum

Pythium mamillatum Meurs
Distr.: Cosmopolitan.
Notes: Fruit and root rots on a wide variety of plants. Description and notes on host range and pathogenicity were provided by Waterhouse & Waterson (C.M.I. Descr. 117. 1966).
Hosts: Ageratum, Ananas, Antirrhinum, Collinsia, Medicago, Pelargonium, Saccharum, Viola

Pythium marsipium Drechs.
Distr.: Central and eastern United States; eastern Asia; Soviet Union.
Host: Nymphaea

Pythium mastophorum Drechs.
Distr.: United States; South America; Australia.
Hosts: Bellis, Calceolaria

Pythium megalacanthum de Bary
Distr.: HI, North America; Europe; New Zealand.
Hosts: Ananas, Chrysanthemum, Lycopersicon, Pyrus

Pythium middletonii Sparrow
Pythium proliferum de Bary non Schenk, hom. illeg.
Distr.: Cosmopolitan.
Notes: Usually considered a saprobe in aquatic habitats, but also isolated from soils and reported to cause damping-off of several plant species.
Hosts: Bromus, Medicago

Pythium monospermum Pringsh.
Pythium gracile Schenk, sensu de Bary
Distr.: Cosmopolitan.
Notes: Isolated from dead insects, soil, and a variety of plants.
Hosts: Hordeum, Panicum, Saccharum, Sporobolus

Pythium myriotylum Drechs.
Distr.: Worldwide in warm regions.
Notes: Pathogenic on a wide variety of plants (Waterhouse & Waterson. C.M.I. Descr. 118. 1966).
Hosts: Antirrhinum, Aponogeton, Arachis, Caladium, Citrullus, Coronilla, Cucumis, Glycine, Ligustrum, Lolium, Lycopersicon, Peperomia, Phaseolus, Robinia, Solanum, Triticum

Pythium nagaii Ito & Tokunaga
Distr.: AR; eastern Asia.
Host: Oryza

Pythium oedochilum Drechs.
Distr.: Cosmopolitan.
Notes: Pathogenic on a wide variety of plants.
Hosts: Bidens, Dahlia, Matthiola

Pythium okanoganense Lipps
Distr.: WA(type).
Notes: Snow rot of wheat (Lipps. Mycologia 72:1127. 1980).
Host: Triticum

Pythium oligandrum Drechs.
Distr.: Cosmopolitan.
Notes: Waterhouse & Waterson (C.M.I. Descr. 119. 1966) provided a complete description.
Hosts: Beta, Cucumis, Iberis, Lathyrus, Lycopersicon, Phaseolus, Pisum, Rheum, Triticum

Pythium ostracodes Drechs.
Distr.: FL, TX; Japan.
Hosts: Gossypium, Triticum

Pythium palingenes Drechs.
Distr.: VA.
Hosts: Ambrosia, Prunella

Pythium paroecandrum Drechs.
Distr.: Cosmopolitan.
Notes: Isolated from soil and from a wide variety of plant species.
Hosts: Allium, Chrysanthemum, Daucus, Impatiens, Medicago, Sanguinaria, Zea

Pythium periilum Drechs.
Distr.: Widespread on sugarcane.
Hosts: Agropyron, Saccharum

Pythium periplocum Drechs.
Distr.: Cosmopolitan.
Notes: Isolated from a wide variety of plants.
Hosts: Citrullus, Cucumis

Pythium perniciosum Serbinow
Distr.: Widespread.
Notes: Pathogenic on a variety of crop plants.
Hosts: Bromus, Euphorbia, Fragaria

Pythium polymastum Drechs.
Distr.: North America; Europe.
Hosts: Brassica, Lactuca

Pythium polytylum Drechs.
Distr.: United States.
Hosts: Prunella, Spinacia

Pythium pulchrum Minden
Pythium epigynum Höhn.

Distr.: Northern hemisphere.
Notes: Saprobic on insects and plant debris, isolated from roots of a variety of plants.
Hosts: Antirrhinum, Medicago, Pyrus, Zea
Pythium rostratum E.J. Butler
Pythium diameson Sideris
Distr.: Cosmopolitan.
Notes: Common in soil and isolated from roots of a wide variety of plants. Generally considered to be nonpathogenic or weakly parasitic.
Hosts: Allium, Ananas, Antirrhinum, Avena, Cajanus, Canavalia, Citrus, Fragaria, Helianthus, Hordeum, Ipomoea, Lactuca, Medicago, Oryza, Panicum, Phaseolus, Saccharum, Solanum, Spinacia, Vicia, Zea
Pythium salpingophorum Drechs.
Distr.: Northern hemisphere, rare.
Hosts: Lycopersicon, Pisum, Spinacia
Pythium scleroteichum Drechs.
Distr.: United States(warmer regions).
Notes: Associated with mottle necrosis of sweet potato.
Host: Ipomoea
Pythium spinosum Sawada
Distr.: Cosmopolitan.
Notes: On a variety of plants.
Hosts: Allium, Asparagus, Carya, Gardenia, Lactuca, Vaccinium, Vicia
Pythium splendens H. Braun
Distr.: Worldwide.
Notes: Waterhouse & Waterson (C.M.I. Descr. 120. 1966).
Hosts: Adiantum, Aeschynanthus, Agave, Aglaonema, Alsophila, Anacardium, Ananas, Anthurium, Antirrhinum, Aphelandra, Aralia, Araucaria, Ardisia, Arecaceae, Aristolochia, Artocarpus, Asparagus, Begonia, Berberis, Bougainvillea, Bucida, Buxus, Cajanus, Caladium, Calathea, Calendula, Callistemon, Callistephus, Camellia, Canavalia, Cattleya, Chamaedorea, Chlorophytum, Chorisia, Chrysalidocarpus, Chrysanthemum, Cinnamomum, Cleyera, Coccoloba, Codiaeum, Columnea, Cornus, Cupressus, Dieffenbachia, Dietes, Dizygotheca, Dracaena, Epipremnum, Fatshedera, Fatsia, Ficus, Fittonia, Gardenia, Geranium, Gossypium, Guaiacum, Hedera, Helianthus, Howeia, Hoya, Illicium, Ipomoea, Ixora, Jasminum, Juniperus, Leea, Ligustrum, Lilium, Liriope, Magnolia, Malus, Maranta, Medicago, Murraya, Nandina, Nautilocalyx, Nephrolepis, Nephthytis, Nolina, Ophiopogon, Orchidaceae, Passiflora, Pelargonium, Pellionia, Peperomia, Philodendron, Pilea, Pinus, Piper, Pittosporum, Plectranthus, Podocarpus, Polypodium, Polyscias, Pteris, Punica, Pyrus, Rheum, Rhododendron, Rosa, Saccharum, Saintpaulia, Schefflera, Serresia, Soleirolia, Solenostemon, Spathiphyllum, Spathoglottis, Stangeria, Strelitzia, Swietenia, Syngonium, Tecomaria, Thuja, Trevesia, Tripogandra, Triticum, Viburnum, Vicia, Vigna, Yucca, Zea
Pythium sylvaticum W.A. Campbell & J.W. Hendrix
Distr.: Cosmopolitan.
Notes: Often misdetermined as *P. debaryanum* (Campbell & Hendrix. Mycologia 59:274. 1967).
Hosts: Allium, Daucus, Fragaria, Juniperus, Lactuca, Triticum
Pythium tardicrescens Vanterpool
Distr.: Widespread in the northern hemisphere.
Notes: Pathogenic on roots of grasses.
Hosts: Avena, Echinochloa
Pythium torulosum Coker & F. Patterson
Distr.: Worldwide on a variety of hosts.
Hosts: Allium, Ananas, Triticum
Pythium tracheiphilum Matta
Distr.: Widespread in Europe and North America.
Notes: Associated with lettuce, but other vegetable crops were found to be susceptible in pathogenicity tests.
Host: Lactuca
Pythium ultimum Trow
Distr.: Cosmopolitan.
Notes: Pathogenic on a wide variety of plants.
Hosts: Aglaonema, Agropyron, Agrostis, Allium, Aloe, Andropogon, Antirrhinum, Apium, Arachis, Atropa, Avena, Begonia, Beta, Brassica, Brassolaeliocattleya, Bromus, Calceolaria, Calendula, Callistephus, Capsicum, Carica, Catalpa, Cattleya, Chamaecyparis, Chilopsis, Chrysanthemum, Cicer, Citrofortunella, Citrullus, Citrus, Cucumis, Cucurbita, Cupressus, Cymbidium, Cynodon, Dactylis, Dahlia, Delphinium, Dendrobium, Dianthus, Dimorphotheca, Echinochloa, Elymus, Epidendrum, Eragrostis, Euphorbia, Fragaria, Gaillardia, Glycine, Gossypium, Gypsophila, Hordeum, Ipomoea, Laelia, Laeliocattleya, Lathyrus, Lilium, Lupinus, Lycopersicon, Maclura, Medicago, Melilotus, Nicotiana, Oenothera, Oncidium, Orchidaceae, Panicum, Parthenium, Pelargonium, Persea, Phaseolus, Phragmites, Picea, Pinus, Pisum, Poa, Prunus, Pseudotsuga, Pyrus, Ranunculus, Rheum, Rhododendron, Ricinus, Saccharum, Saintpaulia, Schizanthus, Secale, Senecio, Sequoiadendron, Shepherdia, Sinningia, Solanum, Sorghum, Spinacia, Tagetes, Triticum, Tulipa, Vanda, Vigna, Viola, Zea
Pythium ultimum Trow var. ***sporangiiferum*** Drechs.
Distr.: Temperate regions.
Host: Triticum
Pythium uncinulatum Plaats-Niterink & Blok
Distr.: AZ; The Netherlands.
Notes: Pathogenic on lettuce (Plaats-Niterink & Blok. Neth. J. Pl. Pathol. 84:135. 1978).
Host: Lactuca
Pythium vanterpoolii V. Kouyeas & H. Kouyeas
Distr.: North America; Europe; Malaysia.
Host: Agrostis
Pythium vexans de Bary
Pythium complectens H. Braun
Distr.: Worldwide on a variety of hosts.
Hosts: Allium, Ananas, Antirrhinum, Capsicum, Carica, Citrus, Delphinium, Dianthus, Lycopersicon, Matthiola, Medicago, Metrosideros, Pelargonium, Persea, Phaseolus, Pisum, Pyrus, Ricinus, Saccharum, Solenostemon, Spinacia, Zea
Pythium violae Chesters & C.J. Hickman
Distr.: North America; Europe.
Hosts: Lactuca, Medicago
Pythium volutum Vanterpool & Truscott
Distr.: Widespread in the northern hemisphere.
Notes: Pathogenic on many grasses, fruits, and ornamentals.
Host: Triticum

Quasiconcha Barr & Blackwell. Loculoascomycetes, Dothideales

Notes: Monotypic.
Quasiconcha reticulata Barr & Blackwell
Chalara sp. (anamorph)
Distr.: AZ, TX.
Notes: On seeds and roots of conifers (Barr & Blackwell. Mycologia 72:1224. 1980; Blackwell & Gilbertson. Mycologia 77:50. 1985).
Hosts: Juniperus, Pinus, Thuja

Questieria G. Arnaud. Loculoascomycetes, Dothideales

Notes: This genus is considered a synonym of *Schiffnerula*, but not all names have been redisposed.
Questieria euphorbiae G. Arnaud
Distr.: HI(type).
Host: Euphorbia

Rabenhorstia Fr. Coelomycetes

Rabenhorstia rudis (Fr.) Fr.
Sphaeria rudis Fr.
Distr.: Eastern United States; Europe; Great Britain.
Host: Laburnum
Rabenhorstia tiliae Fr.
Hercospora tiliae (Pers.:Fr.) Fr. ex Tul. & C. Tul. (teleomorph)
Distr.: Central and eastern United States; Europe; Great Britain.
Notes: On dead branches (Sutton. The Coelomycetes. p. 161. 1980).
Host: Tilia

Racodium Fr., nom. cons. Agonomycetes

Racodium resinae Schwein.
Myxotrichella resinae (Fr.:Fr.) Jaap
Pycnostysanus resinae Lindau

Sorocybe resinae (Fr.) Fr.
Distr.: Temperate North America; Europe.
Notes: On resin and bark of conifers (Hughes. Canad. J. Bot. 46:1100. 1968). Parbery (Austral. J. Bot. 17:349. 1969) suggested that Hughes (Canad. J. Bot. 36:805. 1958) was incorrect in listing *Hormodendron resinae* as a synonym of *Sorocybe resinae*.
Host: Picea

Radulodon Ryvarden. Basidiomycetes, Aphyllophorales

Notes: Ryvarden (Canad. J. Bot. 50:2073. 1972).
Radulodon americanus Ryvarden
Distr.: Northern United States and adjacent Canada.
Notes: On wood. Reports of *R. casearium* from North America, especially on *Populus* spp., probably refer to *R. americanus*.
Hosts: Betula, Populus
Radulodon casearium (Morg.) Ryvarden
Hydnum casearium Morg.
Distr.: OH(type).
Notes: This species is probably known only from the type.
Host: Carya

Ramaricium J. Eriksson. Basidiomycetes, Aphyllophorales

Phlyctibasidium Jülich
Notes: Ginns (Bot. Not. 132:93. 1979).
Ramaricium alboflavescens (Ellis & Everh.) Ginns
Corticium alboflavescens Ellis & Everh.
Coniophora alboflavescens (Ellis & Everh.) Höhn. & Litsch.
Distr.: Appalachian Mountains; Canada (Ontario).
Notes: On wood.
Host: Thuja
Ramaricium flavomarginatum (Burt) Ginns
Coniophora flavomarginata Burt
Distr.: WA; Canada(Vancouver Island).
Notes: On wood and bark.
Host: Quercus
Ramaricium polyporoideum (Berk. & M.A. Curtis) Ginns
Corticium polyporoideum Berk. & M.A. Curtis in Berk.
Coniophora polyporoidea (Berk. & M.A. Curtis) Burt
Phlyctibasidium polyporoideum (Berk. & M.A. Curtis) Jülich
Trechispora polyporoidea (Berk. & M.A. Curtis) Liberta
Distr.: North and South America.
Notes: On wood, woody debris, needles, and living plant parts.
Hosts: Pinus, Quercus, Robinia, Vitis

Ramichloridium Stahel ex De Hoog. Hyphomycetes

Notes: De Hoog (Stud. Mycol. 15:59. 1977).
Ramichloridium indicum (Subramanian) De Hoog
Chloridium indicum Subramanian
Distr.: FL; England; India.
Notes: On dead leaves.
Host: Eucalyptus

Ramularia Unger non Roussel, hom. illeg. Hyphomycetes

Notes: Gunnerbeck (Svensk Bot. Tidskr. 61:126. 1967); Poelt & Fritz-Schroeder (Mitt. Naturwiss. Vereines Steiermark 13:79. 1983).
Ramularia acalyphae Tharp
Distr.: TX(type).
Notes: Tharp (Mycologia 9:120. 1917).
Host: Acalypha
Ramularia actaeae Ellis & Holw.
Distr.: Central North America.
Host: Actaea
Ramularia adoxae (Rabenh.) P. Karst.
Fusidium adoxae Rabenh.
Distr.: Rocky Mountains; Europe.
Notes: On leaves.
Host: Adoxa
Ramularia aequivoca (Ces.) Sacc.
Fusisporium aequivoca Ces.
Distr.: Temperate North America; Europe.
Host: Ranunculus
Ramularia agoseridis Ellis & Everh.
Distr.: Western United States.
Host: Agoseris
Ramularia agrestis Sacc.
Ramularia deflectens Bres.
Distr.: OR; Europe.
Host: Viola
Ramularia alba (Dowson) Nannf.
Cladosporium album Dowson
Hyalodendron album Diddens
Ramularia deusta (Fuckel) Karakulin f. *odorati* K. Baker, W.C. Snyder & L.H. Davis, nom. inval.
Distr.: North America; Europe.
Notes: White blight of sweet pea (Ingham. C.M.I. Descr. 869. 1986).
Host: Lathyrus
Ramularia albomaculata Peck
Distr.: NY.
Host: Carya
Ramularia alismatis Fautrey
Didymaria alismatis (Oudem.) J.J. Davis
Septoria alismatis Oudem.
Distr.: Reported from an unknown locality in the United States; Europe.
Hosts: Alisma, Sagittaria
Ramularia alnicola Cooke
Distr.: ID; Europe(type).
Host: Alnus
Ramularia anagallidis Lindr.
Distr.: WI; Europe(type).
Host: Veronica
Ramularia andromedae Ellis & G. Martin
Distr.: NJ.
Hosts: Leucothoe, Lyonia
Ramularia angustata Peck
Distr.: Eastern United States.
Host: Rhododendron
Ramularia anomala Peck
Distr.: NE(type), eastern United States.
Notes: Leaf spot (Wilson. C.M.I. Descr. 858. 1986).
Hosts: Fagopyrum, Polygonum
Ramularia aquatilis Peck
Doassansiopsis occulta (H. Hoffm.) Dietel (teleomorph)
Distr.: North America; Europe.
Host: Potamogeton
Ramularia archangelicae Lindr.
Ramularia angelicae Höhn.
Ramularia grantii Dearn.
Distr.: Temperate North America; Europe.
Notes: Savile (Canad. J. Bot. 43:574. 1965).
Host: Angelica
Ramularia arisaematis Ellis & Dearn.
Distr.: WI; Canada(Ontario, type).
Host: Arisaema
Ramularia artemisiae J.J. Davis
Distr.: NY, WI(type).
Host: Artemisia
Ramularia arvensis Sacc.
Distr.: Temperate North America; Europe; northern Africa.
Notes: Leaf spot.
Hosts: Horkelia, Potentilla
Ramularia asteris (W. Phillips & Plowr.) Bubák
Fusidium asteris W. Phillips & Plowr.
Distr.: Temperate North America; Europe.
Notes: Leaf spot.
Host: Aster
Ramularia astragali Ellis & Holw.
Didymaria astragali (Ellis & Holw.) Sacc.
Distr.: Midwestern United States.
Notes: Leaf spot (Ingham. C.M.I. Descr. 861. 1986).
Host: Astragalus

Ramularia barbareae **Peck**
Distr.: NY(type), temperate North America; Europe.
Host: Barbarea
Ramularia beticola **Fautrey & Lambotte**
Distr.: Temperate North America; Europe.
Host: Beta
Ramularia bistortae **Fuckel**
Ovularia bistortae (Fuckel) Sacc.
Distr.: WY; Europe.
Notes: Wilson (C.M.I. Descr. 854. 1986).
Host: Polygonum
Ramularia brevipes **Ellis & Everh.**
Distr.: AL(type), southeastern United States.
Host: Monarda
Ramularia brunellae **Ellis & Everh.**
Distr.: Central and eastern United States.
Host: Prunella
Ramularia brunnea **Peck**
Ramularia tulasnei Sacc.
Mycosphaerella fragariae (Tul.) Lindau (teleomorph)
Distr.: Cosmopolitan.
Notes: Leaf spot or leaf blight of strawberries. Sivanesan & Holliday (C.M.I. Descr. 708. 1981).
Hosts: Fragaria, Tussilago
Ramularia canadensis **Ellis & Everh.**
Septocylindrium canadense (Ellis & Everh.) Pound & Clements
Distr.: Central United States; Canada(Ontario, type).
Host: Carex
Ramularia carniformis **(Ellis & Tracy) Sherb.**
Fusarium carniformis Ellis & Tracy
Distr.: MS.
Host: Tripsacum
Ramularia castillejae **Ellis & Everh.**
Ramularia coccinea Dearn. & Bisby
Distr.: Western North America.
Notes: Savile (Canad. J. Bot. 46:465. 1968).
Host: Castilleja
Ramularia celastri **Ellis & G. Martin**
Distr.: PA(type), central and eastern United States.
Host: Celastrus
Ramularia celtidis **Ellis & Kellerm.**
Cercosporella celtidis (Ellis & Kellerm.) J.J. Davis
Distr.: Central and eastern United States.
Host: Celtis
Ramularia centranthi **Brunaud**
Distr.: North America; France(type); temperate Europe.
Hosts: Centranthus, Valeriana
Ramularia chimaphilae **H.C. Greene**
Distr.: WI(type).
Host: Chimaphila
Ramularia chrysopsidis **Dearn.**
Distr.: NY.
Host: Chrysopsis
Ramularia cichorii **Dearn. & House**
Distr.: NY.
Host: Cichorium
Ramularia cilinodis **J.J. Davis**
Distr.: Central and eastern United States.
Host: Polygonum
Ramularia claytoniae **W.B. Cooke**
Distr.: CA.
Hosts: Claytonia, Montia
Ramularia clematidis **Dearn. & Barth.**
Distr.: MT(type), northern United States.
Host: Clematis
Ramularia coalescens **(J.J. Davis) Pirozynski**
Cercospora coalescens J.J. Davis
Distr.: Western North America.
Notes: On living leaves (Pirozynski. Fungi Canadenses 23. 1974).
Host: Ribes
Ramularia collo-cygni **Sutton & Waller**
Ophiocladium hordei Cavara
Ovularia hordei (Cavara) R. Sprague
Distr.: Temperate North America; Mexico; Europe.
Notes: Sutton & Waller (Trans. Brit. Mycol. Soc. 90:57. 1988).
Hosts: Bromus, Festuca, Glyceria, Leucopoa, Lolium, Phalaris
Ramularia crepidis **Ellis & Everh.**
Distr.: NM.
Host: Crepis
Ramularia crypta **Cooke**
Distr.: SC(type).
Host: Quercus
Ramularia cyclamanicola **Trel.**
Cladosporium cyclaminis L. Massey & Tilford
Distr.: North America(greenhouses).
Notes: Cyclamen stunt disease.
Host: Cyclamen
Ramularia cyclaminicola **Trel.**
Distr.: Central and eastern United States.
Host: Cyclamen
Ramularia cylindriopsis **Peck**
Distr.: NY.
Host: Lyonia
Ramularia cynarae **Sacc.**
Distr.: CA; southern Europe.
Hosts: Cirsium, Cynara
Ramularia decipiens **Ellis & Everh.**
Distr.: Central United States; Europe; Asia; Africa.
Notes: Leaf spot (Wilson. C.M.I. Descr. 852. 1986).
Host: Rumex
Ramularia delphinii **Jaap**
Distr.: Temperate North America; Europe.
Host: Delphinium
Ramularia desmodii **Cooke**
Distr.: SC(type), central and eastern United States.
Host: Desmodium
Ramularia destructiva **W. Phillips & Plowr.**
Ovularia destructiva (W. Phillips & Plowr.) Massee
Distr.: Eastern North America; Europe.
Hosts: Comptonia, Myrica
Ramularia destruens **Peck**
Distr.: NY(Adirondack Mountains).
Host: Sorbus
Ramularia deusta **(Fuckel) Karakulin**
Ovularia deusta (Fuckel) Sacc.
Ramularia lathyri W.B. Cooke & C.G. Shaw
Ramularia roseola Bubák & Vleugel
Melanospora multiformis (G.H. Martin & V. Charles) Arx & E. Müller (teleomorph)
Distr.: Temperate North America; Europe.
Notes: Baker et al. (Mycologia 42:403. 1950) questioned the teleomorph connection. Ingham (C.M.I. Descr. 868. 1986).
Host: Lathyrus
Ramularia deusta **(Fuckel) K. Baker, W.C. Snyder, & W.H. Davis f. *latifolii* K. Baker, W.C. Snyder, & W.H. Davis, nom. inval.**
Distr.: CA(type).
Notes: Published without a latin diagnosis.
Host: Lathyrus
Ramularia didyma **Unger**
Didymaria didyma (Unger) J. Schröt.
Cercospora calthae Cooke
Ramularia calthae Lindr.
Ramularia ranunculi Peck
Distr.: Temperate regions.
Notes: Chupp (A monograph of the fungus genus *Cercospora*. p. 462. 1954).
Hosts: Anemone, Caltha, Ranunculus
Ramularia diervillae **Peck**
Distr.: Northeastern North America.
Hosts: Diervilla, Weigela
Ramularia dioscoreae **Ellis & Everh.**
Distr.: WI.
Host: Dioscorea
Ramularia dispar **J.J. Davis**
Distr.: Central and eastern United States.
Host: Eupatorium
Ramularia dubia **Riess**
Cercospora dubia (Riess) G. Wint. non Speg., hom. illeg.
Distr.: Temperate northern hemisphere.
Notes: Leaf spot of Chenopodiaceae.

Hosts: Atriplex, Chenopodium
Ramularia eamesii Dearn. & House
Distr.: NY.
Host: Valeriana
Ramularia effusa Peck
Distr.: NY(type), eastern United States.
Hosts: Gaylussacia, Vaccinium
Ramularia equiseti C. Massal.
Distr.: WI; Italy(type).
Host: Equisetum
Ramularia euonymi Ellis & Kellerm.
Septocylindrium euonymi (Ellis & Kellerm.) Pound & Clements
Distr.: United States.
Host: Euonymus
Ramularia filaris Fresen.
Distr.: Temperate North America; Europe.
Notes: On Asteraceae.
Hosts: Aster, Senecio
Ramularia fragariae Peck
Distr.: NY(type).
Notes: Leaf spot of strawberry.
Host: Fragaria
Ramularia fraxinea J.J. Davis
Distr.: FL, WI(type).
Host: Fraxinus
Ramularia gei (Eliasson) Lindr.
Ovularia gei Eliasson
Cercospora gei Bubák non Fuckel, hom. illeg.
Cercospora gei Fuckel
Ramularia gei (Eliasson) Höhn.
Distr.: Central United States; temperate Europe.
Host: Geum
Ramularia geranii (Westend.) Fuckel
Distr.: Southern and western United States; Europe.
Host: Geranium
Ramularia giliae R. Sprague
Distr.: OR.
Hosts: Linanthus, Microsteris, Phlox
Ramularia glauca Ellis & Everh.
Distr.: Western United States.
Host: Sambucus
Ramularia gossypii (Speg.) Cif.
Cercosporella gossypii Speg.
Septocylindrium gossypii (Speg.) Subramanian
Ramularia areola Atk.
Septocylindrium areola (Atk.) Pound & Clements
Symphyosira areola (Atk.) Sawada
Mycosphaerella areola J. Ehrlich & F.A. Wolf (teleomorph)
Distr.: Worldwide in cotton-growing regions.
Notes: Areolate mildew of cotton (Mulder & Holliday. C.M.I. Descr. 520. 1976).
Hosts: Gossypium, Sida
Ramularia gracilipes J.J. Davis
Distr.: WI(type).
Host: Cornus
Ramularia graminicola Peck
Distr.: NY, WI.
Hosts: Agrostis, Poa
Ramularia grindeliae Ellis & Kellerm.
Distr.: Central and western United States.
Host: Grindelia
Ramularia hedericola Heald & F.A. Wolf
Distr.: TX(type).
Host: Hedera
Ramularia helianthi Ellis & Everh.
Distr.: Western United States.
Hosts: Helianthella, Helianthus
Ramularia heraclei (Oudem.) Sacc.
Distr.: Temperate North America; Europe.
Notes: Leaf spot.
Hosts: Heracleum, Lomatium, Oxypolis
Ramularia hypochoeridis Magnusson
Distr.: WA; Germany(type).
Host: Hypochaeris
Ramularia impatientis Peck
Distr.: NY(type), central and eastern United States.
Host: Impatiens
Ramularia ionophila J.J. Davis
Distr.: Central and western United States.
Host: Viola
Ramularia ipomoeae F. Stevens
Distr.: HI(type).
Host: Ipomoea
Ramularia ivae Dearn. & Barth. in Dearn.
Distr.: WY(type).
Host: Iva
Ramularia junci Peck
Distr.: NY.
Host: Juncus
Ramularia lactea (Desmaz.) Sacc.
Fusisporium lacteum Desmaz.
Ramularia violae Fuckel
Distr.: Temperate North America; Europe.
Notes: Leaf spot.
Host: Viola
Ramularia lanceolata Dearn. & House
Distr.: CA, MA, NY(type).
Host: Plantago
Ramularia lappulae (J.J. Davis) J.J. Davis
Ovularia asperifolia Sacc. var. *lappulae* J.J. Davis
Distr.: TX, WI(type).
Hosts: Cynoglossum, Hackelia
Ramularia lethalis Ellis & Everh.
Distr.: Northeastern North America.
Host: Acer
Ramularia lineola Peck
Distr.: NY(type), central and eastern United States.
Host: Taraxacum
Ramularia liriodendri Ellis & Everh.
Distr.: AL, DE(type).
Host: Liriodendron
Ramularia lonicerae Voglino
Distr.: WY; Italy(type).
Host: Lonicera
Ramularia lophanthi Ellis & Everh.
Distr.: Western United States.
Host: Agastache
Ramularia lucidae J.J. Davis
Distr.: WI(type).
Host: Salix
Ramularia lupini J.J. Davis
Distr.: Western United States.
Host: Lupinus
Ramularia lysimachiae Thuem.
Distr.: Temperate North America; Europe.
Host: Lysimachia
Ramularia macrospora Fresen.
Distr.: North America(questionable); temperate Europe.
Notes: *Ramularia macrospora* Fresen. in the strict sense appears to be restricted to *Campanula* spp.
Hosts: Aster, Campanula, Conyza, Erigeron, Hieracium, Malus, Rosa, Rumex
Ramularia macrospora Fresen. var. ***asteris*** Trel.
Distr.: IA, WI(type).
Host: Aster
Ramularia meliloti Ellis & Everh.
Distr.: CA; Great Britain.
Host: Melilotus
Ramularia menthicola Sacc.
Ramularia menthae Sacc. non Thuem., hom. illeg.
Distr.: North America; temperate Europe.
Host: Mentha
Ramularia microlepiae F. Stevens
Distr.: HI.
Host: Microlepia
Ramularia mimuli Ellis & Everh.
Distr.: Temperate United States.
Host: Mimulus
Ramularia mitellae Peck
Distr.: Central and eastern United States.

Hosts: Heuchera, Mitella
Ramularia mitellae Peck var. ***heucherae*** Dearn.
Distr.: SD, WA.
Host: Heuchera
Ramularia momordicae Heald & F.A. Wolf
Distr.: TX(type).
Host: Momordica
Ramularia monilioides (Ellis & G. Martin) Ellis & Everh.
Ovularia monilioides Ellis & G. Martin
Distr.: MA(type), NY.
Notes: Sometimes listed as a synonym of *Ovularia destructiva* (Phillip & Plowr.) Masse (Cash. U.S.D.A. Spec. Publ. 2:350. 1954).
Host: Myrica
Ramularia montana Speg.
Cercospora montana (Speg.) Sacc.
Ramularia artenii Sacc.
Ramularia cercosporoides Ellis & Everh.
Cercospora epilobii A. Schneid. in Thuem.
Ramularia karstenii Sacc.
Ramularia punctiformis (Schlechtend.) Höhn. non Sacc., hom. illeg.
Distr.: Temperate regions.
Notes: Savile (Canad. J. Bot. 40:1395. 1962).
Hosts: Boisduvalia, Epilobium, Potentilla
Ramularia multiplex Peck
Distr.: Central and eastern United States.
Host: Vaccinium
Ramularia necator Massee
Distr.: FL; Jamaica(type).
Host: Cocos
Ramularia nemopanthis G.W. Clinton & Peck in Peck
Distr.: NY(type).
Host: Nemopanthus
Ramularia nephrolepis F. Stevens
Distr.: HI(type).
Host: Nephrolepis
Ramularia nivosa (Ellis & Everh.) W.B. Cooke & C.G. Shaw
Cercosporella nivosa Ellis & Everh.
Distr.: Pacific Northwest.
Notes: Cooke & Shaw (Lloydia 15:127. 1952); Savile (Canad. J. Bot. 46: 467. 1968).
Host: Penstemon
Ramularia nymphaearum (Allesch.) Ramsb.
Gloeosporium nymphaearum Allesch.
Ovularia nymphaearum (Allesch.) Allesch.
Ovulariella nymphaearum (Allesch.) Bubák & Kab.
Distr.: North America; Europe; India.
Host: Nymphaea
Ramularia obducens Thuem.
Distr.: Western United States; Europe.
Host: Pedicularis
Ramularia occidentalis Ellis & Kellerm.
Distr.: Central and eastern United States.
Host: Rumex
Ramularia orogeniae W.B. Cooke & C.G. Shaw
Distr.: WA(type).
Host: Orogenia
Ramularia orontii Ellis & G. Martin
Distr.: NJ(type).
Host: Orontium
Ramularia ovularioides H.C. Greene
Distr.: WI(type).
Host: Anemone
Ramularia oxalidis Farl.
Distr.: Temperate North America.
Host: Oxalis
Ramularia parietariae Pass.
Distr.: North America(questionable); temperate Europe.
Host: Parietaria
Ramularia parva H.C. Greene
Distr.: WI(type).
Host: Hieracium
Ramularia pentstemonis W.B. Cooke
Distr.: CA(type).
Host: Penstemon
Ramularia phaceliae Bonar
Distr.: CA(type).
Host: Phacelia
Ramularia philadelphi Sacc.
Distr.: Temperate United States; Europe.
Host: Philadelphus
Ramularia polemoniae W.B. Cooke & C.G. Shaw
Distr.: WA.
Host: Polemonium
Ramularia pratensis Sacc.
Distr.: North America; Europe.
Notes: Wilson (C.M.I. Descr. 853. 1986).
Host: Rumex
Ramularia primulae Thuem.
Cercospora primulae Fautrey
Cercosporella primulae Allesch.
Ovularia primulara (P. Karst.) Sacc.
Ramularia primulara P. Karst.
Distr.: North America; Europe; New Zealand.
Host: Primula
Ramularia prini Peck
Distr.: Central and eastern United States.
Host: Ilex
Ramularia pruinosa Speg.
Distr.: North America; Europe.
Host: Senecio
Ramularia psoraleae Ellis & Everh.
Distr.: Central and western United States.
Notes: Ingham (C.M.I. Descr. 867. 1986).
Hosts: Psoralea, Psoralidium
Ramularia punctiformis Sacc.
Distr.: AK.
Hosts: Epilobium, Potentilla
Ramularia pusilla Unger
Ovularia pusilla (Unger) Sacc.
Ovularia pulchella (Ces.) Sacc.
Ramularia pulchella Ces.
Distr.: Temperate North America; Europe.
Notes: Sutton & Waller (Trans. Brit. Mycol. Soc. 90:55. 1988); Savile (Canad. J. Bot. 35:203. 1957).
Hosts: Agropyron, Agrostis, Alopecurus, Arctagrostis, Arrhenatherum, Bromus, Calamagrostis, Deschampsia, Elymus, Festuca, Glyceria, Hierochloe, Holcus, Hordeum, Leucopoa, Lolium, Melica, Muhlenbergia, Phalaris, Phleum, Poa, Puccinellia, Trisetum, Vulpia
Ramularia ranunculi-lyallii Dearn. & Barth.
Distr.: Northwestern North America; Canada(British Columbia, type).
Host: Ranunculus
Ramularia repens Ellis & Everh.
Distr.: Northcentral and northeastern North America.
Host: Aralia
Ramularia reticulata Ellis & Everh.
Distr.: Temperate North America.
Hosts: Ligusticum, Lomatium, Osmorhiza
Ramularia rhabdospora (Berk. & Broome) Nannf. in S. Lundell & Nannf.
Cylindrosporium rhabdosporum Berk. & Broome
Ramularia peckii Sacc.
Ramularia plantaginis Ellis & G. Martin
Distr.: Temperate North America; Europe.
Host: Plantago
Ramularia rhei Allesch.
Distr.: North America; temperate Europe.
Host: Rheum
Ramularia rigidula (Delacr.) Nannf. in S. Lundell & Nannf.
Ovularia rigidula Delacr.
Distr.: Central North America; central Europe; Soviet Union.
Notes: Leaf spot (Wilson. C.M.I. Descr. 855. 1986).
Host: Polygonum
Ramularia rosea (Fuckel) Sacc.
Fusidium roseum Fuckel
Distr.: Temperate North America; Europe.
Host: Salix
Ramularia rubella (Bonord.) Nannf. in S. Lundell & Nannf.
Ovularia rubella (Bonord.) Sacc.
Ovularia canaegricola (Henn.) Henn.

Ramularia circumfusa Ellis & Everh.
Ovularia monosporia (Westend.) Pound & Clements
Oidium monosporium Westend.
Ovularia obliqua (Cooke) Oudem.
Ramularia obliqua (Cooke) Oudem.
Ovularia obovata (Fuckel) Sacc.
Ramularia obovata Fuckel
Distr.: Cosmopolitan.
Notes: Savile (Canad. J. Bot. 35:203. 1957); Wilson (C.M.I. Descr. 851. 1986).
Host: Rumex
Ramularia rudbeckiae Peck
Distr.: Temperate North America.
Hosts: Ratibida, Rudbeckia
Ramularia rufomaculans Peck
Septocylindrium rufomaculans (Peck) Pound & Clements
Distr.: Temperate North America.
Notes: Leaf spot (Wilson. C.M.I. Descr. 856. 1986).
Host: Polygonum
Ramularia rumicis Kalchbr. & Cooke
Distr.: AK; South Africa(type).
Host: Rumex
Ramularia salviicola Tharp
Distr.: TX.
Host: Salvia
Ramularia sambucina Sacc.
Distr.: Temperate North America; Europe.
Host: Sambucus
Ramularia saximontanensis Solheim
Distr.: WY.
Notes: Solheim (Univ. Wyoming Publ. 10:45. 1943).
Host: Clematis
Ramularia schulzeri Bäumler
Distr.: NE, NY, TX; Hungary(type).
Notes: Reports in the United States are questionable.
Host: Lotus
Ramularia senecionis (Berk. & Broome) Sacc.
Cylindrosporium senecionis Berk. & Broome
Distr.: Temperate North America; Europe.
Host: Senecio
Ramularia senecionis (Berk. & Broome) Sacc. var. ***carniolica*** Jaap
Distr.: CA; Austria(type).
Host: Senecio
Ramularia serotina Ellis & Everh.
Distr.: Temperate North America.
Hosts: Euthamia, Solidago
Ramularia sheldonii Trott.
Ramularia delphinii Dearn. & House non Jaap, hom. illeg.
Distr.: Western United States.
Host: Delphinium
Ramularia sidalceae Ellis & Everh.
Distr.: Western North America; Canada(British Columbia).
Hosts: Iliamna, Sidalcea
Ramularia silvestris Sacc.
Distr.: UT; Europe.
Host: Dipsacus
Ramularia smilacinae J.J. Davis
Distr.: Western North America.
Notes: Savile (Mycologia 53:50. 1961).
Host: Smilacina
Ramularia solani Sherb.
Distr.: NY(type).
Host: Solanum
Ramularia sphaerioides Ellis & Everh.
Distr.: WA(type).
Host: Thermopsis
Ramularia spiraeae Peck
Distr.: Northeastern United States.
Host: Physocarpus
Ramularia stachydis (Pass.) C. Massal.
Fusidium stachydis Pass.
Distr.: North America; central and southern Europe.
Host: Stachys
Ramularia stolonifera Ellis & Everh.
Septocylindrium stolonifera (Ellis & Everh.) Pound & Clements
Distr.: WA, northeastern North America.
Host: Cornus
Ramularia subrufa Ellis & Holw.
Distr.: IA, eastern United States.
Host: Smilax
Ramularia subsanguinea (Ellis & Everh.) Karakulin
Cercospora subsanguinea Ellis & Everh.
Ramularia subsanguinea (Ellis & Everh.) Savile non (Ellis & Everh.) Karakulin, hom. illeg.
Ramularia rubicunda Bres.
Distr.: Temperate North America.
Notes: Savile (Canad. J. Bot. 35:205. 1957).
Host: Maianthemum
Ramularia synthyridis W.B. Cooke in W.B. Cooke & C.G. Shaw
Distr.: WA(type).
Host: Synthyris
Ramularia tanaceti Lind
Distr.: North America(questionable); temperate Europe.
Host: Tanacetum
Ramularia taraxaci P. Karst.
Distr.: North America; temperate Europe; New Zealand.
Hosts: Hieracium, Taraxacum
Ramularia tenuis J.J. Davis
Distr.: Central United States.
Notes: Deighton (Trans. Brit. Mycol. Soc. 61:117. 1973).
Host: Solidago
Ramularia trautvetteriae C.G. Shaw & R. Sprague
Distr.: Pacific Northwest.
Host: Trautvetteria
Ramularia trifolii Jaap
Distr.: WY; Europe(type).
Notes: Ingham (C.M.I. Descr. 866. 1986).
Host: Trifolium
Ramularia trillii W.B. Cooke & C.G. Shaw
Distr.: ID.
Host: Trillium
Ramularia ulmariae Cooke
Distr.: Temperate North America; Europe.
Hosts: Aruncus, Spiraea
Ramularia umbrina J.J. Davis
Distr.: WI.
Host: Diervilla
Ramularia urticae Ces.
Septocylindrium urticae (Ces.) Subramanian
Mycosphaerella superflua (Auersw.) Petr. (teleomorph)
Distr.: Temperate North America; Europe; India.
Notes: On leaves.
Host: Urtica
Ramularia vaccinii Peck
Distr.: Eastern North America.
Host: Vaccinium
Ramularia vagnerae Barth.
Distr.: CA, CO.
Host: Smilacina
Ramularia valerianae (Speg.) Sacc.
Cylindrosporium valerianae Speg.
Distr.: North America; Europe.
Host: Valeriana
Ramularia vallisumbrosae Cavara
Distr.: Western North America(coastal); Europe.
Notes: White mold (Gregory. Trans. Brit. Mycol. Soc. 23:24. 1939).
Host: Narcissus
Ramularia vancouveriae (Ellis & Everh.) R. Sprague
Ovularia vancouveriae Ellis & Everh.
Distr.: Pacific Northwest.
Host: Vancouveria
Ramularia variabilis Fuckel
Distr.: North America; Europe.
Notes: Savile (Canad. J. Bot. 46:467. 1968) provided comments on North American collections.
Hosts: Digitalis, Verbascum
Ramularia variata J.J. Davis
Distr.: WI.
Hosts: Mentha, Monarda

Ramularia variegata Ellis & Holw.
Distr.: MN.
Hosts: Luina, Petasites
Ramularia veronicae Fuckel
Ovularia veronicae (Fuckel) Sacc.
Distr.: Temperate North America; Europe.
Notes: Savile (Canad. J. Bot. 46:467. 1968) provided a complete synonymy and comments on North American specimens.
Hosts: Veronica, Veronicastrum
Ramularia viburni Ellis & Everh.
Distr.: WI(type), eastern United States.
Host: Viburnum
Ramularia waldsteiniae Ellis & J.J. Davis
Distr.: WI(type).
Host: Waldsteinia
Ramularia wisconsina H.C. Greene
Distr.: WI(type).
Host: Erechtites

Ramulispora Miura. Hyphomycetes

Notes: Rawla (Trans. Brit. Mycol. Soc. 60:283. 1973).
Ramulispora bromi (R. Sprague) R. Sprague
Cercospora bromi R. Sprague
Ansatospora bromi (R. Sprague) R. Sprague
Centrospora bromi (R. Sprague) A.G. Newhall
Cercosporina bromi (R. Sprague) R. Sprague
Distr.: Pacific Northwest.
Host: Bromus
Ramulispora sorghi (Ellis & Everh.) Olive & Lefebvre in Olive et al.
Septorella sorghi Ellis & Everh.
Distr.: Worldwide.
Notes: Sooty blotch or leaf stripe (Anahosur. C.M.I. Descr. 585. 1978).
Hosts: [Agrostis], Sorghum

Rasutoria Barr. Loculoascomycetes, Dothideales

Notes: Barr (Mycotaxon 29:501. 1987).
Rasutoria abietis (Dearn.) Barr
Dimerosporium abietis Dearn.
Epipolaeum abietis (Dearn.) Shoemaker
Distr.: Northern North America.
Notes: Shoemaker (Canad. J. Bot. 43:631. 1965 as *E. abietis*).
Host: Abies
Rasutoria pseudotsugae (V. Miller & Bonar) Barr
Dimeriella pseudotsugae V. Miller & Bonar
Epipolaeum pseudotsugae (V. Miller & Bonar) Shoemaker
Distr.: Northern North America.
Notes: Shoemaker (Canad. J. Bot. 43:631. 1965 as *E. pseudotsugae*).
Host: Pseudotsuga
Rasutoria tsugae (Dearn.) Barr
Dimerosporium tsugae Dearn.
Dimeriella tsugae (Dearn.) Petr.
Epipolaeum tsugae (Dearn.) Shoemaker
Eudimeriolum tsugae (Dearn.) M.L. Farr
Epipolaeum terrieri (Petr.) Shoemaker
Distr.: North America.
Notes: On conifers. Farr (Mycologia 76:801. 1984) reviewed the synonymy of this species. She considerd the four species in *Rasutoria* to be one species, *Eudimeriolum tsugae*.
Hosts: Abies, Tsuga

Ravenelia Berk. Basidiomycetes, Uredinales

Notes: Cummins (Rust fungi on legumes and composites in North America. 1978).
Ravenelia annulata W.H. Long
Distr.: FL.
Notes: Autoecious.
Host: Lysiloma
Ravenelia arizonica Ellis & Everh.
Distr.: Southwestern United States; Mexico.
Notes: Autoecious.
Host: Prosopis
Ravenelia australis Dietel & Neger
Distr.: TX; Mexico; South America.
Notes: Autoecious.
Host: Acacia
Ravenelia cassiaecola Atk.
Distr.: Eastern United States; West Indies.
Notes: Autoecious.
Host: Cassia
Ravenelia epiphylla (Schwein.) Dietel
Sphaeria epiphylla Schwein.
Distr.: Central and eastern United States; Mexico; El Salvador.
Notes: Autoecious.
Host: Tephrosia
Ravenelia expansa Dietel & Holw.
Ravenelia igualica Arth.
Distr.: TX; Mexico; Guatemala.
Notes: Autoecious.
Host: Acacia
Ravenelia floridana Cummins & J.W. Baxter
Distr.: FL.
Notes: Autoecious. Reported as *R. pithecolobii* Arth. in the older literature (Cummins & Baxter. Mycotaxon 2:199. 1975).
Host: Pithecellobium
Ravenelia fragrans W.H. Long
Distr.: Southwestern United States; Mexico.
Notes: Autoecious.
Host: Mimosa
Ravenelia fragrans W.H. Long var. **evernia** (Syd.) J.W. Baxter
Ravenelia evernia Syd. in Syd. & Mitter
Ravenelia dysocarpae W.H. Long & Goodd.
Distr.: Southwestern United States; Mexico; India.
Notes: Autoecious.
Host: Mimosa
Ravenelia gracilis Arth.
Distr.: TX; Mexico.
Notes: Autoecious.
Host: Pithecellobium
Ravenelia hieronymii Speg.
Distr.: TX; South America.
Notes: According to Cummins (Rust fungi on legumes and composites in North America. p. 340. 1978), this fungus does not occur in North America. The aecial stage of *Spumula heteromorpha* J.W. Baxter may have been confused with this species.
Host: Acacia
Ravenelia holwayi Dietel
Distr.: Southwestern United States; Mexico.
Notes: Autoecious.
Host: Prosopis
Ravenelia humphreyana Henn.
Distr.: FL, TX; West Indies; Central America.
Notes: Autoecious.
Hosts: Caesalpinia, Delonix
Ravenelia indigoferae Tranzschel in Dietel
Distr.: AZ; Central and South America; Africa; China; The Philippines.
Notes: Autoecious.
Host: Indigofera
Ravenelia laevis Dietel & Holw. in Holw.
Distr.: TX; Mexico.
Notes: Autoecious.
Host: Indigofera
Ravenelia leucaenae W.H. Long
Distr.: TX; Mexico.
Notes: Autoecious.
Host: Leucaena
Ravenelia lysilomae Arth.
Distr.: FL; Central America.
Host: Lysiloma
Ravenelia mesilliana Ellis & Barth. in Ellis & Everh.
Ravenelia cassiae-covesii W.H. Long & Goodd. in W.H. Long
Distr.: Southwestern United States; Mexico.
Notes: Autoecious.
Hosts: Cassia, Senna
Ravenelia opaca Dietel

Ravenelia indica Berk. f. *opaca* Seym. & Earle, nom. nud.
Distr.: IL.
Notes: Autoecious.
Host: Gleditsia
Ravenelia piscidiae W.H. Long
Ravenelia piscidiae W.H. Long subsp. *rugosa* J. Hennen & Cummins
Distr.: FL; Mexico; Cuba.
Notes: Autoecious.
Host: Piscidia
Ravenelia pithecolobii Arth.
Distr.: FL(questionable); Mexico; West Indies.
Notes: Autoecious.
Host: Pithecellobium
Ravenelia pringlei Cummins
Ravenelia versatilis Auct. non Dietel
Distr.: Southwestern United States; Mexico.
Notes: Autoecious.
Host: Acacia
Ravenelia roemerianae W.H. Long
Distr.: TX; Mexico.
Notes: Autoecious.
Host: Acacia
Ravenelia siderocarpi W.H. Long
Distr.: TX.
Notes: Autoecious.
Host: Pithecellobium
Ravenelia spegazziniana Lindquist
Ravenelia siliquae W.H. Long
Distr.: HI, TX; West Indies; Central and South America.
Notes: Autoecious. Only urediniospores known from North America and Hawaii.
Host: Acacia
Ravenelia spinulosa Dietel & Holw. in Holw. var. ***papillifera*** (Syd. & P. Syd.) Cummins & J.W. Baxter
Ravenelia papillifera Syd. & P. Syd.
Distr.: Southwestern United States; Bahamas; Mexico.
Notes: Autoecious.
Host: Senna
Ravenelia subtortuosae W.H. Long
Distr.: TX.
Notes: Autoecious.
Host: Acacia
Ravenelia texensis Dietel
Ravenelia gooddingii W.H. Long
Ravenelia reticulata W.H. Long
Ravenelia texanus Ellis & B.T. Galloway in H.S. Jennings
Distr.: Southwestern United States; Mexico.
Notes: Autoecious; on Fabaceae.
Hosts: Acacia, Calliandra, Desmanthus
Ravenelia texensis Dietel in Engl. & Prantl var. ***morongiae*** (W.H. Long) Cummins
Ravenelia morongiae W.H. Long
Distr.: TX, Mexico.
Notes: Autoecious.
Host: Schrankia
Ravenelia thornberiana W.H. Long
Distr.: Southwestern United States; Mexico.
Notes: Autoecious.
Host: Acacia

Rebentischia P. Karst. Loculoascomycetes, Dothideales

Notes: Barr (Mycotaxon 12:141. 1980).
Rebentischia massalongii (Mont.) Sacc.
Rebentischia pomiformis P. Karst.
Distr.: Northeastern North America; Europe.
Notes: On branches and trunks of various woody plants.
Hosts: Abies, Acer, Ilex, Robinia, Rubus, Ulmus

Rectipilus Agerer. Basidiomycetes, Aphyllophorales

Notes: Agerer (Persoonia 7:413. 1973).
Rectipilus fasciculatus (Pers.:Fr.) Agerer
Solenia fasciculata Pers.:Fr.
Distr.: Northern hemisphere.
Hosts: Alnus, Betula, Ostrya, Pinus, Quercus, Sabal
Rectipilus idahoensis (W.B. Cooke) Agerer
Solenia idahoensis W.B. Cooke
Distr.: ID.
Host: Pinus

Redbia Deighton & Pirozynski. Hyphomycetes

Redbia elegans Pirozynski & C.S. Hodges
Distr.: SC(type).
Notes: Pirozynski & Hodges (Canad. J. Bot. 51:166. 1973).
Host: Persea

Reessia C. Fisch. Hyphochytridiomycetes, Hyphochytriales

Notes: According to Sparrow (Aquatic Phycomycetes: p. 764. 1960), this is an imperfectly known genus of the Hyphochytriales.
Reessia amoeboides C. Fisch
Distr.: MN; Germany(type).
Notes: Parasitic on epidermal cells.
Host: Lemna

Repetobasidium J. Eriksson. Basidiomycetes, Aphyllophorales

Repetobasidium mirificum J. Eriksson
Distr.: Temperate northern hemisphere; Africa.
Notes: On wood.
Host: Abies

Requienella Fabre. Loculoascomycetes, Pyrenulales

Requienella seminuda (Pers.:Fr.) Boise
Trematosphaeria seminuda (Pers.:Fr.) Fuckel
Acrocordiella occulta (Romell) O. Eriksson
Hymenopleella occulta (Romell) Shoemaker & P.M. LeClair
Trematosphaeria subcollapsa (Ellis & Everh.) Ellis & Everh.
Distr.: Eastern North America; Europe.
Notes: On limbs and bark of hardwoods.
Hosts: Fraxinus, Nyssa, Quercus

Resinicium Parmasto. Basidiomycetes, Aphyllophorales

Resinicium bicolor (Albertini & Schwein.:Fr.) Parmasto
Odontia bicolor (Albertini & Schwein.:Fr.) Quél.
Hydnum balsameum Peck
Hydnum serratum Peck
Hydnum subtile Fr.:Fr.
Distr.: Temperate northern hemisphere; India; New Zealand.
Notes: On wood.
Hosts: Abies, Larix, Oxydendrum, Picea, Pinus, Populus, Pseudotsuga, Thuja, Tsuga
Resinicium chiricahuaensis R.L. Gilbertson & Budington
Distr.: Southwestern North America.
Notes: On wood of conifers (Gilbertson & Budington. Mycologia 62:673. 1970).
Hosts: Picea, Pinus, Pseudotsuga
Resinicium furfuraceum (Bres.) Parmasto
Corticium furfuraceum Bres.
Corticium subpallidulum Litsch.
Distr.: Eastern North America(rare), western North America(common); Europe.
Notes: On wood of conifers.
Hosts: Abies, Larix, Picea, Pinus, Pseudotsuga, Thuja, Tsuga

Resinomycena Redhead & Singer. Basidiomycetes, Agaricales

Notes: Redhead & Singer (Mycotaxon 13:150. 1981).
Resinomycena rhododendri (Peck) Redhead & Singer
Marasmius decurrens Peck non Mont., hom. illeg.
Distr.: Eastern North America.
Notes: On branches and debris of hardwoods.
Host: Quercus
Resinomycena saccharifera (Berk. & Broome) Redhead subsp. ***kalalochensis*** (A.H. Sm.) Redhead
Mycena kalalochensis A.H. Sm.
Resinomycena kalalochensis (A.H. Sm.) Redhead & Singer
Distr.: Northwestern North America.
Notes: On debris in wetlands (Redhead. Canad. J. Bot. 62:1850. 1984).
Host: Alnus

Resupinatus (Nees) S.F. Gray. Basidiomycetes, Agaricales

Resupinatus applicatus (Batsch:Fr.) S.F. Gray
Pleurotus applicatus (Batsch:Fr.) P. Kumm.
Distr.: Cosmopolitan.
Notes: On woody debris.
Hosts: Fouquieria, Fraxinus, Haplopappus, Heteromeles, Juniperus, Pinus, Platanus, Quercus

Reticularia Bull. Myxomycetes

Reticularia lycoperdon Bull.
Distr.: Cosmopolitan.
Host: Eucalyptus

Rhabdocline Syd. Discomycetes, Rhytismatales

Notes: Parker & Reid (Canad. J. Bot. 47:1533. 1969).
Rhabdocline parkeri M.A. Sherwood, J.K. Stone & Carroll
Meria parkeri M.A. Sherwood, J.K. Stone & Carroll (anamorph)
Distr.: OR.
Notes: On recently abscised needles of Douglas fir (Sherwood-Pike et al. Canad. J. Bot. 64:1851. 1986).
Host: Pseudotsuga
Rhabdocline pseudotsugae Syd.
Distr.: Northeastern and northwestern United States; Europe.
Notes: Needlecast in plantations (Millar & Minter. C.M.I. Descr. 651. 1980).
Host: Pseudotsuga
Rhabdocline pseudotsugae Syd. subsp. ***epiphylla*** A.K. Parker & J. Reid
Distr.: Pacific Northwest.
Notes: On leaves of douglas fir.
Host: Pseudotsuga
Rhabdocline weirii A.K. Parker & J. Reid
Rhabdogloeum pseudotsugae Syd. (anamorph)
Distr.: Northwestern North America.
Notes: Parker & Reid (Canad. J. Bot. 47:1541. 1969) cited evidence for the connection with *Rhabdogloeum pseudotsugae*.
Host: Pseudotsuga
Rhabdocline weirii A.K. Parker & J. Reid subsp. ***oblonga*** A.K. Parker & J. Reid
Distr.: Northeastern United States.
Notes: On one-year needles.
Host: Pseudotsuga
Rhabdocline weirii A.K. Parker & J. Reid subsp. ***obovata*** A.K. Parker & J. Reid
Distr.: Northwestern North America.
Notes: On two-year and older needles.
Host: Pseudotsuga

Rhabdogloeopsis Petr. Coelomycetes

Rhabdogloeopsis balsameae (J.J. Davis) Petr.
Gloeosporium balsameae J.J. Davis
Kabatiella balsameae (J.J. Davis) Arx
Rhabdogloeum abietinum Dearn.
Sarcotrochila balsameae (J.J. Davis) Korf (teleomorph)
Distr.: Northern North America.
Notes: On needles (Sutton. The Coelomycetes. p. 226. 1980; Nag Raj & Morgan-Jones. Canad. J. Bot. 51:565. 1973).
Host: Abies

Rhabdogloeum Syd. Coelomycetes

Rhabdogloeum pseudotsugae Syd.
Rhabdocline weirii A.K. Parker & J. Reid (teleomorph)
Distr.: MT; western Canada.
Notes: Nag Raj & Morgan-Jones (Canad. J. Bot. 51:565. 1973).
Host: Pseudotsuga

Rhabdospora (Durieu & Mont.) Mont., nom. cons. Coelomycetes

Rhabdospora allantoidea (Berk. & M.A. Curtis) Sacc.
Septoria allantoidea Berk. & M.A. Curtis
Distr.: PA.
Host: Medicago
Rhabdospora arctii Kauffm.
Distr.: MI.
Host: Arctium
Rhabdospora aristata Dearn. & Barth.
Distr.: MT.
Host: Heracleum
Rhabdospora avicenniae Kohlmeyer & E. Kohlmeyer
Distr.: Subtropical and tropical coastal regions of the Atlantic Ocean.
Notes: Kohlmeyer & Kohlmeyer (Marine mycology. p. 534. 1979).
Host: Avicennia
Rhabdospora baculum (W. Gerard in Cooke) Grove
Macrophoma baculum (W. Gerard in Cooke) Berl. & Voglino
Sphaeropsis baculum W. Gerard in Cooke
Distr.: NY(type).
Host: Catalpa
Rhabdospora breviuscula (Berk. & M.A. Curtis) Sacc.
Septoria breviuscula Berk. & M.A. Curtis
Distr.: SC.
Hosts: Linnaea, Robinia
Rhabdospora camptospora Sacc. & Scalia
Distr.: AK(type).
Host: Anemone
Rhabdospora clarkeana Sacc.
Distr.: NY.
Host: Aquilegia
Rhabdospora conii Lambotte & Fautrey
Distr.: WA; Europe.
Host: Ligusticum
Rhabdospora continua (Berk. & M.A. Curtis) Sacc.
Septoria continua Berk. & M.A. Curtis
Distr.: Eastern United States.
Host: Plantago
Rhabdospora ellisii Sacc. & Syd.
Rhabdospora artemisiae Ellis & Everh. non Trail, hom. illeg.
Distr.: CO.
Host: Artemisia
Rhabdospora fragariae Atk.
Distr.: Eastern United States.
Host: Fragaria
Rhabdospora galiorum (Ellis) Kuntze non (Ellis) Dearn. & House, hom. illeg.
Septoria galiorum Ellis
Distr.: Eastern United States.
Host: Galium
Rhabdospora gutierreziae Tracy & Earle in Greene
Distr.: CO.
Host: Gutierrezia
Rhabdospora hedeomina (Peck) Sacc.
Septoria hedeomina Peck
Distr.: United States.
Host: Hedeoma
Rhabdospora heraclei Earle

Distr.: NV.
Host: Heracleum
Rhabdospora hibiscicola (Schwein.) Starb.
Sphaeria hibisicola Schwein.
Distr.: Reported from an unknown locality in the United States.
Host: Hibiscus
Rhabdospora hyperici H.C. Greene
Distr.: WI.
Host: Hypericum
Rhabdospora ilicigena Fairm.
Distr.: NY.
Host: Ilex
Rhabdospora inaequalis Sacc. & Roum.
Distr.: AK; Europe.
Host: Sorbus
Rhabdospora interrupta (Berk. & M.A. Curtis) Sacc.
Septoria interrupta Berk. & M.A. Curtis
Distr.: AK.
Host: Viburnum
Rhabdospora jasmini (Schwein.) Cooke
Distr.: PA.
Host: Jasminum
Rhabdospora juglandis (Schwein.) Sacc.
Septoria juglandis Schwein. in Berk. & M.A. Curtis
Distr.: PA.
Host: Juglans
Rhabdospora junci (Desmaz.) Allesch.
Septoria junci Desmaz.
Distr.: ID; Europe.
Host: Juncus
Rhabdospora kalmiarum (Schwein.) Ellis & Everh.
Sphaeria kalmiarum Schwein.
Distr.: Eastern United States.
Notes: This species was first published as "*kalmiae*".
Host: Kalmia
Rhabdospora kellermanii Ellis & G. Martin
Distr.: OH.
Host: Mimulus
Rhabdospora lactucarum (Schwein.) Starb.
Sphaeria lactucarum Schwein.
Distr.: MT.
Host: Lactuca
Rhabdospora lonicerae (Cooke & Ellis) Sacc.
Cryptosporium lonicerae Cooke & Ellis
Distr.: NJ.
Host: Lonicera
Rhabdospora maculans Sacc. & Berl. in Sacc. non (Berk. & M.A. Curtis) Sacc., hom. illeg.
Distr.: SC; Europe.
Notes: The validity of this report is questionable.
Host: Alnus
Rhabdospora mirabilissima (Peck) Dearn. & House
Septoria mirabilissima Peck
Distr.: NY(type).
Host: Pinus
Rhabdospora mueggenburgii (Pirotta) Sacc.
Septoria mueggenburgii Pirotta
Distr.: LA; Europe.
Host: Vitis
Rhabdospora oxycocci Shear
Distr.: Eastern United States.
Host: Vaccinium
Rhabdospora physostegiae Peck
Distr.: NY.
Host: Dracocephalum
Rhabdospora pinea P. Karst.
Distr.: WI; Europe.
Host: Pinus
Rhabdospora pini (Berk. & M.A. Curtis) Sacc.
Cryptosporium pini Berk. & M.A. Curtis
Distr.: Reported from an unknown locality in the United States.
Host: Pinus
Rhabdospora pittospori F. Stevens & P.A. Young in F. Stevens
Distr.: HI.
Host: Pittosporum
Rhabdospora pleosporoides Sacc.
Distr.: Western United States; Europe.
Hosts: Heracleum, Lupinus
Rhabdospora pleosporoides Sacc. var. **drabae** Wehmeyer
Distr.: WY.
Host: Draba
Rhabdospora polygoni Dearn. & House
Distr.: NY(type).
Host: Polygonum
Rhabdospora ribicola (Berk. & M.A. Curtis) Sacc.
Septoria ribicola Berk. & M.A. Curtis
Distr.: WI.
Host: Ribes
Rhabdospora rudis P. Karst.
Distr.: OH; Brazil.
Host: Laburnum
Rhabdospora sabalensis Cooke
Distr.: Southern United States.
Host: Sabal
Rhabdospora saponariae Bres. & Sacc. in Berl. & Bres.
Distr.: NY; Europe.
Host: Saponaria
Rhabdospora solidaginis (Cooke & Ellis) Sacc.
Cryptosporium solidaginis Cooke & Ellis
Distr.: United States.
Hosts: Lactuca, Solidago
Rhabdospora sphaerelloides (Ellis & Kellerm.) Sacc. & Syd.
Septoria sphaerelloides Ellis & Kellerm.
Distr.: Central and western United States.
Host: Hypericum
Rhabdospora subgrisea Peck
Distr.: MI, NY(type).
Host: Solidago
Rhabdospora translucens Fairm.
Distr.: NY(type).
Host: Campsis
Rhabdospora verruciformis (Berk. & M.A. Curtis) Sacc.
Septoria verruciformis Berk. & M.A. Curtis
Distr.: AL.
Host: Cephalanthus
Rhabdospora xanthii Peck
Distr.: NY(type), TX.
Host: Xanthium

Rhamphospora D.D. Cunn. Basidiomycetes, Ustilaginales

Rhamphospora nymphaeae D.D. Cunn.
Entyloma nymphaeae (D.D. Cunn.) Setch.
Distr.: Eastern United States; Asia; Europe.
Hosts: Nuphar, Nymphaea

Rhinocladiella Nannf. in Melin & Nannf. Hyphomycetes

Notes: De Hoog (Stud. Mycol. 15. 1977).
Rhinocladiella atrovirens Nannf. in Melin & Nannf.
Exophiala sp. (synanamorph)
Distr.: Temperate regions.
Notes: Ellis (Dematiaceous Hyphomycetes. p. 248. 1971).
Host: Pinus

Rhinocladium Sacc. & Ém. Marchal. Hyphomycetes

Rhinocladium dingleyae S.J. Hughes
Distr.: WI; New Zealand(type).
Notes: On woody debris (Hughes. New Zealand J. Bot. 18:171. 1980).
Hosts: Populus, Ulmus

Rhizidiocystis Sideris. Chytridiomycetes, Chytridiales

Rhizidiocystis ananasi Sideris
Distr.: HI(type).
Host: Ananas

Rhizina Fr.:Fr. Discomycetes, Pezizales

Rhizina undulata Fr.:Fr.
Rhizina inflata (Schaeff.) Quél. non (Schaeff.) P. Karst., hom. illeg.
Rhizina inflata (Schaeff.) P. Karst.
Distr.: Temperate northern hemisphere to Africa.
Notes: On roots and debris of conifers. Often prevalent after fires (Gremmen. Eur. J. Forest Pathol. 1:1. 1971; Booth & Gibson. C.M.I. Descr. 324. 1972; Ginns. Fungi Canadenses 16. 1974).
Hosts: Abies, Larix, Picea, Pinus, Pseudotsuga, Tsuga

Rhizocalyx Petr. Discomycetes, Helotiales

Rhizocalyx abietis Petr.
Rhizothyrium abietis Naoumoff (anamorph)
Distr.: Temperate regions.
Notes: Smerlis (Canad. J. Bot. 45:1715. 1967).
Host: Abies

Rhizoctonia DC. Agonomycetes

Notes: Ogoshi et al. (J. Fac. Agric., Hokkaido Univ., Japan 61:244. 1983); Oniki et al. (Trans. Mycol. Soc. Japan 26:189. 1985); Tu & Kimbrough (Bot. Gaz. 139:454. 1978). Moore (Mycotaxon 29:91. 1987) has recently transferred many *Rhizoctonia* spp. to segregate genera; however, until more evidence is presented, these genera will not be accepted here.
Rhizoctonia borealis J. Curtis, nom. nud.
Distr.: WI(type).
Notes: This name was published without a latin description.
Hosts: Goodyera, Spiranthes
Rhizoctonia carotae Rader
Distr.: Central and eastern North America; Europe.
Notes: Crater rot of carrot in cold storage (Mordue. C.M.I. Descr. 408. 1974).
Host: Daucus
Rhizoctonia cerealis Van der Hoeven
Ceratorhiza cerealis (Van der Hoeven) R.T. Moore
Ceratobasidium cereale D. Murray & L.L. Burpee (teleomorph)
Distr.: Temperate regions.
Notes: Sharp eyespot of wheat and other small grains, yellow patch of turfgrass and other diseases of grasses. This species is similar to *Rhizoctonia solani* (Martin & Lucas. Phytopathology 74:170. 1984).
Hosts: Agrostis, Festuca, Poa, Triticum, Zoysia
Rhizoctonia crocorum (Pers.:Fr.) DC.
Rhizoctonia violacea Tul. & C. Tul.
Helicobasidium brebissonii (Desmaz.) Donk (teleomorph)
Distr.: Widespread.
Notes: Violet root rot (Boerema. Neth. J. Pl. Pathol. 93. Suppl. 1:9. 1987).
Hosts: Acer, Ambrosia, Ampelopsis, Catalpa, Daucus, Fraxinus, Helianthus, Iresine, Malvaviscus, Medicago, Parthenocissus, Phytolacca, Ratibida, Rhus, Rivina, Sambucus, Solanum, Trifolium, Verbesina, Viguiera, Viola
Rhizoctonia fragariae Hussain & W.E. McKeen
Ceratobasidium sp. (teleomorph)
Distr.: CA, WV; Canada(Ontario).
Host: Fragaria
Rhizoctonia lanuginosa N. Bernard
Distr.: WI.
Hosts: Calopogon, Goodyera
Rhizoctonia leguminicola Gough & E.S. Elliott
Distr.: Central and eastern United States.
Notes: Blackpatch of red clover and other legumes (Gough & Elliott. West Virginia Agr. Exp. Sta. Bull. 387T. 1956). This is probably a synonym of *R. solani*.
Host: Medicago
Rhizoctonia monilioides J. Curtis, nom. inval.
Distr.: WI(type).
Notes: This name was published without a latin description.
Hosts: Habenaria, Orchidaceae, Pogonia, Spiranthes
Rhizoctonia neottiae (H. Wolff) Burgeff
Distr.: WI.
Host: Aplectrum
Rhizoctonia oryzae Ryker & Gooch
Moniliopsis oryzae (Ryker & Gooch) R.T. Moore
Distr.: Southern United States; eastern Asia.
Notes: Sheath spot and root rot of rice and other grasses.
Hosts: Avena, Bromus, Hordeum, Oryza, Schizachyrium, Stenotaphrum, Triticum
Rhizoctonia oryzae-sativae (Sawada) Mordue
Sclerotium oryzae-sativae Sawada
Distr.: CA; most common in the rice-growing regions of Asia.
Notes: Disease of rice (Mordue. C.M.I. Descr. 409. 1974). The teleomorph was described by Gunnell & Webster (Mycologia 79:737. 1987).
Host: Oryza
Rhizoctonia ramicola G.F. Weber & D.A. Roberts
Ceratorhiza ramicola (G.F. Weber & D.A. Roberts) R.T. Moore
Ceratobasidium ramicola Tu, D.A. Roberts & Kimbrough (teleomorph)
Distr.: FL.
Notes: Threadblight disease.
Hosts: Buxus, Carissa, Cinnamomum, Elaeagnus, Erythrina, Euonymus, Feijoa, Ilex, Lagerstroemia, Ligustrum, Myrica, Pittosporum, Rhaphiolepis, Rhododendron
Rhizoctonia repens N. Bernard
Epulorhiza repens (N. Bernard) R.T. Moore
Tulasnella calospora (Boud.) Juel (teleomorph)
Distr.: NH, WI.
Hosts: Calopogon, Goodyera, Habenaria, Orchidaceae, Pogonia, Spiranthes
Rhizoctonia robusta Burgeff
Distr.: WI; Germany(type).
Hosts: Cypripedium, Habenaria
Rhizoctonia sclerotica Burgeff
Distr.: WI.
Hosts: Cypripedium, Habenaria, Spiranthes
Rhizoctonia solani Kühn
Moniliopsis solani Kühn
Rhizoctonia macrosclerotia J. Matz
Rhizoctonia microsclerotia J. Matz
Thanatephorus cucumeris (A.B. Frank) Donk (teleomorph)
Distr.: Cosmopolitan.
Notes: Adams & Butler (Phytopathology 69:629. 1979); Mordue (C.M.I. Descr. 406. 1974 as *Thanatephorus cucumeris*).
Hosts: Abelia, Abelmoschus, Acacia, Acalypha, Acer, Achillea, Acoelorrhaphe, Aconitum, Acorus, Actinidia, Adiantum, Aechmea, Aeschynanthus, Agapanthus, Ageratum, Aglaonema, Agropyron, Agrostis, Ajuga, Alcea, Aleurites, Allamanda, Allium, Alocasia, Aloe, Alsophila, Alternanthera, Alyssum, Amaranthus, Ambrosia, Ammophila, Anchusa, Andropogon, Anethum, Annona, Anthemis, Anthoxanthum, Anthurium, Antirrhinum, Aphelandra, Apium, Aquilegia, Arabis, Arachis, Aralia, Araucaria, Ardisia, Arecaceae, Aristida, Armoracia, Asclepias, Asparagus, Asplenium, Aster, Atropa, Aucuba, Avena, Axonopus, Bassia, Bauhinia, Begonia, Berberis, Beta, Boerhavia, Bougainvillea, Bouteloua, Brassica, Brassolaeliocattleya, Bromus, Brunfelsia, Bucida, Buddleja, Buxus, Caesalpinia, Caladium, Calamagrostis, Calathea, Calendula, Callistemon, Callistephus, Calluna, Calopogonium, Camellia, Campanula, Capsella, Capsicum, Caragana, Carica, Carissa, Carthamus, Carya, Caryota, Cassia, Casuarina, Catalpa, Catharanthus, Cattleya, Cedrus, Celosia, Celtis, Centaurea, Centrosema, Cephalanthus, Cerastium, Cereus, Chamaecyparis, Chamaedorea, Cheiranthus, Chenopodium, Chilopsis, Chionanthus, Chlorophytum, Chrysalidocarpus, Chrysanthemum, Cicer, Cichorium, Cinnamomum, Cissus, Citrofortunella, Citrullus, Citrus, Clarkia, Clusia, Coccoloba, Cocos, Codiaeum, Coffea, Columnea, Convolvulus, Cordyline, Coreopsis, Cornus, Cortaderia, Cosmos, Cotoneaster, Crassula, Crinum, Crossandra, Crotalaria, Ctenanthe, Cucumis, Cucurbita, Cuphea, Cupressus, Cyamopsis, Cycas, Cyclamen, Cydonia, Cymbidium, Cynara, Cynodon, Cynoglossum, Cyperus, Dactylis, Dahlia, Daphne, Datura, Daucus, Davallia,

Delonix, Delphinium, Dendrobium, Deschampsia, Desmodium, Dianthus, Dichondra, Dieffenbachia, Dietes, Digitalis, Digitaria, Dimorphotheca, Diodia, Diospyros, Dizygotheca, Doryopteris, Dracaena, Echeveria, Echinochloa, Elaeagnus, Elettaria, Eleusine, Elymus, Epidendrum, Epipremnum, Episcia, Equisetum, Eragrostis, Eranthemum, Eremochloa, Eriobotrya, Eriophyllum, Erodium, Erysimum, Eucalyptus, Eucharis, Eugenia, Euonymus, Eupatorium, Euphorbia, Exacum, Fagopyrum, Fatshedera, Fatsia, Feijoa, Festuca, Ficus, Firmiana, Fittonia, Flindersia, Foeniculum, Fragaria, Fraxinus, Gaillardia, Galphimia, Gardenia, Gazania, Gelsemium, Gerbera, Gladiolus, Gloriosa, Glycine, Gnaphalium, Gossypium, Grevillea, Gynura, Gypsophila, Hedera, Helianthus, Heliconia, Hemerocallis, Hemigraphis, Hibiscus, Hierochloe, Hippeastrum, Holcus, Hordeum, Hosta, Howeia, Hoya, Hydrangea, Hydrastis, Hygrophila, Iberis, Ilex, Impatiens, Indigofera, Ipomoea, Iresine, Iris, Ixora, Jasminum, Juniperus, Kalanchoe, Koeleria, Kopsia, Kummerowia, Lablab, Lactuca, Laeliocattleya, Lagenaria, Lagerstroemia, Lamium, Lantana, Larix, Lathyrus, Lavatera, Leea, Lepidium, Lespedeza, Leucophyllum, Leucothoe, Ligustrum, Lilium, Limonium, Linaria, Linum, Liquidambar, Liriodendron, Liriope, Litchi, Lobelia, Lobularia, Lolium, Lonicera, Lotus, Lupinus, Lychnis, Lycopersicon, Lysimachia, Lythrum, Macadamia, Maclura, Macroptilium, Macrotyloma, Magnolia, Malpighia, Malus, Malva, Mandevilla, Mangifera, Manihot, Maranta, Matthiola, Medicago, Melaleuca, Melia, Melilotus, Mentha, Mentzelia, Michelia, Mikania, Mirabilis, Mollugo, Monstera, Mucuna, Muhlenbergia, Murraya, Musa, Myrica, Myrsine, Nandina, Narcissus, Nasturtium, Nematanthus, Neonotonia, Nepeta, Nephrolepis, Nerium, Nicotiana, Nyssa, Ochrosia, Ocimum, Oenothera, Oncidium, Ophiopogon, Opuntia, Orchidaceae, Oreocallis, Ornithopus, Oryza, Oryzopsis, Ostrya, Oxalis, Paeonia, Panax, Pandanus, Panicum, Papaver, Parthenium, Parthenocissus, Paspalum, Passiflora, Pastinaca, Pelargonium, Pellaea, Pellionia, Pennisetum, Pentas, Peperomia, Persea, Petroselinum, Petunia, Phalaenopsis, Phalaris, Phanerophlebia, Phaseolus, Philodendron, Phleum, Phlox, Phoenix, Photinia, Phytolacca, Picea, Pilea, Pinus, Piqueria, Pistacia, Pisum, Pittosporum, Plantago, Platanus, Platycerium, Platycodon, Plectranthus, Poa, Podocarpus, Polianthes, Polygonum, Polypodium, Polyscias, Poncirus, Populus, Portulaca, Potamogeton, Potentilla, Primula, Prunus, Pseudotsuga, Pseudovigna, Psidium, Pteris, Ptilimnium, Pueraria, Punica, Pyracantha, Pyrus, Quercus, Ranunculus, Raphanus, Redfieldia, Reseda, Rhaphiolepis, Rhapis, Rheum, Rhododendron, Rhodoleia, Richardia, Ricinus, Robinia, Rosa, Roystonea, Rubus, Ruellia, Rumex, Rumohra, Russelia, Sabal, Saccharum, Sagina, Saintpaulia, Salix, Salvia, Sambucus, Sanchezia, Sansevieria, Sarracenia, Saururus, Saxifraga, Scaevola, Schefflera, Schizanthus, Schlumbergera, Sciadopitys, Scutellaria, Secale, Sedum, Selaginella, Senecio, Senna, Setaria, Shepherdia, Sicana, Siderasis, Silene, Silphium, Sinapis, Sinningia, Sitanion, Solanum, Soleirolia, Solenostemon, Sophora, Sorghum, Spathiphyllum, Spinacia, Sporobolus, Stellaria, Stenotaphrum, Stipa, Strelitzia, Striga, Stromanthe, Synadenium, Syngonium, Syringa, Tabebuia, Tagetes, Taraxacum, Taxus, Tecomaria, Tephrosia, Teramnus, Thuja, Thymus, Torreya, Trachelospermum, Trachymene, Tradescantia, Tragopogon, Tribulus, Trichocereus, Tridens, Trifolium, Triticum, Tropaeolum, Tsuga, Tulipa, Ulmus, Vaccinium, Valeriana, Vanda, Veitchia, Verbena, Vernonia, Veronica, Veronicastrum, Viburnum, Vicia, Vigna, Vinca, Viola, Vitis, Vriesea, Vulpia, Westringia, Zamia, Zantedeschia, Zea, Zebrina, Zingiber, Zinnia, Zoysia

Rhizoctonia sphacelati Burgeff
Rhizoctonia gracilis Burgeff
Distr.: WI.
Host: Habenaria

Rhizoctonia stahlii Burgeff
Distr.: WI.
Hosts: Habenaria, Spiranthes

Rhizoctonia subtilis Burgeff
Distr.: WI; Germany(type).
Hosts: Cypripedium, Habenaria

Rhizoctonia subtilis Burgeff var. ***nigra*** J. Curtis, nom. inval.
Distr.: WI(type).
Notes: This name was published without a latin description.
Host: Spiranthes

Rhizoctonia tuliparum (Kleb.) Whetzel & J.M. Arthur
Sclerotium tuliparum Kleb.
Distr.: Temperate regions.
Notes: Grey bulb rot (Mordue. C.M.I. Descr. 407. 1974).
Hosts: Iris, Lilium, Tulipa

Rhizoctonia zeae Voorhees
Moniliopsis zeae (Voorhees) R.T. Moore
Waitea circinata Warcup & Talbot (teleomorph)
Distr.: Temperate regions.
Notes: Sclerotial rot of corn. Isolated from various plants.
Hosts: Agrostis, Avena, Beta, Cucumis, Daucus, Festuca, Glycine, Helianthus, Pennisetum, Phaseolus, Raphanus, Triticum, Zea

Rhizodiscina Hafellner. Loculoascomycetes, Patellariales

Notes: Hafellner (Beih. Nova Hedwigia 62:195. 1979).

Rhizodiscina lignyota (Fr.:Fr.) Hafellner
Buellia lignyota (Fr.:Fr.) E. Müller in E. Müller & Arx
Karschia lignyota (Fr.:Fr.) Sacc.
Karschia sphaeroides Ellis & Everh.
Distr.: North America; Africa; Europe.
Hosts: Acer, Betula, Juniperus, Metrosideros, Pinus, Vaccinium

Rhizogene Syd. & P. Syd. Loculoascomycetes, Dothideales

Rhizogene impressa (Ellis & Everh.) Barr
Karschia impressa Ellis & Everh.
Lasiobotrys symphoricarpi Syd. & P. Syd.
Rhizogene symphoricarpi (Syd. & P. Syd.) Syd. & P. Syd.
Distr.: North America.
Notes: On living twigs and leaves (Barr. Contr. Univ. Michigan Herb. 9:612. 1972; Hafellner. Nova Hedwigia 27:903. 1976).
Host: Symphoricarpos

Rhizomucor (Lucet & Costantin) Wehmer & Vuill. Zygomycetes, Mucorales

Rhizomucor miehei (Cooney & R. Emerson) M.A.A. Schipper
Mucor miehei Cooney & R. Emerson
Distr.: Widespread.
Notes: Isolated from soil, animals, and various organic substrates (Schipper. Stud. Mycol. 17:58. 1978; Lunn. C.M.I. Descr. 528. 1977 as *Mucor miehei*).
Host: Helianthus

Rhizomucor pusillus (Lindt) M.A.A. Schipper
Mucor pusillus Lindt
Distr.: Cosmopolitan.
Notes: Lunn (C.M.I. Descr. 527. 1977).
Hosts: Arachis, Carya, Zea

Rhizophydium Schenk. Chytridiomycetes, Chytridiales

Notes: Barr (Canad. J. Bot. 47:991. 1969).

Rhizophydium annulatum Sparrow
Distr.: FL(type).
Host: Pinus

Rhizophydium graminis Ledingham
Distr.: Northeastern North America.
Notes: Parasitic on roots of grasses.
Hosts: Panicum, Triticum

Rhizophydium pollinis-pini (A. Braun) Zopf
Distr.: Widespread.
Notes: On pine pollen in water.
Host: Pseudotsuga

Rhizopus Ehrenb. Zygomycetes, Mucorales

Notes: Schipper (Stud. Mycol. 25:1. 1984); Schipper & Stalpers (Stud. Mycol. 25:20. 1984).

Rhizopus arrhizus A. Fischer
Rhizopus nodosus Namyslowski
Rhizopus oryzae Went & Prinsen Geerligs
Rhizopus tritici K. Saito

Distr.: Cosmopolitan.
Notes: Soft rot of fruit and vegetables. Also isolated from soil, animals and various organic substrates. Ellis (Mycologia 77:243. 1985) proposed the synonymy accepted here. Lunn (C.M.I. Descr. 525. 1977 as *Rhizopus oryzae*).
Hosts: Beta, Carya, Cucumis, Daucus, Helianthus, Ipomoea, Phaseolus, Prunus, Vitis, Zea
Rhizopus circinans Tiegh.
Distr.: Widespread.
Notes: Isolated from various organic substrates.
Host: Prunus
Rhizopus fusiformis M.L. Dawson & Povah
Distr.: IL(type).
Notes: Rot of rutabaga.
Host: Brassica
Rhizopus microsporus Tiegh.
Distr.: Widespread.
Notes: Isolated from soil and various organic substrates (Lunn. C.M.I. Descr. 523. 1977).
Host: Zea
Rhizopus microsporus Tiegh. var. ***rhizopodiformis*** (Cohn) M.A.A. Schipper
Rhizopus rhizopodiformis (Cohn) Zopf in Schenk
Distr.: Temperate regions.
Notes: Isolated from sawdust, stored grains, and various organic substrates (Lunn. C.M.I. Descr. 522. 1977 as *Rhizopus rhizopodiformis*).
Host: Zea
Rhizopus necans Massee
Distr.: WA.
Notes: No account of this name could be located.
Host: Tulipa
Rhizopus stolonifer (Ehrenb.:Fr.) Vuill.
Rhizopus nigricans Ehrenb.
Distr.: Cosmopolitan.
Notes: Soft rot of fruits and vegetables (Lunn. C.M.I. Descr. 524. 1977).
Hosts: Allium, Amaryllis, Ananas, Arachis, Brassica, Caladium, Capsicum, Carica, Carya, Chrysanthemum, Citrullus, Citrus, Colocasia, Cucumis, Cucurbita, Daucus, Dioscorea, Diospyros, Euphorbia, Ficus, Fragaria, Glycine, Gossypium, Helianthus, Ipomoea, Lycopersicon, Malus, Medicago, Musa, Narcissus, Persea, Phaseolus, Pisum, Prunus, Psidium, Pyrus, Rubus, Solanum, Trifolium, Trigonella, Triticum, Tulipa, Vaccinium, Vicia, Vitis, Zea

Rhizosphaera L. Mangin & Har. Coelomycetes

Rhizosphaera kalkhoffii Bubák
Sclerophoma pini (Desmaz.) Höhn.
Sphaeronaema pini Desmaz.
Distr.: Temperate North America; Europe.
Notes: Needle blight of Pinaceae (Diamandis & Minter. C.M.I. Descr. 656. 1980).
Hosts: Picea, Pinus
Rhizosphaera pini (Corda) Maubl.
Rhizosphaera abietis L. Mangin & Har.
Distr.: Circumboreal.
Notes: Needle blight of Pinaceae (Diamandis & Minter. C.M.I. Descr. 657. 1980).
Host: Abies

Rhizothyrium Naoumoff. Coelomycetes

Rhizothyrium abietis Naoumoff
Rhizocalyx abietis Petr. (teleomorph)
Distr.: Temperate regions.
Notes: Smerlis (Canad. J. Bot. 45:1715. 1967).
Host: Abies

Rhodocybella T.J. Baroni & R. Petersen. Basidiomycetes, Agaricales

Rhodocybella rhododendri T.J. Baroni & R. Petersen
Distr.: NC.
Notes: On living stems (Baroni & Petersen. Mycologia 79:358. 1987).
Host: Rhododendron

Rhodosticta Woronichin in Tranzschel & Serebrianikow. Coelomycetes

Rhodosticta quercina J.C. Carter
Distr.: IL.
Notes: Oak dieback (Carter. Illinois Nat. Hist. Survey Bull. 21:223. 1941).
Hosts: Liquidambar, Prunus, Quercus

Rhodotorula F.C. Harrison. Hyphomycetes

Notes: Phaff & Ahearn in Lodder (The yeasts. p. 1187. 1970).
Rhodotorula glutinis (Fresen.) F.C. Harrison
Rhodotorula glutinis (Fresen.) F.C. Harrison var. *rubescens* (K. Saito) Lodder
Distr.: Widespread.
Notes: Isolated from air, soil, sea water, marine life, and various kinds of organic debris.
Hosts: Trifolium, Triticum
Rhodotorula rubra (R. Demme) Lodder
Saccharomyces ruber R. Demme
Rhodotorula mucilaginosa (A. Jörgensen) F.C. Harrison
Distr.: Terrestrial and marine habitats.
Notes: Isolated from humans, beer, flowers, and various kinds of organic debris.
Host: Trifolium

Rhodotus Maire. Basidiomycetes, Agaricales

Rhodotus palmatus (Bull.:Fr.) Maire
Pleurotus pubescens Peck
Distr.: Eastern North America; northern Africa; Europe.
Notes: On wood of hardwoods (Pegler & Young. Kew Bull. 30:30. 1975).
Hosts: Acer, Aesculus, Liriodendron, Tilia, Ulmus

Rhopalidium Mont. Hyphomycetes

Notes: This genus is considered a synonym of *Alternaria*, but not all names have been redisposed.
Rhopalidium cercosporelloides Dearn.
Distr.: ID.
Notes: On living leaves.
Hosts: Holodiscus, Spiraea

Rhopaloconidium Petr. Hyphomycetes

Rhopaloconidium asiminae (Ellis & Morg.) Petr.
Phloeospora asiminae Ellis & Morg.
Distr.: Eastern United States.
Host: Asimina

Rhopographus Nitschke ex Fuckel. Loculoascomycetes, Dothideales

Rhopographus anceps Petr.
Distr.: HI(type).
Host: Pteris
Rhopographus filicinus (Fr.:Fr.) Nitschke ex Fuckel
Rhopographus pteridis (Sowerby) G. Wint.
Acremonium sp. (anamorph)
Distr.: North America; temperate Europe; the Himalayas.
Notes: On fern stems (Obrist. Phytopathol. Z. 35:367. 1959).
Host: Pteridium
Rhopographus zeae Pat.
Distr.: GA; Ecuador(type).
Host: Zea

Rhynchophoma P. Karst. Pyrenomycetes, Ophiostomatales

Notes: This genus is considered a synonym of *Ceratocystis*, but not all names have been redisposed..
Rhynchophoma raduloides Sacc. & Scalia
Distr.: AK(type).
Host: Ribes

Rhynchosphaeria (Sacc.) Berl. Pyrenomycetes, Sphaeriales

Rhynchosphaeria shearii Petr.
Distr.: HI.
Hosts: Physostegia, Pipturus

Rhynchosporina Arx. Coelomycetes

Rhynchosporina meinersii (R. Sprague) Arx
Gloeosporium meinersii R. Sprague
Distr.: Western United States.
Hosts: Phleum, Poa
Rhynchosporina tridentis R. Sprague & C.T. Rogerson
Distr.: KS.
Notes: Sprague & Rogerson (Mycologia 50:634. 1958).
Host: Tridens

Rhynchosporium Heinsen ex A.B. Frank. Hyphomycetes

Notes: Most *Rhynchosporium* species have been transferred to *Microdochium*.
Rhynchosporium orthosporum R.M. Caldwell
Distr.: Central and western North America; Great Britain.
Notes: Scald or white leaf stripe of Poaceae.
Hosts: Agropyron, Agrostis, Alopecurus, Calamagrostis, Dactylis, Elymus, Festuca, Lolium, Poa
Rhynchosporium secalis (Oudem.) J.J. Davis
Marssonia secalis Oudem.
Distr.: Temperate regions.
Notes: Barley and rye leaf blotch and leaf scald (Owen. C.M.I. Descr. 387. 1973).
Hosts: Agropyron, Agrostis, Bouteloua, Bromus, Dactylis, Danthonia, Elymus, Hordeum, Leymus, Lolium, Phalaris, Phleum, Poa, Quercus, Secale

Rhynchostoma P. Karst. Pyrenomycetes, Trichosphaeriales

Rhynchostoma exasperans P. Karst.
Distr.: CA; Finland(type).
Host: Lupinus
Rhynchostoma tinctum (Ellis & Everh.) Berl. & Voglino
Ceratostoma tinctum Ellis & Everh.
Distr.: Pacific Northwest.
Hosts: Acer, Quercus

Rhytidenglerula Höhn. Loculoascomycetes, Dothideales

Rhytidenglerula carnea (Ellis & G. Martin) Höhn.
Asterina carnea Ellis & G. Martin
Asterella carnea (Ellis & G. Martin) Sacc.
Englerula carnea (Ellis & G. Martin) Höhn.
Distr.: FL.
Host: Persea

Rhytidhysteron Speg. Loculoascomycetes, Patellariales

Notes: Samuels & Müller (Sydowia 32:277. 1979).
Rhytidhysteron hysterinum (J.L. Dufour) Samuels & E. Müller
Triblidium hysterinum J.L. Dufour
Eutryblidiella hysterina (J.L. Dufour) Petr.
Tryblidiella hysterina (J.L. Dufour) Shear
Aposphaeria sp. (synanamorph)
Diplodia sp. (synanamorph)
Distr.: Temperate northern hemisphere.
Notes: On woody plants.
Host: Ilex
Rhytidhysteron rufulum (Spreng.:Fr.) Speg.
Triblidium rufulum (Spreng.:Fr.) Ellis & Everh.
Tryblidiella rufula (Spreng.:Fr.) Sacc.
Tryblidiella fusca (Ellis & Everh.) Rehm
Rhytidhysterium prosopidis Peck
Aposphaeria sp. (synanamorph)
Diplodia sp. (synanamorph)
Distr.: Subtropical and tropical regions.
Notes: On woody plants.
Hosts: Albizia, Ampelopsis, Casuarina, Celtis, Citrus, Dichrostachys, Diospyros, Elaeagnus, Ficus, Grevillea, Hibiscus, Ilex, Lespedeza, Leucaena, Maclura, Magnolia, Mangifera, Morus, Nyssa, Oxydendrum, Pistacia, Platanus, Poncirus, Prosopis, Pyrus, Quercus, Rhus, Robinia, Senna, Vitis

Rhytidiella Zalasky. Loculoascomycetes, Dothideales

Rhytidiella baranyayi Funk & Zalasky
Distr.: AK, CA, western Canada.
Notes: Causes cork-bark disease of aspen (Funk & Zalasky. Canad. J. Bot. 53:752. 1975).
Host: Populus

Rhytisma Fr.:Fr. Discomycetes, Rhytismatales

Rhytisma acerinum (Pers.:Fr.) Fr.
Melasmia acerina Lév. (anamorph)
Distr.: Temperate northern hemisphere.
Notes: Tar spot (Cannon & Minter. C.M.I. Descr. 791. 1984).
Host: Acer
Rhytisma andromedae (Pers.:Fr.) Fr.
Placuntium andromedae (Pers.:Fr.) Höhn.
Melasmia sp. (anamorph)
Distr.: Circumboreal.
Notes: On leaves of Ericaceae. Eriksson (Symb. Bot. Upsal. 19:56. 1970).
Hosts: Andromeda, Kalmia, Leucothoe, Lyonia, Pieris
Rhytisma arbuti W. Phillips in Vize
Distr.: AK, TX, northwestern United States.
Notes: On Ericaceae.
Hosts: Arbutus, Menziesia, Rhododendron
Rhytisma asteris Fr.
Xyloma asteris Schwein.
Distr.: WI.
Host: Aster
Rhytisma concavum Ellis & Kellerm.
Distr.: AL, MA, OH, WI.
Host: Ilex
Rhytisma decolorans Fr.
Rhytisma andromedae-ligustrinae (Schwein.) G.W. Wils. & Seaver
Xyloma andromedae-ligustrinae Schwein.
Distr.: Eastern United States.
Hosts: Leucothoe, Lyonia
Rhytisma ilicincola Fr.
Distr.: Midwestern and southeastern United States.
Host: Ilex
Rhytisma liriodendri Wallr.
Distr.: Southeastern United States; Europe.
Host: Liriodendron
Rhytisma prini (Schwein.:Fr.) Fr.
Rhytisma ilicis-canadensis Schwein.
Distr.: Temperate North America.
Notes: Tar spot of Aquifoliaceae.
Hosts: Ilex, Nemopanthus
Rhytisma punctatum (Pers.:Fr.) Fr.
Melasmia punctata Sacc. & Roum. (anamorph)

Distr.: Temperate North America; Europe.
Host: Acer
Rhytisma salicinum (Pers.:Fr.) Fr.
Melasmia salicina Lév. (anamorph)
Distr.: Temperate northern hemisphere.
Notes: On leaves.
Hosts: Alnus, Salix
Rhytisma solidaginis Schwein.
Distr.: LA, WI, northeastern United States.
Hosts: Euthamia, Solidago
Rhytisma tostum Berk. & M.A. Curtis
Distr.: AL, FL.
Host: Quercus
Rhytisma ulmi Fr.:Fr.
Distr.: FL, MN, northeastern United States.
Host: Ulmus
Rhytisma vaccinii (Schwein.:Fr.) Fr.
Distr.: Temperate North America.
Notes: Tar spot of Ericaceae.
Hosts: Arctostaphylos, Gaylussacia, Lyonia, Oxydendrum, Vaccinium
Rhytisma velatum Fr.:Fr.
Xyloma velatum Schwein.
Distr.: Southeastern United States.
Host: Ilex
Rhytisma vitis Schwein.
Distr.: NC, PA, WI.
Host: Vitis

Rigidoporus Murrill. Basidiomycetes, Aphyllophorales

Notes: Gilbertson & Ryvarden (North American Polypores 2:693. 1984).
Rigidoporus crocatus (Pat.) Ryvarden
Poria nigrescens Bres.
Rigidoporus nigrescens (Bres.) Donk
Distr.: Temperate northern hemisphere.
Notes: On wood.
Hosts: Abies, Betula, Fagus, Ilex, Liquidambar, Pinus, Thuja
Rigidoporus lineatus (Pers.) Ryvarden
Polyporus lineatus Pers.
Polyporus rigidus Lév.
Polyporus zonalis Berk.
Rigidoporus zonalis (Berk.) Imazeki
Distr.: Southeastern United States; Africa; Asia, warmer parts.
Notes: On hardwoods (Pegler & Waterston. C.M.I. Descr. 200. 1968 as *R. zonalis*).
Hosts: Carya, Celtis, Liquidambar, Liriodendron, Nyssa, Platanus, Populus, Quercus
Rigidoporus microporus (Sw.:Fr.) Overeem
Polyporus microporus (Sw.:Fr.) Fr.
Polyporus lignosus Klotzsch
Rigidoporus lignosus (Klotzsch) Imazeki
Distr.: Subtropical America, Africa and Asia.
Notes: On wood (Pegler & Waterston. C.M.I. Descr. 198. 1968).
Host: Citrus
Rigidoporus ulmarius (Sowerby:Fr.) Imazeki in Ito
Fomes ulmarius (Sowerby:Fr.) Gill.
Fomitopsis ulmaria (Sowerby:Fr.) Bondartsev & Singer
Fomes geotropus (Cooke) Cooke
Distr.: Southeastern United States; Europe; Africa.
Notes: On dead wood or wounds of living trees (Pegler & Waterston. C.M.I. Descr. 199. 1968).
Hosts: Acer, Carya, Celtis, Diospyros, Fraxinus, Liquidambar, Magnolia, Nyssa, Persea, Populus, Quercus, Taxodium, Tilia, Ulmus
Rigidoporus vinctus (Berk.) Ryvarden
Polyporus vinctus Berk.
Poria vincta (Berk.) Cooke
Distr.: Subtropical and tropical regions.
Notes: On hardwoods.
Hosts: Magnolia, [Pinus]

Robergea Desmaz. Discomycetes, Ostropales

Notes: Sherwood (Mycotaxon 5:96. 1977).
Robergea albicedrae (Heald & F.A. Wolf) Sacc. & Traverso
Cyanospora albicedrae Heald & F.A. Wolf
Distr.: TX; Mexico.
Notes: White blight of mountain white cedar.
Host: Juniperus
Robergea nigra M.A. Sherwood
Distr.: OR.
Host: Lonicera
Robergea obliqua M.A. Sherwood
Distr.: CA.
Host: Penstemon

Robillarda Sacc., nom. cons. Coelomycetes

Robillarda rhizophorae Kohlmeyer
Distr.: HI, subtropical and tropical coastal regions of the Americas.
Notes: Kohlmeyer & Kohlmeyer (Marine mycology. p. 536. 1979).
Host: Rhizophora

Roesleria Thuem. & Pass. Discomycetes, Caliciales

Roesleria pallida (Pers.) Sacc.
Coniocybe pallida (Pers.) Fr.
Roesleria hypogaea Thuem. & Pass.
Distr.: Temperate regions.
Notes: On bark, especially of dead roots, often well below soil level. Dennis (British Ascomycetes. p. 116. 1978) provides a modern description and an explanation of the disposition of this unusual fungus.
Hosts: Malus, Vitis

Rogersiomyces J.L. Crane & Schoknecht. Basidiomycetes, Filobasidiales

Rogersiomyces okefenokeensis J.L. Crane & Schoknecht
Distr.: GA.
Notes: On submerged leaves (Crane & Schoknecht. Amer. J. Bot. 65:902. 1978).
Host: Taxodium

Romellia Berl. Pyrenomycetes, Calosphaeriales

Notes: Barr (Mycologia 77:559. 1985).
Romellia cornina (Peck) Barr
Valsa cornina Peck
Calospora allantospora Ellis & Everh.
Eutypella venusta (Ellis) Sacc.
Valsa venusta Ellis
Distr.: Eastern North America.
Notes: On limbs of hardwoods.
Hosts: Cornus, Robinia
Romellia tympanoides Barr
Distr.: MA.
Host: Prunus

Rosellinia De Not. Pyrenomycetes, Xylariales

Rosellinia albolanata Ellis & Everh.
Distr.: MT(type).
Host: Salix
Rosellinia aquila (Fr.:Fr.) De Not.
Nodulisporium sp. (anamorph)
Distr.: Temperate northern hemisphere.
Notes: Root rot of mulberry; also on wood of many hardwoods (Rogers & Stiers. Canad. J. Bot. 52:5. 1974).
Hosts: Acer, Alnus, Asclepias, Berchemia, Carpinus, Carya, Cornus, Fraxinus, Lonicera, Magnolia, Malus, Morus, Ostrya, Picea, Pinus, Platanus, Prunus, Pyrus, Quercus, Robinia, Salix, Sambucus, Vitis
Rosellinia caryae Bonar

Dothichiza caryae Bonar (anamorph)
Distr.: MI(type).
Notes: Canker on young trees (Bonar. Phytopathology 12:384. 1922).
Host: Carya
Rosellinia corticium (Schwein.) Sacc.
Distr.: Eastern United States; northwestern Himalayas (Dargan & Thind. Mycologia 71:1016. 1979).
Hosts: Nyssa, Quercus
Rosellinia deerata (M.A. Curtis & Ellis) Sacc.
Distr.: NJ.
Host: Juniperus
Rosellinia desmazieresii (Berk. & Broome) Sacc.
Distr.: NC; Germany; Great Britain; Italy.
Notes: Barrett & Payne (Trans. Brit. Mycol. Soc. 78:566. 1982) reported this organism causing ring-dying of *Salix repens*. Francis (Sydowia 38:75. 1985).
Host: Betula
Rosellinia diathrausta (Rehm) L.E. Petrini
Hypoxylon diathrauston Rehm
Creosphaeria pinea Petr.
Distr.: Western North America; Europe; Japan.
Notes: On limbs of conifers.
Hosts: Abies, Pinus, Pseudotsuga, Tsuga
Rosellinia glandiformis Ellis & Everh.
Distr.: Southeastern United States.
Notes: On hardwoods.
Hosts: Acer, Quercus
Rosellinia herpotrichioides Hepting & R.W. Davidson
Distr.: Temperate North America; France.
Notes: Needle blight of conifers (Hepting & R.W. Davidson. Phytopathology 22:307. 1937).
Hosts: Picea, Pseudotsuga, Tsuga
Rosellinia limoniispora Ellis & Everh.
Sporothrix sp. (anamorph)
Distr.: Northwestern North America; Europe.
Notes: Petrini & Petrini (Sydowia 38:229. 1985) isolated this species as an endophyte and described it in culture.
Host: Pinus
Rosellinia mammiformis (Pers.:Fr.) Ces. & De Not.
Distr.: Temperate regions; northwest Himalayas.
Notes: Dennis (British Ascomycetes. p. 321. 1978).
Host: Quercus
Rosellinia medullaris (Wallr.) Ces. & De Not.
Distr.: ID, IN; Germany; Himalayas.
Notes: Dargan & Thind (Mycologia 71:1021. 1979).
Hosts: Cercis, Rosa
Rosellinia necatrix Prill.
Dematophora necatrix R. Hartig (anamorph)
Distr.: Cosmopolitan.
Notes: White root rot (Sivanesan & Holliday. C.M.I. Descr. 352. 1972). Francis (Sydowia 38:75. 1985) discussed the problems of identifying this species.
Hosts: Castanea, Fragaria, Hypericum, Ligustrum, Malus, Medicago, Osmanthus, Persea, Prunus, Pyrus, Rubus, Viburnum, Vitis
Rosellinia obliquata (Sommerf.) Sacc.
Distr.: MT; Lapland(type).
Host: Pinus
Rosellinia obtusissima (Berk. & M.A. Curtis) Sacc.
Distr.: Eastern United States.
Host: Liriodendron
Rosellinia ovalis Ellis & Everh.
Distr.: Western United States.
Hosts: Artemisia, Chrysothamnus, Populus
Rosellinia parasitica Ellis & Everh.
Distr.: Western United States.
Hosts: Ribes, Symphoricarpos
Rosellinia rhyncospora Cooke & Harkn.
Distr.: CA(type).
Host: Sambucus
Rosellinia riminicola Rehm
Distr.: Western United States; Europe.
Hosts: Arctostaphylos, Haplopappus, Populus
Rosellinia rosarum (Niessl) Rehm
Distr.: ND, NM; Czechoslovakia(type).
Host: Rosa
Rosellinia spinosa Harkn.
Distr.: CA(type).
Host: Eucalyptus
Rosellinia subcompressa Ellis & Everh.
Distr.: SD(type).
Host: Populus
Rosellinia subiculata (Schwein.:Fr.) Sacc.
Distr.: Southeastern United States.
Notes: Rogers (Mycotaxon 23:434. 1985) described the anamorph.
Hosts: Castanea, Citrus, Liriodendron, Quercus
Rosellinia sublimbata (Durieu & Mont.) Pass.
Distr.: HI, MD; Algeria; India; Portugal.
Notes: Dargan & Thind (Mycologia 71:1016. 1979).
Hosts: Arundinaria, Pandanus
Rosellinia subsimilis Sacc.
Distr.: CA; Finland(type).
Host: Arctostaphylos
Rosellinia thelena (Fr.) Rabenh.
Nodulisporium sp. (anamorph)
Distr.: Temperate regions; Himalayas (Dargan & Thind. Mycologia 71:1010. 1979).
Notes: Reported to sporulate on roots, this organism has been isolated as an endophyte by Petrini & Petrini (Sydowia 38:230. 1985).
Hosts: Picea, Pinus
Rosellinia trichota (Cooke & Ellis) Ellis & Everh.
Rosellinia abietina Fuckel var. *trichota* (Cooke & Ellis) Sacc.
Distr.: GA, NJ.
Host: Pinus
Rosellinia umbrinella (Berk. & M.A. Curtis) Sacc.
Distr.: SC.
Host: Eupatorium
Rosellinia weiriana Sacc.
Distr.: WA(type).
Host: Picea

Rosenscheldia Speg. Loculoascomycetes, Dothideales

Rosenscheldia abundans (Dobrozrakova) Petr.
Gibberidea abundans (Dobrozrakova) Shear
Distr.: Temperate North America; Europe.
Host: Prunella

Rutola J.L. Crane & Schoknecht. Hyphomycetes

Notes: Crane & Schknecht (Canad. J. Bot. 55:3013. 1977).
Rutola graminis (Desmaz.) J.L. Crane & Schoknecht
Torula graminis Desmaz.
Torula caricina Ellis & Dearn.
Distr.: Temperate North America; Europe.
Notes: On Poaceae and Cyperaceae.
Host: Bambusa

Rutstroemia P. Karst. Discomycetes, Helotiales

Rutstroemia bolaris (Batsch ex Pers.:Fr.) Rehm
Calycina bolaris (Batsch ex Pers.:Fr.) Seaver
Ciboria bolaris (Batsch ex Pers.:Fr.) Fuckel
Phialophora sp. (anamorph)
Distr.: Eastern North America; Europe.
Notes: On hardwood twigs. The anamorph of this species is discussed by Berthet (Bull. Soc. Mycol. France 80:140. 1964).
Hosts: Carpinus, Castanea
Rutstroemia renispora (Ellis) W.L. White
Helotium renisporum Ellis
Ciboria renispora (Ellis) Sacc.
Hymenoscypha renispora (Ellis) W. Phillips
Distr.: Eastern United States.
Host: Nyssa

Saccardia Cooke. Loculoascomycetes, Dothideales

Saccardia quercina Cooke
Distr.: Warmer regions.
Notes: On living leaves (Arx. Persoonia 2:463. 1963).
Host: Quercus

Saccardoella Speg. Pyrenomycetes, Sphaeriales

Saccardoella transylvanica (Rehm) Berl.
Zignoella transylvanica Rehm
Distr.: OK; Hungary(type).
Host: Quercus

Saccharomyces Meyen ex E. Hans. Hemiascomycetes, Endomycetales

Saccharomyces zopfii Artari
Distr.: LA.
Notes: Sugar deterioration. No account could be found of this species and this disease is not listed in modern references.
Host: Saccharum

Saccoblastia A. Möller. Basidiomycetes, Auriculariales

Saccoblastia farinacea (Höhn.) Donk
Helicogloea farinacea (Höhn.) D.P. Rogers in G.W. Martin
Distr.: North America; Europe.
Notes: On wood. Ginns (Fungi Canadenses 286. 1984 as *Helicogloea farinacea*).
Host: Quercus

Sacidium Nees in Kunze & J.C. Schmidt. Coelomycetes

Notes: This genus is considered a synonym of *Pilobolus* (Sutton. Mycol. Pap. 141:179. 1977), but not all names have been redisposed.
Sacidium ulmi-gallae Kellerm. & Swingle
Distr.: Central United States.
Host: Ulmus

Sarcinella Sacc. Hyphomycetes

Notes: Hughes (Canad. J. Bot. 61:1727. 1983).
Sarcinella copallina (Ellis & Everh.) S.J. Hughes
Stemphylium copallinum Ellis & Everh.
Distr.: Southeastern United States.
Notes: On leaves.
Host: Rhus
Sarcinella heterospora Sacc.
Schiffnerula pulchra (Sacc.) Petr. (teleomorph)
Questieriella sp. (synanamorph)
Distr.: Central and eastern North America; Europe.
Notes: On leaves.
Hosts: [Cornus], [Corylus], Fraxinus, [Philadelphus], [Vernonia]
Sarcinella quadrata (Cooke) S.J. Hughes
Stemphylium quadratum (Cooke) Sacc.
Questieriella sp. (synanamorph)
Distr.: Southeastern United States.
Notes: On leaves.
Host: Fraxinus
Sarcinella questieri (Desmaz.) G. Arnaud
Questieriella sp. (synanamorph)
Schiffnerula sp. (teleomorph)
Distr.: Central and eastern North America; Europe.
Notes: On leaves.
Host: Cornus
Sarcinella vernoniae (Dearn. & Barth.) S.J. Hughes
Stigmella vernoniae Dearn. & Barth.
Piricauda vernoniae (Dearn. & Barth.) R.T. Moore
Questieriella sp. (synanamorph)
Distr.: Southeastern United States.
Host: Vernonia

Sarcodontia Schulzer. Basidiomycetes, Aphyllophorales

Sarcodontia setosa (Pers.) Donk
Hydnum setosum Pers.
Mycoacia setosa (Pers.) Donk
Oxydontia setosa (Pers.) L.W. Miller
Hydnum amplissimum Berk. & M.A. Curtis in Berk.
Distr.: Northern hemisphere.
Hosts: Crataegus, Malus, Prunus

Sarcophoma Höhn. Coelomycetes

Sarcophoma eriogoni (Ellis & Everh.) Arx
Gloeosporium eriogoni Ellis & Everh.
Distr.: CO(type).
Notes: Arx (Biblioth. Mycol. 24:121. 1970).
Host: Eriogonum

Sarcoscypha (Fr.) Boud. Discomycetes, Pezizales

Sarcoscypha occidentalis (Schwein.) Sacc.
Plectania occidentalis (Schwein.) Seaver
Geopyxis hesperidea (Cooke & Peck) Sacc.
Distr.: North America(east of the Rocky Mountains); Caribbean; Central America, high elevations.
Notes: On sticks, twigs, and fragments of dead wood (Denison. Mycologia 64:615. 1972).
Hosts: Carya, Juglans

Sarcosoma Casp. in Rehm. Discomycetes, Pezizales

Sarcosoma godronioides Rick
Bulgaria godronioides (Rick) Seaver
Distr.: HI, possibly CA; Brazil(type).
Host: Scaevola
Sarcosoma latahensis Paden & Tylutki
Verticicladium sp. (anamorph)
Distr.: ID, WA.
Notes: On rotting wood (Paden & Tylutki. Mycologia 61:686. 1969.)
Host: Paxistima

Sarcosphaera Auersw. Discomycetes, Pezizales

Sarcosphaera coronaria (Jacq.) J. Schröt.
Sarcosphaera crassa (Santi) Pouzar
Distr.: Temperate regions.
Notes: On soil, duff, or litter in conifer and hardwood forests.
Host: Abies

Sarcotrochila Höhn. Discomycetes, Helotiales

Sarcotrochila balsameae (J.J. Davis) Korf
Phacidium balsameae J.J. Davis
Stegopezizella balsameae (J.J. Davis) Syd. in Syd. & Petr.
Rhabdogloeopsis balsameae (J.J. Davis) Petr. (anamorph)
Distr.: Northern North America.
Notes: On needles.
Hosts: Abies, Pseudotsuga
Sarcotrochila piniperda (Rehm) Korf
Naevia piniperda Rehm
Phacidium expansum J.J. Davis
Distr.: WI(type).
Host: Picea

Sarea Fr. Discomycetes, Lecanorales

Notes: Hawksworth & Sherwood (Canad. J. Bot. 59:357. 1981).
Sarea difformis (Fr.:Fr.) Fr.
Tromera difformis (Fr.:Fr.) Tuck.
Retinocyclus abietis (H. Crouan) Groves
Tympanis abietis H. Crouan
Epithyrium resinae (Sacc. & Berl.) Sacc.
Distr.: Temperate northern hemisphere as far south as Morocco.
Hosts: Chamaecyparis, Juniperus, Pseudotsuga
Sarea resinae (Fr.:Fr.) Kuntze
Biatorella resinae (Fr.:Fr.) Th. Fr.
Pycnidiella resinae (Fr.:Fr.) Höhn. (anamorph)
Distr.: Temperate northern hemisphere; Kenya.
Notes: Hawksworth (C.M.I. Descr. 677. 1980).
Host: Pinus

Sarocladium W. Gams & D. Hawksworth. Hyphomycetes

Sarocladium oryzae (Sawada)W. Gams & D. Hawksworth
Acrocylindrium oryzae Sawada
Distr.: LA; eastern Asia; India.
Notes: Sheath rot of rice (Brady. C.M.I. Descr. 673. 1980).
Host: Oryza

Satchmopsis Sutton & C.S. Hodges in Sutton. Coelomycetes

Satchmopsis brasiliensis Sutton & C.S. Hodges in Sutton
Distr.: North and South America; Asia; Australia.
Notes: On fallen leaves (Sutton & Pascoe. Trans. Brit. Mycol. Soc. 88:177. 1987).
Host: Eucalyptus

Scaphidium Clements. Coelomycetes

Notes: Sutton (The Coelomycetes. p. 296. 1980).
Scaphidium boutelouae Clements
Distr.: Central and western United States.
Notes: Narrow brown spot.
Host: Bouteloua

Schiffnerula Höhn. Loculoascomycetes, Dothideales

Schiffnerula pulchra (Sacc.) Petr.
Dimerosporium pulchrum (Sacc.) Sacc.
Sarcinella heterospora Sacc. (synanamorph)
Questieriella sp. (synanamorph)
Distr.: Central and eastern North America; Europe.
Notes: On leaves (Hughes. Canad. J. Bot. 61:1738. 1983).
Hosts: Cornus, Fraxinus, Trachelospermum

Schizonella J. Schröt. in Cohn. Basidiomycetes, Ustilaginales

Schizonella melanogramma (DC.) J. Schröt. in Cohn
Schizonella pusilla (Cooke & Peck) Cif.
Distr.: Widespread.
Notes: On Cyperaceae. Vánky (Symb. Bot. Upsal. 24:101. 1985).
Host: Carex

Schizoparme Shear. Pyrenomycetes, Hypocreales

Schizoparme straminea Shear
Coniella castaneicola (Ellis & Everh.) Sutton (anamorph)
Distr.: Eastern North America; Europe.
Notes: Fruit rot of strawberries and on foliage of broadleafed trees. The anamorph has been reported as *Pilidiella quercicola* (Oudem.) Petr., but this taxon is a synonym of *Coniella castaneicola* (Sutton. The Coelomycetes. p. 420. 1980).
Hosts: Eucalyptus, Rosa

Schizophyllum Fr.:Fr. Basidiomycetes, Aphyllophorales

Schizophyllum commune Fr.:Fr.
Schizophyllum radiatum (Sw.) Fr.
Distr.: Cosmopolitan.
Notes: Usually saprotrophic on wood, but sometimes parasitic on various woody and herbaceous hosts. Variation in sporocarp morphology, especially in the tropics, has led to the recognition of several segregate species and forms. Ginns (Fungi Canadenses 42. 1974).
Hosts: Acacia, Acer, Ailanthus, Albizia, Alnus, Baccharis, Bambusa, Betula, Caesalpinia, Carpinus, Carya, Castanea, Catalpa, Ceanothus, Cedrela, Cercis, Citrus, Diospyros, Fagus, Ficus, Fouquieria, Fraxinus, Gleditsia, Gossypium, Helianthus, Ilex, Ipomoea, Juglans, Liquidambar, Liriodendron, Malus, Melia, Morus, Nyssa, Phyllostachys, Picea, Pinus, Pistacia, Poaceae, Poncirus, Populus, Prosopis, Prunus, Pseudotsuga, Pyrus, Quercus, Rhus, Ricinus, Sabal, Saccharum, Salix, Sassafras, Schinus, Sequoia, Sorghum, Thuja, Tilia, Tsuga, Ulmus, Umbellularia, Vitis

Schizopora Velen. emend. Donk. Basidiomycetes, Aphyllophorales

Notes: Gilbertson & Ryvarden (North American Polypores 2:703. 1987).
Schizopora apacheriensis (R.L. Gilbertson & E.R. Canfield) R.L. Gilbertson & Ryvarden
Poria apacheriensis R.L. Gilbertson & E.R. Canfield
Distr.: Southern United States.
Notes: On wood.
Hosts: Agave, Buddleja, Carnegiea, Chilopsis, Cowania, Haplopappus, Juglans, Juniperus, Mortonia, Prosopis, Quercus
Schizopora paradoxa (Schrad.:Fr.) Donk
Irpex paradoxus (Schrad.:Fr.) Fr.
Poria versipora (Pers.) Romell
Polyporus versiporus Pers.
Distr.: Widespread.
Notes: On wood.
Hosts: Acer, Alnus, Betula, Carpinus, Carya, Castanea, Cornus, Eucalyptus, Fagus, Holodiscus, Ilex, Juglans, Juniperus, Larix, Liquidambar, Lithocarpus, Lyonia, Magnolia, Nyssa, Phoenix, Pinus, Populus, Prunus, Pseudotsuga, Quercus, Sequoia, Thuja, Vitis

Schizothyrella Thuem. Coelomycetes

Schizothyrella fraxini Ellis & Everh.
Distr.: KS.
Host: Fraxinus

Schizothyrium Desmaz. Loculoascomycetes, Dothideales

Notes: Arx & Müller (Stud. Mycol. 9:27. 1975); Müller & Arx (Beitr. Kryptogamenfl. Schweiz 11(2):200. 1962).
Schizothyrium acuum Bubák
Distr.: CA; Austria(type).
Notes: On needles of pines.
Host: Pinus
Schizothyrium discoideum (Ellis & G. Martin) Arx
Asterina discoidea Ellis & G. Martin
Asterella discoidea (Ellis & G. Martin) Sacc.
Microthyriella discoidea (Ellis & G. Martin) Theiss.
Distr.: North America.
Notes: On leaves.
Hosts: Osmanthus, Quercus
Schizothyrium perexiguum (Roberge) Höhn.
Phacidium perexiguum Roberge
Microthyriella perexigua (Roberge) Höhn.
Microthyriella rubi Petr.
Distr.: North America; Europe.
Notes: On cuticle of living and dead stems of various plants.
Hosts: Acer, Brassolaeliocattleya, Cattleya, Dendrobium, Dianthus, Epidendrum, Malus, Orchidaceae, Salix, Vanda
Schizothyrium pomi (Mont. & Fr.) Arx

Leptothyrium pomi (Mont. & Fr.) Sacc.
Asterina gaultheriae M.A. Curtis in Ellis
Epipeltis gaultheriae (M.A. Curtis) Theiss.
Schizothyrium gaultheriae (M.A. Curtis) Höhn.
Microthyriella paludosa C. Booth
Aulographum reticulatum W. Phillips & Harkn.
Zygophiala jamaicensis E. Mason (anamorph)
Distr.: Cosmopolitan.
Notes: This pathogen occurs on at least 78 species in 36 families of flowering plants (Baker et al. Phytopathology 67:580. 1977).
Hosts: Acer, Arctostaphylos, Cinnamomum, Citrus, Cydonia, Diospyros, Gaultheria, Malus, Poncirus, Prunus, Pyrus, Rubus, Sambucus, Sassafras, Smilax, Vaccinium, Vitis, Zanthoxylum
Schizothyrium rufulum (Berk. & M.A. Curtis) Arx
Rhytisma rufulum Berk. & M.A. Curtis
Eremotheca rufula (Berk. & M.A. Curtis) Theiss. & Syd.
Distr.: FL; tropical regions.
Notes: On living and dying leaves of numerous tropical plants.
Host: Jasminum

Schizoxylon Pers. Discomycetes, Ostropales

Notes: Sherwood (Mycotaxon 5:108. 1977; Mycotaxon 6:221. 1977).
Schizoxylon albo-atrum Rehm
Distr.: North America; Europe.
Hosts: Juglans, Pyrus, Salix
Schizoxylon berkeleyanum (Durieu & Lév.) Fuckel
Schizoxylon decipiens (P. Karst.) Sacc.
Schizoxylon occidentale Ellis & Everh.
Distr.: Temperate North America; northern Africa; Europe; Pakistan.
Notes: On a variety of herbaceous and occasionally woody plants.
Hosts: Apocynum, Fraxinus, Helianthus, Populus
Schizoxylon compositum Ellis & Everh.
Schizoxylon demateoides Rehm
Distr.: Central and eastern North America.
Hosts: Acer, Anredera, Carya, Cornus, Fraxinus, Populus, Prunus, Salix, Symphoricarpos
Schizoxylon decipiens (P. Karst.) Sacc. var. ***symphoricarpi*** Rehm
Distr.: ND(type).
Notes: The type variety is considered a synonym of *Schizoxylon berkeleyanum* but this variety has not been treated in recent monographs.
Host: Symphoricarpos
Schizoxylon insigne (De Not.) Rehm
Distr.: North America; Europe.
Notes: Sherwood (Mycotaxon 5:197. 1977) was unable to characterize this species, suggesting that it may be *Stictis mollis* or *Schizoxylon alboatrum*.
Hosts: Adenostoma, Amelanchier, Carya, Corylus, Crataegus, Fraxinus, Hamamelis, Holodiscus, Ilex, Juglans, Lantana, Laurocerasus, Platanus, Salix, Solidago, Tilia, Viburnum, Vitis
Schizoxylon ligustri (Schwein.) M.A. Sherwood
Stictis ligustri Schwein.
Schizoxylon bethelii (Ellis & Everh.) Cash
Stictis glaucoma Berk. & M.A. Curtis
Distr.: North America; Europe.
Hosts: Abrus, Bigelowia, Ligustrum, Viburnum
Schizoxylon melleum (Dearn. & House) M.A. Sherwood
Ostropa mellea Dearn. & House
Distr.: Northeastern United States.
Host: Platanus
Schizoxylon microstomum Ellis & Everh.
Distr.: NJ.
Host: Lyonia
Schizoxylon spiraeae M.A. Sherwood
Distr.: NH.
Host: Spiraea

Scirrhia Nitschke ex Fuckel. Loculoascomycetes, Dothideales

Notes: Barr (Contr. Univ. Michigan Herb. 9:563. 1972).
Scirrhia concaviuscula (Ellis & Everh.) Barr
Dothidea concaviuscula Ellis & Everh.
Dothidella concaviuscula (Ellis & Everh.) Theiss. & Syd.
Plowrightia concaviuscula (Ellis & Everh.) Sacc.
Distr.: NJ, NY.
Host: Magnolia
Scirrhia conigena (Peck) Barr
Mycosphaerella conigena (Peck) House
Sphaerella canadensis Ellis & Everh.
Mycosphaerella thujae Petr. in Syd.
Distr.: Temperate North America.
Notes: On leaves and cones.
Host: Thuja
Scirrhia insculpta (Wallr.) Barr
Dothidea insculpta Wallr.
Dothidella insculpta (Wallr.) Theiss. & Syd.
Systremma insculpta (Wallr.) H. Hess & E. Müller
Plowrightia insculpta (Wallr.) Sacc.
Asteromellopsis insculpta H. Hess & E. Müller (anamorph)
Distr.: Western United States; Europe.
Host: Clematis
Scirrhia lophodermioides Ellis & Everh.
Distr.: HI, PR.
Notes: On living leaves.
Hosts: Bambusa, Phyllostachys, Poaceae
Scirrhia osmundae (Peck & G.W. Clinton) Arx in E. Müller & Arx
Dothidea osmundae Peck & G.W. Clinton
Dothidella osmundae (Peck & G.W. Clinton) Sacc.
Metameris osmundae (Peck & G.W. Clinton) Arx & E. Müller
Scirrhophragma osmundae (Peck & G.W. Clinton) Obrist
Distr.: NY.
Notes: Holm & Holm (Bot. Not. 131:112. 1978).
Host: Osmunda
Scirrhia ostiolata Ellis & B.T. Galloway
Distr.: TX(type).
Host: Cyperus
Scirrhia pluristriae R. Sprague
Distr.: WA.
Host: Muhlenbergia

Scleroderris (Fr.) Bonord. Discomycetes, Helotiales

Notes: This genus is considered a synonym of *Godronia* (Groves. Canad. J. Bot. 43:1195. 1965), but not all names have been redisposed.
Scleroderris eucalypti (Cooke & Harkn.) Sacc.
Dermea eucalypti Cooke & Harkn.
Distr.: CA.
Notes: The generic placement of this species is uncertain.
Host: Eucalyptus

Sclerophoma Höhn. Coelomycetes

Sclerophoma eustomonis Taubenhaus & Ezekiel
Distr.: TX.
Notes: Stem blight of Texas bluebells.
Host: Eustoma
Sclerophoma pityella (Sacc.) Died.
Phoma pityella Sacc.
Distr.: Eastern United States; Europe.
Host: Pinus
Sclerophoma pythiophila (Corda) Höhn.
Dothichiza pythiophila (Corda) Petr.
Phoma acicola (Lév.) Sacc.
Phoma pinicola (Zopf) Sacc.
Phoma strobiligena Desmaz.
Hormonema dematioides Lagerberg & Melin (synanamorph)
Sydowia polyspora (Bref. & Tavel) E. Müller (teleomorph)
Distr.: Cosmopolitan.
Notes: Cankering, needle blight and die-back of conifers (Sutton. The Coelomycetes. p. 486. 1980).
Hosts: Calocedrus, Juniperus, Picea, Pinus, Pseudotsuga
Sclerophoma salicis Died.
Distr.: CA; Germany(type).

Host: Salix

Sclerophthora Thirumalachar, C.G. Shaw, & Narasimhan. Oomycetes, Peronosporales

Sclerophthora macrospora (Sacc.) Thirumalachar, C.G. Shaw & Narasimhan
Sclerospora macrospora Sacc.
Phytophthora macrospora (Sacc.) Ito & I. Tanaka
Distr.: Northern hemisphere; Australia.
Notes: On at least 46 genera of the Poaceae and *Cyperus*. Thirumalachar et al. (Bull. Torrey Bot. Club 80:299. 1953).
Hosts: Agrohordeum, Agropyron, Agrostis, Avena, Bromus, Digitaria, Eleusine, Elymus, Eragrostis, Festuca, Hordeum, Lolium, Muhlenbergia, Oryza, Panicum, Phalaris, Phleum, Poa, Schedonnardus, Secale, Setaria, Sorghum, Sporobolus, Stenotaphrum, Triticum, Zea

Scleropleella Höhn. Loculoascomycetes, Dothideales

Scleropleella hyperborea (Fuckel) L. Holm
Pleospora hyperborea Fuckel
Leptosphaeria hyperborea (Fuckel) Berl. & Voglino
Leptosphaerulina hyperborea (Fuckel) Barr
Distr.: Subarctic region.
Notes: On woody dicotyledonous hosts. Barr (Contr. Univ. Michigan Herb. 9:542. 1972); Holm (Svensk Bot. Tidskr. 69:155. 1975).
Host: Arctostaphylos

Scleropycnium Heald & I. Lewis. Coelomycetes

Notes: Sutton (Mycol. Pap. 141:184. 1977) suggested that this may be a synonym of *Phomopsis*.
Scleropycnium aureum Heald & I. Lewis
Distr.: Southwestern United States.
Notes: Blight of mesquite.
Host: Prosopis

Sclerospora J. Schröt. Oomycetes, Peronosporales

Sclerospora farlowii D. Griffiths
Distr.: Southwestern United States.
Hosts: Chloris, Cynodon
Sclerospora graminicola (Sacc.) J. Schröt.
Peronospora graminicola Sacc.
Distr.: Widespread on all continents except Australia.
Notes: Downy mildew of grasses, green ear of pearl millet (Francis & Williams. C.M.I. Descr. 770. 1983).
Hosts: Alopecurus, Panicum, Setaria, Sorghum, Zea

Sclerotinia Fuckel, nom. cons. Discomycetes, Helotiales

Notes: Kohn (Mycotaxon 9:365. 1979; Phytopathology 69:881. 1979).
Sclerotinia bulborum (Wakk.) Rehm
Distr.: WA; Europe.
Notes: Black slime disease of bulbs (Dennis. Mycol. Pap. 62:147. 1962).
Hosts: Hyacinthus, Iris, Muscari, Narcissus
Sclerotinia homoeocarpa F.T. Bennett
Rhizoctonia monteithiana F.T. Bennett (anamorph)
Distr.: Temperate regions.
Notes: Dollar spot of turf (Mordue. C.M.I. Descr. 618. 1979). Although this species does not belong to *Sclerotinia*, it has not been placed in an appropriate genus (Kohn. Mycotaxon 9:413. 1979).
Hosts: Agrostis, Cynodon, Cyperus, Eremochloa, Festuca, Lolium, Paspalum, Poa, Stenotaphrum, Zoysia
Sclerotinia minor Jagger
Sclerotinia intermedia Ramsey
Sclerotinia sativa Drayton & Groves
Distr.: Temperate regions.
Hosts: Antirrhinum, Apium, Arachis, Brassica, Daucus, Helianthus, Ipomoea, Lactuca, Lycopersicon, Melilotus, Nicotiana, Parthenium, Phaseolus, Solanum, Tragopogon, Tulipa
Sclerotinia sclerotiorum (Lib.) de Bary
Whetzelinia sclerotiorum (Lib.) Korf & Dumont
Sclerotinia libertiana Fuckel
Distr.: Worldwide, most common in cool moist regions.
Notes: Mordue & Holliday (C.M.I. Descr. 513. 1976).
Hosts: Abelmoschus, Aconitum, Alcea, Allium, Amaranthus, Ambrosia, Anemone, Anethum, Antirrhinum, Apium, Aquilegia, Arachis, Aralia, Ardisia, Asarum, Asparagus, Aster, Begonia, Bellis, Beta, Brassica, Calceolaria, Calendula, Camellia, Campanula, Cannabis, Capsicum, Carthamus, Carum, Centaurea, Chenopodium, Chrysanthemum, Cichorium, Cirsium, Citrullus, Citrus, Cnicus, Coreopsis, Crambe, Cucumis, Cucurbita, Cynara, Cynoglossum, Dahlia, Daucus, Delphinium, Dicentra, Digitalis, Dimorphotheca, Dipsacus, Dracocephalum, Epipremnum, Erechtites, Erigeron, Euonymus, Euphorbia, Fagopyrum, Ficus, Foeniculum, Forsythia, Fragaria, Galinsoga, Gerbera, Glycine, Gnaphalium, Gynura, Hedera, Helianthus, Hemigraphis, Hibiscus, Iliamna, Iris, Iva, Lactuca, Lamium, Lavatera, Lens, Lepidium, Liatris, Lilium, Linaria, Linum, Lobularia, Lotus, Lupinus, Lycopersicon, Malus, Malvaviscus, Matthiola, Medicago, Melilotus, Mentha, Mertensia, Moluccella, Myosotis, Nasturtium, Nicandra, Nicotiana, Osteospermum, Paeonia, Panax, Papaver, Parthenium, Pastinaca, Persea, Petroselinum, Petunia, Phaseolus, Philodendron, Pimpinella, Pisum, Proboscidea, Prunus, Pyrus, Ranunculus, Raphanus, Ricinus, Rorippa, Rudbeckia, Rumohra, Scabiosa, Schefflera, Schizanthus, Senecio, Setaria, Sinapis, Sinningia, Solanum, Solenostemon, Sonchus, Sorghum, Spinacia, Stachys, Stellaria, Stokesia, Syringa, Tagetes, Taraxacum, Thlaspi, Tragopogon, Trifolium, Tulipa, Urtica, Vicia, Vigna, Vitis, Youngia, Zinnia
Sclerotinia spermophila M. Noble
Botrytis anthophila Bondartsev (anamorph)
Distr.: Northwestern North America; Europe.
Notes: According to Kohn (Mycotaxon 9:427. 1979), the generic position of this species is uncertain.
Host: Trifolium
Sclerotinia trifoliorum Eriks.
Distr.: Worldwide on forage legumes.
Hosts: Agropyron, Coronilla, [Dactylis], Lens, Lotus, Medicago, Melilotus, [Plantago], Trifolium, Vicia
Sclerotinia veratri Cash & R.W. Davidson
Distr.: CO(type).
Notes: Kohn (Mycotaxon 9:436. 1979) suggested that this species belongs in *Botryotinia*.
Host: Veratrum

Sclerotiomyces Woronichin. Hyphomycetes

Notes: This genus is probably a synonym of *Sclerotium*.
Sclerotiomyces colchicus Woronichin
Distr.: WI; Europe(type).
Notes: On leaves.
Hosts: Acer, Corylus, Geum, Phryma, Polymnia, Ranunculus, Solidago, Viola, Zanthoxylum

Sclerotium Tode:Fr. Agonomycetes

Notes: Fungi are placed in this genus because they produce only sterile mycelium and sclerotia.
Sclerotium caladii Schwein.:Fr.
Distr.: NC(type).
Host: Peltandra
Sclerotium cepivorum Berk.
Stromatinia cepivorum (Berk.) Whetzel
Distr.: Cosmopolitan.
Notes: White rot of *Allium* spp. and other liliaceous plants (Mordue. C.M.I. Descr. 512. 1976).
Host: Allium
Sclerotium deciduum J.J. Davis
Ceratobasidium anceps (Bres. & Syd.) H. Jacks. (teleomorph)
Distr.: WI.
Hosts: Adiantum, Aralia, Aster, Bidens, Fragaria, Hydrophyllum, Lysimachia, Matteuccia, Mitella, Pteridium, Ranunculus, Rubus,

Silphium, Solidago, Trientalis

Sclerotium desmodii Thuem.

Distr.: SC(type).

Host: Desmodium

Sclerotium durum Pers.:Fr.

Distr.: Western United States; Europe(type).

Hosts: Rhododendron, Veratrum

Sclerotium fulvum Fr.

Distr.: MT; Europe(type).

Notes: This fungus may be the sclerotial state of a *Typhula* but the association has not been confirmed.

Host: Sisymbrium

Sclerotium globuliferum J.J. Davis, nom. illeg.

Distr.: WI(type).

Notes: This name was never validly published. The fungus is probably a smut.

Host: Glyceria

Sclerotium hydrophilum Sacc. in Rothert

Distr.: Widespread.

Notes: On various aquatic and semi-aquatic plants. This species apparently has basidiomycetous affinities (Punter et al. Mycologia 76:722. 1984).

Hosts: Alisma, Echinochloa, Glyceria, Nuphar, Nymphaea, Oryza, Sagittaria, Trapa, Typha, Zizania, Zizaniopsis

Sclerotium mendax Sacc.

Distr.: NY, WI; Europe.

Host: Solidago

Sclerotium oryzae Cattaneo

Magnaporthe salvinii (Cattaneo) R. Krause & R.K. Webster (teleomorph)

Nakataea sigmoidea (Cavara) K. Hara (synanamorph)

Distr.: Worldwide in rice-growing regions.

Notes: Stem rot of rice (Krause & Webster. Mycologia 64:103. 1972).

Hosts: Echinochloa, Glyceria, Oryza, Zizania

Sclerotium oryzae Ryker & Gooch var. ***irregulare*** L. Roger

Distr.: LA; Asia.

Notes: Stem rot of rice (Shahjahan & Rush. Pl. Dis. Reporter 63:220. 1979).

Host: Oryza

Sclerotium rhizodes Auersw.

Distr.: Temperate northern hemisphere.

Notes: Leaf blight or root rot of Poaceae.

Hosts: Agropyron, Agrostis, Bromus, Calamagrostis, Glyceria, Hordeum, Phalaris, Phleum, Poa, Sphenopholis

Sclerotium rolfsii Sacc.

Athelia rolfsii (Curzi) Tu & Kimbrough (teleomorph)

Distr.: Cosmopolitan.

Notes: Leaf blight, stem canker, damping-off, crown and root rot on a wide variety of hosts (Mordue. C.M.I. Descr. 410. 1974).

Hosts: Abelmoschus, Acnida, Ageratum, Aglaonema, Agropyron, Agrostis, Ajuga, Alcea, Aleurites, Allium, Aloe, Alternanthera, Amaranthus, Amaryllis, Ambrosia, Anacardium, Anemone, Antirrhinum, Aphelandra, Apium, Aquilegia, Arachis, Araucaria, Arctostaphylos, Arctotis, Ardisia, Aristolochia, Arrhenatherum, Asparagus, Asplenium, Aster, Aucuba, Avena, Azara, Barbarea, Barleria, Begonia, Berberis, Beta, Brassica, Buddleja, Cajanus, Caladium, Calendula, Callistephus, Campanula, Canna, Cannabis, Capsicum, Carica, Casuarina, Catalpa, Cattleya, Centaurea, Cerastium, Chlorophytum, Chrysanthemum, Chrysogonum, Cichorium, Citrullus, Citrus, Cnicus, Codiaeum, Colocasia, Convallaria, Conyza, Coreopsis, Cornus, Coronilla, Cosmos, Crataegus, Crotalaria, Cucumis, Cucurbita, Cyamopsis, Cynara, Cynodon, Cynoglossum, Dahlia, Daphne, Datura, Daucus, Davallia, Delphinium, Descurainia, Desmodium, Dianthus, Dicentra, Dichondra, Dichorisandra, Dieffenbachia, Digitalis, Dimorphotheca, Diodia, Dipsacus, Dracaena, Dracocephalum, Dryopteris, Duranta, Dyschoriste, Echinops, Elaeagnus, Eleusine, Emilia, Epilobium, Epipremnum, Erechtites, Erigeron, Ernodea, Erodium, Eschscholzia, Eucharis, Eupatorium, Euphorbia, Exacum, Festuca, Ficus, Fittonia, Forsythia, Fragaria, Gardenia, Gerbera, Gilia, Gladiolus, Gloriosa, Glycine, Gnaphalium, Gossypium, Helianthus, Helichrysum, Hemerocallis, Hemigraphis, Hibiscus, Hippeastrum, Hordeum, Hosta, Hydrangea, Hydrilla, Hyptis, Iberis, Impatiens, Indigofera, Ipomoea, Iris, Jasminum, Juglans, Kalanchoe, Kummerowia, Lablab, Lactuca, Lantana, Lathyrus, Lavatera, Lespedeza, Liatris, Ligularia, Ligustrum, Lilium, Limonium, Linaria, Linum, Lippia, Liriope, Lolium, Lotus, Lupinus, Lycopersicon, Lycopus, Lysimachia, Malus, Matthiola, Medicago, Melilotus, Mitchella, Mollugo, Monarda, Mucuna, Myrtus, Narcissus, Nasturtium, Neomarica, Nepeta, Nicotiana, Nymphaea, Onobrychis, Orchidaceae, Ornithogalum, Oryza, Oxalis, Paeonia, Parthenium, Passiflora, Pellionia, Pennisetum, Penstemon, Pentas, Peperomia, Persea, Petunia, Phalaenopsis, Phaseolus, Philodendron, Phlox, Phyla, Physalis, Physostegia, Pilea, Piper, Pisum, Pittosporum, Plantago, Platycerium, Poa, Polianthes, Poncirus, Proboscidea, Prunella, Prunus, Ranunculus, Rhaphiolepis, Rheum, Richardia, Ricinus, Rosa, Rudbeckia, Saccharum, Salvia, Sansevieria, Sarracenia, Scabiosa, Schefflera, Sechium, Sedum, Senecio, Senna, Sida, Silene, Sinningia, Solanum, Sorghum, Spathiphyllum, Stokesia, Striga, Syngonium, Syringa, Tagetes, Taraxacum, Thuja, Thunbergia, Tithonia, Torreya, Tragopogon, Trifolium, Triticum, Tulipa, Vanda, Verbesina, Vernonia, Veronica, Vicia, Vigna, Vinca, Viola, Vitis, Washingtonia, Xantheranthemum, Xanthosoma, Yucca, Zantedeschia, Zea, Zebrina, Zephyranthes, Zingiber, Zinnia

Sclerotium rolfsii Sacc. var. ***delphinii*** (Welch) Boerema & Hamers

Sclerotium delphinii Welch

Distr.: Widespread.

Notes: Stalk rot of bulbs and various other plants (Moore. Diseases of bulbs. p. 131. 1979; Boerema & Hamers. Neth. J. Pl. Pathol. 94:7. 1988).

Hosts: Aconitum, Ajuga, Antirrhinum, Daucus, Delphinium, Dracocephalum, Hosta, Iris, Linum, Lobelia, Lysimachia, Narcissus, Piqueria, Potentilla, Scilla, Trillium, Valeriana, Viola

Sclerotium sacidioides Speg.

Distr.: VA(type).

Host: Arundinaria

Sclerotium sulcatum Roberge in Desmaz.

Claviceps caricina D. Griffiths

Myriosclerotinia sulcata (Whetzel) Buchwald, nom. illeg. (teleomorph)

Distr.: Eastern and western United States; Europe.

Notes: This species has not been well characterized nor has the teleomorph been adequately named (Kohn. Mycotaxon 9:428. 1979).

Host: Carex

Sclerotium varium Pers.:Fr.

Distr.: AK; Europe.

Host: Heracleum

Sclerotium zizaniae J.J. Davis, nom. illeg.

Distr.: MD, WI(type).

Notes: Punter et al. (Mycologia 76:730. 1984) explained the status of this name which was never validly published.

Host: Zizania

Scolecobasidium E. Abbott. Hyphomycetes

Scolecobasidium dendroides Pirozynski & C.S. Hodges

Distr.: SC.

Notes: On dead, fallen leaves (Ellis. More dematiaceous Hyphomycetes. p. 192. 1976).

Host: Persea

Scolecobasidium echinophilum (C. Massal.) Sutton

Dactylaria echinophila C. Massal.

Distr.: HI; England; Italy.

Notes: On woody to leathery plant parts (Sutton. Trans. Brit. Mycol. Soc. 61:424. 1973).

Host: Eucalyptus

Scolecobasidium humicola Barron & L.V. Busch

Distr.: North America; India.

Hosts: Acacia, Eichhornia, Juncus, Metrosideros

Scolecobasidium terreum E. Abbott

Distr.: North America; India.

Host: Cheirodendron

Scolecobasidium variabile Barron & L.V. Busch

Distr.: North America; India.

Host: Acacia

Scoleconectria Seaver. Pyrenomycetes, Hypocreales

Scoleconectria cucurbitula (Tode:Fr.) C. Booth
Chilonectria cucurbitula (Tode:Fr.) Sacc.
Nectria cucurbitula (Tode:Fr.) Fr.
Nectria cylindrospora Sollm.
Ophionectria cylindrospora (Sollm.) Berl. & Voglino
Ophionectria scolecospora Bref. & Tavel
Scoleconectria scolecospora (Bref. & Tavel) Seaver
Zythiostroma pinastri (P. Karst.) Höhn. (anamorph)
Distr.: Circumboreal.
Notes: On bark and leaf scars of conifers (Booth. Mycol. Pap. 73. 1959; Rossman. Mycol. Pap. 150:26. 1983). Some older reports refer to *Nectria fuckeliana* C. Booth.
Hosts: Abies, Larix, Picea, Pinus

Scolecosporiella Petr. Coelomycetes

Notes: Sutton (The Coelomycetes. p. 83. 1980).
Scolecosporiella sisyrinchii (Ellis & Everh.) Sutton
Kellermania sisyrinchii Ellis & Everh.
Brencklea sisyrinchii (Ellis & Everh.) Petr.
Distr.: Central and western United States.
Host: Sisyrinchium
Scolecosporiella typhae (Oudem.) Petr.
Hendersonia typhae Oudem.
Scolicosporium typhae (Oudem.) Höhn.
Phaeosphaeria typharum (Desmaz.) L. Holm (teleomorph)
Distr.: Temperate North America; Europe; India.
Notes: Sutton (The Coelomycetes. p. 83. 1980).
Host: Typha

Scolicosporium Lib. in Roum. Coelomycetes

Scolicosporium coryli Dearn. & House
Distr.: NY(type).
Host: Corylus
Scolicosporium macrosporium (Berk.) Sutton
Coryneum macrosporium Berk.
Scolicosporium fagi Lib. in Sacc.
Asteromassaria macrospora (Desmaz.) Höhn. (teleomorph)
Distr.: United States; Europe.
Notes: On twigs, usually of *Fagus* (Spooner & Kirk. Trans. Brit. Mycol. Soc. 78:251. 1982; Sutton. The Coelomycetes. p. 193. 1980).
Host: Castanea
Scolicosporium pedicellatum Dearn. & Overh.
Distr.: PA.
Host: Malus

Scolicotrichum Kunze in Kunze & J.C. Schmidt:Fr. Hyphomycetes

Notes: Hughes (Canad. J. Bot. 36:831. 1958) considered this to be a dubious genus, based on an examination of the presumed type specimen. *Scolecotrichum* is an orthographic variant.
Scolicotrichum maculicola Ellis & Kellerm.
Distr.: Central and western North America.
Hosts: Arundo, Phragmites
Scolicotrichum typhae Ellis & Everh.
Distr.: KS(type).
Host: Typha

Scopinella Lév. in C. d'Orb. Pyrenomycetes, Sordariales

Scopinella caulincola (Fuckel) Malloch
Ceratostoma caulincola Fuckel
Melanospora caulincola (Fuckel) Arx & E. Müller
Distr.: North America; Europe.
Host: Alnus
Scopinella solani (Zukal) Malloch
Melanospora solani Zukal
Melanospora poae D. Griffiths
Distr.: Probably widespread in temperate regions (Cannon & Hawksworth. Bot. J. Linn. Soc. 84:141. 1982).
Notes: On decaying plant material (Malloch. Fungi Canadenses 82. 1976).
Host: Poa

Scopulariopsis Bainier. Hyphomycetes

Notes: Morton & Smith (Mycol. Pap. 86. 1963); Ellis (Dematiaceous Hyphomycetes. p. 326. 1971)
Scopulariopsis asperula (Sacc.) S.J. Hughes
Torula asperula Sacc.
Distr.: Widespread.
Hosts: Alnus, Pseudotsuga
Scopulariopsis brevicaulis (Sacc.) Bainier
Distr.: Cosmopolitan.
Notes: Isolated from various substrates including air, soil, animals, wood and various plants. This species is used for the detection of arsenic. Onions (C.M.I. Descr. 100. 1966).
Hosts: Alnus, Helianthus, Pseudotsuga, Zea
Scopulariopsis brumptii Salvanet-Duval
Marssoniella grisea (G. Sm.) G. Sm.
Distr.: Cosmopolitan.
Notes: Isolated from air, soil, straw, and various plants.
Hosts: Alnus, Helianthus, Zea
Scopulariopsis candida (Guég.) Vuill.
Distr.: Widespread.
Notes: Isolated from cheese, human lesions and a variety of substrates.
Host: Helianthus
Scopulariopsis koningii (Oudem.) Vuill.
Distr.: Widespread.
Host: Carya
Scopulariopsis stercoraria (Link) S.J. Hughes
Distr.: OR; Europe.
Notes: According to Morton & Smith (Mycol. Pap. 86:89. 1963) this species is unidentifiable.
Host: Trifolium

Scorias Fr.:Fr. Loculoascomycetes, Dothideales

Scorias spongiosa (Schwein.:Fr.) Fr.
Distr.: Eastern North America; Europe.
Notes: Especially common on *Fagus* and *Alnus*. Reynolds (Mycotaxon 8:430. 1979) reviewed the species and rejected the anamorph *Polychaeton* suggested by Hughes (Mycologia 68:802. 1976).
Hosts: Alnus, Corylus, Fagus, Fragaria, Pinus, Viola

Scoriomyces Ellis & Sacc. Hyphomycetes

Scoriomyces andersonii Ellis & B.T. Galloway
Distr.: MT(type).
Host: Pinus

Scutellinia (Cooke) Lambotte. Discomycetes, Pezizales

Notes: These species occur on soil and have been reported in association with vascular plants. Denison (Mycologia 51:605. 1959).
Scutellinia erinaceus (Schwein.:Fr.) Kuntze
Humaria erinacea (Schwein.:Fr.) Kanouse
Trichophaea erinaceus (Schwein.:Fr.) Le Gal
Distr.: Western hemisphere.
Notes: on well-rotted, deciduous wood.
Hosts: Carya, Populus, Tilia
Scutellinia hirta (Schumach.:Fr.) Kuntze
Lachnea hirta (Schumach.:Fr.) Gill.
Distr.: HI; Europe(type).
Hosts: Alnus, Mangifera, Populus
Scutellinia pennsylvanica (Seaver) Denison
Melastiza pennsylvanica Seaver
Melastiziella pennsylvanica (Seaver) Svrček
Distr.: Temperate regions.
Notes: On rotting wood, rarely on adjacent soil.
Host: Fagus

Scutellinia scutellata (L.:Fr.) Lambotte
Humaria scutellata (L.:Fr.) Fuckel
Lachnea scutellata (L.:Fr.) Gill.
Patella scutellata (L.:Fr.) Morg.
Distr.: Temperate northern hemisphere.
Notes: Usually on rotting wood.
Hosts: Abies, Acer, Aleurites, Alnus, Betula, Castanea, Fagus, Pseudotsuga, Quercus, Salix, Tsuga

Scytalidium Pesante. Hyphomycetes

Scytalidium lignicola Pesante
Distr.: Widespread.
Notes: Isolated from wood, roots and soil (Ellis. Dematiaceous Hyphomycetes. p. 28. 1971).
Hosts: Alnus, Pinus, Pseudotsuga

Scytinostroma Donk. Basidiomycetes, Aphyllophorales

Scytinostroma arachnoideum (Peck) R.L. Gilbertson
Distr.: North America.
Notes: On wood.
Hosts: Picea, Pinus, Pseudotsuga, Tsuga
Scytinostroma fallax Burdsall & Nakasone
Distr.: Southeastern United States.
Notes: Burdsall & Nakasone (Mycologia 73:454. 1981).
Host: Pinus
Scytinostroma galactinum (Fr.) Donk
Corticium galactinum (Fr.) Burt
Distr.: Temperate northern hemisphere.
Notes: Root, butt, and collar rots of conifers and hardwoods. Nakasone and Micales (Mycologia 80:546. 1988) have separated this species complex into two species distinguished by substrate and isozyme profiles.
Hosts: Abies, Acer, Baptisia, Betula, Ceanothus, Clethra, Cornus, Exochorda, Fagus, Hibiscus, Ilex, Iris, Jasminum, Kalmia, Larix, Malus, Nerium, Paeonia, Picea, Pinus, Populus, Prunus, Pseudotsuga, Quercus, Rhus, Rubus, Silene, Spiraea, Thuja, Tsuga, Viburnum
Scytinostroma ochroleucum (Bres. & Torrend) Donk
Gloeocystidium ochroleucum Bres. & Torrend in Torrend
Corticium abeuns Burt
Distr.: North America; South Africa; Asia; Europe.
Notes: On wood.
Hosts: Pinus, Pseudotsuga, Tsuga
Scytinostroma portentosum (Berk. & M.A. Curtis in Berk.) Donk
Corticium portentosum Berk. & M.A. Curtis in Berk.
Distr.: North America.
Notes: On hardwoods. The European form is *S. aluta* Lanquetin.
Hosts: Acacia, Acer, Aleurites, Alnus, Betula, Bougainvillea, Carya, Crossosoma, Gleditsia, Juniperus, Mortonia, Prosopis, Pseudotsuga, Quercus, Sambucus, Sapindus

Scytinostromella Parmasto. Basidiomycetes, Aphyllophorales

Scytinostromella heterogenea (Bourd. & Galzin) Parmasto
Peniophora heterogenea Bourd. & Galzin
Gloeocystidiellum heterogeneum (Bourd. & Galzin) Donk
Distr.: Northern hemisphere.
Notes: On wood.
Host: Pinus
Scytinostromella nannfeldtii (J. Eriksson) G. Freeman & R. Petersen
Gloeocystidiellum nannfeldtii J. Eriksson
Distr.: Northern hemisphere.
Notes: On wood.
Host: Picea

Seaverinia Whetzel. Discomycetes, Helotiales

Seaverinia geranii (Seaver & W.T. Horne) Whetzel
Sclerotinia geranii Seaver & W.T. Horne
Distr.: Eastern North America.
Notes: Hennbert (Persoonia 7:193. 1973) redescribed the anamorph.
Host: Geranium

Sebacina Tul. Basidiomycetes, Tremellales

Sebacina epigaea (Berk. & Broome) Bourd. & Galzin
Distr.: Temperate northern hemisphere.
Notes: On wood, bark and plant debris.
Hosts: Betula, Elaeagnus, Fraxinus, Juniperus, Mimosa, Populus, Quercus
Sebacina helvelloides (Schwein.) Burt
Distr.: Temperate northern hemisphere.
Notes: Encrusting woody and herbaceous stems, leaves and plant debris.
Hosts: Fagus, Fraxinus, Kalmia, Ostrya, Picea, Prunus, Quercus
Sebacina incrustans (Pers.:Fr.) Tul.
Distr.: Temperate northern hemisphere.
Notes: Encrusting woody and herbaceous stems, leaves, and plant debris.
Hosts: Acer, Adiantum, Aster, Cornus, Corylus, Elephantopus, Erigeron, Glechoma, Hypericum, Juniperus, Nyssa, Onoclea, Poa, Polystichum, Potentilla, Quercus, Thelypteris, Tilia, Viola, Vitis
Sebacina sublilacina G.W. Martin
Exidiopsis sublilacina (G.W. Martin) Ervin
Distr.: GA, IA(type).
Notes: According to Wells (Mycologia 53:367. 1961) this species is probably related to *Myxarium*.
Hosts: Liriodendron, Ulmus

Seimatosporium Corda. Coelomycetes

Notes: Shoemaker & Müller (Canad. J. Bot. 42:403. 1964); Sutton (The Coelomycetes. p. 279. 1980).
Seimatosporium anomalum (Harkn.) Shoemaker
Pestalotia anomala Harkn.
Cryptostictis anomala (Harkn.) Guba
Distr.: CA(type).
Notes: On twigs.
Host: Eriogonum
Seimatosporium arbuti (Bonar) Shoemaker
Cryptostictis arbuti (Bonar) Zeller
Distr.: Western United States.
Hosts: Arbutus, Arctostaphylos, Gaultheria, Kalmia, Ledum
Seimatosporium berckmansii (Milbrath) Sutton
Coryneum berckmansii Milbrath
Distr.: OR.
Notes: On leaves.
Hosts: Cupressus, Thuja
Seimatosporium caninum (Brunaud) Sutton
Hendersonia canina Brunaud
Distr.: NY; Europe.
Host: Rosa
Seimatosporium cassiopes (Rostr.) Sutton
Coryneum cassiopes Rostr.
Coryneum sorbi Peck
Distr.: Widespread in Europe and North America.
Host: Sorbus
Seimatosporium caudatum (G. Preuss) Shoemaker
Sporocadus caudatus G. Preuss
Cryptostictis caudata (G. Preuss) Sacc.
Coryneum foliorum (Fuckel) Petr.
Hendersonia foliorum Fuckel
Cryptostictis ludibunda Vestergr.
Distr.: Widespread.
Notes: On leaves and stems of various dicotyledons.
Hosts: Crataegus, Primula, Pyrus, Rubus, Thuja
Seimatosporium discosioides (Ellis & Everh.) Shoemaker
Pestalotia discosioides Ellis & Everh.
Monochaetia discosioides (Ellis & Everh.) Sacc.
Coryneum microstictum Berk. & Broome var. *foliae* Dearn. & Overh.
Amphichaeta rosicola H.C. Greene
Distr.: Eastern North America.
Notes: Leaf spot.
Hosts: Geum, Pyrus, Rosa

Seimatosporium foliicola (Berk.) Shoemaker
Podisoma foliicola Berk.
Hendersonia foliicola (Berk.) Fuckel
Distr.: Eastern United States; Europe; possibly Great Britain.
Notes: On needles.
Host: Pinus
Seimatosporium kriegerianum (Bres.) Morgan-Jones & Sutton in Sutton
Pestalotia kriegeriana Bres.
Hyaloceras kriegeriana (Bres.) Died.
Monoceras kriegerianum (Bres.) Guba
Discosia bubakii Kab.
Paradidymella tosta (Berk. & Broome) Petr. (teleomorph)
Distr.: Cosmopolitan.
Host: Epilobium
Seimatosporium lichenicola (Corda) Shoemaker & E. Müller
Hendersonia crataegi Brenckle
Hendersonia eucalypti Cooke & Harkn.
Coryneum foliicola Fuckel
Coryneum microstictum Berk. & Broome
Hendersonia platypus Ellis & Everh.
Discostroma corticola (Fuckel) I. Brockmann (teleomorph)
Distr.: Temperate North America; Europe.
Notes: On limbs of woody plants (Pirozynski & Shoemaker. Canad. J. Bot. 48:2199. 1970).
Hosts: Crataegus, Eucalyptus, Ledum, Malus, Prunus, Pyrus, Rhododendron, Rosa, Rubus, Vaccinium
Seimatosporium mariae (G.W. Clinton) Shoemaker
Cryptostictis mariae (G.W. Clinton) Sacc.
Distr.: Eastern and western United States.
Host: Rhododendron
Seimatosporium parasiticum (Dearn. & House) Shoemaker
Pestalotia monochaetoidea Sacc. & Ellis var. *parasitica* Dearn. & House
Cryptostictis parasitica (Dearn. & House) Guba
Distr.: NY.
Host: Physocarpus
Seimatosporium pestalozzioides (Sacc.) Sutton
Coryneum pestalozzioides Sacc.
Coryneum tumoricola Peck
Distr.: Eastern North America; Europe.
Host: Ulmus
Seimatosporium pezizoides (Ellis & Everh.) Sutton
Coryneum pezizoides Ellis & Everh.
Distr.: Western United States.
Notes: On bark.
Host: Salix
Seimatosporium rhododendri (Schwein.) Pirozynski & Shoemaker
Coryneum rhododendri Schwein.
Coryneum triseptatum Peck
Distr.: Eastern United States.
Notes: According to Pirozynski & Shoemaker (Canad. J. Bot. 48:2199. 1970), reports on hosts other than *Rhododendron maximum* are either misdeterminations or are in need of verification.
Hosts: [Kalmia], [Ledum], [Quercus], Rhododendron, [Vaccinium]
Seimatosporium rosae Corda in Sturm
Monochaetia compta (Sacc.) Allesch.
Pestalotia compta Sacc.
Hyaloceras comptum (Sacc.) Died.
Cryptostictis cynosbati (Fuckel) Sacc.
Hendersonia cynosbati Fuckel
Hyaloceras depazeoides (G. Otth) Died.
Pestalotia depazeoides G. Otth
Discostroma rosae I. Brockmann (teleomorph)
Distr.: Widespread in Europe and North America.
Notes: On leaves, overwintered fruits, and dead branches (Brockmann. Sydowia 28:324. 1975).
Host: Rosa
Seimatosporium vaccinii (Fuckel) B. Eriksson
Coryneum vaccinii Fuckel
Coryneum microstictum Berk. & Broome var. *staphyleae* W.H. Davis
Coryneum rhododendri Schwein. var. *fusoideum* Dearn., nom. inval.
Distr.: Temperate regions.
Host: Staphylea

Seiridium Nees:Fr. Coelomycetes
Notes: Sutton (The Coelomycetes. p. 274. 1980).
Seiridium breviaristata (Tracy & Earle) Sutton
Pestalotia breviaristata Tracy & Earle
Monochaetia breviaristata (Tracy & Earle) Sacc.
Distr.: MS, NY.
Host: Campsis
Seiridium cardinale (W. Wagener) Sutton & I. Gibson
Coryneum cardinale W. Wagener
Distr.: Temperate regions.
Notes: Canker of Cupressaceae (Sutton & Gibson. C.M.I. Descr. 326. 1972).
Hosts: Calocedrus, Chamaecyparis, Cupressus, Juniperus, Thuja
Seiridium castaneae (Berk. & M.A. Curtis ex Sacc.) Sutton
Coryneum decipiens Berk. & M.A. Curtis f. *castanae* Berk. & M.A. Curtis ex Sacc.
Distr.: VA(type).
Notes: On bark.
Host: Castanea
Seiridium juniperi (Allesch.) Sutton
Pestalotia juniperi Allesch.
Distr.: OR; Europe.
Hosts: Calocedrus, Thuja
Seiridium unicorne (Cooke & Ellis) Sutton
Pestalotia unicornis Cooke & Ellis
Monochaetia unicornis (Cooke & Ellis) Sacc.
Cryptostictis cupressi Guba
Monochaetia mali (Ellis & Everh.) Sacc.
Pestalotia mali Ellis & Everh.
Pestalotia uniseta Tracy & Earle
Lepteutypa cupressi (Nattras, C. Booth, & Sutton) H.J. Swart (teleomorph)
Distr.: Cosmopolitan.
Notes: According to Swart (Trans. Brit. Mycol. Soc. 61:71. 1973) this species causes canker of Cupressaceae. However, Boesewinkel (Trans. Brit. Mycol. Soc. 80:544. 1983) delineated the three species causing cypress canker and accepted the broad host range of *S. unicorne*. Booth & Gibson (C.M.I. Descr. 325. 1972 as *Rhynchosphaeria cupressi*).
Hosts: Celtis, Chamaecyparis, Cupressus, Juniperus, Malus, Vitis

Selenophoma Maire. Coelomycetes
Notes: Most species of *Selenophoma*, now *Pseudoseptoria*, are restricted to grasses.
Selenophoma boltoniae (Dearn. & Barth.) Sivanesan
Macrophoma boltoniae Dearn. & Barth.
Distr.: Central United States.
Host: Boltonia
Selenophoma donacis (Pass.) R. Sprague & A.G. Johnson var. **linearis** R. Sprague & A.G. Johnson
Distr.: Northwestern United States.
Notes: Although the type variety has been transferred to *Pseudoseptoria*, this variety has not been studied.
Hosts: Agropyron, Vulpia
Selenophoma drabae (Fuckel) Petr.
Phoma drabae Fuckel
Rhabdospora drabae (Fuckel) Berl. & Voglino
Distr.: Boreal regions.
Host: Carex
Selenophoma juncea (Mont.) Arx
Sacidium junceum Mont.
Selenophomopsis juncea (Mont.) Petr.
Guignardia cytisi (Fuckel) Arx & E. Müller
Distr.: United States; Europe.
Notes: Sutton (The Coelomycetes. p. 374. 19800.
Host: Cytisus
Selenophoma linicola Vanterpool
Distr.: CA; Canada(Saskatchewan,type).
Notes: Described from *Linum* (Vanterpool. Mycologia 39:341. 1947).
Host: Iliamna
Selenophoma maculicola Wehmeyer
Distr.: WY.
Host: Cymopterus

Selenosporella G. Arnaud ex MacGarvie. Hyphomycetes

Selenosporella verticillata Sutton & C.S. Hodges
Distr.: HI.
Notes: Sutton & Hodges (Nova Hedwigia 29:602. 1978).
Host: Eucalyptus

Septobasidium Pat., nom. cons. Basidiomycetes, Septobasidiales

Notes: Species described by Couch (The genus *Septobasidium*. 1938) were not validly published because of the lack of latin diagnoses. Most species occur on scale insects, but are often reported on their plant hosts.
Septobasidium castaneum Burt
Distr.: Southeastern United States.
Hosts: Acer, Camellia, Carya, Cornus, Crataegus, Fagus, Ilex, Magnolia, Myrica, Quercus
Septobasidium cokeri Couch, nom. nud.
Distr.: Eastern United States.
Hosts: Acer, Carya, Cornus, Fagus, Pyracantha, Quercus
Septobasidium fumigatum Burt
Distr.: Southeastern United States; Brazil.
Hosts: Acer, Coccoloba, Cornus, Nyssa, Tilia
Septobasidium pilosum Boedijn & B.A. Steinmann
Johncouchia mangiferae (Batista) S.J. Hughes & Cavalcanti (anamorph)
Distr.: Southern United States to Brazil; Java; Taiwan.
Notes: Parasitic on scale insects on limbs and leaves.
Host: Mangifera
Septobasidium pseudopedicellatum Burt
Distr.: Southeastern United States.
Hosts: Acer, Aesculus, Aleurites, Betula, Camellia, Carpinus, Carya, Castanopsis, Chaenomeles, Citrofortunella, Citrus, Cornus, Crataegus, Ficus, Fraxinus, Ilex, Leucothoe, Liquidambar, Macadamia, Magnolia, Malus, Mangifera, Myrica, Nyssa, Ostrya, Persea, Phoradendron, Poncirus, Prunus, Pyrus, Quercus, Smilax, Taxodium, Ulmus

Septocylindrium Bonord. ex Sacc. Hyphomycetes

Septocylindrium aromaticum Sacc.
Ramularia aromatica (Sacc.) Höhn.
Cylindrosporium acori Peck
Distr.: Central and eastern North America; Europe.
Notes: Jenkins & Viégas (Pl. Dis. Reporter 45:145. 1961).
Host: Acorus
Septocylindrium concomitans (Ellis & Holw.) Halst. in Seym. & Earle
Ramularia concomitans Ellis & Holw.
Distr.: Central and eastern North America.
Host: Bidens
Septocylindrium hydrophylli E.Y. Daniels in Tehon & E.Y. Daniels
Distr.: IL(type).
Host: Hydrophyllum
Septocylindrium ranunculi Peck
Distr.: North America; Europe; India.
Notes: On leaves.
Host: Ranunculus
Septocylindrium scirpinum Peck
Distr.: NY(type).
Host: Scirpus

Septocyta Petr. Coelomycetes

Septocyta ruborum (Lib.) Petr.
Ascochyta ruborum Lib.
Septoria ruborum (Lib.) Westend.
Rhabdospora ramealis (Roberge ex Desmaz.) Sacc.
Rhabdospora rubi Ellis in G. Martin
Distr.: Temperate North America; Europe.
Notes: On stems (Punithalingam. C.M.I. Descr. 667. 1980).
Host: Rubus

Septogloeum Sacc. Coelomycetes

Notes: Sutton & Pollack (Mycopath. Mycol. Appl. 52:331. 1974) redescribed this genus restricting it to two species. More than 120 species have been described in *Septogloeum* which should be referred elsewhere.
Septogloeum angelicae (Cooke) Sacc.
Gloeosporium angelicae Cooke
Distr.: SC(type).
Host: Angelica
Septogloeum arachidis Racib.
Distr.: HI; Java.
Host: Arachis
Septogloeum bartholomaei (Peck) Wollenweb.
Fusarium bartholomaei Peck
Trichofusarium bartholomaei (Peck) Sacc.
Distr.: KS.
Host: Sorghastrum
Septogloeum celtidis Dearn.
Distr.: NY.
Host: Celtis
Septogloeum lupini Ellis & Everh.
Distr.: Central and western United States.
Host: Lupinus
Septogloeum maculans Harkn.
Distr.: CA.
Host: Salix
Septogloeum parasiticum Kauffm. & Dearn.
Distr.: MI.
Host: Ulmus
Septogloeum profusum (Ellis & Everh.) Sacc.
Gloeosporium profusum Ellis & Everh.
Distr.: Central and eastern United States.
Hosts: Corylus, Ulmus
Septogloeum querceum J.J. Davis
Phyllosticta quercea J.J. Davis (synanamorph)
Distr.: WI.
Host: Quercus
Septogloeum rhopaloideum Dearn. & Bisby
Distr.: Western North America.
Host: Populus
Septogloeum salicinum (Peck) Sacc.
Gloeosporium salicinum Peck
Distr.: United States.
Host: Salix
Septogloeum salicis-fendlerianae Dearn. & Barth.
Distr.: Western United States.
Host: Salix
Septogloeum schizonoti Dearn.
Distr.: Temperate North America.
Host: Holodiscus
Septogloeum spartinae (Ellis & Everh.) Wollenweb. & Reinking
Fusarium spartinae Ellis & Everh.
Distr.: CA.
Host: Spartina
Septogloeum subnudum J.J. Davis
Distr.: IL, WI.
Host: Smilax
Septogloeum thomasianum (Sacc.) Höhn. in Lind
Gloeosporium thomasianum Sacc.
Marssonia thomasiana (Sacc.) Sacc.
Marssonina thomasiana (Sacc.) Magnus
Distr.: North America; Europe.
Notes: Sutton (The Coelomycetes. p. 613. 1980).
Hosts: Celastrus, Euonymus
Septogloeum variegatum Ellis, nom. nud.
Distr.: CA.
Host: Alnus

Septomyxa Sacc. Coelomycetes

Notes: This genus is considered a synonym of *Diplodina*, but not all names have been redisposed.
Septomyxa affinis Wollenweb.
Fusarium affine Sherb. non Lambotte & Fautrey, hom. illeg.
Distr.: Central and eastern United States.

Hosts: Brassica, Nicotiana, Solanum
Septomyxa grisea Dearn. & House
Distr.: NY(type).
Host: Salix

Septonema Corda. Hyphomycetes

Septonema fasciculare (Corda) S.J. Hughes
Helminthosporium fasciculare Corda
Dendryphion pini Höhn.
Cladosporium rectum G. Preuss
Distr.: Temperate North America; Europe.
Notes: Usually on the inner side of bark on dead logs and stumps of conifers.
Host: Abies
Septonema griseofulvum Ellis & Everh.
Distr.: MT(type).
Host: Populus
Septonema secedens Corda
Distr.: Eastern United States(rare); Europe.
Notes: On dead wood.
Hosts: Ceanothus, Juncus
Septonema subramosum Ellis & Everh.
Distr.: CA, OR.
Host: Alnus

Septopatella Petr. Coelomycetes

Notes: Monotypic (Dyko & Sutton. Canad. J. Bot. 57:370. 1979).
Septopatella septata (Jaap) Petr.
Pseudocenangium septatum Jaap
Distr.: CA; Europe.
Host: Pinus

Septoria Sacc., nom. cons. Coelomycetes

Notes: Jørstad (Skr. Norske Vidensk.-Akad. Oslo, Mat.-Naturvidensk. Kl. 22:1. 1965; 24:1. 1967); Radulescu et al. (Septoriozele din Romania. 1973); Sprague (Oregon State Monogr. Stud. Bot. 1. 1950); Teterevnikova-Babayan (Fungi of the genus *Septoria* in the U.S.S.R. 1987).
Septoria abortiva (Ellis & Kellerm.) Tehon & E.Y. Daniels
Phyllosticta abortiva Ellis & Kellerm.
Distr.: Central United States.
Host: Menispermum
Septoria aceris (Lib.) Berk. & Broome
Phloeospora aceris (Lib.) Sacc.
Distr.: Widespread in Europe and North America.
Host: Acer
Septoria aceris-macrophylli Peck
Distr.: OR, WA(type).
Host: Acer
Septoria acetosella Dearn. & House
Distr.: NY.
Host: Oxalis
Septoria achilleae H.C. Greene
Distr.: WI.
Host: Achillea
Septoria aciculosa Ellis & Everh.
Distr.: Central and eastern North America.
Host: Fragaria
Septoria adenocauli Ellis & Everh.
Distr.: ID.
Host: Adenocaulon
Septoria aegopodii Desmaz.
Sphaeria aegopodii Fr. (teleomorph)
Distr.: Central and eastern United States.
Host: Osmorhiza
Septoria agrimoniae Roum., nom. nud.
Distr.: Central and eastern United States; Europe.
Notes: The exsiccati specimen lacks a description.
Host: Agrimonia
Septoria agrimoniae-eupatoriae E. Bommer & M. Rousseau
Distr.: ID, WI; Belgium.
Host: Agrimonia
Septoria agropyrina Lobik
Distr.: United States; Asia; Europe.
Notes: Leaf spot and blotch (Sprague. Diseases of cereals and grasses in North America. p. 215. 1950).
Hosts: Agrohordeum, Agropyron, Elymus, Leymus
Septoria agrosticola R. Sprague
Distr.: Northwestern United States.
Host: Agrostis
Septoria alabamensis Atk., nom. nud.
Distr.: AL.
Host: Glechoma
Septoria alba Ellis & Everh.
Distr.: IL, KS.
Host: Silphium
Septoria albicans Ellis & Everh.
Distr.: MA, WI.
Host: Saxifraga
Septoria albopunctata Cooke
Distr.: Southern United States.
Hosts: Rhododendron, Vaccinium
Septoria allardii J.A. Stevenson & Pollack
Distr.: VA.
Host: Veratrum
Septoria alni Sacc.
Distr.: Temperate northern hemisphere.
Host: Alnus
Septoria alnifolia Ellis & Everh.
Distr.: Temperate North America.
Host: Alnus
Septoria ambrosiicola Speg.
Distr.: TX; Argentina(type).
Host: Ambrosia
Septoria ammodeniae Dearn.
Distr.: AK.
Host: Honkenya
Septoria ampelina Berk. & M.A. Curtis
Distr.: Central and eastern United States; Europe; Asia.
Host: Vitis
Septoria anagallidis Ellis & Halst., nom. nud.
Distr.: NJ, TX.
Notes: Published without a description.
Host: Anagallis
Septoria andropogonis J.J. Davis
Distr.: Central and western United States.
Notes: Leaf spot.
Hosts: Andropogon, Bouteloua, Schizachyrium, Sorghastrum, Sphenopholis, Sporobolus, Stipa
Septoria andropogonis J.J. Davis f. ***sporobolicola*** R. Sprague
Distr.: Central and western United States.
Notes: Leaf spot and blotch.
Hosts: Sporobolus, Stipa
Septoria andropogonis J.J. Davis var. ***sorghastri*** H.C. Greene & R. Sprague
Distr.: Central United States.
Notes: Leaf spot.
Host: Sorghastrum
Septoria anemones Desmaz.
Distr.: Central and eastern United States.
Host: Anemone
Septoria angularis Dearn. & Barth.
Septoria angularis Tharp non Dearn. & Barth., hom. illeg.
Septoria tharpiana Trott.
Distr.: TX(*sensu* Tharp), WI(*sensu* Dearness).
Notes: Greene (Amer. Midl. Naturalist 39:446. 1948) discussed the difficulty of distinguishing the various species reported from *Solidago* and related plants.
Hosts: Aster, Euthamia, Solidago
Septoria angustifolia Ellis & Everh.
Distr.: ID, NC, NJ, WI, WV.
Host: Kalmia
Septoria angustissima Peck
Distr.: MS.
Host: Maclura
Septoria anthoxanthina Gonz. Frag.
Distr.: OR; Spain(type).

Host: Anthoxanthum
Septoria antirrhinorum Tharp
Distr.: TX(type).
Host: Maurandya
Septoria anychiae H.C. Greene
Distr.: WI.
Host: Paronychia
Septoria apiicola Speg.
Septoria apii F. Chester
Septoria apii-graveolentis Dorogin
Septoria petroselini (Lib.) Desmaz. var. *apii* Briosi & Cavara
Distr.: Temperate regions.
Notes: Leaf spot of celery (Sheridan. Trans. Brit. Mycol. Soc. 51:207. 1968; Sutton & Waterston. C.M.I. Descr. 88. 1966).
Host: Apium
Septoria aquilegiae Ellis & Kellerm. in G. Martin
Distr.: Central and eastern United States.
Host: Aquilegia
Septoria arabidis Sacc.
Distr.: CO, TX, WY.
Host: Arabis
Septoria arabidis Sacc. var. ***macrospora*** Solheim
Distr.: CO, WY.
Host: Arabis
Septoria araliae Ellis & Everh.
Distr.: CA, WI.
Host: Panax
Septoria arctica Berk. & M.A. Curtis
Distr.: AK, OR; Soviet Union(Bering Straits).
Notes: Leaf spot.
Host: Calamagrostis
Septoria arechavaletae G. Wint.
Distr.: VA; Uruguay.
Notes: Leaf spot (Sprague. Diseases of cereals and grasses in North America. p. 220. 1950).
Host: Panicum
Septoria argophylla Ellis & Kellerm. in G. Martin
Distr.: ID, KS, OR.
Hosts: Pediomelum, Psoralea
Septoria aromatica Kab. & Bubák
Distr.: CA, CO; Europe(type).
Host: Ligusticum
Septoria artemisiae Unamuno
Distr.: WA; Spain.
Host: Artemisia
Septoria arundinacea Sacc.
Distr.: Temperate regions.
Host: Deschampsia
Septoria asaricola Allesch.
Distr.: Western United States; Europe.
Host: Asarum
Septoria asclepiadicola Ellis & Everh.
Distr.: Central and eastern United States.
Host: Asclepias
Septoria asclepiadicola Ellis & Everh. f. ***syriaca*** Dearn.
Distr.: MN, PA.
Host: Asclepias
Septoria asiatica Speg.
Distr.: FL; Chile.
Host: Centella
Septoria asiminae Ellis & Everh., nom. nud.
Distr.: TX.
Notes: Listed in Seymour (Host index of the fungi of North America. p. 324. 1929) without a description.
Host: Asimina
Septoria asplenii Ellis & Everh.
Distr.: MI; Europe; Asia.
Host: Athyrium
Septoria astericola Ellis & Everh.
Distr.: Eastern United States; Europe; Asia.
Host: Aster
Septoria asterina Tharp
Distr.: OK, TX, WY.
Host: Aster
Septoria astragali Roberge ex Desmaz.
Phloeospora reticulata Ellis & Everh.
Distr.: Temperate regions.
Hosts: Lathyrus, Vicia
Septoria astragalicola Peck
Distr.: Western United States; Europe; Asia.
Host: Astragalus
Septoria astragalina Hollós
Distr.: CO; Hungary(type).
Host: Astragalus
Septoria atropurpurea Peck
Distr.: Temperate regions.
Hosts: Aster, Callistephus, Euonymus, Solidago
Septoria azaleae Voglino
Distr.: Temperate regions.
Host: Rhododendron
Septoria azaleae-indicae Maubl.
Distr.: OR; Brazil(type).
Host: Rhododendron
Septoria bacilligera G. Wint.
Distr.: Central United States.
Hosts: Ambrosia, Solidago
Septoria baptisiae Cooke
Distr.: SC, TX.
Host: Baptisia
Septoria bataticola Taubenhaus
Distr.: HI, central and eastern United States.
Host: Ipomoea
Septoria baudysiana Sacc.
Distr.: Western United States; Europe(type).
Host: Carex
Septoria betae Westend.
Distr.: Temperate regions.
Host: Beta
Septoria betulae Pass. non (Lib.) Westend. ex Sacc., hom. illeg.
Septoria betulicola Peck
Septoria betulina Pass.
Septoria carpinea J.J. Davis
Distr.: Temperate northern hemisphere.
Hosts: Betula, Carpinus
Septoria betulae-odoratae Bubák & Vleugel
Cylindrosporium utahense Solheim
Distr.: Temperate North America; northern Europe.
Host: Betula
Septoria bidentis Sacc.
Distr.: Temperate regions.
Host: Bidens
Septoria blasdalei Sacc. & P. Syd.
Septoria variegata Ellis & Everh. non Vize in Cooke, hom. illeg.
Distr.: Western United States.
Host: Rhamnus
Septoria brassicae Ellis & Everh. in Millsp.
Distr.: WV.
Host: Brassica
Septoria brencklei Sacc.
Distr.: Central and western United States.
Host: Echinocystis
Septoria brevis House
Distr.: NY.
Host: Solidago
Septoria britannica Trott.
Septoria taraxaci J.W. Ellis non Hollós, hom. illeg.
Distr.: WI; Europe; Asia.
Host: Taraxacum
Septoria bromi Sacc.
Distr.: Temperate regions.
Notes: Leaf spot (Bissett. Fungi Canadenses 241. 1983).
Hosts: Bromus, Hystrix
Septoria bromi Sacc. var. ***phalaricola*** R. Sprague
Distr.: OR; WA.
Host: Phalaris
Septoria brunellae Ellis & Holw.
Distr.: United States; Europe.
Hosts: Lactuca, Monarda, Prunella
Septoria bumeliae L.E. Miles
Distr.: Southcentral United States.

Host: Sideroxylon
Septoria cacaliae Ellis & Kellerm.
Distr.: Central and southeastern United States.
Hosts: Cacalia, Senecio
Septoria cacticola Henn.
Distr.: TX; Germany(type).
Host: Cereus
Septoria calamagrostidis (Lib.) Sacc.
Distr.: Worldwide.
Notes: Leaf spot. Sprague (Diseases of cereals and grasses in North America. p. 227. 1950).
Hosts: Agrostis, Calamagrostis, Trisetum
Septoria calamagrostidis (Lib.) Sacc. f. sp. ***koeleriae*** (Cocc. & Morini) R. Sprague
Distr.: Western United States; Europe; Asia.
Notes: Blotch.
Host: Koeleria
Septoria calamovilfae Petr.
Distr.: Central and western United States.
Notes: Leaf spot.
Host: Calamovilfa
Septoria callistephi Gloyer
Distr.: United States; Europe; Asia.
Host: Callistephus
Septoria calypsonis Povah
Distr.: MI.
Host: Calopogon
Septoria campanulae (Lév.) Sacc.
Distr.: Temperate regions.
Host: Campanula
Septoria canadensis Peck
Distr.: United States.
Host: Cornus
Septoria canavaliae Lyon
Distr.: HI.
Host: Canavalia
Septoria cannabina Peck
Distr.: NY; Europe; Asia.
Host: Cannabis
Septoria caraganae Henn.
Distr.: Temperate regions.
Host: Caragana
Septoria caricina Brunaud
Distr.: WI; Europe.
Host: Carex
Septoria caricinella Sacc. & Roum.
Distr.: Central and western United States; Europe; Asia.
Host: Carex
Septoria caricis Pass.
Distr.: Central and western United States; Europe; Asia.
Host: Carex
Septoria carnea Ellis & Everh.
Distr.: SD; Canada(type).
Host: Carex
Septoria carthami Murashkinskij
Distr.: North America; Africa; Asia; Europe.
Notes: Punithalingam (C.M.I. Descr. 669. 1980).
Host: Carthamus
Septoria caryae Ellis & Everh.
Distr.: DE(type), MI, TX.
Host: Carya
Septoria cassiicola Kellerm. & Swingle
Distr.: KS, TX, WI.
Hosts: Cassia, Senna
Septoria ceanothi Dearn.
Distr.: ID, WA.
Host: Ceanothus
Septoria cenchrina J.J. Davis
Distr.: Central United States.
Notes: Leaf spot.
Host: Cenchrus
Septoria centaureicola Brunaud
Distr.: OR; Europe; Asia.
Host: Centaurea
Septoria centaureicola Brunaud var. ***brevispora*** Peck
Distr.: NY, IL, OK.
Host: Centaurea
Septoria cephalanthi Ellis & Kellerm.
Distr.: FL, KS, WI.
Hosts: Cephalanthus, Genipa
Septoria cerastii Roberge ex Desmaz.
Distr.: United States; Europe; Asia.
Host: Cerastium
Septoria ceuthosporioides (Cooke & Harkn.) Sacc.
Distr.: CA.
Host: Eucalyptus
Septoria chaenorrhini H.C. Greene
Distr.: WI.
Host: Chaenorrhinum
Septoria chamaecisti Vestergr.
Distr.: WA; Sweden(type).
Host: Helianthemum
Septoria chamissonis Sacc. & Scalia
Distr.: AK.
Host: Eriophorum
Septoria chanousiana Ferraris
Distr.: Western United States; Europe.
Host: Luzula
Septoria chelidonii (Lib.) Desmaz.
Ascochyta chelidonii Lib.
Distr.: TX; Europe.
Notes: Jørstad (Skr. Norske Vidensk.-Akad. Oslo, Mat.-Naturvidensk. Kl. 22:45. 1965).
Host: Chelidonium
Septoria chionanthi Cooke
Distr.: SC, TX.
Host: Chionanthus
Septoria chlorantha Dearn. & House, nom. nud.
Distr.: NY(type).
Notes: A poorly known species (Parmelee, Canad. J. Bot. 36:876. 1958).
Host: Pyrola
Septoria chlorogali (Cooke & Harkn.) R. Sprague
Rhabdospora chlorogali Cooke & Harkn.
Distr.: CA, OR.
Hosts: Camassia, Chlorogalum
Septoria chrysanthemella Sacc.
Septoria rostrupii Sacc. & Syd.
Distr.: Cosmopolitan.
Notes: Punithalingam (C.M.I. Descr. 137. 1967).
Host: Chrysanthemum
Septoria chrysanthemi Halst. in Seym. & Earle
Distr.: Temperate North America.
Host: Chrysanthemum
Septoria circinata Ellis & Everh.
Distr.: WA(type).
Host: Acer
Septoria cirrhosa G. Wint.
Distr.: MO.
Host: Staphylea
Septoria cirsii Niessl
Distr.: Temperate regions.
Host: Cirsium
Septoria citri Pass.
Distr.: CA, TX; Europe; Asia.
Host: Citrus
Septoria citrulli Ellis & Everh.
Distr.: NJ, NY.
Host: Citrullus
Septoria clematidis Roberge in Desmaz.
Septoria jackmanii Ellis & Everh.
Distr.: North America; Europe.
Host: Clematis
Septoria clementsii H.C. Greene
Septoria macrostoma Clements, nom. nud.
Distr.: WI.
Notes: According to Greene (Trans. Wisconsin Acad. Sci. 56:278. 1967) this is the same as *Septoria macrostoma* Clements which was published in an exsiccati set without a description.
Host: Aralia

Septoria clermontiae F. Stevens & P.A. Young
Distr.: HI.
Host: Clermontia
Septoria cocoina Ellis & Everh.
Distr.: MO.
Host: Syagrus
Septoria collinsiae H.W. Anderson
Distr.: IL.
Host: Collinsia
Septoria commersoniana Speg.
Distr.: WY; Argentina(type).
Host: Cerastium
Septoria commonsii Ellis & Everh.
Distr.: DE, WI.
Host: Cirsium
Septoria consocia Peck
Distr.: IN, MI, WI.
Host: Polygala
Septoria conspicua Ellis & G. Martin in G. Martin
Distr.: United States; Europe.
Host: Lysimachia
Septoria convolvuli Desmaz.
Septoria flagellaris Desmaz.
Septoria septulata Beach
Distr.: United States; Europe.
Hosts: Calystegia, Convolvulus, Ipomoea
Septoria coptis Berk. & M.A. Curtis
Mycosphaerella coptis (Schwein.) House (teleomorph)
Distr.: Northeastern North America.
Notes: This name was originally published as *coptidis* (Sivanesan. The bitunicate Ascomycetes. p. 239. 1984).
Host: Coptis
Septoria coreopsidis J.J. Davis
Distr.: IA, TX, WI.
Host: Coreopsis
Septoria cornicola Desmaz.
Septoria corni-maris Sacc.
Distr.: North America; Europe; Asia.
Host: Cornus
Septoria cornicola Desmaz. var. ***ampla*** H.C. Greene
Distr.: WI.
Host: Cornus
Septoria corydalis Ellis & J.J. Davis
Distr.: TX, WI.
Host: Corydalis
Septoria crepidis Westend.
Distr.: WI; Europe; Asia.
Host: Crepis
Septoria cruciatae Roberge ex Desmaz.
Septoria aparines Ellis & Kellerm.
Distr.: United States; Europe; Asia.
Host: Galium
Septoria cryptotaeniae Ellis & Rau in G. Martin
Distr.: IA, MN, PA(type), TX.
Hosts: Asclepias, Cryptotaenia
Septoria cucurbitacearum Sacc.
Distr.: Central and eastern North America; eastern Asia; Australia; western Europe.
Notes: Leaf spot of Cucurbitaceae. Punithalingam (C.M.I. Descr. 740. 1982).
Hosts: Citrullus, Cucumis, Cucurbita
Septoria cunillae Tehon
Distr.: IL.
Host: Cunila
Septoria curvispora Ellis & Everh., nom. nud.
Distr.: CA(type).
Host: Acer
Septoria cyani Hollós
Distr.: OK; Europe.
Host: Centaurea
Septoria cylindrica Ellis & Everh.
Distr.: MA, VA.
Host: Anemone
Septoria cynodontis Fuckel
Distr.: Southern United States, HI; Europe; Asia.
Host: Cynodon
Septoria cyperi Ellis & Everh.
Distr.: AL, FL.
Host: Cyperus
Septoria dalibardae Peck
Distr.: Central and eastern United States.
Host: Dalibarda
Septoria darlingtoniae Dearn. & House
Distr.: OR.
Host: Darlingtonia
Septoria darrowii Zeller
Septoria rubi Westend. var. *brevispora* Sacc.
Distr.: WI.
Notes: *Mycosphaerella rubi* Roark or *Sphaerella ligea* Sacc. may be the telemorph (Zeller. Phytopathology 27:1000. 1937).
Host: Rubus
Septoria davisii Sacc.
Septoria canadensis Ellis & J.J. Davis non Peck, hom. illeg.
Distr.: WI.
Hosts: Cornus, Solidago
Septoria dearnessii Ellis & Everh.
Distr.: AR, NC, TX, WI.
Host: Angelica
Septoria delphinella Sacc.
Distr.: Central United States; Europe; Asia.
Host: Delphinium
Septoria dentariae Peck
Distr.: Central and eastern United States.
Hosts: Cardamine, Dentaria
Septoria destruens Desmaz., nom. nud.
Distr.: CA, NY.
Host: Sidalcea
Septoria dianthi Desmaz.
Distr.: Widespread.
Notes: Radulescu et. al (Septoriozele din Romania. p. 75. 1973).
Host: Dianthus
Septoria dianthicola Sacc.
Distr.: WI; Europe; Asia.
Host: Dianthus
Septoria didyma Fuckel
Distr.: WI; Europe; Asia.
Host: Salix
Septoria didyma Fuckel f. ***santonensis*** Pass. in Brunaud
Distr.: WI; France.
Host: Salix
Septoria diervillae Ellis & Everh.
Distr.: Central and eastern United States.
Host: Diervilla
Septoria difformis G.W. Clinton & Peck
Distr.: ME, NY(type), WI.
Host: Vaccinium
Septoria digitalidis Pass.
Distr.: WI; Europe; Asia.
Host: Besseya
Septoria digitarivora R. Sprague
Distr.: WV.
Notes: Leaf spot.
Host: Digitaria
Septoria dimera Sacc.
Distr.: Central United States; Europe; Asia.
Host: Silene
Septoria dodecatheonis J.J. Davis
Distr.: WI.
Host: Dodecatheon
Septoria dracocephali Thuem.
Distr.: TX, WI; Soviet Union(type).
Host: Dracocephalum
Septoria dryina Cooke in Ravenel
Distr.: OK, SC.
Host: Quercus
Septoria echinocystis Ellis & Everh.
Distr.: CA.
Hosts: Echinocystis, Marah
Septoria elaeagni (Chev.) Desmaz.
Septoria argyraea Sacc.

Distr.: Temperate regions.
Host: Elaeagnus
Septoria eleospora Sacc.
Distr.: SC, TX; Italy(type).
Host: Chionanthus
Septoria ellisiana Sacc. & Syd.
Septoria chimaphilae Ellis & Everh. in G. Martin
Septoria pyrolae Ellis & G. Martin
Distr.: North America.
Notes: Parmelee (Canad. J. Bot. 36:876. 1958).
Hosts: Chimaphila, Pyrola
Septoria elymi Ellis & Everh.
Distr.: North America; Central Asia; China; France; Germany.
Notes: Speckled leaf blotch (Bissett. Fungi Canadenses 242. 1983).
Hosts: Agropyron, Elymus, Lolium
Septoria elymi Ellis & Everh. f. **elymina** (J.J. Davis) R. Sprague
Septoria bromi Sacc. var. *elymina* J.J. Davis
Septoria agropyri Ellis & Everh.
Distr.: Central and western United States.
Hosts: Agropyron, Elymus
Septoria emaculata Peck & G.W. Clinton
Distr.: NY, WI.
Notes: Probably closely related to *Septoria fulvescens* Sacc. (Jørstad. Skr. Norske Vidensk.-Akad. Oslo, Mat.-Naturvidensk. Kl. 22:41. 1965)
Host: Lathyrus
Septoria epilobii Westend.
Distr.: Temperate regions.
Host: Epilobium
Septoria erechtitis Ellis & Everh.
Distr.: United States.
Host: Erechtites
Septoria erigerontis Peck
Septoria erigerontis Berk. & M.A. Curtis non Peck, hom. illeg.
Distr.: Temperate North America; Europe.
Hosts: Conyza, Erigeron
Septoria erigerontis Berk. & M.A. Curtis non Peck, hom. illeg. f. **boltoniae** J.H. Webber
Distr.: Central United States.
Host: Boltonia
Septoria eriobotryae Maffei
Distr.: FL; Italy(type).
Host: Eriobotrya
Septoria eryngicola Oudem. & Sacc.
Distr.: Temperate regions.
Host: Eryngium
Septoria eupatorii Roberge & Desmaz.
Distr.: Central United States; Europe; Asia.
Host: Eupatorium
Septoria eupatoriicola Tehon
Distr.: IL.
Host: Eupatorium
Septoria evonymella Pass.
Distr.: NJ, SC.
Host: Euonymus
Septoria evonymi Rabenh.
Distr.: Temperate regions.
Host: Euonymus
Septoria exotica Speg.
Distr.: CA, HI, WI; Argentina; England; Germany.
Hosts: Hebe, Veronica
Septoria expansa Niessl
Distr.: Temperate regions.
Host: Geranium
Septoria fairmanii Ellis & Everh.
Distr.: NY(type).
Host: Alcea
Septoria farfaricola Dearn., nom. inval.
Distr.: TN.
Host: Tussilago
Septoria ficarioides Peck
Distr.: NE.
Host: Ranunculus
Septoria fici-indicae Voglino
Distr.: TX; Italy(type).
Host: Opuntia
Septoria fimbristylis Ellis & B.T. Galloway, nom. nud.
Distr.: MO.
Host: Fimbristylis
Septoria flagellifera Ellis & Everh.
Distr.: Midwestern United States; Europe; Asia.
Host: Pisum
Septoria floridae Tehon & E.Y. Daniels
Distr.: IL.
Host: Cornus
Septoria fragariae (Lib.) Desmaz.
Septogloeum potentillae Allesch.
Distr.: Temperate regions.
Hosts: Fragaria, Potentilla
Septoria fructigena Berk. & M.A. Curtis
Distr.: SC.
Host: Passiflora
Septoria fumosa Peck
Distr.: Central and eastern United States; Europe.
Hosts: Aster, Euthamia, Solidago
Septoria fusca Peck
Distr.: NY.
Host: Artemisia
Septoria gaillardiae Ellis & Everh.
Distr.: Central United States.
Host: Gaillardia
Septoria galeopsidis Westend.
Distr.: North America; Europe.
Host: Galeopsis
Septoria gallarum Ellis & Everh.
Distr.: NY.
Host: Solidago
Septoria gaurina Ellis & Kellerm.
Distr.: Central and western United States.
Hosts: Gaura, Oenothera, Zauschneria
Septoria gei Roberge & Desmaz.
Distr.: United States; Europe; Asia.
Host: Geum
Septoria gentianoides Dearn. & House
Septoria gentianae Dearn. & House non Thuem., hom. illeg.
Distr.: NY.
Host: Gentiana
Septoria gilletiana Sacc.
Distr.: GA, LA; Europe.
Host: Castanea
Septoria glabra Ellis & Everh.
Distr.: IN.
Host: Aesculus
Septoria gladioli Pass.
Distr.: Temperate regions.
Host: Gladiolus
Septoria glycericola R. Sprague
Distr.: United States.
Notes: Buff leaf spot.
Hosts: Glyceria, Puccinellia
Septoria glycines Hemmi
Distr.: Central and eastern North America; eastern Asia.
Notes: Brown spot of soybean (Punithalingam & Holliiday. C.M.I. Descr. 339:1972).
Hosts: Amphicarpaea, Glycine
Septoria glycyrrhizae Ellis & Kellerm.
Distr.: KS, WA.
Host: Glycyrrhiza
Septoria gossypina Cooke in Ravenel
Distr.: MS, SC.
Host: Gossypium
Septoria gouldiae F. Stevens & P.A. Young
Distr.: HI.
Host: Gouldia
Septoria graminum Desmaz.
Distr.: Temperate regions.
Notes: According to Sprague (Mycologia 30:672. 1938) this species does not occur in the U.S. These reports most likely should be referred to *S. tritici*.
Hosts: Agrostis, Calamagrostis, Panicum, Poa, Setaria

Septoria grindeliae Ellis & Barth.
Distr.: Central and western United States.
Host: Grindelia
Septoria grindeliicola H.C. Greene
Distr.: WI.
Host: Grindelia
Septoria gryllii Sacc.
Distr.: North America; Europe; Asia.
Notes: Sprague (Oregon State Monogr. Stud. Bot. 6:104. 1944) suggested that reports for this fungus from Alaska are *S. calamagrostidis*.
Host: Agrostis
Septoria hawaiiensis F. Stevens & Plunkett
Distr.: HI.
Host: Gouldia
Septoria hedeomae Dearn. & House
Distr.: WI.
Host: Hedeoma
Septoria helenii Ellis & Everh.
Distr.: Central and eastern United States; Europe; Asia.
Hosts: Gutierrezia, Helenium
Septoria helianthi Ellis & Kellerm.
Distr.: Temperate regions.
Notes: Leaf spot (Holliday & Punithalingam. C.M.I. Descr. 276. 1970).
Hosts: [Echinacea], Helianthus, [Heliopsis]
Septoria helianthicola Cooke & Harkn.
Rhabdospora helianthicola (Cooke & Harkn.) Sacc.
Distr.: CA.
Host: Helianthus
Septoria heliopsidis Ellis & Dearn.
Distr.: WI.
Host: Heliopsis
Septoria hepaticicola (Duby) Jørst.
Septoria hepaticae Desmaz.
Distr.: Temperate regions.
Host: Hepatica
Septoria heucherae Pass.
Distr.: IN.
Notes: The place of publication of this name could not be located.
Host: Heuchera
Septoria hicoriae Tharp
Distr.: TX.
Host: Carya
Septoria hieracicola Dearn. & House
Distr.: NY, WA, WI.
Host: Hieracium
Septoria hippocastani Berk. & Broome
Septoria aesculicola (Fr.) Westend.
Distr.: Temperate regions.
Host: Aesculus
Septoria hosackiae Harkn.
Distr.: CA.
Host: Lotus
Septoria humuli Westend.
Ascochyta humuli Lasch
Phloeospora humuli Westend.
Distr.: Temperate northern hemisphere.
Notes: Leaf spot of hops (Punithalingam. C.M.I. Descr. 829. 1985).
Host: Humulus
Septoria hyalina Ellis & Everh.
Distr.: Northwestern United States.
Host: Viola
Septoria hydrangeae Bizz.
Distr.: Central and eastern United States; Europe; Asia.
Host: Hydrangea
Septoria hydrocotyles Desmaz.
Distr.: United States; Europe.
Hosts: Centella, Hydrocotyle
Septoria hydrophylli Ellis & Dearn.
Distr.: NY; Canada(type).
Host: Hydrophyllum
Septoria hypoxis Overh.
Distr.: PA.
Host: Hypoxis
Septoria ilicifolia Cooke & Ellis
Distr.: FL, NJ(type).
Host: Ilex
Septoria incarnata Ellis & Barth.
Distr.: IN, KS.
Host: Asclepias
Septoria inclusa H.C. Greene
Distr.: WI.
Host: Parthenium
Septoria inconspicua Berk. & M.A. Curtis
Distr.: United States; Europe; Asia.
Host: Plantago
Septoria infuscans (Ellis & Everh.) R. Sprague
Cylindrosporium infuscans Ellis & Everh.
Distr.: Western United States.
Notes: Brown leaf blotch.
Hosts: Agropyron, Elymus
Septoria infuscata G. Wint.
Distr.: MI, MO(type), WI.
Host: Ratibida
Septoria intermedia Ellis & Everh.
Distr.: WI.
Host: Solidago
Septoria inulae Sacc. & Speg.
Distr.: IL; Europe; Asia.
Host: Helianthus
Septoria iresines Speg.
Distr.: OK(type).
Host: Iresine
Septoria ivicola Ellis & Everh., nom. nud.
Distr.: TX.
Notes: This name was listed in Seymour (Host index of the fungi of North America. p. 665. 1929). According to Cash (U.S.D.A. Spec. Publ. 2:374. 1952) this is the only place where it appears.
Host: Iva
Septoria jacucella R. Sprague in Cooke
Distr.: Western United States.
Notes: Western brome speckle. Sprague (Oregon State Monogr. Stud. Bot. 6:119. 1944) redescribed this species with the spelling "*S. jaculella*".
Host: Bromus
Septoria jamesii Pammel & H. Hume in H. Hume
Distr.: CO(type), MT.
Host: Stellaria
Septoria jatrophae Heald & F.A. Wolf
Distr.: Southern United States.
Host: Cnidoscolus
Septoria jonesii Pammel & H. Hume, nom. nud.
Distr.: CA.
Notes: Distributed in an exsiccati set. No other information on this name was located.
Host: Stellaria
Septoria jussiaeae Ellis & Everh.
Distr.: Southern United States.
Host: Ludwigia
Septoria kellermaniana Thuem. in Kellerm.
Distr.: KS(type).
Host: Vitis
Septoria krigiae Ellis & Everh.
Distr.: Central and eastern United States.
Host: Krigia
Septoria lactucae Pass.
Septoria lactucae Peck non Pass., hom. illeg.
Septoria consimilis Ellis & G. Martin
Septoria ludowiciana Ellis & Everh.
Distr.: Widespread.
Notes: Jørstad (Skr. Norske Vidensk.-Akad. Oslo, Mat.-Naturvidensk. Kl. 22:21. 1965); Punithalingam & Holliday (C.M.I. Descr. 335. 1972).
Hosts: Agoseris, Lactuca
Septoria lactucicola Ellis & G. Martin
Distr.: United States.
Notes: Jørstad (Skr. Norske Vidensk.-Akad. Oslo, Mat.-Naturvidensk. Kl. 22: 21. 1965).
Host: Lactuca
Septoria lamii Pass. var. **brevior** H.C. Greene

Distr.: WI.
Notes: Type variety considered a synonym of *Septoria lamiicola* Sacc. (Jørstad. Skr. Norske Vidensk.-Akad. Oslo, Mat.-Naturvidensk. Kl. 22:39. 1965).
Host: Leonurus
Septoria lamiicola Sacc.
Septoria lamii Sacc.
Septoria lamii Pass.
Distr.: Temperate regions.
Hosts: Lamium, Leonurus
Septoria lanaria Fairm.
Distr.: NY, WI.
Host: Antennaria
Septoria lapparum Sacc.
Distr.: Central and eastern United States; Europe; Asia.
Host: Arctium
Septoria lathyri Ellis & Everh.
Distr.: NJ.
Host: Lathyrus
Septoria lavandulae Desmaz.
Distr.: OH, OK: Europe.
Host: Lavandula
Septoria lepachydis Ellis & Everh.
Distr.: KS, WI.
Hosts: Echinacea, Ratibida
Septoria lepidiicola Ellis & G. Martin
Distr.: Central and eastern United States.
Host: Lepidium
Septoria leptostachya Ellis & Kellerm.
Distr.: Central and eastern United States; Asia.
Host: Phryma
Septoria leucanthemi Sacc. & Speg.
Septoria cercosporoides Trail
Septoria macrosporia Dearn.
Distr.: North and South America; Africa; Asia; Europe.
Notes: Leaf spot or leaf blotch. Sutton (The Coelomycetes. p. 92. 1980).
Host: Chrysanthemum
Septoria liatridis Ellis & J.J. Davis
Distr.: Midwestern United States.
Host: Liatris
Septoria ligustici Guba
Distr.: MA.
Host: Ligusticum
Septoria limonum Pass.
Distr.: CA; Italy.
Host: Citrus
Septoria linariae H.C. Greene
Distr.: WI.
Host: Linaria
Septoria lineolata Sacc. & Speg.
Distr.: United States; France; Italy; The Netherlands.
Host: Carex
Septoria linicola (Speg.) Garassini
Mycosphaerella linicola Naumov (teleomorph)
Distr.: Worldwide in flax-growing regions.
Notes: Pasmo disease of flax (Sivanesan & Holliday. C.M.I. Descr. 709. 1981).
Host: Linum
Septoria liquidambaris Cooke & Ellis
Distr.: Central and eastern United States.
Hosts: Hamamelis, Liquidambar
Septoria lithospermi J.J. Davis & H.C. Greene in H.C. Greene, nom. nud.
Distr.: Used as a herbarium name by Davis (Trans. Wisconsin Acad. Sci. 24:290. 1929).
Hosts: Lithospermum, Onosmodium
Septoria littorea Sacc.
Distr.: Central United States; Europe; Asia.
Host: Apocynum
Septoria lobeliae Peck
Distr.: Central and eastern United States; Europe; Asia.
Host: Lobelia
Septoria lobeliae-syphiliticae Henn.
Distr.: NY; Europe.
Host: Lobelia
Septoria loligena R. Sprague
Distr.: CA.
Notes: Brown leaf spot.
Host: Lolium
Septoria longispora Overh. non Shear, hom. illeg.
Distr.: PA.
Host: Aquilegia
Septoria lophanthi G. Wint.
Distr.: Central and eastern United States.
Host: Agastache
Septoria ludwigiae Cooke
Distr.: United States.
Host: Ludwigia
Septoria lupini Harkn.
Distr.: Western United States; Europe.
Host: Lupinus
Septoria lupinicola Dearn.
Distr.: North America.
Notes: Dearness (Mycologia 8:103. 1916) reported that this species was similar to *Septoria pisi* Westendorp.
Host: Lupinus
Septoria lupulina Ellis & Kellerm.
Distr.: KS; Europe; Asia.
Notes: May be a synonym of *Septoria humuli* Westendorp (Punithalingam. C.M.I. Descr. 829. 1985).
Host: Humulus
Septoria lychnidis Desmaz.
Septoria lychnidis Desmaz. var. *pusilla* Trail
Septoria melandrii Pass.
Distr.: Temperate regions.
Hosts: Agrostemma, Lychnis, Silene
Septoria lycopersici Speg.
Distr.: Worldwide in tomato-growing regions.
Notes: Leaf spot of tomato (Sutton & Waterston. C.M.I. Descr. 89. 1966).
Hosts: [Datura], Lycopersicon, [Solanum]
Septoria lycopi Pass.
Distr.: ID, WI; Europe.
Host: Lycopus
Septoria lysimachiae (Lib.) Westend.
Distr.: Temperate regions.
Host: Lysimachia
Septoria lythrina Peck
Distr.: FL, KS, NY, WI.
Host: Lythrum
Septoria macropoda Pass.
Septoria annua Ellis & Everh.
Septoria poae-trivialis Cocc.
Distr.: Temperate regions.
Notes: Leaf blotch.
Host: Poa
Septoria macropoda Pass. f. ***sitadakaensis*** R. Sprague
Distr.: AK.
Host: Poa
Septoria macropoda Pass. var. ***grandis*** R. Sprague
Distr.: Western United States.
Notes: Leaf spot.
Host: Poa
Septoria macropoda Pass. var. ***septulata*** (Gonz. Frag.) R. Sprague
Distr.: Central and western United States; Europe; Asia.
Notes: Leaf blotch.
Host: Poa
Septoria maculifera Sacc.
Septoria maculosa W. Gerard non Lév., hom. illeg.
Distr.: Eastern United States.
Notes: Saccardo believed that *Septoria maculosa* W. Gerard was a later homonym of *Septoria maculosa* Lév. and thus proposed *Septoria maculifera* for Gerard's name. Gerard's name may not be a later homonym if Léveille's name is not considered validly published as an exsiccati specimen. This could not be determined.
Host: Cuphea
Septoria magnoliae Cooke
Distr.: Southern United States; Europe; Asia.

Host: Magnolia
Septoria malvicola Ellis & G. Martin
Distr.: Eastern North America.
Hosts: Alcea, Althaea, Malva
Septoria margaritaceae Peck
Distr.: United States.
Host: Anaphalis
Septoria marginata Heald & F.A. Wolf
Distr.: CA, TX(type).
Host: Acer
Septoria mariae-wilsonii G.W. Clinton
Distr.: NY(type), eastern and central United States.
Notes: The place of publication of this name could not be located.
Host: Chelone
Septoria matricariae Hollós
Distr.: WI; Europe; Asia.
Host: Anthemis
Septoria megarrhizae Ellis & Everh.
Distr.: CA, OR, WA.
Host: Marah
Septoria melanophthalmi Berk. & M.A. Curtis
Distr.: SC(type).
Host: Vigna
Septoria melicae Pass.
Distr.: Temperate regions.
Notes: Leaf blotch.
Hosts: Melica, Schizachne
Septoria menthae Oudem.
Septoria menthicola Sacc. & Letendre
Distr.: Temperate regions.
Hosts: [Blephilia], Mentha, [Monarda]
Septoria mentzeliae Ellis & Kellerm.
Distr.: KS, WA.
Host: Mentzelia
Septoria menyanthis (Lib.) Desmaz.
Distr.: WI; Europe; Asia.
Host: Menyanthes
Septoria micropuncta Ellis & Everh.
Distr.: DC, VA.
Host: Osmorhiza
Septoria microspora Ellis non Speg., hom. illeg.
Distr.: IN(type).
Notes: Sprague (Oregon State Monogr. Stud. Bot. 6:81. 1944) suggested that this is a microspore stage of *S. passerinii*.
Host: Hystrix
Septoria mikanii G. Wint.
Distr.: CT, TX; Brazil(type).
Host: Mikania
Septoria mimuli Ellis & Kellerm.
Septoria mimuli G. Wint.
Distr.: United States.
Notes: Winter (J. Mycol. 1:155. 1885) reported that his *Septoria mimuli* was identical to that of Ellis & Kellerman.
Host: Mimulus
Septoria minuta J. Schröt.
Distr.: Western United States; Europe; Asia.
Host: Luzula
Septoria mississippiensis R. Sprague
Distr.: Central United States.
Notes: Leaf spot.
Host: Muhlenbergia
Septoria mitellae Ellis & Everh.
Distr.: MI, WI.
Host: Mitella
Septoria mollereana Bres. & Roum.
Distr.: HI; St. Thomas(type).
Host: Canavalia
Septoria mollisia Dearn. & House
Distr.: NY, WI.
Host: Antennaria
Septoria mortolensis Penz. & Sacc.
Distr.: Temperate regions.
Notes: On Myrtaceae.
Host: Eucalyptus
Septoria munroae Ellis & Barth.
Distr.: Central United States.
Host: Munroa
Septoria musiva Peck
Mycosphaerella populorum G.E. Thompson (teleomorph)
Distr.: Eastern North America.
Notes: On leaves (Sivanesan. The bitunicate Ascomycetes. p. 247. 1984).
Host: Populus
Septoria myricae Ellis & Everh. nom. nud. non Trail, hom. illeg.
Distr.: Eastern United States; England.
Notes: The reports from the U.S. are all for the Ellis & Everhart name, which is a later homonym of *Septoria myricae* Trail. Grove (British stem and leaf fungi. Vol. I. p. 393. 1935) suggested that these are probably the same species.
Host: Myrica
Septoria myricata Dearn. & House
Distr.: NY.
Host: Myrica
Septoria nabali Berk. & M.A. Curtis
Distr.: Central and eastern United States.
Host: Prenanthes
Septoria narcissi Pass.
Distr.: MS; Italy(type).
Host: Narcissus
Septoria narvisiana Sacc.
Distr.: United States; Europe.
Notes: Saccardo (Michelia 1:196. 1878) first described this fungus as *Septoria holoschoeni* (Mont.) Sacc., but later decided it was a different fungus. This fungus may have been associated with *Leptosphaeria micropogonis* Sacc.
Host: Scirpus
Septoria neglecta Earle
Distr.: AL, MS.
Host: Quercus
Septoria nematospora J.J. Davis
Distr.: United States.
Host: Carex
Septoria nematospora J.J. Davis f. **aggregata** R. Sprague
Distr.: AR.
Host: Carex
Septoria nepetae Ellis & Everh.
Distr.: TX, WI.
Host: Nepeta
Septoria niphostoma Berk. & M.A. Curtis
Distr.: Eastern United States.
Host: Magnolia
Septoria noctiflorae Ellis & Kellerm. in G. Martin
Distr.: Central and eastern United States.
Hosts: Saponaria, Silene
Septoria noli-tangeris W. Gerard
Distr.: United States; Europe; Asia.
Host: Impatiens
Septoria obesa Syd.
Distr.: FL; Europe.
Notes: Considered by some (Punithalingam. C.M.I. Descr. 139. 1967; Punithalingam & Wheeler. Trans. Brit. Mycol. Soc. 48:436 1965) as distinct and by others (Jørstad. Skr. Norske Vidensk.-Akad. Oslo, Mat.-Naturvidensk. Kl. 22:22 1965) as a synonym of *Septoria leucanthemi* Sacc. & Speg.
Host: Chrysanthemum
Septoria occidentalis Ellis & Everh., nom. nud.
Distr.: WA.
Notes: Only known from a specimen in BPI.
Host: Osmorhiza
Septoria ocellata (Lév.) Sacc.
Distr.: OR; France; Italy.
Host: Quercus
Septoria oedospora Clements, nom. nud.
Distr.: CO, WY.
Notes: Published in an exsiccati without a description.
Host: Symphoricarpos
Septoria oenanthis Ellis & Everh.
Distr.: WA; Europe.
Host: Oenanthe
Septoria oenotherae Westend.

Distr.: Temperate North America; Europe.
Hosts: Calylophus, Camissonia, Gaura, Ludwigia, Oenothera
Septoria oleandrina Sacc.
Distr.: Southern United States; Europe; Asia.
Host: Nerium
Septoria orchidearum Westend.
Distr.: Temperate regions.
Host: Tofieldia
Septoria ornithogali Pass.
Distr.: Central and eastern United States; Europe; Asia.
Host: Ornithogalum
Septoria oryzae Cattaneo
Distr.: Warm temperate and tropical regions.
Host: Oryza
Septoria ostryae Peck
Septoria corylina Peck
Cylindrosporium dearnessii Ellis & Everh.
Phloeospora dearnessii (Ellis & Everh.) Höhn.
Gloeosporium rostratum Ellis & Everh.
Distr.: Central and eastern North America.
Notes: Leaf spot of Betulaceae.
Hosts: Alnus, Carpinus, Corylus, Ostrya
Septoria oudemansii Sacc.
Distr.: United States; Europe.
Notes: Leaf spot.
Hosts: Alopecurus, Hierochloe, Poa, Setaria
Septoria oxalidis Lind
Distr.: Central and western United States; Europe.
Host: Oxalis
Septoria oxyacanthae Kunze & Schumach.:Fr.
Septoria crataegi Desmaz. ex J. Kickx fil.
Distr.: Central United States; Europe.
Host: Crataegus
Septoria pachysandrae Dearn., nom. inval.
Distr.: FL, TN.
Host: Pachysandra
Septoria pachyspora Ellis & Holw.
Distr.: Central United States.
Host: Zanthoxylum
Septoria pacifica R. Sprague
Distr.: Western United States.
Notes: Leaf blotch.
Host: Elymus
Septoria paeoniae Westend.
Distr.: United States; Europe; Asia.
Host: Paeonia
Septoria pallidula Dearn. & House
Distr.: NY.
Host: Hydrocotyle
Septoria parietariae J.J. Davis
Distr.: Midwestern United States.
Host: Parietaria
Septoria parvimaculans H.C. Greene
Distr.: WI.
Host: Kuhnia
Septoria passerinii Sacc.
Distr.: Probably widespread.
Notes: Barley leaf speckle (Bissett. Fungi Canadenses 243. 1983). Mostly on *Hordeum* spp.
Hosts: Elymus, Hordeum, Hystrix, Sitanion
Septoria pastinacae Westend.
Cylindrosporium pastinacae (Westend.) Lind
Rhabdospora pastinacina (Sacc.) Allesch.
Distr.: TX, UT, WI.
Hosts: Heracleum, Pastinaca
Septoria paupera Ellis in G. Martin
Distr.: NJ.
Host: Helianthus
Septoria peckii Sacc.
Cryptosporium scirpi Peck
Distr.: NY.
Notes: This name is based on *Cryptosporium scirpi* Peck which if transferred by Saccardo to *Septoria* would have been a later homonym.
Host: Scirpus
Septoria pentstemonicola Ellis & Everh.
Distr.: SD, WI.
Host: Penstemon
Septoria pentstemonis Ellis & Everh.
Distr.: United States.
Host: Penstemon
Septoria peraphylli Peck
Distr.: UT.
Host: Peraphyllum
Septoria peregrina Sacc.
Distr.: ND.
Host: Conringia
Septoria pertusa Heald & F.A. Wolf
Distr.: FL, TX; Europe; Asia.
Notes: Leaf spot.
Host: Sorghum
Septoria petroselini (Lib.) Desmaz.
Distr.: Temperate northern hemisphere.
Notes: Leaf spot of Apiaceae.
Hosts: Conioselinum, Petroselinum
Septoria petroselini (Lib.) Desmaz. subsp. ***treleasiana*** Sacc. & Scalia
Distr.: AK.
Host: Conioselinum
Septoria philadelphi Ellis & Everh.
Distr.: Central and western United States; Europe; Asia.
Host: Philadelphus
Septoria phlogis Sacc. & Speg.
Septoria divaricatae Ellis & Everh.
Septoria drummondii Ellis & Everh.
Distr.: Temperate North America; Europe.
Host: Phlox
Septoria phlyctanoides Berk. & M.A. Curtis
Distr.: Central and eastern United States.
Hosts: Clerodendrum, Phytolacca
Septoria photiniae Berk. & M.A. Curtis
Distr.: CA, NC, PA.
Hosts: Heteromeles, Photinia
Septoria physostegiae Ellis & Everh.
Distr.: IL, WI.
Hosts: Dracocephalum, Physostegia
Septoria pileae Thuem.
Distr.: Central and eastern United States.
Host: Pilea
Septoria pinicola Dearn.
Distr.: NC, VA.
Notes: Considered by Dearness (Mycologia 20:237. 1928) to be similar to *Septoria spadicea* Patterson & Charles.
Host: Pinus
Septoria pisi Westend.
Distr.: Temperate North America; Europe.
Hosts: Pisum, Vicia
Septoria plantaginis (Ces.) Sacc.
Septoria plantaginea Pass.
Distr.: Temperate regions.
Host: Plantago
Septoria plantaginis-majoris (Sacc.) Nannf.
Septoria plantaginea Pass. var. *plantaginis-majoris* Sacc.
Distr.: WI; Europe; Asia.
Host: Plantago
Septoria platanifolia Cooke in Ravenel
Distr.: Central and eastern United States.
Host: Platanus
Septoria pleosporoides Sacc.
Distr.: TX; Germany(type).
Host: Rumex
Septoria plucheae Guba
Distr.: FL, MA.
Host: Pluchea
Septoria podophyllina Peck
Distr.: Central and eastern United States.
Host: Podophyllum
Septoria polaris P. Karst.
Distr.: Central United States; Scandinavia.
Host: Ranunculus

Septoria polemonii Thuem.
Septoria polemoniicola Ellis & G. Martin
Distr.: Temperate regions.
Host: Polemonium
Septoria poliomela Syd.
Distr.: Western United States.
Notes: Leaf spot.
Host: Deschampsia
Septoria polita J.J. Davis
Distr.: ID, WI.
Host: Carex
Septoria polygalae Peck & G.W. Clinton
Distr.: NY.
Host: Polygala
Septoria polygonorum Desmaz.
Septoria persicariae O'Gara
Septoria polygonicola Lasch ex Sacc.
Septoria polygonina Thuem.
Distr.: Temperate northern hemisphere.
Host: Polygonum
Septoria populi Desmaz.
Gloeosporium stenosporum Ellis & Kellerm.
Mycosphaerella populi (Auersw.) J. Schröt. in Cohn (teleomorph)
Distr.: Temperate North America; Europe.
Host: Populus
Septoria populicola Peck
Mycosphaerella populicola G.E. Thompson (teleomorph)
Distr.: Temperate North America.
Notes: Thompson (Phytopathology 31:241. 1941).
Host: Populus
Septoria poseyi R. Sprague
Distr.: OR.
Hosts: Celtis, Mertensia
Septoria prenanthis Ellis & Everh.
Distr.: WI.
Host: Prenanthes
Septoria primulae Buckn. in Cooke
Distr.: FL; England.
Host: Primula
Septoria pruni Ellis
Distr.: Central and western United States.
Host: Prunus
Septoria psammophila Sacc.
Distr.: ND.
Hosts: Astragalus, Psoralidium
Septoria psilostega Ellis & G. Martin
Distr.: WI.
Host: Galium
Septoria pteleae Ellis & Everh.
Distr.: Central United States.
Host: Ptelea
Septoria pteridis Peck
Distr.: NY(type).
Host: Pteridium
Septoria pulchella Berk. & M.A. Curtis
Distr.: AL(type).
Host: Lyonia
Septoria punctoidea P. Karst.
Distr.: WI; Europe; Asia.
Host: Carex
Septoria punicea J.J. Davis
Phloeospora anemones Ellis & Kellerm. in G. Martin
Distr.: KS.
Hosts: Anemone, Aster
Septoria purpurascens Ellis & G. Martin in G. Martin
Distr.: United States.
Host: Potentilla
Septoria purpureo-cincta Ellis & Everh.
Distr.: KS(type).
Host: Prunus
Septoria pyricola (Desmaz.) Desmaz.
Septoria pyri Castagne
Mycosphaerella pyri (Auersw.) Boerema (teleomorph)
Distr.: Temperate northern hemisphere.
Notes: Leaf-fleck of pear (Boerema. Neth. J. Pl. Pathol. 76:166. 1970).
Hosts: Malus, Pyrus
Septoria querceti Thuem.
Mycosphaerella psilospora Gilman & Wadley (teleomorph)
Distr.: Eastern North America; Europe.
Notes: Leaf spot.
Host: Quercus
Septoria quercicola (Desmaz.) Sacc.
Septoria quercina Fautrey
Distr.: Temperate regions.
Host: Quercus
Septoria quercicola Sacc. var. **cinerea** Ellis & Everh.
Distr.: FL.
Host: Quercus
Septoria quercus Fuckel
Distr.: Temperate regions.
Host: Quercus
Septoria quinqueseptata R. Sprague
Distr.: WA.
Hosts: Koeleria, Sphenopholis
Septoria radiculae Dearn.
Distr.: NY.
Host: Rorippa
Septoria ranunculacearum Lév.
Distr.: LA, MS; France; Italy; Soviet Union.
Host: Ranunculus
Septoria ravenelii Thuem.
Distr.: SC.
Host: Prunus
Septoria rectae H.C. Greene
Distr.: WI.
Host: Potentilla
Septoria recurvata Ellis & Halst., nom. nud.
Distr.: IN.
Notes: Cited in a list of fungi in Indiana by Underwood (Proc. Indiana Acad. Sci. 1894:149. 1895).
Host: Trillium
Septoria repanda Bubák
Distr.: WI; Czechoslovakia(type).
Host: Erysimum
Septoria rhabdocarpa Ellis & Barth.
Distr.: CA, KS(type).
Hosts: Heteromeles, Populus, Salix, Salvia
Septoria rhamni-catharticae Ces.
Distr.: CA; Europe; Asia.
Host: Rhamnus
Septoria rhapontici Thuem.
Distr.: IA; Soviet Union(type).
Host: Rheum
Septoria rhododendri Cooke
Distr.: ME, NC, WA.
Host: Rhododendron
Septoria rhoina Berk. & M.A. Curtis
Mycosphaerella coggygriae Zerova (teleomorph)
Distr.: Temperate regions.
Hosts: Cotinus, Rhus
Septoria ribis (Lib.) Desmaz.
Ascochyta ribis Lib.
Septoria aurea Ellis & Everh.
Septoria grossulariae (Lib.) Westend.
Septoria sibirica Thuem.
Phyllosticta grossulariae Sacc. (synanamorph)
Mycosphaerella ribis Feltgen (teleomorph)
Distr.: Temperate northern hemisphere.
Notes: Leaf spot.
Host: Ribes
Septoria riparia Pass.
Distr.: Central and western United States; Europe; Asia.
Host: Carex
Septoria riviniae Pat.
Distr.: TX; Guadeloupe(type).
Host: Rivina
Septoria rollandiae F. Stevens & P.A. Young
Distr.: HI.
Host: Rollandia
Septoria rosae Desmaz.

Sphaerulina rehmiana Jaap (teleomorph)
Distr.: Widespread.
Notes: Jørstad (Skr. Norske Vidensk.-Akad. Oslo, Mat.-Naturvidensk. Kl. 22: 57. 1965); Boerema (Neth. J. Pl. Pathol. 69:76. 1963)
Host: Rosa
Septoria rubi Westend.
Distr.: Temperate northern hemisphere.
Notes: Leaf spot.
Host: Rubus
Septoria rubi Westend. var. **pallida** Ellis & Holw.
Distr.: LA, MN.
Host: Rubus
Septoria rudbeckiae Ellis & Halst.
Distr.: United States.
Hosts: Ratibida, Rudbeckia
Septoria rudbeckiae Ellis & Halst. var. **oaklandica** Sacc.
Distr.: NY; Canada.
Host: Rudbeckia
Septoria rumicis Trail
Septoria rumicicola Allesch.
Septoria rumicis Ellis non Trail, hom. illeg.
Distr.: Temperate regions.
Host: Rumex
Septoria saccharina Ellis & Everh.
Distr.: MA.
Host: Acer
Septoria salicicola (Fr.) Sacc.
Distr.: Temperate regions.
Notes: Jørstad (Skr. Norske Vidensk.-Akad. Oslo, Mat.-Naturvidensk. Kl. 22:63. 1965).
Host: Salix
Septoria salicifolia (Trel.) Ellis & Everh.
Pleospora salicifolia (Trel.) Petr.
Septoria salicifolia (Trel.) Berl. & Voglino non (Trel.) Ellis & Everh., hom. illeg.
Distr.: FL, TX; Europe.
Host: Spiraea
Septoria salicinia Peck
Distr.: NY, WI.
Host: Salix
Septoria salicis Westend.
Distr.: Temperate regions.
Host: Salix
Septoria salviae-pratensis Pass.
Distr.: HI.
Host: Salvia
Septoria samarae-macrophylli Dearn. & Barth. in Dearn.
Distr.: WA(type).
Host: Acer
Septoria sambucina Peck
Distr.: United States; Europe; Asia.
Host: Sambucus
Septoria sanguinea Dearn.
Distr.: OR, WA; Canada.
Host: Ribes
Septoria saniculae Ellis & Everh.
Distr.: Central United States.
Host: Sanicula
Septoria sarcobati Dearn. & House
Distr.: MT.
Host: Sarcobatus
Septoria saxifragae Pass.
Distr.: WA, WY; Europe.
Host: Heuchera
Septoria scorzonellae W.B. Cooke & C.G. Shaw
Distr.: WA.
Host: Microseris
Septoria scrophulariae Peck
Distr.: United States.
Host: Scrophularia
Septoria scrophulariae Peck var. **agalinis** H.C. Greene
Distr.: WI.
Host: Agalinis
Septoria scutellariae Thuem.
Distr.: North America; Europe.
Host: Scutellaria
Septoria secalis Prill. & Delacr.
Phoma secalina Jacz.
Septoria secalina (Jacz.) Sacc.
Distr.: Temperate North America; Europe.
Notes: Leaf blotch of rye.
Hosts: Secale, Stipa
Septoria secalis Prill. & Delacr. var. **stipae** R. Sprague
Distr.: Central and western United States.
Hosts: Agrostis, Stipa
Septoria sedi (Lib.) Westend.
Ascochyta sedi Lib.
Septoria sedicola Peck
Distr.: Central and eastern United States; Europe; Asia.
Host: Sedum
Septoria selenophomoides Cash & A.M. Watson
Distr.: FL; Central and South America; England; Italy.
Notes: Cash & Watson (Mycologia 47:740. 1055).
Hosts: Cattleya, Cymbidium, Dendrobium, Odontoglossum, Oncidium, Phalaenopsis
Septoria senecionis Westend.
Distr.: CA; Europe; Asia.
Host: Senecio
Septoria senecionis-aurei J.J. Davis
Distr.: WI.
Host: Senecio
Septoria serpentaria Ellis & G. Martin
Distr.: FL.
Hosts: Olea, Quercus
Septoria shastensis Bonar & W.B. Cooke
Distr.: CA.
Hosts: Aster, Machaeranthera
Septoria sheareana Sacc. & Trott.
Septoria longispora Shear non Overh., Voglino, I. Miyake & Bondartsev, hom. illeg.
Distr.: NJ.
Host: Vaccinium
Septoria shepherdiae (Sacc.) Dearn.
Cylindrosporium shepherdiae Sacc.
Distr.: North America.
Hosts: Elaeagnus, Shepherdia
Septoria siberica Thuem.
Distr.: WY; Soviet Union(type).
Host: Ribes
Septoria sicyi Peck
Distr.: United States.
Hosts: Echinocystis, Sicyos
Septoria sigmoidea Ellis & Everh.
Distr.: IA, WI.
Notes: Sprague (Mycologia 38:55. 1946) suggested this is similar to *Wojnowicia hirta*.
Host: Panicum
Septoria signalensis Solheim
Distr.: ID, WY.
Host: Symphoricarpos
Septoria sii Roberge ex Desmaz.
Distr.: Temperate regions.
Hosts: Berula, Chaerophyllum, Cicuta, Sium
Septoria silenes Westend.
Distr.: Temperate regions.
Notes: Species on *Silene* and *Melandryum* are in need of further study (Jørstad. Skr. Norske Vidensk.-Akad. Oslo, Mat.-Naturvidensk. Kl. 22:16. 1965).
Host: Silene
Septoria silenicola Ellis & G. Martin
Distr.: United States; Europe.
Host: Silene
Septoria silphii Ellis & Everh.
Distr.: IA, TX, WI.
Host: Silphium
Septoria silybi Pass.
Distr.: CA; Italy(type).
Host: Silybum
Septoria sisymbrii Ellis
Ascochyta sisymbrii Ellis & Kellerm.

Distr.: Temperate North America.
Hosts: Descurainia, Sisymbrium
Septoria sitchensis Dearn.
Distr.: ID, OR, WA.
Host: Sorbus
Septoria smardosii Moesz
Distr.: FL; Latvia(type).
Host: Amaranthus
Septoria smilacinae Ellis & G. Martin
Distr.: Central United States.
Hosts: Smilacina, Smilax
Septoria smilacis Ellis & Everh.
Distr.: WV.
Host: Smilax
Septoria socia Pass.
Distr.: WI; Czechoslovakia; Great Britain; Italy; Pakistan; Soviet Union; Spain.
Notes: Leaf spot.
Host: Chrysanthemum
Septoria solanina Speg.
Distr.: CA; Argentina(type).
Host: Solanum
Septoria solidaginicola Peck
Distr.: Temperate northern hemisphere.
Hosts: Aster, Solidago
Septoria solidaginis Thuem. non Cooke in Ravenel, hom. illeg.
Distr.: Central United States.
Hosts: Aster, Solidago
Septoria solitaris Ellis & Everh.
Distr.: CA, NC, OR, TX.
Host: Rhododendron
Septoria sonchifolia Cooke
Distr.: HI, North America; Europe; India.
Host: Sonchus
Septoria sorbi Lasch ex Sacc.
Distr.: IA; Europe; Asia.
Host: Sorbus
Septoria spadicea F. Patterson & V. Charles
Distr.: Northeastern United States.
Host: Pinus
Septoria spartinae (Trel.) R. Sprague
Ascochyta spartinae Trel.
Distr.: Central and western United States; Asia.
Notes: Leaf spot.
Hosts: Beckmannia, Spartina
Septoria speculariae Berk. & M.A. Curtis
Distr.: Central and eastern United States.
Hosts: Legousia, Triodanis
Septoria specularina Sacc.
Septoria speculariae Berk. & M.A. Curtis, nom. illeg.
Distr.: SC.
Notes: Berkeley & Curtis described *Septoria speculariae* two times from different collections. Saccardo erected a legitimate name for the later homonym.
Host: Legousia
Septoria spiculispora Ellis & Everh.
Distr.: DE.
Host: Euonymus
Septoria spiculosa Ellis & Holw.
Distr.: Central and eastern United States.
Host: Symplocarpus
Septoria spraguei Uecker & J.M. Krupinsky
Distr.: ND, SD.
Notes: Uecker & Krupinsky (Mycotaxon 14:273. 1982).
Hosts: Elymus, Leymus
Septoria stachydis Roberge in Desmaz.
Distr.: North America; Europe.
Host: Stachys
Septoria stellariae Roberge & Desmaz.
Distr.: Temperate regions.
Host: Stellaria
Septoria stigma Berk. & M.A. Curtis
Distr.: AL.
Host: Symplocos
Septoria stipina Died. in Syd., P. Syd. & E.J. Butler
Distr.: OR; Kashmir(type).
Notes: Leaf spot. Sprague (Diseases of cereals and grasses in North America. p. 255. 1950)
Host: Stipa
Septoria streptopidis Peck
Distr.: Western United States.
Hosts: Disporum, Streptopus
Septoria submaculata G. Wint.
Distr.: MO.
Host: Fraxinus
Septoria symphoricarpi Ellis & Everh.
Distr.: Northwestern United States.
Host: Symphoricarpos
Septoria symploci Ellis & G. Martin
Distr.: Southern United States.
Host: Symplocos
Septoria tageticola Changsri & G.F. Weber
Distr.: FL.
Host: Tagetes
Septoria tandilensis R. Sprague
Distr.: United States; Argentina.
Notes: Leaf spot.
Hosts: Dichanthelium, Digitaria, Panicum
Septoria tecomae Ellis & Everh.
Distr.: OK, TX, WV.
Host: Campsis
Septoria tecomaxochitl Tehon & G.L. Stout
Distr.: IL(type).
Host: Campsis
Septoria tellimae W.B. Cooke & C.G. Shaw
Distr.: WA.
Host: Lithophragma
Septoria tenella Cooke & Ellis
Distr.: Central and western United States.
Notes: Leaf blotch.
Hosts: Festuca, Vulpia
Septoria tenuis Dearn. & House
Distr.: NY, WI.
Host: Carex
Septoria tenuissima G. Wint.
Distr.: KS, MO.
Host: Boehmeria
Septoria thalictri Ellis & Everh.
Distr.: KS.
Host: Thalictrum
Septoria thlaspii H.C. Greene
Distr.: WI.
Host: Thlaspi
Septoria tinctoria Dearn. & House non Brunaud, hom. illeg.
Distr.: AR.
Host: Symplocos
Septoria tosevi Bubák
Distr.: CO; Bulgaria(type).
Host: Gentiana
Septoria tradescantiae (Ellis & Kellerm.) J.J. Davis
Cylindrosporium tradescantiae Ellis & Kellerm.
Distr.: Temperate North America.
Host: Tradescantia
Septoria trautvetteriae Ellis & Everh.
Distr.: WV.
Host: Trautvetteria
Septoria trichostematis Peck
Distr.: NY.
Host: Trichostema
Septoria trientalis (Lasch) Sacc.
Septoria increscens Peck
Distr.: United States; Europe; Asia.
Host: Trientalis
Septoria trifolii Ellis
Distr.: NJ, OR, WA.
Host: Trifolium
Septoria trillii Peck
Distr.: Central and eastern United States.
Host: Trillium
Septoria triseti Speg.

Distr.: Northwestern United States; Argentina.
Notes: Sprague (Diseases of cereals and grasses in North America. p. 261. 1950).
Host: Agrostis
Septoria tritici Roberge in Desmaz.
Mycosphaerella graminicola (Fuckel) J. Schröt. in Cohn (teleomorph)
Distr.: Worldwide.
Notes: Leaf spot or speckled blotch of wheat (Sutton & Waterson. C.M.I. Descr. 90. 1966; Sivanesan. The bitunicate Ascomycetes. p. 240. 1984; Bissett. Fungi Canadenses 244. 1983).
Hosts: Secale, Triticum
Septoria tritici Roberge in Desmaz. f. **avenae** (Desmaz.) R. Sprague
Distr.: Temperate regions.
Notes: Speckled leaf blotch.
Host: Avena
Septoria tritici Roberge in Desmaz. f. **holci** R. Sprague
Distr.: Northwestern United States; Scotland.
Notes: Speckled leaf blotch.
Host: Holcus
Septoria tritici Roberge in Desmaz. var. **lolicola** R. Sprague & A.G. Johnson
Distr.: Northwestern United States; Europe.
Notes: Speckled leaf blotch.
Host: Lolium
Septoria ulmariae Oudem.
Distr.: CT, WI; Europe.
Host: Filipendula
Septoria umatillensis R. Sprague
Distr.: OR.
Host: Psoralidium
Septoria unedinis Roberge in Desmaz.
Distr.: OR; Europe.
Host: Arbutus
Septoria unicolor G. Wint.
Distr.: Central and western United States.
Hosts: Lactuca, Taraxacum
Septoria urticae Roberge ex Desmaz.
Distr.: Temperate regions.
Hosts: Laportea, Urtica
Septoria urticaria Tharp
Distr.: TX.
Host: Urtica
Septoria vaccinii Ellis & Everh., nom. nud.
Distr.: MA.
Notes: This name is not in Cash's list of fungi named by Ellis (U.S.D.A. Spec. Publ. 2. 1952). There is a specimen in BPI with this name.
Host: Vaccinium
Septoria valerianae Sacc. & Fautrey
Distr.: WI; France(type).
Host: Valeriana
Septoria valerianellae L.E. Miles
Distr.: Southcentral United States.
Host: Valerianella
Septoria verbascicola Berk. & M.A. Curtis
Distr.: Temperate North America.
Host: Verbascum
Septoria verbenae Roberge in Desmaz.
Distr.: Temperate North America; Europe.
Host: Verbena
Septoria veronicae Roberge in Desmaz.
Distr.: North America; Europe.
Hosts: Veronica, Veronicastrum
Septoria vestita Berk. & M.A. Curtis
Distr.: PA.
Host: Cucurbita
Septoria viburni Westend.
Distr.: WI; Europe.
Host: Viburnum
Septoria viciicola Jørst.
Septoria viciae (Westend.) Pass.
Distr.: Temperate regions.
Notes: The reports of *Septoria viciae* could be referable to this species or to *Ascochyta viciae* Lib.
Host: Vicia
Septoria vincae Desmaz.
Distr.: Eastern United States; Europe; Asia.
Host: Vinca
Septoria violacea H.C. Greene
Distr.: WI.
Host: Oxalis
Septoria violae-palustris Died.
Septoria violae Westend. non Rabenh., hom. illeg.
Distr.: Temperate northern hemisphere.
Host: Viola
Septoria virgaureae (Lib.) Desmaz.
Distr.: Temperate regions.
Host: Solidago
Septoria viridi-tingens M.A. Curtis in Peck
Distr.: Northeastern North America.
Host: Allium
Septoria viticola Berk. & M.A. Curtis
Sacidium viticola (Berk. & M.A. Curtis) Cooke
Distr.: SC.
Host: Vitis
Septoria waldsteiniae Peck & G.W. Clinton
Distr.: Central and eastern United States.
Host: Waldsteinia
Septoria weiriana Sacc.
Septoria boycei Dearn.
Distr.: Northern North America.
Hosts: Alnus, Betula
Septoria wisconsina H.C. Greene
Distr.: WI.
Host: Astragalus
Septoria wistariae Tharp non Brunaud, hom. illeg.
Distr.: OK, TX.
Host: Wisteria
Septoria wyethiae Harkn.
Distr.: CA, UT.
Host: Wyethia
Septoria xanthifolia Ellis & Kellerm.
Distr.: United States; Europe.
Host: Iva
Septoria xanthii Desmaz.
Distr.: Temperate regions.
Host: Xanthium
Septoria xylostei Sacc. & G. Wint.
Distr.: Temperate regions.
Host: Lonicera
Septoria yuccae (Schwein.) Sacc.
Distr.: FL, PA.
Host: Yucca
Septoria zeae G.L. Stout
Distr.: IL; Europe; Asia.
Host: Zea
Septoria zeicola G.L. Stout
Distr.: IL(type).
Host: Zea
Septoria zeina G.L. Stout
Distr.: IL(type); Europe.
Host: Zea
Septoria ziziae Ellis & Everh., nom. nud.
Distr.: ND(type).
Notes: This name is in Brenckle (Mycologia 10:220. 1918).
Hosts: Thaspium, Zizia

Septosporium Corda. Hyphomycetes

Septosporium scyphophorum Cooke & Harkn.
Distr.: CA.
Host: Eucalyptus

Septotinia Whetzel ex Groves & M.E. Elliott. Discomycetes, Helotiales

Notes: Sutton (Mycologia 72:208. 1980).
Septotinia podophyllina Whetzel ex Groves & M.E. Elliott
Septotis podophyllina (Ellis & Everh.) Arx (anamorph)

Distr.: Central and eastern United States; Europe(introduced).
Notes: Leaf blight of mayapple. Recently this fungus is reported to cause a dieback of *Populus, Prunus* and *Salix* in Europe (Gremmen. Mycotaxon 28:255. 1987).
Host: Podophyllum
Septotinia populiperda A.M. Waterman & Cash ex Sutton
Septotis populiperda (Moesz & Smarods) Sutton (anamorph)
Distr.: Eastern North America; Europe.
Notes: Leaf blotch of poplar (Waterman & Cash. Mycologia 42:374. 1950).
Host: Populus

Septotis Buchwald ex Arx. Coelomycetes

Notes: Sutton (Mycologia 72:208. 1980).
Septotis podophyllina (Ellis & Everh.) Arx
Gloeosporium podophyllinum Ellis & Everh.
Septogloeum podophyllinum (Ellis & Everh.) Sacc.
Septotinia podophyllina Whetzel ex Groves & M.E. Elliott (teleomorph)
Distr.: Central and eastern United States; introduced into Europe.
Notes: Leaf blight of mayapple. Recently reported to cause a dieback of *Populus, Prunus* and *Salix* in Europe (Gremmen. Mycotaxon 28:255. 1987).
Host: Podophyllum

Serpula (Pers.) S.F. Gray. Basidiomycetes, Aphyllophorales

Serpula hexagonoides (Burt) W.B. Cooke
Merulius hexagonoides Burt
Distr.: CA.
Notes: On wood.
Host: Sequoia
Serpula himantioides (Fr.:Fr.) P. Karst.
Merulius himantioides Fr.:Fr.
Serpula americana (Burt) W.B. Cooke
Merulius americanus Burt
Merulius brassicaefolius Schwein.:Fr.
Distr.: Widespread.
Notes: On wood.
Hosts: Abies, Picea, Pinus, Populus, Pseudotsuga, Quercus, Tsuga
Serpula lacrimans (Wulfen:Fr.) J. Schröt.
Boletus lacrimans Wulfen in Jacq.:Fr.
Merulius lacrimans (Wulfen:Fr.) Fr.
Merulius terrestris (Peck) Burt
Merulius vastator Tode:Fr.
Distr.: Widespread.
Notes: One of the most destructive decay fungi. Causes "dry rot" of conifer wood in service; also in forests.
Hosts: Abies, Pinus, Pseudotsuga, Thuja, Tsuga

Servazziella J. Reid & C. Booth. Pyrenomycetes, Amphisphaeriales

Servazziella longispora (Servazzi) J. Reid & C. Booth
Cryptospora longispora Servazzi
Distr.: FL; Australia; Italy; New Zealand; Norfolk Island.
Notes: The report from Florida of "*Cytospora longispora*" Servazzi probably refers to this species. Reid & Booth (Canad. J. Bot. 65:1334. 1987).
Host: Araucaria

Sesquicillium W. Gams. Hyphomycetes

Sesquicillium buxi (J.C. Schmidt in Link)W. Gams
Fusidium buxi J.C. Schmidt in Link
Paecilomyces buxi (J.C. Schmidt in Link) Bezerra
Verticillium buxi (J.C. Schmidt in Link) Auersw. & Fleischhack
Nectriella coronata Juel (teleomorph)
Distr.: North America; Europe.
Notes: The connection with *Nectriella coronata* Juel is tentative (Gams. Acta Bot. Neerl. 17:455. 1968).
Host: Buxus
Sesquicillium candelabrum (Bonord.) W. Gams
Verticillium candelabrum Bonord.
Distr.: Temperate regions.
Notes: On seeds and decaying wood, isolated from soil (Domsch et al. Compendium of soil fungi. p. 732. 1980).
Host: Betula

Setosphaeria K.J. Leonard & E.G. Suggs. Loculoascomycetes, Dothideales

Notes: Sivanesan (Mycol. Pap. 158:201. 1987).
Setosphaeria holmii (Luttrell) K.J. Leonard & E.G. Suggs
Trichometasphaeria holmii Luttrell
Keissleriella holmii (Luttrell) Arx
Exserohilum holmii (Luttrell) K.J. Leonard & E.G. Suggs (anamorph)
Distr.: United States; Africa; India.
Notes: Leaf blight of grasses (Sivanesan. C.M.I. Descr. 885. 1986).
Host: Dactyloctenium
Setosphaeria prolata K.J. Leonard & E.G. Suggs
Exserohilum prolatum K.J. Leonard & E.G. Suggs (anamorph)
Distr.: Southeastern United States; Central America; eastern Africa.
Notes: Leaf spot of grasses (Sivanesan. C.M.I. Descr. 888. 1986).
Host: Zea
Setosphaeria turcica (Luttrell) K.J. Leonard & E.G. Suggs
Trichometasphaeria turcica Luttrell
Keissleriella turcica (Luttrell) Arx
Trichometasphaeria turcica Luttrell f. sp. *sorghi* Bergquist & Masias
Trichometasphaeria turcica Luttrell f. sp. *zeae* Bergquist & Masias
Exserohilum turcicum (Pass.) K.J. Leonard & E.G. Suggs (anamorph)
Distr.: Cosmopolitan.
Notes: Northern leaf blight of maize and sorghum (Ellis & Holliday. C.M.I. Descr. 304. 1971). Cultural variability and differential pathogenicity exists.
Hosts: Sorghum, Zea

Seuratia Pat. Loculoascomycetes, Arthoniales

Atichia Flot.
Notes: Meeker (Canad. J. Bot. 53:2485. 1975).
Seuratia globifera (Ellis & Everh.) Meeker
Asterina globifera Ellis & Everh.
Atichia globifera Petr.
Euthrypton globiferum (Ellis & Everh.) Theiss.
Phycopsis africana Syd.
Atichia treubii Höhn.
Phycopsis treubii (Höhn.) Boedijn
Myriophysella chilensis Speg. (anamorph)
Distr.: Tropical regions.
Notes: On numerous trees and shrubs.
Hosts: Grevillea, Micranthemum, Psidium
Seuratia maunauluana Meeker
Distr.: HI.
Host: Metrosideros
Seuratia millardetii (Racib.) Meeker
Atichia millardetii Racib.
Distr.: HI; tropical regions.
Notes: On numerous hosts.
Hosts: Coffea, Gardenia, Metrosideros, Passiflora, Pipturus, Rapanea, Scaevola, Schinus, Smilax, Straussia, Syzygium, Wikstroemia

Seynesia Sacc. Pyrenomycetes, Amphisphaeriales

Seynesia serrulata (Ellis & G. Martin) Petr.
Didymosphaeria serrulata Ellis & G. Martin
Distr.: FL.
Host: Serenoa

Seynesiella G. Arnaud. Loculoascomycetes, Dothideales

Notes: Müller & Arx (Beitr. Kryptogamenfl. Schweiz 11(2):427. 1962).
Seynesiella exigua Barr
Distr.: Northeastern United States and Canada.
Notes: On dead leaves (Bigelow & Barr. Rhodora 71:196. 1969).
Host: Juniperus
Seynesiella juniperi (Desmaz.) G. Arnaud
Dothidea juniperi Desmaz.
Asterina juniperi (Desmaz.) Jacz. in G. Boyer & Jacz.
Gibbera juniperi (Desmaz.) Auersw.
Microthyrium juniperi (Desmaz.) Sacc.
Seynesia juniperi (Desmaz.) Höhn.
Stigmatea juniperi (Desmaz.) G. Wint. in Rabenh.
Stigmatea alpina Speg. in Thuem.
Distr.: North America; Europe.
Notes: On living leaves. Barr (Canad. J. Bot. 46:860. 1968).
Host: Juniperus
Seynesiella sequoiae (Cooke & Harkn.) E. Müller in E. Müller & Arx
Dothidea sequoiae Cooke & Harkn.
Stigmatea sequoiae (Cooke & Harkn.) Sacc.
Distr.: CA.
Hosts: Calocedrus, Cupressus, Juniperus

Seynesiopeltis F. Stevens & Ryan. Loculoascomycetes, Dothideales

Distr.: Monotypic.
Seynesiopeltis tetraplasandrae F. Stevens & Ryan
Distr.: HI(type).
Host: Tetraplasandra

Shearia Petr. Coelomycetes

Shearia fusa (Berk. & M.A. Curtis) Barr
Sporidesmium fusus Berk. & M.A. Curtis
Shearia formosa (Ellis & Everh.) Petr.
Stegonsporium formosum Ellis & Everh.
Camarosporium magnoliae Shear
Shearia magnoliae Petr.
Pleomassaria maxima Ellis & Everh. (teleomorph)
Distr.: United States.
Notes: On branches (Barr. Mycotaxon 15:373. 1982).
Host: Magnolia

Sigmoidea J.L. Crane. Hyphomycetes

Sigmoidea marina J.M. Haythorn & E.B.G. Jones in J.M. Haythorn, E.B.G. Jones & J.L. Harrison
Distr.: Temperate marine habitats.
Notes: Haythorn et al. (Trans. Brit. Mycol. Soc. 74:615. 1980).
Host: Zostera

Simocybe P. Karst. Basidiomycetes, Agaricales

Simocybe atomacea (Murrill) Singer
Naucoria atomacea (Murrill) Murrill
Distr.: FL(type).
Host: Liquidambar
Simocybe haustellaris (Fr.:Fr.) Watling
Crepidotus haustellaris (Fr.:Fr.) P. Kumm.
Distr.: North America, scattered reports; Europe.
Hosts: Populus, Quercus
Simocybe tiliophila (Peck) Singer
Naucoria tiliophila (Peck) Singer
Distr.: NY.
Host: Tilia

Sirexcipula Bubák. Coelomycetes

Sirexcipula wyomingensis Wehmeyer
Distr.: Western United States.
Hosts: Aquilegia, Arabis, Castilleja, Hedysarum, Lupinus, Saxifraga

Sirobasidium Lagerh. & Pat. Basidiomycetes, Tremellales

Sirobasidium brefeldianum A. Möller
Distr.: Warmer parts of eastern North America to South America.
Notes: On wood.
Host: Mimosa

Sirococcus G. Preuss. Coelomycetes

Sirococcus americanus Sacc.
Distr.: WA.
Host: Heracleum
Sirococcus clavigignenti-juglandacearum N.B. Nair, Kostichka & Kuntz
Distr.: Eastern and midwest United States.
Notes: Canker of butternut (Nair et al. Mycologia 71:641. 1979).
Host: Juglans
Sirococcus conigenus (DC.) P. Cannon & Minter
Ascochyta piniperda Lindau in Engl. & Prantl
Ascochyta strobilina (Corda) Wollenweb.
Sporonema strobilinum Desmaz.
Sirococcus strobilinus G. Preuss
Distr.: Temperate northern hemisphere.
Notes: On cones and needles of Pinaceae. Cannon & Minter (Taxon 32:577. 1983); Sutton (The Coelomycetes. p. 602. 1980 as *S. strobilinus*).
Hosts: Calocedrus, Picea, Pinus, Tsuga
Sirococcus halesiae Ellis & Everh.
Distr.: WV.
Host: Halesia
Sirococcus phlei Tehon & E.Y. Daniels
Distr.: IL.
Host: Phleum

Sirodothis Clements. Coelomycetes

Sirodothis citri Batista in Batista & Vittal
Distr.: FL.
Host: Citrus
Sirodothis populnea (Thuem.) Sutton & Funk
Dothiorella populnea Thuem.
Sirodothis populi Clements
Tympanis spermatiospora (Nyl.) Nyl. (teleomorph)
Distr.: North America; Australia.
Notes: On branches of poplars (Sutton & Funk. Canad. J. Bot. 53:524. 1975; Sutton. The Coelomycetes. p. 559. 1980).
Host: Populus

Sirogloea Petr. Coelomycetes

Sirogloea euonymi Petr.
Distr.: CA; Europe.
Host: Euonymus

Siroplacodium Petr. Coelomycetes

Siroplacodium shastense (R. Sprague & W.B. Cooke) Petr.
Placosphaeria shastensis R. Sprague & W.B. Cooke
Distr.: CA.
Notes: Sutton (The Coelomycetes. p. 566. 1980).
Host: Monardella

Siroscyphellina Petr. Coelomycetes

Siroscyphellina lupini Wehmeyer
Distr.: WA.
Notes: On dead stems (Wehmeyer. Sydowia 6:440. 1952).
Host: Lupinus

Sirosporium Bubák & Serebrianikow. Hyphomycetes

Notes: Ellis (More dematiaceous Hyphomycetes. p. 299. 1976).
Sirosporium antenniforme (Berk. & M.A. Curtis) Bubák &

Serbinow
Macrosporium antenniforme Berk. & M.A. Curtis
Alternaria antenniforme (Berk. & M.A. Curtis) J.A. Elliott, nom. inval.
Gyroceras divergens Peck
Distr.: Eastern North America.
Notes: On leaves (Ellis. Dematiaceous Hyphomycetes. p. 288. 1971).
Host: Celtis
Sirosporium beaumontii (Sacc.) M.B. Ellis
Helminthosporium beaumontii Sacc.
Helminthosporium dubium Berk. & M.A. Curtis non Schwein., hom. illeg.
Distr.: Southeastern United States
Notes: Ellis (Mycol. Pap. 87:11. 1963).
Host: Viburnum
Sirosporium diffusum (Heald & F.A. Wolf) Deighton
Clasterosporium diffusum Heald & F.A. Wolf
Cercospora fusca Rands
Distr.: Southern United States; Malawi.
Notes: Brown leaf spot of pecan.
Host: Carya
Sirosporium diospyri (Thuem.) Deighton
Cercospora diospyri Thuem.
Distr.: Central and southeastern United States.
Notes: Brown leaf spot.
Host: Diospyros

Sirothyriella Höhn. Coelomycetes

Notes: Sutton (Mycol. Pap. 141:235. 1977).
Sirothyriella pini-austriacae (Roum. & Fautrey) Sutton
Leptostroma pinastri Desmaz.
Distr.: WI; Europe.
Notes: Minter (Canad. J. Bot. 58:916. 1980).
Host: Pinus

Sistotrema Pers.:Fr. Basidiomycetes, Aphyllophorales

Notes: Gilbertson & Ryvarden (North American Polypores 2:709. 1987).
Sistotrema brinkmannii (Bres.) J. Eriksson
Odontia brinkmannii Bres.
Grandinia brinkmannii (Bres.) Bourd. & Galzin
Trechispora brinkmannii (Bres.) D.P. Rogers & H. Jacks.
Corticium coronilla Höhn.
Distr.: Widespread.
Notes: Usually on wood, sometimes on plant debris and basidiomata. This species represents several biological species which are similar morphologically (Bigg. Mycologia 29:686. 1937; Hallberg. Mycotaxon 21:389. 1984; Lemke. Mycologia 61:57. 1969; Ullrich. Mycologia 65:1234. 1973).
Hosts: Abies, Acer, Alnus, Betula, Liriodendron, Pinus, Populus, Quercus, Salix, Tsuga
Sistotrema confluens Pers.:Fr.
Distr.: Temperate northern hemisphere.
Notes: On wood, litter and plant debris.
Hosts: Pinus, Populus
Sistotrema coroniferum (Höhn. & Litsch.) Donk
Trechispora coronifera (Höhn. & Litsch.) D.P. Rogers & H. Jacks.
Distr.: Temperate northern hemisphere.
Notes: On wood.
Hosts: Picea, Populus, Quercus
Sistotrema diademiferum (Bourd. & Galzin) Donk
Trechispora diademifera (Bourd. & Galzin) D.P. Rogers
Distr.: North and Central America; Europe.
Notes: On wood and plant debris.
Hosts: Picea, Populus
Sistotrema hirschii (Donk) Donk
Trechispora hirschii (Donk) D.P. Rogers
Distr.: Central and western United States; Europe.
Notes: On hardwoods.
Hosts: Quercus, Salix
Sistotrema muscicola (Pers.) S. Lundell in S. Lundell & Nannf.
Trechispora muscicola (Pers.) D.P. Rogers
Distr.: Temperate northern hemisphere.
Notes: On wood and plant debris.
Host: Populus
Sistotrema raduloides (P. Karst.) Donk
Trechispora raduloides (P. Karst.) D.P. Rogers
Distr.: Temperate northern hemisphere.
Notes: On wood.
Hosts: Picea, Populus, Tilia, Tsuga
Sistotrema subtrigonospermum D.P. Rogers
Trechispora subtrigonosperma (D.P. Rogers) D.P. Rogers & H. Jacks.
Distr.: North, Central, and South America.
Notes: On wood.
Hosts: Abies, Pinus, Populus

Sistotremastrum J. Eriksson. Basidiomycetes, Aphyllophorales

Sistotremastrum suecicum Litsch. ex J. Eriksson
Distr.: Temperate northern hemisphere.
Notes: On wood, usually of conifers.
Host: Populus

Skeletocutis Kotlaba & Pouzar. Basidiomycetes, Aphyllophorales

Incrustoporia Dománski
Notes: Gilbertson & Ryvarden (North American Polypores 2:711. 1987).
Skeletocutis alutacea (J. Lowe) J. Keller
Poria alutacea J. Lowe in Overh. & J. Lowe
Incrustoporia alutacea (J. Lowe) D. Reid
Distr.: Widespread.
Notes: On wood.
Hosts: Alnus, Picea, Tsuga
Skeletocutis amorpha (Fr.:Fr.) Kotlaba & Pouzar
Polyporus amorphus Fr.:Fr.
Distr.: North America; Europe; New Zealand.
Notes: On wood of conifers.
Hosts: Abies, Larix, Picea, Pinus, Pseudotsuga, Sequoia, Tsuga
Skeletocutis nivea (Jungh.) J. Keller
Incrustoporia nivea (Jungh.) Ryvarden
Polyporus semipileatus Peck
Tyromyces semipileatus (Peck) Murrill
Distr.: Temperate northern hemisphere.
Notes: On wood.
Hosts: Abies, Acer, Alnus, Betula, Fagus, Gaultheria, Hamamelis, Larix, Pinus, Populus, Pseudotsuga, Quercus, Thuja
Skeletocutis stellae (Pilát) J. Keller
Poria stellae Pilát
Distr.: Temperate northern hemisphere.
Notes: On conifers.
Hosts: Abies, Picea, Pinus
Skeletocutis subincarnata (Peck) J. Keller
Incrustoporia subincarnata (Peck) Dománski
Poria subincarnata (Peck) Murrill
Distr.: Temperate northern hemisphere.
Notes: On conifers.
Hosts: Abies, Picea, Pinus, Prosopis, Pseudotsuga

Soleella Darker. Discomycetes, Rhytismatales

Soleella striiformis (Darker) Darker
Bifusella striiformis Darker
Distr.: CA.
Notes: Cannon & Minter (C.M.I. Descr. 800. 1984).
Host: Pinus

Sonderhenia eucalypticola (A.R. Davis) H.J. Swart & J. Walker
Hendersonia eucalypticola A.R. Davis
Hendersonia fraserae Hansf.

Distr.: CA; Brazil; Australia.
Notes: Swart & Walker (Trans. Brit. Mycol. Soc. 90:640. 1988).
Host: Eucalyptus

Sordaria Ces. & De Not., nom. cons. Pyrenomycetes, Sordariales

Notes: Lundqvist (Symb. Bot. Upsal. 20:1. 1972).
Sordaria destruens (Shear) Hawker
Anthostomella destruens Shear
Melanospora destruens (Shear) Shear & B.O. Dodge
Distr.: North America; Europe.
Hosts: Rhododendron, Vaccinium
Sordaria fimicola (Roberge ex Desmaz.) Ces. & De Not.
Distr.: Cosmopolitan.
Notes: Usually on dung, also on a variety of hosts.
Hosts: Glycine, Helianthus, Platanus, Pseudotsuga, Zea
Sordaria minima Sacc. & Speg.
Distr.: Cosmopolitan.
Notes: Usually on dung.
Host: Pseudotsuga

Sorosphaera J. Schröt. Plasmodiophoromycetes, Plasmodiophorales

Notes: Karling (The Plasmodiophorales. p. 52. 1968).
Sorosphaera radicalis Cooke
Distr.: CA; England(type).
Notes: Parasitic in the root hairs of various plants.
Host: Poa
Sorosphaera veronicae (J. Schröt.) J. Schröt.
Tuburcinia veronicae J. Schröt.
Distr.: Widespread.
Host: Veronica

Sorosporium Rudolphi. Basidiomycetes, Ustilaginales

Notes: Vánky (Symb. Bot. Upsal. 24:108. 1985; Cryptog. Stud. 1:100. 1987) restricted this genus to one species occurring on Caryophyllaceae, but not all names have been redisposed. Species on grasses probably belong in the genus *Sporisorium*.
Sorosporium confusum H. Jacks.
Distr.: United States.
Host: Aristida
Sorosporium contortum D. Griffiths
Distr.: Southwestern United States; Mexico.
Host: Heteropogon
Sorosporium ellisii G. Wint.
Distr.: North America.
Hosts: Andropogon, Heteropogon, Schizachyrium
Sorosporium everhartii Ellis & B.T. Galloway
Distr.: Central and eastern United States.
Hosts: Andropogon, Schizachyrium
Sorosporium mixtum (Massee) McAlpine
Tilletia mixta Massee
Sorosporium eriochloae D. Griffiths
Distr.: Southwestern United States.
Host: Eriochloa
Sorosporium ovarium D. Griffiths
Distr.: FL; Mexico.
Host: Brachiaria
Sorosporium paranensis Hirschhorn
Distr.: HI; South America.
Host: Axonopus
Sorosporium paspali-thunbergii (Henn.) Ito
Ustilago paspali-thunbergii Henn.
Sorosporium paspali McAlpine
Distr.: Subtropical and tropical regions.
Notes: Zundel (The Ustilaginales of the world. p. 69. 1953).
Host: Paspalum
Sorosporium provinciale (Ellis & B.T. Galloway) G.P. Clinton
Sorosporium ellisii G. Wint. var. *provincialis* Ellis & B.T. Galloway
Distr.: Central United States.
Host: Andropogon
Sorosporium rhynchosporae Henn.
Distr.: MS; Brazil(type).
Host: Rhynchospora
Sorosporium saponariae Rudolphi
Distr.: Widespread.
Notes: On Caryophyllaceae.
Hosts: Arenaria, Cerastium, Silene, Stellaria
Sorosporium texanum Zundel
Distr.: TX(type).
Host: Pennisetum

Spadicoides S.J. Hughes. Hyphomycetes

Spadicoides catenulata Wang & Sutton
Distr.: NY.
Notes: On rotten wood of hardwoods.
Hosts: Acer, Ulmus
Spadicoides grovei Ellis
Distr.: Northern North America; England.
Notes: On decaying wood (Hughes. Mycol. Pap. 93:12. 1963; Hughes. Fungi Canadenses 7. 1973).
Host: Fagus
Spadicoides obovata (Cooke & Ellis) S.J. Hughes
Acrothecium obovatum Cooke & Ellis
Distr.: Northeastern North America; New Zealand.
Notes: On decaying wood (Hughes. Mycol. Pap. 93:6. 1963; Hughes. Fungi Canadenses 6. 1973).
Hosts: Castanea, Magnolia

Sparassis Fr.:Fr. Basidiomycetes, Aphyllophorales

Sparassis crispa (Wulfen:Fr.) Fr.
Clavaria crispa Wulfen in Jacq.
Sparassis radicata Weir
Distr.: Widespread in temperate regions.
Notes: Burdsall & Miller (Mycotaxon 31:591. 1988).
Hosts: Abies, Carex, Larix, Picea, Pinus, Pseudotsuga

Spathulariopsis Maas G. Discomycetes, Helotiales

Notes: Monotypic (Maas Geesteranus. Proc. Koninkl. Nederl. Akad. Wetens., Ser. C, 75:254. 1972).
Spathulariopsis velutipes (Cooke & Farl.) Maas G.
Spathularia velutipes Cooke & Farl.
Distr.: Central and eastern United States.
Host: Betula

Spegazzinia Sacc. Hyphomycetes

Notes: Ellis (Dematiaceous Hyphomycetes. p. 565. 1971); Ellis (More dematiaceous Hyphomycetes. p. 474. 1976).
Spegazzinia parkeri Sivasithamparam
Distr.: FL; Australia(type).
Host: Acer
Spegazzinia tessarthra (Berk. & M.A. Curtis) Sacc.
Distr.: Warm-temperate to tropical regions.
Notes: Saprobic on plant debris and in soil.
Hosts: Juncus, Zea

Speiropsis Tubaki. Hyphomycetes

Speiropsis pedatospora Tubaki
Distr.: HI; Japan; Sri Lanka.
Notes: On leaves of *Eriobotrya* and Myrtaceae (Ellis. More dematiaceous Hyphomycetes. p. 363. 1976).
Host: Eucalyptus

Spermospora R. Sprague. Hyphomycetes

Spermospora avenae (R. Sprague & A.G. Johnson) R. Sprague
Pseudodiscosia avenae R. Sprague & A.G. Johnson
Distr.: Northwestern North America; Turkey.
Notes: Red leather leaf.
Hosts: Arrhenatherum, Avena
Spermospora ciliata (R. Sprague) Deighton
Spermospora subulata R. Sprague f. *ciliata* R. Sprague
Distr.: Northwestern North America.
Hosts: Festuca, Poa
Spermospora subulata (R. Sprague) R. Sprague
Cercosporella subulata R. Sprague
Cercospora subulata (R. Sprague) R. Sprague
Distr.: Western United States.
Hosts: Agrostis, Bromus, Calamagrostis, Deschampsia, Festuca, Melica, Trisetum

Sphaceloma de Bary. Coelomycetes

Sphaceloma ampelinum de Bary
Gloeosporium ampelophagum (Pass.) Sacc.
Elsinoe ampelina Shear (teleomorph)
Distr.: Worldwide in grape-growing regions.
Notes: Anthracnose, bird's-eye rot of grapes (Sivanesan & Critchett. C.M.I. Descr. 439. 1974).
Host: Vitis
Sphaceloma araliae Jenk.
Distr.: FL, MD, MO.
Hosts: Aralia, Dizygotheca
Sphaceloma bucidae A.M. Watson
Distr.: FL(type).
Host: Bucida
Sphaceloma cercocarpi Bitancourt & Jenk. in Jenk., Bitancourt & Pollack
Distr.: CA(type).
Notes: Sutton & Pollack (Mycologia 65:1128. 1973).
Host: Cercocarpus
Sphaceloma choisyae A.M. Watson & Jenk.
Distr.: AZ(type).
Host: Choisya
Sphaceloma crataegi (Dearn. & Barth.) Jenk.
Gloeosporium crataegi Dearn. & Barth.
Melanophora crataegi (Dearn. & Barth.) Arx
Distr.: WA.
Host: Crataegus
Sphaceloma euonymi-japonici Kurosawa & Katsuki
Elsinoe euonymi-japonici Jenk. & Bitancourt (teleomorph)
Distr.: FL; Japan.
Host: Euonymus
Sphaceloma fawcettii Jenk.
Elsinoe fawcettii Bitancourt & Jenk. (teleomorph)
Distr.: Worldwide in citrus-growing regions.
Notes: Common citrus scab (Sivanesan & Critchett. C.M.I. Descr. 438. 1974).
Hosts: Citrus, Fortunella
Sphaceloma floridense Bitancourt & Jenk.
Distr.: FL.
Host: Galactia
Sphaceloma hederae Bitancourt & Jenk. in Jenk., Bitancourt, & Pollack
Distr.: CA, southeastern United States; Brazil.
Host: Hedera
Sphaceloma lippiae Baines & Cummins
Distr.: FL, IN.
Hosts: Lippia, Phyla
Sphaceloma magnoliae Jenk. & J.H. Miller
Elsinoe magnoliae J.H. Miller & Jenk. (teleomorph)
Distr.: Southern United States.
Host: Magnolia
Sphaceloma mangiferae Bitancourt & Jenk.
Elsinoe mangiferae Bitancourt & Jenk. (teleomorph)
Distr.: FL; Caribbean.
Host: Mangifera
Sphaceloma mattirolianum (Sacc. & D. Sacc.) Jenk.
Illosporium mattirolianum Sacc. & D. Sacc.
Elsinoe mattirolianum G. Arnaud & Bitancourt (teleomorph)
Distr.: CA; Italy.
Host: Arbutus
Sphaceloma menthae Jenk.
Distr.: IN, MI, MD.
Host: Mentha
Sphaceloma morindae Bitancourt & Jenk.
Distr.: FL; Colombia.
Host: Morinda
Sphaceloma murrayae Grods. & Jenk.
Distr.: WI, Pacific coastal United States; Argentina; Latvia; New Zealand.
Host: Salix
Sphaceloma necator (Ellis & Everh.) Jenk. & Shear
Gloeosporium necator Ellis & Everh.
Gloeosporium venetum Speg.
Elsinoe veneta (Burkholder) Jenk. (teleomorph)
Distr.: Temperate regions.
Notes: Cane spot or anthracnose of raspberry (Sivanesan & Critchett. C.M.I. Descr. 484. 1974).
Host: Rubus
Sphaceloma oleandri Thirumalachar
Distr.: FL; India.
Host: Nerium
Sphaceloma penstemonis A.M. Watson & Jenk.
Distr.: NY(type).
Host: Penstemon
Sphaceloma perseae Jenk.
Distr.: Southern United States; West Indies; Central and South America.
Notes: Spot anthracnose of avocado.
Host: Persea
Sphaceloma plantaginis Jenk. & Bitancourt
Distr.: Northeastern and midwestern United States; Taiwan.
Host: Plantago
Sphaceloma poinsettiae Jenk. & Ruehle
Distr.: Subtropical and tropical America; South Pacific.
Host: Euphorbia
Sphaceloma psidii Bitancourt & Jenk.
Distr.: FL; Brazil.
Hosts: Feijoa, Psidium
Sphaceloma punicae Bitancourt & Jenk.
Distr.: FL; South America.
Host: Punica
Sphaceloma ribis Jenk. & Bitancourt in Jenk.,Bitancourt, & Pollack
Distr.: WA.
Host: Ribes
Sphaceloma rosarum (Pass.) Jenk.
Phyllosticta rosarum Pass.
Gloeosporium rosicola Dearn. & Barth.
Distr.: Temperate regions.
Notes: Rose anthracnose.
Host: Rosa
Sphaceloma spondiadis Bitancourt & Jenk.
Elsinoe spondiadis A.M. Watson & Jenk. (teleomorph)
Distr.: FL; Honduras; Brazil.
Host: Spondias
Sphaceloma symphoricarpi Barrus & Horsfall
Distr.: United States.
Host: Symphoricarpos
Sphaceloma venezuelanum Bitancourt & Jenk.
Distr.: FL.
Notes: A doubtful species. No other record of this name was found.
Host: Codiaeum
Sphaceloma viburni Jenk. & Bitancourt in Jenk., Bitancourt, & Pollack
Distr.: Pacific coastal states, southern Atlantic coastal states.
Host: Viburnum
Sphaceloma violae Jenk.
Gloeosporium violae Berk. & Broome
Distr.: Central and eastern North America.
Notes: Spot anthracnose or scab of violet.
Host: Viola

Sphacelotheca de Bary. Basidiomycetes, Ustilaginales

Notes: Langdon & Fullerton (Mycotaxon 6:421. 1978).
Sphacelotheca andropogonis-hirtifolii (Henn.) G.P. Clinton
Ustilago andropogonis-hirtifolii Henn.
Sphacelotheca sorghastri Zundel
Distr.: Southern United States; Mexico.
Notes: On Poaceae.
Hosts: Andropogon, Bothriochloa, Sorghastrum
Sphacelotheca cordobensis (Speg.) H. Jacks.
Sphacelotheca panici-leucophaei (Bref.) G.P. Clinton
Sphacelotheca paspali-notati (Henn.) G.P. Clinton
Distr.: Southern United States; Central America.
Notes: On Poaceae.
Hosts: Digitaria, Paspalum
Sphacelotheca digitariae (Kunze) G.P. Clinton
Ustilago digitariae (Kunze) G. Wint.
Distr.: Widespread.
Host: Digitaria
Sphacelotheca diplospora (Ellis & Everh.) G.P. Clinton
Ustilago diplospora Ellis & Everh.
Distr.: Central and southern United States; Mexico.
Notes: On Poaceae.
Hosts: Digitaria, Echinochloa
Sphacelotheca magnusiana (A. Fisch. v. Waldh.) Cif.
Tilletia magnusiana A. Fisch. v. Waldh.
Distr.: Warm regions.
Notes: Ciferri (Trans. Brit. Mycol. Soc. 18:262. 1933) suggested that this is an older name for *Sphacelotheca pamparum*. Later authors have not agreed.
Host: Setaria
Sphacelotheca monilifera (Ellis & Everh.) G.P. Clinton
Ustilago monilifera Ellis & Everh.
Distr.: Subtropical and tropical regions.
Notes: On Poaceae.
Hosts: Andropogon, Heteropogon
Sphacelotheca nealii (Ellis & F.W. Anderson) G.P. Clinton
Distr.: FL; Mexico.
Host: Heteropogon
Sphacelotheca occidentalis (Seym.) G.P. Clinton
Sorosporium ellisii G. Wint. var. *occidentalis* Seym.
Ustilago andropogonis Kellerm. & Sw.
Distr.: United States.
Notes: This species must be transferred to *Sporisporium*.
Hosts: Andropogon, Heteropogon, Schizachyrium
Sphacelotheca pamparum (Speg.) G.P. Clinton
Ustilago pamparum Speg.
Distr.: Subtropical and tropical regions.
Notes: On Poaceae. Ciferri (Trans. Brit. Mycol. Soc. 18:262. 1933) suggested that *S. magnusiana* is an earlier name for this species. Later authors have not agreed.
Host: Pennisetum
Sphacelotheca seymouriana G.P. Clinton
Distr.: Southeastern United States.
Hosts: Andropogon, Schizachyrium
Sphacelotheca utahensis Zundel
Distr.: UT(type).
Host: Sporobolus

Sphaerella Rabenh. non Sommerf., hom. illeg. Loculoascomycetes, Dothideales

Notes: This name is a later homonym. Although replaced by *Mycosphaerella*, not all names have been redisposed.
Sphaerella alni-viridis De Not.
Distr.: AK; Italy(type).
Host: Alnus
Sphaerella californica Cooke & Harkn.
Distr.: AK, CA.
Host: Agrostis
Sphaerella graminum Sacc. & Scalia
Distr.: AK(type).
Host: Poa
Sphaerella grossulariae (Fr.:Fr.) Auersw. var. ***salicella*** Sacc. & Scalia
Distr.: AK(type).
Notes: Although the type variety has been transferred to *Mycosphaerella*, this name is still in the outdated genus *Sphaerella*.
Host: Salix
Sphaerella paulula Cooke
Distr.: FL(type).
Host: Zea
Sphaerella platani Ellis & G. Martin
Distr.: KS(type).
Host: Platanus
Sphaerella sesbaniae Ellis & Everh.
Distr.: LA(type).
Host: Sesbania
Sphaerella umbellulariae Cooke & Harkn.
Distr.: CA.
Notes: Barr (Contr. Univ. Michigan Herb. 9:587. 1972).
Host: Umbellularia

Sphaerellopsis Cooke. Coelomycetes

Darluca Castagne
Notes: Sutton (The Coelomycetes. p. 470. 1980).
Sphaerellopsis filum (Biv.-Bern. ex Fr.) Sutton
Phoma filum Biv.-Bern. ex Fr.
Darluca filum (Biv.-Bern. ex Fr.) Berk.
Diplodina quercuum (Cooke) Tracy & Earle
Sphaerellopsis quercuum Cooke
Eudarluca caricis (Fr.) O. Eriksson (teleomorph)
Distr.: Cosmopolitan.
Notes: Hyperparasitic on Uredinales.
Hosts: Arctagrostis, Poa, Quercus

Sphaeridium Fresen. Hyphomycetes

Sphaeridium luteum Höhn.
Distr.: CA; Austria(type).
Host: Abies

Sphaerobasidium Oberwinkler. Basidiomycetes, Aphyllophorales

Sphaerobasidium minutum (J. Eriksson) Oberwinkler
Xenasma minutum J. Eriksson
Paullicorticium minutum (J. Eriksson) Liberta
Distr.: Temperate northern hemisphere.
Notes: On wood.
Hosts: Abies, Picea, Pinus, Tsuga

Sphaerobolus Tode:Pers. Basidiomycetes, Sclerodermatales

Sphaerobolus stellatus Tode:Pers.
Distr.: Temperate regions.
Notes: On dung, wood, or plant debris.
Hosts: Aglaonema, Dracaena, Fraxinus, Pseudotsuga, Salix

Sphaerodothis (Sacc. & Syd.) Shear. Pyrenomycetes, Phyllachorales

Sphaerodothis neowashingtoniae Shear
Phaeochora neowashingtoniae (Shear) Theiss. & Syd.
Distr.: CA.
Host: Washingtonia
Sphaerodothis pirifera (Speg.) Theiss. & Syd.
Phyllachora pirifera Speg.
Distr.: FL; Brazil(type).
Notes: On leaves.
Host: Verbesina

Sphaerognomonia Potebnia. Pyrenomycetes, Diaporthales

Sphaerognomonia carpinea (Fr.:Fr.) Potebnia
Laestadia carpinea (Fr.:Fr.) Sacc.
Laestadia castanicola Ellis & Everh.
Laestadia orientalis Ellis & Everh.
Distr.: Eastern North America; Europe.
Notes: On overwintered leaves of various hardwoods.
Hosts: Acer, Alnus, Betula, Carpinus, Castanea, Corylus, Ostrya
Sphaerognomonia papillostoma (Dearn. & House) Barr
Gnomonia papillostoma Dearn. & House
Distr.: Northeastern North America.
Notes: On overwintered leaves.
Host: Spiraea

Sphaerographium Sacc. Coelomycetes

Sphaerographium abditum Sacc. & Scalia
Distr.: AK(type).
Host: Heracleum
Sphaerographium avenaceum Fairm. in Millsp. & L. Nutt.
Distr.: CA.
Host: Avena
Sphaerographium niveum Dearn. & House
Distr.: NY.
Host: Rhamnus

Sphaeromyces Mont. Hyphomycetes

Notes: This genus is considered a synonym of *Aspergillus*, but not all names have been redisposed.
Sphaeromyces delphinii Peck
Distr.: UT(type).
Host: Delphinium

Sphaeronaema Fr.:Fr. Coelomycetes

Sphaeronaema acerinum Peck
Lagynodella acerina (Peck) Petr.
Naemosphaera acerina (Peck) Höhn.
Sphaeronaema nigripes Ellis
Dermea acerina (Peck) Rehm (teleomorph)
Distr.: Temperate North America.
Notes: Groves (Mycologia 38:409. 1946).
Hosts: Acer, Amelanchier
Sphaeronaema canum Ellis & Everh.
Distr.: KS(type), PA.
Notes: On dead branches.
Host: Acer
Sphaeronaema clethrincola Ellis
Chondropodiella clethrincola (Ellis) Höhn.
Godronia urceolata (Ellis) Höhn. (teleomorph)
Distr.: Eastern North America.
Host: Clethra
Sphaeronaema eucalypti Cooke & Harkn.
Distr.: CA.
Host: Eucalyptus
Sphaeronaema longirostre Clements
Distr.: NE.
Host: Crataegus
Sphaeronaema minutulum D. Sacc.
Distr.: NY; Italy(type).
Host: Sedum
Sphaeronaema pallidum Peck
Excipulina pallida (Peck) Dearn. & House
Distr.: NY(type).
Host: Sorbus
Sphaeronaema physocarpi Ellis & Everh.
Distr.: Eastern United States.
Host: Physocarpus
Sphaeronaema pithyum Sacc.
Distr.: NY; Europe(type).
Notes: On *Pinus*, possibly the anamorph of a *Cenangium*.
Host: Pinus
Sphaeronaema pomorum Shear
Distr.: NJ.
Host: Vaccinium
Sphaeronaema rhoidis Berk. & M.A. Curtis
Distr.: SC(type).
Host: Rhus

Sphaeronaemella P. Karst. Pyrenomycetes, Amphisphaeriales

Notes: According to Cannon & Hawksworth (Bot. J. Linn. Soc. 84:149. 1982), this genus is monotypic with only the fungicolous species *S. helvellae*. Several names have not been transferred elsewhere.
Sphaeronaemella carnea Ellis & Everh.
Distr.: NY(type).
Host: Fraxinus
Sphaeronaemella rosae Ellis & Everh.
Distr.: Eastern United States.
Host: Rosa

Sphaerophoma Petr. Coelomycetes

Sphaerophoma brencklei Petr.
Distr.: ND.
Host: Helianthus

Sphaerophragmium Magnus. Basidiomycetes, Uredinales

Sphaerophragmium acaciae (Cooke) Magnus
Distr.: Subtropical and tropical regions.
Notes: Autoecious.
Host: Albizia

Sphaeropsis Sacc., nom. cons. Coelomycetes

Macrophoma (Sacc.) Berl. & Voglino
Sphaeropsis abietis Povah
Distr.: MI(type).
Notes: Povah (Pap. Michigan Acad. Sci. 13:182. 1931).
Host: Abies
Sphaeropsis alni Cooke & Ellis
Distr.: SC(type), WA.
Host: Alnus
Sphaeropsis amelanchieris Dearn.
Distr.: NY.
Host: Amelanchier
Sphaeropsis amorphae Ellis & Barth.
Distr.: KS(type), NY.
Host: Amorpha
Sphaeropsis amplispora Dearn. & Barth.
Distr.: KS.
Host: Acer
Sphaeropsis begoniicola Ellis & Everh.
Distr.: AL.
Host: Begonia
Sphaeropsis betulae Cooke
Distr.: NY.
Host: Betula
Sphaeropsis betulae Cooke var. ***foliicola*** J.J. Davis
Distr.: WI.
Host: Betula
Sphaeropsis cercidis Dearn. & House non Brunaud, hom. illeg.
Distr.: KS.
Host: Cercis
Sphaeropsis cornicola Dearn.
Distr.: MD.
Host: Cornus
Sphaeropsis coryli Ellis & Everh.
Distr.: Eastern United States.
Host: Corylus
Sphaeropsis diervillae Fairm.
Distr.: NY.
Host: Diervilla

Sphaeropsis filamentosa Cooke
Distr.: SC.
Host: Yucca
Sphaeropsis foliicola (Berl. & Roum.) Sacc.
Sphaeria foliicola Berl. & Roum.
Distr.: Temperate North America.
Host: Crataegus
Sphaeropsis gleditschiae Cooke
Distr.: SC.
Host: Gleditsia
Sphaeropsis gouldiae F. Stevens & Plunkett
Distr.: HI.
Hosts: Gouldia, Leucaena, Nerium
Sphaeropsis hyalina Berk. & M.A. Curtis
Diplodia infuscans Ellis & Everh.
Diplodia rhizogena Ellis & Barth.
Distr.: United States.
Host: Fraxinus
Sphaeropsis ipomoeae Ellis & Everh.
Distr.: WV.
Host: Ipomoea
Sphaeropsis lappae Ellis & Everh.
Distr.: NY(type), TX.
Host: Arctium
Sphaeropsis lupini Cooke & Harkn.
Distr.: CA.
Host: Lupinus
Sphaeropsis lycii Dearn. & Barth.
Distr.: MO.
Host: Lycium
Sphaeropsis macrospermum Cooke & Harkn.
Distr.: CA.
Host: Eucalyptus
Sphaeropsis malorum Berk.
Sphaeropsis biformis Peck
Sphaeropsis cerasina Peck
Sphaeropsis druparum (Schwein.) Cooke
Sphaeropsis fertilis Peck
Sphaeropsis maclurae Cooke
Sphaeropsis nubilosa Ellis & Barth.
Sphaeropsis pennsylvanica Berk. & M.A. Curtis
Sphaeropsis phlei Ellis & Everh.
Sphaeropsis rosarum Cooke & Ellis
Sphaeropsis rubicola Cooke & Ellis
Botryosphaeria obtusa (Schwein.) Shoemaker (teleomorph)
Distr.: Temperate regions.
Notes: Stevens (Mycologia 25:536. 1933) provided the complete synonymy for the *Sphaeropsis* anamorph of *Botryosphaeria obtusa*. Shoemaker (Canad. J. Bot. 42:1298. 1964) reviewed the problems with a name for the anamorph.
Hosts: Ficus, Fraxinus, Gossypium, Juglans, Maclura, Malus, Phleum, Pinus, Prosopis, Prunus, Pyrus, Rosa, Rubus
Sphaeropsis menispermi Peck
Distr.: Central and eastern United States.
Host: Menispermum
Sphaeropsis microscopica Tassi
Distr.: CA.
Host: Castilleja
Sphaeropsis nebelina Fairm.
Distr.: CA.
Host: Galium
Sphaeropsis negundinis Tehon & E.Y. Daniels
Distr.: IL(type).
Host: Acer
Sphaeropsis pallidula Fairm.
Distr.: Central and eastern United States.
Host: Carya
Sphaeropsis peckii Sacc.
Distr.: Central United States.
Hosts: Amorpha, Prunus
Sphaeropsis pericarpii (Schwein.) Ellis & Everh.
Sphaeria pericarpii Schwein.
Distr.: Central and eastern United States.
Host: Carya
Sphaeropsis physocarpi Ellis & Everh.
Distr.: WV.
Host: Physocarpus
Sphaeropsis profundae Tehon & E.Y. Daniels
Distr.: United States.
Host: Fraxinus
Sphaeropsis rhamni Cooke
Distr.: OK.
Host: Rhamnus
Sphaeropsis rhodocarpa Fairm.
Distr.: NY.
Host: Rosa
Sphaeropsis sabalicola Ellis & G.W. Carver
Distr.: AL.
Host: Sabal
Sphaeropsis salviae Hollós
Distr.: MS; Hungary(type).
Host: Salvia
Sphaeropsis sambucina (Cooke) Sacc.
Macroplodia sambucina Cooke
Distr.: Central and western United States.
Notes: Petrak & Sydow (Repert. Spec. Nov. Regni Veg. 62:87. 1926) stated that this species may be identical with *Lasmeniella alpina* (Ellis & Everh.) Petr. & Syd.
Host: Sambucus
Sphaeropsis sapinea (Fr.:Fr.) Dyko & Sutton in Sutton
Macrophoma sapinea (Fr.:Fr.) Petr.
Sphaeropsis ellisii Sacc.
Diplodia pinea (Desmaz.) J. Kickx fil.
Macrophoma pinea (Desmaz.) Syd. & Petr.
Distr.: Cosmopolitan.
Notes: On needles, shoots, and bark of conifers (Punithalingam & Waterson. C.M.I. Descr. 273. 1970; Sutton. The Coelomycetes. p. 120. 1980).
Hosts: Cedrus, Juniperus, Picea, Pinus, Pseudotsuga
Sphaeropsis sphaerelloides Ellis & Everh.
Distr.: OH.
Host: Rumex
Sphaeropsis sphaerospora Peck
Distr.: NY.
Host: Asclepias
Sphaeropsis staphyleae Brunaud
Distr.: Northeastern United States.
Host: Staphylea
Sphaeropsis sumachi (Schwein.) Cooke & Ellis
Sphaeria sumachi Schwein.
Diplodia rhois Sacc.
Distr.: Central and eastern United States; Europe.
Host: Rhus
Sphaeropsis syringae (Fr.) Peck & G.W. Clinton
Sphaeria syringae Fr.
Distr.: Central and eastern United States.
Host: Syringa
Sphaeropsis thermopsidis (Ellis & B.T. Galloway) Kuntze
Phoma thermopsidis Ellis & B.T. Galloway
Distr.: MT.
Host: Thermopsis
Sphaeropsis tumefaciens Hedges
Distr.: Subtropical and tropical regions.
Notes: Knot of *Citrus* spp. (Holliday & Punithalingam. C.M.I. Descr. 278. 1970).
Hosts: Bauhinia, Callistemon, Carissa, Cinnamomum, Citrofortunella, Citrus, Cupressus, Eucalyptus, Eugenia, Ilex, Ixora, Jatropha, Myrica, Nerium, Pittosporum, Poncirus, Portlandia, Pyracantha, Schinus, Wisteria
Sphaeropsis ulmicola Ellis & Everh.
Botryodiplodia ulmicola (Ellis & Everh.) Buisman
Botryodiplodia hypodermia (Sacc.) Petr.
Distr.: Central and eastern United States.
Host: Ulmus
Sphaeropsis vaccinii W.B. Cooke & C.G. Shaw
Distr.: WA.
Host: Vaccinium
Sphaeropsis visci (Fr.) Sacc.
Botryosphaerostroma visci (DC.) Petr. in Petr. & Syd. non (Sollm.) Petr., hom. illeg.

Botryosphaerostroma visci (Sollm.) Petr.
Distr.: CA, FL; Africa; Europe.
Notes: Sutton (The Coelomycetes. p. 120. 1980).
Host: Phoradendron
Sphaeropsis viscosa Cooke & Ellis in Guba, nom. nud.
Distr.: MA(type).
Host: Aronia

Sphaerosporium Schwein. Hyphomycetes

Sphaerosporium lignatile Schwein.
Distr.: DE, IA, PA.
Notes: On dead wood (Damon & Downing. Mycologia 46:216. 1954).
Host: Fraxinus

Sphaerotheca Lév., nom. cons. Pyrenomycetes, Erysiphales

Notes: Braun (Beih. Nova Hedwigia 89. 1987).
Sphaerotheca alpina S. Blumer
Distr.: Temperate regions.
Notes: On Saxifragaceae. Previously reported from the U.S. as *Sphaerotheca humuli* var. *fuligineae*.
Hosts: Heuchera, Tellima
Sphaerotheca aphanis (Wallr.) U. Braun
Sphaerotheca alchemillae (Grev.) Junell
Distr.: Temperate regions.
Host: Physocarpus
Sphaerotheca collomiae U. Braun
Distr.: North America.
Notes: On Polemoniaceae.
Host: Collomia
Sphaerotheca epilobii (Wallr.) Sacc.
Oidium epilobii (Corda) Lindau in Rabenh. (anamorph)
Distr.: Circumboreal.
Notes: Specimens determined as *S. macularis* (Wallroth:Fr.) Magnus on *Epilobium* appear to be this species (Parmelee. Canad. J. Bot. 55:1967. 1977).
Host: Epilobium
Sphaerotheca fuliginea (Schlechtend.:Fr.) Pollacci
Sphaerotheca humuli (DC.) Burrill var. *fuliginea* (Schlechtend.:Fr.) E.S. Salmon
Sphaerotheca macularis (Wallr.:Fr.) W.B. Cooke var. *fuliginea* (Schlechtend.:Fr.) E.S. Salmon
Sphaerotheca castagnei Lév.
Distr.: Cosmopolitan.
Notes: Powdery mildew of Asteraceae, Cucurbitaceae, and Scrophulariaceae. Junell (Svensk Bot. Tidskr. 60:365. 1966) recognized 12 segregate species. Kapoor (C.M.I. Descr. 159. 1967).
Hosts: Acer, Agastache, Agoseris, Apargidium, Arnica, Aruncus, Astragalus, Aureolaria, Bensoniella, Bidens, Boykinia, Castilleja, Citrullus, Collinsia, Collomia, Conyza, Cucumis, Cucurbita, Delphinium, Epilobium, Erechtites, Erigeron, Fragaria, Gaillardia, Geranium, Haplopappus, Heuchera, Hieracium, Humulus, Hydrophyllum, Kalanchoe, Lactuca, Lagenaria, Lasthenia, Malus, Matricaria, Microsteris, Mitella, Oxytropis, Pedicularis, Phacelia, Physalis, Prenanthes, Prunella, Ribes, Rosa, Rosmarinus, Saxifraga, Senecio, Shepherdia, Solidago, Sonchus, Spiraea, Synthyris, Taraxacum, Tellima, Tiarella, Vernonia, Veronica, Veronicastrum, Viola
Sphaerotheca macularis (Wallr.:Fr.) Lind
Sphaerotheca humuli (DC.) Burrill
Distr.: Cosmopolitan.
Notes: Powdery mildew of hops and strawberry (Mukerji. C.M.I. Descr. 188; Parmelee. Fungi Canadenses 63. 1975).
Hosts: Agastache, Agoseris, Agrimonia, Anthemis, Arnica, Astragalus, Besseya, Bidens, Castilleja, Cercocarpus, Chamaebatia, Collinsia, Collomia, Conyza, Coreopsis, Crassula, Crepis, Delphinium, Epilobium, Erechtites, Erigeron, Eriophyllum, Filipendula, Forsellesia, Fragaria, Gaillardia, Geranium, Geum, Gilia, Heuchera, Hieracium, Humulus, Hydrophyllum, Lactuca, Linanthus, Madia, Malus, Matricaria, Mentha, Microseris, Microsteris, Mitella, Navarretia, Nemophila, Pedicularis, Phlox, Physocarpus, Plantago, Polemonium, Potentilla, Prunella, Pyrus, Ranunculus, Rhus, Ribes, Rosa, Rubus, Sambucus, Sanguisorba, Saxifraga, Senecio, Shepherdia, Spiraea, Stachys, Tamarix, Taraxacum, Veronica, Vicia, Viola, Wyethia
Sphaerotheca mors-uvae (Schwein.) Berk. & M.A. Curtis
Distr.: Temperate northern hemisphere.
Notes: Powdery mildew of gooseberry (Purnell & Sivanesan. C.M.I. Descr. 254. 1970).
Hosts: Heuchera, Ribes, Tolmiea
Sphaerotheca pannosa (Wallr.: Fr.) Lév.
Oidium leucoconium Desmaz. (anamorph)
Distr.: Cosmopolitan.
Notes: Powdery mildew of rose (Mukerji. C.M.I. Descr. 189. 1968).
Hosts: Lycium, Photinia, Prunus, Rhododendron, Rhus, Rosa
Sphaerotheca pannosa (Wallr.:Fr.) Lév. var. ***persicae*** Woronichin
Distr.: Widespread.
Host: Prunus
Sphaerotheca pannosa (Wallr.:Fr.) Lév. var. ***rosae*** Woronichin
Distr.: North America; Europe.
Notes: Bender & Coyier (Phytopathology 74:100. 1984).
Host: Rosa
Sphaerotheca phytoptophila Kellerm. & Swingle
Distr.: Central and northeastern North America.
Notes: On hackberry. Associated with eriophiid mites in witches' broom.
Host: Celtis
Sphaerotheca shepherdiae U. Braun
Distr.: Northern temperate regions.
Notes: Braun (Zentralbl. Mikrobiol. 140:237. 1985).
Host: Shepherdia
Sphaerotheca sparsa U. Braun
Distr.: North America.
Notes: On Asclepiadaceae.
Host: Asclepias
Sphaerotheca violae U. Braun
Distr.: Central and western North America.
Host: Viola

Sphaerulina Sacc. Loculoascomycetes, Dothideales

Notes: Barr (Contr. Univ. Michigan Herb. 9:605. 1972).
Sphaerulina acori Dearn. & House
Distr.: NY(type).
Host: Acorus
Sphaerulina cibotii F. Stevens & Guba
Distr.: HI.
Notes: Petrak (Sydowia 7:398. 1953).
Host: Cibotium
Sphaerulina conflicta (Cooke) Barr
Ceuthocarpon conflictum (Cooke) Berl.
Linospora conflicta (Cooke) Sacc.
Distr.: Western North America.
Host: Lithocarpus
Sphaerulina divergens Rehm
Distr.: ND(type).
Host: Elymus
Sphaerulina ipomoeae F. Stevens
Distr.: HI(type).
Host: Ipomoea
Sphaerulina marginata (Wallr.) Kirschst.
Leptosphaeria marginata (Wallr.) Niessl
Distr.: AK; Europe.
Host: Pyrola
Sphaerulina myriadea (DC.:Fr.) Sacc.
Sphaerella myriadea (DC.:Fr.) Auersw.
Distr.: Temperate northern hemisphere.
Notes: On leaves of Fagaceae.
Hosts: Castanopsis, Quercus
Sphaerulina pallens J.J. Davis
Distr.: WI(type).
Host: Carex
Sphaerulina pedicellata T.W. Johnson
Distr.: RI.
Notes: According to Kohlmeyer & Kohlmeyer (Marine mycology. p.

552. 1979), this is a confused name; most specimens are actually *Buergenerula spartinae*.
Host: Spartina
Sphaerulina potebniae Sacc.
Distr.: ID(type).
Host: Pyrus
Sphaerulina rubi Demaree & M.S. Wilcox
Cylindrosporium rubi Ellis & Morg. (anamorph)
Distr.: Central and eastern North America; Europe.
Notes: On leaves.
Host: Rubus
Sphaerulina salicina Syd.
Distr.: ND(type).
Host: Salix
Sphaerulina tarda (Harkn.) Barr
Ophiodothis tarda Harkn.
Ophiocarpella tarda (Harkn.) Theiss. & Syd.
Ophiodothella tarda (Harkn.) Höhn.
Distr.: Western North America.
Host: Rhus

Sphenospora Dietel. Basidiomycetes, Uredinales

Sphenospora kevorkianii Linder
Uredo nigropuncta Henn. (anamorph)
Distr.: FL; West Indies; Central America.
Notes: Autoecious.
Hosts: Cattleya, Epidendrum, Orchidaceae
Sphenospora mera Cummins
Distr.: FL; Venezuela.
Notes: Autoecious.
Host: Orchidaceae
Sphenospora saphena Cummins
Distr.: FL.
Notes: Autoecious.
Host: Oncidium

Spilocaea Fr.:Fr. Hyphomycetes

Spilocaea nebulosa (Ellis & Everh.) S.J. Hughes & Pirozynski
Dicoccum nebulosum Ellis & Everh.
Distr.: WI(type).
Notes: Hughes & Pirozynski (Canad. J. Bot. 50:2530. 1972).
Host: Fraxinus
Spilocaea oleaginea (Castagne) S.J. Hughes
Cycloconium oleaginum Castagne
Distr.: CA; the Mediterranean region.
Notes: On living leaves (Ellis. Dematiaceous Hyphomycetes. p. 143. 1971).
Host: Olea
Spilocaea photinicola (R.L. McClain) M.B. Ellis
Fusicladium photinicola R.L. McClain
Distr.: Southwestern United States.
Notes: Scab disease on leaves and berries (Ellis. More dematiaceous Hyphomycetes. p. 112. 1976).
Hosts: Heteromeles, Photinia
Spilocaea pomi Fr.:Fr.
Cladosporium dendriticum Wallr.
Fusicladium dendriticum (Wallr.) Fuckel
Coniosporium mali Dearn. & W.R. Foster
Helminthosporium pyrorum Lib.
Venturia inaequalis (Cooke) G. Wint. (teleomorph)
Distr.: Temperate regions.
Notes: Applescab. On a variety of plants in the Rosaceae (Corlett. Fungi Canadenses 35. 1974; Sivanesan. The bitunicate Ascomycetes. p. 615. 1984; Sivanesan & Waller. C.M.I. Descr. 401. 1974).
Hosts: Eriobotrya, Malus
Spilocaea pyracanthae (G. Otth) Arx
Passalora pyracanthae G. Otth
Fusicladium pyracanthae (G. Otth) Vien.-Bourg.
Fusicladium pyrorum (Lib.) Fuckel var. *pyracanthae* (G. Otth) Rostr.
Fusicladium eriobotryae (Cavara) Cavara
Distr.: Widespread in Europe and North America.
Notes: On leaves and fruits (Ellis. More dematiaceous Hyphomycetes. p. 112. 1976).
Hosts: Eriobotrya, Pyracantha

Spiniger Stalpers. Hyphomycetes

Notes: Anamorphs of a variety of Aphyllophorales. Stalpers (Proc. Koninkl. Nederl. Akad. Wetens., Ser. C, 77:402. 1974).
Spiniger meineckellum (A. Olson) Stalpers
Cunninghamella meineckella A. Olson
Oedocephalum lineatum Bakshi
Heterobasidion annosum (Fr.:Fr.) Bref. (teleomorph)
Distr.: Temperate regions.
Host: Pinus

Splanchnonema Corda. Loculoascomycetes, Dothideales

Stigmatomassaria Munk
Notes: Barr (Mycotaxon 15:349. 1980).
Splanchnonema ampullaceum (Pers.:Fr.) Shoemaker & LeClair
Sphaeria ampullacea Pers.:Fr.
Massaria curreyi C. Tul. & Tul.
Massariella curreyi (C. Tul. & Tul.) Sacc.
Massariella curreyi (C. Tul. & Tul.) Sacc. var. *americana* Peck
Distr.: North America; Europe.
Notes: On branches.
Host: Tilia
Splanchnonema argus (Berk. & Broome) Kuntze
Sphaeria argus Berk. & Broome
Massaria argus (Berk. & Broome) Fresen.
Splanchnonema niessliana (Rehm) Kuntze
Myxocyclus polycistis (Berk. & Broome) Sacc. (anamorph)
Distr.: Eastern North America; Europe.
Notes: On branches.
Host: Betula
Splanchnonema clandestinum Barr
Distr.: PA(type).
Notes: On twigs.
Host: Magnolia
Splanchnonema foedans (Fr.:Fr.) Kuntze
Sphaeria foedans Fr.:Fr.
Massaria foedans (Fr.:Fr.) Fr.
Distr.: North America; Europe.
Notes: On branches of deciduous trees.
Hosts: Celtis, Ulmus
Splanchnonema hicoria Barr
Distr.: Eastern North America.
Notes: On branches.
Host: Carya
Splanchnonema loricatum (C. Tul. & Tul.) Barr
Massaria loricata C. Tul. & Tul.
Distr.: North America; Europe.
Notes: On branches.
Host: Fagus
Splanchnonema melanterum (Ellis & Everh.) Barr
Amphisphaeria melantera Ellis & Everh.
Distr.: Western North America.
Notes: On branches.
Host: Quercus
Splanchnonema platani (Ces.) Barr
Massaria platani Ces.
Splanchnonema atroinquinans (Berk. & M.A. Curtis) Kuntze
Macrodiplodiopsis desmazieresii (Mont.) Petr. (anamorph)
Distr.: North America; Europe.
Notes: On branches.
Host: Platanus
Splanchnonema pupula (Fr.:Fr.) Kuntze
Sphaeria pupula Fr.:Fr.
Massaria pupula (Fr.:Fr.) C. Tul. & Tul.
Stigmatomassaria pupula (Fr.:Fr.) Munk
Stegonsporium pyriforme (Hoffm.:Fr.) Corda (anamorph)
Distr.: North America; Europe.
Notes: On branches of trees.

Hosts: Acer, Corylus
Splanchnonema scoriadeum **(Fr.:Fr.) Barr**
Sphaeria scoriadea Fr.:Fr.
Pteridiospora scoriadea (Fr.:Fr.) Dennis
Distr.: Eastern North America; Europe.
Notes: On branches.
Host: Betula
Splanchnonema semitectum **(Berk. & M.A. Curtis) Kuntze**
Massaria semitecta (Berk. & M.A. Curtis) Sacc.
Distr.: Eastern North America.
Notes: On branches.
Host: Platanus
Splanchnonema sporadicum **(Ellis & Everh.) Barr**
Melanomma sporadicum Ellis & Everh.
Distr.: Eastern North America.
Notes: On wood of deciduous trees.
Hosts: Platanus, Quercus

Spogotteria Dyko, Sutton & Roquebert. Coelomycetes

Spogotteria magnoliae **(Sacc.) Dyko, Sutton & Roquebert**
Protostegia magnoliae Sacc.
Eustegia magnoliae Ravenel, nom. nud.
Distr.: Southern United States.
Notes: On decaying leaves (Dyko et al. Mycologia 71:926. 1979).
Host: Magnolia

Spongipellis Pat. Basidiomycetes, Aphyllophorales

Notes: Gilbertson & Ryvarden (North American Polypores 2:722. 1987).
Spongipellis delectans **(Peck) Murrill**
Polyporus delectans Peck
Tyromyces delectans (Peck) J. Lowe
Distr.: Temperate northern hemisphere.
Notes: On dead hardwoods, occasionally on wounds of living trees.
Hosts: Acer, Alnus, Carya, Fagus, Juglans, Quercus, Ulmus
Spongipellis pachyodon **(Pers.) Kotlaba & Pouzar**
Irpex pachyodon (Pers.) Quél.
Irpex mollis Berk. & M.A. Curtis
Distr.: Temperate northern hemisphere.
Notes: On wood.
Hosts: Liquidambar, Quercus, Salix
Spongipellis spumeus **(Sowerby:Fr.) Pat.**
Polyporus spumeus (Sowerby:Fr.) Fr.
Tyromyces spumeus (Sowerby:Fr.) Imazeki
Spongipellis occidentalis Murrill
Polyporus whetstonei C.G. Lloyd
Distr.: Temperate northern hemisphere.
Notes: On hardwoods.
Hosts: Acer, Aesculus, Castanea, Fagus, Fraxinus, Juglans, Magnolia, Malus, Populus, Tilia, Tsuga, Ulmus
Spongipellis unicolor **(Schwein.) Murrill**
Tyromyces unicolor (Schwein.) J. Lowe
Polyporus obtusus Berk.
Polyporus tomentoso-quercinus A.E. Johnson
Distr.: North America.
Notes: On trunks and branches of living hardwoods, especially *Quercus* spp.
Hosts: Acer, Betula, Fagus, Quercus, Robinia, Salix

Spongospora Brunch. Plasmodiophoromycetes, Plasmodiophorales

Notes: Karling (The Plasmodiophorales. p. 68. 1968).
Spongospora subterranea **(Wallr.) Lagerh.**
Distr.: Cosmopolitan.
Notes: Powdery scab of potatoes (Hims & Preece. C.M.I. Descr. 477. 1975).
Hosts: Lycopersicon, Solanum
Spongospora subterranea **(Wallr.) Lagerh. f. sp. *nasturtii* J.A. Tomlinson**
Distr.: Eastern United States; Europe.
Notes: Crook root of watercress.
Host: Nasturtium

Sporidesmium Link:Fr. Hyphomycetes

Notes: Ellis (Mycol. Pap. 70:16. 1958).
Sporidesmium capsularum **Thuem.**
Distr.: OK.
Notes: The place of publication of this name could not be found.
Host: Catalpa
Sporidesmium carrii **Morgan-Jones**
Distr.: AL.
Notes: On dead twigs associated with *Helminthosporium velutinum* (Morgan-Jones. Mycotaxon 6:357. 1977).
Host: Buxus
Sporidesmium ellipticum **Cooke**
Sirodesmium ellipticum (Cooke) Sacc.
Distr.: SC(type).
Host: Magnolia
Sporidesmium folliculatum **(Corda) E. Mason & S.J. Hughes**
Helminthosporium folliculatum Corda
Distr.: WV; Europe.
Notes: On dead wood of various trees (Ellis. Dematiaceous Hyphomycetes. p. 118. 1971).
Host: Zea
Sporidesmium fragilissimum **(Berk. & M.A. Curtis) M.B. Ellis**
Helminthosporium fragilissimum Berk. & M.A. Curtis in Berk.
Clasterosporium obclavatum (Cooke) Sacc.
Sporidesmium obclavatum Cooke
Distr.: Southeastern United States.
Notes: Ellis (More dematiaceous Hyphomycetes. p. 77. 1976).
Hosts: Smilax, Torreya
Sporidesmium hormiscioides **Corda**
Clasterosporium caespitulosum Ellis & Everh.
Clasterosporium vermiculatum Cooke
Distr.: Northern North America; Europe.
Notes: On decaying wood (Hughes & Illman. Fungi Canadenses 58. 1974).
Host: Acer
Sporidesmium hysterioideum **Cooke & Ellis**
Distr.: DE, NJ.
Notes: According to Hughes (Canad. J. Bot. 36:809. 1958), this is a coelomycete.
Host: Juniperus
Sporidesmium induratum **Cooke**
Distr.: CA.
Host: Arctostaphylos
Sporidesmium inflatum **(Berk. & Ravenel) M.B. Ellis**
Helminthosporium inflatum Berk. & Ravenel
Helminthosporium collabendum Cooke
Helminthosporium episphaericum Cooke & Peck
Clasterosporium herculeum Ellis
Clasterosporium sigmoideum Ellis & Everh.
Helminthosporium subfuscum Berk. & M.A. Curtis in Berk.
Distr.: Eastern North America; Africa.
Notes: Ellis (Mycol. Pap. 70:70. 1958).
Hosts: Castanea, Myrica, Rhus
Sporidesmium irregularis **Cooke**
Distr.: SC(type).
Host: Eupatorium
Sporidesmium maclurae **Thuem.**
Distr.: Central and southeastern United States.
Host: Maclura
Sporidesmium macrosporoides **Ellis & B.T. Galloway**
Distr.: MT(type).
Host: Betula
Sporidesmium punctiphyllum **Cooke**
Distr.: SC(type).
Host: Magnolia
Sporidesmium rauii **Ellis & Harkn.**
Distr.: PA(type).
Notes: Moore (Mycologia 51:738. 1959) stated that this may be an immature *Berkleasmium*.
Host: Vitis

Sporidesmium rude Ellis
Distr.: FL, NJ.
Host: Magnolia
Sporidesmium spiraeacola Cooke
Distr.: SC(type).
Host: Spiraea
Sporidesmium stictophyllum Cooke
Sirodesmium stictophyllum (Cooke) Sacc.
Distr.: SC(type).
Host: Magnolia
Sporidesmium tabacinum Ellis & Everh.
Distr.: MT(type).
Host: Populus
Sporidesmium taxodii J.L. Crane
Distr.: IL.
Host: Taxodium
Sporidesmium velutinum Cooke
Distr.: FL(type).
Host: Persea

Sporidiobolus Nyland. Basidiomycetes, Ustilaginales

Notes: Species in this genus have anamorphs in *Sporobolomyces* Kluyver & Niel. Bandoni (Canad. J. Bot. 49:683. 1971).
Sporidiobolus johnsonii Nyland
Sporobolomyces sp. (anamorph)
Distr.: WA(type).
Host: Rubus
Sporidiobolus pararoseus J. Fell & Tallman
Sporobolomyces pararoseus H. Olson & B. Hammer, nom. nud.
Sporobolomyces shibatanus (Okunuki) Verona & Cif. (anamorph)
Distr.: KS.
Hosts: Althaea, Crataegus, Euonymus, Helianthus, Malus, Plantago, Stachys, Taraxacum
Sporidiobolus salmonicolor J. Fell & Tallman
Sporobolomyces odorus Derx
Sporobolomyces salmonicolor (B. Fischer & Brebeck) Kluyver & Niel (anamorph)
Distr.: KS.
Host: Althaea

Sporisorium Ehrenb. ex Link. Basidiomycetes, Ustilaginales

Sporisorium andropogonis (Opiz) K. Vánky
Sphacelotheca andropogonis (Opiz) Bubák
Ustilago cylindrica Peck
Sphacelotheca ischaemi (Fuckel) G.P. Clinton
Ustilago ischaemi Fuckel
Distr.: Widespread.
Notes: On Poaceae.
Hosts: Andropogon, Bothriochloa, Heteropogon, Schizachyrium
Sporisorium cenchri (Lagerh.) K. Vánky
Ustilago cenchri Lagerh.
Sorosporium cenchri Henn.
Ustilago syntherismae Peck non (Schwein.) Peck, hom. illeg.
Distr.: Cosmopolitan.
Notes: On Poaceae (Vánky. Symb. Bot. Upsal. 24:114. 1985).
Hosts: Andropogon, Cenchrus, Cynodon, Digitaria, Panicum
Sporisorium consanguineum (Ellis & Everh.) K. Vánky
Sorosporium consanguineum Ellis & Everh.
Distr.: United States; Australia.
Notes: Vánky (Mycotaxon 31:402. 1988).
Hosts: Aristida, Poa
Sporisorium cruentum (Kühn) K. Vánky
Ustilago cruenta Kühn
Sphacelotheca cruenta (Kühn) A.A. Potter
Sphacelotheca chrysopogonis G.P. Clinton
Sphacelotheca holci H. Jacks.
Distr.: Widespread.
Notes: Ainsworth (C.M.I. Descr. 71. 1965 as *Sphacelotheca cruenta*).
Hosts: Sorghastrum, Sorghum
Sporisorium destruens (Schlechtend.) K. Vánky
Sphacelotheca destruens (Schlechtend.) J.A. Stevenson & A.G. Johnson
Ustilago destruens (Schlechtend.) Rabenh. in Rabenh.
Sphacelotheca panici-miliacei (Pers.) Bubák
Ustilago panici-miliacei (Pers.) G. Wint. in Rabenh.
Distr.: Widespread.
Notes: On Poaceae (Ainsworth. C.M.I. Descr. 72. 1965 as *Sphacelotheca destruens*).
Host: Panicum
Sporisorium holci-sorghi (Rivolta) K. Vánky
Sphacelotheca reiliana (Kühn) G.P. Clinton
Ustilago reiliana Kühn
Sorosporium reilianum (Kühn) McAlpine
Distr.: Cosmopolitan.
Notes: Ainsworth (C.M.I. Descr. 73. 1965 as *Sphacelotheca reiliana*).
Hosts: Hackelochloa, Sorghum, Zea
Sporisorium montaniensis (Ellis & Holw.) K. Vánky
Ustilago montaniensis Ellis & Holw. in Ellis & Everh.
Sphacelotheca montaniensis (Ellis & Holw.) G.P. Clinton
Sphacelotheca strangulans (Issatchenko) G.P. Clinton
Distr.: Widespread.
Notes: On Poaceae.
Hosts: Eragrostis, Muhlenbergia
Sporisorium neglectum (Niessl) K. Vánky
Ustilago neglecta Niessl
Ustilago boutelouae Kellerm. & Swingle
Ustilago panici-glauci (Wallr.) G. Wint. in Rabenh.
Distr.: Widespread.
Hosts: Bouteloua, Buchloe, Setaria
Sporisorium sorghi Link in Willd.
Cintractia sorghi (Link) Hirschhorn
Sphacelotheca sorghi (Link) G.P. Clinton
Ustilago sorghi Pass. in Thuem.
Distr.: Cosmopolitan.
Notes: Ainsworth (C.M.I. Descr. 74. 1965 as *Sphacelotheca sorghi*).
Host: Sorghum

Sporobolomyces Kluyver & Niel. Hyphomycetes

Notes: Species in this genus of yeasts have teleomorphs in *Sporidiobolus* Nyland.
Sporobolomyces albo-rubescens Derx
Distr.: KS; Europe.
Hosts: Althaea, Chrysanthemum
Sporobolomyces roseus Kluyver & Niel
Sporobolomyces boleticola C. Ramirez
Sporobolomyces miniatus Yamazaki & Fujii
Sporobolomyces salmoneus Derx
Sporobolomyces tenuis Kluyver & Niel
Distr.: Widespread.
Notes: Known from a variety of habitats, including living and dead plant parts.
Host: Acer

Sporomega Corda. Discomycetes, Rhytismatales

Notes: The separation of this genus from *Coccomyces* is discussed by Eriksson (Symb. Bot. Upsal. 19:14. 1970).
Sporomega degenerans (Fr.) Corda
Hysterium degenerans Fr.
Clithris degenerans (Fr.) Rehm
Distr.: Northern temperate regions.
Hosts: Chamaedaphne, Vaccinium

Sporonema Desmaz. Coelomycetes

Sporonema camelliae Earle
Distr.: Central and eastern United States.
Notes: According to Sutton (The Coelomycetes. p. 585. 1980) this species belongs in *Coleophoma*.
Host: Camellia
Sporonema fraxini Limber

Distr.: Northwestern United States.
Host: Fraxinus
Sporonema ilicis Earle
Distr.: AL.
Host: Ilex
Sporonema phacidioides Desmaz.
Ascochyta medicaginis Fuckel
Phyllosticta medicaginis (Fuckel) Sacc.
Gloeosporium morianum Sacc.
Leptotrochila medicaginis (Fuckel) H. Schüepp (teleomorph)
Distr.: Cosmopolitan.
Notes: Yellow leaf blotch of alfalfa (Sutton. The Coelomycetes. p. 585. 1980).
Hosts: Medicago, Melilotus
Sporonema punctiforme Fuckel
Distr.: Western North America; Europe.
Notes: Limber (Mycologia 47:399. 1955) suggested that this species is closer to *Placosphaeria* than to *Sporonema*.
Host: Galium
Sporonema trifolii Greene
Distr.: Central and eastern United States.
Host: Trifolium

Sporophlyctidium Sparrow. Chytridiomycetes, Chytridiales

Sporophlyctidium neustonicum Sparrow
Distr.: FL(type).
Host: Pinus

Sporormiella Ellis & Everh. Loculoascomycetes, Dothideales

Sporormiella minima (Auersw.) Ahmed & Cain
Sporormia minima Auersw.
Preusia minima (Auersw.) Arx
Philocopra multifera (Berk. & Ravenel) Sacc.
Distr.: Cosmopolitan.
Notes: Usually on dung (Ahmed & Cain. Canad. J. Bot. 50:450. 1972).
Host: Glycine

Sporothrix L. Hektoen & C.F. Perkins ex J. Nicot & F. Mariat. Hyphomycetes

Notes: De Hoog (Stud. Mycol. 7:12. 1974).
Sporothrix subannulata Livingston & R.W. Davidson
Ophiostoma suannulatum Livingston & R.W. Davidson (teleomorph)
Distr.: ID(type).
Notes: Livingston & Davidson (Mycologia 79:145. 1987).
Host: Abies

Sporotrichum Link:Fr. Hyphomycetes

Notes: Stalpers (Stud. Mycol. 24:1. 1984).
Sporotrichum bombacinum Link
Distr.: DE; Europe.
Notes: A doubtful name.
Host: Pinus
Sporotrichum columbiense R. Sprague
Distr.: Western United States.
Notes: A doubtful name, possibly based on a *Cryptococcus* sp.
Host: Agropyron
Sporotrichum laxum Nees:Fr.
Distr.: HI; Europe.
Notes: A doubtful name.
Host: Acacia
Sporotrichum pruinosum Gilman & E. Abbott
Chrysosporium pruinosum (Gilman & E. Abbott) J.W. Carmichael
Sporotrichum pulverulentum Novobranova
Phanerochaete chrysosporium Burdsall in Burdsall & Eslyn (teleomorph)
Distr.: Widespread.
Host: Cucumis

Stachybotryella Ellis & Barth. Hyphomycetes

Stachybotryella repens Ellis & Barth.
Distr.: TX(type).
Host: Verbesina

Stachybotrys Corda. Hyphomycetes

Notes: Ellis (Dematiaceous Hyphomycetes. p. 540. 1971); Jong & Davis (Mycotaxon 3:409. 1976).
Stachybotrys chartarum (Ehrenb.) S.J. Hughes
Stachybotrys atra Corda
Stachybotrys atrogrisea Ellis & Everh.
Sporocybe lobulata Berk.
Stachybotrys lobulata (Berk.) Berk.
Stachybotrys scabra Cooke & Harkn.
Distr.: Cosmopolitan.
Notes: An ubiquitous saprobe; produces mycotoxins responsible for stachybotryotoxicosis.
Hosts: Acacia, Gossypium, Juncus, Lycopersicon, Metrosideros, Phaseolus, Tragopogon
Stachybotrys cylindrospora C.N. Jensen
Distr.: North temperate region; Israel.
Notes: On dead stems and isolated from soil.
Host: Juncus
Stachybotrys elegans (Pidoplichko)W. Gams
Hyalobotrys elegans Pidoplichko
Stachybotrys aurantia Barron
Stachybotrys bisbyi (Srinivasan) Barron
Stachybotrys sacchari (Srinivasan) Barron
Distr.: North America; Africa; Asia; Europe.
Notes: In soil.
Host: Bambusa
Stachybotrys kampalensis Hansf.
Distr.: Southeastern United States; West Indies; Africa; Asia.
Notes: Saprobic on plant debris.
Hosts: Hydrangea, Juncus
Stachybotrys nephrospora Hansf.
Distr.: Widespread in tropical regions.
Notes: On dead wood of trees and on herbaceous plants.
Host: Juncus
Stachybotrys zeae Morgan-Jones & G.W. Karr
Distr.: AL.
Notes: Morgan-Jones & Karr (Mycotaxon 4:510. 1976).
Host: Zea

Stachycoremium Seifert. Hyphomycetes

Stachycoremium parvulum (Cooke & Ellis) Seifert
Stilbum parvulum Cooke & Ellis
Distr.: Northeastern and southern United States; Japan.
Notes: On decaying wood, usually associated with carbonaceous pyrenomycetes (Seifert. Mycologia 77:987. 1985).
Host: Acer

Stachylidium Link. Hyphomycetes

Stachylidium bicolor Link:Fr.
Spondylocladium tenellum Peck
Stachylidium verticillatum (Pers.) S.J. Hughes
Distr.: Subtropical and tropical regions.
Notes: On dead stems and twigs of various plants (Ellis. Dematiaceous Hyphomycetes. p. 538. 1971).
Hosts: Collinsonia, Juncus

Stagonospora (Sacc.) Sacc., nom. cons. Coelomycetes

Notes: Sutton (The Coelomycetes. p. 106. 1980).
Stagonospora achlydis (Dearn.) R. Sprague
Ascochyta achlydis Dearn.
Distr.: Pacific Northwest.
Host: Achlys
Stagonospora albescens J.J. Davis

Distr.: Central and western United States.
Host: Carex
Stagonospora ambrosiae Savile
Distr.: ND(type).
Host: Ambrosia
Stagonospora andropogonis Castellani & E.G. Germano
Distr.: OK(type).
Host: Andropogon
Stagonospora anglica Cunnell
Distr.: Western United States; Great Britain.
Host: Carex
Stagonospora apocyni (Peck) J.J. Davis
Septogloeum apocyni Peck
Stagonospora apocyni (Peck) Arx non (Peck) J.J. Davis, hom. illeg.
Distr.: Central and eastern United States.
Host: Apocynum
Stagonospora aquatica Sacc. var. ***luzulicola*** Sacc. & Scalia
Distr.: Temperate Europe; North America.
Host: Luzula
Stagonospora arenaria (Sacc.) Sacc.
Distr.: North America; temperate Europe.
Notes: Purple-brown blotch of grasses.
Hosts: Agropyron, Bouteloua, Calamagrostis, Cinna, Dactylis, Elymus, Leymus, Lolium, Oryzopsis, Schizachyrium, Scolochloa
Stagonospora astericola (J.J. Davis) H.C. Greene
Asteromella astericola J.J. Davis
Distr.: WI.
Notes: Greene (Farlowia 1:569. 1944; Trans. Wisconsin Acad. Sci. 50:160. 1961).
Hosts: Aster, Solidago
Stagonospora atriplicis (Westend.) Lind
Phyllosticta atriplicis Westend.
Septoria atriplicis (Westend.) Fuckel
Distr.: North America; temperate Europe.
Hosts: Atriplex, Chenopodium
Stagonospora avenae (A.B. Frank) Bissett
Septoria avenae A.B. Frank
Distr.: Cosmopolitan.
Notes: Speckled blotch of oats and grasses (Bissett. Fungi Canadenses 239. 1982).
Hosts: Agropyron, Agrostis, Arrhenatherum, Avena, Calamagrostis, Elymus, Festuca, Glyceria, Hordeum, Oryzopsis, Puccinellia, Schizachne, Scolochloa, Sitanion, Trisetum, Triticum
Stagonospora avenae Bissett f. sp. ***triticea*** T. Johnson
Septoria avenae A.B. Frank f. sp. *triticea* T. Johnson
Phaeosphaeria avenaria (G.F. Weber) O. Eriksson f. sp. *triticea* T. Johnson (teleomorph)
Distr.: Temperate regions.
Notes: A physiological form (Bissett. Fungi Canadenses 239. 1982).
Hosts: Agrohordeum, Agropyron, Elymus, Hordeum, Phalaris, Triticum
Stagonospora baccharidis Ellis
Distr.: TX, VA(type).
Host: Baccharis
Stagonospora baptisiae (Ellis & Everh.) J.J. Davis
Marssonia baptisiae Ellis & Everh.
Marssonina baptisiae (Ellis & Everh.) Magnus
Distr.: Central United States.
Hosts: Baptisia, Lathyrus
Stagonospora bellunensis (Speg.) Jørst.
Stagonospora agrostidis Syd. f. *angusta* R. Sprague
Distr.: Central and western United States; Argentina; Italy.
Hosts: Agrostis, Stipa
Stagonospora berberidina Sacc.
Distr.: NY(type).
Host: Berberis
Stagonospora biseptata H.C. Greene
Distr.: WI.
Host: Carex
Stagonospora brachyelytri H.C. Greene
Stagonospora rubi-albida Dearn. & House, nom. inval.
Distr.: Central and eastern United States.
Host: Brachyelytrum
Stagonospora brevispora (Ellis & J.J. Davis) Castellani & E.G. Germano
Septoria brevispora Ellis & J.J. Davis
Distr.: WI(type).
Host: Bromus
Stagonospora bromi A.L. Sm. & Ramsb.
Distr.: Temperate North America; Europe.
Notes: Purple-brown blotch.
Host: Bromus
Stagonospora californica Petr.
Distr.: CA(type).
Host: Juncus
Stagonospora calystegiae (Westend.) Grove
Septoria calystegiae Westend.
Distr.: United States; North Africa; Cyprus; Europe.
Hosts: Calystegia, Convolvulus
Stagonospora canadensis Greene
Distr.: WI.
Host: Anemone
Stagonospora caricinella Brunaud
Distr.: North America; temperate Europe.
Notes: On leaves.
Hosts: Carex, Juncus
Stagonospora caricis (Oudem.) Sacc.
Hendersonia caricis Oudem.
Distr.: MA, OK; Europe.
Host: Carex
Stagonospora celtidis (Ellis & G. Martin) H.C. Greene
Phloeospora celtidis Ellis & G. Martin
Distr.: Central United States.
Host: Celtis
Stagonospora chenopodii House
Distr.: ND.
Host: Chenopodium
Stagonospora cirsii J.J. Davis
Distr.: WI.
Hosts: Carduus, Cirsium
Stagonospora convolvuli Dearn. & House
Distr.: NY(type), WI.
Hosts: Calystegia, Convolvulus
Stagonospora cryptotaeniae Greene
Distr.: WI.
Host: Cryptotaenia
Stagonospora culmicola (Sacc.) Castellani & E.G. Germano
Hendersonia culmicola Sacc.
Hendersonia simplex J. Schröt.
Distr.: Temperate North America; Europe.
Notes: Leaf rot of Poaceae.
Hosts: Agropyron, Agrostis, Bromus, Calamagrostis, Deschampsia, Elymus, Festuca, Juncus, Koeleria, Melica, Oryzopsis, Poa, Schizachne, Sporobolus, Trisetum, Vulpia
Stagonospora curtisii (Berk.) Sacc.
Hendersonia curtisii Berk.
Distr.: North America; Europe; Great Britain.
Hosts: Amaryllis, Chlorophytum, Crinum, Eucharis, Hippeastrum, Hymenocallis, Leucojum, Lycoris, Narcissus, Nerine, Sternbergia, Vallota, Zephyranthes
Stagonospora curvula E. Bommer, M. Rousseau, & Sacc.
Distr.: TX; Europe.
Host: Panicum
Stagonospora cylindrocarpa (Ellis & Everh.) Sacc.
Hendersonia cylindrocarpa Ellis & Everh.
Distr.: WA.
Host: Brodiaea
Stagonospora cyperi Ellis & Tracy
Distr.: MS(type).
Host: Cyperus
Stagonospora cypericola Greene non Hollós, hom. illeg.
Distr.: WI.
Host: Cyperus
Stagonospora cytisporoides Ellis & Barth.
Distr.: KS(type).
Host: Amorpha
Stagonospora desmodii Ellis & Everh.
Distr.: FL(type), TX.
Host: Desmodium
Stagonospora eriophorella (Sacc.) Lind

Septoria eriophorella Sacc. & Scalia
Distr.: AK; Norway.
Host: Eriophorum
Stagonospora erythrinae F. Stevens & P.A. Young
Distr.: HI.
Host: Erythrina
Stagonospora foliicola (Bres.) Bubák
Stagonospora intermixta (Cooke) Sacc.
Distr.: Temperate North America; Europe; Mediterranean region.
Hosts: Agrostis, Calamagrostis, Cinna, Glyceria, Phalaris, Poa
Stagonospora gigantea Heald & F.A. Wolf
Distr.: Southern United States.
Hosts: Agave, Yucca
Stagonospora glycericola R. Sprague
Stagonospora sporobolicola R. Sprague
Distr.: Central and western North America; Argentina.
Host: Glyceria
Stagonospora graminum Sacc. & Scalia
Distr.: AK.
Host: Deschampsia
Stagonospora heleocharidis Trail var. ***caricina*** Sacc. & Scalia
Distr.: AK.
Host: Carex
Stagonospora herbarum (Cooke & Harkn.) Sacc.
Hypocenia arundinacea Cooke & Harkn.
Distr.: CA.
Host: Aster
Stagonospora heterospora Sacc.
Distr.: NY(type).
Host: Parthenocissus
Stagonospora humuli-americani Fairm.
Distr.: NM.
Host: Humulus
Stagonospora ischaemi (Sacc.) Sacc.
Hendersonia ischaemi Sacc.
Distr.: Central United States; Italy.
Host: Andropogon
Stagonospora lactucicola Greene
Distr.: WI.
Host: Lactuca
Stagonospora linearis Peck
Distr.: IL(type).
Host: Carya
Stagonospora luzulae Hollós
Distr.: WI; Hungary.
Host: Carex
Stagonospora macromaculans Greene
Distr.: WI.
Host: Celtis
Stagonospora maculata Castellani & E.G. Germano
Stagonospora maculata (Grove) R. Sprague, hom. illeg.
Stagonospora subseriata (Desmaz.) Sacc. var. *maculata* Grove, nom. nud.
Distr.: Northeastern United States; Europe; New Zealand.
Host: Dactylis
Stagonospora mariae R. Sprague
Distr.: WA.
Hosts: Agrostis, Bromus, Trisetum
Stagonospora maritima Syd.
Distr.: TX; Germany.
Host: Rottboellia
Stagonospora meliloti (Lasch) Petr.
Gloeosporium americanum Ellis & Everh.
Gloeosporium medicaginis Ellis & Kellerm.
Septogloeum medicaginis (Roberge & Desmaz.) Ellis & Everh.
Septoria medicaginis Roberge & Desmaz.
Gloeosporium trifolii Peck
Phoma meliloti Allesch. (synanamorph)
Leptosphaeria pratensis Sacc. & Briard (teleomorph)
Distr.: Temperate North America; Europe.
Notes: Leaf spot and root rot of alfalfa (Sivanesan. The bitunicate Ascomycetes. p. 479. 1984).
Hosts: Medicago, Melilotus, Trifolium, Vicia
Stagonospora montagnei Castellani & E.G. Germano
Ascospora arundinis Mont.
Stagonospora graminella (Sacc.) Sacc.
Hendersonia gramniella Sacc.
Distr.: North and South America; Europe.
Hosts: Muhlenbergia, Phragmites
Stagonospora myricae Ellis & Everh.
Distr.: NJ(type).
Host: Myrica
Stagonospora nitida (Ellis & Everh.) Sacc.
Hendersonia nitida Ellis & Everh.
Distr.: HI.
Host: Myrsine
Stagonospora nodorum (Berk.) Castellani & E.G. Germano
Depazea nodorum Berk.
Hendersonia nodorum (Berk.) Petr.
Septoria nodorum (Berk.) Berk. in Berk. & Broome
Stagonospora hennebergii (Kühn) Petr. & Syd.
Phaeosphaeria nodorum (E. Müller) Hedjaroude (teleomorph)
Distr.: Cosmopolitan.
Notes: Glume blotch of wheat and other grasses (Sutton & Waterson. C.M.I. Descr. 86. 1966; Bissett. Fungi Canadenses 240. 1982).
Hosts: Agropyron, Alopecurus, Bromus, Cinna, Deschampsia, Elymus, Festuca, Glyceria, Hordeum, Hystrix, Melica, Poa, Secale, Stipa, Triticum
Stagonospora nyssaecola Fairm.
Distr.: NY.
Host: Nyssa
Stagonospora paludosa (Sacc. & Speg.) Sacc.
Hendersonia paludosa Sacc. & Speg.
Distr.: ID; Great Britain.
Host: Carex
Stagonospora petasitidis Ellis & Everh.
Distr.: Temperate North America.
Host: Petasites
Stagonospora phaseoli Dearn. & Barth.
Distr.: TN, TX.
Hosts: Phaseolus, Vigna
Stagonospora phragmitis (Oudem.) Leuchtmann
Septoriella phragmitis Oudem.
Distr.: IN; Europe.
Notes: Leuchtmann (Sydowia 27:139. 1984).
Host: Phragmites
Stagonospora physocarpi Ellis & Everh.
Distr.: WV(type).
Host: Physocarpus
Stagonospora pini Grove
Distr.: IL, TX; Great Britain.
Host: Juniperus
Stagonospora polytaeniae Greene
Distr.: WI.
Hosts: Pastinaca, Polytaenia
Stagonospora pulsatillae Vestergr.
Distr.: AK; Sweden.
Host: Anemone
Stagonospora recedens (C. Massal.) F.R. Jones & J.L. Weimer
Pleospora trifolii Cavara var. *recedens* C. Massal.
Distr.: Central and eastern United States.
Host: Trifolium
Stagonospora sacchari Lo & Ling
Distr.: HI; Taiwan.
Notes: Sivanesan (C.M.I. Descr. 776. 1983).
Host: Saccharum
Stagonospora samarorum (Desmaz.) Boerema
Ascochyta sambuci Sacc.
Distr.: Widespread in temperate regions.
Notes: Boerema & Dorenbosch (Stud. Mycol. 3:40. 1973).
Hosts: Salix, Sambucus
Stagonospora scirpi Greene
Distr.: IL.
Host: Scirpus
Stagonospora scirpini Ellis & Everh. in Langl., nom. nud.
Distr.: LA.
Host: Scirpus
Stagonospora septorioides Ellis & Everh.
Distr.: MS(type).
Host: Quercus

Stagonospora simplicior Sacc. & Briard
Stagonospora simplicior Sacc. & Briard f. *andropogonis* Sacc.
Distr.: Temperate North America; western Africa; Europe.
Hosts: Agropyron, Andropogon, Arundinaria, Calamagrostis, Calamovilfa, Carex, Dichanthelium, Elymus, Festuca, Panicum, Phragmites, Schizachyrium, Sorghastrum, Stipa, Stiporyzopsis
Stagonospora sparganii (Fuckel) Sacc.
Distr.: WI; Europe(type).
Host: Sparganium
Stagonospora spartinicola R. Sprague
Distr.: Central and eastern United States.
Host: Spartina
Stagonospora spiculisperma H.C. Greene
Distr.: WI.
Host: Stipa
Stagonospora strictae Ellis & Everh.
Distr.: KS(type).
Host: Carex
Stagonospora subchlorinoflavida R. Sprague & C.T. Rogerson
Distr.: KS.
Notes: Sprague & Rogerson (Mycologia 50: 640. 1958).
Host: Bouteloua
Stagonospora subseriata (Desmaz.) Sacc.
Distr.: North America; temperate Europe.
Notes: Leaf rot of grasses.
Hosts: [Carex], Danthonia, Deschampsia, Elymus, Phleum, Poa, Sporobolus
Stagonospora tephrosiae H.C. Greene
Distr.: WI(type).
Host: Tephrosia
Stagonospora tetramera J.J. Davis
Distr.: WI(type).
Host: Carex
Stagonospora thaspii (Ellis & Everh.) H.C. Greene
Ascochyta thaspii Ellis & Everh.
Distr.: Northeastern North America.
Hosts: Cicuta, Osmorhiza, Oxypolis, Sanicula, Thaspium, Zizia
Stagonospora tridentis R. Sprague & C.T. Rogerson
Distr.: KS.
Notes: Sprague & Rogerson (Mycologia 50:640. 1958).
Host: Tridens
Stagonospora trifidae H.C. Greene
Distr.: WI(type).
Host: Ambrosia
Stagonospora typhoidearum (Desmaz.) Sacc.
Hendersonia typhoidearum Desmaz.
Distr.: Central and southern United States; Great Britain.
Host: Typha
Stagonospora verbenae H.C. Greene
Distr.: WI.
Host: Verbena
Stagonospora vexatula (Sacc.) Sacc.
Hendersonia vexulata Sacc.
Distr.: Temperate North America; Europe.
Hosts: Deschampsia, Phragmites, Poa
Stagonospora virens Ellis & G. Martin
Distr.: FL(type).
Host: Quercus
Stagonospora zonata J.J. Davis
Distr.: WI.
Host: Asclepias

Stagonosporopsis Died. Coelomycetes

Stagonosporopsis hortensis (Sacc. & Malbr.) Petr.
Ascochyta boltshauseri Sacc. in Boltshauser
Distr.: Temperate regions.
Notes: Leaf spot disease of beans. Boerema and Verhoeven (Neth. J. Pl. Pathol. 85:174. 1979) and Boerema et al. (Z. Pflanzenkrankh. Pflanzenschutz 88:606. 1981) provided information on distinguishing this species from closely related species in *Ascochyta* and *Phoma*.
Host: Phaseolus

Stamnaria Fuckel. Discomycetes, Helotiales

Notes: Carpenter (Mem. New York Bot. Gard. 33:20. 1981).
Stamnaria americana Massee & Morg.
Distr.: North America.
Host: Equisetum
Stamnaria persoonii (Moug.) Fuckel
Stamnaria equiseti (Hoffm.) Sacc.
Titaeospora equiseti (Desmaz.) Vassiljevsky (anamorph)
Distr.: Temperate regions.
Host: Equisetum
Stamnaria thujae Seaver
Distr.: WI(type).
Host: Thuja

Stauronema (Sacc.) Syd., P. Syd. & E.J. Butler. Coelomycetes

Dinemasporium Lév.
Stauronema cruciferum (Ellis) Syd., P. Syd. & E.J. Butler
Dinemasporium cruciferum Ellis
Dinemasporium bicristatum Cooke
Distr.: SC; India; Pakistan.
Notes: Sutton (The Coelomycetes. p. 466. 1980).
Host: Zea

Steccherícium D. Reid. Basidiomycetes, Aphyllophorales

Steccherícium seriatum (C.G. Lloyd) Maas G.
Steccherinum westii Murrill
Distr.: Subtropical North America.
Notes: On hardwoods.
Host: Quercus

Steccherinum S.F. Gray. Basidiomycetes, Aphyllophorales

Steccherinum alaskense Lindsey & R.L. Gilbertson
Distr.: AK.
Host: Alnus
Steccherinum ciliolatum (Berk. & M.A. Curtis) R.L. Gilbertson & Budington
Hydnum ciliolatum Berk. & M.A. Curtis
Odontia ciliolata (Berk. & M.A. Curtis) L.W. Miller
Distr.: Temperate northern hemisphere.
Notes: On hardwoods.
Hosts: Abies, Fagus, Malus, Picea, Populus, Pseudotsuga, Quercus
Steccherinum fimbriatum (Pers.:Fr.) J. Eriksson
Odontia fimbriata (Pers.:Fr.) Fr.
Distr.: Northern hemisphere.
Hosts: Abies, Acer, Alnus, Amelanchier, Castanea, Liriodendron, Lithocarpus, Ostrya, Pinus, Populus, Quercus, Salix, Umbellularia
Steccherinum laeticolor (Berk. & M.A. Curtis) Banker
Mycoleptodon laeticolor (Berk. & M.A. Curtis) Pat.
Hyphodontia setulosa (Berk. & M.A. Curtis) Maas G.
Steccherinum setulosum (Berk. & M.A. Curtis) L.W. Miller
Distr.: Temperate northern hemisphere.
Notes: On wood.
Hosts: Acer, Celtis, Pinus, Platanus, Prunus, Quercus, Salix, Ulmus
Steccherinum ochraceum (Pers.:Fr.) S.F. Gray
Hydnum ochraceum Pers.:Fr.
Distr.: Temperate northern hemisphere.
Notes: On wood, usually of hardwoods.
Hosts: Acer, Alnus, Betula, Carpinus, Carya, Castanea, Cupressus, Juglans, Juniperus, Liquidambar, Malus, Nyssa, Ostrya, Picea, Populus, Prunus, Pseudotsuga, Quercus, Rhus, Salix
Steccherinum oreophilum Lindsey & R.L. Gilbertson
Distr.: North America.
Notes: On hardwoods.
Host: Populus
Steccherinum peckii Banker
Distr.: NY.
Notes: On wood.
Host: Acer

Steccherinum subcrinale (Peck) Ryvarden
Odontia subcrinale (Peck) R.L. Gilbertson
Distr.: Northern hemisphere.
Notes: On wood.
Hosts: Acer, Pinus, Tsuga
Steccherinum tenue Burdsall & Nakasone
Distr.: Southeastern United States.
Notes: On hardwoods.
Host: Acer
Steccherinum vagum Burdsall & Nakasone
Distr.: TN.
Host: Liriodendron

Stegonsporiopsis Van Warmelo & Sutton. Coelomycetes

Stegonsporiopsis cenangioides (Ellis & Rothr.) Van Warmelo & Sutton
Stegonsporium cenangioides Ellis & Rothr. in Ellis & Everh.
Camarographium abietis (M. Wilson & R.B. Anderson) Grove
Camarosporium abietis M. Wilson & R.B. Anderson
Distr.: Northeastern North America; Great Britain.
Notes: Associated with dieback (Van Warmelo & Sutton. Mycol. Pap. 145:17. 1981).
Host: Abies

Stegonsporium Corda in Opiz. Coelomycetes

Notes: Van Warmelo & Sutton (Mycol. Pap. 145:4. 1981) amended the generic concept to include only two species. Also spelled "*Steganosporium*".
Stegonsporium acerinum Peck
Distr.: Northeastern North America.
Notes: Bissett (Fungi Canadenses 102. 1977 as *Steganosporium acerinum*).
Host: Acer
Stegonsporium mori (Nomura) Sacc. & Trott.
Coryneum mori Nomura
Distr.: Northwestern United States; Japan(type).
Host: Morus
Stegonsporium pyriforme (Hoffm.:Fr.) Corda
Stilbospora ovata Pers.
Stegonsporium ovatum (Pers.) S.J. Hughes
Splanchnonema pupula (Fr.:Fr.) Kuntze (teleomorph)
Distr.: Eastern North America; Europe.
Notes: Usually restricted to *Acer* spp. in North America but occurs on other hardwoods in Europe. Shoemaker & Kokko (Fungi Canadenses 103. 1977 as *Steganosporium ovatum*).
Hosts: Acer, Betula
Stegonsporium tuberculiforme (Ellis & Everh.) Wehmeyer
Sporidesmium tuberculiforme Ellis & Everh.
Distr.: Rocky Mountains.
Notes: According to Van Warmelo & Sutton (Mycol. Pap. 145:39. 1981) this may be a *Trimmatostroma*.
Host: Sambucus

Stegophora Syd. & P. Syd. Pyrenomycetes, Diaporthales

Notes: Barr (Mycol. Mem. 7:70. 1978).
Stegophora ulmea (Schwein.:Fr.) Syd. & P. Syd.
Sphaeria ulmea Schwein.:Fr.
Dothidella ulmea (Schwein.:Fr.) Ellis & Everh.
Lambro ulmea (Schwein.:Fr.) E. Müller in E. Müller & Arx
Gnomonia ulmea (Schwein.:Fr.) Thuem.
Asteroma ulmeum (L.E. Miles) Sutton (anamorph)
Distr.: North America.
Notes: On leaves.
Host: Ulmus

Stemonitis Roth. Myxomycetes

Stemonitis fusca Roth
Distr.: Cosmopolitan.
Hosts: Ipomoea, Quercus
Stemonitis herbatica Peck
Distr.: North America; Africa; Europe; Fiji.
Host: Fragaria
Stemonitis pallida Wingate
Distr.: Central and eastern United States; Europe; Japan; Malay Peninsula.
Host: Quercus
Stemonitis splendens Rostafinski
Stemonitis fenestrata (Rex) T. Macbr.
Distr.: Cosmopolitan.
Host: Pinus

Stemphylium Wallr. Hyphomycetes

Notes: Simmons (Mycologia 59:67. 1967; Mycologia 61:1. 1969; Sydowia 38:284. 1985).
Stemphylium arnyi Reifschneider
Distr.: WI.
Host: Smilacina
Stemphylium bolickii Sobers & C.P. Seymour
Distr.: FL.
Notes: Leaf spots on a variety of hosts (Sobers & Seymour. Phytopathology 53:1444. 1963).
Hosts: Crassula, Echeveria, Kalanchoe, Sedum
Stemphylium botryosum Wallr.
Macrosporium commune Rabenh.
Pleospora tarda E. Simmons (teleomorph)
Distr.: Cosmopolitan.
Notes: Saprobic or rarely weakly parasitic on a wide variety of plants. Booth & Pirozynski (C.M.I. Descr. 150. 1967) and Simmons (Mycologia 61:6 1967) provided traditional accounts. Simmons (Sydowia 38:284. 1985) reinterpreted the anamorph-teleomorph connection.
Hosts: Acacia, Agropyron, Allium, Apium, Asparagus, Aster, Beta, Brassica, Capsicum, Chrysothamnus, Cicuta, Cucurbita, Daucus, Dianthus, Festuca, Gnaphalium, Helianthus, Heracleum, Impatiens, Jacquemontia, Lactuca, Lens, Lupinus, Lycopersicon, Medicago, Melilotus, Petroselinum, Petunia, Phaseolus, Phleum, Phlox, Pisum, Pyrus, Raphanus, Secale, Setaria, Spinacia, Trifolium, Trigonella, Vicia, Vinca, Yucca, Zigadenus
Stemphylium botryosum Wallr. var. **tragopogonis** M.B. Linn
Distr.: NY.
Notes: Leaf spot of cultivated salsify (Linn. Phytopathology 32:150. 1942).
Host: Tragopogon
Stemphylium callistephi K. Baker & L.H. Davis
Distr.: CA.
Notes: Leaf lesions (Ellis. More dematiaceous Hyphomycetes. p. 148. 1976).
Host: Callistephus
Stemphylium citri F. Patterson & V. Charles
Distr.: AZ(type).
Notes: According to Simmons (pers. comm.), the type specimen has shown this to be an *Alternaria*.
Host: Citrus
Stemphylium congestum G.A. Newton
Distr.: WA.
Notes: Storage rot of apples (Newton. Phytopathology 18:576. 1928). According to Simmons (pers. comm.) this is a species of *Ulocladium*, possibly a synonym of *U. consortiale*.
Host: Malus
Stemphylium congestum G.A. Newton var. **minor** Ruehle
Distr.: WA.
Notes: Isolated from apple fruits in cold storage (Mycologia 22:304. 1930). According to Simmons (pers. comm.) this is a *Ulocladium*.
Host: Malus
Stemphylium elasticae F. Patterson
Distr.: DC(type).
Host: Ficus
Stemphylium globuliferum (Vestergr.) E. Simmons
Macrosporium globuliferum Vestergr.

Distr.: Temperate regions.
Notes: Leaf spots on legumes and other plants (Simmons. Mycologia 61:14. 1969).
Hosts: Commelina, Trifolium
Stemphylium graminis (Corda) Bonord.
Soredospora graminis Corda
Distr.: WA; Germany(type).
Notes: According to Simmons (pers. comm.) the report from Washington on apple is probably a *Ulocladium*.
Host: Malus
Stemphylium lancipes (Ellis & Everh.) E. Simmons
Alternaria lancipes Ellis & Everh.
Distr.: Southcentral United States.
Hosts: Aquilegia, Argemone
Stemphylium loti J.H. Graham
Distr.: Central and eastern United States.
Notes: Leaf spot of birdsfoot trefoil (Graham. Phytopathology 43:577. 1953).
Host: Lotus
Stemphylium lycopersici (Enjoji) W. Yamamoto
Thyrospora lycopersici Enjoji
Stemphylium floridanum Hannon & G.F. Weber
Distr.: Widespread in tropical regions.
Notes: On *Lycopersicon* but also occurring on other plants (Ellis. Dematiaceous Hyphomycetes. p. 167. 1971; Ellis & Gibson. C.M.I. Descr. 471. 1975).
Hosts: Antirrhinum, Callistephus, Capsicum, Carica, Chrysanthemum, Dianthus, Juncus, Lycopersicon, Salvia, Solanum, Stokesia
Stemphylium majusculum E. Simmons
Pleospora sp. (teleomorph)
Distr.: NY.
Notes: On decaying leaves (Simmons. Mycologia 61:16. 1969).
Host: Lathyrus
Stemphylium polymorphum (Corda) Bonord.
Sporidesmium polymorphum Corda
Distr.: ME; Europe(type).
Notes: According to Simmons (pers. comm.) this is a *Monodictys*.
Host: Pisum
Stemphylium sarciniforme (Cavara) Wiltshire
Macrosporium sarciniforme Cavara
Thyrospora sarciniforme (Cavara) Tehon & E.Y. Daniels
Distr.: Cosmopolitan.
Notes: Target or pepper spot of clover (Booth. C.M.I. Descr. 671. 1980; Corlett et al. Fungi Canadenses 233. 1982).
Hosts: Melilotus, Trifolium
Stemphylium solani G.F. Weber
Distr.: Cosmopolitan.
Notes: Gray leaf spot of potato and tomato (Ellis & Gibson. C.M.I. Descr. 472. 1975).
Hosts: Capsicum, Gossypium, Lupinus, Lycopersicon, Physalis, Solanum
Stemphylium trifolii J.H. Graham
Distr.: NH, OR, and PA.
Notes: Leaf disease of white clover (Graham. Phytopathology 47:213. 1957).
Host: Trifolium
Stemphylium tritici F. Patterson
Distr.: OK, TX.
Notes: On living leaves and ovaries of wheat. According to Simmons (pers. comm.) this is an *Alternaria*.
Host: Triticum
Stemphylium vesicarium (Wallr.) E. Simmons
Macrosporium parasiticum Thuem.
Stemphylium parasiticum (Thuem.) J.A. Elliott ex A.W. Henry
Pleospora allii (Rabenh.) Ces. & De Not. (teleomorph)
Distr.: North and South America; Africa; Europe.
Notes: On a wide variety of herbaceous plants (Ellis. Dematiaceous Hyphomycetes. p. 167. 1971).
Hosts: Allium, Asparagus, Eichhornia, Juncus, Triticum

Stenella Syd. Hyphomycetes

Notes: Ellis (More dematiaceous Hyphomycetes. p. 307. 1976).
Stenella canavaliae (Syd. & P. Syd.) Deighton
Cercospora canavaliae Syd. & P. Syd.
Distr.: Subtropical and tropical regions.
Host: Canavalia
Stenella lythri (Westend.) J.L. Mulder
Cladosporium lythri Westend.
Cercospora lythri (Westend.) Niessl
Cercospora sanguinea Fuckel
Distr.: WI; Europe.
Host: Lythrum

Stenellopsis Morgan-Jones. Hyphomycetes

Stenellopsis magnoliae (Weedon) Morgan-Jones
Heterosporium magnoliae Weedon
Distr.: Southeastern United States.
Notes: Morgan-Jones (Mycotaxon 10:405. 1980).
Host: Magnolia

Stenocarpella Syd. Coelomycetes

Notes: Sutton (The Coelomycetes. p. 431. 1980).
Stenocarpella macrospora (Earle) Sutton
Diplodia macrospora Earle
Macrodiplodia zeae (Schwein.) Petr. & Syd. var. *macrospora* (Earle) Petr. & Syd.
Stenocarpella zeae Syd.
Distr.: Cosmopolitan.
Notes: On stems, leaves, and seeds (Sutton & Waterston. C.M.I. Descr. 83. 1966 as *Diplodia macrospora*).
Host: Zea
Stenocarpella maydis (Berk.) Sutton
Diplodia zeae (Schwein.) Lév.
Macrodiplodia zeae (Schwein.) Petr. & Syd.
Diplodia zeae-maydis Mekhtijeva
Distr.: Cosmopolitan.
Notes: On seeds and culms.
Host: Zea

Stephanoma Wallr. Hyphomycetes

Stephanoma phaeospora E.E. Butler & A. McCain
Distr.: CA.
Notes: Isolated from the root of an orchid (Butler & McCain. Mycologia 60:956. 1968).
Host: Cattleya

Stephanosporium Dal Vesco. Hyphomycetes

Stephanosporium cereale (Thuem.) H.J. Swart
Sporotrichum cerealis Thuem.
Distr.: Widespread.
Notes: On dead leaves and culms of grasses, isolated from air, paper, soil and textiles (Ellis. Dematiaceous Hyphomycetes. p. 36. 1971).
Host: Pseudotsuga

Stereum J. Hill ex Pers. Basidiomycetes, Aphyllophorales

Haematostereum Pouzar
Notes: Chamuris (Mycol. Mem. 14. 1988).
Stereum gausapatum (Fr.:Fr.) Fr.
Haematostereum gausapatum (Fr.:Fr.) Pouzar
Stereum plicatum (Peck) C.G. Lloyd
Stereum spadiceum (Fr.:Fr.) Fr.
Stereum spadiceum (Fr.:Fr.) Fr. var. *plicatum* Peck
Distr.: Widespread.
Notes: Heart rot of living trees. On wood of *Quercus* species, rarely on *Castanea* species. Reports on other hardwoods are in question as this species has been confused with *S. complicatum* and *S. hirsutum*.
Hosts: [Acer], [Alnus], [Baccharis], [Carpinus], Castanea, [Lithocarpus], [Lyonothamnus], [Malus], [Pseudotsuga], Quercus
Stereum hirsutum (Willd.:Fr.) S.F. Gray
Stereum complicatum (Fr.:Fr.) Fr.
Stereum rameale (Schwein.) Burt
Stereum styracifluum (Schwein.:Fr.) Fr.

Stereum subtomentosum Pouzar
Distr.: Widespread.
Notes: On wood, usually of hardwoods.
Hosts: Acacia, Acer, Adenostoma, Aesculus, Alnus, Amelanchier, Arbutus, Arctostaphylos, Betula, Calocedrus, Carpinus, Carya, Castanea, Catalpa, Cephalanthus, Cercocarpus, Corylus, Eucalyptus, Fagus, Hamamelis, Heteromeles, Ilex, Juniperus, Kalmia, Larix, Ligustrum, Liquidambar, Liriodendron, Lithocarpus, Lyonothamnus, Magnolia, Malus, Myrica, Nyssa, Ostrya, Physocarpus, Picea, Platanus, Populus, Prunus, Pseudotsuga, Pyrus, Quercus, Rhus, Rosa, Salix, Schinus, Sequoia, Tilia, Ulmus, Umbellularia, Vitis
Stereum ochraceo-flavum (Schwein.) Ellis
Stereum striatum (Fr.:Fr.) Fr. var. *ochraceo-flavum* (Schwein.) A.L. Welden
Stereum sulphuratum Berk. & Ravenel
Stereum vellereum Berk. & Ravenel
Distr.: Widespread.
Notes: On branches. The name *Stereum rameale* (Berk.) Massee has been used for this species, especially in Europe.
Hosts: Adenostoma, Ailanthus, Alnus, Amelanchier, Andromeda, Aronia, Betula, Castanea, Cercis, Corylus, Eucalyptus, Fagus, Liquidambar, Myrica, Nicotiana, Prunus, Quercus, Rhus, Salix, Symphoricarpos, Vaccinium
Stereum ostrea (Blume & Nees:Fr.) Fr.
Stereum fasciatum (Schwein.) Fr.
Stereum lobatum (Kunze:Fr.) Fr.
Distr.: North and South America; Africa; tropical Asia; Australasia.
Notes: On hardwoods.
Hosts: Acer, Alnus, Arbutus, Betula, Carpinus, Carya, Castanea, Catalpa, Celtis, Fagus, Fraxinus, Juglans, Liquidambar, Liriodendron, Myrica, Nyssa, Platanus, Populus, Quercus, Sequoia, Tilia, Ulmus
Stereum rugosum Pers.:Fr.
Haematostereum rugosum (Pers.:Fr.) Pouzar
Distr.: North America, rare; Asia; Australia; Europe, common.
Notes: Only a few verified reports of *S. rugosum* exist for the United States, with a dozen or so reports from eastern Canada.
Hosts: Alnus, Rhododendron, Salix
Stereum sanguinolentum (Albertini & Schwein.:Fr.) Fr.
Haematostereum sanguinolentum (Albertini & Schwein.:Fr.) Pouzar
Stereum balsameum Peck
Distr.: Widespread.
Notes: On conifers. Red heart rot of living trees.
Hosts: Abies, Alnus, Amelanchier, Larix, Picea, Pinus, Pseudotsuga, Thuja, Tsuga
Stereum striatum (Fr.:Fr.) Fr.
Stereum sericeum (Schwein.) Sacc.
Distr.: Central and eastern North America.
Notes: On wood of *Carpinus*, rarely on other hardwoods.
Hosts: Alnus, Betula, Carpinus, Castanea, Fagus, Ilex, Liquidambar, Nyssa, Prunus, Quercus, Rhus

Stevensonula Petr. Coelomycetes

Stevensonula ciliata Petr.
Distr.: FL; New Caledonia.
Notes: Sutton (Nova Hedwigia 26:1. 1975; The Coelomycetes. p. 157. 1980).
Host: Baccharis

Sthughesia Barr. Loculoascomycetes, Dothideales

Sthughesia juniperi (Dearn.) Barr
Dimerium juniperi Dearn.
Xenomeris juniperi (Dearn.) Barr & E. Müller
Xenomeris hemispherica E. Müller
Distr.: Western North America.
Notes: On needles. Barr (Mycotaxon 29:503. 1987).
Host: Juniperus

Stictis Pers.:Fr. Discomycetes, Ostropales

Notes: Sherwood (Mycotaxon 5:142. 1977).
Stictis arundinacea Pers.
Distr.: North America; temperate Europe.
Notes: On leaves and culms of grasses.
Host: Andropogon
Stictis bicolor (Ellis & Everh.) Starb.
Schizoxylon bicolor Ellis & Everh.
Distr.: Western North America.
Host: Populus
Stictis chrysopsis Ellis & Everh.
Distr.: Western North America.
Notes: On a variety of woody plants.
Hosts: Arctostaphylos, Ceanothus, Corylus, Rosa, Salix
Stictis dryophila Cooke & Ellis
Distr.: NJ.
Notes: Not a *Stictis* (Sherwood. Mycotaxon 5:174. 1977).
Host: Quercus
Stictis filicina Peck
Distr.: NY.
Notes: According to Sherwood (Mycotaxon 5:181. 1977) this species is a *Karstenia*.
Host: Osmunda
Stictis fusca Ellis & Barth.
Distr.: North America.
Host: Symphoricarpos
Stictis hawaiiensis Cash
Distr.: Subtropical and tropical regions.
Notes: On herbaceous stems.
Host: Rubus
Stictis hydrangeae Schwein.
Distr.: North America; temperate Europe.
Notes: On dead twigs.
Host: Hydrangea
Stictis lanuginicincta Fairm.
Distr.: CA(type).
Host: Marrubium
Stictis lupini W. Phillips & Harkn.
Stictis valdiviensis Speg.
Distr.: CA; Chile.
Notes: Sherwood (Mycotaxon 5:207. 1977).
Host: Lupinus
Stictis maydis Henn.
Distr.: GA; Germany.
Notes: The identity of this species is unclear (Sherwood. Mycotaxon 5:209. 1977).
Hosts: Andropogon, Tripsacum
Stictis mollis Pers.
Distr.: Western North America; Africa; Europe.
Notes: On bark and decorticated wood.
Host: Picea
Stictis monilifera W. Phillips & Harkn.
Stictis araucariae W. Phillips & Harkn.
Stictis megarrhizae W. Phillips & Harkn.
Distr.: CA.
Notes: On herbaceous stems, woody capsules, and conifer needles.
Hosts: Araucaria, Marah, Pittosporum
Stictis ostropoides M.A. Sherwood
Distr.: Western North America.
Hosts: Lonicera, Scrophularia
Stictis pupula Fr.
Schmitzomia pupula (Fr.) Fr.
Distr.: OR; Europe.
Notes: This name cannot be characterized based on the type specimen. Sherwood (Mycotaxon 5:229. 1977) proposed that the name be abandoned.
Host: Umbellularia
Stictis pustulata Ellis
Stictis helicotricha Ellis & Everh.
Distr.: North America; temperate Europe.
Notes: On herbaceous stems and small twigs.
Hosts: Arundinaria, Eremochloa, Vinca
Stictis radiata Pers.:Fr.
Stictis annulata Cooke & W. Phillips
Distr.: Cosmopolitan.

Notes: On bark, wood, and plant debris.
Hosts: Abutilon, Acer, Alnus, Arbutus, Arctostaphylos, Artemisia, Baccharis, Carya, Ceanothus, Cibotium, Conyza, Cornus, Corylus, Erechtites, Garrya, Gaultheria, Heteromeles, Ligustrum, Liriodendron, Lonicera, Maclura, Melia, Physocarpus, Picea, Pinus, Populus, Pseudotsuga, Pteridium, Pueraria, Pyrus, Rhamnus, Rhododendron, Rhus, Ribes, Rubus, Rumex, Sadleria, Salix, Sambucus, Sequoia, Smilax, Solidago, Sorghum, Spiraea, Umbellularia, Zea
Stictis sagaretiae **Cash**
Distr.: FL; South America.
Host: Sageretia
Stictis schizoxyloides **Ellis & Everh.**
Stictis curtispora Dearn. & Bisby
Stictis edwiniae Clements
Coccomyces salicinus (Ellis & Everh.) Sacc.
Distr.: Western North America; Sweden.
Notes: On wood and bark.
Hosts: Ephedra, Fraxinus, Salix
Stictis serenoae **Cash**
Distr.: FL(type).
Notes: Sherwood (Mycotaxon 5:325. 1977) suggested that this species may be a synonym of *Propolis quadrifida*.
Host: Serenoa
Stictis serpentaria **Ellis & Everh.**
Distr.: Western North America.
Notes: On wood.
Hosts: Salix, Sambucus
Stictis sphaeroboloidea **(Schwein.) Berk. in Ellis & Everh.**
Distr.: Eastern North America.
Hosts: Cornus, Populus, Rosa
Stictis stellata **Schwein.**
Distr.: Southeastern United States.
Notes: On herbaceous stems, occasionally on wood.
Hosts: Arundinaria, Phytolacca, Zea
Stictis stellata **Wallr. var.** ***philippensis*** **Rehm**
Distr.: HI; The Philippines.
Host: Verbena
Stictis stigma **Cooke & Ellis**
Distr.: Eastern North America.
Host: Cornus
Stictis viticola **M.A. Sherwood**
Distr.: AL.
Host: Vitis

Stictopatella Höhn. Coelomycetes

Stictopatella iridis **Tehon**
Distr.: IA(type).
Host: Iris

Stigmatea Fr. Loculoascomycetes, Dothideales

Stigmatea polygonorum **(Fr.:Fr.)Fr.**
Dothidea polygoni Fr.:Fr.
Distr.: CA; Europe.
Host: Polygonum
Stigmatea pyrolae **J. Schröt.**
Distr.: NY.
Host: Chimaphila
Stigmatea ranunculi **Fr.**
Distr.: CA; Europe.
Host: Ranunculus
Stigmatea sclerotidea **Cooke**
Distr.: GA, SC.
Host: Arundinaria

Stigmatolemma Kalchbr. Basidiomycetes, Agaricales

Stigmatolemma poriaeforme **(Pers.:Fr.) W.B. Cooke**
Porotheleum poriaeforme (Pers.:Fr.) W.B. Cooke
Solenia poriaeformis (Pers.:Fr.) Fr.
Distr.: North and South America; South Africa; Europe; New Zealand.
Notes: On wood.
Hosts: Juglans, Platanus, Populus, Quercus, Vitis
Stigmatolemma taxi **(Lév.) Donk**
Cyphella taxi Lév.
Cyphella cupuliformis Berk. & Ravenel in Berk.
Rhodocyphella cupuliformis (Berk. & Ravenel) W.B. Cooke
Distr.: Southern United States.
Hosts: Juniperus, Robinia

Stigmatophragmia Tehon & G.L. Stout. Loculoascomycetes, Dothideales

Stigmatophragmia sassafrasicola **Tehon & G.L. Stout**
Distr.: IL(type).
Host: Sassafras

Stigmella Lév. in Demidoff. Coelomycetes

Stigmella effigurata **(Schwein.) S.J. Hughes**
Coryneum effiguratum Schwein.
Dicoccum dryophyllum Corda
Sporidesmium foliicola Desmaz.
Distr.: Temperate regions.
Notes: On fallen leaves (Sutton. The Coelomycetes. p. 80. 1980).
Host: Quercus

Stigmina Sacc. Hyphomycetes

Notes: Ellis (Dematiaceous Hyphomycetes. p. 144. 1971; More dematiaceous Hyphomycetes. p. 114. 1976).
Stigmina asperula **(Lombard & R.W. Davidson) Sutton**
Coryneum asperulum Lombard & R.W. Davidson
Distr.: AL.
Host: Cupressus
Stigmina beaucarneae **Pollack & D.F. Farr**
Distr.: TX; Mexico.
Notes: On living leaves (Pollack & Farr. Mycotaxon 8:194. 1979).
Host: Nolina
Stigmina carpophila **(Lév.) M.B. Ellis**
Clasterosporium carpophilum (Lév.) Aderhold
Coryneum carpophilum (Lév.) Jauch
Coryneum beyerinckii Oudem.
Distr.: Temperate and warm-temperate regions.
Notes: Shot-hole disease (Ellis. Mycol. Pap. 72:56. 1958).
Hosts: Cleome, Prunus
Stigmina compacta **(Sacc.) M.B. Ellis**
Stegonsporium compactum Sacc.
Thyrostroma compactum (Sacc.) Höhn.
Exosporium ulmi Eriks.
Distr.: North America; Europe.
Notes: On twigs.
Hosts: Sambucus, Ulmus
Stigmina concentrica **(Cooke & Ellis) Deighton**
Cercospora concentrica Cooke & Ellis
Cylindrosporium angustifoliae Ellis & Kellerm.
Mycosphaerella deightonii Morelet (teleomorph)
Distr.: NJ(type), North America; Europe.
Notes: Leaf spot (Deighton. Trans. Brit. Mycol. Soc. 61:107. 1973).
Host: Yucca
Stigmina cornicola **(Ellis & Everh.) Sutton**
Coryneum cornicola Ellis & Everh.
Distr.: Eastern North America.
Notes: On limbs.
Host: Cornus
Stigmina deflectens **(P. Karst.) M.B. Ellis**
Exosporium deflectens P. Karst.
Distr.: SD; Europe.
Notes: On leaves.
Host: Juniperus
Stigmina dothideoides **(Ellis & Everh.) M.B. Ellis**
Clasterosporium dothideoides Ellis & Everh.
Sciniatosporium dothideoides (Ellis & Everh.) Morgan-Jones
Distr.: Northwestern United States.

Notes: Shoemaker & Egger (Fungi Canadenses 212. 1982).
Hosts: Artemisia, Shepherdia
Stigmina earlei Deighton
Cercospora thermopsidis Earle non Harkn.
Distr.: Central and western North America.
Notes: Leaf spot.
Host: Thermopsis
Stigmina ellisii M.B. Ellis
Clasterosporium pulvinatum Ellis & Everh.
Distr.: CO.
Host: Chrysothamnus
Stigmina glomerulosa (Sacc.) S.J. Hughes
Exosporium glomerulosum (Sacc.) Höhn.
Exosporium bethelii Dearn.
Coryneum juniperinum Ellis
Distr.: North America; Europe.
Notes: On needles.
Hosts: Cupressus, Juniperus, Thuja
Stigmina heteromelis (Harkn.) Deighton
Cercospora heteromelis Harkn.
Distr.: CA.
Notes: Deighton (Trans. Brit. Mycol. Soc. 61:113. 1973).
Hosts: Heteromeles, Photinia
Stigmina juniperina (Georgescu & Badea) M.B. Ellis
Cercospora juniperina Georgescu & Badea
Sciniatosporium juniperina (Georgescu & Badea) Morgan-Jones
Distr.: Northeastern North America; Europe.
Notes: Needlecast. Often confused with *Cercospora sequoiae* Ellis & Everh. var. *juniperina* Ellis (Hodges. Mycologia 54:66. 1962).
Host: Juniperus
Stigmina liquidambaris (Tharp) Morgan-Jones & Kendrick
Exosporium liquidambaris Tharp
Distr.: TX.
Notes: Morgan-Jones & Kendrick (Canad. J. Bot. 50:1817. 1972).
Host: Liquidambar
Stigmina liriodendri Ellis & Everh.
Distr.: MS(type).
Host: Liriodendron
Stigmina negundinis (Berk. & M.A. Curtis) M.B. Ellis
Coryneum negundinis Berk. & M.A. Curtis in Berk.
Coryneum negundinis Ellis & Everh.
Sciniatosporium negundinis (Berk. & M.A. Curtis) Morgan-Jones
Thyrostromella negundinis (Berk. & M.A. Curtis) S.J. Hughes
Coryneum septosporioides Sacc. & Syd.
Helminthosporium subcuticulare Ellis & Everh.
Distr.: Temperate North America.
Hosts: Acer, Sambucus
Stigmina pallida (Ellis & Everh.) M.B. Ellis
Exosporium pallidum Ellis & Everh.
Coryneum cydoniae Dearn. & House
Coryneum rhoinum (Dearn. & Barth.) S.J. Hughes
Exosporium rhoinum Dearn. & House
Distr.: Central and eastern North America; Europe.
Hosts: Chaenomeles, Rhus
Stigmina palmivora (Sacc. in Trel.) S.J. Hughes
Exosporium palmivorum Sacc. in Trel.
Distr.: Widespread in tropical regions.
Notes: On leaves of various palms.
Hosts: Acoelorrhaphe, Arecaceae, Butia, Caryota, Howeia, Phoenix, Rhapis, Roystonea, Sabal, Syagrus, Thrinax, Veitchia, Washingtonia
Stigmina pedunculata (Ellis & Everh.) M.B. Ellis
Brachysporium pedunculatum Ellis & Everh.
Exosporium pedunculatum (Ellis & Everh.) W.B. Cooke
Coryneum pedunculatum (Ellis & Everh.) S.J. Hughes
Thyrostromella pedunculatum (Ellis & Everh.) S.J. Hughes
Distr.: Western North America.
Notes: On twigs.
Host: Sambucus
Stigmina platani (Fuckel) Sacc.
Stigmella platani Fuckel
Distr.: Central and eastern North America; Europe.
Notes: Leaf blight.
Host: Platanus
Stigmina platani-racemosae (Dearn. & Barth.) S.J. Hughes
Stigmella platani-racemosae Dearn. & Barth.
Sciniatosporium platani-racemosae (Dearn. & Barth.) Morgan-Jones
Distr.: CA(type).
Host: Platanus
Stigmina robusta (Cooke & Ellis) Sutton
Arthrobotryum robustum Cooke & Ellis
Distr.: Widespread in northern North America.
Notes: On bark of deciduous trees (Hughes. Fungi Canadenses 167. 1980; Sutton. Mycol. Pap. 132:117. 1973).
Host: Acer
Stigmina sambucina (Ellis & Everh.) Sutton
Coryneum sambucinum Ellis & Everh.
Exosporium sambucinum Tracy & Earle
Sciniatosporium sambucinum (Ellis & Everh.) Morgan-Jones
Thyrostromella sambucinum (Ellis & Everh.) S.J. Hughes
Distr.: Temperate North America.
Notes: On limbs.
Host: Sambucus
Stigmina sycina (Thuem.) Sutton
Sporidesmium sycinum Thuem.
Sciniatosporium sycinum (Thuem.) Morgan-Jones
Macrosporium puccinioides Ellis & F.W. Anderson
Distr.: CO, MT; Europe.
Notes: On dead twigs (Morgan-Jones. Canad. J. Bot. 49:1005. 1971; Sutton. Trans. Brit. Mycol. Soc. 58:166. 1972).
Hosts: Artemisia, Lathyrus
Stigmina tharpii Morgan-Jones & Kendrick
Exosporium platanorum Tharp
Distr.: TX.
Host: Platanus
Stigmina thermopsidis Harkn.
Sciniatosporium thermopsidis (Harkn.) Morgan-Jones
Distr.: CA.
Host: Thermopsis
Stigmina thujina (Dearn.) Sutton
Coryneum thujinum Dearn.
Sciniatosporium thujinum (Dearn.) Morgan-Jones
Distr.: Western North America.
Hosts: Chamaecyparis, Thuja
Stigmina tinea (Sacc.) Ellis
Cercospora tinea Sacc.
Cercoseptoria tinea (Sacc.) Deighton
Distr.: LA; Europe.
Host: Viburnum
Stigmina trimera (Sacc.) Sutton
Coryneum trimerum Sacc.
Sciniatosporium trimerum (Sacc.) Morgan-Jones
Thyrostromella trimerum (Sacc.) Syd. in Syd. & Petr.
Distr.: Western United States.
Host: Robinia
Stigmina utahensis (Sacc.) Sutton
Stegonsporium utahense Sacc.
Sciniatosporium utahense (Sacc.) Morgan-Jones
Distr.: UT.
Host: Chrysothamnus

Stigmochora Theiss. & Syd. Pyrenomycetes, Phyllachorales

Stigmochora variegata Barr & C.S. Hodges
Distr.: HI(type).
Notes: Barr & Hodges (Mycologia 79:785. 1987).
Host: Pisonia

Stigmopeltis Syd. Coelomycetes

Stigmopeltis graminicola Wehmeyer
Distr.: WA(type).
Host: Elymus

Stilbella Lindau in Engl. & Prantl, nom. cons. Hyphomycetes

Notes: Seifert (Stud. Mycol. 27. 1985).
Stilbella acerina Overh., nom. nud.

Distr.: TN(type).
Notes: Published without a latin diagnosis. Probably a *Phialographium* sp.
Host: Acer
Stilbella aciculosa (Ellis & Everh.) Seifert
Stilbum aciculosum Ellis & Everh.
Distr.: Cosmopolitan.
Notes: On a variety of organic substrates.
Host: Orontium
Stilbella aleuriata (Berk. & M.A. Curtis) Seifert
Stilbum aleuriatum Berk. & M.A. Curtis
Isaria aggregata Cooke & Massee
Stilbum cinnabarinum Mont.
Stilbum corallinum Cooke & Massee
Nectria macrostoma Berk. & M.A. Curtis (teleomorph)
Distr.: Southern United States; subtropical and tropical regions.
Notes: On bark and decayed wood.
Host: Acer
Stilbella annulata (Berk. & M.A. Curtis) Seifert
Stilbum annulatum Berk. & M.A. Curtis
Blistum musae Seifert
Stilbella proliferans F. Stevens
Distr.: Cosmopolitan, more frequent in tropical regions.
Notes: On various kinds of organic debris.
Hosts: Brassica, Pastinaca
Stilbella aterrima (Welw. & Curr.) Seifert
Distr.: Subtropical and tropical regions.
Notes: On bark.
Host: Carya
Stilbella clavulata (Mont.) Seifert
Graphium clavulatum (Mont.) Sacc.
Nectria gracilipes (Tul. & C. Tul.) Wollenweb. (teleomorph)
Distr.: Temperate(rare), subtropical and tropical regions.
Notes: On bark and wood.
Host: Melia
Stilbella fimetaria (Pers.) Lindau
Stilbella erythrocephala (Ditmar) Lindau in Engl. & Prantl
Distr.: Cosmopolitan.
Notes: On dung.
Hosts: Albizia, Eucalyptus
Stilbella flavipes (Peck) Seifert
Stilbum flavipes Peck
Dendrostilbella hanlinii T. Hammill & J. Shipman
Distr.: Central and eastern North America; Asia.
Notes: On decaying angiosperms, often in association with other fungi; isolated from soil.
Hosts: Alnus, Rubus
Stilbella fusca (Sacc.) Seifert
Atractium fuscum Sacc.
Stilbum madidum Peck
Distr.: Temperate northern hemisphere.
Notes: On bark and decaying wood.
Host: Betula

Stilbospora Pers. ex Mérat. Coelomycetes

Stilbospora macrosperma Pers. ex Mérat
Distr.: FL; Europe.
Notes: On woody branches (Sutton. The Coelomycetes. p. 343. 1980).
Host: Smilax
Stilbospora pinicola Berk. & M.A. Curtis
Distr.: MS, OR.
Host: Pinus

Stilbotulasnella Oberwinkler & Bandoni in Bandoni & Oberwinkler. Basidiomycetes, Tulasnellales

Notes: Bandoni & Oberwinkler (Canad. J. Bot. 60:1875. 1982).
Stilbotulasnella conidiophora Bandoni & Oberwinkler
Distr.: HI.
Notes: On palm leaves.
Host: Arecaceae

Stilbum Tode. Basidiomycetes, Auriculariales

Notes: Although this genus is in the basidiomycete order Auriculariales, several species belonging in the ascomycetous anamorph genus *Stilbella* have not been transferred.
Stilbum rhois Berk. & M.A. Curtis
Sporocybe rhois (Berk. & M.A. Curtis) Sacc.
Amphiporthe aculeans (Schwein.) Barr (teleomorph)
Distr.: Eastern United States.
Notes: On dead branches and twigs. According to Seifert (Stud. Mycol. 27:206. 1985) the generic affinities of these species are unclear.
Hosts: Diospyros, Rhus
Stilbum sphaerocephalum Massee
Distr.: CA; England(type).
Notes: According to Seifert (Stud. Mycol. 27:207. 1985) this may be a synonym of *Stilbella aciculosa*.
Host: Baccharis

Stomatogene Theiss. Loculoascomycetes, Dothideales

Stomatogene agaves (Ellis & Everh.) Theiss.
Asterina agaves Ellis & Everh.
Dimerosporium agavectona Pat. & Har.
Distr.: CA, subtropical and tropical North America.
Notes: Barr (Contr. Univ. Michigan Herb. 9:537. 1972).
Host: Yucca

Stomiopeltis Theiss. Loculoascomycetes, Dothideales

Notes: Müller & Arx (Beitr. Kryptogamenfl. Schweiz 11(2):545. 1962).
Stomiopeltis cassiae Mendoza in F. Stevens & Manter
Clypeolum exiguum Syd.
Distr.: ID; South America.
Notes: On various host families.
Host: Pinus
Stomiopeltis polyloculata Luttrell
Distr.: GA.
Notes: Müller & Arx (Beitr. Kryptogamenfl. Schweiz 11(2):546. 1962) listed *Clypeolum japonicum* Hino & Katumoto as a possible synonym.
Host: Arundinaria
Stomiopeltis suttoniae (Mendoza) Luttrell
Stomiopeltella suttoniae Mendoza
Distr.: HI.
Notes: Luttrell (Mycologia 38:572. 1946).
Host: Myrsine
Stomiopeltis versicolor (Desmaz.) Arx
Actinonema rubi Fuckel
Asterella rubi (Fuckel) Höhn.
Distr.: Cosmopolitan.
Notes: On living leaves and twigs of *Rubus* and other plants.
Host: Rubus

Strasseria Bres. & Sacc. in Strass. Coelomycetes

Strasseria geniculata (Berk. & Broome) Höhn.
Sphaeropsis geniculata Berk. & Broome
Distr.: United States; Australia; Europe.
Notes: Parmelee & Cauchon (Canad. J. Bot. 57:1660. 1979); Sutton (The Coelomycetes. p. 478. 1980).
Hosts: Crataegus, Lycopodium, Pyrola, Vaccinium
Strasseria nigra Dearn. in Parmelee
Distr.: Northeastern United States.
Hosts: Lycopodium, Pyrola
Strasseria oxycocci (Shear) Shear
Plagiorhabdus oxycocci Shear
Distr.: MA, WA.
Notes: Schwarz & Boone (Pl. Dis. 67:31. 1983).
Host: Vaccinium

Streptobotrys Hennebert. Hyphomycetes

Notes: Hennebert (Persoonia 7:191. 1973).
Streptobotrys streptothrix (Cooke & Ellis) Hennebert
Botrytis streptothrix (Cooke & Ellis) Sacc.
Distr.: Northeastern United States.
Hosts: Arisaema, Hydrastis, Orontium, Symplocarpus

Streptotinia Whetzel. Discomycetes, Helotiales

Streptotinia arisaemae Whetzel
Streptobotrys arisaemae Hennebert (anamorph)
Distr.: NY.
Notes: The teleomorph was described by Whetzel (Mycologia 37:684. 1945).
Host: Arisaema

Strickeria Koerb. Loculoascomycetes, Dothideales

Notes: Many of the names in this genus are now placed in *Teichospora*.
Strickeria catalinae Fairm.
Distr.: CA(type).
Host: Lyonothamnus
Strickeria inclusa Petr.
Distr.: ID(type).
Host: Salix
Strickeria kelseyi (Ellis & Everh.) Barr
Cucurbitaria kelseyi Ellis & Everh.
Distr.: MT(type).
Notes: Barr (Mycotaxon 29:505. 1987).
Host: Philadelphus
Strickeria mammoides (Ellis & Everh.) Kuntze
Teichospora mammoides Ellis & Everh.
Distr.: MT, OR.
Host: Sarcobatus
Strickeria vaga (Rehm) G. Wint.
Teichospora vaga Rehm
Dictyotrichiella vaga (Rehm) Remler
Distr.: WA; European Alps.
Host: Rhododendron

Strigopodia Batista in Batista, Maia, & Viala. Loculoascomycetes, Dothideales

Notes: Hughes (Canad. J. Bot. 46:1099. 1968).
Strigopodia batistae S.J. Hughes
Antennatula sp. (synanamorph)
Capnophialophora sp. (synanamorph)
Distr.: Pacific Northwest.
Notes: On resin and bark of conifers.
Hosts: Larix, Pseudotsuga
Strigopodia resinae (Sacc. & Bres.) S.J. Hughes
Capnodium resinae Sacc. & Bres.
Limacinia resinae (Sacc. & Bres.) Sacc. & Bres.
Helminthosporium resinaceum Cooke (anamorph conidial)
Racodium resinae Fr. (anamorph mycelial)
Distr.: Temperate North America; Europe.
Notes: On resin and bark of conifers.
Hosts: Abies, Picea
Strigopodia spongiosa (Barr) Barr
Capnodium spongiosum Barr
Capnophaeum spongiosum (Barr) Batista & Cif.
Antennatula sp. (synanamorph)
Capnophialophora sp. (synanamorph)
Distr.: Pacific Northwest.
Notes: Barr (Canad. J. Bot. 33:511. 1955; Contr. Univ. Michigan Herb. 9:622. 1972).
Host: Pseudotsuga

Strobilurus Singer. Basidiomycetes, Agaricales

Strobilurus conigenoides (Ellis) Singer
Collybia conigenoides (Ellis) Sacc.
Pseudohiatula conigenoides (Ellis) Singer
Marasmius bombycirrhizus Berk. & Cooke
Distr.: Southeastern United States.
Notes: On inflorescences.
Host: Magnolia
Strobilurus occidentalis V. Wells & P. Kempton
Distr.: Western North America.
Notes: On cones and cone scales.
Host: Picea
Strobilurus trullisatus (Murrill) J.W. Lennox
Marasmius trullisatus (Murrill) Singer
Strobilurus kemptonae V. Wells in V. Wells & P. Kempton
Distr.: Northwestern North America.
Notes: On buried cones of conifers.
Host: Pseudotsuga

Stromatinia (Boud.) Boud. Discomycetes, Helotiales

Stromatinia gladioli (Drayton) Whetzel
Sclerotinia gladioli Drayton
Distr.: Temperate regions.
Notes: Dry rot of corms (Darvas. Phytophylactica 16:255. 1984).
Hosts: Crocus, Freesia, Gladiolus
Stromatinia narcissi Drayton & Groves
Sclerotium ambiguum Duby var. *narcissi* Sacc. (anamorph)
Distr.: Temperate North America; Europe.
Notes: On bulbs (Drayton & Groves. Mycologia 44:119. 1952).
Hosts: Narcissus, Zephyranthes
Stromatinia panacis (Rankin) L.M. Kohn
Sclerotinia panacis Rankin
Distr.: Eastern North America.
Notes: The name *Sclerotinia smilacinae* E.J. Durand has been misapplied to this species.
Host: Panax
Stromatinia rapulum (Bull. ex Mérat:Fr.) Boud.
Sclerotinia smilacinae E.J. Durand
Stromatinia smilacinae (E.J. Durand) Whetzel
Distr.: Eastern North America; Europe.
Notes: On rhizomes.
Host: Polygonatum

Stromatocrea W.B. Cooke. Hyphomycetes

Stromatocrea cerebriforme W.B. Cooke
Hypocreopsis lichenoides (Tode:Fr.) Seaver (teleomorph)
Distr.: North America; Europe.
Notes: Associated with *Hymenochaete* spp. on wood.
Host: Acer

Stromatocyphella W.B. Cooke. Basidiomycetes, Agaricales

Stromatocyphella conglobata (Burt) W.B. Cooke
Cyphella conglobata Burt
Distr.: North America.
Notes: On dead, still-attached branches of hardwoods, especially *Alnus* spp.
Host: Alnus

Strumella Fr. Hyphomycetes

Strumella acaciae Cooke & Harkn.
Distr.: CA.
Host: Acacia
Strumella sacchari Cooke
Distr.: HI; Australia(type).
Host: Saccharum
Strumella simmondsiae Bonar
Distr.: Southwestern United States.

Notes: On living leaves, calyxes, and peduncles (Bonar. Mycologia 34:190. 1942).
Host: Simmondsia

Stypella A. Möller. Basidiomycetes, Tremellales

Stypella minor A. Möller
Sebacina sphaerospora Bourd. & Galzin
Distr.: Warm regions of the western hemisphere.
Notes: On wood.
Host: Pseudotsuga

Subulicystidium Parmasto. Basidiomycetes, Aphyllophorales

Subulicystidium longisporum (Pat.) Parmasto
Peniophora longispora (Pat.) Höhn.
Peniophora asperipilata Burt
Distr.: Temperate northern hemisphere.
Notes: On wood.
Hosts: Abies, Alnus, Fraxinus, Liriodendron, Populus, Quercus

Suillosporium Pouzar. Basidiomycetes, Aphyllophorales

Suillosporium cystidiatum (D.P. Rogers) Pouzar
Pellicularia cystidiata D.P. Rogers
Distr.: Temperate northern hemisphere.
Notes: On wood.
Host: Tsuga

Sydowia Bres. Loculoascomycetes, Dothideales

Notes: Barr (Contr. Univ. Michigan Herb. 9:570. 1972).
Sydowia dothideoides Dearn. & Barth.
Dothiora polyspora Shear & R.W. Davidson
Dothichiza sp. (anamorph)
Distr.: Temperate North America.
Notes: On branches and twigs.
Hosts: Populus, Salix
Sydowia lepargyrea Dearn. ex Barr
Distr.: Western North America.
Host: Shepherdia
Sydowia polyspora (Bref. & Tavel) E. Müller
Dothidea polyspora Bref. & Tavel
Hormonema dematioides Lagerberg & Melin (synanamorph)
Sclerophoma pithyophila (Corda) Höhn. (synanamorph)
Distr.: Worldwide on conifers.
Notes: Sutton & Waterson (C.M.I. Descr. 228. 1970).
Host: Abies
Sydowia pruni Barr
Distr.: Western North America.
Host: Prunus

Sydowiella Petr. Pyrenomycetes, Diaporthales

Sydowiella depressula (P. Karst.) Barr
Gnomonia depressula P. Karst.
Distr.: North America; northern Europe.
Notes: On limbs.
Host: Rubus
Sydowiella fenestrans (Duby) Petr.
Didymosphaeria fenestrans (Duby) G. Wint.
Gnomonia fenestrans (Duby) Sacc.
Didymella epilobii (Fuckel) Sacc.
Distr.: Circumboreal.
Notes: On overwintered stems.
Host: Epilobium

Syncephalastrum J. Schröt. Zygomycetes, Mucorales

Syncephalastrum racemosum Cohn ex J. Schröt.
Distr.: Widespread.
Notes: Isolated from soil, dung, and various organic substrates (Benjamin. Mycologia 58:5. 1966).
Hosts: Carya, Juncus, Zea

Synchytrium de Bary & Woronin. Chytridiomycetes, Chytridiales

Notes: Karling (*Synchytrium*. 1964).
Synchytrium anemones (DC.:Fr.) Woronin
Distr.: Temperate regions.
Host: Anemone
Synchytrium anomalum J. Schröt.
Distr.: North America; West Indies; Europe.
Hosts: Adoxa, Ranunculus, Rumex
Synchytrium asari Arth. & Holw. in Arth. et al.
Distr.: Northcentral United States.
Host: Asarum
Synchytrium asterum M.T. Cook
Distr.: LA.
Host: Aster
Synchytrium aureum J. Schröt.
Distr.: Cosmopolitan.
Notes: Recorded on at least 123 genera of vascular plants.
Hosts: Acalypha, Artemisia, Caltha, Cardamine, Clintonia, Delphinium, Fragaria, Geum, Hydrocotyle, Lactuca, Ligusticum, Lithophragma, Lysimachia, Parnassia, Petasites, Potentilla, Prenanthes, Ranunculus, Rubus, Rudbeckia, Saxifraga, Sedum, Senecio, Trientalis, Valerianella, Viola
Synchytrium australe Speg.
Synchytrium modioliensis M.T. Cook
Distr.: Eastern United States; Argentina; Australia.
Hosts: Callirhoe, Modiola
Synchytrium bignoniae M.T. Cook
Distr.: LA.
Host: Bignonia
Synchytrium bonaerense Speg.
Synchytrium hydrocotyles M.T. Cook
Distr.: Southeastern United States; Argentina.
Host: Hydrocotyle
Synchytrium callicarpae M.T. Cook
Distr.: LA.
Host: Callicarpa
Synchytrium callirrhoeae Karling
Distr.: Central United States.
Host: Callirhoe
Synchytrium cardiospermi M.T. Cook
Distr.: LA.
Host: Cardiospermum
Synchytrium caricis Tracy & Earle in M.E. Jones
Distr.: UT(type).
Notes: A doubtful species.
Host: Carex
Synchytrium carpini M.T. Cook
Distr.: LA.
Host: Carpinus
Synchytrium cellulare J.J. Davis
Synchytrium pulverum J.J. Davis
Distr.: WI.
Hosts: Boehmeria, Lycopus, Pycnanthemum
Synchytrium cerastii M.T. Cook
Distr.: Southeastern United States.
Host: Cerastium
Synchytrium chamaedryoidis Karling
Synchytrium urticae M.T. Cook non Sorokin, hom. illeg.
Distr.: Southeastern United States.
Host: Urtica
Synchytrium cinnamomeum J.J. Davis
Distr.: WI.
Host: Ranunculus
Synchytrium clematidis M.T. Cook

Distr.: LA.
Host: Clematis
Synchytrium cocculi M.T. Cook
Distr.: LA.
Host: Cocculus
Synchytrium corni M.T. Cook
Distr.: LA.
Host: Cornus
Synchytrium davisii Karling
Distr.: WI.
Notes: Previously reported as *S. aureum* Schröt.
Host: Rubus
Synchytrium decipiens (Farl.) Farl.
Uredo aecidioides Peck non DC., hom. illeg.
Distr.: Central and eastern North America; West Indies; eastern Asia; Europe; India.
Notes: The name *S. aecidioides* (Peck) Lagerh. has been misapplied to this species.
Hosts: Amphicarpaea, Apios
Synchytrium duchesneae M.T. Cook
Distr.: LA, possibly SC.
Host: Duchesnea
Synchytrium echinulatum Karling
Distr.: CA.
Host: Clarkia
Synchytrium endobioticum (Schilberszky) Percival
Chrysophlyctis endobioticum Schilberszky
Synchytrium solani Massee
Distr.: Cosmopolitan.
Notes: Potato wart disease, potato blackwart. Walker (C.M.I. Descr. 755. 1983).
Hosts: Lycopersicon, Solanum
Synchytrium epilobii Karling
Distr.: Western United States.
Host: Epilobium
Synchytrium erechtitis M.T. Cook
Distr.: LA.
Host: Erechtites
Synchytrium eremocarpae Karling
Distr.: CA(type).
Host: Eremocarpus
Synchytrium erigerontis M.T. Cook
Distr.: LA.
Host: Erigeron
Synchytrium fragariae Zeller & L. Campbell
Distr.: WA.
Host: Fragaria
Synchytrium fraxini M.T. Cook
Distr.: LA.
Host: Fraxinus
Synchytrium fulgens J. Schröt.
Synchytrium brownii Karling
Synchytrium oenotherae de Bary
Distr.: North and South America; Europe; Japan.
Hosts: Camissonia, Clarkia, Gaura, Oenothera
Synchytrium geranii Clendenin
Distr.: Southeastern United States.
Host: Geranium
Synchytrium globosum J. Schröt.
Distr.: Eastern United States; Europe.
Hosts: Duchesnea, Veronica
Synchytrium gonolobi M.T. Cook
Distr.: LA.
Host: Gonolobus
Synchytrium graminicola R. Sprague
Distr.: Pacific Northwest.
Notes: A doubtful species.
Hosts: Agropyron, Agrostis, Anthoxanthum, Bromus, Dactylis, Elymus, Festuca, Poa
Synchytrium holwayi Farl.
Distr.: Central United States.
Host: Monarda
Synchytrium ilicicola M.T. Cook
Distr.: LA.
Host: Ilex
Synchytrium impatientis M.T. Cook
Distr.: LA.
Host: Impatiens
Synchytrium innominatum Farl.
Distr.: CA.
Hosts: Malacothrix, Microseris
Synchytrium lacunosum Karling
Distr.: WY(type).
Hosts: Arabis, Draba
Synchytrium lagerheimii Karling
Synchytrium aecidioides Lagerh. in Pat. & Lagerh., nom. inval.
Distr.: Ecuador.
Notes: Old reports of *S. aecidioides* (Peck) Lagerh. may refer to *S. decipiens* (Farl.) Farl.
Hosts: Amphicarpaea, Desmodium
Synchytrium lepidii M.T. Cook
Distr.: CO, LA.
Host: Lepidium
Synchytrium linariae Karling
Distr.: FL.
Host: Linaria
Synchytrium liquidambaris M.T. Cook
Distr.: LA.
Host: Liquidambar
Synchytrium lithophragmae Karling
Distr.: Pacific Northwest.
Host: Lithophragma
Synchytrium lythrii M.T. Cook
Distr.: LA.
Host: Lythrum
Synchytrium macrosporum Karling
Distr.: TX.
Notes: Red gall disease (Walker. C.M.I. Descr. 757. 1983).
Host: Xanthium
Synchytrium marrubii G. Tobl.
Distr.: MA.
Host: Marrubium
Synchytrium mitchellae M.T. Cook
Distr.: LA.
Host: Mitchella
Synchytrium myosotidis Kühn
Distr.: North America; Europe.
Hosts: Cryptantha, Pectocarya
Synchytrium namae Karling
Distr.: TX(type).
Host: Nama
Synchytrium nigrescens J.J. Davis
Distr.: WI.
Host: Aster
Synchytrium nitidum Ellis & B.T. Galloway ex Karling
Distr.: LA(type).
Host: Triodanis
Synchytrium papillatum Farl.
Distr.: Western United States; Canary Islands; Mediterranean region.
Host: Erodium
Synchytrium parksii Karling
Distr.: CA.
Host: Trifolium
Synchytrium parthenocissi M.T. Cook
Distr.: LA.
Host: Parthenocissus
Synchytrium perforatum Ellis
Distr.: Central and western North America.
Notes: Distortion and thickening of leaves, petioles, and stems of Valerianaceae.
Host: Valeriana
Synchytrium pilificum F. Thomas
Pycnochytrium pilificum (F. Thomas) J. Schröt.
Distr.: North America; Europe.
Notes: On stems, leaves, petioles, and runners of Rosaceae.
Host: Potentilla
Synchytrium plantagineum Sacc. & Speg.
Distr.: MO; Italy.
Host: Plantago
Synchytrium polemonii Ellis & Everh. ex Karling

Distr.: WA(type).
Hosts: Microsteris, Polemonium
Synchytrium polygoni M.T. Cook
Distr.: LA.
Host: Polygonum
Synchytrium potentillae (J. Schröt.) Lagerh.
Distr.: Temperate North America; Europe.
Host: Potentilla
Synchytrium ranunculi M.T. Cook
Distr.: LA.
Host: Ranunculus
Synchytrium rugulosum Dietel
Distr.: CA, UT.
Notes: On leaves, petioles, and stems of Onagraceae and other plants.
Host: Penstemon
Synchytrium salviae Karling
Distr.: CA(type).
Host: Salvia
Synchytrium sambuci M.T. Cook
Distr.: LA.
Host: Sambucus
Synchytrium scirpi J.J. Davis
Distr.: Northeastern North America.
Host: Scirpus
Synchytrium shawii Karling
Distr.: SD(type).
Host: Androsace
Synchytrium smilacis M.T. Cook
Distr.: LA.
Host: Smilax
Synchytrium stachydis M.T. Cook
Distr.: Southeastern United States.
Host: Stachys
Synchytrium stellariae Fuckel
Synchytrium chiltonii M.T. Cook
Distr.: Eastern North America; Europe.
Hosts: Cerastium, Stellaria
Synchytrium taraxaci de Bary & Woronin
Distr.: Cosmopolitan.
Hosts: Engelmannia, Taraxacum
Synchytrium tecomae M.T. Cook
Distr.: LA.
Host: Campsis
Synchytrium texanum Karling
Distr.: TX.
Notes: Karling (Mycologia 52:21. 1960).
Host: Plantago
Synchytrium tillaeae Karling
Distr.: CA.
Host: Tillaea
Synchytrium trachelospermi M.T. Cook
Distr.: LA.
Host: Trachelospermum
Synchytrium ulmi M.T. Cook
Distr.: LA.
Host: Ulmus
Synchytrium vaccinii F. Thomas
Distr.: Northeastern North America.
Notes: Red rust or red gall disease of cranberry (Karling. Sydowia 10:149. 1958).
Hosts: Arctostaphylos, Chamaedaphne, Clethra, Gaylussacia, Kalmia, Ledum, Rhododendron, Vaccinium
Synchytrium variabilum Karling
Distr.: CA.
Host: Ranunculus
Synchytrium violae M.T. Cook
Distr.: LA.
Host: Viola

Syspastospora P. Cannon & D. Hawksworth. Pyrenomycetes, Sordariales

Notes: Cannon & Hawksworth (Bot. J. Linn. Soc. 84:115. 1982).
Syspastospora parasitica (Tul.) P. Cannon & D. Hawksworth
Sphaeronema parasitica Tul.
Melanospora parasitica (Tul.) Tul. & C. Tul.
Ceratostoma biparasiticum Ellis & Everh.
Distr.: Worldwide, most common in tropical regions.
Notes: Parasitic on moniliaceous hyphomycetes on plants.
Host: Euphorbia

Taeniolella S.J. Hughes. Hyphomycetes.

Notes: Ellis (More dematiaceous Hyphomycetes. p. 55. 1976).
Taeniolella alta (Ehrenb.:Pers.) S.J. Hughes
Torula alnea Peck
Septonema dichaenoides Peck & G.W. Clinton in Peck
Distr.: Central United States; Canada.
Notes: On woody substrates (Hughes. Fungi Canadenses 184. 1980).
Hosts: Alnus, Quercus
Taeniolella breviuscula (Berk. & M.A. Curtis) S.J. Hughes
Septonema breviusculum Berk. & M.A. Curtis in Berk.
Torula opaca Cooke
Distr.: Eastern United States.
Host: Acer
Taeniolella multiplex (Berk. & M.A. Curtis) S.J. Hughes
Septonema multiplex Berk. & M.A. Curtis
Distr.: NC, SC.
Notes: Described from oak limbs, reported on leaves of *Eucalyptus*.
Host: Eucalyptus
Taeniolella muricata (Ellis & Everh.) S.J. Hughes
Dendryphion muricatum Ellis & Everh.
Dendryphion brunneum Dearn. & Barth. in Dearn.
Helminthosporium repente Dearn. & Barth. in Dearn.
Distr.: Temperate North America.
Notes: Saprobic on wood and bark.
Hosts: Acer, Prunus
Taeniolella rudis (Sacc.) S.J. Hughes
Dendryphion curtipes Ellis & Barth.
Distr.: Central and eastern North America.
Notes: On woody substrates (Hughes. Fungi Canadenses 185. 1980).
Host: Pinus
Taeniolella stilbospora (Corda) S.J. Hughes
Torula stilbospora Corda
Hormiscium stilbosporum (Corda) Sacc.
Distr.: AK, MT; Europe.
Notes: Common on dead branches of *Salix*, also on *Alnus*, *Corylus*, and *Populus*, staining the wood bright reddish purple (Ellis. Dematiaceous Hyphomycetes. p. 94. 1971).
Hosts: [Pinus], Salix
Taeniolella subsessilis (Ellis & Everh.) S.J. Hughes
Dendryphion subsessile Ellis & Everh.
Distr.: KS(type).
Host: Smilax

Taeniolina M.B. Ellis. Hyphomycetes

Taeniolina centaurii (Fuckel) M.B. Ellis
Torula centaurii Fuckel
Hormiscium centaurii (Fuckel) Sacc.
Distr.: CA; Europe.
Notes: On leaves (Ellis. More dematiaceous Hyphomycetes. p. 62. 1976).
Host: Centaurium

Talaromyces C.R. Benjamin. Pyrenomycetes, Eurotiales

Notes: Pitt (The genus *Penicillium* and its teleomorphic states *Eupenicillium* and *Talaromyces*. 1979).
Talaromyces bacillisporus (M.E. Swift) C.R. Benjamin
Penicillium bacillisporium M.E. Swift
Distr.: Widespread.
Host: Begonia
Talaromyces flavus (Klöcker) A.C. Stolk & R.A. Samson
Talaromyces vermiculatus (P.A. Dang.) C.R. Benjamin
Penicillium dangeardii J.I. Pitt (anamorph)
Distr.: Cosmopolitan.

Notes: In soil and on plant debris.
Hosts: Carya, Populus, Selaginella
Talaromyces luteus (Zukal) C.R. Benjamin
Penicillium luteum Zukal
Distr.: Worldwide, but rare.
Hosts: Alnus, Gossypium, Zea
Talaromyces stipitatus (Thom) C.R. Benjamin
Penicillium stipitatum Thom
Distr.: Widespread.
Host: Zea
Talaromyces thermophilus A.C. Stolk
Penicillium dupontii Griffon & Maubl.
Talaromyces dupontii (Griffon & Maubl.) R. Emerson in C.L. Fergus
Distr.: Worldwide, but rare.
Notes: Isolated from decaying plant materials.
Host: Arachis
Talaromyces trachyspermus (Shear) A.C. Stolk & R.A. Samson
Arachniotus trachyspermus Shear
Distr.: Widespread.
Notes: Isolated from soil and various kinds of debris.
Hosts: Carya, Vaccinium

Tapesia (Pers.) Fuckel. Discomycetes, Helotiales

Notes: Aebi (Nova Hedwigia 23:49. 1972).
Tapesia cinerella Rehm
Distr.: WA; Europe.
Notes: Seaver (North Amer. cup fungi. [Inoperculates]. p. 194. 1951) considered this name to be a synonym of *T. fusca* but Aebi (Nova Hedwigia 23:88. 1972) disagreed.
Host: Gaultheria
Tapesia fusca (Pers.) Fuckel
Mollisia fusca (Pers.) P. Karst.
Phialea fusca (Pers.) Gill.
Distr.: Temperate regions.
Notes: On woody debris. LeGal & Mangenot (Rev. Mycol. 26:308. 1961) recognized this species in *Mollisia*.
Hosts: Acer, Alnus, Arctostaphylos, Betula, Ceanothus, Rhododendron, Salix, Shepherdia, Vaccinium
Tapesia lividofusca (Fr.) Rehm
Mollisia lividofusca (Fr.) Gill.
Niptera lividofusca (Fr.) Fuckel
Distr.: CA, IL, NE; Europe.
Notes: On decaying woody plant parts. LeGal & Mangenot (Rev. Mycol. 25:160. 1960) recognized this species in *Mollisia*.
Host: Pinus
Tapesia lonicerae Cash
Distr.: CO(type).
Host: Lonicera
Tapesia melaleucoides Rehm var. ***vaccinii*** Rehm
Distr.: CA, MI; Europe(type).
Notes: The type variety is considered a synonym of *T. lividofusca*.
Host: Arctostaphylos
Tapesia mollisioides (Schwein.) Sacc.
Distr.: Eastern United States, CA.
Hosts: Alnus, Salix
Tapesia ribicola Cash
Distr.: CO.
Host: Ribes
Tapesia rosae (Pers.:Fr.) Fuckel
Tapesia corni Fuckel
Distr.: Temperate regions.
Host: Rosa

Taphrina Fr.:Fr. Hemiascomycetes, Taphrinales

Notes: Mix (Univ. Kansas Sci. Bull. 33:3. 1949; Trans. Kansas Acad. Sci. 57:55. 1954).
Taphrina aceris Mix
Exoascus aceris Dearn. & Barth. non Linh., hom. illeg.
Taphrina bartholomaei (Dearn. & Barth.) Mix
Distr.: Pacific Northwest.
Host: Acer
Taphrina aesculi (F. Patterson) Giesenh.
Distr.: CA.
Host: Aesculus
Taphrina alni (Berk. & Broome) Gjaerum
Taphrina alni-incanae (Kühn) Sadebeck
Taphrina amentorum (Sadebeck) Rostr.
Distr.: AK, GA; Europe; Japan.
Notes: On catkins.
Host: Alnus
Taphrina amelanchieri Mix
Distr.: CA(type).
Host: Amelanchier
Taphrina americana Mix
Distr.: Temperate North America.
Notes: Witches' broom.
Host: Betula
Taphrina armeniacae Georgescu & Badea
Distr.: Roumania.
Notes: Witches' broom.
Host: Prunus
Taphrina australis (Atk.) Giesenh.
Distr.: Eastern North America.
Host: Carpinus
Taphrina bacteriosperma Johans.
Distr.: Circumboreal.
Host: Betula
Taphrina boycei Mix
Distr.: NV; Canada(British Columbia).
Host: Betula
Taphrina bullata (Berk.) Tul.
Distr.: WA; Europe.
Hosts: Malus, Pyrus
Taphrina caerulescens (Desmaz. & Mont.) Tul.
Distr.: Temperate North America; northern Africa; Europe.
Notes: Leaf blister.
Hosts: [Ostrya], Quercus, [Rhus]
Taphrina californica Mix
Distr.: Western United States; Mexico.
Host: Dryopteris
Taphrina carnea Johans.
Distr.: North America; temperate Europe.
Notes: Leaf curl.
Host: Betula
Taphrina carveri Jenk.
Distr.: AL, MO.
Notes: Possibly conspecific with *T. dearnessii*.
Host: Acer
Taphrina castanopsidis Jenk.
Distr.: CA.
Host: Castanopsis
Taphrina communis (Sadebeck) Giesenh.
Exoascus communis Sadebeck
Taphrina longipes (Atk.) Giesenh.
Exoascus mirabilis Atk.
Taphrina mirabilis (Atk.) Giesenh.
Taphrina mirabilis (Atk.) Giesenh. var. *tortilis* Atk.
Taphrina rhizipes (Atk.) Giesenh.
Distr.: Central and eastern North America.
Notes: Bladder plum or plum pocket.
Host: Prunus
Taphrina confusa (Atk.) Giesenh.
Distr.: Temperate North America.
Host: Prunus
Taphrina coryli Nishida
Distr.: Eastern North America; Japan.
Host: Corylus
Taphrina cystopteridis Mix
Distr.: Central United States.
Host: Cystopteris
Taphrina darkeri Mix
Distr.: OR(type).
Host: Acer
Taphrina dearnessii Jenk.
Distr.: Eastern North America.

Notes: Possibly conspecific with *T. letifera*.
Host: Acer
Taphrina deformans (Berk.) Tul.
Exoascus deformans (Berk.) Fuckel
Distr.: Cosmopolitan.
Notes: Peach leaf curl (Booth. C.M.I. Descr. 711. 1981).
Host: Prunus
Taphrina farlowii Sadebeck
Exoascus farlowii (Sadebeck) Sadebeck
Taphrina varia (Atk.) Mix
Exoascus varius Atk.
Distr.: Eastern North America.
Notes: Leaf curl.
Host: Prunus
Taphrina faulliana Mix
Distr.: Pacific Northwest.
Host: Polystichum
Taphrina filicina Rostr. ex Johans.
Distr.: Eastern North America; northern Europe.
Hosts: Dryopteris, Onoclea
Taphrina flava Farl.
Distr.: Eastern North America.
Notes: Yellow leaf blister.
Host: Betula
Taphrina flavorubra W.W. Ray
Distr.: Temperate North America.
Host: Prunus
Taphrina flectans Mix
Distr.: Western North America.
Notes: Leaf curl and witches' broom.
Host: Prunus
Taphrina fusca Giesenh.
Distr.: Eastern North America; southern Europe.
Host: Dryopteris
Taphrina gracilis Mix
Distr.: NY.
Host: Dryopteris
Taphrina hiratsukae Nishida
Distr.: Central and eastern North America; Japan; Soviet Union.
Hosts: Matteuccia, Onoclea
Taphrina japonica Kusano
Taphrina macrophylla W.W. Ray
Distr.: Northwestern North America; Japan.
Notes: Witches' broom.
Hosts: Alnus, Amelanchier
Taphrina jenkinsoniana Mix
Distr.: Western United States.
Host: Prunus
Taphrina johansonii Sadebeck
Distr.: Temperate North America; Europe; Japan.
Host: Populus
Taphrina letifera (Peck) Sacc.
Ascomyces letifera Peck
Distr.: Eastern United States.
Host: Acer
Taphrina lutescens Rostr.
Distr.: Northeastern North America; Europe.
Hosts: Dryopteris, Thelypteris
Taphrina nana Johans.
Distr.: North America(rare); Europe.
Notes: Yellowing of leaves or witches' brooms.
Host: Betula
Taphrina occidentalis W.W. Ray
Distr.: Western North America.
Notes: On catkins.
Host: Alnus
Taphrina polystichi Mix
Distr.: Eastern North America.
Host: Polystichum
Taphrina populi-salicis Mix
Distr.: Pacific Northwest.
Notes: Leaf spot.
Hosts: Populus, Salix
Taphrina populina (Fr.:Fr.) Fr.
Taphrina aurea (Pers.) Fr.
Distr.: Temperate northern hemisphere.
Hosts: Alnus, Populus
Taphrina potentillae (Farl.) Johans.
Taphrina tormentillae Rostr.
Distr.: Eastern North America; Europe.
Host: Potentilla
Taphrina pruni Tul.
Exoascus pruni (Tul.) Fuckel
Taphrina insititiae (Sadebeck) Johans.
Distr.: Temperate northern hemisphere.
Notes: Bladder or pocket plums (Booth. C.M.I. Descr. 713. 1981).
Host: Prunus
Taphrina pruni-subcordatae (Zeller) Mix
Distr.: Western North America.
Host: Prunus
Taphrina purpurescens B.L. Robinson
Distr.: Central and eastern North America; Europe.
Notes: Leaf curl.
Host: Rhus
Taphrina rhizophora Johans.
Distr.: North America; Europe.
Notes: On catkins.
Host: Populus
Taphrina robinsoniana Giesenh.
Taphrina rugosa W.W. Ray
Distr.: Eastern North America.
Notes: On catkins.
Host: Alnus
Taphrina sacchari Jenk.
Distr.: Central and eastern North America.
Notes: Jenkins (J. Wash. Acad. Sci. 28:350. 1938) accepted this taxon as distinct from *T. acericola* because of differences in host relationships and geographic distribution.
Host: Acer
Taphrina thomasii Mix
Distr.: CA.
Host: Prunus
Taphrina tosquinetii (Westend.) Magnus
Distr.: Northeastern North America; Europe.
Notes: Witches' broom.
Host: Alnus
Taphrina ulmi (Fuckel) Johans.
Distr.: Eastern North America; Europe.
Notes: On leaves.
Host: Ulmus
Taphrina virginica Sadebeck
Distr.: Eastern North America.
Host: Ostrya
Taphrina wiesneri (Rathay) Mix
Exoascus cerasi (Fuckel) Sadebeck
Taphrina cerasi (Fuckel) Sadebeck
Distr.: Cosmopolitan.
Notes: Witches' broom and leaf curl of cherry and apricot (Booth. C.M.I. Descr. 712. 1981).
Hosts: Acer, Prunus

Tapinella E.J. Gilbert. Basidiomycetes, Agaricales

Tapinella panuoides (Batsch:Fr.) E.J. Gilbert
Paxillus panuoides (Batsch:Fr.) Fr.
Distr.: Temperate northern hemisphere.
Notes: Lignicolous; brown rot.
Hosts: Abies, Picea, Pinus

Teichospora Fuckel. Loculoascomycetes, Dothideales

Teichospora aberrans Rehm
Distr.: ND(type).
Host: Salix
Teichospora aridophila Peck
Teichospora xerophila Peck in Sacc.
Distr.: AZ, CA.

Notes: Barr et al. (New York State Mus. Bull. 459:9. 1986).
Host: Yucca
Teichospora cercocarpi (Earle) Fairm.
Distr.: Southwestern United States.
Host: Cercocarpus
Teichospora fulgurata Ellis & Everh.
Strickeria fulgurata (Ellis & Everh.) Kuntze
Distr.: KS, ND.
Host: Populus
Teichospora gregaria Ellis & Everh.
Strickeria gregaria (Ellis & Everh.) Kuntze
Distr.: ND; Canada(Ontario).
Host: Fraxinus
Teichospora helenae Ellis & Everh.
Strickeria helenae (Ellis & Everh.) Kuntze
Distr.: Central and western United States; Soviet Union.
Notes: Popushoi (Mycoflora of the fruit trees of U.S.S.R. p. 173. 1971) included this species and mistakenly attributed the transfer in *Strickeria* to himself.
Hosts: Populus, Prunus, Salix
Teichospora mammoides Ellis & Everh. var. ***opuntiae*** Dearn. & Barth.
Distr.: CA(type).
Host: Opuntia
Teichospora megastega Ellis & Everh.
Strickeria megastega (Ellis & Everh.) Kuntze
Distr.: Pacific Northwest.
Hosts: Acer, Ephedra, Haplopappus, Salix
Teichospora nautica Ellis & Everh.
Strickeria nautica (Ellis & Everh.) Kuntze
Distr.: MT(type).
Host: Populus
Teichospora negundinis Ellis & Everh.
Distr.: CO(type).
Host: Acer
Teichospora opuntiae Ellis & Everh.
Distr.: CO(type).
Host: Opuntia
Teichospora patellarioides Sacc.
Strickeria patellarioides (Sacc.) P. Larsen
Distr.: WA; Iceland; Italy(type).
Host: Populus
Teichospora pomiformis P. Karst.
Distr.: CA; Finland(type).
Host: Haplopappus
Teichospora populi (Earle) Sacc. & D. Sacc.
Strickeria populi Earle
Distr.: CO, NM.
Host: Populus
Teichospora pruni-americanae Rehm
Distr.: ND(type).
Host: Prunus
Teichospora solitaria (Ellis) Ellis & Everh.
Cucurbitaria solitaria Ellis
Gibberidea solitaria (Ellis) Kuntze
Strickeria solitaria (Ellis) Kuntze
Distr.: Western United States.
Hosts: Artemisia, Haplopappus, Symphoricarpos
Teichospora symphoricarpi (Tracy & Earle) Sacc. & D. Sacc.
Strickeria symphoricarpi Tracy & Earle
Distr.: CO, WA.
Host: Sambucus
Teichospora trabicola Fuckel
Distr.: OK; Europe(type).
Host: Sapindus
Teichospora umbonata Ellis & Everh.
Strickeria umbonata (Ellis & Everh.) Kuntze
Distr.: MT, SD.
Host: Symphoricarpos
Teichospora variabilis Ellis & Everh.
Strickeria variabilis (Ellis & Everh.) Kuntze
Distr.: MT; Soviet Union.
Host: Artemisia

Teichosporella (Sacc.) Sacc. Loculoascomycetes, Dothideales

Teichosporella dura (Fuckel) Sacc.
Teichospora dura (Fuckel) Fuckel
Strickeria fendlerae Earle
Teichospora fendlerae (Earle) Sacc. & D. Sacc.
Teichospora inflata (Ellis) Ellis & Everh.
Zignoella inflata (Ellis) Sacc.
Teichosporella montanae Ellis & Everh.
Teichospora papillosa Ellis & Everh.
Distr.: Widespread in temperate regions.
Notes: Barr (Mycologia 73:988. 1981).
Hosts: Fendlera, Salix
Teichosporella lonicerina Fairm.
Distr.: CA(type).
Host: Lonicera

Telimena Racib. Pyrenomycetes, Phyllachorales

Telimena rhoina (Ellis & Everh.) Theiss. & Syd.
Distr.: CA(type).
Host: Rhus

Telimenella Petr. Pyrenomycetes, Phyllachorales

Telimenella phacidioidea Barr
Distr.: MA.
Notes: Barr (Mycologia 69:956. 1977).
Host: Smilacina

Teratosperma Syd. Hyphomycetes

Teratosperma cornigerum (Ellis & Everh.) M.B. Ellis
Clasterosporium cornigerum Ellis & Everh.
Distr.: Eastern North America.
Notes: Ellis (Dematiaceous Hyphomycetes. p. 134. 1971).
Host: Carpinus

Testicularia L.J. Klotz. Basidiomycetes, Ustilaginales

Testicularia cyperi Klotzsch
Distr.: North and South America.
Notes: On Cyperaceae (Vánky. Illustrated genera of smut fungi. p. 108. 1987).
Hosts: Cyperus, Rhynchospora

Tetracladium De Wild. Hyphomycetes

Tetracladium marchalianum De Wild.
Distr.: Aquatic habitats in temperate regions.
Notes: On submerged, decaying leaves and in foam (Nilsson. Symb. Bot. Upsal. 18:105. 1964).
Hosts: Acer, Quercus

Tetracoccosporium Szabó. Hyphomycetes

Tetracoccosporium paxianum Szabó
Distr.: Worldwide in soil and on dung.
Notes: Ellis (Dematiaceous Hyphomycetes. p. 82. 1971).
Host: Trifolium

Tetraploa Berk. & Broome. Hyphomycetes

Notes: Ellis (Dematiaceous Hyphomycetes. p. 51. 1971).
Tetraploa aristata Berk. & Broome
Distr.: Subtropical and tropical regions.
Notes: On leaf bases and stems just above the soil (Ellis. Dematiaceous Hyphomycetes. p. 52. 1971).
Hosts: Ammophila, Juncus, Pinus, Sesbania
Tetraploa divergens Tracy & Earle
Distr.: MS, NC.

Host: Panicum

Tetraploa ellisii Cooke in Cooke & Ellis
Distr.: North and South America; Africa.
Notes: On grasses.
Host: Andropogon

Tetraploa scabra Harkn.
Distr.: CA(type).
Host: Scirpus

Thallospora L. Olive. Hyphomycetes

Thallospora aspera L. Olive
Distr.: LA, WI.
Notes: A systemic and obligate parasite (Olive. Mycologia 40:11. 1948).
Host: Veronica

Thamnidium Link. Zygomycetes, Mucorales

Thamnidium elegans Link:Fr.
Distr.: Widespread.
Notes: Isolated from soil, dung, and various organic substrates (O'Donnell. Zygomycetes in culture. p. 196. 1979).
Host: Zea

Thanatephorus Donk. Basidiomycetes, Tulasnellales

Thanatephorus cucumeris (A.B. Frank) Donk
Corticium areolatum Stahel
Pellicularia filamentosa (Pat.) D.P. Rogers
Pellicularia filamentosa (Pat.) D.P. Rogers f. *sasakii* (Shirai) Exner
Pellicularia filamentosa (Pat.) D.P. Rogers f. *solani* (Kühn) Exner
Ceratobasidium filamentosum (Pat.) Olive
Hypochnus filamentosus Pat. in Pat. & Lagerh.
Corticium praticola Kotila
Thanatephorus praticola (Kotila) Talbot
Corticium sasakii (Shirai) Matsumoto
Hypochnus sasakii Shirai
Pellicularia sasakii (Shirai) Ito
Botryobasidium solani (Prill. & Delacr.) Donk
Corticium solani Prill. & Delacr.
Corticium vagum sensu Burt, pro parte
Rhizoctonia solani Kühn (anamorph)
Distr.: Cosmopolitan.
Notes: Pathogenic on a number of cultivated plants. Usually attacks underground portions of stems or roots, causing stem and root rots. Also causes damping-off of seedlings and cuttings and rotting of foliage and fruits (Mordue. C.M.I. Descr. 406. 1974). Some authors prefer to recognize some of the synonyms listed above as distinct species; we accept a broad concept until a stable system of classification is established. *Thanatephorus cucumeris* is composed of a number of "anastamosis groups" which are often used in establishing the identity of a given isolate.
Hosts: Agrohordeum, Agropyron, Agrostis, Aleurites, Allium, Alnus, Alopecurus, Alternanthera, Alysicarpus, Amaranthus, Ambrosia, Anthemis, Apium, Apocynum, Arachis, Armoracia, Arrhenatherum, Aster, Avena, Axonopus, Beta, Bouteloua, Brassica, Bromus, Buchloe, Calamagrostis, Calystegia, Capsella, Capsicum, Caragana, Cenchrus, Cerastium, Chenopodium, Chilopsis, Chimaphila, Chrysanthemum, Cirsium, Citrullus, Commelina, Crotalaria, Cucumis, Cynodon, Cyperus, Dactylis, Dahlia, Danthonia, Daucus, Deschampsia, Dianthus, Dichanthelium, Dichondra, Digitaria, Echinochloa, Eleusine, Elymus, Eragrostis, Erigeron, Euphorbia, Fagopyrum, Festuca, Fragaria, Geranium, Gladiolus, Glycine, Gossypium, Helianthus, Hibiscus, Hierochloe, Holcus, Hordeum, Hydrastis, Ipomoea, Iris, Koeleria, Kummerowia, Lactuca, Lathyrus, Lepidium, Lespedeza, Leymus, Lilium, Lolium, Lotus, Lycopersicon, Malva, Medicago, Melilotus, Mentha, Mucuna, Muhlenbergia, Nicotiana, Oryza, Oryzopsis, Paeonia, Panax, Panicum, Papaver, Parthenocissus, Paspalum, Pastinaca, Pelargonium, Pennisetum, Petunia, Phalaris, Phaseolus, Phleum, Physalis, Picea, Pinus, Pisum, Plantago, Poa, Polygonum, Populus, Prenanthes, Raphanus, Redfieldia, Rheum, Robinia, Rubus, Saccharum, Schedonnardus, Secale, Setaria, Sinapis, Sitanion, Solanum, Solidago, Sorghum, Spinacia, Sporobolus, Stellaria, Stenotaphrum, Stipa, Strelitzia, Taraxacum, Tragopogon, Trifolium, Trisetum, Triticum, Tulipa, Ulmus, Veronica, Vicia, Vigna, Vulpia, Xanthium, Zea

Thecaphora Fingerh. Basidiomycetes, Ustilaginales

Notes: Vánky (Symb. Bot. Upsal. 24:122. 1985).

Thecaphora californica (Harkn.) G.P. Clinton
Sorosporium californicum Harkn.
Distr.: Western United States.
Notes: On Asteraceae.
Hosts: Arnica, Grindelia

Thecaphora cuneata (Schofield) G.P. Clinton
Sorosporium cuneatum Schofield
Sorosporium solidaginis Ellis & Everh.
Distr.: Central and western United States.
Notes: On Asteraceae.
Hosts: Grindelia, Solidago

Thecaphora deformans Durieu & Mont. in Tul. & C. Tul.
Sorosporium astragali Peck
Distr.: Widespread.
Notes: On Fabaceae.
Hosts: Astragalus, Desmodium, Lathyrus, Lotus, Lupinus, Trifolium, Vicia

Thecaphora iresine (J.A. Elliott) H. Jacks.
Sorosporium iresine J.A. Elliott
Distr.: IN.
Notes: On Amaranthaceae.
Host: Iresine

Thecaphora pilulaeformis Berk. & M.A. Curtis
Sorosporium bigloviae D. Griffiths
Tolyposporium davidsonii Dietel & Holw.
Distr.: Southwestern United States.
Host: Haplopappus

Thecaphora rhynchosporae G. Fisch.
Distr.: MS(type).
Host: Rhynchospora

Thecaphora seminis-convolvuli (Desmaz.) Ito
Uredo seminis-convolvuli Desmaz.
Thecaphora hyalina Fingerh.
Distr.: Western United States.
Notes: On Convolvulaceae. According to Vánky (Symb. Bot. Upsal. 24:126. 1985), this species is known only from Europe and Asia.
Host: Convolvulus

Thecaphora thornberi D. Griffiths
Distr.: AZ(type).
Host: Tidestromia

Thecaphora trailii Cooke
Schizonella subtrifida Ellis & Everh.
Distr.: Widespread.
Notes: On Asteraceae.
Host: Cirsium

Thedgonia Sutton. Hyphomycetes

Thedgonia ligustrina (Boerema) Sutton
Cercospora ligustrina Boerema
Distr.: KS, WA; England; The Netherlands(type).
Notes: Sutton (Trans. Brit. Mycol. Soc. 61:428. 1973).
Host: Ligustrum

Thelephora Ehrh. Basidiomycetes, Aphyllophorales

Thelephora albidobrunnea Schwein.
Stereum micheneri Berk. & M.A. Curtis
Thelephora odorifera Peck
Stereum spongiosum Massee
Distr.: North America.
Notes: On the ground, encrusts twigs, stems, etc.
Hosts: Acer, Amelanchier, Carpinus, Fraxinus, Juniperus, Quercus, Viburnum

Thelephora caryophyllea Fr.:Fr.
Distr.: Temperate northern hemisphere.
Notes: According to Corner (Beih. Nova Hedwigia 27:46. 1968) this species inhabits "sandy ground in coniferous woods." There are, however, several reports indicating seedling girdle and habitation of wood or bark.
Hosts: Betula, Larix, Salix
Thelephora cuticularis Berk.
Distr.: Central and eastern United States; England.
Notes: On bark at base of trees, on fallen twigs.
Host: Juniperus
Thelephora terrestris Ehrh.:Fr.
Thelephora crustosa C.G. Lloyd
Thelephora laciniata Fr.:Fr.
Thelephora minor Velen.
Thelephora rhipidium Velen.
Thelephora tristis Sacc.
Distr.: Widespread.
Notes: Mycorrhizal; often a weed in seedling nurseries. A number of forms and varieties have been described but are not accepted here (Corner. Beih. Nova Hedwigia 27:87. 1968).
Hosts: Abies, Larix, Picea, Pinus, Pseudotsuga, Quercus, Ribes, Spiraea, Thuja, Vaccinium

Theleporus Fr. Basidiomycetes, Aphyllophorales

Theleporus ajovalliensis R.L. Gilbertson & Blackwell
Distr.: AZ(type).
Notes: On wood of ocotillo (Gilbertson & Blackwell. Mycotaxon 15:249. 1982).
Host: Fouquieria

Thermoascus Miehe. Plectomycetes, Gymnoascales

Thermoascus aurantiacus Miehe
Distr.: Cosmopolitan.
Notes: Thermophilic; isolated from decomposing plant materials.
Host: Arachis
Thermoascus crustaceus (Apinis & Chesters) A.C. Stolk
Dactylomyces crustaceus Apinis & Chesters
Distr.: North America; Europe.
Notes: Isolated from debris and stored products in the Poaceae.
Host: Helianthus

Thermomyces Tsiklinsky. Hyphomycetes

Notes: Ellis (Dematiaceous Hyphomycetes. p. 58. 1971).
Thermomyces lanuginosus Tsiklinsky
Humicola lanuginosus (Griffon & Maubl.) Bunce
Distr.: Worldwide, most common in tropical regions.
Notes: Isolated from mouldy hay, leaf litter, soil, humans and various substrates in thermophilic habitats.
Hosts: Arachis, Carya, Gossypium, Helianthus, Metrosideros
Thermomyces stellatus (Bunce) Apinis
Humicola stellatus Bunce
Distr.: GA; Great Britain.
Notes: Isolated from mouldy hay and sunflower seeds.
Host: Helianthus

Therrya Sacc. Discomycetes, Rhytismatales

Notes: Reid & Cain (Canad. J. Bot. 39:1117. 1961).
Therrya fuckelii (Rehm) Kujala
Distr.: North America; temperate Europe.
Notes: On dying and dead limbs (Kokko & Shoemaker. Fungi Canadenses 48. 1974).
Host: Pinus
Therrya pini (Albertini & Schwein.) Höhn.
Coccophacidium pini (Albertini & Schwein.) Rehm
Colpomella pini Höhn. (anamorph)
Distr.: This species apparently occurs only on pines in Europe. Reports from North America are probably misidentifications of *Therrya fuckelii*.
Host: Pinus

Thielavia Zopf. Pyrenomycetes, Sordariales

Notes: Arx (Stud. Mycol. 8:5. 1975).
Thielavia basicola Zopf
Distr.: North America; Asia.
Notes: Most records probably refer to *T. terricola* (Gilman & Abbott) Emmons.
Hosts: Antirrhinum, Cucumis, Cytisus, Glycine, Gossypium, Nicotiana, Pisum, Robinia

Thielaviopsis Went. Hyphomycetes

Thielaviopsis basicola (Berk. & Broome) Ferraris
Trichocladium basicola (Berk. & Broome) J.W. Carmichael
Chalara elegans Nag Raj & Kendrick (synanamorph)
Distr.: Cosmopolitan.
Notes: An aggressive pathogen of various plants, commonly referred to as black root rot. This name is applicable only to the endoarthric state, while *Chalara elegans* is the name for the phialidic state. Subramanian (C.M.I. Descr. 170. 1968).
Hosts: Abelmoschus, Allium, Antirrhinum, Arachis, Armoracia, Atropa, Begonia, Catalpa, Cicer, Citrullus, Citrus, Cyclamen, Datura, Epipremnum, Euphorbia, Gerbera, Glycine, Gossypium, Ilex, Lathyrus, Lens, Linaria, Linum, Lupinus, Lycopersicon, Medicago, Nicotiana, Oxalis, Paeonia, Panax, Pelargonium, Phaseolus, Phlox, Pisum, Senecio, Sesamum, Syringa, Trifolium, Ulmus, Verbena, Vicia, Vigna, Viola

Thozetella Kuntze. Hyphomycetes

Thozetella cristata Pirozynski & C.S. Hodges
Distr.: NC, SC.
Notes: On fallen leaves (Pirozynski & Hodges. Canad. J. Bot. 51:168. 1973).
Hosts: Myrica, Persea

Thyridaria Sacc. Loculoascomycetes, Dothideales

Thyridaria californica Rehm
Distr.: CA.
Hosts: Purshia, Umbellularia
Thyridaria comptoniae (Ellis & Everh.) Berl. & Voglino
Diatrype comptoniae Ellis & Everh.
Pseudovalsa comptoniae (Ellis & Everh.) Ellis
Distr.: NJ.
Host: Comptonia
Thyridaria koae Petr.
Distr.: HI.
Host: Acacia
Thyridaria rubronotatum (Berk. & Broome) Sacc.
Distr.: ID; Europe.
Notes: On woody debris (Wehmeyer. Lloydia 4:250. 1941).
Hosts: Acer, Pinus
Thyridaria sambucina (Peck) Wehmeyer
Valsa sambucina Peck
Pseudovalsa sambucina (Peck) Sacc.
Distr.: Temperate North America.
Host: Sambucus
Thyridaria sorbi Dearn. & House, nom. inval.
Distr.: NY.
Host: Sorbus
Thyridaria texensis (Ellis & Everh.) Berl. & Voglino
Diatrype texensis Ellis & Everh.
Pseudovalsa texensis (Ellis & Everh.) Ellis
Distr.: Southeastern United States.
Host: Magnolia
Thyridaria viticola (Ellis & Everh.) Wehmeyer
Pseudovalsa viticola Ellis & Everh.
Aglaospora viticola (Ellis & Everh.) Farl.
Distr.: KS.

Host: Vitis

Thyridium Nitschke. Pyrenomycetes, Sphaeriales

Thyridium tuberculatum Rehm
Distr.: CA(type).
Host: Ceanothus
Thyridium vestitum (Fr.:Fr.) Fuckel
Fenestella vestita (Fr.:Fr.) Sacc.
Mycothyridium vestitum (Fr.:Fr.) E. Müller
Fenestella amorpha Ellis & Everh.
Fenestella rostrata (Fuckel) Sacc.
Thyridium rostratum Fuckel
Fenestella ulmicola Ellis & Everh.
Pleurocytospora vestita Petr. (anamorph)
Distr.: North America; Europe; New Zealand.
Notes: Saprophytic or weakly parastic on numerous hardwood trees and shrubs (Barr. Mycotaxon 18:152. 1983).
Hosts: Ailanthus, Berberis, Caragana, Ulmus
Thyridium vitis Ellis & Everh.
Distr.: KS(type).
Host: Vitis

Thyronectria Sacc. Pyrenomycetes, Hypocreales

Notes: Seeler (J. Arnold Arboretum 21:429. 1940).
Thyronectria austroamericana (Speg.) Seeler
Pleonectria austroamericana Speg.
Pleonectria denigrata G. Wint.
Thyronectria denigrata (G. Wint.) Seaver
Nectria sphaerospora Ellis & Everh.
Thyronectria sphaerospora (Ellis & Everh.) Seaver
Gyrostroma austroamericana G. Wint. (anamorph)
Distr.: North America; Brazil, Paraguay.
Notes: Bark canker of honey locust (Bedker & Wingfield. Trans. Brit. Mycol. Soc. 81:179. 1983).
Hosts: Albizia, Fraxinus, Gleditsia
Thyronectria balsamea (Cooke & Peck) Seeler
Nectria balsamea Cooke & Peck
Calonectria balsamea (Cooke & Peck) Sacc.
Scoleconectria balsmea (Cooke & Peck) Seaver
Gyrostroma sp. (anamorph)
Distr.: Northern North America.
Notes: On dead branches.
Hosts: Abies, Pinus
Thyronectria berolinensis (Sacc.) Seaver
Pleonectria berolinensis Sacc.
Nectria fenestrata Berk. & M.A. Curtis
Pleonectria fenestrata (Berk. & M.A. Curtis) Berl. & Voglino
Dendrodochium berolinense Wollenweb. (anamorph)
Distr.: United States; Europe.
Host: Ribes
Thyronectria chlorinella (Cooke) Seeler
Nectria chlorinella Cooke
Distr.: Southeastern United States.
Host: Ulmus
Thyronectria chrysogramma Ellis & Everh.
Distr.: North America.
Host: Ulmus
Thyronectria lonicerae Seeler
Distr.: Central United States.
Notes: On dead branches.
Hosts: Lonicera, Symphoricarpos
Thyronectria pyrrhochlora (Auersw.) Sacc.
Distr.: Eastern North America; Europe.
Hosts: Acer, Celtis, Diospyros, Rhus
Thyronectria xanthoxyli (Peck) Ellis & Everh.
Distr.: North America.
Hosts: Rhus, Zanthoxylum

Thysanophora Kendrick. Hyphomycetes

Thysanophora longispora Kendrick
Distr.: FL; Canada(Ontario, Quebec).
Notes: On recently-fallen needles (Kendrick. Canad. J. Bot. 39:826. 1961).
Host: Eichhornia

Tiarospora Sacc. & Él. Marchal. Coelomycetes

Tiarospora perforans (Roberge) Höhn.
Sphaeria perforans Roberge
Distr.: OR; Australia; Great Britain.
Notes: On grass culms (Sutton. The Coelomycetes. p. 61. 1980).
Hosts: Ammophila, Elymus

Tiarosporella Höhn. Coelomycetes

Notes: Sutton (The Coelomycetes. p. 58. 1980).
Tiarosporella abietis H.S. Whitney, J. Reid, & Pirozynski
Darkera abietis H.S. Whitney, J. Reid, & Pirozynski (teleomorph)
Distr.: Northern North America; Europe.
Notes: Whitney et al. (Canad. J. Bot. 53:3055. 1975).
Host: Abies
Tiarosporella paludosa (Sacc. & Fiori) Höhn.
Neottiospora arenaria Syd.
Distr.: MI, WI; Europe.
Notes: Nag Raj (Canad. J. Bot. 51:2470. 1973).
Hosts: Aristida, Calamovilfa, Carex, Sporobolus
Tiarosporella parca (Berk. & Broome) H.S. Whitney, J. Reid, & Pirozynski
Macrophoma parca (Berk. & Broome) Berl. & Voglino
Darkera parca H.S. Whitney, J. Reid, & Pirozynski (teleomorph)
Distr.: Western North America; Europe.
Host: Abies

Tilachlidium G. Preuss. Hyphomycetes

Tilachlidium brachiatum (Batsch:Fr.) Petch in E.A. Ellis
Clavaria brachiatum Batsch:Fr.
Isaria citrina Pers.
Pseudonectria tilachlidii W. Gams (teleomorph)
Distr.: Widespread in Europe and North America.
Notes: Although reported on pine, this species usually occurs on other fungi (Gams. Persoonia 8:329. 1975).
Host: Pinus

Tilletia Tul. & C. Tul. Basidiomycetes, Ustilaginales

Notes: Duran & Fischer (The genus *Tilletia*. 1961).
Tilletia asperifolia Ellis & Everh.
Tilletia eremophila Speg.
Distr.: Widespread.
Notes: On grasses.
Host: Muhlenbergia
Tilletia asperifolioides G. Fisch.
Distr.: Western United States.
Notes: On grasses.
Host: Muhlenbergia
Tilletia barclayana (Bref.) Sacc. & Syd. in Sacc.
Neovossia barclayana Bref.
Neovossia horrida (Takah.) Padwick & A. Khan
Tilletia horrida Takah.
Tilletia pennisetina Syd.
Tilletia pulcherrima Ellis & B.T. Galloway in G.P. Clinton
Distr.: Widespread.
Notes: On grasses. Some authors (Tullis & Johnson. Mycologia 44:773. 1952) prefer to place this species in *Neovossia*. Duran & Ainsworth (C.M.I. Descr. 75. 1965).
Hosts: Digitaria, Oryza, Panicum
Tilletia bromi (Brockm.) Brockm.
Ustilago bromi Brockm.
Tilletia bromi-tectorum Urries

Tilletia fusca Ellis & Everh.
Tilletia guyotiana Har.
Distr.: Widespread.
Notes: On grasses. Duran & Fischer (The genus *Tilletia*. p. 66. 1961) recognized this species as *T. fusca*.
Hosts: Bromus, Festuca, Poa, Vulpia
Tilletia buchloeana Kellerm. & Swingle
Distr.: North America.
Notes: On grasses.
Host: Buchloe
Tilletia caries (DC.) Tul. & C. Tul.
Tilletia tritici (Bjerk.) G. Wint. in Rabenh.
Distr.: Widespread.
Notes: On Poaceae (Vánky. Illustrated genera of smut fungi. p. 112. 1987; Mordue & Waller. C.M.I. Descr. 719. 1981).
Hosts: Agropyron, Arrhenatherum, Bromus, Festuca, Secale, Triticum
Tilletia cerebrina Ellis & Everh.
Tilletia airae A. Blytt
Distr.: Widespread.
Notes: On grasses.
Host: Deschampsia
Tilletia controversa Kühn in Rabenh.
Tilletia brevifaciens G. Fisch.
Distr.: Widespread.
Notes: On grasses (Waller & Mordue. C.M.I. Descr. 746. 1981; Parmelee. Fungi Canadenses 33. 1974).
Hosts: Aegilops, Agropyron, Arrhenatherum, Bromus, Elymus, Festuca, Hordeum, Koeleria, Lolium, Secale, Triticum
Tilletia corona Lams.-Scrib. in Tracy & Earle
Neovossia corona (Lams.-Scrib.) Massee
Distr.: United States.
Notes: On grasses.
Host: Leersia
Tilletia earlei D. Griffiths
Distr.: ND, SD.
Notes: On grasses.
Hosts: Agropyron, Elymus
Tilletia elymi Dietel & Holw.
Distr.: Northern hemisphere.
Notes: On grasses.
Host: Elymus
Tilletia eragrostidis G.P. Clinton & Ricker
Distr.: Southeastern United States.
Host: Eragrostis
Tilletia holci (Westend.) J. Schröt. in Cohn
Tilletia anthoxanthi A. Blytt
Tilletia rauwenhoffii A. Fisch. v. Waldh.
Distr.: Temperate northern hemisphere.
Notes: On Poaceae (Waller & Mordue. C.M.I. Descr. 747. 1983).
Hosts: Anthoxanthum, Holcus
Tilletia hyalospora Massee
Tilletia wilcoxiana D. Griffiths
Distr.: Widespread.
Notes: On grasses.
Host: Stipa
Tilletia laevis Kühn in Rabenh.
Tilletia foetens (Berk. & M.A. Curtis) J. Schröt. in Cohn
Ustilago foetens Berk. & M.A. Curtis
Tilletia foetida (Wallr.) Liro
Distr.: Widespread.
Notes: On Poaceae (Mordue & Waller. C.M.I. Descr. 720. 1981 as *T. foetida*). This combination of *T. foetida* was based on an illegitimate name (Mordue & Ainsworth. Mycol. Pap. 154:37. 1984).
Hosts: Agropyron, Elymus, Hordeum, Secale, Triticum
Tilletia maclaganii (Berk.) G.P. Clinton
Tilletia rotundata (Arth.) Massee
Ustilago rotundata Arth.
Distr.: Widespread.
Notes: On grasses.
Host: Panicum
Tilletia montana Ellis & Everh.
Distr.: Western United States.
Notes: On grasses.
Host: Muhlenbergia
Tilletia oklahomae Zundel
Distr.: Central United States.
Notes: On grasses.
Host: Aristida
Tilletia pallida G. Fisch.
Distr.: Widespread.
Notes: On grasses.
Host: Agrostis
Tilletia redfieldiae G.P. Clinton
Distr.: Central United States.
Notes: On grasses.
Host: Redfieldia
Tilletia rugispora Ellis
Distr.: Widespread.
Notes: On grasses.
Host: Paspalum
Tilletia sphaerococca (Rabenh.) A. Fisch. v.
Tilletia decipiens (Pers.) Körn.
Distr.: North America(rare); Europe.
Notes: On grasses (Mordue. C.M.I. Descr. 805. 1984).
Host: Agrostis
Tilletia sterilis Ule
Tilletia scrobiculata G. Fisch.
Distr.: Western United States.
Notes: On grasses.
Host: Poa
Tilletia texana W.H. Long in G.P. Clinton
Tilletia festuca-octoflorana Zundel
Distr.: Widespread.
Notes: On grasses.
Hosts: Festuca, Hordeum, Vulpia
Tilletia youngii G.P. Clinton & Zundel
Distr.: Southeastern United States.
Notes: On grasses.
Host: Alopecurus

Tilletiopsis Derx. Hyphomycetes

Tilletiopsis minor Nyland
Distr.: Western North America.
Notes: On leaves.
Hosts: Acer, Chrysanthemum, Taraxacum
Tilletiopsis washingtonensis Nyland
Distr.: Western North America.
Notes: On leaves.
Hosts: Acer, Althaea, Chrysanthemum, Helianthus, Rosa, Sedum, Stachys

Tinctoporellus Ryvarden. Basidiomycetes, Aphyllophorales

Notes: Gilbertson & Ryvarden (North American Polypores 2:729. 1987).
Tinctoporellus epimiltinus (Berk. & Broome) Ryvarden
Poria epimiltinus Berk. & Broome
Poria borbonica Pat.
Distr.: Pantropical, extending into warmer parts of North America.
Notes: On hardwoods.
Hosts: Liquidambar, Quercus

Titaeospora Bubák. Coelomycetes

Titaeospora equiseti (Desmaz.) Vassiljevsky
Septoria equiseti Desmaz. in Moug.
Gloeosporium equiseti Ellis & Everh.
Septogloeum equiseti Tehon non (Ellis & Everh.) Died., hom. illeg.
Septoria detospora Sacc.
Titaeospora detospora (Sacc.) Bubák
Stamnaria persoonii (Moug.) Fuckel (teleomorph)
Distr.: Widespread in the northern hemisphere.
Notes: On stems and leaves (Sutton. The Coelomycetes. p. 607. 1980).
Hosts: Equisetum, [Ilex], [Lagerstroemia]

Titanella Syd. & P. Syd. Loculoascomycetes, Pyrenulales

Pleamphisphaeria Höhn.
Notes: Barr (Mycotaxon 9:30. 1979).
Titanella macrospora (Speg.) Barr
Lophiostoma macrospora Speg.
Platystomum macrospora (Speg.) Chesters & A. Bell
Distr.: NJ; Italy(type).
Notes: On wood.
Host: Quercus
Titanella pelorospora (Dearn.) Barr
Amphisphaeria pelorospora Dearn.
Distr.: Eastern North America.
Host: Quercus

Togninia Berl. Pyrenomycetes, Calosphaeriales

Togninia minima (Tul. & C. Tul.) Berl.
Calosphaeria alnicola Ellis & Everh.
Distr.: Eastern North America; Europe.
Hosts: Berberis, Ligustrum

Tolyposporella Atk. Basidiomycetes, Ustilaginales

Notes: This genus was reviewed by Thirumalachar et al. (Mycologia 59:389. 1967) who concluded that only the type species belongs to the genus, but not all names have been redisposed.
Tolyposporella brunkii (Ellis & B.T. Galloway) G.P. Clinton
Distr.: PR, TX; Mexico.
Notes: Thirumalachar et al. (Mycologia 59:394. 1967) suggested that this species belongs in *Entyloma*.
Hosts: Andropogon, Bothriochloa
Tolyposporella chrysopogonis Atk.
Distr.: Central and southern United States.
Host: Sorghastrum

Tolyposporium Woronin ex J. Schröt. Basidiomycetes, Ustilaginales

Tolyposporium aterrimum (Tul. & C. Tul.) Dietel
Thecaphora aterrima Tul. & C. Tul.
Sorosporium atrum Peck
Distr.: North America; Asia; Europe.
Notes: On Cyperaceae.
Host: Carex
Tolyposporium junci (J. Schröt.) Woronin
Distr.: Northwestern United States.
Host: Juncus
Tolyposporium montiae (Rostr.) Rostr.
Sorosporium montiae Rostr.
Distr.: Western United States; Europe.
Notes: Fischer (Manual of the North American smut fungi. p. 126. 1953) suggested that this fungus is not a smut. Lindeberg (Symb. Bot. Upsal. 16:77. 1959).
Host: Montia

Tomentella Pat. Basidiomycetes, Aphyllophorales

Notes: Larsen (Mycol. Mem. 4. 1974).
Tomentella angulospora M. Larsen
Distr.: NY.
Notes: On wood.
Host: Acer
Tomentella asperula (P. Karst.) Höhn. & Litsch.
Distr.: Northern hemisphere.
Notes: On wood and plant debris.
Hosts: Juniperus, Populus, Thuja
Tomentella atrorubra (Peck) Bourd. & Galzin
Zygodesmus atroruber Peck
Hypochnus atroruber (Peck) Burt
Distr.: North America.
Notes: On wood.
Hosts: Acer, Castanea, Cedrus, Picea, Pinus, Populus, Quercus, Tsuga
Tomentella avellanea (Burt) Bourd. & Galzin
Hypochnus avellaneus Burt
Distr.: Temperate northern hemisphere.
Notes: On wood of conifers and litter.
Hosts: Abies, Picea, Pinus, Pseudotsuga
Tomentella bicolor (Atk. & Burt) Bourd. & Galzin
Hypochnus bicolor Atk. & Burt
Distr.: Eastern North America.
Notes: On wood.
Host: Acer
Tomentella botryoides (Schwein.) Bourd. & Galzin
Hypochnus botryoides (Schwein.) Burt
Zygodesmus bicolor Cooke & Ellis
Distr.: Northern hemisphere.
Notes: On wood.
Hosts: Acer, Betula, Castanea, Fagus, Liquidambar, Pinus, Populus, Quercus, Sabal, Thuja, Tsuga, Ulmus
Tomentella bresadolae (Brinkmann) Bourd. & Galzin
Hypochnus bresadolae Brinkmann in Bres.
Distr.: Temperate northern hemisphere.
Notes: On wood.
Hosts: Abies, Acer, Juniperus, Picea, Pinus
Tomentella brunneorufa M. Larsen
Distr.: Eastern United States.
Notes: On wood.
Hosts: Liriodendron, Pinus, Robinia
Tomentella bryophila (Pers.) M. Larsen
Tomentella pallidofulva (Peck) Litsch.
Hypochnus pallidofulvus (Peck) Burt
Tomentella pseudoferruginea Skovst.
Tomentella subferruginea (Burt) Donk
Hypochnus subferrugineus Burt
Distr.: Temperate northern hemisphere.
Notes: On wood.
Hosts: Abies, Acer, Alnus, Betula, Castanea, Fagus, Picea, Pinus, Populus, Quercus, Robinia, Thuja, Tsuga
Tomentella calcicola (Bourd. & Galzin) M. Larsen
Caldesiella calcicola (Bourd. & Galzin) M. Christiansen
Distr.: Northern hemisphere.
Notes: On wood.
Hosts: Acer, Alnus, Populus
Tomentella carbonaria M. Larsen
Distr.: NM.
Notes: On wood.
Host: Pinus
Tomentella chlorina (Massee) G. Cunn.
Caldesiella viridis (Albertini & Schwein.:Fr.) Pat.
Odontia viridis (Albertini & Schwein.:Fr.) Quél.
Tomentella viridis (Berk.) G. Cunn.
Distr.: Widespread.
Notes: On wood.
Hosts: Abies, Acer, Alnus, Fraxinus, Juglans, Larix, Picea, Pinus, Platanus, Populus, Quercus
Tomentella cinerascens (P. Karst.) Höhn. & Litsch.
Hypochnus cinerascens P. Karst.
Distr.: Temperate northern hemisphere.
Notes: On wood.
Hosts: Abies, Alnus, Betula, Picea, Populus
Tomentella clavigera Litsch. in Svrček
Distr.: Northern hemisphere.
Notes: On wood.
Host: Pinus
Tomentella coerulea (Bres.) Höhn. & Litsch.
Tomentella papillata Höhn. & Litsch.
Distr.: Temperate northern hemisphere.
Notes: On wood and litter.
Hosts: Acer, Pinus, Populus, Prosopis, Quercus, Tsuga
Tomentella crinalis (Fr.) M. Larsen
Caldesiella ferruginosa (Pers.:Fr.) Sacc.
Distr.: Temperate northern hemisphere.
Notes: On wood.
Hosts: Abies, Acer, Alnus, Fraxinus, Pinus, Populus, Quercus, Tsuga

Tomentella epigaea (Burt) M. Larsen
Distr.: Eastern North America.
Notes: On wood and plant debris.
Host: Acer
Tomentella ferruginea (Pers.:Fr.) Pat.
Hypochnus ferrugineus (Pers.:Fr.) Fr.
Grandinia coriaria Peck
Tomentella coriaria (Peck) Bourd. & Galzin
Hypochnus coriarius (Peck) Burt
Tomentella fusca (Pers.:Fr.) J. Schröt. in Cohn
Distr.: North America; Australasia; Eurasia.
Notes: On wood.
Hosts: Abies, Acer, Alnus, Betula, Fagus, Juglans, Juniperus, Liriodendron, Pinus, Platanus, Populus, Pseudotsuga, Quercus, Tilia, Yucca
Tomentella ferruginella Bourd. & Galzin
Distr.: Northern hemisphere.
Notes: On wood and soil.
Host: Populus
Tomentella fuscoferruginosa (Bres.) Litsch.
Distr.: Northern hemisphere.
Notes: On wood.
Hosts: Acer, Pinus, Populus, Quercus
Tomentella galzinii Bourd.
Distr.: Northern hemisphere.
Notes: On wood.
Hosts: Fagus, Quercus
Tomentella griseo-umbrina Litsch. in S. Lundell
Distr.: Northern hemisphere.
Notes: On wood.
Hosts: Pinus, Populus, Tsuga
Tomentella griseoviolacea Litsch.
Distr.: Northern hemisphere.
Notes: On wood.
Hosts: Abies, Liriodendron, Populus, Quercus
Tomentella jaapii (Bres.) Bourd. & Galzin, nom. nud.
Distr.: MN; France(type).
Host: Betula
Tomentella lateritia Pat.
Distr.: Temperate northern hemisphere.
Notes: On wood.
Hosts: Acer, Betula, Fagus, Picea, Pinus, Platanus, Populus, Prunus, Pseudotsuga, Quercus, Robinia, Tsuga, Ulmus
Tomentella molybdaea Bourd. & Galzin
Distr.: Northern hemisphere.
Notes: On wood.
Hosts: Abies, Populus, Quercus
Tomentella muricata (Ellis & Everh.) Wakef.
Distr.: Eastern United States.
Notes: On wood.
Hosts: Magnolia, Pinus, Quercus
Tomentella neobourdotii M. Larsen
Distr.: Temperate northern hemisphere.
Notes: On wood.
Hosts: Abies, Acer, Alnus, Betula, Carpinus, Fagus, Larix, Pinus, Populus, Quercus, Tsuga
Tomentella nitellina Bourd. & Galzin
Distr.: Western North America; France.
Notes: On wood.
Host: Picea
Tomentella ochracea (Sacc.) M. Larsen
Tomentella microspora (P. Karst.) Höhn. & Litsch.
Tomentella sparsa (Burt) Bourd. & Galzin
Hypochnus sparsus Burt
Distr.: Temperate northern hemisphere.
Notes: On wood, litter, and soil.
Hosts: Acer, Castanea, Larix, Picea, Pinus
Tomentella olivascens (Berk. & M.A. Curtis) Bourd. & Galzin
Hypochnus olivascens (Berk. & M.A. Curtis) Burt
Distr.: North America.
Notes: On wood.
Hosts: Acer, Betula, Castanea, Fagus, Picea, Pinus, Populus, Prunus, Quercus, Tsuga
Tomentella pilatii Litsch.
Distr.: Northern hemisphere.
Notes: On wood.
Host: Populus
Tomentella pilosa (Burt) Bourd. & Galzin
Hypochnus pilosus Burt
Distr.: North America; Australasia; Eurasia.
Notes: On wood.
Hosts: Acer, Betula, Liriodendron, Pinus, Platanus, Populus, Pseudotsuga, Quercus, Ulmus
Tomentella pirolae (Ellis & Halst.) M. Larsen
Distr.: NJ.
Notes: On wood.
Host: Pyrola
Tomentella punicea (Albertini & Schwein.:Fr.) J. Schröt. in Cohn
Hypochnus granulosus (Peck) Burt
Zygodesmus granulosus Peck
Distr.: North America; Australasia; Eurasia.
Notes: On wood.
Hosts: Acer, Betula, Fraxinus, Pinus, Populus, Quercus, Robinia, Thuja, Tsuga, Ulmus, Zea
Tomentella ramosissima (Berk. & M.A. Curtis) Wakef.
Tomentella fuliginea (Burt) Bourd. & Galzin
Hypochnus fuligineus Burt
Distr.: Temperate northern hemisphere.
Notes: On wood.
Hosts: Abies, Acer, Alnus, Larix, Picea, Pinus, Populus, Pseudotsuga, Quercus, Thuja
Tomentella rubiginosa (Bres.) Maire
Hypochnus rubiginosus Bres.
Tomentella subrubiginosa Litsch.
Distr.: Northern hemisphere.
Notes: On wood.
Hosts: Carex, Carya, Cymophyllus, Fagus, Fraxinus, Liriodendron, Picea, Pinus, Populus, Prunus, Quercus, Rhododendron, Thuja, Tilia
Tomentella ruttneri Litsch.
Distr.: Temperate northern hemisphere.
Notes: On wood and litter.
Hosts: Abies, Acer, Betula, Fagus, Magnolia, Picea, Pinus, Populus, Tsuga
Tomentella subcinerascens Litsch.
Distr.: NC; France; Germany.
Notes: On wood.
Host: Acer
Tomentella subclavigera Litsch.
Distr.: Northern hemisphere.
Notes: On wood.
Host: Betula
Tomentella sublilacina (Ellis & Holw.) Wakef.
Zygodesmus sublilacinus Ellis & Holw. in Arth. et al.
Distr.: Temperate and subtropical northern hemisphere.
Notes: On wood.
Hosts: Abies, Acer, Alnus, Betula, Castanea, Fagus, Juglans, Lycopodium, Picea, Pinus, Populus, Quercus, Thuja, Tsuga
Tomentella subtestacea Bourd. & Galzin
Distr.: Northern hemisphere.
Notes: On wood.
Host: Quercus
Tomentella subvinosa (Burt) Bourd. & Galzin
Hypochnus subvinosus Burt
Distr.: North America.
Notes: On wood.
Hosts: Acer, Betula, Fagus, Picea, Populus, Sassafras, Tilia
Tomentella terrestris (Berk. & Broome) M. Larsen
Tomentella badiofusca Bourd. & Galzin
Tomentella umbrinella Bourd. & Galzin
Distr.: Temperate northern hemisphere.
Notes: On wood.
Hosts: Abies, Acer, Castanea, Picea, Pinus, Populus, Quercus, Tsuga
Tomentella umbrinospora M. Larsen
Zygodesmus rubiginosus Peck
Distr.: Northern hemisphere.
Notes: On wood.
Hosts: Acer, Pinus, Populus, Quercus, Tilia, Ulmus
Tomentella violaceofusca (Sacc.) M. Larsen
Hypochnus spiniferus Burt
Tomentella trachychaetes (Ellis & Everh.) M. Larsen

Zygodesmus trachychaetes Ellis & Everh.
Distr.: Temperate northern hemisphere.
Notes: On wood.
Hosts: Acer, Betula, Fraxinus, Larix, Magnolia, Picea, Pinus, Populus, Quercus, Thuja
Tomentella viridescens (Bres. & Torrend) Bourd. & Galzin
Hypochnus viridescens Bres. & Torrend
Distr.: Temperate northern hemisphere.
Notes: On wood.
Hosts: Abies, Acer, Juglans, Pinus, Platanus, Populus
Tomentella viridula Bourd. & Galzin
Distr.: Temperate northern hemisphere.
Notes: On wood.
Hosts: Acer, Pinus, Platanus, Populus

Tomentellastrum (Bourd. & Galzin) Svrček. Basidiomycetes, Aphyllophorales

Tomentellastrum badium (Link) M. Larsen
Tomentella badia (Link) Stalpers
Tomentella atroviolacea Litsch.
Tomentella fimbriata M. Christiansen
Tomentellastrum floridanum (Ellis & Everh.) M. Larsen
Distr.: Northern hemisphere.
Notes: On wood.
Hosts: Alnus, Betula, Fagus, Pinus, Populus
Tomentellastrum brunneofirmum (M. Larsen) M. Larsen
Tomentella brunneofirma M. Larsen
Distr.: AZ.
Notes: On wood.
Host: Quercus
Tomentellastrum montanensis (M. Larsen) M. Larsen
Tomentella montanensis M. Larsen
Distr.: Western United States.
Notes: On wood.
Hosts: Alnus, Populus

Tomentellina Höhn. & Litsch. Basidiomycetes, Aphyllophorales

Tomentellina fibrosa (Berk. & M.A. Curtis) M. Larsen
Kneiffiella fibrosa (Berk. & M.A. Curtis) M. Larsen
Hypochnus canadensis Burt
Tomentella ferruginosa Höhn. & Litsch.
Hypochnus ferruginosus (Höhn. & Litsch.) Burt
Distr.: Temperate northern hemisphere.
Notes: On wood.
Hosts: Larix, Picea, Pinus, Populus, Pseudotsuga

Tomentellopsis Hjortstam. Basidiomycetes, Aphyllophorales

Notes: Hjorstem (Svensk Bot. Tidskr. 68:51. 1974).
Tomentellopsis echinospora (Ellis) Hjortstam
Hypochnus echinosporus (Ellis) Burt
Hypochnus pennsylvanicus Overh.
Distr.: Temperate northern hemisphere.
Notes: On conifers.
Host: Abies
Tomentellopsis zygodesmoides (Ellis) Hjortstam
Hypochnus zygodesmoides (Ellis) Burt
Distr.: Temperate northern hemisphere.
Notes: On conifers.
Host: Pinus

Topospora Fr. Coelomycetes

Topospora uberiformis (Fr.) Fr.
Sphaeria uberiformis Fr.
Distr.: North America; Europe.
Notes: On twigs (Sutton. The Coelomycetes. p. 416. 1980).
Host: Ribes

Torpedospora S.P. Meyers. Pyrenomycetes, Phyllachorales

Notes: Kohlmeyer & Kohlmeyer (Marine mycology. p. 372. 1979).
Torpedospora radiata S.P. Meyers
Distr.: In oceans throughout the world.
Notes: On intertidal wood, bark, test panels and other woody substrates.
Host: Cocos

Torula Pers.:Fr. Hyphomycetes

Torula conglutinata Corda
Distr.: OR; Germany(type).
Host: Pseudotsuga
Torula fuckelii Sacc.
Distr.: DE; Germany(type).
Host: Salix
Torula glutinosa Cooke & Harkn.
Distr.: CA.
Host: Eriodictyon
Torula granulosa Bubák & Serebrianikow
Distr.: CA; Soviet Union(type).
Host: Erechtites
Torula herbarum (Pers.:Fr.) Link
Distr.: Cosmopolitan, most common in temperate regions.
Notes: On plant debris, isolated from air and soil.
Hosts: Alnus, Arceuthobium, Bambusa, Carya, Impatiens, Juncus, Mesembryanthemum, Pinus, Yucca
Torula ligniperda (Willk.) Sacc.
Xenodochus ligniperda Willk.
Distr.: Eastern United States; Germany(type).
Notes: Hughes (Canad. J. Bot. 37:859. 1959) suggested that some North American records of this fungus may be *Trichocladium canadense*.
Hosts: Betula, [Liquidambar], [Liriodendron], [Tilia]
Torula marginata L.W.R. Jackson
Distr.: Southeastern United States.
Notes: Associated with brown patches on roots (Jackson. Phytopathology 35:97. 1945).
Host: Pinus
Torula sulphurea G. Preuss
Oospora sulphurea (G. Preuss) Sacc. & Voglino
Distr.: IA; Europe(type).
Host: Zea

Toxosporium Vuill. Hyphomycetes

Toxosporium camptospermum (Peck) Maubl.
Pestalotia camptospermum Peck
Monochaetia camptospermum (Peck) Sacc.
Toxosporium abietinum Vuill.
Pestalotia mycophaga Vuill.
Coryneum pinicola (Dearn.) Guba
Monochaetia pinicola Dearn.
Distr.: Temperate North America; Europe.
Notes: On conifer needles.
Hosts: Abies, Pinus
Toxosporium effusum (Peck) Sutton
Coryneum effusum Peck
Distr.: KS.
Host: Populus

Trabutia Sacc. & Roum. Pyrenomycetes, Phyllachorales

Trabutia erythrospora (Berk. & M.A. Curtis) Theiss. & Syd.
Rhytisma erythrospora Berk. & M.A. Curtis
Distr.: Southwestern United States.
Host: Quercus
Trabutia quercina (Rudolphi ex Fr.) Sacc. & Roum.
Baeumleria quercina (G. Arnaud) Petr. (anamorph)
Distr.: Southeastern United States; northern Africa; southern Europe.
Notes: Tar spot of oaks.
Host: Quercus

Trabutia rigida Berk. & Mont.
Distr.: FL.
Host: Quercus

Tracya Syd. & P. Syd. Basidiomycetes, Ustilaginales

Tracya lemnae (Setch.) Syd. & P. Syd.
Distr.: North America; Europe.
Notes: Vánky (Symb. Bot. Upsal. 24:148. 1985).
Host: Spirodela

Tracylla (Sacc.) Tassi. Coelomycetes

Notes: Nag Raj (Canad. J. Bot. 53:2435. 1975) discussed this genus erroneously spelling it as *Tracyella*.
Tracylla spartinae (Peck) Tassi
Leptothyrium spartinae Peck
Mitrorhizopeltis ciliata Batsch & A.R. Cavaliere
Distr.: Subtropical and tropical regions.
Hosts: Distichlis, Spartina

Trametes Fr. Basidiomycetes, Aphyllophorales

Notes: Gilbertson & Ryvarden (North American Polypores 2:731. 1987).
Trametes cervina (Schwein.) Bres.
Coriolellus cervinus (Schwein.) Kotlaba & Pouzar
Coriolus cervinus (Schwein.) Bondartsev
Polyporus cervinus (Schwein.) Steud.
Polyporus molliusculus Berk.
Distr.: Temperate northern hemisphere.
Notes: On hardwoods. In North America (Overholts. The Polyporaceae of the United States, Alaska and Canada. p. 328. 1953), this species was formerly known as *Polyporus biformis* (Fr. in Klotzsch) Fr., but *P. biformis* is the correct name for *P. pargamenus*.
Hosts: Platanus, Populus, Quercus, Salix
Trametes conchifer (Schwein.:Fr.) Pilát
Polyporus conchifer (Schwein.:Fr.) Fr.
Poronidulus conchifer (Schwein.:Fr.) Murrill
Boletus virgineus Schwein.
Distr.: Central and eastern North America.
Notes: On hardwoods, especially *Ulmus* spp.
Hosts: Acer, Ostrya, Ulmus
Trametes cubensis (Mont.) Sacc.
Polyporus cubensis Mont.
Distr.: Neotropical and subtropical regions.
Notes: On wood.
Hosts: Liquidambar, Quercus, Sabal
Trametes drummondii (Klotzsch) Ryvarden
Polyporus drummondii Klotzsch
Distr.: FL, LA.
Notes: On wood.
Host: Taxodium
Trametes ectypus (Berk. & M.A. Curtis) R.L. Gilbertson & Ryvarden
Polyporus ectypus Berk. & M.A. Curtis
Coriolus ectypus (Berk. & M.A. Curtis) Pat.
Distr.: Southeastern United States.
Notes: On hardwoods.
Hosts: Acer, Fraxinus
Trametes elegans (Spreng.:Fr.)Fr.
Daedalea elegans Spreng.:Fr.
Lenzites elegans (Spreng.:Fr.) Pat.
Polyporus aesculi Fr.:Fr.
Daedalea ambigua Berk.
Daedalea palisoti Fr.:Fr.
Daedalea repanda Pers.
Distr.: Central and eastern North America; Australasia; pantropical.
Notes: On hardwoods.
Hosts: Acer, Carya, Celtis, Diospyros, Fagus, Gleditsia, Magnolia, Planera, Platanus, Quercus, Salix, Ulmus
Trametes hirsuta (Wulfen:Fr.) Quél.
Polyporus hirsutus (Wulfen:Fr.)Fr.
Coriolus hirsutus (Wulfen:Fr.) Quél.
Polystictus hirsutus (Wulfen:Fr.) Fr.
Coriolus nigromarginatus (Schwein.) Murrill
Polyporus vellereus Berk.
Distr.: Temperate northern hemisphere.
Notes: Several forms and varieties have been described.
Hosts: Abies, Acer, Ailanthus, Albizia, Alnus, Betula, Carpinus, Carya, Castanea, Castanopsis, Catalpa, Celtis, Cercis, Cercocarpus, Citrus, Corylus, Dizygotheca, Eucalyptus, Euonymus, Fagus, Forsythia, Fraxinus, Ginkgo, Gleditsia, Holodiscus, Ilex, Juglans, Juniperus, Liquidambar, Liriodendron, Magnolia, Malus, Myrica, Nyssa, Ostrya, Persea, Philadelphus, Picea, Pinus, Platanus, Populus, Prunus, Pseudotsuga, Pyrus, Quercus, Rhus, Ribes, Robinia, Salix, Sassafras, Sideroxylon, Sorbus, Taxodium, Thuja, Tilia, Tsuga, Ulmus, Vitis
Trametes maxima (Mont.) A. David & Rajchenberg
Irpex maximus Mont.
Coriolus maximus (Mont.) Murrill
Cerrena maxima (Mont.) L. Hans.
Polyporus maximus (Mont.) Overh.
Distr.: Neotropical and subtropical regions.
Notes: On wood.
Host: Ulmus
Trametes membranacea (Sw.:Fr.) Kreisel
Polyporus tenuis (Sacc.) Overh.
Distr.: Southern United States.
Notes: On wood.
Host: Taxodium
Trametes pavonia (Hook.) Ryvarden
Polyporus pavonius (Hook.) Fr.
Distr.: Gulf Coast to Argentina.
Notes: On wood.
Host: Salix
Trametes pubescens (Schumach.:Fr.) Pilát
Coriolus pubescens (Schumach.:Fr.) Quél.
Polyporus pubescens (Schumach.:Fr.) Fr.
Trametes pusilla C.G. Lloyd
Coriolus velutinus (Fr.:Fr.) Quél.
Polyporus velutinus Fr.:Fr.
Distr.: Temperate northern hemisphere.
Notes: On hardwoods.
Hosts: Acer, Alnus, Betula, Carpinus, Castanea, Fagus, Liriodendron, Magnolia, Malus, Picea, Populus, Prunus, Quercus, Salix, Sorbus, Thuja, Tsuga
Trametes suaveolens (L.:Fr.) Fr.
Polyporus suaveolens (L.:Fr.) Fr.
Distr.: Circumboreal.
Notes: On hardwoods.
Hosts: Populus, Salix
Trametes subectypus (Murrill) R.L. Gilbertson & Ryvarden
Coriolus subectypus Murrill
Polyporus subectypus (Murrill) Bres.
Distr.: Southeastern United States.
Notes: On wood.
Hosts: Ficus, Liquidambar, Nyssa, Quercus
Trametes subnivosa Murrill
Distr.: Southcentral and southeastern United States.
Notes: On wood.
Host: Quercus
Trametes trogii Berk. in Trog
Funalia trogii (Berk.) Bondartsev & Singer
Trametes salicina Bres.
Distr.: Temperate northern hemisphere.
Notes: On hardwoods.
Hosts: Acer, Aesculus, Populus, Quercus, Salix
Trametes versicolor (L.:Fr.) Pilát
Polyporus versicolor (L.:Fr.) Fr.
Coriolus versicolor (L.:Fr.) Quél.
Polystictus versicolor (L.:Fr.) Fr.
Distr.: Cosmopolitan.
Notes: Several varieties based on variation in gross morphology have been described.
Hosts: Abies, Acer, Actinidia, Ailanthus, Albizia, Alnus, Arbutus, Betula, Calocedrus, Carpinus, Carya, Castanea, Castanopsis, Catalpa, Celtis, Cercis, Cercocarpus, Citrus, Cornus, Corylus, Crataegus, Cupressus, Diospyros, Eucalyptus, Fagus, Fraxinus, Ginkgo,

Gleditsia, Hydrangea, Ilex, Juglans, Juniperus, Kalmia, Lagerstroemia, Larix, Ligustrum, Liquidambar, Liriodendron, Malus, Melia, Nyssa, Ostrya, Oxydendrum, Persea, Philadelphus, Photinia, Picea, Pinus, Platanus, Populus, Prunus, Pseudotsuga, Psidium, Pyracantha, Pyrus, Quercus, Rhododendron, Rhus, Salix, Sambucus, Sassafras, Schinus, Sequoia, Sorbus, Syringa, Taxodium, Thuja, Tilia, Tsuga, Ulmus, Umbellularia, Vitis

Trametes villosa (Fr.:Fr.) Kreisel
Polyporus villosus Fr.:Fr.
Coriolus pinsitus (Fr.:Fr.) Pat.
Trametes pinsitus (Fr.:Fr.) Fidalgo & M. Fidalgo
Distr.: Widespread.
Notes: On wood.
Hosts: Acer, Chamaecyparis, Fraxinus, Gleditsia, Juniperus, Ligustrum, Prosopis

Tranzschelia Arth. Basidiomycetes, Uredinales

Notes: Bennell & Henderson (Trans. Brit. Mycol. Soc. 71:271. 1978).

Tranzschelia anemones (Pers.:Pers.) Nannf. in S. Lundell & Nannf.
Tranzschelia fusca Dietel
Polythelis pulsatillae Arth.
Tranzschelia suffusca (Holw.) Arth.
Distr.: Temperate northern hemisphere.
Notes: Autoecious.
Host: Anemone

Tranzschelia arthurii Tranzschel & Litv.
Distr.: Central North America.
Notes: Autoecious.
Host: Prunus

Tranzschelia cohaesa (W.H. Long) Arth.
Tranzschelia tucsonensis (Arth.) Dietel
Distr.: Southwestern United States.
Notes: Autoecious.
Host: Anemone

Tranzschelia discolor (Fuckel) Tranzschel & Litv.
Tranzschelia pruni-spinosae (Pers.:Pers.) Dietel var. *discolor* (Fuckel) Dunegan
Tranzschelia persicae (Pers.) Arth.
Tranzschelia punctata (Pers.) Arth.
Distr.: Cosmopolitan.
Notes: Heteroecious; aecial host - *Anemone* spp.; telial host - *Prunus* spp. Bolkan et al. (Pl. Dis. 69:485. 1985) established physiologic strains of this species.
Hosts: Anemone, Prunus

Tranzschelia discolor (Fuckel) Tranzschel & Litv. f. sp. ***dulcis*** Bolkan, Ogawa, Michailides & Kable
Distr.: CA.
Notes: Established for strains that attack almonds (Bolkan et al. Pl. Dis. 69:485. 1985).
Host: Prunus

Tranzschelia discolor (Fuckel) Tranzschel & Litv. f. sp. ***persica*** Bolkan, Ogawa, Michaelides & Kable
Distr.: CA(type).
Notes: Established for strains that attack peaches (Bolkan et al. Pl. Dis. 69:485. 1985).
Host: Prunus

Tranzschelia pruni-spinosae (Pers.:Pers.) Dietel
Puccinia pruni-spinosae Pers.:Pers.
Distr.: North America; temperate Europe.
Notes: Heteroecious; aecial host - Ranunculaceae; telial host - *Prunus* spp.
Hosts: Anemone, [Armeniaca], Hepatica, Prunus, Thalictrum

Tranzschelia thalictri (Chev.) Dietel
Polythelis thalictri Chev.
Distr.: Temperate northern hemisphere.
Notes: Autoecious.
Host: Thalictrum

Tranzschelia viornae (Arth.) Arth.
Distr.: TX.
Notes: Autoecious.
Host: Clematis

Trechispora P. Karst. emend. Liberta. Basidiomycetes, Aphyllophorales

Cristella Pat. emend. Donk
Notes: Liberta (Canad. J. Bot. 51:1871. 1973).

Trechispora alnicola (Bourd. & Galzin) Liberta
Grandinia alnicola Bourd. & Galzin
Distr.: Temperate northern hemisphere.
Notes: On hardwoods.
Host: Poa

Trechispora amianthina (Bourd. & Galzin) Liberta
Corticium amianthinum Bourd. & Galzin
Corticium crustulinum Burt non Bres., hom. illeg.
Distr.: Northern hemisphere.
Notes: On wood and soil.
Host: Pinus

Trechispora cohaerens (Schwein.) Jülich & Stalpers
Corticium confine Bourd. & Galzin
Trechispora confinis (Bourd. & Galzin) Liberta
Trechispora submicrospora (Litsch.) Liberta
Corticium submicrosporum Litsch.
Distr.: Temperate northern hemisphere.
Notes: On wood and plant debris.
Hosts: Abies, Alnus, Picea, Pinus, Populus, Quercus

Trechispora farinacea (Pers.:Fr.) Liberta
Hydnum farinaceum Pers.:Fr.
Cristella farinacea (Pers.:Fr.) Donk
Grandinia farinacea (Pers.:Fr.) Bourd. & Galzin
Odontia farinacea (Pers.:Fr.) Cooke & Quél. non Ces., hom. illeg.
Corticium araneosum (Höhn. & Litsch.) Bourd. & Galzin
Tomentella araneosum Höhn. & Litsch.
Trechispora caucasia (Parmasto) Liberta
Trechispora sphaerospora (Maire) Parmasto
Corticium sphaerosporum (Maire) Höhn. & Litsch.
Corticium submutabile Höhn. & Litsch.
Trechispora submutabilis (Höhn. & Litsch.) Parmasto
Distr.: Temperate northern hemisphere.
Notes: On well-rotted wood.
Hosts: Acer, Fraxinus, Juniperus, Malus, Phoenix, Picea, Pinus, Platanus, Populus, Pseudotsuga, Quercus, Tsuga

Trechispora fastidiosa (Pers.:Fr.) Liberta
Corticium fastidiosum (Pers.:Fr.) P. Karst.
Odontia alliacea (Weinm.) Fr.
Distr.: Eastern North America; Europe.
Notes: On wood and plant debris.
Hosts: Populus, Quercus, Salix, Tsuga

Trechispora lunata (Bourd. & Galzin) Jülich
Grandinia lunata Bourd. & Galzin
Athelopsis lunata (Bourd. & Galzin) Parmasto
Distr.: AZ; Europe.
Notes: On wood. Although Bourdot & Galzin (Hymenomycetes de France. p. 410. 1928) attributed this species to Romell, they were proposing the first validly published name.
Hosts: Pinus, Pseudotsuga

Trechispora microspora (P. Karst.) Liberta
Grandinia microspora P. Karst.
Athelia microspora (P. Karst.) R.L. Gilbertson
Corticium subnullum Burt
Distr.: Temperate northern hemisphere.
Notes: On wood and plant debris.
Hosts: Picea, Pinus

Trechispora mollusca (Pers.:Fr.) Liberta
Polyporus molluscus (Pers.:Fr.) Fr.
Poria mollusca (Pers.:Fr.) Cooke
Phlebiella candidissima (Schwein.) W.B. Cooke
Poria candidissima (Schwein.) Cooke
Trechispora candidissima (Schwein.) Bondartsev & Singer
Polyporus candidissimus Schwein.
Porotheleum subtile (Schrad.:Fr.) Fr.
Polyporus subtilis (Schrad.:Fr.) Fr.
Poria subtilis (Schrad.:Fr.) Bres.
Distr.: Widespread.
Notes: On wood.
Hosts: Abies, Acer, Alnus, Larix, Magnolia, Picea, Pinus, Populus, Pseudotsuga, Quercus, Thuja, Tsuga, Umbellularia

Trechispora pallidoaurantiaca R.L. Gilbertson & Budington
Distr.: North America.
Notes: On wood, usually of conifers (Gilbertson & Budington. Mycologia 62:677. 1970).
Hosts: Pinus, Populus
Trechispora praefocata (Bourd. & Galzin) Liberta
Corticium suffocatum Bourd. & Galzin
Distr.: AZ; Europe.
Notes: On wood and plant debris.
Host: Pinus
Trechispora sphaerocystis Burdsall & R.L. Gilbertson
Distr.: AZ.
Notes: On wood (Burdsall & Gilbertson. Mycotaxon 15:338. 1982).
Host: Juglans
Trechispora variseptata Burdsall & R.L. Gilbertson
Distr.: AZ.
Notes: On wood (Burdsall & Gilbertson. Mycotaxon 15:339. 1982).
Host: Fraxinus

Trematosphaeria Fuckel. Loculoascomycetes, Dothideales

Notes: Boise (Mycologia 75:666. 1983; Mycologia 77:230. 1985).
Trematosphaeria cariosa (Fairm.) Boise
Caryospora cariosa Fairm.
Distr.: NY.
Notes: On dead wood.
Host: Fagus
Trematosphaeria confusa (Garov.) Boise & D. Hawksworth in Boise
Verrucaria confusa Garov.
Lophiostoma langloisii Ellis & Everh.
Melanomma langloisii (Ellis & Everh.) Berl.
Distr.: Temperate regions.
Notes: On woody plants.
Hosts: Quercus, Salix
Trematosphaeria cornina Dearn., nom. nud.
Distr.: KY.
Notes: The original publication lacks a latin description.
Host: Cornus
Trematosphaeria pertusa (Pers.:Fr.) Fuckel
Sphaeria pertusa Pers.:Fr.
Melanomma pertusa (Pers.:Fr.) Sacc.
Trematosphaeria fraxini Ellis & Everh. non Richon, hom. illeg.
Distr.: Temperate regions.
Notes: On dead wood and bark.
Hosts: Fagus, Fraxinus, Populus

Tremella Pers.:Fr., nom. cons. Basidiomycetes, Tremellales

Notes: A number of species are mycoparasites, and are therefore not included here.
Tremella coalescens Olive
Distr.: LA.
Notes: On wood.
Host: Quercus
Tremella fimbriata Pers.:Fr.
Tremella foliacea Pers.:Fr.
Tremella frondosa Fr.:Fr.
Distr.: Widespread.
Notes: On wood. Lowy (Flora Neotropica 6:103. 1971).
Hosts: Acer, Alnus, Quercus
Tremella fuciformis Berk.
Distr.: Warmer parts of the United States, south into South America.
Notes: On wood.
Host: Quercus
Tremella mesenterica Retz.:Fr.
Tremella lutescens Pers.:Fr.
Distr.: Widespread.
Notes: On wood. Wong et al. (Mycologia 77:36. 1985).
Hosts: Acer, Alnus, Amelanchier, Castanea, Corylus, Fraxinus, Heteromeles, Liquidambar, Menziesia, Populus, Prunus, Quercus, Salix, Umbellularia

Tretopileus B.O. Dodge. Hyphomycetes

Tretopileus sphaerophorus (Berk. & M.A. Curtis) S.J. Hughes & Deighton in Deighton
Monotospora sphaerophora Berk. & M.A. Curtis
Tretopileus opuntiae B.O. Dodge
Distr.: FL; West Indies; Africa.
Notes: Deighton (Mycol. Pap. 78:2. 1960).
Host: Opuntia

Triblidium Rebent. Discomycetes, Ostropales

Notes: Also spelled "*Tryblidium*". Pirozynski & Cain (Canad. J. Bot. 44:657. 1966).
Triblidium alpinum (Hazsl.) Rehm
Distr.: OR; Europe.
Host: Vaccinium
Triblidium caliciiforme Rebent.
Blitridium caliciiforme (Rebent.) De Not.
Distr.: North America; Europe.
Host: Acer
Triblidium fatiscens Kauffm.
Distr.: OR(type).
Host: Amelanchier
Triblidium rufulum (Spreng.:Fr.) Ellis & Everh. var. **_microsporum_** Ellis & Everh.
Tryblidiella rufula (Spreng.:Fr.) Sacc. var. *microspora* (Ellis & Everh.) Rehm
Distr.: SC(type).
Host: Morus
Triblidium turgidulum W. Phillips & Harkn.
Distr.: CA(type).
Host: Keckiella

Trichaptum Murrill. Basidiomycetes, Aphyllophorales

Hirschioporus Donk
Notes: Gilbertson & Ryvarden (North American Polypores 2:767. 1987).
Trichaptum abietinum (Dickson:Fr.) Ryvarden
Coriolus abietinus (Dickson:Fr.) Quél.
Hirschioporus abietinus (Dickson:Fr.) Donk
Polystictus abietinus (Dickson:Fr.) Cooke
Polyporus abietinus (Dickson:Fr.) Fr.
Distr.: Circumboreal.
Notes: On wood of conifers. Many records of this species are actually *T. fuscoviolaceum* or *T. laricinum*. See notes under those species.
Hosts: Abies, Calocedrus, Chamaecyparis, Fagus, Juniperus, Larix, Picea, Pinus, Pseudotsuga, Sequoia, Taxodium, Thuja, Tsuga
Trichaptum biforme (Fr.) Ryvarden
Polyporus biformis Fr. in Klotzsch
Polystictus biformis (Fr.) Fr.
Coriolus pargamenus (Fr.) Pat.
Hirschioporus pargamenus (Fr.) Bondartsev & Singer
Polystictus pargamenus (Fr.) Fr.
Trametes pargamenus (Fr.) Kotlaba & Pouzar
Trichaptum pargamenus (Fr.) G. Cunn.
Polyporus sartwellii Berk. & M.A. Curtis in Berk.
Distr.: Widespread.
Notes: On wood, usually hardwoods.
Hosts: Acer, Alnus, Amelanchier, Arbutus, Betula, Carya, Castanea, Celtis, Cytisus, Diospyros, Fagus, Fraxinus, Gleditsia, Ilex, Juglans, Liquidambar, Nyssa, Picea, Pinus, Populus, Prunus, Pseudotsuga, Pyrus, Quercus, Rhododendron, Rhus, Robinia, Salix, Tilia, Tsuga
Trichaptum byssogenum (Jungh.) Ryvarden
Hirschioporus versatilis (Berk.) Imazeki
Polyporus versatilis (Berk.) Romell
Trametes versatilis Berk.
Distr.: Warmer parts of North America; Africa.
Notes: On wood.
Hosts: Juniperus, Pinus, Prosopis, Quercus, Taxodium
Trichaptum fuscoviolaceum (Ehrenb.:Fr.) Ryvarden
Hirschioporus fuscoviolaceus (Ehrenb.:Fr.) Donk

Irpex fuscoviolaceus (Ehrenb.:Fr.) Fr.
Distr.: Circumboreal.
Notes: On wood of conifers. Earlier North American treatments of *Trichaptum abietinum* (Overholts. The Polyporaceae of the United States, Alaska and Canada. p. 33. 1953) were broad, including the species concept of *T. fuscoviolaceum*. Hence, many reports of *T. abietinum* are actually *T. fuscoviolaceum*. Macrae (Canad. J. Bot. 45:1371. 1967); Magasi (Mem. New York Bot. Gard. 28:163. 1976).
Hosts: Abies, Pinus, Tsuga
Trichaptum laricinum (P. Karst.) Ryvarden
Lenzites laricinus P. Karst.
Hirschioporus laricinus (P. Karst.) Teramoto
Lenzites abietis C.G. Lloyd
Distr.: Circumboreal.
Notes: On wood of conifers. Earlier North American treatments of *Trichaptum abietinum* (Overholts. The Polyporaceae of the United States, Alaska and Canada. p. 333. 1953) were broad, including the species concept of *T. laricinum*. Hence, many reports of *T. abietinum* are actually *T. laricinum*. Macrae (Canad. J. Bot. 45:1371. 1967); Magasi (Mem. New York Bot. Gard. 28:163. 1976).
Hosts: Abies, Picea, Pinus, Pseudotsuga
Trichaptum sector (Ehrenb.:Fr.) Kreisel
Coriolus sector (Ehrenb.:Fr.) Pat.
Hirschioporus sector (Ehrenb.:Fr.) S. Teng
Polyporus sector (Ehrenb.:Fr.) Fr.
Polyporus floridanus Berk.
Distr.: Southcentral and southeastern United States.
Notes: On wood.
Hosts: Acer, Carya, Chamaecyparis, Liquidambar, Pseudotsuga, Quercus
Trichaptum subchartaceum (Murrill) Ryvarden
Hirschioporus subchartaceus (Murrill) Bondartsev & Singer
Polyporus subchartaceus (Murrill) Overh.
Distr.: North America.
Notes: On wood, especially of *Populus* spp.
Hosts: Betula, Populus

Trichia Haller. Myxomycetes

Trichia decipiens (Pers.) T. Macbr.
Trichia fallax Pers.
Distr.: Cosmopolitan.
Host: Salix
Trichia favoginea (Batsch) Pers.
Distr.: Temperate regions.
Host: Quercus

Trichobelonium (Sacc.) Rehm. Discomycetes, Helotiales

Trichobelonium rhabdocarpum (Ellis) Dennis
Erinella rhabdocarpa (Ellis) Sacc.
Distr.: NJ.
Notes: On dead twigs (Dennis. Kew Bull. 17:361. 1963).
Host: Comptonia

Trichocladium C. Harz. Hyphomycetes

Trichocladium asperum C. Harz
Distr.: Cosmopolitan.
Notes: In soil and on plant debris.
Host: Alnus

Trichoconis Clements. Hyphomycetes

Trichoconis caudata (Appel & Strunk) Clements
Pyricularia caudata Appel & Strunk
Distr.: Africa.
Notes: Reports from the United States are undoubtedly misidentifications of *Alternaria padwickii* (= *Trichoconis padwickii*) as explained by Ou (Rice diseases. p. 234. 1985).
Host: Oryza

Trichoderma Pers.:Fr. Hyphomycetes

Notes: Rifai (Mycol. Pap. 116:1. 1969); Bissett (Canad. J. Bot. 62. 924. 1984).
Trichoderma citrinoviride Bissett
Distr.: Temperate North America; New Guinea.
Notes: Isolated from wood, soil, and various organic and inorganic substrates.
Host: Pseudotsuga
Trichoderma hamatum (Bonord.) Bainier
Hypocrea sp. (teleomorph)
Distr.: Cosmopolitan.
Notes: In soil.
Hosts: Acer, Pseudotsuga
Trichoderma harzianum Rifai
Hypocrea sp. (teleomorph)
Distr.: Cosmopolitan.
Hosts: Pinus, Pseudotsuga
Trichoderma koningii Oudem.
Hypocrea ceramica Ellis & Everh. (teleomorph)
Distr.: Cosmopolitan.
Notes: In soil.
Hosts: Acer, Carya, Ipomoea, Platanus, Pyrus, Zea
Trichoderma longibrachiatum Rifai
Trichoderma reesei E. Simmons in H.E. Bigelow & E. Simmons
Distr.: Temperate regions; Solomon Islands.
Notes: Isolated from various kinds of organic matter.
Host: Pseudotsuga
Trichoderma polysporum (Link) Rifai
Sporotrichum polysporum Link
Distr.: Widespread.
Notes: Isolated from soil and various kinds of plant debris (Domsch et al. Compendium of soil fungi. p. 801. 1980).
Host: Pseudotsuga
Trichoderma saturnisporum T. Hammill
Distr.: GA, OR.
Notes: Isolated from forest soil and wood (Hammill. Mycologia 62:112. 1970).
Host: Pseudotsuga
Trichoderma viride Pers.:Fr.
Trichoderma lignorum Tode
Hypocrea sp. (teleomorph)
Distr.: Cosmopolitan.
Notes: On a wide variety of organic substrates.
Hosts: Abies, Acacia, Acer, Allium, Alnus, Ananas, Arachis, Arceuthobium, Arctostaphylos, Betula, Cheirodendron, Citrus, Cucumis, Fagus, Gossypium, Hordeum, Juncus, Liriodendron, Metrosideros, Narcissus, Oryza, Pinus, Pseudotsuga, Quercus, Rhizophora, Rhododendron, Saccharum, Sambucus, Sassafras, Trifolium, Triticum, Zea

Trichodothis Theiss. & Syd. Loculoascomycetes, Dothideales

Trichodothis comata (Berk. & Ravenel) Theiss. & Syd.
Asterina comata Berk. & Ravenel
Distr.: Southeastern United States.
Notes: On living leaves.
Host: Magnolia

Tricholomopsis Singer. Basidiomycetes, Agaricales

Tricholomopsis bella A.H. Sm.
Distr.: Western North America.
Hosts: Abies, Pseudotsuga
Tricholomopsis decora (Fr.:Fr.) Singer
Clitocybe decora (Fr.:Fr.) Gill.
Distr.: Circumboreal.
Notes: On wood of conifers.
Hosts: Abies, Populus, Taxodium, Tsuga
Tricholomopsis radicata (Peck) Singer
Pleurotus umbonatus Peck
Distr.: Northeastern North America.
Host: Pseudotsuga

Tricholomopsis rutilans (Schaeff.:Fr.) Singer
Tricholoma rutilans (Schaeff.:Fr.) P. Kumm.
Distr.: Temperate northern hemisphere.
Notes: On conifer wood.
Hosts: Abies, Picea, Pinus, Pseudotsuga
Tricholomopsis sulfureoides (Peck) Singer
Distr.: Northeastern North America.
Notes: On wood of conifers.
Host: Tsuga
Tricholomopsis thompsoniana (Murrill) A.H. Sm.
Distr.: Temperate North America.
Notes: On conifer wood.
Host: Pinus

Trichomerium Speg. Loculoascomycetes, Dothideales

Trichomerium grandisporum (Ellis & G. Martin) Batista & Cif.
Capnodium grandisporum Ellis & G. Martin
Distr.: FL.
Notes: Reynolds (Mycotaxon 14:192. 1982).
Hosts: Gelsemium, Myrica

Trichopeltis Speg. Loculoascomycetes, Dothideales

Trichopeltis rhyacoides F. Stevens
Distr.: HI.
Host: Alyxia

Trichopeziza Fuckel. Discomycetes, Helotiales

Notes: This genus is considered a synonym of *Belonidium*, but not all names have been redisposed.
Trichopeziza agrostina (Peck) Sacc.
Lachnella agrostina (Peck) Seaver
Distr.: NY(type).
Host: Calamagrostis

Trichopezizella Raitviir. Discomycetes, Helotiales

Notes: Haines (Mycologia 66:213. 1974).
Trichopezizella barbata (Kunze:Fr.) Raitviir
Dasyscyphus barbatus (Kunze:Fr.) Massee
Distr.: Temperate North America; Europe.
Notes: On woody stems.
Hosts: Myrica, Oplopanax, Spiraea
Trichopezizella horridula (Desmaz.) Raitviir
Trichopeziza horridula (Desmaz.) Sacc.
Lachnella insularis Speg.
Distr.: North and South America; temperate Europe.
Host: Bromus
Trichopezizella nidulus (J.C. Schmidt & Kunze) Raitviir var. ***onocleae*** J. Haines
Distr.: NY.
Host: Onoclea
Trichopezizella nidulus (J.C. Schmidt & Kunze) Raitviir var. ***setigera*** (W. Phillips) J. Haines
Lachnum setigerum (W. Phillips) Rehm
Distr.: Western United States; Central Asia.
Hosts: Arnica, Oplopanax, Polygonum, Veratrum
Trichopezizella relicina (Fr.:Fr.) Raitviir
Dasyscyphus longipilus (Peck) Sacc.
Distr.: Temperate northern hemisphere.
Notes: On herbaceous and woody plants.
Hosts: Delphinium, Eupatorium, Sambucus
Trichopezizella relicina (Fr.) Raitviir var. ***macrospora*** (Raitviir) J. Haines
Trichopezizella macrospora Raitviir
Distr.: Western United States; Soviet Union.
Hosts: Aconitum, Arnica, Heracleum, Populus, Veratrum

Trichoscyphella Nannf. Discomycetes, Helotiales

Notes: This genus is considered a synonym of *Lachnellula*, but not all names have been redisposed.
Trichoscyphella tenuipilosa Cash
Distr.: WA.
Host: Pinus

Trichoseptoria Cavara. Coelomycetes

Trichoseptoria fructigena Maubl.
Distr.: NY; France(type).
Notes: Soft rot of apples.
Host: Malus

Trichosphaeria Fuckel. Pyrenomycetes, Trichosphaeriales

Trichosphaeria cupressina Syd.
Distr.: NJ(type).
Host: Chamaecyparis
Trichosphaeria pilosa (Pers.:Fr.) Fuckel
Sphaeria pilosa Pers.:Fr.
Distr.: GA; Europe; Great Britain.
Host: Quercus

Trichosporum Fr. Hyphomycetes

Notes: Hughes (Canad. J. Bot. 36:821. 1958) indicated that this was an illegitimate name based on the type of *Acladium* Link. "*Trichosporium*" is a commonly used orthographic varient.
Trichosporum falcatae Dearn. & Barth.
Distr.: TX.
Host: Amphicarpaea
Trichosporum pinicola Dearn. & Barth.
Distr.: ID(type).
Host: Pinus
Trichosporum symbioticum E. Wright
Distr.: CA.
Notes: Associated with *Scolytus ventralis* (Wright. J. Agric. Res. 50:52 5.1935).
Host: Abies

Trichothallus F. Stevens. Hyphomycetes

Trichothallus hawaiiensis F. Stevens
Plokamidomyces sp. (synanamorph)
Trichopeltheca stevensii S.J. Hughes (teleomorph)
Distr.: HI, PR.
Notes: Sooty mold on leaves.
Hosts: Alyxia, Broussaisia, Cibotium, Cladium, Clermontia, Elaphoglossum, Freycinetia, Metrosideros, Pelea, Phyllostegia, Psychotria, Rubus, Scaevola, Smilax

Trichothecium Link:Fr. Hyphomycetes

Notes: Rifai & Cooke (Trans. Brit. Mycol. Soc. 49:147. 1966).
Trichothecium roseum (Pers.:Fr.) Link
Cephalothecium roseum Corda
Distr.: Cosmopolitan.
Notes: On a wide variety of organic substrates.
Hosts: Alnus, Apium, Arachis, Arbutus, Cannabis, Carya, Coffea, Cucumis, Cucurbita, Cydonia, Daucus, Ficus, Gossypium, Helianthus, Ilex, Juglans, Lycopersicon, Malus, Pelea, Phaseolus, Pinus, Platanus, Prunus, Pseudotsuga, Pteris, Pyrus, Quercus, Rhododendron, Solanum, Sorghum, Triticum, Vaccinium, Vitis, Zea

Trichothyrina (Petr.) Petr. Loculoascomycetes, Dothideales

Trichothyrina alpestris Petr.
Microthyrium culmigenum Syd.
Distr.: AK; Europe.
Notes: On various grasses.

Hosts: Agrostis, Deschampsia, Festuca, Phleum, Poa
Trichothyrina tracyi (Peck) Barr in Barr et al.
Asterula tracyi Peck
Distr.: MS.
Notes: Barr et al. (New York State Mus. Bull. 459:46. 1986).
Hosts: Borreria, Spermacoce

Trichothyrium Speg. Loculoascomycetes, Dothideales

Trichothyrium reptans (Berk. & M.A. Curtis) S.J. Hughes
Asterina reptans Berk. & M.A. Curtis
Trichopeltis reptans (Berk. & M.A. Curtis) Speg.
Distr.: Tropical regions.
Notes: Müller & Arx (Beitr. Kryptogamenfl. Schweiz 11(2):55. 1962); Farr (Acta Amaz., Suppl. 15:33. 1985).
Hosts: Alyxia, Cheirodendron, Cladium, Clermontia, Cyanea, Elaeocarpus, Gouldia, Kadua, Metrosideros, Myrsine, Pelea, Peperomia, Psidium, Psychotria, Smilax, Vaccinium

Tricladium Ingold. Hyphomycetes

Tricladium angulatum Ingold
Distr.: Aquatic habitats in temperate regions.
Notes: On decaying, submerged debris.
Host: Quercus

Trimmatostroma Corda. Hyphomycetes

Trimmatostroma americana Thuem.
Distr.: United States.
Notes: On rotting wood.
Hosts: Prunus, Salix
Trimmatostroma brencklei Sacc.
Distr.: ND.
Host: Rosa
Trimmatostroma eriodictyonis (Dearn. & Barth.) M.B. Ellis
Coniothecium eriodictyonis Dearn. & Barth.
Distr.: CA.
Host: Eriodictyon
Trimmatostroma salicis Corda
Distr.: ND; Europe(type).
Notes: On twigs and branches (Ellis. Dematiaceous Hyphomycetes. p. 41. 1971).
Host: Salix

Triphragmium Link in Willd. Basidiomycetes, Uredinales

Triphragmium ulmariae (R. Hedw.) Link
Distr.: Widespread in the northern hemisphere.
Notes: Autoecious.
Host: Filipendula

Tripospermum Speg. Hyphomycetes

Tripospermum myrti (Lind) S.J. Hughes
Triposporium myrti Lind
Distr.: NC, WI; Europe.
Notes: On twigs and leaves of Myrtaceae and various plants (Ellis. Dematiaceous Hyphomycetes. p. 86. 1971).
Hosts: Acer, Malus

Triposporium Corda. Hyphomycetes

Triposporium elegans Corda
Distr.: AK; Europe.
Notes: On dead wood and herbaceous stems (Ellis. Dematiaceous Hyphomycetes. p. 137. 1971).
Host: Sambucus

Triscelophorus Ingold. Hyphomycetes

Triscelophorus monosporus Ingold
Distr.: Tropical regions.
Notes: On decaying, submerged debris.
Hosts: Acacia, Ilex, Psidium, Quercus

Tritirachium Limber. Hyphomycetes

Notes: De Hoog (Stud. Mycol. 1:19. 1972).
Tritirachium dependens Limber
Distr.: Widespread.
Notes: On roots and bulbs of monocotyledonous plants.
Hosts: Gladiolus, Lilium
Tritirachium oryzae (Vincens) De Hoog
Beauveria oryzae Vincens
Tritirachium roseum Van Beyma
Distr.: Worldwide, most common in tropical regions.
Notes: Isolated from various organic substrates.
Host: Zea

Trochila Fr. Discomycetes, Helotiales

Notes: Greenhalgh & Morgan-Jones (Trans. Brit. Mycol. Soc. 47:311. 1964).
Trochila craterium (DC.) Fr.
Distr.: AK; Europe.
Host: Cassiope
Trochila ilicina (Nees:Fr.) Greenhalgh & Morgan-Jones
Trochila ilicis (Chev.) H. Crouan
Distr.: Temperate regions.
Host: Ilex

Troposporella P. Karst. Hyphomycetes

Troposporella fumosa P. Karst.
Distr.: HI; Europe.
Notes: On bark (Ellis. Dematiaceous Hyphomycetes. p. 85. 1971).
Host: Eucalyptus

Troposporium Harkn. Hyphomycetes

Troposporium album Harkn.
Distr.: CA.
Notes: On dead stems, wood, and bark (Peek & Solheim. Mycologia 50:851. 1958).
Host: Corylus

Trullula Ces. Hyphomycetes

Notes: Sutton (The Coelomycetes. p. 34. 1980).
Trullula junci Cooke & Harkn.
Distr.: CA.
Host: Juncus

Truncatella Steyaert. Coelomycetes

Truncatella angustata (Pers.) S.J. Hughes
Pestalotia lignicola Cooke
Pestalotia truncata Lév.
Truncatella truncata (Lév.) Steyaert
Distr.: Cosmopolitan.
Notes: Sutton (The Coelomycetes. p. 267. 1980).
Hosts: Eucalyptus, Picea, Rhododendron, Rubus, Salix

Tryblidaria (Sacc.) Sacc. Loculoascomycetes, Patellariales

Tryblidaria fenestrata (Cooke & Peck) Barr
Patellaria fenestrata Cooke & Peck
Blitridium fenestratum (Cooke & Peck) Sacc.
Triblidium fenestratum (Cooke & Peck) Kuntze
Distr.: NY(type).
Notes: Barr et al. (New York State Mus. Bull. 459:22. 1986).
Host: Populus

Tryblidaria washingtonensis Cash
Distr.: WA(type).
Host: Chamaecyparis

Tryblidiella Sacc. Loculoascomycetes, Patellariales

Notes: This genus is considered a synonym of *Rhytidhysteron* (Samuels & Müller. Sydowia 32:277. 1979), but not all names have been redisposed.
Tryblidiella clavispora (Peck) Berl. & Voglino
Triblidium clavispora Peck
Patellaria clavispora (Peck) Sacc.
Distr.: NY(type).
Notes: Muthappa (Mycologia 62:98. 1970) provided a detailed study of the morphology of this species.
Hosts: Populus, Salix
Tryblidiella nigrocinnabarina (Schwein.) Rehm
Distr.: AL(type).
Host: Morus

Tryblidiopsis P. Karst. Discomycetes, Helotiales

Tryblidiopsis pinastri (Pers.:Fr.) P. Karst.
Tympanis pinastri (Pers.:Fr.) Tul. & C. Tul.
Tympanis buchsii (Henn.) Rehm
Tryblidiopycnis pinastri Höhn. (anamorph)
Distr.: Circumboreal.
Notes: On dead and dying branches of conifers, especially common on *Picea* spp. The name *Tympanis pinastri* has been applied indiscriminately to many *Tympanis*-like fungi on conifers. Groves (Canad. J. Bot. 30:642. 1952).
Hosts: Abies, Picea, Pinus, Pseudotsuga

Tubakia Sutton. Coelomycetes

Tubakia dryina (Sacc.) Sutton
Leptothyrium dryinum Sacc.
Actinopelte dryina (Sacc.) Höhn.
Actinopelte americana Höhn.
Leptothyrium castanicola Ellis & Everh.
Actinothyrium gloeosporioides Tehon
Leptothyrella liquidambaris Tehon & G.L. Stout
Pirostoma nyssae Tehon
Distr.: Temperate North America; Europe.
Notes: Glawe & Crane (Mycotaxon 29:101. 1987).
Hosts: Acer, Carya, Castanea, Cercis, Fraxinus, Liquidambar, Nyssa, Quercus, Rhus, Sassafras, Ulmus

Tubaria (Wm.G. Sm.) Gill. Basidiomycetes, Agaricales

Tubaria pellucida (Bull.:Fr.) Gill.
Distr.: Temperate northern hemisphere.
Notes: On woody debris.
Hosts: Rosa, Salix

Tubercularia Tode, nom. cons. Hyphomycetes

Notes: Seifert (Stud. Mycol. 27:95. 1985).
Tubercularia atomospora Sacc.
Distr.: MT, WA.
Host: Picea
Tubercularia fici Edgerton
Distr.: Southeastern United States.
Notes: Cankers around fruit scars.
Host: Ficus
Tubercularia grayana (Sacc. & Ellis) Seifert
Dendrostilbella ulmi Dearn.
Nectria canadensis Ellis & Everh. (teleomorph)
Distr.: Northeastern North America.
Notes: On wood.
Host: Ulmus
Tubercularia hymenuloides (Sacc.) Höhn.
Dendrodochium hymenuloides Sacc.
Myxosporium hymenuloides (Sacc.) Höhn.
Distr.: OK; Europe(type).
Notes: Weindlmayr (Sydowia 18:31. 1964).
Host: Ulmus
Tubercularia lateritia (Berk.) Seifert
Dendrostilbella bonarii E.F. Morris
Polycephalum subaurantiacum Peck
Nectria pseudotrichia Berk. & M.A. Curtis (teleomorph)
Distr.: Subtropical and tropical regions.
Notes: On bark and wood.
Host: Carya
Tubercularia menispermi Fr.
Distr.: DE; Europe(type).
Host: Menispermum
Tubercularia nigricans (Bull.) Link
Distr.: Central and eastern United States; Europe.
Hosts: Ampelopsis, Fraxinus, Ulmus
Tubercularia smaragdina (Albertini & Schwein.:Fr.) Seifert
Graphium smaragdina (Albertini & Schwein.:Fr.) Sacc.
Distr.: Northeastern North America; Japan, probably more widely distributed.
Notes: On decorticated wood.
Host: Acer
Tubercularia ulmea J.C. Carter
Distr.: IL.
Notes: Canker and dieback of Siberian elm (Carter. Phytopathology 37:246. 1947).
Hosts: Elaeagnus, Rhamnus, Ulmus
Tubercularia vulgaris Tode:Fr.
Tubercularia ailanthi Cooke
Tubercularia hamata Ellis & Everh.
Tubercularia miniata Earle
Tubercularia sambuci Corda
Nectria cinnabarina (Tode:Fr.) Fr. (teleomorph)
Distr.: Cosmopolitan, most common in temperate regions.
Hosts: Acer, Ailanthus, Albizia, Alnus, Celtis, Fagus, Ficus, Hibiscus, Liquidambar, Quercus, Robinia, Salix, Sambucus, Ulmus, Vitis

Tuberculariella Höhn. Coelomycetes

Notes: This genus is considered a synonym of *Cryptosporiopsis*, but not all names have been redisposed.
Tuberculariella ips J.G. Leach, L.W. Orr, & C. Christensen
Distr.: MN(type).
Notes: Blue stain of timber; associated with bark beetles.
Host: Pinus

Tuberculina Sacc. Hyphomycetes

Notes: Gardner et al. (Mycologia 71:848. 1979).
Tuberculina maxima Rostr.
Distr.: Widespread.
Notes: Associated with various rust fungi, particularly the spermatial and/or aecial stages of various *Cronartium* spp.
Host: Pinus

Tubeufia Penz. & Sacc. Loculoascomycetes, Dothideales

Notes: Rossman (Mycol. Pap. 157:37. 1987).
Tubeufia cylindrothecia (Seaver) Höhn.
Ophionectria cylindrothecia Seaver
Distr.: Cosmopolitan.
Notes: Often on monocotyledonous debris, associated with other fungi.
Host: Zea
Tubeufia paludosa (P.L. Crouan & H. Crouan) A.Y. Rossman
Distr.: Temperate regions.
Notes: On rotten wood, often in association with other fungi.
Host: Zea
Tubeufia pezizula (Berk. & M.A. Curtis) Barr
Herpotrichia pezizula (Berk. & M.A. Curtis) Ellis & Everh.

Lasiosphaeria pezizula (Berk. & M.A. Curtis) Sacc.
Thaxteriella pezizula (Berk. & M.A. Curtis) Petr.
Thaxteriella corticola Petr.
Melanomma helicophilum (Cooke) Sacc.
Helicoma muelleri Corda (anamorph)
Distr.: HI, eastern North America; West Indies; South America; Africa; Europe.
Notes: Barr (Mycotaxon 12:157. 1980), Pirozynski (Mycol. Pap. 129:26. 1972).
Hosts: Acer, Arecaceae, Cornus, Diospyros, Fagus, Liquidambar, Liriodendron, Nyssa, Parthenocissus, Pinus, Prunus, Quercus, Salix, Vitis
Tubeufia scopula (Cooke & Peck) Barr
Acanthostigma scopula (Cooke & Peck) Peck
Lasiosphaeria scopula (Cooke & Peck) Sacc.
Helicosporium aureum (Corda) Linder (anamorph)
Distr.: Temperate regions.
Notes: On decorticated coniferous wood. Barr (Mycotaxon 12:164. 1980).
Host: Pinus

Tubulicium Oberwinkler. Basidiomycetes, Aphyllophorales

Tubulixenasma Parmasto
Notes: Oberwinkler (Sydowia 19:53. 1965).
Tubulicium capitatum (D.P. Rogers & Boquiren) Burdsall & Nakasone
Epithele capitata D.P. Rogers & Boquiren in Boquiren
Distr.: North America.
Notes: On wood (Burdsall & Nakasone. Mycotaxon 17:265. 1982).
Host: Phoenix
Tubulicium clematidis (Bourd. & Galzin) Oberwinkler
Peniophora clematidis Bourd. & Galzin
Litschauerella clematidis (Bourd. & Galzin) J. Eriksson & Ryvarden
Xenasma clematidis (Bourd. & Galzin) Liberta
Peniophora abietis (Bourd. & Galzin) Sartory & Maire
Hypochnus albus Burt
Distr.: Temperate northern hemisphere.
Notes: On wood.
Host: Juniperus
Tubulicium vermifera (Bourd.) Oberwinkler
Peniophora vermifera Bourd.
Epithele vermifera (Bourd.) Boquiren
Tubulicrinis vermifera (Bourd.) M. Christiansen
Tubulixenasma vermiferum (Bourd.) Parmasto
Xenasma veriferum (Bourd.) Liberta
Distr.: Widespread.
Notes: On wood.
Host: Gaultheria

Tubulicrinis Donk. Basidiomycetes, Aphyllophorales

Tubulicrinis accedens (Bourd. & Galzin) Donk
Peniophora accedens Bourd. & Galzin
Distr.: Cooler parts of temperate northern hemisphere.
Notes: On wood.
Hosts: Pinus, Pseudotsuga, Tsuga
Tubulicrinis angustus (D.P. Rogers & Weresub) Donk
Peniophora angusta D.P. Rogers & Weresub in Weresub
Distr.: Temperate northern hemisphere, cooler parts.
Notes: On wood.
Hosts: Fagus, Pinus, Quercus, Tsuga
Tubulicrinis calothrix (Pat.) Donk
Peniophora calothrix (Pat.) D.P. Rogers & H. Jacks.
Distr.: Temperate northern hemisphere.
Notes: On wood.
Hosts: Pinus, Pseudotsuga, Quercus, Tsuga
Tubulicrinis chaetophorus (Höhn.) Donk
Peniophora chaetophora (Höhn.) Höhn. & Litsch.
Peniophora dissoluta Overh.
Distr.: Temperate northern hemisphere.
Notes: On wood.
Hosts: Pinus, Populus, Tsuga
Tubulicrinis glebulosus (Bres.) Donk
Peniophora glebulosa Bres.
Peniophora gracillima Ellis & Everh. ex D.P. Rogers & H. Jacks.
Tubulicrinis gracillimus (Ellis & Everh. ex D.P. Rogers & H. Jacks.) G. Cunn.
Distr.: Widespread.
Notes: On conifers.
Hosts: Picea, Pinus, Pseudotsuga, Thuja, Tsuga
Tubulicrinis hamatus (H. Jacks.) Donk
Peniophora hamata H. Jacks.
Distr.: Northern hemisphere.
Notes: On wood.
Host: Pseudotsuga
Tubulicrinis inornatus (H. Jacks. & D.P. Rogers) Donk
Peniophora inornata H. Jacks. & D.P. Rogers in H. Jacks.
Distr.: Temperate northern hemisphere.
Notes: On wood.
Hosts: Pinus, Pseudotsuga
Tubulicrinis medius (Bourd. & Galzin) Oberwinkler
Peniophora media Bourd. & Galzin
Distr.: Northern hemisphere.
Notes: On wood of conifers.
Host: Juniperus
Tubulicrinis prominens (H. Jacks. & Dearden) Donk
Peniophora prominens H. Jacks. & Dearden
Distr.: Western North America.
Notes: On wood.
Host: Pinus
Tubulicrinis propinquus (Bourd. & Galzin) Donk
Peniophora propinqua Bourd. & Galzin
Distr.: Temperate northern hemisphere.
Notes: On wood of conifers.
Hosts: Abies, Picea
Tubulicrinis sceptriferus (H. Jacks. & Weresub) Donk
Peniophora sceptrifera H. Jacks. & Weresub in Weresub
Distr.: Northern hemisphere.
Notes: On wood.
Host: Tsuga
Tubulicrinis subulatus (Bourd. & Galzin) Donk
Peniophora subulata Bourd. & Galzin
Distr.: Northern hemisphere.
Notes: On wood of conifers.
Hosts: Abies, Picea, Pinus

Tuburcinia Fr. Basidiomycetes, Ustilaginales

Notes: This name is considered a synonym of *Urocystis*, but not all names have been redisposed.
Tuburcinia clintoniae Kom.
Distr.: Northwestern North America; northeastern Asia.
Hosts: Smilacina, Streptopus

Tulasnella J. Schröt. Basidiomycetes, Tulasnellales

Tulasnella albida Bourd. & Galzin
Distr.: Eastern United States; Europe.
Notes: On hardwoods.
Host: Quercus
Tulasnella allantospora Wakef. & A. Pearson
Distr.: Northern hemisphere.
Notes: On wood.
Hosts: Pinus, Quercus
Tulasnella bifrons Bourd. & Galzin
Distr.: Northern hemisphere.
Notes: On wood.
Hosts: Acer, Alnus, Betula, Quercus, Ulmus
Tulasnella calospora (Boud.) Juel
Gloeotulasnella calospora (Boud.) D.P. Rogers
Distr.: Temperate northern hemisphere.
Notes: On wood.
Hosts: Pinus, Quercus

Tulasnella eichleriana Bres.
Tulasnella lactea Bourd. & Galzin
Tulasnella microspora Wakef. & A. Pearson
Distr.: Northern hemisphere.
Notes: On wood.
Hosts: Betula, Populus, Quercus, Salix
Tulasnella fuscoviolacea Bres.
Distr.: Northern hemisphere, cooler parts.
Notes: On wood, usually of conifers.
Hosts: Abies, Picea, Pinus, Tsuga
Tulasnella hyalina Höhn. & Litsch. in Sacc.
Gloeotulasnella hyalina Höhn. & Litsch. in D.P. Rogers
Gloeotulasnella metachroa Bourd. & Galzin
Distr.: Northern hemisphere.
Notes: On wood.
Host: Populus
Tulasnella pinicola Bres.
Gloeotulasnella pinicola (Bres.) D.P. Rogers
Distr.: Northern hemisphere.
Notes: On wood.
Hosts: Betula, Carya, Liriodendron, Mimosa, Pinus, Quercus, Salix
Tulasnella pruinosa Bourd. & Galzin
Distr.: Temperate northern hemisphere.
Notes: On wood.
Hosts: Acer, Betula, Carpinus, Picea, Salix, Ulmus
Tulasnella traumatica (Bourd. & Galzin) Sacc. & Trott.
Gloeotulasnella traumatica Bourd. & Galzin
Gloeotulasnella opalea D.P. Rogers
Distr.: Northern hemisphere.
Notes: On wood.
Host: Populus
Tulasnella violacea (Johan-Olsen in Bref.) Juel
Distr.: Temperate northern hemisphere.
Notes: On wood and herbaceous plants.
Host: Quercus
Tulasnella violea (Quél.) Bourd. & Galzin
Distr.: Widespread.
Notes: On wood, plant and fungal debris.
Hosts: Acer, Carya, Fagus, Fraxinus, Liriodendron, Nyssa, Picea, Pinus, Quercus

Tunicago Sutton & Pollack. Coelomycetes

Tunicago uniolae Sutton & Pollack
Distr.: FL.
Notes: Sutton & Pollack (Canad. J. Bot. 55:326. 1977).
Host: Uniola

Tympanis Tode:Fr. Discomycetes, Helotiales

Notes: Groves (Canad. J. Bot. 55:571. 1952).
Tympanis alnea (Pers.:Fr.) Fr.
Sirodothis inversa (Fr.:Fr.) Sutton & Funk (anamorph)
Distr.: Circumboreal.
Notes: Especially common on Betulaceae and *Populus* but occasionally on other hardwoods.
Hosts: Alnus, Malus, Tilia
Tympanis amelanchieris Groves
Distr.: ID; Canada(Ontario).
Host: Amelanchier
Tympanis cephalanthi Dearn. & House
Dendrophoma cephalanthi Peck (anamorph)
Distr.: NY.
Notes: According to Groves (Canad. J. Bot. 30:639. 1952), the type specimen is too poor to allow an accurate determination of the species.
Host: Cephalanthus
Tympanis confusa Nyl.
Distr.: Temperate regions.
Host: Pinus
Tympanis conspersa Fr.
Distr.: Temperate regions.
Host: Malus
Tympanis hypopodia Nyl.
Distr.: Temperate regions.
Notes: Oulette & Pirozynski (Canad. J. Bot. 52:1901. 1974).
Hosts: Pinus, Tsuga
Tympanis laricina (Fuckel) Sacc.
Cenangium laricinum Fuckel
Tympanis abietina Groves
Tympanis rehmiana Groves
Sirodothis sp. (anamorph)
Distr.: Circumboreal on conifers.
Hosts: Abies, Larix, Pinus
Tympanis montanensis (Seaver) Groves
Godronia montanensis Seaver
Distr.: MT.
Host: Artemisia
Tympanis neopithya Ouellette & Pirozynski
Distr.: North America; Europe; Japan.
Host: Pinus
Tympanis oxydendri Ellis & Everh.
Distr.: GA, WV(type).
Host: Oxydendrum
Tympanis spermatiospora (Nyl.) Nyl.
Sirodothis populnea (Thuem.) Sutton & Funk (anamorph)
Distr.: North temperate regions.
Notes: Ouellette & Pirozynski (Canad. J. Bot. 52:1902. 1974).
Host: Populus
Tympanis truncatula (Pers.:Fr.) Rehm
Tympanis acericola Groves in Wehmeyer
Tympanis hansbroughiana Groves
Patellaria pithya P. Karst.
Tympanis pithya (P. Karst.) P. Karst.
Tympanis sorbi Groves
Distr.: Northern North America.
Notes: On hardwoods and conifers.
Hosts: Abies, Acer, Pinus, Pseudotsuga

Typhula Fr.:Fr., emend. Berthier. Basidiomycetes, Aphyllophorales

Notes: Berthier (Bull. Mens. Soc. Linn. Sec. Bot. Lyon, num. spec. 1976).
Typhula athyrii Remsberg
Distr.: Widespread.
Host: Athyrium
Typhula elegans (Berk. & M.A. Curtis) Corner
Pistillaria elegans Berk. & M.A. Curtis
Distr.: IN, SC.
Hosts: Campsis, Viburnum
Typhula filicina Peck
Distr.: NY, VA.
Host: Osmunda
Typhula gyrans Fr.:Fr., sensu Remsberg
Distr.: NY; Europe.
Notes: Berthier (Bull. Mens. Soc. Linn. Sec. Bot. Lyon, num. spec. p. 147. 1976) considered this species under the heading of "*Typhula setipes*" *sensu lato*.
Hosts: Prunus, Rubus
Typhula hyalina (Quél.) Berthier
Pistillina hyalina Quél.
Distr.: Tropical regions.
Host: Aleurites
Typhula idahoensis Remsberg
Distr.: Northwestern United States and adjacent Canada; possibly in northern Europe.
Notes: Some authors prefer to consider *T. idahoensis* 1949 a synonym of *T. ishikariensis* 1930, thus some hosts listed here pertain to the latter species.
Hosts: Agropyron, Bromus, Deschampsia, Elymus, Hordeum, Poa, Secale, Stipa, Trifolium, Triticum
Typhula incarnata Fr.
Typhula elegantula P. Karst.
Typhula itoana Imai
Distr.: Northern hemisphere, cooler parts.
Hosts: Agropyron, Agrostis, Avena, Bromus, Calamagrostis, Dactylis, Elymus, Hordeum, Poa, Secale, Stipa, Triticum
Typhula ishikariensis Imai

Distr.: Northwestern United States and adjacent Canada.
Notes: See note under *T. idahoensis*.
Host: Triticum
Typhula latissima Remsberg
Distr.: NY(type).
Host: Typha
Typhula micans (Fr.) Berthier
Pistillaria micans Fr.
Distr.: Widespread.
Notes: On fallen leaves.
Host: Hypericum
Typhula phacorrhiza (Reichard:Fr.) Fr.
Clavaria phacorrhiza Reichard:Fr.
Typhula complanata de Bary
Typhula juncea (Fr.) P. Karst.
Sclerotium complanatum Tode (synanamorph)
Sclerotium scutellatum Albertini & Schwein. (synanamorph)
Distr.: Eastern and western United States; Europe; Asia.
Notes: On old leaves, petioles, herbaceous stems, and grasses.
Hosts: Acer, Alnus, Aster, Dactylis, Solidago, Zea
Typhula phacorrhiza (Reichard:Fr.) Fr. var. **heterogenea** Berthier
Distr.: ID(type).
Host: Alnus
Typhula setipes (Grev.) Berthier
Pistillaria setipes Grev.
Distr.: North America(questionable); Europe.
Notes: Saprobic on leaves of deciduous trees.
Host: Acer
Typhula spathulata (Corner) Berthier
Clavaria spathulata Peck
Pistillaria spathulata Corner non Peck, hom. illeg.
Distr.: North America.
Host: Carya
Typhula sphaeroidea Remsberg
Distr.: North America.
Host: Rubus
Typhula uncialis (Grev.) Berthier
Typhula falcata P. Karst.
Clavaria typhuloides Peck
Pistillaria typhuloides (Peck) Burt
Distr.: Temperate northern hemisphere.
Notes: On herbaceous plants.
Host: Epilobium
Typhula variabilis Riess
Typhula intermedia Appel & Laubert
Distr.: Temperate regions.
Notes: On rotting leaves, petioles, and herbaceous stems.
Hosts: Acer, Apium, Daucus, Fagopyrum, Populus, Trifolium, Zea
Typhula viburni Remsberg
Distr.: Eastern North America.
Notes: Berthier (Bull. Mens. Soc. Linn. Sec. Bot. Lyon, num. spec. p. 147. 1976) considered this species to come under the heading of "*Typhula setipes*" *sensu lato*.
Host: Viburnum
Typhula viticola (Peck) Berthier
Pistillaria viticola Peck
Distr.: NY(type).
Host: Vitis

Typhulochaeta Ito & K. Hara. Pyrenomycetes, Erysiphales

Typhulochaeta couchii (Solheim, Eboh, & McHenry) U. Braun
Typhulochaeta japonica Ito & K. Hara var. *couchii* Solheim, Eboh, & McHenry
Distr.: AZ.
Notes: Powdery mildew of oak (Solheim et al. J. Elisha Mitchell Sci. Soc. 84:236. 1968; Braun. Beih. Nova Hedwigia 89:552. 1987).
Host: Quercus

Tyromyces P. Karst. Basidiomycetes, Aphyllophorales

Notes: Some species have been redisposed in *Postia*. Gilbertson & Ryvarden (North American Polypores 2:782. 1987).
Tyromyces calkinsii Murrill
Polyporus calkinsii (Murrill) Sacc. & Trott.
Distr.: Southeastern United States.
Notes: On dead wood and from wounds of living trees.
Hosts: Liquidambar, Magnolia
Tyromyces chioneus (Fr.:Fr.) P. Karst.
Polyporus chioneus Fr.:Fr.
Polyporus albellus Peck
Tyromyces albellus (Peck) Bondartsev & Singer
Distr.: Widespread.
Notes: On hardwoods.
Hosts: Abies, Acer, Ailanthus, Alnus, Betula, Castanea, Cornus, Corylus, Fagus, Fraxinus, Ginkgo, Larix, Liriodendron, Malus, Pinus, Populus, Prunus, Pseudotsuga, Quercus, Salix, Thuja, Ulmus
Tyromyces destructor (Schrad.:Fr.) Bondartsev & Singer, nom. conf.
Polyporus destructor (Schrad.:Fr.) Fr.
Distr.: Western United States; Europe.
Notes: A *nomen confusum* (Ryvarden. The Polyporaceae of North Europe. 1:77. 1976).
Hosts: Alnus, Picea
Tyromyces fissilis (Berk. & M.A. Curtis) Donk
Polyporus fissilis Berk. & M.A. Curtis
Polyporus albosordescens Romell
Polyporus fuscomutans C.G. Lloyd
Distr.: Eastern North America; Europe.
Notes: On hardwoods.
Hosts: Acer, Carya, Castanea, Diospyros, Fraxinus, Liquidambar, Malus, Quercus
Tyromyces fumidiceps Atk.
Polyporus fumidiceps (Atk.) Sacc. & Trott.
Polyporus smaragdinus C.G. Lloyd
Distr.: Eastern North America; Europe.
Notes: On hardwoods, especially in areas subject to flooding.
Hosts: Acer, Salix
Tyromyces galactinus (Berk.) J. Lowe
Polyporus galactinus Berk.
Polyporus iowensis C.G. Lloyd
Distr.: Eastern North America.
Notes: On hardwoods.
Hosts: Acer, Alnus, Castanea, Liquidambar, Malus, Nyssa, Platanus, Populus, Prunus, Pseudotsuga, Quercus, Salix, Tilia, Ulmus
Tyromyces humeanus (Murrill) J. Lowe
Trametes humeana Murrill
Distr.: FL.
Host: Quercus
Tyromyces palmarum Murrill
Distr.: FL.
Host: Arecaceae
Tyromyces pseudolacteus Murrill
Distr.: FL.
Host: Quercus
Tyromyces subgiganteus (Berk. & M.A. Curtis) Ryvarden
Polyporus subgiganteus Berk. & M.A. Curtis
Polyporus spumeus (Sowerby:Fr.) Fr. var. *malicola* C.G. Lloyd
Tyromyces spumeus (Sowerby:Fr.) Imazeki var. *malicola* (C.G. Lloyd) J. Lowe
Distr.: Eastern North America.
Notes: On wood.
Hosts: Malus, Populus

Uberispora Pirozynski & C.S. Hodges. Hyphomycetes

Uberispora simplex (Ichin.) Pirozynski & C.S. Hodges
Arachnophora simplex Ichin.
Distr.: SC.
Notes: On fallen leaves (Ellis. More dematiaceous Hyphomycetes. p. 108. 1976).
Host: Persea

Uleomyces Henn. Loculoascomycetes, Dothideales

Uleomyces sandicensis (Ellis & Everh.) Arx
Curreya sandicensis Ellis & Everh.
Hyalocurreya sandicensis (Ellis & Everh.) Theiss. & Syd.
Distr.: HI.
Notes: On living leaves (Arx. Persoonia 2:456. 1963).
Host: Alphitonia

Ulocladium G. Preuss. Hyphomycetes

Notes: Simmons (Mycologia 59:67. 1967).
Ulocladium atrum G. Preuss
Stemphylium atrum (G. Preuss) Sacc.
Distr.: Cosmopolitan.
Notes: On various plant materials and in soil.
Host: Prunus
Ulocladium botrytis G. Preuss
Stemphylium botryosum Wallr. var. *ulocladium* Sacc.
Distr.: Widespread.
Notes: On dead herbaceous plants, rotten wood, paper, textiles, and isolated from soil (Ellis. Dematiaceous Hyphomycetes. p. 498. 1971).
Hosts: Alnus, Pseudotsuga, Sphaeralcea
Ulocladium chartarum (G. Preuss) E. Simmons
Alternaria chartarum G. Preuss
Distr.: Cosmopolitan.
Notes: On a variety of organic substrates.
Hosts: Alnus, Prunus, Pseudotsuga
Ulocladium consortiale (Thuem.) E. Simmons
Alternaria consortiale (Thuem.) Groves & S.J. Hughes
Stemphylium consortiale (Thuem.) Groves & Skolko
Stemphylium ilicis Tengwall
Distr.: North America; Europe; Middle East.
Notes: On plant materials and in soil.
Hosts: Brassica, Cucumis, Cucurbita, Daucus, Lonicera, Lycopersicon, Malus, Medicago, Pisum, Pyrus, Solanum, Tetragonia, Trifolium, Triticum, Vicia
Ulocladium lanuginosum (C. Harz) E. Simmons
Stemphylium lanuginosum C. Harz
Alternaria lanuginosa (C. Harz) Sacc.
Pseudostemphylium lanuginosum (C. Harz) Subramanian
Alternaria harzii Joly
Distr.: Widespread.
Host: Zea

Umbelopsis R.E. Amos & H.L. Barnett. Hyphomycetes

Umbelopsis versiformis R.E. Amos & H.L. Barnett
Distr.: WV.
Notes: Isolated from roots and soil near roots of deciduous trees (Amos & Barnett. Mycologia 58:807. 1966).
Hosts: Acer, Liriodendron, Quercus

Uncinula Lév. Pyrenomycetes, Erysiphales

Notes: Braun (Beih. Nova Hedwigia 89. 1987).
Uncinula adunca (Wallr.:Fr.) Lév.
Uncinula salicis (DC.) G. Wint. in Rabenh.
Distr.: Temperate northern hemisphere.
Notes: Powdery mildew of willows and poplars.
Hosts: Populus, Salix
Uncinula bicornis (Wallr.:Fr.) Lév.
Sawadaia bicornis (Wallr.:Fr.) Miyabe in Homma
Uncinula aceris (DC.) Sacc.
Oidium aceris Rabenh. (anamorph)
Distr.: Cosmopolitan.
Notes: Powdery mildew of maples (Mukerji. C.M.I. Descr. 190. 1968).
Host: Acer
Uncinula circinata Cooke & Peck
Distr.: Eastern North America.
Notes: Most common on leaves of *Acer rubrum* and *A. spicatum*.
Hosts: Acer, Sapindus
Uncinula clintonii Peck
Distr.: Eastern North America.
Notes: Powdery mildew of basswood.
Host: Tilia
Uncinula geniculata W. Gerard
Distr.: Central and eastern United States.
Notes: Powdery mildew of mulberry.
Host: Morus
Uncinula macrospora Peck
Uncinula intermedia Berk. & M.A. Curtis
Distr.: Eastern North America.
Notes: Powdery mildew of elm.
Hosts: Celtis, Ostrya, Ulmus
Uncinula necator (Schwein.) Burrill
Uncinula spiralis Berk. & M.A. Curtis
Uncinula subfusca Berk. & M.A. Curtis
Oidium tuckeri Berk. (anamorph)
Distr.: Cosmopolitan.
Notes: Powdery mildew of Vitaceae (Kapoor. C.M.I. Descr. 160. 1967).
Hosts: Ampelopsis, Parthenocissus, Vitis
Uncinula parvula Cooke & Peck
Uncinula torreyi W. Gerard
Distr.: Central and eastern United States.
Notes: Powdery mildew of hackberry.
Hosts: Celtis, Ulmus

Uncinuliella Zheng & Chen. Pyrenomycetes, Erysiphales

Uncinuliella australiana (McAlpine) Zheng & Chen
Uncinula australiana McAlpine
Distr.: Temperate regions.
Host: Lagerstroemia
Uncinuliella flexuosa (Peck) U. Braun
Uncinula flexuosa Peck
Distr.: North America.
Notes: Powdery mildew of horsechestnut.
Host: Aesculus

Unguiculella Höhn. Discomycetes, Helotiales

Notes: This genus is considered a synonym of *Hyalopeziza*, but not all names have been redisposed. Korf & Kohn (Mycotaxon 10:503. 1980).
Unguiculella uncinata (W. Phillips) Dennis
Dasyscyphus uncinata (W. Phillips) Sacc.
Distr.: CA.
Notes: Dennis (Kew Bull. 17:372. 1963).
Host: Pinus

Uniseta Ciccarone. Coelomycetes

Uniseta flagellifera (Ellis & Everh.) Ciccarone
Pestalotia flagellifera Ellis & Everh.
Neobarclaya flagellifera (Ellis & Everh.) Kuntze
Cryptodiaporthe aubertii (Westend.) Wehmeyer var. *comptoniae* (Schwein.) Wehmeyer (teleomorph)
Distr.: Eastern North America.
Notes: Sutton (The Coelomycetes. p. 261. 1980).
Host: Comptonia

Urceolella Boud. Discomycetes, Helotiales

Notes: Raschle (Sydowia 29:170. 1977).
Urceolella amphipila Huhtinen
Distr.: VT(type).
Notes: Huhtinen (Mycotaxon 29:274. 1987).
Host: Struthiopteris
Urceolella tuberculiformis (Ellis & Everh.) Dennis
Dasyscyphus tuberculiformis Ellis & Everh.
Distr.: CO; Italy.
Notes: On herbaceous stems.
Host: Aquilegia

Uredinopsis Magnus. Basidiomycetes, Uredinales

Uredinopsis acrostichoides (Faull) Hiratsuka
Uredinopsis longimucronata Faull var. *acrostichoides* Faull
Distr.: Northeastern North America.
Notes: Heteroecious; aecial host - unknown; telial host - *Athyrium acrostichoides*.
Host: Athyrium
Uredinopsis americana P. Syd. & Syd.
Uredinopsis mirabilis Magnus
Distr.: Central and eastern North America; introduced into England.
Notes: Heteroecious; aecial host - *Abies* spp.; telial host - *Onoclea sensibilis*.
Hosts: Abies, Onoclea, [Woodwardia]
Uredinopsis arthurii Faull
Distr.: Eastern North America.
Notes: Heteroecious; aecial host - unknown; telial host - *Woodwardia virginica*.
Host: Woodwardia
Uredinopsis atkinsonii Magnus
Distr.: Central and eastern North America.
Notes: Heteroecious; aecial host - *Abies balsamea*; telial host - Polypodiaceae.
Host: Thelypteris
Uredinopsis ceratophora Faull
Distr.: Northcentral and northeastern North America.
Notes: Heteroecious; aecial host - *Abies balsamea*; telial host - *Cystopteris bulbifera*.
Host: Cystopteris
Uredinopsis copelandii P. Syd. & Syd.
Distr.: CA.
Notes: Heteroecious; aecial host - unknown; telial host - *Athyrium cyclosorum*.
Host: Athyrium
Uredinopsis glabra Faull
Distr.: Southwestern United States; Mexico.
Notes: Heteroecious; aecial host - unknown; telial host - Polypodiaceae.
Host: Cystopteris
Uredinopsis hashiokai Hiratsuka
Uredinopsis aspera Faull
Distr.: HI, northwestern North America; Taiwan.
Notes: Heteroecious; aecial host - *Abies* spp.; telial host - *Pteridium aquilinum*.
Host: Pteridium
Uredinopsis longimucronata Faull
Uredinopsis longimucronata Faull f. *cyclosora* Faull
Distr.: Temperate North America.
Notes: Heteroecious; aecial host - *Abies* spp.; telial host - *Athyrium filix-femina*.
Hosts: Abies, Athyrium
Uredinopsis maculata (Faull) Hiratsuka
Uredinopsis arthurii Faull var. *maculata* Faull
Distr.: Eastern United States.
Notes: Heteroecious; aecial host - unknown; telial host - *Woodwardia areolata*.
Host: Woodwardia
Uredinopsis osmundae Magnus
Distr.: Central and Eastern North America; Bermuda.
Notes: Heteroecious; aecial host - *Abies* spp. telial host - *Osmunda* spp.
Hosts: Abies, Osmunda
Uredinopsis phegopteridis Arth.
Distr.: Northern North America.
Notes: Heteroecious; aecial host - *Abies* spp.; telial host - Polypodiaceae.
Hosts: Abies, Dryopteris, Gymnocarpium
Uredinopsis pteridis Dietel & Holw. in Dietel
Uredinopsis macrosperma Magnus
Distr.: Western hemisphere; East Asia; Japan.
Notes: Heteroecious; aecial host - *Abies* spp.; telial host - *Pteridium aquilinum* and its varieties.
Hosts: Abies, Pteridium, [Pteris]
Uredinopsis struthiopteridis Störmer ex Dietel
Distr.: North America; Europe; Japan.
Notes: Heteroecious; aecial host - *Abies* spp.; telial host - Polypodiaceae.
Hosts: Abies, Athyrium, Matteuccia
Uredinopsis virginiana Faull
Distr.: Eastern United States.
Notes: Heteroecious; aecial host - unknown; telial host - *Pteridium aquilinum* var. *pseudocaudatum*.
Host: Pteridium

Uredo Pers.:Pers. Basidiomycetes, Uredinales

Uredo acuta (Faull) Hiratsuka
Milesia acuta Faull
Distr.: CA.
Notes: This species may belong to *Milesina*.
Host: Woodwardia
Uredo apacheca R.S. Peterson
Distr.: NM.
Host: Juniperus
Uredo behnickiana Henn.
Distr.: FL, NJ, OH(greenhouses); tropical regions.
Notes: On Orchidaceae.
Hosts: Cattleya, Epidendrum, Laelia, Oncidium, Orchidaceae
Uredo beringiana (Tranzschel) Parmelee
Pucciniastrum alaskanum Mains, nom. illeg.
Uredo alaskanum (Mains) Cummins
Distr.: Northern North America (Bering Straits).
Notes: *Pucciniastrum alaskanum* is based on a specimen with uredinia. Parmelee (Fungi Canadenses 297. 1985).
Host: Gentiana
Uredo cephalanthi Arth.
Distr.: FL; Cuba.
Host: Cephalanthus
Uredo coccolobae Henn.
Uredo uviferae P. Syd. & Syd.
Distr.: FL, PR.
Host: Coccoloba
Uredo contraria Arth.
Distr.: CA.
Host: Phacelia
Uredo cupheae Henn.
Distr.: FL, PR.
Host: Ginoria
Uredo cupressicola R.S. Peterson
Distr.: Southwestern United States; Central America.
Host: Cupressus
Uredo epidendri Henn.
Distr.: FL; tropical regions.
Hosts: Epidendrum, Oncidium
Uredo ericae A. Naumann
Pucciniastrum ericae (A. Naumann) Cummins, nom. illeg.
Distr.: CA; Europe.
Notes: Telia unknown.
Host: Erica
Uredo floridana Syd.
Distr.: FL; Mexico; Cuba.
Host: Mentzelia
Uredo goodyerae Tranzschel
Pucciniastrum goodyerae (Tranzschel) Arth., nom. illeg.
Distr.: Western North America; Europe.
Notes: Telia unknown.
Host: Goodyera
Uredo guaynabensis F. Kern & Whetzel
Distr.: FL, PR.
Host: Ludwigia
Uredo hameliae Arth.
Distr.: FL, PR.
Notes: On Rubiaceae.
Host: Hamelia
Uredo hawaiiensis Arth. in F. Stevens
Distr.: HI.
Host: Carex
Uredo hoffmanseggiae (W.H. Long) Cummins

Ravenelia hoffmanseggiae W.H. Long, nom. illeg.
Distr.: TX.
Notes: Telia unknown.
Host: Hoffmanseggia
Uredo ignava Arth.
Puccinia ignava Arth., nom. illeg.
Distr.: FL, PR.
Notes: Telia unknown.
Hosts: Bambusa, Poaceae
Uredo laeticolor Arth.
Distr.: FL.
Hosts: Ipomoea, Merremia
Uredo maculans Pat. & Gaillard
Puccinia mogiphanis Arth. (teleomorph)
Distr.: FL(questionable).
Notes: On Amaranthaceae.
Host: Alternanthera
Uredo mckinleyensis Cummins
Distr.: AK(type).
Host: Salix
Uredo mexicana Cummins
Distr.: AZ, TX; Mexico.
Host: Rhus
Uredo nicotianae Anastasia, Sacc. & Splendore in Anastasia & Splendore
Distr.: CA; Italy(type).
Host: Nicotiana
Uredo nigropuncta Henn.
Uredo guacae Mayor
Sphenospora kevorkianii Linder (teleomorph)
Distr.: FL; West Indies; Central America.
Notes: Autoecious.
Hosts: Cattleya, Epidendrum, Odontoglossum, Orchidaceae
Uredo nociviola H. Jacks. & Holw. in H. Jacks.
Distr.: FL; West Indies.
Host: Cyperus
Uredo oncidii Henn.
Distr.: CA, HI.
Host: Oncidium
Uredo phoradendri H. Jacks.
Distr.: AZ, CA, OR.
Hosts: Phoradendron, Salix
Uredo piscariae Cummins
Distr.: OR.
Host: Eremocarpus
Uredo posita J.J. Davis ex Arth.
Distr.: FL, LA.
Notes: The type specimen does not contain a rust, thus this name is spurious.
Host: [Aeschynomene]
Uredo sphacelicola Dietel & Holw. in Dietel
Distr.: CA(type).
Host: Sphacele
Uredo spirostachydis Arth.
Distr.: AZ(type).
Host: Allenrolfea
Uredo stevensii Arth. in F. Stevens
Distr.: HI.
Notes: Hennen & Hodges (Mycologia 73:1122. 1981).
Host: Euphorbia
Uredo wikstroemiae Hiratsuka
Distr.: HI.
Notes: Telia unknown.
Host: Wikstroemia

Urnula Fr. Discomycetes, Pezizales

Notes: Dissing (Mycologia 73:272. 1981).
Urnula craterium (Schwein.:Fr.) Fr.
Conoplea globosa (Schwein.:Fr.) S.J. Hughes (anamorph)
Distr.: Eastern North America; Europe.
Hosts: Carpinus, Quercus, Thuja

Urocystis Rabenh. ex Fuckel, nom. cons. Basidiomycetes, Ustilaginales

Urocystis agropyri (G. Preuss) J. Schröt.
Distr.: Widespread.
Notes: On Poaceae (Mordue & Waller. C.M.I. Descr. 716. 1981).
Hosts: Agrohordeum, Agropyron, Agrostis, Alopecurus, Bromus, Calamagrostis, Dactylis, Elymus, Festuca, Glyceria, Hordeum, Koeleria, Leucopoa, Melica, Phleum, Poa, Sitanion, Trisetum, Triticum
Urocystis alaskana Zundel
Distr.: AK.
Host: Boykinia
Urocystis anemones (Pers.:Pers.) G. Wint. in Rabenh.
Distr.: Northern hemisphere.
Notes: On Ranunculaceae (Mordue. C.M.I. Descr. 806. 1984).
Hosts: Aconitum, Anemone, Anemonella, Hepatica, Ranunculus, Trautvetteria, Trollius
Urocystis carcinodes (Berk. & M.A. Curtis) A. Fisch. v. Waldh.
Distr.: Widespread in the northern hemisphere.
Notes: On Ranunculaceae.
Hosts: Aconitum, Actaea, Cimicifuga, Clematis
Urocystis colchici (Schlechtend.) Rabenh.
Distr.: Temperate northern hemisphere.
Notes: On Liliaceae.
Hosts: Allium, Camassia, Colchicum, Polygonatum, Smilacina
Urocystis coralloides Rostr.
Distr.: Northern hemisphere.
Notes: On Brassicaceae.
Host: Descurainia
Urocystis erythronii G.P. Clinton
Distr.: Central and eastern North America.
Notes: On Liliaceae.
Host: Erythronium
Urocystis fischeri Körn. ex G. Wint. in Rabenh.
Distr.: Northern hemisphere.
Host: Carex
Urocystis flowersii Garrett
Distr.: UT(type).
Host: Zigadenus
Urocystis fraseri G.P. Clinton & Zundel
Sorosporium granulosum Ellis & Tracy
Distr.: Central and western United States; Mexico.
Host: Stipa
Urocystis giliae Speg.
Distr.: WY; Argentina(type).
Host: Gilia
Urocystis gladiolicola Ainsworth
Distr.: Northern hemisphere.
Notes: On cultivated *Gladiolus* spp. (Mordue. C.M.I. Descr. 807. 1984).
Host: Gladiolus
Urocystis granulosa G.P. Clinton
Distr.: ID(type).
Host: Stipa
Urocystis heucherae Garrett
Distr.: UT(type).
Host: Heuchera
Urocystis hypoxidis Thaxt.
Distr.: Eastern United States; South America; Australia.
Host: Hypoxis
Urocystis junci Lagerh.
Distr.: Northern hemisphere.
Host: Juncus
Urocystis kmetiana Magnus
Tuburcinia kmetiana (Magnus) Liro
Distr.: North America; Asia; Europe.
Notes: On Violaceae (Vánky. Symb. Bot. Upsal. 24:167. 1985).
Host: Viola
Urocystis lithophragmae Garrett
Distr.: UT.
Hosts: Heuchera, Lithophragma
Urocystis magica Pass. in Thuem.
Urocystis cepulae Frost
Distr.: Cosmopolitan.
Notes: Mulder & Holliday (C.M.I. Descr. 298. 1971).

Host: Allium
Urocystis occulta (Wallr.) Rabenh. ex Fuckel
Distr.: Widespread.
Notes: Mordue (C.M.I. Descr. 808. 1984).
Host: Secale
Urocystis sorosporioides Körn. ex A. Fisch. v. Waldh.
Sorosporium sorosporioides Körn. ex W.B. Cooke, nom. illeg.
Urocystis sorosporioides Körn. in Fuckel, nom. nud.
Distr.: Northern hemisphere.
Notes: On Ranunculaceae.
Hosts: Aconitum, Anemone, Anemonella, Aquilegia, Delphinium, Ranunculus, Thalictrum
Urocystis trientalis (Berk. & Broome) Lindeberg
Tuburcinia trientalis Berk. & Broome
Distr.: North America; Asia; Europe.
Notes: On Primulaceae (Vánky. Symb. Bot. Upsal. 24:184. 1985).
Host: Trientalis
Urocystis trillii G. Fisch.
Distr.: Western North America; Canada(Quebec); Japan.
Notes: On Liliaceae.
Host: Trillium
Urocystis tritici Körn.
Tuburcinia tritici (Körn.) Liro
Distr.: Cosmopolitan.
Notes: On Poaceae (Vánky. Symb. Bot. Upsal. 24:185. 1985).
Host: Triticum
Urocystis violae (Sowerby) A. Fisch. v. Waldh.
Distr.: Northern hemisphere.
Host: Viola

Urohendersonia Speg. Coelomycetes

Urohendersonia stipae H.C. Greene
Distr.: WI(type); Australia.
Notes: Nag Raj & Kendrick (Canad. J. Bot. 49:1861. 1971).
Host: Stipa

Uromyces (Link) Unger, nom. cons. Basidiomycetes, Uredinales

Notes: Hiratsuke (Rep. Tottori Mycol. Inst. 10:1. 1973).
Uromyces abbreviatus Arth.
Distr.: Western United States.
Notes: Autoecious.
Host: Psoralea
Uromyces acuminatus Arth.
Uromyces acuminatus Arth. var. *magnatus* (Arth.) J.J. Davis
Uromyces acuminatus Arth. var. *polemonii* (Arth.) J.J. Davis
Uromyces acuminatus Arth. var. *spartinae* (Farl.) Arth.
Uromyces acuminatus Arth. var. *steironematis* (Arth.) J.J. Davis
Nigredo polemonii Arth.
Uromyces spartinae Farl.
Aecidium magnatum Arth. (anamorph)
Distr.: Temperate North America.
Notes: Heteroecious; aecial host - several families of angiosperms; telial host - *Spartina* spp. Parmelee & de Carteret (Fungi Canadenses 203. 1981).
Hosts: Collomia, Dodecatheon, Lysimachia, Maianthemum, Phlox, Polemonium, Polygonatum, Smilacina, Spartina, Spergularia, Uvularia
Uromyces aemulus Arth.
Distr.: Western North America.
Notes: Autoecious.
Host: Allium
Uromyces agnatus Arth.
Distr.: Southeastern United States.
Notes: Autoecious.
Host: Jatropha
Uromyces alyxiae Arth. in F. Stevens
Distr.: HI.
Notes: Autoecious.
Host: Alyxia
Uromyces americanus Speg.
Distr.: North and South America.
Notes: Heteroecious; aecial host - Apiaceae; telial host - *Scirpus* spp.
Hosts: Hydrocotyle, Oenanthe, Scirpus
Uromyces americanus Speg. var. ***californicus*** Savile
Distr.: CA.
Notes: Heteroecious; aecial host - *Oenanthe sarmentosa*; telial host - *Scirpus californicus*.
Host: Scirpus
Uromyces amoenus Syd. & P. Syd.
Distr.: Western North America.
Notes: Autoecious.
Host: Anaphalis
Uromyces amphidymus P. Syd. & Syd.
Uromyces glyceriae Arth.
Distr.: Central and eastern United States.
Notes: Aecia unknown; telial host - *Glyceria* spp.
Host: Glyceria
Uromyces andropogonis Tracy
Distr.: Central and eastern United States.
Notes: Heteroecious; aecial host - *Viola* spp.; telial host - *Andropogon* spp.
Hosts: Andropogon, Schizachyrium, [Sorghum], Viola
Uromyces anthyllidis J. Schröt.
Distr.: TX; Europe; Japan.
Notes: Heteroecious; aecial host - *Euphorbia*; telial host - Fabaceae.
Host: Medicago
Uromyces apiosporus Hazsl.
Uromyces nevadensis Harkn.
Distr.: CA, NV; Europe.
Notes: Autoecious.
Host: Primula
Uromyces appendiculatus (Pers.:Pers.) Unger
Nigredo appendiculata Arth.
Uromyces phaseoli (Pers.) G. Wint.
Uromyces phaseoli (Pers.) G. Wint. var. *strophostylis* Arth.
Distr.: Cosmopolitan.
Notes: Autoecious; on Fabaceae (Laundon & Waterston. C.M.I. Descr. 57. 1965).
Hosts: Amphicarpaea, [Glycine], Macroptilium, Phaseolus, Strophostyles, [Vigna]
Uromyces archerianus Arth. & Fromme
Distr.: NM; Mexico; Africa.
Notes: Aecia unknown. Telial host - Poaceae.
Host: Chloris
Uromyces argutus F. Kern
Distr.: FL.
Notes: Aecia unknown; telial host - *Spartina alterniflora*.
Host: Spartina
Uromyces ari-triphylli (Schwein.) Seeler in Guba & Seeler
Puccinia ari-triphylli Schwein.
Uromyces caladii Farl. in Ellis
Distr.: North America; Bermuda; Mexico; northern Europe(on cultivated plants).
Notes: On Araceae.
Hosts: Arisaema, Peltandra, Zantedeschia
Uromyces aristidae Ellis & Everh.
Distr.: NM.
Notes: Aecia unknown; telial host - on Poaceae.
Host: Aristida
Uromyces armeriae (Schlechtend.) Arth.
Uromyces limonii (DC.) Lév. var. *armeriae* Schlechtend.
Distr.: Western North America; Argentina; northern Africa; Eurasia; New Zealand.
Notes: Autoecious.
Host: Armeria
Uromyces armeriae (Schlechtend.) Arth. subsp. ***pacificus*** Savile & Conners
Distr.: Western North America.
Notes: Autoecious. Savile (Mycologia 43:190. 1951).
Host: Armeria
Uromyces asclepiadis Cooke
Uromyces howei Peck
Distr.: Central and eastern North America; West Indies; Central and South America.
Notes: Autoecious; on Asclepiadaceae.
Hosts: Asclepias, Calotropis, Matelea, Vincetoxicum

Uromyces aureus Dietel & Holw. in Dietel
Uromyces chlorogali Dietel & Holw. in Dietel
Distr.: CA.
Notes: Autoecious; on Liliaceae.
Hosts: Allium, Chlorogalum
Uromyces beckmanniae H. Jacks.
Distr.: OR.
Notes: Aecia unknown; telial host - *Beckmannia* spp.
Host: Beckmannia
Uromyces beloperones G. Laundon
Uredo beloperones Arth.
Distr.: Southwestern United States; Mexico.
Notes: Autoecious.
Host: Beloperone
Uromyces betae J. Kickx fil.
Distr.: Worldwide in beet-growing regions.
Notes: Autoecious. Punithalingam (C.M.I. Descr. 177. 1968).
Host: Beta
Uromyces bicolor Ellis ex Coville
Uromyces aterrimus Dietel & Holw. in Dietel
Distr.: Temperate North America.
Notes: Autoecious.
Host: Allium
Uromyces bidenticola Arth.
Distr.: Southern United States; Central and South America; Africa; Asia.
Notes: Autoecious.
Host: Bidens
Uromyces bidentis Lagerh. in Pat. & Lagerh.
Distr.: Southern United States; West Indies; Central and South America.
Notes: Autoecious.
Host: Bidens
Uromyces brodiaeae Ellis & Harkn.
Distr.: Western United States.
Hosts: Brodiaea, Leucocoryne, Triteleia
Uromyces cestri Bertero ex Mont. in Gay
Uromyces cestri Lév., nom. nud.
Distr.: FL.
Notes: Autoecious.
Host: Cestrum
Uromyces ciceris-arietini Jacz. in Boyer & Jacz.
Distr.: SD; Mexico; Central America; Europe; Africa; western Asia to India.
Host: Cicer
Uromyces claytoniae Cooke & Peck in Peck
Distr.: NY.
Notes: Autoecious; on Portulacaceae.
Host: Claytonia
Uromyces coloradensis Ellis & Everh.
Uromyces coloradensis Ellis & Everh. var. *campester* Arth.
Uromyces coloradensis Ellis & Everh. var. *maritimus* Arth.
Uromyces coloradensis Ellis & Everh. var. *montanus* Arth.
Uromycopsis porosa (Peck) Arth.
Distr.: Temperate North America.
Notes: Autoecious. Cummins (Rust fungi on legumes and composites in North America. p. 241. 1978) did not consider these varieties to be distinct.
Hosts: Pediomelum, Vicia
Uromyces columbianus Mayor
Distr.: FL; West Indies; Central and South America.
Notes: Autoecious.
Host: Melanthera
Uromyces coluteae Arth.
Distr.: KS.
Notes: According to Cummins (Rust fungi on legumes and composites in North America. p. 246. 1978), this name was probably based on an introduction of *Uromyces laburni* (DC.) Otth.
Host: [Colutea]
Uromyces commelinae Cooke
Distr.: Cosmopolitan.
Notes: Aecial host - *Cissus* spp.; telial host - Commelinaceae.
Hosts: Cissus, Commelina, Tradescantia, Zebrina
Uromyces compactus Peck
Distr.: Southwestern United States; Mexico.
Notes: Autoecious.
Hosts: Aster, Baccharis, Boltonia
Uromyces coordinatus Arth.
Distr.: Southwestern United States.
Notes: Autoecious.
Host: Euphorbia
Uromyces coronatus Miyabe & Nishida ex Dietel
Distr.: FL; Asia.
Notes: Aecia unknown.
Host: Zizaniopsis
Uromyces costaricensis Syd.
Distr.: Southern United States; Mexico; West Indies; South America.
Notes: Aecia unknown; telial host - *Lasiacis* spp.
Host: Lasiacis
Uromyces dactylidis G. Otth
Distr.: Temperate regions.
Notes: Heteroecious; aecial host - *Ranunculus* spp.; telial host - Poaceae.
Hosts: Alopecurus, Dactylis, Festuca, Poa, Ranunculus
Uromyces dactylidis G. Otth var. **poae** (Rabenh.) Cummins
Uromyces poae Rabenh.
Uromyces alopecuri C.P. Seymour
Aecidium ficariae Pers. (anamorph)
Distr.: Temperate regions.
Notes: Heteroecious; aecial host - Ranunculaceae; telial host - Poaceae.
Hosts: Alopecurus, Ranunculus
Uromyces dianthi (Pers.:Pers.) Niessl
Uromyces caryophyllinus G. Wint.
Distr.: Worldwide, common in greenhouses.
Notes: Heteroecious; aecial host - *Euphorbia* spp.; telial host - Caryophyllaceae.
Host: Dianthus
Uromyces dictyosperma Ellis & Everh.
Distr.: Central and western United States.
Notes: Autoecious; on Euphorbiaceae.
Hosts: Euphorbia, Jatropha
Uromyces dolicholi Arth.
Distr.: Southern United States; West Indies; Central and South America.
Notes: Autoecious; on Fabaceae. The rust on *Cajanus* may be distinct as *Uredo cajani* (Hennen et al. Arq. Biol. 49 [suppl. 1]:110. 1982). Sivanesan (C.M.I. Descr. 269. 1970).
Hosts: Cajanus, Rhynchosia
Uromyces elegans Lagerh.
Distr.: Southeastern United States.
Notes: Autoecious.
Host: Trifolium
Uromyces eleocharidis Arth.
Nigredo eleocharidis (Arth.) Arth.
Distr.: Northcentral United States.
Notes: Aecia unknown; telial host - *Eleocharis*.
Host: Eleocharis
Uromyces epicampis Dietel & Holw. in Holw.
Distr.: Southwestern United States; Central and South America.
Notes: Autoecious.
Host: Muhlenbergia
Uromyces eragrostidis Tracy
Distr.: Warmer regions of the western hemisphere; Africa; Australia; India; Middle East.
Notes: Heteroecious; aecial host - *Anthericum* spp.; telial host - Poaceae.
Hosts: Anthericum, Eragrostis
Uromyces eugentianae Cummins
Uromyces gentianae Arth.
Distr.: Temperate North America.
Notes: Autoecious. Parmelee & de Carteret (Fungi Canadenses 298. 1985).
Hosts: Gentiana, Gentianella
Uromyces euphorbiae Cooke & Peck in Peck
Uromyces proeminens Pass. in Rabenh. var. *euphorbiicola* (Tranzschel) Arth.
Nigredo proeminens (Pass. in Rabenh.) Arth.
Uromyces proeminens Pass. in Rabenh.
Uromyces proeminens Pass. in Rabenh. var. *myristica* (Berk. & M.A.

Curtis) Arth.
Uromyces proeminens Pass. in Rabenh. var. *poinsettiae* (Tranzschel) Arth.
Distr.: Cosmopolitan.
Notes: Autoecious; on Euphorbiaceae.
Hosts: Chamaesyce, Euphorbia, Gymnanthes
Uromyces euphorbiae-corniculati Jordi
Uromyces loti A. Blytt
Uromyces striatus J. Schröt. var. *loti* A. Blytt
Distr.: North America; Africa; Europe; Japan.
Notes: Autoecious; on Fabaceae. Hiratsuka (Rep. Tottori Mycol. Inst. 10:49. 1973).
Host: Lotus
Uromyces fabae (Grev.) Fuckel
Puccinia fabae Grev.
Nigredo fabae (de Bary) Arth.
Uromyces vicae-fabae P. Karst. non (Pers.:Pers.) J. Schröt., hom. illeg.
Uromyces viciae-fabae (Pers.:Pers.) J. Schröt., nom. illeg.
Distr.: Cosmopolitan.
Notes: Autoecious; on Fabaceae. Boerema & Verhoeven (Neth. J. Pl. Pathol. 85:177. 1979) indicated that the commonly used name *Uromyces viciae-fabae* was based only on the uredinial stage and therefore is an illegitimate name. Laundon & Waterston (C.M.I. Descr. 60. 1965).
Hosts: Lathyrus, Pisum, Vicia
Uromyces fragilipes Tranzschel
Uromyces jacksonii Arth. & Fromme
Uromyces mysticus Arth.
Distr.: Western United States; Asia Minor; southern Europe.
Notes: Heteroecious; aecial host - probably Liliaceae; telial host - Poaceae.
Hosts: Agrostis, Deschampsia, Hordeum, Vulpia
Uromyces galii-californici Linder
Distr.: CA.
Notes: Autoecious.
Host: Galium
Uromyces geranii (DC.) Lév.
Distr.: Temperate northern hemisphere.
Notes: Autoecious. Sivanesan (C.M.I. Descr. 270. 1970).
Host: Geranium
Uromyces giganteus Speg.
Uromyces chenopodii J. Schröt.
Distr.: FL; Europe.
Notes: Autoecious.
Hosts: Chenopodium, Suaeda
Uromyces glycyrrhizae Magnus
Distr.: Central and western North America; northern Africa; China; Europe.
Notes: Autoecious.
Host: Glycyrrhiza
Uromyces graminicola Burrill
Distr.: Central and eastern United States; Central America.
Notes: Heteroecious; aecial host - Euphorbiaceae; telial host - *Panicum* spp.
Hosts: Aegilops, Panicum, Sebastiania, Stillingia
Uromyces halstedii De Toni in Sacc.
Distr.: United States; Japan.
Notes: Heteroecious; aecial host - *Trillium* spp.; telial host - *Leersia* and *Brachyelytrum* spp.
Hosts: Brachyelytrum, Leersia, Trillium
Uromyces hedysari-obscuri (DC.) Carestia & Picc.
Uromyces astragalicola Henn.
Distr.: Western North America.
Notes: Autoecious.
Hosts: [Astragalus], Hedysarum
Uromyces hedysari-paniculati (Schwein.) Farl. in Ellis
Uromyces desmodii Cooke
Distr.: Central and eastern North America; Central and South America.
Notes: Autoecious.
Host: Desmodium
Uromyces heterodermus Syd.
Distr.: Western United States, Rocky Mountain region.
Notes: Autoecious; on Liliaceae.
Host: Erythronium
Uromyces holwayi Lagerh.
Uromyces lilii G.W. Clinton in Peck non Kunze, hom. illeg.
Distr.: Temperate North America; Asia.
Notes: Autoecious.
Host: Lilium
Uromyces hordeinus (Arth.) Barth.
Uromyces hordei Tracy
Distr.: Eastern United States to CO.
Notes: Heteroecious; aecial host - *Nothoscordum bivalve*; telial host - Poaceae.
Hosts: Hordeum, [Nolina], Nothoscordum, Vulpia
Uromyces houstoniatus J. Sheld.
Distr.: Central and eastern United States.
Notes: Heteroecious; aecial host - Rubiaceae; telial host - Iridaceae.
Hosts: Hedyotis, Houstonia, Sisyrinchium
Uromyces hyalinus Peck
Uromyces sophorae Peck
Distr.: Central and western United States; Mexico.
Notes: Autoecious.
Host: Sophora
Uromyces inaequialtus Lasch ex Rabenh.
Uromyces silenes Fuckel
Distr.: Temperate northern hemisphere.
Notes: Autoecious.
Hosts: [Arenaria], Silene
Uromyces indigoferae Dietel & Holw. in Holw.
Distr.: Southern United States; Central and South America.
Notes: Autoecious.
Host: Indigofera
Uromyces intricatus Cooke
Nigredo intricata (Cooke) Arth.
Uromyces chorizanthis Ellis & Harkn.
Uromyces eriogoni Ellis & Harkn.
Distr.: Western United States.
Notes: Autoecious; on Polygonaceae.
Hosts: Chorizanthe, Eriogonum
Uromyces jonesii Peck
Distr.: Western North America.
Notes: Autoecious.
Host: Ranunculus
Uromyces junci (Desmaz.) Tul.
Nigredo junci (Desmaz.) Arth.
Distr.: North and South America; Europe.
Notes: Heteroecious; aecial host - Asteraceae; telial host - Juncaceae.
Hosts: Ambrosia, Arnica, Aster, Bahia, Cirsium, Eriophyllum, Grindelia, Helianthus, Jaumea, Juncus
Uromyces junci-effusi P. Syd. & Syd.
Distr.: Temperate North America.
Notes: Aecia unknown; telial host - *Juncus* spp.
Host: Juncus
Uromyces krameriae W.H. Long in Arth.
Distr.: FL, TX.
Notes: Autoecious.
Host: Krameria
Uromyces lapponicus Lagerh.
Uromyces carneus Har.
Distr.: Circumboreal, extending south in the western hemisphere to northern Mexico.
Notes: Autoecious. Savile & Parmelee (Fungi Canadenses 26. 1974) recognized var. *oxytropis*.
Hosts: Astragalus, Oxytropis
Uromyces lespedezae-procumbentis (Schwein.) M.A. Curtis
Uromyces lespedezae Peck
Distr.: Central and eastern North America.
Notes: Autoecious.
Host: Lespedeza
Uromyces limonii (DC.) Lév.
Distr.: North America; Europe.
Notes: Autoecious. Savile & Conners (Mycologia 43:191. 1951).
Host: Limonium
Uromyces limonii-caroliniani Savile & Conners
Distr.: Eastern North America.
Notes: Autoecious. This species was previously considered conspecific with *U. limonii* (Savile & Conners. Mycologia 43:186. 1951).
Host: Limonium

Uromyces lineolatus (Desmaz.) J. Schröt. in Rabenh.
Distr.: North America; Bermuda; Europe.
Notes: Heteroecious; aecial host - Apiaceae; telial host - *Scirpus* spp.
Hosts: Daucus, Scirpus
Uromyces lineolatus (Desmaz.) J. Schröt. subsp. ***nearcticus*** Savile
Nigredo scirpi Arth.
Uromyces scirpi Burrill
Aecidium ligustici Ellis & Everh. (anamorph)
Distr.: Central and eastern North America.
Notes: Savile (Canad. J. Bot. 50:2593. 1972).
Hosts: Cicuta, Daucus, Hydrocotyle, Ligusticum, Oenanthe, Rhynchospora, Scirpus, Sium
Uromyces lupini Berk. & M.A. Curtis
Distr.: Western North America; Mexico.
Notes: Autoecious.
Host: Lupinus
Uromyces lycoctoni (Kalchbr.) Trott.
Distr.: Northwestern United States; Europe.
Notes: Autoecious; on Ranunculaceae.
Host: Aconitum
Uromyces major Arth.
Distr.: TX; Mexico.
Notes: Aecia unknown; telial host - *Muhlenbergia* spp.
Host: Muhlenbergia
Uromyces martinii Farl.
Distr.: FL.
Notes: Autoecious.
Hosts: [Bidens], Melanthera
Uromyces mayorii Tranzschel ex Mayor
Distr.: Central and southern United States; South America.
Notes: Autoecious.
Host: Euphorbia
Uromyces mexicanus Dietel & Holw. in Holw.
Distr.: Southwestern United States; Mexico.
Notes: Autoecious.
Host: Desmodium
Uromyces minimus J.J. Davis
Distr.: Northern North America.
Notes: Aecia unknown; telial host - *Muhlenbergia* spp.
Host: Muhlenbergia
Uromyces minor J. Schröt. in Cohn
Distr.: Temperate northern hemisphere.
Notes: Autoecious.
Host: Trifolium
Uromyces minutus Dietel ex Atk.
Distr.: Central and eastern United States.
Notes: Aecia unknown; telial host - *Carex* spp.
Host: Carex
Uromyces miurae P. Syd. & Syd.
Distr.: Northwestern North America; northeastern Asia.
Notes: Autoecious.
Host: Fritillaria
Uromyces muscari (Duby) Lév.
Uromyces scillarum (Grev. ex Berk.) G. Wint. in Rabenh.
Distr.: CA; Canada(British Columbia); northern Africa; Asia; Europe; New Zealand.
Notes: Autoecious; on Liliaceae. Not all workers accept this synonymy (Anikster. Trans. Brit. Mycol. Soc. 89:327. 1987).
Host: Scilla
Uromyces necopinus Cummins
Distr.: CT, NY.
Notes: Autoecious. This species was segregated from *U. affinis* by Cummins (Bull. Torrey Bot. Club 68:46. 1941).
Host: Hypoxis
Uromyces novissimus Speg.
Uromyces hellerianus Arth.
Distr.: FL; Central and South America.
Notes: Autoecious.
Host: Melothria
Uromyces oaxacanus Dietel & Holw. in Holw.
Distr.: AZ.
Notes: Autoecious.
Host: Jatropha
Uromyces oblongisporus Ellis & Everh.
Distr.: WY.
Notes: Autoecious.
Host: Artemisia
Uromyces occidentalis Dietel
Distr.: Western United States; Mexico.
Notes: Heteroecious; aecial host - *Euphorbia* spp.; telial host - *Lupinus* spp.
Hosts: Euphorbia, Lupinus
Uromyces peckianus Farl.
Distr.: East and west coasts of United States and Canada.
Notes: Heteroecious; aecial host - Chenopodiaceae; telial host - *Distichlis spicata*.
Hosts: [Aristida], Atriplex, Chenopodium, Distichlis, Hedyotis, [Plantago], Salicornia
Uromyces peckianus Farl. subsp. ***rugulosus*** Savile
Distr.: Northern North America.
Notes: Heteroecious; aecial host - Chenopodiacae; telial host - *Distichlis*.
Host: Distichlis
Uromyces perigynius Halst.
Nigredo perigynia (Halst.) Arth.
Distr.: North America.
Notes: Heteroecious; aecial host - Asteraceae; telial host - *Carex* spp.
Hosts: Carex, Ratibida, Rudbeckia, Solidago
Uromyces phacae-frigidae (Wahlenberg) Har.
Distr.: Circumboreal.
Notes: Autoecious. Savile & Parmelee (Fungi Canadenses 25. 1974).
Host: Astragalus
Uromyces plumbarius Peck
Nigredo plumbaria (Peck) Arth.
Distr.: Temperate North America.
Notes: Autoecious; on Onagraceae.
Hosts: Camissonia, Gaura, Oenothera
Uromyces polygoni-avicularíae (Pers.:Pers.) P. Karst.
Nigredo polygoni (Fuckel) Arth.
Uromyces polygoni Fuckel, nom. illeg.
Distr.: Cosmopolitan.
Notes: Autoecious. Savile (Fungi Canadenses 314. 1987).
Host: Polygonum
Uromyces pontederiae W. Gerard
Distr.: Central and eastern United States; South America.
Notes: Autoecious.
Host: Pontederia
Uromyces primaverilis Speg.
Distr.: IL to TX; Mexico; South America.
Notes: Autoecious.
Hosts: Nolina, Nothoscordum
Uromyces primaverilis Speg. subsp. ***nothoscordi*** (Syd.) Savile
Uromyces affinis G. Wint.
Distr.: CT, FL, MO, MS; West Indies; Central and South America.
Notes: Autoecious.
Host: Hypoxis
Uromyces probus Arth.
Distr.: Northeastern WA to central UT.
Notes: Autoecious.
Host: Sisyrinchium
Uromyces psoraleae Peck
Distr.: Central and western North America.
Notes: Autoecious.
Hosts: Pediomelum, Psoralea, Psoralidium
Uromyces psoraleae Peck var. ***argophyllae*** (Seym.) Arth.
Distr.: Central and western United States.
Notes: Autoecious.
Hosts: Pediomelum, Psoralidium
Uromyces punctatus J. Schröt.
Nigredo punctata Arth.
Distr.: Central and western United States; Central America; Africa; Europe.
Notes: Heteroecious; aecial host - *Euphorbia* spp.; telial host - Fabaceae. Savile (Fungi Canadenses 24. 1974).
Hosts: Astragalus, Lotus, Oxytropis
Uromyces rhynchosporae Ellis
Distr.: HI, eastern United States; West Indies; South America.
Notes: Aecial host - unknown; telial host - *Rhynchospora* spp.
Host: Rhynchospora

Uromyces rickerianus Arth.
Distr.: ID to UT.
Notes: Autoecious.
Hosts: Polygonum, Rumex
Uromyces rudbeckiae Arth. & Holw. in Arth.
Distr.: North America; Japan.
Notes: Autoecious.
Host: Rudbeckia
Uromyces ruelliae Holw.
Distr.: AZ, CA; Mexico.
Notes: Autoecious.
Hosts: Beloperone, Justicia
Uromyces seditiosus F. Kern
Distr.: Central and eastern United States.
Notes: Heteroecious; aecial host - *Plantago* spp.; telial host - *Aristida* spp.
Hosts: Aristida, Plantago
Uromyces setariae-italicae Yoshino
Uromyces leptodermus Syd. & P. Syd. in Syd. & E.J. Butler
Distr.: Warm regions.
Notes: Heteroecious; aecial host - *Cordia* spp.; telial host - Poaceae.
Hosts: Eriochloa, Panicum, Setaria
Uromyces shearianus Arth.
Aecidium atriplicis Shear (anamorph)
Distr.: Southwestern United States.
Host: Atriplex
Uromyces silphii Arth.
Nigredo silphii (Arth.) Arth.
Distr.: Western hemisphere.
Notes: Heteroecious; aecial host - Asteraceae; telial host - *Juncus* spp.
Hosts: Aster, Helianthus, Juncus, Silphium
Uromyces sommerfeltii Hyl., Jørst. & Nannf.
Uromyces solidaginis (N.F. Sommer) Niessl
Distr.: Western North America; northeastern Asia; northern Europe.
Notes: Autoecious.
Host: Solidago
Uromyces sparganii Cooke & Peck
Uromyces pyriformis Cooke in Peck
Distr.: Central and eastern North America.
Notes: Heteroecious; aecial host - *Hypericum* spp.; telial host - *Sparangium* spp.
Hosts: [Acorus], Sparganium
Uromyces speciosus Holw.
Distr.: CO, NM.
Notes: Autoecious.
Host: Frasera
Uromyces spermacoces (Schwein.) M.A. Curtis
Distr.: Central and eastern United States.
Notes: Autoecious.
Host: Diodia
Uromyces sporoboli Ellis & Everh.
Distr.: Central United States.
Notes: Heteroecious; aecial host - *Allium* spp.; telial host - *Sporobolus* spp.
Hosts: Allium, Sporobolus
Uromyces sporobolicola Lindquist
Uromyces bravensis Cummins
Uredo egenula Arth.
Distr.: TX; Mexico; Dominican Republic; Argentina.
Notes: Aecia unknown; telial host - *Sporobolus* spp.
Host: Sporobolus
Uromyces striatus J. Schröt.
Uromyces medicaginis Pass.
Uromyces oblongus Vize
Uromyces striatus J. Schröt. var. *medicaginis* (Pass.) Arth.
Distr.: Cosmopolitan.
Notes: Heteroecious; aecial host - *Euphorbia* spp.; telial host - Fabaceae (Laundon & Waterston. C.M.I. Descr. 59. 1965).
Hosts: Medicago, Melilotus, Trifolium
Uromyces suksdorfii Dietel & Holw. in Dietel
Distr.: Western United States to NM.
Notes: Autoecious.
Hosts: Lychnis, Silene
Uromyces tenuicutis McAlpine
Uromyces ignobilis (Syd.) Arth.
Distr.: Circumboreal in warmer regions.
Notes: Aecia unknown.
Hosts: Muhlenbergia, Stenotaphrum
Uromyces tranzschelii Syd. ex Tranzschel
Distr.: Western United States.
Notes: Autoecious.
Host: Euphorbia
Uromyces trifolii (R. Hedw.) Fuckel
Nigredo trifolii (R. Hedw.) Arth.
Uromyces nerviphilus (Grognot) Hotson
Distr.: Temperate regions.
Notes: Autoecious. The rusts on *Trifolium* have not been well segregated; many specimens are misidentified.
Host: Trifolium
Uromyces trifolii-repentis Liro ex Liro
Uromyces trifolii (Hedw.) Fuckel var. *trifolii-repentis* (Liro ex Liro) Arth.
Uromyces hybridi W.H. Davis
Uromyces trifolii-hybridi H. Paul
Distr.: Cosmopolitan.
Notes: Autoecious. The name *U. trifolii* has been misapplied to this species for many years.
Host: Trifolium
Uromyces trifolii-repentis Liro ex Liro var. ***fallens*** (Arth.) Cummins
Uromyces fallens (Arth.) Barth.
Uromyces trifolii (R. Hedw.) Fuckel var. *fallens* (Arth.) Arth.
Distr.: Worldwide in red clover-growing regions.
Notes: Autoecious.
Host: Trifolium
Uromyces triquetrus Cooke
Uromyces hyperici M.A. Curtis
Uromyces hyperici-frondosi (Schwein.) Arth.
Distr.: North and South America; Taiwan.
Notes: Autoecious; on Clusiaceae.
Host: Hypericum
Uromyces uniporulus F. Kern
Distr.: Eastern United States.
Notes: Aecia unknown.
Host: Carex
Uromyces unitus Peck
Distr.: WA; Canada (British Columbia). For this species and all of its subspecies, see Savile & Parmelee (Mycologia 48:580. 1956).
Notes: Autoecious.
Hosts: Calyptridium, Lewisia
Uromyces unitus Peck subsp. ***montanensis*** Savile
Distr.: MT.
Notes: Autoecious.
Host: Lewisia
Uromyces unitus Peck subsp. ***pacificus*** Savile
Distr.: CA, OR.
Notes: Autoecious.
Host: Calyptridium
Uromyces unitus Peck subsp. ***spragueae*** (Harkn.) Savile
Uromyces spragueae Harkn.
Distr.: CA, ID, MT, NV, OR, WY.
Notes: Autoecious.
Hosts: Calyptridium, Lewisia, Spraguea
Uromyces valens F. Kern
Distr.: IN, WI.
Notes: Aecia unknown.
Host: Carex
Uromyces verruculosus J. Schröt.
Distr.: IN, MI, NY; Europe.
Notes: Autoecious.
Host: Silene
Uromyces vignae Barclay
Uromyces phaseoli (Pers.) G. Wint. var. *vignae* (Barclay) Arth.
Distr.: Worldwide in cowpea-growing regions.
Notes: Autoecious.
Hosts: [Vicia], Vigna
Uromyces vossiae Barclay
Uromyces superfluus P. Syd. & Syd.
Distr.: MO(questionable); India.
Notes: Autoecious.
Host: Panicum

Uromyces zygadeni Peck
Distr.: Central and western United States.
Notes: Autoecious.
Host: Zigadenus

Uropyxis J. Schröt. Basidiomycetes, Uredinales

Uropyxis amorphae (M.A. Curtis) J. Schröt.
Puccinia amorphae M.A. Curtis
Distr.: North America.
Notes: Autoecious.
Hosts: Amorpha, Parryella
Uropyxis daleae (Dietel & Holw.) Magnus
Distr.: Southwestern United States; Central and South America.
Notes: Autoecious.
Hosts: Dalea, Eysenhardtia, Marina
Uropyxis daleae (Dietel & Holw.) Magnus var. ***eysenhardtiae*** (Dietel & Holw.) J.W. Baxter
Uropyxis eysenhardtiae (Dietel & Holw.) Magnus
Distr.: AZ; Mexico.
Notes: Autoecious.
Hosts: Dalea, Eysenhardtia, Marina
Uropyxis holwayi (Arth.) Arth.
Distr.: AZ to Central America.
Notes: Autoecious.
Host: Eysenhardtia
Uropyxis petalostemonis (Farl.) De Toni
Uropyxis affinis Arth.
Distr.: Western North America.
Notes: Autoecious.
Host: Dalea

Urosporella Atk. Pyrenomycetes, Amphisphaeriales

Notes: Barr (Mycologia 58:690. 1966).
Urosporella alabamensis (Ellis & Everh.) Barr
Ceriospora alabamensis Ellis & Everh.
Urosporella americana Atk.
Distr.: Southeastern United States.
Hosts: Cassia, Lespedeza
Urosporella magnoliae (Ellis & Everh.) Barr
Dilophia magnoliae Ellis & Everh.
Distr.: WV(type).
Host: Magnolia

Ustacystis Zundel. Basidiomycetes, Ustilaginales

Ustacystis waldsteiniae (Peck) Zundel
Urocystis waldsteiniae Peck
Whetzelia waldsteiniae (Peck) Zundel
Urocystis gei Ellis & Everh.
Distr.: Temperate North America; Europe.
Notes: On herbaceous Rosaceae. Vánky (Illustrated genera of smut fungi. p. 124. 1987).
Hosts: Geum, Waldsteinia

Ustilaginoidea Bref. Hyphomycetes

Ustilaginoidea setariae Bref.
Distr.: LA; Brazil(type).
Host: Setaria
Ustilaginoidea virens (Cooke) Takah.
Distr.: Worldwide in rice-growing regions.
Notes: False smut of rice (Mulder & Holliday. C.M.I. Descr. 299. 1971).
Hosts: Oryza, Zea

Ustilago (Pers.) Roussel. Basidiomycetes, Ustilaginales

Notes: Vánky (Symb. Bot. Upsal. 24:188. 1985).
Ustilago aegopogonis Henn.
Distr.: Southwestern United States and adjacent Mexico.
Notes: On Poaceae.
Host: Hilaria
Ustilago affinis Ellis & Everh.
Distr.: Southeastern United States.
Notes: On Poaceae.
Host: Stenotaphrum
Ustilago affinis Ellis & Everh. var. ***hilariae*** G. Fisch. & Hirschhorn
Ustilago hilariae Henn. non Ellis & Tracy, hom. illeg.
Distr.: Southern United States; Mexico.
Host: Hilaria
Ustilago alsineae G.P. Clinton & Zundel
Distr.: Northwestern United States.
Host: Stellaria
Ustilago anomala J. Kunze ex G. Wint. in Rabenh.
Distr.: Widespread.
Notes: On Polygonaceae.
Host: Polygonum
Ustilago arthurii H. Hume
Distr.: Central and western United States.
Notes: On Poaceae.
Hosts: Glyceria, Scolochloa
Ustilago avenae (Pers.) Rostr.
Ustilago nigra Tapke
Ustilago perennans Rostr.
Distr.: Temperate northern hemisphere.
Notes: On Poaceae (Punithalingam & Waterston. C.M.I. Descr. 279. 1970).
Hosts: Arrhenatherum, Avena, Hordeum
Ustilago bethelii Zundel
Distr.: Western United States.
Host: Muhlenbergia
Ustilago bistortarum (DC.) Körn.
Sphacelotheca borealis (G.P. Clinton) Schellenb.
Ustilago candollei Tul. & C. Tul.
Sphacelotheca inflorescentiae (Trel.) Jaap
Distr.: Temperate northern hemisphere.
Host: Polygonum
Ustilago bosniaca Beck
Ustilago punctata G.P. Clinton
Distr.: North America; Asia; Europe.
Notes: On Polygonaceae (Vánky. Symb. Bot. Upsal. 24:197. 1985).
Host: Polygonum
Ustilago buchloes Ellis & Tracy
Ustilago coloradensis Zundel
Ustilago hieronymi J. Schröt.
Ustilago pseudohieronymi Zundel
Distr.: Central and western United States; Mexico.
Notes: On Poaceae.
Hosts: Bouteloua, Buchloe, Erioneuron, Muhlenbergia
Ustilago bullata Berk. in Hook.
Ustilago bromivora (Tul.) A. Fisch. v. Waldh.
Ustilago holwayi Dietel
Ustilago lorentziana Thuem.
Distr.: Widespread.
Notes: On Poaceae (Mordue & Waller. C.M.I. Descr. 718. 1981).
Hosts: Agrohordeum, Agropyron, Bromus, Elymus, Festuca, Hordeum, Sitanion
Ustilago bullata Berk. in Hook. var. ***macrospora*** (Farl.) G. Fisch.
Distr.: Central United States.
Host: Bromus
Ustilago calamagrostidis (Fuckel) G.P. Clinton
Distr.: Temperate northern hemisphere.
Host: Calamagrostis
Ustilago calandriniae G.P. Clinton
Distr.: CA.
Hosts: Calandrinia, Lewisia
Ustilago calcara D. Griffiths
Distr.: NM(type).
Host: Bouteloua

Ustilago chloridicola Henn.
Distr.: Western United States.
Host: Chloris
Ustilago claytoniae Shear
Distr.: Northwestern United States.
Host: Montia
Ustilago clintoniana Cif.
Ustilago violaceae (Pers.) Roussel var. *major* G.P. Clinton
Distr.: WA(type).
Host: Silene
Ustilago coicis Bref.
Distr.: Eastern United States(on imported seed).
Host: Coix
Ustilago commelinae (Kom.) Zundel
Tilletia commelinae Kom.
Distr.: NY; Asia; Soviet Union.
Notes: Duran & Safeeulla (Mycologia 60:231. 1968).
Host: Commelina
Ustilago compacta G. Fisch.
Distr.: OR(type).
Host: Lolium
Ustilago crameri Körn. in Fuckel
Distr.: Widespread.
Notes: Ainsworth (C.M.I. Descr. 78. 1965).
Host: Setaria
Ustilago cynodontis (Henn.) Henn.
Ustilago cynodontis (Pass.) Curzi in Curzi & Barbaini
Distr.: Cosmopolitan.
Notes: Mulder & Holliday (C.M.I. Descr. 297. 1971).
Host: Cynodon
Ustilago davisii Liro
Ustilago longissima (Sowerby ex Schlechtend.) Meyen var. *macrospora* J.J. Davis, nom. nud.
Distr.: Northern hemisphere.
Host: Glyceria
Ustilago duriaeana Tul. & C. Tul.
Distr.: North America; Asia; Europe.
Notes: On Caryophyllaceae (Vánky. Symb. Bot. Upsal. 24:206. 1985).
Hosts: Cerastium, Stellaria
Ustilago echinata J. Schröt.
Distr.: Temperate northern hemisphere.
Host: Phalaris
Ustilago elegans D. Griffiths
Distr.: Southwestern United States.
Host: Chloris
Ustilago enneapogonis G. Fisch.
Distr.: AZ(type).
Host: Enneapogon
Ustilago eriocauli (Massee) G.P. Clinton
Distr.: Northeastern United States.
Host: Eriocaulon
Ustilago esculenta Henn.
Distr.: DC (greenhouses); Asia.
Notes: Terrell & Batra (Econ. Bot. 36:274. 1982) reviewed the uses of this edible smut and the potential for infection in the United States.
Host: Zizania
Ustilago gayophyti Harkn.
Distr.: Western United States.
Host: Gayophytum
Ustilago grandis Fr.
Distr.: CA; Asia; Europe.
Host: Phragmites
Ustilago halophila Speg.
Distr.: Western North America.
Host: Distichlis
Ustilago halophiloides G. Fisch.
Distr.: OR(type).
Host: Distichlis
Ustilago heterogena Henn.
Ustilago togata Liro
Distr.: North America.
Notes: On Poaceae.
Hosts: Leptochloa, Panicum
Ustilago heufleri Fuckel
Ustilago erythronii G.P. Clinton in Peck
Distr.: Temperate northern hemisphere.
Notes: On Liliaceae.
Host: Erythronium
Ustilago hilariicola G. Fisch.
Distr.: Southwestern United States.
Host: Hilaria
Ustilago hordei (Pers.) Lagerh.
Ustilago avenae (Pers.) Rostr. var. *levis* Kellerm. & Swingle
Ustilago levis (Kellerm. & Swingle) Magnus
Distr.: Widespread.
Notes: On Poaceae (Waller & Mordue. C.M.I. Descr. 749. 1983).
Hosts: Agropyron, Avena, Elymus, Hordeum, Secale
Ustilago hyalino-bipolaris G. Fisch.
Distr.: Southwestern United States.
Host: Muhlenbergia
Ustilago hydropiperis (Schumach.) J. Schröt.
Sphacelotheca hydropiperis (Schumach.) de Bary
Distr.: Cosmopolitan.
Hosts: Bistorta, Polygonum
Ustilago hypodytes (Schlechtend.) Fr.
Ustilago funalis Ellis & Everh.
Ustilago nummularia Speg.
Ustilago spegazzinii Hirschhorn
Ustilago spegazzinii Hirschhorn var. *agrestis* G. Fisch. & Hirschhorn
Ustilago sporoboli Ellis & Everh.
Distr.: Widespread.
Notes: On Poaceae (Mordue. C.M.I. Descr. 809. 1984).
Hosts: Agropyron, Elymus, Elysitanion, Hilaria, Hordeum, Leucopoa, Leymus, Melica, Oryzopsis, Phalaris, Phleum, Poa, Puccinellia, Secale, Sitanion, Sporobolus, Stipa
Ustilago jacksonii Zundel & A.A. Dunlap
Distr.: Western United States.
Host: Stipa
Ustilago kuehneana R. Wolff
Ustilago rumicis (Berk.) G.P. Clinton
Ustilago utriculosa Tul. & C. Tul. var. *rumicis* Berk.
Distr.: SC; Europe.
Notes: Vánky (Symb. Bot. Upsal. 24:218. 1985).
Host: Rumex
Ustilago longissima (Sowerby) Meyen
Distr.: Widespread.
Hosts: Glyceria, Puccinellia
Ustilago luzulae Sacc.
Cintractia luzulae (Sacc.) G.P. Clinton
Distr.: North America; Europe.
Host: Luzula
Ustilago lycuroides D. Griffiths
Distr.: Southwestern United States.
Host: Lycurus
Ustilago mexicana Ellis & Everh.
Distr.: CA; Mexico.
Notes: On Poaceae.
Host: Muhlenbergia
Ustilago minima Arth.
Ustilago stipae Cif.
Distr.: North and South America; Asia; Europe.
Notes: On Poaceae.
Hosts: Oryzopsis, Stipa
Ustilago minor Norton
Distr.: Central and western United States.
Notes: On Poaceae.
Hosts: Bouteloua, Muhlenbergia
Ustilago montagnei Tul. & C. Tul.
Cintractia montagnei (Tul. & C. Tul.) Magnus
Distr.: Widespread.
Host: Rhynchospora
Ustilago montagnei Tul. & C. Tul. var. **major** Desmaz.
Cintractia major (Desmaz.) Liro
Cintractia montagnei (Tul. & C. Tul.) Magnus var. *major* Desmaz.
Distr.: Widespread.
Host: Rhynchospora
Ustilago muhlenbergiae Henn.
Distr.: Southwestern United States.
Host: Muhlenbergia
Ustilago mulfordiana Ellis & Everh.

Distr.: Southern Canada to Mexico.
Hosts: Festuca, Vulpia
Ustilago ornata Tracy & Earle
Distr.: Southeastern United States.
Host: Leptochloa
Ustilago oxalidis Ellis & Tracy
Distr.: Widespread.
Host: Oxalis
Ustilago panici-proliferi Henn.
Distr.: KY, HI; Mexico.
Host: Panicum
Ustilago parlatorei A. Fisch. v. Waldh.
Distr.: North and South America; Europe.
Host: Rumex
Ustilago penniseti Rabenh.
Distr.: VA; Asia(type).
Host: Pennisetum
Ustilago piperi G.P. Clinton
Distr.: Northern hemisphere.
Host: Polygonum
Ustilago residua G.P. Clinton
Distr.: Temperate North America.
Host: Danthonia
Ustilago schroeteriana Henn.
Distr.: Subtropical and tropical regions.
Host: Paspalum
Ustilago scitaminea Syd. & P. Syd.
Distr.: Subtropical and tropical regions.
Notes: Sugarcane smut. This disease is becoming increasingly important in sugarcane-growing regions (Ainsworth. C.M.I. Descr. 80. 1965). Hoy et al. (Pl. Dis. 70:59. 1986).
Host: Saccharum
Ustilago segetum (Bull.:Pers.) Roussel
Ustilago kolleri Wille
Distr.: Widespread.
Notes: On Poaceae.
Host: Avena
Ustilago serpens (P. Karst.) Lindeberg
Ustilago aculeata (Ule) Liro
Ustilago elymicola Syd.
Ustilago macrospora Desmaz.
Distr.: Temperate northern hemisphere.
Notes: On Poaceae.
Hosts: Agropyron, Bromus, Elymus, Phalaris
Ustilago shastensis Zundel in W.B. Cooke
Distr.: CA.
Host: Polygonum
Ustilago shiraiana Henn.
Distr.: Widespread, generally in warmer regions.
Notes: On bamboo.
Hosts: Bambusa, Phyllostachys, Poaceae
Ustilago sieglingiae Ricker
Distr.: Central and eastern United States.
Host: Triplasis
Ustilago sparsa Underw.
Distr.: Southeastern United States; India; Japan.
Host: Dactyloctenium
Ustilago speculariae J.A. Stevenson
Distr.: OK, TX.
Notes: On Campanulaceae.
Host: Triodanis
Ustilago spermophora Berk. & M.A. Curtis ex De Toni in Sacc.
Distr.: Widespread.
Host: Eragrostis
Ustilago striiformis (Westend.) Niessl
Tilletia striiformis (Westend.) Sacc.
Distr.: Widespread.
Notes: On Poaceae (Mordue & Waller. C.M.I. Descr. 717. 1981).
Hosts: Agrohordeum, Agropyron, Agrostis, Alopecurus, Ammophila, Andropogon, Arctagrostis, Beckmannia, Bromus, Calamagrostis, Dactylis, Danthonia, Deschampsia, Elymus, Festuca, Holcus, Hordeum, Hystrix, Koeleria, Leucopoa, Leymus, Lolium, Melica, Muhlenbergia, Phalaris, Phleum, Poa, Puccinellia, Sitanion, Sporobolus, Trisetum
Ustilago syntherismae (Schwein.) Peck
Caeoma syntherismae Schwein.
Sorosporium syntherismae (Schwein.) Farl. in Farl. & Seym.
Ustilago rabenhorstiana Kühn
Distr.: Widespread.
Notes: Vánky (Symb. Bot. Upsal. 24:243. 1985).
Host: Digitaria
Ustilago tenuispora Cif.
Distr.: FL; South America.
Notes: On Polygonaceae.
Host: Polygonum
Ustilago trebouxii Syd. & P. Syd.
Ustilago sitanii G. Fisch.
Ustilago underwoodii Zundel
Distr.: Temperate northern hemisphere.
Notes: On Poaceae.
Hosts: Agrohordeum, Distichlis, Elymus, Elysitanion, Hordeum, Panicum, Poa, Puccinellia, Sitanion, Stipa
Ustilago trichophora (Link) Körn.
Cintractia crus-galli (Tracy & Earle) Magnus
Ustilago crus-galli Tracy & Earle
Ustilago sphaerogena Burrill in Sacc.
Distr.: Widespread.
Host: Echinochloa
Ustilago tricuspidis Ellis & B.T. Galloway
Distr.: Central and southern United States.
Host: Tridens
Ustilago triplasidis Ellis & Everh.
Distr.: Southern United States.
Host: Triplasis
Ustilago tritici (Pers.) Rostr.
Ustilago nuda (C.N. Jensen) Rostr., nom. nud.
Distr.: Widespread.
Notes: On Poaceae (Punithalingam & Waterston. C.M.I. Descr. 280. 1970 as *Ustilago nuda*).
Hosts: Agropyron, Hordeum, Lolium, Secale, Triticum
Ustilago turcomanica Tranzschel ex K. Vánky
Distr.: WA(imported grass seed); Soviet Union.
Notes: Fischer & Meiners. (Mycologia 44:207. 1952).
Host: Eremopyrum
Ustilago uniolae Ellis & Everh.
Distr.: Southeastern United States.
Hosts: Chasmanthium, Uniola
Ustilago utriculosa (Nees) Unger
Distr.: North America; Great Britain.
Host: Polygonum
Ustilago vaillantii Tul. & C. Tul.
Distr.: Widespread.
Notes: On Liliaceae (Mordue. C.M.I. Descr. 810. 1984).
Hosts: Hyacinthus, Muscari, Scilla
Ustilago vilfae G. Wint.
Ustilago hilariae Ellis & Tracy
Tilletia subfusca H. Hume
Distr.: United States.
Notes: On Poaceae.
Hosts: Hilaria, Sporobolus
Ustilago vinosa Tul. & C. Tul.
Distr.: Northern hemisphere, cooler parts.
Notes: On Polygonaceae.
Host: Oxyria
Ustilago violacea (Pers.:Pers.) Roussel
Ustilago antherarum (DC.) Fr.
Ustilago stellariae (Sowerby) Liro
Ustilago violacea (Pers.:Pers.) Roussel var. *stellariae* (Sowerby) Savile
Distr.: Worldwide, except Australia.
Notes: On Caryophyllaceae (Waller & Mordue. C.M.I. Descr. 750. 1983).
Hosts: Arenaria, Cerastium, Dianthus, Lychnis, Silene
Ustilago vuijckii Oudem. & M. Beijerinck
Distr.: North America; Asia; Australia; Europe.
Notes: On Juncaceae (Vánky. Symb. Bot. Upsal. 24:255. 1985).
Host: Luzula
Ustilago williamsii (D. Griffiths) Lavrov
Distr.: Widespread.
Notes: On Poaceae.
Hosts: Hordeum, Oryzopsis, Stipa, Stiporyzopsis

Ustilago zeae (Beckm.) Unger
Lycoperdon zeae Beckm.
Ustilago maydis (DC.) Corda
Distr.: Widespread.
Notes: On *Zea* and related genera. Widely known as *U. maydis*. The earliest name for this fungus is *Lycoperdon zeae* Beckm. 1768 (Mordue & Ainsworth. Mycol. Pap. 154:61. 1984). Vánky (Symb. Bot. Upsal. 24:12. 1985) rejected this name stating that *U. zeae* is preoccupied by *U. zeae* (Link) Unger. Schweinitz treated *zeae* in *Uredo*, not *Ustilgo*; hence the correct name is *U. zeae* (Ainsworth. C.M.I. Descr. 79. 1965 as *Ustilago maydis*).
Host: Zea

Ustulina Tul. & C. Tul. Pyrenomycetes, Xylariales
Notes: Often considered a synonym of *Hypoxylon*.
Ustulina deusta (Hoffm.:Fr.) Lind
Hypoxylon deustum (Hoffm.:Fr.) Grev.
Hypoxylon ustulatum Bull.
Ustulina vulgaris Tul. & C. Tul.
Nodulisporium sp. (anamorph)
Distr.: Cosmopolitan.
Notes: Hawksworth (C.M.I. Descr. 360. 1972).
Hosts: Acer, Aleurites, Alnus, Citrus, Fagus, Ilex, Liriodendron, Nyssa, Poncirus, Quercus, Tilia, Ulmus
Ustulina linearis Rehm
Distr.: KY; Germany(type).
Host: Fagus

Uthatobasidium Donk. Basidiomycetes, Tulasnellales
Uthatobasidium fusisporum (J. Schröt.) Donk
Coniophora vaga Burt
Distr.: Temperate northern hemisphere.
Notes: Saprobic. See notes under *Botryobasidium flavescens*.
Host: Pinus

Valsa Fr., nom. cons. prop. Pyrenomycetes, Diaporthales
Notes: Spielman (Canad. J. Bot. 63:1355. 1985) discussed the *Valsa* species on hardwoods in North America.
Valsa abietis (Fr.:Fr.) Fr.
Cytospora abietis Sacc. (anamorph)
Distr.: Circumboreal.
Notes: Cankers on conifers.
Hosts: Abies, Larix, Pinus, Pseudotsuga, Thuja, Tsuga
Valsa ambiens (Pers.:Fr.) Fr.
Valsa ambiens (Pers.:Fr.) Fr. f. *eleagni* Rehm
Valsa amorphae Ellis & Everh.
Valsa clausa Cooke & Ellis
Valsa laurina Cooke & Ellis
Valsa menispermi Ellis & Holw.
Cryptodiaporthe myinda (Cooke & Ellis) Wehmeyer
Valsa myinda Cooke & Ellis
Valsa pauperata Cooke & Ellis
Valsa ribicola Ellis & Everh.
Valsa symphoricarpi Rehm
Cytospora leucosperma (Pers.:Fr.) Fr. (anamorph)
Distr.: Eastern North America; temperate Europe.
Hosts: Acer, Amorpha, Asimina, Betula, Carya, Celastrus, Celtis, Clethra, Comptonia, Cornus, Corylus, Crataegus, Elaeagnus, Fagus, Fraxinus, Hamamelis, Juglans, Larix, Lyonia, Magnolia, Malus, Menispermum, Morus, Nyssa, Ostrya, Physocarpus, Populus, Pyrus, Quercus, Ribes, Rosa, Rubus, Salix, Sambucus, Sassafras, Shepherdia, Symphoricarpos, Tilia, Ulmus, Viburnum
Valsa ambiens (Pers.:Fr.) Fr. f. ***crataegi*** Rehm
Distr.: ND(type).
Host: Crataegus
Valsa ambiens (Pers.:Fr.) Fr. f. ***rosae*** Rehm
Distr.: ND(type).
Host: Rosa
Valsa ambiens (Pers.:Fr.) Fr. subsp. ***leucostomoides*** (Peck) Spielman
Valsa leucostomoides Peck
Valsa saccharina Rehm
Distr.: Temperate North America; Europe.
Hosts: Acer, Fraxinus, Malus, Pyrus
Valsa ampelopsidis Ellis
Distr.: NJ(type).
Notes: Spielman (Canad. J. Bot. 63:1367. 1985) stated that this is a species of *Melanconis*.
Host: Parthenocissus
Valsa auerswaldii Nitschke
Cytospora personata Fr. (anamorph)
Distr.: WA; Europe.
Host: Prunus
Valsa betulina Nitschke
Distr.: MA, NY; Europe.
Host: Betula
Valsa boreella P. Karst.
Distr.: AK, western North America; Finland(type).
Host: Salix
Valsa castanicola Berk. & M.A. Curtis
Fenestella castanicola (Berk. & M.A. Curtis) Sacc.
Distr.: VA(type).
Notes: An isotype specimen was found to lack peritheical contents (Spielman. Canad. J. Bot. 63:1368. 1985).
Host: Castanea
Valsa cenisia De Not.
Distr.: Central and eastern North America; Europe.
Host: Juniperus
Valsa ceratosperma (Tode:Fr.) Maire
Valsa americana Berk. & M.A. Curtis
Valsa caryigena Berk. & M.A. Curtis
Valsa ceratophora Tul. & C. Tul.
Valsa chlorodisca Cooke & Ellis
Valsa clavigera Dearn. & Barth.
Valsa conspurcata (Schwein.) Cooke
Valsa decorticans (Fr.:Fr.) Fr.
Valsa delicatula Cooke & Ellis
Valsa diospyri Ellis & Everh.
Valsa etherialis Ellis & Everh.
Valsa gossypina Cooke
Valsa leiphaemoides Berk. & M.A. Curtis
Valsa lutescens Ellis
Eutypella maclurae (Cooke & Ellis) Ellis & Everh.
Valsa maclurae Cooke & Ellis
Valsa macrocarpa Ellis & Everh.
Valsa minutella Peck
Valsa morigena Berk. & M.A. Curtis
Valsa multiplex Cooke & Ellis
Valsa nyssae Cooke & Ellis
Valsa rubi Peck
Eutypella rugiella (Cooke & Ellis) Sacc.
Valsa rugiella Cooke & Ellis
Cytospora sacculus (Schwein.) Gvritischvili (anamorph)
Distr.: Cosmopolitan.
Hosts: Acer, Alnus, Amelanchier, Amorpha, Arbutus, Aronia, Betula, Bignonia, Callicarpa, Campsis, Carpinus, Carya, Castanea, Celtis, Cornus, Crataegus, Diospyros, Gaylussacia, Gleditsia, Gossypium, Holodiscus, Ilex, Juglans, Leucothoe, Liquidambar, Liriodendron, Lonicera, Lyonia, Maclura, Magnolia, Malus, Morus, Myrica, Nyssa, Ostrya, Oxydendrum, Populus, Prunus, Pyrus, Quercus, Rhododendron, Rhus, Ribes, Rosa, Rubus, Salix, Sassafras, Tilia, Ulmus, Umbellularia, Vaccinium, Vitis
Valsa chlorina Pat.
Eutypella cocos Ferd. & Winge
Distr.: Tropical regions.
Host: Cocos
Valsa collicula (M. Wormsk.) Cooke
Distr.: Eastern United States; Europe.
Host: Pinus
Valsa coronata (Hoffm.:Fr.) Fr.
Distr.: North America(questionable); Europe.
Hosts: Castanopsis, Cornus, Quercus
Valsa cypri (Tul.) Tul. & C. Tul.

Valsa chionanthi Ellis & Everh.
Valsa fraxinina Peck
Cytospora pruinosa (Fr.:Fr.) Sacc. (anamorph)
Distr.: North America; Europe.
Notes: On hardwoods.
Hosts: Chionanthus, Fraxinus
Valsa eucalypti Cooke & Harkn.
Distr.: CA(type), FL.
Notes: Spielman (Canad. J. Bot. 63:1370. 1985) stated that this is a *Leucostoma*.
Host: Eucalyptus
Valsa fallax Nitschke
Cytospora corni Westend. (anamorph)
Distr.: AK; Europe.
Host: Cornus
Valsa friesii (Duby) Fuckel
Valsa juniperina Cooke
Cytospora pinastri Fr.:Fr. (anamorph)
Distr.: Circumboreal on conifers.
Host: Juniperus
Valsa germanica Nitschke
Cytospora germanica Sacc. (anamorph)
Distr.: MT; Europe.
Host: Salix
Valsa holodiscina Fairm.
Distr.: CA(type).
Host: Holodiscus
Valsa intermedia Nitschke
Cytospora intermedia Sacc. (anamorph)
Distr.: NC; Europe.
Host: Quercus
Valsa lavaterae Cooke & Harkn.
Distr.: CA(type).
Notes: Spielman (Canad. J. Bot. 63:1372. 1985) stated that this is a species of *Eutypella*.
Host: Lavatera
Valsa leucopsis Ellis & Everh.
Distr.: Temperate regions.
Notes: Spielman (Canad. J. Bot. 63:1372. 1985) stated that this is a species of *Leucostoma*.
Host: Fraxinus
Valsa liquidambaris (Schwein.) Cooke
Distr.: Eastern United States.
Notes: Spielman (Canad. J. Bot. 63:1372. 1985) stated that this is a species of *Calosphaeria*.
Host: Liquidambar
Valsa lupini Cooke & Harkn.
Distr.: CA(type).
Notes: Spielman (Canad. J. Bot. 63:1372. 1985) stated that this fungus belongs in *Eutypella*.
Host: Lupinus
Valsa melanodiscus G. Otth
Valsa alni Peck
Valsa truncata Cooke & Peck
Cytospora umbrina (Bonord.) Sacc. (anamorph)
Distr.: Temperate North America; Europe.
Hosts: Alnus, Betula
Valsa microstoma (Pers.:Fr.) Fr.
Cytospora microstoma Sacc. (anamorph)
Distr.: GA; Europe.
Host: Rhododendron
Valsa modesta (Schwein.) Cooke
Distr.: IA, PA.
Notes: Spielman (Canad. J. Bot. 63:1373. 1985) stated that this fungus probably belongs in *Eutypella*.
Host: Ulmus
Valsa opulifoliae Peck
Distr.: NY, WV.
Notes: Spielman (Canad. J. Bot. 63:1373. 1985) stated that this is a species of *Diaporthe*.
Host: Physocarpus
Valsa pallida Ellis
Distr.: NY(type). Spielman (Canad. J. Bot. 63:1374. 1985) was unable to determine the identity of this species.
Host: Salix
Valsa papyriferae (Schwein.) Cooke
Valsella papyriferae (Schwein.) Berl. & Voglino
Distr.: OR, PA(type).
Notes: Spielman (Canad. J. Bot. 63:1374. 1985) stated that this is a species of *Diatrypella*.
Host: Malus
Valsa pini (Albertini & Schwein.) Fr.
Cytospora pini Desmaz. (anamorph)
Distr.: North America; Europe.
Hosts: Abies, Pinus
Valsa polyspora Nitschke
Valsella polyspora (Nitschke) Sacc.
Distr.: WA; Germany(type).
Host: Betula
Valsa ribesia P. Karst.
Distr.: CA; Finland(type).
Host: Ribes
Valsa salicina (Pers.:Fr.) Fr.
Valsa salicina (Pers.:Fr.) Fr. var. *tetraspora* Fr.
Valsa socialis Ellis & Everh.
Cytospora fugax (Bull.:Fr.) Fr. (anamorph)
Distr.: North America; Europe.
Host: Salix
Valsa sordida Nitschke
Cytospora chrysosperma (Pers.:Fr.) Fr. (anamorph)
Distr.: Circumboreal.
Notes: Mostly on *Populus* and *Salix* spp.
Hosts: Acer, Betula, Populus, Salix, Sambucus, Sorbus, Ulmus
Valsa subclypeata Cooke & Peck
Distr.: NY, VA.
Notes: Spielman (Canad. J. Bot. 63:1376. 1985) stated that this is a species of *Leucostoma*.
Host: Sassafras
Valsa superficialis Nitschke
Distr.: MI; Germany(type).
Host: Pinus
Valsa thujae Peck
Distr.: NY.
Host: Thuja
Valsa vitigera Cooke
Distr.: IA; Europe(type).
Hosts: Parthenocissus, Vitis
Valsa weiriana Petr.
Distr.: ID(type).
Host: Thuja

Valsaria Ces. & De Not. Pyrenomycetes, Diaporthales

Valsaria bignoniae (Schwein.) Cooke
Valsa bignoniae (Schwein.) M.A. Curtis
Distr.: PA(type).
Notes: Spielman (Canad. J. Bot. 63:1368. 1985) examined the type collection but was unable to determine the disposition of this species, suggesting that it belonged either in *Melanconis* or was a synonym of *Valsa ceratosperma*.
Host: Campsis
Valsaria celtidis (Cooke) Sacc.
Valsa celtidis Cooke
Distr.: GA, SC.
Notes: Spielman (Canad. J. Bot. 63:1368. 1985) stated that this species is correctly placed in *Valsaria*.
Host: Celtis
Valsaria diospyri (Schwein.:Fr.) De Not.
Distr.: SC; Italy.
Host: Diospyros
Valsaria eucalypti (Kalchbr. & Cooke) Sacc.
Distr.: CA; southern Africa(type).
Host: Eucalyptus
Valsaria exasperans (W. Gerard) Ellis & Everh.
Diatrype exasperans W. Gerard
Distr.: Eastern United States.
Hosts: Acer, Prunus, Quercus, Ulmus
Valsaria gemmata (Berk. & Ravenel) Ellis & Everh.

Hypoxylon gemmatum Berk. & Ravenel
Distr.: GA, SC.
Host: Pinus
Valsaria insitiva (Tode:Fr.) Ces. & De Not.
Diatrype insitiva (Tode:Fr.) Fr.
Valsaria aethiops (Cooke & Ellis) Sacc.
Valsaria rudis (P. Karst. & Har.) Theiss. & Syd.
Distr.: North America; Europe.
Notes: On various woody substrates.
Hosts: Aesculus, Amorpha, Carpinus, Carya, Celastrus, Fraxinus, Gleditsia, Hibiscus, Magnolia, Malus, Morus, Quercus
Valsaria moroides (Cooke & Peck) Sacc. var. **acerina** Rehm
Distr.: ND(type).
Host: Acer
Valsaria nigrofacta (Cooke & Ellis) Sacc.
Valsa nigrofacta Cooke & Ellis
Distr.: NJ(type).
Notes: Spielman (Canad. J. Bot. 63:1373. 1985) examined the type specimen and concluded that this species belonged in *Valsaria*.
Host: Sassafras
Valsaria notarisii (Mont.) Sacc.
Valsa notarisii Mont.
Distr.: GA, SC; Algeria(type).
Host: Gleditsia
Valsaria peckii (E. Howe) Sacc.
Valsa peckii E. Howe
Distr.: NJ, NY.
Notes: Spielman (Canad. J. Bot. 63:1374. 1985) suggested placing this species in *Endoxylina* but the transfer has not been made.
Hosts: Kalmia, Vaccinium
Valsaria phoradendri (Berk. & M.A. Curtis) Berl. & Voglino
Distr.: Reported from an unknown locality in the United States.
Host: Phoradendron
Valsaria rubricosa (Fr.:Fr.) Sacc.
Valsaria hypoxyloides (Rehm) Rehm
Distr.: North America; West Indies; Central America; Europe.
Notes: On dead wood.
Hosts: Acer, Carpinus, Carya, Castanea, Liriodendron, Magnolia, Melia, Morus, Paulownia, Quercus, Sorbus, Wisteria

Valsella Fuckel. Pyrenomycetes, Diaporthales

Valsella alnicola Berl.
Distr.: CA; Italy(type).
Host: Alnus
Valsella furva (P. Karst.) Sacc.
Distr.: WA; Finland(type).
Host: Alnus
Valsella laschii (Nitschke) Sacc.
Valsa laschii Nitschke
Distr.: GA; Germany(type).
Hosts: Amelanchier, Carpinus, Liquidambar, Liriodendron
Valsella melastoma (Fr.) Fuckel
Valsa melastoma Fr.
Distr.: Western United States; Europe.
Host: Malus
Valsella nigroannulata Fuckel
Distr.: Temperate North America; Europe.
Notes: On limbs.
Host: Salix

Valseutypella Höhn. Pyrenomycetes, Diaporthales

Valseutypella tristicha (De Not.) Höhn.
Diatrype tristicha De Not.
Distr.: Temperate North America; Europe.
Notes: On limbs.
Host: Rosa

Vararia P. Karst. Basidiomycetes, Aphyllophorales

Notes: Also see the segregate genus *Dichostereum* Pilát. Gilbertson (Pap. Michigan Acad. Sci. 50:161. 1965); Boidin & Lanquetin (Bull. Soc. Mycol. France 91:457. 1975).
Vararia athabascensis R.L. Gilbertson
Distr.: North America.
Notes: On wood.
Hosts: Picea, Pinus, Populus
Vararia boreale Pouzar
Distr.: Temperate northern hemisphere.
Notes: On conifers. Generally reported as *Grandinia granulosa* or *Vararia granulosa*, this species probably belongs in *Dichostereum*, (Boidin & Lanquetin. Mycotaxon 6:284. 1977). Jülich (Int. J. Mycol. Lichenol. 1:27. 1982); Pouzar (Ceska Mykol. 36:72. 1982).
Hosts: Abies, Crataegus, Larix, Picea, Pinus, Pseudotsuga
Vararia gomezii Boidin & Lanquetin
Distr.: Gulf Coast Region; Africa.
Notes: On dead and living branches.
Host: Juniperus
Vararia investiens (Schwein.) P. Karst.
Corticium investiens (Schwein.) Bres.
Distr.: Widespread.
Notes: On wood.
Hosts: Abies, Acer, Castanea, Fagus, Larix, Paxistima, Picea, Pinus, Populus, Quercus, Tsuga
Vararia pectinata (Burt) D.P. Rogers & H. Jacks.
Corticium pectinatum Burt
Distr.: Eastern United States.
Notes: On hardwoods.
Host: Vitis
Vararia phyllophila (Massee) D.P. Rogers & H. Jacks.
Peniophora phyllophila Massee
Distr.: Southern United States to Central America.
Notes: On twigs and leaves.
Hosts: Pinus, Populus, Quercus
Vararia racemosa (Burt) D.P. Rogers & H. Jacks.
Corticium racemosum Burt
Distr.: North America.
Notes: On wood.
Hosts: Abies, Larix, Picea, Pinus, Populus, Pseudotsuga, Thuja
Vararia tropica A.L. Welden
Distr.: AZ, PR.
Notes: On wood.
Host: Prosopis

Vargamyces S. Toth. Hyphomycetes

Vargamyces aquaticus (I.A. Dudka) S. Toth
Camposporium aquaticum I.A. Dudka
Distr.: IL; Europe.
Notes: On decaying leaves of deciduous trees in aquatic habitats (Toth. Acta Bot. Acad. Sci. Hung. 25:403. 1979).
Hosts: Acer, Quercus

Varicosporina S.P. Meyers & Kohlmeyer. Hyphomycetes

Notes: Kohlmeyer & Kohlmeyer (Marine mycology. p. 471. 1979).
Varicosporina ramulosa S.P. Meyers & Kohlmeyer
Distr.: Widespread in marine habitats.
Host: Zostera

Velutarina Korf. Discomycetes, Helotiales

Velutarina rufo-olivacea (Albertini & Schwein.) Korf
Velutaria rufo-olivacea (Albertini & Schwein.) Fuckel
Dermatea lobata Ellis
Trichopeziza umbilicata (Berk. & M.A. Curtis) Sacc.
Distr.: Temperate regions.
Notes: On dead twigs and branches.
Hosts: Arbutus, Liquidambar, Rhododendron, Rosa, Rubus

Veluticeps (Cooke) Pat. Basidiomycetes, Aphyllophorales

Veluticeps berkeleyi (Berk. & M.A. Curtis) Cooke
Veluticeps fusca C.J. Humphrey & W.H. Long in Burt
Distr.: Central and western United States.
Hosts: Abies, Pinus

Venturia Sacc., nom. cons. Loculoascomycetes, Dothideales

Notes: Barr (Canad. J. Bot. 46:799. 1968); Sivanesan (Biblioth. Mycol. 59:1. 1977).
Venturia acerina Plakidas ex Barr
Cladosporium humile J.J. Davis (anamorph)
Distr.: Northeastern North America.
Notes: The anamorph causes spots on living leaves of *Acer* spp. while the teleomorph forms on overwintered dead leaves.
Host: Acer
Venturia adusta (Fuckel) E. Müller
Sphaerella adusta Fuckel
Asterina chamaenerii (Rostr.) Ellis & Everh.
Sphaerella effusa Sacc. & Syd.
Distr.: Temperate and arctic regions.
Notes: On dead leaves (Corlett & Egger. Fungi Canadenses 194. 1981).
Host: Epilobium
Venturia anthophila Pass.
Distr.: MD, MN; Italy(type).
Host: Hydrangea
Venturia applanata Ellis & G. Martin
Distr.: FL.
Notes: Possibly a species of *Chaetothyrina* (Barr. Canad. J. Bot. 46:858. 1968).
Host: Magnolia
Venturia asteromorpha (Lib.) E. Müller
Euryachora epilobii (Fr.:Fr.) Höhn.
Spilosticta epilobii (Fr.:Fr.) Petr.
Distr.: Temperate North America; Europe.
Host: Epilobium
Venturia carpophila E.E. Fisher
Cladosporium carpophilum Thuem. (anamorph)
Distr.: Widespread.
Notes: Scab and freckle (Sivanesan. C.M.I. Descr. 402. 1974; Sivanesan. The bitunicate Ascomycetes. p. 609. 1984).
Host: Cheirodendron
Venturia chlorospora (Ces.) P. Karst.
Distr.: North America; temperate Europe.
Notes: On leaves (Corlett. Fungi Canadenses 225. 1982).
Host: Salix
Venturia coprosmae (F. Stevens) E. Müller
Actinodothidopsis coprosmae F. Stevens
Distr.: HI.
Host: Coprosma
Venturia crataegi Aderhold
Venturia inaequalis (Cooke) G. Wint. f. sp. *crataegi* R. Menon
Fusicladium crataegi Aderhold (anamorph)
Distr.: FL; Europe.
Notes: Sivanesan (The bitunicate Ascomycetes. p. 612. 1984).
Host: Crataegus
Venturia ditricha (Fr.:Fr.) P. Karst.
Venturia maculans Peck
Fusicladium betulae (Roberge & Desmaz.) Aderhold (anamorph)
Distr.: Temperate North America; Europe.
Notes: On leaves.
Host: Betula
Venturia erysipheoides Ellis & Everh.
Distr.: LA(type); WI.
Notes: Possibly a species of *Niesslia* (Barr. Canad. J. Bot. 46:858. 1968).
Hosts: Panicum, Schizachyrium
Venturia formosa Ellis & G. Martin
Distr.: FL.
Notes: Possibly a species of *Johansonia* (Barr. Canad. J. Bot. 46:858. 1968).
Host: Osmanthus
Venturia fraxini (Fr.) Aderhold
Venturia ditricha (Fr.) P. Karst. var. *fraxinicola* Rehm
Spilocaea fraxini (Aderhold) Sivanesan (anamorph)
Distr.: FL; Europe.
Notes: On petioles and leaves. Withering and premature leaf fall (Sivanesan. The bitunicate Ascomycetes. p. 614. 1984).
Host: Fraxinus
Venturia geranii (Fr.:Fr.) G. Wint. in Rabenh.
Dothidea geranii Fr.:Fr.
Coleroa geranii (Fr.:Fr.)Traverso
Stigmatea geranii (Fr.:Fr.) Fr.
Stigmatea confertissima Fuckel
Distr.: CA, MD; Europe.
Notes: Additional synonyms were listed by Müller & Arx (Beitr. Kryptogamenfl. Schweiz 11(2):410. 1962) and Sivanesan (Biblioth. Mycol. 59:68. 1977).
Host: Geranium
Venturia inaequalis (Cooke) G. Wint.
Venturia inaequalis (Cooke) G. Wint. var. *cinerascens* (Fuckel) Aderhold
Spilocaea pomi Fr.:Fr. (anamorph)
Distr.: Worldwide in apple-growing regions.
Notes: Scab or black spot of apple, reported from a variety of plants in the Rosaceae (Corlett. Fungi Canadenses 35. 1974; Sivanesan. The bitunicate Ascomycetes. p. 615. 1984; Sivanesan & Waller. C.M.I. Descr. 401. 1974).
Hosts: Crataegus, Eriobotrya, Malus, Pyracantha, Pyrus, Sorbus
Venturia kunzei Sacc. var. ***ramicola*** Sacc. & Scalia
Distr.: AK(type).
Notes: Although the type variety is considered a synonym of *Coleroa chaetomium*, this variety has not been treated.
Host: Rubus
Venturia liriodendri Hanlin
Distr.: GA.
Notes: On living leaves.
Host: Liriodendron
Venturia macularis (Fr.:Fr.) E. Müller & Arx
Mycosphaerella macularis (Fr.:Fr.) J. Schröt. in Cohn
Mycosphaerella orbicularis (Peck) House
Didymosphaeria populifolia Ellis & Everh.
Venturia tremulae Aderhold
Pollaccia radiosa (Lib.) Baldacci & Cif. (anamorph)
Distr.: Temperate North America; Europe.
Notes: Shoot tip or scab (Sivanesan. C.M.I. Descr. 403. 1974). Morelet (Crypt. Mycol. 6:101. 1985) concluded that *V. tremulae* is the teleomorph of *Pollaccia radiosa*.
Host: Populus
Venturia nebulosa Ellis & Everh.
Distr.: AL.
Notes: Probably a species of *Epipolaeum*.
Host: Eragrostis
Venturia palustris Sacc., E. Bommer & M. Rousseau
Distr.: WI; Europe(type).
Host: Potentilla
Venturia populina (Vuill.) L. Fabricius in Hollrung
Didymosphaeria populina Vuill.
Pollaccia elegans Servazzi (anamorph)
Distr.: Temperate North America; Europe.
Notes: Leaf and shoot blight (Sivanesan. C.M.I. Descr. 483. 1976).
Host: Populus
Venturia potentillae (Wallr.:Fr.) Cooke
Dothidea potentillae Wallr.:Fr.
Chaetomium potentillae (Wallr.:Fr.) Wallr.
Coleroa potentillae (Wallr.:Fr.) G. Wint. in Rabenh.
Stigmatea potentillae (Wallr.:Fr.) Fr.
Distr.: North America; northern Europe.
Notes: On living and overwintered leaves and stalks.
Host: Potentilla
Venturia pyrina Aderhold
Fusicladium pyrorum (Lib.) Fuckel (anamorph)
Distr.: Temperate regions.
Notes: Scab or black spot of pear (Corlett. Fungi Canadenses 36. 1974; Sivanesan & Waller. C.M.I. Descr. 404. 1974 both as *V. pirina*).
Hosts: Cydonia, Pyrus
Venturia rhamni (Bonar) Barr
Phaeosphaerella rhamni Bonar

Distr.: CA.
Host: Rhamnus
Venturia rumicis (Desmaz.) G. Wint.
Mycosphaerella rumicis (Desmaz.) Grove
Mycosphaerella stromatoidea Dearn.
Distr.: North America; temperate Europe.
Notes: On leaves (Sivanesan. C.M.I. Descr. 405. 1974; Corlett & Egger. Fungi Canadenses 181. 1980).
Host: Rumex
Venturia sabalicola Ellis & Everh.
Distr.: LA(type).
Notes: Barr (Canad. J. Bot. 46:862. 1968) stated that this is a species of *Niesslia*, probably *N. exilis*.
Host: Sabal
Venturia saliciperda J. Nüesch
Pollaccia saliciperda (Allesch. & Tub.) Arx (anamorph)
Distr.: Northeastern North America; Europe.
Notes: Barr (Canad. J. Bot. 46:810. 1968; Corlett (Fungi Canadenses 247. 1983).
Host: Salix
Venturia subcutanea Dearn.
Mycosphaerella minor (P. Karst.) Johans. var. *reticulata* Dearn.
Distr.: Northern North America.
Notes: On leaves (Corlett & Egger. Fungi Canadenses 224. 1982).
Host: Salix
Venturia syringae (Syd.) Barr
Phaeosphaerella syringae Syd.
Distr.: MA; Germany.
Host: Syringa
Venturia systema-solare (Fuckel) G. Wint. in Rabenh.
Venturia clintonii Peck
Distr.: Northeastern North America.
Notes: On leaves.
Host: Cornus
Venturia weiriana (Petr.) Barr
Phaeosphaerella weiriana Petr. in Syd. & Petr.
Distr.: CO, WY.
Host: Yucca

Vermicularia Tode. Coelomycetes

Notes: This genus is considered a synonym of *Colletotrichum* (Sutton. The Coelomycetes. p. 523. 1980), but not all names have been redisposed.
Vermicularia beneficiens Peck
Distr.: NY.
Host: Sedum
Vermicularia cacti (Schwein.) Starb.
Distr.: Southern United States.
Host: Opuntia
Vermicularia coptina Peck
Distr.: NY, WI.
Host: Coptis
Vermicularia dematium (Pers.) Fr. f. ***minor*** Sacc.
Distr.: ND.
Host: Taraxacum
Vermicularia ipomoearum Schwein.
Distr.: Eastern United States.
Host: Ipomoea
Vermicularia phlogina Fairm.
Distr.: Central and eastern United States.
Host: Phlox
Vermicularia polygoni-virginica Schwein.
Distr.: PA, SD.
Host: Rheum
Vermicularia rectispora Cooke
Distr.: SC.
Host: Melia

Verpatinia Whetzel & Drayton. Discomycetes, Helotiales

Verpatinia calthicola Whetzel
Distr.: NY.
Host: Caltha

Verticicladium G. Preuss. Hyphomycetes

Verticicladium effusum Earle
Distr.: FL.
Notes: On decaying leaves.
Host: Coccoloba
Verticicladium pulvereum Peck & G.W. Clinton in Peck
Distr.: NY(type), WA.
Notes: According to Hughes (Mycol. Pap. 43:14. 1951) this species may be a *Nodulisporium*.
Hosts: Cornus, Salix

Verticillium Nees. Hyphomycetes

Notes: Gams (*Cephalosporium*-artige Schimmelpilze [Hyphomycetes]. 1971); Gams (Neth. J. Pl. Pathol. 88:57. 1982).
Verticillium albo-atrum Reinke & Berthier
Verticillium albo-atrum Reinke & Berthier var. *caespitosum* Wollenweb.
Verticillium albo-atrum Reinke & Berthier var. *tuberosum* B.A. Rudolph
Distr.: Cosmopolitan, most common in temperate regions.
Notes: Often confused with *V. dahliae*. Holliday & Talboys (C.M.I. Descr. 255. 1970).
Hosts: Abelmoschus, Abutilon, Acer, Aconitum, Aesculus, Ailanthus, Amaranthus, Antirrhinum, Apium, Arachis, Aralia, Armoracia, Aster, Atropa, Aucuba, Begonia, Berberis, Beta, Brassica, Buxus, Calceolaria, Callirhoe, Callistephus, Capsicum, Carthamus, Castanea, Catalpa, Ceanothus, Centaurea, Ceratonia, Cercis, Chrysanthemum, Cicer, Cinnamomum, Cistus, Citrullus, Cladrastis, Clarkia, Coreopsis, Cotinus, Cucumis, Cucurbita, Dahlia, Daphne, Delphinium, Digitalis, Dimorphotheca, Diospyros, Elaeagnus, Erica, Erigeron, Eschscholzia, Fragaria, Fraxinus, Fremontodendron, Fuchsia, Gerbera, Gossypium, Gymnocladus, Hedeoma, Helichrysum, Heliotropium, Humulus, Impatiens, Ipomoea, Koelreuteria, Lens, Liatris, Ligustrum, Liriodendron, Lycopersicon, Magnolia, Mangifera, Matthiola, Medicago, Mentha, Nicotiana, Nyssa, Olea, Osmanthus, Paeonia, Panax, Papaver, Parthenium, Pelargonium, Persea, Phellodendron, Phlox, Physalis, Pistacia, Pittosporum, Polemonium, Portulaca, Prunus, Quercus, Raphanus, Reseda, Rhaphiolepis, Rheum, Rhododendron, Rhus, Robinia, Rosa, Rubus, Rudbeckia, Salpiglossis, Sambucus, Sassafras, Schinus, Senecio, Sesamum, Sisymbrium, Solanum, Sophora, Tagetes, Tetragonia, Tilia, Tragopogon, Trifolium, Ulmus, Viburnum, Vicia, Vigna, Xanthium
Verticillium albo-atrum Reinke & Berthier var. ***menthae*** Ray Nelson
Distr.: MI(type).
Host: Mentha
Verticillium dahliae Kleb.
Verticillium ovatum G.H. Berkeley & A.B. Jackson
Distr.: Cosmopolitan, most common in temperate and subtropical regions.
Notes: Distinguished from *V. albo-atrum* by the formation of microsclerotia in culture, absence of growth at >30 degrees C, and conidiophores with paler bases. Hawksworth & Talboys (C.M.I. Descr. 256. 1970).
Hosts: Abutilon, Actinidia, Annona, Beta, Carya, Citrullus, Cucumis, Dianthus, Fragaria, Glycine, Gossypium, Helianthus, Hordeum, Humulus, Lycopersicon, Mentha, Olea, Persea, Pistacia, Prunus, Pyrus, Ribes, Rosa, Rubus, Solanum, Tagetes, Vitis
Verticillium dichotomum Ellis & Everh.
Distr.: NJ(type).
Host: Trifolium
Verticillium heterocladium Penz.
Distr.: FL; Italy(type).
Host: Euonymus
Verticillium lecanii (A. Zimmerm.) Viégas
Cephalosporium lecanii A. Zimmerm.
Distr.: Cosmopolitan.
Notes: A common entomogenous and fungicolous hyphomycete, frequently isolated from soil and plant materials (Brady. C.M.I. Descr. 610. 1979).

Hosts: Acer, Citrus, Mangifera
Verticillium lycopersici Pritchard & Porte
Distr.: UT, eastern United States(type).
Notes: This is probably a synonym of *V. dahliae* or *V. albo-atrum*.
Hosts: Lycopersicon, Solanum
Verticillium nigrescens Pethybr.
Distr.: Cosmopolitan.
Notes: Saprophytic or weakly pathogenic, isolated from various kinds of plant material (Domsch et al. Compendium of soil fungi. p. 841. 1980; Hawksworth. C.M.I. Descr. 257. 1970).
Hosts: Glycine, Gossypium, Helianthus, Mentha, Prunus
Verticillium psalliotae Treschow
Distr.: Widespread.
Notes: Brown-spot disease in cultivated mushrooms, isolated from various organic substrates (Brady & Waller. C.M.I. Descr. 497. 1976; Domsch et al. Compendium of soil fungi. p. 844. 1980).
Host: Acer
Verticillium rhizophagum Tehon & E.Y. Daniels nom. nud.
Distr.: OH(type).
Host: Ulmus
Verticillium tenerum (Pers.:Fr.) Link
Verticillium cinnabarinum (Corda) Reinke & Berthier
Acrostalagmus fulvus Berk. & Broome
Verticillium lateritium (Ehrenb.) Rabenh.
Nectria inventa Pethybr. (teleomorph)
Distr.: Cosmopolitan.
Notes: On organic debris (Domsch et al. Compendium of soil fungi. p. 501. 1980).
Hosts: Sambucus, Solanum, Tulipa, Zea
Verticillium tenuissimum Corda
Distr.: HI; Europe.
Notes: On dead wood and bark of various trees (Ellis. Dematiaceous Hyphomycetes. p. 536. 1971).
Host: Eucalyptus
Verticillium theobromae (Turconi) E. Mason & S.J. Hughes
Stachylidium theobromae Turconi
Distr.: Widespread in tropical regions.
Notes: Associated with cigar-end rot of banana (Hughes. Mycol. Pap. 45:10. 1951; Stover. Banana, plantain and abaca diseases. p. 134.1972; Hawksworth & Holliday. C.M.I. Descr. 259. 1970).
Host: Musa
Verticillium tricorpus I. Isaac
Distr.: WA; England(type).
Notes: Isaac (Trans. Brit. Mycol. Soc. 36:193. 1953) distinguished this species from other common *Verticillium* spp. Hawksworth (C.M.I. Descr. 260. 1970).
Hosts: Cucumis, Mentha, Solanum

Vertixore V. Miller & Bonar. Loculoascomycetes, Dothideales

Distr.: CA.
Notes: This genus is considered a synonym of *Aithaloderma*, but not all names have been redisposed.
Vertixore atronitidum V. Miller & Bonar
Distr.: CA.
Hosts: Ceanothus, Heteromeles, Quercus, Umbellularia

Vesiculomyces Hagström. Basidiomycetes, Aphyllophorales

Vesiculomyces citrinus (Pers.) Hagström
Gloeocystidiellum citrinum (Pers.) Donk
Corticium alutaceum (Schrad.) Bres.
Corticium radiosum (Fr.) Fr.
Gloeocystidiellum radiosum (Fr.) Boidin
Distr.: Northern hemisphere.
Notes: On wood. If *C. radiosum* is a sanctioned name, than this epithet should replace *V. citrinus*.
Hosts: Abies, Alnus, Picea, Pinus, Populus, Salix, Thuja

Vialaea Sacc. Pyrenomycetes, Amphisphaeriales

Vialaea insculpta (Fr.:Fr.) Sacc.
Boydia insculpta (Oudem.) Grove
Boydia remuliformis A.L. Sm.
Distr.: Northwestern North America; Europe.
Host: Ilex
Vialaea minutella Petr.
Distr.: HI; tropical regions.
Host: Mangifera

Vibrissea Fr. Discomycetes, Helotiales

Vibrissea truncorum Albertini & Schwein.:Fr.
Distr.: Eastern North America.
Notes: On submerged wood.
Hosts: Acer, Fagus, Liquidambar, Quercus, Tilia

Virgaria Nees. Hyphomycetes

Virgaria nigra (Link) Nees
Distr.: Subtropical and tropical regions.
Notes: On wood and bark of various trees (Ellis. Dematiaceous Hyphomycetes. p. 212. 1971).
Host: Juncus

Virgariella S.J. Hughes. Hyphomycetes

Virgariella fusca (Cooke) S.J. Hughes
Haplaria fusca Cooke
Virgaria fusca (Cooke) Sacc.
Virgaria olivacea Ellis & Everh.
Distr.: NJ, SC.
Host: Quercus

Virgella Darker. Discomycetes, Rhytismatales

Virgella robusta (Tub.) Darker
Hypoderma robustum Tub.
Distr.: Northern North America.
Notes: Needle blight.
Host: Abies

Viscomacula R. Sprague. Hyphomycetes

Viscomacula aenea R. Sprague
Distr.: CO(type).
Notes: On living leaves (Sprague. Mycologia 42:758. 1950).
Host: Poa

Volutella Fr., nom. cons. Hyphomycetes

Notes: Samuels (Mycologia 69:255. 1977).
Volutella acalyphae Atk.
Distr.: AL.
Host: Acalypha
Volutella aciculosa (Ellis & Harkn.) Sacc.
Hymenula aciculosa Ellis & Harkn.
Distr.: CA(type).
Host: Pinus
Volutella albo-pila Boud.
Distr.: United States; France(type).
Notes: According to Tulloch (Mycol. Pap. 130:37. 1972) this could be *Myrothecium leucotrichum*.
Host: Phalaenopsis
Volutella bartholomaei Ellis & Everh.
Distr.: KS.
Host: Sporobolus
Volutella buxi (DC.) Berk.
Pseudonectria rousseliana (Mont.) Seaver (teleomorph)
Distr.: Temperate regions.
Notes: Canker of boxwood.
Host: Buxus

Volutella caricicola L.E. Miles
Distr.: MS.
Host: Carex
Volutella caryogena Fairm.
Distr.: NY.
Host: Carya
Volutella colletotrichoides J.E. Chilton
Distr.: IA(type).
Hosts: Lotus, Medicago, Trifolium
Volutella coronata Cooke & Harkn.
Distr.: CA.
Host: Eucalyptus
Volutella cylindrospora (Berk. & M.A. Curtis) Sacc.
Psilonia cylindrospora Berk. & M.A. Curtis
Distr.: LA, SC.
Host: Tillandsia
Volutella diaphana Ellis
Distr.: NJ.
Host: Orontium
Volutella flexuosa Cooke & Ellis
Distr.: NJ.
Host: Salvia
Volutella mellea J.F. Clark
Distr.: NY(greenhouse).
Host: Pandanus
Volutella occidentalis Ellis & F.W. Anderson
Distr.: MT.
Host: Astragalus
Volutella oxyspora Atk.
Distr.: AL.
Host: Beta
Volutella pachysandrae W.G. Hutchinson
Distr.: Eastern United States.
Notes: Leaf blight. Samuels (Mycologia 69:256. 1977) described the differences between this species and *V. pachysandricola*.
Host: Pachysandra
Volutella phlogina Ellis & Langl., nom. nud.
Distr.: LA.
Host: Phlox
Volutella tecticola Atk.
Distr.: AL.
Host: Arundinaria
Volutella vanillae (Delacr.) Petch & Ragunathan
Botryosphaeria vanillae (Stoneman) Petch & Ragunathan (teleomorph)
Distr.: Subtropical and tropical America; Sri Lanka.
Hosts: Orchidaceae, Vanilla
Volutella vincae Fairm.
Distr.: NY, PA.
Host: Vinca

Volvariella Speg. Basidiomycetes, Agaricales

Notes: Shaffer (Mycologia 49:545. 1957).
Volvariella bombycina (Schaeff.:Fr.) Singer
Volvaria bombycina Schaeff.:Fr.
Distr.: Temperate northern hemisphere.
Notes: On trunks of living and dead hardwoods, especially common on *Acer* spp.
Hosts: Acer, Liquidambar, Populus, Quercus, Salix
Volvariella bombycina (Schaeff.:Fr.) Singer var. ***flaviceps*** (Murrill) R. Shaffer
Volvaria flaviceps Murrill
Distr.: FL(type).
Host: Magnolia
Volvariella lepiotospora Singer
Distr.: FL.
Host: Quercus

Wallemia Johan-Olsen. Hyphomycetes

Wallemia sebi (Fr.) Arx
Sporendonema sebi Fr.
Torula sacchari Corda
Distr.: Cosmopolitan.
Notes: Isolated from foodstuffs, air, hay, textiles and humans (Ellis. Dematiaceous Hyphomycetes. p. 38. 1971).
Hosts: Arachis, Carya

Waydora Sutton in Sutton, V.G. Rao & D.N. Mhaskar. Coelomycetes

Waydora typica (Rodway) Sutton
Pulvinaria typica Rodway
Distr.: Southern United States; South America; Australia; Tasmania.
Notes: On capsules (Sutton. The Coelomycetes. p. 556. 1980).
Host: Eucalyptus

Wettsteinina Höhn. Loculoascomycetes, Dothideales

Notes: Shoemaker & Babcock (Canad. J. Bot. 65:373. 1987).
Wettsteinina dryadis (Rostr.) Petr.
Massarina dryadis Rostr.
Distr.: Circumboreal.
Notes: On leaves.
Host: Dryas
Wettsteinina duplex Shoemaker & Babcock
Distr.: WA.
Notes: On dead stems of various herbaceous plants.
Hosts: Anemone, Erigeron, Lupinus, Valeriana
Wettsteinina eucarpa (P. Karst.) E. Müller & Arx
Distr.: Alpine and arctic North America; Eurasia.
Host: Polygonum
Wettsteinina gentianae (Wehmeyer) Barr
Sphaerulina gentianae Wehmeyer
Distr.: WY.
Notes: On herbaceous stems.
Host: Gentiana
Wettsteinina macrospora (Wehmeyer) Petr.
Eriosphaeria macrospora Wehmeyer
Distr.: WA.
Notes: On dead stems of Asteraceae and other herbaceous plants.
Hosts: Anemone, Erigeron
Wettsteinina niesslii E. Müller
Leptosphaeria gigaspora Niessl
Distr.: IA; Europe.
Hosts: Poa, Scirpus

Wiesneriomyces Koord. Hyphomycetes

Wiesneriomyces laurinus (Tassi) P.M. Kirk
Volutellaria laurina Tassi
Wiesneriomyces javanicus Koord.
Distr.: Cosmopolitan.
Notes: On leaf litter. Kirk (Trans. Brit. Mycol. Soc. 82:748. 1984).
Hosts: Eucalyptus, Quercus, Schefflera

Winterella Kuntze. Pyrenomycetes

Cryptospora Tul. & C. Tul.
Ophiovalsa Petr.
Disculina Höhn.
Notes: Reid & Booth (Canad. J. Bot. 65:1320. 1987).
Winterella albofusca (Cooke & Ellis) J. Reid & C. Booth
Valsa albofusca Cooke & Ellis
Cryptosporella albofusca (Cooke & Ellis) Sacc.
Ophiovalsa trichispora (Cooke & Peck) Barr
Valsa trichispora Cooke & Peck
Distr.: Eastern United States.
Host: Quercus
Winterella alnicola (Höhn.) J. Reid & C. Booth
Distr.: North America; Europe.
Host: Corylus

Winterella aurantiaca (Wehmeyer) J. Reid & C. Booth
Distr.: North America.
Host: Alnus
Winterella betulae (Tul. & C. Tul.) Kuntze
Cryptospora betulae Tul. & C. Tul.
Ophiovalsa betulae (Tul. & C. Tul.) Petr.
Cryptospora tomentella (Peck) Berl. & Voglino
Valsa tomentella Peck
Distr.: Temperate northern hemisphere.
Notes: On limbs.
Host: Betula
Winterella cinctula (Cooke & Peck) Kuntze
Cryptospora cinctula (Cooke & Peck) Sacc.
Distr.: Eastern United States.
Host: Castanea
Winterella femoralis (Peck) Kuntze
Valsa femoralis Peck
Cryptospora femoralis (Peck) Sacc.
Ophiovalsa femoralis (Peck) Petr.
Cryptospora humeralis Dearn. & House
Distr.: Eastern North America; Japan.
Notes: On limbs.
Hosts: Alnus, Betula
Winterella suffusa (Fr.:Fr.) Kuntze
Cryptospora suffusa (Fr.:Fr.) Tul. & C. Tul.
Ophiovalsa suffusa (Fr.:Fr.) Petr.
Disculina vulgaris (Fr.:Fr.) Sutton (anamorph)
Distr.: Temperate northern hemisphere.
Notes: On limbs.
Host: Alnus
Winterella wehmeyeriana J. Reid & C. Booth
Distr.: North America.
Host: Tilia

Winteria Sacc. Pyrenomycetes, Hypocreales
Notes: See Hawksworth & Sherwood (Taxon 30:347. 1981).
Winteria caerulea (Ellis & Everh.) Berl. & Voglino
Distr.: Pacific Northwest.
Host: Pseudotsuga

Winterina Sacc. Pyrenomycetes, Sordariales
Notes: Although this generic name is excluded from the Ascomycetes (Eriksson & Hawksworth. Systema Ascomycetum 5:309. 1986), these species have not been transferred to other genera.
Winterina lobata (Tracy & Earle) Sacc. & Syd.
Winteria lobata Tracy & Earle
Distr.: Southeastern United States.
Host: Ilex

Wojnowicia Sacc. Coelomycetes
Wojnowicia hirta Sacc.
Hendersonia hirta J. Schröt.
Hendersonia crastophila Sacc.
Hendersonia graminis McAlpine
Wojnowicia graminis (McAlpine) Sacc. & D. Sacc.
Wojnowicia tenella Pat.
Distr.: Cosmopolitan.
Notes: Foot rot and root rot of grasses (Punithalingam. C.M.I. Descr. 773. 1983).
Hosts: Agropyron, Avena, Bouteloua, Bromus, Calamagrostis, Calamovilfa, Cinna, Dactylis, Distichlis, Elymus, Festuca, Hordeum, Koeleria, Muhlenbergia, Oryzopsis, Panicum, Phleum, Poa, Secale, Sorghastrum, Spartina, Sporobolus, Stipa, Triticum, Vulpia

Wolfiporia Ryvarden & R.L. Gilbertson. Basidiomycetes, Aphyllophorales
Notes: Gilbertson & Ryvarden (North American Polypores 2:798. 1987).
Wolfiporia cocos (F.A. Wolf) Ryvarden & R.L. Gilbertson
Poria cocos F.A. Wolf
Macrohyporia cocos (F.A. Wolf) I. Johansen & Ryvarden
Merulius albus Burt
Daedalea extensa Peck
Macrohyporia extensa (Peck) Ginns & J. Lowe
Wolfiporia extensa (Peck) Ginns
Poria magnahypha Overh., nom. inval.
Sclerotium cocos Schwein.:Fr. (anamorph)
Distr.: North America.
Notes: On wood; usually found only in the sclerotial state or isolated from wood decay. Ginns & Lowe (Canad. J. Bot. 61:1672. 1983.) provided notes on author citations.
Hosts: Abies, Acer, Betula, Citrus, Diospyros, Eucalyptus, Fagus, Fraxinus, Juniperus, Liriodendron, Magnolia, Pinus, Poncirus, Populus, Prunus, Quercus, Rhus, Sabal, Sassafras, Taxodium, Tsuga, Ulmus, Zea
Wolfiporia dilatohypha Ryvarden & R.L. Gilbertson
Poria inflata Overh., nom. inval.
Distr.: Central and Eastern North America; Bolivia.
Notes: On hardwoods.
Hosts: Prunus, Quercus
Wolfiporia sulphurea (Burt) Ginns
Merulius sulphureus Burt
Byssomerulius sulphureus (Burt) Lindsey in R.L. Gilbertson et al.
Macrohyporia sulphurea (Burt) Ginns & J. Lowe
Distr.: Northern hemisphere.
Notes: On wood.
Hosts: Olneya, Prosopis

Wolkia Ramsb. Ascomycetes
Wolkia decolorans (Van der Wolk) Ramsb.
Protascus decolorans Van der Wolk
Distr.: AR, CA; Java(type).
Notes: Yellow grain of rice (Ou. Rice diseases. p. 332. 1985). The fungus is an ascomycete of uncertain disposition.
Host: Oryza

Wrightoporia Pouzar. Basidiomycetes, Aphyllophorales
Notes: Gilbertson & Ryvarden (North American Polypores 2:802. 1987).
Wrightoporia avellanea (Bres.) Pouzar
Poria avellanea Bres.
Poria calkinsii Murrill
Poria heteromorpha Murrill
Distr.: Southern United States; Africa; neotropical regions.
Notes: On wood.
Host: Sabal
Wrightoporia cylindrospora Ryvarden
Distr.: MD.
Notes: On wood.
Host: Quercus
Wrightoporia lenta (Overh. & J. Lowe) Pouzar
Poria lenta Overh. & J. Lowe
Distr.: North America.
Notes: On wood.
Hosts: Pinus, Tsuga
Wrightoporia subrutilans (Murrill) Ryvarden
Polyporus abieticola Overh., nom. inval.
Poria illudens Overh. & J. Lowe
Tyromyces illudens (Overh. & J. Lowe) J. Lowe
Distr.: Worldwide, not common.
Notes: On wood.
Hosts: Abies, Tsuga

Xenasma Donk. Basidiomycetes, Aphyllophorales
Notes: Liberta (Mycologia 52:884. 1960).
Xenasma praeteritum (H. Jacks.) Donk
Peniophora praeterita H. Jacks.
Distr.: Northern hemisphere.
Notes: On hardwoods.
Host: Acer

Xenasma pulverulentum (Litsch.) Donk
Corticium pulverulentum Litsch.
Distr.: Northern hemisphere.
Notes: On wood.
Host: Tsuga
Xenasma rimicola (P. Karst.) Donk
Distr.: Widespread.
Notes: On dead wood.
Hosts: Acer, Quercus

Xenodochus Schlechtend. Basidiomycetes, Uredinales
Xenodochus carbonarius Schlechtend.
Distr.: Circumboreal.
Notes: Autoecious.
Host: Sanguisorba
Xenodochus minor Arth.
Distr.: Circumboreal.
Notes: Autoecious.
Host: Sanguisorba

Xenomeris Syd. Loculoascomycetes, Dothideales
Xenomeris abietis Barr
Sclerophoma xenomeria Funk (anamorph)
Distr.: Pacific Northwest.
Notes: Die-back of conifers (Funk & Shoemaker. Mycologia 63:567. 1971; Barr. Canad. J. Bot. 46: 842. 1968).
Hosts: Abies, Pseudotsuga

Xenosperma Oberwinkler. Basidiomycetes, Aphyllophorales
Xenosperma ludibundum (D.P. Rogers & Liberta) Oberwinkler
Xenasma ludibundum D.P. Rogers & Liberta
Distr.: Northeastern and northwestern United States.
Notes: On wood.
Hosts: Chamaecyparis, Quercus
Xenosperma murrillii R.L. Gilbertson & Blackwell
Distr.: FL.
Notes: On limbs.
Host: Juniperus

Xenosporium Penz. & Sacc. Hyphomycetes
Xenosporium berkeleyi (M.A. Curtis) Pirozynski
Helicoma berkeleyi M.A. Curtis
Xenosporella berkeleyi (M.A. Curtis) Linder
Distr.: Subtropical and tropical regions.
Notes: On decaying, mostly woody substrates (Ellis. Dematiaceous Hyphomycetes. p. 132. 1971).
Hosts: Aralia, Arecaceae, Betula, Carpinus, Cornus, Fraxinus, Hibiscus, Platanus, Quercus

Xenotypa Petr. Pyrenomycetes, Diaporthales
Notes: Barr (Mycol. Mem. 7:73. 1978).
Xenotypa aterrima (Fr.:Fr.) Petr.
Radulum aterrimum (Fr.:Fr.) Fr.
Distr.: Temperate North America; Europe.
Notes: On twigs and limbs (Pirozynski. Canad. J. Bot. 52:2129. 1974).
Host: Malus

Xeromphalina Kühner & Maire. Basidiomycetes, Agaricales
Notes: Miller (Mycologia 60:156. 1968).
Xeromphalina campanella (Batsch:Fr.) Kühner & Maire
Omphalia campanella (Batsch:Fr.) P. Kumm.
Distr.: Circumboreal.
Notes: On logs and woody debris of conifers.
Hosts: Abies, Picea, Pinus, Prunus, Pseudotsuga, Sequoia, Thuja, Tsuga
Xeromphalina cauticinalis (Bull.) Fr.
Marasmius tomentosipes Peck
Distr.: Northern hemisphere.
Notes: On conifer needles (Redhead. Canad. J. Bot. 66:491. 1988).
Host: Pinus
Xeromphalina fraxinophila A.H. Sm.
Distr.: CO to AK, MI; Canada(Quebec).
Notes: On leaf litter and twigs of hardwoods.
Host: Fraxinus
Xeromphalina fulvipes (Murrill) A.H. Sm.
Collybia fulvipes Murrill
Distr.: Western North America.
Host: Pseudotsuga
Xeromphalina kauffmanii A.H. Sm.
Distr.: Eastern North America.
Notes: On logs and woody debris of hardwoods, frequently oak.
Hosts: Betula, Quercus
Xeromphalina orickiana (A.H. Sm.) Singer
Omphalina orickiana A.H. Sm.
Distr.: CA.
Notes: On redwood logs.
Host: Sequoia

Xerotrema M.A. Sherwood & Coppins. Discomycetes, Ostropales
Notes: Sherwood & Coppins (Notes Roy. Bot. Gard. Edinburgh 38:367. 1980).
Xerotrema megalosporum M.A. Sherwood & Coppins
Distr.: Temperate North America; Europe.
Notes: On decorticated wood.
Hosts: Pinus, Pseudotsuga

Xerula Maire. Basidiomycetes, Agaricales
Xerula americana Dörfelt
Oudemansiella longipes (Bull.) M. Moser var. *americana* D.H. Mitchel & A.H. Sm., nom. inval.
Distr.: Western North America.
Host: Populus
Xerula radicata (Relhan:Fr.) Dörfelt
Collybia radicata (Relhan:Fr.) P. Kumm.
Oudemansiella radicata (Relhan:Fr.) Singer
Distr.: Eastern United States; Eurasia.
Notes: On buried wood. Redhead et al. (Mycotaxon 30:357. 1987) differentiated various taxa which have been considered to be this species in North America. The report on *Pinus* is suspect.
Host: [Pinus]

Xylaria J. Hill ex Schrank, nom. cons. Pyrenomycetes, Xylariales
Notes: Rogers (Mycotaxon 26:85. 1986) presented a key to species in the United States. Although for New Zealand, many North American species are included in Rogers & Samuels (New Zealand J. Bot. 24:615. 1986).
Xylaria acuta Peck
Distr.: Northeastern and midwestern United States.
Notes: Often misdetermined as *X. digitata* (L.:Fr.) Grev. (Rogers. Mycologia 76:23. 1984).
Hosts: Acer, Quercus
Xylaria anisopleura (Mont.) Mont.
Xylosphaera anisopleura (Mont.) Dennis
Penzigia globosa (Spreng.:Fr.) Rehm
Hypoxylon globosum (Spreng.:Fr.) Fr.
Distr.: HI; neotropical regions; New Zealand.
Notes: On dead wood.
Host: Metrosideros
Xylaria apiculata Cooke
Xylaria venulosa Speg.
Distr.: Subtropical and tropical regions.
Host: Solanum
Xylaria arbuscula Sacc.

Distr.: Warm-temperate to tropical regions.
Notes: On wood.
Host: Cephalanthus
Xylaria bulbosa (Pers.) Berk. & Broome
Sphaeria bulbosa Pers.
Distr.: HI, North America; South Africa; Europe.
Notes: On wood and plant debris.
Hosts: Acer, Carya, Castanea, Fagus, Quercus
Xylaria carpophila (Pers.:Fr.) Fr.
Distr.: Eastern North America; Europe.
Notes: Usually on fruits (Rogers. Canad. J. Bot. 57:941. 1970)
Hosts: Crataegus, Fraxinus, Gleditsia, Liquidambar, Magnolia, Morus, Robinia
Xylaria corniformis (Fr.) Fr.
Distr.: Eastern United States
Notes: A confused name; American reports may refer to *Xylaria curta* Fr., *X. bulbosa* (Pers.) Nitschke, or *X. cubensis* (Mont.) Fr. (Rogers. Mycologia 75:457. 1983). Laessoe (Mycotaxon 30:81. 1987).
Hosts: Carpinus, Fagus, Quercus
Xylaria cornu-damae (Schwein.) Fr.
Distr.: Temperate North America.
Notes: Rogers (Mycologia 76:27. 1984).
Hosts: Acer, Alnus, Betula, Pinus, Quercus, Robinia
Xylaria cubensis (Mont.) Fr.
Hypoxylon cubense Mont.
Distr.: Cosmopolitan.
Notes: On dead wood (Rogers. Mycologia 76:912. 1984).
Hosts: Acer, Magnolia, Quercus
Xylaria curta Fr.
Xylaria rhopaloides Mont. in Starb.
Distr.: Cosmopolitan.
Notes: On wood.
Hosts: Acacia, Aleurites, Betula, Fagus, Malus, Tsuga
Xylaria digitata (L.:Fr.) Grev.
Distr.: Cosmopolitan.
Notes: The occurrence of this species in the United States is uncertain. Many specimens determined as *X. digitata* are *X. cornu-damae* (Schwein.) Fr. or *X. acuta* Peck (Rogers. Mycologia 76:25. 1984).
Hosts: Crataegus, Fagus, Populus, Pseudotsuga, Quercus
Xylaria filiformis (Albertini & Schwein.:Fr.) Fr.
Distr.: PR, eastern North America; Europe.
Notes: On leaves and herbaceous stems.
Host: Ilex
Xylaria herculea (Miq.) Cooke
Distr.: Subtropical and tropical America.
Host: Quercus
Xylaria hypoxylon (L.:Fr.) Grev.
Distr.: Cosmopolitan.
Notes: On dead wood.
Hosts: Acer, Castanea, Ceanothus, Chamaecyparis, Corylus, Crataegus, Fouquieria, Holodiscus, Pinus, Quercus, Ribes, Robinia, Salix, Tsuga, Ulmus
Xylaria ianthino-velutina (Mont.) Fr.
Distr.: Tropical regions.
Notes: Usually on fruits of Fabaceae. Records on *Magnolia* from the United States are probably *X. magnoliae* (Rogers. Canad. J. Bot. 57:941. 1979).
Host: Magnolia
Xylaria longiana Rehm
Distr.: Eastern United States.
Notes: On wood.
Hosts: Ginkgo, Malus, Prunus, Quercus, Robinia
Xylaria longipes Nitschke
Xylaria corniformis (Fr.:Fr.) Fr. var. *macrospora* Bres. in Theiss.
Distr.: Northern hemisphere.
Notes: On hardwoods.
Hosts: Acer, Fagus, Tsuga
Xylaria magnoliae J.D. Rogers
Distr.: Southeastern United States.
Notes: On fruits.
Host: Magnolia
Xylaria mali Fromme
Distr.: Central and eastern United States.
Notes: Black root rot of apples (Fromme. Virginia Agr. Exp. Sta. Tech. Bull. 34. 1928; Rogers. Mycologia 76:30. 1984).
Hosts: Acer, Gleditsia, Malus, Prunus, Pyrus, Ulmus
Xylaria multipartita J.H. Miller & L.W. Nielsen
Distr.: NC.
Host: Poa
Xylaria multiplex (Kunze) Fr.
Distr.: Subtropical and tropical regions.
Host: Quercus
Xylaria oxyacanthae Tul. & C. Tul.
Distr.: Temperate North America; Europe.
Notes: On fallen fruits (Stowell & Rogers. Mycotaxon 17:433. 1983).
Hosts: Carya, Crataegus
Xylaria persicaria (Schwein.:Fr.) Berk. & M.A. Curtis
Distr.: Eastern United States; China.
Notes: On fruits.
Host: Liquidambar
Xylaria poitei (Lév.) Fr.
Distr.: Southern United States; Central and South America; Africa.
Notes: This species was thoroughly studied by Rogers & Callan (Mycotaxon 26:287. 1986).
Hosts: Quercus, Sabal
Xylaria polymorpha (Pers.:Fr.) Grev.
Xylosphaera polymorpha (Pers.:Fr.) Dumort.
Distr.: Cosmopolitan.
Notes: White rot of hardwoods (Rogers & Callan. Mycologia 78:391. 1986; Sivanesan & Holliday. C.M.I. Descr. 355. 1972).
Hosts: Acer, Betula, Citrus, Fagus, Malus, Poncirus, Quercus, Robinia, Sassafras, Ulmus, Vitis
Xylaria psamathos Boise
Distr.: NC.
Host: Spartina
Xylaria schweinitzii Berk. & M.A. Curtis
Distr.: HI, subtropical and tropical America.
Host: Aleurites
Xylaria subterranea (Schwein.) Sacc.
Distr.: Temperate North America.
Notes: The status of this taxon is uncertain (J.D. Rogers. pers. comm.).
Host: Alnus

Xylobolus P. Karst. Basidiomycetes, Aphyllophorales

Xylobolus frustulatus (Pers.:Fr.) P. Karst.
Stereum frustulatum (Pers.:Fr.) Fr.
Distr.: North America, most common in the East; Asia; Australia; Europe.
Notes: On wood.
Hosts: Carya, Castanea, Pinus, Pseudotsuga, Quercus
Xylobolus subpileatus (Berk. & M.A. Curtis) Boidin
Stereum subpileatum Berk. & M.A. Curtis
Stereum frustulatum (Pers.:Fr.) Fuckel var. *subpileatum* (Berk. & M.A. Curtis) A.L. Welden
Stereum insigne Bres.
Stereum sepium Burt
Distr.: Warmer parts of eastern North America; Central and South America; Europe; Asia.
Notes: On wood of *Quercus*, rarely on other hardwoods.
Hosts: Liquidambar, Platanus, Quercus, Taxodium, Ulmus

Xylocoremium J.D. Rogers. Hyphomycetes

Xylocoremium flabelliforme (Schwein.:Fr.) J.D. Rogers
Xylaria cubensis (Mont.) Fr. (teleomorph)
Distr.: Worldwide, most common in warm regions.
Notes: On rotting wood (Rogers. Mycologia 76:913. 1984).
Host: Magnolia

Xylogramma Wallr. Discomycetes, Helotiales

Xylogramma nigerrima (Ellis & Everh.) Rehm
Briardia nigerrima Ellis & Everh.
Distr.: CA(type).
Host: Rhus

Xylopezia Höhn. Loculoascomycetes, Dothideales

Xylopezia biseptata M.A. Sherwood
Distr.: OR.
Hosts: Chrysothamnus, Juniperus
Xylopezia hemisphaerica (Fr.:Fr.) M.A. Sherwood
Stictis hemisphaerica Fr.:Fr.
Distr.: Temperate northern hemisphere.
Notes: On decorticated wood.
Hosts: Abies, Ceanothus, Pinus, Vaccinium

Xyloschizon Syd. Loculoascomycetes, Dothideales

Notes: This genus is considered a synonym of *Hysteroglonium*, but not all names have been redisposed.
Xyloschizon atratum Syd.
Distr.: ID(type).
Host: Crataegus
Xyloschizon weirianum Syd.
Distr.: ID(type).
Host: Tsuga

Yoshinagella Höhn. Loculoascomycetes, Dothideales

Notes: This genus is considered a synonym of *Gibberidea*, but not all names have been redisposed.
Yoshinagella polymorpha Lyon var. ***pauciseta*** F. Stevens
Distr.: HI(type).
Notes: Although the type variety has been transferred to *Coccodiella*, this variety has not been redisposed.
Host: Cibotium

Zalerion R.T. Moore & S.P. Meyers. Hyphomycetes

Zalerion varium Anastasiou
Distr.: Widespread in marine habitats.
Notes: On intertidal and drifting wood and woody debris, submerged leaves and seedlings of marine plants (Kohlmeyer & Kohlmeyer. Marine mycology. p. 508. 1979).
Host: Juncus

Zasmidium Fr. Hyphomycetes

Zasmidium cellare (Pers.) Fr.
Racodium cellare Pers.
Cladosporium cellare (Pers.) Schanderl
Rhinocladiella cellaris (Pers.) Schol-Schwarz
Distr.: HI; Europe.
Notes: Commonly found on brickwork and adjacent timber in wine cellars (De Hoog. Stud. Mycol. 15:42. 1977).
Hosts: Cheirodendron, Metrosideros

Zignoella Sacc. Pyrenomycetes, Trichosphaeriales

Zignoella ovoidea (Fr.) Sacc.
Distr.: Temperate regions.
Notes: On dead wood.
Host: Oplopanax
Zignoella roripae Rehm
Distr.: ND(type).
Host: Rorippa

Zoellneria Velen. Discomycetes, Helotiales

Zoellneria callochaetes (Ellis & Everh.) Dennis
Lachnella callochaetes Ellis & Everh.
Dasyscyphus callochaetes (Ellis & Everh.) Sacc.
Distr.: NJ.
Notes: On fallen leaves (Dennis. Kew Bull. 17:333. 1963).
Host: Myrica

Zopfia Rabenh. Loculoascomycetes, Dothideales

Notes: Hawksworth (Canad. J. Bot. 57:91. 1979); Hawksworth & Booth (Mycol. Pap. 135:1. 1974).
Zopfia albiziae M.L. Farr
Distr.: TX(type).
Host: Albizia
Zopfia rhizophila Rabenh.
Distr.: Temperate regions.
Notes: On decaying root and rhizomes (Eriksson. Opera Bot. 60:171. 1981).
Host: Asparagus

Zukalia Sacc. Loculoascomycetes, Dothideales

Notes: This genus is considered a synonym of *Chaetothyrium* (Arx & Müller. Stud. Mycol. 9:117. 1975), but not all names have been redisposed.
Zukalia purpurea (Ellis & G. Martin)Theiss.
Asterina purpurea Ellis & G. Martin
Asteridium purpureum (Ellis & G. Martin) Sacc.
Asterina oleina Cooke
Distr.: FL, GA.
Hosts: Olea, Osmanthus

Zygophiala E. Mason. Hyphomycetes

Zygophiala jamaicensis E. Mason
Schizothyrium pomi (Mont. & Fr.) Arx (teleomorph)
Distr.: Cosmopolitan.
Notes: Flyspeck of apple fruit, greasy blotch of carnation leaves, and parasitic on a wide range of other plants (Baker et al. Phytopathology 67:580. 1977).
Hosts: Dianthus, Ficus, Malus

Zygosaccharomyces B.T. Barker. Hemiascomycetes, Endomycetales

Zygosaccharomyces microellipsodes (Osterwalder) Yarrow
Saccharomyces microellipsodes Osterwalder
Distr.: Widespread.
Notes: Isolated from soil as well as various kinds of organic debris (Arx et al. Stud. Mycol. 14:12. 1977).
Host: Triticum

Zygosporium Mont. Hyphomycetes

Notes: Ellis (Dematiaceous Hyphomycetes. p. 322. 1971).
Zygosporium gibbum (Sacc., M. Rousseau, & E. Bommer) S.J. Hughes
Clasterosporium gibbum Sacc., M. Rousseau, & E. Bommer
Distr.: Subtropical and tropical regions.
Notes: On dead leaves of various plants.
Hosts: Acacia, Eucalyptus, Juncus, Metrosideros, Podocarpus
Zygosporium masonii S.J. Hughes
Distr.: Subtropical and tropical regions.
Notes: Saprobic on plant debris.
Hosts: Juncus, Magnolia
Zygosporium oscheoides Mont.
Distr.: Subtropical and tropical regions.
Notes: Saprobic on plant debris.
Hosts: Musa, Pandanus, Podocarpus, Yucca

Zythia Fr. Coelomycetes

Zythia aurantiaca (Peck) Sacc.
Sphaeronaema aurantiacum Peck
Distr.: Northeastern United States.
Host: Cornus
Zythia fragariae Laibach

Gloeosporium fragariae G. Arnaud
Gnomonia comari P. Karst. (teleomorph)
Distr.: Temperate regions.
Notes: On leaves and stems of Rosaceae.
Host: Fragaria
Zythia rhoina Ellis & Everh.
Distr.: NJ.
Host: Rhus

Literature Cited

1. Aa, H.A. van der. 1971. *Macroventuria*, a new genus of the Venturiaceae. Persoonia 6: 359-363.
2. Aa, H.A. van der. 1973. Studies in *Phyllosticta* I. Stud. Mycol. 5: 1-110.
3. Abawi, G.S., D.C. Crosier, and A.C. Cobb. 1977. Pod-flecking of snap beans caused by *Alternaria alternata*. Pl. Dis. Reporter 61: 901-905.
4. Abawi, G.S., and S.B. Martin. 1985. *Rhizoctonia* foliar blight of cabbage in New York state. Pl. Dis. 69: 158-161.
5. Abbas, H.K., and C.J. Mirocha. 1986. Survival of *Fusarium graminearum* on corn stored at low temperature. Pl. Dis. 70: 78.
6. Abel, G.H. 1977. Brown stem rot of soybean - *Cephalosporium gregatum*. Rev. Plant Pathology 56: 1065-1077.
7. Abercrombie, K., and H.C. Finch. 1976. Powdery mildew of carrot in California. Pl. Dis. Reporter 60: 780-781.
8. Abrahamson, D.S., and D.R. Bergdahl. 1987. Needle weights as a measure of infection by *Uredinopsis mirabilis* and *Chrysomyxa weirii*. Pl. Dis. 71: 270-271.
9. Acedo, G.N., and G.N. Agrios. 1970. *Helminthosporium* blister canker of pear. Pl. Dis. Reporter 54: 296-299.
10. Adams, D.H., and L.F. Roth. 1969. Intraspecific competition among genotypes of *Fomes cajanderi* decaying young-growth Douglas-fir. Forest Sci. 15: 327-331.
11. Adams, E.B., and R.F. Line. 1984. Biology of *Puccinia chondrillina* in Washington. Phytopathology 74: 742-745.
12. Adams, G.C., and E.E. Butler. 1983. Influence of nutrition on the formation of basidia and basidiospores in *Thanatephorus cucumeris*. Phytopathology 73: 147-151.
13. Adams, G.C., Jr., W.D. Gubler, and R.G. Grogan. 1987. Seedling diseases of muskmelon and mixed melons in California caused by *Fusarium equiseti*. Pl. Dis. 71: 370-374.
14. Adams, P.B., R.D. Lumsden, and C.J. Tate. 1974. *Galinosoga parviflora*: A new host for *Whetzelinia sclerotiorum*. Pl. Dis. Reporter 58: 700-701.
15. Adams, P.B., B.H. Marose, and E.M. Dutky. 1983. Cocklebur: a new host for several *Sclerotinia* species. Pl. Dis. 67: 484-485.
16. Adams, R.E., and S.E. Tamburo. 1957. The West Virginia spot-rot complex of apple in 1956. Pl. Dis. Reporter 41: 760-765.
17. Adaskaveg, J.E., and R.L. Gilbertson. 1986. Cultural studies and genetics of sexuality of *Ganoderma lucidum* and *G. tsugae* in relation to the taxonomy of the *G. lucidum* complex. Mycologia 78: 694-705.
18. Adaskaveg, J.E., and R.L. Gilbertson. 1987. Vegetative incompatibility between intraspecific dikaryotic pairings of *Ganoderma lucidum* and *G. tsugae*. Mycologia 79: 603-613.
19. Afanasiev, M.M. 1961. Occurrence of late blight disease of potatoes in Montana. Pl. Dis. Reporter 45: 314.
20. Affeltranger, C.E., P.A. Dugar, P.H. Peacher, and M.J. Weiss. 1976. Elm mortality caused by *Verticillium* and *Cephalosporium* wilts and Dutch elm disease on the Vicksburg National Military Park. Pl. Dis. Reporter 60: 895.
21. Affeltranger, C.E., D.A. Starkey, H.N. Wallace, and S.A. Covington. 1985. Dutch elm disease (*Ceratocystis ulmi*) in Louisiana. Pl. Dis. 69: 268.
22. Aho, P.E., and G.M. Filip. 1982. Incidence of wounding and *Echinodontium tinctorium* infections in advanced white fir regeneration. Canad. J. Forest Res. 12: 705-708.
23. Alcorn, J.L., and J.A.G. Irwin. 1987. *Acrocalymma medicaginis* gen. et. sp. nov. causing root and crown rot of *Medicago sativa* in Australia. Trans. Brit. Mycol. Soc. 88: 163-167.
24. Aldwinckle, H.S., F.J. Polach, W.T. Molin, and R.C. Pearson. 1975. Pathogenicity of *Phytophthora cactorum* isolates from New York apple trees and other sources. Phytopathology 65: 989-994.
25. Alexander, H.M., J.V. Groth, and A.P. Roelfs. 1985. Virulence changes in *Uromyces appendiculatus* after five asexual generations on a partially resistant cultivar of *Phaseolus vulgaris*. Phytopathology 75: 449-453.
26. Alexander, S.A., and J.M. Skelly. 1974. A comparison of isolation methods for determining the incidence of *Fomes annosus* in living loblolly pine. Eur. J. Forest Pathol. 4: 33-38.
27. Alexopoulos, C.J. 1950. A new species of *Gelasinospora*. Mycologia 42: 723-734.
28. Alexopoulos, C.J., and D. Cation. 1952. *Gnomonia fragaria* in Michigan. Mycologia 44: 221-223.
29. Alfieri, S.A. Jr. 1970. *Bougainvillea* blight, a new disease caused by *Phytophthora parasitica*. Phytopathology 60: 1806-1808.
30. Alfieri, S.A., Jr., and N.E. El-Gholl. 1985. *Colletotrichum elasticae* causing leaf spots on fig in Florida. Pl. Dis. 69: 268.
31. Alfieri, S.A., Jr., N.E. El-Gholl, and M.L. Campbell. 1984. *Nectriella* (*Kutilakesa*) *pironii*, a pathogen of fig plants. Proc. Florida State Hort. Soc. 97: 325-327.
32. Alfieri, S.A., Jr., and A.E. Graham. 1972. *Cylindrocladium* leaf spot of *Magnolia grandiflora* in Florida. Proc. Florida State Hort. Soc. 85: 354-356.
33. Alfieri, S.A., Jr., and J.F. Knauss. 1970. Southern blight of *Schefflera*. Proc. Florida State Hort. Soc. 83: 432-435.
34. Alfieri, S.A., Jr., J.F. Knauss, and C. Wehlburg. 1979. A stem gall- and canker-inciting fungus, new to the United States. Pl. Dis. Reporter 63: 1016-1020.
35. Alfieri, S.A., Jr., K.R. Langdon, C. Wehlburg, and J.W. Kimbrough. 1984. Index of Plant Diseases in Florida. Florida Dept. of Agric. & Consumer Services, Div. of Plant Industry. Bull. No. 11 (Revised). 389 p.
36. Alfieri, S.A., Jr., R.G. Linderman, R.H. Morrison, and E.K. Sobers. 1972. Comparative pathogenicity of *Calonectria theae* and *Cylindrocladium scoparium* to leaves and roots of azalea. Phytopathology 62: 647-650.
37. Alfieri, S.A., Jr., A.P. Martinez, and C. Wehlburg. 1967. Stem and needle blight of Florida torreya, *Torreya taxifolia* Arn. Proc. Florida State Hort. Soc. 80: 428-431.
38. Alfieri, S.A., Jr., and J.W. Miller. 1971. Basal stem and root rot of Christmas cactus caused by *Phytophthora parasitica*. Phytopathology 61: 804-806.
39. Alfieri, S.A., Jr., W.H. Ridings, D.J. Mitchell, and N.E. El-Gholl. 1987. Leaf blight of dogwood (*Cornus florida*) caused by *Phytophthora parasitica*. Pl. Dis. 71: 555-556.
40. Alfieri, S.A., Jr., C.P. Seymour, and J.C. Denmark. 1972. Aerial blight of *Carissa grandiflora* caused by *Rhizoctonia solani*. Pl. Dis. Reporter 56: 511-514.
41. Alfieri, S.A., Jr., and C. Wehlburg. 1969. *Cephalosporium* leaf spot of *Syngonium podophyllum* Schott. Proc. Florida State Hort. Soc. 82: 366-368.
42. Alford, D.M., J.B. Sinclair, and T.P. Pirone. 1961. Spread and distribution of camellia flower blight in Louisiana. Pl. Dis. Reporter 45: 856-857.
43. Allen, R.M. 1954. Occurrence of *Verticillium* wilt in Arizona. Pl. Dis. Reporter 38: 36.
44. Allen, S.J., G.L. Barnes, and J.L. Caddel. 1985. Symptomatology and ecology of alfalfa anthracnose in Oklahoma. Pl. Dis. 69: 248-251.

45. Allison, J.L. 1950. Pasture and forage legume and grass diseases in North Carolina in 1949. Pl. Dis. Reporter 34: 205.
46. Allison, J.L., G.A. Hemerick, and R.R. Nelson. 1956. *Pleospora herbarum* and *Pleosphaerulina hyalospora*, two pathogenic Ascomycetes of forage legumes. Pl. Dis. Reporter 40: 377-379.
47. Allison, J.L., H.D. Wells, and J.O. Andes. 1950. Diseases reduce vetch seed crop in Tennessee in 1950. Pl. Dis. Reporter 34: 321.
48. Allison, J.L., J.C. Wells, and H.D. Wells. 1952. Pasture and forage legume and grass diseases in North Carolina in 1950 and 1951. Pl. Dis. Reporter 36: 60.
49. Altschuler, S., and W.M. Brown. 1983. Plant pests of Colorado: An annotated inventory. Ext. Service, Colorado State University. 84 p.
50. Ames, R.W. 1952. *Gloeosporium ulmicolum* reported on fruit of rock elm and variegated English elm. Pl. Dis. Reporter 36: 301.
51. Amos, R.E., Jr., and H.L. Barnett. 1966. *Umbelopsis versiformis*, a new genus and species of the imperfects. Mycologia 58: 805-808.
52. Anastasiou, C.J. 1961. Fungi from salt lakes. I.A new species of *Clavariopsis*. Mycologia 53: 11-16.
53. Anderson, B., and D.G. White. 1987. Fungi associated with cornstalks in Illinois in 1982 and 1983. Pl. Dis. 71: 135-137.
54. Anderson, C.G., R.A. Young, and E.T. Palm. 1952. *Curvularia* on gladiolus in Oregon. Pl. Dis. Reporter 36: 474.
55. Anderson, G.W. 1968. The occurrence of eastern and western gall rust on a single jack pine tree. Pl. Dis. Reporter 52: 328-329.
56. Anderson, G.W., and R.L. Anderson. 1968. Relationship between density of quaking aspen and incidence of *Hypoxylon* canker. Forest Sci. 14: 107-112.
57. Anderson, G.W., and D.W. French. 1965. Western gall rust in the Lake States. Forest Sci. 11: 139-141.
58. Anderson, G.W., and A.L. Schipper Jr. 1975. Variation among isolates of *Hypoxylon mammatum*. Eur. J. Forest Pathol. 5: 216-224.
59. Anderson, J.B., and R.C. Ullrich. 1979. Biological species of *Armillaria mellea* in North America. Mycologia 71: 402-414.
60. Anderson, J.P. 1952. The Uredinales of Alaska and adjacent parts of Canada. Iowa State Coll. J. Sci. 26: 507-526.
61. Anderson, N.A. 1960. Stalactiform rust on jack pine in Minnesota. Forest Sci. 6: 40-41.
62. Anderson, N.A., and R.L. Anderson. 1980. Leaf and shoot blight of aspen caused by *Venturia macularis* in nort hern Minnesota. Pl. Dis. 64: 558-559.
63. Anderson, N.A., S.S. Wang, and J.W. Schwandt. 1973. The *Pleurotus ostreatus-sapidus* species complex. Mycologia 65: 28-35.
64. Anderson, P.J. 1919. Index to American species of *Phyllosticta*. Mycologia 11: 66-79.
65. Anderson, R.L. 1968. Oak wilt, *Ceratocystis fagacearum*, identified in Platte County in Missouri. Pl. Dis. Reporter 52: 811.
66. Anderson, R.L. 1974. Oak wilt fungus, *Ceratocystis fagacearum*, identified in Clay and Knox counties in Missouri. Pl. Dis. Reporter 58: 92.
67. Anderson, R.L., and W.H. Hoffard. 1978. *Fusarium* canker-ambrosia beetle complex on tulip poplar in Ohio. Pl. Dis. Reporter 62: 751.
68. Anderson, R.L., P.N. Joranson, and D.W. Einspahr. 1960. *Hypoxylon* canker on European aspen. Pl. Dis. Reporter 44: 132.
69. Anderson, R.L., W. Ritter, and C.R. Witmer. 1976. *Fomes annosus* found on eastern white pine in Iowa. Pl. Dis. Reporter 60: 981.
70. Anderson, R.L., D.D. Skilling, and D.G. Mosher. 1977. *Naemacyclus* needlecast of scotch pine found in Michigan. Pl. Dis. Reporter 61: 422.
71. Anderson, R.N., and H.L. Walker. 1985. *Colletotrichum coccoides*: A pathogen of eastern black nightshade *Solanum ptycanthum*). Weed Sci. 33: 902-905.
72. Andes, J.O. 1955. Plant disease loss estimates in 1954 for Tennessee. Pl. Dis. Reporter 39: 280-282.
73. Andes, J.O. 1956. Plant disease loss estimates in 1955 for Tennessee. Pl. Dis. Reporter 40: 162-163.
74. Andrews, E.A. 1950. Stem blight of soybeans in Michigan. Pl. Dis. Reporter 34: 214.
75. Andrews, E.A. 1950. Melting-out of bent grass turf in Michigan. Pl. Dis. Reporter 34: 214.
76. Andrews, E.A. 1950. *Septoria* on cereals in Michigan during 1949. Pl. Dis. Reporter 34: 214-215.
77. Andrews, E.A., and M.D. Harrison. 1959. *Cronartium comandrae* in Wyoming. Pl. Dis. Reporter 43: 418-419.
78. Andrews, E.A., and R.L. Janes. 1950. Stalk rot of corn in Michigan during 1949. Pl. Dis. Reporter 34: 207-208.
79. Andrews, E., and W. Klomparens. 1952. Yellow spot of winter wheat in Michigan in 1951. Pl. Dis. Reporter 36: 10-11.
80. Andrews, S.R., and W.E. Eslyn. 1960. Sooty-bark canker of aspen in New Mexico. Pl. Dis. Reporter 44: 373.
81. Anonymous. 1931-1970. California Fungi. (Exsiccati set). Nos. 1 - 1325.
82. Anonymous. 1960. Index of Plant Diseases in the United States. U.S. Dept. Agric. Handbook. No. 165. Washington, D.C. 531 p.
83. Anonymous. 1979. USDA Cooperative Plant Pest Report 4(42) Nov. 30, 1979 p. 819
84. Anonymous. 1983. Plant Pathology Specimen Report, Div. of Plant Industry, Gainesville.
85. Anzalone, L., and A.G. Plakidas. 1957. *Cercospora* leafspot of *Photinia serrulata*. Mycologia 49: 412-416.
86. Apple, J.L., and D.F. Crossan. 1954. *Pythium* stem rot and wilt of pepper. Pl. Dis. Reporter 38: 555-556.
87. Applegate, H.W. 1971. Annosus root rot mortality in once-thinned loblolly pine plantations in Tennessee. Pl. Dis. Reporter 55: 625-627.
88. Applegate, H.W., and T. Seay. 1971. Two new hosts for *Comandra* blister rust. Pl. Dis. Reporter 55: 697.
89. Aragaki, M., and R.R. Bergquist. 1971. Occurrence of *Helminthosporium maydis* in Hawaii. Pl. Dis. Reporter 55: 392-393.
90. Aragaki, M., and S. Goto. 1958. Mango anthracnose control in Hawaii. Pl. Dis. Reporter 42: 474-475.
91. Aragaki, M., and M. Ishii. 1960. A spadix rot of *Anthurium* in Hawaii. Pl. Dis. Reporter 44: 865-867.
92. Aragaki, M., W.S. Kimoto, and J.Y. Uchida. 1981. Limitations of hot water treatment in the control of *Phytophthora* fruit rot of papaya. Pl. Dis. 65: 744-745.
93. Aragaki, M., F.F. Laemmlen, and W.T. Nishijima. 1972. Collar rot of koa caused by *Calonectria crotalariae*. Pl. Dis. Reporter 56: 73-74.
94. Aragaki, M., and J.Y. Uchida. 1978. A new papaya fruit rot in Hawaii caused by *Phytophthora capsici*. Pl. Dis. Reporter 62: 765-768.
95. Aragaki, M., and J.Y. Uchida. 1980. Foliar stage of *Phytophthora* blight of macadamia. Pl. Dis. 64: 483-484.
96. Ark, P.A., and T.A. DeWolfe. 1951. *Phytophthora* rot of *Peperomia*. Pl. Dis. Reporter 35: 46-47.
97. Ark, P.A., and N.A. MacLean. 1951. *Botrytis* spot and blight of tuberoses in California. Pl. Dis. Reporter 35: 45-46.
98. Ark, P.A., and W.C. Snyder. 1951. Black spot of vanda orchids. Pl. Dis. Reporter 35: 43-44.
99. Armolik, N., J.G. Dickson, and A.D. Dickson. 1956. Deterioration of barley in storage by microorganisms. Phytopathology 46: 457-461.
100. Armstrong, G.M. 1951. *Phytophthora cactorum* on cowpeas in South Carolina. Pl. Dis. Reporter 35: 418.
101. Armstrong, G.M. 1954. New locations for *Fusarium* wilt of sesame. Pl. Dis. Reporter 38: 57.
102. Armstrong, G.M., and J.K. Armstrong. 1963. *Fusarium* wilt of bean in South Carolina and some host relations of the bean *Fusarium*. Pl. Dis. Reporter 47: 1088-1091.
103. Arndt, C.H. 1950. Boll rots of cotton in South Carolina in 1949. Pl. Dis. Reporter 34: 176.
104. Arndt, C.H. 1954. *Helminthosporium gossypii* on cotton in South Carolina. Pl. Dis. Reporter 38: 728-729.
105. Arneson, E., and D.L. Stiers. 1977. *Cephalosporium gramineum*: a seedborne pathogen. Pl. Dis. Reporter 61: 619-621.
106. Arny, D.C., E.W. Hanson, G.L. Worf, E.S. Oplinger, and W.H. Hughes. 1975. Powdery mildew on soybean in Wisconsin. Pl. Dis. Reporter 59: 288-290.
107. Arny, D.C., G.L. Worf, R.W. Ahrens, and M.F. Lindsey. 1970. Yellow leaf blight of maize in Wisconsin: its history and the reactions of inbreds and crosses to the inciting fungus (*Phyllosticta* sp.). Pl. Dis. Reporter 54: 281-285.
108. Arsdel, E.P., Van Van. 1972. Some cankers on oaks in Texas. Pl. Dis. Reporter 56: 300-304.
109. Arsdel, E.P., Van, Van, T.W. Jares, and R.F. Ginn. 1971. *Cercospora* leaf spot epidemics on sugarbeets in the Texas-New Mexico high plains. Pl. Dis. Reporter 55: 1073-1077.
110. Arsdel, E.P., van, van, and R.S. Halliwell. 1970. Progress in

research on live oak decline. Pl. Dis. Reporter 54: 669-672.
111. Arthur, J.C. 1934. Manual of the Rusts in the United States and Canada. Purdue Research Foundation. Lafayette, Ind. 438 p.
112. Artman, J.D. 1973. Eastern white pine - a new host for *Fusarium lateritium* f. *pini*. Pl. Dis. Reporter 57: 182-184.
113. Arx, J.A. von. 1963. Revision der zu *Gloeosporium* gestellten pilze. Nachträge und berichtigungen. Proc. Nederl. Akad. Wetens. Ser. C. 66: 174-182.
114. Ashqwort, L.J., and S.A. Gaona. 1982. Evaluation of clear polyethylene mulch for controlling *Verticillium* wilt in established pistachio nut groves. Phytopathology 72: 243-246.
115. Ashworth, L.J., and G. Zimmerman. 1976. *Verticillium* wilt of the pistachio nut tree: occurrence in California and control by soil fumigation. Phytopathology 66: 1449-1451.
116. Atilano, R.A. 1982. *Phytophthora* bud rot of Washingtonia palm. Pl. Dis. 66: 517-519.
117. Atilano, R.A. 1983. *Alternaria* leaf spot of *Schefflera arboricola*. Pl. Dis. 67: 64-66.
118. Atilano, R.A. 1983. A foliar blight of Ming aralia caused by *Alternaria panax*. Pl. Dis. 67: 224-226.
119. Atkins, I.M. 1950. Diseases of small grain in Texas in 1949. Pl. Dis. Reporter 34: 40-42.
120. Atkins, J.G. 1950. *Helminthosporium* on Kentucky 31 fescue in Louisiana. Pl. Dis. Reporter 34: 157.
121. Atkins, J.G. 1956. An outbreak of *Piricularia* on rice in 1955. Pl. Dis. Reporter 40: 372-373.
122. Atkins, J.G. 1957. First report of the Bakanae disease of rice for U.S.A. and preliminary studies on growth stimulation by cultural filtrates. Pl. Dis. Reporter 41: 860-862.
123. Atkins, J.G. 1972. An unreported *Helminthosporium* disease of rice for the United States. Pl. Dis. Reporter 56: 26.
124. Atkins, J.G. Jr. 1951. *Helminthosporium victoriae* as a leaf-spotting pathogen on oats. Phytopathology 41: 300-301.
125. Atkins, J.G. Jr. 1951. Observations on cucumber anthracnose in Louisiana. Pl. Dis. Reporter 35: 336.
126. Aycock, R., and C.H. Arndt. 1951. Notes on the incidence of the diseases of cotton in South Carolina in 1950. Pl. Dis. Reporter 35: 204-206.
127. Ayers, J.E., and P.E. Nelson. 1972. *Fusarium* species associated with corn stalk rot in Pennsylvania in 1970. Phytopathology 62: 493.
128. BPI. National Fungus Collections. U.S. Dept. Agric., Beltsville MD.
129. Babadoost, M., and R.L. Gabrielson. 1979. Pathogens causing *Alternaria* diseases of *Brassica* seed crops in Western Washington. Pl. Dis. Reporter 63: 815-820.
130. Babadoost, M., and T.T. Hebert. 1984. Incidence of *Septoria nodorum* in wheat seed and its effects on plant growth and grain yield. Pl. Dis. 68: 125-129.
131. Bache-Wiig, S. 1940. Contributions to the life history of a systemic fungous parasite, *Cryptomycina pteridis*. Mycologia 32: 214-250.
132. Bachelder, S., and E.R. Orton. 1962. *Botrytis* inflorescence blight on American holly in New Jersey. Pl. Dis. Reporter 46: 320.
133. Bachi, P.R., and J.L. Peterson. 1985. Enhancement of *Sphaeropsis sapinea* stem invaision of pines by water deficits. Pl. Dis. 69: 798-799.
134. Backus, M.P., and G.S. Bryan. 1953. *Penicillium* within ovules of *Zamia*. Mycologia 45: 616-619.
135. Bagga, H.S. 1970. Fungi associated with cotton boll rot in the Yazoo-Mississippi delta, 1966 - 1968. Pl. Dis. Reporter 54: 796-798.
136. Bagyanarayana, G., and P. Ramachar. 1984. New rusts on the genus *Populus*. Curr. Sci. 53: 863-865.
137. Bain, D.C. 1960. *Cercospora* leaf spot of tung in Mississippi. Pl. Dis. Reporter 44: 190-191.
138. Bain, D.C. 1964. *Sclerotinia* blight on nutgrass in Mississippi. Pl. Dis. Reporter 48: 742.
139. Bain, D.C. 1966. *Puccinia zoysiae* in Mississippi. Pl. Dis. Reporter 50: 770.
140. Bair, W., and J.E. Ayers. 1986. Variability in isolates of *Cercospora zeae-maydis*. Phytopathology 76: 129-132.
141. Baker, F.A., and J.G. O'Brien. 1985. Virulence of *Ceratocystis ulmi* from wilting Siberian elms in Minnesota. Pl. Dis. 69: 361.
142. Baker, F.A., S.V. Thomson, B.M. Tkacz, and E.L. Hobbs. 1986. Dutch elm disease (*Ceratocystis ulmi*) in Utah. Pl. Dis. 70: 694.
143. Baker, G.E. 1946. Addenda to the genera *Helicogloea* and *Physalacria*. Mycologia 38: 630-638.
144. Baker, G.E., P.H. Dunn, and W.S. Sakai. 1979. Fungus communities associated with leaf surfaces of endemic vascular plants in Hawaii. Mycologia 71: 272-292.
145. Baker, K.F. 1953. Recent epidemics of downy mildew of rose. Pl. Dis. Reporter 37: 331-339.
146. Baker, K.F. 1953. *Chalaropsis* root rot of rose. Pl. Dis. Reporter 37: 430-433.
147. Baker, K.F., and L.H. Davis. 1950. *Heterosporium* disease of nasturtium and its control. Phytopathology 40: 553-566.
148. Baker, K.F., and L.H. Davis. 1950. *Stemphylium* leaf spot of China aster. Mycologia 42: 477-486.
149. Baker, K.F., and L.H. Davis. 1950. Some diseases of ornamental plants in California caused by species of *Alternaria* and *Stemphylium*. Pl. Dis. Reporter 34: 403-413.
150. Baker, K.F., L.H. Davis, R.D. Durbin, and W.C. Snyder. 1977. Greasy blotch of carnation and flyspeck of apple: diseases caused by *Zygophiala jamaicensis*. Phytopathology 67: 580-588.
151. Baker, K.F., and N.A. MacLean. 1950. Powdery mildew of snapdragon on the Pacific Coast. Pl. Dis. Reporter 34: 183-185.
152. Baker, K.F., W.C. Snyder, and L.H. Davis. 1950. *Ramularia* leaf spots of *Lathyrus odoratus* and *L. latifolius*. Mycologia 42: 403-422.
153. Bald, J.G., and B.B. Markley. 1954. *Curvularia* disease of gladiolus in California. Pl. Dis. Reporter 38: 796.
154. Bandoni, R.J. 1955. A new species of *Helicobasidium*. Mycologia 47: 918-919.
155. Bandoni, R.J. 1956. A preliminary survey of the genus *Platygloea*. Mycologia 48: 821-840.
156. Bandoni, R.J. 1979. *Fibulobasidium*: A new genus in the Sirobasidiaceae. Canad. J. Bot. 57: 264-268.
157. Bandoni, R.J., and F. Oberwinkler. 1982. *Stilbotulasnella*: A new genus in the Tulasnellaceae. Canad. J. Bot. 60: 1875-1879.
158. Baniecki, J.F., and R.J. Young. 1974. New hosts of *Cristulariella pyramidalis*. Pl. Dis. Reporter 58: 421-423.
159. Banker, H.J. 1912. Type studies in the Hydnaceae II. The genus *Steccherinum*. Mycologia 4: 309-318.
160. Barbe, G.D., W.O. McCartney, and D.Y. Rosenberg. 1966. *Gymnosporangium confusum* found in California. Pl. Dis. Reporter 50: 768-769.
161. Bardin, R. 1951. Unusual occurrence of ascosporic infection of tomatoes by *Sclerotinia sclerotiorum*. Pl. Dis. Reporter 35: 246.
162. Barksdale, T.H. 1972. Resistance in tomato to six anthracnose fungi. Phytopathology 62: 660-663.
163. Barmore, C.R., and G.E. Brown. 1985. Influence of ethylene on increased susceptibility of oranges to *Diplodia natalensis*. Pl. Dis. 69: 228-230.
164. Barnard, E.L. 1984. Occurrence, impact, and fungicidal control of girdling stem cankers caused by *Cylindrocladium scoparium* on eucalyptus seedlings in a south Florida nursery. Pl. Dis. 68: 471-473.
165. Barnard, E.L., and G.M. Blakeslee. 1980. Pitch canker of slash pine seedlings: a new disease in forest tree nurseries. Pl. Dis. 64: 695-696.
166. Barnard, E.L., G.M. Blakeslee, J.T. English, S.W. Oak, and R.L. Anderson. 1985. Pathogenic fungi associated with sand pine root disease in Florida. Pl. Dis. 69: 196-199.
167. Barnard, E.L., and P.F. Cannon. 1987. A new species of *Monascus* from pine tissues in Florida. Mycologia 79: 479-484.
168. Barnard, E.L., T. Geary, J.T. English, and S.P. Gilly. 1987. Basal cankers and coppice failure of *Eucalyptus grandis*. Pl. Dis. 71: 358-361.
169. Barnes, E.H., and H.C. Tweedie. 1964. *Fusicoccum* canker of highbush blueberry in Michigan. Pl. Dis. Reporter 48: 687-689.
170. Barnes, L.W., and R.D. Martyn. 1985. *Fusarium* crown and root rot of greenhouse tomatoes in Texas. Pl. Dis. 69: 1100.
171. Barnett, H.L. 1957. *Hypoxylon punctulatum* and its conidial stage on dead oak trees and in culture. Mycologia 49: 588-595.
172. Barnett, H.L., and V.G. Lilly. 1955. The effects of humidity, temperture and carbon dioxide on sporulation of *Choanephora cucurbitarum*. Mycologia 47: 26-29.
173. Baroni, T.J., and R.H. Petersen. 1987. *Rhodocybella*: A new genus in the Entolomataceae. Mycologia 79: 358-361.
174. Barr, J.T. 1960. Isolation of *Olpidium brassicae* from roots of lettuce showing big-vein symptoms. Pl. Dis. Reporter 44: 617.

175. Barr, M.E. 1964. The genus *Pseudomassaria* in North America. Mycologia 56: 841-862.
176. Barr, M.E. 1966. Observations on *Urosporella*. Mycologia 58: 690-693.
177. Barr, M.E. 1968. The Venturiaceae in North America. Canad. J. Bot. 46: 799-864.
178. Barr, M.E. 1970. Some amerosporous ascomycetes on Ericaceae and Empetraceae. Mycologia 62: 377-394.
179. Barr, M.E. 1972. Preliminary studies on the Dothideales in temperate North America. Contr. Univ. Michigan Herb. 9: 523-638.
180. Barr, M.E. 1974. The *Cucurbitaria sorbi* Karsten species complex. Rep. Kevo Subarctic Res. Stat. 11: 12-15.
181. Barr, M.E. 1975. *Pestalosphaeria*, a new genus in the Amphisphaeriaceae. Mycologia 67: 187-194.
182. Barr, M.E. 1976. *Hypoxylon grandineum*: a Loculoascomycete. Mycotaxon 3: 325-329.
183. Barr, M.E. 1977. *Magnaporthe*, *Telimenella*, and *Hyponectria* (Physosporellaceae). Mycologia 69: 952-966.
184. Barr, M.E. 1978. The Diaporthales in North America with emphasis on *Gnomonia* and its segregates. Mycologia Memoir 7: 1-232.
185. Barr, M.E. 1979. On the Massariaceae in North America. Mycotaxon 9: 17-37.
186. Barr, M.E. 1980. On the family Tubeufiaceae (Pleosporales). Mycotaxon 12: 137-167.
187. Barr, M.E. 1982. *Leptosphaeria sepalorum*. Mycotaxon 15: 345-348.
188. Barr, M.E. 1982. On the Pleomassariaceae (Pleosporales) in North America. Mycotaxon 15: 349-383.
189. Barr, M.E. 1984. *Herpotrichia* and its segregates. Mycotaxon 20: 1-38.
190. Barr, M.E. 1985. Notes on the Calosphaeriales. Mycologia 77: 549-565.
191. Barr, M.E. 1987. The genus *Trabutia* and its position in the Ascomycetes. Mycologia 79: 188-192.
192. Barr, M.E., and M. Blackwell. 1980. A new genus in the Lophiaceae. Mycologia 72: 1224-1227.
193. Barr, M.E., and C.S. Hodges Jr. 1987. Hawaiian forest fungi. VIII. New species in *Gnomoniella* and *Stigmo chora*. Mycologia 79: 782-786.
194. Barr, M.E., and C.T. Rogerson. 1983. Two new species of Loculoascomycetes. Mycotaxon 17: 247-252.
195. Barr, M.E., C.T. Rogerson, S.J. Smith, and J.H. Haines. 1986. An annotated catalog of the pyrenomycetes described by Charles H. Peck. Bull. New York State Mus. 459: 1-74.
196. Barrows-Broaddus, J., and L.D. Dwinell. 1985. Branch dieback and cone and seed infection caused by *Fusarium moniliforme* var. *subglutinans* in a loblolly pine seed orchard in South Carolina. Phytopathology 75: 1104-1108.
197. Bartlett, A.B., R.S. Halliwell, and I.C. MacSwan. 1963. Undescribed plant diseases and diseases not previously reported for Oregon during 1961-1963. Pl. Dis. Reporter 47: 762.
198. Bassi, A., Jr., and M.J. Goode. 1978. *Fusarium oxysporum* f. sp. *spinaciae* seedborne in spinach. Pl. Dis. Reporter 62: 203-205.
199. Bates, M.L., and M.E. Stanghellini. 1984. Root rot of hydroponically grown spinach caused by *Pythium aphanider matum* and *P. dissotocum*. Pl. Dis. 68: 989-991.
200. Batista, A.C. 1957. Novos generos e epecies de fungos imperfeitos. Revista Biol. 1: 97-112.
201. Batista, A.C., and A.F. Vital. 1957. Contribuicao ao estudo dos fungos Sphaeropsidales. Anais Soc. Biol. Pernambuco 15: 413-427.
202. Batra, L.R. 1960. The species of *Ciborinia* pathogenic to *Salix*, *Magnolia*, and *Quercus*. Amer. J. Bot. 47: 819-827.
203. Batra, L.R. 1967. Ambrosia fungi: a taxonomic revision, and nutritional studies of some species. Mycologia 59: 976-1017.
204. Batra, L.R. 1973. Nematosporaceae (Hemiascomycetidae): Taxonomy, pathogenicity, distribution, and vector relationships. U.S. Dept. Agric., Agric. Res. Serv. Techn. Bull. 1469: 1-71.
205. Batra, L.R. 1974. *Armillaria mellea* on flowering dogwood (*Cornus florida*). Pl. Dis. Reporter 58: 719-721.
206. Batra, L.R. 1979. First authenticated North American record of *Monilinia fructigena*, with notes on related species. Mycotaxon 8: 476-484.
207. Batra, L.R. 1983. *Monilinia vaccinii-corymbosi* (Sclerotiniaceae): its biology on blueberry and comparison with related species. Mycologia 75: 131-152.
208. Batra, L.R., and R.P. Korf. 1959. The species of *Ciborinia* pathogenic to herbaceous angiosperms. Amer. J. Bot. 46: 441-450.
209. Batra, L.R., and R.W. Lichtwardt. 1962. Red stain of *Acer negundo*. Mycologia 54: 91-96.
210. Batra, L.R., and J.M. Staley. 1960. Occurrence of the sclerotial state of *Ciborinia candolleana* (Lév.) Whet. in the United States of America. Pl. Dis. Reporter 44: 430.
211. Baudoin, A.B.A.M. 1986. First report of *Dichotomophthora indica* on common purslane in Virginia. Pl. Dis. 70: 352.
212. Bauer, A.H. 1951. Black shank on tobacco in Pennsylvania. Pl. Dis. Reporter 35: 56.
213. Baxter, D.V. 1940. Some resupinate polypores from the region of the Great Lakes. XI. Pap. Michigan Acad. Sci. 25: 145-170.
214. Baxter, D.V. 1942. Some resupinate polypores from the region of the Great Lakes. XIII. Pap. Michigan Acad. Sci. 27: 139-161.
215. Baxter, J.W. 1955. Diseases of forage legumes in Iowa in 1954. Pl. Dis. Reporter 39: 239.
216. Baxter, J.W. 1955. *Stemphylium* leaf spot of birdsfoot trefoil in Iowa. Pl. Dis. Reporter 39: 659.
217. Baxter, J.W. 1956. Forage legume diseases in Iowa in 1955. Pl. Dis. Reporter 40: 217-218.
218. Baxter, J.W. 1957. The genus *Cumminsiella*. Mycologia 49: 864-873.
219. Baxter, J.W. 1959. A monograph of the genus *Uropyxis*. Mycologia 51: 210-226.
220. Baxter, J.W. 1959. Notes on labiate rusts. Mycologia 51: 598-601.
221. Baxter, J.W. 1961. North American species of *Puccinia* on *Hyptis*. Mycologia 53: 17-24.
222. Baxter, J.W. 1966. New records of North American rust fungi. Mycologia 58: 971-973.
223. Baxter, J.W., and G.B. Cummins. 1951. A monograph of species of *Puccinia* occurring on *Salvia* in North America. Lloydia 14: 201-230.
224. Baxter, L.W., and J.A. Berly. 1956. Flower blight of Camellia in South Carolina. Pl. Dis. Reporter 40: 831.
225. Baxter, L.W., and A.G. Plakidas. 1954. Dieback and canker of *Camellias* caused by *Glomerella cingulata*. Phytopathology 44: 129-133.
226. Beagle, J.E., J.F. Rissler, and J.G. Kantzes. 1982. *Phytophthora* root rot of soybeans in Maryland. Pl. Dis. 66: 371-372.
227. Bean, G.A., and R. Schwartz. 1961. A severe epidemic of *Helminthosporium* brown spot disease on cultiva ted wild rice in northern Minnesota. Pl. Dis. Reporter 45: 901.
228. Bean, G.A., and R.D. Wilcoxson. 1964. *Helminthosporium* leaf spot of bluegrass. Phytopathology 54: 1065-1070.
229. Beardslee, H.C. 1934. New and interesting fungi. Mycologia 26: 253-260.
230. Bedker, P.J., R.A. Blanchette, and D.W. French. 1982. *Nectria cinnabarina*: the cause of a canker disease of honey locust in Minnesota. Pl. Dis. 66: 1067-1070.
231. Bedker, P.J., and M.J. Wingfield. 1983. Taxonomy of three canker-causing fungi of honey locust in the United States. Trans. Brit. Mycol. Soc. 81: 179-183.
232. Bega, R.V. 1974. *Phytophthora cinnamomi*: Its distribution and possible role in ohia decline on the island of Hawaii. Pl. Dis. Reporter 58: 1069-1073.
233. Bega, R.V. 1978. *Phomopsis lokoyae* outbreak in a California forest nursery. Pl. Dis. Reporter 62: 567-569.
234. Bega, R.V. 1979. Heart and root rot fungi associated with deterioration of *Acacia koa* on the island of Hawaii. Pl. Dis. Reporter 63: 682-684.
235. Bega, R.V., R.S. Smith Jr., A.P. Martinez, and C.J. Davis. 1978. Severe damage to *Pinus radiata* and *P. pinaster* by *Diplodia pinea* and *Lophodermium* spp. on Molokai and Lanai in Hawaii. Pl. Dis. Reporter 62: 329-331.
236. Beil, J.A., and K.J. Kessler Jr. 1979. Sapstreak disease of sugar maple found in New York state. Pl. Dis. Reporter 63: 436.
237. Beisel, M., F.F. Hendrix Jr., and T.E. Starkey. 1984. Natural inoculation of apple buds by *Botryosphaeria obtusa*. Phytopathology 74: 335-338.
238. Bell, D.K., H. Harris, and H.D. Wells. 1973. *Rhizoctonia* blight of grain sorghum foliage. Pl. Dis. Reporter 57: 549-550.
239. Bell, D.K., B.J. Locke, and S.S. Thompson. 1973. The status of

Cylindrocladium black rot of peanut in Georgia since its discovery in 1965. Pl. Dis. Reporter 57: 90 -94.
240. Bender, C.L., and D.L. Coyier. 1984. Isolation and identification of races of *Sphaerotheca pannosa* var. *rosae*. Phytopathology 74: 100-103.
241. Bender, C.L., and D.L. Coyier. 1986. Pathogenic variation in Oregon populations of *Sphaerotheca pannosa* var. *rosae*. Pl. Dis. 70: 383-385.
242. Benson, D.M., and C.L. Campbell. 1985. Spatial pattern of *Phytophthora* root rot and dieback of azalea in container-grown nursery stock. Pl. Dis. 69: 1049-1054.
243. Benson, D.M., R.K. Jones, and K.R. Barker. 1982. Disease loss assessment for azalea, rhododendron, and Japanese holly in North Carolina nurseries. Pl. Dis. 66: 125-128.
244. Beraha, L., D.B. Towner, and T.H. Camp. 1976. Stem gumming and blight and stem-end rot of Texas cantaloups caused by *Diplodia natalensis*. Pl. Dis. Reporter 60: 420-422.
245. Beraha, L., and W.R. Wright. 1973. A new anthracnose of strawberry caused by *Colletotrichum dematium*. Pl. Dis. Reporter 57: 445-448.
246. Bergdahl, D.R., T. Dyrakacz, and A. Gotlieb. 1979. *Gremmeniella abietina* on red pine in New Hampshire. Pl. Dis. Reporter 63: 995-996.
247. Bergdahl, D.R., and D.W. French. 1975. *Endocronartium harknessii* on *Pinus mugo* in Minnesota. Pl. Dis. Reporter 59: 433-434.
248. Bergdahl, D.R., and D.W. French. 1976. Epidemiology of comandra rust on jack pine and comandra in Minnesota. Canad. J. Forest Res. 6: 326-334.
249. Bergdahl, D.R., and D.W. French. 1976. Swiss needlecast of Douglas-fir in Minnesota. Pl. Dis. Reporter 60: 148.
250. Bergdahl, D.R., and D.W. French. 1985. Association of wood decay fungi with decline and mortality of apple trees in Minnesota. Pl. Dis. 69: 887-890.
251. Bergdahl, D.R., S.S. Halik, and D.L.K. Smeltzer. 1985. *Mycosphaerella laricina* on European larch in Vermont. Pl. Dis. 69: 83.
252. Bergdahl, D.R., and D.L.K. Smeltzer. 1983. *Chrysomyxa weirii* on *Picea pungens*. Pl. Dis. 67: 918.
253. Bergquist, R.R. 1973. *Colletotrichum graminicola* on *Sorghum bicolor* in Hawaii. Pl. Dis. Reporter 57: 272-275.
254. Bergquist, R.R., and J.W. Lorbeer. 1972. Apothecial production, compatibility and sex in *Botryotinia squamosa*. Mycologia 64: 1270-1281.
255. Berkeley, M.J. 1874. Notices of North American fungi. Grevillea 3: 1-17.
256. Berkely, M.J. 1875. Notices of North American fungi. Grevillea 3: 97-112.
257. Berkenkamp, B.B. 1961. A powdery mildew on safflower. Pl. Dis. Reporter 45: 826.
258. Bernhardt, E.A., and J.M. Duniway. 1984. Root and stem rot of parrotfeather (*Myriophyllum brasiliense*) caused by *Pythium carolinianum*. Pl. Dis. 68: 999-1003.
259. Bernhardt, E.A., and J.M. Duniway. 1985. A new *Papulaspora* species from the pondweeds *Potamogeton nodosus* and *P. crispus* in California. Canad. J. Bot. 63: 429-431.
260. Bernhardt, E.A., and J.M. Duniway. 1986. Decay of pondweed and *Hydrilla hibernacula* by fungi. J. Aquat. Plant Manage. 24: 20-24.
261. Berry, C.R., and G.E. Thompson. 1961. Extension of range and a new host for *Cristulariella pyramidalis*. Pl. Dis. Reporter 45: 152.
262. Berry, F.H. 1976. *Phlebia chrysocrea* causes a heart rot of oaks and other hardwoods. Pl. Dis. Reporter 60: 308-311.
263. Berry, F.H., N.F. Rogers, and O.J. Dooling. 1961. Shortleaf pine, a new host for comandra blister rust. Pl. Dis. Reporter 45: 576.
264. Berthier, J. 1976. Monographie des *Typhula* Fr., *Pistillaria* Fr. et genres voisins. Bull. Soc. Linn. Lyon n.spec.: 1-213.
265. Bertrand, P.F., I.C. MacSwan, R.L. Rackham, and B.J. Moore. 1977. An outbreak of side rot in Bosc pears in Oregon. Pl. Dis. Reporter 61: 890-893.
266. Bessette, A.E., and R.L. Homola. 1986. *Lentinus adhaerens*, a species new to North America. Mycologia 78: 296-298.
267. Bessey, E.A. 1953. Notes on the genus *Camptomeris*, Fungi Imperfecti. Mycologia 45: 364-390.
268. Beute, M.K., D.M. Porter, and B.A. Hadley. 1975. *Sclerotinia* blight of peanut in North Carolina and Virginia and its chemical control. Pl. Dis. Reporter 59: 697-701.
269. Bigelow, H.E., and M.E. Barr. 1960. Contributions to the fungus flora of northeastern North America. Rhodora 62: 186-198.
270. Bigelow, H.E., and M.E. Barr. 1962. Contribution to the fungus flora of northeastern North America. II. Rhodora 64: 126-137.
271. Bigelow, H.E., and M.E. Barr. 1963. Contribution to the fungus flora of northeastern North America. III. Rhodora 65: 289-309.
272. Bigelow, H.E., and M.E. Barr. 1969. Contribution to the fungus flora of northeastern North America. V. Rhodora 71: 177-203.
273. Bigelow, H.E., and L.R. Hesler. 1960. *Clitocybe* in Tennessee and North Carolina. J. Elisha Mitchell Sci. Soc. 76: 155-167.
274. Bigelow, H.E., O.K. Miller Jr., and H.D. Thiers. 1976. A new species of *Omphalotus*. Mycotaxon 3: 363-372.
275. Bigelow, H.E., and A.H. Smith. 1966. Additional species of *Clitocybe* from Michigan. Michigan Bot. 5: 218-226.
276. Birchfield, W., J.L. Smith, A. Martinez, and E.P. Matherly. 1957. Chinese evergreen plants rejected because of glebal masses of *Sphaerobolus stellatus* on foliage. Pl. Dis. Reporter 41: 537-539.
277. Bird, L.S. 1951. *Ascochyta gossypii* in Texas. Pl. Dis. Reporter 35: 557.
278. Birrenkott, G.L., A. Mengistu, and C.R. Grau. 1984. First report of charcoal rot caused by *Macrophomina phaseolina* on soybean in Wisconsin. Pl. Dis. 68: 628.
279. Bissett, J. 1979. Coelomycetes on Liliales: *Dothiorella smilacina* and *Stagonospora smilacis*. Canad. J. Bot. 57: 2071-2081.
280. Bissett, J. 1979. Coelomycetes on Liliales: The genus *Phyllosticta*. Canad. J. Bot. 57: 2082-2095.
281. Bissett, J. 1986. *Discochora yuccae* sp. nov. with *Phyllosticta* and *Leptodothiorella* synanamorphs. Canad. J. Bot. 64: 1720-1726.
282. Black, L.L., and D.A. Slack. 1967. Dead-arm disease of grapes in Arkansas and West Virginia. Pl. Dis. Reporter 51: 1038-1039.
283. Blackwell, M., and R.L. Gilbertson. 1985. *Quasiconcha reticulata* and its anamorph from conifer roots. Mycologia 77: 50-54.
284. Blackwell, M., and R.L. Gilbertson. 1985. A new species of *Inonotus* (Aphyllophorales, Hymenochaetaceae) on oak in Louisiana. Mycotaxon 23: 285-290.
285. Blain, W.L. 1931. A list of diseases of economic plants in Alabama. Mycologia 23: 300-304.
286. Blaker, N.S., and J.D. MacDonald. 1986. The role of salinity in the development of *Phytophthora* root rot of *Citrus*. Phytopathology 76: 970-975.
287. Blakeslee, G.M. 1980. Residual naval stores stumps as reservoirs of inoculum for infection of slash pines by *Phaeolus schweinitzii*. Pl. Dis. 64: 167.
288. Blakeslee, G.M., S.H. Kratka, R.A. Schmidt, and C.S. Moses. 1978. Sporodochia of the pitch canker fungus (*Fusarium moniliforme* var. *subglutinans*) as found in diseased slash pine in Florida. Pl. Dis. Reporter 62 : 656-657.
289. Blakeslee, G.M., and S.W. Oak. 1979. Significant mortality associated with pitch canker infection of slash pine in Florida. Pl. Dis. Reporter 63: 1023-1025.
290. Blakeslee, G.M., S.W. Oak, and S.H. Kratka. 1980. Shoot dieback of planted sand pine caused by *Fusarium moniliforme* var. *subglutinans*. Pl. Dis. 64: 703-704.
291. Blanchette, R.A. 1982. *Phellinus pini* decay associated with sweetfern rust in sapwood of jack pine. Canad. J. Forest Res. 12: 304-310.
292. Blanchette, R.A. 1982. Decay and canker formation by *Phellinus pini* in white and balsam fir. Canad. J. Forest Res. 12: 538-544.
293. Blanchette, R.A. 1984. Selective delignification of eastern hemlock by *Ganoderma tsugae*. Phytopathology 74: 153-160.
294. Blanchette, R.A. 1984. Manganese accumulation in wood decayed by white rot fungi. Phytopathology 74: 725-730.
295. Blank, L.M. 1965. *Fusarium* wilt of cotton in Arizona. Pl. Dis. Reporter 49: 179.
296. Blazquez, C.H. 1969. Occurrence of gray leafspot on peppers in Florida. Pl. Dis. Reporter 53: 756.
297. Blazquez, C.H. 1972. Target spot of tomato. Pl. Dis. Reporter 56: 243-245.
298. Blazquez, C.H. 1976. A powdery mildew of chilli caused by *Oidiopsis sp.*. Phytopathology 66: 1155-1157.
299. Blazquez, C.H. 1977. A blight of tomatoes caused by *Corynespora cassiicola*. Pl. Dis. Reporter 61: 1002-1006.
300. Blazquez, C.H. 1983. Net spot of cucumbers. Pl. Dis. 67: 534-536.
301. Blazquez, C.H., and S.A. Alfieri Jr. 1974. *Cercospora* leaf mold of

tomato. Phytopathology 64: 443-445.
302. Bliss, D.E. 1941. A new species of *Ceratostromella* on the date palm. Mycologia 33: 468-482.
303. Blodgett, E.C., and L.F. Semler. 1970. Apple scab in the Yakima Valley of Washington. Pl. Dis. Reporter 54: 63-65.
304. Blume, M.C., and G.E. Harman. 1979. *Thielaviopsis basicola*: A component of the pea root rot complex in New York State. Phytopathology 69: 785-788.
305. Bockus, W.W., and M.M. Claassen. 1985. Effect of lime and sulfur application to low-pH soil on incidence of *Cephalosporium* stripe in winter wheat. Pl. Dis. 69: 576-578.
306. Boerema, G.H., M.M.J. Dorenbosch, and H.A. van Kesteren. 1971. Remarks on species of *Phoma* referred to *Peyronellaea*-III. Persoonia 6: 171-177.
307. Boerema, G.H., and H.A. van Kesteren. 1981. Nomenclatural notes on some species of *Phoma* sect. *Plenodomus*. Persoonia 11: 317-331.
308. Boewe, G.H. 1938. *Naucoria* on small grains in Illinois. Phytopathology 28: 852-855.
309. Boewe, G.H. 1953. Early season diseases of grain crops and alfalfa present an unusual picture in southern Illinois. Pl. Dis. Reporter 37: 411-412.
310. Boewe, G.H. 1958. Downy mildew on cucurbits in Illinois in 1957. Pl. Dis. Reporter 42: 554.
311. Boewe, G.H. 1963. Host plants of charcoal rot disease in Illinois. Pl. Dis. Reporter 47: 753-755.
312. Boewe, G.H. 1964. Some plant diseases new to Illinois. Pl. Dis. Reporter 48: 866-870.
313. Boewe, G.H., and W.L. Klarman. 1957. New diseases for Illinois. Pl. Dis. Reporter 41: 904.
314. Bogart, F. van de. 1976. The genus *Coprinus* in western North America, Part I: Section *Coprinus*. Mycotaxon 4: 233-275.
315. Bohn, G.W., K. Mets, J.R. Breece, and T.W. Whitaker. 1964. An unusual downy mildew epiphytotic on muskmelon in Imperial Valley, California. Pl. Dis. Reporter 48: 258.
316. Boise, J. 1982. A new graminicolous species of *Xylaria*. Mycologia 74: 467-471.
317. Boise, J. 1983. On *Trematosphaeria circinans* and reinstatement of the genus *Byssothecium*. Mycologia 75: 666-669.
318. Boise, J. 1985. An amended description of *Trematosphaeria*. Mycologia 77: 230-237.
319. Boise, J. 1985. New combinations in the Pleomassariaceae and the Massarinaceae. Mycotaxon 22: 477-482.
320. Boise, J. 1986. Requienellaceae, a new family of Loculoascomycetes. Mycologia 78: 37-41.
321. Bolick, J.H. 1959. *Cercospora antirrhini* found in Florida. Pl. Dis. Reporter 43: 511.
322. Bolkan, H.A., J.M. Ogawa, T.J. Michailides, and P.F. Kable. 1985. Physiological specialization in *Tranzschelia discolor*. Pl. Dis. 69: 485-486.
323. Bolkan, H.A., J.M. Ogawa, and H.R. Teranishi. 1984. Shoot blight of pistachio caused by *Botrytis cinerea*. Pl. Dis. 68: 163-165.
324. Bollenbacher, K., and P.B. Marsh. 1954. A preliminary note on a fluorescent-fiber condition in raw cotton. Pl. Dis. Reporter 38: 375-379.
325. Bonar, L. 1942. Studies on some California fungi. II. Mycologia 34: 180-192.
326. Bonar, L. 1951. Two new fungi on *Torreya*. Mycologia 43: 62-66.
327. Bonar, L. 1962. *Stegopezizella balsameae* and *Gloeosporium balsamaea*. Mycologia 54: 395-399.
328. Bonar, L. 1964. *Polyporus oleaea* on olive in California. Pl. Dis. Reporter 48: 70.
329. Bonar, L. 1965. Studies on some California fungi IV. Mycologia 57: 379-396.
330. Bonar, L. 1966. A new *Diaporthopsis* causing brooming in *Baccharis*. Amer. J. Bot. 53: 181-184.
331. Bonar, L. 1967. Studies on Deuteromycetes. Mycologia 59: 596-599.
332. Bonde, R. 1955. A survey of late blight of potatoes in Aroostook County, Maine August 26 -- September 10, 1954. Pl. Dis. Reporter 39: 86-87.
333. Bonman, J.M., P.A. Delwiche, R.L. Gabrielson, and P.H. Williams. 1980. *Leptosphaeria maculans* on cabbage in Wisconsin. Pl. Dis. 64: 326.
334. Bonman, J.M., and R.L. Gabrielson. 1981. Localized infections of siliques and seed of cabbage by *Phoma lingam* . Pl. Dis. 65: 868-869.
335. Boosalis, M.G., and R.I. Hamilton. 1957. Root and stem rot of soybean caused by *Corynespora cassiicola* (Berk. & Curt.) Wei. Pl. Dis. Reporter 41: 696-698.
336. Boosalis, M.G., and A.L. Scharen. 1960. The susceptibility of pigweed to *Rhizoctonia solani* in irrigated fields of western Nebraska. Pl. Dis. Reporter 44: 815-818.
337. Booth, J.A., and S.M. Alcorn. 1959. Seedling rot of *Carnegiea gigantea* (Engelm.) Britt. & Rose caused by *Fusarium spp.*. Pl. Dis. Reporter 43: 1038-1041.
338. Boothroyd, C.W. 1951. A new leaf spot of *Celastrus scandens* L., the climbing bittersweet. Mycologia 43: 373-375.
339. Boothroyd, C.W. 1951. Pink rot of potato found in New York state. Pl. Dis. Reporter 35: 55.
340. Boothroyd, C.W., H.J. Otto, and J.M. Barnes. 1955. Stalk rot of corn in New York, 1954. Pl. Dis. Reporter 39: 380.
341. Boquiren, D.T. 1971. The genus *Epithele*. Mycologia 63: 937-957.
342. Borders, H.I. 1953. Preliminary report on fungicide tests for control of downy mildew of cabbage seedlings in south Georgia. Pl. Dis. Reporter 37: 363-364.
343. Bostock, R.M., and M.A. Doster. 1985. Association of *Phytophthora syringae* with pruning wound cankers of almond trees. Pl. Dis. 69: 568-571.
344. Bowden, R.L., and D.J. Eschen. 1986. First report of wild rice diseases in Idaho. Pl. Dis. 70: 800.
345. Bowden, R.L., H.S. Fenwick, L.J. Smith, and J.M. Kraft. 1985. Root rot (*Aphanomyces euteiches*) epidemic in northern Idaho spring peas. Pl. Dis. 69: 451.
346. Bowden, R.L., M.K. Kardin, J.A. Percich, and L.J. Nickelson. 1984. Anthracnose of wild rice. Pl. Dis. 68: 68-69.
347. Bowden, R.L., M.V. Wiese, J.E. Crock, and D.L. Auld. 1985. Root rot of chickpeas and lentils caused by *Thielaviopsis basicola*. Pl. Dis. 69: 1089-1091.
348. Boyce, J.S., and A.A. Graves. 1966. *Monochaetia* canker on Arizona cypress in Georgia and South Carolina. Pl. Dis. Reporter 50: 482-483.
349. Boyce, J.S., and C.F. Speers. 1960. Oak dieback in Virginia in 1959. Pl. Dis. Reporter 44: 351.
350. Boyce, J.S. Jr. 1950. A note on *Leptostroma hedgcockii* Dearn. Pl. Dis. Reporter 34: 228-229.
351. Boyce, J.S. Jr. 1952. Records of flowering dogwood spot anthracnose in North and South Carolina. Pl. Dis. Reporter 36: 296.
352. Boyce, J.S. Jr. 1954. Mat formation by the oak wilt fungus on felled versus standing trees. Pl. Dis. Reporter 38: 676-677.
353. Boyce, J.S. Jr. 1958. Twig blight of eastern white pine caused by *Monochaetia pinicola*. Phytopathology 48: 516-517.
354. Boyce, J.S. Jr. 1959. Brown spot needle blight on Eastern white pine. Pl. Dis. Reporter 43: 420.
355. Boyce, J.S. Jr. 1963. Red root and butt rot in a Georgia slash pine plantation. Pl. Dis. Reporter 47: 572-573.
356. Boyd, O.C. 1950. Some noteworthy diseases of vegetable crops in Massachusetts this year. Pl. Dis. Reporter 34: 339-340.
357. Boyd, P.E., and J. Kohlmeyer. 1982. The influence of temperature on the seasonal and geographic distribution of three marine fungi. Mycologia 74: 894-902.
358. Boyle, A.M. 1955. Three diseases of *Lippia canescens* new to Arizona. Pl. Dis. Reporter 39: 425-428.
359. Boyle, A.M. 1963. Pathogenicity of an isolate of *Verticillium albo-atrum* from olive. Phytopathology 53: 242.
360. Bradford, B., J.M. Skelly, and S.A. Alexander. 1978. Incidence and severity of Annosus root rot in loblolly pine plantations in Virginia. Eur. J. Forest Pathol. 8: 135-145.
361. Brandt, W.H. 1962. Decay resistance of wood from some wilt-infected species of oak. Forest Sci. 8: 334-335.
362. Braselton, J.P., and F.T. Short. 1985. Karyotypic analysis of *Plasmodiophora diplantherae*. Mycologia 77: 940-945.
363. Braun, U. 1982. Descriptions of new species and combinations in *Microsphaera* and *Erysiphe*. Mycotaxon 14: 369-374.
364. Braun, U. 1982. Descriptions of new species and combinations in *Microsphaera* and *Erysiphe* (II). Mycotaxon 15: 121-137.
365. Braun, U. 1982. Taxonomic notes on some powdery mildews. Mycotaxon 15: 138-154.
366. Braun, U. 1983. Descriptions of new species and combinations in

Microsphaera and *Erysiphe* (III). Mycotaxon 16: 417-424.
367. Braun, U. 1983. Descriptions of new species and combinations in *Microsphaera* and *Erysiphe*. IV. Mycotaxon 18: 113-129.
368. Braun, U. 1984. Taxonomic notes on some powdery mildews (IV). Mycotaxon 20: 483-489.
369. Braun, U. 1984. Descriptions of new species and combinations in *Microsphaeria* and *Erysiphe* (VI). Mycotaxon 20: 491-498.
370. Braun, U. 1985. Miscellaneous notes on the genus *Sphaerotheca*. I. Zentralbl. Mikrobiol. 140: 161-170.
371. Braun, U. 1985. Miscellaneous notes on the genus *Sphaerotheca*. II. Zentralbl. Mikrobiol. 140: 237-246.
372. Braverman, S.W. 1959. *Stemphylium* species on comfrey. Pl. Dis. Reporter 43: 1050.
373. Braverman, S.W. 1961. Three additional bromegrass species as hosts of *Helminthosporium sorokinianum* (*H. sativum*). Pl. Dis. Reporter 45: 199.
374. Braverman, S.W. 1966. *Curvularia geniculata* on *Sambucus chinensis*. Pl. Dis. Reporter 50: 488-489.
375. Braverman, S.W. 1976. A leafspot of *Lotus corniculatus* and *Lotus* spp. caused by *Myrothecium verrucaria*. Pl. Dis. Reporter 60: 1065-1068.
376. Braverman, S.W., and J.H. Graham. 1960. *Helminthosporium dictyoides* and related species on forage grasses. Phytopathology 50: 691-695.
377. Brenckle, J.F. 1917. North Dakota fungi - I. Mycologia 9: 275-293.
378. Brenckle, J.F. 1918. North Dakota fungi - II. Mycologia 10: 199-221.
379. Brener, W.D., E.C. Setliff, and R.L. Norgren. 1974. *Sclerophoma pythiophila* associated with a tip dieback of juniper in Wisconsin. Pl. Dis. Reporter 58: 653-657.
380. Brentzel, W.E. 1950. Stem rust on wheat in North Dakota. Pl. Dis. Reporter 34: 340-341.
381. Bresadola, G. 1925. New species of fungi. Mycologia 17: 68-77.
382. Bretz, T.W. 1964. Needle blight of eastern redcedar in Missouri. Pl. Dis. Reporter 48: 418.
383. Bretz, T.W., and W.G. Long. 1950. Oak wilt fungus isolated from Chinese chestnut. Pl. Dis. Reporter 34: 291.
384. Brewster, M.S. 1952. A new species of *Physoderma*. Mycologia 44: 97-100.
385. Bridges, J.R., and T.J. Perry. 1987. *Ceratocystiopsis ranaculosus sp. nov.* associated with the southern pine beetle. Mycologia 79: 630-633.
386. Bridgmon, G.H. 1970. Dutch elm disease identified in Wyoming. Pl. Dis. Reporter 54: 547.
387. Bridgmon, G.H., and B.J. Kolp. 1959. Stripe rust in Wyoming in 1958. Pl. Dis. Reporter 43: 163-164.
388. Brigham, R.D. 1961. Castorbean diseases in Texas in 1960. Pl. Dis. Reporter 45: 124-125.
389. Bristow, P.R., and T.D. Wyllie. 1986. *Macrophomina phaseolina*, another cause of the twin-stem abnormality disease of soybean. Pl. Dis. 70: 1152-1153.
390. Brittingham, R.L., and M.J. O'Brien. 1978. Occurrence of *Phyllosticta cryptomeriae* on *Cryptomeria japonica* in the United States. Pl. Dis. Reporter 62: 152-153.
391. Britton, K.O., and F.F. Hendrix. 1982. Three species of *Botryosphaeria* cause peach tree gummosis in Georgia. Pl. Dis. 66: 1120-1121.
392. Britton, K.O., and F.F. Hendrix. 1986. Population dynamics of *Botryosphaeria* spp. in peach gummosis cankers. Pl. Dis. 70: 134-136.
393. Britton, M.P., and D.P. Rogers. 1963. *Olpidium brassicae* and *Polymyxa graminis* in roots of creeping bent in golf putting greens. Mycologia 55: 758-763.
394. Britton, M.P., and D.P. Rogers. 1966. *Puccinia lobata* on hollyhock in Illinois. Mycologia 58: 804-805.
395. Broscious, S.C., and J.A. Frank. 1986. Effects of crop management practices on common root rot of winter wheat. Pl. Dis. 70: 857-859.
396. Brosten, B.S., and D.C. Sanda. 1986. Field trials of *Sclerotinia sclerotiorum* to control Canada thistle (*Cirsium arvense*). Weed Sci. 34: 377-380.
397. Brotzman, H.G., and R.L. Gilbertson. 1967. *Tyromyces undosus* in North America. Mycopathol. Mycologia Appl. 33: 33-42.
398. Brown, D.H. 1970. Recent discoveries extend distribution range of two destructive diseases of limber pine in southeastern Montana. Pl. Dis. Reporter 54: 441.
399. Brown, D.H. 1978. Extension of the known distribution of *Cronartium ribicola* and *Arceuthobium cyanocarpum* on limber pine in Wyoming. Pl. Dis. Reporter 62: 905.
400. Brown, E.A., and F.F. Hendrix. 1981. Pathogenicity and histopathology of *Botryosphaeria dothidea* on apple stems. Phytopathology 71: 375-379.
401. Brown, E.A., and R.T. Hanlin. 1982. *Beniowskia sphaeroidea* blight on knotroot bristlegrass in Georgia. Pl. Dis. 66: 1197-1198.
402. Brown, E.M., and T.B. Sutton. 1986. Control of sooty blotch and flyspeck of apple with Captan, Mancozeb,and Mancozeb combined with Dinocap in dilute and concentrate applications. Pl. Dis. 70: 281-284.
403. Brown, G.E., and W.C. Wilson. 1968. Mode of entry of *Diplodia natalensis* and *Phomopsis citri* into Florida oranges. Phytopathology 58: 736-739.
404. Brown, H.D. 1971. *Fome annosus* found in Iowa. Pl. Dis. Reporter 55: 317-318.
405. Brown, H.D., D.R. Kucera, and P.H. Peacher. 1969. Dutch elm disease detected in Alabama. Pl. Dis. Reporter 53: 307.
406. Brown, J.G. 1953. A new species of *Hysterographium*. Mycologia 45: 964-967.
407. Brown, J.G., and A.M. Boyle. 1950. Canaigre leaf spot. Pl. Dis. Reporter 34: 178.
408. Brown, J.G., and A.M. Boyle. 1952. Safflower rust appears in Arizona. Pl. Dis. Reporter 36: 325-326.
409. Brown, L.G., and G. Morgan-Jones. 1977. Notes on hyphomycetes. XX. "*Cercospora*-complex" fungi of *Cassia* and *Psoralea*. Mycotaxon 6: 261-276.
410. Brown, W.M. Jr. 1970. A leaf spot of English holly caused by a species of *Sclerophoma*. Phytopathology 60: 1144-1145.
411. Brownell, K.H., and R.W. Schneider. 1983. *Fusarium* hypocotyl rot of sugar pine in California forest nurseries. Pl. Dis. 67: 105-107.
412. Broyles, J.W., and C.O. Grogan. 1962. Crazy top of corn in Mississippi. Pl. Dis. Reporter 46: 8.
413. Bruckart, W.L., and J.W. Lorbeer. 1982. *Pythium* species pathogenic to onion seedlings grown on organic soils in New York. Phytopathology 72: 469-475.
414. Bruehl, G.W. 1955. Barley adaptation in relation to *Pythium* root rot. Phytopathology 45: 97-103.
415. Bruehl, G.W. 1957. *Cephalosporium* stripe disease of wheat. Phytopathology 47: 641-649.
416. Bruehl, G.W., and R. Machtmes. 1980. Cultural variation within *Typhula idahoensis* and *T. ishikariensis* and the species concept. Phytopathology 70: 867-871.
417. Bruehl, G.W., R. Machtmes, and T. Murray. 1985. Tolerance in *Pseudocercosporella herpotrichoides* to benzimidazole fungicides in Washington. Pl. Dis. 69: 360.
418. Bruehl, G.W., T.D. Murray, and R.E. Allan. 1986. Resistance of winter wheats to *Cephalosporium* stripe in the field. Pl. Dis. 70: 314-316.
419. Bruton, B.D., R.W. Toler, and D.L. Blasingame. 1981. Downy mildew on St. Augustinegrass in Mississippi. Pl. Dis. 65: 925.
420. Bruton, B.D., R.W. Toler, and M.P. Grisham. 1986. Preventative and curative control of downy mildew of St. Augustinegrass by metalaxyl. Pl. Dis. 70: 413-415.
421. Buchanan, P.K. 1987. A reappraisal of *Ascochytula* and *Ascochytella* (Coleomycetes). Mycol. Pap. 156: 1-83.
422. Buckley, P.M., T.D. Wyllie, and J.E. DeVay. 1969. Fine structure of conidia and conidium formation in *Verticillium albo-atrum* and *V. nigrescens*. Mycologia 61: 240-250.
423. Buddenhagen, I.W., and R.A. Young. 1957. A leaf and twig disease of English holly caused by *Phytophthora ilicis* n. sp. Phytopathology 47: 95-101.
424. Buddenhagen, I., and R.A. Young. 1954. A new *Phytophthora* leaf and twig blight of English holly in Oregon. Pl. Dis. Reporter 38: 723-724.
425. Bugbee, W.M. 1967. *Verticillium nigrescens* from cotton. Pl. Dis. Reporter 51: 760-761.
426. Bugbee, W.M. 1979. Resistance to sugar beet storage rot pathogens. Phytopathology 69: 1250-1252.
427. Bugbee, W.M., and H.M. El-Nashaar. 1983. A newly recognized symptom of sugar beet root infection caused by *Phoma betae*. Pl. Dis. 67: 101-102.
428. Bugbee, W.M., and G.E. Nielsen. 1978. *Penicillium cyclopium* and *Penicillium funiculosum* as sugarbeet storage rot pathogens. Pl.

Dis. Reporter 62: 953-954.
429. Bullen, S., and R.E. Wood. 1979. *Fomes annosus* on pacific madrone. Pl. Dis. Reporter 63: 844.
430. Burdsall, H.H. Jr. 1971. Notes on some lignicolous basidiomycetes of the southeastern United States. J. Elisha Mitchell Sci. Soc. 87: 239-245.
431. Burdsall, H.H. Jr. 1976. A new species of *Laeticorticium* (Aphyllophorales, Corticiaceae) from the southern Appalachians. J. Elisha Mitchell Sci. Soc. 91: 243-245.
432. Burdsall, H.H. Jr. 1984. The genus *Candelabrochaete* (Corticiaceae) in North America and a note on *Peniophora mexicana*. Mycotaxon 19: 389-395.
433. Burdsall, H.H. Jr. 1985. A contribution to the taxonomy of the genus *Phanerochaete*. Mycologia Memoir 10.: 1-165.
434. Burdsall, H.H. Jr. 1986. *Platygloea acanthophysa*, a new species with single sterigmate basid ia and acanthophyses. Mycotaxon 27: 499-502.
435. Burdsall, H.H., Jr., and W.E. Eslyn. 1974. A new *Phanerochaete* with a *Chrysosporium* imperfect state. Mycotaxon 1: 123-133.
436. Burdsall, H.H., Jr., and G.W. Freeman. 1981. New species of *Gloeocystidiellum* (Corticiaceae) from the southeastern United States. Syst. Bot. 6: 422-434.
437. Burdsall, H.H., Jr., and R.L. Gilbertson. 1974. A new species of *Platygloea* occurring on *Peniophora tamaricicola* in Arizona. Mycologia 66: 702-706.
438. Burdsall, H.H., Jr., and R.L. Gilbertson. 1982. New species of Corticiaceae (Basidiomycotina, Aphyllophorales) from Arizona. Mycotaxon 15: 333-340.
439. Burdsall, H.H., Jr., and M.J. Larsen. 1974. *Lazulinospora*, a new genus of Corticiaceae, and a note on *Tomentel la atrocyanea*. Mycologia 66: 96-100.
440. Burdsall, H.H., Jr., and F.F. Lombard. 1976. The genus *Gloeodontia* in North America. Mem. New York Bot. Gard. 28: 16-31.
441. Burdsall, H.H., Jr., and O.K. Miller Jr. 1975. A reevaluation of *Panellus* and *Dictyopanus* (Agaricales). Beih. Nova Hedwigia 51: 79-91.
442. Burdsall, H.H., Jr., O.K. Miller Jr., and K.A. Nishijima. 1978. Morphological and mating system studies of a new taxon of *Hericium* (Aphyllophorales, Hericiaceae) from the southern Appalachians. Mycotaxon 7: 1-9.
443. Burdsall, H.H., Jr., and K.K. Nakasone. 1981. New or little known lignicolous Aphyllophorales (Basidiomycotina) from southeastern United States. Mycologia 73: 454-476.
444. Burdsall, H.H., Jr., and K.K. Nakasone. 1983. Species of effused Aphyllophorales (Basidiomycotina) from the southeastern United States. Mycotaxon 17: 253-268.
445. Burk, W.R., and R.E. Rex. 1974. *Polyporus squamosus* in Utah. Mycotaxon 1: 135-136.
446. Burke, D.W. 1957. Basidial stage of *Pellicularia filamentosa* on beans. Pl. Dis. Reporter 41: 533.
447. Burke, D.W., J.C. Gomes, and W.G. Foeppel. 1957. Observations on *Sclerotinia* wilt of beans in northeastern Colorado. Pl. Dis. Reporter 41: 72-73.
448. Burkholder, W.H. 1957 . Pea bean rust in New York state. Pl. Dis. Reporter 41: 1036.
449. Burleigh, J.R., W. Wood, and P. Sebesta. 1982. Growth retardation of plantation-grown sugar pine by *Lophodermella arcuata*. Pl. Dis. 66: 680-682.
450. Burnett, H.C. 1958. A new leaf spot disease of mistletoe in Florida. Pl. Dis. Reporter 42: 272.
451. Burnett, H.C., and J.R. McFarlin. 1960. *Cercospora bunchosiae* a new leafspot disease of Barbados cherry. Pl. Dis. Reporter 44: 505.
452. Burnett, H.C., D.P.H. Tucker, and W.H. Ridings. 1974. *Phytophthora* root and stem rot of milkweed vine. Pl. Dis. Reporter 58: 355-357.
453. Burnett, H.C., and H. Van Pelt. 1962. Powdery mildew on strawberry in Florida. Pl. Dis. Reporter 46: 348.
454. Burns, E.E., and M.C. Shurtleff. 1973. Observations of *Physoderma maydis* in Illinois: Effects of tillage practices in field corn. Pl. Dis. Reporter 57: 630-633.
455. Burr, T.J., J.E. Hunter, J.M. Ogawa, and G.S. Abawi. 1978. A root rot of apple caused by *Rhizoctonia solani* in New York nurseries. Pl. Dis. Reporter 62: 476-479.
456. Burt, E.A. 1914. The Thelephoraceae of North America. III. *Craterellus borealis* and *Cyphella*. Ann. Missouri Bot. Gard. 1: 357-382.
457. Burt, E.A. 1915. The Thelephoraceae of North America. IV. *Exobasidium*. Ann. Missouri Bot. Gard. 2: 627-658.
458. Burt, E.A. 1916. The Thelephoraceae of North America. VI. *Hypochnus*. Ann. Missouri Bot. Gard. 3: 203-241.
459. Burt, E.A. 1922. The North American species of *Clavaria* with illustrations of the type specimens. Ann. Missouri Bot. Gard. 9: 1-78.
460. Burt, E.A. 1926. The Thelephoraceae of North America. XV. *Corticium*. Ann. Missouri Bot. Gard. 13: 173-304.
461. Burton, C.L. 1968. Southern corn leaf blight on sweet corn ears in transit. Pl. Dis. Reporter 52: 847-851.
462. Burton, C.L., and W.R. Wright. 1969. Sour rot disease of peaches on the market. Pl. Dis. Reporter 53: 580-582.
463. Bushong, J.W., and J.W. Gerdemann. 1959. Root rot of alfalfa caused by *Phytophthora cryptogea* in Illinois. Pl. Dis. Reporter 43: 1178-1183.
464. Butler, E.E. 1952. A new species of *Mucor*. Mycologia 44: 561-563.
465. Butler, E.E. 1959. Fungi and rots in California canning tomatoes. Pl. Dis. Reporter 43: 187-192.
466. Butler, E.E., and D.H. Hall. 1966. Occurrence in California of wart disease of subterranean clover caused by *Physoderma trifolii*. Pl. Dis. Reporter 50: 796-797.
467. Butler, E.E., K.A. Kimble, and D. Zaiger. 1958. Occurrence of tomato fruit pox in California. Pl. Dis. Reporter 42: 850-851.
468. Butler, E.E., and A.H. McCain. 1968. A new species of *Stephanoma*. Mycologia 60: 955-959.
469. Bynum, H.H., and D.R. Miller. 1964. *Elytroderma deformans* found on knobcone pine in California. Pl. Dis. Reporter 48: 828.
470. Bynum, H.H., and D.R. Miller. 1969. A new needle disease of hard pines in Oregon and California. Pl. Dis. Reporter 53: 232-234.
471. Byrnes, K.J., and R.B. Carroll. 1986. Fungi causing stalk rot of conventional-tillage and no-tillage corn in Delaware. Pl. Dis. 70: 238-239.
472. Byther, R.S. 1971. Black rot of sugarcane cuttings in Hawaii. Pl. Dis. Reporter 55: 7-9.
473. Byther, R.S., and G.W. Steiner. 1972. Four sugarcane seedling diseases in Hawaii: causal agents, control, and a selective medium for isolation. Phytopathology 62: 120-124.
474. Byther, R.S., and G.W. Steiner. 1974. Unusual smut symptoms on sugarcane in Hawaii. Pl. Dis. Reporter 58: 401-405.
475. Byther, R.S., G.W. Steiner, S.L. Ladd, and D.J. Heinz. 1973. Smut disease of sugarcane in Hawaii. Pl. Dis. Reporter 57: 966-968.
476. Calavan, E.C., and J.M. Wallace. 1954. *Hendersonula toruloidea* Nattrass on *Citrus* in California. Phytopathology 44: 635-639.
477. Calvert, O.H., A.S. Foudin, H.C. Minor, and G.F. Krause. 1985. *Fusarium moniliforme* colonization of corn ears in Missouri. Pl. Dis. 69: 988-990.
478. Calvert, O.H., W.P. Sappenfield, R.D. Hicks, and T.D. Wyllie. 1964. The *Cercospora-Alternaria* leaf blight complex of cotton in Missou ri. Pl. Dis. Reporter 48: 466-467.
479. Campana, R.J. 1954. The present status of Dutch elm disease in Illinois. Pl. Dis. Reporter 38: 356-358.
480. Campana, R.J., and J.C. Carter. 1957. The current status of Dutch elm disease in Illinois. Pl. Dis. Reporter 41: 636-639.
481. Campana, R., and M. Rosinski. 1962. Association of *Cladosporium* with needle cast of larch and spruce in the United States. Pl. Dis. Reporter 46: 265-266.
482. Campbell, C.L. 1980. Root rot of Ladino clover induced by *Codinaea fertilis*. Pl. Dis. 64: 959-960.
483. Campbell, C.L. 1982. Cultural characteristics and host range of *Codinaea fertilis*. Phytopathology 72: 501-504.
484. Campbell, R.N. 1960. Some sap-stain fungi found in Minnesota. Pl. Dis. Reporter 44: 625-628.
485. Campbell, R.N., and A.S. Greathead. 1978. *Pseudocercosporella* white spot of crucifers in California. Pl. Dis. Reporter 62: 1066-1068.
486. Campbell, R.N., A.S. Greathead, D.F. Myers, and G.J. Boer. 1985. Factors related to control of clubroot of crucifers in the Salinas Valley of California. Phytopathology 75: 665-670.
487. Campbell, R.N., and R.W. Scheuerman. 1979. *Oidiopsis taurica* on processing tomatoes in the central valley of California. Pl. Dis. Reporter 63: 1087.
488. Campbell, R.N., V.H. Schweers, and D.H. Hall. 1982. Corky root

of tomato in California caused by *Pyrenochaeta lycopersici* and control by soil fumigation. Pl. Dis. 66: 657-661.
489. Campbell, W.A., and G.H. Hepting. 1954. *Fomes annosus* on slash pine. Pl. Dis. Reporter 38: 217.
490. Campbell, W.A., and J.H. Miller. 1952. Windthrow of root-rotted oak shade trees. Pl. Dis. Reporter 36: 490.
491. Campbell, W.A., J.H. Miller, and G.E. Thompson. 1950. Notes on some wood-decaying fungi of Georgia. Pl. Dis. Reporter 34: 128-134.
492. Campbell, W.A., and A.F. Verrall. 1963. *Phytophthora cinnamomi* associated with Lawson cypress mortality in Louisiana. Pl. Dis. Reporter 47: 808.
493. Canaday, C.H., D.G. Helsel, and T.D. Wyllie. 1986. Effects of herbicide-induced stress on root colonization of soybeans by *Macrophomina phaseolina*. Pl. Dis. 70: 863-866.
494. Cappellini, R.A. 1957. Forage crop diseases in New Jersey -- 1956. Pl. Dis. Reporter 41: 123.
495. Cappellini, R.A. 1959. Brown spot of soybeans in New Jersey. Pl. Dis. Reporter 43: 287.
496. Cappellini, R.A., and S. Lund. 1957. Small grain diseases in New Jersey, 1955-56. Pl. Dis. Reporter 41: 117.
497. Cardina, J., R.H. Littrell, and L.J. Stowell. 1987. Bioherbicide for Florida beggarweed. United States Patent number 4,643,756. 6 p.
498. Carey, A.C., E.A. Miller, G.T. Geballe, P.M. Wargo, W.H. Smith, and T.G. Siccama. 1984. *Armillaria mellea* and decline of red spruce. Pl. Dis. 68: 794-795.
499. Carling, D.E., and R.H. Leiner. 1986. Isolation and characterization of *Rhizoctonia solani* and binucleate *R. solani*-like fungi from aerial stems and subterranean organs of potato plants. Phytopathology 76: 725-729.
500. Carlson, M.L. 1987. Effects of two foliar pathogens on seed yield of sunflower. Pl. Dis. 71: 549-551.
501. Carmody, B.E., M.E. Miller, and M.P. Grisham. 1985. A technique to screen muskmelons for resistance to *Alternaria* leaf blight. Pl. Dis. 69: 426-428.
502. Caroselli, N.E. 1957. Juniper blight and progress on its control. Pl. Dis. Reporter 41: 216-218.
503. Carpenter, J.B. 1967. *Aspergillus* species associated with dieback of citrus trees. Pl. Dis. Reporter 51: 523-525.
504. Carpenter, S.E. 1976. Taxonomy, morphology and ontogeny of *Gelatinodiscus flavidus*. Mycotaxon 3: 209-232.
505. Carpenter, S.E. 1981. Monograph of *Crocicreas* (Ascomycetes, Helotiales, Leotiaceae). Mem. New York Bot. Gard. 33: 1-290.
506. Carpenter, T.R., W.D. Wilbur, and W. Young. 1962. Mexican palm trees apparently killed by *Penicillium vermoeseni*. Pl. Dis. Reporter 46: 750.
507. Carris, L.M. 1987. *Vaccinium* fungi: *Corynespora arctespora*, comb. nov. Mycotaxon 30: 127-132.
508. Carroll, G., and O. Petrini. 1983. Patterns of substrate utilization by some fungal endophytes from coniferous foliage. Mycologia 75: 53-63.
509. Carroll, R.B., A.L. Morehart, and M. Stuart. 1976. *Phomopsis* canker of Russian-olive in Delaware. Pl. Dis. Reporter 60: 787-788.
510. Carroll, R.B., and M. Sasser. 1974. An outbreak of potato pink rot in Delaware. Pl. Dis. Reporter 58: 443-445.
511. Carson, M.L. 1981. New race of *Plasmopara halstedii* virulent on resistant sunflowers in South Dakota. Pl. Dis. 65: 842-843.
512. Carson, M.L. 1985. First report of eyespot (*Kabatiella zeae*) of corn in South Dakota. Pl. Dis. 69: 177.
513. Carter, J.C. 1950. Oak wilt in Illinois. Pl. Dis. Reporter 34: 81-82.
514. Carter, J.C. 1951. Dutch elm disease in Illinois. Pl. Dis. Reporter 35: 56.
515. Carter, J.C., and F.O. Dodd. 1969. Discovery and distribution of *Fusicoccum* canker of Russian olive in Illinois. Pl. Dis. Reporter 53: 392.
516. Carter, J.C., and C.M. Sacamano. 1967. *Fusicoccum* canker, a new disease of Russian olive. Mycologia 59: 535-537.
517. Carter, J.C., and N.B. Wysong. 1951. Isolation of the oak wilt fungus from swamp white oak. Pl. Dis. Reporter 35: 173-174.
518. Carter, W.W. 1979. Importance of *Macrophomina phaseolina* in vine decline and fruit rot of cantaloup in South Texas. Pl. Dis. Reporter 63: 927-930.
519. Carter, W.W. 1979. Corky dry rot of cantaloup caused by *Fusarium roseum 'semitectum'*. Pl. Dis. Reporter 63: 1080-1084.
520. Carver, R.B., M.C. Rush, and G.D. Lindberg. 1972. An epiphytotic of ryegrass blast in Louisiana. Pl. Dis. Reporter 56: 157-159.
521. Cash, E.K. 1936. Some new Colorado Discomycetes. Mycologia 28: 297-306.
522. Cash, E.K. 1943. Some new or rare Florida Discomycetes and Hysteriales. Mycologia 35: 595-603.
523. Cash, E.K. 1945. Some new species of fungi on *Libocedrus*. Mycologia 37: 311-317.
524. Cash, E.K. 1951. Some new species of Discomycetes from Mount Shasta. Mycologia 43: 229-234.
525. Cash, E.K. 1952-1953. A record of the fungi named by J.B. Ellis. USDA Division of Mycology and Disease and Survey. Special Publication No. 2. Parts. I & II. 518 p.
526. Cash, E.K. 1953. A checklist of Alaskan fungi. Pl. Dis. Reporter Suppl. 219: 1-70.
527. Cash, E.K. 1958. Some new Discomycetes from California. Mycologia 50: 642-656.
528. Cash, E.K., and R.W. Davidson. 1940. Some new species of ascomycetes on coniferous hosts. Mycologia 32: 728-735.
529. Cash, E.K., and A.M. Waterman. 1957. A new species of *Plagiostoma* associated with a leaf disease of hybrid aspens. Mycologia 49: 756-760.
530. Castellani, E., and G. Germano. 1977. Le Stagonosporae graminicole. Ann. Fac. Sci. Agrar. Univ. Torino 10: 1-135.
531. Cejp, K. 1931. Notes on Iowa species of the genus *Irpex*. Mycologia 23: 130-133.
532. Ceponis, M.J. 1966. Occurence of *Geotrichum candidum* in western melons on the New York market. Pl. Dis. Reporter 50: 222-224.
533. Ceponis, M.J., and J.E. Butterfield. 1974. Market losses in Florida cucumbers and bell peppers in metropolitan New York. Pl. Dis. Reporter 58: 558-560.
534. Ceponis, M.J., and R.A. Cappellini. 1976. Relative importance of *Alternaria* rot in postharvest disease development in New Jersey blueberries. Pl. Dis. Reporter 60: 569-571.
535. Ceponis, M.J., and B.A. Friedman. 1955. The occurrence of *Phomopsis* on tomatoes in New York. Pl. Dis. Reporter 39: 989-990.
536. Ceponis, M.J., J. Kaufman, and J.E. Butterfield. 1969. Moldy carpels in delicious apples on the greater New York market. Pl. Dis. Reporter 53: 136-138.
537. Ceponis, M.J., J. Kaufman, and W.H. Tietjen. 1973. Decays of New Jersey blueberries in marketing channels. Pl. Dis. Reporter 57: 487-488.
538. Chacko, R.J., and A.D. Partridge. 1976. Decay fungi isolated from conifers of northern Idaho. Pl. Dis. Reporter 60: 960-963.
539. Chacko, R.J., and J.D. Rogers. 1981. Cultural characteristics of some species of *Xylaria*. Mycologia 73: 415-428.
540. Chamberlain, D.W. 1951. *Sclerotinia* stem rot of soybeans. Pl. Dis. Reporter 35: 490-491.
541. Chamuris, G.P. 1986. The *Cystostereum pini-canadense* complex in North America. Mycologia 78: 380-390.
542. Chamuris, G.P. 1987. Notes on stereoid fungi I. The genus *Dendrophora*, stat. nov., and *Peniophora malenconii* subsp. *americana*, subsp. nov. ("Stereum heterosporum"). Mycotaxon 28: 543-552.
543. Chamuris, G.P. 1988. The nonstipitate stereoid fungi in the northeastern United States and adjacent Canada. Mycologia Memoir 14:
544. Chamuris, G.P., and S.P. Falk. 1987. The population structure of *Peniophora rufa* in an aspen plantation. Mycologia 79: 451-457.
545. Chaney, D.H., Moller W.J., and S.M. Gotan. 1973. Silver leaf of peach in California. Pl. Dis. Reporter 57: 192.
546. Changsri, W., and G.F. Weber. 1958. Leafspot of marigold, *Tagetes erecta*, caused by *Septoria tageticola* n. sp. Phytopathology 48: 561-565.
547. Chapman, R.A. 1953. Powdery mildew on burley tobacco in Kentucky. Pl. Dis. Reporter 37: 528.
548. Chappelka, A.H., III, and R.A. Schmidt. 1983. The phenology of inoculum production by *Cronartium quercuum* f. sp. *fusiforme* on slash pine in north Florida and south Georgia. Forest Sci. 29: 253-262.
549. Chappelka, A.H., III, R.A. Schmidt, and H.D. Patterson. 1984. Dynamics of aeciospore inoculum production by *Cronartium*

quercuum f. sp. *fusiforme* on slash pine. Forest Sci. 30: 787-792.
550. Charles, V.K. 1942. A *Laschia* on cabbage palmetto. Mycologia 34: 235-240.
551. Charles, V.K., and A.E. Jenkins. 1914. A fungous disease of hemp. J. Agric. Res. 3: 81-84.
552. Charudattan, R., and K.E. Conway. 1975. Comparison of *Uredo eichhorniae*, the water-hyacinth rust, and *Urom yces pontederiae*. Mycologia 67: 653-657.
553. Charudattan, R., and K.E. Conway. 1976. *Mycoleptodiscus terrestris* leaf spot on waterhyacinth. Pl. Dis. Reporter 60: 77-80.
554. Charudattan, R., B.D. Perkins, and R.C. Littell. 1978. Effects of fungi and bacteria on the decline of arthropod-damaged waterhyacinth (*Eichornia crassipes*) in Florida. Weed Sci. 26: 101-107.
555. Chase, A.R. 1978. New fungus associated with vascular wilt of shasta daisy. California Agric. 32: 21.
556. Chase, A.R. 1981. *Phytophthora* stem rot of *Aphelandra squarrosa*. Pl. Dis. 65: 921-922.
557. Chase, A.R. 1982. Stem rot and shattering of Easter cactus caused by *Drechslera cactivora*. Pl. Dis. 66: 602-603.
558. Chase, A.R. 1982. Dematiaceous leaf spots of *Chrysalidocarpus lutescens* and other pal ms in Florida. Pl. Dis. 66: 697-699.
559. Chase, A.R. 1982. *Corynespora* leaf spot of *Aeschynanthus pulcher* and related plants. Pl. Dis. 66: 739-740.
560. Chase, A.R. 1982. *Alternaria* leaf spot of *Calathea* spp. Pl. Dis. 66: 935-954.
561. Chase, A.R. 1984. Leaf spot disease of *Ficus benjamina* caused by *Corynespora cassiicola*. Pl. Dis. 68: 251.
562. Chase, A.R., and N.E. El-Gholl. 1982. Stem rot, cutting rot, and leaf spot of *Dieffenbachia maculata* 'Perfection' incited by *Fusarium solani*. Pl. Dis. 66: 595-598.
563. Chase, A.R., and D.E. Munnecke. 1978. *Pythium* root rot and stunting of *Peperomia obtusifolia* var. *variegata*. Pl. Dis. Reporter 62: 314-315.
564. Chase, A.R., and D.E. Munnecke. 1980. Shasta daisy vascular wilt incited by *Acremonium strictum*. Phytopathology 70: 834-838.
565. Chase, A.R., and R.T. Poole. 1987. Effects of potting medium pH and air temperature on severity of *Cyli ndrocladium* root and petiole rot of *Spathiphyllum* sp. Pl. Dis. 71: 509-511.
566. Chase, A.R., and G.W. Simone. 1986. *Corynespora* bract spot of *Euphorbia pulcherrima* in Florida. Pl. Dis. 70: 1074.
567. Chase, T.E., and R.C. Ullrich. 1983. Sexuality, distribution, and dispersal of *Heterobasidion annosum* in pine plantations of Vermont. Mycologia 75: 825-831.
568. Chase, T.E., and R.C. Ullrich. 1985. Fruiting and genetic analysis of *Heterobasidion annosum*. Mycologia 77: 464-471.
569. Chastagner, G.A. 1983. *Narcissus* fire: Prevalence, epidemiology, and control in western Washington. Pl. Dis. 67: 1384-1386.
570. Chau, K.F., and A.M. Alvarez. 1983. Postharvest fruit rot of papaya caused by *Stemphylium lycopersici*. Pl. Dis. 67: 1279-1281.
571. Cheo, P.C. 1953. Varietal differences in susceptibility of sweet potato to black rot fungus. Phytopathology 43: 78-81.
572. Cheo, P.C. 1976. *Arctotheca calendula*, a new host of *Sclerotium rolfsii* in southern California. Pl. Dis. Reporter 60: 26-27.
573. Chiarappa, L. 1959. Wood decay of the grapevine and its ralationship with black measles disease. Phytopathology 49: 510-519.
574. Childs, J.F.L. 1953. Concentric canker and wood rot of *Citrus* associated with *Fomes applanatus* in Florida. Phytopathology 43: 99-100.
575. Childs, J.F.L., and E. West. 1953. Butt rot of queen palms in Florida associated with *Ganoderma sulcatum*. Pl. Dis. Reporter 37: 632-633.
576. Chilton, J.E. 1954. *Volutella* species on alfalfa. Mycologia 46: 800-809.
577. Chilton, J.E. 1957. Sesame (*Sesamum indicum*), a host for *Verticillium albo-atrum* Reinke & Berth. Pl. Dis. Reporter 41: 803.
578. Cho, J.J. 1977. Shoot and flower blight of *Leucospermum cordifolium* incited by abenomyl-tolerant strain of *Botrytis cinerea*. Phytopathology 67: 124-127.
579. Cho, J.J. 1981. *Phytophthora* root rot of *Banksia*: Host range and chemical control. Pl. Dis. 65: 830-833.
580. Chupp, C. 1950. Early development of rhubarb leaf spot in New York. Pl. Dis. Reporter 34: 240.
581. Chupp, C. 1953. Monograph of the genus *Cercospora*. Published by the author. 667 p.
582. Claflin, L.E., and O.J. Dooling. 1973. Dutch elm disease in Montana. Pl. Dis. Reporter 57: 701.
583. Clark, C.A. 1980. End rot, surface rot, and stem lesions caused on sweet potato by *Fusarium solani*. Phytopathology 70: 109-112.
584. Clark, C.A., and J.W. Lorbeer. 1973. Symptomatology, etiology, and histopathology of *Botrytis* brown stain of onion. Phytopathology 63: 1231-1235.
585. Clayton, C.N. 1951. *Botrytis* leaf spot of apple in North Carolina. Pl. Dis. Reporter 35: 237.
586. Clayton, C.N., and F.A. Haasis. 1964. Blueberry root-rot caused by *Phytophthora cinnamomi* in North Carolina. Pl. Dis. Reporter 48: 460-461.
587. Clayton, C.N., and W.H. Ridings. 1970. Grape rust, *Physopella ampelopsidis*, on *Vitis rotundifolia* in Nor th Carolina. Phytopathology 60: 1022-1023.
588. Clinton, G.P. 1901. Two new smuts on *Eriocaulon septangulare* . Rhodora 3: 79-82.
589. Cobb, F.W., G.W. Slaughter, D.L. Rowney, and C.J. DeMars. 1982. Rate of spread of *Ceratocystis wageneri* in ponderosa pine stands in the central Sierra Nevada. Phytopathology 72: 1359-1362.
590. Cody, Y.S., and O.C. Maloy. 1984. *Cylindrocephalum* rot of mung bean sprouts. Pl. Dis. 68: 304-305.
591. Coffey, M.D. 1987. *Phytophthora* root rot of avocado. Pl. Dis. 71: 1046-1052.
592. Cohen, M. 1963. Infection of lychee and peach seedlings with cultures of *Clitocybe tabescens*. Phytopathology 53: 358-359.
593. Cohen, M., and A.S. Muller. 1957. New host records for *Elsinoe lepagei* Bitanc. & Jenkins. Pl. Dis. Reporter 41: 540.
594. Coker, W.C., W.C., and H.C. Beardslee, H.C. 1921. The Collybias of North Carolina. J. Elisha Mitchell Sci. Soc. 37: 83-107.
595. Cole, G.T. 1983. *Graphiola phoenicis*: A taxonomic enigma. Mycologia 75: 93-116.
596. Cole, H., and G.W. McKee. 1967. A storage decay of crownvetch (*Coronilla varia*) "crowns" caused by *Fusarium roseum* and its control by fungicide dip treatments. Pl. Dis. Reporter 51: 820-822.
597. Cole, J.R. 1950. *Articularia* on pecan in Texas. Pl. Dis. Reporter 34: 414.
598. Cole, J.R. 1953. Diseases causing partial or total losses to pecan trees or nuts in the southeastern states where no control measures are employed. Pl. Dis. Reporter 37: 510.
599. Cole, J.R. 1958. Pecan scab (*Fusicladium effusum* Wint.) is causing damage to susceptible pecan varieties in Georgia this year. Pl. Dis. Reporter 42: 897.
600. Cole, J.R. 1965. Liver spot disease of pecan extends its range into Georgia. Pl. Dis. Reporter 49: 148.
601. Cole, J.R. 1968. Dieback affecting pecan trees in southwest Georgia and north Florida. Pl. Dis. Reporter 52: 947-949.
602. Cole, J.R., and A.C. Gossard. 1956. Stuart pecan found to be susceptible to scab in Mississippi. Pl. Dis. Reporter 40: 156.
603. Comstock, J.C., T.L. Tew, and S.A. Ferreira. 1982. Sugarcane rust in Hawaii. Pl. Dis. 66: 1193-1194.
604. Cones, W.L. 1967. Oak wilt mats on white oak in West Virginia. Pl. Dis. Reporter 51: 430-431.
605. Constantinescu, O. 1975. Revision of *Cercospora* species (Hyphomycetes) parasitic on *Psoralea*. Mycotaxon 3: 119-125.
606. Constantinescu, O. 1979. Revision of *Bremiella* (Peronosporales). Trans. Brit. Mycol. Soc. 72: 510-515.
607. Constantinescu, O. 1984. Taxonomic revision of *Septoria*-like fungi parasitic on Betulaceae. Trans. Brit. Mycol. Soc. 83: 383-398.
608. Converse, R.H. 1953. *Articularia* and *Microstroma* on pecan in Oklahoma. Pl. Dis. Reporter 37: 511-512.
609. Converse, R.H. 1967. Physiologic races of *Phytophthora fragariae* on strawberry in California, Oregon, and Washington. Phytopathology 57: 173-177.
610. Converse, R.H., and C.D. Schwartze. 1968. A root rot of red raspberry caused by *Phytophthora erythroseptica*. Phytopathology 58: 56-59.
611. Converse, R.H., E.H. Varney, and J.G. Kantzes. 1966. Occurrence of *Phytophthora fragariae* race A-5 in Maryland and New Jersey strawberry fields. Pl. Dis. Reporter 50: 351-352.
612. Conway, K.E. 1976. *Cercospora rodmanii*, a new pathogen of water hyacinth with biological control potential. Canad. J. Bot. 54: 1079-1083.
613. Conway, K.E. 1976. Evaluation of *Cercospora rodmanii* as a

biological control of water hyacinths. Phytopathology 66: 914-917.
614. Conway, K.E. 1978. A new species of *Cercospora* from *Limnobium spongia*. Trans. Brit. Mycol. Soc. 71: 521-523.
615. Conway, K.E. 1987. Purple spot (*Pleospora herbarum*) of asparagus in Oklahoma. Pl. Dis. 71: 376.
616. Conway, K.E., T.E. Freeman, and R. Charudattan. 1974. The Fungal Flora of Waterhyacinth in Florida. Florida Agric. Exp. Sta. Journal Ser. No. 5519. Pub. No. 30. 11 p.
617. Conway, K.E., and J.W. Kimbrough. 1975. A new *Doratomyces* from waterhyacinth. Mycotaxon 2: 127-131.
618. Conway, K.E., and L.S. Morrison. 1983. Diseases and decay fungi in windbreaks in Oklahoma. Pl. Dis. 67: 289-291.
619. Conway, K.E., J.S. Motes, B. Bostian, C.G. Fisher, and P.L. Claypool. 1987. *Cercospora* blight development on asparagus fern and effects of fungicides on disease severity and yield. Pl. Dis. 71: 254-259.
620. Conway, K.E., and E. Williams Jr. 1986. *Typhula*-like snow mold on wheat in Oklahoma. Pl. Dis. 70: 169-170.
621. Cook, A.A. 1955. Occurrence of a *Pleospora* species on castor bean leaves. Pl. Dis. Reporter 39: 184.
622. Cook, A.A. 1955. Charcoal rot of castor bean in the United States. Pl. Dis. Reporter 39: 233-235.
623. Cook, A.A. 1955. *Sclerotinia sclerotiorum* on castorbean in Florida. Pl. Dis. Reporter 39: 585-586.
624. Cook, G.E., J.R. Steadman, and E.D. Kerr. 1972. Epidemiology of white mold in western Nebraska. Phytopathology 62: 1107.
625. Cook, M.T. 1951. Species of *Synchytrium* in Louisiana. VI. Two new species on *Impatiens* and *Smilax*. Mycologia 43: 103-107.
626. Cook, M.T. 1952. Species of *Synchytrium* in Louisiana. VII. A new species on *Urtica chamaedryoides*. Mycologia 44: 827-828.
627. Cook, M.T. 1953. Species of *Synchytrium* in Louisiana. VIII. New species. Mycologia 45: 101-114.
628. Cook, R.J. 1967. *Gibberella avenacea* sp. n., perfect stage of *Fusarium roseum* f. sp. *cerealis* 'Avenaceum'. Phytopathology 57: 732-736.
629. Cook, R.J., and G.W. Bruehl. 1968. Ecology and possible significance of perithecia of *Calonectria nival is* in the Pacific Northwest. Phytopathology 58: 702-703.
630. Cook, R.J., D. Huber, R.L. Powelson, and G.W. Bruehl. 1968. Occurrence of take-all in wheat in the Pacific Northwest. Pl. Dis. Reporter 52: 716-718.
631. Cooke, J.C. 1969. Morphology of *Chaetomium funicolum*. Mycologia 61: 1060-1065.
632. Cooke, J.C. 1970. Morphology of *Chaetomium trilaterale*. Mycologia 62: 282-288.
633. Cooke, J.C., and M.E. Barr. 1983. *Plagiostoma solidaginis*, a new species on *Solidago*. Mycotaxon 18: 87-90.
634. Cooke, M.C. 1878. Ravenel's American fungi. Grevillea 6: 129-146.
635. Cooke, M.C., and J.B. Ellis. 1876. New Jersey fungi. Grevillea 5: 30-35.
636. Cooke, M.C., and H.W. Harkness. 1886. Californian fungi. Grevillea 13: 111-114.
637. Cooke, M.C., and W.H. Harkness. 1884. Californian fungi. Grevillea 12: 92-97.
638. Cooke, W.B. 1939-1951. Mycobiota of North America. (Exsiccati set, some subtitled Mycobiota of Mt. Shasta.) Nos. 1 - 450.
639. Cooke, W.B. 1949. Western fungi - I. Mycologia 41: 601-622.
640. Cooke, W.B. 1951. The genus *Cytidia*. Mycologia 43: 196-210.
641. Cooke, W.B. 1952. Western fungi - II. Mycologia 44: 245-261.
642. Cooke, W.B. 1955. Fungi of Mount Shasta. Sydowia 9: 94-215.
643. Cooke, W.B. 1956. The genus *Phlebia*. Mycologia 48: 386-405.
644. Cooke, W.B. 1957. The genera *Serpula* and *Meruliporia*. Mycologia 49: 197-225.
645. Cooke, W.B. 1957. The Porotheleaceae: *Porotheleum*. Mycologia 49: 680-693.
646. Cooke, W.B. 1961. The Cyphellaceous Fungi: A study in the Porotheleaceae. Beih. Sydowia 4: 1-144.
647. Cooke, W.B. 1967. The 1961 Indiana foray. Mycologia 59: 375-381.
648. Cooke, W.B. 1967. The 1962 Oregon foray. Mycologia 59: 519-524.
649. Cooke, W.B. 1968. The 1964 Colorado foray. Mycologia 60: 1093-1098.
650. Cooke, W.B. 1969. The 1965 Illinois foray. Mycologia 61: 817-822.
651. Cooke, W.B. 1970. The 1966 Maryland foray. Mycologia 62: 844-851.
652. Cooke, W.B. 1971. The 1967 foray in Texas. Mycologia 63: 1063-1067.
653. Cooke, W.B. 1973. The 1968 Ohio foray. Mycologia 65: 192-198.
654. Cooke, W.B. 1975. The 1970 Indiana foray. Mycologia 67: 1065-1071.
655. Cooke, W.B. 1975. The 1972 Minnesota foray. Mycologia 67: 1205-1211.
656. Cooke, W.B. 1977. The 1973 Massachusetts foray. Mycologia 69: 1226-1231.
657. Cooke, W.B. 1979. The 1975 Oregon foray and workshop. Mycologia 71: 1078-1081.
658. Cooke, W.B. 1980. The 1976 Louisiana foray. Mycologia 72: 1047-1053.
659. Cooke, W.B. 1982. The 1978 Georgia foray. Mycologia 74: 526-531.
660. Cooke, W.B. 1983. The 1979 Oklahoma foray. Mycologia 75: 752-755.
661. Cooke, W.B. 1985. Fungi of Lassen Volcanic National Park. Cooperative National Park Resources Unit, University of California at Davis. Tech. Rep. 21. 251 p.
662. Cooke, W.B. 1985. The 1980 Arizona foray. Mycologia 77: 168-171.
663. Cooke, W.B. 1986. The 1981 Indiana foray. Mycologia 78: 321-323.
664. Cooke, W.B., M.E. Barr, and H.E. Bigelow. 1967. The 1963 Massachusetts foray. Mycologia 59: 910-917.
665. Cooke, W.B., and J.L. Lowe. 1964. The 1955 Michigan foray. Mycologia 56: 602-607.
666. Cooke, W.B., and C.G. Shaw. 1952. Western fungi. III. Mycologia 44: 795-812.
667. Cooley, J.S. 1948. Natural infection of replanted apple trees by white root rot fungus. Phytopathology 38: 110-113.
668. Cooley, J.S. 1951. Control of white root rot with chloropicrin in a border of ornamental plants. Phytopathology 41: 379-380.
669. Cooley, J.S., and R.W. Davidson. 1940. A white rot of apple trees caused by *Corticium galactinum*. Phytopathology 30: 139-148.
670. Cooley, S.J. 1984. *Meria laricis* on nursery seedlings of western larch in Washington. Pl. Dis. 68: 826.
671. Cooperman, C.J., S.F. Jenkins, and C.W. Averre. 1986. Overwintering and aerobiology of *Cercospora asparagi* in North Carolina. Pl. Dis. 70: 392-394.
672. Coplin, D.L., A.F. Schmitthenner, and W.L. Bauerle. 1980. Root rot of lettuce incited by *Pythium polymastum*. Pl. Dis. 64: 63-66.
673. Cordas, D.I., and R. Duran. 1976. *Clintamra*, a new smut genus. Mycologia 68: 1239-1245.
674. Cordell, C.E., H.W. Applegate, and P.J. Huffman. 1969. Pond pine - A new host for *Comandra* blister rust in eastern Tennessee. Pl. Dis. Reporter 53: 694.
675. Cordell, C.E., and J.S. Astin. 1965. A new host for *Fomes annosus*, *Polyporus schweinitzii* and *Fomes pini*. Pl. Dis. Reporter 49: 360.
676. Cordell, C.E., A.S. Juttner, and W.J. Stambaugh. 1971. *Cylindrocladium floridanum* causes severe mortality of seedling yellow-poplar in a North Carolina nursery. Pl. Dis. Reporter 55: 700-702.
677. Cordell, C.E., and M. Matuszewski. 1974. *Cylindrocladium scoparium* - damaging black walnut seedlings in Kent ucky nurseries. Pl. Dis. Reporter 58: 188-189.
678. Cordell, C.E., and S.J. Rowan. 1975. *Cylindrocladium scoparium* infection in a natural sweetgum stand. Pl. Dis. Reporter 59: 775-776.
679. Cordell, C.E., D.D. Skilling, and J.W. Benzie. 1968. Susceptibility of three pine species to *Scleroderris lagerbergii* in upper Michigan. Pl. Dis. Reporter 52: 37-39.
680. Cordell, C.E., W.J. Stambaugh, C.E. Affeltranger, and J.L. Knighten. 1970. *Rhododendron* and mountain laurel, new hosts of *Fomes annosus* in western North Carolina. Pl. Dis. Reporter 54: 560.
681. Corlett, M. 1981. A taxonomic survey of some species of *Didymella* and *Didymella*-like species. Canad. J. Bot. 59: 2016-2042.
682. Corlett, M., and K.N. Egger. 224. *Venturia subcutanea*. Fungi Canadenses 224: 1-2.
683. Corlett, M., and K.N. Egger. 1980. *Venturia rumicis*. Fungi Canadenses 181: 1-2.

684. Corlett, M., and J.C. Krug. 1984. *Bertia moriformis* and its varieties. Canad. J. Bot. 62: 2561-2569.
685. Corlett, M., and I.A. MacLatchy. 1986. *Pithoascus intermedius*. Fungi Canadenses 305: 1-2.
686. Correa, F.J., and A.W. Saettler. 1987. Angular leaf spot of red kidney beans in Michigan. Pl. Dis. 71: 915-918.
687. Correll, J.C. 1986. Powdery mildew of cotton caused by *Oidiopsis taurica* in California. Pl. Dis. 70: 259.
688. Correll, J.C., T.R. Gordon, and V.J. Elliott. 1987. Host range, specificity, and biometrical measurements of *Leveillula taurica* in California. Pl. Dis. 71: 248-251.
689. Corsini, D.L., J.R. Davis, and J.J. Pavek. 1985. Stability of resistance of potato to strains of *Verticillium dahliae* from different vegetative compatibility groups. Pl. Dis. 69: 980-982.
690. Cotter, H.V.T., and R.O. Blanchard. 1981. Identification of the two *Nectria* taxa causing bole cankers on Amer ican beech. Pl. Dis. 65: 332-334.
691. Cotter, H.V.T., and R.O. Blanchard. 1982. The fungal flora of bark of *Fagus grandifolia*. Mycologia 74: 836-843.
692. Cotty, P.J. 1986. *Alternaria tagetica* on marigold in New Jersey. Pl. Dis. 70: 1159.
693. Cotty, P.J. 1987. Modulation of sporulation of *Alternaria tagetica* by carbon dioxide. Mycologia 79: 508-513.
694. Cotty, P.J., and S.M. Alcorn. 1984. *Alternaria raphani* on turnip in Arizona. Pl. Dis. 68: 732.
695. Covey, R.P. 1967. *Botryosphaeria* canker on apple in Washington. Pl. Dis. Reporter 51: 593-594.
696. Covey, R.P. 1985. Calyx end rot (*Sclerotinia sclerotiorum*) of apples in Washington. Pl. Dis. 69: 726.
697. Covey, R.P., H.J. Larsen, T.J. Fitzgerald, and M.A. Dilley. 1981. *Coriolus versicolor* infection of young apple trees in Washington state. Pl. Dis. 65: 280.
698. Cowling, W.A., and D.G. Gilchrist. 1982. Expression of pathogen virulence and host resistance during infection of alfalfa with *Stemphylium botryosum*. Phytopathology 72: 36-42.
699. Cowling, W.A., D.G. Gilchrist, and J.H. Graham. 1981. Biotypes of *Stemphylium botryosum* on alfalfa in North America. Phytopathology 71: 679-684.
700. Cox, D.E. 1976. A new Homobasidiomycete with anomalous basidia. Mycologia 68: 481-510.
701. Cox, E.A., W.J. Manning, and F.J. Campbell. 1969. *Cylindrocarpon radicicola* found associated with wilt of azalea. Pl. Dis. Reporter 53: 620.
702. Cox, R.S. 1951. A preliminary report on etiological and control studies of damping-off and root-rot in the conifer seedbed in Delaware. Pl. Dis. Reporter 35: 374-378.
703. Cox, R.S. 1953. Control of pyracantha scab. Pl. Dis. Reporter 37: 7-10.
704. Cox, R.S. 1953. Crown canker on a field planting of multiflora rose cuttings. Pl. Dis. Reporter 37: 447.
705. Cox, R.S. 1955. A preliminary report on diseases of lettuce in the Everglades and the ir control. Pl. Dis. Reporter 39: 421-422.
706. Cox, R.S. 1956. *Verticillium* wilt of eggplant in south Florida. Pl. Dis. Reporter 40: 583.
707. Cox, R.S. 1957. Anthracnose, an undescribed disease of celery in the Everglades. Pl. Dis. Reporter 41: 790-793.
708. Cox, R.S. 1969. *Cylindrocladium scoparium* on azalea in south Florida. Pl. Dis. Reporter 53: 139.
709. Cox, R.S. 1970. Diseases of statice in south Florida. Pl. Dis. Reporter 54: 759.
710. Cox, R.S. 1971. Flower blast of statice in south Florida. Pl. Dis. Reporter 55: 848.
711. Cox, R.S. 1972. *Sclerotium rolfsii* on statice. Pl. Dis. Reporter 56: 656-657.
712. Craighead, F.C., and J.C. Nelson. 1960. Oak wilt in Pennsylvania. J. Forest. (Washington) 58: 872-881.
713. Crall, J.M. 1950. Soybean diseases in Iowa in 1949. Pl. Dis. Reporter 34: 96-97.
714. Crall, J.M. 1950. Diseases of forage legumes in Iowa in 1949. Pl. Dis. Reporter 34: 191-192.
715. Crall, J.M. 1951. Soybean diseases in Iowa in 1950. Pl. Dis. Reporter 35: 320-321.
716. Crall, J.M. 1951. Diseases on red clover and alfalfa in Iowa in 1950. Pl. Dis. Reporter 35: 322.
717. Crall, J.M. 1952. Soybean diseases in Iowa in 1951. Pl. Dis. Reporter 36: 302.
718. Crane, J.L. 1971. Illinois fungi. I. *Chaetosticta*. Canad. J. Bot. 49: 31-34.
719. Crane, J.L. 1971. Illinois fungi II. A new species of *Phialocephala*. Trans. Brit. Mycol. Soc. 56: 160-163.
720. Crane, J.L. 1972. Illinois fungi. III. *Dendrosporium lobatum* and *Sporidesmium taxodii* sp. nov. Trans. Brit. Mycol. Soc. 58: 423-426.
721. Crane, J.L., and J.D. Schoknecht. 1975. Revisions of *Torula* species. *Torula brachiata*, *T. maculans* and *T. resinae* reexamined. Mycologia 67: 666-671.
722. Crane, J.L., and J.D. Schoknecht. 1978. *Rogersiomyces*, a new genus in the Filobasidiaceae (Homobasidiomycetes) from an aquatic habitat. Amer. J. Bot. 65: 902-906.
723. Crawford, J.L. 1970. Reduction in yield of cotton caused by parasitic diseases in 1969. Pl. Dis. Reporter 54: 324.
724. Crawley, D.K., H.L. Walker, and J.A. Riley. 1985. Interaction of *Alernaria macrospora* and *Fusarium lateritium* on spurred anoda. Pl. Dis. 69: 977-979.
725. Creager, D.B. 1962. A new *Cercospora* on *Rhizophora mangle*. Mycologia 54: 536-539.
726. Creelman, D.W. 1956. Spot anthracnose of linden. Mycologia 48: 552-557.
727. Crist, C.R., and D.F. Schoeneweiss. 1975. The influence of controlled stresses on susceptibility of European white birch stems to attack by *Botryosphaeria dothidea*. Phytopathology 65: 369-373.
728. Croghan, C.F., and K. Robbins. 1986. Cankers caused by *Botryodiplodia gallae* associated with oak sprout mortality in Michigan. Pl. Dis. 70: 76-77.
729. Crombrink, J.C., N.F. Sommer, R.H. Tyler, and R.J. Fortlage. 1976. Postharvest *Phomopsis* rot of apple fruits. Pl. Dis. Reporter 60: 1060-1064.
730. Crosier, W.F., G.T. Nash, and D.C. Crosier. 1970. *Ustilago nuda* on leaves of "Larker" barley. Pl. Dis. Reporter 54: 927-929.
731. Crosier, W.F., J.J. Natti, and D.C. Crosier. 1967. Occurrence of *Rhizoctonia* spp. in seeds of small grains and other plants. Phytopathology 57: 808.
732. Crosier, W.F., and E.C. Waters. 1959. *Fusarium graminearum* and other fungi in seedstocks of small grains. Pl. Dis. Reporter 43: 1013-1015.
733. Crossan, D.F., F.A. Haasis, and D.E. Ellis. 1954. *Phytophthora* blight of summer squash. Pl. Dis. Reporter 38: 557-559.
734. Crossan, D.F., P.J. Lloyd, R.A. Hyre, and J.W. Heuberger. 1957. Control of downy mildew of lima bean. Pl. Dis. Reporter 41: 156-159.
735. Crowe, F., D. Starkey, and V. Lengkeek. 1982. Honeylocust canker in Kansas caused by *Thyronectria austro-americana*. Pl. Dis. 66: 155-158.
736. Crozier, J.A., and C.W. Boothroyd. 1959. Tomato a new suscept of *Gibberella zeae* (Schw.) Petch. Pl. Dis. Reporter 43: 446-447.
737. Culp, T.W., and C.A. Thomas. 1964. *Alternaria* and *Corynespora* blights of sesame in Mississippi. Pl. Dis. Reporter 48: 608-609.
738. Cummins, G.B. 1952. Uredinales from various regions. Bull. Torrey Bot. Club 79: 212-234.
739. Cummins, G.B. 1956. Nomenclatural changes for some North American Uredinales. Mycologia 48: 601-608.
740. Cummins, G.B. 1962. Records of south Texas rust fungi. Proc. Indiana Acad. Sci. 72: 267-269.
741. Cummins, G.B. 1963. Life cycles of southwestern rust fungi. Mycologia 55: 73-78.
742. Cummins, G.B. 1964. Uredinales of the Big Bend National Park and adjacent areas of Texas. Southw. Naturalist 8: 181-195.
743. Cummins, G.B. 1971. The Rust Fungi of Cereals, Grasses and Bamboos. Springer-Verlag. New York. 570 p.
744. Cummins, G.B. 1975. The status of three western rust fungi. Mycologia 67: 175-177.
745. Cummins, G.B. 1977. Nomenclatural changes and new species in the Uredinales. Mycotaxon 5: 398-408.
746. Cummins, G.B. 1978. Rust Fungi of Legumes and Composites in North America. University of Arizona Press, Tuscon. 424 p.
747. Cummins, G.B. 1979. Annotated, illustrated, host index of Sonoran Desert rust fungi. Mycotaxon 10: 1-20.
748. Cummins, G.B. 1984. Two new rust fungi (Uredinales). Mycotaxon 20: 617-618.
749. Cummins, G.B., and J.W. Baxter. 1962. Nomenclature, life histories, and records of North American Uredinales. Madroño 16:

201-203.
750. Cummins, G.B., and J.W. Baxter. 1975. New species of *Ravenelia* (Uredinales). Mycotaxon 2: 195-203.
751. Cummins, G.B., and J.W. Baxter. 1976. New species and combinations of *Ravenelia*. Mem. New York Bot. Gard. 28: 38-40.
752. Cummins, G.B., M.P. Britton, and J.W. Baxter. 1969. The autoecious species of *Puccinia* on North American Eupatoriae. Mycologia 61: 924-944.
753. Cummins, G.B., and H.C. Greene. 1961. The rust fungi of *Muhlenbergia*, *Sporobolus*, and related genera. Brittonia 13: 271-285.
754. Cummins, G.B., and H.C. Greene. 1966. A review of the grass rust fungi that have uredial paraphyses and aecia on *Berberis-Mahonia*. Mycologia 58: 702-721.
755. Cummins, G.B., and S.M. Husain. 1966. The rust fungi of the genus *Aristida*. Bull. Torrey Bot. Club 93: 56-67.
756. Cunfer, B.M. 1978. The incidence of *Septoria nodorum* in wheat seed. Phytopathology 68: 832-835.
757. Cunfer, B.M., and J. Youmans. 1983. *Septoria nodorum* on barley and relationships among isolates from several hosts. Phytopathology 73: 911-914.
758. Cunningham, J.L. 1974. A new *Gyrothrix* in culture and a key to species. Mycologia 66: 122-129.
759. Curl, E.A., and H.A. Weaver. 1958. Diseases of forage corps under sprinkler irrigation in the southeast. Pl. Dis. Reporter 42: 637-644.
760. Czabator, F.J., J.M. Staley, and G.A. Snow. 1971. Extensive southern pine needle blight during 1970-1971, and associated fungi. Pl. Dis. Reporter 55: 764-766.
761. Daines, R.H., C.M. Haenseler, E. Brennan, and I. Leone. 1960. Rusty spot of peach and its control in New Jersey. Pl. Dis. Reporter 44: 20-22.
762. Daines, R.H., and J.L. Peterson. 1976. The occurrence and control of *Phomopsis* fruit rot of peach. Pl. Dis. Reporter 60: 141-143.
763. Dale, J.L. 1958. Two rusts previously unreported from Arkansas. Pl. Dis. Reporter 42: 402.
764. Dale, J.L. 1963. An ergot fungus on *Panicum anceps*. Pl. Dis. Reporter 47: 190-191.
765. Dale, J.L. 1963. Corn anthracnose. Pl. Dis. Reporter 47: 245-249.
766. Dale, J.L. 1967. A blight disease of *Zoysia*. Pl. Dis. Reporter 51: 376.
767. Dale, J.L. 1978. Atypical symptoms of *Rhizoctonia* infection on zoysia. Pl. Dis. Reporter 62: 645-647.
768. Dale, J.L., and J.P. Fulton. 1957. Severe loss from strawberry leaf spot in Arkansas in 1957. Pl. Dis. Reporter 41: 681-682.
769. Dale, J.L., and G.E. Templeton. 1960. Downy mildew of small grains and grasses in Arkansas. Pl. Dis. Reporter 44: 206-207.
770. Daly, J.M., J.G. Hennen, E.A. Lungren, and J.R. Vaughn. 1957. Short notes concerning stripe rust in other states. Pl. Dis. Reporter 41: 962.
771. Damicone, J.P., and W.J. Manning. 1985. Frequency and pathogenicity of *Fusarium* spp. isolated from first-year asparagus grown from transplants. Pl. Dis. 69: 413-416.
772. Daniel, J.T., G.E. Templeton, R.J. Smith Jr., and W.T. Fox. 1973. Biological control of northern joinvetch in rice with an endemic fungal disease. Weed Sci. 21: 303-307.
773. Darby, J.F. 1955. A progress report on gray mold and ghost spot of tomatoes and their control. Pl. Dis. Reporter 39: 91-97.
774. Darker, G.D. 1932. Contributions from the Arnold Arboretum of Harvard Univ.
775. Darley, E.F., and W.D. Wilbur. 1953. *Phytophthora* trunk rot of Washington palms. Phytopathology 43: 469-470.
776. Daughtrey, M.L., and C.R. Hibben. 1983. A new disease of Northeastern dogwoods. Phytopathology 73: 365.
777. Davidson, R.W. 1935. Fungi causing stain in logs and lumber in the southern states including five new species. J. Agric. Res. 50: 789-807.
778. Davidson, R.W. 1942. Some additional species of *Ceratostomella* in the United States. Mycologia 34: 650-662.
779. Davidson, R.W. 1944. Two American hardwood species of *Endoconidiophora* described as new. Mycologia 36: 300-306.
780. Davidson, R.W. 1950. *Urnula craterium* is possibly the perfect stage of *Strumella coryneoidea*. Mycologia 42: 735-742.
781. Davidson, R.W. 1950. A western white pine twig canker. Pl. Dis. Reporter 34: 99.
782. Davidson, R.W. 1953. Two common lumber-staining fungi in the western United States. Mycologia 45: 579-586.
783. Davidson, R.W. 1955. Wood-staining fungi associated with bark beetles in Engelmann spruce in Colorado. Mycologia 47: 58-67.
784. Davidson, R.W. 1958. Additional species of Ophiostomataceae from Colorado. Mycologia 50: 661-670.
785. Davidson, R.W. 1966. New species of *Ceratocystis* from conifers. Mycopathol. Mycologia Appl. 28: 273-286.
786. Davidson, R.W. 1971. New species of *Ceratocystis*. Mycologia 63: 5-15.
787. Davidson, R.W. 1976. Sapwood staining fungi from two tree species. Mem. New York Bot. Gard. 28: 45-49.
788. Davidson, R.W. 1978. Staining fungi associated with *Dendroctonus adjunctus* in pines. Mycologia 70: 35-40.
789. Davidson, R.W. 1978. A new species of *Ceratocystis* on *Endothia parasitica* canker of American chestnut. Mycologia 70: 856-858.
790. Davidson, R.W. 1979. A *Ceratocystis* associated with an ambrosia beetle in *Dendroctonus* -killed pines. Mycologia 71: 1085-1089.
791. Davidson, R.W. 1985. Brown and red stain in ponderosa pine lumber. Mycologia 77: 494-496.
792. Davidson, R.W., and W.A. Campbell. 1943. Decay in merchantable black cherry on the Allegheny National Forest. Phytopathology 33: 965-985.
793. Davidson, R.W., and W.A. Campbell. 1954. *Poria cocos*, a widely distributed wood-rotting fungus. Mycologia 46: 234-237.
794. Davidson, R.W., and E.K. Cash. 1956. A *Cenangium* associated with sooty-bark canker of aspen. Phytopathology 46: 34-36.
795. Davidson, R.W., C.M. Christensen, and E.F. Darley. 1946. *Polyporus guttulatus* and *Ptychogaster rubescens*. Mycologia 38: 652-663.
796. Davidson, R.W., and T.E. Hinds. 1956. *Hypoxylon* canker of aspen in Colorado. Pl. Dis. Reporter 40: 157.
797. Davidson, R.W., and T.E. Hinds. 1958. Unusual fungi associated with decay in some forest trees in Colorado. Phytopathology 48: 216-218.
798. Davidson, R.W., T.E. Hinds, and E.R. Toole. 1964. Two new species of *Ceratocystis* from hardwoods. Mycologia 56: 793-798.
799. Davidson, R.W., and E.G. Kuhlman. 1978. A species of *Ceratocystis* closely associated with *Endothia* cankers on American chestnut in eastern United States. Mycologia 70: 853-855.
800. Davidson, R.W., P.L. Lentz, and H.H. McKay. 1960. The fungus causing pecky cypress. Mycologia 52: 260-279.
801. Davidson, R.W., F.F. Lombard, and W.A. Campbell. 1959. A white heart rot of *Fraxinus* caused by *Fomes johnsonianus* (Murr.) Lowe. Pl. Dis. Reporter 43: 1148-1149.
802. Davidson, R.W., and R.C. Lorenz. 1938. Species of *Eutypella* and *Schizoxylon* associated with cankers of maple. Phytopathology 28: 733-745.
803. Davidson, R.W., and R.F. Patton. 1961. *Paxillus atrotomentosus* causes brown root rot in dead jack pine in plantations in Wisconsin. Pl. Dis. Reporter 45: 836-838.
804. Davidson, R.W., and R.C. Robinson-Jeffrey. 1965. New records of *Ceratocystis europhioides* and *C. huntii* with *Verticicladiella* imperfect stages from conifers. Mycologia 57: 488-490.
805. Davidson, R.W., E.R. Toole, and W.A. Campbell. 1959. A preliminary note on the cause of "pecky" cypress. Pl. Dis. Reporter 43: 806-808.
806. Davis, D.D., G.N. Mason, and P.H. Peacher. 1970. Dutch elm disease detected in Texas. Pl. Dis. Reporter 54: 931.
807. Davis, J.J. 1907. A new species of *Protomyces*. J. Mycol. 13: 188-189.
808. Davis, J.J. 1915. Notes on parasitic fungi in Wisconsin. II. Trans. Wisconsin Acad. Sci. 18: 93-109.
809. Davis, J.J. 1919. Notes on parasitic fungi in Wisconsin IV-VI. Trans. Wisconsin Acad. Sci. 19: 671-727.
810. Davis, J.J. 1922. Notes on parasitic fungi in Wisconsin. VII. Trans. Wisconsin Acad. Sci. 20: 399-411, 413-431.
811. Davis, J.J. 1924. Notes on parasitic fungi in Wisconsin - IX. Trans. Wisconsin Acad. Sci. 21: 251-302.
812. Davis, J.R., and D.O. Everson. 1986. Relation of *Verticillium dahliae* in soil and potato tissue, irrigation method, and N-fertility to *Verticillium* wilt of potato. Phytopathology 76: 730-736.
813. Davis, L.H. 1952. The *Heterosporium* disease of California poppy. Mycologia 44: 366-376.
814. Davis, R.M., C.J. Farrald, and D. Davila. 1987. *Botryodiplodia* trunk lesions in Texas citrus. Pl. Dis. 71: 848-849.

815. Davis, T.C. 1962. Four new hosts of *Cristulariella pyramidalis*. Phytopathology 52: 1217.
816. Davis, T.C. 1963. Zonate leaf spots on various hosts caused by *Cristulariella pyramidalis*. Pl. Dis. Reporter 47: 983-985.
817. Davis, T.C., W.D. Kelley, and J.F. Goggans. 1972. *Curvularia intermedius* associated with seedling tip blight of Arizona cypress and eastern red cedar. Pl. Dis. Reporter 56: 192.
818. Davis, W.H. 1935. Twig blight of the American bladder nut caused by *Hypomyces ipomoeae*. Mycologia 27: 527-542.
819. Davison, A.D. 1958. Plant diseases of economic crops occurring in Wyoming during 1958. Pl. Dis. Reporter 42: 1409-1410.
820. Davison, A.D. 1972. Factors affecting development of madrone canker. Pl. Dis. Reporter 56: 50-52.
821. Davison, A.D., and R.M. Davidson. 1973. *Apioporthe* and *Monochaetia* cankers reported in western Washington. Pl. Dis. Reporter 57: 522-523.
822. Daykin, M.E., and R.D. Milholland. 1984. Ripe rot of muscadine grape caused by *Colletotrichum gloeosporioides* and its control. Phytopathology 74: 710-714.
823. Daykin, M.E., and R.D. Milholland. 1984. Histopathology of ripe rot caused by *Colletotrichum gloeosporioides* on muscadine grape. Phytopathology 74: 1339-1341.
824. De, Hoog G.S. 1974. The genera *Blastobotrys*, *Sporothrix*, *Calcarisporium* and *Calcarisporiella* gen. nov. Stud. Mycol. 7: 1-84.
825. DeVay, J.E., R.P. Covey, and P.N. Nair. 1957. Corn diseases and their importance in Minnesota in 1956. Pl. Dis. Reporter 41: 505-507.
826. DeVay, J.E., R.W. Davidson, and W.J. Moller. 1968. New species of *Ceratocystis* associated with bark injuries on deciduous fruit trees. Mycologia 60: 635-641.
827. DeVay, J.E., F.L. Lukezic, H. English, E.E. Trujillo, and W.J. Moller. 1968. *Ceratocystis* canker of deciduous fruit trees. Phytopathology 58: 949-954.
828. DeVay, J.E., S.L. Sinden, F.L. Lukezic, L.F. Werenfels, and P.A. Backman. 1968. *Poria* root and crown rot of cherry trees. Phytopathology 58: 1239-1241.
829. DeWolfe, T.A., E.C. Calavan, and J.R. Sufficool. 1954. *Pythium aphanidermatum*, the causal agent of damping-off in some citrus seed beds. Pl. Dis. Reporter 38: 632-633.
830. Dearness, J. 1916. New or noteworthy species of fungi. Mycologia 8: 98-107.
831. Dearness, J. 1917. New or noteworthy North American fungi. Mycologia 9: 345-364.
832. Dearness, J. 1924. New and noteworthy fungi - III. Mycologia 16: 143-176.
833. Dearness, J. 1926. New and noteworthy fungi - IV. Mycologia 18: 236-255.
834. Dearness, J. 1928. New and noteworthy fungi - V. Mycologia 20: 235-246.
835. Dearness, J. 1929. New and noteworthy fungi - VI. Mycologia 21: 326-332.
836. Dearness, J. 1936. The Mycological Society of America summer foray. Mycologia 28: 98-102.
837. Dearness, J. 1941. New species of Tennessee fungi. Mycologia 33: 360-366.
838. Deer, H.J., and T.W. Melder. 1970. Brown-spot resistance in longleaf pine. Forest Sci. 16: 204-209.
839. Deighton, F.C. 1967. Studies on *Cercospora* and allied genera. II. Mycol. Pap. 112: 1-80.
840. Deighton, F.C. 1973. Five North American *Cercospora*-like fungi. Trans. Brit. Mycol. Soc. 61: 107-120.
841. Deighton, F.C. 1976. Studies on *Cercospora* and allied genera. VI. *Pseudocercospora* Speg., *Pantospora* Cif. and *Cercoseptoria* Petr. Mycol. Pap. 140: 1-168.
842. Deighton, F.C. 1976. Brown leaf mould of *Capsicum* caused by *Phaeoramularia capsicicola*. Trans. Brit. Mycol. Soc. 67: 140-142.
843. Deighton, F.C. 1979. Studies on *Cercospora* and allied genera. VII. New species and redispositions. Mycol. Pap. 144: 1-56.
844. Deighton, F.C. 1983. Studies on *Cerospora* and allied genera. VIII. Futher notes on *Cercoseptoria* and some species and redispositions. Mycol. Pap. 151: 1-13.
845. Deighton, F.C. 1986. Misidentification of *Cercospora effusa*. Trans. Brit. Mycol. Soc. 86: 637-641.
846. Delserone, L.M., and H. Cole Jr. 1987. Effects of planting date on development of net blotch epidemics in winter barley in Pennsylvania. Pl. Dis. 71: 438-441.
847. Delwiche, P.A., C.R. Grau, E.B. Holub, and J.B. Perry. 1987. Characterization of *Aphanomyces euteiches* isolates recovered from alfalfa in Wisconsin. Pl. Dis. 71: 155-161.
848. Denison, W.C. 1959. Some species of the genus *Scutellinia*. Mycologia 51: 605-635.
849. Dennis, R.W.G. 1963. A redisposition of some fungi ascribed to the *Hyaloscyphaceae*. Kew Bull. 17: 319-380.
850. Dennis, R.W.G. 1964. Remarks on the genus *Hymenoscyphus* S.F. Gray, with observations on sundry species referred by Saccardo and others to the genera *Helotium*, *Pezizella* or *Phialea*. Persoonia 3: 29-80.
851. Derie, M.L., R.L. Bowden, and K.D. Kephart. 1985. *Ascochyta rabiei* on chickpeas in Idaho. Pl. Dis. 69: 268.
852. Dernoeden, P.H., and N.R. O'Neill. 1983. Occurrence of *Gaeumannomyces* patch disease in Maryland and growth and pathogenicity of the causal agent. Pl. Dis. 67: 528-532.
853. Desjardin, D.E. 1985. New marasmioid fungi from California. Mycologia 77: 894-902.
854. Desjardin, D.E., and R.E. Halling. 1987. California collybias. I. *Collybia bakerensis*: a common snowbank agaric. Mycotaxon 29: 321-327.
855. Desjardin, D.E., and S.A. Redhead. 1987. *Marasmius salalis*, a new Pacific coast North American species. Mycotaxon 29: 307-308.
856. Desjardin Dennis E. 1987. New and noteworthy marasmioid fungi from California. Mycologia 79: 123-134
857. Dewey, W.G., and L.J. Tyler. 1956. Dwarf bunt on ryegrass in New York. Pl. Dis. Reporter 40: 508.
858. Dewey, W.G., and J.A. Hoffman. 1975. Susceptibility of barley to *Tilletia controversa*. Phytopathology 65: 654-657.
859. Dewey, W.G., and D.C. Tingey. 1958. A new grass host for dwarf bunt. Pl. Dis. Reporter 42: 18-19.
860. Dharne, C.G. 1964. Taxonomic investigations on the discomycetous genus *Lachnellula* Karst. Phytopathol. Z. 53: 101-144.
861. DiCosmo, F. 1978. A revision of *Corniculariella*. Canad. J. Bot. 56: 1665-1690.
862. DiCosmo, F., S. Berch, and B. Kendrick. 1983. *Cylindrotrichum*, *Chaetopsis*, and two new genera of Hyphomycetes, *Kylindria* and *Xenokylindria*. Mycologia 75: 949-973.
863. DiCosmo, F., T.R. Nag Raj, and B. Kendrick. 1983. Prodromus for a revision of the Phacidiaceae and related anamorphs. Canad. J. Bot. 61: 31-44.
864. DiCosmo, F., T.R. Nag Raj, and W.B. Kendrick. 1984. A revision of the Phacidiaceae and related anamorphs. Mycotaxon 21: 1-234.
865. Dickens, L.E., N. Oshima, and S.R. Andrews. 1970. Distribution of Dutch elm disease in Colorado - - 1969. Pl. Dis. Reporter 54: 266-267.
866. Dickson, J.G. 1952. General corn rust and leaf blight infection in Wisconsin. Pl. Dis. Reporter 36: 350.
867. Diehl, W.W. 1951. A *Microstroma* on peach. Pl. Dis. Reporter 35: 497.
868. Diehl, W.W. 1952. Powdery mildew of kenaf in Florida. Pl. Dis. Reporter 36: 52.
869. Diehl, W.W. 1953. On *Hyalodothis caricis*. Mycologia 45: 587-591.
870. Diehl, W.W. 1955. *Valsaria megalospora* on red alder. Pl. Dis. Reporter 39: 334.
871. Diener, U.L. 1960. The mycoflora of peanuts in storage. Phytopathology 50: 220-223.
872. Dietrich, R.A., R.A. Blanchette, C.F. Croghan, and S.O. Phillips. 1985. The distribution of *Endocronartium harknessii* and *Cronartium quercuum* on jack pine in Minnesota. Canad. J. Forest Res. 15: 1045-1048.
873. Dietz, S.M., and G.W. Fischer. 1970. *Melanotaenium oxalidis* sp. nov. on *Oxalis oregana* from Washington state. Mycologia 62: 402-405.
874. Digby, S., and R.D. Goos. 1987. Morphology, development and taxonomy of *Loramyces*. Mycologia 79(6): 821-831.
875. Dillard, H.R. 1987. Characterization of isolates of *Rhizoctonia solani* from lima beans grown in New York state. Phytopathology 77: 748-751.
876. Dillard, H.R., and R.G. Grogan. 1985. Influence of green manure crops and lettuce on sclerotial populations of *Sclerotinia minor*. Pl. Dis. 69: 579-582.
877. Dillard, H.R., and J.E. Hunter. 1986. Association of common

ragweed with *Sclerotinia* rot of cabbage in New York state. Pl. Dis. 70: 26-28.
878. Diller, J.D., and R.W. Davidson. 1950. *Cucurbitaria elongata* associated with a canker of honeylocust in Ohio. Pl. Dis. Reporter 34: 224.
879. Dilley, M., and R.P. Covey Jr. 1980. Survey of wood decay and associated Hymenomycetes in central Washington apple orchards. Pl. Dis. 64: 560-561.
880. Dilley, M.A., and R.P. Covey Jr. 1981. Association of *Coriolus versicolor* with a dieback disease of apple trees in Washington state. Pl. Dis. 65: 77-78.
881. Dimitman, J.E. 1962. Some *Phytophthora* diseases of ornamental plants in Southern California. Phytopathology 52: 359-361.
882. Dimock, A.W., R.E. McCoy, and J.E. Knauss. 1968. *Pelargonium* rust, a new geranium disease in New York State. Bull. New York State Flower Growers 268: 1-3.
883. Dochinger, L.S. 1967. Occurrence of poplar cankers caused by *Fusarium solani* in Iowa. Pl. Dis. Reporter 51: 900-903.
884. Dodd, J.L. 1972. The genus *Clavicorona*. Mycologia 64: 737-773.
885. Dolezal, W.E., and F.H. Tainter. 1979. Phenology of *Comandra* blister rust in Arkansas. Phytopathology 69: 41-44.
886. Donald, P.A., and W.M. Bugbee. 1986. First report of *Leptosphaeria lindquistii* (sexual stage of *Phoma macdonaldii*) on sunflower in North Dakota and Minnesota. Pl. Dis. 70: 352.
887. Dooling, O.J. 1961. Oak wilt identified in Texas. Pl. Dis. Reporter 45: 749.
888. Dooling, O.J., and P.H. Peacher. 1964. Oak wilt distribution extended in Arkansas and Oklahoma. Pl. Dis. Reporter 48: 913.
889. Doucette, C.F. 1953. *Botrytis polyblastis* on narcissus in the Pacific Northwest. Pl. Dis. Reporter 37: 556-557.
890. Douglas, D.R. 1972. Occurrence of *Ulocladium consortiale* as an associate fungus of *Alternaria solani* in potato tuber blight. Pl. Dis. Reporter 56: 308-309.
891. Doyle, A.F. 1978. Some secondary metabolites from *Nectria* species. Mycologia 70: 355-362.
892. Drake, C.R. 1958. Diseases of birdsfoot trefoil in six southeastern states in 1956 and 1957. Pl. Dis. Reporter 42: 145-146.
893. Drake, C.R. 1960. Southern blight of birdsfoot trefoil (*Lotus corniculatus*). Pl. Dis. Reporter 44: 115-116.
894. Drake, C.R. 1970. Powdery mildew of peach in Virginia. Pl. Dis. Reporter 54: 686-688.
895. Drayton, F.L., and J.W. Groves. 1952. *Stromatinia narcissi*, a new, sexually dimorphic discomycete. Mycologia 44: 119-139.
896. Drechsler, C. 1952. Bean stem rot in Maryland and Delaware caused by several *Pythium* species. Pl. Dis. Reporter 36: 13.
897. Driver, C.H. 1950. Observations on the "flowering dogwood spot anthracnose" in Atlanta, Georgia. Pl. Dis. Reporter 34: 212.
898. Driver, C.H. 1951. A *Phyllosticta* leaf-spot on Burford holly in Georgia. Pl. Dis. Reporter 35: 412.
899. Driver, C.H. 1952. *Physalospora ilicis* on rotundifolia holly in Georgia. Pl. Dis. Reporter 36: 355.
900. Driver, C.H. 1956. A note on the southern blight of tomato. Pl. Dis. Reporter 40: 259.
901. Driver, C.H., and T.R. Dell. 1961. *Fomes annosus* root-rot in slash pine plantations of the eastern Gulf coast states. Pl. Dis. Reporter 45: 38-40.
902. Driver, C.H., and T.R. Dell. 1961. Observations on *Fomes annosus* root-rot in natural stands of loblolly and shortleaf pine. Pl. Dis. Reporter 45: 352-353.
903. Driver, C.H., and T.R. Dell. 1961. *Fomes annosus* causes severe damage to thinned slash pine plantations in south Georgia. Pl. Dis. Reporter 45: 660.
904. Driver, C.H., and J.H. Ginns Jr. 1969. Ecology of slash pine stumps: Fungal colonization and infection by *Fomes annosus*. Forest Sci. 15: 2-10.
905. Driver, C.H., and A.G. Plakidas. 1953. Spot anthracnose on flowering dogwood in Louisiana. Pl. Dis. Reporter 37: 448.
906. Driver, C.H., and R.E. Wood. 1968. Occurrence of *Fomes annosus* in intensively managed young-growth western hemlock stands. Pl. Dis. Reporter 52: 370-372.
907. Dubin, H.J., and H. English. 1975. Effects of temperature, relative humidity, and desiccation on germina tion of *Nectria galligena* conidia. Mycologia 67: 83-88.
908. Dubin, H.J., and R.C. McCrum. 1975. *Phytophthora cactorum* associated with crown rot of apples in Maine. Pl. Dis. Reporter 59: 365-366.
909. Dumont, K.P. 1971. Sclerotiniaceae II. *Lambertella*. Mem. New York Bot. Gard. 22: 1-178.
910. Dumont, K.P. 1974. Sclerotiniaceae VI. *Pseudociboria*. Mycologia 66: 706-712.
911. Dumont, K.P. 1975. Sclerotiniaceae X. *Ciboriella*, a taxonomic synonym of *Lanzia*. Mycologia 67: 569-585.
912. Dunbar, D.M., and G.R. Stephens. 1975. Association of twolined chestnut borer and shoestring fungus with mortality of defoliated oak in Connecticut. Forest Sci. 21: 169-174.
913. Duncan, E.G. 1972. Microevolution in *Auricularia polytricha*. Mycologia 64: 394-404.
914. Duncan, E.G., and J.A. MacDonald. 1967. Micro-evolution in *Auricularia auricula*. Mycologia 59: 803-818.
915. Dunleavey, J. 1955. Soybean diseases in Iowa in 1954. Pl. Dis. Reporter 39: 169-170.
916. Dunleavy, J. 1954. Soybean diseases in Iowa in 1953. Pl. Dis. Reporter 38: 89-90.
917. Dunleavy, J.M. 1976. A survey of powdery mildew of soybean in central Iowa. Pl. Dis. Reporter 60: 675-677.
918. Dunleavy, J.M., J.W. Keck, K.S. Gobelman-Werner, and M.M. Thompson. 1984. Prevalence of soybean downy mildew in Iowa. Pl. Dis. 68: 778-779.
919. Dunleavy, J., and R.C. Lambe. 1967. Incidence of brown stem rot of soybeans in Iowa. Pl. Dis. Reporter 51: 438-441.
920. Duran, R. 1982. The species of *Thecaphora* on Compositae. Canad. J. Bot. 60: 1512-1522.
921. Duran, R., and G.W. Fischer. 1961. The genus *Tilletia*. Washington State University. Pullman, Washington. 138 p.
922. Duran, R., and K.M. Safeeulla. 1968. Aspects of teliospore germination in some North American smut fungi. Mycologia 60: 231-243.
923. Durbin, R.D., L.H. Davis, and K.F. Baker. 1955. A *Helminthosporium* stem rot of cacti. Phytopathology 45: 509-512.
924. Durbin, R.D., and W.C. Snyder. 1953. Ecology and hosts of flyspeck of apple in California. Phytopathology 43: 586.
925. Durbin, R.D., and W.C. Snyder. 1953. Flyspeck of apple in California. Pl. Dis. Reporter 37: 166.
926. Dutky, E.M., and R.D. Lumsden. 1986. Damping-off, root rot, and wilt of crownvetch caused by *Pythium myriotylum*. Canad. J. Plant Pathol. 8: 107-108.
927. Dyko, B.J., and B.C. Sutton. 1979. A revision of *Linodochium*, *Pseudocenangium*, *Septopatella*, and *Siroscyphella*. Canad. J. Bot. 57: 370-385.
928. Dyko, B.J., and B.C. Sutton. 1979. New and interesting dematiaceous Hyphomycetes from Florida. Mycotaxon 8: 119-124.
929. Earhart, R.W. 1950. Small grain diseases in the southeastern coastal plain, 1949-50. Pl. Dis. Reporter 34: 257-258.
930. Earhart, R.W. 1952. Oat downy mildew found in Florida. Pl. Dis. Reporter 36: 338.
931. Earhart, R.W. 1952. Small grain diseases of the southeastern coastal plain, 1951-52. Pl. Dis. Reporter 36: 420-422.
932. Earhart, R.W. 1955. South Carolina small grain diseases -- 1954-55. Pl. Dis. Reporter 39: 947-948.
933. Earhart, R.W. 1957. Small grain diseases of South Carolina, 1956-57. Pl. Dis. Reporter 41: 863-870.
934. Earhart, R.W., and D.D. Morey. 1953. *Septoria nodorum* attacking wheats in the southeastern coastal plain. Pl. Dis. Reporter 37: 310.
935. Eddins, A.H. 1952. Distribution of cabbage yellows in Florida in 1952. Pl. Dis. Reporter 36: 337-338.
936. Eglitis, M., C.J. Gould, and F. Johnson. 1966. Fungi found on Ericaceae in the Pacific coastal area. Wash. State Agric. Exp. Sta. Bull. 675: 1-21.
937. El-Gholl, N.E. 1985. *Fusarium lateritium* causing needle spots on *Torreya taxifolia* in Florida. Pl. Dis. 69: 905.
938. El-Gholl, N.E., C.L. Schoulties, and S.A. Alfieri Jr. 1983. Homothallism in *Calonectria theae*. Mycologia 75: 162-163.
939. El-Goorani, M.A., and N.F. Sommer. 1979. Suppression of postharvest plant pathogenic fungi by carbon monoxide. Phytopathology 69: 834-838.
940. Eleuterius, L.N. 1970. Observations on *Claviceps purpurea* on *Spartina alterniflora*. Gulf Res. Rep. 3(1): 105-109.
941. Eleuterius, L.N., and S.P. Meyers. 1974. *Claviceps purpurea* on *Spartina* in coastal marshes. Mycologia 66: 978-986.
942. Ellerbrock, L.A., and J.W. Lorbeer. 1977. Etiology and control of

onion flower blight. Phytopathology 67: 155-159.
943. Ellett, C.W. 1953. *Rhizoctonia* blight of *Fittonia*. Pl. Dis. Reporter 37: 373-374.
944. Ellett, C.W. 1953. Rhizome rot of geranium. Pl. Dis. Reporter 37: 374.
945. Ellett, C.W. 1956. *Curvularia* blight of *Ophioglossum* in Ohio. Pl. Dis. Reporter 40: 756.
946. Ellett, C.W. 1957. Diseases not previously reported in Ohio. Pl. Dis. Reporter 41: 369-371.
947. Ellett, C.W. 1959. *Ascochyta hordei* on barley in Ohio. Pl. Dis. Reporter 43: 835.
948. Ellett, C.W. 1963. New records of powdery mildews. Pl. Dis. Reporter 47: 289-291.
949. Ellett, C.W. 1964. Wheat and oat diseases in Ohio in 1963. Pl. Dis. Reporter 48: 487-488.
950. Ellett, C.W. 1966. Host range of the Erysiphaceae of Ohio. Ohio J. Sci. 66: 570-581.
951. Ellett, C.W. 1985. Leaf blight (*Didymosporina aceris*) of Norway maple in Connecticut and Pennsylvania. Pl. Dis. 69: 451.
952. Elliott, E.S. 1954. Notes on forage plant diseases observed in northern West Virginia during 1953. Pl. Dis. Reporter 38: 279-281.
953. Elliott, E.S. 1955. Forage plant diseases observed in West Virginia during 1954. Pl. Dis. Reporter 39: 318-321.
954. Elliott, E.S. 1955. Notes on diseases of cereals and ornamentals recorded in West Virginia during 1953 and 1954. Pl. Dis. Reporter 39: 332-333.
955. Elliott, E.S. 1959. A report of grass diseases observed in West Virginia. Pl. Dis. Reporter 43: 461-463.
956. Ellis, D.E. 1951. Noteworthy diseases of cucurbits in North Carolina in 1949 and 1950. Pl. Dis. Reporter 35: 91-93.
957. Ellis, D.E. 1952. *Ascochyta* leaf spot of bean in North Carolina. Pl. Dis. Reporter 36: 12.
958. Ellis, J.B. 1874. New species of fungi, found at Newfield, N.J. Bull. Torrey Bot. Club 5: 45-47.
959. Ellis, J.B., and B.M. Everhart. 1878-1898. North American Fungi. Centuries I-XXXVI. (Exsiccati Series, set in BPI).
960. Ellis, J.B., and B.M. Everhart. 1885. New species of fungi. J. Mycol. 1: 89-93.
961. Ellis, J.B., and B.M. Everhart. 1885. The North American species of *Gloeosporium*. J. Mycol. 1: 109-120.
962. Ellis, J.B., and B.M. Everhart. 1887. Additions to *Cercospora*, *Gloeosporium* and *Cylindrosporium*. J. Mycol. 3: 13-22.
963. Ellis, J.B., and B.M. Everhart. 1887. New species of fungi from various localities. J. Mycol. 3: 127-130.
964. Ellis, J.B., and B.M. Everhart. 1888. New species of fungi from various localities. J. Mycol. 4: 44-46.
965. Ellis, J.B., and B.M. Everhart. 1888. New species of fungi from various localities. J. Mycol. 4: 97-107.
966. Ellis, J.B., and B.M. Everhart. 1889. New and rare species of North American fungi. J. Mycol. 5: 145-157.
967. Ellis, J.B., and B.M. Everhart. 1892. The North American Pyrenomycetes. William H. Cloyd Press. 793 p.
968. Ellis, J.B., and B.M. Everhart. 1895. New species of fungi from various localities. Proc. Acad. Nat. Sci. Philadelphia 1895: 413-441.
969. Ellis, J.B., and B.M. Everhart. 1900. New species of fungi from various localities. Bull. Torrey Bot. Club 27: 571-578.
970. Ellis, J.B., and B.M. Everhart. 1902. New species of fungi from various localities. J. Mycol. 8: 11-19.
971. Ellis, J.B., and B.M. Everhart. 1904. New species of fungi from various localities. J. Mycol. 10: 167-170.
972. Ellis, J.B., and H.W. Harkness. 1884. New Californian fungi. Bull. Calif. Acad. Sci. 1: 26-29.
973. Ellis, J.B., and W.A. Kellerman. 1887. New Kansas fungi. J. Mycol. 3: 102-105.
974. Ellis, J.B., and G. Martin. 1885. New Florida fungi. J. Mycol. 1: 97-101.
975. Ellis, J.J., G.A. Bennett, and C.W. Hesseltine. 1973. Utilization of carbon compounds by *Helminthosporium maydis* and *H. carbonum*. Mycologia 65: 539-547.
976. Ellis, M.A., and M.A. Bulger. 1986. Anthracnose fruit rot (*Colletotrichum gloeosporioides*) of strawberry in Ohio. Pl. Dis. 70: 475.
977. Ellis, M.A., and C.W. Ellett. 1981. Late leaf rust on Heritage red raspberry in Ohio. Pl. Dis. 65: 924.
978. Ellis, M.A., and G.G. Grove. 1983. Leather rot in Ohio strawberries. Pl. Dis. 67: 549.
979. Ellis, M.A., G.A. Kuter, and L.L. Wilson. 1984. Fungi that cause cane cankers on thornless blackberry in Ohio. Pl. Dis. 68: 812-815.
980. Ellis, M.B. 1961. Dematiaceous hyphomycetes. II. Mycol. Pap. 79: 1-23.
981. Ellis, M.B. 1961. Dematiaceous hyphomycetes. III. Mycol. Pap. 82: 1-55.
982. Ellis, M.B. 1971. Dematicaeous Hyphomycetes. Commonwealth Mycological Institute, Kew. 608 p.
983. Elmer, O.H., I.J. Shields, and C.T. Rogerson. 1953. Oak wilt in seven Kansas counties. Pl. Dis. Reporter 37: 44.
984. Elmer, W.H., and M.L. Lacy. 1984. Fusarium yellows (*F. oxysporium* f. sp. *apii*) of celery in Michigan. Pl. Dis. 68: 537.
985. Elmer, W.H., M.L. Lacy, and S. Honma. 1986. Evaluations of celery germ plasm for resistance to *Fusarium oxysporum* f. sp. *apii* race 2 in Michigan. Pl. Dis. 70: 416-419.
986. Elrod, J.M. 1958. Black stem and leaf spot disease of alfalfa and crimson clover. Pl. Dis. Reporter 42: 279.
987. Emmons, C.W., G.B. Cummins, and W.B. Cooke. 1960. The 1958 foray of the Mycological Society of America. Mycologia 52: 808-817.
988. Endo, R.M. 1961. Turfgrass diseases in southern California. Pl. Dis. Reporter 45: 869-873.
989. Endo, R.M., H.D. Ohr, and E.M. Krausman. 1985. *Leptosphaeria korrae*, a cause of the spring dead spot disease of Bermudagrass in California. Pl. Dis. 69: 235-237.
990. Engelhard, A.W. 1974. Crown rot and wilt of baby's breath incited by *Phytophthora parasitica*. Phytopathology 64: 767.
991. Engelhard, A.W. 1974. A serious new crown rot and wilt of baby's breath (*Gypsophilapaniculata*) incited by *Phytophthora parasitica*. Pl. Dis. Reporter 58: 669-672.
992. Engelhard, A.W., and J.C. Carter. 1956. Isolations of *Verticillium albo-atrum* from woody hosts in Illinois, 1945-1955. Pl. Dis. Reporter 40: 459-462.
993. Englehard, A.W., G.L. Crane, and H.C. Mellinger. 1976. Stem rot, a new disease on *Chrysanthemum* incited by *Fusarium solani*. Pl. Dis. Reporter 60: 437-441.
994. Englerth, G.H., J.S. Boyce Jr., and E.R. Roth. 1956. Longevity of the oak wilt fungus in red oak lumber. Forest Sci. 2: 2-6.
995. English, H., and J.R. Davis. 1965. Apricot dieback fungus found on western choke-cherry. Pl. Dis. Reporter 49: 178.
996. English, H., J.R. Davis, and J.E. DeVay. 1975. Relationship of *Botryosphaeria dothidea* and *Hendersonula toruloide a* to a canker disease of almond. Phytopathology 65: 114-122.
997. Epps, J.M., B.D. Raskopf, and J.O. Andes. 1954. Cotton wilt damage in Tennessee. Pl. Dis. Reporter 38: 304-308.
998. Epps, W.M. 1956. Gummy stem blight and other diseases on cucurbits in South Carolina in the 1955 fall season. Pl. Dis. Reporter 40: 439-440.
999. Erwin, D.C. 1950. *Phytophthora* root rot of safflower in Nebraska caused by *Phytophthora drechsleri*. Pl. Dis. Reporter 34: 306.
1000. Erwin, D.C. 1956. Important diseases of alfalfa in southern California. Pl. Dis. Reporter 40: 380-383.
1001. Erwin, D.C. 1958. *Fusarium lateritium* f. *ciceri*, incitant of *Fusarium* wilt of *Cicer arietinum*. Phytopathology 48: 498-501.
1002. Erwin, D.C. 1958. *Verticillium* wilt of *Cicer arietinum* in southern California. Pl. Dis. Reporter 42: 1111.
1003. Eslyn, W.E. 1960. New records of forest fungi in the Southwest. Mycologia 52: 381-387.
1004. Eslyn, W.E. 1962. Basidiomycetes and decay in silver maple stands of central Iowa. Forest Sci. 8: 263-276.
1005. Eslyn, W.E., and R.W. Davidson. 1976. Some wood-staining fungi from pulpwood chips. Mem. New York Bot. Gard. 28: 50-57.
1006. Eslyn, W.E., and F.F. Lombard. 1983. Decay in mine timbers. Part II. Basidiomycetes associated with decay of coal mine timbers. Forest Prod. J. 33: 19-23.
1007. Eslyn, W.E., and F.F. Lombard. 1984. Fungi associated with decayed wood in stored willow and cottonwood logs. Mycologia 76: 548-550.
1008. Evans, T.A., and C.T. Stephens. 1984. First report in Michigan of the teleomorph of *Stemphylium vesicarium*, causal agent of purple spot of asparagus. Pl. Dis. 68: 1099.
1009. Everts, K.L., H.F. Schwartz, N.D. Epsky, and J.L. Capinera. 1985. Effects of maggots and wounding on occurrence of *Fusarium* basal rot of onions in Colorado. Pl. Dis. 69: 878-882.

1010. Exner, B. 1953. Comparative studies of four *Rhizoctonias* occurring in Louisiana. Mycologia 45: 698-719.
1011. Eyal, Z., J.L. Peterson, and C.O. Johnston. 1964. The occurrence of *Puccinia recondita* Rob. ex Desm. race 53 in New Jersey. Pl. Dis. Reporter 48: 653.
1012. Fairman, C.E. 1913. Notes on new species of fungi from various localities. Mycologia 5: 245-248.
1013. Fairman, C.E. 1918. New or noteworthy Ascomycetes and lower fungi from New Mexico. Mycologia 10: 239-264.
1014. Fakir, G.A., R.E. Welty, and E.B. Cowling. 1971. Prevalence and pathogenicity of fungi associated with achenes of sycamore in the field and in storage. Phytopathology 61: 660-668.
1015. Falk, B.W., and J.E. Duffus. 1977. The first report of *Polymyxa betae* in the western hemisphere. Pl. Dis. Reporter 61: 492-494.
1016. Farmer, L.J., A. Firstman, and J. Betz. 1971. A firm head rot of broccoli. Pl. Dis. Reporter 55: 1136.
1017. Farr, E.R., O.K. Miller Jr., and D.F. Farr. 1977. Biosystematic studies in the genus *Pholiota*, stirps *Adiposa*. Canad. J. Bot. 55: 1167-1180.
1018. Farr, M.L. 1966. The didymosporous dimeriaceous fungi described from leaves of Gramineae. Mycologia 58: 221-248.
1019. Farr, M.L., and H.T. Horner Jr. 1968. Fungi on *Selaginella*. Nova Hedwigia 15: 239-287
1020. Farr, M.L., and E.S. Luttrell. 1979. A new species of *Eudimeriolum* from Georgia. Mycologia 71: 1269-1274.
1021. Farr, M.L., J.D. Schoknecht, and J.L. Crane. 1985. A Costa Rican black mildew found in Everglades National Park, Florida. Canad. J. Bot. 63: 1983-1986.
1022. Farrar, L.L. 1953. Johnson grass as a possible wild host of *Sclerospora* disease of sugarcane. Pl. Dis. Reporter 37: 446.
1023. Farrar, L.L., and U.R. Gore. 1957. Diseases of small grains observed in Georgia during the 1956-57 season. Pl. Dis. Reporter 41: 986-987.
1024. Farrar, L.L., and S.V. Stacy. 1958. Diseases of small grains observed in Georgia during the 1957-58 season. Pl. Dis. Reporter 42: 1262-1267.
1025. Feichtenberger, E., G.A. Zentmyer, and J.A. Menge. 1984. Identity of *Phytophthora* isolated from milkweed vine. Phytopathology 74: 50-55.
1026. Felix, E.L. 1952. Charcoal rot of Irish potato in Tennessee. Pl. Dis. Reporter 36: 369.
1027. Felix, E.L. 1953. Additional hosts of tomato buckeye rot fungus in Tennessee. Pl. Dis. Reporter 37: 527.
1028. Felix, E.L. 1955. Notes on some plant diseases in Tennessee. Pl. Dis. Reporter 39: 275-276.
1029. Felix, E.L. 1955. Some tree diseases in Tennessee. Pl. Dis. Reporter 39: 882.
1030. Felix, E.L. 1958. A *Phytophthora* blight and root rot of strawberry. Pl. Dis. Reporter 42: 818-819.
1031. Felix, E.L. 1958. *Phytophthora* blight of strawberry. Pl. Dis. Reporter 42: 841-842.
1032. Fell, J.W., and I.L. Hunter. 1979. Fungi associated with the decomposition of the black rush, *Juncus roemerianus*, in south Florida. Mycologia 71: 322-342.
1033. Fell, J.W., and I.M. Master. 1975. Phycomycetes (*Phytophthora* spp. nov. and *Pythium* sp. nov.) associated with degrading mangrove (*Rhizophora mangle*) leaves. Canad. J. Bot. 53: 2908-2922.
1034. Fenne, S.B., R.G. Henderson, T.J. Smith, and W.C. White. 1950. Alfalfa-clover disease survey in Virginia. Pl. Dis. Reporter 34: 204-205.
1035. Fenwick, H.S. 1969. Diseases of Austrian winter peas in Idaho. Pl. Dis. Reporter 53: 918-920.
1036. Fergus, C.L. 1950. An epiphytotic of white oak anthracnose in central Pennsylvania. Pl. Dis. Reporter 34: 292-294.
1037. Fergus, C.L. 1951. *Strumella* canker on bur oak in Pennsylvania. Phytopathology 41: 101-103.
1038. Fergus, C.L. 1952. Oak wilt found affecting chestnut oak in Pennsylvania. Pl. Dis. Reporter 36: 386.
1039. Fergus, C.L. 1954. An epiphytotic of *Phyllosticta* leaf spot of maple. Pl. Dis. Reporter 38: 678-679.
1040. Fergus, C.L. 1957. *Myriothecium roridum* on *Gardenia*. Mycologia 49: 124-127.
1041. Fergus, C.L. 1957. *Cylindrosporium* leafspot of *Spiraea*. Mycologia 49: 262-268.
1042. Fergus, C.L. 1962. *Glomosporium leptideum* found in North America. Mycologia 54: 322-324.
1043. Fergus, C.L. 1973. *Peridoxylon petersii* on oak in Pennsylvania. Mycologia 65: 1199-1200.
1044. Fergus, C.L., and C.L. Morris. 1950. Oak wilt in Pennsylvania. Pl. Dis. Reporter 34: 291.
1045. Fergus, C.L., W.M. Sharp, and E. Decker. 1956. The role of plant diseases in wildlife food plants: a smut of dotted smartweed. Pl. Dis. Reporter 40: 1017-1018.
1046. Fergus, C.L., and W.L. Yount. 1953. Scrub oak susceptible to oak wilt. Pl. Dis. Reporter 37: 567.
1047. Ferguson, M.W. 1987. Epidemic of *Phytophthora* root rot of soybeans in South Dakota. Pl. Dis. 71: 850.
1048. Ferguson, M.W., C.L. Lay, and P.D. Evenson. 1987. Effect of pasmo disease on flower production and yield components of flax. Phytopathology 77: 805-808.
1049. Fernandez, J.A., and M.S. McShane. 1980. *Cephalosporium* stripe of winter wheat in Wyoming. Pl. Dis. 64: 1117.
1050. Ferrant, N.P., and R.B. Carroll. 1981. *Fusarium* wilt of soybean in Delaware. Pl. Dis. 65: 596-599.
1051. Ferrell, G.T., and R.S. Smith. 1976. Indicators of *Fomes annosus* root decay and bark beetle susceptibili ty in sapling white fir. Forest Sci. 22: 365-369.
1052. Ferris, J.M., D. Neely, and E.B. Himelick. 1957. Tree leaf diseases in Illinois during 1957. Pl. Dis. Reporter 41: 1051-1052.
1053. Fezer, K.D., and H.G. Johnson. 1960. A barley loose smut epidemic in Minnesota in 1959. Pl. Dis. Reporter 44: 71.
1054. Fidalgo, O. 1958. Studies on *Ptychogaster rubescens* Boud. the chlamydosporiferous form of *Polyporus guttulatus* Pk. Mycologia 50: 831-836.
1055. Filer, T.H., and F.I. McCracken. 1969. *Clitocybe tabescens* associated with decline and death of Chinese elm and water oak. Pl. Dis. Reporter 53: 840.
1056. Filer, T.H., F.I. McCracken, and E.R. Toole. 1968. *Cephalosporium* wilt of elm in lower Mississippi valley. Pl. Dis. Reporter 52: 170-171.
1057. Filer, T.H. Jr. 1964. Outbreak of cankers on plantation-grown cottonwoods in Mississippi. Pl. Dis. Reporter 48: 588.
1058. Filer, T.H. Jr. 1966. Red thread found on Bermuda grass. Pl. Dis. Reporter 50: 525-526.
1059. Filer, T.H. Jr. 1969. Sycamore canker caused by *Botryodiplodia theobromae*. Phytopathology 59: 76-78.
1060. Filer, T.H. Jr. 1969. A new species of *Pteridiospora*. Mycologia 61: 167-169.
1061. Filer, T.H., F.I. McCracken, C.A. Mohn, and W.K. Randall. 1971. *Septoria* canker on nursery stock of *Populus deltoides*. Pl. Dis. Reporter 55: 460-463.
1062. Filip, G.M. 1977. An *Armillaria* epiphytotic on the Winema National Forest, Oregon. Pl. Dis. Reporter 61: 708-711.
1063. Filip, G.M. 1977. Crown mortality of ponderosa pine caused by *Cronartium comandrae*. Pl. Dis. Reporter 61: 1083-1085.
1064. Filip, G.M. 1978. Incense cedar, a new host of *Armillaria mellea* in Oregon. Pl. Dis. Reporter 62: 280.
1065. Filip, G.M. 1979. Root disease in Douglas-fir plantations is associated with infected stumps. Pl. Dis. Reporter 63: 580-583.
1066. Filip, G.M. 1984. Dwarf mistletoe and *Cytospora* canker decrease grand fir growth in central Oregon. Forest Sci. 30: 1071-1079.
1067. Filip, G.M., and D.J. Goheen. 1982. Tree mortality caused by root pathogen complex in Deschutes National Forest, Oregon. Pl. Dis. 66: 240-243.
1068. Filip, G.M., and D.J. Goheen. 1984. Root diseases cause severe mortality in white and grand fir stands of the Pacific Northwest. Forest Sci. 30: 138-142.
1069. Filip, G.M., J.S. Hadfield, and C. Schmitt. 1979. Branch mortality of true firs in west-central Oregon associated with dwarf mistletoe and canker fungi. Pl. Dis. Reporter 63: 189-193.
1070. Filip, G.M., A.M. Kanaskie, and S.J. Frankel. 1984. Substantial decay in Pacific silver fir caused by *Hericium abietis*. Pl. Dis. 68: 992-993.
1071. Filip, G.M., and L.F. Roth. 1977. Stump injections with soil fumigants to eradicate *Armillaria mellea* from young-growth ponderosa pine killed by root rot. Canad. J. Forest Res. 7: 226-231.
1072. Filip, M.G., and C.L. Schmitt. 1979. Susceptibility of native conifers to laminated root rot east of the Cascade range in Oregon and Washington. Forest Sci. 25: 261-165.

1073. Fink, H.C. 1958. Two unusual apple fruit rots in Pennsylvania. Pl. Dis. Reporter 42: 156.
1074. Fischer, E. 1922. Weitere Beiträge zur kenntnis der gattung *Graphiola*. Ann. Mycol. 20: 228-237.
1075. Fischer, G.W. 1944. The blind-seed disease of ryegrass (*Lolium* spp.) in Oregon. Phytopathology 34: 934-935.
1076. Fischer, G.W. 1951. A local, winter-time epidemic of blister smut, *Etyloma crastophilum* , on Kentucky bluegrass at Pullman, Washington. Pl. Dis. Reporter 35: 88.
1077. Fischer, G.W. 1953. Manual of the North American Smut Fungi. Ronald Press, NY. 343 p.
1078. Fischer, G.W., and R. Duran. 1956. Some new records of dwarf bunt of wheat and rye. Phytopathology 46: 12.
1079. Fischer, G.W., and J.P. Meiners. 1952. *Ustilago turcomanica* Tranzsch. in the United States. Mycologia 44: 207-212.
1080. Fisher, F.E. 1967. *Cladosporium* leaf spot of citrus in Florida. Pl. Dis. Reporter 51: 1070.
1081. Fisher, F.E. 1970. *Gloeosporium* rot of citrus fruit in Florida. Pl. Dis. Reporter 54: 173-175.
1082. Flowers, R.A., and A.W. Johnson. 1973. Occurrence of race 1 of *Phytophthora parasitica* var. *nicotianae* in Georgia. Pl. Dis. Reporter 57: 832-833.
1083. Folsom, D. 1954. Strawberry fruit rot in Maine caused by *Gnomonia fructicola*. Pl. Dis. Reporter 38: 796-797.
1084. Folsom, D., and R. Bonde. 1950. Black-dot disease on potato tubers in Maine. Pl. Dis. Reporter 34: 208.
1085. Folsom, D., and R. Bonde. 1950. Skin-spot disease on potato tubers in Maine. Pl. Dis. Reporter 34: 209.
1086. Forbes, G.A., D.C. Collins, G.N. Odvody, and R.A. Frederiksen. 1985. A seedling epiphytotic of sorghum in south Texas caused by *Pythium arrhenomanes*. Pl. Dis. 69: 726.
1087. Ford, R.E. 1959. *Fusarium* blight of birdsfoot trefoil. Phytopathology 49: 481-486.
1088. Ford, R.E. 1962. *Ramularia schulzeri* on birdsfoot trefoil in New York. Pl. Dis. Reporter 46: 655-656.
1089. Ford, R.E., and W.A. Haglund. 1963. *Botrytis cinerea* blight of peas associated with senescent blossoms in northwestern Washington. Pl. Dis. Reporter 47: 483-485.
1090. Forer, L.B., and J.L. Longenecker. 1973. A *Verticillium* wilt of green and white ash. Phytopathology 63: 201.
1091. Forer, L.B., F.L. Lukezic, and V.R. Wagner. 1973. Anthracnose of crownvetch caused by *Colletotrichum destructivum*. Pl. Dis. Reporter 57: 104-106.
1092. Foudin, A.S., and O.H. Calvert. 1982. *Schizophyllum commune* as a possible mycotoxin producer in association with sorghum grain. Mycologia 74: 1041-1043.
1093. Foudin, A.S., and O.H. Calvert. 1982. New head-scab of tall fescue in United States caused by *Fusarium heterosporum*. Pl. Dis. 66: 866.
1094. Fowler, M.E. 1951. Surveys for oak wilt. Pl. Dis. Reporter 35: 112-118.
1095. Fowler, M.E. 1952. Oak wilt surveys in 1951. Pl. Dis. Reporter 36: 162-165.
1096. Fowler, M.E., and F.H. Berry. 1958. Blossom-end rot of Chinese chestnuts. Pl. Dis. Reporter 42: 91-96.
1097. Fowler, M.E., and J.A. Stevenson. 1944. A canker and some decay fungi on mimosa. Phytopathology 34: 985-987.
1098. Francis, S.M. 1975. *Anthostomella* Sacc. (Part I). Mycol. Pap. 139: 1-97.
1099. Frederick, R.D., L. Zang, W. Merrill, and T.A. Tattar. 1980. *Naemacyclus minor* needlecast of Scots pine in Massachusetts. Pl. Dis. 64: 1034.
1100. Frederiksen, R.A., J. Amador, B.L. Jones, and L. Reyes. 1969. Distribution, symptoms and economic loss from downy mildew caused by *Sclerospora sorghi* in grain sorghum in Texas. Pl. Dis. Reporter 53: 995-998.
1101. Frederiksen, R.A., R.W. Berry, and J.H. Foster. 1976. Head smut of maize in Texas. Pl. Dis. Reporter 60: 610-611.
1102. Frederiksen, R.A., and A.J. Bockholt. 1969. *Sclerospora sorghi*, a pathogen of corn in Texas. Pl. Dis. Reporter 53: 566-569.
1103. Frederiksen, R.A., D.T. Rosenow, L. Reyes, and L.K. Edmunds. 1969. A new race of *Sphacelotheca reiliana* in south Texas. Pl. Dis. Reporter 53: 171-172.
1104. Freeman, T.E. 1957. A new *Helminthosporium* disease of bermuda grass. Pl. Dis. Reporter 41: 389-391.
1105. Freeman, T.E. 1959. A leafspot of St. Augustine grass caused by *Cercospora fusimaculans*. Phytopathology 49: 160-161.
1106. Freeman, T.E. 1965. Rust of *Zoysia* spp. in Florida. Pl. Dis. Reporter 49: 382.
1107. Freeman, T.E. 1970. A seed disorder of Bermuda grass caused by *Helminthosporium spiciferum*. Pl. Dis. Reporter 54: 358-359.
1108. Freeman, T.E., and R. Charudattan. 1974. Occurrence of *Cercospora piaropi* on water hyacinth in Florida. Pl. Dis. Reporter 58: 277-278.
1109. Freeman, T.E., R. Charudattan, R.E. Cullen, and E. Addor. 1982. *Rhizoctonia* blight on waterhyacinth in the United States. Pl. Dis. 66: 861-862.
1110. French, A.M. 1987. California plant disease host index. Part 1: Fruit and nuts. California Department of Food and Agriculture. 39 p.
1111. French, A.M., and C.W. Nichols. 1954. *Sclerotium rolfsii* on *Philodendron cordatum* in southern California. Pl. Dis. Reporter 38: 530.
1112. French, D.W. 1985. First report of chestnut blight in Minnesota. Pl. Dis. 69: 905.
1113. French, D.W., and C.M. Christensen. 1950. Oak wilt in Minnesota. Pl. Dis. Reporter 34: 82.
1114. French, D.W., and M.G. Marinos. 1970. Dutch elm disease in Minnesota, 1961 - 1970. Pl. Dis. Reporter 54: 706-707.
1115. French, D.W., and F.W. Perpich. 1974. Cankers and decay in red oaks caused by *Fomes robustus*. Phytopathology 64: 1148-1149.
1116. French, D.W., D.L.K. Smeltzer, and N.A. Anderson. 1983. Effect of stem rust fungi on 45-yr-old jack pine. Pl. Dis. 67: 26-28.
1117. French, W.J. 1968. *Eutypella* canker on species of *Acer* in New York State. Dissertation Abstr. 28 B: 3960.
1118. French, W.J. 1972. *Cristulariella pyramidalis* in Florida: an extension of range and new hosts. Pl. Dis. Reporter 56: 135-138.
1119. French, W.J., and S.B. Silverborg. 1967. *Scleroderris* canker of red pine in New York state plantations. Pl. Dis. Reporter 51: 108-109.
1120. Friend, R.J., and D.M. Boone. 1967. *Dasyscyphus virgineus* on dead cranberry stems. Mycologia 59: 533-534.
1121. Friend, R.J., and D.M. Boone. 1968. *Diaporthe vaccinii* associated with dieback of cranberry in Wisconsin. Pl. Dis. Reporter 52: 341-344.
1122. Froelich, R.C., C.S. Hodges Jr., and S.S. Sackett. 1978. Prescribed burning reduces severity of *Annosus* root rot in the south. Forest Sci. 24: 93-100.
1123. Frosheiser, F.I. 1967. *Phytophthora* root rot of alfalfa in Minnesota. Pl. Dis. Reporter 51: 679-681.
1124. Frosheiser, F.I. 1969. *Phytophthora* root rot of alfalfa in the upper midwest. Pl. Dis. Reporter 53: 595-597.
1125. Frosheiser, F.I., and E.H. Hijano. 1985. *Verticillium* wilt (*Verticillium albo-atrum*) of alfalfa in Minnesota. Pl. Dis. 69: 268.
1126. Frosheiser, F.I., R.H. Morrison, and R.E. Welty. 1981. Anthracnose of alfalfa observed in Minnesota. Pl. Dis. 65: 844-845.
1127. Fuller, W.H., and A.G. Norman. 1944. The nature of the flora on field-retting hemp. Soil Sci. Soc. Amer. Proc. 9: 101-104.
1128. Fuller, W.H., and A.G. Norman. 1946. The retting of hemp. Iowa Agric. Exp. Sta. Res. Bull. 344: 928-944.
1129. Fulton, R.H. 1958. New or unusual small fruit diseases and disease-like occurrences in Michigan. Pl. Dis. Reporter 42: 71-73.
1130. Funk, A. 1972. *Sirococcus* shoot-blight of western hemlock in British Columbia and Alaska. Pl. Dis. Reporter 56: 645-647.
1131. Funk, A. 1986. Two new Discomycetes on *Pinus*. Mycotaxon 27: 283-288.
1132. Funk, A., and R.A. Shoemaker. 1971. *Xenomeris abietis* associated with dieback of Douglas fir and western hemlock. Mycologia 63: 567-574.
1133. Futrell, M.C. 1957. Wheat stripe rust epiphytotic in Texas in 1957. Pl. Dis. Reporter 41: 955-957.
1134. Futrell, M.C., and I.M. Atkins. 1954. Diseases of small grains in Texas in 1953. Pl. Dis. Reporter 38: 167-168.
1135. Futrell, M.C., I.M. Atkins, and C.D. Hobbs. 1959. Unusual occurrence of small-grain diseases in Texas in 1957 and 1958. Pl. Dis. Reporter 43: 777-781.
1136. Futrell, M.C., and R.A. Frederiksen. 1970. Distribution of sorghum downy mildew (*Sclerospora sorghi*) in the U.S.A. Pl. Dis. Reporter 54: 311-314.
1137. Futrell, M.C., K.A. Lahr, K.B. Porter, and I.M. Atkins. 1959. Second stripe rust epiphytotic in Texas hits wheat crop in 1958. Pl.

Dis. Reporter 43: 165-167.
1138. Gabrielson, R.L. 1964. Cucumber scab in western Washington. Pl. Dis. Reporter 48: 801-802.
1139. Gadoury, D.M., and W.E. MacHardy. 1985. Negative geotropism in *Venturia inaequalis*. Phytopathology 75: 856-859.
1140. Gaines, E.F., and F.J. Stevenson. 1923. Occurrence of bunt in rye. Phytopathology 13: 210-215.
1141. Gairola, C., and D. Powell. 1970. *Cytospora* peach canker in Illinois. Pl. Dis. Reporter 54: 832-835.
1142. Garber, R.H., and G.A. Paxman. 1963. *Fusarium* wilt of cotton in California. Pl. Dis. Reporter 47: 398-400.
1143. Gardner, D.E. 1979. Occurrence of fuchsia rust in Hawaii. Pl. Dis. Reporter 63: 136-137.
1144. Gardner, D.E. 1980. *Acacia koa* seedling wilt caused by *Fusarium oxysporum* f. sp. *koae*, f. sp. nov. Phytopathology 70: 594-597.
1145. Gardner, D.E. 1981. Rust on *Commelina diffusa* in Hawaii. Pl. Dis. 65: 690-691.
1146. Gardner, D.E. 1982. Septoria leaf spot on *Canavalia kauensis*, a native Hawaiian bean. Pl. Dis. 66: 263-264.
1147. Gardner, D.E. 1984. Kikuyugrass rust caused by *Phakopsora apoda* in Hawaii. Pl. Dis. 68: 826.
1148. Gardner, D.E. 1985. Red leaf disease of native Hawaiian *Vaccinium* species caused by *Exobasidium vaccinii*. Pl. Dis. 69: 805-806.
1149. Gardner, D.E. 1985. Lemongrass rust caused by *Puccinia nakanishikii* in Hawaii. Pl. Dis. 69: 1100.
1150. Gardner, D.E., and C.S. Hodges Jr. 1985. Spore surface morphology of Hawaiian *Acacia* rust fungi. Mycologia 77: 575-586.
1151. Gardner, D.E., and C.S. Hodges Jr. 1986. Hawaiian forest fungi. VII. A new species of *Elsinoe* on native *Vaccinium*. Mycologia 78: 506-508.
1152. Gardner, D.E., and A.P. Martinez. 1985. Occurrence of canna rust (*Puccinia thaliae*) in Hawaii. Pl. Dis. 69: 1101.
1153. Gardner, M.W., and C.E. Yarwood. 1950. Downy mildew of carnation in California. Pl. Dis. Reporter 34: 270.
1154. Gardner, M.W., C.E. Yarwood, and T. Duafala. 1972. Oak mildews. Pl. Dis. Reporter 56: 313-317.
1155. Gardner, M.W., C.E. Yarwood, and R.D. Raabe. 1970. Unreported powdery mildews. IV. Pl. Dis. Reporter 54: 399-402.
1156. Garren, K.H., and D.M. Porter. 1970. Quiescent endocarpic floral communities in cured mature peanuts from Virginia and Puerto Rico. Phytopathology 60: 1635-1638.
1157. Garren, K.H., D.M. Porter, and A.H. Allison. 1971. *Cylindrocladium* black rot of peanuts in Virginia. Pl. Dis. Reporter 55: 419-421.
1158. Gaskin, T.A., and M.P. Britton. 1956. A new host for *Sclerophthora macrospora*. Pl. Dis. Reporter 40: 830.
1159. Gay, J.D., and S.M. McCarter. 1968. Stem rot of snap bean in southern Georgia caused by *Pythium myriotylum*. Pl. Dis. Reporter 52: 416.
1160. Geils, B.W., and W.R. Jacobi. 1984. Incidence and severity of comandra blister rust on lodgepole pine in north northwestern Wyoming. Pl. Dis. 68: 1049-1051.
1161. Gerdemann, J.W. 1953. An undescribed fungus causing a root rot of red clover and other Leguminosae. Mycologia 45: 548-554.
1162. Gerdemann, J.W. 1954. The association of *Diaporthe phaseolorum* var. *sojae* with root and basal stem rot of soybean. Pl. Dis. Reporter 38: 742-743.
1163. Gerdemann, J.W. 1955. Occurrence of *Polymyxa graminis* in red clover roots. Pl. Dis. Reporter 39: 859.
1164. Gerlach, W.W.P., H.A.J. Hoitink, and C.W. Ellett. 1974. Shoot blight and stem dieback of *Pieris japonica* caused by *Phytoph thora citricola*, *P. citrophthora* and *Botryosphaeria dothidea*. Phytopathology 64: 1368-1370.
1165. Gerten, D.M., and R.L. Bowden. 1985. First report of a rot of *Lithops* spp. caused by *Fusarium oxysporum*. Pl. Dis. 69: 811.
1166. Gessner, R.V. 1977. Seasonal occurrence and distribution of fungi associated with *Spartina alterniflora* from a Rhode Island estuary. Mycologia 69: 477-491.
1167. Gessner, R.V., and R.D. Goos. 1973. Fungi from decomposing *Spartina alterniflora*. Canad. J. Bot. 51: 51-55.
1168. Gessner, R.V., and R.D. Goos. 1973. Fungi from *Spartina alterniflora* in Rhode Island. Mycologia 65: 1296-1301.
1169. Gessner, R.V., and J. Kohlmeyer. 1976. Geographical distribution and taxonomy of fungi from salt marsh *Spartina*. Canad. J. Bot. 54: 2023-2037.
1170. Gessner, R.V., and B.J. Lamore. 1978. Fungi from Nantucket salt marshes and beaches. Rhodora 80: 581-586.
1171. Ghaffar, A., and G.A. Zentmeyer. 1968. *Macrophomina phaseoli* on some new weed hosts in California. Pl. Dis. Reporter 52: 223.
1172. Gibbs, J.N. 1980. Survival of *Ceratocystis fagacearum* in branches of trees killed by oak wilt in Minnesota. Eur. J. Forest Pathol. 10: 218-224.
1173. Gilbertson, R.L. 1962. Resupinate hydnaceous fungi of North America. I.Type studies of species described by Peck. Mycologia 54: 658-677.
1174. Gilbertson, R.L. 1964. Resupinate hydnaceous fungi of North America. III. Additional type studies. Pap. Michigan Acad. Sci. 49: 15-25.
1175. Gilbertson, R.L. 1965. Resupinate hydnaceous fungi of North America V. Type studies of species described by Berkeley and Curtis. Mycologia 57: 845-871.
1176. Gilbertson, R.L. 1969. A new species of *Inonotus* on southwestern sycamores. Southw. Naturalist 14: 123-125.
1177. Gilbertson, R.L. 1970. A new *Vararia* from western North America. Madroño 20: 282-287.
1178. Gilbertson, R.L. 1973. Notes on some corticioid lignicolous fungi associated with snowbanks in southern Arizona. Persoonia 7: 171-182.
1179. Gilbertson, R.L. 1974. Fungi that decay ponderosa pine. Univ. Arizona Press. 197 p.
1180. Gilbertson, R.L. 1975. *Peniophora tamaricicola* in North America. Mycotaxon 2: 143-150.
1181. Gilbertson, R.L. 1976. The genus *Inonotus* (Aphyllophorales: Hymenochaetaceae) in Arizona. Mem. New York Bot. Gard. 28: 67-85.
1182. Gilbertson, R.L. 1979. The genus *Phellinus* (Aphyllophorales: Hymenochaetaceae) in western North America. Mycotaxon 9: 51-89.
1183. Gilbertson, R.L., and M. Blackwell. 1982. *Theleporus ajovalliensis* (Aphyllophorales: Corticiaceae), a new wood-rotting fungus on Ocotillo in the Sonoran Desert. Mycotaxon 15: 249-253.
1184. Gilbertson, R.L., and M. Blackwell. 1984. Two new Basidiomycetes on living live oak in the southeast and Gulf Coast region. Mycotaxon 20: 85-93.
1185. Gilbertson, R.L., and M. Blackwell. 1985. Notes on wood-rotting fungi on junipers in the Gulf coast region. Mycotaxon 24: 325-348.
1186. Gilbertson, R.L., and M. Blackwell. 1987. Notes on wood-rotting fungi on junipers in the Gulf Coast region II. Mycotaxon 28: 369-402.
1187. Gilbertson, R.L., W.M. Brown Jr., and E.G. Ruppel. 1985. Prevalence and virulence of *Fusarium* spp. associated with stalk rot of corn in Colorado. Pl. Dis. 69: 1065-1068.
1188. Gilbertson, R.L., and A.B. Budington. 1970. Three new species of wood-rotting fungi in the Corticiaceae. Mycologia 62: 673-678.
1189. Gilbertson, R.L., and H.H. Burdsall Jr. 1972. *Phellinus torulosus* in North America. Mycologia 64: 1258-1269.
1190. Gilbertson, R.L., H.H. Burdsall Jr., and M.J. Larsen. 1975. Notes on wood-rotting hymenomycetes in New Mexico. Southw. Naturalist 19: 347-360.
1191. Gilbertson, R.L., H.H. Burdsall, H.H., and E. Canfield. 1976. Fungi that decay mesquite in southern Arizona. Mycotaxon 3: 487-551.
1192. Gilbertson, R.L., and E.R. Canfield. 1972. *Poria carnegiea* and decay of saguaro cactus in Arizona. Mycologia 64: 1300-1311.
1193. Gilbertson, R.L., and E.R. Canfield. 1973. A new *Poria* from southern Arizona. Mycologia 65: 1117-1124.
1194. Gilbertson, R.L., E.R. Canfield, and G.B. Cummins. 1972. Notes on fungi from the L.N. Goodding Research Natural Area. Arizona Acad. Sci. 7: 129-138.
1195. Gilbertson, R.L., G.B. Cummins, and E.D. Darnall. 1979. Indexes to W.G. Solheim's Mycoflora Saximontanensis Exsiccata. Mycotaxon 10: 49-92.
1196. Gilbertson, R.L., and J.P. Lindsey. 1975. Basidiomycetes that decay junipers in Arizona. Great Basin Naturalist 35: 288-304.
1197. Gilbertson, R.L., and F.F. Lombard. 1976. Wood-rotting Basidiomycetes - - Itasca State Park annotated list. J. Minnesota Acad. Sci. 42: 25-31.
1198. Gilbertson, R.L., F.F. Lombard, and T.E. Hinds. 1968. *Veluticeps berkeleyi* and its decay of pine in North America. Mycologia 60:

29-41.
1199. Gilbertson, R.L., and J.L. Lowe. 1970. *Tyromyces graminicola* in North America. Mycologia 62: 699-706.
1200. Gilbertson, R.L., W.J. Manning, and D.N. Ferro. 1985. Association of the asparagus miner with stem rot caused in asparagus by *Fusarium* species. Phytopathology 75: 1188-1191.
1201. Gilbertson, R.L., and K.J. Martin. 1976. *Polyporus coronadensis*, a new species from Arizona. Mycologia 68: 1116-1120.
1202. Gilbertson, R.L., K.J. Martin, and J.P. Lindsey. 1974. Annotated check list and host index for Arizona wood-rotting fungi. Univ. Arizona Agric. Exp. Sta. Techn. Bull. 209: 1-48.
1203. Gilbertson, R.L., and J. McHenry. 1969. Check list and host index for Arizona rust fungi. Univ. Arizona Agric. Exp. Sta. Techn. Bull. 186: 1-40.
1204. Gilbertson, R.L., and L. Ryvarden. 1986. North American Polypores. Vol. 1. *Abortiporus - Lindtneria* Fungiflora, Oslo. 433 p.
1205. Gilbertson, R.L., and L. Ryvarden. 1987. North American Polypores Vol. 1. *Megasporoporia - Wrightoporia* Fungiflora, Oslo. 434-885 pp.
1206. Gill, D.L. 1960. A stem and branch rot of snapdragon. Pl. Dis. Reporter 44: 946-947.
1207. Gill, D.L. 1967. *Fusarium* wilt infection of apparently healthy mimosa trees. Pl. Dis. Reporter 51: 148-150.
1208. Gill, D.L. 1970. Pathogenic *Pythium* from irrigation ponds. Pl. Dis. Reporter 54: 1077-1079.
1209. Gill, D.L. 1971. *Centrospora acerina* carried by pansy seed. Pl. Dis. Reporter 55: 731-732.
1210. Gill, D.L. 1977. Downy mildew of roses in Georgia. Pl. Dis. Reporter 61: 230-231.
1211. Gill, D.L. 1977. Root and crown rot of shore juniper (*Juniperus conferta*): cause and control. Pl. Dis. Reporter 61: 584-587.
1212. Gill, D.L. 1979. Powdery mildew and *Cylindrocladium* attack *Leucothoe axillaris*. Pl. Dis. Reporter 63: 358-359.
1213. Gill, D.L., S.A. Alfieri Jr., and E.K. Sobers. 1971. A new leaf disease of *Ilex* spp. caused by *Cylindrocladium avesiculatum* sp. nov. Phytopathology 61: 58-60.
1214. Gill, D.L., and T.J. Ratcliffe. 1967. *Centrospora acerina* on pansy in Georgia. Pl. Dis. Reporter 51: 971-972.
1215. Gill, D.L., and J. Ridley. 1954. An outbreak of *Camellia* flower blight in Georgia. Pl. Dis. Reporter 38: 673.
1216. Gill, D.L., and E.K. Sobers. 1970. A new species of *Cylindrocladium* on *Ilex*. Phytopathology 60: 583.
1217. Gill, H.S. 1975. An unusual outbreak of *Whetzelinia* rot of African daisies in California. Pl. Dis. Reporter 59: 82-83.
1218. Gill, H.S. 1977. Powdery mildew of ocotillo (*Fouquieria splendens*) in California. Pl. Dis. Reporter 61: 562-563.
1219. Gill, H.S., O.K. Ribeiro, and G.A. Zentmyer. 1977. *Phytophthora* blight of periwinkles in the Coachella Valley of California. Pl. Dis. Reporter 61: 560-561.
1220. Gill, H.S., G.A. Zentmeyer, O.K. Ribeiro, and L.J. Klure. 1976. A *Phytophthora* disease of African daisies (*Osteospermum* spp.) in California. Pl. Dis. Reporter 60: 647-649.
1221. Gill, L.S. 1952. A new host for *Septogoleum gillii*. Pl. Dis. Reporter 36: 300.
1222. Gilliam, M.S. 1975. New North American species of *Marasmius*. Mycologia 67: 817-844.
1223. Gilliam, M.S. 1976. The genus *Marasmius* in the northeastern United States and adjacent Canada. Mycotaxon 4: 1-144.
1224. Gilman, J.C. 1952. The annual foray of the Mycological Society of America, 1951. Mycologia 44: 716-719.
1225. Gilman, J.C., and L.H. Tiffany. 1952. Iowa Ascomycetes I. The Xylariaceae. Iowa State Coll. J. Sci. 26: 455-482.
1226. Gilman, J.C., L.H. Tiffany, and R.M. Lewis. 1957. Iowa Ascomycetes II. Diaporthaceae: Valseae. Iowa State Coll. J. Sci. 31: 623-647.
1227. Gilman, J.C., L.H. Tiffany, and R.M. Lewis. 1959. Iowa Ascomycetes III. Diaporthaceae: Diaportheae. Iowa State Coll. J. Sci. 33: 325-393.
1228. Gilman, J.C., L.H. Tiffany, and R.W. Lichtwardt. 1957. Fungi new to Iowa. Proc. Iowa Acad. Sci. 64: 85-92.
1229. Gilman, J.C., and B.N. Wadley. 1952. The ascigerous stage of *Septoria querceti* Thuem. Mycologia 44: 216-220.
1230. Gilmer, R.M. 1960. A blight of ground cherry and Russian almond seedlings caused by *Glo eosporium fructigenum* Berk. Pl. Dis. Reporter 44: 395.
1231. Gilreath, J.P., and J.B. Jones. 1985. White rust epiphytotic on *Amaranthus hybridus* in south Florida incited by *Albugo bliti*. Pl. Dis. 69: 542.
1232. Ginns, J. 1978. *Leucogyrophana* (Aphyllophorales): identification of species. Canad. J. Bot. 56: 1953-1973.
1233. Ginns, J. 1979. The genus *Ramaricium* (Gomphaceae). Bot. Notiser 132: 93-102.
1234. Ginns, J. 1982. *Steccherinum oreophilum*: cultural characters and occurrence in North America. Mycologia 74: 20-25.
1235. Ginns, J. 1982. *Cerinomyces ceraceus* sp. nov. and the similar *C. grandinioides* and *C. lagerheimii*. Canad. J. Bot. 60: 519-524.
1236. Ginns, J. 1982. The wood-inhabiting fungus, *Aleurodiscus dendroideus*, sp. nov., and the distinctions between *A. grantii* and *A. amorphus*. Canad. Field-Naturalist 96: 131-138.
1237. Ginns, J. 1985. *Hericium* in North America: Cultural characteristics and mating behavior. Canad. J. Bot. 63: 1551-1563.
1238. Ginns, J.H. 1970. The genus *Merulius* III. Species of *Merulium* and *Phlebia* proposed by Schweinitz and Peck. Mycologia 62: 238-255.
1239. Ginns, J.H. 1976. *Merulius*: s.s. and s.l., taxonomic disposition and identification of species. Canad. J. Bot. 54: 100-167.
1240. Ginns, J.H., and R.P. True. 1967. Butt rot in yellow-poplar seedling-sprout stands. Forest Sci. 13: 440-447.
1241. Ginns, J., and J.L. Lowe. 1983. *Macrohyporia extensa* and its synonym *Poria cocos*. Canad. J. Bot. 61: 1672-1679.
1242. Glawe, D.A. 1983. Observations on the anamorph of *Eutypella parasitica*. Mycologia 75: 742-743.
1243. Glawe, D.A. 1983. Observations on the anamorph of *Diatrypella frostii*. Mycologia 75: 913-915.
1244. Glawe, D.A. 1984. *Cryptosphaeria pullmanensis*, a new species from Washington state. Mycologia 76: 166-169.
1245. Glawe, D.A. 1985. The pleomorphic asexual state of *Valsaria insitiva* in artificial culture. Mycologia 77: 62-71.
1246. Glawe, D.A., and J.L. Crane. 1987. Illinois fungi XIII. *Tubakia dryina*. Mycotaxon 29: 101-112.
1247. Glawe, D.A., M.A. Dilley, and W.J. Moller. 1983. Isolation and identification of *Eutypa armeniacae* from *Malus domestica* in Washington state. Mycotaxon 18: 315-318.
1248. Glawe, D.A., and J.D. Rogers. 1982. Observations on the anamorphs of six species of *Diatrype* and *Diatrypella*. Canad. J. Bot. 60: 245-251.
1249. Glawe, D.A., and J.D. Rogers. 1982. Observations of the anamorphs of six species of *Eutypa* and *Eutypella*. Mycotaxon 14: 334-346.
1250. Glawe, D.A., and J.D. Rogers. 1986. Conidial states of some species of Diatrypaceae and Xylariaceae. Canad. J. Bot. 64: 1493-1498.
1251. Glawe, D.A., C.B. Skotland, and W.J. Moller. 1982. Isolation and identification of *Eutypa armeniacae* from diseased grapevines in Washington state. Mycotaxon 16: 123-132.
1252. Glover, L.B., and P.M. Halisky. 1973. Disease problems in coastal vegetation in New Jersey. Phytopathology 63: 201.
1253. Godfrey, G.H. 1952. Foliage diseases of cucurbits in the lower Rio Grande Valley of Texas in 1951. Pl. Dis. Reporter 36: 69.
1254. Godfrey, G.H. 1954. Cantaloupe downy mildew in the lower Rio Grande Valley of Texas and its relation to relative humidity. Pl. Dis. Reporter 38: 616-619.
1255. Godfrey, G.H., and A.L. Harrison. 1955. Tomato nailhead spot in the lower Rio Grande Valley. Pl. Dis. Reporter 39: 185-186.
1256. Goheen, D.J., and F.W. Cobb Jr. 1978. Occurrence of *Verticicladiella wagenerii* and its perfect state, *Ceratocystis wageneri* sp. nov., in insect galleries. Phytopathology 68: 1192-1195.
1257. Goheen, D.J., F.W. Cobb Jr., and G.N. McKibbin. 1978. Influence of soil moisture on infection of ponderosa pine by *Verticicladiella wagenerii*. Phytopathology 68: 913-916.
1258. Goheen, D.J., and G.M. Filip. 1980. Root pathogen complexes in Pacific Northwest forests. Pl. Dis. 64: 793-794.
1259. Goheen, D.J., and E.M. Hansen. 1978. Black stain root disease in Oregon and Washington. Pl. Dis. Reporter 62: 1098-1102.
1260. Goldstein, D., and R.L. Gilbertson. 1981. Cultural morphology and sexuality of *Inonotus arizonicus*. Mycologia 73: 167-180.
1261. Golsen, E.W., and H.L. Rubin. 1964. *Cercospora jujubae* Chowdhury - - a new report for the United States. Pl. Dis. Reporter 48: 914.
1262. Gooch, F.S. 1966. The occurrence of *Camellia* flower blight in

southwest Louisiana. Pl. Dis. Reporter 50: 525.
1263. Goode, M.J. 1958. Physiological specialization in *Colletotrichum lagenarium*. Phytopathology 48: 79-83.
1264. Goos, R.D. 1969. Conidium ontogeny in *Cacumisporium capitulatum*. Mycologia 61: 52-56.
1265. Goos, R.D. 1970. A new genus of the Hyphomycetes from Hawaii. Mycologia 62: 171-175.
1266. Goos, R.D. 1975. Production of secondary conidia by *Helicoma proliferans*. Mycologia 67: 888-891.
1267. Goos, R.D. 1977. An addition to the bird's nest fungi of the Hawaiian Islands. Canad. J. Bot. 55: 2761-2763.
1268. Goos, R.D. 1978. Field and laboratory studies of meliolaceous fungi in Hawaii. Mycologia 70: 995-1006.
1269. Goos, R.D. 1980. Some helicosporous fungi from Hawaii. Mycologia 72: 595-610.
1270. Goos, R.D. 1985. A review of the anamorph genus *Helicomyces*. Mycologia 77: 606-618.
1271. Goos, R.D. 1986. A review of the anamorph genus *Helicoma*. Mycologia 78: 744-761.
1272. Goos, R.D., and J.H. Anderson. 1972. The Meliolaceae of Hawaii. Sydowia 26: 73-80.
1273. Gordon, C.C., and C.G. Shaw. 1960. A new genus of the Meliolaceae on the roots of *Solanum* species. Mycologia 52: 327-333.
1274. Gordon, D.E., and L.F. Roth. 1976. Root grafting of Port-Orford-cedar - An infection route of root rot. Forest Sci. 22: 276-278.
1275. Gore, U.R., D.D. Morrey, A.R. Brown, and J.H. Miller. 1956. Diseases of small grains in Georgia - 1955. Pl. Dis. Reporter 40: 224.
1276. Goree, H. 1969. Occurrence of *Cryptostroma corticale* in Northwestern United States. Pl. Dis. Reporter 53: 87.
1277. Goree, H. 1972. *Stictis serpentaria* in western United States. Mycologia 64: 624-628.
1278. Goth, R.W., and W.J. Zaumeyer. 1963. Occurrence of *Phyllosticta* leaf spot in beans in 1963. Pl. Dis. Reporter 47: 1079.
1279. Gotlieb, A.R., A.K. Watson, and C. Poirier. 1987. First report of *Colletotrichum coccodes* on velvetleaf. Pl. Dis. 71: 281.
1280. Goto, S., and M.B. Moore. 1952. Some oat diseases in Minnesota 1951. Pl. Dis. Reporter 36: 69-70.
1281. Gottwald, T.R. 1982. Taxonomy of the pecan scab fungus *Cladosporium caryigenum*. Mycologia 74: 382-390.
1282. Gottwald, T.R., and H.R. Cameron. 1979. Studies in the morphology and life history of *Anisogramma anomala*. Mycologia 71: 1107-1126.
1283. Gough, F.J., and M.E. McDaniel. 1969. *Zoysia* rust in Texas. Pl. Dis. Reporter 53: 232.
1284. Gough, F.J., and M.E. McDaniel. 1974. Occurrence of oat leaf blotch in Texas in 1973. Pl. Dis. Reporter 58: 80-81.
1285. Gough, F.J., N.D. Williams, and W.E. Brentzel. 1959. Occurrence of stripe rust in North Dakota in 1958. Pl. Dis. Reporter 43: 169-171.
1286. Gould, C.J. 1957. Turf diseases in western Washington in 1955 and 1956. Pl. Dis. Reporter 41: 344-347.
1287. Gould, C.J. Jr. 1945. The parasitism of *Glomerularia lonicerae* (Pk.) D. & H. in *Lonicera* species. Iowa State Coll. J. Sci. 19: 301-331.
1288. Gould, C.J., M. Eglitis, and C.C. Doughty. 1955. European Rhododendron rust (*Chrysomyxa ledi var. rhododendri*) in the United States. Pl. Dis. Reporter 39: 781-782.
1289. Gould, C.J., M. Eglitis, and N.A. MacLean. 1958. *Botrytis hyacinthi* in the United States. Pl. Dis. Reporter 42: 534-535.
1290. Gould, C.J., R.L. Goss, and M. Eglitis. 1961. *Ophiobolus* patch disease of turf in western Washington. Pl. Dis. Reporter 45: 296-297.
1291. Graham, J.H. 1951. *Stemphylium* leafspot and stem canker of birdsfoot trefoil. Pl. Dis. Reporter 35: 489.
1292. Graham, J.H. 1954. Dying of orchard grass in southern Pennsylvania. Pl. Dis. Reporter 38: 729.
1293. Graham, J.H. 1955. *Helminthosporium* leaf streak of timothy. Phytopathology 45: 227-228.
1294. Graham, J.H. 1955. A disease of orchardgrass caused by *Pleospora phaeocomes*. Phytopathology 45: 633-634.
1295. Graham, J.H. 1957. A *Stemphylium* disease of Ladino white clover. Phytopathology 47: 213-215.
1296. Graham, J.H., and K.E. Zeiders. 1957. Diseases of crown vetch. Pl. Dis. Reporter 41: 925.
1297. Graham, J.H., and E.S. Luttrell. 1961. Species of *Leptosphaerulina* on forage plants. Phytopathology 51: 680-693.
1298. Graham, J.H., R.N. Peaden, and D.W. Evans. 1977. *Verticillium* wilt of alfalfa found in the United States. Pl. Dis. Reporter 61: 337-340.
1299. Graham, J.H., D.L. Stuteville, F.I. Frosheiser, and D.C. Erwin. 1979. A Compendium of Alfalfa Diseases. Disease Compendia Series. American Phytopathological Society. 65 p.
1300. Graham, S.O., and J.W. Strobel. 1958. The incidence of anthracnose fungi on ornamental foliage plants in Washington state greenhouses. Pl. Dis. Reporter 42: 1294-1296.
1301. Grand, L.F. 1974. *Phytophthora cinnamomi* root rot of Fraser fir in North Carolina. Pl. Dis. Reporter 58: 318-320.
1302. Grand, L.F. 1978. New hosts of *Cristulariella pyramidalis* in North Carolina. Pl. Dis. Reporter 62: 841-842.
1303. Grand, L.F. 1985. North Carolina Plant Disease Index. North Carolina Agric. Res. Serv. Tech. Bull. 240: 1-157.
1304. Grand, L.F. 1987. Teleomorph of *Sphaerotheca fuliginea* on cucurbits in North Carolina. Mycologia 79: 484-486.
1305. Grand, L.F. 1987. Perithecia of *Sphaerotheca fuliginea* on cucurbits in North Carolina. Pl. Dis. 1987: 761.
1306. Grand, L.F., and C.A. Doggett. 1973. Oak wilt in eastern North Carolina. Pl. Dis. Reporter 57: 325.
1307. Grand, L.F., J.A. Menge, and J.J. Bond. 1975. Partial checklist of fungi from Highlands, North Carolina and vicinity. J. Elisha Mitchell Sci. Soc. 91: 221-229.
1308. Grau, C.R., and H.L. Bissonnette. 1974. *Whetzelinia* stem rot of soybean in Minnesota. Pl. Dis. Reporter 58: 693-695.
1309. Grau, C.R., P.A. Delwiche, R.L. Norgren, T.E. O'Connell, and D.P. Maxwell. 1981. *Verticillium* wilt of alfalfa in Wisconsin. Pl. Dis. 65: 843-844.
1310. Grau, C.R., and K.J. Leonard. 1978. *Zea mays*, a new host for *Ligniera junci* (Plasmodiophorales). Mycologia 70: 41-46.
1311. Graves, A.A., and W. Witcher. 1971. *Monochaetia* canker of Arizona cypress and redcedar in South Carolina. Pl. Dis. Reporter 55: 810-813.
1312. Graves, A.H. 1914. Parasitism in *Hymenochaete agglutinans*. Mycologia 6: 279-284.
1313. Gray, F.A., W.H. Bohl, and R.H. Abernethy. 1983. *Phytophthora* root rot of alfalfa in Wyoming. Pl. Dis. 67: 291-294.
1314. Gray, F.A., R.L. Haaland, E.M. Clark, and D.M. Ball. 1980. Diseases of alfalfa in Alabama. Pl. Dis. 64: 1015-1017.
1315. Gray, F.A., R. Rodriguez-Kabana, and J.R. Adams. 1980. *Neocosmospora* stem rot of soybeans in Alabama. Pl. Dis. 64: 321-322.
1316. Gray, F.A., and D.A. Roth. 1982. *Verticillium* wilt of alfalfa in Wyoming. Pl. Dis. 66: 1080.
1317. Gray, F.A., and D.S. Wofford. 1987. *Phytophthora* seed rot and seedling blight of sainfoin (*Onobrychis viciifolia*). Pl. Dis. 71: 74-78.
1318. Gray, L.E. 1978. *Mycoleptodiscus terrestris* root rot of soybeans. Pl. Dis. Reporter 62: 72-73.
1319. Green, D.J., W.C. Shortle, and A.L. Shigo. 1981. Compartmentalization of discolored and decayed wood in red maple branch stubs. Forest Sci. 27: 519-522.
1320. Green, H.J., and A. Kelman. 1964. Distribution of the Dutch elm disease in North Carolina in 1963. Pl. Dis. Reporter 48: 578-580.
1321. Green, R.J. 1961. *Septoria* leafspot disease of Scotch spearmint. Pl. Dis. Reporter 45: 696.
1322. Green, R.J., R.W. Samson, and C. Fordyce. 1961. *Verticillium* wilt of potato in northern Indiana. Pl. Dis. Reporter 45: 30.
1323. Green, R.J. Jr. 1958. "Deep plowing" for controlling *Verticillium* wilt of mint in muck soils. Phytopathology 48: 575-577.
1324. Green, R.J. Jr. 1961. Stolon decay of commercial species of mint in Indiana. Pl. Dis. Reporter 45: 288.
1325. Green, R.J. Jr. 1977. Dieback of black walnut seedlings caused by *Phomopsis elaeagni*. Pl. Dis. Reporter 61: 582-584.
1326. Green, R.J. Jr. 1985. Control of a molding-root rot complex of black walnut seedlings in storage. Pl. Dis. 69: 398-400.
1327. Green, R.J., Jr., and R.G. Pratt. 1970. Root rot of black walnut seedlings caused by *Phytophthora citricola*. Pl. Dis. Reporter 54: 583-585.
1328. Green, V.E. Jr. 1958. Observations on fungus diseases of rice in Florida 1951-1957. Pl. Dis. Reporter 42: 624-628.

1329. Greene, H.C. 1940. Notes on Wisconsin parasitic fungi. I. Trans. Wisconsin Acad. Sci. 32: 77-83.
1330. Greene, H.C. 1942. Notes on Wisconsin parasitic fungi. II. Trans. Wisconsin Acad. Sci. 34: 83-98.
1331. Greene, H.C. 1944. Notes on Wisconsin parasitic fungi. IV. Farlowia 1: 569-581.
1332. Greene, H.C. 1944. Notes on Wisconsin parasitic fungi. III. Trans. Wisconsin Acad. Sci. 35: 113-135.
1333. Greene, H.C. 1944. Notes on Wisconsin parasitic fungi. V. Trans. Wisconsin Acad. Sci. 36: 225-268.
1334. Greene, H.C. 1944. Notes on parasitic Wisconsin fungi. VI. Trans. Wisconsin Acad. Sci. 36: 245-268.
1335. Greene, H.C. 1945. Notes on Wisconsin parasitic fungi. VII. Amer. Midl. Naturalist 34: 258-270.
1336. Greene, H.C. 1946. Notes on Wisconsin parasitic fungi. VIII. Trans. Wisconsin Acad. Sci. 38: 219-234.
1337. Greene, H.C. 1946. Notes on Wisconsin parasitic fungi. IX. Trans. Wisconsin Acad. Sci. 38: 235-248.
1338. Greene, H.C. 1948. Notes on Wisconsin parasitic fungi. X. Amer. Midl. Naturalist 39: 444-456.
1339. Greene, H.C. 1949. Notes on Wisconsin parasitic fungi. XI. Amer. Midl. Naturalist 41: 714-725.
1340. Greene, H.C. 1949. Notes on Wisconsin parasitic fungi. XII. Amer. Midl. Naturalist 41: 726-739.
1341. Greene, H.C. 1949. Notes on Wisconsin parasitic fungi. XIII. Amer. Midl. Naturalist 41: 740-758.
1342. Greene, H.C. 1950. Notes on Wisconsin parasitic fungi. XIV. Amer. Midl. Naturalist 44: 630-642.
1343. Greene, H.C. 1952. Notes on Wisconsin parasitic fungi. XV. Amer. Midl. Naturalist 48: 37-54.
1344. Greene, H.C. 1952. Notes on Wisconsin parasitic fungi. XVI. Amer. Midl. Naturalist 48: 741-757.
1345. Greene, H.C. 1952. Notes on Wisconsin parasitic fungi. XVII. Trans. Wisconsin Acad. Sci. 41: 117-128.
1346. Greene, H.C. 1953. Notes on Wisconsin parasitic fungi. XIX. Amer. Midl. Naturalist 50: 501-508.
1347. Greene, H.C. 1953. Notes on Wisconsin parasitic fungi. XVIII. Trans. Wisconsin Acad. Sci. 42: 69-81.
1348. Greene, H.C. 1954. Notes on Wisconsin parasitic fungi. XX. Trans. Wisconsin Acad. Sci. 43: 165-181.
1349. Greene, H.C. 1955. Notes on Wisconsin parasitic fungi. XXI. Trans. Wisconsin Acad. Sci. 44: 29-43.
1350. Greene, H.C. 1956. Notes on Wisconsin parasitic fungi. XXII. Trans. Wisconsin Acad. Sci. 45: 177-191.
1351. Greene, H.C. 1957. Notes on Wisconsin parasitic fungi. XXIII. Trans. Wisconsin Acad. Sci. 46: 141-158.
1352. Greene, H.C. 1958. Notes on Wisconsin parasitic fungi. XXIV. Trans. Wisconsin Acad. Sci. 47: 99-117.
1353. Greene, H.C. 1958. Notes on Wisconsin parasitic fungi. XXV. Trans. Wisconsin Acad. Sci. 47: 119-129.
1354. Greene, H.C. 1960. Notes on Wisconsin parasitic fungi. XXVI. Trans. Wisconsin Acad. Sci. 49: 85-111.
1355. Greene, H.C. 1961. Notes on Wisconsin parasitic fungi. XXVII. Trans. Wisconsin Acad. Sci. 50: 141-161.
1356. Greene, H.C. 1962. Notes on Wisconsin parasitic fungi. XXVIII. Trans. Wisconsin Acad. Sci. 51: 57-78.
1357. Greene, H.C. 1963. Notes on Wisconsin parasitic fungi. XXIX. Trans. Wisconsin Acad. Sci. 52: 229-253.
1358. Greene, H.C. 1964. Notes on parasitic fungi. XXX. Trans. Wisconsin Acad. Sci. 53: 177-196.
1359. Greene, H.C. 1964. Notes on Wisconsin parasitic fungi. XXXI. Trans. Wisconsin Acad. Sci. 53: 197-215.
1360. Greene, H.C. 1966. Notes on Wisconsin parasitic fungi. XXXII. Trans. Wisconsin Acad. Sci. 55: 147-166.
1361. Greene, H.C. 1968. Notes on Wisconsin parasitic fungi. XXXIII. Trans. Wisconsin Acad. Sci. 56: 263-280.
1362. Greene, H.C., and G.B. Cummins. 1958. A synopsis of the Uredinales which parasitize grasses of the genera *Stipa* and *Nasella*. Mycologia 50: 6-36.
1363. Grey, W.E., D.E. Mathre, J.A. Hoffmann, R.L. Powelson, and J.A. Fernandez. 1986. Importance of seedborne *Tilletia controversa* for infection of winter wheat and its relationship to international commerce. Pl. Dis. 70: 122-125.
1364. Griffin, H.D. 1968. The genus *Ceratocystis* in Ontario. Canad. J. Bot. 46: 689-718.
1365. Grisham, M.P., and N.A. Anderson. 1983. Pathogenicity and host specificity of *Rhizoctonia solani* isolated from carrots. Phytopathology 73: 1564-1569.
1366. Grogan, R.G., and K.A. Kimble. 1959. The association of *Fusarium* wilt with the asparagus decline and replant problem in California. Phytopathology 49: 122-125.
1367. Grogan, R.G., K.A. Kimble, and I. Misaghi. 1975. A stem canker disease of tomato caused by *Alternaria alternata* f.sp. *lycopersici*. Phytopathology 65: 880-886.
1368. Grogan, R.G., and W.C. Snyder. 1952. The occurrence and pathological effects of *Stemphylium radicinum* on carrots in California. Phytopathology 42: 215-218.
1369. Gross, D.C., and L.D. Leach. 1973. Stalk blight of sugarbeet seed crops caused by *Fusarium oxysporum* f. sp. *betae*. Phytopathology 63: 1216.
1370. Gross, H.L., and T.C. Weidensaul. 1967. *Phacidiopycnis pseudotsugae* associated with bleeding cankers on hemlock. Pl. Dis. Reporter 51: 807-808.
1371. Grove, G.G., and R.N. Campbell. 1987. Host range and survival in soil of *Pyrenochaeta lycopersici*. Pl. Dis. 71: 806-809.
1372. Grove, G.G., L.V. Madden, and M.A. Ellis. 1985. Splash dispersal of *Phytophthora cactorum* from infected strawberry fruit. Phytopathology 75: 611-615.
1373. Grove, G.G., L.V. Madden, and M.A. Ellis. 1985. Influence of temperature and wetness duration on sporulation of *Phytophthora cactorum* on infected strawberry fruit. Phytopathology 75: 700-703.
1374. Grove, G.G., and R.G. Roberts. 1987. Survival of *Podosphaera oxyacanthae* on senescent cherry leaves. Phytopathology 77: 1239.
1375. Grove, M.D. 1976. Fusarial disease of asparagus. Proc. Amer. Phytopathol. Soc. 3: 317.
1376. Groves, J.W. 1938. *Dermatea acerina* and *Pezicula acericola*. Mycologia 30: 416-430.
1377. Groves, J.W. 1939. Some *Pezicula* species and their conidial stages. Canad. J. Res., Sect. C, Bot. Sci. 17: 125-143.
1378. Groves, J.W. 1941. *Pezicula carnea* and *Pezicula subcarnea*. Mycologia 33: 510-522.
1379. Groves, J.W. 1946. North American species of *Dermea*. Mycologia 38: 351-431.
1380. Groves, J.W. 1952. The genus *Tympanis*. Canad. J. Bot. 30: 571-651.
1381. Groves, J.W. 1954. The genus *Durandiella*. Canad. J. Bot. 32: 116-144.
1382. Groves, J.W. 1965. The genus *Godronia*. Canad. J. Bot. 43: 1195-1276.
1383. Groves, J.W., and C.A. Bowerman. 1955. The species of *Ciborinia* on *Populus*. Canad. J. Bot. 33: 577-590.
1384. Groves, J.W., and S.C. Hoare. 1954. Notes on fungi from northern Canada. I. Hypocreales and Discomycetes. Canad. Field-Naturalist 68: 1-8.
1385. Groves, J.W., and D.E. Wells. 1956. The genus *Retinocyclus*. Mycologia 48: 865-871.
1386. Gruenhagen, R.H., and C. Fordyce. 1963. *Osmanthus ilicifolius*, a new host of *Verticillium albo-atrum*. Pl. Dis. Reporter 47: 688.
1387. Grybauskas, A.P. 1986. First report of soybean naturally infected with *Leptosphaerulina briosiana*. Pl. Dis. 70: 1159.
1388. Grybauskas, A.P., and E.M. Dutky. 1987. First report of *Verticillium* wilt of alfalfa in Maryland. Pl. Dis. 71: 376.
1389. Guba, E., and J. Stevenson. 1963. Fungus and nematode inhabitants and diseases of holly (*Ilex*). Bull. Exp. Sta. Univ. Massachusetts 530: 1-43.
1390. Guba, E.F. 1937. The fungi of Nantucket. Century I. Rhodora 39: 367-376.
1391. Guba, E.F. 1939. List of second hundred fungi of Nantucket. Rhodora 41: 508-520.
1392. Guba, E.F. 1953. Large leaf spot and canker of peach caused by the fungus *Fusicoccum amygdali* Delacr. Pl. Dis. Reporter 37: 560-564.
1393. Guba, E.F. 1961. Monograph of *Monochaetia* and *Pestalotia*. Harvard University Press, Cambridge. 342 p.
1394. Guba, E.F., and E.V. Seeler Jr. 1942. Third list of fungi of Nantucket. Rhodora 44: 167-175.
1395. Gubler, W.D., and R.G. Grogan. 1976. *Fusarium* wilt of muskmelon in the San Joaquin Valley of California. Pl. Dis. Reporter 60: 742-744.
1396. Gubler, W.D., R.G. Grogan, and A.S. Greathead. 1978. Wilt of

cucumber caused by *Verticillium albo-atrum* in California. Pl. Dis. Reporter 62: 786-788.
1397. Gubler, W.D., J.J. Marois, A.M. Bledsoe, and L.J. Bettiga. 1987. Control of *Botrytis* bunch rot of grape with canopy management. Pl. Dis. 71: 599-601.
1398. Gudauskas, R.T., and S.M. McCarter. 1966. Occurrence of rust on *Zoysia* species in Alabama. Pl. Dis. Reporter 50: 885.
1399. Gudauskas, R.T., and N.E. McGlohon. 1964. *Sclerotinia* blight of bahia grass in Alabama. Pl. Dis. Reporter 48: 418.
1400. Gudauskas, R.T., D.H. Teem, and G. Morgan-Jones. 1977. Anthracnose of *Cassia occidentalis* caused by *Colletotrichum dematium* f. *truncata*. Pl. Dis. Reporter 61: 468-470.
1401. Gulya, T.J., Jr., C.A. Martinson, and L.H. Tiffany. 1979. Ear-rotting fungi associated with opaque-2 maize. Pl. Dis. Reporter 63: 370-373.
1402. Gulya, T.J., J.J. Ooka, and M.K. Mancl. 1985. Diseases of cultivated sunflower in Hawaii. Pl. Dis. 69: 542.
1403. Gunnell, P.S., and R.K. Webster. 1984. Aggregate sheath spot of rice in California. Pl. Dis. 68: 529-531.
1404. Guthrie, J.W. 1958. Early blight of potatoes in southeastern Idaho. Pl. Dis. Reporter 42: 246.
1405. Guthrie, J.W. 1959. *Fusarium eumartii* wilt and tuber necrosis in Idaho. Pl. Dis. Reporter 43: 423.
1406. Guyton, T.L. 1952. An unusual occurrence of oak wilt in Pennsylvania. Pl. Dis. Reporter 36: 386.
1407. Haenseler, C.M. 1924. Okra wilt. New Jersey Agric. Exp. Sta. Dis. Notes 2(7): 4.
1408. Haenseler, C.M. 1952. *Botrytis* on kenaf in New Jersey. Pl. Dis. Reporter 36: 336.
1409. Hagedorn, D.J. 1974. Recent pea anthracnose and downy mildew epiphytotics in Wisconsin. Pl. Dis. Reporter 58: 226-229.
1410. Hagedorn, D.J., and E.K. Wade. 1974. Bean rust and angular leaf spot in Wisconsin. Pl. Dis. Reporter 58: 330-332.
1411. Hahn, G.G. 1940. Distribution and hosts of cedar blight in the United States. Pl. Dis. Reporter 24: 52-57.
1412. Hahn, G.G. 1943. Taxonomy, distribution, and pathology of *Phomopsis occulta* and *P. juniperovora*. Mycologia 35: 112-129.
1413. Hahn, G.G. 1957. A new species of *Phacidiella* causing the so-called *Phomopsis* disease of conifers. Mycologia 49: 226-239.
1414. Hahn, G.G. 1957. *Phacidiopycnis* (*Phomopsis*) canker and dieback of conifers. Pl. Dis. Reporter 41: 623-633.
1415. Hahn, G.G., and H.G. Eno. 1956. Fungus association with birch "dieback" and its significance. Pl. Dis. Reporter 40: 71-79.
1416. Haines, J.H. 1974. Notes on the genus *Trichopezizella* with descriptions of new taxa. Mycologia 66: 213-241.
1417. Haines, J.H. 1980. Studies in the Hyaloscyphaceae I: Some species of *Dasyscyphus* on tropical ferns. Mycotaxon 11: 189-216.
1418. Haines, J.H., and K.H. McKnight, K.H. 1977. Notes on two American Hyaloscyphaceae on aspen. Mycotaxon 5: 423-431.
1419. Halber, M. 1963. *Botrytis* sp. on Douglas fir seedlings. Pl. Dis. Reporter 47: 556.
1420. Halfon-Meiri, A., and M.M. Kulik. 1977. *Septoria nodorum* infection of wheat seeds produced in Pennsylvania. Pl. Dis. Reporter 61: 867-869.
1421. Halisky, P.M., D.G. Smeltzer, and B.R. Houston. 1959. Head smut of sudan grass and sorghum in California. Pl. Dis. Reporter 43: 1084-1090.
1422. Halliwell, R.S., and J.D. Johnson. 1964. Seed rot and seedling tip necrosis of peach caused by a species of *Penicillium*. Pl. Dis. Reporter 48: 136-138.
1423. Halliwell, R.S., and I.C. MacSwan. 1962. Undescribed plant diseases and new hosts recorded in Oregon during 1960-1961. Pl. Dis. Reporter 46: 30.
1424. Halliwell, R.S., and P.W. Miller. 1962. Kernel-spot of filberts found in Oregon. Pl. Dis. Reporter 46: 220.
1425. Halma, F.F., and K.W. Opitz. 1952. Observations on foot rot of orange in California. Pl. Dis. Reporter 36: 413.
1426. Halpin, J.E., and S.M. McCarter. 1961. Fungi associated with white clover stolons in selected areas of the southeast during mid-summer, 1959. Pl. Dis. Reporter 45: 298-299.
1427. Halsted, B.D., and D.G. Fairchild. 1891. Sweet-potato black rot. J. Mycol. 7: 1-11.
1428. Hamm, P.B., and E.M. Hansen. 1981. Host specificity of *Phytophthora megasperma* from Douglas fir, soybean, and alfalfa. Phytopathology 71: 65-68.
1429. Hamm, P.B., and E.M. Hansen. 1983. *Phytophthora pseudotsugae*, a new species causing root rot of Douglas-fir. Canad. J. Bot. 61: 2626-2631.
1430. Hamm, P.B., and E.M. Hanson. 1982. Pathogenicity of *Phytophthora* species to Pacific Northwest conifers. Eur. J. Forest Pathol. 12: 167-174.
1431. Hamm, P.B., A. Kanaskie, P. Morgan, and S.J. Cooley. 1987. Root necrosis caused by *Fusarium oxysporum* on 2-year-old Douglas-fir seedlings in Oregon. Pl. Dis. 71: 651.
1432. Hamm, P.B., and P.A. Koepsell. 1984. *Phytophthora* root rot of cabbage and cauliflower in Oregon. Pl. Dis. 68: 533-535.
1433. Hammer, W., and G. Chastagner. 1987. Factors affecting the production of pseudothecia of *Leptosphaeria korrae* in vitro. Phytopathology 77: 1239.
1434. Hammill, T.M. 1972. Fine structure of annellophores. V. *Stegonosporium pyriforme*. Mycologia 64: 654-657.
1435. Hampson, M.C., and W.A. Sinclair. 1973. Xylem disfunction in peach caused by *Cytospora leucostoma*. Phytopathology 63: 676-681.
1436. Hancock, J.G. 1972. Root rot of cotton caused by *Pythium splendens*. Pl. Dis. Reporter 56: 973-975.
1437. Hancock, J.G. 1985. Fungal infection of feeder rootlets of alfalfa. Phytopathology 75: 1112-1120.
1438. Hanlin, R.T. 1963. A revision of the Ascomycetes of Georgia. Georgia Agric. Exp. Sta. Mimeo Ser. n.s. 175: 1-65.
1439. Hanlin, R.T. 1966. Host index to the Basidiomycetes of Georgia. Georgia Agric. Exp. Sta. Mimeo Ser. n.s. 260: 1-30.
1440. Hanlin, R.T. 1966. Powdery mildew on mock orange in Georgia. Pl. Dis. Reporter 50: 306-307.
1441. Hanlin, R.T. 1969. The identity of *Phyllactinia* on winged elm in Georgia. Pl. Dis. Reporter 53: 395-397.
1442. Hanlin, R.T. 1982. Conidiogenesis in *Spiniger meineckellus*. Mycologia 74: 236-241.
1443. Hanlin, R.T. 1984. *Ustilago sieglingiae* on purple sandgrass in Georgia. Mycologia 76: 1118-1120.
1444. Hanlin, R.T. 1987. *Venturia liriodendri* sp. nov., associated with a leafspot disease of *Liriodendron*. Mycologia 79: 464-467.
1445. Hanlin, R.T. 1987. *Beniowskia* and its synonym *Clathrotrichum*. Mycotaxon 28: 219-231.
1446. Hanlin, R.T., L.L. Foudin, Y. Berisford, S.U. Glover, J.P. Jones, and L.H. Huang. 1978. Plant disease index for maize in the United States, Part I: Host index. Agric. Exp. Sta., University of Georgia, Research Report 277: 1-62.
1447. Hanlin, R.T., and F.A. Pokorny. 1985. First report of *Taphrina* leaf blister of willow oak in Georgia. Pl. Dis. 69: 83.
1448. Hannon, C.I., and G.F. Weber. 1955. A leaf spot of tomato caused by *Stemphylium floridanum* sp. nov. Phytopathology 45: 11-16.
1449. Hansbrough, J.R. 1934. Occurrence and parasitism of *Aleurodiscus amorphus* in North America. J. Forest. (Washington) 32: 452-458.
1450. Hansen, E.M. 1978. Incidence of *Verticicladiella wagenerii* and *Phellinus weirii* in Douglas-fir adjacent to and away from roads in western Oregon. Pl. Dis. Reporter 62: 179-181.
1451. Hansen, H.N., and R.V. Bega. 1955. *Botrytis* rot of *Caladium* tubers. Pl. Dis. Reporter 39: 283.
1452. Hansford, C.G. 1961. The Meliolinae, a monograph. Beih. Sydowia 2: 1-806.
1453. Hansford, C.G. 1962. The Meliolineae supplement. Sydowia 16: 302-323.
1454. Hanson, E.W. 1953. Relative prevalence and severity of the diseases of forage legumes in Wisconsin, 1946-1952. Pl. Dis. Reporter 37: 467-472.
1455. Hanson, E.W. 1963. *Fusarium* head blight of sorghum and head smut of sudan grass in Wisconsin in 1962. Pl. Dis. Reporter 47: 232.
1456. Hanson, E.W. 1963. Violet root rot of alfalfa in Wisconsin. Pl. Dis. Reporter 47: 306-307.
1457. Harding, P.R. Jr. 1968. *Candida krusei* as a primary pathogen in lemons. Pl. Dis. Reporter 52: 433-435.
1458. Hardison, J.H. 1957. Record of blind seed disease control in Oregon. Pl. Dis. Reporter 41: 34-41.
1459. Hardison, J.R. 1954. New grass host records and life history observations of dwarf bunt in eastern Oregon. Pl. Dis. Reporter 38: 345-347.

1460. Hardison, J.R. 1955. Crown wart of *Lotus uliginosus* found in Oregon. Pl. Dis. Reporter 39: 749-750.
1461. Hardison, J.R. 1962. Susceptibility of Gramineae to *Gloeotinia temulenta*. Mycologia 54: 201-216.
1462. Hardison, J.R. 1963. Incidence and control of *Tilletia controversa* in perennial grasses. Phytopathology 53: 579-585.
1463. Hardison, J.R., and M.E. Corden. 1952. A serious outbreak of dwarf bunt in tall oat grass in Oregon. Pl. Dis. Reporter 36: 343-344.
1464. Hardison, J.R., H.S. Fenwick, and J.P. Meiners. 1955. Additional grass hosts and a revised host list for dwarf bunt. Pl. Dis. Reporter 39: 685-687.
1465. Hardison, J.R., and J.P. Meiners. 1956. New host records for dwarf bunt in the Pacific Northwest. Pl. Dis. Reporter 40: 1058-1060.
1466. Hare, W.W. 1949. Tip blight of garden pea. J. Agric. Res. 78: 311-324.
1467. Hare, W.W. 1954. A *Cercospora* on winter pea. Pl. Dis. Reporter 38: 781-782.
1468. Harkness, H.W. 1884. New species of Californian fungi. Bull. Calif. Acad. Sci. 1: 29-47.
1469. Harrington, T.C. 1980. Release of airborne basidiospores from the pouch fungus, *Cryptoporus volvatus*. Mycologia 72: 926-936.
1470. Harrington, T.C., and F.W. Cobb Jr. 1984. *Verticillium albo-atrum* on *Ceonothus* in a California forest. Pl. Dis. 68: 1012.
1471. Harrington, T.C., and F.W. Cobb Jr. 1986. Varieties of *Verticicladiella wageneri*. Mycologia 78: 562-567.
1472. Harrington, T.C., and F.W. Cobb Jr. 1987. *Leptographium wageneri* var. *pseudotsugae*, var. nov., cause of black stain root disease on douglas-fir. Mycotaxon 30: 501-507.
1473. Harrington, T.C., C. Reinhart, D.A. Thornburgh, and F.W. Cobb Jr. 1983. Association of black-stain root disease with precommercial thinning of Douglas-fir. Forest Sci. 29: 12-14.
1474. Harris, H.B., C.D. Fisher, and G. Sowell Jr. 1971. Head smut of sorghum in Georgia. Pl. Dis. Reporter 55: 312-313.
1475. Harris, H.B., U.R. Gore, and E.S. Luttrell. 1952. Take-all of wheat in Georgia. Pl. Dis. Reporter 36: 391.
1476. Harrison, A.L. 1952. Plant disease notes from Texas. Pl. Dis. Reporter 36: 491.
1477. Harrison, A.L. 1967. Some observations on peanut leaf rust and *Cercospora* leaf spots in Texas. Pl. Dis. Reporter 51: 687-689.
1478. Hart, J.H., D.G. Mosher, and R. Ajewole. 1978. Occurrence of *Endocronartium harknessii* and *Cronartium quercuum* on scots and jack pine in Michigan's lower peninsula. Pl. Dis. Reporter 62: 779-782.
1479. Hart, J.H., W.E. Wallner, M.R. Caris, and G.K. Dennis. 1967. Increase in Dutch elm disease associated with summer trimming. Pl. Dis. Reporter 51: 476-479.
1480. Hart, L.P., D.W. Fulbright, J.L. Clayton, and A.V. Ravenscroft. 1984. Occurrence of *Septoria nodorum* blotch and *S. tritici* blotch in Michigan winter wheat. Pl. Dis. 68: 418-420.
1481. Harter, L.L. 1911. A new species of *Alternaria*. Mycologia 3: 154-155.
1482. Hartman, G.L., J.B. Manandhar, and J.B. Sinclair. 1986. Incidence of *Colletotrichum* spp. on soybeans and weeds in Illinois and pathogenicity of *Colletotrichum truncatum*. Pl. Dis. 70: 780-782.
1483. Hartmann, G.C. 1961. *Alternaria* leaf spot of *Pieris japonica*. Pl. Dis. Reporter 45: 938-939.
1484. Hartowicz, L.E., H. Knutson, A. Paulsen, B.J. Eaton, and E. Eshbaugh. 1971. Possible biological control of wild hemp. North Central Weed Control Conf. Proc. 26: 69.
1485. Harvey, G.M. 1976. Epiphytology of a needle cast fungus, *Lophodermella morbida* in Ponderosa pine plantations in western Oregon. Forest Sci. 22: 223-230.
1486. Harvey, J.M. 1951. An anthracnose disease of *Umbellularia californica*. Madroño 11: 162-171.
1487. Harvey, J.M. 1955. A method of forecasting decay in California storage grapes. Phytopathology 45: 229-232.
1488. Hasija, S.K., and F.T. Wolf. 1969. A new species of *Cintractia* on *Fuirena* (Cyperaceae). Mycologia 61: 680-682.
1489. Hassan, S.F. 1956. Pathogenicity of root rotting fungi of oats. Pl. Dis. Reporter 40: 890-897.
1490. Haugen, L.G., R.D. Wilcoxson, and J.S. Baumer. 1985. Effects of fungicides and cultivar genotypes on populations of *Septoria* spp. on spring wheat in Minnesota. Pl. Dis. 69: 162-163.
1491. Havens, O. 1984. Phytosanitary advisory. Plant pests of phytosanitary significance to importing countries and states. Fourth Edition. Department of food and agriculture California. 122 p.
1492. Hawk, J.A., R.B. Carroll, and F.M. Latterell. 1985. First report of gray leaf spot (*Cercospora zeae-maydis*) of corn in Delaware. Pl. Dis. 69: 361.
1493. Hawksworth, D.L. 1979. Ascospore sculpturing and generic concepts in the Testudinaceae (syn. Zopfiaceae). Canad. J. Bot. 57: 91-99.
1494. Hawksworth, D.L., and I.A.S. Gibson. 1976. *Phialophora parasitica* associated with disease conditions in various trees. Trans. Brit. Mycol. Soc. 66: 427-431.
1495. Hawksworth, D.L., and M.A. Sherwood. 1981. A reassessment of three widespread resinicolous Discomycetes. Canad. J. Bot. 59: 357-372.
1496. Hawksworth, F.G. 1953. The pycnial stage of *Cronartium filamentosum*. Pl. Dis. Reporter 37: 246.
1497. Hawksworth, F.G. 1953. *Phoradendron juniperinum*, a new host for *Uredo phoradendri*. Pl. Dis. Reporter 37: 258.
1498. Hawksworth, F.G., and T.E. Hinds. 1960. *Cytospora* canker of Engelmann spruce in Colorado. Pl. Dis. Reporter 44: 72.
1499. Hawksworth, F.G., and J.L. Mielke. 1962. Witches' broom of gambel oak associated with *Articularia quercina* var. *minor*. Phytopathology 52: 451-454.
1500. Hawksworth, F.G., and J.M. Staley. 1968. *Rhizosphaera kalkhoffii* on spruce in Arizona. Pl. Dis. Reporter 52: 804-805.
1501. Hawksworth, F.G., and J.M. Staley. 1978. *Rhizosphaera kalkhoffii* on spruce brooms. Pl. Dis. Reporter 62: 446.
1502. Haygood, R.A., C.H. Graves, and W.H. Ridings. 1986. *Phytophthora* root rot and stem canker of peach trees in Mississippi. Pl. Dis. 70: 866-868.
1503. Headland, J.K., G.J. Griffin, R.J. Stipes, and J.R. Elkins. 1976. Severity of natural *Endothia parasitica* infection of Chinese chestnut. Pl. Dis. Reporter 60: 426-429.
1504. Heagle, A.S., and D.W. French. 1972. *Peridoxylon petersii* on oak in Minnesota. Mycologia 64: 1349-1351.
1505. Heald, F.D., and F.A. Wolf. 1911. New species of Texas fungi. Mycologia 3: 5-22.
1506. Hebert, T.T., and C.N. Clayton. 1963. Pomegranate fruit rot caused by *Coniella granati*. Pl. Dis. Reporter 47: 222-223.
1507. Hecht-Poinar, E.I., J.C. Britton, and J.R. Parmeter Jr. 1981. Dieback of oaks in California. Pl. Dis. 65: 281.
1508. Hecht-Poinar, E.I., and J.R. Parmeter. 1986. *Cryptocline cinerescens* and *Discula quercina* causing twig blight of oaks in California. Pl. Dis. 70: 800.
1509. Hedgcock, G.G. 1929. The large leaf spot of chestnut and oak associated with *Monochaetia desmazierii*. Mycologia 21: 324-325.
1510. Heggestad, H.E., and E.E. Clayton. 1950. Occurrence of black shank and other diseases on burley tobacco in east Tennessee, in 1949. Pl. Dis. Reporter 34: 23-24.
1511. Heimann, M.F., and G.L. Worf. 1987. Basal stem canker of poinsettia caused by *Fusarium solani* in Wisconsin. Pl. Dis. 71: 469.
1512. Heinis, J.L. 1960. *Taphrina pruni-subcordatae* (Zeller) Mix, causing witches'-broom on Italian and French prunes. Pl. Dis. Reporter 44: 630-631.
1513. Helser, L.R. 1945 . Notes on southern Appalachian fungi. VI. J. Tennessee Acad. Sci. 20: 233-238.
1514. Helton, A.W., and R. Dilbeck. 1985. A *Fusarium* root rot problem in Idaho apple orchards. Pl. Dis. 69: 1101.
1515. Henderson, D.M. 1961. *Glomospora* and *Glomopsis*. Notes Roy. Bot. Gard. Edinburgh 23: 497-502.
1516. Hendrix, F.F., and W.M. Powell. 1968. Nematode and *Pythium* species associated with feeder root necrosis of pecan trees in Georgia. Pl. Dis. Reporter 52: 334-335.
1517. Hendrix, F.F., Jr., and W.A. Campbell. 1971. A new species of *Pythium* with spiny oogonia and large chlamydospores. Mycologia 63: 978-982.
1518. Hennebert, G.L., and J.W. Groves. 1963. Three new species of *Botryotinia* on Ranunculaceae. Canad. J. Bot. 41: 341-370.
1519. Hennen, J.F. 1965. The species of *Uromyces* parasitic on the grass tribe Andropogoneae. Mycologia 57: 104-113.
1520. Hennen, J.F. 1967. New records, taxa, and notes on high altitude, west American rust fungi. Mycologia 59: 264-273.

1521. Hennen, J.F., and G.B. Cummins. 1956. Uredinales parasitizing grasses of the tribe Chlorideae. Mycologia 48: 126-162.
1522. Hennen, J.F., and C.S. Hodges Jr. 1981. Hawaiian forest fungi. II. Species of *Puccinia* and *Uredo* on *Euphorbia*. Mycologia 73: 1116-1122.
1523. Hennen, J.F., and M. Komanetsky. 1959. Stripe rust in South Dakota in 1958. Pl. Dis. Reporter 43: 168-169.
1524. Hennon, P.E., and E.M. Hansen. 1987. Nuclear behavior of *Phellinus arctostaphyli*, *P. igniarius*, and *P. tremulae*. Mycologia 79: 501-507.
1525. Henry, B.W. 1951. Oak leaf blister severe in south Mississippi. Pl. Dis. Reporter 35: 384.
1526. Hepperly, P.R., B.L. Kirkpatrick, and J.B. Sinclair. 1980. *Abutilon theophrasti*: wild host for three fungal parasites of soybean. Phytopathology 70: 307-310.
1527. Hepting, G.H. 1944. Sapstreak, a new killing disease of sugar maple. Phytopathology 34: 1069-1076.
1528. Hepting, G.H. 1950. Leaf diseases of dogwood severe this year in Appalachians. Pl. Dis. Reporter 34: 227.
1529. Hepting, G.H. 1951. A new stem rust on Virginia pine. Pl. Dis. Reporter 35: 335.
1530. Hepting, G.H. 1957. A rust on Virginia pine and *Buckleya*. Mycologia 49: 896-899.
1531. Hepting, G.H., and G.B. Cummins. 1952. A new species of *Peridermium* on Virginia pine. Phytopathology 42: 115-116.
1532. Hepting, G.H., and R.W. Davidson. 1937. A leaf and twig disease of hemlock caused by a new species of *Rosellinia*. Phytopathology 27: 305-310.
1533. Hepting, G.H., and E.R. Toole. 1950. Some southeastern tree diseases - 1948 and 1949. Pl. Dis. Reporter 34: 135-137.
1534. Hepton, A., and E.J. Anderson. 1968. Interfruitlet corking of pineapple fruit, a new disease in Hawaii. Phytopathology 58: 74-78.
1535. Hermanides-Nijhof, E.J. 1977. *Aureobasidium* and allied genera. Stud. Mycol. 15: 141-177.
1536. Herr, L.J., and P.E. Lipps. 1982. *Alternaria helianthi* on sunflower in Ohio. Pl. Dis. 66: 509-512.
1537. Herr, L.J., P.E. Lipps, and B.L. Watters. 1983. *Diaporthe* stem canker of sunflower. Pl. Dis. 67: 911-913.
1538. Hesler, L.R. 1945. Notes on southern Appalachian fungi. V. J. Tennessee Acad. Sci. 20: 233-238.
1539. Hesler, L.R. 1951. Notes on southern Appalachian fungi, IX. J. Tennessee Acad. Sci. 26: 4-14.
1540. Hesler, L.R. 1954. Notes on southern Appalachian fungi. XI. J. Tennessee Acad. Sci. 29: 205-219.
1541. Hesler, L.R. 1957. Notes on southeastern Agaricales. I. J. Tennessee Acad. Sci. 32: 198-207.
1542. Hesler, L.R. 1968. North American species of *Gymnopilus*. Mycologia Memoir 3: 1-117.
1543. Hesler, L.R. 1975. New species of *Crepidotus*. Beih. Nova Hedwigia 51: 133-138.
1544. Hesler, L.R., and A.H. Smith. 1965. North American species of *Crepidotus*. New York: Hafner Publ. Co.
1545. Hess, C.E., and D.S. Welch. 1954. *Chalaropsis thielavioides* Peyronel found on evergreen grafting stock. Pl. Dis. Reporter 38: 415-416.
1546. Hessburg, P.F., and E.M. Hansen. 1986. Mechanisms of inter tree transmission of *Ceratocystis wageneri* in young Douglas-fir. Canad. J. Forest Res. 16: 1250-1254.
1547. Hesseltine, C.W., R.J. Bothast, and O.L. Shotwell. 1975. Aflatoxin occurrence in some white corn under loan, 1971: IV. Mold flora. Mycologia 67: 392-408.
1548. Hewitt, W.B. 1951. Grape dead-arm control. Pl. Dis. Reporter 35: 142-143.
1549. Heye, C.C., and J.H. Andrews. 1983. Antagonism of *Athelia bombacina* and *Chaetomium globosum* to the apple scab pathogen, *Venturia inaequalis*. Phytopathology 73: 650-654.
1550. Hibben, C.R., and M.L. Daughtrey. 1988. Dogwood anthracnose in Northeastern United States. Pl. Dis. 72: 199-203.
1551. Higgins, B.B. 1917. A disease of pecan catkins. Phytopathology 7: 42-45.
1552. Highberg, L.M., and J.M. Ogawa. 1986. Yield reduction in almond related to incidence of shot-hole disease. Pl. Dis. 70: 825-828.
1553. Highley, L., and T.A. Tattar. 1985. *Leptographium terebrantis* and black turpentine beetles associated with blue stain and mortality of black and scots pines on Cape Cod, Massachusetts. Pl. Dis. 69: 528-530.
1554. Hilber, O., R. Hilber, and O.K. Miller Jr. 1987. Fungi of the Appalachian Mountains. Mycotaxon 30: 269-288.
1555. Hildebrand, E.M. 1954. Jasmine gall produced by a new species of *Phoma*. Pl. Dis. Reporter 38: 564-565.
1556. Hildebrand, E.M., and C. Drechsler. 1963. An unusual root injury on Jersey-type sweetpotatoes in 1962. Pl. Dis. Reporter 47: 562-563.
1557. Hill, J.P., J.A. Fernandez, and M.S. McShane. 1983. Fungi associated with common root rot of winter wheat in Colorado and Wyoming. Pl. Dis. 67: 795-797.
1558. Himelick, E.B. 1969. Tree and shrub hosts of *Verticillium albo-atrum*. Illinois Nat. Hist. Survey, Biol. Notes 66: 1-8.
1559. Himelick, E.B., and D. Neely. 1960. Juniper hosts of cedar-apple and cedar-hawthorn rust. Pl. Dis. Reporter 44: 109-112.
1560. Hinds, T.E. 1964. Distribution of aspen cankers in Colorado. Pl. Dis. Reporter 48: 610-614.
1561. Hinds, T.E. 1981. *Cryptosphaeria* canker and *Libertella* decay of aspen. Phytopathology 71: 1137-1145.
1562. Hinds, T.E., and G.W. Anderson. 1970. Some *Ceratocystis* spp. and a *Cenangium* found on Minnesota aspen. Pl. Dis. Reporter 54: 460-461.
1563. Hinds, T.E., and R.W. Davidson. 1967. A new species of *Ceratocystis* on aspen. Mycologia 59: 1102-1106.
1564. Hinds, T.E., and R.W. Davidson. 1972. *Ceratocystis* species associated with the aspen ambrosia beetle. Mycologia 64: 405-409.
1565. Hinds, T.E., and R.W. Davidson. 1975. Two new species of *Ceratocystis*. Mycologia 67: 715-721.
1566. Hinds, T.E., R.W. Davidson, and F.F. Lombard. 1968. Decay of Engelmann spruce by *Lentinellus montanus* in Colorado. Pl. Dis. Reporter 52: 820-821.
1567. Hinds, T.E., and T.H. Laurent. 1978. Common aspen diseases found in Alaska. Pl. Dis. Reporter 62: 972-975.
1568. Hinds, T.E., and M.G. Ryan. 1985. Expansion of sooty-bark and *Ceratocystis* cankers on aspen. Pl. Dis. 69: 842-844.
1569. Hinds, T.E., W.D. Shepperd, and T.H. Laurent. 1985. *Dothiora polyspora* on aspen. Pl. Dis. 69: 1100.
1570. Hine, R.B. 1961. The role of fungi in the peach replant problem. Pl. Dis. Reporter 45: 462-465.
1571. Hine, R.B. 1961. Trunk canker of macadamia in Hawaii caused by *Phytophthora cinnamomi* Rands. Pl. Dis. Reporter 45: 868.
1572. Hine, R.B., F.A. Gray, and M.H. Schonhorst. 1972. *Phytophthora* root rot of alfalfa in Arizona. Pl. Dis. Reporter 56: 472-473.
1573. Hine, R.B., and A.H. McCain. 1984. *Verticillium* wilt of *Tagetes* spp. Pl. Dis. 68: 1098.
1574. Hine, R.B., and J.E. Wheeler. 1970. The occurrence of some previously unreported diseases in Arizona. Pl. Dis. Reporter 54: 179-180.
1575. Hiratsuka, N. 1958. Revision of taxonomy of the Pucciniastreae. Tokyo, Japan 167 p.
1576. Hirrel, M.C., and T.L. Kirkpatrick. 1986. First report of stem canker on soybeans in Arkansas. Pl. Dis. 70: 78.
1577. Hjortstam, K., and L. Ryvarden. 1981. Studies in tropical Corticiaceae (Basidiomycetes) III. Two new species of *Laxitextum*. Mycotaxon 13: 35-40.
1578. Ho, H.H. 1986. Notes on the heterothallic behavior of *Phytophthora megasperma* from alfalfa. Mycologia 78: 306-309.
1579. Ho, H.H. 1986. *Pythium dimorphum* from *Rhododendron*. Mycopathologia 93: 141-145.
1580. Ho, H.H., and S.C. Jong. 1986. A comparison between *Phytophthora cryptogea* and *P. drechsleri*. Mycotaxon 27: 289-319.
1581. Ho, W.-C., and I.E. Melhus. 1938. *Pythium graminicolum* on barley in Iowa. Phytopathology 28: 9.
1582. Hobbs, E.L. 1963. *Stemphylium* petal blight of carnations. Pl. Dis. Reporter 47: 917-919.
1583. Hobbs, S.D., and A.D. Partridge. 1979. Wood decays, root rots, and stand composition along an elevation gradient. Forest Sci. 25: 31-42.
1584. Hobbs, T.W., A.F. Schmitthenner, C.W. Ellett, and R.E. Hite. 1981. Top dieback of soybean caused by *Diaporthe phaseolorum* var. *cauliv ora*. Pl. Dis. 65: 618-620.
1585. Hobbs, T.W., A.F. Schmitthenner, and G.A. Kuter. 1985. A new *Phomopsis* species from soybean. Mycologia 77: 535-544.
1586. Hodges, C.F. 1969. Additional varieties of *Agrostis palustris*

infected by *Ustilago striiformis* var. *agrostidis*. Pl. Dis. Reporter 53: 298-299.
1587. Hodges, C.F., and W.M. Blaine. 1974. Midsummer occurrence of *Helminthosporium sorokinianum* leaf spot on selected cultivars of *Poa pratensis*. Pl. Dis. Reporter 58: 448-450.
1588. Hodges, C.F., W.M. Blaine, and P.W. Robinson. 1975. Severity of *Sclerotinia homoeocarpa* blight on various cultivars of fine-leaved fescues. Pl. Dis. Reporter 59: 12-14.
1589. Hodges, C.F., and L.W. Coleman. 1985. *Pythium*-induced root dysfunction of secondary roots of *Agrostis palustris*. Pl. Dis. 69: 336-340.
1590. Hodges, C.S. 1961. New hosts for *Cercospora thujina* Plakidas. Pl. Dis. Reporter 45: 745.
1591. Hodges, C.S. 1962. Comparison of four similar fungi from *Juniperus* and related conifers. Mycologia 54: 62-69.
1592. Hodges, C.S. 1981. Hawaiian forest fungi. I. A new species of *Phyllachora* on the tree fern *Cibotium glaucum*. Mycologia 73: 748-754.
1593. Hodges, C.S. 1982. Needle blight of Port-Orford-cedar caused by *Stigmina thujina* in Hawaii. Pl. Dis. 66: 867-869.
1594. Hodges, C.S. 1983. Pine mortality in Hawaii associated with *Botryosphaeria dothidea*. Pl. Dis. 67: 555-556.
1595. Hodges, C.S. Jr. 1969. A new species of *Phyllachora* on *Persea*. Mycologia 61: 838-840.
1596. Hodges, C.S. Jr. 1984. Hawaiian forest fungi. IV. Rusts on endemic *Acacia* species. Mycologia 76: 332-349.
1597. Hodges, C.S. Jr. 1985. Hawaiian forest fungi VI. A new species of *Brasiliomyces* on *Sapindus oahuensis*. Mycologia 77: 977-981.
1598. Hodges, C.S., and M.E. Barr. 1971. A new species of *Pseudomassaria* and its *Beltraniella* conidial state. Mycologia 63: 562-566.
1599. Hodges, C.S., T.F. Geary, and C.E. Cordell. 1979. The occurrence of *Diaporthe cubensis* on *Eucalyptus* in Florida, Hawaii, and Puerto Rico. Pl. Dis. Reporter 63: 216-220.
1600. Hodges, C.S., and F.A. Haasis. 1962. Etiology of *Cercospora* leaf spot of *Magnolia*. Mycologia 54: 448-454.
1601. Hoes, J.A., G.W. Bruehl, and C.G. Shaw. 1965. A new *Pseudostemphylium*. Mycologia 57: 904-912.
1602. Hofmeister, F.M., and R. Charudattan. 1987. *Pseudocercospora nigricans*, a pathogen of sicklepod (*Cassia obtusifolia*) with biocontrol potential. Pl. Dis. 71: 44-46.
1603. Holcomb, G.E. 1976. Daylily leaf-streak in Louisiana. Pl. Dis. Reporter 60: 232-233.
1604. Holcomb, G.E. 1976. Powdery mildew, a new disease of *Ligustrum japonicum*. Pl. Dis. Reporter 60: 346-347.
1605. Holcomb, G.E., and A.A. Antonopoulos. 1976. *Alternaria alternantherae*: A new species found on alligatorweed. Mycologia 68: 1125-1129.
1606. Holcomb, G.E., K.S. Derrick, R.B. Carver, and R.W. Toler. 1973. Downy mildew and rust found on St. Augustinegrass in Louisiana. Pl. Dis. Reporter 57: 16.
1607. Holcomb, G.E., and J.P. Jones. 1984. Hosts of the leaf spot fungus *Tubakia dryina* in Louisiana. Pl. Dis. 68: 251.
1608. Holcomb, G.E., and J.P. Jones. 1986. Occurrence of *Myriogenospora atramentosa* on centipedegrass in Louisiana. Pl. Dis. 70: 259.
1609. Holcomb, G.E., and B.E. Newman. 1970. *Alternaria circinans* and other fungal pathogens on *Crambeabyssinica* in Louisiana. Pl. Dis. Reporter 54: 28.
1610. Holdeman, Q.L., and T.W. Graham. 1952. *Ascochyta* leaf spot in tobacco plant beds in South Carolina. Pl. Dis. Reporter 36: 8.
1611. Holley, R.C., and B.D. Nelson. 1986. Effect of plant population and inoculum density on incidence of *Sclerotinia* wilt of sunflower. Phytopathology 76: 71-74.
1612. Hollier, C.A., and S.B. King. 1985. Effects of temperature and relative humidity on germinability and infectivity of *Puccinia polysora* uredospores. Pl. Dis. 69: 937-939.
1613. Hollis, J.P. 1960. Occurrence of hadromycosis in Louisiana cotton. Pl. Dis. Reporter 44: 840.
1614. Holmes, F.W. 1957. Recorded Dutch elm disease distribution in North America as of 1956. Pl. Dis. Reporter 41: 634-635.
1615. Holmes, F.W. 1958. Recorded Dutch elm disease distribution in North America as of 1957. Pl. Dis. Reporter 42: 1299-1300.
1616. Holmes, F.W. 1961. Recorded Dutch elm disease distribution in North America as of 1959. Pl. Dis. Reporter 45: 74-75.
1617. Holton, C.S. 1953. Observations and experiments on snow mold of winter wheat in Washingt on state. Pl. Dis. Reporter 37: 354-359.
1618. Holtzmann, O.V. 1963. Raceme blight of macadamia in Hawaii. Pl. Dis. Reporter 47: 416-417.
1619. Holtzmann, O.V., and M. Aragaki. 1966. Susceptibility of acerola to *Cercospora* leaf spot. Phytopathology 56: 1114-1115.
1620. Honda, Y., and M. Aragaki. 1983. Light-dependence for fruiting body formation and its inheritance in *Phoma caricae-papayae*. Mycologia 75: 22-29.
1621. Honey, E.E. 1942. *Monilinia amelanchieris*. Mycologia 34: 575-578.
1622. Hoog, G.S., de, and G. Morgan-Jones. 1978. Notes on Hyphomycetes. XXIII. *Paraphaeoisaria alabamensis* gen. et sp. nov. Mycotaxon 7: 133-138.
1623. Hooker, A.L. 1961. Occurrence of *Puccinia polysora* in Illinois. Pl. Dis. Reporter 45: 236.
1624. Hooker, A.L., A. Mesterhazy, D.R. Smith, and S.M. Lim. 1973. A new *Helminthosporium* leaf blight of corn in the northern corn belt. Pl. Dis. Reporter 57: 195-198.
1625. Hooker, A.L., and D.G. White. 1976. Prevalence of corn stalk rot fungi in Illinois. Pl. Dis. Reporter 60: 1032-1034.
1626. Hoppe, P.E. 1953. Northern corn leaf blight in Wisconsin in 1952. Pl. Dis. Reporter 37: 94.
1627. Horn, B.W. 1984. Fertile stromata of *Camarops petersii* in culture. Mycologia 76: 956-959.
1628. Horn, N.L. 1952. *Diplodina* on peach in Louisiana. Pl. Dis. Reporter 36: 351.
1629. Horn, N.L., and P.L. Hawthorne. 1954. A new fruit rot of peach. Phytopathology 44: 134-136.
1630. Horn, N.L., J.B. Sinclair, and R.T. Brown. 1958. The occurrence of *Oospora* rot of citrus in Louisiana. Pl. Dis. Reporter 42: 264-265.
1631. Horn, N.L., W.F. Wilson, and M. Giamalva. 1957. Seed and insect transmission of cucumber anthracnose. Pl. Dis. Reporter 41: 69-71.
1632. Horne, C.W., and R.S. Halliwell. 1964. Oak wilt in Texas. Pl. Dis. Reporter 48: 419.
1633. Horner, C.E. 1952. Diseases of peppermint in Oregon in 1951. Pl. Dis. Reporter 36: 245.
1634. Horner, C.E. 1971. Rhizome and stem rot of peppermint caused by *Phoma strasseri*. Pl. Dis. Reporter 55: 814-816.
1635. Horner, W.E., S.A. Alexander, and K.J. Lewis. 1987. Colonization patterns of *Verticicladiella procera* in Scots and eastern white pine and associated resin-soaking, reduced sapwood moisture content, and reduced needle water potential. Phytopathology 77: 557-560.
1636. Horst, R.K. 1983. Compendium of rose diseases. The American Phytopathological Society 50 p.
1637. Horst, R.K., and H.A.J. Hoitink. 1968. Occurrence of *Cylindrocladium* blights on nursery crops and control with fungicide 1991 on *Azalea*. Pl. Dis. Reporter 52: 615-617.
1638. Hosford, R.M. Jr. 1961. *Aphanomyces raphani* Kendrick in radishes in Oregon. Pl. Dis. Reporter 45: 482.
1639. Hosford, R.M. Jr. 1971. *Platyspora pentamera* in the Great Plains on wheat. Mycologia 63: 668-669.
1640. Hosford, R.M. Jr. 1978. Effects of wetting period on resistance to leaf spotting of wheat, barley, and rye by *Leptosphaeria herpotrichoides*. Phytopathology 68: 591-594.
1641. Hotchkiss, E.S., and L.W. Baxter Jr. 1983. Pathogenicity of *Alternaria tagetica* on *Tagetes*. Pl. Dis. 67: 1288-1290.
1642. Hotson, J.W. 1927. A new species of *Exobasidium*. Phytopathology 17: 207-216.
1643. Houston, B.R., and P.F. Knowles. 1953. Studies on *Fusarium* wilt of flax. Phytopathology 43: 491-495.
1644. Houston, D.R. 1969. Basal canker of white pine. Forest Sci. 15: 66-83.
1645. Houston, D.R. 1971. Discoloration and decay in red maple and yellow birch: Reduction through wound treatment. Forest Sci. 17: 402-406.
1646. Houston, D.R., and K.D. Fisher. 1964. Sapstreak disease of sugar maple found in the northeast. Pl. Dis. Reporter 48: 788.
1647. Howard, C.M. 1972. A strawberry fruit rot caused by *Colletotrichum fragariae*. Phytopathology 62: 600-602.
1648. Howard, C.M. 1972. Occurrence of strawberry leaf blotch,

Gnomonia fructicola, in Florida. Pl. Dis. Reporter 56: 998-999.
1649. Howard, C.M., and E.E. Albregts. 1973. A strawberry fruit rot caused by *Dendrophoma obscurans*. Phytopathology 63: 419-421.
1650. Howard, C.M., and E.E. Albregts. 1973. *Cassia obtusifolia*, a possible reservoir for inoculum of *Colletotr ichum fragariae*. Phytopathology 63: 533-534.
1651. Howard, C.M., and E.E. Albregts. 1973. A strawberry fruit rot caused by *Pestalotia longisetula*. Phytopathology 63: 862-863.
1652. Howard, C.M., and E.E. Albregts. 1973. A strawberry fruit rot caused by *Alternaria tenuissima*. Phytopathology 63: 938-939.
1653. Howard, C.M., and E.E. Albregts. 1982. Cleistothecia of *Sphaerotheca macularis* on strawberry plants in Florida. Pl. Dis. 66: 261-262.
1654. Howard, C.M., and E.E. Albregts. 1982. Outbreak of *Verticillium* wilt of strawberries in Central Florida. Pl. Dis. 66: 856-857.
1655. Howard, C.M., and E.E. Albregts. 1983. Black leaf spot phase of strawberry anthracnose caused by *Colletotri chum gloeosporioides* (=*C. fragariae*). Pl. Dis. 67: 1144-1146.
1656. Howard, C.M., and E.E. Albregts. 1984. Anthracnose of strawberry fruit caused by *Glomerella cingulata* in Florida. Pl. Dis. 68: 824-825.
1657. Hoy, J.W., C.A. Hollier, D.B. Fontenot, and L.B. Grelen. 1986. Incidence of sugarcane smut in Louisiana and its effect on yield. Pl. Dis. 70: 59-60.
1658. Hsi, C.H. 1954. Dryland root rot complex of winter wheat in eastern New Mexico. Pl. Dis. Reporter 38: 270-274.
1659. Hsi, C.H. 1959. New records of plant diseases in New Mexico. Pl. Dis. Reporter 43: 595.
1660. Hsi, C.H. 1956. Stalk rots of sorghum in eastern New Mexico. Pl. Dis. Reporter 40: 369-371.
1661. Huang, L.H., and R.T. Hanlin. 1975. Fungi occurring in freshly harvested and in-market pecans. Mycologia 67: 689-700.
1662. Huber, D.M. 1981. Incidence and severity of take-all of wheat in Indiana. Pl. Dis. 65: 734-737.
1663. Huber, D.M., and A.M. Finley. 1959. *Gliocladium*, a causal agent in the bean root rot complex in Idaho. Pl. Dis. Reporter 43: 626-628.
1664. Huber, D.M., T.S. Lee, M.A. Ross, and T.S. Abney. 1987. Amelioration of tan spot-infected wheat with nitrogen. Pl. Dis. 71: 49-50.
1665. Hubert, E.E. 1953. Studies of *Leptographium* isolated from western white pine. Phytopathology 43: 637-641.
1666. Hubert, F.P., and W.H. Wheeler. 1950. Disease survey of domestic bulbs 1948-49. Pl. Dis. Reporter 34: 53-55.
1667. Hudler, G.W., M.T. Banik, and S.G. Miller. 1987. Unusual epidemic of tar spot on Norway maple in upstate New York. Pl. Dis. 71: 65-68.
1668. Hudler, G.W., and N. Oshima. 1976. The occurrence and distribution of *Thyronectria austro-americana* on honeylocust in Colorado. Pl. Dis. Reporter 60: 920-922.
1669. Huff, D.R., D. Zagory, and L. Wu. 1987. Report of buffalograss bunt (*Tilletia buchloeana*) in Oklahoma. Pl. Dis. 71: 651.
1670. Huftalen, C.S., and G.C. Bergstrom. 1986. First report of *Ascochyta* leaf spot caused by *Ascochyta hordei* var. *americana* on barley in New York. Pl. Dis. 70: 1074.
1671. Hughes, S.J. 1951. Studies on micro-fungi. III *Mastigosporium*, *Camposporium*, and *Ceratophorum*. Mycol. Pap. 2: 1-43.
1672. Hughes, S.J. 1960. Microfungi *V. conoplea* Pers. and *Exosporium* Link. Canad. J. Bot. 38: 659-696.
1673. Hughes, S.J. 1967. New Zealand Fungi 9. *Ophiocapnocoma* with *Hormiokrypsis* and *Capno phialophora* states. New Zealand J. Bot. 5: 117-133.
1674. Hughes, S.J. 1968. *Strigopodia*. Canad. J. Bot. 46: 1099-1107.
1675. Hughes, S.J. 1973. *Spadicoides obovata*. Fungi Canadenses 6: 1-2.
1676. Hughes, S.J. 1973. *Spadicoides grovei*. Fungi Canadenses 7: 1-2.
1677. Hughes, S.J. 1978. *Endophragmiella biseptata*. Fungi Canadenses 125: 1-2.
1678. Hughes, S.J. 1980. *Stigmina robusta*. Fungi Canadenses 167: 1-2.
1679. Hughes, S.J. 1980. *Taeniolella alta*. Fungi Canadenses 184: 1-2.
1680. Hughes, S.J. 1980. *Taeniolella rudis*. Fungi Canadenses 185: 1-2.
1681. Hughes, S.J. 1983. Five species of *Sarcinella* from North America, with notes on *Questieriella* n. gen., *Mitteriella*, *Endophragmiopsis*, *Schiffnerula*, and *Clypeolella*. Canad. J. Bot. 61: 1727-1767.
1682. Hughes, S.J. 1983. New Zealand fungi 32. *Janetia capnophila* sp. nov. and some allies. New Zealand J. Bot. 21: 177-182.
1683. Hughes, S.J., and W.A. Cavalcanti. 1983. *Janetia mangiferae* n.sp., and the *Johncouchia* n.gen. anamorph of *Septobasidium pilosum* from Brasil. Canad. J. Bot. 61: 2224-2230.
1684. Hughes, S.J., and K.E. Conway. 1978. *Botryosporium madrasense*. Canad. J. Bot. 56: 2405-2407.
1685. Hughes, S.J., and J.C. Cooke. 1979. *Pseudospiropes nodosus*. Fungi Canadenses 144: 1-2.
1686. Hughes, S.J., and W.I. Illman. 1974. *Sporidesmium hormiscioides*. Fungi Canadenses 58: 1-2.
1687. Hughes, S.J., and G.P. White. 1983. *Bactrodesmium betulicola*. Fungi Canadenses 251: 1-2.
1688. Huhtinen, S. 1987. Three new species, and the histochemical delimitation of genera in the glassy-haired Hyaloscyphaceae. Mycotaxon 29: 267-283.
1689. Humphrey, H.B., and A.G. Johnson. 1916. Observations on the occurrence of *Puccinia glumarum* in the United States. Phytopathology 6: 96-97.
1690. Hunst, P.L., F.M. Latterell, and A.E. Rossi. 1986. Variation in double-stranded RNA from isolates of *Pyricularia oryzae*. Phytopathology 76: 674-678.
1691. Hunt, J. 1956. Taxonomy of the genus *Ceratocystis*. Lloydia 19: 1-58.
1692. Hunt, J., and E. Wright. 1957. Needle-cast of Sitka spruce in Oregon. Pl. Dis. Reporter 41: 650.
1693. Hunt, R.S. 1969. *Pinus sabiniana* and *Pinus muricata* as hosts for *Diplodia pinea* in California. Pl. Dis. Reporter 53: 675-677.
1694. Hunter, B.B., and H.L. Barnett. 1978. Growth and sporulation of species and isolates of *Cylindrocladium* in culture. Mycologia 70: 614-635.
1695. Hunter, J.E., R.K. Kunimoto, and K.G. Rohrbach. 1971. *Phytophthora* blight, a new disease of *Macadamia*. Phytopathology 61: 1130-1133.
1696. Hurd, B., and M.P. Grisham. 1983. *Rhizoctonia* spp. associated with brown patch of Saint Augustinegras s. Phytopathology 73: 1661-1665.
1697. Hutchison, J.A., and J.F. Schafer. 1968. A *Pyrenophora* - *Cochliobolus* (*Helminthosporium*) disease complex in barley. Pl. Dis. Reporter 52: 395.
1698. Hwang, S.-W. 1955. Compatibility and variability in *Merulius americanus*. Mycologia 47: 317-328.
1699. Hyre, R. 1951. A new severe case of potato wart in Pennsylvania. Pl. Dis. Reporter 35: 432.
1700. Illman, W.I. 1964. *Gliocladium roseum* (Link) Bainier on declining and dead stems of *Psilotum nudum*. Pl. Dis. Reporter 48: 68.
1701. Inglis, D.A. 1980. Contamination of asparagus seed by *Fusarium oxysporium* f. sp. *asparagi* and *Fusarium moniliforme*. Pl. Dis. 64: 74-76.
1702. Inglis, D.A., and O.C. Maloy. 1983. Scab caused by *Gibberella zeae* occurring on irrigated wheat in east ern Washington. Pl. Dis. 67: 827-828.
1703. Inman, R.E., and J.L. Weihing. 1959. Cyprex: a superior control for shothole disease of chokecherry. Pl. Dis. Reporter 43: 536-539.
1704. Ito, J.S., and M. Aragaki. 1977. *Botrytis* blossom blight of *Dendrobium*. Phytopathology 67: 820-824.
1705. Ivanoff, S.S. 1963. The cause of spikelet drop of oats. Pl. Dis. Reporter 47: 206-207.
1706. Ivanoff, S.S., D.H. Bowman, and P.G. Rothman. 1958. Oat diseases in Mississippi. Pl. Dis. Reporter 42: 521-523.
1707. Jülich, W. 1972. Monographie der Athelieae (Corticiaceae, Basidiomycetes). Willdenowia 7: 1-283.
1708. Jackson, C.R. 1960. *Cercospora* blight of snapdragon. Phytopathology 50: 190-192.
1709. Jackson, C.R. 1960. Crown rots and *Botrytis* flower blight of statice. Pl. Dis. Reporter 44: 643-645.
1710. Jackson, C.R. 1961. *Cercospora* leafspot of statice. Phytopathology 51: 129-130.
1711. Jackson, C.R. 1961. A flower rot symptom associated with *Stemphylium* ray speck of *Chrysanthemum*. Pl. Dis. Reporter 45: 798.
1712. Jackson, C.R. 1962. *Penicillium* core rot of gladiolus. Phytopathology 52: 794-797.
1713. Jackson, C.R., and E.G. Simmons. 1968. Blight of *Richardia scabra* caused by *Alternaria multirostrata* sp. n. Phytopathology 58:

1139-1142.
1714. Jackson, H.S. 1948. Studies of Canadian Thelephoraceae I.Some new species of *Peniophora*. Canad. J. Res., Sect. C, Bot. Sci. 26: 128-139.
1715. Jackson, H.S. 1948. Studies of Canadian Thelephoraceae II. Some new species of *Corticium*. Canad. J. Res., Sect. C, Bot. Sci. 26: 143-157.
1716. Jackson, H.S., and E.R. Dearden. 1951. Studies of North American Thelephoraceae. I. Some new western species of *Peniophora*. Mycologia 43: 54-61.
1717. Jackson, L.F. 1981. Distribution and severity of peanut leafspot in Florida. Phytopathology 71: 324-328.
1718. Jackson, N., and J.T. Campanini. 1975. A dieback of *Euonymus alatus* caused by *Whetzelinia sclerotiorum*. Pl. Dis. Reporter 59: 310-311.
1719. Jackson, N., and J.M. Fenstermacher. 1973. A patch disease of two bentgrasses caused by *Drechslera gigantea*. Pl. Dis. Reporter 57: 84-85.
1720. Jackson, N, and P.J. Landschoot. 1986. *Gaeumannomyces cylindrosporus* associated with diseased turfgrass in Rhode Island. Phytopathology 76: 654.
1721. Jacobi, W.R. 1984. Optimal conditions for in vitro growth, asexual spore release, and germination of *Thyronectria austro-americana*. Phytopathology 74: 566-569.
1722. Jacobi, W.R., E.B. Cowling, N.D. Cost, and J.P. McClure. 1980. Disease losses in North Carolina forests: I. Losses in softwoods, 1973-1974. Pl. Dis. Reporter 64: 573-576.
1723. James, R.L. 1979. *Fomes annosus* on white fir in Colorado. Pl. Dis. Reporter 63: 129-130.
1724. James, R.L., and D.J. Goheen. 1981. Conifer mortality associated with root disease and insects in Colorado. Pl. Dis. 65: 506-507.
1725. Janson, B.F. 1952. *Fusarium* root rot and wilt of dill. Phytopathology 42: 152-155.
1726. Jarvis, W.R., and R.A. Shoemaker. 1978. Taxonomic status of *Fusarium oxysporum* causing foot and root rot of tomato. Phytopathology 68: 1679-1680.
1727. Jeffers, W.F. 1952. Downy mildew of snapdragon found in Maryland. Pl. Dis. Reporter 36: 211.
1728. Jeffers, W.F. 1952. Observations on occurrence of azalea petal blight in Maryland. Pl. Dis. Reporter 36: 301.
1729. Jeffers, W.F., C.E. Cox, R.A. Jehle, L.O. Weaver, O.D. Morgan, and J.E. Moore. 1951. Notes on occurrence of some plant diseases in Maryland during 1950. Pl. Dis. Reporter 35: 385-386.
1730. Jehle, R.A. 1951. Potato wart in Maryland. Pl. Dis. Reporter 35: 432.
1731. Jehle, R.A., and A.E. Jenkins. 1950. Supplemental survey for spot anthracnose of flowering dogwood in Mary land and Virginia. Pl. Dis. Reporter 34: 225-226.
1732. Jehle, R.A., and A.E. Jenkins. 1951. First observations in 1951 of spot anthracnose on flowering dogwood in Maryland. Pl. Dis. Reporter 35: 277.
1733. Jehle, R.A., and A.E. Jenkins. 1951. Two instances of wild violets recently infected by *Sphaceloma violae*. Pl. Dis. Reporter 35: 414.
1734. Jehle, R.A., and A.E. Jenkins. 1952. Northernmost records of flowering dogwood spot anthracnose in Maryland. Pl. Dis. Reporter 36: 110.
1735. Jehle, R.A., and A.E. Jenkins. 1952. Known distribution of flowering dogwood spot anthracnose in Maryland, May 1952. Pl. Dis. Reporter 36: 296-298.
1736. Jehle, R.A., A.E. Jenkins, and W.F. Jeffers. 1951. Spot anthracnoses known in Maryland and their distribution by counties. Pl. Dis. Reporter 35: 194-199.
1737. Jehle, R.A., A.E. Jenkins, and W.F. Jeffers. 1953. Additional records of spot anthracnoses in Maryland. Pl. Dis. Reporter 37: 371-372.
1738. Jehle, R.A., A.E. Jenkins, K.W. Kreitlow, and H.S. Sherwin. 1952. An outbreak of *Phyllosticta* canker and leaf spot of soybeans in Maryland. Pl. Dis. Reporter 36: 155-158.
1739. Jenkins, A.E. 1918. Brown canker of roses, caused by *Diaporthe umbrina*. J. Agric. Res. 15: 593-600.
1740. Jenkins, A.E. 1940. *Taphrina carveri* recently discovered in Missouri. Mycologia 32: 266-267.
1741. Jenkins, A.E. 1951. Records of rose anthracnose in Rhode Island. Pl. Dis. Reporter 35: 511.
1742. Jenkins, A.E. 1952. Spot anthracnose of flowering dogwood found in the Nassawanda Forest, Maryland, and isolation of the pathogen. Pl. Dis. Reporter 36: 294.
1743. Jenkins, A.E. 1958. Sugar maple leaf blister in New York. Pl. Dis. Reporter 42: 1091.
1744. Jenkins, A.E. 1960. A new record of violet scab (*Spaceloma violae*). Pl. Dis. Reporter 44: 596.
1745. Jenkins, A.E., and A.A. Bitancourt. 1940-1963. Myrangiales Selecti Exsiccati. Nos. 1-550.
1746. Jenkins, A.E., and A.A. Bitancourt. 1955. Notas sobre as anthracnoses, maculados e assuntos correlatos. V. Espécimens escolhidos de *Elsinoe* e *Sphaceloma*. Biologico (Sao Paulo) 21: 217-221.
1747. Jenkins, A.E., and A.A. Bitancourt. 1957. Studies in the Myriangiales. V. Elsinoaceae on evergreen euonymus, rose and English ivy. Mycologia 49: 95-101.
1748. Jenkins, A.E., and R.A. Jehle. 1950. *Sphaceloma viburni* destructive to snowball in Maryland. Pl. Dis. Reporter 34: 350-351.
1749. Jenkins, A.E., and R.A. Jehle. 1951. An anthracnose of bittersweet in Maryland. Pl. Dis. Reporter 35 : 413-414.
1750. Jenkins, A.E., and W.W. Ray. 1940. A new host for *Taphrina dearnessii* and geographic distribution of *Taphrina* on North American maples. Mycologia 32: 408-414.
1751. Jenkins, S.F., P.D. Dukes, and S.S. Thompson. 1963. Flue-cured tobacco diseases in Georgia, 1959-1962. Pl. Dis. Reporter 47: 96-98.
1752. Jenkins, S.F., and L.H. Person. 1973. *Ascochyta* blight of cucurbits caused by *Ascochyta phaseolorum*. Phytopathology 63: 444.
1753. Jenkins, S.F., Jr., and T.C. Wehner. 1983. Occurrence of *Fusarium oxysporum* f.sp. *cucumerinum* on greenhouse- grown *Cucumis sativus* seed stocks in North Carolina. Pl. Dis. 67: 1024-1025.
1754. Jenkins, S.F., J. Andreas, D.C. Sanders, and R.S. Gurkin. 1986. Powdery mildew caused by *Erysiphe heraclei* on carrot in North Carolina. Pl. Dis. 70: 892.
1755. Jenkins, W.A. 1952. Notes on early season pathology of fluecured tobacco in southside Virginia and portions of the old belt in North Carolina. Pl. Dis. Reporter 36: 278.
1756. Jenkins, W.A. 1952. Charcoal rot on flue-cured tobacco in Virginia. Pl. Dis. Reporter 36: 368-369.
1757. Jenkins, W.A. 1952. *Fusarium* wilt in flue-cured tobacco in Virginia. Pl. Dis. Reporter 36: 391.
1758. Jenkins, W.A. 1954. Outbreak of *Pythium* rot in newly set fluecured tobacco in Virginia. Pl. Dis. Reporter 38: 421.
1759. Jensen, J.D. 1984. *Melanconis marginalis* from northern Idaho. Mycotaxon 20: 275-281.
1760. Johns, R.M. 1957. A new species of *Physoderma* on *Dulichium*. Mycologia 49: 298-299.
1761. Johnson, A.G., and B. Koehler. 1953. Take-all disease of wheat in Illinois. Pl. Dis. Reporter 37: 148-149.
1762. Johnson, C.S., P.M. Phipps, and M.K. Beute. 1986. *Cercospora* leaf spot management decisions: uses of a correlation between rainfall and disease severity to evaluate the Virginia leaf spot advisory. Phytopathology 76: 860-863.
1763. Johnson, D.A. 1986. Two components of slow-rusting in asparagus infected with *Puccinia asparagi*. Phytopathology 76: 208-211.
1764. Johnson, D.A. 1987. First report in Washington state of the teleomorph of *Stemphylium vesicarium*, the causal agent of purple spot of asparagus. Pl. Dis. 71: 192.
1765. Johnson, D.A., and W.L. Anliker. 1985. Effect of downy mildew epidemics on the seasonal carryover of initial inoculum in hop yards. Pl. Dis. 69: 140-142.
1766. Johnson, D.A., C.M. Christensen, and T.H. King. 1975. *Galerina hypnorum* fruiting on cattail and in pure culture. Mycologia 67: 879-882.
1767. Johnson, D.A., and T.H. King. 1976. A leaf spot disease of three genera of aquatic plants in Minnesota. Pl. Dis. Reporter 60: 726-730.
1768. Johnson, D.A., and J.D. Lunden. 1984. First report of purple spot (*Stemphylium vesicarium*)of asparagus in Washington. Pl. Dis. 68: 1099.
1769. Johnson, D.A., and J.D. Lunden. 1986. Effects of wounding and wetting duration on infection of asparagus by *Stemphylium vesicarium*. Pl. Dis. 70: 419-420.

1770. Johnson, D.A., C.B. Skotland, and J.R. Alldredge. 1983. Weather factors affecting downy mildew epidemics of hops in the Yakima Valley of Washington. Phytopathology 73: 490-493.

1771. Johnson, D.A., E.L. Stewart, and T.H. King. 1976. A *Sclerotium* species associated with water lilies in Minnesota. Pl. Dis. Reporter 60: 807-808.

1772. Johnson, D.W. 1979. Growth and development of *Commandra* rust cankers on young lodgepole pine. Pl. Dis. Reporter 63: 916-918.

1773. Johnson, D.W., and J.E. Kuntz. 1979. *Eutypella* canker of maple: ascospore discharge and dissemination. Phytopathology 69: 130-135.

1774. Johnson, D.W., T.D. Landis, and L.S. Gillman. 1976. Rocky mountain juniper, a new host of *Armillariella mellea* in Colorado. Pl. Dis. Reporter 60: 886.

1775. Johnson, D.W., G.W. Peterson, and R.D. Dorset. 1985. *Diplodia* tip blight of ponderosa pine in the black hills of South Dakota. Pl. Dis. 69: 136-137.

1776. Johnson, E.M., and R.A. Chapman. 1958. Unusual occurrence of certain plant diseases in Kentucky in 1958. Pl. Dis. Reporter 42: 1411-1413.

1777. Johnson, E.M., R.A. Chapman, L. Henson, and W.D. Valleau. 1963. Occurrence of certain plant diseases in Kentucky in 1962. Pl. Dis. Reporter 47: 144-146.

1778. Johnson, E.M., R.A. Chapman, and W.D. Valleau. 1960. Occurrence of certain plant diseases in Kentucky in 1959. Pl. Dis. Reporter 44: 159-161.

1779. Johnson, E.M., R.A. Chapman, W.D. Valleau, and G.A. Everette. 1961. Occurrence of certain plant diseases in Kentucky in 1960. Pl. Dis. Reporter 45: 140-141.

1780. Johnson, E.M., C.C. Litton, and R.A. Reinert. 1964. Occurrence of certain plant diseases in Kentucky in 1963. Pl. Dis. Reporter 48: 111-113.

1781. Johnson, G.T. 1953. The Petersham foray. Mycologia 45: 794-800.

1782. Johnson, G.T. 1959. The Trypetheliaceae of Mississippi. Mycologia 51: 741-750.

1783. Johnson, H.W., and J.P. Jones. 1961. Other legumes prove susceptible to a powdery mildew of *Psoralea tenax*. Pl. Dis. Reporter 45: 542.

1784. Johnson, H.W., and J.P. Jones. 1962. Purple stain of guar. Phytopathology 52: 269-272.

1785. Johnson, H.W., and B.L. Keeling. 1969. Pathogenicity of *Phytophthora megasperma* isolated from subterranean clover roots. Phytopathology 59: 1279-1283.

1786. Johnson, H.W., and B.L. Keeling. 1969. Pathogenicity of *Phytophthora parasitica* isolated from regal white clover roots. Pl. Dis. Reporter 53: 446-450.

1787. Johnson, H.W., and R.A. Kilpatrick. 1953. Soybean diseases in Mississippi in 1951-52. Pl. Dis. Reporter 37: 154-155.

1788. Johnson, L.E.B., F.I. Frosheiser, and R.D. Wilcoxson. 1982. Interaction between *Fusarium oxysporum* f. sp. *medicaginis* and *Corynebacterium insidiosum* in alfalfa. Phytopathology 72: 517-522.

1789. Johnson, T.W., and H.S. Gold. 1957. *Didymosamarospora*, a new genus of fungi from fresh and marine waters. J. Elisha Mitchell Sci. Soc. 73: 103-108.

1790. Johnson, T.W. Jr. 1954. *Piricularia oryzae* on *Eremochloa ophiuroides* in Florida. Pl. Dis. Reporter 38: 796.

1791. Johnson, T.W. Jr. 1958. Marine fungi. IV. *Lulworthia* and *Ceriosporopsis*. Mycologia 50: 151-163.

1792. Johnson, T.W., Jr., M.H. Goforth, R.W. Powell Jr., and N.D. Haar. 1957. Observations on yellow-poplar dieback. Forest Sci. 3: 84-89.

1793. Johnson, T.W.,Jr.Jr. 1956. Marine fungi. I. *Leptosphaeria* and *Pleospora*. Mycologia 48: 495-505.

1794. Johnston, S.A., J.K. Springer, and G.D. Lewis. 1979. *Fusarium moniliforme* as a cause of stem and crown rot of asparagus and its association with asparagus decline. Phytopathology 69: 778-780.

1795. Jokela, J.J., J.D. Paxton, and E.J. Zegar. 1976. *Marssonina* leaf spot and rust on eastern cottonwood. Pl. Dis. Reporter 60: 1020-1024.

1796. Jones, A.L. 1963. Occurrence of *Nectria* twig blight of apple in western New York state in 1962. Pl. Dis. Reporter 47: 538-540.

1797. Jones, A.L., and W.G. Fields. 1972. An apple fruit rot caused by a *Hormiscium* sp. Pl. Dis. Reporter 56: 275.

1798. Jones, B.L., and J. Amador. 1969. Downy mildew, a new disease of St. Augustine grass. Pl. Dis. Reporter 53: 852-854.

1799. Jones, C., G.J. Griffin, and J.R. Elkins. 1980. Association of climatic stress with blight on Chinese chestnut in the eastern United States. Pl. Dis. 64: 1001-1004.

1800. Jones, E.D. 1960. Aecial stage of bean rust found in New York state. Pl. Dis. Reporter 44: 809.

1801. Jones, F.R. 1952. *Botrytis* on flowers of alfalfa, sweetclover and red clover. Pl. Dis. Reporter 36: 61.

1802. Jones, F.R., and J.L. Weimer. 1940. Three anthracnoses of alfalfa. Pl. Dis. Reporter 24: 30-31.

1803. Jones, J.P. 1959. Purple stain of soybean seeds incited by several *Cercospora* species. Phytopathology 49: 430-432.

1804. Jones, J.P. 1981. Nuclear behavior in Hypocreales: *Calonectria crotalariae*. Mycologia 73: 923-930.

1805. Jones, J.P., and G.E. Holcomb. 1978. Conidium ontogeny and cytology of *Tubakia dryina* from Louisiana hardwoods. Mycologia 70: 1212-1216.

1806. Jones, J.P., and H.W. Johnson. 1969. Lupine, a new host for *Phytophthora megasperma* var. *sojae*. Phytopathology 59: 504-507.

1807. Jones, R.K., C.J. Krass, and R.J. Sava. 1978. Isolation of *Ceratocystis ulmi* from 14-year-old annual rings of Eng lish elm in California. Pl. Dis. Reporter 62: 994-995.

1808. Jones, T.W. 1961. First report of pine mortality caused by *Fomes annosus* root rot in Ohio. Pl. Dis. Reporter 45: 980.

1809. Jones, T.W., and T.W. Bretz. 1958. First report of tree mortality from *Fomes annosus* root rot in Missouri. Pl. Dis. Reporter 42: 988.

1810. Jong, S.C., and C.R. Benjamin. 1971. North American species of *Nummularia*. Mycologia 63: 862-876.

1811. Jong, S.C., and J.D. Rogers. 1972. Illustrations and descriptions of conidial states of some *Hypoxylon* species. Wash. Agric. Exp. Sta. Techn. Bull. 71: 1-51.

1812. Jons, V.L. 1980. Crazy top of corn in North Dakota. Pl. Dis. 64: 103-104.

1813. Jons, V.L., B. Nelson, and M. McMullen. 1986. Downy mildew (*Sclerophthora macrospora*) of wheat, barley, and oats in North Dakota. Pl. Dis. 70: 892.

1814. Jordan, E.G., J.B. Manandhar, P.N. Thapliyal, and J.B. Sinclair. 1986. Factors affecting soybean seed quality in Illinois. Pl. Dis. 70: 246-248.

1815. Josephson, L.M., and D.A. Reid. 1950. Small grain diseases in Kentucky in 1949. Pl. Dis. Reporter 34: 24-26.

1816. Joyner, B.G., R.E. Partyka, and P.O. Larsen. 1977. *Rhizoctonia* brown patch of Kentucky bluegrass. Pl. Dis. Reporter 61: 749-752.

1817. Judd, R.W., Jr., and J.L. Peterson. 1971. Leaf spot of garden and sultan balsam caused by *Stemphylium*. Pl. Dis. Reporter 55: 374-375.

1818. Judd, R.W., Jr., and J.L. Peterson. 1974. *Cercospora omphakodes* infection and disease development in *Phlox divaricata*. Phytopathology 64: 1108-1111.

Jülich W. See 1707

1819. Juska, F.V., and K.W. Kreitlow. 1972. Conditions affecting the development and perpetuation of *Zoysia* rust. Pl. Dis. Reporter 56: 227-229.

1820. Juzwik, J., and D.W. French. 1986. Relationship between nitidulids and *Ceratocystis fagacearum* during late summer and autumn in Minnesota. Pl. Dis. 70: 424-426.

1821. Juzwik, J., D.W. French, and T.E. Hinds. 1986. *Encoelia pruinosa* on *Populus tremuloides* in Minnesota: Occurrence, pathogenicity, and comparison with Colorado isolates. Canad. J. Bot. 64: 2728-2731.

1822. Juzwik, J., W.T. Nishijima, and T.E. Hinds. 1978. Survey of aspen cankers in Colorado. Pl. Dis. Reporter 62: 906-910.

1823. Kable, P.F., P. Fliegel, and K.G. Parker. 1967. Cytospora canker on sweet cherry in New York state: association with winter injury and pathogenicity to other species. Pl. Dis. Reporter 51: 155-157.

1824. Kable, P.F., and K.G. Parker. 1963. The occurrence of the imperfect stage of *Monilinia laxa* on *Prunus cerasus* var. *austera* in New York state. Pl. Dis. Reporter 1963: 1104.

1825. Kagen, S.L. 1981. *Aspergillus*: An inhalable contaminant of marijuana. New England J. Medicine 304: 483-484.

1826. Kagen, S.L., V.P. Kurup, P.G. Sohnie, and J.N. Fink. 1983. Marijuana smoking and fungal sensitization. J. Allergy Clinical Immunol. 71: 389-392.

1827. Kais, A.G., C.E. Cordell, and C.E. Affeltranger. 1986. Benomyl root treatment controls brown-spot disease on longleaf pine in the southern United States. Forest Sci. 32: 506-510.
1828. Kais, A.G., and C.H. Walkinshaw. 1964. White-spored strain of *Cronartium fusiforme*. Pl. Dis. Reporter 48: 686.
1829. Kaiser, W.J., and R.M. Hannan. 1986. Incidence of seedborne *Ascochyta lentis* in lentil germplasm. Phytopathology 76: 355-360.
1830. Kaiser, W.J., and R.M. Hannan. 1987. Seed-treatment fungicides for control of seedborne *Ascochyta lentis* on lentil. Pl. Dis. 71: 58-62.
1831. Kaiser, W.J., and R.M. Hannan. 1987. First report of *Mycosphaerella rabiei* on chickpeas in the Western hemisphere. Pl. Dis. 71: 192.
1832. Kalb, D.W., and R.L. Millar. 1986. Dispersal of *Verticillium albo-atrum* by the fungus gnat (*Bradysia impatiens*). Pl. Dis. 70: 752-753.
1833. Kanouse, B.B. 1941. New and unusual species of Discomycetes. Mycologia 33: 461-467.
1834. Kanouse, B.B. 1947. A survey of the discomycete flora of the Olympic National Park and adjacent areas. Mycologia 39: 635-689.
1835. Kao, C.J. 1956. The cytology of *Xenogloea eriophori*. Mycologia 48: 288-301.
1836. Kardin, M.K., R.L. Bowden, J.A. Percich, and L.J. Nickelson. 1982. Zonate eyespot on wild rice caused by *Drechslera gigantea*. Pl. Dis. 66: 737-739.
1837. Karling, J.S. 1953. *Synchytrium chamaedryoidis*. Mycologia 45: 976-977.
1838. Karling, J.S. 1954. *Synchytrium brownii*, a new species with sexual reproduction. Sydowia 8: 27-30.
1839. Karling, J.S. 1955. *Synchytrium ranunculi* Cook. Mycologia 47: 130-139.
1840. Karling, J.S. 1955. Resting spore germination in *Synchytrium australe* in relation to its classification. Mycologia 47: 185-192.
1841. Karling, J.S. 1955. Additional species of *Synchytrium* with prosori. Bull. Torrey Bot. Club 82: 446-462.
1842. Karling, J.S. 1956. Development of *Synchytrium innominatum* in relation to other similar species. Mycologia 48: 534-546.
1843. Karling, J.S. 1956. *Synchytrium fulgens* in relation to other species on onagraceous hosts. Amer. J. Bot. 43: 61-69.
1844. Karling, J.S. 1956. Unrecorded hosts and species of *Physoderma*. Bull. Torrey Bot. Club 83: 292-299.
1845. Karling, J.S. 1956. New and unidentified species of *Synchytrium*. II. Bull. Torrey Bot. Club 83: 415-420.
1846. Karling, J.S. 1956. Undescribed species of *Synchytrium*. Mycologia 48: 83-98.
1847. Karling, J.S. 1956. New and unidentified species of *Synchytrium* I. Sydowia 10: 18-37.
1848. Karling, J.S. 1956. New and unidentified species of *Synchytrium*. - III. Sydowia 10: 243-254.
1849. Karling, J.S. 1957. New and unidentified species of *Synchytrium*. V. Mycologia 49: 740-755.
1850. Karling, J.S. 1958. *Synchytrium callirrhoeae* n. sp. Amer. J. Bot. 45: 327-330.
1851. Karling, J.S. 1960. *Synchytrium texanum* sp. nov. Mycologia 52: 21-29.
1852. Karling, J.S. 1960. *Synchytrium pilificum* in America. Mycologia 52: 436-443.
1853. Kauffman, C.H. 1917. Unreported Michigan fungi for 1915 and 1916, with an index to the hosts and substrata of basidiomycetes. Ann. Rep. Michigan Acad. Sci. 19: 145-157.
1854. Keenan, J.G., H.D. Moore, N. Oshima, and L.E. Jenkins. 1974. Effect of bean root rot on dryland pinto bean production in southwestern Colorado. Pl. Dis. Reporter 58: 890-892.
1855. Keener, P.D. 1953. Unusual numbers of sporophores of a wood-rotting fungus on trunks of *Populus tremuloides* var. *aurea*. Pl. Dis. Reporter 37: 164-165.
1856. Keener, P.D. 1953. A species of *Gymnosporangium* on quince, pear and hawthorn in Arizona. Pl. Dis. Reporter 37: 235.
1857. Keener, P.D. 1954. *Cladosporium aecidiicola* Thuem. and *Tuberculina persicina* (Ditm.) Sacc. associated with *Puccinia conspicua* (Arth.) Mains on *Helenium hoopesii* A. Gray in Arizona. Pl. Dis. Reporter 38: 690-694.
1858. Keener, P.D. 1959. An early record of pear rust in Arizona. Pl. Dis. Reporter 43: 287.
1859. Keener, P.D. 1959. Sporophore size in *Fomes pinicola* (Swartz ex Fries) Cooke. Pl. Dis. Reporter 43: 966-967.
1860. Keener, P.D. 1960. Peach "shot hole" in Arizona --1959. Pl. Dis. Reporter 44: 629.
1861. Keil, H.L., and R.A. Wilson. 1961. Powdery mildew on peach. Pl. Dis. Reporter 45: 10-11.
1862. Keim, R. 1977. Foliage blight of periwinkle in southern California. Pl. Dis. Reporter 61: 182-184.
1863. Keim, R. 1979. Pruning would dieback of oleander. Pl. Dis. Reporter 63: 499-501.
1864. Keim, R., L.J. Klure, and G.A. Zentmyer. 1979. Collar rot of Washington palms in containers. Pl. Dis. Reporter 63: 718-720.
1865. Keim, R., T. Mock, and H. Guenther. 1976. *Phytophthora* crown rot of toyon. Pl. Dis. Reporter 60: 88-90.
1866. Keim, R., G.A. Zentmyer, and L.J. Klure. 1976. *Phytophthora palmivora* on ivy in California and its control with Pyroxychlor. Pl. Dis. Reporter 60: 632-633.
1867. Keller, J.R., and H.S. Potter. 1954. *Thielaviopsis* associated with root rot of some ornamental plants. Pl. Dis. Reporter 38: 354-355.
1868. Keller, J.R., and J.B. Shanks. 1955. *Poinsettia* root rot. Phytopathology 45: 552-558.
1869. Kelman, A. 1963. Dutch elm disease in North Carolina. Pl. Dis. Reporter 47: 154.
1870. Kelman, A., C.S. Hodges, and H.R. Garriss. 1960. Needle blight of redcedar, *Juniperus virginiana* L. Pl. Dis. Reporter 44: 527-531.
1871. Kendrick, J.B., Jr., and J.T. Middleton. 1950. Gray mold of lima beans. Phytopathology 40: 228-234.
1872. Kendrick, J.B., and W.C. Snyder. 1942. *Fusarium* wilt of radish. Phytopathology 32: 1031-1033.
1873. Kenerley, C.M., and R.I. Bruck. 1981. *Phytophthora* root rot of balsam fir and Norway spruce in North Carolina. Pl. Dis. 65: 614-615.
1874. Kennedy, B., J. Orf, W.C. Stienstra, and J. Wiersma. 1986. Outbreak of powdery mildew of soybean in Minnesota in 1985. Pl. Dis. 70: 694.
1875. Kern, F.D. 1960. Additional notes on the rusts of Pennsylvania. Proc. Pennsylvania Acad. Sci. 34: 45-49.
1876. Kern, F.D. 1964. Notes on *Gymnosporangium gracile* Pat. Pl. Dis. Reporter 48: 93-94.
1877. Kern, F.D. 1973. A revised taxonomic account of *Gymnosporangium*. Pennsylvania State Univ. Press, University Park and London. 134 p.
1878. Kern, H. 1955. Taxonomic studies in the genus *Leucostoma*. Pap. Michigan Acad. Sci. 40: 9-21.
1879. Kernkamp, M.F., and J.W. Gibler. 1951. Diseases of soybeans new to Minnesota. Pl. Dis. Reporter 35: 509-510.
1880. Kernkamp, M.F., R. Kroll, and W.C. Woodruff. 1976. Diseases of cultivated wild rice in Minnesota. Pl. Dis. Reporter 60: 771-775.
1881. Kessler, K.J., and R.L. Anderson. 1960. *Ceratocystis coerulescens* on sugar maple in the Lake States. Pl. Dis. Reporter 44: 348-350.
1882. Kessler, K.J. Jr. 1970. A survey of diseases affecting yellow birch seedlings. Pl. Dis. Reporter 54: 16-18.
1883. Kessler, K.J. Jr. 1974. An apparent symbiosis between *Fusarium* fungi and ambrosia beetles causes canker on black walnut stems. Pl. Dis. Reporter 58: 1044-1047.
1884. Kessler, K.J. Jr. 1984. *Mycosphaerella juglandis*, causal agent of a leaf spot of *Juglans nigra*. Mycologia 76: 362-366.
1885. Kessler, K.J. Jr. 1985. *Mycosphaerella* leaf spot of black walnut. Pl. Dis. 69: 1092-1094.
1886. Kessler, K.J., Jr., and E.M. Hayes. 1984. Extension of range of *Grovesinia pyramidalis* on black walnut. Pl. Dis. 68: 918.
1887. Keyworth, W.G. 1952. *Verticillium* wilt of potatoes in Connecticut 1951. Pl. Dis. Reporter 36: 16-17.
1888. Kidambi, R.D., A.L. Lutz, and N.K., van Alfen van. 1985. Disease progress and epidemiology of crown rot of spring barley in Utah. Canad. J. Plant Pathol. 7: 233-237.
1889. Kilpatrick, R.A. 1955. Soybean diseases in the delta area of Mississippi in 1954. Pl. Dis. Reporter 39: 578-579.
1890. Kilpatrick, R.A. 1956. Diseases of forage crops in New England and New York in 1956. Pl. Dis. Reporter 40: 1054-1057.
1891. Kilpatrick, R.A. 1958. *Curvularia* leaf blight of clovers and its causal agent, *Curvularia trifolii*. Phytopathology 48: 513-515.
1892. Kilpatrick, R.A. 1958. Fungi isolated from Ladino white clover seeds (*Trifolium repens* L.) in New Hampshire. Pl. Dis. Reporter 42: 142-144.
1893. Kilpatrick, R.A. 1959. Fungi associated with red and white

clovers in New Hampshire. Pl. Dis. Reporter 43: 1111-1113.

1894. Kilpatrick, R.A. 1961. Floral infection of Ladino white clover, incited by *Curvularia trifolii*. Pl. Dis. Reporter 45: 286-287.

1895. Kilpatrick, R.A. 1961. Diseases of forage grasses and legumes previously unreported from New Hampshire. Pl. Dis. Reporter 45: 926-927.

1896. Kilpatrick, R.A. 1977. Symptoms and control of the peat mold, *Chromelosporium fulvum*, on *Crambe abyssinica*. Pl. Dis. Reporter 61: 93-95.

1897. Kilpatrick, R.A., and G.M. Dunn. 1956. Late season diseases on forage crops in New Hampshire in 1955. Pl. Dis. Reporter 40: 384-386.

1898. Kilpatrick, R.A., E.W. Hanson, and J.G. Dickson. 1954. Root and crown rots of red clover in Wisconsin and the relative prevalence of associated fungi. Phytopathology 44: 252-259.

1899. Kilpatrick, R.A., and H.W. Johnson. 1953. Fungi isolated from soybean plants at Stoneville, Mississippi, in 1951-52. Pl. Dis. Reporter 37: 98-100.

1900. Kilpatrick, R.A., and E.S. Luttrell. 1967. An undescribed species of *Curvularia* pathogenic to *Dichanthium annulatum*. Mycologia 59: 888-892.

1901. Kilpatrick, R.A., A.E. Rich, and J.G. Conklin. 1961. *Juncus effusus*, a new host for *Epichloe typhina*. Pl. Dis. Reporter 45: 899.

1902. Kilpatrick, R.A., F.A. Uecker, and G.A. White. 1975. Powdery mildew, a new disease of Stokes aster. Pl. Dis. Reporter 59: 795.

1903. Kimmey, J.W., and P.C. Lightle. 1955. Fungi associated with cull in redwood. Forest Sci. 1: 104-110.

1904. Kimmey, J.W., and J.A. Stevenson. 1957. A forest disease survey of Alaska. Pl. Dis. Reporter Suppl. 247: 86-98.

1905. King, C., and C.T. Rogerson. 1951. Tomato late blight in Kansas. Pl. Dis. Reporter 35: 120.

1906. Kingsland, G.C. 1963. *Cercospora* leaf blight of corn: a case history of a local epiphyto tic in South Carolina. Pl. Dis. Reporter 47: 724-725.

1907. Kingsland, G.C., and A.E. Rich. 1961. *Fusarium solani* f. *pisi*, a cause of pea root rot in New Hampshire. Pl. Dis. Reporter 45: 845.

1908. Kingsolver, C.H. 1950. Downy mildew on barley in Missouri. Pl. Dis. Reporter 34: 29-30.

1909. Kirby, R.S. 1950. Powdery mildew on African violet in Pennsylvania. Pl. Dis. Reporter 34: 197.

1910. Kirby, R.S. 1950. Some diseases on small grains in Pennsylvania. Pl. Dis. Reporter 34: 198.

1911. Kirk, P.M. 1982. New or interesting microfungi VII. Two new Hyphomycetes from *Pericopsis angolensis* leaf litter. Mycologia 74: 872-876.

1912. Kistler, B.R., W. Merrill, J. Longenecker, and V. Wagner. 1977. *Fabrella* needle cast of hemlock: A new record for Pennsylvania. Pl. Dis. Reporter 61: 863.

1913. Klarman, W.L. 1968. *Coniothyrium* canker on ornamental pear. Pl. Dis. Reporter 52: 792.

1914. Klemmer, H.W., and R.Y. Nakano. 1964. Distribution and pathogenicity of *Phytophthora* and *Pythium* in pineapple soils of Hawaii. Pl. Dis. Reporter 48: 848-852.

1915. Kliejunas, J.T. 1976. Bleeding canker of Norfolk Island pine in Hawaii. Pl. Dis. Reporter 60: 84-87.

1916. Kliejunas, J.T., and D.H. Adams. 1981. *Phytophthora* root rot of Port-Orford-cedar in California. Pl. Dis. 65: 446-447.

1917. Kliejunas, J.T., J.R. Allison, A.H. McCain, and R.S. Smith Jr. 1985. *Phoma* blight of fir and Douglas-fir seedlings in a California nursery. Pl. Dis. 69: 773-775.

1918. Kliejunas, J.T., and W.H. Ko. 1973. Root rot of ohia (*Metrosideros collina* subsp. *polymorpha*) caused by *Phytophthora cinnamomi*. Pl. Dis. Reporter 57: 383-384.

1919. Kliejunas, J.T., and W.H. Ko. 1975. The occurrence of *Pythium vexans* in Hawaii and its relation to ohia decline. Pl. Dis. Reporter 59: 392-395.

1920. Kliejunas, J.T., and W.H. Ko. 1976. Association of *Phytophthora cinnamomi* with *Ohia* decline on the island of Hawaii. Phytopathology 66: 116-121.

1921. Kliejunas, J.T., and J.E. Kuntz. 1974. Microorganisms associated with *Eutypella parasitica* in *Acer saccharum* and *A. rubrum*. Canad. J. Forest Res. 4: 207-212.

1922. Kliejunas, J.T., R.F. Scharpf, and R.S. Smith. 1979. *Teichospora obducens* and a *Pleospora* species on *Korthalsella* in Hawaii. Pl. Dis. Reporter 63: 1060-1062.

1923. Klisiewicz, J.M. 1977. Effect of flooding and temperature on incidence and severity of safflower seedling rust and viability of *Puccinia carthami* teliospores. Phytopathology 67: 787-790.

1924. Klisiewicz, J.M. 1977. Identity and relative virulence of some heterothallic *Phytophthora* species associated with root and stem rot of safflower. Phytopathology 67: 1174-1177.

1925. Klisiewicz, J.M., and B.H. Beard. 1976. Diseases of sunflower in California. Pl. Dis. Reporter 60: 298-301.

1926. Klisiewicz, J.M., S.L. Clement, and R.F. Norris. 1983. Black stem: A fungal disease of common purslane in California. Pl. Dis. 67: 1162.

1927. Kmetz, K.T., A.F. Schmitthenner, and C.W. Ellet. 1978. Soybean seed decay: Prevalence of infection and symptom expression caused by *Phomopsis* sp., *Diaporthe phaseolorum* var. *sojae*, and *D. phaseolorum* var. *caulivora*. Phytopathology 68: 836-840.

1928. Kmetz, K., C.W. Ellett, and A.F. Schmitthenner. 1974. Isolation of seedborne *Diaporthe phaseolorum* and *Phomopsis* from immature soybean plants. Pl. Dis. Reporter 58: 978-982.

1929. Knauss, J.F. 1971. *Rhizoctonia* blight of 'Florida ruffle' fern and its control. Pl. Dis. Reporter 55: 614-616.

1930. Knauss, J.F. 1972. Description and control of *Pythium* root rot on two foliage plant species. Pl. Dis. Reporter 56: 211-215.

1931. Knauss, J.F. 1972. Foliar blight of *Dionaea muscipula* incited by *Colletotrichum gloeosporioides*. Pl. Dis. Reporter 56: 391-393.

1932. Knauss, J.F. 1975. Description and control of *Fusarium* tuber rot of *Caladium*. Pl. Dis. Reporter 59: 975-979.

1933. Knutson, D.M., and A.S. Hutchins. 1979. *Wallrothiella arceuthobii* infecting *Arceuthobium douglasii*: culture and field inoculation. Mycologia 71: 821-828.

1934. Ko, W.H., J.E. Hunter, and R.K. Kunimoto. 1973. *Rhizoctonia* disease of Queensland maple seedlings. Pl. Dis. Reporter 57: 907-909.

1935. Ko, W.H., R.K. Kunimoto, and I. Maedo. 1977. Root decay caused by *Kretzschmaria clavus*: its relation to macadamia decline. Phytopathology 67: 18-21.

1936. Ko, W.H., R.K. Kunimoto, and W.T. Nishijima. 1982. Fruit rot of guava caused by *Phytophthora citricola*. Pl. Dis. 66: 854-855.

1937. Ko, W.H., J.Y. Uchida, R.K. Kunimoto, and M. Aragaki. 1981. Collar rot and leaf spot of leea caused by *Calonectria crotalariae*. Pl. Dis. 65: 621.

1938. Kobayasi, Y., N. Hiratsuka, Y. Otani, K. Tubaki, S. Udagawa, and M. Soneda. 1969. The second report on the mycological flora of the Alaskan Arctic. Bull. Natl. Sci. Mus. 12: 311-430.

1939. Kobriger, K., and D.J. Hagedorn. 1984. Additional *Pythium* species associated with the bean root rot complex in Wisconsin's central sands. Pl. Dis. 68: 595-596.

1940. Koehler, B., and G.H. Boewe. 1957. Causes of corn stalk rot in Illinois. Pl. Dis. Reporter 41: 501-504.

1941. Kohlmeyer, J. 1964. A new marine ascomycete from wood. Mycologia 56: 770-774.

1942. Kohlmeyer, J. 1969. Marine fungi of Hawaii including the new genus *Helicascus*. Canad. J. Bot. 47: 1469-1487.

1943. Kohlmeyer, J., and E. Kohlmeyer. 1971. Marine fungi from tropical America and Africa. Mycologia 63: 831-861.

1944. Kohlmeyer, J., and E. Kohlmeyer. 1974. Distribution of *Epichloe typhina* (Ascomycetes) and its parasitic fly. Mycologia 66: 77-86.

1945. Kohlmeyer, J., and E. Kohlmeyer. 1979. Marine Mycology - The Higher Fungi. Academic Press, NY. 690 p.

1946. Kohn, L.M. 1979. A monographic revision of the genus *Sclerotinia*. Mycotaxon 9: 365-444.

1947. Koike, H. 1980. Rust of sugarcane in Louisiana: a first report. Pl. Dis. 64: 226.

1948. Koike, H., D. Fontenot, K. Damann, and R. Schlub. 1981. Smut of sugarcane in Louisiana. Pl. Dis. 65: 1018.

1949. Koike, S.T. 1987. First report of avocado fruit rot caused by *Phytophthora citricola* in California. Pl. Dis. 71: 1045.

1950. Kommedahl, T., K.K. Sabet, P.M. Burnes, and C.E. Windels. 1987. Occurrence and pathogenicity of *Fusarium proliferatum* on corn in Minnesota. Pl. Dis. 71: 281.

1951. Kommedahl, T., and C.E. Windels. 1977. *Fusarium* stalk rot and common smut in cornfields of southern Minnesota in 1976. Pl. Dis. Reporter 61: 259-261.

1952. Kontaxis, D.G. 1977. Powdery mildew on watermelon fruit in Imperial Valley, California. Pl. Dis. Reporter 61: 397-398.

1953. Kontaxis, D.G. 1977. Epiphytotic of rust on asparagus in the

Imperial Valley, California. Pl. Dis. Reporter 61: 503.
1954. Kontaxis, D.G. 1979. Cleistothecia of cucurbit powdery mildew in California -- a new record. Pl. Dis. Reporter 63: 278.
1955. Kontaxis, D.G., and P. Guerrero. 1978. Downy mildew epiphytotic on broccoli - a new disease in Imperial Valley, California. Pl. Dis. Reporter 62: 170-171.
1956. Kontaxis, D.G., H. Meister, and R.K. Sharma. 1974. Powdery mildew epiphytotic on sugarbeets. Pl. Dis. Reporter 58: 904-905.
1957. Kontaxis, D.G., and A.F. Van Maren. 1978. Powdery mildew on tomato - a new disease in the United States. Pl. Dis. Reporter 62: 892-893.
1958. Korf, R.P. 1954. Discomyceteae Exsiccatae. Fasc. I (nos. 1 - 15).
1959. Korf, R.P., R.C. Pearson, W.Y. Zhuang, and B. Dubos. 1986. *Pseudopezicula* (Helotiales, Peziculoideae) a new discomycete genus for pathogens causing an angular leaf scorch disease of grapes ("rot brenner"). Mycotaxon 26: 457-471.
1960. Korf, R.P, and S.C. Gruff. 1978. Discomycetes exsiccati, Fasc. II & III. Mycotaxon 7: 185-203.
1961. Koucheki, H.K. 1973. Observations on the etiology of a disease of *Pelargonium hortorum*. Pl. Dis. Reporter 57: 284-288.
1962. Kozelnicky, G.M. 1969. First report of stripe smut on bluegrass in Georgia. Pl. Dis. Reporter 53: 580.
1963. Kozelnicky, G.M., and W.N. Garrett. 1966. The occurrence of *Zoysia* rust in Georgia. Pl. Dis. Reporter 50: 839.
1964. Kraft, J.M., and D.W. Burke. 1971. *Pythium ultimum* as a root pathogen on beans and peas in Washington. Pl. Dis. Reporter 55: 1056-1060.
1965. Kraft, J.M., F.J. Muehlbauer, R.J. Cook, and F.M. Entemann. 1974. The reappearance of common wilt of peas in eastern Washington. Pl. Dis. Reporter 58: 62-64.
1966. Kramer, C.L. 1958. A new genus in the Protomycetaceae. Mycologia 50: 916-926.
1967. Kramer, C.L. 1960. Morphological development and nuclear behavior in the genus *Taphrina* . Mycologia 52: 295-320.
1968. Kramer, C.L., and H.E. Thompson. 1961. Dutch elm disease in Kansas in 1960. Pl. Dis. Reporter 45: 309.
1969. Kraus, H., and W. Witcher. 1977. Survey of pine pitch canker in South Carolina. Pl. Dis. Reporter 61: 976-978.
1970. Kraus, J.F., and D.L. Hunt. 1971. Inherent variation of resistance to *Hypoderma lethale* in slash and loblolly pines. Forest Sci. 17: 143-144.
1971. Krause, R.A., and R.K. Webster. 1972. The morphology, taxonomy, and sexuality of the rice stem rot fungus, *Magnaporthe salvinii* (*Leptosphaeria salvinii*). Mycologia 64: 103-114.
1972. Krausz, J.P., and J.D. Caldwell. 1987. *Cylindrocladium* root rot of kiwifruit. Pl. Dis. 71: 374-375.
1973. Krausz, J.P., and B.A. Fortnum. 1983. An epiphytotic of *Diaporthe* stem canker of soybean in South Carolina. Pl. Dis. 67: 1128-1129.
1974. Krebill, R.G. 1964. Blister rust found on limber pine in northern Wasatch Mountains. Pl. Dis. Reporter 48: 532.
1975. Krebill, R.G. 1967. *Gymnosporangium clavariiforme* in witches'-brooms of *Juniperus communis*. Pl. Dis. Reporter 51: 143-147.
1976. Krebill, R.G. 1968. *Chrysomyxa ledi glandulosi* in the Uinta Mountains. Proc. Utah Acad. Sci. 45: 319-320.
1977. Krebill, R.G. 1968. *Cronartium comandrae* in the Rocky Mountain States. U.S. Forest Serv. Res. Pap. INT 50: 1-28.
1978. Krebill, R.G. 1970. Autoecious gall rusts of pines in southern Michigan and New York. Pl. Dis. Reporter 54: 853-855.
1979. Krebill, R.G., and W.G. Ziller. 1968. *Hyalopsora aspidiotus* on fir in the west. Pl. Dis. Reporter 52: 336.
1980. Kreitlow, K.W., and A.T. Bleak. 1964. *Podosporiella verticillata*, a soil-borne pathogen of some western Gramineae. Phytopathology 54: 353-357.
1981. Kreitlow, K.W., F.V. Juska, and R.T. Haard. 1965. A rust on *Zoysia japonica* new to North America. Pl. Dis. Reporter 49: 185.
1982. Kreitlow, K.W., and H.S. Yu. 1955. *Curvularia* leaf blight of red clover. Pl. Dis. Reporter 39: 181-182.
1983. Kreitlow, K.W., and H.S. Yu. 1955. A leaf spot of *Lotus uliginosus* caused by *Cercospora loti*. Pl. Dis. Reporter 39: 236-238.
1984. Krommedahl, T., C.E. Windels, and H.G. Johnson. 1974. Corn stalk rot survey methods and results in Minnesota in 1973. Pl. Dis. Reporter 58: 363-366.
1985. Kropp, B.R., and M.E. Corden. 1986. Morphology and taxonomy of *Pachnocybe ferruginea*. Mycologia 78: 334-342.
1986. Krupinsky, J.M. 1981. *Botryodiplodia hypodermia* and *Tubercularia ulmea* in cankers on Siberian elm in northern Great Plains windbreaks. Pl. Dis. 65: 677-678.
1987. Krupinsky, J.M. 1982. Observations on the host range of isolates of *Pyrenophora trichostoma*. Canad. J. Plant Pathol. 4: 42-46.
1988. Krupinsky, J.M. 1986. *Pyrenophora tritici-repentis*, *P. bromi*, and *Leptosphaeria nodorum* on *Bromus inermis* in the northern Great Plains. Pl. Dis. 70: 61-64.
1989. Krupinsky, J.M. 1986. Cankers on Russian olive in windbreaks in North Dakota and South Dakota. Pl. Dis. 70: 981.
1990. Krupinsky, J.M., and J.D. Berdahl. 1984. *Septoria spraguei*, *Pyrenophora trichostoma*, and *Cochliobolus sativus* incidence on Russian wildrye grass leaves and *S. spraguei* host range. Pl. Dis. 68: 13-16.
1991. Krupinsky, J.M., and R.J. Loring. 1983. An *Alternaria* sp. on leafy spurge (*Euphorbia esula*). Weed Sci. 31: 86-88.
1992. Kucharek, T.A. 1973. Stalk rot of corn caused by *Helminthosporium rostratum*. Phytopathology 63: 1336-1338.
1993. Kuhlman, E.G., and S. Cade. 1985. Pitch canker disease of loblolly and pond pines in North Carolina plantations. Pl. Dis. 69: 175-176.
1994. Kuhlman, E.G., and F.F. Hendrix. 1963. *Phytophthora* root rot of Fraser Fir. Pl. Dis. Reporter 47: 552-553.
1995. Kukkonen, I. 1961. The smuts on the genus *Cintractia* parasitizing *Kobresia* species. Canad. J. Bot. 39: 155-164.
1996. Kukkonen, I. 1963. Taxonomic studies on the genus *Anthracoidea* (Ustilaginales). Ann. Bot. Soc. Zool.-Bot. Fenn. "Vanamo" 34(3): 1-122.
1997. Kukkonen, I. 1964. Taxonomic studies on the species of the section Echinosporae of *Anthracoidea*. Ann. Bot. Fenn. 1: 161-177.
1998. Kulhavy, D.L., R.J. Chacko, and A.D. Partridge. 1978. Some decay and disease fungi isolated from western white pine in northern Idaho. Pl. Dis. Reporter 62: 332-336.
1999. Kulik, M.M., and P.R. Thomison. 1985. First report of soybean stem canker in Maryland. Pl. Dis. 69: 811.
2000. Kunimoto, R.K., M. Aragaki, J.E. Hunter, and W.H. Ko. 1976. *Phytophthora capsici*, corrected name for the cause of Phytophthora blight of macadamia racemes. Phytopathology 66: 546-548.
2001. Kunimoto, R.K., and W.H. Ko. 1984. Brown rot of guava fruit in Hawaii caused by *Botryosphaeria berengeriana*. Pl. Dis. 68: 918.
2002. Kunwar, I.K., J.B. Manandhar, and J.B. Sinclair. 1986. Histopathology of soybean seeds infected with *Alternaria alternata*. Phytopathology 76: 543-546.
2003. Kurtzman, C.P., L.J. Wickerham, and C.W. Hesseltine. 1970. Yeasts from wheat and flour. Mycologia 62: 542-547.
2004. Kuss, F.R., and W.N. Harnish. 1972. *Alternaria* species associated with an apple leaf spot. Pl. Dis. Reporter 56: 721.
2005. Kuter, G.A. 1986. Microfungal populations associated with the decomposition of sugar maple leaf litter. Mycologia 78: 114-126.
2006. Lackner, A.L., and S.A. Alexander. 1982. Occurrence and pathogenicity of *Verticicladiella procera* in Christm as tree plantations in Virginia. Pl. Dis. 66: 211-212.
2007. Lackner, A.L., and S.A. Alexander. 1984. Incidence and development of *Verticicladiella procera* in Virginia Christmas plantations. Pl. Dis. 68: 210-212.
2008. Lacy, M.L. 1977. Influence of chemical treatments on stand establishment in asparagus. Proc. Amer. Phytopathol. Soc. 4: 151.
2009. Laemmlen, F.F., and M. Aragaki. 1971. Collar rot of papaya caused by *Calonectria* sp. Pl. Dis. Reporter 55: 743-745.
2010. Laemmlen, F.F., and R.M. Endo. 1985. Powdery mildew (*Oidiopsis taurica*) on onion in California. Pl. Dis. 69: 451.
2011. Laemmlen, F., and R.V. Bega. 1974. Hosts of *Armillaria mellea* in Hawaii. Pl. Dis. Reporter 58: 102-103.
2012. Lambe, R.C. 1960. Diseases of cut greenery in Oregon. Pl. Dis. Reporter 44: 718-720.
2013. Lambe, R.C., and J.M. Dunleavy. 1967. Soybean root and stem rot caused by *Rhizoctonia solani*, 1966. Pl. Dis. Reporter 51: 872.
2014. Lambe, R.C., and W.H. Wills. 1975. Decline of English boxwood in Virginia. Pl. Dis. Reporter 59: 105-108.
2015. Lambe, R.C., and W.H. Wills. 1977. Further incidence of black root rot of Japanese holly. Proc. Amer. Phytopathol. Soc. 4: 217.
2016. Lambe, R.C., and W.H. Wills. 1980. Distribution of dieback associated with *Thielaviopsis* black root rot of Japanese holly. Pl.

Dis. 64: 956.

2017. Lambe, R.C., and W.H. Wills. 1983. Root rot of Japanese umbrella pine, *Sciadopitys verticillata*, caused by *Phytophthora cinnamomi*. Pl. Dis. 67: 698.

2018. Lamey, H.A. 1970. *Pyricularia oryzae* on rice seed in the United States. Pl. Dis. Reporter 54: 931-935.

2019. Lancaster, F.R., and A.F. Rumph. 1951. Oak wilt in western Pennsylvania. Pl. Dis. Reporter 35: 383.

2020. Landis, T.D. 1977. Oak wilt detected in five new counties in Kansas and Nebraska. Pl. Dis. Reporter 61: 188-189.

2021. Landis, T.D., and A.K. Evans. 1974. A relationship between *Fomes applanatus* and aspen windthrow. Pl. Dis. Reporter 58: 110-113.

2022. Landis, T.D., and L.B. Helburg. 1976. Black stain root disease of pinyon pine in Colorado. Pl. Dis. Reporter 60: 713-717.

2023. Landis, W.R., and J.H. Hart. 1967. Cankers of ornamental crabapples associated with *Physalospora obtusa* and other microorganisms. Pl. Dis. Reporter 51: 230-234.

2024. Landschoot, P.J., and N. Jackson. 1987. The *Phialophora* state of a *Magnaporthe* sp. causes summer patch disease of *Poa pratensis* L. and *P. annua* L. Phytopathology 77: 119.

2025. Langenberg, W.G., and E.D. Kerr. 1982. *Polymyxa betae* in Nebraska. Pl. Dis. 66: 862.

2026. Large, J.R., and J.R. Cole. 1964. Karathane-controlled powdery mildew on the Curtis and Pabst varieties of pecan in Florida and Georgia. Phytopathology 54: 1174.

2027. Large, J.R., J.H. Painter, and W.A. Lewis. 1950. Thread blight in tung orchards and its control. Phytopathology 40: 453-459.

2028. Larsen, H.J., R.P. Covey, and W.R. Fischer. 1980. A red spot fruit blemish in apricots. Phytopathology 70: 139-142.

2029. Larsen, H.J., M.G. Boosalis, and E.D. Kerr. 1985. Temporary depression of *Rhizoctonia solani* field populations by soil amendment with *Laetisaria arvalis*. Pl. Dis. 69: 347-350.

2030. Larsen, M.J. 1966. *Tomentella* and related genera in North America II. Studies of nomenclatural types of species described by Peck. Mycologia 58: 597-613.

2031. Larsen, M.J. 1967. *Tomentella* and related genera in North America III. New species of *Tomentella* and *Pseudotomentella*. Canad. J. Bot. 45: 1297-1307.

2032. Larsen, M.J. 1967. *Tomentella* and related genera in North America V. New North American records of tomentelloid fungi. Mycopathol. Mycologia Appl. 32: 37-67.

2033. Larsen, M.J. 1968. Notes on tomentelloid fungi I. *Tomentella bourdotii* in North America. Mycologia 60: 1178-1184.

2034. Larsen, M.J. 1968. Tomentelloid fungi of North America. State Univ. New York Coll. Forestry Tech. Pub. 93: 1-157.

2035. Larsen, M.J. 1969. Notes on tomentelloid fungi. II. *Tomentella besadolae* and *T. ruttnerii*. Mycologia 61: 670-679.

2036. Larsen, M.J. 1970. *Tomentella* and related genera in North America VI. Some synonmy and additional new records. Mycologia 62: 256-271.

2037. Larsen, M.J. 1971. Notes on tomentelloid fungi III. New species of *Pseudotomentella*. Bull. Torrey Bot. Club 98: 38-41.

2038. Larsen, M.J. 1974. Some notes on *Pseudotomentella*. Mycologia 66: 165-168.

2039. Larsen, M.J. 1974. A contribution to the taxonomy of the genus *Tomentella*. Mycologia Memoir 4: 1-145.

2040. Larsen, M.J. 1975. Some new North American species of *Tomentella*. Beih. Nova Hedwigia 51: 171-175.

2041. Larsen, M.J. 1983. Notes on tomentelloid fungi V.Additional new species of *Pseudotomentella*. Mycologia 75: 556-562.

2042. Larsen, M.J. 1986. *Lindtneria thujatsugina* sp. nov. (Stephanosporales, Stephanosporace ae) and notes on other resupinate Basidiomycetes with ornamented basidiospores. Nova Hedwigia 43: 255-267.

2043. Larsen, M.J., and R.L. Gilbertson. 1977. Studies in *Laeticorticium* (Aphyllophorales, Corticiaceae) and related genera. Norw. J. Bot. 24: 99-121.

2044. Larsen, M.J., and R.L. Gilbertson. 1978. *Laeticorticium lombardiae* (Aphyllophorales, Corticiaceae): A newly recognized segregate from the *L. roseum* complex. Mycologia 70: 206-208.

2045. Larsen, M.J., M.F. Jurgensen, and A.E. Harvey. 1978. N2 fixation associated with wood decayed by some common fungi in western Montana. Canad. J. Forest Res. 8: 341-345.

2046. Larsen, M.J., M.F. Jurgensen, A.E. Harvey, and J.C. Ward. 1978. Dinitrogen fixation associated with sporophores of *Fomitopsis pinico la*, *Fomes fomentarius*, and *Echinodontium tinctorium*. Mycologia 70: 1217-1222.

2047. Larsen, M.J., and F.F. Lombard. 1976. *Phellinus fragrans* sp. nov. (Aphyllophorales, Hymenochaetaceae) associated with a white rot of maple. Mem. New York Bot. Gard. 28: 131-140.

2048. Larsen, M.J., and F.F. Lombard. 1983. *Fibroporia angulopora*, a new species (Aphyllophorales, Polyporaceae) associated with brown-rot of *Pseudotsuga menziesii* residue in western Oregon. Mycologia 75: 623-627.

2049. Larsen, M.J., F.F. Lombard, and P.E. Aho. 1979. A new variety of *Phellinus pini* associated with cankers and decay in white firs in southwestern Oregon and northern California. Canad. J. Forest Res. 9: 31-38.

2050. Larsen, M.J., F.F. Lombard, and C.S. Hodges. 1985. Hawaiian forest fungi V.A new species of *Phellinus* (Hymenochaetaceae) causing decay of*Casuarina* and *Acacia*. Mycologia 77: 345-352.

2051. Larsen, M.J., and K.K. Nakasone. 1984. Additional new taxa of *Laeticorticium* (Aphyllophorales, Corticiaceae). Mycologia 76: 528-532.

2052. Larsen, P.O., A.K. Hagan, B.G. Joyner, and D.A. Spilker. 1981. Leaf blight and crown rot on creeping bentgrass, a new disease caused by *Drechslera catenaria*. Pl. Dis. 65: 79-81.

2053. Laskaris, T. 1950. The *Diplodina* disease of *Delphinium*. Phytopathology 40: 615-626.

2054. Latham, A.J. 1970. Development of apple fruit rot and basidiocarp formation by *Schizoph yllum commune*. Phytopathology 60: 596-598.

2055. Latham, A.J. 1970. Liver spot disease of pecan in Alabama. Pl. Dis. Reporter 54: 408.

2056. Latham, A.J. 1972. Some hosts of *Cristulariella pyramidalis* in pecan orchard hedgerows. Pl. Dis. Reporter 56: 176-177.

2057. Latham, A.J., and H.J. Amling. 1968. A new leafspot disease of pecan incited by *Cristulariella pyramidalis* in Alabama and Georgia. Pl. Dis. Reporter 52: 747.

2058. Laundon, G.F. 1963. Rust fungi 1: On Acanthaceae. Mycol. Pap. 89: 1-89.

2059. Laursen, G.A., and J.F. Ammirati. 1982. Arctic and Alpine Mycology Univ. Washington Press. 559 p.

2060. Laut, J.G., L.B. Helburg, and M.E. Schomaker. 1974. Dutch elm disease in Colorado - 1973. Pl. Dis. Reporter 58: 571-572.

2061. Leach, C.M. 1958. Sclerotia of *Typhula idahoensis* found mixed with Idaho-grown seed of *Trifolium pratense*. Pl. Dis. Reporter 42: 383.

2062. Leach, C.M. 1958. A disease of dodder caused by the fungus *Colletotrichum destructivum*. Pl. Dis. Reporter 42: 827-829.

2063. Leach, C.M. 1959. A survey of root deterioration of *Medicago sativa* in Oregon. Pl. Dis. Reporter 43: 622-625.

2064. Leach, C.M., and M. Tulloch. 1971. *Pithomyces chartarum*, a mycotoxin-producing fungus, isolated from seed and fruit in Oregon. Mycologia 63: 1086- 1089.

2065. Leach, C.M., and M. Tulloch. 1972. World-wide occurrence of the suspected mycotoxin producing fungus *Drechslera biseptata* with grass seed. Mycologia 64: 1357-1359.

2066. Leaphart, C.D. 1955. Preliminary observations on a current outbreak of western gall rust (*Cronartium coleosporioides*). Pl. Dis. Reporter 39: 314-315.

2067. Leaphart, C.D. 1960. A root stain disease of eastern white pine. Pl. Dis. Reporter 44: 704-706.

2068. Leath, K.T. 1978. Crownwart of alfalfa in Pennsylvania. Pl. Dis. Reporter 62: 621-623.

2069. Leath, K.T., and J.E. Baylor. 1975. *Phytophthora* root rot of alfalfa in Pennsylvania. Pl. Dis. Reporter 59: 972-974.

2070. Leath, S., and R.B. Carroll. 1982. Powdery mildew on soybean in Delaware. Pl. Dis. 66: 70-71.

2071. Leath, S., and K.J. Leonard. 1986. Distribution of race 3 of *Cochliobolus carbonus* on corn in North Carolina. Pl. Dis. 70: 800.

2072. Leathers, C.R. 1960. Downy mildew on watermelons in Arizona, a first report. Pl. Dis. Reporter 44: 372.

2073. Leathers, C.R., and A.H. Smith. 1967. Two new species of clavarioid fungi. Mycologia 59: 456-462.

2074. Lebeau, J.B., and C.E. Logsdon. 1958. Snow mold of forage crops in Alaska and Yukon. Phytopathology 48: 148-150.

2075. Lee, B.K.H., and G.E. Baker. 1973. Fungi associated with the roots of red mangrove, *Rhizophora mangle*. Mycologia 65: 894-906.

2076. Lee, F.N. 1980. Number, viability, and buoyancy of *Rhizoctonia solani* sclerotia in Arkansas rice fields. Pl. Dis. 64: 298-300.
2077. Lee, T.A., Jr., and R.W. Toler. 1972. St. Augustine rust in Texas. Pl. Dis. Reporter 56: 630-632.
2078. Lefebvre, C.L. 1950. Observations on plant diseases in Alaska. Pl. Dis. Reporter 34: 3-4.
2079. Lehman, S.G. 1923. Pod and stem blight of soybean. Ann. Missouri Bot. Gard. 10: 111-178.
2080. Lehman, S.G. 1951. *Curvularia* leaf blight of Ladino clover. Pl. Dis. Reporter 35: 79-80.
2081. Lehman, S.G., H. Murakishi, and J.H. Graham. 1951. A leaf spot of soybean caused by *Sclerotium rolfsii*. Pl. Dis. Reporter 35: 167-168.
2082. Lehman, S.G., and F.A. Wolf. 1926. Soy-bean anthracnose. J. Agric. Res. 33: 381-390.
2083. Lemke, P.A. 1964. The genus *Aleurodiscus* (sensu stricto) in North America. Canad. J. Bot. 42: 213-282.
2084. Lemke, P.A. 1964. The genus *Aleurodiscus* (sensu lato) in North America. Canad. J. Bot. 42: 723-768.
2085. Lengkeek, V.H., and T. Sim. 1979. An outbreak of sorghum downy mildew in Kansas. Pl. Dis. Reporter 63: 905-907.
2086. Lenné, J.M., and R.M. Sonoda. 1978. Occurrence of *Colletotrichum dematium* f. sp. *truncata* on *Stylosanthes* spp. Pl. Dis. Reporter 62: 641-644.
2087. Lentz, P.L. 1951. *Labrella coryli* (Desm. & Rab.)Sacc. on filberts in North America. Pl. Dis. Reporter 35: 437-438.
2088. Lentz, P.L. 1955. A scurf of sugar beets caused by *Cylindrocladium scoparium*. Pl. Dis. Reporter 39: 654-655.
2089. Lentz, P.L. 1955. *Microstroma tonellianum* on sweet cherries. Pl. Dis. Reporter 39: 697.
2090. Lentz, P.L., and H.H. Burdsall Jr. 1973. *Scytinostroma galactinum* as a pathogen of woody plants. Mycopathol. Mycologia Appl. 49: 289-305.
2091. Lentz, P.L., C.E. Turner, L.W. Robertson, and W.A. Gentner. 1974. First North American record for *Cercospora cannabina*, with notes on the identification of *C. cannibina* and *C. cannabis*. Pl. Dis. Reporter 58: 165-168.
2092. Lentz, P.L., and L.O. Weaver. 1955. A disease of fig (*Ficus carica*). Pl. Dis. Reporter 39: 822-823.
2093. Lentz, P.W. 1958. *Septoria* spot of plum fruits. Pl. Dis. Reporter 42: 1123-1124.
2094. Lenz, L.W. 1941. A new smut from Louisiana. Mycologia 33: 155-157.
2095. Leonard, K.J. 1974. Foliar pathogens of corn in North Carolina. Pl. Dis. Reporter 58: 532-534.
2096. Leonard, K.J., and S. Leath. 1986. Evidence that race 1 of *Setosphaeria turcica* caused the 1985 northern corn leaf blight epidemic in North Carolina. Pl. Dis. 70: 981.
2097. Leonard, K.J., and E.G. Suggs. 1974. *Setosphaeria prolata*, the ascigerous state of *Exserohilum prolatum* . Mycologia 66: 281-297.
2098. Leonard, K.J., and D.L. Thompson. 1969. Corn stalk rot fungi in North Carolina. Pl. Dis. Reporter 53: 718-720.
2099. Leppik, E.E. 1960. *Cercospora traversiana* and some other pathogens of fenugreek new to North America. Pl. Dis. Reporter 44: 40-44.
2100. Leppik, E.E. 1960. *Melilotus italica*, a new host for *Uromyces striatus*. Pl. Dis. Reporter 44: 184-185.
2101. Leppik, E.E. 1965. *Plasmopara halstedii* and other diseases on *Dimorphotheca*. Pl. Dis. Reporter 49: 940-942.
2102. Levine, M.N., and R.C. Hildreth. 1957. A natural occurrence of the aecial stage of *Puccinia rubigo-vera* var. *tritici* in the United States. Phytopathology 47: 110-112.
2103. Lewis, G.D. 1957. White rot of onion in New York. Pl. Dis. Reporter 41: 814.
2104. Lewis, G.D. 1962. *Stemphylium* leaf spot of parsley. Phytopathology 52: 925.
2105. Lewis, J.A. 1973. Effect of mineral salts on *Aphanomyces euteiches* and *Aphanomyces* root rot of peas. Phytopathology 63: 989-993.
2106. Lewis, R. Jr. 1985. Temperature tolerance and survival of *Ceratocystis fagacearum* in Texas. Pl. Dis. 69: 443-444.
2107. Leyendecker, P.J. 1950. 1949 Plant disease survey for New Mexico. Pl. Dis. Reporter 34: 39-40.
2108. Leyendecker, P.J. 1954. Reduction in yield of cotton caused by diseases in 1953. Pl. Dis. Reporter 38: 223-225.
2109. Leyendecker, P.J. Jr. 1951. 1950 plant disease survey for New Mexico. Pl. Dis. Reporter 35: 168-169.
2110. Leyendecker, P.J. Jr. 1952. Root rot of cotton caused by *Thielaviopsis basicola* discovered in the upper Rio Grande Valley of Texas. Pl. Dis. Reporter 36: 53.
2111. Leyendecker, P.J. Jr. 1952. *Cytospora* canker of apple in New Mexico. Pl. Dis. Reporter 36: 276-277.
2112. Leyendecker, P.J., and R.M. Nakayama. 1956. 1955 plant disease summary for New Mexico. Pl. Dis. Reporter 40: 159-161.
2113. Leyendecker, P.J., and R.M. Nakayama. 1957. 1956 Plant disease summary for New Mexico. Pl. Dis. Reporter 41: 53-54.
2114. Leyendecker, P.J., A.L. Smith, W.E. Cooper, and L. Lett. 1957. Reduction in yield of cotton caused by diseases in 1956. Pl. Dis. Reporter 41 : 124-127.
2115. Liberta, A.E. 1960. A taxonomic analysis of section Athele of the genus *Corticium*. I. Genus *Xenasma*. Mycologia 52:: 884-914.
2116. Liberta, A.E. 1965. Notes on Wisconsin resupinate basidiomycetes. Mycologia 57: 459-464.
2117. Liberta, A.E. 1968. Descriptions of the nomenclatural types of Peniophoras described by Burt. Mycologia 60: 827-857.
2118. Liberta, A.E. 1969. Descriptions of the nomenclatural types of Corticiums described by Burt. Nova Hedwigia 18: 215-233.
2119. Liberta, A.E., and G.H. Boewe. 1960. A new species of *Cercospora* on *Acer saccharinum*. Mycologia 52: 345-347.
2120. Libonati-Barnes, S.D., and S.A. Redhead. 1984. *Panellus longinquus* subsp. *pacificus* a new west coast North Amer ican agaric associated with red alder. Mycotaxon 20: 205-212.
2121. Lichtwardt, R.W., G.L. Barron, and L.H. Tiffany. 1958. Mold flora associated with shelled corn in Iowa. Iowa State Coll. J. Sci. 33: 1-11.
2122. Lightle, P.C. 1960. *Fomes annosus* root rot of loblolly pine. Pl. Dis. Reporter 44: 423.
2123. Lightle, P.C., and F.G. Hawksworth. 1965. New hosts for broom-causing fungi in the southwest. Pl. Dis. Reporter 49: 417-418.
2124. Lim, S.M., and T. Hymonitz. 1987. Reactions of perennial wild species of genus *Glycine* to *Septoria glycines*. Pl. Dis. 71: 891-893.
2125. Limber, D.P. 1950. *Ophiostoma* on *Narcissus* bulbs. Phytopathology 40: 493-496.
2126. Limber, D.P. 1951. New records of *Coniothyrium hellebori* in the United States. Pl. Dis. Reporter 35: 277.
2127. Limber, D.P. 1954. Outbreak of *Fraxinus* rust in Massachusettes. Pl. Dis. Reporter 38: 696.
2128. Limber, D.P. 1955. Studies in the genus *Sporonema*. Mycologia 47: 389-402.
2129. Lin, Y.-S., and R.J. Cook. 1977. Root rot of lentils caused by *Fusarium roseum* 'avenaceum'. Pl. Dis. Reporter 61: 752-755.
2130. Linder, D.H. 1937. Mycological Society of America, the summer foray, September 3-5, 1936. Mycologia 29: 365-375.
2131. Linder, D.H. 1942. A contribution towards a monograph of the genus *Oidium* (Fungi Imperfecti). Lloydia 5: 165-207.
2132. Linderman, R.G. 1972. Occurence of azalea petal blight in Connecticut. Pl. Dis. Reporter 56: 1101-1102.
2133. Linderman, R.G., and F. Zeitoun. 1977. *Phytophthora cinnamomi* causing root rot and wilt of nursery-grown native western azalea and salal. Pl. Dis. Reporter 61: 1045-1048.
2134. Lindsey, J., and R.L. Gilbertson. 1975. Wood-inhabiting Homobasidiomycetes on saguaro in Arizona. Mycotaxon 2: 83-103.
2135. Lindsey, J.P. 1985. Basidiomycetes that decay gambel oak in southwestern Colorado. Mycotaxon 22: 327-362.
2136. Lindsey, J.P. 1986. Basidiomycetes that decay Gambel oak in southwestern Colorado: II. Mycotaxon 25: 67-83.
2137. Lindsey, J.P. 1986. Basidiomycetes that decay gambel oak in southwestern Colorado: III. Mycotaxon 27: 325-345.
2138. Lindsey, J.P. 1987. A new species of *Aleurodiscus* from Colorado. Mycotaxon 30: 433-437.
2139. Lindsey, J.P., and R.L. Gilbertson. 1977. A new *Steccherinum* (Aphyllophorales, Steccherinaceae) on quaking aspen. Mycologia 69: 193-197.
2140. Lindsey, J.P., and R.L. Gilbertson. 1977. Some aspects of bipolar heterothallism and other cultural characters of *Poria carnegiea*. Mycologia 69: 761-772.
2141. Lindsey, J.P., and R.L. Gilbertson. 1977. New species of corticioid fungi on quaking aspen. Mycotaxon 5: 311-319.
2142. Lindsey, J.P., and R.L. Gilbertson. 1978. Basidiomycetes that

decay aspen in North America. Bibliotheca Mycologica 63: 1-406.
2143. Lindsey, J.P., and R.L. Gilbertson. 1979. A new *Steccherinum* (Aphyllophorales, Steccherinaceae) from Alaska. Mycologia 71: 1264-1267.
2144. Lindsey, J.P., and R.L. Gilbertson. 1983. Notes on Basidiomycetes that decay bristlecone pine. Mycotaxon 18: 541-559.
2145. Linford, M.B., and H.H. McKinney. 1954. Occurrence of *Polymyxa graminis* in roots of small grains in the United States. Pl. Dis. Reporter 38: 711-713.
2146. Ling, L. 1950. Studies in the genus *Cintractia*. I. *C. montagnei* and related species. Mycologia 42: 503-513.
2147. Ling, L. 1950. Studies in the genus *Cintractia*. II. *C. axicola* and related species. Mycologia 42: 646-653.
2148. Ling, L. 1951. Studies in the genus *Cintractia*. III. *C. leucoderma* and related species. Mycologia 43: 310-318.
2149. Lipps, P.E. 1980. A new species of *Pythium* isolated from wheat beneath snow in Washington. Mycologia 72: 1127-1133.
2150. Lipps, P.E. 1987. Gray leaf spot epiphytotic in Ohio corn. Pl. Dis. 71: 281.
2151. Lipps, P.E., and G.W. Bruehl. 1978. Snow rot of winter wheat in Washington. Phytopathology 68: 1120-1127.
2152. Lipps, P.E., and L.J. Herr. 1982. Etiology of *Rhizoctonia cerealis* in sharp eyespot of wheat. Phytopathology 72: 1574-1577.
2153. Lipps, P.E., and R.M. Riedel. 1984. *Macrophomina phaseolina* causing charcoal rot of soybean in Ohio. Pl. Dis. 68: 536.
2154. Lipscomb, H.A., and W. Witcher. 1965. Canker of tulip poplar caused by *Fusarium solani*. Pl. Dis. Reporter 49: 507-508.
2155. Livingston, J.E. 1950. Small grain diseases in Nebraska in 1950. Pl. Dis. Reporter 34: 232.
2156. Livingston, W.H., and R.W. Davidson. 1987. *Ophiostoma subannulatum*, a new fungal species pathogenic to grand fir roots. Mycologia 79: 144-147.
2157. Livingston, W.H., and M.J. Wingfield. 1982. First report of *Verticicladiella procera* on pines in Minnesota. Pl. Dis. 66: 260-261.
2158. Llewellyn, G.C., and C.E. O'Rear. 1977. Examination of fungal growth and aflatoxin production on marihuana. Mycopathologia 62: 109-112.
2159. Lloyd, P.J., and D.F. Crossan. 1958. Anthracnose of pansy and violet caused by *Colletotrichum violae-tricoloris*. Pl. Dis. Reporter 42: 86-90.
2160. Lockwood, J.L., D.L. Yoder, and N.A. Smith. 1970. *Thielaviopsis basicola* root rot of soybeans in Michigan. Pl. Dis. Reporter 54: 849-850.
2161. Lockwood, J.L., D.J. deZeeuw, A.L. Andersen, and D.J. Hagedorn. 1957. Pea diseases in Michigan, 1955 and 1956. Pl. Dis. Reporter 41: 478-480.
2162. Logsdon, C.E. 1955. Notes on plant disease incidence in the Matanuska and Tanana Valleys, Alaska - 1953 and 1954. Pl. Dis. Reporter 39: 274.
2163. Logsdon, C.E., and G. Strobel. 1960. Additional records of vegetable diseases in Alaska. Pl. Dis. Reporter 44: 92-93.
2164. Lombard, F.F. 1983. A cultural study of *Piptoporus soloniensis* (Aphyllophorales, Polyporaceae). Mycologia 75: 723-727.
2165. Lombard, F.F., H.H. Burdsall Jr., and R.L. Gilbertson. 1975. Taxonomy of *Corticium chrysocreas* and *Phlebia livida*. Mycologia 67: 495-510.
2166. Lombard, F.F., R.W. Davidson, and J.L. Lowe. 1960. Cultural characteristics of *Fomes ulmarius* and *Poria ambigua*. Mycologia 52: 280-294.
2167. Lombard, F.F., and R.L. Gilbertson. 1965. Studies on some western Porias with negative or weak oxidase reactions. Mycologia 57: 43-76.
2168. Lombard, F.F., and R.L. Gilbertson. 1966. *Poria luteoalba* and some related species in North America. Mycologia 58: 827-845.
2169. Lombard, F.F., and M.J. Larsen. 1985. *Phellinus bicuspidatus* (Hymenochaetales, Hymenochaetaceae), a new species associated with a white sap rot of oak in Louisiana. Mycologia 77: 55-61.
2170. Long, E.A., and J.W. Hilty. 1985. Occurrence of *Acremonium coenophialum* in tall fescue in Tennessee. Pl. Dis. 69: 467-468.
2171. Long, W.H. 1913. Three undescribed heart-rots of hardwood trees, especially oak. J. Agric. Res. 1: 109-128.
2172. Longenecker, J., B.R. Kistler, and W. Merrill. 1975. Major conifer disease occurrences and new records for Pennsylvania. Pl. Dis. Reporter 59: 702-704.
2173. Loring, L.B. 1957. A new disease on western flowering dogwood. Pl. Dis. Reporter 41: 810.
2174. Loring, L.B. 1957. *Phytophthora cinnamomi* Rands apparently parasitic on two juniper varieties. Pl. Dis. Reporter 41: 815.
2175. Loring, L.B., and H. Nicolaison. 1955. Powdery mildew on *Tolmiea menziesii*. Pl. Dis. Reporter 39: 334.
2176. Loring, L.B., and L.F. Roth. 1964. *Pucciniastrum epilobii* on *Fuschia* in Oregon. Pl. Dis. Reporter 48: 99.
2177. Loring, L.B., and H. Smithson. 1970. Twig dieback of *Abies* saplings. Pl. Dis. Reporter 54: 700.
2178. Lorio, P.L. Jr. 1966. *Phytophthora cinnamomi* and *Pythium* species associated with loblol ly pine decline in Louisiana. Pl. Dis. Reporter 50: 596-596.
2179. Lowe, J.L. 1975. Polyporaceae of North America. The genus *Tyromyces*. Mycotaxon 2: 1-82.
2180. Lowe, J.L., and R.L. Gilbertson. 1961. Synopsis of the Polyporaceae of the Western United States and Canada. Mycologia 53: 474-511.
2181. Lowen, R., and K.P. Dumont. 1984. Taxonomy and nomenclature in the genus *Calycellina* (Hyaloscyphaceae). Mycologia 76: 1003-1023.
2182. Lowy, B., and W.B. Cooke. 1965. The 1960 Louisiana foray. Mycologia 57: 478-483.
2183. Lucas, L.T. 1976. *Sclerotium rolfsii* on bentgrass in North Carolina. Pl. Dis. Reporter 60: 820-822.
2184. Lucas, L.T., T.B. Warren, W.W. Woodhouse Jr., and E.D. Seneca. 1971. *Marasmius* blight, a new disease of American beachgrass. Pl. Dis. Reporter 55: 582-585.
2185. Luck-Allen, E.R. 1963. The genus *Basidiodendron*. Canad. J. Bot. 41: 1025-1052.
2186. Luepschen, N.S., J.E. Hetherington, F.J. Stahl, and K.E. Mowrer. 1979. *Cytospora* canker of peach trees in Colorado: survey of incidence, canker location and apparent infection courts. Pl. Dis. Reporter 63: 685-687.
2187. Luke, H.H., and D.T. Sechler. 1963. Rye anthracnose. Pl. Dis. Reporter 47: 936-937.
2188. Luke, H.H., A.T. Wallace, and W.H. Chapman. 1957. A new disease symptom incited by the oat leaf blotch pathogen, *Helminthosporium avenae*. Pl. Dis. Reporter 41: 109-110.
2189. Luley, C.J., L.H. Tiffany, and H.S. McNabb Jr. 1987. In vitro production of *Mycosphaerella populorum* ascomata. Mycologia 79: 654-658.
2190. Lunsford, J.N., and J.A. Spencer. 1976. A variety of *Collecephalus hemerocalli* pathogenic on *Hemerocallis*. Phytopathology 66: 245-248.
2191. Lutey, R.W., and R.P. Covey. 1959. Crown rust on barley in Minnesota. Pl. Dis. Reporter 43: 1287.
2192. Luther, B.S., and S.A. Redhead. 1981. *Crepidotus cinnabarinus* in North America. Mycotaxon 12: 417-430.
2193. Luttrell, E.S. 1940. Tar spot of American holly. Bull. Torrey Bot. Club 67: 692-704.
2194. Luttrell, E.S. 1950. Grain sorghum diseases in Georgia -- 1949. Pl. Dis. Reporter 34: 45-52.
2195. Luttrell, E.S. 1950. *Botryosphaeria* stem canker of elm. Pl. Dis. Reporter 34: 138-139.
2196. Luttrell, E.S. 1951. Diseases of tall fescue grass in Georgia. Pl. Dis. Reporter 35: 83-87.
2197. Luttrell, E.S. 1951. Diseases of guar in Georgia. Pl. Dis. Reporter 35: 166.
2198. Luttrell, E.S. 1952. Diseases of brome grasses in Georgia. Pl. Dis. Reporter 36: 283-286.
2199. Luttrell, E.S. 1953. *Melanconium* leaf and stem fleck of grapes. Phytopathology 43: 347-348.
2200. Luttrell, E.S. 1954. An undescribed species of *Piricularia* on sedges. Mycologia 46: 810-814.
2201. Luttrell, E.S. 1954. Diseases of pearl millet in Georgia. Pl. Dis. Reporter 38: 507-514.
2202. Luttrell, E.S. 1960. The morphology of an undescribed species of *Dothiora*. Mycologia 52: 64-79.
2203. Luttrell, E.S. 1963. A *Trichometasphaeria* perfect stage for a *Helminthosporium* causing leaf blight of *Dactyloctenium*. Phytopathology 53: 281-285.
2204. Luttrell, E.S. 1976. Ovarian infection of *Sporobolus poiretii* by *Bipolaris ravenelii*. Phytopathology 66: 260-268.
2205. Luttrell, E.S. 1977. The disease cycle and fungus-host

relationships in dallisgrass ergot. Phytopathology 67: 1461-1468.
2206. Luttrell, E.S., and M.E. Barr. 1978. Additional species of *Pododimeria* (Loculascomycetes). Amer. J. Bot. 65: 251-257.
2207. Luttrell, E.S., and L.W. Boyle. 1960. Leaf spot of peanut in Georgia caused by *Leptosphaerulina arachidicola*. Pl. Dis. Reporter 44: 609-611.
2208. Luttrell, E.S., H.B. Harris, and H.D. Wells. 1974. *Bipolaris* leaf blight of *Panicum fasciculatum*: Effects of host age and photoperiod on susceptibility. Phytopathology 64: 476-480.
2209. Luttrell, E.S., and J.L. Weimer. 1952. *Macrophomina* stem canker and ashy stem blight of cowpea. Pl. Dis. Reporter 36: 194-195.
2210. Lutz, A., and J. Menge. 1986. Breaking winter dormancy of *Phytophthora parasitica* propagules using heat shock. Mycologia 78: 148-150.
2211. Luz, W.C., da, and G.C. Bergstrom. 1985. *Septoria avenae* spot as an additional component of the fungal leaf spot syndrome of spring wheat in New York. Pl. Dis. 69: 724-725.
2212. Luz, W.C., da, and G.C. Bergstrom. 1986. Distribution, prevalence, and severity of fungal foliar diseases of spring wheat in New York in 1984 and 1985. Pl. Dis. 70: 842-847.
2213. Lyle, J.A. 1950. Diseases of small grains in Alabama. Pl. Dis. Reporter 34: 318-320.
2214. Maas, J.L. 1969. *Idriella* root rot of strawberry: a new disease in Maryland. Pl. Dis. Reporter 53: 494-496.
2215. Maas, J.L. 1971. *Hyalotia pistacina*, a new species, and a note of *Pestalotia thuja* . Mycologia 63: 663-667.
2216. Maas, J.L. 1978. Anthracnose of strawberry fruit in Maryland. Pl. Dis. Reporter 62: 488-492.
2217. Maas, J.L., F.G. Pollack, and F.A. Uecker. 1979. Morphology and development of *Pilidiella quercicola*. Mycologia 71: 92-102.
2218. Maas, J.L., and R.L. Powelson. 1972. Growth and sporulation of *Botrytis convoluta* with various carbon and nitrogen sources. Mycologia 64: 897-903.
2219. Maas, J.L., and D.E. Stuntz. 1969. Mycoecology of serpentine soil. Mycologia 61: 1106-1116.
2220. Maas, J.L., and F.A. Uecker. 1984. *Botryosphaeria dothidea* cane canker of thornless blackberry. Pl. Dis. 68: 720-726.
2221. Maas, J.L., T. van der Zwet, and G. Madden. 1971. A severe *Septoria* leaf spot of pistachio nut trees new to the United States. Pl. Dis. Reporter 55: 72-76.
2222. MacDonald, J.D., and E.E. Butler. 1981. *Cylindrocarpon* root rot of tulip poplar. Pl. Dis. 65: 154-157.
2223. MacDonald, J.D., J.R. Hartman, and J.D. Shapiro. 1984. Pathogens of ice plant in California. Pl. Dis. 68: 965-967.
2224. MacLean, N.A., and K.F. Baker. 1950. Powdery mildew on outdoor snapdragons in California. Pl. Dis. Reporter 34: 346.
2225. Macrae, R. 1955. Cultural and interfertility studies in *Aporpium caryae*. Mycologia 47: 812-820.
2226. Macrae, R., and K. Aoshima. 1966. *Hirschioporus* (*Lenzites*) *laricinus* and its synonyms: *L. abietis, L. ambigua, L. pinicola*. Mycologia 58: 912-926.
2227. Maffel, H.M., and H.L. Morton. 1983. *Phomopsis* canker of Russian-olive in southeastern Michigan. Pl. Dis. 67: 964-965.
2228. Mahoney, M.J., and T.A. Tattar. 1980. Identification, etiology, and control of *Euonymus fortunei* anthracnose caused by *Colletotrichum gloeosporiodes*. Pl. Dis. 64: 854-856.
2229. Mahr, S.E.R., W.R. Stevenson, and L. Sequeira. 1986. Control of bottom rot of head lettuce with iprodione. Pl. Dis. 70: 506-509.
2230. Mains, E.B., and H.S. Jackson. 1923. Aecial stages of the leaf-rust of barley, *Puccinia simplex*, and of rye, *P. dispersa* in the United States. Phytopathology 13: 49-50.
2231. Malca, I.M., and J.H. Owen. 1957. The gray-leaf-spot disease of St. Augustine grass. Pl. Dis. Reporter 41: 871-875.
2232. Malloch, D., and C.T. Rogerson. 1977. *Pulveria*, a new genus of Xylariaceae (Ascomycetes). Canad. J. Bot. 55: 1505-1509.
2233. Malm, N.R., and D.C.H. Hsi. 1964. Occurrence of rust on grain sorghum hybrids and varieties in New Mexico. Pl. Dis. Reporter 48: 381.
2234. Maloy, O.C. 1968. Alfalfa anthracnose in Washington. Pl. Dis. Reporter 52: 983.
2235. Maloy, O.C. 1968. Decay fungi in young grand fir. Pl. Dis. Reporter 52: 489-492.
2236. Maloy, O.C., and F.R. Matthews. 1960. Southern cone rust: distribution and control. Pl. Dis. Reporter 44: 36-39.
2237. Maloy, O.C., C.B. Skotland, and H. Willson. 1974. *Fusarium* wilt of muskmelon in Washington. Pl. Dis. Reporter 58: 10-12.
2238. Manandhar, J.B., G.L. Hartman, and J.B. Sinclair. 1986. *Colletotrichum destructivum*, the anamorph of *Glomerella glycines*. Phytopathology 76: 282-285.
2239. Maneval, W.E. 1948. A supplementary list of Missouri fungi. Univ. Missouri Stud., Sci. Ser. 22: 1-65.
2240. Manion, P.D., and D.W. French. 1967. *Nectria galligena* and *Ceratocystis fimbriata* cankers of aspen in Minnesota. Forest Sci. 13: 23-28.
2241. Manion, P.D., and F.A. Valentine. 1971. Diseases of trembling aspen in the Adirondack region of New York. Pl. Dis. Reporter 55: 662-665.
2242. Mankin, C.J. 1969. Diseases of grasses and cereals in South Dakota. Agric. Exp. Sta., South Dakota State Univ., Tech. Bull. 35: 1-27.
2243. Mankin, C.J. 1969. Fungous diseases on non-grass plants in South Dakota. Agric. Exp. Sta., South Dakota State Univ., Tech. Bull. 36: 1-28.
2244. Manning, W.J. 1980. Relationship of *Rhizoctonia solani* and *Colletotrichum coccodes* to basal stem canker of tomato. Pl. Dis. 64: 76-78.
2245. Manning, W.J., P.M. Papia, and E.A. Cox. 1971. Fusaria associated with wilted tomato plants in Massachusetts. Pl. Dis. Reporter 55: 687-691.
2246. Manning, W.J., and P.M. Vardaro. 1974. Brown root rot of tomato in Massachusetts. Pl. Dis. Reporter 58: 483-486.
2247. Mark, W.R., F.G. Hawksworth, and N. Oshima. 1976. Resin disease: A new disease of lodgepole pine dwarf mistletoe. Canad. J. Forest Res. 6: 415-424.
2248. Marks, G.C., J.G. Berbee, and A.J. Riker. 1965. *Colletotrichum* shoot blight of poplar. Forest Sci. 11: 204-215.
2249. Marlatt, R.B. 1958. Onion *Fusarium* basal rot in Arizona. Pl. Dis. Reporter 42: 667-668.
2250. Marlatt, R.B. 1970. Isolation, inoculation, temperature relations and culture of a *Cercospora* pathogenic to *Ficus elastica* "decora". Pl. Dis. Reporter 54: 199-202.
2251. Marlatt, R.B., R.M. Allen, and R.T. McKittrick. 1959. Control of cantaloup powdery mildew in Arizona with fungicides. Pl. Dis. Reporter 43: 67-69.
2252. Marlatt, R.B., and S. Goldweber. 1969. *Verticillium* wilt of avocado (*Persea americana*) in Florida. Pl. Dis. Reporter 53: 583.
2253. Marlatt, R.B., and J.W. Kimbrough. 1979. *Puccinia psidii* on *Pimenta dioica* in south Florida. Pl. Dis. Reporter 63: 510-512.
2254. Marlatt, R.B., and J.F. Knauss. 1974. A leaf spot of *Aechmea fasciata*. Phytopathology 64: 768-769.
2255. Marlatt, R.B., and J.F. Knauss. 1974. A new leaf disease of *Aechmea fasciata* caused by *Helminthosporium rostratum*. Pl. Dis. Reporter 58: 445-448.
2256. Marlatt, R.B., R.J. Knight, and S. Goldweber. 1970. *Verticillium* wilt of mango (*Mangifera indica*) in Florida. Pl. Dis. Reporter 54: 569-571.
2257. Marlatt, R.B., K. Pohronezny, and R. McSorley. 1983. Field survey of Tahiti lime, *Citrus latifolia*, for algal disease, melanose, and greasy spot in southern Florida. Pl. Dis. 67: 946-949.
2258. Marlatt, R.B., and W.H. Ridings. 1974. *Sphaeropsis* gall of bottlebrush tree, *Callistemon viminalis*, a new host. Phytopathology 64: 1001-1003.
2259. Marlatt, R.B., and W.H. Ridings. 1976. *Sphaeropsis* disease of *Carissa grandiflora*. Pl. Dis. Reporter 60: 842-843.
2260. Marlatt, R.B., and W.H. Ridings. 1979. *Sphaeropsis* gall of *Schinus terebinthifolius*, a new host. Pl. Dis. Reporter 63: 786-787.
2261. Marousky, F.J., and P.P.Q. Wildt de. 1982. Postharvest decay in Florida leatherleaf fern. Pl. Dis. 66: 1029-1031.
2262. Marshall, B.H. Jr. 1952. Cane blight of forsythia in Maryland. Pl. Dis. Reporter 36: 440-441.
2263. Marshall, D. 1987. First report of take-all of wheat in Texas. Pl. Dis. 71: 850.
2264. Marshall, H.G., and R.D. Schein. 1960. Occurrence of snow mold, *Typhula itoana* Imai, on winter oats in Pennsylvania. Pl. Dis. Reporter 44: 894-895.
2265. Martin, K.J., and R.L. Gilbertson. 1973. The mating system and some other cultural aspects of *Veluticeps berkeleyi*. Mycologia 65: 548-557.
2266. Martin, K.J., and R.L. Gilbertson. 1976. Cultural and other

morphological studies of *Sparassis radicata* and related species. Mycologia 68: 622-639.
2267. Martin, K.J., and R.L. Gilbertson. 1977. Synopsis of wood-rotting fungi on spruce in North America: I. Mycotaxon 6: 43-77.
2268. Martin, K.J., and R.L. Gilbertson. 1978. Decay of Douglas-fir by *Sparassis radicata* in Arizona. Phytopathology 68: 149-154.
2269. Martin, K.J., and R.L. Gilbertson. 1978. Synopsis of wood-rotting fungi on spruce in North America: II. Mycotaxon 7: 337-356.
2270. Martin, K.J., and R.L. Gilbertson. 1980. Synopsis of wood rotting-fungi on spruce in North America: III. Mycotaxon 10: 479-501.
2271. Martin, S.B., and L.T. Lucas. 1984. Characterization and pathogenicity of *Rhizoctonia* spp. and binuclea te *Rhizoctonia*-like fungi from turfgrasses in North Carolina. Phytopathology 74: 170-178.
2272. Martinez, A.P. 1961. *Sphaerophragmium* rust discovered in Florida on *Albizia lebbeck* (L.) Benth. Pl. Dis. Reporter 45: 560-561.
2273. Martyn, R.D. 1985. An aggressive race of *Fusarium oxysporum* f. sp. *niveum* new to the United States. Pl. Dis. 69: 1007.
2274. Martyn, R.D., L.W. Barnes, and J. Amador. 1987. *Fusarium* wilt (*F. oxysporum* f. sp. *melonis* Race O) of muskmelon in Texas. Pl. Dis. 71: 469.
2275. Martyn, R.D., L.W. Barnes, and J. Amador. 1987. *Fusarium* yellows of celery in Texas. Pl. Dis. 71: 651.
2276. Martyn, R.D., and T.K. Hartz. 1986. Use of soil solarization to control *Fusarium* wilt of watermelon. Pl. Dis. 70: 762-766.
2277. Mason, D.L. 1973. Host-parasite relations in the *Gibberidea* disease of *Helianthus strumosus*. Mycologia 65: 1158-1170.
2278. Mason, D.L., and M.P. Backus. 1969. Host-parasite relations in spot anthracnose of *Desmodium*. Mycologia 61: 1124-1141.
2279. Matheis, W. 1977. Über einige *Dasyscyphus*-Arten auf blättern von *Vaccinium*. Sydowia 29: 237-244.
2280. Mathre, D.E. 1986. Occurrence of *Verticillium dahliae* on barley. Pl. Dis. 70: 981.
2281. Mathre, D.E., and R.H. Johnston. 1972. *Alternaria dianthi* leafspot of cowcockle in Montana. Pl. Dis. Reporter 56: 728.
2282. Mathre, D.E., R.H. Johnston, and R. Engel. 1986. Effect of seed treatment with triadimenol on severity of take-all of spring wheat caused by *Gaeumannomyces graminis* var. *tritici*. Pl. Dis. 70: 749-751.
2283. Matsumoto, T., D. Showers, D. Luscher, D. Higuera, and C. Krass. 1984. Kernel smut of rice. Calif. Dis. 3: 23-44.
2284. Matthews, F.D. 1969. *Uromyces costaricensis* -- A new report for the continental United States. Pl. Dis. Reporter 53: 213.
2285. Matthews, F.D. 1969. *Pucciniosira pallidula* - a new report for the continental United States. Pl. Dis. Reporter 53: 570.
2286. Matthews, F.D. 1969. *Uredo hameliae* - A new report for the continental United States. Pl. Dis. Reporter 53: 656.
2287. Matthews, F.D. 1969. *Phyllachora domingensis* and *Phyllachora simplex* - - New reports for the continental United States. Pl. Dis. Reporter 53: 820.
2288. Matuszewski, M. 1973. *Comandra* blister rust in Kentucky. Pl. Dis. Reporter 57: 17-18.
2289. Matyac, C.A., and T. Kommedahl. 1985. Factors affecting development of head smut caused by *Sphacelotheca reiliana* on corn. Phytopathology 75: 577-581.
2290. Matyac, C.A., and T. Kommedahl. 1985. Occurrence of chlorotic spots on corn seedlings infected with *Sphacelotheca reiliana* and their use in evaluation of head smut resistance. Pl. Dis. 69: 251-254.
2291. May, C., and R.W. Davidson. 1960. *Endothia parasitica* associated with a canker of live oak. Pl. Dis. Reporter 44: 754.
2292. McAlpine, R.G. 1961. *Hypoxylon tinctor* associated with a canker on American sycamore trees in Georgia. Pl. Dis. Reporter 45: 196-198.
2293. McBeath, J.H. 1985. Pink snow mold on winter cereals and lawn grasses in Alaska. Pl. Dis. 69: 722-723.
2294. McBeath, J.H. 1986. Spruce needle rust epidemic in Alaska caused by *Chrysomyxa ledicola*. Pl. Dis. 70: 801.
2295. McCain, A.H. 1961. *Gymnosporangium fuscum* found on pears in California. Pl. Dis. Reporter 45: 151.
2296. McCain, A.H. 1963. New or unusual hosts of *Verticillium* albo-*atrum*. Pl. Dis. Reporter 47: 233.
2297. McCain, A.H. 1964. *Dematophora* root rot of *Hypericum* in California. Pl. Dis. Reporter 48: 139.
2298. McCain, A.H., and R.F. Scharpf. 1986. *Phytophthora* shoot blight and canker disease of *Abies* spp. Pl. Dis. 70: 1036-1037.
2299. McCain, A.H., and E.E. Trujillo. 1967. Plant diseases previously unreported from Hawaii. Pl. Dis. Reporter 51: 1051-1053.
2300. McCain, J.W., and J.F. Hennen. 1982. Is the taxonomy of *Berberis* and *Mahonia* (*Berberidaceae*) support ed by their rust pathogens *Cumminsiella santa* sp. nov. and other *C umminsiella* species (*Uredinales*)?. Syst. Bot. 7: 48-59.
2301. McCarter, S.M., and S.J. Kays. 1984. Diseases limiting production of Jerusalem artichokes in Georgia. Pl. Dis. 68: 299-302.
2302. McCarter, S.M., and T.J. Ratcliffe. 1977. Incidence of major diseases on tomato transplants produced in Georgia. Pl. Dis. Reporter 61: 129-131.
2303. McCarter, S.M., R.W. Roncadori, and J.L. Crawford. 1970. Microorganisms associated with cotton boll rots in Georgia. Pl. Dis. Reporter 54: 586-590.
2304. McCartney, W.O. 1967. An unusual occurrence of eye rot of apple in California due to *Nectria galligena*. Pl. Dis. Reporter 51: 278-281.
2305. McClain, L.A. 1966. *Uredo ignava* Arthur - - A new report for the United States. Pl. Dis. Reporter 50: 112.
2306. McClain, L.A., and H.L. Rubin. 1965. *Septoria eriobotyrae* Maffei: A first report for the United States. Pl. Dis. Reporter 49: 1016.
2307. McClure, T.T. 1951. *Fusarium* foot rot of sweet-potato sprouts. Phytopathology 41: 72-77.
2308. McCoy, R.E., D.J. Mitchell, and F.J. Subirats. 1976. A new disease of *Pilea* caused by *Phytophthora parasitica*. Pl. Dis. Reporter 60: 680-681.
2309. McCracken, F.I. 1978. Spore release of some decay fungi of southern hardwoods. Canad. J. Bot. 56: 426-431.
2310. McCracken, F.I., and E.C. Burkhardt. 1977. Destruction of sycamore by canker stain in the midsouth. Pl. Dis. Reporter 61: 984-986.
2311. McCracken, F.I., and T.H. Filer Jr. 1970. *Polyporus zonalis* associated with stem breakage of hackberry in Alabama. Pl. Dis. Reporter 54: 322-323.
2312. McCracken, F.I., and E.R. Toole. 1974. Felling infected oaks in natural stands reduces dissemination of *Polyporus hispidus* spores. Phytopathology 64: 265-266.
2313. McDowell, J., and W. Merrill. 1985. *Rhabdocline* taxa in Pennsylvania. Pl. Dis. 69: 714-715.
2314. McGee, D.C. 1978. The severity of red leaf disease caused by *Exobasidium vaccinii* in lowbush blueberries in Maine. Pl. Dis. Reporter 62: 486-487.
2315. McGinnis, M.R. 1971. Selected aspects of the biology of *Hyalospora polypodii* on *Cystopteris fragilis*. Mycologia 63: 277-282.
2316. McGlohon, N.E. 1959. Survey of fungi associated with white clover stolons. Pl. Dis. Reporter 43: 22-24.
2317. McGrath, W.T., A.J. Prey, and F.S. Morse. 1972. Brown spot needle blight of scotch pine Christmas trees in Wisconsin: sporulation and control. Pl. Dis. Reporter 56: 99-102.
2318. McGuire, J.M. 1941. The species of *Sebacina* (Tremellales) of temperate North America. Lloydia 4: 1-43.
2319. McKay, H.H. 1959. Cultural basis for maintaining *Polyporus cinnabarinus* and *Polyporus sanguineus* as two distinct species. Mycologia 51: 465-473.
2320. McKay, H.H. 1967. Cultural studies of *Polyporus meliae* and two similar species. Mycologia 59: 1050-1058.
2321. McKeen, C.G. 1952. Studies of Canadian Thelephoraceae IX. A cultural and taxonomic study of three species of *Peniophora*. Canad. J. Bot. 30: 764-787.
2322. McLaughlin, M.R., J.S. Mignucci, and G.M. Milbrath. 1977. *Microsphaera diffusa*, the perfect stage of the soybean powdery mild ew pathogen. Phytopathology 67: 726-729.
2323. McLean, D.M. 1951. *Sclerotinia* stem rot and *Ramularia* leaf spot of dandelion grown for seed in Skagit County, Washington. Pl. Dis. Reporter 35: 162-163.
2324. McLean, D.M. 1957. *Sclerotinia* on strawberry fruit in northwest Washington. Pl. Dis. Reporter 41: 1057.
2325. McLean, D.M., and B. Sleeth. 1961. *Myrothecium* rind rot of

cantaloup. Pl. Dis. Reporter 45: 728-729.
2326. McMillan, R.T. Jr. 1972. *Choanephora* wet-rot of pole beans. Pl. Dis. Reporter 56: 967-968.
2327. McMillan, R.T. Jr. 1972. *Helminthosporium maydis* on T cytoplasm sweet corn in south Florida. Pl. Dis. Reporter 56: 969-970.
2328. McMillan, R.T. Jr. 1986. *Guignardia citricarpa* a cause of black spot on mango foliage in Florida. J. Phytopathology 117: 260-264.
2329. McNabb, R.F.R. 1973. Taxonomic studies in the Dacrymycetaceae VIII. *Dacrymyces* Nees ex Fries. New Zealand J. Bot. 11: 461-524.
2330. McPartland, J.M. 1983. Fungal pathogens of *Cannabis sativa* in central Illinois. Phytopathology 73: 797.
2331. McPartland, J.M. 1983. *Phomopsis ganjae* sp. nov. on *Cannabis sativa*. Mycotaxon 18: 527-530.
2332. McRitchie, J.J. 1973. Wilt of variegated *Pyracantha* caused by *Fusarium oxysporum*. Phytopathology 63: 445.
2333. McRitchie, J.J. 1973. Pathogenicity and control of *Fusarium oxysporum* wilt of variegated *Pyracantha*. Pl. Dis. Reporter 57: 389-391.
2334. McRitchie, J.J., J.W. Kimbrough, and A.W. Engelhard. 1973. *Itersonilia* petal blight of *Chrysanthemum* in Florida. Pl. Dis. Reporter 57: 181-182.
2335. McWhorter, F.P. 1950. Rust on garden peas in Oregon. Pl. Dis. Reporter 34: 269.
2336. McWhorter, F.P., and C.J. Anderson. 1951. Diseases in Croft lily plantings, 1949-50. Pl. Dis. Reporter 35: 106-108.
2337. Meeker, J.A. 1975. Revision of the Seuratiaceae. II. Taxonomy and nomenclature of *Seuratia*. Canad. J. Bot. 53: 2483-2496.
2338. Mehrlich, F.P., and H.M. Fitzpatrick. 1935. *Dichotomophthora portulacae*, a pathogen of *Portulaca oleracea*. Mycologia 27: 543-550.
2339. Meiners, J.P. 1950. Bean diseases in southern Idaho in 1949. Pl. Dis. Reporter 34: 14.
2340. Meiners, J.P. 1952. A new race of head smut on Bromar variety of mountain brome. Pl. Dis. Reporter 36: 166.
2341. Meiners, J.P. 1954. Grass diseases on new hosts in the Pullman nursery unit of the Soil Conservation Nurseries. Pl. Dis. Reporter 38: 277-278.
2342. Meiners, J.P., and G.W. Fischer. 1957. Some new grass smut records from the western states. III. Mycologia 49: 767-771.
2343. Meiners, J.P., and J.R. Hardison. 1957. New host records for dwarf bunt in the Pacific Northwest. II. Pl. Dis. Reporter 41: 983-985.
2344. Meiners, J.P., J.T. Waldher, J.R. Hardison, and H.S. Fenwick. 1956. Additions to the host range of dwarf bunt. Pl. Dis. Reporter 40: 26-27.
2345. Melchers, L.E. 1956. Fungi associated with Kansas hybrid seed corn. Pl. Dis. Reporter 40: 500-506.
2346. Melouk, H.A., and C.E. Horner. 1974. *Verticillium nigrescens* from peppermint. Phytopathology 64: 1267-1268.
2347. Mengistu, A., and C.R. Grau. 1986. Variation in morphological, cultural, and pathological characteristics of *Phialophora gregata* and *Acremonium* sp. recovered from soybean in Wisconsin. Pl. Dis. 70: 1005-1009.
2348. Menzies, J.D. 1950. *Erysiphe cichoracearum* DC. as a parasite of potatoes. Pl. Dis. Reporter 34: 140-141.
2349. Menzies, J.D. 1954. Plant disease observations in the irrigated areas of central Washington during 1953. Pl. Dis. Reporter 38: 314-315.
2350. Menzies, J.D. 1955. A *Penicillium* crown rot of asparagus. Phytopathology 45: 527-530.
2351. Meredith, D.S. 1969. Fungal diseases of bananas in Hawaii. Pl. Dis. Reporter 53: 63-66.
2352. Meredith, D.S., M. Aragaki, and O.V. Holtzmann. 1968. *Deightoniella torulosa* on bananas in Hawaii. Pl. Dis. Reporter 52: 405.
2353. Meredith, D.S., and J.S. Lawrence. 1969. Black leaf streak disease of bananas (*Mycosphaerella fijiensis*): symptoms of disease in Hawaii, and notes of the conidial state of the causal fungus. Trans. Brit. Mycol. Soc. 52: 459-476.
2354. Merrill, W. 1972. Occurrence of *Endocronartium harknessii* in Pennsylvania. Pl. Dis. Reporter 56: 1058.
2355. Merrill, W., and B.R. Kistler. 1974. *Naemacyclus* needlecast of scots pine epidemic in Pennsylvania. Pl. Dis. Reporter 58: 287-288.
2356. Merrill, W., and B.R. Kistler. 1978. Accelerated development of *Rhizosphaera* needlecast of blue spruce in Pennsylvania. Pl. Dis. Reporter 62: 34-35.
2357. Merrill, W., and B.R. Kistler. 1978. Needlecast and weeds interact to cause branch mortality in Scots pine. Pl. Dis. Reporter 62: 200-202.
2358. Merrill, W., and J. Longenecker. 1973. Swiss needlecast of Douglas-fir in Pennsylvania. Pl. Dis. Reporter 57: 984.
2359. Merrill, W., K. McCall, and L. Zang. 1981. *Fusarium* root rot of Douglas-fir and Fraser fir seedlings in Pennsylvania. Pl. Dis. 65: 913-914.
2360. Merrill, W., N. Wenner, and B. Towers. 1986. Resistance of *Pinus ponderosa* to *Endocronartium harknessii* in Pennsylvania. Pl. Dis. 70: 317-320.
2361. Merrill, W., L. Zang, K. Bowen, and B.R. Kistler. 1979. *Naemacyclus minor* needlecast of *Pinus nigra* in Pennsylvania. Pl. Dis. Reporter 63: 994.
2362. Meyer, R. 1978. A *Cladosporium* sooty mold and its insect associates in central California. Pl. Dis. Reporter 62: 382-385.
2363. Meyer, S.L.F., and E.S. Luttrell. 1986. Ascoma morphology of *Pseudopeziza trifolii* forma specialis *medicaginis-sativae* (Dermateaceae) on alfalfa. Mycologia 78: 529-542.
2364. Meyer, W.A., and J.B. Sinclair. 1970. *Pyrenochaeta terrestris*, a root pathogen on creeping bentgrass. Pl. Dis. Reporter 54: 506-507.
2365. Micales, J.A., and R.J. Stipes. 1986. The differentiation of *Endothia* and *Cryphonectria* species by exposure to selected fungitoxicants. Mycotaxon 26: 99-117.
2366. Micales, J.A., R.J. Stipes, and M.R. Bonde. 1987. On the conspecificity of *Endothia eugeniae* and *Cryphonectria cubensis*. Mycologia 79: 707-720.
2367. Michaels, E., and G.A. Chastagner. 1984. Distribution, severity, and impact of Swiss needle cast in Douglas-fir Christmas trees in western Washington and Oregon. Pl. Dis. 68: 939-942.
2368. Michailides, T.J., J.M. Ogawa, and D.C. Opgenorth. 1987. Shift of *Monilinia* spp. and distribution of isolates sensitive and resistant to benomyl in California prune and apricot orchards. Pl. Dis. 71: 893-896.
2369. Michailides, T.J., J.M. Ogawa, and P.L. Sholberg. 1987. Chemical control of fungi causing decay of fresh prunes during storage. Pl. Dis. 71: 14-17.
2370. Michailides, T.J., and R.A. Spotts. 1986. Mating types of *Mucor piriformis* isolated from soil and pear fruit in Oregon orchards (on the life history of *Mucor piriformis*). Mycologia 78: 766-770.
2371. Middleton, J.T., J.B. Kendrick Jr., and B.J. Hall. 1952. Parasitic and nonparasitic diseases of celery in southern California. Pl. Dis. Reporter 36: 188-193.
2372. Mielke, M.E. 1981. Pathogenicity of *Verticicladiella penicillata* (Grosm.) Kendrick to northern Idaho conifers. Forest Sci. 27: 103-110.
2373. Mielke, M.E., C. Haynes, and W.L. MacDonald. 1982. Beech scale and *Nectria galligena* on beech in the Monongahela National Forest, West Virginia. Pl. Dis. 66: 851-852.
2374. Mihail, J.D., and S.M. Alcorn. 1984. Powdery mildew (*Leveillula taurica*) on native and cultivated plants in Arizona. Pl. Dis. 68: 625-626.
2375. Mihail, J.D., and S.M. Alcorn. 1986. *Macrophomina phaseolina* from guar in Arizona. Canad. J. Bot. 64: 11-12.
2376. Mihail, J.D., S.M. Alcorn, and P.J. Thrapp. 1985. First report of *Pythium aphanidermatum* infecting guayule. Pl. Dis. 69: 177.
2377. Mildenhall, J.P., R.G. Pratt, P.H. Williams, and J.E. Mitchell. 1971. *Pythium* brown root and forking of muck-grown carrots. Pl. Dis. Reporter 55: 536-540.
2378. Mildenhall, J.P., and P.H. Williams. 1970. *Rhizoctonia* crown rot and cavity spot of muck-grown carrots. Phytopathology 60: 887-890.
2379. Milholland, R.D. 1973. A leaf spot disease of highbush blueberry caused by *Alternaria tenuissima*. Phytopathology 63: 1395-1397.
2380. Milholland, R.D. 1974. Blueberry stem canker and dieback caused by *Gloeosporium minus*. Phytopathology 64: 727-730.
2381. Milholland, R.D. 1974. Stem and root rot of blueberry caused by *Calonectria crotalariae*. Phytopathology 64: 831-834.
2382. Milholland, R.D. 1977. *Cercospora* stem blotch disease of rabbiteye blueberry. Phytopathology 67: 816-819.
2383. Milholland, R.D. 1982. Blueberry twig blight caused by

Phomopsis vaccinii. Pl. Dis. 66: 1034-1036.
2384. Milholland, R.D., and G.J. Galletta. 1967. Relative susceptibility of blueberry cultivars to *Phytophthora cinnamomi*. Pl. Dis. Reporter 51: 998-1001.
2385. Miller, C.E., R.W. Martin, and D.P. Dylewski. 1985. The ultrastructure of plasmodia, sporangia, and cytosori of *Ligniera verrucosa* (Plasmodiophoromycetes). Canad. J. Bot. 63: 263-273.
2386. Miller, D.J., D.M. Knutson, and R.D. Tocher. 1983. Chemical brown staining of Douglas-fir sapwood. Forest Prod. J. 33: 44-48.
2387. Miller, D.L., and A.D. Partridge. 1973. Fungus associations in root rots of grand fir. Pl. Dis. Reporter 57: 346-348.
2388. Miller, D.R. 1968. *Bifusella linearis* found on both limber and whitebark pine in California. Pl. Dis. Reporter 52: 305.
2389. Miller, D.R. 1968. White pine blister rust found on foxtail pine in northwestern California. Pl. Dis. Reporter 52: 391-392.
2390. Miller, D.R. 1968. *Furcaspora pinicola* found on needles of white fir. Pl. Dis. Reporter 52: 486.
2391. Miller, D.R. 1969. *Lophodermium durilabrum* found on foxtail pine in California. Pl. Dis. Reporter 53: 271.
2392. Miller, D.R. 1972. *Dermea pseudotsugae* found to be killing young Douglas-fir in northern California. Pl. Dis. Reporter 56: 843.
2393. Miller, D.R. 1974. *Sydowia polyspora* found on white fir twigs in California. Pl. Dis. Reporter 58: 94-95.
2394. Miller, D.R. 1976. A needlecast found on bristlecone fir in California. Pl. Dis. Reporter 60: 54-55.
2395. Miller, D.R., and K.M. Swain. 1969. *Rhabdocline* needle cast found on bigcone Douglas-fir. Pl. Dis. Reporter 53: 270.
2396. Miller, D.R., and E.I. Veirs. 1968. *Verticicladiella wagenerii* found on Douglas-fir in ElDorado County, California. Pl. Dis. Reporter 52: 393.
2397. Miller, J.D., and J.J. Christensen. 1959. Occurrence of stripe rust in Minnesota in 1958. Pl. Dis. Reporter 43: 159.
2398. Miller, J.D., and H.D. Wells. 1984. First report of powdery mildew, *Erysiphe* spp., on arrowleaf clover in Georgia. Pl. Dis. 68: 351.
2399. Miller, J.H. 1941. Georgia Pyrenomycetes. II. Mycologia 33: 74-81.
2400. Miller, J.H. 1951. Apple root rot in Georgia. Pl. Dis. Reporter 35: 465.
2401. Miller, J.H. 1952. Spot anthracnose (*Elsinoe corni*) on *Cornus florida* in Georgia in 1952. Pl. Dis. Reporter 36: 295-296.
2402. Miller, J.H. 1957. *Elsinoe* on southern red oak. Mycologia 49: 277-279.
2403. Miller, J.H. 1961. A monograph of the world species of *Hypoxylon*. University of Georgia Press, Athens
2404. Miller, J.H., A. Brown, and J.R. Johnson. 1952. Notes on small grain diseases in Georgia for 1951-1952. Pl. Dis. Reporter 36: 287.
2405. Miller, J.H., W.A. Campbell, and G.E. Thompson. 1954. Diseases and insects affecting the commonly planted trees and shrubs in Georgia. Pl. Dis. Reporter 38: 362-369.
2406. Miller, J.H., U.R. Gore, A.R. Brown, and D.D. Morey. 1955. Small grain diseases in Georgia in 1953-54. Pl. Dis. Reporter 39: 17-19.
2407. Miller, J.H., and A.E. Jenkins. 1955. A new speices of *Elsinoe* on southern *Magnolia*. Mycologia 47: 104-108.
2408. Miller, J.H., and L.W. Nielsen. 1957. A new species of *Xylaria*. Mycologia 49: 112-114.
2409. Miller, J.H., and J. Taylor. 1952. *Cladosporium* leaf spot of pepper in Georgia. Pl. Dis. Reporter 36: 440.
2410. Miller, J.H., and G.E. Thompson. 1940. Georgia Pyrenomycetes I. Mycologia 32: 1-15.
2411. Miller, J.W. 1971. Tan leaf spot of *Rhoeo discolor* caused by *Curvularia eragrostidis*. Pl. Dis. Reporter 55: 38-40.
2412. Miller, J.W., and S.A. Alfieri Jr. 1974. Leaf spot of *Ligustrum sinense* caused by *Corynespora cassiicola* and its control. Phytopathology 64: 255-256.
2413. Miller, J.W., and J.J. McRitchie. 1973. *Cylindrocladium scoparium*, causal agent of a new disease of silver buttonwood and its control. Pl. Dis. Reporter 57: 500-503.
2414. Miller, M.E., and J.M. Amador. 1985. An epidemic of downy mildew (*Peronospora destructor*) of onion in Texas. Pl. Dis. 69: 268.
2415. Miller, M.E., R.A. Taber, and J.M. Amador. 1978. *Stemphylium* blight of onion in south Texas. Pl. Dis. Reporter 62: 851-853.
2416. Miller, O.K. 1982. Higher fungi in Alaskan subarctic tundra and taiga plant communities. Pp. 123-143. In: Arctic and Alpine Mycology. Eds., G.A. Larson and J.F. Ammirati. Univ. Washington Press, Seattle. 559 p.
2417. Miller, O.K. Jr. 1965. Three new species of lignicolous agarics in the Tricholomataceae. Mycologia 57: 933-945.
2418. Miller, O.K. Jr. 1969. A new species of *Pleurotus* with a coremioid imperfect stage. Mycologia 61: 887-893.
2419. Miller, O.K., Jr., and R.L. Gilbertson. 1969. Notes on Homobasidiomycetes from northern Canada and Alaska. Mycologia 61: 840-844.
2420. Miller, O.K., Jr., G.A. Laursen, and D.F. Farr. 1982. Notes on Agaricales from Arctic tundra in Alaska. Mycologia 74: 576-591.
2421. Miller, O.K., Jr., and D.L. Manning. 1976. Distribution of the lignicolous Tricholomataceae in the southern Appalachians. Pp. 307-344. In: Distributional History of the Biota of the Southern Appalachians Part IV. Algae and Fungi, Biogeography, Systematics, and Ecology. Eds., B.C. Parker, and M.K. Roane. University Press of Virginia, Charlottesville. 416 p.
2422. Miller, O.K., Jr., J.G. Palmer, and L.S. Gillman. 1980. The fruiting and development of *Rhodotus palmatus* in culture. Mycotaxon 11: 409-419.
2423. Miller, O.K., Jr., and L. Stewart. 1971. The genus *Lentinellus*. Mycologia 63: 333-369.
2424. Miller, P.W. 1950. Nut diseases in the Pacific Northwest in 1949. Pl. Dis. Reporter 34: 38-39.
2425. Miller, P.W. 1951. Nut diseases in Oregon in 1950. Pl. Dis. Reporter 35: 145.
2426. Miller, P.W. 1952. Nut diseases in Oregon in 1951. Pl. Dis. Reporter 36: 142.
2427. Miller, P.W. 1953. Nut diseases in Oregon in 1952. Pl. Dis. Reporter 37: 18.
2428. Miller, P.W. 1954. Nut diseases in Oregon in 1953. Pl. Dis. Reporter 38: 80.
2429. Miller, P.W. 1955. Nut diseases in Oregon in 1954. Pl. Dis. Reporter 39: 68.
2430. Miller, P.W. 1955. Nut diseases in Oregon in 1955. Pl. Dis. Reporter 39: 929-930.
2431. Miller, P.W. 1957. Nut diseases in Oregon in 1957. Pl. Dis. Reporter 41: 1057-1058.
2432. Miller, S.B., and L.W. Baxter Jr. 1970. Dieback in azaleas caused by *Phomopsis* species. Phytopathology 60: 387-388.
2433. Miller, T. 1972. Fusiform rust in planted slash pines: influence of site preparation and spacing. Forest Sci. 18: 70-75.
2434. Miller, V.M., and L. Bonar. 1941. A study of the Perisporiaceae, Capnodiaceae, and some other sooty molds from California. Univ. Calif. Publ. Bot. 19: 405-428.
2435. Miller, W.A., and K.W. Roy. 1982. Mycoflora of soybean leaves, pods, and seeds in Mississippi. Canad. J. Bot. 60: 2716-2723.
2436. Miller-Weeks, M., and D. Stark. 1983. European larch canker in Maine. Pl. Dis. 67: 448.
2437. Millikan, D.F., and T.W. Bretz. 1955. *Fusicladium* on Pyracantha. Pl. Dis. Reporter 39: 658.
2438. Millikan, D.F., and J.E. Smith Jr. 1955. Foot rot of *Pothos*, a disease caused by *Rhizoctonia*. Pl. Dis. Reporter 39: 240-241.
2439. Millspaugh, C.F., and L.W. Nuttall. 1923. Flora of Santa Catalina Island. Field Mus. Nat. Hist., Bot. Ser. 5: 314-357.
2440. Mims, C.W. 1982. Ultrastructure of the haustorial apparatus of *Exobasidium camelliae*. Mycologia 74: 188-200.
2441. Mims, F., D.M. Benson, and R.K. Jones. 1981. Susceptibility of Leucothoe, hybrid rhododendron, and azalea to *Cyli ndrocladium scoparium* and *C. theae*. Pl. Dis. 65: 353-354.
2442. Minter, D.W. 1980. *Leptostroma* on pine needles. Canad. J. Bot. 58: 906-917.
2443. Minter, D.W. 1981. *Lophodermium* on pines. Mycol. Pap. 147: 1-54.
2444. Mircetich, S.M., R.N. Campbell, and M.E. Matheron. 1977. *Phytophthora* trunk canker of coast live oak and cork oak trees in California. Pl. Dis. Reporter 61: 66-70.
2445. Mircetich, S.M., and M.E. Matheron. 1976. *Phytophthora* root and crown rot of cherry trees. Phytopathology 66: 549-558.
2446. Mircetich, S.M., and M.E. Matheron. 1983. *Phytophthora* root and crown rot of walnut trees. Phytopathology 73: 1481-1488.
2447. Mircetich, S.M., and G.A. Zentmeyer. 1964. *Rhizoctonia* seed and root rot of avocado. Phytopathology 54: 211-213.
2448. Mirocha, C.J., and E.E. Wilson. 1961. Hull rot disease of

almonds. Phytopathology 51: 843-847.
2449. Mislivec, P.B., and J. Tuite. 1970. Species of *Penicillium* occurring in freshly-harvested and in stored dent corn kernels. Mycologia 62: 67-74.
2450. Mitchel, D.H., and A.H. Smith. 1978. Notes on Colorado fungi III: New and interesting mushrooms from the aspen zone. Mycologia 70: 1040-1063.
2451. Mitchell, D.J., J.O. Strandberg, and M.E. Kannwischer. 1978. Root and stem rot of watercress (*Nasturtium officinale*) caused by *Phytophthora cryptogea*. Pl. Dis. Reporter 62: 599-600.
2452. Mitchell, J.K. 1986. *Dichotomophthora portulacae* causing black stem rot on common pursla ne in Texas. Pl. Dis. 70: 603.
2453. Mix, A.J. 1936. The genus *Taphrina*. II: a list of valid species. Bull. Univ. Kansas 37: 151-176.
2454. Mix, A.J. 1949. A monograph of the genus *Taphrina*. Univ. Kansas Sci. Bull. 33: 3-167.
2455. Mix, A.J. 1953. Differentiation of species of *Taphrina* in culture. Utilization of nitrogen compounds. Mycologia 45: 649-670.
2456. Mix, A.J. 1954. Report of the 1952 foray. Mycologia 46: 112-123.
2457. Mix, A.J. 1954. Additions and emendations to a monograph of the genus *Taphrina*. Trans. Kansas Acad. Sci. 57: 55-65.
2458. Mix, A.J., and R.L. McGregor. 1950. Notes on Kansas fungi, 1949. Trans. Kansas Acad. Sci. 53: 340-343.
2459. Moller, W.J., A.N. Kasimatis, and J.J. Kissler. 1974. A dying arm disease of grape in California. Pl. Dis. Reporter 58: 869-871.
2460. Moller, W.J., D.E. Ramos, and W.R. Hildreth. 1971. Apricot pathogen associated with *Ceanothus* limb dieback in California. Pl. Dis. Reporter 55: 1006-1008.
2461. Monoson, H.L. 1974. The species of *Triphragmium, Nyssopsora*, and *Triphragmiopsis*. Mycopathol. Mycologia Appl. 52: 115-131.
2462. Moody, M.M., R.C. Wharton, N. Schnaper, and S.C. Schimpff. 1982. Do water pipes prevent transmission of fungi from contaminated marijuana?. New England J. Medicine 306: 1492-1493.
2463. Moore, E.J. 1965. Ontogeny of gelatinous fungi. Mycologia 57: 114-130.
2464. Moore, R.T. 1954. Three new species of Helicosporae. Mycologia 46: 89-92.
2465. Moore, R.T. 1955. Index to the Helicosporae. Mycologia 47: 90-103.
2466. Moore, R.T. 1957. Index to the Helicosporae: Addenda. Mycologia 49: 580-587.
2467. Moore, R.T. 1958. Deuteromycetes I: The *Sporidesmium* complex. Mycologia 50: 681-692.
2468. Moore, R.T. 1959. The genus *Berkleasmium*. Mycologia 51: 734-739.
2469. Moore, W.D. 1956. Powdery mildew of pepper in Florida. Pl. Dis. Reporter 40: 756.
2470. Moore, W.F. 1970. Origin and spread of southern corn leaf blight in 1970. Pl. Dis. Reporter 54: 1104-1108.
2471. Moorman, G.W., and R.A. Klemmer. 1980. *Fusarium oxysporum* causes basal stem rot of *Zygocactus truncatus*. Pl. Dis. 64: 1118-1119.
2472. Morgan, F.L., and H.W. Johnson. 1965. *Phytophthora* root and crown rot of vetch. Pl. Dis. Reporter 49: 84-88.
2473. Morgan, O.D. 1964. The occurrence of a *Sclerotinia* on *Vicia villosa* in Maryland. Pl. Dis. Reporter 48: 696-697.
2474. Morgan, O.D. 1971. A study of four weed hosts of *Sclerotinia* species in alfalfa fields. Pl. Dis. Reporter 55: 1087-1089.
2475. Morgan, O.D. 1957. Mimosa wilt spreading in southern Maryland. Pl. Dis. Reporter 41: 51-52.
2476. Morgan, O.D., and J.G. Kantzes. 1971. Observations of *Colletotrichum graminicola* on T corn and blends in Maryland. Pl. Dis. Reporter 55: 755.
2477. Morgan, O.D., J.G. Kantzes, L.O. Weaver, and W.F. Jeffers. 1963. A *Rhizoctonia* leaf blight of strawberry. Pl. Dis. Reporter 47: 463-465.
2478. Morgan, P. 1983. *Fusarium* stem rot of Douglas-fir seedlings. Pl. Dis. 67: 441-442.
2479. Morgan, P.D., and C.H. Driver. 1972. *Rhizina* root rot of Douglas-fir seedlings planted on burned sites in Washington. Pl. Dis. Reporter 56: 407-409.
2480. Morgan, P.D., E.K. Wallin, and C.H. Driver. 1974. Occurrence of *Rhizina* root rot in an old-growth conifer stand in the Pacific northwest. Pl. Dis. Reporter 58: 492-494.
2481. Morgan-Jones, G. 1973. Genera Coelomycetarum. VII. *Cryptocline* Petrak. Canad. J. Bot. 51: 309-325.
2482. Morgan-Jones, G. 1974. Fungi of Alabama. I. Dematiaceous Hyphomycetes. J. Alabama Acad. Sci. 45: 144-155.
2483. Morgan-Jones, G. 1975. Notes on Hyphomycetes. VIII. *Lylea*, a new genus. Mycotaxon 3: 129-132.
2484. Morgan-Jones, G. 1976. Fungi of Alabama. III. Dematiaceous Hyphomycetes. J. Alabama Acad. Sci. 47: 38-48.
2485. Morgan-Jones, G. 1976. Notes on Hyphomycetes. X. *Codinaeopsis* gen. nov. Mycotaxon 4: 166-170.
2486. Morgan-Jones, G. 1977. Fungi of Alabama. VI. Dematiaceous hyphomycetes. J. Alabama Acad. Sci. 48: 26-41.
2487. Morgan-Jones, G. 1977. Notes on Hyphomycetes. XVII. A new species of *Pseudospiropes*. Mycotaxon 5: 481-483.
2488. Morgan-Jones, G. 1977. Notes on Hyphomycetes. XVIII. *Chaetoblastophorum ingramii* gen. et sp. nov. and *Cylindrotrichum oblongisporum*. Mycotaxon 5: 484-490.
2489. Morgan-Jones, G. 1977. Notes on Hyphomycetes. XIX. *Cladosporium leprosum* sp. nov. and *Cladosporium nigrellum*. Mycotaxon 6: 1-5.
2490. Morgan-Jones, G. 1977. Notes on Hyphomycetes. XXI. *Sporidesmium carrii* sp. nov. Mycotaxon 6: 356-358.
2491. Morgan-Jones, G. 1978. Notes on Hyphomycetes. XXII. *Phaeoisariopsis bambusicola* sp. nov. Mycotaxon 7: 130-132.
2492. Morgan-Jones, G. 1978. Notes on Hyphomycetes. XXIV. *Phialophora pinicola* sp. nov. and *Phialophora bubakii*. Mycotaxon 7: 327-332.
2493. Morgan-Jones, G. 1979. Notes on Hyphomycetes XXX. On three species of *Idriella*. Mycotaxon 8: 402-410.
2494. Morgan-Jones, G. 1979. Notes on Hyphomycetes. XXXI. *Chaetopsina auburnensis* sp. nov. Mycotaxon 8: 411-416.
2495. Morgan-Jones, G. 1980. Notes on Hyphomycetes. XXXV. *Stenellopsis* gen. nov. Mycotaxon 10: 405-408.
2496. Morgan-Jones, G., and L.G. Brown. 1976. Notes on Hyphomycetes. XIII. Concerning two species of *Phaeoisariopsis*. Mycotaxon 4: 493-497.
2497. Morgan-Jones, G., and G.W. Karr. 1976. Fungi of Alabama. V. Dematiaceous Hyphomycetes. J. Alabama Acad. Sci. 47: 247-262.
2498. Morgan-Jones, G., and G.W. Karr Jr. 1976. Notes on Hyphomycetes. IX. A new species of *Curvularia*. Mycotaxon 3: 559-563.
2499. Morgan-Jones, G., and G.W. Karr Jr. 1976. Notes on Hyphomycetes. XVI. A new species of *Stachybotrys*. Mycotaxon 4: 510-512.
2500. Morgan-Jones, G., and B. Kendrick. 1972. Notes on Hyphomycetes. III. Redispostions of six species of *Exosporium*. Canad. J. Bot. 50: 1817-1824.
2501. Morgan-Jones, G., T.R. Nag Raj, and B. Kendrick. 1972. Genera coelomycetarum. V. *Alpakesa* and *Bartalinia*. Canad. J. Bot. 50: 877-882.
2502. Morgan-Jones, G., T.R. Nag Raj, and B. Kendrick. 1972. Genera coelomycetarum. VI. *Kellermania*. Canad. J. Bot. 50: 1641-1648.
2503. Morgan-Jones, G., and R.C. Sinclair. 1978. Fungi of Alabama. VII. Dematiaceous hyphomycetes. J. Alabama Acad. Sci. 49: 1-15.
2504. Morgan-Jones, G., and J.F. White. 1983. Studies in the genus *Phoma*. I. *Phoma americana* sp. nov. Mycotaxon 16: 403-413.
2505. Morgan-Jones, G., and J.F. White Jr. 1987. Notes on Coelomycetes. II. Concerning the *Fusicoccum* anamorph of *Botryosphaeria ribis*. Mycotaxon 30: 117-125.
2506. Morgan-Jones, G., and J.F. White Jr. 1987. Notes on Coelomycetes. III. Concerning *Microsphaeropsis concentrica* : morphology and ultrastructure. Mycotaxon 30: 177-187.
2507. Morgan-Jones, G., and E.A. Wiggins. 1974. Fungi of Alabama II. Dematiaceous Hyphomycetes. J. Alabama Acad. Sci. 45: 201-211.
2508. Morris, C.L. 1953. Chemical control of *Hypoderma lethale* on pitch pine. Pl. Dis. Reporter 37: 368-370.
2509. Morris, E.F. 1962. Somes species of *Harpographium*. Amer. Midl. Naturalist 68: 319-324.
2510. Morris, E.F., and D.E. Finley. 1967. Two new genera of stilbellaceous fungi. Amer. Midl. Naturalist 77: 200-204.
2511. Morrison, R.H. 1982. *Drechslera nobleae* on *Lolium multifolrum* in North America. Mycologia 74: 391-397.
2512. Morrison, R.H., and T.H. King. 1971. Stem rot of wildrice in Minnesota. Pl. Dis. Reporter 55: 498-500.
2513. Morton, H.L., and D.W. French. 1970. Attraction toward and

penetration of *Polyporus dryophilus* var. *vulpinus* basidiospores by hyphae of the same species. Mycologia 62: 714-720.
2514. Morton, H.L., and R. Miller. 1977. *Rhabdocline* needle cast in the Lake States. Pl. Dis. Reporter 61: 801-802.
2515. Morton, H.L., and R.F. Patton. 1970. Swiss needle cast of Douglas-fir in the Lake States. Pl. Dis. Reporter 54: 612-616.
2516. Moseman, J.G., and T.T. Hebert. 1950. Small grain diseases in North Carolina in 1949-50. Pl. Dis. Reporter 34: 392-394.
2517. Moss, M.A. 1987. Resistance to metalaxyl in the *Pseudoperonospora cubensis* population causing downy mildew of cucumber in south Florida. Pl. Dis. 71: 1045.
2518. Moyer, J.W., C.L. Campbell, and C.W. Averre. 1982. Stem canker of sweet potato induced by *Fusarium solani*. Pl. Dis. 66: 65-66.
2519. Muchovej, J.J., and H.B. Couch. 1987. Colonization of bentgrass turf by *Curvularia lunata* after leaf clip ping and heat stress. Pl. Dis. 71: 873-875.
2520. Mueller, J.D., B.J. Shortt, and J.B. Sinclair. 1985. Effects of cropping history, cultivar, and sampling date of the internal fungi of soybean roots. Pl. Dis. 69: 520-523.
2521. Mueller, J.D., and J.B. Sinclair. 1986. Occurrence and role of *Gliocladium roseum* in field-grown soybeans in Illinois. Trans. Brit. Mycol. Soc. 86: 677-680.
2522. Mueller, W.C., N. Jackson, and J.M. Fenstermacher. 1974. Occurrence of *Sclerophthora macrospora* in turfgrass affected with yellow tuft. Pl. Dis. Reporter 58: 848-850.
2523. Mullen, J.M., A.K. Hagan, and G. Morgan-Jones. 1985. Bradford pear (*Pyrus calleryana*), a new host of *Botryosphaeria dothidea*. Pl. Dis. 69: 726.
2524. Muller, A.S. 1956. *Elsinoe* and *Spaceloma* in Florida. Pl. Dis. Reporter 40: 256-257.
2525. Muller, A.S., A.E. Jenkins, and A.A. Bitancourt. 1954. *Elsinoe* and *Sphaceloma* in Florida. Pl. Dis. Reporter 38: 599-601.
2526. Mulrean, E.N., and A.H. McCain. 1981. *Rhizopus* soft rot of *Euphorbia trigona*. Pl. Dis. 65: 159-161.
2527. Munnecke, D.E. 1956. *Alternaria* leaf spot of geranium. Pl. Dis. Reporter 40: 452-454.
2528. Munnecke, D.E., and J.L. Bricker. 1976. *Phytophthora* root rot of *Pinus radiata* in Christmas tree plantings. Pl. Dis. Reporter 60: 928-932.
2529. Munnecke, D.E., and P.A. Chandler. 1953. Some diseases of variegated peperomia. Pl. Dis. Reporter 37: 434-435.
2530. Murakishi, H.H. 1951. Purple seed stain of soybean. Phytopathology 41: 305-318.
2531. Murakishi, H.H., and H.S. Potter. 1958. Fungicidal control of *Botrytis* leaf rot of hothouse rhubarb. Pl. Dis. Reporter 42: 1316-1318.
2532. Murdoch, C.L., F.F. Laemmlen, and P.E. Parvin. 1973. Response of Kentucky bluegrass and perennial ryegrass cultivars to rust (*Puccinia striiformis* and *P. coronata*) in Hawaii. Pl. Dis. Reporter 57: 215-216.
2533. Murrill, W.A. 1919. Illustrations of fungi - XXX. Mycologia 11: 101-103.
2534. Murrill, W.A. 1919. Some described species of *Poria*. Mycologia 11: 231-244.
2535. Murrill, W.A. 1919. Collecting fungi in Virginia. Mycologia 11: 277-279.
2536. Murrill, W.A. 1920. Fungi from Hedgcock. Mycologia 12: 41-42.
2537. Murrill, W.A. 1920. Light-colored resupinate Polypores - I. Mycologia 12: 77-92.
2538. Murrill, W.A. 1920. Light-colored resupinate Polypores - II. Mycologia 12: 299-308.
2539. Murrill, W.A. 1938. New Florida polypores. Bull. Torrey Bot. Club 65: 647-661.
2540. Murrill, W.A. 1940. Additions to Florida fungi - II. Bull. Torrey Bot. Club 67: 57-66.
2541. Murrill, W.A. 1941. Some Florida novelties. Mycologia 33: 279-287.
2542. Murrill, W.A. 1951. *Campanella* in Florida. Mycologia 43: 247.
2543. Muthappa, B.N. 1970. Morphology of *Tryblidiella clavispora*. Mycologia 62: 98-106.
2544. Nag Raj, T.R. 1973. Genera coelomycetum. X. *Ellisiella, Samukuta,* and *Sakireeta*. Canad. J. Bot. 51: 2463-2472.
2545. Nag Raj, T.R. 1977. Miscellaneous microfungi. II. Canad. J. Bot. 55: 757-765.
2546. Nag Raj, T.R. 1983. Genera coelomycetum. XXI. *Strasseria* and two new anamorph-genera, *Apostrasseria* and *Nothostrasseria*. Canad. J. Bot. 61: 1-30.
2547. Nag Raj, T.R. 1985. Redisposals and redescriptions in the *Monochaetia-Seiridium*, *Pestalotia-Pestalotiopsis* complexes. II. *Pestalotiopsis besseyii* (Guba) comb. nov., and *Pestalosphaeria varia* sp. nov. Mycotaxon 22: 52-63.
2548. Nag Raj, T.R., and F. DiCosmo. 1981. A monograph of *Harknessia* and *Mastigosporella*, with notes on associated teleomorphs. Bibliotheca Mycologica 80: 1-62.
2549. Nag Raj, T.R., and B. Kendrick. 1972. Genera Coelomycetarum. III. *Pestalozziella*. Canad. J. Bot. 50: 607-617.
2550. Nag Raj, T.R., and B. Kendrick. 1975. A monograph of *Chalara* and allied genera. Univ. Waterloo Press, Waterloo, Ontario. 200 p.
2551. Nagel, C.M. 1952. Wheat rust in South Dakota in late June. Pl. Dis. Reporter 36: 338.
2552. Nair, V.M.G., C.J. Kostichka, and J.E. Kuntz. 1979. *Sirococcus clavigignenti-juglandacearum*: An undescribed species causing canker on butternut. Mycologia 71: 641-646.
2553. Nakasone, K.K. 1982. Cultural and morphological studies of *Gloeocystidiellum porosum* and *Gloeocystidium clavuligerum*. Mycotaxon 14: 316-324.
2554. Nakasone, K.K. 1983. Cultural and morphological studies on *Cystostereum australe* (Cortic iaceae), a new species from southeastern U.S.A. and Costa Rica. Mycotaxon 17: 269-274.
2555. Nakasone, K.K. 1984. Taxonomy of *Crustoderma* (Aphyllophorales, Corticiaceae). Mycologia 76: 40-50.
2556. Nakasone, K.K., H.H. Burdsall Jr., and L.A. Noll. 1982. Species of *Phlebia* section *Leptocystidiophlebia* (Aphyllophorales, Corticiaceae) in North America. Mycotaxon 14: 3-12.
2557. Nakasone, K.K., and W.E. Eslyn. 1981. A new species, *Phlebia brevispora*, a cause of internal decay in utility poles. Mycologia 73: 803-810.
2558. Nakasone, K.K., and R.L. Gilbertson. 1978 . Cultural and other studies of fungi that decay Ocotillo in Arizona. Mycologia 70: 266-299.
2559. Nakasone, K.K., and R.L. Gilbertson. 1982. Three brown-rot fungi in the Corticiaceae. Mycologia 74: 599-606.
2560. Namuco, L.O., W.R. Coffman, G.C. Bergstrom, and M.E. Sorrells. 1987. Virulence spectrum of the *Erysiphe graminis* f. sp. *tritici* population in New York. Pl. Dis. 71: 539-541.
2561. Nannfeldt, J.A. 1972. *Camarops* Karst. (Sphaeriales - Boliniaceae). With special regard to its European species. Svensk Bot. Tidskr. 66: 335-376.
2562. Nannfeldt, J.A. 1977. The species of Anthracoidea (Ustilaginales) on *Carex* subgen. *Vignea* with special regard to the Nordic species. Bot. Notiser 130: 351-375.
2563. Nannfeldt, J.A. 1983. *Nimbomollisia* and *Discocurtisia*: Two new genera of mollisioid Discomycetes. Mycologia 75: 292-310.
2564. Natour, R.M., and H.N. Miller. 1960. Stem rot of *Dracaena sanderiana*. Phytopathology 50: 648.
2565. Natural, M.P., R.A. Frederiksen, and D.T. Rosenow. 1982. *Acremonium* wilt of sorghum. Pl. Dis. 66: 863-865.
2566. Neal, D.C. 1950. *Verticillium* wilt of cotton in Louisiana in 1950. Pl. Dis. Reporter 34: 353.
2567. Neely, D. 1958. Diseases of woody hosts new to Illinois. Pl. Dis. Reporter 42: 711-712.
2568. Neely, D. 1959. New and unusual leaf disease fungi for Illinois. Pl. Dis. Reporter 43: 498-499.
2569. Neely, D. 1968. Bleeding necrosis of sweet gum in Illinois and Indiana. Pl. Dis. Reporter 52: 223-225.
2570. Neely, D. 1971. Additional *Aesculus* species and subspecies susceptible to leaf blotch. Pl. Dis. Reporter 55: 37-38.
2571. Neely, D., and W.M. Black. 1976. Anthracnose of black walnuts in the midwest. Pl. Dis. Reporter 60: 519-521.
2572. Neely, D., and R.A. Evers. 1976. New host records for *Cristulariella pyramidalis*. Pl. Dis. Reporter 60: 590-593.
2573. Neely, D., and E.B. Himelick. 1961 . Additional juniper hosts of cedar-apple and cedar-hawthorn rusts. Pl. Dis. Reporter 45: 351.
2574. Neely, D., and E.B. Himelick. 1963. *Aesculus* species susceptible to leaf blotch. Pl. Dis. Reporter 47: 170.
2575. Neely, D., R. Phares, and B. Weber. 1976. *Cristulariella* leaf spot associated with defoliation of black walnut plantations in Illinois. Pl. Dis. Reporter 60: 587-590.
2576. Neergaard, P., and A.G. Newhall. 1951. Notes on the physiology

and pathogenicity of *Centrospora acerina* (Hartig) Newhall. Phytopathology 41: 1021-1033.
2577. Neilsen, L.W., and J.W. Moyer. 1979. A *Fusarium* root rot of sweetpotatoes. Pl. Dis. Reporter 63: 400-404.
2578. Nelson, E.E., B. Goldfarb, and W.G. Thies. 1987. *Trichoderma* species from fumigated Douglas-fir roots decayed by *Phellinus weirii*. Mycologia 79: 370-374.
2579. Nelson, E.E., and W.G. Thies. 1985. Colonization of *Phellinus weirii*-infected stumps by *Trichoderma viride*. Eur. J. Forest Pathol. 15: 425-430.
2580. Nelson, L.R., D.D. Morey, and A.R. Brown. 1972. Severe leaf rust on wheat in Georgia. Pl. Dis. Reporter 56: 534-535.
2581. Nelson, L.R., D.D. Morey, and A.R. Brown. 1974. Wheat cultivar responses to severe glume blotch in Georgia. Pl. Dis. Reporter 58: 21-23.
2582. Nelson, P.E. 1957. Pathogenicity of *Idriella lunata* on strawberry. Phytopathology 47: 438-443.
2583. Nelson, P.E., B.W. Pennypacker, T.A. Toussoun, and R.K. Horst. 1975. *Fusarium* stub dieback of carnation. Phytopathology 65: 575-581.
2584. Nelson, P.E., and S. Wilhelm. 1956. An undescribed fungus causing a root rot of strawberry. Mycologia 48: 547-551.
2585. Nelson, P.E., C.E. Williamson, and W.D. McClellan. 1956. *Ovulinia* flower spot of azalea found on Long Island. Pl. Dis. Reporter 40 : 1115.
2586. Nelson, R.R. 1956. A new disease of corn caused by *Curvularia maculans*. Pl. Dis. Reporter 40: 210-211.
2587. Nelson, R.R. 1960. *Cochliobolus victoriae*, the perfect stage of *Helminthosporium victoriae*. Phytopathology 50: 774-775.
2588. Nelson, R.R., and J.C. Osborne. 1956. The relative prevalence and geographic distribution of fungi associated with moldy corn in eastern North Carolina in 1955. Pl. Dis. Reporter 40: 225-227.
2589. Nelson, R., and M. Mooar. 1952. Life history of the *Stemphylium* that causes red leaf spot of *Gladiolus*. Phytopathology 42: 363-366.
2590. Nemec, S., and H. Sanders. 1970. *Pythium* species associated with strawberry root necrosis in southern Illinois. Pl. Dis. Reporter 54: 49-51.
2591. Neuhauser, K.S., and R.L. Gilbertson. 1971. Some aspects of bipolar heterothallism in *Fomes cajanderi*. Mycologia 63: 722-735.
2592. Newell, S.Y., J.D. Miller, and R.D. Fallon. 1987. Ergosterol content of salt-marsh fungi: effect of growth conditions and mycelial age. Mycologia 79: 688-695.
2593. Newhall, A.G. 1944. A serious storage rot of celery caused by the fungus *Ansatospora macrospora* n. gen. Phytopathology 34: 92-105.
2594. Nicholls, T.H. 1970. *Lophodermium pinastri* outbreak in Lake States forest nurseries. Pl. Dis. Reporter 54: 731-733.
2595. Nicholls, T.H., and G.W. Hudler. 1971. *Dothistroma pini* on *Pinus nigra* in Minnesota. Pl. Dis. Reporter 55: 1040.
2596. Nicholls, T.H., and G.W. Hudler. 1972. Red pine - - a new host for brown spot (*Scirrhia acicola*). Pl. Dis. Reporter 56: 712-713.
2597. Nicholls, T.H., A.J. Prey, and D.D. Skilling. 1974. *Rhizosphaera kalkhoffii* damages blue spruce Christmas tree plantations. Pl. Dis. Reporter 58: 1094-1096.
2598. Nicholls, T.H., D.D. Skilling, and G.W. Hudler. 1973. *Scirrhia acicola* in Scotch pine Christmas tree plantations. Pl. Dis. Reporter 57: 55-59.
2599. Nichols, C.W., and E.E. Wilson. 1956. An outbreak of European canker in California. Pl. Dis. Reporter 40: 952-953.
2600. Nichols, L.P. 1962. A disease of snow-in-summer, *Cypress euphorbia*, and Corsican sandwo rt caused by a species of *Rhizoctonia*. Pl. Dis. Reporter 46: 172.
2601. Nichols, L.P., and L.B. Forer. 1972. Geranium rust discovered in Pennsylvania. Pl. Dis. Reporter 56: 759.
2602. Nichols, L.P., and D.H. Petersen. 1969. Occurrence of the apple scab fungus on twigs of flowering crab apples and one cultivar of commercial apple. Pl. Dis. Reporter 53: 974.
2603. Nicholson, R.L., L.F. Bauman, and H.L. Warren. 1976. Association of *Fusarium moniliforme* with brown midrib maize. Pl. Dis. Reporter 60: 908-910.
2604. Nickell, L.G. 1952. A species of *Ganoderma* probably pathogenic to sassafras in New York. Pl. Dis. Reporter 36: 28-29.
2605. Nickerson, N.L. 1984. A previously unreported disease of cranberries caused by *Exobasidium perenne* sp. nov. Canad. J. Plant Pathol. 6: 218-220.
2606. Niehaus, M.H. 1968. Relative amounts of stripe and flag smut on merion Kentucky bluegrass. Pl. Dis. Reporter 52: 633-634.
2607. Niemczyk, H.D., and R.L. Kiesling. 1961. Observations on red clover diseases in Michigan. Pl. Dis. Reporter 45: 698-699.
2608. Niemelä, T., and L. Ryvarden. 1975. Studies in the Aphyllophorales of Africa. IV: *Antrodia juniperina*, new for East Africa. Trans. Brit. Mycol. Soc. 65: 427-432.
2609. Nishijima, W.T., and M. Aragaki. 1975. Shoot blights of ohia and koa caused by *Calonectria theae*. Pl. Dis. Reporter 59: 883-885.
2610. Niyo, K.A., H.S. McNabb Jr., and L.H. Tiffany. 1986. Ultrastructure of the ascocarps, asci and ascospores of *Mycosphaerel la populorum*. Mycologia 78: 202-212.
2611. Nobles, M.K. 1967. Conspecificity of *Basidioradulum* (*Radulum*) *radula* and *Corticium hydnans*. Mycologia 59: 192-209.
2612. Norton, D.C. 1954. *Fusarium* root rot of guayule. Pl. Dis. Reporter 38: 500-501.
2613. Norton, D.C., and R. Behrens. 1956. *Ganoderma zonatum* associated with dying mesquite. Pl. Dis. Reporter 40: 253-254.
2614. Norton, D.C., and F.A. Frank. 1953. Charcoal rot (caused by *Sclerotium bataticola* Taub.) on guayule in southwest Texas in 1951-1952. Pl. Dis. Reporter 37: 41-43.
2615. Norton, D.C., and D.W. Rosberg. 1954. Watermelon fruit rot, caused by *Phytophthora parasitica*, in Texas. Pl. Dis. Reporter 38: 854.
2616. Nusbaum, C.J. 1951. The occurrence of tobacco anthracnose in North Carolina. Pl. Dis. Reporter 35: 276.
2617. Nutter, F.W., Jr., C.G. Warren, O.S. Wells, and W.E. MacHardy. 1978. *Fusarium* foot and root rot of tomato in New Hampshire. Pl. Dis. Reporter 62: 976-978.
2618. Nyland, G. 1949. Studies of some unusual Heterobasidiomycetes from Washington State. Mycologia 41: 686-701.
2619. Nyland, G. 1950. The genus *Tilletiopsis*. Mycologia 42: 487-496.
2620. Nyvall, R.F. 1979. Field Crop Diseases Handbook. AVI Pub. Co., Westport CT. 436 p.
2621. O'Brien, J.T. 1973. *Sirococcus* shoot blight of red pine. Pl. Dis. Reporter 57: 246-247.
2622. O'Brien, J.T., and D.A. Stark. 1979. *Scleroderris* canker found in Maine. Pl. Dis. Reporter 63: 194.
2623. O'Neill, N.R., and M.C. Rush. 1982. Etiology of sorghum sheath blight and pathogen virulence on rice. Pl. Dis. 66: 1115-1118.
2624. O'Neill, N.R., M.C. Rush, N.L. Horn, and R.B. Carver. 1977. Aerial blight of soybeans caused by *Rhizoctonia solani*. Pl. Dis. Reporter 61: 713-717.
2625. Oak, S.W., and R.D. Dorset. 1983. *Phomopsis* canker of European black alder found in Kentucky seed-production areas. Pl. Dis. 67: 691-693.
2626. Oberwinkler, F., and R. Bandoni. 1984. *Herpobasidium* and allied genera. Trans. Brit. Mycol. Soc. 83: 639-658.
2627. Oberwinkler, F., and R.J. Bandoni. 1982. *Atractogloea*: a new genus in the Hoehnelomycetaceae (Heterobasidiomycetes). Mycologia 74: 634-639.
2628. Oberwinkler, F., and K. Wells. 1985. A new species of *Herpobasidium* from Idaho. Mycologia 77: 265-271.
2629. Obrero, F.P., E.E. Trujillo, and M. Aragaki. 1968. *Phoma* leaf spot of tomato and potato in Hawaii. Pl. Dis. Reporter 52: 946-947.
2630. Odvody, G.N., and L.D. Dunkle. 1975. Occurrence of *Helminthosporium sorghicola* and other minor pathogens of sorghum in Nebraska. Pl. Dis. Reporter 59: 120-122.
2631. Odvody, G.N., L.D. Dunkle, and M.G. Boosalis. 1973. The occurrence of sooty stripe of sorghum in Nebraska. Pl. Dis. Reporter 57: 681-683.
2632. Odvody, G.N., L.D. Dunkle, and L.K. Edmunds. 1974. Zonate leaf spot in the northern sorghum belt. Pl. Dis. Reporter 58: 267-268.
2633. Ogawa, J.M. 1954. The occurrence of *Hendersonula toruloidea* Nattrass on *Populus* species in California. Pl. Dis. Reporter 38: 238.
2634. Ogawa, J.M., H.A. Bolkan, and W.H. Krueger. 1984. Outbreak of *Tranzschelia discolor* causing almond rust disease in northern California. Pl. Dis. 68: 351.
2635. Ogawa, J.M., W.H. English, and E.E. Wilson. 1954. Survey for brown rot of stone fruits in California. Pl. Dis. Reporter 38: 254-257.
2636. Ogawa, J.M., C.W. Nichols, and H. English. 1955. Almond scab. Pl. Dis. Reporter 39: 504-508.

2637. Ohman, J.H., K.J. Kessler Jr., and G.C. Meyer. 1969. Control of *Cryptostroma corticale* on stored sugar maple pulpwood. Phytopathology 59: 871-873.
2638. Ohr, H.D., F.G. Pollack, and B.F. Ingber. 1977. The occurrence of *Alternaria macrospora* on *Anoda cristata* in Mississippi. Pl. Dis. Reporter 61: 208-209.
2639. Olexa, M.T., and T.E. Freeman. 1978. A gall disease of red mangrove caused by *Cylindrocarpon didymum*. Pl. Dis. Reporter 62: 283-286.
2640. Olive, L.S. 1944. New or rare Heterobasidiomycetes from North Carolina - I. J. Elisha Mitchell Sci. Soc. 60: 17-26.
2641. Olive, L.S. 1945. A new *Dacrymyces*-like parasite of *Arundinaria*. Mycologia 37: 543-552.
2642. Olive, L.S. 1947. Notes on the Tremellales of Georgia. Mycologia 39: 90-108.
2643. Olive, L.S. 1951. Taxonomic notes on Louisiana fungi. III. Additions to the Tremellales. Mycologia 43: 677-690.
2644. Olive, L.S. 1952. A new species of *Cystobasidium* from New Jersey. Mycologia 44: 564-569.
2645. Olive, L.S. 1954. New or noteworthy species of Tremellales from the southern Appalachia ns - III. Bull. Torrey Bot. Club 81: 329-339.
2646. Olsen, M.W., and M.E. Stanghellini. 1981. *Mycosphaerella melonis* on greenhouse cucumber. Pl. Dis. 65: 157-159.
2647. Ono, Y. 1985. Host range, geographic distribution, and life cycle of *Puccinia grum osa* (Uredinales). Trans. Mycol. Soc. Japan 26: 493-497.
2648. Ooka, J.J. 1980. Guava fruit rot caused by *Rhizopus stolonifer* in Hawaii. Pl. Dis. Reporter 64: 412-413.
2649. Ooka, J.J., and J.Y. Uchida. 1982. *Cylindrocladium* root and crown rot of alfalfa in Hawaii. Pl. Dis. 66: 947-948.
2650. Ooka, J.J., and M.S. Zuber. 1980. Aflatoxin-producing *Aspergillus flavus* in Hawaii. Pl. Dis. 64: 550-551.
2651. Orellana, R.G. 1959. *Botrytis* leaf blight of *Ricinus communis* L. Pl. Dis. Reporter 43: 363-364.
2652. Orellana, R.G. 1961. Leaf spot of sesame caused by *Cylindrosporium sesami*. Phytopathology 51: 89-92.
2653. Orellana, R.G. 1971. *Fusarium* wilt of sunflower, *Helianthus annuus*: first report. Pl. Dis. Reporter 55: 1124-1125.
2654. Orellana, R.G., and W.K. Bailey. 1964. Blight of peanuts in Maryland and Virginia in 1963. Pl. Dis. Reporter 48: 519-521.
2655. Orellana, R.G., and J.E. Bear. 1963. Rust on sunflower introductions at Beltsville, Maryland. Pl. Dis. Reporter 47: 45.
2656. Oshima, N., L.E. Dickens, and B.F. Counter. 1969. Incidences of *Pythium* wilt of snap beans in Colorado. Pl. Dis. Reporter 53: 766.
2657. Oshima, N., W.J. Henderson, and L.E. Dickens. 1960. The occurrence of spinach rust in Colorado. Pl. Dis. Reporter 44: 828-829.
2658. Ostazeski, S.A. 1959. A *Curvularia* leaf spot of Alyce clover. Pl. Dis. Reporter 43: 350-351.
2659. Ostazeski, S.A. 1967. An undescribed fungus associated with a root and crown rot of birdsfo ot trefoil (*Lotus corniculatus*). Mycologia 59: 970-975.
2660. Ostazeski, S.A., J.H. Elgin Jr., and J.E. McMurtrey III. 1979. Occurrence of anthracnose on formerly anthracnose-resistant 'Arc' alfalfa. Pl. Dis. Reporter 63: 734-736.
2661. Ostazeski, S.A., and J.J. Murray. 1973. *Helminthosporium triseptatum*, a major pathogen of some cultivars of Kentucky bluegrass (*Poa pratensis*). Phytopathology 63: 804.
2662. Ostazeski, S.A., and H.D. Wells. 1960. A *Phomopsis* stem blight of yellow lupine (*Lupinus luteus* L.). Pl. Dis. Reporter 44: 66-67.
2663. Ostrofsky, A., and W.D. Ostrofsky. 1984. *Kabatina juniperi* associated with branch tip dieback of eastern red cedar in Maine. Pl. Dis. 68: 351.
2664. Ostrofsky, A., and G.W. Peterson. 1977. Occurrence of *Kabatina juniperi* on *Juniperus virginiana* in eastern Nebraska. Pl. Dis. Reporter 61: 512-513.
2665. Ostrofsky, A., and G.W. Peterson. 1981. Etiologic and cultural studies of *Kabatina juniperi*. Pl. Dis. 65: 908-910.
2666. Ostry, M.E. 1986. Association of *Parkerella populi* with declining hybrid aspen in Wis consin. Canad. J. Bot. 64: 1834-1835.
2667. Ostry, M.E., T.H. Nicholls, and M.A. Palmer. 1985. Needlecast of western larch seedlings caused by *Mycosphaerella laricina*. Pl. Dis. 69: 542.
2668. Otta, J.D., and F.L. Bode. 1972. Siberian elm canker in South Dakota. Pl. Dis. Reporter 56: 572-574.
2669. Ouellette, G.B., and L.P. Magasi. 1966. *Lophomerum*, a new genus of Hypodermataceae. Mycologia 58: 275-280.
2670. Ouellette, G.B., and K.A. Pirozynski. 1974. Reassessment of *Tympanis* based on types of ascospore germination within asci. Canad. J. Bot. 52: 1889-1911.
2671. Overholts, L.O. 1921. Some New Hampshire fungi. Mycologia 13: 24-37.
2672. Overholts, L.O. 1925. Mycological notes for 1923. Mycologia 17: 108-112.
2673. Overholts, L.O. 1933. Mycological notes for 1930-32. Mycologia 25: 418-430.
2674. Overholts, L.O. 1934. Mycological notes for 1933. Mycologia 26: 502-515.
2675. Overholts, L.O. 1938. Notes on fungi from the lower Mississippi Valley. Bull. Torrey Bot. Club 65: 167-180.
2676. Overholts, L.O. 1938. Mycological notes for 1934-1935. Mycologia 30: 269-279.
2677. Overholts, L.O. 1940. Mycological notes for 1936-38. Mycologia 32: 251-263.
2678. Overholts, L.O. 1942. Polyporaceae of Pennsylvania. The genus *Poria*. Pennsylvania State Coll. Agric. Exp. Sta. Bull. 418: 1-64.
2679. Overholts, L.O. 1943. Mycological notes for 1939-40. Mycologia 35: 243-257.
2680. Overholts, L.O. 1952. New species of polypores. Mycologia 44: 224-227.
2681. Overholts, L.O. 1953. The Polyporaceae of the United States, Alaska and Canada. Univ. Michigan Press. 466 p.
2682. Overholts, L.O., and J.L. Lowe. 1946. New species of *Poria*. Mycologia 38: 202-212.
2683. Owen, J.H., and D. Connell. 1953. Cucumber diseases in Sumter County, Florida. Pl. Dis. Reporter 37: 327.
2684. Owen, J.H., J.H. Miller, and W.A. Campbell. 1960. *Clitocybe* mushroom root rot in Georgia. Pl. Dis. Reporter 44: 89-91.
2685. Pady, S.M. 1958. Dutch elm disease in Kansas. Pl. Dis. Reporter 42: 402.
2686. Pady, S.M. 1974. Sporobolomycetaceae in Kansas. Mycologia 66: 333-338.
2687. Pady, S.M., and C.O. Johnson. 1958. Cereal rust epidemiology and aerobiology in Kansas in 1957. Pl. Dis. Reporter 42: 726-733.
2688. Pady, S.M., and C.O. Johnston. 1959. Stripe rust in Kansas in 1958. Pl. Dis. Reporter 43: 159-163.
2689. Pady, S.M., C.O. Johnston, and C.T. Rogerson. 1957. Stipe rust of wheat in Kansas in 1957. Pl. Dis. Reporter 41: 959-961.
2690. Palmer, M.A., and T.H. Nicholls. 1985. Shoot blight and collar rot of *Pinus resinosa* caused by *Sphaeropsis sapinea* in forest tree nurseries. Pl. Dis. 69: 739-740.
2691. Palmer, M.A., T.H. Nicholls, and C.F. Croghan. 1986. Fungicidal control of shoot blight caused by *Sphaeropsis sapinea* on red pine nursery seedlings. Pl. Dis. 70: 194-196.
2692. Palmer, M.A., M.E. Ostry, K. Robbins, and T.H. Nicholls. 1986. Occurrence and development of *Mycosphaerella laricina* on larch in the north central United States. Pl. Dis. 70: 921-923.
2693. Palmiter, D.H. 1951. A blossom end rot of apples in New York caused by *Botrytis*. Pl. Dis. Reporter 35: 435-436.
2694. Pantidou, M.E., and G.D. Darker. 1963. The species of *Didymascella* on *Juniperus*. Mycologia 55: 415-420.
2695. Park, J.Y., and R. Sprague. 1953. Studies on some *Selenophoma* species on Gramineae. Mycologia 45: 260-275.
2696. Parker, A.K., and J. Reid. 1969. The genus *Rhabdocline* Syd. Canad. J. Bot. 47: 1533-1545.
2697. Parker, M.A. 1987. Pathogen impact on sexual vs. asexual reproductive success in *Arisaema triphyllum*. Amer. J. Bot. 74: 1758-1763.
2698. Parker, P.E., and D.C. Ramsdell. 1977. Epidemiology and chemical control of *Phomopsis* canker of highbush blueberry. Phytopathology 67: 1481-1484.
2699. Parmelee, J. 1958. Some foliicolous fungi of the Pyrolaceae. Canad. J. Bot. 38: 865-881.
2700. Parmelee, J.A. 1974. *Gymnosporangium connersii*. Fungi Canadenses 28: 1-2.
2701. Parmelee, J.A. 1976. *Puccinia balsamorhizae*. Fungi Canadenses 89: 1-2.
2702. Parmelee, J.A. 1977. *Puccinia xanthii*. Fungi Canadenses 99: 1-2.
2703. Parmelee, J.A. 1980. *Puccinia schedonnardi*. Fungi Canadenses

172: 1-2.
2704. Parmelee, J.A. 1981. *Puccinia limosae*. Fungi Canadenses 207: 1-2.
2705. Parmelee, J.A. 1981. *Puccinia ortonii*. Fungi Canadenses 208: 1-2.
2706. Parmelee, J.A. 1986. *Frommeella tormentillae*. Fungi Canadenses 306: 1-2.
2707. Parmelee, J.A. 1986. *Xenodochus carbonarius*. Fungi Canadenses 308: 1-2.
2708. Parmelee, J.A. 1986. The autoecious species of *Puccinia* on Polemoniaceae in North America. Mycologia 78: 454-468.
2709. Parmelee, J.A. 1987. *Puccinia iridis*. Fungi Canadenses 320: 1-2.
2710. Parmelee, J.A., and R. Cauchon. 1979. *Strasseria* on Pinaceae in Canada. Canad. J. Bot. 57: 1660-1662.
2711. Parmelee, J.A., and M.E. Elliott. 1974. *Pileolaria brevipes*. Fungi Canadenses 50: 1-2:
2712. Parmelee, J.A., and D.B.O. Savile. 1954. Life history and relationship of the rusts of *Sparganium* and *Acorus*. Mycologia 46: 823-836.
2713. Parmelee, J.A., and D.B.O. Savile. 1981. Autoecious species of *Puccinia* on Cichorieae in North America. Canad. J. Bot. 59: 1078-1101.
2714. Parmelee, J.A., and D.B.O. Savile. 1984. *Puccinia ustalis*. Fungi Canadenses 268: 1-2.
2715. Parmelee, J.A., and P.M. deCarteret. 1984. *Puccinia areolata*. Fungi Canadenses 269: 1-2.
2716. Parmelee, J.A., and P.M. deCarteret. 1984. *Puccinia calthae*. Fungi Canadenses 270: 1-2.
2717. Parmelee, J.A., and P.M. deCarteret. 1985. *Puccinia haleniae*. Fungi Canadenses 296: 1-2.
2718. Parmelee, J.A., and P.M. deCarteret. 1985. *Uromyces eugentianae*. Fungi Canadenses 298: 1-2.
2719. Parmeter, J.R., Jr., J.R. Hood, and R.F. Scharpf. 1959. *Colletotrichum* blight of dwarf mistletoe. Phytopathology 49: 812-815.
2720. Parris, G.K. 1959. A revised host index of Mississippi plant diseases. Mississippi State Univ., Botany Dept. Misc. Pub. 1: 1-146.
2721. Parris, G.K. 1966. Death of honeylocust in Mississippi associated with *Polyporus lucidus* Fr. Pl. Dis. Reporter 50: 243-244.
2722. Partridge, A.D., and D.L. Miller. 1972. Dutch elm disease continues to move east in the Idaho - Oregon area. Pl. Dis. Reporter 56: 939.
2723. Partridge, A.D., and H.N. Ward. 1969. Dutch elm disease in Idaho - 1968. Pl. Dis. Reporter 53: 140-141.
2724. Partridge, A.D., and H.N. Ward. 1971. Dutch elm disease crosses treeless rangeland in Idaho. Pl. Dis. Reporter 55: 1064.
2725. Partridge, A.D., and L.C. Weir. 1974. Dutch elm disease moves into Oregon from Idaho. Pl. Dis. Reporter 58: 75-76.
2726. Partridge, J.E., and B.L. Doupnik. 1979. Occurrence of sorghum downy mildew on shattercane and sorghum in Nebraska. Pl. Dis. Reporter 63: 154-155.
2727. Patterson, C.L., and R.G. Grogan. 1985. Differences in epidemiology and control of lettuce drop caused by *Sclerotinia minor* and *S. sclerotiorum*. Pl. Dis. 69: 766-770.
2728. Patton, R.F., and R.N. Spear. 1983. Needle cast of European larch caused by *Mycosphaerella laricina* in Wisconsin and Iowa. Pl. Dis. 67: 1149-1153.
2729. Pavgi, M.S., and A.L. Flangas. 1959. Occurrence of southern corn rust in Wisconsin. Pl. Dis. Reporter 43: 1239-1240.
2730. Paxton, J.D., and D.P. Rogers. 1974. Powdery mildew of soybeans. Mycologia 66: 894-896.
2731. Payne, G.A., and K.J. Leonard. 1985. *Stenocarpella macrospora* on corn in North Carolina. Pl. Dis. 69: 613.
2732. Peacher, P.H. 1969. Live oak in Mississippi infected by *Endothia parasitica*. Pl. Dis. Reporter 53: 304.
2733. Peacher, P.H., and C.A. Bower. 1969. Status of Dutch elm disease in Oklahoma. Pl. Dis. Reporter 53: 768.
2734. Peacher, P.H., M.J. Weiss, and J.F. Wolf. 1975. Southward spread of oak wilt remains static. Pl. Dis. Reporter 59: 303-304.
2735. Pearson, R.C., and D.H. Hall. 1973. Ripe fruit rot of tomato caused by *Pythium ultimum* and *Pythium aphanidermatum*. Pl. Dis. Reporter 57: 1066-1069.
2736. Peck, C.H. 1872. Plants found growing spontaneously in the state and not before reported. Ann. Rep. New York State Mus. 24: 56-103.
2737. Peck, C.H. 1884. Report of the Botanist. Annual Rep. New York State Mus. 35: 125-164.
2738. Peck, C.H. 1909. New species of fungi. Bull. Torrey Bot. Club 36: 329-339.
2739. Peck, C.H. 1912. State Botanist Report for 1911. Bull. New York State Mus. 157: 1-116.
2740. Peek, C.A., and W.G. Solheim. 1958. The hyphomycetous genera of H.W. Harkness and the ascomycetous genus *Cleistosoma* Harkn. Mycologia 50: 844-861.
2741. Pepper, E.H. 1959. Northwestern apple tree anthracnose reported from Michigan. Pl. Dis. Reporter 43: 920-921.
2742. Perez, J.M., and T.E. Summers. 1963. A *Botrytis* disease of kenaf. Pl. Dis. Reporter 47: 200-201.
2743. Perkins, J.M., and A.L. Hooker. 1979. Effects of anthracnose stalk rot on corn yields in Illinois. Pl. Dis. Reporter 63: 26-30.
2744. Perry, R.S. 1970. *Chalaropsis thielavioides*, a new pathogen on poinsettias. Pl. Dis. Reporter 54: 451-452.
2745. Person, L.H., and J.G. Gaines. 1955. The occurrence of anthracnose on tobacco in Georgia. Pl. Dis. Reporter 39: 597.
2746. Persons, T.D. 1954. Destructive outbreak of loose smut in a Georgia wheat field. Pl. Dis. Reporter 38: 422.
2747. Petersen, D.H. 1960. Hymenomycetous fungi except *Polyporus* associated with wood decay of living peach trees in South Carolina. Pl. Dis. Reporter 44: 891-893.
2748. Petersen, D.H. 1961. *Polyporus* spp. associated with wood decay of living peach trees in South Carolina. Pl. Dis. Reporter 45: 186-188.
2749. Petersen, D.H., and G. KenKnight. 1950. *Prunus angustifolia*, a hitherto unreported host of *Glomerella cing ulata*. Pl. Dis. Reporter 34: 352.
2750. Petersen, R.H. 1963. Aquatic Hyphomycetes from North America III. Phialosporae and miscellaneous species. Mycologia 55: 570-581.
2751. Petersen, R.H. 1974. Notes on clavarioid fungi. XIV. Cultures of *Lentaria byssiseda*. Mycologia 66: 530-532.
2752. Peterson, G.W. 1964. Dutch elm disease in Nebraska. Pl. Dis. Reporter 48: 781.
2753. Peterson, G.W. 1973. Infection of Austrian and Ponderosa Pines by *Dothistroma pini* in eastern Nebraska. Phytopathology 63: 1060-1063.
2754. Peterson, G.W. 1977. Epidemiology and control of a blight of *Juniperus virginiana* caused by *Cercospora sequoiae* var. *juniperi*. Phytopathology 67: 234-238.
2755. Peterson, G.W., J.L. Stewart, and W.G. Willis. 1968. Oak wilt distribution in Nebraska and Kansas. Pl. Dis. Reporter 52: 357-358.
2756. Peterson, G.W., and J.A. Walla. 1978. Development of *Dothistroma pini* upon and within needles of Austrian and ponderosa pines in eastern Nebraska. Phytopathology 68: 1422-1430.
2757. Peterson, G.W., and D.S. Wysong. 1968. *Diplodia* tip blight of pines in the central Great Plains: damage and control. Pl. Dis. Reporter 52: 359-360.
2758. Peterson, G.W., and D.S. Wysong. 1968. *Cercospora* blight of junipers: damage and control. Pl. Dis. Reporter 52: 361-362.
2759. Peterson, J.L. 1967. Root and crown rot of perennial phlox caused by *Thielaviopsis basicola*. Pl. Dis. Reporter 51: 436-437.
2760. Peterson, J.L. 1967. Leaf spot of annual phlox caused by *Stemphylium botryosum*. Pl. Dis. Reporter 51: 741-742.
2761. Peterson, J.L. 1970. Diseases of holly in the eastern United States. Amer. Hort. Mag. 49: 231-233.
2762. Peterson, J.L., and S.H. Davis. 1963. Chrysanthemum flower rot caused by *Fusarium tricinctum* f. *poae* (Pk.) Snyd. & Hans. Pl. Dis. Reporter 47: 722-723.
2763. Peterson, J.L., S.H. Davis Jr., and P.V.V. Weber. 1978. The occurrence of *Puccinia horiana* on chrysanthemum in New Jersey. Pl. Dis. Reporter 62: 357-360.
2764. Peterson, R. 1967. Cypress rusts in California and Baja California. Madroño 19: 47-54.
2765. Peterson, R.S. 1959. The *Cronartium coleosporioides* complex in the Black Hills. Pl. Dis. Reporter 43: 1227-1228.
2766. Peterson, R.S. 1959. The *Cronartium coleosporioides* complex in the Black Hills. Pl. Dis. Reporter 43: 1227-1228.
2767. Peterson, R.S. 1961. Notes on western rust fungi I. *Chrysomyxa*. Mycologia 53: 427-431.
2768. Peterson, R.S. 1962. Notes on western rust fungi. II.

Pucciniaceae. Mycologia 54: 389-394.
2769. Peterson, R.S. 1965. Notes on western rust fungi. IV. Mycologia 57: 465-471.
2770. Peterson, R.S. 1967. The *Peridermium* species on pine stems. Bull. Torrey Bot. Club 94: 511-542.
2771. Peterson, R.S. 1967. Studies of juniper rusts in the west. Madroño 19: 79-91.
2772. Peterson, R.S. 1972. On *Coleosporium crowellii* (Uredinales). Pl. Dis. Reporter 56: 474-475.
2773. Peterson, R.S. 1982. Rust fungi (Uredinales) on Cupressaceae. Mycologia 74: 903-910.
2774. Peterson, R.S., and E.A. Cronin. 1967. Non-rosaceous hosts of *Phragmidium* (Uredinales). Pl. Dis. Reporter 51: 766-767.
2775. Petrak, F. 1947. Kleine Beiträge zur Pilzflora von Australien und Polynesien. Sydowia 1: 264-276.
2776. Petrak, F. 1950. Fungi Beltsvillenses. Sydowia 4: 402-408.
2777. Petrak, F. 1951. Fungi Beltsvillenses. II. Sydowia 5: 230-247.
2778. Petrak, F. 1952. *Stevensonula* n. gen., eine neue Gattung der Parasphaerosideen. Sydowia 6: 1-4.
2779. Petrak, F. 1952. Fungi Beltsvillenses. III. Sydowia 6: 5-16.
2780. Petrak, F. 1952. Fungi Beltsvillenses IV. Sydowia 6: 352-360.
2781. Petrak, F. 1952. Ein Beitrag zur Pilzflora von Hawaii. Sydowia 6: 363-371.
2782. Petrak, F. 1953. Ein Beitrag zur Pilzflora Floridas. Sydowia 7: 103-116.
2783. Petrak, F. 1953. Fungi Beltsvillenses. V. Sydowia 7: 121-132.
2784. Petrak, F. 1953. Beiträge zur Pilzflora von Hawaii. Sydowia 7: 381-409.
2785. Petrak, F. 1955. Fungi Beltsvillenses. VI. Sydowia 9: 542-546.
2786. Petrini, L.E., and J.D. Rogers. 1986. A summary of the *Hypoxylon serpens* complex. Mycotaxon 26: 401-436.
2787. Petrini, O., and G. Carroll. 1981. Endophytic fungi in foliage of some Cupressaceae in Oregon. Canad. J. Bot. 59: 629-636.
2788. Pettit, R.E., R.A. Taber, and A.L. Harrison. 1973. *Ascochyta* web-blotch of peanuts. Phytopathology 63: 447.
2789. Pfender, W.F., and D.J. Hagedorn. 1982. *Aphanomyces euteiches* f. sp. *phaseoli*, a causal agent of bean root and hypocotyl rot. Phytopathology 72: 306-310.
2790. Pfleger, F.L. 1974. *Festuca rubra*, a new host species of *Puccinia pygmaea*. Pl. Dis. Reporter 58: 1028-1029.
2791. Pfleger, F.L., and E.K. Vaughn. 1972. *Pyrenochaeta terrestris* as a naturally occurring component of the soil in Oregon. Pl. Dis. Reporter 56: 180-182.
2792. Phaff, H.J., M. Miranda, W.T. Starmer, J. Tredick, and J.S.F. Barker. 1986. *Clavispora opuntiae*, a new heterothallic yeast occurring in necrotic tissue of *Opuntia* species. Int. J. Systematic Bacteriol. 36: 372-379.
2793. Phaff, H.J., W.T. Starmer, M. Miranda, and M.W. Miller. 1978. *Pichia heedii*, a new species of yeast indigenous to necrotic cacti in the North American Sonoran Desert. Int. J. Systematic Bacteriol. 28: 326-331.
2794. Phaff, H.J., W.T. Starmer, J. Tredick, and M. Miranda. 1985. *Pichia deserticola* and *Candida deserticola*, two new species of yeasts associated with necrotic stems of cacti. Int. J. Systematic Bacteriol. 35: 211-216.
2795. Phelps, W.R. 1974. Live oak in North Carolina and Florida infected by *Endothia parasitica*. Pl. Dis. Reporter 58: 596-598.
2796. Phelps, W.R. 1974. Evaluation of fusiform rust incidence on loblolly and slash pine in the south. Pl. Dis. Reporter 58: 1137-1141.
2797. Phillips, D.J., and W.J. Lipton. 1974. Systemic downy mildew found in California head lettuce. Pl. Dis. Reporter 58: 118-119.
2798. Phillips, D.J., B. Mackey, W.R. Ellis, and T.N. Hansen. 1979. Occurrence and interaction of *Aspergillus flavus* with other fungi on almonds. Phytopathology 69: 829-831.
2799. Phillips, D.V., and H.R. Boerma. 1981. *Cercospora sojina* race 5: a threat to soybeans in the southeastern United States. Phytopathology 71: 334-336.
2800. Phillips, D.V., and J.P. Ross. 1968. Incidence of brown stem rot of soybeans in North Carolina. Pl. Dis. Reporter 52: 895-896.
2801. Phillips, W., and H.W. Harkness. 1884. Fungi of California. Bull. Calif. Acad. Sci. 1: 21-25.
2802. Phipps, P.M. 1985. Web blotch of peanut in Virginia. Pl. Dis. 69: 1097-1099.
2803. Phipps, P.M., and D.M. Porter. 1982. *Sclerotinia* blight of soybean caused by *Sclerotinia minor* and *Sclerotinia sclerotiorum*. Pl. Dis. 66: 163-165.
2804. Phipps, P.M., and R.J. Stipes. 1976. Histopathology of mimosa infected with *Fusarium oxysporum* f. sp. *perniciosum*. Phytopathology 66: 839-843.
2805. Pieczarka, D.J., and J.W. Lorbeer. 1975. Microorganisms associated with bottom rot of lettuce grown on organic soil in New York state. Phytopathology 65: 16-21.
2806. Pierce, L., and A.H. McCain. 1982. Stem rot and wilt of *Exacum affine* caused by *Nectria haematococca*. Pl. Dis. 66: 161-163.
2807. Pierce, L., and A.H. McCain. 1986. Rot of oleander cuttings caused by *Calonectria crotalariae*. Pl. Dis. 70: 352.
2808. Piirto, D.D., J.R. Parmeter Jr., and F.W. Cobb Jr. 1974. *Fomes annosus* in giant sequoia. Pl. Dis. Reporter 58: 478.
2809. Piirto, D.D., J.R. Parmeter, and W.W. Wilcox. 1977. *Poria incrassata* in giant sequoia. Pl. Dis. Reporter 61: 50.
2810. Pinckard, J.A., and G. Guidroz. 1968. A parasitic *Phytophthora* boll rot of cotton found in Louisiana. Pl. Dis. Reporter 52: 780-781.
2811. Pinckard, J.A., and G.F. Guidroz. 1973. A boll rot of cotton caused by *Phytophthora parasitica*. Phytopathology 63: 896-899.
2812. Pinckard, J.A., and W.J. Luke. 1967. *Pellicularia filamentosa*, a primary cause of cotton boll rot in Louisiana. Pl. Dis. Reporter 51: 67-70.
2813. Pinkard, J.A. 1964. *Pellicularia filamentosa* a common saprophyte on mature cotton stems in Louisiana. Phytopathology 54: 626.
2814. Pirone, P.P., and J.C. Carter. 1966. *Phomopsis acerina* on Norway Maple. Mycologia 58: 325-328.
2815. Pirone, T.P. 1963. Some woody ornamental hosts of *Clitocybe tabescens* in Louisiana. Pl. Dis. Reporter 47: 721.
2816. Pirozynski, K.A. 1974. *Cercospora handelii*. Fungi Canadenses 22: 1-2.
2817. Pirozynski, K.A. 1974. *Ramularia coalescens*. Fungi Canadenses 23: 1-2.
2818. Pirozynski, K.A., and C.S. Hodges Jr. 1973. New hyphomycetes from South Carolina. Canad. J. Bot. 51: 157-173.
2819. Pirozynski, K.A., and R.A. Shoemaker. 1970. Some Asterinaceae and Meliolaceae on conifers in Canada. Canad. J. Bot. 48: 1321-1328.
2820. Pirozynski, K.A., and R.A. Shoemaker. 1970. *Seimatosporium* leaf spot of *Ledum* and *Rhododendron*. Canad. J. Bot. 48: 2199-2203.
2821. Plaats-Niterink, A.J. van der. 1981. Monograph of the genus *Pythium*. Stud. Mycol. 21: 1-242.
2822. Plakidas, A.G. 1942. *Venturia acerina*, the perfect stage of *Cladosporium humile*. Mycologia 34: 27-37.
2823. Plakidas, A.G. 1950. Camellia flower blight in Louisiana. Pl. Dis. Reporter 34: 188.
2824. Plakidas, A.G. 1950. Rose rust (*Phragmidium americanum*?) in Louisiana. Pl. Dis. Reporter 34: 197.
2825. Plakidas, A.G. 1952. White rust of spinach in Louisiana. Pl. Dis. Reporter 36: 211.
2826. Plakidas, A.G. 1954. Spot anthracnose of Chinese holly. Mycologia 46: 346-353.
2827. Plakidas, A.G. 1955. Spot anthracnose of Chinese holly. Pl. Dis. Reporter 39: 243-244.
2828. Plakidas, A.G. 1955. *Camellia* petal blight in Louisiana in 1955. Pl. Dis. Reporter 39: 519.
2829. Plakidas, A.G. 1955. *Clitocybe tabescens* on camellia. Pl. Dis. Reporter 39: 519.
2830. Plakidas, A.G. 1956. *Cercospora* leaf spot of *Abelia*. Mycologia 48: 382-385.
2831. Plakidas, A.G. 1957. *Camellia* petal blight found in Texas. Pl. Dis. Reporter 41: 547.
2832. Plakidas, A.G. 1957. New or unusual plant diseases in Louisiana. Pl. Dis. Reporter 41: 643-645.
2833. Plakidas, A.G. 1959. *Cercospora theae* on *Camellia sasanqua* and *Thea sinensis*. Pl. Dis. Reporter 43: 668-669.
2834. Plakidas, A.G. 1960. Angular leaf spot of *Magnolia*. Mycologia 52: 255-259.
2835. Ploetz, R.C. 1980. Crown and root rot of gloxinia and other gesneriads caused by *Phytophthora parasitica*. Pl. Dis. 64: 487-490.
2836. Plowright, C.B., and H.W. Harkness. 1884. New species of California fungi. Bull. Calif. Acad. Sci. 1: 26.
2837. Polach, F.J., and G.S. Abawi. 1975. The occurence and biology of

Botryotinia fuckeliana on beans in New York. Phytopathology 65: 657-660.
2838. Polach, F.J., and R.K. Webster. 1972. Identification of strains and inheritance of pathogenicity in *Phytophthora capsici*. Phytopathology 62: 20-26.
2839. Politis, D.J., A.K. Watson, and W.L. Bruckart. 1984. Susceptibility of musk thistle and related composites to *Puccinia carduorum*. Phytopathology 74: 687-691.
2840. Pollack, F.G. 1947. Two additions to the fungi imperfecti. Mycologia 39: 617-621.
2841. Pollack, F.G. 1971. *Cercospora uromycestri*, hyperparasite of rust on *Cestrum diurnum*. Mycologia 63: 689-693.
2842. Pollack, F.G., and C.W. Ellett. 1974. *Mycocentrospora verrucosa*, the cause of foliar shot-hole of *Euonymus*. Mycologia 66: 170-173.
2843. Pollack, F.G., and D.F. Farr. 1979. A new species of *Stigmina* on *Beaucarnea* (Liliaceae). Mycotaxon 8: 193-200.
2844. Pollack, F.G., and F.D. Matthews. 1976. *Deightoniella argemonensis*, a new fungus on Mexican pricklepoppy, associated with *Cercosporidium guanicense*. Mycologia 68: 1093-1097.
2845. Pollack, F.G., and J.A. Stevenson. 1973. A fungal pathogen of *Broussonetia papyrifera* collected by George Washington Carver. Pl. Dis. Reporter 57: 296-297.
2846. Pollack, F.G., and B.C. Sutton. 1972. *Colletogloeum fouquieriae*, a new fungus on ocotillo. Mycologia 64: 702-707.
2847. Pollack, F.G., and F.A. Uecker. 1974. *Monosporascus cannonballus* an unusual ascomycete in cantaloupe roots. Mycologia 66: 346-349.
2848. Pollack, F.G., and H.E. Waterworth. 1969. A leafspot disease of kenaf in Maryland associated with *Cristulariella pyramidalis*. Pl. Dis. Reporter 53: 810-812.
2849. Pomerleau, R., and D.E. Etheridge. 1961. A bluestain in balsam fir. Mycologia 53: 155-170.
2850. Pon, D.S., C.G. Schmitt, and C.H. Kingsolver. 1959. An *Acremonium* associated with stem rust pustules. Pl. Dis. Reporter 43: 173-174.
2851. Pool, R.A.F., and P.W. Hiatt. 1976. Dutch elm disease in California. Pl. Dis. Reporter 60: 166-167.
2852. Pool, R.A.F., and E.L. Sharp. 1968. Distribution of *Cephalosporium* stripe disease of wheat in Montana. Pl. Dis. Reporter 52: 818-819.
2853. Poole, D. 1954. Red stem gall of castorbean caused by *Synchytrium* in Texas. Pl. Dis. Reporter 38: 728.
2854. Poole, D.D. 1954. Disease observations on castorbeans in 1953. Pl. Dis. Reporter 38: 218-220.
2855. Poole, D.D. 1956. Aerial stem rot of sesame caused by *Helminthosporium sesami* in Texas. Pl. Dis. Reporter 40: 235.
2856. Poole, R.F. 1922. A new fruit rot of tomatoes. Bot. Gaz. 74: 210-214.
2857. Pope, W.K., E.L. Sharp, and H.S. Fenwick. 1963. Stripe rust of wheat in the Pacific Northwest in 1962. Pl. Dis. Reporter 47: 554-555.
2858. Porter, D.M., and M.K. Beute. 1974. *Sclerotinia* blight of peanuts. Phytopathology 64: 263-264.
2859. Porter, D.M., F.S. Wright, and N.L. Powell. 1987. Effects of sprinkler irrigation on peanut diseases in Virginia. Pl. Dis. 71: 512-515.
2860. Pottle, W.H., and A.L. Shigo. 1975. Treatment of wounds on *Acer rubrum* with *Trichoderma viride*. Eur. J. Forest Pathol. 5: 274-279.
2861. Pound, G.S., and D.L. Fowler. 1951. A *Fusarium* wilt of radish in Wisconsin. Phytopathology 41: 30.
2862. Povah, A.H. 1931. Fungi of Rock River, Michigan. Pap. Michigan Acad. Sci. 13: 173-189.
2863. Powell, D., and R. Naik. 1971. An apple twig canker in Illinois. Pl. Dis. Reporter 55: 96.
2864. Powers, H.R., and J.S. Boyce. 1961. *Fomes annosus* on slash pine in the southeast. Pl. Dis. Reporter 45: 306-307.
2865. Powers, H.R. Jr. 1969. Virginia pine, a new host for *Cronartium comandrae*. Pl. Dis. Reporter 53: 320.
2866. Powers, H.R., G.H. Hepting, and W.A. Stegall Jr. 1967. *Comandra* rust on loblolly pine in eastern Tennessee. Pl. Dis. Reporter 51: 4-8.
2867. Pratt, R.G. 1981. Morphology, pathogenicity, and host range of *Phytophthora megasperma*, *P. erythroseptica*, and *P. parasitica* from arrowleaf clover. Phytopathology 71: 276-282.
2868. Pratt, R.G. 1982. A new vascular wilt disease caused in crimson clover by *Fusarium oxysporum*. Phytopathology 72: 622-627.
2869. Pratt, R.G. 1984. A new *Cercospora* leaf and stem disease of subterranean clover. Phytopathology 74: 1152-1156.
2870. Pratt, R.G., L.F. Roth, E.M. Hansen, and W.D. Ostrofsky. 1976. Identity and pathogenicity of species of *Phytophthora* causing root rot of Douglas-fir in the Pacific Northwest. Phytopathology 66: 710-714.
2871. Presley, J.T. 1954. Root diseases of cotton observed in Mexico and California. Pl. Dis. Reporter 38: 529.
2872. Preston, D.A. 1945. Host index of Oklahoma plant diseases. Oklahoma Agric. Mechan. College Agric. Exp. Sta. Tech. Bull. T-21: 1-168.
2873. Preston, D.A. 1947. Host index of Oklahoma plant diseases, supplement, 1947. Oklahoma Agric. Mechan. College Agric. Exp. Sta. Tech. Bull. (suppl.) T-21: 1-39.
2874. Preston, D.A., and L. Dosdall. 1955. Minnesota Plant Diseases. USDA Plant Disease Epidemics and Identification Section. Special Publication No. 8. 184 p.
2875. Prey, A.J., and F.S. Morse. 1971. Brown spot needle blight of Scotch pine Christmas trees in Wisconsin. Pl. Dis. Reporter 55: 648-649.
2876. Pronos, J., and R.F. Patton. 1977. *Armillaria* root rot of red pine planted on oak sites in Wisconsin. Pl. Dis. Reporter 61: 955-958.
2877. Pronos, J., and R.F. Patton. 1978. Penetration and colonization of oak roots by *Armillaria mellea* in Wisconsin. Eur. J. Forest Pathol. 8: 259-267.
2878. Provvidenti, R., and W.T. Schroeder. 1959. Foliage infection of tomato and eggplant by *Verticillium*. Pl. Dis. Reporter 43: 821-826.
2879. Prusso, D.C., and K. Wells. 1967. *Sporobolomyces roseus*. I. Ultrastructure. Mycologia 59: 337-348.
2880. Pscheidt, J.W., and W.R. Stevenson. 1986. Comparison of forecasting methods for control of potato early blight in Wisconsin. Pl. Dis. 70: 915-920.
2881. Puhalla, J.E. 1979. Classification of isolates of *Verticillium dahliae* based on heterokaryon incompatibility. Phytopathology 69: 1186-1189.
2882. Pultz, L.M. 1950. Occurrence of *Ovulara canaegricola* P. Henn. on canaigre (*Rumex hymenosepalus* Torr.). Pl. Dis. Reporter 34: 157.
2883. Punithalingham, E. 1981. Microfungi from cereals and grasses. II. Nova Hedwigia 34: 67-95.
2884. Punja, Z.K. 1986. Progression of root rot on processing carrots due to *Sclerotium rolfsii* and the relationship of disease incidence to inoculum density. Canad. J. Plant Pathol. 8: 297-304.
2885. Punja, Z.K., J.D. Carter, G.M. Campbell, and E.L. Rossell. 1986. Effects of calcium and nitrogen fertilizers, fungicides, and tillage practices on incidence of *Sclerotium rolfsii* on processing carrots. Pl. Dis. 70: 819-824.
2886. Punja, Z.K., R.G. Grogan, and G.C. Adams Jr. 1982. Influence of nutrition, environment, and the isolate, on basidiocarp formation, development, and structure in *Athelia (Sclerotium) rolfsii*. Mycologia 74: 917-926.
2887. Punja, Z.K., R.G. Grogan, and T. Unruh. 1982. Comparative control of *Sclerotium rolfsii* on golf greens in northern California with fungicides, inorganic salts, and *Trichoderma* spp. Pl. Dis. 66: 1125-1128.
2888. Punter, D., J. Reid, and A.A. Hopkin. 1984. Notes on sclerotium-forming fungi from *Zizania aquatica* (wildrice) and other hosts. Mycologia 76: 722-732.
2889. Purcell, S.L., D.J. Phillips, and B.E. Mackey. 1980. Distribution of *Aspergillus flavus* and other fungi in several almond-growing areas of California. Phytopathology 70: 926-929.
2890. Purdy, L.H., S.V. Krupa, and J.L. Dean. 1985. Introduction of sugarcane rust into the Americas and its spread to Florida. Pl. Dis. 69: 689-693.
2891. Raabe, R.D. 1955. Additional woody hosts of *Armillaria mellea* in California. Pl. Dis. Reporter 39: 658-659.
2892. Raabe, R.D. 1957. Perithecia of *Erysiphe polygoni* on Iceland poppy. Pl. Dis. Reporter 41: 694-695.
2893. Raabe, R.D. 1958. Some previously unreported non-woody hosts of *Armillaria mellea* in California. Pl. Dis. Reporter 42: 1025.
2894. Raabe, R.D. 1959. *Clasterosporium* shot-hole of the Catalina, or Island cherry. Phytopathology 49: 116-117.
2895. Raabe, R.D. 1959. Additional unrecorded hosts of *Armillaria*

mellea. Pl. Dis. Reporter 43: 1270.
2896. Raabe, R.D. 1964. Cleistothecia of a powdery mildew on coral bells. Pl. Dis. Reporter 48: 416.
2897. Raabe, R.D. 1965. Some previously unreported hosts of *Armillaria mellea* in California. Pl. Dis. Reporter 49: 812.
2898. Raabe, R.D. 1966. Variation of *Armillaria mellea* in culture. Phytopathology 56: 1241-1244.
2899. Raabe, R.D. 1966. Check list of plant diseases previously unreported in Hawaii. Pl. Dis. Reporter 50: 411-414.
2900. Raabe, R.D. 1967. Some previously unreported hosts of *Armillaria mellea* in California, II. Pl. Dis. Reporter 51: 390.
2901. Raabe, R.D. 1985. Association of soil particles with seed and three pathogens of chickpea in California. Pl. Dis. 69: 238-239.
2902. Raabe, R.D., and S.M. Alcorn. 1968. *Armillaria* root and stem rot of prickly pear cactus. Phytopathology 58: 1036-1037.
2903. Raabe, R.D., I.L. Conners, and A.P. Martinez. 1981. Checklist of Plant Diseases in Hawaii. Hawaii Institute of Tropical Agriculture and Human Resources. College of Tropical Agriculture and Human Resources, University of Hawaii. Information Text Series no. 22. 313 p.
2904. Raabe, R.D., and H.N. Hansen. 1955. *Entomosporium* leaf spot of *Rhaphiolepis*. Phytopathology 45: 55.
2905. Raabe, R.D., and F.F. Hendrix Jr. 1966. *Sclerotinia* blight of *Gloxinia*. Mycologia 58: 161-166.
2906. Raabe, R.D., P.J. Lert, and A.H. McCain. 1968. *Armillaria* stem rot of carnation. Pl. Dis. Reporter 52: 415.
2907. Raabe, R.D., and E.E. Trujillo. 1963. *Armillaria mellea* in Hawaii. Pl. Dis. Reporter 47: 776.
2908. Raabe, R.D., and S. Wilhelm. 1958. *Verticillium* wilt of garden stock (*Matthiola incana*). Phytopathology 48: 610-613.
2909. Rader, W.E. 1948. *Helminthosporium portulacae* a new pathogen of *Portulaca oleracea* L. Mycologia 40: 342-346.
2910. Ramachar, P., and G.B. Cummins. 1963. The species of *Uromyces* on the tribe Paniceae. Mycopathol. Mycologia Appl. 19: 49-61.
2911. Ramaley, A.W. 1987. A new rust on Leguminosae. Mycotaxon 28: 361-363.
2912. Ramirez-Villupadua, J., R.M. Endo, P. Bosland, and P.H. Williams. 1985. A new race of *Fusarium oxysporum* f. sp. *conglutinans* that attacks cabbage with type A resistance. Pl. Dis. 69: 612-613.
2913. Ramsey, G.B., and B.C. Heiberg. 1952. *Cladosporium* rot of peppers. Pl. Dis. Reporter 36: 481-482.
2914. Ramsey, G.B., M.A. Smith, L. Beraha, and W.R. Wright. 1959. *Pellicularia rolfsii* on Mexican and Texas watermelons. Pl. Dis. Reporter 43: 1031.
2915. Ramsey, G.B., M.A. Smith, and B.C. Heiberg. 1951. Anthracnose of peaches. Phytopathology 41: 447-455.
2916. Ramsey, G.B., M.A. Smith, and W.R. Wright. 1952. Orange rust, a serious market disease of Colorado spinach in 1952. Pl. Dis. Reporter 36: 323-324.
2917. Ramsey, G.R., and J.W. Lorbeer. 1986. Flower blight and scape girdling of onion grown for seed production in New York. Phytopathology 76: 599-603.
2918. Rand, F.V. 1911. A pecan leaf-blotch. Phytopathology 1: 133-138.
2919. Randall, H., C.M. Nagel, and L.S. Wood. 1968. Dutch elm disease found in South Dakota. Pl. Dis. Reporter 52: 349.
2920. Rane, K.K., A.E. Dorrance, and F.W. Holmes. 1985. *Endocronartium harknessii* on Scots pine in Massachusettes. Pl. Dis. 69: 177.
2921. Rane, K.K., and T.A. Tattar. 1987. Pathogenicity of blue-stain fungi associated with *Dendroctonus terebrans*. Pl. Dis. 71: 879-883.
2922. Raniere, L.C. 1961. Observations on new or unusual diseases of highbush blueberry. Pl. Dis. Reporter 45: 844.
2923. Rankin, H.W. 1953. Peanut rust in Georgia. Pl. Dis. Reporter 37: 528.
2924. Raper, K.B., W.B. Cooke, and R.A. Taft. 1954. The 1950 foray of the Mycological Society of America. Mycologia 46: 670-679.
2925. Rappaz, F. 1987. Taxonomy and nomenclature of the octosporous Diatrypaceae. Myc. Helvetica 2: 285-648.
2926. Ray, W.W. 1953. Leaf rust of Merion bluegrass in Nebraska. Pl. Dis. Reporter 37: 578.
2927. Ray, W.W. 1954. Unusual or new occurrences of fungus pathogens on grasses in Nebraska. Pl. Dis. Reporter 38: 583-587.
2928. Raymond, F.L., J.L. Etchells, T.A. Bell, and P.M. Masley. 1959. Filamentous fungi from blossoms, ovaries, and fruit of pickling cucumbers. Mycologia 51: 492-511.
2929. Redhead, S.A. 1975. The genus *Cristulariella*. Canad. J. Bot. 53: 700-707.
2930. Redhead, S.A. 1980. *Marasmiellus filopes*. Fungi Canadenses 179: 1-2.
2931. Redhead, S.A. 1981. Agaricales on wetland Monocotyledoneae in Canada. Canad. J. Bot. 59: 574-589.
2932. Redhead, S.A. 1982. *Marasmius copelandi*. Fungi Canadenses 214: 1-2.
2933. Redhead, S.A. 1984. Two fern-associated mushrooms, *Mycena lohwagii* and *M. pterigena*, in Canada. Naturaliste Canad. 111: 439-442.
2934. Redhead, S.A. 1984. Mycological observations, 4-12: on *Kuehneromyces, Stropharia, Marasmius, Mycena, Geopetalum, Omphalopsis, Phaeomarasmius, Naucoria* and *Prunulus*. Sydowia 37: 246-270.
2935. Redhead, S.A. 1984. Mycological observations 13-14: On *Hypsizygus* and *Tricholoma*. Trans. Mycol. Soc. Japan 25: 1-9.
2936. Redhead, S.A. 1986. Mycological observations 15-16: On *Omphalia* and *Pleurotus*. Mycologia 78: 522-528.
2937. Redhead, S.A., and J.H. Ginns. 1985. A reappraisal of agaric genera associated with brown rots of wood. Trans. Mycol. Soc. Japan 26: 349-381.
2938. Redhead, S.A., and P. Kroeger. 1984. *Melanotus textilis*, a new fabric- and wood-inhabiting North American agaric. Mycologia 76: 868-872.
2939. Redhead, S.A., O.K. Miller Jr., R. Watling, and E. Ohenoja. 1982. *Marasmius epidryas*. Fungi Canadenses 213: 1-2.
2940. Redhead, S.A., and G.P. White. 1985. *Mycopappus*, a new genus of leaf pathogens, and two parasitic *Anguillospora* species. Canad. J. Bot. 63: 1429-1435.
2941. Reed, H.E. 1953. Occurrence of arborvitae blight in Tennessee. Pl. Dis. Reporter 37: 258-259.
2942. Reed, H.E. 1955. Occurrence of Dutch elm disease in Tennessee. Pl. Dis. Reporter 39: 882.
2943. Reed, H.E., and H.L. Bruer. 1956. Status of Dutch elm disease in Knox County, Tennessee. Pl. Dis. Reporter 40: 756.
2944. Rehm, H. 1909. Ascomycetes novi. 1. Ascomycetes Americae borealis. Ann. Mycol. 7: 531-542.
2945. Reid, D.A. 1963. Notes on some fungi of Michigan -'Cyphellaceae'. Persoonia 3: 97-154.
2946. Reid, D.A. 1970. New or interesting records of British Hymenomycetes, IV. Trans. Brit. Mycol. Soc. 55: 413-441.
2947. Reid, J., and C. Booth. 1967. The Diaporthaceae. I. *Ditopellopsis* gen. nov., *Ditopellina* gen. nov., *Phragmoporthe* and *Ditopella*. Canad. J. Bot. 45: 1479-1487.
2948. Reid, J., and C. Booth. 1987. *Winterella*, the correct name for *Cryptospora* and *Ophiovalsa*. Canad. J. Bot. 65: 1320-1342.
2949. Reid, J., and R.F. Cain. 1960. Additional Diaporthales on twigs from Ontario. Can. J. Botany 38: 945-950.
2950. Reid, J., and R.F. Cain. 1961. The genus *Therrya*. Canad. J. Bot. 39: 1117-1129.
2951. Reid, J., and R.F. Cain. 1962. Studies on the organisms associated with "snow-blight" of conifers in North America. II. Some species of the genera *Phacidium*, *Lophophacidium*, *Sarcotrichila*, and *Hemiphacidium*. Mycologia 54: 481-497.
2952. Reid, J., and R.F. Cain. 1962. Studies on the organisms associated with "snow-blight" of conifers in North America. I. A new genus of the Helotiales. Mycologia 54: 194-200.
2953. Reifschneider, F.J.B. 1979. A new species of *Stemphylium* parasitic on *Smilacina stellata*. Canad. J. Bot. 57: 1484-1485.
2954. Reilly, C.C., and W.R. Okie. 1982. Distribution in the southeastern United States of peach tree fungal gummosis caused by *Botryosphaeria dothidea*. Pl. Dis. 66: 158-161.
2955. Remsberg, R.E. 1940. Studies in the genus *Typhula*. Mycologia 32: 52-96.
2956. Renfro, B.L. 1963. *Helminthosporium sorokinianum*: a cause of black stem of forage legumes. Pl. Dis. Reporter 47: 292-293.
2957. Renfro, B.L., F.I. Frosheiser, and R.D. Wilcoxson. 1960. Diseases of forage legumes in Minnesota. Pl. Dis. Reporter 44: 314-316.
2958. Rexrode, C.O., and A.C. Lincoln. 1965. Distribution of oak wilt. Pl. Dis. Reporter 49: 1007-1010.
2959. Reyes, L., D.T. Rosenow, R.W. Berry, and M.C. Futrell. 1964. Downy mildew and head smut diseases of sorghum in Texas. Pl.

Dis. Reporter 48: 249-253.
2960. Reynolds, J.E. 1964. *Gliocladium* disease of palm in Dade County,Florida. Pl. Dis. Reporter 48: 718-720.
2961. Rhoads, A.F. 1974. Leaf-streak of daylily observed in Pennsylvania. Pl. Dis. Reporter 58: 102.
2962. Rhoads, A.S. 1917. Some new or little known hosts for wood-destroying fungi. Phytopathology 7: 46-48.
2963. Rhoads, A.S. 1954. *Clitocybe* root rot found widespread and destructive in Georgia and South Carolina peach orchards. Pl. Dis. Reporter 38: 42-46.
2964. Rhodes, L.H., and D.K. Myers. 1986. Severity of spring black stem on alfalfa cultivars in Ohio. Pl. Dis. 70: 746-748.
2965. Ribeiro, O.K., and L.L. Black. 1971. *Rhizoctonia fragariae*: a mycorrhizal and pathogenic fungus of straw berry plants. Pl. Dis. Reporter 55: 599-603.
2966. Ribeiro, O.K., D.C. Erwin, and R.A. Khan. 1978. A new high-temperature *Phytophthora* pathogenic to roots of alfalfa. Phytopathology 68: 155-161.
2967. Rich, A.E. 1955. Die-back of asparagus associated with *Botrytis cinerea* in New Hampshire. Pl. Dis. Reporter 39: 305-306.
2968. Rich, S. 1957. *Cladosporium fulvum* on field grown tomatoes in Connecticut. Pl. Dis. Reporter 41: 1058.
2969. Rich, S. 1957. Field infection of radish roots with *Peronospora parasitica*. Pl. Dis. Reporter 41: 1058-1059.
2970. Rich, S. 1958. *Rhizoctonia* root rot on stored horseradishes in Connecticut. Pl. Dis. Reporter 42: 554.
2971. Rich, S. 1958. Brown-spot of celery in Connecticut. Pl. Dis. Reporter 42: 1091-1092.
2972. Rich, S. 1959. Chlorosis of spinach associated with *Olpidium brassicae*. Pl. Dis. Reporter 43: 118.
2973. Rich, S. 1959. *Aphanomyces raphani* on red globe radish in Connecticut. Pl. Dis. Reporter 43: 834.
2974. Rich, S., and E.M. Stoddard. 1950. Connecticut disease notes. Pl. Dis. Reporter 34: 240.
2975. Rich, S., G.S. Walton, and R.W. Judd. 1977. Unusual outbreaks of vegetable diseases in Connecticut during 1976. Pl. Dis. Reporter 61: 668-669.
2976. Richeson, M.L. 1981. Etiology of a late season wilt in *Helianthus annuus*. Pl. Dis. 65: 1019-1021.
2977. Ricker, M.D. 1987. *Fusarium* crown and root rot of tomato in Pennsylvania. Pl. Dis. 71: 469.
2978. Ridings, W.H., and F.W. Zettler. 1973. Aphanomyces blight of amazon sword plants. Phytopathology 63: 289-295.
2979. Ries, S.M., and D.J. Royse. 1977. Rusty spot of peach in Illinois. Pl. Dis. Reporter 61: 317-318.
2980. Riffle, J.W., and T.E. Hinds. 1969. *Hypoxylon* canker of aspen in New Mexico and Arizona. Pl. Dis. Reporter 53: 454.
2981. Riffle, J.W., A.K. Myatt, and R.L. Davis. 1985. Incidence of *Phellinus robineae* in black locust plantings in Oklahoma. Pl. Dis. 69: 116-118.
2982. Riffle, J.W., W.D. Ostrofsky, and R.L. James. 1981. *Fomes fraxinophilus* on green ash in Nebraska windbreaks. Pl. Dis. 65: 667-669.
2983. Riffle, J.W., and G.W. Peterson. 1986. *Thyronectria* canker of honeylocust: influence of temperature and wound age on disease development. Phytopathology 76: 313-316.
2984. Riffle, J.W., E.M. Sharon, and M.O. Harrell. 1984. Incidence of *Fomes fraxinophilus* on green ash in Nebraska woodlands. Pl. Dis. 68: 322-324.
2985. Risley, J.H., and S.B. Silverborg. 1958. *Stereum sanguinolentum* on living Norway spruce following pruning. Phytopathology 48: 337-338.
2986. Rittenburg, L.H., and F.F. Hendrix Jr. 1983. Peach fruit rots caused by *Botryosphaeria* spp. and *Glomerella cingulata*. Pl. Dis. 67: 449-450.
2987. Roane, C.W. 1950. Observations on corn diseases in Virginia from 1947 to 1950. Pl. Dis. Reporter 34: 394-396.
2988. Roane, C.W. 1960. New or unusual diseases of cereal crops in Virginia. Pl. Dis. Reporter 44: 696.
2989. Roane, C.W., and S.B. Fenne. 1955. Some new plant disease records for Virginia. Pl. Dis. Reporter 39: 695-696.
2990. Roane, C.W., R.L. Harrison, and C.F. Genter. 1974. Observations on gray leaf spot of maize in Virginia. Pl. Dis. Reporter 58: 456-459.
2991. Roane, C.W., and P.H. Massey Jr. 1956. An outbreak of tomato late blight in plastic greenhouses. Pl. Dis. Reporter 40: 313.
2992. Roane, C.W., M.K. Roane, and T.M. Starling. 1974. *Ascochyta* species on barley and wheat in Virginia. Pl. Dis. Reporter 58: 455-456.
2993. Roane, C.W., and T.M. Starling. 1958. Miscellaneous notes on small grain diseases in Virginia. Pl. Dis. Reporter 42: 1268-1271.
2994. Roane, C.W., and T.M. Starling. 1976. *Cephalosporium* stripe of wheat in Virginia. Pl. Dis. Reporter 60: 345.
2995. Roane, M.K., and F.R. Fosberg. 1983. A new pyrenomycete associated with *Metrosideros collina* subspecies *polymorpha* (Myrtaceae). Mycologia 75: 163-166.
2996. Roane, M.K., R.J. Stipes, P.M. Phipps, and O.K. Miller Jr. 1974. *Endothia gyrosa*, causal pathogen of pin oak blight. Mycologia 66: 1042-1047.
2997. Robert, A.L. 1956. *Helminthosporium maydis* on sweet corn ears in Florida. Pl. Dis. Reporter 40: 991-995.
2998. Roberts, D.A., K.D. Fezer, and C.S. Ramamurthi. 1956. Diseases of forage crops in New York, 1955. Pl. Dis. Reporter 40: 219-220.
2999. Roberts, D.A., K.D. Fezer, and R.T. Sherwood. 1954. Diseases of forage crops in New York, 1953. Pl. Dis. Reporter 38: 30-31.
3000. Roberts, D.A., R.E. Ford, C.H. Ward, and D.T. Smith. 1959. *Colletotrichum* anthracnose of alfalfa in New York. Pl. Dis. Reporter 43: 352.
3001. Roberts, D.A., E.S. Horner, G.M. Prine, and F.A. Johnson. 1984. An epidemic of Lepto leaf spot in alfalfa caused by *Leptosphaerulina briosiana*. Pl. Dis. 68: 732.
3002. Roberts, D.A., A.S. Jensen, and G.F. Weber. 1971. Blight of Spanish moss. Pl. Dis. Reporter 55: 390-392.
3003. Roberts, D.A., and R.P. Murphy. 1955. *Puccinia dactylidis* on orchard grass in New York. Pl. Dis. Reporter 39: 881-882.
3004. Roberts, D.A., R.T. Sherwood, K.D. Fezer, and C.S. Ramamurthi. 1955. Diseases of forage crops in New York, 1954. Pl. Dis. Reporter 39: 316-317.
3005. Roberts, D.A., and C.H. Ward. 1956. Diseases of forage crops before the first harvest of hay in New York, 1956. Pl. Dis. Reporter 40: 807-809.
3006. Roberts, R.G., J.A. Robertson, and R.T. Hanlin. 1984. *Ascotricha xylina*: Its occurrence, morphology, and typification. Mycologia 76: 963-968.
3007. Roberts, R.G., J.A. Robertson, and R.T. Hanlin. 1986. Fungi occurring in the achenes of sunflower (*Helianthus annuus*). Canad. J. Bot. 64: 1964-1971.
3008. Robinson-Jeffrey, R.C., and R.W. Davidson. 1968. Three new *Europhium* species with *Verticicladiella* imperfect states. Canad. J. Bot. 46: 1523-1527.
3009. Roelfs, A.P., and D.L. Long. 1980. Analysis of recent oat stem rust epidemics. Phytopathology 70: 436-440.
3010. Rogers, C.E., T.E. Thompson, and D.E. Zimmer. 1978. *Rhizopus* head rot of sunflower: etiology and severity in the south ern plains. Pl. Dis. Reporter 62: 769-771.
3011. Rogers, D.P. 1944. The genera *Trechispora* and *Galzinia*. Mycologia 36: 70-103.
3012. Rogers, D.P., and G.W. Martin. 1955. The genus *Mylittopsis*. Mycologia 47: 891-894.
3013. Rogers, J.D. 1966. Notes on *Hypoxylon granadense* var. *macrospora* from Washington state. Mycologia 58: 978-982.
3014. Rogers, J.D. 1979. *Xylaria magnoliae* sp. nov. and comments on several other fruit-inhabiting species. Canad. J. Bot. 57: 941-945.
3015. Rogers, J.D. 1983. *Xylaria bulbosa*, *Xylaria curta*, and *Xylaria longipes* in continental United States. Mycologia 75: 457-467.
3016. Rogers, J.D. 1984. *Xylaria cubensis* and its anamorph *Xylocoremium flabelliforme*, *Xylaria allantoidea*, and *Xylaria poitei* in continental United States. Mycologia 76: 912-923.
3017. Rogers, J.D. 1984. *Xylaria acuta*, *Xylaria cornu-damae*, and *Xylaria mali* in continental United States. Mycologia 76: 23-33.
3018. Rogers, J.D., and B.E. Callan. 1986. *Xylaria polymorpha* and its allies in continental United States. Mycologia 78: 391-400.
3019. Rogers, J.D., and D.A. Glawe. 1983. *Diatrype whitmanensis* sp. nov. and the anamorphs of *Diatrype bulla ta* and *Eutypella sorbi*. Mycotaxon 18: 73-80.
3020. Rogers, J.D., and A.F. Noskowiak. 1976. Brown sapwood stain of ponderosa pine caused by *Cytospora* sp.: Cultural and histological aspects. Phytopathology 66: 25-27.
3021. Rogerson, C.T. 1956. Diseases of grasses in Kansas: 1953-55. Pl. Dis. Reporter 40: 388-397.

3022. Rogerson, C.T. 1957. *Verticillium* wilt in Kansas. Pl. Dis. Reporter 41: 1053-1054.
3023. Rogerson, C.T. 1958. Diseases of grasses in Kansas: 1956-1957. Pl. Dis. Reporter 42: 346-353.
3024. Rogerson, C.T., and C.L. King. 1954. Stem rust of Merion bluegrass in Kansas. Pl. Dis. Reporter 38: 57.
3025. Romanko, R.R., J.M. Ogawa, C.B. Skotland, C.E. Horner, H. Eismann, G.E. Miller Jr., W. Hogue, H. Goschie, and S.N. Brooks. 1964. Hop downy mildew - A symposium. Modern Brewery Age 66: 45-52.
3026. Rosen, H.R. 1950. Oat diseases in Arkansas 1948-49. Pl. Dis. Reporter 34: 43-44.
3027. Rosen, H.R. 1951. Oat diseases in Arkansas 1949-50. Pl. Dis. Reporter 35: 149-151.
3028. Rosen, H.R. 1952. Oat diseases in Arkansas 1950-51. Pl. Dis. Reporter 36: 153-154.
3029. Rosen, H.R. 1957. Epidemics of powdery mildews in Arkansas in 1956 with special referen ce to powdery mildew on winter oats. Pl. Dis. Reporter 41: 330-336.
3030. Rosenberger, D.A., and T.J. Burr. 1982. Fruit decays of peach and apple caused by *Phomopsis mali*. Pl. Dis. 66: 1073-1075.
3031. Rosenthal, G.A. 1963. A new host of *Fomes everhartii*. Pl. Dis. Reporter 47: 152-153.
3032. Ross, E.W. 1963. A pathogenic *Fusicoccum* sp. associated with ash dieback cankers. Pl. Dis. Reporter 47: 20-21.
3033. Ross, E.W. 1966. Incidence of *Polyporus tomentosus* in slash pine plantations in the southeastern United States. Pl. Dis. Reporter 50: 527.
3034. Ross, E.W. 1967. Association of *Cylindrocladium scoparium* with mortality in a 27-year-old yellow-poplar plantation. Pl. Dis. Reporter 51: 38-39.
3035. Ross, E.W. 1968. Duration of stump susceptibility of loblolly pine to infection by *Fomes annosus*. Forest Sci. 14: 206-211.
3036. Ross, E.W. 1968. Sand pine, a new host of *Fomes annosus*. Pl. Dis. Reporter 52: 635.
3037. Ross, E.W. 1971. *Diplodia theobromae* and *Ceratocystis fimbriata* f. *platani* found in silage sycamore plantings. Pl. Dis. Reporter 55: 741-743.
3038. Ross, J.P. 1954. *Mastigosporium album* Riess on *Alopecurus pratensis* in New York. Pl. Dis. Reporter 38: 607-608.
3039. Ross, J.P., and T.J. Smith. 1963. Brown stem rot of soybean in North Carolina and Virginia. Pl. Dis. Reporter 47: 329.
3040. Ross, W.D. 1976. Fungi associated with root diseases of aspen in Wyoming. Canad. J. Bot. 54: 734-744.
3041. Rossman, A.Y. 1977. The genus *Ophionectria* (Euascomycetes, Hypocreales). Mycologia 69: 355-391.
3042. Rossman, A.Y., and K.C. Lu. 1980. Filamentous fungi associated with leaf surfaces of red alder and Douglas-fir seedlings in western Oregon. Mycotaxon 10: 369-371.
3043. Rossman, A.Y., and H.E. Wilcox. 1985. *Microthelia acerina*, a new fungus associated with roots of sugar maple. Mycologia 77: 162-165.
3044. Roth, D.A., and G.J. Griffin. 1981. *Cylindrocladium* root rot of black walnut seedlings and inoculum pattern in nursery soil. Canad. J. Plant Pathol. 3: 1-5.
3045. Roth, E.R. 1950. Cankers and decay of oak associated with *Irpex mollis*. Pl. Dis. Reporter 34: 347-348.
3046. Roth, E.R. 1952. Roots of living *Pinus rigida* decayed by *Fomes annosus*. Pl. Dis. Reporter 36: 330.
3047. Roth, E., G.H. Hepting, and E.R. Toole. 1959. Sapstreak disease of sugar maple and yellow-poplar in North Carolina. Phytopathology 49: 549.
3048. Roth, L.F. 1959. Perennial infection of ponderosa pine by *Elytroderma deformans*. Forest Sci. 5: 182-191.
3049. Roth, L.F., and G.F. Kuhlman. 1966. *Phytophthora cinnamomi*, an unlikely threat to Douglas-fir forestry. Forest Sci. 12: 147-159.
3050. Rothman, P.G., and D.H. Bowman. 1961. Downy mildew on small grains and two other grasses in Mississippi, 1961. Pl. Dis. Reporter 45: 764-765.
3051. Rothman, P.G., and J.S. Hursh. 1967. Observations on powdery mildew of winter oats. Pl. Dis. Reporter 51: 30-32.
3052. Rothrock, C.S., and B.M. Cunfer. 1984. Prevalence of take-all caused by *Gaeumannomyces graminis* var. *tritici* on wheat in Georgia. Pl. Dis. 68: 351.
3053. Rothrock, C.S., T.W. Hobbs, and D.V. Phillips. 1985. Effects of tillage and cropping system on incidence and severity of southern stem canker of soybean. Phytopathology 75: 1156-1159.
3054. Rowe, R.C. 1975. Powdery mildew of potatoes in Ohio. Pl. Dis. Reporter 59: 330-331.
3055. Rowe, R.C., M.K. Beute, and J.C. Wells. 1973. *Cylindrocladium* black rot of peanuts in North Carolina - 1972. Pl. Dis. Reporter 57: 387-389.
3056. Rowe, R.C., M.K. Beute, J.C. Wells, and J.C. Wynne. 1974. Incidence and control of *Cylindrocladium* black rot of peanuts in North Carolina during 1973. Pl. Dis. Reporter 58: 348-352.
3057. Rowe, R.C., J.R. Davis, M.L. Powelson, and D.I. Rouse. 1987. Potato early dying: causal agents and management strategies. Pl. Dis. 71: 482-489.
3058. Rowe, R.C., and A.F. Schmitthenner. 1977. Potato pink rot in Ohio caused by *Phytophthora erythroseptica* and *P. cryptogea*. Pl. Dis. Reporter 61: 807-810.
3059. Roy, K.W. 1982. Seedling diseases caused in soybean by species of *Colletotrichum* and *Glomerella*. Phytopathology 72: 1093-1096.
3060. Roy, K.W., and F.M. Bourland. 1982. Epidemiological and mycofloral relationships in cotton seedling disease in Mississippi. Phytopathology 72: 868-872.
3061. Royle, D.J., and C.J. Hickman. 1963. *Phytophthora cinnamomi* on highbush blueberry. Pl. Dis. Reporter 47: 266-268.
3062. Royse, D.J., L.V. Gregory, J.E. Ayers, and H. Cole Jr. 1980. Powdery mildew of wheat: Relation of yield components to disease severity. Canad. J. Plant Pathol. 2: 131-136.
3063. Royse, D.J., and S.M. Ries. 1976. A cortical decay of petioles and stolons of strawberry caused by *Pestalotia longisetula*. Pl. Dis. Reporter 60: 901-902.
3064. Royse, D.J., and S.M. Ries. 1978. The influence of fungi isolated from peach twigs on the pathogenicity of *Cytospora cincta*. Phytopathology 68: 603-607.
3065. Rubin, H.L. 1965. *Cercospora plumeraie* Chupp - a new report for the United States. Pl. Dis. Reporter 49: 729.
3066. Ruhl, G.E., and J.W. McCain. 1986. First report of rust caused by *Frommeella duchesneae* on false strawberry in Indiana. Pl. Dis. 70: 475.
3067. Rumbold, C.T. 1931. Two blue-staining fungi associated with bark-beetle infestation of pines. J. Agric. Res. 43: 847-873.
3068. Rumbold, C.T. 1936. Three blue-staining fungi, including two new species, associated with bark beetles. J. Agric. Res. 52: 419-437.
3069. Rumbold, C.T. 1941. A blue stain fungus, *Ceratostomella montium* n. sp., and some yeasts associated with two species of *Dendroctonus*. J. Agric. Res. 62: 589-601.
3070. Rupe, J.C., M.R. Siegel, and J.R. Hartman. 1982. Influence of environment and plant maturity on gray leaf spot of corn caused by *Cercospora zeae-maydis*. Phytopatho!ogy 72: 1587-1591.
3071. Ruppel, E.G. 1969. Diseases of sugarbeet in Arizona. Pl. Dis. Reporter 53: 56-58.
3072. Ruppel, E.G., F.J. Hills, and D.L. Mumford. 1975. Epidemiological observations on the sugarbeet powdery mildew epiphytotic in western U.S.A. in 1974. Pl. Dis. Reporter 59: 283-286.
3073. Rusden, P.L. 1965. Dutch elm disease in Atlanta, Georgia. Pl. Dis. Reporter 49: 801.
3074. Rush, C.M. 1987. Root rot of sugar beet caused by *Pythium deliense* in the Texas panhandle. Pl. Dis. 71: 469.
3075. Rush, C.M., T.J. Gerik, and C.M. Kenerley. 1985. Atypical disease symptoms associated with *Phymatotrichum* root rot of cotton. Pl. Dis. 69: 534-537.
3076. Rush, M.C. 1973. Leaf scald of rice observed in Louisiana. Pl. Dis. Reporter 57: 715-716.
3077. Russell, T.E. 1975 . Plantago wilt. Phytopathology 65: 359-360.
3078. Rycyk, F., Jr., and D. Sharpe. 1984. Infection of tall fescue seed in Missouri by the endophyte *Epichloe typhina*. Pl. Dis. 68: 1099.
3079. Ryvarden, L. 1979. *Porogramme* and related genera. Trans. Brit. Mycol. Soc. 73: 9-19.
3080. Saccardo, P.A. 1908. Notae mycologicae. Ann. Mycol. 6: 553-569.
3081. Saccardo, P.A. 1915. Fungi noveboracenses. Ann. Mycol. 13: 115-138.
3082. Saccardo, P.A. 1920. Notae mycologicae, Ser. XXIX. Microcmycetes Dakotenses et Utahenses a Doct. J.F. Brenckle lecti et communicati. Mycologia 12: 199-205.
3083. Salogga, D.S., and J.F. Ammirati. 1983. Discula species

associated with anthracnose of dogwood in the Pacific Northwest. Pl. Dis. 67: 1290.
3084. Samuels, G.J., C.T. Rogerson, A.Y. Rossman, and J.D. Smith. 1984. *Nectria tubercularifformis*, *Nectriella muelleri*, *Nectriella* sp., and *Hyponectria sceptri*: low-temperature tolerant, alpine-boreal fungal antagonists. Canad. J. Bot. 62: 1896-1903.
3085. Samuels, Gary. 1976. A revision of the fungi formerly classified as *Nectria* subgenus *Hyphonectria*. Mem. New York Bot. Gard. 26: 1-126.
3086. Sanderlin, R.S., and L.J. Grauke. 1987. Vein spot disease caused by *Gnomonia nerviseda* on pecan in Illinois. Pl. Dis. 71: 1146.
3087. Sanderlin, R.S., A.S. Hunt, and D.K. Babcock. 1983. Epidemiology and control of vein spot disease of pecan caused by *Gno monia nerviseda*. Pl. Dis. 67: 1209-1213.
3088. Sanderson, P.G., and G.L. Worf. 1986. *Phomopsis* and *Sirococcus* shoot blights of Colorado blue spruce in Wisconsin. Pl. Dis. 70: 1159.
3089. Sauer, D.B., C.L. Storey, and D.E. Walker. 1984. Fungal populations in U.S. farm-stored grain and their relationship to moisture, storage time, regions, and insect infestation. Phytopathology 74: 1050-1053.
3090. Savage, E.F., J.H. Weinberger, E.S. Luttrell, and A.S. Rhoads. 1953. Clitocybe root rot - a disease of economic importance in Georgia peach orchards. Pl. Dis. Reporter 37: 269-270.
3091. Savile, D.B.O. 1965. *Puccinia karelica* and species delimitation in the Uredinales. Canad. J. Bot. 43: 231-238.
3092. Savile, D.B.O. 1946. *Entyloma fuscum* and related smuts attacking Papaveraceae. Canad. J. Res., Sect. C, Bot. Sci. 24: 109-114.
3093. Savile, D.B.O. 1952. A study of the species of *Cintractia* on *Carex*, *Kobresia*, and *Scirpus* in North America. Canad. J. Bot. 30: 410-435.
3094. Savile, D.B.O. 1954. Taxonomy, phylogeny, host relationship, and phytogeography of the microcyclic rusts of Saxifragaceae. Canad. J. Bot. 32: 400-425.
3095. Savile, D.B.O. 1959. Two little-known Ascomycetes attacking Filicales. Mycologia 51: 296-298.
3096. Savile, D.B.O. 1959. Notes on *Exobasidium*. Canad. J. Bot. 37: 641-656.
3097. Savile, D.B.O. 1961. Some fungal parasites of Liliaceae. Mycologia 53: 31-52.
3098. Savile, D.B.O. 1962. Some fungal parasites of Onagraceae. Canad. J. Bot. 40: 1385-1398.
3099. Savile, D.B.O. 1964. Geographic variation and gene flow in *Puccinia cruciferarum*. Mycologia 56: 240-248.
3100. Savile, D.B.O. 1967. Evolution and relationships of North American *Pedicularis* rusts and their hosts. Canad. J. Bot. 45: 1093-1103.
3101. Savile, D.B.O. 1972. Some rusts of *Scirpus* and allied genera. Canad. J. Bot. 50: 2579-2596.
3102. Savile, D.B.O. 1974. *Phragmidium potentillae*. Fungi Canadenses 41: 1-2.
3103. Savile, D.B.O. 1974. *Puccinia holboellii*. Fungi Canadenses 47: 1-2.
3104. Savile, D.B.O. 1974. *Phragmidium fusiforme* var. *novi-boreale*. Fungi Canadenses 54: 1-2.
3105. Savile, D.B.O. 1974. *Puccinia canadensis*. Fungi Canadenses 56: 1-2.
3106. Savile, D.B.O. 1975. *Puccinia fergussonii*. Fungi Canadenses 64: 1-2.
3107. Savile, D.B.O. 1975. *Puccinia ornatula*. Fungi Canadenses 77: 1-2.
3108. Savile, D.B.O. 1975. *Phragmidium occidentale*. Fungi Canadenses 80: 1-2.
3109. Savile, D.B.O. 1975. *Mundkurella mossii*, a smut of *Aralia nudicaulis*. Mycologia 67: 273-279.
3110. Savile, D.B.O. 1976. Notes on some parasitic fungi from southern British Columbia, southwest Alberta, and adjacent United States. Canad. J. Bot. 54: 971-975.
3111. Savile, D.B.O. 1976. *Phragmidium ivesiae* and its allies in North America. Canad. J. Bot. 54: 1690-1696.
3112. Savile, D.B.O. 1977. *Puccinia conglomerata*. Fungi Canadenses 110: 1-2.
3113. Savile, D.B.O. 1978. *Puccinia arnicalis*. Fungi Canadenses 111: 1-2.
3114. Savile, D.B.O., and I.L. Conners. 1951. The rusts of *Armeria* and *Limonium* in North America. Mycologia 43: 186-195.
3115. Savile, D.B.O., and J.A. Parmelee. 1956. Some fungal parasites of Portulacaceae. Mycologia 48: 573-590.
3116. Scardaci, S.C., and R.K. Webster. 1982. Common root rot of cereals in California. Pl. Dis. 66: 31-34.
3117. Schafer, J.F., and R.M. Caldwell. 1953. Powdery mildew of small grains in Indiana in 1953. Pl. Dis. Reporter 37: 569.
3118. Schafer, J.F., and D.L. Long. 1986. Evidence for local source of *Puccinia recondita* on wheat in Pennsylvania. Pl. Dis. 70: 892.
3119. Scharen, A.L., J.W. Bergman, and E.E. Burns. 1976. Leaf diseases of winter wheat in Montana and losses from them in 1975. Pl. Dis. Reporter 60: 686-690.
3120. Scharpf, R.F. 1969. *Cytospora abietis* associated with dwarf mistletoe on true firs in California. Phytopathology 59: 1657-1658.
3121. Scharpf, R.F. 1986. Effect of a foliage disease caused by *Lirula abietis-concoloris* on growth of white fir in California. Pl. Dis. 70: 13-14.
3122. Scharpf, R.F., and F.G. Hawksworth. 1966. Hosts and distribution of *Uredo phoradendri*. Mycologia 58: 811-812.
3123. Scharpf, R.F., J. Staley, and F.G. Hawksworth. 1970. A needle cast, the first known disease of bristlecone fir in California. Pl. Dis. Reporter 54: 275-277.
3124. Scheffer, R.P., and W.J. Haney. 1956. Causes and control of root rot in Michigan greenhouses. Pl. Dis. Reporter 40: 570-579.
3125. Scheffer, R.P., and R.R. Nelson. 1967. Geographical distribution and prevalence of *Helminthosporium victoriae*. Pl. Dis. Reporter 51: 110-111.
3126. Scheifele, G.L., and R.R. Nelson. 1969. The occurrence of *Phyllosticta* leaf spot of corn in Pennsylvania. Pl. Dis. Reporter 53: 186-189.
3127. Schenck, N.C. 1960. Watermelon disease incidence in central Florida, 1931-1959. Pl. Dis. Reporter 44: 556-558.
3128. Schenck, N.C. 1961. Three uncommon watermelon fruit rots in Florida. Pl. Dis. Reporter 45: 841-843.
3129. Schenck, N.C. 1970. Perithecia of *Cochliobolus heterostrophus* on corn leaves in Florida. Pl. Dis. Reporter 54: 1127-1128.
3130. Schenck, N.C. 1972. *Phaeosphaeria herpotricha* on southern corn leaf blight-infected plants in Florida. Pl. Dis. Reporter 56: 276.
3131. Schenck, N.C., and R.A. Kinloch. 1974. Pathogenic fungi, parasitic nematodes, and endomycorrhizal fungi associated with soybean roots in Florida. Pl. Dis. Reporter 58: 169-173.
3132. Schiller, C.T., and J.B. Sinclair. 1979. Soybean seed lot contamination by *Melanopsichium pennsylvanicum* smut galls. Phytopathology 69: 605-608.
3133. Schipper, A.L., Jr., and D.H. Dawson. 1974. Poplar leaf rust - - a problem in maximum wood fiber production. Pl. Dis. Reporter 58: 721-723.
3134. Schmidt, E.L., and C.M. Christensen. 1977. *Strobilurus kemptonae* on cones of *Pinus strobus*. Mycologia 69: 428-430.
3135. Schmidt, E.L., and M.G. Dietz. 1985. *Arthrographis cuboidea* causing pink stain of sodium pentachlorophen oxide-treated red oak. Mycologia 77: 316-318.
3136. Schmidt, E.L., and D.W. French. 1976. *Aureobasidium pullulans* on wood shingles. Forest Prod. J. 26: 34-37.
3137. Schmidt, R.A., R.C. Holley, M.C. Klapproth, and T. Miller. 1986. Temporal and spatial patterns of fusiform rust epidemics in young plantations of susceptible and resistant slash and loblolly pines. Pl. Dis. 70: 661-666.
3138. Schmidt, R.A., and E.M. Underhill. 1974. Incidence and impact of pitch canker in slash pine plantations in Florida. Pl. Dis. Reporter 58: 451-454.
3139. Schmidtling, R.C., and C.H. Walkinshaw. 1985. Fusiform rust infection of loblolly pines that survived resistance screening and of their progeny. Pl. Dis. 69: 491-493.
3140. Schnathorst, W.C., and A.C. Goheen. 1977. A wilt disease of grapevines (*Vitis vinifera*) in California caused by *Verticillium dahliae*. Pl. Dis. Reporter 61: 909-913.
3141. Schnathorst, W.C., and D.E. Mathre. 1966. Host range and differentiation of a severe form of *Verticillium albo-atrum* in cotton. Phytopathology 56: 1155-1161.
3142. Schnathorst, W.C., T.A. Reeve, and D. Fogle. 1975. *Verticillium dahliae* strains in cotton in the Pahrump Valley, Nevada. Pl. Dis. Reporter 59: 863-865.
3143. Schneider, C.L., and G.J. Hogaboam. 1977. New occurrences of powdery mildew and curly top on sugarbeet in Michigan in 1975.

Pl. Dis. Reporter 61: 88-89.
3144. Schneider, R.W., O.D. Dhingra, J.F. Nicholson, and J.B. Sinclair. 1974. *Colletotrichum truncatum* borne within the seedcoat of soybean. Phytopathology 64: 154-155.
3145. Schneider, R.W., R.G. Grogan, and K.A. Kimble. 1978. *Colletotrichum* root rot of greenhouse tomatoes in California. Pl. Dis. Reporter 62: 969-971.
3146. Schoeneweiss, D.F. 1965. *Fusicoccum* canker on mountain ash in Illinois. Pl. Dis. Reporter 49: 251-252.
3147. Schoeneweiss, D.F. 1966. *Cytospora* canker on thornless honey locust trees. Pl. Dis. Reporter 50: 13.
3148. Schoeneweiss, D.F. 1974. *Tubercularia ulmea* canker of tall hedge: influence of freezing stress on disease susceptibility. Pl. Dis. Reporter 58: 937-941.
3149. Schoknecht, J.D., and J.L. Crane. 1977. Revision of *Torula* and *Hormiscium* species. *Torula occulta, T. diversa, T. elasticae, T. bigemina* and *Hormiscium condensatum* reexamined. Mycologia 69: 533-546.
3150. Schoulties, C.L., N.E. El-Gholl, and S.A. Alfieri Jr. 1982. *Cylindrocladium spathiphylli* sp. nov. Mycotaxon 16: 265-272.
3151. Schreiber, L.R. 1964. Stem canker and die-back of *Rhododendron* caused by *Botryosphaeria ribis* Gross. & Dug. Pl. Dis. Reporter 48: 207-210.
3152. Schreiber, L.R., and L.S. Dochinger. 1967. *Fusarium* canker on paper mulberry (*Broussonetia papyrifera*). Pl. Dis. Reporter 51: 531-532.
3153. Schreiber, L.R., C. Fordyce Jr., and R.J. Green Jr. 1961. *Verticillium* wilt of saucer magnolia, *Magnolia soulangeana*. Pl. Dis. Reporter 45: 108.
3154. Schreiber, L.R., and R.J. Green. 1959. Die-back and root rot disease of *Taxus* spp. in Indiana. Pl. Dis. Reporter 43: 814-817.
3155. Schreiber, L.R., and C.L. Wilson. 1969. Occurrence of Dutch elm disease in North Dakota. Pl. Dis. Reporter 53: 994.
3156. Schroeder, D.B., and D.W. French. 1961. Dutch elm disease found in Minnesota. Pl. Dis. Reporter 45: 749.
3157. Schroeder, H.W. 1964. Grain discoloration in Belle Patna rice. Pl. Dis. Reporter 48: 288-291.
3158. Schroeder, W.T., and R. Provvidenti. 1961. *Rhizoctonia* fruit rot of processing tomatoes. Pl. Dis. Reporter 45: 160-163.
3159. Schuster, M.L., and D.W. Christiansen. 1952. A foot and root disease of safflower caused by *Puccinia carthami* Cda. Phytopathology 42: 211-212.
3160. Schwarz, M.R., and D.M. Boone. 1983. Black rot of cranberry caused by *Strasseria oxycocci*. Pl. Dis. 67: 31-32.
3161. Schwarz, M.R., and D.M. Boone. 1985. Effect of wounding on incidence of black rot of cranberry in Wisconsin. Pl. Dis. 69: 225-227.
3162. Schwinghamer, E.A. 1955. A form of crown rust occurring on *Agropyron* sp. in North Dakota. Pl. Dis. Reporter 39: 322-324.
3163. Schwinghamer, E.A. 1956. Physiologic specialization in *Colletotrichum linicolum*. Phytopathology 46: 300-305.
3164. Sciumbato, G.L., and J.A. Pinckard. 1974. *Alternaria macrospora* leaf spot of cotton in Louisiana in 1972. Pl. Dis. Reporter 58: 201-202.
3165. Sears, R.G., R.L. Ditterline, and D.E. Mathre. 1975. Crown and root rotting organisms affecting sainfoin (*Onobrychis viciifolia*) in Montana. Pl. Dis. Reporter 59: 423-426.
3166. Seaver, F.J. 1909. The Hypocreales of North America - II. Mycologia 1: 177-207.
3167. Seaver, F.J. 1928-1951. The North American cup-fungi. Vol. 1 (Operculates). 248 p. Vol. 2 (Inoperculates). 428 p. Published by the author. New York.
3168. Seaver, F.J., and J. Velazquez. 1933. *Dermea* and *Pezicula*. Mycologia 25: 139-148.
3169. Sechler, D., and H.H. Luke. 1967. Stand loss of small grains in Florida. Pl. Dis. Reporter 51: 919-922.
3170. Sechler, D., and H.H. Luke. 1967. Anthracnose of rye in Florida. Pl. Dis. Reporter 51: 923-925.
3171. Seeler, E.V. Jr. 1939. *Thyronectria denigrata* (Winter) Seaver, the cause of disease in *Gleditsia*. J. Arnold Arbor. 20: 114-115.
3172. Seeler, E.V. Jr. 1940. Two diseases of *Gleditsia* caused by a species of *Thyronectria*. J. Arnold Arbor. 21: 405-427.
3173. Seeler, E.V. Jr. 1940. A monographic study of the genus *Thyronectria*. J. Arnold Arbor. 21: 429-460.
3174. Segall, R. 1951. Brown spot disease of celery found in New York. Pl. Dis. Reporter 35: 164.
3175. Segall, R.H. 1953. Onion blast or leaf spotting caused by species of *Botrytis*. Phytopathology 43: 483.
3176. Seifert, K.A. 1983. Decay of wood by the Dacrymycetales. Mycologia 75: 1011-1018.
3177. Seifert, K.A. 1985. *Stachycoremium*, a new genus synnematous Hyphomycetes. Mycologia 77: 987-990.
3178. Seifert, K.A. 1985. A monograph of *Stilbella* and some allied Hyphomycetes. Stud. Mycol. 27: 1-235.
3179. Seifert, K.A. 1987. *Charomyces*, a new genus of Hyphomycetes with thallic-arthric conidia. Canad. J. Bot. 65: 230-232.
3180. Seifert, K.A., and S.E. Carpenter. 1987. *Bisporella resinicola* comb. nov. and its *Eustilbum* anamorph. Canad. J. Bot. 65: 1262-1267.
3181. Seifert, K.A., G. Okada, and V. Rao. 1987. The synnematous hyphomycete *Dictyocatenulata alba*. Mycologia 79: 459-462.
3182. Self, R.L., and C.H. Driver. 1956. A *Physalospora* disease of lawn grass. Pl. Dis. Reporter 40: 509.
3183. Semeniuk, G. 1983. Association of *Trematosphaeria circinans* with crown and root rot of alfalfa in South Dakota. Mycologia 75: 744-747.
3184. Semeniuk, G., and C.J. Mankin. 1966. Additional hosts of *Sclerophthora macrospora* in South Dakota. Phytopathology 56: 351.
3185. Semeniuk, G., and E.F. Vestal. 1952. Leaf diseases of corn in Iowa in 1951. Pl. Dis. Reporter 36: 171-177.
3186. Semer, C.R., and B.C. Raju. 1985. *Phytophthora* leaf blight of *Hibiscus*, a new disease caused by *Phytophthora parasitica*. Pl. Dis. 69: 1005-1006.
3187. Setliff, E.C. 1972. The taxonomy and morphology of *Poria vincta*. Mycologia 64: 689-701.
3188. Setliff, E.C., J.A. Sullivan, and J.H. Thompson. 1975. *Scleroderris lagerbergii* in large red and scots pine trees in New York. Pl. Dis. Reporter 59: 380-381.
3189. Setliff, E.C., and E.K. Wade. 1973. *Stereum purpureum* associated with sudden decline and death of apple trees in Wisconsin. Pl. Dis. Reporter 57: 473-474.
3190. Seymour, C.P., and J.W. Miller. 1974. The eradication of two potentially dangerous rusts, *Physopella palle scens* and *P. zeae*, from Florida. Florida State Horticultural Society 84: 124-126.
3191. Seymour, E.K., and F.T. McFarland. 1921. Loss from rye ergot. Phytopathology 11: 285-289.
3192. Shaffer, R.L. 1957. *Volvariella* in North America. Mycologia 49: 545-579.
3193. Shaffer, R.L., and C.T. Rogerson. 1952. Notes on the fleshy fungi of Kansas. Trans. Kansas Acad. Sci. 55: 282-286.
3194. Shahin, E.A., and L.E. Claflin. 1978. The occurrence and distribution of *Sirococcus* shootblight of spruce in Kansas. Pl. Dis. Reporter 62: 648-650.
3195. Shahin, E.A., and L.E. Claflin. 1980. Twig blight of Douglas-fir: a new disease caused by *Dothiorella dothidea*. Pl. Dis. 64: 47-50.
3196. Shahjahan, A.K.M., Z. Harahap, and M.C. Rush. 1977. Sheath rot of rice caused by *Acrocylindrium oryzae* in Louisiana. Pl. Dis. Reporter 61: 307-310.
3197. Shahjahan, A.K.M., and M.C. Rush. 1979. Occurrence and distribution of diseases caused by sclerotial fungi on rice in Louisiana. Pl. Dis. Reporter 63: 220-223.
3198. Shane, W.W., J.S. Baumer, and S.G. Sederstrom. 1981. *Alternaria helianthi*: a pathogen of sunflower new to Minnesota. Pl. Dis. 65: 269-271.
3199. Shane, W.W., and T.B. Sutton. 1981. The *Glomerella cingulata* perfect stage and apple bitter rot. Pl. Dis. 65: 135-137.
3200. Sharp, E.L. 1959. Two previously unreported fungi on cereals in Montana. Pl. Dis. Reporter 43: 12-13.
3201. Shaw, C.G. 1951. New species of the Peronosporaceae. Mycologia 43: 445-455.
3202. Shaw, C.G. 1973. Host fungus index for the Pacific Northwest - I. Hosts. Wash. State Agric. Exp. Sta. Bull. 765: 1-121.
3203. Shaw, C.G., and C.D. Leaphart. 1960. Two serious foliage diseases of western white pine in the Inland Empire. Pl. Dis. Reporter 44: 655-659.
3204. Shaw, C.G., and B.D. Thyr. 1962. Observations on diseases of coniferous foliage in the Inland Empire. Phytopathology 52: 365.
3205. Shaw, C.G., and W.D. Yerkes. 1952. Additions to the downy mildews (Peronosporaceae) of the Pacific Northwest. Northw. Sci.

26: 19-21.
3206. Shay, J.R. 1953. Southern blight on apple nursery stock in Indiana. Pl. Dis. Reporter 37: 121.
3207. Shea, K.R. 1964. *Rosellinia herpotrichioides* on sitka spruce seedlings in Washington. Pl. Dis. Reporter 48: 512-513.
3208. Shea, K.R., and J.H. Rediske. 1964. *Schizophyllum commune* Fr. isolated from stored Douglas-fir cones. Pl. Dis. Reporter 48: 234.
3209. Shear, C.L. 1902. Mycological notes and new species. Bull. Torrey Bot. Club 29: 449-457.
3210. Shear, C.L. 1923. Life histories and undescribed genera and species of fungi. Mycologia 15: 120-131.
3211. Shear, C.L., and R.W. Davidson. 1944. An undescribed *Corticium* with conidia. Mycologia 36: 294-299.
3212. Shear, C.L., and B.O. Dodge. 1921. The life history and identity of *Patellina fragariae*, *Leptothyrium macrothecium*, and *Peziza oenotherae*. Mycologia 13: 135-170.
3213. Shear, C.L., N.E. Stevens, and M.S. Wilcox. 1925. *Botryosphaeria* and *Physalospora* in the eastern United States. Mycologia 17: 98-107.
3214. Shearer, C.A. 1982. *Polyschema bicellularis*, a new species from llinois. Mycotaxon 14: 91-93.
3215. Shearer, C.A., and J.L. Crane. 1971. Fungi of the Chesapeake Bay and its tributaries. I. Patuxent River. Mycologia 63: 237-260.
3216. Shearer, C.A., and J.L. Crane. 1976. Illinois fungi no. 7. *Rogersia annelidica* gen. & sp. nov. an aquatic hyphomycete colonizing leaves in the Sangamon River. Mycologia 68: 946-950.
3217. Shearer, C.A., D.L. Hawksworth, and S. von Bodman. 1980. *Arnium apiculatum* (Sordariaceae s. lat.) rediscovered. Mycotaxon 11: 165-172.
3218. Shearer, C.A., and L.C. Lane. 1983. Comparison of three techniques for the study of aquatic hyphomycete communities. Mycologia 75: 498-508.
3219. Shefelbine, P.A., D.E. Mathre, and G. Carlson. 1986. Effects of chloride fertilizer and systemic fungicide seed treatments on common root rot of barley. Pl. Dis. 70: 639-642.
3220. Sherbakoff, C.D. 1928. An examination of Fusaria in the herbarium of the Pathological Collections, Bureau of Plant Industry, U.S.D.A. Phytopathology 18: 148.
3221. Sherbakoff, C.D. 1928. *Washingtonia* palm leaf spot due to *Cylindrocladium macrosporum* n.sp. Phytopathology 18: 219-225.
3222. Sherbakoff, C.D. 1953. Fusaria associated with *Citrus* feeder roots in Florida. Phytopathology 43: 395-397.
3223. Sherwin, H.S. 1952. Target spot on soybeans in Virginia. Pl. Dis. Reporter 36: 491.
3224. Sherwood, M. 1974. *Myriophacidium*, a new genus of Rhytismataceae from western Oregon. Mycologia 66: 690-692.
3225. Sherwood, M.A. 1977. The ostropalean fungi. Mycotaxon 5: 1-277.
3226. Sherwood, M.A. 1977. The Ostropalean fungi II: *Schizoxylon*, with notes on *Stictis*, *Acarosporina*, *Coccopeziza*, and *Carestiella*. Mycotaxon 6: 215-260.
3227. Sherwood, M.A. 1978. New Ostropales from the collections of the Farlow Herbarium. Occas. Pap. Farlow Herb. 13: 39-47.
3228. Sherwood, M.A. 1979. Phacidiales Exsiccati. Decades I - III. Mycotaxon 10: 241-245.
3229. Sherwood, M.A. 1980. Taxonomic studies in the Phacidiales: The genus *Coccomyces* (Rhytismataceae). Occas. Pap. Farlow Herb. 15: 1-120.
3230. Sherwood, M., and G. Carroll. 1974. Fungal succession on needles and young twigs of old-growth Douglas fir. Mycologia 66: 499-506.
3231. Sherwood-Pike, M. 1985. New and Unusual Ascomycetes from the Western United States. Sydowia 38: 267-277.
3232. Sherwood-Pike, M.A. 1987. The Ostropalean fungi III: The Odontotremataceae. Mycotaxon 28: 137-177.
3233. Sherwood-Pike, M.A., and J. Boise. 1986. Studies in lignicolous Ascomycetes: *Xylopeziza* and *Mycowintera*. Brittonia 38: 35-44.
3234. Sherwood-Pike, M., J.K. Stone, and G.C. Carroll. 1986. *Rhabdocline parkeri*, a ubiquitous foliar endophyte of Douglas-fir. Canad. J. Bot. 64: 1849-1855.
3235. Shew, H.D., and D.M. Benson. 1981. Fraser fir root rot induced by *Phytophthora citricola*. Pl. Dis. 65: 688-689.
3236. Shew, H.D., and C.E. Main. 1985. *Rhizoctonia* leaf spot of flue-cured tobacco in North Carolina. Pl. Dis. 69: 901-903.
3237. Shields, I.J. 1951. Oak wilt in Kansas. Pl. Dis. Reporter 35: 119.
3238. Shields, I.J. 1952. Oak wilt survey in Kansas. Pl. Dis. Reporter 36: 68.
3239. Shigo, A.L. 1958. Fungi isolated from oak-wilt trees and their effects on *Ceratocystis fagacearum*. Mycologia 50: 757-769.
3240. Shigo, A.L. 1965. Decay and discoloration in sprout red maple. Phytopathology 55: 957-962.
3241. Shigo, A.L. 1969. How *Poria obliqua* and *Polyporus glomeratus* incite cankers. Phytopathology 59: 1164-1165.
3242. Shigo, A.L. 1972. Successions of microorganisms and patterns of discoloration and decay after wounding in red oak and white oak. Phytopathology 62: 256-259.
3243. Shigo, A.L. 1979. Compartmentalization of decay associated with *Heterobasidion annosum* in roots of *Pinus resinosa*. Eur. J. Forest Pathol. 9: 341-347.
3244. Shih, S.C., E.S. Elliott, and C. Veatch. 1959. A study of *Erysiphe graminis* races on barley in West Virginia. Pl. Dis. Reporter 43: 1282-1283.
3245. Shoemaker, R., and C. Babcock. 1987. Additions to *Nodulosphaeria*. Canad. J. Bot. 65: 1921-1926.
3246. Shoemaker, R.A. 1963. Generic correlations and concepts: *Griphosphaerioma* and *Labridella*. Canad. J. Bot. 41: 1419-1423.
3247. Shoemaker, R.A. 1965. Revision of some *Dimeriella* and *Dimerosporium* parasites of conifers. Canad. J. Bot. 43: 631-639.
3248. Shoemaker, R.A. 1976. Canadian and some extralimital *Ophiobolus* species. Canad. J. Bot. 54: 2365-2404.
3249. Shoemaker, R.A. 1984. Canadian and some extralimital *Leptosphaeria* species. Canad. J. Bot. 62: 2688-2729.
3250. Shoemaker, R.A. 1984. Canadian and some extralimital *Nodulosphaeria* and *Entodesmium* species. Canad. J. Bot. 62: 2730-2753.
3251. Shoemaker, R.A., and C.E. Babcock. 1985. Canadian and some extralimital *Paraphaeosphaeria* species. Canad. J. Bot. 63: 1284-1291.
3252. Shoemaker, R.A., and C.E. Babcock. 1987. *Wettsteinina*. Canad. J. Bot. 65: 373-405.
3253. Shoemaker, R.A., and K.N. Egger. 1982. *Stigmina dothideoides*. Fungi Canadenses 212: 1-2.
3254. Shoemaker, R.A., and K.N. Egger. 1982. *Endoxylina allantospora*. Fungi Canadenses 227: 1-2.
3255. Shoemaker, R.A., and K.N. Egger. 1982. *Pseudopeziza singularis*. Fungi Canadenses No. 229: 1-2.
3256. Shoemaker, R.A., and E.G. Kokko. 1977. *Aglaospora profusa*. Fungi Canadenses 101: 1-2.
3257. Shoemaker, R.A., and E.G. Kokko. 1977. *Massaria lantanae*. Fungi Canadenses 105: 1-2.
3258. Shoemaker, R.A., and E.G. Kokko. 1977. *Massaria zanthoxyli*. Fungi Canadenses 106: 1-2.
3259. Shoemaker, R.A., and E. Müller. 1963. Generic correlations and concepts: *Broomella* and *Pestalotia*. Canad. J. Bot. 41: 1235-1243.
3260. Shoemaker, R.A., and A.E. Straby. 1965. *Pestalotia langloisii* on gardenias. Pl. Dis. Reporter 49: 832-833.
3261. Shoemaker, R.A., and L.J. Tyler. 1954. *Cercosporella* foot rot of winter wheat in New York. Pl. Dis. Reporter 38: 710.
3262. Shortt, B.J., J.B. Sinclair, C.G. Helm, M.R. Jeffords, and M. Kogan. 1982. Soybean seed quality losses associated with bean leaf beetles and *Alternaria tenuissima*. Phytopathology 72: 615-618.
3263. Shriner, D.S., and L.F. Grand. 1974. *Rhizosphaera kalkhoffii* on blue spruce in North Carolina. Pl. Dis. Reporter 58: 439.
3264. Siang, W.N. 1954. Observations on *Tilletia cerebrina*. Mycologia 46: 238-244.
3265. Sickinger, S.M., C.R. Grau, and R.G. Harvey. 1987. *Verticillium* wilt of velvetleaf (*Abutilon theophrasti*). Pl. Dis. 71: 415-418.
3266. Silber, G., and H.E. Heggestad. 1960. Black shank of tobacco in Pennsylvania and some observations on the pathogen. Pl. Dis. Reporter 44: 303-305.
3267. Sim, T. IV. 1980. Sorghum downy mildew in Kansas in 1979. Pl. Dis. 64: 499.
3268. Sim, T. IV. 1980. Southern rust of corn recognized in Kansas. Pl. Dis. 64: 500.
3269. Simmons, E.G. 1969. Perfect states of *Stemphylium*. Mycologia 61: 1-26.
3270. Simmons, E.G. 1971. *Helminthosporium allii* as type of a new genus. Mycologia 63: 380-386.
3271. Simmons, E.G. 1986. *Alternaria* themes and variations (14-16). Mycotaxon 25: 195-202.

3272. Simmons, E.G. 1986. *Alternaria* themes and variations (17-21). Mycotaxon 25: 203-216.
3273. Simmons, E.G. 1986. *Gibbago*, a new phaeodictyoconidial genus of Hyphomycetes. Mycotaxon 27: 107-111.
3274. Simms, H.R. 1967. On the ecology of *Herpotrichia nigra*. Mycologia 59: 902-909.
3275. Simone, G.W., and D.D. Brunk. 1983. New leaf spot disease of *Calathea* and *Maranta* spp. incited by *Dr echslera setariae*. Pl. Dis. 67: 1160-1161.
3276. Simpson, M.E., and P.B. Marsh. 1971. The geographical distribution of *Aspergillus flavus* boll rot in the U.S. cotton crop of 1970. Pl. Dis. Reporter 55: 510-514.
3277. Simpson, W.R. 1967. Rust on fence row asparagus. Pl. Dis. Reporter 51: 542-543.
3278. Sinclair, J.B. 1964. Clubroot of rutabaga in Louisiana. Pl. Dis. Reporter 48: 574.
3279. Sinclair, J.B., N.L. Horn, and W.J. Martin. 1959. New or unusual occurrence of certain diseases in Louisiana. Pl. Dis. Reporter 43: 947-948.
3280. Sinclair, J.B., N.L. Horn, and E.C. Tims. 1958. Unusual occurrence of certain diseases in Louisiana. Pl. Dis. Reporter 42: 984-985.
3281. Sinclair, J.B., and M.N. Shatla. 1962. *Stemphyllium* leaf spot of cotton reported from Louisiana. Pl. Dis. Reporter 46: 744.
3282. Sinclair, W.A., D.P. Cowles, and S.M. Hee. 1975. *Fusarium* root rot of Douglas-fir seedlings: Suppression by soil fumigation, fertility management and inoculation with spores of the fungal symbiont *Laccaria laccata*. Forest Sci. 21: 390-399.
3283. Sinclair, W.A., and G.W. Hudler. 1980. Tree and shrub pathogens new or noteworthy in New York state. Pl. Dis. 64: 590-592.
3284. Singer, R. 1945. New and interesting species of Basidiomycetes. Mycologia 37: 425-439.
3285. Singer, R. 1956. Contribution towards a monograph of the genus *Pluteus*. Trans. Brit. Mycol. Soc. 39: 145-232.
3286. Singer, R. 1960. *Pluteus lilacinus*. Mycologia 52: 337-338.
3287. Singer, R. 1973. The genera *Marasmiellus*, *Crepidotus*, and *Simocybe* in the neotropics. Beih. Nova Hedwigia 44: 1-517.
3288. Singer, R. 1976. Marasmieae (Basidiomycetes-Tricholomataceae). Flora Neotropica 17: 1-347.
3289. Singer, R. 1982. *Hydropus* (Basidiomycetes-Tricholomataceae-Mycenae). Flora Neotropica 32: 1-153.
3290. Singer, R., L.T. Lucas, and T.B. Warren. 1973. The *Marasmius*-blight fungus. Mycologia 65: 468-473.
3291. Singer, R. R. 1942. A monographic study of the genera *Crinipellis* and *Chaetocalathus*. Lilloa 8: 441-534.
3292. Singer, R. R. 1955. New and interesting species of basidiomycetes. IV. Mycologia 47: 763-777.
3293. Siradhana, B.S., C.W. Ellett, and A.F. Schmitthenner. 1968. Crown rot of *Peperomia*. Pl. Dis. Reporter 52: 244.
3294. Sitterly, W.R. 1960. *Physalospora obtusa* (Cke.) Winter on South Carolina live oaks. Pl. Dis. Reporter 44: 533.
3295. Sitterly, W.R., and W.M. Epps. 1958. Lima bean scab found in South Carolina. Pl. Dis. Reporter 42: 1309.
3296. Sivanesan, A. 1971. The genus *Herpotrichia* Fuckel. Mycol. Pap. 127: 1-37.
3297. Sivanesan, A. 1984. *Acantharia*, *Gibbera*, and their anamorphs. Trans. Brit. Mycol. Soc. 82: 507-529.
3298. Sivanesan, A., and J.M. Waller. 1986. Sugarcane diseases. Commonwealth Mycological Institute. Phytopathological Paper No. 29. 88 p.
3299. Skelly, J.M. 1972. *Dothiostroma pini* found infecting Scotch pine, *Pinus sylvestris*,a first report. Phytopathology 62: 671.
3300. Skelly, J.M. 1976. Western gall rust found in Virginia. Pl. Dis. Reporter 60: 222-223.
3301. Skelly, J.M., R.B. Carroll, and F.L. Lukezic. 1968. A new disease of crownvetch, *Coronilla varia*, caused by *Sclerotinia trifoliorum*. Pl. Dis. Reporter 52: 882-886.
3302. Skilling, D.D. 1970. *Scleroderris* canker on red and jack pine in Minnesota. Pl. Dis. Reporter 54: 132.
3303. Skilling, D.D. 1979. *Rhabdocline* needlecast of Douglas-fir found in Wisconsin. Pl. Dis. Reporter 63: 626.
3304. Skilling, D.D., and T.H. Nicholls. 1971. *Lophodermium pinastri* -- a new disease problem in Scotch pine Chris tmas tree plantations. Pl. Dis. Reporter 55: 1116-1117.
3305. Skilling, D.D., B.S. Schneider, and J.A. Sullivan. 1977. *Scleroderris* canker on Austrian and ponderosa pine in New York. Pl. Dis. Reporter 61: 707-708.
3306. Skolko, A.J., and J.W. Groves. 1953. Notes on seed-borne fungi VII. *Chaetomium*. Canad. J. Bot. 31: 779-809.
3307. Skotland, C.B. 1955. A *Phytophthora* damping-off disease of soybean. Pl. Dis. Reporter 39: 682-683.
3308. Skotland, C.B. 1971. Pathogenic and nonpathogenic *Verticillium* species from south central Washington. Phytopathology 61: 435-436.
3309. Skotland, C.B., and J.D. Menzies. 1957. Two peppermint diseases found in the Yakima Valley of Washington. Pl. Dis. Reporter 41: 493.
3310. Skoyen, I.O., R.T. Lewellen, and J.S. McFarlane. 1975. Effect of powdery mildew on sugarbeet production in the Salinas Valley of California. Pl. Dis. Reporter 59: 506-510.
3311. Slack, D.A. 1954. Powdery mildew of apple destructive in northwest Arkansas. Pl. Dis. Reporter 38: 607.
3312. Sleeth, B. 1951. *Diplodia* crown rot of alfalfa. Pl. Dis. Reporter 35: 50-51.
3313. Sleeth, B. 1956. Crown necrosis of bells-of-Ireland. Pl. Dis. Reporter 40: 80.
3314. Sleeth, B. 1958. *Cladosporium* on melons in south Texas. Pl. Dis. Reporter 42: 1104.
3315. Slysh, A.R. 1960. The genus *Peniophora* in New York state and adjacent regions. State Univ. New York Coll. Forestry Tech. Pub. 83: 1-95.
3316. Smalley, E.B., and H.N. Hansen. 1962. *Penicillium* decay of garlic. Phytopathology 52: 666-678.
3317. Smalley, G.W., and R.L. Scheer. 1963. Black root rot in Florida sandhills. Pl. Dis. Reporter 47: 669-671.
3318. Smart, G.C. Jr. 1962. Peanut rust in Virginia. Pl. Dis. Reporter 46: 65.
3319. Smiley, E.T., J.J. Kielbaso, and T.J. Proffer. 1985. *Cytospora* canker on *Acer platanoides* in southeastern Michigan. Pl. Dis. 69: 360.
3320. Smiley, J.H., and R.A. Flowers. 1967. Occurrence of tobacco diseases in Kentucky in 1966. Pl. Dis. Reporter 51: 866-867.
3321. Smiley, J.H., and R.A. Flowers. 1968. Tobacco diseases in Kentucky in 1967. Pl. Dis. Reporter 52: 589.
3322. Smiley, J.H., R.A. Flowers, and R.A. Chapman. 1968. Occurrence of blue mold on tobacco in Kentucky in 1967. Pl. Dis. Reporter 52: 204.
3323. Smiley, J.H., R.A. Flowers, and C.Y.Y. Yang. 1968. Occurrence of brown spot on tobacco in Kentucky in 1966. Pl. Dis. Reporter 52: 71.
3324. Smiley, R.W., and M.C. Fowler. 1984. *Leptosphaeria korrae* and *Phialophora graminicola* associated with *Fusarium* blight syndrome of *Poa pratensis* in New York. Pl. Dis. 68: 440-442.
3325. Smiley, R.W., M.C. Fowler, and R.C. O'Knefski. 1985. Arsenate herbicide stress and incidence of summer patch on Kentucky bluegrass turfs. Pl. Dis. 69: 44-48.
3326. Smith, A.H. 1947. North American species of *Mycena*. Univ. Michigan Stud., Sci. Ser. 17: 1-521.
3327. Smith, A.H. 1951. The North American species of *Naematoloma*. Mycologia 43: 467-521.
3328. Smith, A.H. 1952. New and rare Agarics from the Douglas Lake Region and Tahquamenon Fal ls State Park, Michigan, and an account of the North American species of *Xeromphalina*. Pap. Michigan Acad. Sci. 38: 53-87.
3329. Smith, A.H. 1953. New species of *Galerina* from North America. Mycologia 45: 892-925.
3330. Smith, A.H. 1956. Additional new or unusual North American Agarics. Beih. Sydowia 1: 46-61.
3331. Smith, A.H. 1966. New and noteworthy higher fungi from Michigan. Michigan Bot. 5: 18-25.
3332. Smith, A.H., and L.R. Hesler. 1968. The North American Species of *Pholiota*. Hafner Pub. Co. New York. 392 p.
3333. Smith, A.H., and R. Singer. 1955. New species of *Galerina*. Mycologia 47: 557-596.
3334. Smith, A.H., and R. Singer. 1958. New species of *Galerina*. Mycologia 50: 469-489.
3335. Smith, A.H., and D.E. Stuntz. 1950. New or noteworthy fungi from Mt. Rainier National Park. Mycologia 42: 80-134.

3336. Smith, A.H., and D.E. Stuntz. 1958. Studies on the genus *Pluteus* I. Redescriptions of american species based on a study of type specimens. Lloydia 21: 115-136.
3337. Smith, A.L. 1950. *Ascochyta* seedling blight of cotton in Alabama in 1950. Pl. Dis. Reporter 34: 233-235.
3338. Smith, A.L., and C. Wilson. 1951. *Verticillium* wilt of cotton in Alabama. Pl. Dis. Reporter 35: 165.
3339. Smith, B.J., and L.L. Black. 1986. First report of *Colletotrichum acutatum* on strawberry in the United States. Pl. Dis. 70: 1074.
3340. Smith, D.H. 1972. *Arachis hypogaea*: A new host of *Cristulariella pyramidalis*. Pl. Dis. Reporter 56: 796-797.
3341. Smith, H.E., A.L. Smith, W.E. Cooper, and L. Lett. 1959. Reduction in yield of cotton caused by diseases in 1958. Pl. Dis. Reporter 43: 368-370.
3342. Smith, J.D. 1971. *Pleospora* stem eyespot of fescues in Oregon and the *Didymella* per fect stage of the pathogen. Pl. Dis. Reporter 55: 63-67.
3343. Smith, L.J. 1978. Wheat leaf rust in Aberdeen, Idaho in 1976. Pl. Dis. Reporter 62: 74.
3344. Smith, M.A. 1951. Anthracnose on Michigan peaches. Pl. Dis. Reporter 35: 276.
3345. Smith, M.A. 1963. Apple rot caused by *Pyrenochaeta mali* n. sp. Phytopathology 53: 589-591.
3346. Smith, M.A., and W.R. Wright. 1951. Natural infection of peach fruit by *Diplodia*. Pl. Dis. Reporter 35: 222.
3347. Smith, R.L. 1971. Chinese gooseberry, a new host for *Armillaria mellea*. Pl. Dis. Reporter 55: 1099-1100.
3348. Smith, R.S., and R.V. Bega. 1964. *Macrophomina phaseoli* in the forest tree nurseries of California. Pl. Dis. Reporter 48: 206.
3349. Smith, R.S., R.V. Bega, and J. Tarry. 1966. Additional hosts of *Fomes annosus* in California. Pl. Dis. Reporter 50: 181.
3350. Smith, R.S. Jr. 1966. *Rosellinia* needle blight of Douglas-fir in California. Pl. Dis. Reporter 50: 249-250.
3351. Smith, R.S. Jr. 1972. Needle casts of high-altitude white pines in California. Pl. Dis. Reporter 56: 102-103.
3352. Smith, R.S. Jr. 1973. *Sirococcus* tip dieback of *Pinus* spp. in California forest nurseries. Pl. Dis. Reporter 57: 69-73.
3353. Smith, R.S., Jr., A.H. McCain, and M.D. Srago. 1973. Control of *Botrytis* storage rot of giant sequoia seedlings. Pl. Dis. Reporter 57: 67-69.
3354. Smith, R.W., J.W. Lorbeer, and A.A. Abd-Elrazik. 1985. Reappearance and control of onion downy mildew epidemics in New York. Pl. Dis. 69: 703-706.
3355. Smith, T.E. 1960. Occurrence of *Verticillium* wilt on Russian olive. Pl. Dis. Reporter 44: 72.
3356. Smith, T.E. 1960. Observations on cotton rust (*Puccinia stakmanii*) under severe disease conditions. Pl. Dis. Reporter 44: 77-79.
3357. Smith, T.E. 1960. Occurrence of *Verticillium* wilt on peanuts. Pl. Dis. Reporter 44: 435.
3358. Smith, W.L., H.E. Moline, and K.S. Johnson. 1979. Studies with *Mucor* species causing postharvest decay of fresh produce. Phytopathology 69: 865-869.
3359. Smith, W.L., Jr., and M.H. Haller. 1951. Peach anthracnose found in West Virginia. Pl. Dis. Reporter 35: 465.
3360. Smyly, W.B., and T.H. Filer Jr. 1977. *Cylindrocladium scoparium* associated with root rot and mortality of cherry bark oak seedlings. Pl. Dis. Reporter 61: 577-579.
3361. Snider, R.D., and C.L. Kramer. 1974. An electrophoretic protein analysis and numerical taxonomic study of the genus *Taphrina*. Mycologia 66: 754-772.
3362. Snow, G.A., J.W. Beland, and F.J. Czabator. 1974. Formosan sweetgum susceptible to North American *Endothia gyrosa*. Phytopathology 64: 602-605.
3363. Snow, G.A., D.O. Wells, and G.L. Switzer. 1986. Fusiform rust gradient in a loblolly pine plantation. Forest Sci. 32: 372-376.
3364. Snow, J.P., and J.C. Mertely. 1979. A boll rot of cotton caused by *Colletotrichum capsici*. Pl. Dis. Reporter 63: 626-627.
3365. Snyder, W.C., R. Bardin, and R.G. Grogan. 1952. Powdery mildew of lettuce reappears in Salinas Valley. Pl. Dis. Reporter 36: 321-322.
3366. Snyder, W.C., H.N. Hansen, and S. Wilhelm. 1950. New hosts of *Verticillium albo-atrum*. Pl. Dis. Reporter 34: 26-27.
3367. Sobers, E.K. 1966. A leaf spot disease of azalea and hydrangea caused by *Corynespora cassiicola*. Phytopathology 56: 455-457.
3368. Sobers, E.K. 1966. A new form and two new hosts of *Puccinia iridis* in Florida. Phytopathology 56: 804-806.
3369. Sobers, E.K. 1967. The perfect stage of *Coniothyrium concentricum* on leaves of *Yucca aloifolia*. Phytopathology 57: 234-235.
3370. Sobers, E.K. 1969. *Calonectria floridana* sp. nov., the perfect state of *Cylindrocladi um floridanum*. Phytopathology 59: 364-366.
3371. Sobers, E.K., and R.S. Cox. 1973. Anthracnose of statice in southern Florida. Phytopathology 63: 193-194.
3372. Sobers, E.K., and A.P. Martinez. 1966. A leaf spot of *Bougainvillea* caused by *Cercospora bougainvilleae*. Phytopathology 56: 128-130.
3373. Sobers, E.K., and A.G. Plakidas. 1962. Colletotrichums associated with lily bulbs. Phytopathology 52: 884-887.
3374. Sobers, E.K., and C.P. Seymour. 1963. *Stemphylium* leafspot of *Echeveria*, *Kalanchoe*, and *Sedum*. Phytopathology 53: 1443-1446.
3375. Sobers, E.K., and C.P. Seymour. 1967. *Cylindrocladium floridanum* sp. n. associated with decline of peach trees in Florida. Phytopathology 57: 389-393.
3376. Solheim, W.G. 1949. Studies on Rocky Mountain fungi - I. Mycologia 41: 623-631.
3377. Solheim, W.G. 1958. *Puccinia poae-nemoralis* Otth on *Phleum pratense* L. and other grass hosts in Wyoming. Pl. Dis. Reporter 42: 533.
3378. Solheim, W.G. 1979. Mycoflora Saximontanensis Exsiccata Centum XVI. Mycotaxon 8: 385-394.
3379. Solheim, W.G., and G.B. Cummins. 1979. Mycoflora Saximontanensis Exsiccata Centum XVII. Mycotaxon 8: 395-401.
3380. Solheim, W.G., D.O. Eboh, and J. McHenry. 1968. Powdery mildew of oak caused by a species of *Typhulochaeta*. J. Elisha Mitchell Sci. Soc. 84: 236-238.
3381. Solheim, W.G., and R.L. Gilbertson. 1973. *Peronospora* species in Arizona. Mycopathol. Mycologia Appl. 49: 153-159.
3382. Sommer, N.F. 1955. Sunburn predisposes walnut trees to branch-wilt. Phytopathology 45: 607-613.
3383. Sommer, N.F., and L. Beraha. 1975. *Diaporthe actinidiae*, a new species causing stem-end rot of Chinese gooseberry fruits. Mycologia 67: 650-653.
3384. Sommer, N.F., J.R. Buchanan, R.J. Fortlage, and B.E. Bearden. 1985. Relation of floral infection to *Botrytis* blossom-end rot of pears in storage. Pl. Dis. 69: 340-343.
3385. Sonoda, R.M. 1973. Occurrence of a *Pythium* disease in virgin sandy soils of south Florida associated with a new method of field seedling tomatoes. Pl. Dis. Reporter 57: 260-261.
3386. Sonoda, R.M. 1973. *Pythium* fruit rot of tomatoes. Pl. Dis. Reporter 57: 638-639.
3387. Sonoda, R.M. 1973. Incidence of *Colletotrichum* leaf spot and stem canker on introductions and selections of *Stylosanthes humilis*. Pl. Dis. Reporter 57: 747-749.
3388. Sonoda, R.M. 1974. *Helminthosporium hawaiiense* on rhodesgrass in Florida. Pl. Dis. Reporter 58: 490.
3389. Sonoda, R.M. 1976. The occurrence of a *Fusarium* root rot of tomatoes in south Florida. Pl. Dis. Reporter 60: 271-274.
3390. Sonoda, R.M. 1980. *Rhizopus* soft rot of seedless greenhouse cucumbers. Pl. Dis. 64: 296-297.
3391. Sonoda, R.M. 1980. Reduction of forage yield of siratro by *Rhizoctonia solani* foliar blight. Pl. Dis. 64: 667.
3392. Sonoda, R.M. 1985. Foliar blight of *Cynodon* spp. forage caused by *Rhizoctonia solani* in Florida. Pl. Dis. 69: 811.
3393. Sonoda, R.M., and J.M. Ogawa. 1968. *Phytophthora* crown and root rot of hops in California. Pl. Dis. Reporter 52: 76-78.
3394. Southam, C.M., and J. Ehrlich. 1950. Etiology of some sap rots of western red cedar poles. Phytopathology 40: 439-444.
3395. Sowell, G., and H.B. Harris. 1969. Downy mildew of sorghum in Georgia. Pl. Dis. Reporter 53: 4.
3396. Sowell, G., Jr., and R.P. Korf. 1960. An emendation of the genus *Itersonilia* based on studies of morphology and pathogenicity. Mycologia 52: 934-945.
3397. Sparrow, F.K. 1950. Annual foray, 1948. Mycologia 42: 191-195.
3398. Sparrow, F.K. 1956. Observations on chytridiaceous parasites of phanerogams. V. The occur rence of *Physoderma butomi* and *P. vagans* in the United States. Mycologia 48: 765-766.
3399. Sparrow, F.K. 1957. Observations on chytridiaceous parasites of phanerogams VII. A *Physoderma* on *Lycopus americanus*. Amer. J. Bot. 44: 661-665.

3400. Sparrow, F.K. 1961. Observations on chytridaceous parasites of phanerogams X. Notes on some species of *Physoderma* occuring in the Douglas Lake Region of Northern Michigan. Pap. Michigan Acad. Sci. 46: 183-194.
3401. Sparrow, F.K. 1968. *Physoderma hydrocotylidis* and other interesting Phycomycetes from California. J. Elisha Mitchell Sci. Soc. 84: 62-68.
3402. Sparrow, F.K. 1970. Observations on chytridiaceous parasites of phanerogams XVIII. A *Physoderma* on *Juncus pelocarpus*. Arch. Mikrobiol. 70: 104-109.
3403. Sparrow, F.K. 1976. *Phlyctochytrium lackeyi*, a chytrid of pine pollen in Florida. Mycologia 68: 1233-1234.
3404. Sparrow, F.K. 1977. Two new uniporous species of *Rhizophydium*. Canad. J. Bot. 55: 1501-1504.
3405. Sparrow, F.K. 1978. Two neustonic fungi. Mycologia 70: 140-149.
3406. Sparrow, F.K. 1979. Observations on chytridiaceous parasites of phanerogams. XXVIII. Physodermas on *Sium suave*. Mycologia 71: 423-433.
3407. Sparrow, F.K. Jr. 1933. Observations on operculate Chytridiaceous fungi collected in the vicinity of Ithaca, N.Y. Amer. J. Bot. 20: 63-77.
3408. Sparrow, F.K., J.E. Griffin, and R.M. Johns. 1961. Observations on chytridiaceous parasites of phanerogams. XI. A *Physoderma* on *Agropyron repens*. Amer. J. Bot. 48: 850-858.
3409. Sparrow, F.K., and R.M. Johns. 1970. Observations on chytridiaceous parasites of phanerogams XVII. Notes on a *Physoderma* parasitic on *Asclepias incarnata*. Arch. Mikrobiol. 70: 72-81.
3410. Sparrow, F.K., and M.S. Saunders. 1974. Observations on chytridiaceous parasites of phanerogams XXII. A *Physoderma* on *Ranunculus septentrionalis* Poir. Arch. Mikrobiol. 100: 41-50.
3411. Specht, L.P., G.J. Griffin, J.J. Reilly, and D.A. Komm. 1987. Inoculum densities of *Thielaviopsis basicola* in tobacco fields and the role of black root rot in tobacco stunting in Virginia. Pl. Dis. 71: 876-879.
3412. Spencer, J.A. 1968. A new leaf-streak disease of *Hemerocallis* spp. Pl. Dis. Reporter 52: 751-753.
3413. Spencer, J.A. 1973. A new *Cercospora* leaf spot of *Kalanchoe*. Phytopathology 63: 449.
3414. Spencer, S., and D.M. Benson. 1981. Root rot of *Acuba japonica* caused by *Phytophthora cinnamomi* and *P. citricola* and suppressed with bark media. Pl. Dis. 65: 918-921.
3415. Spielman, L.J. 1985. A monograph of *Valsa* on hardwoods in North America. Canad. J. Bot. 63: 1355-1378.
3416. Spilker, D.A., and P.O. Larsen. 1985. Characterization and host range of *Drechslera catenaria*, the pathogen of leaf blight and crown rot of creeping bentgrass. Pl. Dis. 69: 331-333.
3417. Spotts, R.A. 1985. Environmental factors affecting conidial survival of five pear decay fungi. Pl. Dis. 69: 391-392.
3418. Spotts, R.A., and L.A. Cervantes. 1985. Effects of heat treatments on populations of four fruit decay fungi in sodium ortho phenylphenate solutions. Pl. Dis. 69: 574-576.
3419. Spotts, R.A., and L.A. Cervantes. 1986. Populations, pathogenicity, and benomyl resistance of *Botrytis* spp., *Penicillium* spp., and *Mucor piriformis* in packinghouses. Pl. Dis. 70: 106-108.
3420. Spotts, R.A., J.A. Traquair, and B.B. Peters. 1981. d'Angou pear decay caused by a low temperature Basidiomycete. Pl. Dis. 65: 151-153.
3421. Sprague, R. 1937. New or noteworthy parasitic species of fungi in Oregon. Mycologia 29: 426-433.
3422. Sprague, R. 1938. Gill fungi associated with the roots of cereals. Phytopathology 28: 78-79.
3423. Sprague, R. 1950. Diseases of Cereals and Grasses in North America. Ronald Press. 538 p.
3424. Sprague, R. 1950. Some leafspot fungi on western Gramineae. V. Mycologia 42: 758-771.
3425. Sprague, R. 1951. Some leafspot fungi on western gramineae - VI. Mycologia 43: 549-569.
3426. Sprague, R. 1954. Some leafspot fungi on western Gramineae. VII. Mycologia 46: 76-88.
3427. Sprague, R. 1954. Some fungi on Alaskan species of *Carex*. Res. Stud. State Coll. Wash. 22: 161-169.
3428. Sprague, R. 1955. Some leafspot fungi on western Gramineae - VIII. Mycologia 47: 249-262.
3429. Sprague, R. 1955. Some leafspot fungi on western Gramineae. IX. Mycologia 47: 835-845.
3430. Sprague, R. 1955. Fungi occurring on Gramineae in the State College of Washington Herbarium. Northw. Sci. 29: 109-119.
3431. Sprague, R. 1955. A check list of fungi of glacier bay, Alaska. Res. Stud. State Coll. Wash. 23: 202-224.
3432. Sprague, R. 1956. Some leafspot fungi on western Gramineae. X. Mycologia 48: 741-756.
3433. Sprague, R. 1957. Some leafspot fungi on western Gramineae. XI. Mycologia 49: 837-853.
3434. Sprague, R. 1957. Fungi isolated from roots and crowns of pear trees. Pl. Dis. Reporter 41: 74-76.
3435. Sprague, R. 1958. Some leafspot fungi on western Gramineae - XII. Mycologia 50: 814-830.
3436. Sprague, R. 1958. Some fungi on western species of Cyperaceae. Res. Stud. State Coll. Wash. 26: 74-84.
3437. Sprague, R. 1959. Some root-rotting fungi isolated from western grasses. Pl. Dis. Reporter 43: 871.
3438. Sprague, R. 1960. Some leafspot fungi on western Gramineae - XIII. Mycologia 52: 357-377.
3439. Sprague, R. 1960. Some leafspot fungi on western Gramineae - XIV. Mycologia 52: 698-718.
3440. Sprague, R. 1961. Some fungi on western Juncaceae. Res. Stud. State Coll. Wash. 29: 77-83.
3441. Sprague, R. 1962. Some leafspot fungi on western Gramineae - XVI. Mycologia 54: 593-610.
3442. Sprague, R. 1962. Some leafspot fungi on western Gramineae. XV. Mycologia 54: 44-61.
3443. Sprague, R. 1962. Some fungi on western species of Cyperaceae. II. Res. Stud. State Coll. Wash. 30: 45-63.
3444. Sprague, R. 1962. Some leafspot fungi on western Juncaceae - II. Res. Stud. State Coll. Wash. 30: 169-173.
3445. Sprague, R. 1965. An undescribed *Heterosporium* on *Petunia*. Mycologia 57: 657-659.
3446. Sprague, R., and P.M. Halisky. 1955. Take-all in the Olympic National Park. Pl. Dis. Reporter 39: 989.
3447. Sprague, R., and H.H.V. Hord. 1950. The reappearance of silver leaf in the apple orchards of Washington state. Pl. Dis. Reporter 34: 414.
3448. Sprague, R., and H.H.V. Hord. 1952. The reappearance of silver leaf in the apple orchards of Washington state. Pl. Dis. Reporter 36: 30.
3449. Sprague, R., and A.G. Johnson. 1950. *Ascochyta* leaf spots of cereals and grasses in the United States. Mycologia 42: 523-553.
3450. Sprague, R., and A.G. Johnson. 1950. Species of *Selenophoma* on North American grasses. Oregon State College Monographs 10: 1-43.
3451. Sprague, R., and C.T. Rogerson. 1958. Some leafspot fungi on Kansas Gramineae. Mycologia 50: 634-641.
3452. Sproston, T., Jr., and W.W. Scott. 1954. *Valsa leucostomoides*, the cause of decay and discoloration in tapped sugar maples. Phytopathology 44: 12-13.
3453. Stack, R.W. 1980. Disease progression in common root rot of spring wheat and barley. Canad. J. Plant Pathol. 2: 187-193.
3454. Staffeldt, E.E. 1959. *Thielaviopsis basicola*, a part of the cotton (*Gossypium hirsutum*) seedling disease complex in New Mexico. Pl. Dis. Reporter 43: 506-508.
3455. Staffeldt, E.E., and P.J. Leyendecker Jr. 1955. *Verticillium* wilt on two vegetable hosts in southern New Mexico. Pl. Dis. Reporter 39: 589-590.
3456. Staley, J.M. 1964. A new *Lophodermium* on ponderosa pine. Mycologia 56: 757-762.
3457. Staley, J.M. 1964. A survey of coniferous foliage diseases (other than rusts) in Colorado. Pl. Dis. Reporter 48: 562-563.
3458. Staley, J.M. 1965. Decline and mortality of red and scarlet oaks. Forest Sci. 11: 2-17.
3459. Staley, J.M. 1965. Scleroderris canker of Subalpine Fir in Colorado. Pl. Dis. Reporter 49: 882.
3460. Staley, J.M., and H.H. Bynum. 1972. A new *Lophodermella* on *Pinus ponderosa* and *P. attenuata*. Mycologia 64: 722-726.
3461. Staley, J.M., and F.G. Hawksworth. 1967. *Bifusella crepidiformis* on Engelmann spruce. Pl. Dis. Reporter 51: 791-792.
3462. Stalpers, J.A., and I. Vlug. 1983. *Confistulina*, the anamorph of *Fistulina hepatica*. Canad. J. Bot. 61: 1660-1666.
3463. Stambaugh, W.J., and W.C. Bramble. 1952. The occurrence of *Rhabdocline* needle cast of Douglas-fir in Pennsylvania. Pl. Dis.

Reporter 36: 292-293.
3464. Standish, E.D., J.D. MacDonald, and W.A. Humphrey. 1982. *Phytophthora* root and crown rot of junipers in California. Pl. Dis. 66: 925-928.
3465. Stanghellini, M.E., P., von Bretzel von, M.W. Olsen, and W.C. Kronland. 1982. Root rot of sugar beets caused by *Pythium deliense*. Pl. Dis. 66: 857-858.
3466. Stanghellini, M.E., and W.C. Kronland. 1977. Root rot of mature sugarbeets by *Rhizopus arrhizus*. Pl. Dis. Reporter 61: 255-256.
3467. Stanghellini, M.E., and W.C. Kronland. 1982. Root rot of chicory caused by *Phymatotrichum omnivorum* and *Phytophthora cryptogea*. Pl. Dis. 66: 262-263.
3468. Stanton, T.R. 1952. *Septoria* black stem of oats abundant in most of the north central states. Pl. Dis. Reporter 36: 349.
3469. Starmer, W.T., H.J. Phaff, M. Miranda, and M.W. Miller. 1978. *Pichia cactophila*, a new species of yeast found in decaying tissue of cacti. Int. J. Systematic Bacteriol. 28: 318-325.
3470. Starmer, W.T., H.J. Phaff, M. Miranda, and M.W. Miller. 1978. *Pichia amethionina*, a new heterothallic yeast associated with the decaying stems of cereoid cacti. Int. J. Systematic Bacteriol. 28: 433-441.
3471. Stathis, P.D., and A.G. Plakidas. 1959. *Entomosporium* leaf spot of *Photinia glabra* and *Photinia serrulata*. Phytopathology 49: 361-365.
3472. Statler, G.D., and M.A. McVey. 1987. Partial resistance to *Uromyces appendiculatus* in dry edible beans. Phytopathology 77: 1101-1103.
3473. Stavely, J.R. 1984. Pathogenic specialization in *Uromyces phaseoli* in the United States and rust resistance in beans. Pl. Dis. 68: 95-99.
3474. Stavely, J.R., and R.R. Kincaid. 1969. Occurrence of *Phyllosticta* leaf spot on Florida cigar-wrapper tobacco. Pl. Dis. Reporter 53: 837-839.
3475. Steadman, J.R., E.D. Kerr, and R.F. Mumm. 1975. Root rot of bean in Nebraska: primary pathogen and yield loss appraisal. Pl. Dis. Reporter 59: 305-308.
3476. Steenland, A., and E. Burke. 1950. *Typhula* sp. pathogenic on mint. Pl. Dis. Reporter 34: 322.
3477. Stelzer, H.E., R.B. Polk, and W.Q. Loegering. 1983. Rapid method of rating forest trees for general resistance to foliar pathogens. Forest Sci. 29: 215-221.
3478. Sterne, R.E. 1982. *Phytophthora* root rot of blueberry in Arkansas. Pl. Dis. 66: 604-605.
3479. Sterne, R.E., and J.P. Jones. 1978. Sharp eyespot of wheat in Arkansas caused by *Rhizoctonia solani*. Pl. Dis. Reporter 62: 56-60.
3480. Stevens, F.L. 1925. Hawaiian Fungi. Bernice P. Bishop Museum Bulletin No. 19. 189 p.
3481. Stevenson, J.A. 1945. Powdery mildew of mesquite. Pl. Dis. Reporter 29: 214-215.
3482. Stevenson, J.A. 1949. Mycological Society of America report of the 1947 foray. Mycologia 41: 207-215.
3483. Stevenson, J.A. 1969. Additions to the Meliolineae of the United States. Friesia 9: 245-252.
3484. Stevenson, J.A. 1975. Fungi of Puerto Rico and the American Virgin Islands Braun-Brumfield, Inc. 743 p.
3485. Stevenson, J.A., and W.A. Archer. 1950. *Fomes* root-rot of *Ribes* in New Mexico. Pl. Dis. Reporter 34: 212.
3486. Stevenson, W.R., G.E. Evans, and T.H. Barksdale. 1978. Evaluation of tomato breeding lines for resistance to fruit anthracnose. Pl. Dis. Reporter 62: 937-940.
3487. Stevenson, W.R., R.J. Green, and G.B. Bergeson. 1976. Occurrence and control of potato black dot root rot in Indiana. Pl. Dis. Reporter 60: 248-251.
3488. Stewart, D.M. 1951. Heart rot of black ash in Minnesota. Phytopathology 41: 569-570.
3489. Stewart, E.L., and M. Corden. 1979. *Periconiella verrucosa* sp. nov. (Dematiaceous Hyphomycetes) from Douglas-fir wood. Mycotaxon 8: 302-306.
3490. Stewart, E.L., and H.G. Johnson. 1977. *Frommea obtusa* var. *duchesneae* on *Duchesnea indica* (Aztec Indian berry) in Michigan and Minnesota. Pl. Dis. Reporter 61: 999.
3491. Stewart, F.C. 1936. The persistence of *Hypholoma incertum* about tree stumps. Mycologia 28: 445-450.
3492. Stipes, R.J. 1970. Comparative mycelial protein and enzyme patterns in four species of *Ceratocystis*. Mycologia 62: 987-995.
3493. Stipes, R.J., and T.C. Davis. 1972. Diseases of landscape trees observed in Virginia. Pl. Dis. Reporter 56: 108-111.
3494. Stipes, R.J., and C.R. Drake. 1972. Reciprocal pathogenicity of *Glomerella cingulata* isolated from *Betula pendula* and *Malus sylvestris*. Phytopathology 62: 1366-1368.
3495. Stipes, R.J., G.H. Emert, and R.D. Brown Jr. 1982. Differentiation of *Endothia gyrosa* and *Endothia parasitica* by disc electrophoresis of intramycelial enzymes and proteins. Mycologia 74: 138-141.
3496. Stipes, R.J., and P.M. Phipps. 1971. A species of *Endothia* associated with a canker disease of pin oak (*Quercus palustris*) in Virginia. Pl. Dis. Reporter 55: 467-469.
3497. Stipes, R.J., and P.M. Phipps. 1975. *Fusarium oxysporum* f. sp. *perniciosum* on *Fusarium*-wilted mimosa trees. Phytopathology 65: 188-190.
3498. Stoddard, E.M. 1957. A *Fusarium* rot of geraniums and its control. Pl. Dis. Reporter 41: 536.
3499. Stone, W.J., and T.W. Culp. 1959. Effects of diseases on castorbeans in Mississippi. Pl. Dis. Reporter 43: 827-829.
3500. Stoner, W.N. 1951. A list of some of the plant diseases occurring in Everglades region of Florida during the 1949-1950 season. Pl. Dis. Reporter 35: 170-172.
3501. Stoner, W.N. 1952. A fungicidal control for powdery mildew, *Leveillula taurica* (Lev.) Arn., of kenaf, *Hibiscus cannabinus* L. in south Florida. Pl. Dis. Reporter 36: 302-303.
3502. Stoner, W.N., and W.D. Hogan. 1950. Gray mold on tomato causes crop loss at Fort Pierce, Florida. Pl. Dis. Reporter 34: 210.
3503. Stoner, W.N., and W.D. Hogan. 1950. *Verticillium* wilt of eggplant observed in Palm Beach and Martin Cou nties, Florida. Pl. Dis. Reporter 34: 213.
3504. Stoner, W.N., F.V. Stevenson, W.G. Genung, W.H. Thames Jr., C.C. Seale, E.O. Gangstad, and J.B. Pate. 1952. Preliminary reports of some of the disease and pest problems of kenaf, *Hibiscus cannabinus* L., in south Florida. Pl. Dis. Reporter 36: 121-126.
3505. Stowell, E.A., and M.P. Backus. 1966. Morphology and cytology of *Diplocarpon maculatum* on *Crataegus*. I.The *Entomosporium* stage. Mycologia 58: 949-960.
3506. Stowell, E.A., and J.D. Rogers. 1983. Studies on *Xylaria oxyacanthae*. Mycotaxon 17: 433-444.
3507. Strandberg, J.O. 1968. Occurrence of *Spongospora subterranea* f. sp. *nasturii* causing crook rot of watercress in Florida and Pennsylvania. Pl. Dis. Reporter 52: 859.
3508. Stranghellini, M.E., and W.C. Kronland. 1986. Yield loss in hydroponically grown lettuce attributed to subclinical infection of feeder rootlets by *Pythium dissotocum*. Pl. Dis. 70: 1053-1056.
3509. Strausbaugh, C.A., and O.C. Maloy. 1986. *Fusarium* scab of irrigated wheat in central Washington. Pl. Dis. 70: 1104-1106.
3510. Stretch, A.W. 1967. Occurrence and control of *Glomerella cingulata* on highbush blueberry. Pl. Dis. Reporter 51: 401-404.
3511. Strider, D.L. 1973. Damping-off of statice caused by *Botrytis cinerea* and its control. Pl. Dis. Reporter 57: 969-971.
3512. Strider, D.L. 1976. Increased prevalence of powdery mildew of azalea and rhododendron in North Carolina. Pl. Dis. Reporter 60: 149-151.
3513. Strider, D.L. 1985. *Fusarium* wilt of *Chrysanthemum*: cultivar susceptibility and chemical control. Pl. Dis. 69: 564-568.
3514. Strider, D.L., and T.T.L. Chi. 1984. Damping-off of portulaca caused by *Helminthosporium* (*Bipolaris*) *portulacae* in North Carolina. Pl. Dis. 68: 826.
3515. Strider, D.L., and R.K. Jones. 1978. Rust of fuchsia in North Carolina. Pl. Dis. Reporter 62: 745-746.
3516. Strobel, G.A., and F. Sugawara. 1986. The pathogenicity of *Ceratocystis montia* to lodgepole pine. Canad. J. Bot. 64: 113-116.
3517. Stromberg, E.L., W.C. Stienstra, T. Kommedahl, C.A. Matyac, C.E. Windels, and J.L. Geadelmann. 1984. Smut expression and resistance of corn to *Sphacelotheca reiliana* in Minnesota. Pl. Dis. 68: 880-884.
3518. Strong, F.C. 1951. Oak wilt found in Michigan. Pl. Dis. Reporter 35: 383.
3519. Strong, F.C. 1951. Known distribution of oak wilt in Michigan. Pl. Dis. Reporter 35: 557.
3520. Strong, F.C. 1960. Control of hawthorn leaf blight. Pl. Dis. Reporter 44: 396-398.
3521. Strong, F.C., and W.A. Lemmien. 1964. *Fomes annosus* in southwestern Michigan. Pl. Dis. Reporter 48: 110.

3522. Strong, M.C. 1961. *Rhizoctonia* stem canker of tomatoes. Pl. Dis. Reporter 45: 392.
3523. Stroube, W.H. 1954. *Puccinia oahuensis* on *Digitaria Ischaemum* and *D. sanguinalis* in Louisiana. Pl. Dis. Reporter 38: 120.
3524. Stroube, W.H. 1954. Host range of the *Rhizoctonia* aerial blight fungus in Louisiana. Pl. Dis. Reporter 38: 789-790.
3525. Struble, F.B., and L.S. Morrison. 1957. Blackberry anthracnose control and its relation to yield. Pl. Dis. Reporter 41: 766-769.
3526. Stuckey, I.H. 1961. Leaf drop of *Taxus*. Pl. Dis. Reporter 45: 527-529.
3527. Stuntz, D.E., and C.E. Seliskar. 1943. A stem canker of Dogwood and Madrona. Mycologia 35: 207-221.
3528. Sturgeon, R.V. Jr. 1982. Peanut disease loss estimates for major peanut producing states in the United States for 1981. Proc. Amer. Peanut Res. Educ. Soc. 14: 86-87.
3529. Stuteville, D.L., W.G. Willis, J.A.H. Christensen, and T. Sim. 1986. First report of *Verticillium* wilt of alfalfa in Kansas. Pl. Dis. 70: 475.
3530. Subirats, F.J., and R.L. Self. 1971. *Gloeosporium* canker and dieback of Burford holly in Alabama. Pl. Dis. Reporter 55: 424.
3531. Subirats, F.J., and R.L. Self. 1971. A new petal blight of camellias incited by *Pestalotia* sp. Pl. Dis. Reporter 55: 697-700.
3532. Subirats, F.J., and R.L. Self. 1972. *Fusarium* blight of centipede grass. Pl. Dis. Reporter 56: 42-44.
3533. Subirats, F.J., and R.L. Self. 1972. A flower rot disease of camellias. Pl. Dis. Reporter 56: 919-920.
3534. Subirats, F.J., and R.L. Self. 1973. New disease complex on camellias. Phytopathology 63: 449.
3535. Subirats, F.J., and R.L. Self. 1973. *Santolina* blight caused by *Alternaria tenuissima*. Pl. Dis. Reporter 57: 1072.
3536. Suit, R.F., and E.P. DuCharme. 1971. Cause and control of pink pitting on grapefruit. Pl. Dis. Reporter 55: 923-926.
3537. Summers, T.E. 1952. Downey mildew (*Sclerospora macrospora*) of oats in the south. Pl. Dis. Reporter 36: 347-348.
3538. Summers, T.E., C.R. Adair, and T.R. Stanton. 1951. Downy mildew of oats found in Arkansas, Mississippi, and Georgia in 1951. Pl. Dis. Reporter 35: 510.
3539. Summers, T.E., and D.H. Bowman. 1953. The cereal rusts and other diseases of small grains in Mississippi. Pl. Dis. Reporter 37: 142-147.
3540. Sumner, D.R. 1971. Effect of leafspots on quality of turnip greens grown for processing. Pl. Dis. Reporter 55: 540-543.
3541. Sumner, D.R., and D.K. Bell. 1982. Root diseases induced in corn by *Rhizoctonia solani* and *Rhizoctonia zeae*. Phytopathology 72: 86-91.
3542. Sumner, D.R., and J.D. Gay. 1984. Basal rot of onion caused by *Fusarium oxysporium* f. sp. *cepae* in Georgia. Pl. Dis. 68: 450.
3543. Sumstine, D.R. 1949. The Albert Commons collection of fungi in the herbarium of the Academy of Natural Sciences in Philadelphia. Mycologia 41: 11-23.
3544. Susuri, L., D.J. Hagedorn, and R.E. Rand. 1982. *Alternaria* blight of pea. Pl. Dis. 66: 328-330.
3545. Sutton, B.C. 1963. Coelomycetes. II *Neobarclaya, Mycohypallage, Bleptosporium,* and *Cryptostictis*. Mycol. Pap. 88: 1-50.
3546. Sutton, B.C. 1973. Hyphomycetes from Manitoba and Saskatchewan, Canada. Mycol. Pap. 132: 1-143.
3547. Sutton, B.C. 1975. *Coelomycetes.* V. *Coryneum*. Mycol. Pap. 138: 1-224.
3548. Sutton, B.C. 1978. New and interesting Hyphomycetes from Tampa, Florida. Mycologia 70: 784-801.
3549. Sutton, B.C. 1980. The Coelomycetes. Commonwealth Mycological Institute 696 p.
3550. Sutton, B.C. 1980. The Coelomycetes. Commonweath Mycological Institute. Kew. 696 p.
3551. Sutton, B.C. 1980. *Septotinia* and *Septotis*. Mycologia 72: 208-213.
3552. Sutton, B.C., and A. Funk. 1975. Conidial states of some *Pragmopora* and *Tympanis* species. Canad. J. Bot. 53: 521-526.
3553. Sutton, B.C., and C.S. Hodges. 1978. *Eucalyptus* microfungi. *Chaetendophragmiopsis* gen. nov. and other hyphomycetes. Nova Hedwigia 29: 593-607.
3554. Sutton, B.C., and C.S. Hodges Jr. 1983. Hawaiian forest fungi. III. A new species, *Gloeocoryneum hawaiiense*, on *Acacia koa*. Mycologia 75: 280-284.
3555. Sutton, B.C., and F.G Pollack. 1974. Microfungi on *Cerococarpus*. Mycopathol. Mycologia Appl. 52: 331-351.
3556. Sutton, B.C., and F.G. Pollack. 1973. *Gloeosporium cercocarpi* and *Sphaceloma cercocarpi*. Mycologia 65: 1125-1134.
3557. Sutton, B.C., and F.G. Pollack. 1977. *Tunicago uniolae* gen. et sp. nov., a pycnidial fungus with unusual conidia. Canad. J. Bot. 55: 326-330.
3558. Sutton, T.B., E.M. Brown, and D.J. Hawthorne. 1987. Biology and epidemiology of *Mycosphaerella pomi*, cause of Brooks fruit spot of apple. Phytopathology 77: 431-437.
3559. Sutton, T.B., D.W. Hayne, W.T., Jr. Sullivan Jr., J.F. Nardacci, and D.E. Klimstra. 1981. Causes of apple tree death in Henderson County, North Carolina. Pl. Dis. 65: 330-332.
3560. Sutton, T.B., and A.L. Jones. 1979. Analysis of factors affecting dispersal of *Podosphaera leucotricha* conidia. Phytopathology 69: 380-383.
3561. Sutton, T.B., and W.W. Shane. 1983. Epidemiology of the perfect stage of *Glomerella cingulata* on apples. Phytopathology 73: 1179-1183.
3562. Swank, G. Jr. 1951. Tomato rot due to a strain of *Glomerella cingulata*. Pl. Dis. Reporter 35: 329.
3563. Swank, G. Jr. 1951. *Alternaria* leaf spot and dieback of snap bean: a new disease in central Florida. Pl. Dis. Reporter 35: 330-332.
3564. Swift, M.E. 1932. *Phoma conidiogena* on box[wood]. Mycologia 24: 199-206.
3565. Sydow, H., and P. Sydow. 1907. Novae fungorum species. IV. Ann. Mycol. 5: 338-340.
3566. Sykes, E.E., and D. Porter. 1973. Nutritional studies of *Labyrinthula* sp. Mycologia 65: 1302-1311.
3567. Sylvia, D.M., and T.A. Tattar. 1978. Electrical resistance properties of tree tissues in cankers incited by *Endothia parasitica* and *Nectria galligena*. Canad. J. Forest Res. 8: 162-167.
3568. Taber, R.A., and R.E. Pettit. 1975. Occurrence of thermophilic microorganisms in peanuts and peanut soil. Mycologia 67: 157-161.
3569. Taber, R.A., W.A. Taber, R.E. Pettit, and C.W. Horne. 1978. *Beniowskia* blight on bristlegrass in Texas. Pl. Dis. Reporter 62: 157-160.
3570. Tachibana, H. 1968. *Rhizoctonia solani* root rot epidemic of soybeans in central Iowa 1967. Pl. Dis. Reporter 52: 613-614.
3571. Tachibana, H., A.H. Epstein, R.F. Nyvall, and R.A. Musselman. 1975. *Phytophthora* root rot of soybean in Iowa: observations, trends, and control. Pl. Dis. Reporter 59: 994-998.
3572. Tainter, F.H. 1986. Growth, sporulation, and mucilage production by *Ceratocystis fagacearum* at high temperatures. Pl. Dis. 70: 339-342.
3573. Tainter, F.H., and W.D. Gubler. 1974. Effect of invasion by *Hypoxylon* and other microorganisms on carbohydrate reserves of oak-wilted trees. Forest Sci. 20: 337-342.
3574. Takacs, D.J., J.A. Payne, W.L. Tedders, E.J. Wehunt, and B.D. Horton. 1970. *Clitocybe* root rot of pecans. Pl. Dis. Reporter 54: 620-621.
3575. Tattar, T.A., W.C. Shortle, W.C., and A.E. Rich. 1971. Sequence of microorganisms and changes in constituents associated with discoloration and decay of sugar maples infected with *Fomes connatus*. Phytopathology 61: 556-558.
3576. Tattar, T.A., H.B. Teillon, and E.B. Walker. 1973. *Scleroderris* canker on red and scots pine in Vermont. Pl. Dis. Reporter 57: 338.
3577. Taylor, C.F., and R.D. Berger. 1966. Apple fruit rot in Pennsylvania caused by *Phialophora malorum*. Pl. Dis. Reporter 50: 701.
3578. Taylor, G.S. 1969. A black mildew, of the genus *Diporotheca*, found on roots of *Nicotiana tabacum* in Connecticut. Pl. Dis. Reporter 53: 85-86.
3579. Taylor, G.S., and R.E.B. Moore. 1979. A canker of red maples associated with oviposition by the narrow-winged tree cricket. Phytopathology 69: 236-239.
3580. Taylor, J. 1953. Black pox on apples in Georgia. Pl. Dis. Reporter 37: 446-447.
3581. Taylor, J. 1959. The distinctive nature of some apple disease conditions in Georgia. Pl. Dis. Reporter 43: 654-657.
3582. Taylor, J. 1963. An apple leaf spot caused by *Helminthosporium papulosum*. Pl. Dis. Reporter 47: 1105-1106.
3583. Taylor, N.L., and L. Henson. 1955. Crown wart of alfalfa in Kentucky. Pl. Dis. Reporter 39: 697.

3584. Tegethoff, A.C. 1973. Known distribution of *Fomes annosus* in the intermountain region. Pl. Dis. Reporter 57: 407-410.
3585. Tehon, L.R. 1937. Notes on the parasitic fungi of Illinois - IV. Mycologia 29: 434-446.
3586. Tehon, L.R. 1939. New species and taxonomic changes in the Hypodermataceae. Mycologia 31: 674-692.
3587. Tehon, L.R. 1948. Notes on the parasitic fungi of Illinois. Mycologia 40: 314-327.
3588. Tehon, L.R., and G.H. Boewe. 1939. Charcoal rot in Illinois. Pl. Dis. Reporter 23: 312-321.
3589. Tehon, L.R., and E. Daniels. 1925. Notes on the parasitic fungi of Illinois - II. Mycologia 17: 240-249.
3590. Tehon, L.R., and E.Y. Daniels. 1927. Notes on the parasitic fungi of Illinois. III. Mycologia 19: 110-129.
3591. Tehon, L.R., and G.L. Stout. 1929. Notes on the parasitic fungi of Illinois - IV. Mycologia 21: 180-196.
3592. Teixeira, A.R., and D.P. Rogers. 1955. *Aporpium*, a polyporoid genus of the Tremellaceae. Mycologia 47: 408-415.
3593. Templeton, G.E. 1962. A list of the smut fungi of Arkansas. Proc. Arkansas Acad. Sci. 16: 17-18.
3594. Templeton, G.E. 1984. University of Arkansas. Plant Pathology Herbarium.
3595. Templeton, G.E., T.H. Johnston, and S.E. Henry. 1961. Downy mildew of rice in Arkansas. Pl. Dis. Reporter 45: 95.
3596. Teunisson, D.J., H.H. Hall, and L.J. Wickerham. 1960. *Hansenula angusta*, an excellent species for demonstration of the coexistence of haploid and diploid cells in a homothallic yeast. Mycologia 52: 184-188.
3597. Tharp, B.C. 1917. Texas parasitic fungi. Mycologia 9: 105-124.
3598. Thiers, H.D. 1959. The agaric flora of Texas. III. New taxa of brown- and black-spored agarics. Mycologia 51: 529-540.
3599. Thies, W.G. 1979. Frequency of pine stump colonization by *Fomes annosus* on the Fremont National Forest. Pl. Dis. Reporter 63: 542-545.
3600. Thies, W.G. 1983. Determination of growth reduction in Douglas-fir infected by *Phellinus weirii*. Forest Sci. 29: 305-315.
3601. Thies, W.G., and E.E. Nelson. 1982. Control of *Phellinus weirii* in Douglas-fir stumps by the fumigants chloropicrin, allyl alcohol, Vapam and Vorlex. Canad. J. Forest Res. 12: 528-532.
3602. Thirumalachar, M.J., M.D. Whitehead, and M.J. O'Brien. 1967. Studies in the genus *Tolyposporella*. Mycologia 59: 389-396.
3603. Thomas, C.A. 1951. Anthracnose of *Digitalis*. Phytopathology 41: 997-1000.
3604. Thomas, C.A. 1959. *Fusarium* infection of safflower roots on acid soil. Pl. Dis. Reporter 43: 416-417.
3605. Thomas, C.A., and G.C. Papavizas. 1965. Susceptibility of sesame and castorbean to *Thielaviopsis basicola*. Pl. Dis. Reporter 49: 256.
3606. Thomas, C.A., and P.R. Thomison. 1986. First report of race 1 of *Phytophthora megasperma* f. sp. *glycinea* on soybean in Delaware. Pl. Dis. 70: 78.
3607. Thomas, C.E. 1977. Influence of dew on downy mildew of cantaloups in south Texas. Phytopathology 67: 1368-1369.
3608. Thomas, C.S., and J.H. Hart. 1986. Site factors associated with *Nectria* canker on black walnut in Michigan. Pl. Dis. 70: 1117-1121.
3609. Thomas, C.S., J.H. Hart, and C.E. Cress. 1984. Severity of *Endocronartium harknessii* in two provenance stands of *Pinus ponderosa* in Michigan. Pl. Dis. 68: 681-683.
3610. Thomas, J.E., and C.A. Bower. 1962. The occurrence of Dutch elm disease in Oklahoma. Pl. Dis. Reporter 46: 463.
3611. Thomas, W.D. Jr. 1950. New shade tree diseases in Colorado. Pl. Dis. Reporter 34: 83.
3612. Thompkins, C.M. 1959. Leaf rot of poinsettia cuttings caused by *Rhizoctonia solani* and its control. Pl. Dis. Reporter 43: 1036-1037.
3613. Thompson, D.C., and S.F. Jenkins. 1985. Pictorial assessment key to determine fungicide concentrations that control anthracnose development on cucumber cultivars with varying resistance levels. Pl. Dis. 69: 833-836.
3614. Thompson, G.E. 1951. *Microsphaera diffusa* on blue lupine. Pl. Dis. Reporter 35: 221.
3615. Thompson, G.E. 1956. A rust disease of the smoke-tree new to North America. Pl. Dis. Reporter 40: 1015-1016.
3616. Thompson, G.E. 1960. Rust of the smoke-tree. Mycologia 52: 321-326.
3617. Thompson, G.E. 1963. Decay of oaks caused by *Hypoxylon atropunctatum*. Pl. Dis. Reporter 47: 202-205.
3618. Thompson, G.E., and B.R. Murray. 1961. Distribution and control of *Cercospora thujina* on Arizona cypress in Georgia. Pl. Dis. Reporter 45: 731-733.
3619. Thompson, J.H., and T.A. Tattar. 1973. *Rhizina undulata* associated with disease of 80-year-old red spruce in Vermont. Pl. Dis. Reporter 57: 394-396.
3620. Thompson, S.V., and C.R. Bull. 1983. Utah Plant Diseases. Cooperative Extension Service, Utah State Univ., Plant Dis. Control 13: 1-6.
3621. Thompson, S.V., and M.C. Mehdy. 1978. Occurrence of *Aspergillus flavus* in pistachio nuts prior to harvest. Phytopathology 68: 1112-1114.
3622. Thomson, S.V., and W.B. Jones. 1981. An epiphytotic of *Leveillula taurica* on tomatoes in Utah. Pl. Dis. 65: 518-519.
3623. Thorn, R.G. 1986. The "*Pleurotus silvanus*" complex. Mycotaxon 25: 27-66.
3624. Thorn, W.A., and G.A. Zentmyer. 1954. Hosts of *Phytophthora cinnamomi* Rands. Pl. Dis. Reporter 38: 47-52.
3625. Thyr, B.D., and C.G. Shaw. 1964. Identity of the fungus causing red band disease on pines. Mycologia 56: 103-109.
3626. Thyr, B.D., T.K. Stadniuk, and J. Gallian. 1977. *Phoma medicaginis* and *Sclerotinia trifoliorum* on alfalfa in western Nevada. Pl. Dis. Reporter 61: 166-167.
3627. Tidwell, T.E., K.L. Kosta, and J.C. Henderson. 1984. Root rot of western swordfern caused by *Phytophthora cinnamomi* in California. Pl. Dis. 68: 536.
3628. Tiffany, L.H. 1955. Fungus leaf spots of brome grass in Iowa. Iowa State Coll. J. Sci. 30: 21-32.
3629. Tiffany, L.H., and J.C. Gilman. 1965. Iowa Ascomycetes IV. Diatrypaceae. Iowa State Coll. J. Sci. 40: 121-161.
3630. Tiffany, L.H., and J.H. Mathre. 1961. A new species of *Elsinoe* on *Panicum virgatum*. Mycologia 53: 600-604.
3631. Tillman, R.W., and G.A. Bean. 1970. Tentative identification and influence of sterols and dimethyl sulfox ide (DMSO) on the growth and survival of *Fusarium roseum*. Mycologia 62: 428-436.
3632. Timmer, L.W. 1982. Host range and host colonization, temperature effects, and dispersal of *Fusarium oxysporum* f. sp. *citri*. Phytopathology 72: 698-702.
3633. Timmer, L.W., S.M. Garnsey, G.R. Grimm, N.E. El-Gholl, and C.L. Schoulties. 1979. Wilt and dieback of Mexican lime caused by *Fusarium oxysporum*. Phytopathology 69: 730-734.
3634. Timonin, M.I., and R.L. Self. 1955. *Cylindrocladium scoparium* Morgan on azaleas and other ornamentals. Pl. Dis. Reporter 39: 860-863.
3635. Tims, E.C. 1950. Onion and shallot downy mildew in Louisiana. Pl. Dis. Reporter 34: 380-382.
3636. Tims, E.C. 1953. Pink root of shallot, *Allium ascalonicum*. Pl. Dis. Reporter 37: 533-537.
3637. Tims, E.C. 1963. Recurrence of onion white rot in Louisiana. Pl. Dis. Reporter 47: 72.
3638. Tims, E.C. 1963. An unusual combination of pests on shallot. Pl. Dis. Reporter 47: 72.
3639. Tims, E.C. 1963. *Corticium salmonicolor* in the United States. Pl. Dis. Reporter 47: 1055-1059.
3640. Tims, E.C., P.J. Mills, and B. Exner. 1954. Thread-blight (*Pellicularia koleroga*) in Louisiana. Pl. Dis. Reporter 38: 634-637.
3641. Tims, E.C., and L.S. Olive. 1948. Two interesting leaf spots of fig. Phytopathology 38: 707-715.
3642. Tippett, J.T., and A.L. Shigo. 1980. Barrier zone anatomy in red pine roots invaded by *Heterobasidion annosum*. Canad. J. Forest Res. 10: 224-232.
3643. Tippett, J.T., and A.L. Shigo. 1981. Barrier to decay in conifer roots. Eur. J. Forest Pathol. 11: 51-59.
3644. Tisdale, W.B. 1952. What's new in plant pathology. Pl. Dis. Reporter 36: 208-210.
3645. Tisserat, N. 1987. Stem canker of black walnut caused by *Fusarium solani* in Kansas. Pl. Dis. 71: 557.
3646. Tisserat, N., and J.E. Kuntz. 1984. Butternut canker: Development on individual trees and increase withina plantation. Pl. Dis. 68: 613-616.
3647. Todd, E.H. 1957. Rust of St. Augustine grass in Florida. Pl. Dis. Reporter 41: 650.
3648. Todd, E.H. 1960. *Elsinoe* disease of sugarcane in Florida. Pl. Dis. Reporter 44: 153.
3649. Todd, E.H. 1962. Target blotch of sugarcane in Florida. Pl. Dis.

Reporter 46: 486.
3650. Todhunter, M.N., and W.F. Beineke. 1984. Effect of anthracnose on growth of grafted black walnut. Pl. Dis. 68: 203-204.
3651. Toler, R.W., S.S. Thompson, and J.M. Barber. 1963. Cowpea (southern pea) diseases in Georgia, 1961-1962. Pl. Dis. Reporter 47: 746-747.
3652. Tompkins, C.M., and H.N. Hansen. 1950. Flower blight of *Stephanotis floribunda*, caused by *Botrytis elliptica* and its control. Phytopathology 40: 780-781.
3653. Tompkins, C.M., and W.C. Snyder. 1972. *Cyclamen* wilt in California and its control. Pl. Dis. Reporter 56: 493-497.
3654. Tompkins, C.M., and C.M. Tucker. 1950. Rhizome rot of white calla caused by *Phytophthora erythroseptica*. Phytopathology 40: 712-714.
3655. Tompkins, C.M., and C.M. Tucker. 1951. Stem and root rot of *Daphne odora* caused by *Phytophthora parasitica*. Phytopathology 41: 654-656.
3656. Tontyaporn, S., and G.E. Templeton. 1971. Rust on smoke-tree in Arkansas. Pl. Dis. Reporter 55: 226-227.
3657. Toole, E.R. 1951. Bleeding canker of maple in North Carolina and Tennessee. Pl. Dis. Reporter 35: 119.
3658. Toole, E.R. 1955. Red stain of boxelder. Pl. Dis. Reporter 39: 66-67.
3659. Toole, E.R. 1955. Twig blight and canker of red mulberry. Pl. Dis. Reporter 39: 657.
3660. Toole, E.R. 1957. Twig canker of sweetgum. Pl. Dis. Reporter 41: 808-809.
3661. Toole, E.R. 1960. Root rot of white oak in Arkansas. Pl. Dis. Reporter 44: 783.
3662. Toole, E.R. 1961. New sycamore canker. Pl. Dis. Reporter 45: 78.
3663. Toole, E.R. 1963. Sweetgum lesion caused by *Botryosphaeria ribis*. Pl. Dis. Reporter 47: 229-231.
3664. Toole, E.R., and J.S. Boyce Jr. 1952. *Fomes annosus* on Atlantic white cedar. Pl. Dis. Reporter 36: 330.
3665. Toole, E.R., and T.H. Filer. 1965. *Colletotrichum gloeosporioides* on dogwood. Pl. Dis. Reporter 49: 351.
3666. Toole, E.R., and B.J. Huckenpahler. 1954. Yellow-poplar dieback. Pl. Dis. Reporter 38: 786-788.
3667. Toole, E.R., and P.C. Lightle. 1960. Status of persimmon wilt, 1959. Pl. Dis. Reporter 44: 45.
3668. Torgeson, D.C. 1952. Observations of *Verticillium* wilt of geranium in Oregon. Pl. Dis. Reporter 36: 51.
3669. Tortolero, O., and L. Sequeira. 1978. A vascular wilt and leaf blight disease of lettuce in Wisconsin caused by a new strain of *Pythium tracheiphilum*. Pl. Dis. Reporter 62: 616-620.
3670. Towers, B. 1977. The occurrence of *Verticicladiella procera* in Pennsylvania: 1976. Pl. Dis. Reporter 61: 477.
3671. Towers, B., B.A. Ammerman, J.F. Connor, and A.W. Reinke. 1974. 1973 Beech bark disease status in Pennsylvania. Pl. Dis. Reporter 58: 718-719.
3672. Towers, B., E.M. Blumenthal, R. Acciavatti, M. Mielke, C. Haynes, and R. Dekker. 1986. Expanded range of *Cytospora* canker of black cherry. Pl. Dis. 70: 259.
3673. Towers, B., L. Forer, V. Wagner, W. Merrill, and N. Wenner. 1984. Needle blight caused by *Dothistroma pini* in Austrian and red pines in Pennsylvania. Pl. Dis. 68: 450.
3674. Towers, B., M. Matuszewski, and J. Schoenholz. 1971. Beech bark disease in Pennsylvania. Pl. Dis. Reporter 55: 591-593.
3675. Tracy, S.M., and F.S. Earle. 1895. New species of parasitic fungi. Bull. Torrey Bot. Club 22: 174-179.
3676. Traquair, J.A., J.B. Lebeau, J.E. Moffat, and M. Kokko. 1983. Northern distribution of LTB snow mold in Canada. Canad. Plant Dis. Surv. 63: 1-2.
3677. Trese, A.T., C.L. Burton, and D.C. Ramsdell. 1980. *Eutypa armeniacae* in Michigan vineyards: ascospore production and survival, host infection, and fungal growth at low temperatures. Phytopathology 70: 788-793.
3678. Treshow, M., and O.S. Cannon. 1961. Powdery mildew of potato in Utah. Pl. Dis. Reporter 45: 354-355.
3679. Trione, E.J., and L.F. Roth. 1957. Aerial infection of *Chamaecyparis* by *Phytophthora lateralis*. Pl. Dis. Reporter 41: 211-215.
3680. Trolinger, J.C., E.S. Elliott, and R.J. Young. 1978. Host range of *Cristulariella pyramidalis*. Pl. Dis. Reporter 62: 710-714.
3681. Troutman, J.L., and J.C. Matejka. 1972. Downy mildew of small grains and sorghum in Arizona. Pl. Dis. Reporter 56: 773-774.
3682. True, R.P., F.W. Craig, and H.L. Barnett. 1951. Oak wilt found in West Virginia. Pl. Dis. Reporter 35: 382.
3683. Trujillo, E.E. 1962. Occurrence of the bean root rot fungus, *Fusarium solani* f. *phaseoli*, in Hawaii. Pl. Dis. Reporter 46: 819.
3684. Trujillo, E.E., A.M. Alvarez, and D.N. Swindale. 1975. *Phytophthora* leaf spot of ti. Pl. Dis. Reporter 59: 452-453.
3685. Trujillo, E.E., and S. Goto. 1963. The occurrence of Sigatoka disease of bananas in the Hawaiian Islands. Pl. Dis. Reporter 47: 362-363.
3686. Trujillo, E.E., F.M. Latterell, and A.E. Rossi. 1986. *Colletotrichum gloeosporioides*, a possible biological control agent for *Clidemia hirta* in Hawaiian forests. Pl. Dis. 70: 974-976.
3687. Trujillo, E.E., and F.P. Obrero. 1969. Anthracnose of papaya leaves caused by *Colletotrichum gloeosporioides*. Pl. Dis. Reporter 53: 323-325.
3688. Tsao, P.H., and S.D., van Gundy van. 1962. *Thielaviopsis basicola* as a citrus root pathogen. Phytopathology 52: 781-786.
3689. Tu, C.C., D.A. Roberts, and J.W. Kimbrough. 1969. Hyphal fusion, nuclear condition, and perfect stages of three species of *Rhizoctonia*. Mycologia 61: 775-783.
3690. Tu, J.C. 1985. Pink pod rot of bean caused by *Trichothecium roseum*. Canad. J. Plant Pathol. 7: 55-57.
3691. Tucker, C.M., and J.A. Milbrath. 1942. Root rot of *Chamaecyparis* caused by a species of *Phytophthora*. Mycologia 34: 94-103.
3692. Tucker, D.H., Jr., and N.E. McGlohon. 1985. First report of *Monilinia* leaf and shoot blight of peach in Georgia. Pl. Dis. 69: 811.
3693. Tuite, J. 1961. Fungi isolated from unstored corn seed in Indiana in 1956-1958. Pl. Dis. Reporter 45: 212-215.
3694. Tuite, J., and R.W. Caldwell. 1971. Infection of corn seed with *Helminthosporium maydis* and other fungi in 1970. Pl. Dis. Reporter 55: 387-389.
3695. Tullis, E.C., and A.G. Johnson. 1952. Synonymy of *Tilletia horrida* and *Neovossia barclayana*. Mycologia 44: 773-788.
3696. Turner, S.K., A. Kwiatkowski, P.K. Fay, and D.C. Sands. 1986. Factors affecting germination of teliospores of *Puccinia obtegens*. Pl. Dis. 70: 390-391.
3697. Tweedy, B.G., and D. Powell. 1962. Cork rot of apples and its causal organism, a pathogenic strain of *Alternaria mali*. Phytopathology 52: 1073-1079.
3698. Tweedy, B., and D. Powell. 1958. Charcoal rot on strawberry in Illinois. Pl. Dis. Reporter 42: 107.
3699. Tyler, D.J., and L.E. Dickens. 1957. *Cephalosporium* leaf stripe disease of winter wheat. Pl. Dis. Reporter 41: 384.
3700. Uchida, J.Y., and M. Aragaki. 1978. Leaf and stem blight of ivy caused by *Phytophthora palmivora*. Pl. Dis. Reporter 62: 699-701.
3701. Uchida, J.Y., and M. Aragaki. 1979. *Ceratocystis* blight of *Syngonium podophyllum*. Pl. Dis. Reporter 63: 1053-1056.
3702. Uchida, J.Y., and M. Aragaki. 1982. Acremonium leaf spot of *Syngonium*: nomenclature of the causal organism and chemical control. Pl. Dis. 66: 421-423.
3703. Uchida, J.Y., M. Aragaki, and M.A. Yoshimura. 1984. *Alternaria* leaf spots of *Brassaia actinophylla*, *Dizygotheca elegantissima*, and *Tupidanthus calyptratus*. Pl. Dis. 68: 447-449.
3704. Uecker, F.A., and J.M. Krupinsky. 1982. *Septoria spraguei*, a new pathogen on leaves of *Elymus*. Mycotaxon 14: 273-279.
3705. Uecker, F.A., and M.M. Kulik. 1986. *Pseudorobillarda sojae*, a new pycnidial coelomycete from soybean stems. Mycologia 78: 449-453.
3706. Ueki, R., and C.W. Smith. 1973. The genus *Crepidotus* in Hawaii. Canad. J. Bot. 51: 1251-1254.
3707. Ullstrup, A.J. 1950. Corn diseases in Indiana in 1949. Pl. Dis. Reporter 34: 98-99.
3708. Ullstrup, A.J. 1955. Crazy top of some wild grasses and the occurrence of the sporangial stage of the pathogen. Pl. Dis. Reporter 39: 839-841.
3709. Ullstrup, A.J. 1964. A nonpigmented form of *Gibberella roseum* forma *cerealis* on corn in Indiana. Mycologia 56: 110-113.
3710. Ullstrup, A.J., and F.A. Laviolette. 1959. Diseases of corn and of sorghum species in Indiana in 1958. Pl. Dis. Reporter 43: 334-336.
3711. Ullstrup, A.J., E.B. Smalley, G.L. Worf, and R.W. Ahrens. 1969. Eyespot: a serious disease of corn in the USA. Phytopathology 59: 1054.

3712. Ullstrup, A.J., and M.H. Sun. 1969. The prevalence of crazy top of corn in 1968. Pl. Dis. Reporter 53: 246-250.
3713. Underwood, L.M. 1897. Some new fungi, chiefly from Alabama. Bull. Torrey Bot. Club 24: 81-86.
3714. Upadhyay, H.P. 1981. A monograph of *Ceratocystis* and *Ceratocystiopsis*. Univ. Georgia Press, Athens. 176 p.
3715. Urban, Z. 1973. The autoecious species of *Puccinia* on Vernonieae in North America. Acta Univ. Carol. - Biol. 1971: 1-84.
3716. Uyemoto, J.K., W.J. Moller, and A.C. Goheen. 1976. Isolation of *Eutypa armeniacae* from grapevines in New York and inoculation to apricot. Pl. Dis. Reporter 60: 684-686.
3717. VPI. Virginia Polytechnic Institute and State University. Blacksburg, VA. Herbarium.
3718. Valleau, W.D. 1933. Anthracnoses of legumes in Kentucky. Pl. Dis. Reporter 17: 44.
3719. Valleau, W.D. 1950. Some diseases observed on legumes and grasses in Kentucky. Pl. Dis. Reporter 34: 197-198.
3720. Valleau, W.D. 1950. Tobacco plant bed diseases in Kentucky, 1950. Pl. Dis. Reporter 34: 210-211.
3721. Valleau, W.D. 1950. Some diseases of forage legumes in Kentucky. Pl. Dis. Reporter 34: 213-214.
3722. Valverde, A.R., and G.E. Templeton. 1984. Leaf gall of *Torilis japonica* caused by *Protomyces macrosporus* in Arkansas. Pl. Dis. 68: 716-717.
3723. Vanev, S.G. 1982. *Discosia baarnensis* sp. nov. Trans. Brit. Mycol. Soc. 79: 569-571.
3724. Vanky, K. 1981. Two new genera of Ustilaginales: *Nannfeldtiomyces* and *Pseudodoassansia*, and a survey of allied genera. Sydowia 34: 167-178.
3725. Vann, S.R., and R.A. Taber. 1985. *Annellophora* leaf spot of date palm in Texas. Pl. Dis. 69: 903-904.
3726. Vanterpool, T.C., and R. Sprague. 1942. *Pythium arrhenomanes* on cereals and grasses in the Northern Great Plains. Phytopathology 32: 327-328.
3727. Vargo, R.H., and J.S. Baumer. 1986. *Fusarium sporotrichioides* as a pathogen of spring wheat. Pl. Dis. 70: 629-631.
3728. Vartanian, V.G., and R.M. Endo. 1985. Survival of *Phytophthora infestans* in seeds extracted from infected tomato fruits. Phytopathology 75: 375-378.
3729. Vartanian, V.G., and R.M. Endo. 1985. Overwintering hosts, compatibility types, and races of *Phytophthora infestans* on tomato in southern California. Pl. Dis. 69: 516-519.
3730. Vaughn, E.K., F. Johnson, R.E. Fitzpatrick, and R. Stace-Smith. 1951. Diseases observed on bramble fruits in the Pacific Northwest. Pl. Dis. Reporter 35: 34-37.
3731. Venette, J.R., and R.W. Stack. 1987. First report of rust (*Uromyces ciceris-arientini*) on garbanzo (chickpea) in the United States. Pl. Dis. 71: 101.
3732. Verrall, A.F. 1939. Relative importance and seasonal prevalence of wood-staining fungi in the southern United States. Phytopathology 29: 1031-1051.
3733. Verrall, A.F., E.R. Toole, and P.C. Lightle. 1959. Oak wilt in Oklahoma and Arkansas. Pl. Dis. Reporter 43: 1288.
3734. Vest, G., J.L. Peterson, and J.R. Justin. 1969. *Phytophthora* rot of soybeans in New Jersey. Pl. Dis. Reporter 53: 556-557.
3735. Vestal, E.F., and G. Semeniuk. 1951. *Puccinia sorghi* and *Helminthosporium turcicum* on corn in Iowa during 1950. Pl. Dis. Reporter 35: 212-214.
3736. Villalón, B. 1982. Sugarcane smut in lower Rio Grande Valley of south Texas. Pl. Dis. 66: 605-606.
3737. Vogler, D.R., B.B. Kinloch, F.W. Cobb Jr., and T.L. Popenuck. 1987. Isozyme patterns in *Endocronartium harknessii* from several pine hosts and stands in California. Phytopathology 77: 1242.
3738. Vogler, D.R., and J. Pronos. 1984. First report of *Elytroderma deformans* infecting coulter pine in California. Pl. Dis. 68: 918.
3739. Volin, R.B., and L. Ramos. 1981. The association of brown root rot resistance with yield components and root weight among tomato selections grown in infested and non-infested soil in Dade County, Florida. Proc. Florida State Hort. Soc. 93: 224-226.
3740. Wadsworth, D.F. 1973. Research on the nature and control of peanut diseases in Oklahoma. Oklahoma Argic. Exp. Sta. Rep. P-683: 1-17.
3741. Wadsworth, D.F. 1979. *Sclerotinia* blight of peanuts in Oklahoma and occurrence of the sex ual stage of the pathogen. Peanut Sci. 6: 77-79.
3742. Wadsworth, D.F., and H.C. Young Jr. 1960. Spring dead spot of bermudagrass. Pl. Dis. Reporter 44: 516-518.
3743. Wagener, W.W., and J.L. Mielke. 1961. A staining-fungus root disease of ponderosa, jeffrey, and pinyon pines. Pl. Dis. Reporter 45: 831-835.
3744. Waggoner, P.E., and G.S. Taylor. 1955. Tobacco blue mold epiphytotics in the field. Pl. Dis. Reporter 39: 79-85.
3745. Wagner, D.T. 1969. A monocentric, holocarpic fungus in *Lemna minor* L. Nova Hedwigia 18: 203-208.
3746. Wakefield, E.M. 1960. Some species of *Tomentella* from North America. Mycologia 52: 919-933.
3747. Wakeley, P.C. 1970. Thirty-year effects of uncontrolled brown spot on planted longleaf pine. Forest Sci. 16: 197-202.
3748. Walker, E.A. 1944. *Fusarium* root rot of parsley in New Jersey. Pl. Dis. Reporter 28: 807.
3749. Walker, E.A., C.T. Rogerson, and A.E. Jenkins. 1952. Additional collections of plantain scab and violet scab from several North Central states. Pl. Dis. Reporter 36: 331-332.
3750. Walker, H.L. 1981. *Fusarium lateritium*: A pathogen of spurred anoda (*Anoda cristata*), prickly sida (*Sida spinosa*), and velvetleaf (*Abutilon theophras ti*). Weed Sci. 29: 629-631.
3751. Walker, H.L. 1982. Seedling blight of sicklepod caused by *Alternaria cassiae*. Pl. Dis. 66: 426-428.
3752. Walker, H.L., and C.D. Boyette. 1986. Influence of sequential dew periods on biocontrol of sicklepod (*Cassia obtusifolia*) by *Alternaria cassiae*. Pl. Dis. 70: 962-963.
3753. Walker, H.L., and J.C. White. 1979. *Curvularia cymbopogonis*, a pathogen of itchgrass (*Rottboellia exaltata*) in southern Louisiana. Pl. Dis. Reporter 63: 642-644.
3754. Walker, J. 1980. *Gaeumannomyces*, *Linocarpon*, *Ophiobolus* and several other genera of scolecospored ascomycetes and *Phialophora* conidial states, with a note on hyphopodia. Mycotaxon 11: 1-129.
3755. Walker, J.T. 1971. *Dothiorella*, associated with cankers on Katsura-tree, causes wood discoloration in seedling trees. Pl. Dis. Reporter 55: 893-895.
3756. Walker, J.T., and E.F. Earhart. 1962. Medlar is host for *Gymnosporangium clavipes*. Pl. Dis. Reporter 46: 293.
3757. Wall, G.C., L.K. Mughogho, R.A. Frederiksen, and G.N. Odvody. 1987. Foliar disease of *Sorghum* species caused by *Cercospora fusimaculans*. Pl. Dis. 71: 759-760.
3758. Walla, J.A. 1979. *Diplodia pinea* found in North Dakota. Pl. Dis. Reporter 63: 464.
3759. Walla, J.A. 1984. Incidence of *Phellinus punctatus* on living woody plants in North Dakota. Pl. Dis. 68: 252-253.
3760. Walla, J.A. 1984. Needle blight on spruce in North Dakota caused by *Lirula macrospora*. Pl. Dis. 68: 1016.
3761. Walla, J.A., and J.W. Riffle. 1981. *Fomes fraxinophilus* on green ash in North Dakota windbreaks. Pl. Dis. 65: 669-670.
3762. Walla, J.A., and R.W. Stack. 1979. Western gall rust in North Dakota. Pl. Dis. Reporter 63: 432-433.
3763. Wallace, D.B., W.E. MacHardy, and E.M. Meader. 1976. Control of mummy berry of highbush blueberry in New Hampshire. Pl. Dis. Reporter 60: 97-101.
3764. Wallace, H.N., and P.H. Peacher. 1970. Chinkapin in Louisiana infected by *Endothia parasitica*. Pl. Dis. Reporter 54: 713.
3765. Walters, H.J. 1980. Soybean leaf blight caused by *Cercospora kikuchii*. Pl. Dis. 64: 961-962.
3766. Walters, H.J., and K.F. Martin. 1981. *Phyllosticta sojaecola* on pods of soybeans in Arkansas. Pl. Dis. 65: 161-162.
3767. Walters, J.W., and N.R. Walters. 1977. *Verticicladiella wagenerii* in the southwest. Pl. Dis. Reporter 61: 419.
3768. Walters, R.S., and A.L. Shigo. 1978. Discolouration and decay associated with paraformaldehyde-treated tapholes in sugar maple. Canad. J. Forest Res. 8: 54-60.
3769. Walterscheidt, M.J., and E.P. VanArsdel. 1976. Distribution of fusiform rust on slash and loblolly pines in Texas. Pl. Dis. Reporter 60: 718-720.
3770. Walton, G.S., and S. Rich. 1974. Serious and unusual plant diseases in Connecticut in 1973. Pl. Dis. Reporter 58: 428-429.
3771. Wang, C.J.K., and B.C. Sutton. 1982. New and rare lignicolous Hyphomycetes. Mycologia 74: 489-500.
3772. Wang, C.J.K., and B.C. Sutton. 1984. Two new synnematous genera: *Chalarodendron* and *Paradidymobotryum*. Mycologia 76:

569-572.
3773. Wang, S.-Y., and J.A. Pinckard. 1972. Some biochemical factors associated with the infection of cotton fruit by *Diplodia gossypina*. Phytopathology 62: 460-465.
3774. Wang, S.S., and N.A. Anderson. 1972. A genetic analysis of sporocarp production in *Pleurotus sapidus*. Mycologia 64: 521-528.
3775. Wargo, P.M. 1977. *Armillaria mellea* and *Agrilus bilineatus* and mortality of defoliated oak trees. Forest Sci. 23: 485-492.
3776. Warmelo,, K.T. von, K.T. von, and B.C. Sutton. 1981. Coelomycetes VII. *Stegonsporium*. Mycol. Pap. 145: 1-45.
3777. Warren, H.L., R.L. Nicholson, A.J. Ullstrup, and E.G. Sharvelle. 1973. Observations of *Colletotrichum graminicola* of sweet corn in Indiana. Pl. Dis. Reporter 57: 143-144.
3778. Warren, H.L., D.H. Scott, and R.L. Nicholson. 1974. Occurrence of sorghum downy mildew on maize in Indiana. Pl. Dis. Reporter 58: 430-432.
3779. Warren, T.B., and L.T. Lucas. 1973. Histopathology of *Marasmius* blight of American beachgrass. Phytopathology 63: 725-728.
3780. Waterman, A.M. 1941. Diseases of shade and ornamental trees: Annotated list of specimens received in 1940 at the New Haven Office, Division of Forest Pathology. Pl. Dis. Reporter 25: 181-186.
3781. Waterman, A.M. 1955. The relation of *Valsa kunzei* to cankers on conifers. Phytopathology 45: 686-692.
3782. Waterman, A.M., and E.K. Cash. 1950. Leaf blotch of poplar caused by a new species of *Septotinia*. Mycologia 42: 374-384.
3783. Waterman, A.M., and R.P. Marshall. 1947. A new species of *Cristulariella* associated with a leaf spot of maple. Mycologia 39: 690-698.
3784. Waterworth, H.E., J.J. Higgins, and A.M. Golden. 1967. *Euphorbia lagascae*, a new host of *Sclerotium rolfsii* and a root-knot nematode, *Meloidogyne hapla*. Pl. Dis. Reporter 51: 164-166.
3785. Watkins, J.E., G.N. Odvody, M.G. Boosalis, and J.E. Partridge. 1978. An epidemic of tan spot of wheat in Nebraska. Pl. Dis. Reporter 62: 132-134.
3786. Watling, R., and O.K. Miller, O.K. 1973. A *Coprinus* from white rot of sugar maple. Notes Royal Bot. Gard. Edinburgh 32: 445-447.
3787. Watson, A.J. 1950. Fungi associated with camellia flowers. Pl. Dis. Reporter 34: 186-187.
3788. Watson, A.J., and A.E. Jenkins. 1969. New species of *Elsinoe* and *Sphaceloma*. Mycologia 61: 273-280.
3789. Watson, A.K., and K. Brunetti. 1984. *Puccinia carduorum* on *Carduus tenuiflorus* in California. Pl. Dis. 68: 1003-1005.
3790. Wean, R.E. 1937. The parasitism of *Polyporus schweinitzii* on seedling *Pinus strobus*. Phytopathology 27: 1124-1142.
3791. Weaver, D.J. 1974. A gummosis disease of peach trees caused by *Botryosphaeria dothidea*. Phytopathology 64: 1429-1432.
3792. Weaver, L.O. 1951. The constriction disease of peach in Maryland. Pl. Dis. Reporter 35: 144.
3793. Weaver, L.O., and W.F. Jeffers. 1952. Oak wilt in western Maryland. Pl. Dis. Reporter 36: 28.
3794. Weaver, L.O., and W.J. Zaumeyer. 1956. Angular leaf spot of bean found in Maryland. Pl. Dis. Reporter 40: 1092.
3795. Weaver, M.G., and R.L. Shaffer. 1969. Records of the higher fungi of Minnesota. J. Minnesota Acad. Sci. 35: 123-128.
3796. Webb, R.E., and G. Libman. 1963. An epidemic of *Verticillium* wilt in tomato in the metropolitan area of Washington, D.C. in 1962. Pl. Dis. Reporter 47: 81-82.
3797. Webb, R.R., and S.E. Lindow. 1987. Influence of environment and variation in host susceptibility on a disease of bracken fern caused by *Ascochyta pteridis*. Phytopathology 77: 1144-1147.
3798. Webb, R.S. 1983. Seed capsule abortion and twig dieback of *Eucalyptus camaldulensis* in south Florida induced by *Botryosphaeria ribis*. Pl. Dis. 67: 108-109.
3799. Weber, G.F. 1923. III. *Septoria* disease of rye, barley and certain grasses. Phytopathology 13: 1-23.
3800. Weber, G.F. 1933. Occurrence and pathogenicity of *Nematospora* spp. in Florida. Phytopathology 23: 384-388.
3801. Weber, G.F. 1952. *Elsinoe corni* Jenkins & Bitanc., on flowering dogwood in Gainesville, Florida. Pl. Dis. Reporter 36: 294-295.
3802. Weber, G.F. 1963. *Pellicularia* spp. pathogenic on sorghum and pearl millet in Florida. Pl. Dis. Reporter 47: 654-656.
3803. Weber, G.F., and J.H. Owen. 1951. Watermelon diseases in the Gainsville area of Florida for the 1951 season. Pl. Dis. Reporter 35: 355.
3804. Weber, G.F., and D.A. Roberts. 1951. Silky threadblight of *Elaeagnus pungens* caused by *Rhizoctonia ramicola* n. sp. Phytopathology 41: 615-621.
3805. Webster, R.K., D.H. Hall, C.M. Wick, and R.A. Krause. 1971. Distribution and occurrence of *Sclerotium oryzae* on rice in California. Pl. Dis. Reporter 55: 757-759.
3806. Wehlburg, C. 1970. *Pelargonium* rust (*Puccinia pelargonii-zonalis*) in Florida. Pl. Dis. Reporter 54: 827-828.
3807. Wehmeyer, L.E. 1933. The genus *Diaporthe* Nitschke and its segregates. Univ. Michigan Stud., Sci. Ser. 9: 1-349.
3808. Wehmeyer, L.E. 1936. Cultural life histories of *Melanconis* and *Pseudovalsa*. II. Mycologia 28: 528-541.
3809. Wehmeyer, L.E. 1941. A revision of *Melanconis*, *Pseudovalsa*, *Prosthecium* and *Titania*. Univ. Michigan Stud., Sci. Ser. 14: 1-161.
3810. Wehmeyer, L.E. 1946. Studies on some fungi of northwestern Wyoming III. *Pleospora* and *Leptosphaeria*. Lloydia 9: 203-240.
3811. Wehmeyer, L.E. 1946. Studies of some fungi from northwestern Wyoming. I. Pyrenomycetes. Mycologia 38: 144-170.
3812. Wehmeyer, L.E. 1946. Studies on some fungi from northwestern Wyoming. II. Fungi Imperfecti. Mycologia 38: 306-330.
3813. Wehmeyer, L.E. 1951. Studies in the genus *Pleospora*. IV. Mycologia 43: 570-589.
3814. Wehmeyer, L.E. 1952. The genera *Leptosphaeria*, *Pleospora*, and *Clathrospora* in Mt. Ranier National Park. Mycologia 44: 621-655.
3815. Wehmeyer, L.E. 1952. Some pyrenomycetous fungi from Mt. Rainier National Park. Sydowia 6: 412-428.
3816. Wehmeyer, L.E. 1952. Some fungi imperfecti from Mt. Rainier National Park. Sydowia 6: 433-442.
3817. Wehmeyer, L.E. 1953. Studies in the genus *Pleospora*. V. Mycologia 45: 391-414.
3818. Wehmeyer, L.E. 1954. Studies in the genus *Clathrospora*. Mycologia 46: 498-522.
3819. Wehmeyer, L.E. 1966. Development of the ascocarp in *Cryptomycina pteridis*. Mycologia 58: 752-760.
3820. Weidemann, G.J., D.M. Boone, and H.H. Burdsall Jr. 1982. Taxonomy of *Phyllosticta vaccinii* (Coelomycetes) and a new name for the true anamorph of *Botryosphaeria vaccinii* (Dothideales, Dothioraceae). Mycologia 74: 59-65.
3821. Weidensaul, T.C., and F.A. Wood. 1974. Analysis of a maple canker epidemic in Pennsylvania. Phytopathology 64: 1024-1027.
3822. Weigle, C.G. 1956. Powdery mildew (*Podosphaera tridactyla*) on almond. Pl. Dis. Reporter 40: 584.
3823. Weimer, J.L. 1951. *Ascochyta* canker of blue lupine. Pl. Dis. Reporter 35: 81-82.
3824. Weimer, J.L. 1952. *Pythium* root and stem rot of lupines. Pl. Dis. Reporter 36: 279-282.
3825. Weimer, J.L., and E.S. Luttrell. 1948. Angular leaf spot of kudzu by a new species of *Mycosphaerella*. Phytopathology 38: 348-358.
3826. Weingartner, D.P. 1985. First observation of potato late blight in fall potatoes grown in northeast Florida. Pl. Dis. 69: 1101.
3827. Weingartner, D.P., D.W. Dickson, and J.D. Dilbeck. 1974. Early dying disease on potatoes in north Florida. Pl. Dis. Reporter 58: 374-378.
3828. Weingartner, D.P., and E.J. Klos. 1974. Fungi associated with blueberry stems in Michigan. Pl. Dis. Reporter 58: 180-181.
3829. Weingartner, D.P., and E.J. Klos. 1975. Etiology and symptomatology of canker and dieback diseases on highbush blueberries caused by *Godronia (Fusicoccum) cassandrae* and *Diaporthe (Phomopsis) vaccinii*. Phytopathology 65: 105-110.
3830. Weingartner, D.P., and E.J. Klos. 1975. Histopathology of blueberry stems naturally infected with *Godronia cassandrae*. Phytopathology 65: 1327-1328.
3831. Weir, J.R. 1914. Notes on wood destroying fungi which grow on both coniferous and deciduous trees. I. Phytopathology 4: 271-276.
3832. Weir, J.R. 1917. Montana forest tree fungi - I. Polyporaceae. Mycologia 9: 129-137.
3833. Weise, M.V., N.A. Smith, A.H. Ellingboe, A.V. Ravenscroft, and J.L. Clayton. 1973. Wheat diseases in Michigan. Pl. Dis. Reporter 57: 187-189.
3834. Weiss, F. 1948. Checklist revision. Pl. Dis. Reporter 32: 490.
3835. Weiss, M.J., and J.W. Riffle. 1971. *Armillaria* root rot in a ponderosa pine plantation in New Mexico. Pl. Dis. Reporter 55: 823-824.

3836. Welch, A.W., Jr., S.F. Jenkins Jr., and C.W. Averre. 1975. *Trichothecium* fruit rot on greenhouse tomatoes in North Carolina. Pl. Dis. Reporter 59: 255-257.

3837. Welch, D.S., and E.L. Stone. 1953. *Fomes annosus* (Fr.) Cke. in coniferous plantations in New York state. Pl. Dis. Reporter 37: 247-248.

3838. Weller, D.M., R.J. Cook, G. MacNish, E.N. Bassett, R.L. Powelson, and R.R. Petersen. 1986. *Rhizoctonia* root rot of small grains favored by reduced tillage in the Pacific northwest. Pl. Dis. 70: 70-73.

3839. Wells, D.G. 1956. Downy mildew on wheat and barley. Pl. Dis. Reporter 40: 258.

3840. Wells, H.D. 1953. Diseases of big trefoil in Georgia. Pl. Dis. Reporter 37: 521.

3841. Wells, H.D. 1956. Top rot of pearl millet caused by *Fusarium moniliforme*. Pl. Dis. Reporter 40: 387.

3842. Wells, H.D. 1959. Annual ryegrass, *Lolium multiflorum*, host for *Sclerotium rolfsii*. Pl. Dis. Reporter 43: 834.

3843. Wells, H.D., G.W. Burton, and J.F. Hennen. 1973. *Puccinia substriata* var. *indica* on pearl millet in the southeast. Pl. Dis. Reporter 57: 262.

3844. Wells, H.D., G.W. Burton, and D.K. Ourecky. 1963. *Tolyposporium* smut, a new disease on pearl millet, *Pennisetum glaucum*, in the United States. Pl. Dis. Reporter 47: 16-19.

3845. Wells, H.D., and I. Forbes. 1963. Anthracnose of *Desmodium* in Georgia. Pl. Dis. Reporter 47: 837-839.

3846. Wells, H.D., I. Forbes, and J.R. Edwardson. 1961. The *Stemphyllium* leaf spot complex on blue lupine. Pl. Dis. Reporter 45: 725-727.

3847. Wells, H.D., and I. Forbes Jr. 1971. Anthracnose on *Dolichos biflorus* in Georgia. Pl. Dis. Reporter 55: 10-11.

3848. Wells, H.D., I. Forbes, T.E. Webb, and J.R. Edwardson. 1956. Two *Stemphylium* diseases of blue lupine. Pl. Dis. Reporter 40: 803-806.

3849. Wells, H.D., N.A. Minton, and E.D. Donnelly. 1969. *Lespedeza cuneata*, a new host for *Sclerotium rolfsii*. Pl. Dis. Reporter 53: 528.

3850. Wells, J.C. 1962. Peanut rust in North Carolina. Pl. Dis. Reporter 46: 65.

3851. Wells, J.M. 1977. Sour rot of peaches caused by *Monilia implicata* and *Geotrichum candidum*. Phytopathology 67: 404-408.

3852. Wells, K. 1975. Studies of some Tremellaceae. V.A new genus, *Efibulobasidium*. Mycologia 67: 147-156.

3853. Wells, K., and G. Wong. 1985. Interfertility and comparative morphological studies of *Exidiopsis plumbescens* from the west coast. Mycologia 77: 285-299.

3854. Wells, V.L., and P.E. Kempton. 1971. Studies on the fleshy fungi of Alaska. V. The genus *Strobilurus* with notes on extralimital species. Mycologia 63: 370-379.

3855. Welty, R.E. 1977. *Lamium amplexicaule* (henbit): A new host for *Sclerotinia sclerotiorum* (*trifoliorum*). Pl. Dis. Reporter 61: 508-510.

3856. Welty, R.E., M.D. Azevedo, and K.L. Cook. 1986. Detecting viable *Acremonium* endophytes in leaf sheaths and meristems of tall fescue and perennial ryegrass. Pl. Dis. 70: 431-435.

3857. Welty, R.E., and T.H. Busbice. 1976. *Phytophthora* root rot of alfalfa in North Carolina. Pl. Dis. Reporter 60: 14-18.

3858. Wenner, N., and W. Merrill. 1984. New conifer hosts for *Fusarium* root rot in Pennsylvania. Pl. Dis. 68: 536.

3859. Weresub, L.K. 1953. Studies of Canadian Thelephoraceae X. Some species of *Peniophora*, section *Tubuliferae*. Canad. J. Bot. 31: 760-778.

3860. Weresub, L.K., and S. Gibson. 1960. *Stereum pini* in North America. Canad. J. Bot. 38: 833-867.

3861. West, E., M. Cohen, and L.C. Knorr. 1954. Brown rot of citrus fruit on the tree in Florida. Pl. Dis. Reporter 38: 120-121.

3862. Wester, H.V., R.W. Davidson, and M.E. Fowler. 1950. Cankers of linden and redbud. Pl. Dis. Reporter 34: 219-223.

3863. Westerlund, F.V., Jr., R.N. Campbell, and K.A. Kimble. 1974. Fungal root rots and wilt of chickpea in California. Phytopathology 64: 432-436.

3864. Wheeler, J.E., and A.M. Boyle. 1971. Identification of four *Phytophthora* isolates previously unreported from Arizona. Phytopathology 61: 1293-1296.

3865. Whetzel, H.H. 1945. A synopsis of the genera and species of the Sclerotiniaceae, a family of stromatic inoperculate Discomycetes. Mycologia 37: 648-714.

3866. White, B.L., and W. Merrill. 1969. Pathological anatomy of *Abies balsamea* infected with *Melampsorella caryophyllacearum*. Phytopathology 59: 1238-1242.

3867. White, D.G., B.J. Jacobsen, and A.L. Hooker. 1978. Occurrence of sorghum downy mildew in Illinois. Pl. Dis. Reporter 62: 720.

3868. White, D.G., J. Sleesman, and C.W. Ellett. 1973. *Mycocentrospora* leaf spot of *Euonymus*. Pl. Dis. Reporter 57: 807-808.

3869. White, J.F., Jr., and G.T. Cole. 1985. Endophyte-host associations in forage grasses. II. Taxonomic observations on the endophyte of *Festuca arundinacea*. Mycologia 77: 483-486.

3870. White, J.F., Jr., and G. Morgan-Jones. 1987. Endophyte-host associations in forage grasses. VII. *Acremonium chisosum*, a new species isolated from *Stipa eminens* in Texas. Mycotaxon 28: 179-189.

3871. White, J.F., Jr., and G. Morgan-Jones. 1987. Studies in the genus *Phoma*. VII. Concerning *Phoma glomerata*. Mycotaxon 28: 437-445.

3872. White, J.F., Jr., and G. Morgan-Jones. 1987. Endophyte-host associations in forage grasses. IX. Concerning *Acremonium typhinum*, the anamorph of *Epichloe typhina*. Mycotaxon 29: 489-500.

3873. White, J.F., Jr., and G. Morgan-Jones. 1987. Endophyte-host associations in forage grasses. X. Cultural studies on some species of *Acremonium* sect. *Albo-lanosa*, including a new species, *A. starrii*. Mycotaxon 30: 87-95.

3874. White, W.L. 1942. Studies in the genus *Helotium*. I. A review of the species described by Peck. Mycologia 34: 154-179.

3875. White, W.L., and H.H. Whetzel. 1938. Pleomorphic life cycles in a new genus of the Helotiaceae. Mycologia 30: 187-203.

3876. Whitehead, M.D. 1954. Downy mildew of corn in Texas. Pl. Dis. Reporter 38: 896.

3877. Whitehead, M.D. 1958. Pathology and pathological histology of downy mildew, *Sclerophthora macrospora*, on six graminicolous hosts. Phytopathology 48: 485-493.

3878. Whitehead, M.D., and J.G. Dickson. 1952. Pathology, morphology and nuclear cycle of two species of *Pyrenophora*. Mycologia 44: 747-758.

3879. Whiteside, J.O. 1972. Histopathology of citrus greasy spot and identification of the causal fungus. Phytopathology 62: 260-263.

3880. Whiteside, J.O. 1976. A newly recorded *Alternaria*-induced brown spot disease on Dancy tangerines in Florida. Pl. Dis. Reporter 60: 326-329.

3881. Whiteside, J.O. 1986. Semiselective media for the isolation of *Elsinoe fawcettii* from cit rus scab pustules. Pl. Dis. 70: 204-206.

3882. Whitney, E.D. 1971. The first confirmable occurrence of *Urophlyctis leproides* on sugarb eet in North America. Pl. Dis. Reporter 55: 30-32.

3883. Whitney, H.S., R.J. Bandoni, and F. Oberwinkler. 1987. *Entomocorticium dendroctoni* gen. et sp. nov. (Basidiomycotina), a possible nutritional symbiote of the mountain pine beetle in lodgepole pine in British Columbia. Canad. J. Bot. 65: 95-102.

3884. Whitney, H.S., J. Reid, and K.A. Pirozynski. 1975. Some new fungi associated with needle blight of conifers. Canad. J. Bot. 53: 3051-3063.

3885. Whitney, N.G. 1986 . First report of *Amphobotrys ricini* infecting *Caperonia palustris* in the United States. Pl. Dis. 70: 892.

3886. Whitney, N.G., and G.R. Bowers Jr. 1985. Stem canker of soybean in Texas. Pl. Dis. 69: 361.

3887. Wick, R.L., K.K. Rane, and D.P. Sutton. 1987. Two new ornamental hosts for *Phytophthora cactorum*: *Trachymeme caerulea* and *Gypsophila paniculata*. Pl. Dis. 71: 281.

3888. Wicker, E.F. 1965. A *Phomopsis* canker on western larch. Pl. Dis. Reporter 49: 102-105.

3889. Wicker, E.F., and C.G. Shaw. 1968. Fungal parasites of dwarf mistletoes. Mycologia 60: 372-383.

3890. Wicklow-Howard, M.C., and J. Skujins. 1980. Infection of Engelmann-spruce seed by *Geniculodendron pyriforme* in Western North America. Mycologia 72: 406-410.

3891. Widin, K.D., and A.L. Schipper. 1980. Epidemiology of *Melampsora medusae* leaf rust of poplars in the north central United States. Canad. J. Forest Res. 10: 257-263.

3892. Widin, K.D., and A.L. Schipper Jr. 1981. Effect of *Melampsora medusae* leaf rust infection on yield of hybrid poplars in the north-central United States. Eur. J. Forest Pathol. 11: 438-448.

3893. Wiese, M.V. 1987. Compendium of wheat diseases. (Second edition) The American Phytopathological Society 112 p.
3894. Wilhelm, S. 1957. *Rhizoctonia* bud rot of strawberry. Pl. Dis. Reporter 41: 941-944.
3895. Wilhelm, S., G. Nyland, and F. Charles. 1954. *Verticillium* wilt of sweet cherry in California. Pl. Dis. Reporter 38: 438.
3896. Wilhelm, S., R.D. Raabe, and E.B. Smalley. 1955. Some previously unrecorded hosts of *Verticillium albo-atrum* in California. Pl. Dis. Reporter 39: 693-694.
3897. Wilhelm, S., R.D. Raabe, and H.E. Thomas. 1954. *Dematophora* root and crown rot of strawberry. Pl. Dis. Reporter 38: 72-73.
3898. Wilhelm, S., and E.E. Stevenson. 1955. *Verticillium* wilt of melons. Pl. Dis. Reporter 39: 881.
3899. Wilhelm, S., and H.E. Thomas. 1950. *Verticillium* wilt of bramble fruits with specialreference to *Rubus ursinus* derivatives. Phytopathology 40: 1103-1110.
3900. Wilkinson, H.T. 1987. Association of *Trechispora alnicola* with yellow ring disease of *Poa pratensis*. Canad. J. Bot. 65: 150-153.
3901. Wilkinson, H.T., and R.L. Millar. 1981. *Phytophthora* root rot of alfalfa in central New York. Pl. Dis. 65: 127-129.
3902. Wilks, D.S., P.L. Gersper, and F.W. Cobb Jr. 1985. Association of soil moisture with spread of *Ceratocystis wageneri* in ponderosa pine disease centers. Pl. Dis. 69: 206-208.
3903. Williams, A.S. 1964. Twist disease of orchardgrass in Virginia. Pl. Dis. Reporter 48: 119.
3904. Williams, A.S., and J.W. Herron. 1974. Occurrence of downy mildew of sorghum in Kentucky. Pl. Dis. Reporter 58: 90-91.
3905. Williams, E., Jr., and F.J. Gough. 1984. Bunt of wheat caused by *Tilletia caries* widespread in southwestern Oklahoma in 1983. Pl. Dis. 68: 537.
3906. Williams, H.E., and H.R. Cameron. 1956. Silver-leaf of Montmorency sour cherry in Oregon. Pl. Dis. Reporter 40: 954-956.
3907. Williams, L., and S.C. Hayne. 1982. Index of plant diseases in West Virginia. Plant Pest Control Division, West Virginia Dept. Agric., Charleston. 115 p.
3908. Williams, P.H., and E.K. Wade. 1967. Scab of celeriac. Pl. Dis. Reporter 51: 427-429.
3909. Williams, R.E., and S. Hagland. 1976. *Fomes annosus* in roots of mature ponderosa pine in western Montana. Pl. Dis. Reporter 60: 101-103.
3910. Williams, R.L., and S.C. Haynes. 1982. Index of Plant Diseases in West Virginia (addenda). Plant Pest Control Division, West Virginia Dept. Agric., Charleston. 1 p.
3911. Williams, R.L., and S.C. Haynes. 1982. Index of Plant Diseases in West Virginia (addenda). Plant Pest Control Division, West Virginia Dept. Agric., Charleston. 1 p.
3912. Willis, W.G., C.L. Kramer, and H.E. Thompson. 1963. Dutch elm disease in Kansas in 1962. Pl. Dis. Reporter 47: 443.
3913. Willis, W.G., C.L. Kramer, and H.E. Thompson. 1964. Dutch elm disease in Kansas in 1963. Pl. Dis. Reporter 48: 324.
3914. Willis, W.G., and O.D. Shively. 1974. *Cephalosporium* stripe of winter wheat and barley in Kansas. Pl. Dis. Reporter 58: 566-567.
3915. Wills, W.H. 1953. A large specimen of *Polyporus berkeleyi*. Mycologia 45: 144-145.
3916. Wilson, C.L. 1959. The Columbian timber beetle and associated fungi in white oak. Forest Sci. 5: 114-127.
3917. Wilson, C.L. 1961. An undescribed blight disease of Arizona cypress. Pl. Dis. Reporter 45: 96-98.
3918. Wilson, C.L. 1961. Epidemics of pine needle rust in Arkansas. Pl. Dis. Reporter 45: 957.
3919. Wilson, C.L. 1963. *Fomes annosus* root rot of shortleaf, loblolly, and slash pine in Arkansas. Pl. Dis. Reporter 47: 328.
3920. Wilson, C.L. 1963. Dutch elm disease survey in Arkansas in 1962. Pl. Dis. Reporter 47: 442.
3921. Wilson, C.L. 1968. Presence of two tree vascular diseases confirmed in Arkansas. Pl. Dis. Reporter 52: 79.
3922. Wilson, C.L., M.C. McDaniel, and C.P. Seymour. 1961. Dutch elm disease in Arkansas. Pl. Dis. Reporter 45: 900.
3923. Wilson, C.L., M.C. Tucker, and M.C. McDaniel. 1967. Dutch elm disease distribution in Arkansas. Pl. Dis. Reporter 51: 71.
3924. Wilson, C.L., M.C. Tucker, and J.V. Tiner. 1964. Oak wilt in Arkansas (1950-1963). Pl. Dis. Reporter 48: 370-372.
3925. Wilson, V.E. 1956. 1955 bean disease survey in southern Idaho. Pl. Dis. Reporter 40: 312.
3926. Wilson, V.E. 1971. Downy mildew on peas in northern Idaho and eastern Washington in 1970. Pl. Dis. Reporter 55: 730.
3927. Wilson, V.E., and J. Brandsberg. 1965. Fungi isolated from diseased lentil seedlings in 1963 - 1964. Pl. Dis. Reporter 49: 660-662.
3928. Winstead, N.N., and H.R. Garriss. 1960. Control of cabbage clubroot in North Carolina. Pl. Dis. Reporter 44: 14-18.
3929. Winstead, N.N., L.H. Person, and R.D. Riggs. 1957. *Cercospora* leaf spot on cucumbers in North Carolina. Pl. Dis. Reporter 41: 794.
3930. Winstead, N.N., D.L. Strider, and S.F. Jenkins. 1959. Scab disease on cantaloupes in North Carolina. Pl. Dis. Reporter 43: 424.
3931. Winstead, N.N., D.L. Strider, and L.H. Person. 1960. Vegetable diseases in North Carolina during 1958 and 1959. Pl. Dis. Reporter 44: 491-495.
3932. Winstead, N.N., J.C. Wells, and W.W. Reid. 1958. *Botrytis* in pepper seedbeds and on young plants in the field. Pl. Dis. Reporter 42: 981-982.
3933. Witcher, W. 1969. Oak wilt in South Carolina. Pl. Dis. Reporter 53: 916.
3934. Witcher, W., and E.I. Zehr. 1971. Dutch elm disease in South Carolina. Pl. Dis. Reporter 55: 158.
3935. Witcosky, J.J., and E.M. Hansen. 1985. Root-colonizing insects recovered from Douglas-fir in various stages of decline due to black-stain root disease. Phytopathology 75: 399-402.
3936. Wolf, F.A. 1926. Brown leaf spot of leather leaf fern. J. Elisha Mitchell Sci. Soc. 42: 55-62.
3937. Wolf, F.A. 1927. Pomegranate blotch. J. Agric. Res. 35: 465-469.
3938. Wolf, F.A. 1927. Blight of asparagus fern. J. Elisha Mitchell Sci. Soc. 43: 91-95.
3939. Wolf, F.A. 1938. Life histories of two leaf-inhabiting fungi on sycamore. Mycologia 30: 54-63.
3940. Wolf, F.A. 1940. A leafspot fungus on *Nyssa*. Mycologia 32: 331-335.
3941. Wolf, F.A. 1970. A large sporophore of *Polyporus berkeleyi*. Mycologia 62: 198-200.
3942. Wolf, F.A., and A.R. Cavaliere. 1965. Two new species of leafblight fungi of *Kalmia latifolia*. Mycologia 57: 576-582.
3943. Wolf, F.A., and R.W. Davidson. 1941. Life cycle of *Piggotia fraxini*, causing leaf disease of ash. Mycologia 33: 526-539.
3944. Wolf, F.A., K.H. Garren, and J.K. Miller. 1938. Fungi of the Duke Forest. Bull. School Forest. Duke Univ. 2: 1-122.
3945. Wolf, F.T., and F.A. Wolf. 1952. Pathology of *Camellia* leaves infected by *Exobasidium camelliae* var. *gracilis* Shirai. Phytopathology 42: 147-149.
3946. Wolfe, R.D. 1970. Status of *Commandra* blister rust in Arkansas. Pl. Dis. Reporter 54: 471-474.
3947. Wolfe, R.D., and C.L. Hatch. 1969. Dutch elm disease detected in Mississippi. Pl. Dis. Reporter 53: 3.
3948. Wong, G.J., and K. Wells. 1987. Comparative morphology, compatibility, and interfertility of *Auricularia cornea, A. polytricha*, and *A. tenuis*. Mycologia 79: 847-856.
3949. Wong, G.J., K. Wells, and R.J. Bandoni. 1985. Interfertility and comparative morphological studies of *Tremella mesenterica*. Mycologia 77: 36-49.
3950. Wood, F.A. 1964. *Ceratocystis* canker of aspen in Pennsylvania. Pl. Dis. Reporter 48: 441.
3951. Wood, F.A., and D.W. French. 1963. *Ceratocystis fimbriata*, the cause of a stem canker of quaking aspen. Forest Sci. 9: 232-235.
3952. Woodard, K.E., and R.M. Jones. 1985. First report of *Sclerotium rolfsii* and *Rhizoctonia solani* on sain foin. Pl. Dis. 69: 905.
3953. Woodward, R.W., C.S. Holton, and O.A. Vogel. 1952. Dwarf bunt in rye. Pl. Dis. Reporter 36: 434.
3954. Worf, G.L., J.S. Stewart, and R.C. Avenius. 1986. Necrotic ring spot disease of turfgrass in Wisconsin. Pl. Dis. 70: 453-458.
3955. Worrall, J.J., J.C. Correll, and A.H. McCain. 1986. Pathogenicity and teleomorph-anamorph connection of *Botryosphaeria dothidea* on *Sequoiadendron giganteum* and *Sequoia sempervirens*. Pl. Dis. 70: 757-759.
3956. Wright, E. 1957. *Cytospora* cankers of cottonwood. Pl. Dis. Reporter 41: 892-893.
3957. Wright, E. 1957. *Cytospora* canker of Rocky Mountain Douglas-fir. Pl. Dis. Reporter 41: 811-813.
3958. Wright, W.R., L. Beraha, and M.A. Smith. 1964. Leather rot of

Louisiana strawberries on the Chicago market in 1964. Pl. Dis. Reporter 48: 747-748.
3959. Wright, W.R., and B.A. Billeter. 1974. "Red kernel" disease of sweet corn on the retail market. Pl. Dis. Reporter 58: 1065-1066.
3960. Wright, W.R., M.A. Smith, and L. Beraha. 1964. Sour rot of carrots. Pl. Dis. Reporter 48: 837-838.
3961. Wright, W.R., M.A. Smith, and G.B. Ramsey. 1954. *Diplodina* rot of South Carolina peaches. Pl. Dis. Reporter 38: 436-437.
3962. Wright, W.R., M.A. Smith, G.B. Ramsey, and L. Beraha. 1960. *Gloeosporium* rot of strawberry fruit. Pl. Dis. Reporter 44: 212-213.
3963. Wright, W.R., M.A. Smith, G.B. Ramsey, and L. Beraha. 1960. Observations on pustular spot on peaches. Pl. Dis. Reporter 44: 424-425.
3964. Wright,, E. E. 1942. *Cytospora abietis*, the cause of a canker of true firs in California and Nevada. J. Agric. Res. 65: 143-153.
3965. Wymore, L.A., and J.W. Lorbeer. 1986. Timing fungicidal control of *Sclerotinia minor* causing drop of lettuce grown on organic soils in New York. Pl. Dis. 70: 1134-1138.
3966. Wysong, D.S., and W.G. Willis. 1968. Recorded distribution of dutch elm disease west of the Mississippi River as of 1967. Pl. Dis. Reporter 52: 652-653.
3967. Yamamoto, B.T., and M. Aragaki. 1983. Etiology and control of seedling blight of *Brassaia actinophylla* caused by *Pythium splendens* in Hawaii. Pl. Dis. 67: 396-399.
3968. Yandle, O.D., and E.R. Roth. 1971. Ratio estimation of fusiform rust incidence in southern pine plantations. Forest Sci. 17: 135-147.
3969. Yang, S.M., and T. Gulya. 1986. Prevalent races of *Puccinia helianthi* in cultivated sunflower on the Texas high plains. Pl. Dis. 70: 603.
3970. Yang, S.M., J.B. Morris, P.W. Unger, and T.E. Thompson. 1979. *Rhizopus* head rot of cultivated sunflower in Texas. Pl. Dis. Reporter 63: 833-835.
3971. Yang, S.M., and D.F. Owen. 1982. Symptomatology and detection of *Macrophomina phaseolina* in sunflower plants parasitized by *Cylindrocopturus adspersus* larvae. Phytopathology 72: 819-821.
3972. Yang, S., R.W. Berry, E.S. Luttrell, and T. Vongkaysone. 1984. A new sunflower disease in Texas caused by *Diaporthe helianthi*. Pl. Dis. 68: 254-255.
3973. Yarwood, C.E. 1951. Hop cankers. Pl. Dis. Reporter 35: 361-363.
3974. Yarwood, C.E. 1969. Sunflower, a new host of *Coleosporium madia*. Pl. Dis. Reporter 53: 648.
3975. Yarwood, C.E., and M.W. Gardner. 1964. Unreported powdery mildews III. Pl. Dis. Reporter 48: 310.
3976. Yarwood, C.E., and M.W. Gardner. 1970. Leaf surface and leaf hairs affect length of conidiophores of Erysiphaceae. Mycologia 62: 707-713.
3977. Yarwood, C.E., and M.W. Gardner. 1972. Ascospore discharge by *Erysiphe trina*. Mycologia 64: 799-805.
3978. Yarwood, C.E., and S. Wilhelm. 1951. Downy mildew on rose in California. Pl. Dis. Reporter 35: 56-57.
3979. Yeh, C.C., and J.B. Sinclair. 1980. Effect of *Chaetomium cupreum* on seed germination and antagonism to other seedborne fungi of soybean. Pl. Dis. 64: 468-470.
3980. Yerkes, W.D., Jr., C.G. Shaw., and G.W. Fischer. 1951. Spinach smut, *Entyloma ellisii* Halsted, a new record for Washington state. Phytopathology 41: 854-858.
3981. Yohem, K.H., G.B. Cummins, and R.L. Gilbertson. 1985. Revised list and host index of Arizona rust fungi. Mycotaxon 22: 451-468.
3982. Yoshimura, M.A., J.Y. Uchida, and M. Aragaki. 1985. Etiology and control of poinsettia blight caused by *Phytophthora nicotianae* var. *parasitica* and *P. drechsleri*. Pl. Dis. 69: 511-513.
3983. Yoshimura, M.A., J.Y. Uchida, and M. Aragaki. 1986. Etiology and control of *Alternaria* blight of poinsettia. Pl. Dis. 70: 73-75.
3984. Young, D.J., and S.M. Alcorn. 1981. A *Coniothyrium* sp. causing leaf blight of jojoba. Mycologia 73: 822-832.
3985. Young, D.J., and S.M. Alcorn. 1982. Soilborne pathogens of *Euphorbia lathyris*: *Macrophomina phaseolina*, *Pythium aphanidermatum*, and *Rhizoctonia solani*. Pl. Dis. 66: 236-238.
3986. Young, D.J., and S.M. Alcorn. 1984. Latent infection of *Euphorbia lathyris* and weeds by *Macrophomina phaseolina* and propagule populations in Arizona field soil. Pl. Dis. 68: 587-589.
3987. Young, H.C., Jr., and L.E. Browder. 1957. Occurrence of stripe rust in Oklahoma in 1957. Pl. Dis. Reporter 41: 958-959.
3988. Young, P.A. 1950. First epidemic of tomato late blight in central east Texas. Pl. Dis. Reporter 34: 338-339.
3989. Young, R.A. 1954. Dissemination of plant pathogens on nursery and ornamental plants. Pl. Dis. Reporter 38: 417-420.
3990. Young, T.R., and T.A. Kucharek. 1977. Succession of fungal communities in roots and stalks of hybrid field corn grown in Florida. Pl. Dis. Reporter 61: 76-80.
3991. Young, W.J. 1962. *Tuberculina maxima* on white pine in Michigan. Pl. Dis. Reporter 46: 750.
3992. Yuen, G.Y., M.N. Schroth, and A.H. McCain. 1985. Reduction of *Fusarium* wilt of carnation with suppressive soils and antagonistic bacteria. Pl. Dis. 69: 1071-1075.
3993. Yuen, J.E., and J.W. Lorbeer. 1987. Natural and experimental production of oospores of *Bremia lactucae* in lettuce in New York. Pl. Dis. 71: 63-64.
3994. Zabel, R.A. 1976. Basidiocarp development in *Inonotus obliquus* and its inhibition by stem treatments. Forest Sci. 22: 431-437.
3995. Zabel, R.A., A.M. Kenderes, and F.F. Lombard. 1980. Fungi associated with decay in treated Douglas-fir transmission poles in the northeastern United States. Forest Prod. J. 30: 51-56.
3996. Zabel, R.A., F.F. Lombard, C.-J. Wang, and F. Terracina. 1985. Fungi associated with decay in treated southern pine utility poles in the eastern United States. Wood Fiber Sci. 17: 75-91.
3997. Zak, B. 1950. Observations on *Cronartium strobilinum* in Georgia. Pl. Dis. Reporter 34: 349.
Zeeuw D.J. du. See 4030
3998. Zeiders, K.E. 1976. A new disease of reed canarygrass caused by *Helminthosporium catenarium*. Pl. Dis. Reporter 60: 556-560.
3999. Zeiders, K.E. 1979. A *Septoria* disease of reed canarygrass in Pennsylvania. Pl. Dis. Reporter 63: 796-800.
4000. Zeiders, K.E. 1980. A variable-spored isolate of *Drechslera dactylidis* pathogenic on orchardgrass and corn. Pl. Dis. 64: 211-213.
4001. Zeiders, K.E. 1982. Leaf spots of big bluestem, little bluestem, and indiangrass caused by *Ascochyta brachypodii*. Pl. Dis. 66: 502-505.
4002. Zeiders, K.E. 1984. *Helminthosporium* spot blotch of switchgrass in Pennsylvania. Pl. Dis. 68: 120-122.
4003. Zeiders, K.E. 1985. First report of rust caused by *Uromyces* species on birdsfoot trefoil in the United States. Pl. Dis. 69: 727.
4004. Zeiders, K.E. 1987. Leaf spot of indiangrass caused by *Colletotrichum caudatum*. Pl. Dis. 71: 348-350.
4005. Zeiders, K.E., and R.T. Sherwood. 1986. First report of rust caused by *Puccinia recondita* on smooth bromegrass. Pl. Dis. 70: 801.
4006. Zeller, S.M. 1922. Contributions to our knowledge of Oregon fungi - I. Mycologia 14: 173-199.
4007. Zeller, S.M. 1929. Contribution to our knowledge of Oregon fungi - III. Mycologia 21: 97-111.
4008. Zentmyer, G.A. 1960. *Phytophthora* canker of macadamia trees in California. Pl. Dis. Reporter 44: 819.
4009. Zentmyer, G.A., L. Jefferson, C.J. Hickman, and Y. Chang-Ho. 1974. Studies of *Phytophthora citricola*, isolated from *Persea americana*. Mycologia 66: 830-845.
4010. Zentmyer, G.A., and D.E. Munnecke. 1952. *Phytophthora* root rot of nursery stock. Pl. Dis. Reporter 36: 211-212.
4011. Zentmyer, G.A., E.C. Pond, and L.J. Klure. 1979. A new *Phytophthora* disease of ornamental flannel bush (*Fremontia californica*). Pl. Dis. Reporter 63: 721-724.
4012. Ziller, W.G. 1954. Studies of western tree rusts I.A new cone rust on sitka spruce. Canad. J. Bot. 32: 432-439.
4013. Zimmer, D.E. 1961. Powdery mildew on safflower. Pl. Dis. Reporter 45: 944.
4014. Zimmer, D.E. 1971. A serious outbreak of downy mildew in the principal sunflower production area of the United States. Pl. Dis. Reporter 55: 11-12.
4015. Zimmer, D.E. 1973. Basal gall of sunflowers incited by *Plasmopara halstedii*. Pl. Dis. Reporter 57: 647-649.
4016. Zimmer, D.E., J.M. Klisiewicz, and C.A. Thomas. 1963. *Alternaria* leaf spot and other diseases of safflower in 1962. Pl. Dis. Reporter 47: 643.
4017. Zimmer, D.E., and C.A. Thomas. 1967. *Rhizoctonia* blight of safflower and varietal resistance. Phytopathology 57: 946-949.

4018. Zink, F.W., and P.G. Smith. 1958. A second physiologic race of spinach downy mildew. Pl. Dis. Reporter 42: 818.
4019. Zitter, T.A., and H.C. Wien. 1984. Outbreak of *Alternaria alternata* causing fruit rot of tomato in upstate New York. Pl. Dis. 68: 628.
4020. Zuckerman, B.M. 1959. *Fusicoccum* canker of the highbush blueberry in Massachusetts. Pl. Dis. Reporter 43: 803.
4021. Zuckerman, B.M. 1960. Studies of two blueberry stem diseases recently found in eastern Massachusetts. Pl. Dis. Reporter 44: 409-415.
4022. Zuckerman, B.M. 1960. Fungi collected from blueberry stems in Massachusetts. Pl. Dis. Reporter 44: 416.
4023. Zummo, N. 1980. Sugarcane rust in Mississippi. Pl. Dis. 64: 706.
4024. Zummo, N. 1986. Red spot (*Helminthosporium rostratum*) of sweet sorghum and sugarcane, a new disease resembling anthracnose and red rot. Pl. Dis. 70: 800.
4025. Zummo, N. 1987. Occurrence of target leaf spot (*Bipolaris sorghicola*) on sorghum in Mississippi. Pl. Dis. 71: 1045.
4026. Zundel, G.L. 1951. Notes on the Ustilaginales of the world. V. Mycologia 43: 267-270.
4027. Zwet, T. van der. 1963. Occurrence of *Cercospora websteri* on tung in the United States. Pl. Dis. Reporter 47: 70-71.
4028. Zwet, T. van der, I.L. Forbes, and R.J. Steib. 1960. Studies on the *Phytophthora* rot of sugarcane seed pieces in Louisia na. Pl. Dis. Reporter 44: 519-523.
4029. Zwet, T. van der, and W.W. Kilby. 1969. A new leaf spot of Chinese chestnut in Mississippi. Pl. Dis. Reporter 53: 621-622.
4030. de Zeeuw, D.J. 1954. Pythium root rot of parsley. Phytopathology 44: 486.

Host Index

Abelia 123
Abelmoschus 289
Abies 326
Abronia 308
Abrus 187
Abutilon 289
Acacia 187
Acaena 460
Acalypha 182
Acanthaceae 7
Acanthospermum 53
Acer 9
Aceraceae 9
Achillea 53
Achimenes 243
Achlys 91
Acnida 21
Acoelorrhaphe 46
Acokanthera 33
Aconitum 451
Acorus 38
Acrostichum 445
Actaea 452
Actinidia 17
Actinidiaceae 17
Adenium 33
Adenocaulon 54
Adenostoma 460
Adiantum 445
Adoxa 17
Adoxaceae 17
Aechmea 116
Aegilops 363
Aeonium 145
Aeschynanthus 243
Aeschynomene 188
Aesculus 249
Agalinis 515
Agapanthus 271
Agastache 262
Agavaceae 17
Agave 17
Agdestis 326
Ageratina 54
Ageratum 54
Aglaomorpha 445
Aglaonema 38
Agoseris 54
Agrimonia 460
Agrohordeum 363
Agropogon 363
Agropyron 363
Agrositanion 366
Agrostemma 129
Agrostis 366
Ailanthus 522
Aira 368
Aizoaceae 20
Ajuga 263
Akebia 268
Albizia 188
Alcea 290
Aletris 272
Aleurites 182
Alisma 20
Alismataceae 20
Allamanda 33
Allenrolfea 134
Allionia 308
Allium 272
Alnus 92
Alocasia 38
Aloe 273
Alopecurus 368
Aloysia 545
Alphitonia 458
Alpinia 553
Alsophila 445
Alstonia 33
Alternanthera 21
Althaea 290
Alysicarpus 189
Alyssum 108
Alyxia 33
Amaranthaceae 21
Amaranthus 22
Amaryllis 273
Ambrosia 54
Amelanchier 461
Ammannia 284
Ammophila 369
Amorpha 189
Amorphophallus 38
Ampelopsis 549
Amphicarpaea 189
Amsinckia 106
Amsonia 33
Anacardiaceae 23
Anacardium 23
Anagallis 442
Ananas 116
Anaphalis 55
Anchusa 106
Andromeda 170
Andropogon 369
Androsace 442
Anemone 452
Anemonella 453
Anemopsis 513
Anethum 26
Angelica 26
Angelonia 515
Angiopteris 445
Anisacanthus 7
Annona 26
Annonaceae 26
Anoda 290
Anredera 90
Antennaria 55
Anthaenantia 370
Anthemis 56
Anthericum 273
Anthoxanthum 370
Anthurium 38
Anthyllis 189
Antidesma 183
Antigonon 436
Antirrhinum 515
Apargidium 56
Apera 370
Aphanes 462
Aphelandra 7
Apiaceae 26
Apios 189
Apium 27
Aplectrum 320
Apocynaceae 33
Apocynum 33
Aponogeton 36
Aponogetonaceae 36
Aquifoliaceae 36
Aquilegia 453
Arabidopsis 108
Arabis 108
Araceae 38
Arachis 189
Aralia 42
Araliaceae 42
Araucaria 45
Araucariaceae 45
Arbutus 170
Arceuthobium 549
Archontophoenix 46
Arctagrostis 370
Arctium 56
Arctophila 370
Arctostaphylos 170
Arctotis 56
Ardisia 302
Arecaceae 45
Arenaria 129
Arenga 46
Argemone 324
Argyreia 141
Argyroxiphium 56
Argythamnia 183
Arisaema 39
Aristida 371
Aristolochia 51
Aristolochiaceae 51
Armeniaca 462
Armeria 362
Armoracia 109
Arnica 56
Aronia 462
Arrhenatherum 371
Artemisia 57
Arthraxon 372
Artocarpus 298
Aruncus 462
Arundinaria 372
Arundo 372
Asarum 51
Asclepiadaceae 51
Asclepias 51
Asimina 26
Asparagus 274
Aspidistra 274
Asplenium 445
Aster 58
Asteraceae 53
Astilbe 513
Astragalus 190
Astrophytum 118
Athrotaxis 534
Athyrium 445
Atriplex 134
Atropa 524
Aucuba 143
Aureolaria 516
Avena 372
Averrhoa 323
Avicennia 545
Axonopus 374
Azara 240
Baccaurea 183
Baccharis 60
Bactris 46
Bahia 60
Baileya 60
Balaka 46
Balsaminaceae 90
Balsamorhiza 60
Bambusa 374
Banksia 444
Baptisia 191
Barbarea 109
Barleria 7
Barringtonia 271
Basella 90
Basellaceae 90
Bassia 135
Bauhinia 192
Baumea 158
Beckmannia 374
Begonia 90
Begoniaceae 90
Belamcanda 253
Bellis 61
Beloperone 7
Benincasa 146
Bensoniella 513
Berberidaceae 91
Berberis 91
Berchemia 458
Bergenia 513
Berteroa 109
Bertholletia 271
Berula 27
Besseya 516
Beta 135
Betula 97
Betulaceae 92
Bidens 61
Bigelowia 61
Bignonia 103
Bignoniaceae 103
Billbergia 116
Bischofia 183
Biserrula 192
Bismarckia 46
Bistorta 436
Bixa 105
Bixaceae 105
Blechnum 446
Blechum 7
Blepharoneuron 375
Blephilia 263
Blighia 511
Boehmeria 543
Boerhavia 308
Boisduvalia 316
Boltonia 61
Bombacaceae 105
Bombax 105
Bonamia 141
Boraginaceae 106
Borago 106
Borreria 490
Borrichia 61
Boschniakia 323
Bothriochloa 375
Botrychium 446
Bouea 23
Bougainvillea 308
Bouteloua 375
Bouvardia 490
Boykinia 513
Brachiaria 376
Brachychiton 532
Brachyelytrum 376
Brachypodium 376
Brasenia 118
Brassica 109
Brassicaceae 108
Brassolaeliocattleya 320
Breynia 183
Brickellia 62
Briza 376
Brodiaea 274
Bromelia 116
Bromeliaceae 116
Bromus 376
Broussaisia 250
Broussonetia 298
Browallia 524
Brunfelsia 524
Brunnichia 436
Bryonopsis 146
Buchloe 379

Buchnera 516
Bucida 139
Buckleya 510
Buddleja 284
Bumelia 512
Bupleurum 27
Bursera 117
Burseraceae 117
Butia 46
Butomaceae 117
Butomus 117
Buxaceae 117
Buxus 117
Byrsonima 289
Cabombaceae 118
Cacalia 62
Cactaceae 118
Cactus 118
Caesalpinia 192
Cajanus 192
Cakile 111
Caladium 39
Calamagrostis 379
Calamovilfa 380
Calamus 46
Calandrinia 441
Calathea 295
Calceolaria 516
Calendula 62
Calla 39
Calliandra 192
Callicarpa 545
Callirhoe 290
Callisia 140
Callistemon 303
Callistephus 62
Callitrichaceae 121
Callitriche 121
Callitris 151
Calluna 172
Calocedrus 151
Calochortus 274
Calophyllum 138
Calopogon 320
Calopogonium 192
Calothamnus 303
Calotropis 52
Caltha 453
Calycadenia 62
Calycanthaceae 121
Calycanthus 121
Calycocarpum 297
Calylophus 316
Calyptranthes 303
Calyptridium 441
Calyptrocalyx 46
Calystegia 141
Camassia 275
Camelina 111
Camellia 535
Camissonia 316
Camoensia 192
Campanula 121
Campanulaceae 121
Campsis 103
Camptosorus 446
Camptotheca 309
Canavalia 193
Canistrum 116
Canna 123
Cannabidaceae 122
Cannabis 122
Cannaceae 123
Canotia 133
Caperonia 183
Capparidaceae 123
Capparis 123
Caprifoliaceae 123
Capsella 111
Capsicum 524
Caragana 193
Cardamine 111
Cardiospermum 511
Carduus 62
Carex 158
Carica 128
Caricaceae 128
Carissa 34
Carlowrightia 7
Carnegiea 118
Carpinus 101
Carpobrotus 20
Carthamus 63
Carum 27
Carya 255
Caryophyllaceae 129
Caryota 46
Casimiroa 494
Cassia 193
Cassiope 172
Castanea 222
Castanopsis 224
Castanospermum 193
Castela 523
Castilleja 516
Casuarina 132
Casuarinaceae 132
Catabrosa 380
Catalpa 104
Catha 133
Catharanthus 34
Cattleya 320
Caulophyllum 92
Ceanothus 458
Cecropia 133
Cecropiaceae 133
Cedrela 296
Cedrus 332
Ceiba 106
Celastraceae 133
Celastrus 133
Celosia 22
Celtis 539
Cenchrus 380
Centaurea 63
Centaurium 241
Centella 27
Centranthus 544
Centrosema 193
Cephalanthus 490
Cephalocereus 118
Cephalotaxaceae 134
Cephalotaxus 134
Cerastium 129
Ceratoides 136
Ceratonia 193
Ceratopteris 446
Cercidiphyllaceae 134
Cercidiphyllum 134
Cercidium 194
Cercis 194
Cercocarpus 462
Cercropiaceae 134
Cereus 119
Cestrum 525
Chaenactis 63
Chaenomeles 462
Chaenorrhinum 517
Chaerophyllum 27
Chamaebatia 463
Chamaebatiaria 463
Chamaecyparis 151
Chamaedaphne 172
Chamaedorea 46
Chamaesyce 183
Chambeyronia 47
Chasmanthium 381
Cheilanthes 446
Cheiranthus 111
Cheirodendron 42
Chelidonium 324
Chelone 517
Chenopodiaceae 134
Chenopodium 136
Chilopsis 104
Chimaphila 451
Chimonanthus 121
Chiococca 491
Chionanthus 310
Chloris 381
Chlorogalum 275
Chlorophytum 275
Choisya 494
Chondrilla 63
Chorisia 106
Chorizanthe 436
Chrysalidocarpus 47
Chrysanthemum 63
Chrysobalanaceae 138
Chrysobalanus 138
Chrysogonum 64
Chrysophyllum 512
Chrysopsis 64
Chrysothamnus 64
Chrysothemis 243
Cibotium 446
Cicer 194
Cichorium 65
Cicuta 27
Cimicifuga 454
Cinchona 491
Cinna 381
Cinnamomum 268
Circaea 316
Cirsium 65
Cissus 549
Cistaceae 138
Cistus 138
Citrofortunella 494
Citrullus 146
Citrus 494
Cladium 163
Cladrastis 194
Claoxylon 183
Clarkia 316
Clausena 497
Claytonia 441
Clematis 454
Cleome 123
Clermontia 121
Clerodendrum 545
Clethra 138
Clethraceae 138
Cleyera 536
Clidemia 296
Cliftonia 166
Clinopodium 263
Clintonia 275
Clitoria 194
Clivia 275
Clusia 138
Clusiaceae 138
Cnicus 66
Cnidoscolus 183
Coccocypselum 491
Coccoloba 437
Coccothrinax 47
Cocculus 297
Cochlearia 111
Cochliostema 140
Cochlospermum 105
Cocos 47
Codiaeum 183
Coelopleurum 28
Coffea 491
Coix 382
Colchicum 275
Collinsia 517
Collinsonia 263
Collomia 434
Colocasia 39
Colubrina 459
Columnea 243
Colutea 195
Comandra 510
Combretaceae 139
Commelina 140
Commelinaceae 140
Comptonia 302
Condalia 459
Conioselinum 28
Conium 28
Conocarpus 139
Conringia 112
Convallaria 275
Convolvulaceae 141
Convolvulus 141
Conyza 66
Coprosma 491
Coptis 454
Corallorrhiza 320
Cordia 106
Cordyline 17
Coreopsis 66
Coriandrum 28
Cornaceae 143
Cornus 143
Coronilla 195
Coronopus 112
Cortaderia 382
Corydalis 325
Corylus 102
Coryphantha 119
Cosmos 66
Costus 553
Cotinus 23
Cotoneaster 463
Coursetia 195
Cowania 463
Cracca 195
Crambe 112
Crassula 145
Crassulaceae 145
Crataegus 463
Crepis 67
Crescentia 104
Cressa 141
Crinum 275
Crocus 253
Crossandra 7
Crossopetalum 133
Crossosoma 146
Crossosomataceae 146
Crotalaria 195
Croton 183
Crotonopsis 183
Cryptantha 106
Cryptanthus 116
Cryptbergia 117
Cryptocarya 269
Cryptocereus 119
Cryptocoryne 39
Cryptogramma 446
Cryptomeria 534
Cryptotaenia 28
Ctenanthe 296
Ctenium 382
Cucumis 147
Cucurbita 149
Cucurbitaceae 146
Cunila 263
Cunninghamia 534
Cupaniopsis 511
Cuphea 284
Cupressaceae 151
Cupressus 152
Curcuma 553
Cuscuta 141
Cussonia 42
Cyamopsis 195
Cyanea 121
Cyanotis 140
Cyathea 446
Cycadaceae 158
Cycas 158
Cycladenia 34
Cyclamen 443
Cycloloma 137
Cydista 104
Cydonia 465
Cymbalaria 517
Cymbidium 320
Cymbopogon 382
Cymodoceaceae 158
Cymophyllus 163
Cymopterus 28
Cynanchum 52
Cynara 67
Cynodon 382
Cynoglossum 106
Cynosurus 383
Cyperaceae 158
Cyperus 163
Cyphia 121
Cyphomandra 525
Cypripedium 321
Cyrilla 166
Cyrillaceae 166
Cyrtandra 243
Cyrtanthus 275
Cystopteris 446
Cytisus 195
Dactylis 383
Dactyloctenium 384
Dahlia 67
Dalbergia 196
Dalea 196
Dalibarda 465
Danthonia 384
Daphne 536
Darlingtonia 512
Dasylirion 18
Datisca 166
Datiscaceae 166
Datura 525
Daucus 28
Davallia 446
Decodon 285
Decumaria 250
Delonix 196
Delphinium 455
Dendrobium 321
Dendromecon 325
Dennstaedtia 446
Dentaria 112

Deschampsia 384
Descurainia 112
Desmanthus 196
Desmodium 196
Deutzia 250
Dianella 275
Dianthera 7
Dianthus 130
Diapensia 166
Diapensiaceae 166
Diarrhena 385
Dicentra 240
Dichanthelium 385
Dichanthium 385
Dichelostemma 275
Dichondra 141
Dichorisandra 140
Dichrostachys 197
Dicliptera 7
Dicranopteris 447
Dictamnus 497
Didymopanax 42
Dieffenbachia 39
Diervilla 123
Dietes 253
Digitalis 517
Digitaria 385
Dimorphotheca 68
Diodia 491
Dionaea 167
Dioon 552
Dioscorea 167
Dioscoreaceae 167
Diospyros 167
Dipsacaceae 167
Dipsacus 167
Dipterocarpaceae 167
Dirca 537
Disporum 276
Distichlis 386
Dizygotheca 43
Dodecatheon 443
Dodonaea 511
Dombeya 532
Doronicum 68
Doryopteris 447
Dovyalis 240
Downingia 121
Draba 112
Dracaena 18
Dracocephalum 263
Dracopis 68
Droseraceae 167
Dryas 465
Dryopteris 447
Dubautia 68
Duboisia 526
Duchesnea 465
Dulichium 164
Dupontia 386
Duranta 545
Dyschoriste 7
Dyssodia 68
Ebenaceae 167
Echeveria 145
Echinacea 68
Echinocactus 119
Echinocereus 119
Echinochloa 386
Echinocystis 150
Echinodorus 21
Echinops 68
Echinopsis 119
Echium 106
Eclipta 68
Edithcolea 52
Eichhornia 441
Elaeagnaceae 168
Elaeagnus 168
Elaeis 47
Elaeocarpaceae 169
Elaeocarpus 169
Elaphoglossum 447
Eleocharis 164
Elephantopus 68
Elettaria 553
Eleusine 387
Eleutherococcus 43
Ellisia 252
Elmera 513
Elodea 251
Elymus 387
Elysitanion 390
Emilia 68
Empetraceae 169
Empetrum 169
Encelia 68
Encephalartos 552
Engelmannia 68
Enneapogon 390
Enterolobium 197
Epacridaceae 169
Ephedra 169
Ephedraceae 169
Epidendrum 321
Epigaea 172
Epilobium 317
Epiphyllum 119
Epipremnum 39
Episcia 244
Equisetaceae 169
Equisetum 169
Eragrostis 390
Eranthemum 7
Erechtites 69
Eremocarpus 184
Eremochloa 391
Eremopyrum 391
Eremurus 276
Erianthus 391
Eriastrum 434
Erica 172
Ericaceae 170
Erigenia 29
Erigeron 69
Eriobotrya 465
Eriocaulaceae 182
Eriocaulon 182
Eriochloa 391
Eriodictyon 252
Eriogonum 437
Erioneuron 391
Eriophorum 164
Eriophyllum 70
Ernodea 491
Erodium 242
Eryngium 29
Erysimum 112
Erythrina 197
Erythronium 276
Escallonia 245
Eschscholzia 325
Espostoa 119
Eucalyptus 303
Eucharis 276
Eugenia 305
Euonymus 133
Eupatorium 70
Euphorbia 184
Euphorbiaceae 182
Euphrasia 517
Euryops 71
Eustoma 241
Euterpe 47
Euthamia 71
Evolvulus 141
Exacum 241
Exochorda 466
Exothea 511
Eysenhardtia 197
Fabaceae 187
Fabiana 526
Fagaceae 222
Fagopyrum 437
Fagus 224
Fatshedera 43
Fatsia 43
Feijoa 305
Felicia 71
Fendlera 250
Ferocactus 119
Festuca 391
Ficus 298
Filago 71
Filicium 511
Filipendula 466
Fimbristylis 164
Firmiana 532
Fittonia 7
Flacourtia 240
Flacourtiaceae 240
Flindersia 497
Floerkea 283
Foeniculum 29
Fontanesia 310
Forestiera 311
Forsellesia 146
Forsythia 311
Fortunella 497
Fouquieria 240
Fouquieriaceae 240
Fragaria 466
Frasera 241
Fraxinus 311
Freesia 253
Fremontodendron 532
Freycinetia 324
Fritillaria 276
Froelichia 22
Fuchsia 318
Fuirena 164
Fumariaceae 240
Furcraea 18
Gahnia 164
Gaillardia 71
Galactia 197
Galanthus 276
Galax 166
Galeopsis 263
Galinsoga 71
Galium 491
Galphimia 289
Garcinia 139
Gardenia 492
Garrya 241
Garryaceae 241
Gasteria 276
Gastridium 393
Gaultheria 172
Gaura 318
Gaylussacia 173
Gayophytum 318
Gazania 71
Gelsemium 284
Genipa 492
Genista 197
Gentiana 241
Gentianaceae 241
Gentianella 242
Geocaulon 510
Geonoma 47
Geraea 71
Geraniaceae 242
Geranium 242
Gerbera 71
Gesneriaceae 243
Geum 467
Gilia 434
Gillenia 468
Ginkgo 244
Ginkgoaceae 244
Ginoria 285
Gladiolus 253
Glaux 443
Glechoma 263
Gleditsia 198
Glehnia 29
Gleichenia 447
Gloriosa 276
Gloxinia 244
Glyceria 393
Glycine 198
Glycyrrhiza 200
Gnaphalium 71
Gomphrena 22
Gonolobus 52
Goodeniaceae 244
Goodyera 321
Gordonia 536
Gossypium 290
Gouania 459
Gouldia 492
Gratiola 517
Grevillea 444
Grindelia 72
Gronophyllum 47
Grossulariaceae 245
Guaiacum 553
Guardiola 72
Guettarda 492
Guilleminea 22
Gunnera 247
Gunneraceae 247
Gutierrezia 72
Guzmania 117
Gymnanthes 185
Gymnocarpium 447
Gymnocladus 200
Gymnopogon 394
Gynura 72
Gypsophila 130
Habenaria 321
Hackelia 107
Hackelochloa 394
Haemanthus 276
Haemodoraceae 247
Hakea 444
Halenia 242
Halesia 533
Halodule 158
Haloragaceae 247
Haloragidaceae 247
Hamamelidaceae 247
Hamamelis 247
Hamelia 493
Haplopappus 72
Harpullia 511
Haworthia 276
Hebe 517
Hedeoma 263
Hedera 43
Hedychium 553
Hedyotis 493
Hedysarum 200
Helenium 73
Helianthella 73
Helianthemum 138
Helianthus 73
Helichrysum 75
Heliconia 249
Heliconiaceae 249
Helictotrichon 394
Heliopsis 75
Heliotropium 107
Helipterum 76
Helleborus 455
Hemarthria 394
Hemerocallis 276
Hemigraphis 8
Hemizonia 76
Hepatica 455
Heracleum 29
Herissantia 292
Hesperis 113
Hesperochiron 252
Hesperochloa 394
Heteranthera 441
Heterocodon 121
Heteromeles 468
Heteropogon 394
Heterotheca 76
Heuchera 513
Hevea 185
Hibiscadelphus 292
Hibiscus 292
Hicriopteris 447
Hieracium 76
Hierochloe 394
Hilaria 395
Hippeastrum 277
Hippocastanaceae 249
Hoffmanseggia 200
Hohenbergia 117
Holcus 395
Holodiscus 468
Homalocladium 438
Homalomena 40
Honkenya 130
Hordeum 395
Horkelia 468
Hosta 277
Houstonia 493
Hovenia 459
Howeia 48
Hoya 53
Hudsonia 138
Huernia 53
Humulus 123
Hura 185
Hyacinthus 277
Hybanthus 547
Hydnophytum 493
Hydrangea 250
Hydrangeaceae 250
Hydrastis 456
Hydrilla 251
Hydrocharitaceae 251
Hydrocotyle 29
Hydrolea 252
Hydrophyllaceae 252
Hydrophyllum 252
Hygrophila 8
Hylocereus 119
Hymenocallis 277
Hymenoclea 76

Hymenopappus 76
Hymenoxys 76
Hyophorbe 48
Hyoscyamus 526
Hyparrhenia 397
Hypericum 139
Hyphaene 48
Hypochaeris 76
Hypoxis 277
Hyptis 263
Hystrix 397
Iberis 113
Idesia 240
Ilex 36
Iliamna 293
Illiciaceae 253
Illicium 253
Impatiens 90
Imperata 397
Indigofera 200
Inula 76
Ipomoea 142
Iresine 22
Iridaceae 253
Iris 254
Isanthus 264
Isopyrum 456
Itea 245
Iva 77
Ivesia 468
Ixia 255
Ixiolirion 277
Ixora 493
Jacaranda 105
Jacquemontia 143
Jacquinia 536
Jamesia 251
Jasminum 313
Jatropha 185
Jaumea 77
Jeffersonia 92
Juglandaceae 255
Juglans 258
Juncaceae 260
Juncaginaceae 262
Juncus 260
Junellia 545
Juniperus 153
Justicia 8
Kadua 493
Kalanchoe 145
Kalmia 173
Keckiella 518
Kelloggia 493
Kerria 469
Kigelia 105
Kniphofia 277
Kobresia 165
Koeleria 397
Koelreuteria 511
Kohleria 244
Kokia 294
Kolkwitzia 124
Kopsia 34
Korthalsella 549
Krameria 262
Krameriaceae 262
Krigia 77
Krugiodendron 459
Kuhnia 77
Kummerowia 200
Lablab 201
Labordia 284
Laburnum 201
Laccospadix 48
Lachnanthes 247
Lactuca 77
Laelia 322
Laeliocattleya 322
Lagenaria 150
Lagerstroemia 285
Lagophylla 78
Lagotis 518
Laguncularia 139
Lagurus 398
Lamarckia 398
Lamiaceae 262
Lamium 264
Lampranthus 20
Lansium 297
Lantana 545
Laportea 543
Lappula 107
Lapsana 78
Lardizabalaceae 268
Larix 332
Larrea 553
Lasiacis 398
Lasianthaea 78
Lasiocroton 186
Lasthenia 78
Lathyrus 201
Lauraceae 268
Laurocerasus 469
Laurus 269
Lavandula 264
Lavatera 294
Layia 79
Lechea 138
Lecythidaceae 271
Ledum 174
Leea 271
Leeaceae 271
Leersia 398
Legousia 121
Leiophyllum 174
Lemaireocereus 119
Lemna 271
Lemnaceae 271
Lens 202
Leonotis 264
Leontodon 79
Leonurus 264
Lepidium 113
Leptarrhena 513
Leptochloa 398
Leptodactylon 435
Leptospermum 305
Lespedeza 202
Lesquerella 113
Leucaena 203
Leucocoryne 277
Leucocrinum 277
Leucojum 277
Leucophyllum 518
Leucophysalis 526
Leucopoa 398
Leucospermum 444
Leucothoe 174
Lewisia 442
Leymus 399
Liatris 79
Libocedrus 156
Licania 138
Ligularia 79
Ligusticum 30
Ligustrum 314
Liliaceae 271
Lilium 277
Limnanthaceae 283
Limnanthes 283
Limnobium 251
Limnodea 399
Limonia 497
Limonium 362
Limosella 518
Linaceae 283
Linanthus 435
Linaria 518
Lindera 269
Linnaea 124
Linum 283
Lippia 546
Liquidambar 247
Liriodendron 285
Liriope 279
Litchi 511
Lithocarpus 227
Lithophragma 513
Lithops 20
Lithospermum 107
Litsea 269
Livistona 48
Lloydia 279
Loasaceae 283
Lobelia 122
Lobularia 113
Loganiaceae 284
Lolium 399
Lomatium 30
Lonchocarpus 203
Lonicera 124
Lophocereus 119
Lotus 203
Ludwigia 318
Luffa 150
Luina 79
Lunaria 113
Lupinus 203
Luzula 262
Lychnis 131
Lycium 526
Lycopersicon 526
Lycopodiaceae 284
Lycopodium 284
Lycopus 264
Lycoris 279
Lycurus 400
Lygodesmia 79
Lygodium 447
Lyonia 174
Lyonothamnus 469
Lysichiton 40
Lysiloma 205
Lysimachia 443
Lythraceae 284
Lythrum 285
Maackia 205
Macadamia 444
Macfadyena 105
Machaeranthera 79
Machaerina 165
Maclura 300
Macroptilium 205
Macrotyloma 206
Macrozamia 552
Madia 79
Magnolia 287
Magnoliaceae 285
Mahonia 92
Maianthemum 279
Malachra 294
Malacothamnus 294
Malacothrix 79
Mallotus 186
Malpighia 289
Malpighiaceae 289
Malus 469
Malva 294
Malvaceae 289
Malvastrum 294
Malvaviscus 294
Malvella 294
Mammillaria 119
Mandevilla 34
Manfreda 18
Mangifera 23
Manihot 186
Manilkara 512
Marah 150
Maranta 296
Marantaceae 295
Marina 206
Marrubium 264
Marsilea 447
Matelea 53
Matricaria 79
Matteuccia 447
Matthiola 114
Maurandya 518
Maytenus 134
Meconopsis 325
Medeola 279
Medicago 206
Medinilla 296
Melaleuca 305
Melanthera 80
Melastoma 296
Melastomataceae 296
Melia 297
Meliaceae 296
Melica 400
Melicoccus 511
Melilotus 207
Melinis 400
Melissa 264
Melothria 150
Menispermaceae 297
Menispermum 297
Mentha 264
Mentzelia 283
Menyanthaceae 298
Menyanthes 298
Menziesia 175
Merremia 143
Mertensia 107
Meryta 43
Mesembryanthemum 20
Mespilus 473
Metasequoia 534
Metrosideros 306
Michelia 289
Miconia 296
Micranthemum 518
Microlepia 447
Microseris 80
Microsorium 447
Microsteris 435
Mikania 80
Milium 400
Mimosa 208
Mimulus 518
Mimusops 512
Mirabilis 308
Mitchella 493
Mitella 514
Mitreola 284
Modiola 294
Molinia 400
Molluginaceae 298
Mollugo 298
Moluccella 265
Momordica 150
Monanthochloe 400
Monarda 265
Monardella 265
Moneses 451
Monimiaceae 298
Monodora 26
Monolena 296
Monolepis 137
Monotropa 298
Monotropaceae 298
Monstera 40
Montia 442
Moraceae 298
Morinda 493
Morrenia 53
Mortonia 134
Morus 300
Mucuna 208
Muhlenbergia 400
Munroa 402
Muntingia 169
Murraya 497
Musa 301
Musaceae 301
Muscari 279
Musineon 31
Myoporaceae 301
Myoporum 301
Myosotis 107
Myrciaria 307
Myrica 302
Myricaceae 302
Myriophyllum 247
Myrsinaceae 302
Myrsine 303
Myrtaceae 303
Myrtus 307
Nama 252
Nandina 92
Napaea 294
Narcissus 279
Nasturtium 114
Nautilocalyx 244
Navarretia 435
Nectandra 269
Neillia 473
Nelumbo 308
Nelumboaceae 308
Nematanthus 244
Nemopanthus 38
Nemophila 252
Nengella 48
Neomarica 255
Neonotonia 208
Neoregelia 117
Nepenthaceae 308
Nepenthes 308
Nepeta 265
Nephelium 511
Nephrolepis 447
Nephrophyllidium 298
Nephthytis 40
Nerine 280
Nerium 34
Nicandra 528
Nicotiana 528
Nidularium 117
Nolina 18
Nopalea 120
Noronhia 315
Nothochelone 518
Notholaena 448

Nothoscordum 280
Nuphar 309
Nyctaginaceae 308
Nymphaea 309
Nymphaeaceae 309
Nymphoides 298
Nyssa 309
Nyssaceae 309
Ochrosia 35
Ocimum 266
Ocotea 269
Odontoglossum 322
Oemleria 473
Oenanthe 31
Oenothera 319
Olea 315
Oleaceae 310
Olneya 208
Omphalodes 108
Onagraceae 316
Oncidium 322
Oncoba 240
Onobrychis 209
Onoclea 448
Onosmodium 108
Ophioglossum 448
Ophiopogon 280
Oplismenus 402
Oplopanax 43
Opuntia 120
Orchidaceae 320
Orchis 322
Oreocallis 445
Origanum 266
Ornithogalum 280
Ornithopus 209
Orobanchaceae 323
Orobanche 323
Orogenia 31
Orontium 40
Orthocarpus 518
Oryza 402
Oryzopsis 403
Osmanthus 315
Osmorhiza 31
Osmunda 448
Osteomeles 473
Osteospermum 80
Ostrya 102
Oxalidaceae 323
Oxalis 323
Oxydendrum 175
Oxypolis 31
Oxyria 438
Oxytropis 209
Pachira 106
Pachycarpus 53
Pachypodium 35
Pachyrhizus 209
Pachysandra 118
Pachystachys 8
Paederia 493
Paeonia 323
Paeoniaceae 323
Panax 44
Pandanaceae 324
Pandanus 324
Pandorea 105
Panicum 403
Papaver 325
Papaveraceae 324
Pappophorum 405
Paralinospadix 48
Parietaria 543
Parkinsonia 209
Parnassia 514
Paronychia 131
Parrya 114
Parryella 209
Parthenice 80
Parthenium 80
Parthenocissus 550
Paspalum 405
Passiflora 325
Passifloraceae 325
Pastinaca 31
Paulownia 519
Pavonia 294
Paxistima 134
Pectocarya 108
Pedaliaceae 326
Pedicularis 519
Pedilanthus 186
Pediomelum 209
Pelargonium 243
Pelea 497
Pellaea 448
Pellionia 543
Peltandra 40
Pennisetum 406
Penstemon 519
Pentas 493
Penthorum 514
Peperomia 359
Peraphyllum 473
Perideridia 31
Perilla 266
Perovskia 266
Perrottetia 134
Persea 269
Petasites 80
Petradoria 80
Petroselinum 32
Petunia 529
Peumus 298
Phacelia 252
Phalaenopsis 322
Phalaris 407
Phanerophlebia 448
Phaseolus 209
Phellodendron 497
Philadelphus 251
Philibertia 53
Philodendron 40
Phleum 407
Phlomis 266
Phlox 435
Phoenicaulis 114
Phoenix 48
Pholidocarpus 49
Phoradendron 549
Phormium 19
Photinia 473
Phragmites 408
Phryma 546
Phyla 546
Phyllodoce 175
Phyllostachys 409
Phyllostegia 266
Physalis 529
Physocarpus 473
Physostegia 266
Phytolacca 326
Phytolaccaceae 326
Picea 334
Picramnia 523
Pieris 175
Pilea 543
Pimenta 307
Pimpinella 32
Pinaceae 326
Pinaropappus 80
Pinckneya 493
Pinus 339
Piper 359
Piperaceae 359
Piptanthus 211
Piptochaetium 409
Pipturus 544
Piqueria 81
Piscidia 211
Pisonia 308
Pistacia 23
Pisum 211
Pithecellobium 212
Pittosporaceae 359
Pittosporum 359
Pityrogramma 448
Plagiobothrys 108
Planera 540
Plantaginaceae 359
Plantago 359
Platanaceae 360
Platanus 360
Platycerium 448
Platycodon 122
Platydesma 498
Plectranthus 266
Plectritis 544
Pleuropogon 409
Pluchea 81
Plumbaginaceae 362
Plumbago 362
Plumeria 35
Poa 409
Poaceae 362
Podocarpaceae 434
Podocarpus 434
Podophyllum 92
Pogonia 322
Polemoniaceae 434
Polemonium 436
Polianthes 19
Polygala 436
Polygalaceae 436
Polygonaceae 436
Polygonatum 280
Polygonella 438
Polygonum 438
Polymnia 81
Polypodium 449
Polypogon 413
Polyscias 44
Polystichum 449
Polytaenia 32
Poncirus 498
Pongamia 212
Pontederia 441
Pontederiaceae 441
Populus 498
Porophyllum 81
Portlandia 494
Portulaca 442
Portulacaceae 441
Portulacaria 442
Potamogeton 442
Potamogetonaceae 442
Potentilla 474
Pothos 41
Pourouma 134
Pouteria 512
Prenanthes 81
Primula 443
Primulaceae 442
Pritchardia 49
Proboscidea 326
Proserpinaca 247
Prosopis 212
Protea 445
Proteaceae 444
Prunella 266
Prunus 475
Pseudanamomis 307
Pseuderanthemum 8
Pseudobombax 106
Pseudocymopterus 32
Pseudolarix 352
Pseudopanax 44
Pseudotsuga 352
Pseudovigna 213
Psidium 307
Psilostrophe 81
Psilotaceae 445
Psilotum 445
Psoralea 213
Psoralidium 213
Psychotria 494
Ptelea 498
Pteridium 449
Pteris 450
Pterocarpus 213
Pterocarya 260
Pterygota 532
Ptilimnium 32
Ptychosperma 49
Puccinellia 413
Pueraria 213
Punica 450
Punicaceae 450
Purshia 481
Puya 117
Pycnanthemum 266
Pyracantha 481
Pyrola 451
Pyrolaceae 451
Pyrostegia 105
Pyrrhopappus 81
Pyrularia 510
Pyrus 482
Quercus 227
Quisqualis 139
Rafinesquia 81
Railliardia 81
Randia 494
Ranunculaceae 451
Ranunculus 456
Rapanea 303
Raphanus 114
Ratibida 81
Ravenala 532
Redfieldia 413
Reseda 458
Resedaceae 458
Rhabdadenia 35
Rhamnaceae 458
Rhamnus 459
Rhaphiolepis 483
Rhapis 49
Rheum 439
Rhexia 296
Rhinanthus 520
Rhipsalidopsis 120
Rhizophora 460
Rhizophoraceae 460
Rhododendron 176
Rhodoleia 249
Rhodotypos 484
Rhus 24
Rhynchelytrum 413
Rhynchosia 214
Rhynchospora 165
Ribes 245
Richardia 494
Ricinus 186
Rivina 326
Robinia 214
Rockia 309
Rohdea 280
Rollandia 122
Rollinia 26
Romanzoffia 253
Rorippa 115
Rosa 484
Rosaceae 460
Rosmarinus 266
Rottboellia 413
Roystonea 49
Rubiaceae 490
Rubus 486
Rudbeckia 82
Ruellia 8
Rumex 440
Rumohra 450
Russelia 520
Rutaceae 494
Sabal 49
Sabatia 242
Saccharum 413
Sadleria 450
Sageretia 460
Sagina 131
Sagittaria 21
Saintpaulia 244
Salicaceae 498
Salicornia 137
Salix 505
Salpiglossis 529
Salsola 137
Salvia 267
Sambucus 125
Samolus 444
Sanchezia 8
Sanguinaria 325
Sanguisorba 488
Sanicula 32
Sansevieria 19
Santalaceae 510
Santalum 510
Santolina 82
Sapindaceae 511
Sapindus 511
Sapium 186
Saponaria 131
Sapotaceae 512
Sarcobatus 137
Sarcococca 118
Sarcostemma 53
Sarracenia 512
Sarraceniaceae 512
Sassafras 270
Satureja 267
Saururaceae 513
Saururus 513
Saussurea 82
Saxifraga 514
Saxifragaceae 513
Scabiosa 167
Scaevola 244
Schedonnardus 415
Schefflera 44
Scheuchzeria 515
Scheuchzeriaceae 515
Schinus 25
Schizachne 415
Schizachyrium 415

Schizanthus 529
Schlumbergera 120
Schrankia 215
Sciadopitys 534
Scilla 280
Scirpus 165
Scleria 166
Sclerochloa 415
Scleropogon 415
Scolochloa 415
Scoparia 520
Scorzonera 82
Scrophularia 521
Scrophulariaceae 515
Scutellaria 267
Sebastiania 186
Secale 416
Sechium 150
Sedum 146
Selaginella 522
Selaginellaceae 522
Selenicereus 120
Sempervivum 146
Senecio 82
Senna 215
Sequoia 534
Sequoiadendron 535
Serenoa 50
Sericocarpus 83
Serjania 511
Serresia 547
Sesamum 326
Sesbania 215
Sesuvium 20
Setaria 417
Severinia 498
Shepherdia 169
Shorea 167
Shortia 166
Sibara 115
Sibbaldia 489
Sicana 150
Sicyos 150
Sida 294
Sidalcea 295
Siderasis 140
Sideroxylon 512
Silene 131
Silphium 83
Silybum 84
Simarouba 523
Simaroubaceae 522
Simmondsia 523
Simmondsiaceae 523
Sinapis 115
Sinningia 244
Sisymbrium 115
Sisyrinchium 255
Sitanion 418
Sium 32
Smallanthus 84
Smelowskia 115
Smilacaceae 523
Smilacina 280
Smilax 523
Solanaceae 524
Solanum 530
Soleirolia 544
Solenostemon 267
Solidago 84
Sonchus 86
Sonerila 296
Sophora 215
Sorbaria 489
Sorbus 489
Sorghastrum 418
Sorghum 419
Sparganiaceae 532
Sparganium 532
Spartina 420
Spartium 216
Spathiphyllum 41
Spathodea 105
Spathoglottis 322
Spergula 132
Spergularia 132
Spermacoce 494
Sphacele 268
Sphaeralcea 295
Sphenomeris 450
Sphenopholis 421
Sphenosciadium 32
Spigelia 284
Spinacia 137
Spiraea 489
Spiranthes 322
Spirodela 271
Spondias 26
Sporobolus 421
Spraguea 442
Stachys 268
Stachytarpheta 546
Stangeria 450
Stanleya 115
Stapelia 53
Staphylea 532
Staphyleaceae 532
Steirodiscus 86
Stelechocarpus 26
Stellaria 132
Stenanthium 281
Stenocarpus 445
Stenosiphon 319
Stenotaphrum 422
Stephanomeria 86
Stephanotis 53
Sterculiaceae 532
Sternbergia 281
Stigmaphyllon 289
Stillingia 187
Stipa 422
Stiporyzopsis 424
Stokesia 86
Straussia 494
Strelitzia 532
Strelitziaceae 532
Streptanthus 115
Streptocarpus 244
Streptopus 281
Striga 521
Strobilanthes 8
Stromanthe 296
Strongylodon 216
Strophostyles 216
Struthiopteris 450
Stylosanthes 216
Styphelia 169
Styracaceae 533
Styrax 533
Suaeda 138
Sullivantia 515
Swertia 242
Swietenia 297
Syagrus 50
Symphoricarpos 126
Symphytum 108
Symplocaceae 533
Symplocarpus 41
Symplocos 533
Synadenium 187
Syngonium 41
Synsepalum 512
Synthyris 521
Syringa 315
Syzygium 307
Tabebuia 105
Tabernaemontana 35
Tacca 533
Taccaceae 533
Taeniatherum 424
Taenidia 32
Tagetes 86
Talinum 442
Tamaricaceae 533
Tamarix 533
Tanacetum 87
Tapeinocheilos 553
Taraxacum 87
Tauschia 32
Taxaceae 533
Taxodiaceae 534
Taxodium 535
Taxus 533
Tecoma 105
Tecomaria 105
Tectaria 450
Tectona 546
Tellima 515
Tephrosia 216
Teramnus 216
Terminalia 139
Ternstroemia 536
Tetradymia 87
Tetragonia 20
Tetramerium 8
Tetrapanax 44
Tetraplasandra 44
Tetrastigma 550
Tetrazygia 296
Teucrium 268
Thalassia 252
Thalia 296
Thalictrum 457
Thaspium 33
Thaumatophyllum 41
Thea 536
Theaceae 535
Thelypodium 115
Thelypteris 450
Theobroma 532
Theophrastaceae 536
Thermopsis 216
Thespesia 295
Thevetia 35
Thibaudia 179
Thlaspi 115
Thrinax 50
Thuja 156
Thujopsis 158
Thunbergia 9
Thymelaeaceae 536
Thymus 268
Thysanocarpus 116
Tiarella 515
Tibouchina 296
Tidestromia 23
Tigridia 255
Tilia 537
Tiliaceae 537
Tillaea 146
Tillandsia 117
Tipuana 216
Tithonia 87
Tofieldia 281
Tolmiea 515
Tonella 521
Toona 297
Torenia 521
Torilis 33
Torreya 534
Tournefortia 108
Trachelospermum 35
Trachycarpus 50
Trachymene 33
Tradescantia 140
Tragia 187
Tragopogon 87
Trapa 538
Trapaceae 538
Trautvetteria 457
Trema 540
Trevesia 45
Trianthema 20
Tribulus 553
Trichocereus 121
Trichosanthes 151
Trichostema 268
Tridax 88
Tridens 424
Trientalis 444
Trifolium 216
Triglochin 262
Trigonella 219
Trillium 281
Triodanis 122
Triodia 424
Triosteum 127
Triphasia 498
Triplasis 424
Tripogandra 140
Tripsacum 424
Trisetum 425
Tristania 307
Triteleia 281
Triticum 425
Tritonia 255
Triumfetta 538
Trollius 458
Tropaeolaceae 538
Tropaeolum 538
Tsuga 355
Tulbaghia 281
Tulipa 281
Tupidanthus 45
Turnera 538
Turneraceae 538
Tussilago 88
Typha 538
Typhaceae 538
Ulmaceae 539
Ulmus 540
Umbellularia 271
Ungnadia 512
Uniola 428
Urena 295
Urtica 544
Urticaceae 543
Uvularia 282
Vaccaria 132
Vaccinium 179
Valeriana 544
Valerianaceae 544
Valerianella 545
Vallota 282
Vancouveria 92
Vanda 322
Vanilla 323
Vauquelinia 490
Veitchia 51
Veratrum 282
Verbascum 521
Verbena 546
Verbenaceae 545
Verbesina 88
Vernonia 88
Veronica 521
Veronicastrum 522
Vetiveria 428
Viburnum 127
Vicia 219
Victoria 309
Vigna 220
Viguiera 88
Vinca 35
Vincetoxicum 53
Viola 547
Violaceae 547
Viscaceae 549
Viscum 549
Vitaceae 549
Vitex 547
Vitis 550
Vriesea 117
Vulpia 428
Waldsteinia 490
Washingtonia 51
Watsonia 255
Wedelia 89
Weigela 128
Westringia 268
Whipplea 251
Wikstroemia 537
Wisteria 221
Woodsia 450
Woodwardia 450
Wyethia 89
Xantheranthemum 9
Xanthium 89
Xanthorhiza 458
Xanthosoma 41
Xerophyllum 282
Xylosma 240
Xyridaceae 552
Xyris 552
Youngia 89
Yucca 19
Zamia 552
Zamiaceae 552
Zantedeschia 41
Zanthoxylum 498
Zauschneria 319
Zea 428
Zebrina 140
Zelkova 543
Zenobia 182
Zephyranthes 282
Zexmenia 89
Zigadenus 283
Zingiber 553
Zingiberaceae 553
Zinnia 89
Zizania 433
Zizaniopsis 433
Zizia 33
Ziziphus 460
Zornia 222
Zostera 553
Zosteraceae 553
Zoysia 433
Zygophyllaceae 553

Host Common Name Index

Aaron's-beard - *Hypericum calycinum* 139
abele - *Populus* x *alba* 498
abelia, glossy - *Abelia* x *grandiflora* 123
absinthe - *Artemisia absinthium* 57
absinthium - *Artemisia absinthium* 57
acacia
  blackwood acacia - *Acacia melanoxylon* 187
  cat's-claw acacia - *Acacia greggii* 187
  cultivated acacia - *Acacia cyclopis* 187
  false acacia - *Robinia pseudoacacia* 214
  mescat acacia - *Acacia constricta* 187
  prairie acacia - *Acacia angustissima* 187
  sweet acacia - *Acacia farnesiana* 187
aceituna - *Simarouba glauca* 523
acerola - *Malpighia glabra* 289
achiote - *Bixa orellana* 105
achira - *Canna edulis* 123
aconite - *Aconitum napellus* 451
Adam's-needle - *Yucca filamentosa, Y. smalliana* 19
Adam-and-Eve - *Aplectrum hyemale* 320
adder's-tongue
  yellow adder's-tongue - *Erythronium americanum* 276
agapanthus, blue - *Agapanthus africanus* 271
ageratum - *Ageratum houstonianum* 54
  hardy ageratum - *Eupatorium coelestinum* 70
agrimony - *Agrimonia eupatoria* 460
aguerroot - *Aletris farinosa* 272
ailanthus - *Ailanthus altissima* 522
airplant - *Kalanchoe pinnata* 146
  South American airplant - *Kalanchoe fedtschenkoi* 146
akebia, five-leaf - *Akebia quinata* 268
akee - *Blighia sapida* 511
akulikuli - *Lampranthus glomeratus* 20
albizia, silktree - *Albizia julibrissin* 188
alder
  American green alder - *Alnus crispa* 92
  black alder - *Alnus glutinosa* 92
  European alder - *Alnus glutinosa* 92
  European green alder - *Alnus viridis* 92
  green alder - *Alnus crispa, A. sinuata* 92
  hazel alder - *Alnus rugosa, A. serrulata* 92
  Japanese alder - *Alnus japonica* 92
  mountain alder - *Alnus crispa, A. tenuifolia* 92
  New Mexican alder - *Alnus oblongifolia* 92
  Oregon alder - *Alnus rubra* 92
  red alder - *Alnus rubra* 92
  Sierra alder - *Alnus rhombifolia* 92
  Sitka alder - *Alnus sinuata* 92
  smooth alder - *Alnus rugosa* 92
  speckled alder - *Alnus incana, A. rugosa* 92
  thinleaf alder - *Alnus tenuifolia* 92
  white alder - *Alnus incana, A. rhombifolia* 92
alehoof - *Glechoma hederacea* 263
alexanders - *Angelica atropurpurea* 26
  golden alexanders - *Zizia aurea* 33
alfalfa - *Medicago sativa* 206
  yellow-flowered alfalfa - *Medicago sativa* subsp. *falcata* 206
alfilaria - *Erodium cicutarium* 242
algarroba - *Prosopis chilensis, P. juliflora, P. pallida* 212
alimau - *Citrus macrophylla* 494
alkanet - *Anchusa officinalis* 106
alkekengi - *Physalis alkekengi* 529
allamanda - *Allamanda cathartica* 33
  bush allamanda - *Allamanda neriifolia* 33
allgood - *Chenopodium bonus-henricus* 136
alligator-pear - *Persea americana* 269
alligator-weed - *Alternanthera philoxeroides* 21
allscale - *Atriplex polycarpa* 135
allspice - *Pimenta dioica* 307
  Carolina allspice - *Calycanthus floridus* 121
  wild allspice - *Lindera benzoin* 269
almond - *Prunus dulcis* 475
  dwarf Russian almond - *Prunus tenella* 475
  flowering almond - *Prunus triloba* 475
aloe
  Barbados aloe - *Aloe barbadensis* 273
  bitter aloe - *Aloe barbadensis* 273
  Curacao aloe - *Aloe barbadensis* 273
  false aloe - *Manfreda virginica* 18
  kamniedood aloe - *Aloe variegata* 273
  medicinal aloe - *Aloe barbadensis* 273
  tiger aloe - *Aloe variegata* 273
  true aloe - *Aloe barbadensis* 273
althea, shrub - *Hibiscus syriacus* 292
aluminum-plant - *Pilea cadierei* 543
alumroot - *Geranium maculatum* 242
alyce-clover - *Alysicarpus vaginalis* 189
alyssum
  golden-tuft alyssum - *Alyssum saxatile* 108
  hoary alyssum - *Berteroa incana* 109
  sweet alyssum - *Lobularia maritima* 113
amaranth
  Chinese amaranth - *Amaranthus tricolor* 22
  Ganges amaranth - *Amaranthus tricolor* 22
  globe amaranth - *Gomphrena globosa* 22
  green amaranth - *Amaranthus hybridus, A. retroflexus, A. viridis* 22
  prostrate amaranth - *Amaranthus albus, A. blitoides, A. graecizans* 22
  slender amaranth - *Amaranthus gracilis, A. viridis* 22
  spiny amaranth - *Amaranthus spinosus* 22
amatungulu - *Carissa macrocarpa* 34
Amazon-vine - *Stigmaphyllon ciliatum* 289
amberbell - *Erythronium americanum* 276
amole - *Chlorogalum pomeridianum* 275
ampelopsis, heartleaf - *Ampelopsis cordata* 549
anacahuita - *Cordia boissieri* 106
andromeda
  Japanese andromeda - *Pieris japonica* 175
  mountain andromeda - *Pieris floribunda* 175
anemone
  candle anemone - *Anemone cylindrica* 452
  European wood anemone - *Anemone nemorosa* 452
  Japanese anemone - *Anemone hupehensis* var. *japonica* 452
  long-headed anemone - *Anemone cylindrica* 452
  meadow anemone - *Anemone canadensis* 452
  poppy anemone - *Anemone coronaria* 452
  rue anemone - *Anemonella thalictroides* 453
  wood anemone - *Anemone quinquefolia* 452
angel's-tears - *Soleirolia soleirolii* 544
angel's-trumpet - *Datura inoxia* 525
angelica
  filmy angelica - *Angelica triquinata* 26
  great angelica - *Angelica atropurpurea* 26
  purple-stemmed angelica - *Angelica atropurpurea* 26
angelica-tree - *Aralia spinosa* 42
angle-pod - *Cynanchum laeve* 52
anise - *Pimpinella anisum* 32
  purple anise - *Illicium floridanum* 253
anise-hyssop - *Agastache foeniculum* 262
aniseroot - *Osmorhiza longistylis* 31
annatto - *Bixa orellana* 105
anoda, spurred - *Anoda cristata* 290
anthericum, walking - *Chlorophytum comosum* 275
Appalachian-tea - *Ilex glabra* 36, *Viburnum cassinoides*
apple - *Malus sylvestris* 469
  Adam's apple - *Tabernaemontana divaricata* 35
  Chinese apple - *Malus prunifolia* 469
  common apple - *Malus domestica* 469
  Indian apple - *Datura inoxia* 525
  Mexican apple - *Casimiroa edulis* 494
  paradise apple - *Malus pumila, M. sylvestris* 469
  plum-leaved apple - *Malus prunifolia* 469
  wild apple - *Malus sylvestris* 469
apple-of-Peru - *Nicandra physalodes* 528
apricot - *Prunus armeniaca* 475
apricotvine - *Passiflora incarnata* 325
aquacate - *Persea americana* 269
aralia
  Balfour aralia - *Polyscias balfouriana* 44
  geranium-leaf aralia - *Polyscias guilfoylei* 44
  ming aralia - *Polyscias fruticosa* 44
aramina - *Urena lobata* 295

arborvitae
American arborvitae - *Thuja occidentalis* 156
false arborvitae - *Thujopsis dolabrata* 158
giant arborvitae - *Thuja plicata* 156
hiba arborvitae - *Thujopsis dolabrata* 158
Oriental arborvitae - *Thuja orientalis* 156
arbutus, trailing - *Epigaea repens* 172
ardisia - *Ardisia crispa* 302
Arizona-giant - *Carnegiea gigantea* 118
aromo - *Azara microphylla* 240
arrow-feather - *Aristida purpurascens* 371
arrowhead - *Sagittaria lancifolia* 21
Chinese arrowhead - *Sagittaria sagittifolia* 21
coastal arrowhead - *Sagittaria graminea* 21
common arrowhead - *Sagittaria latifolia* 21
giant arrowhead - *Sagittaria montevidensis* 21
old-world arrowhead - *Sagittaria sagittifolia* 21
slender arrowhead - *Sagittaria graminea* 21
arrowhead-vine - *Syngonium podophyllum* 41
arrowroot - *Maranta arundinacea* 296
Florida arrowroot - *Zamia pumila* 553
Queensland arrowroot - *Canna edulis* 123
Spanish arrowroot - *Canna edulis* 123
arrowwood - *Viburnum acerifolium* 127
downy-leaved arrowwood - *Viburnum rafinesquianum* 127
smooth arrowwood - *Viburnum recognitum* 127
artichoke - *Cynara cardunculus* 67
globe artichoke - *Cynara cardunculus* 67
artillery-plant - *Pilea microphylla* 543
artillery-weed - *Pilea microphylla* 543
arum
arrow arum - *Peltandra virginica* 40
dragon arum - *Arisaema dracontium* 39
ivy arum - *Epipremnum aureum* 40
water arum - *Calla palustris* 39
arum-lily - *Zantedeschia aethiopica* 41
ash
black ash - *Fraxinus nigra* 311
blue ash - *Fraxinus quadrangulata* 311
Carolina ash - *Fraxinus caroliniana* 311
European ash - *Fraxinus excelsior* 311
green ash - *Fraxinus pennsylvanica* 311
Oregon ash - *Fraxinus latifolia* 311
pop ash - *Fraxinus caroliniana* 311
red ash - *Fraxinus pennsylvanica* 311
velvet ash - *Fraxinus velutina* 311
water ash - *Fraxinus caroliniana* 311
white ash - *Fraxinus americana* 311
asparagus
common asparagus - *Asparagus officinalis* 274
Cossack asparagus - *Typha latifolia* 538
cultivated asparagus - *Asparagus officinalis* var. *altilis* 274
fern asparagus - *Asparagus setaceus* 274
garden asparagus - *Asparagus officinalis* 274
Smilax asparagus - *Asparagus asparagoides* 274
Sprenger asparagus - *Asparagus densiflorus* 274
asparagus-fern - *Asparagus plumosus* 274
aspen
bigtooth aspen - *Populus grandidentata* 499
large-toothed aspen - *Populus grandidentata* 499
quaking aspen - *Populus tremuloides* 499
trembling aspen - *Populus tremuloides* 499
aster
annual aster - *Callistephus chinensis* 62
bluewood aster - *Aster cordifolius* 58
China aster - *Callistephus chinensis* 62
heath aster - *Aster ericoides* 58
New England aster - *Aster novae-angliae* 58
rock aster - *Aster alpinus* 58
Stokes' aster - *Stokesia laevis* 86
white upland aster - *Aster ptarmicoides* 58
whitewood aster - *Aster divaricatus* 58
ati - *Calophyllum inophyllum* 138
aubergine - *Solanum melongena, S. melongena* var. *esculentum* 530
aucuba, Japanese - *Aucuba japonica* 143
auricula-tree - *Calotropis procera* 52
Australian-cedar - *Toona ciliata* var. *australis* 297
Australian-laurel - *Pittosporum tobira* 359
Australian-nut - *Macadamia integrifolia* 444
Australian-pepper - *Schinus molle* 25
avens - *Geum aleppicum* 467
mountain avens - *Dryas octopetala, D. octopetala* subsp. *alaskensis* 465
purple avens - *Geum rivale* 467
water avens - *Geum rivale* 467
avocado - *Persea americana* 269
azalea
Chinese azalea - *Rhododendron molle* 176
clammy azalea - *Rhododendron viscosum* 176
flame azalea - *Rhododendron calendulaceum* 176
hiryu azalea - *Rhododendron obtusum* 176
hoary azalea - *Rhododendron canescens* 176
Japanese azalea - *Rhododendron japonicum* 176
macranthum azalea - *Rhododendron indicum* 176
pink-shell azalea - *Rhododendron vaseyi* 176
plum-leaved azalea - *Rhododendron prunifolium* 176
pontic azalea - *Rhododendron luteum* 176
smooth azalea - *Rhododendron arborescens* 176
snow azalea - *Rhododendron mucronatum* 176
swamp azalea - *Rhododendron viscosum* 176
sweet azalea - *Rhododendron arborescens* 176
torch azalea - *Rhododendron kaempferi* 176
western azalea - *Rhododendron occidentale* 176
white swamp azalea - *Rhododendron viscosum* 176
yellow azalea - *Rhododendron calendulaceum* 176
baby's-breath - *Gypsophila elegans, G. paniculata* 130
baby's-tears - *Soleirolia soleirolii* 544
baby-blue-eyes - *Nemophila menziesii* 252
baby-pepper - *Rivina humilis* 326
baby-rubberplant - *Peperomia obtusifolia* 359
baccharis
dwarf baccharis - *Baccharis pilularis* 60
sticky baccharis - *Baccharis glutinosa* 60
bachelor's-button - *Centaurea cyanus* 63
yellow bachelor's-button - *Polygala lutea* 436
bagflower - *Clerodendrum thomsoniae* 545
bald-cypress - *Taxodium distichum* 535
ball-moss - *Tillandsia recurvata* 117
ball-nightshade - *Solanum carolinense* 530
balloonvine - *Cardiospermum halicacabum* 511
balm
bee balm - *Melissa officinalis* 264, *Monarda didyma* 265
common balm - *Melissa officinalis* 264
lemon balm - *Melissa officinalis* 264
molucca balm - *Moluccella laevis* 265
spotted bee balm - *Monarda punctata* 265
sweet balm - *Melissa officinalis* 264
balmony - *Chelone glabra* 517
balsam
garden balsam - *Impatiens balsamina* 90
he balsam - *Picea rubens* 334
rose balsam - *Impatiens balsamina* 90
Zanzibar balsam - *Impatiens wallerana* 90
balsam-apple - *Clusia rosea* 138, *Momordica balsamina* 150
wild balsam-apple - *Echinocystis lobata* 150
balsam-pear - *Momordica charantia* 150
balsamo-cor-de-carne - *Justicia carnea* 8
bamboo
black bamboo - *Phyllostachys nigra* 409
canebreak bamboo - *Arundinaria gigantea* 372
common bamboo - *Bambusa vulgaris* 374
fishpole bamboo - *Phyllostachys aurea* 409
golden bamboo - *Phyllostachys aurea* 409
hardy timber bamboo - *Phyllostachys bambusoides* 409
hedge bamboo - *Bambusa multiplex* 374
Japenese timber bamboo - *Phyllostachys bambusoides* 409
bamboo-vine - *Smilax laurifolia* 523
banana
dwarf banana - *Musa acuminata, M. nana* 301
edible banana - *Musa acuminata* 301
plantain cooking banana - *Musa* × *paradisiaca* 301
banana-shrub - *Michelia figo* 289
band-plant - *Vinca major* 35
baneberry
red baneberry - *Actaea rubra* 452
white baneberry - *Actaea alba* 452
banyan, Indian - *Ficus benghalensis* 298
Barbados-cherry - *Malpighia glabra* 289
Barbados-nut - *Jatropha curcas* 185
barbe-de-capuchin - *Cichorium intybus* 65
barberry
Allegheny barberry - *Berberis canadensis* 91
American barberry - *Berberis canadensis* 91
blue barberry - *Berberis aquifolium* 91
Colorado barberry - *Berberis fendleri* 91
common barberry - *Berberis vulgaris* 91
creeping barberry - *Berberis repens* 91
European barberry - *Berberis vulgaris* 91
holly barberry - *Berberis aquifolium* 91
Japanese barberry - *Berberis thunbergii* 91
red barberry - *Berberis haematocarpa* 91
barley
bulbous barley - *Hordeum bulbosum* 395
common barley - *Hordeum vulgare* 395
foxtail barley - *Hordeum jubatum* 395
little barley - *Hordeum pusillum* 395
mouse barley - *Hordeum murinum* 395
Nepal barley - *Hordeum vulgare* 395
short-awned barley - *Hordeum brevisubulatum* 395
squirrel-tail barley - *Hordeum jubatum* 395
wall barley - *Hordeum murinum* 395

barometer-bush - *Leucophyllum frutescens* 518
barroom-plant - *Aspidistra elatior* 274
basil
common basil - *Ocimum basilicum* 266
sweet basil - *Ocimum basilicum* 266
basket-flower - *Centaurea americana* 63
basket-of-gold - *Alyssum saxatile* 108
basketflower - *Hymenocallis narcissiflora* 277
basketvine, scarlet - *Aeschynanthus pulcher* 243
basswood
American basswood - *Tilia americana* 537
white basswood - *Tilia heterophylla* 537
bastard-willow - *Rhus lancea* 24
batino - *Alstonia macrophylla* 33
Batoko-plum - *Flacourtia indica* 240
bauhinia
nasturtium bauhinia - *Bauhinia galpinii* 192
red bauhinia - *Bauhinia galpinii* 192
bay - *Laurus nobilis* 269
bull bay - *Magnolia grandiflora* 287
California bay - *Umbellularia californica* 271
sweet bay - *Magnolia virginiana* 287
bayberry, California - *Myrica californica* 302
baygall-bush - *Ilex coriacea* 36
be-still-tree - *Thevetia peruviana* 35
beadtree
Japanese beadtree - *Melia azedarach* 297
Syrian beadtree - *Melia azedarach* 297
beakrush
brown beakrush - *Rhynchospora fusca* 165
nodding beakrush - *Rhynchospora inexpansa* 165
white beakrush - *Rhynchospora alba* 165
bean
algarroba bean - *Ceratonia siliqua* 193
American potato bean - *Apios americana* 189
asparagus bean - *Vigna unguiculata* subsp. *sesquipedalis* 221
azuki bean - *Vigna angularis* 220
bonavist bean - *Lablab purpureus* 201
broad bean - *Vicia faba* 219
bush bean - *Phaseolus vulgaris* var. *humilis* 209
butter bean - *Phaseolus lunatus* 209
Carolina bean - *Phaseolus lunatus* 209
castor bean - *Ricinus communis* 186
Cherokee bean - *Erythrina herbacea* 197
civet bean - *Phaseolus lunatus* 209
cluster bean - *Cyamopsis tetragonoloba* 195
coral bean - *Erythrina herbacea* 197
Dutch-case-knife bean - *Phaseolus coccineus* 209
fava bean - *Vicia faba* 219
French bean - *Phaseolus vulgaris* 209
garden bean - *Phaseolus lunatus, P. vulgaris* 209
giant stock bean - *Canavalia ensiformis* 193
green bean - *Phaseolus vulgaris* 209
haricot bean - *Phaseolus vulgaris* 209
horse bean - *Canavalia ensiformis* 193, *Vicia faba* 219
hyacinth bean - *Lablab purpureus* 201
Indian bean - *Catalpa speciosa* 104
jack bean - *Canavalia ensiformis* 193
jequirity bean - *Abrus precatorius* 187
jumbie bean - *Leucaena leucocephala* 203
kidney bean - *Phaseolus vulgaris* 209
lablab bean - *Lablab purpureus* 201
lima bean - *Phaseolus lunatus* 209
locust bean - *Ceratonia siliqua* 193
mung bean - *Vigna radiata* 220
navy bean - *Phaseolus vulgaris* 209
precatory bean - *Abrus precatorius* 187
purple bean - *Macroptilium atropurpureum* 205
runner bean - *Phaseolus vulgaris* 209
salad bean - *Phaseolus vulgaris* 209
scarlet runner bean - *Phaseolus coccineus* 209
sewee bean - *Phaseolus lunatus* 209
sieva bean - *Phaseolus lunatus* 209
snap bean - *Phaseolus vulgaris* 209
string bean - *Phaseolus vulgaris* 209
sword bean - *Canavalia ensiformis, C. gladiata* 193
Tepary bean - *Phaseolus acutifolius* 209
wax bean - *Phaseolus vulgaris* 209
wild bean - *Apios americana* 189
Windsor bean - *Vicia faba* 219
wolf bean - *Lupinus albus* 203
wonder bean - *Canavalia ensiformis* 193
yard-long bean - *Vigna unguiculata* 220, *V. unguiculata* subsp. *sesquipedalis* 221
bear's-foot - *Aconitum napellus* 451
bear's-grape - *Arctostaphylos uva-ursi* 171
bearberry - *Arctostaphylos uva-ursi* 171, *Rhamnus purshiana* 459
alpine bearberry - *Arctostaphylos alpina* 170
black bearberry - *Arctostaphylos alpina* 170
common bearberry - *Arctostaphylos uva-ursi* 171
beardstem, bushy - *Andropogon glomeratus* 369
beauty-of-the-night - *Mirabilis jalapa* 308
beautyberry
American beautyberry - *Callicarpa americana* 545
purple beautyberry - *Callicarpa dichotoma* 545
beautybush - *Kolkwitzia amabilis* 124
bedstraw
catchweed bedstraw - *Galium aparine* 491
northern bedstraw - *Galium boreale* 491
sweet-scented bedstraw - *Galium triflorum* 491
beech
American beech - *Fagus grandifolia* 224
blue beech - *Carpinus caroliniana* 101
European beech - *Fagus sylvatica* 224
water beech - *Carpinus caroliniana* 101
beefplant - *Iresine herbstii* 22
beefsteak-geranium - *Begonia* × *rex-cultorum* 90, *Saxifraga stolonifera* 514
beefsteak-plant - *Acalypha wilkesiana* 182, *Iresine herbstii* 22
beefwood-tree - *Casuarina equisetifolia* 132
beeplant, yellow - *Cleome lutea* 123
beet
common beet - *Beta vulgaris* 135
garden beet - *Beta vulgaris, B. vulgaris* subsp. *vulgaris* 135
sea beet - *Beta vulgaris* 135
spinach beet - *Beta vulgaris* subsp. *cicla* 135
sugar beet - *Beta vulgaris* subsp. *vulgaris* 135
wild beet - *Saxifraga pensylvanica* 514
beetroot - *Beta vulgaris, B. vulgaris* subsp. *vulgaris* 135
beggarticks - *Bidens bipinnata* 61
devil's beggarticks - *Bidens frondosa* 61
hairy beggarticks - *Bidens pilosa* 61
nodding beggarticks - *Bidens cernua* 61
tall beggarticks - *Bidens vulgata* 61
beggarweed - *Desmodium canadense, D. tortuosum* 196
Florida beggarweed - *Desmodium tortuosum* 196
begonia
hybrid tuberous begonia - *Begonia* x *tuberhybrida* 90
rex begonia - *Begonia* x *rex-cultorum* 90
wild begonia - *Rumex venosus* 440
belladonna - *Atropa belladonna* 524
cape belladonna - *Amaryllis belladonna* 273
belle-apple - *Passiflora laurifolia* 325
bellflower
American bellflower - *Campanula americana* 121
creeping bellflower - *Campanula rapunculoides* 121
marsh bellflower - *Campanula aparinoides* 121
rover bellflower - *Campanula rapunculoides* 121
tall bellflower - *Campanula americana* 121
tussock bellflower - *Campanula carpatica* 121
willow bellflower - *Campanula persicifolia* 121
bells, Roanoke - *Mertensia virginica* 107
bells-of-Ireland - *Moluccella laevis* 265
Benjamin-bush - *Lindera benzoin* 269
Benjamin-tree - *Ficus benjamina* 298
bergamot, wild - *Monarda fistulosa* 265
berry
baked-apple berry - *Rubus chamaemorus* 486
crimson berry - *Rubus arcticus* 486
Himalaya berry - *Rubus procerus* 486
miraculous berry - *Synsepalum dulcificum* 512
Panama berry - *Muntingia calabura* 169
bigarade - *Citrus aurantium* 494
bignay - *Antidesma bunius* 183
bigtree - *Sequoiadendron giganteum* 535
bilberry - *Vaccinium myrtillus* 179
bog bilberry - *Vaccinium uliginosum* 179
dwarf bilberry - *Vaccinium caespitosum* 179
early sweet bilberry - *Vaccinium vacillans* 179
low bilberry - *Vaccinium vacillans* 179
bilsted - *Liquidambar styraciflua* 247
bimble-box - *Eucalyptus populnea* 303
bindweed
black bindweed - *Polygonum convolvulus* 438
corn bindweed - *Polygonum convolvulus* 438
field bindweed - *Convolvulus arvensis* 141
hedge bindweed - *Calystegia sepium* 141
Japanese bindweed - *Calystegia hederacea* 141
low bindweed - *Calystegia spithamaea* 141
small bindweed - *Convolvulus arvensis* 141
bine - *Humulus lupulus* 123
birch
black birch - *Betula lenta, B. nigra* 97
bog birch - *Betula glandulosa* 97
canoe birch - *Betula papyrifera* 97
cherry birch - *Betula lenta* 97
dwarf birch - *Betula glandulosa* 97
European white birch - *Betula pendula* 97
fire birch - *Betula populifolia* 97
gray birch - *Betula alleghaniensis, B. populifolia* 97

Kenai birch - *Betula papyrifera* var. *kenaica* 97
low birch - *Betula pumila* 97
mahogany birch - *Betula lenta* 97
mountain birch - *Betula occidentalis* 97
old-field birch - *Betula populifolia* 97
paper birch - *Betula papyrifera* 97
red birch - *Betula nigra* 97
river birch - *Betula nigra* 97
Sandberg's birch - *Betula* × *sandbergii* 97
swamp birch - *Betula pumila* 97
sweet birch - *Betula lenta* 97
water birch - *Betula occidentalis* 97
West Indian birch - *Bursera simaruba* 117
white birch - *Betula papyrifera*, *B. populifolia* 97
yellow birch - *Betula alleghaniensis* 97
bird's-eyes - *Gilia tricolor* 434
bird-catcher-tree - *Pisonia umbellifera* 308
bird-of-paradise - *Caesalpinia gilliesii* 192, *Strelitzia reginae* 532
queen's bird-of-paradise - *Strelitzia reginae* 532
bird-on-the-wing - *Polygala paucifolia* 436
biriba - *Rollinia pulchrinervis* 26
bishop's-cap - *Astrophytum myriostigma*, *A. ornatum* 118
bishop's-hood - *Astrophytum myriostigma* 118
bistort - *Polygonum bistortoides* 438
alpine bistort - *Polygonum viviparum* 438
bitter-bloom - *Sabatia angularis* 242
bitterbush - *Picramnia pentandra* 523
bitteroot - *Lewisia rediviva* 442
bittersweet
American bittersweet - *Celastrus scandens* 133
climbing bittersweet - *Celastrus scandens* 133
false bittersweet - *Celastrus scandens* 133
Oriental bittersweet - *Celastrus orbiculata* 133
shrubby bittersweet - *Celastrus scandens* 133
bitterweed - *Helenium amarum* 73
bitterwood - *Simarouba glauca* 523
black-alder - *Ilex verticillata* 36, *Viburnum nudum* 127
black-eyed-Susan - *Rudbeckia hirta* 82
black-geranium - *Heuchera americana* 513
black-olive - *Bucida buceras* 139
black-thorn - *Crataegus calpodendron* 463
blackbead - *Pithecellobium guadalupense*, *P. unguis-cati* 212
ebony blackbead - *Pithecellobium flexicaule* 212
blackbean - *Castanospermum australe* 193
blackberry
Allegheny blackberry - *Rubus allegheniensis* 486
arctic blackberry - *Rubus arcticus* 486
California blackberry - *Rubus ursinus* 486
common blackberry - *Rubus allegheniensis* 486
crimson blackberry - *Rubus arcticus* 486
cutleaf blackberry - *Rubus laciniatus* 486
evergreen blackberry - *Rubus laciniatus* 486
Pacific blackberry - *Rubus ursinus* 486
parsley-leaved blackberry - *Rubus laciniatus* 486
running blackberry - *Rubus hispidus* 486
sand blackberry - *Rubus cuneifolius* 486
sow-teat blackberry - *Rubus allegheniensis* 486
swamp blackberry - *Rubus hispidus* 486
blackberry-lily - *Belamcanda chinensis* 253
blackbutt - *Eucalyptus pilularis* 303
blackcap - *Rubus occidentalis* 486
blackening-plant - *Hibiscus rosa-sinensis* 292
blackplum - *Syzygium cumini* 307
blackroot - *Veronicastrum virginicum* 522
blackthorn - *Prunus spinosa* 475
West Indian blackthorn - *Acacia farnesiana* 187
blackwood - *Avicennia germinans* 545
Australian blackwood - *Acacia melanoxylon* 187
bladderfern
berry bladderfern - *Cystopteris bulbifera* 446
bulblet bladderfern - *Cystopteris bulbifera* 446
bladdernut, American - *Staphylea trifolia* 532
bladderpod, Gordon - *Lesquerella gordonii* 113
blaspheme-vine - *Smilax laurifolia* 523
blazing-star - *Liatris spicata* 79
bleeding-heart - *Dicentra spectabilis* 240
tropical bleeding-heart - *Clerodendrum thomsoniae* 545
western bleeding-heart - *Dicentra formosa* 240
blimbing - *Averrhoa carambola* 323
blonde-Lilian - *Erythronium albidum* 276
bloodberry - *Rivina humilis* 326
bloodroot - *Sanguinaria canadensis* 325
bloodwort, burnet - *Sanguisorba officinalis* 488
blue-buttons - *Vinca major* 35
blue-devil - *Echium vulgare* 106
blue-dicks - *Dichelostemma pulchellum* 275
blue-eyed-Mary - *Collinsia verna* 517
blue-sailors - *Cichorium intybus* 65
blue-stem, little - *Schizachyrium scoparium* 415
blue-tangle - *Gaylussacia frondosa* 173
bluebell - *Campanula rotundifolia* 121
bluebells, Virginia - *Mertensia virginica* 107
blueberry
creeping blueberry - *Vaccinium crassifolium* 179
Elliott's blueberry - *Vaccinium elliottii* 179
highbush blueberry - *Vaccinium corymbosum* 179
late sweet blueberry - *Vaccinium angustifolium* 179
low blueberry - *Vaccinium vacillans* 179
low sweet blueberry - *Vaccinium angustifolium* 179
lowbush blueberry - *Vaccinium angustifolium* 179
mountain blueberry - *Vaccinium membranaceum* 179
rabbit's-eye blueberry - *Vaccinium ashei* 179
sourtop blueberry - *Vaccinium myrtilloides* 179
swamp blueberry - *Vaccinium corymbosum* 179
sweet blueberry - *Vaccinium vacillans* 179
velvetleaf blueberry - *Vaccinium myrtilloides* 179
western blueberry - *Vaccinium occidentale* 179
blueblossum - *Ceanothus thyrsiflorus* 458
bluebonnet
Texas bluebonnet - *Lupinus subcarnosus*, *L. texensis* 204
bluebottle - *Centaurea cyanus* 63
bluebrush - *Ceanothus thyrsiflorus* 458
bluejoint - *Calamagrostis canadensis* 379
bluelips - *Collinsia grandiflora* 517
bluestem - *Andropogon glomeratus* 369, *Schizachyrium scoparium* 415
big bluestem - *Andropogon gerardii* 369
cane bluestem - *Bothriochloa barbinodis* 375
Diaz bluestem - *Dichanthium annulatum* 385
King Ranch bluestem - *Bothriochloa ischaemum* 375
sand bluestem - *Andropogon hallii* 369
silver bluestem - *Bothriochloa saccharoides* 375
Turkestan bluestem - *Bothriochloa ischaemum* 375
yellow bluestem - *Bothriochloa ischaemum* 375
bluet, mountain - *Centaurea montana* 63
bluevine - *Cynanchum laeve* 52
blueweed - *Echium vulgare* 106
bluewood, Mexican - *Condalia mexicana* 459
boat-lily - *Tradescantia spathacea* 140
bog-bean - *Menyanthes trifoliata* 298
bog-candle - *Habenaria dilatata* 321
bolbonac - *Lunaria annua* 113
boldo - *Peumus boldus* 298
boneset - *Symphytum officinale* 108
blue boneset - *Eupatorium coelestinum* 70
common boneset - *Eupatorium perfoliatum* 70
false boneset - *Kuhnia eupatorioides* 77
upland boneset - *Eupatorium sessilifolium* 70
borage - *Borago officinalis* 106
borecole - *Brassica oleracea* var. *acephala* 109
borracho, sapote - *Pouteria campechiana* 512
bottle, scented - *Habenaria dilatata* 321
bottle-tree, flame - *Brachychiton acerifolius* 532
bottlebrush - *Hystrix patula* 397
crimson bottlebrush - *Callistemon citrinus* 303
one-sided bottlebrush - *Calothamnus sanguineus* 303
weeping bottlebrush - *Callistemon viminalis* 303
bouncing-bet - *Saponaria officinalis* 131
bow-wood - *Maclura pomifera* 300
bowman's-root - *Gillenia trifoliata* 468, *Veronicastrum virginicum* 522
box-leaf, myrtle - *Paxistima myrsinites* 134
boxwood
African boxwood - *Myrsine africana* 303
common boxwood - *Buxus sempervirens* 117
Japanese boxwood - *Buxus microphylla* var. *japonica* 117
Korean boxwood - *Buxus microphylla* var. *koreana* 117
littleleaf boxwood - *Buxus microphylla* 117
Oregon boxwood - *Paxistima myrsinites* 134
Turkish boxwood - *Buxus sempervirens* 117
boysenberry - *Rubus loganobaccus* 486
brake
Cretan brake - *Pteris cretica* 450
pasture brake - *Pteridium aquilinum* 449
sword brake - *Pteris ensiformis* 450
braschette - *Brassica oleracea* var. *acephala* 109
Brazil-cherry - *Eugenia uniflora* 305
bread-and-cheeses - *Pithecellobium unguis-cati* 212

breadfruit - *Artocarpus altilis* 298, *Pandanus odoratissimus* 324
breadfruit-vine - *Monstera deliciosa* 40
breadroot - *Pediomelum esculentum* 209
bridal-wreath - *Spiraea prunifolia* 489, *Spiraea* × *vanhouttei* 490
bride's-bonnet - *Clintonia uniflora* 275
brier
  Austrian brier - *Rosa foetida* 484
  Jackson brier - *Smilax lanceolata* 523
  sensitive brier - *Schrankia nuttallii* 215
bright-eyes - *Catharanthus roseus* 34
brisbane-box - *Tristania conferta* 307
broccoli - *Brassica oleracea* var. *botrytis* 109
  asparagus broccoli - *Brassica oleracea* var. *italica* 109
  Italian broccoli - *Brassica oleracea* var. *italica* 109
  sprouting broccoli - *Brassica oleracea* var. *italica* 109
brodiaea
  golden brodiaea - *Triteleia ixioides* 281
  harvest brodiaea - *Brodiaea coronaria* 274
brome
  barren brome - *Bromus sterilis* 377
  California brome - *Bromus carinatus* 376
  Canada brome - *Bromus purgans* 376
  compact brome - *Bromus madritensis* 376
  downy brome - *Bromus tectorum* 377
  drooping brome - *Bromus tectorum* 377
  field brome - *Bromus arvensis* 376
  foxtail brome - *Bromus rubens* 376
  fringe brome - *Bromus ciliatus* 376
  Japanese brome - *Bromus japonicus* 376
  Madrid brome - *Bromus madritensis* 376
  meadow brome - *Bromus erectus* 376
  mountain brome - *Bromus carinatus* 376
  nodding brome - *Bromus anomalus* 376
  poverty brome - *Bromus sterilis* 377
  rattlesnake brome - *Bromus brizaeformis* 376
  red brome - *Bromus rubens* 376
  ripgut brome - *Bromus rigidus* 376
  rye brome - *Bromus secalinus* 376
  sand brome - *Bromus arenarius* 376
  smooth brome - *Bromus inermis* 376
  soft brome - *Bromus mollis* 376
  Spanish brome - *Bromus madritensis* 376
  sterile brome - *Bromus sterilis* 377
  upright brome - *Bromus erectus* 376
brooklime, American - *Veronica americana* 521
broom-sedge - *Andropogon virginicus* 369
broombrush - *Hypericum prolificum* 139
broomcorn - *Panicum miliaceum* 403, *Sorghum bicolor* 419
broomweed - *Sida acuta* 294
brown-eyed-Susan - *Rudbeckia triloba* 82
browntop - *Agrostis capillaris* 366
buchu, Indian - *Myrtus communis* 307
buck-bean - *Menyanthes trifoliata* 298
buckbrush - *Ceanothus cuneatus* 458
buckeye
  bottlebrush buckeye - *Aesculus parviflora* 249
  California buckeye - *Aesculus californica* 249
  Ohio buckeye - *Aesculus glabra* 249
  red buckeye - *Aesculus pavia* 249
  sweet buckeye - *Aesculus octandra* 249
  Texas buckeye - *Aesculus glabra* var. *arguta* 249
  yellow buckeye - *Aesculus octandra* 249
buckhorn - *Lycopodium clavatum* 284, *Osmunda cinnamomea* 448
buckroot - *Psoralea canescens* 213
buckthorn
  alder buckthorn - *Rhamnus frangula* 459
  alder-leaved buckthorn - *Rhamnus alnifolia* 459
  California buckthorn - *Rhamnus californica* 459
  Carolina buckthorn - *Rhamnus caroliniana* 459
  cascara buckthorn - *Rhamnus purshiana* 459
  common buckthorn - *Rhamnus cathartica* 459
  false buckthorn - *Sideroxylon lanuginosa* 512
  glossy buckthorn - *Rhamnus frangula* 459
  hollyleaf buckthorn - *Rhamnus crocea* subsp. *ilicifolia* 459
  lame-leaved buckthorn - *Rhamnus lanceolata* 459
  red-berried buckthorn - *Rhamnus crocea* subsp. *ilicifolia* 459
  southern buckthorn - *Sideroxylon lycioides* 512
  tough buckthorn - *Sideroxylon tenax* 512
  yellow buckthorn - *Rhamnus caroliniana* 459
buckwheat
  California buckwheat - *Eriogonum fasciculatum* 437
  climbing false buckwheat - *Polygonum scandens* 438
  common buckwheat - *Fagopyrum esculentum* 437
  notch-seeded buckwheat - *Fagopyrum esculentum* 437
  wild buckwheat - *Polygonum convolvulus* 438
buckwheat-bush - *Cliftonia monophylla* 166
buckwheat-tree - *Cliftonia monophylla* 166
buckwheat-vine - *Brunnichia cirrhosa* 436
buddleja, orange-eye - *Buddleja davidii* 284
buffaloberry
  russet buffaloberry - *Shepherdia canadensis* 169
  silver buffaloberry - *Shepherdia argentea* 169
buffalobur - *Solanum rostratum* 530
buffaloweed - *Ambrosia trifida* 55
bugler, scarlet - *Penstemon centranthifolius* 519
bugleweed
  carpet bugleweed - *Ajuga reptans* 263
  creeping bugleweed - *Ajuga reptans* 263
bugloss - *Anchusa officinalis* 106
  viper's bugloss - *Echium vulgare* 106
Bukhara-clover - *Melilotus alba* 207
bull-nettle - *Solanum elaeagnifolium* 530
bullbrier - *Smilax rotundifolia* 523
bullnettle - *Cnidoscolus stimulosus* 183
  Texas bullnettle - *Cnidoscolus texanus* 183
bullock's-heart - *Annona reticulata* 26
bullrush - *Typha latifolia* 538
bulrush
  American bulrush - *Scirpus americanus* 165
  American great bulrush - *Scirpus validus* 165
  California bulrush - *Scirpus californicus* 165
  dark green bulrush - *Scirpus atrovirens* 165
  great bulrush - *Scirpus lacustris* 165
  hard-stem bulrush - *Scirpus acutus* 165
  river bulrush - *Scirpus fluviatilis* 165
  small bulrush - *Typha angustifolia* 538
  southern bulrush - *Scirpus californicus* 165
  woolgrass bulrush - *Scirpus cyperinus* 165
bumelia, gum - *Sideroxylon lanuginosa* 512
bunch-evergreen - *Lycopodium obscurum* 284
bunch-moss - *Tillandsia recurvata* 117
bunchberry - *Cornus canadensis* 143
bundleflower, velvet - *Desmanthus velutinus* 196
bur
  Sacramento bur - *Triumfetta semitriloba* 538
  skeleton-leaf bur - *Ambrosia tomentosa* 55
bur-cucumber - *Sicyos angulatus* 150
bur-marigold - *Bidens pilosa* 61
bur-sage, skeleton-leaf - *Ambrosia tomentosa* 55
burclover
  California burclover - *Medicago polymorpha* 206
  spotted burclover - *Medicago arabica* 206
burdock
  common burdock - *Arctium minus* 56
  edible burdock - *Arctium lappa* 56
  great burdock - *Arctium lappa* 56
  lesser burdock - *Arctium minus* 56
burnet
  American burnet - *Sanguisorba canadensis* 488
  Canadian burnet - *Sanguisorba canadensis* 488
  great burnet - *Sanguisorba officinalis* 488
  Sitka burnet - *Sanguisorba stipulata* 488
burning-bush - *Dictamnus albus* 497, *Euonymus atropurpureus* 133
  western burning-bush - *Euonymus occidentalis* 133
burnut - *Tribulus terrestris* 553
burnweed - *Erechtites hieracifolia* 69
  American burnweed - *Erechtites hieracifolia* 69
burro-weed - *Haplopappus tenuisectus* 73
bursting-heart - *Euonymus americanus* 133
bushclover, shrub - *Lespedeza bicolor* 202
bushman's-poison - *Acokanthera oppositifolia* 33
bushpoppy - *Dendromecon rigida* 325
busy-lizzy - *Impatiens wallerana* 90
butter-and-eggs - *Linaria vulgaris* 518
butter-daisy - *Ranunculus repens* 456
butterbough - *Exothea paniculata* 511
buttercup
  bulbous buttercup - *Ranunculus bulbosus* 456
  celery-leaved buttercup - *Ranunculus sceleratus* 456
  creeping buttercup - *Ranunculus repens* 456
  early buttercup - *Ranunculus fascicularis* 456
  meadow buttercup - *Ranunculus acris* 456
  Persian buttercup - *Ranunculus asiaticus* 456
  small-flowered buttercup - *Ranunculus abortivus* 456
  swamp buttercup - *Ranunculus septentrionalis* 456
  tall buttercup - *Ranunculus acris* 456
  turban buttercup - *Ranunculus asiaticus* 456
  yellow water buttercup - *Ranunculus flabellaris* 456
butterfly-lily - *Hedychium coronarium* 553
butterfly-vine - *Stigmaphyllon ciliatum* 289
butterfly-weed - *Asclepias tuberosa* 51
butternut - *Juglans cinerea* 258

butterprint - *Abutilon theophrastii* 289
butterweed - *Conyza canadensis* 66
button-mangrove - *Conocarpus erectus* 139
button-snakeroot - *Eryngium aquaticum* 29
buttonbush - *Cephalanthus occidentalis* 490
common buttonbush - *Cephalanthus occidentalis* 490
buttonwood - *Conocarpus erectus* 139, *Platanus occidentalis* 360
silver buttonwood - *Conocarpus erectus* var. *sericeus* 139
white buttonwood - *Laguncularia racemosa* 139
cabbage - *Brassica oleracea*, *B. oleracea* var. *capitata* 109
celery cabbage - *Brassica pekinensis* 109
Chinese cabbage - *Brassica pekinensis* 109
flowering cabbage - *Brassica oleracea* var. *acephala* 109
John's cabbage - *Hydrophyllum virginianum* 252
Shantung cabbage - *Brassica pekinensis* 109
wild cabbage - *Brassica oleracea* 109
cabbage-tree - *Cussonia spicata* 42
cacao - *Theobroma cacao* 532
cactus
Christmas cactus - *Schlumbergera bridgesii* 120
claw cactus - *Schlumbergera truncata* 120
cochineal cactus - *Nopalea cochenillifera* 120
crab cactus - *Schlumbergera truncata* 120
Easter cactus - *Rhipsalidopsis gaertneri* 120
giant cactus - *Carnegiea gigantea* 118
jumping cactus - *Opuntia fulgida* 120
lady-of-Mexico cactus - *Mammillaria hahniana* 119
new old-man cactus - *Espostoa lanata* 119
organ-pipe cactus - *Lemaireocereus thurberi* 119
Peruvian apple cactus - *Cereus peruvianus* 119
Peruvian old-man cactus - *Espostoa lanata* 119
Peruvian snowball cactus - *Espostoa lanata* 119
prickly-pear cactus - *Opuntia stricta* var. *dillenii* 120
spineless cactus - *Opuntia ficus-indica* 120
star cactus - *Astrophytum ornatum* 118
Thanksgiving cactus - *Schlumbergera truncata* 120
torch cactus - *Trichocereus spachianus* 121
vine cactus - *Fouquieria splendens* 240
whisker cactus - *Lophocereus schottii* 119
white torch cactus - *Trichocereus spachianus* 121
yoke cactus - *Schlumbergera truncata* 120
cadillo - *Urena lobata* 295
caesarweed - *Urena lobata* 295
cafta - *Catha edulis* 133
cajeput - *Melaleuca leucadendra* 305
cajeput-tree - *Melaleuca quinquenervia* 305
caju - *Anacardium occidentale* 23
calabash - *Crescentia cujete* 104
calabazilla - *Cucurbita foetidissima* 149
calabur - *Muntingia calabura* 169
caladium, fancy-leaved - *Caladium* x *hortulanum* 39
calamondin - x *Citrofortunella mitis* 494
calamus - *Acorus calamus* 38
calico-plant - *Alternanthera ficoidea* cv. 'bettzichiana' 21
calicobush - *Kalmia latifolia* 173
California-fern - *Conium maculatum* 28
California-nutmeg - *Torreya californica* 534
California-olive - *Umbellularia californica* 271
calla
florist's calla - *Zantedeschia aethiopica* 41
garden calla - *Zantedeschia aethiopica* 41
wild calla - *Calla palustris* 39
calla-lily
common calla-lily - *Zantedeschia aethiopica* 41
golden calla-lily - *Zantedeschia elliottiana* 41
pink calla-lily - *Zantedeschia rehmannii* 41
red calla-lily - *Zantedeschia rehmannii* 41
white calla-lily - *Zantedeschia aethiopica* 41
yellow calla-lily - *Zantedeschia elliottiana* 41
calliopsis - *Coreopsis tinctoria* 66
caltrop - *Trapa natans* 538
camas
common camas - *Camassia quamash* 275
white camas - *Zigadenus elegans* 283
camasey - *Clidemia hirta* 296
camass, Eastern - *Camassia scilloides* 275
camellia
common camellia - *Camellia japonica* 535
sasanqua camellia - *Camellia sasanqua* 535
camosh - *Camassia quamash* 275
camphor-tree - *Cinnamomum camphora* 268
campion
bladder campion - *Silene cucubalus* 131
morning campion - *Silene dioica* 131
moss campion - *Silene acaulis* 131
red campion - *Silene dioica* 131
rose campion - *Lychnis coronaria* 131
starry campion - *Silene stellata* 131
white campion - *Silene alba* 131
cana-brava - *Aristida longespica* 371, *Arundo donax* 372
canaigre - *Rumex hymenosepalus* 440
candleberry - *Myrica cerifera* 302
candleberry-tree - *Aleurites moluccana* 182
candlenut-tree - *Aleurites moluccana* 182
candlewood - *Fouquieria splendens* 240
candytuft
globe candytuft - *Iberis umbellata* 113
rocket candytuft - *Iberis amara* 113
candyweed - *Polygala lutea* 436
cane
giant cane - *Arundinaria gigantea* subsp. *tecta* 372
large cane - *Arundinaria gigantea* 372
southern cane - *Arundinaria gigantea* 372
switch cane - *Arundinaria gigantea* subsp. *tecta* 372
cane-apples - *Arbutus unedo* 170
canistel - *Pouteria campechiana* 512
cankerbrake - *Polystichum acrostichoides* 449
cankerroot - *Coptis groenlandica* 454
canna
common garden canna - *Canna* x *generalis* 123
edible canna - *Canna edulis* 123
cantaloupe - *Cucumis melo* var. *cantalupensis* 147
Canterbury-bells - *Campanula medium* 121
caouthchouc-tree - *Hevea brasiliensis* 185
caper-tree, Jamaica - *Capparis cynophallophora* 123
capeweed - *Phyla nodiflora* 546
caramba - *Averrhoa carambola* 323
carambola - *Averrhoa carambola* 323
caraway - *Carum carvii* 27
false caraway - *Perideridia gairdneri* 31
Indian caraway - *Perideridia gairdneri* 31
cardamom - *Elettaria cardomomum* 553
Ceylon cardamom - *Elettaria cardomomum* 553
Malabar cardamom - *Elettaria cardomomum* 553
cardinal-climber - *Ipomoea quamoclit* 142
cardinal-flower - *Lobelia cardinalis*, *L. siphilitica* 122
cardinal-spear - *Erythrina herbacea* 197
cardo-del-valle - *Centaurea americana* 63
cardoon - *Cynara cardunculus* 67
carnation - *Dianthus caryophyllus* 130
carob - *Ceratonia siliqua* 193
carpenter's-square - *Scrophularia marilandica* 521
carpenter-weed - *Prunella vulgaris* 266
carpet-plant, Corsican - *Soleirolia soleirolii* 544
carpetweed - *Mollugo verticillata* 298
carrionflower - *Smilax herbacea* 523
carrizo - *Arundo donax* 372, *Phragmites australis* 408
carrot - *Daucus carota* subsp. *sativus* 28
wild carrot - *Daucus carota* 28
cart-track-plant - *Plantago major* 359
casabanana - *Sicana odorifera* 150
cashew - *Anacardium occidentale* 23
cassandra - *Chamaedaphne calyculata* 172
cassava - *Manihot esculenta* 186
bitter cassava - *Manihot esculenta* 186
cassia
feathery cassia - *Cassia artemisioides* 193
purging cassia - *Cassia fistula* 193
cassie - *Acacia farnesiana* 187
cassina - *Ilex cassine*, *I. vomitoria* 36
castor-oil-plant - *Ricinus communis* 186
casuarina, horsetail - *Casuarina equisetifolia* 132
cat's-breeches - *Hydrophyllum capitatum* 252
cat's-claw - *Pithecellobium unguis-cati* 212, *Schrankia nuttallii* 215
Gregg cat's-claw - *Acacia greggii* 187
long-flowered cat's-claw - *Acacia greggii* 187
cat's-claw-vine - *Macfadyena unguis-cati* 105
cat's-ear, smooth - *Hypochaeris glabra* 76
catalpa
common catalpa - *Catalpa bignonioides* 104
Indian bean catalpa - *Catalpa bignonioides* 104
northern catalpa - *Catalpa speciosa* 104
southern catalpa - *Catalpa bignonioides* 104
western catalpa - *Catalpa speciosa* 104
catawba - *Catalpa bignonioides* 104
catawba-tree - *Catalpa speciosa* 104
catberry - *Nemopanthus mucronatus* 38
catbrier - *Smilax glauca*, *S. rotundifolia* 523
catchfly
French catchfly - *Silene gallica* 131
garden catchfly - *Silene armeria* 131
German catchfly - *Lychnis viscaria* 131
night-flowering catchfly - *Silene noctiflora* 131
sleepy catchfly - *Silene antirrhina* 131
sweet-William catchfly - *Silene armeria* 131
catclaw - *Acacia tortuosa* 187
Texas catclaw - *Acacia wrightii* 187
catgut - *Tephrosia virginiana* 216
catjang - *Vigna unguiculata* subsp. *cylindrica* 220
catmint - *Nepeta cataria* 265
catnip - *Nepeta cataria* 265

cattail
common cattail - *Typha latifolia* 538
narrowleaf cattail - *Typha angustifolia* 538
cauliflower - *Brassica oleracea* var. *botrytis* 109
cauliflower-ears - *Crassula argentea* 145
ceanothus
carmel ceanothus - *Ceanothus griseus* 458
Catalina ceanothus - *Ceanothus arboreus* 458
felt-leaf ceanothus - *Ceanothus arboreus* 458
Jersey tea ceanothus - *Ceanothus americanus* 458
Monterey ceanothus - *Ceanothus rigidus* 458
Point Reyes ceanothus - *Ceanothus gloriosus* 458
Yankee point ceanothus - *Ceanothus griseus* var. *horizontalis* 458
cedar
Atlantic white cedar - *Chamaecyparis thyoides* 151
atlas cedar - *Cedrus atlantica* 332
California incense cedar - *Calocedrus decurrens* 151
Colorado red cedar - *Juniperus scopulorum* 153
creeping cedar - *Juniperus horizontalis* 153
eastern red cedar - *Juniperus virginiana* 153
giant cedar - *Thuja plicata* 156
Japanese cedar - *Cryptomeria japonica* 534
Ozark white cedar - *Juniperus ashei* 153
Port Orford cedar - *Chamaecyparis lawsoniana* 151
red cedar - *Juniperus virginiana* 153
Russian cedar - *Pinus cembra* 339
Southern red cedar - *Juniperus silicicola* 153
southern white cedar - *Chamaecyparis thyoides* 151
swamp white cedar - *Chamaecyparis thyoides* 151
western red cedar - *Thuja plicata* 156
cedar-of-Goa - *Cupressus lusitanica* 152
cedrela - *Cedrela sinensis* 296
ceiba - *Ceiba pentandra* 106
celandine
greater celandine - *Chelidonium majus* 324
lesser celandine - *Ranunculus ficaria* 456
small celandine - *Ranunculus ficaria* 456
celeriac - *Apium graveolens* var. *rapaceum* 27
celery - *Apium graveolens, A. graveolens* var. *dulce* 27
knob celery - *Apium graveolens* var. *rapaceum* 27
turnip-rooted celery - *Apium graveolens* var. *rapaceum* 27
cenicilla - *Sesuvium portulacastrum* 20
cenizo - *Atriplex canescens* 134, *Leucophyllum frutescens* 518, *Tetrazygia elaeagnoides* 296
centaury - *Centaurium erythraea* 241
centipede-plant - *Homalocladium platycladum* 438
century-plant - *Agave americana* 17
American century-plant - *Agave americana* 17
cereus, night-blooming - *Hylocereus undatus* 119
ceriman - *Monstera deliciosa* 40
cestrum, day-blooming - *Cestrum diurnum* 525
chamise - *Adenostoma fasciculatum* 460
chamiza - *Atriplex canescens* 134
chamomile
corn chamomile - *Anthemis arvensis* 56
stinking chamomile - *Anthemis cotula* 56
chaparral-broom - *Baccharis pilularis* 60
chard - *Beta vulgaris* subsp. *cicla* 135
Swiss chard - *Beta vulgaris, B. vulgaris* subsp. *cicla* 135
charity - *Polemonium caeruleum* subsp. *villosum* 436
charlock - *Sinapis arvensis* 115
jointed charlock - *Raphanus raphanistrum* 114
chaste-tree - *Vitex agnus-castus* 547
chat - *Catha edulis* 133
chayote - *Sechium edule* 150
cheat - *Bromus secalinus* 376
checkerberry - *Gaultheria procumbens* 172
checkerbloom - *Sidalcea malviflora* 295
cheeses - *Malva neglecta* 294
cheeseweed - *Malva parviflora* 294
cherimalla - *Annona cherimola* 26
cherimoya - *Annona cherimola* 26
cherry
Bessey cherry - *Prunus besseyi* 475
bird cherry - *Prunus avium, P. pensylvanica* 475
bitter cherry - *Prunus emarginata* 475
Catalina Island cherry - *Prunus lyonii* 475
dwarf cherry - *Prunus pumila* 475
European bird cherry - *Prunus padus* 475
European dwarf cherry - *Prunus fruticosa* 475
European ground cherry - *Prunus fruticosa* 475
evergreen cherry - *Prunus ilicifolia* 475
fire cherry - *Prunus pensylvanica* 475
hollyleaf cherry - *Prunus ilicifolia* 475
Indian cherry - *Rhamnus caroliniana* 459
Jamaica cherry - *Muntingia calabura* 169
Japanese flowering cherry - *Prunus serrulata* 475
mahaleb cherry - *Prunus mahaleb* 475
mazzard cherry - *Prunus avium* 475
morello cherry - *Prunus cerasus* var. *austera* 475
Oregon cherry - *Prunus emarginata* 475
oriental cherry - *Prunus serrulata* 475
perfumed cherry - *Prunus mahaleb* 475
pie cherry - *Prunus cerasus* 475
pin cherry - *Prunus pensylvanica* 475
rum cherry - *Prunus serotina* 475
sand cherry - *Prunus pumila* 475
sour cherry - *Prunus cerasus* 475
Spanish cherry - *Mimusops elengi* 512
St.-Lucie cherry - *Prunus mahaleb* 475
sweet cherry - *Prunus avium* 475
western sand cherry - *Prunus besseyi* 475
wild black cherry - *Prunus serotina* 475
wild cherry - *Prunus ilicifolia* 475
wild red cherry - *Prunus pensylvanica* 475
cherry-laurel
Caroline cherry-laurel - *Prunus caroliniana* 475
English cherry-laurel - *Prunus laurocerasus* 475
Portuguese cherry-laurel - *Prunus lusitanica* 475
cherry-pie - *Heliotropium arborescens* 107
chervil, white - *Cryptotaenia canadensis* 28
chess - *Bromus secalinus* 376
Australian chess - *Bromus arenarius* 376
Chilean chess - *Bromus trinii* 377
downy chess - *Bromus tectorum* 377
foxtail chess - *Bromus rubens* 376
hairy chess - *Bromus commutatus* 376
Japanese chess - *Bromus japonicus* 376
rattlesnake chess - *Bromus brizaeformis* 376
soft chess - *Bromus mollis* 376
chestnut
American chestnut - *Castanea dentata* 222
Chinese chestnut - *Castanea mollissima* 222
Eurasian chestnut - *Castanea sativa* 222
European chestnut - *Castanea sativa* 222
Japanese chestnut - *Castanea crenata* 222
Moreton Bay chestnut - *Castanospermum australe* 193
Spanish chestnut - *Castanea sativa* 222
chestnut-vine - *Tetrastigma voinieranum* 550
chicken-claws - *Salicornia europaea* 137
chicken-gizzard - *Iresine herbstii* 22
chickpea - *Cicer arietinum* 194
chickweed
common chickweed - *Stellaria media* 132
field chickweed - *Cerastium arvense* 129
forked chickweed - *Paronychia canadensis* 131
mouse-ear chickweed - *Cerastium vulgatum* 129
sea chickweed - *Honkenya peploides* 130
chicle - *Manilkara zapota* 512
chicory, common - *Cichorium intybus* 65
chicot - *Gymnocladus dioica* 200
chicozapote - *Manilkara zapota* 512
chigger-flower - *Asclepias tuberosa* 51
chin-chin - *Azara microphylla* 240
China-box - *Murraya paniculata* 497
China-root - *Smilax tamnoides* 523
China-rose - *Hibiscus rosa-sinensis* 292
China-tree - *Melia azedarach* 297
chinaberry - *Melia azedarach* 297
Chinese-laurel - *Antidesma bunius* 183
chinquapin
Allegheny chinquapin - *Castanea pumila* 222
bush chinquapin - *Castanopsis sempervirens* 224
California chinquapin - *Castanopsis sempervirens* 224
giant chinquapin - *Castanopsis chrysophylla* 224
golden chinquapin - *Castanopsis chrysophylla* 224
Sierra chinquapin - *Castanopsis sempervirens* 224
water chinquapin - *Nelumbo lutea* 308
chirimoya - *Annona cherimola* 26
chittamwood - *Cotinus obovatus* 23, *Sideroxylon lanuginosa* 512
chive, Chinese - *Allium tuberosum* 272
chives - *Allium schoenoprasum* 272
chocolate, Indian - *Geum rivale* 467
chocolate-root - *Geum rivale* 467
chocolate-vine - *Akebia quinata* 268
chokeberry
black chokeberry - *Aronia melanocarpa* 462
purple chokeberry - *Malus floribunda* 469
red chokeberry - *Aronia arbutifolia* 462
chokecherry - *Prunus virginiana* 475
black chokecherry - *Prunus virginiana* var. *melanocarpa* 475
southwestern chokecherry - *Prunus serotina* subsp. *virens* 475
choyote - *Sechium edule* 150
Christ-plant - *Euphorbia milii* 184
Christ-thorn - *Euphorbia milii* 184
Christmas-berry - *Lycium carolinianum* 526

Christmas-berry-tree - *Schinus terebinthifolius* 25
Christmas-candle - *Senna alata* 215
Christmas-flower - *Euphorbia pulcherrima* 184
Christmas-green - *Lycopodium complanatum* 284
Christmas-rose - *Helleborus niger* 455
Christmas-star - *Euphorbia pulcherrima* 184
Christmasberry - *Heteromeles arbutifolia* 468
christophine - *Sechium edule* 150
chrysanthemum
  corn chrysanthemum - *Chrysanthemum segetum* 63
  daisy chrysanthemum - *Chrysanthemum maximum* 63
  florist's chrysanthemum - *Chrysanthemum* x *morifolium* 63
  Max chrysanthemum - *Chrysanthemum maximum* 63
chufa - *Cyperus esculentus* 163
chuparosa - *Anisacanthus thurberi* 7
ciboule - *Allium fistulosum* 272
cicely
  hairy sweet cicely - *Osmorhiza claytonii* 31
  woolly sweet cicely - *Osmorhiza claytonii* 31
cigar-flower - *Cuphea ignea* 284
cigar-tree - *Catalpa speciosa* 104
cilantro - *Coriandrum sativum* 28
cimarron, tabacco - *Nicotiana repanda* 528
cineraria, florist's - *Senecio cruentus* 82
cinnamon - *Cinnamomum verum* 268
  Ceylon cinnamon - *Cinnamomum verum* 268
cinnamon-jasmine - *Hedychium coronarium* 553
cinquefoil
  bush cinquefoil - *Potentilla fruticosa* 474
  common cinquefoil - *Potentilla canadensis* 474
  hoary cinquefoil - *Potentilla argentea* 474
  old-field cinquefoil - *Potentilla simplex* 474
  rough cinquefoil - *Potentilla norvegica* 474
  shrubby cinquefoil - *Potentilla fruticosa* 474
  silverweed cinquefoil - *Potentilla anserina* 474
  silvery cinquefoil - *Potentilla argentea* 474
  sulphur cinquefoil - *Potentilla recta* 474
  tall cinquefoil - *Potentilla arguta* 474
  three-toothed cinquefoil - *Potentilla tridentata* 474
  upright cinquefoil - *Potentilla recta* 474
cistus, St.-Vincent - *Cistus palhinhai* 138
citron - *Citrus medica* 494
citronella - *Collinsonia canadensis* 263
cive - *Allium schoenoprasum* 272
clearweed - *Pilea pumila* 544
cleavers - *Galium aparine* 491
cleome
  golden cleome - *Cleome lutea* 123
  yellow cleome - *Cleome lutea* 123
cliff-green - *Paxistima canbyi* 134
cliffbrake
  fragile cliffbrake - *Cryptogramma stelleri* 446
  New Zealand cliffbrake - *Pellaea rotundifolia* 448
  slender cliffbrake - *Cryptogramma stelleri* 446
  smooth cliffbrake - *Pellaea glabella* 448
cliffbush - *Jamesia americana* 251
climbing-beauty - *Aeschynanthus pulcher* 243
clintonia, yellow - *Clintonia borealis* 275
cloudberry - *Rubus chamaemorus* 486
clover
  alsike clover - *Trifolium hybridum* 217
  arrowleaf clover - *Trifolium vesiculosum* 217
  bigflower clover - *Trifolium michelianum* 217
  Carolina clover - *Trifolium carolinianum* 216
  cow hop clover - *Trifolium procumbens* 217
  crimson clover - *Trifolium incarnatum* 217
  holy clover - *Onobrychis viciifolia* 209
  hop clover - *Trifolium agrarium* 216
  Hubam clover - *Melilotus alba* var. *annua* 207
  Italian clover - *Trifolium incarnatum* 217
  Kenya clover - *Trifolium semipilosum* 217
  Kenya wild white clover - *Trifolium semipilosum* 217
  ladino clover - *Trifolium repens* 217
  old-field clover - *Trifolium arvense* 216
  one-leaf clover - *Alysicarpus vaginalis* 189
  Persian clover - *Trifolium resupinatum* 217
  rabbitfoot clover - *Trifolium arvense* 216
  red clover - *Trifolium pratense* 217
  reversed clover - *Trifolium resupinatum* 217
  seaside clover - *Trifolium wormskioldii* 217
  Sierra clover - *Trifolium wormskioldii* 217
  small hop clover - *Trifolium procumbens* 217
  stone clover - *Trifolium arvense* 216
  strawberry clover - *Carthamus tinctorius* 63, *Trifolium fragiferum* 217
  subterranean clover - *Trifolium subterraneum* 217
  white clover - *Trifolium repens* 217
  whitetip clover - *Trifolium variegatum* 217
  yellow clover - *Trifolium agrarium* 216, *T. procumbens* 217
  zigzag clover - *Trifolium medium* 217
club-moss
  bristly club-moss - *Lycopodium annotinum* 284
  common club-moss - *Lycopodium clavatum* 284
  interrupted club-moss - *Lycopodium annotinum* 284
  running club-moss - *Lycopodium clavatum* 284
  shining club-moss - *Lycopodium lucidulum* 284
  stiff club-moss - *Lycopodium annotinum* 284
  tree club-moss - *Lycopodium obscurum* 284
coach-whip - *Fouquieria splendens* 240
cochineal-plant - *Nopalea cochenillifera* 120
cock's-foot - *Dactylis glomerata* 383
cockle - *Vaccaria hispanica* 132
  corn cockle - *Agrostemma githago* 129
  purple cockle - *Agrostemma githago* 129
  white cockle - *Silene alba* 131
cocklebur - *Xanthium strumarium* var. *canadense* 89
  common cocklebur - *Xanthium strumarium* 89
  spiny cocklebur - *Xanthium spinosum* 89
cockscomb - *Celosia argentea* var. *cristata* 22
cockspur - *Crataegus crus-galli* 463
cockspur-thorn - *Crataegus crus-galli* 463
coco-plum - *Chrysobalanus icaco* 138
coco-yam - *Colocasia esculenta* 39
coffee
  arabica coffee - *Coffea arabica* 491
  common coffee - *Coffea arabica* 491
  wild coffee - *Colubrina arborescens* 459
coffee-tree - *Polyscias guilfoylei* 44
  Kentucky coffee-tree - *Gymnocladus dioica* 200
coffeeberry - *Rhamnus californica* 459
cohosh, white - *Actaea alba* 452
cole - *Brassica oleracea* var. *acephala* 109
  southern cole - *Brassica juncea* var. *crispifolia* 109
colewort - *Brassica oleracea* var. *acephala* 109, *Crambe abyssinica* 112
colicroot, yellow - *Aletris aurea* 272
collards - *Brassica oleracea* var. *acephala* 109
coltsfoot - *Tussilago farfara* 88
columbine
  European columbine - *Aquilegia vulgaris* 453
  wild columbine - *Aquilegia canadensis* 453
colza - *Brassica napus*, *B. napus* var. *napus* 109
comfrey
  blue comfrey - *Symphytum* x *uplandicum* 108
  common comfrey - *Symphytum officinale* 108
  Quaker comfrey - *Symphytum* x *uplandicum* 108
  Russian comfrey - *Symphytum* x *uplandicum* 108
common-threesquare - *Scirpus americanus* 165
compass-plant - *Silphium laciniatum* 84
comptie - *Zamia pumila* 553
conchito - *Macroptilium atropurpureum* 205
condalia, Mexican - *Condalia mexicana* 459
coneflower - *Dracopis amplexicaulis* 68, *Rudbeckia laciniata* 82
confederate-jasmine - *Trachelospermum jasminoides* 35
consumption-weed - *Baccharis halimifolia* 60
convulsion-root - *Monotropa uniflora* 298
cool-tankard - *Borago officinalis* 106
coolwort - *Mitella diphylla* 514, *Pilea pumila* 544
coontie - *Zamia pumila* 553
copa-tree - *Ailanthus altissima* 522
copey - *Clusia rosea* 138
copperleaf
  painted copperleaf - *Acalypha wilkesiana* 182
  Virginia copperleaf - *Acalypha virginica* 182
coral-bead-plant - *Abrus precatorius* 187
coral-evergreen - *Lycopodium clavatum* 284
coralbeads - *Cocculus carolinus* 297
coralbells - *Heuchera sanguinea* 513
coralberry - *Ardisia crenata* 302, *Symphoricarpos orbiculatus* 126, *Symphoricarpos* × *chenaultii* 126
coralplant - *Russelia equisetiformis* 520
coralroot
  large coralroot - *Corallorrhiza maculata* 320
  spotted coralroot - *Corallorrhiza maculata* 320
coraltree, cockspur - *Erythrina crista-galli* 197
corazon - *Annona reticulata* 26
coreopsis, plains - *Coreopsis tinctoria* 66
coriander - *Coriandrum sativum* 28
corn
  Dent corn - *Zea mays* 428
  field corn - *Zea mays* 428
  flint corn - *Zea mays* 428
  pod corn - *Zea mays* 428

sweet corn - *Zea mays* 428
corn-lily - *Clintonia borealis* 275
corn-plant - *Dracaena fragrans* cv. 'massangeana' 18
cornel
dwarf cornel - *Cornus canadensis* 143
small roughleaf cornel - *Cornus asperifolia* 143
cornflower - *Centaurea cyanus* 63
cornsalad, beaked - *Valerianella radiata* 545
coroa - *Sicana odorifera* 150
corpse-plant - *Monotropa uniflora* 298
Corsican-curse - *Soleirolia soleirolii* 544
cosmetic-bark-tree - *Murraya paniculata* 497
cosmos
common cosmos - *Cosmos bipinnatus* 66
orange cosmos - *Cosmos sulphureus* 67
yellow cosmos - *Cosmos sulphureus* 67
cotton
American-Egyptian cotton - *Gossypium barbadense* 290
Arizona wild cotton - *Gossypium thurberi* 291
Ceylon cotton - *Gossypium arboreum* 290
extra-long-staple cotton - *Gossypium barbadense* 290
Levant cotton - *Gossypium herbaceum* 290
Sea Island cotton - *Gossypium barbadense* 290
tree cotton - *Gossypium arboreum, G. barbadense* 290
upland cotton - *Gossypium hirsutum* 291
wild cotton - *Hibiscus moscheutos* 292
cotton-gum - *Nyssa aquatica* 309
cottontop, Arizona - *Digitaria californica* 385
cottonwood
black cottonwood - *Populus heterophylla, P. trichocarpa* 499
eastern cottonwood - *Populus deltoides* 499
Fremont cottonwood - *Populus fremontii* 499
plains cottonwood - *Populus deltoides* subsp. *monilifera* 499
southern cottonwood - *Populus deltoides* 499
swamp cottonwood - *Populus heterophylla* 499
couch, bearded - *Agropyron trachycaulum* 363
council-tree - *Ficus altissima* 298
country-spinach - *Basella alba* 90
country-walnut - *Aleurites moluccana* 182
cow-herb - *Vaccaria hispanica* 132
cow-itch - *Campsis radicans* 103, *Rhus radicans* 24
cow-parsnip, American - *Heracleum sphondylium* subsp. *montanum* 29
cowbane - *Oxypolis rigidior* 31
spotted cowbane - *Cicuta maculata* 27
cowberry - *Vaccinium vitis-idaea* 179, *Viburnum lentago* 127
cowpea - *Vigna unguiculata* 220
cowslip - *Caltha palustris* 453, *Primula veris* 443
American cowslip - *Dodecatheon conjugens, D. meadia* 443
Virginia cowslip - *Mertensia virginica* 107
coyote-brush - *Baccharis pilularis* 60
crab's-eye - *Abrus precatorius* 187
crabapple - *Malus sylvestris* 469
American crabapple - *Malus angustifolia, M. coronaria* 469
Chinese flowering crabapple - *Malus spectabilis* 469
garland crabapple - *Malus coronaria* 469
Japanese flowering crabapple - *Malus floribunda* 469
Oregon crabapple - *Malus fusca* 469
prairie crabapple - *Malus ioensis* 469
showy crabapple - *Malus floribunda* 469
Siberian crabapple - *Malus baccata* 469
southern wild crabapple - *Malus angustifolia* 469
sweet-scented crabapple - *Malus coronaria* 469
Toringo crabapple - *Malus sieboldii* 469
wild crabapple - *Malus ioensis* 469
wild sweet crabapple - *Malus coronaria* 469
crabwood - *Gymnanthes lucida* 185
crackerberry - *Cornus canadensis* 143
crakeberry - *Empetrum nigrum* 169
crambe - *Crambe abyssinica* 112
crampbark - *Viburnum trilobum* 127
cranberry - *Vaccinium vitis-idaea* 179
American cranberry - *Vaccinium macrocarpon* 179
cultivated cranberry - *Vaccinium macrocarpon* 179
highbush cranberry - *Viburnum trilobum* 127
hog cranberry - *Arctostaphylos uva-ursi* 171
large cranberry - *Vaccinium macrocarpon* 179
mountain cranberry - *Vaccinium vitis-idaea* var. *minus* 179
rock cranberry - *Vaccinium vitis-idaea* var. *minus* 179
small cranberry - *Vaccinium oxycoccos* 179
cranberry-bush - *Viburnum trilobum* 127
European cranberry-bush - *Viburnum opulus* 127
crane-lily - *Strelitzia reginae* 532
craneflower - *Strelitzia reginae* 532
cranesbill
spotted cranesbill - *Geranium maculatum* 242
crape-gardenia - *Tabernaemontana divaricata* 35
crape-jasmine - *Tabernaemontana divaricata* 35
crape-myrtle - *Lagerstroemia indica* 285
crazyweed
Lambert crazyweed - *Oxytropis lambertii* 209
creambush - *Holodiscus discolor* 468
creashak - *Arctostaphylos uva-ursi* 171
creeper
carmel creeper - *Ceanothus griseus* var. *horizontalis* 458
golden Ceylon creeper - *Epipremnum aureum* 40
Point Reyes creeper - *Ceanothus gloriosus* 458
Rangoon creeper - *Quisqualis indica* 139
Virginia creeper - *Parthenocissus quinquefolia* 550
creeping-Charlie - *Lysimachia nummularia* 443, *Pilea nummulariifolia* 544, *Wedelia trilobata* 89
creeping-Jennie - *Lysimachia nummularia* 443
creeping-sailor - *Saxifraga stolonifera* 514
creosote-bush - *Larrea mexicana* 553
cress
field cress - *Lepidium campestre* 113
field penny cress - *Thlaspi arvense* 115
garden cress - *Lepidium sativum* 113
Indian cress - *Tropaeolum majus* 538
marsh cress - *Rorippa islandica* 115
mouse-ear cress - *Arabidopsis thaliana* 108
rock cress - *Arabis virginica* 108
upland cress - *Barbarea vulgaris* 109
winter cress - *Barbarea vulgaris* 109
crinum, southern swamp - *Crinum americanum* 275
crista-de-galo - *Celosia argentea* var. *cristata* 22
crocus
autumn crocus - *Colchicum autumnale* 275
wild crocus - *Anemone nuttalliana, A. patens* 452
cross-vine - *Bignonia capreolata* 103
crotalaria
showy crotalaria - *Crotalaria spectabilis* 195
smooth crotalaria - *Crotalaria pallida* 195
striped crotalaria - *Crotalaria pallida* 195
croton
garden croton - *Codiaeum variegatum* 183
woolly croton - *Croton capitatus* 183
crow-corn - *Aletris farinosa* 272
crowberry, black - *Empetrum nigrum* 169
crowfoot
cursed crowfoot - *Ranunculus sceleratus* 456
European crowfoot - *Aquilegia vulgaris* 453
garden crowfoot - *Aquilegia vulgaris* 453
small-flowered crowfoot - *Ranunculus abortivus* 456
yellow water crowfoot - *Ranunculus flabellaris* 456
crown-of-thorns - *Euphorbia milii* 184
crownflower - *Calotropis gigantea* 52
small crownflower - *Calotropis procera* 52
crownplant - *Calotropis gigantea* 52
crownvetch - *Coronilla varia* 195
trailing crownvetch - *Coronilla varia* 195
cry-baby-tree - *Erythrina crista-galli* 197
crypthanthus, pink - *Cryptanthus bromelioides* 116
cryptomeria, common - *Cryptomeria japonica* 534
crystal-tea - *Ledum decumbens* 174
cuckold - *Arctium lappa* 56
cuckoo-flower - *Lychnis flos-cuculi* 131
cucumber - *Cucumis sativus* 147
bitter cucumber - *Momordica charantia* 150
mock cucumber - *Echinocystis lobata* 150
prickly cucumber - *Echinocystis lobata* 150
wild cucumber - *Echinocystis lobata* 150
cucumber-tree
large-leaved cucumber-tree - *Magnolia macrophylla* 287
yellow cucumber-tree - *Magnolia cordata* 287
cucumbertree - *Magnolia acuminata* 287
cudweed - *Artemisia gnaphalodes* 57, *Gnaphalium purpureum* var. *spathulatum* 71
fragrant cudweed - *Gnaphalium obtusifolium* 71
low cudweed - *Gnaphalium uliginosum* 71
Culver's-root - *Veronicastrum virginicum* 522
cup-plant
Indian cup-plant - *Sarracenia purpurea* 512
cupey - *Clusia rosea* 138
cuphea, clammy - *Cuphea viscosissima* 284
curlewberry - *Empetrum nigrum* 169
currant
alpine currant - *Ribes alpinum* 245
American black currant - *Ribes americanum* 245
black currant - *Ribes nigrum* 245
bristly black currant - *Ribes lacustre* 245

buffalo currant - *Ribes aureum, R. odoratum* 245
California black currant - *Ribes bracteosum* 245
Chaparral currant - *Ribes malvaceum* 245
European black currant - *Ribes nigrum* 245
fetid currant - *Ribes glandulosum* 245
golden currant - *Ribes aureum* 245
Hudson Bay currant - *Ribes hudsonianum* 245
Indian currant - *Symphoricarpos orbiculatus, Symphoricarpos × chenaultii* 126
Missouri currant - *Ribes aureum, R. odoratum* 245
mountain currant - *Ribes alpinum* 245
northern black currant - *Ribes hudsonianum* 245
northern red currant - *Ribes rubrum* 245
red-flowered currant - *Ribes sanguineum* 245
Sierra currant - *Ribes nevadense* 245
skunk currant - *Ribes glandulosum* 245
squaw currant - *Ribes cereum* 245
stink currant - *Ribes bracteosum* 245
swamp red currant - *Ribes triste* 245
wax currant - *Ribes cereum* 245
white currant - *Ribes rubrum* 245
white-flowered currant - *Ribes cereum* 245
wild black currant - *Ribes americanum* 245
curtain-plant - *Kalanchoe pinnata* 146
curua - *Sicana odorifera* 150
curubá - *Sicana odorifera* 150
custard-apple - *Annona cherimola, A. reticulata, A. squamosa* 26
cyclamen, florist's - *Cyclamen persicum* 443
cyperus
awned cyperus - *Cyperus aristatus* 163
globose cyperus - *Cyperus ovularis* 163
marsh cyperus - *Cyperus pseudovegetus* 163
pine-barren cyperus - *Cyperus cylindricus* 163
red-rooted cyperus - *Cyperus erythrorhizos* 163
shining cyperus - *Cyperus rivularis* 163
slender cyperus - *Cyperus filiculmis* 163
straw-colored cyperus - *Cyperus strigosus* 163
cypress
Arizona cypress - *Cupressus arizonica* 152
Hinoki cypress - *Chamaecyparis obtusa* 151
Hinoki false cypress - *Chamaecyparis obtusa* 151
Italian cypress - *Cupressus sempervirens* 152
Japanese false cypress - *Chamaecyparis obtusa* 151
Lawson cypress - *Chamaecyparis lawsoniana* 151
Mendocino cypress - *Cupressus pygmaea* 152
Mexican cypress - *Cupressus lusitanica* 152
modoc cypress - *Cupressus bakeri* 152
Monterey cypress - *Cupressus macrocarpa* 152
pond cypress - *Taxodium distichum* var. *imbricarium* 535
Portuguese cypress - *Cupressus lusitanica* 152
rough-barked Arizona cypress - *Cupressus arizonica* 152
Sargent cypress - *Cupressus sargentii* 152
Sawara cypress - *Chamaecyparis pisifera* 151
cypress-vine - *Ipomoea quamoclit* 142
daffodil
common daffodil - *Narcissus pseudonarcissus* 279
fall daffodil - *Sternbergia lutea* 281
Peruvian daffodil - *Hymenocallis narcissiflora* 277
dagger-plant - *Yucca aloifolia* 19
daggerfern - *Polystichum acrostichoides* 449
dahl - *Cajanus cajan* 192
dahoon - *Ilex cassine* 36
daikon - *Raphanus sativus* cv. 'longipinnatus' 114
daisy
African daisy - *Gerbera jamesonii* 71
Barberton daisy - *Gerbera jamesonii* 71
blue-eyed African daisy - *Arctotis stoechadifolia* 56
butter daisy - *Verbesina encelioides* 88
Clanwilliam daisy - *Euryops speciosissimus* 71
Dahlberg daisy - *Dyssodia tenuiloba* 68
Engelmann daisy - *Engelmannia pinnatifida* 68
English daisy - *Bellis perennis* 61
goldencrown daisy - *Verbesina encelioides* 88
kingfisher daisy - *Felicia bergeriana* 71
mountain daisy - *Arenaria groenlandica* 129
oxeye daisy - *Chrysanthemum leucanthemum* 63
painted daisy - *Chrysanthemum coccineum* 63
Paris daisy - *Chrysanthemum frutescens* 63
Shasta daisy - *Chrysanthemum maximum* 63
Transvaal daisy - *Gerbera jamesonii* 71
Vanstadens River daisy - *Osteospermum ecklonis* 80
veldt daisy - *Gerbera jamesonii* 71
white daisy - *Chrysanthemum frutescens* 63
dalea, black - *Dalea frutescens* 196
dame's-rocket - *Hesperis matronalis* 113
dame's-violet - *Hesperis matronalis* 113
damson, bitter - *Simarouba glauca* 523
dandelion
Carolina false dandelion - *Pyrrhopappus carolinianus* 81
common dandelion - *Taraxacum officinale* 87
false dandelion - *Hypochaeris radicata* 76
redseed dandelion - *Taraxacum erythrospermum* 87
Russian dandelion - *Taraxacum kok-saghyz* 87
dangleberry - *Gaylussacia frondosa* 173
daphne
February daphne - *Daphne mezereum* 536
rose daphne - *Daphne cneorum* 536
winter daphne - *Daphne odora* 536
darnel - *Lolium temulentum* 399
dasheen - *Colocasia esculenta* 39
date, Chinese - *Ziziphus jujuba* 460
date-plum - *Diospyros kaki, D. virginiana* 167
datil - *Yucca baccata* 19
datura
Hindu datura - *Datura metel* 525
sacred datura - *Datura inoxia* 525
dawnflower, blue - *Ipomoea indica* 142
dayflower - *Commelina communis* 140
creeping dayflower - *Commelina diffusa* 140
Virginia dayflower - *Commelina virginica* 140
daylily
fulvous daylily - *Hemerocallis fulva* 276
orange daylily - *Hemerocallis fulva* 276
tawny daylily - *Hemerocallis fulva* 276
deerberry - *Vaccinium stamineum* 179
deerbrush - *Ceanothus integerrimus* 458
deerbush - *Ceanothus integerrimus* 458
deerfoot - *Achlys triphylla* 91
deernut - *Simmondsia chinensis* 523
deertongue - *Panicum clandestinum* 403
deervetch - *Lotus purshianus* 203
deodar - *Cedrus deodara* 332
desert-candle - *Eremurus robustus* 276
desert-honeysuckle - *Anisacanthus thurberi* 7
desert-plume - *Stanleya pinnata* 115
desert-rose - *Adenium obesum* 33
Sturt's desert-rose - *Gossypium sturtianum* 291
desert-thorn - *Lycium pallidum* 526
desert-willow - *Chilopsis linearis* 104
devil's-backbone - *Pedilanthus tithymaloides* 186
devil's-claw - *Proboscidea louisianica* 326
devil's-club - *Oplopanax horridum* 43
devil's-darning-needle - *Clematis virginiana* 454
devil's-grandmother - *Elephantopus tomentosus* 68
devil's-paintbrush - *Hieracium aurantiacum* 76
devil's-plague - *Daucus carota* 28
devil's-shoestrings - *Viburnum alnifolium* 127
devil's-walkingstick - *Aralia spinosa* 42
devilwood - *Osmanthus americanus* 315
dewberry
American dewberry - *Rubus flagellaris* 486
northern dewberry - *Rubus flagellaris* 486
Pacific dewberry - *Rubus ursinus* 486
southern dewberry - *Rubus trivialis* 486
swamp dewberry - *Rubus hispidus* 486
dewdrop - *Dalibarda repens* 465
garden dewdrop - *Duranta repens* 545
golden dewdrop - *Duranta repens* 545
digitalis - *Digitalis lanata* 517
dill - *Anethum graveolens* 26
dilly, wild - *Manilkara jaimiqui* subsp. *emarginata* 512
dittany - *Dictamnus albus* 497
American dittany - *Cunila origanoides* 263
common dittany - *Cunila origanoides* 263
divine-flower - *Dianthus caryophyllus* 130
dock
bitter dock - *Rumex obtusifolius* 440
blunt-leaved dock - *Rumex obtusifolius* 440
broadleaf dock - *Rumex obtusifolius* 440
curly dock - *Rumex crispus* 440
patience dock - *Rumex patientia* 440
purple-wan dock - *Brasenia schreberi* 118
red-veined dock - *Rumex obtusifolius* 440
sour dock - *Rumex acetosa* 440
spinach dock - *Rumex patientia* 440
tanner's dock - *Rumex hymenosepalus* 440
yellow dock - *Rumex crispus* 440
dockmackie - *Viburnum acerifolium* 127
dodder
clover dodder - *Cuscuta epithymum* 141
lespedeza dodder - *Cuscuta pentagona* 141
swamp dodder - *Cuscuta gronovii* 141
dog-bramble - *Ribes cynosbati* 245
dog-hobble - *Leucothoe axillaris* var. *editorum* 174, *Viburnum alnifolium* 127
dogbane
common dogbane - *Apocynum androsaemifolium* 33
hemp dogbane - *Apocynum cannabinum* 33

prairie dogbane - *Apocynum sibiricum* 33
spreading dogbane - *Apocynum androsaemifolium* 33
dogberry - *Cornus sanguinea* 143, *Ribes cynosbati* 245, *Sorbus americana* 489, *Viburnum alnifolium* 127
dogbrier - *Rosa canina* 484
dogfennel - *Eupatorium capillifolium* 70
common dogfennel - *Anthemis cotula* 56
dogtail, crested - *Cynosurus cristatus* 383
dogtooth-violet, white - *Erythronium albidum* 276
dogwood
American dogwood - *Cornus sericea* 143
blood-twig dogwood - *Cornus sanguinea* 143
brown dogwood - *Cornus glabrata* 143
creek dogwood - *Cornus* × *californica* 143
false dogwood - *Sapindus saponaria* 511
flowering dogwood - *Cornus florida* 143
gray dogwood - *Cornus racemosa* 143
Jamaican dogwood - *Piscidia piscipula* 211
mountain dogwood - *Cornus nuttallii* 143
Pacific dogwood - *Cornus nuttallii* 143
pagoda dogwood - *Cornus alternifolia* 143
panicled dogwood - *Cornus racemosa* 143
red osier dogwood - *Cornus sericea* 143
roughleaf dogwood - *Cornus drummondii* 143
round-leaved dogwood - *Cornus rugosa* 143
silky dogwood - *Cornus amomum, C. obliqua* 143
stiff dogwood - *Cornus stricta* 143
West Indian dogwood - *Piscidia piscipula* 211
White Mountains dogwood - *Viburnum alnifolium* 127
doll's-eyes - *Actaea alba* 452
dollar-plant - *Crassula argentea* 145
doum, Egyptian - *Hyphaene thebaica* 48
dove's-dung - *Ornithogalum umbellatum* 280
doveweed - *Eremocarpus setigerus* 184
dracaena
gold-dust dracaena - *Dracaena surculosa* 18
golden dracaena - *Dracaena aurea* 18
spotted dracaena - *Dracaena surculosa* 18
dragon-tree - *Dracaena draco* 18
dragonroot - *Arisaema dracontium, A. triphyllum* 39
drake - *Avena fatua* 372
dropseed
annual dropseed - *Sporobolus neglectus* 421
giant dropseed - *Sporobolus giganteus* 421
hairy dropseed - *Blepharoneuron tricholepis* 375
meadow dropseed - *Sporobolus asper* var. *hookeri* 421
pineywood's dropseed - *Sporobolus junceus* 421
poverty dropseed - *Sporobolus vaginiflorus* 421
prairie dropseed - *Sporobolus heterolepis* 421
sand dropseed - *Sporobolus cryptandrus* 421
spike dropseed - *Sporobolus contractus* 421
dropwort - *Filipendula vulgaris* 466
duck-potato - *Sagittaria latifolia* 21
duckweed
common duckweed - *Lemna minor* 271
giant duckweed - *Spirodela polyrhiza* 271
great duckweed - *Spirodela polyrhiza* 271
lesser duckweed - *Lemna minor* 271
dumb-cotton - *Calotropis procera* 52
dumbcane - *Dieffenbachia seguine* 39
spotted dumbcane - *Dieffenbachia maculata* 39
durra - *Sorghum bicolor* 419
dusty-miller - *Centaurea cineraria* 63, *Lychnis coronaria* 131
Dutchman's-breeches - *Dicentra cucullaria* 240
Dutchman's-pipe - *Aristolochia durior* 51
dyeroot - *Lachnanthes caroliniana* 247
earthnut - *Arachis hypogaea* 189
eddo - *Colocasia esculenta* 39
egg-fruit-tree - *Pouteria campechiana* 512
eggplant - *Solanum melongena, S. melongena* var. *esculentum* 530
fruited eggplant - *Solanum integrifolium* 530
scarlet eggplant - *Solanum integrifolium* 530
tomato fruited eggplant - *Solanum integrifolium* 530
eglantine - *Rosa eglanteria* 484
Egyptian-pea - *Cicer arietinum* 194
einkorn - *Triticum monococcum* 425
elaeagnus
autumn elaeagnus - *Elaeagnus umbellata* 168
thorny elaeagnus - *Elaeagnus pungens* 168
elder
American elder - *Sambucus canadensis* 125
blackbead elder - *Sambucus melanocarpa* 125
blue elder - *Sambucus caerulea* 125
box elder - *Acer negundo* 9
California box elder - *Acer negundo* subsp. *californicum* 9
European elder - *Sambucus nigra* 125
European red elder - *Sambucus racemosa* 125
Florida elder - *Sambucus simpsonii* 125
Pacific coast red elder - *Sambucus callicarpa* 125
red-berried elder - *Sambucus racemosa* subsp. *pubens* 125
sweet elder - *Sambucus canadensis* 125
yellow elder - *Tecoma stans* 105
elderberry
American elderberry - *Sambucus canadensis* 125
American red elderberry - *Sambucus racemosa* subsp. *pubens* 125
blue elderberry - *Sambucus caerulea* 125
red elderberry - *Sambucus callicarpa* 125
western elderberry - *Sambucus caerulea* 125
elecampane - *Inula helenium* 76
elephant's-ear - *Caladium bicolor* 39
elephant's-ear-plant - *Colocasia esculenta* 39
elephant-apple - *Limonia acidissima* 497
elephant-bush - *Portulacaria afra* 442
elephant-climber - *Argyreia nervosa* 141
elephant-foot-tree - *Nolina tuberculata* 18
elephant-heads - *Pedicularis groenlandica* 519
eleven-o'clock - *Portulaca grandiflora* 442
elfin-herb - *Cuphea hyssopifolia* 284
elk-clover - *Aralia californica* 42
elm
American elm - *Ulmus americana* 540
cedar elm - *Ulmus crassifolia* 540
Chinese elm - *Ulmus parvifolia, U. pumila* 540
cork elm - *Ulmus thomasii* 540
dwarf elm - *Ulmus pumila* 540
English elm - *Ulmus procera* 540
Japanese elm - *Ulmus davidiana* var. *japonica* 540
leatherleaf elm - *Ulmus parvifolia* 540
Moline elm - *Ulmus americana* cv. 'moline' 540
red elm - *Ulmus rubra, U. serotina* 540
rock elm - *Ulmus thomasii* 540
Scotch elm - *Ulmus glabra* 540
September elm - *Ulmus serotina* 540
Siberian elm - *Ulmus pumila* 540
slippery elm - *Ulmus rubra* 540
Wahoo elm - *Ulmus alata* 540
water elm - *Ulmus americana* 540
white elm - *Ulmus americana* 540
winged elm - *Ulmus alata* 540
Wych elm - *Ulmus glabra* 540
elodea - *Elodea canadensis* 251
emmer - *Triticum dicoccon* 425
wild emmer - *Triticum dicoccoides* 425
empress-candleplant - *Senna alata* 215
endive - *Cichorium endivia* 65
ephedra
Chinese ephedra - *Ephedra sinica* 169
gray ephedra - *Ephedra nevadensis* 169
green ephedra - *Ephedra viridis* 169
longleaf ephedra - *Ephedra trifurca* 169
Nevada ephedra - *Ephedra nevadensis* 169
eryngo - *Eryngium aquaticum* 29
escarole - *Cichorium endivia* 65
eschalot - *Allium cepa* 272
esparcet - *Onobrychis viciifolia* 209
estafiata - *Artemisia frigida* 57
estragon - *Artemisia dracunculus* 57
eucalyptus, spiral - *Eucalyptus cinerea* 303
euonymus
climbing euonymus - *Euonymus fortunei* 133
evergreen euonymus - *Euonymus japonica* 133
euphorbia, Abyssinian - *Euphorbia trigona* 184
evening-campion - *Silene alba* 131
evening-primrose
common evening-primrose - *Oenothera biennis* 319
desert evening-primrose - *Oenothera deltoides* 319
evergreen
African evergreen - *Syngonium podophyllum* 41
Belgian evergreen - *Dracaena sanderana* 18
Chinese evergreen - *Aglaonema modestum, A. simplex* 38
everlasting - *Gnaphalium obtusifolium* 71
pearly everlasting - *Anaphalis margaritacea* 55
sweet everlasting - *Gnaphalium obtusifolium* 71
ezo-negi - *Allium schoenoprasum* 272
fairy-bells - *Disporum hookeri* 276
fairy-duster - *Calliandra eriophylla* 192
fall-dandelion - *Leontodon autumnalis* 79
fameflower - *Talinum paniculatum* 442
fanweed - *Thlaspi arvense* 115
farkleberry - *Rhododendron arboreum* 176
fat-hen - *Chenopodium bonus-henricus* 136
fatsia, Japanese - *Fatsia japonica* 43
feather-geranium - *Chenopodium botrys* 136
feather-plume - *Dalea formosa* 196
featherbells - *Stenanthium gramineum* 281
feijoa - *Feijoa sellowiana* 305
felonherb - *Artemisia vulgaris* 57

feng-hsiang-shu - *Liquidambar formosana* 247
fennel - *Foeniculum vulgare* 29
  Florence fennel - *Foeniculum vulgare* 29
fenugreek - *Trigonella foenum-graecum* 219
fern
  American maidenhair fern - *Adiantum pedatum* 445
  American wall fern - *Polypodium virginianum* 449
  Australian tree fern - *Alsophila australis* 445, *Cyathea cooperi* 446
  ball fern - *Davallia trichomanoides* 446
  bear's-paw fern - *Aglaomorpha meyeniana* 445
  bird's-nest fern - *Asplenium nidus* 445
  blond tree fern - *Cibotium splendens* 446
  boulder fern - *Dennstaedtia punctilobula* 446
  bracken fern - *Pteridium aquilinum* 449
  brittle fern - *Cystopteris fragilis* 446
  brittle maidenhair fern - *Adiantum tenerum* 445
  button fern - *Pellaea rotundifolia* 448
  California gold-bark fern - *Pityrogramma triangularis* 448
  Christmas fern - *Polystichum acrostichoides* 449
  cinnamon fern - *Osmunda cinnamomea* 448
  coffee fern - *Pellaea andromedifolia* 448
  common staghorn fern - *Platycerium bifurcatum* 448
  crested fern - *Microsorium punctatum* 447
  delta maidenhair fern - *Adiantum raddianum* cv. 'decorum' 445
  European chain fern - *Woodwardia radicans* 450
  European parsley fern - *Cryptogramma crispa* 446
  fan maidenhair fern - *Adiantum tenerum* 445
  fancy fern - *Dryopteris austriaca* var. *spinulosa* 447
  five-finger fern - *Adiantum pedatum* 445
  floating water fern - *Ceratopteris thalictroides* 446
  florist's fern - *Dryopteris austriaca* var. *spinulosa* 447
  flowering fern - *Osmunda regalis* 448
  fragile fern - *Cystopteris fragilis* 446
  giant chain fern - *Woodwardia fimbriata* 450
  halberd fern - *Tectaria heracleifolia* 450
  hand fern - *Doryopteris pedata* 447
  Hawaiian tree fern - *Cibotium chamissoi*, *C. splendens* 446
  hay-scented fern - *Dennstaedtia punctilobula* 446
  interrupted fern - *Osmunda claytoniana* 448
  iron fern - *Rumohra adiantiformis* 450
  Jamaica gold fern - *Pityrogramma sulphurea* 448
  Japanese climbing fern - *Lygodium japonicum* 447
  lace fern - *Sphenomeris chusana* 450
  leather fern - *Acrostichum aureum* 445, *Rumohra adiantiformis* 450
  licorice fern - *Polypodium glycyrrhiza* 449
  man tree fern - *Cibotium splendens* 446
  marsh fern - *Thelypteris palustris* 450
  meadow fern - *Thelypteris palustris* 450
  mountain parsley fern - *Cryptogramma crispa* 446
  nettle chain fern - *Woodwardia areolata* 450
  New York fern - *Thelypteris noveboracensis* 450
  northern maidenhair fern - *Adiantum pedatum* 445
  oak fern - *Gymnocarpium dryopteris* 447
  ostrich fern - *Matteuccia struthiopteris* 447
  Pacific Christmas fern - *Polystichum munitum* 449
  rattlesnake fern - *Botrychium virginianum* 446
  resurrection fern - *Polypodium polypodioides* 449
  ribbon fern - *Polypodium phyllitidis* 449
  royal fern - *Osmunda regalis* 448
  savannah fern - *Gleichenia linearis* 447
  sensitive fern - *Onoclea sensibilis* 448
  silver fern - *Pityrogramma calomelanos* 448
  snuffbox fern - *Thelypteris palustris* 450
  spinulose wood fern - *Dryopteris austriaca* var. *spinulosa* 447
  squirrel-foot fern - *Davallia trichomanoides* 446
  strap fern - *Polypodium phyllitidis* 449
  sword fern - *Nephrolepis biserrata* 447
  Virginia chain fern - *Woodwardia virginica* 450
  walking fern - *Camptosorus rhizophyllus* 446
  water fern - *Ceratopteris thalictroides* 446
  waxy cloak fern - *Notholaena sinuata* 448
  West Indian tree fern - *Cyathea arborea* 446
fern-palm - *Cycas circinalis* 158
  Japanese fern-palm - *Cycas revoluta* 158
fescue
  Arizona fescue - *Festuca arizonica* 391
  bluebunch fescue - *Festuca idahoensis* 391
  Foxtail fescue - *Vulpia myuros* 428
  green-leaf fescue - *Festuca viridula* 391
  hair fescue - *Festuca capillata* 391
  hard fescue - *Festuca ovina* var. *duriuscula* 391
  Idaho fescue - *Festuca idahoensis* 391
  meadow fescue - *Festuca pratensis* 391
  nodding fescue - *Festuca obtusa* 391
  rattail fescue - *Vulpia myuros* 428
  red fescue - *Festuca rubra* 391
  rough fescue - *Festuca scabrella* 391
  shade fescue - *Festuca rubra* var. *heterophylla* 391
  sheep fescue - *Festuca ovina* 391
  slender fescue - *Festuca capillata* 391
  squirreltail fescue - *Vulpia bromoides* 428
  tall fescue - *Festuca arundinacea* 391
  Thurber fescue - *Festuca thurberi* 391
  western fescue - *Festuca occidentalis* 391
festuca, chewing - *Festuca rubra* 391
fetterbush - *Lyonia lucida* 174, *Pieris floribunda* 175
feverbush - *Lindera benzoin* 269
feverfew - *Chrysanthemum parthenium* 63
  American feverfew - *Parthenium integrifolium* 80
fevertree - *Pinckneya pubens* 493
feverwort - *Triosteum perfoliatum* 127
ficus
  climbing ficus - *Ficus pumila* 299
  creeping ficus - *Ficus pumila* 299
  fiddleleaf ficus - *Ficus lyrata* 299
  mistletoe ficus - *Ficus deltoidea* 298
fiddleheads - *Osmunda cinnamomea* 448
fieldbalm - *Glechoma hederacea* 263
fieldcress, yellow - *Rorippa sylvestris* 115
fig
  banyan fig - *Ficus benghalensis* 298
  common fig - *Ficus carica* 298
  Florida strangler fig - *Ficus aurea* 298
  golden fig - *Ficus aurea* 298
  India rubber fig - *Ficus elastica* 298
  Indian fig - *Opuntia ficus-indica* 120
  Indian laurel fig - *Ficus microcarpa* 299
  Java fig - *Ficus benjamina* 298
  weeping fig - *Ficus benjamina* 298
fig-tree, East Indian - *Ficus benghalensis* 298
filaree
  redstemmed filaree - *Erodium cicutarium* 242
  whitestem filaree - *Erodium moschatum* 242
filbert
  American filbert - *Corylus americana* 102
  beaked filbert - *Corylus cornuta* 102
  European filbert - *Corylus avellana* 102
  giant filbert - *Corylus maxima* 102
finger-of-God - *Aechmea orlandiana* 116
fingercups - *Tellima grandiflora* 515
finocchio - *Foeniculum vulgare* 29
fir
  alpine fir - *Abies lasiocarpa* 326
  balsam fir - *Abies balsamea* 326
  bristlecone fir - *Abies bracteata* 326
  California red fir - *Abies magnifica* 326
  Cascade fir - *Abies amabilis* 326
  cork fir - *Abies lasiocarpa* var. *arizonica* 326
  douglas fir - *Pseudotsuga menziesii* 352
  grand fir - *Abies grandis* 326
  Greek fir - *Abies cephalonica* 326
  lowland white fir - *Abies grandis* 326
  noble fir - *Abies procera* 326
  red fir - *Abies magnifica* 326
  Santa Lucia fir - *Abies bracteata* 326
  silver fir - *Abies alba* 326
  subalpine fir - *Abies lasiocarpa* 326
  white fir - *Abies amabilis*, *A. concolor* 326
fir-balsam - *Abies balsamea* 326
fire-chalice - *Zauschneria californica* 319
fire-dragon - *Acalypha wilkesiana* 182
fire-plant, Mexican - *Euphorbia heterophylla* 184
fire-thorn, narrowleaf - *Pyracantha angustifolia* 481
firebush - *Hamelia patens* 493
firecracker-flower - *Crossandra infundibuliformis* 7
firecracker-plant - *Cuphea ignea* 284
firefinger, marsh - *Potentilla palustris* 474
firethorn, scarlet - *Pyracantha coccinea* 481
fireweed - *Epilobium angustifolium* 317, *Erechtites hieracifolia* 69
firewheel - *Gaillardia pulchella* 71
firewheel-tree - *Stenocarpus sinuatus* 445
fish-poison-tree - *Piscidia piscipula* 211
fistula, purging - *Cassia fistula* 193
fitsroot - *Monotropa uniflora* 298
fittonia
  silver fittonia - *Fittonia verschaffeltii* 7
  white-leaved fittonia - *Fittonia verschaffeltii* 7
flag - *Iris × germanica* 254
  larger blue flag - *Iris versicolor* 254
  poison flag - *Iris versicolor* 254
  southern blue flag - *Iris virginica* var. *schrevei* 254
  western blue flag - *Iris missouriensis* 254
flagroot - *Acorus calamus* 38
flamboyant-tree - *Delonix regia* 196

flame-of-the-forest - *Spathodea campanulata* 105
flame-of-the-woods - *Ixora coccinea* 493
flame-tree - *Delonix regia* 196
Australian flame-tree - *Brachychiton acerifolius* 532
flameflower - *Kniphofia uvaria* 277, *Pyrostegia venusta* 105
flamegold - *Koelreuteria elegans* 511
flameleaf, Mexican - *Euphorbia pulcherrima* 184
flamevine - *Pyrostegia venusta* 105
Mexican flamevine - *Senecio chenopodioides* 82
flaming-trumpet - *Pyrostegia venusta* 105
flamingo-lily - *Anthurium andraeanum* 38
flamingo-plant - *Justicia carnea* 8
flannel-bush - *Fremontodendron californica* 532
flannelplant - *Verbascum thapsus* 521
flaver - *Avena fatua* 372
flawn - *Zoysia matrella* 433
flax
common flax - *Linum usitatissimum* 283
New Zealand flax - *Phormium tenax* 19
perennial flax - *Linum perenne* subsp. *lewisii* 283
prairie flax - *Linum perenne* subsp. *lewisii* 283
woodland flax - *Linum virginianum* 283
fleabane
annual fleabane - *Erigeron annuus* 69
daisy fleabane - *Erigeron annuus* 69
marsh fleabane - *Pluchea camphorata* 81
fleece-flower - *Polygonum cuspidatum* 438
fleur-de-lis - *Iris* x *germanica* 254
flixweed - *Descurainia sophia* 112
floppers - *Kalanchoe pinnata* 146
floss-silk-tree - *Chorisia speciosa* 106
flower-of-an-hour - *Hibiscus trionum* 292
flowering-fence - *Caesalpinia pulcherrima* 192
flowering-maple - *Abutilon pictum* 289
flowering-rush - *Butomus umbellatus* 117
flowering-willow - *Chilopsis linearis* 104
flowering-wintergreen - *Polygala paucifolia* 436
flowers-of-love - *Tabernaemontana divaricata* 35
foamflower - *Tiarella cordifolia* 515
forget-me-not - *Myosotis scorpioides* 108
Chinese forget-me-not - *Cynoglossum amabile* 106
garden forget-me-not - *Myosotis sylvatica* 108
fountainbush - *Russelia equisetiformis* 520
fountainplant - *Russelia equisetiformis* 520
four-o'clock
Colorado four-o'clock - *Mirabilis multiflora* 308
common four-o'clock - *Mirabilis jalapa* 308
trailing four-o'clock - *Allionia incarnata* 308
fox's-brush - *Centranthus ruber* 544
foxberry - *Vaccinium vitis-idaea* 179
foxglove - *Digitalis purpurea* 517
common foxglove - *Digitalis purpurea* 517
downy false foxglove - *Aureolaria virginica* 516
false foxglove - *Aureolaria pedicularia* 516
Grecian foxglove - *Digitalis lanata* 517
straw foxglove - *Digitalis lutea* 517
foxtail
alpine foxtail - *Alopecurus alpinus* 368
bristly foxtail - *Setaria verticillata* 417
creeping foxtail - *Alopecurus arundinaceus* 368
giant foxtail - *Setaria faberi* 417
green foxtail - *Setaria viridis* 417
meadow foxtail - *Alopecurus pratensis* 368
reed foxtail - *Alopecurus arundinaceus* 368
short-awn foxtail - *Alopecurus aequalis* 368
water foxtail - *Alopecurus geniculatus* 368
yellow foxtail - *Setaria glauca, S. pumila* 417
framboise - *Rubus idaeus* 486
fraxinella - *Dictamnus albus* 497
fremontia, California - *Fremontodendron californica* 532
frenchweed - *Thlaspi arvense* 115
friar's-cap - *Aconitum napellus* 451
friendship-plant - *Pilea involucrata* 543
frijol - *Phaseolus vulgaris* 209
frijolito - *Sophora secundiflora* 215
fringetree - *Chionanthus virginicus* 310
fritillary, narrow-leaved - *Fritillaria lanceolata* 276
frogfruit - *Phyla nodiflora* 546
northern frogfruit - *Phyla lanceolata* 546
frostwort - *Helianthemum canadense* 138
fruit-salad-plant - *Monstera deliciosa* 40
fuchsia, California - *Zauschneria californica* 319
funnel-creeper - *Macfadyena unguis-cati* 105
galinsoga, small-flowered - *Galinsoga parviflora* 71
gallberry - *Ilex glabra* 36
large gallberry - *Ilex coriacea* 36
sweet gallberry - *Ilex coriacea* 36
galleta, big - *Hilaria rigida* 395
garbanzo - *Cicer arietinum* 194
gardenia, common - *Gardenia jasminoides* 492
garget - *Phytolacca americana* 326
garland-flower - *Daphne cneorum* 536, *Hedychium coronarium* 553
garlic - *Allium sativum* 272
crow garlic - *Allium vineale* 272
field garlic - *Allium vineale* 272
meadow garlic - *Allium canadense* 272
oriental garlic - *Allium tuberosum* 272
society garlic - *Tulbaghia violacea* 281
stag's garlic - *Allium vineale* 272
wild garlic - *Allium canadense, A. vineale* 272
garlicvine - *Cydista aequinoctialis* 105
gas-plant - *Dictamnus albus* 497
gean - *Prunus avium* 475
genipap - *Genipa americana* 492
genipe - *Melicoccus bijugatus* 511
gentian
alpine gentian - *Gentiana newberryi* 241
blind gentian - *Gentiana clausa* 241
bottle gentian - *Gentiana andrewsii, G. clausa* 241
closed gentian - *Gentiana andrewsii, G. clausa, G. linearis* 241
fringed gentian - *Gentiana crinita* 241
Mendocino gentian - *Gentiana setigera* 241
narrow-leaved gentian - *Gentiana linearis* 241
soapwort gentian - *Gentiana saponaria* 241
geranium
bedding geranium - *Pelargonium* × *hortorum* 243
blood-red geranium - *Geranium sanguineum* 242
Carolina geranium - *Geranium carolinianum* 242
cutleaf geranium - *Geranium dissectum* 242
dove's-foot geranium - *Geranium molle* 242
fancy geranium - *Pelargonium* × *domesticum* 243
fish geranium - *Pelargonium* × *hortorum* 243
hanging geranium - *Pelargonium peltatum* 243
horseshoe geranium - *Pelargonium* × *hortorum* 243
house geranium - *Pelargonium* × *hortorum* 243
ivy geranium - *Pelargonium peltatum* 243
Lady Washington geranium - *Pelargonium* × *domesticum* 243
Martha Washington geranium - *Pelargonium* × *domesticum* 243
pansy-flowered geranium - *Pelargonium* × *domesticum* 243
pole-cat geranium - *Lantana montevidensis* 546
regal geranium - *Pelargonium* × *domesticum* 243
rose geranium - *Pelargonium graveolens* 243
show geranium - *Pelargonium* × *domesticum* 243
spotted geranium - *Geranium maculatum* 242
sweet-scented geranium - *Pelargonium graveolens* 243
wild geranium - *Geranium maculatum* 242
zonal geranium - *Pelargonium* × *hortorum* 243
germander, American - *Teucrium canadense* 268
gherkin
bur gherkin - *Cucumis anguria* 147
West Indian gherkin - *Cucumis anguria* 147
ghostweed - *Euphorbia marginata* 184
gilia, granite - *Leptodactylon pungens* 435
gill-over-the-ground - *Glechoma hederacea* 263
gillyflower - *Matthiola incana* 114
ginberbread-tree - *Hyphaene thebaica* 48
ginger
buterfly ginger - *Hedychium coronarium* 553
Canton ginger - *Zingiber officinale* 553
common ginger - *Zingiber officinale* 553
Kahili ginger - *Hedychium gardneranum* 553
red ginger - *Alpinia purpurata* 553
shell ginger - *Alpinia zerumbet* 553
spiral ginger - *Costus malortieanus* 553
true ginger - *Zingiber officinale* 553
white ginger - *Hedychium coronarium* 553
zerumbet ginger - *Zingiber zerumbet* 553
ginger-lily - *Hedychium coronarium* 553
ginkgo - *Ginkgo biloba* 244
ginseng - *Panax quinquefolius* 44
American ginseng - *Panax quinquefolius* 44
dwarf ginseng - *Panax trifolius* 44
girasole - *Helianthus tuberosus* 73
gladiola, garden - *Gladiolus* x *hortulanus* 253
gladiolus, water - *Butomus umbellatus* 117
glasswort, starry - *Cerastium arvense* 129
globeflower - *Kerria japonica* 469
spreading globeflower - *Trollius laxus* 458
glory-bower, bleeding - *Clerodendrum thomsoniae* 545
glory-bush - *Tibouchina urvilleana* 296
glory-of-the-sun - *Leucocoryne ixioides* 277
glory-tree - *Clerodendrum thomsoniae* 545

glorytree, purple - *Tibouchina urvilleana* 296
gloxinia
Brazilian gloxinia - *Sinningia speciosa* 244
violet-slipper gloxinia - *Sinningia speciosa* 244
goat's-rue - *Tephrosia virginiana* 216
goatnut - *Simmondsia chinensis* 523
goatsbeard - *Aruncus sylvester* 462
gobo - *Arctium lappa* 56
gold-and-silver-flower - *Lonicera japonica* 124
gold-dust - *Alyssum saxatile* 108
gold-moss - *Sedum acre* 146
goldapple - *Lycopersicon esculentum* 526
golden-buttons - *Tanacetum vulgare* 87
golden-carpet - *Sedum acre* 146
golden-chain - *Laburnum anagyroides* 201
golden-column - *Trichocereus spachianus* 121
golden-creeper - *Stigmaphyllon ciliatum* 289
golden-fleece - *Dyssodia tenuiloba* 68
golden-glow - *Rudbeckia laciniata* 82
golden-rain - *Cassia fistula* 193
golden-shower - *Cassia fistula* 193, *Pyrostegia venusta* 105
golden-slipper - *Cypripedium calceolus* var. *pubescens* 321
golden-trumpet - *Allamanda cathartica* 33
golden-vine, Brazilian - *Stigmaphyllon ciliatum* 289
goldenclub - *Orontium aquaticum* 40
goldeneye - *Viguiera annua* 88
goldenflower - *Hypericum calycinum* 139
goldenrod
blue-stem goldenrod - *Solidago caesia* 84
Canada goldenrod - *Solidago canadensis* 84
Leavenworth goldenrod - *Solidago leavenworthii* 84
seaside goldenrod - *Solidago sempervirens* 84
sweet goldenrod - *Solidago odora* 84
tall goldenrod - *Solidago canadensis* var. *scabra* 84
white goldenrod - *Solidago bicolor* 84
wreath goldenrod - *Solidago caesia* 84
goldenseal - *Hydrastis canadensis* 456
goldentop - *Lamarckia aurea* 398
goldfields - *Lasthenia chrysostoma* 78
goober - *Arachis hypogaea* 189
good-King-Henry - *Chenopodium bonus-henricus* 136
good-luck-leaf - *Kalanchoe pinnata* 146
good-luck-plant - *Cordyline terminalis* 17
goose-tansy - *Potentilla anserina* 474
gooseberry
Barbados gooseberry - *Physalis peruviana* 529
canyon gooseberry - *Ribes menziesii* 245
Cape gooseberry - *Physalis peruviana* 529
Chinese gooseberry - *Actinidia chinensis* 17
country gooseberry - *Averrhoa carambola* 323
dwarf cape gooseberry - *Physalis pruinosa* 529
English gooseberry - *Ribes uva-crispa* 245
European gooseberry - *Ribes grossularis*, *R. uva-crispa* 245
fuchsia-flowered gooseberry - *Ribes speciosum* 245
gummy gooseberry - *Ribes lobbii* 245
hairystem gooseberry - *Ribes hirtellum* 245
hawthorn-leaved gooseberry - *Ribes oxyacanthoides* 245
Missouri gooseberry - *Ribes missouriense*, *R. setosum* 245
pasture gooseberry - *Ribes cynosbati* 245
prickly gooseberry - *Ribes cynosbati* 245
Sierra gooseberry - *Ribes roezlii* 245
swamp gooseberry - *Ribes lacustre* 245
goosefoot
nettleleaf goosefoot - *Chenopodium murale* 136
perennial goosefoot - *Chenopodium bonus-henricus* 136
red goosefoot - *Chenopodium rubrum* 136
white goosefoot - *Chenopodium album* 136
gourd
ash gourd - *Benincasa hispida* 146
bitter gourd - *Momordica charantia* 150
bottle gourd - *Lagenaria siceraria* 150
calabash gourd - *Lagenaria siceraria* 150
Chinese fuzzy gourd - *Benincasa hispida* 146
club gourd - *Trichosanthes anguina* 151
dishcloth gourd - *Luffa aegyptiaca* 150
goarberry gourd - *Cucumis anguria* 147
gooseberry gourd - *Cucumis anguria* 147
Missouri gourd - *Cucurbita foetidissima* 149
snake gourd - *Trichosanthes anguina* 151
sponge gourd - *Luffa aegyptiaca* 150
Texas gourd - *Cucurbita texana* 149
towel gourd - *Luffa acutangula* 150
viper's gourd - *Trichosanthes anguina* 151
wax gourd - *Benincasa hispida* 146
white gourd - *Benincasa hispida* 146
white-flowered gourd - *Lagenaria siceraria* 150
yellow-flowered gourd - *Cucurbita pepo* var. *ovifera* 149
governor's-plum - *Flacourtia indica* 240
gow - *Lycium chinense* 526
gow-kee - *Lycium chinense* 526
gram
Bengal gram - *Cicer arietinum* 194
black gram - *Vigna mungo* 220
golden gram - *Vigna radiata* 220
green gram - *Vigna radiata* 220
red gram - *Cajanus cajan* 192
yellow gram - *Cicer arietinum* 194
grama
black grama - *Bouteloua eriopoda* 375
blue grama - *Bouteloua gracilis* 375
hairy grama - *Bouteloua hirsuta* 375
mat grama - *Bouteloua simplex* 375
needle grama - *Bouteloua aristidoides* 375
purple grama - *Bouteloua radicosa* 375
Rothrock grama - *Bouteloua rothrockii* 375
side-oats grama - *Bouteloua curtipendula* 375
six-weeks grama - *Bouteloua barbata* 375
granadilla
giant granadilla - *Passiflora quadrangularis* 325
purple granadilla - *Passiflora edulis* 325
yellow granadilla - *Passiflora laurifolia* 325
granjeno - *Celtis pallida* 539
grape
American grape - *Vitis labrusca* 550
bird grape - *Vitis munsoniana* 550
bullace grape - *Vitis rotundifolia* 550
bunch grape - *Vitis aestivalis* 550
bush grape - *Vitis rupestris* 550
California wild grape - *Vitis californica* 550
canyon grape - *Vitis arizonica* 550
cat grape - *Vitis palmata* 550
catbird grape - *Vitis palmata* 550
chicken grape - *Vitis vulpina* 550
European grape - *Vitis vinifera* 550
fox grape - *Vitis labrusca* 550
frost grape - *Vitis riparia*, *V. vulpina* 550
grayback grape - *Vitis cinerea* 550
mountain grape - *Vitis rupestris* 550
muscadine grape - *Vitis rotundifolia* 550
pigeon grape - *Vitis aestivalis* 550
possum grape - *Vitis baileyana* 550
red grape - *Vitis palmata* 550
riverbank grape - *Vitis riparia* 550
rock grape - *Vitis rupestris* 550
sand grape - *Vitis rupestris* 550
skunk grape - *Vitis labrusca* 550
southern fox grape - *Vitis rotundifolia* 550
Spanish grape - *Vitis berlandieri* 550
stunt grape - *Vitis labrusca* 550
sugar grape - *Vitis rupestris* 550
summer grape - *Vitis aestivalis* 550
Western wild grape - *Vitis californica* 550
wine grape - *Vitis vinifera* 550
winter grape - *Vitis berlandieri*, *V. vulpina* 550
grape-hyacinth, common - *Muscari botryoides* 279
grapefruit - *Citrus* x *paradisi* 495
grass
Alaska oniongrass - *Melica subulata* 400
aleppo grass - *Sorghum halepense* 419
alkali bluegrass - *Poa juncifolia* 409
alkali cordgrass - *Spartina gracilis* 420
alkali-grass - *Zigadenus elegans* 283
alkaligrass - *Puccinellia distans* 413
alpine bluegrass - *Poa alpina* 409
American beachgrass - *Ammophila breviligulata* 369
American dunegrass - *Elymus mollis* 387
American mannagrass - *Glyceria grandis* 393
American sloughgrass - *Beckmannia syzigachne* 375
annual beardgrass - *Polypogon monspeliensis* 413
annual bluegrass - *Poa annua* 409
annual hairgrass - *Deschampsia danthonioides* 384
annual ryegrass - *Lolium multiflorum* 399
Aparejo grass - *Muhlenbergia utilis* 401
arctic bluegrass - *Poa arctica* 409
Arizona three-awn grass - *Aristida arizonica* 371
Australian ryegrass - *Lolium multiflorum*, *L. multiflorum* cv. 'gulf' 399
autumn bentgrass - *Agrostis perennans* 366
awnless bromegrass - *Bromus inermis* 376
Bahia grass - *Paspalum notatum* 405
Baker wheatgrass - *Agropyron bakeri* 363
barb goat grass - *Aegilops triuncialis* 363
barnyard grass - *Echinochloa crusgalli* 386
basket grass - *Oplismenus hirtellus* 402
bayonet-grass - *Scirpus maritimus* 165
bear-grass - *Yucca filamentosa*, *Y. smalliana* 19
bearded wheatgrass - *Agropyron* × *subsecundum* var. *andinum* 363
beardless wheatgrass - *Agropyron spicatum* 363
beggartick grass - *Aristida orcuttiana* 371
Bengal grass - *Setaria italica* 417
bent-awn plumegrass - *Erianthus contortus* 391
bermudagrass - *Cynodon dactylon* 382
big bluegrass - *Poa ampla* 409
big cordgrass - *Spartina cynosuroides* 420
Bigelow bluegrass - *Poa bigelovii* 409
billion-dollar grass - *Echinochloa crusgalli* var. *frumentacea* 386
birdseed grass - *Phalaris canariensis* 407

black oatgrass - *Stipa avenacea* 422
blackseed needlegrass - *Stipa avenacea* 422
bluebunch wheatgrass - *Agropyron spicatum* 363
Boer lovegrass - *Eragrostis curvula* 390
bog bluegrass - *Poa leptocoma* 409
brahman grass - *Dichanthium annulatum* 385
broadleaf signalgrass - *Brachiaria platyphylla* 376
broom beardgrass - *Schizachyrium scoparium* 415
broom wiregrass - *Schizachyrium scoparium* 415
brown bentgrass - *Agrostis canina, A. perennans* 366
brown plumegrass - *Erianthus brevibarbis* 391
buffalograss - *Buchloe dactyloides* 379
bulb panicgrass - *Panicum bulbosum* 403
bulbous bluegrass - *Poa bulbosa* 409
bulbous canarygrass - *Phalaris aquatica* 407
bullgrass - *Muhlenbergia emersleyi* 400
bunchgrass - *Schizachyrium scoparium* 415
bur bristlegrass - *Setaria verticillata* 417
burgrass - *Cenchrus incertus* 381
California needlegrass - *Stipa pulchra* 423
California oatgrass - *Danthonia californica* 384
California sweetgrass - *Hierochloe occidentalis* 394
Canada bluegrass - *Poa compressa* 409
canarygrass - *Phalaris canariensis* 407
Canby bluegrass - *Poa canbyi* 409
caribgrass - *Eriochloa polystachya* 391
carpetgrass - *Axonopus affinis* 374
catchfly grass - *Leersia lenticularis* 398
centipede grass - *Eremochloa ophiuroides* 391
cheatgrass - *Agropyron* sp. 363, *Bromus secalinus* 376, *B. tectorum* 377
China-grass - *Boehmeria nivea* 543
colonial bentgrass - *Agrostis capillaris* 366
Columbia needlegrass - *Stipa columbiana* 423
couchgrass - *Agropyron repens* 363
creeping bentgrass - *Agrostis stolonifera* 366
crowfoot grass - *Dactyloctenium aegyptium* 384
curly mesquitegrass - *Hilaria belangeri* 395
Cusick bluegrass - *Poa cusickii* 409
Dallis grass - *Paspalum dilatatum* 405
deergrass - *Muhlenbergia rigens* 401
desert needlegrass - *Stipa speciosa* 423
desert saltgrass - *Distichlis stricta* 386
devilgrass - *Cynodon dactylon* 382
downy oatgrass - *Danthonia sericea* 384
dune bentgrass - *Agrostis pallens* 366
dwarf meadow grass - *Poa annua* 409
eastern gamagrass - *Tripsacum dactyloides* 424
eastern mannagrass - *Glyceria septentrionalis* 393
eelgrass - *Zostera marina* 553
elephant grass - *Pennisetum purpureum* 406
Elliott's beardgrass - *Andropogon elliottii* 369
English bluegrass - *Festuca pratensis* 391
English ryegrass - *Lolium perenne* 399
European beachgrass - *Ammophila arenaria* 369
European dunegrass - *Leymus arenarius* 399
fairway crested wheatgrass - *Agropyron cristatum* 363
fall panicgrass - *Panicum dichotomiflorum* 403
false buffalograss - *Munroa squarrosa* 402
false wheatgrass - *Leymus chinensis* 399
feather bunchgrass - *Stipa viridula* 423
feather fingergrass - *Chloris virgata* 381
fevergrass - *Cymbopogon citratus* 382
Florida gamagrass - *Tripsacum floridanum* 424
foothill needlegrass - *Stipa lepida* 423
fowl bluegrass - *Poa palustris* 409
fowl mannagrass - *Glyceria striata* 393
foxtail bromegrass - *Bromus rubens* 376
freshwater cordgrass - *Spartina pectinata* 420
galleta grass - *Hilaria jamesii* 395
gallowgrass - *Cannabis sativa* 122
Geyer oniongrass - *Melica geyeri* 400
giant cutgrass - *Zizaniopsis miliacea* 433
goat grass - *Aegilops* sp. 363
goose-grass - *Potentilla anserina* 474
goosegrass - *Eleusine indica* 387
green needlegrass - *Stipa viridula* 423
Grisebach bristlegrass - *Setaria grisebachii* 417
Guinea grass - *Panicum maximum* 403
hairy crabgrass - *Digitaria sanguinalis* 385
Harding grass - *Phalaris stenoptera* 407
hedgehog grass - *Cenchrus echinatus* 381
herd's grass - *Phleum pratense* 407
India lovegrass - *Eragrostis pilosa* 390
Indian grass - *Sorghastrum nutans* 418
inland bluegrass - *Poa interior* 409
intermediate wheatgrass - *Agropyron intermedium* 363
Italian ryegrass - *Lolium multiflorum, L. multiflorum* cv. 'gulf' 399
itchgrass - *Rottboellia cochinchinensis* 413
Japanese carpet grass - *Zoysia matrella* 433
Japanese false bromegrass - *Brachypodium pinnatum* 376
Japanese lawngrass - *Zoysia japonica* 433
Johnson grass - *Sorghum halepense* 419
jointed goatgrass - *Aegilops cylindrica* 363
June grass - *Koeleria nitida* 397
jungle ricegrass - *Echinochloa colonum, E. crusgalli* 386
Kentucky bluegrass - *Poa pratensis* 409
kikuyu grass - *Pennisetum clandestinum* 406
Kleberg grass - *Dichanthium annulatum* 385
knot grass - *Paspalum distichum* 405
knotroot bristlegrass - *Setaria geniculata* 417
Korean grass - *Zoysia japonica* 433
lacegrass - *Eragrostis capillaris* 390
large crabgrass - *Digitaria sanguinalis* 385
lazy-man's grass - *Eremochloa ophiuroides* 391
Lehmann lovegrass - *Eragrostis lehmanniana* 390
Leiberg bluegrass - *Poa leibergii* 409
Lemmon needlegrass - *Stipa lemmonii* 423
lemongrass - *Cymbopogon citratus* 382
Letterman needlegrass - *Stipa lettermanii* 423
limpo grass - *Hemarthria altissima* 394
little lovegrass - *Eragrostis poaeoides* 390
little quaking-grass - *Briza minor* 376
little ricegrass - *Oryzopsis exigua* 403
littleseed canarygrass - *Phalaris minor* 407
littleseed ricegrass - *Oryzopsis micrantha* 403
longtongue mutton grass - *Poa longiligula* 409
low speargrass - *Poa annua* 409
lyme grass - *Leymus arenarius* 399
lyme ryegrass - *Lolium perenne* 399
Manila grass - *Zoysia matrella* 433
marram grass - *Ammophila arenaria* 369
Mascarene grass - *Zoysia tenuifolia* 434
mat-grass - *Phyla nodiflora* 546
May grass - *Phalaris caroliniana* 407
means-grass - *Sorghum halepense* 419
Mexican lovegrass - *Eragrostis mexicana* 390
molasses grass - *Melinis minutiflora* 400
moor grass - *Molinia caerulea* 400
mountain hairgrass - *Deschampsia atropurpurea* 384
mutton bluegrass - *Poa fendleriana* 409, *Setaria faberi* 417
napier grass - *Pennisetum purpureum* 406
Natal grass - *Rhynchelytrum repens* 413
needlegrass - *Stipa sibirica* 423
Nevada bluegrass - *Poa nevadensis* 409
New Mexican feathergrass - *Stipa neomexicana* 423
nodding semaphore-grass - *Pleuropogon refractus* 409
northern mannagrass - *Glyceria borealis* 393
northern reedgrass - *Calamagrostis neglecta* 379
nutgrass - *Cyperus rotundus* 163
Nuttall alkaligrass - *Puccinellia nuttalliana* 413
one-spike oatgrass - *Danthonia unispicata* 384
oniongrass - *Melica bulbosa* 400
orangegrass - *Hypericum gentianoides* 139
orchardgrass - *Dactylis glomerata* 383
Oregon bentgrass - *Agrostis oregonensis* 366
Pacific bluegrass - *Poa gracillima* 409
Pacific reedgrass - *Calamagrostis nutkaensis* 379
pampasgrass - *Cortaderia selloana* 382
pangola grass - *Digitaria eriantha* subsp. *pentzii* 385
panicgrass - *Panicum anceps* 403
Pará grass - *Panicum purpurascens* 403
Parry oatgrass - *Danthonia parryi* 384
peppergrass - *Lepidium virginicum* 113
perennial quaking-grass - *Briza media* 376
perennial ryegrass - *Lolium perenne* 399
pin-grass - *Erodium cicutarium* 242
pine bluegrass - *Poa scabrella* 409
pinegrass - *Calamagrostis rubescens* 379
pinyon ricegrass - *Piptochaetium fimbriatum* 409
plains bluegrass - *Poa arida* 409
plains bristlegrass - *Setaria macrostachya* 417
plains lovegrass - *Eragrostis intermedia* 390
plains reedgrass - *Calamagrostis montanensis* 379
porcupine grass - *Stipa spartea* 423
poverty oatgrass - *Danthonia spicata* 384
poverty-grass - *Hudsonia tomentosa* 138
prairie bayonet-grass - *Scirpus maritimus* 165
prairie beardgrass - *Schizachyrium scoparium* 415
prairie cordgrass - *Spartina pectinata* 420

prairie wedge grass - *Sphenopholis obtusata* 421
Pringle needlegrass - *Stipa pringlei* 423
pubescent wheatgrass - *Agropyron intermedium* 363
pudding-grass - *Hedeoma pulegioides* 263
purple lovegrass - *Eragrostis spectabilis* 390
purple needlegrass - *Stipa pulchra* 423
purple oniongrass - *Melica spectabilis* 400
purple reedgrass - *Calamagrostis purpurascens* 379
purple-eyed grass - *Sisyrinchium douglasii* 255
quackgrass - *Agropyron repens* 363
rabbit's-tail grass - *Lagurus ovatus* 398
rabbitfoot grass - *Polypogon monspeliensis* 413
rattlesnake mannagrass - *Glyceria canadensis* 393
reed canarygrass - *Phalaris arundinacea* 407
rescue grass - *Bromus catharticus* 376
Rhode Island bentgrass - *Agrostis capillaris* 366
Rhodes grass - *Chloris gayana* 381
ribgrass - *Plantago lanceolata* 359
rice cutgrass - *Leersia oryzoides* 398
Richardson needlegrass - *Stipa richardsonii* 423
ring grass - *Muhlenbergia torreyi* 401
ringed beardgrass - *Dichanthium annulatum* 385
ripgut bromegrass - *Bromus rigidus* 376
ripgut grass - *Bromus rigidus* 376
ripplegrass - *Plantago lanceolata* 359
rough bentgrass - *Agrostis scabra* 366
roughstalk bluegrass - *Poa palustris, P. trivialis* 409
russet cottongrass - *Eriophorum russeolum* 164
salt marsh cordgrass - *Spartina alterniflora* 420
salt meadow cordgrass - *Spartina patens* 420
sand lovegrass - *Eragrostis trichodes* 390
Sandberg bluegrass - *Poa sandbergii, P. secunda* 409
sawgrass - *Cladium jamaicense* 163
scratchgrass - *Muhlenbergia asperifolia* 400
Scribner needlegrass - *Stipa scribneri* 423
Scribner reedgrass - *Calamagrostis scribneri* 379
scurvygrass - *Cochlearia officinalis* 111, *Crambe maritima* 112
scutch grass - *Cynodon dactylon* 382
seashore saltgrass - *Distichlis spicata* 386
serpent-grass - *Polygonum viviparum* 438
sheathed cottongrass - *Eriophorum vaginatum* 164
Siberian wheatgrass - *Agropyron fragile* subsp. *sibiricum* 363
silk grass - *Oryzopsis hymenoides* 403
silk-grass - *Chrysopsis graminifolia* 64, *Yucca filamentosa* 19
silver beardgrass - *Bothriochloa saccharoides* 375
silver plumegrass - *Saccharum alopecuroideum* 413
six-weeks grass - *Poa annua* 409
skyline bluegrass - *Poa epilis* 409
sleepy grass - *Stipa robusta* 423
slender bluegrass - *Poa gracillima* 409
slender false bromegrass - *Brachypodium sylvaticum* 376
slender hairgrass - *Deschampsia elongata* 384
slender wheatgrass - *Agropyron trachycaulum, A. trachycaulum* var. *glaucum* 363
slough grass - *Spartina pectinata* 420
smilograss - *Oryzopsis miliacea* 403
smooth cordgrass - *Spartina alterniflora* 420
smooth crabgrass - *Digitaria ischaemum* 385
sourgrass - *Paspalum conjugatum* 405
Spanish bromegrass - *Bromus madritensis* 376
speargrass - *Poa pratensis* 409
spidergrass - *Aristida ternipes* 371
St.-Augustine grass - *Stenotaphrum secundatum* 422
standard crested wheatgrass - *Agropyron desertorum* 363
stargrass - *Hypoxis hirsuta* 277
stinkgrass - *Eragrostis cilianensis* 390
Sudan grass - *Sorghum* × *drummondii* 419
sugarcane plumegrass - *Erianthus giganteus* 391
sweet vernalgrass - *Anthoxanthum odoratum* 370
sweetgrass - *Hierochloe odorata* 394
switchgrass - *Panicum virgatum* 404
tall mannagrass - *Glyceria elata* 393
tall oatgrass - *Arrhenatherum elatius* 371
tall wheatgrass - *Agropyron elongatum* 363
Terrell grass - *Lolium perenne* 399
Texas bluegrass - *Poa arachnifera* 409
Texas needlegrass - *Stipa leucotricha* 423
thick-spike wheatgrass - *Agropyron dasystachyum* 363
thin grass - *Agrostis diegoensis* 366
Thurber needlegrass - *Stipa thurberiana* 423
timber oatgrass - *Danthonia intermedia* 384
timberline bluegrass - *Poa rupicola* 409
tobosa grass - *Hilaria mutica* 395
toothache grass - *Ctenium aromaticum* 382
torpedograss - *Panicum repens* 403
tropical carpetgrass - *Axonopus compressus* 374
tuber oatgrass - *Arrhenatherum elatius* 371
tufted hairgrass - *Deschampsia caespitosa* 384
tufted lovegrass - *Eragrostis pectinacea* 390
tumble windmill-grass - *Chloris verticillata* 381
tumblegrass - *Schedonnardus paniculatus* 415
upland bentgrass - *Agrostis perennans* 366
Vasey grass - *Paspalum urvillei* 405
velvet bentgrass - *Agrostis canina* 366
velvetgrass - *Holcus lanatus* 395
vine mesquitegrass - *Panicum obtusum* 403
Virginia cottongrass - *Eriophorum virginicum* 164
water mannagrass - *Glyceria fluitans* 393
water stargrass - *Heteranthera dubia* 441
weeping alkaligrass - *Puccinellia distans* 413
weeping lovegrass - *Eragrostis curvula* 390
West Indian lemongrass - *Cymbopogon citratus* 382
western needlegrass - *Stipa occidentalis* 423
western wheatgrass - *Elymus smithii* 387
Wheeler bluegrass - *Poa nervosa* 409
windmill-grass - *Chloris verticillata* 381
wiregrass - *Eleusine indica* 387, *Poa compressa* 409
witch grass - *Panicum capillare* 403
wood bluegrass - *Poa nemoralis* 409
woolgrass - *Scirpus congdonii* 165
woolly fingergrass - *Digitaria eriantha* subsp. *pentzii* 385
yardgrass - *Eleusine indica* 387
yellow nutgrass - *Cyperus esculentus* 163
zoysia grass - *Zoysia matrella* 433
grass-widow - *Sisyrinchium douglasii* 255
grassnut - *Arachis hypogaea* 189, *Brodiaea laxa* 274
grassy-rush - *Butomus umbellatus* 117
graybeard - *Tillandsia usneoides* 117
greasewood - *Adenostoma fasciculatum* 460, *Salvia apiana* 267, *Sarcobatus vermiculatus* 137
green-almond - *Pistacia vera* 23
green-dragon - *Arisaema dracontium* 39
green-onion, bracted - *Habenaria viridis* var. *bracteata* 321
greenbrier
bristly greenbrier - *Smilax hispida* 523
cat greenbrier - *Smilax glauca* 523
common greenbrier - *Smilax rotundifolia* 523
laurel-leaved greenbrier - *Smilax laurifolia* 523
gromwell, western - *Lithospermum ruderale* 107
ground-cedar - *Lycopodium complanatum, L. tristachyum* 284
ground-cherry, Peruvian - *Physalis peruviana* 529
ground-hemlock - *Taxus canadensis* 533
ground-pine - *Lycopodium tristachyum* 284
flat-branch ground-pine - *Lycopodium obscurum* 284
round branch ground-pine - *Lycopodium obscurum* var. *dendroideum* 284
groundcherry
clammy groundcherry - *Physalis heterophylla* 529
downy groundcherry - *Physalis pubescens* 529
hairy groundcherry - *Physalis pruinosa* 529
purple groundcherry - *Physalis lobata* 529
strawberry groundcherry - *Physalis alkekengi* 529
groundivy - *Glechoma hederacea* 263
groundnut - *Apios americana* 189, *Arachis hypogaea* 189, *Panax trifolius* 44
groundsel
balsam groundsel - *Senecio pauperculus* 83
common groundsel - *Senecio vulgaris* 83
groundsel-bush - *Baccharis halimifolia* 60
groundsel-tree - *Baccharis halimifolia* 60
grouseberry - *Vaccinium scoparium* 179, *Viburnum trilobum* 127
gruya - *Canna edulis* 123
guar - *Cyamopsis tetragonoloba* 195
guava
apple guava - *Psidium guajava* 307
cattley guava - *Psidium littorale* var. *longipes* 307
common guava - *Psidium guajava* 307
pineapple guava - *Feijoa sellowiana* 305
purple strawberry guava - *Psidium littorale* var. *longipes* 307
yellow guava - *Psidium guajava* 307
guaymochil - *Pithecellobium dulce* 212
guayule - *Parthenium argentatum* 80
guelder-rose - *Viburnum opulus* 127

gueules-noires - *Gaylussacia baccata* 173
Guiana-chestnut - *Pachira aquatica* 106
gum
black gum - *Nyssa sylvatica* 309
cabbage gum - *Eucalyptus amplifolia, E. pauciflora* 303
forest red gum - *Eucalyptus tereticornis* 303
Formosan gum - *Liquidambar formosana* 247
gray gum - *Eucalyptus punctata* 303
Lehmann's gum - *Eucalyptus lehmannii* 303
lemon-scented gum - *Eucalyptus citriodora* 303
manna gum - *Eucalyptus viminalis* 303
Murray red gum - *Eucalyptus camaldulensis* 303
red gum - *Eucalyptus calophylla* 303, *Liquidambar styraciflua* 247
Red River gum - *Eucalyptus camaldulensis* 303
rose gum - *Eucalyptus grandis* 303
silver-dollar gum - *Eucalyptus polyanthemos* 303
silver-leaved mountain gum - *Eucalyptus pulverulenta* 303
snow gum - *Eucalyptus niphophila* 303
sour gum - *Nyssa sylvatica* 309
Sydney blue gum - *Eucalyptus saligna* 303
Tasmanian blue gum - *Eucalyptus globulus* 303
gum-elemi - *Bursera simaruba* 117
gumbo - *Abelmoschus esculentus* 289, *Bursera simaruba* 117
gumbo-limbo - *Bursera simaruba* 117
gumweed, curly-cup - *Grindelia squarrosa* 72
gypsyweed - *Veronica officinalis* 521
hackberry
common hackberry - *Celtis occidentalis* 539
desert hackberry - *Celtis pallida* 539
Mississippi hackberry - *Celtis laevigata* 539
netleaf hackberry - *Celtis reticulata* 539
southern hackberry - *Celtis laevigata* 539
hackmatack - *Larix laricina* 332, *Populus balsamifera* 499
hagberry - *Prunus padus* 475
hagbrier - *Smilax hispida* 523
halapepe - *Dracaena aurea* 18
hapuu - *Cibotium splendens* 446
hapuu-ii - *Cibotium chamissoi* 446
harakeke - *Phormium tenax* 19
harbinger-of-spring - *Erigenia bulbosa* 29
hardhack - *Spiraea tomentosa* 489
golden hardhack - *Potentilla fruticosa* 474
hare's-tail - *Lagurus ovatus* 398
harebell
common harebell - *Campanula rotundifolia* 121
southern harebell - *Campanula divaricata* 121
harlock - *Arctium lappa* 56
hartshorn-plant - *Anemone nuttalliana* 452
haw
black haw - *Viburnum lentago, V. prunifolium* 127
possum haw - *Ilex decidua* 36, *Viburnum nudum* 127
rusty black haw - *Viburnum rufidulum* 127
Southern black haw - *Viburnum rufidulum* 127
swamp haw - *Viburnum cassinoides, V. nudum* 127
sweet haw - *Viburnum prunifolium* 127
hawkbit - *Leontodon autumnalis* 79
hawksbeard
narrowleaf hawksbeard - *Crepis tectorum* 67
smooth hawksbeard - *Crepis capillaris* 67
Western hawksbeard - *Crepis occidentalis* 67
hawkweed
king-devil hawkweed - *Hieracium piloselloides* 76
narrow-leaved hawkweed - *Hieracium umbellatum* 76
orange hawkweed - *Hieracium aurantiacum* 76
hawthorn - *Crataegus aestivalis* 463
black hawthorn - *Crataegus douglasii* 463
downy hawthorn - *Crataegus mollis* 463
English hawthorn - *Crataegus laevigata, C. monogyna* 463
green hawthorn - *Crataegus viridis* 463
Indian hawthorn - *Rhaphiolepis indica* 483
hazel, western - *Corylus cornuta* var. *californica* 102
hazelnut
American hazelnut - *Corylus americana* 102
beaked hazelnut - *Corylus cornuta* 102
California hazelnut - *Corylus cornuta* var. *californica* 102
European hazelnut - *Corylus avellana* 102
heal-all - *Prunella vulgaris* 266
heart-of-Jesus - *Caladium bicolor* 39
heartnut - *Juglans cordiformis* 258
hearts-ease - *Polygonum persicaria* 438, *Viola tricolor* 547
heath, spring - *Erica carnea* 172
heather - *Calluna vulgaris* 172
false heather - *Cuphea hyssopifolia* 284
red heather - *Phyllodoce breweri* 175
snow heather - *Erica carnea* 172
heavenly-bamboo - *Nandina domestica* 92
heliotrope - *Heliotropium arborescens* 107
garden heliotrope - *Valeriana officinalis* 545
salt heliotrope - *Heliotropium curassavicum* 107
seaside heliotrope - *Heliotropium curassavicum* 107
hellebore, white - *Veratrum viride* 282
hellfetter - *Smilax hispida, S. tamnoides* 523
helmut-flower - *Aconitum napellus* 451
hemlock
Canada hemlock - *Tsuga canadensis* 355
Carolina hemlock - *Tsuga caroliniana* 355
mountain hemlock - *Tsuga mertensiana* 355
northern Japanese hemlock - *Tsuga diversifolia* 355
western hemlock - *Tsuga heterophylla* 355
hemp
bog hemp - *Boehmeria cylindrica* 543
bowstring hemp - *Calotropis gigantea* 52
Ceylon bowstring hemp - *Sansevieria zeylanica* 19
Colorado River hemp - *Sesbania macrocarpa* 215
Deccan hemp - *Hibiscus cannabinus* 292
Indian hemp - *Apocynum cannabinum* 33, *Hibiscus cannabinus* 292
New Zealand hemp - *Phormium tenax* 19
sisal hemp - *Agave sisalana* 17
soft hemp - *Cannabis sativa* 122
true hemp - *Cannabis sativa* 122
hemp-nettle - *Galeopsis tetrahit* 263
hemp-plant - *Agave sisalana* 17
hemp-tree - *Vitex agnus-castus* 547
hempweed, climbing - *Mikania scandens* 80
henbane - *Hyoscyamus niger* 526
black henbane - *Hyoscyamus niger* 526
henbit - *Lamium amplexicaule* 264
hens-and-chickens - *Sempervivum tectorum* 146
herb, healing - *Symphytum officinale* 108
herb-patience - *Rumex patientia* 440
herb-robert - *Geranium robertianum* 242
herb-sophia - *Descurainia sophia* 112
Hercules-club - *Aralia spinosa* 42, *Zanthoxylum clava-herculis* 498
hibiscus - *Hibiscus moscheutos* 292
Chinese hibiscus - *Hibiscus rosa-sinensis* 292
Hawaiian hibiscus - *Hibiscus rosa-sinensis* 292
sea hibiscus - *Hibiscus tiliaceus* 292
hickory
bitternut hickory - *Carya cordiformis* 255
black hickory - *Carya texana* 255
broom hickory - *Carya glabra* 255
mockernut hickory - *Carya tomentosa* 255
pale hickory - *Carya pallida* 255
pignut hickory - *Carya glabra* 255
sand hickory - *Carya pallida* 255
shagbark hickory - *Carya ovata* 255
shellbark hickory - *Carya laciniosa, C. ovata* 255
small-fruited hickory - *Carya glabra* 255
swamp hickory - *Carya cordiformis* 255
water hickory - *Carya aquatica* 255
white-heart hickory - *Carya tomentosa* 255
hiedra - *Rhus toxicodendron* 24
hills-of-snow - *Hydrangea arborescens* 250
himegurumi - *Juglans cordiformis* 258
hobblebush - *Viburnum alnifolium* 127
hog-peanut - *Amphicarpaea bracteata* 189
pitcher's hog-peanut - *Amphicarpaea bracteata* var. *comosa* 189
hog-plum - *Spondias purpurea* 26
hogweed - *Conyza canadensis* 66
holly
American holly - *Ilex opaca* 36
box-leaved holly - *Ilex crenata* 36
Chinese holly - *Ilex cornuta* 36
dahoon holly - *Ilex cassine* 36
dwarf holly - *Malpighia coccigera* 289
English holly - *Ilex aquifolium* 36
European holly - *Ilex aquifolium* 36
false holly - *Osmanthus heterophyllus* 315
horned holly - *Ilex cornuta* 36
Japanese holly - *Ilex crenata* 36
large-leaved holly - *Ilex montana* 36
miniature holly - *Malpighia coccigera* 289
mountain holly - *Ilex montana* 36, *Prunus ilicifolia* 475
Oregon holly - *Ilex aquifolium* 36
Singapore holly - *Malpighia coccigera* 289
West Indian holly - *Leea coccinea* 271
hollyfern
giant hollyfern - *Polystichum munitum* 449
house hollyfern - *Phanerophlebia falcata* 448
Japanese hollyfern - *Phanerophlebia falcata* 448
mountain hollyfern - *Polystichum lonchitis* 449
northern hollyfern - *Polystichum lonchitis* 449
hollyhock - *Alcea ficifolia, A. rosea* 290
desert hollyhock - *Sphaeralcea ambigua* 295
sea hollyhock - *Hibiscus moscheutos* subsp. *palustris* 292

honesty-plant - *Lunaria annua* 113
honewort - *Cryptotaenia canadensis* 28
honeyberry - *Melicoccus bijugatus* 511
honeyplant - *Hoya carnosa* 53
honeyshuck - *Gleditsia triacanthos* 198
honeysuckle - *Aquilegia canadensis* 453
American honeysuckle - *Lonicera canadensis* 124
bush honeysuckle - *Diervilla lonicera* 123
Cape honeysuckle - *Tecomaria capensis* 105
coral honeysuckle - *Lonicera sempervirens* 124
fly honeysuckle - *Lonicera canadensis* 124
grape honeysuckle - *Lonicera prolifera* 124
hairy honeysuckle - *Lonicera hirsuta* 124
Jamaica honeysuckle - *Passiflora laurifolia* 325
Japanese honeysuckle - *Lonicera japonica* 124
limber honeysuckle - *Lonicera dioica* 124
Morrow honeysuckle - *Lonicera morrowii* 124
pink honeysuckle - *Lonicera hispidula* 124
swamp honeysuckle - *Rhododendron viscosum* 176
swamp-fly honeysuckle - *Lonicera oblongifolia* 124
Tatarian honeysuckle - *Lonicera tatarica* 124
trumpet honeysuckle - *Campsis radicans* 103, *Lonicera sempervirens* 124
yellow honeysuckle - *Lonicera flava* 124
honeyvine - *Cynanchum laeve* 52
Honolulu-queen - *Hylocereus undatus* 119
hop
common hop - *Humulus lupulus* 123
European hop - *Humulus lupulus* 123
Japanese hop - *Humulus japonicus* 123
hop-clover - *Medicago lupulina* 206
hop-plant, false - *Justicia brandegeana* 8
hophornbeam, American - *Ostrya virginiana* 102
horehound
common horehound - *Marrubium vulgare* 264
white horehound - *Marrubium vulgare* 264
horn-of-plenty - *Datura metel* 525
hornbeam
American hornbeam - *Carpinus caroliniana* 101
European hornbeam - *Carpinus betulus* 101
horse-bean - *Parkinsonia aculeata* 209
horse-nettle
silver horse-nettle - *Solanum elaeagnifolium* 530
white horse-nettle - *Solanum elaeagnifolium* 530
horse-sugar - *Symplocos tinctoria* 533
horsebrier - *Smilax rotundifolia* 523
horsechestnut
California horsechestnut - *Aesculus californica* 249
common horsechestnut - *Aesculus hippocastanum* 249
dwarf horsechestnut - *Aesculus parviflora* 249
European horsechestnut - *Aesculus hippocastanum* 249
Japanese horsechestnut - *Aesculus turbinata* 250
red horsechestnut - *Aesculus* × *carnea*, *A.* × *carnea* cv. 'briotii' 249
horsegram - *Macrotyloma uniflorum* 206
horsemint - *Monarda punctata* 265
sweet horsemint - *Cunila origanoides* 263
horsenettle - *Solanum carolinense* 530
horseradish - *Armoracia rusticana* 109
horsetail, field - *Equisetum arvense* 169
horsetail-tree - *Casuarina equisetifolia* 132
horseweed - *Conyza canadensis* 66
hortensia - *Hydrangea macrophylla* 250
hottentot-fig - *Carpobrotus edulis* 20
hound's-tongue - *Cynoglossum officinale* 106
houseleek - *Sempervivum tectorum* 146
huamuchil - *Pithecellobium dulce* 212
huckleberry
black huckleberry - *Gaylussacia baccata* 173
blue huckleberry - *Vaccinium membranaceum* 179
box huckleberry - *Gaylussacia brachycera* 173
California huckleberry - *Vaccinium ovatum* 179
dwarf huckleberry - *Gaylussacia dumosa*, *G. frondosa* 173
evergreen huckleberry - *Vaccinium ovatum* 179
he huckleberry - *Cyrilla racemiflora* 166, *Lyonia ligustrina* 174
little-leaf huckleberry - *Vaccinium scoparium* 179
red huckleberry - *Vaccinium parvifolium* 179
shot huckleberry - *Vaccinium ovatum* 179
squaw huckleberry - *Vaccinium stamineum* 179
thin-leaf huckleberry - *Vaccinium membranaceum* 179
tree huckleberry - *Rhododendron arboreum* 176
huesito - *Malpighia glabra* 289
huisache - *Acacia farnesiana* 187
hunter's-robe - *Epipremnum aureum* 40
huntsman's-cup - *Sarracenia purpurea* 512
huntsman's-horn - *Sarracenia flava* 512
hurricane-plant - *Monstera deliciosa* 40
hyacinth
common hyacinth - *Hyacinthus orientalis* 277
Dutch hyacinth - *Hyacinthus orientalis* 277
garden hyacinth - *Hyacinthus orientalis* 277
water hyacinth - *Eichhornia crassipes* 441
wild hyacinth - *Dichelostemma pulchellum* 275
hydrangea
climbing hydrangea - *Decumaria barbara*, *Hydrangea anomala* subsp. *petiolaris* 250
French hydrangea - *Hydrangea macrophylla* 250
house hydrangea - *Hydrangea macrophylla* 250
panicle hydrangea - *Hydrangea paniculata* 250
peegee hydrangea - *Hydrangea paniculata* cv. 'grandiflora' 250
smooth hydrangea - *Hydrangea arborescens* 250
wild hydrangea - *Hydrangea arborescens* 250, *Rumex venosus* 440
hydrilla - *Hydrilla verticillata* 251
hyssop
blue giant hyssop - *Agastache foeniculum* 262
fennel giant hyssop - *Agastache foeniculum* 262
fragrant giant hyssop - *Agastache foeniculum* 262
nettle-leaf giant hyssop - *Agastache urticifolia* 262
purple giant hyssop - *Agastache scrophulariifolia* 262
yellow giant hyssop - *Agastache nepetoides* 262
icaco - *Chrysobalanus icaco* 138
ice-plant - *Mesembryanthemum crystallinum* 20
iigiri-tree - *Idesia polycarpa* 240
Inca-wheat - *Amaranthus caudatus* 22
inchplant - *Zebrina pendula* 140
fern-leaf inchplant - *Tripogandra multiflora* 140
Indian-cup - *Silphium perfoliatum* 84
Indian-paint - *Chenopodium capitatum* 136, *Lithospermum canescens* 107
Indian-physic - *Gillenia trifoliata* 468
Indian-salad - *Hydrophyllum virginianum* 252
Indian-shoe, yellow - *Cypripedium calceolus* var. *pubescens* 321
Indian-shot - *Canna indica* 123
indigo
bastard indigo - *Amorpha fruticosa* 189
blue wild indigo - *Baptisia australis* 191
fragrant false indigo - *Amorpha nana* 189
hairy indigo - *Indigofera hirsuta* 200
indigo-bush, false - *Amorpha fruticosa* 189
indigo-squill - *Camassia scilloides* 275
inkberry - *Ilex glabra* 36, *Randia aculeata* 494
innocence - *Collinsia verna* 517
insect-flower
Dalmatian insect-flower - *Chrysanthemum cinerariifolium* 63
Persian insect-flower - *Chrysanthemum coccineum* 63
inside-out-flower - *Vancouveria planipetala* 92
ipecac
American ipecac - *Gillenia stipulata* 468
false ipecac - *Asclepias curassavica* 51
ipil - *Leucaena leucocephala* 203
ipomoea, star - *Ipomoea coccinea* 142
iris
African iris - *Dietes vegeta* 253
copper iris - *Iris fulva* 254
dwarf-crested iris - *Iris cristata* 254
English iris - *Iris xiphioides* 254
Japanese iris - *Iris kaempferi* 254
red iris - *Iris fulva* 254
Spanish iris - *Iris xiphium* 254
wild iris - *Iris versicolor* 254
Irish-moss - *Soleirolia soleirolii* 544
ironbark, red - *Eucalyptus sideroxylon* 303
ironweed - *Vernonia noveboracensis* 88
desert ironweed - *Olneya tesota* 208
tall ironweed - *Vernonia gigantea* 88
western ironweed - *Vernonia baldwinii* 88
ironwood - *Cliftonia monophylla* 166, *Cyrilla racemiflora* 166, *Sideroxylon lycioides* 512
South Sea ironwood - *Casuarina equisetifolia* 132
islay - *Prunus ilicifolia* 475
itchweed - *Veratrum viride* 282
ivy - *Kalmia latifolia* 173
American ivy - *Parthenocissus quinquefolia* 550
Boston ivy - *Parthenocissus tricuspidata* 550
cape ivy - *Senecio mikanioides* 83
coliseum ivy - *Cymbalaria muralis* 517
devil's ivy - *Epipremnum aureum* 40
English ivy - *Hedera helix* 43

five-leaved ivy - *Parthenocissus quinquefolia* 550
German ivy - *Senecio mikanioides* 83
grape ivy - *Cissus rhombifolia* 549
Japanese ivy - *Parthenocissus tricuspidata* 550
Kenilworth ivy - *Cymbalaria muralis* 517
marine ivy - *Cissus incisa* 549
parlor ivy - *Senecio mikanioides* 83
Solomon Island ivy - *Epipremnum aureum* 40
switch ivy - *Leucothoe axillaris* var. *editorum* 174
water ivy - *Senecio mikanioides* 83
ivybush - *Kalmia latifolia* 173
ixora, scarlet - *Ixora coccinea* 493
izote - *Yucca elephantipes* 19
jaboncillo - *Sapindus saponaria, S. saponaria* var. *drummondii* 511
jaboticaba - *Myrciaria cauliflora* 307
Jack-go-to-bed-at-noon - *Tragopogon pratensis* 87
jack-in-the-pulpit - *Arisaema triphyllum* 39
jackfruit - *Artocarpus heterophyllus* 298
Jacob's-coat - *Acalypha wilkesiana* 182
Jacob's-ladder - *Polemonium caeruleum* subsp. *villosum* 436
Jacob's-staff - *Fouquieria splendens* 240
jadeplant
Chinese jadeplant - *Crassula arborescens* 145
silver jadeplant - *Crassula arborescens* 145
jadetree - *Crassula argentea* 145
jadevine - *Strongylodon macrobotrys* 216
jambolan - *Syzygium cumini* 307
jambolan-plum - *Syzygium cumini* 307
jambool - *Syzygium cumini* 307
jambos, pomerac - *Syzygium malaccense* 307
jambu - *Syzygium cumini* 307
Jamestown-weed - *Datura stramonium* 525
Japanese-clover - *Lespedeza striata* 202
Japanese-lantern - *Physalis alkekengi* 529
Japanese-laurel - *Aucuba japonica* 143
Japanese-moss - *Soleirolia soleirolii* 544
Japanese-rose - *Kerria japonica* 469
Jarrah - *Eucalyptus marginata* 303
jarvil, sweet - *Osmorhiza claytonii* 31
jasmine
angel-wing jasmine - *Jasminum nitidum* 313
Arabian jasmine - *Jasminum sambac* 313
Cape jasmine - *Gardenia jasminoides* 492
Carolina jasmine - *Gelsemium sempervirens* 284
Confederate jasmine - *Jasminum nitidum* 313
star jasmine - *Jasminum multiflorum, J. nitidum* 313
windmill jasmine - *Jasminum nitidum* 313
winter jasmine - *Jasminum nudiflorum* 313
jaundice-berry - *Berberis vulgaris* 91
Java-plum - *Syzygium cumini* 307
javillo - *Hura crepitans* 185
Jerusalem-artichoke - *Helianthus tuberosus* 73
Jerusalem-cherry - *Solanum pseudocapsicum* 530
false Jerusalem-cherry - *Solanum capsicastrum* 530
Jerusalem-cross - *Lychnis chalcedonica* 131
Jerusalem-oak - *Chenopodium botrys* 136
Jerusalem-thorn - *Parkinsonia aculeata* 209
jessamine
day jessamine - *Cestrum diurnum* 525
night jessamine - *Cestrum nocturnum* 525
orange jessamine - *Murraya paniculata* 497
yellow jessamine - *Gelsemium sempervirens* 284
jesuit-nut - *Trapa natans* 538
jetbead, black - *Rhodotypos scandens* 484
jewels-of-Opar - *Talinum paniculatum* 442
jewelweed - *Impatiens capensis* 90
jicama - *Pachyrhizus erosus* 209
jimsonweed - *Datura stramonium* 525
Job's-tears - *Coix lacryma-jobi* 382
jocote - *Spondias purpurea* 26
Joe-Pye-weed - *Eupatorium maculatum, E. purpureum* 70
green-stemmed Joe-Pye-weed - *Eupatorium purpureum* 70
sweet Joe-Pye-weed - *Eupatorium purpureum* 70
John-go-to-bed-at-noon - *Tragopogon pratensis* 87
Johnny-jump-up - *Viola tricolor* 547
joint-vetch - *Aeschynomene americana* 188
jointwood - *Cassia javanica* var. *indochinensis* 193
jojoba - *Simmondsia chinensis* 523
jonquil - *Narcissus jonquilla* 279
Joseph's-coat - *Amaranthus tricolor* 22
Joshua-tree - *Yucca brevifolia* 19
joyweed, shoofly - *Alternanthera ficoidea* cv. 'amoena' 21
jujube
Chinese jujube - *Ziziphus jujuba* 460
common jujube - *Ziziphus jujuba* 460
cottony jujube - *Ziziphus mauritiana* 460
Indian jujube - *Ziziphus mauritiana* 460
juniper
alligator juniper - *Juniperus deppeana* var. *pachyphloea* 153
Ashe juniper - *Juniperus ashei* 153
California juniper - *Juniperus occidentalis* 153
cherrystone juniper - *Juniperus monosperma* 153
Chinese juniper - *Juniperus chinensis* 153
common juniper - *Juniperus communis* 153
creeping juniper - *Juniperus horizontalis* 153
creeping savin juniper - *Juniperus horizontalis* 153
Greek juniper - *Juniperus excelsa, J. excelsa* cv. 'stricta' 153
mountain juniper - *Juniperus communis* var. *saxatilis* 153
one-seed juniper - *Juniperus monosperma* 153
Pinchot juniper - *Juniperus pinchotii* 153
prickly juniper - *Juniperus oxycedrus* 153
redberry juniper - *Juniperus pinchotii* 153
Rocky Mountain juniper - *Juniperus scopulorum* 153
Sargent juniper - *Juniperus chinensis* var. *sargentii* 153
shore juniper - *Juniperus conferta* 153
Sierra juniper - *Juniperus occidentalis* 153
sweet-fruited juniper - *Juniperus deppeana* var. *pachyphloea* 153
Utah juniper - *Juniperus osteosperma* 153
junta-de-cobra-pintada - *Justicia brandegeana* 8
Jupiter's-beard - *Centranthus ruber* 544
jute
bastard jute - *Hibiscus cannabinus* 292
bimli jute - *Hibiscus cannabinus* 292
China jute - *Abutilon theophrastii* 289
Congo jute - *Urena lobata* 295
Cuba jute - *Sida rhombifolia* 295
kaffir - *Sorghum bicolor* 419
kaffir-corn - *Sorghum bicolor* 419
kai-apple - *Dovyalis caffra* 240
kalanchoe
Christmas-tree kalanchoe - *Kalanchoe laciniata* 146
fig-tree kalanchoe - *Kalanchoe laciniata* 146
kale - *Brassica oleracea, B. oleracea* var. *acephala* 109
sea kale - *Crambe maritima* 112
kalmia, bog - *Kalmia polifolia* 173
kalo - *Colocasia esculenta* 39
kangkong - *Ipomoea aquatica* 142
kapok - *Ceiba pentandra* 106
kapundung - *Baccaurea racemosa* 183
karanda - *Carissa carandas* 34
karee - *Rhus lancea* 24
karo - *Pittosporum crassifolium* 359
karoo-tree - *Rhus lancea* 24
karum-tree - *Pongamia pinnata* 212
kat - *Catha edulis* 133
kat-sola - *Aeschynomene indica* 188
katsura-tree - *Cercidiphyllum japonicum* 134
kau-apple - *Dovyalis caffra* 240
kawa'ii - *Ilex anomala* f. *sandwicensis* 36
keg-fig - *Diospyros kaki* 167
kei-apple - *Dovyalis caffra* 240
kenaf - *Hibiscus cannabinus* 292
kepau - *Bougainvillea umbellifera* 308
kerria - *Kerria japonica* 469
khat - *Catha edulis* 133
khus-khus - *Vetiveria zizanioides* 428
kidney-vetch - *Anthyllis vulneraria* 189
kidneyleaf - *Ranunculus abortivus* 456
kidneywood - *Eysenhardtia polystachya* 197
kidneywort - *Baccharis pilularis* 60
king's-crown - *Justicia carnea* 8
king's-cup - *Caltha palustris* 453
king's-devil - *Hieracium aurantiacum* 76
king's-spear - *Eremurus robustus* 276
king-of-the-meadow - *Thalictrum polygamum* 457
kingnut - *Carya laciniosa* 255
kingpalm - *Archontophoenix alexandrae* 46
kinnikinick - *Arctostaphylos uva-ursi* 171
kino - *Coccoloba uvifera* 437
kiri-tree - *Paulownia tomentosa* 519
kiss-me-over-the-garden-gate - *Polygonum orientale* 438
kittul-tree - *Caryota urens* 46
kiwi - *Actinidia chinensis* 17
klamathweed - *Hypericum perforatum* 139
knapweed
creeping knapweed - *Centaurea repens* 63
Russian knapweed - *Centaurea repens* 63
knesheneka - *Rubus stellatus* 486
knotweed
bushy knotweed - *Polygonum ramosissimum* 438
erect knotweed - *Polygonum erectum* 438
Japanese knotweed - *Polygonum cuspidatum* 438
prostrate knotweed - *Polygonum aviculare* 438
koa - *Acacia koa* 187
kochia - *Bassia scoparia* 135
kohlrabi - *Brassica oleracea, B. oleracea* var. *gongylodes* 109
kokio - *Kokia drynarioides* 294
kopiko-tea - *Psychotria kuduana* 494
korakan - *Eleusine coracana* 387
korari - *Phormium tenax* 19
kudzu - *Pueraria lobata* 213
tropical kudzu - *Pueraria phaseoloides* 213

kumquat
Hong Kong kumquat - *Fortunella hindsii* 497
oval kumquat - *Fortunella margarita* 497
kurumaba-zakuro-so - *Mollugo verticillata* 298
la-kwa - *Momordica charantia* 150
laburnum, Indian - *Cassia fistula* 193
lace-fern - *Asparagus setaceus* 274
laceflower, blue - *Trachymene coerulea* 33
ladder, Jacob's - *Smilax herbacea* 523
lady's-eardrops - *Brunnichia cirrhosa* 436
lady's-earrings - *Impatiens capensis* 90
lady's-finger - *Abelmoschus esculentus* 289
lady's-fingers - *Anthyllis vulneraria* 189
lady's-slipper
large lady's-slipper - *Cypripedium calceolus* var. *pubescens* 321
showy lady's-slipper - *Cypripedium reginae* 321
small white lady's-slipper - *Cypripedium candidum* 321
lady's-sorrel, creeping - *Oxalis corniculata* 323
lady's-thumb - *Polygonum persicaria* 438
lady-of-the-night - *Brunfelsia americana* 524
ladyfern - *Athyrium felix-femina* 445
northern ladyfern - *Athyrium felix-femina* var. *michauxii* 445
lamb's-quarters - *Chenopodium album* 136
lambkill - *Kalmia angustifolia* 173
lancewood - *Pseudopanax crassifolia* 44
langsat - *Lansium domesticum* 297
lantana - *Lantana camara* 545
trailing lantana - *Lantana montevidensis* 546
weeping lantana - *Lantana montevidensis* 546
lanternplant, Chinese - *Physalis alkekengi* 529
larch
American larch - *Larix laricina* 332
black larch - *Larix laricina* 332
European larch - *Larix decidua* 332
Japanese larch - *Larix kaempferi* 332
western larch - *Larix occidentalis* 332
larkspur
bouquet larkspur - *Delphinium grandiflorum* 455
candle larkspur - *Delphinium elatum* 455
dwarf larkspur - *Delphinium tricorne* 455
field larkspur - *Delphinium consolida* 455
rocket larkspur - *Delphinium ajacis* 455
scarlet larkspur - *Delphinium cardinale* 455
lasiandra - *Tibouchina urvilleana* 296
laurel - *Laurus nobilis* 269
Alexandrian laurel - *Calophyllum inophyllum* 138
black laurel - *Gordonia lasianthus* 536
bog laurel - *Kalmia polifolia* 173
California laurel - *Umbellularia californica* 271
drooping laurel - *Leucothoe axillaris* var. *editorum* 174
dwarf laurel - *Kalmia angustifolia* 173
great laurel - *Rhododendron maximum* 176
Grecian laurel - *Laurus nobilis* 269
Indian laurel - *Calophyllum inophyllum* 138
mountain laurel - *Kalmia latifolia* 173
pale laurel - *Kalmia polifolia* 173
pig laurel - *Kalmia angustifolia* 173
purple laurel - *Rhododendron catawbiense* 176
sheep laurel - *Kalmia angustifolia* 173
laureltree - *Persea borbonia* 269
laurelwood - *Calophyllum inophyllum* 138
laurustinus - *Viburnum tinus* 127
lavender - *Lavandula angustifolia* 264
desert lavender - *Hyptis emoryi* 263
English lavender - *Lavandula angustifolia* 264
lavender-cotton - *Santolina chamaecyparissus* 82
lavender-scallops - *Kalanchoe fedtschenkoi* 146
lawnleaf - *Dichondra carolinensis* 141
leadplant - *Amorpha canescens* 189
leadtree - *Leucaena leucocephala* 203
leadwort, Cape - *Plumbago auriculata* 362
leather-flower - *Clematis viorna* 454
leatherflower - *Clematis versicolor*, *C. virginiana* 454
leatherleaf - *Chamaedaphne calyculata* 172
leatherroot - *Psoralea macrostachya* 213
leatherwood - *Cyrilla racemiflora* 166, *Dirca palustris* 537
southern leatherwood - *Cyrilla racemiflora* 166
lebbek-tree - *Albizia lebbeck* 188
leechee - *Litchi chinensis* 511
leek - *Allium porrum* 272
lady's leek - *Allium cernuum* 272
meadow leek - *Allium canadense* 272
rose leek - *Allium canadense* 272
stone leek - *Allium fistulosum* 272
wild leek - *Allium tricoccum* 272
lehua, 'ohi'a - *Metrosideros polymorpha* 306
lemandarin - *Citrus* x *limonia* 494
lemon - *Citrus limon* 494
rough lemon - *Citrus jambhiri* 494
lemonade-berry - *Rhus integrifolia* 24
lentil - *Lens culinaris* 202
lentisco - *Rhus virens* 24
leopard's-bane - *Doronicum plantagineum* 68
leopard-flower - *Belamcanda chinensis* 253
lespedeza
bicolor lespedeza - *Lespedeza bicolor* 202
Chinese lespedeza - *Lespedeza cuneata* 202
common lespedeza - *Lespedeza striata* 202
Japanese lespedeza - *Lespedeza striata* 202
Korean lespedeza - *Kummerowia stipulacea* 200
roundhead lespedeza - *Lespedeza capitata* 202
Sericea lespedeza - *Lespedeza cuneata* 202
striate lespedeza - *Lespedeza striata* 202
lettuce - *Lactuca sativa* 77
blue lettuce - *Lactuca tatarica* subsp. *pulchella* 77
Canada wild lettuce - *Lactuca canadensis* 77
Cos lettuce - *Lactuca sativa* var. *longifolia* 77
garden lettuce - *Lactuca sativa* 77
grassleaf lettuce - *Lactuca graminifolia* 77
head lettuce - *Lactuca sativa* var. *capitata* 77
Indian lettuce - *Montia linearis* 442
prickly lettuce - *Lactuca serriola* 77
Romaine lettuce - *Lactuca sativa* var. *longifolia* 77
tall lettuce - *Lactuca canadensis* 77
leucaena - *Leucaena leucocephala* 203
leverwood - *Ostrya virginiana* 102
licorice
American licorice - *Glycyrrhiza lepidota* 200
chinook licorice - *Lupinus littoralis* 204
Indian licorice - *Abrus precatorius* 187
wild licorice - *Glycyrrhiza lepidota* 200
licorice-vine - *Abrus precatorius* 187
life-of-man - *Aralia racemosa* 42
life-plant - *Kalanchoe pinnata* 146
lignum, holy-wood - *Guaiacum sanctum* 553
lignum-vitae - *Guaiacum sanctum* 553
lilac
Amur lilac - *Syringa amurensis* 315
Chinese lilac - *Syringa* × *chinensis* 315
common lilac - *Syringa vulgaris* 315
Hungarian lilac - *Syringa josikaea* 315
Indian lilac - *Melia azedarach* 297
Persian lilac - *Melia azedarach* 297, *Syringa* × *persica* 315
wild lilac - *Ceanothus sanguineus* 458
lily - *Agapanthus africanus* 271
African blood lily - *Haemanthus coccineus* 276
African lily - *Agapanthus africanus* 271
Amazon lily - *Eucharis grandiflora* 276
American turk's-cap lily - *Lilium superbum* 278
atamasco lily - *Zephyranthes atamasco* 282
avalanche lily - *Erythronium grandiflorum* 276
belladonna lily - *Amaryllis belladonna* 273
Bermuda Easter lily - *Lilium longiflorum* var. *eximium* 278
blood lily - *Haemanthus coccineus* 276
bluebead lily - *Clintonia borealis* 275
Canada lily - *Lilium canadense* 277
candlestick lily - *Lilium pensylvanicum*, *Lilium* × *hollandicum* 278
Caucasian lily - *Lilium monadelphum* 278
checker lily - *Fritillaria lanceolata* 276
Columbia lily - *Lilium columbianum* 278
common torch lily - *Kniphofia uvaria* 277
Easter lily - *Lilium longiflorum* var. *eximium* 278
eucharist lily - *Eucharis grandiflora* 276
fragrant plantain lily - *Hosta plantaginea* 277
gloriosa lily - *Gloriosa rothschildiana* 276
gold-banded lily - *Lilium auratum* 277
golden-rayed lily - *Lilium auratum* 277
Guernsey lily - *Nerine sarniensis* 280
Japanese turk's-cap lily - *Lilium hansonii* 278
Kamchatka lily - *Fritillaria camschatcensis* 276
leopard lily - *Lilium pardalinum* 278
madonna lily - *Eucharis grandiflora* 276, *Lilium candidum* 277
magic lily - *Lycoris squamigera* 279
Martagon lily - *Lilium martagon* 278
Michigan lily - *Lilium michiganense* 278
mountain lily - *Lilium auratum* 277
naked-lady lily - *Amaryllis belladonna* 273
Nankeen lily - *Lilium* × *testaceum* 278
orange-cup lily - *Lilium philadelphicum* 278
Oregon lily - *Lilium columbianum* 278
panther lily - *Lilium pardalinum* 278
pond lily - *Nymphaea odorata* 309
regal lily - *Lilium regale* 278
resurrection lily - *Lycoris squamigera* 279
royal lily - *Lilium regale* 278
Scarboro lily - *Vallota speciosa* 282
showy Japanese lily - *Lilium speciosum* 278
showy lily - *Lilium speciosum* 278
Siberian lily - *Ixiolirion tataricum* 277
swamp lily - *Lilium superbum* 278
tartar lily - *Ixiolirion tataricum* 277
tiger lily - *Lilium lancifolium* 278
trout lily - *Erythronium americanum* 276

turban lily - *Lilium martagon* 278
turk's-cap lily - *Lilium martagon* 278
Washington lily - *Lilium washingtonianum* 278
western orange-cup lily - *Lilium philadelphicum* var. *andinum* 278
white-trumpet lily - *Lilium longiflorum* 278
wild orange-red lily - *Lilium philadelphicum* 278
wild yellow lily - *Lilium canadense* 277
wood lily - *Lilium philadelphicum* 278
yellow-bell lily - *Lilium canadense* 277
lily-of-the-Amazon - *Eucharis grandiflora* 276
lily-of-the-Nile - *Agapanthus africanus* 271
lily-of-the-valley - *Convallaria majalis* 275, *Smilacina stellata* 280
lily-of-the-valley-bush - *Pieris japonica* 175
lily-royal - *Lilium superbum* 278
lilyturf, big blue - *Liriope muscari* 279
lime - *Citrus aurantifolia* 494
mandarin lime - *Citrus* x *limonia* 494
Persian lime - *Citrus latifolia* 494
Rangpur lime - *Citrus* x *limonia* 494
Spanish lime - *Melicoccus bijugatus* 511
wild lime - *Zanthoxylum fagara* 498
limeberry - *Triphasia trifolia* 498
linden
American linden - *Tilia americana* 537
European linden - *Tilia cordata* 537
littleleaf linden - *Tilia cordata* 537
ling - *Trapa natans* 538
lingberry - *Vaccinium vitis-idaea* var. *minus* 179
lingen - *Vaccinium vitis-idaea* var. *minus* 179
lingenberry - *Vaccinium vitis-idaea* var. *minus* 179
linkleaf - *Schlumbergera truncata* 120
lion's-beard - *Anemone nuttalliana* 452
lipstick-plant - *Aeschynanthus pulcher* 243
lipstick-tree - *Bixa orellana* 105
litchi - *Litchi chinensis* 511
live-forever - *Sedum telephium*, *Sempervivum tectorum* 146
lizard-plant - *Tetrastigma voinieranum* 550
lizard-tail - *Saururus cernuus* 513
lobelia
blue lobelia - *Lobelia siphilitica* 122
edging lobelia - *Lobelia erinus* 122
great lobelia - *Lobelia siphilitica* 122
pale-spike lobelia - *Lobelia spicata* 122
loblolly-bay - *Gordonia lasianthus* 536
lobsterplant - *Euphorbia pulcherrima* 184
loco
Lambert loco - *Oxytropis lambertii* 209
purple loco - *Oxytropis lambertii* 209
woolly loco - *Astragalus mollissimus* 191
locoweed - *Oxytropis lambertii* 209
woolly locoweed - *Astragalus mollissimus* 191
locust
black locust - *Robinia pseudoacacia* 214
bristly locust - *Robinia hispida* 214
honey locust - *Gleditsia triacanthos* 198
Japanese honey locust - *Gleditsia japonica* 198
moss locust - *Robinia hispida* 214
sweet locust - *Gleditsia triacanthos* 198
yellow locust - *Robinia pseudoacacia* 214
loganberry - *Rubus loganobaccus* 486
lomatium, slender-fruited - *Lomatium leptocarpum* 30
London-pride - *Lychnis chalcedonica* 131
London-rocket - *Sisymbrium irio* 115
loofah - *Luffa aegyptiaca* 150
smooth loofah - *Luffa aegyptiaca* 150
looking-glass-plant - *Coprosma repens* 491
loosestrife
garden loosestrife - *Lysimachia vulgaris* 443
purple loosestrife - *Lythrum salicaria* 285
spiked loosestrife - *Lythrum salicaria* 285
swamp loosestrife - *Decodon verticillatus* 285, *Lysimachia terrestris* 443
tufted loosestrife - *Lysimachia thyrsiflora* 443
winged loosestrife - *Lythrum alatum* 285
loquat - *Eriobotrya japonica* 465
Lord's-candlestick - *Yucca gloriosa* 19
lotus
American lotus - *Nelumbo lutea* 308
East Indian lotus - *Nelumbo nucifera* 308
Hindu lotus - *Nelumbo nucifera* 308
sacred lotus - *Nelumbo nucifera* 308
lousewort
common lousewort - *Pedicularis canadensis* 519
love-in-a-mist - *Passiflora foetida* 325
love-lies-bleeding - *Amaranthus caudatus* 22
love-plant, Mexican - *Kalanchoe pinnata* 146
loveapple - *Lycopersicon esculentum* 526
lucerne - *Medicago sativa* 206
Brazilian lucerne - *Stylosanthes guianensis* 216
luckynut - *Thevetia peruviana* 35
luffa - *Luffa aegyptiaca* 150
angled luffa - *Luffa acutangula* 150
lupine
Bentham's annual lupine - *Lupinus benthamii* 203
bicolored lupine - *Lupinus bicolor* 203
blue lupine - *Lupinus pilosus* 204
Egyptian lupine - *Lupinus albus* 203
European blue lupine - *Lupinus angustifolius* 203
European yellow lupine - *Lupinus luteus* 204
field lupine - *Lupinus albus* 203
Nootka lupine - *Lupinus nootkatensis* 204
riverbank lupine - *Lupinus rivularis* 204
seashore lupine - *Lupinus littoralis* 204
sickle-keeled lupine - *Lupinus albicaulis* 203
sundial lupine - *Lupinus perennis* 204
Texas lupine - *Lupinus subcarnosus* 204
tree lupine - *Lupinus arboreus* 203
Washington lupine - *Lupinus polyphyllus* 204
white lupine - *Lupinus albus* 203
wild lupine - *Lupinus perennis* 204
woolly-leaved lupine - *Lupinus leucophyllus* 204
lychee - *Litchi chinensis* 511
lychnis
evening lychnis - *Silene alba* 131
scarlet lychnis - *Lychnis chalcedonica* 131
lycopod, dwarf - *Selaginella rupestris* 522
lysiloma, Bahama - *Lysiloma bahamensis* 205
maackia, Amur - *Maackia amurensis* 205
Madagascar-plum - *Flacourtia indica* 240
madake - *Phyllostachys bambusoides* 409
madapple - *Solanum melongena* var. *esculentum* 530
madar - *Calotropis gigantea*, *C. procera* 52
Madeira-nut - *Juglans regia* 259
madia, common - *Madia elegans* 79
madia-oilplant - *Madia sativa* 79
madrone - *Arbutus menziesii* 170
Pacific madrone - *Arbutus menziesii* 170
madwort
golden-tuft madwort - *Alyssum saxatile* 108
rock madwort - *Alyssum saxatile* 108
magnolia
Chinese magnolia - *Magnolia* x *soulangiana* 287
Fraser magnolia - *Magnolia fraseri* 287
great-leaved magnolia - *Magnolia macrophylla* 287
mountain magnolia - *Magnolia fraseri* 287
saucer magnolia - *Magnolia* x *soulangiana* 287
southern magnolia - *Magnolia grandiflora* 287
star magnolia - *Magnolia stellata* 287
sweetbay magnolia - *Magnolia virginiana* 287
umbrella magnolia - *Magnolia tripetala* 287
maguey - *Agave americana* 17, *Furcraea selloa* 18
mahoe - *Hibiscus tiliaceus* 292
mahogany
Florida mahogany - *Persea borbonia* 269
Spanish mahogany - *Swietenia mahagoni* 297
West Indian mahogany - *Swietenia mahagoni* 297
mahonia
Cascades mahonia - *Berberis nervosa* 91
cluster mahonia - *Berberis pinnata* 91
Fremont mahonia - *Berberis fremontii* 91
holly mahonia - *Berberis aquifolium* 91
maidencane - *Panicum hemitomon* 403
maidenhair-berry - *Gaultheria hispidula* 172
maidenhair-tree - *Ginkgo biloba* 244
maize - *Zea mays* 428
water maize - *Victoria amazonica* 309
majagua - *Hibiscus tiliaceus* 292
Malabar-plum - *Syzygium jambos* 307
Malabar-spinach - *Basella alba* 90
Malay-apple - *Syzygium malaccense* 307
male-blueberry - *Lyonia ligustrina* 174
malka - *Rubus chamaemorus* 486
mallee, gooseberry - *Eucalyptus calycogona* 303
mallow
alkali mallow - *Malvella leprosa* 294
bur mallow - *Urena lobata* 295
common mallow - *Malva neglecta* 294
common rose mallow - *Hibiscus moscheutos* 292
desert mallow - *Sphaeralcea ambigua* 295
Drummond wax mallow - *Malvaviscus arboreus* var. *drummondii* 294
great rose mallow - *Hibiscus grandiflorus* 292
Halberd-leaved mallow - *Hibiscus laevis* 292
hemp-leaved mallow - *Althaea cannabina* 290
Indian mallow - *Abutilon incanum*, *A. theophrastii* 289
little mallow - *Malva parviflora* 294
marsh mallow - *Hibiscus moscheutos* subsp. *palustris* 292
prairie mallow - *Sphaeralcea coccinea* 295
prickly mallow - *Sida spinosa* 295
red false mallow - *Sphaeralcea coccinea* 295
swamp rose mallow - *Hibiscus moscheutos*, *H. moscheutos* subsp. *palustris* 292
Texas mallow - *Malvaviscus arboreus* var. *drummondii* 294
tree mallow - *Lavatera arborea* 294
Venice mallow - *Hibiscus trionum* 292
wax mallow - *Malvaviscus arboreus* 294

woolly rose mallow - *Hibiscus lasiocarpus* 292
mallow-rose - *Hibiscus moscheutos* subsp. *palustris* 292
malojilla - *Eriochloa polystachya* 391
Maltese-cross - *Lychnis chalcedonica* 131
mamane - *Sophora chrysophylla* 215
mamey-colorado - *Pouteria sapota* 512
mamoncillo - *Melicoccus bijugatus* 511
man-in-a-boat - *Tradescantia spathacea* 140
man-of-the-earth - *Ipomoea leptophylla, I. pandurata* 142
man-root - *Ipomoea pandurata* 142
maní - *Arachis hypogaea* 189
mandarin
rose mandarin - *Streptopus roseus* 281
white mandarin - *Streptopus amplexifolius* 281
mandrake - *Podophyllum peltatum* 92
mangel - *Beta macrorrhiza, B. vulgaris, B. vulgaris* subsp. *vulgaris* 135
mango - *Mangifera indica* 23
mangold - *Beta vulgaris, B. vulgaris* subsp. *vulgaris* 135
mangosteen - *Garcinia mangostana* 139
mangrove - *Rhizophora mangle* 460
American mangrove - *Rhizophora mangle* 460
black mangrove - *Avicennia germinans* 545
red mangrove - *Rhizophora mangle* 460
white mangrove - *Laguncularia racemosa* 139
manila - *Pithecellobium dulce* 212
manioc - *Manihot esculenta* 186
manroot - *Ipomoea leptophylla* 142
manuka - *Leptospermum scoparium* 305
manzanita
Bragg manzanita - *Arctostaphylos nummularia* 170
dune manzanita - *Arctostaphylos pumila* 171
Eastwood manzanita - *Arctostaphylos glandulosa* 170
greenleaf manzanita - *Arctostaphylos patula* 171
hairy manzanita - *Arctostaphylos columbiana* 170
Mexican manzanita - *Arctostaphylos pungens* 171
Monterey manzanita - *Arctostaphylos hookeri* 170
paintleaf manzanita - *Arctostaphylos pungens* 171
Parry manzanita - *Arctostaphylos manzanita* 170
pine-mat manzanita - *Arctostaphylos nevadensis* 170
whiteleaf manzanita - *Arctostaphylos viscida* 171
woolly manzanita - *Arctostaphylos tomentosa* 171
maple
Amur maple - *Acer ginnala* 9
ash-leaved maple - *Acer negundo* 9
bigleaf maple - *Acer macrophyllum* 9
bigtooth maple - *Acer saccharum* subsp. *grandidentatum* 9
black maple - *Acer nigrum* 9
black sugar maple - *Acer nigrum* 9
canyon maple - *Acer macrophyllum* 9
chalk maple - *Acer saccharum* subsp. *leucoderme* 9
Douglas maple - *Acer glabrum* subsp. *douglasii* 9
dwarf maple - *Acer glabrum* 9
field maple - *Acer campestre* 9
Florida maple - *Acer barbatum* 9
hard maple - *Acer saccharum* 9
hedge maple - *Acer campestre* 9
Japanese maple - *Acer palmatum* 9
mono maple - *Acer truncatum* subsp. *mono* 9
moose maple - *Acer spicatum* 9
mountain maple - *Acer spicatum* 9
Norway maple - *Acer platanoides* 9
Oregon maple - *Acer macrophyllum* 9
Pennsylvania maple - *Acer pensylvanicum* 9
red maple - *Acer rubrum, A. rubrum* var. *rubrum* 9
river maple - *Acer saccharinum* 9
rock maple - *Acer nigrum, A. saccharum* 9
Rocky Mountain maple - *Acer glabrum* 9
scarlet maple - *Acer rubrum* 9
Shantung maple - *Acer truncatum* 9
silver maple - *Acer saccharinum* 9
soft maple - *Acer rubrum, A. saccharinum* 9
southern sugar maple - *Acer barbatum* 9
striped maple - *Acer pensylvanicum* 9
sugar maple - *Acer saccharum, A. saccharum* subsp. *grandidentatum* 9
swamp maple - *Acer rubrum* 9
sycamore maple - *Acer pseudoplatanus* 9
trident maple - *Acer rubrum* var. *trilobum* 9
vine maple - *Acer circinatum* 9
white maple - *Acer saccharinum* 9
whitebark maple - *Acer saccharum* subsp. *leucoderme* 9
maranon - *Anacardium occidentale* 23
marbleseed - *Onosmodium molle* 108
marguerite - *Chrysanthemum frutescens* 63
white marguerite - *Chrysanthemum frutescens* 63
marigold
African marigold - *Tagetes erecta* 86
Aztec marigold - *Tagetes erecta* 86
big marigold - *Tagetes erecta* 86
blue-eyed cape marigold - *Dimorphotheca sinuata* 68
corn marigold - *Chrysanthemum segetum* 63
French marigold - *Tagetes patula* 86
marsh marigold - *Caltha palustris* 453
pot marigold - *Calendula officinalis* 62
marijuana - *Cannabis sativa* 122
marine-vine - *Cissus incisa* 549
mariposa
desert mariposa - *Calochortus kennedyi* 274
green-banded mariposa - *Calochortus macrocarpus* 274
marjoram - *Origanum vulgare* 266
annual marjoram - *Origanum majorana* 266
pot marjoram - *Origanum vulgare* 266
sweet marjoram - *Origanum majorana* 266
wild marjoram - *Origanum vulgare* 266
markry - *Rhus radicans* 24
marlberry - *Ardisia escallonioides* 302
marlock, white-leaved - *Eucalyptus tetragona* 303
marmalade-fruit - *Pouteria sapota* 512
marmelade - *Genipa americana* 492
marmelade-box - *Genipa americana* 492
marri - *Eucalyptus calophylla* 303
marrow - *Cucurbita pepo* 149
wild marrow - *Cucurbita texana* 149
marsh-elder - *Iva xanthifolia* 77
marsh-trefoil - *Menyanthes trifoliata* 298
marvel-of-Peru - *Mirabilis jalapa* 308
masterwort - *Angelica atropurpurea* 26, *Heracleum sphondylium* subsp. *montanum* 29
mastic, false - *Sideroxylon foetidissimum* 512
mastic-tree, Peruvian - *Schinus molle* 25
match-me-if-you-can - *Acalypha wilkesiana* 182
mathers - *Vaccinium ovalifolium* 179
matrimony-vine
barbary matrimony-vine - *Lycium barbarum* 526
Chinese matrimony-vine - *Lycium chinense* 526
May-blob - *Caltha palustris* 453
mayapple - *Podophyllum peltatum* 92
mayflower - *Epigaea repens* 172
mayweed - *Anthemis cotula* 56
meadow-bright - *Caltha palustris* 453
meadow-fern - *Myrica gale* 302
meadow-hyacinth - *Camassia scilloides* 275
meadow-parsnip - *Zizia aurea* 33
meadow-saffron - *Colchicum autumnale* 275
meadow-weed - *Ruellia tuberosa* 8
meadowrue
alpine meadowrue - *Thalictrum alpinum* 457
early meadowrue - *Thalictrum dioicum* 457
purple meadowrue - *Thalictrum dasycarpum* 457
tall meadowrue - *Thalictrum polygamum* 457
meadowsweet - *Spiraea alba, S. latifolia* 489
mealberry - *Arctostaphylos uva-ursi* 171
medic
black medic - *Medicago lupulina* 206
spotted medic - *Medicago arabica* 206
medlar - *Mespilus germanica* 473, *Mimusops elengi* 512
Japanese medlar - *Eriobotrya japonica* 465
meeting-houses - *Aquilegia canadensis* 453
melic
California melic - *Melica imperfecta* 400
false melic - *Schizachne purpurascens* 415
Harford melic - *Melica harfordii* 400
Porter melic - *Melica porteri* 400
Smith melic - *Melica smithii* 400
three-flower melic - *Melica nitens* 400
two-flower melic - *Melica mutica* 400
melilot
white melilot - *Melilotus alba* 207
yellow melilot - *Melilotus officinalis* 207
melist - *Melilotus officinalis* 207
melon - *Cucumis melo* 147
casaba melon - *Cucumis melo* var. *inodorus* 147
Chinese preserving melon - *Benincasa hispida* 146
Chinese winter melon - *Benincasa hispida* 146
honeydew melon - *Cucumis melo* var. *inodorus* 147
netted melon - *Cucumis melo* var. *reticulatus* 147
nutmeg melon - *Cucumis melo* var. *reticulatus* 147
Persian melon - *Cucumis melo* var. *reticulatus* 147
winter melon - *Cucumis melo* var. *inodorus* 147
melon-tree - *Carica papaya* 128
melongene - *Solanum melongena* var. *esculentum* 530
menow-weed - *Ruellia tuberosa* 8
menteng - *Baccaurea racemosa* 183
mercury - *Chenopodium bonus-henricus* 136, *Rhus radicans* 24

mescal - *Agave parryi* 17
mescal-bean - *Sophora secundiflora* 215
mesquite - *Prosopis chilensis, P. glandulosa, P. glandulosa* var. *torreyana, P. juliflora* 212
velvet mesquite - *Prosopis velutina* 212
western honey mesquite - *Prosopis glandulosa* var. *torreyana* 212
mesquitilla - *Calliandra eriophylla* 192
metel - *Datura metel* 525
Mexican-bamboo - *Polygonum cuspidatum* 438
Mexican-buckeye - *Ungnadia speciosa* 512
Mexican-clover - *Richardia scabra* 494
mezereon - *Daphne mezereum* 536
mignonette, common - *Reseda odorata* 458
mile-tree - *Casuarina equisetifolia* 132
milfoil - *Achillea millefolium* 53
milk-tree, African - *Euphorbia trigona* 184
milkbush, American - *Synadenium grantii* 187
milkvetch - *Astragalus adsurgens* var. *robustior* 190, *A. falcatus* 190, *A. missouriensis* 191
cicer milkvetch - *Astragalus cicer* 190
flexile milkvetch - *Astragalus flexuosus* 190
milkweed
bloodflower milkweed - *Asclepias curassavica* 51
common milkweed - *Asclepias syriaca* 51
eastern whorled milkweed - *Asclepias verticillata* 51
giant milkweed - *Calotropis gigantea* 52
horsetail milkweed - *Asclepias verticillata* 51
purple milkweed - *Asclepias purpurascens* 51
swamp milkweed - *Asclepias incarnata* 51
milkwort, yellow - *Polygala lutea* 436
millet
African millet - *Eleusine coracana* 387
broom millet - *Panicum miliaceum* 403
browntop millet - *Panicum fasciculatum* 403
bulrush millet - *Pennisetum glaucum* 406
common millet - *Panicum miliaceum* 403
Egyptian millet - *Sorghum halepense* 419
finger millet - *Eleusine coracana* 387
foxtail millet - *Setaria italica* 417
German millet - *Setaria italica* 417
hog millet - *Panicum miliaceum* 403
Hungarian millet - *Setaria italica* 417
Italian millet - *Setaria italica* 417
Japanese millet - *Echinochloa crusgalli* var. *frumentacea* 386, *Setaria italica* 417
pearl millet - *Pennisetum glaucum* 406
proso millet - *Panicum miliaceum* 403
Sanwa millet - *Echinochloa crusgalli* var. *frumentacea* 386
Texas millet - *Panicum texanum* 404
milo - *Sorghum bicolor* 419
mimosa - *Albizia julibrissin* 188
Texas mimosa - *Acacia greggii* 187
mind-your-own-business-plant - *Soleirolia soleirolii* 544
miner's-lettuce - *Montia perfoliata* 442
minniebush - *Menziesia pilosa* 175
mint
bergamot mint - *Mentha* × *piperita* var. *citrata* 264
corn mint - *Mentha arvensis* 264
coyote mint - *Monardella villosa* 265
dotted mint - *Monarda punctata* 265
field mint - *Mentha arvensis* 264
Japanese mint - *Mentha arvensis* 264
lemon mint - *Mentha* × *piperita* var. *citrata* 264, *Monarda citriodora* 265
mountain mint - *Monardella odoratissima* 265
red mint - *Mentha* × *gentilis* 264
stone mint - *Cunila origanoides* 263
water mint - *Mentha aquatica* 264
wood mint - *Blephilia hirsuta* 263
miracle-leaf - *Kalanchoe pinnata* 146
mirasol - *Helianthus annuus* 73
mirror-plant - *Coprosma repens* 491
missey-moosey - *Sorbus americana* 489
mistflower - *Eupatorium coelestinum* 70
mistletoe, American - *Phoradendron serotinum* 549
mock-mesquite - *Calliandra eriophylla* 192
mock-orange - *Prunus caroliniana* 475, *Sideroxylon lycioides* 512, *Styrax americana* 533
house blooming mock-orange - *Pittosporum tobira* 359
sweet mock-orange - *Philadelphus coronarius* 251
mock-plane - *Acer pseudoplatanus* 9
mock-strawberry - *Duchesnea indica* 465
moleplant - *Euphorbia heterophylla, E. lathyris* 184
molle - *Schinus molle* 25
mombin
purple mombin - *Spondias purpurea* 26
red mombin - *Spondias purpurea* 26
money-plant - *Lunaria annua* 113
money-tree - *Eucalyptus pulverulenta* 303
moneywort - *Lysimachia nummularia* 443
monk's-hood - *Astrophytum myriostigma* 118
ornamental monk's-hood - *Astrophytum ornatum* 118
monk's-hood-vine - *Ampelopsis aconitifolia* 549
monk's-pepper-tree - *Vitex agnus-castus* 547
monkey's-dinner-bell - *Hura crepitans* 185
monkey-flower
Allegheny monkey-flower - *Mimulus ringens* 518
common monkey-flower - *Mimulus guttatus* 518
scarlet monkey-flower - *Mimulus cardinalis* 518
monkey-pistol - *Hura crepitans* 185
monkey-plant - *Ruellia makoyana* 8
monkey-puzzle-tree - *Araucaria araucana* 45
monkeynut - *Arachis hypogaea* 189
monkshood
azure monkshood - *Aconitum carmichaelii* 451
garden monkshood - *Aconitum napellus* 451
monox - *Empetrum nigrum* 169
monstera - *Monstera deliciosa* 40
moonbeam - *Tabernaemontana divaricata* 35
moonflower - *Ipomoea alba* 142
bush moonflower - *Ipomoea leptophylla* 142
moonseed
Carolina moonseed - *Cocculus carolinus* 297
red moonseed - *Cocculus carolinus* 297
moorberry - *Vaccinium uliginosum* 179
moort, round-leaved - *Eucalyptus platypus* 303
mooseberry - *Viburnum edule, V. pauciflorum* 127
moosewood - *Acer pensylvanicum* 9, *Dirca palustris* 537, *Viburnum alnifolium* 127
morning-glory
beach morning-glory - *Ipomoea pes-caprae* 142
big-root morning-glory - *Ipomoea pandurata* 142
bush morning-glory - *Ipomoea leptophylla* 142
common morning-glory - *Ipomoea purpurea* 142
ivy-leaf morning-glory - *Ipomoea hederacea* 142
red morning-glory - *Ipomoea coccinea* 142
swamp morning-glory - *Ipomoea aquatica* 142
tall morning-glory - *Ipomoea purpurea* 142
white-edge morning-glory - *Ipomoea nil* 142
wild morning-glory - *Calystegia sepium* 141
woolly morning-glory - *Argyreia nervosa* 141
mosaic-plant - *Fittonia verschaffeltii* 7
moschatel - *Adoxa moschatellina* 17
Moses-in-a-boat - *Tradescantia spathacea* 140
Moses-in-the-bulrushes - *Tradescantia spathacea* 140
Moses-on-a-raft - *Tradescantia spathacea* 140
mosquito-bills - *Dodecatheon hendersonii* 443
mother-in-law's-tongue - *Sansevieria trifasciata* var. *laurentii* 19
mother-of-thousands - *Saxifraga stolonifera* 514
mother-of-thyme - *Thymus serpyllum* 268
motherwort - *Leonurus cardiaca* 264
mountain-ash
American mountain-ash - *Sorbus americana* 489
black mountain-ash - *Eucalyptus sieber* 303
European mountain-ash - *Sorbus aucuparia* 489
Pacific mountain-ash - *Sorbus sitchensis* 489
mountain-box - *Arctostaphylos uva-ursi* 171
mountain-ebony - *Bauhinia hookeri, B. variegata* 192
mountain-glory - *Holodiscus dumosus* 468
mountain-grape - *Berberis aquifolium* 91
mountain-lilac, Catalina - *Ceanothus arboreus* 458
mountain-lover - *Paxistima canbyi* 134
mountain-mahogany - *Betula lenta* 97, *Cercocarpus montanus* 462
birchlead mountain-mahogany - *Cercocarpus betuloides* 462
mountain-misery - *Chamaebatia foliolosa* 463
mountain-pride - *Penstemon newberryi* 519
mountain-sorrel - *Oxyria digyna* 438
mountain-sweet - *Ceanothus americanus* 458
mountain-tea - *Gaultheria procumbens* 172
mourning-bride - *Scabiosa atropurpurea* 167
moxie-plum - *Gaultheria hispidula* 172
mu - *Aleurites montana* 182
mu-oil-tree - *Aleurites montana* 182
mu-tree - *Aleurites montana* 182
mud-baby, Texas - *Echinodorus cordifolius* 21
mudar - *Calotropis gigantea* 52
small mudar - *Calotropis procera* 52
muguet - *Maianthemum canadense* 279
mugwort - *Artemisia vulgaris* 57
western mugwort - *Artemisia gnaphalodes, A. ludoviciana* 57
muhly
alkali muhly - *Muhlenbergia asperifolia* 400
bush muhly - *Muhlenbergia porteri* 401
creeping muhly - *Muhlenbergia repens* 401
foxtail muhly - *Muhlenbergia andina* 400
Mexican muhly - *Muhlenbergia mexicana* 401

mountain muhly - *Muhlenbergia montana* 401
New Mexican muhly - *Muhlenbergia pauciflora* 401
plains muhly - *Muhlenbergia cuspidata* 400
pull-up muhly - *Muhlenbergia filiformis* 400
slimstem muhly - *Muhlenbergia filiculmis* 400
spike muhly - *Muhlenbergia wrightii* 401
mulberry
American mulberry - *Morus rubra* 300
black mulberry - *Morus nigra* 300
French mulberry - *Callicarpa americana* 545
Indian mulberry - *Morinda citrifolia* 493
red mulberry - *Morus rubra* 300
silkworm mulberry - *Morus alba* var. *multicaulis* 300
white mulberry - *Morus alba* 300
mule-fat - *Baccharis viminea* 60
mullein - *Verbascum thapsus* 521
common mullein - *Verbascum thapsus* 521
moth mullein - *Verbascum blattaria* 521
turkey mullein - *Eremocarpus setigerus* 184
white moth mullein - *Verbascum blattaria* var. *albiflora* 521
mullet, Indian - *Oryzopsis hymenoides* 403
mum - *Chrysanthemum* x *morifolium* 63
musclewood - *Carpinus caroliniana* 101
musk, wild - *Erodium cicutarium* 242
musk-clover - *Erodium moschatum* 242
muskflower - *Mimulus moschatus* 518
muskmelon - *Cucumis melo* var. *reticulatus* 147
muskplant - *Mimulus moschatus* 518
muskrat-weed - *Thalictrum polygamum* 457
muskroot - *Adoxa moschatellina* 17
musquash-root - *Cicuta maculata* 27
mustard
black mustard - *Brassica nigra* 109
brown mustard - *Brassica juncea, B. juncea* var. *crispifolia* 109
curled mustard - *Brassica juncea* var. *crispifolia* 109
field mustard - *Brassica rapa* 109
hare's-ear mustard - *Conringia orientalis* 112
hedge mustard - *Sisymbrium altissimum, S. officinale* 115
Indian mustard - *Brassica juncea, B. juncea* var. *crispifolia* 109
leaf mustard - *Brassica juncea* var. *crispifolia* 109
mithridate mustard - *Thlaspi arvense* 115
tansy mustard - *Descurainia pinnata, D. sophia* 112
treacle mustard - *Erysimum repandum* 112
tumble mustard - *Sisymbrium altissimum* 115
white mustard - *Sinapis alba* 115
wormseed mustard - *Erysimum cheiranthoides* 112
mustard-greens - *Brassica juncea, B. juncea* var. *crispifolia* 109
mustard-tree - *Nicotiana glauca* 528
myrtle - *Cyrilla racemiflora* 166, *Myrtus communis* 307
bog myrtle - *Myrica gale* 302
cape myrtle - *Myrsine africana* 303
Greek myrtle - *Myrtus communis* 307
Oregon myrtle - *Umbellularia californica* 271
running myrtle - *Vinca minor* 35
Swedish myrtle - *Myrtus communis* 307
wax myrtle - *Myrica cerifera* 302
nagoonberry - *Rubus stellatus* 486
nail-rod - *Typha latifolia* 538
naio - *Myoporum sandwicense* 301
nance - *Byrsonima crassifolia* 289
nannyberry - *Viburnum lentago, V. prunifolium* 127
nap-at-noon - *Ornithogalum umbellatum* 280
narcissus - *Narcissus poeticus* 279
poet's narcissus - *Narcissus poeticus* 279
polyanthus narcissus - *Narcissus tazetta* 279
trumpet narcissus - *Narcissus pseudonarcissus* 279
naseberry - *Manilkara zapota* 512
nasturtium
garden nasturtium - *Tropaeolum majus* 538
tall nasturtium - *Tropaeolum majus* 538
natal-plum - *Carissa macrocarpa* 34
Nebraska-fern - *Conium maculatum* 28
neckweed - *Veronica peregrina* 521
nectarine - *Prunus persica* var. *nucipersica* 475
needle-and-thread - *Stipa comata* 423
needle-palm - *Yucca filamentosa* 19
negro-coffee - *Senna occidentalis* 215
nephthytis - *Syngonium podophyllum* 41
nerveplant - *Fittonia verschaffeltii* 7
silver nerveplant - *Fittonia verschaffeltii* 7
nerveroot - *Cypripedium calceolus* var. *pubescens* 321
netbrush - *Calothamnus sanguineus* 303
nettle
burning nettle - *Urtica urens* 544
Carolina nettle - *Solanum carolinense* 530
dog nettle - *Urtica urens* 544
European nettle - *Urtica dioica* 544
false nettle - *Boehmeria cylindrica* 543
silver-leaved nettle - *Solanum elaeagnifolium* 530
stinging nettle - *Urtica dioica* 544
nettle-tree - *Celtis occidentalis* 539
nicker-tree - *Gymnocladus dioica* 200
nightshade
bittersweet nightshade - *Solanum dulcamara* 530
black nightshade - *Solanum nigrum* 530
common nightshade - *Solanum nigrum* 530
deadly nightshade - *Atropa belladonna* 524, *Solanum dulcamara* 530
poisonous nightshade - *Solanum dulcamara* 530
silver-leaved nightshade - *Solanum elaeagnifolium* 530
stinking nightshade - *Hyoscyamus niger* 526
nimblewill - *Muhlenbergia schreberi* 401
ninebark - *Physocarpus opulifolius* 473
nira - *Allium tuberosum* 272
nispero - *Manilkara zapota* 512
Noah's-ark - *Cypripedium calceolus* var. *pubescens* 321
Nogal - *Juglans major* 258
nogal - *Juglans microcarpa* 258
none-so-pretty - *Silene armeria* 131
nonesuch - *Medicago lupulina* 206
nose-bleed - *Achillea millefolium* 53
nosegay - *Plumeria rubra* 35
num-num, big - *Carissa macrocarpa* 34
nut, macadamia - *Macadamia integrifolia* 444
nutgall-tree - *Rhus chinensis* 24
nutmeg
African nutmeg - *Monodora myristica* 26
Calabash nutmeg - *Monodora myristica* 26
Jamaica nutmeg - *Monodora myristica* 26
nutsedge - *Cyperus rotundus* 163
purple nutsedge - *Cyperus rotundus* 163
oak
basket oak - *Quercus prinus* 227
bear oak - *Quercus ilicifolia* 227
black oak - *Quercus velutina* 227
blackjack oak - *Quercus marilandica* 227
blue oak - *Quercus douglasii* 227
bluejack oak - *Quercus incana* 227
bur oak - *Quercus macrocarpa* 227
California black oak - *Quercus kelloggii* 227
California field oak - *Quercus agrifolia* 227
California live oak - *Quercus agrifolia* 227
California scrub oak - *Quercus dumosa* 227
California white oak - *Quercus lobata* 227
canyon oak - *Quercus chrysolepis* 227
Catesby's oak - *Quercus laevis* 227
chestnut oak - *Quercus prinus* 227
chinquapin oak - *Quercus prinoides* 227
coast live oak - *Quercus agrifolia* 227
cork oak - *Quercus suber* 227
deer oak - *Quercus sadleriana* 227
dwarf chestnut oak - *Quercus prinoides* 227
Emory oak - *Quercus emoryi* 227
Engelmann oak - *Quercus engelmannii* 227
English oak - *Quercus robur* 227
European turkey oak - *Quercus cerris* 227
Gambel oak - *Quercus gambelii* 227
high-ground willow oak - *Quercus incana* 227
holly oak - *Quercus ilex* 227
Holm oak - *Quercus ilex* 227
huckleberry oak - *Quercus vaccinifolia* 227
interior live oak - *Quercus wislizenii* 227
island oak - *Quercus tomentella* 227
jack oak - *Quercus ellipsoidalis, Q. marilandica* 227
Kellogg oak - *Quercus kelloggii* 227
laurel oak - *Quercus imbricaria* 227
laurel-leaved oak - *Quercus laurifolia* 227
leather oak - *Quercus durata* 227
MacDonald oak - *Quercus macdonaldii* 227
maul oak - *Quercus chrysolepis* 227
Mongolian oak - *Quercus mongolica* 227
mossy-cup oak - *Quercus macrocarpa* 227
northern pin oak - *Quercus ellipsoidalis* 227
northern red oak - *Quercus rubra* 227
Oregon oak - *Quercus garryana* 227
overcup oak - *Quercus lyrata* 227
pin oak - *Quercus palustris* 227
possum oak - *Quercus nigra* 227
post oak - *Quercus stellata* 227
rock chestnut oak - *Quercus prinus* 227
Rocky Mountain scrub oak - *Quercus undulata* 227
sawtooth oak - *Quercus acutissima* 227
scarlet oak - *Quercus coccinea* 227
scrub oak - *Quercus ilicifolia* 227
shingle oak - *Quercus imbricaria* 227
Shumard's red oak - *Quercus shumardii* 227
southern live oak - *Quercus virginiana* 227
southern red oak - *Quercus falcata* 227
Spanish oak - *Quercus palustris* 227
Spanish red oak - *Quercus falcata* 227
swamp chestnut oak - *Quercus prinus* 227
swamp post oak - *Quercus lyrata* 227
swamp white oak - *Quercus bicolor* 227
tanbark oak - *Lithocarpus densiflora* 227
Texas red oak - *Quercus texana* 227
truffle oak - *Quercus robur* 227
turkey oak - *Quercus incana, Q. laevis* 227
valley oak - *Quercus lobata* 227
water oak - *Quercus nigra* 227

western oak - *Quercus garryana* 227
white oak - *Quercus alba* 227
willow oak - *Quercus phellos* 227
yellow chestnut oak - *Quercus muhlenbergii* 227
yellow-barked oak - *Quercus velutina* 227
oat
animated oat - *Avena sterilis* 372
bristle oat - *Avena strigosa* 372
common oat - *Avena sativa* 372
naked oat - *Avena nuda* 372
sand oat - *Avena strigosa* 372
slender oat - *Avena barbata* 372
small oat - *Avena strigosa* 372
Tartarian oat - *Avena fatua* 372
oats
potato oats - *Avena fatua* 372
red oats - *Avena fatua* 372
wild oats - *Avena fatua* 372, *Chasmanthium latifolium* 381
obedience-plant - *Maranta arundinacea* 296
oceanspray - *Holodiscus discolor* 468
ocffee, wild robusta - *Coffea canephora* 491
Oconee-bells - *Shortia galacifolia* 166
ocotillo - *Fouquieria splendens* 240
octopus-tree - *Schefflera actinophylla* 44
oil-tree
China wood oil-tree - *Aleurites fordii* 182
poonga oil-tree - *Pongamia pinnata* 212
tung oil-tree - *Aleurites fordii* 182
oilcloth-flower - *Anthurium andraeanum* 38
oiticica - *Licania rigida* 138
okra - *Abelmoschus esculentus* 289
old-man - *Artemisia abrotanum* 57
old-man's-beard - *Chionanthus virginicus* 310
oleander - *Nerium oleander* 34
yellow oleander - *Thevetia peruviana* 35
oleaster - *Elaeagnus angustifolia* 168
olive
American olive - *Osmanthus americanus* 315
Chinese olive - *Osmanthus heterophyllus* 315
common olive - *Olea europaea* 315
desert olive - *Forestiera neomexicana* 311
fragrant olive - *Osmanthus fragrans* 315
holly olive - *Osmanthus heterophyllus* 315
Russian olive - *Elaeagnus angustifolia* 168
sweet olive - *Osmanthus fragrans* 315
tea olive - *Osmanthus fragrans* 315
wild olive - *Elaeagnus angustifolia* 168, *Nyssa aquatica* 309, *Osmanthus americanus* 315
one-leaf-clover - *Macroptilium lathyroides* 205
onigurumi - *Juglans sieboldiana* 259
onion - *Allium cepa* 272
bunching onion - *Allium fistulosum* 272
Egyptian onion - *Allium cepa* 272
Japanese bunching onion - *Allium fistulosum* 272
multiplier onion - *Allium cepa* 272
nodding onion - *Allium cernuum* 272
potato onion - *Allium cepa* 272
prairie onion - *Allium stellatum* 272
Spanish onion - *Allium fistulosum* 272
spring onion - *Allium fistulosum* 272
swamp onion - *Allium validum* 272
top onion - *Allium cepa* var. *viviparum* 272
tree onion - *Allium cepa* 272
two-bladed onion - *Allium fistulosum* 272
Welsh onion - *Allium fistulosum* 272
wild onion - *Allium canadense*, *A. cernuum* 272
opiuma - *Pithecellobium dulce* 212
oppossumwood - *Halesia carolina* 533
orache
halberdleaf orache - *Atriplex patula* subsp. *hastata* 135
red orache - *Atriplex rosea* 135
spreading orache - *Atriplex patula* 135
orange
bitter orange - *Citrus aurantium* 494
Chinese box orange - *Severinia buxifolia* 498
hardy orange - *Poncirus trifoliata* 498
king orange - *Citrus reticulata* 495
Mandarin orange - *Citrus reticulata* 495
Panama orange - x *Citrofortunella mitis* 494
Satsuma orange - *Citrus reticulata*, *C. unshiu* 495
Seville orange - *Citrus aurantium* 494
sour orange - *Citrus aurantium* 494
sweet orange - *Citrus sinensis* 495
trifoliate orange - *Poncirus trifoliata* 498
orange-jasmine - *Murraya paniculata* 497
orangeroot - *Hydrastis canadensis* 456
orchid
leaf white orchid - *Habenaria dilatata* 321
northern green orchid - *Habenaria hyperborea* 321
slender bog orchid - *Habenaria leucostachys* 321
tall white bog orchid - *Habenaria dilatata* 321
orchid-vine - *Stigmaphyllon ciliatum* 289
orchidtree - *Bauhinia variegata* 192
Oregon-grape - *Berberis aquifolium* 91
organy - *Origanum vulgare* 266
origano - *Origanum vulgare* 266
orpine - *Sedum telephium* 146
osage-orange - *Maclura pomifera* 300
osier
common osier - *Salix viminalis* 506
green osier - *Cornus alternifolia* 143
purple osier - *Salix purpurea* 506
western osier - *Cornus occidentalis* 143
osmanthus, holly - *Osmanthus heterophyllus* 315
osoberry - *Oemleria cerasiformis* 473
ostrich-plume - *Brassica juncea* var. *crispifolia* 109
otaheite-walnut - *Aleurites moluccana* 182
oudo - *Aralia cordata* 42
our-lord's-candle - *Yucca whipplei* 19
oxslip - *Primula elatior* 443
oyster-plant - *Tragopogon porrifolius* 87
oysterplant - *Tradescantia spathacea* 140
black oysterplant - *Scorzonera hispanica* 82
Spanish oysterplant - *Scorzonera hispanica* 82
pacaya - *Chamaedorea elegans* 46
pachysandra
Alleghany pachysandra - *Pachysandra procumbens* 118
Japanese pachysandra - *Pachysandra terminalis* 118
padauk - *Pterocarpus indicus* 213
padouk - *Pterocarpus indicus* 213
pagoda-tree - *Plumeria rubra* f. *acutifolia* 35
Japanese pagoda-tree - *Sophora japonica* 215
paintbrush
Flora's paintbrush - *Emilia javanica* 68
Indian paintbrush - *Asclepias tuberosa* 51, *Castilleja affinis* 516
scarlet paintbrush - *Castilleja coccinea* 516
painted-cup
seaside painted-cup - *Castilleja latifolia* 516
woolly painted-cup - *Castilleja foliolosa* 516
painted-leaf - *Euphorbia pulcherrima* 184
painted-tongue - *Salpiglossis sinuata* 529
paintleaf - *Euphorbia heterophylla* 184
paintroot - *Lachnanthes caroliniana* 247
pak-choi - *Brassica chinensis* 109
pakai - *Amaranthus gracilis* 22
palm
Alexandra palm - *Archontophoenix alexandrae* 46
areca palm - *Chrysalidocarpus lutescens* 47
areng palm - *Arenga pinnata* 46
assai palm - *Euterpe edulis* 47
bamboo palm - *Chamaedorea erumpens* 46, *Rhapis excelsa* 49
Belmore palm - *Howeia belmoreana* 48
black-fiber palm - *Arenga pinnata* 46
Burmese fishtail palm - *Caryota mitis* 46
butterfly palm - *Chrysalidocarpus lutescens* 47
California fan palm - *Washingtonia filifera* 51
California palm - *Washingtonia filifera* 51
California Washington palm - *Washingtonia filifera* 51
Canary Island date palm - *Phoenix canariensis* 48
cane palm - *Chrysalidocarpus lutescens* 47
Chinese windmill palm - *Trachycarpus fortunei* 50
Christmas palm - *Veitchia merrillii* 51
cluster palm - *Ptychosperma macarthurii* 49
clustered fishtail palm - *Caryota mitis* 46
coconut palm - *Cocos nucifera* 47
curly palm - *Howeia belmoreana* 48
date palm - *Phoenix dactylifera* 48
desert fan palm - *Washingtonia filifera* 51
desert palm - *Washingtonia robusta* 51
doum palm - *Hyphaene thebaica* 48
dwarf date palm - *Phoenix roebelenii* 48
everglades palm - *Acoelorrhaphe wrightii* 46
fishtail palm - *Caryota urens* 46
Florida cabbage palm - *Sabal palmetto* 49
Florida royal palm - *Roystonea elata* 49
Forster palm - *Howeia forsteriana* 48
Forster sentry palm - *Howeia forsteriana* 48
gingerbread palm - *Hyphaene thebaica* 48
golden cane palm - *Chrysalidocarpus lutescens* 47
golden feather palm - *Chrysalidocarpus lutescens* 47
gomuti palm - *Arenga pinnata* 46
good-luck palm - *Chamaedorea elegans* 46
hemp palm - *Trachycarpus fortunei* 50
hurricane palm - *Ptychosperma macarthurii* 49
jaggery palm - *Caryota urens* 46
kentia palm - *Howeia forsteriana* 48
lady palm - *Rhapis excelsa* 49
Macarthur feather palm - *Ptychosperma macarthurii* 49
Madagascar palm - *Chrysalidocarpus lutescens* 47
Manila palm - *Veitchia merrillii* 51
Merrill palm - *Veitchia merrillii* 51
Mexican Washington palm - *Washingtonia robusta* 51
miniature date palm - *Phoenix roebelenii* 48
miniature fan palm - *Rhapis excelsa* 49
northern Bangalow palm - *Archontophoenix alexandrae* 46
parlor palm - *Chamaedorea elegans* 46

Paurotis palm - *Acoelorrhaphe wrightii* 46
petticoat palm - *Washingtonia filifera* 51
pupunha palm - *Syagrus inajai* 50
pygmy date palm - *Phoenix roebelenii* 48
queen palm - *Syagrus romanzoffianum* 50
rattan palm - *Rhapis excelsa* 49
Roebelin palm - *Phoenix roebelenii* 48
sago palm - *Caryota urens* 46
saw cabbage palm - *Acoelorrhaphe wrightii* 46
Senegal date palm - *Phoenix reclinata* 48
silver saw palm - *Acoelorrhaphe wrightii* 46
slender lady palm - *Rhapis excelsa* 49
South American jelly palm - *Butia capitata* 46
sugar palm - *Arenga pinnata* 46
thatch palm - *Howeia forsteriana* 48
thatch-leaf palm - *Howeia forsteriana* 48
thread palm - *Washingtonia robusta* 51
toddy palm - *Caryota urens* 46
tufted fishtail palm - *Caryota mitis* 46
wild date palm - *Phoenix reclinata* 48
windmill palm - *Trachycarpus fortunei* 50
wine palm - *Caryota urens* 46
yellow butterfly palm - *Chrysalidocarpus lutescens* 47
yellow palm - *Chrysalidocarpus lutescens* 47
palm-lily - *Yucca gloriosa* 19
palmella - *Yucca elata* 19
palmetto
bluestem palmetto - *Sabal minor* 49
bush palmetto - *Sabal minor* 49
cabbage palmetto - *Sabal palmetto* 49
dwarf palmetto - *Sabal minor* 49
Etonia palmetto - *Sabal etonia* 49
Mexican palmetto - *Sabal mexicana* 49
Rio Grande palmetto - *Sabal mexicana* 49
saw palmetto - *Serenoa repens* 50
scrub palmetto - *Sabal etonia* 49, *S. minor* 49, *Serenoa repens* 50
Texas palmetto - *Sabal mexicana* 49
Victoria palmetto - *Sabal mexicana* 49
palo-verde - *Parkinsonia florida* 209
Mexican palo-verde - *Parkinsonia aculeata* 209
palta - *Persea americana* 269
panamiga - *Pilea involucrata* 543
panda-bear-plant - *Kalanchoe tomentosa* 146
pandang - *Pandanus odoratissimus* 324
pandanus-palm - *Pandanus tectorius* 324
pansy - *Viola tricolor* 547
bedding pansy - *Viola cornuta* 547
European field pansy - *Viola arvensis* 547
European wild pansy - *Viola tricolor* 547
field pansy - *Viola rafinesquii, V. tricolor* 547
miniature pansy - *Viola tricolor* 547
tufted pansy - *Viola cornuta* 547
papala - *Bougainvillea umbellifera* 308
papaya - *Carica papaya* 128
paper-mulberry - *Broussonetia papyrifera* 298
paperbark-tree - *Melaleuca quinquenervia* 305
paperflower - *Bougainvillea glabra* 308
paperplant - *Fatsia japonica* 43
glossy-leaved paperplant - *Fatsia japonica* 43
para-para - *Pisonia umbellifera* 308
paradise-flower - *Caesalpinia pulcherrima* 192
paradise-plant - *Justicia carnea* 8
paradise-tree - *Melia azedarach* 297, *Simarouba glauca* 523
parakeet-flower - *Heliconia psittacorum* 249
parasol-tree, Chinese - *Firmiana simplex* 532
parilla, yellow - *Menispermum canadense* 297
parrot's-feather - *Myriophyllum aquaticum* 247
parrot's-flower - *Heliconia psittacorum* 249
parrot's-plantain - *Heliconia psittacorum* 249
parrotleaf - *Alternanthera ficoidea* cv. 'amoena' 21
parsley - *Coriandrum sativum* 28, *Petroselinum crispum* 32
Chinese parsley - *Coriandrum sativum* 28
parsnip - *Pastinaca sativa* 31
bladder parsnip - *Lomatium utriculatum* 30
buck parsnip - *Lomatium triternatum* 30
pestle parsnip - *Lomatium nudicaule* 30
wild parsnip - *Pastinaca sativa* 31
partridge-breast - *Aloe variegata* 273
partridgeberry - *Mitchella repens* 493
partridgepea, sensitive - *Cassia procumbens* 193
paspalum
brownseed paspalum - *Paspalum plicatulum* 405
bull paspalum - *Paspalum boscianum* 405
Combs paspalum - *Paspalum almum* 405
field paspalum - *Paspalum laeve* 405
fringeleaf paspalum - *Paspalum ciliatifolium* 405
sour paspalum - *Paspalum conjugatum* 405
pasque-flower - *Anemone nuttalliana* 452
American pasque-flower - *Anemone patens* 452
passionflower
blue passionflower - *Passiflora caerulea* 325
maypop passionflower - *Passiflora incarnata* 325
wild passionflower - *Passiflora incarnata* 325
passionfruit - *Passiflora edulis* 325
patience-plant - *Impatiens wallerana* 90
patient-lucy - *Impatiens wallerana* 90
paulownia, royal - *Paulownia tomentosa* 519
pawpaw - *Asimina triloba* 26, *Carica papaya* 128
pe-tsai - *Brassica pekinensis* 109
pea
Angola pea - *Cajanus cajan* 192
Austrian winter pea - *Lathyrus hirsutus* 201
beach pea - *Lathyrus japonicus, L. littoralis* 201
black-eyed pea - *Vigna unguiculata* 220
butterfly pea - *Clitoria ternatea* 194
Caley pea - *Lathyrus hirsutus* 201
Canada pea - *Vicia cracca* 219
catjang pea - *Cajanus cajan* 192
common pea - *Pisum sativum* 211
Congo pea - *Cajanus cajan* 192
cow pea - *Vigna unguiculata* 220
crowder pea - *Vigna unguiculata* 220
edible-podded pea - *Pisum sativum* var. *macrocarpon* 211
English pea - *Pisum sativum* 211
everlasting pea - *Lathyrus latifolius, L. sylvestris* 201
field pea - *Pisum sativum* var. *arvense* 211
flat pea - *Lathyrus sylvestris* 201
garden pea - *Pisum sativum* 211
heath pea - *Lathyrus japonicus* 201
Jerusalem pea - *Vigna unguiculata* subsp. *cylindrica* 220
love pea - *Abrus precatorius* 187
marble pea - *Vigna unguiculata* subsp. *cylindrica* 220
no-eye pea - *Cajanus cajan* 192
partridge pea - *Cassia fasciculata* 193
perennial pea - *Lathyrus latifolius, L. sylvestris* 201
pigeon pea - *Cajanus cajan* 192
rabbit's pea - *Tephrosia virginiana* 216
rosary pea - *Abrus precatorius* 187
seaside pea - *Lathyrus japonicus* 201
singletary pea - *Lathyrus hirsutus* 201
snow pea - *Pisum sativum* var. *macrocarpon* 211
southern pea - *Vigna unguiculata* 220
sugar pea - *Pisum sativum* var. *macrocarpon* 211
sweet pea - *Lathyrus odoratus* 201
Tangier pea - *Lathyrus tingitanus* 201
pea-tree, Siberian - *Caragana arborescens* 193
peace-in-the-home - *Soleirolia soleirolii* 544
peach - *Prunus persica* 475
Chinese wild peach - *Prunus davidiana* 475
desert peach - *Prunus andersonii* 475
peach-bells - *Campanula persicifolia* 121
peanut, common - *Arachis hypogaea* 189
pear
Asian pear - *Pyrus pyrifolia* 482
Bradford pear - *Pyrus calleryana* 482
Callery pear - *Pyrus calleryana* 482
Chinese pear - *Pyrus pyrifolia* 482
common pear - *Pyrus communis* 482
Japanese pear - *Pyrus pyrifolia* 482
oriental pear - *Pyrus pyrifolia* 482
sand pear - *Pyrus pyrifolia* 482
pear-thorn - *Crataegus calpodendron* 463
pearlbush, common - *Exochorda racemosa* 466
pearlwort, bird's-eye - *Sagina procumbens* 131
pearthorn - *Crataegus laevigata* 463
peatree - *Sesbania macrocarpa* 215
peavine
marsh peavine - *Lathyrus palustris* 201
rough peavine - *Lathyrus hirsutus* 201
wing-stemmed wild peavine - *Lathyrus palustris* 201
pecan - *Carya illinoensis* 255
bitter pecan - *Carya aquatica* 255
pegwood - *Cornus sanguinea* 143
pellionia, satin - *Pellionia pulchra* 543
pellitory, Pennsylvania - *Parietaria pensylvanica* 543
pen-wiper - *Kalanchoe marmorata* 146
pennisetum, Chinese - *Pennisetum alopecuroides* 406
pennyflower - *Lunaria annua* 113
pennyroyal
American pennyroyal - *Hedeoma pulegioides* 263
bastard pennyroyal - *Trichostema dichotomum* 268
mock pennyroyal - *Hedeoma pulegioides* 263
pennywort - *Cymbalaria muralis* 517
Asiatic pennywort - *Centella asiatica* 27
water pennywort - *Hydrocotyle umbellata* 30
peony
Chinese peony - *Paeonia lactiflora* 323
common garden peony - *Paeonia lactiflora* 323
tree peony - *Paeonia suffruticosa* 323
peperomia
emerald-ripple peperomia - *Peperomia caperata* 359
green-ripple peperomia - *Peperomia caperata* 359
little-fantasy peperomia - *Peperomia caperata* 359

oval-leaf peperomia - *Peperomia obtusifolia, P. obtusifolia* cv. 'variegata' 359
pepper
bell pepper - *Capsicum annuum* 524, *C. frutescens* cv. 'grossum' 525
black pepper - *Piper nigrum* 359
Cayenne pepper - *Capsicum annuum* 524
Celebes pepper - *Piper ornatum* 359
chili pepper - *Capsicum annuum* 524
common garden pepper - *Capsicum annuum* 524
green pepper - *Capsicum annuum* 524
mango pepper - *Capsicum annuum* 524
paprika pepper - *Capsicum annuum* 524
sweet pepper - *Capsicum frutescens* cv. 'grossum' 525
tabasco pepper - *Capsicum frutescens* 524
white pepper - *Piper nigrum* 359
wild pepper - *Vitex agnus-castus* 547
pepper-and-salt - *Erigenia bulbosa* 29
pepper-face - *Peperomia obtusifolia* 359
pepper-tree
Brazilian pepper-tree - *Schinus terebinthifolius* 25
California pepper-tree - *Schinus molle* 25
Peruvian pepper-tree - *Schinus molle* 25
pepperbush, sweet - *Clethra alnifolia* 138
pepperidge - *Nyssa sylvatica* 309
peppermint - *Mentha* x *piperita* 264
pepperplant - *Piper nigrum* 359
peppervine - *Ampelopsis arborea* 549
pepperweed
field pepperweed - *Lepidium campestre* 113
Virginia pepperweed - *Lepidium virginicum* 113
pepperwood - *Umbellularia californica* 271, *Zanthoxylum clava-herculis* 498
perilla - *Perilla frutescens* 266
periquito-saracura - *Alternanthera philoxeroides* 21
periwinkle
bigleaf periwinkle - *Vinca major* 35
common periwinkle - *Vinca minor* 35
greater periwinkle - *Vinca major* 35
lesser periwinkle - *Vinca minor* 35
Madagascar periwinkle - *Catharanthus roseus* 34
rose periwinkle - *Catharanthus roseus* 34
perpetua - *Gomphrena globosa* 22
persimmon
American persimmon - *Diospyros virginiana* 167
black persimmon - *Diospyros texana* 167
common persimmon - *Diospyros virginiana* 167
Japanese persimmon - *Diospyros kaki* 167
kaki persimmon - *Diospyros kaki* 167
Mexican persimmon - *Diospyros texana* 167
Texas persimmon - *Diospyros texana* 167
perunkila - *Carissa carandas* 34
petty-morel - *Aralia racemosa* 42
petunia
common garden petunia - *Petunia* x *hybrida* 529
large white petunia - *Petunia axillaris* 529
seaside petunia - *Petunia parviflora* 529
whitemoon petunia - *Petunia axillaris* 529
wild petunia - *Petunia parviflora* 529
pheasant's-eye - *Narcissus poeticus* 279
pheasant's-wings - *Aloe variegata* 273
philodendron
cutleaf philodendron - *Monstera deliciosa* 40
heart-leaf philodendron - *Philodendron cordatum, P. scandens* 40
variegated philodendron - *Epipremnum aureum* 40
velvet-leaf philodendron - *Philodendron scandens* f. *micans* 40
phlox
annual phlox - *Phlox drummondii* 435
blue phlox - *Phlox divaricata* 435
Drummond phlox - *Phlox drummondii* 435
fall phlox - *Phlox paniculata* 435
meadow phlox - *Phlox maculata* 435
perennial phlox - *Phlox paniculata* 435
smooth phlox - *Phlox glaberrima* 435
summer perennial phlox - *Phlox paniculata* 435
thick-leaf phlox - *Phlox carolina* 435
Phoenix-tree - *Firmiana simplex* 532
photinia, Japanese - *Photinia glabra* 473
pichi - *Fabiana imbricata* 526
pichi-pichi - *Fabiana imbricata* 526
pickaback-plant - *Tolmiea menziesii* 515
pickerelweed - *Pontederia cordata* 441
lanceleaf pickerelweed - *Pontederia cordata* var. *lancifolia* 441
piemacker - *Abutilon theophrastii* 289
pieplant - *Rheum rhabarbarum* 439
pig-lily - *Zantedeschia aethiopica* 41
pigeon-berry - *Phytolacca americana* 326
pigeon-foot - *Salicornia europaea* 137
pigeon-plum - *Coccoloba diversifolia* 437
pigeonberry - *Duranta repens* 545
pigeonwings, Asian - *Clitoria ternatea* 194
piggyback-plant - *Tolmiea menziesii* 515
pignut - *Carya cordiformis* 255
pigweed - *Amaranthus hybridus, A. retroflexus* 22
prostrate pigweed - *Amaranthus blitoides* 22
redroot pigweed - *Amaranthus retroflexus* 22
rough pigweed - *Amaranthus retroflexus* 22
smooth pigweed - *Amaranthus hybridus* 22
spiny pigweed - *Amaranthus spinosus* 22
spreading pigweed - *Amaranthus graecizans*
tumble pigweed - *Amaranthus albus* 22
winged pigweed - *Cycloloma atriplicifolium* 137
pilea, watermelon - *Pilea cadierei* 543
pilewort - *Ranunculus ficaria* 456
pimbina - *Viburnum trilobum* 127
pimento - *Capsicum annuum* 524, *Pimenta dioica* 307
pimpernel
scarlet pimpernel - *Anagallis arvensis* 442
yellow pimpernel - *Taenidia integerrima* 32
pin-clover - *Erodium cicutarium* 242
pincushions - *Scabiosa atropurpurea* 167
pindar - *Arachis hypogaea* 189
pine
Aleppo pine - *Pinus halepensis* 339
Apache pine - *Pinus engelmannii* 339
Australian pine - *Araucaria heterophylla* 45, *Casuarina equisetifolia* 132
Austrian pine - *Pinus nigra* 339
beach pine - *Pinus contorta* 339
Bhutan pine - *Pinus wallichiana* 339
big-cone pine - *Pinus coulteri* 339
bishop pine - *Pinus muricata* 339
black pine - *Pinus nigra* 339
blue pine - *Pinus wallichiana* 339
Brazilian pine - *Araucaria angustifolia* 45
bristlecone pine - *Pinus aristata* 339
Canary Island pine - *Pinus canariensis* 339
Caribbean pine - *Pinus caribaea* 339
cedar pine - *Pinus glabra* 339
Chilean pine - *Araucaria araucana* 45
cluster pine - *Pinus pinaster* 339
Corsican pine - *Pinus nigra* 339
cow's-tail pine - *Cephalotaxus harringtonia* var. *drupacea* 134
Cuban pine - *Pinus caribaea* 339
digger pine - *Pinus sabiniana* 339
eastern white pine - *Pinus strobus* 339
foxtail pine - *Pinus balfouriana* 339
frankincense pine - *Pinus taeda* 339
Georgia pine - *Pinus palustris* 339
giant pine - *Pinus lambertiana* 339
gray pine - *Pinus banksiana* 339
hickory pine - *Pinus aristata* 339
Himalayan white pine - *Pinus wallichiana* 339
house pine - *Araucaria heterophylla* 45
Italian stone pine - *Pinus pinea* 339
Jack pine - *Pinus banksiana* 339
Japanese black pine - *Pinus thunbergiana* 339
Japanese red pine - *Pinus densiflora* 339
Jeffrey pine - *Pinus ponderosa* var. *jeffreyi* 339
Jersey pine - *Pinus virginiana* 339
Jerusalem pine - *Pinus halepensis* 339
knobcone pine - *Pinus attenuata* 339
limber pine - *Pinus flexilis* 339
loblolly pine - *Pinus taeda* 339
lodgepole pine - *Pinus contorta, P. contorta* var. *latifolia* 339
long-tag pine - *Pinus echinata* 339
longleaf pine - *Pinus palustris* 339
maritime pine - *Pinus pinaster* 339
Mexican stone pine - *Pinus cembroides* 339
Mexican yellow pine - *Pinus patula* 339
Monterey pine - *Pinus radiata* 339
New Caledonia pine - *Araucaria columnaris* 45
Norfolk Island pine - *Araucaria heterophylla* 45
Norway pine - *Pinus resinosa* 339
nut pine - *Pinus monophylla* 339
old-field pine - *Pinus taeda* 339
Paraná pine - *Araucaria angustifolia* 45
Parry pinyon pine - *Pinus quadrifolia* 339
pinyon pine - *Pinus edulis* 339
pitch pine - *Pinus rigida* 339
pond pine - *Pinus serotina* 339
ponderosa pine - *Pinus ponderosa* 339
poverty pine - *Pinus virginiana* 339
prickly pine - *Pinus pungens* 339
red pine - *Pinus resinosa* 339
Rocky Mountain yellow pine - *Pinus ponderosa* var. *scopulorum* 339
rough-barked Mexican pine - *Pinus montezumae* 339
Scotch pine - *Pinus sylvestris* 339
Scots pine - *Pinus sylvestris* 339
scrub pine - *Pinus banksiana, P. virginiana* 339
shore pine - *Pinus contorta* 339
shortleaf pine - *Pinus echinata* 339
single-leaf pinyon pine - *Pinus monophylla* 339
slash pine - *Pinus elliottii* 339
soledad pine - *Pinus torreyana* 339
southern yellow pine - *Pinus palustris* 339
spruce pine - *Pinus glabra, P. virginiana* 339
stone pine - *Pinus monophylla* 339

sugar pine - *Pinus lambertiana* 339
Swiss mountain pine - *Pinus mugo* 339
Swiss stone pine - *Pinus cembra* 339
table mountain pine - *Pinus pungens* 339
Torrey pine - *Pinus torreyana* 339
two-leaved nut pine - *Pinus edulis* 339
umbrella pine - *Pinus pinea* 339
Virginia pine - *Pinus virginiana* 339
western white pine - *Pinus monticola* 339
western yellow pine - *Pinus ponderosa* 339
white-barked pine - *Pinus albicaulis* 339
yellow pine - *Pinus echinata* 339
pine-sap - *Monotropa uniflora* 298
pineapple - *Ananas comosus* 116
pineapple-shrub - *Calycanthus floridus* 121
pineapple-weed - *Matricaria matricarioides* 79
pineweed - *Hypericum gentianoides* 139
piney - *Paeonia officinalis* 323
pink
California Indian pink - *Silene californica* 131
clove pink - *Dianthus caryophyllus* 130
cottage pink - *Dianthus plumarius* 130
cushion pink - *Silene acaulis* 131
Deptford pink - *Dianthus armeria* 130
garden pink - *Dianthus plumarius* 130
grass pink - *Dianthus armeria, D. plumarius* 130
ground pink - *Phlox subulata* 435
Indian pink - *Lobelia cardinalis* 122, *Spigelia marilandica* 284
maiden pink - *Dianthus deltoides* 130
moss pink - *Phlox subulata* 435
mullein pink - *Lychnis coronaria* 131
rainbow pink - *Dianthus chinensis* 130
rose pink - *Sabatia angularis* 242
wild pink - *Silene caroliniana* subsp. *pensylvanica* 131
pink-and-white-shower - *Cassia javanica* var. *indochinensis* 193
pinwheel-flower - *Tabernaemontana divaricata* 35
pinxter, Florida - *Rhododendron canescens* 176
pinxterbloom - *Rhododendron nudiflorum* 176
pipe, Indian - *Monotropa uniflora* 298
pipe-plant - *Aeschynanthus pulcher* 243
pipevine - *Aristolochia durior* 51
piprage - *Berberis vulgaris* 91
pipsissewa - *Chimaphila umbellata* var. *cisatlantica* 451
pirul - *Schinus molle* 25
pistachio, Chinese - *Pistacia chinensis* 23
pistachio-nut - *Pistacia vera* 23
pitanga - *Eugenia uniflora* 305
pitcher-sage - *Sphacele calycina* 268
pitcherplant
California pitcherplant - *Darlingtonia californica* 512
common pitcherplant - *Sarracenia purpurea* 512
southern pitcherplant - *Sarracenia purpurea* 512
sweet pitcherplant - *Sarracenia purpurea* 512
yellow pitcherplant - *Sarracenia flava* 512
pittosporum, Japanese - *Pittosporum tobira* 359
plama-christi - *Ricinus communis* 186
planertree - *Planera aquatica* 540
planetree - *Platanus occidentalis* 360
London planetree - *Platanus* x *acerifolia* 360
oriental planetree - *Platanus orientalis* 360
plant
mother-in-law plant - *Dieffenbachia seguine* 39
urn plant - *Aechmea fasciata* 116
plantain - *Musa acuminata* 301
blackseed plantain - *Plantago rugelii* 360
bracted plantain - *Plantago aristata* 359
broadleaf plantain - *Plantago major* 359
buckhorn plantain - *Plantago lanceolata* 359
common plantain - *Plantago major* 359
English plantain - *Plantago lanceolata* 359
great Indian plantain - *Cacalia muhlenbergii* 62
great plantain - *Plantago major* 359
narrow-leaved plantain - *Plantago lanceolata* 359
pale Indian plantain - *Cacalia atriplicifolia* 62
paleseed plantain - *Plantago virginica* 360
poor-Robin's plantain - *Hieracium venosum* 76
robin's plantain - *Erigeron pulchellus* 69
Rugel's plantain - *Plantago rugelii* 360
platterleaf - *Coccoloba uvifera* 437
pleroma - *Tibouchina urvilleana* 296
plum
Alleghany plum - *Prunus alleghaniensis* 475
American plum - *Prunus americana* 475
apricot plum - *Prunus simonii* 475
August plum - *Prunus americana* 475
beach plum - *Prunus maritima* 475
bullace plum - *Prunus domestica* subsp. *insititia* 475
Canada plum - *Prunus nigra* 475
cherry plum - *Prunus cerasifera* 475
chickasaw plum - *Prunus angustifolia* 475
common plum - *Prunus domestica* 475
Damson plum - *Prunus domestica* subsp. *insititia* 475
European plum - *Prunus domestica* 475
flatwoods plum - *Prunus umbellata* 475
garden plum - *Prunus domestica* 475
goose plum - *Prunus americana* 475
hog plum - *Prunus americana, P. umbellata* 475
hortulan plum - *Prunus hortulana* 475
Japanese plum - *Eriobotrya japonica* 465, *Prunus salicina* 475
Klamath plum - *Prunus subcordata* 475
marmalade plum - *Pouteria sapota* 512
Mexican plum - *Prunus mexicana* 475
Myrobalan plum - *Prunus cerasifera* 475
Pacific plum - *Prunus subcordata* 475
prune plum - *Prunus domestica* 475
sand plum - *Prunus angustifolia* 475
shore plum - *Prunus maritima* 475
Sierra plum - *Prunus subcordata* 475
Simon plum - *Prunus simonii* 475
sloe plum - *Prunus umbellata* 475
Spanish plum - *Spondias purpurea* 26
wild plum - *Prunus americana* 475
wild-goose plum - *Prunus hortulana, P. munsoniana* 475
plume, Brazilian - *Justicia carnea* 8
plume-flower - *Justicia carnea* 8
plume-plant - *Justicia carnea* 8
plush-plant - *Kalanchoe tomentosa* 146
pocan - *Phytolacca americana* 326
podocarpus, yew - *Podocarpus macrophyllus* 434
poha - *Physalis peruviana* 529
poinciana
dwarf poinciana - *Caesalpinia pulcherrima* 192
royal poinciana - *Delonix regia* 196
poinsettia - *Euphorbia pulcherrima* 184
annual poinsettia - *Euphorbia heterophylla* 184
Japanese poinsettia - *Euphorbia heterophylla* 184, *Pedilanthus tithymaloides* 186
poison-dogwood - *Rhus vernix* 24
poison-elder - *Rhus vernix* 24
poison-hemlock - *Conium maculatum* 28
poison-ivy - *Rhus radicans, R. toxicodendron* 24
poison-oak - *Rhus toxicodendron* 24
Pacific poison-oak - *Rhus diversiloba* 24
Western poison-oak - *Rhus diversiloba* 24
poisonberry - *Solanum nigrum* 530
poisonwood - *Gymnanthes lucida* 185
poke
Indian poke - *Veratrum viride* 282
Virginian poke - *Phytolacca americana* 326
pokeberry - *Phytolacca americana* 326
poker-plant - *Kniphofia uvaria* 277
pokeweed - *Phytolacca americana* 326
polecat-bush - *Rhus aromatica* 24
polecat-weed - *Symplocarpus foetidus* 41
pollyana-vine - *Soleirolia soleirolii* 544
polyanthus - *Primula* x *polyantha* 443
polygala, fringed - *Polygala paucifolia* 436
polypody
California polypody - *Polypodium californicum* 449
rock polypody - *Polypodium virginianum* 449
western polypody - *Polypodium hesperium* 449
pomegranate - *Punica granatum* 450
dwarf pomegranate - *Punica granatum* cv. 'nana' 450
pomme-de-liane - *Passiflora laurifolia* 325
pondnuts - *Nelumbo lutea* 308
pondspice - *Litsea aestivalis* 269
pondweed
American pondweed - *Potamogeton nodosus* 442
curlyleaf pondweed - *Potamogeton crispus* 442
sago pondweed - *Potamogeton pectinatus* 442
pongam - *Pongamia pinnata* 212
ponytail - *Nolina tuberculata* 18
poorman's-weatherglass - *Anagallis arvensis* 442
popcorn - *Zea mays* 428
popinac - *Acacia farnesiana* 187
poplar
balsam poplar - *Populus balsamifera* 499
black poplar - *Populus nigra* 499, *P. nigra* cv. 'italica'
Chinese white poplar - *Populus tomentosa* 499
downy poplar - *Populus heterophylla* 499
Lombardy poplar - *Populus nigra* cv. 'italica' 499
necklace poplar - *Populus deltoides* 499
silver-leaved poplar - *Populus* × *alba* 498
Simon poplar - *Populus simonii* 499
Western balsam poplar - *Populus trichocarpa* 499
white poplar - *Populus* × *alba* 498
poppy
arctic poppy - *Papaver nudicaule* 325
blue poppy - *Meconopsis betonicifolia* 325

California poppy - *Eschscholzia californica* 325
corn poppy - *Papaver rhoeas* 325
field poppy - *Papaver rhoeas* 325
Flanders poppy - *Papaver rhoeas* 325
Iceland poppy - *Papaver nudicaule* 325
opium poppy - *Papaver somniferum* 325
oriental poppy - *Papaver orientale* 325
porcelain-lily, pink - *Alpinia zerumbet* 553
porch-vine - *Wedelia trilobata* 89
possum-apple - *Diospyros virginiana* 167
possum-grape - *Cissus incisa* 549
possum-haw - *Viburnum acerifolium* 127
possumwood - *Diospyros virginiana* 167
potato - *Solanum tuberosum* 530
white Irish potato - *Solanum tuberosum* 530
pothos - *Epipremnum aureum* 40
poverty-weed - *Iva axillaris* 77
powderpuff, red - *Calliandra haematocephala* 192
prairie-clover, white - *Dalea candida* 196
prairie-dock - *Silphium terebinthinaceum* 84
prairie-gentian - *Eustoma russellianum* 241
prairie-mimosa - *Desmanthus illinoensis* 196
prairie-potato - *Pediomelum esculentum* 209
prairie-smoke - *Anemone nuttalliana, A. patens* 452
prairie-tea - *Croton monanthogynus* 183
prairie-turnip - *Pediomelum esculentum* 209
prayer-beads - *Abrus precatorius* 187
prayer-plant - *Maranta leuconeura* 296
pretty-face - *Triteleia ixioides* 281
pricklepoppy, crested - *Argemone platyceras* 324
prickleweed - *Desmanthus illinoensis* 196
prickly-ash - *Aralia spinosa* 42
prickly-pear
Lindheimer prickly-pear - *Opuntia lindheimeri* 120
Texas prickly-pear - *Opuntia lindheimeri* 120
prickly-phlox - *Leptodactylon californicum* 435
pricklyash
lime pricklyash - *Zanthoxylum fagara* 498
northern pricklyash - *Zanthoxylum americanum* 498
southern pricklyash - *Zanthoxylum clava-herculis* 498
pricklypoppy, Mexican - *Argemone mexicana* 324
pride-of-Barbados - *Caesalpinia pulcherrima* 192
pride-of-Bolivia - *Tipuana tipu* 216
pride-of-China - *Melia azedarach* 297
pride-of-India - *Melia azedarach* 297
primrose
baby primrose - *Primula malacoides* 443
Chinese primrose - *Primula sinensis* 443
English primrose - *Primula vulgaris* 444
fairy primrose - *Primula malacoides* 443
German primrose - *Primula obconica, P. obconica* cv. 'gigantea' 443
poison primrose - *Primula obconica* cv. 'gigantea' 443
spike primrose - *Boisduvalia densiflora* 316
white evening primrose - *Oenothera speciosa* 319
primrose-willow - *Ludwigia octovalvis* 318
prince's-feather - *Polygonum orientale* 438
prince's-pipe - *Chimaphila umbellata* var. *cisatlantica* 451
princess-feather - *Polygonum orientale* 438
princess-flower - *Tibouchina urvilleana* 296
princess-pine - *Lycopodium obscurum* var. *dendroideum* 284
princess-tree - *Paulownia tomentosa* 519
princess-vine - *Cissus sicyoides* 549
privet
Amur privet - *Ligustrum amurense* 314
California privet - *Ligustrum obtusifolium, L. ovalifolium* 314
Chinese privet - *Ligustrum lucidum, L. sinense* 314
common privet - *Ligustrum vulgare* 314
glossy privet - *Ligustrum lucidum* 314
Japanese privet - *Ligustrum japonicum* 314
Nepal privet - *Ligustrum lucidum* 314
swamp privet - *Forestiera acuminata* 311
wax-leaf privet - *Ligustrum japonicum, L. lucidum* 314
proboscis-flower - *Proboscidea louisianica* 326
protea
giant protea - *Protea cynaroides* 445
king protea - *Protea cynaroides* 445
provision-tree - *Pachira aquatica* 106
psoralea, scurfy - *Psoralidium tenuiflorum* 213
puccoon, hoary - *Lithospermum canescens* 107
pudding-berry - *Cornus canadensis* 143
pudding-pipe-tree - *Cassia fistula* 193
pummelo - *Citrus maxima* 494
pumpkin - *Cucumis sativus* 147, *Cucurbita moschata* 149, *C. pepo* 149
bush pumpkin - *Cucurbita pepo* var. *melopepo* 149
Canada pumpkin - *Cucurbita moschata* 149
fetid wild pumpkin - *Cucurbita foetidissima* 149
white pumpkin - *Benincasa hispida* 146
puncture-vine - *Tribulus terrestris* 553
punk-tree - *Melaleuca quinquenervia* 305
purple-passionvine - *Gynura aurantiaca* 72
purpletop - *Tridens flavus* 424
purret - *Allium porrum* 272
purslane
common purslane - *Portulaca oleracea* 442
horse purslane - *Trianthema portulacastrum* 20
sea purslane - *Sesuvium portulacastrum* 20
Siberian purslane - *Montia sibirica* 442
winter purslane - *Montia perfoliata* 442
pussy-ears - *Calochortus tolmiei* 275, *Kalanchoe tomentosa* 146
pussy-paws - *Calyptridium umbellatum* 441
puttyroot - *Aplectrum hyemale* 320
pyrethrum - *Chrysanthemum cinerariifolium* 63
common pyrethrum - *Chrysanthemum coccineum* 63
Dalmatian pyrethrum - *Chrysanthemum cinerariifolium* 63
pyrola
arctic pyrola - *Pyrola glandiflora* 451
one-flowered pyrola - *Moneses uniflora* 451
one-sided pyrola - *Pyrola secunda* 451
pink pyrola - *Pyrola asarifolia* 451
pysicnut - *Jatropha curcas* 185
qat - *Catha edulis* 133
quamash - *Camassia quamash* 275
quarter-vine - *Bignonia capreolata* 103
queen's-delight - *Stillingia sylvatica* 187
queen's-lace - *Daucus carota* 28
queen's-root - *Stillingia sylvatica* 187
Queen-Anne's-lace - *Daucus carota* 28
queen-of-the-meadow - *Filipendula ulmaria* 466
queen-of-the-night - *Hylocereus undatus* 119
queen-of-the-prairie - *Filipendula rubra* 466
queencup - *Clintonia uniflora* 275
Queensland-nut - *Macadamia integrifolia* 444
small-fruited Queensland-nut - *Macadamia tetraphylla* 444
quercitron - *Quercus velutina* 227
quick-set-thorn - *Crataegus laevigata* 463
quick-silverweed - *Thalictrum dioicum* 457
quickbeam - *Sorbus aucuparia* 489
quickweed - *Galinsoga parviflora* 71
quince
Chinese quince - *Chaenomeles sinensis* 463
common quince - *Cydonia oblonga* 465
flowering quince - *Chaenomeles japonica* 462
Japanese quince - *Chaenomeles lagenaria* 463
quinine - *Cinchona officinalis* 491
quiverleaf - *Populus tremuloides* 499
rabbit's-foot - *Maranta leuconeura* var. *kerchoviana* 296
rabbit's-tracks - *Maranta leuconeura* var. *kerchoviana* 296
rabbitbrush
Douglas rabbitbrush - *Chrysothamnus viscidiflorus* 64
rubber rabbitbrush - *Chrysothamnus nauseosus* 64
raccoonberry - *Podophyllum peltatum* 92
radiator-plant, American - *Peperomia obtusifolia* 359
radish - *Raphanus sativus* 114
Chinese radish - *Raphanus sativus* cv. 'longipinnatus' 114
wild radish - *Raphanus raphanistrum* 114
ragged-robin - *Lychnis flos-cuculi* 131
ragi - *Eleusine coracana* 387
ragweed
common ragweed - *Ambrosia artemisiifolia* 54
giant ragweed - *Ambrosia trifida* 55
western ragweed - *Ambrosia psilostachya* 54
ragwort - *Senecio jacobaea* 82
tansy ragwort - *Senecio jacobaea* 82
rain-tree - *Albizia saman* 188
rainbow-star - *Cryptanthus bromelioides* var. *tricolor* 117
rainbow-vine - *Pellionia pulchra* 543
raisin-tree, Japanese - *Hovenia dulcis* 459
rambutan - *Nephelium lappaceum* 511
ramie - *Boehmeria nivea* 543
ramontchi - *Flacourtia indica* 240
ramp - *Allium tricoccum* 272
rampion, German - *Oenothera biennis* 319
rams-horn - *Proboscidea louisianica* 326
rapanea, Guiana - *Rapanea guianensis* 303
rape - *Brassica napus, B. napus* var. *napus* 109
bird rape - *Brassica rapa* 109
California rape - *Sinapis arvensis* 115
raspberry
American red raspberry - *Rubus strigosus* 486
black raspberry - *Rubus occidentalis* 486
dwarf red raspberry - *Rubus pubescens* 486
European red raspberry - *Rubus idaeus* 486
Mauritius raspberry - *Rubus rosifolius* 486
New Mexico raspberry - *Rubus neomexicanus* 486
purple flowering raspberry - *Rubus odoratus* 486
red raspberry - *Rubus idaeus* 486
Rocky Mountain raspberry - *Rubus deliciosus* 486

rattan
dwarf ground rattan - *Rhapis excelsa* 49
rattan-vine - *Berchemia scandens* 458
rattlebox - *Ludwigia alternifolia* 318, *Rhinanthus crista-galli* 520
rattlesnake-master - *Eryngium yuccifolium* 29, *Manfreda virginica* 18
rattlesnake-weed - *Hieracium venosum* 76
red-bead-plant - *Abrus precatorius* 187
red-buglevine - *Aeschynanthus pulcher* 243
red-cole - *Armoracia rusticana* 109
red-ivy - *Hemigraphis alternata* 8
red-robin - *Geranium robertianum* 242
red-white-and-blue-flower - *Cuphea ignea* 284
red-willow - *Cornus amomum* 143
redbay - *Persea borbonia* 269
swamp redbay - *Persea borbonia* 269
redberry - *Rhamnus crocea* 459
redbird-cactus - *Pedilanthus tithymaloides* 186
redbird-flower - *Pedilanthus tithymaloides* 186
redbud
Chinese redbud - *Cercis chinensis* 194
eastern redbud - *Cercis canadensis* 194
western redbud - *Cercis occidentalis* 194
redflame-ivy - *Hemigraphis alternata* 8
redhead - *Asclepias curassavica* 51
redmaids - *Calandrinia ciliata* 441
redroot - *Amaranthus retroflexus* 22, *Lachnanthes caroliniana* 247
redscale - *Atriplex rosea* 135
redtop - *Agrostis gigantea* 366
Idaho redtop - *Agrostis idahoensis* 366
Ross redtop - *Agrostis rossae* 366
spike redtop - *Agrostis exarata* 366
Thurber redtop - *Agrostis thurberiana* 366
redwood - *Sequoia sempervirens* 534
coast redwood - *Sequoia sempervirens* 534
dawn redwood - *Metasequoia glyptostroboides* 534
giant redwood - *Sequoiadendron giganteum* 535
Madeira redwood - *Swietenia mahagoni* 297
reed
common reed - *Phragmites australis* 408
giant reed - *Arundo donax* 372
resin-bush - *Viguiera stenoloba* 88
retaima - *Parkinsonia aculeata* 209
rhapis
fern rhapis - *Rhapis excelsa* 49
rhododendron
Canadian rhododendron - *Rhododendron canadense* 176
Carolina rhododendron - *Rhododendron carolinianum* 176
Catawba rhododendron - *Rhododendron catawbiense* 176
great rhododendron - *Rhododendron maximum* 176
honey-bell rhododendron - *Rhododendron campylocarpum* 176
Indian rhododendron - *Melastoma malabathricum* 296
Lapland rhododendron - *Rhododendron lapponicum* 176
Piedmont rhododendron - *Rhododendron minus* 176
rosebay rhododendron - *Rhododendron maximum* 176
silvery rhododendron - *Rhododendron grande* 176
Smirnow rhododendron - *Rhododendron smirnowii* 176
tree rhododendron - *Rhododendron arboreum* 176
West Coast rhododendron - *Rhododendron macrophyllum* 176
rhodora - *Rhododendron canadense* 176
rhubarb - *Rheum rhabarbarum* 439
Chinese rhubarb - *Rheum officinale* 439
monk's rhubarb - *Rumex alpinus, R. patientia* 440
mountain rhubarb - *Rumex alpinus* 440
ribbon-bush - *Homalocladium platycladum* 438
ribbon-cactus - *Pedilanthus tithymaloides* 186
ribbon-plant - *Chlorophytum comosum* 275
rice - *Oryza sativa* 402
annual wild rice - *Zizania aquatica* 433
cultivated northern wild rice - *Zizania palustris* 433
eastern wild rice - *Zizania aquatica* 433
Indian rice - *Oryzopsis hymenoides* 403, *Zizania aquatica* 433
Manchurian wild rice - *Zizania latifolia* 433
Russian wild rice - *Elymus junceus* 387
upland rice - *Oryza sativa* 402
rice-paper-plant, Chinese - *Tetrapanax papyriferum* 44
rice-tree, Formosa - *Fatsia japonica* 43
richweed - *Collinsonia canadensis* 263, *Pilea pumila* 544
ringworm-bush - *Senna alata* 215
river-beauty - *Epilobium latifolium* 317
robin-run-away - *Dalibarda repens* 465
rock-harlequin - *Corydalis sempervirens* 325
rocket, prairie - *Erysimum asperum* 112
Roman-candle - *Yucca gloriosa* 19
root
pleurisy root - *Asclepias tuberosa* 51
ropebark - *Dirca palustris* 537
rose
Arkansas rose - *Rosa arkansana* 484
baby rose - *Rosa multiflora* 484
Bengal rose - *Rosa chinensis* 484
Burnet rose - *Rosa spinosissima* 484
cabbage rose - *Rosa centifolia* 484
California wild rose - *Rosa californica* 484
Cherokee rose - *Rosa laevigata* 484
China rose - *Rosa chinensis* 484
climbing rose - *Rosa setigera* 484
cluster rose - *Rosa pisocarpa* 484
dog rose - *Rosa canina* 484
French rose - *Rosa gallica* 484
Japanese rose - *Rosa rugosa* 484
Macartney rose - *Rosa bracteata* 484
mallow rose - *Hibiscus moscheutos* 292
Manettii rose - *Rosa × noisettiana* cv. 'manettii'
memorial rose - *Rosa wichuraiana* 484
multiflora rose - *Rosa multiflora* 484
nootka rose - *Rosa nutkana* 484
nutka rose - *Rosa nutkana* 484
pasture rose - *Rosa carolina* 484
prairie rose - *Rosa arkansana, R. setigera* 484
prickly rose - *Rosa acicularis* 484
rugosa rose - *Rosa rugosa* 484
Scotch rose - *Rosa spinosissima* 484
swamp rose - *Rosa palustris* 484
tea rose - *Rosa odorata* 484
Turkestan rose - *Rosa rugosa* 484
wood rose - *Rosa gymnocarpa* 484
Wood's rose - *Rosa woodsii* 484
rose-acacia - *Robinia hispida* 214
rose-mallow, soldier - *Hibiscus laevis* 292
rose-moss - *Portulaca grandiflora* 442
rose-of-China - *Hibiscus rosa-sinensis* 292
rose-of-Sharon - *Hibiscus syriacus* 292, *Hypericum calycinum* 139
roseapple - *Syzygium jambos, S. malaccense* 307
large-fruited roseapple - *Syzygium malaccense* 307
rosebay - *Nerium oleander* 34
broadleaved rosebay - *Tabernaemontana divaricata* 35
California rosebay - *Rhododendron macrophyllum* 176
East Indian rosebay - *Tabernaemontana divaricata* 35
Lapland rosebay - *Rhododendron lapponicum* 176
mountain rosebay - *Rhododendron catawbiense* 176
roselle - *Hibiscus sabdariffa* 292
rosemary - *Rosmarinus officinalis* 266
wild rosemary - *Ledum decumbens* 174
rosewood - *Tipuana tipu* 216
Burmese rosewood - *Pterocarpus indicus* 213
Indian rosewood - *Dalbergia sissoo* 196
rosin-brush - *Baccharis sarothroides* 60
rougeplant - *Rivina humilis* 326
roundwood - *Sorbus americana* 489
rowan - *Sorbus aucuparia* 489
royal-red-bugler - *Aeschynanthus pulcher* 243
royal-vineplant - *Gynura aurantiaca* 72
rubber, Assam - *Ficus elastica* 298
rubber-tree - *Hevea brasiliensis* 185, *Schefflera actinophylla* 44
Pará rubber-tree - *Hevea brasiliensis* 185
rubberplant - *Ficus elastica* 298
Chinese rubberplant - *Crassula argentea* 145
creeping rubberplant - *Ficus pumila* 299
dwarf rubberplant - *Crassula argentea* 145
Japanese rubberplant - *Crassula argentea* 145
mistletoe rubberplant - *Ficus deltoidea* 298
small-leaved rubberplant - *Ficus benjamina* 298
rubbertree, Indian - *Ficus elastica* 298
runaway-robin - *Glechoma hederacea* 263
running-evergreen - *Lycopodium complanatum* 284
running-pop - *Passiflora foetida* 325
rush
Baltic rush - *Juncus balticus* 260
bog rush - *Juncus effusus* 260
chairmakers rush - *Scirpus americanus* 165
horned rush - *Rhynchospora corniculata* 165
Japanese mat rush - *Juncus effusus* 260
mat rush - *Scirpus lacustris* 165
ovoid spike rush - *Eleocharis ovata* 164
path rush - *Juncus tenuis* 260
salt rush - *Juncus lesueurii* 260
slender rush - *Juncus tenuis* 260
slender spike rush - *Eleocharis acicularis* 164
soft rush - *Juncus effusus* 260
toad rush - *Juncus bufonius* 260
tufted club rush - *Scirpus caespitosus* 165
wire rush - *Juncus balticus* 260
Russian-thistle - *Salsola australis, S. iberica* 137
rustyleaf - *Menziesia ferruginea* 175
rutabaga - *Brassica napus, B. napus* var. *napobrassica* 109
rye
common rye - *Secale cereale* 416
mountain rye - *Secale montanum* 416

sabicu - *Lysiloma latisiliqua* 205
sacahuista - *Nolina microcarpa* 18
sacaton, alkali - *Sporobolus airoides* 421
sacred-bamboo - *Nandina domestica* 92
safflower - *Carthamus tinctorius* 63
saffron
  bastard saffron - *Carthamus tinctorius* 63
  false saffron - *Carthamus tinctorius* 63
saffron-spike - *Aphelandra squarrosa* 7
sage
  autumn sage - *Salvia greggii* 267
  black sage - *Salvia mellifera* 267
  blue sage - *Eranthemum pulchellum* 7, *Salvia azurea* 267
  common sage - *Salvia officinalis* 267
  creeping sage - *Salvia sonomensis* 267
  garden sage - *Salvia officinalis* 267
  gray-ball sage - *Salvia dorrii* 267
  Jerusalem sage - *Phlomis fruticosa* 266
  mealy-cup sage - *Salvia farinacea* 267
  pitcher sage - *Salvia spathacea* 267
  scarlet sage - *Salvia coccinea, S. splendens* 267
  Texas sage - *Salvia coccinea* 267
  white sage - *Artemisia gnaphalodes* 57, *A. ludoviciana* 57, *Salvia apiana* 267
  wood sage - *Teucrium canadense* 268
sage-tree - *Vitex agnus-castus* 547
sagebrush
  basin sagebrush - *Artemisia tridentata* 57
  big sagebrush - *Artemisia tridentata* 57
  California sagebrush - *Artemisia californica* 57
  common sagebrush - *Artemisia tridentata* 57
  fringed sagebrush - *Artemisia frigida* 57
  Louisiana sagebrush - *Artemisia ludoviciana* 57
  low sagebrush - *Artemisia arbuscula* 57
  silver sagebrush - *Artemisia cana* 57
sago, queen - *Cycas circinalis* 158
sago-cycas - *Zamia pumila* 553
sago-palm - *Cycas circinalis, C. revoluta* 158
  Japanese sago-palm - *Cycas revoluta* 158
sagrada, cascara - *Rhamnus purshiana* 459
saguaro - *Carnegiea gigantea* 118
  giant saguaro - *Carnegiea gigantea* 118
sahuaro - *Carnegiea gigantea* 118
sailor-caps - *Dodecatheon hendersonii* 443
sainfoin - *Onobrychis viciifolia* 209
salal - *Gaultheria shallon* 172
saligot - *Trapa natans* 538
salmonberry - *Rubus chamaemorus, R. parviflorus, R. spectabilis* 486
salsify - *Tragopogon porrifolius* 87
  black salsify - *Scorzonera hispanica* 82
  meadow salsify - *Tragopogon pratensis* 87
  Spanish salsify - *Scorzonera hispanica* 82
  western salsify - *Tragopogon dubius* 87
saltbush
  Australian saltbush - *Atriplex semibaccata* 135
  berry saltbush - *Atriplex semibaccata* 135
  desert saltbush - *Atriplex polycarpa* 135
  four-wing saltbush - *Atriplex canescens* 134
  spiny saltbush - *Atriplex confertifolia* 134
saltwort - *Salsola iberica* 137
saman - *Albizia saman* 188
samphire - *Salicornia europaea* 137
sand-jack - *Quercus incana* 227
sandberry - *Arctostaphylos uva-ursi* 171
sandbox-tree - *Hura crepitans* 185
sandbur
  coast sandbur - *Cenchrus incertus* 381
  dune sandbur - *Cenchrus tribuloides* 381
  field sandbur - *Cenchrus incertus* 381
  southern sandbur - *Cenchrus echinatus* 381
sandlewood, bastard - *Myoporum sandwicense* 301
sandreed, prairie - *Calamovilfa longifolia* 380
sandspurry, red - *Spergularia rubra* 132
sandverbena, yellow - *Abronia latifolia* 308
sandvine - *Cynanchum laeve* 52
sandwort
  mountain sandwort - *Arenaria groenlandica* 129
  rock sandwort - *Arenaria stricta* 129
  seabeach sandwort - *Honkenya peploides* 130
sanguinary - *Achillea millefolium* 53
sanicle
  purple sanicle - *Sanicula bipinnatifida* 32
  white sanicle - *Eupatorium rugosum* 70
santolina, green - *Santolina virens* 82
sapodilla - *Manilkara zapota* 512
  wild sapodilla - *Manilkara jaimiqui* subsp. *emarginata* 512
sapota - *Pouteria sapota* 512
sapote - *Pouteria sapota* 512
  mammee sapote - *Pouteria sapota* 512
  white sapote - *Casimiroa edulis* 494
sapote-amarillo - *Pouteria campechiana* 512
sapotilla - *Manilkara zapota* 512
sarsaparilla
  bristly sarsaparilla - *Aralia hispida* 42
  wild sarsaparilla - *Aralia nudicaulis* 42
sassafras, white - *Sassafras albidum* 270
satinflower - *Clarkia amoena* 316
satinwood - *Murraya paniculata* 497
sausage-tree - *Kigelia africana* 105
savin - *Juniperus sabina, J. sabina* cv. 'tamariscifolia' 153
sawbrier - *Smilax glauca* 523
saxifraga, swamp - *Saxifraga pensylvanica* 514
saxifrage
  alpine-brook saxifrage - *Saxifraga rivularis* 514
  early saxifrage - *Saxifraga virginiensis* 514
scabious
  sweet scabious - *Erigeron annuus* 69, *Scabiosa atropurpurea* 167
scarlet-lightening - *Lychnis chalcedonica* 131
scarletbush - *Hamelia patens* 493
schnittlauch - *Allium schoenoprasum* 272
scoke - *Phytolacca americana* 326
Scotch-attorney - *Clusia rosea* 138
Scotch-broom - *Cytisus scoparius* 195
scouring-rush
  common scouring-rush - *Equisetum hyemale* 169
  variegated scouring-rush - *Equisetum variegatum* 169
screw-pine
  common screw-pine - *Pandanus utilis* 324
  thatch screw-pine - *Pandanus tectorius* 324
  Veitch screw-pine - *Pandanus veitchii* 324
scupernong - *Vitis rotundifolia* 550
sea-ash - *Zanthoxylum clava-herculis* 498
sea-fig - *Carpobrotus chilensis* 20
sea-grape - *Coccoloba uvifera* 437
sea-lovage - *Ligusticum scothicum* 30
sea-milkwort - *Glaux maritima* 443
sea-myrtle - *Baccharis halimifolia* 60
sea-oats - *Uniola paniculata* 428
sea-purslane - *Honkenya peploides* 130
sea-rocket, American - *Cakile edentula* 111
sedge
  awn-fruited sedge - *Carex stipata* 160
  black sedge - *Carex atratiformis* 159
  black-edged sedge - *Carex nigromarginata* var. *elliptica* 159
  bladder sedge - *Carex intumescens* 159
  blunt-broom sedge - *Carex tribuloides* 160
  bristle-stalked sedge - *Carex leptalea* 159
  bristly sedge - *Carex comosa* 159
  browned sedge - *Carex adusta* 158
  brownish sedge - *Carex brunnescens* 159
  creeping sedge - *Carex chordorrhiza* 159
  crested sedge - *Carex cristatella* 159
  drooping wood sedge - *Carex arctata* 158
  dry-spiked sedge - *Carex siccata* 160
  elk sedge - *Carex geyeri* 159
  false nut sedge - *Cyperus strigosus* 163
  few-flowered sedge - *Carex pauciflora* 159
  few-seeded sedge - *Carex oligosperma* 159
  field sedge - *Carex conoidea* 159
  fox sedge - *Carex vulpinoidea* 160
  foxtail sedge - *Carex alopecoidea* 158
  Fraser's sedge - *Cymophyllus fraseri* 163
  fringe sedge - *Carex crinita* 159
  golden-fruited sedge - *Carex aurea* 159
  graceful sedge - *Carex gracillima* 159
  green sedge - *Carex oederi* 159
  hay sedge - *Carex foenea* 159
  heavy sedge - *Carex gravida* 159
  hillside sedge - *Carex siccata* 160
  hop sedge - *Carex lupulina* 159
  hop-like sedge - *Carex lupuliformis* 159
  inland sedge - *Carex interior* 159
  lenticular sedge - *Carex lenticularis* 159
  lesser prickly sedge - *Carex muricata* 159
  little prickly sedge - *Carex sterilis* 160
  long-awned arctic sedge - *Carex podocarpa* 159
  long-stalked sedge - *Carex pedunculata* 159
  loose-flowered alpine sedge - *Carex rariflora* 160
  loose-flowered sedge - *Carex laxiflora* 159
  low northern sedge - *Carex concinnoides* 159
  marsh straw sedge - *Carex tenera* 160
  mud sedge - *Carex limosa* 159
  northern clustered sedge - *Carex arcta* 158
  northern sedge - *Carex deflexa* 159
  oval-headed sedge - *Carex cephalophora* 159
  plantain-leaved sedge - *Carex plantaginea* 159
  pointed-broom sedge - *Carex scoparia* 160
  porcupine sedge - *Carex hystricina* 159
  retrorse sedge - *Carex retrorsa* 160
  riverbank sedge - *Carex riparia* 160
  rough sedge - *Carex scabrata* 160
  russet sedge - *Carex saxatilis* 160
  sallow sedge - *Carex lurida* 159
  silvery sedge - *Carex canescens* 159
  stellate sedge - *Carex rosea* 160
  straw sedge - *Carex straminea* 160
  summer sedge - *Carex aestivalis* 158
  thicket sedge - *Carex platyphylla* 159
  thin leaved sedge - *Carex cephaloidea* 159
  threadleaf sedge - *Carex filifolia* 159
  three-fruited sedge - *Carex trisperma* 160
  tussock sedge - *Carex stricta* 160
  twisted sedge - *Carex torta* 160
  umbrella sedge - *Cyperus alternifolius* 163
  velvet sedge - *Carex vestita* 160
  water sedge - *Carex aquatilis* 158
  weak arctic sedge - *Carex supina* 160
  white bear sedge - *Carex albursina* 158
  wood sedge - *Carex tetanica* 160
  woolly sedge - *Carex lanuginosa* 159
  yellow nut sedge - *Cyperus esculentus* 163
seedbox - *Ludwigia alternifolia* 318

seep-willow - *Baccharis glutinosa* 60
sego-lily - *Calochortus nuttallii* 274
self-heal - *Prunella vulgaris* 266
seminole-bread - *Zamia pumila* 553
senecio, bush - *Senecio douglasii* 82
senita - *Lophocereus schottii* 119
senna
  Alexandrian senna - *Senna alexandrina* 215
  bladder senna - *Colutea arborescens* 195
  candlestick senna - *Senna alata* 215
  coffee senna - *Senna occidentalis* 215
  desert senna - *Senna covesii* 215
  Indian senna - *Senna alexandrina* 215
  sickle senna - *Cassia tora* 193
  wild senna - *Senna marilandica* 215
  wormwood senna - *Cassia artemisioides* 193
sensitive-plant, wild - *Cassia nictitans* 193
sequoia, giant - *Sequoiadendron giganteum* 535
serpent-cucumber - *Trichosanthes anguina* 151
serradella - *Ornithopus sativus* 209
serviceberry
  Allegheny serviceberry - *Amelanchier laevis* 461
  cluster serviceberry - *Amelanchier pumila* 461
  Cusick serviceberry - *Amelanchier alnifolia* var. *cusickii* 461
  downy serviceberry - *Amelanchier arborea*, *A. canadensis* 461
  Saskatoon serviceberry - *Amelanchier alnifolia* 461
  western serviceberry - *Amelanchier alnifolia* 461
sesame - *Sesamum indicum* 326
sevenbark - *Hydrangea arborescens* 250
shaddock - *Citrus maxima* 494
shadscale - *Atriplex confertifolia* 134
shallon - *Gaultheria shallon* 172
shallot - *Allium ascalonicum*, *A. cepa* 272
shamrock
  Irish shamrock - *Oxalis acetosella* 323, *Trifolium procumbens* 217
shaving-brush-tree - *Pseudobombax ellipticum* 106
Shawnee-salad - *Hydrophyllum virginianum* 252
she-oak
  drooping she-oak - *Casuarina stricta* 132
  river she-oak - *Casuarina cunninghamiana* 132
sheep-fat - *Atriplex confertifolia* 134
sheepberry - *Viburnum lentago*, *V. prunifolium* 127
shellflower - *Alpinia zerumbet* 553, *Moluccella laevis* 265
shepherd's-clock - *Anagallis arvensis* 442
shepherd's-purse - *Capsella bursa-pastoris* 111
shieldfern, marginal - *Dryopteris marginalis* 447
shinleaf - *Pyrola elliptica* 451
  one-flowered shinleaf - *Moneses uniflora* 451
  white-veined shinleaf - *Pyrola picta* 451
shittimwood - *Halesia carolina* 533, *Sideroxylon lanuginosa* 512
shoo-fly-plant - *Nicandra physalodes* 528
shooting-star - *Dodecatheon conjugens* 443
  alpine shooting-star - *Dodecatheon alpinum* 443
shower-tree - *Cassia artemisioides* 193
shrimp-bush - *Justicia brandegeana* 8
shrimp-plant - *Justicia brandegeana* 8
sicklepod - *Cassia tora* 193, *Senna obtusifolia* 215
sida
  alkali sida - *Malvella leprosa* 294
  arrowleaf sida - *Sida rhombifolia* 295
  prickly sida - *Sida spinosa* 295
  southern sida - *Sida acuta* 294
side-saddle-flower - *Sarracenia purpurea* 512
silk-cotton-tree - *Ceiba pentandra* 106
  red silk-cotton-tree - *Bombax ceiba* 105
  white silk-cotton-tree - *Ceiba pentandra* 106
silk-oak - *Grevillea robusta* 444
silk-plant, Chinese - *Boehmeria nivea* 543
silver-dollar - *Crassula arborescens* 145, *Lunaria annua* 113
silver-dollar-tree - *Eucalyptus cinerea* 303
silver-rod - *Solidago bicolor* 84
silver-threads - *Fittonia verschaffeltii* 7
silver-tree - *Conocarpus erectus* var. *sericeus* 139
silverberry - *Elaeagnus angustifolia*, *E. commutata* 168
silverbush - *Convolvulus cneorum* 141, *Sophora tomentosa* 215
silverfir, Pacific - *Abies amabilis* 326
silverling - *Baccharis halimifolia* 60
silvernet-plant - *Fittonia verschaffeltii* 7
simpler's-joy - *Verbena hastata* 546
sing-kwa - *Luffa acutangula* 150
siratro - *Macroptilium atropurpureum* 205
siris-tree - *Albizia lebbeck* 188
sisal - *Agave sisalana* 17
sissoo - *Dalbergia sissoo* 196
sisu - *Dalbergia sissoo* 196
skeels - *Sophora davidii* 215
skullcap, mad-dog - *Scutellaria lateriflora* 267
skunk-cabbage - *Symplocarpus foetidus* 41
  western skunk-cabbage - *Lysichiton americanum* 40
  yellow skunk-cabbage - *Lysichiton americanum* 40
skunkbush - *Rhus trilobata* 24
skyflower - *Duranta repens* 545
  Brazilian skyflower - *Duranta repens* 545
slipper-flower - *Pedilanthus tithymaloides* 186
slipper-plant - *Pedilanthus tithymaloides* 186
sloe - *Prunus alleghaniensis*, *P. americana*, *P. spinosa* 475
smallwood - *Helianthus microcephalus* 73
smartweed
  common smartweed - *Polygonum hydropiper* 438
  dotted smartweed - *Polygonum punctatum* 438
  marshpepper smartweed - *Polygonum hydropiper* 438
  Pennsylvania smartweed - *Polygonum pensylvanicum* 438
  swamp smartweed - *Polygonum coccineum* 438
  water smartweed - *Polygonum amphibium* 438
smoke-tree
  American smoke-tree - *Cotinus obovatus* 23
  common smoke-tree - *Cotinus coggygria* 23
smokebush - *Cotinus coggygria* 23
smokeweed - *Eupatorium maculatum* 70
snailseed - *Coccoloba diversifolia* 437, *Cocculus carolinus* 297
snake-plant - *Sansevieria trifasciata* var. *laurentii* 19
snake-root, button - *Eryngium yuccifolium* 29
snakeberry - *Actaea rubra* 452
snakehead - *Chelone glabra* 517
snakeroot - *Asarum canadense* 51
  black snakeroot - *Cimicifuga racemosa* 454
  broom snakeroot - *Gutierrezia sarothrae* 72
  Seneca snakeroot - *Polygala senega* 436
  Virginia snakeroot - *Aristolochia serpentaria* 51
  white snakeroot - *Eupatorium rugosum* 70
snakeweed - *Polygonum bistortoides* 438
  broom snakeweed - *Gutierrezia sarothrae* 72
snapdragon
  common snapdragon - *Antirrhinum majus* 515
  garden snapdragon - *Antirrhinum majus* 515
  pale snapdragon - *Impatiens pallida* 90
  wild snapdragon - *Linaria vulgaris* 518
sneezeweed - *Achillea ptarmica* 53, *Helenium autumnale* 73,, *H. microcephalum* 73
  bitter sneezeweed - *Helenium amarum* 73
sneezewort - *Achillea ptarmica* 53
snow-in-summer - *Cerastium tomentosum* 129
snow-on-the-mountain - *Euphorbia marginata* 184
snowball, wild - *Ceanothus americanus* 458
snowball-bush - *Viburnum opulus*, *V. opulus* var. *roseum* 127
snowbell
  American snowbell - *Styrax americana* 533
  bigleaf snowbell - *Styrax grandifolia* 533
snowberry - *Chiococca alba* 491
  common snowberry - *Symphoricarpos albus*, *S. albus* var. *laevigatus* 126
  creeping snowberry - *Gaultheria hispidula* 172
snowbush - *Ceanothus cordulatus* 458
soap-plant - *Chlorogalum pomeridianum* 275
soap-tree, Hawaiian - *Sapindus oahuensis* 511
soapberry - *Shepherdia canadensis* 169
  southern soapberry - *Sapindus saponaria* 511
  western soapberry - *Sapindus saponaria* var. *drummondii* 511
  wingleaf soapberry - *Sapindus saponaria* 511
soapbush - *Clidemia hirta* 296
soaptree - *Yucca elata* 19
soapweed - *Yucca elata*, *Y. glauca* 19
soapwell - *Yucca glauca* 19
soapwort - *Saponaria officinalis* 131
  cow soapwort - *Vaccaria hispanica* 132
soft-flag - *Typha angustifolia* 538
sojabean - *Glycine max* 198
soldier's-cap - *Aconitum napellus* 451
Solomon, fat - *Smilacina racemosa* 280
Solomon's-seal, two-leaved - *Maianthemum canadense* 279
Solomon's-zigzag - *Smilacina racemosa* 280
sophora
  silky sophora - *Sophora sericea* 215
  vetchleaf sophora - *Sophora davidii* 215
sorghum - *Sorghum bicolor* 419
  grain sorghum - *Sorghum bicolor* 419
  sweet sorghum - *Sorghum bicolor* 419
sorrel
  common sorrel - *Rumex acetosella* 440
  European wood sorrel - *Oxalis acetosella* 323
  garden sorrel - *Rumex acetosa* 440
  Indian sorrel - *Hibiscus sabdariffa* 292
  Jamaican sorrel - *Hibiscus sabdariffa* 292
  red sorrel - *Rumex acetosella* 440
  sheep sorrel - *Rumex acetosella* 440

sorreltree - *Oxydendrum arboreum* 175
sour-clover - *Melilotus indica* 207
sour-greens - *Rumex venosus* 440
sourberry - *Rhus integrifolia* 24
sourwood - *Oxydendrum arboreum* 175
southernwood - *Artemisia abrotanum* 57
sowthistle
  annual sowthistle - *Sonchus oleraceus* 86
  common sowthistle - *Sonchus oleraceus* 86
  perennial sowthistle - *Sonchus arvensis* 86
  spiny sowthistle - *Sonchus asper* 86
  spiny-leaved sowthistle - *Sonchus asper* 86
soybean - *Glycine max* 198
Spanish-bayonet - *Yucca aloifolia, Y. baccata* 19
Spanish-broom - *Spartium junceum* 216
Spanish-dagger - *Yucca gloriosa* 19
Spanish-moss - *Tillandsia usneoides* 117
Spanish-needles - *Bidens bipinnata, B. pilosa* 61
spatterdock, common - *Nuphar advena* 309
spearmint - *Mentha spicata* 264
  Scotch spearmint - *Mentha* x *gentilis* 264
speedwell
  beach speedwell - *Veronica longifolia* 521
  common speedwell - *Veronica officinalis* 521
  corn speedwell - *Veronica arvensis* 521
  marsh speedwell - *Veronica scutellata* 521
  purslane speedwell - *Veronica peregrina* 521
  showy speedwell - *Hebe speciosa* 517
  spike speedwell - *Veronica spicata* 521
  thyme-leaved speedwell - *Veronica serpyllifolia* 521
spelt - *Triticum spelta* 425
spice, Indian - *Vitex agnus-castus* 547
spice-lily - *Manfreda maculosa* 18
spiceberry - *Ardisia crenata* 302
spicebush - *Lindera benzoin* 269
spider-ivy - *Chlorophytum comosum* 275
spider-plant - *Chlorophytum comosum* 275
spiderflower, spiny - *Cleome spinosa* 123
spiderwort
  common spiderwort - *Tradescantia virginiana* 140
  purple-leaved spiderwort - *Tradescantia spathacea* 140
spike-moss, rock - *Selaginella rupestris* 522
spikenard - *Aralia cordata* 42
  American spikenard - *Aralia racemosa* 42
  false spikenard - *Smilacina racemosa* 280
spinach - *Spinacia oleracea* 137
  Indian spinach - *Basella alba* 90
  New Zealand spinach - *Tetragonia tetragonioides* 20
  wild spinach - *Chenopodium bonus-henricus* 136
spindle-tree
  European spindle-tree - *Euonymus europaeus* 133
  Japanese spindle-tree - *Euonymus japonica* 133
  winged spindle-tree - *Euonymus alatus* 133
spiraea, rock - *Holodiscus dumosus* 468
spiral-flag - *Costus malortieanus* 553
spirea
  Japanese spirea - *Spiraea japonica* 489
  rock spirea - *Holodiscus discolor* 468
spleenwort, silvery - *Athyrium thelypteroides* 445
sponge, vegetable - *Luffa aegyptiaca* 150
spongetree - *Acacia farnesiana* 187
spoonflower - *Dasylirion wheeleri* 18, *Peltandra sagittifolia* 40
spoonwood - *Kalmia latifolia* 173
spotted-hemlock - *Conium maculatum* 28
sprangletop
  green sprangletop - *Leptochloa dubia* 398
  red sprangletop - *Leptochloa filiformis* 398
spring-beauty - *Claytonia megarrhiza, C. virginica* 441
spring-gold - *Lomatium utriculatum* 30
spring-snowflake - *Leucojum vernum* 277
springvetch - *Vicia sativa* 219
sprouting-leaf - *Kalanchoe pinnata* 146
sprouts, brussel - *Brassica oleracea* var. *gemmifera* 109
spruce
  Alberta spruce - *Picea glauca* var. *albertiana* 334
  big-cone spruce - *Pseudotsuga macrocarpa* 352
  black spruce - *Picea mariana* 334
  bog spruce - *Picea mariana* 334
  cat spruce - *Picea glauca* 334
  Colorado blue spruce - *Picea pungens* 334
  Colorado spruce - *Picea pungens* 334
  double spruce - *Picea mariana* 334
  Engelmann spruce - *Picea engelmannii* 334
  Norway spruce - *Picea abies* 334
  red spruce - *Picea rubens* 334
  Sitka spruce - *Picea sitchensis* 334
  white spruce - *Picea glauca* 334
spurge
  Alleghany spurge - *Pachysandra procumbens* 118
  caper spurge - *Euphorbia lathyris* 184
  cypress spurge - *Euphorbia cyparissias* 184
  flowering spurge - *Euphorbia corollata* 184
  garden spurge - *Euphorbia hirta* 184
  gopher spurge - *Euphorbia lathyris* 184
  Japanese spurge - *Pachysandra terminalis* 118
  leafy spurge - *Euphorbia esula* 184
  myrtle spurge - *Euphorbia lathyris* 184
  nodding spurge - *Euphorbia maculata* 184
  painted spurge - *Euphorbia heterophylla* 184
  petty spurge - *Euphorbia peplus* 184
  round-leaved spurge - *Euphorbia serpens* 184
  seaside spurge - *Euphorbia polygonifolia* 184
  spotted spurge - *Euphorbia maculata* 184
  thyme-leaved spurge - *Euphorbia serpyllifolia* 184
  toothed spurge - *Euphorbia dentata* 184
  tramp's spurge - *Euphorbia corollata* 184
spurry, corn - *Spergula arvensis* 132
squarenut - *Carya tomentosa* 255
squash
  autumn squash - *Cucurbita maxima* 149
  bush squash - *Cucurbita pepo* var. *melopepo* 149
  crookneck squash - *Cucurbita moschata* 149
  summer squash - *Cucurbita pepo* 149
  winter squash - *Cucurbita maxima, C. moschata, C. pepo* 149
squashberry - *Viburnum edule* 127
squaw-apple - *Peraphyllum ramosissimum* 473
squawberry - *Mitchella repens* 493
squawbush - *Condalia spathulata* 459
squawroot - *Perideridia gairdneri* 31
squill
  blue-flowered squill - *Scilla hispanica* 280
  Siberian squill - *Scilla sibirica* 280
squirrel-corn - *Dicentra canadensis* 240
squirreltail - *Sitanion hystrix* 418
St.-Andrew's-cross - *Hypericum hypericoides* 139
St.-Catharine's-lace - *Eriogonum giganteum* 437
St.-John's-bread - *Ceratonia siliqua* 193
St.-John's-wort
  Blue Ridge St.-John's-wort - *Hypericum mitchellianum* 139
  common St.-John's-wort - *Hypericum perforatum* 139
  creeping St.-John's-wort - *Hypericum calycinum* 139
  kalms St.-John's-wort - *Hypericum kalmianum* 139
  marsh St.-John's-wort - *Hypericum virginicum* 139
  shrubby St.-John's-wort - *Hypericum prolificum* 139
  spotted St.-John's-wort - *Hypericum punctatum* 139
St.-Peter's-wort - *Hypericum stans* 139
staff-vine - *Celastrus scandens* 133
stagbush - *Viburnum prunifolium* 127
staggerbrush - *Lyonia mariana* 174
staghorn-evergreen - *Lycopodium clavatum* 284
star-cluster - *Pentas lanceolata* 493
  Egyptian star-cluster - *Pentas lanceolata* 493
star-cucumber - *Sicyos angulatus* 150
star-glory - *Ipomoea quamoclit* 142
star-jasmine - *Trachelospermum jasminoides* 35
star-of-Bethlehem - *Ornithogalum umbellatum* 280
star-tulip, Sierra - *Calochortus nudus* 274
starfish-plant - *Cryptanthus acaulis* 116
starflower - *Smilacina stellata* 280, *Trientalis borealis* 444
starfruit - *Averrhoa carambola* 323
starleaf - *Schefflera actinophylla* 44
starwort
  common starwort - *Stellaria longipes* 132
  lesser starwort - *Stellaria graminea* 132
  little starwort - *Stellaria graminea* 132
steeplebush - *Spiraea tomentosa* 489
stepladder-plant - *Costus malortieanus* 553
stevia - *Piqueria trinervia* 81
stickseed - *Lappula redowskii* 107
sticky-cockle - *Silene noctiflora* 131
stinking-ash - *Ptelea trifoliata* 498
stinking-Benjamin - *Trillium erectum* 281
stinking-cedar - *Torreya taxifolia* 534
stinkweed - *Thlaspi arvense* 115
stitchwort - *Stellaria media* 132
  lesser stitchwort - *Stellaria graminea* 132
stock
  Brampton's stock - *Matthiola incana* 114
  common stock - *Matthiola incana* 114
  evening-scented stock - *Matthiola bicornis* 114
  imperial stock - *Matthiola incana* 114
  ten-weeks stock - *Matthiola incana* var. *annua* 114
stoneroot - *Collinsonia canadensis* 263
strand-wheat - *Lolium perenne* 399
strangler-vine - *Morrenia odorata* 53
strawberry
  alpine strawberry - *Fragaria vesca* 466
  barren strawberry - *Waldsteinia fragarioides* 490
  beach strawberry - *Fragaria chiloensis* 466
  Chilean strawberry - *Fragaria chiloensis* 466
  chiloe strawberry - *Fragaria chiloensis* 466

cultivated strawberry - *Fragaria × ananassa* 466
European strawberry - *Fragaria vesca* 466
false strawberry - *Duchesnea indica* 465
garden strawberry - *Fragaria × ananassa* 466
Indian strawberry - *Duchesnea indica* 465
sow-teat strawberry - *Fragaria vesca* 466
Virginia strawberry - *Fragaria virginiana* 466
wild strawberry - *Fragaria virginiana* 466
woodland strawberry - *Fragaria vesca, F. vesca* subsp. *americana* 466
strawberry-begonia - *Saxifraga stolonifera* 514
strawberry-blite - *Chenopodium capitatum* 136
strawberry-bush - *Euonymus americanus* 133
strawberry-geranium - *Saxifraga stolonifera* 514
strawberry-shrub - *Calycanthus floridus* 121
strawberry-tree - *Arbutus unedo* 170
strawflower - *Helichrysum bracteatum* 75
stringybark
mealy stringybark - *Eucalyptus cinerea* 303
messmate stringybark - *Eucalyptus obliqua* 303
stylo
Brazilian stylo - *Stylosanthes guianensis* 216
Caribbean stylo - *Stylosanthes hamata* 216
Nigerian stylo - *Stylosanthes erecta* 216
Townsville stylo - *Stylosanthes humilis* 216
stypticweed - *Senna occidentalis* 215
subclover - *Trifolium subterraneum* 217
succory - *Cichorium intybus* 65
sugar-apple - *Annona squamosa* 26
sugar-scoop - *Tiarella unifoliata* 515
sugar-tree - *Acer barbatum* 9
sugarbeet - *Beta vulgaris* 135
sugarberry - *Celtis laevigata, C. occidentalis* 539
sugarcane - *Saccharum officinarum* 413
sulphur-flower - *Eriogonum umbellatum* 437
sultana - *Impatiens wallerana* 90
sumac
Chinese sumac - *Rhus chinensis* 24
dwarf sumac - *Rhus copallina* 24
evergreen sumac - *Rhus sempervirens, R. virens* 24
flame-tree sumac - *Rhus copallina* 24
fragrant sumac - *Rhus aromatica* 24
laurel sumac - *Rhus laurina* 24
lemonade sumac - *Rhus integrifolia* 24
mountain sumac - *Rhus copallina* 24
poison sumac - *Rhus vernix* 24
scarlet sumac - *Rhus glabra* 24
shining sumac - *Rhus copallina* 24
smooth sumac - *Rhus glabra* 24
staghorn sumac - *Rhus typhina* 24
sugar sumac - *Rhus ovata* 24
sugarbush sumac - *Rhus ovata* 24
swamp sumac - *Rhus vernix* 24
sweet-scented sumac - *Rhus aromatica* 24
tobacco sumac - *Rhus virens* 24
velvet sumac - *Rhus typhina* 24
Virginian sumac - *Rhus typhina* 24
wing-rib sumac - *Rhus copallina* 24
winged sumac - *Rhus copallina* 24
summer-azalea - *Pelargonium* x *domesticum* 243
summer-cypress - *Bassia scoparia* 135
summer-haw - *Crataegus flava* 463
summer-lilac - *Buddleja davidii* 284
summer-snowflake - *Ornithogalum umbellatum* 280
summer-sweet - *Clethra alnifolia* 138
summerberry - *Viburnum trilobum* 127
sun-rose - *Helianthemum nummularium* 138
sunchoke - *Helianthus tuberosus* 73
suncup - *Camissonia bistorta, C. subacaulis* 316
sundrops - *Oenothera fruticosa* 319
sunflower
ashy sunflower - *Helianthus mollis* 73
beach sunflower - *Helianthus debilis* 73
cheerful sunflower - *Helianthus × laetiflorus* 73
common sunflower - *Helianthus annuus* 73
dark-eye sunflower - *Helianthus atrorubens* 73
dark-red sunflower - *Helianthus atrorubens* 73
different-leaf sunflower - *Helianthus heterophyllus* 73
divergent sunflower - *Helianthus divaricatus* 73
giant sunflower - *Helianthus decapetalus, H. giganteus* 73
hairy sunflower - *Helianthus heterophyllus* 73
hairywood sunflower - *Helianthus atrorubens* 73
Hopi sunflower - *Helianthus annuus* 73
Kansas sunflower - *Helianthus petiolaris* 73
Maximilian's sunflower - *Helianthus maximilianii* 73
narrow-leaf sunflower - *Helianthus angustifolius* 73
Nuttall's sunflower - *Helianthus nuttallii* 73
Oregon sunflower - *Balsamorhiza sagittata* 61
petioled sunflower - *Helianthus petiolaris* 73
prairie sunflower - *Helianthus petiolaris* 73
rough sunflower - *Helianthus hirsutus* 73
Sawtooth sunflower - *Helianthus grosseserratus* 73
showy sunflower - *Helianthus × laetiflorus* 73
silver-leaved sunflower - *Helianthus argophyllus* 73
small-headed sunflower - *Helianthus microcephalus* 73
soft sunflower - *Helianthus mollis* 73
stiff sunflower - *Helianthus rigidus* 73
stiff-haired sunflower - *Helianthus hirsutus* 73
swamp sunflower - *Helianthus angustifolius* 73
swollen sunflower - *Helianthus strumosus* 73
ten-petals sunflower - *Helianthus decapetalus* 73
thin-leaf sunflower - *Helianthus decapetalus* 73
weak sunflower - *Helianthus debilis* 73
western sunflower - *Helianthus occidentalis* 73
woodland sunflower - *Helianthus decapetalus, H. strumosus, H. tuberosus* 73
sunplant - *Portulaca grandiflora* 442
supplejack - *Berchemia scandens* 458
Alabama supplejack - *Berchemia scandens* 458
Surinam-cherry - *Eugenia uniflora* 305
swamp-lily - *Saururus cernuus* 513
swamp-mahogany - *Eucalyptus robusta* 303
swamp-rose - *Hibiscus moscheutos* subsp. *palustris* 292
swan-potato - *Sagittaria sagittifolia* 21
swede - *Brassica napus* var. *napobrassica* 109
sweet-bay - *Persea borbonia* 269
sweet-bells - *Leucothoe racemosa* 174
sweet-clover
Indian sweet-clover - *Melilotus indica* 207
white sweet-clover - *Melilotus alba* 207
yellow sweet-clover - *Melilotus officinalis* 207
sweet-fern - *Comptonia peregrina* 302
sweet-flag - *Acorus calamus* 38
sweet-hurts - *Vaccinium angustifolium* 179
sweet-potato-tree - *Manihot esculenta* 186
sweet-rocket - *Hesperis matronalis* 113
sweet-William - *Dianthus barbatus* 130
wild sweet-William - *Phlox divaricata, P. maculata* 435
sweetbay - *Laurus nobilis* 269
sweetbrier - *Rosa eglanteria* 484
sweetfern - *Comptonia peregrina* var. *asplenifolia* 302
sweetgale - *Myrica gale* 302
sweetgum, American - *Liquidambar styraciflua* 247
sweetleaf
common sweetleaf - *Symplocos tinctoria* 533
sapphire-berry sweetleaf - *Symplocos tinctoria* 533
sweetpotato - *Ipomoea batatas* 142
sweetpotato-vine, wild - *Ipomoea pandurata* 142
sweetshrub, California - *Calycanthus occidentalis* 121
sweetsop - *Annona squamosa* 26
sweetspire - *Itea virginica* 245
Swiss-cheese-plant - *Monstera deliciosa* 40
swordfern, western - *Polystichum munitum* 449
sycamore
American sycamore - *Platanus occidentalis* 360
eastern sycamore - *Platanus occidentalis* 360
tacamahac - *Populus balsamifera* 499
talewort - *Borago officinalis* 106
tallow-tree - *Sapium sebiferum* 186
Chinese tallow-tree - *Sapium sebiferum* 186
tamanu - *Calophyllum inophyllum* 138
tamarack - *Larix laricina* 332
tamarind, wild - *Lysiloma latisiliqua* 205
tamarisk, athel - *Tamarix aphylla* 533
tamarix, Kashgar - *Tamarix hispida* 533
tamono - *Calophyllum inophyllum* 138
tampala - *Amaranthus tricolor* 22
tangelfoot - *Viburnum alnifolium* 127
tangelo - *Citrus* x *tangelo* 495
tangerine - *Citrus reticulata* 495
tanglehead - *Heteropogon contortus* 394
sweet tanglehead - *Heteropogon melanocarpus* 394
tansy, common - *Tanacetum vulgare* 87
tapa-cloth-tree - *Broussonetia papyrifera* 298
tapeworm-plant - *Homalocladium platycladum* 438
tapioca-plant - *Manihot esculenta* 186
tare - *Vicia sativa* 219
taro - *Colocasia esculenta* 39
Chinese taro - *Alocasia cucullata* 38
taro-vine - *Epipremnum aureum* 40
tarragon - *Artemisia dracunculus* 57
tarweed - *Cuphea viscosissima* 284
Chilean tarweed - *Madia sativa* 79
coast tarweed - *Madia sativa* 79
common tarweed - *Madia elegans* 79

tarwi - *Lupinus mutabilis* 204
tasi - *Morrenia odorata* 53
tassel-flower - *Amaranthus caudatus* 22, *Brickellia grandiflora* 62, *Emilia javanica* 68
red tassel-flower - *Emilia sonchifolia* 68
tassel-hyacinth - *Muscari comosum* 279
tassel-rue - *Trautvetteria carolinensis* 457
tassel-white - *Itea virginica* 245
tea
Abyssinian tea - *Catha edulis* 133
Arabian tea - *Catha edulis* 133
Labrador tea - *Ledum groenlandicum* 174
Mexican tea - *Chenopodium ambrosioides* 136
New Jersey tea - *Ceanothus americanus* 458
Oswego tea - *Monarda didyma* 265
Somali tea - *Catha edulis* 133
Spanish tea - *Chenopodium ambrosioides* 136
tea-plant - *Camellia sinensis* 535, *Viburnum lentago* 127
tea-tree
broad-leaved tea-tree - *Melaleuca quinquenervia* 305
New Zealand tea-tree - *Leptospermum scoparium* 305
river tea-tree - *Melaleuca leucadendra* 305
swamp tea-tree - *Melaleuca quinquenervia* 305
weeping tea-tree - *Melaleuca leucadendra* 305
teaberry - *Gaultheria procumbens* 172, *Viburnum cassinoides*
teak - *Tectona grandis* 546
tearthumb, arrow-leaved - *Polygonum sagittatum* 438
teasel
common teasel - *Dipsacus sylvestris* 167
wild teasel - *Dipsacus sylvestris* 167
teddy-bear-plant - *Cyanotis kewensis* 140
teddy-bear-vine - *Cyanotis kewensis* 140
ten-commandments - *Maranta leuconeura* 296
teosinte - *Zea mays* subsp. *mexicana* 428
perennial teosinte - *Zea perennis* 428
terongan - *Solanum torvum* 530
Texas-ebony - *Pithecellobium flexicaule* 212
thimbleberry - *Rubus occidentalis*, *R. odoratus*, *R. parviflorus* 486
thimbleweed - *Anemone cylindrica*, *A. virginiana* 452
thistle
blessed thistle - *Cnicus benedictus* 66, *Silybum marianum* 84
bull thistle - *Cirsium vulgare* 65
Canada thistle - *Cirsium arvense* 65
holy thistle - *Silybum marianum* 84
milk thistle - *Silybum marianum* 84
plumeless thistle - *Carduus acanthoides* 62
St.-Mary's thistle - *Silybum marianum* 84
thornless thistle - *Centaurea americana* 63
thorn, Washington - *Crataegus phaenopyrum* 463
thornapple
common thornapple - *Datura stramonium* 525
downy thornapple - *Datura inoxia*, *D. metel* 525
Stramonium thornapple - *Datura stramonium* 525
thousand-mothers - *Tolmiea menziesii* 515
thousand-seal - *Achillea millefolium* 53
three-awn
pineland three-awn - *Aristida stricta* 371
poverty three-awn - *Aristida divaricata* 371
prairie three-awn - *Aristida adscensionis* 371
purple three-awn - *Aristida purpurea* 371
red three-awn - *Aristida purpurea* 371
Reverchon three-awn - *Aristida purpurea* 371
six-weeks three-awn - *Aristida adscensionis* 371
Wooton three-awn - *Aristida pansa* 371
three-men-in-a-boat - *Tradescantia spathacea* 140
three-seeded-mercury - *Acalypha virginica* 182
thyme
common thyme - *Thymus vulgaris* 268
creeping thyme - *Thymus serpyllum* 268
garden thyme - *Thymus vulgaris* 268
lemon thyme - *Thymus serpyllum* 268
Spanish thyme - *Lippia micromera* 546
wild thyme - *Thymus serpyllum* 268
ti - *Cordyline terminalis* 17
ti-es - *Pouteria campechiana* 512
ti-palm - *Cordyline terminalis* 17
tibisee - *Lasiacis divaricata* 398
tigerflower - *Tigridia pavonia* 255
tigernut - *Cyperus esculentus* 163
timothy
alpine cat's timothy - *Phleum alpinum* 407
alpine timothy - *Phleum alpinum* 407
common timothy - *Phleum pratense* 407
mountain timothy - *Phleum pratense* 407
tinker's-weed - *Triosteum perfoliatum* 127
tipu-tree - *Tipuana tipu* 216
tisswood - *Persea borbonia* 269
titi - *Oxydendrum arboreum* 175
black titi - *Cliftonia monophylla*, *Cyrilla racemiflora* 166
red titi - *Cyrilla racemiflora* 166
toadflax
blue toadflax - *Linaria canadensis* 518
old-field toadflax - *Linaria canadensis* 518
yellow toadflax - *Linaria vulgaris* 518
tobacco - *Nicotiana tabacum* 528
desert tobacco - *Nicotiana trigonophylla* 528
fiddle-leaf tobacco - *Nicotiana repanda* 528
flowering tobacco - *Nicotiana alata* 528
Indian tobacco - *Lobelia inflata* 122
jasmine tobacco - *Nicotiana alata* 528
rabbit tobacco - *Gnaphalium obtusifolium* 71
tree tobacco - *Nicotiana glauca* 528
wild tobacco - *Nicotiana repanda* 528
winged tobacco - *Nicotiana alata* 528
tobacco-root - *Valeriana edulis* 545
tobacco-weed - *Elephantopus tomentosus* 68
tolguacha - *Datura inoxia* 525
tomato - *Lycopersicon esculentum* 526
cherry tomato - *Lycopersicon esculentum* var. *cerasiforme* 526, *Physalis peruviana* 529
currant tomato - *Lycopersicon pimpinellifolium* 526
gooseberry tomato - *Physalis peruviana* 529
pear tomato - *Lycopersicon esculentum* var. *pyriforme* 526
strawberry tomato - *Physalis peruviana*, *P. pruinosa*, *P. pubescens* 529
tree tomato - *Cyphomandra betacea* 525
tomato-tree - *Cyphomandra betacea* 525
toog - *Bischofia trifoliata* 183
touch-me-not - *Impatiens noli-tangere* 90
pale touch-me-not - *Impatiens pallida* 90
spotted touch-me-not - *Impatiens capensis* 90
tous-les-mois - *Canna edulis* 123
townhall-clock - *Adoxa moschatellina* 17
trailing-evergreen - *Lycopodium complanatum* 284
trapa-nut - *Trapa natans* 538
traveler's-palm - *Ravenala madagascariensis* 532
treacleberry - *Smilacina racemosa* 280
tread-softly - *Cnidoscolus texanus* 183
treasure-flower - *Gazania rigens* 71
tree
salamander tree - *Antidesma bunius* 183
service tree - *Sorbus domestica* 489
toothache tree - *Zanthoxylum americanum* 498
varnish tree - *Koelreuteria paniculata* 511
tree-of-gold - *Tabebuia caraiba* 105
tree-of-heaven - *Ailanthus altissima* 522
tree-of-kings - *Cordyline terminalis* 17
treebine, Venezuela - *Cissus rhombifolia* 549
treepoppy - *Dendromecon rigida* 325
trefoil
big trefoil - *Lotus uliginosus* 203
bird's-foot trefoil - *Lotus corniculatus* 203
prairie trefoil - *Lotus purshianus* 203
small-flowered birdsfoot trefoil - *Lotus micranthus* 203
trillium, purple - *Trillium erectum* 281
triplet-lily - *Brodiaea laxa* 274
trisetum
nodding trisetum - *Trisetum cernuum* 425
spike trisetum - *Trisetum spicatum* 425
tall trisetum - *Trisetum canescens* 425
Wolf's trisetum - *Trisetum wolfii* 425
trompillo - *Solanum elaeagnifolium* 530
tropic-laurel - *Ficus benjamina* 298
trumpet, cat's-claw - *Macfadyena unguis-cati* 105
trumpet-creeper - *Campsis radicans* 103
trumpet-flower - *Tecoma stans* 105
evening trumpet-flower - *Gelsemium sempervirens* 284
trumpet-leaf - *Sarracenia flava* 512
trumpet-lily - *Zantedeschia aethiopica* 41
trumpet-tree
Paraguayan trumpet-tree - *Tabebuia caraiba* 105
silver trumpet-tree - *Tabebuia caraiba* 105
trumpets - *Sarracenia flava* 512
yellow trumpets - *Sarracenia alata* 512
trumpetvine - *Campsis radicans* 103
trumpetweed - *Eupatorium fistulosum* 70
ts'ai, kui - *Allium tuberosum* 272
tsuru-na - *Tetragonia tetragonioides* 20
tuber-root - *Asclepias tuberosa* 51
tuberose - *Polianthes tuberosa* 19
tuckahoe - *Peltandra virginica* 40
tule - *Scirpus acutus* 165
tulip-poplar - *Liriodendron tulipifera* 285
tuliptree - *Liriodendron tulipifera* 285
African tuliptree - *Spathodea campanulata* 105
tumbleweed - *Amaranthus albus* 22, *A. graecizans* , *Salsola australis* 137, *S. iberica* 137
tumeric - *Hydrastis canadensis* 456
tung - *Aleurites montana* 182
tunka - *Benincasa hispida* 146
tupelo
large tupelo - *Nyssa aquatica* 309
Ogeechee lime tupelo - *Nyssa ogeche* 309

swamp tupelo - *Nyssa sylvatica* var. *biflora* 309
upland tupelo - *Nyssa sylvatica* 309
water tupelo - *Nyssa aquatica* 309
Turk's-cap - *Aconitum napellus* 451, *Malvaviscus arboreus* 294, *M. arboreus* var. *drummondii* 294
turkey's-beard - *Xerophyllum asphodeloides* 282
turkey-tangle - *Phyla nodiflora* 546
turkeyberry - *Solanum torvum* 530
turkeyfoot - *Andropogon gerardii, A. hallii* 369
turnip - *Brassica rapa* 109
Indian turnip - *Arisaema triphyllum* 39, *Pediomelum esculentum* 209
swedish turnip - *Brassica napus* var. *napobrassica* 109
turtlehead - *Chelone glabra* 517
twinberry - *Lonicera involucrata* 124
twinleaf - *Jeffersonia diphylla* 92
twistwood - *Viburnum lantana* 127
udo - *Aralia cordata* 42
umbilroot - *Cypripedium calceolus* var. *pubescens* 321
umbrella-palm - *Cyperus alternifolius* 163
umbrella-pine, Japanese - *Sciadopitys verticillata* 534
umbrella-plant - *Cyperus alternifolius* 163
umbrella-tree - *Magnolia tripetala* 287
Australian umbrella-tree - *Schefflera actinophylla* 44
ear-leaved umbrella-tree - *Magnolia fraseri* 287
Queensland umbrella-tree - *Schefflera actinophylla* 44
umbrella-trumpets - *Sarracenia flava* 512
umkokolo - *Dovyalis caffra* 240
unguentine-cactus - *Aloe barbadensis* 273
unicorn-flower, common - *Proboscidea louisianica* 326
unicornroot - *Aletris farinosa* 272
unionleaf - *Gaultheria shallon* 172
urd - *Vigna mungo* 220
valerian
American valerian - *Cypripedium calceolus* var. *pubescens* 321
common valerian - *Valeriana officinalis* 545
edible valerian - *Valeriana edulis* 545
Greek valerian - *Polemonium caeruleum* subsp. *villosum* 436
red valerian - *Centranthus ruber* 544
vanilla - *Vanilla planifolia* 323
vanillaleaf - *Achlys triphylla* 91
varnish-tree - *Ailanthus altissima* 522, *Aleurites moluccana* 182
Japanese varnish-tree - *Firmiana simplex* 532
vase-vine - *Clematis viorna* 454
vegetable-oyster - *Tragopogon porrifolius* 87
veludo-branco - *Celosia argentea* 22
velvet-plant, trailing - *Ruellia makoyana* 8
velvetbean - *Mucuna pruriens* var. *utilis* 208
Bengal velvetbean - *Mucuna pruriens* var. *utilis* 208
cowage velvetbean - *Mucuna pruriens* var. *utilis* 208
Florida velvetbean - *Mucuna pruriens* var. *utilis* 208
Mauritius velvetbean - *Mucuna pruriens* var. *utilis* 208
Yokohama velvetbean - *Mucuna pruriens* var. *utilis* 208
velvetleaf - *Abutilon theophrastii* 289
velvetplant - *Gynura aurantiaca* 72, *Verbascum thapsus* 521
Venetian-sumac - *Cotinus coggygria* 23
Venus's-flytrap - *Dionaea muscipula* 167
Venus's-shoe - *Cypripedium calceolus* var. *pubescens* 321
verbena
blue verbena - *Verbena hastata* 546
clump verbena - *Verbena canadensis* 546
garden verbena - *Verbena* x *hybrida* 546
rose verbena - *Verbena canadensis* 546
white verbena - *Verbena urticifolia* 546
woolly verbena - *Verbena stricta* 546
vervain
blue vervain - *Verbena hastata* 546
creeping vervain - *Verbena canadensis* 546
Dakota vervain - *Verbena bipinnatifida* 546
hoary vervain - *Verbena stricta* 546
prostrate vervain - *Verbena bracteata* 546
rose vervain - *Verbena canadensis* 546
white vervain - *Verbena urticifolia* 546
vetch
bird vetch - *Vicia cracca* 219
blackpod vetch - *Vicia angustifolia* 219
boreal vetch - *Vicia cracca* 219
common vetch - *Vicia angustifolia, V. sativa* 219
cow vetch - *Vicia cracca* 219
hairy vetch - *Vicia villosa* 219
Hungarian vetch - *Vicia pannonica* 219
large Russian vetch - *Vicia villosa* 219
purple vetch - *Vicia benghalensis* 219
Sitka vetch - *Vicia gigantea* 219
tufted vetch - *Vicia cracca* 219
winter vetch - *Vicia villosa* 219
vetiver - *Vetiveria zizanioides* 428
viburnum
arrowwood viburnum - *Viburnum dentatum* 127
doublefile viburnum - *Viburnum plicatum* var. *tomentosum* 127
Linden viburnum - *Viburnum dilatatum* 127
maple-leaved viburnum - *Viburnum acerifolium* 127
naked viburnum - *Viburnum nudum* 127
sweet viburnum - *Viburnum lentago, V. odoratissimum* 127
vine
cinnamon vine - *Dioscorea batatas* 167
railroad vine - *Ipomoea pes-caprae* 142
vine-spinach - *Basella alba* 90
vinegar-pear - *Passiflora laurifolia* 325
vinegar-tree - *Rhus glabra* 24
viola - *Viola cornuta* 547
violet
Alaska violet - *Viola langsdorfii* 547
alpine marsh violet - *Viola palustris* 547
American dog violet - *Viola conspersa* 547
arrow-leaved violet - *Viola sagittata* 547
birdsfoot violet - *Viola pedata* 547
Canada violet - *Viola canadensis* 547
common African violet - *Saintpaulia ionantha* 244
cream violet - *Viola striata* 547
downy yellow violet - *Viola pubescens* 547
early blue violet - *Viola palmata* 547
English violet - *Viola odorata* 547
field violet - *Viola arvensis* 547
florist's violet - *Viola odorata* 547
German violet - *Exacum affine* 241
great-spurred violet - *Viola selkirkii* 547
halberb-leaved violet - *Viola hastata* 547
hook-spur violet - *Viola adunca* 547
horned violet - *Viola cornuta* 547
lance-leaved violet - *Viola lanceolata* 547
large-leaved white violet - *Viola incognita* 547
larkspur violet - *Viola pedatifida* 547
long-spurred violet - *Viola rostrata* 547
marsh blue violet - *Viola cucullata* 547
Missouri violet - *Viola missouriensis* 547
northern bog violet - *Viola nephrophylla* 547
northern white violet - *Viola renifolia* 547
pale violet - *Viola striata* 547
Persian violet - *Exacum affine* 241
Philippine violet - *Barleria cristata* 7
primrose-leaved violet - *Viola primulifolia* 547
purple prairie violet - *Viola pedatifida* 547
smooth yellow violet - *Viola pubescens* 547
stream violet - *Viola glabella* 547
stripped violet - *Viola striata* 547
sweet violet - *Viola odorata* 547
sweet white violet - *Viola blanda* 547
tall white violet - *Viola canadensis* 547
two-eyed violet - *Viola ocellata* 547
western dog violet - *Viola adunca* 547
western round-leaved violet - *Viola orbiculata* 547
western sweet white violet - *Viola macloskeyi* 547
wild violet - *Viola tricolor* 547
woolly blue violet - *Viola sororia* 547
yellow prairie violet - *Viola nuttallii* 547
yellow wood violet - *Viola lobata* 547
violettas - *Anoda cristata* 290
virgilia - *Cladrastis lutea* 194
virgin's-bower - *Clematis virginiana* 454
wahoo - *Euonymus atropurpureus* 133
eastern wahoo - *Euonymus atropurpureus* 133
wake-robin, Virginian - *Peltandra virginica* 40
wallflower - *Cheiranthus cheiri* 111
bushy wallflower - *Erysimum repandum* 112
coast wallflower - *Erysimum capitatum* 112
English wallflower - *Cheiranthus cheiri* 111
western wallflower - *Erysimum asperum* 112
walnut
Arizona walnut - *Juglans major* 258
black walnut - *Juglans nigra* 259
English walnut - *Juglans regia* 259
Hind's walnut - *Juglans hindsii* 258
Indian walnut - *Aleurites moluccana* 182
Japanese walnut - *Juglans sieboldiana* 259
little walnut - *Juglans microcarpa* 258
Northern California black walnut - *Juglans hindsii* 258
Persian walnut - *Juglans regia* 259
river walnut - *Juglans microcarpa* 258
Southern California walnut - *Juglans californica* 258
Texas walnut - *Juglans microcarpa* 258
white walnut - *Juglans cinerea* 258
wampi - *Clausena lansium* 497
wandering-jew - *Tradescantia fluminensis, Zebrina pendula* 140
wapato - *Sagittaria cuneata, S. latifolia* 21
warrigal-cabbage - *Tetragonia tetragonioides* 20
warrior, Indian - *Pedicularis densiflora* 519
warron - *Cucurbita pepo* 149
watches - *Sarracenia flava* 512
water-ash - *Ptelea trifoliata* 498
water-chestnut - *Pachira aquatica* 106, *Trapa natans* 538

water-dragon - *Calla palustris* 39, *Saururus cernuus* 513
water-elm - *Planera aquatica* 540
water-flaxseed - *Spirodela polyrhiza* 271
water-hemlock - *Cicuta maculata* 27
spotted water-hemlock - *Cicuta maculata* 27
water-hemp - *Acnida tamariscina* 21
water-lily
Amazon water-lily - *Victoria amazonica* 309
American water-lily - *Nymphaea odorata* 309
European white water-lily - *Nymphaea alba* 309
fragrant water-lily - *Nymphaea odorata* 309
magnolia water-lily - *Nymphaea tuberosa* 309
royal water-lily - *Victoria amazonica* 309
tuberous water-lily - *Nymphaea tuberosa* 309
yellow water-lily - *Nuphar luteum* 309
water-motie - *Baccharis glutinosa* 60
water-oleander - *Decodon verticillatus* 285
water-parsnip - *Sium suave* 32
water-pepper, mild - *Polygonum hydropiperoides* 438
water-plantain, large-flowered - *Alisma plantago-aquatica* 21
water-platter, Amazon - *Victoria amazonica* 309
water-purslane - *Ludwigia palustris* 318
water-spinach - *Ipomoea aquatica* 142
water-willow - *Baccharis glutinosa* 60, *Decodon verticillatus* 285
water-wisteria - *Hygrophila difformis* 8
watercress - *Nasturtium officinale* 114
waterfeather - *Myriophyllum aquaticum* 247
waterleaf, Virginia - *Hydrophyllum virginianum* 252
watermelon - *Citrullus lanatus* 147
Chinese watermelon - *Benincasa hispida* 146
watermelon-begonia, trailing - *Pellionia daveauana* 543
watermilfoil, Eurasian - *Myriophyllum spicatum* 247
waterprimrose, winged - *Ludwigia decurrens* 318
watershield - *Brasenia schreberi* 118
waterweed - *Elodea canadensis* 251
wattle, coastal - *Acacia cyclopis* 187
wax-myrtle
California wax-myrtle - *Myrica californica* 302
Pacific wax-myrtle - *Myrica californica* 302
wax-tree, white - *Ligustrum lucidum* 314
waxberry - *Myrica cerifera* 302, *Symphoricarpos albus* 126, *S. albus* var. *laevigatus* 126
waxflower - *Tabernaemontana divaricata* 35
waxgourd - *Benincasa hispida* 146
waxplant - *Hoya carnosa* 53
waxweed, blue - *Cuphea viscosissima* 284
waxwork - *Celastrus scandens* 133
wayfaring-tree - *Viburnum lantana* 127
American wayfaring-tree - *Viburnum alnifolium* 127
weather-plant - *Abrus precatorius* 187
weather-vine - *Abrus precatorius* 187
weaver's-broom - *Spartium junceum* 216
wedelia, trailing - *Wedelia trilobata* 89
weigela, pink - *Weigela florida* 128
wheat
Alaska wheat - *Triticum turgidum* 425
bread wheat - *Triticum aestivum* 425
common wheat - *Triticum aestivum* 425
cone wheat - *Triticum turgidum* 425
durum wheat - *Triticum durum* 425
English wheat - *Triticum turgidum* 425
Mediterranean wheat - *Triticum turgidum* 425
Polish wheat - *Triticum polonicum* 425
poulard wheat - *Triticum turgidum* 425
rivet wheat - *Triticum turgidum* 425
wheel-of-fire - *Stenocarpus sinuatus* 445
whinberry - *Vaccinium myrtillus* 179
whippoorwill-shoe - *Cypripedium calceolus* var. *pubescens* 321
whisk-fern - *Psilotum nudum* 445
whistlewood - *Acer pensylvanicum* 9
white-cedar, northern - *Thuja occidentalis* 156
white-man's-foot - *Plantago major* 359
whitebrush - *Aloysia gratissima* 545
whitetop - *Erigeron annuus* 69
whiteweed - *Ageratum conyzoides* 54, *Chrysanthemum frutescens* 63
whitewood - *Liriodendron tulipifera* 285, *Tabebuia riparia* 105, *Tilia americana* 537
whitten-tree - *Viburnum opulus* 127
whortleberry - *Vaccinium corymbosum*, *V. myrtillus* 179
wickup - *Epilobium angustifolium* 317
wicky - *Kalmia angustifolia* 173
wicopy - *Dirca palustris* 537
widdy - *Potentilla fruticosa* 474
widow's-frill - *Silene stellata* 131
widow's-tears - *Tradescantia virginiana* 140
wigtree - *Cotinus coggygria* 23
wild-beet - *Amaranthus hybridus*, *A. retroflexus* 22
wild-cocoa - *Pachira aquatica* 106
wild-coffee - *Polyscias guilfoylei* 44, *Triosteum aurantiacum* 127, *T. perfoliatum* 127
wild-hyacinth - *Camassia scilloides* 275
wild-jalap - *Podophyllum peltatum* 92
wild-lemon - *Podophyllum peltatum* 92
wild-licorice - *Abrus precatorius* 187
wild-lily-of-the-valley - *Maianthemum canadense* 279, *Pyrola elliptica* 451, *P. rotundifolia* 451
wild-oats - *Uvularia sessilifolia* 282
wild-okra - *Viola palmata* 547
wild-olive - *Halesia carolina* 533
wild-orange - *Prunus caroliniana* 475
wild-pineapple - *Tillandsia fasciculata* 117
wild-potato - *Chlorogalum pomeridianum* 275
wild-raisin - *Viburnum cassinoides* 127
wild-rhubarb - *Rumex hymenosepalus* 440
wild-rice, Chinese - *Leymus chinensis* 399
wild-rye
Altai wild-rye - *Leymus angustus* 399
basin wild-rye - *Elymus cinereus* 387
beardless wild-rye - *Elymus triticoides* 387
blue wild-rye - *Elymus glaucus* 387
Canada wild-rye - *Elymus canadensis* 387
giant wild-rye - *Elymus cinereus* 387
hairy wild-rye - *Elymus villosus* 387
mammoth wild-rye - *Leymus racemosus* 399
Salina wild-rye - *Elymus salina* 387
Siberian wild-rye - *Elymus sibiricus* 387
Virginia wild-rye - *Elymus virginicus* 387
wild-tuberose - *Manfreda maculosa* 18
wild-water-lemon - *Passiflora foetida* 325
willow
Alaska bog willow - *Salix fuscescens* 505
arctic willow - *Salix arctica* 505
Barkley willow - *Salix barclayi* 505
barren-ground willow - *Salix brachycarpa* subsp. *niphoclada* 505
basket willow - *Salix purpurea*, *S. viminalis* 506
bay willow - *Salix pentandra* 506
bay-leaved willow - *Salix pentandra* 506
Bebb willow - *Salix bebbiana* 505
black willow - *Capparis cynophallophora* 123, *Salix nigra* 506
brittle willow - *Salix fragilis* 505
Chamisso willow - *Salix chamissonis* 505
crack willow - *Salix fragilis* 505
diamond-leaf willow - *Salix planifolia* 506
ditchbank willow - *Salix exigua* 505
feltleaf willow - *Salix alaxensis* 505
golden willow - *Salix alba* var. *vitellina* 505
gray willow - *Salix humilis* 505
grayleaf willow - *Salix glauca* 505
heart-leaved willow - *Salix cordata* 505
Hooker willow - *Salix hookeriana* 505
large pussy willow - *Salix discolor* 505
laurel willow - *Salix pentandra* 506
least willow - *Salix rotundifolia* 506
littletree willow - *Salix arbusculoides* 505
low blueberry willow - *Salix myrtillifolia* 505
mountain willow - *Salix scouleriana* 506
netleaf willow - *Salix reticulata* 506
oval-leaf willow - *Salix ovalifolia*, *S. stolonifera* 506
Pacific willow - *Salix lasiandra* 505
park willow - *Salix monticola* 505
peach-leaved willow - *Salix amygdaloides* 505
polished willow - *Salix laevigata* 505
prairie willow - *Salix humilis* 505
purple willow - *Salix purpurea* 506
red willow - *Salix laevigata* 505
Richardson willow - *Salix lanata* subsp. *richardsonii* 505
sandbar willow - *Salix exigua* 505
Scouler willow - *Salix scouleriana* 506
silky willow - *Salix sericea* 506
silver willow - *Salix candida* 505
Sitka willow - *Salix sitchensis* 506
skeleton-leaf willow - *Salix petiolaris* 506
small pussy willow - *Salix humilis* 505
tall blueberry willow - *Salix novaeangliae* 506
undergreen willow - *Salix commutata* 505
Virginia willow - *Itea virginica* 245
weeping willow - *Salix babylonica* 505
white willow - *Salix alba* 505
willowherb
great willowherb - *Epilobium angustifolium* 317
wineplant - *Rheum rhabarbarum* 439
wingnut, Chinese - *Pterocarya stenoptera* 260
wingstem - *Verbesina alternifolia* 88
winter-cherry - *Physalis alkekengi*, *P. peruviana* 529
winter-fern - *Conium maculatum* 28
winterberry - *Ilex glabra* 36
common winterberry - *Ilex verticillata* 36
mountain winterberry - *Ilex montana* 36
smooth winterberry - *Ilex laevigata* 36
winterfat - *Ceratoides lanata* 136
wintergreen - *Gaultheria procumbens* 172
alpine wintergreen - *Gaultheria humifusa* 172
arctic wintergreen - *Pyrola glandiflora* 451
one-sided wintergreen - *Pyrola secunda* 451
pink wintergreen - *Pyrola asarifolia* 451

spotted wintergreen - *Chimaphila maculata* 451
wintersweet - *Chimonanthus praecox* 121
wiss, blue - *Teramnus labialis* 216
wisteria
Chinese wisteria - *Wisteria sinensis* 221
Japanese wisteria - *Wisteria floribunda* 221
wisteria-tree, scarlet - *Sesbania punicea* 215
witch-hazel - *Hamamelis virginiana* 247
witch-hobble - *Viburnum alnifolium* 127
witchweed - *Striga asiatica* 521
tobacco witchweed - *Striga gesnerioides* 521
withe-rod - *Viburnum cassinoides* 127
smooth withe-rod - *Viburnum nudum* 127
witloof - *Cichorium intybus* 65
wolf's-claws - *Lycopodium clavatum* 284
wolf's-milk - *Euphorbia esula* 184
wolfbane, garden - *Aconitum napellus* 451
wolfberry - *Symphoricarpos occidentalis* 126
wonder-bulb - *Colchicum autumnale* 275
wonder-tree - *Ricinus communis* 186
wonkapin - *Nelumbo lutea* 308
wood-vamp - *Decumaria barbara* 250
woodapple - *Limonia acidissima* 497
Indian woodapple - *Limonia acidissima* 497
woodbetony - *Pedicularis canadensis* 519
woodbine - *Clematis virginiana* 454, *Parthenocissus quinquefolia* 550
woodfern
coastal woodfern - *Dryopteris arguta* 447
leather woodfern - *Dryopteris marginalis* 447
woodnymph - *Moneses uniflora* 451
woodreed
drooping woodreed - *Cinna latifolia* 381
stout woodreed - *Cinna arundinacea* 381
woodrush
common woodrush - *Luzula campestris* 262
field woodrush - *Luzula campestris* 262
millet woodrush - *Luzula parviflora* 262
woodsia
Oregon woodsia - *Woodsia oregana* 450
Rocky Mountain woodsia - *Woodsia scopulina* 450
smooth woodsia - *Woodsia glabella* 450
woodsorrel
creeping woodsorrel - *Oxalis corniculata* 323
red woodsorrel - *Oxalis oregana* 323
violet woodsorrel - *Oxalis violacea* 323
woolflower - *Celosia argentea* 22
wormseed - *Chenopodium ambrosioides* 136
American wormseed - *Chenopodium ambrosioides* 136
wormwood
common wormwood - *Artemisia absinthium* 57
Louisiana wormwood - *Artemisia ludoviciana* 57
Roman wormwood - *Corydalis sempervirens* 325
southern wormwood - *Artemisia abrotanum* 57
woundwort - *Anthyllis vulneraria* 189
wysteria - *Colchicum autumnale* 275
yam - *Ipomoea batatas* 142
Chinese yam - *Dioscorea batatas* 167
wild yam - *Dioscorea villosa* 167
yambean - *Pachyrhizus erosus* 209
yanquapin - *Nelumbo lutea* 308
yarrow, common - *Achillea millefolium* 53
yate, bushy - *Eucalyptus lehmannii* 303
yaupon - *Ilex cassine, I. vomitoria* 36
Yedda-hawthorn - *Rhaphiolepis umbellata* 483
yellow-alder - *Turnera ulmifolia* 538
yellow-bells - *Tecoma stans* 105
yellow-cedar
Alaska yellow-cedar - *Chamaecyparis nootkatensis* 151
Nootka yellow-cedar - *Chamaecyparis nootkatensis* 151
yellow-fruited-thorn - *Crataegus flava* 463
yellow-gowan - *Ranunculus repens* 456
yellow-ironweed - *Verbesina alternifolia* 88
yellow-poplar - *Liriodendron tulipifera* 285
yellow-rocket - *Barbarea vulgaris* 109
yellow-sage - *Lantana camara* 545
yellow-trefoil - *Medicago lupulina* 206
yellow-wood - *Cladrastis lutea* 194, *Symplocos tinctoria* 533
yellowberry - *Rubus chamaemorus* 486
yellowcress, creeping - *Rorippa sylvestris* 115
yellowroot, shrub - *Xanthorhiza simplicissima* 458
yerba-buena - *Satureja douglasii* 267
yerba-de-selva - *Whipplea modesta* 251
yew
American yew - *Taxus canadensis* 533
Canada yew - *Taxus canadensis* 533
English yew - *Taxus baccata* 533
Harrington plum yew - *Cephalotaxus harringtonia* 134
Japanese plum yew - *Cephalotaxus harringtonia* var. *drupacea* 134
Japanese yew - *Taxus cuspidata* 533
Pacific yew - *Taxus brevifolia* 533
plum fruited yew - *Cephalotaxus harringtonia* var. *drupacea* 134
western yew - *Taxus brevifolia* 533
Yorkshire-fog - *Holcus lanatus* 395
youth-on-age - *Tolmiea menziesii* 515
yuca - *Manihot esculenta* 186
yucca
aloe yucca - *Yucca aloifolia* 19
banana yucca - *Yucca baccata* 19
blue yucca - *Yucca baccata* 19
spineless yucca - *Yucca elephantipes* 19
spoonleaf yucca - *Yucca filamentosa* 19
twisted-leaf yucca - *Yucca rupicola* 19
weakleaf yucca - *Yucca flaccida* 19
zapote-blanco - *Casimiroa edulis* 494
zebraplant - *Aphelandra squarrosa* 7
zelkova
elmleaf zelkova - *Zelkova carpinifolia* 543
Japanese zelkova - *Zelkova serrata* 543
saw-leaf zelkova - *Zelkova serrata* 543
zephyr-lily - *Zephyranthes candida* 282
zinnia - *Zinnia violacea* 89
zit-kwa - *Benincasa hispida* 146
zoysia - *Zoysia japonica* 433

Fungus Index

abbreviata, Pleospora = **P. anthyllidis** var. **abbreviata** 884
abbreviatum, Glonium (Hysterium) 720
abbreviatus, Uromyces 1009
abdita, Physalospora (Anthostomella) 876
abditum, Sphaerographium 966
abeliae, Cercospora 605
abelmoschi, Ascochyta 573
abelmoschi, Pseudocercospora (Cercospora) 893
aberrans, Dothiorella 688
aberrans, Nimbomollisia (Belonium) 816
aberrans, Puccinia 900
aberrans, Teichospora 987
abeuns, Corticium = **Scytinostroma ochroleucum** 944
abieti-capraearum, Melampsora 781
abieticola, Grandinia (Hyphodontia) 722
abieticola, Grovesiella (Godronia, Scleroderris) 724
abieticola, Pholiota 851
abieticola, Polyporus = **Wrightoporia subrutilans** 1024
abietina, Ascocalyx (Gremmeniella, Lagerbergia, Scleroderris) 573
abietina, Botryosphaeria (Cryptosporina, Physalospora) 592
abietina, Columnocystis 640
abietina, Conoplea 643
abietina, Cryptocline 651
abietina, Gloniella 719
abietina, Micropera = **Foveostroma abietinum** 706
abietina, Oncospora 819
abietina, Patinella (Patellaria) 827
abietina, Phomopsis (Phoma) 857
abietina, (Sporocybe, Verticicladiella) = **Leptographium abietinum** 759
abietina, Tympanis = **T. laricina** 1004
abietina var. trichota, Rosellinia = **R. trichota** 937
abietinellum, Nothophacidium (Phacidium) 817
abietinum, Foveostroma (Gelatinosporium) 706
abietinum, Fusicoccum = **Phomopsis abietina** 857
abietinum, Leptographium 759
abietinum, Phaeophacidium 840
abietinum, Rhabdogloeum = **Rhabdogloeopsis balsameae** 930
abietinum, Stereum = **Columnocystis abietina** 640
abietinum, Toxosporium = **T. camptospermum** 995
abietinum, Trichaptum 998
abietinus, Dacrymyces = **D. stillatus** 663
abietinus, (Coriolus, Hirschioporus, Polyporus, Polystictus) = **Trichaptum abietinum** 998
abietis, Aleurodiscus 560
abietis, Apiosporium (Meliola) 569
abietis, (Camarographium, Camarosporium) = **Stegonsporiopsis cenangioides** 976
abietis, Cenangium = **C. ferruginosum** 602
abietis, Cytospora 661
abietis, Darkera 664
abietis, Delphinella (Mycosphaerella, Rehmiellopsis) 666
abietis, Helminthosporium = **Hyphosoma lumbricoideum** 742
abietis, Hericium (Hydnum) 734
abietis, Heyderia (Mitrula) 736
abietis, Isthmiella (Bifusella) 750
abietis, Lachnellula (Dasyscyphus) 753
abietis, Lenzites = **Trichaptum laricinum** 999
abietis, Peniophora = **Tubulicium clematidis** 1003
abietis, Phacidium 838
abietis, Pholiota = **P. limonella** 851
abietis, Rasutoria (Dimerosporium, Epipolaeum) 928
abietis, Rhizocalyx 932
abietis, Rhizosphaera = **R. pini** 934
abietis, Rhizothyrium 934
abietis, Sphaeropsis 966
abietis, Tiarosporella 991
abietis, (Retinocyclus, Tympanis) = **Sarea difformis** 939
abietis, Valsa 1017
abietis, Xenomeris 1025
abietis-canadensis, Melampsora (Caeoma) 781
abietis-concoloris, Lirula (Hypodermella) 769
abiocarpa, Ceratocystis = **Ophiostoma abiocarpum** 821
abiocarpum, Ophiostoma 821
abnormale, Camarosporium 599
abnormalis, Hendersonia = **Camarosporium abnormale** 599
abnorme, Herpobasidium 735
abnorme, Lachnum (Peziza) 754
abnormis, Calosphaeria 598
abnormis, Puccinia 900
abortiva, Septoria (Phyllosticta) 947
abortivus, Polyporus = **Abortiporus biennis** 555
abramsianus, Fomes = **Phellinus robustus** 846
abramsii, Puccinia 900
abroniae, Ascochyta = **A. oxybaphi** 576
abroniae, Heterosporium 735
abrotani, Diplodia 679
abrupta, Botryosphaeria 592
abrupta, Puccinia 900
abrupta, Valsa = **Leucostoma translucens** 767
abscondens, Pleurotus = **Ossicaulis lignatilis** 823
abscondita, Pleospora 883
absinthii, (Cercospora, Helminthosporium) = **Mycovellosiella ferruginea** 807
absinthii, Puccinia = **P. tanaceti** 915
absinthii, Pyrenopeziza 918
abundans, Clitocybula (Collybia) 633
abundans, Rosenscheldia (Gibberidea) 937
abutilonis, Alternaria (Macrosporium) 560
abutilonis, Cercospora 605
abutilonis, Orbilia 822
abutilonis, Schizoxylon = **Acarosporina hormospora** 556
acaciae, Endoraecium 694
acaciae, Mycosphaerella (Sphaerella) 799
acaciae, Phyllachora 862
acaciae, Sphaerophragmium 966
acaciae, Strumella 982
acaciae-confusae, Meliola = **M. koae** 785
acacicola, Lophodermium 771
acalyphae, Cercospora 605
acalyphae, Plasmopara (Rhysotheca) 882
acalyphae, Ramularia 923
acalyphae, Volutella 1022
acalypharum, Cercospora = **C. acalyphae** 605
acanthicum, Pythium 920
acanthophoron, Pythium 920
acanthophysa, Platygloea 882
acanthospermi, Phyllosticta 864
acanthostroma, Nitschkia (Scortechinia) 816
accedens, Tubulicrinis (Peniophora) 1003
acericola, Diplodia 679
acericola, Pezicula (Dermatea) 837
acericola, Phyllosticta = **P. minima** 870
acericola, Pleomassaria 883
acericola, Tympanis = **T. truncatula** 1004
acerina, Botryodiplodia 591
acerina, Ciboria 627
acerina, Cryptosporella 652
acerina, Dendrothele 667
acerina, Dermea (Tympanis) 669
acerina, Diaporthe (Valsa) 670
acerina, Diplodina (Discella, Marssonina) 682
acerina, Kirschsteiniothelia (Microthelia) 752
acerina, Libertella 768
acerina, Melasmia 784
acerina, Mollisina (Pezizella, Urceolella) 794
acerina, Mycocentrospora (Centrospora, Cercospora) 799
acerina, (Lagynodella, Naemosphaera) = **Sphaeronaema acerinum** 966
acerina, Odontia = **Hyphoderma setigerum** 741
acerina, Phlebia = **P. rufa** 848
acerina, Phomopsis 857
acerina, (Haplosporella, Sphaeropsis) = **Aplosporella clintonii** 570

acerina, Stilbella 980
acerina, Trichopeziza = **Dasyscyphus acerinus** 664
acerina, Venturia 1020
acerinum, Aleurocorticium = **Dendrothele acerina** 667
acerinum, Cryptodiaporthe 651
acerinum, Cylindrosporium 658
acerinum, Gloeosporium = **Diplodina acerina** 682
acerinum, Hysterographium = **H. fraxini** 747
acerinum, Myxosporium 810
acerinum, Pilidium (Leptothyrium) 880
acerinum, Rhytisma 935
acerinum, Septogloeum = **Diplodina acerina** 682
acerinum, Sphaeronaema 966
acerinum, Stegonsporium 976
acerinum, Stereum = **Dendrothele acerina** 667
acerinus, Aleurodiscus = **Dendrothele acerina** 667
acerinus, Coccomyces = **Pilidium acerinum** 880
acerinus, Dasyscyphus 664
aceris, Didymosporina 678
aceris, Gloeosporium 717
aceris, Leptothyrella 765
aceris, Mycopappus (Cercosporella) 799
aceris, Septoria (Phloeospora) 947
aceris, Taphrina (Exoascus) 986
aceris, Uncinula = **U. bicornis** 1006
aceris-macrophylli, Septoria 947
acerophila, Dictyoporthe 675
acerophila, (Gnomonia, Gnomoniopsis) = **Plagiostoma acerophilum** 881
acerophilum, Plagiostoma 881
acerva, Dothidea 687
acervalis, Gibberella 714
acervata, Asteridiella (Irene, Meliola) 580
acervata, Collybia 640
acervata, Mycosphaerella (Diaporthe, Diatrype) 799
acervulatum, Hysterium = **H. pulicare** 747
acetosae, Puccinia 900
acetosella, Septoria 947
acetosellae, Cercospora 605
acetosellae, Phyllosticta 864
acetosellae var. maculosa, Cercospora = **C. peckiana** 614
acharii, Eutypa = **E. maura** 700
achilleae, Phomopsis 857
achilleae, Septoria 947
achlydis, Stagonospora (Ascochyta) 972
achlyicola, Ascochyta 573
acicola, Amphisphaeria = **Neopeckia coulteri** 815
acicola, Cenangium (Encoelia) 602
acicola, Cylindrosporium 658
acicola, Desmazierella 670
acicola, Lecanosticta (Cryptosporium, Septoria) 758
acicola, Lembosia (Aulographum) 758
acicola, Mycosphaerella (Sphaerella) 800
acicola, Mytilinidion 809
acicola, (Dothidea, Oligostroma, Scirrhia, Systremma) = **Mycosphaerella dearnessii** 801
acicola, Phoma = **Sclerophoma pythiophila** 940
acicularum, Phialea (Helotium) 847
aciculina, Cercospora 605
aciculispora, Myriogenospora 809
aciculosa, Septoria 947
aciculosa, Stilbella 981
aciculosa, Volutella (Hymenula) 1022
aciculosum, Stilbum = **Stilbella aciculosa** 981
aclada, Botrytis 594
acnidae, Cercospora 605
aconiti-elymi, Puccinia = **P. recondita** 912
aconiti-rubrae, Puccinia = **P. recondita** 912
acori, Cylindrosporium = **Septocylindrium aromaticum** 946
acori, Phaeosphaeria 841
acori, Sphaerulina 968
acremonioides, Harzia 730
acremonium, Cephalosporium = **Acremonium strictum** 557
acrocystis, Prosthecium (Melanconis, Valsa) 892
acrophila, Puccinia 900
acroptili, Puccinia 900
acrostichoides, Uredinopsis 1007
actaeae, Ascochyta (Actinonema, Marssonia, Marssonina) 573
actaeae, Ramularia 923
actinidiae, Diaporthe 670
acuaria, Chalara 625
aculeans, Amphiporthe (Cryptodiaporthe, Valsa) 563
aculeata, Ustilago = **U. serpens** 1016
aculeatum, Penicillium 827
acuminata, Alatospora 559
acuminata, Duplicaria (Bifusella) 690
acuminata, Phragmopyxis 861
acuminatum, Fusarium 707
acuminatum, Hysterium = **H. angustatum** 747
acuminatum var. alpinum, Hysterium = **H. angustatum** 747
acuminatus, Ophiobolus (Sphaeria) 819
acuminatus, Uromyces 1009
acuminatus var. magnatus, Uromyces = **Uromyces acuminatus** 1009
acuminatus var. polemonii, Uromyces = **Uromyces acuminatus** 1009
acuminatus var. spartinae, Uromyces = **Uromyces acuminatus** 1009
acuminatus var. steironematis, Uromyces = **Uromyces acuminatus** 1009
acuta, Discosia = **Heteropatella umbilicata** 735
acuta, Leptosphaeria (Pleospora) 760
acuta, Uredo (Milesia) 1007
acuta, Xylaria 1025
acutatum, Colletotrichum 637
acuum, Cenangium = **C. acicola** 602
acuum, Phoma 851
acuum, Schizothyrium 939
adenocauli, Septoria 947
adenocaulonis, Coleosporium 635
adhaerens, Neolentinus (Lentinus) 814
adhaerens, Pythium 920
adiposa, Ceratocystis (Ceratostomella, Endoconidiophora, Ophiostoma) 603
adiposa, Pholiota 851
adjuncti, Ophiostoma (Ceratocystis) 821
admirabilis, Polyporus 889
adoxae, Phyllosticta 864
adoxae, Puccinia 900
adoxae, Ramularia (Fusidium) 923
adunca, Marssonia 777
adunca, Uncinula 1006
adusta, Bjerkandera 590
adusta, Cercospora 605
adusta, Dothidella 687
adusta, Pestalotia 835
adusta, Phloeospora 849
adusta, Sebacina = **Exidiopsis fuliginea** 702
adusta, Venturia (Sphaerella) 1020
adustum, Mycorrhaphium (Steccherinum) 799
adustum, Mystrosporium = **Bipolaris iridis** 588
adustus, Polyporus = **Bjerkandera adusta** 590
advena, Botryosphaeria = **B. melanops** 592
adzamethica, Ascochyta = **Phoma arachidicola** 852
aecidiicola, Cladosporium 629
aecidioides, Melampsora = **M. populnea** 781
aecidioides, Synchytrium = **S. lagerheimii** 984
aecidioides, Uredo = **Synchytrium decipiens** 984
aegerita, Agrocybe (Pholiota) 559
aegopodii, Septoria 947
aegopogonis, Ustilago 1014
aemulans, Peniophora = **P. incarnata** 829
aemulans, Puccinia = **P. enceliae** 906
aemulus, Uromyces 1009
aenea, Viscomacula 1022
aeneus, Berteromyces = **Cercosporidium cassiae** 620
aequalis, Torula = **Bispora betulina** 589
aequilinearis, Eutypella (Sphaeria) 700
aequivoca, Ramularia (Fusisporium) 923
aeruginascens, Clypeopycnis 633
aeruginosa, Cercospora = **C. bacilligera** 606
aeruginosa, Chlorociboria 625
aeruginosum, Chlorosplenium = **Chlorociboria aeruginosa** 625
aesculi, Aplosporella 569
aesculi, Cryptodiaporthe 651
aesculi, Diplodia 679
aesculi, Diplodina (Septomyxa) 682
aesculi, Guignardia (Botryosphaeria, Laestadia) 724
aesculi, Mycosphaerella 800
aesculi, Polyporus = **Trametes elegans** 996
aesculi, Taphrina 986
aesculicola, Septoria = **S. hippocastani** 952
aesculina, Cercospora 605
aesculina, Eutypella = **E. scoparia** 701
aestivale, Poria = **Antrodiella americana** 568
aethiops, Kirschsteiniothelia (Amphisphaeria) 752
aethiops, Valsaria = **V. insitiva** 1019
affine, Dendrodochium 666
affine, Fusarium = **Hymenula affinis** 740
affine, Gloeosporium = **Colletotrichum gloeosporioides** 638
affine, Mastigonetron = **Harknessia liquidambaris** 729
affine, (Gloeosporium, Phyllosticta) = **Discula sassafras** 685
affinis, Cintractia 628
affinis, Claudopus 631
affinis, Curvularia 655
affinis, Doassansia = **Doassansiopsis intermedia** 686
affinis, Harknessia = **H. liquidambaris** 729
affinis, Hymenula 740
affinis, Lasiobotrys = **L. lonicerae** 756
affinis, Pestalotia = **Monochaetia monochaeta** 795
affinis, Phanerochaete (Peniophora) 843
affinis, Septomyxa 946
affinis, Uromyces = **U. primaverilis** subsp. **nothoscordi** 1012
affinis, Uropyxis = **U. petalostemonis** 1014
affinis, Ustilago 1014
affinis var. **hilariae, Ustilago** 1014

affinissimum, Helotium = **Lachnum pygmaeum** 755
afflata, Cercospora = **C. pteleae** 615
afflata, Diatrype = **Graphostroma platystoma** 723
afflata, Xyloma = **Ectostroma afflatum** 691
afflatum, Ectostroma 691
africana, Phycopsis = **Seuratia globifera** 960
africana, Sebacina = **Hyphoderma praetermissum** 741
agapanthi, Phoma (Sphaeropsis) 851
agassizii, Lachnellula (Dasyscyphus, Lachnella) 754
agavectona, Dimerosporium = **Stomatogene agaves** 981
agavei, Laetisaria 756
agaves, Colletotrichum = **C. gloeosporioides** 638
agaves, Coniothecium 642
agaves, Gloeosporium = **Colletotrichum crassipes** 637
agaves, Stomatogene (Asterina) 981
agavicola, Dothiorella 688
agerati, (Cercospora, Ragnhildiana) = **Mycovellosiella perfoliati** 808
ageratinae, Aecidium 558
ageratoides, Cercospora 605
agglutinans, Hymenochaete 739
aggregata, Erysiphe (Erysiphella) 697
aggregata, Harknessia 729
aggregata, Isaria = **Stilbella aleuriata** 981
aggregata, Mucronella 797
aggregata, Mycosphaerella 800
aggregatum, Belonium 587
agminalis, Phaeosphaeria (Leptosphaeria) 841
agnatus, Uromyces 1009
agnita, Leptosphaeria 760
agnita, Puccinia 900
agnita var. erigerontis, Leptosphaeria = **L. erigerontis** 761
agnostoica, Cercospora 605
agoseridis, Puccinia 900
agoseridis, Ramularia 923
agrariella var. vaccinii, Psilocybe 899
agrestis, Ramularia 923
agrifolia, Phyllosticta 864
agrifolia, Pseudomassaria (Physalospora) 896
agrimoniae, Puccinia = **P. laterípes** 909
agrimoniae, Pucciniastrum 916
agrimoniae, Septoria 947
agrimoniae-eupatoriae, Septoria 947
agropyri, Apiocarpella = **Ascochyta agropyri-repentis** 573
agropyri, Puccinia = **P. recondita** 912
agropyri, Septoria = **S. elymi f. elymina** 951
agropyri, Urocystis 1008
agropyrina, Ascochyta (Ascochytula) 573
agropyrina, Puccinia = **P. recondita** 912
agropyrina, Septoria 947
agropyri-repentis, Ascochyta 573
agropyri-repentis, Hendersonia 732
agrosticola, Septoria 947
agrostidis, Cercospora 605
agrostidis, Didymella = **Herpotrichia macrotricha** 735
agrostidis, Pseudorobillarda (Robillarda) 898
agrostidis f. angusta, Stagonospora = **S. bellunensis** 973
agrostina, Trichopeziza (Lachnella) 1000
agrostis, Cheilaria (Labrella) 625
ahmadii, Helminthosporium 731
ailanthi, Aplosporella (Sphaeropsis) 569
ailanthi, Botryodiplodia (Diplodia) 591
ailanthi, Cucurbitaria 654
ailanthi, Cytospora = **C. sacculus** 662
ailanthi, Glabrocyphella 715
ailanthi, Gloeosporium = **Colletotrichum gloeosporioides** 638
ailanthi, Guignardia (Laestadia) 724
ailanthi, Mycosphaerella (Sphaerella) 800
ailanthi, Phomopsis (Phoma) 857
ailanthi, Phyllosticta 864
ailanthi, Tubercularia = **T. vulgaris** 1002
airae, Phaeoseptoria 841
airae, Tilletia = **T. cerebrina** 992
ajovalliensis, Theleporus 990
akebiae, Diplodia 679
akebiae, Phoma 851
alabamae, Pachykytospora (Polyporus, Poria) 825
alabamensis, Cercospora = **C. ipomoeae** 611
alabamensis, Crepidotus 647
alabamensis, Micropeltis 790
alabamensis, Septoria 947
alabamensis, Urosporella (Ceriospora) 1014
alachuana, Ceriporia (Poria) 621
alachuanus, Marasmius = **M. cohortalis** var. **alachuanus** 777
alascensis, Acremoniella 556
alascensis, Muirella (Gloeocercospora) 797
alaskana, Trametes = **Antrodia heteromorpha** 567
alaskana, Urocystis 1008
alaskanum, Endophyllum 694
alaskanum, Phragmidium 860
alaskanum, (Pucciniastrum, Uredo) = **Uredo beringiana** 1007
alaskanus, Coriolellus = **Antrodia heteromorpha** 567
alaskense, Steccherinum 975
alaskensis, Euantennaria (Aithalomyces, Limacinia, Morfea) 698
alba, Bullera 596
alba, Ceratocystiopsis (Ceratocystis) 603
alba, Dictyocatenulata 675
alba, Peniophora = **Fibricium rude** 704
alba, Ramularia 923
alba, Septoria 947
albaniensis, Septoria = **Cylindrosporium salicinum** 659
albellus, (Polyporus, Tyromyces) = **Tyromyces chioneus** 1005
albertensis, Melampsora = **M. medusae** 781
albescens, Crepidotus (Geopetalum) 647
albescens, Didymosphaeria = **D. futilis** 678
albescens, Sphaeropsis = **Aplosporella clintonii** 570
albescens, Stagonospora 972
albicans, Candida (Oidium) 600
albicans, Hyphodontia 742
albicans, Septoria 947
albicedrae, Robergea (Cyanospora) 936
albiceps, Cytospora 661
albicolor, Mycena 797
albida, Antrodia 567
albida, Chrysomyxa = **Kuehneola uredinis** 753
albida, Hemimycena 732
albida, Patella = **Humaria hemisphaerica** 738
albida, Phlebia 848
albida, Tulasnella 1003
albidobrunnea, Thelephora 989
albido-maculans, Cercospora = **C. ricinella** 616
albido-tomentosus, Panus = **Hohenbuehelia atrocaerulea** 737
albidula, Mycena 797
albidum, Phacidium = **Propolis phacidioides** 892
albidus, Coccomyces = **Propolis phacidioides** 892
albidus, Cryptococcus 651
albidus, Gymnopus = **Hemimycena albida** 732
albipellucida, Poria = **Ceriporiopsis rivulosa** 621
albissima, Cyphella = **Calathella eruciformis** 597
albissima, Mycena 797
albiziae, Zopfia 1027
alboalutacea, Psathyrella 893
alboatrum, Cenangium 602
albo-atrum, Schizoxylon 940
albo-atrum, Verticillium 1021
albo-atrum var. caespitosum, Verticillium = **Verticillium albo-atrum** 1021
albo-atrum var. **menthae, Verticillium** 1021
albo-atrum var. tuberosum, Verticillium = **Verticillium albo-atrum** 1021
albobadia, Dendrophora (Peniophora) 667
albobadium, Stereum = **Dendrophora albobadia** 667
albobrunnea, Antrodia (Poria) 567
albobrunneus, Polyporus = **Antrodia albobrunnea** 567
albocarnea, Cyphella = **Calathella eruciformis** 597
albocincta, Phlyctema 850
albocincta, Valsa = **Diaporthe acerina** 670
albocincta, Zignoella = **Exarmidium morthieri** 702
albocinctum, Hypoxylon 744
albocitrinus, Dasyscyphus = **Lachnum virginellum** 755
albocremea, Conohypha 643
albocremeum, (Corticium, Hyphoderma) = **Conohypha albocremea** 643
albocrenulata, Pholiota 851
alboflavescens, Ramaricium (Coniophora, Corticium) 923
albofusca, Winterella (Cryptosporella, Valsa) 1023
alboglaucum, Coronicium 645
albo-glaucum, (Corticium, Xenasma) = **Coronicium alboglaucum** 645
albogrisea, Collybia = **Marasmius albogriseus** 776
albogriseus, Marasmius 776
albolabra, Dasyscyphus (Lachnella) 664
albolanata, Rosellinia 936
albolanatus, Pleurotus 886
albolutescens, Anomoporia (Poria) 565
alboluteum, Myxosporium 810
alboluteus, Pycnoporellus (Aurantioporellus, Phaeolus, Polyporus) 917
albomaculans, Cercosporella = **Pseudocercosporella capsellae** 895
albomaculans, Dendrophoma (Sphaeria) 667
albomaculata, Ramularia 923
albomarginata, Peniophora = **Dendrophora albobadia** 667
albomyces, Melanocarpus (Myriococcum, Thielavia) 783
albo-pila, Volutella 1022
albopileata var. subaurata, Lachnella = **Dasyscyphus subauratus** 665
albopileatus, Dasyscyphus 664
albopruinosa, Diatrype 673
albopruinosa var. salicina, Diatrype = **Eutypella canodisca** 700
albopunctata, Leptosphaeria 760

albopunctata, Septoria 947
alborosea, Poria = **Physisporinus vitreus** 876
albo-rosella, Isariopsis 749
albo-rubescens, Sporobolomyces 971
albosordescens, Polyporus = **Tyromyces fissilis** 1005
albostramineus, Hypochnus = **Hypochnicium eichleri** 743
albostramineus, Meruliopsis (Ceraceomerulius, Merulius) 787
alboviolascens, Lachnella (Cyphella, Cyphellopsis) 753
albovirens, Helotium = **Claussenomyces prasinula** 631
alboviride, (Mycoacia, Oxydontia) = **Kavinia alboviridis** 752
alboviridis, Kavinia 752
alboviridis, (Peziza, Trichopeziza) = **Eriopezia microspora** 697
albugo, Peniophora = **Hyphoderma praetermissum** 741
albula, Peniophora = **Cylindrobasidium torrendii** 656
albulensis, Puccinia 900
albulensis subsp. cascadensis, Puccinia = **P. cascadensis** 903
albulum, (Corticium, Cylindrobasidium) = **Cylindrobasidium torrendii** 656
album, Acrospermum 557
album, (Cladosporium, Hyalodendron) = **Ramularia alba** 923
album, Cylindrocarpon (Fusarium) 656
album, Graphium (Ceratopodium) 723
album, Mastigosporium 780
album, Microstroma (Fusisporum) 791
album, Troposporium 1001
album var. majus, Cylindrocarpon = **Cylindrocarpon album** 656
albuminum, (Calycina, Helotium) = **Pezizella vulgaris** 838
albus, Hypochnus = **Tubulicium clematidis** 1003
albus, Merulius = **Wolfiporia cocos** 1024
albus, Sporobolomyces = **Bullera alba** 596
albuscorticis, Marasmiellus = **M. candidus** 776
alcalina, Mycena 797
alchemillae, Sphaerotheca = **S. aphanis** 968
alcides, Phyllosticta 864
aletridis, Gloeosporium = **Colletotrichum gloeosporioides** 638
aletridis, Puccinia 900
aleuriata, Stilbella 981
aleuriatum, Stilbum = **Stilbella aleuriata** 981
aleurina, (Eutypella, Valsa) = **Eutypella platani** 701
aleuriticum, Gloeosporium = **Colletotrichum gloeosporioides** 638
aleuritis, Mycosphaerella 800
aleuritis, Pseudocercospora (Cercospora, Cercosporina) 893
alfalfae, Physoderma (Cladochytrium, Urophlyctis) 876
aliena, Diaporthe 670
alienum, Basidioradulum = **Hyphoderma litschaueri** 741
alismatis, Cercospora 605
alismatis, Doassansia 686
alismatis, Pleospora = **Massariosphaeria rubelloides** 779
alismatis, Ramularia (Didymaria, Septoria) 923
allantoidea, Rhabdospora (Septoria) 930
allantoideum, Gloeosporium = **Phlyctema vagabunda** 850
allantospora, Calospora = **Romellia cornina** 936
allantospora, Endoxylina (Valsaria) 695
allantospora, Phanerochaete (Grandiniella) 843
allantospora, Tulasnella 1003
allantosporum, Gloeosporium = **Phlyctema vagabunda** 850
allardii, Septoria 947
alleghaniensis, Diaporthe 670
allegheniensis, Phyllosticta 864
allenii, (Aecidium, Puccinia) = **Puccinia caricis-shepherdiae** 902
allescheri, Peniophora = **Hyphoderma mutatum** 741
allescherianum, Gloeosporium = **Colletotrichum gloeosporioides** 638
alli, Alternaria = **A. porri** 562
alliacea, Dendrothele 667
alliacea, Odontia = **Trechispora fastidiosa** 997
alliaceum, Aleurocorticium = **Dendrothele alliacea** 667
alliaceus, Aleurodiscus = **Dendrothele alliacea** 667
alliaceus, Aspergillus 578
alliariifoliae, Phyllosticta 864
allicina, Mycosphaerella (Sphaerella) 800
allicola, Phoma 851
allii, Botrytis = **B. aclada** 594
allii, Cladosporium (Heterosporium) 629
allii, Embellisia (Helminthosporium) 693
allii, Phyllosticta 864
allii, Pleospora (Sphaeria) 883
allii, Puccinia 900
allii var. allii-porri, Heterosporium = **Cladosporium allii** 629
allii-porri, Cladosporium = **C. allii** 629
allostoma, (Eutypa, Valsa) = **Eutypella juglandicola** 700
alnea, Byssosphaeria (Cucurbitaria, Gibberidea, Massarina) 596
alnea, Cylindrosporella = **Asteroma alneum** 581
alnea, Melasmia = **Leptothyrium alneum** 765
alnea, Phomopsis (Phoma) 857
alnea, Phyllosticta 864
alnea, Torula = **Taeniolella alta** 985
alnea, Tympanis 1004
alneum, Asteroma (Gloeosporium, Ophiodothis) 581
alneum, Lachnum 754
alneum, Leptothyrium 765
alneum, Plagiostoma 881
alni, Ciboria (Sclerotinia) 627
alni, Cylindrosporium 658
alni, Ditopellopsis (Mamiania) 686
alni, Gnomonia 720
alni, Melampsoridium = **M. betulinum** 781
alni, Melanconis 781
alni, Microsphaera = **M. penicillata** 790
alni, Monostichella (Gloeosporium) 796
alni, Mycopappus (Cercosporella) 799
alni, Passalora (Cercospora) 826
alni, Pezicula 837
alni, Platychora (Dothidella) 882
alni, Septoria 947
alni, Sphaeropsis 966
alni, Taphrina 986
alni, Valsa = **V. melanodiscus** 1018
alni var. calocladophora, Microsphaera = **M. calocladophora** 790
alni var. cinnamomi, Microsphaera = **M. cinnamomi** 790
alni var. extensa, Microsphaera = **M. extensa** 790
alni var. ludens, Microsphaera = **M. ludens** 790
alni var. marginalis, Melanconis = **M. marginalis** 782
alni var. vaccinii, Microsphaera = **M. vaccinii** 791
alnicola, Aplosporella (Sphaeropsis) 569
alnicola, Calosphaeria = **Togninia minima** 993
alnicola, Crepidotus 647
alnicola, Crinipellis = **Phaeomarasmius erinaceus** 840
alnicola, Gloeosporium = **Asteroma alneum** 581
alnicola, Karstenula 752
alnicola, Mycena 797
alnicola, Mycosphaerella 800
alnicola, Pholiota (Flammula) 851
alnicola, Ramularia 923
alnicola, Trechispora (Grandinia) 997
alnicola, Valsella 1019
alnicola, Winterella 1023
alniella, Phialea = **Hymenoscyphus alniellus** 740
alniellus, Hymenoscyphus 740
alnifolia, Septoria 947
alnifraga, Eutypella = **E. cerviculata** 700
alni-incanae, Taphrina = **T. alni** 986
alniphila, Pholiota (Naucoria) 851
alni-rubrae, Diplodia = **D. sarmentorum** 681
alniseda, Hyaloscypha 738
alni-viridis, Sphaerella 965
alnivorus, Coprinus 644
aloes, Leptosphaeria = **Paraphaeosphaeria obtusispora** 826
alopallonella, Humicola 738
alopecuri, Dilophospora 679
alopecuri, Fusicladium 711
alopecuri, Uromyces = **U. dactylidis** var. **poae** 1010
alopecuri var. airae, Septoria = **Phaeoseptoria airae** 841
aloysii-sabaudiae, Patinella 827
alpestre, Bostrichonema = **B. polygonia** 590
alpestre, Melanomma 783
alpestris, Trichothyrina 1000
alpigenum, Lophiostoma (Lophiotrema) 770
alpina, Antennulariella 565
alpina, Antrodia (Poria) 567
alpina, Apiosporella 569
alpina, Cyathicula = **Crocicreas cyathoideum** var. **cacaliae** 648
alpina, Guepiniopsis 724
alpina, Kellermania (Heteropatella) 752
alpina, Lasmeniella (Aplosporella) 757
alpina, Mycosphaerella 800
alpina, Physoderma 876
alpina, Protoventuria (Antennularia, Venturia) 893
alpina, Sphaerotheca 968
alpina, Stigmatea = **Seynesiella juniperi** 961
alpinicola, Phyllosticta 864
alpinum, Cenangium 602
alpinum, Dimerium = **Antennulariella alpina** 565
alpinum, Lophodermium 771
alpinum, Triblidium 998
alpinus, Heterotextus = **Guepiniopsis alpina** 724
alsineae, Ustilago 1014
alsinearum, Peronospora 832
alsines, Entyloma 695
alta, Peronospora 832

alta, Taeniolella 985
altera, Mycosphaerella 800
alternans, Puccinia = **P. recondita** 912
alternantherae, Alternaria 561
alternantherae, Cercospora 605
alternarina, Pyrenophora = **Lewia scrophulariae** 767
alternata, Alternaria 561
althaeina, Ascochyta = **Phoma exigua** 853
althaeina, Cercospora 605
althaeina, Phyllosticta = **Phoma exigua** 853
althaeina var. praecincta, Cercospora = **C. praecincta** 615
altissimorum, Puccinia 900
altissimorum var. **horriduli, Puccinia** 900
altissimus, Phaeostalagmus 842
alutacea, (Hyphodontia, Odontia) = **Grandinia stenospora** 723
alutacea, Skeletocutis (Incrustoporia, Poria) 962
alutaceum, Corticium = **Vesiculomyces citrinus** 1022
alutaceum, Hydnum = **Grandinia stenospora** 723
alutaria, Grandinia (Hyphodontia, Peniophora) 722
alutipes, Chloroscypha (Kriegeria) 625
alveolaris, (Favolus, Polyporus) = **Polyporus mori** 889
alyxiae, Amazonia 563
alyxiae, Elsinoe 692
alyxiae, Guignardia 724
alyxiae, Meliola = **M. moerenhoutiana** 785
alyxiae, Uromyces 1009
amara, Postia 890
amaranthi, Albugo 559
amaranthi, Alternaria 561
amaranthi, Gloeosporium = **Colletotrichum gloeosporioides** 638
amaranthi, Peronospora = **P. farinosa** 833
amaranthi, Phoma 851
amaranthi, Phyllosticta 864
amaranthicola, Gloeosporium 717
amaranthicola, Phoma 851
amarissimus, Gymnopilus 725
amarus, Crepidotus 647
amarus, (Fomes, Oligoporus, Polyporus, Tyromyces) = **Postia amara** 890
amaryllidis, Cercospora 605
ambiens, Cytospora = **C. leucosperma** 662
ambiens, Helicoma (Helicosporium) 730
ambiens, Valsa 1017
ambiens f. **crataegi, Valsa** 1017
ambiens f. eleagni, Valsa = **Valsa ambiens** 1017
ambiens subsp. **leucostomoides, Valsa** 1017
ambiens f. **rosae, Valsa** 1017
ambigua, Botryosphaeria = **B. quercuum** 593
ambigua, Columnocystis 640
ambigua, Daedalea = **Trametes elegans** 996
ambigua, Diaporthe = **D. eres** 671
ambigua, Phomopsis (Phoma) 857
ambigua, Pleospora = **P. penicillus** 885
ambigua, Poria = **Oxyporus latemarginatus** 824
ambigua, Puccinia = **P. difformis** 905
ambigua, Sphaeropsis = **Phaeocytostroma ambiguum** 839
ambiguum, Helicoma = **Helicomyces ambiguus** 731
ambiguum, Melogramma = **Botryosphaeria quercuum** 593
ambiguum, Phaeocytostroma 839
ambiguum, Stereum = **Columnocystis ambigua** 640
ambiguus, Helicomyces 731
ambiguus, Meruliopsis (Byssomerulius, Merulius) 787
ambiguus, Mucor = **M. circinelloides** 796
ambrosia, Ceratocystis = **Ophiostoma piliferum** 821
ambrosiae, Erysiphe = **E. cichoracearum** var. **latispora** 697
ambrosiae, Mycovellosiella (Ormathodium) 807
ambrosiae, Oidium 818
ambrosiae, Phyllachora (Dothidea) 862
ambrosiae, Phyllosticta 864
ambrosiae, Stagonospora 973
ambrosiaecola, Metasphaeria 787
ambrosiaecola var. **praetans, Metasphaeria** 787
ambrosiicola, Septoria 947
ambrosioides, Phyllosticta 864
amelanchieri, Taphrina 986
amelanchieris, Diplodia 680
amelanchieris, Monilinia 794
amelanchieris, Pleospora 884
amelanchieris, Sphaeropsis 966
amelanchieris, Tympanis 1004
amentacea, Ciboria = **C. caucus** 627
amenti, Crocicreas (Ciboria, Helotium, Pezizella) 648
amenticola, Mollisia 793
amentorum, Taphrina = **T. alni** 986
americana, Actinopelte = **Tubakia dryina** 1002
americana, Anthracoidea (Cintractia) 566
americana, Antrodiella 568
americana, Asteridiella 580
americana, (Cadophora, Phialophora) = **Phialophora verrucosa** 847
americana, Choanephora = **C. cucurbitarum** 626
americana, Ciboria (Rutstroemia) 627
americana, Diatrype = **D. albopruinosa** 673
americana, Didymochaeta = **D. fraserae** 677
americana, Erysiphe 697
americana, Fracchiaea = **Nitschkia broomeiana** 816
americana, Harknessia 729
americana, Leptosphaerulina (Pleospora) 764
americana, Melampsora = **M. abieti-capraearum** 781
americana, Peronospora 832
americana, Phoma 851
americana, Phyllachora 862
americana, Serpula = **S. himantioides** 960
americana, Stamnaria 975
americana, Taphrina 986
americana, Trametes = **Gloeophyllum odoratum** 717
americana, Trimmatostroma 1001
americana, Uredinopsis 1007
americana, Urosporella = **U. alabamensis** 1014
americana, Valsa = **V. ceratosperma** 1017
americana, Xerula 1025
americana f. **ostryae, Diatrype** 673
americana f. **quercus, Diatrype** 673
americanum, Cladosporium 629
americanum, Gloeosporium = **Stagonospora meliloti** 974
americanum, Hericium 734
americanum, Melanconium 782
americanum, (Ostreichnion, Ostreion) = **Ostreichnion sassafras** 823
americanum, Phragmidium 860
americanum, Pocillum 887
americanum, Pucciniastrum 916
americanus, Merulius = **Serpula himantioides** 960
americanus, Radulodon 923
americanus, Sirococcus 961
americanus, Uromyces 1009
americanus var. **californicus, Uromyces** 1009
amethionina var. **pachycereana, Pichia** 879
amianthina, Trechispora 997
amianthinum, Corticium = **Trechispora amianthina** 997
amica, Otthia 823
amicta, Phyllosticta 864
ammanniae, Cercospora 606
ammodeniae, Septoria 947
ammophilae, Mollisina 794
ammophilae, Phyllachora 862
ammophilina, Puccinia = **P. pygmaea** var. **ammophilina** 912
amoeboides, Reessia 929
amoena, Peniophora = **Hyphoderma amoenum** 740
amoena var. petiolorum, Gnomoniella = **Gnomonia petiolorum** 721
amoenum, Hyphoderma 740
amoenus, Uromyces 1009
amorpha, Fenestella = **Thyridium vestitum** 991
amorpha, Skeletocutis 962
amorphae, Cucurbitaria 654
amorphae, Cytospora = **C. sacculus** 662
amorphae, Diaporthe 670
amorphae, Diplodia 680
amorphae, Eutypella = **E. tetraploa** 701
amorphae, Sphaeropsis 966
amorphae, Uropyxis (Puccinia) 1014
amorphae, Valsa = **V. ambiens** 1017
amorphus, Aleurodiscus 560
amorphus, Polyporus = **Skeletocutis amorpha** 962
ampelicida, Phyllosticta 864
ampelina, Ascochyta (Pseudodiplodia) 573
ampelina, Cryptovalsa 654
ampelina, Diplodia 680
ampelina, Elsinoe 692
ampelina, Micropera 790
ampelina, Phoma (Phomopsis) 851
ampelina, Septoria 947
ampelinum, Coniothyrium 642
ampelinum, Sphaceloma 964
ampelophaga, Briosia 595
ampelophagum, Gloeosporium = **Sphaceloma ampelinum** 964
ampelopsidis, Cercospora 606
ampelopsidis, Diaporthe 670
ampelopsidis, Macrophoma (Sphaeropsis) 773
ampelopsidis, Phloeospora (Gloeosporium, Phloeospora, Rhabdospora, Septogloeum, Septoria) 849
ampelopsidis, Phyllosticta = **P. ampelicida** 864
ampelopsidis, Physopella 877
ampelopsidis, Valsa 1017
amphidymus, Uromyces 1009
amphigena, Puccinia 900
amphimelas, Charomyces 625
amphipila, Urceolella 1006
amphispilusa, Puccinia 900
amphitricha, Meliola 784
ampla, Cyphella = **Cytidia flocculenta** 661
ampla, Davisomycella (Hypodermella, Lophodermium) 666
ampla, Phialea = **Hymenoscyphus conscriptus** 740

ampla, Pyrenophora = **Pleospora anthyllidis var. abbreviata** 884
amplectens, (Inonotus, Polyporus) = **Phylloporia frutica** 864
amplispora, Pleospora 884
amplispora, Sphaeropsis 966
amplissimum, Hydnum = **Sarcodontia setosa** 938
amplum, Helotium = **Hymenoscyphus conscriptus** 740
ampullacea, Sphaeria = **Splanchnonema ampullaceum** 969
ampullaceum, Splanchnonema 969
amstelodami, Aspergillus = **A. hollandicus** 578
amstelodami, Eurotium 699
amygdali, Fusicoccum 711
amygdalina, Aplosporella 569
amygdalinum, Anthostoma 565
amygdalinum, Gloeosporium = **Colletotrichum gloeosporioides** 638
amygdalinus, Polyporus = **P. virgatus** 889
amygdalosporus, Crepidotus 647
amylacea, Amyloathelia 564
amylaceus, Aleurodiscus = **Amyloathelia amylacea** 564
amylodextrinoidea, Perenniporia 830
amylohypha, Perenniporia = **P. narymica** 830
anacardiacearum, Phyllosticta 864
anagallidis, Peronospora = **P. candida** 832
anagallidis, Ramularia 923
anagallidis, Septoria 947
analogum, Hypochnicium (Corticium) 742
ananasi, Rhizidiocystis 932
anandrum, Pythium 920
anasaziense, Hyphoderma 740
anceps, Ceratobasidium (Corticium, Tulasnella) 603
anceps, Phoma 851
anceps, Polyporus = **Dichomitus squalens** 675
anceps, Rhopographus 934
ancilis, Discina 684
andersonii, Dothidotthia (Gibbera) 688
andersonii, Inonotus (Mucronoporus, Poria) 748
andersonii, Pestalozziella 837
andersonii, Phlyctema 850
andersonii, Phragmidium 860
andersonii, Scoriomyces 943
andersonii, (Mycosphaerella, Sphaerella) = **Didymella douglasii** 676
andina, Puccinia 900
andinus, Protomyces 892
andrewsii, Asteromella 582
andrewsii, Mycosphaerella 800
andromedae, Clithris (Sporomega) 633
andromedae, Diplodina 682
andromedae, Durandiella (Cenangium) 691
andromedae, Exobasidium = **E. vaccinii** 703
andromedae, Extrawettsteinina (Leptosphaeria, Wettsteinina) 704
andromedae, Gibbera (Coleroa) 714
andromedae, Lembosiopsis (Lembosia) 758
andromedae, Mycosphaerella = **Botryosphaeria vaccinii** 593
andromedae, Phyllosticta 864
andromedae, Ramularia 923
andromedae, Rhytisma (Placuntium) 935
andromedae var. cassiopes, Wettsteinina = **Lophiostoma sieversiae** 770
andromedae-ligustrinae, (Rhytisma, Xyloma) = **Rhytisma decolorans** 935
andropogoni, Cylindrosporium 658
andropogonis, (Belonium, Hysteropezizella) = **Nimbomollisia aberrans** 816
andropogonis, Cercoseptoria 605
andropogonis, Cerebella (Epicoccum) 621
andropogonis, Clypeoporthe (Gnomonia) 633
andropogonis, Didymella 676
andropogonis, Didymosphaeria (Microthelia) 678
andropogonis, Leptostromella 765
andropogonis, Phyllachora = **P. americana** 862
andropogonis, Puccinia 900
andropogonis, Septoria 947
andropogonis, Sporisorium (Sphacelotheca) 971
andropogonis, Stagonospora 973
andropogonis, Uromyces 1009
andropogonis, Ustilago = **Sphacelotheca occidentalis** 965
andropogonis var. gerardiae, Puccinia = **Puccinia andropogonis** 900
andropogonis var. melampyri, Puccinia = **Puccinia andropogonis** 900
andropogonis var. micropuncta, Puccinia = **Puccinia andropogonis** 900
andropogonis var. onobrychidis, Puccinia = **Puccinia andropogonis** 900
andropogonis var. oxalidis, Puccinia = **Puccinia andropogonis** 900
andropogonis var. pentstemonis, Puccinia = **Puccinia andropogonis** 900
andropogonis var. polygalina, Puccinia = **Puccinia andropogonis** 900
andropogonis var. pustulata, Puccinia = **Puccinia andropogonis** 900
andropogonis var. sorghastri, Septoria 947
andropogonis f. sporobolicola, Septoria 947
andropogonis var. xanthoxyli, Puccinia = **Puccinia andropogonis** 900
andropogonis-hirti, Puccinia 900
andropogonis-hirtifolii, Sphacelotheca (Ustilago) 965
andropogonivora, Phyllosticta 864
androsaceus, Marasmius 776
aneimiae, Desmella 670
aneirina, Ceriporiopsis (Poria) 621
anemone, Phoma 851
anemones, Ascochyta = **A. infuscans** 575
anemones, Phloeospora = **Septoria punicea** 956
anemones, Phyllosticta = **P. anemonicola** 864
anemones, Septoria 947
anemones, Synchytrium 983
anemones, Tranzschelia 997
anemones, Urocystis 1008
anemones-virginianae, Puccinia 900
anemonicola, Phyllosticta 864
anethi, Cercospora = **Cercosporidium punctum** 620
anethi, Cercosporella 619
anethi, Phoma 851
anethiana, Didymaria 676
angelicae, Cercospora 606
angelicae, Cercosporidium (Fusicladium) 620
angelicae, Leptostromella 765
angelicae, Phyllosticta 864
angelicae, Puccinia 900
angelicae, Ramularia = **R. archangelicae** 923
angelicae, Septogloeum (Gloeosporium) 946
anglica, Stagonospora 973
angraeci, Cercospora 606
anguillidus, Ophiobolus (Sphaeria) 819
anguioides, Fusarium = **F. avenaceum** 707
angulare, Diatrype = **Dothivalsaria megalospora** 689
angularis, Septoria (Septoria) 947
angulata, Cercospora 606
angulata, Mycosphaerella 800
angulata, Phyllactinia 863
angulata, Puccinia 900
angulatum, Tricladium 1001
angulopora, Postia (Fibroporia) 890
angulosa, (Eutypella, Valsa) = **Eutypella sorbi** 701
angulosa f. fraxini, Eutypella 700
angulosa f. negundinis, Eutypella 700
angulospora, Tomentella 993
angusiana, Periconiella 831
angusta, Hansenula = **H. polymorpha** 728
angusta, Peniophora = **Tubulicrinis angustus** 1003
angustata, Hohenbuehelia 736
angustata, Puccinia 900
angustata, Ramularia 923
angustata, Truncatella 1001
angustata var. angustatoides, Puccinia = **P. angustatoides** 900
angustata var. eriophori, Puccinia = **P. eriophori** 906
angustatoides, Puccinia 900
angustatum, Hysterium 747
angustatum, Pythium 920
angustatus, Panus = **Hohenbuehelia angustata** 736
angusticollis, Ceratocystis 603
angustifolia, Septoria 947
angustifoliae, Cylindrosporium = **Stigmina concentrica** 979
angustiforme, Aulographum = **Echidnodella angustiformis** 691
angustiformis, Echidnodella (Lembosia, Morenoella) 691
angustilabrum var. crenatum, Lophiostoma 770
angustior, Monilia 794
angustiovoidea, Alternaria 561
angustispora, Idriella 748
angustisporum, Exobasidium 702
angustissima, Septoria 947
angustum, Cylindrocarpon 656
angustum, Fusarium = **F. oxysporum** 708
angustum, Melanconium 782
angustus, Tubulicrinis 1003
anisacanthi, Puccinia 900
anisometra, Metasphaeria (Sphaeria) 787
anisopleura, Xylaria (Xylosphaera) 1025
annae, Metasphaeria 787
annelidica, Filosporella (Rogersia) 705
annexa, Coronophora (Cryptosphaerella) 645
annosa, Fomitopsis = **Heterobasidion annosum** 735
annosum, Heterobasidion 735
annosus, Fomes = **Heterobasidion annosum** 735
annua, Septoria = **S. macropoda** 953
annulans, Diatrypella 674
annularis, Cytospora = **C. pruinosa** 662
annularis, Ganoderma (Fomes) 713
annulata, Cercospora 606
annulata, Cheiromycella = **Eversia subopaca** 701
annulata, Cytospora 661
annulata, Mycosphaerella (Sphaerella) 800
annulata, Pestalotiopsis (Pestalotia) 836
annulata, Ravenelia 928
annulata, Stictis = **S. radiata** 978
annulata, Stilbella 981

annulatum, Hypoxylon = **H. truncatum** 746
annulatum, Rhizophydium 933
annulatum, Stilbum = **Stilbella annulata** 981
annuum, Fusarium 707
anomala, Anisogramma (Apioporthe, Cryptosporella) 564
anomala, Burrillia = **Nannfeldtiomyces anomalus** 811
anomala, Cercospora 606
anomala, Cyphellopsis (Cyphella, Solenia) 660
anomala, Hansenula 728
anomala, Kellermania (Discella, Discula) 752
anomala, Limacinula (Asterina, Chaetasterina, Phaeosaccardinula) 768
anomala, (Cryptostictis, Pestalotia) = **Seimatosporium anomalum** 944
anomala, Phialea = **Lanzia longipes** 756
anomala, Puccinia = **P. hordei** 908
anomala, Ramularia 923
anomala, Ustilago 1014
anomalum, Seimatosporium 944
anomalum, Synchytrium 983
anomalus, Henningsomyces = **Cyphellopsis anomala** 660
anomalus, Merismodes = **Cyphellopsis anomala** 660
anomalus, Nannfeldtiomyces 811
anomia, Aglaospora = **A. profusa** 559
anserinae, Phyllosticta 864
ansosa, Pellicularia = **Botryobasidium ansosum** 591
ansosum, Botryobasidium 591
antennariae, Globulina 715
antennariae, Phoma 851
antennariae, Phyllosticta 864
antennatum, Camposporium 600
antenniforme, Sirosporium (Alternaria, Macrosporium) 961
anthaenantiae, Cerebella = **C. andropogonis** 621
anthelmintica, Cercospora = **C. beticola** 606
anthelmintica, Leptosphaeria (Metasphaeria, Sphaeria) 760
antherarum, Ustilago = **U. violacea** 1016
anthophila, Botrytis 594
anthophila, Venturia 1020
anthophilum, Sporotrichum = **Fusarium poae** 710
anthoxanthi, Tilletia = **T. holci** 992
anthoxanthina, Septoria 947
anthoxella, Phyllosticta 864
anthrisci, Ophiobolus 819
anthyllidis, Pleospora 884
anthyllidis, Uromyces 1009
anthyllidis var. **abbreviata, Pleospora** 884
antibiotica, Verticicladiella = **Leptographium antibioticum** 759
antibioticum, Leptographium 759
antigoni, Phaeoramularia (Cercospora) 840
antipus, Phaeoramularia (Cercospora) 840
antiqua, Monodictys 795
antiquum, Sporidesmium = **Monodictys antiqua** 795
antirrhini, Cercospora 606
antirrhini, Colletotrichum = **C. gloeosporioides** 638
antirrhini, Peronospora 832
antirrhini, Phyllosticta 864
antirrhini, Puccinia 900
antirrhinorum, Septoria 948
anychiae, Septoria 948
aorista, Diaporthe = **D. linearis** 671
apacheca, Uredo 1007
apacheriensis, Schizopora (Poria) 939
aparines, (Leptosphaeria, Pleospora) = **Leptosphaeria galiorum** 762
aparines, Septoria = **S. cruciatae** 950
apatela, Valsa = **Phragmodiaporthe caryae** 861
apatosa, (Valsa, Valsaria) = **Massariovalsa sudans** 780
aphanidermatum, Pythium (Nematosporangium, Rheosporangium) 920
aphanis, Sphaerotheca 968
aphyophyllicum, Myriophacidium 809
apicale, (Acrothecium, Brachysporium, Helminthosporium) = **Brachysporium nigrum** 595
apicalis, Marssonina (Marssonia) 777
apicalis, Phyllosticta 864
apiculata, Atropellis 584
apiculata, Delitschia = **Arnium apiculatum** 572
apiculata, Xylaria 1025
apiculatum, Arnium 572
apiculatum, Gonatobotryum (Haplographium) 722
apiculatum, Lophodermium 771
apiculatus, Aleurodiscus = **Acanthophysium mirabile** 556
apiculosa, Diaporthopsis (Diaporthe) 672
apii, Cercospora 606
apii, Phyllosticta 865
apii, Septoria = **S. apiicola** 948
apii var. angelicae, Cercospora = **C. angelicae** 606
apii var. pastinacae, Cercospora = **C. pastinacae** 614
apii var. petroselini, Cercospora = **Cercosporidium punctum** 620
apii var. selini-gmelini, Cercospora = **C. selini-gmelini** 616
apii f. sp. **clerodendri, Cercospora** 606
apiicola, Phoma 852
apiicola, Septoria 948
apiifoliae, Cercospora = **C. crataegi** 608
apii-graveolentis, Septoria = **S. apiicola** 948
apiocarpum, Melanconium 782
apiospora, Apioporthe (Diaporthe) 569
apiosporoides, Metasphaeria 787
apiosporus, Uromyces 1009
aplectri, Fusicladium 711
aplectri, Phyllosticta 865
aplopappi, Dasyscyphus 664
apobasidialis, Chionosphaera 625
apoclastosporum, Lophionema = **Glyphium corrugatum** 720
apocrypta, Diaporthe 670
apocrypta, Kabatiella = **Aureobasidium apocryptum** 584
apocrypta, Melanconis 781
apocrypta, Phoma 852
apocrypta, Puccinia = **P. recondita** 912
apocryptum, Aureobasidium (Gloeosporium, Microstroma) 584
apocryptum var. ramicola, Gloeosporium = **Colletotrichum gloeosporioides** 638
apocynaceum, Coleosporium 635
apocyni, Botryosphaeria (Laestadia) 592
apocyni, Cercospora (Cercosporella) 606
apocyni, Cylindrosporium 658
apocyni, Nectria 811
apocyni, Phyllosticta 865
apocyni, Stagonospora (Septogloeum, Stagonospora) 973
apoda, Phakopsora (Puccinia) 843
aposphaerioides, Phoma = **P. cava** 852
appalachianensis, Cercophora 605
appalachianensis, Crepidotus 647
appalachianum, Cronartium (Peridermium) 649
appalachiensis, Corticium (Laeticorticium) 645
appendicularis, Papulaspora 825
appendiculata, (Diaporthe, Melanconis) = **Prosthecium appendiculatum** 892
appendiculata, Nigredo = **Uromyces appendiculatus** 1009
appendiculatum, Lophiostoma = **L. caulium** 770
appendiculatum, Prospodium (Puccinia) 892
appendiculatum, Prosthecium 892
appendiculatus, Uromyces 1009
applanata, Didymella (Didymosphaeria) 676
applanata, Kirschsteiniella (Amphisphaeria, Microthelia, Trematosphaeria) 752
applanata, Mycosphaerella (Sphaerella) 800
applanata, Nectria = **N. purtonii** 813
applanata, Venturia 1020
applanatum, Coniothecium 642
applanatum, Ganoderma 713
applanatus, (Fomes, Polyporus) = **Ganoderma applanatum** 713
applanatus var. **globigera, Crepidotus** 647
applicatus, Resupinatus (Pleurotus) 930
approximans, Hohenbuehelia 737
aquaeductuum, Fusarium (Selenosporium) 707
aquaeductuum var. dimerum, Fusarium = **Microdochium dimerum** 789
aquatica, Bacillispora 585
aquatica, Clavariopsis 632
aquatica, Lemonniera 758
aquatica, Mycosphaerella 800
aquatica, Pestalotiopsis (Pestalotia) 836
aquatica, Pleospora = **Pyrenophora scirpi** 919
aquatica var. **luzulicola, Stagonospora** 973
aquaticum, Camposporium = **Vargamyces aquaticus** 1019
aquaticus, Heliscus = **H. lugdunensis** 731
aquaticus, Vargamyces 1019
aquatilis, Colletotrichum 637
aquatilis, Ramularia 923
aquifoliae, Pseudonectria (Hymenoscyphus, Peziza, Pezizella) 897
aquifolii, Gloeosporium 717
aquifolii, Nectria 811
aquila, Rosellinia 936
aquilegiae, Ascochyta (Ascochyta, Phyllosticta) 573
aquilegiae, Cercospora 606
aquilegiae, Septoria 948
aquilegiae, Sphaerella = **Mycosphaerella aquilegiae-jonesii** 800
aquilegiae-jonesii, Mycosphaerella 800
aquilegicola, Phyllosticta 865
aquilina, Cyathicula = **Mollisia pteridina** 793
arabicola, Belonium 587
arabidis, Septoria 948
arabidis var. **macrospora, Septoria** 948
arachidicola, Cercospora 606
arachidicola, Leptosphaerulina = **L. crassiasca** 764
arachidicola, Mycosphaerella = **M. arachidis** 800
arachidicola, Phoma 852
arachidis, Cercospora = **Phaeoisariopsis personata** 840
arachidis, Mycosphaerella 800
arachidis, Puccinia 900
arachidis, Septogloeum 946
arachidis var. **offuscata, Puccinia** 900

arachidis-hypogaea, Phyllosticta 865
arachispora, Amphinema 563
arachnoidea, Athelia 583
arachnoidea, Cercosporella 619
arachnoidea, Mycosphaerella 800
arachnoidea, Peniophora = **Phanerochaete sordida** 844
arachnoideum, Corticium = **Athelia arachnoidea** 583
arachnoideum, Scytinostroma 944
araiosporum, Pythium = **P. dissotocum** 920
araliae, Alternaria 561
araliae, Aplosporella 569
araliae, Appendiculella (Amphitrichum, Irene, Meliola) 571
araliae, Diaporthe 670
araliae, Phyllosticta = **P. everhartii** 868
araliae, Puccinia 901
araliae, Septoria 948
araliae, Sphaceloma 964
araneosum, (Corticium, Tomentella) = **Trechispora farinacea** 997
araucaria, Clonostachys 633
araucaria var. **rosea, Clonostachys** 633
araucariae, Cryptosporella (Melanops) 652
araucariae, Phyllosticta 865
araucariae, Stictis = **S. monilifera** 978
arborescens, Peronospora 832
arborescentis, Cercospora = **C. hydrangeae** 611
arboriae, Cercospora 606
arbuscula, Allomyces 560
arbuscula, Xylaria 1025
arbusculus, Phaeostalagmus 843
arbuti, Phacidium 838
arbuti, Rhytisma 935
arbuti, Seimatosporium (Cryptostictis) 944
arbuticola, Didymosporium 678
arbuticola, Micromphale 789
arbuticola, Mycosphaerella 800
arbutifoliae, Mycosphaerella 800
arbutifoliae, Phyllosticta 865
arceuthobii, Caliciopsis (Wallrothiella) 597
archangelicae, Ramularia 923
archeri, Phomopsis 857
archerianus, Uromyces 1009
arctespora, Corynespora 646
arctesporum, Helminthosporium = **Corynespora arctespora** 646
arcti-ambrosiae, Cercospora 606
arctica, Calosphaeria 598
arctica, Cintractia = **Orphanomyces arcticus** 823
arctica, Discocainia 684
arctica, Labrella = **Leptothyrium arcticum** 765
arctica, Laestadia = **Mycosphaerella tassiana** var. **arctica** 806
arctica, Melampsora 781
arctica, Pleospora = **P. heleocharidis** var. **arctica** 884
arctica, Septoria 948
arcticoceanorum, Piricauda = **Monodictys pelagica** 795
arcticum, Cenangium = **Discocainia arctica** 684
arcticum, Leptothyrium 765
arcticum, Pucciniastrum 916
arcticum var. americanum, Pucciniastrum = **P. americanum** 916
arcticus, Orphanomyces 823
arctii, Cercospora = **C. arcti-ambrosiae** 606
arctii, Diaporthe 670
arctii, Rhabdospora 930
arctii var. **achilleae, Diaporthe** 670
arctostaphyli, Arachnopeziza = **A. aurata** 571
arctostaphyli, Botryosphaeria 592
arctostaphyli, Cercospora = **C. gaultheriae** 610
arctostaphyli, Chrysomyxa 626
arctostaphyli, Coccomyces 634
arctostaphyli, Coniothyrium (Sphaeropsis) 642
arctostaphyli, Exobasidium = **E. vaccinii** var. **arctostaphyli** 703
arctostaphyli, Gloeosporium 717
arctostaphyli, Harknessia 729
arctostaphyli, Hysterographium 747
arctostaphyli, (Phacidium, Sphaeropezia) = **Eupropolella vaccinii** 699
arctostaphyli, Phellinus (Fomes) 844
arctostaphyli, Placosphaeria 881
arctostaphyli, Venturia = **Protoventuria alpina** 893
arcuata, Lophodermella (Hypodermella) 771
arcuata, Phlyctema 850
arcularium, Aecidium 558
arcularius, Polyporus 889
arecae, Phomopsis 857
arechavaletae, Puccinia 901
arechavaletae, Septoria 948
arenaria, Dendryphiella (Scolecobasidium) 668
arenaria, Neottiospora = **Tiarosporella paludosa** 991
arenaria, Stagonospora 973
arenaria var. **macrospora, Didymosphaeria** 678
arenariae, Puccinia 901
arenarium, Acrothecium = **Curvularia inaequalis** 655
areola, Mycosphaerella 800
areola, (Ramularia, Septocylindrium, Symphyosira) = **Ramularia gossypii** 925
areolata, Puccinia 901
areolatum, Corticium = **Thanatephorus cucumeris** 989
argemonensis, Deightoniella 666
argemones, Septoria = **Gloeosporium argemonis** 717
argemonis, Gloeosporium 717
argentata, Puccinia 901
argentatum, Corticium 645
argentina, Meliola 784
argentina var. africana, Meliola = **M. cyperi** 784
argentina var. **hawaiiensis, Meliola** 784
argentinensis, Leptosphaerulina (Pleosphaerulina) 764
argillacea, Peniophora = **Hyphoderma argillaceum** 740
argillacea, Phyllosticta (Ascochyta) 865
argillacea, Poria = **Aporpium caryae** 571
argillaceum, Fusarium = **F. ventricosum** 711
argillaceum, Hyphoderma 740
argillaceum, Hypoxylon 744
argillaceus, Polyporus = **Aporpium caryae** 571
argophylla, Septoria 948
argus, Splanchnonema (Massaria, Sphaeria) 969
arguta, Grandinia (Hyphodontia, Odontia) 722
argutus, Uromyces 1009
argyraea, Phyllosticta 865
argyraea, Septoria = **S. elaeagni** 950
argyrospora, Pleospora 884
argyrostigma, Phomatospora 857
argythamniae, Cercospora 606
ariae, Cercospora 606
ariae, Dermea 669
ariaefoliae, Phloeosporella (Cylindrosporium, Phloeospora) 850
aricola, Phyllosticta 865
arida, Coniophora 641
arida, Discella 683
arida, Hymenochaete 739
arida, Lachnellula (Lachnella, Trichoscyphella) 754
arida, Phyllosticta 865
arida var. **suffocata, Coniophora** 641
aridophila, Teichospora 987
aridula, Gorgoniceps 722
aridulum, Apostemidium = **Gorgoniceps aridula** 722
aridum, Coleosporium = **C. arizonicum** 635
aridum, Gloeosporium = **Discula fraxinea** 685
aridus, Dasyscyphus = **Lachnellula arida** 754
arisaemae, Streptotinia 982
arisaematis, Colletotrichum 637
arisaematis, Ramularia 923
aristata, Diaporthella 672
aristata, Rhabdospora 930
aristata, Tetraploa 988
aristidae, Balansia (Dothichloe) 586
aristidae, Dothidella 687
aristidae, Euryachora 699
aristidae, Puccinia 901
aristidae, Uromyces 1009
aristidae var. **chaetariae, Puccinia** 901
aristolochiae, Ovularia 823
aristolochiae, Phyllosticta 865
aristospora, Pezizella = **Ceratophacidium aristosporum** 604
aristosporum, Ceratophacidium 604
aristosporum, Pythium 920
ari-triphylli, Uromyces (Puccinia) 1009
arizonica, Atropellis = **A. piniphila** 584
arizonica, Bipolaris 588
arizonica, Cucurbitaria 654
arizonica, Leucogyrophana 767
arizonica, Phanerochaete (Grandiniella) 843
arizonica, Ravenelia 928
arizonicum, Coleosporium 635
arizonicum, Cronartium 649
arizonicum, Hyphoderma = **Cylindrobasidium laeve** 656
arizonicus, Inonotus 748
armeniacae, Eutypa 699
armeniacae, Taphrina 986
armeniacus, Byssomerulius = **Meruliopsis hirtellus** 787
armeniacus, Merulius = **Meruliopsis albostramineus** 787
armeriae, Phyllosticta 865
armeriae, Uromyces 1009
armeriae subsp. **pacificus, Uromyces** 1009
armillatus, Gymnopilus 725
armoraciae, Ascochyta 573
armoraciae, Cercospora 606
armoraciae, Entylomella (Entyloma, Ramularia) 696
arnicae, Phyllosticta 865
arnicale, Entyloma (Entylomella) 695
arnicalis, Puccinia 901
arnicalis, Ramularia = **Entyloma calendulae** 695
arnoldiae, Phomopsis 857
arnyi, Stemphylium 976
aromatica, Ramularia = **Septocylindrium aromaticum** 946
aromatica, Septoria 948
aromaticum, Cladosporium 629

aromaticum, Septocylindrium 946
aroniae, Cylindrosporium 658
aronici, Mycosphaerella 800
arrhenomaes, Nematosporangium = **Pythium arrhenomanes** 920
arrhenomanes, Pythium 920
arrhizus, Rhizopus 933
artemisiae, Cercosporidium = **Mycovellosiella ferruginea** 807
artemisiae, Cladosporium 629
artemisiae, Cylindrosporium 658
artemisiae, Gibberidea = **Leptosphaeria preandina** 763
artemisiae, Leptosphaeria 760
artemisiae, Pyrenopeziza 918
artemisiae, Ramularia 923
artemisiae, Rhabdospora = **R. ellisii** 930
artemisiae, Septoria 948
artemisiae, Systremma = **Dothidea puccinioides** 687
artemisiae var. **solidaginis, Pyrenopeziza** 918
artemisiae-norvegicae, Puccinia 901
artenii, Ramularia = **R. montana** 926
arthrinioides, Cladosporium 629
arthrophyma, Leptosphaeria (Gibberidea) 760
arthrosporioides, Fusarium 707
arthuriana, Physalospora = **Phyllachora ambrosiae** 862
arthurii, Peronospora 832
arthurii, Tranzschelia 997
arthurii, Uredinopsis 1007
arthurii, Ustilago 1014
arthurii var. maculata, Uredinopsis = **U. maculata** 1007
artocarpi, Colletotrichum = **C. gloeosporioides** 638
artocarpi, Mycosphaerella 800
artocarpi, Phyllosticta 865
artocreas, Discosia 684
artocreas var. **viticola, Discosia** 684
artotrogus, Pythium = **P. hydnosporum** 921
artotrogus var. macracanthum, Pythium = **P. hydnosporum** 921
arunci, Leptosphaeria 760
arundinacea, Deightoniella 666
arundinacea, Hendersonia (Sphaeria) 732
arundinacea, Hypocenia = **Stagonospora herbarum** 974
arundinacea, Massarina (Leptosphaeria) 779
arundinacea, Septoria 948
arundinacea, Stictis 978
arundinaceum, (Helminthosporium, Napicladium) = **Deightoniella arundinacea** 666
arundinaceum, Lophodermium 771
arundinaceum var. alpinum, Lophodermium = **L. alpinum** 771
arundinaceum var. culmigenum, Lophodermium = **L. culmigenum** 771
arundinaceum var. **gramineum, Lophodermium** 771
arundinariae, Aulographum 584
arundinariae, Botryosphaeria 592
arundinariae, Diatrype (Eutypella) 673
arundinariae, Didymosphaeria (Microthelia) 678
arundinariae, Discocurtisia (Belonium, Pyrenopeziza, Scleroderris) 684
arundinariae, Eupropolella (Sphaeropezia) 699
arundinariae, Lachnum (Lachnella) 754
arundinariae, Mycosphaerella 800
arundinariae, Phyllachora = **P. shiraiana** 863
arundinariae, Puccinia 901
arundinis, Arthrinium (Papularia) 572
arundinis, Ascospora = **Stagonospora montagnei** 974
arundinis, Lophiostoma 770
arvensis, Heptameria (Leptosphaeria) 734
arvensis, Ramularia 923
arxii, Phyllosticta 865
asari, Synchytrium 983
asaricola, Septoria 948
asarifolia, Laestadia 755
asarina, Puccinia 901
aschersoniana, Ciboria (Sclerotinia) 627
asclepiadea, Diplodia 680
asclepiadea, Ditopella 686
asclepiadea, Phoma 852
asclepiadeum var. quercuum, Cronartium = **C. quercuum** 649
asclepiadeum var. thesii, Cronartium = **C. comandrae** 649
asclepiadicola, Ascochyta 573
asclepiadicola, Septoria 948
asclepiadicola f. **syriaca, Septoria** 948
asclepiadis, Ascochyta 573
asclepiadis, Asperisporium (Passalora, Scolicotrichum) 579
asclepiadis, Belaina 587
asclepiadis, Cercospora 606
asclepiadis, Leptosphaeria 760
asclepiadis, Uromyces 1009
asclepiodorae, Cercospora 606
ascyrinum, Fusicladium 711
asiatica, Poria = **Postia sericeomollis** 891
asiatica, Septoria 948
asiaticum, Gymnosporangium 726
asiminae, Aplosporella (Sphaeropsis) 569
asiminae, Dothiorella 688
asiminae, Mycocentrospora (Centrospora, Cercospora) 799
asiminae, Mycosphaerella 800
asiminae, Phyllosticta 865
asiminae, Rhopaloconidium (Phloeospora) 934
asiminae, Septoria 948
asparagi, Cercospora 606
asparagi, Confertopeltis 641
asparagi, Diplodia 680
asparagi, Hendersonia 732
asparagi, Leptosphaeria = **Massarina gloeospora** 779
asparagi, Metasphaeria 787
asparagi, Phomopsis (Phoma) 857
asparagi, Puccinia 901
asparagina, Ascochyta (Ascochytula) 574
asparagina, Leptosphaeria 760
asparagina, Phoma 852
aspera, Anthracoidea (Cintractia) 566
aspera, Chaetochalara 623
aspera, (Grandinia, Hyphodontia) = **Grandinia granulosa** 722
aspera, Papulaspora 825
aspera, Peniophora = **Hyphoderma setigerum** 741
aspera, Puccinia = **P. pazschkei** var. **heterisiae** 911
aspera, Thallospora 989
aspera, Uredinopsis = **U. hashiokai** 1007
aspera, (Eutypa, Valsa) = **Eutypa polycocca** 700
asperatum, Heterosporium 735
asperifolia, Tilletia 991
asperifolia var. lappulae, Ovularia = **Ramularia lappulae** 925
asperifolii var. **lappulae, Ovularia** 823
asperifolioides, Tilletia 991
asperifoliorum, Erysiphe = **E. cynoglossi** 698
asperior, Puccinia 901
asperipilata, Peniophora = **Subulicystidium longisporum** 983
asperula, Scopulariopsis (Torula) 943
asperula, Stigmina 979
asperula, Tomentella 993
asperulispora, (Asterina, Englerulaster) = **Asterina orbicularis** 581
asperulum, Coryneum = **Stigmina asperula** 979
asperum, Lophiostoma (Lophiosphaera) 770
asperum, Trichocladium 999
aspidii, Asterina (Asterella, Opeasterina) 580
aspidii, Incrupila (Dasyscyphus, Peziza) 748
aspidiicola, (Helotium, Mollisia, Pezizella) = **Psilachnum chrysostigmum** 899
aspidiotus, Hyalopsora (Melampsorella) 738
aspidistrae, Ascochyta 574
aspidistrae, Labrella 753
asplenii, Cercospora 606
asplenii, Septoria 948
asprata, Cyptotrama (Armillaria, Xerulina) 660
asprellae, Cercospora 606
assimile, Hyphoderma 740
assimilis, Peniophora = **Hyphoderma assimile** 740
asterata, Cercospora 606
astericola, Cladosporium 629
astericola, Exophoma 703
astericola, Leptosphaeria 760
astericola, Leptothyrium (Phyllosticta) 765
astericola, Macrophoma (Phoma) 774
astericola, Septoria 948
astericola, Stagonospora (Asteromella) 973
asterigena, Phyllachora 862
asterina, Septoria 948
asterinoides, Amazonia 563
asterinoides, Chaetothyrina (Venturia) 624
asterinoides, Mycosphaerella (Sphaerella) 800
asterinosporum, (Cenangium, Myriangium) = **Myriangium superficialis** 808
asterinum, Coniothyrium 642
asterinum, Epicoccum = **E. nigrum** 696
asterinum, Helminthosporium 731
asteris, (Ascochyta, Phyllosticta) = **Phoma exigua** 853
asteris, Asteromella 582
asteris, Leptosphaeria = **L. planiuscula** 763
asteris, Ovularia 823
asteris, Puccinia = **P. cnici-oleracei** 903
asteris, Ramularia (Fusidium) 923
asteris, Rhytisma (Xyloma) 935
asteriscus, Phomopsis (Phoma) 857
asteromorpha, Venturia 1020
asterophora, Asterina 580
asterospermum, Asterosporium 582
asterostoma, Diatrype 673
aster-sericeanum, Entyloma = **E. polysporum** 696
asterum, Coleosporium 636
asterum, Erysiphe = **E. cichoracearum** 697
asterum, Puccinia = **P. dioicae** 905
asterum, Synchytrium 983
astragali, Cercospora (Cercospora) 606
astragali, Napicladium 811
astragali, Phoma 852
astragali, Phyllosticta 865
astragali, Polystigma (Physalospora) 889
astragali, Ramularia (Didymaria) 923
astragali, Septoria 948

astragali, Sorosporium = **Thecaphora deformans** 989
astragalicola, Septoria 948
astragalicola, Uromyces = **U. hedysari-obscuri** 1011
astragalina, Septoria 948
asymmetrica, Pleospora = **P. rainierensis** 885
aterrima, Chaetostroma (Psilonia) 624
aterrima, Corynespora 646
aterrima, Stilbella 981
aterrima, Thecaphora = **Tolyposporium aterrimum** 993
aterrima, Xenotypa 1025
aterrimum, Chaetomium 623
aterrimum, Mystrosporium = **Corynespora aterrima** 646
aterrimum, Radulum = **Xenotypa aterrima** 1025
aterrimum, Tolyposporium 993
aterrimus, Uromyces = **U. bicolor** 1010
athabascensis, Vararia 1019
athaeae, Colletotrichum = **C. malvarum** 639
athyrii, Typhula 1004
atkinsonii, Meliola 784
atkinsonii, Peniophora = **Amylostereum chailletii** 564
atkinsonii, Scoleconectria = **Pseudotrichia mutabilis** 899
atkinsonii, Seynesia = **Myiocopron freycinetiae** 808
atkinsonii, Uredinopsis 1007
atlanticus, Marasmius 777
atomacea, Simocybe (Naucoria) 961
atomaespora, Valsa = **Eutypella scoparia** 701
atomaria, Hyaloscypha = **H. stevensonii** 738
atomata, Phyllosticta 865
atomatoides, Psathyrella (Psilocybe) 893
atomospora, Tubercularia 1002
atra, Acremoniella = **Harzia acremonioides** 730
atra, Cercospora = **C. fuligniosa** 610
atra, Chaetomella = **Amerosporium atrum** 563
atra, Dendryphiopsis 668
atra, Phillipsiella 848
atra, Puccinia = **P. esclavensis** var. **panicophila** 906
atra, Sebacina = **Exidiopsis molybdea** 702
atra, Stachybotrys = **S. chartarum** 972
atra, Streptothrix = **Conoplea juniperi** 644
atramentaria, Dothidea 687
atramentaria, Venturia = **Pyrenobotrys conferta** 918
atramentarium, Colletotrichum = **C. coccodes** 637
atramentarius, Coprinus 644
atramentosa, Myriogenospora (Dothichloe, Hypocrea) 809
atramentosum, Penicillium 827
atrans, Alternaria 561
atrata, Diplodia 680
atrata, Mollisia (Pyrenopeziza) 793
atrata, Naematelia = **Exidia nucleata** 702
atrata, Oliveonia = **O. subviolacea** 818
atrata, Patellaria = **Lecanidion atratum** 758
atrata var. **liliaceara, Mollisia** 793
atratae, Anthracoidea (Cintractia) 566
atratum, (Ceratobasidium, Corticium) = **Oliveonia subviolacea** 818
atratum, Lecanidion 758
atratum, Xyloschizon 1027
atricapillus, Pluteus 887
atriella, Mollisia 793
atriellum, Cladosporium 629
atriplicis, Stagonospora (Phyllosticta, Septoria) 973
atroapicula, Acrodictys 557
atrobarbum, Chaetosphaeria (Acanthostigma) 624
atrobrunneum, Haplographium (Penicillium) 729
atrocaerulea, Hohenbuehelia 737
atrocephalum, Arthrobotryum (Stilbum) 573
atrocincta, Phoma 852
atrocinerea, Coniophora = **C. olivacea** 641
atrocinerea, Mollisia = **Pyrenopeziza polygoni** 919
atrocoerulea, Diplodia 680
atrocyanea, Durella 691
atrofusca, Nectria 811
atrofusca, Patellaria = **Lecanidion simile** 758
atrofusca, Pseudotomentella 898
atrofusca, Puccinia 901
atrogrisea, Cercospora 606
atrogrisea, Stachybotrys = **S. chartarum** 972
atroinquinans, Splanchnonema = **S. platani** 969
atromaculans, Cercospora = **Phaeoisariopsis nigricans** 840
atromarginalis, Pseudocercospora (Cercospora) 893
atronitidum, Vertixore 1022
atro-olivaceum, Helminthosporium = **Pithomyces atro-olivaceus** 880
atro-olivaceus, Pithomyces 880
atropae, Cercospora 606
atropuncta, Diaporthe = **Phragmodiaporthe tiliacea** 861
atropuncta, Puccinia 901
atropunctatum, Hypoxylon (Albocrustum, Numulariola) 745
atropurpurascens, Punctularia 917
atropurpurea, Lachnella = **Dasyscyphus atropurpureus** 664
atropurpurea, Septoria 948
atropurpureum, Cenangium 602
atropurpureum, Hypoxylon 745
atropurpureum, Lachnum = **Dasyscyphus atropurpureus** 664
atropurpureum, Sporotrichum = **Fusarium moniliforme** 708
atropurpureus, Dasyscyphus 664
atropurpureus, Merulius = **Meruliopsis albostramineus** 787
atroruber, (Hypochnus, Zygodesmus) = **Tomentella atrorubra** 993
atrorubra, Tomentella 993
atrosporum, Chaetomium = **C. megalocarpum** 623
atrotomentosus, Paxillus 827
atroviolacea, Tomentella = **Tomentellastrum badium** 995
atrovirens, Byssocorticium (Corticium) 596
atrovirens, Claussenomyces (Corynella) 631
atrovirens, Merulius = **Leucogyrophana pinastri** 767
atrovirens, Rhinocladiella 931
atrovirens, Spondylocladium = **Helminthosporium solani** 732
atroviride, Hypoxylon = **Camarops ohiensis** 599
atroviridis, Patinella 827
atrum, Amerosporium 563
atrum, Dendryphion = **Dendryphiopsis atra** 668
atrum, Melanconium 782
atrum, Sorosporium = **Tolyposporium aterrimum** 993
atrum, Ulocladium (Stemphylium) 1006
attenuata, Kellermania = **Piptarthron yuccae** 880
attenuata, Lulworthia 773
attenuata, Poria = **Junghuhnia nitida** 751
attenuatus, Polyporus = **Junghuhnia nitida** 751
aubertii var. **comptoniae, Cryptodiaporthe**
auburnensis, Chaetopsina 624
aucubae, Pestalotia = **P. subcuticularis** 835
aucubae, Phomopsis (Phoma) 857
aucubae, Phyllosticta 865
aucubicola, Phyllosticta 865
aucupariae, Lophodermium 771
audibertiae, Mycosphaerella (Sphaerella) 800
auerswaldii, Valsa 1017
aulica, Leptosphaeria (Metasphaeria, Sphaeria) 760
aurantia, Cercospora = **C. penzigii** 615
aurantia, Physalospora = **Polystigma astragali** 889
aurantia, Stachybotrys = **S. elegans** 972
aurantiaca, Leptophyma (Ascomycetella) 760
aurantiaca, Nectria (Sphaerostilbe) 811
aurantiaca, Peniophora 829
aurantiaca, Pezizella 838
aurantiaca, Phialophora = **P. hoffmannii** 847
aurantiaca, Winterella 1024
aurantiaca, Zythia 1027
aurantiacum, Fusarium = **F. oxysporum** 708
aurantiacum, Gymnosporangium = **G. cornutum** 726
aurantiacum, Nematogonum = **N. ferrugineum** 814
aurantiacum, Sphaeronaema = **Zythia aurantiaca** 1027
aurantiacus, Marasmius = **M. corrugatus** var. **aurantiacus** 777
aurantiacus, Polyporus = **Pycnoporellus fulgens** 917
aurantiacus, Thermoascus 990
aurantiicola, Phyllosticta 865
aurantiogriseum, Penicillium 827
aurantiorugosus, Pluteus 887
aurantiorum, (Paradiplodia, Pseudodiplodia) = **Botryodiplodia lecanidion** 591
aurantius, Aleurodiscus 560
aurantius, Hypomyces 744
aurata, Arachnopeziza 571
aurata, Pseudotrichia = **P. mutabilis** 899
aurea, Auriporia (Poria) 585
aurea, Ceratocystis = **Ophiostoma aureum** 821
aurea, Cryptosporella = **Melanconis xanthostroma** 782
aurea, Mycoacia 798
aurea, Mycosphaerella = **M. ribis** 805
aurea, Ocellaria = **Pezicula ocellata** 837
aurea, (Clathrospora, Pleospora) = **Clathrospora pentamera** 631
aurea, Plicatura = **Pseudomerulius aureus** 897
aurea, Septoria = **S. ribis** 956
aurea, Taphrina = **T. populina** 987
aurelia, Arachnopeziza 571
aureo-fulva, Nectria 811
aureum, Cenangium 602
aureum, Chaetomium 623
aureum, Eustilbum 699
aureum, Gliocladium = **G. roseum** 715
aureum, Helicosporium 731
aureum, Ophiostoma (Europhium) 821
aureum, Scleropycnium 941

aureum, Synchytrium 983
aureus, Pseudomerulius (Merulius) 897
aureus, Uromyces 1010
auriconum, Ozonium 824
auricula, Auricularia 585
auricula-judae, Auricularia = **A. auricula** 585
auricularis, Auricularia = **A. auricula** 585
auripuncta, Hyponectria 744
auripunctum, Laestadia = **Hyponectria auripuncta** 744
auriscalpium, (Auriscalpium, Hydnum) = **Auriscalpium vulgare** 585
aurivella, Pholiota = **P. adiposa** 851
aurivelloides, Pholiota 851
austinii, Mycena 797
australasica, Physoderma 877
australe, Cystostereum 660
australe, Entyloma 695
australe, Ganoderma 713
australe, Lophodermium 771
australe, Synchytrium 983
australiae, Diplodia 680
australiana, Uncinuliella (Uncinula) 1006
australiensis, Bipolaris (Drechslera, Helminthosporium) 588
australiensis, Leptosphaeria (Metasphaeria) 760
australis, Fomes = **Ganoderma australe** 713
australis, Leptosphaerulina = **L. trifolii** 764
australis, Pezicula 837
australis, Plasmopara (Peronospora) 882
australis, Ravenelia 928
australis, Taphrina 986
austrina, Monodictys 795
austro-americana, Diaporthe 670
austroamericana, Gyrostroma 727
austroamericana, Thyronectria (Pleonectria) 991
austro-americanum, Melanopsichium 783
austro-occidentale, Mycoacia 798
autosteira, Gelasinospora 713
autumnale, Lophomerum (Lophodermium) 773
autumnalis, Galerina (Pholiota) 712
autumnalis, Meliola = **Appendiculella calostroma** 571
avellanea, Clavicorona = **C. piperata** 632
avellanea, Coniophora = **Cerocorticium confluens** 622
avellanea, Mycena = **M. intertexta** 798
avellanea, Phanerochaete 843
avellanea, Tomentella 993
avellanea, Wrightoporia (Poria) 1024
avellaneum, Cladosporium 629
avellaneum, Corticium = **Phanerochaete avellanea** 843
avellaneus, Hypochnus = **Tomentella avellanea** 993
avenacea, Drechslera 689
avenacea, Gibberella 714
avenaceum, Fusarium (Fusisporium) 707
avenaceum, Helminthosporium = **Drechslera avenacea** 689
avenaceum, Sphaerographium 966
avenae, Ascochyta (Ascochytella, Pseudodiplodia) 574
avenae, Drechslera (Helminthosporium) 689
avenae, Heterosporium 736
avenae, Leptothyrium 765
avenae, Pestalotia 835
avenae, Pyrenophora 919
avenae, Spermospora (Pseudodiscosia) 964
avenae, Stagonospora (Septoria) 973
avenae, Ustilago 1014
avenae var. levis, Ustilago = **U. hordei** 1015
avenae f. sp. **triticea, Stagonospora** (Septoria) 973
avenaria, Phaeosphaeria (Leptosphaeria) 841
avenaria f. sp. **triticea, Phaeosphaeria** (Leptosphaeria) 842
avenophila, Phyllosticta 865
averrhoae, Cercospora 606
avesiculatum, Cylindrocladium 657
avicennae, Cercospora 606
avicennae, Macrosporium 775
avicenniae, Leptosphaeria 760
avicenniae, Rhabdospora 930
avicularis, Cercospora 606
avicularis, Ovularia 823
avicularis var. sagittati, Cercospora = **C. polygonacea** 615
avocense, Aecidium 558
avocensis, Puccinia 901
awamorii, Aspergillus = **A. niger** var. **awamorii** 579
axicola, Cintractia 628
axillatum, Leptoxyphium (Capnodium, Polychaeton) 767
azaleae, Colletotrichum = **C. gloeosporioides** 638
azaleae, Colpoma 640
azaleae, Dacrymyces = **D. capitatus** 663
azaleae, Dendrophoma 667
azaleae, Exobasidium = **E. vaccinii** 703
azaleae, Monilinia 794
azaleae, Ovulinia (Sclerotinia) 824
azaleae, Pycnostysanus (Briosia, Periconia, Sporocybe) 917
azaleae, Septoria 948
azaleae-indicae, Septoria 948
azedarachis, Phyllosticta 865
azedarachtae, Diatrype (Eutypella) 673
baarnensis, Discosia 684
baarnensis, Pestalotia 835
baboquivariensis, Lindtneria (Poria) 768
babylonica, Ascochyta 574
baccata, Gibberella 714
baccata, Pleospora = **Comoclathris permunda** 640
baccatum, Macrosporium 775
baccharidicola, Aplosporella (Sphaeropsis) 570
baccharidis, Cercospora 606
baccharidis, Diatrype = **D. prominens** 673
baccharidis, (Dimerosporium, Meliola) = **Lasiostemma melioloides** 757
baccharidis, Phyllosticta 865
baccharidis, Placosphaeria 881
baccharidis, Puccinia 901
baccharidis, Stagonospora 973
baccharidis, Systremma = **Dothidea puccinioides** 687
bacillata, Acerbia (Ophioceras) 556
bacillifera, Mycosphaerella (Sphaerella) 800
bacilliformis, Phoma 852
bacilligera, Cercospora 606
bacilligera, Septoria 948
bacilligera f. alnobetulae, Passalora = **P. alni** 826
bacilligera var. alnobetulae, Passalora = **P. alni** 826
bacilligerum, Fusarium = **Cercospora bacilligera** 606
bacillisporium, Penicillium = **Talaromyces bacillisporus** 985
bacillisporum, Acremonium (Paecilomyces) 556
bacillisporus, Talaromyces 985
bacterioides, Phyllosticta 865
bacteriophila, Phoma 852
bacteriosperma, Taphrina 986
baculiferum, Lophodermium 771
baculorubrense, Hyphoderma 740
baculum, Rhabdospora (Macrophoma, Sphaeropsis) 930
badia, Naucoria = **Phaeomarasmius erinaceus** 840
badia, Tomentella = **Tomentellastrum badium** 995
badio-ferruginea, Hymenochaete 739
badiofusca, Tomentella = **T. terrestris** 994
badium, Tomentellastrum 995
badius, Phellinus (Fomes) 844
badius, Polyporus 889
bainii, Nectria = **N. ochroleuca** 813
bakerae, Crepidotus 647
bakerensis, Collybia 640
bakerensis, Pholiota 851
bakeri, Clathrospora 631
bakeri, Diaporthe 670
bakeri, Hysterographium 747
bakeri, Mycosphaerella = **M. tassiana** 806
bakeriana, Phoma 852
balansae, Hypoxylon = **H. caries** 745
balansae var. ildefonsiae, Seynesia = **Asterina ildefonsiae** 581
baldensis, Phyllosticta 865
ballotaeflorae, Puccinia 901
balloui, Polyporus = **Abortiporus biennis** 555
ballouii, Echinodontium (Steccherinum) 691
balsamea, Dermea 669
balsamea, Postia 890
balsamea, Thyronectria (Calonectria, Nectria) 991
balsameae, Delphinella (Rehmiellopsis) 666
balsameae, Rhabdogloeopsis (Gloeosporium, Kabatiella) 930
balsameae, Sarcotrochila (Phacidium, Stegopezizella) 938
balsameum, Cenangium = **Dermea balsamea** 669
balsameum, Hydnum = **Resinicium bicolor** 929
balsameum, Stereum = **S. sanguinolentum** 978
balsameum var. abietinum, Cenangium = **Dermea balsamea** 669
balsameus, (Oligoporus, Polyporus, Tyromyces) = **Postia balsamea** 890
balsamicola, Adelopus = **Phaeocryptopus nudus** 839
balsamicola, Dimerium (Dimeriella, Dimerosporium, Meliola) 679
balsamicola, Phacidium (Potebniamyces) 838
balsamorhizae, Cylindrosporium 658
balsamorhizae, Mycosphaerella (Sphaerella) 800
balsamorhizae, Phloeospora 849
balsamorhizae, Pleospora = **P. chlamydospora** 884
balsamorhizae, Puccinia 901
balsmea, Scoleconectria = **Thyronectria balsamea** 991
bambusae, Cylindrosporium 658
bambusae, Diplodia 680
bambusae, Guignardia 724
bambusae, Idriella 748
bambusae, Konenia 753
bambusae, Lageniformia 756
bambusae, Placostroma (Scirrhodothis) 881
bambusicola, Phaeoisariopsis 839
bananae, Glabrocyphella (Cyphella) 715
baptisiae, Anthostomella 565

baptisiae, Aplosporella 570
baptisiae, Cladosporium 629
baptisiae, Coniothyrium (Sphaeropsis) 642
baptisiae, Phoma 852
baptisiae, Septoria 948
baptisiae, Stagonospora (Marssonia, Marssonina) 973
baptisiicola, Mycosphaerella (Sphaerella) 800
baranyayi, Rhytidiella 935
barba-jovis, Grandinia (Hyphodontia, Odontia) 722
barba-jovis subsp. abieticola, Odontia = **Grandinia abieticola** 722
barbarea, Cercospora 606
barbareae, Ramularia 924
barbata, Trichopezizella 1000
barbatus, Dasyscyphus = **Trichopezizella barbata** 1000
barbirostris, Calosphaeria 598
barclayana, Tilletia (Neovossia) 991
bardanae, Puccinia = **P. calcitrapae** var. **bardanae** 902
barringtoniae, Dothiorella (Phoma) 688
barringtoniae, Gloeosporium = **Colletotrichum gloeosporioides** 638
barssii, Phyllosticta 865
bartholomaei, Coniothyrium 642
bartholomaei, Gloeosporidiella (Gloeosporium) 717
bartholomaei, Hypoxylon 745
bartholomaei, Puccinia = **P. chloridis** 903
bartholomaei, Septogloeum (Fusarium, Trichofusarium) 946
bartholomaei, Taphrina = **T. aceris** 986
bartholomaei, Volutella 1022
bartholomei, Cercospora = **C. toxicodendri** 618
basicola, Thielavia 990
basicola, Thielaviopsis (Trichocladium) 990
basilaris, (Polyporus, Tyromyces) = **Postia balsamea** 890
bataanensis, Pleospora 884
batatae, Pestalotia 835
batatae, Phomopsis = **P. phaseoli** 859
batatas, (Depazea, Phyllosticta) = **Phomopsis ipomeae-batatas** 858
batatas, Elsinoe 692
batatatis, Diaporthe = **D. phaseolorum** 672
batatatis, Fusarium = **F. oxysporum** f. sp. **batatas** 708
bataticola, (Rhizoctonia, Sclerotium) = **Macrophomina phaseolina** 775
bataticola, Septoria 948
batesiana, Puccinia 901
batesii, Aecidium 558
batistae, Strigopodia 982
baudysiana, Septoria 948
bauhiniae, Pseudocercospora (Cercospora) 893
baumeae, Aulographella 584
beaucarneae, Stigmina 979
beaumontii, Phyllachora (Dothidea) 862
beaumontii, Sirosporium (Helminthosporium) 962
beckhausii, Diaporthe 670
beckmanniae, Physoderma 877
beckmanniae, Uromyces 1010
begoniae, Cercospora 606
begoniicola, Sphaeropsis 966
behnickiana, Uredo 1007
bella, Tricholomopsis 999
bellulus, Gymnopilus 725
bellunensis, Stagonospora 973
bellus, Helicomyces 731
bellus, Meruliopsis (Merulius) 787
bellynckii, Cercospora (Cladosporium) 606
belonosporum, Odontotrema = **Coccomycetella richardsonii** 635
beloperones, Uromyces (Uredo) 1010
beltsvillensis, Cryptosporella 653
beneficiens, Vermicularia 1021
benesuada, Mollisia (Pyrenopeziza) 793
benzoina, Trametes = **Ischnoderma resinosum** 750
benzoinus, Polyporus = **Ischnoderma resinosum** 750
berberidina, Diplodia 680
berberidina, Stagonospora 973
berberidis, Cucurbitaria 654
berberidis, Dothidea (Dothidella, Plowrightia, Sphaeria) 687
berberidis, Kabatiella (Gloeosporium) 751
berberidis, Laestadia 755
berberidis, Leptosphaeria 760
berberidis, Mycosphaerella 800
berberidis, Pestalotia = **P. pestalozzioides** 835
berberidis, Phragmodothis (Montagnella) 861
berberidis, Phyllosticta 865
berberidis, Pyrenochaeta (Phoma) 918
berberina, Phomopsis (Phoma) 857
berchemiae, Massarina 779
berchemiae, (Eutypella, Valsa) = **Eutypella goniostoma** 700
berckmansii, Seimatosporium (Coryneum) 944
berengeriana, Botryosphaeria = **B. dothidea** 592
berengeriana, Dothiorella 688
berenice, Acanthostigma 556
berenice, Bothrodiscus (Fusisporium) 590
bergii, Poria = **Antrodia vaillantii** 568
beringiana, Uredo 1007
berkeleyanum, Camarosporium (Hendersonia) 599
berkeleyanum, Schizoxylon 940
berkeleyi, Bondarzewia (Polyporus) 590
berkeleyi, Cercospora = **C. seymouriana** 617
berkeleyi, Corticium = **Hyphoderma setigerum** 741
berkeleyi, Daedalea = **Gloeophyllum mexicanum** 717
berkeleyi, Hapalocystis 728
berkeleyi, Mycosphaerella 800
berkeleyi, Phomatospora 857
berkeleyi, Veluticeps 1020
berkeleyi, Xenosporium (Helicoma, Xenosporella) 1025
berlandieri, Coniothyrium 642
berlesiana, Diaporthe = **D. syngenesia** 672
bermudianum, Gymnosporangium 726
berolinense, Dendrodochium 666
berolinensis, Phyllosticta 865
berolinensis, Thyronectria (Pleonectria) 991
bertholletianum, Cephalosporium 602
bertholletianum, Phomopsis 857
bertii, Acanthophysium (Aleurodiscus) 555
besseyi, Septoria = **Piggotia fraxini** 879
besseyii, Pestalotiopsis (Pestalotia) 836
betae, Heterosporium 736
betae, Phoma (Gloeosporium, Phyllosticta) 852
betae, Polymyxa 889
betae, Septoria 948
betae, Uromyces 1010
bethelii, Aecidium 558
bethelii, Exosporium = **Stigmina glomerulosa** 980
bethelii, Gymnosporangium 726
bethelii, Schizoxylon = **S. ligustri** 940
bethelii, Ustilago 1014
beticola, Cercospora 606
beticola, Ramularia 924
betulae, Ciboria 627
betulae, Cylindrosporium 658
betulae, Erysiphe = **Phyllactinia guttata** 864
betulae, Septoria 948
betulae, Sphaeropsis 966
betulae, Winterella (Cryptospora, Ophiovalsa) 1024
betulae var. **foliicola, Sphaeropsis** 966
betulae-luteae, Gloeosporium = **Asteroma microspermum** 582
betulae-odoratae, Septoria 948
betulae-papyriferae, Cylindrosporella (Gloeosporium) 658
betularum, Cryptocline (Gloeosporidium, Gloeosporium) 651
betulicola, Gloeosporium = **Discula betulina** 685
betulicola, Septoria = **S. betulae** 948
betuliginum, Hysterium = **H. pulicare** 747
betulina, Atopospora (Dothidea, Dothidella, Euryachora, Phyllachora, Rehmiodothis) 584
betulina, Bispora 589
betulina, Diatrypella (Diatrype) 674
betulina, Discula 685
betulina, Fomitiporella = **Phellinus laevigatus** 845
betulina, Gelatinosporis = **Gelatinosporium betulinum** 713
betulina, Hyponectria 744
betulina, Lenzites 759
betulina, Melampsora = **Melampsoridium betulinum** 781
betulina, Phyllosticta 865
betulina, Septoria = **S. betulae** 948
betulina, Valsa 1017
betulinum, Asterosporium 582
betulinum, Coryneum 646
betulinum, Gelatinosporium 713
betulinum, Gloeosporium = **Discula betulina** 685
betulinum, Melampsoridium 781
betulinum, Melanconium 782
betulinus, Piptoporus (Polyporus) 880
beyerinckii, Coryneum = **Stigmina carpophila** 979
bicellularis, Polyschema 889
bicolor, Capitotricha (Dasyscyphus, Erinella, Lachnum) 600
bicolor, Cercospora 606
bicolor, Chaetocalathus (Lachnella) 622
bicolor, Dermea (Tympanis) 669
bicolor, Gloeosporium = **Cryptocline cinerescens** 651
bicolor, Hypoxylon = **H. rubiginosum** 746
bicolor, Laxitextum (Stereum) 757
bicolor, Melanconium 782
bicolor, Ophiostoma (Ceratocystis) 821
bicolor, Phyllosticta = **P. variabilis** 874
bicolor, Piloderma (Athelia, Corticium) 880
bicolor, Resinicium (Odontia) 929
bicolor, Septonema = **Coniosporium olivaceum** 642
bicolor, Stachylidium 972
bicolor, Stictis (Schizoxylon) 978
bicolor, Tomentella (Hypochnus) 993
bicolor, Uromyces 1010
bicolor, Zygodesmus = **Tomentella botryoides** 993

bicolor var. candidum, Melanconium = **M. candidum** 782
bicolor var. **indicus, Dasyscyphus** 664
bicorne, Arthrinium = **A. cuspidatum** 572
bicorne, Triposporium = **Ceratosporella bicornis** 604
bicornis, Ceratosporella 604
bicornis, Uncinula (Sawadaia) 1006
bicristatum, Dinemasporium = **Stauronema cruciferum** 975
bicuspidata, Leptosphaeria (Sphaeria) 760
bicuspidatus, Phellinus 844
bidentata, Meliola 784
bidenticola, Uromyces 1010
bidentis, Cercospora 606
bidentis, Septoria 948
bidentis, Uromyces 1010
bidwellii, Guignardia (Botryosphaeria, Physalospora) 724
bidwellii f. **muscadinii, Guignardia** 725
bidwellii f. **parthenocissi, Guignardia** 725
biennis, Abortiporus (Heteroporus, Polyporus) 555
bifida, Theclospora = **Emericella variecolor** 693
biforme, Trichaptum 998
biformis, Cercospora 607
biformis, Guepinia = **Dacryopinax elegans** 663
biformis, Phyllosticta 865
biformis, (Polyporus, Polystictus) = **Trichaptum biforme** 998
biformis, Sphaeropsis = **S. malorum** 967
bifoveolata, Metasphaeria 788
bifrons, (Ciborinia, Sclerotinia) = **Ciborinia whetzelii** 627
bifrons, Tulasnella 1003
bifurcata, Botrytis = **Amphobotrys ricini** 564
bigeloviae, Cylindrocolla 658
bigeloviae, Dothidea 687
bigelowii, Melampsora = **M. paradoxa** 781
bigemina, Torula = **Eversia subopaca** 701
bigeminum, Cryptocoryneum = **Eversia subopaca** 701
biglobosa, Diaporthe 670
bigloviae, Sorosporium = **Thecaphora pilulaeformis** 989
bignoniacearum, Meliola 784
bignoniae, Asterella (Asterina) 579
bignoniae, Leptostromella 765
bignoniae, Synchytrium 983
bignoniae, Valsaria (Valsa) 1018
biguttulata, Ascochyta 574
biloculare, Phragmidium 860
bina, Dicarpella (Disperma, Physalospora) 674
binale, Torula = **Bispora betulina** 589
binoculata, Diaporthe 670
biocellata, Puccinia 901
biparasiticum, Ceratostoma = **Syspastospora parasitica** 985
bipartitus, Coccomyces (Schizoxylon) 634
biporula, Puccinia 901
bisbyi, Stachybotrys = **S. elegans** 972
bischofiae, Cercospora 607
biseptata, Drechslera 689
biseptata, Endophragmiella (Monotospora) 694
biseptata, Stagonospora 973
biseptata, Xylopezia 1027
biseptatum, Gymnosporangium 726
biseptatum, Helminthosporium = **Drechslera biseptata** 689
bismarckii, Phoma = **P. pomorum** 855
bispora, Athelia = **A. arachnoidea** 583
bisporum, Corticium = **Athelia arachnoidea** 583
bistortae, Pseudorhytisma (Pseudopeziza) 898
bistortae, Puccinia 901
bistortae, Ramularia (Ovularia) 924
bistortarum, Ustilago 1014
biuncinata, Podosphaera 888
bixae, Cercospora 607
bixae, Crossopsora 650
bixae, Oidium 818
blakei, Hypoxylon = **H. mammatum** 745
blasdaleanum, Gymnosporangium = **G. libocedri** 727
blasdalei, Hadrotrichum 727
blasdalei, Puccinia 901
blasdalei, Septoria 948
blattaria, Conocybe (Pholiotina) 643
blepharospora, Keissleriella 752
blephiliae, Cercoseptoria 605
blephiliae, Cercospora 607
bletiae, Coleosporium 636
bletiae, Colletotrichum = **C. gloeosporioides** 638
bliti, Albugo (Cystopus) 559
bliti, Cercospora = **Pseudocercospora rubi** 895
bliti, Hendersonia 733
bloxami, Cercospora = **C. brassicicola** 607
bloxamii, Dactylospora (Buellia, Karschia, Patellaria) 663
blumencronii, Discosia 684
blyttiana, Puccinia 901
boehmeriae, Aecidium 558
boehmeriae, Cercospora (Cercospora) 607
boehmeriicola, Phyllosticta 865
boerhaviae, Ascochyta = **A. oxybaphi** 576
boerhaviae, Cercospora = **C. furfurella** 610
boewei, Endophragmiella (Endophragmia) 694
bohemica, Ascochyta (Stagonospora) 574
bohemica, Rehmiellopsis = **Delphinella abietis** 666
bolaris, Rutstroemia (Calycina, Ciboria) 937
boleticola, Sporobolomyces = **S. roseus** 971
bolickii, Stemphylium 976
bolleana, Mycosphaerella (Sphaerella) 800
bolleana, Pseudocercospora (Cercospora, Cercosporidium, Septosporium) 894
bolleyana, Puccinia 901
bolleyi, Microdochium (Aureobasidium, Gloeosporium) 789
boltoniae, Discosphaerina (Guignardia) 684
boltoniae, Phoma 852
boltoniae, Selenophoma (Macrophoma) 945
boltshauseri, Ascochyta = **Stagonosporopsis hortensis** 975
bombacina, Athelia 583
bombacinum, Sporotrichum 972
bombycina, Anomoporia (Poria) 565
bombycina, Volvariella (Volvaria) 1023
bombycina var. **flaviceps, Volvariella** 1023
bombycinum, Hypochnicium (Corticium) 742
bombycirrhizus, Marasmius = **Strobilurus conigenoides** 982
bonaerense, Synchytrium 983
bonariensis, Phyllachora 862
bonarii, Dendrostilbella = **Tubercularia lateritia** 1002
bonarii, Janetia (Sporidesmium) 750
bonnayae, Microdiplodia 788
boothii, Endophragmiella (Endophragmia) 694
borbonica, Poria = **Tinctoporellus epimiltinus** 992
boreale, Kabatiella (Gloeosporium, Microstroma) 751
boreale, Lasiobelonium 756
boreale, Phragmidium 860
boreale, Vararia 1019
borealis, Ceraceomyces (Athelia, Merulius) 603
borealis, Climacocystis (Abortiporus, Polyporus, Spongipellis, Tyromyces) 632
borealis, Dasyscyphus = **Lasiobelonium boreale** 756
borealis, Ephelis 696
borealis, Hainesia 728
borealis, Hymenochaete 739
borealis, Leptosphaeria 761
borealis, Mazzantia (Heteropera, Physalospora) 780
borealis, Mycena 797
borealis, Mycosphaerella 800
borealis, Myriosclerotinia (Sclerotinia) 809
borealis, Rhizoctonia 932
borealis, Sphacelotheca = **Ustilago bistortarum** 1014
boreella, Valsa 1017
borneensis, Phanerochaete = **P. filamentosa** 843
borreriae, Pseudocercospora (Cercospora) 894
borrichiae, Aecidium 558
bosniaca, Ustilago 1014
bostrychodes, Chaetomium 623
botrya, Dothiorella 688
botrychii, Mycosphaerella 800
botryoides, Tomentella (Hypochnus) 993
botryosum, Acanthophysium = **Aleurobotrys botryosus** 559
botryosum, Botryobasidium 591
botryosum, Stemphylium 976
botryosum var. **tragopogonis, Stemphylium** 976
botryosum var. ulocladium, Stemphylium = **Ulocladium botrytis** 1006
botryosus, Aleurobotrys (Aleurodiscus) 559
botrys, Hypoxylon = **H. rubiginosum** 746
botrytis, Ulocladium 1006
bougainvilleae, Cercosporidium (Cercospora) 620
boutelouae, Ascochyta 574
boutelouae, Cercospora 607
boutelouae, Cercosporella 619
boutelouae, Cochliobolus 635
boutelouae, Phyllachora = **P. cynodontis** 862
boutelouae, Puccinia 901
boutelouae, Scaphidium 939
boutelouae, Ustilago = **Sporisorium neglectum** 971
bouvardiae, Aecidium 558
bouvardiae, Puccinia = **P. anisacanthi** 900
boycei, Foveostroma (Brunchorstia, Cryptosporium, Micropera) 706
boycei, Phacidiopycnis (Phomopsis) 838
boycei, Septoria = **S. weiriana** 959
boycei, Taphrina 986
brachiata, Cercospora 607
brachiata, Papulaspora (Torula) 825
brachiatum, Tilachlidium (Clavaria) 991
brachyascus, Ophiobolus (Rhaphidospora) 819
brachycladia, Lemonniera = **L. terrestris** 758
brachyelytri, Cladosporium 629
brachyelytri, Stagonospora 973
brachypoda, Cercospora 607
brachypodii, Ascochyta 574
brachypodii, Puccinia 901
brachypodii var. **arrhenatheri, Puccinia** 902

brachypodii var. **poae-nemoralis, Puccinia** 902
brachypodii-phoenicoidis var. **chisosana, Puccinia** 902
brachypodii-phoenicoidis var. **davisii, Puccinia** 902
brachypus, Cercospora 607
brachyspora, Curvularia 655
brachysporum, Hypoderma = **Meloderma desmazieresii** 786
brachysporum, Microstroma = **M. juglandis** 791
brachystoma, Anthostomella (Xylosphaeria) 565
bracteosa, Marssonina 777
bramleyi, Ciboriopsis = **Moellerodiscus tenuistipes** 792
brandegei, Puccinia 902
brasiliense, Chaetomium 623
brasiliensis, Favolus = **Polyporus tenuiculus** 889
brasiliensis, (Fistulina, Pseudofistulina) = **Pseudofistulina radicata** 896
brasiliensis, Mycoleptodiscus 799
brasiliensis, Satchmopsis 939
brassicae, Alternaria (Macrosporium) 561
brassicae, Asteromella 582
brassicae, Cercosporella = **Pseudocercosporella capsellae** 895
brassicae, Cladosporium (Cladotrichum) 629
brassicae, Nectria 811
brassicae, Olpidium 818
brassicae, Plasmodiophora 881
brassicae, Plasmopara 882
brassicae, Septoria 948
brassicae var. microspora, Alternaria = **A. brassicicola** 561
brassicae f. phaseoli, Alternaria = **A. tenuissima** 562
brassicaefolius, Merulius = **Serpula himantioides** 960
brassicae-rapae, Ascochyta = **A. cheiranthi** 574
brassicicola, Alternaria (Helminthosporium) 561
brassicicola, Cercospora 607
brassicicola, Mycosphaerella 800
brassicicola, Phyllosticta = **Asteromella brassicae** 582
bravensis, Uromyces = **U. sporobolicola** 1013
brebissonii, Helicobasidium 730
brefeldianum, Sirobasidium 961
brenckleana, Discula (Macrophoma) 685
brenckleana, Phaeosolenia 841
brenckleanum, Lophiostoma = **L. caulium** var. **congregatum** 770
brencklei, Clypeoporthella (Diaporthopsis) 633
brencklei, Phaeotrype = **Diatrype albopruinosa** 673
brencklei, Septoria 948
brencklei, Sphaerophoma 966
brencklei, Trimmatostroma 1001
bresadolae, Ascochyta 574
bresadolae, Hyphoderma = **H. litschaueri** 741
bresadolae, Phoma 852
bresadolae, (Ceriporia, Poria) = **Ceriporia purpurea** 621
bresadolae, Tomentella (Hypochnus) 993
bresadoleana, Myriogenospora 809
breviaristata, Seiridium (Monochaetia, Pestalotia) 945
brevicaulis, Scopulariopsis 943
brevicolla, Ophiostoma 821
brevicollis, Ceratocystis = **Ophiostoma brevicolla** 821
brevicompactum, Penicillium 827
brevifaciens, Tilletia = **T. controversa** 992
brevifolia, Phyllachora = **P. luteo-maculata** 863
brevipes, Cladosporium 629
brevipes, Fusicladium 711
brevipes, Pestalotia 835
brevipes, Pileolaria (Uromyces) 879
brevipes, Ramularia 924
brevis, Humicola (Acremoniella, Monotospora) 738
brevis, Lembosia (Lembosiopsis) 758
brevis, Septoria 948
brevisaccata, Gloniopsis = **G. praelonga** 720
breviseta, Grandinia (Hyphodontia) 722
breviseta, Pestalotia 835
brevispina, Cylindrosporium 658
brevispora, Ceuthospora 622
brevispora, Phlebia 848
brevispora, Stagonospora (Septoria) 973
breviuscula, Rhabdospora (Septoria) 930
breviuscula, Taeniolella 985
breviusculum, Septonema = **Taeniolella breviuscula** 985
briareus, Cercospora 607
brickelliae, Puccinia = **P. kuhniae** var. **brickelliae** 908
bridgesii, Phyllosticta 865
bridgii, Pholiota 851
brightonensis, Leptosphaeria 761
brinkmannii, Sistotrema (Grandinia, Odontia, Trechispora) 962
briosiana, (Leptosphaerulina, Pleosphaerulina, Pseudoplea) = **Leptosphaerulina trifolii** 764
britannica, Septoria 948
brizae, Drechslera (Bipolaris, Helminthosporium) 689
brodiaeae, Uromyces 1010
bromeliacearum, Colletotrichum 637
bromi, Apiocarpella = **Ascochyta macrospora** 576
bromi, Drechslera (Helminthosporium) 689
bromi, Pyrenophora (Pleospora) 919
bromi, Ramulispora (Ansatospora, Centrospora, Cercospora, Cercosporina) 928
bromi, Septoria 948
bromi, Stagonospora 973
bromi, Tilletia (Ustilago) 991
bromi var. elymina, Septoria = **S. elymi** f. **elymina** 951
bromi var. **phalaricola, Septoria** 948
bromi var. phalaridis, Septoria = **Phaeoseptoria phalaridis** 841
bromicola, Puccinia = **P. recondita** 912
bromigena, Pseudoseptoria (Selenophoma, Septoria) 898
bromi-maximi, Puccinia = **P. recondita** 912
bromi-tectorum, Tilletia = **T. bromi** 991
bromivora, Phyllosticta 865
bromivora, Ustilago = **U. bullata** 1014
broomeiana, Nitschkia 816
broomeiana, Nummularia (Hypoxylon, Jongiella) 817
broussonetiae, Cercospora 607
brownii, Ganoderma (Elfvingia) 713
brownii, Mycena = **M. cinerella** 797
brownii, Synchytrium = **S. fulgens** 984
brumalis, Polyporus 889
brumptii, Scopulariopsis 943
brunaudiana var. ribis-aurei, Eutypella = **Diatrype standleyi** 673
brunellae, Ceuthocarpon = **Ophiobolus prunellae** 820
brunellae, Linospora = **Ophiobolus prunellae** 820
brunellae, Phyllosticta 865
brunellae, Ramularia 924
brunellae, Septoria 948
bruneopruinosa, Dermatea = **Pestalopezia brunneopruinosa** 834
brunkii, Cercospora 607
brunkii, Tolyposporella 993
brunnea, Ambrosiella (Monilia) 563
brunnea, Ceratocystis 603
brunnea, Cercospora = **C. prenanthis** 615
brunnea, Cladosporium = **C. brunneolum** 629
brunnea, (Gloeosporium, Marssonia, Marssonina) = **Marssonina tremulae** 778
brunnea, Laestadia (Depazea, Sphaerella) 755
brunnea, Phyllosticta 865
brunnea, Ramularia 924
brunneofirma, Tomentella = **Tomentellastrum brunneofirmum** 995
brunneofirmum, Tomentellastrum 995
brunneola, Didymosphaeria = **D. futilis** 678
brunneolum, Cladosporium 629
brunneo-maculatum, Gloeosporium 717
brunneopruinosa, Pestalopezia 834
brunneopunctata, Erysiphe 697
brunneorufa, Tomentella 993
brunnescens, (Cadophora, Phialophora) = **Phialophora richardsiae** 847
brunneum, Dendryphion = **Taeniolella muricata** 985
brunneum, Gloeosporium = **Marssonina tremulae** 778
brunnichiae, Leptothyrium 765
bryoniae, Didymella 676
bryophila, Tomentella 993
bubakii, Discosia = **Seimatosporium kriegerianum** 945
bubakii, Phialophora (Margarinomyces) 847
buccina, Guepiniopsis 724
buchloeana, Tilletia 992
buchloes, Bipolaris (Drechslera, Helminthosporium) 588
buchloes, Ustilago 1014
buchnerae, Puccinia 902
buchsii, Tympanis = **Tryblidiopsis pinastri** 1002
bucidae, Sphaceloma 964
buddleiae, Phomopsis 857
budingtonii, Hyphoderma 740
bugellensis, (Hyphodontia, Odontia) = **Grandinia pruni** 723
bulbicola, Nectria 811
bulbigenum, Fusarium = **F. oxysporum** 708
bulbigenum var. niveum, Fusarium = **F. oxysporum** f. sp. **niveum** 709
bulbigenum var. tracheiphilum, Fusarium = **F. oxysporum** f. sp. **tracheiphilum** 710
bulbigera, Isariopsis (Ovularia) 749
bulbigerium, Scolicotrichum = **Isariopsis bulbigera** 749
bulbilium, Gliocephalotrichum 715
bulborum, Sclerotinia 941
bulbosa, Clavariopsis 632
bulbosa, Xylaria (Sphaeria) 1026
bullata, Diatrype 673
bullata, Ovularia 823
bullata, Puccinia = **P. longipes** 909
bullata, Taphrina 986
bullata, Ustilago 1014
bullata var. **macrospora, Ustilago** 1014

bullatum, (Sorosporium, Tolyposporium) = **Moesziomyces bullatus** 792
bullatus, Moesziomyces 792
bulliardii, Nummularia = **Biscogniauxia nummularia** 589
bumeliae, Guignardia 725
bumeliae, Meliola 784
bumeliae, Phoma 852
bumeliae, Polystigma 890
bumeliae, Septoria 948
bumeliae, (Mycosphaerella, Sphaerella) = **Mycosphaerella hermione** 802
bumeliifolia, Phyllosticta 865
bunchosiae, Cercospora 607
bupleuri, Leptosphaeria 761
bupleuri, Puccinia 902
burkii, Cylindrosporium 658
burnettii, Puccinia 902
burtii, Exobasidium 702
burtii, Grandinia = **G. crustosa** 722
burtii, Phanerochaete (Peniophora) 843
burtonii, Endomycopsis (Pichia) 694
butleri, Pythium = **P. aphanidermatum** 920
butomi, Physoderma 877
buxbaumii, Anthracoidea 566
buxi, Fusarium = **F. buxicola** 707
buxi, Hyponectria 744
buxi, Sesquicillium (Fusidium, Paecilomyces, Verticillium) 960
buxi, Volutella 1022
buxicola, Fusarium 707
buxifolia, Laestadia = **Hyponectria buxi** 744
byrsina, Coriolopsis 644
byrsinus, Polyporus = **Coriolopsis byrsina** 644
byssicola, Ophiobolus 819
byssina, Athelia = **Piloderma byssinum** 880
byssina, Poria = **Antrodiella romellii** 568
byssincola, Leptosphaeria 761
byssinum, Piloderma (Corticium) 880
byssiseda, Lentaria 758
byssogenum, Trichaptum 998
byssoidea, Botrytis 594
byssoidea, (Coniophora, Peniophora) = **Amphinema byssoides** 563
byssoides, Amphinema 563
byssoides, Periconia (Sporocybe) 831
byssoideus, Merulius = **Leucogyrophana pinastri** 767
caborcensis, Puccinia = **P. franseriae** 906
cacabata, Puccinia 902
cacaina, Phanerochaete (Peniophora) 843
cacaliae, (Cyathicula, Helotium) = **Crocicreas cyathoideum** var. **cacaliae** 648
cacaliae, Phyllosticta 865
cacaliae, Septoria 949
cacaoicola, Diplodia 680
cacti, Anthostomella 565
cacti, Nectriella (Hyponectria) 814
cacti, Phomopsis (Phoma, Phyllosticta) 857
cacti, Vermicularia 1021
cacticola, Septoria 949
cactivora, Bipolaris (Drechslera) 588
cactivorum, Helminthosporium = **Bipolaris cactivora** 588
cactophila, Pichia 879
cactorum, Gloeosporium = **Colletotrichum gloeosporioides** 638
cactorum, Lembosia (Lembosiopsis) 758
cactorum, Phytophthora 877
cactorum var. applanata, Phytophthora = **P. citricola** 878
cacuminispora, Leptosphaeria 761
caducum, Cladosporium 629
caelata, Laestadia 755
caerulea, Winteria 1024
caerulescens, Taphrina 986
caeruleum, Pulcherricium (Corticium) 917
caesia, Postia 890
caesiocinerea, (Bourdotia, Sebacina) = **Basidiodendron caesiocinereum** 586
caesiocinereum, Basidiodendron 586
caesius, Marasmius = **Marasmiellus nigripes** 776
caesius, (Oligoporus, Polyporus, Tyromyces) = **Postia caesia** 890
caespiticia, Mollisia (Pyrenopeziza) 793
caespitosa, Cercospora 607
caespitosa, Dendrophoma 667
caespitosa, Hypocenia 742
caespitosa, Mycosphaerella 801
caespitosa, Rosellinia = **Lopadostoma caespitosum** 769
caespitosum, Capnodium 601
caespitosum, Lopadostoma 769
caespitosum, Microdochium 789
caespitosus, Aspergillus 578
caespitulosum, Clasterosporium = **Sporidesmium hormiscioides** 970
caffra, Pestalotia = **Pestalotiopsis glandicola** 836
cajanderi, Fomitopsis (Fomes) 705
cajani, Colletotrichum = **C. gloeosporioides** 638
cajani, Phyllosticta 865
caladii, Sclerotium 941
caladii, Uromyces = **U. ari-triphylli** 1009
calamagrostidis, Cylindrosporium 658
calamagrostidis, Hyalothyridium 738
calamagrostidis, Mycosphaerella = **M. greenei** 802
calamagrostidis, Phaeoseptoria 841
calamagrostidis, Septoria 949
calamagrostidis, Ustilago 1014
calamagrostidis f. sp. **koeleriae, Septoria** 949
calamagrostis, Phyllachora = **P. ammophilae** 862
calami, Eriosporella (Cryptosporium) 697
calamovilfae, Hendersonia 733
calamovilfae, Septoria 949
calandriniae, Ustilago 1014
calanticariae, Puccinia 902
calatheae, Cercospora 607
calcara, Ustilago 1014
calcea, Exidiopsis (Sebacina) 702
calcicola, Tomentella (Caldesiella) 993
calcitrapae, Puccinia 902
calcitrapae var. **bardanae, Puccinia** 902
calcitrapae var. **centaureae, Puccinia** 902
calderi, Anthracoidea (Cintractia) 566
caleae, Puccinia 902
calendulae, Cercospora 607
calendulae, Entyloma 695
caliciiforme, Triblidium (Blitridium) 998
caliciiformis, Crinula 648
calicioides, Caliciopsis (Hypsotheca) 597
californica, Absidia 555
californica, Ceratocystis = **Ophiostoma longirostellatum** 821
californica, Cercosporella 619
californica, Cintractia = **Anthracoidea caricis** 566
californica, Erysiphe 697
californica, Hypospila = **Gnomonia alni** 720
californica, Leptosphaeria (Sphaeria) 761
californica, Marssonina (Marssonia) 778
californica, Massariovalsa (Melanconis) 780
californica, Patellea 827
californica, Pleosphaerulina 883
californica, Puccinia 902
californica, Pyrenopeziza (Pezizella) 918
californica, Sphaerella 965
californica, Stagonospora 973
californica, Taphrina 986
californica, Thecaphora 989
californica, Thyridaria 990
californica var. **deserticola, Puccinia** 902
californicum, Byssocorticium 596
californicum, Colpoma = **Coccomyces castanopsidis** 634
californicum, Helminthosporium = **Bipolaris sorokiniana** 589
californicum, Heterosporium 736
californicum, Leptothyrium 765
californicum, Peridermium 831
californicum, Propolidium 891
californicum, Sorosporium = **Thecaphora californica** 989
californicus, Baeodromus 585
californiense, Helotium = **Mycena albidula** 797
californiensis, Omphalopsis = **Mycena albidula** 797
calkinsii, Poria = **Wrightoporia avellanea** 1024
calkinsii, (Fomes, Pyropolyporus) = **Phellinus robustus** 846
calkinsii, Tyromyces (Polyporus) 1005
callae, Cercospora 607
callae, Gloeosporium = **Colletotrichum gloeosporioides** 638
callae, Marssonia 777
callaquensis, Puccinia 902
callicarpae, Atractilina = **A. parasitica** 584
callicarpae, Botryosphaeria (Melogramma) 592
callicarpae, Cercospora 607
callicarpae, Coniothyrium 642
callicarpae, Diaporthe = **D. sociabilis** var. **sambuci** 672
callicarpae, Diatrype = **Eutypella juglandicola** 700
callicarpae, Synchytrium 983
callimorpha, Entoleuca = **Hypoxylon mammatum** 745
callirrhoeae, Synchytrium 983
callista, Microsphaeropsis 791
callista, Nitschkia (Cryptosphaerella, Cucurbitaria, Fracchiaea) 816
callistephi, Phomopsis 857
callistephi, Septoria 949
callistephi, Stemphylium 976
callistum, Coniothyrium = **Microsphaeropsis callista** 791
callitriches, Doassansia 686
callochaetes, Zoellneria (Dasyscyphus, Lachnella) 1027
callosa, Poria = **Antrodia serialis** 567
callostroma, Hypoxylon = **H. sassafras** 746
callunae, Pseudophacidium = **P. ledi** 898
caloceps, Pluteus = **P. aurantiorugosus** 887
calochorti, Puccinia 902
calocladophora, Microsphaera 790
calolepioides, Crepidotus = **C. mollis** 647
calolepsis, Crepidotus = **C. kauffmanii** 647
calosphaerioides, Cryptodiaporthe (Diaporthe) 651
calospora, Hendersonia 733
calospora, Tulasnella (Gloeotulasnella) 1003
calostroma, Appendiculella (Irene, Sphaeria) 571
calotheca, Peronospora 832
calothrix, Tubulicrinis (Peniophora) 1003
calotrichum, Corticium = **Phanerochaete sanguinea** 844

calotropidis, Cercospora 607
calthae, Botryotinia 593
calthae, (Cercospora, Ramularia) = **Ramularia didyma** 924
calthae, Pseudopeziza 897
calthae, Puccinia 902
calthicola, Puccinia 902
calthicola, Verpatinia 1021
calvescens, Pleospora (Chaetoplea, Pyrenophora) 884
calycanthi, Aplosporella 570
calycanthi, Botryosphaeria 592
calyciformis, Lachnellula (Dasyscyphus) 754
calycinus, Dasyscyphus = **Lachnellula subtilissima** 754
calyculiforme, Lachnum = **Brunnipila calyculiformis** 595
calyculiformis, Brunnipila (Dasyscyphus) 595
calyculus, Hymenoscyphus 740
calyculus, Nitschkia (Byssosphaeria, Calyculosphaeria, Winterina) 816
calypsonis, Septoria 949
calystegiae, Cercospora 607
calystegiae, Dothidella (Dothidea, Plowrightia) 687
calystegiae, Stagonospora (Septoria) 973
cambivora, Phytophthora 878
cameliae, Physopella (Angiopsora) 877
camelliae, Alternaria (Macrosporium) 561
camelliae, Chaetospermum = **C. carneum** 624
camelliae, Ciborinia (Sclerotinia) 627
camelliae, Colletotrichum = **C. gloeosporioides** 638
camelliae, Cylindrocladiella (Cylindrocladium) 657
camelliae, Cylindrosporium 658
camelliae, Exobasidium 702
camelliae, Guignardia 725
camelliae, Meliola (Fumago) 784
camelliae, Monochaetia = **Pestalotiopsis karstenii** 836
camelliae, Phoma 852
camelliae, Phyllosticta 865
camelliae, Sporonema 971
camelliae var. **gracilis, Exobasidium** 703
camelliaecola, Phyllosticta = **P. camelliae** 865
camembertii, Penicillium 827
campanella, Xeromphalina (Omphalia) 1025
campanula, Lachnella = **Calathella ellisii** 597
campanula, Phialea (Belonioscypha) 847
campanulae, Coleosporium = **C. tussilaginis** 636
campanulae, Mycosphaerella 801
campanulae, Phlyctema 850
campanulae, Puccinia 902
campanulae, Septoria 949
campanulastri, Aecidium 558
campanulata, Drechslera (Angiopoma) 689
campanulatus, Pleurotus = **Hohenbuehelia cyphelliformis** 737
campbellii, Amylosporus (Polyporus) 564
campestris, Dichomitus (Coriolellus, Trametes) 675
camphorae, Diplodia 680
camphorae, Echidnodella (Lembosia, Morenoella) 691
camphorae, Gloeosporium = **Colletotrichum gloeosporioides** 638
campicola, Cylindrosporium 658
camptosori, Cercospora 607
camptospermum, Toxosporium (Monochaetia, Pestalotia) 995
camptospora, Rhabdospora 930
camptostylus, Aphanomyces 568
camptothecae, Cercospora 607
campylosporum, Melogramma 786
campylostyla, Gnomonia = **Plagiostoma campylostylum** 881
campylostylum, Plagiostoma 881
campylostylum var. **mirabilis, Plagiostoma** 881
cana, Ceratocystis (Ceratostomella) 603
cana, (Cercospora, Cercosporella) = **Cercosporella virgaureae** 619
cana var. gracilis, Cercosporella = **C. virgaureae** 619
canadense, Acanthophysium 555
canadense, Amylocorticium (Corticium) 564
canadense, Chlorosplenium = **Holwaya mucida** 737
canadense, Exobasidium 703
canadense, Gloeosporium = **Discula umbrinella** 685
canadense, (Lachnella, Pseudohelotium) = **Crocicreas cyathoideum** 648
canadense, Septocylindrium = **Ramularia canadensis** 924
canadensis, Aleurodiscus = **Acanthophysium canadense** 555
canadensis, Hansenula 728
canadensis, Hypochnus = **Tomentellina fibrosa** 995
canadensis, Nectria (Scoleconectria) 811
canadensis, Peniophora = **Hyphodermopsis polonensis** 742
canadensis, (Polyporus, Tyromyces) = **Antrodiella overholtsii** 568
canadensis, Puccinia 902
canadensis, Ramularia 924
canadensis, Septoria 949
canadensis, Sphaerella = **Scirrhia conigena** 940
canadensis, Stagonospora 973
canaegricola, Ovularia = **Ramularia rubella** 926
canafistulae, Phyllachora 862
canaliculata, Phyllachora (Dothidea) 862
canaliculata, Puccinia 902
canarii, Oudemansiella 823
canavaliae, Enchnoa 693
canavaliae, Fusicoccum 711
canavaliae, Gloeosporium = **Colletotrichum gloeosporioides** 638
canavaliae, Septoria 949
canavaliae, Stenella (Cercospora) 977
cancellatum, Dictydium 675
cancellatum f. anomalum, Dictydium = **Dictydium cancellatum** 675
cancellatum var. purpureum, Dictydium = **Dictydium cancellatum** 675
cancri, Nectria (Hypomyces) 811
candelabrum, Sesquicillium (Verticillium) 960
candicans, Botryobasidium 591
candicans, Oidium (Monilia) 818
candida, Albugo 559
candida, Dendrothele 667
candida, Exidia 702
candida, Nidula (Nidularia) 815
candida, Peronospora 832
candida, Scopulariopsis 943
candida, (Cyphella, Solenia) = **Henningsomyces candidus** 734
candidissima, (Phlebiella, Poria, Trechispora) = **Trechispora mollusca** 997
candidissima, Psathyrella 893
candidissimum, Cheimonophyllum (Pleurotus) 625
candidissimus, Polyporus = **Trechispora mollusca** 997
candidum, (Acanthophysium, Aleurocorticium) = **Dendrothele candida** 667
candidum, Cryptosporium 653
candidum, Cyanodermella (Acrospermum, Cyanoderma) 656
candidum, Geotrichum 714
candidum, Helicobasidium 730
candidum, Melanconium 782
candidum, Stereum = **Dendrothele candida** 667
candidum var. citri-aurantii, Geotrichum = **G. citri-aurantii** 714
candidus, Aleurodiscus = **Dendrothele candida** 667
candidus, Aspergillus 578
candidus, Cystopus = **Albugo candida** 559
candidus, Henningsomyces 734
candidus, Marasmiellus (Marasmius) 776
candidus var. sphaerosporus, Aleurodiscus = **Dendrothele candida** 667
candolleana, Ciborinia (Phialea, Sclerotinia) 627
candollei, Macrophoma (Ludwigiella) 774
candollei, Ustilago = **U. bistortarum** 1014
canescens, Cercospora 607
canescens, Lasiosphaeria (Leptospora) 757
canescens, Penicillium 827
canescens, Phyllosticta 865
canescens, (Aporpium, Poria) = **Aporpium caryae** 571
canfieldii, Corticium (Laeticorticium) 645
canina, Hendersonia = **Seimatosporium caninum** 944
caninum, Seimatosporium 944
cannabina, Phomopsis 857
cannabina, Pseudocercospora (Cercospora) 894
cannabina, Septoria 949
cannabis, Ascochyta (Phyllosticta, Septoria) 574
cannabis, Cercospora 607
cannae, Puccinia = **P. thaliae** 915
cannonballus, Monosporascus 796
canodisca, Eutypella (Valsa) 700
canum, Corticium = **Leptosporomyces galzinii** 764
canum, Ophiostoma = **Ceratocystis cana** 603
canum, Sphaeronaema 966
caperoniae, Pseudocercospora (Helicomina) 894
capillaris, Marasmius 777
capillaris, Mycena 797
capillata, (Eutypella, Valsa) = **Eutypella scoparia** 701
capillifera, Ceratocystis = **Ophiostoma piliferum** 821
capitalensis, Phyllosticta 865
capitata, Ceratocystis = **C. tenella** 604
capitata, Epithele = **Tubulicium capitatum** 1003
capitata, Lachnella = **Incrucipulum capitatum** 748
capitati, Cercospora 607
capitatum, Incrucipulum 748
capitatum, Tubulicium 1003
capitatus, Dacrymyces 663
capitulatum, Cacumisporium (Helminthosporium) 597
capnoides, Hypholoma (Naematoloma) 742
capnostroma, Diatrype = **Eutypella aequilinearis** 700
capreae, Asteroma 581

capreae, Linospora 769
capreolata, Cercospora = **C. duplicata** 609
capsellae, Pseudocercosporella (Cylindrosporium) 895
capsici, Ascochyta = **Phoma exigua** 853
capsici, Cercospora 607
capsici, Colletotrichum (Vermicularia) 637
capsici, Phytophthora 878
capsicicola, Phaeoramularia (Cercospora) 840
capsularum, Gloeosporium 717
capsularum, Sporidesmium 970
capsulata, Hansenula 728
capsulatum, Penicillium 827
capula, Calyptella (Cyphella) 599
caput-ursi, Hericium = **H. americanum** 734
caracallae, Cercospora (Cercospora, Cercosporina) 607
caraganae, Cucurbitaria 654
caraganae, Phloeospora 849
caraganae, Phomopsis 857
caraganae, Septoria 949
caraganae var. elaeagni, Cucurbitaria = **C. elaeagni** 654
carbonacea, Cytospora 661
carbonacea, Discella = **D. microsperma** 683
carbonaria, Griseoporia = **Gloeophyllum carbonarium** 716
carbonaria, (Hexagonia, Poria, Trametes) = **Gloeophyllum carbonarium** 716
carbonaria, Tomentella 993
carbonarium, Gloeophyllum 716
carbonarium, Hydnum = **Mycoacia fuscoatra** 798
carbonarius, Aspergillus 578
carbonarius, Coriolellus = **Gloeophyllum carbonarium** 716
carbonarius, (Polyporus, Tyromyces) = **Postia balsamea** 890
carbonarius, Xenodochus 1025
carbonica, Antrodia (Poria) 567
carbonum, Cochliobolus 635
carbonum, (Drechslera, Helminthosporium) = **Bipolaris zeicola** 589
carcinodes, Urocystis 1008
carcinophthalmum, Stilbum = **Stilbella annulata** 981
cardinale, Seiridium (Coryneum) 945
cardiospermi, Synchytrium 983
carduorum, Puccinia 902
carduorum, Pyrenopeziza (Mollisia) 918
caribaea, Cercospora = **Phaeoramularia manihotis** 841
carica, (Kawakamia, Phytophthora) = **Phytophthora palmivora** 879
caricae, Ascochyta 574
caricae, Asperisporium 579
caricae, Colletotrichum = **C. gloeosporioides** 638
caricae, Mycosphaerella 801
caricae, Oidium 818
caricae, Phoma = **P. caricae-papayae** 852
caricae-papayae, Phoma (Ascochyta) 852
caricae-papayae, Phomopsis 857
caricae-papayae, Phyllosticta 865
caricarum, Asterina 580
cariceti, Ophiobolus (Sphaeria) 819
caricetorum, Cintractia = **Anthracoidea caricis** 566
caricicola, Aposphaeria 571
caricicola, Arthrinium 572
caricicola, Ascochyta 574
caricicola, Cladosporium = **C. herbarum** 630
caricicola, Marasmius = **M. caricis** 777
caricicola, Oidium = **O. caricae** 818
caricicola, Phaeoseptoria (Septoria) 841
caricicola, Phyllosticta 865
caricicola, Volutella 1023
caricina, Cercospora = **C. caricis** 607
caricina, Claviceps = **Sclerotium sulcatum** 942
caricina, Cylindrosporella 658
caricina, Cyphella = **Cellypha goldbachii** 601
caricina, Dothidella 687
caricina, Lophodermina = **Lophodermium caricinum** 771
caricina, Pseudopeziza (Pyrenopeziza) 897
caricina, Puccinia 902
caricina, Septoria 949
caricina, Torula = **Rutola graminis** 937
caricina var. limosae, Puccinia = **P. limosae** 909
caricina var. **uliginosa, Puccinia** 902
caricinella, Calloria 597
caricinella, Phaeosphaeria (Leptosphaeria) 842
caricinella, Septoria 949
caricinella, Stagonospora 973
caricinellum, Helotium = **Calloria caricinella** 597
caricinum, Amerosporium = **A. atrum** 563
caricinum, Clasterosporium 631
caricinum, Leptostroma 764
caricinum, Lophodermium 771
caricis, Anthracoidea (Cintractia, Ustilago) 566
caricis, Ascochyta = **A. caricicola** 574
caricis, Cercospora (Cercospora) 607
caricis, Leptothyrella 765
caricis, Marasmius (Gloiocephala) 777
caricis, Melanodothis (Hyalodothis) 783
caricis, Mycosphaerella (Guignardia, Laestadia) 801
caricis, Phaeoseptoria 841
caricis, Phloeospora 849
caricis, Phoma (Macrophoma) 852
caricis, Phyllachora (Dothidea) 862
caricis, Phyllosticta 866
caricis, Puccinia = **P. caricina** 902
caricis, Septoria 949
caricis, Stagonospora (Hendersonia) 973
caricis, Synchytrium 983
caricis, Uredo = **Puccinia caricina** 902
caricis var. acutarum, Cintractia = **Anthracoidea heterospora** 566
caricis var. caricis-strictae, Puccinia = **P. caricina** 902
caricis var. douglasii, Ustilago = **Anthracoidea douglasii** 566
caricis var. grossulariata, Puccinia = **P. caricina** 902
caricis var. **intermedia, Anthracoidea** (Cintractia) 566
caricis var. quadriporula, Puccinia = **P. caricina** 902
caricis var. uniporula, Puccinia = **P. caricina** 902
caricis var. urticata, Puccinia = **P. caricina** 902
caricis-ampullaceae, Myriosclerotinia (Sclerotinia) 809
caricis-asteris, Puccinia = **P. dioicae** 905
caricis-shepherdiae, Puccinia 902
caries, Hypoxylon 745
caries, Tilletia 992
cariosa, Trematosphaeria (Caryospora) 998
carletonii, Ovularia 823
carnea, Cytosporella 662
carnea, Isariopsis 750
carnea, Pezicula 837
carnea, Rhytidenglerula (Asterella, Asterina, Englerula) 935
carnea, Septoria 949
carnea, Sphaeronaemella 966
carnea, Taphrina 986
carnea, Trametes = **Fomitopsis cajanderi** 705
carnegiana, Puccinia 903
carnegiea, Ceriporiopsis (Poria) 621
carneolum, Lachnum 754
carneolus, Dasyscyphus = **Lachnum carneolum** 754
carneopallens, Polyporus = **Rigidoporus vinctus** 936
carneorosea, Pezizella (Peziza) 838
carneum, Chaetospermum 624
carneum, Coleosporium = **C. vernoniae** 636
carneum var. carpini, Myxosporium = **M. carpini** 810
carneus, Paecilomyces 825
carneus, Uromyces = **U. lapponicus** 1011
carnicolor, Poria = **Postia placenta** 891
carniformis, Ramularia (Fusarium) 924
carnosa, Mycogloea 799
carnosa, Phanerochaete (Peniophora) 843
carolinense, Crustoderma 650
carolinensis, Calycellina 598
carolinensis, Cercospora 607
carolinensis, Mycena 797
carolinensis, Pseudomassaria 896
caroliniana, Dothichiza 686
caroliniana, Mycosphaerella 801
caroliniana, Platygloea 882
carolinianum, Pythium 920
carotae, Alternaria = **A. dauci** 561
carotae, Cercospora 607
carotae, Rhizoctonia 932
carpigenum, Gloeosporium = **Diplodina aesculi** 682
carpinae, Pezicula = **Cryptosporiopsis fasciculata** 653
carpinea, Depazea 669
carpinea, Pezicula 837
carpinea, Septoria = **S. betulae** 948
carpinea, Sphaerognomonia (Laestadia) 966
carpineum, Fusarium 707
carpineum, Xyloma = **Depazea carpinea** 669
carpini, Melampsoridium (Melampsora) 781
carpini, Myxosporium 810
carpini, Phyllosticta 866
carpini, Pleomassaria (Massaria) 883
carpini, Synchytrium 983
carpinicola, (Karstenula, Pleospora) = **Karstenula shepherdiae** 752
carpinigera, Melanconis = **M. chrysostroma** var. **ellisii** 782
carpogena, Marssonina 778
carpogena, Phyllosticta (Phyllostictina) 866
carpogenum, Cephalosporium 602
carpogenum, Gloeosporium = **Marssonina carpogena** 778
carpogenum, Sporotrichum = **Phialophora malorum** 847
carpophila, Stigmina 979
carpophila, (Cintractia, Uredo) = **Anthracoidea heterospora** 566
carpophila, Venturia 1020
carpophila, Xylaria 1026
carpophila var. elynae, Cintractia = **Anthracoidea elynae** 566
carpophila var. **kenaica, Cintractia** 628
carpophila var. verrucosa, Cintractia = **Anthracoidea verrucosa** 567
carpophilum, Cladosporium (Fusicladium) 629

carpophilum, (Clasterosporium, Coryneum) = **Stigmina carpophila** 979
carpophilum, Colletotrichum = **C. dematium** 637
carpophilum, Leptothyrium 765
carposchiza, Phomopsis 857
carrii, Cercospora = **C. elaeagni** 609
carrii, Sporidesmium 970
carthaginensis, Ascochyta 574
carthami, Gloeosporium (Marssonia) 717
carthami, Puccinia = **P. calcitrapae** var. **centaureae** 902
carthami, Septoria 949
carunculoides, Ciboria (Sclerotinia) 627
carveri, Taphrina 986
carveriana, Cercospora 607
caryae, Aporpium (Polyporus, Poria) 571
caryae, Ascochyta 574
caryae, Asteroma (Cylindrosporella, Gloeosporium, Phyllosticta) 581
caryae, Gnomonia 720
caryae, Phragmodiaporthe (Cryptospora) 861
caryae, Rosellinia 936
caryae, Septoria 949
caryae var. pecanae, Gnomonia = **G. nerviseda** 721
caryigena, Cytospora = **C. sacculus** 662
caryigena, Diplodia 680
caryigena, Mycosphaerella 801
caryigena, Pseudocercosporella (Cercoseptoria, Cercosporella) 895
caryigena, Valsa = **V. ceratosperma** 1017
caryigenum, Cladosporium (Fusicladium) 629
caryigenum, Cylindrosporium = **Pseudocercosporella caryigena** 895
caryigenum var. **carpinum, Cladosporium** 629
caryogena, Volutella 1023
caryogenum, Coniothyrium 642
caryophagum, Melanomma 783
caryophyllacearum, Albugo 559
caryophyllacearum, Melampsorella 781
caryophyllea, Guignardia (Sphaerella) 725
caryophyllea, Thelephora 990
caryophylleae, Anthracoidea 566
caryophylli, Mycosphaerella (Sphaerella) 801
caryophyllinus, Uromyces = **U. dianthi** 1010
cascadensis, Puccinia 903
casearium, Radulodon (Hydnum) 923
caseicola, Penicillium = **P. camembertii** 827
casimiroae, Phyllosticta 866
cassandrae, Ascochyta 574
cassandrae, Cenangium 602
cassandrae, (Chrysomyxa, Melampsoropsis) = **Chrysomyxa ledi** var. **cassandrae** 626
cassandrae, Cucurbitaria 654
cassandrae, Dothichiza 686
cassandrae, Exobasidium = **E. vaccinii** 703
cassandrae, Gibbera (Venturia) 714
cassandrae, Godronia 721
cassandrae, Hypoderma (Hypodermopsis) 743
cassandrae, Psilachnum (Helotium, Phialea) 899
cassandrae f. **ribicola, Godronia** 721
cassandrae f. **vaccinii, Godronia** 721
cassavae, Cercospora = **Cercosporidium henningsii** 620
cassiae, Alternaria 561
cassiae, Cercosporidium (Cercospora) 620
cassiae, Phomopsis (Phoma) 857
cassiae, Pleospora 884
cassiae, Stomiopeltis 981
cassiaecola, Leptosphaeria 761
cassiaecola, Ramularia = **Phaeoramularia occidentalis** 841
cassiaecola, Ravenelia 928
cassiae-covesii, Ravenelia = **R. mesilliana** 928
cassiicola, Corynespora (Helminthosporium) 646
cassiicola, Phyllosticta 866
cassiicola, Septoria 949
cassiocarpa, Cercospora = **Cercosporidium cassiocarpum** 620
cassiocarpum, Cercosporidium 620
cassiopes, Phaeospora = **Lophiostoma sieversiae** 770
cassiopes, Seimatosporium (Coryneum) 944
castagnei, Marssonina (Gloeosporium) 778
castagnei, Sphaerotheca = **S. fuliginea** 968
castanae, Ascoconidium 577
castanea, Amphiporthe (Cryptodiaporthe) 563
castanea, Calycellina 598
castanea, Phlebia = **P. rufa** 848
castanea, Phoma 852
castaneae, Botryosphaeria 592
castaneae, Cenangium 602
castaneae, Coryneum = **C. modonium** 646
castaneae, Cylindrosporium 658
castaneae, Leptothyrium (Leptostroma) 765
castaneae, Monodictys 795
castaneae, Phyllosticta 866
castaneae, Phytophthora = **P. katsurae** 878
castaneae, Seiridium 945
castaneae var. **quercus, Leptothyrium** 766
castaneicola, Coniella (Gloeosporium, Pilidiella) 641
castaneicola, Coryneum (Stegonsporium) 646
castaneipes, Galerina 712
castanella, Didymella 676
castaneum, Helotium = **Calycellina castanea** 598
castaneum, Myxosporium 810
castaneum, Septobasidium 946
castaneum var. **quercus, Myxosporium** 810
castanicola, Asteromella (Phyllosticta) 582
castanicola, Dothidella = **Mycosphaerella janus** 803
castanicola, Laestadia = **Sphaerognomonia carpinea** 966
castanicola, Leptothyrium = **Tubakia dryina** 1002
castanicola, Valsa (Fenestella) 1017
castanophilus, Polyporus = **Hapalopilus croceus** 729
castanopsidis, Coccomyces (Godronia) 634
castanopsidis, Coronellaria 645
castanopsidis, (Dothidella, Mycosphaerella) = **Mycosphaerella janus** 803
castanopsidis, Gloeosporium 718
castanopsidis, Taphrina 986
castelae, Meliola 784
castillejae, Didymella 676
castillejae, Leptosphaeria 761
castillejae, Ophiobolus 819
castillejae, Puccinia 903
castillejae, Ramularia 924
castrensis, Leptosphaeria 761
catabrosae, Entyloma = **E. dactylidis** 695
catalina, Didymosphaeria 678
catalinae, Myriangium 808
catalinae, Strickeria 982
catalpae, Alternaria (Macrosporium) 561
catalpae, Ascochyta 574
catalpae, Cercospora 607
catalpae, Didymosphaeria 678
catalpae, Gloeosporium = **Colletotrichum gloeosporioides** 638
catalpae, Phyllosticta 866
catariae, Didymella (Cercidospora) 676
catenaria, Drechslera 689
catenarium, Helminthosporium = **Drechslera catenaria** 689
catenoides, Hyphochytrium 740
catenospora, Phaeoramularia (Cercospora) 840
catenulata, Lylea 773
catenulata, Spadicoides 963
catenulatum, Gliocladium 715
catenulatum, Pythium 920
catesbeyi, Mycosphaerella 801
catesbyana, Phyllachora 862
cattleyae, Gloeosporium 718
cattleyae, Physalospora = **Glomerella cingulata** 719
cattleyicola, Macrophoma = **Colletotrichum gloeosporioides** 638
caucasia, Trechispora = **T. farinacea** 997
caucus, Ciboria 627
caudata, Cryptostictis = **Seimatosporium caudatum** 944
caudata, (Ellisiella, Ellisiellina) = **Colletotrichum caudatum** 637
caudata, Harknessia 729
caudata, Lasiosphaeria (Leptospora) 757
caudata, Trichoconis (Pyricularia) 999
caudatum, Colletotrichum 637
caudatum, Helotium = **Hymenoscyphus caudatus** 740
caudatum, Mastigonetron = **Harknessia caudata** 729
caudatum, Seimatosporium 944
caudatus, Hymenoscyphus 740
caudatus, Sporocadus = **Seimatosporium caudatum** 944
caulicola, Ascochyta = **A. lethalis** 575
caulicola, Colletotrichum = **C. truncatum** 639
caulicola, Heterosporium 736
caulicola, Microsphaera = **M. trifolii** 791
caulicola, Platycarpium 882
caulicola, Puccinia 903
caulincola, Scopinella (Ceratostoma, Melanospora) 943
caulium, Lophiostoma 770
caulium var. **congregatum, Lophiostoma** 770
caulivora, Catosphaeropsis 601
caulivora, Kabatiella = **Aureobasidium caulivorum** 584
caulivorum, Aureobasidium (Gloeosporium) 584
caulophylli, Cercospora 607
caulophylli, Laetinaevia (Calloria, Orbilia) 755
cauticinalis, Xeromphalina 1025
cava, Phoma 852
cavernosa, Metasphaeria (Sphaeria) 788
ceanothi, Camarosporium 599
ceanothi, Cercospora 607
ceanothi, Cryptocline (Gloeosporium) 651
ceanothi, Cucurbitaria 654
ceanothi, Cytospora 661
ceanothi, Diaporthe 670
ceanothi, (Diatrype, Eutypella) = **Diatrype prominens** 673
ceanothi, Didymosphaeria 678
ceanothi, Leptosphaeria (Sphaeria) 761
ceanothi, Macrophoma = **Myxofusicoccum peckianum** 810
ceanothi, Microdiplodia 788

ceanothi, Mycothyridium (Thyridium) 807
ceanothi, Phloeosporella (Cylindrosporium) 850
ceanothi, Phyllosticta 866
ceanothi, Puccinia = **P. tripsaci** 915
ceanothi, Septoria 949
ceanothina, Jattaea (Physalospora) 750
cedricola, Bactrodesmium 585
cedrina, Chloroscypha 626
celastri, Cytospora 661
celastri, Diplodia = **D. propullans** 681
celastri, Dothiorella 688
celastri, Labrella 753
celastri, Microthyriella (Asteridium, Asterina) 791
celastri, Phyllosticta 866
celastri, Ramularia 924
celastrina, Cytospora = **C. leucosperma** 662
celastrina, Diaporthe 670
celastrina, (Diatrype, Valsa) = **Eutypella leprosa** 701
celastrina, Diplodia = **D. propullans** 681
celastrina, Sphaeropsis = **Diplodia propullans** 681
celata, Laestadia = **Pseudomassaria polystigma** 897
cellare, Zasmidium (Cladosporium, Racodium) 1027
cellaris, Rhinocladiella = **Zasmidium cellare** 1027
cellulare, Synchytrium 983
celosiae, Cercospora 607
celtidicola, Macrophoma = **Botryodiplodia celtidis** 591
celtidigena, Diplodia 680
celtidis, Aplosporella (Sphaeropsis) 570
celtidis, Botryodiplodia (Macrophoma) 591
celtidis, Camarosporium 599
celtidis, Cucurbitaria 654
celtidis, Cylindrosporium 658
celtidis, Cytospora = **C. leucosperma** 662
celtidis, Didymella 676
celtidis, Didymosphaeria 678
celtidis, Dothiorella 688
celtidis, Fusarium = **F. lateritium** 708
celtidis, Gloeosporium = **Discula umbrinella** 685
celtidis, Hendersonia (Sporocadus) 733
celtidis, Melanconium 782
celtidis, Phlebia 848
celtidis, Phyllosticta 866
celtidis, Placosphaeria 881
celtidis, Pseudoperonospora 897
celtidis, Ramularia (Cercosporella) 924
celtidis, Septogloeum 946
celtidis, Stagonospora (Phloeospora) 973
celtidis, Valsaria (Valsa) 1018
celtidis-australis, Coniothyrium 642
celtifolia, Hendersonia 733
cembrae, Coccophacidium = **Coccomyces strobi** 634
cenangioides, Lachnella 753
cenangioides, Stegonsporiopsis (Stegonsporium) 976
cenchri, Puccinia 903
cenchri, Sporisorium (Sorosporium, Ustilago) 971
cenchricola, Phyllachora = **P. sphaerosperma** 863
cenchrina, Septoria 949
cenisia, Valsa 1017
cenisia f. littoralis, Cytospora 661
centaureae, Cercospora 607
centaureicola, Septoria 949
centaureicola var. **brevispora, Septoria** 949
centaurii, Taeniolina (Hormiscium, Torula) 985
centranthi, Ramularia 924
centrifuga, Phlebia = **P. albida** 848
centrifugum, Corticium = **Butlerelfia eustacei** 596
cephalanthi, Ascochyta 574
cephalanthi, Cercospora (Ramularia) 607
cephalanthi, Coniothyrium 642
cephalanthi, Dendrophoma 667
cephalanthi, Diatrypella 674
cephalanthi, Godronia (Scleroderris) 721
cephalanthi, Lambertella 756
cephalanthi, Phoma 852
cephalanthi, Phyllosticta 866
cephalanthi, Puccinia = **P. seymouriana** 913
cephalanthi, Septoria 949
cephalanthi, Tympanis 1004
cephalanthi, Uredo 1007
cephalophora, Plasmopara 882
cepivorum, Sclerotium (Stromatinia) 941
cepulae, Urocystis = **U. magica** 1008
ceracellus, (Byssomerulius, Merulius) = **Ceraceomyces serpens** 603
ceraceus, Cerinomyces 621
ceramica, Hypocrea (Chromocrea) 743
ceranata, Eutypella 700
cerasella, Cercospora = **C. circumscissa** 608
cerasella, Mycosphaerella (Sphaerella) 801
cerasi, Dermea (Cenangium, Dermatea) 669
cerasi, (Exoascus, Taphrina) = **Taphrina wiesneri** 987
cerasi, Fusicladium 711
cerasina, Sphaeropsis = **S. malorum** 967
cerasinum, Melanconium 782
cerastii, Cercoseptoria 605
cerastii, Didymobotryopsis 677
cerastii, Melampsorella = **M. caryophyllacearum** 781
cerastii, Pleospora (Pyrenophora) 884
cerastii, Septoria 949
cerastii, Synchytrium 983
cerastiorum, Leptotrochila (Pseudopeziza) 766
cerastiorum var. **arenariae, Pseudopeziza** 897
ceratoniae, Pseudocercospora (Cercospora) 894
ceratophora, Cytospora = **C. sacculus** 662
ceratophora, Uredinopsis 1007
ceratophora, Valsa = **V. ceratosperma** 1017
ceratosperma, Valsa 1017
cercidicola, Cercospora 607
cercidicola, Mycosphaerella (Sphaerella) 801
cercidicola, Phoma 852
cercidicola, Phyllosticta 866
cercidicola var. coremioides, Cercospora = **Cercospora cercidicola** 607
cercidis, Aplosporella 570
cercidis, Cercospora = **C. chionea** 608
cercidis, Cytospora 661
cercidis, Diplodia 680
cercidis, Gloeosporidiella 717
cercidis, Ovularia 824
cercidis, Plasmopara 882
cercidis, Sphaeropsis 966
cercis, Macrophoma 774
cercocarpi, Camarosporellum 599
cercocarpi, Gibbera (Dothidea) 714
cercocarpi, Gloeosporidina (Gloeosporium, Melanophora) 717
cercocarpi, Leptosphaeria 761
cercocarpi, Phloeospora (Septogloeum) 849
cercocarpi, Sphaceloma 964
cercocarpi, Teichospora 988
cercosperma, Heteropatella (Septoria) 735
cercosporelloides, Rhopalidium 934
cercosporoides, Cylindrosporium 658
cercosporoides, Ramularia = **R. montana** 926
cercosporoides, Septoria = **S. leucanthemi** 953
cereale, Colletotrichum 637
cereale, Stephanosporium 977
cerealis, Fusisporium 712
cerealis, Hymenula 740
cerealis, Naucoria 811
cerealis, Rhizoctonia (Ceratorhiza) 932
cerealis, Sporotrichum = **Stephanosporium cereale** 977
cerebella, Coniophora = **C. puteana** 641
cerebriforme, Stromatocrea 982
cerebrina, Tilletia 992
cerebroides, Peridermium = **Endocronartium harknessii** 694
cerebrum, Cronartium = **C. quercuum** 649
cerebrum, Peridermium 831
cerei, Camarosporium (Hendersonia) 599
cerei, (Colletotrichum, Gloeosporium) = **Colletotrichum gloeosporioides** 638
ceriflua, Postia 890
cerifluus, (Polyporus, Tyromyces) = **Postia ceriflua** 890
cerina, Lophodermella (Hypodermella) 771
cerinum, Belonidium = **Dasyscyphus cerinus** 664
cerinus, Dasyscyphus 664
cerussatum, Acanthophysium 555
cerussatus, Aleurodiscus = **Acanthophysium cerussatum** 555
cervicolor, Asterostroma 583
cerviculata, Eutypella (Eutypa, Valsa) 700
cervina, Lachnellula (Erinella, Erinellina) 754
cervina, Trametes 996
cervinus, Pluteus = **P. atricapillus** 887
cervinus, (Coriolellus, Coriolus, Polyporus) = **Trametes cervina** 996
cesatii, Puccinia 903
cespitosus, Marasmius = **Collybia dichrous** 640
cestri, Chrysocyclus (Puccinia) 626
cestri, Uromyces (Uromyces) 1010
ceuthosporioides, Septoria 949
chaenorrhini, Septoria 949
chaetochloae, Meliola = **M. panici** 785
chaetochloae, Puccinia 903
chaetomium, Cercosporidium (Cladosporium) 620
chaetomium, Coleroa 636
chaetomium var. americana, Coleroa = **Hormotheca rubicola** 738
chaetophora, Peniophora = **Tubulicrinis chaetophorus** 1003
chaetophorus, Tubulicrinis 1003
chaetospira, Heteroconium (Septocylindrium, Septonema) 735
chaetosporum, Chaetospermum 624
chaetostroma, (Macbridella, Nectria) = **Byssosphaeria rhodomphala** 596
chailletii, Amylostereum (Stereum) 564
chailletii, Phragmotrichum 862
chamaecisti, Septoria 949
chamaecrista, Cercospora 608
chamaecristae, Aecidium 558
chamaecristae, Aplosporella 570
chamaecyparidis, Didymascella 676
chamaedaphnes, Monostichella (Gloeosporium) 796
chamaedryoidis, Synchytrium 983

chamaehyphon, Pythium 920
chamaenerii, Asterina = **Venturia adusta** 1020
chamaenerii, Phyllosticta 866
chamaesarachae, Puccinia 903
chamaesyces, Peronospora = **P. euphorbiae** 833
chamissonis, Septoria 949
chanousiana, Septoria 949
chaparralum, Erythricium 698
chardonii, Phyllachora = **P. paspalicola** 863
charlesii, Penicillium = **P. fellutanum** 828
chartarum, Pithomyces (Piricauda, Sporidesmium) 880
chartarum, Stachybotrys 972
chartarum, Ulocladium (Alternaria) 1006
chartarum f. stemphylioides, Alternaria = **A. stemphylioides** 562
cheilanthis, Hyalopsora 738
cheiranthi, Alternaria (Helminthosporium, Macrosporium) 561
cheiranthi, Ascochyta 574
cheirodendronis, Asteridiella (Irenina, Meliola) 580
chelidonii, Septoria (Ascochyta) 949
chelones, Erysiphe = **E. galeopsidis** 698
chelonis, Puccinia 903
chenopodii, Cercospora 608
chenopodii, Leptostromella 765
chenopodii, Leptothyrium 766
chenopodii, Mycosphaerella 801
chenopodii, Phloeospora = **Ascochyta hyalospora** 575
chenopodii, Phyllosticta 866
chenopodii, Stagonospora 973
chenopodii, Uromyces = **U. giganteus** 1011
chenopodii-albi, Phyllosticta 866
chenopodiicola, Cercospora = **C. beticola** 606
chenopodii-glauci, Peronospora = **P. farinosa** 833
chenopodinum, Hadrotrichum 727
cherimoliae, Phakopsora 843
chermesinum, Penicillium 827
chesapeakensis, Aniptodera 564
chevalieri, Aspergillus = **A. equitis** 578
chevalieri, Eurotium 699
chevalieri var. **intermedius, Aspergillus** 578
chiltonii, Synchytrium = **S. stellariae** 985
chimaphilae, Mycosphaerella 801
chimaphilae, Ramularia 924
chimaphilae, Septoria = **S. ellisiana** 951
chimaphilina, Mycosphaerella = **M. chimaphilae** 801
chiogenis, Chrysomyxa 626
chionanthi, Cercospora 608
chionanthi, Coniothyrium 642
chionanthi, Cytospora = **C. pruinosa** 662
chionanthi, Dermea 669
chionanthi, Didymosphaeria 678
chionanthi, Phyllosticta 866
chionanthi, Septoria 949
chionanthi, Valsa = **V. cypri** 1017
chionea, Cercospora (Cercosporella) 608
chionea, Hypocrea 743
chioneus, Tyromyces (Polyporus) 1005
chionostoma, Anthostomella 565
chiricahuaensis, Resinicium 929
chisosensis, Puccinia 903
chisosum, Acremonium 556
chlamydospora, Embellisia 693
chlamydospora, Pleospora 884
chlamydosporum, Fusarium 707
chlamydosporum, Pseudostemphylium = **Embellisia chlamydospora** 693
chloes, Graphyllium (Pleospora) 723
chlora, Chlorosplenium 626
chloracae, Puccinia 903
chlorantha, Septoria 949
chlorascens, Lachnella = **Chlorosplenium chlora** 626
chloridicola, Ustilago 1015
chloridis, Puccinia 903
chlorina, Tomentella 993
chlorina, Valsa 1017
chlorinella, Thyronectria (Nectria) 991
chlorodisca, Valsa = **V. ceratosperma** 1017
chlorogali, Mycosphaerella 801
chlorogali, Septoria (Rhabdospora) 949
chlorogali, Uromyces = **U. aureus** 1010
chlorogalli, Pleospora 884
chloromela, Chloroscypha (Kriegeria) 626
chlorospora, Venturia 1020
choisyae, Sphaceloma 964
chondrillina, Puccinia 903
chondrospora, Pseudomassaria 896
chordulata, Pellicularia = **Botryohypochnus chordulatus** 591
chordulatus, Botryohypochnus 591
chorizanthis, Uromyces = **U. intricatus** 1011
chromatica, Poria = **Diplomitoporus crustulinus** 683
chrysanthemella, Septoria 949
chrysanthemi, Alternaria = **A. leucanthemi** 562
chrysanthemi, Cercospora 608
chrysanthemi, Cylindrosporium 658
chrysanthemi, Leptosphaeria = **Ophiobolus ophioboloides** 820
chrysanthemi, Phoma (Ascochyta) 852
chrysanthemi, Phyllosticta 866
chrysanthemi, Puccinia = **P. tanaceti** 915
chrysanthemi, Septoria 949
chrysentera, Lasiosphaeria 757
chrysitis, Penicillium = **P. rugulosum** 829
chrysobalani, Cercospora 608
chrysocoma, (Byssonectria, Nectria) = **Nectria squamulosa** 813
chrysocoma, Orbilia (Calloria) 822
chrysocoma, Peziza = **Guepiniopsis chrysocomus** 724
chrysocomus, Guepiniopsis 724
chrysocrea, Phlebia 848
chrysocreas, Corticium = **Phlebia chrysocrea** 848
chrysogenum, Penicillium 827
chrysogramma, Thyronectria 991
chrysoloma, Phellinus (Polyporus) 844
chrysopepla, (Collybia, Xerulina) = **Cyptotrama asprata** 660
chrysophthalma, Lachnellula = **L. suecica** 754
chrysophylla, Omphalina = **Gerronema chrysophyllum** 714
chrysophyllum, Gerronema 714
chrysopogonis, Sphacelotheca = **Sporisorium cruentum** 971
chrysopogonis, Tolyposporella 993
chrysopsidis, Ramularia 924
chrysopsis, Stictis 978
chrysorhiza, Phanerochaete (Mycoacia) 843
chrysorhizon, Hydnum = **Phanerochaete chrysorhiza** 843
chrysosperma, Cytospora 661
chrysospermus, Dacrymyces 663
chrysospora, (Pleospora, Pyrenophora) = **Pleospora helvetica** 884
chrysosporium, Phanerochaete 843
chrysostigma, Pezizella = **Psilachnum chrysostigmum** 899
chrysostigmum, Psilachnum 899
chrysostroma, Melanconis (Valsa) 782
chrysostroma var. ellisii, Melanconis 782
chrysothamni, Macrophoma 774
chrysothamni, Marssonina 778
chumrungensis, Pseudobeltrania 893
cibotii, Helminthosporium 731
cibotii, Pestalotia 835
cibotii, Phyllachora 862
cibotii, Sphaerulina 968
cibotti, Chalara (Excioconidium) 625
cicatricum, Phoma = **P. macrostoma** 854
ciceris-arietini, Uromyces 1010
cichoracearum, Erysiphe 697
cichoracearum f. galeopsidis, Erysiphe = **E. galeopsidis** 698
cichoracearum var. latispora, Erysiphe 697
cichorii, Cercospora 608
cichorii, Ramularia 924
cichorii-intybi, Cercospora = **C. cichorii** 608
cicutae, Cylindrosporium 658
cicutae, Puccinia 903
ciliare, Incrucipulum (Lachnum) 748
ciliaris, Lachnella = **Incrucipulum ciliare** 748
ciliata, Hyalopeziza (Lachnella, Lachnum, Trichopeziza) 738
ciliata, Lachnellula 754
ciliata, Mitrorhizopeltis = **Tracylla spartinae** 996
ciliata, Spermospora 964
ciliata, Stevensonula 978
ciliata var. eocoronis, Pyrenophora = **Pleospora eocoronis** 884
ciliatospora, Belonioscypha = **Hymenoscyphus scutula** 740
ciliatus, Dasyscyphus = **Lachnellula ciliata** 754
cilinodis, Ramularia 924
ciliolata, Odontia = **Steccherinum ciliolatum** 975
ciliolatum, Steccherinum (Hydnum) 975
cinchonae, Cercospora 608
cincta, Cytospora 661
cincta, Fabraea 704
cincta, Leucostoma (Valsa) 767
cincta, Phoma 852
cinctula, Winterella (Cryptospora) 1024
cinctum, (Colletotrichum, Gloeosporium) = **Colletotrichum gloeosporioides** 638
cinctum, Glomerella = **G. cingulata** 719
cinctum, Melanconium 782
cinctum, Myrothecium 809
cinerariae, Alternaria 561
cinerascens, Lopharia (Stereum) 770
cinerascens, Phomopsis (Phoma) 858
cinerascens, (Polyporus, Poria) = **Diplomitoporus linbladii** 683
cinerascens, Tomentella (Hypochnus) 993
cinerea, Botrytis 594
cinerea, Claviceps 632
cinerea, Maireina (Solenia) 776
cinerea, Mollisia 793
cinerea, Monilia = **M. laxa** 794
cinerea, Ostropa 823
cinerea, Peniophora 829
cinerea, Phyllosticta 866
cinerea, Puccinia = **P. recondita** 912
cinerea, (Bourdotia, Sebacina) = **Basidiodendron cinereum** 586
cinerea f. douglasii, Botrytis = **B. douglasii** 594
cinereipallens, Crepidotus 647
cinerella, Mycena 797
cinerella, Tapesia 986
cinereo-lilacinum, Hypoxylon 745

cinerescens, Cryptocline (Gloeosporium) 651
cinerescens, Phialophora (Verticillium) 847
cinereum, Basidiodendron 586
cinereum, Gloeocoryneum (Coryneum, Leptomelanconium) 716
cinereum, Hypoxylon = **H. hypophlaeum** 745
cinereum, Physarum 876
cinereus, Microascus 788
cingulata, Glomerella 719
cingulata var. orbiculare, Glomerella = **G. lagenaria** 719
cingulata var. vaccinii, Glomerella = **Glomerella cingulata** 719
cingulatum, Gloeosporium = **Colletotrichum gloeosporioides** 638
cinnabarina, Nectria 812
cinnabarina, Phlebia = **P. merismoides** 848
cinnabarina, Trametes = **Pycnoporus cinnabarinus** 917
cinnabarina var. sanguinea, Trametes = **Pycnoporus sanguineus** 917
cinnabarinum, Fusarium = **F. lateritium** 708
cinnabarinum, Stilbum = **Stilbella aleuriata** 981
cinnabarinum, Verticillium = **V. tenerum** 1022
cinnabarinus, Crepidotus 647
cinnabarinus, Pycnoporus (Coriolus, Phellinus, Polyporus, Polystictus) 917
cinnafolia, Mycosphaerella 801
cinnamomea, Hymenochaete 739
cinnamomea, Lazulinospora 757
cinnamomea, Pezicula 837
cinnamomeum, Cephalosporium 602
cinnamomeum, Synchytrium 983
cinnamomeus, Crepidotus 647
cinnamomeus, Irpex = **Hydnochaete olivacea** 738
cinnamomi, Colletotrichum = **C. gloeosporioides** 638
cinnamomi, Elsinoe 692
cinnamomi, Microsphaera 790
cinnamomi, Phytophthora 878
cinnamopurpureum, Eupenicillium 699
cinnamopurpureum, Penicillium = **P. dierckxii** 828
cintum, Fusarium = **Myrothecium cinctum** 809
circaeae, Entyloma = **E. arnicale** 695
circaeae, Puccinia 903
circinans, Alternaria = **A. brassicicola** 561
circinans, Botryodiplodia (Diplodia) 591
circinans, Byssothecium (Trematosphaeria) 596
circinans, Coleroa (Venturia) 636
circinans, Colletotrichum (Vermicularia) 637
circinans, Cylindrosporium 658
circinans, Didymosphaeria 678
circinans, Meliola 784
circinans, Pirostoma (Coniosporium) 880
circinans, Rhizopus 934
circinans var. **rhynchosporae, Meliola** 784
circinata, Gyrothrix 727
circinata, Onnia = **Inonotus circinatus** 748
circinata, Periconia 831
circinata, Septoria 949
circinata, Uncinula 1006
circinatum, Campsotrichum = **Gyrothrix circinata** 727
circinatus, Inonotus (Polyporus, Trametes) 748
circinelloides, Mucor 796
circuligerens, Phyllosticta 866
circumdata, Puccinia = **P. dolosa** var. **circumdata** 905
circumfusa, Ramularia = **R. rubella** 926
circumscissa, Cercospora 608
circumscissa, Phragmodothis (Plowrightia) 861
circumscissa, Phyllosticta 866
circumscriptum, Endophyllum 694
cirrhosa, Endoxyla (Ceratostomella) 695
cirrhosa, Septoria 949
cirrosus, Microascus 788
cirsii, Cercospora 608
cirsii, Phyllosticta 866
cirsii, Puccinia = **P. calcitrapae** var. **centaureae** 902
cirsii, Septoria 949
cirsii, Stagonospora 973
cirsii f. **furcatus, Ophiobolus** 819
cirsii-altissima, Ophiobolus 820
cirsii-lanceolati, Puccinia = **P. cnici** 903
cissi, Mycosyrinx (Ustilago) 807
cissicola, Phyllosticta 866
citreonigrum, Penicillium 828
citreoviride, Penicillium = **P. citreonigrum** 828
citri, Alternaria 561
citri, Capnodium (Meliola, Pleosphaeria) 601
citri, Cladosporium 629
citri, Cylindrocladium (Candelospora) 657
citri, Diaporthe 670
citri, Mycosphaerella 801
citri, Myriangium 808
citri, Phomopsis 858
citri, Septoria 949
citri, Sirodothis 961
citri, Stemphylium 976
citri var. cerasi, Alternaria = **A. tenuissima** 562
citri-aurantii, Geotrichum (Oidium, Oospora) 714
citricarpa, Guignardia 725
citricarpa, Phyllosticta (Phoma, Phyllostictina) 866
citricola, Capnodium (Limacinula) 601
citricola, Eutypella 700
citricola, Phyllosticta 866
citricola, Phytophthora 878
citriforme, Fusarium = **F. tricinctum** 711
citrina, Bisporella (Calycella) 590
citrina, Hypocrea 743
citrina, Isaria = **Tilachlidium brachiatum** 991
citrinoalba, Trichopeziza = **Dasyscyphus citrinoalbus** 665
citrinoalbus, Dasyscyphus 665
citrinomarginata, Mycena 797
citrinoviride, Trichoderma 999
citrinum, Gloeocystidiellum = **Vesiculomyces citrinus** 1022
citrinum, Helotium = **Bisporella citrina** 590
citrinum, Penicillium 828
citrinum, Physarum 876
citrinus, Crepidotus 647
citrinus, Vesiculomyces 1022
citrispora, Physalospora = **Pseudomassaria chondrospora** 896
citrophthora, Phytophthora 878
citrulli, Septoria 949
citrulli, Volutella = **Colletotrichum gloeosporioides** 638
citrullina, Cercospora 608
citrullina, Cribropeltis 648
citrullina, Mycosphaerella = **Didymella bryoniae** 676
citrullina, (Ascochyta, Diplodina, Phyllosticta) = **Phoma cucurbitacearum** 852
cladii, Puccinia 903
cladogamus, Aphanomyces 568
cladosporioides, Cladosporium (Hormodendrum, Penicillium) 629
cladosporioides, Mycocentrospora (Cercospora) 799
cladosporioides f. sp. **pisicola, Cladosporium**
clallamensis, Mycosphaerella 801
clandestina, Brunnipila 595
clandestina, Podosphaera 888
clandestinum, Lachnum = **Brunnipila clandestina** 595
clandestinum, Splanchnonema 969
clandestinus, Dasyscyphus = **Brunnipila clandestina** 595
clarkeana, Rhabdospora 930
clausa, Valsa = **V. ambiens** 1017
clavariiforme, Gymnosporangium 726
clavata, Cercospora (Cercospora) 608
clavata, Hemitrichia 732
clavata, Mycena (Omphalia) 797
clavata, Pestalotia 835
clavatum, Cylindrocladium 657
clavatum, (Helminthosporium, Virgasporium) = **Cercospora clavata** 608
clavatus, Aspergillus 578
clavellosa, Nyssopsora 817
clavellosum, Triphragmium = **Nyssopsora clavellosa** 817
clavicarpa, Cercospora 608
clavicarpa, Leptosphaeria = **Phaeosphaeria clavispora** 842
claviceps, Balansia 586
clavicularis, Mycena 797
claviforme, Penicillium 828
claviger, Ophiobolus 820
clavigera, Ceratocystis = **Ophiostoma clavigerum** 821
clavigera, Leptosphaeria 761
clavigera, Mollisia (Peziza) 793
clavigera, Peniophora = **Hyphoderma clavigerum** 740
clavigera, Tomentella 993
clavigera, Valsa = **V. ceratosperma** 1017
clavigerum, Hyphoderma 740
clavigerum, Ophiostoma (Europhium) 821
clavigignenti-juglandacearum, Sirococcus 961
clavipes, Gymnosporangium 726
clavispora, Isariopsis = **Pseudocercospora vitis** 895
clavispora, Leptosphaeria = **L. helminthospora** 762
clavispora, Patellariopsis 826
clavispora, Pestalotiopsis (Pestalotia) 836
clavispora, Phaeosphaeria (Dothidea) 842
clavispora, Puccinia = **P. virgata** 916
clavispora, Tryblidiella (Patellaria, Triblidium) 1002
clavisporum, (Coryneum, Exosporium) = **Corynespora olivacea** 646
clavisporum, Glonium = **G. lineare** 720
clavisporum, Lecanidion = **Patellariopsis clavispora** 826
clavisporus, Rhopographus = **Phaeosphaeria clavispora** 842
clavula, Balansia = **Balansiopsis pilulaeformis** 586
clavulata, Gnomonia (Didymella) 720
clavulata, Stilbella 981
clavulata, (Eutypella, Valsa) = **Eutypella scoparia** 701

clavulatum, Graphium = **Stilbella clavulata** 981
clavuligera, Asterina 580
clavuligerum, Gloeocystidiellum 716
clavus, Kretzschmaria (Hypoxylon) 753
claytoniae, Peronospora 832
claytoniae, Ramularia 924
claytoniae, Uromyces 1010
claytoniae, Ustilago 1015
claytoniana, Physoderma 877
claytoniana var. sparrowii, Physoderma = **P. australasica** 877
claytoniicola, Puccinia 903
cleistotheca, Massaria 779
clematidina, Phoma (Ascochyta) 852
clematidina var. **thalictri, Ascochyta** 574
clematidis, Aplosporella (Macroplodia) 570
clematidis, Cylindrosporium 658
clematidis, (Dicaeoma, Puccinia) = **Puccinia recondita** 912
clematidis, Didymaria 676
clematidis, Didymosphaeria = **D. futilis** 678
clematidis, Macrosporium 775
clematidis, Phyllosticta 866
clematidis, Pleospora = **P. vitalbae** 886
clematidis, Ramularia 924
clematidis, Septoria 949
clematidis, Synchytrium 983
clematidis, Tubulicium (Litschauerella, Peniophora, Xenasma) 1003
clemensiae, Cercospora 608
clementsii, Didymosphaeria 678
clementsii, Pleospora = **P. dichromatricha** 884
clementsii, Septoria 949
cleomes, Cercospora 608
clermontiae, Amazonia 563
clermontiae, Asterina 580
clermontiae, Calothyriopeltis 598
clermontiae, Clypeolella 633
clermontiae, Septoria 950
clethrae, Ditopellopsis 686
clethrae, Phyllosticta 866
clethricola, Phyllosticta 866
clethrincola, Sphaeronaema (Chondropodiella) 966
clidemiae, Microsphaeropsis 791
cliftoniae, Lembosia (Morenoella) 758
cliftoniae, Pestalotia = **P. versicolor** 836
clintoniae, Tuburcinia 1003
clintoniana, Mycosphaerella (Sphaerella) 801
clintoniana, Ustilago 1015
clintonianum, Entyloma 695
clintonii, Aplosporella (Haplosporella, Sphaeropsis) 570
clintonii, Puccinia 903
clintonii, Uncinula 1006
clintonii, Venturia = **V. systema-solare** 1021
clintonii var. **albae, Puccinia** 903
clintonii var. **bracteosae, Puccinia** 903
clintonii var. **groenlandicae, Puccinia** 903
clintonii var. **racemosae, Puccinia** 903
clitoriae, Pseudocercospora (Cercospora) 894
clitoricarpa, Phoma (Sphaeropsis) 852
clivensis, Kalmusia (Diapleella, Leptosphaeria) 751
clonostachoides, Isaria = **Gliocladium roseum** 715
clymenia, Mycosphaerella 801
clypeata, Dendrophoma 667
clypeata, Pachyella (Peziza) 825
clypeata, Phyllosticta 866
clypeatum, Melanomma 783
clypeosphaerioides, Leptothyrium 766
clypeus, (Diatrype, Nummularia) = **Hypoxylon mediterraneum** 745
cnici, Puccinia 903
cnici-oleracei, Puccinia 903
coalescens, Ramularia (Cercospora) 924
coalescens, Tremella 998
coarctata, Cistella (Trichopeziza) 628
coccicola, Podonectria 888
coccinea, Cytospora 661
coccinea, Nectria 812
coccinea, Ramularia = **R. castillejae** 924
coccinea var. **faginata, Nectria** 812
coccinella, Orbilia (Calloria, Mollisia) 822
coccinellum, Illosporium 748
coccineo-fulva, Phlebia (Peniophora) 848
coccineum, Hypoxylon = **H. fragiforme** 745
coccineus, Pluteus = **P. aurantiorugosus** 887
coccodes, Colletotrichum 637
coccolobae, Pestalotia = **P. versicolor** 836
coccolobae, Phyllosticta 866
coccolobae, Uredo 1007
coccolobii, Cryphonectria (Endothia) 650
coccophila, Nectria = **N. flammea** 812
coccophilum, Fusarium 707
cocculi, Lembosina (Echidnodella, Maurodothella) 758
cocculi, Synchytrium 984
cocculicola, Cercospora 608
cochleariae, Peronospora 832
cochleatus, Lentinellus 759
cochliodes, Chaetomium = **C. globosum** 623
cochlioides, Aphanomyces 568
cochlospermi, Cercospora 608
cockerelliana, Puccinia 903
cocoes, Lasmeniella (Coniothyrium) 757
cocoina, Septoria 950
cocos, Eutypella = **Valsa chlorina** 1017
cocos, Macrosporium 775
cocos, Wolfiporia (Macrohyporia, Poria) 1024
codiaei, Phyllosticta = **Phyllosticta ghaesembillae** 868
coelopleuri, Puccinia 903
coenophialum, Acremonium 556
coerulea, Ceratostomella = **Ophiostoma piliferum** 821
coerulea, Mycosphaerella 801
coerulea, Tomentella 993
coerulescens, Ceratocystis (Endoconidiophora, Ophiostoma) 603
coerulescens f. douglasii, Endoconidiophora = **Ceratocystis coerulescens** 603
coeruleum, Fusarium = **F. solani** var. **coeruleum** 711
coffeae, Cercospora = **C. coffeicola** 608
coffeanum, Colletotrichum 637
coffeicola, Cercospora 608
coffeicola, Massarina (Oraniella) 779
cognata, Poria = **Postia rancida** 891
cognata, Puccinia 903
cohaerens, Hypoxylon 745
cohaerens, Trechispora 997
cohaesa, Tranzschelia 997
cohortalis var. **alachuanus, Marasmius** 777
coicis, Ustilago 1015
cokeri, Dacrymyces 663
cokeri, Exidia 702
cokeri, Septobasidium 946
colchici, Urocystis 1008
colchicus, Sclerotiomyces 941
coleosanthi, Cercospora 608
coleosporioides, Cronartium 649
coleroides, Cercospora 608
collabendum, Helminthosporium = **Sporidesmium inflatum** 970
collabens, Junghuhnia (Polyporus) 751
collabens, Phoma = **P. cava** 852
collabescens, Aposphaeria = **Phoma cava** 852
collapsa, (Hendersonia, Stagonospora) = **Diplodina acerina** 682
collapsa, Pseudomeliola = **Nectria leucorrhodina** 813
collapsus, Ophiobolus 820
collecta, Dothidea = **D. puccinioides** 687
colletotrichoides, Volutella 1023
collicula, Valsa 1017
colliculosa, Trametes = **Pachykytospora tuberculosa** 825
colliculosum, Corticium = **Basidioradulum radula** 586
colligatus, Helicomyces 731
collinsiae, Entyloma = **E. linariae** 696
collinsiae, Puccinia 903
collinsiae, Septoria 950
collinsii, Apiosporina (Dimerosporium) 569
collinsoniae, Leptosphaeria 761
collo-cygni, Ramularia 924
collomiae, Sphaerotheca 968
collybiiformis, Mycena 797
colocasiae, Phytophthora 878
colocasiophila, Phyllosticta 866
coloradense, Peridermium 831
coloradensis, Crepidotus 647
coloradensis, Diplodina (Diplodinula) 682
coloradensis, Flagelloscypha = **Calathella ellisii** 597
coloradensis, Niesslia (Acanthonitschkea) 815
coloradensis, Phyllachora 862
coloradensis, Pleospora 884
coloradensis, Pyrenopeziza 918
coloradensis, Uromyces 1010
coloradensis, Ustilago = **U. buchloes** 1014
coloradensis, Valsaria = **Endoxylina allantospora** 695
coloradensis var. campester, Uromyces = **Uromyces coloradensis** 1010
coloradensis var. maritimus, Uromyces = **Uromyces coloradensis** 1010
coloradensis var. montanus, Uromyces = **Uromyces coloradensis** 1010
coloradoensis, Phoma 852
colorata, Ascochyta = **Marssonina fragariae** 778
colorata, Mycosphaerella 801
coloratum, Myxosporium (Melanconium) 810
coloratum, Pythium 920
colorea, Collybia = **Callistosporium luteo-olivaceum** 597
colorea, Poria = **Perenniporia subacida** 831
colossum, Ganoderma 713
colubrina, Cercosporella = **Entyloma compositarum** 695
columbi, Mycosphaerella 801
columbianus, Uromyces 1010
columbiense, Aecidium 558
columbiense, Sporotrichum 972
columbiensis, Cercospora = **C. tandilensis** 617
columbiensis, Gloeodontia 716
columbiensis, Puccinia = **P. cnici-oleracei** 903
columnaris, Calyptospora = **Pucciniastrum goeppertianum** 916
columnaris, (Cercospora, Pseudocercospora) = **Phaeoisariopsis griseola** 839

coluteae, Camarosporium (Hendersonia) 599
coluteae, Diplodia 680
coluteae, Dothidea = **D. puccinioides** 687
coluteae, Phomopsis (Phoma) 858
coluteae, Uromyces 1010
comandrae, Cercospora 608
comandrae, Cronartium 649
comandrae, Puccinia 903
comari, Cercospora 608
comari, Gnomonia 720
comari, Physoderma (Protomyces) 877
comata, Pleospora (Pyrenophora) 884
comata, Trichodothis (Asterina) 999
comatella, Leptosphaeria = **Massarina gloeospora** 779
combinans, Hydnum = **Hyphodontia stipata** 742
comitatum, Myrioconium 808
comitatum var. **salicarium, Myrioconium** 808
comitialis, Coccomyces = **C. tumidus** 634
commelinae, Colletotrichum = **C. gloeosporioides** 638
commelinae, Illosporium 748
commelinae, Puccinia 903
commelinae, Uromyces 1010
commelinae, Ustilago (Tilletia) 1015
commelinicola, Phyllosticta 866
commersoniana, Septoria 950
commixtus, Claudopus = **Pleurotellus hypnophilus** 886
commonsii, Cercospora 608
commonsii, Phyllosticta 866
commonsii, Septoria 950
commune, Hypoderma 743
commune, Macrosporium = **Stemphylium botryosum** 976
commune, Penicillium = **P. puberulum** 828
commune, Schizophyllum 939
communis, Erysiphe 697
communis, Phomopsis (Phoma) 858
communis, Taphrina (Exoascus) 986
commutata, Puccinia 903
compacta, Alternaria 561
compacta, (Irenopsis, Meliola) = **Irenopsis tehoniana** 749
compacta, Ovularia 824
compacta, Perenniporia (Poria) 830
compacta, Pestalotia = **P. pezizoides** 835
compacta, Pyrenobotrys (Gibbera, Venturia) 918
compacta, Stigmina 979
compacta, Ustilago 1015
compacta, Vermicularia = **Colletotrichum dematium** 637
compactum, Cercosporidium (Cladosporium) 620
compactum, (Stegonsporium, Thyrostroma) = **Stigmina compacta** 979
compactus, Polyporus = **Perenniporia compacta** 830
compactus, Uromyces 1010
complanata, Leptosphaeria (Metasphaeria, Sphaeria) 761
complanata, Phlyctema (Septoria) 850
complanata, Phoma 852
complanata, Typhula = **T. phacorrhiza** 1005
complectens, Pythium = **P. vexans** 922
complens, Completoria 640
complicatula, Mollisia 793
complicatum, Stereum = **S. hirsutum** 977
compositarum, Ascochyta 574
compositarum, Camarosporium (Dichomera) 599
compositarum, Entyloma 695
compositarum, Pleospora = **P. scrophulariae** var. **compositarum** 885
compositarum var. parva, Ascochyta = **Ascochyta compositarum** 574
compositum, Schizoxylon 940
compressa, Leptosphaeria (Ophiobolus) 761
compressa, Physopella (Angiopsora) 877
compressa, Pleospora = **Comoclathris permunda** 640
compressa f. **toxicodendri, Botryodiplodia** 591
compressula, Pyrenopeziza 918
compressum, Acrospermum 557
compressum, Platystomum (Lophidium) 882
compressum var. foliicola, Acrospermum = **A. foliicola** 557
compressum var. **pseudomacrostomum, Platystomum** 882
compta, Cryptosporella (Cryptospora, Valsa) 653
compta, Monochaetia = **Seimatosporium rosae** 945
compta, Pestalotia = **Seimatosporium rosae** 945
comptoniae, Cronartium 649
comptoniae, Cucurbitaria 654
comptoniae, Peridermium 832
comptoniae, Thyridaria (Diatrype, Pseudovalsa) 990
comptum, Hyaloceras = **Seimatosporium rosae** 945
concava, (Ceuthospora, Sclerotiopsis) = **Pilidium concavum** 880
concava, Phyllosticta 866
concaviuscula, Dothiorella 688
concaviuscula, Scirrhia (Dothidea, Dothidella, Plowrightia) 940
concavum, Pilidium 880
concavum, Rhytisma 935
concavus, Dacrymyces = **D. capitatus** 663
concentrica, Daldinia 664
concentrica, Hendersonia 733
concentrica, Microsphaeropsis 791
concentrica, Monochaetia (Pestalotia) 795
concentrica, Pestalosphaeria 834
concentrica, Phlebia 848
concentrica, Phyllosticta 866
concentrica, Stigmina (Cercospora) 979
concentrica, Vermicularia = **Colletotrichum gloeosporioides** 638
concentricum, Aristastoma = **A. oeconomicum** 572
concentricum, Coniothyrium = **Microsphaeropsis concentrica** 791
concentricum, Discogloeum (Exosporium) 684
concentricum, Hypoxylon = **Daldinia concentrica** 664
concentricum, (Pirex, Radulum) = **Phlebia concentrica** 848
conchatus, Phellinus (Fomes, Polyporus) 844
conchifer, Trametes (Polyporus, Poronidulus) 996
concinna, Cercospora = **Cercosporidium galii** 620
concinna, Hercospora = **H. taleola** 734
concinna, Leptosphaeria 761
concinna, Puccinia 903
concinnum, Berkleasmium (Sporidesmium) 587
concinnum, Didymosporium 678
concolor, Diatrype 673
concolor, Fusarium 707
concolor, Lophodermella (Hypodermella) 771
concomitans, Phyllosticta 866
concomitans, Septocylindrium (Ramularia) 946
concors, Mycovellosiella (Cercospora) 807
condaliae, Meliola 784
condensata, Cercospora 608
condensata var. desmanthi, Cercospora = **Cercosporidium desmanthi** 620
condensatum, Cryptocoryneum 651
conferta, Didymaria 676
conferta, (Irenopsis, Meliola) = **Irenopsis tehoniana** 749
conferta, Poria = **Perenniporia ellipsospora** 830
conferta, Puccinia = **P. cnici-oleracei** 903
conferta, Pyrenobotrys (Gibbera, Sphaeria, Stigmatea) 918
confertissima, Phyllosticta 866
confertissima, Stigmatea = **Venturia geranii** 1020
confertula, Nitschkia 816
confertus, Godronia 721
confine, Corticium = **Trechispora cohaerens** 997
confinis, Mycosphaerella = **M. tassiana** 806
confinis, Trechispora = **T. cohaerens** 997
conflicta, Sphaerulina (Linospora) 968
conflictum, Ceuthocarpon = **Sphaerulina conflicta** 968
confluens, Cercospora 608
confluens, Cerocorticium (Corticium, Radulomyces) 622
confluens, Gloeosporium 718
confluens, Melampsora = **M. ribesii-purpureae** 781
confluens, (Byssomerulius, Merulius) = **Meruliopsis corium** 787
confluens, Pyronema = **P. omphalodes** 920
confluens, Sistotrema 962
confraga, Puccinia 904
confragosa, Daedaleopsis (Daedalea) 664
confundens, (Ciborinia, Sclerotinia) = **Ciborinia seaveri** 627
confusa, Ascochyta 574
confusa, Cyphellopsis (Solenia) 660
confusa, Eutypa = **E. limaeformis** 700
confusa, Mycosphaerella 801
confusa, Taphrina 986
confusa, Trematosphaeria (Verrucaria) 998
confusa, Tympanis 1004
confusum, Gymnosporangium 726
confusum, Sorosporium 963
congdonii, Puccinia 904
congenera, Diaporthe = **D. dubia** 671
congesta, Asterina 580
congesta, Botryodiplodia (Diplodia) 591
congesta, Cucurbitaria 654
congesta, Diaporthe = **D. impulsa** 671
congesta, Graphiola 723
congesta, Phyllosticta 866
congestum, Stemphylium 976
congestum var. **minor, Stemphylium** 976
conglobata, Asterina 580
conglobata, Cucurbitaria 654
conglobata, Stromatocyphella (Cyphella) 982
conglobata var. **crataegi, Cucurbitaria** 654
conglobatum, Berkleasmium (Sporidesmium) 587
conglobatus, Polyporus = **Globifomes graveolens** 715
conglomerata, Peronospora 833
conglomerata, Puccinia 904
conglomeratus, Protomyces 892
conglutinans, Fusarium = **F. oxysporum** f. sp. **conglutinans** 709

conglutinans var. betae, Fusarium = **F. oxysporum** f. sp. **spinaciae** 709
conglutinans var. callistephi, Fusarium = **F. oxysporum** f. sp. **callistephi** 708
conglutinata, Heteropatella (Excipula) 735
conglutinata, Torula 995
congregata, Puccinia = **P. heucherae** 907
congregatum, Lophiostoma = **L. caulium** var. **congregatum** 770
congruens, Phyllachora = **P. bonariensis** 862
conica, Lulworthia 773
conica, Macrophoma = **Diplodia rubi** 681
conica, Physalospora = **Botryosphaeria festucae** 592
conicola, Ascochyta 574
conicola, Illosporium 748
conidiogena, Phoma = **P. glomerata** 853
conidiophora, Stilbotulasnella 981
coniferarum, Phacidium (Phacidiella, Potebniamyces) 838
coniferarum, Poria = **Anomoporia bombycina** 565
conigena, Leptostromella 765
conigena, Microdiplodia 788
conigena, Mollisia 793
conigena, Pseudopatellina = **Propolis rhodoleuca** 892
conigena, Scirrhia (Mycosphaerella) 940
conigena, Sphaerella = **Didymella douglasii** 676
conigenoides, Strobilurus (Collybia, Pseudohiatula) 982
conigenum, Corticium = **Punctularia atropurpurascens** 917
conigenum, Cronartium 649
conigenus, Sirococcus 961
conii, Rhabdospora 930
coniothyrium, Diapleella (Leptosphaeria, Melanomma) 670
conjuncta, Choanephora = **C. infundibulifera** 626
conjuncta, (Cucurbidothis, Gibberidea) = **Melanomma conjunctum** 783
conjunctum, Melanomma 783
connata, Fomitopsis = **Oxyporus populinus** 824
connata, Pholiota = **P. limonella** 851
connatus, Fomes = **Oxyporus populinus** 824
connatus, (Flaviporus, Oxyporus) = **Oxyporus populinus** 824
connersii, Gymnosporangium = **G. bethelii** 726
conocarpi, Cercospora 608
conoclinii, Puccinia 904
conoidea, Leptosphaeria 761
conorum, Diaporthe 670
conorum, Phomopsis (Phoma) 858
conradii, Plagiostoma (Diaporthe, Gnomonia) 881
consanguinea, Anthostomella (Sordaria) 565
consanguineum, Belonium 587
consanguineum, Helotium 732
consanguineum, Sporisorium (Sorosporium) 971
conscriptus, Hymenoscyphus 740
conseptata, Eutypella (Valsa) 700
conservans, Cylindrosporium = **C. nuttallii** 659
consessa, Leptosphaeria (Sphaeria) 761
consimile, Corticium = **Athelia arachnoidea** 583
consimilis, Hypocreopsis (Hypocrea) 743
consimilis, Ophiobolus 820
consimilis, Puccinia 904
consimilis, Septoria = **S. lactucae** 952
consobrina, Cercospora = **C. rubrotincta** 616
consobrina, Eutypa (Diatrype) 699
consocia, Septoria 950
consociata, Cercospora 608
consociata, Dermascia = **Lophodermium consociatum** 771
consociata, (Laestadia, Physalospora, Wallrothiella) = **Acanthostigma sequoiae** 556
consociatum, Cylindrosporium 658
consociatum, Lophodermium 771
consors, Mystrosporium 809
consortiale, Ulocladium (Alternaria, Stemphylium) 1006
conspersa, Tympanis 1004
conspersum, Haplotrichum (Oidium) 729
conspicua, Cylindrosporella 658
conspicua, Dothidea (Phragmodothis, Thyridaria) 687
conspicua, Pseudocercospora (Cercospora) 894
conspicua, Puccinia = **P. poarum** 912
conspicua, Septoria 950
conspicuum, Leptothyrium 766
conspurcata, Massaria 779
conspurcata, Valsa = **V. ceratosperma** 1017
constellata, (Eutypella, Valsa) = **Eutypella juglandicola** 700
constricta, Dactylaria 663
constrictum, Scolecobasidium = **Dactylaria constricta** 663
contaminans, Anthostomella 565
contigua, Meliola = **M. palmicola** 785
contigua, Poria = **Phellinus contiguus** 845
contiguum, Corticium = **Intextomyces contiguus** 749
contiguus, Intextomyces 749
contiguus, Phellinus 845
continua, Rhabdospora (Septoria) 930
contorta, Helicogloea 730
contortae, Lecanidiella 758
contortum, Sorosporium 963
contortus, Crepidotus 647
contraria, Uredo 1007
controversa, Tilletia 992
convallariae, Gloeosporium 718
convallariae, Phyllosticta 866
convexula, Mycosphaerella 801
convexula, Phyllosticta 866
convexum, Hemiphacidium (Phacidium) 732
convexum, Myrothecium = **Greeneria uvicola** 724
convoluta, Botryotinia (Sclerotinia) 593
convoluta, Botrytis 594
convolutum, Chaetomium 623
convolutum, Coniothyrium 642
convolvens, Gloeocystidiellum 716
convolvuli, Diplodia 680
convolvuli, Puccinia 904
convolvuli, Septogloeum = **Cylindrosporium convolvulicola** 658
convolvuli, Septoria 950
convolvuli, Stagonospora 973
convolvulicola, Cylindrosporium 658
conwayana, Poria = **Diplomitoporus crustulinus** 683
cookeana, Gloniopsis = **G. praelonga** 720
cookeana, Meliola 784
cookeana var. major, Meliola = **M. lyoni** 785
cookei, Bahusakala 586
cookei, Bipolaris (Helminthosporium) 588
cookei, Clathrospora = **C. multiseptata** 631
cookei, Melanographium 783
cookei, Periconia 831
cookei, Phyllosticta 866
cooperiae, Puccinia 904
coordinatus, Uromyces 1010
coorgica, Cercosporella = **Mycovellosiella perfoliati** 808
copallina, Cercospora = **Pseudocercospora rhoina** 895
copallina, Sarcinella 938
copallinum, Stemphylium = **Sarcinella copallina** 938
copelandii, Marasmius 777
copelandii, Uredinopsis 1007
copelandii var. olidus, Marasmius 777
coprophila, Athelia 583
coprophila, Papulaspora 825
coprophilum, Corticium = **Athelia coprophila** 583
coprophilum, Helicosporangium = **Papulaspora coprophila** 825
coprosmae, Asteridiella 580
coprosmae, Venturia (Actinodothidopsis) 1020
coptina, Vermicularia 1021
coptis, Mycosphaerella (Laestadia, Sphaerella) 801
coptis, Septoria 950
corallinum, Stilbum = **Stilbella aleuriata** 981
coralloides, Hericium 734
coralloides, Urocystis 1008
corallorhizae, Leptosphaeria (Phaeosphaeria) 761
corallorhizae, Phoma 852
cordatae, Cercospora = **C. thaspiicola** 618
cordeanum, Berkleasmium = **B. concinnum** 587
cordiae, Alveolaria 563
cordiae, Puccinia 904
cordilleranum var. minor, Exobasidium 703
cordobensis, Sphacelotheca 965
cordylina, Poria = **Aporpium caryae** 571
cordylines, Cercospora (Cercospora) 608
cordylinophila, Phyllosticta 867
coreopsidis, Cercospora 608
coreopsidis, Cladosporium 629
coreopsidis, Phyllosticta 867
coreopsidis, Septoria 950
coriaceus, Irpex = **Fuscocerrena portoricensis** 711
coriaria, Grandinia = **Tomentella ferruginea** 994
coriaria, Tomentella = **T. ferruginea** 994
coriarius, Hypochnus = **Tomentella ferruginea** 994
corium, Meruliopsis (Byssomerulius, Merulius) 787
cornea, Auricularia (Exidia) 585
cornea, Calocera (Clavaria) 597
cornea, Phlebia = **Crustoderma corneum** 650
corneum, Crustoderma 650
corneum, Sphaeronaema = **Hainesia lythri** 728
corni, Cercospora 608
corni, Cryptodiaporthe (Apioporthe) 651
corni, Cylindrosporium 658
corni, Cytospora 661
corni, Elsinoe 692
corni, Erysiphe = **Microsphaera tortilis** 791
corni, Gloeosporium = **Colletotrichum gloeosporioides** 638
corni, Glomopsis (Glomerularia) 719
corni, Hapalocystis (Melanconiella, Melanconis) 728
corni, Hendersonia = **H. fiedleri** 733
corni, Lachnum (Lachnella) 754
corni, Monilinia (Sclerotinia) 794

corni, Mycosphaerella 801
corni, Myxosporium = **M. everhartii** 810
corni, Pezicula 837
corni, Phomopsis (Phoma) 858
corni, Physalospora = **Botryosphaeria obtusa** 592
corni, Pseudomassaria 896
corni, Synchytrium 984
corni, Tapesia = **T. rosae** 986
cornicola, Anthostomella (Entosordaria) 566
cornicola, Ascochyta 574
cornicola, Calosphaeria (Tognina) 598
cornicola, Cercospora 608
cornicola, Mycosphaerella 801
cornicola, Pezicula 837
cornicola, Phyllosticta 867
cornicola, Placosphaeria 881
cornicola, Septoria 950
cornicola, Sphaeropsis 966
cornicola, Stigmina (Coryneum) 979
cornicola var. **ampla, Septoria** 950
corniculans, Gymnosporangium = **G. nelsonii** 727
corniculata, Peroneutypa 832
corniformis, Xylaria 1026
corniformis var. macrospora, Xylaria = **X. longipes** 1026
cornigerum, Ceratobasidium (Corticium) 603
cornigerum, Teratosperma (Clasterosporium) 988
corni-maris, Lambertella 756
corni-maris, Septoria = **S. cornicola** 950
cornina, Apiosporella = **Pseudomassaria corni** 896
cornina, Cryptosporiopsis (Macrophoma, Sphaeropsis) 653
cornina, Romellia (Valsa) 936
cornina, Trematosphaeria 998
cornispora, Phyllachora = **P. minutissima** 863
cornu-cervae, Labridella 753
cornu-cervae f. longiseta, Pestalotia = **Labridella cornu-cervae** 753
cornu-damae, Xylaria 1026
cornuta, Didymella 676
cornuta, Roestelia = **Gymnosporangium cornutum** 726
cornuti, Phyllosticta 867
cornutum, Gymnosporangium 726
corona, Tilletia (Neovossia) 992
coronadensis, Polyporus 889
coronaria, Marssonina (Ascochyta, Marssonia) 778
coronaria, Sarcosphaera 938
coronariae, Marssonia 777
coronata, Ceratocystis = **C. tenella** 604
coronata, Cilioplea (Pleospora) 627
coronata, Clavaria = **Clavicorona pyxidata** 632
coronata, (Cyathicula, Hymenoscypha) = **Crocicreas coronatum** 648
coronata, Puccinia 904
coronata, Valsa 1017
coronata, Volutella 1023
coronata var. **avenae, Puccinia** 904
coronata var. bromi, Puccinia = **Puccinia coronata** 904
coronata var. calamagrostis, Puccinia = **Puccinia coronata** 904
coronata var. elaeagni, Puccinia = **Puccinia coronata** 904
coronatum, Crocicreas (Helotium) 648
coronatus, Coccomyces 634
coronatus, Uromyces 1010
coronifera, Trechispora = **Sistotrema coroniferum** 962
coroniferum, Sistotrema 962
coronilla, Corticium = **Sistotrema brinkmannii** 962
coronopsora, Puccinia 904
corrugata, (Daedalea, Lenzites) = **Daedaleopsis confragosa** 664
corrugata, Hymenochaete 739
corrugata, Hyphodermella (Grandinia, Odontia) 742
corrugata, Trametes = **Earliella scabrosa** 691
corrugatum, Glyphium (Acrospermum) 720
corrugatus, Paxillus = **Pseudomerulius curtisii** 897
corrugatus var. **aurantiacus, Marasmius** 777
corruge, Corticium = **Cylindrobasidium corrugum** 656
corrugis, Coniophora = **Cylindrobasidium corrugum** 656
corrugum, Cylindrobasidium 656
corticale, Cryptostroma (Coniosporium) 654
corticale, Hapalocystis (Prosthecium) 728
corticale, Lachnum = **Dasyscyphus corticalis** 665
corticalis, Dasyscyphus 665
corticalis, Mycena 797
corticalis, Peniophora = **P. quercina** 830
corticatus, Pleurotus = **P. dryinus** 886
corticioides, (Ceracea, Dacrymyces) = **Arrhytidia involuta** 572
corticioides, Helicobasidium 730
corticis, Botryosphaeria (Physalospora) 592
corticium, Rosellinia 937
corticola, Berkleasmium 587
corticola, Coccodinium (Pleosphaeria) 634
corticola, Cryptosporiopsis (Myxosporium) 653
corticola, Discostroma (Clethridium, Griphosphaeria, Pleosphaerulina) 685
corticola, Kmetia 753
corticola, Massarina (Trematosphaeria) 779
corticola, Mycena = **M. meliigena** 798
corticola, Oxyporus (Poria) 824
corticola, Thaxteriella = **Tubeufia pezizula** 1002
corticola f. **crataegi, Pleosphaerulina** 883
corticola f. **rosae, Pleosphaerulina** 883
corticolor, Hymenochaete 739
cortinicinctus, Claudopus 631
corydalis, Peronospora 833
corydalis, Phyllosticta 867
corydalis, Septoria 950
corylea, Phyllactinia = **P. guttata** 864
corylea var. angulata, Phyllactinia = **P. angulata** 863
coryli, Anguillospora 564
coryli, Diplodia 680
coryli, Eutypella = **E. tetraploa** 701
coryli, Mamianiella (Gnomoniella) 776
coryli, Nectria (Creonectria) 812
coryli, Nematospora 814
coryli, Phyllosticta 867
coryli, Piggotia (Gloeosporium, Labrella, Monostichella) 879
coryli, Scolicosporium 943
coryli, Sphaeropsis 966
coryli, Taphrina 986
coryli var. **spiralis, Mamianiella** (Gnomoniella) 776
corylina, Cercospora 608
corylina, Dothidea = **Endotryblidium insculptum** 694
corylina, Pezicula 837
corylina, Septoria = **S. ostryae** 955
corylophilum, Penicillium 828
corymbata, Phanerochaete 843
corymbatum, Corticium = **Phanerochaete corymbata** 843
corymbifera, Absidia 555
corymbiferum, Penicillium = **P. hirsutum** 828
corymbosa, Diaporthe = **D. tuberculosa** var. **corymbosa** 672
coryneoidea, Monochaetia (Pestalotia) 795
coryneoidea, Strumella = **Conoplea globosa** 644
corynoidea, Hendersonia 733
corynostoma, (Eutypella, Valsa) = **Eutypella scoparia** 701
coryphae, Phoma 852
costaricensis, Uromyces 1010
cotini-coggygriae, Pileolaria 879
cotizensis, Cercospora 608
cotoneastri, Phyllosticta 867
couchii, Typhulochaeta 1005
coulteri, Neopeckia (Herpotrichia, Lasiosphaeria) 815
cowdellii, Diplodia (Diplodiella) 680
crameri, Ustilago 1015
crandallii, Puccinia 904
crassa, Alternaria 561
crassa, Antrodia (Poria) 567
crassa, Peniophora = **Dacryobolus karstenii** 663
crassa, Phanerochaete (Lopharia) 843
crassa, Sarcosphaera = **S. coronaria** 938
crassiasca, Leptosphaerulina (Pleospora) 764
crassipes, Colletotrichum (Gloeosporium) 637
crassipes, Puccinia 904
crassivaginata, (Ceratocystiopsis, Ceratocystis) = **Ophiostoma crassivaginatum** 821
crassivaginatum, Ophiostoma 821
crassoides, Cercospora 608
crassulae, Diplodia 680
crassum, Laxitextum = **Phanerochaete crassa** 843
crassus, Ophiobolus 820
crassus, Physisporus = **Antrodia crassa** 567
crastophila, Didymosphaeria = **D. futilis** 678
crastophila, Hendersonia = **Wojnowicia hirta** 1024
crastophilum, Entyloma = **E. dactylidis** 695
crataegi, Cenangium 602
crataegi, Cercoseptoria (Cylindrosporium, Phloeospora) 605
crataegi, Cercospora 608
crataegi, Cucurbitaria 654
crataegi, Diplodia 680
crataegi, Hendersonia = **Seimatosporium lichenicola** 945
crataegi, Monilia 794
crataegi, Monochaetia = **M. turgida** 795
crataegi, Otthia = **O. spiraeae** 823
crataegi, Phyllosticta 867
crataegi, Septoria = **S. oxyacanthae** 955
crataegi, Sphaceloma (Gloeosporium, Melanophora) 964
crataegi, (Stemphylium, Stigmella) = **Pithomyces quadrata** 881
crataegi, Venturia 1020
crataegicola, (Dermatea, Pezicula) = **Pezicula sepium** 838
crataegicola, Hendersonia 733
crataegicola, Phyllosticta 867
craterium, Trochila 1001
craterium, Urnula 1008

crec'hqueraultii, Lamprospora 756
cremea, Calloria 597
cremea, (Peniophora, Phanerochaete) = **Phanerochaete sordida** 844
cremeicolor, Poria = **Ceriporia xylostromatoides** 621
cremeo-albidum, Corticium (Laeticorticium) 645
cremeum, Helotium = **Calloria cremea** 597
cremeus, Aleurodiscus = **Acanthophysium bertii** 555
cremoricolor, Cerocorticium (Corticium) 622
crenato-marginata, Hyalinia 738
crenatum, (Lophionema, Lophiostoma, Lophiotrema) = **Lophiostoma angustilabrum** var. **crenatum** 770
crepidiformis, Isthmiella (Bifusella) 750
crepidis, Cercoseptoria 605
crepidis, Ramularia 924
crepidis, Septoria 950
crepidis-montanae, Puccinia 904
crepini, Leptosphaeria = **Phaeosphaeria lycopodina** 842
crepiniana, Physalospora = **P. empetri** 876
crepinii, Curvularia 655
crescentum, Cylindrosporium 658
cressae, Puccinia = **P. tuyutensis** 915
cretaceum, Corticium = **Hyphoderma sambuci** 741
crinalis, Tomentella 993
crinella, Dasyscyphus = **Psilachnum crinellum** 899
crinellum, Psilachnum 899
crinitus, Lentinus 759
crinospora, Cercospora 608
crispa, Clithris = **Colpoma crispum** 640
crispa, Plicaturopsis (Plicatura, Trogia) 887
crispa, Sparassis (Clavaria) 963
crispatus, Merulius = **Ceraceomyces serpens** 603
crispellus, Polyporus = **Postia balsamea** 890
crispulum, Trichosporum = **Conoplea fusca** 643
crispum, Colpoma 640
crispus, (Cantharellus, Merulius) = **Plicaturopsis crispa** 887
crispus, Polyporus = **Bjerkandera adusta** 590
cristata, Clavicorona 632
cristata, Costantinella = **C. terrestris** 646
cristata, Thozetella 990
cristulata, Odontia = **Hyphoderma cristulatum** 741
cristulatum, Hyphoderma 741
crocatus, Phellinus (Polyporus) 845
crocatus, Rigidoporus 936
crocea, Allescheriella 560
croceoluteus, Gymnopilus 725
croceum, Hypoxylon 745
croceus, Hapalopilus (Polyporus, Tyromyces) 729
crocipora, Poria = **Ceriporia spissa** 621
crociporus, Polyporus = **Ceriporia spissa** 621
crocitinctus, Inonotus 748
crocopeplum, Hypoxylon 745
crocorum, Rhizoctonia 932
crookwellense, Fusarium 707
crossota, Trichopeziza = **Dennisiodiscus crossotus** 668
crossotus, Dennisiodiscus (Trichodiscus) 668
crotalariae, Calonectria 597
crotalariae, Cercospora = **C. cotizensis** 608
crotalariae, Colletotrichum = **C. gloeosporioides** 638
crotalariae, Cylindrocladium 657
crotalariae, Diaporthe 671
crotalariae, Phaeotrichoconis (Trichoconis) 843
crotocinigenum, Acremonium (Cephalosporium) 556
crotonicola, Cercospora 608
crotonifolia, Cercospora 608
crotonis, Phakopsora (Bubakia, Melampsora, Pucciniastrum) 843
crotonis, Pseudocercospora (Cercospora) 894
crotonophila, Cercospora 608
crowellii, Coleosporium 636
cruciatae, Septoria 950
cruciatus, Asteromyces 582
crucifera, Dasyscyphus = **Incrucipulum sulphurellum** 748
cruciferarum, Albugo = **A. candida** 559
cruciferarum, Cercospora 608
cruciferarum, Mycosphaerella 801
cruciferarum, Puccinia 904
cruciferarum subsp. **borealis, Puccinia** 904
cruciferarum subsp. **wyomingensis, Puccinia** 904
cruciferum, Stauronema (Dinemasporium) 975
cruenta, Mycosphaerella 801
cruenta, Orbilia 822
cruenta, (Phyllosticta, Phyllostictina) = **Phyllosticta convallariae** 866
cruenta, Pseudocercospora (Cercospora) 894
cruenta, Sphaeropsis = **Phyllosticta convallariae** 866
cruenta, (Sphacelotheca, Ustilago) = **Sporisorium cruentum** 971
cruenta var. discincta, Phyllosticta = **P. uvulariae** 874
cruenta var. pallidior, Phyllosticta = **P. pallidior** 871
cruentum, Sporisorium 971
crus-galli, Mycosphaerella (Sphaerella) 801
crus-galli, (Cintractia, Ustilago) = **Ustilago trichophora** 1016
crustacea, Mucilago 796
crustacea, Nitschkia (Coelosphaeria) 816
crustaceus, Thermoascus (Dactylomyces) 990
crustosa, Grandinia (Hyphodontia, Odontia) 722
crustosa, Phyllosticta 867
crustosa, Plasmopara 882
crustosa, Thelephora = **T. terrestris** 990
crustosum, Penicillium 828
crustula, Odontia 817
crustulina, Poria = **Diplomitoporus crustulinus** 683
crustulinum, Corticium = **Trechispora amianthina** 997
crustulinus, Diplomitoporus 683
crypta, Dermea (Scleroderris) 669
crypta, Phyllosticta 867
crypta, Ramularia 924
cryptandri, Puccinia 904
cryptandri var. **luxurians, Puccinia** 904
cryptanthes, Puccinia 904
cryptocarpa, Irenopsis (Irenina, Meliola) 749
cryptogea, Phytophthora 878
cryptogea var. richardiae, Phytophthora = **P. richardiae** 879
cryptomeriae, Monochaetia 795
cryptomeriae, Pestalotia 835
cryptomeriae, Phyllosticta 867
cryptosphaerioides, Didymosphaeria 678
cryptotaeniae, Cercospora 609
cryptotaeniae, Puccinia 904
cryptotaeniae, Septoria 950
cryptotaeniae, Stagonospora 973
crystalligerum, Lachnum 754
crystallina, Dasyscyphella 664
crystallina, Hyalinia (Cyathicula) 738
crystallinum, Helotium = **Hyalinia crystallina** 738
crystallinus, Dasyscyphus = **Dasyscyphella crystallina** 664
crystallophora, Dothidea = **D. puccinioides** 687
cubense, Fusarium = **F. oxysporum** f. sp. **cubense** 709
cubense, Hypoxylon = **Xylaria cubensis** 1026
cubensis, Cryphonectria (Diaporthe) 650
cubensis, Pseudoperonospora (Peronoplasmopara, Peronospora) 897
cubensis, Ptychogaster (Ceriomyces) 899
cubensis, Trametes (Polyporus) 996
cubensis, Xylaria 1026
cuboidea, Arthrographis (Oospora) 573
cucullata, Campanella = **C. junghuhnii** 600
cucullata, Halosphaeria (Remispora) 728
cucullata, Mitrula = **Heyderia abietis** 736
cucullata, Mycena 797
cucullatus, Marasmius = **Mycena cucullata** 797
cucullatus, Pseudofavolus (Favolus) 896
cucumerina, Alternaria 561
cucumerinum, Cladosporium 629
cucumerinum, Macrosporium = **Alternaria cucumerina** 561
cucumerinum, (Gonatobotrys, Rhinotrichum) = **Olpitrichum tenellum** 819
cucumeris, Thanatephorus 989
cucumis, Ascochyta = **Phoma cucurbitacearum** 852
cucurbitacearum, Laestadia 755
cucurbitacearum, Phoma 852
cucurbitacearum, Phyllosticta 867
cucurbitacearum, Septoria 950
cucurbitacearum, Stemphylium = **Leandria momordicae** 758
cucurbitae, Cercospora = **C. citrullina** 608
cucurbitaria, Dermatella 669
cucurbitarum, Choanephora 626
cucurbiticola, Phaeoramularia (Cercospora) 840
cucurbitula, Scoleconectria (Chilonectria, Nectria) 943
culicitella, Scortechinia = **Nitschkia acanthostroma** 816
culmicola, Crocicreas (Belonium, Cyathicula) 648
culmicola, Leptosphaeria = **Phaeosphaeria nigrans** 842
culmicola, Stagonospora (Hendersonia) 973
culmicola var. **minor, Hendersonia** 733
culmifida, Keissleriella (Leptosphaeria, Trichometasphaeria) 752
culmifida, Septoria = **Pseudoseptoria stomaticola** 898
culmifraga, Leptosphaeria = **Phaeosphaeria herpotrichoides** 842
culmigena, Didymella 676
culmigena, Mycena 797
culmigenum, Lophodermium 771
culmigenum, Microthyrium = **Trichothyrina alpestris** 1000
culmigenus, Naemacyclus 810
culmina, Mollisia 793
culmiseda, Hendersonia 733
culmorum, Fusarium 707
culmorum, Hypoxylon 745
culmorum, Phaeosphaeria (Leptosphaeria) 842

culmorum, Pleospora = **Pyrenophora trichostoma** 919
culmorum var. leteius, Fusarium = **Fusarium culmorum** 707
cumminsiana, Erysiphe 697
cumminsii, Melampsora 781
cumminsii, Phyllosticta 867
cumulodentata, Phanerochaete = **P. magnoliae** 844
cumulodentatum, Radulum = **Phanerochaete magnoliae** 844
cuneata, Thecaphora 989
cuneatum, Sorosporium = **Thecaphora cuneata** 989
cuneatus, Oxyporus (Coriolellus, Polyporus) 824
cuneiforme, Fusarium = **F. ventricosum** 711
cuneomaculata, Ascochyta = **A. verbenae** 577
cunillae, Septoria 950
cunninghamianum, Gymnosporangium 726
cunninghamii, Phyllosticta 867
cupheae, Uredo 1007
cupressi, Cryptostictis = **Seiridium unicorne** 945
cupressi, Gymnosporangium = **G. cunninghamianum** 726
cupressi, Macrophoma (Sphaeropsis) 774
cupressi, Physalospora = **Botryosphaeria quercuum** 593
cupressi, Pithya = **P. cupressina** 881
cupressi var. cascadense, Gymnosporangium = **G. cunninghamianum** 726
cupressicola, Uredo 1007
cupressina, Asterina (Asterinella, Venturia) 580
cupressina, Microdiplodia (Diplodia) 788
cupressina, Pithya 881
cupressina, Trichosphaeria 1000
cupressi-thyoides, Lophodermina = **Lophodermium juniperinum** f. **cupressi-thyoidis** 772
cupreum, Chaetomium 623
cupula, (Amphisphaeria, Cainia, Didymosphaeria) = **Amphisphaeria succinea** 564
cupulata, Johansonia (Venturia) 750
cupulata, Melanomma 783
cupulatum, Coryneum = **Corynespora olivacea** 646
cupuliformis, (Cyphella, Rhodocyphella) = **Stigmatolemma taxi** 979
curcubitula, Creonectria = **Nectria fuckeliana** 812
curreyana, Myriosclerotinia 809
curreyi, (Massaria, Massariella) = **Splanchnonema ampullaceum** 969
curreyi var. americana, Massariella = **Splanchnonema ampullaceum** 969
curta, Xylaria 1026
curtipes, Dendryphion = **Taeniolella rudis** 985
curtipes, Puccinia = **P. heucherae** var. **saxifragae** 907
curtisii, Haplotrichum (Acladium, Oidium) 729
curtisii, Helicoma = **H. muelleri** 730
curtisii, Hymenochaete 739
curtisii, Macroderma (Phacidium, Rhytisma) 773
curtisii, Myriangium = **M. duriaei** 808
curtisii, Ostreichnion (Glonium, Hysterium) 823
curtisii, Phyllosticta (Phoma) 867
curtisii, Pluteus = **P. petasatus** 887
curtisii, (Ganoderma, Polyporus) = **Ganoderma lucidum** 713
curtisii, Pseudomerulius (Paxillus) 897
curtisii, Stagonospora (Hendersonia) 973
curtispora, Stictis = **S. schizoxyloides** 979
curvata, Gloniopsis 719
curvatum, Gloeosporium = **Gloeosporidiella ribis** 717
curvatum, (Clasterosporium, Sporidesmium) = **Hirudinaria macrospora** 736
curvicolla, Jattaea (Eutypa, Gnomoniella) 750
curvicolla, Podospora (Sordaria) 888
curviseta, Protoventuria (Antennularia, Venturia) 893
curvisetum, Colletotrichum = **C. gloeosporioides** 638
curvispora, Cryptosporiopsis (Macrophoma) 653
curvispora, Septoria 950
curvula, Flagellospora 705
curvula, Stagonospora 973
curvulum, Acremonium 556
cuscutae, Phomopsis 858
cuspidatum, Arthrinium (Camptoum) 572
cuticularis, Inonotus (Polyporus) 748
cuticularis, Thelephora 990
cuticulosa, Microthyriella (Asterina) 791
cutifractus, (Polyporus, Tyromyces) = **Postia balsamea** 890
cyaneae, Mycosphaerella 801
cyaneum, Penicillium 828
cyani, Puccinia 904
cyani, Septoria 950
cyanococci, Phyllosticta 867
cyanogena, Gibberella (Sphaeria) 714
cyathiformis, Pseudoclitocybe (Clitocybe) 896
cyathoidea, Cyathicula = **Crocicreas cyathoideum** 648
cyathoidea, Phialea = **Crocicreas cyathoideum** 648
cyathoideum, Crocicreas (Helotium) 648
cyathoideum var. **cacaliae, Crocicreas** 648
cyathoideum var. **olympianum, Crocicreas** 649
cyathoideus, Hymenoscyphus = **Crocicreas cyathoideum** 648
cycadina, Ascochyta 574
cycadina, Phyllosticta 867
cycadis, Pestalotia 835
cyclamanicola, Ramularia 924
cyclamenae, Colletotrichum = **C. gloeosporioides** 638
cyclaminicola, Phyllosticta 867
cyclaminicola, Ramularia 924
cyclaminis, Cladosporium = **Ramularia cyclamanicola** 924
cyclaminis, Phyllosticta 867
cyclopium, Penicillium = **P. aurantiogriseum** 827
cyclopium var. echinulatum, Penicillium = **P. echinulatum** 828
cyclops, (Bipolaris, Helminthosporium) = **Drechslera campanulata** 689
cydoniae, Aplosporella (Sphaeropsis) 570
cydoniae, Cercospora (Cercospora) 609
cydoniae, Coniothyrium 642
cydoniae, Coryneum = **Stigmina pallida** 980
cydoniae, Entomosporium = **E. mespili** 695
cydoniae, Hendersonia 733
cydoniae, Macrosporium 775
cydoniae, Myriellina (Gloeosporium, Phyllosticta, Septogloeum) 808
cydoniae, Phoma 852
cydoniae, Physalospora = **Botryosphaeria quercuum** 593
cylindrica, Septoria 950
cylindrica, Ustilago = **Sporisorium andropogonis** 971
cylindricum, Leptothyrium 766
cylindricum, Mastigosporium 780
cylindricum, Microdochium 789
cylindriopsis, Ramularia 924
cylindrocarpa, Stagonospora (Hendersonia) 973
cylindroides, Cylindrocarpon 657
cylindroides, Polyscytalum 889
cylindrophorum, Hypoxylon = **Camarops polysperma** 599
cylindrospermum, Gloeosporium = **Asteroma alneum** 581
cylindrospora, Codinaea 635
cylindrospora, Coleophoma (Macrophoma) 635
cylindrospora, Leptosphaeria 761
cylindrospora, Mycena 797
cylindrospora, (Nectria, Ophionectria) = **Scoleconectria cucurbitula** 943
cylindrospora, Stachybotrys 972
cylindrospora, Volutella (Psilonia) 1023
cylindrospora, Wrightoporia 1024
cylindrosporum, Diplocladium = **Cylindrocladium scoparium** 657
cylindrosporus, Gaeumannomyces 712
cylindrotheca, Phomatospora 857
cylindrothecia, Tubeufia (Ophionectria) 1002
cymatilis, Cremasteria 647
cymbopogonis, Curvularia (Helminthosporium) 655
cymopteri, Puccinia = **P. jonesii** 908
cymosa, Galzinia 713
cynarae, Ramularia 924
cynodonifolii, Lojkania = **Lasiostemma erysipheoides** 757
cynodontis, Bipolaris (Drechslera, Helminthosporium) 588
cynodontis, Leptostromella 765
cynodontis, Phyllachora (Physalospora) 862
cynodontis, Puccinia 904
cynodontis, Septoria 950
cynodontis, Ustilago (Ustilago) 1015
cynoglossi, Cercospora 609
cynoglossi, Erysiphe (Alphitomorpha) 698
cynoglossi, Peronospora 833
cynoglossi, Phoma 852
cynoglossi var. echinospermi, Peronospora = **P. echinospermi** 833
cynosbati, (Cryptostictis, Hendersonia) = **Seimatosporium rosae** 945
cyparissa, Diplodia 680
cyperacearum, Balansia 586
cyperi, Balansia 586
cyperi, Cintractia 628
cyperi, Meliola 784
cyperi, Peronospora (Phytophthora) 833
cyperi, Phyllachora 862
cyperi, Puccinia 904
cyperi, Septoria 950
cyperi, Stagonospora 973
cyperi, Testicularia 988
cypericola, Cercospora 609
cypericola, Stagonospora 973
cyperi-tagetiformis, Puccinia 904
cyphelliformis, Hohenbuehelia (Pleurotus) 737
cypri, Valsa 1017
cypripedii, Ascochyta 574
cypripedii, Cercospora 609

cypripedii, Mycosphaerella 801
cypripedii, Puccinia 904
cyrillae, Aecidium 558
cyrillae, Phyllosticta 867
cyrillicola, Lophodermium 771
cyrtandrae, Asteridiella (Irenina, Meliola) 580
cystidiata, Dacryomitra = **Dacrymyces capitatus** 663
cystidiata, Exidia = **Ductifera sucina** 690
cystidiata, Pellicularia = **Suillosporium cystidiatum** 983
cystidiata, Phlebia = **P. merismoides** 848
cystidiatum, Suillosporium 983
cystidiosus, Pleurotus 886
cystopteridis, Phyllosticta 867
cystopteridis, Taphrina 986
cytisi, Guignardia = **Selenophoma juncea** 945
cytisporoides, Stagonospora 973
cytosporeum, Myxosporium 810
dactylidis, Drechslera 689
dactylidis, Entyloma (Thecaphora) 695
dactylidis, Puccinia = **P. graminis** subsp. **graminicola** 907
dactylidis, Uromyces 1010
dactylidis var. **poae, Uromyces** 1010
dactyliferum, Pythium = **P. irregulare** 921
daemonicola, Cercospora 609
dahliae, Diplodina 682
dahliae, Entyloma 695
dahliae, Mycosphaerella 801
dahliae, Phoma 852
dahliae, Verticillium 1021
dahliaecola, Phyllosticta 867
dakotense, Fusicoccum 711
dakotense, Graphyllium 724
dakotense, Helotium = **Mollisia dakotensis** 793
dakotensis, Diaporthe 671
dakotensis, Diatrype 673
dakotensis, Hyponectria 744
dakotensis, Mollisia (Pezizella) 793
dakotensis, (Clathrospora, Pleospora) = **Graphyllium dakotense** 724
daldiniana, (Diatrype, Eutypa) = **Eutypella leprosa** 701
daleae, Cercospora 609
daleae, Uropyxis 1014
daleae var. **eysenhardtiae, Uropyxis** 1014
dalibardae, Phyllachora (Dothidea) 862
dalibardae, Septoria 950
damae-cornis, Calopposis = **Calocera cornea** 597
dangeardii, Penicillium 828
danthoniae, Lophodermium 771
daphnes, Marssonina (Marssonia) 778
darkeri, Leptosphaeria 761
darkeri, Milesina (Milesia) 792
darkeri, Taphrina 986
darlingtoniae, Septoria 950
darrowii, Septoria 950
dasycarpa, Melanconis = **Macrodiaporthe everhartii** 773
dasylirii, Botryosphaeria (Dothidea, Phyllachora) 592
datiscae, Phomatospora 857
daturae, Cercospora = **Alternaria crassa** 561
daturicola, Cercospora (Cercosporina) 609
dauci, Alternaria 561
davidsoniana, Ciborinia 627
davidsonii, Godronia 721
davidsonii, Tolyposporium = **Thecaphora pilulaeformis** 989
davisii, Cercospora 609
davisii, Entyloma 695
davisii, Gymnosporangium 726
davisii, Hendersonia 733
davisii, Mycosphaerella 801
davisii, Peronospora 833
davisii, Phyllosticta (Gloeosporium, Phyllostictina) 867
davisii, Septoria 950
davisii, Synchytrium 984
davisii, Ustilago 1015
dayi, Puccinia 904
dearnessii, Calonectria = **Nectria decora** 812
dearnessii, Cercosporella = **C. virgaureae** 619
dearnessii, (Cylindrosporium, Phloeospora) = **Septoria ostryae** 955
dearnessii, Hymenoscyphus (Helotium, Phialea) 740
dearnessii, Leptosphaeria 761
dearnessii, Leptothyrium 766
dearnessii, Mycosphaerella 801
dearnessii, Phyllosticta 867
dearnessii, Physalospora 876
dearnessii, Septoria 950
dearnessii, Taphrina 986
debaryana, Puccinia = **P. pulsatillae** 912
debaryanum, Pythium 920
debaryi, Peronospora 833
decastylum, Acanthostigma 556
decedens, Allantoporthe (Diaporthe) 560
decemcellulare, Fusarium 707
deceptivus, Polyporus = **Coltriciella dependens** 640
decidua, Mycosphaerella 801
decidua, Phyllosticta = **Phoma exigua** 853
decidua, Propolis (Stictis) 892
deciduum, Sclerotium 941
decipiens, Athelia (Corticium) 583
decipiens, Burrillia 596
decipiens, Coniothyrium 642
decipiens, Cytospora 661
decipiens, Diatrypella 674
decipiens, Dinemasporium (Excipula) 679
decipiens, Gloeosporium = **Piggotia fraxini** 879
decipiens, Leptostroma 764
decipiens, Mytilinidion (Lophium) 809
decipiens, Ovularia 824
decipiens, Phyllosticta = **P. concentrica** 866
decipiens, Placosphaeria 881
decipiens, Ramularia 924
decipiens, Schizoxylon = **S. berkeleyanum** 940
decipiens, Synchytrium 984
decipiens, Tilletia = **T. sphaerococca** 992
decipiens, Trichia 999
decipiens f. castanae, Coryneum = **Seiridium castaneae** 945
decipiens var. **symphoricarpi, Schizoxylon** 940
decodontis, Cercospora = **C. nesaeae** 614
decolorans, Didymosphaeria = **D. futilis** 678
decolorans, Exobasidium 703
decolorans, Gloeosporium 718
decolorans, Poria = **Physisporinus sanguinolentus** 876
decolorans, Rhytisma 935
decolorans, Wolkia (Protascus) 1024
decolorans var. **americana, Didymosphaeria** 678
decolorata, Pestalotia = **P. unicolor** 836
decora, Nectria 812
decora, Tricholomopsis (Clitocybe) 999
decorahensis, Melanconis (Melanconiella) 782
decorata, Diatrypella 674
decorata, Pholiota 851
decorticans, Diaporthe 671
decorticans, Peniophora 829
decorticans, Valsa = **V. ceratosperma** 1017
decorticata, Lojkania (Amphisphaeria) 769
decorum, Lophodermium 771
decumariae, Cercospora 609
decumbens, Penicillium 828
decumbens, Phialophora (Margarinomyces) 847
decurrens, Crepidotus = **C. cinnabarinus** 647
decurrens, Gymnopilus 725
decurrens, Marasmius = **Resinomycena rhododendri** 930
deerata, Pseudombrophila (Ascophanus, Humaria) 897
deerata, Rosellinia 937
definitum, Hyphoderma (Corticium) 741
deflectens, Ascochytulina (Diplodia) 577
deflectens, Discosia 684
deflectens, Ramularia = **R. agrestis** 923
deflectens, Stigmina (Exosporium) 979
defodiens, (Leptosphaeria, Metasphaeria, Sphaeria) = **Massarina gloeospora** 779
defoliatum, Cylindrosporium 658
deformans, Doassansia 686
deformans, Elytroderma (Hypoderma) 693
deformans, Hapalosphaeria (Paepalopsis) 729
deformans, Insolibasidium (Herpobasidium) 749
deformans, Taphrina (Exoascus) 987
deformans, Thecaphora 989
deformata, (Cenangella, Dermatella) = **Holmiella sabina** 737
deformatum, Cenangium = **Holmiella sabina** 737
deformis, Ascochyta (Diplodia, Diplodina, Pseudodiplodia) 574
degenerans, Sporomega (Clithris, Hysterium) 971
degeneri, Mycosphaerella 802
deglubens, Eichleriella (Radulum) 692
deglubens, Phragmopyxis 861
dehnii, (Fabraea, Mollisia) = **Leptotrochila repanda** 766
delastrei, Diplosporonema (Gloeosporium, Marssonina) 683
delectans, Spongipellis (Polyporus, Tyromyces) 970
delicatella, Mycena = **Hemimycena lactea** 732
delicatellum, Helotium = **Hemimycena lactea** 732
delicatula, Valsa = **V. ceratosperma** 1017
delicatulum, Coleosporium 636
deliense, Pythium 920
deliquescens, Dacrymyces = **D. stillatus** 663
deliquescens, Gliocladium = **G. viride** 715
delitescens, Asterina 581
delphinella, Septoria 950
delphinii, Cercospora 609
delphinii, Didymella 676
delphinii, Diplodina = **Ascochyta aquilegiae** 573
delphinii, Ovularia 824
delphinii, Phyllosticta 867
delphinii, Puccinia 904
delphinii, Ramularia 924
delphinii, Sclerotium = **S. rolfsii** var. **delphinii** 942
delphinii, Sphaeromyces 966
delphiniicola, Mycosphaerella 802

delphiniicola, Phoma 853
dematеoides, Schizoxylon = **S. compositum** 940
dematia, Phaeosaccardinula 841
dematioidea, Drechslera 689
dematioides, Harpocephalum = **Periconia lateralis** 831
dematioideum, Helminthosporium = **Drechslera dematioidea** 689
dematium, Colletotrichum (Vermicularia) 637
dematium f. circinans, Colletotrichum = **C. circinans** 637
dematium f. **ipomoeae, Colletotrichum** 637
dematium f. **minor, Vermicularia** 1021
dematium f. **spinaciae, Colletotrichum** 637
dematium f. truncatum, Colletotrichum = **C. truncatum** 639
demetrioniana, Cercospora 609
demidoffii, Pyrofomes (Fomes, Polyporus) 920
deminuta, (Bourdotia, Sebacina) = **Basidiodendron eyrei** 586
deminutum, Basidiodendron = **B. eyrei** 586
demissa, Monilinia (Sclerotinia) 794
dendritica, Phoma 853
dendriticum, (Cladosporium, Fusicladium) = **Spilocaea pomi** 969
dendrobii, Cercospora 609
dendrobii, Gloeosporium = **Colletotrichum gloeosporioides** 638
dendroctoni, Entomocorticium 695
dendroides, Mycosphaerella 802
dendroides, Scolecobasidium 942
dendromeconis, Pocosphaeria 887
denigrata, (Pleonectria, Thyronectria) = **Thyronectria austroamericana** 991
densipes, Dendrodochium 666
densissima, Cercospora = **C. sidaecola** 617
densissima, Cryptodiaporthe = **Amphiporthe raveneliana** 563
densissima var. spicata, Cryptodiaporthe = **C. petiolophila** 651
densus, Fomes = **Phellinus conchatus** 844
dentariae, Puccinia 904
dentariae, Septoria 950
dentata, Cistella (Cyathicula, Hyaloscypha) 628
dentatum, Clavidisculum = **Cistella dentata** 628
dentatus, Coccomyces 634
denticulata, Ceratocystis 604
depallens, Dacryomitra = **Dacrymyces capitatus** 663
depauperata, Nectria 812
depazeoides, Cercospora 609
depazeoides, (Hyaloceras, Pestalotia) = **Seimatosporium rosae** 945
dependens, Coltriciella (Polyporus) 640
dependens, Phellinus (Fomes, Pyropolyporus) 845
dependens, Tritirachium 1001
depraedans, Cristulariella 648
depressa, Amphisphaeria 564
depressa, Cercospora = **Cercosporidium depressum** 620
depressa, Hyponectria (Guignardia, Mycosphaerella) 744
depressa, Passalora = **Cercosporidium depressum** 620
depressa, Phomopsis 858
depressa, Piggotia 879
depressa, Pyrenophora = **Comoclathris permunda** 640
depresseum f. petroselini, Fusicladium = **Cercosporidium punctum** 620
depresssum, Megacladosporium = **Cercosporidium depressum** 620
depressula, Sydowiella (Gnomonia) 983
depressum, Cercosporidium (Cladosporium, Fusicladium, Scolicotrichum) 620
depressum, Hysterium = **H. insidens** 747
depressum, Myxosporium 810
depressum var. platysporum, Fusicladium = **Cercosporidium punctiforme** 620
depressum var. sii, Fusicladium = **Cercosporidium sii** 620
derridis, Colletotrichum = **C. gloeosporioides** 638
deschampsiae, Mycosphaerella 802
deschampsiae, Puccinia = **P. brachypodii** var. **arrhenatheri** 902
deserticola, Candida 600
deserticola, Hyphoderma 741
deserticola, Pichia 879
desmanthi, Cercosporidium (Cercospora) 620
desmazieresii, Ascochyta 574
desmazieresii, Macrodiplodiopsis 773
desmazieresii, Meloderma (Hypoderma) 786
desmazieresii, Rosellinia 937
desmazierii, Lophiostoma = **L. viridarium** 771
desmazierii, Monochaetia = **M. monochaeta** 795
desmium, (Aecidium, Cerotelium, Phakopsora) = **Phakopsora gossypii** 843
desmodii, Cercospora 609
desmodii, Gloeosporidiella (Gloeosporium) 717
desmodii, Phyllosticta 867
desmodii, Ramularia 924
desmodii, Sclerotium 942
desmodii, Stagonospora 973
desmodii, Uromyces = **U. hedysari-paniculati** 1011
desmodiicola, Cercospora 609
desmosporus, Microascus 788
destructans, Cylindrocarpon (Ramularia) 657
destructiva, Cercospora 609
destructiva, Phoma (Diplodina) 853
destructiva, Ramularia (Ovularia) 924
destructivum, Colletotrichum 637
destructor, Peronospora 833
destructor, Tyromyces (Polyporus) 1005
destruens, Fusicladium 711
destruens, Gnomoniella 721
destruens, Pholiota = **P. populnea** 851
destruens, Phyllosticta 867
destruens, Plenodomus 883
destruens, Ramularia 924
destruens, Septoria 950
destruens, Sordaria (Anthostomella, Melanospora) 963
destruens, Sporisorium (Sphacelotheca, Ustilago) 971
detospora, (Septoria, Titaeospora) = **Titaeospora equiseti** 992
detrusa, Diaporthe 671
deusta, Cryptovalsa (Diatrypella) 654
deusta, Ramularia (Ovularia) 924
deusta, Ustulina 1017
deusta, (Eutypella, Valsa) = **Eutypella scoparia** 701
deusta f. **latifolii, Ramularia**
deusta f. odorati, Ramularia = **R. alba** 923
deustum, Hypoxylon = **Ustulina deusta** 1017
deutziae, Cercospora 609
deutziae, Phyllosticta 867
devexum, Plagiostoma 881
deviata, Xylosphaeria = **Mycothyridium deviatum** 807
deviatum, Mycothyridium 807
diachenii, Phomopsis 858
diademifera, Trechispora = **Sistotrema diademiferum** 962
diademiferum, Sistotrema 962
diameson, Pythium = **P. rostratum** 922
dianallae, Colletotrichum 637
dianellae, Mycosphaerella 802
dianellae, Phaeosphaerella 841
dianellae, Pirostoma 880
diantherae, Cercospora 609
diantherae, Dothideovalsa (Bagnisiella, Eutypa) 687
diantherae, Puccinia = **P. justiciae** 908
dianthi, Alternaria = **A. saponariae** 562
dianthi, Ascochyta 574
dianthi, Fusarium = **F. oxysporum** 708
dianthi, Puccinia = **P. arenariae** 901
dianthi, Septoria 950
dianthi, Uromyces 1010
dianthicola, Alternaria 561
dianthicola, Peronospora 833
dianthicola, Septoria 950
dianthii, Asteroma 581
diaphana, Volutella 1023
diaphana, Zignoella = **Exarmidium diaphanum** 702
diaphanum, Exarmidium 702
diaporthoides, Pleospora = **P. lactucicola** 885
diathrausta, Rosellinia 937
diathrauston, Hypoxylon = **Rosellinia diathrausta** 937
diatrypea, Dothiorella (Phoma, Sphaeropsis) 688
dicentrae, Cerotelium 622
dicentrae, Peronospora 833
dichaenoides, Septonema = **Taeniolella alta** 985
dichaeta, Pestalotiopsis (Pestalotia) 836
dichelostemmae, Puccinia 904
dichondrae, Puccinia 905
dichotomum, Verticillium 1021
dichroa, Poria = **Antrodia albobrunnea** 567
dichromatricha, Pleospora (Pyrenophora) 884
dichrous, Collybia (Marasmius) 640
dichrous, Gloeoporus (Polyporus) 717
dickiei, Metacoleroa (Gibbera, Venturia) 787
dictamni, Phyllosticta 867
dictyoides, Drechslera (Helminthosporium) 689
dictyoides var. phlei, Helminthosporium = **Drechslera phlei** 690
dictyosperma, Uromyces 1010
dictyosporum, Pythium (Nematosporangium) 920
dictyosporus, Dacrymyces 663
didyma, Ramularia (Didymaria) 924
didyma, Septoria 950
didyma f. **santonensis, Septoria** 950
didymospora, Cercospora 609
didymum, Cylindrocarpon (Fusisporium) 657
diedickei, Ascochyta = **A. zeicola** 577
diedickei, Myxosporium 810
dieffenbachiae, Cephalosporium 603
dierckxii, Penicillium 828
diervillae, Cercospora 609
diervillae, Gloeosporium = **Phlyctema vagabunda** 850

diervillae, Microdiplodia 788
diervillae, Phyllosticta 867
diervillae, Ramularia 924
diervillae, Septoria 950
diervillae, Sphaeropsis 966
dietelii, Puccinia 905
diffidens, Diatrype 673
diffissum, Acanthophysium 555
diffissus, Aleurodiscus = **Acanthophysium diffissum** 555
difformis, Cercospora = **C. violae** 618
difformis, Puccinia 905
difformis, Sarea (Tromera) 939
difformis, Septoria 950
diffusa, Byssosphaeria (Herpotrichia, Neopeckia) 596
diffusa, Cercospora 609
diffusa, Microsphaera 790
diffusa f. **elongata, Microsphaera** 790
diffusa var. rhodomphala, Herpotrichia = **Byssosphaeria rhodomphala** 596
diffusum, Sirosporium (Clasterosporium) 962
digitalidis, Septoria 950
digitalis, Phyllosticta 867
digitariae, Phyllosticta 867
digitariae, Sphacelotheca (Ustilago) 965
digitarivora, Septoria 950
digitata, Atelocauda 583
digitata, Periconia 831
digitata, Puccinia = **P. mesnieriana** 910
digitata, Xylaria 1026
digitatum, Penicillium 828
digitatus, Uromyces = **Atelocauda digitata** 583
digressus, Dacrymyces = **D. capitatus** 663
dilatata, Milesina (Milesia) 792
dilatohypha, Wolfiporia 1024
dimera, Septoria 950
dimeriae, Phyllachora 862
dimerum, Microdochium (Fusarium) 789
dimidiata, Torula = **Hendersonula toruloidea** 734
diminuens, Hysteropezizella 747
diminuta, Otthia 823
dimorpha, Diatrype = **Eutypella scoparia** 701
dimorphospora, Endophragmiella (Endophragmia) 694
dimorphospora, Phialocephala 847
dimorphosporum, Exobasidium 703
dimorphum, Pythium 920
dinemasporium, Phomatospora 857
dingleyae, Chaetosphaeria 624
dingleyae, Rhinocladium 931
diodiae, Cercospora 609
diodiae-virginianae, Cercospora 609
dioicae, Puccinia 905
dioscoreae, Colletotrichum = **C. gloeosporioides** 638
dioscoreae, Didymella 676
dioscoreae, Phaeoramularia (Cercospora) 840
dioscoreae, Phoma 853
dioscoreae, Phyllosticta (Phyllostictina, Septoria) 867
dioscoreae, Ramularia 924
diospyri, Acremonium (Cephalosporium) 556
diospyri, Aplosporella (Sphaeropsis) 570
diospyri, Dendrina 666
diospyri, Elsinoe 692
diospyri, (Dothiorella, Gloeosporium, Phyllostictina) = **Phyllosticta arxii** 865
diospyri, Macrophoma 774
diospyri, Phomopsis (Phoma) 858
diospyri, Sirosporium (Cercospora) 962
diospyri, Valsa = **V. ceratosperma** 1017
diospyri, Valsaria 1018
diplachnis, Puccinia 905
diplantherae, Plasmodiophora 881
diploa, Nectria (Calonectria) 812
diplocarpa, Botryodiplodia 591
diplocarpa, (Homostegia, Phyllachora) = **Phyllachora sylvatica** 863
diplodiella, Coniella (Coniothyrium, Phoma) 641
diplodioides, Asterina 581
diplospora, (Clathrospora, Pleospora) = **Comoclathris permunda** 640
diplospora, Sphacelotheca (Ustilago) 965
dipsaci, Peronospora 833
dircae, Phyllosticta 867
disciflorum, Phragmidium = **P. mucronatum** 861
disciforme, Aleurocystidiellum 560
disciforme, Coryneum = **C. umbonatum** 646
disciformis, Aleurodiscus = **Aleurocystidiellum disciforme** 560
disciformis, Diatrype 673
disciformis var. magnoliae, Diatrype = **D. tremellophora** 673
discincola, Nummularia = **Biscogniauxia marginata** 589
discincola, Phyllosticta 867
discincta, Phyllosticta = **P. uvulariae** 874
discoidea, (Asterella, Asterina, Microthyriella) = **Schizothyrium discoideum** 939
discoidea, Balansia = **B. strangulans** f. **discoidea** 586
discoidea, Diatrypella = **D. decorata** 674
discoidea var. alni, Diatrypella = **D. placenta** 674
discoideum, Exobasidium = **E. vaccinii** 703
discoideum, Schizothyrium 939
discolor, Gloeodontia (Irpex) 716
discolor, Tranzschelia 997
discolor f. sp. **dulcis, Tranzschelia** 997
discolor f. sp. **persica, Tranzschelia** 997
discophora, Nectria (Sphaeria) 812
discors, (Leptosphaeria, Passeriniella) = **Leptosphaeria obiones** 763
discors, Pleospora (Sphaeria) 884
discosiiformis, Phoma 853
discosioides, Seimatosporium (Monochaetia, Pestalotia) 944
discospora, Coniochaeta = **C. ligniaria** 641
discostoma, Diatrype = **Graphostroma platystoma** 723
discreta, Nummularia = **Biscogniauxia marginata** 589
dispar, Ramularia 924
dispersa, Cercospora = **C. penicillus** 614
dispersa f. sp. tritici, Puccinia = **P. triticina** 915
dispora, Gnomonia 720
disrupta, Botryosphaeria (Phomatospora, Physalospora) 592
disseminans, Ophiobolus = **O. acuminatus** 819
disseminata, Pestalotiopsis (Pestalotia) 836
disseminata, Psathyrella = **Coprinus disseminatus** 644
disseminatus, Coprinus (Pseudocoprinus) 644
dissepta, Eutypella (Eutypa, Quaternaria, Valsa) 700
dissita, Torula = **Bispora betulina** 589
dissoluta, Peniophora = **Tubulicrinis chaetophorus** 1003
dissotocum, Pythium 920
distans, Hendersonia 733
distans, Phaeomarasmius 840
distegiae, (Gibbera, Otthia) = **Dothidotthia andersonii** 688
distichlidis, Puccinia 905
distincta, Asteromassaria (Massaria) 582
distinctus, Dasyscyphus 665
distorta, Ceratocystis = **Ophiostoma distortum** 821
distorta, Poria = **Antrodiella incrustans** 568
distorta, Puccinia 905
distortum, Ophiostoma 821
distortus, (Abortiporus, Polyporus) = **Abortiporus biennis** 555
distributa, Leptosphaeria (Metasphaeria) 761
ditissima, Cercospora = **C. obesa** 614
ditissima, Nectria 812
ditricha, Venturia 1020
ditricha var. fraxinicola, Venturia = **V. fraxini** 1020
divaricata, Clavicorona 632
divaricatae, Septoria = **S. phlogis** 955
divergens, Gloeosporium = **Discula umbrinella** 685
divergens, Gyroceras = **Sirosporium antenniforme** 961
divergens, Phyllosticta 867
divergens, Phyllostictina 875
divergens, Sphaerulina 968
divergens, Tetraploa 988
diversa, Torula = **Coniosporium diversum** 641
diversisporum, Fusarium = **F. semitectum** var. **majus** 710
diversum, Coniosporium (Sirodesmium) 641
diversum, Penicillium 828
diversum var. aureum, Penicillium = **P. primulinum** 828
dochmia, Puccinia 905
dodecathei, Phyllosticta 867
dodecatheonis, Septoria 950
dodgei, Papulaspora = **P. sepedonioides** 826
dodonaeae, Episphaerella (Dimerina) 697
doellingeriae, Leptothyrium 766
dolabriforme, (Glyphium, Lophium) = **Glyphium elatum** 720
dolichi, Pseudocercospora (Cercospora) 894
dolicholi, Uromyces 1010
dolichospora, Mycosphaerella 802
dolichotrichum, Chaetomium 623
dolioloides, Nodulosphaeria (Leptosphaeria) 816
doliolum, Leptosphaeria 761
doliolum var. conoidea, Leptosphaeria = **L. conoidea** 761
dolominuta, Ceratocystis = **Ceratocystiopsis minuta** 603
dolosa var. **circumdata, Puccinia** 905
domicalis, Eutypella 700
domingensis, Cercoseptoria = **Pseudocercospora subsessilis** 895
domingensis, Coleosporium = **C. plumierae** 636
domingensis, Doassansia = **Doassansiopsis hydrophila** 686
domingensis, Phyllachora 862
donacis, Hendersonia 733
donacis, Leptostroma 765
donacis, Pseudoseptoria (Selenophoma, Septoria) 898
donacis var. **linearis, Selenophoma** 945

donacis var. stomaticola, Selenophoma = **Pseudoseptoria stomaticola** 898
donkii, (Cristella, Trechispora) = **Phlebiella vaga** 849
doronici, Ascochyta 574
dothidea, Botryosphaeria 592
dothidea f. chromogena, Botryosphaeria = **Botryosphaeria dothidea** 592
dothideaspora, Mycomicrothelia (Amphisphaeria) 799
dothideasporum, Melanomma 783
dothideoides, Haplosporella = **Aplosporella ailanthi** 569
dothideoides, Maurodothina (Asteridium, Morenoella) 780
dothideoides, Stigmina (Clasterosporium, Sciniatosporium) 979
dothideoides, Sydowia 983
douglasii, Anthracoidea 566
douglasii, Botrytis 594
douglasii, Didymella 676
douglasii, Phoma 853
douglasii, Puccinia 905
douglasii var. **hoodii, Puccinia** 905
dovrensis var. **leptoticha, Puccinia** 905
dovrensis var. **russa, Puccinia** 905
downingiae, Doassansia 686
drabae, Leptosphaeria (Sphaeria) 761
drabae, Puccinia 905
drabae, Selenophoma (Phoma, Rhabdospora) 945
dracaenae, Colletotrichum = **C. gloeosporioides** 638
dracaenae, Coniothyrium 642
dracaenae, Leptosphaeria 761
dracaenae, Lophodermium (Dermascia) 771
dracaenae, Meliola 784
dracaenae, Phyllosticta 867
dracaenae, Physalospora = **Glomerella cingulata** 719
dracaenatum, Cladosporium 629
dracocephali, Phyllosticta 867
dracocephali, Septoria 950
draconis, Leptosphaeria = **Paraphaeosphaeria obtusispora** 826
draconis, Phyllosticta 867
dracunculina, Puccinia = **P. tanaceti** var. **dracunculina** 915
drechsleri, Ophiobolus 820
drechsleri, Phytophthora 878
drosodes, Bisporella (Allophylaria) 590
drummondii, Septoria = **S. phlogis** 955
drummondii, Trametes (Polyporus) 996
drupacearum, Foveostroma (Micropera) 706
druparum, Sphaeropsis = **S. malorum** 967
dryadeus, Inonotus (Polyporus) 748
dryadis, Melanomma 783
dryadis, Wettsteinina (Massarina) 1023
dryina, Cleistophoma (Macrophoma) 632
dryina, Dendrothele 667
dryina, Peniophora = **Crustoderma dryinum** 650
dryina, Septoria 950
dryina, Tubakia (Actinopelte) 1002
dryinum, Aleurocorticium = **Dendrothele dryina** 667
dryinum, Crustoderma (Corticium) 650
dryinum, Helotium = **Dasyscyphella dryinus** 664
dryinum, Lachnella = **Dasyscyphella dryinus** 664
dryinum, Leptothyrium = **Tubakia dryina** 1002
dryinus, Aleurodiscus = **Dendrothele dryina** 667
dryinus, Dasyscyphella (Dasyscyphus) 664
dryinus, Pleurotus 886
dryinus, Polyporus = **Perenniporia tenuis** 831
dryophila, Collybia 640
dryophila, Leptosphaeria (Sphaerella) 761
dryophila, Stictis 978
dryophilum, Anthostoma = **Lopadostoma sphinctrinum** 769
dryophilus, Inonotus (Polyporus) 748
dryophilus var. vulpinus, Polyporus = **Inonotus rheades** 749
dryophyllum, Dicoccum = **Stigmella effigurata** 979
dryopteridis, Phyllosticta 867
dualis, Polyporus = **Inonotus circinatus** 748
dubautiae, Graphium 723
dubia, Diaporthe 671
dubia, Mycosphaerella = **M. confusa** 801
dubia, Ramularia (Cercospora) 924
dubium, Caeoma = **Melampsora epitea** f. sp. **tsugae** 781
dubium, Helminthosporium = **Sirosporium beaumontii** 962
dubyi, Ceriospora 621
dubyi, Cytospora = **C. pinastri** 662
duchesneae, Frommeella (Kuehneola) 706
duchesneae, Synchytrium 984
duclauxii, Penicillium 828
dulcamarae, Cercospora (Ramularia) 609
dulcamarae, Phoma = **Phomopsis dulcamarina** 858
dulcamarae, Phyllosticta 867
dulcamarina, Phomopsis (Phoma) 858
dulcarmaraecola, Cercospora = **C. dulcamarae** 609
dulichii, Physoderma 877
dulichii, Puccinia 905
dumetorum, Cyphella = **Cellypha goldbachii** 601
duplex, Peniophora = **P. pini** subsp. **duplex** 830
duplex, Wettsteinina 1023
duplicata, Cercospora 609
dupontii, (Penicillium, Talaromyces) = **Talaromyces thermophilus** 986
dura, (Agrocybe, Pholiota) = **Agrocybe molesta** 559
dura, Phoma 853
dura, Teichosporella (Teichospora) 988
durangensis, Puccinia 905
durantae, Cercospora 609
durescens, Fomitopsis (Polyporus) 705
duriaeana, Ustilago 1015
duriaeanum, Epicoccum = **E. nigrum** 696
duriaei, Myriangium 808
duricaulis, Aspergillus 578
durieui, Diatrype 673
durilabrum, Lophodermium 772
durissimum, Leptostroma 765
durum, Sclerotium 942
dysocarpae, Ravenelia = **R. fragrans** var. **evernia** 928
dyssodiae, Puccinia 905
eamesii, Ramularia 925
earleae, Fomitiporia = **Phellinus punctatus** 846
earlei, Fomes = **Pyrofomes demidoffii** 920
earlei, Poria = **Diplomitoporus lenis** 683
earlei, Stigmina 980
earlei, Tilletia 992
earliana, Meliola = **Irenopsis tehoniana** 749
earliana, (Fabraea, Peziza) = **Diplocarpon earlianum** 679
earlianum, Diplocarpon 679
earoleuca, Trichopeziza = **Proliferodiscus earoleucus** 891
earoleucus, Proliferodiscus 891
eatoniae, Puccinia 905
eatoniae var. myosotidis, Puccinia = **Puccinia eatoniae** 905
eatoniae var. ranunculi, Puccinia = **Puccinia eatoniae** 905
eccentrica, Clitocybe 633
eccentrica, Gnomoniella = **Plagiostoma devexum** 881
echeveriae, Puccinia 905
echii, Cercospora 609
echinata, Acantharia (Dimerosporium, Venturia) 555
echinata, Eutypa = **Eutypella scoparia** 701
echinata, Memnoniella (Stachybotrys) 786
echinata, Nyssopsora 817
echinata, Petrakia 837
echinata, Pleosphaeria (Cucurbitaria) 883
echinata, Ustilago 1015
echinatum, Epicoccum = **Petrakia echinata** 837
echinatum, Triphragmium = **Nyssopsora echinata** 817
echinella, Ceratostomella = **Ophiostoma piliferum** 821
echinella, Pyrenophora = **Pleospora calvescens** 884
echinochloae, Cercospora 609
echinochloae, Periconia (Periconiella) 831
echinocystis, Cercospora 609
echinocystis, Hyphoderma 741
echinocystis, Septoria 950
echinodori, Burrillia 596
echinophila, Dactylaria = **Scolecobasidium echinophilum** 942
echinophila, Lanzia (Ciboria, Rutstroemia, Sclerotinia) 756
echinophilum, Scolecobasidium 942
echinophilus, Dasyscyphus 665
echinospermi, Peronospora 833
echinospora, Tomentellopsis 995
echinosporus, Hypochnus = **Tomentellopsis echinospora** 995
echinulata, Cunninghamella 655
echinulatum, Cladosporium (Heterosporium) 629
echinulatum, Eurotium 699
echinulatum, Penicillium 828
echinulatum, Synchytrium 984
echinulatus, Aspergillus 578
ectypus, Trametes (Coriolus, Polyporus) 996
edwiniae, Ovularia 824
edwiniae, Stictis = **S. schizoxyloides** 979
effigurata, Mycosphaerella (Sphaerella) 802
effigurata, Stigmella 979
effiguratum, Coryneum = **Stigmella effigurata** 979
effugiens, Hysteromyxa 747
effusa, Bispora 589
effusa, Cercospora (Didymaria) 609
effusa, Cryptocline 651
effusa, Didymella (Didymosphaeria) 676
effusa, Hendersonia 733
effusa, Peronospora 833
effusa, Pileolaria 880
effusa, Puccinia = **P. violae** 916
effusa, Ramularia 925
effusa, Sphaerella = **Venturia adusta** 1020
effuscata, Vararia = **Dichostereum effuscatum** 675
effuscatum, Dichostereum (Corticium) 675
effusum, Corticium = **Cystostereum murrayi** 660

effusum, (Cladosporium, Fusicladium) = **Cladosporium caryigenum** 629
effusum, Gymnosporangium 726
effusum, Hypoxylon = **H. serpens** var. **effusum** 746
effusum, Melogramma = **Bertiella rhodospila** 587
effusum, Toxosporium (Coryneum) 995
effusum, Verticicladium 1021
effusum var. carpinum, Fusicladium = **Cladosporium caryigenum** var. **carpinum** 629
efibulatum, Corticium (Laeticorticium) 645
egenula, Uredo = **Uromyces sporobolicola** 1013
ehrenbergii, Cylindrocarpon = **C. magnusianum** 657
eichhornius, Doratomyces 686
eichleri, Hypochnicium (Peniophora) 743
eichleriana, Tulasnella 1004
elaeagnellum, Camarosporium 599
elaeagni, Cercospora 609
elaeagni, Cucurbitaria 654
elaeagni, Cytospora 661
elaeagni, Diaporthe = **D. beckhausii** 670
elaeagni, (Fusicoccum, Phomopsis) = **Phomopsis arnoldiae** 857
elaeagni, Septoria 950
elaeagnina, Aplosporella (Sphaeropsis) 570
elaeochroma, Cercospora 609
elaeodes, (Clasterosporium, Torula) = **Polyschema olivacea** 889
elaeospora, Karschia = **Dactylospora stygia** 664
elastica, Leptostromella 765
elasticae, Cercospora 609
elasticae, Chaetospermum = **C. carneum** 624
elasticae, (Colletotrichum, Gloeosporium) = **Colletotrichum gloeosporioides** 638
elasticae, Stemphylium 976
elasticae var. artocarpi, Chaetospermum = **C. carneum** 624
elatina, Melampsorella = **M. caryophyllacearum** 781
elatina, Pseudographis 896
elatior, Rhinocladiella = **Leptodontium elatius** 759
elatius, Leptodontium 759
elatum, Chaetomium 623
elatum, Cladosporium (Cadophora) 629
elatum, Glyphium (Lophium) 720
eleagni, Phyllactinia 864
electrae, Puccinia 905
electrae var. **depressiporosa, Puccinia** 905
electrae var. **robusta, Puccinia** 905
elegans, Actinomucor (Rhizopus) 558
elegans, Aspergillus 578
elegans, Corallomyces = **Nectria mauritiicola** 813
elegans, Cunninghamella 655
elegans, Dacryopinax (Guepinia) 663
elegans, Dictyosporium 676
elegans, Endobotrya 693
elegans, Hohenbuehelia (Pleurotus) 737
elegans, Lobatopedis 769
elegans, Mariannaea (Paecilomyces, Penicillium) 777
elegans, Polyporus = **P. varius** 889
elegans, Redbia 929
elegans, Stachybotrys (Hyalobotrys) 972
elegans, Thamnidium 989
elegans, Trametes (Daedalea, Lenzites) 996
elegans, Triposporium 1001
elegans, Typhula (Pistillaria) 1004
elegans, Uromyces 1010
elegans, Ustilago 1015
elegans var. sorghina, Spicaria = **Mariannaea elegans** 777
elegantissimum, Phacidium 839
elegantula, Conoplea (Botrytis) 643
elegantula, Trichopeziza = **Belonidium elegantulum** 587
elegantula, Typhula = **T. incarnata** 1004
elegantula var. **inflata, Mycena** 797
elegantulum, Belonidium (Helotium, Lachnella) 587
elegantulus, Dasyscyphus = **Belonidium elegantulum** 587
eleocharidis, Anthracoidea 566
eleocharidis, Cercospora 609
eleocharidis, Puccinia 905
eleocharidis, Uromyces (Nigredo) 1010
eleospora, Septoria 951
elephantopi, Cercospora 609
elephantopodis, Coleosporium = **C. vernoniae** 636
elevans, Eutypella (Eutypa) 700
elevata, Microsphaera 790
elinorae, Helicosporium = **H. pannosum** 731
ellipsospora, Perenniporia 830
ellipsosporum, Prosthecium 892
elliptica, Botrytis 594
elliptica, Diatrype = **Pseudovalsa lanciformis** 899
ellipticum, Camarosporium 599
ellipticum, Cylindrocladium 657
ellipticum, Sporidesmium (Sirodesmium) 970
ellisiana, Glabrocyphella 715
ellisiana, Lachnella = **Dasyscyphus ellisianus** 665
ellisiana, Leptosphaeria 761
ellisiana, Monochaetia 795
ellisiana, Perenniporia 830
ellisiana, Phyllosticta 867
ellisiana, Puccinia 905
ellisiana, Septoria 951
ellisianum, Fusicoccum 712
ellisianus, Dasyscyphus 665
ellisianus, Fomes = **Perenniporia ellisiana** 830
ellisianus, Polyporus = **Dichomitus squalens** 675
ellisii, Calathella 597
ellisii, Caliciopsis = **C. calicioides** 597
ellisii, Coniophora = **C. olivacea** 641
ellisii, Dacrymyces = **D. capitatus** 663
ellisii, Dimerosporium = **Protoventuria maculosa** 893
ellisii, Diplodina = **Ascochyta hyalospora** 575
ellisii, Entyloma 696
ellisii, Gloniopsis = **G. curvata** 719
ellisii, Gymnosporangium 726
ellisii, Meliola = **M. niessleana** 785
ellisii, Myxosporium = **Discosporium populeum** 685
ellisii, Phyllosticta 867
ellisii, Pleospora 884
ellisii, Puccinia 905
ellisii, Pyrenopeziza (Niptera) 918
ellisii, Rhabdospora 930
ellisii, Sorosporium 963
ellisii, Sphaeropsis = **S. sapinea** 967
ellisii, Stigmina 980
ellisii, Tetraploa 989
ellisii var. occidentalis, Sorosporium = **Sphacelotheca occidentalis** 965
ellisii var. provincialis, Sorosporium = **S. provinciale** 963
elongata, Cercospora (Cercosporina) 609
elongata, Cucurbitaria 654
elongata, Leptosphaeria 761
elongata, Macrophoma 774
elongata, Phyllosticta 868
elongata, Poria = **Perenniporia narymica** 830
elongatipes, (Hypsizygus, Pleurotus) = **Hypsizygus tessellatus** 747
elongatum, Capnodium 601
elongatum, Hysterographium = **H. fraxini** 747
elymi, Ascochyta = **A. leptospora** 575
elymi, Leptosphaeria 761
elymi, Phaeoseptoria 841
elymi, Pyrenochaeta 918
elymi, Septoria 951
elymi, Tilletia 992
elymi f. **elymina, Septoria** 951
elymicola, Ustilago = **U. serpens** 1016
elymina, Davisiella (Cytodiplospora) 666
elynae, Anthracoidea 566
elynae, Clathrospora (Pleospora) 631
emaculata, Puccinia 905
emaculata, Septoria 951
emarginata, Propolis 892
emergens, Mollisia 793
emergens, Nummularia = **N. broomeiana** 817
emiliae, Puccinia 905
eminens, Cylindrosporium 658
eminens, Phyllosticta 868
empetri, Arwidssonia (Heterosphaeria, Sphaeropezia) 573
empetri, Chrysomyxa 626
empetri, Coleophoma (Septoria) 635
empetri, Didymosphaeria 678
empetri, Phaeangellina (Cenangium) 839
empetri, Physalospora (Botryosphaeria) 876
enalia, Didymosphaeria 678
enceliae, Puccinia 906
endobioticum, Synchytrium (Chrysophlyctis) 984
engadinensis, Wettsteinina = **Nodulosphaeria winteri** 817
engelmanii, Leptographium 759
engelmannii, Lachnum = **Lachnellula arida** 754
enneapogonis, Ustilago 1015
enterochroma, Chloroscypha (Kriegeria, Ombrophila) 626
entospora, Basidiophora (Plasmopara) 586
entylomicola, Phyllosticta 868
eocoronis, Pleospora 884
ephedrae, Peridermium 832
epicampis, Phyllachora 862
epicampis, Puccinia = **P. schedonnardi** 913
epicampis, Uromyces 1010
epicarecta, Leptosphaeria (Sphaeria) 761
epichloe, Balansia (Dothichloe) 586
epichysium, Omphalina (Clitocybe) 819
epicoccoides, Cercospora 609
epicocos, Diplodia 680
epidendri, Gloeosporium = **Colletotrichum gloeosporioides** 638
epidendri, Uredo 1007
epidendronis, Cercospora 609
epidermidis, Didymosphaeria = **D. futilis** 678
epidryas, Marasmius 777
epifagus, Marasmius 777
epigaea, Sebacina 944
epigaea, Tomentella 994
epigaeae, Cercospora 610
epigaeae, Phyllosticta 868
epigaeina, Cercospora = **C. epigaeae** 610

epigynum, Pythium = **P. pulchrum** 921
epileuca, (Massaria, Trematosphaeria) = **Asteromassaria olivaceohirta** 582
epilobii, Cercospora = **Ramularia montana** 926
epilobii, Didymella = **Sydowiella fenestrans** 983
epilobii, Doassansia 686
epilobii, (Euryachora, Spilosticta) = **Venturia asteromorpha** 1020
epilobii, Plasmopara 882
epilobii, Puccinia 906
epilobii, Pucciniastrum 916
epilobii, Septoria 951
epilobii, Sphaerotheca 968
epilobii, Synchytrium 984
epilobii, (Datronia, Trametes) = **Datronia stereoides** 665
epilobii-tetragoni, Puccinia = **P. pulverulenta** 912
epimiltinus, Tinctoporellus (Poria) 992
epipactidis, Cercospora 610
epiphaeum f., Hypoxylon = **H. investiens** var. **epiphaeum** 745
epiphylla, Athelia 583
epiphylla, Gerwasia (Mainsia) 714
epiphylla, Ravenelia (Sphaeria) 928
epiphyllum, Ceratophorum (Coryneum, Sporidesmium) 604
epiphyllum, Cladosporium = **C. herbarum** 630
epiphyllum, Corticium = **Athelia epiphylla** 583
epiphyllum, Helotium = **Hymenoscyphus epiphyllus** 740
epiphyllus, Hymenoscyphus 740
epiphyllus, Marasmius 777
epipora, Orbilia 822
episphaeria, Cenangium 602
episphaeria, Fusarium 707
episphaeria, Hymenochaete 739
episphaeria, Nectria 812
episphaericum, Fusarium = **F. lateritium** 708
episphaericum, Helminthosporium = **Sporidesmium inflatum** 970
epistomium, Phytophthora 878
epitea, Melampsora 781
epitea f. sp. tsugae, Melampsora 781
equiseti, Ascochyta (Phoma) 575
equiseti, Fusarium 707
equiseti, Heterosporium 736
equiseti, Leptosphaeria 761
equiseti, Ramularia 925
equiseti, Stamnaria = **S. persoonii** 975
equiseti, Titaeospora (Gloeosporium, Septogloeum, Septoria) 992
equiseti var. bullatum, Fusarium = **Fusarium equiseti** 707
equitis, Aspergillus 578
eragrostidis, Curvularia 655
eragrostidis, Gloeosporium 718
eragrostidis, Phyllachora 862
eragrostidis, Tilletia 992
eragrostidis, Uromyces 1010
erechtitis, Cercospora 610
erechtitis, Phyllosticta 868
erechtitis, Septoria 951
erechtitis, Synchytrium 984
erecta, Acrodictys 557
erecta, Phlebia = **P. rufa** 848
erectum, Chaetomium 623
erectum, (Macrosporium, Mystrosporium) = **Acrodictys erecta** 557
eremocarpae, Synchytrium 984
eremophila, Coniophora 641
eremophila, Tilletia = **T. asperifolia** 991
eres, Diaporthe 671
erianthi, Cladosporium 629
erianthi, Hendersonia 733
erianthi, Phyllachora = **P. dimeriae** 862
ericaceana, Phomopsis 858
ericae, Uredo (Pucciniastrum) 1007
ericina, Phanerochaete (Peniophora) 843
erigeniae, Puccinia 906
erigeronicola, Dimeriella = **Lasiostemma melioloides** 757
erigerontis, Leptosphaeria 761
erigerontis, Septoria (Septoria) 951
erigerontis, Synchytrium 984
erigerontis f. boltoniae, Septoria
erikssonii, Peniophora 829
erinacea, Humaria = **Scutellinia erinaceus** 943
erinaceus, Hericium (Hydnum) 734
erinaceus, Phaeomarasmius 840
erinaceus, Scutellinia (Trichophaea) 943
erinaceus subsp. **erinaceo-abietis, Hericium** 735
eriobotryae, Cercospora (Cercosporina) 610
eriobotryae, Diplodia 680
eriobotryae, Fusicladium = **Spilocaea pyracanthae** 969
eriobotryae, Phyllosticta 868
eriobotryae, Septoria 951
eriocauli, Moesziomyces (Tolyposporium) 793
eriocauli, Ustilago 1015
eriochloae, Sorosporium = **S. mixtum** 963
eriocryptum, Camarosporium 599
eriodictyonis, Trimmatostroma (Coniothecium) 1001
eriogoni, Camarosporium 600
eriogoni, Cercospora 610
eriogoni, Cytospora 661
eriogoni, Hendersonia 733
eriogoni, Leptostromella 765
eriogoni, Peronospora 833
eriogoni, Phoma 853
eriogoni, Phyllosticta 868
eriogoni, Pyrenophora = **Pleospora herbarum** var. **occidentalis** 884
eriogoni, Sarcophoma (Gloeosporium) 938
eriogoni, Uromyces = **U. intricatus** 1011
eriophila, Mycosphaerella 802
eriophorella, Stagonospora (Septoria) 973
eriophori, Kriegeria (Platygloea, Xenogloea) 753
eriophori, Puccinia 906
eriophori-alpini, Puccinia 906
eriophylli, Puccinia = **P. hieracii** 907
erodii, Peronospora (Pseudoperonospora) 833
erosum, Lophiostoma = **Platystomum compressum** var. **pseudomacrostomum** 882
errabunda, Apiognomonia (Gnomonia) 568
erratica, Fenestella (Cucurbitaria) 704
erratica, Phyllosticta 868
erratica, Physalospora 876
erraticus, Anomothallus 565
erubescens, Nectria (Calonectria) 812
eruciformis, Calathella (Cyphella, Flagelloscypha, Lachnella) 597
erumpens, Colletotrichum = **C. dematium** 637
erumpens, Dendrophora (Peniophora, Stereum) 667
erumpens, Pseudomassaria 896
eryngicola, Septoria 951
eryngii, Cylindrosporium 658
eryngii, Entyloma 696
erysimi, Cercospora 610
erysimi, Phyllosticta 868
erysipheoides, Lasiostemma (Dimeriella, Dimerosporium, Epipolaeum) 757
erysipheoides, Venturia 1020
erysiphina, Mycosphaerella 802
erysiphoides, Asterina = **Phillipsiella atra** 848
erysiphoides, Oidium 818
erysiphoides, Phoma 853
erysiphoides, Phyllosticta 868
erysiphoides f. crotalariae, Oidium 818
erythraeae, Cercospora = **C. sabbatiae** 616
erythrinae, Cercospora 610
erythrinae, Colletotrichum = **C. gloeosporioides** 638
erythrinae, Stagonospora 974
erythrinicola, Cercospora 610
erythrinicola, Phoma 853
erythrocephala, Stilbella = **S. fimetaria** 981
erythrogena, Cercospora 610
erythronii, Ciborinia (Sclerotinia) 627
erythronii, Phyllosticta 868
erythronii, Urocystis 1008
erythronii, Ustilago = **U. heufleri** 1015
erythroseptica, Phytophthora 878
erythrospila, Drechslera 690
erythrospilum, Helminthosporium = **Drechslera erythrospila** 690
erythrospora, Nodulosphaeria 817
erythrospora, Trabutia (Rhytisma) 995
erythrosporus, Ophiobolus = **Nodulosphaeria erythrospora** 817
eschscholtziae, Acroconidiella (Heterosporium) 557
eschscholtziae, Crocicreas (Helotium, Hymenoscyphus, Peziza, Phialea) 649
eschscholtziae, Entyloma 696
eschscholzii, Daldinia 664
esclavensis, Puccinia 906
esclavensis var. panicophila, Puccinia 906
esculenta, Ustilago 1015
estrelti, Phoma 853
ethacetica, Thielaviopsis = **Chalara paradoxa** 625
etherialis, Valsa = **V. ceratosperma** 1017
eucalypti, Aulographina (Aulographum) 584
eucalypti, Bagnisiopsis 585
eucalypti, Cercospora 610
eucalypti, Cryptosporium 653
eucalypti, Diaporthe 671
eucalypti, Diatrype = **D. prominens** 673
eucalypti, Diplodia 680
eucalypti, Dothiorella (Sphaeropsis) 688
eucalypti, Gloeosporium = **Colletotrichum gloeosporioides** 638
eucalypti, Harknessia 729
eucalypti, Hendersonia = **Seimatosporium lichenicola** 945
eucalypti, Heterosporium 736
eucalypti, Hypoderma 743
eucalypti, Hysterium = **H. angustatum** 747
eucalypti, Lembosia 758
eucalypti, Nectria 812
eucalypti, Orbilia (Calloria) 822
eucalypti, Phaeoseptoria 841
eucalypti, Phragmodothidea 861
eucalypti, Phyllachora 862
eucalypti, Pleurophomopsis 886
eucalypti, Scleroderris (Dermea) 940
eucalypti, Sphaeronaema 966
eucalypti, Valsa 1018
eucalypti, Valsaria 1018
eucalypti var. maculicola, Heterosporium = **H. californicum** 736

eucalypticola, Coniella = **C. castaneicola** 641
eucalypticola, Melanconium = **Fairmaniella leprosa** 704
eucalypticola, Sonderhenia (Hendersonia) 962
eucalyptina, Lembosiopsis = **Aulographina eucalypti** 584
eucalyptina, Physalospora 876
eucarpa, Wettsteinina 1023
eucastaneae, Ceratocystis = **Ophiostoma stenoceras** 822
eugeniae, Cercospora (Cercosporina) 610
eugeniae, (Cryptosporella, Endothia) = **Cryphonectria cubensis** 650
eugeniae, Mycosphaerella 802
eugeniae, Phyllosticta 868
eugentianae, Uromyces 1010
eumacrospora, Puccinia 906
eumartii, Fusarium = **F. solani** f. sp. **eumartii** 710
eumorpha, Cyphella = **Calathella eruciformis** 597
eumorpha, Didymella 676
eumorpha, Leptosphaeria (Sphaeria) 761
eunomia, Cryptosphaeria (Valsa) 652
eunomia var. **fraxini, Cryptosphaeria** 652
eunotiaespora, Wallrothiella = **Botryosphaeria quercuum** 593
euomphala, Tympanopsis = **Nitschkia confertula** 816
euonymi, Aplosporella 570
euonymi, Cercospora 610
euonymi, Cytospora 661
euonymi, Diaporthe 671
euonymi, Diplodia 680
euonymi, Microsphaera 790
euonymi, Phyllosticta 868
euonymi, Ramularia (Septocylindrium) 925
euonymi, Sirogloea 961
euonymi-japonici, Elsinoe 692
euonymi-japonici, Oidium 818
euonymi-japonici, Sphaceloma 964
eupatorii, Belonidium (Dasyscyphus, Lachnella, Trichopeziza) 587
eupatorii, Cercospora 610
eupatorii, Coleosporium = **C. steviae** 636
eupatorii, Septoria 951
eupatoriicola, Septoria 951
eupatorii-rotundifolii, Aecidium 558
euphorbiae, Alternaria (Alternaria) 561
euphorbiae, Cercospora 610
euphorbiae, Erysiphe 698
euphorbiae, Melampsora 781
euphorbiae, Melanotaenium (Tilletia) 784
euphorbiae, Microsphaera 790
euphorbiae, Peronospora 833
euphorbiae, Phyllostictina = **Phyllosticta ghaesembillae** 868
euphorbiae, Questieria 922
euphorbiae, (Cercosporidium, Pyricularia, Scolicotrichum) = **Cercosporidium chaetomium** 620
euphorbiae, Uromyces 1010
euphorbiaecola, Cercospora 610
euphorbiaecola var. **tragiae, Cercospora** 610
euphorbiae-corniculati, Uromyces 1011
euphorbiae-gerardianae, Melampsora 781
euphorbiicola, Alternaria 561
euphorbiicola, Phoma 853
eupora, Poria = **Junghuhnia nitida** 751
eupyrena, Didymella 677
eupyrena, Phoma 853
eurhododendri, Diplodina 682
europa, Cercospora = **Cercosporidium cassiocarpum** 620
europhioides, Ceratocystis = **C. piceaperda** 604
eurotiae, Didymella 677
eurotiae, Puccinia 906
euryasca, Monascostroma (Didymosphaeria) 794
euryhalina, Didymosamarospora 678
eustacei, Butlerelfia 596
eustegiaeforme, Belonium = **Discocurtisia arundinariae** 684
eustoma, Phaeosphaeria (Leptosphaeria, Pleospora) 842
eustomae, Cercospora 610
eustomonis, Sclerophoma 940
euteiches, Aphanomyces 568
euteiches f. sp. **phaseoli, Aphanomyces** 568
euteiches f. sp. **pisi, Aphanomyces** 568
eutypaeformis, Diatrypella (Cryptovalsa) 674
eutypoides, Diatrypella 674
eutypoides, Leptosphaeria = **Pleospora calvescens** 884
evadens, Puccinia (Eriosporangium) 906
everhartii, Dothiorella 688
everhartii, Gloeosporium = **Marssonina punctiformis** 778
everhartii, Macrodiaporthe (Melanconis) 773
everhartii, Myxosporium 810
everhartii, Phellinus (Fomes, Mucronoporus, Pyropolyporus) 845
everhartii, Phyllosticta 868
everhartii, Physalospora = **Botryosphaeria obtusa** 592
everhartii, Pseudoseptoria (Selenophoma, Septoria) 898
everhartii, Sorosporium 963
evernia, Ravenelia = **R. fragrans** var. **evernia** 928
evolvens, (Corticium, Cylindrobasidium) = **Cylindrobasidium laeve** 656
evonymella, Septoria 951
evonymi, Septoria 951
exaridum, Lophodermium (Hysterium) 772
exasperans, Cytospora = **C. leucosperma** 662
exasperans, Puccinia 906
exasperans, Rhynchostoma 935
exasperans, Valsaria (Diatrype) 1018
excavata, Phyllosticta 868
excelsa, Ceriporia (Poria) 621
excelsior, Phomopsis 858
excisa, Mycena 797
excursionis, Puccinia 906
exercitalis, Diaporthe = **D. linearis** 671
exigua, Ceratostomella = **Ophiostoma minus** 821
exigua, Didymella (Didymosphaeria) 677
exigua, Eutypella = **E. scoparia** 701
exigua, Mollisia (Peziza, Pezizella) 793
exigua, Phoma 853
exigua, Seynesiella 961
exigua, Zignoella = **Melanopsamma pomiformis** 783
exigua var. **inoxydabilis, Phoma** 853
exiguum, Clypeolum = **Stomiopeltis cassiae** 981
exiguum, Gymnosporangium 726
exiguum, Hypoxylon = **Penzigia frustulosa** 830
exile, Gampsonema 713
exilis, Asteridiella (Irene, Irenina, Meliola) 580
exilis, Cercosporella 619
exilis, Niesslia 815
exilis, Pestalotia = **Gampsonema exile** 713
exima, Peniophora 829
eximia, Pezicula = **P. ocellata** 837
eximia, Phyllosticta 868
exitiosa, Cercospora = **C. microsora** 613
exocarpina, Phoma 853
exotica, Cercospora = **Pseudocercospora nymphaeacea** 895
exotica, Septoria 951
expansa, Donkioporia 686
expansa, Puccinia = **P. glomerata** 907
expansa, Ravenelia 928
expansa, Septoria 951
expansum, Penicillium 828
expansum, Phacidium = **Sarcotrochila piniperda** 938
extensa, (Daedalea, Macrohyporia, Wolfiporia) = **Wolfiporia cocos** 1024
extensa, Microsphaera 790
extensa, Phyllosticta 868
extensicola, Puccinia = **P. dioicae** 905
extensicola var. asteris, Puccinia = **P. dioicae** 905
extensicola var. erigerontis, Puccinia = **P. dioicae** 905
extensicola var. euthamii, Puccinia = **P. dioicae** 905
extensicola var. hieraciata, Puccinia = **P. dioicae** 905
extensicola var. oenotherae, Puccinia = **P. dioicae** 905
extensicola var. phrymae, Puccinia = **P. dioicae** 905
extensicola var. rudbeckiae, Puccinia = **P. dioicae** 905
extensicola var. solidaginis, Puccinia = **P. dulichii** 905
extensicola var. valerianae, Puccinia = **P. dioicae** 905
extensus, Phellinus (Fomes, Polyporus) 845
externa, Anthracoidea (Cintractia) 566
exterum, Gymnosporangium 726
extorre, Cladosporium 629
exudans, Valsa = **Camarops microspora** 599
exutans, Mycosphaerella 802
exutans, Nummularia = **Hypoxylon nummularium** var. **exutans** 745
eyrei, Basidiodendron (Bourdotia, Sebacina) 586
eysenhardtiae, Uropyxis = **U. daleae** var. **eysenhardtiae** 1014
fabae, Cercospora = **C. zonata** 619
fabae, Pythium = **P. irregulare** 921
fabae, Uromyces (Nigredo, Puccinia) 1011
fabaeformis, Aplosporella (Diplodia, Sphaeropsis) 570
fabaeformis var. lignicola, Sphaeropsis = **Aplosporella fabaeformis** 570
fagacearum, Ceratocystis (Endoconidiophora) 604
fagaricola, Cercospora = **C. xanthoxyli** 618
fagi, Coniothyrium 642
fagi, Diaporthe 671
fagi, Discosphaerina (Guignardia) 685
fagi, (Discula, Gloeosporium) = **Discula umbrinella** 685
fagi, Microxiphium (Fumago) 792
fagi, Mycosphaerella 802
fagi, Scolicosporium = **S. macrosporium** 943
fagi var. **longispora, Diaporthe** 671
fagicola, Calosphaeria (Valsa) 598
fagicola, Gloeosporium = **Discula umbrinella** 685
fagicola, Sphaeropsis = **Dothiorella macrospora** 688
faginea, Diaporthe = **D. rudis** 672

faginea, Dichaena = **Ascodichaena rugosa** 578
faginea, Libertella 768
faginea, Phleogena 849
faginea, Phyllosticta 868
faginea, Plicatura = **Plicaturopsis crispa** 887
faginea, Propolis = **Propolomyces versicolor** 892
faginea var. **atra, Propolis** 892
fagineus, Marasmius = **Collybia dichrous** 640
fagineus, Merulius = **Plicaturopsis crispa** 887
fagopyri, Ascochyta = **A. bresadolae** 574
fagopyri, Cercospora 610
fagopyri, Phomopsis 858
fagopyri, Phytophthora = **Pythium helicoides** 921
fairmani, Pseudovalsa = **Phragmodiaporthe caryae** 861
fairmanianum, Cylindrosporium 658
fairmanii, Diplodia 680
fairmanii, Septoria 951
falcata, Chaetasbolisia 622
falcata, Curvularia = **C. senegalensis** 655
falcata, Monostichella 796
falcata, Typhula = **T. uncialis** 1005
falcatae, Trichosporum 1000
falcatum, Acrothecium = **Curvularia senegalensis** 655
falcatum, Colletotrichum 637
falcatum, Cryptosporium = **C. candidum** 653
falcatum, Gloeosporium = **Monostichella falcata** 796
falcatum, Microdochium 789
fallaciosa, Diaporthe = **D. eres** 671
fallax, Curvularia 655
fallax, Mycena 797
fallax, Scytinostroma 944
fallax, Trichia = **T. decipiens** 999
fallax, Valsa 1018
fallens, Uromyces = **U. trifolii-repentis** var. **fallens** 1013
familia, Clitocybula (Collybia) 633
farfaricola, Septoria 951
farinacea, Daedalea = **Fuscocerrena portoricensis** 711
farinacea, Puccinia 906
farinacea, Saccoblastia (Helicogloea) 938
farinacea, Trechispora (Cristella, Grandinia, Odontia) 997
farinacea var. **azurea, Puccinia** 906
farinaceum, Hydnum = **Trechispora farinacea** 997
farinaceus, Irpex = **Fuscocerrena portoricensis** 711
farinosa, Bulbillomyces (Peniophora) 595
farinosa, Cryptosporella (Cryptospora, Valsa) 653
farinosa, Ovulariopsis 824
farinosa, Peronospora 833
farinosa, Pichia 879
farinosa f. sp. betae, Peronospora = **Peronospora farinosa** 833
farinosus, Paecilomyces 825
farlowiana, Macrophoma (Phoma) 774
farlowiana, Pleospora 884
farlowii, Aleurodiscus 560
farlowii, Cintractia 628
farlowii, Coccomyces 634
farlowii, Inonotus (Polyporus) 748
farlowii, Medeolaria 780
farlowii, Melampsora (Necium) 781
farlowii, (Byssomerulius, Merulius) = **Ceraceomyces serpens** 603
farlowii, Sclerospora 941
farlowii, Taphrina (Exoascus) 987
farrae, Maurodothina (Eupelte) 780
fasciatum, Stereum = **S. ostrea** 978
fasciatus, Fomes 705
fasciculare, Hypholoma (Naematoloma) 742
fasciculare, Septonema (Helminthosporium) 947
fascicularis, Encoelia (Phibalis) 693
fasciculata, Alternaria = **A. alternata** 561
fasciculata, Bombardia 590
fasciculata, Cryptosporiopsis (Tubercularia) 653
fasciculata, Cyphella = **Merismodes fasciculatus** 787
fasciculata, Gnomonia (Gnomoniella) 720
fasciculata, Monotospora = **Melanographium cookei** 783
fasciculata, Passalora = **Fusicladium fasciculatum** 711
fasciculata, Solenia = **Rectipilus fasciculatus** 929
fasciculatum, Cladosporium = **C. herbarum** 630
fasciculatum, Fusicladium (Cercosporidium, Scolicotrichum) 711
fasciculatum, Harpographium 730
fasciculatum, Macrosporium = **Alternaria alternata** 561
fasciculatum var. **didymum, Fusicladium** 711
fasciculatus, Fomes 705
fasciculatus, Merismodes 787
fasciculatus, Rectipilus 929
fastidiosa, Calycina = **Helotium fastidiosum** 732
fastidiosa, Trechispora 997
fastidiosum, Corticium = **Trechispora fastidiosa** 997
fastidiosum, Helotium 732
fastigiata, Phialophora (Cadophora) 847
fatiscens, Phyllosticta 868
fatiscens, Triblidium 998
faulliana, Caeoma = **Melampsora medusae** 781
faulliana, Taphrina 987
faullii, Isthmiella (Bifusella) 750
favacea, Diatrypella (Diatrype) 674
favillensis, Phyllosticta 868
favoginea, Trichia 999
fawcettii, Elsinoe 692
fawcettii, Exosporina = **Hendersonula toruloidea** 734
fawcettii, Sphaceloma 964
feei, Fomitopsis (Fomes, Polyporus) 705
fellutanum, Penicillium 828
femoralis, Winterella (Cryptospora, Ophiovalsa, Valsa) 1024
fendlerae, (Strickeria, Teichospora) = **Teichosporella dura** 988
fendleri, Mycosphaerella 802
fenestrans, Sydowiella (Didymosphaeria, Gnomonia) 983
fenestrata, (Nectria, Pleonectria) = **Thyronectria berolinensis** 991
fenestrata, Pyrenophora = **Pleospora tragacanthae** 886
fenestrata, Stemonitis = **S. splendens** 976
fenestrata, Stilbospora = **Kaleidosporium fenestratum** 751
fenestrata, Tryblidaria (Patellaria) 1001
fenestratum, (Blitridium, Triblidium) = **Tryblidaria fenestrata** 1001
fenestratum, Kaleidosporium (Stegonsporium) 751
fennicum, Acanthophysium 555
fennicus, Aleurodiscus = **Acanthophysium fennicum** 555
ferax, Phyllosticta 868
fergusii, Perenniporia 830
fergussonii, Puccinia 906
fermentans, Pichia 879
ferox, Antrodia (Poria) 567
ferrea, Poria = **Phellinus ferreus** 845
ferreus, Phellinus (Polyporus) 845
ferruginea, Ambrosiella (Monilia) 563
ferruginea, Mycovellosiella (Cercospora) 807
ferruginea, Ophiodothella (Linospora) 820
ferruginea, Pachnocybe 825
ferruginea, Tomentella 994
ferruginea var. ambrosiae, Cercospora = **Mycovellosiella ambrosiae** 807
ferruginella, Tomentella 994
ferrugineofusca, Poria = **Phellinus ferrugineofuscus** 845
ferrugineofuscus, Phellinus 845
ferrugineo-velutinus, Phellinus 845
ferrugineum, Ceuthocarpon = **Ophiodothella ferruginea** 820
ferrugineum, Gloeosporium 718
ferrugineum, Hypoxylon = **H. rubiginosum** var. **ferrugineum** 746
ferrugineum, Nematogonum 814
ferrugineus, Hypochnus = **Tomentella ferruginea** 994
ferruginosa, Caldesiella = **Tomentella crinalis** 993
ferruginosa, Poria = **Phellinus ferruginosus** 845
ferruginosa, Tomentella = **Tomentellina fibrosa** 995
ferruginosum, Asterodon 581
ferruginosum, Cenangium 602
ferruginosus, Hypochnus = **Tomentellina fibrosa** 995
ferruginosus, Phellinus (Polyporus) 845
fertile, Circinotrichum 628
fertilis, Dictyochaeta (Codinaea) 675
fertilis, Sphaeropsis = **S. malorum** 967
festucae, Botryosphaeria (Physalospora) 592
festucae, Cercospora 610
festucae, Didymella (Didymosphaeria) 677
festucae, Ophiobolus = **Lophodermium alpinum** 771
festucae, Phaeoseptoria = **P. airae** 841
festucae, Puccinia 906
festucae var. **andropogonis, Phaeoseptoria** 841
festucae var. **muhlenbergiae, Phaeoseptoria** 841
festuca-octoflorana, Tilletia = **T. texana** 992
feurichii, Ceuthospora = **C. foliicola** 622
fibriatum, Gliocladium = **Myrothecium verrucaria** 809
fibricola, (Aposphaeria, Phoma) = **Phoma glomerata** 853
fibrillosus, (Phaeolus, Polyporus) = **Pycnoporellus fulgens** 917
fibrosa, Tomentellina (Kneiffiella) 995
fibulata, Athelia 583
ficariae, Entyloma 696
ficariae, Gloeosporium (Cylindrosporium) 718
ficariae, Peronospora 833
ficarioides, Septoria 951
fici, Cephalosporium = **Acremonium zonatum** 557
fici, Cercospora 610
fici, Cerotelium (Physopella) 622
fici, Eutypella = **E. scoparia** 701
fici, Macrophoma 774

fici, Ophiodothella 820
fici, Ormathodium = **Gonatophragmium mori** 722
fici, Tubercularia 1002
fici-indicae, Septoria 951
ficina, Cercospora 610
ficus, Botryosphaeria (Melogramma) 592
ficuum, Aspergillus 578
ficuum, Phlyctema 850
ficuum, Phyllachora (Trabutia) 862
fiedleri, Hendersonia 733
fiedleri, Leptosphaeria (Cryptosphaeria, Metasphaeria) 761
fijiensis, Mycosphaerella 802
filamentosa, Leptosphaeria 761
filamentosa, Pellicularia = **Thanatephorus cucumeris** 989
filamentosa, Phanerochaete (Peniophora) 843
filamentosa, Sphaeropsis 967
filamentosa f. sasakii, Pellicularia = **Thanatephorus cucumeris** 989
filamentosa f. solani, Pellicularia = **Thanatephorus cucumeris** 989
filamentosum, Ceratobasidium = **Thanatephorus cucumeris** 989
filamentosum, Corticium = **Phanerochaete filamentosa** 843
filamentosum, Cronartium = **C. coleosporioides** 649
filamentosum, Peridermium 832
filamentosus, Hypochnus = **Thanatephorus cucumeris** 989
filaris, Ramularia 925
filicina, Cryptomycina 652
filicina, Leptostromella 765
filicina, Nectria 812
filicina, Phlebiella (Xenasmatella) 849
filicina, Phomopsis 858
filicina, Phyllachora (Trabutiella) 862
filicina, Stictis 978
filicina, Taphrina 987
filicina, Typhula 1004
filicinum, Aphanobasidium = **Phlebiella filicina** 849
filicinum, (Corticium, Xenasma) = **Phlebiella filicina** 849
filicinum, Cryptosporium = **Leptostromella filicina** 765
filicinum, Herpobasidium 735
filicinum, Leptostroma = **Cryptomycina filicina** 652
filicinus, Rhopographus 934
filicum, Mycosphaerella 802
filiforme, Lemonniera 758
filiforme, Lophodermium = **Lirula macrospora** 769
filiformis, Cercospora (Cercosporella) 610
filiformis, Galerina 712
filiformis, Mycena 797
filiformis, Xylaria 1026
filipendulae, Cylindrosporium 658
filispora, Cercospora = **C. longispora** 612
filopes, Marasmiellus (Marasmius) 776
filum, Sphaerellopsis (Darluca, Phoma) 965
fimbriata, Asterina = **A. punctiformis** var. **fimbriata** 581
fimbriata, Ceratocystis (Ceratostomella, Endoconidiophora, Ophiostoma) 604
fimbriata, Entyloma 696
fimbriata, Mamiania (Gnomoniella) 776
fimbriata, Odontia = **Steccherinum fimbriatum** 975
fimbriata, Phyllosticta 868
fimbriata, (Poria, Stromatoscypha) = **Porotheleum fimbriatum** 890
fimbriata, (Lasiostictis, Stictis) = **Naemacyclus fimbriatus** 811
fimbriata, Tomentella = **Tomentellastrum badium** 995
fimbriata, Tremella 998
fimbriata, Venturia = **Lasiostemma fimbriatum** 757
fimbriata f. **platani, Ceratocystis** (Endoconidiophora) 604
fimbriatella, Junghuhnia (Chaetoporus, Poria) 751
fimbriatellus, Polyporus = **Junghuhnia fimbriatella** 751
fimbriatum, Gloeocystidiellum 716
fimbriatum, Lasiostemma 757
fimbriatum, Porotheleum 890
fimbriatum, Steccherinum 975
fimbriatus, Naemacyclus 811
fimbriatus, Wentiomyces = **Lasiostemma fimbriatum** 757
fimbristylidis, Puccinia 906
fimbristylis, Septoria 951
fimetaria, Stilbella 981
fimicola, Acrodictys 557
fimicola, Sordaria 963
findens, Hendersonia (Hendersonula) 733
fingens, Cercospora 610
firma, Fistulina = **Pseudofistulina radicata** 896
firma, Peniophora = **Phanerochaete carnosa** 843
fischeri, Anthracoidea (Cintractia) 566
fischeri, Aspergillus = **A. fischerianus** 578
fischeri, Urocystis 1008
fischerianus, Aspergillus 578
fissiliformis, Perenniporia (Poria) 830
fissilis, Tyromyces (Polyporus) 1005
fissus, Polyporus = **P. badius** 889
flabella, Dothidea 687
flabella, Phyllachora (Sphaeria) 862
flabelliforme, Xylocoremium 1026
flabelliformis, Lentinellus 759
flabellum, Catacauma = **Phyllachora flabella** 862
flabellum, Dacrymyces = **D. chrysospermus** 663
flaccida, Poria = **Ceriporia xylostromatoides** 621
flaccida, Puccinia = **P. abnormis** 900
flagellaris, Cercospora 610
flagellaris, Septoria = **S. convolvuli** 950
flagellata, Monochaetia = **M. kansensis** 795
flagellifera, Cercospora 610
flagellifera, Septoria 951
flagellifera, Uniseta (Neobarclaya, Pestalotia) 1006
flagelloidea, Bipolaris = **Alternaria flagelloideum** 561
flagelloideum, Alternaria (Helminthosporium) 561
flammans, Pholiota 851
flammea, Nectria (Sphaerostilbe) 812
flammea, Perrotia (Helotium, Lachnella) 834
flammea, Pleuroflammula 886
flammeus, Crepidotus = **Pleuroflammula flammea** 886
flammeus, Dasyscyphus = **Perrotia flammea** 834
flava, Lindtneria 768
flava, Taphrina 987
flaveriae, Puccinia = **P. melampodii** 910
flavescens, Botryobasidium (Hypochnus, Pellicularia) 591
flavescens, Corticium = **Athelia teutoburgensis** 583
flavescens, Crustoderma 650
flaviceps, Volvaria = **Volvariella bombycina** var. **flaviceps** 1023
flavida, Poria = **Antrodia radiculosa** 567
flavida, Trichosphaeria = **Niesslia pulchriseta** 815
flavido-alba, Phanerochaete (Peniophora) 844
flavidulum, Lachnum 754
flavidus, Gelatinodiscus 713
flavipes, Aspergillus 578
flavipes, Stilbella (Stilbum) 981
flavobrunnea, Patinella 827
flavofuligineum, Lachnum 754
flavofuligineus, Dasyscyphus = **Lachnum flavofuligineum** 754
flavolivens, Melanotus (Crepidotus) 784
flavomarginata, Coniophora = **Ramaricium flavomarginatum** 923
flavomarginata, Poria = **Phellinus ferrugineo-velutinus** 845
flavomarginatum, Ramaricium 923
flavorubra, Taphrina 987
flavovirens, Diatrype (Eutypa, Valsa) 673
flavovirens, Melanconis 782
flavovirens, Pseudotomentella (Tomentella) 898
flavum, Bactridium 585
flavus, Aspergillus 578
flavus, Dacrymyces = **D. stillatus** 663
flavus, Talaromyces 985
flavus var. oryzae, Aspergillus = **A. oryzae** 579
flectans, Taphrina 987
flexuosa, Uncinuliella (Uncinula) 1006
flexuosa, Volutella 1023
flexuosum, Helminthosporium = **Cladosporium herbarum** 630
flexuosum, Hysterographium 747
flexuosus, Crepidotus 647
flocciferum, Fusarium 707
floccosa, Grandinia (Hyphodontia) 722
flocculenta, Cytidia 661
floerkeae, Entyloma 696
floerkeae, Peronospora 833
floricola, Cercospora 610
florida, Phoma 853
floridae, Septoria 951
floridaensis, Flosculomyces 705
floridana, Calonectria = **C. kyotensis** 597
floridana, Campanella 600
floridana, Camptomeris 600
floridana, Hysterostomella 748
floridana, Lulworthia 773
floridana, Massarina 779
floridana, Ophiodothella 820
floridana, Ravenelia 928
floridana, Uredo 1007
floridana, Xylosphaeria = **Mycothyridium floridanum** 807
floridanum, Cylindrocladium = **C. scoparium** 657
floridanum, Helotium = **Hemimycena albida** 732
floridanum, Mycothyridium 807
floridanum, Myriangium = **M. citri** 808
floridanum, Peridermium = **P. guatemalense** 832
floridanum, Stemphylium = **S. lycopersici** 977
floridanum, Tomentellastrum = **T. badium** 995
floridanus, Monascus 794

floridanus, Polyporus = **Trichaptum sector** 999
floridense, Sphaceloma 964
floridensis, Corticium (Laeticorticium) 645
floridensis, Meliola 784
floriforme, Gymnosporangium 726
floriformis, Postia (Oligoporus, Polyporus, Tyromyces) 890
flosculosorum, Puccinia = **P. hieracii** 907
flowersii, Urocystis 1008
flumeanum, Pleurophragmium (Helminthosporium) 886
fluviatilis, Lophiosphaera 770
foeda, Harknessia = **H. thujina** 729
foedans, Pestalotia 835
foedans, Splanchnonema (Massaria, Sphaeria) 969
foeniculacea subsp. **lupina, Leptosphaeria** 761
foeniculi, Cercospora = **Cercosporidium punctum** 620
foetens, (Tilletia, Ustilago) = **Tilletia laevis** 992
foetida, Tilletia = **T. laevis** 992
foetidum, Micromphale 789
foetidus, Aspergillus 578
foetidus, Marasmius = **Micromphale foetidum** 789
foliacea, Tremella = **T. fimbriata** 998
foliicola, Acrospermum 557
foliicola, Ceuthospora (Cytospora) 622
foliicola, Ciborinia (Sclerotinia) 627
foliicola, Coryneum = **Seimatosporium lichenicola** 945
foliicola, Crinipellis 648
foliicola, Dimorphospora 679
foliicola, Fusicolla 712
foliicola, Gloeosporium = **Colletotrichum gloeosporioides** 638
foliicola, Lasiosphaeria 757
foliicola, Lophodermium 772
foliicola, Pseudomassaria 896
foliicola, Pyrenopeziza 918
foliicola, Seimatosporium (Hendersonia, Podisoma) 945
foliicola, Sphaeropsis (Sphaeria) 967
foliicola, Sporidesmium = **Stigmella effigurata** 979
foliicola, Stagonospora 974
foliorum, Botryosphaeria = **Discochora philoprina** 684
foliorum, (Coryneum, Hendersonia) = **Seimatosporium caudatum** 944
foliorum var. **hamamelidina, Hendersonia** 733
foliorum var. **viburni, Hendersonia** 733
folliculata var. **oxyspora, Leptosphaeria** 761
folliculatum, Sporidesmium (Helminthosporium) 970
fomentarius, Fomes 705
footii, (Capnodium, Microxiphium) = **Microxiphium fagi** 792
foraminulatum, Leptothyrium 766
formosa, Ostreola 823
formosa, Shearia = **S. fusa** 961
formosa, Venturia 1020
formosum, (Hysterium, Hysterographium) = **Ostreola formosa** 823
formosum, Stegonsporium = **Shearia fusa** 961
forsythiae, Alternaria 561
forsythiae, Phyllosticta 868
fouquieriae, Colletogloeum 637
fouquieriae, Hyphoderma 741
foveolaris, Cryptodiscus 651
foveolaris, Phomopsis (Sphaeria) 858
foveolata, Corynespora 646
foveolatum, Helminthosporium = **Corynespora foveolata** 646
fractipes, Abortiporus (Heteroporus, Polyporus) 555
fragariae, Colletotrichum = **C. gloeosporioides** 638
fragariae, Gnomonia 720
fragariae, Kabatia 751
fragariae, Marssonina (Ascochyta, Leptothyrium) 778
fragariae, Mycosphaerella (Sphaerella) 802
fragariae, Peronospora 833
fragariae, Phytophthora 878
fragariae, Ramularia 925
fragariae, Rhabdospora 930
fragariae, Rhizoctonia 932
fragariae, Septoria 951
fragariae, Sphaeronaemella = **Phomopsis obscurans** 859
fragariae, Synchytrium 984
fragariae, Zythia (Gloeosporium) 1027
fragariae f. fructicola, Gnomonia = **G. comari** 720
fragaricola, Phyllosticta 868
fragiforme, Hypoxylon 745
fragile, Dicranidion 675
fragilipes, Uromyces 1011
fragilis, Dentipellis 669
fragilis, Leocarpus 759
fragilis, Mucor 796
fragilis, Panus = **Lentinus strigosus** 759
fragilis, Postia (Oligoporus, Polyporus, Spongipellis, Tyromyces) 890
fragilis, Puccinia 906
fragilis var. **divaricatae, Puccinia** 906
fragilissimum, Sporidesmium (Helminthosporium) 970
fragillima, Mycena 797
fragillissima, (Hydnum, Mycoacia, Odontia) = **Phanerochaete chrysorhiza** 843
fragrans, Phellinus 845
fragrans, Polyporus = **Bjerkandera fumosa** 590
fragrans, Ravenelia 928
fragrans var. **evernia, Ravenelia** 928
frangulae, Diplodia 680
frangulae, Pezicula (Dermatea, Tympanis) 837
frangulae, Pleospora = **P. herbarum** 884
frankiana, Phyllosticta = **Greeneria uvicola** 724
franseriae, Puccinia 906
fraserae, Asteroma 581
fraserae, Cercosporella (Cercospora) 619
fraserae, Cylindrosporium 658
fraserae, Didymochaeta (Ascochyta, Diplodina) 677
fraserae, Hendersonia = **Sonderhenia eucalypticola** 962
fraserae, Marssonina (Marssonia) 778
fraserae, Phyllosticta 868
fraseri, Puccinia 906
fraseri, Urocystis 1008
fraternum, Gymnosporangium 726
fraternum, Poculum (Helotium) 887
fraternus, Hymenoscyphus = **Poculum fraternum** 887
fraxinata, Puccinia = **P. sparganioides** 914
fraxinea, Discula 685
fraxinea, Perenniporia 830
fraxinea, Ramularia 925
fraxinea, Sphaerella = **Mycosphaerella effigurata** 802
fraxineum, Gloeosporium = **Discula fraxinea** 685
fraxineus, Fomes = **Perenniporia fraxinea** 830
fraxini, Ascochyta = **A. syringae** 577
fraxini, Cucurbitaria 654
fraxini, Cylindrosporium (Cercosporella, Phloeochora, Phloeospora, Septoria) 659
fraxini, Diatrypella 674
fraxini, Dothiorella = **D. everhartii** 688
fraxini, Durandiella (Godronia) 691
fraxini, Eutypa = **E. lata** 699
fraxini, Hendersonulina (Hendersonia) 734
fraxini, Heterosporium 736
fraxini, Hysterographium 747
fraxini, Leptosphaeria 761
fraxini, Micropera 790
fraxini, Microsphaera = **M. syringae** 791
fraxini, Mycosphaerella (Sphaerella) 802
fraxini, Phloeosporina (Gloeosporium, Septogloeum) 850
fraxini, Piggotia (Asterostromella, Marssonia, Marssonina) 879
fraxini, Schizothyrella 939
fraxini, Sporonema 971
fraxini, Synchytrium 984
fraxini, Venturia 1020
fraxinicola, Dothiorella 688
fraxinicola, Laestadia (Depazea) 755
fraxinicola, Mycosphaerella 802
fraxinicola, Phyllosticta 868
fraxinicola, (Eutypella, Valsa) = **Eutypella juglandicola** 700
fraxinii, (Sphaerographium, Sphaeronaema) = **Cornicularia spina** 644
fraxinina, Pyrenochaeta 918
fraxinina, Valsa = **V. cypri** 1017
fraxinites, Cercospora 610
fraxinopennsylvanica, Ceratocystis (Calosphaeria) 604
fraxinophila, Perenniporia (Poria) 830
fraxinophila, Xeromphalina 1025
fraxinophilus, (Fomes, Polyporus) = **Perenniporia fraxinophila** 830
fraxinophilus f. ellisianus, Fomes = **Perenniporia ellisiana** 830
fremontiae, Ascochyta 575
frequentans, Penicillium = **P. glabrum** 828
freycinetiae, Mycosphaerella 802
freycinetiae, Myiocopron (Peltella) 808
freycinetiae, Phyllachora 862
friabilis, Diatrype (Diatrypella) 673
friesii, Microsphaera 790
friesii, Valsa 1018
frigidum, Gloeosporium = **Phlyctema vagabunda** 850
fritillariae, Phyllosticta 868
froelichiae, Albugo = **A. bliti** 559
frondicola, (Lachnella, Pirottaea) = **Dasyscyphus pteridis** 665
frondosa, Grifola 724
frondosa, Tremella = **T. fimbriata** 998
frondosus, Polyporus = **Grifola frondosa** 724
frostii, Diatrypella 674
fructicola, Gnomonia = **G. comari** 720
fructicola, Monilinia (Sclerotinia) 794
fructicola, Pleospora = **P. herbarum** 884
fructicum, Mycosphaerella 802
fructigena, Diplodina 682
fructigena, Monilinia (Sclerotinia) 794
fructigena, Septoria 951
fructigena, Trichoseptoria 1000
fructigenum, Fusamen 706
fructigenum, Gloeosporium = **Colletotrichum gloeosporioides** 638

fructigenum, Helotium = **Hymenoscyphus fructigenus** 740
fructigenus, Hymenoscyphus 740
fructuosa, Milesina (Milesia) 792
fructus, Colletotrichum = **C. dematium** 637
frumenti, Diplodia 680
frustulatum, Stereum = **Xylobolus frustulatus** 1026
frustulatum var. subpileatum, Stereum = **Xylobolus subpileatus** 1026
frustulatus, Xylobolus 1026
frustulosa, Penzigia 830
frutica, Phylloporia 864
fruticetorum, Acanthophysium (Aleurodiscus) 555
fruticola, Otthia (Parodiella) 823
fruticum, Polyporus = **Phylloporia frutica** 864
fuchsiae, Cercospora 610
fuchsiae, Didymella 677
fuchsiae, Diplodia 680
fuchsiae, Pucciniastrum = **P. epilobii** 916
fuciforme, Corticium = **Laetisaria fuciformis** 756
fuciformis, Laetisaria (Hypochnus, Phanerochaete) 756
fuciformis, Tremella 998
fuckeliana, Botryotinia (Sclerotinia) 594
fuckeliana, Nectria 812
fuckelii, Coniothyrium 642
fuckelii, Lepteutypa (Massaria) 759
fuckelii, Phaeosphaeria (Leptosphaeria) 842
fuckelii, Therrya 990
fuckelii, Torula 995
fugacissima, Exidiopsis (Sebacina) 702
fugax, Cytospora 661
fugax, Drechslera (Helminthosporium) 690
fugax, Merulius = **Leucogyrophana mollusca** 767
fuirenae, Cintractia 628
fuirenae, Puccinia 906
fujikuroi, Gibberella (Lisea) 714
fujikuroi var. subglutinans, Gibberella = **G. subglutinans** 715
fukushiana, Cercospora 610
fulgens, Pycnoporellus (Hydnum) 917
fulgens, Synchytrium 984
fulgida, Leptosphaeria 762
fulgidus, Ophiobolus = **Leptosphaeria fulgida** 762
fulgurata, Teichospora (Strickeria) 988
fuligena, Pseudocercospora (Cercospora) 894
fuliginea, Catenularia 601
fuliginea, Exidiopsis 702
fuliginea, Phoma 853
fuliginea, Sphaerotheca 968
fuliginea, Tomentella = **T. ramosissima** 994
fuliginella, Mycena 797
fuligineum, Coniothyrium 642
fuligineum, (Greeneria, Melanconium) = **Greeneria uvicola** 724
fuligineus, Hypochnus = **Tomentella ramosissima** 994
fuliginosa, Cercospora = **C. macclatchieana** 613
fuliginosa, Godronia (Scleroderris) 721
fuliginosa, Hymenochaete = **H. fusca** 739
fuliginosum, Ischnoderma = **I. resinosum** 750
fuliginosum, Septosporium = **Helminthosporium velutinum** 732
fuliginosus, Pluteus 887
fuliginosus, Polyporus = **Ischnoderma resinosum** 750
fuligniosa, Cercospora 610
fuligo, Grammothele 722
fullageri, Fomes (Polyporus) 705
fultum, Acrospermum = **Glyphium elatum** 720
fulva, Botrytis = **Chromelosporium fulvum** 626
fulva, Chaetopsina 624
fulva, Cyphella = **Merismodes fasciculatus** 787
fulva, Didymaria 676
fulva, Fulvia 706
fulva, Stictis = **Propolis rhodoleuca** 892
fulvella, Cercospora 610
fulvella, Poria = **Anomoporia bombycina** 565
fulvida, Poria = **Phellinus ferreus** 845
fulvidus, Panus = **Heliocybe sulcata** 731
fulvifibrillosus, Crepidotus = **C. nephrodes** 647
fulvipes, Xeromphalina (Collybia) 1025
fulvosquamosa, Pholiota = **Agaricus subrufescens** 559
fulvotingens, Ionomidotis (Cenangium) 749
fulvum, Chromelosporium 626
fulvum, Cladosporium = **Fulvia fulva** 706
fulvum, Gelatinosporium 713
fulvum, (Gymnosporium, Rhinotrichum) = **Allescheriella crocea** 560
fulvum, Sclerotium 942
fulvus, Acrostalagmus = **Verticillium tenerum** 1022
fulvus, Fomes = **Phellinus pomaceus** 846
fulvus, Panus = **Lentinus velutinus** 759
fumagineum, Cladosporium 629
fumago, Cladosporium = **Fumago salicina** 706
fumidiceps, Tyromyces (Polyporus) 1005
fumigata, Mollisia 793
fumigatum, Septobasidium 946
fumigatus, Aspergillus 578
fumosa, Bjerkandera 590
fumosa, Cercospora = **C. penzigii** 615
fumosa, Cyphella = **Cellypha musaecola** 602
fumosa, Pseudotomentella 899
fumosa, Septoria 951
fumosa, Troposporella 1001
fumosella, Phialea = **Pseudocenangium succineum** 893
fumosum, Helminthosporium (Brachysporium) 731
fumosum, Myxosporium 810
fumosus, Hypochnus = **Phlebiella vaga** 849
fumosus, Polyporus = **Bjerkandera fumosa** 590
funalis, Ustilago = **U. hypodytes** 1015
funerea, Pestalotiopsis (Pestalotia) 836
funerea var. camelliae, Pestalotia = **Pestalotiopsis maculans** 836
funerea var. multiseta, Pestalotia = **P. multiseta** 835
funicola, Chaetomium 623
funiculosum, Penicillium 828
furcata, Meliola 784
furfuracea, Encoelia 693
furfuraceum, Cenangium = **Encoelia furfuracea** 693
furfuraceum, Helotium 732
furfuraceum, Resinicium (Corticium) 929
furfurella, Cercospora 610
furva, Doassansiopsis (Doassansia) 686
furva, Valsella 1019
fusa, Shearia 961
fusariisporum, Exarmidium 702
fusariisporus, Rhopographus = **Exarmidium fusariisporum** 702
fusarioides, Calloria = **C. neglecta** 597
fusarioides, Colletotrichum (Gloeosporium) 637
fusarioides, (Dactylium, Fusarium) = **Fusarium chlamydosporum** 707
fusarioideum, Pseudofusarium = **Fusarium pallidoroseum** 710
fusca, Botrytis (Polyactis) 594
fusca, Cercospora = **Sirosporium diffusum** 962
fusca, Conoplea (Streptothrix) 643
fusca, Hymenochaete 739
fusca, Peniophora = **Hyphoderma argillaceum** 740
fusca, Physalospora = **P. abdita** 876
fusca, Protomyces = **Plasmopara pygmaea** var. **fusca** 882
fusca, Septoria 951
fusca, Serpula = **Cerocorticium fuscum** 622
fusca, Stemonitis 976
fusca, Stictis 978
fusca, Stilbella 981
fusca, Tapesia (Mollisia, Phialea) 986
fusca, Taphrina 987
fusca, Tilletia = **T. bromi** 991
fusca, Tomentella = **T. ferruginea** 994
fusca, Tryblidiella = **Rhytidhysteron rufulum** 935
fusca, Veluticeps = **V. berkeleyi** 1020
fusca, Virgariella (Haplaria, Virgaria) 1022
fuscans, Sphaerella = **Mycosphaerella tulipifera** 807
fuscatum, Cladosporium = **C. herbarum** 630
fuscescens, Hydnoporia = **Hydnochaete olivacea** 738
fuscescens, Lachnum (Dasysyphus) 754
fuscidulum, Coniothyrium 642
fuscidulus, Dasyscyphus = **Lachnum fuscescens** 754
fuscispora, Hendersonia 733
fuscoatra, Humicola 738
fuscoatra, Mycoacia (Odontia) 798
fuscoatra, Patellaria = **Lecanidion simile** 758
fuscoatra, Pyrenopeziza = **Heterosphaeria patella** 735
fuscoatrum, Hydnum = **Mycoacia fuscoatra** 798
fuscoatrum, Lecanidion = **L. simile** 758
fuscoatrum, Steccherinum = **Mycoacia fuscoatra** 798
fuscoferruginosa, Tomentella 994
fuscofloccosum, Lachnum 754
fuscomaculans, Aposphaeria (Plenodomus) 571
fuscomarginata, Peniophora (Phanerochaete) 829
fuscomutans, Polyporus = **Tyromyces fissilis** 1005
fusco-ocula, Mycena 797
fuscopes, Helicomyces = **H. roseus** 731
fuscopurpureum, Hypoxylon = **H. rubiginosum** 746
fuscosanguinea, Lachnellula 754
fuscosanguineus, Dasyscyphus = **Lachnellula fuscosanguinea** 754
fuscosporum, Melogramma = **M. campylosporum** 786
fuscosquamulosus, Gymnopilus 725
fuscostrata, Athelia = **Confertobasidium olivaceo-album** 641
fuscostratum, Corticium = **Confertobasidium olivaceo-album** 641
fuscosuccinea, Auricularia 585
fuscoviolacea, Tulasnella 1004
fuscoviolaceum, Trichaptum 998

fuscoviolaceus, (Hirschioporus, Irpex) = **Trichaptum fuscoviolaceum** 998
fusco-virens, Cercospora 610
fuscum, Atractium = **Stilbella fusca** 981
fuscum, Cerocorticium 622
fuscum, Colletotrichum 637
fuscum, Crocicreas (Belonidium) 649
fuscum, Entyloma 696
fuscum, Gymnosporangium 726
fuscum, Helminthosporium = **Dendryphiella vinosum** 668
fuscum, Hypoxylon 745
fuscum, Stereum = **Laxitextum bicolor** 757
fuscus, Merulius = **Cerocorticium fuscum** 622
fuscus, Polyporus = **Ischnoderma resinosum** 750
fuscus, Radulomyces = **Cerocorticium fuscum** 622
fusicarpa, Phyllachora 862
fusiforme, Cronartium = **Cronartium quercuum** f. sp. **fusiforme** 650
fusiforme, Peridermium 832
fusiforme, Phragmidium 860
fusiforme var. **novi-boreale, Phragmidium** 861
fusiformis, Rhizopus 934
fusiger, Hysterium = **H. insidens** 747
fusimaculans, Phaeoramularia (Cercospora) 840
fusispora, Coniophora 641
fusispora, Diplodina 682
fusispora, Lepteutypa 759
fusispora, Leptosphaeria 762
fusispora, Phoma 853
fusispora, Phyllosticta 868
fusisporum, Mytilidion = **Mytilinidion gemmigenum** 809
fusisporum, Corticium = **Coniophora fusispora** 641
fusisporum, Lophium = **Mytilinidion gemmigenum** 809
fusisporum, Uthatobasidium 1017
fusisporus var. **abietinus, Crepidotus** 647
fusus, Sporidesmium = **Shearia fusa** 961
futilis, Didymosphaeria 678
gaeumannii, Phaeocryptopus (Adelopus) 839
gaillardiae, Septoria 951
galactiae, Cercospora 610
galactina, Guignardia (Laestadia) 725
galactinum, Scytinostroma (Corticium) 944
galactinus, Tyromyces (Polyporus) 1005
galactis, Dimerosporium = **Dimerium langloisii** 679
galactis, Phyllosticta (Phoma) 868
galanthina, Botrytis (Polyactis) 594
galeopsidis, Erysiphe 698
galeopsidis, Septoria 951
galericulata, Cryptodiaporthe 651
galericulata, Mycena 798
galii, Cercosporidium (Cercospora) 620
galii, Marssonia 777
galii, Mazzantia (Physalospora) 780
galii, Melasmia 784
galii, Nectria 812
galii, Placosphaeria 881
galii, Pucciniastrum = **P. guttatum** 916
galii-californici, Uromyces 1011
galiorum, Leptosphaeria (Metasphaeria, Sphaeria) 762
galiorum, Rhabdospora (Septoria) 930
galiorum var. **gnaphaliana, Leptosphaeria** 762
gallae, Botryodiplodia (Dothiorella, Sphaeropsis) 591
gallarum, Basisporium = **Khuskia oryzae** 752
gallarum, Septoria 951
gallica, Coriolopsis (Funalia) 644
gallica, Cristinia 648
gallicola, Botryodiplodia (Macrophoma) 591
gallicola, Phyllosticta 868
galligena, Nectria 812
gallorum, Phyllosticta 868
gallula, Puccinia 906
galzinii, Bourdotia 595
galzinii, Leptosporomyces (Athelia, Corticium) 764
galzinii, Tomentella 994
gambellii, Dictyochora 675
gangliformis, Peronospora = **Bremia lactucae** 595
ganjae, Phomopsis 858
gardeniae, Mycosphaerella (Sphaerella) 802
gardeniae, Phomopsis 858
gardeniorum, Conidioxyphium 641
gardoquiae, Asterella = **Asterina ildefonsiae** 581
garrettiana, Ascochyta 575
garrettianum, Asteroma 581
garrettii, Phyllosticta 868
garryae, Cercospora 610
garryae, Dothichiza 686
garryae, Janetia (Sporidesmium) 750
garryae, Lambertella (Phaeociboria) 756
garryae, Meliola 784
garryae, Phyllosticta 868
gastrinum, Lopadostoma (Anthostoma) 769
gaudefroyi, Pleospora 884
gaultheri, Mycena 798
gaultheriae, (Asterina, Epipeltis, Schizothyrium) = **Schizothyrium pomi** 939
gaultheriae, Cercospora 610
gaultheriae, Coccomyces (Phacidium) 634
gaultheriae, Lachnum (Dasyscyphus, Lachnella) 755
gaultheriae, Leptosphaeria 762
gaultheriae, Mollisia 793
gaultheriae, Mycosphaerella 802
gaultheriae, Phyllosticta 868
gaultheriae f. shallon, Phyllosticta = **Phyllosticta gaultheriae** 868
gaurae, Cercospora 610
gaurae, Pestalotia = **Pestalotiopsis uvicola** 836
gaurina, Septoria 951
gausapatum, Stereum (Haematostereum) 977
gayophyti, Phaeoramularia (Cercospora, Ramularia) 840
gayophyti, Puccinia = **P. pulverulenta** 912
gayophyti, Ustilago 1015
gei, Cylindrosporium 659
gei, Ramularia (Cercospora, Ovularia, Ramularia) 925
gei, Septoria 951
gei, Urocystis = **Ustacystis waldsteiniae** 1014
gelatinosa, Calloriopsis (Coryne) 597
gelatinosa, Ciliospora = **Chaetospermum carneum** 624
gelatinosa, Exidia = **E. recisa** 702
gelatinosa, Hypocrea (Chromocrea) 743
gelatinosa, Pododimeria 888
gelatinospora, Phomatospora 857
gelatinosum, Hormiscium 737
gelatinosum, Pseudohydnum (Tremellodon) 896
gelatinosus, Creopus = **Hypocrea gelatinosa** 743
gelsemii, Phyllosticta 868
gemella, Puccinia 906
gemmata, Valsaria 1018
gemmatum, Hypoxylon = **Valsaria gemmata** 1018
gemmigenum, Mytilinidion 809
geniculata, Conoplea (Botrytis) 643
geniculata, Curvularia 655
geniculata, Strasseria (Sphaeropsis) 981
geniculata, Uncinula 1006
geniculatum, Helminthosporium = **Curvularia geniculata** 655
genista, Phoma 853
gentianae, Cercospora 610
gentianae, Mycosphaerella 802
gentianae, Phoma 853
gentianae, Puccinia 906
gentianae, Septoria = **S. gentianoides** 951
gentianae, Uromyces = **U. eugentianae** 1010
gentianae, Wettsteinina (Sphaerulina) 1023
gentianaecola, Leptothyrium (Depazea) 766
gentianicola, Cercospora 610
gentianicola, Phyllosticta = **Asteromella andrewsii** 582
gentianoides, Septoria 951
geogenium, Hypochnicium (Corticium) 743
geographicum, Xyloma = **Leptothyrium castaneae** 765
georgiana, Dicarpella (Gnomoniella) 674
georgius-fischeri, Colletotrichum 637
georgius-fischeri, Leptosphaeria 762
geotropus, Fomes = **Rigidoporus ulmarius** 936
geranii, Cercospora 610
geranii, Cercosporella 619
geranii, Cylindrosporium 659
geranii, (Dilophospora, Neottiospora) = **Pestalozziella subsessilis** 837
geranii, Phyllosticta 868
geranii, Physalospora 876
geranii, Plasmopara (Peronospora) 882
geranii, Ramularia 925
geranii, Seaverinia (Sclerotinia) 944
geranii, (Hormotheca, Sphaeria) = **Hormotheca robertiani** 737
geranii, Synchytrium 984
geranii, Uromyces 1011
geranii, Venturia (Coleroa, Dothidea, Stigmatea) 1020
geranii-pusilli, Pestalozziella = **P. subsessilis** 837
geranii-silvatici, Puccinia 906
gerardiae, Cercospora 610
gerardiae, Erysiphe = **E. galeopsidis** 698
gerardiana, Gloniopsis = **G. curvata** 719
gerberae, Alternaria 561
gerberae, Ascochyta = **A. doronici** 574
gerberae, Cercospora 611
gerhardtii, Physoderma 877
germanica, Valsa 1018
germinale, Gymnosporangium = **G. clavipes** 726
ghaesembillae, Phyllosticta 868
gibberulosa, Puccinia 906
gibbosa, Brunchorstia = **Foveostroma boycei** 706
gibbum, Zygosporium (Clasterosporium) 1027
gigantasca, Pleospora = **P. longispora** 885
gigantea, Cercospora 611
gigantea, Drechslera 690
gigantea, Phanerochaete (Peniophora, Phlebia, Phlebiopsis) 844
gigantea, Puccinia 906
gigantea, Stagonospora 974
giganteum, Helminthosporium = **Drechslera gigantea** 690

giganteum, (Ceratopodium, Graphium, Stilbum) = **Crinula caliciiformis** 648
giganteus, Uromyces 1011
gigantispora, Puccinia 906
gigantissima, Cintractia = **Anthracoidea limosa** var. **gigantissima** 566
gigaspora, Leptosphaeria = **Wettsteinina niesslii** 1023
gigaspora, Massaria = **M. inquinans** 779
gigaspora, (Gremmenia, Phragmonaevia) = **Phacidium infestans** 839
gigaspora, Pleospora 884
giliae, Peronospora = **P. phlogina** 834
giliae, Puccinia 906
giliae, Ramularia 925
giliae, Urocystis 1008
giliicola, Puccinia 906
gilletiana, Septoria 951
gillii, Cylindrocarpon (Septogloeum) 657
gilvescens, Ceriporiopsis (Poria) 621
gilvidula, Peniophora = **Phanerochaete affinis** 843
gilvoides, Trametes = **Phellinus viticola** 846
gilvus, Phellinus (Hapalopilus, Polyporus) 845
ginkgo, Phyllosticta 868
glabella, Puccinia 906
glabra, Isthmospora 750
glabra, Septoria 951
glabra, Uredinopsis 1007
glabrum, Penicillium 828
gladioli, Botrytis 594
gladioli, Curvularia 655
gladioli, Macrophoma 774
gladioli, Penicillium 828
gladioli, Phyllosticta 868
gladioli, Septoria 951
gladioli, Stromatinia (Sclerotinia) 982
gladiolicola, Urocystis 1008
gladiolorum, Botrytis 594
glaira, Exidiopsis 702
glandicola, Cryptostictis = **Discosia artocreas** 684
glandicola, Pestalotiopsis (Robillarda) 836
glandicola, Phomopsis (Phoma) 858
glandicola, Physalospora = **Botryosphaeria quercuum** 593
glandiformis, Rosellinia 937
glandulosa, Cercospora 611
glandulosa, Dothiorella (Sphaeropsis) 688
glandulosa, Eutypella (Valsa) 700
glandulosa, Exidia 702
glandulosum, Phoma 853
glauca, Monodictys 795
glauca, Mycosphaerella (Sphaerella) 802
glauca, Phyllosticta 868
glauca, Propolis = **Propolidium glaucum** 892
glauca, Ramularia 925
glauca, Streptothrix = **Basidiobotrys pallida** 586
glaucescens, Cercospora = **C. tuberosa** 618
glaucescens, Laestadia (Sphaerella) 755
glaucoma, Stictis = **Schizoxylon ligustri** 940
glaucum, (Coremium, Penicillium) = **Penicillium expansum** 828
glaucum, (Epochnium, Stemphylium) = **Monodictys glauca** 795
glaucum, Propolidium 892
glaucus, Aspergillus 578
glebulosa, Peniophora = **Tubulicrinis glebulosus** 1003
glebulosus, Tubulicrinis 1003
gleditschiae, Aplosporella = **A. gleditschiicola** 570
gleditschiae, Dothiorella 688
gleditschiae, Libertella 768
gleditschiae, Linospora 769
gleditschiae, Melasmia = **M. hypophylla** 784
gleditschiae, Sphaeropsis 967
gleditschiicola, Aplosporella (Sphaeropsis) 570
gleditsiae, Botryosphaeria = **Cucurbitaria recuperata** 654
gleditsiae, Cytospora = **C. sacculus** 662
gleditsiae, Kaskaskia = **Gyrostroma austroamericana** 727
gleicheniae, Helminthosporium 731
gleicheniae, (Dasyscyphus, Lachnella, Lachnum) = **Lachnum flavidulum** 754
gleicheniae, Leptothyrium 766
gliocladioides, Nectria = **N. ochroleuca** 813
globifera, Cytospora 661
globifera, Ovularia = **Hadrotrichum globiferum** 727
globifera, Phyllosticta 868
globifera, Seuratia (Asterina, Atichia) 960
globiferum, Euthrypton = **Seuratia globifera** 960
globiferum, Hadrotrichum 727
globigera, Phyllosticta 868
globosa, Anungitea 568
globosa, Conoplea (Streptothrix) 644
globosa, Harknessia 729
globosa, Penzigia = **Xylaria anisopleura** 1025
globosa, Plectrothrix = **Olpitrichum macrosporum** 819
globosipes, Puccinia 906
globosum, Chaetomium 623
globosum, Gymnosporangium 726
globosum, Hypoxylon = **Xylaria anisopleura** 1025
globosum, Melanconium 782
globosum, Synchytrium 984
globulifera, Peniophora = **Phanerochaete gigantea** 844
globuliferum, Sclerotium 942
globuliferum, Stemphylium (Macrosporium) 976
globuligerum, Tolyposporium = **Moesziomyces bullatus** 792
globuligerus, Moesziomyces = **M. bullatus** 792
gloeodes, Marssonina (Marssonia) 778
gloeospora, Massarina (Leptosphaeria, Trichometasphaeria) 779
gloeosporioides, Actinothyrium = **Tubakia dryina** 1002
gloeosporioides, Cladosporium 630
gloeosporioides, Colletotrichum 638
gloeosporioides f. sp. **aeschynomene, Colletotrichum**
gloeosporus, Dicellomyces 675
glomerata, Phoma 853
glomerata, Podosporiella (Cercospora) 888
glomerata, Puccinia 907
glomerata, (Spilosticta, Venturia) = **Coleroa circinans** 636
glomeratus, Inonotus (Polyporus) 748
glomerula, Mucor = **Actinomucor elegans** 558
glomerulisporum, Stilbum = **Dictyocatenulata alba** 675
glomerulosa, Atichia (Synalissa) 583
glomerulosa, Stigmina 980
glomerulosum, Collema = **Atichia glomerulosa** 583
glomerulosum, Exosporium = **Stigmina glomerulosa** 980
gloniospora, Lophiosphaera 770
glottidii, Gloeosporium = **Colletotrichum gloeosporioides** 638
glottidiicola, Cercospora 611
glumarum, Phoma (Phyllosticta) 853
glumarum, Puccinia = **P. striiformis** 914
glumicola, (Phoma, Phyllosticta) = **Phoma sorghina** 856
glutinis, Rhodotorula 934
glutinis var. rubescens, Rhodotorula = **Rhodotorula glutinis** 934
glutinosa, Torula 995
glutinosum, Metarhizium = **Myrothecium verrucaria** 809
glyceriae, Cylindrosporium 659
glyceriae, Uromyces = **U. amphidymus** 1009
glycericola, Septoria 951
glycericola, Stagonospora 974
glycinea, Phyllosticta = **P. sojicola** 873
glycines, Colletotrichum = **C. truncatum** 639
glycines, Glomerella 719
glycines, Phomopsis = **P. phaseoli** 859
glycines, Pseudocercospora (Cercospora) 894
glycines, Septoria 951
glycosmae, Mycosphaerella 802
glycyrrhizae, Cylindrosporium 659
glycyrrhizae, Septoria 951
glycyrrhizae, Uromyces 1011
gnaphaliaceae, Phaeoisariopsis (Cercospora) 839
gnaphaliatum, Fusicladium 711
gnaphalicola, Cylindrosporium 659
gnaphalii, Cercospora 611
gnaphalii, Puccinia = **P. gnaphaliicola** 907
gnaphalii var. **lanaria, Ophiobolus** 820
gnaphaliicola, Puccinia 907
gnomon, Gnomonia (Gnomoniella) 720
godronioides, Sarcosoma (Bulgaria) 938
goeppertiana, Calyptospora = **Pucciniastrum goeppertianum** 916
goeppertianum, Pucciniastrum 916
goldbachii, Cellypha (Cyphella) 601
gomezii, Vararia 1019
gomphrenae, Alternaria 561
gomphrenae, Cercospora 611
goniostoma, Eutypella (Eutypella) 700
gonolobi, Cercospora 611
gonolobi, Plasmopara (Peronospora) 882
gonolobi, Synchytrium 984
gonytrichodes, Codinaeopsis (Codinaea) 635
gooddingii, Ravenelia = **R. texensis** 929
goodyerae, Uredo (Pucciniastrum) 1007
gordoniae, Cytospora 661
gordoniae, Mycosphaerella (Sphaerella) 802
gordoniae, Phyllosticta 868
gossypii, Alternaria (Macrosporium) 561
gossypii, Ascochyta 575
gossypii, Ashbya (Nematospora) 578
gossypii, Botryodiplodia = **Lasiodiplodia theobromae** 756
gossypii, Colletotrichum 639
gossypii, Glomerella 719
gossypii, Helminthosporium 731
gossypii, Hyponectria 744
gossypii, Phakopsora (Doassansia, Kuehneola, Phakopsora) 843
gossypii, Phlyctema 850
gossypii, Phoma 853
gossypii, Ramularia (Cercosporella, Septocylindrium) 925
gossypina, Antrodia (Poria) 567
gossypina, Ceratocystis = **Ophiostoma stenoceras** 822
gossypina, Cercospora 611
gossypina, Diplodia = **Lasiodiplodia theobromae** 756

gossypina, Mycosphaerella (Sphaerella) 802
gossypina, Pestalozziella 837
gossypina, Phyllosticta 868
gossypina, Septoria 951
gossypina, Valsa = **V. ceratosperma** 1017
gossypina var. robusta, Ceratocystis = **Ophiostoma stenoceras** 822
gossypinum, Ophiostoma = **O. stenoceras** 822
gouldiae, Asterina 581
gouldiae, Dendrophoma 667
gouldiae, Enthallopycnidium 695
gouldiae, Meliola 784
gouldiae, Pluriporus = **Dothidella rubiacearum** 687
gouldiae, Septoria 951
gouldiae, Sphaeropsis 967
gracile, Graphium 723
gracile, Gymnosporangium 726
gracile, (Brachysporium, Helminthosporium, Heterosporium) = **Cladosporium iridis** 630
gracile, Lophodermium = **L. orbiculare** 772
gracile, Melanconium 782
gracile, Peridermium = **Puccinia subnitens** 914
gracile, Phacidium = **Phacidina gracilis** 838
gracile, Phragmidium = **P. rubi-idaei** 861
gracile, Pythium = **P. monospermum** 921
gracilipes, Ciborinia (Sclerotinia) 627
gracilipes, Nectria (Sphaerostilbe) 812
gracilipes, Ramularia 925
gracilis, Ciborinia (Sclerotinia) 627
gracilis, Pestalotia 835
gracilis, Phacidina 838
gracilis, Ravenelia 928
gracilis, Rhizoctonia = **R. sphacelati** 933
gracilis, Taphrina 987
gracillima, Peniophora = **Tubulicrinis glebulosus** 1003
gracillimus, Tubulicrinis = **T. glebulosus** 1003
graebnerianum, Aecidium 558
graminea, Ascochyta (Diplodia, Diplodina) 575
graminea, Drechslera 690
graminea, Herpotrichia 735
graminearum, Fusarium 707
graminearum, Phloeospora 849
graminella, Neophoma (Macrophoma, Phoma) 815
graminella, Puccinia 907
graminella, Stagonospora = **S. montagnei** 974
gramineum, Cephalosporium = **Hymenula cerealis** 740
gramineum, Coniosporium 641
gramineum, Crocicreas 649
gramineum, Helminthosporium = **Drechslera graminea** 690
gramineum, Lophodermium (Hysterium) 772
gramineum, Myrothecium 809
gramineum, Napicladium = **Cercosporidium graminis** 620
graminicola, Amastigosporium = **Mastigosporium rubricosum** 780
graminicola, Ascochyta 575
graminicola, Asterina (Asterella) 581
graminicola, Calonectria = **Monographella nivalis** 795
graminicola, Camarosporium = **Amarenographium metableticum** 563
graminicola, Cercospora = **Cercosporidium graminis** 620
graminicola, Colletotrichum (Vermicularia) 639
graminicola, Didymosphaeria (Microthelia) 678
graminicola, Glomerella 719
graminicola, Phialophora 847
graminicola, Pythium 920
graminicola, Ramularia 925
graminicola, Sclerospora (Peronospora) 941
graminicola, Stigmella = **Cerebella andropogonis** 621
graminicola, Stigmopeltis 980
graminicola, Synchytrium 984
graminicola, Tyromyces = **Amylosporus campbellii** 564
graminicola, Uromyces 1011
graminicola var. brachypodii, Ascochyta = **A. brachypodii** 574
graminicola var. diedickeana, Ascochyta = **A. leptospora** 575
graminicola var. **holci, Ascochyta** 575
graminicolor, Callistosporium = **C. luteo-olivaceum** 597
graminis, Ascospora 578
graminis, Asteroma 581
graminis, Cercosporidium (Cercospora, Passalora, Scolicotrichum) 620
graminis, Dilophospora = **D. alopecuri** 679
graminis, Erysiphe (Blumeria) 698
graminis, Gaeumannomyces (Ophiobolus) 712
graminis, (Hendersonia, Wojnowicia) = **Wojnowicia hirta** 1024
graminis, (Graphyllium, Hysterographium) = **Graphyllium hysterioides** 724
graminis, Leptothyrella 765
graminis, Phoma 853
graminis, Phyllachora 862
graminis, Physoderma (Cladochytrium) 877
graminis, Placosphaeria = **Cheilaria agrostis** 625
graminis, Polymyxa 889
graminis, Puccinia 907
graminis, Pyrenochaeta 918
graminis, Rhizophydium 933
graminis, Rutola (Torula) 937
graminis, Stemphylium (Soredospora) 977
graminis var. agropyri-repentis, Puccinia = **Puccinia graminis** 907
graminis var. **avenae, Gaeumannomyces** (Ophiobolus) 712
graminis var. cynodontis, Phyllachora = **P. cynodontis** 862
graminis subsp. **graminicola, Puccinia** 907
graminis var. stakmanii, Puccinia = **Puccinia graminis** 907
graminis var. **tritici, Gaeumannomyces** 712
graminis var. tritici, Puccinia = **Puccinia graminis** 907
graminis f. sp. **avenae, Erysiphe** 698
graminis f. sp. avenae, Puccinia = **Puccinia graminis** 907
graminis f. sp. **hordei, Erysiphe** 698
graminis f. sp. phlei-pratensis, Puccinia = **Puccinia graminis** subsp. **graminicola** 907
graminis f. sp. poae, Puccinia = **Puccinia graminis** 907
graminis f. sp. **secalis, Erysiphe** 698
graminis f. sp. secalis, Puccinia = **Puccinia graminis** 907
graminis f. sp. **tritici, Erysiphe** 698
graminis f. sp. tritici, Puccinia = **Puccinia graminis** 907
graminum, Acrospermum 557
graminum, Alternaria (Macrosporium) 561
graminum, Cladosporium = **C. herbarum** 630
graminum, Dactylaria (Dactylium) 663
graminum, Dinemasporium = **D. strigosum** 679
graminum, Discula (Gloeosporium) 685
graminum, Fusarium = **F. heterosporum** 708
graminum, Marasmius 777
graminum, Septoria 951
graminum, Sphaerella 965
graminum, Stagonospora 974
graminum f. **caricicola, Cladosporium** 630
graminum var. **strigosulum, Dinemasporium** 679
gramma, Phoma (Sphaeria) 853
grammicospora, Nectria 812
grammodes, Parodiella = **P. hedysari** 826
gramniella, Hendersonia = **Stagonospora montagnei** 974
granati, Coniella 641
grandifoliae, Ophiostoma (Ceratocystis) 821
grandinea, Sphaeropsis = **Aplosporella clintonii** 570
grandineum, Loculohypoxylon (Hypoxylon) 769
grandinioides, Basidiodendron (Bourdotia, Sebacina) 586
grandis, Cytospora = **C. sacculus** 662
grandis, Daldinia 664
grandis, Poria = **Anomoporia albolutescens** 565
grandis, Ustilago 1015
grandispora, Lulworthia 773
grandispora var. apiculata, Lulworthia = **Lulworthia grandispora** 773
grandisporangium, Pythium 921
grandisporum, Trichomerium (Capnodium) 1000
granigena, Gloeotinia (Phialea) 718
grantii, Aleurodiscus 560
grantii, Hendersonia 733
grantii, Ramularia = **R. archangelicae** 923
granulata, (Mycosphaerella, Sphaerella) = **Mycosphaerella oenotherae** 804
granulatum, Corticium = **Phlebia subochracea** 848
granulatum, Epicoccum = **E. nigrum** 696
granulatum, Penicillium 828
granuliformis, Cercospora 611
granulispora, Puccinia 907
granulosa, Amphisphaeria 564
granulosa, Dothichloe = **Balansia strangulans** f. **discoidea** 586
granulosa, Grandinia 722
granulosa, Pholiota (Flammula) 851
granulosa, Torula 995
granulosa, Urocystis 1008
granulosum, Fusarium = **F. avenaceum** 707
granulosum, Sirodesmium = **Coniosporium olivaceum** 642
granulosum, Sorosporium = **Urocystis fraseri** 1008
granulosus, (Hypochnus, Zygodesmus) = **Tomentella punicea** 994
graphidea, Botryosphaeria 592
graphideum, Melogramma = **Botryosphaeria graphidea** 592
graphioides, (Ceratostoma, Chaetoceratostoma) = **Arxiomyces vitis** 573
graphioides, Cercospora 611
graphis, Clithris 633
gratiolae, Entyloma 696
graveolens, Globifomes (Polyporus) 715
graveolens, Pholiota (Flammula) 851
gravidus, Protomyces 892
grayana, Tubercularia 1002
greenei, Ascochyta 575

greenei, Mycosphaerella 802
gregaria, Dothiorella 688
gregaria, Hyponectria 744
gregaria, Monilinia (Sclerotinia) 794
gregaria, Physalospora 876
gregaria, Teichospora (Strickeria) 988
gregarium, Nodulisporium (Stachylidium) 816
gregata, Phialophora 847
gregatum, Cephalosporium = **Phialophora gregata** 847
gregoriana, Meliola 784
grenadense var. **macrospora, Hypoxylon** 745
grenadensis, Phellinus (Pyropolyporus) 845
greschikii, (Corticium, Fibricium, Peniophora) = **Fibricium rude** 704
grevillei, Cistella (Dasyscyphus) 628
grevillii, Nitschkia (Melanopsamma) 816
grewiae, (Hansfordia, Verticicladium) = **Dicyma pulvinata** 676
griffithsii, Parodiella = **P. hedysari** 826
grindeliae, Cercospora 611
grindeliae, Puccinia 907
grindeliae, Ramularia 925
grindeliae, Septoria 952
grindeliae subsp. **riddellii, Puccinia** 907
grindeliicola, Septoria 952
grisea, Athelia = **Leptosporomyces galzinii** 764
grisea, Cercospora 611
grisea, Chaetopsis 624
grisea, Exidiopsis 702
grisea, Haplaria = **Botrytis cinerea** 594
grisea, Marssoniella = **Scopulariopsis brumptii** 943
grisea, Peronospora 833
grisea, Phyllosticta 868
grisea, Pyricularia 919
grisea, Septomyxa 947
grisea, Valsa = **V. ambiens** subsp. **leucostomoides** 1017
griseo-cana, Dendrothele 668
griseo-canum, Aleurocorticium = **Dendrothele griseo-cana** 668
griseo-canus, Aleurodiscus = **Dendrothele griseo-cana** 668
griseofulvum, Penicillium 828
griseofulvum, Septonema 947
griseofuscus, Irpex = **Fuscocerrena portoricensis** 711
griseola, Phaeoisariopsis (Cercospora, Isariopsis) 839
griseolazulina, Nidularia 815
griseopergamacea, Pseudotomentella 899
griseo-umbrina, Tomentella 994
griseoveneta, Pseudotomentella 899
griseoviolacea, Tomentella 994
griseoviridis var. **cascadensis, Mycena** 798
griseum, Cenangium 602
griseum, Colletotrichum = **C. gloeosporioides** 638
griseum, Cylindrosporium 659
griseum, Fusidium 712
griseum, Microdochium 789
griseum, Oidiodendron 818
griseum, Trichothecium = **Pyricularia grisea** 919
grohii, Claviceps 632
grossulariae, Ascochyta (Diplodina, Hendersonia) 575
grossulariae, Cytospora 661
grossulariae, Godronia 721
grossulariae, Marssonina (Marssonia) 778
grossulariae, Microsphaera 790
grossulariae, Mycosphaerella = **M. ribis** 805
grossulariae, Phyllosticta 869
grossulariae, Puccinia = **P. caricina** 902
grossulariae, Septoria = **S. ribis** 956
grossulariae var. **salicella, Sphaerella** 965
grovei, Spadicoides 963
grumata, Didymosphaeria 678
grumosa, Puccinia 907
grumosum, Cladosporium 630
gryllii, Septoria 952
guacae, Uredo = **U. nigropuncta** 1008
guanicense, Cercosporidium (Cladosporium) 620
guaranitica, Nummularia 817
guaraniticum, Entyloma = **E. polysporum** 696
guatemalense, Peridermium 832
guaynabensis, Uredo 1007
gudauskasii, Curvularia 655
guepini, (Pestalotia, Pestalotiopsis) = **Pestalotiopsis maculans** 836
guepini var. **macrotricha, Pestalotiopsis** 836
guepini var. vaccinii, Pestalotia = **P. vaccinii** 836
guestphalica, Niptera 816
guestphalicum, Belonopsis = **Niptera guestphalica** 816
guianensis, Phyllachora = **P. punctum** 863
guillemineae, Puccinia 907
guilliermondii, Candida (Endomyces) 600
gulosa, Puccinia 907
gunnerae, Harknessia 729
gutierreziae, Rhabdospora 930
guttata, Phyllactinia 864
guttatum, Cylindrosporium 659
guttatum, Pucciniastrum 916
guttifera, Aplosporella = **A. tiliacea** 571
guttulata, Cercospora 611
guttulata, Petriella 837
guttulata, Postia 890
guttulatae, Phyllosticta 869
guttulatus, (Oligoporus, Polyporus, Tyromyces) = **Postia guttulata** 890
guttulifera, Peniophora = **Hyphoderma guttuliferum** 741
guttuliferum, Hyphoderma 741
guyotiana, Tilletia = **T. bromi** 991
gymnandrae, Puccinia 907
gymnocladi, Cercospora 611
gymnocladi, Phyllosticta 869
gyneriicola, Phoma 853
gyrans, Typhula 1004
gyrosa, Endothia 694
gyrosa, Endothiella 694
gyrosum, Physarum 876
hadrotrichoides, Bipolaris (Helminthosporium, Septosporium) 588
haedinum, Geopetalum = **Cheimonophyllum candidissimum** 625
haedinus, Pleurotus = **Cheimonophyllum candidissimum** 625
haematococca, Nectria 812
haematococcus, Hypomyces = **Nectria haematococca** 812
haematopoda, Mycena 798
haematopoda var. **cuspidata, Mycena** 798
haematopus, Mycena = **M. haematopoda** 798
hahniana, (Lachnella, Lachnellula, Trichoscyphella) = **Lachnellula occidentalis** 754
haleniae, Cercospora 611
haleniae, Puccinia 907
halesiae, Cercospora 611
halesiae, Cytospora 661
halesiae, Diaporthe = **D. beckhausii** 670
halesiae, Lachnum (Dasyscyphus) 755
halesiae, Lophiotrema 771
halesiae, Phoma 853
halesiae, Polyporus = **Bjerkandera adusta** 590
halesiae, Sirococcus 961
halima, Ceriosporopsis 621
halima, Leptosphaeria 762
halima, Papulaspora 825
haliotrepha, Dactylospora (Buellia) 663
haliotrephus, Kymadiscus = **Dactylospora haliotrepha** 663
halodes, (Bipolaris, Drechslera, Exserohilum, Helminthosporium) = **Exserohilum rostratum** 704
halophila, Ustilago 1015
halophiloides, Ustilago 1015
halseyana, (Eutypella, Valsa) = **Eutypella cerviculata** 700
halstediana, Phyllachora 862
halstedii, Cercospora 611
halstedii, Macrophoma 774
halstedii, Phyllosticta 869
halstedii, Plasmopara (Peronospora) 882
halstedii, Uromyces 1011
hamamelidis, Cercospora (Ramularia) 611
hamamelidis, Dermea (Patellaria) 669
hamamelidis, Diplodia 680
hamamelidis, Graphium (Stilbum) 723
hamamelidis, Leptosphaeria 762
hamamelidis, Phyllosticta (Gloeosporium, Phyllosticta, Phyllostictina) 869
hamata, Halosphaeria (Remispora) 728
hamata, Peniophora = **Tubulicrinis hamatus** 1003
hamata, Tubercularia = **T. vulgaris** 1002
hamatum, Trichoderma 999
hamatus, Tubulicrinis 1003
hameliae, Uredo 1007
hamulatus, Crepidotus 647
handelii, Pseudocercospora (Cercoseptoria, Cercospora) 894
hanlinii, Dendrostilbella = **Stilbella flavipes** 981
hansbroughiana, Tympanis = **T. truncatula** 1004
hansenii, Ascochyta 575
hansenii, Cercospora 611
hansenii, Phloeospora 849
hapalocystis, Pseudovalsa = **Hapalocystis berkeleyi** 728
haplochaeta, Meliolina 786
haraeanum, Gymnosporangium = **G. asiaticum** 726
hariotii, (Corticium, Hyphodontia) = **Hyphoderma sambuci** 741
harknessianum, Gymnosporangium 726
harknessii, Diplodia = **Microdiplodia laurina** 788
harknessii, Endocronartium (Cronartium) 694
harknessii, Peridermium 832
harknessii, Phoma 853
harknessii, Puccinia = **P. hieracii** var. **harknessii** 907
harknessii var. major, Puccinia = **P. hieracii** var. **stephanomeriae** 908
harknessii var. stephanomeriae, Puccinia = **P. hieracii** var. **stephanomeriae** 908
harknessioides, Coniosporium (Gymnosporium) 641
harperi, Pleiopatella 883
harperianum, Myrmaecium 809
harpographoidea, Corniculariella 644
harpographoides, Cornularia = **Corniculariella harpographoidea** 644

harrimanii, Microthyrium 791
harthensis, Mycosphaerella (Sphaerella) 802
hartigii, Pestalotia 835
hartigii, Phellinus (Fomes, Polyporus) 845
harzianum, Trichoderma 999
harzii, Alternaria = **Ulocladium lanuginosum** 1006
hashiokai, Uredinopsis 1007
hastata, Grandinia (Hyphodontia, Peniophora) 722
haustellaris, Simocybe (Crepidotus) 961
havanensis, Cryphonectria (Endothia) 650
hawaiense, Gloeosporium = **Colletotrichum gloeosporioides** 638
hawaiiense, Chaetothyrium 624
hawaiiense, Endoraecium 694
hawaiiense, Gloeocoryneum 716
hawaiiense, Helminthosporium = **Bipolaris hawaiiensis** 588
hawaiiensis, Antennellina 565
hawaiiensis, Bipolaris (Drechslera) 588
hawaiiensis, Dothiorella 688
hawaiiensis, Fomes 705
hawaiiensis, Harknessia 729
hawaiiensis, Meliola 784
hawaiiensis, Metasphaeria 788
hawaiiensis, Mycosphaerella 802
hawaiiensis, Phyllosticta 869
hawaiiensis, Septoria 952
hawaiiensis, Stictis 978
hawaiiensis, Trichothallus 1000
hawaiiensis, Uredo 1007
haydenii, Guignardia (Dothidea, Ophiodothis) 725
haydenii, Placosphaeria 881
haynaldii, Phyllosticta 869
healdii, Phyllosticta 869
hedeomae, Peronospora 833
hedeomae, Septoria 952
hedeomina, Rhabdospora (Septoria) 930
hederae, Aulographum 584
hederae, Calonectria = **C. pyrochroa** 598
hederae, Cercospora 611
hederae, Diplodia 680
hederae, Phyllosticta 869
hederae, Sphaceloma 964
hedericola, Aplosporella (Diplodia, Sphaeropsis) 570
hedericola, Phyllosticta 869
hedericola, Ramularia 925
hedgcockii, Leptostroma 765
hedgcockii, Ploioderma (Hypoderma) 887
hedychii, Mycosphaerella 802
hedysari, Parodiella (Coryneum, Didymosporium) 826
hedysari, Phloeosporella (Septogloeum) 850
hedysari-obscuri, Uromyces 1011
hedysari-paniculati, Uromyces 1011
heedii, Pichia 879
heimii, Gloeocystidiellum 716
helenae, Phyllosticta 869
helenae, Teichospora (Strickeria) 988
helenii, Cercospora 611
helenii, Septoria 952
heleocharidis, Physoderma 877
heleocharidis var. **arctica, Pleospora** 884
heleocharidis var. **caricina, Stagonospora** 974
helianthellae, Cercosporella 619
helianthellae, Puccinia 907
helianthi, Alternaria (Helminthosporium) 561
helianthi, Cercospora 611
helianthi, Coleosporium 636
helianthi, Colletotrichum 639
helianthi, Diaporthe 671
helianthi, Leptosphaeria 762
helianthi, Phyllosticta 869
helianthi, Puccinia 907
helianthi, Ramularia 925
helianthi, Septoria 952
helianthi var. **macromaculans, Colletotrichum** 639
helianthicola, Septoria (Rhabdospora) 952
helianthinficiens, Alternaria 562
helianthorum, Puccinia = **P. helianthi** 907
helicalis, Puccinia 907
helicandrum, Pythium 921
helicoides, Pythium 921
heliconiae, Phyllosticta 869
helicophilum, Melanomma = **Tubeufia pezizula** 1002
helicosporus, Acanthophiobolus (Ophiobolus) 555
helicotricha, Stictis = **S. pustulata** 978
heliomyces, Marasmius 777
heliopsidis, Leptosphaeria (Gibberidea, Montagnella, Rosenscheldia) 762
heliopsidis, Phyllosticta 869
heliopsidis, Puccinia = **P. helianthi** 907
heliopsidis, Septoria 952
helios, Crocicreas (Davincia) 649
heliotropii, Cercospora 611
heliotropii, Cladosporium 630
hellebori, Coniothyrium 642
helleboricola var. **coptidis, Phyllosticta** 869
hellerianus, Uromyces = **U. novissimus** 1012
helminthospora, Leptosphaeria 762
helminthospora, Leptospora = **Acanthophiobolus helicosporus** 555
helotioides, Cenangella 602
helotioides, Pezizella 838
helvelloides, Sebacina 944
helvetica, Cristinia 648
helvetica, Leptosphaeria 762
helvetica, Pleospora 884
helveticum, Hydnum = **Cristinia helvetica** 648
hemerocalli, Collecephalus 636
hemerocalli, Gloeocephalus 716
hemerocalli var. **macrosporum, Collecephalus** 636
hemerocallidis, Cercospora 611
hemerocallidis, Gloeosporium 718
hemicrypta, Balansia 586
hemileucus, Polyporus = **Fomitella supina** 705
hemisphaerica, Humaria 738
hemisphaerica, Puccinia = **P. minussensis** 910
hemisphaerica, Xylopezia (Stictis) 1027
hemispherica, Xenomeris = **Sthughesia juniperi** 978
hemizoniae, Puccinia = **P. nuda** 910
hendersoniae, Clypeosphaeria = **Diapleella coniothyrium** 670
hennebergii, Stagonospora = **S. nodorum** 974
henningsiana, Balansia 586
henningsii, Cercosporidium (Cercospora) 620
henningsii, Phoma 853
hepatica, Confistulina 641
hepatica, Fistulina 705
hepaticae, (Ascochyta, Gloeosporium) = **Ascochyta vodakii** 577
hepaticae, Colletotrichum 639
hepaticae, Septoria = **S. hepaticicola** 952
hepaticicola, Septoria 952
hepaticus, (Ceriomyces, Ptychogaster) = **Confistulina hepatica** 641
heptaseptatum, Cylindrocladium 657
heraclei, (Ascochyta, Cylindrosporium, Septoria) = **Cylindrosporium umbelliferarum** 660
heraclei, Erysiphe 698
heraclei, Phoma 854
heraclei, Phyllachora 862
heraclei, Phyllosticta 869
heraclei, Ramularia 925
heraclei, Rhabdospora 930
herbacea, Diatrypella 674
herbariorum, Eurotium 699
herbarum, Calycina (Helotium, Hymenoscyphus) 599
herbarum, Cladosporium 630
herbarum, Crepidotus = **Pleurotellus hypnophilus** 886
herbarum, Dinemasporium = **Pseudolachnea hispidula** 896
herbarum, Diplodia (Sporocadus) 680
herbarum, Leptostroma (Sclerotium) 765
herbarum, Lophiostoma (Eriosphaeria) 770
herbarum, Pezizellaster 838
herbarum, Phoma 854
herbarum, Pleospora 884
herbarum, Stagonospora 974
herbarum, Torula 995
herbarum, Vermicularia = **Colletotrichum dematium** 637
herbarum var. **citricola, Cladosporium** 630
herbarum var. citrorum, Pleospora = **Pleospora herbarum** 884
herbarum f. medicaginum, Phoma = **P. medicaginis** 855
herbarum var. **occidentalis, Pleospora** 884
herbatica, Stemonitis 976
herbicola, Calosphaeria (Jattaea) 598
herbicola, Eutypella = **E. goniostoma** 700
herbicola, Phoma 854
herculea, Alternaria = **A. brassicae** 561
herculea, Xylaria 1026
herculeum, Clasterosporium = **Sporidesmium inflatum** 970
herculeum, Macrosporium = **Alternaria brassicae** 561
hermione, Mycosphaerella (Sphaerella) 802
herpotricha, Ophiosphaerella (Phaeosphaeria) 820
herpotrichioides, Rosellinia 937
herpotrichoides, Herpotrichia 735
herpotrichoides, Phaeosphaeria (Leptosphaeria, Trematosphaeria) 842
herpotrichoides, Pseudocercosporella (Cercosporella) 895
herpotrichus, Ophiobolus = **Ophiosphaerella herpotricha** 820
herquei, Penicillium 828
hesleri, Aecidium 558
hesleri, Phyllosticta 869
hesperia, Leptosphaeria = **Ophiobolus prunellae** 820
hesperidea, Geopyxis = **Sarcoscypha occidentalis** 938
hesperidearum, Phyllosticta 869
hesperidis, Ascochyta = **A. cheiranthi** 574
heteracantha, (Eutypa, Peroneutypa, Valsa) = **Eutypella scoparia** 701
heterantherae, Membranosorus 786
heterisiae, Puccinia = **P. pazschkei** var. **heterisiae** 911
heterocarpa, Peziza = **Crocicreas nigrescens** 649
heterocladium, Verticillium 1021
heterocystidia, Hyphoderma (Peniophora) 741

heterodermus, Uromyces 1011
heterogena, Ustilago 1015
heterogenea, Fracchiaea = **Nitschkia broomeiana** 816
heterogenea, Scytinostromella (Peniophora) 944
heterogeneum, Gloeocystidiellum = **Scytinostromella heterogenea** 944
heteromeles, (Capnodium, Limacinia, Phaeosaccardinula) = **Limacinula anomala** 768
heteromeles, Phyllosticta 869
heteromeles, Pyrenopeziza 918
heteromelina, Diplodia 680
heteromelina, Physalospora 876
heteromelis, Hadrotrichum 727
heteromelis, Lophodermium (Hypoderma) 772
heteromelis, Phacidium 839
heteromelis, Stigmina (Cercospora) 980
heteromorpha, Antrodia (Lenzites, Trametes) 567
heteromorpha, Phialophora = **Exophiala jeanselmei** var. **heteromorpha** 703
heteromorpha, Poria = **Wrightoporia avellanea** 1024
heteromorphum, Trichosporum = **Exophiala jeanselmei** var. **heteromorpha** 703
heteromorphus, Coriolellus = **Antrodia heteromorpha** 567
heteronemum, Macrosporium (Septonema) 775
heterophragmia, Hendersonia 733
heterophyllae, Coccomyces 634
heterophyllae, Gloeosporium = **Colletotrichum phyllachoroides** 639
heteroseta, Meliola 784
heterosperma, Cercospora = **Mycovellosiella concors** 807
heterospora, Anthracoidea 566
heterospora, Cercospora 611
heterospora, Clathrospora (Pleospora) 631
heterospora, Phaeoramularia 841
heterospora, Phaeosphaeria (Leptosphaeria, Trematosphaeria) 842
heterospora, Puccinia 907
heterospora, Sarcinella 938
heterospora, Stagonospora 974
heterospora var. **simmonsii, Clathrospora** 631
heterosporum, Fusarium 708
heterosporum, Septosporium = **Phaeoramularia heterospora** 841
heterostrophus, Cochliobolus (Ophiobolus) 635
heterothallicum, Pythium 921
heterothecae, Coleosporium = **C. asterum** 636
heucherae, Cercospora 611
heucherae, Colletotrichum = **C. gloeosporioides** 638
heucherae, Mycosphaerella (Dothidella) 802
heucherae, Phyllosticta (Asteromella, Stictochorella) 869
heucherae, Puccinia 907
heucherae, Septoria 952
heucherae, Urocystis 1008
heucherae var. **austroberingiana, Puccinia** 907
heucherae var. **lithophragmae, Puccinia** 907
heucherae var. **saxifragae, Puccinia** 907
heufleri, Ustilago 1015
heugelinianum, Cladosporium 630
heveae, Drechslera (Helminthosporium) 690
heveae, Ophiobolus 820
heveae, Phytophthora 878
hewittiae, Meliola = **M. malacotricha** 785
hexagonoides, Serpula (Merulius) 960
hibernalis, Phytophthora 878
hibernica, Postia 890
hibernicus, Oligoporus = **Postia hibernica** 890
hibisci, Aplosporella (Sphaeria, Sphaeropsis) 570
hibisci, Botryosphaeria (Melogramma) 592
hibisci, Cercospora = **Pseudocercospora abelmoschi** 893
hibisci, Colletotrichum = **C. gloeosporioides** 638
hibisci, Coniothyrium 642
hibisci, Microthyriella 791
hibisci, Phomopsis 858
hibiscicola, Rhabdospora 931
hibiscina, Diplodia 680
hibiscina, Phyllosticta 869
hibisicola, Sphaeria = **Rhabdospora hibiscicola** 931
hicoria, Splanchnonema 969
hicoriae, Cryptodiaporthe (Melanconis) 651
hicoriae, Diaporthe 671
hicoriae, Dothiorella 688
hicoriae, Septoria 952
hiemale, Cerocorticium (Corticium) 622
hiemalis, (Aleurodiscus, Radulomyces) = **Cerocorticium hiemale** 622
hiemalis, (Coccomyces, Higginsia) = **Blumeriella jaapii** 590
hiemalis, Cylindrosporium = **Phloeosporella padi** 850
hiemalis, Mucor 796
hiemalis, Peronospora 833
hiemalis f. silvaticus, Mucor 796
hieracicola, Septoria 952
hieracii, Cercospora 611
hieracii, Phoma 854
hieracii, Phomopsis 858
hieracii, Phyllosticta 869
hieracii, Puccinia 907
hieracii var. **harknessii, Puccinia** 907
hieracii var. **stephanomeriae, Puccinia** 908
hieronymi, Ustilago = **U. buchloes** 1014
hieronymii, Ravenelia 928
higginsianum, Colletotrichum 639
higginsii, Dactylaria (Pyricularia) 663
higginsii, Taphrina = **Mixia osmundae** 792
hilariae, Ustilago = **U. vilfae** 1016
hilariicola, Ustilago 1015
himantia, Kavinia (Clavaria, Mycoacia, Odontia, Oxydontia) 752
himantioides, Serpula (Merulius) 960
hinnuleum, Nodulisporium 816
hippocastani, Nectria 812
hippocastani, Septoria 952
hiratsukae, Taphrina 987
hirschii, Sistotrema (Trechispora) 962
hirsuta, Oramasia 822
hirsuta, Trametes 996
hirsuta var. **ramosa, Oramasia** 822
hirsutum, Gloeophyllum = **G. sepiarium** 717
hirsutum, Penicillium 828
hirsutum, Stereum 977
hirsutus, (Coriolus, Polyporus, Polystictus) = **Trametes hirsuta** 996
hirta, Dendropleella (Massaria) 667
hirta, Scutellinia (Lachnea) 943
hirta, Wojnowicia (Hendersonia) 1024
hirtellus, Meruliopsis (Byssomerulius, Merulius) 787
hirticeps, Collybia = **Crinipellis zonata** 648
hirtus, Jahnoporus (Fomes, Polyporus) 750
hispalensis, Dendrophoma 667
hispida, Lasiosphaeria 757
hispida, Phyllosticta 869
hispida, (Pleospora, Pyrenophora) = **Pleospora penicillus** 885
hispida, Trametes = **Coriolopsis gallica** 644
hispidellus, Polyporus = **Jahnoporus hirtus** 750
hispidula, Cornularia = **Morrisographium hispidulum** 796
hispidula, Periconia (Conoplea) 831
hispidula, Pseudolachnea 896
hispidulum, Dinemasporium = **Pseudolachnea hispidula** 896
hispidulum, Morrisographium (Pseudographium, Sphaeronaema) 796
hispidus, Inonotus (Polyporus) 748
hiulca, Phanerochaete (Peniophora) 844
hochelagae, Diatrype = **Eutypella leprosa** 701
hoffmannii, (Asterosporium, Botryosphaeria) = **Asterosporium asterospermum** 582
hoffmannii, Phialophora (Margarinomyces) 847
hoffmanseggiae, Uredo (Ravenelia) 1007
holboellii, Puccinia = **P. thlaspeos** 915
holci, Cercosporella (Cercospora) 619
holci, Dilophospora = **D. alopecuri** 679
holci, Entyloma = **E. dactylidis** 695
holci, Mycosphaerella 802
holci, Sphacelotheca = **Sporisorium cruentum** 971
holci, Tilletia 992
holcina, Puccinia = **P. hordei** 908
holci-sorghi, Sporisorium 971
hollandicus, Aspergillus 578
holmii, Exserohilum (Drechslera, Helminthosporium) 703
holmii, Setosphaeria (Keissleriella, Trichometasphaeria) 960
holodisci, Diplodia 680
holodiscina, Valsa 1018
holoseparans, Poria = **Physisporinus vitreus** 876
holospirum, Helicobasidium 730
holwayi, Entyloma = **E. calendulae** 695
holwayi, Peridermium 832
holwayi, Ravenelia 928
holwayi, Synchytrium 984
holwayi, Uraecium = **Pucciniastrum vaccinii** 917
holwayi, Uromyces 1011
holwayi, Uropyxis 1014
holwayi, Ustilago = **U. bullata** 1014
homaena, Poria = **Perenniporia subacida** 831
homoeocarpa, Sclerotinia 941
hordei, Ascochyta 575
hordei, Helminthosporium = **Drechslera teres** 690
hordei, Heterosporium 736
hordei, Hormodendrum 737
hordei, (Ophiocladium, Ovularia) = **Ramularia collo-cygni** 924
hordei, Puccinia (Puccinia) 908
hordei, Puccinia = **P. recondita** 912
hordei, Uromyces = **U. hordeinus** 1011
hordei, Ustilago 1015
hordei var. **americana, Ascochyta** 575
hordei f. skagwayensis, Ascochyta = **A. skagwayensis** 577
hordei-murini, Puccinia = **P. hordei** 908
hordeinus, Uromyces 1011
horiana, Puccinia 908

horii, Mycosphaerella 802
horkeliae, Phragmidium 861
hormiscioides, Sporidesmium 970
hormospora, Acarosporina 556
hormosporum, Schizoxylon = **Acarosporina hormospora** 556
horrida, (Neovossia, Tilletia) = **Tilletia barclayana** 991
horridula, Erysiphe = **E. cynoglossi** 698
horridula, Trichopezizella (Trichopeziza) 1000
hortensis, Diplodia 680
hortensis, Stagonosporopsis 975
hortilecta, Hendersonia 733
hortorum, Phyllosticta = **Phoma exigua** 853
hosackiae, Microdiplodia (Phoma) 788
hosackiae, Mycosphaerella (Ditopella, Sphaerella) 803
hosackiae, Phyllosticta 869
hosackiae, Septoria 952
houseana, Phoma 854
housei, Didymosphaeria 678
houstoniae, Cercospora 611
houstoniatus, Uromyces 1011
howei, Uromyces = **U. asclepiadis** 1009
howeianum, Hypoxylon 745
huberi, Puccinia 908
huerfanum, Acremonium 557
hughesiana, Ovularia 824
hullensis, Diatrype = **D. flavovirens** 673
humboldtiana, Melampsora = **M. abieti-capraearum** 781
humeana, Trametes = **Tyromyces humeanus** 1005
humeanus, Tyromyces 1005
humeralis, Cryptospora = **Winterella femoralis** 1024
humicola, Dactylaria 663
humicola, Gilmaniella 715
humicola, Scolecobasidium 942
humile, Cladosporium 630
humile, Podosporium = **Podosporiella glomerata** 888
humilis, Podosporiella = **P. glomerata** 888
humilis, Polyporus = **Abortiporus fractipes** 555
humilis, Poria = **Antrodiella incrustans** 568
humiriae, Asterinella (Seynesia) 581
humphreyana, Ravenelia 928
humuli, Aecidium 558
humuli, Cylindrosporium 659
humuli, Phyllosticta 869
humuli, Pseudoperonospora 897
humuli, Septoria (Ascochyta, Phloeospora) 952
humuli, Sphaerotheca = **S. macularis** 968
humuli var. fuliginea, Sphaerotheca = **S. fuliginea** 968
humuli var. major, Phyllosticta = **Phyllosticta humuli** 869
humuli-americani, Stagonospora 974
huntii, Ophiostoma (Ceratocystis) 821
huricola, Cercospora 611
hyacinthi, Botrytis 594
hyalina, Cercospora 611
hyalina, Hyaloscypha (Lachnella) 738
hyalina, Macrophoma 774
hyalina, Mastigosporella (Harknessia) 780
hyalina, Septoria 952
hyalina, Sphaeropsis 967
hyalina, Thecaphora = **T. seminis-convolvuli** 989
hyalina, Tulasnella (Gloeotulasnella) 1004
hyalina, Typhula (Pistillina) 1004
hyalinellum, Lachnum 755
hyalino-bipolaris, Ustilago 1015
hyalinum, Gymnosporangium 726
hyalinum, Linodochium 769
hyalinum, Melanconium = **Discosporium populeum** 685
hyalinus, Dacrymyces = **Linodochium hyalinum** 769
hyalinus, Uromyces 1011
hyalospora, Ascochyta (Diplodia) 575
hyalospora, Cercospora 611
hyalospora, (Pleosphaerulina, Pleospora) = **Leptosphaerulina americana** 764
hyalospora, Tilletia 992
hyalothecium, Ceratocystis 604
hybridi, Uromyces = **U. trifolii-repentis** 1013
hybridum, Heterosporium 736
hybridum, Macrosporium 775
hydnans, Corticium = **Basidioradulum radula** 586
hydnicola, Hypoxylon = **H. albocinctum** 744
hydnoidea, Peniophora 829
hydnoides, Epithele 697
hydnoides, Hexagonia (Pogonomyces, Polyporus, Trametes) 736
hydnoides, (Odontia, Peniophora, Phlebia, Scopuloides) = **Phanerochaete rimosa** 844
hydnoideum, Aecidium 558
hydnosporum, Pythium 921
hydrangeae, Ascochyta = **Phoma exigua** 853
hydrangeae, Cercospora 611
hydrangeae, Diaporthe = **D. eres** 671
hydrangeae, Hendersonia 733
hydrangeae, Phyllosticta 869
hydrangeae, Pucciniastrum (Thekopsora) 916
hydrangeae, Septoria 952
hydrangeae, Stictis 978
hydrangeana, Cercospora = **C. hydrangeae** 611
hydrangeicola, Cercosporina = **Cercospora hydrangeae** 611
hydrocotyles, Cercospora 611
hydrocotyles, Entyloma 696
hydrocotyles, Puccinia 908
hydrocotyles, Septoria 952
hydrocotyles, Synchytrium = **S. bonaerense** 983
hydrocotylidis, Physoderma 877
hydrophila, Doassansiopsis (Doassansia) 686
hydrophila, Hymenopsis 739
hydrophila, (Hypholoma, Psathyrella) = **Psathyrella piluliformis** 893
hydrophila, Leptosphaeria 762
hydrophilum, Sclerotium 942
hydrophylli, Didymaria (Ramularia) 676
hydrophylli, Gloeosporium (Ascochyta) 718
hydrophylli, Peronospora 833
hydrophylli, Puccinia 908
hydrophylli, Septocylindrium 946
hydrophylli, Septoria 952
hydrophylli subsp. **mertensiae, Puccinia** 908
hydrophylli-virginiani, Ascochyta 575
hydropiperis, (Cercospora, Helminthosporium) = **Cercospora effusa** 609
hydropiperis, Ustilago (Sphacelotheca) 1015
hylodes, Valsa = **Eutypella scoparia** 701
hymenocallidis, Cercospora = **C. pancratii** 614
hymenocallidis, Phyllosticta 869
hymenuloides, Tubercularia (Dendrodochium, Myxosporium) 1002
hyoscyami, Peronospora 833
hyperborea, Sclerophleella (Leptosphaeria, Leptosphaerulina, Pleospora) 941
hyperboreum, Lophophacidium 773
hyperici, Cercospora 611
hyperici, Leptostroma 765
hyperici, Rhabdospora 931
hyperici, Uromyces = **U. triquetrus** 1013
hyperici-frondosi, Uromyces = **U. triquetrus** 1013
hypericina, Mycosphaerella (Sphaerella) 803
hypericorum, Melampsora (Mesopsora) 781
hyperopta, (Diaporthe, Melanconis) = **Melanconis chrysostroma** 782
hyperoxysporum, Fusarium = **F. oxysporum** 708
hyphaenea, Protostroma 893
hypnophilus, Pleurotellus 886
hypocarpa, Hendersonia 733
hypochaeridis, Puccinia 908
hypochoeridis, Ramularia 925
hypodermellae, Mycosphaerella 803
hypodermia, Botryodiplodia = **Sphaeropsis ulmicola** 967
hypodermia, Cryptosporella (Valsa) 653
hypodytes, Ustilago 1015
hypogaea, Roesleria = **R. pallida** 936
hypogeae, Ligniera (Rhizomyxa) 768
hypogynum, Pythium 921
hypophlaeum, Hypoxylon 745
hypophloea, Nummularia (Anthostoma, Diatrype) 817
hypophloia, Cryptendoxyla 650
hypophylla, Melasmia 784
hypophyllum, Cryptocline (Rhabdogloeum) 651
hypophyllum, Leptostroma = **Melasmia hypophylla** 784
hypophyllum, Lophodermium = **L. oxycocci** 772
hypopodia, Tympanis 1004
hypoxidis, Urocystis 1008
hypoxis, Septoria 952
hypoxyloides, Valsaria = **V. rubricosa** 1019
hypoxylon, Atkinsonella (Balansia, Epichloe, Hypocrella) 583
hypoxylon, Xylaria 1026
hypsicola, Mycosphaerella (Sphaerella) 803
hypticola, Cercospora = **Pseudocercospora lycopodis** 894
hyptidis, Puccinia 908
hyptidis-mutabilis, Puccinia 908
hyssopi, Puccinia 908
hysterella, Phoma = **Phyllosticta concentrica** 866
hysterella, Phyllostictina = **Phyllosticta concentrica** 866
hysteriiforme, Exarmidium (Dothidea, Rhopographus) 702
hysteriiformis, Colletotrichum (Vermicularia) 639
hysteriiformis, Monochaetia (Pestalotia) 795
hysterina, Stictis = **Melittosporium hysterinum** 786
hysterina, (Eutryblidiella, Tryblidiella) = **Rhytidhysteron hysterinum** 935
hysterinum, Melittosporium (Xylogramma) 786
hysterinum, Rhytidhysteron (Triblidium) 935
hysterioides, Crocicreas (Cyathicula, Helotium) 649
hysterioides, Gloeosporium = **Colletotrichum gloeosporioides** 638
hysterioides, Graphyllium (Hysterographium, Pleospora) 724
hysterioides, Leptosphaeria = **L. xerophylli** 764

hysterioides, Lophiostoma 770
hysterioides, (Lophodermellina, Lophodermium) = **Lophodermium foliicola** 772
hysterioides var. graminicola, Leptostromella 765
hysterioideum, Coniothyrium = **Microsphaeropsis concentrica** 791
hysterioideum, Sporidesmium 970
hysteroidea, Monostichella 796
hysteroides, Mollisia 793
hysteroideum, Gloeosporium = **Monostichella hysteroidea** 796
hystricina, Ceratostomella 605
hystricina, Cornicurariella 644
hystricinum, (Sphaerographium, Sphaeronaema) = **Cornicurariella hystricina** 644
hystricula, Lachnella = **Dasyscyphus hystriculus** 665
hystriculus, Dasyscyphus 665
hystrix, Lasiosphaeria = **L. canescens** 757
ianthino-velutina, Xylaria 1026
ichthyomethiae, Cercospora 611
idaei, Ascochyta 575
idahoense, Eupenicillium = **E. cinnamopurpureum** 699
idahoensis, Cercosporella 619
idahoensis, Phloeospora 849
idahoensis, Rectipilus (Solenia) 929
idahoensis, Typhula 1004
ignava, Uredo (Puccinia) 1008
igniaria, Periconia 831
igniarius, Phellinus (Fomes) 845
igniarius var. laevigatus, Fomes = **Phellinus laevigatus** 845
igniarius var. nigricans, Fomes = **Phellinus nigricans** 846
igniarius f. tremulae, Fomes = **Phellinus tremulae** 846
ignobilis, Hysteropezizella (Naevia) 747
ignobilis, Mycosphaerella (Sphaerella) 803
ignobilis, Uromyces = **U. tenuicutis** 1013
igualica, Ravenelia = **R. expansa** 928
ii, Cercospora = **Mycovellosiella murina** 808
ildefonsiae, Asterina 581
iliamnae, Kabatiella (Gloeosporium, Microstroma) 751
iliau, Clypeoporthe (Gnomonia) 633
ilicella, Macrophoma (Dothiorella) 774
ilicella, Mycosphaerella (Sphaerella) 803
ilicicola, Cylindrocladium (Candelospora) 657
ilicicola, (Phoma, Phyllosticta, Sphaeropsis) = **Phyllosticta concentrica** 866
ilicicola, Synchytrium 984
ilicifolia, Septoria 952
ilicigena, Microdiplodia 788
ilicigena, Rhabdospora 931
ilicina, Chrysomyxa 626
ilicina, Micropera = **Fusicoccum ilicinum** 712
ilicina, Phoma (Sphaeropsis) 854
ilicina, Trochila 1001
ilicincola, Rhytisma 935
ilicinum, Amerosporium 563
ilicinum, Coniothyrium 642
ilicinum, Dimerosporium 679
ilicinum, Fusicoccum 712
ilicinum, Hypoderma 743
ilicis, Amerodothis = **Botryosphaeria quercuum** 593
ilicis, Cercospora 611
ilicis, Diaporthe 671
ilicis, Elsinoe 692
ilicis, Gloeosporium 718
ilicis, Metasphaeria 788
ilicis, (Lembosia, Morenoella) = **Echidnodella angustiformis** 691
ilicis, Mycosphaerella 803
ilicis, Phillipsiella (Asterina, Henningsiella, Microphyma) 848
ilicis, Phomopsis (Phoma) 858
ilicis, (Discochora, Physalospora) = **Discochora philoprina** 684
ilicis, Phytophthora 878
ilicis, Sporonema 972
ilicis, Stemphylium = **Ulocladium consortiale** 1006
ilicis, Trochila = **T. ilicina** 1001
ilicis-canadensis, Rhytisma = **R. prini** 935
ilicis-opacae, Cercospora 611
illicii, Asteridium 580
illiciicola, Laestadia 755
illiciicola, Lembosia 758
illinii, Chaetopsina (Phialocephala) 624
illinoensis, Cercospora = **C. venturioides** 618
illinoensis, Phyllosticta 869
illinoensis, Plasmopara (Peronospora) 882
illinoisensis, Polyporus = **Inonotus radiatus** 749
illitum, Hypoxylon 745
illosporioides, Fusarium = **F. lateritium** 708
illudens, (Clitocybe, Omphalotus) = **Omphalotus olearius** 819
illudens, (Poria, Tyromyces) = **Wrightoporia subrutilans** 1024
imberbe, Helotium = **Hymenoscyphus imberbis** 740
imberbis, Hymenoscyphus 740
imitans, Melasmia 784
imitans, Phragmidium = **P. rubi-idaei** 861
immersa, Erysiphe 698
immersa, Mycosphaerella 803
immitis, (Polyporus, Tyromyces) = **Postia stiptica** 891
impatiens, Mycosphaerella 803
impatientis, Gloeosporium 718
impatientis, Phomopsis 858
impatientis, Puccinia = **P. recondita** 912
impatientis, Ramularia 925
impatientis, Synchytrium 984
imperfecta, Ascochyta = **Phoma medicaginis** 855
imperfecta, Dactylospora (Buellia, Karschia, Patellaria) 664
imperialis, Phomopsis (Phoma) 858
implexa, Leptospora 764
implexum, Lophiostoma = **Leptospora implexa** 764
implexus, Ophiobolus = **Leptospora implexa** 764
implicata, Monilia 794
implicatua, Monilia = **Acremonium implicatum** 557
implicatum, Acremonium 557
implicatum, Penicillium 828
imposita, Puccinia = **P. substriata** var. **imposita** 914
impressa, Rhizogene (Karschia) 933
improvisa, Melanopsamma 783
impulsa, Diaporthe 671
inaequale, Helminthosporium = **Curvularia inaequalis** 655
inaequalis, Cryptostictis 653
inaequalis, Curvularia 655
inaequalis, Diaporthe 671
inaequalis, Dicranidion = **D. fragile** 675
inaequalis, Rhabdospora 931
inaequalis, Venturia 1020
inaequalis var. cinerascens, Venturia = **Venturia inaequalis** 1020
inaequalis f. sp. crataegi, Venturia = **V. crataegi** 1020
inaequialtus, Uromyces 1011
inanipes, Puccinia 908
incanum, Corticium = **Phlebiella tulasnelloidea** 849
incarnata, Cercospora = **C. clavata** 608
incarnata, Cyathicula = **Crocicreas culmicola** 648
incarnata, Peniophora 829
incarnata, Phlebia 848
incarnata, Septoria 952
incarnata, Typhula 1004
incarnatifolia, Mycena 798
incarnatum, Corticium = **Peniophora incarnata** 829
incarnatus, (Byssomerulius, Merulius) = **Phlebia incarnata** 848
incerta, Psathyrella 893
incertum, Hypholoma = **Psathyrella incerta** 893
inclusa, Puccinia 908
inclusa, Septoria 952
inclusa, Strickeria 982
incomptum, Chaetomium 623
inconspicua, Alternaria 562
inconspicua, Cercosporella (Cylindrosporium) 619
inconspicua, Gloeocercospora 716
inconspicua, Septoria 952
inconspicuum, Asteroma (Gloeosporium) 582
inconspicuum, Belonium 587
inconspicuum, Coleosporium 636
inconspicuum, Fibulobasidium 704
inconspicuum, Gymnosporangium 727
inconspicuum, Helminthosporium = **Exserohilum turcicum** 704
inconspicuum var. buchloes, Helminthosporium = **Bipolaris buchloes** 588
incrassata, Meruliporia (Poria, Serpula) 787
incrassatus, Merulius = **Meruliporia incrassata** 787
increscens, Septoria = **S. trientalis** 958
incrustans, (Amphisphaeria, Microthelia) = **Kirschsteiniothelia aethiops** 752
incrustans, Antrodiella (Poria) 568
incrustans, Dendrothele (Aleurocorticium) 668
incrustans, Sebacina 944
incrustata, Athelia = **A. salicum** 583
incrustatum, Laxitextum 757
incurvata, Bipolaris (Drechslera) 588
incurvatum, Helminthosporium = **Bipolaris incurvata** 588
indianensis, Phyllosticta 869
indica, Beltrania = **B. rhombica** 587
indica, Dichotomophthora = **D. lutea** 675
indica f. opaca, Ravenelia = **R. opaca** 928
indicum, Chaetomium 623
indicum, Ramichloridium (Chloridium) 923
indicus, Mycoleptodiscus 799
indigoferae, Pythium 921
indigoferae, Ravenelia 928
indigoferae, Uromyces 1011
indistincta, Eutypella = **E. goniostoma** 700
indistincta, Mycosphaerella = **M. pteridis** 805
induratum, Sporidesmium 970
inermis, Asteridiella (Irene, Meliola) 580
inermis, Phellinus (Poria) 845
infectoria, Lewia (Alternaria, Pleospora) 767

infestans, Phacidium 839
infestans, Phytophthora 878
infestans var. abietis, Phacidium = **P. abietis** 838
infestans f. sp. **thalictri, Phytophthora** 878
inflata, Coniophora = **C. submembranacea** 641
inflata, Phytophthora 878
inflata, Poria = **Wolfiporia dilatohypha** 1024
inflata, Rhizina = **Rhizina undulata** 932
inflata, (Teichospora, Zignoella) = **Teichosporella dura** 988
inflatum, Helicoma 730
inflatum, Pythium 921
inflatum, Sporidesmium (Helminthosporium) 970
inflorescentiae, Sphacelotheca = **Ustilago bistortarum** 1014
infossa, Phoma 854
infula, Gyromitra 727
infundibulifera, Choanephora 626
infuscans, Ascochyta 575
infuscans, Cercospora = **C. toxicodendri** 618
infuscans, Dendryphiella (Cladosporium) 668
infuscans, Diatrype = **Eutypella quadrifida** 701
infuscans, Diplodia = **Sphaeropsis hyalina** 967
infuscans, Metasphaeria 788
infuscans, Monilochaetes 795
infuscans, Mycosphaerella (Sphaerella) 803
infuscans, Septoria (Cylindrosporium) 952
infuscata, Septoria 952
ingramii, Chaetoblastophorum 622
innominatum, Synchytrium 984
innumera, Phyllosticta 869
innumerabilis, Eutypella = **E. juglandicola** 700
innumerabilis, Phyllosticta 869
innumerabilis, Valsa = **Eutypella juglandicola** 700
innumerella, Mycosphaerella = **M. ranunculi** 805
inornata, Nais 811
inornata, Peniophora = **Tubulicrinis inornatus** 1003
inornatus, Tubulicrinis 1003
inquinans, Bulgaria (Phaeobulgaria) 596
inquinans, Ellisiodothis (Asterina) 692
inquinans, Massaria 779
inquinans, Melanconiopsis 781
inquinans, Pestalotia = **Pestalotiopsis karstenii** 836
inquinans var. , Coniosporium = **Arthrinium phaeospermum** 572
insculpta, Scirrhia (Dothidea, Dothidella, Plowrightia, Systremma) 940
insculpta, Vialaea (Boydia) 1022
insculptum, Endotryblidium (Triblidium) 694
insidens, Cryptostictis (Monochaetia, Pestalotia) 653
insidens, Hysterium 747
insidiosa, Phoma = **P. sorghina** 856
insidiosum, Lophiostoma = **L. caulium** 770
insidiosum f. **salicis, Lophiostoma** 770
insigne, Schizoxylon 940
insigne, Stereum = **Xylobolus subpileatus** 1026
insignis, Gloniopsis = **G. praelonga** 720
insignis, Phaeosphaeria (Leptosphaeria) 842
insipidus, Marasmius 777
insititiae, Taphrina = **T. pruni** 987
insitiva, Cytoplea 661
insitiva, Valsaria (Diatrype) 1019
insitivum, Coniothyrium = **Cytoplea insitiva** 661
insolita, Phanerochaete = **Candelabrochaete langloisii** 600
insperata, Puccinia = **P. variabilis** var. **insperata** 915
instabilis, Ophiobolus 820
insueta, Harknessia 729
insulana, Cercospora 611
insularis, Lachnella = **Trichopezizella horridula** 1000
insularis, Phyllachora = **P. paspalicola** 863
insularis, Stagonospora = **S. bellunensis** 973
insulsum, Aecidium 558
intensa, Asterinella (Asterina) 581
intermedia, Candida 600
intermedia, Curvularia 655
intermedia, Doassansiopsis (Doassansia) 686
intermedia, Milesia = **Milesina fructuosa** 792
intermedia, Pseudographis 896
intermedia, Sclerotinia = **S. minor** 941
intermedia, Septoria 952
intermedia, Typhula = **T. variabilis** 1005
intermedia, Uncinula = **U. macrospora** 1006
intermedia, Valsa 1018
intermedium, Blastodendrion = **Candida intermedia** 600
intermedium, Gloeosporium = **Colletotrichum gloeosporioides** 638
intermedium, Pythium 921
intermedium var. **poinsettiae, Gloeosporium** 718
intermedius, Dacrymyces = **D. stillatus** 663
intermedius, Pithoascus (Microascus) 880
intermissum, Acrodontium 557
intermissum, Lophodermium 772
intermixta, Phyllosticta 869
intermixta, Pleosphaerulina = **Pringsheimia sepincola** 891
intermixta, Puccinia 908
intermixta, Sphaerulina = **Pringsheimia sepincola** 891
intermixta, Stagonospora = **S. foliicola** 974
interna, Melanospora 783
interrupta, Rhabdospora (Septoria) 931
interruptus, Merulius = **Phlebia rufa** 848
interseminata, Dendryphiella = **D. vinosum** 668
interseminatum, Dendryphion = **Dendryphiella vinosum** 668
interseminatum, Helminthosporium = **Dendryphiella vinosum** 668
interseminatum, Heterosporium 736
interstitialis, Cucurbitaria (Teichospora) 654
interstitialis, Cylindrosporium 659
interstitialis, Gymnoconia = **Arthuriomyces peckianus** 573
intertexta, Mycena 798
interveniens, Puccinia 908
intricata, Meliola 785
intricata, Nigredo = **Uromyces intricatus** 1011
intricata, Plochmopeltis (Asterina, Microthyriella) 887
intricatus, Uromyces 1011
inulae, Septoria 952
inusta, Endoxyla 695
invaginata, Puccinia 908
invelata, Puccinia 908
inventa, Nectria 812
inversa, Dothiorella (Sphaeria) 688
inversa, Hohenbuehelia 737
inversa, Pseudomassaria 896
inversus, Pleurotus = **Hohenbuehelia inversa** 737
investiens, Hypoxylon 745
investiens, Vararia (Corticium) 1019
investiens var. **epiphaeum, Hypoxylon** 745
investita, Puccinia 908
involucrum, Corticium = **Basidiodendron eyrei** 586
involuta, Arrhytidia 572
involutus, Dacrymyces = **D. capitatus** 663
iodinus, Cyclomyces (Polyporus) 656
ionophila, Ramularia 925
iowana, Phoma 854
iowensis, Gorgoniceps 722
iowensis, Neovossia 815
iowensis, Polyporus = **Tyromyces galactinus** 1005
ipomeae-batatas, Phomopsis 858
ipomoeae, Cercospora 611
ipomoeae, Coleosporium 636
ipomoeae, Colletotrichum = **C. gloeosporioides** 638
ipomoeae, Graphyllium (Comoclathris, Pleospora) 724
ipomoeae, Meliola = **Meliola malacotricha** 785
ipomoeae, Nectria (Hypomyces) 812
ipomoeae, Phlyctema 850
ipomoeae, Phyllosticta 869
ipomoeae, Ramularia 925
ipomoeae, Sphaeropsis 967
ipomoeae, Sphaerulina 968
ipomoeae-panduranae, Cystopus = **Albugo ipomoeae-panduratae** 559
ipomoeae-panduratae, Albugo (Cystopus) 559
ipomoearum, Vermicularia 1021
ips, Ophiostoma (Ceratocystis, Ceratostomella) 821
ips, Tuberculariella 1002
iresine, Thecaphora (Sorosporium) 989
iresines, Septoria 952
iridicola, Alternaria 562
iridina, Phoma 854
iridis, Ascochyta 575
iridis, Bipolaris (Clasterosporium, Drechslera) 588
iridis, Cercoseptoria (Cylindrosporium) 605
iridis, Cladosporium (Heterosporium) 630
iridis, Didymium 677
iridis, Macrosporium 775
iridis, Mollisia (Micropeziza) 793
iridis, Phyllosticta 869
iridis, Puccinia 908
iridis, Stictopatella 979
iridis f. sp. **australis, Puccinia** 908
iridis f. sp. **floridanum, Puccinia** 908
iridis f. sp. **septentrionalis, Puccinia** 908
irpicinus, Merulius = **Leucogyrophana pinastri** 767
irregulare, Coryneum = **Myxocyclus polycistis** 810
irregulare, Cylindrosporium = **Gloeosporium toxicodendri** 718
irregulare, Entyloma = **E. dactylidis** 695
irregulare, Gloeosporium = **Discula umbrinella** 685
irregulare, Nakatea = **Nakataea sigmoidea** var. **irregulare** 811
irregulare, Oidium = **Ovularia isarioides** 824
irregulare, Pythium 921
irregularis, Peniophora = **Hyphoderma sambuci** 741

irregularis, (Phloeospora, Septoria) = **Gloeosporium toxicodendri** 718
irregularis, Sporidesmium 970
irrequisita, Puccinia = **P. calcitrapae** var. **centaureae** 902
irretitus, Coccomyces 634
isabellina, Clitocybe (Omphalina) 633
isabellina, Mortierella 796
isabellina, Pellicularia = **Botryohypochnus isabellinus** 591
isabellina, Peniophora 829
isabellina, (Poria, Trametes) = **Phellinus viticola** 846
isabellinus, Botryohypochnus (Hypochnus) 591
isabellinus, Phellinus = **P. viticola** 846
isanthi, Cercospora 611
isarioides, Ovularia 824
isariphora, Mycosphaerella (Sphaerella) 803
ischaemi, Ascochyta 575
ischaemi, Sphacelotheca = **Sporisorium andropogonis** 971
ischaemi, Stagonospora (Hendersonia) 974
ischaemi, Ustilago = **Sporisorium andropogonis** 971
ishikariensis, Typhula 1004
islandicum, Penicillium 828
isthmophorum, Epochnium 697
italicum, Penicillium 828
iteae, Phyllosticta 869
ithacensis, Cercospora 611
itihummensis, Dendrothele 668
itoana, Typhula = **T. incarnata** 1004
ivae, Aecidium 558
ivae, Dasyscyphus (Lachnella) 665
ivae, Ramularia 925
ivaecola, Phyllosticta 869
ivesiae, Phragmidium 861
ivesiae subsp. wyomingense, Phragmidium = **P. boreale** 860
ivicola, Microdiplodia (Diplodia) 788
ivicola, Septoria 952
iwayamai, Pythium 921
ixorae, Phyllosticta 869
ixorae, Pseudocercospora (Cercospora) 894
jaapii, Blumeriella (Pseudopeziza) 590
jaapii, Tomentella 994
jackmanii, Septoria = **S. clematidis** 949
jacksonensis, Leptosphaeria 762
jacksonii, Chloroscypha = **C. enterochroma** 626
jacksonii, Peronospora 833
jacksonii, Uromyces = **U. fragilipes** 1011
jacksonii, Ustilago 1015
jacobinae, Cercospora = **C. diantherae** 609
jacquiniana, Cercospora 612
jacucella, Septoria 952
jahnii, Cercospora 612
jamaicensis, Inonotus 748
jamaicensis, Pseudocercospora (Cercospora) 894
jamaicensis, Zygophiala 1027
jambosae, Phyllostictina = **Phyllosticta jambosicola** 869
jambosicola, Phyllosticta 869
jamesii, Septoria 952
janczewskii, Penicillium 828
janiphae, Colletotrichum = **C. gloeosporioides** 638
janthinellum, Penicillium 828
janus, Mycosphaerella (Dothidella, Leptosphaeria) 803
japonica, Drechslera = **D. tuberosa** 690
japonica, Phomopsis (Phoma) 858
japonica, Phyllosticta 869
japonica, Pyrenophora 919
japonica, Taphrina 987
japonica var. couchii, Typhulochaeta = **T. couchii** 1005
japonicum, Coryneum 646
japonicum, Gymnosporangium 727
japonicum, Helminthosporium = **Drechslera tuberosa** 690
japonicum, Penicillium = **P. resticulosum** 829
jasmini, Aplosporella 570
jasmini, Elsinoe 692
jasmini, Lambertella 756
jasmini, Rhabdospora 931
jasminicola, Pseudocercospora (Cercospora) 894
jatrophae, Cercospora 612
jatrophae, Septoria 952
jatropharum, Cercospora 612
jatrophicola, Cercospora 612
jatrophicola, Phakopsora 843
javanicum, Fusarium 708
javanicum var. radicicola, Fusarium = **Fusarium javanicum** 708
javanicus, Dasyscyphus = **D. oncospermatis** 665
javanicus, Wiesneriomyces = **W. laurinus** 1023
jeanselmei, Exophiala (Torula) 703
jeanselmei var. **heteromorpha, Exophiala** 703
jecorinum, Hypoxylon 745
jefferisii, Monochaetia (Pestalotia) 795
jejuna, Phoma 854
jenkinsoniana, Taphrina 987
jensenii, Penicillium 828
johansenii, Didymosphaeria 678
johansonii, Taphrina 987
johnsii, Physoderma 877
johnsonianus, Phellinus (Fomes) 845
johnsonii, Monilinia 794
johnsonii, Sporidiobolus 971
johnstonii, Postia (Oligoporus, Poria, Tyromyces) 890
jollymannii, Myrothecium = **M. leucotrichum** 809
jolyana, Phoma 854
jonesiana, Hendersonia 733
jonesii, Coleosporium 636
jonesii, Phragmidium 861
jonesii, Pseudopeziza = **Leptotrochila medicaginis** 766
jonesii, Puccinia 908
jonesii, Septoria 952
jonesii, Uromyces 1011
jonesii var. cymopteri, Puccinia = **Puccinia jonesii** 908
juddiana, Meliola 785
juglandicola, Cytospora = **C. leucosperma** 662
juglandicola, Eutypella (Valsa) 700
juglandicola var. juglandina, Eutypella = **Eutypella juglandicola** 700
juglandina, Cryptosphaeria = **Eutypella dissepta** 700
juglandina, Phomopsis (Phoma) 858
juglandina, Phyllosticta 869
juglandina, (Eutypella, Valsa) = **Eutypella juglandicola** 700
juglandinum var. **caryae, Anthostoma** 565
juglandis, Actinothecium 558
juglandis, Ascochyta 575
juglandis, Cercospora 612
juglandis, Cylindrosporium 659
juglandis, Dichomera 675
juglandis, Diplodia (Sphaeria) 680
juglandis, Marssonina (Marssonia, Marssoniella) 778
juglandis, Melanconis (Diaporthe) 782
juglandis, Microstroma (Exobasidium) 791
juglandis, Mycosphaerella 803
juglandis, Phyllosticta 869
juglandis, Rhabdospora (Septoria) 931
juglandis var. **caryae, Melanconis** 782
juglandis var. **robustum, Microstroma** 791
juglandis var. **tiliae, Melanconis** 782
jujubae, Cercospora 612
julibrissin, Diplodia 680
juliflora, Phyllosticta 869
juncacearum, Leptostroma 765
juncea, Selenophoma (Selenophomopsis) 945
juncea, Typhula = **T. phacorrhiza** 1005
junceum, Sacidium = **Selenophoma juncea** 945
junci, Cercospora = **C. juncina** 612
junci, Cintractia 628
junci, Claviceps 632
junci, Coniothyrium 642
junci, Ligniera (Sorosphaera) 768
junci, Ophiobolus 820
junci, Phoma 854
junci, Phyllachora (Dothidella, Endodothella) 862
junci, Placosphaeria 881
junci, Ramularia 925
junci, Rhabdospora (Septoria) 931
junci, Tolyposporium 993
junci, Trullula 1001
junci, Urocystis 1008
junci, Uromyces (Nigredo) 1011
juncicola, Clathrospora (Pleospora) 631
juncicola, Crandallia 646
juncicola, Loramyces 773
juncicola, Mycena = **M. culmigena** 797
juncicola, Phaeosphaeria (Leptosphaeria) 842
junci-effusi, Uromyces 1011
juncigena, (Ciboria, Sclerotinia) = **Myriosclerotinia curreyana** 809
juncina, Cercospora 612
juncina, Didymella 677
juncina, Phaeosphaeria (Leptosphaeria) 842
juncinum, Melanconium (Leptothyrium) 782
junciseda, Belonopsis (Mollisia) 587
juncisedum, Belonidium = **Belonopsis junciseda** 587
juncoidea, Tureenia = **Arthrinium cuspidatum** 572
junghuhnii, Campanella 600
juniperi, Amphisphaeria 564
juniperi, Aplosporella (Sphaeropsis) 570
juniperi, (Clithris, Coccomyces) = **Coccomyces petersii** 634
juniperi, Conoplea 644
juniperi, Dothidella = **Mycosphaerella juniperina** 803
juniperi, Grandinia (Hyphodontia) 723
juniperi, Herpotrichia 735
juniperi, Kabatina 751
juniperi, Keithia = **Coccodothis sphaeroidea** 634
juniperi, Lophodermium = **L. juniperinum** 772
juniperi, Mycosphaerella = **M. juniperina** 803
juniperi, Mytilidion = **Mytilinidion tortile** 810
juniperi, Pododimeria (Lojkania) 888
juniperi, Seiridium (Pestalotia) 945

juniperi, Seynesiella (Asterina, Dothidea, Gibbera, Microthyrium, Seynesia, Stigmatea) 961
juniperi, Sthughesia (Dimerium, Xenomeris) 978
juniperi var. **robusta, Conoplea** 644
junipericola, Peniophora 829
juniperina, Antrodia (Daedalea, Poria) 567
juniperina, Botryodiplodia (Macrophoma) 591
juniperina, (Dermatea, Kriegeria) = **Chloroscypha sabinae** 626
juniperina, Lophodermina = **Lophodermium juniperinum** 772
juniperina, Mycosphaerella (Dothidella, Laestadia, Sphaerella) 803
juniperina, Stigmina (Cercospora, Sciniatosporium) 980
juniperina, Valsa = **V. friesii** 1018
juniperinum, (Ceratostoma, Phaeostoma) = **Caliciopsis nigra** 597
juniperinum, Colpoma = **Coccomyces petersii** 634
juniperinum, Coryneum = **Stigmina glomerulosa** 980
juniperinum, Gymnosporangium = **G. cornutum** 726
juniperinum, Lophodermium 772
juniperinum f. **cupressi-thyoidis, Lophodermium** 772
juniperinus, Inonotus 749
juniperinus, Leptosporomyces 764
juniperinus, Marasmiellus 776
juniperinus, (Fomes, Fulvifomes, Polyporus) = **Pyrofomes demidoffii** 920
juniperi-virginianae, Gymnosporangium 727
juniperovora, Phomopsis 858
juruanum, Fusarium 708
jussiaeae, Colletotrichum = **C. gloeosporioides** 638
jussiaeae, Guignardia 725
jussiaeae, Pseudocercospora (Cercospora) 894
jussiaeae, Puccinia 908
jussiaeae, Septoria 952
justiciae, Cercospora 612
justiciae, Puccinia 908
justiciicola, Cercospora 612
juvenescens, Gymnosporangium = **G. nidus-avis** 727
kaduae, Dothidella (Polystomella) 687
kaduae, Meliola 785
kaduae, Mycosphaerella 803
kaernbachii, Puccinia = **P. posadensis** 912
kaki, Cercospora 612
kakombensis, Cryptophiale 652
kalalauensis, Appendiculella 571
kalalochensis, (Mycena, Resinomycena) = **Resinomycena saccharifera** subsp. **kalalochensis**
kalkhoffii, Rhizosphaera 934
kalmiae, Cercospora 612
kalmiae, Diaporthe 671
kalmiae, Gibbera (Coleroa, Venturia) 714
kalmiae, Godronia (Gorgoniceps) 721
kalmiae, (Gibberidea, Leptosphaeria) = **Melanomma rhododendri** 783
kalmiae, Phomopsis 858
kalmiae, Phyllachora (Dothidella) 863
kalmiae, Physalospora 876
kalmiarum, Rhabdospora (Sphaeria) 931
kalmicola, Hendersonia 733
kalmicola, Pestalotia = **P. adusta** 835
kalmicola, Phyllosticta 869
kampalensis, Stachybotrys 972
kanaloanus, Helicascus 730
kaniksuensis, Pseudotomentella 899
kansense, Clasterosporium = **Fusariella kansensis** 706
kansense, Hysterographium = **H. subrugosum** 747
kansensis, Cercospora 612
kansensis, Cilioplea (Pleospora) 628
kansensis, Diplodia = **Holmiella sabina** 737
kansensis, Fusariella 706
kansensis, Griphosphaerioma (Calospora, Cryptospora) 724
kansensis, Monochaetia (Pestalotia) 795
kansensis, Puccinia 908
karelica, Puccinia 908
karelica subsp. **laurentina, Puccinia** 908
karii, Anthracoidea (Cintractia) 566
karstenii, Dacryobolus (Stereum) 663
karstenii, Gloeocystidiellum (Gloeocystidium) 716
karstenii, Hyphoderma 741
karstenii, Mytilinidion = **M. rhenanum** 809
karstenii, Pestalotiopsis (Monochaetia, Pestalotia) 836
karstenii, Phoma 854
karstenii, Pleospora = **P. heleocharidis** var. **arctica** 884
karstenii, Pyrenopeziza 918
karstenii, Ramularia = **R. montana** 926
katsurae, Phytophthora 878
kauaiensis, Meliola 785
kauffmaniana, Pholiota = **P. flammans** 851
kauffmanii, Crepidotus 647
kauffmanii, Neolentinus (Lentinus, Panus) 815
kauffmanii, Peniophora = **Hyphoderma heterocystidia** 741
kauffmanii, Xeromphalina 1025
kawakamii, Phellinus 845
kellermaniana, Septoria 952
kellermanii, Basidiophora 586
kellermanii, Cercospora = **C. althaeina** 605
kellermanii, Leptothyrium 766
kellermanii, Rhabdospora 931
kelloggii, Cylindrosporium 659
kelseyi, (Homostegia, Phragmodothella) = **Discostroma massarina** 685
kelseyi, Strickeria (Cucurbitaria) 982
kemptonae, Strobilurus = **S. trullisatus** 982
kentiae, Colletotrichum = **C. gloeosporioides** 638
kernianum, Gymnosporangium 727
kerriae, Blumeriella (Coccomyces, Higginsia) 590
kerriae, Cylindrosporium 659
kevorkianii, Sphenospora 969
kickxii, Neohendersonia 814
kikuchii, Cercospora (Cercosporina) 612
killianii, Mycosphaerella = **Cymadothea trifolii** 660
kisubiensis var. **peleicola, Meliola** 785
kiusana, Cercospora = **C. violae** 618
kmetiana, Urocystis (Tuburcinia) 1008
koae, Atelocauda (Uromyces) 583
koae, Meliola 785
koae, Mycosphaerella 803
koae, Thyridaria 990
koeleriae, Puccinia = **P. brachypodii** var. **arrhenatheri** 902
koelreuteriae, Diplodia 680
koepkei, Mycovellosiella (Cercospora) 807
kokiae, Diaporthe 671
kokkiae, Libertella 768
koleroga, Pellicularia (Corticium) 827
kolkwitziae, Cercospora 612
kolleri, Ustilago = **U. segetum** 1016
koningii, Scopulariopsis 943
koningii, Trichoderma 999
konseiensis, Cryptodiaporthe = **Pseudovalsella thelebola** 899
koorchalomoides, Starkeyomyces = **Myrothecium verrucaria** 809
korrae, Leptosphaeria 762
krameriae, Uromyces 1011
kravtzevianus, (Parmastomyces, Tyromyces) = **Parmastomyces transmutans** 826
kriegeri, Camarosporium 600
kriegeriana, Marssonina (Marssonia) 778
kriegeriana, (Hyaloceras, Pestalotia) = **Seimatosporium kriegerianum** 945
kriegerianum, Seimatosporium (Monoceras) 945
krigiae, Mycosphaerella (Sphaerella) 803
krigiae, Phoma 854
krigiae, Septoria 952
krugiodendri, Meliola 785
krusei, Candida (Saccharomyces) 600
kuehneana, Ustilago 1015
kuhniae, Ascochyta = **A. doronici** 574
kuhniae, Puccinia 908
kuhniae var. **brickelliae, Puccinia** 908
kuhniae var. **robusta, Puccinia** 909
kunzeana, Leptosphaeria = **Phaeosphaeria typharum** 842
kunzei, Coryneum = **C. umbonatum** 646
kunzei, Cytospora (Leucocytospora) 661
kunzei, Lahmia 756
kunzei, Leucostoma (Valsa) 767
kunzei, Venturia = **Coleroa chaetomium** 636
kunzei var. japonicum, Coryneum = **C. japonicum** 646
kunzei var. **ramicola, Venturia** 1020
kusanoi, Bipolaris (Drechslera, Helminthosporium) 588
kyotensis, Calonectria 597
labiatarum, Pleospora = **P. herbarum** 884
labrosus, Dasyscyphus 665
labruscae, Phyllosticta = **P. ampelicida** 864
laburni, Camarosporium 600
laburni, Cucurbitaria 654
laburni, Pseudocercospora (Cercospora) 894
lacera, Dermascia = **Lophodermium lacerum** 772
lacerata, Pyrenopeziza 918
lacerata, Trametes = **Antrodia heteromorpha** 567
lacerum, Lophodermium 772
lacerum, Phacidium 839
laciniariae, Coleosporium 636
laciniata, Thelephora = **T. terrestris** 990
laciniatum, Hericium = **H. coralloides** 734
lackeyi, Phlyctochytrium 850
lacrimans, Serpula (Boletus, Merulius) 960
lacrimiformia, Crumenulopsis 650
lacrimiformis, Davisomycella (Hypodermella) 666
lactea, Cyphella = **Cellypha goldbachii** 601
lactea, Hemimycena 732
lactea, Nectria = **N. oropensoides** 813
lactea, Ramularia 925
lactea, Tulasnella = **T. eichleriana** 1004
lactescens, Megalocystidium (Corticium, Gloeocystidiellum) 780
lactescens, Sebacina = **Ductifera sucina** 690
lacteum, Fusisporium = **Ramularia lactea** 925
lacteus, Irpex 749
lacteus, Polyporus = **Postia tephroleuca** 891
lactis, Oospora = **Geotrichum candidum** 714

lactis var. parasitica, Oospora = **Geotrichum candidum** 714
lactucae, Asteroma 582
lactucae, Bremia 595
lactucae, Cercospora = **Cercospora longissima** 612
lactucae, Leptothyrium 766
lactucae, Mycosphaerella (Sphaerella) 803
lactucae, Phyllosticta 869
lactucae, Septoria (Septoria) 952
lactucarum, Rhabdospora (Sphaeria) 931
lactucicola, Cylindrosporium 659
lactucicola, Pleospora 885
lactucicola, Septoria 952
lactucicola, Stagonospora 974
lacunosa, Helvella 732
lacunosum, Synchytrium 984
lacustris, Massarina (Leptosphaeria, Metasphaeria) 779
laeliae, Gloeosporium = **Colletotrichum gloeosporioides** 638
laeliae, Phyllosticta 870
laelio-cattleyae, Diplodia 680
laeta, Peniophora = **P. hydnoidea** 829
laeticolor, Nectria = **N. flammea** 812
laeticolor, Steccherinum (Mycoleptodon) 975
laeticolor, Uredo 1008
laetum, Erythricium 698
laevata, Eutypa (Valsa) 699
laeve, Botryobasidium 591
laeve, Crucibulum 650
laeve, Cylindrobasidium (Corticium) 656
laevigata, Peniophora = **Amylostereum laevigatum** 564
laevigata, (Fuscoporia, Poria) = **Phellinus laevigatus** 845
laevigatum, Amylostereum (Corticium) 564
laevigatus, Phellinus (Polyporus) 845
laevis, Lignincola 768
laevis, Ravenelia 928
laevis, Tilletia 992
laevisporum, Oidium (Rhinotrichum) 818
laevisporus, Zygodesmus = **Oidium laevisporum** 818
laeviuscula, Milesina (Milesia) 792
lagenaria, Glomerella 719
lagenariae, Phoma (Sphaeropsis) 854
lagenarium, (Colletotrichum, Gloeosporium) = **Colletotrichum orbiculare** 639
lagenarium var. **musarum, Gloeosporium** 718
lageniformis, Mycosphaerella (Sphaerella) 803
lagerbergii, Scleroderris = **Ascocalyx abietina** 573
lagerheimii, Gibberella 715
lagerheimii, Helicogloea 730
lagerheimii, Synchytrium 984
lagerstroemiae, Cercospora = **C. lythracearum** 613
lagerstroemiae, Erysiphe 698
lagerstroemiae, Phyllosticta 870
lagerstroemiae, Platygloea 882
lagerstroemiae var. **foliicola, Phoma** 854
lagopus var. **sphaerosporus, Coprinus** 644
lamii, Peronospora 833
lamii, Septoria = **Septoria lamiicola** 953
lamii var. **brevior, Septoria** 952
lamiicola, Septoria 953
laminata, Poria = **Phellinus punctatus** 846
lamprocarpa, Erysiphe = **E. cichoracearum** 697
lanaria, Septoria 953
lanata, Cytidia 661
lanata, Hyphodontia 742
lanceolata, (Dothiorella, Macrophoma) = **Macrophoma elongata** 774
lanceolata, Ramularia 925
lanceolata-paraphysa, Pezizella = **Psilachnum lanceolato-paraphysatum** 899
lanceolato-paraphysatum, Psilachnum 899
lanceolatum, (Helminthosporium, Sporidesmium) = **Paratomenticola lanceolatus** 826
lanceolatus, Paratomenticola 826
lanciformis, Pseudovalsa 899
lancipes, Stemphylium (Alternaria) 977
landeghemiae, Phomopsis (Phoma) 858
lanea, Venturia = **Eriosphaeria vermicularia** 697
lanestris, Cystotheca (Sphaerotheca) 661
langloisiana, Fomitiporella = **Phellinus melleoporus** 846
langloisii, Candelabrochaete (Botryobasidium, Pellicularia) 600
langloisii, Caryospora 601
langloisii, Cercospora = **C. duplicata** 609
langloisii, Dimerium (Dimerosporium) 679
langloisii, Diplodia 680
langloisii, Fomes = **Phylloporia ribis** 864
langloisii, Fomitiporia = **Phellinus punctatus** 846
langloisii, Leptothyrella (Asterinula) 765
langloisii, (Lophiostoma, Melanomma) = **Trematosphaeria confusa** 998
langloisii, Pestalotia 835
lannosum, Penicillium = **P. puberulum** 828
lantanae, Asteromella 582
lantanae, Diatrypella 674
lantanae, Godronia (Scleroderris) 721
lantanae, Massaria 779
lantanae, Mycovellosiella (Cercospora) 808
lantanae, Puccinia 909
lantanoides, Phyllosticta = **Phoma viburni** 856
lantanoides, Rhabdospora = **Phoma viburni** 856
lanugine, Aphanostigme 568
lanuginicincta, Stictis 978
lanuginis, Phoma 854
lanuginosa, Alternaria = **Ulocladium lanuginosum** 1006
lanuginosa, Cercospora 612
lanuginosa, Herpotrichia = **Byssosphaeria rhodomphala** 596
lanuginosa, Rhizoctonia 932
lanuginosum, Ulocladium (Pseudostemphylium, Stemphylium) 1006
lanuginosus, Thermomyces (Humicola) 990
lapathicola, Puccinia 909
laplata, Platygloea 882
laporteae, Epistigme (Gloeosporium) 697
lappae, Ascochyta = **A. doronici** 574
lappae, Gloeosporium = **Ascochyta doronici** 574
lappae, Phyllosticta 870
lappae, Sphaeropsis 967
lapparum, Septoria 953
lapponica, Amylocystis 564
lapponica, Gloniella 719
lapponicum, Acanthophysium 556
lapponicum, Entyloma 696
lapponicum, Hysterium = **Gloniella lapponica** 719
lapponicus, Aleurodiscus = **Acanthophysium lapponicum** 556
lapponicus, Polyporus = **Amylocystis lapponica** 564
lapponicus, Uromyces 1011
lappulae, Ramularia 925
lapsanae, Puccinia 909
laricina, Encoeliopsis (Crumenula, Scleroderris) 693
laricina, Mycosphaerella 803
laricina, Pezoloma (Sphagnicola) 838
laricina, Pleospora 885
laricina, Tympanis 1004
laricinum, Cenangium = **Tympanis laricina** 1004
laricinum, Pseudohelotium = **Pezoloma laricina** 838
laricinum, Trichaptum 999
laricinus, (Hirschioporus, Lenzites) = **Trichaptum laricinum** 999
laricis, Coccodinium 634
laricis, Fomes = **Fomitopsis officinalis** 706
laricis, Heterosporium 736
laricis, Hypodermella 744
laricis, Lachnellula 754
laricis, Lophodermium = **L. pinastri** 772
laricis, Meria 787
laricis var. octospora, Hypodermella = **Hypodermella laricis** 744
laschii, Puccinia = **P. calcitrapae** var. **centaureae** 902
laschii, Valsella (Valsa) 1019
lasiocarpae, Anthracoidea 566
lasiostoma, Valsa = **Calosphaeria barbirostris** 598
lata, Eutypa 699
lata var. **aceri, Eutypa** 700
lataensis, Stromatocyphella = **Calathella eruciformis** 597
latahensis, Sarcosoma 938
latebrosa, Pyrenopeziza (Lachnella, Trichopeziza) 918
latemarginata, Poria = **Oxyporus latemarginatus** 824
latemarginatus, Oxyporus 824
latens, Cercospora 612
lateralis, Periconia 831
lateralis, Phytophthora 878
lateralis, (Aspergillus, Theclospora) = **Aspergillus versicolor** 579
lateripes, Puccinia 909
lateritia, Cercospora 612
lateritia, Diatrype = **Melogramma campylosporum** 786
lateritia, Nectria (Stilbonectria) 813
lateritia, Nummularia = **Hypoxylon rubiginosum** 746
lateritia, Odontia 817
lateritia, Puccinia 909
lateritia, Sphaerostilbe = **Nectria pseudotrichia** 813
lateritia, Tomentella 994
lateritia, Tubercularia 1002
lateritium, Anthracophyllum 567
lateritium, Fusarium 708
lateritium, Verticillium = **V. tenerum** 1022
lateritium var. fructigenum, Fusarium = **Fusarium lateritium** 708
lateritium var. mori, Fusarium = **Fusarium lateritium** f. sp. **mori** 708
lateritium f. sp. ciceri, Fusarium = **F. oxysporum** f. sp. **ciceris** 709
lateritium f. sp. crotalariae, Fusarium = **F. udum** f. sp. **crotalariae** 711
lateritium f. sp. **mori, Fusarium** 708
lateritium f. sp. **pini, Fusarium** 708
lathamii, Aplosporella 570
lathyri, Ascochyta 575
lathyri, Cercospora 612
lathyri, Fusarium = **F. solani** 710

lathyri, Physoderma (Urophlyctis) 877
lathyri, Ramularia = **R. deusta** 924
lathyri, Septoria 953
lathyrii, Robillarda = **Pseudorobillarda agrostidis** 898
lathyrina, Cercospora 612
lathyrinum, Fusicladium (Dicoccum) 711
lathyri-palustris, Peronospora 833
laticeps, Galerina 712
latifolia, Phyllosticta 870
latifolii, Plasmopara 882
latifolius, Crepidotus 647
latifolius, Pluteus 887
latispora, Protoventuria (Antennularia) 893
latissima, Typhula 1005
latitans, Ceuthospora (Dothidea, Dothiopsis, Dothiorella, Phyllachora) 622
latitans, Chaetoporellus (Poria) 624
latitans, Physalospora = **Phyllachora eucalypti** 862
laurentiana, Puccinia 909
laurentii, Cryptococcus (Torula) 651
lauricola, Kernella (Kernia) 752
laurina, Microdiplodia (Diplodia) 788
laurina, Valsa = **V. ambiens** 1017
laurina, Volutellaria = **Wiesneriomyces laurinus** 1023
laurinus, Wiesneriomyces 1023
laurocerasi, Pestalotia 835
laurocerasi, Phyllosticta 870
laurocerasi, Pyrenopeziza (Trochila) 918
laurus, Hypoxylon 745
lavandulae, Septoria 953
lavaterae, Phomopsis (Phoma) 858
lavaterae, Valsa 1018
laxa, Athelia (Peniophora) 583
laxa, Coniophora = **C. puteana** 641
laxa, Cytospora 662
laxa, Isariopsis = **Phaeoisariopsis griseola** 839
laxa, Monilia 794
laxa, Monilinia (Sclerotinia) 794
laxa, Pleospora = **P. longispora** 885
laxum, Asterostroma 583
laxum, Graphium = **Phaeoisariopsis griseola** 839
laxum, Sporotrichum 972
leaiana, Mycena (Collybia) 798
leaianus, Prunulus = **Mycena leaiana** 798
lebiseyi, Cryptodiaporthe 651
lebiseyi, Phomopsis 858
lecanidion, Botryodiplodia (Metadiplodia, Pseudopatella) 591
lecanidium, Nothopatella = **Botryodiplodia lecanidion** 591
lecanii, Verticillium (Cephalosporium) 1021
lecanora, Pleospora (Leptosphaeria) 885
lecheae, Cercospora 612
lecideola, Patellaria (Durella) 826
lecomtei, Lentinus = **L. strigosus** 759
ledi, Ascochyta 575
ledi, Chrysomyxa 626
ledi, Elsinoe (Aulographum) 692
ledi, Monilinia (Sclerotinia) 794
ledi, Pseudophacidium 898
ledi var. **cassandrae, Chrysomyxa** 626
ledi var. glandulosi, Chrysomyxa = **Chrysomyxa ledi** 626
ledi var. **rhododendri, Chrysomyxa** 626
ledicola, Chrysomyxa 626
leemingii, Clypeolella (Asterina) 633
leersiae, Bipolaris (Drechslera, Helminthosporium) 588
leersiae, Cladotrichum 631
leersiae, Leptosphaeria = **L. leersiana** 762
leersiae, (Metasphaeria, Sphaerella) = **Phaeosphaeria nigrans** 842
leersiana, Leptosphaeria 762
leguminicola, Rhizoctonia 932
leguminis-cytisi, (Mycosphaerella, Sphaerella) = **Didymella leguminis-cytisis** 677
leguminis-cytisis, Didymella 677
leguminum, Discella 683
leguminum, Fusisporium 712
leguminum, Gloeosporium = **Colletotrichum gloeosporioides** 638
leguminum, Macrosporium 775
leguminum, Pleospora = **P. herbarum** 884
leiostega, Metasphaeria = **Discostroma corticola** 685
leiphaemia var. raveneliana, Diaporthe = **Amphiporthe raveneliana** 563
leiphaemoides, Cryptospora 652
leiphaemoides, Valsa = **V. ceratosperma** 1017
lelebae, Haplobasidion 729
lemnae, Tracya 996
lemnischea, Cercospora = **C. mikaniae** 613
lenis, Diplomitoporus (Poria) 683
lenta, Hypocrea = **H. schweinitzii** 743
lenta, Wrightoporia (Poria) 1024
lenticellicola, Durandiella 691
lentis, Ascochyta 575
lentisci, Phyllosticta (Phoma) 870
lentum, Helotium = **Moellerodiscus lentus** 792
lentus, Moellerodiscus 792
leonensis, Curvularia = **C. pallescens** 655
leonina, Hymenochaete 739
leoninum, Hyphoderma 741
leonis, Propolis (Stictis) 892
leonotidicola, Puccinia 909
leonotidis, Cercospora 612
leonotidis, Puccinia = **P. leonotidicola** 909
leonuri, Ascochyta 575
leonuri, Cercospora 612
lepachydis, Physalospora = **Phyllachora ambrosiae** 862
lepachydis, Septoria 953
lepagei, Elsinoe 692
lepargyrea, Sydowia 983
lepida, Peniophora = **P. aurantiaca** 829
lepideus, Neolentinus (Lentinus) 815
lepidigena, Asterina = **Asteridium lepidigenum** 580
lepidigenoides, Asteridium (Asterina) 580
lepidigenum, Asteridium 580
lepidii, Peronospora = **P. parasitica** 834
lepidii, Synchytrium 984
lepidiicola, Pleospora = **P. herbarum** var. **occidentalis** 884
lepidiicola, Septoria 953
lepidii-sativi, Peronospora = **P. parasitica** 834
lepidii-virginici, Peronospora = **P. parasitica** 834
lepigoni, Albugo = **A. caryophyllacearum** 559
lepiniana, Asterina 581
lepiotospora, Volvariella 1023
lepistemonis, Meliola = **M. malacotricha** 785
leporinae, Hysteropezizella = **Pyrenopeziza karstenii** 918
leporinus, Polyporus = **Inonotus circinatus** 748
leprogena, Pestalotia 835
leproides, Physoderma (Urophlyctis) 877
leprosa, Eutypella (Valsa) 701
leprosa, Fairmaniella 704
leprosa, Phoma 854
leprosum, Coniothyrium = **Fairmaniella leprosa** 704
leptandrae, Cercospora 612
leptasca, Phomatospora (Cryptospora, Cryptosporella, Flageoletia, Valsa) 857
leptidea, Meliola = **M. heteroseta** 784
leptidea var. major, Meliola = **M. heteroseta** 784
leptideum, Glomosporium (Tolyposporium) 719
leptideus, Coccomyces 634
leptochloae, Puccinia 909
leptoclada, Meliola = **M. heteroseta** 784
leptodactyli, Diplodia 680
leptodermus, Uromyces = **U. setariae-italicae** 1013
leptographioides, Ophiostoma (Ceratocystis, Ceratostomella) 821
leptosperma, Cercospora (Cercoseptoria) 612
leptosperma, (Holwaya, Patellaria) = **Holwaya mucida** 737
leptosperma, Peronospora 833
leptosperma, Physalospora = **Mycosphaerella tassiana** 806
leptospermum, Cenangium 602
leptospermum, (Cercosporella, Cylindrosporium, Septoriopsis) = **Cercospora leptosperma** 612
leptospora, Ascochyta 575
leptospora, Mycosphaerella (Sphaerella) 803
leptospora, Phoma 854
leptostachya, Septoria 953
leptostromiforme, Cryptosporium = **Phomopsis leptostromiformis** 858
leptostromiformis, Phomopsis 858
leptostyla, Gnomonia 720
leptothecium, (Glyphium, Lophium) = **Glyphium corrugatum** 720
leptothyrioides, Cylindrosporella (Gloeosporium) 658
lespedezae, Cercospora = **C. latens** 612
lespedezae, Phyllachora (Sphaeria) 863
lespedezae, Uromyces = **U. lespedezae-procumbentis** 1011
lespedezae-procumbentis, Uromyces 1011
lesquereuxii, Hysterographium = **H. mori** 747
lethale, Ploioderma (Hypoderma) 887
lethalis, Ascochyta 575
lethalis, Didymella (Mycosphaerella) 677
lethalis, Leptosphaeria 762
lethalis, Ramularia 925
letifera, Taphrina (Ascomyces) 987
letiferum, Cladosporium 630
leucaenae, Ravenelia 928
leucanthemi, Alternaria 562
leucanthemi, Septoria 953
leucelenes, Hendersonia 733
leucelenes, Microdiplodia 788
leucelenes, Pyrenophora 919
leucella, Calycellina (Peziza) 598
leucellum, Helotium = **Calycellina leucella** 598
leucobasis, Anthostomella 566
leucobryophila, Lindtneria (Trechispora) 768
leucobryophilum, Corticium = **Lindtneria leucobryophila** 768
leucocarpa, Ceratocystis 604
leucocephalum, Craterium 647
leucoderma, Cintractia 628
leucoderma var. striata, Cintractia = **C. striata** 628
leucodermis, Pyrenopeziza 919

leucomallela, Postia 890
leucomallelus, (Oligoporus, Tyromyces) = **Postia leucomallela** 890
leucophaea, Eichleriella (Exidiopsis) 692
leucophaeum, Physarum 876
leucophaeus, Fomes = **Ganoderma applanatum** 713
leucopodia, Diachea 670
leucopsis, Valsa 1018
leucorrhodina, Nectria (Peziza) 813
leucosperma, Cytospora 662
leucospila, Elsinoe 692
leucospila, Ophiodothella (Linospora) 820
leucospongia, Postia (Oligoporus, Polyporus, Spongiporus, Tyromyces) 890
leucosticta, Cercospora = **C. meliae** 613
leucosticta, Cytospora 662
leucostigma, Orbilia (Calloria, Mollisia) 822
leucostoma, Cytospora 662
leucostoma, Valsa = **Leucostoma persoonii** 767
leucostomoides, Valsa = **V. ambiens** subsp. **leucostomoides** 1017
leucostyla, Bipolaris = **B. nodulosa** 588
leucostylum, Helminthosporium = **Bipolaris nodulosa** 588
leucothoes, Pestalotiopsis (Pestalotia) 836
leucothoes, Phyllosticta = **P. terminalis** 874
leucothoes, Pseudocercospora (Cercospora) 894
leucothoes, Pseudomassaria (Guignardia, Laestadia, Mycosphaerella) 896
leucothoes var. **gaultheriae, Pseudomassaria** 896
leucothoicola, Clithris 633
leucothoicola, Colpoma 640
leucothrix, Coniophora = **C. olivacea** 641
leucotricha, Excipula = **Myrothecium leucotrichum** 809
leucotricha, Podosphaera (Sphaerotheca) 888
leucotrichum, Myrothecium 809
leucoxanthum, Megalocystidium (Corticium, Gloeocystidiellum) 780
levata, Puccinia 909
leve, Ostropella (Xenolophium) 823
leveillei, Puccinia 909
leveilliana, Eichleriella 692
leveillianum, Corticium = **Eichleriella leveilliana** 692
levieri, Fusicladium 711
levis, Pleurotus (Lentinus, Panus) 886
levis, Puccinia 909
levis, Ustilago = **U. hordei** 1015
levitum, Microthecium 791
liatridis, Phomopsis 858
liatridis, Phyllosticta 870
liatridis, Puccinia = **P. poarum** 912
liatridis, Septoria 953
liberta, Puccinia 909
libertiana, Cryptostictis 653
libertiana, Flagelloscypha (Cyphella, Lachnella) 705
libertiana, Sclerotinia = **S. sclerotiorum** 941
libertum, Aecidium 558
libocedri, Camaropycnis 599
libocedri, Dermea 669
libocedri, Gymnosporangium 727
libocedri, Parksia 826
licatensis, Leptosphaeria 762
lichenalis, Arthothelium (Pleospora) 572
lichenicola, Seimatosporium 945
licnoides, Polyporus = **Phellinus gilvus** 845
ligea, Mycosphaerella 803
lignatile, Sphaerosporium 968
lignatilis, Ossicaulis (Pleurotus) 823
ligni, Mollisia 793
ligniaria, Coniochaeta (Rosellinia) 641
lignicola, Kuehneromyces 753
lignicola, Pestalotia = **Truncatella angustata** 1001
lignicola, Scytalidium 944
ligniperda, Torula (Xenodochus) 995
lignitalis, Hyalotrochophora (Everhartia) 738
lignorum, Massarinula 779
lignorum, Trichoderma = **T. viride** 999
lignosus, (Polyporus, Rigidoporus) = **Rigidoporus microporus** 936
lignyota, Cryptosphaeria (Diatrype) 652
lignyota, Rhizodiscina (Buellia, Karschia) 933
ligula, Clavariadelphus 632
ligulicola, Didymella (Mycosphaerella) 677
ligustici, Puccinia 909
ligustici, Septoria 953
ligusticicola, Cercospora 612
ligustri, Pseudocercospora (Cercospora) 894
ligustri, Schizoxylon (Stictis) 940
ligustricola, Cercospora 612
ligustrina, Diaporthe 671
ligustrina, Phoma 854
ligustrina, Thedgonia (Cercospora) 989
ligustri-vulgaris, Phomopsis 859
lilacea, Phlebia = **P. umbrata** 849
lilacifolia, Mycena 798
lilacina, Cercospora = **Mycovellosiella murina** 808
lilacino-fuscum, (Corticium, Stereum) = **Dendrocorticium roseocarneum** 666
lilacinum, Penicillium = **Paecilomyces lilacinus** 825
lilacinus, Paecilomyces 825
lilacinus, Pluteus 887
lilacis, Cercospora 612
lilascens, Phlebia (Corticium) 848
liliacearum, Colletotrichum = **Colletotrichum dematium** 637
liliacearum, Puccinia 909
liliacearum, Vermicularia = **Colletotrichum dematium** 637
lilii, Cercosporella = **C. inconspicua** 619
lilii, Colletotrichum 639
lilii, Uromyces = **U. holwayi** 1011
limaeformis, Eutypa 700
limbata, Septoplaca = **Piptarthron yuccae** 880
limbatum, Piptarthron = **P. yuccae** 880
limetticola, Gloeosporium = **Colletotrichum gloeosporioides** 638
limitata, Cintractia 628
limitata, Davisomycella (Hypodermella) 666
limitata, Dothichloe = **Balansia epichloe** 586
limitata, (Phoma, Phyllosticta) = **Phoma macrostoma** 854
limnanthemi, Burrillia 596
limnobii, Cercospora 612
limnodeae, Puccinia 909
limonella, Pholiota 851
limonia, Peniophora = **Phanerochaete sanguinea** 844
limonii, Uromyces 1011
limonii var. armeriae, Uromyces = **U. armeriae** 1009
limonii-caroliniani, Uromyces 1011
limoniiforme, Coniosporium 642
limoniispora, Rosellinia 937
limonum, Septoria 953
limosa, Anthracoidea (Cintractia) 566
limosa var. **gigantissima, Anthracoidea** (Cintractia) 566
limosa var. **minor, Anthracoidea** (Cintractia) 567
limosae, Puccinia 909
limosellae, Doassansia (Burrillia, Protomyces) 686
limpidum, Helicoma 730
limpidus, Pleurotus = **P. ostreatus** 886
linariae, Entyloma 696
linariae, Heterosphaeria = **H. patella** 735
linariae, Peronospora 833
linariae, Phoma 854
linariae, Septoria 953
linariae, Synchytrium 984
linariae var. gratiolae, Entyloma = **E. gratiolae** 696
linariae var. veronicae, Entyloma = **Entyloma linariae** 696
linbladii, Diplomitoporus 683
lindebergiae, Anthracoidea (Cintractia) 567
lindemuthiana, Glomerella 719
lindemuthianum, Colletotrichum 639
linderae, Aplosporella (Sphaeropsis) 570
linderae, Camarosporium 600
linderae, Diplodia (Microdiplodia) 680
linderae, Dothidea 687
linderae, Hendersonia 733
linderae, Isariopsis (Exosporium, Graphium) 750
linderae, Phyllosticta 870
linderae, Rosellinia = **Hypoxylon sassafras** 746
linderae, (Eutypella, Valsa) = **Eutypella goniostoma** 700
lindericola, Phyllosticta 870
lindquistii, Leptosphaeria 762
lindrothii, Puccinia = **P. jonesii** 908
lineare, Glonium (Hysterium) 720
lineare, Hadrotrichum 727
lineare, Lophodermium = **Bifusella linearis** 588
neariformis, Diaporthe (Apioporthe) 671
linearis, Aplosporella (Sphaeropsis) 570
linearis, Bifusella 588
linearis, Diaporthe 671
linearis, Diatrype = **D. prominens** 673
linearis, Phoma 854
linearis, Stagonospora 974
linearis, Ustulina 1017
linearis f. **salicis, Diatrype** 673
linearispora, Phoma 854
lineata, Grammothele 722
lineata, Phoma = **P. linearis** 854
lineata, Scirrhia = **Mycosphaerella scirrhoides** 806
lineata, Sphaeropsis = **Aplosporella clintonii** 570
lineatum, Entyloma 696
lineatum, Oedocephalum = **Spiniger meineckellum** 969
lineatus, Rigidoporus (Polyporus) 936
lineola, Colletotrichum = **C. dematium** 637
lineola, Ramularia 925
lineola var. **halepense, Colletotrichum** 639
lineolans, Hendersonia (Sphaeria) 733
lineolata, Gloniopsis 720
lineolata, Mycosphaerella 803
lineolata, Phoma 854
lineolata, Septoria 953
lineolatum, Hysterium = **Gloniopsis lineolata** 720
lineolatum, Melanconium = **Papularia vinosa** 825
lineolatus, Uromyces 1012

lineolatus subsp. **nearcticus, Uromyces** 1012
lingam, Phoma (Plenodomus) 854
linhartiana, Mycosphaerella 803
lini, Aureobasidium (Kabatiella, Polyspora) 584
lini, Cercospora 612
lini, Colletotrichum (Colletotrichum, Gloeosporium) 639
lini, Fusarium = **F. oxysporum** f. sp. **lini** 709
lini, Melampsora 781
lini, Peronospora 833
lini var. liniperda, Melampsora = **Melampsora lini** 781
linicola, Colletotrichum 639
linicola, Mycosphaerella 803
linicola, Phlyctema 850
linicola, Selenophoma 945
linicola, Septoria 953
linkii, Puccinia 909
linnaeae, Gnomonia 721
linnaeae, Godronia 721
linnaeae, Halbaniella 728
linorum, Mycosphaerella = **M. linicola** 803
linteus, Phellinus (Polyporus) 845
lippiae, Cercospora 612
lippiae, Cylindrosporium 659
lippiae, Meliola 785
lippiae, Prospodium (Puccinia) 892
lippiae, Sphaceloma 964
liquidambaris, Aplosporella (Sphaeropsis) 570
liquidambaris, Cercospora (Cercospora) 612
liquidambaris, Cryptodiaporthe 651
liquidambaris, Harknessia (Monochaetia, Seiridium) 729
liquidambaris, Leptothyrella = **Tubakia dryina** 1002
liquidambaris, Septoria 953
liquidambaris, Stigmina (Exosporium) 980
liquidambaris, Synchytrium 984
liquidambaris, Valsa 1018
liquiritiae, Gymnopilus 725
lirella, Phomopsis (Phoma) 859
liriodendri, Asteroma 582
liriodendri, Cylindrocarpon 657
liriodendri, Diplodia 680
liriodendri, Ectostroma 691
liriodendri, Erysiphe = **Phyllactinia guttata** 864
liriodendri, Gloeosporium = **Colletotrichum gloeosporioides** 638
liriodendri, Leptothyrium 766
liriodendri, Mycosphaerella 803
liriodendri, Myxosporium 810
liriodendri, Phaeoisariopsis (Cercospora) 840
liriodendri, Phyllosticta = **P. liriodendrica** 870
liriodendri, Ramularia 925
liriodendri, Rhytisma 935
liriodendri, Stigmina 980
liriodendri, Venturia 1020
liriodendrica, Phyllosticta 870
liriopes, Cercospora 612
lithocarpi, Limacinia (Scolecobonaria) 768
lithocarpi, Mollisia 793
lithocarpi, Protopeltis 893
lithophragmae, Entyloma 696
lithophragmae, Puccinia = **P. heucherae** var. **lithophragmae** 907
lithophragmae, Synchytrium 984
lithophragmae, Urocystis 1008
lithospermi, Calloria (Niptera) 597
lithospermi, Cercospora 612
lithospermi, Puccinia 909
lithospermi, Septoria 953
litigiosa, Leptopeltis 760
litigiosum, Leptothyrium (Leptostroma) 766
litigiosum, Microthyrium = **Leptopeltis litigiosa** 760
litschaueri, Hyphoderma (Corticium) 741
littorea, Septoria 953
liturae, Lembosina (Aulographum, Echidnodes, Lembosia) 758
livida, Pezicula (Dermea, Durella) 837
livida, Phlebia 848
livida, Phyllosticta 870
lividocaeruleum, Acanthophysium (Corticium, Gloeocystidiellum) 556
lividocaeruleus, Aleurodiscus = **Acanthophysium lividocaeruleum** 556
lividofusca, Tapesia (Mollisia, Niptera) 986
lividum, Corticium = **Phlebia livida** 848
lividum, Mycothyridium (Thyridium) 807
lividum, Penicillium 828
lobata, Dermatea = **Velutarina rufo-olivacea** 1019
lobata, Godronia (Scleroderris) 721
lobata, Puccinia 909
lobata, Winterina (Winteria) 1024
lobatum, Dendrosporium 667
lobatum, Ganoderma 713
lobatum, Stereum = **S. ostrea** 978
lobatus, (Fomes, Polyporus) = **Ganoderma lobatum** 713
lobeliae, Asterina (Parasterina) 581
lobeliae, Cercospora 612
lobeliae, Colletotrichum = **C. crassipes** 637
lobeliae, Entyloma 696
lobeliae, Meliola 785
lobeliae, Puccinia 909
lobeliae, Septoria 953
lobeliae-cardinalis, Mycovellosiella 808
lobeliae-syphiliticae, Septoria 953
lobeliicola, Cercospora = **Mycovellosiella lobeliae-cardinalis** 808
lobulata, (Sporocybe, Stachybotrys) = **Stachybotrys chartarum** 972
loculata, Daldinia 664
lohwagii, Mycena 798
lokoyae, Diaporthe 671
lokoyae, Phomopsis 859
loligena, Septoria 953
lolii, Diplodina = **Ascochyta desmazieresii** 574
lolii, Ovularia 824
lomatiae, Phoma 854
lomatii, Ascochyta 575
lombardiae, Corticium (Laeticorticium) 645
longestipitatum, Coryneum 646
longiana, Puccinia 909
longiana, Xylaria 1026
longiaristata, Pestalotia 835
longibrachiatum, Trichoderma 999
longicolla, Phomopsis 859
longimucronata, Uredinopsis 1007
longimucronata var. acrostichoides, Uredinopsis = **U. acrostichoides** 1007
longimucronata f. cyclosora, Uredinopsis = **Uredinopsis longimucronata** 1007
longinquus subsp. **pacificus, Panellus** 825
longipedicellata, Leptosphaeria 762
longipes, Alternaria 562
longipes, Cercospora 612
longipes, Chalara (Cylindrosporium) 625
longipes, Dendrophoma (Phoma, Phomopsis) 667
longipes, Harknessia = **H. uromycoides** 730
longipes, Hendersonia = **Pestalotia pezizoides** 835
longipes, Lanzia (Phialea) 756
longipes, Pseudovalsa 899
longipes, Puccinia 909
longipes, Taphrina = **T. communis** 986
longipes, Xylaria 1026
longipes var. americana, Oudemansiella = **Xerula americana** 1025
longipes var. **brevipes, Puccinia** 909
longipilus, Dasyscyphus = **Trichopezizella relicina** 1000
longirostellata, Ceratocystis = **Ophiostoma longirostellatum** 821
longirostellatum, Ophiostoma 821
longirostre, Sphaeronaema 966
longirostris, Eutypella = **E. scoparia** 701
longirostris, Microascus 788
longisclerotialis, Myriosclerotinia (Sclerotinia) 809
longiseta, Mycena 798
longisetula, Pestalotia 835
longisetum, Dinemasporium = **D. strigosum** 679
longispora, Cercospora (Cercoseptoria) 612
longispora, Diplodia 681
longispora, Erinella = **Lachnum abnorme** 754
longispora, Helicogloea 730
longispora, Macrophoma 774
longispora, Peniophora = **Subulicystidium longisporum** 983
longispora, Phyllosticta 870
longispora, Pleospora 885
longispora, Septoria 953
longispora, Servazziella (Cryptospora) 960
longispora, Thysanophora 991
longisporium, Cylindrosporium = **Cercospora longispora** 612
longisporum, Myxosporium 810
longisporum, Subulicystidium 983
longisporus, Dasyscyphus = **Lachnum abnorme** 754
longissima, Alternaria 562
longissima, Anguillospora 564
longissima, Cercospora 612
longissima, Dothidella (Sphaeria) 687
longissima, Mycosphaerella 803
longissima, Phoma 854
longissima, Ustilago 1015
longissima var. macrospora, Ustilago = **U. davisii** 1015
longissimum, Fusarium = **Anguillospora longissima** 564
longissimum, Pleurothyrium (Leptostroma) 886
longistipitatum, Coryneum 646
longistriatus, Pluteus 887
longitudinalis, Cucurbitaria 654
lonicera, Teichosporella = **Julella vitrispora** 750
lonicerae, Cercospora 612
lonicerae, Diplodia 681
lonicerae, Gloniopsis = **G. praelonga** 720
lonicerae, Hyponectria (Guignardia) 744
lonicerae, Kabatia (Marssonia, Marssonina) 751
lonicerae, Lachnella = **Dasyscyphus corticalis** 665
lonicerae, Lasiobotrys 756
lonicerae, Libertella 768
lonicerae, Ophiobolus 820
lonicerae, Phyllosticta = **P. vulgaris** 875
lonicerae, Ramularia 925
lonicerae, Rhabdospora (Cryptosporium) 931

lonicerae, Systremma = **Dothidea sambuci** 687
lonicerae, Tapesia 986
lonicerae, Thyronectria 991
lonicerae var. **involucratae, Kabatia** 751
lonicerae-canadensis, Ascochyta 575
lonicerina, Teichosporella 988
lophanthi, Ascochyta 575
lophanthi, Peronospora 833
lophanthi, Phoma 854
lophanthi, Ramularia 925
lophanthi, Septoria 953
lophanthi var. **lycopina, Ascochyta** 575
lophanthi var. **moldavicae, Peronospora** 833
lophanthi var. **osmophila, Ascochyta** 575
lophodermioides, Scirrhia 940
lophospora, Lophiosphaera (Didymella, Didymosphaeria) 770
lophospora var. acetosellae, Didymella = **Lophiosphaera lophospora** 770
lorentziana, Ustilago = **U. bullata** 1014
loricata, Massaria = **Splanchnonema loricatum** 969
loricatum, Splanchnonema 969
loti, Cercospora 613
loti, Stemphylium 977
loti, Uromyces = **U. euphorbiae-corniculati** 1011
lotophaga, Ovularia 824
lotorum, Peronospora 833
lotorus, Pseudospiropes 898
louisianae, Mycosphaerella 803
lucens, Morenoina (Aulographum, Lembosia) 796
luchuensis, Aspergillus = **A. niger** var. **awamorii** 579
lucidae, Colletotrichum 639
lucidae, Ramularia 925
lucidum, Ganoderma 713
lucidus, Polyporus = **Ganoderma lucidum** 713
lucidus var. zonatus, Polyporus = **Ganoderma zonatum** 713
luctuosa, Phaeosphaeria (Leptosphaeria) 842
lucumae, Achrotelium 556
ludens, Microsphaera 790
ludibunda, Cryptostictis = **Seimatosporium caudatum** 944
ludibunda, Cytosporina 662
ludibunda, Puccinia = **P. caricina** 902
ludibunda, (Eutypa, Valsa) = **Eutypella leprosa** 701
ludibundum, Xenosperma (Xenasma) 1025
ludoviciana, Anthostomella 566
ludoviciana, Fuscoporella = **Phellinus laevigatus** 845
ludoviciana, Phlebia (Peniophora) 848
ludoviciana, Phyllosticta 870
ludovicianae, Puccinia 909
ludovicianum, Hyphoderma = **Phlebia ludoviciana** 848
ludovicianus, Inonotus (Polyporus) 749
ludowiciana, Septoria = **S. lactucae** 952
ludwigiae, Cercospora 613
ludwigiae, Phyllosticta 870
ludwigiae, Septoria 953
lugdunensis, Heliscus 731
lumbricoides, Cercospora 613
lumbricoides, Helicosporium 731
lumbricoideum, Hyphosoma (Helminthosporium) 742
luna, Chaetoderma (Peniophora) 623
lunariae, Helminthosporium 731
lunariae, Leptothyrium 766
lunasporum, Cryptosporium 653
lunata, Apostrasseria (Ceuthospora) 571
lunata, Curvularia 655
lunata, Idriella 748
lunata, Schizotrichella = **Colletotrichum dematium** 637
lunata, Septoria = **Pseudoseptoria stomaticola** 898
lunata, Trechispora (Athelopsis, Grandinia) 997
lunata var. **aeria, Curvularia** 655
lunatum, Microdochium (Fusarium, Gloeosporium) 789
lunatus, Cochliobolus (Pseudocochliobolus) 635
lundbergii, Leptographium 759
lundellii, Hypochnicium 743
lundellii, Phellinus 845
lunelliana, Phloeospora (Septoria) 849
lupini, Cryptosporium 653
lupini, Cylindrosporium 659
lupini, Diaporthe 671
lupini, Didymella 677
lupini, Diplodia 681
lupini, Hadrotrichum = **H. globiferum** 727
lupini, Hendersonia 733
lupini, Leptosphaeria (Trematosphaeria) 762
lupini, Leptostroma 765
lupini, Macrophoma 774
lupini, Ovularia 824
lupini, Phoma (Asteromella) 854
lupini, Phyllosticta 870
lupini, Placosphaeria (Stictochorella) 881
lupini, Pseudocercospora (Cercospora, Chaetotrichum, Septosporium) 894
lupini, Ramularia 925
lupini, Septogloeum 946
lupini, Septoria 953
lupini, Siroscyphellina 961
lupini, Sphaeropsis 967
lupini, Stictis 978
lupini, Uromyces 1012
lupini, Valsa 1018
lupinicola, Cercospora 613
lupinicola, Gloeosporium 718
lupinicola, Ovularia 824
lupinicola, Phoma 854
lupinicola, Puccinia = **P. thelypodii** 915
lupinicola, Septoria 953
lupulina, Septoria 953
luridum, Megalocystidium (Corticium, Gloeocystidiellum) 780
lusatiae, Humaria (Peziza) 738
lutea, Dichotomophthora (Dactylaria) 675
lutea, Pleospora 885
luteoalba, Junghuhnia (Poria) 751
luteofimbriata, Poria = **Antrodia radiculosa** 567
luteofuscum, Callistosporium = **C. luteo-olivaceum** 597
luteo-maculata, Phyllachora 863
luteo-niger, Aspergillus 578
luteo-olivaceum, Callistosporium 597
luteopallens, Mycena 798
luteophylla, Clitocybe = **Gerronema chrysophyllum** 714
luteorubella, Orbilia (Calloria) 822
luteovirescens, Lanzia (Ciboria, Rutstroemia) 756
luteoviridis, Gymnopilus 725
lutescens, (Coccomyces, Higginsia) = **Blumeriella jaapii** 590
lutescens, Cylindrosporium = **Phloeosporella padi** 850
lutescens, Hymenoscyphus (Helotium) 740
lutescens, Taphrina 987
lutescens, Tremella = **T. mesenterica** 998
lutescens, Valsa = **V. ceratosperma** 1017
luteum, Myxosporium 810
luteum, Penicillium = **Talaromyces luteus** 986
luteum, Sphaeridium 965
luteus, Talaromyces 986
luttrellii, Physalacria 875
luxuriosa, Puccinia = **P. cryptandri** var. **luxurians** 904
luzulae, Phomatospora 857
luzulae, Stagonospora 974
luzulae, Ustilago (Cintractia) 1015
luzulicola, Ascochyta 576
lychnidis, Cercospora 613
lychnidis, Leptothyrium 766
lychnidis, Phyllosticta 870
lychnidis, Septoria 953
lychnidis var. pusilla, Septoria = **Septoria lychnidis** 953
lycii, Aecidium 558
lycii, Cercospora 613
lycii, Phyllosticta 870
lycii, Sphaeropsis 967
lycoctoni, Uromyces 1012
lycoperdon, Reticularia 930
lycopersici, Fusarium = **F. oxysporum** f. sp. **lycopersici** 709
lycopersici, Nematospora 814
lycopersici, Phoma (Ascochyta, Diplodina, Phoma) 854
lycopersici, Phyllosticta = **Phoma destructiva** 853
lycopersici, Pleospora = **P. herbarum** 884
lycopersici, Pyrenochaeta 918
lycopersici, Septoria 953
lycopersici, Stemphylium (Thyrospora) 977
lycopersici, Verticillium 1022
lycopi, Cercospora = **Pseudocercospora lycopodis** 894
lycopi, Physoderma 877
lycopi, Septoria 953
lycopodii, Ceuthospora 622
lycopodii, Mollisia 793
lycopodii, Mycosphaerella 803
lycopodiicola, Leptosphaeria = **Phaeosphaeria fuckelii** 842
lycopodina, Phaeosphaeria (Leptosphaeria) 842
lycopodina, Pseudomassaria (Mycosphaerella, Pseudapiospora) 897
lycopodina, (Neottiospora, Strasseria) = **Apostrasseria lunata** 571
lycopodis, Pseudocercospora (Cercospora) 894
lycuroides, Ustilago 1015
lyndonvillae, Camarosporium 600
lyndonvillae, Sphaeropsis = **Aplosporella hibisci** 570
lyngei, Diplodina 682
lyoni, Meliola 785
lyoniae, Cercospora 613
lysichitonis, Ombrophila 819
lysilomae, Ravenelia 928
lysimachiae, Cladosporium 630
lysimachiae, Phyllosticta 870
lysimachiae, Pseudocercospora (Cercospora) 894
lysimachiae, Ramularia 925
lysimachiae, Septoria 953
lythracearum, Cercospora 613
lythracearum, Mycosphaerella 803
lythri, Hainesia 728
lythri, Leptosphaeria 762

lythri, Pezizella = **Discohainesia oenotherae** 684
lythri, Stenella (Cercospora, Cladosporium) 977
lythrii, Synchytrium 984
lythrina, Septoria 953
mabae, Asterinella 581
mabae, Echidnodella (Lembosia) 691
macclatchieana, Cercospora 613
macdonaldii, Phoma 854
machaerinae, Helicoma 730
maclaganii, Tilletia 992
maclurae, Cercospora 613
maclurae, Cytospora = **C. sacculus** 662
maclurae, Leptosphaeria (Metasphaeria) 762
maclurae, Mycosphaerella (Sphaerella) 803
maclurae, Mycovellosiella (Ovularia) 808
maclurae, Phyllosticta 870
maclurae, Sphaeropsis = **S. malorum** 967
maclurae, Sporidesmium 970
maclurae, (Eutypella, Valsa) = **Valsa ceratosperma** 1017
macounii, Belonidium = **Godronia spiraeae** 721
macounii, Cryptodiaporthe 651
macounii, Diatrype = **D. bullata** 673
macrocarpa, Valsa = **V. ceratosperma** 1017
macrocarpum, Cladosporium 630
macrocarpum, Helminthosporium 731
macrocarpum var. **caudatum, Helminthosporium** 731
macrocephala, Cirrenalia (Helicoma) 628
macrochaeta, Amerosporium 563
macrochaeta, Cercospora = **C. polytricha** 615
macrocystidiata, Phanerochaete = **P. chrysosporium** 843
macrocystidiatum, Acanthophysium 556
macrocystidiatus, Aleurodiscus = **Acanthophysium macrocystidiatum** 556
macrodens, Dendrothele (Aleurocorticium, Aleurodiscus) 668
macroguttata, Cercospora 613
macroguttata, Phyllosticta 870
macromaculans, Cercospora 613
macromaculans, Stagonospora 974
macrophylla, Taphrina = **T. japonica** 987
macropoda, Septoria 953
macropoda var. **grandis, Septoria** 953
macropoda var. **septulata, Septoria** 953
macropoda f. **sitadakaensis, Septoria** 953
macropus, Gymnosporangium = **G. juniperi-virginianae** 727
macrosclerotia, Rhizoctonia = **R. solani** 932
macrosperma, Melanconis = **Prosthecium ellipsosporum** 892
macrosperma, Stilbospora 981
macrosperma, Uredinopsis = **U. pteridis** 1007
macrospermum, Cryptosporium 653
macrospermum, Sphaeropsis 967
macrospinosa, Periconia 831
macrospora, Alternaria 562
macrospora, Ascochyta (Apiocarpella, Apiosporella) 576
macrospora, Asteromassaria (Massaria) 582
macrospora, Athelia = **A. teutoburgensis** 583
macrospora, (Ansatospora, Cercospora) = **Mycocentrospora acerina** 799
macrospora, Dimerosporiopsis 679
macrospora, Dothiorella 688
macrospora, Eichleriella (Exidiopsis, Sebacina) 692
macrospora, Eutryblidiella (Tryblidiella) 699
macrospora, Hirudinaria 736
macrospora, Leptosphaeria 762
macrospora, Lirula 769
macrospora, Melittosporiella 786
macrospora, (Didymellina, Mycosphaerella, Mycosphaerella) = **Mycosphaerella macrospora** 803
macrospora, Oidiopsis 818
macrospora, Pellionella 827
macrospora, Phomopsis 859
macrospora, Phyllosticta = **P. liriodendrica** 870
macrospora, Pleospora = **Pyrenophora schroeteri** 919
macrospora, Puccinia = **P. eumacrospora** 906
macrospora, Ramularia 925
macrospora, Sclerophthora (Phytophthora, Sclerospora) 941
macrospora, (Septoria, Stagonospora) = **Piptarthron macrosporum** 880
macrospora, Stenocarpella (Diplodia) 977
macrospora, Titanella (Lophiostoma, Platystomum) 993
macrospora, Trichopezizella = **T. relicina** var. **macrospora** 1000
macrospora, Uncinula 1006
macrospora, Ustilago = **U. serpens** 1016
macrospora, Wettsteinina (Eriosphaeria) 1023
macrospora var. **asteris, Ramularia** 925
macrosporia, Septoria = **S. leucanthemi** 953
macrosporium, Glonium = **Ostreichnion curtisii** 823
macrosporium, Scolicosporium (Coryneum) 943
macrosporoidea, Acrospeira = **Monodictys castaneae** 795
macrosporoides, Sporidesmium 970
macrosporum, Chaetotrichum 625
macrosporum, Cylindrocladium = **C. pteridis** 657
macrosporum, Fusicoccum = **Dothiorella macrospora** 688
macrosporum, Glutinium = **Cryptosporiopsis pruinosa** 653
macrosporum, Hysterium 747
macrosporum, Lophodermium = **Lirula macrospora** 769
macrosporum, Olpitrichum (Oidium, Rhinotrichum) 819
macrosporum, Phlyctochytrium 850
macrosporum, Piptarthron 880
macrosporum, Synchytrium 984
macrosporus, Protomyces 892
macrostoma, Phoma 854
macrostoma, Septoria = **S. clementsii** 949
macrostomoides, Lophiostoma 770
macrotricha, Herpotrichia 735
macrotricha, Pestalotia = **Pestalotiopsis guepini** var. **macrotricha** 836
macrotricha, Pestalotiopsis = **P. guepini** var. **macrotricha** 836
maculans, Cercospora = **C. asclepiadis** 606
maculans, Curvularia = **C. eragrostidis** 655
maculans, Leptosphaeria 762
maculans, Periconia (Torula) 831
maculans, Pestalotiopsis (Hendersonia, Pestalotia, Sporocadus) 836
maculans, Phloeospora (Cercosporella) 849
maculans, Phyllosticta 870
maculans, Rhabdospora 931
maculans, Septogloeum 946
maculans, (Phoma, Sphaeropsis) = **Phyllosticta curtisii** 867
maculans, Uredo 1008
maculans, Venturia = **V. ditricha** 1020
maculare, Corticium 645
maculare, Lophodermium (Hysterium) 772
maculare, Physoderma (Cladochytrium) 877
macularis, Mycomicrothelia 799
macularis, Phoma 854
macularis, Sphaerotheca 968
macularis, Venturia (Mycosphaerella) 1020
macularis var. fuliginea, Sphaerotheca = **S. fuliginea** 968
maculata, Dendrothele 668
maculata, Diplodia 681
maculata, Fabraea = **Diplocarpon mespili** 679
maculata, Phoma (Sphaeropsis) 854
maculata, Psathyrella (Hypholoma) 893
maculata, Stagonospora (Stagonospora) 974
maculata, Uredinopsis 1007
maculatum, Aleurocorticium = **Dendrothele maculata** 668
maculatum, Diplocarpon = **D. mespili** 679
maculatum, Entomosporium = **E. mespili** 695
maculatum, Heterosporium 736
maculatum, Macrosporium 775
maculatum var. cydoniae, Entomosporium = **E. mespili** 695
maculatum var. domesticum, Entomosporium = **E. mespili** 695
maculicola, Discosia 684
maculicola, (Gonatobotrys, Gonatobotryum) = **Gonatobotryum apiculatum** 722
maculicola, Phaeoramularia (Torula) 841
maculicola, Phyllosticta 870
maculicola, Scolicotrichum 943
maculicola, Selenophoma 945
maculifera, Septoria 953
maculiformans, Pestalotiopsis (Pestalotia) 836
maculiforme, Circinotrichum 628
maculiformis, Macrophoma 774
maculiformis, Phyllosticta 870
maculiformis, (Mycosphaerella, Sphaerella) = **Mycosphaerella punctiformis** 805
maculiformis var. hippocastani, Mycosphaerella = **M. aesculi** 800
maculosa, Calycina 599
maculosa, Protoventuria (Antennularia, Meliola) 893
maculosa, Septoria = **S. maculifera** 953
maculosum, Helotium = **Calycina maculosa** 599
maculum, Hypoxylon 745
madiae, (Coleosporium, Stichopsora) = **Coleosporium pacificum** 636
madiae, Entyloma = **E. compositarum** 695
madidum, Stilbum = **Stilbella fusca** 981
madisonensis, Ascochyta 576
madisonensis, Cercospora 613
madisonensis, Colletotrichum 639
madrasense, Botryosporium 593
madronicola, Mycena 798
maesae, Cercospora 613
magica, Urocystis 1008
magna, Glomerella 719
magnacystidiata, Hyphodontia 742
magnahypha, Peniophora = **Candelabrochaete langloisii** 600
magnahypha, Poria = **Wolfiporia cocos** 1024
magnifica, Microsphaera 790
magnifica, Pleospora 885
magnifructa, Macrophoma (Dothiorella, Phoma) 774
magnispora, (Cryptodiaporthe, Valsa) = **Macrodiaporthe everhartii** 773

magnisporum, Haplotrichum (Oidium) 729
magnisporum, Phaeangium 839
magnisporus, Marasmius = **Marasmiellus candidus** 776
magnoliae, Anthostomella = **A. unguiculata** 566
magnoliae, (Camarosporium, Shearia) = **Shearia fusa** 961
magnoliae, Cenangium (Dermatella) 602
magnoliae, Cercosporidium (Cercospora, Isariopsis, Phaeoisariopsis) 620
magnoliae, Diaporthe 671
magnoliae, Didymosphaeria 678
magnoliae, Dothidasteromella (Dimerosporium) 687
magnoliae, Elsinoe 692
magnoliae, Exophoma 703
magnoliae, Hendersonia 733
magnoliae, Hyponectria (Guignardia, Laestadia, Mycosphaerella, Sphaerella) 744
magnoliae, Marasmius 777
magnoliae, (Massaria, Pleomassaria) = **Pleomassaria maxima** 883
magnoliae, Melanconium 782
magnoliae, Meliola 785
magnoliae, Nectria 813
magnoliae, Phanerochaete (Radulum) 844
magnoliae, Phyllosticta 870
magnoliae, Plagiostoma (Gnomonia) 881
magnoliae, Septoria 953
magnoliae, Sphaceloma 964
magnoliae, Spogotteria (Eustegia, Protostegia) 970
magnoliae, Stenellopsis (Heterosporium) 977
magnoliae, Urosporella (Dilophia) 1014
magnoliae, Xylaria 1026
magnoliicola, Phomopsis (Phoma) 859
magnosporum, Hysterium 747
magnum, Harpographium = **Morrisographium persicae** 796
magnum, Melanconium 782
magnusiana, Pleospora 885
magnusiana, Puccinia 909
magnusiana, Ramularia = **Cylindrocarpon magnusianum** 657
magnusiana, Sphacelotheca (Tilletia) 965
magnusianum, Colletotrichum = **C. malvarum** 639
magnusianum, Cylindrocarpon (Septocylindrium) 657
mahoniana, (Phoma, Phyllosticta) = **Phyllosticta mahoniicola** 870
mahoniicola, Phyllosticta 870
maianthemi, Phyllosticta 870
majalis, Ascochyta 576
majanthemi, Phoma 854
majanthemi, Puccinia = **P. sessilis** 913
major, Aposphaeria 571
major, Ceratocystis = **C. adiposa** 603
major, Cintractia = **Ustilago montagnei** var. **major** 1015
major, Kellermania = **K. anomala** 752
major, Uromyces 1012
major, Urophlyctis = **Physoderma majus** 877
majus, Physoderma (Cladochytrium) 877
majuscula, Phomopsis 859
majusculum, Hypoxylon = **H. fragiforme** 745
majusculum, Odontotrema 818
majusculum, Stemphylium 977
malachrae, Cercospora 613
malacotricha, Coniochaeta 641
malacotricha, Meliola 785
malayensis, Cercospora 613
malbranchei, Cyphella = **Cellypha goldbachii** 601
malenconii subsp. **americana, Peniophora** 829
mali, Alternaria 562
mali, Ascochyta 576
mali, Botrytis 594
mali, Coniosporium = **Spilocaea pomi** 969
mali, Endomycopsis (Endomyces) 694
mali, Hendersonia 733
mali, Monochaetia = **Seiridium unicorne** 945
mali, Oospora 819
mali, Pestalotia = **Seiridium unicorne** 945
mali, (Phoma, Phyllosticta) = **Phoma macrostoma** 854
mali, Phomopsis 859
mali, Pleospora 885
mali, Pseudocercospora (Cercospora) 894
mali, Pyrenochaeta 918
mali, Xylaria 1026
mali var. **fructus, Dothiorella** 689
malicola, Antrodia (Coriolellus, Trametes) 567
malicola var. **macropoda, Pholiota** 851
malicorticis, (Cryptosporiopsis, Gloeosporium) = **Cryptosporiopsis curvispora** 653
malicorticis, Pezicula (Neofabraea) 837
malifoliorum, Illosporium 748
malkoffii, Cercospora 613
malkoffii, Phyllosticta 870
malleola, Daldinia = **Hypoxylon thouarsianum** 746
malleolus, Hypoxylon = **H. thouarsianum** 746
malli, Fusarium = **F. solani** 710
mallotii, Cercospora 613
malorum, Cladosporium 630
malorum, Diplodia 681
malorum, Phialophora (Sporotrichum) 847
malorum, Physalospora = **Botryosphaeria stevensii** 593
malorum, Sphaeropsis 967
malorum, Sphaeropsis = **Diplodia mutila** 681
malvacearum, Phomopsis = **P. lavaterae** 858
malvacearum, Puccinia 909
malvarum, Cercospora 613
malvarum, Colletotrichum (Steirochaete) 639
malvastri, Puccinia = **P. sherardiana** 914
malvavisci, Phyllosticta 870
malvicola, Cercospora 613
malvicola, Kuehneola (Uredo) 753
malvicola, Septoria 954
mamaonis, Cercospora 613
mamillana, Clypeosphaeria 633
mamillana, Diplodia 681
mamillariae, Phoma (Sphaeropsis) 855
mamillaris, Macrophoma 774
mamillaris, Phoma 855
mamillatum, Pythium 921
mammatum, Hypoxylon 745
mammiformis, Rosellinia 937
mammillaris, Sphaeropsis = **Phoma mamillaris** 855
mammoidea, Nectria = **N. discophora** 812
mammoides, Strickeria (Teichospora) 982
mammoides var. **opuntiae, Teichospora** 988
manca, Asteridiella (Irenina, Meliola) 580
manca var. tenuis, Meliola = **Asteridiella manca** 580
mangiferae, Chaetothyrium 624
mangiferae, Elsinoe 692
mangiferae, Gloeosporium = **Colletotrichum gloeosporioides** 638
mangiferae, Johncouchia 750
mangiferae, Pestalotiopsis (Pestalotia) 836
mangiferae, Phaeosphaerella 841
mangiferae, Phyllostictina = **Phyllosticta anacardiacearum** 864
mangiferae, Sphaceloma 964
mangifolia, Pestalotia 835
mangini, Aspergillus 578
manihotis, Phaeoramularia (Ragnhildiana) 841
manihotis, (Cercospora, Septogloeum) = **Cercosporidium henningsii** 620
manitobana, Cercospora = **Cercosporidium manitobanum** 620
manitobanum, Cercosporidium 620
manitobiensis, (Didymosphaeria, Microthelia) = **Hormotheca rubicola** 738
manshurica, Peronospora 833
mansonii, Capronia (Dictyotrichiella) 601
mansonii, Exophiala (Microsporum, Rhinocladiella) 703
mappa, Postia (Poria) 891
mappus, Oligoporus = **Postia mappa** 891
marantaceae, Phyllosticta 870
marchalianum, Tetracladium 988
marconii, Botryosphaeria 592
marconii, Dendrophoma 667
marcyensis, Phaeosphaeria (Leptosphaeria) 842
margaritaceae, Septoria 954
marginalis, Melanconis (Diaporthe) 782
marginalis, Milesina (Milesia) 792
marginata, Ascochyta 576
marginata, Biscogniauxia 589
marginata, Galerina (Pholiota) 712
marginata, Maireina (Cyphella) 776
marginata, Septoria 954
marginata, Sphaerulina (Leptosphaeria) 968
marginata, Torula 995
marginata, Trichopeziza = **Lachnum fuscescens** 754
marginatum, Hypoxylon = **H. truncatum** 746
marginella, Mycena = **Hydropus marginellus** 739
marginella, Poria = **Phellinus ferrugineofuscus** 845
marginellus, Hydropus 739
marginellus, Polyporus = **Phellinus ferrugineofuscus** 845
mariae, Phoma 855
mariae, Seimatosporium (Cryptostictis) 945
mariae, Stagonospora 974
mariae-wilsoniae, Puccinia 909
mariae-wilsoniae var. **montiae, Puccinia** 909
mariae-wilsonii, Septoria 954
marianae, Puccinia 909
marianum, Crustoderma 650
marianum, Gloeocystidiellum 716
marina, Leptosphaeria 762
marina, Sigmoidea 961
maritima, Alternaria 562
maritima, Cercospora 613
maritima, Corollospora 645
maritima, Leptosphaeria = **L. neomaritima** 762
maritima, Stagonospora 974
marmorata, Mylittopsis (Tremella) 808
marmoratus, Fomes = **F. fasciatus** 705
marmoreus, (Agaricus, Hypsizygus) = **Hypsizygus tessellatus** 747
marquandii, Paecilomyces 825
marrubii, Cercospora (Cercospora) 613
marrubii, Coniothyrium 642

marrubii, Synchytrium 984
marsiliae, Physoderma 877
marsipium, Pythium 921
martensii, Penicillium = **P. aurantiogriseum** 827
martiana, (Peniophora, Phlebia) = **Phlebia coccineo-fulva** 848
martianoffiana, (Doassansia, Doassansiopsis) = **Doassansiopsis hydrophila** 686
martianum, Corticium = **Phlebia coccineo-fulva** 848
martii, Erysiphe = **Microsphaera trifolii** 791
martii var. phaseoli, Fusarium = **F. solani** 710
martii var. pisi, Fusarium = **F. solani** f. sp. **pisi** 710
martinae, Coccomyces = **Propolis emarginata** 892
martini, Saccardia = **Phaeosaccardinula martinii** 841
martiniana, Meliola (Irenopsis) 785
martinii, Coprinus 644
martinii, Marssonina (Gloeosporium, Marssonia) 778
martinii, Phaeosaccardinula 841
martinii, Uromyces 1012
martiorum, Clitocybe 633
marylandica, Puccinia 910
marylandica, Xylosphaeria = **Mycothyridium marylandicum** 807
marylandicum, Ceratostoma 605
marylandicum, Mycothyridium 807
masonii, Zygosporium 1027
massalis, Puccinia 910
massalongii, Rebentischia 929
massariana, Leucocytospora (Cytospora) 767
massariana, Leucostoma (Valsa) 767
massarina, Discostroma (Clethridium, Metasphaeria) 685
massarioides, Didymosphaeria 678
mastophorum, Pythium 921
mastrucata, Hohenbuehelia 737
mastrucatus, Pleurotus = **Hohenbuehelia mastrucata** 737
matildae, Pestalotia 835
matricariae, Septoria 954
mattirolianum, Elsinoe 692
mattirolianum, Sphaceloma (Illosporium) 964
maunauluana, Seuratia 960
maura, Diplodia = **D. malorum** 681
maura, Eutypa (Diatrype, Valsa) 700
maura, Pestalotia 835
mauritanica, Nummularia 817
mauritiicola, Nectria (Corallomyces) 813
maxima, Phyllosticta (Phyllostictina) 870
maxima, Pleomassaria 883
maxima, Trametes (Cerrena) 996
maxima, Tuberculina 1002
maximum, Leptothyrium 766
maximum, Rhytisma = **Cryptomyces maximus** 652
maximus, Crepidotus 647
maximus, Cryptomyces 652
maximus, (Coriolus, Irpex, Polyporus) = **Trametes maxima** 996
maydicum, Clasterosporium = **Pithomyces maydicus** 880
maydicus, Pithomyces 880
maydis, Ascochyta 576
maydis, Bipolaris (Drechslera, Helminthosporium) 588
maydis, Diplodia (Sphaeria) 681
maydis, Leptosphaeria 762
maydis, Macrosporium = **Alternaria alternata** 561
maydis, Phyllosticta 870
maydis, Physoderma 877
maydis, Stenocarpella 977
maydis, Stictis 978
maydis, Ustilago = **U. zeae** 1017
mayor, Glomerella = **G. cingulata** 719
mayorii, Peronospora 833
mayorii, Uromyces 1012
mcclatchieana, Puccinia 910
mckinleyensis, Uredo 1008
medeolae, Phyllosticta 870
media, Peniophora = **Tubulicrinis medius** 1003
media, Phoma 855
media, Pleospora = **P. penicillus** 885
media var. obtusa, Pleospora = **Lewia scrophulariae** 767
medicaginis, (Ascochyta, Phyllosticta) = **Sporonema phacidioides** 972
medicaginis, Cercospora 613
medicaginis, (Gloeosporium, Septogloeum, Septoria) = **Stagonospora meliloti** 974
medicaginis, Leptotrochila (Pyrenopeziza) 766
medicaginis, Phoma 855
medicaginis, Placosphaeria 881
medicaginis, Pseudopeziza 897
medicaginis, Pseudoplea = **Leptosphaerulina trifolii** 764
medicaginis, Uromyces = **U. striatus** 1013
medicaginis var. pinodella, Phoma = **P. pinodella** 855
medioburiense, Hyphoderma 741
medioburiensis, Peniophora = **Hyphoderma medioburiense** 741
mediosetigera, Halosphaeria (Halosphaeriopsis) 728
mediosetigera var. grandispora, Halosphaeria = **Halosphaeria mediosetigera** 728
mediterraneum, Hypoxylon 745
mediterraneum var. **microspora, Hypoxylon** 745
medium, Asterostroma 583
medium, Botryobasidium 591
medius, Tubulicrinis 1003
medulla-panis, Perenniporia (Poria) 830
medullaris, Rosellinia 937
medusa, Davisomycella (Hypodermella) 666
medusa, Lulworthia (Linocarpon, Ophiobolus) 773
medusae, Melampsora 781
medusae f. bromi, Ophiobolus = **Ophiosphaerella herpotricha** 820
medusaea, Diaporthe = **D. rudis** 672
megalacanthum, Pythium 921
megalobrunnea, Ceratocystis = **Ophiostoma megalobrunneum** 821
megalobrunneum, Ophiostoma 821
megalocarpum, Chaetomium 623
megalocystis, Galerina 712
megaloma, (Elfvingia, Fomes, Ganoderma) = **Ganoderma applanatum** 713
megalopotamica, Cercospora 613
megalospora, Diaporthe 671
megalospora, Diplodia 681
megalospora, Dothivalsaria (Massariovalsa, Myrmaecium, Valsaria) 689
megalospora, Monilinia (Sclerotinia) 794
megalosporum, Xerotrema 1025
megarrhizae, Didymella 677
megarrhizae, Phloeospora 849
megarrhizae, Phoma 855
megarrhizae, Septoria 954
megarrhizae, Stictis = **S. monilifera** 978
megasperma, Bremiella 595
megasperma, Phytophthora 878
megasperma var. sojae, Phytophthora = **Phytophthora megasperma** f. sp. **glycinea** 878
megasperma f. sp. **glycinea, Phytophthora** 878
megasperma f. sp. **medicaginis, Phytophthora** 879
megaspermum, Coryneum 646
megastega, Teichospora (Strickeria) 988
megastoma, Diatrype = **Eutypella cerviculata** 700
megastoma, Physalospora = **Polystigma astragali** 889
meibomiae, Phyllosticta 870
meineckella, Cunninghamella = **Spiniger meineckellum** 969
meineckellum, Spiniger 969
meinersii, Rhynchosporina (Gloeosporium) 935
meinersii var. alpina, Gloeosporium = **Rhynchosporina meinersii** 935
mela, Eutypa 700
melaena, Pseudoplectania (Melascypha) 898
melaleuca, Cercospora = **C. desmodiicola** 609
melaleuca, Diplonaevia 683
melaleuca, Hypocrea = **H. schweinitzii** 743
melaleuca, Mollisia (Niptera, Patella) 793
melaleuca, Phoma (Phomopsis) 855
melaleuca, Phyllosticta 870
melaleuca var. epiphyllum, Lophodermium = **L. phloxii** 772
melaleucoides var. **vaccinii, Tapesia** 986
melaleucum, Lophodermium 772
melaleucum var. epiphyllum, Lophodermium = **Lophodermium melaleucum** 772
melampodii, Puccinia 910
melanconioides, Puccinia 910
melanconioides, Sphaeropsis = **Aplosporella ailanthi** 569
melandrii, Septoria = **S. lychnidis** 953
melanocephala, Puccinia 910
melanochaeta, Phaeoisariopsis (Cercospora) 840
melanochlorum, Fusarium (Fusisporium) 708
melanodiscus, Cytospora = **C. umbrina** 662
melanodiscus, Valsa 1018
melanogramma, Schizonella 939
melanophaea, Niptera 816
melanophthalmi, Septoria 954
melanoplaca, Phyllachora 863
melanoplaca, Phyllosticta 870
melanops, Botryosphaeria 592
melanopus, Acrothecium (Dactylium) 557
melanopus, Polyporus 889
melanosporum, Crocicreas (Belonium, Scelobelonium) 649
melanosticta, Anthostomella 566
melanotes var. **symphoricarpi, Anthostoma** 565
melanoxanthus, Endocalyx 693
melantera, Amphisphaeria = **Splanchnonema melanterum** 969
melantera, Sphaeria = **Diapleella coniothyrium** 670
melanterum, Splanchnonema 969
melaspileoides, Karschia 752
melastoma, Plectania (Bulgaria) 883
melastoma, Valsella (Valsa) 1019
melathroa, Botryosphaeria 592

meliae, (Botryodiplodia, Diplodia) = **Diplodia langloisii** 680
meliae, Cercospora 613
meliae, Fomitopsis (Fomes, Polyporus) 706
meliae, Melogramma 786
meliae, Phyllosticta 870
melicae, Septoria 954
meliigena, Mycena 798
meliloti, Entyloma 696
meliloti, (Ascochyta, Gloeosporium) = **Ascochyta lethalis** 575
meliloti, Plenodomus = **Phoma sclerotioides** 856
meliloti, Pleospora = **P. herbarum** 884
meliloti, Pseudopeziza = **P. trifolii** 897
meliloti, Ramularia 925
meliloti, Stagonospora 974
melinii, Phialophora (Cadophora) 847
melioloides, Chaetomium = **C. indicum** 623
melioloides, Lasiostemma (Asterina, Dimeriella, Wentiomyces) 757
melioloides, Passalora 826
mellea, Armillaria (Armillariella) 572
mellea, Cyphellopsis (Cyphella) 660
mellea, Ostropa = **Schizoxylon melleum** 940
mellea, Volutella 1023
melleopora, Fomitiporella = **Phellinus melleoporus** 846
melleoporus, Phellinus 846
melleum, Gloeosporium = **Phoma exigua** 853
melleum, Schizoxylon 940
melleus, Gymnopilus 725
melleus, Merismodes = **Cyphellopsis mellea** 660
mellifera, Puccinia 910
mellispora, Leptosphaeria 762
melongenae, Cercospora 613
melongenae, Gloeosporium = **Colletotrichum gloeosporioides** 638
melonis, Ascochyta = **Phoma cucurbitacearum** 852
melonis, Cercospora = **Corynespora cassiicola** 646
melonis, (Didymella, Mycosphaerella) = **Didymella bryoniae** 676
melonis, Marssonina 778
membranacea, Trametes 996
mendax, Sclerotium 942
menezesiana, Pestalotia 835
menispermi, Diaporthe = **D. beckhausii** 670
menispermi, Entyloma 696
menispermi, Phaeoisariopsis (Cercospora, Cercosporiopsis) 840
menispermi, Phomopsis (Phoma) 859
menispermi, Sphaeropsis 967
menispermi, Tubercularia 1002
menispermi, Valsa = **V. ambiens** 1017
menispermicola, Phyllosticta 870
menthae, Physoderma 877
menthae, Puccinia 910
menthae, Ramularia = **R. menthicola** 925
menthae, Septoria 954
menthae, Sphaceloma 964
menthicola, Cercospora 613
menthicola, Ramularia 925
menthicola, Septoria = **S. menthae** 954
mentzeliae, Phyllosticta 870
mentzeliae, Septoria 954
menyanthicola, Ascochyta 576
menyanthis, Ascochyta = **A. menyanthicola** 576
menyanthis, Physoderma 877
menyanthis, Septoria 954
menziesiae, Melasmia 784
mera, Sphenospora 969
meridionalis, Mycoacia 799
merismoides, Fusarium 708
merismoides, Phlebia (Merulius) 848
merremiae, Meliola = **M. malacotricha** 785
merrillii, Phellinus (Pyropolyporus) 846
mertensiae, Puccinia = **P. hydrophylli** subsp. **mertensiae** 908
mertensianae, Coccomyces 634
merulioidea, Phlebia = **P. rufa** 848
mesadeniae, Aecidium 558
mesaverdensis, Aleurodiscus 560
mesenterica, Tremella 998
mesilliana, Ravenelia 928
mesnieriana, Puccinia 910
mesoedema, Leptosphaeria = **Heptameria obesa** 734
mesomajalis, Puccinia (Micropuccinia) 910
mesosporus, Marasmiellus 776
mespili, Diplocarpon 679
mespili, Entomosporium 695
mespili var. cydoniae, Entomosporium = **Entomosporium mespili** 695
metableticum, Amarenographium (Camarographium, Camarosporium) 563
metachroa, Gloeotulasnella = **Tulasnella hyalina** 1004
metata, Mycena 798
metrosideri, Calothyriopeltis 598
metrosideri, Endothia (Diaporthopsis) 694
metrosideri, Mycosphaerella 803
mexicana, Ephelis 696
mexicana, Lenzites = **Gloeophyllum mexicanum** 717
mexicana, Phakopsora (Bubakia) 843
mexicana, Uredo 1008
mexicana, Ustilago 1015
mexicanum, Aecidium 558
mexicanum, Gloeophyllum 717
mexicanus, Uromyces 1012
mexicensis, Puccinia 910
mezerei, Gloeosporium = **Marssonina daphnes** 778
micaceus, Coprinus 644
micans, Typhula (Pistillaria) 1005
michelianum, Microthyrium 791
micheneri, Costantinella (Botrytis) 646
micheneri, Lentinellus 759
micheneri, Pestalotia = **Pestalotiopsis dichaeta** 836
micheneri, Stereum = **Thelephora albidobrunnea** 989
michotii, Paraphaeosphaeria (Leptosphaeria) 826
micrantha, Puccinia 910
microcarpa, Eutypella = **E. scoparia** 701
microcarpa, Massarina (Massaria) 779
microcephala, Claviceps = **C. purpurea** 632
microellipsodes, Zygosaccharomyces (Saccharomyces) 1027
microlepiae, Ramularia 925
microplaca, Nummularia = **Hypoxylon microplacum** 745
microplacum, Hypoxylon 745
microporus, Rigidoporus (Polyporus) 936
micropuncta, Eutypa = **Graphostroma platystoma** 723
micropuncta, Phyllosticta 870
micropuncta, Septoria 954
micropus, Bipolaris (Helminthosporium) 588
microsclerotia, Rhizoctonia = **R. solani** 932
microsclerotium, Corticium 645
microscopica, Cookella 644
microscopica, Phaeosphaeria (Leptosphaeria) 842
microscopica, Sphaeropsis 967
microscopica var. culmorum, Phaeosphaeria = **P. culmorum** 842
microscopicum, Coniothyrium 642
microscopicum, Helicoma (Helicosporium) 730
microscopicum, Microthyrium 791
microsora, Cercospora 613
microsora, Mycosphaerella 803
microsora, Puccinia 910
microsperma, Calosphaeria (Circinostoma) 598
microsperma, Discella 683
microsperma, Discula 685
microsperma, Gyrothrix 727
microsperma, Phoma = **P. karstenii** 854
microsperma, Septoria = **Asteroma microspermum** 582
microspermum, Asteroma 582
microspermum, Circinotrichum = **Gyrothrix microsperma** 727
microspilum, Cylindrosporium 659
microspora, Acarosporina (Schizoxylon) 556
microspora, Camarops (Anthostoma, Diatrype) 599
microspora, Ceratocystis (Ceratostomella, Ophiostoma) 604
microspora, Ceratostomella 605
microspora, (Coryne, Ombrophila) = **Neobulgaria pura** 814
microspora, Cytospora 662
microspora, Dendrothele 668
microspora, Diplodia = **Microdiplodia microsporella** 788
microspora, Eriopezia (Lachnella) 697
microspora, Helotiella 732
microspora, Pestalotia 835
microspora, Phoma 855
microspora, Poria = **Postia placenta** 891
microspora, Puccinia 910
microspora, Septoria 954
microspora, Tomentella = **T. ochracea** 994
microspora, Trechispora (Athelia, Grandinia) 997
microspora, Tulasnella = **T. eichleriana** 1004
microspora, (Eutypella, Valsa) = **Eutypella tetraploa** 701
microsporella, Microdiplodia (Diplodia) 788
microsporella, Phoma 855
microsporum, Aleurocorticium = **Dendrothele microspora** 668
microsporum, Entyloma 696
microsporum, Hormodendrum 737
microsporum, Leptographium 759
microsporus, Rhizopus 934
microsporus var. rhizopodiformis, Rhizopus 934
microstega, Diatrype = **D. prominens** 673
microsticta, Kabatiella = **Aureobasidium microstictum** 584
microstictum, Aureobasidium 584
microstictum, Coryneum = **Seimatosporium lichenicola** 945
microstictum var. foliae, Coryneum = **Seimatosporium discosioides** 944
microstictum var. staphyleae, Coryneum = **Seimatosporium vaccinii** 945
microstigma, Cercospora = **C. caricis** 607
microstoma, Valsa 1018
microstomum, Schizoxylon 940
microtheca, Jattaea (Calosphaeria) 750
miczynskii, Penicillium 828
middletonii, Pythium 921

miehei, Rhizomucor (Mucor) 933
mikaniacola, Cercospora 613
mikaniae, Cercospora 613
mikanii, Septoria 954
millardetii, Seuratia (Atichia) 960
millefolii, Leptosphaeria (Pleospora) 762
millefolii, Puccinia = **P. cnici-oleracei** 903
millepunctata, Cryptosphaeria = **C. eunomia** 652
millepunctata, Mycosphaerella (Guignardia, Laestadia, Phyllachora) 804
milleri, Dasyscyphus = **Lachnum arundinariae** 754
milliaria, Eutypa = **E. lata** 699
mimuli, Cercospora 613
mimuli, Didymolepta (Apiosporella) 677
mimuli, Macrophoma 774
mimuli, Microdiplodia 788
mimuli, Ramularia 925
mimuli, Septoria (Septoria) 954
minax, Mycovellosiella (Ramularia) 808
miniata, Belonioscypha = **Crocicreas panici** 649
miniata, Peniophora = **Phanerochaete sanguinea** 844
miniata, Tubercularia = **T. vulgaris** 1002
miniatus, Lachnaster 753
miniatus, Sporobolomyces = **S. roseus** 971
minima, Ascochyta 576
minima, Caryospora 601
minima, Cercospora = **Pseudocercospora mali** 894
minima, Ceuthospora 622
minima, Discosia = **Mycoleptodiscus minimus** 799
minima, Dothidella 687
minima, Phyllosticta 870
minima, Physalospora 876
minima, Sordaria 963
minima, Sporormiella (Preusia, Sporormia) 972
minima, Togninia 993
minima, Ustilago 1015
minimum, Chaetomium = **C. aureum** 623
minimum, Cryptosporium 653
minimum, Gloeosporium = **Ascochyta minima** 576
minimus, Mycoleptodiscus 799
minimus, Uromyces 1012
minnsiae, Corticium (Aleurodiscus, Laeticorticium) 645
minor, Anthostomella 566
minor, Ascochyta (Apiocarpella) 576
minor, Aspergillus = **Aspergillus glaucus** 578
minor, Asteromassaria (Caryospora) 582
minor, Botryosphaeria (Melanops) 592
minor, (Ceratocystis, Ceratostomella) = **Ophiostoma minus** 821
minor, Cintractia = **C. peribebuyensis** 628
minor, Dacrymyces 663
minor, Dictyosporium = **Eversia subopaca** 701
minor, Dothiorella 689
minor, Mycosphaerella 804
minor, Naemacyclus = **Cyclaneusma minus** 656
minor, Phloeosporina = **P. fraxini** 850
minor, Phyllosticta 870
minor, Podospora (Pleurage, Sordaria) 888
minor, Sclerotinia 941
minor, Stypella 983
minor, Thelephora = **T. terrestris** 990
minor, Tilletiopsis 992
minor, Uromyces 1012
minor, Ustilago 1015
minor, Xenodochus 1025
minor var. reticulata, Mycosphaerella = **Venturia subcutanea** 1021
minus, Cyclaneusma 656
minus, Cylindrosporium = **Phloeosporina fraxini** 850
minus, Gloeosporium 718
minus, Odontotrema 818
minus, Ophiostoma 821
minus f. salicella, Odontotrema = **O. majusculum** 818
minusculoides, (Oligoporus, Polyporus, Tyromyces) = **Postia ceriflua** 890
minussensis, Puccinia 910
minuta, Ceratocystiopsis (Ceratocystis) 603
minuta, Cercospora (Cercoseptoria) 613
minuta, Cytospora 662
minuta, Diaporthe = **D. ceanothi** 670
minuta, Eutypella 701
minuta, Extrawettsteinina 704
minuta, Guepiniopsis = **Dacrymyces minutus** 663
minuta, Microdiplodia (Diplodia) 789
minuta, Pezizella 838
minuta, Phoma 855
minuta, Phorcys = **Didymosphaeria clementsii** 678
minuta, Puccinia 910
minuta, Pyrenochaeta 918
minuta, Pyrenopeziza 919
minuta, Septoria 954
minuta-bicolor, Ceratocystiopsis (Ceratocystis) 603
minutaspora, Phyllosticta 870
minutella, Physalospora = **Phomatospora berkeleyi** 857
minutella, Valsa = **V. ceratosperma** 1017
minutella, Vialaea 1022
minutispora, Diatrypella 674
minutissima, Flagelloscypha (Cyphella) 705
minutissima, Periconia 831
minutissima, Phyllachora (Isothea) 863
minutissima, Phyllosticta 870
minutissima, Puccinia 910
minutissimum, Helicoma 730
minutulum, Aecidium 558
minutulum, Asperisporium (Fusicladium) 579
minutulum, Sphaeronaema 966
minutum, Coleosporium 636
minutum, Melanomma 783
minutum, Sphaerobasidium (Paullicorticium, Xenasma) 965
minutus, Dacrymyces 663
minutus, Marasmius 777
minutus, Pleurotus (Crepidopus) 886
minutus, Uromyces 1012
mirabile, Acanthophysium 556
mirabilis, Aleurodiscus = **Acanthophysium mirabile** 556
mirabilis, Cercospora 613
mirabilis, Cercosporella 619
mirabilis, (Exoascus, Taphrina) = **Taphrina communis** 986
mirabilis, (Gnomonia, Gnomoniella) = **Plagiostoma campylostylum** var. **mirabilis** 881
mirabilis, Hobsonia (Helicomyces) 736
mirabilis, Kabatia 751
mirabilis, Lirula (Hypodermella) 769
mirabilis, Mycotribulus 807
mirabilis, Uredinopsis = **U. americana** 1007
mirabilis var. **oblongifoliae, Kabatia** 751
mirabilis var. tortilis, Taphrina = **T. communis** 986
mirabilissima, Cumminsiella (Puccinia) 655
mirabilissima, Rhabdospora (Septoria) 931
mirifica, Puccinia 910
mirificum, Repetobasidium 929
miscanthi, Pseudocercospora = **Mycovellosiella koepkei** 807
mississippiensis, Cercospora 613
mississippiensis, Septoria 954
missouriense, Gyrostroma 727
missouriense, Melanopsichium = **M. pennsylvanicum** 783
missouriensis, Ascochyta 576
missouriensis, Cercospora 613
missouriensis, Diatrypella 674
missouriensis, Nectria (Pleonectria, Thyronectria) 813
missouriensis, Phomopsis 859
mitchellae, Meliola 785
mitchellae, Microthyrium 792
mitchellae, Synchytrium 984
mitellae, Cercospora 613
mitellae, Phyllosticta 871
mitellae, Ramularia 925
mitellae, Septoria 954
mitellae var. **heucherae, Ramularia** 926
mitis, Panellus 825
miurae, Uromyces 1012
mixta, Puccinia = **P. allii** 900
mixta, Tilletia = **Sorosporium mixtum** 963
mixtum, Sorosporium 963
miyabeana, Physalospora = **Glomerella cingulata** 719
miyabeanus, Cochliobolus (Ophiobolus) 635
miyakei, Phyllosticta 871
modesta, Nodulosphaeria (Leptosphaeria, Sphaeria) 817
modesta, Pleosphaeria 883
modesta, Valsa 1018
modestum, Aecidium 558
modiolae, Cercospora 613
modioliensis, Synchytrium = **S. australe** 983
modonia, Pseudovalsa (Melanconis) 899
modonium, Coryneum 646
moerenhoutiana, Meliola 785
mohavensis, Hyponectria 744
molesta, Agrocybe 559
molinea, Carpenterella 601
molitoris, Lophodermium 772
molle, Cladosporium 630
molle, Helminthosporium = **Dendryphiella vinosum** 668
molle, Hypochniciellum 742
mollereana, Septoria 954
molleriana, Cercospora 613
molleriana, Macrophoma 774
molleriana, Mycosphaerella 804
molleriana, Peniophora = **Phanerochaete ravenelii** 844
mollerianum, Coniothecium 642
mollis, Aposphaeria 571
mollis, Crepidotus 647
mollis, Datronia (Trametes) 665
mollis, Irpex = **Spongipellis pachyodon** 970
mollis, Leptoporus (Polyporus, Tyromyces) 760
mollis, Meliolina (Meliola) 786
mollis, (Leucogyrophana, Peniophora) = **Hypochniciellum molle** 742
mollis, Stictis 978
mollis var. **cystidiosus, Crepidotus** 647
mollisia, Septoria 954
mollisioides, Tapesia 986
molliuscula, Dermea 669
molliusculus, Polyporus = **Trametes cervina** 996

mollugínicola, Cercospora = **C. molluginis** 613
molluginis, Cercospora (Cercospora) 613
molluginis, Phyllosticta 871
mollusca, Leucogyrophana (Serpula) 767
mollusca, Trechispora (Poria) 997
molluscus, Merulius = **Leucogyrophana mollusca** 767
molluscus, Polyporus = **Trechispora mollusca** 997
molokaiensis, Puccinia 910
molybdaea, Tomentella 994
molybdea, Exidiopsis (Sebacina) 702
molybdea, Pseudotomentella 899
momordicae, Leandria 758
momordicae, Ramularia 926
monadelphus, Clitocybe = **Armillaria tabescens** 572
monanthochloes, Phyllachora 863
monardae, Cladosporium 630
monardae, Phyllosticta 871
monardae, Puccinia 910
monardellae, Phyllosticta (Asteromella) 871
monesis, Chrysomyxa 626
monilifera, Acarosporina 556
monilifera, Sphacelotheca (Ustilago) 965
monilifera, Stictis 978
moniliferum, Schizoxylon = **Acarosporina monilifera** 556
moniliforme, Fusarium 708
moniliforme, Gibberella = **G. fujikuroi** 714
moniliforme, Ophiostoma = **Ceratocystis moniliformis** 604
moniliforme var. subglutinans, Fusarium = **F. subglutinans** 711
moniliformis, Ceratocystis (Ceratostomella, Endoconidiophora) 604
monilioides, Ramularia (Ovularia) 926
monilioides, Rhizoctonia 932
monilipes, Helicoma 730
monilispora, (Leptosphaeria, Pleospora) = **Paraphaeosphaeria michotii** 826
monoceras, Exserohilum (Bipolaris, Helminthosporium) 703
monochaeta, Monochaetia (Pestalotia) 795
monochaeta var. libertiana, (Monochaetia, Pestalozzia) = **Cryptostictis libertiana** 653
monochaetoidea, Pestalotia = **Monochaetia ellisiana** 795
monochaetoidea var. parasitica, Pestalotia = **Seimatosporium parasiticum** 945
monoica, Puccinia 910
monoicae, Pseudocercospora (Cercospora) 894
monospermum, Pythium 921
monosporia, Ovularia = **Ramularia rubella** 926
monosporium, Oidium = **Ramularia rubella** 926
monosporum, Exobasidium 703
monosporus, Triscelophorus 1001
montagnei, Apiospora 569
montagnei, Hendersonia = **Piptarthron macrosporum** 880
montagnei, Stagonospora 974
montagnei, Ustilago (Cintractia) 1015
montagnei var. **major, Ustilago** (Cintractia) 1015
montana, Bondarzewia 590
montana, Davisomycella (Hypodermella) 666
montana, Leucogyrophana 767
montana, Peniophora = **Hyphoderma amoenum** 740
montana, Pleospora 885
montana, Poria = **Diplomitoporus lenis** 683
montana, Ramularia (Cercospora) 926
montana, Tilletia 992
montanae, Teichosporella = **T. dura** 988
montanensis, Crepidotus 647
montanensis, Phomopsis 859
montanensis, Puccinia 910
montanensis, Tomentellastrum (Tomentella) 995
montanensis, Tympanis (Godronia) 1004
montaniensis, Broomella (Keissleria, Lophiostoma) 595
montaniensis, Dothidea = **D. bigeloviae** 687
montaniensis, Sporisorium (Sphacelotheca, Ustilago) 971
montanum, Confertobasidium 640
montanus, Lentinellus 759
montanus, Merulius = **Leucogyrophana montana** 767
montanus, Polyporus = **Bondarzewia montana** 590
montellica, Pestalotiopsis (Pestalotia) 836
montenegrinum, Cylindrosporium 659
montevidensis, Aspergillus = **A. hollandicus** 578
montia, (Ceratocystis, Ceratostomella) = **Ophiostoma ips** 821
montiae, Tolyposporium (Sorosporium) 993
monticola, Leptosphaeria 762
monticola, Melampsora 781
monticola, Poria = **Postia placenta** 891
monticola, Sebacina = **Eichleriella macrospora** 692
monticulosa, Valsa = **Eutypella goniostoma** 700
montis-rainieri, Entyloma = **E. lapponicum** 696
montivaga, Lophodermella (Hypodermella) 771
montivaga f. concolor, Hypodermella = **Lophodermella concolor** 771
montivagum, Phragmidium 861
moorei, Pestalotia 835
moravica, Pleospora (Pyrenophora) 885
moravicum, Microthyrium = **M. microscopicum** 791
morbida, Colpoma (Clithris) 640
morbida, Lophodermella 771
morbidum, Triblidium = **Colpoma morbida** 640
morbosa, Apiosporina (Plowrightia) 569
morbosa, Meliola 785
morbosum, Dibotryon = **Apiosporina morbosa** 569
moreniana, Puccinia 910
morganii, Helicoma 730
morganii, Hypoxylon 745
morganii, Trametes = **Pachykytospora tuberculosa** 825
mori, Aplosporella (Sphaeropsis) 570
mori, Cercosporella 619
mori, Dermea 669
mori, Dothiorella 689
mori, Gonatophragmium 722
mori, Hysterographium 747
mori, Mycosphaerella 804
mori, Polyporus 889
mori, (Cylindrosporium, Phloeospora, Septoria) = **Phloeospora maculans** 849
mori, Stegonsporium (Coryneum) 976
morianum, Gloeosporium = **Sporonema phacidioides** 972
moricarpa, Fracchiaea = **Nitschkia broomeiana** 816
moricola, Cercospora 613
moricola, Cristulariella 648
moricola, Gibberella = **G. baccata** 714
moricola, Phloeospora (Septoria) 849
moricola, Phomopsis (Phoma) 859
moricola, Phyllosticta 871
moriforme, Berkleasmium (Sporidesmium) 587
moriforme subsp. corticola, Sporidesmium = **Berkleasmium corticola** 587
moriformis, Bertia 587
moriformis var. **latispora, Bertia** 587
morigena, Valsa = **V. ceratosperma** 1017
morindae, (Ainsworthia, Phaeosaccardinula) = **Limacinula theae** 768
morindae, Sphaceloma 964
morindicola, Cercospora 613
moroides, Endoxylina (Diatrype, Valsaria) 695
moroides var. **acerina, Valsaria** 1019
morongiae, Cercospora 614
morongiae, Ravenelia = **R. texensis** var. **morongiae** 929
mororum, Phoma = **P. macrostoma** 854
morsei, Hypoxylon = **H. mammatum** 745
mors-uvae, Sphaerotheca 968
morthieri, Exarmidium (Trematosphaeria, Trematostoma) 702
morthieri, Lachnum (Lachnella) 755
morthieriana, Leptosphaeria = **Nodulosphaeria winteri** 817
mortolensis, Septoria 954
mortonii, Phyllosticta 871
mortuosum, Anthostoma = **Anthostomella tomicoides** 566
mosigii, Atichia = **A. glomerulosa** 583
mossii, Mundkurella 797
mouiensis, Hendersonia 733
mounceae, Streptothrix = **Conoplea fusca** 643
mucedo, Mucor 796
mucida, Ceriporiopsis (Poria) 621
mucida, Holwaya (Ditiola) 737
mucidula, Pseudotomentella 899
mucidulus, Hypochnus = **Pseudotomentella mucidula** 899
mucidum, Hydnum = **Cristinia gallica** 648
mucilaginosa, Rhodotorula = **R. rubra** 934
mucor, Mycena 798
mucronatum, Phragmidium 861
mucunae, Phyllosticta 871
mueggenburgii, Rhabdospora (Septoria) 931
muelleri, Helicoma (Helicosporium) 730
muenscheri, Lagenidium 756
muhlenbergiae, Cercospora 614
muhlenbergiae, Clypeostagonospora 634
muhlenbergiae, Leptosphaeria 762
muhlenbergiae, Mycosphaerella (Sphaerella) 804
muhlenbergiae, Phloeospora 849
muhlenbergiae, Puccinia = **P. schedonnardi** 913
muhlenbergiae, Robillarda = **Pseudorobillarda agrostidis** 898
muhlenbergiae, Ustilago 1015
muirensis, Leptosphaeria 762
mulfordiana, Ustilago 1015
mulgedii, Phyllosticta = **Phoma exigua** 853
mulleri, Elsinoe 692
multiannulata, (Ceratocystis, Ceratostomella) = **Ophiostoma multiannulatum** 821
multiannulatum, Ophiostoma 821
multicolor, Penicillium = **P. sclerotiorum** 829
multifera, Philocopra = **Sporormiella minima** 972

multiforme, Hypoxylon 745
multiformis, Melanospora (Erostrotheca) 784
multimaculans, Phloeospora 849
multipartita, Xylaria 1026
multiplex, Ramularia 926
multiplex, Taeniolella (Septonema) 985
multiplex, Valsa = **V. ceratosperma** 1017
multiplex, Xylaria 1026
multiporum, Gymnosporangium 727
multipuncta, Hysteropezizella (Pyrenopeziza) 747
multipunctoidea, Pyrenopeziza 919
multirostrata, Alternaria 562
multiseptata, Clathrospora 631
multiseptata, Kellermania = **Piptarthron macrosporum** 880
multiseptata, Leptosphaeria 762
multiseptata, (Limacinia, Ophiocapnocoma) = **Ophiocapnocoma phloiophilia** 820
multiseta, Pestalotia 835
multispora, Beltrania = **B. rhombica** 587
munda, (Athelia, Peniophora) = **Leptosporomyces mundus** 764
mundus, Leptosporomyces 764
munroae, Septoria 954
muntingiae, Pseudocercospora (Cercospora) 894
munzii, Inonotus (Polyporus) 749
muricata, Discostroma (Strickeria, Teichospora) 685
muricata, Leptosphaeria 762
muricata, Taeniolella 985
muricata, Tomentella 994
muricatum, Dendryphion = **Taeniolella muricata** 985
muriculata, Botryosphaeria (Melanops) 592
murina, Mycovellosiella (Cercospora) 808
murorum, Chaetomium 623
murorum var. **felina, Gliomastix** 715
murrayae, Sphaceloma 964
murrayi, Cystostereum (Stereum) 660
murrillii, Phlebia 848
murrillii, Xenosperma 1025
musae, Blistum = **Stilbella annulata** 981
musae, Colletotrichum (Myxosporium) 639
musae, Cordana (Scolicotrichum) 644
musae, Cyphella = **Cellypha musaecola** 602
musae, Guignardia (Guignardia) 725
musae, Macrophoma = **Phyllosticta musarum** 871
musae, Mycosphaerella 804
musae, Phoma = **P. jolyana** 854
musae, Phyllosticta 871
musae, Pseudocercospora (Cercospora) 894
musaecola, Cellypha 602
musarum, Chaetothyrina (Chaetothyrium) 624
musarum, Gloeosporium = **Colletotrichum musae** 639
musarum, Phyllosticta (Phyllostictina, Sphaeropsis) 871
muscae, Circinella 628
muscari, Uromyces 1012
muscicola, Asterostroma 583
muscicola, Sistotrema (Trechispora) 962
musenii, Puccinia 910
musicola, Mycosphaerella 804
musicola, Phyllosticta 871
musiva, Septoria 954
mustialaensis, Hypochnopsis (Coniophora) 743
mutabile, Corticium = **Fibulomyces mutabilis** 705
mutabilis, Fibulomyces (Athelia) 705
mutabilis, Kuehneromyces (Pholiota) 753
mutabilis, Polyporus = **Microporellus obovatus** 790
mutabilis, Pseudotrichia (Herpotrichia) 899
mutabilis, Puccinia 910
mutans, Hapalopilus (Polyporus, Poria) 729
mutata, Peniophora = **Hyphoderma mutatum** 741
mutatum, Hyphoderma (Corticium) 741
mutica, Ellisiella = **Colletotrichum graminicola** 639
mutila, Diplodia 681
mutila, Physalospora = **Botryosphaeria stevensii** 593
myceliosa, Anomoporia (Poria) 565
mycophaga, Pestalotia = **Toxosporium camptospermum** 995
myinda, (Cryptodiaporthe, Valsa) = **Valsa ambiens** 1017
myosotidis, Peronospora 833
myosotidis, Plasmopara 882
myosotidis, Synchytrium 984
myosura, Baeospora (Collybia) 585
myriadea, Sphaerulina (Sphaerella) 968
myriangioides, (Hypoxylon, Thyridaria) = **Melogramma campylosporum** 786
myricae, Anthostomella = **Coniochaeta pulveracea** 641
myricae, Asterella 580
myricae, Cercospora = **C. penicillus** 614
myricae, Gnomonia (Apiognomonia) 721
myricae, Leptosphaeria 762
myricae, Massarina (Metasphaeria) 779
myricae, Mycosphaerella 804
myricae, Pestalotia = **P. stellata** 835
myricae, Phyllosticta 871
myricae, Septoria 954
myricae, Stagonospora 974
myricae, Stictis = **Propolis phacidioides** 892
myricae, Tremella = **Exidia glandulosa** 702
myricata, Septoria 954
myriospora, Orbilia (Calloria) 822
myriosporum, Cladosporium 630
myriotylum, Pythium 921
myrothecioides, Phaeodochium 839
myrti, Tripospermum (Triposporium) 1001
myrticola, Pseudocercospora (Cercospora) 894
myrtilli, Exobasidium 703
myrtilli, Gibbera (Venturia) 714
myrtilli, Pucciniastrum = **P. vaccinii** 917
myrtillina, Pezicula 837
mysticus, Uromyces = **U. fragilipes** 1011
mytilinellum, Mytilinidion (Lophium) 809
nabali, Septoria 954
nagaii, Pythium 921
nakanishikii, Puccinia 910
nalgiovense, Penicillium = **P. jensenii** 828
namae, Cercospora (Cercosporina) 614
namae, Synchytrium 984
nana, Didymosphaeria 678
nana, Taphrina 987
nana var. **brachyspora, Didymosphaeria** 678
nandinae, Cercospora 614
nandinae, Phyllosticta 871
nannfeldtii, Endophragmia = **Endophragmiella biseptata** 694
nannfeldtii, Plectania 883
nannfeldtii, Scytinostromella (Gloeocystidiellum) 944
nanum, Dendryphion (Helminthosporium) 668
napi, Phyllosticta = **Asteromella brassicae** 582
narbonensis, Peronospora 833
narcissi, Ophiostoma (Ceratocystis) 821
narcissi, Septoria 954
narcissi, Stromatinia 982
narcissicola, Botryotinia (Sclerotinia) 594
narcissicola, Botrytis 594
nardi, Lachnum 755
narvisiana, Septoria 954
narymica, Perenniporia 830
nashii, Meliola 785
nasturtii, Cercospora 614
natalensis, Diplodia = **Lasiodiplodia theobromae** 756
natans, Systremma = **Dothidea sambuci** 687
naucoria, Psathyrella 893
naucorioides, Gymnopilus 725
nautica, Teichospora (Strickeria) 988
navicularis, Hemibeltrania 732
naviculasporum, Helotium = **Hymenoscyphus caudatus** 740
nealii, Sphacelotheca 965
nebelina, Sphaeropsis 967
nebulosa, Phoma 855
nebulosa, Phyllosticta 871
nebulosa, Spilocaea 969
nebulosa, Venturia 1020
nebulosum, Dicoccum = **Spilocaea nebulosa** 969
necans, Ascochyta (Gloeosporium) 576
necans, Myxosporium = **M. pruni** 810
necans, Rhizopus 934
necator, Ramularia 926
necator, Sphaceloma (Gloeosporium) 964
necator, Uncinula 1006
necatrix, Phoma 855
necatrix, Rosellinia 937
necopinus, Uromyces 1012
neglecta, Calloria 597
neglecta, Microsphaera 790
neglecta, Pestalotia 835
neglecta, Septoria 954
neglecta, Ustilago = **Sporisorium neglectum** 971
neglectum, Epicoccum = **E. nigrum** 696
neglectum, Sporisorium 971
negundi, Fusarium = **F. heterosporum** 708
negundicola, Myxosporium 810
negundicola var. **ramicola, Phoma** 855
negundinicola, Alternaria (Macrosporium) 562
negundinicola, Phoma 855
negundinis, Aplosporella (Haplosporella) 570
negundinis, Ascochyta 576
negundinis, Asteridiella 580
negundinis, Cercospora 614
negundinis, Coniothyrium 642
negundinis, Cucurbitaria = **C. ailanthi** 654
negundinis, Cylindrosporium 659
negundinis, Cytospora = **C. annulata** 661
negundinis, Epicoccum = **E. nigrum** 696
negundinis, Macrophoma 774
negundinis, Phacidium 839
negundinis, Phyllosticta 871
negundinis, Piggotia 879
negundinis, Sphaeropsis 967
negundinis, Stigmina (Coryneum, Sciniatosporium, Thyrostromella) 980
negundinis, Teichospora 988
neilliae, Aplosporella (Sphaeropsis) 570
neilliae, Marssonina (Marssonia) 778
nelsonii, Gymnosporangium 727
nelumbii, Alternaria 562
nelumbii, Dothiorella 689
nelumbonis, Cercospora = **Pseudocercospora nymphaeacea** 895

nematospora, Septoria 954
nematospora f. **aggregata, Septoria** 954
nematosporum, Helicosporium = **H. pannosum** 731
nemopanthis, Durandiella (Godronia, Tympanis) 691
nemopanthis, Godroniopsis 722
nemopanthis, Ramularia 926
nemophilae, Peronospora 833
nemoseridis, Mycosphaerella 804
neobourdotii, Tomentella 994
neoceras, Fusarium = **F. subglutinans** 711
neocoronata, Puccinia 910
neomaritima, Leptosphaeria 762
neomexicana, Microsphaera 790
neomexicana, Phyllosticta 871
neomexicana, Plowrightia 887
neomexicanum, Byssocorticium 596
neopithya, Tympanis 1004
neottiae, Rhizoctonia 932
neowashingtoniae, Sphaerodothis (Phaeochora) 965
neowashingtonianae, Euryachora 699
nepetae, Ascochyta = **A. leonuri** 575
nepetae, Cercospora 614
nepetae, Septoria 954
nepetae, Teichospora = **Cilioplea coronata** 627
nepheloides, Cercospora = **C. eustomae** 610
nephrodes, Crepidotus 647
nephrolepis, Glomerella 719
nephrolepis, Ramularia 926
nephrophyllidii, Puccinia = **P. areolata** 901
nephrospora, Stachybotrys 972
neriella, Pseudocercospora (Cercoseptoria, Cercospora) 894
nerii, Aplosporella 570
nerii, Botryodiplodia 591
nerii, Diplodia 681
nerii, Macrosporium 775
nerii, Phyllosticta 871
nerii-indici, Cercospora = **Pseudocercospora neriella** 894
nervalis, Pestalotia = **Monochaetia monochaeta** 795
nervata, Lirula (Hypodermella) 769
nervicola, Macrophoma 774
nerviphilus, Uromyces = **U. trifolii** 1013
nerviseda, Gnomonia 721
nervisedum, Leptothyrium 766
nervisequia, (Phyllachora, Sphaeria) = **Phyllachora americana** 862
nervisequia var. **conspicua, Lirula** 769
nervisequum, Gloeosporium = **Discula umbrinella** 685
nesaeae, Cercospora 614
nesaeae, Phyllosticta 871
nespori, Grandinia (Hyphodontia, Odontia) 723
neuhoffii, Athelia 583
neurophila, Lyonella 773
neuroterigallicola, Phyllosticta 871
neustonicum, Sporophlyctidium 972
nevadensis, Uromyces = **U. apiosporus** 1009
nicotiana, Phyllosticta 871
nicotianae, Ascochyta = **Phoma exigua** 853
nicotianae, Cercospora 614
nicotianae, Oospora (Toruloidea) 819
nicotianae, Phomopsis 859
nicotianae, Phytophthora 879
nicotianae, Uredo 1008
nicotianae var. **parasitica, Phytophthora** 879
nidicola, Pithoascus (Microascus) 880
nidulans, Aspergillus = **A. nidulellus** 579
nidulans, Emericella 693
nidulans, Meliola 785
nidulans, Phyllotopsis (Claudopus) 875
nidulans, (Hapalopilus, Polyporus) = **Hapalopilus rutilans** 729
nidulans var. germanica, Meliola = **M. niessleana** 785
nidulellus, Aspergillus 579
nidulus var. **onocleae, Trichopezizella** 1000
nidulus var. **setigera, Trichopezizella** 1000
nidus-avis, Gymnosporangium 727
niessleana, Leptosphaeria = **Entodesmium niessleanum** 695
niessleana, Meliola 785
niessleanum, Entodesmium 695
niessliana, Splanchnonema = **S. argus** 969
niesslii, Ophiobolus 820
niesslii, Wettsteinina 1023
nigella, Phillipsiella 848
nigellum, Microphyma = **Phillipsiella nigella** 848
niger, Aspergillus 579
niger, Nigroporus = **Melanoporia nigra** 783
niger, Pleurotus = **Hohenbuehelia nigra** 737
niger, Polyporus = **Melanoporia nigra** 783
niger var. **awamorii, Aspergillus** 579
niger var. **phoenicis, Aspergillus** 579
nigerrima, Diatrype = **Eutypella aequilinearis** 700
nigerrima, (Karschia, Patellaria) = **Dactylospora stygia** 664
nigerrima, Xylogramma (Briardia) 1026
nigerristigma, Mycosphaerella (Sphaerella) 804
nigra, Caliciopsis (Lagenula) 597
nigra, Ceratocystis = **Ophiostoma nigrum** 821
nigra, Comatricha 640
nigra, Herpotrichia = **H. juniperi** 735
nigra, Hohenbuehelia 737
nigra, Melanoporia (Poria) 783
nigra, Pseudotomentella (Tomentella) 899
nigra, Robergea 936
nigra, Strasseria 981
nigra, Ustilago = **U. avenae** 1014
nigra, Virgaria 1022
nigrans, Phaeosphaeria (Leptosphaeria) 842
nigredo, Mycosphaerella (Sphaerella) 804
nigrella, Paxina = **Plectania nannfeldtii** 883
nigrella, Pseudoplectania 898
nigrellum, Cladosporium 630
nigrescens, Crocicreas (Helotium, Phialea) 649
nigrescens, Dendrodochium 666
nigrescens, Dendrophoma 667
nigrescens, Phyllosticta 871
nigrescens, (Poria, Rigidoporus) = **Rigidoporus crocatus** 936
nigrescens, Synchytrium 984
nigrescens, Verticillium 1022
nigri, Cercospora = **Pseudocercospora atromarginalis** 893
nigricans, Aureobasidium (Kabatiella, Protocoronospora) 584
nigricans, Claviceps 632
nigricans, Dothichloe = **Balansia strangulans** 586
nigricans, Marssonina (Marssonia) 778
nigricans, Penicillium = **P. janczewskii** 828
nigricans, Phaeoisariopsis (Cercospora, Pseudocercospora) 840
nigricans, Phellinus (Fomes) 846
nigricans, Rhizopus = **R. stolonifer** 934
nigricans, Tubercularia 1002
nigricans var. **inculta, Leptosphaeria** 762
nigricantia, Pleospora 885
nigricantium, Macrosporium 775
nigriceps, Periconia (Sporocybe) 831
nigrificans, Didymella 677
nigripes, Galerina 712
nigripes, Marasmiellus (Marasmius) 776
nigripes, Sphaeronaema = **S. acerinum** 966
nigrita, Mycosphaerella 804
nigritella, Mollisia 793
nigro-angulata, Diatrypella (Sphaeria) 674
nigro-annulata, Anthostomella 566
nigroannulata, Valsella 1019
nigrocarpa, Ceratocystis = **Ophiostoma nigrocarpum** 821
nigrocarpum, Ophiostoma 821
nigrocinnabarina, Tryblidiella 1002
nigro-clypeata, Ophiobolus 820
nigrocyanea, Patellaria 826
nigrofacta, Valsaria (Valsa) 1019
nigrofusca, Phialea = **Crocicreas nigrofuscum** 649
nigrofuscum, Crocicreas 649
nigrofuscum var. **allantosporum, Crocicreas** 649
nigrofusum, Helotium = **Crocicreas nigrofuscum** 649
nigrolimitatus, Phellinus (Fomes, Polyporus) 846
nigromaculans, Mycosphaerella 804
nigromaculans, Phyllosticta 871
nigromaculatus f. anthrisci, Ophiobolus = **O. anthrisci** 819
nigromarginatus, Coriolus = **Trametes hirsuta** 996
nigropuncta, Uredo 1008
nigrospora, Chapeckia (Melanconis) 625
nigrosporum, Dimerium 679
nigrotecta, Anthostomella (Sphaeria) 566
nigrum, Brachysporium (Cladotrichum) 595
nigrum, Colletotrichum = **C. gloeosporioides** 638
nigrum, Epicoccum 696
nigrum, Gliocladium = **Myrothecium roridum** 809
nigrum, Ophiostoma 821
nigrum, Phacidium 839
nimbosum, Dimerosporium 679
niphostoma, Septoria 954
nitellina, Tomentella 994
nitens, Gymnoconia (Kunkelia) 725
nitens, Lophodermium (Lophodermina) 772
nitens, Phycomyces 862
nitida, Cryptosphaeria = **Hormotheca robertiani** 737
nitida, Junghuhnia (Poria) 751
nitida, Stagonospora (Hendersonia) 974
nitida, (Eutypa, Valsa) = **Eutypa laevata** 699
nitidum, Actidium (Glonium) 557
nitidum, Hypoderma 743
nitidum, Myxosporium 810
nitidum, Synchytrium 984
nitschkei, Diatrypella (Cryptovalsa) 674
nivale, Microdochium (Fusarium) 789
nivalis, Gerlachia = **Microdochium nivale** 789
nivalis, Monographella (Calonectria, Griphosphaeria, Micronectriella) 795
nivalis, Pleospora 885
nivea, Byssochlamys 596
nivea, Dasyscyphella = **Lachnum niveum** 755
nivea, Leucostoma (Valsa) 767
nivea, (Metulodontia, Peniophora) = **Hyphoderma karstenii** 741
nivea, Plasmopara = **P. crustosa** 882
nivea, Plicatura 887

nivea, Pseudocercosporella (Cercosporella) 895
nivea, Skeletocutis (Incrustoporia) 962
niveocremeum, Paullicorticium 827
niveopurpureus, Cryptodiscus (Stictis) 651
niveum, Cyclaneusma 656
niveum, Fusarium = **F. oxysporum** f. sp. **niveum** 709
niveum, Gloeosporium 718
niveum, Lachnum (Lachnum) 755
niveum, Sphaerographium 966
niveus, Dasyscyphus = **Lachnum niveum** 755
niveus, Merulius = **Plicatura nivea** 887
niveus, Naemacyclus = **Cyclaneusma niveum** 656
nivosa, Dendrothele 668
nivosa, Ramularia (Cercosporella) 926
nivosum, (Acanthophysium, Aleurocorticium) = **Dendrothele nivosa** 668
nivosus, Aleurodiscus = **Dendrothele nivosa** 668
njegusensis, Pleospora 885
nobile, Macrosporium = **Alternaria saponariae** 562
nobilissimus, Oxyporus (Fomes) 824
nobleae, Drechslera 690
nociviola, Uredo 1008
noctiflorae, Septoria 954
nodorum, Phaeosphaeria (Leptosphaeria) 842
nodorum, Stagonospora (Depazea, Hendersonia, Septoria) 974
nodosa, Puccinia 910
nodosum, (Helminthosporium, Pleurophragmium) = **Pseudospiropes nodosus** 898
nodosum, Myxomycidium = **Gloeomucro nodosus** 716
nodosus, Gloeomucro 716
nodosus, Pseudospiropes 898
nodosus, Rhizopus = **R. arrhizus** 933
nodulosa, Bipolaris (Drechslera) 588
nodulosa, Calopposis = **Calocera cornea** 597
nodulosum, Dendryphion = **Dendryphiella vinosum** 668
nodulosum, Helminthosporium = **Bipolaris nodulosa** 588
nodulosus, Cochliobolus 635
noelii, Phragmopyxis 861
nolinae, Alpakesa (Bartalinia) 560
nolinae, Clintamra (Melanotaenium, Tolyposporella) 633
noli-tangeris, Septoria 954
nootkatense, Gymnosporangium 727
norvegica, Leptosphaeria 762
notabile, Cerocorticium (Corticium) 622
notarisii, Valsaria (Valsa) 1019
notata, Poria = **Ceriporiopsis subvermispora** 622
notatum, Hypoxylon 745
notatum, Penicillium = **P. chrysogenum** 827
nova-caesariense, Hysterographium = **H. flexuosum** 747
novae-hollandiae, Aplosporella (Diplodia, Sphaeropsis) 570
novae-zealandiae, Meliolina 786
noveboracensis, Cercospora 614
noveboracensis, Leptosphaeria 762
novissimus, Uromyces 1012
nowakowskii, Cladochytrium = **C. replicatum** 629
nubigenum, Nivatogastrium (Secotium) 816
nubilosa, Pseudostegia 898
nubilosa, Sphaeropsis = **S. malorum** 967
nubilosum, Cryptosporium = **Pseudostegia nubilosa** 898
nubilosum, Dendryphion 668
nucicola, Hysterographium (Hysterium) 747
nucicola, Marasmius = **Mycena luteopallens** 798
nucicola, Pestalotia = **P. microspora** 835
nucifoedum, Coniosporium 642
nucinata, Pyrenochaeta 918
nucis, Dothiorella 689
nucis-hicoriae, Didymella 677
nucleata, Exidia (Naematelia) 702
nuda, Coccodiella (Bagnisiopsis, Yoshinagella) 634
nuda, Dacryomitra = **Ditiola radicata** 686
nuda, Lojkania (Amphisphaeria) 769
nuda, Peniophora 829
nuda, Puccinia 910
nuda, Ustilago = **U. tritici** 1016
nudiflorae, Cercospora 614
nudum, Coccostroma = **Coccodiella nuda** 634
nudus, Phaeocryptopus (Adelopus) 839
numerosa, Macrophoma 774
numerospora, Phyllosticta 871
nummularia, Biscogniauxia 589
nummularia, Ustilago = **U. hypodytes** 1015
nummularia var. **pseudopachyloma, Biscogniauxia** 589
nummularia var. **rumpens, Biscogniauxia** 589
nummularium, Hypoxylon = **Biscogniauxia nummularia** 589
nummularium var. **exutans, Hypoxylon** 745
nummularium var. **macrosporum, Hypoxylon** 746
nummularium var. pseudopachyloma, Hypoxylon = **Biscogniauxia nummularia** var. **pseudopachyloma** 589
nummularium var. rumpens, Hypoxylon = **Biscogniauxia nummularia** var. **rumpens** 589
nutans, Arcyria 571
nutans, Physarum 876
nutans var. leucophaeum, Physarum = **P. leucophaeum** 876
nuttalliana, Phyllachora = **P. sylvatica** 863
nuttallii, Cylindrosporium (Septogloeum) 659
nuttallii, Hypoxylon = **H. rubiginosum** 746
nymphaeacea, Phyllosticta 871
nymphaeacea, Pseudocercospora (Cercoseptoria, Cercospora) 895
nymphaeae, Cercospora = **Pseudocercospora nymphaeacea** 895
nymphaeae, Rhamphospora (Entyloma) 931
nymphaearum, Dichotomophthoropsis (Helicoceras) 675
nymphaearum, Ramularia (Gloeosporium, Ovularia, Ovulariella) 926
nymphaeicola, Phyllosticta 871
nyssae, Aplopsora 569
nyssae, Cercospora 614
nyssae, Cytospora = **C. leucosperma** 662
nyssae, Isothea 750
nyssae, Mastigosporella 780
nyssae, Phyllosticta 871
nyssae, Pirostoma = **Tubakia dryina** 1002
nyssae, Valsa = **V. ceratosperma** 1017
nyssaecola, Stagonospora 974
nyssaegena, Ciboria (Peziza, Sclerotinia) 627
nyssae-sylvaticae, Cercospora 614
nyssagena, Melanconis = **Massariovalsa sudans** 780
nyssicarpa, Phoma 855
nyssicola, Ceratostomella 605
nyssicola, Crepidotus (Pleuropus) 647
nyssicola, Mycosphaerella (Sphaerella) 804
oahuensis, Puccinia 910
oakesiae, Phyllosticta = **Macrophoma smilacina** 774
oakesii, Aleurodiscus 560
oaxacanus, Uromyces 1012
obclavata, Corynespora 646
obclavatum, (Clasterosporium, Sporidesmium) = **Sporidesmium fragilissimum** 970
obducens, Cucurbitaria (Strickeria, Teichospora) 654
obducens, Plasmopara (Peronospora) 882
obducens, Ramularia 926
obducta, Osteina 823
obductum, Oidium (Acrosporium) 818
obductus, Oligoporus = **Osteina obducta** 823
obesa, Cercospora 614
obesa, Heptameria (Sphaeria) 734
obesa, Septoria 954
obiones, Camarosporium = **C. roumeguerii** 600
obiones, Leptosphaeria (Pleospora) 763
obliqua, (Ovularia, Ramularia) = **Ramularia rubella** 926
obliqua, Poria = **Inonotus obliquus** 749
obliqua, Puccinia 910
obliqua, Robergea 936
obliquata, Rosellinia 937
obliquus, Inonotus (Polyporus) 749
oblita, Phomopsis 859
oblonga, Chaetomella 623
oblonga, Phomopsis 859
oblonga, Physarella 876
oblonga var. major, Chaetomella = **Chaetomella oblonga** 623
oblongata, Macrophoma 774
oblongispora, Chaetoplea (Teichospora) 623
oblongisporum, Mytilinidion 809
oblongisporus, Uromyces 1012
oblongospora, Lachnella = **Dasyscyphus oblongosporus** 665
oblongosporus, Dasyscyphus 665
oblongum, Melanconium 782
oblongum, Paradidymobotryum 826
oblongum, Phoma = **Phomopsis oblonga** 859
oblongus, Uromyces = **U. striatus** 1013
obovata, Peronospora 833
obovata, (Ovularia, Ramularia) = **Ramularia rubella** 926
obovata, Spadicoides 963
obovatum, Acrothecium = **Spadicoides obovata** 963
obovatus, Microporellus (Polyporus) 790
obscura, (Cadophora, Phialophora) = **Phialophora bubakii** 847
obscura, (Ceratocystis, Ceratostomella) = **Ophiostoma obscurum** 821
obscura, Cercospora 614
obscura, Didymella 677
obscura, Peniophora = **P. cinerea** 829
obscura, Pseudodoassansia (Doassansia) 896
obscura, Puccinia 910
obscurans, Phomopsis (Dendrophoma, Phoma, Phyllosticta) 859
obscurum, Ophiostoma 821
obscurus, Crepidotus 647
obstipa, Fusariella 706
obstipum, Dendryphion = **Fusariella obstipa** 706
obtecta, Balansia 586
obtecta, Puccinia 911

obtegens, Gloeosporium = **Cryptomycella pteridis** 652
obtegens, Puccinia = **P. punctiformis** 912
obtusa, Botryosphaeria (Physalospora) 592
obtusa, Frommea = **Phragmidium potentillae** 861
obtusa, Montagnula (Pleospora, Teichospora) 796
obtusa, Pseudoseptoria (Selenophoma) 898
obtusa var. duchesneae, Frommea = **Frommeella duchesneae** 706
obtusiforme, Hyphoderma 741
obtusispora, Paraphaeosphaeria (Leptosphaeria) 826
obtusisporum, Cylindrocarpon (Fusarium) 657
obtusissima, Rosellinia 937
obtusus, Polyporus = **Spongipellis unicolor** 970
occidentale, Acanthostigma = **Nematostoma occidentalis** 814
occidentale, Coleosporium = **C. tussilaginis** 636
occidentale, Cronartium 649
occidentale, Hypoxylon = **H. thouarsianum** 746
occidentale, Melanomma (Cucurbitaria) 783
occidentale, Peridermium 832
occidentale, Phragmidium 861
occidentale, Schizoxylon = **S. berkeleyanum** 940
occidentale var. tetonense, Melanomma = **Leptosphaeria tetonensis** 763
occidentalis, Albugo 559
occidentalis, Arcyria (Lachnobolus) 572
occidentalis, Crepidotus 647
occidentalis, Cyanodiscus 656
occidentalis, Daldinia 664
occidentalis, Galzinia 713
occidentalis, Lachnellula (Dasyscyphus) 754
occidentalis, Leptosphaeria 763
occidentalis, Melampsora 781
occidentalis, Nematostoma (Venturia) 814
occidentalis, Phaeoramularia (Cercospora, Cercosporina) 841
occidentalis, Phellinus (Fomes) 846
occidentalis, Pholiota = **P. alniphila** 851
occidentalis, Ramularia 926
occidentalis, Sarcoscypha (Plectania) 938
occidentalis, Septoria 954
occidentalis, Sphacelotheca 965
occidentalis, Spongipellis = **S. spumeus** 970
occidentalis, Strobilurus 982
occidentalis, Taphrina 987
occidentalis, Tryblidiopsis = **Holmiella sabina** 737
occidentalis, Uromyces 1012
occidentalis, Volutella 1023
occidentalis var. cassiocarpa, Cercospora = **Cercosporidium cassiocarpum** 620
occulta, Aglaospora = **Massaria pyri** 779
occulta, Bispora (Torula) 589
occulta, (Calloria, Orbilia) = **Orbilia cruenta** 822
occulta, Doassansiopsis (Doassansia) 686
occulta, (Acrocordiella, Hymenopleella) = **Requienella seminuda** 929
occulta, Macrodiaporthe (Melanconis) 773
occulta, Phomopsis (Phoma) 859
occulta, Phyllosticta 871
occulta, Urocystis 1009
occulta var. farlowii, Doassansia = **Doassansiopsis occulta** 686
occultata, Cucurbitaria 654
ocellata, Cercoseptoria 605
ocellata, Pezicula (Ocellaria) 837
ocellata, Septoria 954
ochracea, Cercospora = **Mycovellosiella lobeliae-cardinalis** 808
ochracea, Peniophora = **P. nuda** 829
ochracea, Solenia = **Merismodes ochraceus** 787
ochracea, Tomentella 994
ochraceo-flavum, Stereum 978
ochraceofulva, Phlebia 848
ochraceum, Gloeocystidiellum 716
ochraceum, Gloeosporium = **Colletotrichum gloeosporioides** 638
ochraceum, Steccherinum (Hydnum) 975
ochraceus, Aspergillus 579
ochraceus, Crepidotus 647
ochraceus, Merismodes 787
ochrofarctum, Corticium = **Hyphoderma pallidum** 741
ochroleuca, Cyphella = **Cellypha goldbachii** 601
ochroleuca, Discella (Gloeosporium, Marssonina, Septoria) 683
ochroleuca, Nectria 813
ochroleucum, Asterostroma 583
ochroleucum, Scytinostroma (Gloeocystidium) 944
ochrotinctellus, (Coriolus, Polyporus) = **Antrodiella versicutis** 568
octoseptata, Nodulosphaeria (Leptosphaeria) 817
oculata, Cercospora 614
oculatum, Cylindrosporium 659
odontoglossi, Cercospora 614
odora, Antrodia (Poria) 567
odora, Leptosphaeria 763
odorata, Peniophora = **Chaetoderma luna** 623
odorata, Trametes = **Gloeophyllum odoratum** 717
odoratum, Gloeophyllum 717
odoratus, Osmoporus = **Gloeophyllum odoratum** 717
odoratus, Polyporus = **Gloeophyllum odoratum** 717
odorifera, Thelephora = **T. albidobrunnea** 989
odorus, Polyporus = **Antrodia odora** 567
odorus, Sporobolomyces = **Sporidiobolus salmonicolor** 971
oeconomicum, Aristastoma (Amerosporium) 572
oedochilum, Pythium 921
oedospora, Septoria 954
oenanthis, Septoria 954
oenotherae, Cercospora 614
oenotherae, Cryptostictis (Pestalotia) 653
oenotherae, Discohainesia (Peziza, Pezizella) 684
oenotherae, (Dothiorella, Macrophoma) = **Macrophoma oenotherae-biennis** 774
oenotherae, Mycosphaerella (Sphaerella) 804
oenotherae, Phomopsis (Myxosporium) 859
oenotherae, Puccinia 911
oenotherae, Septoria 954
oenotherae, Synchytrium = **S. fulgens** 984
oenotherae var. heteranthae, Puccinia = **Puccinia oenotherae** 911
oenotherae-biennis, Macrophoma 774
oenotherae-sinuatae, Cercospora 614
officinale, Cylindrosporium 659
officinalis, Diplodia 681
officinalis, Fomitopsis (Fomes, Laricifomes) 706
offuscata, Nectria = **N. cinnabarina** 812
offuscata, Puccinia = **P. arachidis** var. **offuscata** 900
ogilviensis, Leptosphaeria 763
ohiana, Asteridiella 580
ohianus, Amazonia = **Asteridiella ohiana** 580
ohiense, Hypoxylon = **Camarops ohiensis** 599
ohiensis, Calyptronectria (Strickeria, Teichospora) 599
ohiensis, Camarops 599
ohiensis, Perenniporia (Fomes, Fomitopsis, Trametes, Truncospora) 830
okanoganense, Pythium 921
okefenokeensis, Rogersiomyces 936
oklahomae, Tilletia 992
okra, Phoma 855
oleae, Coleophoma (Macrophoma) 635
oleae, Gloeosporium = **Colletotrichum gloeosporioides** 638
oleae, Lembosia (Lembosiopsis) 758
oleae, Phyllosticta 871
oleae, Polyporus = **Abortiporus biennis** 555
oleagina, Poria = **Antrodia sordida** 568
oleaginea, Spilocaea 969
oleaginum, Cycloconium = **Spilocaea oleaginea** 969
oleandri, Gloeosporium = **Colletotrichum gloeosporioides** 638
oleandri, Pestalotia 835
oleandri, Sphaceloma 964
oleandrina, Septoria 955
olearius, Omphalotus 819
oleina, Asterina = **Zukalia purpurea** 1027
oleina, Mycosphaerella 804
oleosa, Calloria = **Hyponectria sceptri** 744
oleracea, Alternaria = **A. brassicicola** 561
oleracea, Antrodia (Poria) 567
oleracea, Phoma = **P. herbarum** 854
oleracea var. dipsaci, Phoma = **P. herbarum** 854
olericola, Leptosphaeria 763
olidus, Marasmius = **M. copelandii** var. **olidus** 777
oligandrum, Pythium 921
oligasca, Pleospora = **P. potentillae** 885
oligospora, Arthrobotrys 573
oligosporum, Helminthosporium = **Corynespora olivacea** 646
oligostachyae, Pleospora = **P. pellita** var. **oligostachyae** 885
olindensis, Mycosphaerella 804
olivacea, Ceratocystis = **Ophiostoma olivaceum** 821
olivacea, Cercospora = **Mycovellosiella ferruginea** 807
olivacea, Coniophora (Coniophorella) 641
olivacea, Conoplea 644
olivacea, Corynespora 646
olivacea, Dermea 669
olivacea, (Farysia, Ustilago) = **Farysia thuemenii** 704
olivacea, Hydnochaete 738
olivacea, Hypocrea = **H. schweinitzii** 743
olivacea, Libertella 768
olivacea, Massaria = **Asteromassaria olivaceohirta** 582
olivacea, Microsphaeropsis 791
olivacea, Nodulosphaeria (Leptosphaeria) 817
olivacea, Phialophora 847
olivacea, Polyschema (Clasterosporium) 889
olivacea, Tympanis = **Claussenomyces olivaceus** 631

olivacea, Virgaria = **Virgariella fusca** 1022
olivaceapinii, Ceratocystis 604
olivaceo-album, Confertobasidium 641
olivaceo-album var. montanum, Confertobasidium = **C. montanum** 640
olivaceohirta, Asteromassaria (Massaria) 582
olivaceum, Chaetomium = **C. globosum** 623
olivaceum, Circinotrichum (Helicotrichum) 628
olivaceum, Clasterosporium = **Corynespora olivacea** 646
olivaceum, Coniosporium 642
olivaceum, Coniothyrium = **Microsphaeropsis olivacea** 791
olivaceum, Helicoma 730
olivaceum, Helminthosporium = **Cercospora seymouriana** 617
olivaceum, Ophiostoma 821
olivaceum, Trichocladium = **Coniosporium diversum** 641
olivaceum f. celtidis-australis, Coniothyrium = **C. celtidis-australis** 642
olivaceus, Claussenomyces (Retinocyclus, Tromera) 631
olivaceus, Ophiobolus = **Nodulosphaeria olivacea** 817
olivarum, Gloeosporium = **Colletotrichum gloeosporioides** 638
olivascens, Leucogyrophana (Coniophora, Corticium) 767
olivascens, Omphalotus 819
olivascens, Tomentella (Hypochnus) 994
olivinoviride, Penicillium = **P. viridicatum** 829
olla, Cyathus 656
ollare, (Chromelosporium, Dematium) = **Chromelosporium fulvum** 626
olympiana, Galerina 712
olympiana, Phialea = **Crocicreas cyathoideum var. olympianum** 649
olympiana, Pholiota 851
olympianum, (Hypholoma, Naematoloma) = **Pholiota olympiana** 851
olympica, Phyllosticta 871
omnivora, Phanerochaete 844
omnivora, Phymatotrichopsis 875
omnivorum, Colletotrichum = **C. dematium** 637
omnivorum, Hydnum = **Phanerochaete omnivora** 844
omnivorum, (Ozonium, Phymatotrichum) = **Phymatotrichopsis omnivora** 875
omphacodes, Cercospora 614
omphalodes, (Lentinellus, Lentinus) = **Lentinellus micheneri** 759
omphalodes, Pyronema 920
onagrae, Leptosphaeria = **L. ellisiana** 761
oncidii, Botryodiplodia (Macrophoma) 591
oncidii, Gloeosporium = **Colletotrichum gloeosporioides** 638
oncidii, Uredo 1008
oncidii-sphacelati, Phoma 855
oncospermatis, Dasyscyphus 665
oncostoma, Diaporthe 671
oncostoma, Phomopsis (Phoma) 859
ontariensis, (Anthostoma, Eutypa) = **Cryptosphaeria subcutanea** 652
ontariensis, Diaporthe = **D. dubia** 671
oospermum, Coniothyrium = **C. ovispermum** 643
oosporum, Brachysporium (Helminthosporium) 595
ootheca, Mycosphaerella 804
ootheca, Pleurostoma (Coronophora, Neoarcangelia) 886
opaca, Diatrypella 674
opaca, Doassansia 686
opaca, Phyllosticta 871
opaca, Ravenelia 928
opaca, Torula = **Taeniolella breviuscula** 985
opacum, Berkleasmium (Sporidesmium) 587
opacum, Dictyosporium = **D. elegans** 676
opacus, Marasmiellus 776
opalea, Gloeotulasnella = **Tulasnella traumatica** 1004
opalea, (Exidiopsis, Sebacina) = **Exidiopsis glaira** 702
operculata, Endoxyla 695
operculata, Licea (Orcadella) 768
operculata, Mycosphaerella (Sphaerella) 804
ophioboloides, Ophiobolus (Leptosphaeria) 820
ophiobolus, Bulgaria = **Holwaya mucida** 737
ophiopogonis f. graminum, Leptosphaeria 763
ophites, Phomopsis (Phoma) 859
opulasteris, Phyllosticta 871
opulenta, (Asterella, Asterina) = **Asterina orbicularis** 581
opuli, Cercospora 614
opulifoliae, Valsa 1018
opuntiae, Clavispora 632
opuntiae, Crustoderma 650
opuntiae, Diplodia 681
opuntiae, Dothiorella 689
opuntiae, Gloeosporium = **Colletotrichum gloeosporioides** 638
opuntiae, Hendersonia 733
opuntiae, Hysterographium 747
opuntiae, Leptodermella 759
opuntiae, Leptosphaeria 763
opuntiae, Monographella (Mycosphaerella, Sphaerella) 795
opuntiae, Teichospora 988
opuntiae, Tretopileus = **T. sphaerophorus** 998
opunticola, Macrophoma (Phoma) 774
oraemaris, Leptosphaeria 763
orbicula, Phyllosticta 871
orbicula, Puccinia 911
orbicula, Valsa = **Leucostoma translucens** 767
orbicula, Venturia = **Protoventuria quercina** 893
orbiculare, Colletotrichum (Gloeosporium) 639
orbiculare, Lophodermium 772
orbiculare, Radulum = **Basidioradulum radula** 586
orbicularis, Asterina (Englerulaster) 581
orbicularis, Mycosphaerella = **Venturia macularis** 1020
orbicularis, Phyllosticta 871
orbiculata, Phyllachora (Dothidea, Sphaeria) 863
orchidearum, Colletotrichum = **C. gloeosporioides** 638
orchidearum, Physalospora = **Glomerella cingulata** 719
orchidearum, Septoria 955
oregonense, Ganoderma 713
oregonense, (Geopetalum, Pleurotus) = **Cheimonophyllum candidissimum** 625
oregonense, Hypoxylon = **H. vogesiacum** 746
oregonense, Odontotrema 818
oregonensis, Diatrype (Eutypella) 673
oregonensis, Didymosphaeria 678
oregonensis, Gymnopilus 725
oregonensis, Lachnella = **Calathella ellisii** 597
oregonensis, Pholiota 851
oregonensis, Polyporus = **Ganoderma oregonense** 713
oreophila, Leptosphaeria 763
oreophilum, Steccherinum 975
orickiana, Xeromphalina (Omphalina) 1025
oricostata, Encoeliopsis (Cenangella) 693
orientalis, Laestadia = **Sphaerognomonia carpinea** 966
ornamentale, Peridermium 832
ornamentata, Neocosmospora = **N. vasinfecta** 814
ornata, Chaetosphaeria 624
ornata, Puccinia 911
ornata, Ustilago 1016
ornatula, Puccinia 911
orni, Coniothyrium 643
ornithogali, Mycosphaerella (Didymellina) 804
ornithogali, Septoria 955
ornithogali var. allii-porri, Heterosporium = **Cladosporium allii** 629
orobella, Phyllosticta 871
orogeniae, Ramularia 926
orontii, Phyllosticta 871
orontii, Physalospora 876
orontii, Ramularia 926
oropensoides, Nectria 813
orthoceras, Fusarium = **F. oxysporum** 708
orthoceras var. gladioli, Fusarium = **F. oxysporum** f. sp. **gladioli** 709
orthoceras var. longius, Fusarium = **F. oxysporum** 708
orthoceras var. pisi, Fusarium = **F. oxysporum** f. sp. **pisi** 709
orthogramma, Leptosphaeria (Sphaeria) 763
orthosporum, Rhynchosporium 935
orthosticha, Phoma 855
ortonii, Puccinia 911
oryzae, Ascochyta 576
oryzae, Aspergillus (Eurotium) 579
oryzae, Bipolaris (Drechslera, Helminthosporium) 588
oryzae, Cercospora 614
oryzae, Diplodia 681
oryzae, Entyloma 696
oryzae, Helicoceras 730
oryzae, Khuskia (Monotospora) 752
oryzae, Microdochium (Rhynchosporium) 789
oryzae, Mycosphaerella 804
oryzae, Nigrospora 816
oryzae, Ophiobolus = **Ophiosphaerella herpotricha** 820
oryzae, Pleosphaerulina = **Leptosphaerulina trifolii** 764
oryzae, Pyrenochaeta 918
oryzae, Pyricularia 919
oryzae, Pythium = **P. dissotocum** 920
oryzae, Rhizoctonia (Moniliopsis) 932
oryzae, Rhizopus = **R. arrhizus** 933
oryzae, Sarocladium (Acrocylindrium) 939
oryzae, Sclerotium 942
oryzae, Septoria 955
oryzae, Tritirachium (Beauveria) 1001
oryzae var. irregulare, Sclerotium 942
oryzae-sativae, Balansia 586
oryzae-sativae, Rhizoctonia (Sclerotium) 932
oryzinus, Ophiobolus = **Gaeumannomyces graminis** 712
oryzopsidis, Phyllachora 863
oscheoides, Zygosporium 1027
osmanthi, Aulacostroma 584
osmanthi, Calothyriella 598
osmanthi, Calothyrium 598

osmanthi, Meliola 785
osmanthi, Phyllosticta 871
osmanthi var. **hawaiiensis, Meliola** 785
osmanthicola, Meliola 785
osmanthina, Meliola 785
osmaroniae, (Ascochyta, Gloeosporium) = **Ascochyta idaei** 575
osmorrhizae, Cercospora 614
osmorrhizae, Phloeospora (Septoria) 849
osmorrhizae, Puccinia = **P. pimpinellae** 911
osmundae, Ascochyta 576
osmundae, Calycellina (Peziza, Trichopeziza) 598
osmundae, (Cryptomycina, Hysterium) = **Cryptomycina filicina** 652
osmundae, Mixia (Gloeosporium, Taphrina) 792
osmundae, Mycosyrinx 807
osmundae, Scirrhia (Dothidea, Dothidella, Metameris, Scirrhophragma) 940
osmundae, Uredinopsis 1007
osoyoosensis, Puccinia 911
osseus, Polyporus = **Osteina obducta** 823
osteospora, Phyllosticta 871
ostiolata, Scirrhia 940
ostracodes, Pythium 921
ostrea, Stereum 978
ostreatus, Pleurotus 886
ostropoides, Stictis 978
ostryae, Melanconis 782
ostryae, Septoria 955
otophila, Oospora 819
oudemansii, Diplodina = **Ascochyta grossulariae** 575
oudemansii, Hendersonia 733
oudemansii, Puccinia 911
oudemansii, Septoria 955
ovalifolii, Phyllosticta 871
ovalis, Rosellinia 937
ovarium, Sorosporium 963
ovata, Stilbospora = **Stegonsporium pyriforme** 976
ovatum, Stegonsporium = **S. pyriforme** 976
ovatum, Verticillium = **V. dahliae** 1021
overholtsii, Antrodiella 568
overholtsii, Diplomitoporus (Poria) 683
overholtsii, Mycena 798
overholtsii, Peniophora = **P. nuda** 829
ovina, Lasiosphaeria 757
ovispermum, Coniothyrium 643
ovoidea, Zignoella 1027
ovoideus, Leptosporomyces 764
ovularioides, Ramularia 926
owensii, (Irpex, Radulum) = **Phlebia concentrica** 848
owensii, Phyllosticta 871
oxalicum, Penicillium 828
oxalidiphila, Cercospora 614
oxalidis, Cercospora 614
oxalidis, Melanotaenium 784
oxalidis, Phyllosticta 871
oxalidis, Puccinia 911
oxalidis, Ramularia 926
oxalidis, Septoria 955
oxalidis, Ustilago 1016
oxalina, Phomopsis (Phyllachora) 859
oxyacanthae, Podosphaera = **P. clandestina** 888
oxyacanthae, Septoria 955
oxyacanthae, Xylaria 1026
oxyacanthae var. tridactyla, Podosphaera = **P. tridactyla** 888
oxybaphi, Ascochyta 576
oxybaphi, Cercospora 614
oxybaphi, Heterosporium 736
oxybaphi, Peronospora 834
oxycocci, Cladosporium 630
oxycocci, Eupropolella (Naevia) 699
oxycocci, Lophodermium (Hysterium) 772
oxycocci, Monilinia (Sclerotinia) 794
oxycocci, Mycosphaerella = **Botryosphaeria vaccinii** 593
oxycocci, Rhabdospora 931
oxycocci, Sporonema = **Coleophoma empetri** 635
oxycocci, Strasseria (Plagiorhabdus) 981
oxycocci var. hypophyllum, Lophodermium = **Lophodermium oxycocci** 772
oxydendri, Cercospora (Cercospora) 614
oxydendri, Phyllosticta 871
oxydendri, Pseudomassaria (Venturia) 897
oxydendri, (Fusarium, Rhabdospora) = **Fusarium aquaeductuum** 707
oxydendri, Tympanis 1004
oxyriae, Puccinia 911
oxyspora, Diaporthe 671
oxyspora, Phomopsis 859
oxyspora, (Lunospora, Septoria) = **Pseudoseptoria donacis** 898
oxyspora, Volutella 1023
oxysporum ssp. lycopersici, Fusarium = **Fusarium oxysporum** f. sp. **lycopersici** 709
oxysporum, Cladosporium 630
oxysporum, Fusarium 708
oxysporum, Septogloeum = **Cheilaria agrostis** 625
oxysporum var. aurantiacum, Fusarium = **Fusarium oxysporum** 708
oxysporum var. cubense, Fusarium = **Fusarium oxysporum** f. sp. **cubense** 709
oxysporum var. orthoceras, Fusarium = **Fusarium oxysporum** 708
oxysporum var. redolens, Fusarium = **F. redolens** 710
oxysporum f. sp. **apii, Fusarium** 708
oxysporum f. sp. **asparagi, Fusarium** 708
oxysporum f. sp. barbati, Fusarium = **Fusarium oxysporum** f. sp. **dianthi** 709
oxysporum f. sp. **batatas, Fusarium** 708
oxysporum f. sp. betae, Fusarium = **Fusarium oxysporum** f. sp. **spinaciae** 709
oxysporum f. sp. **callistephi, Fusarium** 708
oxysporum f. sp. **cannabis, Fusarium** 708
oxysporum f. sp. **carthami, Fusarium** 708
oxysporum f. sp. **cattleyae, Fusarium** 709
oxysporum f. sp. **cepae, Fusarium** 709
oxysporum f. sp. **chrysanthemi, Fusarium** 709
oxysporum f. sp. **ciceris, Fusarium** 709
oxysporum f. sp. **citri, Fusarium** 709
oxysporum f. sp. **conglutinans, Fusarium** 709
oxysporum f. sp. **cubense, Fusarium** 709
oxysporum f. sp. **cucumerinum, Fusarium** 709
oxysporum f. sp. **cyclaminis, Fusarium** 709
oxysporum f. sp. **delphinii, Fusarium** 709
oxysporum f. sp. **dianthi, Fusarium** 709
oxysporum f. sp. **gladioli, Fusarium** 709
oxysporum f. sp. **koae, Fusarium** 709
oxysporum f. sp. **lilii, Fusarium** 709
oxysporum f. sp. **lini, Fusarium** 709
oxysporum f. sp. **lycopersici, Fusarium** 709
oxysporum f. sp. **mathioli, Fusarium** 709
oxysporum f. sp. **medicaginis, Fusarium** 709
oxysporum f. sp. **melonis, Fusarium** 709
oxysporum f. sp. **narcissi, Fusarium** 709
oxysporum f. sp. **nicotianae, Fusarium** 709
oxysporum f. sp. **niveum, Fusarium** 709
oxysporum f. sp. **perniciosum, Fusarium** 709
oxysporum f. sp. **phaseoli, Fusarium** 709
oxysporum f. sp. **pisi, Fusarium** 709
oxysporum f. sp. **radici-lupini, Fusarium** 709
oxysporum f. sp. **radicis-lycopersici, Fusarium** 709
oxysporum f. sp. **raphani, Fusarium** 709
oxysporum f. sp. **rhois, Fusarium** 709
oxysporum f. sp. **spinaciae, Fusarium** 709
oxysporum f. sp. **tracheiphilum, Fusarium** 710
oxysporum f. sp. **vanillae, Fusarium** 710
oxysporum f. sp. **vasinfectum, Fusarium** 710
oxysporum f. sp. **zingiberi, Fusarium** 710
oxystoma, Physalospora = **Botryosphaeria festucae** 592
oxytropidis, Phloeospora (Septoria) 850
pachyasca, Mycosphaerella 804
pachycheiles, Polyporus = **Antrodiella semisupina** 568
pachyderma, Cintractia 628
pachydermus, Protomyces 893
pachyodon, Spongipellis (Irpex) 970
pachypus, Cercospora 614
pachysandrae, Macrophoma 774
pachysandrae, Phyllosticta 871
pachysandrae, Septoria 955
pachysandrae, Volutella 1023
pachysandricola, Pseudonectria 897
pachyspora, Cercospora = **C. callae** 607
pachyspora, Monochaetia = **M. kansensis** 795
pachyspora, Septoria 955
pachysterigmata, Dendrothele 668
pachysterigmatum, Aleurocorticium = **Dendrothele pachysterigmata** 668
pachystimae, Mycosphaerella 804
pachystimae, Phyllosticta 871
pacifica, Leptosphaeria 763
pacifica, Puccinia 911
pacifica, Septoria 955
pacificensis, Hypoderma 743
pacificum, Coleosporium 636
pacificus, Gymnopilus 725
pacificus, Polyporus = **Climacocystis borealis** 632
padi, Dermea 669
padi, Diaporthe = **D. decorticans** 671
padi, Monilinia 794
padi, Phloeosporella (Ascochyta, Cylindrosporium) 850
padi var. **patria, Diaporthe** 671
padina, Phomopsis (Phoma) 859
paederiae, Coleosporium 636
paeoniae, Botrytis 594
paeoniae, Cercospora 614
paeoniae, Cladosporium 630
paeoniae, Cryptostictis 653
paeoniae, Monochaetia (Pestalotia) 795
paeoniae, Septoria 955
pagana, Puccinia 911
palingenes, Pythium 921
palisoti, Daedalea = **Trametes elegans** 996
palitans, Penicillium = **P. viridicatum** 829
paliuri, Ascochyta 576
pallens, Merulius = **Phlebia rufa** 848
pallens, Phoma 855
pallens, Phyllosticta 871
pallens, Sphaerulina 968
pallescens, Curvularia 655
pallescens, Dichostereum (Vararia) 675
pallescens, Physopella (Angiopsora) 877
palliatum, Camarosporium 600
pallida, Basidiobotrys (Botrytis) 586
pallida, Botryodiplodia 591
pallida, Ceratocystis = **Ceratocystiopsis minuta-bicolor** 603

pallida, Cercospora = **C. duplicata** 609
pallida, Cytospora 662
pallida, Excipulina = **Sphaeronaema pallidum** 966
pallida, Fistulina = **Pseudofistulina radicata** 896
pallida, Hendersonia (Stagonopsis) 733
pallida, Melanconis (Melanconiella) 782
pallida, Nectria = **N. ochroleuca** 813
pallida, Pestalotia = **Monochaetia monochaeta** 795
pallida, Peziza = **P. repanda** 838
pallida, Phialea = **Crocicreas cyathoideum** 648
pallida, Phomopsis (Libertella) 859
pallida, Phyllosticta = **P. nebulosa** 871
pallida, Roesleria (Coniocybe) 936
pallida, Stemonitis 976
pallida, Stictis = **Cryptodiscus pallidus** 652
pallida, Stigmina 980
pallida, Tilletia 992
pallida, Uredo = **Physopella pallescens** 877
pallida, Valsa 1018
pallidior, Phyllosticta (Phyllostictina) 871
pallidispora, Galerina 712
pallidispora, Hendersonia 733
pallidissima, Puccinia 911
pallidoaurantiaca, Trechispora 998
pallidocephalus, Marasmius 777
pallido-cervinus, Polyporus = **Hapalopilus rutilans** 729
pallidofulva, Tomentella = **T. bryophila** 993
pallidofulvus, Hypochnus = **Tomentella bryophila** 993
pallidoroseum, Fusarium (Fusisporium) 710
pallido-roseum, Fusisporium 712
pallidospermum, Coniothyrium 643
pallidula, Grandinia (Hyphodontia, Peniophora) 723
pallidula, Nectria 813
pallidula, Pucciniosira 917
pallidula, Septoria 955
pallidula, Sphaeropsis 967
pallidum, Exosporium = **Stigmina pallida** 980
pallidum, Hyphoderma (Corticium) 741
pallidum, Hypoxylon = **H. notatum** 745
pallidum, Illosporium 748
pallidum, Melanconium 782
pallidum, Sphaeronaema 966
pallidus, Cerinomyces 621
pallidus, Cryptodiscus 652
pallidus, Dacrymyces = **D. cokeri** 663
palmarum, Botryodiplodia (Macrophoma, Phoma, Sphaeropsis, Striodiplodia) 591
palmarum, Cytospora 662
palmarum, Melanconium 782
palmarum, Pestalotiopsis (Pestalotia) 836
palmarum, Tyromyces 1005
palmata, Calocera = **C. cornea** 597
palmata, Tremella = **Dacrymyces chrysospermus** 663
palmatus, Dacrymyces = **D. chrysospermus** 663
palmatus, Rhodotus 934
palmeri, Puccinia 911
palmetta, Metasphaeria (Sphaeria) 788
palmetto, Linocarpon (Linospora) 769
palmetto, Phyllosticta 871
palmicola, Delortia 666
palmicola, Meliola 785
palmicola, Phomopsis (Phoma) 859
palmicola, Phyllachora (Catacauma) 863
palmicolus, Phellinus (Polyporus) 846
palmigenum, Helicoma 730
palmivora, Phytophthora 879
palmivora, Stigmina 980
palmivorum, Exosporium = **Stigmina palmivora** 980
palousensis, Bipolaris 588
paludosa, Microthyriella = **Schizothyrium pomi** 939
paludosa, Mitrula 792
paludosa, Periconia = **P. digitata** 831
paludosa, Stagonospora (Hendersonia) 974
paludosa, Tiarosporella 991
paludosa, Tubeufia 1002
paludosum, Colletotrichum (Gloeosporium) 639
palustris, Cercosporella 619
palustris, Fomitopsis (Polyporus, Tyromyces) 706
palustris, Phyllosticta 872
palustris, Physoderma 877
palustris, Venturia 1020
palustrium, Cercospora 614
pammelii, Phyllachora = **P. sylvatica** 863
pammelii, Puccinia = **P. emaculata** 905
pamparum, Sphacelotheca (Ustilago) 965
pampeana, Melanospora = **M. zamiae** 784
panacheum, Helicosporium 731
panacicola, Ramularia = **Cylindrocarpon destructans** 657
panacis, Stromatinia (Sclerotinia) 982
panamensis, Aspergillus 579
panattoniana, (Marssonia, Marssonina) = **Microdochium panattonianum** 789
panattonianum, Microdochium 789
panax, Alternaria 562
pancratii, Cercospora 614
pandani, Linocarpon (Linospora) 769
pandani, Macrophoma 774
pandani, Melanconium 782
pandani, Phomatospora (Physalospora) 857
pandani, Schizochora = **Oxydothis pandanicola** 824
pandanicola, Coniothyrium 643
pandanicola, Oxydothis (Didymella) 824
pandanicola, Phyllosticta 872
panicea, Badhamia 585
paniceae, Anthracoidea 567
panici, (Cercospora, Cercosporina) = **Phaeoramularia fusimaculans** 840
panici, Cerebella = **C. andropogonis** 621
panici, Crocicreas (Phialea) 649
panici, Elsinoe 692
panici, Leptostromella 765
panici, Meliola 785
panici, Phloeospora 850
panici, Phyllachora = **P. punctum** 863
panici, Phyllosticta 872
panici, Puccinia = **P. emaculata** 905
panicicola, Hendersonia 733
panici-glauci, Ustilago = **Sporisorium neglectum** 971
panici-miliacei, Cercospora = **Phaeoramularia fusimaculans** 840
panici-miliacei, (Sphacelotheca, Ustilago) = **Sporisorium destruens** 971
panici-proliferi, Ustilago 1016
panicophila, Puccinia = **P. esclavensis** var. **panicophila** 906
panicorum, Metasphaeria 788
paniculata, Macrophoma = **Cryptosporiopsis cornina** 653
panicum, Sphaerella = **Metasphaeria panicorum** 788
pannicola, Chaetosphaeria 624
pannocincta, Ceriporiopsis (Poria) 621
pannocinctus, Gloeoporus = **Ceriporiopsis pannocincta** 621
pannosa, Drepanospora = **Helicosporium pannosum** 731
pannosa, Sphaerotheca 968
pannosa var. **persicae, Sphaerotheca** 968
pannosa var. **rosae, Sphaerotheca** 968
pannosum, Helicosporium 731
panuoides, Tapinella (Paxillus) 987
papagoensis, Coprinus 644
papakurae, Circinotrichum 628
papaveri, Cercospora = **C. papavericola** 614
papavericola, Cercospora 614
papayae, Cercospora 614
papayae, Colletotrichum = **C. gloeosporioides** 638
papillata, Tomentella = **T. coerulea** 993
papillatum, Hypoxylon 746
papillatum, Synchytrium 984
papillifera, Ravenelia = **R. spinulosa** var. **papillifera** 929
papillosa, Cercospora 614
papillosa, Odontia = **Grandinia nespori** 723
papillosa, Teichospora = **Teichosporella dura** 988
papillostoma, Sphaerognomonia (Gnomonia) 966
papulatum, Gloeosporium 718
papulosum, Helminthosporium 731
papulosum f. **pyri, Helminthosporium** 731
papyracea, Pachykytospora (Poria) 825
papyraceus, Hypomyces (Hypocrea) 744
papyriferae, Valsa (Valsella) 1018
papyrina, Lopharia 770
papyrinum, Stereum = **Lopharia papyrina** 770
paradoxa, Ceratocystis (Ceratostomella, Endoconidiophora) 604
paradoxa, Chalara (Thielaviopsis) 625
paradoxa, Melampsora 781
paradoxa, Orbilia 822
paradoxa, Schizopora 939
paradoxica, Puccinia 911
paradoxum, Ophiostoma = **Ceratocystis paradoxa** 604
paradoxus, Irpex = **Schizopora paradoxa** 939
paraguayensis, Parodiella = **P. hedysari** 826
parallela, Apiospora (Sphaerella) 569
parallela, Endoxyla 695
paranensis, Sorosporium 963
paraphysaria, Diplodia 681
parapsilosis, Candida (Monilia) 600
pararoseus, Sporidiobolus (Sporobolomyces) 971
parasitans, Hydnum = **Steccherinum laeticolor** 975
parasitica, Ascochyta = **Phoma exigua** 853
parasitica, Atractilina 584
parasitica, Clitocybe = **Armillaria tabescens** 572
parasitica, Cryphonectria (Diaporthe, Endothia) 650
parasitica, Cryptostictis = **Seimatosporium parasiticum** 945
parasitica, Eutypella 701
parasitica, Herpotrichia (Acanthostigma, Trichosphaeria) 735
parasitica, Peronospora 834
parasitica, Phialophora 847
parasitica, Phytophthora = **P. nicotianae** var. **parasitica** 879
parasitica, Pyricularia 919
parasitica, Rosellinia 937
parasitica, Syspastospora (Melanospora, Sphaeronema) 985

parasitica var. nicotianae, Phytophthora = **P. nicotianae** 879
parasiticum, Camarosporium 600
parasiticum, (Macrosporium, Stemphylium) = **Stemphylium vesicarium** 977
parasiticum, Seimatosporium 945
parasiticum, Septogloeum 946
parasiticus, Aspergillus 579
parca, Puccinia 911
parca, Tiarosporella (Macrophoma) 991
pardalota, Diaporthe 671
pardalota, Mycosphaerella 804
pargamenus, (Coriolus, Hirschioporus, Polyporus, Polystictus, Trametes, Trichaptum) = **Trichaptum biforme** 998
parietalis, Herpotrichia = **Byssosphaeria diffusa** 596
parietariae, Ramularia 926
parietariae, Septoria 955
parkerae, Puccinia 911
parkeri, Meria 787
parkeri, Rhabdocline 930
parkeri, Spegazzinia 963
parkinsoniae, Asteroma 582
parkinsoniae, Cylindrosporium 659
parkinsoniae, Phyllosticta 872
parksii, Cistella 628
parksii, Nipterella (Belonidium, Belonium) 816
parksii, Poria = **Byssoporia terrestris** 596
parksii, Synchytrium 984
parlatorei, Ustilago 1016
parnassiae, Puccinia 911
paroecandrum, Pythium 921
paronychia, Mucor 797
paroselae, Puccinia 911
parryi, Dothidella (Endothia) 687
parthenices, Puccinia = **P. invelata** 908
parthenii, Puccinia 911
partheniphila, Cercospora 614
parthenocissi, Elsinoe 692
parthenocissi, Synchytrium 984
parva, Codinaea 635
parva, Cylindrocladiella 657
parva, Ramularia 926
parvicapsa, Chaetosphaeria 624
parvifolii, Exobasidium = **E. vaccinii** 703
parvimaculans, Cercospora = **C. stomatica** 617
parvimaculans, Septoria 955
parvula, Uncinula 1006
parvulum, Crucibulum 650
parvulum, Glonium = **G. abbreviatum** 720
parvulum, Helminthosporium 731
parvulum, Leptothyrium 766
parvulum, Stachycoremium (Stilbum) 972
parvulus, Coccomyces 634
parvum, Cylindrocladium = **Cylindrocladiella parva** 657
parvum, Entyloma 696
paspali, Ascochyta (Ascochytula) 576
paspali, Cercospora = **C. setariae** 617
paspali, Cerebella = **C. andropogonis** 621
paspali, Claviceps 632
paspali, Massariothea (Apocoryneum, Coryneum) 780
paspali, Myriogenospora = **M. atramentosa** 809
paspali, Neottiosporina (Stagonospora) 815
paspali, Phaeosphaerella 841
paspali, Puccinia = **P. levis** 909
paspali, Sorosporium = **S. paspali-thunbergii** 963
paspalicola, Phyllachora 863
paspali-notati, Sphacelotheca = **S. cordobensis** 965
paspali-thunbergii, Sorosporium (Ustilago) 963
passaloroides, Cercospora (Cylindrosporium) 614
passerinii, Septoria 955
passiflorae, Alternaria 562
passiflorae, Cercospora 614
passiflorae, Colletotrichum = **C. gloeosporioides** 638
pastinacae, Cercospora 614
pastinacae, Cercosporella (Ramularia) 619
pastinacae, Cyphellopycnis = **Phomopsis diachenii** 858
pastinacae, Septoria (Cylindrosporium) 955
pastinacina, Cercospora = **Cercosporella pastinacae** 619
pastinacina, Rhabdospora = **Septoria pastinacae** 955
patella, Heterosphaeria 735
patella, Hypocrea 743
patellarioides, Amerosporium = **Chaetomella oblonga** 623
patellarioides, Teichospora (Strickeria) 988
patelliforme, Stereum = **Cytidia patelliformis** 661
patelliformis, Cytidia 661
patelloides, (Asterina, Microthyriella) = **Phillipsiella atra** 848
patellum, Dinemasporium = **Pseudolachnea hispidula** 896
patens, Melogramma 786
patouillardii, Inonotus 749
pattersoniana, Puccinia 911
patula, Lachnella = **Dasyscyphus patulus** 665
patulus, Dasyscyphus 665
patzcuarensis, Pileolaria 880
paucilamellata, Mycena 798
paucispora, Cryptosporella (Cryptospora, Valsa) 653
paucisporula, Ascochyta 576
paucitricha, Pleospora (Pyrenophora) 885
paulowniae, Ascochyta 576
paulowniae, Phyllosticta 872
paulula, Diaporthe (Cryptodiaporthe, Valsa) 671
paulula, Sphaerella 965
paupera, Septoria 955
pauperata, Valsa = **V. ambiens** 1017
paupercula, Asterina (Asterella) 581
paupercula, Microdiplodia (Diplodia) 789
paupercula, Phyllosticta 872
paurida, Nitschkia = **N. broomeiana** 816
paviae, Phyllosticta = **P. sphaeropsoidea** 873
pavonia, Trametes 996
pavoninum, Haplobasidion = **H. thalictri** 729
pavonius, Polyporus = **Trametes pavonia** 996
paxianum, Tetracoccosporium 988
paxilloides, Crepidotus 647
payettensis, Crepidotus 647
pazschkei, Puccinia 911
pazschkei var. **heterisiae, Puccinia** 911
pazschkei var. **tricuspidatae, Puccinia** 911
pearsonii, Paullicorticium 827
peckiana, Cercospora 614
peckiana, Curreya 655
peckiana, Dermea 669
peckiana, Macrophoma = **Colletotrichum gloeosporioides** 638
peckiana, Monilia 794
peckiana, (Gymnoconia, Puccinia) = **Arthuriomyces peckianus** 573
peckiana var. angustior, Monilia = **M. angustior** 794
peckianum, Cenangium = **Dermea peckiana** 669
peckianum, Myxofusicoccum 810
peckianum, Phragmidium 861
peckianus, Albatrellus (Polyporus) 559
peckianus, Arthuriomyces 573
peckianus, Uromyces 1012
peckianus subsp. **rugulosus, Uromyces** 1012
peckii, Colletotrichum = **C. dematium** 637
peckii, Delphinella (Mycosphaerella) 666
peckii, Diaporthe 671
peckii, Exobasidium = **E. vaccinii** 703
peckii, Puccinia = **P. dioicae** 905
peckii, Ramularia = **R. rhabdospora** 926
peckii, Septoria 955
peckii, Sphaeropsis 967
peckii, Steccherinum 975
peckii, Valsaria (Valsa) 1019
pectinata, Vararia 1019
pectinatum, Corticium = **Vararia pectinata** 1019
pedatospora, Speiropsis 963
pedatum, Ploioderma (Hypoderma) 887
pedicellata, Drechslera = **Exserohilum pedicellatum** 703
pedicellata, Galzinia 713
pedicellata, Sphaerulina 968
pedicellatum, Exserohilum (Helminthosporium) 703
pedicellatum, Scolicosporium 943
pedicularidis, Diplodina 682
pedicularidis, Phyllosticta 872
pedicularis, Nectriella = **Hyponectria sceptri** 744
pedicularis, Phoma 855
pediculata, Gyrothrix 727
pedunculata, Exidia = **Dacrymyces pedunculatus** 663
pedunculata, Stigmina 980
pedunculatum, (Brachysporium, Coryneum, Exosporium, Thyrostromella) = **Stigmina pedunculata** 980
pedunculatus, Dacrymyces 663
pedunculi, Phoma 855
peglionii, Alternaria 562
pelagica, Leptosphaeria 763
pelagica, Monodictys (Piricauda) 795
pelagica, Pleospora 885
pelagica, Speira = **Dictyosporium pelagicum** 676
pelagicum, Dictyosporium 676
pelargonii-zonalis, Puccinia 911
peleae, Gloeosporium = **Colletotrichum gloeosporioides** 638
peleae, Hexagonella 736
peleae, Meliola 785
pellicula, Chaetophoma 623
pelliculare, Corticium = **Fibulomyces mutabilis** 705
pellicularis, Athelia = **Fibulomyces mutabilis** 705
pelliculosa, Asterina = **A. lepiniana** 581
pelliculosa, Limacinia = **Capnodium pelliculosum** 601
pelliculosum, Capnodium 601
pellita var. bardanae, Pleospora = **P. lecanora** 885
pellita var. **oligostachyae, Pleospora** 885
pellucida, Ascochyta 576
pellucida, Oxyporus 824
pellucida, Tubaria 1002
pelorospora, Titanella (Amphisphaeria) 993
pendulus, Porodisculus 890

penetrans, Gymnopilus (Flammula) 725
penicillariae, Tolyposporium = **Moesziomyces bullatus** 792
penicillata, (Ceratocystis, Ceratostomella) = **Ophiostoma penicillatum** 821
penicillata, Isariopsis = **Atractilina parasitica** 584
penicillata, Microsphaera (Erysiphe) 790
penicillata, Verticicladiella = **Leptographium penicillatum** 760
penicillata var. extensa, Microsphaera = **M. extensa** 790
penicillata var. opuli, Cercospora = **C. opuli** 614
penicillata var. vaccinii, Microsphaera = **M. vaccinii** 791
penicillatum, Leptographium 760
penicillatum, Ophiostoma 821
penicillatus, Aleurodiscus 560
penicillioides, Aspergillus 579
penicillioides, Graphium 723
penicillus, Cercospora 614
penicillus, Pleospora (Sphaeria) 885
penniseti, Cercospora 614
penniseti, Curvularia (Acrothecium) 655
penniseti, Puccinia = **P. substriata** 914
penniseti, Ustilago 1016
pennisetina, Tilletia = **T. barclayana** 991
pennsylvanica, Diaporthe 672
pennsylvanica, Peronospora 834
pennsylvanica, Phoma 855
pennsylvanica, Phyllosticta 872
pennsylvanica, Scutellinia (Melastiza, Melastiziella) 943
pennsylvanica, Sphaeropsis = **S. malorum** 967
pennsylvanicum, Melanopsichium 783
pennsylvanicum var. besseyanum, Melanopsichium = **Melanopsichium pennsylvanicum** 783
pennsylvanicum var. caulicola, Melanopsichium = **Melanopsichium pennsylvanicum** 783
pennsylvanicus, Hypochnus = **Tomentellopsis echinospora** 995
penstemonis, Coniothyrium 643
penstemonis, Sphaceloma 964
pentamera, Clathrospora (Platyspora, Pleospora) 631
pentstemonicola, Gloeosporium 718
pentstemonicola, Phyllosticta 872
pentstemonicola, Septoria 955
pentstemonii, Ascochytella 577
pentstemonis, Cercospora 614
pentstemonis, Mycosphaerella 804
pentstemonis, Puccinia 911
pentstemonis, Ramularia 926
pentstemonis, Septoria 955
penzigii, Cercospora 615
perangusta, Galerina 712
peraphylli, Septoria 955
perdelicata, Postia 891
perdelicatus, (Oligoporus, Polyporus, Tyromyces) = **Postia perdelicata** 891
pereffusa, (Fomitiporia, Poria) = **Phellinus laevigatus** 845
peregrina, Cercospora 615
peregrina, Pestalotia = **P. foedans** 835
peregrina, Septoria 955
peregrinum, Colletotrichum = **C. gloeosporioides** 638
perelegans, Haglundia 728
perelegans, Helicoma 730
perennans, Cryptosporiopsis = **C. curvispora** 653
perennans, Gloeosporium = **Cryptosporiopsis curvispora** 653
perennans, Neofabraea = **Pezicula malicorticis** 837
perennans, Ustilago = **U. avenae** 1014
perenne, Exobasidium 703
perenne, Porotheleum 890
perexigua, Microthyriella = **Schizothyrium perexiguum** 939
perexigua, Peniophora 829
perexiguum, Schizothyrium (Phacidium) 939
perextensa, Poria = **Anomoporia myceliosa** 565
perfoliati, Mycovellosiella (Cercospora) 808
perforans, Marssonia = **Microdochium panattonianum** 789
perforans, Micromphale (Marasmius) 789
perforans, Tiarospora (Sphaeria) 991
perforata, Chaetosticta (Chaetomella) 624
perforatum, Hypoxylon = **H. rubiginosum** 746
perforatum, Synchytrium 984
peribebuyense, Sporotrichum = **Beniowskia sphaeroidea** 587
peribebuyensis, Cintractia 628
pericarpii, Sphaeropsis (Sphaeria) 967
pericarpium, Cladosporium 630
periclymeni, Kabatia (Colletotrichella, Labrella, Leptothyrium) 751
periclymeni, Phaeoramularia (Cercospora) 841
peridermiospora, Puccinia = **P. sparganioides** 914
perigynia, Nigredo = **Uromyces perigynius** 1012
perigynius, Uromyces 1012
perigynosum, Pythium = **P. dissotocum** 920
periilum, Pythium 921
periplocum, Pythium 921
perisporioides, Dothidea = **Parodiella hedysari** 826
perisporioides, Parodiella = **P. hedysari** 826
peristeriae, Cercospora 615
perlata, Discina = **D. ancilis** 684
permunda, Comoclathris (Clathrospora, Platyspora, Pleospora) 640
perniciosa, Cercospora = **C. cephalanthi** 607
perniciosa, Diaporthe 672
perniciosum, Fusarium = **F. oxysporum** f. sp. **perniciosum** 709
perniciosum, Pythium 921
perparva, (Mycosphaerella, Sphaerella) = **Mycosphaerella harthensis** 802
perparvispora, Ceratocystis = **C. microspora** 604
perplexans, Itersonilia 750
perplexum, Hypholoma = **H. sublateritium** 742
perpusila, Microdiplodia 789
perrottetiae, Actinodothis = **Amazonia stevensii** 563
perrottetiae, Asteridiella (Amazonia, Irene) 580
perseae forma setulifera, Meliola = **M. martiniana** 785
perseae, Asteridiella (Irene) 580
perseae, Mycosphaerella 804
perseae, Phyllachora 863
perseae, Phyllosticta = **P. micropuncta** 870
perseae, Sphaceloma 964
persica, Miuraea (Cercospora, Cercosporella) 792
persica, Mycosphaerella = **M. pruni-persicae** 805
persicae, Cytospora 662
persicae, Diplodina 682
persicae, Fusicoccum 712
persicae, Morrisographium (Cornularia, Periconia, Sphaeronaema) 796
persicae, Phoma 855
persicae, Phyllosticta 872
persicae, Tranzschelia = **T. discolor** 997
persicaria, Gilbertella (Choanephora) 715
persicaria, Xylaria 1026
persicariae, Septoria = **S. polygonorum** 956
persicina, Diplodia (Sphaeropsis) 681
persicina, Macrophoma 774
persicinus, Laetiporus (Meripilus, Polyporus) 755
persicum, Fusarium = **Miuraea persica** 792
persistens, Helminthosporium = **Corynespora pruni** 646
personata, Cytospora 662
personata, Mycosphaerella 804
personata, Phaeoisariopsis (Cercospora) 840
personatum, Cercosporidium = **Phaeoisariopsis personata** 840
personatum, Cladosporium = **Phaeoisariopsis personata** 840
persoonii, Leucostoma 767
persoonii, (Polyporus, Trametes) = **Earliella scabrosa** 691
persoonii, Quaternaria = **Eutypella quaternata** 701
persoonii, Stamnaria 975
pertusa, Septoria 955
pertusa, Trematosphaeria (Melanomma, Sphaeria) 998
peruviana, Cylindrocladiella (Cylindrocladium) 657
perversa, Gnomonia = **Plagiostoma alneum** 881
pestalozzioides, Pestalotia (Arthrobotryum) 835
pestalozzioides, Seimatosporium (Coryneum) 945
petalodes, Hohenbuehelia (Pleurotus) 737
petalostemonis, Hendersonia 733
petalostemonis, Uropyxis 1014
petasata, Cyathicula = **Crocicreas cyathoideum** 648
petasatus, Pluteus 887
petasitidis, Stagonospora 974
petersii, Camarops (Hypoxylon, Peridoxylon) 599
petersii, Cercospora (Helminthosporium) 615
petersii, Coccomyces (Hysterium) 634
petersii, Uncinula = **Pleochaeta polychaeta** 883
petiolata, (Macrophoma, Sphaeropsis) = **Discula platani** 685
petiolicola, Lophodermium 772
petiolophila, Cryptodiaporthe (Gnomonia, Gnomoniella) 651
petiolorum, Gnomonia 721
petiolorum, Hendersonia (Stagonospora) 733
petiolorum, Leptothyrium (Leptostroma) 766
petiolorum, Mollisia 793
petiolorum, Poculum (Calycina, Rutstroemia) 887
petrakii, Leptothyrium 766
petrophilum, Corticium = **Lindtneria leucobryophila** 768
petroselini, Cercospora = **Cercosporidium punctum** 620
petroselini, Phyllosticta 872
petroselini, Septoria 955

petroselini var. apii, Septoria = **S. apiicola** 948
petroselini subsp. **treleasiana, Septoria** 955
petuniae, Ascochyta 576
petuniae, Cercospora (Cercosporina) 615
petuniae, Heterosporium 736
peucedani, Pollaccia (Asperisporium, Fusicladium) 888
peziza, Nectria (Neuronectria) 813
peziza, Sporidesmium = **Coniosporium olivaceum** 642
pezizaeformis, Ditiola (Exidia, Femsjonia) 685
pezizoidea, Cyphellathelia 660
pezizoidea, Glomerella 719
pezizoidea, Venturia = **Protoventuria maculosa** 893
pezizoides, Cenangium 602
pezizoides, Dendrodochium 666
pezizoides, Nummularia = **Biscogniauxia repanda** 589
pezizoides, Peniophora = **P. quercina** 830
pezizoides, Pestalotia 835
pezizoides, Phoma (Aposphaeria) 855
pezizoides, Porotheleum = **P. fimbriatum** 890
pezizoides, Seimatosporium (Coryneum) 945
pezizoideum, Corticium = **Cyphellathelia pezizoidea** 660
pezizula, Tubeufia (Herpotrichia, Lasiosphaeria, Thaxteriella) 1002
phacae-frigidae, Uromyces 1012
phaceliae, Cylindrosporium 659
phaceliae, Puccinia 911
phaceliae, Ramularia 926
phacidiella, Phoma (Macrophoma) 855
phacidioidea, Telimenella 988
phacidioides, Propolis (Naemacyclus, Stictis) 892
phacidioides, Sphaeropsis = **Phoma phacidiella** 855
phacidioides, Sporonema 972
phacidioideum, Fusarium = **Foveostroma boycei** 706
phacidiomorpha, Didymella 677
phacorrhiza, Typhula (Clavaria) 1005
phacorrhiza var. **heterogenea, Typhula** 1005
phaeocomes, Pyrenophora (Pleospora) 919
phaeocomoides, (Pleospora, Sphaeria) = **Pleospora penicillus** 885
phaeoconia, Phialea = **Crocicreas phaeoconium** 649
phaeoconium, Crocicreas (Helotium) 649
phaeospermum, Arthrinium 572
phaeospora, Fenestella 704
phaeospora, Pleospora (Pyrenophora) 885
phaeospora, Stephanoma 977
phaeosporium, Helicoma 731
phalaridis, Phaeoseptoria 841
phalaridis, Phyllachora 863
phari-latifolia, Phyllachora = **P. sylvatica** 863
phaseoli, Botryodiplodia = **Macrophomina phaseolina** 775
phaseoli, Cercospora = **Pseudocercospora cruenta** 894
phaseoli, Elsinoe 692
phaseoli, (Macrophoma, Macrophomina) = **Macrophomina phaseolina** 775
phaseoli, Nematospora 814
phaseoli, Phomopsis 859
phaseoli, Phytophthora 879
phaseoli, Stagonospora 974
phaseoli, Uromyces = **U. appendiculatus** 1009
phaseoli var. strophostylis, Uromyces = **U. appendiculatus** 1009
phaseoli var. vignae, Uromyces = **U. vignae** 1013
phaseoliformis, Crepidotus = **C. albescens** 647
phaseolina, Macrophomina (Macrophoma) 775
phaseolina, Phyllosticta 872
phaseolorum, Ascochyta = **Phoma exigua** 853
phaseolorum, Cercospora = **Pseudocercospora cruenta** 894
phaseolorum, Diaporthe 672
phaseolorum, Leptosphaeria 763
phaseolorum var. batatatis, Diaporthe = **Diaporthe phaseolorum** 672
phaseolorum var. caulivora, Diaporthe = **Diaporthe phaseolorum** 672
phaseolorum var. sojae, Diaporthe = **Diaporthe phaseolorum** 672
phegopteridis, Uredinopsis 1007
phiala, Hymenoscyphus (Helotium) 740
philadelphi, Ascochyta 576
philadelphi, Phoma 855
philadelphi, Ramularia 926
philadelphi, Septoria 955
philodendri, Colletotrichum = **C. gloeosporioides** 638
philodendri, Macrophoma 774
philoprina, Discochora (Botryosphaeria, Physalospora) 684
phlebioides, (Peniophora, Phlebia) = **Phlebia subserialis** 848
phlegmaceum, (Belonidium, Belonium) = **Arachnopeziza aurata** 571
phlei, Cladosporium (Heterosporium) 630
phlei, Drechslera 690
phlei, Lophodermium 772
phlei, Macrophoma 774
phlei, Rostrosphaeria = **Botryosphaeria festucae** 592
phlei, Sirococcus 961
phlei, Sphaeropsis = **S. malorum** 967
phleina, Ascochyta 576
phlogina, Cercospora = **C. omphacodes** 614
phlogina, Mycosphaerella 804
phlogina, Peronospora 834
phlogina, Vermicularia 1021
phlogina, Volutella 1023
phlogis, Ascochyta 576
phlogis, Dasyscyphus 665
phlogis, Leptosphaeria 763
phlogis, Pyrenochaeta 918
phlogis, Septoria 955
phlogis var. phlogina, Ascochyta = **Ascochyta phlogis** 576
phloiophila, Perenniporia 830
phloiophilia, Ophiocapnocoma (Limacinia) 820
phloxii, Lophodermium 772
phlyctanoides, Septoria 955
phoeniceum, Penicillium = **P. dierckxii** 828
phoenicis, Annellophora 565
phoenicis, Aspergillus = **A. niger** var. **phoenicis** 579
phoenicis, Graphiola 723
phoenicola, Phomopsis 859
phoenicum, Diplodia (Botryodiplodia) 681
pholiotoides, Lentinus = **Heliocybe sulcata** 731
phomaspora, Cryptodiaporthe (Apioporthe, Valsa) 651
phomiformis, Dothiorella (Phyllosticta) 689
phomoides, Colletotrichum = **C. gloeosporioides** 638
phomoides, Glomerella = **G. cingulata** 719
phomopsis, Dothiorella 689
phoradendri, Exosporium 703
phoradendri, Phyllosticta 872
phoradendri, (Kabatiella, Protocoronospora) = **Colletotrichum gloeosporioides** 638
phoradendri, Uredo 1008
phoradendri, Valsaria 1019
phoradendri f. umbellulariae, Kabatiella = **Aureobasidium umbellulariae** 585
phoradendricola, Asterina 581
phormii, Cercospora 615
phosphorescens, Peniophora = **Phanerochaete velutina** 844
photiniae, Septoria 955
photiniae-serrulatae, Cercospora 615
photinicola, Spilocaea (Fusicladium) 969
phragmitis, Helicosporium 731
phragmitis, Phragmopeltis 861
phragmitis, Pseudographis 896
phragmitis, Puccinia 911
phragmitis, Stagonospora (Septoriella) 974
phyllachoroides, Colletotrichum (Gloeosporium) 639
phyllitidis, Cercospora 615
phyllocacti, Colletotrichum = **C. gloeosporioides** 638
phyllogena, Dendrophoma 667
phyllogenum, Coniothyrium 643
phyllophila, Phialea = **Hymenoscyphus phyllophilus** 740
phyllophila, Vararia (Peniophora) 1019
phyllophilum, Helotium = **Hymenoscyphus phyllophilus** 740
phyllophilus, Hymenoscyphus 740
phyllostachydis, Puccinia 911
phyllostegiae, Asterina 581
phyllostictoides, Phoma = **P. macrostoma** 854
physaleos, Phyllosticta 872
physalicola, Cercospora = **C. physalidis** 615
physalidis, Aecidium 558
physalidis, Cercospora 615
physalidis, Entyloma = **E. australe** 695
physalidis, Leptosphaeria 763
physalidis, Puccinia 911
physocarpi, Cercospora 615
physocarpi, Phyllosticta 872
physocarpi, Sphaeronaema 966
physocarpi, Sphaeropsis 967
physocarpi, Stagonospora 974
physopellae, Phyllosticta 872
physostegiae, Mycosphaerella 804
physostegiae, Puccinia 911
physostegiae, Rhabdospora 931
physostegiae, Septoria 955
phytolaccae, Phomopsis 859
phytolaccae, Phyllosticta 872
phytoptophila, Sphaerotheca 968
piaropi, Cercospora 615
picaceum, Anthostomella (Anthostoma) 566
picea, Phaeostigme = **Dimerium piceum** 679
picea, Phyllachora (Dothidea) 863
piceae, Crinipellis 648
piceae, Dermea 669
piceae, Diedickea 679
piceae, Hyphosoma (Monilia) 742
piceae, Lophodermium 772
piceae, Mollisia 793
piceae, Ophiostoma (Ceratocystis, Ceratostomella) 821
piceae, Peniophora 829

piceaperda, Ceratocystis (Ceratostomella, Grosmannia) 604
piceaperdum, Ophiostoma = **Ceratocystis piceaperda** 604
piceicola, Protodontia 892
piceina, Peniophora = **Cystostereum pini-canadense** 660
piceina, Phoma 855
piceinum, Cystostereum = **C. pini-canadense** 660
piceinum, Dendrocorticium 666
piceinus, Aleurodiscus 560
piceum, Dimerium 679
picipes, Polyporus = **P. badius** 889
pictum, Leptothyrium = **Kabatia periclymeni** 751
pigmentata, Omphalia = **Marasmiellus pigmentatus** 776
pigmentatus, Marasmiellus 776
pilati, (Aporpium, Poria) = **Aporpium caryae** 571
pilatii, Tomentella 994
pileae, Cercospora 615
pileae, Septoria 955
pilifera, (Ceratocystis, Ceratostomella) = **Ophiostoma piliferum** 821
piliferum, Ophiostoma 821
pilificum, Synchytrium (Pycnochytrium) 984
pilosa, Isariopsis = **Morrisographium persicae** 796
pilosa, Peniophora = **Hyphoderma pilosum** 741
pilosa, Tomentella 994
pilosa, Trichosphaeria (Sphaeria) 1000
pilosula, Discella 683
pilosum, Corticium = **Hyphoderma amoenum** 740
pilosum, Helicosporium = **H. aureum** 731
pilosum, Hyphoderma 741
pilosum, Phragmographium = **Morrisographium persicae** 796
pilosum, Septobasidium 946
pilosus, Hypochnus = **Tomentella pilosa** 994
pilosus, Merulius = **Phlebia rufa** 848
pilotae, Polyporus = **Hapalopilus croceus** 729
pilulaeformis, Balansiopsis (Balansia) 586
pilulaeformis, Thecaphora 989
piluliformis, Psathyrella 893
pimeriensis, Columnocystis 640
pimpinellae, Phomopsis (Rhabdospora, Septoria) 859
pimpinellae, Puccinia 911
pinaropappi, Puccinia 911
pinastri, Cytospora 662
pinastri, Diplodia 681
pinastri, Dothiorella 689
pinastri, Leptostroma = **Sirothyriella pini-austriacae** 962
pinastri, Leucogyrophana (Gyrophana, Merulius, Serpula) 767
pinastri, Lophodermium (Hysterium) 772
pinastri, (Mollisia, Phibalis) = **Cenangium acicola** 602
pinastri, Phaeocryptopus (Asterina) 839
pinastri, Tryblidiopsis (Tympanis) 1002
pinastri var. thujae, Hysterium = **H. thujae** 747
pinea, Caliciopsis 597
pinea, Creosphaeria = **Rosellinia diathrausta** 937
pinea, (Diplodia, Macrophoma) = **Sphaeropsis sapinea** 967
pinea, Nectria 813
pinea, Rhabdospora 931
pinguedinis, Macrosporium 775
pini, Aplosporella 570
pini, Bifusella (Hypoderma) 588
pini, Camarosporium (Hendersonia) 600
pini, Capnodium 601
pini, Ceratostomella = **Ophiostoma minus** 821
pini, Coleosporium = **C. pinicola** 636
pini, Corticium (Aleurodiscus, Laeticorticium) 645
pini, Dasyscyphus = **Lachnellula fuscosanguinea** 754
pini, Dendryphion = **Septonema fasciculare** 947
pini, Dermea = **Cenangium ferruginosum** 602
pini, Didymopeltis 677
pini, Discosia 684
pini, Dothistroma = **D. septospora** 689
pini, Lachnellula (Lachnella) 754
pini, Lecanosticta = **L. acicola** 758
pini, Mycosphaerella (Scirrhia) 804
pini, Paschelkiella 826
pini, Phellinus (Fomes, Porodaedalea, Trametes) 846
pini, Phoma = **P. harknessii** 853
pini, Phragmotrichum (Trimmatostroma) 862
pini, Pragmopora 891
pini, Rhabdospora (Cryptosporium) 931
pini, Rhizosphaera 934
pini, (Sclerophoma, Sphaeronaema) = **Rhizosphaera kalkhoffii** 934
pini, Stagonospora 974
pini, Therrya (Coccophacidium) 990
pini, Valsa 1018
pini subsp. **duplex, Peniophora** 830
pini var. lineare, Dothistroma = **D. septospora** var. **lineare** 689
pini-austriacae, Sirothyriella 962
pini-canadense, Cystostereum 660
pini-canadense subsp. **subabruptum, Cystostereum**
pini-canadensis, Radulum = **Cystostereum pini-canadense** 660
pinicola, Appendiculella (Asteridiella, Irenina, Meliola) 571
pinicola, Atropellis 584
pinicola, Bothrodiscus = **B. berenice** 590
pinicola, Coleosporium 636
pinicola, Fomitopsis (Fomes) 706
pinicola, Furcaspora 706
pinicola, Gelatinosporium (Cryptosporium, Micropera) 713
pinicola, Hendersonia 733
pinicola, Hendersonula 734
pinicola, Kalmusia 751
pinicola, Lophodermium = **L. pinastri** 772
pinicola, (Coryneum, Monochaetia) = **Toxosporium camptospermum** 995
pinicola, Neofuckelia 814
pinicola, Phoma = **Sclerophoma pythiophila** 940
pinicola, Septoria 955
pinicola, Stilbospora 981
pinicola, Trichosporum 1000
pinicola, Tulasnella (Gloeotulasnella) 1004
piniperda, Ascochyta = **Sirococcus conigenus** 961
piniperda, Sarcotrochila (Naevia) 938
piniphila, Atropellis 584
piniphilum, Cenangium = **Atropellis piniphila** 584
pinnatifida, Hymenochaete 739
pinnulaecola, Cercospora 615
pinodella, Phoma (Ascochyta) 855
pinodes, Ascochyta 576
pinodes, Mycosphaerella (Didymella) 804
pinophilum, Apiosporium 569
pinophilum, Penicillium 828
pinophyllum, Microthyrium = **M. microscopicum** 791
pinorum var. **graupianum, Leptostroma** 765
pinsapo, Mycosphaerella 804
pinsitus, (Coriolus, Polyporus, Trametes) = **Trametes villosa** 997
piperata, Clavicorona (Clavaria) 632
piperata, Glomerella = **G. cingulata** 719
piperatum, Gloeosporium = **Colletotrichum gloeosporioides** 638
piperi, Puccinia 911
piperi, Ustilago 1016
piperiana, Chrysomyxa (Melampsoropsis) 626
pipericola, Pseudonectria = **Nectria leucorrhodina** 813
piptarthra, Hendersonia = **Piptarthron macrosporum** 880
pipturi, Asteridiella (Irenina) 580
pipturi, Cercospora 615
piri, Elsinoe 692
piricularioides, Cladosporium 630
pirifera, Sphaerodothis (Phyllachora) 965
piriforme, Didymosporium = **Marssonina piriformis** 778
piriformis, Marssonina (Marssonia) 778
piriformis, Mucor 797
pirina, Ascochyta = **Phoma exigua** 853
pirina, Cercospora = **Cercosporella pyrina** 619
pirinum, Oidium 818
pirolae, Tomentella 994
pirolata, Chrysomyxa 627
pironii, Kutilakesa 753
pironii, Nectriella 814
pisa-sativae, Cercospora 615
piscariae, Uredo 1008
piscarium, Penicillium = **P. simplicissimum** 829
piscidiae, Ravenelia 929
piscidiae subsp. rugosa, Ravenelia = **Ravenelia piscidiae** 929
pisi, Ascochyta 576
pisi, Cladosporium 630
pisi, Colletotrichum = **C. gloeosporioides** 638
pisi, Peronospora 834
pisi, Septoria 955
pisicola, Ascochyta = **A. pisi** 576
pisicola, Cladosporium = **C. cladosporioides** f. sp. **pisicola**
pisicola, Fusicladium 711
pisoniae, Echidnodes (Maurodothella) 691
pistacina, Hyalotia 738
pistorica, Puccinia 912
pitcheriana, Phyllosticta 872
pithecolobii, Phyllosticta 872
pithecolobii, Ravenelia 929
pithoides, Nectria 813
pithoideum, Bactrodesmium (Coryneum) 585
pithya, Diaporthe = **D. eres** 671
pithya, (Patellaria, Tympanis) = **Tympanis truncatula** 1004
pithya, Peniophora 830
pithya, Pragmopora (Phacidium) 891
pithya, Pragmopycnis 891
pithyophila, Cucurbidothis (Cucurbitaria) 654

pithyophilum, Hydnum = **Grandinia spathulata** 723
pithyum, Sphaeronaema 966
pittospori, Cercospora 615
pittospori, Diplodia = **D. mutila** 681
pittospori, Mycosphaerella 804
pittospori, Phomopsis (Phoma) 859
pittospori, Rhabdospora 931
pitya, Asteridiella (Irenina) 580
pityella, Sclerophoma (Phoma) 940
placenta, Diatrypella 674
placenta, Postia (Poria) 891
placentus, Oligoporus = **Postia placenta** 891
placentus, Polyporus = **Postia placenta** 891
plagia, Diatrype = **Eutypella goniostoma** 700
plagiopus, Prospodium (Puccinia) 892
planellus, (Coriolus, Polyporus) = **Datronia stereoides** 665
planimi, Pestalotia 835
planispora, Graphyllium (Comoclathris, Platyspora, Pleospora) 724
planiuscula, Leptosphaeria 763
plantaginea, Septoria = **S. plantaginis** 955
plantaginea var. plantaginis-majoris, Septoria = **S. plantaginis-majoris** 955
plantaginella, Ascochyta = **A. sodalis** 577
plantaginella, Cercospora = **C. plantaginis** 615
plantaginella, Phyllosticta 872
plantagineum, Synchytrium 984
plantaginicola, Ascochyta = **A. sodalis** 577
plantaginicola, Mycosphaerella 804
plantaginicola, Phyllosticta 872
plantaginis, Ascochyta = **A. sodalis** 577
plantaginis, Cercospora 615
plantaginis, Eudimeriolum (Stigmatea) 699
plantaginis, Hormotheca (Aphysa, Asterina, Coleroa, Mycosphaerella) 737
plantaginis, Peronospora 834
plantaginis, Phyllosticta 872
plantaginis, Ramularia = **R. rhabdospora** 926
plantaginis, Septoria 955
plantaginis, Sphaceloma 964
plantaginis-majoris, Septoria 955
planum, Hemiphacidium (Phacidium) 732
planus, Polyporus = **Datronia stereoides** 665
platani, Cytospora 662
platani, Discula 685
platani, Eutypella (Valsa) 701
platani, Exosporium 703
platani, Gnomonia = **Apiognomonia veneta** 569
platani, Libertella 768
platani, Phyllosticta 872
platani, Sphaerella 965
platani, Splanchnonema (Massaria) 969
platani, (Hendersonia, Stegonsporium) = **Macrodiplodiopsis desmazieresii** 773
platani, Stigmina (Stigmella) 980
platanicola, Cercospora 615
platanicola, Leptosphaeria (Sphaeria) 763
platanicola, Myxosporium = **Discula umbrinella** 685
platanifolia, Mycosphaerella (Sphaerella) 804
platanifolia, Septoria 955
platani-racemosae, Stigmina (Sciniatosporium, Stigmella) 980
platanoidis, Phyllosticta 872
platanorum, Exosporium = **Stigmina tharpii** 980
platasca, (Diaporthella, Diatrype) = **Diaporthella aristata** 672
platensis, Albugo 559
platycerii, Cercospora 615
platyphylla, Megacollybia (Collybia, Tricholomopsis) 780
platypus, Hendersonia = **Seimatosporium lichenicola** 945
platysperma, Phoma 855
platyspora, (Cercospora, Didymaria) = **Cercosporidium punctiforme** 620
platyspora, Pleospora = **Clathrospora pentamera** 631
platysporioides, Clathrospora 631
platysporum, Mycoenterolobium 799
platystoma, Graphostroma (Diatrype) 723
platystomum, Hypoxylon = **H. stygium** 746
platystroma, Melanconis 782
plenum, Aecidium 558
pleosphaerioides, Pleospora 885
pleosporoides, Leptosphaeria = **Pleospora vitalbae** 886
pleosporoides, Rhabdospora 931
pleosporoides, Septoria 955
pleosporoides var. **drabae, Rhabdospora** 931
pleurostyla, Gnomonia = **Pleuroceras pleurostylum** 886
pleurostylum, Pleuroceras 886
plicata, Ionomidotis (Midotis) 749
plicatulus, Marasmius 777
plicatum, Stereum = **S. gausapatum** 977
plicosa, Mycena 798
plucheae, Phloeospora 850
plucheae, Puccinia = **P. biocellata** 901
plucheae, Septoria 955
plumasensis, Puccinia 912
plumbaginicola, Ascochyta 576
plumbaria, Nigredo = **Uromyces plumbarius** 1012
plumbaria, Puccinia 912
plumbarius, Uromyces 1012
plumbea, Mycena 798
plumbescens, (Exidiopsis, Sebacina) = **Exidiopsis grisea** 702
plumbeum, Physarum = **P. cinereum** 876
plumbeus, Mucor 797
plumeriae, Cercospora 615
plumierae, Coleosporium 636
plumigera, Massaria = **M. lantanae** 779
plumigera var. tetraspora, (Massaria, Massarina) = **Massaria lantanae** 779
pluriannulata, (Ceratocystis, Ceratostomella) = **Ophiostoma pluriannulatum** 822
pluriannulata, Urophlyctis = **Physoderma pluriannulatum** 877
pluriannulatum, Ophiostoma 822
pluriannulatum, Physoderma (Synchytrium) 877
pluriseptata, Pleospora 885
pluristriae, Scirrhia 940
pneumatophorae, Mycosphaerella 804
poae, Drechslera (Helminthosporium) 690
poae, Fusarium (Sporotrichum) 710
poae, (Leptosphaeria, Metasphaeria) = **Keissleriella culmifida** 752
poae, Melanospora = **Scopinella solani** 943
poae, Phaeoseptoria 841
poae, Uromyces = **U. dactylidis** var. **poae** 1010
poae-nemoralis, Puccinia = **P. brachypodii** var. **poae-nemoralis** 902
poae-sudeticae, Puccinia = **P. brachypodii** var. **poae-nemoralis** 902
poae-sudeticae var. airae, Puccinia = **P. brachypodii** var. **poae-nemoralis** 902
poae-trivialis, Septoria = **S. macropoda** 953
poagena, Cercosporella 619
poarum, Puccinia 912
poavora, Coniothyrium 643
pocula, Polyporus = **Porodisculus pendulus** 890
poculiformis, Puccinia = **P. graminis** 907
podlachica, Exidiopsis (Sebacina) 702
podophylli, Cercospora 615
podophylli, Phyllosticta (Phyllostictina) 872
podophylli, Puccinia 912
podophyllina, Phyllosticta 872
podophyllina, Septoria 955
podophyllina, Septotinia 959
podophyllina, Septotis 960
podophyllinum, (Gloeosporium, Septogloeum) = **Septotis podophyllina** 960
poecilospora, Didymellina 677
poikilimera, Discosia 684
poinsettiae, Sphaceloma 964
poitei, Xylaria 1026
polaris, Septoria 955
polemoniae, Ramularia 926
polemonii, Nigredo = **Uromyces acuminatus** 1009
polemonii, Phyllachora 863
polemonii, Puccinia (Micropuccinia) 912
polemonii, Septoria 956
polemonii, Synchytrium 984
polemoniicola, Septoria = **S. polemonii** 956
polifoliae, Mycosphaerella 804
poliomela, Septoria 956
polita, Septoria 956
pollaccii, Colletotrichum = **C. gloeosporioides** 638
pollinaria, Lachnella = **Lachnum pollinarium** 755
pollinarium, Lachnum 755
pollinarius, Dasyscyphus = **Lachnum pollinarium** 755
pollinis-pini, Rhizophydium 933
polonense, Hyphoderma = **Hyphodermopsis polonensis** 742
polonensis, Hyphodermopsis (Peniophora) 742
polonica, Microsphaera = **M. friesii** 790
polyblastis, Botryotinia (Sclerotinia) 594
polyblastis, Botrytis 595
polychaeta, Pleochaeta (Erysiphe, Uncinula, Uncinulopsis) 883
polychaetia, Pestalotia = **Pestalotiopsis funerea** 836
polychromum, Ganoderma = **G. lucidum** 713
polycistis, Myxocyclus (Hendersonia) 810
polycocca, Eutypa (Valsa) 700
polycodii, Monilinia 795
polygalae, Cercospora = **C. grisea** 611
polygalae, Phoma 855
polygalae, Phyllachora (Dothidea) 863
polygalae, Septoria 956
polygonacea, Cercospora 615
polygonacea var. avicularis, Cercospora 615
polygonati, Guignardia 725
polygoni, Chaetoconis (Amphorula, Kellermania) 623
polygoni, Dothidea = **Stigmatea polygonorum** 979
polygoni, Erysiphe 698
polygoni, (Ascochyta, Gloeosporium) = **Ascochyta volubilis** 577
polygoni, Nitschkia 816
polygoni, Peronospora 834
polygoni, Pyrenopeziza (Mollisia) 919
polygoni, Rhabdospora 931
polygoni, Synchytrium 985
polygoni, (Nigredo, Uromyces) = **Uromyces polygoni-avicularíae** 1012
polygonia, Bostrichonema 590

polygonia, Peniophora (Cryptochaete) 830
polygoni-alpini, Puccinia 912
polygoni-amphibii, Puccinia 912
polygoni-amphibii var. convolvuli, Puccinia = **Puccinia polygoni-amphibii** 912
polygoni-amphibii var. persicariae, Puccinia = **Puccinia polygoni-amphibii** 912
polygoni-amphibii var. tovariae, Puccinia = **Puccinia polygoni-amphibii** 912
polygoni-avicularìae, Uromyces 1012
polygonicola, Ascochyta = **A. volubilis** 577
polygonicola, Diplodia 681
polygonicola, Septoria = **S. polygonorum** 956
polygonina, Septoria = **S. polygonorum** 956
polygonium, Corticium = **Peniophora polygonia** 830
polygoni-virginica, Vermicularia 1021
polygonorum, Cercospora = **C. effusa** 609
polygonorum, Phyllosticta 872
polygonorum, Septoria 956
polygonorum, Stigmatea 979
polygramma var. **plantaginis, Phoma** 855
polyloculata, Stomiopeltis 981
polymastum, Pythium 921
polymorpha, Cercospora = **C. malvicola** 613
polymorpha, Coccodiella 634
polymorpha, Hansenula 728
polymorpha, Mycosphaerella = **M. stigmina-platani** 806
polymorpha, Xylaria (Xylosphaera) 1026
polymorpha var. **pauciseta, Yoshinagella** 1027
polymorphon, Pythium = **P. irregulare** 921
polymorphum, (Bagnisiopsis, Coccostroma) = **Coccodiella polymorpha** 634
polymorphum, Gloeosporium = **Colletotrichum gloeosporioides** 638
polymorphum, Stemphylium (Sporidesmium) 977
polyphragmoides, Pyrenophora = **Pleospora anthyllidis** var. **abbreviata** 884
polypoda, Amazonia 563
polypodii, Hyalopsora 738
polypodii, Mycosphaerella 804
polypodophila, (Milesia, Milesina, Uredinopsis) = **Milesina pycnograndis** 792
polyporoidea, (Coniophora, Trechispora) = **Ramaricium polyporoideum** 923
polyporoidea, Solenia = **Henningsomyces candidus** 734
polyporoideum, Ramaricium (Corticium, Phlyctibasidium) 923
polysora, Puccinia 912
polysperma, Camarops (Hypoxylon) 599
polyspora, Dothiora = **Sydowia dothideoides** 983
polyspora, Sphaerulina = **Dothiora wolfii** 688
polyspora, Sydowia (Dothidea) 983
polyspora, Valsa (Valsella) 1018
polysporum, Cladosporium = **Conoplea globosa** 644
polysporum, Entyloma 696
polysporum, Helicoma 731
polysporum, Trichoderma (Sporotrichum) 999
polystichi, Helotium 732
polystichi, Milesia = **Milesina winelandii** 792
polystichi, Taphrina 987
polysticta, Didymosphaeria 678
polystigma, Pseudomassaria (Laestadia, Sphaerognomonia) 897
polytaeniae, Cercospora 615
polytaeniae, Stagonospora 974
polythalama, (Calonectria, Nectria, Scoleconectria) = **Nectria pseudotrichia** 813
polythrincioides, Passalora = **Cercosporidium depressum** 620
polytricha, Auricularia = **A. cornea** 585
polytricha, Cercospora 615
polytricha, Vermicularia = **Colletotrichum truncatum** 639
polytylum, Pythium 921
pomaceus, Phellinus (Fomes, Polyporus) 846
pomi, Mycosphaerella (Sphaerella) 804
pomi, (Aposphaeria, Phoma) = **Phoma macrostoma** 854
pomi, Schizothyrium (Leptothyrium) 939
pomi, Spilocaea 969
pomicola, Alternaria 562
pomiformis, Melanopsamma 783
pomiformis, Rebentischia = **R. massalongii** 929
pomiformis, Teichospora 988
pomigena, Gloeodes (Dothidea, Phyllachora) 716
pomorum, Phoma 855
pomorum, Sphaeronaema 966
ponderosae, Ceratocystis = **Ophiostoma populinum** 822
ponderosae, Davisomycella (Lophodermium) 666
ponderosus, Neolentinus (Lentinus) 815
pongamiae, Mycovellosiella (Cercospora) 808
pontederiae, Cercosporella (Cercospora) 619
pontederiae, Mycosphaerella 805
pontederiae, Uromyces 1012
populea, (Chondroplea, Dothichiza, Phoma) = **Discosporium populeum** 685
populeum, Discosporium 685
populi, Apioplagiostoma (Plagiostoma) 569
populi, (Clasterosporium, Stigmina) = **Pollaccia radiosa** 888
populi, Cornicularìella (Cornularia) 644
populi, Diatrypella 674
populi, Dimerosporium 679
populi, Eutypella 701
populi, Hyponectria (Guignardia) 744
populi, Marssonina (Gloeosporium) 778
populi, Mycosphaerella 805
populi, Parkerella = **Lahmia kunzei** 756
populi, Pleuroceras 886
populi, Septoria 956
populi, Sirodothis = **S. populnea** 961
populi, Teichospora (Strickeria) 988
populi-albae, Drepanopeziza (Pseudopeziza) 690
populicola, Cercospora 615
populicola, Lachnella = **Dasyscyphus populicolus** 665
populicola, Mycosphaerella 805
populicola, Septoria 956
populicolus, Dasyscyphus 665
populifolia, Didymosphaeria = **Venturia macularis** 1020
populifolia, Mycosphaerella (Sphaerella) 805
populina, Calycellina (Pezizella) 599
populina, Ceratocystis = **Ophiostoma populinum** 822
populina, Cercospora = **Pseudocercospora salicina** 895
populina, (Cryptosphaeria, Eutypa) = **Cryptosphaeria lignyota** 652
populina, Dermatella 669
populina, Diplodia 681
populina, Discella 683
populina, Perrotia (Lachnella) 834
populina, Phyllosticta = **P. intermixta** 869
populina, Taphrina 987
populina, Venturia (Didymosphaeria) 1020
populinum, Chaetospermella = **Chaetospermum carneum** 624
populinum, Cylindrosporium 659
populinum, Dicoccum = **Pollaccia radiosa** 888
populinum, Helotium = **Calycellina populina** 599
populinum, Ophiostoma 822
populinus, Oxyporus (Fomes, Rigidoporus) 824
populiperda, Septotinia 960
populi-salicis, Taphrina 987
populnea, Melampsora 781
populnea, Mycosphaerella (Sphaerella) 805
populnea, Peniophora = **Hyphoderma populneum** 741
populnea, Pholiota 851
populnea, Sirodothis (Dothiorella) 961
populneum, Cenangium = **Encoelia fascicularis** 693
populneum, Hyphoderma (Stereum) 741
populneum var. **prunicola, Cenangium** 602
populorum, Drepanopeziza 690
populorum, Mycosphaerella 805
poriaeforme, Stigmatolemma (Porotheleum) 979
poriaeformis, Solenia = **Stigmatolemma poriaeforme** 979
poricola, Fusicladium 711
porinoides, Athelia = **Ceraceomyces serpens** 603
porioides, Cyphella = **Porotheleum fimbriatum** 890
poromera, Puccinia 912
porosa, Uromycopsis = **Uromyces coloradensis** 1010
porosum, Gloeocystidiellum (Corticium) 716
porphyrogenita, Puccinia 912
porphyrogonus, Ophiobolus = **Leptospora rubella** 764
porrecta, Pulveria 917
porrectus, Inonotus (Polyporus) 749
porri, Alternaria (Macrosporium) 562
porri, Puccinia = **P. allii** 900
porrigens, Pleurocybella (Nothopanus, Pleurotus) 886
portentosum, Scytinostroma (Corticium) 944
porteri, Phyllosticta 872
portoricensis, Beltraniella (Ellisiella, Ellisiopsis) 587
portoricensis, Desmotascus = **Botryosphaeria disrupta** 592
portoricensis, Fuscocerrena (Polyporus) 711
portoricensis, Halstedia = **Phyllachora halstediana** 862
portulacae, Albugo (Cystopus) 559
portulacae, Dichotomophthora 675
portulacae, Drechslera (Helminthosporium) 690
posadensis, Puccinia 912
poseyi, Septoria 956
posita, Uredo 1008
potebniae, Sphaerulina 969
potentillae, Cercospora 615
potentillae, Discosia 684
potentillae, Marssonia = **Marssonina fragariae** 778
potentillae, (Marssonina, Phyllosticta) = **Marssonina fragariae** 778
potentillae, Peronospora 834
potentillae, Phragmidium 861
potentillae, Physalospora 876

potentillae, Pleospora 885
potentillae, Pucciniastrum 917
potentillae, Pyrenopeziza (Trochila) 919
potentillae, Septogloeum = **Septoria fragariae** 951
potentillae, Synchytrium 985
potentillae, Taphrina 987
potentillae, Venturia (Chaetomium, Coleroa, Dothidea, Stigmatea) 1020
potentillae var. **helleri, Marssonia** 777
potentillae var. **tormentillae, Marssonina** (Marssonia) 778
potentillica, Phoma 855
pothi, Leptothyrium 766
pothicola, Phyllosticta 872
potteri, Physoderma 877
praecincta, Cercospora 615
praecox, Agrocybe (Pholiota) 559
praecox, Gymnopilus (Flammula) 725
praefloccosus, Gymnopilus 725
praefocata, Trechispora 998
praegracilis, Puccinia 912
praegrandis, Cercospora = **Mycocentrospora acerina** 799
praelatifolius, Crepidotus 647
praelonga, Gloniopsis 720
praelongum, (Hysterium, Hysterographium) = **Gloniopsis praelonga** 720
praerimosus, Fomes = **Phellinus everhartii** 845
praeterita, Peniophora = **Xenasma praeteritum** 1024
praeteritum, Xenasma 1024
praetermissa, Peniophora = **Hyphoderma praetermissum** 741
praetermissum, Hyphoderma 741
praetervisa, Meliola 785
praetervisa, Phyllosticta 872
prasinula, Claussenomyces (Corynella, Peziza) 631
pratensis, Leptosphaeria 763
pratensis, Ramularia 926
praticola, (Corticium, Thanatephorus) = **Thanatephorus cucumeris** 989
preandina, Leptosphaeria (Gibberidea) 763
prenanthis, Cercospora 615
prenanthis, Mycosphaerella (Laestadia) 805
prenanthis, Septoria 956
primaverilis, Uromyces 1012
primaverilis subsp. **nothoscordi, Uromyces** 1012
primulae, Colletotrichum = **C. gloeosporioides** 638
primulae, Mycosphaerella (Stigmatea) 805
primulae, Ramularia (Cercospora, Cercosporella) 926
primulae, Septoria 956
primulara, (Ovularia, Ramularia) = **Ramularia primulae** 926
primulicola, Phyllosticta 872
primulinum, Penicillium 828
princeps, Calosphaeria = **C. pulchella** 598
princeps, Diatrype 673
pringlei, Apodothina (Dothidea, Sphaerodothis) 571
pringlei, Ravenelia 929
prini, Ramularia 926
prini, Rhytisma 935
prinicola, Pyrenopeziza (Mollisia) 919
prinicola, Rosellinia = **Hypoxylon sassafras** 746
prinoides, Echidnodella (Lembosia, Morenoella) 691
pritchardiae, Phomopsis (Macrophoma, Phoma) 859
probabilis, Puccinia = **P. gentianae** 906
probus, Uromyces 1012
procera, Puccinia 912
procera, Verticicladiella = **Leptographium procerum** 760
procerum, Leptographium 760
proeminens, (Nigredo, Uromyces) = **Uromyces euphorbiae** 1010
proeminens var. euphorbiicola, Uromyces = **U. euphorbiae** 1010
proeminens var. myristica, Uromyces = **U. euphorbiae** 1010
proeminens var. poinsettiae, Uromyces = **U. euphorbiae** 1010
profundae, Sphaeropsis 967
profusa, Aglaospora 559
profusa, Menisporopsis 786
profusum, Septogloeum (Gloeosporium) 946
prolata, Drechslera = **Exserohilum prolatum** 703
prolata, Setosphaeria 960
prolatum, Exserohilum 703
proliferans, Cystobasidium 660
proliferans, Helicoma = **Helicosporium viride** 731
proliferans, Stilbella = **S. annulata** 981
proliferatum, Fusarium (Cephalosporium) 710
proliferum, Pythium = **P. middletonii** 921
prolifica, Periconia 831
prolificans, Cercosporella (Cercospora) 619
prolificans, Polyporus = **Trichaptum biforme** 998
prominens, Diatrype 673
prominens, Diatrypella 674
prominens, Didymella 677
prominens, (Hysterium, Hysterographium) = **Hysterographium flexuosum** 747
prominens, Lophiostoma 770
prominens, Tubulicrinis (Peniophora) 1003
propinqua, Mollisia (Calycina) 793
propinqua, Peniophora = **Tubulicrinis propinquus** 1003
propinquum, Helotium = **Mollisia propinqua** 793
propinquus, Tubulicrinis 1003
propullans, Diplodia (Sphaeria) 681
proserpinacae, Puccinia 912
prosopidis, Aplosporella (Sphaeropsis) 570
prosopidis, Cercospora 615
prosopidis, Hypochnicium 743
prosopidis, Pleochaeta (Uncinula) 883
prosopidis, Rhytidhysterium = **Rhytidhysteron rufulum** 935
prosopodium, Napicladium 811
prostii, Hysteropatella (Hysterium) 747
proteispora, Leptosphaeria 763
protracta, Trametes = **Gloeophyllum protractum** 717
protractum, Gloeophyllum 717
protuberans, Leptothyrium 766
protuberata, Curvularia 655
provinciale, Sorosporium 963
proxima, Poria = **Oxyporus pellucida** 824
proximella, Didymella 677
proximus, Oxyporus = **O. pellucida** 824
pruinatoides, Hypoxylon = **H. fuscum** 745
pruinatum, Botryobasidium 591
pruinatum, Hypoxylon = **H. mammatum** 745
pruinopsipes, Stilbum = **Stilbella flavipes** 981
pruinosa, Cryptosporiopsis (Lagynodella) 653
pruinosa, Cytospora (Cytophoma) 662
pruinosa, Dendrophoma (Sphaeria) 667
pruinosa, Encoelia (Cenangium, Dermea, Phibalis) 693
pruinosa, Pezicula 837
pruinosa, Ramularia 926
pruinosa, Tulasnella 1004
pruinosum, Sphaeronaema = **Cryptosporiopsis pruinosa** 653
pruinosum, Sporotrichum (Chrysosporium) 972
prunastri, Dermea (Cenangium) 669
prunastri, Eutypella (Eutypa) 701
prunellae, Ophiobolus (Linospora) 820
pruni, Catenophora 601
pruni, Corynespora (Helminthosporium) 646
pruni, Diaporthe 672
pruni, Diplodia 681
pruni, Grandinia (Hyphoderma, Hyphodontia, Odontia) 723
pruni, Lambertella 756
pruni, Lophiostoma = **L. quadrinucleatum** var. **triseptatum** 770
pruni, Massaria = **M. pyri** 779
pruni, Merulius = **Phlebia rufa** 848
pruni, Microgloeum 789
pruni, Myxosporium (Phoma, Sphaeropsis) 810
pruni, Septoria 956
pruni, Sydowia 983
pruni, Taphrina (Exoascus) 987
pruni-americanae, Teichospora 988
pruni-avium, Phyllosticta = **Phoma pomorum** 855
prunicola, Aureobasidium (Gloeosporium, Kabatiella, Microstroma) 584
prunicola, Cercoseptoria (Cercospora) 605
prunicola, Colletotrichum 639
prunicola, (Depazea, Phoma) = **Phoma pomorum** 855
prunicola, Diaporthe (Valsa) 672
prunicola, Phellinus (Fomitiporia, Poria) 846
prunicola, Phyllosticta = **Phoma pomorum** 855
pruni-persicae, Mycosphaerella 805
pruni-spinosae, Tranzschelia (Puccinia) 997
pruni-spinosae var. discolor, Tranzschelia = **T. discolor** 997
pruni-subcordatae, Taphrina 987
pruni-virginianae, Phyllosticta 872
pruni-yedoensis, Cercospora = **Cercoseptoria prunicola** 605
prunophorae, Coccomyces = **Blumeriella jaapii** 590
prunophorae, Cylindrosporium = **Phloeosporella padi** 850
prunorum, Phomopsis (Phoma) 859
psalliotae, Verticillium 1022
psamathos, Xylaria 1026
psammae, Coniothyrium 643
psammophila, Septoria 956
psederae, Linospora 769
psedericola, Cercospora = **C. ampelopsidis** 606
pseudacaciae, Mycosphaerella (Sphaerella) 805
pseudobifrons, Ciborinia 627
pseudocapsici, Phyllosticta 872
pseudocasei, Penicillium = **P. crustosum** 828
pseudocymopteri, Puccinia 912
pseudodiplodia, Diplodia = **D. malorum** 681
pseudoeurophioides, Ceratocystis = **Ophiostoma penicillatum** 821
pseudoferruginea, Tomentella = **T. bryophila** 993

pseudoflammeus, Crepidotus 647
pseudofloscula, Lemonniera 758
pseudohieronymi, Ustilago = **U. buchloes** 1014
pseudohimantia, (Ascospora, Discosphaerina) = **Guignardia haydenii** 725
pseudolacteus, Tyromyces 1005
pseudomacrocephala, Cirrenalia 628
pseudomacrostomum, Lophiostoma = **Platystomum compressum** var. **pseudomacrostomum** 882
pseudomollusca, Leucogyrophana = **L. mollusca** 767
pseudomolluscus, Merulius = **Leucogyrophana mollusca** 767
pseudopedicellatum, Septobasidium 946
pseudopini, Peniophora 830
pseudoplatani, Phloeospora 850
pseudosepulta, Erysiphe 698
pseudotrichia, Nectria (Megalonectria, Sphaerostilbe, Thyronectria) 813
pseudotropicalis, Candida (Endomyces) 600
pseudotsugae, Caliciopsis 597
pseudotsugae, Ceratostomella = **Ophiostoma minus** 821
pseudotsugae, Claussenomyces (Tympanis) 631
pseudotsugae, Coccomyces 634
pseudotsugae, Corniculariella (Chondropodium) 644
pseudotsugae, Dermea 669
pseudotsugae, Durandiella 691
pseudotsugae, Helminthosporium = **Hyphosoma resinacea** 742
pseudotsugae, Lachnellula (Dasyscyphus, Lachnella) 754
pseudotsugae, Leptothyrium 766
pseudotsugae, Phacidiopycnis (Phomopsis) 838
pseudotsugae, Phytophthora 879
pseudotsugae, Rasutoria (Dimeriella, Epipolaeum) 928
pseudotsugae, Rhabdocline 930
pseudotsugae, Rhabdogloeum 930
pseudotsugae subsp. **epiphylla, Rhabdocline** 930
pseudoversicolor, Peniophora 830
psidii, Cercospora 615
psidii, Limaciniella 768
psidii, Phomopsis 860
psidii, Puccinia 912
psidii, Sphaceloma 964
psidiicola, Phyllosticta (Phyllostictina) 872
psilocaryae, Cintractia 628
psilocybe, Callistosporium = **C. luteo-olivaceum** 597
psilospora, Mycosphaerella 805
psilostega, Septoria 956
psoraleae, Actinonema 558
psoraleae, Cercospora = **Cercosporidium cassiocarpum** 620
psoraleae, Colletotrichum (Gloeosporium) 639
psoraleae, Diplodina (Diplodinula) 682
psoraleae, Phyllosticta (Phoma) 872
psoraleae, Ramularia 926
psoraleae, Uromyces 1012
psoraleae var. **argophyllae, Uromyces** 1012
psoraleae-bituminosae, Cercospora = **Cercosporidium cassiocarpum** 620
psychotriae, Amazonia 563
psychotriae, (Calothyrium, Microthyrium) = **Calothyrium ryanae** 598
psychotriae, Cercospora 615
psychotriae, Meliola 785
psychotriae var. **chiococcae, Meliola** 785
psychotriae var. **labordiae, Amazonia** 563
psychotriae var. straussiae, Amazonia = **A. polypoda** 563
psychromorbidus, Coprinus 644
pteleae, Cercospora 615
pteleae, Phloeospora 850
pteleae, Septoria 956
pteleicola, Phyllosticta 872
pteridicola, Lachnella = **Albotricha washingtonensis** 559
pteridicola, Mycosphaerella 805
pteridina, Mollisia 793
pteridina, Pezizella 838
pteridis, Ascochyta = **A. necans** 576
pteridis, Camarosporium 600
pteridis, Cercospora 615
pteridis, Cryptomycella (Cylindrosporium, Fusidium, Gloeosporium) 652
pteridis, Cryptomycina (Cryptomyces, Phyllachora) 652
pteridis, Cylindrocladium 657
pteridis, Dasyscyphus (Urceolella) 665
pteridis, Hyalopeziza = **Albotricha washingtonensis** 559
pteridis, Leptostroma 765
pteridis, Mycosphaerella (Sphaerella) 805
pteridis, Phyllosticta 872
pteridis, Rhopographus = **R. filicinus** 934
pteridis, Septoria 956
pteridis, Uredinopsis 1007
pterigena, Mycena 798
pubera, Peniophora = **Hyphoderma puberum** 741
puberula, Peziza = **Dasyscyphus puberulus** 665
puberulum, Penicillium 828
puberulus, Dasyscyphus 665
puberum, Hyphoderma 741
pubescens, Phaeoisariopsis (Didymobotryum, Graphium) 840
pubescens, Pleurotus = **Rhodotus palmatus** 934
pubescens, Stereum = **Cytidia flocculenta** 661
pubescens, Trametes (Coriolus, Polyporus) 996
puccinioides, Auerswaldiella (Auerswaldia) 584
puccinioides, Dothidea 687
puccinioides, Macrosporium = **Stigmina sycina** 980
puderi, Pseudocercospora (Cercospora) 895
puerariae, Diplodia 681
puerariicola, Mycosphaerella 805
puerariicola, Pseudocercospora (Cercospora) 895
pugillus, Camarops 599
puiggarii, Asterinella (Asterina) 581
puiggarii, Irene = **Appendiculella calostroma** 571
pulchella, Calosphaeria (Valsa) 598
pulchella, Corollospora 645
pulchella, Phoma (Sphaeropsis) 855
pulchella, Poria = **Perenniporia tenuis** 831
pulchella, Pyrenobotrys (Gibbera, Stigmatea, Venturia) 918
pulchella, (Ovularia, Ramularia) = **Ramularia pusilla** 926
pulchella, Septoria 956
pulchella var. sulfurea, Malbranchea = **M. sulfurea** 776
pulchellicola, Phoma 855
pulchelloidea, Calosphaeria (Valsa) 598
pulchellum, Circinostoma = **Calosphaeria pulchella** 598
pulchellus, Polyporus = **Perenniporia tenuis** 831
pulcherrima, Cytospora = **C. chrysosperma** 661
pulcherrima, Donkia = **Climacodon pulcherrimus** 632
pulcherrima, Metschnikowia 788
pulcherrima, Tilletia = **T. barclayana** 991
pulcherrimae, Cercospora 615
pulcherrimae var. minima, Cercospora = **Cercospora pulcherrimae** 615
pulcherrimum, (Hydnum, Steccherinum) = **Climacodon pulcherrimus** 632
pulcherrimus, Climacodon 632
pulcherripes, Physarum 876
pulchra, Asteromassaria (Massaria) 582
pulchra, Schiffnerula 939
pulchrifolius, Gymnopilus 725
pulchriseta, Niesslia (Lohwagiella, Trichosphaeria) 815
pulchrispora, Macrophoma 774
pulchrum, Botryosporium 593
pulchrum, Cylindrosporium 659
pulchrum, Dimerosporium = **Schiffnerula pulchra** 939
pulchrum, Leptothyrium 766
pulchrum, Pythium 921
pulicare, Hysterium 747
pulicaris, Gibberella 715
pullata, Mollisia 793
pullatum, Helotium = **Mollisia pullata** 793
pullmanensis, Cryptosphaeria 652
pullulans, Aureobasidium (Dematium, Pullularia) 585
pullulans var. lini, Aureobasidium = **A. lini** 584
pulposa, Urophlyctis = **Physoderma pulposum** 877
pulposum, Physoderma 877
pulsatillae, Cercospora = **C. filiformis** 610
pulsatillae, Polythelis = **Tranzschelia anemones** 997
pulsatillae, Puccinia 912
pulsatillae, Stagonospora 974
pulveracea, Coniochaeta (Rosellinia) 641
pulveracea, Lachnellula = **Proliferodiscus pulveraceus** 891
pulveraceus, Proliferodiscus (Dasyscyphus) 891
pulvereum, Verticicladium 1021
pulverulenta, Puccinia 912
pulverulentum, Phacidium = **P. coniferarum** 838
pulverulentum, Sporotrichum = **S. pruinosum** 972
pulverulentum, Xenasma (Corticium) 1025
pulverum, Synchytrium = **S. cellulare** 983
pulvinata, Diatrypella 674
pulvinata, Dicyma (Hansfordia, Polyactis) 676
pulvinata, Eichleriella 692
pulvinatum, Clasterosporium = **Stigmina ellisii** 980
pulvinatum, Sporonema = **Pilidium concavum** 880
pulviniceps, Cryptospora 652
pulvinula, Cercospora 615
pulvinulata, Cercospora = **C. missouriensis** 613
pulviscula, Calycellina (Orbilia) 599
pulviscula, Chaetosphaeria (Zignoella) 624
pulvis-pyrius, Melanomma 783
pulvis-pyrius f. **fraxini, Melanomma** 783

pumilo, Hypoxylon = **H. crocopeplum** 745
puncta, Didymaria 676
punctata, Calycellina 599
punctata, Helicopsis = **Helicoma olivaceum** 730
punctata, Lirula (Hypodermella) 769
punctata, Mycosphaerella 805
punctata, Nigredo = **Uromyces punctatus** 1012
punctata, Phyllosticta = **Phoma viburni** 856
punctata, Pleospora 885
punctata, Poria = **Phellinus punctatus** 846
punctata, Puccinia 912
punctata, Tranzschelia = **T. discolor** 997
punctata, Ustilago = **U. bosniaca** 1014
punctata var. **troglodytes, Puccinia** 912
punctato-striatum, Eriosporangium = **Puccinia baccharidis** 901
punctatum, Rhytisma 935
punctatus, Phellinus (Polyporus) 846
punctatus, Uromyces 1012
punctiforme, Cercosporidium (Fusicladium) 620
punctiforme, Cryptosporium 653
punctiforme, Gloeosporium = **Marssonina punctiformis** 778
punctiforme, Leptothyrium 766
punctiforme, Lophodermium 772
punctiforme, Sporonema 972
punctiformis, Cercospora 615
punctiformis, Dacrymyces = **D. tortus** 663
punctiformis, Marssonina 778
punctiformis, Mycosphaerella (Sphaerella) 805
punctiformis, Phoma (Phyllosticta, Sclerophoma) 855
punctiformis, Placosphaeria (Phyllachora) 881
punctiformis, Puccinia 912
punctiformis, Ramularia 926
punctiformis, Ramularia = **R. montana** 926
punctiformis var. clematidis, Mycosphaerella = **M. hermione** 802
punctiformis var. **fimbriata, Asterina** 581
punctipetiolae, Diplodia 681
punctiphyllum, Sporidesmium 970
punctoidea, Septoria 956
punctulata, Metasphaeria 788
punctulatum, Asperisporium (Scolicotrichum) 579
punctulatum, Cladosporium 630
punctulatum, Hypochnicium (Corticium) 743
punctulatum, Hypoxylon 746
punctum, Aplosporella (Sphaeropsis) 570
punctum, Cercosporidium 620
punctum, Coniothyrium = **C. decipiens** 642
punctum, Phyllachora (Sphaeria) 863
pungens, Diaporthe = **D. eres** 671
punicae, Pseudocercospora (Cercospora) 895
punicae, Sphaceloma 964
punicea, Nectria 813
punicea, Septoria 956
punicea, Tomentella 994
punicea var. ilicis, Nectria = **Nectria punicea** 813
pupula, Splanchnonema (Massaria, Sphaeria, Stigmatomassaria) 969
pupula, Stictis (Schmitzomia) 978
pura, Neobulgaria 814
puritanica, Puccinia 912
puro-alba, Dacrymyces = **D. stillatus** 663
purpurascens, Epicoccum = **E. nigrum** 696
purpurascens, Pezicula (Dermatea) 837
purpurascens, Septoria 956
purpurea, Ceriporia (Poria) 621
purpurea, Claviceps 632
purpurea, Creonectria = **Nectria cinnabarina** 812
purpurea, Harknessiella (Phillipsiella) 730
purpurea, Herpotrichia 735
purpurea, Pseudocercospora (Cercospora) 895
purpurea, Puccinia 912
purpurea, Zukalia (Asterina) 1027
purpurea var. roseo-lilacina, Poria = **Ceriporia purpurea** 621
purpureo-cincta, Septoria 956
purpurescens, Taphrina 987
purpureum, Asteridium = **Zukalia purpurea** 1027
purpureum, Cenangium 602
purpureum, Chondrostereum (Stereum) 626
purpureum, Helicobasidium = **H. brebissonii** 730
purpureus, Monascus 794
purpurogenum, Penicillium 829
purtonii, Nectria 813
pusilla, Kretzschmaria = **K. clavus** 753
pusilla, Microsphaera 790
pusilla, Mycosphaerella 805
pusilla, Niesslia = **N. exilis** 815
pusilla, Peniophora 830
pusilla, Ramularia (Ovularia) 926
pusilla, Schizonella = **S. melanogramma** 939
pusilla, Trametes = **T. pubescens** 996
pusillum, Physarum 876
pusillus, Panellus 825
pusillus, Rhizomucor (Mucor) 933
pusiola, Allophylaria = **Crocicreas gramineum** 649
pustula, Cercospora = **C. ampelopsidis** 606
pustulans, Polyscytalum (Oospora) 889
pustulata, Arrhytidia = **Dacrymyces stillatus** 663
pustulata, (Asterella, Asterina) = **Calothyrium pustulatum** 598
pustulata, Burrillia 596
pustulata, Cryptovalsa = **Diatrypella eutypaeformis** 674
pustulata, Cytospora 662
pustulata, Hendersonulina (Hendersonia) 734
pustulata, Puccinia = **P. andropogonis** 900
pustulata, Stictis 978
pustulatum, Calothyrium (Microthyrium) 598
pustulatum, Coryneum = **C. umbonatum** 646
pustulatum, Lophiostoma = **Thyridaria viticola** 990
pustulatum, Pucciniastrum 917
putaminum, Caryospora 601
puteana, Coniophora 641
putearius, Fomes = **Phellinus nigrolimitatus** 846
putrefaciens, Clasterosporium (Sporidesmium) 631
putrefaciens, Fusicoccum 712
putrefaciens, Phyllosticta 872
putrida, Armillaria 572
puyae, Metasphaeria 788
pycnanthemi, Cercosporella 619
pycnograndis, Milesina (Milesia) 792
pycnospora, Periconia = **P. byssoides** 831
pygmaea, Plasmopara (Peronospora) 882
pygmaea, Puccinia 912
pygmaea var. **ammophilina, Puccinia** 912
pygmaea var. chisosana, Puccinia = **P. brachypodii-phoenicoidis** var. **chisosana** 902
pygmaea var. **fusca, Plasmopara** 882
pygmaeum, Lachnum 755
pygmaeus, Dasyscyphus = **Lachnum pygmaeum** 755
pyracanthae, Cercospora 615
pyracanthae, Spilocaea (Fusicladium, Passalora) 969
pyramidalis, Cristulariella = **C. moricola** 648
pyramidalis, Grovesinia 724
pyrenaica, Mycosphaerella 805
pyrenophora, Dothiora (Dothidea) 688
pyrenophora, Dothiorella = **Dothichiza sorbi** 687
pyrenophora var. **salicis, Dothiorella** 689
pyri, Aspergillus = **A. niger** var. **phoenicis** 579
pyri, Cercospora (Cercosporella) 615
pyri, Hormodendrum 737
pyri, Massaria 779
pyri, Mycosphaerella 805
pyri, Pseudopeziza 897
pyricola, Septoria 956
pyriforme, Cronartium = **C. comandrae** 649
pyriforme, Geniculodendron 714
pyriforme, Lycoperdon 773
pyriforme, Stegonsporium 976
pyriformis, (Hendersonia, Neohendersonia) = **Neohendersonia kickxii** 814
pyriformis, Itersonilia = **I. perplexans** 750
pyriformis, Phyllostictina = **Phyllosticta capitalensis** 865
pyriformis, Uromyces = **U. sparganii** 1013
pyrina, Cercosporella 619
pyrina, Mycosphaerella 805
pyrina, Phyllosticta = **Phoma pomorum** 855
pyrina, Venturia 1020
pyrinum, Coniothyrium 643
pyrinum, Leptographium 760
pyrinus, Marasmius = **M. minutus** 777
pyriospora, Botryosphaeria 593
pyrochroa, Calonectria 598
pyrolae, Chrysomyxa = **C. pirolata** 627
pyrolae, Colletotrichum (Ovularia, Ramularia) 639
pyrolae, Lophodermium 772
pyrolae, Phyllosticta (Phyllostictina) 872
pyrolae, Puccinia 912
pyrolae, Pucciniastrum 917
pyrolae, Septoria = **S. ellisiana** 951
pyrolae, Stigmatea 979
pyrorum, Fusicladium 711
pyrorum, Fusicoccum 712
pyrorum, Helminthosporium = **Spilocaea pomi** 969
pyrorum, Phyllosticta 872
pyrorum var. pyracanthae, Fusicladium = **Spilocaea pyracanthae** 969
pyrostegiae, Cercospora 615
pyrrhochlora, Thyronectria 991
pyrrhocystis, Cryptodiaporthe (Diaporthe) 651
pythiophila, Sclerophoma (Dothichiza) 940
pyxidata, Clavicorona (Clavaria) 632
quadrangulata, Sphaerella = **Mycosphaerella effigurata** 802
quadraspora, Phyllachora 863
quadrata, Albugo = **A. portulacae** 559
quadrata, Pithomyces (Piricauda, Scheleobrachea) 881
quadrata, Sarcinella 938
quadratum, Hypoxylon 746
quadratum, Sporidesmium = **Pithomyces quadrata** 881
quadratum, Stemphylium = **Sarcinella quadrata** 938
quadratus, Coccomyces = **C. leptideus** 634

quadricornuta, Halosphaeria 728
quadrifida, Eutypella 701
quadrinucleatum, Lophiostoma 770
quadrinucleatum var. **triseptatum, Lophiostoma** 770
quadriseptata, Pleospora = **Clathrospora pentamera** 631
quarta, Cercospora 616
quaternata, Eutypella (Eutypa, Quaternaria, Valsa) 701
quercea, Phyllosticta 872
querceti, Lophiosphaera 770
querceti, Septoria 956
querceum, Septogloeum 946
querci, Cercospora 616
querci, Leptostroma 765
quercicola, Elsinoe 692
quercicola, Pilidiella = **Coniella castaneicola** 641
quercicola, Septoria 956
quercicola var. **cinerea, Septoria** 956
quercifolia, Chaetophoma 623
quercifolia, Dicarpella (Physalospora, Pseudomassaria) 674
quercifolia, Irenopsis 749
quercifolia, Propolis 892
quercifoliae, Allosoma = **Phillipsiella atra** 848
quercina var major, Marssonia = **Discella quercina** 683
quercina, Anthostoma (Leptomassaria) 565
quercina, Articularia 573
quercina, (Ascomycetella, Cookella) = **Cookella microscopica** 644
quercina, Chalara 625
quercina, Clithris = **Colpoma quercinum** 640
quercina, Daedalea 664
quercina, Diatrypella 674
quercina, Dichaena = **Polymorphum quercinum** 888
quercina, Diplodia 681
quercina, Discella (Marssonia, Marssonina) 683
quercina, Discula = **D. umbrinella** 685
quercina, Eutypa = **E. mela** 700
quercina, Fumago = **Polychaeton quercinum** 888
quercina, Gnomonia = **Apiognomonia errabunda** 568
quercina, Grandinia (Hyphodontia) 723
quercina, Lembosina (Echidnodes, Lembosia, Morenoella) 758
quercina, Microsphaera (Erysiphe) 791
quercina, Peniophora 830
quercina, Pestalotia = **P. breviseta** 835
quercina, Phoma 855
quercina, Phomopsis 860
quercina, Protoventuria (Antennularia, Phaeodimeriella) 893
quercina, Rhodosticta 934
quercina, Saccardia 938
quercina, Septoria = **S. quercicola** 956
quercina, (Dothiorella, Sphaeropsis) = **Botryodiplodia gallae** 591
quercina, Stictis = **Propolis quercifolia** 892
quercina, Trabutia 995
quercina var. extensa, Microsphaera = **M. extensa** 790
quercina var. **minor, Articularia** 573
quercinum, Aulographum = **Lembosina quercina** 758
quercinum, Colpoma (Cenangium, Hysterium) 640
quercinum, Cylindrosporium 659
quercinum, Fusicoccum = **F. ellisianum** 712
quercinum, Gloeosporium = **Discula umbrinella** 685
quercinum, Polychaeton (Capnodium) 888
quercinum, Polymorphum 888
quercophilus, Marasmius 777
quercus, (Marssonia, Marssonina) = **Marssonina martinii** 778
quercus, Phyllosticta 872
quercus, Septoria 956
quercus-falcatae, Elsinoe 692
quercus-ilicis, Phyllosticta 872
quercus-prini, Phyllosticta 872
quercustris, Inonotus 749
quercuum, Botryosphaeria 593
quercuum, Cronartium 649
quercuum, Diplodina = **Sphaerellopsis filum** 965
quercuum, Gloeosporium 718
quercuum, Sphaerellopsis = **S. filum** 965
quercuum f. sp. **banksianae, Cronartium** 649
quercuum f. sp. **echinatae, Cronartium** 649
quercuum f. sp. **fusiforme, Cronartium** 650
quercuum f. sp. **virginianae, Cronartium** 650
quericfolia, Stictis = **Propolis quercifolia** 892
querna, Beltrania 587
quernea, Godroniopsis (Dichaena) 722
quernum, Gloeosporium = **Discula umbrinella** 685
questieri, Sarcinella 938
quinqueseptata, Gibberella 715
quinqueseptata, Herpotrichia = **Mytilinidion gemmigenum** 809
quinqueseptata, Meliola = **Asteridiella inermis** 580
quinqueseptata, Septoria 956
quinquespora, Meliola = **Asteridiella inermis** 580
quisquilaris, Irpex = **Hydnochaete olivacea** 738
quitensis, Limacinia = **Limacinula anomala** 768
rabenhorstiana, Ustilago = **U. syntherismae** 1016
rabiei, Mycosphaerella 805
rabiei, Phoma (Ascochyta) 856
racemosa, Cercospora = **Mycovellosiella teucrii** 808
racemosa, Vararia 1019
racemosa var. ambrosiae, Cercospora = **Mycovellosiella ambrosiae** 807
racemosum, Corticium = **Vararia racemosa** 1019
racemosum, Syncephalastrum 983
racemosus, Mucor 797
racemula, Ditopellopsis (Diaporthe) 686
raciborskii, Cercospora = **C. nicotianae** 614
radians, Asterina 581
radians, Coprinus 644
radiata, Diatrype = **Eutypella leprosa** 701
radiata, Phlebia = **P. merismoides** 848
radiata, Stictis 978
radiata, Torpedospora 995
radiatum, Boreostereum (Stereum) 590
radiatum, Polynema (Dinemasporium) 889
radiatum, Schizophyllum = **S. commune** 939
radiatus, Coccomyces 634
radiatus, Inonotus (Polyporus) 749
radiatus var. cephalanthi, Polyporus = **Inonotus radiatus** 749
radicalis, Cryphonectria (Endothia) 650
radicalis, Sorosphaera 963
radicans, Dothiorella 689
radicata, Ditiola 686
radicata, Pseudofistulina (Fistulina) 896
radicata, Sparassis = **S. crispa** 963
radicata, Tricholomopsis 999
radicata, Xerula (Collybia, Oudemansiella) 1025
radicatella, Mycena 798
radicatum, Helotium = **Ditiola radicata** 686
radicatus, Dacrymyces = **Ditiola radicata** 686
radicicola, Allantospora 560
radicicola, Ceratocystis (Ceratostomella, Ophiostoma) 604
radicicola, Coniothyrium = **C. celtidis-australis** 642
radicicola, Cylindrocarpon = **C. destructans** 657
radicicola, Diplodia 681
radicicola, Fusarium = **F. javanicum** 708
radicicola, Nectria 813
radicicola var. graminicola, Phialophora = **P. graminicola** 847
radicina, Alternaria (Pseudostemphylium, Thyrospora) 562
radicinum, Stemphylium = **Alternaria radicina** 562
radicis, Asterocystis = **Plasmodiophora brassicae** 881
radiculae, Septoria 956
radiculata, Femsjonia = **Ditiola pezizaeformis** 685
radiculosa, Antrodia (Poria) 567
radiculosus, Polyporus = **Antrodia radiculosa** 567
radiosa, Pollaccia 888
radiosum, (Corticium, Gloeocystidiellum) = **Vesiculomyces citrinus** 1022
radiosum, Fusicladium = **Pollaccia radiosa** 888
radula, Basidioradulum (Hyphoderma) 586
radula, Eutypella = **E. stellulata** 701
radula, (Polyporus, Poria) = **Junghuhnia subfimbriata** 751
raduloides, Phanerochaete = **P. magnoliae** 844
raduloides, Rhynchophoma 935
raduloides, Sistotrema (Trechispora) 962
rafinesquiae, Cercospora 616
rafinesquii, Phyllosticta 872
ragnhildae, Phyllosticta 872
raillardiae, Echidnodella 691
rainierensis, Pleospora 885
raistrickii, Penicillium 829
ralla, Phlebiella 849
rallum, (Aphanobasidium, Corticium, Xenasma) = **Phlebiella ralla** 849
ramanniana var. **angulispora, Mortierella** 796
rambosum, Cladosporium 630
rameale, Stereum = **S. hirsutum** 977
ramealis, Marasmiellus (Marasmius) 776
ramealis, Rhabdospora = **Septocyta ruborum** 946
ramealis var. **californicus, Marasmiellus** 776
ramicola, Ceratobasidium 603
ramicola, Rhizoctonia (Ceratorhiza) 932
ramifera, Chaetopsina 624
ramonae, Didymella 677
ramonae, Microdiplodia 789
ramosa, Absidia = **A. corymbifera** 555
ramosa, Blastocladia 590
ramosa, Gloeosporidiella 717
ramosissima, Tomentella 994
ramosissimum, Haplotrichum (Acladium, Oidium) 729
ramosum, Gloeosporium = **Gloeosporidiella ramosa** 717
ramosum, Hericium = **H. coralloides** 734
ramosus, Crepidotus 647

ramularia, Cercospora = **C. althaeina** 605
ramulariae, Nectria (Neonectria) 813
ramularis, Microdothella (Asterina) 789
ramulicola, Dothidotthia (Leptosphaeria, Neodeightonia) 688
ramulicola, Macrophoma 774
ramulorum, Diplodina 682
ramulosa, Alternaria 562
ramulosa, Varicosporina 1019
ramulosum, Stemphylium = **Alternaria ramulosa** 562
ranaculosus, Ceratocystiopsis 603
rancida, Postia (Poria) 891
rancidus, Oligoporus = **Postia rancida** 891
randii, Elsinoe 692
ranunculacearum, Aecidium 558
ranunculacearum, Septoria 956
ranunculi, Botryotinia 594
ranunculi, Cercospora 616
ranunculi, Cylindrosporium (Fusidium) 659
ranunculi, Entyloma = **E. ficariae** 696
ranunculi, Leptotrochila (Fabraea, Pseudopeziza) 766
ranunculi, Mycosphaerella (Sphaerella) 805
ranunculi, Peronospora 834
ranunculi, Puccinia = **P. blyttiana** 901
ranunculi, Ramularia = **R. didyma** 924
ranunculi, Septocylindrium 946
ranunculi, Stigmatea 979
ranunculi, Synchytrium 985
ranunculi f. thalictri, Cylindrosporium = **C. thalictri** 660
ranunculi-lyallii, Ramularia 926
ranunculina, Doassansia 686
ranunculoides, Claviceps 632
rapaneae, Cercospora 616
raphani, Alternaria 562
raphani, Aphanomyces 568
rapulum, Stromatinia 982
ratibidae, Anthostomella 566
ratibidae, Cercospora 616
raui, Phyllosticta (Macrophoma, Sphaeropsis) 872
rauii, Gnomonia (Didymella) 721
rauii, Sporidesmium 970
raunkaerii, Puccinia 912
raunkiaerii, Leptosporomyces (Athelia) 764
rautensis, Cercospora 616
rauwenhoffii, Tilletia = **T. holci** 992
ravenalae, Poria = **Grammothele fuligo** 722
raveneliana, Amphiporthe (Cryptodiaporthe, Diaporthe) 563
ravenelii, Bipolaris (Drechslera, Helminthosporium) 588
ravenelii, Botryodiplodia 591
ravenelii, Cenangella = **Eutryblidiella viburni** 699
ravenelii, (Broomella, Hypocrea) = **Melogramma campylosporum** 786
ravenelii, Irpex = **Hydnochaete tabacina** 739
ravenelii, Lentinus = **L. tigrinus** 759
ravenelii, Lophodermium 772
ravenelii, Merulius = **Meruliopsis taxicola** 787
ravenelii, Microsphaera 791
ravenelii, Phaeosolenia (Cyphella) 841
ravenelii, Phanerochaete (Peniophora) 844
ravenelii, Septoria 956
recedens, Puccinia 912
recedens, Stagonospora 974
recessa, Kirschsteiniothelia (Melanopsamma) 752
recisa, Exidia 702
recondita, Puccinia 912
recondita f. sp. tritici, Puccinia = **P. triticina** 915
rectae, Septoria 956
rectangularis, Antennaria 565
rectispora, Vermicularia 1021
rectum, Cladosporium = **Septonema fasciculare** 947
recuperata, Cucurbitaria 654
recurvata, Septoria 956
recurvipoda, Meliola 785
recurvus, Mucor 797
recutita, Mycosphaerella 805
redfieldiae, Puccinia 913
redfieldiae, Tilletia 992
redolens, Fusarium 710
reducta, Cercospora = **Pseudocercospora salicina** 895
reesei, Trichoderma = **T. longibrachiatum** 999
reflexa, Galerina (Galerula) 712
reflexa, Phlebia = **Punctularia strigoso-zonata** 917
refracta, Didymella = **D. proximella** 677
regale, Hypoxylon 746
regale var. macrosporum, Hypoxylon = **H. sphaeriostomum** 746
regalis, Cercospora 616
regalis, Pezizella (Peziza) 838
regina, Phoma 856
rehmiana, Pleospora 885
rehmiana, Tympanis = **T. laricina** 1004
rehmii, Curreya = **Discostroma massarina** 685
rehmii, (Curreyella, Discostroma) = **Discostroma massarina** 685
reiliana, (Sphacelotheca, Ustilago) = **Sporisorium holci-sorghi** 971
reilianum, Sorosporium = **Sporisorium holci-sorghi** 971
relicina, Pyrenophora = **P. trichostoma** 919
relicina, Trichopezizella 1000
relicina var. **macrospora, Trichopezizella** 1000
remuliformis, Boydia = **Vialaea insculpta** 1022
renatum, Aecidium 558
reniformis, Macrophoma (Dothiorella) 774
reniformis, Paxillus = **Crepidotus paxilloides** 647
renispora, Rutstroemia (Ciboria, Hymenoscypha) 937
renisporum, Helotium = **Rutstroemia renispora** 937
renouana, Phyllosticta = **P. typhina** 874
repanda, Biscogniauxia (Nummularia) 589
repanda, Daedalea = **Trametes elegans** 996
repanda, Exidia 702
repanda, Leptotrochila (Pseudopeziza) 766
repanda, Peziza (Plicaria) 838
repanda, Septoria 956
repanda f. querceti, Nummularia 817
repandus, Phellinus (Fomes) 846
repens, Absidia 555
repens, Aspergillus = **A. reptans** 579
repens, Cercospora 616
repens, Eurotium 699
repens, Helicoma 731
repens, Phialocephala (Penicillium) 847
repens, Phialophora (Cadophora) 847
repens, Ramularia 926
repens, Rhizoctonia (Epulorhiza) 932
repens, Sphaerostilbe = **Nectria mauritiicola** 813
repens, Stachybotryella 972
repente, Helminthosporium = **Taeniolella muricata** 985
replicatum, Cladochytrium 629
reptans, Aspergillus 579
reptans, Trichothyrium (Asterina, Trichopeltis) 1001
resedae, Cercospora 616
residua, Ustilago 1016
residuum, Aecidium 558
resinacea, Hyphosoma 742
resinaceum, Helmisporium = **Hyphosoma resinacea** 742
resinae, Epithyrium = **Sarea difformis** 939
resinae, Hormoconis (Cladosporium, Hormodendrum) 737
resinae, Racodium (Myxotrichella, Pycnostysanus, Sorocybe) 922
resinae, Sarea (Biatorella) 939
resinae, Strigopodia (Capnodium, Limacinia) 982
resinaria, Lachnellula (Trichoscypha, Trichoscyphella) 754
resinarium, Stilbum = **Eustilbum aureum** 699
resinarius, Dasyscyphus = **Lachnellula resinaria** 754
resinosa, Peniophora = **Crustoderma resinosum** 650
resinosum, Crustoderma (Hyphoderma) 650
resinosum, Ischnoderma 750
resinosus, Marasmius = **Resinomycena rhododendri** 930
resinosus, Polyporus = **Ischnoderma resinosum** 750
resticulosum, Penicillium 829
restrictum, Penicillium 829
restrictus, Aspergillus 579
resupinatus, Leptoporus = **Antrodia gossypina** 567
resurgens, Diplodia (Microdiplodia) 681
retecta, Puccinia 913
reticulata, Athelia = **Piloderma reticulatum** 880
reticulata, Cercospora = **Cercosporella virgaureae** 619
reticulata, Ceriporia (Poria) 621
reticulata, Guignardia (Dothidea) 725
reticulata, Mycocalia (Nidularia) 799
reticulata, Phloeospora = **Septoria astragali** 948
reticulata, Quasiconcha 922
reticulata, Ramularia 926
reticulata, Ravenelia = **R. texensis** 929
reticulatum, Aulographum = **Schizothyrium pomi** 939
reticulatum, Fusarium = **F. heterosporum** 708
reticulatum, Hypoxylon = **H. atropurpureum** 745
reticulatum, Piloderma (Corticium) 880
reticulatum var. negundi, Fusarium = **F. heterosporum** 708
reticulatus, Bolbitius (Pluteolus) 590
retusi, Ceratocystiopsis (Ceratocystis) 603
retusum, Fusarium 710
revolutum, Gloeosporium = **Colletotrichum gloeosporioides** 638
rhabarbarina, Daedalea = **Gloeophyllum mexicanum** 717
rhabdocarpa, Erinella = **Trichobelonium rhabdocarpum** 999
rhabdocarpa, Septoria 956
rhabdocarpum, Trichobelonium 999
rhabdospora, (Marssonia, Marssonina) = **Cylindrosporium populinum** 659

rhabdospora, Ramularia 926
rhabdosporum, Cylindrosporium = **Ramularia rhabdospora** 926
rhaetica subsp. **americana, Puccinia** 913
rhamni, Ascochyta = **A. paliuri** 576
rhamni, Capnodium 601
rhamni, Cercospora 616
rhamni, Cylindrosporium 659
rhamni, Marssonina (Marssonia) 778
rhamni, Phyllosticta 873
rhamni, (Dicaeoma, Puccinia) = **Puccinia coronata** 904
rhamni, Sphaeropsis 967
rhamni, Venturia (Phaeosphaerella) 1020
rhamni-catharticae, Septoria 956
rhamnicola, Phoma 856
rhamnigena, Ovularia 824
rhamnigena, Phyllosticta = **Coniothyrium rhamnigenum** 643
rhamnigenum, Coniothyrium 643
rhaphidospora, (Arachnopeziza, Erinella) = **Gorgoniceps aridula** 722
rhapontici, Cercospora 616
rhapontici, Septoria 956
rheades, Inonotus (Polyporus) 749
rhei, Ascochyta (Phyllosticta) 576
rhei, Ramularia 926
rhenana, Herpotrichia = **H. herpotrichoides** 735
rhenanum, Mytilinidion 809
rhexiae, Colletotrichum 639
rhexiae, Phyllosticta 873
rhipidium, (Favolus, Polyporus) = **Panellus pusillus** 825
rhipidium, Thelephora = **T. terrestris** 990
rhizicola, Helotium 732
rhizipes, Taphrina = **T. communis** 986
rhizodes, Sclerotium 942
rhizogena, Botryosphaeria 593
rhizogena, Diplodia = **Sphaeropsis hyalina** 967
rhizogenum, Helotium = **Lachnum pygmaeum** 755
rhizogenum, Melogramma = **Botryosphaeria rhizogena** 593
rhizomorphae, Anthostomella 566
rhizophagum, Verticillium 1022
rhizophila, Aplosporella = **A. ulmea** var. **rhizophila** 571
rhizophila, Diporotheca 683
rhizophila, Lachnella = **Dasyscyphus corticalis** 665
rhizophila, Zopfia 1027
rhizophilum, Coniosporium (Gymnosporium) 642
rhizophora, Taphrina 987
rhizophorae, Cercospora 616
rhizophorae, Cytospora 662
rhizophorae, Robillarda 936
rhizopodiformis, Rhizopus = **R. microsporus** var. **rhizopodiformis** 934
rhodella, Poria = **Ceriporia viridans** 621
rhodina, Botryosphaeria (Physalospora) 593
rhodocarpa, Phoma 856
rhodocarpa, Sphaeropsis 967
rhodochroa, Peniophora = **Phlebia coccineo-fulva** 848
rhododendri, Ascochyta 576
rhododendri, Cercospora = **Pseudocercospora handelii** 894
rhododendri, Chaetapiospora (Venturia) 622
rhododendri, Chrysomyxa = **C. ledi** var. **rhododendri** 626
rhododendri, Coniothyrium 643
rhododendri, Dendrophoma (Cytospora) 667
rhododendri, Diaporthe = **D. eres** 671
rhododendri, Encoeliopsis (Cenangella) 693
rhododendri, Euantennaria (Aithalomyces) 698
rhododendri, Hendersonia = **H. oudemansii** 733
rhododendri, (Coccomyces, Hysterium, Lophodermium) = **Lophodermium schweinitzii** 772
rhododendri, Melanomma 783
rhododendri, Melasmia 784
rhododendri, Pestalopezia 834
rhododendri, Pestalotia = **Pestalotiopsis sydowiana** 836
rhododendri, Phyllosticta 873
rhododendri, Physalospora 876
rhododendri, Protoventuria (Antennularia) 893
rhododendri, Resinomycena 930
rhododendri, Rhodocybella 934
rhododendri, Seimatosporium (Coryneum) 945
rhododendri, Septoria 956
rhododendri, Sphaerella = **Mycosphaerella clintoniana** 801
rhododendri var. **albimontana, Physalospora** 876
rhododendri var. fusoideum, Coryneum = **Seimatosporium vaccinii** 945
rhododendricola, Pezicula 837
rhododendricola, Phyllosticta 873
rhododendri-flavi, Phyllosticta = **P. berolinensis** 865
rhodoleuca, Phialea (Hymenoscyphus) 847
rhodoleuca, Propolis (Stictis) 892
rhodomphala, Byssosphaeria 596
rhodorae, Botryosphaeria (Guignardia, Laestadia) 593
rhodorae, Phyllosticta (Phoma) 873
rhodospila, Bertiella (Herpotrichia) 587
rhodospiloides, Herpotrichia = **Byssosphaeria rhodomphala** 596
rhodosticta, Herpotrichia = **Byssosphaeria rhodomphala** 596
rhodosticta, Neopeckia = **Byssosphaeria rhodomphala** 596
rhodostoma, Neopeckia = **Byssosphaeria schiedermayeriana** 596
rhodotypi, Ascochyta 576
rhoicola, Phyllosticta 873
rhoidis, Sphaeronaema 966
rhoina, Dothiorella 689
rhoina, Hainesia = **H. lythri** 728
rhoina, Harknessia 729
rhoina, Lachnella = **Perrotia flammea** 834
rhoina, Pseudocercospora (Cercospora) 895
rhoina, Septoria 956
rhoina, Telimena 988
rhoina, Zythia 1028
rhoina var. nigromaculans, Cercospora = **Pseudocercospora rhoina** 895
rhoinum, (Coryneum, Exosporium) = **Stigmina pallida** 980
rhois, Colletogloeum (Coryneum) 637
rhois, Coniothyrium = **C. rhois-radicantis** 643
rhois, Diatrypella (Diatrype) 674
rhois, Diplodia = **Sphaeropsis sumachi** 967
rhois, Godronia 721
rhois, Myxosporium (Naemaspora) 810
rhois, Stilbum (Sporocybe) 981
rhois, Tubercularia = **Hainesia lythri** 728
rhoiseda, Phyllosticta 873
rhois-radicantis, Coniothyrium 643
rhombica, Beltrania 587
rhopaloides, Xylaria = **X. curta** 1026
rhopaloideum, Septogloeum 946
rhuina, Cytospora 662
rhyacoides, Trichopeltis 1000
rhynchosiae, Ascochyta 576
rhynchosiae, Phyllosticta 873
rhynchosporae, Sorosporium 963
rhynchosporae, Thecaphora 989
rhynchosporae, Uromyces 1012
rhyncospora, Rosellinia 937
rhytismoides, Puccinia 913
ribesia, Chapeckia (Pseudovalsa) 625
ribesia, Dothiora (Dothidea, Phragmodothella, Plowrightia) 688
ribesia, Peziza = **Godronia ribis** 721
ribesia, Valsa 1018
ribesicida, Phyllosticta 873
ribesii-purpureae, Melampsora 781
ribicola, Ascochyta = **A. philadelphi** 576
ribicola, Asteroma 582
ribicola, Cercospora 616
ribicola, Coleosporium = **C. jonesii** 636
ribicola, Cronartium 650
ribicola, Gnomonia 721
ribicola, Marssonina (Marssonia) 778
ribicola, Plasmopara 882
ribicola, Rhabdospora (Septoria) 931
ribicola, Tapesia 986
ribicola, Valsa = **V. ambiens** 1017
ribis, Botryosphaeria 593
ribis, Cercospora 616
ribis, Cucurbitaria 654
ribis, Cylindrosporium (Cercoseptoria) 659
ribis, Dothiorella (Podosporium) 689
ribis, Drepanopeziza (Pseudopeziza) 690
ribis, Gloeosporidiella (Gloeosporium, Leptothyrium) 717
ribis, Godronia (Cenangium) 721
ribis, Lophiostoma (Gibberidea) 770
ribis, Mycosphaerella 805
ribis, Nectria = **N. cinnabarina** 812
ribis, Neokeissleria (Ceriospora, Melanconis) 814
ribis, Phylloporia (Fomes, Phellinus, Polyporus) 864
ribis, Phyllosticta = **P. ellisii** 867
ribis, Puccinia 913
ribis, Septoria (Ascochyta) 956
ribis, Sphaceloma 964
ribis f. achromogena, Botryosphaeria = **Botryosphaeria ribis** 593
ribis f. chromogena, Botryosphaeria = **Botryosphaeria ribis** 593
richardiae, Coniothecium (Phoma) 642
richardiae, Curvularia 655
richardiae, Phytophthora 879
richardiaecola, Cercospora 616
richardsiae, Phialophora (Cadophora) 847
richardsoniae, Cercospora = **C. carveriana** 607
richardsoniae, Phyllosticta 873
richardsonii, Coccomycetella 635
richtophensis, Pleospora 885
richtophensis var. **pallida, Pleospora** 885
ricinella, Cercospora 616
ricini, Alternaria 562
ricini, Amphobotrys (Botrytis) 564
ricini, Botryotinia (Sclerotinia) 594
ricini, Cercospora = **C. ricinella** 616
ricini, Didymella (Cercidospora) 677
ricini, Phomopsis 860
rickerianus, Uromyces 1013
rickii, Asterina 581
rickii, Inonotus 749

riddelliae, Gymnoconia = **Puccinia grindeliae** subsp. **riddellii** 907
rigida, Coriolopsis (Trametes) 644
rigida, Trabutia 996
rigidae, Pyrenopeziza (Hysteropezizella) 919
rigidula, Ramularia (Ovularia) 926
rigidum, Graphium (Stilbum) 723
rigidum, Podosporium 888
rigidus, Polyporus = **Rigidoporus lineatus** 936
rigospora, Cercospora = **Pseudocercospora atromarginalis** 893
rimicola, Xenasma 1025
riminicola, Rosellinia 937
rimosa, Diatrypella 674
rimosa, Phanerochaete (Peniophora, Scopuloides) 844
rimosa, Poria = **Diplomitoporus rimosus** 683
rimosissima, Hyphodontia (Odontia) 742
rimosum, Fusarium = **F. merismoides** 708
rimosum, Hyphoderma 741
rimosus, Diplomitoporus 683
rimosus, Irpex = **I. lacteus** 749
rimosus, (Fomes, Polyporus) = **Phellinus robiniae** 846
rimularum, Metasphaeria (Sphaeria) 788
rimulosum, Hydnum = **Hyphodontia stipata** 742
riparia, Septoria 956
ripulae, Puccinia = **P. melampodii** 910
riviniae, Septoria 956
rivulosa, Ceriporiopsis (Poria) 621
rivulosa, Eutypa 700
rixosa, Poria = **Junghuhnia collabens** 751
rizalensis, Meliola = **M. cookeana** 784
roanensis, Chrysomyxa 627
robergei, Monostichella (Gloeosporium) 796
robertiani, Hormotheca (Coleroa, Dothidea, Stigmatea) 737
robertii, Phyllosticta 873
robiniae, Cryptosporium 653
robiniae, Heterosporium 736
robiniae, Phaeoisariopsis (Camptomeris, Fusicladium, Passalora) 840
robiniae, Phellinus (Fomes, Fulvifomes, Pyropolyporus) 846
robiniae, Phloeospora (Ascochyta) 850
robiniae, Phyllosticta 873
robinianus, Marasmius 777
robiniophila, Perenniporia (Trametes) 831
robiniophilus, Polyporus = **Perenniporia robiniophila** 831
robinsoniana, Taphrina 987
robinsonii, Antennularia 565
robusta, Apostrasseria 571
robusta, Ceratocystis = **Ophiostoma robustum** 822
robusta, Curvularia 655
robusta, Diaporthe = **D. dubia** 671
robusta, Poria = **Phellinus robustus** 846
robusta, Rhizoctonia 932
robusta, Stigmina 980
robusta, Virgella 1022
robustum, Arthrobotryum = **Stigmina robusta** 980
robustum, Hypoderma = **Virgella robusta** 1022
robustum, Ophiostoma (Europhium) 822
robustus, Phellinus (Fomes) 846
rockiana, Hendersonia 733
rockii, Clypeoseptoria 633
rodmanii, Cercospora 616
roemerianae, Ravenelia 929
rogleri, Phyllosticta 873
rolfsii, Athelia (Corticium, Pellicularia) 583
rolfsii, Claviceps = **C. paspali** 632
rolfsii, Lembosia 758
rolfsii, Penicillium 829
rolfsii, Sclerotium 942
rolfsii var. **delphinii, Sclerotium** 942
rollandiae, Limaciniopsis 768
rollandiae, Septoria 956
rolleyi, Efibulobasidium (Exidia) 691
romanzoffiae, Puccinia 913
romellii, Antrodiella (Flaviporus, Poria) 568
romellii, Odonticium (Odontia) 817
romellii, Phlebia (Peniophora) 848
roquefortii, Penicillium 829
roridum, Myrothecium 809
roripae, Zignoella 1027
rosae, Cercospora = **C. rosicola** 616
rosae, Cucurbitaria 654
rosae, Diplocarpon 679
rosae, Diplodia = **D. rosarum** 681
rosae, Durandiella (Cenangium) 691
rosae, Gloeosporium 718
rosae, Lachnella = **Flagelloscypha libertiana** 705
rosae, Marssonina (Actinonema, Asteroma) 778
rosae, Myxosporium 810
rosae, Pestalotia 835
rosae, (Dermatea, Pezicula) = **Dermea bicolor** 669
rosae, Phomatospora 857
rosae, Phomopsis (Phoma) 860
rosae, Phyllachora 863
rosae, Phyllosticta 873
rosae, Seimatosporium 945
rosae, Septoria 956
rosae, Sphaeronaemella 966
rosae, Tapesia 986
rosae-acicularis, Phragmidium = **P. fusiforme** 860
rosae-arkansanae, Phragmidium 861
rosae-californicae, Phragmidium 861
rosae-indiananensis, Cercospora = **C. rosicola** 616
rosae-pimpinellifoliae, Phragmidium 861
rosae-setigerae, Phyllosticta 873
rosarium, Macrosporium 775
rosarum, Coniothyrium 643
rosarum, Diplodia 681
rosarum, Elsinoe 692
rosarum, Leptothyrium 766
rosarum, Rosellinia 937
rosarum, Sphaceloma (Phyllosticta) 964
rosarum, Sphaeropsis = **S. malorum** 967
rosea, Fomitopsis 706
rosea, Ramularia 926
rosellinioides, Camarosporium 600
rosellus, Hypomyces 744
rosenwaldii, Monochaetia = **Pestalotia laurocerasi** 835
roseocarneum, Dendrocorticium (Laeticorticium, Laxitextum, Stereum) 666
roseocremeum, Hyphoderma (Corticium) 741
roseola, Ramularia = **R. deusta** 924
roseomaculata, Bjerkandera = **Postia placenta** 891
roseum, Corticium (Laeticorticium) 645
roseum, Fusarium 710
roseum, Fusidium = **Ramularia rosea** 926
roseum, Gibberella = **G. zeae** 715
roseum, Gliocladium (Penicillium) 715
roseum, Trichothecium (Cephalothecium) 1000
roseum, Tritirachium = **T. oryzae** 1001
roseum f. cerealis, Fusarium = **F. avenaceum** 707
roseum f. gibbosum, Fusarium = **F. equiseti** 707
roseum var. viride, Gliocladium = **G. catenulatum** 715
roseum f. sp. cerealis, Gibberella = **G. zeae** 715
roseus, Aleurodiscus = **Corticium roseum** 645
roseus, Fomes = **Fomitopsis rosea** 706
roseus, Helicomyces 731
roseus, Sporobolomyces 971
rosicola, Amphichaeta = **Seimatosporium discosioides** 944
rosicola, Cercospora 616
rosicola, Gloeosporium = **Sphaceloma rosarum** 964
rosicola, Mycosphaerella (Phaeosphaerella) 805
rosicola, Phragmidium 861
rosicola, Phyllosticta 873
rosicola var. undosa, Cercospora = **Cercospora rosicola** 616
rosigena, Cercospora = **C. rosicola** 616
rosigena, Mycosphaerella 805
rostellata, Gnomonia 721
rostraspora, Anthostomella 566
rostrata, (Bipolaris, Drechslera) = **Exserohilum rostratum** 704
rostrata, Endoxyla (Ceratostomella) 695
rostrata, Fenestella = **Thyridium vestitum** 991
rostrata, Leptosphaeria 763
rostratum, Exserohilum (Helminthosporium) 704
rostratum, Gloeosporium = **Septoria ostryae** 955
rostratum, Pythium 922
rostratum, Thyridium = **T. vestitum** 991
rostrocoronata, Ceratocystis = **Ophiostoma rostrocoronatum** 822
rostrocoronatum, Ophiostoma 822
rostrocylindrica, (Ceratocystis, Ceratostomella) = **Ophiostoma rostrocylindricum** 822
rostrocylindricum, Ophiostoma 822
rostrupii, Diplodina = **D. andromedae** 682
rostrupii, Leptostroma 765
rostrupii, (Leptosporopsis, Ophiobolus) = **Ophiobolus prunellae** 820
rostrupii, Septoria = **S. chrysanthemella** 949
rotula, Marasmius 777
rotunda, Lulworthia 773
rotundata, (Tilletia, Ustilago) = **Tilletia maclaganii** 992
roumeguerii, Camarosporium 600
roumeguerii, (Metulodontia, Peniophora, Phlebia, Phlebiopsis) = **Phanerochaete ravenelii** 844
rousseauana, Fabraea = **Pseudopeziza calthae** 897
rousseliana, Pseudonectria 897
rousselii, Hysterographium = **H. mori** 747
rubefaciens, Puccinia 913
rubella, Cercospora 616
rubella, Leptospora (Sphaeria) 764
rubella, Puccinia = **P. phragmitis** 911
rubella, Ramularia (Ovularia) 926
rubelloides, Massariosphaeria (Pleospora) 779
rubellum, Corticium = **Cerocorticium confluens** 622
rubellum, Dendrodochium 667
rubellus, Merulius = **Phlebia incarnata** 848

rubellus, Ophiobolus = **Leptospora rubella** 764
rubens, Poria = **Postia salmonicolor** 891
ruber, Aspergillus = **A. rubrobrunneus** 579
ruber, Monascus 794
ruber, Saccharomyces = **Rhodotorula rubra** 934
rubescens, Cytospora 662
rubescens, Hyponectria (Anisostomula, Laestadia) 744
rubescens, Ptychogaster 899
rubi, (Actinonema, Asterella) = **Stomiopeltis versicolor** 981
rubi, Cercosporella (Fusisporium) 619
rubi, Cylindrosporium 659
rubi, Cyphella = **Cellypha goldbachii** 601
rubi, Cytospora 662
rubi, Diplodia 681
rubi, Gnomonia 721
rubi, Herpotrichia = **H. herpotrichoides** 735
rubi, Hypoderma (Hysterium) 744
rubi, Macrophoma 774
rubi, Microthyriella = **Schizothyrium perexiguum** 939
rubi, Mycosphaerella 805
rubi, Peronospora 834
rubi, Pezicula 837
rubi, (Coccomyces, Phacidium) = **Coleroa chaetomium** 636
rubi, Pseudocercospora (Cercospora, Fusarium, Ramularia) 895
rubi, Pyrenopeziza (Mollisia) 919
rubi, Rhabdospora = **Septocyta ruborum** 946
rubi, Septoria 957
rubi, Sphaerulina 969
rubi, Valsa = **V. ceratosperma** 1017
rubi var. brevispora, Septoria = **S. darrowii** 950
rubi var. **pallida, Septoria** 957
rubiacearum, Dothidella (Polystomella) 687
rubi-albida, Stagonospora = **S. brachyelytri** 973
rubicarpa, Nectria 813
rubicola, Gloeosporium = **Colletotrichum gloeosporioides** 638
rubicola, Glomerella = **G. cingulata** 719
rubicola, Hormotheca (Asterina, Coleroa, Stigmatea) 738
rubicola, Meliola = **Appendiculella calostroma** 571
rubicola, Microthyrium 792
rubicola, Otthia 823
rubicola, Phoma 856
rubicola, Phyllosticta 873
rubicola, Psathyrella 893
rubicola, Sphaeropsis = **S. malorum** 967
rubicola, Venturia = **Glomerella cingulata** 719
rubicunda, Leptosphaeria 763
rubicunda, Massariosphaeria (Pleospora) 779
rubicunda, Ramularia = **R. subsanguinea** 927
rubicunda var. americana, Pleospora = **Massariosphaeria rubelloides** 779
rubicundum, Corticium 645
rubida, Metasphaeria = **Bertiella rhodospila** 587
rubida, Papulaspora 826
rubiginosa, Hymenochaete 739
rubiginosa, Marssonina (Marssonia) 778
rubiginosa, Tomentella 994
rubiginosum, Hypoxylon 746
rubiginosum var. **ferrugineum, Hypoxylon** 746
rubiginosus, Hypochnus = **Tomentella rubiginosa** 994
rubiginosus, Zygodesmus = **Tomentella umbrinospora** 994
rubigo, Cercospora 616
rubigo-vera, Puccinia = **P. recondita** 912
rubigo-vera f. aconiti-elymi, Puccinia = **P. recondita** 912
rubigo-vera f. aconiti-rubrae, Puccinia = **P. recondita** 912
rubigo-vera var. agropyri, Puccinia = **P. recondita** 912
rubigo-vera var. agropyrina, Puccinia = **P. recondita** 912
rubigo-vera f. alternans, Puccinia = **P. recondita** 912
rubigo-vera var. apocrypta, Puccinia = **P. recondita** 912
rubigo-vera f. arthuri, Puccinia = **P. recondita** 912
rubigo-vera f. bromicola, Puccinia = **P. recondita** 912
rubigo-vera var. impatientis, Puccinia = **P. recondita** 912
rubigo-vera f. phaceliae, Puccinia = **P. recondita** 912
rubigo-vera var. secalis, Puccinia = **P. recondita** 912
rubigo-vera f. triseti, Puccinia = **P. recondita** 912
rubigo-vera var. tritici, Puccinia = **P. recondita** 912
rubigo-vera f. sp. holcina, Puccinia = **P. hordei** 908
rubiicola, Lophodermium (Dermascia) 772
rubi-idaei, Phragmidium 861
rubina, Mycosphaerella = **Didymella applanata** 676
rubi-odorati, Phragmidium = **P. occidentale** 861
rubiseda, Phomopsis 860
ruborum, Ascospora 578
ruborum, Phyllosticta 873
ruborum, Septocyta (Ascochyta, Septoria) 946
rubra, Rhodotorula 934
rubricarpa, Nectria 813
rubricosa, Valsaria 1019
rubricosum, Mastigosporium (Fusoma, Mastigosporium) 780
rubrobrunneus, Aspergillus 579
rubrococcinea, Orbilia 822
rubromaculans, Aecidium 558
rubronotatum, Thyridaria 990
rubrostromaticum, Hypoxylon 746
rubrotincta, Cercospora 616
rubrotincta, Leptosphaeria 763
rubrum, Eurotium 699
rubrum, Graphium 723
rubrum, Penicillium = **P. purpurogenum** 829
rudbeckiae, Ascochyta = **A. doronici** 574
rudbeckiae, Cercospora = **C. tabacina** 617
rudbeckiae, Colletotrichum 639
rudbeckiae, Mycosphaerella 805
rudbeckiae, Phyllosticta 873
rudbeckiae, Puccinia = **P. cnici-oleracei** 903
rudbeckiae, Ramularia 927
rudbeckiae, Septoria 957
rudbeckiae, Uromyces 1013
rudbeckiae var. **oaklandica, Septoria** 957
rude, Fibricium (Corticium) 704
rude, Sporidesmium 971
rudi, Lophiostoma = **Platystomum compressum** var. **pseudomacrostomum** 882
rudis, Diaporthe 672
rudis, Panus = **Lentinus strigosus** 759
rudis, Peniophora = **Fibricium rude** 704
rudis, Phomopsis 860
rudis, Pleospora 885
rudis, Rabenhorstia (Sphaeria) 922
rudis, Rhabdospora 931
rudis, Taeniolella 985
rudis, Valsaria = **V. insitiva** 1019
ruelliae, Puccinia = **P. lateripes** 909
ruelliae, Uromyces 1013
ruelliae-parviflorae, Puccinia 913
rufa, Calocera = **C. cornea** 597
rufa, Galiella (Bulgaria) 712
rufa, Haematomyxa = **H. vinosa** 728
rufa, Hypocrea 743
rufa, Nummularia = **Hypoxylon stygium** 746
rufa, Peniophora (Cryptochaete) 830
rufa, Phlebia 848
rufa, Poria = **Meruliopsis taxicola** 787
rufescens, Angelina 564
rufescens, Lanzia (Ciboria) 756
rufescens, Puccinia 913
rufescens, Valsa = **Amphiporthe aculeans** 563
rufesens, Ciboriella = **Lanzia rufescens** 756
rufilabrum, Hypoderma (Hysterium) 744
rufo-fusca, Ciboria 627
rufolateritius, Phaeomarasmius 840
rufo-maculans, Gloeosporium = **Colletotrichum gloeosporioides** 638
rufomaculans, Glomerella = **G. cingulata** 719
rufomaculans, Ramularia (Septocylindrium) 927
rufo-olivacea, Velutarina (Velutaria) 1019
rufosquamulosus, Gymnopilus 725
rufula, Eremotheca = **Schizothyrium rufulum** 940
rufula, Tryblidiella = **Rhytidhysteron rufulum** 935
rufula var. microspora, Tryblidiella = **Triblidium rufulum** var. **microsporum** 998
rufulum, Rhytidhysteron (Triblidium) 935
rufulum, Schizothyrium (Rhytisma) 940
rufulum var. **microsporum, Triblidium** 998
rufum, Sterellum = **Peniophora rufa** 830
rufum, Stereum = **Peniophora rufa** 830
rufus, Merulius = **Phlebia rufa** 848
rugiella, (Eutypella, Valsa) = **Valsa ceratosperma** 1017
rugispora, Echidnodella (Lembosia, Morenoella) 691
rugispora, Hymenochaete = **Columnocystis abietina** 640
rugispora, Tilletia 992
rugisporum, Stereum = **Columnocystis abietina** 640
rugodisca, Dothidea (Homostegia) 687
rugosa, Ascodichaena 578
rugosa, Durandiella (Godronia) 691
rugosa, Taphrina = **T. robinsoniana** 987
rugosiusculum, Stereum = **Chondrostereum purpureum** 626
rugosum, Dichaena = **Ascodichaena rugosa** 578
rugosum, Peridermium 832
rugosum, Stereum (Haematostereum) 978
rugulosa, Discosia 684
rugulosa, Emericella 693
rugulosiceps, Mycena 798
rugulosum, Penicillium 829
rugulosum, Synchytrium 985
rumicicola, Aplosporella (Sphaeropsis) 570
rumicicola, Septoria = **S. rumicis** 957
rumicis, Cercospora 616

rumicis, Gloeosporium 718
rumicis, Kellermania = **Chaetoconis polygoni** 623
rumicis, Peronospora 834
rumicis, Ramularia 927
rumicis, Septoria (Septoria) 957
rumicis, Ustilago = **U. kuehneana** 1015
rumicis, Venturia (Mycosphaerella) 1021
rumpens, Hypoxylon = **Biscogniauxia nummularia** var. **rumpens** 589
runcinata, Discula 685
rupestris, Coniothyrium 643
russa, Puccinia = **P. dovrensis** var. **russa** 905
russa var. leptoticha, Puccinia = **P. dovrensis** var. **leptoticha** 905
russellii, Microsphaera 791
ruthenica, Passalora = **Cercosporidium galii** 620
ruthenicum, (Fusicladium, Scolicotrichum) = **Cercosporidium galii** 620
rutilans, Hapalopilus (Polyporus) 729
rutilans, Tricholomopsis (Tricholoma) 1000
ruttneri, Tomentella 994
ryanae, Calothyrium 598
rydbergii, Puccinia 913
sabal, Catacauma 601
sabal, Melanconium = **M. palmarum** 782
sabalensioides, Anthostomella 566
sabalensis, Favolaschia (Laschia) 704
sabalensis, Oxydothis (Dilophia, Metasphaeria) 824
sabalensis, Rhabdospora 931
sabaleos, Hendersonia 733
sabaleos, Myriangium (Myrianginella) 808
sabali, Mycena 798
sabalicola, Asterina 581
sabalicola, Sphaeropsis 967
sabalicola, Venturia 1021
sabalidis, Meliola 785
sabalidis, Mollisia (Cenangium, Dermatea) 793
sabalina, Eutypella (Valsa) 701
sabalinum, Amerosporium 563
sabalis, Hydropus 739
sabbatiae, Cercospora 616
sabina, Holmiella (Caldesia, Eutryblidiella) 737
sabina, Macroplodia = **Aplosporella juniperi** 570
sabinae, Chloroscypha 626
sabinae, Gymnosporangium = **G. fuscum** 726
sabiniana, Anthostomella 566
saccardioides, Venturia = **Acanthostigma berenice** 556
saccardoi, Asterostomella (Phyllosticta) 583
saccata, Bifusella 588
saccata, Didymosphaeria 678
saccatum, Hypoderma = **Bifusella saccata** 588
sacchari, Bipolaris (Drechslera, Helminthosporium) 588
sacchari, Cytospora 662
sacchari, Dimeriella (Coleroa, Eriosphaeria, Venturia) 679
sacchari, Elsinoe 692
sacchari, Fusarium (Cephalosporium) 710
sacchari, Hormiactella 737
sacchari, Leptosphaeria 763
sacchari, Lophodermium 772
sacchari, Macrophoma 774
sacchari, Marasmius 777
sacchari, Periconia 831
sacchari, Phaeocytostroma (Melanconium, Pleocyta, Trullula) 839
sacchari, Phyllosticta = **Phoma sorghina** 856
sacchari, Stachybotrys = **S. elegans** 972
sacchari, Stagonospora 974
sacchari, Strumella 982
sacchari, Taphrina 987
sacchari, Torula = **Wallemia sebi** 1023
saccharicola, Odontia 817
saccharifera subsp. **kalalochensis, Resinomycena**
sacchariferum, Pseudohelotium 896
saccharina, Exidia 702
saccharina, Nectria = **N. ochroleuca** 813
saccharina, Phyllosticta = **P. minima** 870
saccharina, Septoria 957
saccharina, Valsa = **V. ambiens** subsp. **leucostomoides** 1017
saccharini, Cercospora 616
saccharinum, Cylindrosporium 659
saccharinum, Gloeosporium = **Aureobasidium apocryptum** 584
saccharinum, Melanconium = **Papularia vinosa** 825
sacculus, Cytospora 662
sacidioides, Sclerotium 942
sadleriae, Calycellina (Dasyscyphus) 599
sagaretiae, Stictis 979
sagittariae, Cercospora 616
sagittariae, Doassansia 686
sagittata, Puccinia = **P. ellisiana** 905
sagmatospora, Ceratocystis 604
salalis, Marasmius 777
salebrosa, Byssosphaeria (Amphisphaeria, Trematosphaeria) 596
salicaria, Melanopsamma 783
salicaria subsp. **fallax, Melanopsamma** 783
salicella, Cryptodiaporthe 651
salicellum, Helotium = **Hymenoscyphus salicellus** 740
salicellus, Hymenoscyphus 740
salicicola, Aplosporella (Sphaeropsis) 570
salicicola, Eutypa = **Cryptosphaeria subcutanea** 652
salicicola, Leptothyrium 766
salicicola, Marssonina (Marssonia) 778
salicicola, Phyllosticta 873
salicicola, Septoria 957
salicifolia, Septoria (Pleospora, Septoria) 957
salicifoliae, (Ascochyta, Cylindrosporium, Phloeospora, Septoria) = **Cylindrosporium filipendulae** 658
salicina, Cryptodiaporthe = **C. salicella** 651
salicina, Cytidia 661
salicina, Diplodia 681
salicina, Fumago 706
salicina, Marssonina 778
salicina, Mycosphaerella 805
salicina, Phomopsis (Phoma) 860
salicina, Pleosphaeria 883
salicina, Pseudocercospora (Cercospora) 895
salicina, Septoria = **Cylindrosporium salicinum** 659
salicina, Sphaerulina 969
salicina, Trametes = **T. trogii** 996
salicina, Valsa 1018
salicina var. tetraspora, Valsa = **Valsa salicina** 1018
salicinia, Septoria 957
salicinum, Capnodium (Polychaeton) 601
salicinum, Coryneum 646
salicinum, Cylindrosporium 659
salicinum, (Dematium, Sporotrichum) = **Fumago salicina** 706
salicinum, Melanconium 782
salicinum, Myxosporium 810
salicinum, Rhytisma 936
salicinum, Septogloeum (Gloeosporium) 946
salicinus, Coccomyces = **Stictis schizoxyloides** 979
salicinus, Fomes = **Phellinus ferruginosus** 845
saliciperda, Pollaccia 888
saliciperda, Venturia 1021
saliciperdum, Fusicladium = **Pollaccia saliciperda** 888
salicis, Ascochyta 576
salicis, Cercospora 616
salicis, Discella = **D. microsperma** 683
salicis, Drepanopeziza (Pseudopeziza) 690
salicis, Macrophoma = **Discula brenckleana** 685
salicis, Marssonina (Marssonia) 778
salicis, Monostichella (Gloeosporidiella, Gloeosporium) 796
salicis, Myxofusicoccum 810
salicis, Phyllosticta 873
salicis, Sclerophoma 940
salicis, Septoria 957
salicis, Sphaeropsis = **Aplosporella salicicola** 570
salicis, Trimmatostroma 1001
salicis, Uncinula = **U. adunca** 1006
salicis-fendlerianae, Septogloeum 946
salicis-sitchensis, Cladosporium 630
salicorniae, Mycosphaerella 806
salicum, Aposphaeria = **Phoma pezizoides** 855
salicum, Athelia 583
saligna, Linospora = **L. capreae** 769
salignus, Pleurotus = **P. ostreatus** 886
salina, Dendryphiella (Cercospora) 668
salina, Gnomonia 721
salina, Halosphaeria (Arenariomyces, Remispora) 728
salinum, Scolecobasidium = **Dendryphiella salina** 668
sallei, Hymenochaete 739
salmoneolutea, Phanerochaete 844
salmoneus, Dasyscyphus 665
salmoneus, Sporobolomyces = **S. roseus** 971
salmonicolor, Colletotrichum = **C. fusarioides** 637
salmonicolor, Erythricium (Corticium, Phanerochaete) 698
salmonicolor, Postia (Hapalopilus, Polyporus, Poria) 891
salmonicolor, Sporidiobolus 971
salpingophorum, Pythium 922
salsolae, Leptosphaeria 763
salsolae, Pleospora = **P. herbarum** 884
salsolae, Pyrenophora = **Graphyllium ipomoeae** 724
salviae, Sphaeropsis 967
salviae, Synchytrium 985
salviae-pratensis, Septoria 957
salviicola, Cercospora 616
salviicola, Puccinia 913
salviicola, Ramularia 927
salvinii, Magnaporthe (Leptosphaeria) 776
samarae-macrophylli, Septoria 957
samaricola, Melasmia 784
samarorum, Stagonospora 974
sambuci, Aplosporella (Sphaeropsis) 570
sambuci, Ascochyta = **Stagonospora samarorum** 974
sambuci, Cercospora = **C. depazeoides** 609
sambuci, Coniothyrium = **C. sambucinum** 643
sambuci, Cryptodiscus 652
sambuci, Diaporthe = **D. sociabilis** var. **sambuci** 672
sambuci, Diplodia 681

sambuci, Dothidea 687
sambuci, Heterosporium 736
sambuci, Hyphoderma (Hyphodontia, Lyomyces, Peniophora) 741
sambuci, Leptosphaeria 763
sambuci, Melanomma 783
sambuci, Myxosporium 810
sambuci, Nectriella (Charonectria) 814
sambuci, Phyllosticta = **Phoma exigua** 853
sambuci, Synchytrium 985
sambuci, Tubercularia = **T. vulgaris** 1002
sambucicola, Cytospora = **C. leucosperma** 662
sambucicola, Phyllosticta 873
sambucina, Cercospora = **C. depazeoides** 609
sambucina, Cytospora = **Cytospora leucosperma** 662
sambucina, Didymella = **D. effusa** 676
sambucina, Diplodia 681
sambucina, Dothiora (Sphaerulina) 688
sambucina, Leptosphaeria 763
sambucina, Phomopsis (Phoma) 860
sambucina, Ramularia 927
sambucina, Septoria 957
sambucina, Sphaeropsis (Macroplodia) 967
sambucina, Stigmina 980
sambucina, Thyridaria (Pseudovalsa, Valsa) 990
sambucinum, Coniothyrium 643
sambucinum, (Coryneum, Exosporium, Sciniatosporium, Thyrostromella) = **Stigmina sambucina** 980
sambucinum, Fusarium 710
sambucivora, Diatrype 673
sandicensis, Uleomyces (Curreya, Hyalocurreya) 1006
sandwicense, Chaetotrichum 625
sandwicensis, Ceriospora 621
sandwicensis, Coleophoma 635
sandwicensis, Echidnodes 691
sandwicensis, Meliola 785
sandwicensis, Peltosphaeria 827
sandwicensis var. **gouldiae, Meliola** 785
sandwicensis var. **major, Meliola** 785
sanguinariae, Aureobasidium (Gloeosporium, Kabatiella, Microstroma) 585
sanguinariae, Cercospora 616
sanguinariae, Phyllosticta 873
sanguinea, Cercospora = **Stenella lythri** 977
sanguinea, Irenina = **Appendiculella calostroma** 571
sanguinea, Nectria 813
sanguinea, Phanerochaete (Peniophora) 844
sanguinea, Septoria 957
sanguinea, Trametes = **Pycnoporus sanguineus** 917
sanguinea, (Cumminsiella, Uropyxis) = **Cumminsiella mirabilissima** 655
sanguineus, Pycnoporus (Coriolus, Polyporus, Polystictus) 917
sanguinolenta, Mycena 798
sanguinolenta, Poria = **Physisporinus sanguinolentus** 876
sanguinolentum, Stereum (Haematostereum) 978
sanguinolentus, Physisporinus (Rigidoporus) 876
sanguisorbae, Leptotrochila (Fabraea) 767
saniculae, Cercospora 616
saniculae, Entyloma = **E. eryngii** 696
saniculae, Septoria 957
sansevieriae, Gloeosporium = **Colletotrichum gloeosporioides** 638
saphena, Sphenospora 969
sapidus, Pleurotus = **P. ostreatus** 886
sapindi, (Mycosphaerella, Sphaerella) = **Mycosphaerella effigurata** 802
sapindii, Phyllosticta 873
sapinea, Flammula = **Gymnopilus sapineus** 726
sapinea, Sphaeropsis (Macrophoma) 967
sapineus, Gymnopilus 726
saponariae, Alternaria (Macrosporium) 562
saponariae, Rhabdospora 931
saponariae, Sorosporium 963
sapotae, Maravalia (Scopella) 777
saprophilum, Anthostoma 565
sarciniforme, Stemphylium (Macrosporium, Thyrospora) 977
sarcobati, Eutypella 701
sarcobati, Puccinia = **P. subnitens** 914
sarcobati, Septoria 957
sarcochroum, Fusarium = **F. sambucinum** 710
sarcoides, Ascocoryne (Coryne) 577
sarcoides var. **viridescens, Coryne** 646
sarmentii, Didymosphaeria 678
sarmentorum, Diplodia 681
sarmentorum, Hendersonia 734
sarothamni, Diaporthe 672
sarothamni, Diplodia 681
sarothamni, Phomopsis (Phoma) 860
sarothamni var. **baccharidis, Diaporthe** 672
sarraceniae, Drudeola (Peckia) 690
sarraceniae, Helminthosporium 732
sarraceniae, Mycosphaerella 806
sartwellii, Polyporus = **Trichaptum biforme** 998
sasakii, (Corticium, Hypochnus, Pellicularia) = **Thanatephorus cucumeris** 989
sassafras, Cytospora = **C. leucosperma** 662
sassafras, Diatrypella (Cryptovalsa) 674
sassafras, Diplodia 681
sassafras, Discula (Gloeosporium, Phyllosticta) 685
sassafras, Hypoxylon 746
sassafras, Mycosphaerella (Sphaerella) 806
sassafras, Ostreichnion (Lophium, Mytilidion) 823
sassafras, Pestalotia = **P. gracilis** 835
sassafras, Pleuroceras (Gnomonia) 886
sassafrasicola, Cytospora 662
sassafrasicola, Diplopeltis 683
sassafrasicola, Metasphaeria 788
sassafrasicola, Pseudodictya 896
sassafrasicola, Stigmatophragmia 979
sativa, Sclerotinia = **S. minor** 941
sativum, Colletotrichum = **C. destructivum** 637
sativum, Helminthosporium = **Bipolaris sorokiniana** 589
sativus, Cochliobolus (Ophiobolus) 635
saturnisporum, Trichoderma 999
saubinetii, Gibberella = **G. cyanogena** 714
saururi, Cercospora (Ramularia) 616
saururi, Gloeosporium 718
saururicola, Cercospora = **C. saururi** 616
saxifragae, Asteroma 582
saxifragae, Cercosporella 619
saxifragae, Laestadia 755
saxifragae, Marssonina 778
saxifragae, Peronospora 834
saxifragae, Puccinia = **P. heucherae** var. **saxifragae** 907
saxifragae, Septoria 957
saxifragarum, Phyllosticta 873
saximontanense, Cylindrosporium 659
saximontanensis, Cercospora 616
saximontanensis, Ramularia 927
scaber, Puccinia = **P. substerilis** 914
scabies, Cladosporium = **C. cucumerinum** 629
scabiosae, Guignardia (Laestadia) 725
scabra, Phomopsis (Phoma) 860
scabra, Stachybotrys = **S. chartarum** 972
scabra, Tetraploa 989
scabriseta, Eutypa = **Calosphaeria barbirostris** 598
scabrosa, Earliella (Trametes) 691
scabrosa, Eutypa 700
scabrovillosus, Dasyscyphus 665
scaevolae, Amazonia 563
scaevolae, Asteromella 582
scaevolae, Calothyriopeltis 598
scaevolae, Meliola 785
scaevolae, Mycosphaerella 806
scaevolae, Phyllosticta 873
scaevolae, Pleospora 885
scaevolicola, Meliola (Irene, Irenopsis) 785
scandens, Helicomyces 731
scandica, Puccinia 913
scariolicola, Phyllosticta 873
scelestum, Lophiostoma = **Platystomum compressum** var. **pseudomacrostomum** 882
sceptri, Hyponectria (Nectriella) 744
sceptrifera, Peniophora = **Tubulicrinis sceptriferus** 1003
sceptriferus, Tubulicrinis 1003
schachtii, Peronospora = **P. farinosa** 833
schedonnardi, Puccinia 913
schiedermayeriana, Byssosphaeria (Herpotrichia) 596
schinii, Cercospora 616
schinii, Inonotus = **I. munzii** 749
schixanthi, Gloeosporium = **Colletotrichum gloeosporioides** 638
schizanthi, Colletotrichum 639
schizonoti, Septogloeum 946
schizospora, Dothiora 688
schizosporum, Glyphium 720
schizoxyloides, Stictis 979
schleideni, Peronospora = **P. destructor** 833
schleideniana, Peronospora = **P. destructor** 833
schoenoprasii, (Phaeosphaerella, Sphaerella) = **Mycosphaerella tassiana** 806
schrenkiana, Ceratocystis = **Ophiostoma piliferum** 821
schrenkii, Eichleriella = **E. leucophaea** 692
schroeteri, Pyrenophora 919
schroeteriana, Ustilago 1016
schulzeri, Ramularia 927
schumacheri, Pithoascus (Microascus, Rosellinia) 880
schwarziana, Ovularia = **O. sphaeroidea** 824
schweinitzii, Fusarium = **Colletotrichum crassipes** 637
schweinitzii, Hypocrea 743
schweinitzii, Lophodermium 772
schweinitzii, Phaeolus (Polyporus) 840
schweinitzii, Xylaria 1026
scillarum, Uromyces = **U. muscari** 1012
scirpi, Anthracoidea (Cintractia, Ustilago) 567
scirpi, Cercoseptoria 605
scirpi, Coniothyrium 643
scirpi, Cryptosporium = **Septoria peckii** 955
scirpi, Fusarium = **F. equiseti** 707
scirpi, Phaeoseptoria (Hendersonia) 841
scirpi, Puccinia 913
scirpi, Pyrenophora 919
scirpi, Stagonospora 974
scirpi, Synchytrium 985

scirpi, (Nigredo, Uromyces) = **Uromyces lineolatus** subsp. **nearcticus** 1012
scirpi var. acuminatum, Fusarium = **F. acuminatum** 707
scirpi var. compactum, Fusarium = **F. equiseti** 707
scirpicola, Hendersonia 734
scirpicola, Naucoria = **Psathyrella typhae** 893
scirpicola, Phaeosphaerella 841
scirpina, Cercosporella 619
scirpina, Leptostromella 765
scirpina, Massariosphaeria (Leptosphaeria) 780
scirpina, Mollisia 793
scirpina, Pestalotia 835
scirpini, Stagonospora 974
scirpinum, Hypoderma (Hysterium) 744
scirpinum, Septocylindrium 946
scirpivora, Cylindrosporium 659
scirpoideae, Anthracoidea 567
scirrhoides, Mycosphaerella 806
scirrofaciens, Pestalotia 835
scitaminea, Ustilago 1016
scleriicola, Phyllachora 863
scleriicola, Puccinia 913
sclerodermeus, Fomes = **F. fasciatus** 705
sclerophaeum, Hypoxylon 746
sclerophila, Diaporthopsis 672
scleropogonis, Puccinia 913
scleroteichum, Pythium 922
sclerotica, Rhizoctonia 932
sclerotidea, Stigmatea 979
sclerotigenum, Acremonium (Cephalosporium) 557
sclerotii, Lachnum (Belonidium) 755
sclerotioides, Phoma 856
sclerotiorum, Aspergillus 579
sclerotiorum, Penicillium 829
sclerotiorum, Sclerotinia (Whetzelinia) 941
scolecospora, (Ophionectria, Scoleconectria) = **Scoleconectria cucurbitula** 943
scolecotrichoides, Cercospora = **Cercosporidium compactum** 620
scoparia, Eutypella (Eutypa, Valsa) 701
scoparium, Cylindrocladium 657
scopula, Tubeufia (Acanthostigma, Lasiosphaeria) 1003
scopulina, Dothiorella 689
scoriadea, (Pteridiospora, Sphaeria) = **Splanchnonema scoriadeum** 970
scoriadeum, Splanchnonema 970
scorzonellae, Septoria 957
scorzonerae, Alternaria (Sporidesmium) 562
scrobiculata, Tilletia = **T. sterilis** 992
scrophulariae, Coniothyrium 643
scrophulariae, Cylindrosporium 659
scrophulariae, Lewia (Pleospora) 767
scrophulariae, Septoria 957
scrophulariae var. **agalinis, Septoria** 957
scrophulariae var. **compositarum, Pleospora** 885
scrophulariae var. spinosella, Pleospora = **Montagnula spinosella** 796
scutellare, Corticium = **Athelia scutellaris** 583
scutellariae, Cercospora 616
scutellariae, Septoria 957
scutellaris, Athelia 583
scutellata, Datronia (Fomitopsis) 665
scutellata, Scutellinia (Humaria, Lachnea, Patella) 944
scutellatus, Fomes = **Datronia scutellata** 665
scutula, Coccoidella 634
scutula, Hymenoscyphus (Helotium, Phialea) 740
scyphophorum, Septosporium 959
seaveri, Chloroscypha 626
seaveri, Ciborinia 627
seaveri, Monilia 794
seaveri, Monilinia (Sclerotinia) 795
sebi, Wallemia (Sporendonema) 1023
secalina, Macrophoma 774
secalina, Nectria 813
secalina, (Phoma, Septoria) = **Septoria secalis** 957
secalis, Cercospora 616
secalis, Helminthosporium = **Drechslera tuberosa** 690
secalis, Pyrenophora = **P. japonica** 919
secalis, Rhynchosporium (Marssonia) 935
secalis, Septoria 957
secalis var. **stipae, Septoria** 957
secedens, Septonema 947
sechii, Cercospora 616
sector, Trichaptum (Coriolus, Hirschioporus, Polyporus) 999
sedi, Cercospora 616
sedi, Septoria (Ascochyta) 957
sedicola, Diplodia 681
sedicola, Septoria = **S. sedi** 957
seditiosum, Lophodermium 772
seditiosus, Uromyces 1013
sedoides, Cercospora 616
segetum, Ustilago 1016
segna, Didymella 677
sejuncta, Puccinia = **P. hieracii** 907
selaginellae, Chalara 625
selaginellae, Crucellisporium = **Belaina asclepiadis** 587
selaginellae, Gibbera 714
selaginellae, Morenoina 796
selaginellae, Papulospora 826
selaginellae, Plectosphaera 883
selaginellarum, Cercospora 616
selaginellarum, Heterosporium 736
selenophomoides, Phoma 856
selenophomoides, Septoria 957
selini-gmelini, Cercospora 616
seminalis, Cercospora 616
seminalis, Macrophoma 774
seminata, (Dothidea, Stigmatea) = **Parodiella hedysari** 826
seminicola, (Helotium, Phialea) = **Ciboria alni** 627
seminis-convolvuli, Thecaphora (Uredo) 989
seminuda, Farrowia 704
seminuda, Requienella (Trematosphaeria) 929
seminudum, Chaetomium = **Farrowia seminuda** 704
semipileatus, (Polyporus, Tyromyces) = **Skeletocutis nivea** 962
semisupina, Antrodiella 568
semisupinus, (Flaviporus, Polyporus, Tyromyces) = **Antrodiella semisupina** 568
semitecta, Massaria = **Splanchnonema semitectum** 970
semitectum, (Fusarium, Pseudofusarium) = **Fusarium pallidoroseum** 710
semitectum, Splanchnonema 970
semitectum var. **majus, Fusarium** 710
semitincta, Poria = **Ceriporia tarda** 621
semiustus, Marasmiellus (Collybia, Marasmius) 776
sempervivi, Endophyllum 694
senecionicola, Cercospora 616
senecionis, Alternaria = **A. cinerariae** 561
senecionis, Cercospora = **C. jacquiniana** 612
senecionis, Coleosporium = **C. tussilaginis** 636
senecionis, Gloeosporium 718
senecionis, Leptosphaeria (Pleospora) 763
senecionis, Puccinia 913
senecionis, Ramularia (Cylindrosporium) 927
senecionis, Septoria 957
senecionis var. **carniolica, Ramularia** 927
senecionis-aurei, Septoria 957
senegalense, Brachysporium = **Curvularia senegalensis** 655
senegalensis, Curvularia 655
senegalensis, Microascus 788
sennenis, Marssonina 778
sennensis, Marssonia = **Marssonina sennenis** 778
sentina, Mycosphaerella = **M. pyri** 805
sepalorum, Bricookea (Leptosphaeria, Metasphaeria) 595
separabilima, Junghuhnia (Chaetoporus) 751
separans, Lojkania (Amphisphaeria, Herpotrichia) 769
separans, Peniophora = **P. piceae** 829
separans, Poria = **Oxyporus corticola** 824
sepedonioides, Papulaspora 826
sepedonium, Corynascus (Thielavia) 645
sepelibilis, Anthostomella 566
sepiaria, Lenzites = **Gloeophyllum sepiarium** 717
sepiarium, Gloeophyllum 717
sepincola, Phoma (Sphaeropsis) 856
sepincola, Pringsheimia (Saccothecium) 891
sepincolaeformis, Pseudomassaria (Didymella) 897
sepium, Coniothyrium 643
sepium, Pezicula 838
sepium, Stereum = **Xylobolus subpileatus** 1026
sepium, (Coriolellus, Trametes) = **Antrodia albida** 567
seposita, Phomopsis (Phoma) 860
septata, Lophodermina = **Lophomerum septatum** 773
septata, Septopatella 947
septatissima, Cercospora 616
septatum, Lophomerum (Lophodermium) 773
septatum, Pseudocenangium = **Septopatella septata** 947
septem-septata, Hendersonia 734
septentrionale, Eudimeriolum 699
septentrionale, (Hydnum, Steccherinum) = **Climacodon septentrionalis** 633
septentrionalis, Climacodon 633
septentrionalis, Fibulomyces (Athelia) 705
septentrionalis, Puccinia 913
septica, Fuligo 706
septicus, Pleurotus 887
septocystidia, Candelabrochaete (Peniophora, Phanerochaete) 600
septorioides, Cercospora = **Pseudocercospora rubi** 895
septorioides, Gloeosporium = **Discella quercina** 683
septorioides, Mycosphaerella 806
septorioides, Sphaerella = **Mycosphaerella thalictri** 806
septorioides, Stagonospora 974
septorioides var. major, Gloeosporium = **Discella quercina** 683
septoriopsidis, Cercoseptoria (Septoria) 605
septoriopsis, Cercospora = **Cylindrosporium ribis** 659

septospora, Dothistroma (Cytosporina, Septoria, Septoriella) 689
septospora var. **lineare, Dothistroma** 689
septosporioides, Coryneum = **Stigmina negundinis** 980
septulata, Septoria = **S. convolvuli** 950
sepulta, Cryptosphaeria = **C. eunomia** 652
sepulta, (Eutypa, Eutypella) = **Eutypella scoparia** 701
sepulta, Sphaeropsis = **Aplosporella mori** 570
sepultum, Dendrodochium 667
sequoiae, Acanthostigma 556
sequoiae, Cercospora 617
sequoiae, Clithris 633
sequoiae, Ganoderma = **G. oregonense** 713
sequoiae, Haematomyxa 728
sequoiae, Leptostroma 765
sequoiae, Micromphale 789
sequoiae, Murangium (Cenangium, Scleroderris) 797
sequoiae, Mycosphaerella 806
sequoiae, Postia (Oligoporus, Poria) 891
sequoiae, Seynesiella (Dothidea, Stigmatea) 961
sequoiae, Trametes = **Gloeophyllum carbonarium** 716
sequoiae var. **juniperi, Cercospora** 617
serendipita, Piricauda = **Acrodictys erecta** 557
serenoae, Stictis 979
serialis, Antrodia (Coriolellus, Polyporus, Trametes) 567
serialis, Phyllachora = **P. cynodontis** 862
seriata, Asteromassaria (Massariella) 582
seriata, Cercospora 617
seriata, Dendrothele 668
seriata, Durandiella (Scleroderris) 691
seriata, Otthia = **Nectria atrofusca** 811
seriata, Phorcys 860
seriatum, Aleurocorticium = **Dendrothele seriata** 668
seriatum, Cenangium = **Durandiella seriata** 691
seriatum, Myxosporium 810
seriatum, Steccherinum 975
seriatus, Aleurodiscus = **Dendrothele seriata** 668
sericea, Athelia = **Phlebiella vaga** 849
sericeomollis, Postia (Oligoporus, Polyporus, Poria, Tyromyces) 891
sericeum, Polyscytalum 889
sericeum, Stereum = **S. striatum** 978
seriflua, Mycena = **Hydropus serifluus** 739
serifluus, Hydropus (Galactopus) 739
serinella, Erinella = **Dasyscyphus serinellus** 665
serinellus, Dasyscyphus 665
serjaniae, Meliola 785
serotina, Dothichiza 687
serotina, Phyllosticta 873
serotina, Ramularia 927
serotinum, Entyloma 696
serotinum, Helotium = **Hymenoscyphus serotinus** 740
serotinus, Hymenoscyphus 740
serotinus, Panellus (Pleurotus) 825
serpens, Antrodia = **A. albida** 567
serpens, Ceraceomyces (Byssomerulius, Ceraceomerulius, Merulius) 603
serpens, Ceratocystis 604
serpens, Geniculosporium 714
serpens, Hypoxylon 746
serpens, Trametes = **Datronia mollis** 665
serpens, Ustilago 1016
serpens var. **effusum, Hypoxylon** 746
serpens var. **macrospora, Hypoxylon** 746
serpentaria, Cercospora 617
serpentaria, Septoria 957
serpentaria, Stictis 979
serpentinum, Helicosporium = **H. pannosum** 731
serpula, Hemitrichia 732
serratum, Hydnum = **Resinicium bicolor** 929
serrulata, Mycosphaerella (Sphaerella) 806
serrulata, Seynesia (Didymosphaeria) 960
sesami, Alternaria 562
sesami, Cercoseptoria (Cylindrosporium) 605
sesami, Cercospora 617
sesami, Helminthosporium 732
sesamicola, Cercospora = **Cercoseptoria sesami** 605
sesbaniae, Elsinoe 693
sesbaniae, Physalospora (Phomatospora) 876
sesbaniae, Sphaerella 965
sessile, Ganoderma = **G. lucidum** 713
sessile, Graphium 723
sessilis, Cercospora = **Pseudocercospora salicina** 895
sessilis, Puccinia 913
sessilis var. **minor, Puccinia** 913
setacea, Gnomonia 721
setacea var. caryae, Gnomonia = **G. nerviseda** 721
setacea var. macrospora, Gnomonia = **G. caryae** 720
setariae, Bipolaris (Drechslera, Helminthosporium) 589
setariae, Cercospora 617
setariae, Melasmia 784
setariae, Puccinia 913
setariae, Pyrenochaeta 918
setariae, Ustilaginoidea 1014
setariae-italicae, Uromyces 1013
setariicola, Cercospora = **C. setariae** 617
seticollis, Ceratocystis 604
setigera, Chaetophoma (Chaetophomella) 623
setigera, (Kneiffia, Odontia, Peniophora) = **Hyphoderma setigerum** 741
setigera, Trichopeziza = **Dasyscyphus setigerus** 665
setigerum, Hyphoderma 741
setigerum, Lachnum = **Trichopezizella nidulus** var. **setigera** 1000
setigerus, Dasyscyphus (Lachnum) 665
setipes, Typhula (Pistillaria) 1005
setiseda, Clitocybe 633
setosa, Chaetochalara (Chalara) 623
setosa, Cucurbitaria = **Pleosphaeria echinata** 883
setosa, Pleiochaeta 883
setosa, Sarcodontia (Mycoacia, Oxydontia) 938
setosum, Ceratophorum = **Pleiochaeta setosa** 883
setosum, Hydnum = **Sarcodontia setosa** 938
setosum, Myxotrichum 810
setosus, Brasiliomyces 595
setosus, Gymnoascus = **Myxotrichum setosum** 810
setosus, Trametes = **Phellinus viticola** 846
setulifera, Meliola = **M. martiniana** 785
setulosa, Hyphodontia = **Steccherinum laeticolor** 975
setulosa, Mycena 798
setulosum, (Hydnum, Steccherinum) = **Steccherinum laeticolor** 975
seymouriana, Cercospora 617
seymouriana, Puccinia 913
seymouriana, Sphacelotheca 965
seymourii, Peronospora 834
shastense, Siroplacodium 961
shastensis, Leptosphaeria 763
shastensis, Melomastia 786
shastensis, Mollisia 793
shastensis, Naemosphaera 811
shastensis, Placosphaeria = **Siroplacodium shastense** 961
shastensis, Septoria 957
shastensis, Ustilago 1016
shawii, Mycosphaerella 806
shawii, Synchytrium 985
sheareana, Septoria 957
sheariana, Asteridiella 580
sheariana, Diaporthe 672
shearianus, Uromyces 1013
shearii, Cryptosphaerella 652
shearii, Diplodia 681
shearii, Harknessia 729
shearii, Heterochaete (Sebacina) 735
shearii, Massaria 779
shearii, Otthia 823
shearii, Peniophora = **P. aurantiaca** 829
shearii, Propolidium 892
shearii, Rhynchosphaeria 935
sheldonii, Ramularia 927
shepherdiae, Cucurbitaria (Curreya) 654
shepherdiae, Karstenula (Pleospora) 752
shepherdiae, Septoria (Cylindrosporium) 957
shepherdiae, Sphaerotheca 968
sherardiana, Puccinia 914
sherwoodiae, Phacidium 839
shiraiana, Phyllachora 863
shiraiana, Ustilago 1016
siberica, Septoria 957
sibirica, Septoria = **S. ribis** 956
siccans, Drechslera (Helminthosporium) 690
siccata, Phyllosticta 873
sicula, Oidiopsis 818
sicyi, Septoria 957
sicyicola, Mycosphaerella (Sphaerella) 806
sicyna, Phyllosticta 873
sidae, Leptothyrium 766
sidaecola, Cercospora 617
sidaecola, Mycosphaerella (Sphaerella) 806
sidalceae, Leptosphaerulina 764
sidalceae, Phoma 856
sidalceae, Ramularia 927
sidarum, Mycovellosiella (Ramularia) 808
siderocarpi, Ravenelia 929
sideroxyfoliae, Creothyriella 647
sideroxyli, Meliola 785
sideroxyli, Pauahia 827
sideroxylicola, Meliola 785
sieglingiae, Ustilago 1016
sieversiae, Lophiostoma (Leptosphaeria, Massaria, Wettsteinina) 770
sieversiae, Puccinia 914
sigmoidea, Nakataea (Curvularia, Vakrabeeja) 811
sigmoidea, Pseudovalsa = **P. longipes** 899
sigmoidea, Septoria 957
sigmoidea var. **irregulare, Nakataea** 811
sigmoideum, Clasterosporium = **Sporidesmium inflatum** 970
sigmoideum, Helminthosporium = **Nakataea sigmoidea** 811
sigmoideum var. irregulare, Helminthosporium = **Nakataea sigmoidea** var. **irregulare** 811
sigmoideum var. irregulare, Vakrabeeja = **Nakataea sigmoidea** var. **irregulare** 811

signalensis, Septoria 957
signata, Nummularia (Jongiella) 817
signata, Patellaria 826
signatum, (Blitridium, Triblidium) = **Patellaria signata** 826
sii, Cercosporidium (Cercospora) 620
sii, Septoria 957
silenes, Ascochyta 577
silenes, Diplodina 682
silenes, Peronospora 834
silenes, Phyllosticta 873
silenes, Pleospora = **Comoclathris permunda** 640
silenes, Septoria 957
silenes, Uromyces = **U. inaequialtus** 1011
silenes-acaulis, Phaeosphaeria (Leptosphaeria) 842
silenicola, Septoria 957
siliquae, Ravenelia = **R. spegazziniana** 929
siliquastri, Phyllosticta 873
siliquosum, Helminthosporium 732
silphii, Cercospora 617
silphii, Colletotrichum = **C. dematium** 637
silphii, Puccinia 914
silphii, Septoria 957
silphii, Uromyces (Nigredo) 1013
silphii var. laciniati, Cercospora = **Cercospora silphii** 617
silvaticus, Mucor = **M. hiemalis** f. **silvaticus** 796
silvestris, Ramularia 927
silybi, Septoria 957
simanii, (Poria, Postia) = **Postia hibernica** 890
simile, Lecanidion 758
similis, Oxyporus (Poria) 824
similis, Patellaria = **Lecanidion simile** 758
similis, Pleurotus = **P. ostreatus** 886
similis, Puccinia 914
similispora, Phyllosticta = **Leptothyrium similisporum** 766
similisporum, Leptothyrium 766
simillima, Dermatea = **Pezicula acericola** 837
simillima, Phyllachora (Dothidea) 863
simillima, Puccinia = **P. magnusiana** 909
simillima, Sphaeropsis = **Aplosporella clintonii** 570
simmondsiae, Strumella 982
simmonsii, Clathrospora = **C. heterospora** var. **simmonsii** 631
simmonsii, Cryptocoryneum 651
simmonsii, Leptosphaeria 763
simmonsii, Oncopodium 819
simplex, Anthostoma (Leptomassaria) 565
simplex, Codinaea 635
simplex, Epicoccum = **Arthrinium phaeospermum** 572
simplex, Gliocephalotrichum (Cylindrocladium) 715
simplex, Gonatobotrys 722
simplex, Hendersonia = **Stagonospora culmicola** 973
simplex, Peronospora = **Basidiophora entospora** 586
simplex, Phyllachora 863
simplex, Uberispora (Arachnophora) 1005
simplicibasidium, Laeticorticium 755
simplicior, Stagonospora 975
simplicior f. andropogonis, Stagonospora = **Stagonospora simplicior** 975
simplicissimum, Penicillium 829
simulans, Cercospora (Cercosporidium) 617
simulans, Glonium = **G. lineare** 720
simulans, Hymenochaete = **Amylostereum chailletii** 564
simulans, Phlyctema (Septoria) 850
simulans, Puccinia = **P. cryptandri** 904
simulata, (Ciboriopsis, Peziza) = **Moellerodiscus lentus** 792
simulata, Phaeoisariopsis (Cercospora) 840
singulare, Cenangium = **Encoelia pruinosa** 693
singularis, Endothia 694
singularis, Pseudopeziza (Mollisia) 897
sinuosa, Antrodia (Poria) 567
sinuosa, Phyllosticta 873
sinuosum, Gloeocystidiellum 716
sinuosus, Irpex = **I. lacteus** 749
siparia, Pleomassaria (Massaria, Sphaeria) 883
sirodesmioides, Peyronelia 837
sisymbrii, Septoria (Ascochyta) 957
sisyrinchii, Scolecosporiella (Brencklea, Kellermania) 943
sitanii, Ustilago = **U. trebouxii** 1016
sitchense, Cladosporium 630
sitchensis, Antrodia (Poria) 568
sitchensis, Septoria 958
sitophila, Monilia = **Chrysonilia sitophilia** 627
sitophila, Neurospora 815
sitophilia, Chrysonilia 627
skagwayensis, Ascochyta 577
skagwayensis, Phloeospora 850
slippii, Dothiora 688
smaragdina, Tubercularia (Graphium) 1002
smaragdinus, Polyporus = **Tyromyces fumidiceps** 1005
smardosii, Septoria 958
smilacicola, Hypoxylon 746
smilacicola, Mycosphaerella (Sphaerella) 806
smilacicola, Pyrenopeziza 919
smilacina, Cercospora = **C. smilacis** 617
smilacina, Diplodia 681
smilacina, Discosia 684
smilacina, Macrophoma (Dothiorella, Phoma, Phyllosticta, Sphaeropsis) 774
smilacina, Mycosphaerella (Sphaerella) 806
smilacina, Phragmocapnias 861
smilacina, Plochmopeltidella = **Chaetothyrina musarum** 624
smilacina var. smilacis, Macrophoma 774
smilacinae, Cylindrosporium 659
smilacinae, Guignardia = **Discochora smilacinina** 684
smilacinae, Macrophoma = **M. smilacina** 774
smilacinae, Phyllosticta 873
smilacinae, Ramularia 927
smilacinae, (Sclerotinia, Stromatinia) = **Stromatinia rapulum** 982
smilacinae, Septoria 958
smilacinae-trifoliae, Phyllosticta 873
smilacinina, Discochora (Anthostomella, Botryosphaeria) 684
smilacis, Aplosporella (Sphaeropsis) 570
smilacis, (Ascochyta, Stagonospora) = **Macrophoma smilacina var. smilacis** 774
smilacis, Cercospora 617
smilacis, Colletotrichum = **C. dematium** 637
smilacis, Cylindrosporium 659
smilacis, Cytospora = **C. sacculus** 662
smilacis, Diplodina (Diplodinula) 683
smilacis, Hypoderma (Hypodermopsis) 744
smilacis, (Gloniopsis, Hysterium) = **Gloniopsis curvata** 719
smilacis, Leptostroma 765
smilacis, Leptothyrium 766
smilacis, Myiocopron (Microthyrium, Peltella) 808
smilacis, Pestalotiopsis (Coryneum, Pestalotia) 836
smilacis, Phlyctema 850
smilacis, Phoma 856
smilacis, Phyllosticta = **Macrophoma smilacina** 774
smilacis, Puccinia 914
smilacis, Septoria 958
smilacis, Synchytrium 985
smilacis var. subeffusa, Phyllosticta = **P. subeffusa** 874
smithii, Crepidotus 647
socia, Phoma 856
socia, Septoria 958
sociabilis var. sambuci, Diaporthe 672
socialis, Ascochyta 577
socialis, Valsa = **V. salicina** 1018
sociata, Curvularia 656
sociata, Diaporthe (Valsa) 672
sociatum, Exosporium = **Curvularia sociata** 656
sodalis, Ascochyta 577
sojae, Diaporthe = **D. phaseolorum** 672
sojae, Peronospora = **P. manshurica** 833
sojae, Phomopsis = **P. phaseoli** 859
sojae, Phytophthora = **P. megasperma** f. sp. **glycinea** 878
sojae, Pseudorobillarda 898
sojicola, Phyllosticta 873
sojicola, Pleosphaerulina 883
sojina, Cercospora 617
sojinum, Cercosporidium = **Cercospora sojina** 617
solani, Alternaria (Macrosporium) 562
solani, Cercospora 617
solani, (Botryobasidium, Corticium) = **Thanatephorus cucumeris** 989
solani, Fusarium 710
solani, Helminthosporium 732
solani, Leptosphaeria = **L. corallorhizae** 761
solani, Mycosphaerella (Didymella) 806
solani, Phyllosticta 873
solani, Ramularia 927
solani, Rhizoctonia (Moniliopsis) 932
solani, Scopinella (Melanospora) 943
solani, Stemphylium 977
solani, Synchytrium = **S. endobioticum** 984
solani var. aduncisporum, Fusarium = **Fusarium solani** 710
solani **var. coeruleum, Fusarium** 711
solani var. martii, Fusarium = **Fusarium solani** 710
solani var. striatum, Fusarium = **Fusarium solani** 710
solani f. sp. **cucurbitae, Fusarium** 710
solani f. sp. **eumartii, Fusarium** 710
solani f. sp. **lupini, Fusarium** 710
solani f. sp. **phaseoli, Fusarium** 710
solani f. sp. **pisi, Fusarium** 710
solani f. sp. radicicola, Fusarium = **F. javanicum** 708
solanicola, Cercospora 617
solanicola, Phoma = **P. exigua** 853
solanina, Septoria 958
solani-nigri, Ascochyta = **A. petuniae** 576
solaridiscoidea, Atichia 583
solaris, Bombardia (Sphaeria, Trichosphaeria) 590
solenia, Dasyscyphus (Lachnella, Solenopezia) 665
solheimii, Puccinia 914
solidaginicola, Phyllosticta 873
solidaginicola, Septoria 958

solidaginis, Ascochyta = **A. greenei** 575
solidaginis, Asteroma 582
solidaginis, Cercospora 617
solidaginis, Coleosporium = **C. asterum** 636
solidaginis, Dothidea (Xyloma) 687
solidaginis, Elsinoe 693
solidaginis, Phyllosticta 873
solidaginis, Plagiostoma 881
solidaginis, Rhabdospora (Cryptosporium) 931
solidaginis, Rhytisma 936
solidaginis, Septoria 958
solidaginis, Sorosporium = **Thecaphora cuneata** 989
solidaginis, Uromyces = **U. sommerfeltii** 1013
solidaginis var. **longispora, Phoma** 856
solidaginis-serotinae, Phyllosticta 873
solidaginum, Phyllachora (Sphaeria) 863
solitaria, Phyllosticta 873
solitaria, Teichospora (Cucurbitaria, Gibberidea, Strickeria) 988
solitaris, Septoria 958
solitarium, Colletotrichum = **C. dematium** 637
solitarium, Cylindrosporium 659
solitum, Penicillium = **P. aurantiogriseum** 827
soloniensis, Piptoporus 880
somalensis, Cercospora = **Phaeoramularia occidentalis** 841
sommerfeltii, Uromyces 1013
sonchi, Alternaria 562
sonchi, (Ascochyta, Phyllosticta) = **Phoma exigua** 853
sonchi, Coleosporium = **C. tussilaginis** 636
sonchi-arvensis, Coleosporium = **C. tussilaginis** 636
sonchifolia, Septoria 958
sonorae, Cristinia 648
sonorae, Phellinus 846
sonorica, Puccinia 914
sophorae, Cytospora 662
sophorae, Diplodia 681
sophorae, Phyllosticta 873
sophorae, Uromyces = **U. hyalinus** 1011
sorbariae, Phyllosticta 873
sorbi, Coryneum = **Seimatosporium cassiopes** 944
sorbi, Cucurbitaria 654
sorbi, Dothichiza (Micropera) 687
sorbi, Eutypella (Valsa) 701
sorbi, Graphium 723
sorbi, Pestalotia 835
sorbi, Phyllosticta 873
sorbi, Septoria 958
sorbi, Thyridaria 990
sorbi, Tympanis = **T. truncatula** 1004
sorbicola, Karstenula = **Cucurbitaria sorbi** 654
sordaria, Coniochaeta 641
sordida, Antrodia 568
sordida, Exidiopsis (Sebacina) 702
sordida, Peronospora 834
sordida, Phanerochaete (Peniophora) 844
sordida, Pseudocercospora (Cercospora) 895
sordida, Valsa 1018
sordidum, Colletotrichum = **C. crassipes** 637
sordidum, Corticium = **Phanerochaete sordida** 844
sorghastri, Sphacelotheca = **S. andropogonis-hirtifolii** 965
sorghi, Ascochyta 577
sorghi, Cercospora 617
sorghi, Cerebella = **C. andropogonis** 621
sorghi, Gloeocercospora 716
sorghi, Helminthosporium 732
sorghi, Peronosclerospora (Sclerospora) 832
sorghi, Puccinia 914
sorghi, Ramulispora (Septorella) 928
sorghi, Sporisorium (Cintractia, Sphacelotheca, Ustilago) 971
sorghi var. maydis, Cercospora = **Cercospora sorghi** 617
sorghicola, Bipolaris (Drechslera, Helminthosporium) 589
sorghicola, Macrophoma 775
sorghina, Ascochyta 577
sorghina, Phoma (Phyllosticta) 856
sorokiniana, Bipolaris (Drechslera) 589
sorokinianum, Helminthosporium = **Bipolaris sorokiniana** 589
sororia, Crumenulopsis (Crumenula, Godronia) 650
sororia, Leucogyrophana 767
sororiae, Cercospora = **C. granuliformis** 611
sororius, Merulius = **Leucogyrophana sororia** 767
sorosporioides, Urocystis (Sorosporium, Urocystis) 1009
spadicea, Septoria 958
spadiceum, Stereum = **S. gausapatum** 977
spadiceum var. plicatum, Stereum = **S. gausapatum** 977
spadiceus, Marasmius 777
sparganii, Cylindrosporium (Septoria) 659
sparganii, Stagonospora 975
sparganii, Uromyces 1013
sparganioides, Puccinia 914
sparsa, Ceratocystis = **Ophiostoma sparsum** 822
sparsa, Cercospora 617
sparsa, Cryptovalsa = **Diatrypella eutypaeformis** 674
sparsa, Diaporthe = **D. peckii** 671
sparsa, Erysiphe 698
sparsa, Hysterostomella (Cocconia, Rhytisma) 748
sparsa, Leptosphaeria = **Phaeosphaeria herpotrichoides** 842
sparsa, Peronospora 834
sparsa, Phyllosticta = **P. vaccinii** 874
sparsa, Sphaerotheca 968
sparsa, Thekopsora = **Pucciniastrum sparsum** 917
sparsa, Tomentella = **T. ochracea** 994
sparsa, Ustilago 1016
sparsipila, Vermicularia = **Colletotrichum gloeosporioides** 638
sparsum, Ophiostoma 822
sparsum, Pucciniastrum 917
sparsus, Hypochnus = **Tomentella ochracea** 994
spartiicola, Macrophoma 775
spartinae, Buergenerula 595
spartinae, Cerebella = **C. andropogonis** 621
spartinae, Didymosphaeria = **Leptosphaeria obiones** 763
spartinae, Haligena 728
spartinae, Hendersonia 734
spartinae, Mycosphaerella (Sphaerella) 806
spartinae, Phyllachora 863
spartinae, Septogloeum (Fusarium) 946
spartinae, Septoria (Ascochyta) 958
spartinae, Tracylla (Leptothyrium) 996
spartinae, Uromyces = **U. acuminatus** 1009
spartinicola, Stagonospora 975
spathiphylli, Cylindrocladium 657
spathularia, Dacryopinax (Guepinia) 663
spathulata, Fistulina = **Pseudofistulina radicata** 896
spathulata, Grandinia (Hyphodontia, Odontia) 723
spathulata, Typhula (Clavaria, Pistillaria) 1005
spathulatus, Pleurotus = **Hohenbuehelia petalodes** 737
speciosa, (Aregma, Earlea) = **Phragmidium speciosum** 861
speciosum, Entyloma = **E. dactylidis** 695
speciosum, Gymnosporangium 727
speciosum, Phragmidium 861
speciosus, Uromyces 1013
spectabilis, Entyloma = **Physoderma gerhardtii** 877
spectabilis, Gymnopilus (Pholiota) 726
spectabilis, Leptosphaeria 763
specularіae, Cercospora 617
specularіae, Septoria 958
specularіae, Ustilago 1016
specularina, Septoria 958
spegazziniana, Ravenelia 929
spegazzinii, Cercospora 617
spegazzinii, Puccinia 914
spegazzinii, Ustilago = **U. hypodytes** 1015
spegazzinii var. agrestis, Ustilago = **U. hypodytes** 1015
spergulae, Puccinia = **P. arenariae** 901
spermacoces, Uromyces 1013
spermatiospora, Tympanis 1004
spermoides, Lasiosphaeria 757
spermoides, Phoma 856
spermoides, Phyllosticta 873
spermophila, Sclerotinia 941
spermophora, Ustilago 1016
sphacelati, Rhizoctonia 933
sphacelicola, Uredo 1008
sphaeralceae, Puccinia = **P. sphaeralceoides** 914
sphaeralceoides, Puccinia 914
sphaerelloides, (Didymella, Mycosphaerella) = **Mycosphaerella tassiana** 806
sphaerelloides, Dothidella 687
sphaerelloides, Pestalotia 835
sphaerelloides, Rhabdospora (Septoria) 931
sphaerelloides, Sphaeropsis 967
sphaerellula, Mycosphaerella (Didymella) 806
sphaeriaeformis, Cercospora 617
sphaerica, Conoplea = **C. olivacea** 644
sphaerica, Nigrospora (Trichosporum) 816
sphaericus, Mycoleptodiscus 799
sphaeriiforme, Heterosporium 736
sphaerioides, Lophodermium 772
sphaerioides, Ramularia 927
sphaeriostomum, Hypoxylon 746
sphaeroboloidea, Stictis 979
sphaerocephala, Cytospora = **C. sacculus** 662
sphaerocephalum, Stilbum 981
sphaerococca, Tilletia 992
sphaerocystidiata, Peniophora 830
sphaerocystis, Trechispora 998
sphaerogena, Ustilago = **U. trichophora** 1016
sphaeroidea, Beniowskia 587
sphaeroidea, Coccodothis (Dothidea, Dothidella) 634
sphaeroidea, Cornularia (Sphaeronaema) 645
sphaeroidea, Ovularia 824
sphaeroidea, Typhula 1005
sphaeroides, Karschia = **Rhizodiscina lignyota** 933
sphaeroides, Phaeangium 839

sphaeroideum, Melanconium 782
sphaeroideum, Phacidium 839
sphaerophora, Monotospora = **Tretopileus sphaerophorus** 998
sphaerophoroides, Hymenoscyphus 740
sphaerophorus, Tretopileus 998
sphaeropsispora, Macrophoma 775
sphaeropsoidea, Phyllosticta 873
sphaerosperma, Papularia = **Arthrinium phaeospermum** 572
sphaerosperma, Phyllachora (Sphaerodothis) 863
sphaerosperma var. inquinans, Papularia = **Arthrinium phaeospermum** 572
sphaerospermum, Cladosporium 630
sphaerospora, (Nectria, Thyronectria) = **Thyronectria austroamericana** 991
sphaerospora, Peniophora = **Hypochnicium sphaerosporum** 743
sphaerospora, Phyllachora = **P. sphaerosperma** 863
sphaerospora, Platygloea 882
sphaerospora, Sebacina = **Stypella minor** 983
sphaerospora, Sphaeropsis 967
sphaerospora, Trechispora = **T. farinacea** 997
sphaerosporum, Corticium = **Trechispora farinacea** 997
sphaerosporum, Hypochnicium 743
sphaerosporum, Piloderma 880
sphaerosporus, Crepidotus 647
sphendamnina, Diaporthella 672
sphinctrinum, Lopadostoma 769
spicata, Diaporthe = **Cryptodiaporthe petiolophila** 651
spicifer, Cochliobolus 635
spicifera, Bipolaris (Curvularia, Drechslera) 589
spiciferum, Brachycladium = **Bipolaris spicifera** 589
spiciferum, Helminthosporium = **Bipolaris spicifera** 589
spiculata, Exidia = **E. glandulosa** 702
spiculata, Pezicula 838
spiculiferum, Helminthosporium 732
spiculisperma, Stagonospora 975
spiculispora, Septoria 958
spiculosa, Diaporthe 672
spiculosa, Exidia = **E. glandulosa** 702
spiculosa, Poria = **Phellinus spiculosus** 846
spiculosa, Septoria 958
spiculosa var. saccharina, Tremella = **Exidia saccharina** 702
spiculosus, Phellinus 846
spilomeum, Septonema = **Coniosporium olivaceum** 642
spilosticta, Pseudocercospora (Cercospora) 895
spina, Corniculariella (Cornularia, Sphaeronaema) 644
spinaciae, Cercospora = **C. beticola** 606
spinaciae, Colletotrichum = **C. dematium** f. **spinaciae** 637
spinaciae, Hendersonia (Stagonospora) 734
spinaciae, Phoma = **P. betae** 852
spinaciae, Phyllosticta = **Phoma betae** 852
spinicola, Guignardia 725
spiniferus, Hypochnus = **Tomentella violaceofusca** 994
spiniger, Acanthophysium (Aleurodiscus) 556
spinosa, Absidia 555
spinosa, Eutypa 700
spinosa, Phyllosticta 873
spinosa, Phytophthora 879
spinosa, Rosellinia 937
spinosella, Montagnula (Pleospora) 796
spinosispora, Pteridiospora 899
spinosum, Pythium 922
spinosus, Mucor = **M. plumbeus** 797
spinulosa, Apiospora 569
spinulosa, Heterochaete (Eichleriella) 735
spinulosa var. **papillifera, Ravenelia** 929
spinulosum, Penicillium 829
spinulosum, Radulum = **Heterochaete spinulosa** 735
spinulosus, Aleurodiscus = **Acanthophysium mirabile** 556
spiraeacola, Sporidesmium 971
spiraeae, Cercospora 617
spiraeae, Dermea 669
spiraeae, Godronia (Belonidium, Cenangium) 721
spiraeae, Heterosporium 736
spiraeae, Lophiostoma (Lophiotrema) 770
spiraeae, Otthia (Cucurbitaria) 823
spiraeae, Phomopsis (Phoma) 860
spiraeae, Ramularia 927
spiraeae, Schizoxylon 940
spiraeicola, Cercospora 617
spiraeicola, Lachnum = **Dasyscyphus spiraeicolus** 665
spiraeicola, Microdiplodia (Diplodia) 789
spiraeicola, Phloeospora (Cylindrosporium) 850
spiraeicolus, Dasyscyphus 665
spirale, Chaetomium 623
spiralis, Uncinula = **U. necator** 1006
spirostachydis, Uredo 1008
spissa, Ceriporia (Poria) 621
spissa, Didymaria 676
spissus, Physisporinus = **Ceriporia spissa** 621
splendens, Appendiculella (Irene) 571
splendens, Puccinia 914
splendens, Pythium 922
splendens, Stemonitis 976
splendida, Chaetopsina 624
spleniata, Mycosphaerella 806
spokanense, Coniothyrium 643
spondiadis, Sphaceloma 964
spongiosa, Capnocybe 601
spongiosa, Mucilago = **M. crustacea** 796
spongiosa, Scorias 943
spongiosa, Strigopodia 982
spongiosum, Arthrobotryum = **Capnocybe spongiosa** 601
spongiosum, (Capnodium, Capnophaeum) = **Strigopodia spongiosa** 982
spongiosum, Cladosporium 630
spongiosum, Stereum = **Thelephora albidobrunnea** 989
sporadicum, Splanchnonema (Melanomma) 970
sporoboli, Coleroa (Venturia) 636
sporoboli, Leptosphaeria 763
sporoboli, Macrophoma 775
sporoboli, Placostroma (Scirrhia) 881
sporoboli, Puccinia 914
sporoboli, Uromyces 1013
sporoboli, Ustilago = **U. hypodytes** 1015
sporoboli var. robusta, Puccinia 914
sporobolicola, Stagonospora = **S. glycericola** 974
sporobolicola, Uromyces 1013
sporophleum, Arthrinium 573
sporotrichiella var. sporotrichioides, Fusarium = **F. sporotrichioides** 711
sporotrichiella var. tricintum, Fusarium = **F. tricinctum** 711
sporotrichioides, Fusarium 711
sporotrichioides var. chlamydosporum, Fusarium = **F. chlamydosporum** 707
spragueae, Uromyces = **U. unitus** subsp. **spragueae** 1013
sprgueanum, Entyloma = **Physoderma gerhardtii** 877
spraguei, Fomitopsis (Polyporus, Trametes, Tyromyces) 706
spraguei, Melogramma = **Cucurbidothis pithyophila** 654
spraguei, Mycosphaerella = **M. zizaniae** 807
spraguei, Scolecosporiella = **Ascochyta paspali** 576
spraguei, Septoria 958
spreta, Hymenochaete 739
spretum, Corticium 645
spumeus, Spongipellis (Polyporus, Tyromyces) 970
spumeus var. malicola, (Polyporus, Tyromyces) = **Tyromyces subgiganteus** 1005
spumosa, Pholiota 851
spuria, Micropera 790
spurium, (Ceratostoma, Sphaeronema) = **Micropera spuria** 790
squalens, Dichomitus 675
squalidula, Cercospora 617
squalidus, Merulius = **Serpula himantioides** 960
squamosa, Botryotinia (Sclerotinia) 594
squamosa, Botrytis 595
squamosus, Polyporus (Agaricus) 889
squamulosa, Nectria 813
squamulosum, Lentodium = **Lentinus tigrinus** 759
squarrosa, Pholiota 851
squarroso-adiposa, Pholiota = **P. limonella** 851
squarrosoides, Pholiota 851
squieriae, Aplosporella (Sphaeropsis) 571
stachydis, Cercospora 617
stachydis, Cylindrosporium 660
stachydis, Ramularia (Fusidium) 927
stachydis, Septoria 958
stachydis, Synchytrium 985
stachydis-ciliatae, Ovularia 824
stachytarphetae, Cercospora 617
stahlii, Rhizoctonia 933
stakmanii, Puccinia = **P. cacabata** 902
stalactiforme, Peridermium 832
staleyi, Lophodermium 772
standleyi, Diatrype 673
stanleyellae, Hendersonia 734
staphula, Cucurbitaria 654
staphyleae, Cercospora 617
staphyleae, Eutypella 701
staphyleae, Hendersonia 734
staphyleae, Massarina 779
staphyleae, Metasphaeria = **Discostroma corticola** 685
staphyleae, Phyllosticta 874
staphyleae, Sphaeropsis 967
staphylina, Diaporthe 672
staphylina, Dothiora (Hysterium, Metasphaeria) 688
staphylina, Mycosphaerella (Sphaerella) 806
staphylina, (Otthia, Plowrightia) = **Nectria atrofusca** 811
starbackii, Lophiostoma 771
starbaeckii, Crocicreas (Phialea) 649
starbaeckii, Phyllosticta 874
starrii, Acremonium 557
staticis, Fusicladium 711
staticis, Phyllosticta 874
steckii, Penicillium = **P. citrinum** 828

steironematis, Cylindrosporium 660
steironematis, Leptosphaeria = **Kalmusia clivensis** 751
steironematis, Phyllosticta (Phyllostictina) 874
stellae, Skeletocutis (Poria) 962
stellariae, Leptosphaeria 763
stellariae, Omphalospora (Euryachora, Phyllachora) 819
stellariae, Septoria 958
stellariae, Synchytrium 985
stellariae, Ustilago = **U. violacea** 1016
stellarinearum, Mycosphaerella 806
stellata, Pestalotia 835
stellata, Stictis 979
stellata var. nerii, Pestalotia = **P. oleandri** 835
stellata var. **philippensis, Stictis** 979
stellatum, Glonium 720
stellatum, Myxosporium 810
stellatus, Cheiromyces 625
stellatus, Sphaerobolus 965
stellatus, Thermomyces (Humicola) 990
stellifer, Aspergillus 579
stellifera, Himantia 736
stellulata, Eutypella (Valsa) 701
stemonitis, Doratomyces (Cephalotrichum, Stysanus) 686
stemphylioides, Alternaria 562
stenacra, Drechslera = **D. fugax** 690
stenacrum, Helminthosporium = **Drechslera fugax** 690
stenoceras, Ophiostoma (Ceratocystis, Ceratostomella) 822
stenocyboides, Caliciopsis = **C. pinea** 597
stenodon, Mycoacia = **M. aurea** 798
stenophyllus, Marasmiellus (Marasmius) 776
stenospila, Bipolaris (Drechslera) 589
stenospilum, Helminthosporium = **Bipolaris stenospila** 589
stenospilus, Cochliobolus 635
stenospora, Grandinia 723
stenospora, Marssonina (Gloeosporium) 778
stenospora, Naevia 811
stenospora, Pleospora 885
stenospora, Poria = **Dichomitus squalens** 675
stenosporum, Gloeosporium = **Septoria populi** 956
stenosporum, Leptothyrium 766
stenostomum, Lophiostoma = **L. vigheffulensis** 771
stenotaphri, Meliola = **M. panici** 785
stenotaphri, Puccinia = **P. stenotaphricola** 914
stenotaphricola, Puccinia 914
stenotheca, Metasphaeria (Sphaeria) 788
stephanomeriae, Puccinia = **P. hieracii** var. **stephanomeriae** 908
stercoraria, Scopulariopsis 943
stereoides, Cytidia 661
stereoides, Datronia (Antrodia, Polyporus) 665
sterilis, Tilletia 992
stevensianum, Monochytrium 795
stevensii, Amazonia 563
stevensii, Botryosphaeria 593
stevensii, Clypeosphaeria 633
stevensii, Corticium 645
stevensii, Diplodina 683
stevensii, Uredo 1008
stevensonii, Cercospora 617
stevensonii, Griphosphaeria (Griphosphaerella) 724
stevensonii, Hyaloscypha 738
stevensonii, Pestalotia 835
steviae, Coleosporium 636
stewartii, Diaporthe 672
stictella, Niptera (Mollisia) 816
stictica, Phomopsis 860
stictiformis, Naemacyclus 811
stictispora, Sphaeria = **Ophiobolus stictisporus** 820
stictisporus, Ophiobolus 820
stictochaetophora, Lasiosphaeria = **L. canescens** 757
stictoidea, Nimbomollisia (Mollisia) 816
stictoidea, Pyrenopeziza 919
stictoides, Sphaeropsis = **Harknessia uromycoides** 730
stictophyllum, Sporidesmium (Sirodesmium) 971
stictostoma, Leptosphaeria = **Massarina gloeospora** 779
stictum, Phoma = **Phomopsis stictica** 860
stigma, Diatrype 673
stigma, Septoria 958
stigma, Stictis 979
stigmaoides, Diatrype 673
stigmaticola, Peronospora 834
stigmina-platani, Mycosphaerella 806
stigmoides, Diatrype = **D. stigma** 673
stilbospora, Taeniolella (Torula) 985
stilbosporum, Hormiscium = **Taeniolella stilbospora** 985
stilbostoma, Melanconis 782
stillata, Atractogloea 584
stillatus, Dacrymyces 663
stillingiae, Phyllosticta 874
stillingiae, Pseudocercospora (Cercospora) 895
stipae, Ascochyta 577
stipae, Cladosporium 630
stipae, (Cyathicula, Phialea) = **Crocicreas gramineum** 649
stipae, Puccinia 914
stipae, Urohendersonia 1009
stipae, Ustilago = **U. minima** 1015
stipae-pennatae, Hendersonia 734
stipata, Hyphodontia (Odontia) 742
stipatum, Hydnum = **Hyphodontia stipata** 742
stipina, Septoria 958
stipitata, Dacryomitra = **Ditiola radicata** 686
stipitatum, Penicillium = **Talaromyces stipitatus** 986
stipitatus, Dacrymyces = **D. capitatus** 663
stipitatus, Talaromyces 986
stiptica, Postia 891
stipticus, Panellus (Panus) 825
stipticus, (Oligoporus, Polyporus) = **Postia stiptica** 891
stizolobii, Pseudocercospora (Cercospora) 895
stolonifer, Rhizopus 934
stolonifera, Ramularia (Septocylindrium) 927
stoloniferum, Penicillium = **P. brevicompactum** 827
stomatica, Cercospora 617
stomaticola, Pseudoseptoria (Phyllosticta) 898
stomatophora, (Asterella, Asterina) = **Calothyrium stomatophorum** 598
stomatophorum, Calothyrium 598
straminea, Schizoparme 939
straminella, Macrophoma (Phyllosticta) 775
stramineum, Corticium = **Gloeocystidiellum porosum** 716
straminis, Pleospora 885
stramonii, Dothiorella 689
strangulans, Balansia (Epichloe) 586
strangulans, Sphacelotheca = **Sporisorium montaniensis** 971
strangulans f. discoidea, Balansia 586
strasseri, Phoma 856
stratosa, Peniophora = **Phanerochaete ravenelii** 844
stratosa, Pholiota 851
stratosum, Hypochnicium 743
straussiae, Chaetothyrium 624
straussiae, Meliola 785
streetsii, Tricholomopsis = **Cyptotrama asprata** 660
streptopi, Cercospora 617
streptopidis, Septoria 958
streptothrix, Streptobotrys (Botrytis) 982
striata, Cintractia 628
striata, Lenzites = **Gloeophyllum striatum** 717
striatiformans, Mycosphaerella 806
striatospora, Puccinia = **P. heucherae** var. **saxifragae** 907
striatum, Gloeophyllum 717
striatum, Stereum 978
striatum var. ochraceo-flavum, Stereum = **S. ochraceo-flavum** 978
striatus, Cyathus 656
striatus, Uromyces 1013
striatus var. loti, Uromyces = **U. euphorbiae-corniculati** 1011
striatus var. medicaginis, Uromyces = **Uromyces striatus** 1013
strictae, Stagonospora 975
strictum, Acremonium 557
strigosa, Lasiosphaeria 757
strigoso-zonata, Punctularia (Phaeophlebia, Phlebia) 917
strigoso-zonatus, Merulius = **Punctularia strigoso-zonata** 917
strigosum, Dinemasporium 679
strigosum, Hydnum = **Gloiodon strigosus** 718
strigosus, Gloiodon 718
strigosus, Lentinus 759
strigosus, (Panus, Pleurotus) = **Pleurotus levis** 886
striiformis, Puccinia 914
striiformis, Soleella (Bifusella) 962
striiformis, Ustilago (Tilletia) 1016
strobi, Coccomyces 634
strobi, Phomopsis 860
strobilarum, Glonium (Hysterium) 720
strobiligena, Phoma = **Sclerophoma pythiophila** 940
strobilina, Ascochyta = **Sirococcus conigenus** 961
strobilina, Propolis (Stictis) 892
strobilinum, Cronartium 650
strobilinum, Peridermium 832
strobilinum, Sporonema = **Sirococcus conigenus** 961
strobilinus, Sirococcus = **S. conigenus** 961
stromatica, Didymella 677
stromatica, Puccinia 914
stromaticum, Ceratostoma = **Caliciopsis nigra** 597
stromatis, Cercospora 617
stromatoidea, Mycosphaerella = **Venturia rumicis** 1021
stromatoidea, Stigmina = **Coryneum stromatoideum** 646
stromatoideum, Coryneum (Helminthosporium, Sporidesmium) 646
strumella, Diaporthe 672

strumosa, Dendrothele 668
strumosa, Dichaena 675
strumosum, Aleurocorticium = **Dendrothele strumosa** 668
strumosus, Aleurodiscus = **Dendrothele strumosa** 668
struthanthi, Cercospora 617
struthiopteridis, Uredinopsis 1007
stuckertiana, Cercospora = **C. ipomoeae** 611
stuppea, Lasiosphaeria 757
stygia, Dactylospora (Karschia, Patellaria) 664
stygium, Hypoxylon 746
stylismae, Cercospora 617
stylosanthis, Cercospora = **C. commonsii** 608
stylospora, Pseudovalsa = **Prosthecium stylosporum** 892
stylosporum, Prosthecium 892
styracifluum, Stereum = **S. hirsutum** 977
styracis, Cercospora 617
suaveolens, Puccinia = **P. lapathicola** 909
suaveolens, Trametes (Polyporus) 996
subabrupta, Odontia = **Cystostereum pini-canadense** subsp. **subabruptum**
subacida, Perenniporia (Poria) 831
subalba, Peniophora = **Hyphoderma amoenum** 740
subalbida, Phlebia 848
subalbum, Corticium = **Hyphoderma amoenum** 740
subalpina, Ascochyta 577
subalutacea, Grandinia (Hyphodontia, Peniophora) 723
subalutaceum, Corticium = **Grandinia subalutacea** 723
subangulata, Puccinia 914
subannulata, Sporothrix 972
subannulatum, Ophiostoma 822
subapplanatus, Crepidotus 647
subargillaceus, Claudopus 631
subaurantiacum, Corticium = **Phanerochaete viticola** 844
subaurantiacum, Polycephalum = **Tubercularia lateritia** 1002
subauratus, Dasyscyphus 665
subbadia, Galerina 712
subcaespitosa, Fenestella (Cucurbitaria) 704
subcana, Mycena 798
subcarnea, Pezicula 838
subcarneum, Clathrotrichum = **Beniowskia sphaeroidea** 587
subcartilagineus, (Polyporus, Tyromyces) = **Parmastomyces transmutans** 826
subceracea, Phanerochaete 844
subceraceum, Corticium = **Phanerochaete subceracea** 844
subchartaceum, Trichaptum 999
subchartaceus, (Hirschioporus, Polyporus) = **Trichaptum subchartaceum** 999
subchlorinoflavida, Stagonospora 975
subcinerascens, Tomentella 994
subcinereum, Corticium = **Intextomyces contiguus** 749
subcinereus, Marasmius = **Marasmiellus nigripes** 776
subcircinata, Phoma = **Phomopsis phaseoli** 859
subcircinata, Puccinia = **P. senecionis** 913
subclausum, Amerosporium 563
subclavata, (Clitocybe, Omphalina) = **Gerronema subclavatum** 714
subclavatum, Gerronema 714
subclavigera, Tomentella 994
subclypeata, Valsa 1018
subcoccinea, Nectria = **N. flammea** 812
subcollapsa, Poria = **Ceriporia xylostromatoides** 621
subcollapsa, Trematosphaeria = **Requienella seminuda** 929
subcompressa, Rosellinia 937
subconcava, Nummularia (Cytoplea, Sphaeria) 817
subcongregata, Fracchiaea = **Nitschkia broomeiana** 816
subcongregata, Mycosphaerella (Sphaerella) 806
subcongrua, Diaporthe = **D. dubia** 671
subconica, Leptosphaeria = **L. doliolum** 761
subconica, Macrophoma 775
subconnata, Fracchiaea = **Nitschkia broomeiana** 816
subcoriacea, Cerrenella = **Fuscocerrena portoricensis** 711
subcornea, Mollisia 793
subcoronata, Pellicularia = **Botryobasidium subcoronatum** 591
subcoronatum, Botryobasidium (Corticium) 591
subcorticalis, Enchnoa (Trichosphaeria) 693
subcorticalis, (Dasyscyphella, Erinella) = **Lachnum sclerotii** 755
subcorticium, Phragmidium = **P. mucronatum** 861
subcorticola, Poria = **Ceriporia xylostromatoides** 621
subcremea, Peniophora = **Fibricium rude** 704
subcrinale, Steccherinum (Odontia) 976
subcruentatum, Aleurocystidiellum 560
subcruentatus, Aleurodiscus = **Aleurocystidiellum subcruentatum** 560
subcutanea, Cryptosphaeria (Eutypa, Valsa) 652
subcutanea, Leptosphaeria (Conisphaeria, Metasphaeria, Sphaeria) 763
subcutanea, Venturia 1021
subcuticulare, Helminthosporium = **Stigmina negundinis** 980
subcuticularis, Calosphaeria (Valsa) 598
subcuticularis, Diplodia 682
subcuticularis, Pestalotia 835
subcuticularis, Phyllachora (Dothidea) 863
subcyanea, Asterina 581
subcyaneum, Dictyothyrium = **Asterina subcyanea** 581
subdecora, Puccinia 914
subdentata, Meliola 786
subdryophilus, Gymnopilus 726
subectypus, Trametes (Coriolus, Polyporus) 996
subeffusa, Phyllosticta 874
subelatinum, (Geopetalum, Pleurotus) = **Cheimonophyllum candidissimum** 625
subepipterygia, Mycena 798
suberina, Diplodia 682
suberumpens, Physalospora = **Botryosphaeria dothidea** 592
subferruginea, Tomentella = **T. bryophila** 993
subferrugineus, Hypochnus = **Tomentella bryophila** 993
subfibrillosus, Crepidotus 647
subfimbriata, Junghuhnia 751
subflavidum, Lasiobelonium = **L. boreale** 756
subfulva, Diatrypella 674
subfusca, Tilletia = **Ustilago vilfae** 1016
subfusca, Uncinula = **U. necator** 1006
subfuscoflavida, Poria = **Diplomitoporus linbladii** 683
subfuscum, Helminthosporium = **Sporidesmium inflatum** 970
subfuscum, Hydnum = **Kavinia himantia** 752
subfusispora, Dendrothele 668
subgelatinosa, Cellypha (Cyphella) 602
subgibbosum, Belonium = **Crocicreas culmicola** 648
subgiganteum, Licrostroma (Corticium) 768
subgiganteus, Aleurodiscus = **Licrostroma subgiganteum** 768
subgiganteus, Tyromyces (Polyporus) 1005
subglobosum, Macrosporium 775
subglutinans, Fusarium 711
subglutinans, Gibberella 715
subgranulosum, Coniosporium 642
subgrisea, Rhabdospora 931
subiculata, Rosellinia 937
subiculosa, Amphisphaeria = **Byssosphaeria rhodomphala** 596
subiculosa, Poria = **Inonotus subiculosus** 749
subiculosus, Inonotus 749
subillaqueata, Trechispora = **Hypochniciellum subillaqueatum** 742
subillaqueatum, Hypochniciellum (Corticium, Leucogyrophana) 742
subilludens, Monodelphus = **Omphalotus olearius** 819
subincarnata, Mycena 798
subincarnata, Peniophora = **Amylocorticium subincarnatum** 564
subincarnata, Skeletocutis (Incrustoporia, Poria) 962
subincarnatum, Amylocorticium 564
subinclinata, Mycena 798
subinclusa, Anthracoidea (Cintractia, Ustilago) 567
subinvisibile, Corticium = **Paullicorticium pearsonii** 827
sublaeve, Corticium = **Ceraceomyces sublaevis** 603
sublaevis, Ceraceomyces (Athelia, Peniophora) 603
sublateritia var. **intermedia, Psathyrella** 893
sublateritium, Hypholoma (Naematoloma) 742
sublateritium, Penicillium 829
sublilacina, Sebacina (Exidiopsis) 944
sublilacina, Tomentella 994
sublilacinus, Zygodesmus = **Tomentella sublilacina** 994
sublimbata, Rosellinia 937
sublineola, Colletotrichum = **C. graminicola** 639
sublividula, Mollisia 794
subluteum, Hypoxylon = **H. serpens** 746
subluteus, Crepidotus 648
submaculata, Septoria 958
submagnum, Lophiostoma 771
submembranacea, Coniophora 641
submersa, Lulworthia 773
submicrospora, Trechispora = **T. cohaerens** 997
submicrosporum, Corticium = **Trechispora cohaerens** 997
submollis, Crepidotus 648
submurinus, Polyporus = **Fomitopsis meliae** 706
submutabile, Corticium = **Trechispora farinacea** 997
submutabilis, Trechispora = **T. farinacea** 997
subnidulans, Crepidotus (Claudopus) 648
subnitens, Phlebiella (Aphanobasidium, Corticium, Xenasma) 849
subnitens, Puccinia 914
subnivosa, Trametes 996

subnodosum, Cladosporium 630
subnudum, Septogloeum 946
subnullum, Corticium = **Trechispora microspora** 997
subochracea, Calycella (Lachnella, Trichopeziza) 598
subochracea, Coniophora = **Leucogyrophana olivascens** 767
subochracea, Phlebia (Grandinia, Mycoacia) 848
subopaca, Eversia 701
subopacum, Helminthosporium = **Eversia subopaca** 701
subordinaria, Phomopsis (Phoma) 860
subpallidulum, Corticium = **Resinicium furfuraceum** 929
subpendulus, Oligoporus (Tyromyces) 818
subpileatum, Stereum = **Xylobolus subpileatus** 1026
subpileatus, Xylobolus 1026
subplicosa, Mycena 798
subquaternata f. farinosa, Nectria 813
subradiculosa, Poria = **Antrodia radiculosa** 567
subramosum, Septonema 947
subrosea, Trametes = **Fomitopsis cajanderi** 705
subroseus, Fomes = **Fomitopsis cajanderi** 705
subrubiginosa, Tomentella = **T. rubiginosa** 994
subrufa, Ceriporiopsis (Poria) 622
subrufa, Ramularia 927
subrufescens, Agaricus 559
subrufum, Ceratostoma 605
subrugosum, Hysterographium 747
subrutilans, Wrightoporia 1024
subsanguinea, Ramularia (Cercospora, Ramularia) 927
subseriale, Corticium = **Phlebia subserialis** 848
subserialis, Phlebia (Peniophora) 848
subseriata, Metasphaeria 788
subseriata, Stagonospora 975
subseriata var. maculata, Stagonospora = **S. maculata** 974
subsessile, Cladosporium = **Phaeoramularia maculicola** 841
subsessile, Dendryphion = **Taeniolella subsessilis** 985
subsessilis, Pestalozziella 837
subsessilis, Pseudocercospora (Cercospora) 895
subsessilis, Taeniolella 985
subsimilis, Rosellinia 937
subsimplex, Memnoniella (Stachybotrys) 786
subsimulans, Aecidium 558
subsphaericum, Stemphylium = **Monodictys glauca** 795
substenospora, Meliola = **M. panici** 785
substerilis, Leptosphaeria 763
substerilis, Puccinia 914
substerilis var. **oryzopsidis, Puccinia** 914
substerilis var. **scribneri, Puccinia** 914
substriata, Puccinia 914
substriata var. **imposita, Puccinia** 914
substriata var. **indica, Puccinia** 915
subsulphurea, Peniophora = **Amylocorticium subsulphureum** 564
subsulphureum, Amylocorticium (Corticium) 564
subsynodicus, Marasmius = **Marasmiellus stenophyllus** 776
subtecta, Diplodia 682
subtecta, Microdiplodia = **M. perpusila** 789
subterranea, Spongospora 970
subterranea, Xylaria 1026
subterranea f. sp. **nasturtii, Spongospora** 970
subtestacea, Peniophora = **Hyphoderma setigerum** 741
subtestacea, Tomentella 994
subtestaceum, Hyphoderma = **H. setigerum** 741
subtile, Hydnum = **Resinicium bicolor** 929
subtile, Porotheleum = **Trechispora mollusca** 997
subtilis, Phyllosticta 874
subtilis, Polyporus = **Trechispora mollusca** 997
subtilis, Poria = **Trechispora mollusca** 997
subtilis, Rhizoctonia 933
subtilis var. **nigra, Rhizoctonia** 933
subtilissima, Lachnellula (Lachnella) 754
subtilissimus, Dasyscyphus = **Lachnellula subtilissima** 754
subtomentosum, Stereum = **S. hirsutum** 977
subtortuosae, Ravenelia 929
subtrifida, Schizonella = **Thecaphora trailii** 989
subtrigonosperma, Trechispora = **Sistotrema subtrigonospermum** 962
subtrigonospermum, Sistotrema 962
subtropicus, Gymnopilus 726
subulata, Acrospermoides 557
subulata, Isariopsis 750
subulata, Peniophora = **Tubulicrinis subulatus** 1003
subulata, Spermospora (Cercospora, Cercosporella) 964
subulata f. ciliata, Spermospora = **S. ciliata** 964
subulatus, Tubulicrinis 1003
subvelata, Hysteropezizella = **H. diminuens** 747
subvelata, Phoma 856
subvelutina, Lasiosphaeria = **Acanthostigma decastylum** 556
subvelutipes, Pholiota = **P. limonella** 851
subvermispora, Ceriporiopsis (Poria) 622
subvermisporus, Spongiporus = **Ceriporiopsis subvermispora** 622
subverrucisporus var. **megalosporus, Crepidotus** 648
subvinosa, Tomentella 994
subvinosa, Trametes = **Fomitopsis meliae** 706
subvinosus, Hypochnus = **Tomentella subvinosa** 994
subviolacea, Oliveonia 818
subvitrea, Mycena 798
succina, Perrotia (Dasyscyphus, Lachnella) 834
succinea, Amphisphaeria (Charonectria, Nectriella) 564
succineum, Acanthophysium 556
succineum, Chaetomium 623
succineum, Pseudocenangium 893
succineus, Aleurodiscus = **Acanthophysium succineum** 556
succineus, Dacrymyces = **Pseudocenangium succineum** 893
succineus, Merulius = **Meruliopsis ambiguus** 787
succinosa, Phyllosticta 874
sucina, Ductifera 690
sudans, Dacryobolus (Odontia) 663
sudans, Massariovalsa (Melanconis) 780
suecica, Lachnellula 754
suecica, Phyllosticta 874
suecicum, Sistotremastrum 962
suffocata, Coniophora = **C. arida** var. **suffocata** 641
suffocata, Pestalotia 836
suffocatum, Corticium = **Trechispora praefocata** 998
suffulta, Leptosphaeria 763
suffulta, Phyllactinia = **P. guttata** 864
suffulta var. macrospora, Phyllactinia = **P. guttata** 864
suffusa, Winterella (Cryptospora, Ophiovalsa) 1024
suffusca, Tranzschelia = **T. anemones** 997
suksdorfii, Puccinia 915
suksdorfii, Uromyces 1013
sulcata, Heliocybe 731
sulcata, Helvella = **H. lacunosa** 732
sulcata, Laurilia 757
sulcata, Myriosclerotinia (Sclerotinia) 809
sulcatum, Ganoderma = **G. zonatum** 713
sulcatum, Sclerotium 942
sulcatum, (Echinodontium, Stereum) = **Laurilia sulcata** 757
sulcatus, (Lentinus, Pleurotus) = **Heliocybe sulcata** 731
sulfurea, Malbranchea 776
sulfurella, Ciboria = **Lanzia luteovirescens** 756
sulfurellum, Helotium = **Lanzia luteovirescens** 756
sulfureoides, Tricholomopsis 1000
sulfureo-isabellina, Phlebia = **Cerocorticium sulfureo-isabellinum** 622
sulfureo-isabellinum, Cerocorticium (Corticium) 622
sulfureum, Thermoideum = **Malbranchea sulfurea** 776
sulfurina, Bisporella 590
sulfurinum, Helotium = **Bisporella sulfurina** 590
sulphurata, Nectria (Pseudonectria) 813
sulphuratum, Stereum = **S. ochraceo-flavum** 978
sulphurea, Cristella = **Phlebiella vaga** 849
sulphurea, Epithele 697
sulphurea, Hypocrea 743
sulphurea, Lachnella = **Belonidium sulphureum** 587
sulphurea, Torula (Oospora) 995
sulphurea, Wolfiporia (Macrohyporia) 1024
sulphurellum, Dentocorticium (Grandinia, Hydnum, Laeticorticium) 669
sulphurellum, Incrucipulum 748
sulphurellus, Dasyscyphus = **Incrucipulum sulphurellum** 748
sulphureum, Belonidium (Lachnum) 587
sulphureum, (Corticium, Hyphoderma) = **Phlebiella vaga** 849
sulphureus, Aspergillus 579
sulphureus, Dasyscyphus = **Belonidium sulphureum** 587
sulphureus, Laetiporus (Polyporus) 755
sulphureus, (Byssomerulius, Merulius) = **Wolfiporia sulphurea** 1024
sulphurina, (Peniophora, Phanerochaete) = **Ceraceomyces sulphurinus** 603
sulphurinus, Ceraceomyces 603
sumachi, Sphaeropsis (Sphaeria) 967
sumstinei, Meripilus (Grifola) 787
superficialis, Desmella = **D. aneimiae** 670
superficialis, Myriangium (Fenestella, Melogramma) 808
superficialis, Valsa 1018
superflua, Cercospora = **C. gymnocladi** 611
superflua, Mycosphaerella (Didymella, Didymosphaeria) 806

superflua var. macedonica, Didymella = **D. eupyrena** 677
superfluus, Uromyces = **U. vossiae** 1013
supina, Fomitella 705
supinus, Polyporus = **Fomitella supina** 705
suttoniae, Amazonia (Actinodothis) 563
suttoniae, Asterina 581
suttoniae, Beelia 586
suttoniae, Calothyrium 598
suttoniae, Oligostroma 818
suttoniae, Stomiopeltis (Stomiopeltella) 981
swertiae, Puccinia 915
swieteniae, Pestalotia = **P. sorbi** 835
swieteniae, Phyllachora 863
swieteniae-mahagoni, Nectria 813
swinglei, Peronospora = **P. lamii** 833
sycina, Diplodia 682
sycina, Stigmina 980
sycina var. **syconophila, Diplodia** 682
sycinum, (Sciniatosporium, Sporidesmium) = **Stigmina sycina** 980
sydowiana, Meliola 786
sydowiana, Meliolina 786
sydowiana, Microsphaera 791
sydowiana, Pestalotia 836
sydowiana, Pestalotiopsis 836
sydowiana, Poculum (Ciboria, Ombrophila, Rutstroemia) 887
sydowii, Aspergillus 579
sylvana, Nectria 813
sylvatica, Mycosphaerella = **M. juniperina** 803
sylvatica, Phyllachora 863
sylvaticum, Pythium 922
sylvestris, Poria = **Antrodia sinuosa** 567
symbioticum, Trichosporum 1000
symphoricarpi, (Clethridium, Curreyella) = **Griphosphaerioma kansensis** 724
symphoricarpi, Herpotrichia (Gibberidea) 735
symphoricarpi, (Lasiobotrys, Rhizogene) = **Rhizogene impressa** 933
symphoricarpi, Microsphaera 791
symphoricarpi, Phaeoramularia (Cercospora) 841
symphoricarpi, Phyllosticta 874
symphoricarpi, (Dibotryon, Gibbera, Otthia, Plowrightia, Pseudotthia, Valsaria) = **Dothidotthia andersonii** 688
symphoricarpi, Puccinia 915
symphoricarpi, Pyrenopeziza (Lachnella) 919
symphoricarpi, Septoria 958
symphoricarpi, Sphaceloma 964
symphoricarpi, Teichospora (Strickeria) 988
symphoricarpi, Valsa = **V. ambiens** 1017
symplocarpi, Cercospora 617
symploci, Exobasidium 703
symploci, Leptothyrium (Sacidium) 766
symploci, Septoria 958
sympodialis, Pycnopeziza 917
syngenesia, Diaporthe 672
syntherismae, Ustilago (Caeoma, Sorosporium) 1016
synthyridis, Ramularia 927
syriaca, Phyllosticta 874
syringae, Ascochyta 577
syringae, Dendrophoma 667
syringae, Gloeosporium = **Colletotrichum gloeosporioides** 638
syringae, Heterosporium 736
syringae, Microsphaera 791
syringae, Peniophora = **P. violaceolivida** 830
syringae, Phyllosticta 874
syringae, Phytophthora 879
syringae, Sphaeropsis (Sphaeria) 967
syringae, Venturia (Phaeosphaerella) 1021
syringella, Phyllosticta = **P. syringae** 874
systema-solare, Venturia 1021
tabaci, Phyllosticta 874
tabacina, Alternaria = **A. tenuissima** 562
tabacina, Cercospora 617
tabacina, Dermateopsis (Dermatea, Godronia) 669
tabacina, Hydnochaete 739
tabacina, Hymenochaete 739
tabacina, Peniophora = **Crustoderma dryinum** 650
tabacina, Peronospora 834
tabacinium, Macrosporium = **Alternaria tenuissima** 562
tabacinum, Microdochium (Cephalosporium, Fusarium) 789
tabacinum, Sporidesmium 971
tabacinus, Irpex = **Hydnochaete tabacina** 739
tabernaemontanae, Gloeosporium = **Colletotrichum gloeosporioides** 638
tabernaemontanae, Pseudocercospora 895
tabernaemontane, Cercospora = **Pseudocercospora tabernaemontanae** 895
tabescens, Armillaria (Armillariella, Clitocybe) 572
tabifica, Mycosphaerella 806
taenia, Helicoma 731
tagetica, Alternaria 562
tageticola, Cercospora 617
tageticola, Puccinia 915
tageticola, Septoria 958
taleola, Hercospora (Diaporthe) 734
tamaricicola, Peniophora 830
tamaricis, Botryosphaeria (Bagnisiella, Dothidea) 593
tamaricis, Leptosphaeria (Sphaeria) 763
tamaricis, Pyrenopeziza = **Mollisia ligni** 793
tamarii, Aspergillus 579
tamariscina, Diplodia 682
tanaceti, Ophiobolus (Leptosphaeria) 820
tanaceti, Puccinia 915
tanaceti, Ramularia 927
tanaceti var. **dracunculina, Puccinia** 915
tanakae, Cerotelium 622
tandilensis, Cercospora 617
tandilensis, Septoria 958
tantalina, Asterina 581
taphrinicola, Monochaetia = **M. monochaeta** 795
taphrinicola, Pestalotia = **Monochaetia monochaeta** 795
taraxaci, (Ascochyta, Phyllosticta) = **Ascochyta doronici** 574
taraxaci, Mycosphaerella 806
taraxaci, Puccinia = **P. hieracii** 907
taraxaci, Ramularia 927
taraxaci, Septoria = **S. britannica** 948
taraxaci, Synchytrium 985
tarda, Ceriporia (Poria) 621
tarda, Sphaerulina (Ophiocarpella, Ophiodothella, Ophiodothis) 969
tardicrescens, Pythium 922
tardissima, Puccinia 915
tassiana, Mycosphaerella (Sphaerella) 806
tassiana var. **arctica, Mycosphaerella** 806
tatrensis, Puccinia = **P. sieversiae** 914
taubertiana, Cintractia 628
taurica, Leveillula (Erysiphe) 767
taurica var. macrospora, Oidiopsis = **O. macrospora** 818
tautilla, Hyalopeziza (Lachnella) 738
tautillus, Dasyscyphus = **Hyalopeziza tautilla** 738
taveliana, Karschia = **Dactylospora stygia** 664
taxi, Coleophoma (Xenodomus) 635
taxi, Diplodia 682
taxi, Endophragmiella (Endophragmia) 694
taxi, (Asteridiella, Irenina) = **Asteridiella pitya** 580
taxi, Macrophoma 775
taxi, Mycosphaerella 806
taxi, Phacidium 839
taxi, Phyllosticta = **Phoma macrostoma** 854
taxi, Sphaerulina = **Dothiora taxicola** 688
taxi, Stigmatolemma (Cyphella) 979
taxicola, Cryptocline (Gloeosporium) 651
taxicola, Dothiora (Leptosphaeria, Metasphaeria, Sphaerulina) 688
taxicola, Meruliopsis (Caloporus, Gloeoporus, Merulius, Poria) 787
taxicola, Phacidium 839
taxodii, Fomes = **Phellinus gilvus** 845
taxodii, Hydropus (Mycena, Prunulus) 739
taxodii, Laurilia (Stereum) 757
taxodii, Mycosphaerella 806
taxodii, Peniophora = **Hyphoderma praetermissum** 741
taxodii, Sporidesmium 971
taxodium, Poria = **Ceriporia xylostromatoides** 621
tecomae, Aplosporella (Sphaeropsis) 571
tecomae, Ascochyta 577
tecomae, Diplodiella (Diplodia) 682
tecomae, Mycosphaerella 806
tecomae, Phyllosticta 874
tecomae, Septoria 958
tecomae, Synchytrium 985
tecomaxochitl, Septoria 958
tecta, Phakopsora 843
tecticola, Volutella 1023
tectoniae, Cercospora 617
tehoniana, Irenopsis (Meliola) 749
tellimae, Septoria 958
temulenta, (Gloeotinia, Phialea) = **Gloeotinia granigena** 718
temulentum, Endoconidium 694
tenax, Mycena 798
tenella, Asterina = **A. delitescens** 581
tenella, Ceratocystis 604
tenella, (Gnomonia, Gnomoniella) = **Pleuroceras tenellum** 886
tenella, Hendersonia 734
tenella, Micropera 790
tenella, Peniophora = **Hyphoderma puberum** 741
tenella, Septoria 958
tenella, Wojnowicia = **W. hirta** 1024
tenellum, Olpitrichum (Acladium, Oidium, Rhinotrichum) 819
tenellum, Pleuroceras 886
tenellum, Spondylocladium = **Stachylidium bicolor** 972
tenellus, Ophiobolus = **Leptospora rubella** 764
tenera, Heptameria (Leptosphaeria) 734
tenerrima, Ascochyta 577
tenerrima, Phyllosticta 874
tenerrimum, Asteroma 582
tenerum, Verticillium 1022
tennerrimum var. **erythronii, Asteroma** 582
tennesseensis, Crepidotus 648
tenue, Hyphoderma = **H. praetermissum** 741
tenue, Steccherinum 976
tenuiceps, Mycena 798
tenuiculus, Polyporus 889
tenuicutis, Uromyces 1013
tenuifilum, Helicoma 731

tenuipilosa, Trichoscyphella 1000
tenuiporia, Poria = **Diplomitoporus lenis** 683
tenuis, Alternaria = **A. alternata** 561
tenuis, Auricularia = **A. cornea** 585
tenuis, Cercospora 617
tenuis, Diplodia 682
tenuis, (Fomes, Trametes) = **Phellinus viticola** 846
tenuis, Hymenochaete 739
tenuis, Meliola 786
tenuis, Odontia = **Phlebiella vaga** 849
tenuis, Perenniporia (Poria) 831
tenuis, Phoma 856
tenuis, Polyporus = **Trametes membranacea** 996
tenuis, Puccinia 915
tenuis, Ramularia 927
tenuis, Septoria 958
tenuis, Sporobolomyces = **S. roseus** 971
tenuis var. pulchella, Poria = **Perenniporia tenuis** 831
tenuispora, Discella 683
tenuispora, Ustilago 1016
tenuisporum, Cylindrosporium 660
tenuissima, Alternaria 562
tenuissima, Diatrype = **Graphostroma platystoma** 723
tenuissima, Peniophora = **Hyphoderma puberum** 741
tenuissima, Septoria 958
tenuissima f. sp. **euphorbiae, Alternaria** 563
tenuissimum, Cladosporium 630
tenuissimum, Verticillium 1022
tenuistipes, Moellerodiscus (Ciboria, Ciboriopsis) 792
tephroleuca, Postia 891
tephroleucus, (Oligoporus, Polyporus, Tyromyces) = **Postia tephroleuca** 891
tephrosiae, Cercospora 617
tephrosiae, Stagonospora 975
terebinthinaceae, Coleosporium = **C. helianthi** 636
terebrantis, Leptographium 760
teres, Drechslera (Helminthosporium) 690
teres, Hysterium = **H. pulicare** 747
teres, Pyrenophora (Pleospora) 919
teretiuscula, Ascochyta 577
terminalis, Cercosporella = **Cylindrosporium veratrinum** 660
terminalis, Helicogloea 730
terminalis, Phyllosticta 874
ternateae, Cercospora 618
terrestre, Byssocorticium = **Byssoporia terrestris** 596
terrestre, Heterosporium = **Dactylaria constricta** 663
terrestre, Penicillium = **P. crustosum** 828
terrestre, Stachylidium = **Costantinella terrestris** 646
terrestris, Byssoporia (Poria) 596
terrestris, Costantinella 646
terrestris, Lemonniera 758
terrestris, Merulius = **Serpula lacrimans** 960
terrestris, Mycoleptodiscus (Leptodiscus) 799
terrestris, Phoma (Pyrenochaeta) 856
terrestris, Phytophthora = **P. nicotianae** 879
terrestris, Thelephora 990
terrestris, Tomentella 994
terreum, Scolecobasidium 942
terreus, Aspergillus 579
terricola, (Acremonium, Fusidium) = **Acremonium implicatum** 557
terricola, Aspergillus 579
terricola, Conohypha (Hyphoderma, Peniophora) 643
terricola var. **indicus, Aspergillus** 579
terrieri, Epipolaeum = **Rasutoria tsugae** 928
tertia, Denticularia (Cercospora) 668
tertium, Colletotrichum = **C. gloeosporioides** 638
tessarthra, Spegazzinia 963
tessella, Allantoporthe (Diaporthe) 560
tessellata, Cercospora 618
tessellatus, Hypsizygus 747
tessulata, Athelia = **Ceraceomyces tessulatus** 603
tessulatum, Corticium = **Ceraceomyces tessulatus** 603
tessulatus, Ceraceomyces 603
tethys, Hydronectria 739
tetonensis, Leptosphaeria 763
tetonensis, Pellionella 827
tetonensis, Phyllosticta 874
tetramera, (Bipolaris, Helminthosporium) = **Bipolaris spicifera** 589
tetramera, Stagonospora 975
tetramerii, Puccinia 915
tetramicrospora, Didymascella (Keithia) 676
tetraplasandrae, Seynesiopeltis 961
tetraploa, Eutypella (Valsa) 701
tetraptera, Diaporthe 672
tetrasperma, Gelasinospora 713
tetraspora, Dermatea = **Cenangium tetrasporum** 602
tetraspora, Dothidea = **D. puccinioides** 687
tetraspora, Haematomyxa 728
tetraspora, Linospora 769
tetraspora, Mycosphaerella 806
tetraspora, Pseudographis 896
tetraspora, Pseudopezicula 897
tetrasporum, Cenangium (Phaeangium) 602
tetrorchidii, Amazonia 563
teucrii, Mycovellosiella (Cercospora) 808
teucrii, Phyllosticta 874
teutoburgense, (Corticium, Hyphoderma) = **Athelia teutoburgensis** 583
teutoburgensis, Athelia 583
texana, Cumminsiella 655
texana, Entophlyctis 695
texana, Peniophora = **Phanerochaete flavido-alba** 844
texana, Phyllachora = **P. acaciae** 862
texana, Tilletia 992
texanum, Ozonium 824
texanum, Sorosporium 963
texanum, Synchytrium 985
texanus, Inonotus (Polyporus) 749
texanus, Phellinus (Fomes, Pyropolyporus) 846
texanus, Ravenelia = **R. texensis** 929
texensis, Cercospora 618
texensis, Maireina (Cyphella) 776
texensis, Mycena 798
texensis, Phaeodiscula 839
texensis, Phyllachora 863
texensis, Phyllosticta 874
texensis, Ravenelia 929
texensis, Thyridaria (Diatrype, Pseudovalsa) 990
texensis var. **morongiae, Ravenelia** 929
textilis, Melanotus 784
thaliae, Cercospora 618
thaliae, Puccinia 915
thalictri, Aplosporella (Sphaeropsis) 571
thalictri, Cylindrosporium 660
thalictri, Didymosphaeria 678
thalictri, Diplodia 682
thalictri, Entyloma 696
thalictri, Gloeosporium 718
thalictri, Haplobasidion 729
thalictri, Leptosphaeria 763
thalictri, Mycosphaerella 806
thalictri, Phomopsis 860
thalictri, Phytophthora = **P. infestans** f. sp. **thalictri** 878
thalictri, Pyrenopeziza 919
thalictri, Septoria 958
thalictri, Tranzschelia (Polythelis) 997
thalictricola, Diplodia 682
thalictricola, Sphaerella = **Mycosphaerella thalictri** 806
thalictrina, Phomopsis (Phoma) 860
thapsi, Phoma 856
tharpiana, Septoria = **S. angularis** 947
tharpii, Stigmina 980
thaspii, Cercospora 618
thaspii, Stagonospora (Ascochyta) 975
thaspiicola, Cercospora 618
thaxteri, Graphiola 723
theae, Calonectria 598
theae, Cercospora = **Cercoseptoria ocellata** 605
theae, Cercosporella = **Cylindrocladium crotalariae** 657
theae, Cylindrocladium 657
theae, Limacinula 768
theae, Pseudocercospora (Cercoseptoria, Septoria) 895
theae var. crotalariae, Calonectria = **C. crotalariae** 597
theiodea, Lachnellula 754
theissenii, Mycosphaerella = **Hormotheca plantaginis** 737
thelebola, Pseudovalsella (Melanconis) 899
thelena, Rosellinia 937
thelephoroides, Corticium = **Dichostereum pallescens** 675
thelypodii, Puccinia 915
theobromae, Lasiodiplodia (Botryodiplodia, Diplodia) 756
theobromae, Verticillium (Stachylidium) 1022
thermophilus, Talaromyces 986
thermopsidicola, Phoma 856
thermopsidis, Apiocarpella (Ascochyta) 568
thermopsidis, Cercospora = **Stigmina earlei** 980
thermopsidis, Macrophoma 775
thermopsidis, Phyllosticta 874
thermopsidis, Poliomopsis 888
thermopsidis, Sphaeropsis (Phoma) 967
thermopsidis, Stigmina (Sciniatosporium) 980
therophila, Phomatospora (Phyllachora) 857
thielavioides, Chalara (Chalaropsis) 625
thiersii, Marasmius 777
thistletonia, Pseudomassaria (Physalospora) 897
thlaspeos, Puccinia 915
thlaspiae, Cercospora 618
thlaspii, Septoria 958
thomasiana, Leptosphaeria 764
thomasiana, (Marssonia, Marssonina) = **Septogloeum thomasianum** 946
thomasianum, Septogloeum (Gloeosporium) 946
thomasii, Taphrina 987
thomii, Penicillium 829
thompsoniana, Tricholomopsis 1000
thornberi, Thecaphora 989
thornberiana, Ravenelia 929
thouarsianum, Hypoxylon 746
thuemeniana, Montagnula (Pleospora) 796

thuemenii, Entomosporium = **E. mespili** 695
thuemenii, Fabraea 704
thuemenii, Farysia 704
thujae, Cytospora 662
thujae, Hysterium 747
thujae, Lophiostoma (Navicella) 771
thujae, Lophodermium (Lophodermina) 772
thujae, Microthyrium 792
thujae, Mycosphaerella = **Scirrhia conigena** 940
thujae, Peniophora = **Hyphoderma sambuci** 741
thujae, Pestalotia = **P. thujicola** 836
thujae, Stamnaria 975
thujae, Valsa 1018
thujana, Nectria (Nectriella) 813
thujarum, Mytilinidion (Hysterium) 809
thujatsugina, Lindtneria 768
thujicola, Pestalotia 836
thujina, Caliciopsis (Hypsotheca) 597
thujina, Cercospora = **C. sequoiae** 617
thujina, Cryptosporella 653
thujina, Didymascella (Keithia) 676
thujina, Diplodia 682
thujina, Galerina 712
thujina, Harknessia 729
thujina, Kirschsteiniothelia (Amphisphaeria, Kirschsteiniella) 752
thujina, Mycena = **M. clavata** 797
thujina, Stigmina 980
thujinum, (Coryneum, Sciniatosporium) = **Stigmina thujina** 980
thujinum, Helotium = **Pithya cupressina** 881
thumenii, Gloeosporium = **Colletotrichum gloeosporioides** 638
thuyoidea, Physalospora = **Botryosphaeria quercuum** 593
thyoidea, Diplodia 682
thyoides, Hendersonia 734
tiara, Helminthosporium = **Sporidesmium inflatum** 970
tigrinus, Lentinus (Panus) 759
tiliacea, Aplosporella (Sphaeropsis) 571
tiliacea, Holwaya = **H. mucida** 737
tiliacea, Macrophoma 775
tiliacea, Phragmodiaporthe (Diaporthe, Diatrype, Melanconis) 861
tiliae, Apiognomonia (Gnomonia) 569
tiliae, Aplosporella = **A. tiliacea** 571
tiliae, Asteroma 582
tiliae, Cercospora = **C. microsora** 613
tiliae, Ciboria (Sclerotinia) 627
tiliae, Dendrophoma 667
tiliae, Elsinoe 693
tiliae, Exosporium (Helminthosporium) 703
tiliae, Gloeosporium = **Discula umbrinella** 685
tiliae, Helicosporium = **Helicoma muelleri** 730
tiliae, Hercospora 734
tiliae, Lachnella (Cyphella, Trichopeziza) 753
tiliae, Myxosporium 810
tiliae, Phlyctema 850
tiliae, Phorcys 860
tiliae, Phyllosticta 874
tiliae, Rabenhorstia 922
tiliophila, Simocybe (Naucoria) 961
tillaeae, Synchytrium 985
tillandsiae, Glyphium (Lophium) 720
tillandsiae, Puccinia 915
tilletii, Costantinella = **C. terrestris** 646
tinctor, Hypoxylon 746
tinctor, Nummularia 817
tinctoria, Septoria 958
tinctorium, Echinodontium 691
tinctorius, Fomes = **Echinodontium tinctorium** 691
tinctum, Rhynchostoma (Ceratostoma) 935
tinea, Phyllosticta 874
tinea, Stigmina (Cercoseptoria, Cercospora) 980
tineola, Phyllosticta 874
tineum, Gloeosporium = **Phlyctema vagabunda** 850
tingens, Atropellis 584
tingens, Mycosphaerella 806
tingitaninum, Oidium (Acrosporium) 818
tini, Hendersonia 734
tini, Leptosphaeria 764
tithoniae, Puccinia = **P. enceliae** 906
tithymali, Aecidium 558
tocciaeana, Diatrypella = **D. favacea** 674
togata, Ustilago = **U. heterogena** 1015
togwotiensis, Pleospora 885
tolimensis, Puccinia 915
tomato, Helminthosporium (Brachysporium) 732
tomato, (Alternaria, Macrosporium) = **Alternaria tenuissima** 562
tomentella, (Cryptospora, Valsa) = **Winterella betulae** 1024
tomentellum, Amphinema = **A. byssoides** 563
tomentocincta, Poria = **Perenniporia tenuis** 831
tomentosa, (Coltricia, Onnia) = **Inonotus tomentosus** 749
tomentosa, Pleospora 885
tomentosipes, Marasmius = **Xeromphalina cauticinalis** 1025
tomentoso-quercinus, Polyporus = **Spongipellis unicolor** 970
tomentosus, Inonotus (Polyporus) 749
tomentosus var. circinatus, Polyporus = **Inonotus circinatus** 748
tomicoides, Anthostomella 566
tomipara, Puccinia 915
tonellianum, Microstroma = **Aureobasidium prunicola** 584
tonophilum, Eurotium 699
torae, Cercospora = **Phaeoisariopsis nigricans** 840
tormentillae, Taphrina = **T. potentillae** 987
torrendii, Cylindrobasidium (Peniophora) 656
torreyae, Caeoma 597
torreyi, Uncinula = **U. parvula** 1006
torta, Botrytis = **Conoplea fusca** 643
torta, Cercospora 618
torta, Guepiniopsis = **G. buccina** 724
tortile, Mytilinidion (Hysterium) 810
tortilis, Microsphaera (Erysiphe) 791
tortipes, Cercospora 618
tortus, Dacrymyces 663
toruloidea, Hendersonula 734
toruloides, Macrosporium 775
torulosa, Deightoniella 666
torulosum, (Brachysporium, Helminthosporium) = **Deightoniella torulosa** 666
torulosum, Chaetomium 623
torulosum, Pythium 922
torulosus, Lentinus (Panus) 759
torulosus, Phellinus (Fomes, Polyporus) 846
tosevi, Septoria 958
tosquinetii, Taphrina 987
tosta, Puccinia = **P. schedonnardi** 913
tostum, Rhytisma 936
townei, Melanospora = **M. zamiae** 784
toxica, Dothiorella 689
toxica, Phyllosticta 874
toxicodendri, Cercospora 618
toxicodendri, Gloeosporium (Cylindrosporium, Marssonia, Marssonina, Septoria) 718
toxicodendri, Phyllosticta 874
toxicodendri, (Pileolaria, Uromyces) = **Pileolaria brevipes** 879
trabea, (Lenzites, Trametes) = **Gloeophyllum trabeum** 717
trabeum, Gloeophyllum 717
trabicola, Teichospora 988
tracheiphilum, Fusarium = **F. oxysporum** f. sp. **tracheiphilum** 710
tracheiphilum, Pythium 922
trachelospermi, Synchytrium 985
trachychaetes, (Tomentella, Zygodesmus) = **Tomentella violaceofusca** 994
trachysorum, Gymnosporangium 727
trachyspermus, Talaromyces (Arachniotus) 986
trachytricha, Peniophora = **Hyphoderma setigerum** 741
tracyanum, Aecidium 558
tracyi, (Dothidella, Endodothella, Phyllachora) = **Phyllachora sylvatica** 863
tracyi, Trichothyrina (Asterula) 1001
tradescantiae, Septoria (Cylindrosporium) 958
tragacanthae, Pleospora 886
tragopogonis, Albugo (Cystopus) 559
tragopogonis, Cercospora 618
trailii, Thecaphora 989
tralucida, Omphalia 819
trames, Botryosphaeria (Lasiosphaeria, Trichosphaeria) 593
transformans, (Gymnosporangium, Roestelia) = **Gymnosporangium fraternum** 726
translucens, Cytospora 662
translucens, Leucostoma (Valsa) 767
translucens, Rhabdospora 931
translucida, Dasyscyphus = **Pezizella vulgaris** 838
transluscens, Leptosphaeria = **Paraphaeosphaeria obtusispora** 826
transmutans, Parmastomyces (Polyporus, Tyromyces) 826
transylvanica, Saccardoella (Zignoella) 938
tranzschelii, Uromyces 1013
traumatica, Tulasnella (Gloeotulasnella) 1004
trautvetteriae, Puccinia 915
trautvetteriae, Ramularia 927
trautvetteriae, Septoria 958
traversiana, Cercospora 618
trebouxii, Ustilago 1016
treleasei, Ascochyta 577
treleasei, Discocainia (Atropellis, Godronia, Scleroderris) 684
treleasiana, Puccinia 915
trematis, Cercospora (Ragnhildiana) 618
tremelloides, Exidia 702
tremelloides, Gymnosporangium 727
tremelloides, Pestalotia = **P. pezizoides** 835
tremellophora, Diatrype 673
tremellosus, Phlebia (Merulius) 849
tremulae, Fusicladium 711
tremulae, Marssonina (Leptothyrium) 778
tremulae, Napicladium = **Pollaccia radiosa** 888
tremulae, Phellinus 846
tremulae, Venturia = **V. macularis** 1020

tremulo-aurea, Ceratocystis = **Ophiostoma tremulo-aureum** 822
tremulo-aureum, Ophiostoma 822
tremuloides, Discosporium (Cryptocline, Gloeosporium, Leptomelanconium) 685
treubii, (Atichia, Phycopsis) = **Seuratia globifera** 960
triacanthi, Macrophoma (Phoma) 775
triacanthi, Sphaeropsis = **Aplosporella gleditschiicola** 570
triangularis, Coccomyces (Colpoma) 634
triannulata, Puccinia = **P. mirifica** 910
trianthemae, Albugo 559
trianthemae, Gibbago 714
trichellum, Colletotrichum (Amerosporium) 639
trichispora, (Ophiovalsa, Valsa) = **Winterella albofusca** 1023
trichophila, Cercosporella 619
trichophilum, Cladosporium 631
trichophora, Ustilago 1016
trichosanthis, Macrophoma 775
trichosporus, Ophiobolus = **Leptospora rubella** 764
trichostematis, Septoria 958
trichostoma, (Leptosphaeria, Metasphaeria) = **Massarina gloeospora** 779
trichostoma, Pyrenophora (Pleospora) 919
trichota, Rosellinia 937
trichothecioides, Fusarium 711
tricinctum, Fusarium (Selenosporium) 711
tricinctum f. sp. poae, Fusarium = **F. poae** 710
tricolor var. microspora, Lachnella = **Eriopezia microspora** 697
tricorpus, Verticillium 1022
tricristatum, Polynema 889
tricuspidis, Ustilago 1016
tridactyla, Podosphaera 888
tridentatae, Diplodina 683
tridentis, Rhynchosporina 935
tridentis, Stagonospora 975
trientalis, Septoria 958
trientalis, Urocystis (Tuburcinia) 1009
trifidae, Cercospora = **Mycovellosiella ambrosiae** 807
trifidae, Stagonospora 975
trifidum, Lophidium 770
triflorae, Lasiodiplodia = **L. theobromae** 756
triflori, Cylindrosporium 660
trifolii, Colletotrichum 639
trifolii, Curvularia (Blennoria, Brachysporium) 656
trifolii, Cymadothea (Dothidella, Phyllachora, Plowrightia) 660
trifolii, Didymella 677
trifolii, Gloeosporium = **Stagonospora meliloti** 974
trifolii, Leptosphaerulina (Pseudoplea, Sphaerulina) 764
trifolii, Microsphaera (Erysiphe) 791
trifolii, Olpidium 818
trifolii, Phoma = **P. pinodella** 855
trifolii, Physoderma (Urophlyctis) 877
trifolii, Polythrincium 890
trifolii, Pseudopeziza (Phacidium) 897
trifolii, Ramularia 927
trifolii, Septoria 958
trifolii, Sporonema 972
trifolii, Stemphylium 977
trifolii, Uromyces (Nigredo) 1013
trifolii var. fallens, Uromyces = **U. trifolii-repentis** var. **fallens** 1013
trifolii var. recedens, Pleospora = **Stagonospora recedens** 974
trifolii var. trifolii-repentis, Uromyces = **U. trifolii-repentis** 1013
trifolii f. sp. **gladioli, Curvularia** 656
trifolii f. sp. **medicaginis-sativae, Pseudopeziza**
trifolii-hybridi, Uromyces = **U. trifolii-repentis** 1013
trifolii-repentis, Uromyces 1013
trifolii-repentis var. **fallens, Uromyces** 1013
trifoliorum, Peronospora 834
trifoliorum, Sclerotinia 941
trigonellae, Entyloma = **E. meliloti** 696
trigonosporus, Microascus 788
trilaterale, Chaetomium 623
trillii, Ciborinia 627
trillii, Colletotrichum = **C. gloeosporioides** 638
trillii, Cylindrogloeum (Gloeosporium) 658
trillii, Heterosporium 736
trillii, Phyllosticta 874
trillii, Ramularia 927
trillii, Septoria 958
trillii, Urocystis 1009
triloba, Asteridiella (Irene, Irenina, Meliola) 580
trimera, Stigmina 980
trimeroides, Leptosphaeria 764
trimerum, (Coryneum, Sciniatosporium, Thyrostromella) = **Stigmina trimera** 980
trina, Brasiliomyces (Californiomyces, Erysiphe) 595
trinucleata, Diaporthopsis (Diaporthe) 672
triostei, Aecidium 558
triostei, Cercospora 618
triostei, Cladosporium 631
triostei, Coremium 644
triostei, Cylindrosporium 660
triostei, Gnomonia (Diaporthe) 721
triplasidis, Ustilago 1016
tripsaci, Claviceps 632
tripsaci, Puccinia 915
triquetrus, Uromyces 1013
triseptata, Drechslera 690
triseptatum, Coryneum = **Seimatosporium rhododendri** 945
triseptatum, Helminthosporium = **Drechslera triseptata** 690
triseptatum, Lophiostoma = **L. quadrinucleatum** var. **triseptatum** 770
triseptatum f. **fraxini, Lophiostoma** 771
triseti, Hadrotrichum 727
triseti, Puccinia = **P. recondita** 912
triseti, Septoria 958
trispora, Blakeslea 590
tristicha, Valseutypella (Diatrype) 1019
tristis, Acanthonitschkea (Calyculosphaeria) 555
tristis, Pseudotomentella (Hypochnus, Tomentella) 899
tristis, Thelephora = **T. terrestris** 990
tritici, Ascochyta 577
tritici, Marasmius = **M. graminum** 777
tritici, Phaeosphaeria (Leptosphaeria, Pleospora) 842
tritici, Rhizopus = **R. arrhizus** 933
tritici, Septoria 959
tritici, Stemphylium 977
tritici, Tilletia = **T. caries** 992
tritici, Urocystis (Tuburcinia) 1009
tritici, Ustilago 1016
tritici f. **avenae, Septoria** 959
tritici f. **holci, Septoria** 959
tritici var. **lolicola, Septoria** 959
triticina, Puccinia 915
tritici-repentis, Drechslera (Helminthosporium) 690
tritici-repentis, Pyrenophora (Pleospora) 919
tritici-vulgaris, (Drechslera, Helminthosporium) = **Drechslera tritici-repentis** 690
tritici-vulgaris, Pyrenophora = **P. tritici-repentis** 919
trochiloides, Hymenopsis (Myrothecium) 739
trochus, Melanomma 783
trogii, Trametes (Funalia) 996
trollii, Cercosporella (Didymaria, Ramularia) 619
trollii, Phomopsis 860
trollii, Phyllosticta 874
tropaeoli, Acroconidiella (Heterosporium) 557
tropaeoli, Cercospora 618
tropaeoli, Pleospora 886
tropica, Cephaliophora 602
tropica, Vararia 1019
tropicale, Gloeocystidiellum 716
troximontis, Puccinia 915
trullisatus, Strobilurus (Marasmius) 982
truncata, Cercospora 618
truncata, Cytospora = **C. umbrina** 662
truncata, Nectria 814
truncata, (Pestalotia, Truncatella) = **Truncatella angustata** 1001
truncata, Phialocephala 847
truncata, Valsa = **V. melanodiscus** 1018
truncata var. rubi, Pestalotia = **P. laurocerasi** 835
truncatella, Cercospora 618
truncatula, Marssonina = **Didymosporina aceris** 678
truncatula, Tympanis 1004
truncatulum, Gloeosporium = **Didymosporina aceris** 678
truncatum, Colletotrichum 639
truncatum, Hypoxylon 746
truncatum, Oidiodendron 818
truncatum, Polyscytalum 889
truncicolor, (Ceratocystis, Ophiostoma) = **Ophiostoma penicillatum** 821
truncisedum, Coniothyrium = **C. microscopicum** 642
truncorum, Vibrissea 1022
tryblidioides, Cenangium 602
tryblidioides, Glonium 720
tsugae, Aleurodiscus 560
tsugae, Fabrella (Didymascella) 704
tsugae, Galerina 712
tsugae, Ganoderma (Polyporus) 713
tsugae, Hyphoderma (Corticium) 741
tsugae, Korfia (Phacidium) 753
tsugae, (Delphinella, Mycosphaerella) = **Delphinella peckii** 666
tsugae, Rasutoria (Dimeriella, Dimerosporium, Epipolaeum, Eudimeriolum) 928
tsugina, (Fomitiporia, Poria) = **Phellinus punctatus** 846
tuba, Capnodium 601
tuberculans, Cercospora 618
tuberculans, Heterosporium 736
tuberculans, Myriangium 808
tuberculariformis, Nectria (Creonectria, Hypocrea) 814
tubercularioides, Chaetospermum = **C. chaetosporum** 624
tuberculata, Curvularia 656
tuberculata, Phanerochaete (Grandiniella) 844

tuberculatum, Corticium = **Phanerochaete tuberculata** 844
tuberculatum, Endophyllum 694
tuberculatum, Thyridium 991
tuberculella, Cercospora = **C. calystegiae** 607
tuberculifera, Winteria = **Nitschkia grevillii** 816
tuberculiforme, Cenangium 602
tuberculiforme, Stegonsporium (Sporidesmium) 976
tuberculiformis, Dothidea (Plowrightia, Syncarpella) 687
tuberculiformis, Urceolella (Dasyscyphus) 1006
tuberculosa, Diaporthe 672
tuberculosa, Pachykytospora 825
tuberculosa var. **corymbosa, Diaporthe** 672
tuberculosa var. **pruni, Diaporthe** 672
tuberculosus, Polyporus = **Pachykytospora tuberculosa** 825
tubericola, Diplodia = **Lasiodiplodia theobromae** 756
tubericola, Lasiodiplodia = **L. theobromae** 756
tubericola, Stysanus = **Doratomyces stemonitis** 686
tuberiformis, Hypocrella (Echinodothis, Hypocrea) 743
tuberivora, Phacidiopycnis (Phomopsis) 838
tuberosa, Cercospora 618
tuberosa, Drechslera 690
tuberosa, Phoma = **P. exigua** 853
tuberosa, Phyllosticta 874
tuberosum, Helminthosporium = **Drechslera tuberosa** 690
tubiformis, Gnomoniella (Gnomonia) 721
tubulatum, Gymnosporangium = **G. bethelii** 726
tubulina, Camarops (Bolinia) 599
tucsonensis, Tranzschelia = **T. cohaesa** 997
tucumanensis, Glomerella (Physalospora) 719
tulasnei, Marssonina = **Diplodina acerina** 682
tulasnei, Mycosphaerella 807
tulasnei, Ramularia = **R. brunnea** 924
tulasnelloidea, Phlebiella 849
tulasnelloideum, (Corticium, Xenasma) = **Phlebiella tulasnelloidea** 849
tulipae, Botrytis (Sclerotium) 595
tuliparum, Rhizoctonia (Sclerotium) 933
tulipifera, Mycosphaerella 807
tulipiferae, Myxosporium 810
tulipiferae, Oospora (Alysidium, Toruloidea) 819
tulipiferae, Phoma 856
tulipiferae, (Irpex, Polyporus) = **Irpex lacteus** 749
tumamocensis, Puccinia 915
tumefaciens, Diplodia (Macrophoma) 682
tumefaciens, Leptosphaeria (Montagnella, Syncarpella) 764
tumefaciens, Sphaeropsis 967
tumida, (Diatrype, Eutypella) = **Eutypella leprosa** 701
tumida f. **populi, Diatrype** 673
tumida f. **pruni, Diatrype** 673
tumidipes, Puccinia 915
tumidula, (Eutypella, Valsa) = **Eutypella goniostoma** 700
tumidulum, Leptothyrium = **L. similisporum** 766
tumidum, Lophodermium = **L. aucupariae** 771
tumidus, Coccomyces 634
tumoricola, Coryneum = **Seimatosporium pestalozzioides** 945
tumoricola, Godronia (Scleroderris) 721
tumoricola, Phyllosticta 874
tumulosa, Cytospora = **C. sacculus** 662
tunae, Diplotheca 683
tunicatum, Hypoderma (Hypodermopsis) 744
turbinata, Brachysporiella 595
turbinata, Godronia 722
turbinatum, Mystrosporium = **Brachysporiella turbinata** 595
turcica, (Bipolaris, Drechslera) = **Exserohilum turcicum** 704
turcica, Setosphaeria (Keissleriella, Trichometasphaeria) 960
turcica f. sp. sorghi, Trichometasphaeria = **Setosphaeria turcica** 960
turcica f. sp. zeae, Trichometasphaeria = **Setosphaeria turcica** 960
turcicum, Exserohilum (Helminthosporium) 704
turcomanica, Ustilago 1016
turfosa, Gibberidea = **Melanomma rhododendri** 783
turgida, Monochaetia (Pestalotia) 795
turgidipes, Puccinia 915
turgidulum, Triblidium 998
turgidum, Lopadostoma (Anthostoma) 769
turnerae, Pseudocercospora (Cercospora) 895
turpe, Gloeocystidiellum 716
turrita, Puccinia = **P. pazschkei** var. **tricuspidatae** 911
tussilaginis, Coleosporium 636
tussilaginis, Mycosphaerella 807
tuyutensis, Puccinia 915
tympanoides, Romellia 936
typhae, Cladosporium 631
typhae, Colletotrichum 640
typhae, Didymosphaeria = **D. futilis** 678
typhae, Hymenopsis (Cryptomela, Melanconium) 739
typhae, Mycosphaerella 807
typhae, Pleospora = **P. herbarum** 884
typhae, Psathyrella 893
typhae, Scolecosporiella (Hendersonia, Scolicosporium) 943
typhae, Scolicotrichum 943
typharum, Cladosporium 631
typharum, Phaeosphaeria (Leptosphaeria, Pleospora) 842
typharum, Phoma 856
typhicola, Hyphoderma (Peniophora) 742
typhicola, Phoma 856
typhina, Epichloe 696
typhina, Phyllosticta 874
typhinum, Acremonium 557
typhinum, Lophodermium 773
typhinum, Typhodium = **Epichloe typhina** 696
typhoidearum, Stagonospora (Hendersonia) 975
typhoides, Comatricha 640
typhuloides, (Clavaria, Pistillaria) = **Typhula uncialis** 1005
typica, Waydora (Pulvinaria) 1023
uberiformis, Godronia 722
uberiformis, Topospora (Sphaeria) 995
uda, Mycoacia (Odontia) 799
udagawae, Cryptophiale 652
udicola, Melzericium (Corticium) 786
udum var. crotalariae, Fusarium = **Fusarium udum** f. sp. **crotalariae** 711
udum f. sp. **crotalariae, Fusarium** 711
ulei, Dasyscyphus 665
uliginosa, Puccinia = **P. caricina** var. **uliginosa** 902
ulmaria, Fomitopsis = **Rigidoporus ulmarius** 936
ulmariae, Ramularia 927
ulmariae, Septoria 959
ulmariae, Triphragmium 1001
ulmarium, Lyophyllum = **Hypsizygus ulmarius** 747
ulmarius, Hypsizygus (Pleurotus) 747
ulmarius, Rigidoporus (Fomes) 936
ulmea, Stegophora (Dothidella, Gnomonia, Lambro, Sphaeria) 976
ulmea, Tubercularia 1002
ulmea var. **rhizophila, Aplosporella** 571
ulmeum, Asteroma (Gloeosporium) 582
ulmi, Cercospora = **C. sphaeriaeformis** 617
ulmi, Coniothyrium = **C. celtidis-australis** 642
ulmi, Dendrostilbella = **Tubercularia grayana** 1002
ulmi, Diplodia 682
ulmi, Dothiorella 689
ulmi, Exosporium = **Stigmina compacta** 979
ulmi, Hapalocystis (Prosthecium, Pseudovalsa) 728
ulmi, Massaria 779
ulmi, Mycosphaerella 807
ulmi, Myxosporium (Discella) 810
ulmi, Ophiostoma (Ceratocystis, Ceratostomella) 822
ulmi, Phloeospora (Septoria) 850
ulmi, Rhytisma 936
ulmi, Synchytrium 985
ulmi, Taphrina 987
ulmicola, Cylindrosporium 660
ulmicola, Dictyodesmium (Ceratophorum) 675
ulmicola, Fenestella = **Thyridium vestitum** 991
ulmicola, Gloeosporium 718
ulmicola, Melasmia 784
ulmicola, Mycena 798
ulmicola, Myxosporella 810
ulmicola, Sphaeropsis (Botryodiplodia) 967
ulmi-gallae, Sacidium 938
ulmi-rubrae, Phyllosticta 874
ultimum, Pythium 922
ultimum var. **sporangiiferum, Pythium** 922
umatillensis, Septoria 959
umbellata, Grifola = **Polyporus umbellatus** 889
umbellatus, Polyporus (Dendropolyporus, Polypilus) 889
umbelliferarum, Cylindrosporium 660
umbelliferarum, Neottiospora 815
umbelliferaum, Erysiphe = **E. heraclei** 698
umbellulariae, Aureobasidium (Microstroma) 585
umbellulariae, Cyclodomus 656
umbellulariae, Diplodia 682
umbellulariae, Nectria = **N. peziza** 813
umbellulariae, Phialea = **Crocicreas cyathoideum** 648
umbellulariae, Sphaerella 965
umbilicata, Cucurbitaria 654
umbilicata, Heteropatella 735
umbilicata, Trichopeziza = **Velutarina rufo-olivacea** 1019
umbilicatus, (Lentinellus, Lentinus) = **Lentinellus cochleatus** 759
umbilici, Puccinia 915
umbonata, Caulorhiza (Collybia, Hydropus) 601

umbonata, Teichospora (Strickeria) 988
umbonatum, Camarosporium 600
umbonatum, Coryneum 646
umbonatus, Pleurotus = **Tricholomopsis radicata** 999
umbrata, Cercospora 618
umbrata, Phlebia 849
umbrata var. maculata, Cercospora = **Cercospora umbrata** 618
umbrina, Cryptosporella (Diaporthe) 653
umbrina, Cytospora (Cytosporopsis) 662
umbrina, Poria = **Phellinus ferruginosus** 845
umbrina, Pseudociboria 895
umbrina, Pseudotomentella = **P. tristis** 899
umbrina, Ramularia 927
umbrina, Sebacina = **Exidiopsis grisea** 702
umbrinella, Discula 685
umbrinella, Exidia 702
umbrinella, Rosellinia 937
umbrinella, Tomentella = **T. terrestris** 994
umbrinellum, (Gloeosporidium, Gloeosporium) = **Discula umbrinella** 685
umbrino-fumosa, Phyllosticta 874
umbrinospora, Tomentella 994
umbrinum, Helotium = **Pseudociboria umbrina** 895
umbrinum, Stereum = **Phanerochaete crassa** 843
umbrosum, Eurotium = **E. herbariorum** 699
umbrosus, Aspergillus = **A. glaucus** 578
unamunoi, (Cercospora, Cladosporium, Phaeoramularia) = **Phaeoramularia capsicicola** 840
uncialis, Typhula 1005
uncinata, Lophodermina = **Lophodermium uncinatum** 773
uncinata, Menispora 786
uncinata, Unguiculella (Dasyscyphus) 1006
uncinatum, Ceratophorum (Clasterosporium) 604
uncinatum, Colletotrichum = **C. dematium** 637
uncinatum, Lophodermium 773
uncinulatum, Pythium 922
undata, Poria = **Physisporinus vitreus** 876
undatus, Rigidoporus = **Physisporinus vitreus** 876
underwoodii, Diatrypella 674
underwoodii, Gymnopilus 726
underwoodii, Polyporus = **P. admirabilis** 889
underwoodii, Ustilago = **U. trebouxii** 1016
undosa, Postia 891
undosus, (Oligoporus, Polyporus, Tyromyces) = **Postia undosa** 891
undulata, Diatrype 673
undulata, Rhizina 932
unedinis, Septoria 959
ungeri, Chalara 625
unguicularis, Hohenbuehelia (Pleurotus, Resupinatus) 737
unguiculata, Anthostomella 566
unguis, Aspergillus 579
unica, Puccinia 915
unicolor, Cerrena (Daedalea) 622
unicolor, Peniophora = **Phanerochaete filamentosa** 843
unicolor, Pestalotia 836
unicolor, Septoria 959
unicolor, Spongipellis (Tyromyces) 970
unicorne, Seiridium 945
unicornis, (Monochaetia, Pestalotia) = **Seiridium unicorne** 945
uniforme, Hormiscium = **Cryptocoryneum condensatum** 651
uniformis, Torula = **Cryptocoryneum condensatum** 651
uniolae, Tunicago 1004
uniolae, Ustilago 1016
uniporulus, Uromyces 1013
uniseptata, Alpakesa 560
uniseptata, Calonectria = **C. kyotensis** 597
uniseta, Pestalotia = **Seiridium unicorne** 945
unisetula, Codinaea 635
unispora, Platygloea 882
unita, (Fomitopsis, Poria) = **Perenniporia medulla-panis** 830
unitus, Fomes = **Perenniporia medulla-panis** 830
unitus, Polyporus = **Perenniporia medulla-panis** 830
unitus, Uromyces 1013
unitus subsp. **montanensis, Uromyces** 1013
unitus subsp. **pacificus, Uromyces** 1013
unitus subsp. **spragueae, Uromyces** 1013
universalis, Puccinia = **P. atrofusca** 901
uramensis, Cercospora 618
urbaniana, Phyllachora 863
urbaniana, Puccinia 915
urbanianum, Catacauma = **Phyllachora urbaniana** 863
urceolata, Godronia (Cenangella) 722
urceolatum, Acrospermum 557
urceolus, Godronia 722
urceolus var. confertus, Godronia = **G. confertus** 721
uredinicola, Botrytis = **Beniowskia sphaeroidea** 587
uredinis, Kuehneola 753
urens, Phoma 856
urnula, Monilinia (Sclerotinia) 795
urochloae, Bipolaris (Drechslera, Helminthosporium) 589
uromycestri, Cercospora 618
uromycoides, Harknessia 730
ursinus, Lentinellus (Lentinus) 759
ursinus, Polyporus = **Amylocystis lapponica** 564
urticae, Colletotrichum 640
urticae, Cylindrosporium 660
urticae, Helminthosporium = **Dendryphiella vinosum** 668
urticae, Mollisia 794
urticae, Penicillium = **P. griseofulvum** 828
urticae, Phialea = **Crocicreas cyathoideum** 648
urticae, Pseudoperonospora 897
urticae, Puccinia = **P. caricina** 902
urticae, Ramularia (Septocylindrium) 927
urticae, Septoria 959
urticae, Synchytrium = **S. chamaedryoidis** 983
urticaria, Septoria 959
urticata, Puccinia = **P. caricina** 902
urticina, Trichopeziza = **Cistella grevillei** 628
ustalis, Puccinia 915
ustulatum, Hypoxylon = **Ustulina deusta** 1017
ustus, Aspergillus 579
utahense, Cylindrosporium = **Septoria betulae-odoratae** 948
utahense, (Sciniatosporium, Stegonsporium) = **Stigmina utahensis** 980
utahensis, Aleurodiscus 560
utahensis, Apioporthe (Chorostate, Diaporthe) 569
utahensis, Ascochyta 577
utahensis, Lojkania (Melanopsammina) 769
utahensis, Pleospora = **Comoclathris permunda** 640
utahensis, Sphacelotheca 965
utahensis, Stigmina 980
utriculicola, Cintractia 628
utriculosa, Ustilago 1016
utriculosa var. rumicis, Ustilago = **U. kuehneana** 1015
uvae-ursi, Exobasidium 703
uvicola, Greeneria (Phoma) 724
uvicola, Pestalotiopsis (Pestalotia) 836
uviferae, Uredo = **U. coccolobae** 1007
uvispora, Physalospora = **Glomerella cingulata** 719
uvulariae, Diplodia = **Macrophoma smilacina var. smilacis** 774
uvulariae, Phyllosticta 874
vaccinicola, Coniothyrium 643
vaccinicola, Gloniella 719
vaccinicola, Pestalotia = **P. adusta** 835
vaccinii, Acremonium = **Conoplea fusca** 643
vaccinii, Bifusella 588
vaccinii, Botryosphaeria (Guignardia) 593
vaccinii, Diaporthe 672
vaccinii, Diplodia 682
vaccinii, Dothidella = **Pyrenobotrys conferta** 918
vaccinii, Elsinoe 693
vaccinii, Eupropolella 699
vaccinii, Exobasidium 703
vaccinii, Gibbera 714
vaccinii, Meliola 786
vaccinii, Micropera 790
vaccinii, Microsphaera (Erysiphe) 791
vaccinii, Mycosphaerella (Carlia) 807
vaccinii, Ophiodothella 820
vaccinii, Pestalotia 836
vaccinii, Phacidium 839
vaccinii, Phoma 856
vaccinii, Phomopsis 860
vaccinii, Phyllosticta (Phyllostictina) 874
vaccinii, Physalospora (Acanthorhynchus) 876
vaccinii, Piggotia 879
vaccinii, Pucciniastrum (Thekopsora) 917
vaccinii, Ramularia 927
vaccinii, Rhytisma 936
vaccinii, Seimatosporium (Coryneum) 945
vaccinii, Septoria 959
vaccinii, Sphaeropsis 967
vaccinii, Stictis = **Propolis phacidioides** 892
vaccinii, Synchytrium 985
vaccinii, Venturia = **Niesslia exilis** 815
vaccinii var. **arctostaphyli, Exobasidium** 703
vaccinii var. myrtilli, Exobasidium = **E. myrtilli** 703
vaccinii var. oxycoccini, Phacidium = **Eupropolella vaccinii** 699
vaccinii-corymbosi, Monilinia (Sclerotinia) 795
vaccinii-uliginosi, Exobasidium 703
vaga, Coniophora = **Uthatobasidium fusisporum** 1017
vaga, Pellicularia = **Botryobasidium vagum** 591
vaga, Phlebiella (Grandinia, Odontia, Phlebia, Trechispora) 849
vaga, Strickeria (Dictyotrichiella, Teichospora) 982
vagabunda, Leptosphaeria = **Massarina microcarpa** 779
vagabunda, Phlyctema 850
vagabundum, Lophiostoma (Lophiotrema) 771
vagans, Fumago 706
vagans, (Drechslera, Helminthosporium) = **Drechslera poae** 690

vagans, Leptosphaeria = **Phaeosphaeria caricinella** 842
vagans, Melogramma = **M. campylosporum** 786
vagans, Phaeosphaeria (Pleospora) 842
vagans, Phyllosticta 874
vagans, Physoderma 877
vagans, Puccinia = **P. pulverulenta** 912
vagans var. epilobii-tetragoni, Puccinia = **P. pulverulenta** 912
vagans var. gayophyti, Puccinia = **P. pulverulenta** 912
vaginae, Mycovellosiella (Cercospora) 808
vagnerae, Cylindrosporium 660
vagnerae, Mycosphaerella 807
vagnerae, Phloeospora 850
vagnerae, Ramularia 927
vagum, Aulographum = **A. hederae** 584
vagum, Botryobasidium 591
vagum, Corticium = **Thanatephorus cucumeris** 989
vagum, Steccherinum 976
vagus, Hypochnus = **Phlebiella vaga** 849
vaillantii, Antrodia (Polyporus, Poria) 568
vaillantii, Ustilago 1016
valdiviensis, Stictis = **S. lupini** 978
valens, Uromyces 1013
valerianae, Ascochyta 577
valerianae, Cercoseptoria 605
valerianae, Peronospora 834
valerianae, Ramularia (Cylindrosporium) 927
valerianae, Septoria 959
valerianellae, Aecidium 559
valerianellae, Septoria 959
vallisumbrosae, Ramularia 927
valsoideum, (Gloeosporium, Myxosporium) = **Discula umbrinella** 685
valtellinensis, Heteropatella 735
vancouveriae, Ramularia (Ovularia) 927
vanillae, Botryosphaeria (Gnomoniopsis) 593
vanillae, (Colletotrichum, Gloeosporium) = **Colletotrichum gloeosporioides** 638
vanillae, Volutella 1023
vaninii, Phellinus 846
vanterpoolii, Pythium 922
vaporaria, Poria = **Antrodia sinuosa** 567
vaporarius, Polyporus = **Antrodia sinuosa** 567
varia, Cercospora 618
varia, Pestalosphaeria 834
varia, Taphrina = **T. farlowii** 987
variabile, Cladosporium (Helminthosporium, Heterosporium) 631
variabile, Gloeosporium = **Gloeosporidiella variabilis** 717
variabile, Penicillium 829
variabile, Scolecobasidium 942
variabilis, Crepidotus 648
variabilis, Gloeosporidiella 717
variabilis, Microascus = **M. longirostris** 788
variabilis, Phyllosticta 874
variabilis, Puccinia 915
variabilis, Ramularia 927
variabilis, Teichospora (Strickeria) 988
variabilis, Typhula 1005
variabilis var. **insperata, Puccinia** 915
variabilis var. lapsanae, Puccinia = **P. lapsanae** 909
variabilum, Synchytrium 985
varians, Dasyscyphus = **Lachnum flavidulum** 754
varians, Gliocladium = **G. catenulatum** 715
varians, Hendersonia 734
variata, Ramularia 927
variecolor, Aspergillus = **A. stellifer** 579
variecolor, Emericella 693
variecolor, Trechispora = **Lindtneria leucobryophila** 768
variegata, Ramularia 928
variegata, Septoria = **S. blasdalei** 948
variegata, Stigmochora 980
variegatum, Helminthosporium 732
variegatum, Hypoderma 744
variegatum, Septogloeum 946
variicolor, Cercospora 618
variiformis, Antrodia (Coriolellus, Polyporus, Trametes) 568
variiseptata, Leptosphaeria 764
variispora, Articulophora 573
variisporus, Dacrymyces 663
variospora, (Ceratocystis, Endoconidiophora, Ophiostoma) = **Ceratocystis fimbriata** 604
variotii, Paecilomyces 825
variseptata, Sebacina = **Exidiopsis fuliginea** 702
variseptata, Trechispora 998
varium, Sclerotium 942
varium, Zalerion 1027
varius, Exoascus = **Taphrina farlowii** 987
varius, Polyporus 889
vasinfecta, Neocosmospora 814
vasinfectum, Fusarium = **F. oxysporum** f. sp. **vasinfectum** 710
vasinfectum var. lutulatum, Fusarium = **F. oxysporum** 708
vastator, Merulius = **Serpula lacrimans** 960
vauqueliniae, Gymnosporangium 727
velata, Lophiosphaera 770
velatum, Glonium = **Lophiosphaera velata** 770
velatum, Rhytisma (Xyloma) 936
vellereum, Hypochnicium (Corticium) 743
vellereum, Stereum = **S. ochraceo-flavum** 978
vellereus, Polyporus = **Trametes hirsuta** 996
veluticeps, Hymenochaete = **Veluticeps berkeleyi** 1020
velutina, Cercospora 618
velutina, Eutypa = **E. lata var. aceri** 700
velutina, Phanerochaete (Peniophora) 844
velutina var. quercina, Eutypa = **E. mela** 700
velutinum, Cladosporium 631
velutinum, (Helicoma, Helicosporium) = **Helicoma phaeosporium** 731
velutinum, Helminthosporium 732
velutinum, Sporidesmium 971
velutinus, Lentinus 759
velutinus, (Coriolus, Polyporus) = **Trametes pubescens** 996
velutipes, Flammulina (Collybia) 705
velutipes, Spathulariopsis (Spathularia) 963
venenosa, Phomopsis 860
venensoa, Phoma = **Phomopsis venenosa** 860
veneta, Apiognomonia (Gnomonia) 569
veneta, Elsinoe (Plectodiscella) 693
venetum, Gloeosporium = **Sphaceloma necator** 964
venezuelana var. **floridensis, Meliola** 786
venezuelanum, Sphaceloma 964
venezuelensis, (Pseudohansfordia, Sympodiophora) = **Fusarium pallidoroseum** 710
ventricosa, Harknessia 730
ventricosum, Fusarium 711
ventriosa, Valsa = **Eutypella juglandicola** 700
venturioides, Cercospora 618
venturioides, Dasyscyphus (Lachnella, Lachnum, Pirottaea, Trichopeziza) 665
venulosa, Xylaria = **X. apiculata** 1025
venulosum, Asteroma 582
venusta, (Eutypella, Valsa) = **Romellia cornina** 936
venustula, Puccinia = **P. posadensis** 912
vepris, Cryptodiaporthe (Apioporthe) 651
veratri, Ascochyta 577
veratri, Cercosporella 619
veratri, Diplodia (Ascochytella) 682
veratri, Gloeosporium (Fusoma, Marssonia, Marssonina, Septogloeum) 718
veratri, Laetinaevia 755
veratri, Leptosphaeria 764
veratri, Puccinia 915
veratri, Sclerotinia 941
veratrina, Ascochyta 577
veratrinum, Cylindrosporium 660
veratrinum, Gloeosporium = **Ascochyta veratri** 577
verbasci, Gloniopsis = **G. praelonga** 720
verbascicarpa, Phoma 856
verbascicola, Cercospora 618
verbascicola, Mycosphaerella 807
verbascicola, Phoma 856
verbascicola, Phyllosticta 874
verbascicola, Septoria 959
verbenae, Ascochyta 577
verbenae, Septoria 959
verbenae, Stagonospora 975
verbenae-strictae, Cercospora = **C. verbenicola** 618
verbenicola, Cercospora 618
verbenicola, Phyllosticta 874
verbenicola, Puccinia = **P. vilfae** 916
verbesinae, Phyllosticta 874
verbesinae, Puccinia 916
veriferum, Xenasma = **Tubulicium vermifera** 1003
vermicularia, Eriosphaeria 697
vermicularioides, Colletotrichum = **C. gloeosporioides** 638
vermicularis, Calocera = **C. cornea** 597
vermicularis, Perichaena 831
vermiculatum, Clasterosporium = **Sporidesmium hormiscioides** 970
vermiculatum, Penicillium = **P. dangeardii** 828
vermiculatus, Talaromyces = **T. flavus** 985
vermifera, Tubulicium (Epithele, Peniophora, Tubulicrinis) 1003
vermiferum, Tubulixenasma = **Tubulicium vermifera** 1003
vermiforme, Cylindrosporium = **Anguillospora vermiformis** 564
vermiformis, Anguillospora (Cercoseptoria) 564
vermisporum, (Lophionema, Lophiostoma) = **Ophiobolus vermisporus** 820
vermisporus, Ophiobolus 820
vermoeseni, Gliocladium (Penicillium) 715
vernalis, (Kuehneromyces, Pholiota) = **Kuehneromyces lignicola** 753
vernalis var. aggregata, Erysiphe = **E. aggregata** 697
vernicosa, Daldinia 664
vernicosa, Peniophora = **Phanerochaete flavido-alba** 844
vernicosum, Hypoxylon = **H. truncatum** 746
vernicosus, Cyathus = **C. olla** 656
vernoniae, Cercospora 618
vernoniae, Coleosporium 636
vernoniae, Puccinia = **P. longipes** 909

vernoniae, Sarcinella (Piricauda, Stigmella) 938
vernum, Physarum 876
veronicae, Discogloeum (Gloeosporium, Leptothyrium) 684
veronicae, Entyloma = **E. linariae** 696
veronicae, Ramularia (Ovularia) 928
veronicae, Septoria 959
veronicae, Sorosphaera (Tuburcinia) 963
veronicarum, Puccinia 916
verrucaria, Myrothecium 809
verrucarioides, Phlyctema 850
verruciformis, Diatrypella = **D. favacea** 674
verruciformis, Rhabdospora (Septoria) 931
verrucisporus, Botryohypochnus 591
verrucoides, Diatrype = **D. albopruinosa** 673
verrucosa, Anthracoidea 567
verrucosa, Ligniera 768
verrucosa, Mycocentrospora 799
verrucosa, Nectria (Creonectria) 814
verrucosa, Ostropella 823
verrucosa, Periconiella 831
verrucosa, Phialophora 847
verrucosum, Penicillium 829
verrucosum, Xenolophium = **Ostropella verrucosa** 823
verruculosus, Uromyces 1013
versatilis, Ravenelia = **R. pringlei** 929
versatilis, (Hirschioporus, Polyporus, Trametes) = **Trichaptum byssogenum** 998
versicolor, Ascochyta 577
versicolor, Aspergillus 579
versicolor, Cladosporium = **Mycovellosiella perfoliati** 808
versicolor, Lophodermium 773
versicolor, Pestalotia 836
versicolor, Propolomyces (Propolis, Stictis) 892
versicolor, Puccinia 916
versicolor, Stomiopeltis 981
versicolor, Trametes (Coriolus, Polyporus, Polystictus) 996
versicutis, Antrodiella (Polyporus) 568
versiforme, Stereum = **Dendrophora versiformis** 667
versiformis, Chlorencoelia (Chlorociboria, Midotis) 625
versiformis, Dendrophora (Peniophora) 667
versiformis, Umbelopsis 1006
versipora, Poria = **Schizopora paradoxa** 939
versiporus, Polyporus = **Schizopora paradoxa** 939
versisporum, Linocarpon 769
versisporus, Ophiobolus = **Linocarpon versisporum** 769
versutus, Crepidotus 648
verticillata, (Drechslera, Podosporiella) = **Drechslera campanulata** 689
verticillata, Selenosporella 946
verticillatum, Cladobotryum (Acremonium) 629
verticillatum, Stachylidium = **S. bicolor** 972
verticilloides, Gliocladium = **G. roseum** 715
vertisepta, Puccinia 916
vesca, Ceratocystis = **Ophiostoma olivaceum** 821
vesicarium, Stemphylium 977
vesicatoria, Phyllosticta 874
vesicula, Phytophthora 879
vesiculosa, Odontia = **Hyphoderma setigerum** 741
vesiculosa, Puccinia 916
vesiculosum, Corticium = **Gloeocystidiellum clavuligerum** 716
vesiculosus, Poria = **Oxyporus corticola** 824
vesparium, Metatrichia (Hemitrichia) 788
vestigiale, Glonium 720
vestita, Fenestella = **Thyridium vestitum** 991
vestita, Septoria 959
vestitum, Thyridium (Mycothyridium) 991
vexans, Cercospora 618
vexans, Gloeosporium = **Colletotrichum gloeosporioides** 638
vexans, Phomopsis 860
vexans, Puccinia 916
vexans, Pythium 922
vexata, Belonioscypha = **Crocicreas culmicola** 648
vexata, Cryptosphaeria 652
vexatula, Stagonospora 975
vexulata, Hendersonia = **Stagonospora vexatula** 975
vialis, Lenzites = **Gloeophyllum trabeum** 717
vibrissa, Nia 815
viburni, Coleosporium 636
viburni, Diaporthe 672
viburni, Eutryblidiella (Cenangium) 699
viburni, Lambertella 756
viburni, Phoma (Ascochyta) 856
viburni, Plasmopara 882
viburni, Ramularia 928
viburni, Septoria 959
viburni, Sphaceloma 964
viburni, Typhula 1005
viburni var. **spiraeicola, Diaporthe** 672
viburnicola, Cercospora 618
viburnicola, Durandiella (Godronia) 691
viburnicola, Phyllosticta = **Phoma viburni** 856
vicae-fabae, Uromyces = **U. fabae** 1011
viciae, Ascochyta 577
viciae, Botrytis 595
viciae, Cercospora = **C. zonata** 619
viciae, Colletotrichum = **C. truncatum** 639
viciae, Microdiplodia 789
viciae, Peronospora 834
viciae, Septoria = **S. viciicola** 959
viciae var. **americana, Peronospora** 834
viciae-fabae, Uromyces = **U. fabae** 1011
vicicola, Peronospora 834
viciicola, Septoria 959
vicina, Poria = **Oxyporus corticola** 824
vicinula, (Cryptosphaeria, Eutypa) = **Cryptosphaeria subcutanea** 652
victoriae, Bipolaris (Drechslera, Helminthosporium) 589
viequesensis, Phyllachora 863
vigheffulensis, Lophiostoma (Lophiosphaera) 771
vignae, Cercospora = **Pseudocercospora cruenta** 894
vignae, Chaetoseptoria 624
vignae, Cladosporium 631
vignae, Helminthosporium = **Corynespora cassiicola** 646
vignae, Leptosphaerulina = **L. trifolii** 764
vignae, Uromyces 1013
vignicaulis, Cercospora = **C. canescens** 607
vignicaulis, Glomerella = **G. cingulata** 719
vignicola, (Cercospora, Helminthosporium) = **Corynespora cassiicola** 646
viguierae, Coleosporium 636
vilfae, Puccinia 916
vilfae, Ustilago 1016
villosa, Halocyphina 728
villosa, Lachnella (Cyphella, Peziza) 753
villosa, Phyllosticta 874
villosa, Trametes 997
villosum, Colletotrichum = **C. truncatum** 639
villosus, Polyporus = **Trametes villosa** 997
viminei, Cercospora = **Cercosporella virgaureae** 619
vincae, Diplodia 682
vincae, Macrophoma 775
vincae, Puccinia 916
vincae, Septoria 959
vincae, Volutella 1023
vincae-majoris, Phyllosticta = **Phoma exigua** var. **inoxydabilis** 853
vincetoxici, Cercospora = **C. punctiformis** 615
vincta, Poria = **Rigidoporus vinctus** 936
vinctus, Rigidoporus (Polyporus) 936
vinosa, Haematomyxa 728
vinosa, Orbilia (Calloria, Mollisia) 822
vinosa, Papularia 825
vinosa, Ustilago 1016
vinosum, Dendryphiella (Dendryphion, Helminthosporium) 668
vinosum, Graphiothecium = **Ramularia coalescens** 924
vinosum, Gymnosporium = **Papularia vinosa** 825
vinosus, Nigroporus (Polyporus) 815
violacea, Fuligo = **F. septica** 706
violacea, Rhizoctonia = **R. crocorum** 932
violacea, Septoria 959
violacea, Tulasnella 1004
violacea, Ustilago 1016
violacea var. stellariae, Ustilago = **Ustilago violacea** 1016
violaceae var. major, Ustilago = **U. clintoniana** 1015
violaceofulvus, Panellus (Pleurotus) 825
violaceofusca, Tomentella 994
violaceolivida, Peniophora 830
violae, Alternaria = **A. tenuissima** 562
violae, Ascochyta 577
violae, Cercospora 618
violae, Ciborinia 627
violae, Cryptostictis 653
violae, Cylindrosporium 660
violae, Marssonina 779
violae, Peronospora 834
violae, Phyllosticta 874
violae, Pleosphaerulina 883
violae, Puccinia 916
violae, Pythium 922
violae, Ramularia = **R. lactea** 925
violae, Septoria = **S. violae-palustris** 959
violae, Sphaceloma (Gloeosporium) 964
violae, Sphaerotheca 968
violae, Synchytrium 985
violae, Urocystis 1009
violae-palustris, Septoria 959
violae-rotundifoliae, Colletotrichum = **C. dematium** 637
violae-tricoloris, Cercospora = **C. violae** 618
violae-tricoloris, Colletotrichum = **C. gloeosporioides** 638
violamaculans, Cercospora 618
violarum, Colletotrichum = **C. dematium** 637
violea, Tulasnella 1004
violicola, Ascochyta 577
viornae, Tranzschelia 997
virens, Gliocladium 715
virens, Phyllosticta 874
virens, Stagonospora 975
virens, Ustilaginoidea 1014
virescens, Diatrype 673
virescens, (Ceratocystis, Endoconidiophora) = **Ceratocystis coerulescens** 603

virescens, Eutypella = **Diatrype flavovirens** 673
virgae-aurea, Puccinia 916
virgata, Puccinia 916
virgatus, Polyporus 889
virgaureae, Cercosporella (Cercospora, Ovularia, Ramularia) 619
virgaureae, Septoria 959
virginella, Lachnella = **Lachnum virginellum** 755
virginellum, Lachnum 755
virginellus, Dasyscyphus = **Lachnum virginellum** 755
virgineum, Lachnum 755
virgineus, Boletus = **Trametes conchifer** 996
virgineus, Dasyscyphus = **Lachnum virgineum** 755
virginiana, Diplodia 682
virginiana, Phyllosticta (Phoma) 875
virginiana, Uredinopsis 1007
virginianum, Cylindroxyphium 660
virginica, Leptosphaeria = **L. planiuscula** 763
virginica, Taphrina 987
virgultorum, Anisogramma 565
virgultorum, Helotium = **Hymenoscyphus calyculus** 740
virgultorum, (Hypoderma, Leptostroma) = **Hypoderma rubi** 744
virgultorum var. fructigenum, Helotium = **Hymenoscyphus fructigenus** 740
viridans, Ceriporia (Polyporus, Poria) 621
viridarii, Phomopsis (Phoma) 860
viridarium, Lophiostoma 771
viride, Cylindrosporium = **Phloeosporina fraxini** 850
viride, Gliocladium 715
viride, Helicosporium 731
viride, Hormodendrum (Penicillium) 737
viride, Trichoderma 999
viridella, Phaeosphaeria (Leptosphaeria) 842
viridescens, Tomentella (Hypochnus) 995
viridicatum, Penicillium 829
viridicoma, Pseudotrichia = **P. mutabilis** 899
viridis, (Caldesiella, Odontia, Tomentella) = **Tomentella chlorina** 993
viridis, Phyllosticta 875
viridistroma, Endothia 694
viridi-tingens, Septoria 959
viridula, Cercospora = **C. ipomoeae** 611
viridula, Tomentella 995
viridulum, Cyanoderma (Acrospermum) 656
visci, Echidnodes 691
visci, Meliola 786
visci, Sphaeropsis (Botryosphaerostroma) 967
viscidula, Haligena 728
viscosa, Sphaeropsis 968
viscosum, Melanconium 783
vitalbae, Pleospora (Leptosphaeria, Teichospora) 886
vitata, Puccinia 916
viticis, Cercospora (Cercospora) 618
viticis, Helminthosporium 732
viticola, Cercospora = **Pseudocercospora vitis** 895
viticola, Cryptosporella 653
viticola, Dacrymyces = **D. capitatus** 663
viticola, Dichomera 675
viticola, Diplodia 682
viticola, Fusarium = **F. avenaceum** 707
viticola, Hysterographium = **H. mori** 747
viticola, Macrophoma 775
viticola, Myxosporium 810
viticola, Nectria 814
viticola, Phanerochaete (Peniophora) 844
viticola, Phellinus (Fomes, Polyporus, Poria) 846
viticola, Phomopsis (Fusicoccum) 860
viticola, Phyllosticta = **P. ampelicida** 864
viticola, Plasmopara (Peronospora) 882
viticola, Septoria (Sacidium) 959
viticola, Sphaeropsis = **Aplosporella fabaeformis** 570
viticola, Stictis 979
viticola, Thyridaria (Aglaospora, Pseudovalsa) 990
viticola, Typhula (Pistillaria) 1005
viticola var. ampelopsidis, Phomopsis = **Phomopsis viticola** 860
vitigena, Sphaeropsis = **Aplosporella fabaeformis** 570
vitigera, Valsa 1018
vitis, Arxiomyces (Phaeostoma) 573
vitis, (Aureobasidium, Exobasidium) = **Aureobasidium pullulans** 585
vitis, Cytospora 662
vitis, Phoma 856
vitis, Physopella = **P. ampelopsidis** 877
vitis, Pseudocercospora (Cercospora, Cladosporium) 895
vitis, Pyrenochaeta 918
vitis, Rhytisma 936
vitis, Stigmina = **Phaeoramularia heterospora** 841
vitis, Thyridium 991
vitis, (Eutypella, Valsa) = **Eutypella aequilinearis** 700
vitrea, Poria = **Physisporinus vitreus** 876
vitreus, Physisporinus (Rigidoporus) 876
vitrispora, Julella (Peltosphaeria, Pleospora) 750
vodakii, Ascochyta 577
vogesiaca, Milesina (Milesia) 792
vogesiaca, Pseudoplectania = **P. melaena** 898
vogesiacum, Hypoxylon 746
vogesiacum var. **macrospora, Hypoxylon** 746
vogesiacum var. **microspora, Hypoxylon** 746
volkartiana, Puccinia 916
volubilis, Ascochyta 577
volutum, Pythium 922
volvatus, Cryptoporus (Polyporus) 652
vomitoria, Massaria = **M. inquinans** 779
vossiae, Uromyces 1013
vossii, Phyllosticta = **Kabatia periclymeni** 751
vuijckii, Ustilago 1016
vulgare, Auriscalpium 585
vulgare, Crucibulum = **C. laeve** 650
vulgare, Epicoccum = **E. nigrum** 696
vulgare, Leptothyrium (Leptostroma) 766
vulgare, Preussia (Perisporium) 891
vulgare f. **parryae, Leptothyrium** 766
vulgaris, Ascochyta = **A. tenerrima** 577
vulgaris, Botrytis = **B. cinerea** 594
vulgaris, Diplodia 682
vulgaris, (Gnomonia, Gnomoniella) = **Gnomonia gnomon** 720
vulgaris, Nectria = **N. ochroleuca** 813
vulgaris, Pezizella (Allophylaria, Hyalinia, Mollisia) 838
vulgaris, Phoma 856
vulgaris, Phyllosticta 875
vulgaris, Pithya 881
vulgaris, Pleospora = **Lewia scrophulariae** 767
vulgaris, Tubercularia 1002
vulgaris, Ustulina = **U. deusta** 1017
vulgaris f. **astericola, Pleospora** 886
vulgaris var. **interrupta, Botrytis** 595
vulgaris var. **sanguinella, Pezizella** (Peziza) 838
vulgata, Phyllachora 863
vulpina, Nectria (Peziza) 814
vulpinae, Cercospora 618
vulpinoideae, Puccinia = **P. dioicae** 905
vulpinus, Lentinellus (Lentinus) 759
vulpinus, Polyporus = **Inonotus rheades** 749
vulvatum, Hysterographium = **H. flexuosum** 747
wageneri, Leptographium (Verticicladiella) 760
wageneri, Ophiostoma (Ceratocystis) 822
wageneri var. ponderosa, Verticicladiella = **Leptographium wageneri** var. **ponderosum** 760
wageneri var. **ponderosum, Leptographium** 760
wageneri var. **pseudotsugae, Leptographium** 760
waimeana, Asteridiella 580
wakefieldii, Lazulinospora 758
waksmanii, Penicillium 829
wakullum, Gloeocystidiellum 716
waldsteiniae, Gnomonia 721
waldsteiniae, Puccinia 916
waldsteiniae, Ramularia 928
waldsteiniae, Septoria 959
waldsteiniae, Ustacystis (Urocystis, Whetzelia) 1014
wallowaensis, Pseudosphaeria 898
washingtonensis, Albotricha (Dasyscyphus) 559
washingtonensis, Coriolus = **Oxyporus cuneatus** 824
washingtonensis, Tilletiopsis 992
washingtonensis, Tryblidaria 1002
washingtoniae, Metasphaeria 788
washingtoniae, Mycosphaerella = **M. tassiana** 806
weberi, Hysterostoma 747
websteri, Cercospora 618
wehmeyeri, Didymella 677
wehmeyeri, Leptosphaeria 764
wehmeyeriana, Winterella 1024
weigelae, Diplodia 682
weigelae, Mycosphaerella 807
weigeliae, Phoma 856
weigeliae, Pseudocercospora (Cercospora) 895
weiri, Peniophora = **Crustoderma dryinum** 650
weiriana, Rosellinia 937
weiriana, Septoria 959
weiriana, Valsa 1018
weiriana, Venturia (Phaeosphaerella) 1021
weirianum, (Gloeosporium, Platycarpium) = **Fusamen fructigenum** 706
weirianum, Xyloschizon 1027
weirianus, Phellinus (Fomes) 847
weirii, Acanthophysium (Aleurodiscus) 556
weirii, Chrysomyxa 627
weirii, (Clypeothecium, Exarmidium) = **Exarmidium morthieri** 702
weirii, Hericium = **H. abietis** 734
weirii, Peridermium 832
weirii, Phellinus (Fomitiporia, Inonotus, Poria) 847
weirii, Rhabdocline 930
weirii subsp. **oblonga, Rhabdocline** 930
weirii subsp. **obovata, Rhabdocline** 930
weldenii, Hypoxylon 746
wentii, Aspergillus 579
wentii, Macroventuria 776
wernsdorffiae, Coniothyrium 643

westendorpii, Septoria = **Phyllosticta chenopodii** 866
westii, Daedalea = **Antrodia juniperina** 567
westii, Polyporus = **P. squamosus** 889
westii, Steccherinum = **Stecchericium seriatum** 975
westonii, Helicoma = **H. palmigenum** 730
westonii, Megachytrium 780
whetstonei, Polyporus = **Spongipellis spumeus** 970
whetzelii, Ciborinia (Sclerotinia) 627
whippleae, Peronospora 834
whitmanensis, Diatrype 673
wichuriana, Mycosphaerella 807
wiesneri, Taphrina 987
wikstroemiae, Amazonia 563
wikstroemiae, Didymosphaeria 678
wikstroemiae, Uredo 1008
wilcoxiana, Tilletia = **T. hyalospora** 992
williamsii, Ustilago 1016
willkommii, Lachnellula (Dasyscyphus, Trichoscyphella) 754
wilsonii, Phyllachora = **P. paspalicola** 863
windsoriae, Puccinia 916
winelandii, Milesina 792
winteri, Entyloma 696
winteri, Nodulosphaeria (Leptosphaeria, Wettsteinina) 817
wisconsina, Ascochyta 577
wisconsina, Ramularia 928
wisconsina, Septoria 959
wisconsinensis, Cercospora 618
wisconsinensis, Ciborinia 627
wisconsinensis, Coniothyrium 643
wisconsinensis, Elsinoe 693
wisconsinensis, Phyllosticta 875
wisconsiniense, Entyloma = **E. compositarum** 695
wislizenii, Phyllosticta 875
wistariae, Aplosporella 571
wistariae, Botryosphaeria (Melogramma) 593
wistariae, Hendersonia 734
wistariae, Phomatospora 857
wistariae, Phyllosticta 875
wistariae, Septoria 959
wistariana, Sphaeropsis = **Aplosporella wistariae** 571
wistarianum, Camarosporium 600
wisteriae, Cercospora 618
wittrockii, Phylleutypa (Dothidea, Phyllachora) 864
wolfii, Dothiora 688
wootoniana, Cumminsiella (Uropyxis) 655
wrightii, Octospora (Humaria, Lamprospora) 817
wrightii, (Perisporiopsis, Perisporium, Stevensea) = **Diplotheca tunae** 683
wulfeniae, Phyllosticta 875
wulfeniae, Puccinia 916
wulffiae, Cercospora 618
wyethiae, Didymaria (Marssonia, Marssonina) 676
wyethiae, Puccinia = **P. balsamorhizae** 901
wyethiae, Septoria 959
wyomingense, Entyloma = **E. winteri** 696
wyomingensis, Phoma 856
wyomingensis, Phyllosticta 875
wyomingensis, Sirexcipula 961
xantha, Antrodia (Amyloporia, Poria) 568
xanthicola, Cercospora 618
xanthicola, Mycosphaerella 807
xanthifolia, Septoria 959
xanthifoliae, Puccinia = **P. helianthi** 907
xanthii, Colletotrichum = **C. gloeosporioides** 638
xanthii, Puccinia 916
xanthii, Rhabdospora 931
xanthii, Septoria 959
xanthorrhizae, Phyllosticta 875
xanthostigma, Orbilia (Calloria, Mollisia) 823
xanthostroma, Melanconis (Valsa) 782
xanthoxyli, Cercospora 618
xanthoxyli, Dermea 670
xanthoxyli, Discula 685
xanthoxyli, Macrophoma 775
xanthoxyli, Massaria (Massariella) 779
xanthoxyli, Thyronectria 991
xerophila, Phanerochaete 844
xerophila, Teichospora = **T. aridophila** 987
xerophylli, Asteridium (Asterina) 580
xerophylli, Hendersonia 734
xerophylli, Leptosphaeria (Metasphaeria, Sphaerulina) 764
xerophylli, Leptostroma 765
xerophylli, Mycosphaerella 807
xerophylli, Pleospora = **Comoclathris permunda** 640
ximenesiae, Puccinia 916
xylebori, Ambrosiella 563
xylina, Ascotricha 578
xylita, Cistella 628
xylostei, Depazea = **Kabatia periclymeni** 751
xylostei, Phoma 856
xylostei, Septoria 959
xylostromatoides, Ceriporia (Poria) 621
xyridis, Cercospora 618
xyridis, Cladosporium 631
yamadai, Bipolaris (Helminthosporium) 589
yanagawaensis, Claviceps 632
yogoensis, Pyrenopeziza 919
yosemitana, Puccinia 916
youngii, Tilletia 992
yuccae FILE , Dothidea = **D. conspicua** 687
yuccae, Allantonectria 560
yuccae, Catenophora 601
yuccae, Cenangium 602
yuccae, Cercospora 619
yuccae, Discochora 684
yuccae, (Dothidea, Phaeodothis) = **Dothidea conspicua** 687
yuccae, Mycosphaerella (Sphaerella) 807
yuccae, Pestalozziella 837
yuccae, Piptarthron (Phyllachora) 880
yuccae, Septoria 959
yuccaegena, Phyllosticta = **Macrophoma smilacina** 774
yuccaegenum, Gloeosporium = **Colletotrichum gloeosporioides** 638
yuccifolia, Alpakesa (Neottiospora) 560
yuccigena, Kellermania 752
yuccogena, Didymosphaeria 678
zahlbruckneriana, Pestalotia 836
zamiae, Melanospora 784
zanthoxyli, Phyllosticta 875
zauschneriae, Puccinia = **P. oenotherae** 911
zeae, Ascochyta = **A. ischaemi** 575
zeae, Aureobasidium (Kabatiella) 585
zeae, Botryosphaeria (Physalospora) 593
zeae, Cladosporium 631
zeae, Clasterosporium = **Polyschema olivacea** 889
zeae, Coniothyrium = **C. scirpi** 643
zeae, Dendrophoma 667
zeae, (Diplodia, Macrodiplodia) = **Stenocarpella maydis** 977
zeae, Gibberella 715
zeae, Gonatobotrys 722
zeae, Leptosphaeria = **Paraphaeosphaeria michotii** 826
zeae, Leptothyrium 766
zeae, Macrophoma 775
zeae, Mycosphaerella 807
zeae, Perisporium 832
zeae, Phaeocytosporella = **Phaeocytostroma ambiguum** 839
zeae, Phyllosticta 875
zeae, Physopella (Angiopsora) 877
zeae, Rhizoctonia (Moniliopsis) 933
zeae, Rhopographus 934
zeae, Septoria 959
zeae, Stachybotrys 972
zeae, Stenocarpella = **S. macrospora** 977
zeae, Ustilago (Lycoperdon) 1017
zeae var. macrospora, Macrodiplodia = **Stenocarpella macrospora** 977
zeae-maydis, Cercospora 619
zeae-maydis, Diplodia = **Stenocarpella maydis** 977
zeae-maydis, Mycosphaerella 807
zeae-maydis, Physoderma = **P. maydis** 877
zebrina, Cercospora 619
zeicola, Ascochyta 577
zeicola, Bipolaris (Drechslera, Helminthosporium) 589
zeicola, Lophiosphaera 770
zeicola, Mycosphaerella = **M. zeae** 807
zeicola, Phoma 856
zeicola, Physalospora = **Botryosphaeria festucae** 592
zeicola, Pleosphaerulina = **Leptosphaerulina trifolii** 764
zeicola, Septoria 959
zeina, Septoria 959
zelkowae, Cercospora 619
zelleri, Aleurodiscus = **Peniophora aurantiaca** 829
zelleri, Exidia 702
zelleri, Godronia = **Atropellis pinicola** 584
zexmeniae, Puccinia 916
zingiberi, Cercospora 619
zingiberi, Coniothyrium 643
zingiberis, Phyllosticta 875
zinniae, Alternaria 563
zinniae, Cercospora 619
zizaniae, Claviceps (Spermoedia) 632
zizaniae, Doassansia 686
zizaniae, Mycosphaerella 807
zizaniae, Sclerotium 942
ziziae, Cercospora 619
ziziae, Cylindrosporium = **Cercospora thaspiicola** 618
ziziae, Puccinia 916
ziziae, Septoria 959
ziziphi, Cladosporium 631
zizyphi-vulgaris, Phakopsora 843
zonalis, Polyporus = **Rigidoporus lineatus** 936
zonalis, Rigidoporus = **R. lineatus** 936
zonata, Cercospora 619
zonata, Crinipellis (Collybia) 648
zonata, Junghuhnia (Poria) 751
zonata, Phyllosticta 875
zonata, Stagonospora 975
zonatum, Acremonium (Cephalosporium) 557
zonatum, Ganoderma 713
zonatum, Melanconium = **M. bicolor** 782
zopfii, Saccharomyces 938
zoysiae, Puccinia 916
zygadeni, Uromyces 1014
zygodesmoides, Tomentellopsis (Hypochnus) 995

Authors of Fungal Names

Aa, H.A. van der - **van der Aa**
Abadie, F. - **Abadie**
Abawi, George S. - **Abawi**
Abbott, Ernest Victor - **E. Abbott**
Abe, Seiji - **Abe**
Acedo, Gregoria N. - **Acedo**
Acharius, Erik - **Achar.**
Adam, David Bonar - **D.B. Adam**
Adams, James Fowler - **J.F. Adams**
Adanson, Michel - **Adans.**
Aderhold, Rudolph Ferdinand Theodor - **Aderhold**
Afzelius, Adam - **Afzel.**
Agardh, Carl Adolf - **C. Agardh**
Agerer, Reinhard - **Agerer**
Agnihothrudu, V. - **Agnihothrudu**
Agnihotri, V.P. - **Agnihotri**
Agrios, George N. - **Agrios**
Ahlburg - **Ahlburg**
Ahmad, Sultan - **Ahmad**
Ahmed, S. Iktikhar - **Ahmed**
Ainsworth, Geoffrey Clough - **Ainsworth**
Ajello, Libero - **Ajello**
Albertini, Johannes Baptista von - **Albertini**
Alcorn, John L. - **Alcorn**
Alexopoulos, Constantine J. - **Alexopoulos**
Alfieri, Salvatore A., Jr. - **Alfieri**
Alippi, Hector E. - **Alippi**
Allescher, Andreas - **Allesch.**
Allington, William B. - **Allington**
Almeida, José Verissimo de - **J.V. Almeida**
Al-Musallam, A. - **Al-Musallam**
Alvarez, Garcia - **Alvarez**
Ames, Adeline - **A. Ames**
Ames, Lawrence Marion - **L. Ames**
Ammon, H.U. - **Ammon**
Amos, Raymond E. - **R.E. Amos**
Anastasia, Giuseppe Emilio - **Anastasia**
Anastasiou, C.J. - **Anastasiou**
Anderson, Frederick William - **F.W. Anderson**
Anderson, Harvey Warren - **H.W. Anderson**
Anderson, Paul Johnson - **P.J. Anderson**
Anderson, Redreis B. - **R.B. Anderson**
Andres, Heinrich - **Andres**
Andrus, Charles Frederick - **Andrus**
Antonopoulos, Antonios A. - **A.A. Antonopoulos**
Anzalone, Louis - **Anzalone**
Aoshima, Kiyowo - **Aoshima**
Apinis Arvids E. - **Apinis**
Appel, Friedrich Carl Louis Otto - **Appel**
Aragaki, Minoru - **Aragaki**
Arcangeli, Giovanni - **Arcang.**
Archer, S.A. - **S.A. Archer**
Archer, William Andrew - **W.A. Archer**
Arenberg - **Arenberg**
Armstrong, George Miller - **G.M. Armstrong**
Armstrong, Joanne K. - **J.K. Armstrong**
Arnaud, Gabriel - **G. Arnaud**
Arnaud, Jean Andre Michel - **J. Arnaud**
Arnaud, Madeleine - **M. Arnaud**
Arnold, Ruth Horner - **R.H. Arnold**
Arnolds, E.E.F. - **E. Arnolds**
Arny, Deane C. - **D.C. Arny**
Artari, Alexander Petrovich - **Artari**
Arthur, John M. - **J.M. Arthur**
Arthur, Joseph Charles - **Arth.**
Arx, Josef Adloph von - **Arx**
Asano, Iasmu - **I. Asano**
Ascherson, Paul Fredrich August - **Aschers.**
Ashby, Sidney Francis - **S. Ashby**
Ashford, Bailey K. - **Ashford**
Athow, Kirk L. - **K.L. Athow**
Atienza, J.D. - **Atienza**
Atkinson, George Francis - **Atk.**
Auerswald, Bernhard - **Auersw.**
Awao, Takeyoshi - **Awao**
Ayers, Theodore Thomas - **Ayers**
Babcock, C.E. - **Babcock**
Babington, Charles Cardale - **Bab.**
Badea, M. - **Badea**
Badham, Charles David - **Badham**
Baglietto, Francesco - **Bagl.**
Bagyanarayana, G. - **Bagyanarayana**
Bailey, Floyd Douglas - **F.D. Bailey**
Bailey, Harold Edward - **H.E. Bailey**
Bain, Douglas Cogburn - **D. Bain**
Bain, Henry Franklin - **H. Bain**
Bain, Samuel McCutcheon - **Bain**
Baines, Richard Cecil - **Baines**
Bainier, Georges - **Bainier**
Baker, Charles Fuller - **C. Baker**
Baker, Gladys Elizabeth - **G. Baker**
Baker, Kenneth Frank - **K. Baker**
Baker, Richard Eric Defoe - **R.E.D. Baker**
Baker, Shirley D. - **S.D. Baker**
Bakke, Arthur Lawrence - **Bakke**
Bakshi, B. - **Bakshi**
Balbis, Giovanni-Batista - **Balb.**
Baldacci, Elio - **Baldacci**
Balsamo-Crivelli, Giuseppe Gabriel - **Bals.-Criv.**
Bancroft, Claude Keith - **Bancroft**
Bandoni, Robert J. - **Bandoni**
Banker, Howard James - **Banker**
Baral, Hans Otto - **Baral**
Baranyay, J.A. - **Baranyay**
Barbaini, Maria - **Barbaini**
Barbosa, Maria Antonieta de Freitas - **Barbosa**
Barbour, W. J. - **Barbour**
Barclay, Arthur - **Barclay**
Barghoorn, Elso - **Barghoorn**
Barker, Berthie Thomas Percival - **B.T. Barker**
Barker, J.S.F. - **J. Barker**
Barla, Joseph Hieronymus Jean Baptiste - **Barla**
Barnard, Edward L. - **E. Barnard**
Barnett, Horace L. - **H.L. Barnett**
Baroni, Timothy J. - **T.J. Baroni**
Barr, Margaret E. - **Barr**
Barras, Stanley J. - **S.J. Barras**
Barrett, James Theophilus - **Barrett**
Barron, George L. - **Barron**
Barrus, Mortier F. - **Barrus**
Barthelet, J. - **Barthelet**
Bartholomew, Elam - **Barth.**
Bartlett, Albert W. - **A.W. Bartlett**
Bary, Heinrich Anton de - **de Bary**
Bas, Cornelius - **Bas**
Basgal, W. - **Basgal**
Bataille, Frederic - **Bataille**
Batista, Augusto Chaves - **Batista**
Batra, Lekh Raj - **Batra**
Batsch, August Johann Georg Carl - **Batsch**
Battarra, Giovanni Antonio - **Battarra**
Baudys, Eduard - **Baudys**
Baumgarten, Johann Christian Gottlob - **Baumg.**
Bäumler, Johann Andreas - **Bäumler**
Baxter, Dow Vauter - **D. Baxter**
Baxter, John W. - **J.W. Baxter**
Bayliss-Elliott, Jessie - **Bayliss-Elliott**
Beach, Walter Sturgeon - **Beach**
Beaumont, Albert - **A. Beaumont**
Beauvois, Ambroise Marie Francois Joseph Palisot - **P. Beauv.**
Beccari, C. - **Beccari**
Beck, Ritter von Mannagetta, Günther - **Beck**
Beckmann, Johann - **Beckm.**
Behera - **Behera**
Beijerinck, Martinus Willem - **M. Beijerinck**
Bell, Ann - **A. Bell**
Bell, Durham K. - **D.K. Bell**
Bell, Hugh Philip - **H.P. Bell**
Bellynck, Alexis Adolphe Alexandre - **Bellynck**
Beltrán, Francisco - **Beltrán**
Beltrani, Vito - **Beltrani**
Benjamin, Chester R. - **C.R. Benjamin**
Bennett, Alfred William - **A.W. Benn.**
Bennett, Carlye Wilson - **C.W. Bennett**
Bennett, Frederick Thomas - **F.T. Bennett**
Beraha, Louis - **L. Beraha**
Berenger, Giuseppe Adolfo de - **Berenger**
Berg, Anthony - **A. Berg**
Bergquist, Richard R. - **Bergquist**
Berkeley, Garven Hugh - **G.H. Berkeley**
Berkeley, Miles Joseph - **Berk.**
Berkhout, Christine Marie - **Berkhout**
Berlese, Augusto Napoleone - **Berl.**
Bernard, Charles Jean - **C. Bernard**
Bernard, Georges Eugene - **G. Bernard**
Bernard, Noel - **N. Bernard**
Bernhardt, Elizabeth A. - **Bernhardt**

Bertero, Carlo Guiseppe - **Bertero**
Berthier, Jacques - **Berthier**
Bertoloni, Giuseppe - **G. Bertol.**
Besl, H. - **Besl**
Bessey, Charles Edwin - **Bessey**
Bessey, Ernst Athearn - **E.A. Bessey**
Bestagno, G. - **Bestagno**
Bethel, Ellsworth - **Bethel**
Bezerra, Jose Luiz - **Bezerra**
Bhatt, G.C. - **G.C. Bhatt**
Bier, John Ertel - **Bier**
Biga, M.L.A. - **Biga**
Bigelow, Howard E. - **H.E. Bigelow**
Bilai, Vera Iosifovna - **Bilai**
Billings, John Shaw - **J.S. Billings**
Biourge, Philibert Melchior Joseph Ehislam - **Biourge**
Birge - **Birge**
Birkinshaw, J.H. - **Birkinshaw**
Bisby, Guy Richard - **Bisby**
Bissett, John - **Bissett**
Bitancourt, Agesilau Antonio - **Bitancourt**
Bivona-Bernardi, Antonio de - **Biv.-Bern.**
Bizzozero, Giacomo - **Bizz.**
Bjerkander, Clas - **Bjerk.**
Björling, Karl - **Björling**
Blackwell, Meredith - **Blackwell**
Blasdale, Walter Charles - **Blasdale**
Bliss, Donald Everett - **D.E. Bliss**
Blok, Ida - **Blok**
Bloxam, Andrew - **Bloxam**
Bloxam, Richard Roland - **R. Bloxam**
Blume, Carl Ludwig - **Blume**
Blumer, Samuel - **S. Blumer**
Blytt, Axel Gudbrand - **A. Blytt**
Boedijn, Karl Bernard - **Boedijn**
Boerema, Gerhard H. - **Boerema**
Boesewinkel, Hans J. - **Boesewinkel**
Boewe, Gideon Herman - **Boewe**
Boidin, Jacques - **Boidin**
Boise, Jean R. - **Boise**
Bolick - **Bolick**
Bolkan, Hasan A. - **Bolkan**
Bolle, Pierette Cornelie - **P.C. Bolle**
Bolley, Henry Luke - **Bolley**
Bolton, James - **Bolton**
Boltshauser, Heinrich - **Boltshauser**
Bommer, Elisa Caroline - **E. Bommer**
Bonar, Lee - **Bonar**
Bondartsev, Appolinaris Semenovich - **Bondartsev**
Bondartzeva-Monteverde, Vera Nikolaevna - **Bondartzeva-Monteverde**
Bonorden, Hermann Friedrich - **Bonord.**
Boone, Donald M. - **D.M. Boone**
Booth, Colin C. - **C. Booth**
Boothroyd Carl W. - **Boothroyd**
Boquiren, Daisy - **Boquiren**
Borisov, P.N. - **Borisov**
Born, Gerald L. - **G.L. Born**
Borrer, William J. - **Borrer**
Borzi, Antonino - **Borzi**
Bosc, Louis Auguste Guillaume - **Bosc**
Bose, Sunil Kumar - **Bose**
Boudier, Jean Louis Emile - **Boud.**
Boulanger, Emile - **Boul.**
Bourchier, Robert James - **Bourchier**
Bourdot, Hubert - **Bourd.**
Bowerman, Constance A. - **Bowerman**
Boyd, Emma Sophia - **Boyd**
Boyer, Gaston - **G. Boyer**
Boyer, Léon - **Boyer**
Boyle - **Boyle**
Boyle, John Samuel - **J. Boyle**
Brader, L. - **L. Brader**
Brandza, Marcel Alex - **M. Brandza**
Brasfield, Travis W. - **Brasfield**
Braun, Alexander Carl Heinrich - **A. Braun**
Braun, Harry - **H. Braun**
Braun, Uwe - **U. Braun**
Braverman, Samuel W. - **Braverman**
Brebeck, Carl - **Brebeck**
Breda de Haan, Jacob van - **Breda de Haan**
Brefeld, Julius Oscar - **Bref.**
Bremer Widar - **W. Bremer**
Brenckle, Jacob Frederick - **Brenckle**
Bresadola, Don Giacopo - **Bres.**
Bresinsky, Andreas - **Bresinsky**
Bretz, Theodore Walter - **T.W. Bretz**
Brewster, M.S. - **M.S. Brewster**
Briard, Pierre Alfred - **Briard**
Brick, Carl - **C. Brick**
Bridges, J. Robert - **J.R. Bridges**
Briganti, Vincenzo - **V. Briganti**
Brinkmann, Wilhelm - **Brinkmann**
Briosi, Giovanni - **Briosi**
Brittlebank, Charles C. - **Brittlebank**
Britton, M.P. - **M.P. Britton**
Britton, Nathaniel Lord - **Britton**
Britzelmayr, Max - **Britzelm.**
Brockmann, Ingrid - **I. Brockmann**
Brockmüller, Hans Joachim Heinrich - **Brockm.**
Brodie, Harold J. - **Brodie**
Brondeau, Louis de - **Brond.**
Brongniart, Adolphe Théodore - **Brongn.**
Brooks, Albert Nelson - **A.N. Brooks**
Brooks, Charles - **C. Brooks**
Broome, Christopher Edmund - **Broome**
Brotero, Felix de Avellar - **Brot.**
Brown, Agnes H.S. - **A.H.S. Brown**
Brown, Clair A. - **C.A. Brown**
Brown, James Greenlief - **J. Brown**
Brown, Lawrence G. - **L.G. Brown**
Brown, Nellie Adalesa - **N.A. Brown**
Bruehl, George William - **G.W. Bruehl**
Bruhne, Karl - **Bruhne**
Brun, C. - **C. Brun**
Brunaud, Paul - **Brunaud**
Brunchorst, Jörgen - **Brunch.**
Bruner, Stephen Cole - **Bruner**
Bruschi, Diana - **D. Bruschi**
Bubák, Frantisek - **Bubák**
Buchanan, P.K. - **P.K. Buchanan**
Buchwald, Niels Fabritius - **Buchwald**
Bucknall, Cedric - **Buckn.**
Buddenhagen, Ivan W. - **Buddenhagen**
Buddin, Walter - **Buddin**
Budington, A.B. - **Budington**
Bugnicourt, F. - **Bugnicourt**
Buisman, Christine Johanna - **Buisman**
Buller, Arthur Henry Reginald - **Buller**
Bulliard, Jean Baptiste Francoise - **Bull.**
Bunce, M.E. - **Bunce**
Burch, Katherine B. - **K.B. Burch**
Burdsall, Harold H., Jr. - **Burdsall**
Burgeff, Hans Edmund Nicola - **Burgeff**
Burgess, Lester W. - **L.W. Burgess**
Buriticá, Pablo - **Buriticá**
Burkholder, Walter Hagemeyer - **Burkholder**
Burlingham, Gertrude Simmons - **Burl.**
Burnett, Harry C. - **H.C. Burnett**
Burpee, L.L. - **L.L. Burpee**
Burrill, Thomas Jonathan - **Burrill**
Burt, Catherine C. - **C.C. Burt**
Burt, Edward Angus - **Burt**
Burton, Mary Gwendolyn - **M.G. Burton**
Busch, Lloyd V. - **L.V. Busch**
Büsgen, Maritz - **Büsgen**
Butin, Heinz - **Butin**
Butler, Edward Eugene - **E.E. Butler**
Butler, Edwin John - **E.J. Butler**
Butler, Ellys Theodora - **E.T. Butler**
Byford, William John - **Byford**
Bynum, H.H. - **Bynum**
Caballero Y Segares, Arturo - **Caball.**
Cain, Roy Franklin - **Cain**
Caldwell, Ralph Merrill - **R.M. Caldwell**
Calkins, William Wirt - **Calk.**
Camara, Manuel de Sousa de - **Camara**
Campanile, Giulia - **Campanile**
Campbell, Leo - **L. Campbell**
Campbell, N.J. - **N.J. Campbell**
Campbell, William Andrew - **W.A. Campbell**
Candolle, Augustin Pyramus de - **DC.**
Candoussau, Francoise - **Candoussau**
Canfield, E.R. - **E.R. Canfield**
Cannon, Paul F. - **P. Cannon**
Canonaco, A. - **Canonaco**
Carestia, Antonio - **Carestia**
Carleton, Mark Alfred - **Carleton**
Carmichael, Dugald - **Carmich.**
Carmichael, John W. - **J.W. Carmichael**
Carne, Walter Merryn - **Carne**
Caroselli, Nestor E. - **Caroselli**
Carpenter, Clarence William - **C. Carpenter**
Carpenter, Steven E. - **S. Carpenter**
Carris, Lori M. - **Carris**
Carroll, George C. - **Carroll**
Carter, Carlos Newton - **C.N. Carter**
Carter, James Cedric - **J.C. Carter**
Carter, M.V. - **M.V. Carter**
Carus, Carl Gustav - **Carus**
Carver, George Washington - **G.W. Carver**
Cash, Edith Katherine - **Cash**
Caspary, Johann Xaver Robert - **Casp.**
Castagne, Jean Louis Martin - **Castagne**
Castellani, Aldo - **A. Castellani**
Castellani, Ettore - **Castellani**
Cattaneo, Achille - **Cattaneo**
Caum, Edward Leonard - **Caum**
Cavalcanti, Wlandemir de A. - **Cavalcanti**
Cavaliere, A.R. - **A.R. Cavaliere**
Cavara, Fridiano - **Cavara**
Cejp, Karel - **Cejp**
Cerny, Alois - **Cerny**
Cesati, Vincenzo de - **Ces.**
Chamberlain, Donald William - **D.W. Chamberlain**
Chamuris, George Peter - **Chamuris**
Chang, Ho-shii - **H.S. Chang**
Changsri, W. - **Changsri**
Chárdon Palacios, Carlos Eugenio - **Chárdon**
Charles, Vera K. - **V. Charles**
Chen, Chi-Cheng - **Chen**
Cheney, Gwendolyn M. - **G. Cheney**
Cheo, C.C. - **Cheo**
Chester, Frederick Dixon - **F. Chester**
Chesters, Charles G.C. - **Chesters**
Chevallier, Francois Fulgis - **Chev.**
Child, Marion (Moss) - **Child**
Childs, Leroy - **Childs**
Chilton, John E. - **J.E. Chilton**
Chimba, O. - **Chiba**
Chiu, Wei Fan - **Chiu**
Chivers, Authur Houston - **Chivers**
Chochrjakov, M. - **Chochrjakov**
Choisy, Jacques Denis - **Choisy**
Chowdhury, Shaista Begum - **Chowdhury**
Christensen, Clyde - **C. Christensen**
Christiansen, M.P. - **M. Christiansen**
Christiansen, Mogens Skytte - **M.S. Christiansen**

Chupp, Charles David - **Chupp**
Church, Margaret Brooks - **Church**
Churchland, L.M. - **Churchland**
Ciccarone, Antonio - **Ciccarone**
Ciferri, Raffaele - **Cif.**
Claflin, Larry E. - **Claflin**
Clapp, Grace L. - **Clapp**
Clark, Judson Freeman - **J.F. Clark**
Clausen, Roy Elwood - **R.E. Clausen**
Clements, Frederick Edward - **Clements**
Clendenin, Ida - **Clendenin**
Cline, Molly Niedbalski - **M. Cline**
Cline, Steven D. - **S. Cline**
Clinton, George Perkins - **G.P. Clinton**
Clinton, George William - **G.W. Clinton**
Clusius, Carolus - **Clus.**
Cobb, Fields W., Jr. - **F.W. Cobb**
Cobb, Nathan Augustus - **N.A. Cobb**
Cocconi, Girolamo - **Cocc.**
Cockerell, Theodor Dru Alison - **Cockerell**
Coemans, Henri Eugene Lucien Gaëtan - **Coem.**
Cohen, Sylvan Irving - **S.I. Cohen**
Cohn, Ferdinand Julius - **Cohn**
Coker, William Chambers - **Coker**
Colby, Arthur Samuel - **Colby**
Cole, Gary T. - **G.T. Cole**
Cole, John Rufus - **J.R. Cole**
Coleman, Leslie Charles - **L. Coleman**
Conant, Norman Francis - **Conant**
Conn, Harold Joel - **Conn**
Conners, Ibra Lockwood - **Conners**
Constantinescu, Ovidiu - **O. Constantinescu**
Conway, Kenneth E. - **K. Conway**
Cook, Melville Thurston - **M.T. Cook**
Cooke, John C. - **J.C. Cooke**
Cooke, Mordecai Cubitt - **Cooke**
Cooke, R. James - **R.J. Cooke**
Cooke, William Bridge - **W.B. Cooke**
Cooney, Donald G. - **Cooney**
Coons, George Herbert - **Coons**
Copeland, Edwin Bingham - **Copeland**
Copeland, Herbert F. - **H. Copel.**
Coppins, B.J. - **Coppins**
Corda, August Karl Joseph - **Corda**
Cordas, David I. - **Cordas**
Corden, Malcolm - **M. Corden**
Cordier, Fancois Simon - **Cordier**
Cordley, Arthur Burton - **Cordley**
Corlett, Michael - **Corlett**
Corner, Edred John Henry - **Corner**
Cornu, Marie Maxime - **Cornu**
Costa, C.A. Amaral - **C.A. Costa**
Costa, P. - **P. Costa**
Costantin, Julien Noël - **Costantin**
Couch, John Nathaniel - **Couch**
Coville, Frederick Vernon - **Coville**
Cox, Darrell E. - **Cox**
Cragin, Francis Whittemore - **Cragin**
Cralley, Elza Monroe - **Cralley**
Crandall, Bowen Sinclair - **Crandall**
Crane, J. Leland - **J.L. Crane**
Creager, Don B. - **D.B. Creager**
Creelman, D.W. - **Creelman**
Cribb, A.B. - **A.B. Cribb**
Cribb, Joan W. - **J.W. Cribb**
Crivelli, Paolo Giuseppe - **P.G. Crivelli**
Crompton, J.G. - **J.G. Crompton**
Crosier - **Crosier**
Crouan, Hippolyte Marie - **H. Crouan**
Crouan, Pierre Louis - **P.L. Crouan**
Crowell, Ivan Herrett - **Crowell**
Cruchet - **Cruchet**
Cuboni, Giuseppe - **Cuboni**
Cugini, Gino - **Cugini**
Cummins, George Baker - **Cummins**
Cunnell, G.J. - **Cunnell**
Cunningham, David Douglas - **D.D. Cunn.**
Cunningham, Gordon Herriot - **G. Cunn.**
Cunningham, John L. - **J.L. Cunningham**
Currey, Fredrick - **Curr.**
Currie, James Nimrod - **Currie**
Curtis, John Thomas - **J. Curtis**
Curtis, Moses Ashley - **M.A. Curtis**
Curtis, William - **Curtis**
Curzi, Mario - **Curzi**
Cutter, Victor Macombes - **V.M. Cutter**
D.B. Slope - **D. Slope**
Dade, Harry Authur - **Dade**
Dal Vesco, Giovanna - **Dal Vesco**
Dalbey, Nora Elizabeth - **N. Dalbey**
Dale, W.T. - **W.T. Dale**
Damon, Samuel C. - **Damon**
Dana, Bliss F. - **Dana**
Dangeard, Pierre Augustin - **P.A. Dang.**
Daniels, Eve Ypin - **E.Y. Daniels**
Darker, Grant Dooks - **Darker**
Darling, Louise - **L. Darling**
Das, B.K. - **B.K. Das**
Dastur, Jehangir Fardunji - **Dastur**
David, Alix - **A. David**
Davidson, John G.N. - **J.G.N. Davidson**
Davidson, Ross Wallace - **R.W. Davidson**
Davis, Alva Raymond - **A.R. Davis**
Davis, Benjamin Harold - **B.H. Davis**
Davis, John Jefferson - **J.J. Davis**
Davis, Lily H. - **L.H. Davis**
Davis, William Harold - **W.H. Davis**
Dawson, Marion Lucille - **M.L. Dawson**
De Hoog, G.S. - **De Hoog**
De Notaris, Giuseppe - **De Not.**
De Toni, Giovanni Batista - **De Toni**
De Vries, Gerardus Albertus - **G.A. De Vries**
De Wildeman, Emile Auguste Joseph - **De Wild.**
DeVay, J.E. - **DeVay**
Deacon J.W. - **Deacon**
Dearden, Elizabeth Ruth - **Dearden**
Dearness, John - **Dearn.**
Defago, Genevieve - **Defago**
Deighton, Charles Fredrick - **Deighton**
Delacroix, Edouard Georges - **Delacr.**
Demange, Victor - **Demange**
Demaree, Juan Brewer - **Demaree**
Demidoff, Anatole - **Demidoff**
Demme, R. - **R. Demme**
Denison, William C. - **Denison**
Dennis, Richard William George - **Dennis**
Derx, H.G. - **Derx**
Descals, E. - **E. Descals**
Desjardin, Dennis E. - **Desjardin**
Desmazieres, Jean Baptiste Henri Joseph - **Desmaz.**
Desvaux, Nicaise Auguste - **Desv.**
Dharne, Chandrashekar Ganesh - **Dharne**
DiCosmo, Frank - **DiCosmo**
Dias, Maria Rosália de Sousa - **Dias**
Dickinson, C.H. - **C.H. Dickinson**
Dickson, Jacobus - **Dicks.**
Dickson, James - **Dickson**
Diddens, Harmanna - **Diddens**
Diedicke, Hermann - **Died.**
Diehl, William Webster - **Diehl**
Dierckx, Pere - **Dierckx**
Dietel, Paul - **Dietel**
Dietrich, Albert Gottfried - **A. Dietr.**
Dietz, Samuel M. - **S. Dietz**
Dimock, A.W. - **Dimock**
Dingley, Joan M. - **J.M. Dingley**
Ditmar, L.P. - **Ditmar**
Dixon, Dennis M. - **D.M. Dixon**
Dixon, John R. - **J.R. Dixon**
Dixon, Marjorie - **M. Dixon**
Doassans, E. - **Doassans**
Dobrozrakova, Taisiia Leonidovna - **Dobrozrakova**
Dodge, Bernard Ogilvie - **B.O. Dodge**
Dodge, Carroll William - **C.W. Dodge**
Doi, Y. - **Doi**
Doidge, Ethel Mary - **Doidge**
Dolan, Desmond - **Dolan**
Dománski, Stanislaw - **Dománski**
Don, David - **D. Don**
Donk, Marinus Anton - **Donk**
Dorenbosch, Maria M. J. - **Dorenbosch**
Dörfelt, Heinrich - **Dörfelt**
Dorogin, Georgi Nikolaevich - **Dorogin**
Doroguine, M. - **Doroguine**
Dorsett, Palemon Howard - **Dorsett**
Doty, Maxwell Stanford - **Doty**
Dovaston, H.F. - **Dovaston**
Dowding, Eleanor Silver - **Dowding**
Dowson, Walter John - **Dowson**
Drayton, Frank Lisle - **Drayton**
Drechsler, Charles - **Drechs.**
Dubin, Jesse - **Dubin**
Dubois, Francois Noel Alexandre - **Dubois**
Dubos, Bernadette - **B. Dubos**
Duby, Jean Etienne - **Duby**
Duche, Jacques - **Duche**
Dudka, I.A. - **I.A. Dudka**
Dudley, William Russell - **W.R. Dudley**
Dufour, Jean-Marie Leon - **J.L. Dufour**
Dufour, Leon - **Dufour**
Duggar, Benjamin Minge - **Duggar**
Duke, Maude M. - **M.M. Duke**
Dumont, Kent P. - **Dumont**
Dumortier, Barthelemy Charles Joseph - **Dumort.**
Dunegan, John Clymer - **Dunegan**
Duniway, John M. - **Duniway**
Dunlap, Albert Atkinson - **A.A. Dunlap**
Durán, Rubén - **Durán**
Durand, Elias Judah - **E.J. Durand**
Durieu de Maisonneuve, Michel Charles - **Durieu**
Duss, Pere Antoine - **Duss**
Duvernoy, A. - **Duvernoy**
Dyko, Barbara J. - **Dyko**
Earle, Franklin Summer - **Earle**
Eaton, Amos - **Eaton**
Eboh, Dan - **Eboh**
Eckblad, Finn-Egil - **Eckblad**
Eddy, Emery D. - **E.D. Eddy**
Edgerton, Claude Wilbur - **Edgerton**
Edson, Howard Austin - **Edson**
Edwards, Eric Thomas - **E. Edwards**
Egeland, John - **Egeland**
Egger, Keith N. - **Egger**
Ehrenberg, Christian Gottfried - **Ehrenb.**
Ehrhart, Jakob Friedrich - **Ehrh.**
Ehrlich, John - **J. Ehrlich**
Eidam, Michael Emil Howard - **Eidam**
El-Gholl, Nabih E. - **El-Gholl**
Eliasson, Albin Gottfrid - **Eliasson**
Ellett, C. Wayne - **C.W. Ellett**
Elliott, Edward S. - **E.S. Elliott**
Elliott, John Asbury - **J.A. Elliott**
Elliott, Mary E. - **M.E. Elliott**
Ellis, Don Edwin - **D.E. Ellis**
Ellis, E.A. - **E.A. Ellis**
Ellis, Job Bicknell - **Ellis**
Ellis, John J. - **J.J. Ellis**

Ellis, John William - **J.W. Ellis**
Ellis, Martin B. - **M.B. Ellis**
Emerson, Ralph - **R. Emerson**
Emile-Weil, P. - **Emile-Weil**
Emmons, Chester Wilson - **C.W. Emmons**
Emoto, Yoshikadzu - **Y. Emoto**
Endlicher, Stephan Ladislaus - **Endl.**
Engelhard, Arthur W. - **Engelhard**
Engelmann, Georg - **Engelm.**
Engler, Heinrich Gustav Adolf - **Engl.**
English, William Harley - **English**
Enjoji, Sadao - **Enjoji**
Enlows, Ella Morgan - **Enlows**
Eriksson, Birgitta - **B. Eriksson**
Eriksson, Jakob - **Eriks.**
Eriksson, John - **J. Eriksson**
Eriksson, Ove - **O. Eriksson**
Ervin, Marion D. - **Ervin**
Erwin, Donald C. - **D.C. Erwin**
Eschweiler, Franz Gerhard - **Eschw.**
Eslyn, Wallace E. - **Eslyn**
Essary, Samuel Henry - **Essary**
Etheridge, D.E. - **D.E. Etheridge**
Ettlinger, Leopold - **Ettlinger**
Everhart, Benjamin Matlack - **Everh.**
Exner, Beatrice - **Exner**
Ezekiel, Walter N. - **Ezekiel**
Fabre, Jean Henri Casimir - **Fabre**
Fabricius, Ludwig - **L. Fabricius**
Fahrendorff, Ernst - **Fahrendorff**
Fairman, Charles Edward - **Fairm.**
Fall, Joan - **J. Fall**
Fang, C.T. - **C.T. Fang**
Faris, James A. - **Faris**
Farlow, William Gilson - **Farl.**
Farr, David Frederick - **D.F. Farr**
Farr, Ellen - **E. Farr**
Farr, Marie L. - **M.L. Farr**
Fassatiová, Olga - **Fassatiová**
Faull, Joseph Horace - **Faull**
Fautrey, F. - **Fautrey**
Favre, Jules - **J. Favre**
Fawcett, Howard Samuel - **H. Fawc.**
Fayod, Victor - **Fayod**
Feldman, Genevieve - **G. Feldman**
Fell, J.W. - **J. Fell**
Fell, Jack W. - **Fell**
Feltgen, Johann - **Feltgen**
Fennell, Dorthy Irene - **Fennell**
Ferdinandsen, Carl Christian Frederick - **Ferd.**
Fergus, Charles L. - **C.L. Fergus**
Ferraris, Teodoro - **Ferraris**
Ferry, Rene - **Ferry**
Feuillebois, P.V.A. - **Feuillebois**
Fidalgo, Maria Eneyda Pacheco Kauffman - **M. Fidalgo**
Fidalgo, Oswaldo - **Fidalgo**
Field, Ethel Content - **E.C. Field**
Filer, Theodore H. - **Filer**
Fingerhuth, Carl Anton - **Fingerh.**
Finley, David E. - **Finley**
Fiori, Adriano - **Fiori**
Fisch, Carl - **C. Fisch**
Fischer von Waldheim, Alexander Alexandrovitch - **A. Fisch. v. Waldh.**
Fischer, Alfred - **A. Fischer**
Fischer, Bernhard - **B. Fischer**
Fischer, Eduard - **E. Fisch.**
Fischer, George William - **G. Fisch.**
Fischer, Johann Carl - **J.C. Fisch.**
Fisher, Eileen E. - **E.E. Fisher**
Fisher, Fran E. - **F.E. Fisher**
Fitzpatrick, Harry Morton - **Fitzp.**
Fleischhack - **Fleischhack**
Flentje, N.T. - **Flentje**
Flotow, Julius Christian Gottlieb Ulrich Gustav Georg von - **Flot.**
Foister, Charles Edward - **Foister**
Fosberg, Francis Raymond - **Fosb.**
Foster, A.A. - **A.A. Foster**
Foster, Arthur Crawford - **A.C. Foster**
Foster, Virginia - **V. Foster**
Foster, William Robert - **W.R. Foster**
Fragoso, Romualdo Gonzalez - **Gonz. Frag.**
Francis, Sheila M. - **S. Francis**
Frandsen, N.O. - **Frandsen**
Frank, Albert Bernhard - **A.B. Frank**
Fraser, Elizabeth M. - **E.M. Fraser**
Fraser, William Pollock - **W.P. Fraser**
Freeman, Glenn W. - **G. Freeman**
Fresenius, Johann Baptist Georg Wolfgang - **Fresen.**
Freyer, K. - **Freyer**
Frezzi, Mariano J. - **Frezzi**
Fries, Elias Magnus - **Fr.**
Fries, Theodor Magnus - **Th. Fr.**
Fromme, Fred Denton - **Fromme**
Fron, Georges - **Fron**
Frost, Charles Christopher - **Frost**
Fuckel, Karl Wilhelm Gottlieb Leopold - **Fuckel**
Fujii, H. - **Fujii**
Fukui, Takeji - **Fukui**
Fukushi, Teikichi - **Fukushi**
Funck, Heinrich Christian - **Funck**
Funk, Alvin - **Funk**
Furukawa, Hisahiki - **H. Furukawa**
Futrell, Maurice C. - **M.C. Futrell**
Fyles, Faith - **Fyles**
Gaertner, Joseph - **Gaertn.**
Gaillard, Albert - **Gaillard**
Galea, V.J. - **Galea**
Galloway, Beverly Thomas - **B.T. Galloway**
Galloway, Leslie Douglas - **L.D. Galloway**
Galzin, Amedee - **Galzin**
Gams, Walter - **W. Gams**
Garassini, Luiz A. - **Garassini**
Garcia-Rada, German - **Garcia-Rada**
Gardner, Donald E. - **D.E. Gardner**
Gardner, Max W. - **M.W. Gardner**
Garovaglio, Santo - **Garov.**
Garrett, Albert Osbun - **Garrett**
Gasparrini, Gulielmo - **Gasp.**
Gaudin, L. - **L. Gaudin**
Gäumann, Ernest Albert - **Gäum.**
Gay, Claude - **Gay**
Géneau de Lamarlière, L. - **Géneau de Lamarlière**
Georg, Lucille K. - **L.K. Georg**
Georgescu, C.C. - **Georgescu**
Gerard, William Ruggles - **W. Gerard**
Gerdemann, J.W. - **J.W. Gerdemann**
Gerlach, Walter - **Gerlach**
Germano, E.G. - **E.G. Germano**
Gessner, Robert V. - **R.V. Gessner**
Ghesquière, Jean - **Ghesq.**
Gibson, I.A.S. - **I. Gibson**
Gibson, Shelagh - **S. Gibson**
Gibson, William Hamilton - **W. Gibson**
Giddens, J.E. - **J.E. Giddens**
Giesenhagen, Karl Friedrich Georg - **Giesenh.**
Gilbert, Edward Jean - **E.J. Gilbert**
Gilbertson, Robert L. - **R.L. Gilbertson**
Gilkey, Helen Margaret - **Gilkey**
Gill, D.L. - **D.L. Gill**
Gilles, G. - **Gilles**
Gillet, Claude Casimir - **Gill.**
Gilliam, Martina S. - **Gilliam**
Gillor, Francois Xavier - **Gillot**
Gilman, Joseph C. - **Gilman**
Ginns, James H. - **Ginns**
Gjaerum, Halvar B. - **Gjaerum**
Glawe, Dean A. - **Glawe**
Glick, Perry Aaron - **Glick**
Gloyer, Walter Oscar - **Gloyer**
Gmelin, Johann Friedrich - **J.F. Gmelin**
Godfrey, George Harold - **Godfrey**
Goheen, Donald J. - **D.J. Goheen**
Goidanich, Gabriel - **Goidanich**
Gold, Harvey S. - **H.S. Gold**
Golovin, I.N. - **Golovin**
Gooch, F.S. - **Gooch**
Goodding, Leslie Newton - **Goodd.**
Goos, Roger D. - **Goos**
Gordon, C.C. - **C.C. Gordon**
Gordon, William Laurence - **W.L. Gordon**
Goree, Harold - **H. Goree**
Gorenz, August Mark - **Gorenz**
Gottwald, Tim R. - **Gottwald**
Gough, Francis J. - **Gough**
Gould, Charles Jay, Jr. - **C.J. Gould**
Govindu, Heirehalli Chenniah - **Govindu**
Graff, Paul Weidemeyer - **Graff**
Graham, Joseph H. - **J.H. Graham**
Graves, Arthur Harmount - **Graves**
Gray, Elizabeth - **E. Gray**
Gray, Samuel Fredrick - **S.F. Gray**
Greene, Edward Lee - **Greene**
Greene, Henry Campbell - **H.C. Greene**
Greenhalgh, G.N - **Greenhalgh**
Gregor, Mary J.F. - **M. Gregor**
Gregory, Charles Truman - **C.T. Gregory**
Gregory, Philip Herries - **P. Gregory**
Gremmen, J. - **Gremmen**
Greville, Robert Kaye - **Grev.**
Griffin, H.D. - **H.D. Griffin**
Griffin, M.J. - **M.J. Griffin**
Griffiths, David - **D. Griffiths**
Griffon, Edouard - **Griffon**
Griggs, Robert Fiske - **Griggs**
Grigorieva-Manoilova, O.C. - **Grigorieva-Manoilova**
Grinchenko, A.H. Hertha - **Grinchenko**
Grodsinski, Leon - **Grods.**
Grognot, Camille - **Grognot**
Grosmann, Helene - **Grosmann**
Gross, Henry L. - **H. Gross**
Grossenbacher, John Gasser - **Gross.**
Grout, Abel Joel - **Grout**
Grove, William Bywater - **Grove**
Groves, James Walton - **Groves**
Gruff, Susan - **S. Gruff**
Guba, Emil Frederick - **Guba**
Guéguen, Fernand Pierre Joseph - **Guég.**
Guépin, Jean Pierre - **Guépin**
Guerra, Paul - **Guerra**
Guilliermond, Alexandre - **Guilliermond**
Gunnell, Pamela S. - **P.S. Gunnell**
Gunnell, published with M.B. Ellis - **Gunnell**
Guo, Ying-lan - **Guo**
Güssow, Hans Theodor - **H.T. Güssow**
Guthrie, E.J. - **Guthrie**
Gutsevich, S. A. - **Gutsevich**
Gutteridge, R.J. - **R. Gutteridge**
Guyot, A. Lucien - **Guyot**
Gvritischvili, M.N. - **Gvritischvili**
Haasis, Frank Arling. - **Haasis**
Hafellner, Joseph - **Hafellner**
Hafiz Khan, Abdul - **Hafiz Khan**
Hagedorn, Donald J. - **D.J. Hagedorn**

Hagelstein, Robert - **Hagelst.**
Hagem, Oscar - **Hagem**
Hagström, Elisabeth - **Hagström**
Hahn, Glen Gardener - **Hahn**
Hahn, Gotthold - **G. Hahn**
Haines, John - **J. Haines**
Hall, Constant Johan Jacob van - **J.C. Hall**
Hall, Harlow H. - **H.H. Hall**
Hall, John Galentine - **J.G. Hall**
Hall, Joyce A. - **J.A. Hall**
Hallenberg, Nils - **Hallenberg**
Haller, Albrect von - **Haller**
Hallett, Ian C. - **I.C. Hallett**
Halsted, Bryon David - **Halst.**
Hamers, M.E.C. - **Hamers**
Hamm, Phil B. - **P.B. Hamm**
Hammer, Bernard Wernick - **B. Hammer**
Hammill, Terrence M. - **T. Hammill**
Hampe, Georg Ernst Ludwig - **Hampe**
Hanlin, Richard T. - **Hanlin**
Hanna, M. - **Hanna**
Hannon, Chancellor I. - **Hannon**
Hansbrough, John Raymond - **Hansbrough**
Hansen, Emil Christian - **E. Hans.**
Hansen, Everett M. - **E.M. Hans.**
Hansen, Hans Nicholas - **H.N. Hans.**
Hansen, Lise - **L. Hans.**
Hansford, Clifford Gerald - **Hansf.**
Hara, Kanesuke - **K. Hara**
Hard, Miron Elisha - **Hard**
Hardison, John Robert - **Hardison**
Hariot, Paul Auguste - **Har.**
Harkness, Harvey Wilson - **Harkn.**
Harr, Jost - **J. Harr**
Harrington, Thomas C. - **T.C. Harrington**
Harris, Hubert Andrew - **H.A. Harris**
Harris, Richard C. - **R.C. Harris**
Harrison, Arthur Leslie - **A.L. Harrison**
Harrison, F.C. - **F.C. Harrison**
Harrison, J.L. - **J.L. Harrison**
Harrison, Kenneth Archibald - **K.A. Harrison**
Harter, Leonard Lee - **Harter**
Hartig, Heinrich Julius Adolph Robert - **R. Hartig**
Harting, Pieter - **Harting**
Harvey, Francis Leroy - **Harvey**
Harvey, John M. - **J.M. Harvey**
Harz, Carl(Karl) Otto - **C. Harz**
Hashioka Yoshio - **Hashioka**
Hasija, S.K. - **Hasija**
Hasselbring, Heinrich - **Hasselbring**
Hauerslev, K. - **Hauerslev**
Hawker, Lilian E. - **Hawker**
Hawksworth, David L. - **D. Hawksworth**
Hawley, Henry Cusack Wingfield - **Hawley**
Hawthorne, P.L. - **Hawthorne**
Haythorn, J.M. - **J.M. Haythorn**
Hazslinszky, von Hazslin, Friedrich August - **Hazsl.**
Heald, Frederick De Forest - **Heald**
Hebert, Teddy T. - **T.T. Hebert**
Hedgcock, George Grant - **Hedgc.**
Hedges, Florence - **Hedges**
Hedjaroude, Ghorban-Ali - **Hedjaroude**
Hedwig, Johann - **Hedw.**
Hedwig, Romanus Adolf - **R. Hedw.**
Heim, Roger Jean - **R. Heim**
Hein, Burghard - **B. Hein**
Heinsen, Ernst - **Heinsen**
Hektoen, Ludwig - **L. Hektoen**
Hemmi, Takewo - **Hemmi**
Henderson, Douglas Mackay - **D.M. Henderson**
Hendrix, James W. - **J.W. Hendrix**
Hennebert, Gregoire L. - **Hennebert**
Hennen, Joe F. - **J. Hennen**
Hennings, Ernst Johan - **E. Henn.**
Hennings, Paul Christoph - **Henn.**
Henrard, Paul - **P. Henrard**
Henry, Arthur Wellesley - **A.W. Henry**
Henry, Berch Waldo - **B.W. Henry**
Hepp, Johann Adam Philipp - **Hepp**
Hepting, George Henry - **Hepting**
Herbst, William - **Herbst**
Hermanides-Nijhof, E.J. - **Hermanides-Nijhof**
Herpell, Gustav Jacob - **Herp.**
Herrera, M.P. - **Herrera**
Hesler, Lexemuel Ray - **Hesler**
Hess, Hans - **H. Hess**
Hesse, Rudolph - **R. Hesse**
Hesseltine, Charles W. - **Hesseltine**
Heufler, Ludwig Joseph Ritter von - **Heufl.**
Hickman, C.J. - **C.J. Hickman**
Hicks, Gilbert H. - **G.H. Hicks**
Higgins, Bascombe Britt - **Higgins**
Hilber, Oswald - **O. Hilber**
Hilber, Ruzena - **R. Hilber**
Hildebrand, Alexander Anderson - **A.A. Hildebrand**
Hilitzer, Alfred C. - **Hilitzer**
Hill, John - **J. Hill**
Hinds, Thomas E. - **T.E. Hinds**
Hino, Iwao - **Hino**
Hiratsuka, Naohide - **Hiratsuka**
Hiratsuka, Toshiko - **T. Hiratsuka**
Hiratsuka, Yasuyuki - **Y. Hiratsuka**
Hiroë, Isamu (Matsuura) - **Hiroë**
Hirschhorn, Elisa - **Hirschhorn**
Hitomi, Tsuyoshi - **Hitomi**
Hjortstam, Kurt - **Hjortstam**
Hobbs, Thomas W. - **T.W. Hobbs**
Hochapfel, Heinz - **Hochapfel**
Hodges, Charles S. Jr. - **C.S. Hodges**
Hodgson, Elmer Reed - **E.R. Hodgson**
Hoerl, Ruth A. - **Hoerl**
Hoes, John A. - **J.A. Hoes**
Hoffmann, Georg Franz - **Hoffm.**
Hoffmann, Heinrich Karl Hermann - **H. Hoffm.**
Höhnel, Franz Xaver Rudolf, Ritter von - **Höhn.**
Höhnk, Willy - **W. Höhnk**
Holcomb, Gordon E. - **G.E. Holcomb**
Holliday, Paul - **Holliday**
Hollós, Ladislaus (Lászlo) - **Hollós**
Hollrung, Max - **Hollrung**
Holm, Kerstin - **K. Holm**
Holm, Lennart - **L. Holm**
Holmskjold, Theodor - **Holmsk.**
Holubová-Jechová, Vera - **Holubová-Jechová**
Holway, Edward Willet Dorland - **Holw.**
Homma, Yasu - **Homma**
Hone, Daisy S. - **Hone**
Honey, Edwin Earle - **Honey**
Hooker, William Jackson - **Hook.**
Hoozemans - **Hoozemans**
Hora, Frederich Bayard - **Hora**
Hori, Shotaro - **S. Hori**
Horn, Norman L. - **Horn**
Hornby D. - **D. Hornby**
Horne, Arthur S. - **A.S. Horne**
Horne, William Titus - **W.T. Horne**
Hornemann, Jens Wilken - **Hornem.**
Horner, Harry T., Jr. - **H.T. Horner**
Horsfall, James G. - **Horsfall**
Hotson, Hugh Howison - **H.H. Hotson**
Hotson, John William - **Hotson**
House, Homer Doliver - **House**
Houston, Byron Robinson - **Houston**
Howe, Elliot Calvin - **E. Howe**
Howe, Marshall Avery - **M. Howe**
Huang, Keng Tang - **K. Huang**
Huber, Glenn Anthony - **G.A. Huber**
Hubert, Ernest Everett - **Hubert**
Hudson, H.J. - **H.J. Hudson**
Hudson, William - **Huds.**
Hughes, Gilbert C. - **G.C. Hughes**
Hughes, Stanley J. - **S.J. Hughes**
Hugueney, Robert - **Hugueney**
Huhtinen, Seppo - **Huhtinen**
Hulbary, Robert Louis - **Hulbary**
Hume, Hardrada Harold - **H. Hume**
Humphrey, Clarence John - **C.J. Humphrey**
Humphrey, James Ellis - **J.E. Humphrey**
Hunt, John - **J. Hunt**
Hunt, Nicholas Rex - **N. Hunt**
Hunt, Willis Robert - **W. Hunt**
Hunter, Lillian Mary - **L.M. Hunter**
Husain, Syed M. - **S.M. Husain**
Hussain, S.S. - **Hussain**
Hussey, A.M. - **Hussey**
Hutchinson, Wesley Gillis - **W.G. Hutchinson**
Hütter, Ralf - **Hütter**
Hylander, Nils - **Hyl.**
Ichinoe, Masakatsu - **Ichin.**
Ideta, Arata - **Ideta**
Ikata, Suehiko - **Ikata**
Illman, William I. - **Illman**
Imai, Sanshi - **Imai**
Imazeki, Rokuya - **Imazeki**
Imle, Ernest Paul - **Imle**
Ing, Bruce - **Ing**
Ingold, Cecil Terence - **Ingold**
Ingram, Edwin G. - **E.G. Ingram**
Inui, Tamaki - **Inui**
Inzenga, Giuseppe - **Inz.**
Irwin, John A. G. - **J.A.G. Irwin**
Isaac, Ivor - **I. Isaac**
Issatchenko, V. - **Issatchenko**
Ito, Seiya - **Ito**
Jaap, Otto - **Jaap**
Jackson, Albert Bruce - **A.B. Jackson**
Jackson, C.R. - **C.R. Jackson**
Jackson, Herbert Spencer - **H. Jacks.**
Jackson, Lyle Wendell Redverse - **L.W.R. Jackson**
Jacky, Ernst - **Jacky**
Jacques, Joseph Emile - **J. Jacques**
Jacquin, Nikolaus Joseph, Baron von - **Jacq.**
Jaczewski, Arthur Louis Arthurovic - **Jacz.**
Jagger, Ivan Claude - **Jagger**
Jahn, Eduard Adolf Wilhelm - **E. Jahn**
Jain, Prakash Chandra - **P.C. Jain**
Jamalainen, E.A. - **Jamalainen**
James, Joseph Francis - **J. James**
Jamieson, Clara Octavia - **Jamieson**
Janczewski, Eduard von Glinka - **Jancz.**
Jarvis, William R. - **W.R. Jarvis**
Jauch, Clothile - **Jauch**
Jeffers, Walter F. - **Jeffers**
Jenkins, Anna Eliza - **Jenk.**
Jenkins, S. F., Jr. - **S.F. Jenkins**
Jenkins, Wilbert Armonde - **W.A. Jenkins**
Jennings, Herbert Spencer - **H.S. Jennings**
Jensen, Christian Nephi - **C.N. Jensen**
Joffe, Abraham Z. - **Joffe**
Johan-Olsen, Olav - **Johan-Olsen**
Johansen, Inger - **I. Johansen**
Johanson, Carl Johan - **Johans.**
Johns, Robert M. - **R.M. Johns**
Johnson, Aaron Guy - **A.G. Johnson**

Johnson, Asa Emery - **A.E. Johnson**
Johnson, Edward Marshall - **E.M. Johnson**
Johnson, James - **J. Johnson**
Johnson, L.N. - **L.N. Johnson**
Johnson, Terry W. - **T.W. Johnson**
Johnson, Thorvaldur - **T. Johnson**
Johnston, George - **G. Johnst.**
Johnston, John Robert - **J.R. Johnston**
Johri, Bhavdish Narain - **Johri**
Johnson, Thorvaldur - **T. Johnson**
Joly, Patrick - **Joly**
Jones, E.B. Garreth - **E.B.G. Jones**
Jones, Fred Ruel - **F.R. Jones**
Jones, Leon Killy - **L.K. Jones**
Jones, Lewis Ralph - **L. Jones**
Jones, Marcus Eugene - **M.E. Jones**
Jones, Philip Malory - **P.M. Jones**
Jong, Shung C. - **S.C. Jong**
Jonnes - **Jonnes**
Jordi, Ernst A. - **Jordi**
Jörgensen, Alfred - **A. Jörgensen**
Jorgensen, Carl Adolph - **C.A. Jorgensen**
Jørstad, Ivar - **Jørst.**
Josserand, Marcel - **Josserand**
Juel, Hans Oscar - **Juel**
Jülich, Walter - **Jülich**
Junell, Lena - **Junell**
Junghuhn, Franz Wilhelm - **Jungh.**
Jurair, A.M.M. - **A.M.M. Jurair**
Kabat, Josef Emanuel - **Kab.**
Kable, Phillip F. - **Kable**
Kalchbrenner, Karoly - **Kalchbr.**
Kamat, M.N. - **M.N. Kamat**
Kambly, Paul E. - **Kambly**
Kamienski, T. - **T. Kamienski**
Kanitz, August - **Kanitz**
Kanouse, Bessie Bernice - **Kanouse**
Kapoor, Jagmohan Nath - **J.N. Kapoor**
Kar, A.K. - **A.K. Kar**
Karakulin, Boris Palladiyerrich - **Karakulin**
Karling, John Sidney - **Karling**
Karr, Guy W., Jr. - **G.W. Karr**
Karsten, Petter Adolf - **P. Karst.**
Katsuki, Shigetaka - **Katsuki**
Katsura, Kiichi - **Katsura**
Katumoto, Ken - **Katumoto**
Kauffman, Calvin Henry - **Kauffm.**
Kaufmann, M.J. - **M.J. Kaufmann**
Kavina, Karel - **Kavina**
Kawamura, Eikichi - **E. Kawamura**
Kawamura, Seiichi - **S. Kawamura**
Kaznowski, L. - **Kaznowski**
Keissler, Karl von - **Keissl.**
Keller, Jean - **J. Keller**
Kellerman, William Ashbrook - **Kellerm.**
Kelly, Howard Atwood - **H. Kelly**
Kempton, Phyllis - **P. Kempton**
Kendrick, James Blair - **J.B. Kendrick**
Kendrick, W. Bryce - **Kendrick**
Kennedy, Lorene L. - **L. Kennedy**
Kern, Frank Dunn - **F. Kern**
Keskin, B. - **Keskin**
Kessler, Kenneth J., Jr. - **K.J. Kessler**
Khan, Azmatullah - **A. Khan**
Khazanoff, Amram - **Khazanoff**
Khune, N.N. - **Khune**
Kickx, Jean Jacques, Jr. - **J. Kickx fil.**
Kidd, Mary Nest - **M.N. Kidd**
Kiely, Temple B. - **Kiely**
Kienholz, Jess Reuben - **Kienholz**
Killian, Charles - **C. Killian**
Kilpatrick, R.A. - **Kilpatr.**
Kimbrough, James W. - **Kimbrough**
Kimura, Kojiro - **K. Kimura**
King (published with F. Stevens) - **King**
King, Chalmers Jackson - **C.J. King**
King, Charlotte Maria - **C.M. King**
Kirchner, Emil Otto Oskar - **Kirchn.**
Kirchner, Leopold Anton - **L. Kirchner**
Kirk, Paul M. - **P.M. Kirk**
Kirschstein, Wilhelm - **Kirschst.**
Kita, G. - **Kita**
Klebahn, Heinrich - **Kleb.**
Klisiewicz, John M. - **Klisiewicz**
Klöcker, Albert - **Klöcker**
Klotz, Leo Joseph - **L.J. Klotz**
Klotzsch, Johann Friedrich - **Klotzsch**
Kluyver, Albert Jan - **Kluyver**
Knapp, Josef Armin - **J.A. Knapp**
Knauss, James F. - **J.F. Knauss**
Ko, Wen-Hsiung - **W.H. Ko**
Kobayashi, Takao - **T. Kobayashi**
Kobayasi, Yosio - **Y. Kobayasi**
Kochman, Józef - **Kochman**
Koerber, Gustav Wilhelm - **Koerb.**
Kohlmeyer, Erika - **E. Kohlmeyer**
Kohlmeyer, Jan - **Kohlmeyer**
Kohn, Linda M. - **L.M. Kohn**
Kokko, E.G. - **Kokko**
Komarov, Vladimir Leontjevic - **Kom.**
König - **König**
Koning, C.J. - **Koning**
Konrad, Paul - **Konrad**
Koorders, Sijfert Hendrik - **Koord.**
Korf, Richard Paul - **Korf**
Körnicke, Friedrich August - **Körn.**
Kostichka, C.J. - **Kostichka**
Kotila, John Ernest - **Kotila**
Kotlaba, Frantisek - **Kotlaba**
Kouyeas, H. - **H. Kouyeas**
Kouyeas, V. - **V. Kouyeas**
Kovachevski, C. - **Kovachevski**
Kramer, Charles Lawrence - **C.L. Kramer**
Krause, Raymond A. - **R. Krause**
Kreger-Van Rij, N.J.W. - **Kreger-Van Rij**
Kreisel, H. - **Kreisel**
Krenner, Joseph Andor - **Krenner**
Krieger, Karl Wilhelm - **K. Krieg.**
Krieger, Louis Charles Christopher - **L. Krieg.**
Krieger, Willi - **W. Krieg.**
Krieglsteiner, German J. - **Krieglsteiner**
Kroeger, Paul - **P. Kroeger**
Krombholz, Julius Vincenz von - **Krombh.**
Kropp, Bradley R. - **Kropp**
Krug, John C. - **J. Krug**
Krüger, Friedrich - **Krüger**
Krupinsky, Joseph M. - **J.M. Krupinsky**
Kuan, Ta-Li - **T. Kuan**
Kudriavzev - **Kudriavzev**
Kufferath, Hubert - **Kuff.**
Kühn, Julius Gotthelf - **Kühn**
Kühner, Robert - **Kühner**
Kujala, V. - **Kujala**
Kukkonen, Ilkka - **Kukk.**
Kulik, Martin M. - **Kulik**
Kulkarni, G.S. - **G.S. Kulkarni**
Kummer, Paul - **P. Kumm.**
Kunth, Carl Sigismund - **Kunth**
Kuntz, James Eugene - **Kuntz**
Kuntze, Carl Ernst Otto - **Kuntze**
Kunze, Gustav - **Kunze**
Kunze, Johannes - **J. Kunze**
Kuribayashi, Kazue - **Kuribayashi**
Kurosawa, Eiichi - **Kurosawa**
Kusano, Shunsuke - **Kusano**
Kützing, Friedrich Traugott - **Kütz.**
Kuyper, Th.W. - **Kuyper**
Kvashnina, E.S. - **Kvashnina**
Labrousse, Francis Jean - **Labrousse**
Lacroix, Louis-Sosthene Veyron de - **Lacr.**
Lafferty, Henry A. - **Lafferty**
Lagerberg, Karl Erik Torsten - **Lagerberg**
Lagerheim, Nils Gustaf - **Lagerh.**
Laibach, Friedrich - **Laibach**
Lair, Eugene D. - **Lair**
Lam, Annie - **A. Lam**
Lamarck, Jean Baptiste Antoine Pierre Monnet de - **Lam.**
Lambotte, Jean Baptiste Emile - **Lambotte**
Lamson-Scribner, Frank - **Lams.-Scrib.**
Lange, Jakob Emanuel - **J.E. Lange**
Lange, Johan Martin Christian - **Lange**
Langeron, Maurice - **Langeron**
Langlois, Auguste Barthelemy - **Langl.**
Lanquetin, Paule - **Lanquetin**
Lanzi, Matteo - **Lanzi**
Larsen, Michael J. - **M. Larsen**
Larsen, Poul - **P. Larsen**
Larson, Russell Harold - **R.H. Larson**
Larson, published with F. Stevens - **Larson**
Larsson, Karl-Henrik - **K. Larsson**
Lasch, Wilhelm Gottfried - **Lasch**
Laskaris, Thomas - **Laskaris**
Lasnier, E. - **Lasnier**
Latham, Dennis Harold - **D.H. Latham**
Laubert, (Karl) Richard - **Laubert**
Laundon, Gillian F. - **G. Laundon**
Laurila, Matti - **Laurila**
Lavrov, Nikolai Nicolaevich - **Lavrov**
Laxa, O. - **Laxa**
Lázaro é Ibiza, Blas - **Lázaro**
Le Gal, Marcelle Louise Fernande - **Le Gal**
LeClair, A. - **LeClair**
LeClair, P.M. - **P.M. LeClair**
LeMonnier, Alexandre Alexis George - **A. LeMonnier**
Leach, Julian Gilbert - **J.G. Leach**
Leathers, Chester R. - **Leathers**
Lebert, Hermann - **Lebert**
Lebreton, F. - **Lebreton**
Lechmere, Arthur Eckley - **Lechmere**
Ledingham, George Alek - **Ledingham**
Lee, Robert E. - **R.E. Lee**
Leers, Johann Daniel - **Leers**
Lefebvre, Camille Leon - **Lefebvre**
Lehman, Samuel George - **S.G. Lehman**
Lehmann, Friedrich - **F. Lehm.**
Lehodey, Y. - **Lehodey**
Lehtola, V.B. - **Lehtola**
Leighton, William Allport - **Leight.**
Lemke, Paul A. - **P.A. Lemke**
Lendner, Alfred - **Lendn.**
Lennox, Joanne Williams - **J.W. Lennox**
Lentz, Paul Lewis - **Lentz**
Lenz, Lee Wayne - **L. Lenz**
Leonard, Kurt J. - **K.J. Leonard**
Leonian, Leon Hatchig - **Leonian**
Lessing, Christian Friedrich - **Less.**
Letellier, Jean Baptiste Louis - **Letell.**
Letendre, Jean Baptiste Pierre [L'Abbe] - **Letendre**
Leuchtmann, Adrian - **Leuchtmann**
Léveille, Joseph Henri - **Lév.**
Levier, Emile - **Levier**
Lewis, Isaac McKinney - **I. Lewis**
Lewis, R.M. - **R.M. Lewis**
Leysser, Friedrich Wilhelm von - **Leyss.**
Libert, Marie-Anne - **Lib.**
Liberta, Anthony - **Liberta**
Libonati-Barnes, Susan D. - **S.D. Libonati-Barnes**

Lieneman, Catherine Mary - **Lieneman**
Lim, G. - **G. Lim**
Lima, J. Americo de - **Lima**
Limber, Donald Philips - **Limber**
Lind, Jens Wilhelm August - **Lind**
Lindau, Gustav - **Lindau**
Lindblad, Matts Adolf - **Lindblad**
Lindeberg, B. - **Lindeberg**
Linder, David Hunt - **Linder**
Lindfors, [Karl Magnus] Theodor - **Lindfors**
Lindner, Paul - **Lindner**
Lindquist, Juan Carlos - **Lindquist**
Lindroth, Johan Ivar - **Lindr.**
Lindsey, J. Page - **Lindsey**
Lindstedt, Karl - **Lindst.**
Lindt, W. - **Lindt**
Linford, Maurice Blood - **Linford**
Ling, C.C. - **Ling**
Linhart, Gyorgy - **Linh.**
Link, Johann Heinrich Friedrich - **Link**
Linn, Manson B. - **M.B. Linn**
Linnaeus, Carl - **L.**
Linnemann, Germaine - **Linnemann**
Lipps, P.E. - **Lipps**
Liro, Johan Ivar - **Liro**
Lister, Arthur - **List.**
Lister, Guilielma - **G. List.**
Litschauer, Viktor - **Litsch.**
Littrell, Robert H. - **R.H. Littrell**
Litvinov, Dimitri Ivanovich - **Litv.**
Liu, Xi-jin - **X. Liu**
Livingston, William H. - **Livingston**
Lloyd, Curtis Gates - **C.G. Lloyd**
Lo, T.T. - **Lo**
Lobik, Alexis Iulionovich - **Lobik**
Locquin, Marcel - **Locquin**
Lodder, Jocomina - **Lodder**
Lodha, B.C. - **Lodha**
Loerakker, W.M. - **Loerakker**
Lohde, Georg - **Lohde**
Lohman, Marion Lee - **Lohman**
Lohwag, Heinrich - **H. Lohwag**
Lombard, Frances Foust - **Lombard**
Long, William Henry - **W.H. Long**
Lonnegren, August Valfrid - **Lonnegren**
Loos, C.A. - **C.A. Loos**
Loos, Walter - **W. Loos**
Lorentz, Paul Guther - **Lorentz**
Lorenz, Russell J. - **R.J. Lorenz**
Lowe, Josiah Lincoln - **J. Lowe**
Lowen, Rosalind - **R. Lowen**
Lowy, Bernard - **Lowy**
Lucand, Jean Louis - **Lucand**
Lucet, Adrien - **Lucet**
Luck-Allen, E. Robena - **Luck-Allen**
Lüdi, Rudolf - **Lüdi**
Ludwig, Clinton Albert - **C.A. Ludw.**
Ludwig, Friedrich - **F. Ludw.**
Lundell, Seth - **S. Lundell**
Lundqvist, Nils - **N. Lundqvist**
Lunsford, J.N. - **J.N. Lunsford**
Luttrell, Everett Stanley - **Luttrell**
Lutz, M.L. - **M.L. Lutz**
Luz, Carlos Gomes da - **Luz**
Lyman, George Richard - **Lyman**
Lynge, Bernt Arne - **Lynge**
Lyon, Harold Lloyd - **Lyon**
Maas Geesteranus, Rudolph Arnold - **Maas G.**
Maas, John L. - **J.L. Maas**
MacDonald, J.D. - **J.D. MacDonald**
MacGarvie, Q.D. - **MacGarvie**
MacMillan, Conway - **MacMill.**
MacOwan, Peter - **MacOwan**
Macbride, Thomas Huston - **T. Macbr.**
Macchiati, Luigi - **Macch.**
Mackie, William Wylie - **Mackie**
Mackinnon, Juan E. - **Mackinnon**
Macoun, John - **Macoun**
Maffei, Liro - **Maffei**
Magasi, Laszlo P. - **Magasi**
Magnaghi, Angelo - **Magnaghi**
Magnus, Paul Wilhelm - **Magnus**
Magnusson, Adolf Hugo - **Magnusson**
Maia, M.H. Dalia - **Maia**
Mains, Edwin Butterworth - **Mains**
Maire, Rene Charles Joseph Ernest - **Maire**
Major, Thomas Grant - **T.G. Major**
Malbranche, Alexandre Francois - **Malbr.**
Malencon, Georges Jean Louis - **Malencon**
Malloch, David - **Malloch**
Malme, Gustaf Oskar Andersson - **Malme**
Mandal, M. - **M. Mandal**
Maneval, Willis Edgar - **Maneval**
Mangenot, G. - **G. Mangenot**
Mangin, Louis Alexandre - **L. Mangin**
Mann, Horace, Jr. - **H. Mann**
Manners, J.G. - **Manners**
Manns, Thomas Franklin - **T.F. Manns**
Manter, Harold Winfred - **Manter**
Marasas, W.F.O. - **Marasas**
Marchal, Élie - **Él. Marchal**
Marchal, Émile Jules Joseph - **Ém. Marchal**
Marchionatto, Juan B. - **Marchionatto**
Maresquelle, Henri J. - **Maresq.**
Mariat, Francois - **F. Mariat**
Marignoni, Giuseppi Br. - **Marignoni**
Marshall, Rush Porter - **R.P. Marshall**
Martin, George - **G. Martin**
Martin, George Hamilton - **G.H. Martin**
Martin, George Willard - **G.W. Martin**
Martin, Kenneth J. - **K.J. Martin**
Martin, Philip Michael Dunlop - **P. Martin**
Martinez, Albert P. - **A.P. Martinez**
Martius, Carl Friedrich Phil. Sigm. - **C. Mart.**
Martius, Carl Friedrich Philipp von - **Mart.**
Martius, Heinrich von - **H. Mart.**
Masias, O.R. - **Masias**
Mason, Edmund William - **E. Mason**
Massalongo, Caro Benigno - **C. Massal.**
Massee, George Edward - **Massee**
Massenot, M. - **Massenot**
Massey, Louis Melville - **L. Massey**
Master, I.M. - **Master**
Mathiesen, A. - **Mathiesen**
Mathiesen-Kaarik, Aino - **Mathiesen-Kaarik**
Mathre, Judith H. - **Mathre**
Mathur, S.B. - **Mathur**
Matruchot, Louis - **Matr.**
Matsumoto, Teikichi - **Matsumoto**
Matsushima, Takashi - **T. Matsushima**
Matsuura, Isamu [after 1935 known as I. Hiroe] - **Matsuura**
Matta, Alberto - **Matta**
Matthews, Dawn - **D. Matthews**
Matthews, Velma Dare - **Matthews**
Matuo, Takken - **Matuo**
Matz, Julius - **J. Matz**
Maublanc, Andre - **Maubl.**
May, Curtis - **May**
May, Luiza Cardos - **L.C. May**
Mayor, Eugéne - **Mayor**
Mayr, Heinrich - **Mayr**
McAlpine, Daniel - **McAlpine**
McCain, Arthur H. - **A. McCain**
McCain, John W. - **J. McCain**
McCarthy, G.J.P. - **G. McCarthy**
McClain, Ravenda Ladd - **R.L. McClain**
McClellen, Wilbur Dwight - **McClellen**
McClure, T.T. - **T. McClure**
McColloch, Lacy Porter - **McColloch**
McCulloch, Lucia - **L. McCulloch**
McDougall, Walter Bryon - **W. McDougall**
McGinnis, Michael - **McGinnis**
McGinty, Professor for C.G. Lloyd - **McGinty**
McGuire, James M. - **J.M. McGuire**
McHenry, Jerry - **McHenry**
McIlvaine, Charles - **McIlv.**
McKay, Hazel H. - **McKay**
McKeen, Wilber E. - **W.E. McKeen**
McKenzie, Eric H.C. - **E. McKenzie**
McKinney, Harold Hall - **McKinney**
McKnight, Kent H. - **McKnight**
McLaughlin, J. Harvey - **J.H. McLaughlin**
McMurran, Stockton Mosby - **S.M. McMurran**
McNabb, R.F.R. - **McNabb**
McPartland, John M. - **McPartland**
McWhorter, Frank P. - **McWhorter**
Medlar, Edgar Mathias - **Medlar**
Meehan, Frances - **F. Meehan**
Meeker, Joseph Andrew - **Meeker**
Mehrlich, Ferdinand Paul - **Mehrlich**
Mehrotra, Brahma Swarup - **Mehrotra**
Meier, Fred Campbell - **Meier**
Meinecke, Emilio Pepe Michael - **E. Meinecke**
Mekhtijeva - **Mekhtijeva**
Melhus, Irving E. - **Melhus**
Melin, Johannes Botwid Elis - **Melin**
Melnik, V.A. - **Melnik**
Mendoza, José Miguel - **Mendoza**
Menon, R. - **R. Menon**
Mérat de Vaumartoise, Francois Victor - **Mérat**
Mercer, W.B. - **W.B. Mercer**
Metcalf, Haven - **H. Metcalf**
Métrod, Georges - **Métrod**
Meurs, Abraham - **Meurs**
Meyen, Franz Julius Ferdinand - **Meyen**
Meyer, J.A. - **J.A. Meyer**
Meyers, Samuel P. - **S.P. Meyers**
Mhaskar, D.N. - **D.N. Mhaskar**
Micales, Jessica - **Micales**
Michael, Edmund - **Michael**
Michailides, Themis J. - **Michailides**
Micheli, Pier' Antonio - **P. Mich.**
Middleton, John Tylor - **Middleton**
Miehe, H. - **Miehe**
Migula, Emil Friedrich August Walther - **Mig.**
Mihaljcevic, M. - **M. Mihaljcevic**
Mikawa, Takashi - **Mikawa**
Mikawaki, Y. - **Mikawaki**
Milbraith, David Gallus - **Milbraith**
Milbrath, John A. - **Milbrath**
Miles, Lee Ellis - **L.E. Miles**
Millar, C.S. - **Millar**
Millardet, Pierre Marie Alexis - **Millard.**
Miller, J. Kyle - **J.K. Miller**
Miller, Julian Howell - **J.H. Miller**
Miller, Lee Wallace - **L.W. Miller**
Miller, M.W. - **M.W. Miller**
Miller, Margann - **M. Miller**
Miller, Orson K., Jr. - **O.K. Miller**
Miller, Vera A. Mentzer - **V. Miller**
Millspaugh, Charle Frederick - **Millsp.**
Minden, Max D. von - **Minden**
Minter, David W. - **Minter**
Miquel, Friedrich Anton Wilhelm - **Miq.**
Miranda, Mary - **M. Miranda**
Mirza, F. - **Mirza**
Mitchel, D.H. - **D.H. Mitchel**
Mitchell, John E. - **J.E. Mitchell**

Mitra, Manoranjan - **M. Mitra**
Mitter, Julian H. - **Mitter**
Miura, Michiya - **Miura**
Mix, Arthur Jackson - **Mix**
Miyabe, Kingo - **Miyabe**
Miyake, Ichiro - **I. Miyake**
Miyake, Tsutome - **Miyake**
Miyawaki, K. - **Miyawaki**
Moesz, Gustav - **Moesz**
Mohanty, N.N. - **Mohanty**
Mohr, Charles Theodore - **C. Mohr**
Moller, E. - **Moller**
Möller, Friedrich Alfred Gustav Jobst - **A. Möller**
Moller, William J. - **W.J. Moller**
Molliard, Marin - **Molliard**
Monod, M. - **M. Monod**
Montagne, Jean Pierre Francois Camille - **Mont.**
Moore, Justin Payson - **J.P. Moore**
Moore, Royall T. - **R.T. Moore**
Moore, Willian Dewey - **W.D. Moore**
Morais, J.O. Falcao De - **J. Morais**
Moravec, Jiri - **J. Moravec**
Mordue, J. Elizabeth M. - **Mordue**
Moreau, Claude - **C. Moreau**
Moreau, Ferdnand - **Moreau**
Moreau, Madame (Valentine) - **Mme. Moreau**
Morelet, Michel - **Morelet**
Morgan, Andrew Price - **Morg.**
Morgan-Jones, Gareth - **Morgan-Jones**
Morini, Fausto - **Morini**
Morris, Everett F. - **E.F. Morris**
Morthier, Paul - **Morthier**
Morton, F.J. - **F.J. Morton**
Moser, Meinhard - **M. Moser**
Mougeot, Jean Baptiste - **Moug.**
Mounce, Irene (Stewart) - **Mounce**
Mouton, Victor - **V. Mouton**
Mueller, Ferdinand Jacob Heinrich von - **F.v. Muell.**
Mueller, Wilhelm - **W. Muell. Stettin**
Mühlenberg, Heinrich Ernst - **Mühlenberg**
Muir, John A. - **J.A. Muir**
Mukerji, Krisha Gopal - **Mukerji**
Mukunya, D.M. - **Mukunya**
Mulder, J.L. - **J.L. Mulder**
Müller Argoviensis, Johannes - **Müll. Arg.**
Müller, (Thurgau) Herman - **H. Müll.-Thurg.**
Muller, Albert Stanley - **Muller**
Müller, Emil - **E. Müller**
Müller, Julius Heinrich Hans - **J. Müller**
Münch, Ernst - **Münch**
Mundkur, Bhalchendra Bhavanishankar - **Mundk.**
Munk, Anders - **Munk**
Munn, Moncel Thornton - **Munn**
Muntañola, Maria - **Muntañola**
Muntanola-Cvetkovic, M. - **M. Muntanola-Cvetkovic**
Murashkinskij, Konstantin Eugenevich - **Murashkinskij**
Murphy, Hickman Charles - **Murphy**
Murray, D.I.L. - **D. Murray**
Murrill, William Alfonso - **Murrill**
Mustafee, T.P. - **Mustafee**
Muthappa, B.N. - **Muthappa**
Nag Raj, T.R. - **Nag Raj**
Nagatomo, Isamu - **Nagatomo**
Nagornyi, P. - **Nagornyi**
Nair, N.B. - **N.B. Nair**
Nakasone, Karen K. - **Nakasone**
Nakata, Kakugoro - **Nakata**
Nakazawa, R. - **Nakazawa**
Namyslowski, Boleslaw - **Namyslowski**
Nannenga-Bremekamp, N.E. - **Nann.-Brem.**
Nannfeldt, John Axel - **Nannf.**
Naoumoff, M.N. - **Naoumoff**
Narasimhan, M.J. - **Narasimhan**
Narita, Takeshi - **Narita**
Nascimento, M.L. - **Nascim.**
Nattras, Roland Marshall - **Nattras**
Naumann, Arno - **A. Naumann**
Naumov, Nikolai Alexandrovic - **Naumov**
Nawaschin, Sergius Gawrilowitsch - **Nawaschin**
Nawawi, A. - **Nawawi**
Neal, David Carleton - **D.C. Neal**
Neergaard, Paul - **Neergaard**
Nees von Esenbeck, Christian Gottfried Daniel - **Nees**
Neger, Franz Wilhelm - **Neger**
Nelen, E.S. - **E.S. Nelen**
Nelson, Paul E. - **P.E. Nelson**
Nelson, R.R. - **R.R. Nelson**
Nelson, Ray - **Ray Nelson**
Nestler, Chrétien Géofrey - **Nestl.**
Neuhoff, Walther - **Neuhoff**
Neuman, Julius John - **J.J. Neuman**
Nevodovsky - **Nevodovsky**
Newhall, Allan Goodrich - **A.G. Newhall**
Newton, George A. - **G.A. Newton**
Nickerson, Nancy L. - **N. Nickerson**
Nicot, Jacqueline - **J. Nicot**
Niel, C.B. van - **Niel**
Nielsen, Lowell Wendell - **L.W. Nielsen**
Nielsen, Peter - **P. Nielsen**
Niemelä, Tuomo - **Niemelä**
Niessl von Mayendorf, Gustav - **Niessl**
Nikolaeva, T.A. - **Nikolaeva**
Nilsson, Lennart - **L. Nilsson**
Nilsson, Sven - **S. Nilsson**
Nirenberg, Helgard - **Nirenberg**
Nishida, Toji - **Nishida**
Nishihara, Natsuki - **Nishihara**
Nishijima, Kate A. - **Nishijima**
Nisikado, Yosikazu - **Nisikado**
Nitschke, Theodor Rudolf Joseph - **Nitschke**
Noack, Fritz - **F. Noack**
Noble, Mary - **M. Noble**
Nobles, Mildred Katherine - **Nobles**
Noll, Laurie A. - **Noll**
Nolla, José Antonia Bernabé - **Nolla**
Nomura, Hikotaro - **Nomura**
Norton, John Bitting Smith - **Norton**
Noviello, Carmine - **Noviello**
Novobranova - **Novobranova**
Nowakowski, Leon - **Nowakowski**
Nowell, William - **W. Nowell**
Nüesch, Jakob - **J. Nüesch**
Nuss, I. - **I. Nuss**
Nutman, F.J. - **F.J. Nutman**
Nuttall, Lawrence William - **L. Nutt.**
Nyberg, W. - **Nyberg**
Nyland, George - **Nyland**
Nylander, William - **Nyl.**
O'Brien, Muriel J. - **M. O'Brien**
O'Gara, Patrick Joseph - **O'Gara**
Oberwinkler, Franz - **Oberwinkler**
Obrist, Walter - **Obrist**
Ogawa, Joseph M. - **Ogawa**
Ohtsuki, Torao - **Ohtsuki**
Okunuki, Kazuo - **Okunuki**
Olchowecki, Alexander - **Olchowecki**
Olive, Edgar William - **Olive**
Olive, Lindsay S. - **L. Olive**
Olivier, Dorothea L. - **D. Olivier**
Olson, Alver J. - **A. Olson**
Olson, H.C. - **H. Olson**
Olson, Mary E. - **M.E. Olson**
Onions, Agnes H.S. - **Onions**
Ono, Yoshitaka - **Y. Ono**
Opiz, Philipp Maximilian - **Opiz**
Orbigny, Charle Victor Dessalines d' - **C. d'Orb.**
Orejuela, Carlos Garces - **Orejuela**
Orlicz, Anna - **Orlicz**
Orr, Geoffrey Francis - **G.F. Orr**
Orr, Leslie Wayne - **L.W. Orr**
Orton, Clayton Roberts - **Orton**
Orton, P.D. - **P.D. Orton**
Orton, William Allen - **W. Orton**
Osner, George Adin - **G.A. Osner**
Ostazeski, Stanley A. - **Ostazeski**
Ostenfeld, Carl Emil Hansen - **Ostenf.**
Osterwalder, Adolph - **Osterwalder**
Otani, Yoshio - **Otani**
Otth, Gutav Heinrich - **G. Otth**
Ou, Shu-Huang - **Ou**
Oudemans, Corneille Antoine Jean Abram - **Oudem.**
Ouellette, G.B. - **Ouellette**
Overeem, Casper van - **Overeem**
Overholts, Lee Oras - **Overh.**
Owen, John H. - **J.H. Owen**
Owen, Mary Nest - **M.N. Owen**
Paden, John - **Paden**
Padhye, Arvind A. - **Padhye**
Padwick, Geoffrey Watts - **Padwick**
Palisot de Beauvois, Ambroise Marie Francois Joseph - **P. Beauv.**
Palliser, Helen Letitia - **Palliser**
Palm, Bjorn Torvald - **B. Palm**
Palmer, John T. - **J.T. Palmer**
Palmer, Julius Auboineau - **J.A. Palmer**
Pammel, Louis Hermann - **Pammel**
Pande, A.K. - **Pande**
Pandotra, V.R. - **Pandotra**
Panizzi-Savio, Francesco - **Panizzi**
Pantidou, Maria E. - **Pantidou**
Parbery, Douglas George - **D.G. Parbery**
Parker, Arthur K. - **A.K. Parker**
Parker, Charles Stewart - **C.S. Parker**
Parker, John Bernard - **J.B. Parker**
Parkinson, Verona O. - **V.O. Parkinson**
Parks, Harold Ernest - **Parks**
Parmasto, Erast - **Parmasto**
Parmelee, John A. - **Parmelee**
Passerini, Giovanni - **Pass.**
Patel, M.K. - **M.K. Patel**
Patil, S.D. - **S.D. Patil**
Patoulliard, Narcisse Theophile - **Pat.**
Patterson, Flora Wambaugh - **F. Patterson**
Pauer, G.D. - **G.D. Pauer**
Paul, Arthur Richard - **A.R. Paul**
Paul, Hermann Karl Gustav - **H. Paul**
Paulet, Jean Jacques - **Paul.**
Pavgi, M.S. - **Pavgi**
Paxton, Glen Ernst - **G. Paxton**
Pazschke, Franz Otto - **Pazschke**
Pearson, Arthur Anselm - **A. Pearson**
Pearson, Roger C. - **R.C. Pearson**
Peck, Charles Horton - **Peck**
Peek, Chester A. - **C.A. Peek**
Pegler, David N. - **Pegler**
Peglion, Vittorio - **Peglion**
Pennington, Leigh H. - **Pennington**

Penzig, Albert Julius Otto - **Penz.**
Percival, John - **Percival**
Peredo, H. - **Peredo**
Peres, G.E.P. - **Peres**
Perkins, C.F. - **C.F. Perkins**
Perry, Thelma J. - **T.J. Perry**
Persoon, Christiaan Hendrik - **Pers.**
Pesante, Aldo - **Pesante**
Petch, Thomas - **Petch**
Petersen, Henning Eiler - **H.E. Petersen**
Petersen, L.J. - **L.J. Petersen**
Petersen, Ronald H. - **R. Petersen**
Peterson, Alvah - **A. Peterson**
Peterson, Roger S. - **R.S. Peterson**
Pethybridge, George Herbert - **Pethybr.**
Petrak, Franz - **Petr.**
Petri, Lionello - **Petri**
Petrini, Liliane E. - **L.E. Petrini**
Petrini, Orlando - **Petrini**
Petrov, Marija - **M. Petrov**
Peyronel, Beniamino - **Peyronel**
Pfender, W.F. - **W.F. Pfender**
Phaff, Herman J. - **Phaff**
Phillips, William - **W. Phillips**
Picard, Francois - **F. Picard**
Picbauer, Richard - **Picb.**
Piccone, Antonio - **Picc.**
Pidoplichko, N.M. - **Pidoplichko**
Piemeisel, Frank Joseph - **Piemeisel**
Pierce, Newton Barris - **N. Pierce**
Pignal, Marie-Claire - **Pignal**
Pilát, Albert - **Pilát**
Pirone, Pascal P. - **Pirone**
Pirotta, Pietro Romualdo - **Pirotta**
Pirozynski, Kris A. - **Pirozynski**
Pitard, Charles-Joseph Marie - **Pit.**
Pitt, John I. - **J.I. Pitt**
Pitt, P.I. - **P.I. Pitt**
Plakidas, Antonios George - **Plakidas**
Plessis, J. du - **Plessis**
Plöttner - **Plöttner**
Plowright, Charles Bagge - **Plowr.**
Plunkett, Orda Allen - **Plunkett**
Poiteau, Pierre Antoine - **Poit.**
Pole-Evans, Illtyd Buller - **Pole-Evans**
Politis, Demetrios J. - **Politis**
Pollacci, Gino - **Pollacci**
Pollack, Flora G. - **Pollack**
Pollini, Ciro - **Pollini**
Pomerleau, René - **Pomerleau**
Ponnappa, K.M. - **Ponnappa**
Pontual, Duarte - **Pontual**
Pope, Seth Alison - **S. Pope**
Poradielova, N.N. - **Poradielova**
Porte, William Solomon - **Porte**
Potebnia, Andrej A. - **Potebnia**
Potter, Alden Archibald - **A.A. Potter**
Potter, Michael Cresse - **Potter**
Pound, Roscoe - **Pound**
Pouzar, Zdenek - **Pouzar**
Povah, Alfred Hubert William - **Povah**
Powell, Paul E. - **P.E. Powell**
Prantl, Karl Aton Eugen - **Prantl**
Prasad, Nirankar - **Prasad**
Pratt, Robert G. - **R.G. Pratt**
Presl, Karl Borowag - **K. Presl**
Presley, John Thomas - **Presley**
Preston, Norman C. - **N.C. Preston**
Preuss, (Carl) Gotlieb Traugott - **G. Preuss**
Price, Terence Vincent - **Price**
Prillieux, Edouard Ernest - **Prill.**
Pringsheim, Nathanael - **Pringsh.**
Prinsen-Geerligs, Hendrik Coenraad - **Prinsen Geerligs**
Pritchard, Frederick John - **Pritchard**
Pugh, G.J.F. - **Pugh**
Punithalingam, E. - **Punithalingam**
Puttemans, Arséne - **A. Puttemans**
Putterill, Victor Armsby - **Putterill**
Quélet, Lucien - **Quél.**
Raabe, Robert D. - **Raabe**
Rabenhorst, Gottlob Ludwig - **Rabenh.**
Raciborski, Marjan - **Racib.**
Rader, William Ernst - **Rader**
Radlkofer, Ludwig Adolph Timotheus - **Radlk.**
Raghukumar, S. - **Raghukumar**
Ragunathan, C. - **Ragunathan**
Raillo, Madame A. - **Raillo**
Raistrick, Harold - **Raistrick**
Raitviir, Ain - **Raitviir**
Rajchenberg, Mario - **Rajchenberg**
Rall, Gloria - **Rall**
Ramachar, P. - **Ramachar**
Ramakrishan, K. - **Ramakrishan**
Ramaley, Annette W. - **A.W. Ramaley**
Ramamurthi, C.S. - **Ramamurthi**
Rambelli, Anton - **Rambelli**
Ramirez, C. - **C. Ramirez**
Ramsbottom, John - **Ramsb.**
Ramsey, Glen Blaine - **Ramsey**
Rand, Frederick Vernon - **F.V. Rand**
Rands, Robert Delafield - **Rands**
Rangel, Eugenio dos Santos - **Rangel**
Rankin, William Howard - **Rankin**
Ranojevic, Nikola - **Ranoj.**
Rao, P.N. - **P.N. Rao**
Rao, Vasant Gurunath - **V.G. Rao**
Raper, Kenneth Bryan - **Raper**
Rappaz, Francois - **Rappaz**
Rathay, Emerich - **Rathay**
Ratiani - **Ratiani**
Rattan, Sarjit S. - **S.S. Rattan**
Rau, Eugene Abraham - **Rau**
Raunkiaer, Christen Christiansen - **Raunk.**
Ravaz, Louis Etienne - **Ravaz**
Ravenel, Henry William - **Ravenel**
Ray, William Winfield - **W.W. Ray**
Raymond, F.L. - **F.L. Raymond**
Rayner, Ronald W. - **R.W. Rayner**
Rea, Carleton - **Rea**
Reade, John Moore - **Reade**
Rebentisch, Johann Friedrich - **Rebent.**
Reddick, Donald - **Reddick**
Redhead, Scott A. - **Redhead**
Redlin, Scott Clare - **Redlin**
Reed, Minnie - **M. Reed**
Regel, Eduard [August] von - **Regel**
Rehm, Heinrich (Simon Ludwig Friedrich Felix) - **Rehm**
Reichard, Johann Jakob - **Reichard**
Reichardt, Heinrich Wilhelm - **Reichardt**
Reichenbach, Heinrich Gottlieb Ludwig - **Rchb.**
Reichert, Israel - **Reichert**
Reid, Derek - **D. Reid**
Reid, James - **J. Reid**
Reifschneider, Francisco J.B. - **Reifschneider**
Reinke, Johannes - **Reinke**
Reinking, Otto August - **Reinking**
Reinsch, Paul Friedrich - **Reinsch**
Reisinger, Otto - **O. Reisinger**
Reitsma, J. - **Reitsma**
Relhan, Richard - **Relhan**
Remler, Paula - **Remler**
Remsberg, Ruth Elizabeth - **Remsberg**
Retzius, Anders Jahan - **Retz.**
Rex, George Abraham - **Rex**
Reynolds, Donald R. - **Reynolds**
Richardson, M.J. - **M.J. Richardson**
Richatt, Fernando Mujica - **Richatt**
Richon, Charles Edouard - **Richon**
Rick, Johann - **Rick**
Ricken, Adalbert - **Ricken**
Ricker, Percy Leroy - **Ricker**
Riddle, Lincoln Ware - **Riddle**
Riess, H. - **Riess**
Rifai, Mien A. - **Rifai**
Rimpau, Rudolf H. - **Rimpau**
Rivolta, Sebastiano - **Rivolta**
Roane, Martha K. - **M.K. Roane**
Roark, Eugene Washburn - **Roark**
Robak, Hakan - **Robak**
Roberge, Michael Robert - **Roberge**
Roberts, Daniel A. - **D.A. Roberts**
Roberts, Florence M. - **F.M. Roberts**
Roberts, John William - **Roberts**
Robertson, Noel F. - **N.F. Robertson**
Robin, Charles Philippe - **C.P. Robin**
Robinson, Benjamin Lincoln - **B.L. Robinson**
Robinson-Jeffrey, Robena C. - **Robinson-Jeffrey**
Röder, K. - **Röder**
Rodrigues de Miranda, L. - **Rodrigues de Miranda**
Rodway, Leonard - **Rodway**
Roethert, W. - **Roethert**
Roger, L. - **L. Roger**
Rogers, Donald Phillip - **D.P. Rogers**
Rogerson, Clark Thompson - **C.T. Rogerson**
Roges, Jack D. - **J.D. Rogers**
Rohde, Theo - **T. Rohde**
Röhling, Johann Christoph - **Röhl.**
Roldan, Emiliano F. - **Roldan**
Rolland, Leon - **Rolland**
Romagnesi, Henri Charles Louis - **Romagnesi**
Romell, Lars - **Romell**
Roquebert, Marie-France - **Roquebert**
Rosenbaum, Joseph - **Rosenb.**
Rossman, Amy Yarnell - **A.Y. Rossman**
Rostafinski, T.J. - **Rostafinski**
Rostkovius, Friedrich Wilhelm Gottlieb - **Rostk.**
Rostovzev, Semen Ivanovich - **Rostovzev**
Rostrup, (Frederik Georg) Emil - **Rostr.**
Rostrup, Ove Georg Frederik - **O. Rostr.**
Roth, Albrecht Wilhelm - **Roth**
Rothers, Boris - **Rothers**
Rothert, Wladislaw - **Rothert**
Rothrock, Joseph Trimble - **Rothr.**
Roumeguere, Casimir - **Roum.**
Rousseau, Marietta - **M. Rousseau**
Roussel, Henri Francois Anne de - **Roussel**
Routien, John Broderick - **Routien**
Roux, Celia - **Roux**
Roy, A.J. - **A.J. Roy**
Roze, Ernest - **Roze**
Rudolph, Bert Alexander - **B.A. Rudolph**
Rudolphi, Friedrich (Karl Ludwig) - **F. Rudolphi**
Rudolphi, Karl Asmund - **Rudolphi**
Ruehle, George Dewey - **Ruehle**
Ruhland, Eugene Otto - **Ruhland**
Rühle, George Dewey - **Rühle**
Rumbold, Caroline Thomas - **Rumbold**
Rupprecht, Heinrich - **H. Rupprecht**
Russell, R. C. - **Russell**
Rutgers, Abraham Arnold Loderwijk - **Rutgers**
Ryan, Mary Hilaire - **Ryan**
Ryker, Truman Clifton - **Ryker**
Ryvarden, Leif - **Ryvarden**
Sacamano, C.M. - **Sacamano**

Saccardo, Domenico - **D. Sacc.**
Saccardo, Francesco - **F. Sacc.**
Saccardo, Pier Andrea - **Sacc.**
Saccas, A.M. - **Saccas**
Sadebeck, Richard Emil Benjamin - **Sadebeck**
Sagra, Ramon de la - **Sagra**
Sahni, V.P. - **Sahni**
Saint-Amans, Jean Florimond Boudon de - **Saint-Amans**
Saito, Hidesaku - **H. Saito**
Saito, Kendo - **K. Saito**
Salam, M.A. - **M.A. Salam**
Salkin, Ira F. - **Salkin**
Salmon, Ernst Stanley - **E.S. Salmon**
Salt, G.A. - **G.A. Salt**
Salvanet-Duval - **Salvanet-Duval**
Sampson, Kathleen - **K. Sampson**
Samson, Robert A. - **R.A. Samson**
Samuels, Gary Joseph - **Samuels**
Sanderson, F.R. - **F.R. Sanderson**
Sandhu, Dhanwant K. - **Sandhu**
Sandu-Ville, Constantin - **Sandu**
Sanford, Guthrie Brown - **G.B. Sanford**
Santi, Georgio - **Santi**
Sardina, Juan Rodriguez - **Sardina**
Sarkar, Anjali - **A. Sarkar**
Sartoris, George Bartholomew - **Sartoris**
Sartory, Auguste Theodore - **Sartory**
Sartory, Réne - **R. Sartory**
Sastry, K.S.M. - **K.S. Sastry**
Sato, Kunihiko - **K. Sato**
Sauter, Anton Eleutherius - **Saut.**
Sauvageau, Camille Francois - **Sauv.**
Savile, Douglas Barton Osborne - **Savile**
Savulescu, Trajan - **Savul.**
Sawada, Kaneyoshi - **Sawada**
Scalia, Giuseppe - **Scalia**
Schaeffer, Jacob Christian - **Schaeff.**
Schaffnit, Johannes Martin Ernst Christian Otto - **Schaffnit**
Schanderl, Richard - **Schanderl**
Scharif, G. - **Scharif**
Scheffer, R.J. - **R.J. Scheffer**
Schellenberg, Hans Conrad - **Schellenb.**
Schenck, Bernhard - **B. Schenck**
Schenk, Joseph August - **Schenk**
Scherffel, Aladár - **Scherff.**
Schilberszky, Karoly - **Schilberszky**
Schipper, Maria A. A. - **M.A.A. Schipper**
Schläpfer, E. - **Schläpfer**
Schlechtendahl, Diederich Franz Leonhard von - **Schlechtend.**
Schleicher, Johann Cristoph - **Schleich.**
Schmidt, Johann Carl - **J.C. Schmidt**
Schmiedeknecht, Martin - **Schmiedeknecht**
Schnabl, Johann Nepomuk - **Schnabl**
Schnegg, Hans - **Schegg**
Schneider, Albert - **A. Schneid.**
Schneider, Roswitha - **R. Schneider**
Schneider, Wilhelm Gottlieb - **W.G. Schneid.**
Schofield, Joseph Robert - **Schofield**
Schoknecht, Jean D. - **Schoknecht**
Schol-Schwarz, M. Beatrice - **Schol-Schwarz**
Scholten, G. - **Scholten**
Schoschiaschvili, I. - **Schoschiaschvili**
Schostakowitsch, Wladimir - **Schostakowitsch**
Schoulties, Calvin L. - **Schoulties**
Schrader, Heinrich Adolph - **Schrad.**
Schrank, Franz von Paula von - **Schrank**
Schrantz, Jean-Paul - **Schrantz**
Schrenk, Hermann von - **H. Schrenk**
Schröter, Joseph - **J. Schröt.**
Schubert, Carl - **C. Schub.**
Schüepp, Hannes - **H. Schüepp**
Schultz, Carl Friedrich - **C.F. Schultz**
Schultz, Eugene S. - **E.S. Schultz**
Schulzer von Muggenburg, Stephan - **Schulzer**
Schumacher, (Heinrich) Christian Friedrich - **Schumach.**
Schumacher, Trond - **T. Schumacher**
Schwabe, Samuel Heinrich - **Schwabe**
Schwartz, Ernest Justus - **E.J. Schwartz**
Schwartz, Marie Beatrice - **M.B. Schwartz**
Schwarze, Carl Alois - **C.A. Schwarze**
Schweinitz, Lewis David von - **Schwein.**
Scopoli, Joannes Antonius - **Scop.**
Scott, D.B. - **D.B. Scott**
Seaver, Fred Jay - **Seaver**
Sechet, M. - **Sechet**
Secretan, Louis - **Secr.**
Seeler, Edgar Viguers - **Seeler**
Seifert, Keith A. - **Seifert**
Selby, Augustine Dawson - **A. Selby**
Serbinow, Ivan L'rovich - **Serbinow**
Serebrianikow, J. - **Serebrianikow**
Servazzi, Ottone - **Servazzi**
Setchell, William Albert - **Setch.**
Seymour, Arthur Bliss - **Seym.**
Seymour, C.P. - **C.P. Seymour**
Seynes, Jules de - **De Seyn.**
Shaffer, Robert L. - **R. Shaffer**
Shahin, E.A. - **Shahin**
Sharma, L.C. - **Sharma**
Shaw, Charles Gardener - **C.G. Shaw**
Shaw, Frederick John Freshwater - **F.J. Shaw**
Shear, Cornelius Lott - **Shear**
Shearer, Carol A. - **C.A. Shearer**
Sheldon, John Lewis - **J. Sheld.**
Sherbakoff, Constantine Demetry - **Sherb.**
Sherwin, Helen Shedd - **Sherwin**
Sherwood-Pike, Martha A. - **M.A. Sherwood**
Shipman, Jill S. - **J. Shipman**
Shipton, W.A. - **Shipton**
Shirai, Mitsutaro - **Shirai**
Shoemaker, Robert A. - **Shoemaker**
Shome, S.K. - **Shome**
Shope, Paul Franklin - **Shope**
Sibilia, Cesare - **Sibilia**
Sibthorp, John - **Sibth.**
Sicard, Guillaume - **Sicard**
Sideris, Christos Plutarchos - **Sideris**
Siegler, Eugene Alfred - **Siegler**
Siegmund, Wilhelm - **Siegm.**
Siemaszko, Wincenty - **Siemaszko**
Siggers, Paul Victor - **Siggers**
Sigler, Lynn - **Sigler**
Silva, J. Nery da - **J.N. Silva**
Simmonds, John Howard - **J.H. Simmonds**
Simmons, Emory G. - **E. Simmons**
Singer, Rolf - **Singer**
Singh, B. - **B. Singh**
Singh, R.D. - **R.D. Singh**
Singh, Udai Pratap - **U.P. Singh**
Sivanesan, A. - **Sivanesan**
Sivasithamparam, Krishnapillai - **Sivasithamparam**
Skinner, C.E. - **C.E. Skinner**
Skolko, Arthur John - **Skolko**
Skovsted, Aage Thorsen - **Skovst.**
Slope, D.B. - **D. Slope**
Slysh, Anton R. - **Slysh**
Smalley, Eugene B. - **Smalley**
Smarods, Julius - **Smarods**
Smerlis, E. - **Smerlis**
Smith, Alexander Hanchett - **A.H. Sm.**
Smith, Alexander Martin - **A.M. Sm.**
Smith, Annie Lorrain - **A.L. Sm.**
Smith, Charles Leonhard - **C.L. Sm.**
Smith, Clayton O. - **C.O. Sm.**
Smith, Clifford W. - **C.W. Sm.**
Smith, Donald J. - **D.J. Sm.**
Smith, Elizabeth H. - **E.H. Sm.**
Smith, Erwin Frink - **E.F. Sm.**
Smith, George - **G. Sm.**
Smith, George Worthington - **G.W. Sm.**
Smith, James Edward - **Sm.**
Smith, Jeffrey Drew - **J.D. Sm.**
Smith, M.A. - **M.A. Sm.**
Smith, Ralph Eliot - **R.E. Sm.**
Smith, William Gardner - **Wm.G. Sm.**
Snell, Walter Henry - **Snell**
Snow, Glen A. - **G. Snow**
Snyder, William C. - **W.C. Snyder**
Sobers, Edward K. - **Sobers**
Solheim, Wilhelm Gerhard - **Solheim**
Sollmann, August - **Sollm.**
Sommer, Noel F. - **N.F. Sommer**
Sommerfelt, Soren Christian - **Sommerf.**
Sommier, [Carlo Pietro] Stefano - **Sommier**
Sopp, Olav Johan - **Sopp**
Soppitt, Henry Thomas - **Soppitt**
Sorauer, Paul Carl Moritz - **Sorauer**
Sorokin, Nikolai Vasilevich - **Sorokin**
Sousa de Camara, Manuel de - **Sousa de Carmara**
Southworth, Effie Almira - **Southworth**
Sowerby, James - **Sowerby**
Sparrow, Fredrick Kroeber - **Sparrow**
Spaulding, Perley - **Spauld.**
Speare, Alden True - **Speare**
Speer, Eberhard O. - **E.O. Speer**
Spegazzini, Carlo Luigi - **Speg.**
Spencer, Edwin Rollin - **E.R. Spencer**
Spencer, James A. - **J.A. Spencer**
Spessa, Carolina - **Spessa**
Spieckermann, Albert - **Spieckermann**
Spielman, Linda J. - **Spielman**
Splendore, Achille - **Splendore**
Sprague, Charles James - **C. Sprague**
Sprague, Roderick - **R. Sprague**
Sprée, T. - **Sprée**
Sprengel, Kurt Polycarp Joachim - **Spreng.**
Srinivasan, Kattalaicheri Venkataraman - **Srinivasan**
Stack, Robert W. - **Stack**
Stahel, Gerold - **Stahel**
Stakman, Elvin Charles - **Stakman**
Staley, John M. - **Staley**
Stalpers, Joost A. - **Stalpers**
Stansfield, Olive P. - **Stansfield**
Starbäck, Karl - **Starb.**
Staritz, Richard - **Staritz**
Starmer, William T. - **Starmer**
States, Jack - **J. States**
Staude, Friedrich - **Staude**
Stein, Berthold - **Stein**
Steinmann, B. Alfred - **B.A. Steinmann**
Stelling-Dekker, N.M. - **Dekker**
Steudel, Ernst Gottlieb von - **Steud.**
Stevens, Frank Lincoln - **F. Stevens**
Stevens, Neil Everett - **N. Stevens**
Stevenson, John Albert - **J.A. Stevenson**
Stevenson, William C. - **W.C. Stevenson**
Stewart, Dewey - **D. Stewart**
Stewart, Elwin - **E. Stewart**
Stewart, Fred Carlton - **F.C. Stewart**
Stewart, Vern Bonham - **V.B. Stewart**
Steyaert, René - **Steyaert**
Stipes, Roland Jay - **Stipes**
Stolk, Amelia C. - **A.C. Stolk**
Stoll, O. - **O. Stoll**
Stone, Jeffrey K. - **J.K. Stone**
Stone, Roland Elisha - **R. Stone**
Stoneman, Bertha M. - **Stoneman**

Störmer, [Frederick] Carl [Mülerz] - **Störmer**
Stout, Gilbert Leonidas - **G.L. Stout**
Stover, Robert Harry - **Stover**
Strasser, Pius - **Strass.**
Straus, Adolf Paul Carl - **Straus**
Strauss, Friedrich Karl Joseph von - **F. Strauss**
Stretton, Helena M. - **Stretton**
Strid, Ake - **A. Strid**
Strömbom, Nils Gustaf - **Strömbom**
Strunk, H.F. - **Strunk**
Studer-Steinhäuslin, Bernhard - **Studer**
Stuntz, Daniel Elliot - **D.E. Stuntz**
Sturm, Jacob - **Sturm**
Subramanian, C.V. - **Subramanian**
Subramanian, Lekshminarayanapuram S. - **L. Subramanian**
Suggs, Edna G. - **E.G. Suggs**
Sumstine, David Ross - **Sumstine**
Sung, Sun Huang - **Sung**
Sutherland, George Kenneth - **G.K. Sutherland**
Sutherland-Campbell, H. - **Suth.-Camp.**
Sutton, Brian C. - **Sutton**
Svrček, Mirko - **Svrček**
Swank, George Jr. - **Swank**
Swart, H.J. - **H.J. Swart**
Swartz, Olof Peter - **Sw.**
Swift, Marjorie Elizabeth - **M.E. Swift**
Swingle, Deane Bret - **D. Swingle**
Swingle, Walter Tennyson - **Swingle**
Sydow, Hans - **Syd.**
Sydow, Paul - **P. Syd.**
Szabó, Zoltan von - **Szabó**
Tai, Fan Lan - **F.L. Tai**
Takahashi, Yoshinao - **Takah.**
Takimoto, Seito - **Takimoto**
Talbot, Patrick Henry Brabazon - **Talbot**
Talice, Rudolfo V. - **Talice**
Tallman, A. Statzell - **Tallman**
Tamiya, Hiroshi - **Tamiya**
Tanaka, Ichiro - **I. Tanaka**
Tanda, Seinoske - **S. Tanda**
Tapke, Victor Ferdinand - **Tapke**
Tapy - **Tapy**
Tarr, S. A. J. - **Tarr**
Tassi, Flaminio - **Tassi**
Taubenhaus, Jacob Joseph - **Taubenhaus**
Tavel, Catherine von - **C. Tavel**
Tavel, Rudolf Franz von - **Tavel**
Taylor, John J. - **J.J. Taylor**
Taylor, Thomas - **T. Tayl.**
Tehon, Leo Roy - **Tehon**
Teixeira, Alcides Ribeiro - **Teix.**
Teng, Shu-ch'ün - **S. Teng**
Tengwall, Tor Ake - **Tengwall**
Teramoto, T. - **Teramoto**
Terashita, Takakiyo - **Terashita**
Terra, Paule - **P. Terra**
Terrier, Charles A. - **Terrier**
Teunisson, Dorothea J. - **Teunisson**
Tharp, Benjamin Carroll - **Tharp**
Thaxter, Roland - **Thaxt.**
Theissen, Ferdinand - **Theiss.**
Therry, Jean Joseph - **Therry**
Thiers, Harry Delbert - **Thiers**
Thind, Kartar Singh - **Thind**
Thirumalachar, Mandayam Jeersannidhi - **Thirumalachar**
Thom, Charles - **Thom**
Thomas, Friedrich August Wilhelm - **F. Thomas**
Thomé, Otto Wilhelm - **Thomé**
Thompson, A. - **A. Thompson**
Thompson, Charles Henry - **C.H. Thomps.**
Thompson, George Edward - **G.E. Thompson**
Thore, Jean - **Thore**
Thorn, R. Greg - **R.G. Thorn**
Thümen, Felix Karl Albert Ernst Joachim von - **Thuem.**
Thurston, Henry Winfred - **H.W. Thurston**
Thyr, Bill D. - **Thyr**
Tieghem, Phillippe Edouard Leon van - **Tiegh.**
Tiffany, Lois C. - **L.C. Tiffany**
Tilford, Paul Edward - **Tilford**
Timmer, L.W. - **Timmer**
Timmermans, Adrianna J. - **Timmermans**
Tims, Eugene Chapel - **Tims**
Tiraboschi, Carlo - **Tiraboschi**
Tison, Adrien - **Tison**
Tjallingii-Beukers, G. - **Tjall.-Beukers**
Tobisch, Julius - **Tobisch**
Tobler, Gertrud [Paula] née Wolf - **G. Tobl.**
Tochinai, Yoshihiko - **Tochinai**
Tode, Heinrich Julius - **Tode**
Togashi, Kogo - **Togashi**
Tognini, Filippo - **Tognini**
Tokunaga, Yosio - **Tokunaga**
Tomilin, B.A. - **Tomilin**
Tomlinson, J.A. - **J.A. Tomlinson**
Tomoyasu, Ryokichi - **Tomoyasu**
Toole, Eben Richard - **Toole**
Torkelson, Anna-Elise - **Torkelson**
Toro, Rafael A. - **Toro**
Torrend, Camille - **Torrend**
Torrey, John - **Torr.**
Toth, Sandor - **S. Toth**
Tournefort, Joseph Pitton de - **Tourn.**
Toussoun, T. A. - **T.A. Toussoun**
Traaen, Alf E. - **Traaen**
Trabut, Louis Charles - **Trab.**
Tracy, Samuel Mills - **Tracy**
Trail, James William Helenus - **Trail**
Tranzschel, Woldemar Andrejevich - **Tranzschel**
Traquair, James A. - **J.A. Traquair**
Traverso, Giovanni Battista - **Traverso**
Tredick, Joanne - **Tredick**
Trelease, William - **Trel.**
Trentepohl, Johann Friedrich - **Trentep.**
Treschow, Cecil - **Treschow**
Trevisan de Saint Leon, Vittore [Benedetto Antonio] - **Trevis.**
Trinchieri, Giulio - **Trinchieri**
Trog, Jakob Gabriel - **Trog**
Trotter, Alessandro - **Trott.**
Trow, Albert Howard - **Trow**
Trujillo, Eduardo E. - **E.E. Trujillo**
Truscott, John Henry Lloyd - **Truscott**
Tsiklinsky, P. - **Tsiklinsky**
Tsuda, Mitsuya - **Tsuda**
Tu, C.C. - **Tu**
Tubaki, Keisuke - **Tubaki**
Tubeuf, Carl von - **Tub.**
Tucker, Clarence Mitchell - **Tucker**
Tuckerman, Edward - **Tuck.**
Tulasne, Charles - **C. Tul.**
Tulasne, Louis-Réne - **Tul.**
Tullis, Edgar Cecil - **Tullis**
Tulloch, Margaret C. - **Tulloch**
Tunstall, A.C. - **Tunstall**
Turconi, Malusio - **Turconi**
Turner, Elizabeth M. - **E.M. Turner**
Tylutki, Edward E. - **Tylutki**
Uchida, - **Uchida**
Udagawa, Shun-ichi - **Udagawa**
Uecker, Francis August - **Uecker**
Ueki, Robert - **Ueki**
Ueyama, Akinori - **Ueyama**
Ukkelberg, Harry G. - **Ukkelberg**
Ule, Ernst Heinrich Georg - **Ule**
Ullstrup, Arnold John - **Ullstrup**
Unamuno Yrigoyen, Luis M. - **Unamuno**
Underwood, Lucien Marcus - **Underw.**
Unger, Franz Joseph Andreas Nicolaus - **Unger**
Upadhyay, H.P. - **Upadhyay**
Uppal, Badri Nath - **Uppal**
Urban, Zdenek - **Z. Urban**
Urries, Manuel J. de - **Urries**
Vaartaja, Olli - **Vaartaja**
Vahl, Martin Hendriksen - **Vahl**
Valenta - **Valenta**
Valleau, William Dorney - **Valleau**
Van Beyma, Thoe Kingm, T.H. - **Van Beyma**
Van De Bogart, Fred - **Van De Bogart**
Van Hoof, H.A. - **Van Hoof**
Van Hook, James Mon - **Van Hook**
Van Kesteren, H.A. - **Van Kesteren**
Van Oorschot, Connie A.M. - **Van Oorschot**
Van Warmelo, Konrad T. - **Van Warmelo**
Van Weerdt, L. - **Van Weerdt**
Van der Hoeven, E.P. - **Van der Hoeven**
Van der Plaats-Niterink, A.J. - **Plaats-Niterink**
Van der Wolk, P.C. - **Van der Wolk**
Van der Zwet, Thomas - **Van der Zwet**
Vanev, Simeon G. - **Vanev**
Vánky, Kálmán - **K. Vánky**
Vanterpool, Thomas Clifford - **Vanterpool**
Vasconcelos, C.T. de - **Vasconcelos**
Vassiljevsky, N.I. - **Vassiljevsky**
Vasudeva, R. Sahai - **Vasudeva**
Vegh, I. - **Vegh**
Velenovsky, Josef - **Velen.**
Venkata Ram, C.S. - **Venkata Ram**
Venkataramani, K.S. - **Venkataramani**
Ventenat, Etienne Pierre - **Venten.**
Venturi, Antonio - **A. Venturi**
Verona, O. - **Verona**
Verplancke, Germain - **Verplancke**
Verrall, Arthur F. - **Verrall**
Verwoerd, Len - **Verwoerd**
Vestergren, Jacob Tycho Conrad - **Vestergr.**
Vey, A. - **Vey**
Viala, Pierre - **Viala**
Viégas, Ahmés Pinto - **Viégas**
Viennot-Bourgin, Georges - **Vien.-Bourg.**
Villars, Dominique - **Villars**
Vincens, Francois - **Vincens**
Vital, A. Fernandes - **Vital**
Vittadini, Carlo - **Vittad.**
Vittal, Balumuri Pandu Ranga - **Vittal**
Viviani, Domenico - **Viv.**
Vize, John Edward - **Vize**
Vizioli, José - **Vizioli**
Vleugel, Jens Schanke - **Vleugel**
Vlug, Ijda - **Vlug**
Voglino, Pietro - **Voglino**
Volkart, Albert - **Volkart**
Voorhees, Richard Kenneth - **Voorhees**
Voss, Guglielmo - **G. Voss**
Voss, Wilhelm - **W. Voss**
Vouaux, Léon - **Vouaux**
Vuillemin, Jean Paul - **Vuill.**
Wadley, Bryce N. - **Wadley**
Wagener, Willis Westlake - **W. Wagener**

Wahlenberg, Göran - **Wahlenberg**
Waite, Merton Benway - **M.B. Waite**
Wakefield, Elsie Maud - **Wakef.**
Wakker, Jan Hendrik - **Wakk.**
Walcz, Ilona - **I. Walcz**
Walker, John - **J. Walker**
Walker, John Charles - **J.C. Walker**
Walker, Leva Bella - **L.B. Walker**
Waller, James M. - **Waller**
Waller, S. - **S. Waller**
Wallroth, Karl Friedrich Wilhelm - **Wallr.**
Walter, James M. - **J.M. Walter**
Wang, Chun-Juan Kao - **Wang**
Warburg, Otto - **Warb.**
Warcup, J.H. - **Warcup**
Ware, W.N. - **W.N. Ware**
Watanabe, Tatsuwo - **T. Watanabe**
Waterhouse, Grace Marion - **G.M. Waterhouse**
Waterman, Alma May - **A.M. Waterman**
Waterston, John MacLaren - **Waterston**
Watling, Roy - **Watling**
Watson, Alice Marie Johnson - **A.M. Watson**
Webber, Herbert John - **J.H. Webber**
Weber, George Frederick - **G.F. Weber**
Weberbauer, Otto - **O. Weberb.**
Webster, John - **J. Webster**
Webster, R.E. - **R.E. Webster**
Webster, Robert K. - **R.K. Webster**
Weedon, Amy Gertrude - **Weedon**
Weese, Josef - **Weese**
Wegelin, Heinrich - **H. Wegelin**
Wehmer, Carl - **Wehmer**
Wehmeyer, Lewis Edgar - **Wehmeyer**
Wei, Ching-Chao - **C.T. Wei**
Weidemann, Gregory J. - **G.J. Weidemann**
Weimer, James Leroy - **J.L. Weimer**
Weindlymayr, J. - **J. Weindlymayr**
Weinmann, Johannes Anton - **Weinm.**
Weir, James Robert - **Weir**
Weiss, Freeman Albert - **Weiss**
Welch, Donald Stuart - **Welch**
Welden, Arthur L. - **A.L. Welden**
Welles, Colin Gilchrist - **C. Welles**
Wellman, Frederick Lovejoy - **Wellman**
Wells, Doreen - **D. Wells**
Wells, Kenneth - **K. Wells**
Wells, Virginia - **V. Wells**
Welwitsch, Friedrich Martin Josef - **Welw.**
Went, Friedrich August Ferdinand Christian - **Went**
Werdermann, Erich - **Werderm.**
Weresub, Luella K. - **Weresub**
West, Erdman - **E. West**
Westendorp, Gérard Daniel - **Westend.**
Westerdijk, Johanna - **Westerd.**
Westling, Per Richard - **Westling**
Weston, William Henry, Jr. - **W. Weston**
Wettstein, Richard von - **Wettst.**
Whetzel, Herbert Hice - **Whetzel**
White, Francis Buchanan - **F.B. White**
White, George P. - **G.P. White**
White, James F., Jr. - **J.F. White**
White, Richard P. - **R.P. White**
White, Violetta Susan Elizabeth - **V.S. White**
White, William Lawrence - **W.L. White**
Whitehead, Marvin D. - **M.D. Whitehead**
Whiteside, Jack O. - **Whiteside**
Whitney, Hugh S. - **H.S. Whitney**
Wickerham, Lynford J. - **Wickerham**
Wiehe, P. O. - **Wiehe**
Wiggers, [Heinrich] August - **A. Wigg.**
Wilcox, Edwin Mead - **E.M. Wilcox**
Wilcox, Hugh - **H. Wilcox**
Wilcox, Marguerite Statira - **M.S. Wilcox**
Wiley, Bonnie - **B. Wiley**
Wilhelm, Karl Adolf - **K. Wilh.**
Willdenow, Carl Ludwig - **Willd.**
Wille, Fritz - **Wille**
Willkomm, Heinrich Mariz - **Willk.**
Wilson, Doreen E. - **D.E. Wilson**
Wilson, Edward Elmer - **E.E. Wilson**
Wilson, Guy West - **G.W. Wils.**
Wilson, K.S. - **K.S. Wilson**
Wilson, Malcolm - **M. Wilson**
Wiltshire, Samuel Paul - **Wiltshire**
Wineland, Grace Odel - **Wineland**
Wingard, Samuel Andrew - **Wingard**
Wingate, Harold - **Wingate**
Winge, Ojvind - **Winge**
Wingfield, Michael J. - **M.J. Wingfield**
Winstead, Nash N. - **Winstead**
Winter, Georg - **G. Wint.**
Wolf, Fredrick Adolphus - **F.A. Wolf**
Wolf, Fredrick Taylor - **F.T. Wolf**
Wolff, Heinrich - **H. Wolff**
Wolff, Jacobus Otto Reinhold - **R. Wolff**
Wollenweber, Hans Wilhelm - **Wollenweb.**
Wormskjold, Martin - **M. Wormsk.**
Woronichin, Nikolai Nikolaevich - **Woronichin**
Woronin, Michael Stepanovitch - **Woronin**
Woronow, G.N. - **Woronow**
Wright, Ernst - **E. Wright**
Wright, Jorge Eduardo - **J. Wright**
Wróblewski, Antoni - **Wróbl.**
Wulfen, Franz Xavier von - **Wulfen**
Yaegashi, Hiroshi - **Yaegashi**
Yamada, Gentaro - **G. Yamada**
Yamamoto, Wataro - **W. Yamamoto**
Yamazaki, I. - **Yamazaki**
Yarrow, D. - **Yarrow**
Yasuda, Atsushi - **Yasuda**
Yen, Jomin - **J. Yen**
Yohem, K.H. - **Yohem**
Yokogi, - **Yokogi**
Yokoyama, Tatsuo - **Yokoyama**
Yoshii, Hazime - **Yoshii**
Yoshino, Kiichi - **Yoshino**
Young, Esther - **E. Young**
Young, Paul Allen - **P.A. Young**
Young, Roy A. - **R.A. Young**
Yu, T.F. - **Yu**
Zabriskie, Jeremiah Lott - **Zabr.**
Zahlbruckner, Alexander - **Zahlbr.**
Zak, Bratislav - **Zak**
Zalasky, Harry - **Zalasky**
Zaleski, K. - **Zaleski**
Zambettakis, Charalambos - **Zambettakis**
Zeller, Sanford Myron - **Zeller**
Zentmyer, George Aubrey, Jr. - **Zentmyer**
Zerova, M. - **Zerova**
Zheng, Ru-yong - **Zheng**
Zhuang, Wen-Ying - **Zhuang**
Ziller, Wolf Gunther - **Ziller**
Zimmer, Ramon C. - **Zimmer**
Zimmermann, Albrecht Wilhelm Phillip - **A. Zimmerm.**
Zinssmeister, Carl Luther - **Zinssmeister**
Zobel, Johann Baptista - **J. Zobel**
Zogg, Hans - **Zogg**
Zopf, Friederich Wilhelm - **Zopf**
Zukal, Hugo - **Zukal**
Zundel, George Lorenzo Ingram - **Zundel**